North Pole
ARCTIC OCEAN
ZEMLYA FRANTSA IOSIFA
SVALBARD
NOVAYA ZEMLYA
Mys Chelyuskin
NOVOSIBIRSKIYE OSTROVA
POLUOSTROV TAYMYR
Ostrov Vrangelya
Jan Mayen
BARENTS SEA
NORDKAPP
Karskoye More
N. AMERICA
ICELAND
FAEROE IS.
SCANDINAVIAN PEN.
Beloye More (White Sea)
Onezhskoye Oz.
Ladozhskoye Oz.
URAL MTS.
St. Lawrence
BRITISH ISLES
NORTH SEA
Baltic Sea
SEA OF OKHOTSK
POLUOSTROV KAMCHATKA
Klyuchevskaya (Vol.) 15 584
BERING SEA
ALEUTIAN IS.
EUROPE
ASIA
CASPIAN DEPRESSION
Gora El'brus 18 481
Mt. Blanc 15 771
Corse
Sardegna
BALKAN PEN.
Black Sea
CAUCASUS
Oz. Balkhash
ALTAI MTS.
KHREBET KHANGAY
PLATEAU OF MONGOLIA
GOBI DESERT
PLAINS OF MANCHURIA
SAKHALIN
Mys Lopatka
KURIL'SKIY OSTROVA
HOKKAIDŌ
SEA OF JAPAN
JAPAN TRENCH
KOREAN PEN.
HONSHŪ
Fuji-San (Vol.) 12 388
KYUSHŪ
Sicilia
Etna (Vol.) 11 122
Krete
Cyprus
MEDITERRANEAN SEA
PLATEAU OF ASIA MINOR
Ararat 16 946
Demavend 18 934
PLATEAU OF IRAN
TARIM BASIN
PAMIRS
HINDU KUSH
KUNLUN SHAN
PLATEAU OF TIBET
GREAT PLAIN OF CHINA
Yellow Sea
EAST CHINA SEA
RYŪKYŪ RETTŌ
TAIWAN
BONIN IS.
Everest 29 028
HIMALAYAS
BARKA PLATEAU
LIBYAN DESERT
OASES OF FEZZAN
SAHARA
NUBIAN DESERT
RED SEA
Persian Gulf
G. of Oman
Ras al Hadd
PLATEAU AND PENINSULA OF ARABIA
THAR DESERT
PENINSULA OF INDIA
G. of Cutch
Hsinkao Shan 13 113
HAINAN
LUZON
PHILIPPINES
MARIANA ISLANDS
Guam
MARIANA TRENCH
WAKE
MARSHALL ISLANDS
ARABIAN SEA
DECCAN PLATEAU
BAY OF BENGAL
INDOCHINA PENINSULA
SOUTH CHINA SEA
PHILIPPINES TRENCH
YAP
CAROLINE ISLANDS
Ras Dashan 15 158
PLATEAU OF ABYSSINIA
Gulf of Aden
Socotra
C. GUARDAFUI
LACCADIVE IS.
ANDAMAN ISLANDS
ISTHMUS OF KRA
NICOBAR IS.
G. of Thailand
MALAY PENINSULA
Sulu Sea
MINDANAO
PALAU IS.
GREAT LAKES
ADAMAWA HIGHLANDS
Mt. Cameroon 13 354
C. COMORIN
CEYLON
MALDIVE ISLANDS
Str. of Malacca
Kinabalu 13 455
BORNEO
Celebes Sea
MALAY ARCHIPELAGO
Halmahera
GILBERT IS.
Nauru
CENTRAL AFRICA
Lake Victoria
Kilimanjaro 19 340
Zanzibar
AMIRANTE IS.
CHAGOS ARCH.
DIEGO GARCIA
SUMATRA
EAST INDIES
CELEBES
Moluccas
Jaya Peak 16 500
NEW GUINEA
New Ireland
New Britain
SOLOMON ISLANDS
Lake Tanganyika
ALDABRA IS.
COMORO IS.
C. d'Ambre
INDIAN OCEAN
Java Sea
JAVA
Banda Sea
SUNDA ISLANDS
Flores
Timor
Arafura Sea
Timor Sea
Torres Str.
C. YORK
Gulf of Carpentaria
TUVALU
CHRISTMAS
SUNDA TRENCH
COCOS IS.
GREAT BARRIER REEF
CORAL SEA
NEW HEBRIDES
FIJI IS.
Viti Levu
Lake Nyasa
Mozambique Channel
MADAGASCAR
MASCARENE IS.
Rodrigues
Mauritius
Réunion
C. FRIO
KALAHARI DESERT
NORTH WEST CAPE
GT. SANDY DESERT
AUSTRALIA
WESTERN PLATEAU
GT. VICTORIA DESERT
EASTERN HIGHLANDS
NEW CALEDONIA
Shark Bay
Bala Delagoa
C. Ste. Marie
Mont aux Sources 10 822
GREAT KARROO
C. OF GOOD HOPE
C. AGULHAS
AMSTERDAM
ST. PAUL
C. LEEUWIN
Great Australian Bight
Spencer Gulf
Mt. Kosciusko 7 316
C. HOWE
NORTH CAPE
NORTH ISLAND
NEW ZEALAND
Bass Str.
TASMANIA
SOUTH EAST CAPE
Mt. Cook 12 349
SOUTH ISLAND
Stewart
BOUNTY IS.
ANTIPODES
PRINCE EDWARD IS.
IS. CROZET
IS. DE KERGUELEN
Heard
BOUVETØEN
AUCKLAND IS.
Campbell
MACQUARIE IS.
DAVIS SEA
BALLENY IS.
Enderby Land
ANTARCTICA
WILKES LAND
VICTORIA LAND
Ross Sea
South Pole

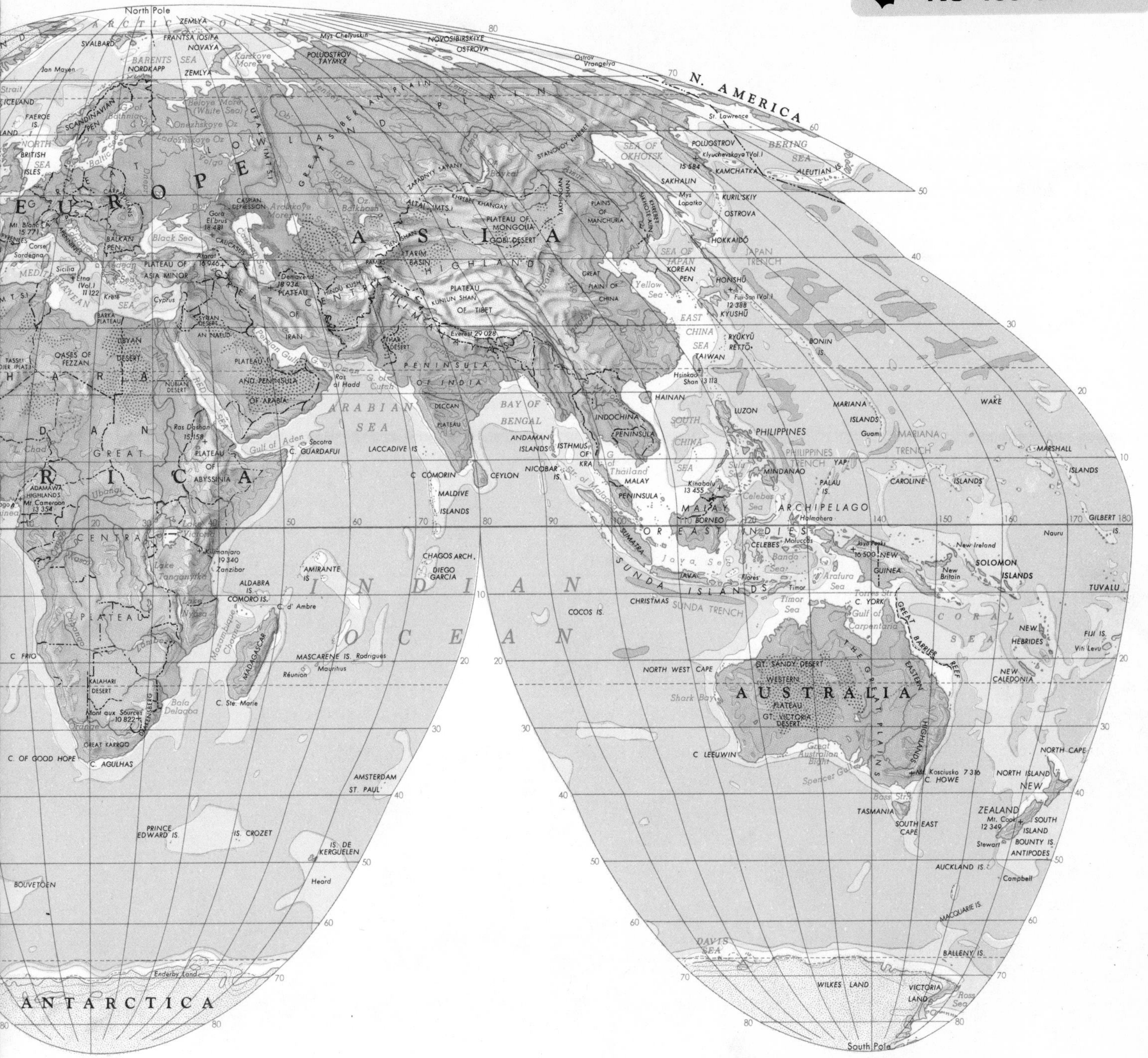

Goode's Homolosine Equal Area Projection

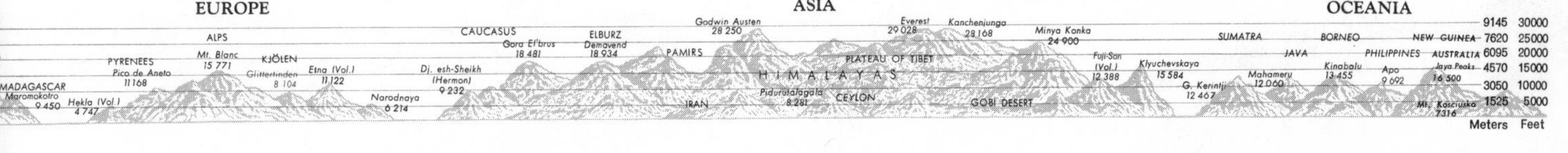

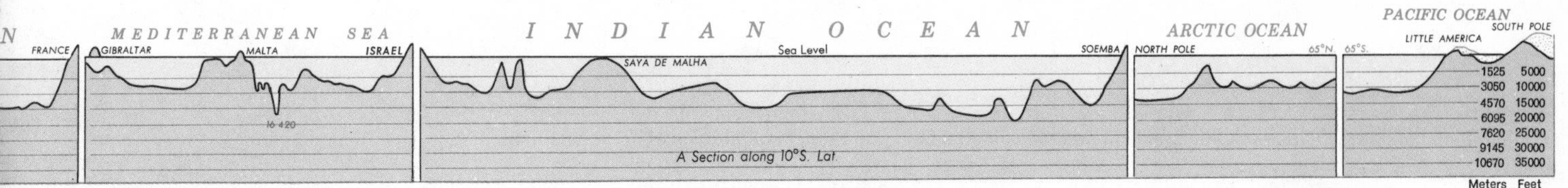

en in feet

THE INTERNATIONAL GEOGRAPHIC ENCYCLOPEDIA AND ATLAS

ENCYCL

THE INTERNATIONAL GEOGRAPHIC OPEDIA AND ATLAS

The Macmillan Press Limited • London

 Houghton Mifflin Company adapted, revised, and supplemented the encyclopedic text entries that originally appeared as geographic entries in *The New Columbia Encyclopedia,* published by Columbia University Press.

ISBN: 0-333-27498-9

Contents

Staff

Editorial Director

Fernando de Mello Vianna

Reference Editor

Kaethe Ellis

Art Director

Geoffrey Hodgkinson

Managing Editor for this Book

Pamela Burton DeVinne

Associate Editors

Mark Boyer
Walter M. Havighurst

Editorial Assistant

Paula Rubenstein

Contributing Editors

Bruce Dale
Glenn Davis
Kathleen Gerard
George D. Griffin
Stephen Krensky
George M. Margolis
Elizabeth Marine
Sarah F. McMahon

Editorial Researchers

Cindy Barlow
Carolyn S. Casagrande
Janet R. Cooper
Joan S. Cooper
Barbara Dullea
Michele Fossier
Roxann M. Fossier
John E. Gutowski
Sally Hehir
Deborah A. Karacozian
Jane G. Kennedy
David Lovler
Pamela J. Martin
Catherine Menand
Robert M. Pollak
Ellen R. Regal
Charles F. Robbart
Terri L. Smith
David Tinsley

Manuscript Coding and Composition

Joanne L. Nichols
Brenda Bregoli Sturtevant

Art

Jacket Design

Geoffrey Hodgkinson

Maps

Geoffrey Hodgkinson
Logan Smith
Sue Wetherill

Map and Research Editor

Carole LaMond

Proofreading

Barbara Collins
Kendra K. Ho
Joyce O'Connor
Doris Scarpello

Foreword

The *International Geographic Encyclopedia and Atlas* is intended to fill the need for a completely up-to-date one-volume encyclopedia of geographic names and terms. It is believed that this unique volume, which has the added benefit of an attractive and current atlas of the world, will provide the information to answer people's geographic inquiries.

The rapid rate of change in international geographic and census data, combined with the growing sense of global interdependence, should ensure the general usefulness of the *International Geographic Encyclopedia and Atlas*. New regions of the world are coming to prominence; new leaders, new countries, and new cities are constantly being mentioned in the news. Although the encyclopedia is primarily intended for English-speaking readers, the wide variety of its coverage of peoples, countries, and cultures makes it a valuable reference for the international market, in keeping with the increased need for knowledge and the increased sophistication of readers in all parts of the world.

The more than 25,000 entries go well beyond location and geographic statistics; they also show how to pronounce entry words and give information on history, economic importance, and educational and cultural institutions.

One of the book's most helpful sources of information is a 64-page atlas of four-color maps, showing major cities and major physical and political features of every region of the world. This map section is followed by a 48-page index to the atlas, giving the page location of all important features. In addition, more than 200 black-and-white maps pinpoint the location of countries, U.S. states, and Canadian provinces.

Population figures are from the most recent sources available and were revised as necessary up to the time the book was printed. Dates for population figures are also given. The appendix section updates major world events to June, 1979.

The editors of *The International Geographic Encyclopedia and Atlas* would like to express appreciation to Columbia University Press for their cooperation and for making the geographic text entries of the *New Columbia Encyclopedia* available for adaptation.

We wish to thank the people involved in our major survey of U.S. counties. All county seats of the United States have been included; information about population, dates of incorporation, manufacturing, and products have been supplied directly by the public officials of these locations.

Thanks are due also to Rocappi, the typesetting division of Lehigh Press, Inc., for their help and assistance in all phases of manuscript composition.

How to Use This Book

The *International Geographic Encyclopedia and Atlas* contains a great wealth of information about the important places and physical features of the world around you. Its more than 25,000 entries give the vital statistics – geographic designation, population, size, and location – plus pertinent information on historical significance, economic importance, governmental organization, and the most interesting places to visit and sights to see.

The following guide will help you to find precisely the information you are seeking in this book.

Main Entry

Main entries in the *International Geographic Encyclopedia and Atlas* appear in boldface type. All new entry words are syllabicated and pronounced. For a full explanation of the pronunciation symbols used, see pages XI and XII.

Entry words are listed in alphabetical order:

town (toun), in the United States. In the New England states the town is the basic unit of local government. Elsewhere in the United States the term signifies either a place incorporated as a town or simply a population center. A township is a geographic division of a county, usually made up of 36 sections, each with an area of 1 sq mi (2.6 sq km). Except in the Middle Atlantic states, townships are seldom units of local government.

Tow·ner (tou'nər), county (1970 pop. 4,645), 1,043 sq mi (2,701.4 sq km), N N.Dak., in a prairie area bordering on Man., Canada, and watered by numerous creeks and streams; formed 1883; co. seat Cando. Wheat, barley, oats, flax, potatoes, and some sunflowers are grown. Dairy products, livestock, and poultry are also important.

Towner, city (1970 pop. 870), seat of McHenry co., N central N.Dak., E of Minot on the Souris River, in a diversified farming area. Dairy products, poultry, and grain are processed, and hay and cattle are shipped.

Towns (tounz), county (1970 pop. 4,565), 166 sq mi (430 sq km), N Ga., bounded on the N by Chatuge Lake and the N.C. border and drained by the Hiawassee River; formed 1856; co. seat Hiawassee. It has agriculture (corn, hay, potatoes, fruit, and livestock), lumbering, and a resort area that is part of Chattahoochee National Forest.

Town·send (toun'zənd), town (1970 pop. 1,371), seat of Broadwater co., W central Mont., on the Missouri River NE of Butte; settled 1883, inc. 1895. Formerly a mining town, it still has some mines but also has ranching and irrigated farming.

Townsend, Mount, 7,260 ft (2,214.3 m) high, SE New South Wales, in the Australian Alps. It is Australia's second-highest peak.

Towns·ville (tounz'vĭl'), city (1973 est. pop. 76,500), NE Queensland, Australia, on Cleveland Bay. It is a major port. Wool, hides, meat, copper, and sugar are the chief exports. Copper and sugar refining, meat packing and freezing, and cement making are other industries.

Entries that are spelled the same are combined only if they have the same geographic designation and are located in the same country. For example, all United States counties named Campbell are combined; all United States cities named Campbell are combined in a separate entry. Individual localities within such combined entries are listed alphabetically by state and are designated by boldface numerals:

Camp·bell (kăm'bəl). **1.** County (1970 pop. 88,704), 151 sq mi (391.1 sq km), N Ky., in a gently rolling upland area in the outer Bluegrass region, bounded on the N and E by the Ohio River, here forming the Ohio border, on the W by the Licking River; formed 1794; co. seats Alexandria and Newport. It is chiefly industrial and residential, with some agriculture (vegetables, corn, alfalfa, tobacco, dairy products, poultry, and livestock). **2.** County (1970 pop. 2,866), 732 sq mi (1,895.9 sq km), N S.Dak., in an agricultural area bounded on the W by the Missouri River; formed 1873; co. seat Mound City. Dairy products, livestock, wheat, corn, and barley are produced. It includes a state game refuge. **3.** County (1970 pop. 26,045), 450 sq mi (1,165.5 sq km), NE Tenn., with the Cumberlands in the NW and bounded on the N by the Ky. border, on the SE by the Clinch River; formed 1806; co. seat Jacksboro. It has bituminous-coal mining, lumbering, and some agriculture (livestock, fruit, tobacco, corn, and hay). **4.** County (1970 pop. 43,319), 524 sq mi (1,357.2 sq km), SW central Va., bounded on the N by the James River, on the S by the Roanoke River; formed 1782; co. seat Rustburg. Agriculture (tobacco, corn, hay, and wheat), dairying, and poultry raising are important. **5.** County (1970 pop. 12,957), 4,755 sq mi (12,315.5 sq km), NE Wyo., in a grain and livestock area bordering on Mont. and watered by the Little Powder and Belle Fourche rivers; formed 1911; co. seat Gillette. It has coal deposits.

Campbell. 1. City (1978 est. pop. 25,400), Santa Clara co., W Calif., in the fertile Santa Clara valley; founded 1885, inc. 1952. A processing center for fruits and vegetables, it has a huge fruit-drying facility. **2.** City (1970 pop. 12,577), Mahoning co., NE Ohio, on the Mahoning River, adjacent to Youngstown; inc. 1908. It has extensive iron and steel works.

Entries that are spelled the same but are different in nature or in national location appear as separate entries, and are listed in the following order.

Ancient geographic entities precede modern ones:

Troy (troi), ancient city made famous by Homer's account of the Trojan War. It is also called Ilion or, in Latin, Ilium. Its site is the mound now named Hissarlik, in Asian Turkey, c.4 mi (6.4 km) from the mouth of the Dardanelles. Heinrich Schliemann identified the site and conducted excavations here beginning in 1871. Nine successive cities or villages have occupied the site, the earliest dating from the Neolithic period. The Troy of the Trojan War was a Phrygian city and the center of a region known as Troas. The culture of the Trojans dates from the Bronze Age.

Troy. 1. City (1970 pop. 11,482), seat of Pike co., SE Ala., on the Conecuh River; inc. 1843. Products include lumber and wood items, textiles, truck bodies, feed, and pecans. **2.** City (1970 pop. 1,047), seat of Doniphan co., extreme NE Kansas, near the Missouri River W of St. Joseph, Mo.; founded 1855, inc. 1860. It is the shipping center of an apple-growing and vinegar-producing region. Indian mounds have been excavated nearby. **3.** City (1978 est. pop. 62,000), Oakland co., SE Mich., a suburb of Detroit; settled 1821, inc. 1955. Its manufactures include automobile and electronic parts. **4.** City (1970 pop. 2,538), seat of Lincoln co., E Mo., NW of St. Louis; settled 1801, laid out 1819, inc. 1852. A state park is nearby. **5.** City (1978 est. pop. 57,400), seat of Rensselaer co., E N.Y., on the east bank of the Hudson River; inc. 1816. It is known for its manufacture of collars and shirts. Other important products are abrasives, auto parts, instruments, railroad supplies, and apparel. Henry Hudson explored (1609) the area near present Troy, and the site was included in the patroonship given to Kiliaen Van Rensselaer by the Dutch West India Company. The town was laid out in 1786. **6.** Town (1970 pop. 2,429), seat of Montgomery co., central N.C., E of Albermarle. It is a trade center and manufactures textiles and wood products. **7.** City (1978 est. pop. 18,500), seat of Miami co., W central Ohio, on the Great Miami River, in a farm area; inc. 1814. Food-processing machinery, motor generators, gummed paper, and tools are manufactured. Growth and industrialization came with the arrival of the Miami and Erie Canal in 1837.

Modern geographic entities are alphabetized according to national location:

Ham·il·ton (hăm'əl-tən), city (1970 pop. 2,060), capital of Bermuda, on Bermuda Island. It is a free port at the head of Great Sound, a lagoon protected by coral reefs. The city is a major tourist resort.

Hamilton, city (1976 pop. 312,003), S Ont., Canada, at the W end of Lake Ontario. Hamilton is an important port, transportation center, and manufacturing city. It is Canada's leading producer of iron and steel; other manufactures include automobiles, heavy machinery, and paper and textile products. The site was settled by United Empire Loyalists in 1778.

Hamilton, city (1976 pop. 87,968), N central North Island, New Zealand, on the Waikato River. Hamilton is the urban center of a densely populated dairy area. It was founded as a military settlement on the site of a deserted Maori village in 1864.

Hamilton, burgh (1976 est. pop. 107,178), Strathclyde region, S central Scotland, at the confluence of the Avon and the Clyde rivers. It is a market town with metal products and other industries. Rudolf Hess landed near Hamilton after his flight from Germany in May, 1941.

Hamilton. 1. Town (1976 est. pop. 4,196), seat of Marion co., NW Ala., near the Miss. border NW of Birmingham, in a timber and cotton area; settled c.1818. **2.** City (1970 pop. 357), seat of Harris co., W Ga., NNE of Columbus, in an agricultural area. It is a summer resort. **3.** Town (1970 pop. 6,373), Essex co., NE Mass., W of Gloucester; settled 1638, set off from Ipswich 1793. **4.** City (1970 pop. 2,499), seat of Ravalli co., W central Mont., on the Bitterroot River SSW of Missoula; inc. 1894. It is in a farming, dairying, and lumbering area. **5.** City (1978 est. pop. 64,400), seat of Butler co., SW Ohio, on the Great Miami River; inc. 1857. A manufacturing center, Hamilton has paper and pulp mills, blast furnaces, and many factories that make a great variety of products, including safes, machinery, chemicals, and pumps and motors. Hamilton was settled on the site of Fort Hamilton, built in 1791. **6.** City (1970 pop. 2,760), seat of Hamilton co., central Texas, W of Waco, in a cotton, wheat, and livestock area. Deposits of clay, limestone, and natural gas are in the vicinity.

Note that for geographic entities within the United States only a state location is given. However, these entries are alphabetized as though the country name appeared in the entry.

Demographic entries (those with a population figure) precede physical geographic entries (for example, rivers, mountains, lakes, or islands):

Fra·ser (frā'zər, -zhər), city (1970 pop. 11,868), Macomb co., SE Mich., a suburb of Detroit; inc. as a village 1894, as a city 1957. Automated machine tools and steel products are manufactured.

Fraser, chief river of British Columbia, Canada, c.850 mi (1,370 km) long, rising in the Rocky Mts., at Yellowhead Pass, near the British Columbia-Alta. line and flowing NW through the Rocky Mt. Trench to Prince George, then S and W to the Strait of Georgia at Vancouver. The Fraser River canyon, which begins at Yale, is noted for its scenery; its mountain walls rise more than 3,000 ft (915 m). The river contains the chief spawning grounds in North America for the Pacific salmon. Logging is important along the upper course. The Fraser delta is the most fertile agricultural region of British Columbia; dairying and truck farming are important.

The Fraser River was discovered by Sir Alexander Mackenzie, the Canadian explorer, who followed its upper course on his expedition (1793) to the Pacific Ocean and takes its name from Simon Fraser, the Canadian explorer and fur trader, who followed (1808) the river to its mouth, establishing fur-trading posts along the way. With the discovery of gold (1859) in the Cariboo dist., on the river's upper reaches, the government built a road to serve the valley and settlement of the region followed.

Demographic entries that are located within the same country but are different in geographic designation are alphabetized so that larger units precede smaller ones:

Wash·ing·ton (wŏsh'ĭng-tən, wôsh'-), state (1975 pop. 3,522,000), 68,192 sq mi (176,617 sq km), including 1,483 sq mi (3,841 sq km) of inland water surface, extreme NW United States, in the Pacific Northwest, admitted 1889 as the 42nd state. Olympia is the capital; Seattle, Spokane, and Tacoma are the largest cities. Washington is bounded on the north by the Canadian province of British Columbia, on the east by Idaho, on the south by Oregon (with the Columbia River marking much of the boundary), and on the west by the Pacific, with Puget Sound in the northwest and two inlets, Grays Harbor and Willapa Bay, farther south.

Washington. 1. County (1970 pop. 16,241), 1,066 sq mi (2,761 sq km), SW Ala., in the coastal plain on the Miss. border, bounded on the E by the Tombigbee River; formed 1800; co. seat Chatom. Its agriculture includes corn, soybeans, cattle, hogs, and cotton. Naval stores and agricultural chemicals are produced. **2.** County (1970 pop. 77,370), 962 sq mi (2,491.6 sq km), NW Ark., in the Ozarks, bordered on the W by the Okla. line, drained by the Illinois and White rivers; formed 1828; co. seat Fayetteville. It has agriculture (livestock, poultry, dairy products, truck crops, fruit, and grain), lumber, limestone deposits, and diversified industry. **3.** County (1970 pop. 5,550), 2,526

Washington. 1. City (1970 pop. 4,094), seat of Wilkes co., NE Ga., NW of Augusta; settled 1773, laid out 1780. It was the colony's temporary capital during the American Revolution. Eli Whitney's workshop is nearby. The last cabinet meeting of the Confederacy was held here in May, 1865. Many ante-bellum houses remain. **2.** City (1975 est. pop. 11,200), seat of Daviess co., SW Ind.; settled 1805, inc. as a city 1871. It has railroad shops and a rubber industry. **3.** City (1970 pop. 6,317), seat of Washington co., SE Iowa, SSW of Iowa City, in a farm area; inc. 1864. **4.** City (1970 pop. 1,584), seat of Washington co., NE Kansas, near the Nebr. border NW of Topeka, in a farm area; inc. 1875. **5.** City (1970 pop. 8,499), Franklin co., E Mo., on the Missouri River WSW of St. Louis; settled c.1822, platted 1837, inc. 1841. It is a farm trade center. Shoes, metal and dairy products, and corncob pipes are made. **6.** Borough (1970 pop. 5,943), Warren co., NW N.J., NE of Phillipsburg on the old Morris Canal; settled 1741, inc. 1868. Plastic containers and furniture are made. **7.** City (1970 pop. 8,961), seat of Beaufort co., E N.C., at the head of the Pamlico estuary SSE of Greenville; founded 1771. It is a market, processing, and shipping center in a tobacco, grain, and livestock area and has clothing and lumber mills. **8.** City (1979 est. pop. 20,400), seat of Washington co., SW Pa., in a bituminous coal region; settled 1769, laid out 1781, inc. as a city 1924. Chief among its many manufactures are glass, steel, and electronic products. **9.** Town (1970 pop. 189), seat of Rappahannock co., N Va., in the E foothills of the Blue Ridge W of Warrenton; platted 1749 by George Washington. Apples are shipped.

Variants and Cross-references

Alternate spellings and names are given if they are in common usage:

Hel·sing·ør (hĕl'sĭng-œr') or **El·si·nore** (ĕl'sə-nôr'), city (1976 est. pop. 55,964), Frederiksborg co., E Denmark, on the Øresund opposite Hälsingborg, Sweden. It is an industrial center, fishing port, and summer resort. Manufactures include ships, machinery, beer, and textiles. Known since the 13th cent., Helsingør is the site of Kronborg castle (1754–85; restored 1925–37), which, although the strongest fortress in Denmark at the time, was taken by the Swedes in 1660. The castle is now a maritime museum and is also used for performances of Shakespeare's *Hamlet,* which is set here.

Cross-references to such names are given only if the words do not fall within close alphabetical order:

El·si·nore (ĕl'sə-nôr'): *see* Helsingør, Denmark.

Entry Information

Each main entry in this book contains a geographic designation, such as state, province, county or parish, city or town, river, island, or mountain. Each demographic entry has the latest population figures available. Figures updated since the last official national census have been given wherever possible:

Fair·born (fâr'bôrn'), city (1978 est. pop.35,300), Greene co., SW Ohio; settled 1799, inc. 1950 with the merging of Osborn and Fairborn. Major employers are Wright State Univ. in nearby Dayton and Wright-Patterson Air Force Base.

Fair·bur·y (fâr'bûr'ē, -bə-rē), city (1970 pop. 5,265), seat of Jefferson co., SE Nebr., near the Kansas border SSW of Lincoln, in a farm area; founded 1869, inc. 1871.

Area figures and length, width, height, etc., are given where required, and appear in both U.S. standard and metric figures. A chart of metric conversions appears on page XV.

Ta·na (tä'nə), river, c.500 mi (805 km) long, rising near Mt. Kenya, central Kenya, E Africa, and flowing E, then S across Kenya to the Indian Ocean. There are hydroelectric plants and irrigation projects in the Tana basin.

Tana or **Tsa·na** (tsä'nə), largest lake of Ethiopia, c.1,400 sq mi (3,625 sq km), S of Gondar. One of the streams feeding it is regarded as the source of the Blue Nile.

Ta·ney (tā'nē), county (1970 pop. 13,023), 615 sq mi (1,592.9 sq km), S Mo., in the Ozarks, drained by the White River, which here forms Lake Taneycomo; formed 1837; co. seat Forsyth. It is a resort and agricultural area, known especially for livestock.

The opening statement of each main entry also contains precise information on geographic location:

Wa·bash (wô'băsh'). **1.** County (1970 pop. 12,841), 222 sq mi (575 sq km), SE Ill., bounded on the E and S by the Wabash River and drained by Bonpas Creek; formed 1824; co. seat Mount Carmel. It has agriculture (corn, wheat, soybeans, and livestock), manufacturing of electrical and sports equipment, clothing, flour, and paper products, and deposits of petroleum, natural gas, and bituminous coal. **2.** County (1970 pop. 35,553), 398 sq mi (1,030.8 sq km), N Ind., drained by the Wabash, Eel, Salamonie, and Mississinewa rivers; formed 1832; co. seat Wabash. It is an agricultural area that produces livestock, soybeans, wheat, and corn. Its manufacturing includes furniture, abrasive products, metal products, and farm machinery. There is timber sawmilling in the county.

Wabash, city (1975 est. pop. 13,300), seat of Wabash co., N central Ind., on the Wabash River; inc. 1849. Wabash is in a fertile area that yields grain, vegetables, and fruit. Rubber products, paperboard, temperature controls, insulation, and electrical parts are manufactured. The U.S. government built (1820) a mill here for the Miami Indians. A treaty concluded (1826) with the Indians opened the way for the first permanent white settlers, who arrived in 1827. Wabash was the world's first electrically lighted city.

Abbreviations

For ease in readability, abbreviations have been used sparingly in this book. For example, geographic directions are abbreviated only within the opening statement of each entry. A complete list of abbreviations appears on page XIII.

Subheads

Within longer entries, such as those for countries, states, or provinces, the subheads *Economy, History, Government,* and *Educational Institutions* are used to indicate where specific information on those subjects can be found.

Charts and Tables

Entries for major countries and U.S. states include a chart of political subdivisions.

Utah is divided into 29 counties:

NAME	COUNTY SEAT	NAME	COUNTY SEAT
Beaver	Beaver	Piute	Junction
Box Elder	Bringham City	Rich	Randolph
Cache	Logan	Salt Lake	Salt Lake City
Carbon	Price	San Juan	Monticello
Daggett	Manila	Sanpete	Manti
Davis	Farmington	Sevier	Richfield
Duchesne	Duchesne	Summit	Coalville
Emery	Castle Dale	Tooele	Tooele
Garfield	Panguitch	Uintah	Vernal
Grand	Moab	Utah	Provo
Iron	Parowan	Wasatch	Heber City
Juab	Nephi	Washington	St. George
Kane	Kanab	Wayne	Loa
Millard	Fillmore	Weber	Ogden
Morgan	Morgan		

In addition, at the entry for national parks and monuments there is a table listing the name, location, date authorized, and special characteristics of all officially designated parks, monuments, historic sites, etc., within the United States.

Locator Maps

All entries for independent countries, U.S. states, and Canadian provinces include locator maps showing the capital, all surrounding geographic entities, and major bodies of water.

Atlas

A four-color world atlas, showing demographic and physical features, appears at the end of this book. The atlas section includes a complete index with geographic coordinates for all features that appear on the maps.

Pronunciation

Syllabication

Entry words are divided into syllables by means of centered dots and, often, hyphens:

Val·pa·rai·so
Lab·ra·dor-Un·ga·va

A word may properly be broken at the end of a line wherever a syllable division is indicated.

In entries consisting of more than one word, words that appear as separate entries are not syllabicated:

A·leu·tian Islands
La·hon·tan, Lake

Island and **lake** are separate entries; **Aleutian** and **Lahontan** are not. Separate entries with the same first element and identical pronunciations are divided only at their first occurrence:

O·neg·a
Onega, Lake

Pronunciation

Many of the entry words in this book are foreign geographic names. Often their spelling gives native speakers and readers of English no clue to their pronunciation. Every effort has been made to indicate these pronunciations adequately and accurately, but it must be remembered that they can only be an approximation, since other languages have sounds that cannot be represented exactly in a respelling system and indeed often cannot be correctly articulated by speakers of English. In addition, dialectal variations cause variations in pronunciation, in English as well as in other languages. In many instances an anglicized pronunciation has been provided.

The pronunciation, enclosed in parentheses, appears immediately after the boldface entry word. Like the entry word, pronunciations are syllabicated, though the syllabication of the pronunciation does not necessarily match that of the entry word. The former follows phonological rules in an effort to show how the word is syllabicated when it is pronounced. The latter represents the established practice of printers and editors in breaking words at the ends of lines.

The set of symbols used in the pronunciations is designed to enable the reader to reproduce a satisfactory pronunciation with no more than a quick reference to the pronunciation key, which explains the symbols used and appears on page XII. All pronunciations given are acceptable in all circumstances. When more than one is given, the first is assumed to be the most common, but the difference in frequency may be insignificant.

Differing pronunciations within an entry are given wherever necessary; these are indicated in parentheses following the forms to which they apply:

Or·e·gon (ôr′ə-gən, -gŏn′, ŏr′-)
Ne·vad·a. 1. (nə-vā′də) County . . . **2.** (nə-văd′ə, -vä′də)

Stress

In this book stress, the relative degree of loudness with which the syllables of a word or phrase are spoken, is indicated in three different ways. The strongest, or primary, stress is marked with a bold mark (**′**):

Qui·to (kē′tō)

The second mark, a lighter stress (′), indicates syllables that are pronounced with less stress than those marked with a primary stress but with stronger stress than unmarked syllables:

O·ber·hau·sen (ō′bər-hou′zən)

An unmarked syllable has the weakest stress in the word.

Words of one syllable show no stress mark at all, since there is no other stress level to which the syllable is compared:

Niles (nīlz)

Pronunciation Key

ă	pat
ā	aid, fey, pay
â	air, care, wear
ä	father
b	bib
ch	church
d	deed
ĕ	pet, pleasure
ē	be, bee, easy, leisure
f	fast, fife, off, phase, rough
g	gag
h	hat
hw	which
ĭ	pit
ī	by, guy, pie
î	dear, deer, fierce, mere
j	judge
k	cat, kick, pique
l	lid, needle
m	am, man, mum
n	no, sudden
ng	thing
ŏ	horrible, pot
ō	go, hoarse, row, toe
ô	alter, caught, for, paw
oi	boy, noise, oil
ou	cow, out
o͝o	took
o͞o	boot, fruit
p	pop
r	roar
s	miss, sauce, see
sh	dish, ship
t	tight
th	path, thin
th	bathe, this
ŭ	cut, rough
û	circle, firm, heard, term, turn, urge, word
v	cave, valve, vine
w	with
y	yes
yo͞o	abuse, use
z	rose, size, xylophone, zebra
zh	garage, pleasure, vision
ə	about, silent, pencil, lemon, circus
ər	butter

Foreign

œ	*French* feu
ü	*French* tu
KH	*Scottish* loch
N	*French* bon

Stress

Primary stress ′
Sic·i·ly (sĭs′ə-lē)

Secondary stress ′
Yel·low·stone (yĕl′ō-stōn′)

Abbreviations

AAA Agricultural Adjustment Agency
Acad. Academy
ACLU American Civil Liberties Union
A.D. anno Domini
AEC Atomic Energy Commission
Ala. Alabama
alt. altitude
Alta. Alberta
A.M. ante meridiem
Apr. April
Ariz. Arizona
Ark. Arkansas
Assn. Association
ASSR Autonomous Soviet Socialist Republic
Aug. August
Ave. Avenue
b. born
BBC British Broadcasting Corporation
B.C. before Christ
C Celsius (centigrade)
c. circa
CAA Civil Aeronautics Administration
Calif. California
CARE Cooperative for American Remittances to Everywhere
cc cubic centimeter(s)
cent. century, centuries
CENTO Central Treaty Organization
CIA Central Intelligence Agency
cm centimeter(s)
co. county
Coll. Collection
Colo. Colorado
COMECON Council for Mutual Economic Assistance
comp. compiled, compiler
Conn. Connecticut
Corp. Corporation
cos. counties
cu cubic
CWA Civil Works Administration
d. died
DAR Daughters of the American Revolution
D.C. District of Columbia
Dec. December
Del. Delaware
dept. department
dist. district
div. division
E east
ECA Economic Cooperation Administration
ECSC European Coal and Steel Community
ed. edition
EDC European Defense Community
EEC European Economic Community
EFTA European Free Trade Association
e.g. exempli gratia (for example)
ENE east-northeast
ERA Emergency Relief Administration
ERP European Recovery Program
ESC Economic and Social Council of the United Nations
ESE east-southeast
est. established; estimated
et al. and others *(et alii)*
F Fahrenheit
FAA Federal Aviation Administration
fac. facsimile
FAO Food and Agriculture Organization of the United Nations
FBI Federal Bureau of Investigation
FCC Federal Communications Commission
Feb. February
FDA Food and Drug Administration
FEPC Fair Employment Practices Committee
fl. floruit (flourished)
Fla. Florida
ft foot, feet
FTC Federal Trade Commission
Ga. Georgia
GMT Greenwich mean time
GNP gross national product
H.M.S. His (Her) Majesty's Ship
hr hour(s)
IADB Inter-American Defense Board
IAU International Astronomical Union
ICAO International Civil Aviation Organization
ICC Interstate Commerce Commission
i.e. id est (that is)
IGY International Geophysical Year
Ill. Illinois
in. inch(es)
inc. incorporated
Ind. Indiana
Inst. Institute, Institution
IRA Irish Republican Army
ITO International Trade Organization
Jan. January
Jr. Junior
K Kelvin
kc kilocycle(s)
kg kilogram(s)
KKK Ku Klux Klan
km kilometer(s)
kw kilowatt(s)
kwh kilowatt hour(s)
Ky. Kentucky
La. Louisiana
lat. latitude
lb pound(s)
Lib. Library
long. longitude
Ltd. Limited
m meter(s)
Man. Manitoba
Mar. March
Mass. Massachusetts
Md. Maryland
MEV million electron volts
mg milligram(s)
mi mile(s)
Mich. Michigan
min minute(s)
Minn. Minnesota
Miss. Mississippi
Mlle Mademoiselle
mm millimeter(s)
Mme Madame
Mo. Missouri
Mont. Montana
mph miles per hour
MS, MSS manuscript(s)
Mt. Mount, Mountain
Mts. mountains
Mus. Museum
N north
NAACP National Association for the Advancement of Colored People
NAM National Association of Manufacturers
NASA National Aeronautics and Space Administration
NATO North Atlantic Treaty Organization
N.B. New Brunswick
N.C. North Carolina
N.Dak. North Dakota
NE northeast
Nebr. Nebraska
Nev. Nevada
N.F. Newfoundland
N.H. New Hampshire
N.J. New Jersey
NLRB National Labor Relations Board
N.Mex. New Mexico
NNE north-northeast
NNW north-northwest
no. number
Nov. November
NRA National Recovery Administration

N.S. Nova Scotia
NW northwest
N.Y. New York
OAS Organization of American States
Oct. October
OECD Organization for Economic Cooperation and Development
OEO Office of Economic Opportunity
OES Office of Economic Stabilization
Okla. Oklahoma
Ont. Ontario
OPA Office of Price Administration
OPEC Organization of Petroleum Exporting Countries
OSS Office of Strategic Services
oz ounce(s)
Pa. Pennsylvania
PAU Pan American Union
P.E.I. Prince Edward Island
pop. population
prov(s). province(s)
pseud. pseudonym
PWA Public Works Administration
Que. Quebec
RAF Royal Air Force
rev. revised
R.I. Rhode Island
ROTC Reserve Officers Training Corps
rpm revolution(s) per minute
RR railroad
RSFSR Russian Soviet Federated Socialist Republic
S south
Sask. Saskatchewan
S.C. South Carolina
S.Dak. South Dakota
SE southeast
SEATO Southeast Asia Treaty Organization
SEC Securities and Exchange Commission
sec second(s)
Sept. September
sq square
Sr. Senior
S.S. Steamship
SSE south-southeast
SSR Soviet Socialist Republic
SSW south-southwest
St. Saint; Street
SW southwest
TASS Telegraphnoye Agentstvo Sovyetskovo Soyuza (Soviet News Agency)
Tenn. Tennessee
tr. translated, translation, translator(s)
TVA Tennessee Valley Authority
UN United Nations
UNESCO United Nations Educational, Scientific, and Cultural Organization
UNICEF United Nations Children's Fund
uninc. unincorporated
Univ. University
UNRRA United Nations Relief and Rehabilitation Administration
U.S. United States
USA United States Army
USAF United States Air Force
USBGN United States Board on Geographic Names
USCG United States Coast Guard
USMC United States Marine Corps
USN United States Navy
USO United Service Organizations
U.S.S. United States Ship
USSR Union of Soviet Socialist Republics
VA Veterans Administration
Va. Virginia
VFW Veterans of Foreign Wars
VISTA Volunteers in Service to America
vol. volume(s)
vs. versus
Vt. Vermont
W west
WAC Women's Army Corps
Wash. Washington
WAVES Women Accepted for Voluntary Emergency Service (United States Women's Naval Reserve)
WCTU Women's Christian Temperance Union
WHO World Health Organization
Wis. Wisconsin
WMO World Meteorological Organization
WNW west-northwest
WPA Work Projects Administration
WSW west-southwest
W.Va. West Virginia
Wyo. Wyoming
yd yard(s)

Guide to the Metric System

Length

Unit	Number of Meters	Approximate U.S. Equivalent
myriameter	10,000	6.2 miles
kilometer	1,000	0.62 mile
hectometer	100	109.36 yards
dekameter	10	32.81 feet
meter	1	39.37 inches
decimeter	0.1	3.94 inches
centimeter	0.01	0.39 inch
millimeter	0.001	0.04 inch

Area

Unit	Number of Square Meters	Approximate U.S. Equivalent
square kilometer	1,000,000	0.3861 square mile
hectare	10,000	2.47 acres
are	100	119.60 square yards
centare	1	10.76 square feet
square centimeter	0.0001	0.155 square inch

Volume

Unit	Number of Cubic Meters	Approximate U.S. Equivalent
dekastere	10	13.10 cubic yards
stere	1	1.31 cubic yards
decistere	0.10	3.53 cubic feet
cubic centimeter	0.000001	0.061 cubic inch

Capacity

Unit	Number of Liters	Cubic	Approximate U.S. Equivalents: Dry	Approximate U.S. Equivalents: Liquid
kiloliter	1,000	1.31 cubic yards		
hectoliter	100	3.53 cubic feet	2.84 bushels	
dekaliter	10	0.35 cubic foot	1.14 pecks	2.64 gallons
liter	1	61.02 cubic inches	0.908 quart	1.057 quarts
deciliter	0.10	6.1 cubic inches	0.18 pint	0.21 pint
centiliter	0.01	0.6 cubic inch		0.338 fluid ounce
milliliter	0.001	0.06 cubic inch		0.27 fluid dram

Mass and Weight

Unit	Number of Grams	Approximate U.S. Equivalent
metric ton	1,000,000	1.1 tons
quintal	100,000	220.46 pounds
kilogram	1,000	2.2046 pounds
hectogram	100	3.527 ounces
dekagram	10	0.353 ounce
gram	1	0.035 ounce
decigram	0.10	1.543 grains
centigram	0.01	0.154 grain
milligram	0.001	0.015 grain

Metric Conversion Chart – Approximations

When You Know	Multiply By	To Find
	Length	
millimeters	0.04	inches
centimeters	0.4	inches
meters	3.3	feet
meters	1.1	yards
kilometers	0.6	miles
inches	2.5	centimeters
feet	30	centimeters
yards	0.9	meters
miles	1.6	kilometers
	Area	
square centimeters	0.16	square inches
square meters	1.2	square yards
square kilometers	0.4	square miles
hectares (10,000m^2)	2.5	acres
square inches	6.5	square centimeters
square feet	0.09	square meters
square yards	0.8	square meters
square miles	2.6	square kilometers
acres	0.4	hectares
	Mass and Weight	
grams	0.035	ounce
kilograms	2.2	pounds
tons (1000kg)	1.1	short tons
ounces	28	grams
pounds	0.45	kilograms
short tons (2000 lb)	0.9	tons
	Volume	
milliliters	0.03	fluid ounces
liters	2.1	pints
liters	1.06	quarts
liters	0.26	gallons
cubic meters	35	cubic feet
cubic meters	1.3	cubic yards
fluid ounces	30	milliliters
pints	0.47	liters
quarts	0.95	liters
gallons	3.8	liters
cubic feet	0.03	cubic meters
cubic yards	0.76	cubic meters
	Temperature (exact)	
Celsius temp.	9/5, +32	Fahrenheit temp.
Fahrenheit temp.	−32, 5/9 x remainder	Celsius temp.
	Speed	
miles per hour	1.6	kilometers per hour
kilometers per hour	0.6	miles per hour

Aa·chen (ä'kʜən), **Aix-la-Cha·pelle** (ěks'lä-shä-pěl'), or **Bad Aa·chen** (bät ä'kʜən), city (1976 pop. 242,453), North Rhine-Westphalia, W West Germany, near the Belgian and Dutch borders. One of the great historic cities of Europe, it is now chiefly important as an industrial center and rail and road junction. Its manufactures include textiles, machinery, rubber goods, metal products, and furniture. Hard coal is mined in the region. The city's hot mineral baths, frequented by the Romans in the 1st cent. A.D., are still used to treat gout, rheumatism, and skin diseases. Charlemagne, who was probably born in Aachen in 742, made the city his northern capital and the leading center of Carolingian civilization. From 936 to 1531, German kings were usually crowned at Aachen. Although it later declined in importance, Aachen remained a free imperial city until it was occupied (1794) by French troops and later annexed (1801) by France. It passed to Prussia in 1815. From 1918 to 1930 the city was occupied by the Allies as a result of Germany's defeat in World War I. During World War II approximately two thirds of Aachen was destroyed by aerial bombardment, and the city was the first major German city to fall (Oct., 1944) to the Allies.

Aals·meer (äls'mār'), town (1976 pop. 20,779), North Holland prov., W central Netherlands, on Westeinder Plassen lake, near Amsterdam. It has one of the largest flower nurseries in Europe.

Aalst (älst), city (1970 pop. 46,659), East Flanders prov., W central Belgium. It is a commercial and industrial center; manufactures include textiles, clothing, and footwear. Known since the 9th cent., Aalst was held by France from 1667 to 1706 and was the capital of Austrian Flanders in the 18th cent.

Aa·rau (ä'rou'), town (1970 pop. 16,881), capital of Aargau canton, N Switzerland, at the foot of the Jura Mts. and on the Aare River. A noted shoe-manufacturing center, it also has factories producing bells, mathematical instruments, and electrical and optical goods.

Aa·re (ä'rə) or **Aar** (är), longest river entirely in Switzerland, 183 mi (294.5 km) long, rising in the Bernese Alps and fed by several glaciers. The upper Aare emerges from dam-impounded Grimsel Lake and flows generally west through Lake Brienz, past Interlaken (where it is canalized), and through Lake Thun, the head of navigation. The Aare continues northwest, flowing through Bern before turning and flowing generally northeast, past Solothurn and Aarau, to join the Rhine River opposite Waldshut, West Germany.

Aar·gau (är'gou'), canton (1970 pop. 433,284), 542 sq mi (1,403.8 sq km), N Switzerland; capital Aarau.

A·ba (ä'bä), city (1969 est. pop. 152,000), SE Nigeria. It is an important regional market, a road and rail hub, and a manufacturing center for textiles, pharmaceuticals, processed palm oil, shoes, plastics, soap, and beer. Originally a small Ibo village, Aba was developed by the British as an administrative center in the early 20th cent.

A·ba·co and Cays (ăb'ə-kō'; kēz, kāz), island group, c.780 sq mi (2,020 sq km), most northerly of the Bahama Islands. The low islands, composed mainly of coral limestone, have native pine forests. Fish and sponges are taken from surrounding waters.

Ab·a·dan (ăb'ə-dăn', ä'bä-dän'), city (1971 est. pop. 281,000), Khuzestan prov., SW Iran, on Abadan Island, in the delta of the Shatt al Arab, at the head of the Persian Gulf. It is the terminus of major oil pipelines and is an important oil refining and shipping center. There is a large petrochemical complex that produces plastics, detergents, and caustic soda. Abadan is the point of origin of a natural-gas pipeline to the USSR. Abadan Island was ceded to Iran by Turkey in 1847. Abadan city was an unimportant village until the discovery (1908) of nearby oil fields.

A·ba·deh (ä'bä-dā'), town (1966 pop. 16,000), Fars prov., S central Iran. It is the trade center for a grain and fruit-growing region. Sesame oil, castor oil, and opium are also produced here. Woodcarving is a local craft.

A·ba·jo (ä'bə-hō'), peak, 11,445 ft (3,490.7 m) high, SE Utah, in the Abajo Mts. close to the Colo. border.

A·ba·kan (ə-bə-kän'), city (1970 pop. 90,000), capital of the Khakass Autonomous Oblast, in S central Siberian USSR, on the Yenisei River. A commercial center on the South Siberian RR, it produces textiles, furniture, foodstuffs, and metal products. It was founded in 1707 as a fortress.

Abakan, river, 350 mi (563.2 km) long, SE Khakass Autonomous Oblast, in central Siberian USSR. It rises in two headstreams in the west part of the Sayan Mts. and continues northeast to the Yenisei River south of the city of Abakan.

A·ba·shir·i (ä-bä'shĭr-ē), city (1970 pop. 43,904), Hokkaido prefecture, E Hokkaido, Japan, on the Sea of Okhotsk and the Abashiri River, lying on the Abashiri plain. It is a fishing center and port.

A·ba·ya, Lake (ə-bī'ə), 485 sq mi (1,256.2 sq km), Great Rift Valley, SW Ethiopia, in E Africa SSW of Addis Ababa. Discovered by an Italian expedition in 1896, the lake has several inhabited islands.

Abbe·ville (äb-vēl'), town (1975 pop. 25,398), Somme dept., N France, in Picardy, on the Somme River. Sugar refining, brewing, and the manufacture of jute and hemp are the chief industries. Abbeville received its commercial charter in 1184 and enjoyed prosperity until the revocation of the Edict of Nantes (1685) caused the Protestants, who constituted the skilled labor, to flee. Although heavily damaged in World War II, the town retains the late Gothic Church of St. Wolfram, with its 13th cent. belfry.

Ab·be·ville (ăb'ē-vĭl'), county (1970 pop. 21,112), 509 sq mi (1,318.3 sq km), NW S.C., bounded on the W by the Savannah River, on the NE by the Saluda River; formed 1785; co. seat Abbeville. It is in a piedmont agricultural area (cotton, oats, and corn), with some textile manufacturing.

Ab·be·ville (ăb'ē-vĭl'). **1.** City (1970 pop. 2,996), seat of Henry co., SE Ala., NNE of Dothan near the Chattahoochee River, in a cotton, corn, and timber area. There are sawmills and cotton gins here. **2.** City (1970 pop. 781), seat of Wilcox co., S central Ga., E of Cordele on the Ocmulgee River, in an agricultural area. Lumber and naval stores are processed. **3.** City (1970 pop. 10,996), seat of Vermilion parish, S La., on the Vermilion River, with access to the Intracoastal Waterway; inc. 1850. It is a trade and processing center for a region of dairies and rice and sugar-cane fields. Abbeville was settled (1843) by descendants of Acadians from Nova Scotia and was laid out like a French town. It grew around the Roman Catholic chapel built in 1845 and preserves much of the early atmosphere in its old buildings. **4.** City (1970 pop. 5,515), seat of Abbeville co., NW S.C., in a piedmont area near the Savannah River WNW of Columbia. Settled by the Huguenots in the 18th cent., it is in an agricultural and dairy-farming region. It was the scene of the last Confederate cabinet meeting held by Jefferson Davis (1865).

Ab·bot·ta·bad (ăb'ə-tä-bäd'), town (1961 pop. 31,036), NE Pakistan. It is a popular health resort c.4,000 ft (1,220 m) above sea level in the Himalaya region. It is also a market town for an agricultural and timber area.

Ab·de·ra (ăb-dîr'ə) or **Av·di·ra** (äv-dîr'ä), town, NE Greece, in Thrace, near the mouth of the Mesta River. It is a small agricultural settlement. Founded c.650 B.C., it was destroyed by the Thracians (c.550 B.C.) and rebuilt (c.500 B.C.) by refugees from Teos. The town passed to Macedon in 352 B.C. and in 198 B.C. became a free city under Roman rule.

A·ben·gou·rou (ä'bĕng-gŏŏ'rŏŏ), town (1964 est. pop. 18,000), E Ivory Coast. It is the commercial center for a region producing cacao, coffee, kola nuts, plantains, yams, manioc, and timber. The French established an administrative post in Abengourou in 1896.

Å·ben·rå (ô'bən-rô'), city (1975 pop. 15,156), capital of Sønderjylland co., S Denmark, at the head of the Åbenrå Fjord. It is a port and the commercial center for a rich agricultural region. The city was chartered in 1335. It was held by Prussia from 1864 to 1920.

A·be·o·ku·ta (ä'bē-ō-kŏŏ'tə, ăb'ē-), city (1969 est. pop. 217,000), SW Nigeria. It is the trade center for an agricultural region producing cacao, kola nuts, and palm products. Manufactures include beer, cement, dyed textiles, and canned foods. Abeokuta was founded in the 1830s. The city repelled attacks by raiders from Dahomey in 1851 and 1864. It came under British protection in 1893.

Ab·er·bro·thock (ăb'ər-brə-thŏk'): *see* Arbroath, Scotland.

Ab·er·carn (ăb'ər-kärn'), urban district (1976 est. pop. 18,370), Monmouthshire, SE Wales. It is in a coal-mining district and has iron and tin-plate works.

Ab·er·dare (ăb-ər-dâr'), urban district (1976 pop. 38,030), Mid Glamorgan co., S Wales. It is in an anthracite and iron-ore region.

Ab·er·deen (ăb'ər-dēn'), city (1976 pop. 210,362), Grampian region, NE Scotland, on the North Sea at the mouth of the Dee River. Formerly the county town of Aberdeenshire, it is Scotland's third-largest city and the only industrial center outside the midland belt. Famous as a herring and whitefish port, it is also known for its granite quarries. Other manufactures are paper, textiles, linen, and wool. There are shipyards, engineering and chemical works, and facilities for agricultural research. Aberdeen became a royal burgh in 1176 and was a leading port for trade with England and the Low Countries as early as the 14th cent. The town was burned by the English in 1336.

Ab·er·deen (ăb'ər-dēn'). **1.** Town (1970 pop. 12,375), Harford co., NE Md., in a farm region; inc. 1892. Just south, on Chesapeake Bay, is Aberdeen Proving Ground, a major research, development, and testing installation and site of the army ordnance center and school. **2.** City (1970 pop. 6,507), seat of Monroe co., NE Miss., on the Tombigbee River NNW of Columbus; inc. 1837. It is in a cotton, dairying, and timber area. A number of ante-bellum houses are here. **3.** City (1978 est. pop. 25,700), seat of Brown co., NE S.Dak.; inc. 1882. The trade and distributing center for a wheat and livestock region, it has flour mills, dairy-processing plants, and a bottling house. Manufactures include fertilizers and feeds, gear boxes, computers, and tools. Northern State College is in the city. **4.** City (1970 pop. 18,489), Grays Harbor co., W Wash., a port of entry on Grays Harbor, at the confluence of the Chehalis and the Wishkah rivers; inc. 1890. With its adjacent twin city, Hoquiam, it has lumbering, fishing, canning, and shipping industries. The two communities are in a region containing some of the world's densest stands of cedar, hemlock, and Douglas fir.

Ab·er·deen·shire (ăb'ər-dēn'shîr', -shər), former county, NE Scotland. Its county town was Aberdeen. Under the Local Government Act of 1973, Aberdeenshire became (1975) part of the Grampian region.

Ab·er·taw·e (ăb'ər-tou'ē): *see* Swansea, Wales.

Ab·er·til·ler·y (ăb'ər-tə-lâr'ē), urban district (1976 pop. 20,550), Gwent, SE Wales. It is located in an area of coal and iron mines.

Ab·e·ryst·wyth (ăb'ə-rĭst'wĭth), municipal borough (1973 pop. 10,900), Dyfed, W Wales, on Cardigan Bay. It is a summer resort and a cultural center. Before the construction of railroads, Aberystwyth was a coastal trade center. The National Library of Wales, which has an outstanding collection of Welsh manuscripts, is here.

Ab·i·djan (ăb-ĭ-jän'), city (1973 est. pop. 408,000), capital of Ivory Coast, a port on the Ebrié Lagoon (an arm of the Gulf of Guinea). Coffee, cacao, timber, pineapples, and plantains are the chief items shipped. Abidjan's major industries are food processing, sawmilling, and the manufacture of textiles, chemicals, beverages, and soap. Abidjan was a small village until the French began to enlarge it in the 1920s. In 1934 it became the capital of France's Ivory Coast colony. The Museum of the Ivory Coast is in Abidjan.

A·bi·ko (ä-bē'kō), city (1976 pop. 76,218), Chiba prefecture, central Honshu, Japan. It is an important railway junction, a resort town, and a residential suburb northeast of Tokyo.

Ab·i·lene (ăb'ə-lēn). **1.** City (1970 pop. 6,661), seat of Dickinson co., central Kansas, on the Smoky Hill River; inc. 1869. It was (1867–71) a railhead for a large cattle-raising region extending southwest into Texas. One of the wildest and toughest cow towns of the Old West, Abilene once had Wild Bill Hickok as its marshal. The city, now a shipping point for a wheat and farm region, has feed and flour mills. Greyhound racing dogs are bred in Abilene, which is the headquarters of the National Greyhound Association. President Dwight D. Eisenhower lived in Abilene in his youth; the Eisenhower Center (completed 1961) includes his family homestead, a museum, the Eisenhower Library, and his grave. **2.** City (1978 est. pop. 99,400), seat of Taylor co., W central Texas; inc. 1882. Buffalo hunters first settled here; the town, which was founded in 1881 with the coming of the railroad, was named after Abilene, Kansas. Abilene grew as a shipping point for cattle ranches and is now the financial, commercial, and educational center of a large part of west Texas. The city's diversified manufactures include electronic, aircraft, and missile components, oil-field equipment, food and dairy products, cottonseed oil, agricultural equipment, clothing, metals, and musical instruments. Agriculture (cattle, sheep, poultry, cotton, and grain sorghums) and minerals (oil, natural gas, stone, sand and gravel, and clays) are important in the economy of the surrounding area. Dyess Air Force Base and the ruins of Fort Phantom Hill, an early army post and stagecoach stop, are nearby.

Ab·ing·don (ăb'ĭng-dən), municipal borough (1976 est. pop. 20,130), Berkshire, S central England; chartered 1556. It is a popular resort and an agricultural center; automobiles and scientific instruments are manufactured. There are ruins of a Benedictine abbey dating from 675.

Abingdon. **1.** City (1970 pop. 3,936), Knox co., W central Ill., S of Galesburg in a farm and bituminous-coal area; inc. 1857. Plumbing supplies and pottery are made. **2.** Resort town (1970 pop. 4,376), seat of Washington co., extreme SW Va., in a farming and dairying region of the Blue Ridge Mts. NE of Bristol; settled c.1765, inc. 1778. Burley tobacco, grain, and livestock are auctioned here, and the town is noted for its handicrafts and the manufacture of lusterware.

Ab·ing·ton (ăb'ĭng-tən). **1.** Town (1970 pop. 12,334), Plymouth co., E Mass.; settled 1668, inc. 1713. Chiefly residential, it has some light industry. **2.** Township (1970 pop. 63,625), Montgomery co., SE Pa., a residential suburb of Philadelphia; settled 1696, inc. 1906.

Ab·i·tib·i Lake (ăb'ə-tĭb'ē), irregularly shaped lake, c.60 mi (95 km) long, SW Que. and E Ont., Canada. It is a popular tourist area and the site of the Abitibi Game Reserve. The Abitibi River drains the lake and flows west and north to the Moose River.

Ab·khaz Autonomous Soviet Socialist Republic (ăb-kăz', äb-ᴋʜäz'), autonomous region (1970 pop. 487,000), 3,300 sq mi (8,547 sq km), SE European USSR, in Georgia, between the Black Sea and the Greater Caucasus; capital Sukhumi.

Ab·o·mey (ăb-ō-mā', ə-bō'mē), town (1970 est. pop. 42,000), S Benin. It is the trade center for an agricultural region where grain and palm products are processed. Abomey was the capital of the kingdom of Dahomey, which was founded in the early 17th cent. and conquered by the French between 1892 and 1894.

A·bra·ham, Plains of (ā'brə-hăm'), field adjoining the upper part of the city of Quebec, Canada. In 1759 the English under Gen. James Wolfe defeated the French under Gen. Louis Montcalm here. The battle decided the last of the French and Indian Wars and led to British supremacy in Canada.

Abraham Lin·coln Birthplace National Historic Site (lĭngk'ən): *see* National Parks and Monuments Table.

A·bran·tes (ə-brän'tĭsh), town (1970 municipal pop. 48,161), Santarem dist., W central Portugal in Ribatejo, on the Tagus River. It is the commercial center of a fruit-growing region. In the Napoleonic Wars the French won the Battle of Abrantes in 1807, but in 1810 they were unable to take the town by siege.

A·bruz·zi (ä-brŏŏt'tsē), region (1971 pop. 1,166,664), 4,167 sq mi (10,792.5 sq km), central Italy, bordering on the Adriatic Sea in the E; capital L'Aquila. Abruzzi is mostly mountainous and is crossed by the Apennines. A generally poor region, Abruzzi has mostly small-scale agriculture and limited, but growing, industry. Tourism is important. Abruzzi was conquered by the Romans in the 4th cent. B.C. Later, it was part of the Lombard duchy of Spoleto (6th–11th cent. A.D.), the Norman kingdom of Sicily (12th-13th cent.), and the kingdom of Naples (13th-19th cent.).

Ab·sa·ro·ka Range (ăb-sə-rō'kə, ăb-sôr'kē), part of the Rocky Mts., c.175 mi (282 km) long, NW Wyo. and S Mont., partly in Yellowstone National Park and rising to 13,140 ft (4,007.7 m).

Ab·se·con (ăb-sē'kən), city (1970 pop. 6,094), Atlantic co., SE N.J., NW of Atlantic City; settled c.1780, inc. 1902.

A·bu Dha·bi (ä'bōō dä'bē, zä'-, *thä'*-), sheikdom (1968 pop. 46,375), c.26,000 sq mi (67,340 sq km), part of the federation of United Arab Emirates, E Arabia, on the Persian Gulf. The sheikdom became a British protectorate in 1892. Oil was discovered here in the early 1960s. It became part of the United Arab Emirates in 1971.

A·bu·ná (ä-bōō'nä), river, with several headstreams SW of Santa Rosa, NE Bolivia, flowing 200 mi (321.8 km) NE through rubber-producing rain forest to the Madeira River in Brazil, forming part of the Bolivian-Brazilian border in its course.

A·bu Qir or **A·bu·kir** (both: ă'bōō-kîr', ə-bōō'kər), village, N Egypt, on a promontory in the Nile River delta. Adm. Horatio Nelson's victory over the French fleet off Abu Qir on Aug. 1, 1798, restored British prestige in the Mediterranean and cut short the French venture in the Middle East begun by Napoleon I.

A·bu-Sim·bel (ä'bōō-sĭm'bəl) or **Ip·sam·bul** (ĭp'säm-bōōl'), village, S Egypt, on the Nile River. Its two temples, hewn (c.1250 B.C.) out of rock cliffs during the reign of Ramses II, were raised over 200 ft (61 m) to avoid the rising waters caused by the construction of the Aswan High Dam.

A·by·dos (ə-bī'dəs), ancient city of Egypt, NW of Thebes. Associated in religion with Osiris, Abydos became the most venerated place in Egypt. It was the favorite burial place for the kings of the earliest dynasties, and later kings such as Seti I and Ramses II continued to build temples and sanctuaries here.

Abydos, ancient town of Phrygia, Asia Minor, on the Asiatic side of the Hellespont opposite Sestos, in present-day Turkey. It was originally a Milesian colony and was the scene of the story of Hero and Leander.

Ab·ys·sin·i·a (ăb'ĭ-sĭn'ē-ə): *see* Ethiopia.

A·ca·di·a (ə-kā'dē-ə), region and former French colony, E Canada, centered on Nova Scotia but including also N.B., P.E.I., and the mainland coast from the Gulf of St. Lawrence S into Maine. The first and chief town, Port Royal (now Annapolis Royal, N.S.), was founded by the sieur de Monts in 1605. The French colony grew to be fairly prosperous, despite periodic attacks by the British, who also claimed the region. During the French and Indian War (1755-63) the British fell upon the peaceful Acadian farms, seized most of the Acadians, and deported them to the more southerly British colonies. The Treaty of Paris (1763) ceded all of Acadia to England.

Acadia, parish (1970 pop. 52,109), 662 sq mi (1,714.6 sq km), S La., bounded on the W by Bayou Nezpique, on the S by Bayou Queue de Tortue; formed 1805; parish seat Crowley. It is crossed by the Gulf Intracoastal Waterway. In a rice-growing region, it also has cotton, corn, sugar cane, and sweet potato farms, lumbering, and oil and natural-gas wells. Cotton ginning and rice milling are done here. Edible frogs are raised.

Acadia National Park, 41,642 acres (16,865 hectares), SE Maine, on the Atlantic coast; est. 1919. The park occupies a major portion of Mount Desert Island, Isle au Haut and several smaller islands, and the southern tip of Schoodic Peninsula. Almost completely surrounded by the sea, the park is characterized by a rugged, glacier-scoured interior with numerous valleys, lakes, and peaks, and a wave-eroded coastline.

Ac·a·pul·co (äk'ə-pōōl'kō), city (1970 pop. 234,866), Guerrero state, S Mexico. A fashionable resort, it has lavish hotels and facilities for deep-sea fishing and skin diving. Its fine natural harbor, surrounded by cliffs and promontories, served as a base for Spaniards exploring the Pacific and later played a key role in trade with the Philippines. Coconuts, beans, and bananas are grown in the area. Near the city, which was founded in 1550, are the archaeological remains of the Ciudad Perdida, estimated to be 2,000 years old. Acapulco has suffered frequent earthquake and hurricane damage.

Ac·ar·na·ni·a (ăk'ər-nā'nē-ə), region of ancient Greece, between the Achelous River and the Ionian Sea. The inhabitants generally sided with Athens, and Athens helped Acarnania to uphold its independence against Corinth and Sparta in the 5th cent. B.C. Later (390-375 B.C.) Sparta controlled the region. The Acarnanians kept some autonomy under the Roman Empire until the Christian era. When the Byzantine Empire broke up (1204), Acarnania passed to Epirus and in 1480 to the Turks.

Ac·co·mac or, officially, **Ac·co·mack** (both: ăk'ə-măk'), county (1970 pop. 29,004), 470 sq mi (1,217.3 sq km), E Va., on the Eastern Shore near its S tip, bounded on the N by Md., on the W by Chesapeake Bay and its arms, Tangier and Pocomoke sounds; formed 1663; co. seat Accomac. Off its east coast lie barrier islands that shelter its bays from the Atlantic. In a productive coastal plain, it has truck farming (potatoes, sweet potatoes, and tomatoes), poultry raising, dairying, and fruit growing. Pine timber, fishing, oyster culture, and crabbing are important. Fish, shellfish, and vegetables are canned and shipped. Many old buildings (some dating from the 17th cent.) remain.

Accomac, town (1970 pop. 373), seat of Accomac co., E Va., on the Eastern Shore, on the N part of Delmarva Peninsula, S of Pocomoke City, Md. It was twice a refuge of Sir William Berkeley during Bacon's Rebellion (1675-76).

Ac·cra (ăk-rä', ăk'rə), city (1970 pop. 564,194), capital of Ghana, a port on the Gulf of Guinea. It is Ghana's communications, transportation, and economic center. The chief manufactures are processed food, beverages, timber and plywood, textiles, clothing, chemicals, and printed materials. Accra originally was the capital of a Ga kingdom. It developed into a sizable town around British and Dutch forts built in the 17th cent. After the completion (1923) of a railroad to the mining and agricultural hinterland, Accra rapidly became the economic center of Ghana. Riots in the city (1948), against high retail prices and European control, led to the rise of Kwame Nkrumah as a popular leader.

Ac·cring·ton (ăk'rĭng-tən), municipal borough (1976 pop. 36,470), Lancashire, NW England. The principal industry is cotton weaving. Textile printing and dyeing and the manufacture of machinery and bricks are also important.

A·chae·a (ə-kē'ə), region of ancient Greece, in the N part of the Peloponnesus on the Gulf of Corinth. It lay between Sicyon and Elis. There the Achaeans supposedly remained when driven from other parts of Greece by the Dorian invasion. The small Achaean cities eventually banded together in the First Achaean League, but were of little significance in Greek politics. Later, however, the Second Achaean League became an important factor. After the downfall of the league, the name Achaea, or Achaia, was given to a Roman province in the Peloponnesus.

Ach·e·lo·us (äk'hĕl-ô'ôs), river: *see* Akhelóos.

Ach·ill (ăk'ĭl), island, 56 sq mi (145 sq km), Co. Mayo, W Republic of Ireland; the largest island of Ireland. The rugged island is barren, and its inhabitants subsist by fishing and farming.

A·ci·re·a·le (ä'chē-rā-ä'lā), city (1971 pop. 47,086), E Sicily, Italy. Beautifully situated on a volcanic plateau near Mt. Etna and near the Ionian Sea, Acireale has been frequented since Roman times for its warm sulfur springs and today is also a commercial center. The city was damaged by earthquakes in 1169 and 1693.

Ack·er·man (ăk'ər-mən), town (1970 pop. 1,502), seat of Choctaw co., central Miss., WSW of Columbus, in an agricultural and timber-growing area. Sawmilling is done.

Ac·o·ma (ăk'ə-mə), pueblo (1970 est. pop. 2,750), alt. c.7,000 ft (2,135 m), Valencia co., W central N.Mex.; founded c.1100-1250. This "sky city" on top of a steep-sided sandstone mesa, 357 ft (108.9 m) high and difficult of access, is considered to be the oldest continuously inhabited community in the United States. The residents are skilled pottery makers. Below the mesa are the cultivated fields and grazing grounds that help support the community.

A·con·ca·gua (ä-kōn-kä'gwä), peak, 22,835 ft (6,964.7 m) high, Mendoza prov., W Argentina, in the Andes, near the Chilean border. It is the highest peak of the Western Hemisphere. The snow-capped Aconcagua was first scaled in 1897.

Aconcagua, river, c.120 mi (193 km) long, Valparaíso prov. and Aconcagua prov., central Chile. Rising at the northwest foot of Mt. Aconcagua, it flows west to the Pacific Ocean northeast of Valparaiso. Tobacco, fruit, grain, and hemp are grown along its banks.

A·çores (ə-zôrz', ā'zôrz): *see* Azores.

Ac·re (ä'krə), state (1970 pop. 216,200), 58,915 sq mi (152,590 sq km),

W Brazil, on the borders of Peru and Bolivia; capital Rio Branco.

A·cre (ä'kər, ä'kər): *see* Akko, Israel.

A·cre (ä'krə), river, rising at the border of Peru and Brazil and flowing NE for c.400 mi (645 km) to join the Purus River at Bôca do Acre. It forms part of the boundary between Brazil and Bolivia.

Ac·ro·ce·rau·ni·an Mountains (ăk'rō-sə-rô'nē-ən): *see* Ceraunian Mountains.

Ac·ti·um (ăk'tē-əm, -shē-), promontory, NW Acarnania, Greece, at the mouth of the Ambracian Gulf. There are vestiges of several temples and an ancient town. At Actium was fought the naval battle (31 B.C.) in which the forces of Octavian (later Augustus) under Agrippa defeated the sea and land forces of Antony and Cleopatra. The battle established Octavian as ruler of Rome.

Ac·ton (ăk'tən), town (1970 pop. 14,770), Middlesex co., E Mass., NW of Boston; settled c.1680, inc. 1735. Among its manufactures are electrical machinery and chemicals. The Acton Minutemen's march to the Battle of Concord is re-enacted in the town annually.

A·cush·net (ə-ko͝osh'nət), town (1970 pop. 7,767), Bristol co., SE Mass., NE of New Bedford; settled c.1660, inc. 1860. The town was devastated in King Philip's War. In the Revolution it was the scene of a skirmish (Sept., 1776) between British and American forces.

Ac·worth (ăk'wərth), city (1970 pop. 3,929), Cobb co., NW Ga. NW of Atlanta on Allatoona Reservoir; inc. 1860. It has hosiery and lumber mills. Kennesaw Mountain National Battlefield Park is nearby.

A·da (ā'də), county (1970 pop. 112,230), 1,140 sq mi (2,952.6 sq km), SW Idaho, in an irrigated area in the Boise River valley and bounded on the S by the Snake River; formed 1864; co. seat Boise. Its agriculture includes hay, sugar beets, fruit, truck crops, and dairy products. There is some manufacturing.

Ada. 1. City (1970 pop. 2,076), seat of Norman co., NW Minn., in the Red River valley NNE of Moorhead; founded 1874, inc. 1881. It is a farm trade center for an area yielding dairy products, potatoes, grain, and sugar beets. **2.** Village (1970 pop. 5,309), Hardin co., NW Ohio, E of Lima, in a farm area; laid out 1853 as Johnstown, inc. 1861 as Ada. Ohio Northern Univ. (1871) is here. **3.** City (1978 est. pop. 14,700), seat of Pontotoc co., S central Okla.; inc. 1904. It is a large cattle market and the center of a rich oil and ranch area. The city is also a center for horse breeding.

A·dair (ə-dâr'). **1.** County (1970 pop. 9,487), 569 sq mi (1,473.7 sq km), SW Iowa, drained by the Middle, North, Thompson, and Nodaway rivers; formed 1851; co. seat Greenfield. In an agricultural region (corn, hogs, and poultry), it has bituminous-coal deposits. **2.** County (1970 pop. 13,037), 393 sq mi (1,017.9 sq km), S Ky., drained by the Green River and several creeks; formed 1801; co. seat Columbia. Its agriculture includes livestock, grain, burley tobacco, poultry, and fruit. It has some mining and timber (sawmills), and there is manufacturing in the county. **3.** County (1970 pop. 22,472), 572 sq mi (1,481.5 sq km), N Mo., drained by the Chariton and Salt rivers; formed 1841; co. seat Kirksville. It is in a grain, cattle, poultry, and coal-mining area. There is miscellaneous light industry and manufacturing. **4.** County (1970 pop. 15,141), 569 sq mi (1,473.7 sq km), E Okla., in the Ozarks, bounded by the Ark. border on the E and drained by the Illinois River and Barre Creek; formed 1907; co. seat Stilwell. It has agriculture (fruit, truck crops, livestock, grain, and berries), some manufacturing, stone and marble quarries, timbering, hunting and fishing, and processing plants for farm products.

A·dak (ā'dăk), island, 289 sq mi (748.5 sq km), off W Alaska in the Andreanof group of the Aleutian Islands. In World War II it was occupied by the Americans in 1942 for use as a military base.

A·da·ma·wa Massif (ä-dä-mä-wä', ăd'ə-mä'wə), plateau, c.26,000 sq mi (67,340 sq km), N central Cameroon and E Nigeria, W central Africa. It is sparsely populated, and grazing is the chief occupation; bauxite is mined here.

Ad·ams (ăd'əmz). **1.** County (1970 pop. 185,789), 1,237 sq mi (3,203.8 sq km), N central Colo., in an irrigated agricultural area drained by the South Platte River; formed 1901; co. seat Brighton. Sugar beets, wheat, and beans are grown, and livestock is raised. **2.** County (1970 pop. 2,877), 1,377 sq mi (3,566.4 sq km), W Idaho, in a mountainous area bounded on the W by the Snake River and the Idaho border and watered by the headstreams of the Weiser and Little Salmon rivers; formed 1911; co. seat Council. The Seven Devils Mts. are in the northwest. Its economy is based on agriculture, mining, and lumber products. Weiser National Forest extends throughout much of the county. **3.** County (1970 pop. 70,861), 866 sq mi (2,243 sq km), W Ill., bounded on the W by the Mississippi and drained by Bear and McKee creeks; formed 1825; co. seat Quincy. In an agricultural area (with livestock, corn, wheat, soybeans, poultry, and dairy products), it has limestone quarries and diversified manufacturing. **4.** County (1970 pop. 26,871), 345 sq mi (893.6 sq km), E Ind., bounded on the E by the Ohio border and drained by the Wabash and St. Marys rivers; formed 1835; co. seat Decatur. In a dairying and farming (livestock, soybeans, grain, poultry, and truck crops) area, it has timber and diversified manufacturing. **5.** County (1970 pop. 6,322), 426 sq mi (1,103.3 sq km), SW Iowa, in a gently rolling prairie region drained by the Nodaway, West Platte, and One Hundred and Two rivers; formed 1851; co. seat Corning. It has agriculture (corn, hogs, cattle, and poultry), bituminous-coal deposits, sand and gravel pits, and limestone quarries. **6.** County (1970 pop. 37,293), 449 sq mi (1,163 sq km), SW Miss., bounded on the W by the Mississippi, on the S by the Homochitto River; formed 1799; co. seat Natchez. There is agriculture (cotton and corn), dairying, lumbering (logging, sawmills), oil and gas fields, and clay and limestone deposits. Its manufactures include paper and clay products. Part of Homochitto National Forest is here. **7.** County (1970 pop. 30,553), 562 sq mi (1,455.6 sq km), S Nebr., in an agricultural area drained by the Little Blue River; formed 1872; co. seat Hastings. Grain, livestock, and poultry are produced. Wood, rubber, and plastics products, farm equipment, and textiles are manufactured. **8.** County (1970 pop. 3,862), 990 sq mi (2,564.1 sq km), SW N.Dak., drained by Cedar Creek; formed 1907; co. seat Hettinger. Some coal mining is done, cattle is raised, and winter and spring wheat and feed crops are grown. **9.** County (1970 pop. 18,957), 588 sq mi (1,523 sq km), S Ohio, bounded on the S by the Ohio River and drained by small creeks; formed 1799; co. seat West Union. It is mainly agricultural (livestock, dairy products, tobacco, and grain), with some manufacturing (wood products and textiles). It includes Serpent Mound State Park. **10.** County (1970 pop. 56,937), 526 sq mi (1,362.3 sq km), S Pa., in an agricultural and fruit-growing area bounded on the S by the Md. border, on the W and NW by South Mt.; formed 1800; co. seat Gettysburg. Its agriculture includes apples, peaches, cherries, and wheat. There are limestone quarries and a canning industry. Textiles, wood, clay, and metal products, and machinery are manufactured. **11.** County (1970 pop. 12,014), 1,895 sq mi (4,908 sq km), SE Wash., in an agricultural region of the Columbia basin bounded on the SE by the Palouse River; formed 1883; co. seat Ritzville. Its agriculture includes wheat, alfalfa, livestock, and dairy products. **12.** County (1970 pop. 9,234), 646 sq mi (1,673.1 sq km), central Wis., drained by tributaries of the Wisconsin River; formed 1848; co. seat Friendship. In a dairying, agricultural, and timber area, it has recreational facilities and a growing industrial base.

Adams, town (1970 pop. 11,772), Berkshire co., NW Mass., in the Berkshires, on the Hoosic River; inc. 1778. Its manufactures include lime products and decorative textiles (made here since 1862). The region attracts summer and winter vacationers. A Society of Friends meeting house (built 1782) is the site of annual Quaker meetings. Susan B. Anthony was born in Adams.

Adams, Mount. 1. Peak, 5,798 ft (1,768.4 m) high, N N.H., in the Presidential Range of the White Mts. **2.** Peak, 12,307 ft (3,753.6 m) high, SW Wash., in the Cascade Range.

Adam's Bridge or **Ra·ma's Bridge** (rä'məz), chain of shoals, c.18 mi (30 km) long, in the Palk Strait between India and Sri Lanka. At high tide it is covered by water. According to Hindu legend, the bridge was built to transport Rama, hero of the *Ramayana,* to the island to rescue his wife from the demon king Ravanna.

Adams National Historic Site: *see* National Parks and Monuments Table.

Adam's Peak, mountain, 7,360 ft (2,244.8 m) high, S central Sri Lanka. It is a sacred mountain, famous as a goal of pilgrimage for Buddhists, Hindus, and Moslems. On its summit is a large flat rock that bears the impression of a gigantic human foot. This stone footprint is regarded as Buddha's by Buddhists, Siva's by Hindus, and Adam's by Moslems, who believe this to be the site of Adam's fall from Paradise.

A·da·na (ä'dä-nä'), city (1975 pop. 467,122), capital of Adana prov., S Turkey, on the Seyhan River. It is the commercial center of a farm region where cotton, grains, and fruits are grown. Manufactures include processed food, cotton textiles, cement, and soap. An ancient city probably founded by the Hittites, Adana was colonized (66 B.C.) by the Romans. The city prospered, then declined, and was revived (A.D. c.782) by Harun ar-Rashid. In the 16th cent. the city passed to the Ottoman Turks.

Adana, Plain of, fertile region along the Mediterranean coast, S central Turkey. It has a subtropical climate and receives rainfall mainly during the autumn and winter months. The plain, traversed and irrigated by the Seyhan River, is a major agricultural region, producing a large variety of crops.

A·da·pa·zar·i (ä-dä'pä-zär'ə), city (1975 pop. 113,411), capital of Adapazarı prov., NW Turkey, on the Sakarya River. It is the trade center for a rich agricultural region where tobacco, sugar beets, and grains are produced. The city's manufactures include refined sugar, farm machinery, textiles, and cement.

Ad·da (äd'dä), river, 194 mi (312.1 km) long, rising in the Rhaetian Alps, N Italy, and flowing SW through Lake Como, then S into the Po River near Cremona. Its upper course furnishes much electric power; the lower river irrigates the Lombard plain.

Ad·dis Ab·a·ba (ăd'ĭs ăb'ə-bə), city (1974 est. pop. 1,161,300), capital of Ethiopia. It is situated at c.8,000 ft (2,440 m) on a well watered plateau surrounded by hills and mountains. Addis Ababa is Ethiopia's largest city and its administrative, transportation, and communications center. It is the main trade center for coffee, the country's chief export, and for tobacco, grains, and hides. The major industries produce food, beverages, processed tobacco, textiles, and shoes. In 1886 the city was chosen by Menelik II as the capital of his kingdom. In 1889 it was made the capital of Ethiopia. Here, in 1896, Italy recognized Ethiopian independence. However, in 1936 Italy captured Addis Ababa and made it the capital of Italian East Africa. The city was recaptured by the Allies in 1941 and returned to Ethiopian rule. Major growth began after 1945.

Ad·di·son (ăd'-ə-sən), county (1970 pop. 24,266), 785 sq mi (2,033.2 sq km), W Vt., bounded on the W by Lake Champlain and on the E by the Green Mts., drained by Otter Creek and the New Haven and White rivers; organized 1787; co. seat Middlebury. In an agricultural and resort region, it has dairy farms and orchards. Wood products and poultry were once important. Today industry is increasing in importance, with the manufacture of plastics, electrical components, and business forms.

Addison. 1. Village (1970 pop. 24,482), Du Page co., NE Ill.; inc. 1884. It has some light manufacturing. **2.** Village (1970 pop. 2,104), Steuben co., S central N.Y., on the Canisteo River W of Elmira, in a dairy region; inc. 1873.

Ad Di·wan·i·yah (äd dē-wän'ē-yä), city (1970 pop. 62,300), S central Iraq, on a branch of the Euphrates River.

A·del (ā-dĕl'). **1.** City (1970 pop. 4,972), seat of Cook co., S Ga., NE of Thomasville; inc. 1889. Settled nearby in 1860 as Puddleville, it was moved to its present site in 1888 to be on the railroad. It is a tobacco market and a trade center for a farm area yielding watermelon, corn, and canned food. It also has lumber milling, and veneers are made. **2.** Town (1975 est. pop. 2,771), seat of Dallas co., S central Iowa, a short distance W of Des Moines; settled 1846, inc. 1887. There is a brick and tile plant here.

Ad·e·laide (ăd'ə-lād'), city (1976 urban agglomeration pop. 900, 379), capital and chief port of South Australia state, S Australia, at the mouth of the Torrens River on Gulf St. Vincent. It has automotive, textile, and other industries. Grains, wool, dairy products, and fruit are exported. Named for the consort of William IV, it was founded in 1836 and is the oldest city in the state. The Univ. of Adelaide (1874) and a natural history museum (1895) are here. The Adelaide Festival of the Arts has been held biennially since 1960.

Adelaide, island, c.68 mi (109.4 km) long, 20 mi (32.2 km) wide, British Antarctic Territory, Antarctica, off the W coast of the Antarctica Peninsula NW of Marguerite Bay. The British explorer John Biscoe discovered the island in 1832.

A·dé·lie Coast (ə-dā'lē), region, E Antarctica, between George V Coast and Wilkes Land. It was discovered by Dumont d'Urville, a French explorer who landed in 1840 to collect rock samples; it was explored by an Australian geologist, Douglas Mawson, from 1911 to 1914. The French claim the area. In 1950 they established meteorological stations here.

A·den (ä'dən, ā'dən), city (1973 est. pop. 264,300), SW Southern Yemen, on the Gulf of Aden near the S entrance to the Red Sea. It is the capital and chief port of Southern Yemen. It has a large oil refinery; the manufacture of soap, cigarettes, and salt is also important here. Aden, a free port since 1850, has been the chief entrepôt and trading center of southern Arabia since ancient times. It enjoyed commercial importance until the discovery (late 15th cent.) of an all-water route around Africa to India. With the opening of the Suez Canal (1869), Aden again became a major trading center; the harbor was deepened to accommodate the largest vessels able to use the canal. Aden's economy suffered from the closing of the canal during the Arab-Israeli wars.

Aden's strategic location and its importance as a commercial center long made it a coveted conquest. Moslem Arabs held the region from the 7th to the 16th cent. The Portuguese failed in an attempt to capture it in 1513, but it fell in 1538 to the Ottoman Turks. At the end of the 18th cent. Aden's importance as a strategic post grew as a result of British policy to contain French expansion in the region. Aden was formally made into a British crown colony in 1935, and the surrounding region (now Southern Yemen) became known as the Aden Protectorate in 1937. Aden was granted a legislative council in 1944 and later received other rights of self-government. In 1963 Aden was joined to the Federation of the Emirates of the South, which then became the Federation of South Arabia. The British-sponsored federation was opposed by nationalists in Aden who feared domination by the tribal states. They conducted terrorist activities against the British and the federation administration, forcing the collapse of the federal government. With the establishment (1967) of the independent country of Southern Yemen, Aden became the capital along with Madinat ash Shab. In 1970 Aden became the country's sole capital.

Aden, Gulf of, W arm of the Arabian Sea, 550 mi (885 km) long, lying between Southern Yemen and the Somali Republic and connected with the Red Sea by the Bab el Mandeb.

A·di·ge (ä'dē-jā), second-longest river of Italy, c.225 mi (360 km) long, rising in the Tyrolean Alps, N Italy. It flows generally south, past Bolzano, Trent, and Verona, to the Po valley, where it turns east to empty into the Adriatic Sea. The Adige is used for irrigation and hydroelectric-power production.

Ad·i·ron·dack Mountains (ăd'ə-rŏn'dăk), circular mountain mass, NE N.Y., between the St. Lawrence valley in the N and the Mohawk valley in the S; rising to 5,344 ft (1,630 m) at Mt. Marcy, the highest point in the state. Geologically a southern extension of the Laurentian Plateau, the Adirondacks are sometimes mistakenly included in the Appalachian system. Composed chiefly of metamorphic rock, the Adirondacks were formed as igneous rocks (mainly granite) intruded upward, doming the earth's surface; subsequent faulting of the earth's crust and surface erosion, particularly by the Pleistocene glaciers, have given the mountains a rugged topography. The glaciers also carved scenic gorges, waterfalls, and numerous lakes. The region is a year-round resort area. Lumbering, once a major occupation in the Adirondacks, declined after a forest preserve was established in 1892. Important mineral products of the mountains include iron ore, titanium, vanadium, and talc.

Ad·mi·ral·ty Inlet (ăd'mər-əl-tē), arm of the Pacific Ocean, NW Wash., entry and northernmost part of Puget Sound, between Whidbey Island and the mainland.

Admiralty Island, off SE Alaska, in the Alexander Archipelago SW of Juneau. The large island (1,664 sq mi/4,309.8 sq km) is separated from the mainland by Stephens Passage. It is mountainous and heavily forested. Wildlife abounds.

Admiralty Islands, group of 40 volcanic islands (1969 pop. 22,035), c.800 sq mi (2,070 sq km), SW Pacific, in the Bismarck Archipelago and part of Papua New Guinea. Copra, pearls, and marine shells are the principal products. Discovered by the Dutch navigator Willem Schouten in 1616, the group became part of German New Guinea in 1884 and an Australian League of Nations mandate in 1920.

Admiralty Range, mountain range, stretching along the N coast of Victoria Land, Antarctica, NW of Ross Sea. Its highest peak is Mt. Sabine (9,859 ft/3,007 m). The Admiralty Range was discovered by Sir James Ross on his 1841 expedition to Antarctica.

A·do (ä'dō), city (1971 est. pop. 190,000), SW Nigeria. Located in a region where rice and yams are grown, the town has rice mills and also manufactures textiles, bricks, tile, and pottery. Ado was the capital of the Yoruba Ekiti state that was probably founded in the 15th cent. It alternated between independence and subjection to Benin until the British gained control in 1894.

A·dour (ä-do͝or'), river, 210 mi (337.9 km) long, rising in the Pyrenees of Gascony, SW France. It flows north and then west in a wide arc and enters the Bay of Biscay near Bayonne.

A·dra (ä'*th*rä), town (1970 pop. 16,283), Almería prov., S Spain, in Andalusia, on the Mediterranean Sea. Adra, a port, is the center of a fertile agricultural region. At the foot of a hill below the present town stood Abdera, founded by Phoenician traders and later a Roman colony. Adra was the last stronghold of the Moors in Spain.

A·dra·no (ə-drä'nō), town (1976 pop. 33,592), E Sicily, Italy, at the foot of Mt. Etna, NW of Catania. It is the commercial center for a region where olives and citrus fruit are grown. Adrano was founded c.400 by Dionysius the Elder near a temple of the god Hadranus. Fierce fighting took place in Adrano during World War II. Of note are the ruins of the town's ancient walls and an imposing 11th cent. Norman castle.

A·dri·an (ā'drē-ən), city (1978 est. pop. 20,600), seat of Lenawee co., SE Mich., on the Raisin River; inc. 1836. It is a trading center for a fertile farm region; its many products include automobile and aircraft parts, metal ware, chemicals, and paper goods.

A·dri·a·no·ple (ā'drē-ə-nō'pəl): *see* Edirne, Turkey.

A·dri·a·tic Sea (ā'drē-ă'tĭk), arm of the Mediterranean Sea, between Italy and the Balkan Peninsula, extending c.500 mi (805 km) from the Gulf of Venice, at its head, SE to the Strait of Otranto, which leads to the Ionian Sea. It is from 58 to 140 mi (93.3–225.3 km) wide, with a maximum depth of c.4,100 ft (1,250 m). The Italian coast (west and north) is low; Yugoslavia and Albania border the irregular eastern shore. The Yugoslavian coast, which is rugged and has many offshore islands and sheltered bays, is a popular tourist resort. Fishing is an important activity in the Adriatic Sea; lobsters, sardines, and tuna are the chief catch.

A·du·wa or **A·do·wa** (both: ä'də-wə), town (1970 est. pop. 16,000), Tigre prov., N Ethiopia. Aduwa was the most important commercial center of Tigre in the 19th cent., but declined in the 1870s as a result of fighting between Ethiopia and Egypt. In 1896 Aduwa was the site of the battle in which Menelik II decisively defeated Italian invaders and forced them out of Ethiopia.

A·dy·ge Autonomous Oblast (ä'də-gā'), administrative division (1970 pop. 386,000), c.2,935 sq mi (7,600 sq km), Krasnodar Kray, SE European USSR, in the N foothills of the Greater Caucasus; capital Maikop.

Ad·zhar Autonomous Soviet Socialist Republic (ə-jär') or **Ad·zhar·i·stan** (ə-jär'ə-stăn'), autonomous region (1970 pop. 310,000), c.1,160 sq mi (3,005 sq km), SE European USSR, on the Black Sea, bordering Turkey on the S; capital Batumi. Colonized by Greek merchants in the 5th and 4th cent. B.C., the region later came under Roman rule and after the 9th cent. A.D. was part of Georgia. The Turks conquered Adzharistan in the late 17th and early 18th cent. and introduced Islam. Acquired by Russia in 1878, the region became an autonomous republic in 1921.

Ae·ga·di·an Isles (ē-gā'dē-ən): *see* Egadi Islands.

Ae·ge·an Sea (ĭ-jē'ən), arm of the Mediterranean Sea, c.400 mi (645 km) long and 200 mi (320 km) wide, off SE Europe between Greece and Turkey. Crete and Rhodes mark its southern limit. Irregular in shape, it is dotted with islands, most of which belong to Greece. The Aegean Sea's greatest depths (more than 6,600 ft/2,013 m) are found off north Crete. The Dardanelles strait connects the Aegean Sea with the Sea of Marmara and the Black Sea. Sardines and sponges taken from the Aegean are economically important; natural gas has been found off northeast Greece.

Ae·gi·na (ĭ-jī'nə) or **Aí·yi·na** (ā'yē-nä), island (1971 pop. 5,704), 32 sq mi (82.9 sq km), off SE Greece, in the Saronic Gulf near Athens. Sponge fishing and farming (figs, almonds, grapes, olives, and peanuts) are the most important occupations. The island, inhabited from late Neolithic times, was named for the mythological figure Aegina. Its culture was influenced by Minoan Crete. Conquered by Dorian Greeks, it grew rapidly as a commercial state and struck the first Greek coins. In 431 B.C. the Athenians, against whom Aegina sided in the Peloponnesian War, expelled the population of the island, and Aegina fell into insignificance. In the 12th cent. it served as a haven for pirates, and the Venetians, in suppressing the outlaws, conquered the island. Albanians settled here in the 16th cent. During the Greek War of Independence the town of Aegina was (1828–29) the capital of Greece.

Ae·gos·pot·a·mos (ē'gəs-pŏt'ə-məs), river of ancient Thrace flowing into the Hellespont. At its mouth in 405 B.C. the culminating battle of the Peloponnesian War occurred, in which Lysander and his Spartan fleet completely destroyed the Athenian fleet.

Ae·o·li·an Islands (ē-ō'lē-ən): *see* Lipari Islands.

Ae·o·lis (ē'ə-lĭs) or **Ae·o·li·a** (ē-ō'lē-ə), ancient region of the W coast of Asia Minor (in present-day Turkey). Aeolis was not a geographic term but a collective term for the cities founded there by the Aeolians, a branch of the Hellenic peoples.

Aet·na (ĕt'nə), volcano: *see* Etna.

Ae·tol·ia (ē-tōl'yə), region of ancient Greece, N of the Gulf of Corinth and the Gulf of Calydon and E of the Achelous River (separating it from Acarnania). The Aetolians, though they had coastal cities, were primarily an inland farming and pastoral people. Aetolia was of little significance in Greek history until the rise of the Aetolian League. After the downfall of that confederation, Aetolia was absorbed by the Romans into Achaea.

Af·ghan·i·stan (ăf-găn'ə-stăn'), republic (1975 est. pop. 19,280,000), 249,999 sq mi (647,497 sq km), S central Asia. The capital is Kabul. Afghanistan is bordered by Iran on the west, by Pakistan on the east and south, and by the USSR on the north. A narrow strip, the Vakhan, extends in the northeast to touch Kashmir and the Sinkiang Uigur Autonomous Region of China. The great mass of the country is mountainous, with ranges fanning out from the towering Hindu Kush (reaching a height of more than 24,000 ft/7,320 m) across the center of the country. In the south, and particularly in the southwest, are great stretches of desert. To the north, between the central mountain chains and the Amu Darya River, which marks part of the boundary with the USSR, are the highlands of Badakhshan (with the finest lapis lazuli in the world), Afghan Turkistan, the Amu Darya plain, and the rich valley of Herat on the Hari Rud River in the northwest corner of the country. The regions thus vary widely, although most of the land is dry.

Economy. Agriculture is the main occupation, but less than 10% of the land is cultivated. There are, however, within the mountain ranges and on their edges many fertile valleys and plains, with fields of wheat, corn, barley, and rice and orchards yielding fine fruits, such as the famous peaches and grapes of Kandahar. The fat-tailed sheep, a staple of Afghan life, supplies skins and wool for clothing and meat and fat for food. Fine horses are the pride of many tribesmen. There are deposits of iron ore, coal, copper, and sulfur; oil and natural-gas fields are found in the north. Industry is still only in the beginning stages. Cotton and other fabrics, cement, and processed agricultural goods are the main products. Fruits and lambskins (karakul) are the main exports.

History. The location of Afghanistan astride the land route to India has enticed conquerors throughout history. But its high mountains, while hindering unity, have helped the hill tribes preserve their independence. The archaeological record of Afghanistan is not clear. Certainly cultures had flourished in the north and east before Darius I (c.500 B.C.) by conquest annexed these areas to the Persian

Empire. Later, Alexander the Great conquered (329-327 B.C.) them on his way to India. Buddhism was introduced from the east by the Yüechi, who founded the Kushan dynasty (early 2nd cent. B.C.), which declined 400 years later.

The Arab conquest of Afghanistan began in the 7th cent. Several short-lived Moslem dynasties were founded. Mahmud of Ghazni, who conquered the lands from Khurasan in Iran to the Punjab in India early in the 11th cent., was the greatest of Afghanistan's rulers. Genghis Khan (c.1220) and Tamerlane (late 14th cent.) were subsequent conquerors of renown. Babur, a descendant of Tamerlane, used Kabul as the base for his conquest of India and the establishment of the Mogul empire in the 16th cent. In the 18th cent. the Persian Nadir Shah extended his rule north of the Hindu Kush. After his death (1747) his lieutenant, Ahmad Shah, an Afghan tribal leader, established the Durani dynasty and a united state covering most of present-day Afghanistan.

The reign of the Durani line ended in 1818, and no predominant ruler emerged until Dost Mohammed became emir in 1826. During his rule Britain and Russia vied for influence in central Asia. Aiming to protect the northern approaches to India, the British tried to replace Dost Mohammed with a former emir whom they controlled. This policy caused the first Afghan War (1838–42). Dost Mohammed was at first deposed but, after an Afghan revolt in Kabul, was restored. In 1857 Dost Mohammed signed an alliance with the British. He died in 1863 and was succeeded, after familial fighting, by his third son, Shere Ali. As the Russians acquired territory bordering on the Amu Darya, Shere Ali and the British quarreled, and the second Afghan War began (1878). Subsequently, the Khyber Pass and other areas were ceded to the British, and after a British envoy was murdered, the British occupied Kabul. Border agreements were reached with Russia (1885 and 1895), British India (the Durand Agreement, 1893), and Persia (1905), although the line with what is now Pakistan remained disputed. The Anglo-Russian agreement of 1907 guaranteed the independence of Afghanistan under British influence in foreign affairs. Despite British pressure, however, Afghanistan remained neutral in World War I. The emir, Amanullah, attempting to free himself of British influence, invaded India (1919). This third Afghan War was ended by the Treaty of Rawalpindi, which gave Afghanistan full control over its foreign relations. The attempts of Amanullah at westernization—including reducing the power of the country's religious leaders and increasing the freedom of its women—provoked opposition that led to his deposition in 1929. A tribal leader, Bacha-i Saqao, held Kabul for a few months until defeated by Amanullah's cousin, Mohammed Nadir Khan, who became King Nadir Shah. He pursued cautious modernization efforts until he was assassinated in 1933. His son Mohammed Zahir Shah continued his efforts.

Afghanistan was neutral in World War II. When British India was partitioned (1947), Afghanistan wanted the Pathans of the North-West Frontier Province to be able to choose whether to join Afghanistan, join Pakistan, or be independent; the Pathans were only offered the choice of joining Pakistan or joining India—they chose the former. Since then relations between Afghanistan and Pakistan have been embittered. In the early 1970s the country was beset by serious economic problems, particularly a severe long-term drought in the center and north. Maintaining that King Mohammed Zahir Shah had mishandled the economic crisis and in addition was stifling political reform, a group of young military officers deposed (July, 1973) the king and proclaimed a republic. Lt. Gen. Mohammed Daud Khan, the former king's cousin and brother-in-law and a former prime minister (1953–63), became president and prime minister. A new constitution was adopted on Feb. 14, 1977, and at the same time Daud was named to a six-year term as president by the Grand Constituent Assembly. In April, 1978, he and a number of other government officials were assassinated and his regime overthrown by Noor Mohammed Taraki, who set up a strongly pro-Soviet government. Since that time the country has been torn by internal strife. Early in 1979, the U.S. ambassador to Afghanistan, Adolph Dubs, was shot to death after being kidnapped by right-wing terrorists.

Af·ri·ca (ăf'rĭ-kə), second-largest continent, c.11,677,240 sq mi (30,244,050 sq km) including adjacent islands. Broad to the north (c.4,600 mi/7,400 km wide), Africa straddles the equator and stretches c.5,000 mi (8,045 km) from Cape Blanc (Tunisia) in the north to Cape Agulhas (South Africa) in the south. It is connected with Asia by the Sinai Peninsula (which is crossed by the Suez Canal) and is bounded on the north by the Mediterranean Sea, on the west and south by the Atlantic Ocean, and on the east and south by the Indian Ocean. The largest offshore island is Madagascar. Most of Africa is a stable, ancient plateau that has been warped into a series of basins, low in the north and west and higher (rising to more than 6,000 ft/1,830 m) in the south and east. The plateau is composed mainly of metamorphic rock that has been overlaid in places by sedimentary rock. The escarpment of the plateau is in close proximity to the coast, thus leaving the continent with a generally narrow coastal plain. The escarpment also forms a barrier of falls and rapids in the lower course of rivers that impedes their use as transportation routes into the interior. North Africa, a region composed mainly of folded sedimentary rock, is geologically more closely related to Europe than to the rest of Africa; the Atlas Mts., which occupy most of the region, are a part of the Alpine mountain system of southern Europe. The entire African continent is surrounded by a narrow continental shelf. The lowest point on the continent is 436 ft (133 m) below sea level in the Qattarah Depression in northwest Egypt; the highest point is Mt. Kibo (19,340 ft/5,898.7 m), a peak of Kilimanjaro in northeast Tanzania.

Geologists now have evidence that Africa formed the center of a large ancestral supercontinent known as Pangaea. It began to break apart in the Jurassic period to form Gondwanaland, from which Africa, the other southern continents, and India were formed. Similar large-scale earth movements are also believed responsible for the formation of the Great Rift Valley of East Africa, the continent's most spectacular land feature. The lava flows of the recent and ancient epochs in the Ethiopian Highlands, and volcanoes farther south, are associated with the rift. A less spectacular rift, the Cameroon Rift, is in the west.

Africa's climatic zones are largely controlled by the continent's location astride the equator and its almost symmetrical extensions into the Northern and Southern hemispheres. Thus, except where altitude exerts a moderating influence on temperature or precipitation (permanently snow-capped peaks are found near the equator), Africa may be divided into six general climatic regions. Areas near the equator and on the windward shores of southeast Madagascar have a tropical rain forest climate, with heavy rain and high temperatures throughout the year. North and south of the rain forest are belts of tropical savanna climate, with high temperatures all year and a seasonal distribution of rain during the summer season. The savanna grades poleward in both hemispheres into a region of semiarid steppe (with limited summer rain) and then into true desert conditions in the extensive Sahara (north) and the smaller Kalahari (south). Belts of semiarid steppe with limited winter rain occur on the poleward sides of the desert regions. At the northern and southern extremities of the continent are narrow belts of Mediterranean type climate with subtropical temperatures and a concentration of rainfall mostly in the autumn and winter months.

African peoples, who account for about 10% of the world's population, are divided into more than 50 different political units and are further fragmented into a larger (and disputed) number of linguistic and cultural groups. The Sahara forms a great ethnic divide. North of it Caucasoids, mostly Arabs along the coast and Berbers, Tuareg, and Tibbu in the interior regions, predominate. The southern (or sub-Saharan) sections of the continent are occupied by a diverse group of predominantly Negroid peoples, mostly Bantu-speaking. Numerous other groups, of mixed and often disputed origin, occupy transitional areas south of the Sahara. In the south are persons of Dutch and British descent and in the northwest are persons of French, Italian, and Spanish descent. Indians are an important minority in many coastal towns of southern and eastern Africa. As a whole, the continent is sparsely populated; about three quarters of its population is rural.

Economy. Africa produces three quarters of the world's cocoa beans and about one third of its groundnuts, but only small percentages of the world's corn, wheat, meat, and eggs. Rare and precious minerals (including most of the world's diamonds) are abundant in the continent's ancient crystalline rocks, which are found mostly to the south and east of a line from the Gulf of Guinea to the Sinai Peninsula; extensive oil, gas, and phosphate deposits occur in sedimentary rocks to the north and west of this general line. Despite Africa's enormous potential for hydroelectric power production, only a small percentage of it has been developed. Africa's fairly regular coastline affords few natural harbors, and the shallowness of coastal waters makes it difficult for large ships to approach the shore; deepwater ports, protected by breakwaters, have been built offshore to facilitate commerce and trade. Major fishing grounds are found over the wider sections of the continental shelf as off northwest, southwest, and southern Africa and northwest Madagascar.

History. Africa's history is long, complex, and only partly known. Man's oldest ancestor, discovered (1959) by Louis S. B. Leaky, the British anthropologist, lived in East Africa's Olduvai Gorge at least 1,750,000 years ago; agriculture, brought from southwest Asia, appears to date from the 6th or 5th millennium B.C. Africa's first civilization began in Egypt in 3400 B.C. Phoenicians established Carthage in the 9th cent. B.C. and probably explored the northwestern coast as far as the Canary Islands by the 1st cent. B.C. Romans conquered Carthage in 146 B.C., controlled northern Africa until the 4th cent. A.D., and, in the 1st or 2nd cent. A.D., were probably the first Europeans to cross the Sahara into tropical Africa. Arabs began their conquest in the 7th cent. and, except in Ethiopia, extended Arabic and the religion of Islam across north Africa and south across the Sahara into the great medieval kingdoms of the western Sudan. The earliest of these kingdoms, which drew their wealth and power from the control of a lucrative trans-Saharan trade in gold, salt, and slaves, was ancient Ghana, already thriving when first recorded by Arabs in the 8th cent.

There are few written accounts of the interior of the continent before 1500, but it appears from available evidence that the original San, Pygmy, and Azanian inhabitants were displaced beginning in the 1st cent. A.D. by the Bantu, a group of black African peoples speaking related languages. The Bantu spread over most of the continent south of the equator, probably from an original homeland in the southern portion of modern Zaire. Prior to 1500 pastoralists moved south until they encountered the various Bantu groups. The Portuguese began to explore the coasts of Africa in the 15th cent. in an attempt to establish a safe route to India and to tap the lucrative gold trade of the Sudan and the east coast trade in gold, slaves, and ivory conducted for centuries by Arabs, Persians, and Indians. In 1488 Bartolomeu Dias rounded the Cape of Good Hope; in 1498 Vasco da Gama reached the east coast and, the following year, India. In the centuries that followed, coastal trading stations were established by European maritime powers; under them the slave trade rapidly expanded. At the same time Ottoman Turks extended their control over northern Africa and the shores of the Red Sea, and the Omani Arabs established suzerainty over the east coast as far south as Cape Delgado. Explorations in the 18th and 19th cent. by Mungo Park, David Livingstone, Henry Stanley, and others uncovered the great natural wealth of the continent. Between 1880 and 1912 all of Africa except Liberia and Ethiopia passed under the control or protection of European powers, the boundaries of the new colonies and protectorates often bearing no relationship to the realities of geography or to the political and social organization of the indigenous population.

In the northwest and west, France acquired French Equatorial Africa, the French Cameroons, Algeria, Morocco, and Tunisia. Other French territories were French Somaliland, French Togoland, Madagascar, and Réunion. The main group of British possessions was in the east and southeast, including Anglo-Egyptian Sudan, British Somaliland, Uganda, Kenya, Tanganyika (after World War I), Zanzibar, Nyasaland, Northern and Southern Rhodesia, Bechuanaland, Basutoland, Swaziland, and South Africa. Gambia, Sierra Leone, the Gold Coast, and Nigeria were British possessions on the west coast. Portugal's African empire was made up of Portuguese Guinea, Angola, and Mozambique, in addition to various enclaves and islands on the west coast. Belgium held the Belgian Congo and, after World War I, Ruanda-Urundi. The Spanish possessions in Africa were the smallest, being composed of Spanish Guinea, Spanish Sahara, Ifni, and the protectorate of Spanish Morocco. The extensive German holdings were lost after World War I and redistributed among the Allies. Italy's empire included Libya, Eritrea, Italian Somaliland, and, briefly after 1936, Ethiopia.

Beginning in 1950, in the face of rising nationalism, the former colonies and protectorates were granted independence by all the European powers except Portugal, which finally agreed to grant its territories independence in 1974. The subsequent history of the continent has been extremely turbulent, complicated by prolonged drought conditions, agitation for native self-rule, and border disputes between many of the newly independent countries. (For detailed information, see individual country entries.)

Af·ton (ăf'tən). **1.** Uninc. city (1970 pop. 24,898), St. Louis co., E Mo., a suburb of St. Louis. **2.** Town (1970 pop. 1,290; alt. c.6,100 ft/1,860.5 m), Lincoln co., W Wyo., near the Idaho line NNW of Jackson; settled 1895 by Mormons, inc. 1902. It is a trade center for the Star Valley, with flour and lumber mills and plants processing dairy goods. There is a Mormon tabernacle here, and nearby is a Univ. of Wyoming agricultural experiment station.

Afton, river, 9 mi (14.5 km) long, Strathclyde region, SW Scotland, flowing N to the Nith River. It is the "sweet Afton" of Robert Burns's poem.

Af·yon·ka·ra·his·ar (äf-yōn'kä'rä-hĭs-är'), city (1975 pop. 60,117), capital of Afyonkarahisar prov., W central Turkey, at an elevation of c.3,500 ft (1,070 m). It is the commercial center of a region where opium poppies and grains are grown. Carpets are manufactured in the city, which is a major rail junction.

A·ga·dès (ä'gə-dĕs'), town (1963 est. pop. 7,100), W central Niger, in the Aïr Mts. A traditional, picturesque town, Agadès is a trade center. Leather and silver handicrafts are made. Tin, tungsten, uranium, and salt are mined nearby. Founded by the 11th cent., Agadès developed mainly because of its location on trans-Saharan caravan routes linking Egypt and Libya with the Lake Chad area.

A·ga·dir (ä'gə-dîr', ăg'ə-), city (1970 est. pop. 34,000), SW Morocco, on the Atlantic Ocean. Agadir has metal-processing industries and exports fruit and vegetables. In 1960 it was almost completely destroyed by an earthquake.

A·ga·na (ə-gä'nyə), city (1970 pop. 2,119), capital of the island of Guam, W Pacific, in the Marianas Islands. Most of the city's economic activities are related to the provision of goods and services to the large U.S. military bases on the island. Completely destroyed in World War II, Agana was subsequently rebuilt.

A·gar·ta·la (ə-gŭr'tə-lə), city (1971 pop. 59,682), capital of Tripura state, NE India, near the Bangladesh border. It is a market town for rice, tea, jute, and oilseed.

Ag·as·siz, Lake (ăg'ə-sē), glacial lake of the Pleistocene epoch, c.700 mi (1,125 km) long, 250 mi (402.3 km) wide, formed by the melting of the continental ice sheet some 10,000 years ago. It covered much of present-day northwest Minn., northeast N.Dak., south Man., and southwest Ont., Canada

Ag·ate Fos·sil Beds National Monument (ăg'ĭt fŏs'əl): *see* National Parks and Monuments Table.

Ag·at·tu (ăg'ə-to͞o'), mountainous island, 85 sq mi (220.2 sq km), part of the Near Island group, SW Alaska, in the W section of the Aleutian Islands SE of Attu Island. Japanese forces occupied Agattu briefly from June to Oct., 1942.

Ag·a·wam (ăg'ə-wäm'), town (1970 pop. 21,717), Hampden co., SW Mass., on the Connecticut River; settled 1636, inc. 1855. Leather goods, machinery, and electronic equipment are produced.

Ag·bo·ville (äg'bō-vēl'), town (1970 est. pop. 18,068), S Ivory Coast. Situated in a forest zone, the town is the market center for a region producing plantains, yams, coffee, cassava, manioc, rice, and timber.

A·gen (ä-zhäN'), town (1975 pop. 34,039), capital of Lot-et-Garonne dept., SW France, on the Garonne River, in Guienne. It is an agricultural market place in the center of a fruit-growing region and an industrial center where food products, clothing, agricultural machinery, bicycles, tiles, drugs, furniture, and musical instruments are manufactured. Originally a Gallic settlement, Agen was a crossroads in Roman times. It became the capital of the county of Agenois under the Carolingians. An episcopal see from the 10th cent., it passed (1154) to England with the rest of Aquitaine. It was reconquered in the Hundred Years' War (1337–1453) and incorporated into the province of Guienne. Among the historic structures are chapels from the 13th and 14th cent.

A·ge·o (ä'gā-ō), city (1975 pop. 150,805), Saitama prefecture, central Honshu, Japan. It is an agricultural and communications center. Raw silk and sake are produced in the city.

Agh·rim (ôg'rĭm, ôKH'-): *see* Aughrim, Republic of Ireland.

A·gin·court (ä'zhäN-ko͞or', ăj'ĭn-kôrt'), village (1968 pop. 276), Pas-de-Calais dept., N France. Here, on Oct. 25, 1415, Henry V of England defeated a much larger French army in the Hundred Years' War (1337–1453). His success, due mainly to the superiority of the masses of English longbow men over the heavily armored French knights, demonstrated the obsolescence of the methods of warfare of the age of chivalry.

Ag·no (äg'nō), river, 128 mi (206 km) long, NW Luzon, Philippines. Originating in the south Mountain prov., it flows south through Pangasinan prov., then turns north to empty into the Lingayen Gulf. The Agno was the site of severe fighting in 1942 and Jan., 1945.

A·gra (ä'grə, ăg'rə), city (1971 pop. 594,858), Uttar Pradesh state, N central India, on the Jumna River. It is noted for its shoes, glass products, handicrafts, carpets, and historic architecture. The present city was established (1566) by Akbar and was for many years a Mogul capital. In the reign of Shah Jahan (1628–58), the magnificent Taj Mahal was built. Agra's importance diminished after the Mogul court moved to Delhi in 1658. During the decline of the Mogul empire, the city frequently changed rulers until 1803, when it was annexed by the British.

A·gri·gen·to (ä'grē-jĕn'tō), city (1975 est. pop. 49,979), capital of Agrigento prov., S Sicily, Italy, on a hill above the Mediterranean Sea. It is an agricultural market and a tourist center. Sulfur, salt, and gypsum are produced. Founded c.580 B.C. by Greek colonists, the city became one of the most prosperous in the Greek world. It was destroyed c.406 B.C. by Carthage but recovered. It fell definitively to Rome in 210 B.C. during the Second Punic War. After the fall of Rome, Agrigento passed to the Byzantines and then to the Arabs (9th cent.) and to the Normans (11th cent.). Of note in the city are the remains of several Doric temples (6th-5th cent. B.C.), Roman ruins, Christian catacombs, and archaeological and art museums.

A·gua (ä'wä, ä'gwä), inactive volcano, 12,310 ft (3,754.6 m) high, S Guatemala. In 1541, after several days of unceasing rain and earthquakes, a wall of water swept down from its slopes, completely destroying Ciudad Vieja. Over 1,000 inhabitants were drowned.

A·gua·dil·la (ä'gwä-dē'yä, ä'wä-), town (1970 pop. 21,031), NW Puerto Rico, a port on Mona Passage. It is the trade center for an agricultural region. Columbus reputedly landed at the site of Aguadilla in 1493.

A·gua Fri·a (ä'gwə frē'ə, ä'wə), river, 120 mi (193.1 km) long, W Ariz. It rises in Yavapai co., east of Prescott, and flows intermittently south to meet the Gila River in Maricopa co. west of Phoenix.

A·guas·ca·lien·tes (ä'gwä-skä-lyĕn'tās, ä'wäs-), city (1974 est. pop. 213,400), capital of Aguascalientes state, central Mexico. The city is a health resort, noted for its mineral waters. Its industries include smelting and the manufacture of textiles. Aguascalientes is built over an ancient, intricate system of tunnels constructed by early, still unidentified, inhabitants. Founded in 1575, the city was long a Spanish outpost against hostile Indians.

A·gul·has, Cape (ə-gŭl'əs), W Cape Province, Republic of South Africa. It is the southernmost point of Africa. Its name refers to the saw-edged reefs and sunken rocks that run out to sea and make navigation hazardous. The meridian of Cape Agulhas, long. 20° E, is used to divide the Atlantic and Indian oceans.

A·gung (ä'gŏŏng), volcanic mountain, 10,308 ft (3,144 m) high, NE Bali, Indonesia. Also known as the Peak of Bali, it last erupted in 1963.

A·gu·san (ä-gōō'sän), river, c.240 mi (385 km) long, rising in the mountains of SE Mindanao, Philippines, and flowing N past Butuan to Butuan Bay. It is navigable for small craft c.160 mi (260 km) upstream. The Agusan valley is very fertile and is one of the Philippines' chief rice-growing regions.

Ah·mad·na·gar or **Ah·med·na·gar** (both: ä'məd-nŭg'ər), city (1971 pop. 117,275), Maharashtra state, W central India. It has textile manufacturing and some light industry. Founded in 1490, it was the capital of a kingdom that lasted until 1600.

Ah·me·da·bad or **Ah·ma·da·bad** (both: ä'mə-də-bäd'), city (1971 pop. 1,588,378), capital of Gujarat state, NW India, NW of Baroda. An industrial center noted for its cotton mills, Ahmedabad is also a transportation hub and a commercial center. Founded in 1412, it fell to Akbar in 1573 and enjoyed great prosperity under the Mogul empire. The British opened a trading post here in 1619; by the early 19th cent. they controlled the city. The cultural center of Gujarat, Ahmedabad has many outstanding mosques, temples, and tombs.

A·ho·me (ä-ō'mā), city (1970 pop. 165,612), Sinaloa state, W Mexico, on the Pacific Ocean. Sugar cane, grains, and cotton are grown in the region. The city also has an important fishing industry, based mainly on shrimp.

A·hua·cha·pán (ä'wä-chä-pän'), city (1974 est. pop. 17,300), W El Salvador, near the Guatemalan border. It is the center of an agricultural region producing coffee, sugar, grain, and fruit. There are thermal springs nearby.

Ah·vaz (ä-väz') or **Ah·waz** (ä-wäz'), city (1972 est. pop. 286,000), SW Iran, on the Karun River. It is an oil center, a transportation hub, and an industrial city that has petrochemical, textile, and food-processing industries. An ancient city, Ahvaz was rebuilt (3rd cent. A.D.) by Ardashir I. It was an important Arab trading center in the 12th and 13th cent. but later declined. The discovery of oil nearby in the early 20th cent. restored the city to its former importance.

Ah·ven·an·maa Islands (ä'vĕn-än-mä') or **Å·land Islands** (ä'lənd, ô'lənd), archipelago (1970 pop. 21,010), 581 sq mi (1,504.8 sq km), in the Baltic Sea between Sweden and Finland, at the entrance of the Gulf of Bothnia. The archipelago consists of about 7,000 islands, but fewer than 100 are inhabited. Shipping, fishing, forestry, farming, and tourism are the chief occupations. The islands, colonized by Swedes, are of strategic importance. With Finland, they were ceded by Sweden to Russia in 1809. At the end of World War I, the islanders sought to join Sweden. The League of Nations in 1921 recognized Finland's sovereignty, but guaranteed the autonomous status of the islands. After the Finnish-Russian War (1939–40) Finland and Russia signed a demilitarization agreement that was renewed after World War II. Under pressure from Russia, Finland's parliament renounced the League guarantee of autonomy in 1951 but accorded the islanders additional rights of self-government.

A·i·e·a (ä'ē-ā'ä), city (1970 pop. 12,560), Honolulu co., Oahu, Hawaii, a residential suburb of Honolulu, on the E shore of Pearl Harbor.

Aigues-Mortes (ĕg-môrt'), town (1968 pop. 3,776), Gard dept., S France, in Languedoc, in the Rhone River delta near the Mediterranean. Founded by Louis IX as a port for his two crusades (1248 and 1270), its canal gradually silted up, and it has long since lost its commercial prosperity. A well-preserved medieval town, its splendid 13th cent. ramparts and fortified tower remain intact.

Ai·jal (ī'jəl), city (1971 pop. 31,740), capital of the union territory of Mizoram, NE India. Situated on a ridge in the Lushai Hills that is 3,500 ft (1,067.5 m) high, Aijal is a trade center.

Ai·ken (ā'kən), county (1970 pop. 91,023), 1,097 sq mi (2,841.2 sq km), W S.C., at the edge of the Piedmont, in the Sand Hills belt, bounded on the W by the Savannah River, on the NE by the North Fork of the Edisto River, and drained by the South Fork of the Edisto; formed 1871; co. seat Aiken. It has agriculture (cotton, corn, fruit, truck, and dairy products), kaolin deposits, textile manufacturing, and resorts. Part of an Atomic Energy Commission installation is in the south.

Aiken, city (1978 est. pop. 14,400), seat of Aiken co., W S.C.; inc. 1835. It is a fashionable resort and polo center located in the midst of sand hills and pine forests. Aiken is also an industrial city, with textile and lumber mills and a large fiberglass plant. There are kaolin mines in the vicinity.

Ain (ăɴ), department (1968 pop. 339,262), 2,222 sq mi (5,755 sq km), E central France, in Burgundy, bordering on Switzerland; capital Bourg-en-Bresse.

Ains·worth (ānz'wərth), city (1970 pop. 2,073), seat of Brown co., N Nebr., NE of Grand Island; settled 1877, inc. 1883. It is a trade center in an irrigated region that processes livestock, poultry, grain, and dairy products.

Ain·tab (īn-täb'): *see* Gaziantep, Turkey.

Ai·oi (ī-oi'), city (1976 est. pop. 42,008), Hyogo prefecture, W Honshu, Japan, on the Inland Sea and Aioi Bay. It is a major port with a good natural harbor and a flourishing shipbuilding industry.

Air·drie (âr'drē), burgh (1974 est. pop. 38,833), Strathclyde region, S central Scotland. Chemicals and electrical and electronic equipment are produced, and there are facilities for electronic research. Airdrie's free library was the first established in Scotland.

Aisne (ān), department (1975 pop. 533,862), 2,849 sq mi (7,378.9 sq km), NE France, in Île-de-France, Picardy, and Champagne, touching the Belgian border; capital Laon.

Aisne, river, 165 mi (265.5 km) long, N France, rising in Meuse dept. and flowing NW and W to join the Oise River near Compiègne. Four battles of World War I were fought along its banks. In Sept.–Oct., 1918, the Germans were defeated by French and American troops.

Ait·kin (ā'kən), county (1970 pop. 11,403), 1,828 sq mi (4,734.5 sq km), E central Minn., in an agricultural and resort area drained by the Mississippi and watered in the SW by part of Mille Lacs Lake; formed 1857; co. seat Aitkin. There are deposits of marl and peat here. Dairy products, potatoes, and meat and poultry dressing are processed, and lumbering is done.

Aitkin, village (1970 pop. 1,553), seat of Aitkin co., central Minn., on the Mississippi N of Mille Lacs Lake and WSW of Duluth; settled 1870. In a farm and resort region, it is a shipping center for dairy products, turkeys, and fruit. A wildlife refuge is nearby.

Aix-en-Pro·vence (ěks′äɴ-prô-väɴs′), city (1975 pop. 110,659), Bouches-du-Rhône dept., in Provence, SE France. It is a commercial center in an area producing olives, grapes, and almonds. Its manufactures include food products, wine-making equipment, and electrical apparatus. Founded (123 B.C.) by the Romans near the site of mineral springs, it has long been a popular spa. A music center since the 11th cent. and a focus of Provençal literature, Aix is a favorite sojourn for painters. A music festival is held each summer.

Aix-la-Cha·pelle (ěks′lä-shä-pěl′): *see* Aachen, West Germany.

Aix-les-Bains (ěks-lā-băɴ′), town (1975 pop. 22,210), Savoie dept., SE France, situated on Lake Bourget at the foot of the Alps. It is a popular resort and spa. The town's alum and sulfur springs have been frequented since Roman times.

Aí·yi·na (ā′yē-nä), island: *see* Aegina.

Ai·zu-Wa·ka·mat·su (ī′zo͞o-wä′kə-mät′so͞o), city (1976 est. pop. 109,859), Fukushima prefecture, N Honshu, Japan. Its major products are wooden items, sake, rice, and persimmons.

A·jac·cio (ä-yät′chō), town (1968 pop. 42,300), capital of Corsica, France, on the Gulf of Ajaccio, an inlet of the Mediterranean. A fortified seaport, it is an important market town, an active industrial center, and a year-round tourist attraction. Its present site was established by Genoese colonists in 1492. Ajaccio was the birthplace of Napoleon I; the house where he was born is preserved. In World War II, Ajaccio was occupied by the Italians until the people successfully revolted (Sept., 1943) with the aid of Free French troops.

A·jan·ta (ə-jŭn′tə), village, Maharashtra state, W central India, in the Ajanta Hills. The famous Ajanta caves, discovered in 1819, contain remarkable examples of Buddhist art. The caves, carved out of the side of a steep ravine, consist of chapels and monasteries dating from c.200 B.C.–A.D. 650 with magnificent frescoes and sculpture depicting scenes from the life of Buddha.

A·jax (ā′jăks), mountain, 10,900 ft (3,324.5 m) high, S section of the Bitterroot Range, on the Mont.–Idaho border.

Aj·man (äj-män′), sheikdom (1968 pop. 4,245), c.100 sq mi (260 sq km), part of the federation of United Arab Emirates, E Arabia, primarily on the Persian Gulf. Oil production in Ajman began in 1964. A former British protectorate, it joined the United Arab Emirates in 1971.

Aj·mer (ăj-mîr′, əj-), city (1971 pop. 262,480), NW India. Founded in the 12th cent., the city is a trade center and has cotton mills and railroad shops. Marble is quarried nearby.

A·jo (ä′hō), uninc. town (1970 pop. 5,881), Pima co., SW Ariz., SW of Gila; settled 1854. One of the oldest mining towns in the state, it still derives its income primarily from copper mining. Ajo is also a health resort. Nearby is Organ Pipe Cactus National Monument, containing unique desert growth.

A·jod·hya (ə-jōd′yə), village, Uttar Pradesh state, N India, on the Gogra River. It is a joint municipality with Faizabad. Ajodhya was the capital of the kingdom of Kosala (7th cent. B.C.). Long associated with Hindu legend, the town is a center of pilgrimage and is one of the seven sites sacred to Hindus.

A·jus·co (ä-ho͞o′skō), extinct volcanic mountain, 12,887 ft (3,930.5 m) high, Federal Dist., central Mexico, SW of Mexico City and N of Cuernavaca. An eruption c.5000 B.C. left a large part of the Cuicuilco pyramid under lava and ash.

A·ka·dem·gor·o·dok (ä-kä-dăm-gôr′ô-dôk), city, W central Siberian USSR, near Novosibirsk. A scientific center begun in 1959, it is the site of 15 institutes of the Soviet Academy of Sciences.

A·kai·shi (ä-kī′shē), mountain, 10,234 ft (3,121.4 m) high, N Shizuoka prefecture, central Honshu, Japan, near the border of Nagano prefecture. It has a wide variety of alpine vegetation.

A·ka·shi (ä′kä′shē), city (1976 est. pop. 239,401), Hyogo prefecture, W Honshu, Japan, on the Akashi Channel. It is a fishing port and industrial center where electrical machinery is produced.

Akh·el·ó·os or **Ach·el·o·us** (both: äkh′ĕl-ô′ôs), river, 137 mi (220.4 km) long, rising in the Pindus Mts., NW Greece, and flowing generally S, traversing many mountain gorges and emptying into the Ionian Sea opposite Kefallinía. It is used for floating logs and is an important source of hydroelectric power. It formed a part of the boundary between ancient Aetolia and Acarnania.

Ak·hi·sar (äk′hə-sär′), city (1975 pop. 53,027), W Turkey. It is in a region where tobacco, cotton, and grapes are grown. The city is noted for its rugs.

Akh·mim (ăkh-mēm′), city (1966 pop. 44,800), E central Egypt, on the Nile. Textiles and handicrafts are produced. The city was long noted for its linen and limestone. The temple of Pan is here.

A·ki·ta (ä-kē′tə), city (1976 est. pop. 266,779), capital of Akita prefecture, NW Honshu, Japan, on the Sea of Japan. An oil-refining center, it is also a large port that exports lumber and rice. It became an important feudal town in the 8th cent.

Ak·kad (ă′kăd, ä′käd), ancient region of Mesopotamia, occupying the N part of later Babylonia; the S part was Sumer. In both regions city-states first appeared in the 4th millennium B.C. Akkad flourished after Sargon began (c.2340 B.C.) to spread wide his conquests, which ranged to the Mediterranean shores. After more than a century the empire declined and was overrun by mountain tribes.

Ak·ko (äk′ō) or **A·cre** (ā′kər, ä′kər), city (1975 est. pop. 35,600), NW Israel, a fishing port on the Bay of Haifa (an arm of the Mediterranean Sea). Its manufactures include iron and steel, chemicals, and textiles. The city was occupied (A.D. 638) by the Arabs, who developed its natural harbor. In 1104 it was captured in the First Crusade and was held by Christians until 1187, when it was taken by Saladin. In the Third Crusade it was won back (1191) by Guy of Lusignan, Richard I of England, and Philip II of France, who gave it to the Knights Hospitalers. For the next century it was the center of the Christian possessions in the Holy Land. Its surrender and virtual destruction by the Saracens in 1291 marked the decline of the Latin Kingdom of Jerusalem and the Crusades. In 1799 Ottoman forces, with the aid of Great Britain, withstood a 61-day siege by Napoleon. The city was taken in 1832 by Ibrahim Pasha for Mohammed Ali of Egypt, but European and Ottoman forces won it back for the Ottoman Empire in 1840. British troops captured the city in 1918. Akko was assigned to the Arabs in the 1948 partition of Palestine, but was captured by Israeli forces later that year.

A·ko·la (ə-kō′lə), town (1971 pop. 168,454), Maharashtra state, W central India, just WSW of Amravati. It is a market town. Cotton and groundnuts are the chief products of the region.

A-k'o-su (ä′kō′so͞o′), town, SW Sinkiang Uigur Autonomous Region, China, on the A-k'o-su River. The center of an oasis at the foot of the Tien Shan, it is a caravan hub on the Old Silk Road. Industries include textile and carpet manufacturing, jade carving, tanning, and metalworking. Iron deposits are in the area.

Ak·pa·tok (ăk′pə-täk′), island, 296 sq mi (766.6 sq km), SE Franklin dist., Canada, in Ungava Bay.

A·kra·nes (ä′krä-nĕs′), town (1970 est. pop. 4,253), SW Iceland, on a peninsula in the Faxaflói. It is a fishing port and industrial center, with a large cement plant.

Ak·ron (ăk′rən). **1.** Town (1970 pop. 1,775), seat of Washington co., NE Colo., in an agricultural region NE of Denver; founded 1882, inc. 1887. Poultry, grain, and dairy products are processed. **2.** Village (1970 pop. 2,863), Erie co., W N.Y., NE of Buffalo, in a farm and dairy region; inc. 1849. Metal products and cement are made here. **3.** City (1978 est. pop. 242,600), seat of Summit co., NE Ohio, on the Cuyahoga River, on the highest point of the Ohio and Erie Canal; inc. 1825. It is a port of entry, an important industrial and transportation center, and the heart of the country's rubber industry. Its manufactures include fishing tackle, plastics, missiles, and heavy machinery. Points of interest include a giant dirigible airdock, one of the world's largest buildings without inner supports, and the John Brown home, where the abolitionist lived from 1844 to 1846.

Ak·sum or **Ax·um** (both: äk-so͞om′), town (1970 est. pop. 12,800), Tigre prov., N Ethiopia. Aksum was the capital of an empire (c.1st–8th cent. A.D.) that controlled much of what is now northern Ethiopia. The Ark of the Covenant is said to have been brought here from Jerusalem and placed in the church of St. Mary of Zion, where Ethiopia's emperors were crowned. There are gigantic carved obelisks dating from pre-Christian times.

Ak·te (äk′tē): *see* Athos, Greece.

Ak·tyu·binsk (äk-tyo͞o′bĭnsk), city (1970 pop. 550,000), capital of Aktyubinsk oblast, Kazakhstan, S European USSR, on the Ilek River.

Aktyubinsk has an important ferroalloy plant and chromium complex. Founded in 1869, the city grew rapidly with the expansion of metallurgical industries during World War II.

A·ku·re (ä-ko͞o'rā), town (1969 est. pop. 82,000), S Nigeria. Timber is cut nearby and processed in Akure. The town is also a cacao marketing center. Akure was a small independent Yoruba kingdom until it was conquered by Benin in the early 19th cent. Great Britain gained control in 1894.

A·ku·rey·ri (ä'kü-rā'rē), city (1970 pop. 10,735), N Iceland, at the head of the Eyjafjörður. The second-largest city of Iceland, it is a fishing, commercial, and industrial center. It was settled c.900 A.D. and chartered in 1786.

A·ku·tan (ə-ko͞o'tən), volcanic island, 127 sq mi (328.9 sq km), Fox Islands, in the Aleutian Islands off Alaska. Akutan Peak, 4,244 ft (1,294.4 m), last erupted in 1946. The island is across Akutan Pass from Unalaska Island.

Al, for some Arabic names beginning thus, see the second element; e.g., for Al Mansurah, *see* Mansurah, Al.

A·la·bam·a (ăl-ə-băm'ə), state (1975 pop. 3,608,000), 51,609 sq mi (133,667 sq km), SE United States, bounded on the N by Tenn., on the E by Ga., on the S by Fla. and the Gulf of Mexico, and on the W by Miss.; admitted as the 22nd state of the Union in 1819; capital Montgomery. Except for the mountainous section in the northeast

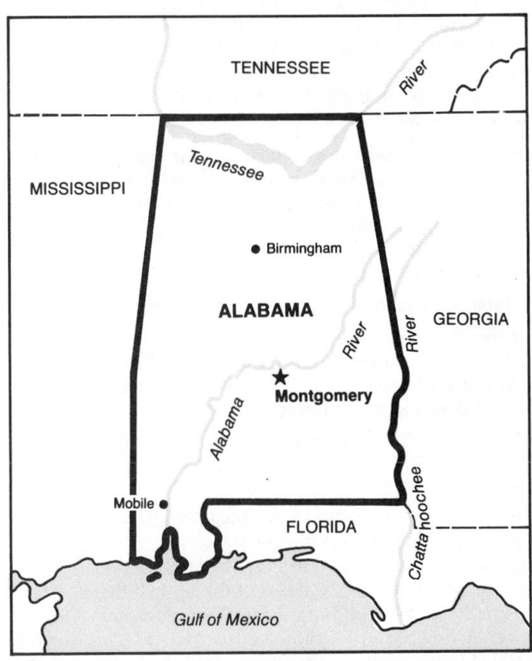

(the southern end of the Cumberland Plateau) the state is a rolling plain with a mean elevation of c.500 ft (155 m) in two geologic regions—the Appalachian Piedmont above the Fall Line and the coastal plain below.

Alabama is divided into 67 counties:

NAME	COUNTY SEAT	NAME	COUNTY SEAT
Autauga	Prattville	Coosa	Rockford
Baldwin	Bay Minette	Covington	Andalusia
Barbour	Clayton	Crenshaw	Luverne
Bibb	Centreville	Cullman	Cullman
Blount	Oneonta	Dale	Ozark
Bullock	Union Springs	Dallas	Selma
Butler	Greenville	De Kalb	Fort Payne
Calhoun	Anniston	Elmore	Wetumpka
Chambers	Lafayette	Escambia	Brewton
Cherokee	Centre	Etowah	Gadsden
Chilton	Clanton	Fayette	Fayette
Choctaw	Butler	Franklin	Russellville
Clarke	Grove Hill	Geneva	Geneva
Clay	Ashland	Greene	Eutaw
Cleburne	Heflin	Hale	Greensboro
Coffee	Elba	Henry	Abbeville
Colbert	Tuscumbia	Houston	Dothan
Conecuh	Evergreen	Jackson	Scottsboro

NAME	COUNTY SEAT	NAME	COUNTY SEAT
Jefferson	Birmingham	Perry	Marion
Lamar	Vernon	Pickens	Carrollton
Lauderdale	Florence	Pike	Troy
Lawrence	Moulton	Randolph	Wedowee
Lee	Opelika	Russell	Phenix City
Limestone	Athens	St. Clair	Ashville
Lowndes	Hayneville	Shelby	Columbiana
Macon	Tuskegee	Sumter	Livingston
Madison	Huntsville	Talladega	Talladega
Marengo	Linden	Tallapoosa	Dadeville
Marion	Hamilton	Tuscaloosa	Tuscaloosa
Marshall	Guntersville	Walker	Jasper
Mobile	Mobile	Washington	Chatom
Monroe	Monroeville	Wilcox	Camden
Montgomery	Montgomery	Winston	Double Springs
Morgan	Decatur		

Economy. Cattle and poultry are Alabama's most valuable agricultural products. Cotton, grown in the Tennessee River valley, is still the chief crop. Other important crops are peanuts, soybeans, and hay. Although about half of Alabama's area is devoted to agriculture, manufacturing accounts for a larger share of the state's income. Where the Tennessee River loops across the north, hydroelectric power from the Tennessee Valley Authority has been increasingly turning an agricultural land into an industrial section. The mineral riches of coal, oil, stone, and iron also contribute to the state's industries. Other major industries produce chemicals, textiles, paper products, and processed foods. In addition Gulf fishing and lumbering add to the wealth of Alabama; pine plywood is also an important product.

History. The Creek, Cherokee, Choctaw, and Chickasaw Indians inhabited the region when Spanish explorers arrived. Cabeza de Vaca visited Alabama in 1528, and Hernando De Soto spent some time in the region in 1540. White settlement was begun in 1702 by the French under the sieur de Bienville in the Mobile area. The region passed (1763) to the British, who were victorious over France and Spain in the French and Indian Wars. At the close of the American Revolution, Great Britain ceded (1783) to the United States all lands east of the Mississippi except the Floridas. The territory of Mississippi, which included parts of present-day Alabama, was set up in 1798, but the land was still largely wilderness. Andrew Jackson decisively defeated the Creek Confederacy at Horseshoe Bend on Mar. 27, 1814. Newcomers settled in the fertile bottomlands and established great cotton plantations based on slave labor. Poorer settlers took over less fertile uplands. The Territory of Alabama was set up in 1817; two years later it became a state.

Alabama broke away from the Union on Jan. 11, 1861. The government of the Confederacy was organized at Montgomery on Feb. 4, 1861. Federal troops held the Tennessee valley after 1862. One of the great naval battles of the war was won by Admiral D. G. Farragut in Mobile Bay in 1864, but most of the state was not occupied in force until 1865. Alabama ratified the Thirteenth Amendment to the U.S. Constitution in 1865, but in 1867 it refused to ratify the Fourteenth Amendment and was placed under military rule. That rule ended the following year when a new state legislature operating under a new constitution approved the Fourteenth Amendment. However, Federal troops did not leave Alabama until 1876. In the Reconstruction era Alabama's government was filled with carpetbaggers and scalawags, and corruption was widespread. Few reforms emerged during the period; but the mining of coal and iron was expanded, marking the rise of industry. Birmingham was founded in 1870, and its first blast furnace began operationsin 1880. The cotton textile industry developed in the 1880s.

Diversification of crops, much advocated in the 20th cent., was accelerated when the boll weevil invaded the cotton fields. The Great Depression caused more farmers to produce subsistence crops and took more land away from the wasting cotton culture. Industrialization was greatly increased during World War II. In 1954 the U.S. Supreme Court handed down a decision ruling racial segregation in public schools unconstitutional.

Government. Alabama's constitution, adopted in 1901, provides for an elected governor, who may not succeed himself in office. The legislature is made up of a 35-member senate and a 105-member house of representatives.

Alabama, river, 315 mi (506.8 km) long, formed in central Ala. by the confluence of the Coosa and Tallapoosa rivers N of Montgomery and flowing SW to Mobile, where it joins the Tombigbee to form the Mobile River.

Al·a·bat (äl'ə-bät'), island (1969 pop. 11,700), 74 sq mi (191.7 sq km), Quezon prov., E of S Luzon, Philippines. Long and narrow, it lies between Calauag Bay on the east and Lopez Bay on the west. Fishing is important.

A·la·chu·a (ə-lăch'-ə-wə), county (1970 pop. 104,764), 916 sq mi (2,372.4 sq km), N Fla., bounded on the N by the Santa Fe River; formed 1824; co. seat Gainesville. It produces corn, vegetables, peanuts, cotton, tobacco, citrus fruit, livestock, and lumber. Limestone, phosphate, and flint are quarried.

Al·a·go·as (ä-lä-gō'əs), state (1970 pop. 1,589,605), 10,707 sq mi (27,731 sq km), NE Brazil, on the Atlantic; capital Maceió.

Al·a·gón (äl'ə-gôn'), river, c.120 mi (195 km) long, Salamanca and Cáceres provs., W Spain. The main tributary of the Tagus River, it originates in a ridge of the Peña de Francia range, then flows southwest into the Tagus northeast of Alcántara.

A·lai or **A·lay** (both: ä'lī'), mountain range, SW Kirghizstan, Central Asian USSR. A western branch of the Tien Shan system, it extends c.200 mi (320 km) west from the Chinese border and rises to c.19,280 ft (5,880 m) in its western portion. The Alai Valley, south of the range, is a fertile elevated (c.9,800 ft/2,990 m) grassland used for grazing; there is irrigated grain cultivation in the west.

A·lais (ä-lā'): *see* Alès, France.

Al·a·jue·la (äl'ä-hwā'lä), city (1976 est. pop. 35,000), capital of Alajuela prov., central Costa Rica. On the central plateau, it is a commercial and agricultural center with sugar, coffee, and lumber industries. It was the national capital in the 1830s.

Al·a·kol (äl'ə-kôl'), salt lake, 803 sq mi (2,079.8 sq km), E Kazakhstan, USSR, near the China border E of Lake Balkhash.

Al·a·ma·gan (ăl'ə-mə-găn'), uninhabited volcanic island, 4 sq mi (10.4 sq km), N Mariana Islands, in the W Pacific NNE of Guam. It was included in the 1920 Japanese mandate. Taken by the United States in Aug., 1945, it is today part of an important strategic link in the U.S. defense network in the Pacific.

Al·a·mance (ăl'ə-măns'), county (1970 pop. 96,362), 434 sq mi (1,124.1 sq km), N central N.C., in the Piedmont and crossed by the Haw River; formed 1849; co. seat Graham. It produces tobacco, corn, hay, and dairy products. Textiles and furniture are made.

Al·a·me·da (ăl'ə-mē'də), county (1970 pop. 1,071,446), 733 sq mi (1,898.5 sq km), W Calif., in a residential and industrial region adjacent to San Francisco; formed 1853; co. seat Oakland. Flowers, plants, grain, hay, fruit, and nuts are grown. Poultry and livestock are raised. It has sand, gravel, and clay pits, stone quarries, and magneite deposits.

Alameda, city (1970 pop. 70,968), Alameda co., W central Calif., on an island just off the E shore of San Francisco Bay; settled 1850, inc. as a city 1884. It is primarily residential, with excellent beaches, parks, and pleasure-boating facilities. The major employer in the city is the Alameda Naval Air Station. The city is connected with the mainland by four bridges and two tunnels.

Al·a·mein, El (ĕl ăl'ə-mān'), or **Al Al·a·mayn** (äl äl'ə-mān'), town, N Egypt, on the Mediterranean Sea. It was the site of a decisive British victory in World War II. In preparation for an attack by German Field Marshal Rommel from Libya (May 26, 1942) the British forces retreated into Egypt and set up a defense line extending 35 mi (56 km) from Alamein south to the Qattara Depression. If this position had fallen, the British might have lost Alexandria and been forced to withdraw from North Africa. In Aug., Gen. Bernard L. Montgomery took command of the 8th Army. The British offensive opened on Oct. 23 with tremendous air and artillery bombardments. Montgomery's forces burst through the German lines and forced a swift Axis retreat out of Egypt, across Libya, and into Tunisia. With the landing on Nov. 7 and 8 of American troops in Algeria the Axis soon suffered (May, 1943) total defeat in North Africa.

Al·a·mo (ăl'ə-mō'). **1.** City (1970 pop. 833), seat of Wheeler co., SE central Ga., SSE of Dublin, in a cotton-growing area. **2.** Town (1970 pop. 2,499), seat of Crockett co., W Tenn., NW of Jackson, in a cotton, grain, livestock, and truck-farm area; inc. 1911. **3.** City (1970 pop. 4,291), Hidalgo co., extreme S Texas, NW of Brownsville, in the irrigated district of the lower Rio Grande valley; inc. 1924. Fruits and vegetables are canned.

Alamo, the, in San Antonio, Texas. Built as a chapel after 1744, it is all that remains of a mission that was founded in 1718 by the Franciscans and later converted into a fortress. In the Texas Revolution, San Antonio was taken by Texas revolutionaries in Dec., 1835, and was lightly garrisoned. When Santa Anna approached with an army of several thousand in Feb., 1836, only some 150 men held the Alamo, and confusion, indifference, and bickering among the insurgents throughout Texas prevented any help from joining them, except for 32 volunteers from Gonzales who slipped through the Mexican lines. The siege, which began Feb. 24, ended with hand-to-hand fighting within the walls on Mar. 6. Most of the garrison died, but the heroic resistance roused fighting anger among Texans, who six weeks later defeated the Mexicans at San Jacinto, crying "Remember the Alamo!" The complex was restored from 1936 to 1939.

Al·a·mo·gor·do (ăl'ə-mə-gôr'dō, -də), city (1978 est. pop. 22,900), seat of Otero co., S N.Mex., near the Sacramento Mts.; inc. 1912. It is a trade center for a large livestock, irrigated farm, timber, and recreational area. Pressure cookers, wearing apparel, and lumber are among its products. Holloman Air Force Base, site of the White Sands Missile Range, where the first atomic bomb was exploded on June 16, 1945, is located in Alamogordo. The city was founded in 1898 with the arrival of the Southern Pacific RR.

Alamo Heights, city (1970 pop. 6,933), Bexar co., S central Texas, a suburb ENE of San Antonio; inc. 1926.

Al·a·mo·sa (ăl'ə-mō'sə), county (1970 pop. 11,422), 720 sq mi (1,864.8 sq km), S Colo., in an irrigated agricultural area watered by the Rio Grande and bounded on the E by the Sangre de Cristo Mts.; formed 1913; co. seat Alamosa. Livestock and potatoes are its agricultural products. It includes part of Great Sand Dunes National Monument and San Isabel National Forest.

Alamosa, city (1970 pop. 6,985), seat of Alamosa co., S central Colo., on the Rio Grande at an elevation of 7,500 ft (2,287.5 m); founded and inc. 1878 with the coming of the railroad. It is an industrial, shipping, and retail center in a prosperous agricultural area yielding potatoes, livestock, and dairy and meat products. There are stockyards, a flour mill, and an oil refinery here.

Å·land Islands (ä'lənd, ô'lənd): *see* Ahvenanmaa Islands.

Al·a·şe·hir (äl'ä-shĕ-hēr'), town (1975 pop. 23,120), W Turkey, ESE of Manisa. It is the trade center for a region where tobacco, fruit, and mineral water are produced. The town is picturesque, with narrow, winding streets and a Byzantine wall.

A·la·shan (ä'lä-shän'), mountain range, W Inner Mongolia, N China, rising to c.12,000 ft (3,660 m). It stretches west of and parallel to the Huang Ho.

A·las·ka (ə-lăs'kə), state (1975 pop. 344,000), 586,400 sq mi (1,518,776 sq km), including 15,335 sq mi (39,718 sq km) of water surface, NW North America; admitted 1959 as the 49th state; capital Juneau. Nearly one fifth the size of the rest of the United States, Alaska is the largest state in the Union but the least populous one. Alaska is at the northwestern extremity of the North American continent, between the Arctic Ocean on the north and the Gulf of Alaska and the Pacific Ocean on the south. It is bounded on the east by Canada (Yukon and British Columbia) and on the west by the Bering Sea, Bering Strait, and Chukchi Sea.

The tip of the Seward Peninsula is only a few miles from Far Eastern USSR; the two are separated by the narrow Bering Strait. Seward Peninsula is chiefly tundra-covered and sparsely inhabited.

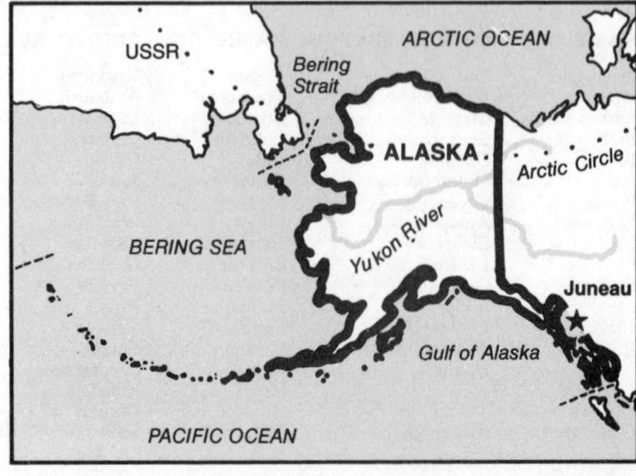

The Bering Strait widens in the north to the Chukchi Sea, which slices into Alaska with Kotzebue Sound; in the south the strait widens to the Bering Sea, which cuts into Alaska with Norton Sound and Bristol Bay. The state again extends toward the USSR in the Alaska Peninsula and the Aleutian Islands, reaching out a total of 1,200 mi (1,930.8 km) and dividing the Bering Sea from the Pacific. The Aleutian Range, which is the spine of the Alaska Peninsula, is continued in the grass-covered, treeless Aleutian Islands.

The southern shore of Alaska is deeply indented by two inlets of the wide Gulf of Alaska, Cook Inlet and Prince William Sound; the Kenai Peninsula between them extends southwest toward Kodiak Island. The narrow Panhandle dips southeast along the coast from the Gulf of Alaska, cutting into British Columbia. It consists of the offshore islands of the Alexander Archipelago and the narrow coast, which rises steeply to the mountains of the Coast Range and the St. Elias Mts.

The state abounds in natural wonders. In the Panhandle, the scenic beauty of the mountains and the rugged fjord-indented coast are augmented by such attractions as the Malaspina Glacier and the acres of blue ice in Glacier Bay National Monument. In the Alaska Range of south-central Alaska stands the highest point in North America, Mt. McKinley. The Alaska Peninsula and the Aleutian Islands have numerous volcanoes.

Alaska is divided into nine boroughs:

NAME	BOROUGH SEAT	NAME	BOROUGH SEAT
Bristol Bay	Naknek	Kenai Peninsula	Soldotna
Fairbanks North Star	Fairbanks	Kodiak Island	Kodiak
Greater Sitka	Sitka	Matanuska-Susitna	Palmer
Haines	Haines	Municipality of Anchorage	Anchorage
Juneau	Juneau		

Economy. Alaska has very little agriculture. Most of Alaska's farms are dairies or poultry ranches, and the state's most valuable farm commodities are dairy products, potatoes, cattle, and eggs. Fishing is a leading industry. Alaska heads the nation in the value of its commercial catch—chiefly salmon, crab, shrimp, halibut, herring, and cod. Its largest manufacturing enterprise is food processing, particularly the freezing and canning of fish. Lumbering and related industries are second.

Mining, principally of petroleum, sand and gravel, natural gas, and coal, is the state's most valuable industry. Alaska leads the nation in the production of platinum, is second in production of antimonies, and is a leading producer of tin. Mercury, uranium, and beryllium are also found. Gold is no longer mined in quantity. Fur-trapping, Alaska's oldest industry, still endures, and pelts are obtained from a great variety of animals. Government—Federal, state, and local—is Alaska's major source of employment. The state's strategic location has generated considerable defense activity, including the establishment of permanent military bases. Oil and natural gas offer the best hope for Alaska's future. The vast discoveries on the Arctic North Slope indicate that that area, along with the offshore deposits in south-central Alaska, may make the state one of the world's greatest petroleum and natural-gas producers. The construction of an 800-mi (1,287.2-km) pipeline from Prudhoe Bay on the Arctic North Slope to the ice-free port of Valdez encountered opposition from ecologists, but was built between 1974 and 1977. Legal disputes and construction problems have curtailed the expected outflow of oil since the completion of the pipeline.

History. Alaska was discovered by Russian explorers. The disastrous voyage of Vitus Bering and Aleksey Chirikov in 1741 climaxed the march of Russian traders across Siberia. The survivors started a rush of fur hunters to the Aleutian Islands. Grigori Shelekhov in 1784 founded the first permanent settlement in Alaska on Kodiak Island and sent (1790) to Alaska Aleksandr Baranov, who became director of the Alaskan activities of the Russian American Company. Sitka was founded in 1799 as his capital. Rivalry for the northwest coast was strong, and British and American trading vessels began to threaten the Russian monopoly. A negotiated settlement (1824) fixed the southern boundary of Alaska at lat. 54°40′ N.

Russian interests in Alaska gradually declined, and in 1867 Alaska was sold to the United States for $7,200,000. The U.S. purchase was accomplished solely through the determined efforts of Secretary of State William H. Seward, and for many years afterward the land was derisively called Seward's Folly. It was not until after the discovery of gold in the Juneau region in 1880 that Alaska was given a governor and a feeble local administration. The great Canadian Klondike strike of 1896 brought a stampede, mainly of Americans, and most of them came through Alaska. The big discoveries in Alaska itself followed—Nome in 1898-99, Fairbanks in 1902. The miners and prospectors (the sourdoughs) took over Alaska, and the era of the rough mining camps reached its height.

The longstanding controversy concerning the boundary between the Alaska Panhandle and British Columbia was settled by negotiation in 1903. A period of rapid building and development began. Mining, requiring heavy financing, passed into the hands of Eastern capitalists. A new era began for Alaska when local government was established in 1912 and it became a U.S. territory (Juneau had officially replaced Sitka as capital in 1900). The building of the Alaska RR from Seward to Fairbanks was commenced with government funds in 1915. Gold mining died out, and the fishing industry became the major enterprise.

Alaska enjoyed its greatest economic boom during World War II. The Alaska Highway was built, supplying a much-needed link with the United States. Japanese troops occupied the Aleutian islands of Attu and Kiska until U.S. forces retook them in 1943. The growth of air travel after the war and the establishment of permanent military bases brightened hopes for Alaska's growth; between 1950 and 1960 the population nearly doubled. In 1959 Alaska was admitted into the Union as a state. On Mar. 27, 1964, the strongest earthquake ever recorded in North America occurred in Alaska, taking approximately 114 lives and causing extensive property damage.

Government. Alaska operates under a constitution drawn up and ratified in 1956 (effective with statehood). Its executive branch is headed by a governor and a secretary of state, both elected (on the same ticket) for four-year terms. Alaska's bicameral legislature has a senate with 20 members elected for four-year terms and a house of representatives with 40 members elected for two years.

Alaska Highway, all-weather graveled road, 1,523 mi (2,450.5 km) long, extending NW from Dawson Creek, British Columbia, to Fairbanks, Alaska. An extension of an existing Canadian road between Dawson Creek and Edmonton, Alta., the Alaska Highway was constructed (Mar.-Sept., 1942) by U.S. troops as a supply route to military forces in Alaska during World War II. It was a significant engineering feat because of the difficulties of terrain and weather. In 1947 the entire highway was opened to unrestricted travel. One of the best routes to Alaska, the highway is open throughout the year.

Alaska North Slope or **Arctic North Slope,** region, N Alaska, between the Arctic Ocean and the Brooks Range. Large petroleum reserves were found here in the late 1960s.

Alaska Range, S central Alaska, rising to the highest mountain in North America, Mt. McKinley (20,320 ft/6,197.6 m). The range divides south-central Alaska from the interior plateau.

A·la-Tau (ä′lä-tou′), several ranges of the Tien Shan system in central Asia. The Dzungarian Ala-Tau, the northernmost and loftiest branch of the Tien Shan, forms part of the USSR-China border. Silver and lead mines and hot springs are found here. The Kungei Ala-Tau lies north of Issyk-Kul, a huge lake in the Tien Shan. The Trans-Ili Ala-Tau, on the Kirghizia-Kazakhstan border, supports intensive, irrigated agriculture.

A·la·tyr (ăl-ə-tĭr′), city (1974 est. pop. 46,000), Chuvash Autonomous Republic, E European USSR, at the confluence of the Sura and Alatyr rivers. Founded in 1552, it is a river port and railroad junction with locomotive and food-processing plants.

A·lay (ä′lī′), mountain range: *see* Alai.

Al Az·i·zi·yah (äl äz′ē-zē′yä) or **Az·i·zi·a** (äz′ĭ-zē′ä), town, NW Libya, near Tripoli. It is a trade center. The hottest recorded temperature on earth, 141°F (60.6°C), was recorded here.

Al·ba·ce·te (ăl′bə-sät′ē), city (1970 pop. 93,233), capital of Albacete prov., SE Spain, in Murcia. Under the Moors, Albacete was a part of the kingdom of Murcia, with which it was incorporated (1269) into Castile. The city now has a modern aspect and is mainly an agricultural center. It is noted for the manufacture of fine knives.

Al·ba Iu·li·a (ăl′bə yo͞o′lyä), town (1969 est. pop. 84,000), W central Rumania, in Transylvania, on the Mureşul River. It is a rail junction and distribution center for a winemaking region, where grain, poultry, and fruit are raised. The town's light manufactures include soap, furniture, and footwear. Alba Iulia is the site of the ancient Apulum, founded by the Romans in the 2nd cent. A.D., and destroyed by Tatars in 1241. It was the seat (16th-17th cent.) of the princes of Transylvania, of a Roman Catholic bishop, and of an Eastern Orthodox metropolitan. From 1599 to 1601, Alba Iulia was the capital of the united principalities of Walachia, Transylvania, and Moldavia. It was the site (1918) of the proclamation of Transylvania's union with Rumania and of the coronation (1922) of King Ferdinand.

Al·ba Lon·ga (ăl'bə lông'gə), city of ancient Latium, in the Alban Hills near Lake Albano, SE of Rome. It was a city before 1100 B.C. and apparently the most powerful in Latium. Legend says that Romulus and Remus were born there, thus making it the mother city of Rome. Possibly Rome was founded from Alba Longa, and certainly the Romans destroyed it (c.600 B.C.). The modern Castel Gandolfo occupies the site.

Al·ba·ni·a (ăl-bā'nē-ə, -bān'yə, ôl-), ancient name of a mountainous region in E Transcaucasia, bordering on the Caspian Sea. It is now within the Azerbaijan SSR and the Dagestan ASSR.

Albania, independent republic (1970 est. pop. 2,100,000), 11,101 sq mi (28,751.6 sq km), SE Europe, on the Adriatic Sea coast of the Balkan

Peninsula, between Yugoslavia on the N and E and Greece on the S. Tiranë is the capital. Albania is rugged and mountainous, except for the fertile Adriatic coast. The coastal climate is typically Mediterranean, with hot dry summers and mild, wet winters. The mountainous interior, especially in the north, has severe winters.

Economy. More than one third of Albania's land is covered by forests and swamps, about one third is pasture, and only about one tenth is cultivated, nearly one half of the cultivated land being given over to vineyards and olive groves. Grains (especially wheat and maize), cotton, tobacco, potatoes, and sugar beets are also grown. Livestock raising is important. Agriculture is socialized in the form of collective and state farms, but small private plots are permitted. Albania is rich in mineral resources, notably oil, lignite, copper, chromium, limestone, salt, and bauxite. Agricultural processing and the manufacture of textiles and cement are among the leading industries; other important products include naphtha, copper, and machinery. Engineering, chemical, and iron and steel plants are being developed, and several hydroelectric stations have been built. All industrial enterprises and mines are nationalized, and the economy is run on the basis of five-year plans.

History. The Albanians are reputedly descendants of Illyrian and Thracian tribes. The coastal towns were ancient Greek colonies, but the interior formed an independent kingdom that reached its height in the 3rd cent. A.D. Albania passed (A.D. 395) from Roman to Byzantine rule, during which northern Albania was invaded (7th cent.) by the Serbs and southern Albania was annexed (9th cent.) by Bulgaria. In succeeding centuries Venetians, Normans, and Serbs vied with the Byzantines and Turks for control of the area. The Ottoman Turks completed their conquest in 1478. Under Turkish rule Islam became the predominant religion. A cultural awakening began in the 19th cent., and Albanian nationalism grew.

During the Balkan Wars (1912-13) and World War I, Albania became a battleground for contending Serbian, Montenegrin, Greek, Italian, Bulgarian, and Austrian forces. These armies retreated, but civil war followed, which lasted until 1925, when Albania was declared a republic under the presidency of Ahmed Zogu. Zogu became King Zog in 1928. Italy, whose political and economic influence in Albania had steadily increased, invaded the country in 1939 and set up a puppet government. Resistance groups waged guerrilla warfare against the occupying Axis armies. In late 1944 Hoxha's partisans seized most of Albania and in 1946 proclaimed Albania a republic with Hoxha as premier.

After Marshal Tito of Yugoslavia broke with Stalin (1948), Albania turned away from Yugoslavia and became a virtual satellite of the USSR. Albania's disapproval of de-Stalinization led in 1961 to a break between Moscow and Tiranë. Chinese influence and economic aid replaced Soviet. In the 1970s Albania re-established relations with Yugoslavia, but Hoxha vowed eternal hostility to the USSR and maintained close ties with China.

Government. The Albanian constitution (adopted 1946) names the People's Assembly as "the highest organ of state power," but in practice the Communist party (officially the Albania Workers' Party) wields complete control. Deputies to the unicameral people's assembly are elected by universal suffrage for four-year terms. The assembly elects a presidium, whose chairman becomes titular head of state. The country's highest executive body, the council of ministers, is appointed by the assembly. The chairman of the council of ministers serves as premier.

Al·ba·ny (ôl'bə-nē), ancient and literary name of Scotland, N of the Firth of Forth and Firth of Clyde.

Albany, town (1971 pop. 12,434), Western Australia, SW Australia. It is a port on Princess Royal Harbour of King George Sound. The town has woolen mills and fish canneries. It was founded in 1826 as a penal colony.

Albany. 1. County (1970 pop. 286,742), 531 sq mi (1,375.3 sq km), E N.Y., bound on the E by the Hudson River, partly on the N by the Mohawk River and the Barge Canal; formed 1683; co. seat Albany. It is principally the seat of the state government. It also has dairying and agriculture (apples and truck farming), stock and poultry raising, and extensive manufacturing. **2.** County (1970 pop. 26,431), 4,248 sq mi (11,002.3 sq km), SE Wyo., in a grain and livestock area bordering on Colo. and watered by the Laramie River and Wheatland Reservoir; formed 1868; co. seat Laramie. Parts of Medicine Bow National Forest and Medicine Bow Mts. are here.

Albany. 1. Residential city (1970 pop. 14,674), Alameda co., W Calif., on the E shore of San Francisco Bay; inc. 1908. A U.S. Dept. of Agriculture research laboratory and a Univ. of California agricultural experiment station are here. **2.** City (1978 est. pop. 72,500), seat of Dougherty co., SW Ga., on the Flint River; inc. 1841. It is the industrial center of a great pecan and peanut area. Among its many industries are peanut and pecan processing, meat packing, and cotton milling. Manufactures include airplanes and airplane parts, farm tools, fertilizers, pharmaceuticals, and paper, wood, cotton, and concrete products. **3.** Town (1970 pop. 2,293), Delaware co., E central Ind., NE of Muncie, in a farm area. Corn and wheat are grown. **4.** City (1970 pop. 1,891), seat of Clinton co., S Ky., SSW of Danville, in the Cumberland foothills near the Tenn. border; settled c.1800. It is in a fishing, timber, coal-mining, and oil region. Its agriculture includes livestock, poultry, and dairy products. Dale Hollow Reservoir is to the southwest. **5.** City (1970 pop. 1,804), seat of Gentry co., NW Mo., on the East Fork of the Grand River NE of St. Joseph, in a livestock, grain, poultry, and soybean region; laid out 1845, inc. 1851. **6.** City (1978 est. pop. 106,600), state capital and seat of Albany co., E N.Y., on the W bank of the Hudson; inc. 1686. A deepwater port of entry, it handles much shipping and is a major transshipment point. The trading center for a large agricultural and resort area, it has oil tanks, breweries, machine shops, foundries, meat-packing houses, and plants making paper items, felt, textiles, chemicals, brushes, and sports equipment. In 1609 Henry Hudson visited the site, and four years later the Dutch built a fur-trading post, called Fort Nassau, on Castle Island. In 1624 several Walloon families began permanent settlement at the Dutch post of Fort Orange, which was renamed Albany when the English took control (1664). Albany was long important as a fur-trading center and was involved in the French and Indian Wars. In 1754 the Albany Congress met here, and after the Revolution the state capital was moved (1797) to Albany from New York City. Albany's trade grew with the development of the state, particularly after the opening of the Champlain and Erie canals in the 1820s. Today it is the seat of the State Univ. of New York at Albany, the Albany Institute of History and Art, and other schools. Among the many old buildings are the Schuyler mansion (1762); Ten Broeck Mansion (1798); and Cherry

Hill (1768), the home of Philip Van Rensselaer and his descendants until 1963. Dominating the city, at the top of State Street hill, is the capitol, built (1867–98) in the French château style. An annual tulip festival is held in the city. In the 1960s a major urban renewal project resulted in the razing of 90 acres (36.5 hectares) in the downtown section for a great complex of state administrative buildings, residences, and parks. **7.** City (1978 est. pop. 24,900), seat of Linn co., NW Oregon, on the Willamette River; inc. 1864. A metallurgical center, it is the seat of a U.S. Bureau of Mines experimental station. The city also has important lumbering and paper and wood-product industries. Other manufactures are packaged meats, frozen foods, mobile homes, and seeds. An annual world championship timber carnival is held here. **8.** City (1970 pop. 1,978), seat of Shackelford co., N central Texas, NE of Abilene. Albany originated as a ranching and salt-working town and is still noted for its Hereford cattle ranching. It is a farm market and shipping center for cattle, and an oil-refining center.

Albany, river, 610 mi (981.5 km) long, rising in Lake St. Joseph, W Ont., Canada, and flowing generally E into James Bay, near Fort Albany. The river was long an important fur-trading route. Gold is found near Lake St. Joseph.

Al Bay·da (äl bā'dä) or **Bei·da** (bā'də), city (1964 pop. 12,799), NE Libya, in the Jabal al Akhdar plateau. Construction of the city began in 1961 on the site of the tomb of Raweifi ibn Thabit, a revered Moslem holy person.

Al·be·marle (ăl'bə-märl'), county (1970 pop. 37,780), 739 sq mi (1,914 sq km), central Va., bounded on the SW by the Rockfish River, on the S by the James River, and drained by the headstreams of the Rivanna River and by the Hardware River; formed 1774; co. seat Charlottesville (an independent city). The west part of the county is in a piedmont region rising to the Blue Ridge Mts. It has state government offices and some large factories manufacturing electrical machinery. Food, textiles, and wood products are made. It is known for its historic estates (notably Monticello), and includes part of Shenandoah National Park.

Albemarle, city (1970 pop. 11,126), seat of Stanly co., central N.C., in the Piedmont region; inc. 1857. A marketing center in an agricultural and aluminum-mining area, Albemarle has poultry-processing and textile and clothing industries.

Albemarle Sound, large inland body of generally fresh water, c.55 mi (90 km) long, from 3 to 14 mi (4.8 to 22.5 km) wide, NE N.C. Shallow and tideless, the sound is separated from the Atlantic Ocean by a long, narrow barrier island. Albemarle Sound forms a vital link in the Intracoastal Waterway.

Al·ber·che (äl-bĕr'kā), river, c.110 mi (177 km) long, central Spain, rising in the Sierra de Gredos N of Arenas de San Pedro in Castile prov. and flowing E through Ávila prov., then SW through the La Sagra plain to the Tagus River E of Talavera de la Reina. It is used for irrigation and hydroelectricity.

Al·ber·ga (ăl-bûr'gə), river, c.350 mi (565 km) long, N South Australia. It rises in the Musgrave Ranges and flows intermittently east, connecting with the Finke River only during the rainy season.

Al·ber·ni (ăl-bûr'nē), city (1970 pop. 4,783), on Vancouver Island, British Columbia, Canada. Port Alberni (1970 pop. 20,063), which adjoins it, is at the head of the Alberni Canal, a 22-mi (35.4-km) inlet of the Pacific, allowing passage to the largest ocean vessels. Both cities were nearly destroyed by fire (Aug., 1947). Today the cities' industries are based on forest products, fisheries, and tourism.

Al·ber·ta (ăl-bûr'tə), province (1975 pop. 1,768,000), 255,285 sq mi (661,188 sq km), including 6,485 sq mi (16,796 sq km) of water surface, W Canada, bounded on the E by Sask., on the N by Mackenzie dist., on the W by British Columbia, and on the S by Mont.; capital Edmonton. It lies on a high plateau, rising on the west to the Continental Divide at the British Columbia border. Alberta has parklike, partly wooded country, and the northern stretches bear thousands of acres of virgin timberland. Endowed with many lakes, streams, and rivers, the province is drained by the Peace, Athabasca, Saskatchewan, Red Deer, St. Mary, Milk, and many other rivers.

Economy. Until recently agriculture was Alberta's basic industry. Grain, especially wheat, is the dominant crop, but farming is becoming increasingly diversified. In the south, large irrigation developments have placed thousands of additional acres under cultivation. In this area is grown a variety of crops, and livestock is raised. Meat packing, flour milling, dairying, and food processing are important industries. Alberta's major industry, since the early 1960s, has been the exploitation of its vast petroleum and other mineral resources. Alberta's coal beds contain about one half of Canada's reserves, while the province leads the country in the production of oil; it is believed to have the richest oil deposits in the world, most notably in the famous tar beds of the Athabasca River. Its sources of natural gas are also among the world's greatest. Other industries include lumbering, textile milling, and the manufacture of iron, steel, and clay products.

History. Alberta was originally part of the territory granted to the Hudson's Bay Company by King Charles II in 1670, and its early history was dominated by the fur trade. In 1794 a Hudson's Bay Company fort was built at the site of present-day Edmonton. Destroyed by Indians in 1807, it was rebuilt 12 years later, and for 50 years it served traders and missionaries within a wide radius. The area remained under the control of the Hudson's Bay Company until 1870, when it was sold to the newly created confederation of Canada. An act of 1882 created four administrative divisions from the Northwest Territories, and one was named Alberta in honor of Queen Victoria's daughter, Princess Louise Alberta, whose husband was then governor general of Canada. The railroad came through in the mid-1880s, opening up the area to ranchers and homesteaders. Alberta became a province in 1905. By then fur trapping was in decline, and agriculture suffered from droughts and crop failures in the 1930s. The discovery (1947) of oil in quantity in the Leduc fields near Edmonton began a new era for the mineral-rich province.

Alberta, Mount, 11,874 ft (3,621.6 m) high, W Alta., Canada, in the Rocky Mts. near the border of British Columbia.

Albert Canal, waterway, c.80 mi (130 km) long, N Belgium, from the Meuse River to the Scheldt River; constructed 1930–39. The canal connects the important industrial region around Liège with the port of Antwerp.

Albert Lea (lē), city (1970 pop. 19,418), seat of Freeborn co., S Minn., near the Iowa border; inc. 1878. It is an important manufacturing and marketing center in a dairy, livestock, and poultry region.

Albert Nyan·za (nyän'zä, nī-ăn'zə) or **Lake Albert,** 2,064 sq mi (5,345.8 sq km), on the Zaire-Uganda border, E central Africa. The lake is c.100 mi (160 km) long and c.19 mi (30 km) wide, with a maximum depth of 168 ft (51.2 m). Lying in the Great Rift Valley, 2,030 ft (619.2 m) above sea level, Albert Nyanza is drained by the Albert Nile, which becomes the Bahr-el-Jebel when it enters the Republic of the Sudan. Albert Nyanza, discovered in 1864 by Sir Samuel Baker, was named for Queen Victoria's consort.

Al·ber·ton (ăl'bər-tən), town (1970 pop. 30,322), Transvaal, NE Republic of South Africa, on the Witwatersrand; founded 1904. It is an industrial center manufacturing cast iron, machine tools, paints, and abrasives.

Al·bert·ville (ăl'bərt-vĭl'), city (1970 pop. 9,963), Marshall co., NE Ala., NNW of Gadsden, in a cotton, truck-farm, and poultry area; inc. 1890. It has textile firms, and lumber is produced.

Al·bert·ville (äl-bĕr-vēl'): *see* Kalemi, Zaire.

Al·bi (ăl-bē′), town (1975 pop. 46,162), capital of Tarn dept., S France, in Languedoc, on the Tarn River. A commercial center in an area yielding coal, salt, and sand, it has glassworks, foundries, and food and textile industries. An old Roman city, it was the center of the Albigensian heresy. The huge Gothic Cathedral of Sainte-Cécile, begun in 1282, resembles a fortress. Other structures include the episcopal palace (13th–15th cent.) and an 11th cent. bridge. The birthplace of Toulouse-Lautrec, Albi has an art museum containing much of his work.

Al·bi·a (ăl′bē-ə), city (1970 pop. 4,151), seat of Monroe co., S Iowa, W of the Des Moines River and SE of the city of Des Moines; inc. 1859. It is a rail junction where farm equipment and concrete blocks are manufactured. Large fields of bituminous coal are in the area.

Al·bi·on (ăl′bē-ən), ancient and literary name of Britain. It is usually restricted to England and is perhaps derived from the Latin *albus*, meaning "white," referring to the chalk cliffs of southern England.

Albion. **1.** City (1970 pop. 1,791), seat of Edwards co., SE Ill., E of Belleville; founded 1818 by English settlers, inc. 1869. It is in an agricultural (livestock, poultry, and fruit) and manufacturing (clothing, flour, and brick) region. Several of the early houses in the area have been preserved. **2.** Town (1970 pop. 1,498), seat of Noble co., NE Ind., NNW of Fort Wayne; laid out 1847. Its agriculture includes livestock, soybeans, grain, poultry, fruit, and dairy products. Furniture, lumber products, tile, and flour are manufactured. Scenic lake resorts are nearby. **3.** Industrial city (1970 pop. 12,112), Calhoun co., S Mich., at the forks of the Kalamazoo River; inc. as a village 1855, as a city 1885. Among its manufactures are iron castings, electronic parts, air conditioners, heaters, bakery ovens, and wire products. Albion College is here; it was established in 1835, and the city developed around it. **4.** City (1970 pop. 2,074), seat of Boone co., E central Nebr., SW of Norfolk on Beaver Creek; settled 1871, inc. 1873. Meat products, poultry, beverages, and grain are processed. It also produces feed and manufactures electronic equipment. **5.** Resort village (1970 pop. 5,122), seat of Orleans co., W N.Y., on the Barge Canal W of Rochester; inc. 1828. It is a shipping center for an agricultural area yielding fruit and vegetables. It manufactures packers' supplies, textiles, feed, and machinery and has stone quarries.

Ål·borg (ôl′bôrg, -bôr), city (1970 com. pop. 154,343), capital of Nordjylland co., N Denmark, on the Limfjørd. It is a major industrial, transportation, and cultural center. Its manufactures include cement, machinery, chemicals, liquor, ships, and textiles. Known in the 11th cent., Ålborg was chartered in 1342.

Al·bu·quer·que (ăl′bə-kûr′kē), city (1978 est. pop. 288,600), seat of Bernalillo co., W central N.Mex., on the upper Rio Grande; inc. 1890. It is an important commercial, industrial, and transportation center serving a rich timber, livestock, and farm area. It has railroad shops, lumber mills, food-processing plants, and a large electronics industry. A major employer is the huge Atomic Energy Commission installation. Spanish settlers arrived in the mid-1600s but were driven out (1680) by the Indians. The old town was founded in 1706; the new town was platted in 1880 in connection with the railroad and grew rapidly, soon enveloping the old town. Albuquerque is a noted health resort with many sanatoriums and hospitals. It is the seat of the Univ. of New Mexico and the headquarters for Cibola National Forest. Tourist attractions in and about the city include the Church of San Felipe de Nerí (1706); the Old Town plaza; numerous museums; caves that contain remains of some of the oldest inhabitants in the western hemisphere; and many Indian pueblos. Coronado State Monument, to the north, is an excavated pueblo near which Coronado camped in 1541.

Al·bur·y (ăl′bĕr′ē, -bə-rē), city (1976 pop. 32,942), New South Wales, SE Australia, on the Murray River at the Victoria border. It is an agricultural market. Among its industries are food processing (including wine) and woolen milling.

Al·ca·lá de He·na·res (ăl′kə-lä′ dā ə-när′əs), town (1970 pop. 59,783), Madrid prov., central Spain, on the Henares River, in New Castile. Leather, soap, and china are produced in the town, which is surrounded by an agricultural district that yields wheat. Called Complutum in Roman times, the town is famous as the former seat of a great university founded in 1508 and transferred in 1836 to Madrid and as the birthplace of Ferdinand I, Catherine of Aragón, and Cervantes. The town was severely damaged in the Spanish civil war.

Alcalá la Re·al (lä rē-äl′), town (1970 pop. 21,349), Jaén prov., S Spain, in Andalusia. It has mineral springs. The town played an important part in the recovery of Granada from the Moors (15th cent.). In 1810 it was the site of a French victory in the Peninsular War.

Al·ca·mo (äl′kä-mō′), city (1971 pop. 41,448), NW Sicily, Italy. It is an agricultural and industrial center and is noted for its white wine. The ruins of the ancient Greek settlement of Segesta are nearby.

Al·cán·ta·ra (ăl-kän′tä-rə), town (1970 pop. 4,636), Cáceres prov., W Spain, in Estremadúra, near the Tagus River. A fine Roman bridge built in honor of Emperor Trajan and the ruins of the convent and church of the Knights of Alcántara are located in the town.

Al·ca·traz (ăl′kə-trăz′), rocky island in San Francisco Bay, W Calif. Discovered by the Spanish in 1769, it came under U.S. control in 1851. The island was used as a U.S. military prison from 1859 until 1933, when it became a Federal prison; the prison was closed in 1963. The island became part of Golden Gate National Recreation Area in 1972.

Al·ca·zar·qui·vir (ăl-käz′ər-kĭ-vîr′), city (1960 pop. 34,035), N Morocco. Near the city on Aug. 4, 1578, the Moroccans soundly defeated the Portuguese led by King Sebastian of Portugal. Abd al-Malik, ruler of Morocco, and King Sebastian died in the fighting. Portugal soon passed (1580) to Philip II of Spain, since there were no direct heirs to the Portuguese throne.

Al·co·a (ăl-kō′ə), industrial city (1970 pop. 7,739), Blount co., E Tenn., S of Knoxville; inc. 1919. It was founded in 1913 by the Aluminum Company of America, which has established large reduction plants here.

Al·co·ba·ça (ăl′kə-bä′sə), town (1970 pop. 4,799), Leiria dist., W central Portugal, in Estremadura. The town became a center of the Cistercians in the reign of Alfonso I, and its abbey (begun 1152) was the greatest of medieval Portugal. The early kings of Portugal are buried in the abbey.

Al·co·na (ăl-kō′nə), county (1970 pop. 7,113), 677 sq mi (1,753.4 sq km), NE Mich., bounded on the E by Lake Huron and drained by the Au Sable and Pine rivers; est. 1840, organized 1869; co. seat Harrisville. It has agriculture (potatoes, grain, corn, and livestock), fisheries, and nurseries. It includes part of Huron National Forest.

Al·corn (ôl′kôrn′), county (1970 pop. 27,179), 405 sq mi (1,049 sq km), NE Miss., bordered on the N by the Tenn. border and drained by the Hatchie and Tuscumbia rivers; formed 1870; co. seat Corinth. It is a processing center for a lumbering and agricultural region yielding cotton, corn, dairy products, soybeans, and livestock. There is some textile manufacturing in the county.

Al·coy (äl-koi′), city (1971 pop. 61,371), Alicante prov., SE Spain, in Valencia, N of Alicante. An important industrial center with manufactures of paper (especially cigarette paper), matches, and textiles, it also has trade in grain, wine, and oil from the surrounding region.

Al·dab·ra (ăl-dăb′rə), main island, 60 sq mi (155.4 sq km), of the Aldabra island group, British Indian Ocean Territory, in the Indian Ocean NW of Malagasy Republic. Among the distinctive fauna found here are giant land tortoises. The Aldabras were first discovered by the Portuguese in 1511. In 1810 Britain assumed control.

Al·dan (äl-dän′), city (1970 est. pop. 19,000), Yakut Autonomous Republic, E Siberian USSR, on the Aldan Plateau. Located on a major north-south highway of the region, it is also in the heart of an important gold-mining area. Valuable mica deposits are nearby.

Al·dan (ôl′dən), borough (1970 pop. 5,001), Delaware co., SE Pa., a suburb SW of Philadelphia; inc. 1893.

Aldan, river, c.1,400 mi (2,255 km) long, rising in the Stanovoy Range, Yakut Autonomous Republic, SE Siberian USSR. It flows north and east, around the Aldan Plateau, before flowing generally northwest to enter the Lena River north of Yakutsk. The Aldan River is navigable c.1,000 mi (1,610 km) upstream. Gold is found in its basin.

Al·der·ney (ôl′dər-nē), island (1971 pop. 1,686), c.3 sq mi (7.8 sq km), in the English Channel, northernmost of the larger Channel Islands. It is separated from the French coast and from other islands by swift tidal races. The island's main crops are potatoes and grains. Tourism is important.

Al·der·shot (ôl′dər-shät′), municipal borough (1971 pop. 33,311), Hampshire, S central England. It is the site of the largest military training center (est. 1854) in Great Britain. The minister of defense appoints most of the borough council.

Al·dridge-Brown·hills (ôl′drĭj-broun′hĭlz′), urban district (1976 est.

pop. 89,370), West Midlands co., central England. It was created in 1966 through the merger of two former districts. Aldridge-Brownhills is residential and has extensive areas of open countryside. Chasewater Pleasure Park, in the northern part of the district, has the largest area of open water in the Midlands.

A·le·do (ə-lē'dō), city (1970 pop. 3,325), seat of Mercer co., NW Ill., SSW of Rock Island; inc. 1885. It is a trade and shipping center in an agricultural (corn, oats, wheat, soybeans, livestock, poultry, and dairy products) and bituminous-coal mining area. It has some manufacturing. Roosevelt Military Academy is here.

Al·ek·san·drov (ăl'ĭk-săn'drəf), city (1976 est. pop. 57,000), Russian Republic, E European USSR. It has radio, textile, and food industries. The city came under the control of the Muscovite princes in 1302. Ivan IV resided (1564-81) in Aleksandrov, where he organized his political police, the Oprichnina. The city is also the site of the first printing establishment in Russia, founded during the reign of Ivan IV, and of the famous Uspenski convent (late 17th cent.).

Al·ek·san·drovsk-Sak·ha·lin·ski (ăl'ĭk-săn'drəfsk-săk'hə-lĭn'skē), city on N Sakhalin island, Far Eastern USSR. A port on the Tatar Strait, it is also a coal-mining center and has lumber and fishing industries. The city was founded in 1881 as a place of exile.

Al·en·çon (ăl-äN-sôN'), town (1975 pop. 33,680), capital of Orne dept., N France, in Normandy, on the Sarthe River WSW of Paris. A commercial center in a fertile farm area, it is particularly noted for its fine lace work, an industry that dates from the 17th cent. The town also has spinning mills, printing plants, sawmills, and quarries. Alençon was heavily damaged in World War II. Among its surviving structures are Notre Dame Church, with windows and a porch from the 16th cent.; the Gothic St. Leonard's Church (completed in 1505); and the Ozé House (15th cent.).

A·len·te·jo (ə-lān-tā'zhōō), historic province, SE Portugal, now divided into Alto (Upper) Alentejo (4,888 sq mi/12,660 sq km) and Baixo (Lower) Alentejo (5,318 sq mi/13,773.6 sq km). The capital of Alto Alentejo is Évora; the capital of Baixo Alentejo is Beja.

Al·e·nu·i·ha·ha (äl'ə-nōō'ē-hä'hä), channel, 26 mi (41.8 km) wide, in the Hawaiian Islands separating Hawaii Island from Maui Island.

A·lep·po (ə-lĕp'ō) or **A·lep** (ə-lĕp'), city (1970 pop. 639,428), capital of Aleppo governorate, NW Syria. It is a commercial center located in a semidesert region where grains, cotton, and fruit are grown. The city is also a transportation hub and a market for wool, hides, and fruit. Manufactures include silk, printed cotton textiles, dried fruits and nuts (especially pistachios), and cement. The city was inhabited perhaps as early as the 6th millennium B.C. In the 14th-13th cent. B.C. it was controlled by the Hittites. Later, Aleppo was a key point on the major caravan route across Syria to Baghdad. From the 9th to the 7th cent. B.C. it was mostly ruled by Assyria and was known as Halman. It was later (6th cent. B.C.) held by the Persians and Seleucids. Seleucus I (d. 280 B.C.) rebuilt much of the city, renaming it Berea. The city's commercial importance was enhanced by the fall of Palmyra in A.D. 272, and by the 4th cent. Aleppo was a major center of Christianity. A flourishing city of the Byzantine Empire, it was taken without a struggle by the Arabs in 638; subsequently, in the late 11th cent., it was captured by the Seljuk Turks. Saladin captured it in 1183, making it his stronghold. The city was held briefly by the Mongols and by Tamerlane (1401); in 1517 the Ottoman Empire annexed Aleppo, which then became a great commercial city. Aleppo's importance declined in the late 19th cent. with the advent of the Suez Canal and other trade routes, but the city revived under French control after World War I and continued to prosper after Syrian independence (1941). Points of interest include the Byzantine citadel (12th cent.) and the Great Mosque (715).

A·lert (ə-lûrt'), settlement (1977 pop. 15), on Ellesmere Island, extreme N Northwest Territories, Canada, on the Arctic Ocean. It is the most northerly permanent settlement in the world. The settlement has a radio and meteorological station and a landing strip operated jointly by Canada and the United States.

A·lès (ä-lĕs'), formerly **A·lais** (ä-lā'), city (1975 pop. 44,245), Gard dept., S France, in Languedoc, at the foot of the Cévennes Mts. NW of Nîmes. Situated in one of the most important coal basins in southeast France, it has iron and steel industries, vehicle-repair facilities, and factories making machinery and hosiery. In the 16th cent. Alais was one of the principal centers of French Protestantism. Several buildings date from the 17th cent.

Al·es·san·dri·a (äl'ə-săn'drē-ə), city (1975 est. pop. 102,910), capital of Alessandria prov., in Piedmont, NW Italy, on the Tanaro River just E of Asti. It is an industrial center and agricultural market. Manufactures include furniture, machinery, and hats. Alessandria was built (c.1168) as a stronghold of the Lombard League and was named for Pope Alexander III. There are two 13th cent. churches and remains of the city's medieval fortifications.

Å·le·sund (ô'lə-sōōn'), city (1977 est. pop. 40,961), Møre og Romsdal co., W Norway, on three islands in the Atlantic Ocean at the mouth of the Storfjord. It is a major commercial and fishing port. Products include clothing, processed fish, and dairy goods.

Al·etsch (äl'ĭch), glacier, 66 sq mi (171 sq km), 16 mi (25.7 km) long and 1.2 mi (1.9 km) wide, S central Switzerland, largest in the Alps. It lies between the Jungfrau and the Aletschhorn, one of the highest (13,721 ft/4,184.9 m) peaks in the Bernese Alps.

A·leu·tian Islands (ə-lōō'shən), chain of rugged, volcanic islands curving c.1,200 mi (1,930 km) W from the tip of the Alaska Peninsula and approaching the Komandorski Islands, USSR. A partially submerged continuation of the Aleutian Range, they separate the Bering Sea from the Pacific Ocean. The Aleutians are composed of four main groups: Fox Islands, nearest to the mainland; Andreanof Islands; Rat Islands; and Near Islands, smallest and westernmost group. The Aleutians have few good harbors, and the numerous reefs make navigation treacherous. Temperatures are relatively moderate, but heavy rains and constant fog make the climate dreary. Almost completely treeless, the islands have a luxuriant growth of grasses, bushes, and sedges. Sheep and reindeer are raised. Hunting and fishing are the main occupations of the Eskimo population.

The Aleutian Islands were discovered in 1741 by Vitus Bering, a Danish explorer employed by Russia. The indigenous Aleuts were exploited by the Russian trappers and traders who, in search of sea otter, seal, and fox fur, established settlements on the islands in the late 18th and early 19th cent. The Aleutian Islands were included in the Alaska purchase in 1867 and at that time became part of the United States. Dutch Harbor, one of the few good harbors, became a transshipping point for Nome in 1900, after the discovery of gold turned Nome into a boom town. During World War II a U.S. naval base was established at Dutch Harbor. In 1942 the Japanese bombed the base and later occupied Attu, Kiska, and Agattu islands. The United States regained the islands in 1943. The Aleutian Islands play an important role in U.S. defense because of their proximity to the USSR.

Aleutian Range, volcanic mountain chain, c.600 mi (965 km) long, SW Alaska, extending W from Anchorage along the Alaska Peninsula and continuing, partly submerged as the Aleutian Islands, to Attu Island. Mt. Redoubt (10,200 ft/3,111 m) is the highest peak.

Aleutian Trench, submarine trough, 26,574 ft (8,105.1 m) deep, S of the W end of the Aleutian Islands, in the N Pacific Ocean.

Al·ex·an·der (ăl'ĭg-zăn'dər). **1.** County (1970 pop. 12,015), 224 sq mi (580.2 sq km), extreme S Ill., bounded on the W and S by the Mississippi, on the SE by the Ohio River, and drained by the Cache River; formed 1819; co. seat Cairo. It is mainly agricultural (cotton, fruit, and corn), with light manufacturing. It includes part of Shawnee National Forest. **2.** County (1970 pop. 19,466), 255 sq mi (660.5 sq km), W central N.C., in the Piedmont, bounded on the S by the Catawba River; formed 1847; co. seat Taylorsville. It has agriculture and timber. Furniture, textiles, and wood and paper products are manufactured, and food is processed.

Alexander I, island, 16,700 sq mi (43,253 sq km), in Bellingshausen Sea off the W coast of the Antarctic Peninsula, British Antarctic Territory. It was named Alexander I Land by its discoverer, F. G. von Bellingshausen, on his 1819-21 expedition. In 1940 a U.S. team proved it was an island.

Alexander Archipelago, island group off SE Alaska. The islands are the exposed tops of the submerged coastal mountains that rise steeply from the Pacific Ocean. Deep, fjordlike channels separate the islands and cut them off from the mainland; the northern part of the Inside Passage threads its way among the islands. All the islands are rugged, densely forested, and have an abundance of wildlife. Lumbering, trapping, fishing, and canning are the main industries. The archipelago was discovered by the Russians in 1741 and was later explored by Britain, Spain, and the United States.

Alexander Bay, town, Cape Prov., NW South Africa, where the Orange River enters the Atlantic Ocean. It is the site of some of the world's richest alluvial diamond deposits.

Alexander City, city (1978 est. pop. 13,200), Tallapoosa co., E central Ala., in a piedmont farm area; inc. 1873. Nearby Martin Dam supplies power for the city's textile mills, foundries, and mobile home manufactures. Horseshoe Bend National Military Park, site of a fierce battle (1814) between Andrew Jackson and the Creek Indians, is nearby.

Al·ex·an·dret·ta (ăl'ĭg-zăn-drĕt'ə): *see* İskenderun, Turkey.

Alexandretta, San·jak of (sän-jäk'), former name of Hatay prov. (1970 pop. 596,201), 2,141 sq mi (5,545.2 sq km), S Turkey. It was awarded to Syria in 1920; in 1936 Turkey complained to the League of Nations that the privileges of the Turkish minority were being infringed. It was given autonomous status in 1937 by an agreement between France and Turkey. Rioting by Turks and Arabs resulted (1938) in the establishment of joint French and Turkish military control. In 1939 France transferred the sanjak to Turkey.

Al·ex·an·dri·a (ăl'ĭg-zăn'drē-ə), city (1976 pop. 2,161,916), N Egypt, on the Mediterranean Sea. It is at the western extremity of the Nile River delta, situated on a narrow isthmus between the sea and Lake Maryut. The city is Egypt's leading port, a commercial and transportation center, and the heart of a major industrial area where refined petroleum, asphalt, cotton textiles, processed food, paper, and plastics are produced. In addition, motor vehicles are assembled and fish are caught.

Alexandria, founded in 332 B.C. by Alexander the Great, became the largest city in the Mediterranean basin and was the greatest center of Hellenistic and Jewish culture. Alexandria had two celebrated royal libraries. The collections at their maximum were said to contain, counting duplicates, c.700,000 rolls. Julius Caesar temporarily occupied (47 B.C.) the city while in pursuit of Pompey, and Octavian (later Augustus) entered it (30 B.C.) after the suicide of Antony and Cleopatra. In the later centuries of Roman rule and under the Byzantine Empire, Alexandria was a center of Christian learning that rivaled Rome and Constantinople. The libraries, however, were gradually destroyed from the time of Caesar's invasion, and suffered especially in A.D. 391, when Theodosius I had pagan temples and other structures razed. When the Moslem Arabs took Alexandria in 642, its prosperity had fallen severely, largely because of a decline in shipping. The Arabs moved the capital of Egypt to Cairo in 969 and Alexandria's decline continued, becoming especially rapid in the 14th cent., when the canal to the Nile silted up. During his Egyptian campaign, Napoleon took the city in 1798, but it fell to the British in 1801. The city gradually regained importance after 1819, when the Mahmudiyah Canal to the Nile was completed. During the 19th cent. many foreigners settled in Alexandria, and in 1907 they made up about 25% of the population. In World War II Alexandria, the chief Allied naval base in the eastern Mediterranean, was bombed by the Germans. The city's foreign population declined during the 20th cent.

Much of ancient Alexandria is covered by modern buildings or is underwater; only a few landmarks are readily accessible, including ruins of the emporium and a granite shaft (88 ft/26.8 m high) called Pompey's Pillar. Nothing remains of the lighthouse on the Pharos (3rd cent. B.C.), which was one of the Seven Wonders of the World. The Greco-Roman Museum in Alexandria houses a vast collection of Coptic, Roman, and Greek art.

Alexandria. 1. City (1970 pop. 5,600), Madison co., E central Ind., N of Anderson; settled c.1821. Canned goods are produced, and rockwool insulation is made. **2.** City (1970 pop. 3,844), a seat of Campbell co., N Ky., SSE of Cincinnati, Ohio. **3.** City (1978 est. pop. 50,200), seat of Rapides parish, central La., on the Red River; inc. 1818. It is a trade, rail, and medical center for a rich agricultural and timber area. Among its many manufactures are valves, lumber, paper, and soaps and cleansers. During the Civil War the city was burned (May, 1864) to the ground by Federal troops. It is the headquarters for Kisatchie National Forest and the seat of a branch of Louisiana State Univ. Several nearby lakes, recreation areas, state parks, and a hot mineral springs resort attract tourists. **4.** City (1975 est. pop. 6,843), seat of Douglas co., W Minn., in a rich farm and timber region surrounded by over 200 lakes; inc. 1877. Its economy is based upon tourism, agriculture, and light manufacturing. The Kensington Rune Stone is on exhibition at a museum here. **5.** City (1970 pop. 598), seat of Hanson co., SE S.Dak., ESE of Mitchell near the James River. Dairy products, building stone, corn, and wheat are produced. **6.** Independent city (1978 est. pop. 108,700), N Va., a port of entry on the Potomac; patented 1657, permanently settled 1730s, inc. 1779. Primarily a residential suburb of Washington, D.C., it also has extensive railroad yards and repair shops, a deepwater port, and a great variety of manufactures, including fertilizers, chemicals, and farm equipment. A number of U.S. government buildings and scientific and engineering research firms are here. George Washington helped lay out the streets in 1749. The city was part of the District of Columbia from 1789 to 1847. Its many historic buildings include Gadsby's Tavern (1752), Carlyle House (1752), Christ Church (1767–73), and Ramsey House (1749–51). Nearby are Mount Vernon; Woodlawn, one of the Washington family estates (made a national shrine in 1949); and Fort Belvoir.

Al·ex·an·droú·po·lis (ăl'ĭg-zăn-dro͞o'pə-ləs), city (1971 pop. 22,995), capital of Evros prefecture, NE Greece, W Thrace, a seaport on the Gulf of Ainos, an inlet of the Aegean Sea. Alexandroúpolis is a commercial center; wheat, cotton, rice, tobacco, salt, and dairy products are traded. It developed from a small fishing village after 1871. The city suffered greatly at the hands of the Bulgarians in both World Wars. It was ceded to Greece in 1919.

Al·fal·fa (ăl-făl'fə), county (1970 pop. 7,224), 884 sq mi (2,289.6 sq km), N. Okla., bounded on the N by the Kansas border and drained by the Medicine Lodge River; formed 1907; co. seat Cherokee. The Great Salt Plains Dam impounds the Salt Fork of the Arkansas River here. The county is mainly agricultural (wheat, barley, cotton, alfalfa, livestock, and poultry), with gas and oil wells.

Al·fi·ós (äl-fyo͞os'), river: *see* Alpheus.

Al·föld (ôl'föld'), great central plain of Hungary extending into N Yugoslavia and W Rumania. Formerly wooded, the Alföld gradually became a steppe region as the 13th cent. Mongol invaders cut down many trees, exposing the soil to dry winds. Grasslands covered most of the Alföld until the late 19th cent., when extensive irrigation and drainage projects transformed it into fertile farmland; grains, hemp, flax, and livestock are now raised.

Al·fred (ăl'frĭd). **1.** Town (1970 pop. 1,211), seat of York co., SW Maine, W of Biddeford; settled 1764, inc. 1794. **2.** Village (1970 pop. 3,804), Allegany co., SW N.Y., SW of Hornell; inc. 1881. Alfred Univ. is here.

Al·fre·ton (ôl'frə-tən), urban district (1976 est. pop. 21,560), Derbyshire, N central England. Its products include clothing, cosmetics, and agricultural machinery.

Al·gar·ve (äl-gär'və), province (1975 est. pop. 276,800), 1,958 sq mi (5,071.2 sq km), extreme S Portugal, coextensive with Faro dist.; capital Faro.

Al·ge·cir·as (ăl'jə-sĭr'əs), city (1970 pop. 81,622), Cádiz prov., S Spain, in Andalusia, on the Bay of Algeciras opposite Gibraltar. A Mediterranean seaport, it has fishing and tourist industries. It was the first Spanish town taken (711) by the Moors. In the naval engagements of July, 1801, near Algeciras, the British defeated the French and Spanish fleets.

Al·ger (ăl'jər), county (1970 pop. 8,568), 904 sq mi (2,341.4 sq km), N Mich., in the Upper Peninsula, bounded on the N by Lake Superior and drained by the Whitefish and Sturgeon rivers; organized 1885; co. seat Munising. Resorts and recreation areas abound in the county, which also depends on dairying, agriculture (livestock, poultry, potatoes, grain, and fruit), timber and lumber products, fish, and some manufacturing. It includes part of Hiawatha National Forest.

Al·ge·ri·a (ăl-jĭr'ē-ə), republic (1974 est. pop. 16,275,000), 919,590 sq mi (2,381,738 sq km), NW Africa, bordering on Mauritania, Spanish Sahara, and Morocco in the W, on the Mediterranean Sea in the N, on Tunisia and Libya in the E, and on Niger and Mali in the S. Algiers is the capital. Algeria falls into two main geographical areas, the northern region and the much larger Saharan, or southern, region. The northern region, which is part of the Maghreb, is made up of four parallel east-west zones: a narrow lowland strip (interspersed with mountains) along the country's Mediterranean coastline; the Tell Atlas Mts., which have a Mediterranean climate and abundant fertile soil; the sparsely populated, semiarid Plateau of the Chotts; and the Saharan Atlas Mts., a broken series of mountain ranges and massifs, also a semiarid area and used chiefly for pasturing livestock. The arid and very sparsely populated Saharan region has an average elevation of c.1,500 ft (460 m), but reaches greater heights in the Ahaggar Mts. in the south, where Algeria's loftiest point, Mt. Tahat (9,850 ft/3,004.3 m), is located. Most of the region is covered with gravel or rocks, with little vegetation; there are also large areas of sand dunes in the north and east.

Economy. About half of Algeria's workers are engaged in farming, but agriculture's contribution to the country's annual domestic product is much less than that of either mining or manufacturing, both of which began their main growth in the mid-1960s. The state plays a leading role in planning the economy and owns many important industrial concerns. Farming is concentrated in the fertile valleys and basins of the north and in the oases of the Sahara. The principal crops are wheat, barley, oats, potatoes, citrus fruit, wine grapes, olives, tomatoes, tobacco, figs, and dates. Large numbers of sheep, poultry, goats, and cattle are raised. Petroleum, found principally in the east Sahara, is Algeria's most important mineral resource and its leading export. Much natural gas is also produced. Other minerals extracted in significant quantities include iron, lead, and copper ores; phosphates; zinc; mercury; antimony; kaolin; salt; and coal. The country's leading manufactures are processed food (notably olive oil), beverages (especially wine), tobacco products, construction materials, chemicals, metals (including steel), refined petroleum, liquefied natural gas, textiles, and clothing. There are small forest-products and fishing industries.

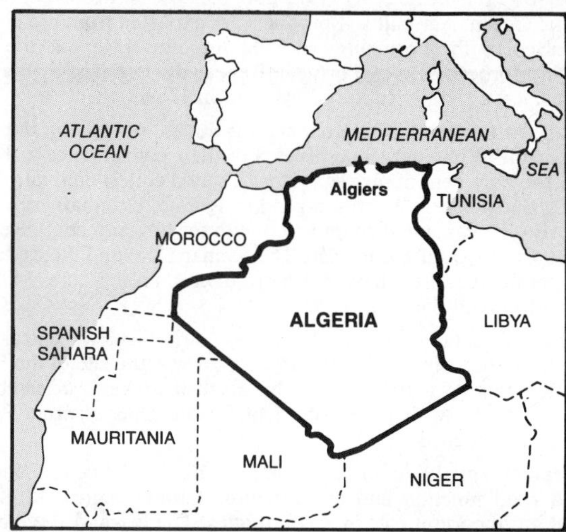

History. The earliest known inhabitants of Algeria were Berber-speaking nomads, who lived in small political units by the 2nd millennium B.C. In the 9th cent. B.C., Carthage was founded in modern-day Tunisia, and Carthaginians eventually established trading posts at Annaba, Skikda, and Algiers. Coastal Algeria was known as Numidia and was usually divided into two kingdoms. The Romans held Numidia by 106 B.C. and also gained control of the Tell Atlas region and part of the Plateau of the Chotts; the rest of present-day Algeria remained under Berber rulers and was outside Roman influence. By the Christian era, Algeria (divided into Numidia and Mauritania Caesariensis) was an integral, albeit relatively unimportant, part of the Roman Empire. However, by the 5th cent. the incursions of Berbers and Saharan tribes and the destruction wreaked by the Vandals in 430-431 marked the end of effective Roman influence.

In the early 6th cent. a temporary veneer of unity and order was forged by the Byzantine Empire, which conquered parts of the North African coast including the region east of Algiers. In the late 7th and early 8th cent. Moslem Arabs conquered Algeria and ousted the Byzantines. In the late 15th cent. Spain expelled the Moslems from its soil and soon thereafter captured the coastal cities of Algeria. Algerians appealed to Turkish pirates for help and, with the aid of the Ottoman Empire, ended Spanish control by the mid-16th cent. Algeria then came under Ottoman rule. The country was at first governed by officials sent from Constantinople, but in 1671 the dey of Algiers, chosen by local civilian, military, and pirate leaders to govern for life and virtually independent of the Ottoman Empire, became head of Algeria. The power of the Ottomans, and later of the deys, did not extend much beyond the Tell Atlas. The coast was a stronghold of pirates. Privateering reached a high point in the 16th and 17th cent. and declined thereafter; there was a temporary increase during the Napoleonic Wars (early 19th cent.). The country was, in addition, a center of the slave trade.

In the 1820s a minor dispute with the French reached a climax that had far-reaching effects: two Algerian merchants had delivered wheat to France in the 1790s but had never been paid for it. The dey unsuccessfully pressed their claim for payment, and, in exasperation, he flicked the French consul in Algeria with a fly whisk during an audience in 1827. To avenge this insult, Charles X of France responded first by instituting a naval blockade of Algeria and then, in June, 1830, by invading the country. The dey capitulated in July, 1830, but most of the country resisted the French. In 1834 the French renewed their drive to occupy Algeria and in 1837 they took Constantine, the last major city to retain its independence.

Colonization by Europeans accelerated after 1848, when Algeria was declared to be French territory. By 1880 persons of European descent numbered about 375,000, and they controlled most of the better farmland. During the 19th cent. Algeria was usually administered under civil departments in Paris, with intermittent periods of military rule. By 1900 the colonists had started large-scale agricultural and industrial enterprises and had built roads, railroads, schools, and hospitals and modernized the cities. Although the official French policy in Algeria was to encourage the Moslems to adapt to European ways and thus to prepare them for full citizenship, very little was done to implement this policy.

After World War I two types of protest groups were started by the Moslems, one movement calling for a fully independent, Moslem-controlled Algeria; the other faction seeking assimilation with France and the equality of Moslems and Europeans in Algeria. The hopes of the nationalists were buoyed by Allied statements during World War II concerning self-determination, but the Moslems' actual status improved little. Despairing of ever gaining meaningful concessions from the colonists or the French government, a radical group of Moslems in 1954 seceded from Messali Hadj's Movement for the Triumph of Democratic Liberties (MTLD), forming the National Liberation Front (FLN). The FLN called for the establishment of an independent Algerian state controlled by the Moslem majority. The MTLD was reorganized into the Algerian Nationalist Movement, which, led by Messali, unsuccessfully competed with—and at times fought against—the FLN. By 1956 the FLN had the support of virtually all Algerian nationalists except Messali, controlled much of the countryside, and was organizing frequent terrorist actions in the cities (especially Algiers). In 1957 the French successfully used massive measures to rid the cities of most of the terrorists, and the FLN was forced to concentrate on guerrilla activities in the rural areas.

By 1960 De Gaulle had come to recognize the inevitability of some form of Algerian independence; the main problem concerned the future status of the almost one million European colonists, many of whom had been born in Algeria. Shortly thereafter, negotiations began and in Mar., 1962, an agreement was signed. The accord provided for an end to the fighting and for the establishment of an independent Algerian state after a transition period. Members of the French army in Algeria, banded together in the Secret Army Organization (OAS), launched a terrorist campaign against Moslems in an attempt to prevent the implementation of the accord. However, in late April their leader, Gen. Raoul Salan, was captured and by late June the army revolt had been ended. Already in April the colonists had begun to leave Algeria in large numbers. As a result of the more than seven years' fighting at least 100,000 Moslem and 10,000 French soldiers had been killed; in addition, many thousands of Moslem civilians and a much smaller number of colonists lost their lives.

On July 1, 1962, the people of Algeria voted almost unanimously for independence and on July 3 France recognized Algeria's sovereignty. As a result of the fighting and of the exodus of colonists, the Algerian economy lay in ruins by mid-1962. Ben Khedda, the moderate leader of the Government of the Algerian Republic (GPRA), formed the initial Algerian government, but in Sept., 1962, he was replaced as prime minister by Ahmed Ben Bella, a leftist radical who had the support of the FLN. In Sept., 1963, Ben Bella was elected president. Ben Bella, who increasingly concentrated power in his hands, followed a left-wing domestic policy that included the confiscation of European-held farms and the nationalization of various parts of the economy. On June 19, 1965, Ben Bella was deposed in a bloodless coup d'état by Houari Boumedienne, his defense minister, who was angered by the army's greatly reduced influence and by the deterioration of the economy. Boumedienne suspended the constitution and established a revolutionary council, of which he was president, to run the country. By the end of 1968 he had a secure hold on Algeria. Algeria gave strong vocal support to the Arabs in the Arab-Israeli wars of 1967 and 1973 and also contributed some soldiers and military supplies (especially aircraft). After an initial slowdown Boumedienne increased the pace of state involvement in the economy. In 1971 he nationalized (with compensation) the French oil and natural-gas companies active in Algeria. By 1972 output had reached record levels, and there was a growing emphasis on the export of liquefied natural gas. After Boumedienne's death (1978) he was succeeded by Nbenjagid Chadli.

Al·giers (ăl-jĭrz'), city (1966 pop. 943,142), capital of Algeria, N Algeria, on the Bay of Algiers of the Mediterranean Sea. It is one of the leading ports of North Africa (wine, citrus fruit, iron ore, cork, and cereals are the major exports), as well as a popular winter resort and a commercial center. Industries include metallurgy, oil refining, automotive construction, machine-building, and the production of chemicals, tobacco, paper, and cement. Founded by the Phoenicians and called Icosium by the Romans, the city disappeared after the fall of the Roman Empire. It was re-established in the late 10th cent. by the Moslems. In 1511 Barbarossa captured Algiers for the Turks. Algiers then became a base for the Moslem fleet that preyed upon Christian commerce in the Mediterranean. As European navies repeatedly attacked Algiers, the city's prosperity declined. French forces captured the port in 1830. Algiers became headquarters for the Allied forces in North Africa in World War II, as well as for Charles de Gaulle's provisional French government. An anti-French uprising in the city in 1954 provided a major spark in the Algerian armed struggle for independence. In May, 1958, Algiers was the principal scene of a revolt that ended the Fourth French Republic and returned De Gaulle to power. During the final months before Algeria won independence (1962), bombings by the French terrorist Organization of the Secret Army (OAS) damaged industrial and communications facilities in Algiers. The city is divided into the newer, French-built sector, with wide boulevards and modern buildings, and the original Moslem quarter, with narrow streets, numerous mosques, and the 16th cent. casbah.

Al·go·na (ăl-gō'nə), city (1970 pop. 6,032), seat of Kossuth co., N Iowa, on the Des Moines River N of Fort Dodge; settled 1854, inc. 1872. It is a rail junction where poultry packing and grain milling are done. There are rendering plants, creameries, machine shops, and concrete works here, with sand pits nearby.

Al·ham·bra (ăl-hăm'brə), city (1978 est. pop. 60,600), Los Angeles co., S Calif., a suburb NE of Los Angeles; inc. 1903. Its many manufactures include aircraft parts, electronic equipment, oil-refinery machinery, air conditioners, and felt products.

Alhambra, extensive group of buildings on a hill overlooking Granada, Spain. They were built chiefly between 1230 and 1354 and they formed a great citadel of the Moorish kings of Spain. After the expulsion of the Moors in 1492, the structures suffered mutilation, but were extensively restored after 1828. The Alhambra is a true expression of the once flourishing Moorish civilization and is the finest example of its architecture in Spain.

Al Hil·lah (ăl hĭl'ə), city (1970 est. pop. 128,800), central Iraq, on a branch of the Euphrates River. It is a port and the main cereal market of the area. It was built (c.1100) largely of material taken from the nearby ruins of ancient Babylon.

Al Hu·day·dah (ăl ho͝o-dā'də): *see* Hodeida, Yemen.

Al Hu·fuf (ăl ho͝o-fo͝of'): *see* Hofuf, Saudi Arabia.

Al·iák·mon (äl-yäk'môn), longest river of Greece, c.200 mi (320 km) long, rising in the mountains near Lake Prespa and flowing SE then NE into the Gulf of Thessaloníki. The river wáters an agricultural region and forms the western portion of the extensive Vardar River delta.

Al·i·ba·tes Flint Quarries and Texas Panhandle Pueblo Culture National Monument (ăl'ə-bät'ēz): *see* National Parks and Monuments Table.

Al·i·can·te (ăl'ə-kăn'tē), city (1975 est. pop. 214,760), capital of Alicante prov., SE Spain, in Valencia. A Mediterranean port and resort, it has exports of wine, oil, cereals, fruit, and esparto from the fertile surrounding region. Textiles and tobacco and clay products are made. The Romans had a naval base on the site. The town was recaptured from the Moors c.1250.

Al·ice (ăl'ĭs), city (1978 est. pop. 20,300), seat of Jim Wells co., S Texas; inc. 1910. Long a cow town at a railroad junction, Alice is still a cattle-shipping center. Oil and natural gas are also important to its economy. Manufactures include oil well equipment and cottonseed oil. Nearby are a wildlife refuge, the King Ranch, and several Gulf Coast resorts.

Alice Springs, town (1975 est. pop. 13,400), Northern Territory, Australia. It lies in a pastoral area near the center of the continent and at the terminus of the Central Australian RR. The town became important as a telegraph station on the overland route from Adelaide to Darwin. Gold, copper, wolfram, and mica are mined in the area.

Al·i·garh (ăl'ĭ-gär'), city (1971 pop. 252,314), Uttar Pradesh state, N central India. An important agricultural trade center, it also has cotton mills. Aligarh is famous chiefly for its university, opened in 1875 as Anglo-Oriental College.

A·ling·sås (ä'lĭng-sōs'), city (1975 est. pop. 23,000), Älvsborg co., SW Sweden, NE of Göteborg; chartered 1619. It is an industrial center. Manufactures include textiles, leather goods, processed food, candy, beer, and metal goods.

Al·i·quip·pa (ăl'ĭ-kwĭp'ə), borough (1978 pop. 20,300), Beaver co., W Pa., in a highly industrialized region along the Ohio River N of Pittsburgh; inc. 1894. Aliquippa grew after the expansion of steel mills in 1909.

Al-Ja·di·da (ăl'jə-dē'də), city (1971 pop. 55,501), W Morocco, on the Atlantic Ocean. Agricultural products are exported from the port. It was seized by the Portuguese in 1502 and after 1541 was the only place in Morocco held by Portugal. Repeatedly besieged by the Moroccans, it was finally captured by them in 1769.

Al Ja·zi·rah (ăl jă-zē'rä), region, central Sudan, occupying the tract between the White and Blue Niles S of their convergence at Khartoum. Development of the region for irrigated cotton cultivation has made it Sudan's leading cotton-producing area. Originally operated by a private company in conjunction with the government, the entire project was nationalized in 1950. The Sannar Dam and the irrigation canals built since 1925 have put more than 1 million acres (405,000 hectares) into cultivation.

Al·ju·bar·ro·ta (äl'jo͞o-bə-rō'tə), village, Leiria dist., W central Portugal, in Beira Litoral. On Aug. 14, 1385, it was the site of the battle in which the Portuguese, aided by English archers, defeated the forces of the Spanish King John I of Castile, thus assuring Portuguese independence.

Al Ka·rak (ăl kär'ək), town (1973 est. pop. 10,000), W central Jordan. It is a road junction and an agricultural trade center. Al Karak played an important role in the Crusades. The brigand Reginald of Châtillon was lord of Al Karak and Montreal when, in 1187, he attacked a caravan led by Sultan Saladin and thus provoked the events leading to the fall of Jerusalem. Al Karak was taken by Saladin in 1188 after a long siege. The Christians were massacred or expelled in 1910.

Alk·maar (ălk'mär'), town (1977 est. pop. 67,554), North Holland prov., NW Netherlands. It is an important market town and has varied industries. The Edam-cheese market is world famous. Alkmaar was chartered in 1254. Its successful defense (1573) against Spanish troops was a turning point in the revolt of the Netherlands.

Al Ku·fah (ăl ko͞o'fə), town (1965 pop. 30,862), S central Iraq. Founded (638) by Caliph Omar I, it was one of the two Moslem centers (the other was Basra) of the early Umayyad caliphs.

Al Kut (ăl ko͞ot'), town (1965 pop. 42,116), SE Iraq, on the Tigris River. It is a port and a market center for grains, dates, fruit, and vegetables. Much of the town was destroyed during World War I.

Al·lah·a·bad (ăl'ə-hə-băd', -bäd'), city (1971 pop. 490,662), Uttar Pradesh state, N central India. On the site of Prayag, an ancient Indo-Aryan holy city, Allahabad is at the junction of two sacred rivers, the Jumna and the Ganges, and is visited by many Hindu pilgrims. The oldest monument is a pillar (c.242 B.C.) with inscriptions from the reign of Asoka. The city was the scene of much fighting in the Indian Mutiny (1857).

Al·la·ma·kee (ăl'ə-mə-kē'), county (1970 pop. 14,968), 639 sq mi (1,655 sq km), extreme NE Iowa, bounded on the N by the Minn. border, on the E by the Wis. border; formed 1847; co. seat Waukon. It is in a prairie area drained by the upper Iowa River. Dairying is important, as are hogs, poultry, corn, and grass seed. There are limestone quarries and iron deposits. Some manufacturing is done.

All-A·mer·i·can Canal (ôl'ə-mĕr'ĭ-kən), 80 mi (128.7 km) long, SE Calif.; part of the Federal irrigation system of the Hoover Dam. Built between 1934 and 1940 across the Colorado Desert, the canal is entirely within the United States and replaces the Inter-California Canal, which passes through Mexico. The Imperial Dam, northeast of Yuma, Ariz., diverts water from the Colorado River into the All-

American Canal, which runs west to Calexico, Calif. This canal system irrigates more than 630,000 acres (255,150 hectares) and has greatly increased crop yield in the area.

Al·le·gan (ăl'ĭ-gən), county (1970 pop. 66,575, 829 sq mi (2,147.1 sq km), SW Mich., bounded on the W by Lake Michigan and drained by the Kalamazoo River; organized 1835; co. seat Allegan. It has agriculture (fruit, dairy products, vegetables, corn, onions, grain, hay, and livestock), fisheries, resorts, and some manufacturing.

Allegan, city (1970 pop. 4,516), seat of Allegan co., SW Mich., on the Kalamazoo River SSW of Grand Rapids, in a farm region; settled 1834, inc. as a village 1838, as a city 1907. Automobile parts and furniture are made.

Al·le·ga·ny (ăl'ə-gā'nē, -gĕn'ē). **1.** County (1970 pop. 84,044), 428 sq mi (1,108.5 sq km), NW Md., bounded on the N by the Pa. border, on the S by the Potomac River; formed 1789; co. seat Cumberland. It has bituminous-coal fields, limestone quarries, sand, gravel, and clay pits, and stands of timber. The fertile agricultural areas of the county produce apples, corn, wheat, and dairy products. Textile, paper, glass, and clay products are manufactured. Green Ridge State Forest and the Cumberland Narrows are here. **2.** County (1970 pop. 46,458), 1,048 sq mi (2,714.3 sq km), W N.Y., bounded on the S by the Pa. border and drained by the Genesee and Canisteo rivers; formed 1806; co. seat Belmont. It has dairying, diversified manufacturing, and a number of colleges. There are resorts on the small lakes here. It includes part of the Oil Spring Indian Reservation.

Al·le·gha·ny (ăl'ə-gā'nē, -gĕn'ē). **1.** County (1970 pop. 8,134), 230 sq mi (595.7 sq km), NW N.C., in the Blue Ridge on the Va. border; formed 1859; co. seat Sparta. Its agriculture includes tobacco, cabbage, dairy products, and beef cattle. Sawmilling is also important. It is included in Yadkin National Forest. **2.** County (1970 pop. 12,461), 451 sq mi (1,168.1 sq km), W Va., in the Alleghenies bounded on the W by the W.Va. border and drained by the Jackson and Cowpasture rivers; formed 1882; co. seat Covington (an independent city). It has agriculture (grain hay, fruit, livestock, and dairy products), iron and coal mining, limestone quarrying, and some manufacturing (especially paper and synthetic textiles). It includes part of Jefferson National Forest.

Al·le·ghe·ny (ăl'ə-gā'nē, -gĕn'ē), county (1970 pop. 1,605,133), 730 sq mi (1,890.7 sq km), W Pa., in an industrial area drained by the Allegheny and Monongahela rivers; formed 1788; co. seat Pittsburgh. It is a center of steel production. Its resources include bituminous coal, natural gas, oil, clay, limestone, and sandstone. Agricultural services, forestry, and fishing are also important. Textiles, food, lumber, and metal products, and electrical equipment are made.

Allegheny, river, 325 mi (523 km) long, rising in N central Pa., and flowing NW into N.Y., then SW through Pa. to the Monongahela River, with which it forms the Ohio River at Pittsburgh. It drains 11,580 sq mi (29,992 sq km). Before the railroad era, the river was an important commercial route and is still used to transport coal and other bulky freight.

Allegheny Mountains, dissected plateau, W part of the Appalachian Mts., extending c.500 mi (805 km) SW from N Pa. to SW Va., rising to c.4,860 ft (1,480 m) at Spruce Knob. The eastern Alleghenies, with a steep escarpment often called the Allegheny Front (c.1,500–1,600 ft/460–490 m high), are more rugged than the western portion, which is a plateau extending into Ohio and Ky. The Alleghenies, formed by the folding of sedimentary rocks, have been subsequently reduced by erosion. The mountains are rich in coal and timber and contain iron ore, petroleum, and natural gas.

Allegheny Portage Railroad National Historic Site: *see* National Parks and Monuments Table.

Al·len (ăl'ən). **1.** County (1970 pop. 280,455), 671 sq mi (1,737.9 sq km), NE Ind., bounded on the E by the Ohio border and drained by the Maumee, St. Joseph, and St. Marys rivers; formed 1823; co. seat Fort Wayne. Farming (grain, corn, and soybeans), stock raising, and dairying are important. **2.** County (1970 pop. 15,043), 505 sq mi (1,308 sq km), SE Kansas, in a rolling prairie area drained by the Neosho River; formed 1855; co. seat Iola. It has stock and grain raising, dairying, and oil and gas fields. Its manufacturing includes some primary metal industries. **3.** County (1970 pop. 12,598), 364 sq mi (942.8 sq km), S Ky., bounded on the S by the Tenn. border, on the NE by the Barren River, and drained by Trammel Fork; formed 1815; co. seat Scottsville. Livestock, grain, burley tobacco, fruit, poultry, and dairy products are processed. It also has oil wells, sawmills, and some manufacturing. **4.** Parish (1970 pop. 20,794), 755 sq mi (1,955.5 sq km)., SW La., drained by the Calcasieu River; formed 1910; parish seat Oberlin. It has pulp, paper, and lumber mills, rice and feed mills, and cotton gins. Its agriculture includes rice, corn, cotton, sweet potatoes, citrus fruit, livestock, and poultry. **5.** County (1970 pop. 111,144), 410 sq mi (1,061.9 sq km), W Ohio, intersected by the Ottawa and Auglaize rivers; formed 1831; co. seat Lima. It has diversified agriculture (livestock, grain, poultry, and soybeans), oil wells, and limestone quarries. Auto parts, machine tools, textiles, cigars, and lumber, wood, concrete, gypsum, and plaster products are manufactured.

Allen, Bog of, area of several peat bogs, c.375 sq mi (970 sq km), with patches of cultivable land, in the central lowlands, E Republic of Ireland. It is a source of fuel and contains peat-fired electrical generating stations.

Al·len·dale (ăl'ən-dāl'), county (1970 pop. 9,783), 418 sq mi (1,082.6 sq km), SW S.C., bounded on the W by the Savannah River and drained by the Coosawhatchie River; formed 1919; co. seat Allendale. Its agriculture includes truck crops and livestock.

Allendale. 1. Borough (1970 pop. 6,240), Bergen co., NE N.J., NNE of Paterson; settled 1740, inc. 1894. **2.** Town (1970 pop. 3,620), seat of Allendale co., SW S.C., near the Savannah River SSW of Columbia; settled in the mid-18th cent., inc. 1873. It is an agricultural area noted for watermelons, cotton, and lumber. Veneer, beverages, and confectionery are produced.

Allen Park, city (1978 est. pop. 38,500), Wayne co., SE Mich., a suburb of Detroit; inc. as a village 1927, as a city 1957. Its manufactures include automobiles, tires, liquor, bread, and potato chips. The area was settled in the early 1800s.

Al·len·town (ăl'ən-toun'), city (1978 est. pop. 104,900), seat of Lehigh co., E Pa., on the Lehigh River; founded 1762, inc. as a borough 1811, as a city 1867. Allentown, situated in the agricultural Lehigh valley and in the Pennsylvania Dutch region, is an industrial and commercial city. Cement, truck and bus bodies, clothing, machinery, small appliances, transistors, tubes, air-reduction equipment, gas-generating equipment, pneumatic loading machinery, and beer are the major products. In the city are Muhlenberg College and a campus of Pennsylvania State Univ. The Liberty Bell was brought here (1777) for safekeeping during the Revolutionary War. Points of interest include the Zion Reformed Church (where the Liberty Bell was kept) and an art museum.

Al·lep·pey (ə-lĕp'ē), town (1971 pop. 160,166), Kerala state, SW India. It is a district administrative center and port on the Arabian Sea. Copra, coir, rubber, and spices are its chief exports. Fishing is a major industry.

Al·ler (äl'ər), river, 131 mi (210.8 km) long, NW Germany. It rises near Magdeburg, East Germany, then continues north and northwest to the Weser River, near Bremen, West Germany.

Al·li·a (ăl'ē-ə), small river, in Latium, central Italy. Near its confluence with the Tiber, north of Rome, the Gauls defeated the Romans in 390 B.C.

Al·li·ance (ə-lī'əns). **1.** City (1970 pop. 6,862), seat of Box Butte co., NW Nebr., in the Great Plains region, NE of Scottsbluff; founded c.1888. A shipping and trade center for a large grain and livestock area, it has railroad repair shops and a seed-potato testing ground. Farm machinery, electronic equipment, and dairy and poultry goods are among its products. **2.** City (1978 est. pop. 25,600), Mahoning and Stark cos., NE Ohio, on the Mahoning River, in a farm area; inc. 1854. It is an industrial, distributing, and rail center, with manufactures of steel, heavy machinery, electric tubing, chinaware, and farm, railroad, and industrial equipment. Mount Union College and Clarke Observatory are here.

Al·lier (ăl-yā'), department (1968 pop. 386,533), 2,829 sq mi (7,327.1 sq km), central France, in Bourbonnais; capital Moulins.

Allier, river, rising in central France, Lozère dept., in the Cévennes. It flows c.255 mi (410 km) northwest past Vichy and Moulins to join the Loire River below Nevers.

Al·li·son (ăl'ə-sən), town (1970 pop. 1,071), seat of Butler co., NE central Iowa, NW of Waterloo. Dairy products and poultry are produced. A limestone quarry and a sand and gravel pit are nearby.

Al·lo·a (ăl'ə-wə), burgh (1974 est. pop. 13,498), Central region, central Scotland, on the Firth of Forth. Coal mining, brewing, and bottle making are the principal industries.

Al·ma (ăl'mə), city (1976 pop. 25,638), S central Que., Canada, on the Saguenay River. In 1954 its name was shortened from St. Joseph d'Alma. There are granite quarries in the region, and the town has pulp and paper and aluminum plants.

Alma. 1. City (1970 pop. 3,756), seat of Bacon co., SE Ga., N of Waycross, in a coastal-plain farm area; inc. 1906. Tobacco, cotton, peanuts, pulpwood, and naval stores are produced. **2.** City (1970 pop. 905), seat of Wabaunsee co., E central Kansas, on a small affluent of the Kansas River W of Topeka, in a grain, poultry, and livestock area. **3.** City (1970 pop. 9,611), Gratiot co., S Mich., N of Lansing; settled 1853, inc. as a village 1872, as a city 1905. It has oil refineries, sugar-processing plants, and factories producing trailers and automobile parts. Alma College (1886) is here. Several Indian mounds are in the vicinity. **4.** City (1970 pop. 1,299), seat of Harlan co., S Nebr., SW of Grand Island, on the Republican River near the Kansas border; founded 1871. It is a trade center for an agricultural region yielding dairy and poultry produce, grain, and livestock. **5.** City (1970 pop. 956), seat of Buffalo co., W Wis., on the Mississippi SW of Eau Claire; settled c.1852, inc. 1885. It is noted for dairy products and timber. A Federal dam here was completed in 1935.

Al·ma-A·ta (ăl'mə-ə-tä'), city (1971 est. pop. 871,000), capital of the Kazakh SSR, Central Asian USSR, in the foothills of the Trans-Ili Ala-Tau. A terminus of the Turkistan-Siberia RR, Alma-Ata is an industrial and cultural center. Leading industries include motion-picture production, fruit canning, meat packing, tobacco processing, and the repair of railroad equipment. The city was founded in 1854 as a Russian fort and trade center.

Al·ma·dén (äl'mə-dān'), town (1970 pop. 10,774), Ciudad Real prov., central Spain, in New Castile. It is the center of one of the richest mercury-mining regions in the world. The mines have been exploited since Roman times.

Al Mah·di·yah (ăl mə-dē'yə) or **Mah·di·a** (mə-dē'ə), town (1975 pop. 21,711), E Tunisia, on the Mediterranean Sea. It is a fishing port where olive oil and handicrafts are marketed. The town was founded in 912 on the site of Phoenician and Roman colonies.

Al Ma·nam·ah (ăl mə-năm'ə), town (1971 pop. 89,112), capital of Bahrain, on the Persian Gulf. It has oil refineries and light industries and is a free port.

Al·me·lo (äl'mə-lō'), city (1977 est. pop. 62,501), Overijssel prov., E Netherlands. It is a manufacturing center and has a large textile industry.

Al·me·rí·a (ăl'mə-rē'ə), city (1975 est. pop. 120,072), capital of Almería prov., SE Spain, in Andalusia, on the Gulf of Almería. A busy Mediterranean port, it exports the celebrated grapes of the region, other fruits, esparto, as well as iron and other minerals mined nearby. The city has refineries and processing plants and light industries. Probably founded by Phoenicians, Almería flourished from the 13th to the 15th cent. as the outlet of the Moorish kingdom of Granada. It fell to the Christians in 1489. There is a Moorish fort, now in ruins, and a Gothic cathedral.

Al·mi·ran·te Brown (ăl'mə-rän'tē broun'), city (1970 pop. 245,017), Buenos Aires prov., E Argentina. It was settled in 1873 by families fleeing a yellow fever epidemic in the city of Buenos Aires.

A·lor (ä'lôr, ä'lôr), largest island, 810 sq mi (2,097.9 sq km) of the Alor Islands, 1,126 sq mi (2,916.3 sq km), E Lesser Sunda Island group, Indonesia, N of Timor in the S Flores Sea.

A·lor Se·tar (ä'lôr sĕ-tär') or **Alor Star** (stär), city (1971 pop. 66,179), capital of Kedah state, Malaysia, central Malay Peninsula. It is a major center for trade in rice and rubber.

Al·pe·na (ăl-pē'nə), county (1970 pop. 30,708), 565 sq mi (1,463.4 sq km), NE Mich., bounded on the E by Lake Huron and drained by Thunder Bay River and its affluents; organized 1840 as Anamickee co., renamed 1843; co. seat Alpena. In a dairy and farm area (livestock, potatoes, grain, and raspberries), it has limestone quarries, a lumber industry, and fisheries. Its manufacturing includes wood products, paper, cement, and iron and steel. It is a summer and winter resort and has a state forest and hunting area.

Alpena, city (1970 pop. 13,805), seat of Alpena co., N Mich., on Thunder Bay, an arm of Lake Huron; inc. 1871. Limestone quarried nearby is used to make cement. Other products include hardboard, paper, machinery, and automobile parts. Alpena lies in a year-round resort area and has an annual winter carnival.

Alpes-de-Hautes-Pro·vence (ălp'də-ōt'prə-väɴs'), department (1975 pop. 112,178), 2,681 sq mi (6,943.8 sq km), SE France; formerly Basses-Alpes dept.; capital Digne.

Alpes-Mar·i·times (ălp'mär-ĭ-tēm'), department (1975 pop. 816,681), 1,658 sq mi (4,294.2 sq km), SE France, bounded by Italy on the E and the Mediterranean Sea on the S and surrounding the independent principality of Monaco; capital Nice.

Al·phe·us (ăl-fē'əs) or **Al·fi·ós** (äl-fyōōs'), river, c.70 mi (112 km) long, rising in the Taygetus Mts., S Greece. The longest river in the Peloponnesus, it flows northwest through gorges, past Olympia, and onto the Olympia plains before entering the Ionian Sea.

Al·pine (ăl'pīn'), county (1970 pop. 484), 723 sq mi (1,872.6 sq km), E Calif., along the crest of the Sierra Nevada S of Lake Tahoe, bounded on the NE by the Nev. border and drained by the Mokelumne River, a headstream of the Stanislaus River, and by forks of the Carson River; formed 1864; co. seat Markleeville. It has an important mining industry and stands of timber. Its agriculture includes cattle and sheep raising and dairying. Alpine and Blue lakes and mineral springs attract hunting, fishing, and camping tourists.

Alpine, city (1970 pop. 5,971), seat of Brewster co., W Texas, in the mountains N of the Big Bend of the Rio Grande, SE of El Paso; founded in 1882 with the coming of the railroad. It is a railroad junction and shipping point for cattle and sheep. Sul Ross State College is here. The mountain scenery, with state parks and dude ranches, attracts many visitors. Big Bend National Park is nearby.

Alps (ălps), mountain system of S central Europe, c.500 mi (805 km) long and c.100 mi (160 km) wide, curving in an arc from the Riviera coast on the Mediterranean Sea, along the borders of N Italy and adjacent regions of SE France, Switzerland, S West Germany, and Austria, and into NW Yugoslavia. Cut by numerous gaps and passes, the Alps do not form a complete climatic or strategic barrier, as is evidenced by the similarities of air, people, and animals on either side of the system. Geologically, the Alps were formed during the Oligocene and Miocene epochs. Permanently snowcapped peaks rise above the snowline—located between 8,000 ft and 10,000 ft (2,440-3,050 m)—and glaciers (the longest being Aletsch glacier) form the headwaters of many Alpine rivers. Glaciation was more extensive during the Pleistocene epoch and carved a distinctive mountain landscape of arêtes, cirques, matterhorns, U-shaped and hanging valleys, and long moraine-blocked lakes. Below the snowline is a treeless zone of pastures that have for generations been used for the summer grazing of goats and cattle. Agriculture is confined to the valleys and foothills, with fruit growing and viticulture on some sunny slopes. Hydroelectric power, used for industries in the mountains and in nearby regions, is generated from the many waterfalls and swift-flowing rivers. Tourism, based on the scenic attractions of the Alps and the mountaineering and winter sports they provide, is a major source of income.

Al·pu·jar·ras, Las (läs ăl'pōō-här'äs), mountainous region, Granada and Almería provs., Andalusia, S Spain. Situated between the Mediterranean Sea and the Sierra Nevada region, the Alpujarras region has forests, pastures, and large valleys. It was the site of Moorish uprisings between 1492 and 1571.

Al Qayr·a·wan (äl kī'rə-wän') or **Kair·ouan** (kĕr-wän'), city (1975 pop. 54,546), NE Tunisia. It is a sacred city of Islam. Founded in 670, it was the seat of Arab governors in West Africa until 800 and was the first capital (909-21) of the Fatimids. When the city was ruined (1057) by invaders, it was supplanted by Tunis.

Als (äls), island, 121 sq mi (313.4 sq km), Sønderjylland co., S Denmark, in the Lille Baelt, separated from the mainland by the narrow Als Sund. Farming (particularly of apples and grain), fishing, and manufacturing (especially of motor vehicle parts) are the main occupations. The island was held by Germany from 1864 to 1920.

Al·sace (ăl-săs', -sās'), region and former province, E France. It is separated from West Germany by a part of the Rhine River. Alsace is rich agriculturally, geologically, and industrially. Textiles and wines (notably Riesling) are produced here.

Of Celtic origin, Alsace became part of the Roman province of Upper Germany. It fell to the Alemanni (5th cent.) and to the Franks (496). The Treaty of Verdun (843) included it in Lotharingia; the Treaty of Mersen (870) put it in the kingdom of the East Franks (later Germany). The 10 chief cities of Alsace gained (13th cent.) virtual independence as free imperial cities. Alsace became a center of the Reformation (although the rural areas remained generally Catholic). The Peace of Westphalia (1648) transferred all Hapsburg lands in Alsace to France. Lower Alsace was conquered (1680–97) by Louis XIV of France; the Treaty of Ryswick (1697) confirmed French possession. In 1871, as a result of the Franco-Prussian War, all Alsace (except Belfort) was annexed by Germany. With part of Lorraine, it formed the "imperial land" of Alsace-Lorraine, held in common by all the German states. The return of Alsace-Lorraine became the chief rallying force for French nationalism and was a major cause of the armaments race that led to World War I. France's recovery (1918) of this territory was confirmed by the Treaty of Versailles (1919). In 1940 German troops occupied Alsace; a large part of the population had already been evacuated to central France. French and American troops recovered (Jan., 1945) Alsace for France and were generally hailed as liberators.

Al·sip (ăl'səp), village (1975 est. pop. 16,108), Cook co., NE Ill., a suburb of Chicago; inc. 1927.

Al·ta Cal·i·for·nia (ăl'tə kăl'ə-fôr'nyə, -fôr'nē-ə), term used by the Spanish to refer to their possessions along the entire Pacific coast N of what is now the Mexican state of Baja California. California was often represented on maps as an island some 3,000 mi (4,800 km) long until the 18th cent. explorations of the Jesuit father Eusebio Kino proved conclusively that the southern part of the area was a peninsula and the rest of it mainland. Thereafter the peninsula came to be called Baja (Lower) and the mainland Alta (Upper) California.

Al·ta·de·na (ăl'tə-dē'nə), uninc. residential city (1978 est. pop. 39,400), Los Angeles co., S Calif., just N of Pasadena, on the slopes of the San Gabriel Mts., in an orange and avocado area; founded 1887.

Al·tai or **Al·tay** (both: ăl'tī'), geologically complex mountain system of central Asia; largely in the Gorno-Altai Autonomous Oblast, and in Kazakh SSR, but extending into W Mongolia and into N China. The Soviet Altai are bounded by the Sayan Mts. in the west, the Mongolian Altai in the south, and the Tannu-Ola range in the east. Meltwater from more than 230 sq mi (595.7 sq km) of glaciers feeds many rivers. Lake Teletskoye, with an area of 90 sq mi (233.1 sq km) and a depth of 1,066 ft (325.1 m), is the largest of the Altai's more than 3,000 lakes. Rich deposits of gold, silver, mercury, iron, lead, zinc, and copper are found in the mountains, especially in eastern Kazakhstan. Dense forests on the lower slopes are used for timber. Bears, martens, musk deer, and mountain goats inhabit the mountains. The first Russians entered the area in the 17th cent., settled in the foothills, and mined silver. In the late 19th cent., piedmont agriculture replaced mining as the main occupation. After the Soviet takeover in the early 20th cent., the area became both an important farming and mining region.

Al·tai Kray (ăl'tī' krī), administrative division (1970 pop. 2,766,000), c.102,400 sq mi (265,215 sq km), S central Siberian USSR; capital Barnaul.

Al·ta·mont (ăl'tə-mŏnt). **1.** City (1975 pop. 2,130), Effingham co., SE central Ill., SE of Springfield, in a farm area; inc. as a village 1872, as a city 1901. It is a railroad junction and a shipping point. Clothing and egg cases are manufactured. **2.** Uninc. town (1970 pop. 15,746), Klamath co., S Oregon, a suburb of Klamath Falls. **3.** Town (1970 pop. 546), seat of Grundy co., S central Tenn., in the Cumberlands NW of Chattanooga. It lies in a lumbering area.

Al·ta·monte Springs (ăl'tə-mŏnt), city (1975 est. pop. 18,000), Seminole co., E central Fla., N of Orlando.

Al·ta·mu·ra (äl'tə-moŏr'ə), city (1976 est. pop. 48,128), Apulia, S Italy. It is a commercial and agricultural center.

Alt·dorf (ält'dôrf'), town (1977 est. pop. 8,500), capital of Uri canton, central Switzerland. Cables and rubber goods are manufactured. Altdorf was the scene of the legendary exploits of William Tell.

Al·te·el·va (äl'tə-ĕl'və), river, c.120 mi (195 km) long, N Norway. It originates on the border of Finland and flows north into Alteelva Fjord, an inlet of the Arctic Ocean. Alteelva Fjord served as a secret station for the German fleet during World War II.

Al·ten·burg (äl'tən-boŏrg'), city (1974 est. pop. 51,193), Leipzig district, S East Germany, on the Pleisse River S of Leipzig. Manufactures include sewing machines, machine tools, textiles, and playing cards. Lignite is mined nearby. Built on the site of early 9th cent. Slavic fortifications, Altenburg became an important trade center and was made an imperial city in the 12th cent. It formally passed in 1329 to the house of Wettin and later (1603–72, 1826–1918) was the capital of the duchy of Saxe-Altenburg.

al·ti·pla·no (äl'tē-plä'nō), high plateau (alt. c.12,000 ft/3,660 m) in the Andes Mts., c.65,000 sq mi (168,350 sq km), W Bolivia, extending into S Peru. The altiplano is a sediment-filled depression between the Cordillera Oriental and the Cordillera Occidental. Its lowest point is occupied by Lake Titicaca, the largest high-altitude lake in the world. The sparsely vegetated region receives little precipitation and has several large salt flats. Potatoes and hardy grains are the principal crops here. Mining is the chief industry in the mineral-rich plateau.

Al·ton (ôl'tən). **1.** City (1978 pop. 33,800), Madison co., SW Ill., on bluffs of the Mississippi River above its confluence with the Missouri; inc. 1837. Alton is a shipping and industrial center, with machine shops, foundries, oil refineries, and a large bottle-making plant. Among its many other manufactures are food products, building materials, and ammunition and explosives. Lewis and Clark built their first camp and spent the winter of 1803–04 just south of what is now Alton. The town was laid out in 1817. During the Civil War it grew as a main supply point for the Union armies. The Principia (at Elsah) and a state park are nearby. **2.** City (1970 pop. 715), seat of Oregon co., S Mo., E of West Plains. It is a trade center in an agricultural area.

Al·too·na (ăl-too͞'nə). **1.** Town (1974 pop. 4,151), Polk co., central Iowa, ENE of Des Moines. It has a corn cannery. **2.** Industrial city (1978 est. pop. 57,400), Blair co., central Pa., on the E slopes of the Allegheny Mts. E of Pittsburgh; settled c.1769, laid out (1849) by the Pennsylvania RR as a switching point for locomotives preparing to cross the Alleghenies; inc. as a city 1868. It is still a major railroad center with huge construction and repair shops. The city's great variety of manufactures include foundry products, machinery, electrical equipment, paper items, shoes, clothing, and textiles.

Al·trinc·ham (ôl'trĭng-əm), municipal borough (1971 pop. 40,752), Greater Manchester, W central England. A suburb of Manchester, it has a textile-printing industry and engineering works and is also noted for its market gardens. The town's growth was stimulated by the construction of the Bridgewater Canal in 1760.

Al·tu·ras (ăl-toŏr'əs), city (1970 pop. 2,799), seat of Modoc co., extreme NE Calif., on the Pit River SE of Klamath Falls, Oregon; settled 1869, inc. 1901. It is a shipping and trade center for a stock-raising, farming, and lumbering region.

Al·tus (ăl'təs), city (1978 est. pop. 26,700), seat of Jackson co., SW Okla.; founded c.1892, inc. 1901. The city's agricultural products include cotton, wheat, and cattle. Altus Air Force Base, a large training facility, also contributes to the economy. Wichita Mountain Wildlife Refuge is nearby.

Al U·bay·yid (äl o͝o-bā-yĭd') or **El O·beid** (ĕl ō-bād'), city (1973 pop. 90,060), central Sudan. It is a rail terminus, a road and camel caravan junction, and the end of a pilgrim road from Nigeria. Al Ubayyid is also a trade and transshipment point. Founded by the Turko-Egyptian pashas in 1821, it fell to the Mahdists in 1883 and was destroyed. Its reconstruction followed the fall of the Mahdist empire in 1898.

Al·um Rock (ăl'əm), uninc. town (1970 pop. 18,355), Santa Clara co., W central Calif., a suburb NE of San Jose; formed in 1956 as a planned residential community. The town has a large park with many mineral springs.

Al·va (ăl'və), city (1970 pop. 7,440), seat of Woods co., NW Okla., on the Salt Fork of the Arkansas River NW of Enid; settled 1893. It is the commercial and processing center of a large wheat, dairy farming, and livestock area. Northwestern State College is here.

Al·vend or **El·vend** (both: ĕl-vĕnd', ĕl'vĕnd'), mountain, c.11,600 ft (3,540 m) high, W Iran. It bears cuneiform inscriptions of Darius and Xerxes.

Al·vin (ăl'vən), city (1970 pop. 10,671), Brazoria co., S Texas; inc. 1893. The city is chiefly residential, and many of its citizens work in Houston or at the nearby Lyndon B. Johnson Space Center. There is a petrochemical industry in the city.

Al·war (ŭl'wər), city (1971 pop. 100,378), Rajasthan state, N central India. Alwar is a market for grain, oilseed, cotton, and marble. There are textile and oilseed mills, iron foundries, and chemical and porcelain factories.

Am·a·de·us (ăm'ə-dē'əs), salt lake, 340 sq mi (880.6 sq km), SW Northern Territory, Australia. It is c.90 mi (145 km) from east to west, but is usually nearly dry.

Am·a·dor (ăm′ə-dôr′), county (1970 pop. 11,821), 593 sq mi (1,535.9 sq km), central Calif., extending from Sacramento Valley on the W to the Sierra Nevada on the E, bounded on the N by the Mokelumne River, on the S by the Cosumnes River; formed 1854; co. seat Jackson. There is lode-gold mining, and clay, sand, and gravel quarrying. Its agriculture includes cattle and sheep grazing, dairying, poultry raising, and some farming. It is a lumber area, lying partly in the El Dorado National Forest. The Mother Lode gold-mining region, made famous in the tales of Mark Twain and Bret Harte, centers around Jackson, where many old gold camps survive.

A·ma·ga·sa·ki (ä′mä-gä-sä′kē), city (1976 est. pop. 544,291), Hyogo prefecture, S Honshu, Japan, a port on Osaka Bay. An important industrial center, with iron and steel factories, chemical plants, and textile mills, it has a 16th cent. castle.

A·ma·ger (ăm′ə-gər), island (1965 pop. 177,818), 25 sq mi (64.8 sq km), Copenhagen co., E Denmark, in the Øresund. Northern Amager is occupied by a part of Copenhagen city that has important shipbuilding and harbor facilities. Southern Amager includes fishing ports, beach resorts, and farms.

A·ma·gi (ä-mä′gē), city (1975 pop. 42,725), Fukuoka prefecture, N Kyushu, Japan. It is an agricultural center and railway terminus. Textiles are produced in the city.

A·ma·ku·sa Islands (ä-mä′kə-sä), archipelago, c.340 sq mi (880 sq km), Kumamoto and Kagoshima prefectures, in the East China Sea, off W Kyushu, Japan. There are about 70 islands in the group. The interior of the islands is rugged; the coastal lowlands are fertile. Rice, fish, porcelain, and coal are the principal products.

A·mal·fi (ə-mäl′fē), town (1975 est. pop. 6,453), in Campania, S Italy, a small fishing port on the Gulf of Sorrento. Built on a mountain slope, it is also a picturesque seaside resort. According to legend, Amalfi was founded by the Romans; it later became (9th cent. A.D.) an early Italian maritime republic. It rivaled Pisa, Venice, and Genoa in wealth and power. Amalfi reached its zenith in the 11th cent. Thereafter it declined rapidly; it was captured (1131) by the Normans and sacked (1135, 1137) by the Pisans. In 1343 a storm destroyed much of the town.

A·ma·mi (ə-mäm′ē), island group, 498 sq mi (1,289.8 sq km), N Ryukyu Islands, Japan, NE of Okinawa, between the Philippine Sea on the E and the East China Sea on the W. The islands produce sweet potatoes, sugar cane, and silk. Tuna fishing is also important.

Am·a·pá (ăm′ə-pä′), federal territory (1975 est. pop. 142,100), 53,013 sq mi (137,304 sq km), extreme N Brazil, bounded on the N by French Guiana and the Atlantic Ocean; capital Macapá.

A·ma·ra (ə-mär′ə), town (1965 pop. 64,847), SE Iraq, on the Tigris River. A marketplace for dates and grains, it was taken by the British during the Mesopotamian campaign in 1915.

Am·a·ra·pu·ra (ăm-ə-rä′pə-rä, äm′-), town (1970 est. pop. 11,268), Mandalay division, central Burma, on the Irrawaddy River. It is a silk-weaving center and has varied handicraft industries. Amarapura was founded in 1782 and was twice (1783-1823 and 1837-60) the capital of Burma.

Am·a·ra·va·ti (ăm′ə-rä′və-tē, äm′-), ancient ruined city, Andhra Pradesh state, SE India, near the mouth of the Kistna River. The former capital of the Buddhist Andhra kingdom, it is a well-known archaeological site.

Am·a·ril·lo (ăm′ə-rĭl′ō, -rĭl′ə), city (1978 est. pop. 140,500), seat of Potter co., N Texas; inc. 1899. A commercial, banking, and industrial center of the Texas Panhandle, Amarillo is situated in the midst of treeless plains that are swept by summer dust storms and winter blizzards. The city grew after the coming of the railroad in 1887, and at the turn of the century it was a market for wheat farmers. After the discovery of gas (1918) and oil (1921), Amarillo mushroomed into an industrial city. In addition to oil and gas, the city's economy is based on cattle ranching, meat packing, flour milling, zinc smelting, as well as the production of helicopters, synthetic rubber, and farm and dairy items. Nearby are Amarillo Air Force Base and an atomic energy project.

Am·a·zon (ăm′ə-zŏn′, -zən), world's second-longest river, c.3,900 mi (6,275 km) long, formed by the junction in N Peru of two major headstreams, the Ucayili and the shorter Marañón. It flows across northern Brazil before entering the Atlantic Ocean near Belém. The Amazon carries more water than any other river in the world. For most of its course the river has an average depth of c.150 ft (45 m); ships with a draft of 14 ft (4.3 m) can reach Iquitos, Peru, c.2,300 mi (3,700 km) from the sea. The drainage basin is enormous (c.2,500,000 sq mi/6,475,000 sq km; c.35% of South America), gathering waters from both hemispheres and covering not only most of northern Brazil but also parts of Bolivia, Peru, Ecuador, Colombia, and Venezuela. In the lowlands stretching east from the Andes is the largest rain forest in the world—a wet, green land rich in plant life. Geologically, the Amazon basin is a sediment-filled structural depression between crystalline highlands of Brazil and Guiana. The river bed (1-8 mi/1.6-12.9 km wide) is in a wide flood plain that is up to 30 mi (48.3 km) wide. For much of its course the Amazon wanders in a maze of brownish channels amid countless islands. Its headstreams, however, arise cold and clear in the heights of the Andes. They descend northward before turning east to join and form the Amazon. Below the Xingú River the river reaches its delta, with many islands formed by alluvial deposit and submergence of the land. Around the largest of these, Marajó, the river splits into two large streams. The northern stream is the principal outlet and threads its way around many islands. The southern channel is called the Pará River. The awesome tidal bore (up to 12 ft/3.7 m high) of the Amazon travels c.500 mi (805 km) upstream. The river's immense silt-laden discharge is visible far out to sea.

The Amazon was probably first seen in 1500 by the Spanish commander Vicente Yáñez Pinzón, who explored the lower part. Real exploration of the river came with the voyage of the Spanish explorer Francisco de Orellana in 1540-41; his fanciful stories of female warriors gave the river its name. In 1637-38 the Portuguese explorer Pedro Teixeira led the voyage upstream that definitively opened the Amazon to world knowledge. The river continued to be of enormous importance to explorers and naturalists, among them Charles Darwin and Louis Agassiz. The valley was largely left to its sparse Indian inhabitants until the mid-19th cent., when steamship service was regularly established on the river and when some settlements were made. The area still remains largely unpopulated and undeveloped, yielding small quantities of forest products (rubber, timber, vegetable oils, Brazil nuts, and medicinal plants) and cacao. Oil and manganese resources are exploited near Manaus. In the 1960s the Amazon region began experiencing increased economic development brought on by tax incentives and construction of the Trans-Amazon Highway, the Belém-Brasília Highway, and two rail lines.

Am·a·zo·nas (ăm′ə-zō′nəs), state (1975 est. pop. 1,089,700), 604,032 sq mi (1,564,443 sq km), NW Brazil; capital Manaus.

Am·ba·la (əm-bäl′ə), town (1971 pop. 83,633), Haryana state, N central India. It is a military station and a transportation center. Automobile parts, pharmaceuticals, scientific instruments, iron products, porcelain, and glassware are manufactured.

Am·ba·to (äm-bä′tō), city (1974 pop. 77,052), capital of Tungurahua prov., central Ecuador, in a high Andean valley. A major commercial and transportation center, Ambato is noted for the variety of fruit grown in its outskirts. Sugar cane, grains, and cotton are also raised, and hides are processed. Picturesque Ambato is a favorite resort. The city has been frequently damaged by volcanic eruptions and earthquakes and in 1949 was almost totally destroyed.

Am·berg (äm′bĕrg′), city (1974 est. pop. 47,432), Bavaria, SE West Germany, on the Vils River, near Czechoslovakia. Its manufactures include precision instruments, machinery, blast furnaces, plastics, and porcelain. Nearby are large iron mines known since the Middle Ages. Until 1810 Amberg was capital of the Upper Palatinate. St. Martin's church (15th cent.) and the town hall (14th-16th cent.) are the city's outstanding buildings.

Am·bler (ăm′blər), suburban borough (1970 pop. 7,800), Montgomery co., SE Pa., N of Philadelphia; settled 1728, inc. 1888. It was a major supply center for George Washington's army during the Revolution. Asbestos, metal products, pharmaceuticals, and chemicals are manufactured. There are quarries and large nurseries in the vicinity, and a school of horticulture is here.

Am·boise (äɴ-bwäz′), town (1968 pop. 8,899), Indre-et-Loire dept., N central France, in Touraine, on the Loire. It is a wine and wool market, and its manufactures include precision instruments, shoes, sporting goods, pharmaceuticals, and film and radio equipment. The town is chiefly famous for its Gothic chateau. Leonardo da Vinci, who probably worked on it, is said to be buried in its chapel.

Am·bon (ăm′bän′), island, c.1,800 sq mi (4,660 sq km), E Indonesia, one of the Moluccas, in the Banda Sea. It is mountainous, well watered, and fertile. Maize and sago are produced, and hunting and fishing supplement the diet. Nutmeg and cloves, once grown in abundance, are produced in limited quantities, and copra is exported. The island was discovered (1512) by the Portuguese, who made it a religious and military headquarters. It was captured by the Dutch in 1605. An English settlement here was destroyed (1623) by the Dutch in what is called the Ambon massacre. Ambon was temporarily under British rule from 1796 to 1802 and again from 1810 to 1814. The town was the site of a major Dutch naval base captured (1942) by the Japanese in World War II, and it was the scene (1950) of a revolt against the Indonesian government during the short-lived South Moluccan Republic.

Am·bra·cia (ăm-brā'shə): *see* Arta, Greece.

Am·bridge (ăm'brĭj), industrial borough (1978 est. pop. 9,500), Beaver co., W Pa., on the Ohio River; inc. 1905. Founded by and named for the American Bridge Co. in 1901, it is still the home of the bridge company and of one of the largest structural steel plants in the world. Manufactures include steel, foundry and machine-shop products, and electrical equipment. On the northwest edge of town are 17 restored buildings and homes from the old village of Economy, a communistic colony established by members of the Harmony Society in 1825. The most successful of the society's communities, it thrived until 1906.

Am·brim or **Am·brym** (both: ăm'brĭm'), island (1967 pop. 4,246), c.230 sq mi (595.7 sq km), NE part of New Hebrides, SW Pacific Ocean. The island has an active volcano, Mt. Benbow (3,720 ft/1,134.6 m), that erupted in 1913 causing great destruction.

Am·chit·ka (ăm-chĭt'kə), island, 40 mi (64.4 km) long, off W Alaska; one of the Aleutian Islands. It was selected in 1967 by the Atomic Energy Commission as the site for underground tests of nuclear weapons. In 1971 the use of Amchitka for the detonation of atomic devices without specific Presidential approval was banned. The first test, sanctioned by President Richard Nixon, was made on Nov. 6, 1971.

Am·e·land (äm'ə-länt'), island (1970 est. pop. 2,899), 22 sq mi (57 sq km), W Frisian Island group, Friesland prov., N Netherlands. The North Sea is to the north, and the Waddenzee divides the island from the mainland on the south. The island is noted for truck farming and horse breeding.

A·me·lia (ə-mēl'yə), county (1970 pop. 7,592), 366 sq mi (947.9 sq km), central Va., bounded on the N and E by the Appomattox River and on the SE by Namozinc Creek; formed 1735; co. seat Amelia Court House. It has agriculture (mainly tobacco, and also corn, wheat, and hay), dairying, livestock raising, and extensive lumbering and lumber milling. Textile and wood products are manufactured. There are feldspar, mica, and pegmatite deposits in the county.

Amelia Court House or **Amelia**, village (1970 pop. c.700), seat of Amelia co., central Va., WSW of Richmond, in a timber and agricultural area.

A·me·ni·a (ə-mē'nē-ə), uninc. town (1970 pop. 7,842), Dutchess co., SE N.Y., NE of Poughkeepsie near the Conn. border, in a farm and dairy region; founded 1788. Thomas Lake Harris had his Brotherhood of the New Life settlement (1863–67) here.

A·mer·i·ca (ə-mĕr'ə-kə), the lands of the Western Hemisphere—North America, Central America (sometimes called Middle America), and South America. In English, "America" and "American" are frequently used to refer only to the United States. Martin Waldseemüller was the first to use the name (1507).

A·mer·i·can (ə-mĕr'ə-kən), river, 30 mi (48.3 km) long, rising in N central Calif. in the Sierra Nevada near Lake Tahoe and flowing SW into the Sacramento River. Two dams on the river, regulating its flow and generating hydroelectric power, are part of the Central Valley project. The discovery of gold at Sutter's Mill along the river in 1848 led to the California gold rush of 1849.

American Falls, city (1976 pop. 3,403), seat of Power co., SE Idaho. It is an important wheat-shipping center near the Great American Falls dam and reservoir on the Snake River. The city grew after the arrival of the railroad in 1892. It was moved to its present site after construction in 1927 of the dam, the reservoir of which inundated the former site. A trout hatchery is nearby.

American Fork, city (1975 est. pop. 10,462), Utah co., N central Utah, on American Fork Creek in the valley of Utah Lake, S of Salt Lake City; settled 1850 by Mormons; inc. 1853. A poultry-raising center in an irrigated area, it is served by the Provo River project.

American Geographical Society (AGS), oldest geographic society in the United States, founded 1852 in New York City. Its purpose is to advance the science of geography through discussion and publication. The society has the largest private geographic library in the Western Hemisphere. Its archives contain many rare maps and globes, historic letters, and artifacts from explorations. The society is noted for its support of scientific research and exploration, for its research facilities (extensively used by the Federal government during the 1919 Paris Peace Conference and again during World War II), and for its cartographic work. The *Geographical Review* is its quarterly journal.

American Samoa, uninc. territory of the United States (1970 pop. 27,159), 76 sq mi (196.8 sq km), comprising the E half of the Samoa island chain in the South Pacific. Pago Pago, the capital, is on Tutuila Island. Most of the islands are mountainous, heavily wooded, and surrounded by coral reefs. Subsistence agriculture and the export of

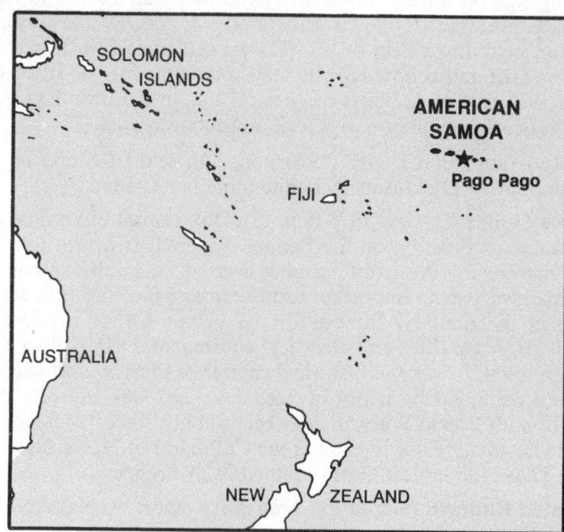

canned fish, copra, cocoa, and handicrafts became the mainstays of the economy after the U.S. naval base at Pago Pago was closed down in 1951. American Samoa was defined by a treaty in 1899 between the United States, Great Britain, and Germany, which gave the United States control of all Samoan islands east of long. 171° W.

A·mer·i·cus (ə-mĕr'əkəs), city (1978 est. pop. 15,200), seat of Sumter co., SW Ga.; inc. 1855. It is a manufacturing city, a livestock market, and a processing center for the area's timber, crops (peanuts, corn, cotton), and minerals (kaolin and bauxite).

A·mers·foort (ä'mərs-fōrt'), city (1977 est. pop. 87,203), Utrecht prov., central Netherlands. It is a transportation and manufacturing center. Points of interest include a 14th cent. water gate, the 15th cent. Gate of Our Lady, and the old town, which has medieval houses.

Ames (āmz), city (1978 est. pop. 45,200), Story co., central Iowa, on the Skunk River; inc. 1870. Its chief manufactures are electronic equipment and water-analysis and water-treatment equipment. Iowa State Univ. of Science and Technology is located in Ames. The National Disease Laboratory is also here.

Ames·bur·y (āmz'bĕr'ē), rural district (1971 pop. 27,611), Wiltshire, S central England. There are British remains that predate the Roman occupation. Amesbury Abbey, where Queen Guinevere of Arthurian legend is believed to have died, was founded in 980. Stonehenge, the chief megalithic monument in Britain, is nearby.

Amesbury, town (1975 est. pop. 13,923), Essex co., NE Mass., on the Merrimack River; inc. 1668. Rubber, metal, and vinyl products are manufactured. John Greenleaf Whittier lived here most of his life, and his house is preserved.

Am·gun (äm'go͞o'), river, 490 mi (788.4 km) long, Khabarovsk Kray, Far Eastern USSR. It rises in the Bureya Mts. and flows northeast to the Amur River above Nikolayevsk. Its upper course has yielded much gold.

Am·herst (ăm'ərst), town (1976 pop. 10,263), Cumberland co., N central N.S., Canada. Amherst is an industrial center. Its products include steel, aircraft parts, clothing, luggage, and insulating materials. Nearby are salt beds.

Amherst, county (1970 pop. 26,072), 467 sq km (1,209.5 sq km), W central Va., in the Piedmont bounded on the NW by the Blue Ridge and on the SW, S, and SE by the James River; formed 1761; co. seat Amherst. It is chiefly agricultural (cattle, corn, tobacco, apples, and peaches), with some mining and manufacturing (textiles, lumber, and stone, clay, and glass products). Iron and steel foundries are in the vicinity. It includes part of George Washington National Forest.

Amherst. 1. Town (1978 est. pop. 24,700), Hampshire co., W Mass., in a fertile farm area; inc. 1759. Named for Lord Jeffrey Amherst, it is a lovely, tree-lined college town. Emily Dickinson was born and lived here all her life. Helen Hunt Jackson was also born here, and Ray Stannard Baker, Robert Frost, and Noah Webster lived in the town. It is the seat of the Univ. of Massachusetts and Amherst College. **2.** Town (1975 est. pop. 6,847), Hillsborough co., S N.H., SW of Manchester; settled c.1733, inc. 1760. It was granted by Massachusetts in 1728 to soldiers of King Philip's War and their heirs. Horace Greeley was born here. Franklin Pierce was married in the Robert Means House (built 1785). **3.** City (1795 est. pop. 9,980), Lorain co., N Ohio, near Lake Erie NW of Elyria; settled c.1810. It is a fruit-growing area and is noted for its sandstone quarries. **4.** Town (1970 pop. 1,108), seat of Amherst co., central Va., in the Blue Ridge foothills NNE of Lynchburg, in a fruit and tobacco area.

Am·i·don (ăm'ə-dŏn'), city (1970 pop. 54), seat of Slope co., SW N.Dak., SW of Dickinson. A lignite mine is nearby.

A·miens (ä-myäN'), city (1975 pop. 131,476), capital of Somme dept., N France, in Picardy, on the Somme River. It is a rail hub and a large market for the truck farming carried on in the surrounding marshlands. Also an important textile center (since the 16th cent.), it has been particularly famous for its velvet. Other products are chemicals, soap, tires, and electrical equipment. Originally a Gallo-Roman town, it was the historical capital of Picardy and was overrun and occupied by many invaders. Amiens was severely devastated in both World Wars and has been rebuilt since 1945, largely in the medieval style. Of interest is the Cathedral of Notre Dame (begun c.1220), the largest Gothic cathedral in France.

Am·i·stad National Recreation Area (ăm'ə-stăd): *see* National Parks and Monuments Table.

A·mite (ā-mēt'), county (1970 pop. 13,763), 729 sq mi (1,888.1 sq km), SW Miss., bordered on the S by La. and drained by the Homochitto, Tangipahoa, Amite, and Tickfaw rivers; formed 1809; co. seat Liberty. It has agriculture (cotton, corn, and truck farms), dairying and logging, and lumber and plywood mills. It includes part of Homochitto National Forest.

Amite, town (1970 pop. 3,593), seat of Tangipahoa parish, SE La., NW of Lake Pontchartrain; settled 1836, inc. 1861 as Amite City. This cotton, corn, and lumber area ships truck produce and strawberries.

Am·i·ty·ville (ăm'ə-tē-vĭl'), residential village (1975 est. pop. 8,153), Suffolk co., SE N.Y., on the S shore of Long Island E of New York City; settled 1780, inc. 1894. It is a summer resort.

Am·man (ă-män'), city (1971 pop. 520,720), capital of Jordan, N central Jordan, on the Zarqa River. Jordan's largest city and industrial and commercial heart, it is also a transportation hub, especially for pilgrims en route to Mecca. Amman, which is built on a series of hills and valleys, is noted for its locally quarried colored marble. Industries include the manufacture of textiles, leather and leather goods, cement, marble, tiles, flour, and tobacco products. On a site occupied since prehistoric times, it was conquered by King David in the 11th cent. B.C. but regained independence under Solomon. The city was taken by Assyria in the 8th cent. B.C. and by Antiochus III c.218 B.C. Ptolemy II Philadelphus named it Philadelphia, by which it was known throughout the Roman and Byzantine periods. After the Arab conquest of 635, the city, which then became known as Amman, experienced a steady decline; it was only a small village when Emir Abdullah (later king) made it the capital of newly created Trans-Jordan in 1921. Growth was particularly rapid after World War II, when Amman absorbed refugees from Palestine. The city is the site of the Univ. of Jordan (est. 1962). Historical monuments include a Roman amphitheater (1st cent. B.C.), remains of a temple that was probably built by Hercules, and some tombs and a section of wall that date to the 9th or 8th cent. B.C.

Am·ne Ma·chin or **Am·ne Ma·chin Shan** (ăm'nē mə-jĭn' shän), spur of the Kunlun Mt. system, SE Tsinghai prov., W central China. Its highest peak rises to 23,490 ft (7,164.5 m).

A·mol (ä-mōl'), city (1966 pop. 40,076), Mazanderan prov., N Iran, near the Caspian Sea. It is an agricultural trade center. Amol was a provincial capital under the Abbassides in the 9th cent.

A·mor·gos (ə-môr'gəs), island (1961 pop. 2,096), 47 sq mi (121.7 sq km), one of the Cyclades Greece, in the S Aegean Sea. Its maximum height is 2,695 ft (822 m). Its products include wine, vegetables, figs, and tobacco.

Am·o·ry (ām'ə-rē), city (1975 est. pop. 7,374), Monroe co., NE Miss., N of Columbus near the Tombigbee River; founded 1887. It is a railroad, shipping, and trade center for a cotton and livestock area.

A·moy (ä-moi'), city (1970 est. pop. 400,000), S Fukien prov., China, on the SW shore of Amoy Island. It has an excellent natural harbor and is connected to the mainland by a railroad (built 1957) that crosses on a dike. Fishing, shipbuilding, and food processing are the major industries; machine tools and chemicals are also manufactured. Opposite Amoy proper, across the inner harbor, is the island of Ku-lang Hsü, the former foreign settlement and a fine residential section. Amoy was one of the earliest seats of European commerce in China, with Portuguese (16th cent.) and Dutch (17th cent.) establishments. It was captured (1841) by the British in the Opium War and became a treaty port in 1842.

Am·phip·o·lis (ăm-fĭ'pə-lĭs), ancient city of Macedonia, on the Struma River near the sea and NE of later Thessaloniki. Athenian colonists were driven out (c.464 B.C.) by Thracians, but a colony was established in 437 B.C. This colony was captured by Sparta in 424 B.C. After it was returned to Athens in 421 B.C., it had virtual independence until captured (357 B.C.) by Philip II of Macedon. He had promised to restore it to Athens, and his retention of Amphipolis was a major cause of the war with Athens. It was the capital (168–148 B.C.) of Macedonia Prima, one of the Roman republics.

Am·ra·va·ti (əm'rä'və-tē), town (1971 pop. 193,800), Maharashtra state, central India. The town is the site of the Great Stupa (c.A.D. 200) of the Andhra Dynasty.

Am·rit·sar (əm'rĭt'sər), city (1971 pop. 407,628), Punjab state, NW India. It is a trade and industrial city where carpets, fabrics of goat hair, and handicrafts are made. The center of the Sikh religion, Amritsar was founded in 1577. The Golden Temple (refurbished 1802), is especially sacred to Sikhs. The city was the center of a Sikh empire in the early 19th cent., and modern Sikh nationalism was founded here. The Amritsar massacre took place in Apr., 1919; hundreds of Indian nationalists were killed and thousands wounded when they were fired upon by the troops under British control.

Am·stel·veen (äm'stəl-vān'), town (1977 est. pop. 70,924), North Holland prov., W Netherlands, a suburb of Amsterdam.

Am·ster·dam (ăm'stər-dăm'), city (1977 est. pop. 740,650), constitutional capital of the Netherlands, North Holland prov., W Netherlands, on the IJ, an inlet of the IJsselmeer. The city derives its name from the fact that it is situated where the small, bifurcated Amstel River (which empties into the IJ) is joined by a sluice dam (originally built c.1240). A major port, Amsterdam is also the seat of one of the world's chief stock exchanges and a center of the diamond-cutting industry. Its manufactures include food products, clothing, printed materials, and metal goods. Amsterdam is connected with the North Sea by the North Sea Canal (opened in 1876), which can accommodate large ocean-going vessels, and by the older North Holland Canal (opened 1824). The Amsterdam-Rhine Canal connects the city with the Rhine delta. Because of the underlying soft ground, Amsterdam is built on wooden and concrete piles. The city is cut by about 40 concentric and radial canals that are flanked by streets and crossed by some 400 bridges. The many old and picturesque houses along the canals, once patrician dwellings, are now mostly offices and warehouses.

Amsterdam was chartered c.1300 and in 1369 joined the Hanseatic League. Having accepted the Reformation, the people of Amsterdam in 1578 expelled their pro-Spanish magistrates and joined the rebellious Netherland provinces. The commercial decline of Antwerp and Ghent and a large influx of refugees from all nations contributed to the rapid growth of Amsterdam after the late 16th cent. The Peace of Westphalia (1648) further stimulated the growth of Amsterdam at the expense of the Spanish Netherlands. Amsterdam reached its apex as an intellectual and artistic center in the 17th cent. The city was captured by the French in 1795 and became the capital of the Kingdom of the Netherlands. The constitution of 1814 made it the capital of the Netherlands. During World War II, Amsterdam was occupied by German troops from 1940 to 1945 and suffered severe hardship, including famine.

Outstanding buildings include the Oude Kerk, built in 1334; the weighhouse (15th cent.); the Nieuwthe Kerk (15th–17th cent.); the city hall (16th cent.); and the Beguinage, or almshouses, of the 17th cent. Rembrandt and the other Dutch masters are best represented in the world-famous Rijks Museum, founded in 1808 by Louis Bonaparte. Among the many other notable museums are the municipal museum (with a magnificent Van Gogh collection) and Rembrandt's house. The Univ. of Amsterdam, which was founded as an academy in 1632 and achieved university status in 1876, is the largest center of learning in the Netherlands.

Amsterdam, city (1978 est. pop. 24,000), Montgomery co., E central N.Y., on the Mohawk River; inc. 1885. It is an industrial city where carpets, rugs, clothing, and novelties are manufactured. The area was settled in 1783 and was named Amsterdam because many of the early settlers were from the Netherlands.

Amsterdam or **New Amsterdam,** volcanic island, 18 sq mi (46.6 sq km), S Indian Ocean, part of the French Southern and Antarctic Territories off the SE coast of Africa. It is mountainous, rising to 2,760 ft (841.8 m), and supports sea lions and herds of wild cattle. Anton Van Diemen, a Dutchman, named the island in 1633. France claimed it in 1843.

A·mu Dar·ya (ä-mo͞o' där'yə), river, c.1,600 mi (2,575 km) long, formed by the junction of the Vakhsh and Pandj rivers, which rise in the Pamir Mts. of central Asia. It flows generally northwest, marking much of the USSR-Afghanistan border before flowing through the Kara-Kum desert of Turkmenistan and Uzbekistan, Central Asian USSR, and entering the south Aral Sea through a large delta. The river drains c.180,000 sq mi (466,200 sq km). It flows swiftly until it reaches the Kara-Kum where its course braids into several channels. The Amu Darya is rich in fish and provides water for irrigation. In ancient times it was called the Oxus and figured importantly in the history of Persia and in the campaigns of Alexander the Great.

A·mund·sen Gulf (ä'mən-sən), c.250 mi (400 km), long, inlet of the Arctic Ocean, Northwest Territories, Canada. It lies between the Mackenzie dist. mainland on the south and south Banks Island on the north, opening on the Beaufort Sea to the west. Amundsen Gulf was first navigated completely by Roald Amundsen during his 1903–1906 expedition.

Amundsen Sea, arm of the S Pacific Ocean, off the coast of Marie Byrd Land, Antarctica, between Cape Dart and Thurston Peninsula. Amundsen Sea was explored and named by a Norwegian, Nils Larsen, in 1929.

A·mur (ä-mo͝or'), river, c.1,800 mi (2,895 km) long, formed by the confluence of the Shilka and Argun rivers, NE Asia, at the Soviet-Chinese border. The Amur-Shilka-Onon system is c.2,700 mi (4,345 km) long. The Amur flows generally southeast, forming for more than 1,000 mi (1,609 km) the border between the Soviet Union and China, then northeast through the Far Eastern USSR before entering the Tatar Strait opposite Sakhalin Island. One of the chief waterways of Asia, the Amur is navigable for small craft for its entire length during the ice-free season.

An·a·con·da (ăn'ə-kŏn'də), city (1978 est. pop. 14,100), seat of Deer Lodge co., SW Mont.; inc. 1887. Marcus Daly chose this place (1883) to build the smelter for the Anaconda Copper Mining Company and in the 1890s tried unsuccessfully to make it the state capital. The present high-stacked smelter (585 ft/178.4 m high), one of the largest in the world, dominates the life of the city and produces copper, zinc, and manganese.

An·a·cor·tes (ăn'ə-kôr'tĭs), city (1975 est. pop. 8,647), Skagit co., NW Wash., on Fidalgo Island, on the Strait of Georgia SSW of Bellingham; settled 1860, platted 1876, inc. 1890. It is a port of entry and a shipping and processing center for a lumbering, farming, and fishing region. There are large oil refineries here.

An·a·cos·ti·a (ăn'ə-kôs'tē-ə), river, rising near Blandensburg, Md., and flowing c.12 mi (19 km) SW before entering the Potomac at Washington, D.C. Part of Washington's harbor lies along its lower course. Much of the river is bordered by parks.

An·a·dar·ko (ăn'ə-där'kō), city (1970 pop. 6,682), seat of Caddo co., W central Okla., on the Washita River SW of Oklahoma City; founded 1901. It is the trade center of a rich agricultural area producing cotton, grain, alfalfa, livestock, and dairy products. Oil from nearby fields is processed. The city lies in an area once inhabited by the Delaware Indians. An Indian exposition is held annually, and the Southern Plains Indian Arts and Crafts Museum is here.

An·a·dyr (än'ə-dĭr'), river, c.695 mi (1,120 km) long, rising on the Anadyr Plateau, NE Far Eastern USSR, and flowing S then E into Anadyr Bay, an inlet of the Bering Sea. The Anadyr basin, a lowland between the Anadyr and Koryak ranges, is mostly covered by tundra. There are coal and gold deposits near the river's mouth.

An·a·heim (ăn'ə-hīm'), city (1978 est. pop. 203,000), Orange co., S Calif., SE of Los Angeles; inc. 1870. Anaheim was founded by Germans in 1857 as an experiment in communal living. Lying in an area of citrus fruit and walnut groves, the city is an important industrial center and one of the great tourist and convention centers in the United States. Among the city's manufactures are electronic equipment, guidance systems, paper converters, metal fabricators, greeting cards, and processed foods. Disneyland (opened in 1955) is here.

An·a·huac (ăn'ə-wăk'), resort and fishing city (1970 pop. 1,881), seat of Chambers co., on Galveston Bay, SE Texas, at the mouth of the Trinity River E of Houston, in a fishing, rice, cattle, lumber, and oil area; inc. after 1940. It was formerly a Mexican military post and a prominent port of early Texas. It was the scene of two clashes (1832, 1835) in which Anglo-American settlers sought to release William B. Travis, thus precipitating the Texan War of Independence.

An·am·bas (ə-näm'bəs), island group, 260 sq mi (673.4 sq km), Indonesia, in the S China Sea, between W Borneo and the SE Malay Peninsula NE of Singapore. Fishing, copra, and lumbering are important in the islands.

An·a·mo·sa (ăn'ə-mō'sə), city (1970 pop. 4,389), seat of Jones co., E central Iowa, SW of Dubuque on the Wapsipinicon River; inc. 1856. The area is noted for work clothes, poultry and eggs, dairy products, and concrete blocks. Limestone quarries are nearby. Grant Wood was born near Anamosa.

An·an (än'än'), city (1970 pop. 58,467), Tokushima prefecture, E Shikoku, Japan, on the Kii Channel. It is a fishing port and agricultural center.

An·a·to·li·a (ăn'ə-tō'lē-ə), Asian part of Turkey, usually synonymous with Asia Minor.

A·nau (ə-nou'): *see* Annau, USSR.

An·ch'ing (än'chĭng') or **An·king** (än'kĭng'), city (1970 est. pop. 160,000), SW Anhwei prov., China. A port and trading center on the Yangtze River, it was capital of the province until 1949.

An·chor·age (ăng'kə-rĭj), city (1978 est. pop. 178,600), borough seat of the Municipality of Anchorage, S central Alaska, a port at the head of Cook Inlet; inc. 1920. It is the largest city in the state, the administrative and commercial heart of south-central and western Alaska, one of the nation's key defense centers, and a vital transportation hub. Adjacent to the city are two huge U.S. military bases, Fort Richardson and Elmendorf Air Force Base. Anchorage is also a focus for the state's oil, coal, and natural gas industries. Tourism also contributes to the city's economy. Anchorage was founded (1915) as construction headquarters for the Alaska RR and grew as a railroad town. It also became a fishing center, a market and supply point for gold-mining regions to the north, and the metropolis for the coal mining and farming of the Matanuska valley. World War II brought the establishment of the large military bases and the enormous growth of air and rail traffic. The city suffered severe damage in the 1964 earthquake. Points of interest include Earthquake Park and several notable museums.

An·co·na (äng-kô'nə), city (1975 est. pop. 107,829), capital of Ancona prov., chief city of Marche region, central Italy, on a promontory in the Adriatic Sea. It is a leading Adriatic port and an industrial and commercial center. Manufactures include ships, musical instruments, and refined sugar. There is a fishing industry and an annual fish fair. Late in the 4th cent. B.C., Greeks from Syracuse took refuge in Ancona. The city prospered under the Romans, and its harbor was enlarged (2nd cent. A.D.) by Emperor Trajan. In the 9th cent. it became a semi-independent maritime republic under the nominal rule of the popes, to whose direct control it passed in 1532. The city was badly damaged in World War II. Noteworthy buildings include the Romanesque Cathedral of San Ciriaco (11th-13th cent.) and the Venetian-Gothic Merchants' Loggia (15th cent.).

An·da·lu·sia (ăn'də-lo͞o'zhə, -shə) or *Spanish* **An·da·lu·cí·a** (än'dä-lo͞o-*thē*'ä), region (1970 pop. 5,971,277), 33,675 sq mi (87,218.3 sq km), S Spain, on the Mediterranean Sea, the Strait of Gibraltar, and the Atlantic Ocean. Spain's largest and most populous region, it covers all of southern Spain.

An·da·lu·sia (ăn-də-lo͞o'shə), city (1975 est. pop. 10,128), seat of Covington co., S Ala., in a farming and forestry area; inc. 1844. Its manufactures include processed peanuts and pecans, meat products, textiles, lumber, and plywood.

An·da·man and Nicobar Islands (ăn'də-mən; nĭk'ə-bär'), union territory (1971 pop. 115,113), India, in the Bay of Bengal. Port Blair, in the Andamans, is the capital. Comprising the Andaman Islands (2,508 sq mi/6,495.7 sq km) in the south and the Nicobar Islands (707 sq mi/1,831.1 sq km) in the north, the territory chiefly exports tropical products and lumber. Known to Europeans since the 7th cent. A.D., the Andamans, consisting of more than 200 islands, were the site of a British penal colony from 1858 to 1945. The Nicobars, which comprise 19 small islands, are separated from the Andamans by a channel that is 90 mi (144.8 km) wide. The Nicobars became a British possession in 1869.

An·delys, Les (lā'zäɴ-dlē'), town (1968 pop. 6,292), in Eure dept., N France, Normandy, on the Seine. The twin communities of Le Grand Andely and Le Petit Andely form a commercial center, with metalworks, glassworks, and silk and leather industries. On the border between Normandy and Vexin, it was of considerable strategic importance in the Middle Ages. The impressive Château Gaillard was built (1197) by Richard I of England.

An·der·lecht (än'dər-lĕKHT'), commune (1976 est. pop. 99,942), Brabant prov., central Belgium, on the Charleroi-Brussels Canal, an industrial and residential suburb of Brussels. Erasmus lived (1517-21) in Anderlecht, and his house is now a museum.

An·der·nach (än'dər-näk'), city (1974 est. pop. 27,153), Rhineland-Palatinate, W West Germany, a port on the Rhine River. Its manufactures include chemicals, steel, wood products, and construction materials. Drusus founded a Roman frontier garrison here about A.D. 12. From 1167 to 1801 the city belonged to the archbishopric of Cologne. In 1815 it passed to Prussia. Andernach has a Romanesque church (13th cent.), a 16th cent. town hall, and parts of its medieval city wall.

An·der·son (ăn'dər-sən). **1.** County (1970 pop. 8,501), 577 sq mi (1,494.4 sq km), E Kansas, in a gently rolling agricultural area watered by Pottawatomie Creek; formed 1855; co. seat Garnett. Its agriculture includes livestock, dairy products, and grain. Food processing is done. It also has oil and natural-gas fields. **2.** County (1970 pop. 9,358), 206 sq mi (533.5 sq km), central Ky., in the Bluegrass bounded on the E by the Kentucky River, on the SW by Beech Fork, and drained by the Salt River; formed 1827; co. seat Lawrenceburg. It is in a gently rolling upland agricultural area (producing burley tobacco, dairy products, poultry, horses, corn, and hay), with some manufacturing. **3.** County (1970 pop. 105,474), 776 sq mi (2,009.8 sq km), NW S.C., bounded on the W by the Tugaloo and Savannah rivers, on the E by the Saluda River, and drained by the Seneca River; formed 1826; co. seat Anderson. It has an extensive textile-mill industry. Its agriculture includes cotton, grain, fruit, and vegetables, and it has diverse manufacturing (clothing, wood products, paper, chemicals, rubber and plastic products, glassware, aluminum, and textile machinery). **4.** County (1970 pop. 60,300), 338 sq mi (875.4 sq km), E Tenn., in the Cumberlands and the Great Appalachian Valley and drained by the Clinch River; formed 1801; co. seat Clinton. Bituminous-coal mining, lumbering, livestock, raising, dairying, and farming are done here. It includes Norris Dam and Reservoir and nuclear research facilities at Oak Ridge. **5.** County (1970 pop. 27,789), 1,072 sq mi (2,776 sq km), E Texas, bounded on the W by the Trinity River, on the E by the Neches River; formed 1846; co. seat Palestine. It has agriculture (truck crops, fruit, legumes, cotton, oats, peanuts, cattle, hogs, poultry, and dairy products) and oil, natural-gas, asphalt, clay, and coal deposits. Wood and metal products, auto parts, and clothing are manufactured. Hunting and fishing attract tourists.

Anderson. 1. City (1975 est. pop. 6,158), Shasta co., N central Calif., SSE of Redding near the Sacramento River, in a timber and dairy area; founded 1872, inc. 1956. Lumber products are made. **2.** City (1978 est. pop. 68,600), seat of Madison co., E central Ind., on the White River; inc. as a town 1838, as a city 1865. It is a manufacturing center in a rich farm area; its products include automotive parts, steel tools, and corrugated paper boxes. The city's industrial growth began with the discovery of natural gas in 1887. The automotive industry was established in 1901. Nearby Mounds State Park has numerous prehistoric mounds. **3.** City (1978 est. pop. 29,000), seat of Anderson co., NW S.C.; settled in the 17th cent., inc. 1828. The commercial center of a farming and livestock area, its manufactures include textiles, fiberglass products, and sewing machines. **4.** Town (1970 pop. c.300), seat of Grimes co., E central Texas, NE of Navasota. It is a trade and shipping point in an agricultural area.

Anderson, river, c.465 mi (750 km) long, rising in several lakes in NW Mackenzie dist., Northwest Territories, Canada. It meanders north and west before receiving the Carnwath River and flowing north to Liverpool Bay, an arm of the Arctic Ocean.

An·der·son·ville (ăn'dər-sən-vĭl'), village (1970 pop. 274), Sumter co., SW Ga., near Americus; inc. 1881. In Andersonville Prison, tens of thousands of Union soldiers were confined during the Civil War under conditions so bad that more than 12,000 soldiers died. It is now a national historic site. Andersonville National Cemetery, nearby, contains more than 13,000 soldiers' graves.

An·des (ăn'dēz), mountain system, more than 5,000 mi (8,045 km) long, W South America. The ranges run generally parallel to the Pacific coast and extend from Tierra del Fuego northward, across the equator, as the backbone of the entire continent. The Falkland Islands are a continuation of the Andes, and evidence shows that the system is continued in Antarctica. A geologically young system, the Andes were originally uplifted in the Cretaceous and Tertiary periods. They are still rising; volcanoes and earthquakes are common. The folded ranges are discontinuous—merging and bifurcating within the system—but as a whole they form one of the world's most important mountain masses. They are loftier than any other mountains except the Himalayas, with many snow-capped peaks more than 22,000 ft (6,710 m) high.

Far south in Tierra del Fuego, the mountains run east and west, then turn north between Argentina and Chile. The westernmost of the mountains run into the sea, lining the coast of southern Chile with islands. In the Patagonian Andes are high, glacier-fed lakes in both Argentina and Chile. The highest range of the Andes is on the central and northern Argentine-Chilean border; Aconcagua, the highest mountain of the Western Hemisphere, is here. In north Chile subandean ranges enclose the high, cold Desert of Atacama. The central Andes broaden out in Bolivia and Peru in multiple ranges (c.400 mi/645 km wide) with high plateau country (the altiplano) and many high intermontane valleys, where the great civilization of the Inca had its home. The western or coastal range in Peru has lofty peaks and is crossed by the highest railroad of the Andes. The ranges approach each other again in Ecuador, where the northern Andes begin. Between two volcanic cordilleras are rich intermontane basins. In Colombia the Andes divide again, the western range running between the coast and the Cauca River, the central between the Cauca and the Magdalena rivers, and the eastern running north parallel to the Magdalena River, then stretching out on the coast into Venezuela. The Andes continue in some of the islands of the West Indies, and in Panama Andean spurs connect with the mountains of Central America and thus with the Sierra Madre and the Rocky Mts.

An·dhra Pra·desh (än'drə prə-dāsh'), state (1971 pop. 43,502,708), 106,052 sq mi (274,675 sq km), SE India, on the Bay of Bengal; capital Hyderabad.

An·di·zhan (ăn'dĭ-zhăn'), city (1970 pop. 188,000), capital of Andizhan oblast, Uzbekistan, Central Asian USSR, in the Fergana Valley, on the Andizhan-Say River. It is an industrial center in an irrigated area that produces cotton and silk.

And·khui (änd-kho͞o'ē), city (1967 pop. 30,000), N Afghanistan, near the USSR border. Wool is its chief product, and it has a noted trade in fruits and karakul skins. Andkhui is also known for its handwoven rugs. Legend attributes the city's founding to Alexander the Great (4th cent. B.C.). A Russo-Afghan boundary commission assigned it to Afghanistan in 1885.

An·dong (än'do͝ong'), city (1970 est. pop. 76,000), E South Korea. It is a railroad junction and commercial center in an agricultural area where rice, hemp, cotton, and tobacco are grown.

An·dor·ra (ăn-dôr'ə), state (1970 est. pop. 21,000), 191 sq mi (494.7 sq km), high in the E Pyrenees between France and Spain, under the joint suzerainty of the president of France and the bishop of Urgel (Spain). In actuality Andorra is independent; it pays homage to France and Spain through nominal yearly gifts. Andorra comprises

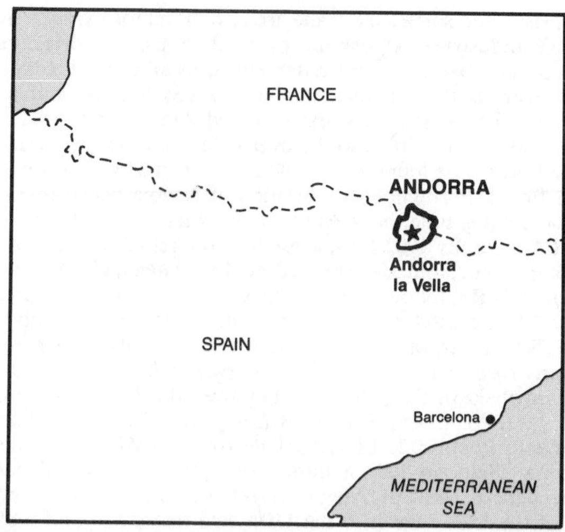

several high mountain valleys that are generally poor in soil but support large flocks of sheep. Livestock raising, the traditional source of livelihood, is being supplemented by a growing tourist trade. It also has iron and lead deposits, marble quarries, and extensive pine forests.

An·do·ver (ăn'dō-vər), town (1975 est. pop. 26,468), Essex co., NE Mass.; inc. 1646. Chiefly a textile producer in the 19th cent., Andover now makes toiletries, electronic equipment, chemicals, rubber products, and other items. Phillips Andover Academy (1778) is here. The Addison Gallery of American Art and the Robert S. Peabody Foundation archaeological museum are on the campus. Harriet Beecher Stowe lived in the town and is buried here.

An·doy (än'doi), island, 188 sq mi (486.9 sq km), Nordland co., off the NW coast of N Norway, in the Norwegian Sea. Fishing for herring, salmon, and cod is the main industry.

An·drew (ăn'drōō), county (1970 pop. 11,913), 430 sq mi (1,113.7 sq km), NW Mo., bounded on the W by the Nodaway River, on the SW by the Missouri River, and drained by the One Hundred and Two and Little Platte rivers; formed 1841; co. seat Savannah. It is in an agricultural area producing corn, oats, wheat, livestock, and poultry.

Andrew John·son National Historic Site (jŏn'sən): *see* National Parks and Monuments Table.

An·drews (ăn'drōōz), county (1970 pop. 10,372), 1,504 sq mi (3,895.4 sq km), W Texas, bounded on the W by the N.Mex. border, on the S by the Llano Estacado; formed 1876; co. seat Andrews. It has agriculture (hogs, sheep, poultry, cattle, and cotton) and mineral resources. It is a leading oil-producing county.

Andrews. 1. Town (1970 pop. 2,879), Georgetown and Williamsburg cos., E S.C., NNE of Charleston; inc. 1909. It makes textiles and wood products and processes minerals. **2.** City (1970 pop. 8,625), seat of Andrews co., W Texas, in the S Llano Estacado NW of Midland. Originally a prairie cow town, it grew rapidly in the 1950s when great quantities of oil began to be produced. Cattle-ranching and truck farming are still important.

An·dri·a (än'drē-ə), city (1975 est. pop. 81,646), in Apulia, S Italy. It is an agricultural market, handling wine, olives, and almonds. Andria was founded in the 11th cent. It was a favorite residence of Emperor Frederick II, who built (13th cent.) nearby the imposing Castel del Monte with eight round towers.

Án·dros (ăn'drəs), island (1971 pop. 10,457), 146 sq mi (378.1 sq km), SE Greece, in the Aegean Sea, northernmost and second largest of the Cyclades. The island produces silk, wine, and lemons and has manganese deposits. Colonized by Athens in the 5th cent. B.C., Ándros rebelled in 410 B.C., became a free state, and later passed successively to Macedon, Pergamum, and Rome. Seized (1204) from the Byzantines by Venice and made a principality, it remained almost entirely under Venetian rule until its conquest (1514) by the Turks. In 1829 it passed to Greece.

An·dros·cog·gin (ăn'drə-skä'gən), county (1970 pop. 91,279), 478 sq mi (1,238 sq km), SW Maine; formed 1854; co. seat Auburn. The "Twin Cities" of Lewiston and Auburn, on the Androscoggin River, are the center of the state's shoe and textile industries. Among the various other industries of the county are pulp and paper, wood, rubber and plastic, and metal products. It is in a rich lumber, agricultural, and dairy area, and fruit and vegetables are canned and shipped. Several lakes nearby offer recreational facilities.

An·dú·jar (än-dōō'här'), city (1970 pop. 25,962), Jaén prov., S Spain, in Andalusia, on the Guadalquivir River. Its pottery and water-cooling jars made of porous stone are famous.

A·ne·to, Pi·co de (pē'kō dā ə-nä'tō), peak, 11,168 ft (3,406.2 m) high, NE Spain, in the Maladetta near the French border. It is the highest peak of the Pyrenees.

An·ga·ra (äng'gə-rä'), river, c.1,150 mi (1,850 km) long, SE Siberian USSR, the outlet of Lake Baykal. After leaving the southwestern end of Lake Baykal, it flows north past Irkutsk and Bratsk, then turns west after receiving the Ilim River and flows into the Yenisei River near Strelka. The Angara is navigable between Irkutsk and Bratsk; below Bratsk there are many rapids. At Bratsk is a large dam with one of the world's largest hydroelectric power plants. Iron, coal, and gold deposits are found in the Angara basin.

Angara Shield (shēld): *see* Siberian Platform.

An·gaur (äng-our'), small coral island (c.3 sq mi/7.8 sq km wide and c.25 mi/40 km long), at the S end of the Palau Islands, W Pacific Ocean. It is noted for its chalk cliffs and phosphate deposits. American forces took over the island in a siege lasting from Sept. 17 to Oct. 13, 1944.

Án·gel de la Gar·da (äng'hĕl dā lə gwärd'ə), island, 50 mi (80.5 km) long, in upper Gulf of California, NW Mexico, off the NE coast of Baja California. Barren and uninhabited, it is a fishing base.

An·gel Fall (ān'jəl), waterfall, 3,212 ft (979.7 m) high, SE Venezuela, in the Guiana Highlands. It is the highest uninterrupted waterfall in the world.

Äng·el·holm (ĕng'əl-hôlm'), city (1975 est. pop. 15,100), Kristianstad co., SW Sweden, on Skälderviken Bay (an arm of the Kattegat); chartered 1516. It is a beach resort and has tanneries.

An·ge·li·na (ăn'jə-lē'nə), county (1970 pop. 49,349), 804 sq mi (2,082.4 sq km), E Texas, bounded on the N and E by the Angelina River, on the W and S by the Neches River; formed 1846; co. seat Lufkin. It has agriculture (truck crops, peanuts, fruit, corn, potatoes, sugar cane, and dairy products) and deposits of oil, natural gas, clay, lignite, and iron. Some lumbering is done. Parts of Angelina and Davy Crockett national forests are here.

Angel Island, largest island in San Francisco Bay, W Calif. Discovered by the Spanish in 1775, it came under U.S. control in 1851. The U.S. army used the island as a base from 1863 to 1946; in 1952 a radar and missile site was established. Part of the island is now a state park.

Ång·er·ma·näl·ven (ông'ər-mä-nĕl'vən), river, c.280 mi (450 km) long, rising in Västerbotten prov., W central Sweden, and flowing generally SE through narrow lakes to the Gulf of Bothnia at Härnösand. The river is used to float logs downstream to sawmills. There are numerous hydroelectric power plants on the river.

An·gers (äɴ-zhā'), city (1975 pop. 137,587), capital of Maine-et-Loire dept., W France, in Anjou, on the Maine River. A business and trade center, it is known for its wine. It also has glassworks, printing plants, and factories making electronic and photographic equipment, textiles, food, paper products, and tiles. On its outskirts are the largest slate quarries in France. Of pre-Roman origin, Angers became the seat (870–1204) of the powerful counts of Anjou and the historical capital of the province. There is a fine cathedral (12th–13th cent.) and a museum containing 14th cent. tapestries. The 13th cent. castle was damaged in World War II.

Ang·kor (ăng'kôr'), site of several capitals of the Khmer Empire, N of Tônlé Sap, NW Cambodia, for five and a half centuries the heart of the empire. Extending over an area of 40 sq mi (103.6 sq km), the ruins contain some of the most imposing monuments in the world. The first capital was founded c.900 and was centered around the pyramidal temple of Phnom Bak Kheng. To the southeast of the original capital, a new temple complex, Angkor Wat, was created under Suryavarman II (1113–50). Planned as a sepulcher and a monument to the divinity of the monarch, it is probably the largest religious structure in the world. The architecture of Angkor Wat, derived from the stupa form, is enormously impressive, but the most remarkable feature of the temple compound is its sculptural ornament, covering thousands of feet of wall space. The decoration is in the form of low relief of impeccable craftsmanship, illustrating scenes from the legends of Vishnu and Krishna, with some historical events. In 1777 Angkor was sacked by the Chams, and Angkor Wat fell into ruins. Jayavarman VII (1181–c.1218) established a new capital, Angkor Thom, north of Phnom Bak Kheng. The buildings of an already existing city were used as residential palaces and governmental buildings; an excellent system of moats and canals was constructed. The central tower of the temple bears a giant image of Buddha, which has been interpreted as the incarnation of Jayavarman VII. Surrounding the main structure is a forest of more than 50 smaller towers studded with multiple heads of the king as a Buddhist god. Angkor was raided in the 14th and 15th cent. by the Thai. It was abandoned in 1434 for Phnom Penh. Overgrown by the jungle, the ruins were discovered by the French in 1861. Many of the monuments were subsequently restored to their former glory, but were later damaged during the Vietnam conflict.

An·gle·sey or **An·gle·sea** (both: ăng'gəl-sē), island (1976 est. pop. 64,500), 275 sq mi (712.3 sq km), Gwynedd, NW Wales. It is a region of low, rolling hills. The principal industries are agriculture and stock raising. Two bridges over the Menai Strait connect the island to the mainland. Anglesey is said to have been the last refuge of the druids from the Romans in Britain.

An·gle·ton (ăng'gəl-tən), city (1970 pop. 9,770), seat of Brazoria co., S Texas, SSW of Houston; settled 1896, inc. 1912. It is a rail center in an oil-producing and agricultural (cotton, rice, cattle, and truck crops) region that once boasted large plantations. There is also manufacturing in the area.

Ang·mags·sa·lik (äng-mäg'sä-lĭk), settlement and trading post (1975 est. pop. 938), E Greenland, on the Denmark Strait just S of the Arctic Circle. It was founded in 1894. Its radio-meteorological station (est. 1925) is the oldest on Greenland.

An·go·la (ăng-gō'lə), officially People's Republic of Angola, (1977 est. pop. 7,205,000), including the exclave of Cabinda, 481,351 sq mi (1,246,700 sq km), SW Africa. Luanda is the capital. Angola is bounded by the Atlantic Ocean on the west, by Zaire on the north and northeast, by Zambia on the east, and by Namibia on the south. The Bié Plateau, which forms the central region of the territory, has an average altitude of 6,000 ft (1,830 m). Rising abruptly from the

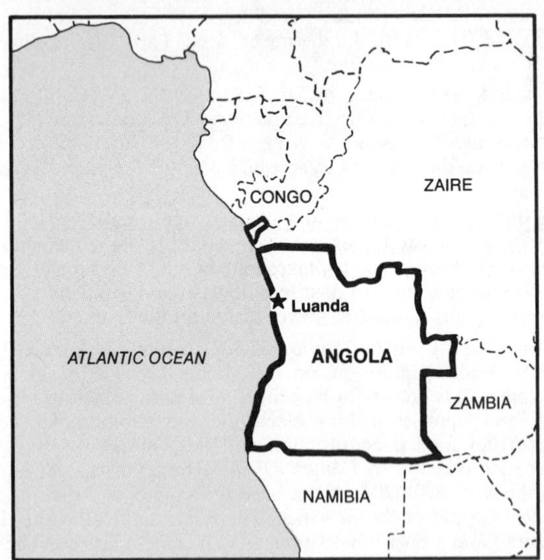

coastal lowland, the plateau slopes gently eastward toward the Congo and Zambesi river basins and forms one of Africa's major watersheds. The uneven topography of the plateau has resulted in the formation of numerous rapids and waterfalls, which are used for the production of hydroelectric power. The northeast has densely forested valleys that yield hardwoods, and palm trees are cultivated along a narrow coastal strip.

Economy. Diamond mining is the principal industry. Oil has been produced and refined near Luanda since the 1950s, and the exploitation of large reserves off Cabinda began in 1968. Angola's important deposits of copper, iron, and manganese ores remain largely undeveloped. Livestock, notably sheep and goats, is raised in much of the savanna region. Coffee is the most important cash crop. Maize, sisal, and some sugar cane are also raised for export. Among Angola's industries are fishing, railroad shops, foundries, cereal mills, fish and palm-oil processing plants, meat and fish canneries, and enterprises that manufacture jute, cotton textiles, and paper.

History. The Portuguese first explored Angola in the late 15th cent. Although they failed to discover gold and other precious metals, the Portuguese found in Angola a source of slaves for their colony in Brazil. Portuguese colonization of Angola began in 1575 at Luanda. By this time the Mbundu dynasty had established itself in central Angola. By 1902 Portuguese troops finally broke the back of the Mbundu kingdom and captured the Bié Plateau. Construction of the Benguela RR followed, and white settlers arrived in the Angolan highlands. The modern development of Angola began only after World War II. Inspired by nationalist movements elsewhere, the Angolans rose in revolt in 1961. The two major liberation movements involved in the 1961 uprisings were the Union of Angolan Peoples (UPA), led by Holden Roberto, and the presently ruling Movimento Popular de Libertação de Angola (MPLA), headed by Agostinho Neto. In 1966 a third group, União Nacional para la Independência Total de Angola (UNITA), emerged as a splinter party of the UPA. By 1970 the UPA had disbanded and Roberto assumed leadership of the National Front for the Liberation of Angola (FNLA). As a result of the guerrilla warfare, Portugal was forced to keep more than 50,000 troops in Angola by the early 1970s. In 1972 the Portuguese national assembly changed Angola's status from an overseas province to an "autonomous state," with internal sovereignty. After the overthrow of the Portuguese government in 1974, the new regime moved rapidly toward granting independence to Angola. At the time of the proclamation of independence (Nov. 11, 1975), three rival guerrilla movements were entrenched in different parts of the country. In the north were the forces of the FNLA (National Front for the Liberation of Angola), and in the extreme south, UNITA was predominant. In central Angola, principally around the capital of Luanda, was the army of the Marxist-oriented MPLA. By mid-1975 the battle seemed to have been won by the pro-Western FNLA and UNITA, both of which had received material aid from the United States and direct military assistance from South African troops, who had entered the conflict from Namibia. But in late 1975, largely through the intervention of some 13,000 Cuban soldiers, the tide turned in favor of the MPLA. By Feb., 1976, the MPLA had triumphed.

Angola. 1. Resort city (1970 pop. 5,117), seat of Steuben co., extreme NE Ind., N of Fort Wayne, in an area of hills and lakes; platted 1838, inc. as a town 1886, as a city 1906. It manufactures die castings and produces flour. Tri-State College is here, and Pokagon State Park is nearby. **2.** Village (1970 pop. 2,676), Erie co., W N.Y., near Lake Erie SW of Buffalo, in a farm area; inc. 1873. It is in a hilly lake region with some manufacturing (feed, brick, and automobile parts). Canned goods are processed.

An·gou·lême (äɴ-go͞o-lĕm'), city (1975 pop. 47,221), capital of Charente dept., W France, on the Charente River. A former river port, it is now a major road and rail center. Its paper industry dates from the 15th cent., and it has copper foundries, and plants making electric motors, soap, and shoes. Ceded (1360) to England, it was reconquered (1373) by Charles V. Its remarkable Cathedral of St. Pierre was begun c.1110.

An·gou·mois (äɴ-go͞o-mwä'), region and former province, W France. In the region is the Charente valley, with its excellent vineyards; the brandy made from their grapes is named for Cognac, the chief distillery center. In pre-Roman times the region was occupied by Gallic peoples. Part of the kingdom of Aquitaine under Charlemagne's empire, Angoumois was united with the French crown in 1307. Under the Treaty of Brétigny (1360) Angoumois, was recognized as English territory, but in 1371 it became a fief of the dukes of Berry. When Francis I, formerly count of Angoulême, became king in 1515, Angoumois was definitively incorporated into the French crown lands.

An·gra do He·ro·ís·mo (äng'grə do͝o ĕ'ro͞o-ēzh'mo͞o), town (1970 pop. 13,795), capital of Angra do Heroísmo dist., Portugal, in the Azores, on Terceira Island. It is a port and was until 1832 capital of the Azores. There is an old castle in the town.

An·gren (ən-gryĕn'), city (1969 est. pop. 94,000), Uzbekistan, Central Asian USSR. The largest lignite-mining center in Soviet Central Asia, it was developed during and after World War II.

An·guil·la (ăng-gwĭl'ə), island (1974 pop. 6,519), 35 sq mi (90.7 sq km), British West Indies, one of the Leeward Islands. Salt mining, fishing, and stock raising are the mainstays of the economy. In 1967 the British possessions of Anguilla, St. Kitts, Nevis, and Sombrero were united in the self-governing state of St. Kitts-Nevis-Anguilla. Anguilla seceded in 1967, returned to British colonial rule in 1971, and became a self-governing colony in 1976.

An·guille, Cape (ăng-gwĭl'), most westerly point of N.F., Canada, in the SW. It is northwest of Port aux Basques and south of the entrance to St. George Bay.

An·gus (ăng'gəs), former county, NE Scotland. The county town was Forfar. Under the Local Government Act of 1973, Angus became part of the Tayside region.

An·halt (än'hält), former state, central Germany, surrounded by the former Prussian provinces of Saxony and Brandenburg and located in the Halle and Magdeburg dists. of East Germany.

An·hwei (än'hwā') or **An-hui** (än'hwē'), province (1968 est. pop. 35,000,000), c.55,000 sq mi (142,450 sq km), E central China; capital Ho-fei. The northern half, within the north China plain, is cold in winter and dry throughout the year. It has a single harvest annually, the chief crops being wheat, kaoliang, corn, soybeans, and cotton. The southern half, through which the Yangtze River flows, is mountainous and has a relatively moist, warm climate. It is a major rice-producing region. Wheat, sweet potatoes, cotton, barley, and tobacco are also grown, and tea is produced in the southeast. Fish culture is important. Coal and iron are abundant throughout the province.

A·ni (ä'nē), ancient ruined city of Asia Minor, now in NE Turkey, SE of Kars. An ancient settlement, the city became the capital of Armenia in the 10th cent. It was often besieged by invaders and was finally destroyed by an earthquake in the 14th cent. There are notable ruins of a cathedral (built 989-1001) and several churches (11th-13th cent.), and remnants of a double wall.

An·i·ak·chak (ăn'ē-ăk'chăk), volcano, 4,420 ft (1,348.1 m) high, in the Aleutian Range, SW Alaska. Its crater is c.6 mi (9.7 km) in diameter. It was thought to be extinct until it erupted on May 1, 1931.

An·jo (än-jō'), city (1970 pop. 94,307), Aichi prefecture, S central Honshu, Japan. It is an agricultural and poultry center with cotton textile mills and food canneries. There are agricultural and forestry schools in the city.

An·jou (äɴ-zhōō'), region and former province, W France. A fertile lowland, Anjou is chiefly an agricultural area with excellent vineyards. Occupied by a Gallic people, the region was conquered by Caesar. Anjou fell to the Franks in the 5th cent. and became a countship under Charlemagne in the 9th cent. By the 10th cent. it was in the hands of the first line of the counts of Anjou. Fulk (d. 1143), after protracted wars with Henry I of England over the possession of Maine, married his son Geoffrey Plantagenet to Henry's daughter Matilda. Geoffrey ruled Anjou (1129-51) and conquered Normandy, of which he was crowned duke in 1144. His son, later Henry II of England, married Eleanor of Aquitaine and with her inheritance ruled most of western France. When Henry II's grandson, Arthur I, duke of Brittany, rebelled against his uncle, John of England, he won the support of Philip II of France. After Arthur's death, Philip II seized (1204) all Anjou. In 1246 Louis IX of France gave Anjou in appanage to his brother Charles, count of Provence. Charles II of Naples gave Anjou as dowry to his daughter Margaret when she married Charles of Valois, son of Philip III of France. When their son became (1328) King Philip VI of France, Anjou was again reunited to the French crown. It was definitively annexed to France in 1487. The region was devastated during the Wars of Religion (1562-98). During the French Revolution the rising of the Vendée, the Royalist revolt against the revolution, occurred in Anjou.

An·jou·an (än'jōō-än'), island, 89 sq mi (230.5 sq km), Comoro Islands, Mozambique Channel, Indian Ocean, off the NW coast of the Malagasy Republic. Its principal products include copra, vanilla, and coffee.

An·ka·ra (ăng'kə-rə), city (1975 pop. 1,698,542), capital of the Republic of Turkey and Ankara prov., W central Turkey, at an elevation of c.3,000 ft (915 m). It is an administrative, commercial, and cultural center. Grains, vegetables, and fruit are grown nearby. Manufactures of the city include food products, farm machinery, and cement. Known in ancient times as Ancyra and later as Angora, the city was an important commercial center in Hittite times (18th cent. B.C.). In the 1st cent. A.D. it became the capital of a Roman province. The city was conquered by the Ottoman Turks in the mid-14th cent., and in 1402 Tamerlane defeated and captured Sultan Beyazid I here. In the late 19th cent. Ankara declined and by the early 20th cent. was a small town known only for the production of mohair. In 1920, Kemal Atatürk made the city the seat of his Turkish Nationalist government. In 1923 it replaced İstanbul as the capital of Turkey. The city grew rapidly after the 1920s. There are few historic remains; Ankara's leading modern monument is the Atatürk mausoleum, completed in 1953. The city has numerous museums.

An·king (än'kĭng'): *see* An-ch'ing, China.

Ann, Cape (ăn), NE Mass., N of Massachusetts Bay. It is noted for its old fishing villages, resorts, and artists' colonies.

An·na·ba (ə-nä'bə), formerly **Bône** (bōn), city (1968 est. pop. 223,000), capital of Annaba dept., extreme NE Algeria, a port on the Mediterranean Sea. The city is an important commercial and industrial center. The large El Hadjar ironworks constitutes the chief industry; others include chemical manufacturing, food canning, cork production, and railway construction. Founded by the Carthaginians, it was called Hippo Regius by the Romans and was a center of early Christianity. The city was captured by the Vandals in 431. After the Arab conquest of Algeria in the 7th cent., Annaba became an important Moslem city and port. Spanish forces occupied it in the 16th cent.

An Na·jaf (än nä'jäf), city (1970 pop. 179,200), S central Iraq, on a lake near the Euphrates River. The tomb of Ali, son-in-law of Mohammed, is an object of pilgrimage by Shiite Moslems and a starting point for the pilgrimage to Mecca.

An·na·ka (än-nä'kä), city (1976 pop. 41,581), Gumma prefecture, central Honshu, Japan. It is an agricultural and tourist center, noted for its mineral springs.

An·nam (ə-năm', ăn'ăm), historical region, (c.58,000 sq mi (150,220 sq km) and former state, in central Vietnam, SE Asia. The region extended nearly 800 mi (1,290 km) along the South China Sea between Tonkin, in northern Vietnam, and Cochin China, which is situated in the southern section. The ridge of the Annamese Cordillera separated north and central Annam from Laos on the west; the ridge then swung southeastward and ran along the coast of southern Annam, which included the plateaus that stretched to the borders of Cambodia and Cochin China. The narrow coastal plains of north and central Annam were interrupted by spurs of mountains that almost reached the sea.

The origins of the Annamese state can be traced to the peoples of the Red River valley. These people fell under Chinese rule as the result of a Han invasion in 111 B.C. The region, to which the Chinese gave the name Annam ("Pacified South"; a name resented by the people), comprised all of what later became northern Annam and Tonkin. Southern Annam was occupied by the kingdom of the Chams, or Champa, from the late 2nd cent. A.D. In 939 the Annamese drove out the Chinese and maintained their independence, except for one brief period of Chinese reoccupation (1407-28), until their conquest by the French in the 19th cent. A long series of wars against the Chams ended in 1471 when the Chams were defeated and the Annamese kingdom was extended southward to the vicinity of Da Nang. French military operations began in 1858 and resulted in the seizure of southern Vietnam (Cochin China) and the establishment of protectorates (by 1884) over northern Vietnam (Tonkin) and central Vietnam (Annam). In 1887 Annam became part of the Union of Indochina. In World War II, Indochina was occupied by the Japanese, who set up the autonomous state of Vietnam, comprising Tonkin, Annam, and Cochin China. After the war Annamese and Tonkinese nationalists demanded independence for the state of Vietnam, and the region was plunged into a bloody conflict.

An·na·mese Cor·dil·ler·a (ăn'ə-mēz', -mēs' kôr'dĭl-yâr'ə, kôr-dĭl'ər-ə), principal mountain range of Southeast Asia, extending c.700 mi (1,125 km) from N central Laos SE to S central Vietnam. The range forms the divide between rivers draining into the Mekong basin and those flowing into the South China Sea and has a steep eastern face and a gently sloping western section.

An·nap·o·lis (ə-năp'ə-lĭs), city (1978 est. pop. 33,300), state capital and seat of Anne Arundel co., central Md., on the S bank of the Severn River. Annapolis is a port of entry and the business and shipping center for the fruit and vegetable farmers of southern Md. Local industries include the packaging of seafood and the manufacture of small boats and plastics. Annapolis was settled in 1649 by

Puritans fleeing Virginia. Hostility between the Puritans and the Roman Catholic governors resulted in the Battle of the Severn River in 1655, in which the Puritans successfully revolted, only to lose control after the Restoration in England. In 1694 it became the provincial capital of Md. and was named Annapolis for Princess (later Queen) Anne of England. During the 1700s the city prospered, largely because of its tobacco exports and trade with the West Indies and Europe; it rapidly became an important social and commercial center for the colonies. In 1783–84, Annapolis served as the capital of the United States. The city was the site of the Annapolis Convention (1786), which led to the Federal Constitutional Convention. Still standing is the statehouse where George Washington resigned as commander in chief of the Continental Army in 1783 and where the treaty that ended the Revolutionary War was ratified in 1784. Other notable landmarks are the Old Treasury (c.1695), the oldest original building in the state; the library (1737); St. John's College; and St. Anne's Church (1858–59). Annapolis is also the site of the U.S. Naval Academy, founded in 1845.

Annapolis, river, c.75 mi (120 km) long, rising in W N.S., Canada, and flowing SW past Annapolis Royal to Annapolis Basin, an arm of the Bay of Fundy. The entrance to the basin, bordered by cliffs 500 ft (152.5 m) high, is known as Digby Gut.

Annapolis Roy•al (roi'əl), town (1971 pop. 758), W N.S., Canada, on the Annapolis River. Founded as Port Royal in 1605, the settlement was destroyed (1613) by English colonists but was rebuilt by the French. The fort changed hands between the French and the English five times from 1605 to 1710, when it capitulated to a force of New Englanders. The name was then changed in honor of Queen Anne. Annapolis Royal was the capital of Nova Scotia from 1713 to 1749. Fort Anne Historic National Park includes the ruins of the fort.

An•na•pur•na (ăn'ə-po͝or'nə), massif of the Himalayas, N central Nepal, forming a ridge 35 mi (56.3 km) long, including two of the highest peaks in the world. It rises to Annapurna I (26,502 ft/8,083.1 m) in the west and Annapurna II (26,041 ft/7,942.5 m) in the east. Annapurna I was first climbed in 1950 by a French expedition.

Ann Ar•bor (ăn är'bər), city (1978 est. pop. 104,500), seat of Washtenaw co., S Mich., on the Huron River; inc. 1851. It is a research and educational center, with a large number of government and industrial research and development firms and the Univ. of Michigan. Products include lasers, computers, hospital and laboratory equipment, scientific instruments, automotive parts, and precision machinery. There are Indian mounds in the region.

An Na•si•ri•yah (ăn nä'sĭ-rī'yə), city (1970 est. pop. 62,400), SE Iraq, on the Euphrates River. It is the center of a date-growing region. Founded in 1870, the city was captured by the British in 1915. Nearby are the ruins of Ur.

An•nau or **A•nau** (both: ə-nou'), village, Central Asian USSR, in Turkmenistan, SE of Ashkhabad, near the Iranian border. It has a 15th cent. mosque, a citadel, ancient burial mounds, and other remains. Traces of habitation dating back to c.3000 B.C. were discovered here in 1903.

Anne A•run•del (ăn ə-rŭn'dəl), county (1970 pop. 298,042), 417 sq mi (1,080 sq km), central Md., bounded on the E by the Chesapeake Bay, on the N and NE by the Patapsco River, and on the W by the Patuxent River; formed 1650; co. seat Annapolis. It has truck farms, a seafood industry (fresh and frozen fish), pine timbering (sawmills, and furniture and paper products), and many resorts. Tobacco, poultry, livestock, and dairy products are produced. Residential suburbs of Baltimore are in the north.

Anne•cy (än-sē'), town (1975 est. pop. 103,543), capital of Haute-Savoie dept., SE France, in Savoy in the N Alps, on Lake Annecy. A popular tourist resort, it also has printing plants and factories making jewelry and wood and leather products. St. Francis of Sales, who was born in Annecy, was bishop from 1602 to 1622. The city has many fine churches, monasteries, and seminaries.

An•nis•ton (ăn'ĭs-tən), city (1970 pop. 31,533), seat of Calhoun co., NE Ala., in a mining region of the Appalachian foothills; inc. 1873. Its many varied manufactures include soil pipes, textiles, microwave ovens, factory-built homes, and vaccines. Founded (1872) as an iron-manufacturing company town, it was opened to the public in 1883. A local landmark is the Episcopal Church of St. Michael and All Angels (1887). Fort McClellan is nearby.

A•no•ka (ə-nō'kə), county (1970 pop. 154,712), 425 sq mi (1,100.8 sq km), E Minn., in an agricultural area bounded on the SW by the Mississippi and drained by the Rum River; formed 1857; co. seat Anoka. It has deposits of marl and peat. Dairy products, truck crops, livestock, grain, and poultry are produced. Fruit, vegetables, and oil are processed, and paper, furniture, and metal manufactured.

Anoka, city (1970 pop. 13,579), seat of Anoka co., E Minn., on the Mississippi at its confluence with the Rum River; inc. 1878. Originally a trading post and lumber town, it grew as a farm trade center. Ammunition and metal products are among its manufactures.

Ans•bach (äns'bäKH), city (1974 pop. 39,673), capital of Middle Franconia, Bavaria, S West Germany, on the Rezat River WSW of Nuremberg. Its manufactures include machine tools, electrical products, and chemicals. The city developed around an 8th cent. Benedictine abbey. Ansbach passed to Prussia in 1791 and to Bavaria in 1806. Noteworthy buildings include the 12th cent. Romanesque Church of St. Gumbertus, which was redone in baroque style in the 18th cent., and an 18th cent. castle.

An-shan or **An•shan** (both: än'shän'), city (1970 est. pop. 1,500,000), central Liaoning prov., China, on a branch of the South Manchurian RR. Its huge integrated iron and steel complex is the largest in China and one of the ten largest in the world. Many varieties of steel and steel products (including rails and cables) are produced. Other manufactures in An-shan include chemicals, tractors, machinery, alarm clocks, and cement.

An-shun or **An•shun** (both: än'sho͞on'), town, W central Kweichow prov., SW China. A flourishing town during the opium traffic days, it is known for its green tea. Other industries include sugar refining and machine building. Coal deposits are here.

An•son (ăn'sən), county (1970 pop. 23,488), 533 sq mi (1,380.5 sq km), S N.C., bounded on the S by the S.C. border, on the E by the Pee Dee River, and on the N by the Rocky River; formed 1749; co. seat Wadesboro. In a piedmont region, it has agriculture (cotton, corn, hay, fruit, and dairy products), pine and oak timber, and some manufacturing (dairy and wood products and textiles).

Anson. 1. Town (1970 pop. 2,168), Somerset co., central Maine, on the Kennebec River W of Skowhegan; inc. 1798. Wood products are made. **2.** City (1970 pop. 2,615), seat of Jones co., W central Texas, NNW of Abilene; settled c.1880, inc. 1904. Formerly a cow town, today it is primarily a shipping point for a cotton and cattle-ranching area, with cotton gins and miscellaneous industries. Oil wells and a refinery are in the area. To the southeast are the ruins of Fort Phantom Hill, a frontier outpost built in 1851.

An•so•ni•a (ăn-sō'nĭ-ə, -sōn'yə), city (1970 pop. 21,160), New Haven co., SW Conn., on the Naugatuck River; inc. as a city 1893. Its manufactures include brass and copper products, iron castings, foundry products, plastics, and electronic devices. Settled in 1651, Ansonia was founded (1844) as an industrial community by Anson G. Phelps, a metals merchant and philanthropist. Ansonia's historical landmarks include the birthplace of David Humphreys, who accepted Gen. Charles Cornwallis's sword in surrender after the Yorktown Campaign (1781), and Pork Hollow, where Revolutionary patriots hid food stores from British raiders.

An•ta•kya (än'tä-kyä'): *see* Antioch, Turkey.

An•tal•ya (än'täl-yä'), city (1975 pop. 130,759), capital of Antalya prov., SW Turkey, a seaport on the Mediterranean Sea. Its manufactures include textiles and ships. Nearby are deposits of chrome and manganese. Founded in the 2nd cent. B.C. by Attalus II, king of Pergamum, the city passed under the control of the Seljuk Turks in the 13th cent. and in the 15th cent. was annexed by the Ottoman Empire. Situated on a steep cliff, Antalya is a picturesque city surrounded by an old wall. Nearby are numerous ancient ruins.

Ant•arc•ti•ca (ănt'ärk'tĭ-kə, -är'tĭ-kə), the fifth-largest continent, c.5,500,000 sq mi (14,245,000 sq km), asymmetrically centered on the South Pole and almost entirely within the Antarctic Circle. It consists of two major regions: West Antarctica, a mountainous archipelago that includes the Antarctic Peninsula, and East Antarctica, geologically a continental shield. They are joined into a single continental mass by an ice cap thousands of feet thick. Masses of ice often break off from the seaward margins of the ice cap, floating away as icebergs. Where the outward ice is channeled into ice streams (zones of more rapid flowage), great floating ice tongues project into the sea; where mountains retard outward movement, the flow is channeled into great valley glaciers. The two major coastal indentations are the Ross Sea, facing the Pacific Ocean, and the Weddell Sea, facing the Atlantic Ocean. At the head of each sea are great ice shelves. Mostly afloat, these shelves are from 600 to 4,000 ft (183-1,220 m) thick. They move steadily toward the sea, fed by valley glaciers, ice streams, and surface snow accumulations. Smaller ice shelves are found all along the coast. Except for mountain ranges, much of East Antarctica's rock surface is near sea level; however, the continent's domed, snow-covered glacial surface rises to about 13,000 ft (3,965 m). In West Antarctica the subglacial relief suggests mountainous islands or submerged ranges separated by deep sounds beneath the ice cover. Less than 5% of Antarctica is free of ice; these areas include mountain peaks, small coastal areas, and islands.

The Transantarctic Mts., which extend from the east side of the Filchner Ice Shelf in the Weddell Sea to the western portal of the Ross Sea, form the inner margin of East Antarctica. Primarily formed by block faulting, the lower slopes have a complex structure of late Precambrian and early Paleozoic metamorphic rocks. These are overlaid by essentially horizontal sedimentary rock, ranging in age from the Devonian period to the early Jurassic, which are similar to rocks found in Australia, southern Africa, and eastern South America; coal-bearing Permian strata are also found here. Distinctive plant, insect, fish, and animal fossils in the Triassic and Jurassic strata strongly indicate that the continents of the Southern Hemisphere are parts of a hypothetical supercontinent, Gondwanaland, which broke up in the late Mesozoic era. In the early 1970s fossil finds and geological studies gave further support to the theory of continental drift. These studies concluded that Antarctica has been frozen for at least 20 million years—not 7 million years as previously thought—and that a tropical environment existed there 250 million years ago. The ice-drowned, mountainous archipelago of West Antarctica is related to the Andes Mts. of South America and is structurally connected to them by way of the Antarctic Peninsula and the Scotia Arc (South Georgia and the South Orkney and south Sandwich islands). The complex structure consists of highly folded metasedimentary strata from Paleozoic to Pliocene epochs, marked by much volcanism. A variety of mineral deposits have been discovered in Antarctica, but the extent of the deposits is largely unknown.

Antarctica is surrounded by the world's stormiest seas. A belt of pack ice surrounds the continent; only a few areas are ice-free at the end of most summers. The physical boundary most widely accepted today for the antarctic region is the Antarctic Convergence, a zone c.25 mi (40 km) wide encircling the earth roughly between the 50th and 60th parallels of latitude. Within this zone the colder and denser north-flowing antarctic surface waters sink beneath warmer and saltier subantarctic waters.

Antarctic climate is characterized by low temperature, high wind velocities, and frequent blizzards. Rapidly changing weather is typical of coastal locations, where temperatures for the warmest month average around freezing. Winter minimums drop as low as −40°F (−40°C). High altitude and continuous darkness in winter combine to make the interior of Antarctica the coldest place on earth. Summer temperatures are unlikely to be warmer than 0°F (−18°C); winter mean temperatures are −70°F (−57°C) and lower. The lowest temperature ever recorded on earth was −126.9°F (−88.3°C) at Vostok, a Russian station.

There is no native human population in Antarctica, nor are there any large land animals. Life that depends completely on the land is limited to microscopic life in summer meltwater ponds, tiny wingless insects living in patches of moss and lichens, and two types of flowering plants (both in the Antarctic Peninsula). Birds and seals that spend part of their time on land are dependent on the surrounding sea for food. Antarctic waters are rich in plankton, which serves as food for krill—small shrimplike crustaceans that are the principal food of baleen whales, crabeater seals, Adélie penguins, and several kinds of fish. Fur and elephant seals were the basis for 19th cent. commercial activity in Antarctica. In the 20th cent., commercial interest shifted to baleen whales. Whaling has been declining since the peak year of 1930-31 when the Norwegians dominated the industry; since 1967 only the Japanese and the Russians have continued whaling. The baleen whales that spend the summer in a zone up to 300 mi (480 km) north of the pack ice are now in danger of extinction.

History. Although there was for centuries a tradition that another land lay south of the known world, attempts to find it were defeated by the ice. The English explorer Capt. James Cook did not see the continent as he circumnavigated the world during his second voyage, but he was the first to cross the Antarctic Circle. British and U.S. seal hunters followed him to South Georgia, an island in the South Atlantic. In 1819 the British mariner William Smith discovered the South Shetland Islands. Searching for rookeries, sealers explored the coastal and offshore regions of the Antarctic Peninsula. Most notable were the British captains James Weddell, George Powell, and Robert Fildes and the Americans Nathaniel B. Palmer, Benjamin Pendleton, Robert Johnson, and John Davis. Davis made the first landing on the Antarctic continent (Feb. 7, 1821). First to spend the winter in Antarctica, on King George Island in 1821, were 11 men from the wrecked British vessel *Lord Mellville.* John Biscoe, a British navigator, circumnavigated Antarctica from 1830 to 1832, sighting Enderby Land in 1831 and exploring the western side of the Antarctic Peninsula in 1832. Four naval exploring expeditions visited Antarctica in the first half of the 19th cent. Capt. T. T. Bellingshausen was the leader of a Russian expedition that circumnavigated Antarctica (1819-21). He apparently was the first to see (1820) the part of the continent that is now called Queen Maud Land. Admiral J. S. C. Dumont d'Urville led a French expedition to the Pacific Ocean that made two visits to Antarctica. He explored in the area of the Antarctic Peninsula in 1838 and in 1840 discovered Clarie Coast and Adélie Coast in East Antarctica. In 1840, Lt. Charles Wilkes, leader of the U.S. Exploring Expedition to the Pacific (1838-42), sailed along the coast of East Antarctica for 1,500 mi (2,413.5 km), sighting land at nine points. British Capt. James C. Ross commanded two vessels on an expedition (1841-43) that discovered Victoria Land, the Ross Sea, and the Ross Ice Shelf and explored and mapped the western approaches of the Weddell Sea.

In the 1890s, after a half-century of neglect, a period of extensive Antarctic exploration began during which 16 exploring expeditions from nine countries visited the continent. For the first time, many of them were financed by private individuals and sponsored by scientific societies. Notable among them was a British expedition led by C. E. Borchgrevink, the first to establish a base for wintering on the continent (Cape Adare, 1899) and the first to make sledge journeys. Exploration in the Ross Sea area during this period was characterized by long inland journeys. Four British expeditions had bases on Ross Island at McMurdo Sound. British Capt. R. F. Scott headed two expeditions (1901-4 and 1910-13), E. H. Shackleton led another expedition (1907-9), and A. E. Mackintosh headed the Ross Sea party of Shackleton's unsuccessful Trans-Antarctic Expedition (1914-17). Roald Amundsen, a Norwegian, set up his base at the Bay of Whales, an indentation in the front of the Ross Ice Shelf, and a Japanese expedition (1911-12) was ship-based. The British expeditions carried out extensive exploration and scientific investigation of Victoria Land. Shackleton sledged to within 97 mi (156.1 km) of the South Pole (Jan., 1909), but it was Amundsen who reached the Pole first, on Dec. 14, 1911. Scott reached it on Jan. 17, 1912, but he and four companions perished on the return trip. The Weddell Sea border of East Antarctica was seen first (1904) by W. S. Bruce, leader of the Scottish National Antarctic Expeditions, and it was later explored by the German expedition of Dr. Wilhelm Filchner, discoverer of the Filchner Ice Shelf. A German and an Australasian expedition also explored East Antarctica during the early 20th cent.

In the period following World War I, scientific and technological advances were applied to further antarctic exploration. The first airplane flight in Antarctica (Nov. 26, 1928) was by Sir Hubert Wilkins, an Australian who later flew down the eastern side of the Antarctic Peninsula. However, it was U.S. explorer Richard E. Byrd who most successfully coordinated radios, tractors, airplanes, and aerial cameras for the purposes of exploration. On his first expedition Byrd established his base, Little America, near the site of Amundsen's old base at the Bay of Whales. From Little America he made the first flight over the South Pole on Nov. 29, 1929. On this expedition Marie Byrd Land was discovered and explored from the air. On his second expedition (1933-35) Byrd successfully integrated flights with long sledge and tractor journeys in a more complete exploration of Marie Byrd Land. In 1929-30 three other expeditions were also using aircraft for short flights over the coast. In Nov., 1935, U.S. explorer Lincoln Ellsworth made the first transantarctic flight, from the tip of the Antarctic Peninsula to the Bay of Whales, landing four times en route. The British Graham Land Expedition explored the Antarctic Peninsula by sea, air, and dog team from 1935 to 1937, using a different base each winter. Germany made a calculatedly spectacular effort at aerial surveying when two aircraft flying from a catapult ship photographed approximately 135,000 sq mi (349,650 sq km) of Queen Maud Land. In 1925-26 the Norwegians introduced pelagic whaling with factory ships that could operate in the open sea. Between 1927 and 1937, Lars Christensen led an extensive program of aerial exploration and mapping of the coast of East Antarctica from the Weddell Sea to the Shackleton Ice Shelf on Queen

Mary Coast. Also allied to whaling were the investigations in physical oceanography, marine biology, and coastal mapping carried out by the Discovery Committee of the British Colonial Office from 1925 to 1939.

The 1930s were a period of international rivalry in Antarctica, and the map was cut into wedgelike territorial claims that often overlapped. Although the U.S. government did not make a claim nor recognize those of other nations, it supported Antarctic exploration. The U.S. Antarctic Service Expedition (1939–41), directed by Byrd, introduced the concept of continuously occupied bases. The onset of World War II forced the evacuation of the bases, but before the war ended Great Britain, in order to offset claims by Chile and Argentina, had established permanent bases on the Antarctic Peninsula and off-lying islands.

Interest in Antarctica intensified after the war, and several governments established permanent agencies to direct Antarctic affairs. From 1945 to 1957 the U.S. navy conducted Operation Highjump, the largest expedition ever sent to Antarctica. It involved c.5,000 men, 13 ships including 2 icebreakers, 6 seaplanes, 6 airplanes, 2 small amphibian planes, and 3 helicopters. About 60% of the coastline was photographed, of which about 25% was reported as sighted for the first time. Much of the interior bordering the Ross Ice Shelf was also photographed.

After World War II, most expeditions were again government-financed. The Ronne Antarctic Research Expedition (1947–48), led by Finn Ronne, was the last privately sponsored U.S. expedition. Using the old U.S. Antarctic Service Expedition Base on Stonington Island, Ronne closed the unexplored gap at the head of the Weddell Sea. Some work was done as a joint effort with the British party that also had a base on the island. A portent of the international cooperation soon to follow, the Norwegian-British-Swedish Antarctic Expedition was organized by the respective governments and scientific societies for exploration and scientific investigation in Queen Maud Land. In a cooperative program during the International Geophysical Year (IGY)—July 1, 1957 to Dec. 31, 1958—12 nations maintained 65 stations and operational facilities in Antarctica. World data centers were established to collect and organize information and make it available to all scientists. The more difficult logistical problems of establishing inland bases were undertaken by the United States and the USSR. The American effort, beginning in 1955–56, was carried out by Naval Task Force 43 (Operation Deep Freeze). A major base of operations was built on Ross Island, and an airfield was established on the ice. Five other U.S. stations were established, including one at the South Pole that was entirely supplied by air. The Russians concentrated on East Antarctica, building a station on the Queen Mary Coast and two relay stations and three bases inland. There were 14 British stations, 8 Argentine stations, and 6 Chilean stations. France reoccupied the station opened in 1950 on the Adélie Coast and set up another inland near the Magnetic South Pole. Australia, Belgium, Japan, Norway, South Africa, and New Zealand also participated and occupied either insular or coastal sites.

The success of the IGY effort led to the signing (1959) of the Antarctic Treaty by representatives of the 12 nations that had been involved in the IGY. The treaty applied to the area south of lat. 60° S, exclusive of the high seas, and provided for cooperation and freedom of movement for scientific investigation as well as for the exchange of observers and scientific data. It prohibited military operations, nuclear explosions, and the disposal of radioactive wastes.

Ant·arc·tic Circle (ănt-ärk'tĭk, -är'tĭk), imaginary circle on the surface of the earth at lat. 66½° S, or 23½° N of the South Pole. It marks the southernmost point at which the sun can be seen during the summer solstice (about June 22) and the northernmost point of the southern polar regions at which the midnight sun is visible.

Antarctic Peninsula, glaciated mountain region of W Antarctica, extending c.1,200 mi (1,930 km) N toward South America. It is surrounded by numerous islands, including the South Shetlands and the Palmer Archipelago. The tip of the peninsula, 670 mi (1,078 km) from Cape Horn, is Antarctica's farthest point from the South Pole. The continent's only flowering plants are found here. The northwest coast of the peninsula was mapped by the British navigator James Bransfield in Jan., 1820, and was explored by sealers in 1820–21. First considered to be part of the continent, the peninsula was later (1928) thought to be a group of islands; the John Rymill expedition (1934–37) proved its peninsularity. Under different names, it was claimed successively by the United States (1820), Great Britain (1832), Argentina (1940) and Chile (1942). In 1964, by international agreement, the entire feature was called the Antarctic Peninsula.

An·te·lope (ăn'tə-lōp'), county (1970 pop. 9,047), 853 sq mi (2,209.3 sq km), NNE Nebr., in an agricultural region drained by the Elkhorn River; formed 1871; co. seat Neligh. It is the one of the largest irrigated areas in the Midwest and grows corn, soybeans, and alfalfa.

An·te·que·ra (än'tā-kā'rä), city (1970 pop. 40,908), Málaga prov., S Spain, in Andalusia. It is the center of a fertile agricultural region. The Cueva de Menga, a large prehistoric burial chamber, possibly Celtic, was discovered in the vicinity in 1842.

An·tho·ny (ăn'thə-nē). **1.** City (1970 pop. 2,653), seat of Harper co., S Kansas, SW of Wichita near the Okla. border; laid out 1878, inc. 1879. It is a trade center for a grain and livestock area. Flour milling is done. **2.** Town (1970 pop. 2,154), El Paso co., extreme W Texas, NNW of El Paso on the N.Mex. border; inc. 1952.

An·tibes (äɴ-tēb'), resort town (1975 pop. 55,960), in Alpes-Maritimes dept., SE France, on the Riviera. It is a seaport and the center of a great flower-growing region; a school of horticulture is here. Nearby is the fashionable resort Cap d'Antibes. The town was founded as a Greek colony in the 4th cent. B.C. A fortified port, it still has the 16th cent. Fort Carré. Also of interest is a Grimaldi château (14th and 16th cent.) housing a museum that includes numerous works by Picasso. Roman ruins are to the south.

An·ti·cos·ti (ăn'tə-kŏs'tē, -kô'stē), low, flat island (1971 pop. 419), 135 mi (217.2 km) long and 10 to 30 mi (16.1 to 48.3 km) wide, E Que., Canada, at the head of the Gulf of St. Lawrence. The island was discovered by Jacques Cartier in 1534. Louis XIV granted it to Louis Jolliet as a reward for his discovery of the Mississippi. Jolliet's heirs held it until 1763, when it was annexed to Newfoundland (then a separate colony). It was returned to Canada in 1774 and has been privately owned since 1895. Lumbering for pulpwood is the chief occupation on the island.

An·tie·tam (ăn-tē'təm), village (1970 pop. 300), Washington co., W Md., N of Harpers Ferry, W.Va., on the Potomac River at the mouth of Antietam Creek. On its banks just southeast of Sharpsburg and northeast of Antietam is Antietam National Battlefield Site, commemorating the Civil War Battle (Sept. 17, 1862) of Antietam, often called the Battle of Sharpsburg in the South. It was the bloodiest single day's fighting during the war, with heavy losses on both sides. As a result, Gen. Robert E. Lee abandoned his first invasion of Union territory. Antietam National Cemetery is nearby.

An·ti·go (ăn'tĭ-gō'), city (1970 pop. 9,005), seat of Langlade co., NE central Wis., NE of Wausau, in an agricultural and forested region; settled 1876, inc. 1885. Dairy and wood products are manufactured. The county historical society has a museum here.

An·ti·gua (ăn-tē'gwə) or **Antigua Gua·te·ma·la** (gwä'tə-mä'lə), town (1973 pop. 26,631), S central Guatemala. Founded in 1542 by surviviors from nearby Ciudad Veija, which had been destroyed by flood and earthquake, Antigua became the capital of Spanish Guatemala. In the 17th cent. it flourished as one of the richest capitals of the New World, rivaling Lima and Mexico City; by the 18th cent. its population had increased to c.100,000. In 1773 two earthquakes leveled the city. The Spanish captain general ordered (1776) the removal of the capital to a plain supposedly free from earthquakes and there founded Guatemala city. Antigua is now a major tourist center with many fine Spanish colonial buildings. It is also the commercial center of a rich coffee-growing region.

An·ti·gua (ăn-tē'gə, -gwə), island (1975 pop. 69,700), 108 sq mi (279.7 sq km), British West Indies, in the Leeward Islands. Saint John's is the capital. With its dependencies of Barbuda and Redonda, Antigua is an associated state of Great Britain and enjoys full internal self-government, with the British responsible for foreign affairs and defense. Hilly, with a much indented coast, Antigua has farms that grow mainly sugar cane and cotton. Tourism is a major industry. Discovered by Columbus in 1493, Antigua was named for a Spanish church in Seville. Unsuccessful Spanish and French settlements on the island were followed by a fruitful British effort in 1632, when sugar cane was introduced from St. Kitts. After a brief French occupation in 1666, Antigua passed permanently to Britain. The abolition of slavery in 1834 hurt the sugar industry; in the early 19th cent. cotton was introduced.

An·ti-Leb·a·non (ăn'tĭ-lĕb'ə-nən), mountain range between Syria and Lebanon, rising to Mt. Hermon, 9,232 ft (2,815.8 m) high. Once noted for its forests of oak, pine, cypress, and juniper, the range is now largely barren and stony.

An·til·les (ăn-tĭl'ēz): *see* West Indies.

An·ti·och (ăn'tē-ŏk), ancient town of Phrygia, near the Pisidian border. The site is north of the present-day Antalya, Turkey. It was founded by Seleucus I and became a center of Hellenistic influence.

Antioch or **An·ta·kya** (än'tä-kyä'), city (1970 pop. 66,400), capital of Hatay prov., S Turkey, on the Orontes River, near the Mediterranean Sea. It is the trade center for a farm region where grains, cotton, grapes, olives, and vegetables are grown. The city's few manufactures include processed foods, textiles, and leather goods.

Antioch was founded c.300 B.C. by Seleucus I and named for his father Antiochus, a Macedonian general. Situated at the crossing of north-south and east-west trade routes, the city soon became a rich commercial center. Antioch was occupied by Pompey in 64 B.C. and quickly became an important Roman military, commercial, and cultural center. Peter and Paul preached here, and it was in Antioch that the followers of Jesus were first called Christians. In 526 Antioch suffered a severe earthquake, and in 540 it was captured by Persia. In 637 the city was conquered by the Arabs. Nicephorus II reconquered it (969) for the Byzantine Empire, but in 1085 it fell, through treason, to the Seljuk Turks. The army of the First Crusade captured Antioch in 1098, after a half-year siege. In 1268 the Mamelukes captured and sacked the city; it was further damaged by Tamerlane in 1401. In 1516 Antioch, by then an unimportant city, was taken by the Ottoman Empire. The city was held (1832–40) by Egypt, and in 1872 it was badly disrupted by an earthquake. After World War I, Antioch was incorporated into the French Syria League of Nations mandate. In 1939 it was restored to Turkey as part of the Sanjak of Alexandretta. Modern Antioch occupies only a fraction of the area of the ancient city, most of which is buried under alluvial deposits. Numerous important archaeological finds have been made in and near Antioch. They include the Great Chalice of Antioch, held by some to be the Holy Grail, and splendid mosaics (1st–6th cent. A.D.), which are mostly copies of lost paintings.

Antioch. **1.** City (1970 pop. 28,060), Contra Costa co., W Calif., on the San Joaquin River near the mouth of the Sacramento; inc. 1872. It is a processing and shipping center for the agricultural products of the fertile islands in the delta area between the rivers. **2.** Summer resort village (1970 pop. 3,189), Lake co., NE Ill., near the Wis. border NNW of Chicago, in a lake region and farming area; settled 1836, inc. 1857. Chinaware and metal furniture are made. Chain-O-Lakes State Park is nearby.

an·tip·o·des (ăn-tĭp'ə-dēz'), places diametrically opposite on the globe. Thus antipodes must be separated by half the circumference of the earth (180°), and one must be as far north as the other is south of the equator; midnight at one is noonday at the other. For example, New Amsterdam and St. Paul, small islands nearly midway between southern Africa and Australia, are more nearly antipodal to Washington, D.C., than is any other land.

Antipodes, rocky uninhabited islands, 24 sq mi (62.2 sq km), South Pacific, c.450 mi (725 km) SE of New Zealand, to which they belong. The Antipodes were discovered by British seamen in 1800 and are so named because they are diametrically opposite Greenwich, England.

An·ti·sa·na (än'tə-sä'nä), active volcano, 18,885 ft (5,760 m), N central Ecuador, in the Andes SE of Quito.

Ant·lers (ănt'lərz), town (1970 pop. 2,685), seat of Pushmataha co., SE Okla., near the Kiamichi River S of McAlester; inc. 1903. It is a lumber-milling town, with miscellaneous industries (wood products, canned foods, and machine-shop products).

An·to·fa·gas·ta (än'tō-fə-gä'stä), city (1970 pop. 126,252), capital of Antofagasta prov., N Chile, a port on the Pacific Ocean. Antofagasta was founded by Chileans in 1870 to exploit nitrates in the Desert of Atacama, then under Bolivian administration. Its occupation by Chilean troops in 1879 sparked the War of the Pacific, and after the war the city and province were ceded to Chile. Antofagasta has depended primarily on nitrates and copper exports, and its economy has often been affected by sharp fluctuations in world demands. The city is an international commercial center and a major industrial hub with large foundries and ore refineries.

An·ton·gil Bay (äN-tōN-zhēl'), bay, 50 mi (80.5 km) long and 25 mi (40.3 km) wide, arm of the Indian Ocean, meeting the NE coast of Malagasy Republic. The French established a temporary settlement here in 1642.

An·trim (ăn'trĭm), county (1971 pop. 355,716), 1,098 sq mi (2,843.8 sq km), NE Northern Ireland; co. town Belfast.

Antrim, town (1969 est. pop. 33,980), Co. Antrim, E Northern Ireland, on Lough Neagh. An agricultural market in a flax-growing region, it has linen and lumber mills. The round tower here dates from c.900.

Antrim, county (1970 pop. 12,612), 477 sq mi (1,235.4 sq km), NW Mich., bounded on the W by Grand Traverse Bay and intersected by Torch Lake; formed 1840, name changed from Meegisee in 1843; co. seat Bellaire. Its agriculture includes livestock, dairy products, poultry, fruit, truck crops, potatoes, beans, and alfalfa. It has lumber and metal industries. A resort county, it includes Elk, Round, Intermediate, and Bellaire lakes.

An·tsi·ra·ne (än'tsə-rä'nə), *see* Diégo-Suarez, Malagasy Republic.

An·tu·co (än-to͞o'kō'), inactive volcano 9,060 ft (2,763.3 m), part of the Andes Range, S Chile, near the border of Argentina.

An·tung (än'do͝ong'): *see* Tan-tung, China.

Ant·werp (ăn'twûrp), or *French* **An·vers** (äN-vâr'), province (1975 pop. 1,559,269), 1,104 sq mi (2,859.4 sq km), N Belgium, bordering on the Netherlands in the N; capital Antwerp.

Antwerp, or *French* **Anvers,** city (1975 est. pop. 209,200), capital of Antwerp prov., N Belgium, on the Scheldt River. Antwerp is one of the busiest ports in Europe, a commercial, industrial, and financial center, and a rail junction. The city is linked with industrial eastern Belgium by the Albert Canal. Manufactures of Antwerp and its surrounding region include refined petroleum, petrochemicals, dyes, photographic supplies, motor vehicles, leather goods, and processed food. In addition, the city is a major international center of the diamond trade and industry, has large shipyards, and is the seat of the world's first stock exchange (founded 1460).

Antwerp was a small trading center by the early 8th cent. It was destroyed by the Normans in 836, but by the 11th cent. it was a fairly important port. The city was chartered in 1291. Antwerp was held (13th to mid-14th cent.) by Brabant and then became an early seat of the counts of Flanders. In the 15th cent. it rose to prominence as Bruges and Ghent declined. By the middle of the 16th cent. Antwerp was Europe's chief commercial and financial center. The city's prosperity suffered greatly in 1576 when it was sacked and about 6,000 of its inhabitants killed by Spanish troops and again in 1584-85 when the city was captured by the Spanish after a 14-month siege. Under the Peace of Westphalia (1648), the Scheldt was closed to navigation (as a means of favoring Amsterdam), and Antwerp declined rapidly. The city revived with the opening of the Scheldt by the French in 1795 and with the expansion of its port facilities by Napoleon I. The incorporation (1815) of Belgium in the Netherlands again hindered Antwerp's economic development; in 1839 the Netherlands secured the right to collect tolls on Scheldt shipping. The expansion of Antwerp as a major modern port dates only from 1863, when Belgium ended Dutch restrictions on traffic on the Scheldt. Antwerp was seriously damaged in World War I when it was captured (Oct., 1914) by the Germans after a 12-day siege. In World War II, it was again taken (May, 1940) by the Germans, who bombarded it heavily after it had been recaptured (Sept., 1944) by the Allies.

The artistic fame of Antwerp dates from the rule (15th cent.) of Philip the Good of Burgundy, who founded an academy of painting here. The painters Quentin Massys and Peter Paul Rubens resided in the city, and Sir Anthony Van Dyke was born here. Among Antwerp's many splendid structures are the large Gothic Cathedral of Notre Dame (14th-16th cent.), with a spire c.400 ft (122 m) high; the churches of St. James (containing the tomb of Rubens) and of St. Paul (both 16th cent.); the Renaissance-style city hall (mid-16th cent.); Rubens's house (now a museum); and old guildhalls lining the Groote Markt, or marketplace. Antwerp is the site of a famous zoological garden and a noted school of music.

A·nu·ra·dha·pur·a (ə-no͞o'rä-də-po͞o'rə), town (1972 est. pop. 33,000), capital of the North Central prov., Sri Lanka, on the Aruvi River NNE of Colombo. Rice plantations and vegetable gardens surround the town, which is famous chiefly for its vast Buddhist ruins and as a pilgrimage center. Founded in 437 B.C., it was the capital of a Sinhalese kingdom and a Buddhist center until the 8th cent. A.D., when, after a Tamil invasion, it was abandoned. Ruins include several colossal stupas (some larger than the pyramids of Egypt), a temple hewn from rock, and the Brazen Palace (so called from its metal roof). A sacred bo tree at Anuradhapura was grown from a slip of the tree at Bodh Gaya, India, under which Buddha reputedly attained enlightenment.

An-yang or **An·yang** (both: än'yäng'), city (1970 est. pop. 225,000), N

Honan prov., China, on the Peking-Canton RR, in a cotton-growing area. It has textile mills, coal mines, and a medium-sized iron and steel complex. An-yang was once a capital of the Shang dynasty and one of the earliest centers of Chinese civilization. Excavations, begun here in 1928, have revealed a rich Bronze Age culture.

An·zhe·ro-Sud·zhensk (ən-zhĕ'rə-sōōd'zhĕnsk), city (1976 est. pop. 104,000), SW Siberian USSR, on the Trans-Siberian RR. One of the oldest and largest coal-mining centers of the Kuznetsk Basin, the city was developed as a source of coal for the railroad. Mining equipment, chemicals, and pharmaceuticals are manufactured.

An·zi·o (ăn'zĭ-ō), town (1976 est. pop. 26,136), in Latium, central Italy, on the Tyrrhenian Sea. It is a seaside resort with a fishing industry. A Volscian town, it was captured by Rome in 341 B.C. and became a favorite resort of the Romans. Nero and Caligula were born here. Anzio declined in the Middle Ages, but it revived c.1700 and became a residence of the popes. During World War II, Allied troops landed (Jan., 1944) at Anzio and nearby Nettuno to draw German forces from Cassino, thus effecting a breakthrough (May, 1944) to Rome.

A·o·mo·ri (ä-ō-môr'ē),city (1976 est. pop. 270,012), capital of Aomori prefecture, extreme N Honshu, Japan, on Aomori Bay, an inlet of Mutsu Bay. First opened to foreign trade in 1906, Aomori is now the chief port of northern Honshu. Rice, textiles, lumber, fish and tobacco are exported. A modern city, it was rebuilt after a disastrous fire in 1910 and again after severe air raids in 1945.

A·o·sta (ä-ŏs'tä), city (1976 pop. 39,026), capital of Valle d'Aosta and of Aosta prov., NW Italy, near the junction of the Great and Little St. Bernard passes. Aosta is an industrial and tourist center. Manufactures include iron and steel, aluminum, and chemicals. Emperor Augustus here founded (c.25 B.C.) a colony called Augusta Praetoria, on the site of an older settlement. In the 11th cent. Aosta was given as a fief to Count Humbert I, the founder of the Savoy dynasty. Roman remains in Aosta include walls and gates, a majestic triumphal arch honoring Augustus, a theater, and an amphitheater. There is also a fine cathedral (12th-19th cent.).

Aosta, Valle d', region (1973 est. pop. 111,802), 1,260 sq mi (3,263.4 sq km), NW Italy, bordering on France in the W and on Switzerland in the N; capital Aosta. A high Alpine country, the Valle d'Aosta includes the Italian slopes of Mont Blanc, the Matterhorn, and Monte Rosa; its highest peak is the Gran Paradiso. Farming is the main occupation; cereals and grapes are grown, and dairy cattle are raised. Iron and steel and textiles are the leading manufactures, and there are major hydroelectric facilities. Under the Italian constitution of 1947 Valle d'Aosta was made a region with considerable autonomy. The feudal system long prevailed in the region, and more than 70 castles are still standing.

A·pach·e (ə-păch'ē), county (1970 pop. 32,304), 11,174 sq mi (28,940.7 sq km), NE corner of Ariz., in a mountain area bordering on N.Mex and Utah and touching Colo., and crossed by the Zuni, Little Colorado, and Puerco rivers; formed 1879; co. seat St. Johns. A small dam on the Little Colorado provides water for an irrigated area producing alfalfa, truck crops, and grain. It has livestock and dairying, lumbering (logging and sawmills), and metal mining. The Navajo Indian Reservation occupies the northern half of the county. Canyon de Chelly National Monument and part of the Petrified Forest National Monument are in the west.

Ap·a·lach·i·co·la (ăp'ə-lăch'ə-kō'lə), city (1970 pop. 3,102), seat of Franklin co., NW Fla., SW of Tallahassee; founded 1821 as West Point, inc. 1827, renamed 1831. It lies on Apalachicola Bay and is a port of entry and an important center for oysters, shrimp, and sport fishing. Lumber and naval stores are shipped. Its position after 1830 as one of the country's largest cotton-shipping ports was ended during the Civil War.

Apalachicola, river, NW Fla., formed by the junction of the Chattahoochee and Flint rivers in Lake Seminole at the Ga. border. The river flows 112 mi (180.2 km) south to Apalachicola Bay, an arm of the Gulf of Mexico, and is navigable throughout its length.

A·pa·po·ris (ä'pä-pōr'ēs), river, S central Colombia, rising SE of the Cordillera Macarena and flowing c.500 mi (800 km) S to join the Japurá River on the Brazilian border.

A·par·ri (ä-pär'rē), city (1969 est. pop. 45,700), Cagayan prov., on N Luzon, the Philippines. Situated on the mouth of the Cagayan River on the Babuyan Channel, it is the port for the rich Cagayan valley, the Philippines's leading tobacco-producing area.

A·pel·doorn (ä'pəl-dōrn), city (1977 est. pop. 135,251), Gelderland prov., central Netherlands. Its varied manufactures include paper and paint. The city is a transportation center and attracts many tourists. Nearby is Het Loo, a royal palace.

Ap·en·nines (ăp'ə-nīnz), mountain system, running the entire length of the Italian peninsula. It extends south c.840 mi (1,350 km) from the Cadibona Pass in Liguria, NW Italy, where the Apennines join with the Ligurian Alps, to the Strait of Messina. The mountains of Sicily are a southwest continuation of the system. The Apennines are widest (c.80 mi/130 km) in the central section, which also contains the highest peaks, rising to 9,560 ft (2,915.8 m). The central and southern Apennines have mineral springs, crater lakes, fumaroles, and volcanoes (two, Vesuvius and Etna, are still active). The southern section also experiences many earthquakes. The north and central Apennines are rich in a great variety of minerals.

A·pi·a (ä-pē'ə), town (1971 est. pop. 30,593), capital of Western Samoa, on the N coast of Upolu Island. It is the economic, social, and political center of Western Samoa. Through its harbor bananas, copra, and cocoa are exported. Robert Louis Stevenson is buried on a hill overlooking the city.

A·po, Mount (ä'pō), active volcano, 9,692 ft (2,955.6 m) high, on S Mindanao island, the Philippines. It is the highest peak of the islands. Mt. Apo has a snow-capped appearance but is actually covered with white sulfur.

Ap·ol·lo·ni·a (ăp'ə-lō'nē-ə), name of several ancient Greek towns. The most important was a port in Illyria on the Adriatic. It was founded by Corinthians and was later a Greek and a Roman intellectual center. Among the other towns of this name, there was one in Thrace on the Aegean (a town famous for a large statue of Apollo) and one in northern Sicily.

A·pos·tle Islands (ə-pŏs'əl), group of more than 20 wooded islands, in Lake Superior, off N Wis. Noted for their wave-eroded cliffs and abundant wildlife, the islands are visited by tourists and hunters. The islands, along with an 11 mi (17.7 km) strip of the adjacent shoreline, make up Apostle Islands National Lakeshore.

Ap·pa·la·chi·an Mountains (ăp'ə-lā'chən, -chē-ən, -lăch'ən), mountain system of E North America, extending in a broad belt c.1,600 mi (2,575 km) SW from the St. Lawrence valley in Que., Canada, to the Gulf coastal plain in Ala. The Appalachian Mts., much-eroded remnants of a great mountain mass formed by folding, consist largely of sedimentary rocks. In general the eastern portions are more rugged than the western, which are mainly of horizontal rock structure. Mt. Mitchell (6,684 ft/2,038.6 m) in the Black Mts. is the highest peak. The Great Appalachian Valley is a chain of lowlands extending along most of the system's length; it has long been an important north-south highway and is one of the most fertile areas in the eastern United States. The Appalachians themselves are rich in mineral resources, including coal, iron, petroleum, and natural gas. The scenic ranges also abound in resorts and recreation areas. Crossed by few passes, the Appalachians, especially their central portion, were a barrier to early westward expansion and played an important role in U.S. history; major east-west routes followed river valleys and gaps.

Appalachian Trail, world's longest continuous hiking path, 2,050 mi (3,298.5 km) long, passing through 14 states, E United States. Completed in 1937, the trail extends along the ridges of the Appalachian Mts. from Mt. Katahdin in Me. to Springer Mt. in Ga. It passes through eight national forests and two national parks, but the greatest part of its length is on private property. Hiking and trail clubs maintain shelters and campsites along the path. The trail was designated a national scenic trail in 1968.

Ap·pa·noose (ăp'ə-nōōs), county (1970 pop. 15,007), 523 sq mi (1,345.6 sq km), S Iowa, on the Mo. border, drained by the Chariton River; formed 1843; co. seat Centerville. It lies in a prairie agricultural region (with hogs, cattle, poultry, and corn, soybean, and hay farming) and has bituminous-coal mines and limestone quarries. There is some industry (wood products, plastics, primary metal industries, and metal products).

Ap·pen·zell (äp'ən-tsĕl), canton, NE Switzerland. It became a canton in 1513, and in 1597 it was split into two independent half cantons. Ausser-Rhoden, or Outer Rhodes (1970 pop. 49,023), 94 sq mi (243.5 sq km), with its capital at Herisau, accepted the Reformation; Inner-Rhoden, or Inner Rhodes (1970 pop. 13,124), 67 sq mi (173.5 sq km), with its capital at the town of Appenzell (1970 pop. 5,217), remained Catholic.

Ap·pi·an Way (ăp′ē-ən), most famous of the Roman Roads, built (312 B.C.) under Appius Claudius Caecus. It connected Rome with Capua and was later extended to Benevento, Taranto, and Brindisi. It was the chief highway to Greece and the East. Its total length was more than 350 mi (563 km). The substantial construction of cemented stone blocks has preserved it to the present. On the first stretch of road out of Rome are interesting tombs and the Church of St. Sebastian with its catacombs.

Ap·ple·ton (ăp′əl-tən), city (1975 est. pop. 59,500), seat of Outagamie co., E Wis., on the Fox River near its exit from the N end of Lake Winnebago, in a dairying and stock-raising region; inc. 1857. Waterfalls provide power for the city's industries, which produce paper, wood, metal, concrete, knitted goods, and dairy products. Appleton had the nation's first hydroelectric plant (1882) and the state's first electric streetcar (1886). The city is the seat of Lawrence Univ.

Ap·pling (ăp′lĭng), county (1970 pop. 12,726), 513 sq mi (1,328.7 sq km), SE Ga., bounded on the NE by the Altamaha River and on the SW by the Little Satilla River; formed 1818; co. seat Baxley. It is in a coastal plain agricultural area yielding tobacco, corn, melons, peanuts, and livestock. Lumber and naval stores are processed.

Appling, division (1970 pop. 2,724), seat of Columbia co., E Ga., WNW of Augusta. The division contains the small town of Appling (1970 pop. c.100). It is in an agricultural region.

Ap·po·mat·tox (ăp′ə-măt′əks), county (1970 pop. 9,784), 343 sq mi (888.4 sq km), central Va., bounded on the NW by the James River and drained by the Appomattox River; formed 1845; co. seat Appomattox. It is mainly agricultural (tobacco, hay, livestock, and dairy products) and has a large lumber industry. Fishing, mining, and manufacturing (textiles and furniture) are also important.

Appomattox, town (1970 pop. 1,400), seat of Appomattox co., central Va.; inc. 1925. Confederate general Robert E. Lee surrendered to Union general Ulysses S. Grant at nearby Appomattox Courthouse on April 9, 1865. The surrender marked the virtual end of the war, as the remaining Confederate armies, on hearing of Lee's act, followed suit. The site of the surrender is a national historical park.

A·pra Harbor (ä′prä) or **Port Apra,** port on the W coast of the island of Guam, W Pacific, in the Mariana Islands. The only good harbor on the island, it is a port of entry closed to foreign vessels except by permit. There is a large U.S. naval base here.

Ap·she·ron (əp-shĭ-rôn′), peninsula, c.40 mi (65 km) long, extending into the Caspian Sea, E Azerbaijan, SW USSR. It is a dry, hilly area at the eastern end of the Caucasus Mts. and is underlain by rich oil-bearing rock strata. The oil industry was developed in the 1870s, although the presence of oil was known long before. The peninsula, with its Baku oil fields, was once the USSR's chief oil-producing region but now accounts for only a small portion of Soviet production. Natural-gas wells, salt lakes, mineral springs, and mud volcanoes are also found on Apsheron.

A·pu·li·a (ə-pyo͞o′lē-ə), region (1973 est. pop. 3,674,408), 7,469 sq mi (19,344.7 sq km), S Italy, bordering on the Adriatic Sea in the E and the Strait of Otranto and Gulf of Taranto in the S; capital Bari. Its southern portion forms the heel of the Italian "boot." Apulia is mostly a plain; its low coast, however, is broken by a mountainous peninsula in the north, and there are mountains in the north-central part of the region. Farming is the chief occupation, but industry is expanding. There are saltworks in the north and bauxite mines in the south. Fishing is pursued in the Adriatic and in the Gulf of Taranto. The region was settled by several Italic peoples and by Greek colonists before it was conquered (4th cent. B.C.) by Rome. After the fall of Rome, Apulia was held successively by the Goths, the Lombards, and the Byzantines. In the 11th cent. it was conquered by the Normans. After the Norman conquest of Sicily (late 11th cent.), Apulia became a province, first of the kingdom of Sicily, then of the kingdom of Naples. The coast later was occupied at times by the Turks and by the Venetians. In 1861 the region joined Italy.

A·pu·re (ä-po͞o′rā), river, c.500 mi (805 km) long, rising in the Andes, N Colombia, and flowing E across W central Venezuela to the Orinoco River. It drains much of the western portion of the Orinoco basin and is navigable by river steamers for c.400 mi (645 km) during the rainy season.

A·pu·rí·mac (ä′po͞o-rē′mäk), river, c.550 mi (885 km) long, rising in the Andes, S Peru. It flows generally northwest in a narrow valley to join the Urubamba River and form the Ucayali, which is one of the main headstreams of the Amazon.

A·qa·ba (ä′kä-bä), town (1973 est. pop. 15,000), SW Jordan, at the head of the Gulf of Aqaba, on the border with Israel. Phosphates are the chief export. Aqaba is also a popular winter seaside resort. Since at least 1000 B.C., a port has existed continuously on the site to handle trade between Palestine and Syria. Occupied and fortified by the Crusaders in 1115, Aqaba was retaken by Saladin in 1187. During the 19th cent. the town became a staging point on the pilgrim route to Mecca. T. E. Lawrence captured Aqaba for the Allies in World War I; it later became part of the Hejaz but was ceded to Trans-Jordan in 1924.

Aqaba, Gulf of, N arm of the Red Sea, 118 mi (190 km) long and 10 mi (16.1 km) wide, between the Sinai and Arabian peninsulas. It is a part of the Great Rift Valley. The gulf, which is entered through the Straits of Tiran, has played a major role in the tensions between Israel and the Arab states bordering it.

A·qui·le·ia (ä′kwē-lĕ′yä), town (1971 pop. 1,938), in Friuli-Venezia Giulia, NE Italy, near the Adriatic Sea. Founded in 181 B.C. by the Romans, it was a stronghold against the barbarians and a trade center. Later, the town was destroyed several times by invaders, notably by Attila (A.D. 452).

Aq·ui·taine (ăk′wə-tān), former duchy and kingdom in SW France. Julius Caesar conquered the Aquitani, an Iberian people of southwestern Gaul, in 56 B.C. The province that he created occupied the territory between the Garonne River and the Pyrenees; under Roman rule it was extended northward and eastward almost as far as the Loire River. It was occupied (5th cent.) by the Visigoths, who were defeated (507) by the Frankish ruler Clovis I. In the chaotic strife among Clovis's successors, much of Aquitaine escaped Frankish control. After the 7th cent., the area north of the Garonne was considered Aquitaine proper. In 781 Charlemagne made Aquitaine into a kingdom for his son Louis, who later (838) granted it to his youngest son, Charles the Bald. A struggle for control ensued between Charles and the Aquitanians (840–52; 862–65). During this period Aquitaine was subject to attacks by both Normans and Moslems. Although Charles was the eventual victor, the repeated invasions, combined with the civil wars, weakened Carolingian control. Charles's successors were forced to recognize the hereditary rights of a number of independent noble families, and during the 10th cent. royal influence virtually disappeared. In the 11th cent. the dukes of Aquitaine expanded at the expense of their weaker neighbors, and the new duchy of Aquitaine was one of the most powerful states in western Europe. The marriage (1137) of Eleanor of Aquitaine to the French king Louis VII joined Aquitaine to France. Eleanor's subsequent marriage to Henry II, duke of Normandy, who became king of England in 1154, initiated a long struggle between France and England for possession of Aquitaine. In the Hundred Years' War France finally recovered all of Aquitaine. After its recovery, Aquitaine was constituted as the French province of Guienne, a name that had been used interchangeably with Aquitaine for many years.

A·ra·bah or **A·ra·ba** (both: ä′rä-bä, ăr′ə-bə), depression, on the Israel-Jordan border, extending c.100 mi (160 km) from the Dead Sea S to the Gulf of Aqaba. Limestone, salt, and potash are mined near the Dead Sea.

A·ra·bat, Tongue of (ŭ-rŭ-bät′; tŭng), peninsula, c.70 mi (115 km) long, NE Crimean Oblast, European USSR. It divides the Sea of Azov from the Sivash Sea. It is narrow and sandy, with saltworks on its southern tip.

A·ra·bi·a (ə-rā′bĭ-ə), peninsula (1970 est. pop. 17,000,000), c.1,000,000 sq mi (2,590,000 sq km), SW Asia. It is bordered on the west by the Gulf of Aqaba and the Red Sea, on the south by the Gulf of Aden and the Arabian Sea, on the east by the Gulf of Oman and the Persian Gulf, and on the north by Iraq and Jordan. Politically, Arabia consists of Saudi Arabia (the largest and most populous state), Yemen, Southern Yemen, Oman, the United Arab Emirates, Qatar, Bahrain, Kuwait, and several neutral zones. Arabia is mainly a great plateau of ancient crystalline rock, covered with limestone and sandstone. It rises steeply from the narrow Red Sea coastal plain, achieving its greatest height (c.12,000 ft/3,660 m) in southwest Arabia, and slopes gently east to the Persian Gulf. The coastal mountains catch the little moisture carried by the dry winds that cross Arabia, making the interior so arid that there is not a single perennial stream and large areas lack water. The basin-shaped interior consists of alternating steppe and desert landscape. The coastal lands, however, are much more humid than the interior; fog and dew are common.

Economy. Because of their dependence on isolated sources of water, about four fifths of the Arabian population is sedentary, concentrated around oases, notably in the Nejd (central Arabia) and the Hejaz (along the northeast coast of the Red Sea). Agriculture is the main occupation, with dates, grains, and fruits the chief crops; it is extensive and varied (coffee, grains, and fruits) only in southwest Arabia, particularly Yemen, where high coastal mountains intercept the moist southwest monsoon winds during the summer. Pastoral nomads raise goats and sheep. Until the mid-20th cent., when oil was discovered in eastern Arabia, the peninsula's main exports were hides, wool, coffee, spices, and the famed, highly bred Arabian horses. Arabia has an estimated one third of the earth's petroleum reserves; Kuwait and Saudi Arabia are among the world's leading producers. Modern technology and the huge wealth generated by oil resources have profoundly altered traditional life in Arabia

History. Archaeological evidence points to very early trade between Yemen and the northeast African coast. However, little is definitely known of Arabian history in the period preceding the oldest inscriptions discovered—those dating from c.1000 B.C. No ancient power ever attempted the complete conquest of Arabia, because of the formidable obstacles of crossing the deserts, though the Romans, the Ethiopians, and the Sassanids of Persia each invaded the peninsula with some success.

Arabia was briefly unified after the founding of Islam by Mohammed in the 7th cent. His dynamic faith, furthered by his successors, reconciled the warring Arab tribes and soon sent them out on a career of conquest. They subjugated northern Africa and southwestern Asia and gained control of Spain and southern France. They were stopped in the west by the Frankish ruler Charles Martel in 732 and in the east by the Byzantine Empire c.750. However, the tremendous territorial expansion of Islam diluted its exclusively Arabic character. Because of the need for a more convenient administrative center, the seat of the caliphate was transferred from Medina to Damascus; Arabia was again left without political cohesion, and independent emirates arose in Yemen, Oman, and elsewhere.

After the discovery of the route to India around the Cape of Good Hope in 1498, European powers were attracted to Arabia as a site for trading bases. The Portuguese seized Oman in 1508 but were driven out in 1659 by the Ottoman Empire, which attempted, never with complete success, to control all Arabia. Great Britain established a physical presence in Arabia in 1799 and in 1839 gained control of Aden from the Ottoman Empire. In 1853 Britain and the Arabian sheiks signed the Perpetual Maritime Truce by which the Arabs agreed not to harass British shipping in the Arabian Sea and recognized Britain as the dominant power in the Persian Gulf.

Arab nationalist opposition to the Ottoman Turks was aroused in the mid-19th cent. by a rekindling of the Wahabi, a reform movement within Islam; it waned toward the end of the century. Just before World War I, Ibn Saud revived the Wahabi, and during the war he signed a military pact with Britain against the Turks. His strongest rival, Husein ibn Ali of the influential Hashemite family, led a successful revolt against the Turks in the Hejaz and set up an independent state there. After the war, however, the Saud family prevailed in a violent struggle against Husein and other Arab families and founded (1925) Saudi Arabia, which absorbed the state in the Hejaz.

Between the World Wars, Britain was the dominant foreign power in Arabia, holding protectorates over the Arab sheikdoms. The post-World War II era witnessed a gradual decline of Britain's presence, culminating in the withdrawal of British military forces east of Suez in the late 1960s. Both the United States and the USSR sought to fill the vacuum created by Britain's withdrawal from the oil-rich, strategically important peninsula, but by the late 1970s the Arab nations had successfully asserted their independence.

A·ra·bi·an Desert (ə-rā'bē-ən) or **East·ern Desert** (ēs'tərn), c.86,000 sq mi (222,740 sq km), E Egypt, bordered by the Nile valley in the W and the Red Sea and the Gulf of Suez in the E. It extends along most of Egypt's eastern border and merges into the Nubian Desert in the south. Since ancient times Egypt has used the porphyry, granite, and sandstone found in the desert mountains as building materials. Oil is produced in the north.

Arabian Sea, NW part of the Indian Ocean, lying between Arabia and India. The Gulf of Aden, extended by the Red Sea, and the Gulf of Oman, extended by the Persian Gulf, are its principal arms. The submarine Carlsberg ridge, southeast of Socotra Island, is the sea's southern boundary. The Arabian Sea has long been an important trade route between India and the West.

A·ra·ca·ju (ä'rə-kə-zhōō'), city (1970 pop. 183,333), capital of Sergipe state, E central Brazil, a port on the Sergipe River near the Atlantic Ocean. Mainly a commercial center, the city has cotton-spinning and weaving industries. Aracaju was founded in 1855.

A·rad (ä-räd'), city (1977 est. pop. 171,110), W Rumania, in the Banat, on the Mureşul River, near the Hungarian border. It is an important railroad junction and a leading regional commercial and industrial center. Distilling, sawmilling, and the manufacture of textiles, machine tools, locomotives, electrical goods, and leather products are the chief industries. Long (c.1551–1685) under Turkish rule, Arad passed to the Hungarians, who made it the headquarters of their insurrection against the Hapsburg Empire (1848–49). In 1920 Arad became part of Rumania. The 18th cent. citadel was built by Empress Maria Theresa.

A·ra·fat (ŭ-rŏ-făt') or **A·ra·fa** (ŭ-rŏ-fă'), granite hill, Saudi Arabia, near Mecca. The hill was an ancient pagan sanctuary and is shrouded in many legends. It is now a site for prayers during the annual pilgrimage to Mecca. Atop the hill is a minaret, reached by broad stone steps.

A·ra·fu·ra Sea (ä'rə-fōō'rə), shallow part of the Pacific Ocean, between the Timor and Coral seas, separating Australia from New Guinea. It contains several islands of Indonesia.

Ar·a·gats, Mount (ăr'ə-găts'), extinct volcano, 13,435 ft (4,097.7 m) high, N Armenia, S European USSR, in the Lesser Caucasus. It is the highest peak in Armenia.

Ar·a·gón (âr'ə-gŏn), region (1974 pop. 1,192,387), 18,382 sq mi (47,609.4 sq km), and former kingdom, NE Spain, bordered on the N by France. Aragón includes the southern slopes of the Pyrenees, where the mountains reach their greatest height; a central plain drained by the Ebro River; and the western fringe of the central plateau of Spain. Much of the region is sparsely populated and desertlike. Irrigation works, started by the Moors, were resumed in the 16th cent. In the oases and irrigated areas cereals, grapes, olives, and sugar beets are grown. Sheep are raised throughout Aragón, and cattle in the Pyrenees. There are iron, sulfur, and lignite deposits, but sugar refining is the only important industry.

First settled by the Romans, the area was conquered by Visigoths in the late 5th cent. and by Moslems in the early 8th cent. Carolingians pushed out the Moslems (c.850), and Aragón came under the rule of Navarre. After 1035 the area was organized as the kingdom of Aragón. In 1076 Aragón annexed Navarre, and in 1137 it became united with Catalonia. Both regions preserved their own cortes, laws, languages, and customs and evolved along separate lines; their historical differences at times caused great friction. With the expansion of the house of Aragón, the name Aragón came to signify a confederation of its Spanish possessions (Aragón, Catalonia, Majorca, and Valencia) and several French fiefs. In the bitter struggles (12th–15th cent.) between kings and nobles, the nobles gained more and more privileges until Peter IV defeated them in 1348. United with Castile after 1479 through the marriage of Ferdinand V with Isabella, Aragón preserved its cortes and its city privileges. These, however, were gradually limited by the centralizing policies of the Spanish monarchy, and in 1716 Philip V abolished most of the remaining political privileges to punish the Aragonese for siding with Archduke Charles (later Emperor Charles VI) in the War of the Spanish Succession.

A·ra·guai·a (ă'rə-gwī'ə), river, c.1,300 mi (2,090 km) long, rising in the Serra des Araras, at the border of Goiás and Mato Grosso states, S central Brazil. It flows generally northward into the Tocantins River, forming most of the border between Goiás and the states of Mato Grosso and Pará. Diamonds are washed along its upper tributaries. There are numerous falls on the Araguaía. The island of Bananal (c.200 mi/320 km long; 35 mi/56 km wide), separating the river into two arms, is one of the largest freshwater islands in the world. It is also a national park. The Araguaía region has been made accessible by new highways.

A·rak (ä-räk'), city (1966 pop. 71,925), Tehran prov., W central Iran. A center for agricultural trade as well as for road and rail, the city is also known for its rugs and carpets. It was founded c.1800.

Ar·al Sea (ăr'əl), inland sea and the world's fourth-largest lake, c.26,000 sq mi (67,340 sq km), SW Kazakhstan and NW Uzbekistan, Central Asian USSR, E of the Caspian Sea. It is c.260 mi (420 km) long and c.175 mi (280 km) wide. Generally very shallow, making navigation possible only in a small area, it attains a maximum depth of c.220 ft (70 m). The Aral Sea is fed by the Syr Darya and Amu Darya rivers but has no outlet. Because of its geologically recent separation from the Caspian Sea, the Aral Sea's water is only slightly saline. There are many small islands in the sea. The sparse population of the region engages in fishing (carp, perch, and pike). Sodium and magnesium sulfate are mined along the shore.

Ar·an Islands (âr'ən), 18 sq mi (46.6 sq km), Co. Galway, W Republic of Ireland, in Galway Bay. The islands are barren, and living is primitive; farming and fishing are important. There are many early Christian and prehistoric remains.

A·ran·juez (ä'räng-hwāth'), town (1970 pop. 29,548), Madrid prov., central Spain, in New Castile, on the Tagus River. A market town (the region is known for asparagus and strawberries; horses are bred), it was once a royal residence.

A·ran·sas (ə-răn'səs), county (1970 pop. 8,902), 276 sq mi (714.8 sq km), S Texas, on the Gulf Coast, here indented by Aransas, Copano, and St. Charles bays, and traversed by the Gulf Intracoastal Waterway; formed 1871; co. seat Rockport. Cattle and poultry raising, truck farming, dairying, and fishing are important. There are also oil and natural-gas wells. The county includes Aransas National Wildlife Refuge, which is noted for its whooping cranes.

Aransas Pass, city (1970 pop. 5,813), Aransas, Nueces, and San Patricio cos., S Texas, NE of Corpus Christi; settled 1890, inc. 1910. A port on the Gulf of Mexico, it has a deepwater ship channel that goes through Harbor Island and through the pass between St. Joseph Island and Mustang Island, where Port Aransas stands. A sea wall protects the city, which is a resort and fishing center. The city has a carbon-black plant and ships oil. A hurricane (1961) caused great damage.

A·ra·o (ä-rä'ō), city (1975 pop. 58,296), Kumamoto prefecture, W Kyushu, Japan, on Ariake Bay. It is a port and produces cement, chemicals, fertilizers, and plastics.

A·rap·a·hoe (ə-răp'ə-hō'), county (1970 pop. 162,142), 827 sq mi (2,142 sq km), NE central Colo., bordering Denver and bounded on the W by the South Platte River; formed 1861; co. seat Littleton. It is in an agricultural area yielding dairy products and wheat and has petroleum and natural-gas wells and lumber. There is diversified manufacturing (furniture, paper, plastics, concrete, primary metal industries, and fabricated metal products) in the county.

Arapahoe Peak, mountain, 13,506 ft (4,119.3 m) high, N central Colo., in the Front Range of the Rocky Mts. It has a large glacier.

Ar·a·rat (ăr'ə-răt), name of two mountains, Little Ararat (12,877 ft/3,927.5 m) and Great Ararat (16,946 ft/5,168.5 m), E Turkey, near the Iranian and Soviet borders. It is the traditional resting place of Noah's ark.

A·ras (ä-räs'), river, c.600 mi (965 km) long, rising in the Transcaucasian Mts., NE Turkey. It flows generally east, forming parts of the Turkey-USSR and USSR-Iran borders, before entering the Azerbaijan Republic, USSR, where it joins the Kura River. Much of its upper and middle courses are rapid and tumultuous, and its waters are used for irrigation. The Aras was called Araxes in ancient times.

A·rau·ca (ə-rou'kä), river, rising in the Cordillera Oriental E of Bucaramanga, N central Colombia, and flowing E c.500 mi (805 km). It forms part of the Venezuela-Colombia boundary and empties into the Orinoco River.

A·ra·val·li (ə-rä'vəl-lī), mountain range, central and S Rajasthan, NW India. The range, c.300 mi (485 km) long, separates the north upland region from the Thar Desert on the west. The average height of the system varies from 1,000 ft (305 m) to 3,000 ft (915 m). Leopards and tigers are found in the forests, but the Aravallis are generally bare and sparsely populated.

Ar·be·la (är-bē'lə), town of ancient Assyria. Its name is sometimes given to the battle fought at Gaugamela, some 60 mi (100 km) away, in which Alexander the Great defeated (331 B.C.) Darius III.

Ar·bo·ga (är-bō'gä), town (1970 pop. 11,932), Västmanland co., S Sweden, on the Arboga River, near Lake Hjälmaren. It is a transportation and tourist center. Manufactures include metal goods and processed food. Of great importance in the Middle Ages, Arboga was the site of several parliaments, including Sweden's first (1435).

Ar·bon (ȧr-bôɴ), town (1970 pop. 13,122) Thurgau canton, NE Switzerland, on the Lake of Constance. It has an automobile factory and machine works. Originally a Celtic town, then a Roman settlement, Arbon has historic ruins and remains of neolithic pile dwellings.

Ar·broath (är-brōth') or **Ab·er·bro·thock** (ăb'ər-brə-thŏk'), burgh (1974 pop. 23,207), Tayside region, E central Scotland, on the North Sea at the mouth of the Brothock River. A seaport, it is known for its smoked haddock, shipbuilding, and the processing of flax and jute. There are engineering works, breweries, an iron foundry, and diverse small industries. Arbroath Abbey was founded by William the Lion c.1178 and contains his tomb.

Ar·buck·le Mountains (är'bŭk-əl), range of low, rolling hills, rising c.700 ft (215 m) above the prairie, S Okla. The hills are remnant of mountains formed in the Precambrian era. Interesting geological formations have resulted from the varying erosion rates of the different rock types found in the area. Arbuckle National Recreation Area surrounds Lake of the Arbuckles, a 2,350 acre (951.8 hectare) reservoir formed behind Arbuckle Dam.

Ar·bu·tus (är-byo͞o'təs), uninc. town (1975 est. pop. 24,100), Baltimore co., NE Md., a suburb of Baltimore.

Ar·ca·di·a (är-kā'dē-ə), region of ancient Greece, in the middle of the Peloponnesus, without a seaboard, and surrounded and dissected by mountains. The Arcadians, relatively isolated from the rest of the world, lived a proverbially simple and natural life. The independent mountaineers periodically fought against Spartan power.

Arcadia. 1. City (1975 pop. 45,300), Los Angeles co., S Calif., a residential suburb of Los Angeles, at the foot of the San Gabriel Mts.; inc. 1903. The city has electronic, aerospace, optical, and camera industries. Santa Anita Racetrack and an arboretum are in Arcadia. **2.** City (1970 pop. 5,658), seat of De Soto co., SW central Fla., on the Peace River SE of Tampa; inc. 1887. It is a trade and shipping center for an extensive cattle and citrus-fruit area, and has fruit-packing houses and canneries. **3.** Town (1970 pop. 2,970), seat of Bienville parish, N La., W of Ruston. It has a cottonseed-oil mill, a cotton gin, and lumber mills. There are natural-gas wells and salt deposits in the area.

Ar·ca·ta (är-kā'tə), city (1970 pop. 10,900), Humboldt co., N Calif., NNE of Eureka on Humboldt Bay; founded 1850, inc. 1858. Lumbering is its major industry. Humboldt State College is here.

Arc Dome (ärk), mountain, 11,775 ft (3,591.4 m) high, in the Toiyabe Range, central Nev.

Arch·an·gel (ärk'ān'jəl): *see* Arkhangelsk, USSR.

Arch·bald (ärch'bôld), mining and industrial borough (1970 pop. 6,118), Lackawanna co., NE Pa., on the Lackawanna River NE of Scranton; settled 1831, inc. 1877. Anthracite-coal mines are here. Clothing and textiles are made.

Ar·cher (är'chər), county (1970 pop. 5,759), 917 sq mi (2,375 sq km), central N Texas, drained by the West Fork of the Trinity River and the Little Wichita and Wichita rivers; formed 1858; co. seat Archer City. It has agriculture (wheat, oats, grain sorghums, corn, cotton, cattle, dairy products, and poultry) and oil and natural-gas fields.

Archer City, city (1970 pop. 1,722), seat of Archer co., N Texas, S of Wichita Falls; settled c.1880, inc. 1910. Founded as a cow town, it is a trade center for an oil-producing, cattle-ranching, and poultry, grain, and dairy area.

Arch·es National Park (ärch'ĭz), 82,953 acres (33,596 hectares), E Utah; est. 1971. Located in red-rock country and overlooking the gorge of the Colorado River, this area contains a vast and unusual array of natural rock formations.

Ar·chi·pel·a·go (är'kĭ-pĕl'ə-gō'), ancient name of the Aegean Sea, later applied to the numerous islands it contains. The word now designates any cluster of islands.

Ar·chi·pié·la·go de Co·lón (är'chē-pyā'lā-gō' dā kō-lŏn'): *see* Galápagos Islands.

Ar·chu·le·ta (är'chə-lē'tə), county (1970 pop. 2,733), 1,364 sq mi (3,532.8 sq km), S Colo., on the N.Mex. border, drained by the San Juan River and bounded on the E by the San Juan Mts.; formed 1885; co. seat Pagosa Springs. It lies in an agricultural area engaged primarily in raising sheep. Mining and lumbering are also done.

Ar·co (är'kō), town (1970 pop. 1,244), seat of Butte co., SE central Idaho, NW of Pocatello, in a mountainous region. It is a trade center for an agricultural (grain and livestock) and mining (silver and lead) area. Craters of the Moon National Monument is nearby.

Ar·co·le (är'kō-lā), village (1971 pop. 4,009), Venetia, N Italy. Here, in Nov., 1796, Napoleon Bonaparte defeated the Austrians in a three-day battle.

Ar·cos de la Fron·te·ra (är'kōs dā lä frōn-tā'rä), town (1970 pop. 25,966), Cádiz prov., S Spain, in Andalusia, on a rocky hill above the Guadalete River. A Gothic church and the palace of the duke of Arcos are at the summit. Wine and olive oil are produced.

Ar·cot (är'kŏt), town (1971 pop. 30,230), Tamil Nadu state, SE India, on the Palar River. It is an agricultural market and has a weaving industry. It became the capital of the Nawab of Carnatic in 1712. Arcot was the first important fortified town captured (1751) by Robert Clive in the British-French struggle for southern India.

Arc·tic Archipelago (ärk'tĭk, är'tĭk), group of more than 50 large islands, Franklin dist., Northwest Territories, N Canada, in the Arctic Ocean. Tundra and permanent ice cover the islands, on which oil and coal have been discovered.

Arc·tic Circle, imaginary circle on the surface of the earth at lat. 66½° N, or 23½° S of the North Pole. It marks the northernmost point at which the sun can be seen during the winter solstice (about Dec. 22) and the southernmost point of the northern polar regions at which the midnight sun is visible.

Arctic North Slope or **Arctic Slope**: *see* Alaska North Slope.

Arctic Ocean, the smallest ocean, c.5,400,000 sq mi (13,986,000 sq km), located entirely within the Arctic Circle and occupying the region around the North Pole. Once called the Frozen Ocean, it is covered with ice (2-14 ft/.6-4.3 m thick) throughout the year except in fringe areas. Nearly landlocked, the Arctic Ocean is bordered by Greenland, Canada, Alaska, the USSR, and Norway. The Bering Strait connects it with the Pacific Ocean and the Greenland Sea is the chief link with the Atlantic Ocean. The floor of the Arctic Ocean is divided by three submarine ridges—Alpha Ridge, Lomonosov Ridge, and the Arctic Mid-Oceanic Ridge; other submarine ridges, such as the Faeroe-Icelandic Ridge, act to separate the Arctic Ocean from the Atlantic.

The Arctic Ocean has the widest continental shelf of all the oceans; it extends c.750 mi (1,205 km) seaward from Siberia and encloses a deep oval basin (average depth 12,000 ft/3,660 m) that stretches between Svalbard Island and the Bering Strait; east of Greenland the ring of the continental shelf is broken by the Greenland Sea. The greatest depth (17,850 ft/5,444.3 m) in the Arctic Ocean is found just north of the Chukchi Sea. Since the Arctic's connection with the Pacific Ocean is narrow and very shallow, its principal exchange of water is therefore with the Atlantic Ocean through the Greenland Sea. This outflow creates the cold East Greenland Current, which flows south along the coast of east Greenland. A weaker current goes through Smith Sound and Baffin Bay and is known as the Labrador Current. Another weak current flows out of Bering Strait. The water that does not flow out by the Greenland Sea seems to be deflected by north Greenland and forms the current that gives rise to a circular current in the Arctic basin itself. This circular current causes the relatively light ice of the Siberian seas, which contrasts with the heavy-pressure ice phenomenon off Greenland and Ellesmere Island.

The drift of ice southward and westward has been noted and utilized by explorers. Some of the ice pack remains in the Arctic basin, and some, carried out by the East Greenland Current, melts before going far enough south to reach the regular Atlantic shipping lanes; the icebergs that harass ships are generally brought from the fjords of west Greenland by the Labrador Current. The cold Arctic currents give the shores of northeast North America and northeast Asia a much colder climate than the northwest shores of Europe and North America, which are warmed by the North Atlantic Drift and the Japan Current. It was long thought that no life could exist in the Arctic; however, despite drifting ice, ice packs, vast ice floes, and winter temperatures to −60°F (−51°C), there are hares, polar bears, seals, gulls, and guillemots as far north as 88° and plankton in all Arctic waters.

The Arctic basin was almost wholly unexplored until the Amundsen-Ellsworth flight over it in 1926. Arctic research was stimulated when it was recognized that the shortest air routes between the great cities of the Northern Hemisphere cross the Arctic Ocean. Improved technology has also facilitated research, and detailed knowledge of drifts and ice floes, water depths, and the ocean floor has been vastly increased. One fact of great potential importance is now being studied—the Arctic Ocean is warming. Recorded temperatures, glacial regressions, and the appearance of observed species of fish in larger numbers, at higher latitudes, at earlier seasons, and for long periods prove that over the decades a "climatic improvement" has taken place. Similar changes have been reported in subarctic latitudes. Whether the warming is a phase in a cycle or a permanent development cannot yet be said.

Arctic Red River, c.310 mi (500 km) long, rising in the Mackenzie Mts. of Mackenzie dist., W Northwest Territories, Canada, and flowing generally NW to the Mackenzie River. At its mouth is a post of the Royal Canadian Mounted Police.

arctic regions or **the Arctic,** northernmost area of the earth, centered on the North Pole. The arctic regions are usually defined by the irregular and shifting 50°F (10°C) July isotherm that closely corresponds to the northern limit of tree growth and that varies both north and south of the Arctic Circle. The regions therefore include the Arctic Ocean; the northern reaches of Canada, Alaska, the USSR, Norway, and the Atlantic Ocean; Svalbard; most of Iceland; Greenland; and the Bering Sea. In the center of the arctic regions is a large basin occupied by the Arctic Ocean. Surface features vary from low coastal plains (swampy in summer) to high ice plateaus and glaciated mountains. Tundras, extensive flat and poorly drained lowlands, dominate the regions. The most notable highlands are the Brooks Range of Alaska, the Innuitians of the Canadian Arctic Archipelago, the Urals, and the mountains of eastern USSR. Greenland, the world's largest island, is a high plateau covered by a vast ice sheet except in the coastal regions.

The climate of the Arctic, classified as polar, is characterized by long, cold winters and short, cool summers. Polar climate may be further subdivided into tundra climate (the warmest month of which has an average temperature below 50°F/10°C but above 32°F/0°C) and ice cap climate (all months average below 32°F/0°C and there is a permanent snow cover). Precipitation, almost entirely in the form of snow, is very low, with the annual average accumulations less than 20 in. (50.8 cm). Persistent winds whip up fallen snow to create the illusion of constant snowfall. Regions abutting the Atlantic and Pacific oceans have generally warmer temperatures and heavier snowfalls than the colder and drier interior areas. However, except for along its fringe, the Arctic Ocean remains frozen throughout the year. Great seasonal changes in the length of days and nights are experienced north of the Arctic Circle, ranging from 24 hours of constant daylight (midnight sun) or darkness at the Arctic Circle to six months of daylight or darkness at the North Pole.

Vegetation in the Arctic, limited to regions having a tundra climate, flourishes during the short spring and summer seasons. The tundra's restrictive environment for plant life increases northward, with dwarf trees giving way to grasses (mainly mosses, lichen, sedges, and some flowering plants). There are about 20 species of land animals in the Arctic, including squirrel, wolf, fox, moose, caribou, reindeer, polar bear, and musk ox, and about six species of aquatic mammals such as the walrus, seal, and whale. Most of the species are year-round inhabitants of the Arctic, migrating to the southern margins as winter approaches. Some of the species, especially the fur-bearing ones, are in danger of extinction. A variety of fish is found in arctic seas, rivers, and lakes. The Arctic's bird population increases tremendously each spring with the arrival of migratory birds. During the short warm season, a large number of insects breed in the marshlands of the tundra.

In parts of the Arctic natural resources abound, but many known reserves are too inaccessible to exploit. The arctic region of the USSR is the most developed and is a vast storehouse of mineral wealth, including deposits of nickel, copper, coal, gold, uranium, tungsten, and diamonds. The North American Arctic yields uranium, copper, nickel, iron, natural gas, and oil. The arctic region of Europe (including western USSR) benefits from good overland links with southern areas and ship routes that are open throughout the year. The arctic regions of the Asian USSR and North America depend on isolated overland routes, summertime ship routes, and air transportation.

The Arctic is one of the world's most sparsely populated areas. Its inhabitants, basically of the Mongoloid race, are thought to be descendants of a people who migrated northward from central Asia after the Ice Age and subsequently spread west into Europe and east into North America. Because of their common background and the general lack of contact with other peoples, arctic peoples have strikingly similar physical characteristics and cultures, especially in such things as clothing, tools, techniques, and social organization. Once totally nomadic, they are now largely sedentary or seminomadic. Hunting, fishing, reindeer herding, and indigenous arts and crafts are the chief activities.

History. The first explorers in the arctic regions were the Vikings.

Much later the search for the Northwest Passage and the Northeast Passage to reach the Orient from Europe spurred exploration to the north. This activity began in the 16th cent. and continued in the 17th, but the hardships suffered and the negative results obtained by early explorers caused interest to wane. The fur traders in Canada did not begin serious explorations across the tundras until the latter part of the 18th cent. After 1815, British naval officers took up the challenge of the Arctic. The disappearance of John Franklin on his expedition between 1845 and 1848 gave rise to more than 40 searching parties. Although Franklin was not found, a great deal of knowledge was learned about the Arctic as a result, including the general outline of Canada's arctic coast. The Northeast Passage was finally navigated in 1879 by Nils A. E. Nordenskjöld. Although Fridtjof Nansen, drifting with his vessel *Fram* in the ice (1893-96), failed to reach the North Pole, he added enormously to the knowledge of the Arctic Ocean. The race to be first at the North Pole was won by Robert E. Peary in 1909.

Air exploration of the regions began with the tragic balloon attempt of S. A. Andree in 1897. In 1926 Richard E. Byrd and Floyd Bennett flew over the North Pole. The use of the "great circle" route for world air travel increased the importance of arctic regions, while new agricultural and other uses of arctic and subarctic regions led to many projects for development, especially by the USSR. In 1937 and 1938 many field expeditions were sent out by British, Danish, Norwegian, Soviet, Canadian, and American groups to learn more about the Arctic. The Soviet group under Ivan Papinin set down and wintered on an ice floe near the North Pole and drifted with the current for 274 days. Valuable hydrological, meteorological, and magnetic observations were made; by the time they were taken off the floe, the group had drifted 19° of latitude and 58° of longitude.

Before World War II, the USSR had established many meteorological and radio stations in the arctic regions. Soviet activity in practical exploitation of resources also pointed the way to the development of arctic regions. During World War II, interest in transporting supplies gave rise to considerable study of arctic conditions. In 1946 the Canadian army undertook a project that had as one of its objects the testing of new machines (notably the snowmobile) for use in developing arctic regions. There was also a strong impulse to develop Alaska and northern Canada, but no consolidated effort, like that of the Soviets, to take the natives into partnership for a full-scale development of the regions. In 1955, as part of joint U.S.-Canadian defense, construction was begun on a c.3,000-mi (4,830-km) radar network (the Distant Early Warning line, commonly called the DEW line) stretching from Alaska to Baffin Island and, subsequently, across Greenland.

With the continuing development of northern regions, the Arctic is assuming greater importance in the world. During the International Geophysical Year (1957-58) more than 300 arctic stations were established by the northern countries interested in the arctic regions. In 1958 the *Nautilus*, a U.S. navy atomic-powered submarine, became the first ship to cross beneath the North Pole. Two years later the submarine *Skate* became the first to surface at the Pole. The discovery of oil on the Alaska North Slope (1968) and on Canada's Ellesmere Island (1972) led to a great effort to find new oil fields along the edges of the continents. In the summer of 1969 the SS *Manhattan*, a specially designed oil tanker with icebreaker and oceanographic research vessel features, successfully sailed from Philadelphia to Alaska by way of the Northwest Passage in the first attempt to bring commercial shipping into the region. Practically all parts of the Arctic have now been photographed and scanned (by remote sensing devices) from aircraft and satellites.

Ar·da (är'dä), river, 180 mi (289.6 km) long, S Bulgaria and Turkey. It rises in the Rhodope Mts. and flows east through Bulgaria to the Maritsa River in northwest Turkey.

Ar·de·bil (är'də-bēl'), town (1971 est. pop. 88,000), NW Iran, near the USSR border. It is a market center for a fertile agricultural region. Carpets and rugs are produced in the town. Ardebil was probably founded in the 5th cent. A.D. It became (10th cent.) the capital of Azerbaijan but was soon superseded by Tabriz. In 1220 it was destroyed by the Mongols. The town was occupied by the Turks in 1725 and by the Russians in 1828. Its fine library was taken to St. Petersburg by the Russians.

Ar·dèche (är-dĕsh'), department (1975 pop. 257,065), 2,132 sq mi (5,521.9 sq km), S France, in Vivarais; capital Privas.

Ar·den (är'dən), uninc. city (1975 est. pop. 50,700), Sacramento co., N central Calif.

Arden, Forest of, well-wooded area, formerly very extensive, in Warwickshire, central England. It is the setting for Shakespeare's *As You Like It*.

Ar·dennes (är-dĕn'), department (1975 est. pop. 309,306), 2,014 sq mi (5,216.3 sq km), NE France, in Champagne; capital Charleville-Mézières.

Ardennes, wooded plateau, from 1,600 to 2,300 ft (488-701.5 m) high, in SE Belgium, N Luxembourg, and Ardennes dept., N France, E and S of the Meuse River. The plateau is cut into wild crags and ravines by rapid rivers. Agriculture and cattle raising are the main occupations of this sparsely populated region. A traditional battleground, the Ardennes saw heavy fighting in both World Wars, notably in the Battle of the Bulge (Dec., 1944-Jan., 1945).

Ard·more (ärd'môr). **1.** City (1975 est. pop. 21,915), seat of Carter co., S Okla.; inc. 1898. It is the commercial center of a rich oil and farm area. Its industries include oil refining, cotton and food processing, and the manufacture of tires, telephone equipment, and electronic and plastic parts. The Goddard Center for the Visual and Performing Arts, the Southern Oklahoma Area Vocational-Technical Center, and Carter Seminary for Indian children are in Ardmore. **2.** Uninc. residential area (1970 pop. 5,801), Montgomery co., SE Pa., a suburb W of Philadelphia; settled c.1800. Motor vehicles are manufactured here.

Ard·ros·san (ärd-rŏs'ən), burgh (1971 pop. 10,569), Strathclyde region, SW Scotland, on the Firth of Clyde. A resort and seaport, it has an oil refinery, coal mines, and shipyards. Ardrossan Castle was razed by Oliver Cromwell's troops.

Ar·e·ci·bo (är'ə-sē'bō), city (1970 pop. 35,484), N Puerto Rico, a port on the Atlantic Ocean at the mouth of the Rio Grande de Arecibo. It is the commercial and industrial center of a region producing coffee, tobacco, sugar cane, and pineapples.

Ar·e·nac (ăr'ə-năk), county (1970 pop. 11,149), 368 sq mi (953.1 sq km), E Mich., bounded on the SE by Saginaw Bay and drained by the Rifle, Pine, and Au Gres rivers; formed 1831; co. seat Standish. Its agriculture includes livestock, dairy products, potatoes, corn, sugar beets, and grain. It has a lumbering industry, and wood products and automobile parts are made. The county has some commercial fishing, and there are hunting, fishing, and bathing resorts.

A·ren·dal (ä'rən-däl'), city (1977 est. pop. 11,709), capital of Aust-Agder co., SE Norway, a port on the Skagerrak. Manufactures include forest products and electric light bulbs. Chartered in 1723, Arendal has had one of Norway's largest merchant fleets since 1880.

Ar·e·op·a·gus (ăr'ē-ŏp'ə-gəs), rocky hill, 370 ft (112.9 m) high, NW of the Acropolis of Athens, famous as the sacred meeting place of the prime council of Athens.

A·re·qui·pa (ä'rā-kē'pä), city (1970 est. pop. 194,700), alt. c.7,800 ft (2,380 m), capital of Arequipa dept., S Peru, on the Chili River at the foot of El Misti. One of Peru's largest cities, it is the commercial center of southern Peru and northern Bolivia. Leather goods, textiles, and foodstuffs are the chief products. Alpaca wool is graded, sorted, and shipped out through the port of Mollendo. Founded in 1540 on the site of an Inca town, Arequipa stands on an oasis in an arid plain and grows crops for local consumption. Although the city was almost totally destroyed by an earthquake in 1868, its lovely examples of Spanish colonial architecture have been restored.

A·rez·zo (ä-rĕt'tsō), city (1975 est. pop. 91,299), capital of Arezzo prov., Tuscany, central Italy. It is an agricultural trade center and has machine and textile industries. Arezzo was an Etruscan town, later became a Roman military station and colony, and was made (11th cent.) a free commune. Siding with the Ghibellines, it was defeated (1289) by Florence, to which it passed definitively in 1384. The city retains much of its medieval character. Noteworthy buildings include the Gothic cathedral (1286-1510); the Gothic Church of San Francesco (14th cent.); the Romanesque Church of Santa Maria della Pieve (1330); and Bruni Palace (15th cent.), which now houses an art gallery and museum.

Ar·gen·teuil (är'zhäɴ-tœ'yə), city (1975 est. pop. 102,530), Val-d'Oise dept., N France, on the Seine, a suburb of Paris. It has important metalworks and factories making furniture, railroad and airplane parts, and chemicals. It is also famous for its asparagus and grapes. It grew around a convent founded in the 7th cent. The convent (later a monastery) was destroyed in the French Revolution; a famous relic, the Seamless Tunic, said to have been worn by Christ, was given by

Charlemagne to the convent and is now enshrined in Saint-Denis Basilica (1866).

Ar·gen·ti·na (är'jən-tē'nə), republic (1970 pop. 23,364,443), 1,072,157 sq mi (2,776,887 sq km), S South America; capital Buenos Aires. It is the second-largest nation of South America. Argentina is triangular in shape and stretches c.2,300 mi (3,700 km) from its broad northern region near the Tropic of Capricorn to southern Tierra del Fuego, an island shared with Chile. On the northeast, Argentina fronts on the Río de la Plata, which separates Argentina from southern Uruguay; its tributaries also act as international boundaries—the Uruguay River, with west Uruguay and south Brazil, and the Paraná and Pilcomayo rivers, with Paraguay. The northwest boundary with Bolivia lies in the Gran Chaco and the Andes. The western boundary with Chile follows the crestline of the Andes. The Atlantic

Ocean borders Argentina on the east; there, off south Argentina, are the Falkland Islands, and the South Georgia, South Sandwich, and South Orkney islands, all claimed by Argentina but administered by Great Britain. Argentina also claims a sector of Antarctica. Argentina may be divided into six geographic regions—the Paraná Plateau (an extension of the highlands of southern Brazil), the Gran Chaco, the Pampa (also known as Pampas), the Monte (an arid region in the rain shadow of the Andes), Patagonia, and the Andes Mts. The climate varies from subtropical in the north to cold and windswept in the south, with temperate and dry areas found throughout much of the country.

Argentina is divided into a federal district, 22 provinces, and one territory:

NAME	CAPITAL	NAME	CAPITAL
Buenos Aires	La Plata	Misiones	Posadas
Catamarca	Catamarca	Neuquén	Neuquén
Chaco	Resistencia	Río Negro	Viedma
Chubut	Rawson	Salta	Salta
Córdoba	Córdoba	San Juan	San Juan
Corrientes	Corrientes	San Luis	San Luis
Entre Ríos		Santa Cruz	Río Gallegos
Federal District	Paraná	Santa Fe	Santa Fe
Formosa	Formosa	Santiago del Estero	Santiago del Estero
Jujuy	Jujuy		
La Pampa	Santa Rosa	Tierra del Fuego Territory	Ushuaia
La Rioja	La Rioja		
Mendoza	Mendoza	Tucumán	Tucumán

Economy. Argentina's economy is based on agriculture, with grains and livestock (cattle and sheep) the bulwark of its wealth. As an exporter of wheat, corn, flax, oats, beef, mutton, hides, and wool, Argentina has traditionally rivaled the United States, Canada, and Australia. Argentina is the world's largest source of tannin and linseed oil. Since the 1930s there has been a great rise in production in areas besides the Pampa, especially in the oases of the Monte and the irrigated valleys of northern Patagonia. Argentina is nearly self-sufficient in its agricultural needs. Domestic oil and gas production supplies most of the nation's energy. The large coal field of south Patagonia has low-grade coal. All mining operations in the country have been under federal control since 1954. Developed after World War I and protected by a strong nationalistic policy, Argentine industry has made the country virtually self-sufficient in the production of consumer goods and many types of machinery. Food processing (in particular meat packing, flour milling, and canning) is the chief manufacturing industry of Argentina; leather goods and textiles are also major products.

History. The Europeans probably first arrived in 1502 during the voyage of Amerigo Vespucci. The search for a Southwest Passage to the Orient brought Juan Díaz de Solís to the Río de la Plata in 1516. Ferdinand Magellan entered (1520) the estuary, and Sebastian Cabot ascended (1536) the Paraná and Paraguay rivers. Pedro de Mendoza in 1536 founded the first settlement of the present Buenos Aires, but Indian attacks forced abandonment of the settlement. Buenos Aires was refounded in 1580. It achieved (1617) a sort of semi-independence under the viceroyalty of Peru. The mercantilist system, however, severely hampered commerce, and smuggling, especially with Portuguese traders in Brazil, became an accepted profession. While the cities of present west and northwest Argentina grew by supplying the mining towns of the Andes, Buenos Aires was threatened by Portuguese competition. By the 18th cent., cattle (introduced to the Pampa in the 1550s) roamed wild throughout the Pampa in large herds and were hunted by gauchos for their skins and fat.

In 1776 Buenos Aires was made a free port and the capital of a viceroyalty that included present-day Argentina, Uruguay, Paraguay, and (briefly) Bolivia. From this combination grew the idea of a Greater Argentina to include all the Río de la Plata countries. In 1806 the British took the city after the Spanish viceroy fled. An Argentine militia force ended the British occupation and beat off a renewed attack. On May 25, 1810 (May 25 is the Argentine national holiday), revolutionists, acting nominally in favor of the Bourbons dethroned by Napoleon, deposed the viceroy, and the government was controlled by a junta. The result was war against the royalists. On July 9, 1816, a congress in Tucumán proclaimed the independence of the United Provinces of La Plata. Uruguay and Paraguay went their own ways, however, despite hopes of reunion.

In Argentina, a struggle ensued between those wanting to unify the country and those unwilling to be dominated by Buenos Aires. Independence was followed by virtually permanent civil war, with countless coups d'état by regional, social, or political factions. Argentinians united to help the Uruguayans repel Brazilian conquerors in the Battle of Ituzaingó (1827), which led to the independence of Uruguay. The internal conflict was, however, soon resumed and was not even quelled when Juan Manuel de Rosas established a dictatorship that lasted until 1852. Ironically, this federalist leader did more than the unitarians to unify the country. Rosas was overthrown by Justo José de Urquiza, who called a constituent assembly at Santa Fe. A constitution was adopted (1853) based on the principles enunciated by Juan Bautista Alberdi.

In 1880 federalism triumphed, and Gen. Julio A. Roca became president (1880–86); Buenos Aires remained the capital, but the federal district was set up, and Buenos Aires prov. was given La Plata as its capital. Argentina flourished during Roca's administration. The conquest of the Indians by Gen. Roca (1878–79) had made colonization in the south and the southwest possible. Already the Pampa had begun to undergo its agricultural transformation. Establishment of refrigerating plants for meat made expansion of commerce possible. The British not only became the prime consumers of Argentine products but also invested substantially in the construction of factories, public utilities, and railroads (which were nationalized in 1948). Efforts to end the power of the great landowners, however, were not genuinely successful, and the military tradition continued to play a significant part in politics. The second administration of Roca (1898–1904) was marked by recovery from the crises of the intervening years; a serious boundary dispute with Chile was settled (1902), and perpetual peace between the two nations was symbolized in the Christ of the Andes.

Even before World War I, in which Argentina maintained neutrality, the wealthy nation had begun to act as spokesman for the rights and interests of Latin America as a whole. Internal problems, however, remained vexing. A number of factions controlled Argentina between 1910 and World War II. A "palace revolt" in 1944 brought to power a group of army colonels, chief among them Juan Perón. After four years of pro-Axis "neutrality" Argentina belatedly

(Mar., 1945) entered World War II on the side of the Allies. Perón, an admirer of Mussolini, established a type of popular dictatorship new to Latin America, based initially on support from the army, reactionaries, nationalists, and some clerical groups. His regime was marked by curtailment of freedom of speech, confiscation of liberal newspapers, imprisonment of political opponents, and transition to a one-party state. His second wife, the popular Eva Duarte de Perón, helped him gain the support of the trade unions, thereafter the main foundation of Perón's political power. In 1949 a new constitution permitted Perón to succeed himself as president; the Peronista political party was established the same year. Perón inaugurated a program of industrial development—which by the early 1950s was severely hampered by the lack of power resources and machine tools—supplemented by social welfare programs. Perón also placed the sale and export of wheat and beef under government control, thus undermining the political and economic power of the rural oligarchs. Agricultural production, long the chief source of revenue, dropped sharply, and the economy faltered. The Roman Catholic church, alienated by the reversal of close church-state relations, excommunicated Perón, and, finally, in 1955, Perón was ousted by a military coup.

In 1957 Argentina reverted to the constitution of 1853 as modified up to 1898. In 1958 Dr. Arturo Frondizi was elected president. Faced with the economic and fiscal crisis inherited from Perón, Frondizi, with U.S. advice and the promise of financial aid, initiated a program of austerity to "stabilize" the economy and check inflation. Leftists, as well as Peronistas, who still commanded strong popular support, criticized the plan because the burden lay most heavily on the working and lower middle classes. Frondizi later fell into disfavor with the military because of his leniency toward the regime of Fidel Castro in Cuba and toward Peronistas at home. Outraged by the resurgence of Peronista strength in the congressional elections of 1962, the military arrested Frondizi.

In 1963, after months of political crisis and control by the military, presidential elections were held. The Peronista and Communist parties were banned before the election, and many persons were placed under arrest. Following the election as president of the moderate liberal Dr. Arturo Illía, many political prisoners were released, and relative political stability returned. The new president was faced, however, with serious economic depression and with the difficult problem of reintegrating the Peronist forces into Argentine political life. In 1964 an attempt by Perón to return from Spain was thwarted when Perón was turned back at Rio de Janeiro by Brazilian authorities. In elections in 1965 and 1966 the Peronists showed that they remained the strongest political force in the country; unwilling to tolerate another Peronist resurgence, a junta of military leaders, supported by business interests, seized power (1966). The new government dissolved the legislature, banned all political parties, and exercised unofficial press censorship. The national universities were also placed under government control. Widespread opposition to the rigid rule grew, and an antigovernment campaign developed. Faced with labor and student unrest, the military named a new president (1970). Economic problems and terrorist activities increased.

An active program for economic growth, distribution of wealth, and political stability was matched by a call for national elections and a civilian government led to the return of Perón to Argentina in 1972. After failing to achieve unity among the various Peronist groups, Perón declined the nomination from his supporters to run for president in the March, 1973, elections, but when new elections were held in Sept., 1973, Perón was elected president and his third wife, Isabel Martínez Perón, vice president. Perón died in July, 1974, and, as provided for in the constitution, was succeeded as president by his widow. During her term in office, economic problems and labor unrest grew and divisions in the Peronista party worsened. In March, 1976, she was overthrown by a three-man military junta headed by Jorge Videla.

Ar·ghan·dab (är'gän-däb'), river, rising in E Afghanistan and flowing c.250 mi (400 km) SW past Kandahar to the Helmand River.

Ar·gonne (är'gŏn'), region of the Paris basin, NE France, in Champagne and Lorraine, a hilly and woody district. Here, in 1792, the French repulsed the Prussians. The sector was a battleground throughout World War I. In the Allied victory drive (Sept.-Nov., 1918), the Meuse-Argonne sector was carried by the Americans.

Ar·gos (är'gŏs, -gəs), city of ancient Greece, in NE Peloponnesus, 3 mi (4.8 km) inland from the Gulf of Argos, near the modern Nauplia. It was occupied from the early Bronze Age. Argos was the center of Argolis and in the 7th cent. B.C. dominated much of the Peloponnesus. For centuries it was one of the most powerful Greek cities, struggling with Sparta and rivaling Athens and Corinth. Much of Argos's power disappeared after Sparta took (c.494 B.C.) the city. In 146 it was occupied by Rome, under whose rule trade flourished. There is a small modern town called Argos on the site.

Ar·gun (är'go͞on'), river, 950 mi (1,528.6 km) long, rising in the Great Khingan Mts., Heilungkiang prov., NE China, and flowing W to the USSR border, then NE along the USSR-China frontier, where it joins the Shilka River to form the Amur. Corn, grains, and sugar beets are grown in the fertile valley. Silver, lead, and coal are found along the river banks.

Ar·gyll·shire (är-gīl'shîr', -shər) or **Ar·gyll** (är-gīl'), former county, W central Scotland. Its county town was Inveraray. Argyllshire was settled by Celts from Ireland in the 6th cent. Under the Local Government Act of 1973, Argyllshire was divided between the new Highland and Strathclyde regions.

År·hus (ôr'ho͞os), city (1976 est. pop. 246,355), capital of Århus co., central Denmark, on Århus Bay, an arm of the Kattegat. It is a commercial, industrial, and shipping center. Manufactures include beer, textiles, machinery, processed food, locomotives, and tobacco products. First mentioned in the mid-10th cent., Århus is one of the oldest cities in Denmark. It developed rapidly after it became an episcopal see in the 11th cent. The city declined after the Reformation (16th cent.) but recovered its prosperity in the 18th cent. Noteworthy buildings include the Cathedral of St. Clemens (12th cent.) and the town hall (1942), made of Norwegian marble.

Ar·i·a·na (ăr'ē-ā'nə, -ăn'ə), general name for the E provinces of the ancient Persian Empire. It was used to mean the regions south of the Oxus (modern Amu Darya) River.

A·ri·ca (ä-rē'kä), city (1970 pop. 92,394), N Chile, on the Pacific Ocean, just S of the Peruvian border at the N limit of the Desert of Atacama. The district of Arica is now a free zone where both Chile and Peru maintain customs houses. The city is a resort and a port through which the mineral exports (chiefly copper, tin, and sulfur) of both countries are shipped.

A·riège (ä-ryĕzh'), department (1975 pop. 137,857), 1,888 sq mi (4,890 sq km), SW France, in Languedoc, bounded by Spain and Andorra; capital Foix.

Ariège, river, 106 mi (170.6 km) long, Ariège and Haute-Garonne depts., S France, flowing NNW from the E Pyrenees to the Garonne River S of Toulouse. There are hydroelectric plants along its course.

A·ri·nos (ə-rē'nəs, -no͞os), river, c.400 mi (645 km) long, rising in central Mato Grosso state, Brazil, and flowing N into the Juruena River. Considered the longest headstream of the Tapajós River, it is interrupted by rapids.

A·ri·pua·nã (ä'rē-pwə-năɴ'), river, rising in N Mato Grosso state, W central Brazil, and flowing c.400 mi (645 km) N to the Madeira River NE of Manicoré. It receives the Roosevelt River and is navigable in its lower course.

A·rish, Al (ăl ă-rēsh'), town (1970 est. pop. 43,000), NE Egypt, in the Sinai peninsula, on the Mediterranean Sea. It is a fishing port and administrative center. In 1800, during the venture of Napoleon I in Egypt, the French signed a convention agreeing to evacuate the country. The British did not ratify the convention, and fighting resumed. Israeli troops briefly held Al Arish during the 1956 Arab-Israeli war and occupied the town in the 1967 war.

Ar·i·zo·na (ăr'ə-zō'nə), state (1970 pop. 1,770,900), 113,909 sq mi (295,024 sq km), SW United States, admitted as the 48th state of the Union in 1912. The capital and largest city is Phoenix. Arizona is bounded on the north by Utah; on the west, where the Colorado River forms the border, by southern Nev. and Calif.; on the south by Mexico; and on the east by N.Mex.

In northern Arizona are the Colorado Plateau, an area of dry plains more than 4,000 ft (1,220 m) high, and deep canyons, including the famous Grand Canyon cut out by the Colorado River. Along the Little Colorado River, which runs northwest through the plateau to join the Colorado, are the Painted Desert and Petrified Forest National Park. South of the Grand Canyon are the San Francisco Peaks, including Humphreys Peak, the highest point in the state. The southern edge of the Colorado Plateau is marked by an escarpment called Mogollon Rim. The southern half of the state has desert basins broken up by mountains with rocky peaks and extending northwest to southeast across central Arizona. To the south, the Gila River, a major tributary of the Colorado, flows west across the entire state. This area has desert plains separated by mountain chains running north and south; in the west the plains fall to the relatively low altitude of c.140 ft (43 m) in the region around Yuma. Precipitation in most of the state is low, and much of Arizona's history has been shaped by the inadequate water supply. Since the early 20th cent. massive irrigation projects have been built in Arizona's valleys.

National and state forests attract millions of tourists yearly. Tourism is bolstered in the north by the Grand Canyon, the Painted Desert, the Petrified Forest, meteor craters, ancient Indian ruins, and the Navaho and Hopi Indian reservations that cover nearly all of the state's northeast quadrant. Southeast Arizona's warm, dry climate, often recommended for people in ill health, also attracts a large tourist trade. Between 1940 and 1960 Arizona's population increased more than 100% and between 1960 and 1970 it increased another 36%. The mountainous, arid north has not shared the population growth of the southern sections of the state. In the 1960s the population included some 85,000 Indians.

Arizona is divided into 14 counties:

NAME	COUNTY SEAT	NAME	COUNTY SEAT
Apache	St. Johns	Mohave	Kingman
Cochise	Bisbee	Navajo	Holbrook
Coconino	Flagstaff	Pima	Tucson
Gila	Globe	Pinal	Florence
Graham	Safford	Santa Cruz	Nogales
Greenlee	Clifton	Yavapai	Prescott
Maricopa	Phoenix	Yuma	Yuma

Economy. The state's principal crops are cotton, hay, lettuce, and sorghum. Cattle, calves, and dairy products are also important. The state's major industries produce machinery, food products, and primary metals. Copper is the state's most valuable mineral; Arizona produces over half of all copper mined in the United States. Other leading mineral resources are molybdenum, sand and gravel, and cement. The mountains in the north and central regions have 3,180,000 acres (1,287,900 hectares) of commercial forests, chiefly ponderosa pines and other firs, which support the state's lumber and building-materials industries. Arizona's Indians produce many fine handicrafts, including leather goods, woven items, pottery, and the famous silver and turquoise jewelry of the Navahos.

History. Indians probably lived in the region as early as 25,000 B.C. Pueblo Indians flourished in Arizona between the 11th and 14th cent. and built many of the cliff dwellings that still stand. Apache and Navaho Indians came to the area in c.1300 from Canada. Probably the first Spanish explorer to enter Arizona (c.1536) was Cabeza de Vaca. Francisco Vasquez de Coronado led an expedition from Mexico in 1540, but the region was later neglected by the Spanish. Father Eusebio Kino converted the Indians of southern Arizona, founded missions (1692-1700), and introduced cattle and sheep raising. The Arizona region came under Mexican control following the Mexican war of independence from Spain (1810-21). In the Treaty of Guadalupe Hidalgo (1848), Mexico relinquished control of the area north of the Gila River to the United States. This area became part of the U.S. Territory of New Mexico in 1850. The United States bought the area between the Gila River and the present-day southern boundary of Arizona from Mexico in the Gadsden Purchase (1853).

Arizona was organized as a separate territory in 1863. By the 1870s mining, especially of copper and silver, was flourishing. The capital, earlier at Prescott, then Tucson, was moved to Phoenix in 1889. The region was held precariously by U.S. soldiers during intermittent warfare (1861-86) with the Apache Indians, who were led by Cochise and later Geronimo. Ranching, which had foundered under the Apache attacks on livestock, thrived after their defeat.

In 1912 Arizona, still a raw frontier territory, attained statehood. Its constitution created a storm, with such "radical" political features as initiative, referendum, and judicial recall. With the opening of the Roosevelt Dam (1911), a federally financed project, massive irrigation projects began to transform Arizona's valleys. During World War II defense industries were established; manufacturing, notably electronic industries, continued to develop after the war.

Government. The state's constitution provides for an elected governor and bicameral legislature, with a 30-member senate and a 60-member house of representatives. The governor and members of the legislature serve two-year terms.

Educational Institutions. Arizona's major schools include the Univ. of Arizona, at Tucson; Arizona State Univ., at Tempe; Northern Arizona Univ., at Flagstaff; and several private institutions.

Ar·ka·del·phi·a (är′kə-dĕl′fĭ-ə), city (1970 pop. 9,841), seat of Clark co., SW Ark., on the Ouachita River S of Hot Springs, in a cotton-growing and timber area; settled c.1810, inc. 1857. Flour, lumber and wood products, clothing, and aluminum products are made.

Ar·kan·sas (är′kən-sô′), state (1975 pop. 2,089,000), 53,104 sq mi (137,539 sq km), central and SW United States, admitted as the 25th state of the Union in 1836. The capital and largest city is Little Rock. On the east the Mississippi River separates Arkansas from Tenn. and Miss. The state is bounded on the north by Mo., on the west by Okla. and a part of Texas, and on the south by La.

The Arkansas River flows southeast across the state between the Ozark plateaus and the Ouachita Mts. and runs down to the south-

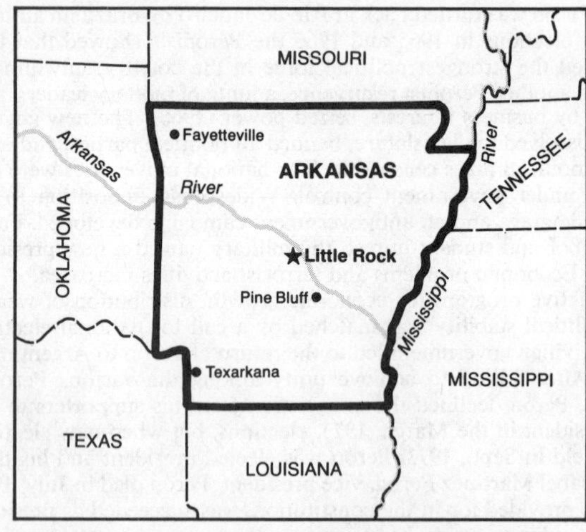

ern and eastern plains to empty into the Mississippi. The other rivers of the state also flow generally southeast or south to the Mississippi. The climate of Arkansas is marked by long, hot summers and mild winters. The state's many lakes and streams and its abundant wildlife provide excellent hunting and fishing and bring thousands of sportsmen annually. Mineral springs also attract many visitors to Arkansas, where tourism is an important industry.

Arkansas is divided into 75 counties:

NAME	COUNTY SEAT	NAME	COUNTY SEAT
Arkansas	De Witt and Stuttgart	Baxter	Mountain Home
Ashley	Hamburg	Benton	Bentonville

NAME	COUNTY SEAT	NAME	COUNTY SEAT
Boone	Harrison	Little River	Ashdown
Bradley	Warren	Logan	Booneville and Paris
Calhoun	Hampton	Lonoke	Lonoke
Carroll	Berryville and Eureka Springs	Madison	Huntsville
		Marion	Yellville
Chicot	Lake Village	Miller	Texarkana
Clark	Arkadelphia	Mississippi	Blytheville and Osceola
Clay	Corning and Piggott	Monroe	Clarendon
Cleburne	Heber Springs	Montgomery	Mount Ida
Cleveland	Rison	Nevada	Prescott
Columbia	Magnolia	Newton	Jasper
Conway	Morrilton	Ouachita	Camden
Craighead	Jonesboro and Lake City	Perry	Perryville
		Phillips	Helena
Crawford	Van Buren	Pike	Murfreesboro
Crittenden	Marion	Poinsett	Harrisburg
Cross	Wynne	Polk	Mena
Dallas	Fordyce	Pope	Russellville
Desha	Arkansas City	Prairie	Des Arc and De Valls Bluff
Drew	Monticello		
Faulkner	Conway	Pulaski	Little Rock
Franklin	Charleston and Ozark	Randolph	Pocahontas
Fulton	Salem	Saint Francis	Forrest City
Garland	Hot Springs	Saline	Benton
Grant	Sheridan	Scott	Waldron
Greene	Paragould	Searcy	Marshall
Hempstead	Hope	Sebastian	Fort Smith and Greenwood
Hot Spring	Malvern		
Howard	Nashville	Sevier	De Queen
Independence	Batesville	Sharp	Evening Shade and Hardy
Izard	Melbourne		
Jackson	Newport	Stone	Mountain View
Jefferson	Pine Bluff	Union	El Dorado
Johnson	Clarksville	Van Buren	Clinton
Lafayette	Lewisville	Washington	Fayetteville
Lawrence	Powhatan and Walnut Ridge	White	Searcy
		Woodruff	Augusta
Lee	Marianna	Yell	Danville and Dardanelle
Lincoln	Star City		

Economy. A major cotton-producing state in the 19th cent., Arkansas has since diversified its agricultural production and overall economy. Cotton is still an important crop, but it ranked second in value, below soybeans, in 1971. Rice is also important. Livestock (including chickens, cattle, and calves) and dairy products almost equal crops as a source of farm income. Important mineral products are petroleum, bromine and bromine compounds, natural gas, and bauxite. Arkansas is the nation's leading bauxite producer. Lumbering is important in this heavily wooded state. Arkansas's major manufactures are food products; electrical equipment; paper, lumber, and wood products; furniture; and fixtures. The state also has a fast-growing chemical industry.

History. A people known as the Bluff Dwellers, who inhabited caves, probably lived in the Arkansas area before 500. They were followed by the Mound Builders. The first white men to arrive (1541-42) were probably members of the Spanish expedition under Hernando De Soto. Later the French explorers Jacques Marquette and Louis Jolliet came south along the Mississippi to the mouth of the Arkansas River. In 1686 La Salle's lieutenant, Henri de Tonti, established Arkansas Post, the first white settlement in the Arkansas area. La Salle claimed the Mississippi valley for France, and the region became part of the French territory of Louisiana. The French ceded the Louisiana territory to Spain in 1762 but regained it again before it passed to the United States with the Louisiana Purchase (1803). Arkansas became part of the Territory of Missouri in 1812. In 1819 it was made a separate territory, and in 1836 a state with Little Rock as the capital.

The cotton boom of 1818 brought the first large wave of settlers, and the Southern plantation system fixed itself in the alluvial plains of southern and eastern Arkansas. After some hesitation Arkansas seceded (May 6, 1861) from the Union. In the Civil War, Confederate defeats led to Union occupation of northern Arkansas. In Sept., 1863, Federal troops entered Little Rock, where a Unionist convention in Jan., 1864, set up a government that repudiated secession and abolished slavery. Arkansas was not readmitted to the Union until 1868, when a new constitution gave Negroes the right to vote and hold office.

Because of high cotton prices and the failure to give the freed Negroes any economic status, the broken plantation system was replaced by sharecropping and farm tenancy. The lives of the people of the Ozarks remained largely unchanged. In the late 19th cent., as railroad construction proceeded, Arkansas's population grew substantially, and bauxite and lumbering industries developed. Oil was discovered near El Dorado in 1921. With the fortunes of the state pegged to the price of cotton, the depression of the early 1930s struck hard. Impoverished farmers migrated west to California as "Arkies"—like the "Okies" from neighboring Oklahoma. World War II created a boom for new industries in the state, notably the processing of bauxite into aluminum. In 1957 Governor Orval Faubus of Arkansas became a center of attention when he resisted the attempted desegregation of public schools in Little Rock.

Government. The state constitution (1874) provides for an elected governor and bicameral legislature, with a 35-member senate and a 100-member house of representatives. The governor and representatives serve two-year terms; senators serve for four years.

Educational Institutions. Among those prominent in the state are the Univ. of Arkansas, at Fayetteville; Arkansas State Univ., near Jonesboro; Hendrix College and the State College of Arkansas, at Conway; Ouachita Baptist College and Henderson State College, at Arkadelphia; the College of the Ozarks, at Clarksville; Arkansas College, at Batesville; and Harding College, at Searcy.

Ar·kan·sas (är'kən-sô'), county (1970 pop. 23,347), 1,035 sq mi (2,680.7 sq km), SE Ark., bounded on the S by the Arkansas River, on the SE by the Mississippi River, and drained by the White River and Bayou Meto; formed 1813; co. seats De Witt and Stuttgart. It has agriculture (cotton, rice, corn, oats, soybeans, hay, lespedeza, and livestock), lumbering, rice and cotton processing, and commercial fishing. Pecans are shipped. There is duck and quail hunting in the area.

Ar·kan·sas (är'kən-sô', är-kăn'zəs), river, c.1,450 mi (2,335 km) long, rising in the Rocky Mts., central Colo., and flowing generally SE across the plains to the Mississippi River, SE Ark. It drains 160,500 sq mi (415,695 sq km). The upper course of the Arkansas River has many rapids and flows through Royal Gorge, one of the deepest canyons in the United States. More than 25 dams on the river provide flood control, power, and irrigation. During the warm months, because of its extensive use for irrigation, the middle course of the Arkansas is reduced to a trickle. The Arkansas River Navigation System, opened in 1971, makes the river navigable to Tulsa, Okla., c.500 mi (805 km) upstream. The Spanish explorers Coronado and De Soto probably traveled along portions of the river in the 1540s. In 1806 Zebulon Pike, an American army officer, explored the river's upper reaches in Colorado. The Arkansas River was an important trade and travel route in the 19th cent.

Ar·kan·sas City. 1. (är'kən-sô) Town (1970 pop. 615), seat of Desha co., SE Ark., SE of Pine Bluff on the Mississippi; founded c.1873. **2.** (är-kăn'zəs) City (1970 pop. 13,216), Cowley co., S Kansas, at the confluence of the Arkansas and the Walnut rivers, near the Okla. border; inc. 1872. Located in an agricultural and oil region, it has oil refineries, flour mills, and meat-packing plants. Arkansas City was the starting point for the "run" (1893) of thousands of homesteaders into the Cherokee Strip.

Ar·kan·sas Post (är'kən-sô), community on the Arkansas River, SE Ark. Founded by the French in 1686 as a trading post, it is the oldest white settlement in the state; it became the capital of Arkansas Territory in 1819. Once an important port, Arkansas Post was a Confederate stronghold during the Civil War until it was captured by Union troops in 1863. Arkansas Post National Memorial is here.

Ark·han·gelsk (ŭr-KHän'gəlysk) or **Arch·an·gel** (ärk'ān'jəl), city (1970 pop. 343,000), NW European USSR, on the Northern Dvina near its mouth at the White Sea. Although icebound much of the year, it is a leading Soviet port and can generally be made usable by icebreakers. Timber and wood products make up the bulk of the exports. The city has factories producing pulp and paper, turpentine, resin, cellulose, building materials, and prefabricated houses. Fishing and shipbuilding are also major industries. Once the site of a Norse settlement, the city was founded (1584) as Novo-Kholmogory; it was renamed (1613) for the monastery of the Archangel Michael (which still stands). Arkhangelsk was Russia's principal port until the founding of St. Petersburg in 1703; it regained importance after the rail line to Moscow was completed in 1898. A supply port during World War I, Arkhangelsk was occupied from 1918 to 1920 by Allied forces (including Americans) and by the White Army; it served as their base for unsuccessful campaigns against the Bolsheviks. During World War II, U.S. and British shipments landed at Arkhangelsk. The city has a maritime school (1771), a regional museum (1859), and institutes of forestry and medicine.

Arl·berg (ärl'bûrg), pass, 5,946 ft (1,813.5 m) high, W Austria, beside

Arlberg peak, on the boundary between Tyrol and Vorarlberg. The Arlberg region forms the water divide between rivers flowing to the North Sea and those flowing into the Black Sea. The Arlberg Tunnel (built 1880–84) is one of the world's longest (6.2 mi/10 km) railway tunnels. The Arlberg district is a noted winter sports center.

Arles (ärl), kingdom, formed in 933, when Rudolf II, king of Transjurane Burgundy, united the kingdom of Provence or Cisjurane Burgundy to his lands and established his capital at Arles. Holy Roman Emperor Conrad II annexed the kingdom to the Holy Roman Empire in 1034. In 1378 Holy Roman Emperor Charles IV ceded the realm to the dauphin (later King Charles VI of France), and the kingdom for all practical purposes ceased to exist.

Arles, city (1975 pop. 50,059), Bouches-du-Rhône dept., S central France, in Provence, on the Rhône River delta. Arles is an important railroad and industrial center with varied manufactures. It was a flourishing Roman town and the metropolis of Gaul in the late Roman Empire. Constantine I convoked (314) a synod at Arles that condemned Donatism; Constantine II was born here. Arles became (879) the capital of Provence and (933) of the kingdom of Arles. In the 12th cent. it became a free city. Arles retained its special status until the French Revolution. Among its noteworthy attractions are a Roman arena (2nd cent. A.D.), now used for bullfights; a Roman theater (1st or 2nd cent. A.D.); the Church of St. Trophime (11th-15th cent.; formerly a cathedral); and the Museon Arlaten, a museum of Provençal culture and folklore. Arles has attracted many painters, notably Van Gogh and Gauguin.

Ar·ling·ton (är'lĭng-tən), county (1970 pop. 174,284), 26 sq mi (67.3 sq km), N Va., across the Potomac River from Washington, D.C. A residential suburb of Washington, the county is governed as a single unit. Most of the residents are employed by the U.S. government. Arlington was ceded to the Federal government in 1790 and was part of the District of Columbia until 1847.

Arlington. 1. Town (1970 pop. 53,534), Middlesex co., E Mass., a residential suburb of Boston; settled c.1630 as Menotomy, inc. as West Cambridge 1807, renamed Arlington 1867. Menotomy was the scene of fierce fighting after the Lexington and Concord battles in 1775. Some 17th cent. buildings remain. **2.** Town (1970 pop. 11,203), Dutchess co., SE N.Y., a suburb E of Poughkeepsie. It is the seat of Vassar College. Nearby is a state park. **3.** Industrial city (1975 est. pop. 111,500), Tarrant co., N Texas, midway between Dallas and Fort Worth; inc. 1896. The center of a rapidly growing area, it has a huge industrial park with its own railroad. There are steel and iron works, and other industries that produce automobile parts, cans and containers, rubber items, mobile homes, electronic equipment, oilfield equipment, aircraft and parts, insecticides, and paving and road equipment. Six Flags Over Texas (an amusement park) and the Pecan Bowl are located here. **4.** Uninc. city (1970 pop. 174,284), seat of and coextensive with Arlington co., N Va.

Arlington Heights, village (1976 pop. 71,012), Cook and Lake cos., NE Ill., a residential suburb of Chicago; founded 1836, inc. 1887. Its manufactures include heating and air-conditioning equipment, electronic components, radioactive drugs, and office supplies. Arlington Park Racetrack and a missile base are in the village.

Arlington House National Memorial, 3 acres (1.2 hectares), NE Va., in Arlington National Cemetery; est. 1955. Formerly called the Custis-Lee Mansion, it is a memorial to the Confederate Gen. Robert E. Lee. Arlington House was the home of Lee, inherited by his wife, the daughter of George Washington Parke Custis. It was abandoned by the Lees early in the Civil War and was later used as headquarters for the Union army. The estate was confiscated for nonpayment of taxes, and c.200 acres (80 hectares) were set aside for a national cemetery in 1864.

Arlington National Cemetery, 420 acres (170.1 hectares), N Va., across the Potomac River from Washington, D.C.; est. 1864. More than 60,000 American war dead, as well as notable Americans including William Howard Taft, John F. Kennedy, Gen. John J. Pershing, and Adm. Robert E. Peary, are interred here. Burial in Arlington is limited to active, retired, and former members of the armed forces, Medal of Honor recipients, high-ranking Federal government officials, and their dependents. There are commemorative monuments, including the Tomb of the Unknown Soldier.

Ar·lon (är-lôɴ'), town (1977 pop. 22,066), capital of Luxembourg prov., SE Belgium, near the border with Luxembourg. A strategic point since Roman times, the town has suffered numerous attacks in its history. It has Roman ruins and a picturesque marketplace.

Ar·magh (är-mä'), county (1971 pop. 133,196), 489 sq mi (1,266.5 sq km), S Northern Ireland; co. town Armagh.

Armagh, urban district (1971 pop. 12,297), county town of Co. Armagh, S Northern Ireland. Textiles are produced here. Armagh has been the ecclesiastical capital of all Ireland since the 5th cent., when St. Patrick founded his church here. Besides its Roman Catholic and Protestant cathedrals, the town contains an observatory and St. Patrick Diocesan College. Armagh suffered several Danish raids; it was destroyed by Shane O'Neill in 1566 and was burned in 1642.

Ar·ma·gnac (är'mən-yăk'), region and former county, SW France, in Gascony. Armagnac is famous for the brandy bearing the same name. The counts of Armagnac originated in the 10th cent. as vassals of the dukes of Gascony. Their power reached its height in the early 15th cent. Margaret of Angoulême married the last count of Armagnac, who died without issue. Armagnac eventually passed to her second husband, Henri d'Albret, king of Navarre, whose grandson became King Henry IV. Henry added Armagnac to the royal domain in 1607.

Ar·ma·vir (är'mə-vîr'), city (1976 pop. 158,000), Krasnodar Kray, SE European USSR, on the Kuban River. An important railroad junction, it has machine and tool plants. Armavir was founded in 1848.

Ar·me·ni·a (är-mē'nē-ə), region and former kingdom of Asia Minor. Greater Armenia lies east of the Euphrates River, and Little, or Lesser, Armenia is west of the river. Armenia is generally understood to include northeast Turkey, the Armenian Soviet Socialist Republic, and parts of Iranian Azerbaijan. It thus forms a continuation of the Anatolian plateau. Mt. Ararat, the highest point, is in Turkey, as are the sources of the Euphrates, Tigris, and Aras rivers and Lake Van.

According to tradition, the kingdom was founded by Haig, or Haik, a descendant of Noah. Modern scholars, however, believe that the Armenians crossed the Euphrates and came into Asia Minor in the 8th cent. B.C. They intermarried with the indigenous peoples and formed a homogeneous nation by the 6th cent. B.C. This state was a Persian satrapy from the late 6th cent. B.C. to the late 4th cent. B.C. Conquered (330 B.C.) by Alexander the Great, it became after his death part of the Syrian kingdom of Seleucus I and his descendants. After the Roman victory over the Seleucids in 190 B.C., the Armenians declared (189 B.C.) their independence. The imperialistic ambitions of King Tigranes led to war with Rome, and Tiridates, a Parthian prince, was confirmed as king of Armenia by Nero in A.D. 66.

Christianity was introduced early; Armenia is reckoned the oldest Christian state. In the 3rd cent. A.D. Ardashir I, founder of the Sassanid, came to power in Persia and overran Armenia. The persecution of Christians created innumerable martyrs and kindled nationalism among the Armenians. Attempts at independence were short-lived, as Armenia was the constant prey of Persians, Byzantines, White Huns, Khazars, and Arabs. From 886 to 1046 the kingdom enjoyed autonomy under native rulers; it was then reconquered by the Byzantines, who promptly lost it to the Seljuk Turks following the Byzantine defeat at the Battle of Manzikert in 1071. With the Mongol invasion of the mid-13th cent., many Armenians pushed westward, establishing in Cilicia the kingdom of Little Armenia. Shortly afterward (1386–94) the Mongol conqueror Tamerlane seized Greater Armenia and massacred a large part of the population. By the 15th cent. the Ottoman Turks invaded Armenia and by the 16th cent. held all of it. Under Ottoman rule the Armenians, although often persecuted and always discriminated against, nevertheless acquired a vital economic role, many of them serving as merchants and financiers. Eastern Armenia was chronically disputed between Turkey and Persia. It was from Persia that Russia, in 1828, acquired the present Armenian SSR.

The Armenian people underwent one of the worst trials in their history between 1894 and 1915 when a systematic plan for their extermination was put into action under Ottoman Sultan Abd al-Hamid II and was sporadically resumed, notably in 1915, when the Armenians were accused of aiding the Russian invaders during World War I. The Armenians rose in revolt at Van, which they held until relieved by Russian troops. The Treaty of Brest-Litovsk (1918) between Soviet Russia and Germany made Russian Armenia an independent republic under German auspices. It was superseded by the Treaty of Sèvres (1920), which created an independent Greater Armenia, comprising both the Turkish and the Soviet Russian parts.

In the same year, however, the Communists gained control of Russian Armenia and proclaimed it a Soviet republic, and in 1921 the Russo-Turkish Treaty established the present boundaries, thus ending Armenian independence.

Ar·me·nia (är-mā′nyä), city (1968 est. pop. 142,200), W central Colombia. Located in a fertile agricultural region (especially for coffee and cattle), Armenia is an industrial center and a transportation hub.

Ar·me·ni·an Soviet Socialist Republic (är-mē′nē-ən), constituent republic (1976 pop. 2,834,000), 11,500 sq mi (29,785 sq km), SE European USSR, in the S Caucasus; capital Yerevan. Armenia is bounded by Turkey on the west, the Azerbaijan Republic on the east, Iran on the south, and the Georgian Republic on the north. The region is one of extinct volcanoes and rugged mountains. Lake Sevan supports the important fishing industry and is a source of hydroelectric power. Armenia is rich in mineral resources, notably copper but also molybdenum, zinc, lead, iron, pyrite, manganese, gold, chromite, and mercury. Food processing, nonferrous metallurgy, and the manufacture of chemicals, electrical equipment, machinery, textiles, automobiles, and the famous cognacs and wines are the republic's major industries. Wine grapes and other fruits, wheat, barley, potatoes, and sugar beets are the major food crops; cotton and tobacco are the foremost industrial crops. Armenia was acquired by Russia from Persia in 1828. After the 1917 Bolshevik Revolution, Armenia became an independent republic. In 1920 it was occupied by the Red Army and proclaimed a Soviet republic. Two years later, Armenia, Azerbaijan, and Georgia were combined to form the Transcaucasian SFSR. In 1936 Armenia, like Azerbaijan and Georgia, became a separate constituent republic.

Ar·men·tières (är′mən-tyâr′), town (1975 pop. 58,000), Nord dept., N France, in Flanders, on the Lys River. It has foundries, boiler works, breweries, and a large textile industry. It became known through the World War I song "Mademoiselle from Armentières."

Ar·mor·i·ca (är-môr′ə-kə, -mŏr′-), ancient name for the NW part of France, especially Brittany.

Ar·mour (är′mĕr), city (1970 pop. 925), seat of Douglas co., SE S. Dak., NW of Yankton; founded 1886. It is a shipping center for grain, livestock, and poultry.

Arm·strong (ärm′strông′). **1.** County (1970 pop. 75,590), 660 sq mi (1,709.4 sq km), W Pa., bounded on the SW by the Kiskiminetas River and drained by the Allegheny River; formed 1800; co. seat Kittanning. It has bituminous-coal, oil, and gas deposits. Beverages, textiles, and glass, clay, wood, and metal products are manufactured. **2.** County (1970 pop. 1,895), 907 sq mi (2,349.1 sq km), extreme N Texas, in the high plains of the Panhandle; formed 1876; co. seat Claude. In a cattle-ranching area, it has some agriculture (wheat, grain, sorghums, dairy products, and poultry).

Ar·nett (är-nĕt′), town (1970 pop. 711), seat of Ellis co., NW Okla., SW of Woodward, in an agricultural area producing wheat, sorghums, livestock, and poultry. Dairy products are made.

Arn·hem (ärn′hĕm), city (1977 pop. 282,000), capital of Gelderland prov., E Netherlands, a port on the Lower Rhine. It is an industrial and transportation center. Textiles, electrical equipment, and metal goods are manufactured. During World War II British airborne troops suffered (Sept., 1944) a serious defeat here.

Arn·hem Land (är′nəm), region, 31,200 sq mi (80,808 sq km), N Northern Territory, Australia, on a wide peninsula W of the Gulf of Carpentaria. It contains an aboriginal reservation. Bauxite is mined in the area.

Ar·no (är′nō), river, c.150 mi (240 km) long, rising in the N Apennines, Tuscany, central Italy, and flowing S to Arezzo where it turns NW, then proceeds generally W, through Florence and Pisa, to empty into the Ligurian Sea.

Ar·nold (är′nəld). **1.** Community (1970 pop. 17,381), Jefferson co., E central Mo., just S of St. Louis. **2.** City (1970 pop. 8,174), Westmoreland co., SW Pa., on the Allegheny River adjoining New Kensington and NE of Pittsburgh; inc. as a borough 1895, as a city 1939. Glass is manufactured.

Arn·stadt (ärn′shtät), city (1974 pop. 29,462), Erfurt district, SW East Germany, on the Gera River. Gloves, shoes, and machinery are manufactured, and fluorspar and manganese are mined nearby. Arnstadt passed to the counts of Schwarzburg in the 14th cent. and later was the capital of the principality of Schwarzburg-Sonderhausen.

A·roe Islands (ä′rōō): *see* Aru Islands.

A·roo·stook (ə-rōōs′tŏŏk), county (1970 pop. 94,078), 6,821 sq mi (17,666.4 sq km), N Maine, bordering on Que. and N.B., Canada, the northernmost point of the E United States; formed 1839; co. seat Houlton. In one of the nation's richest agricultural areas, it produces c.15% of the U.S. potato crop. It has a food-processing industry (frozen fruit and vegetables) and a lumber industry (logging, sawmills, wood products, and paper). There are manganese deposits in the area, and farm machinery and fertilizers are manufactured. Large areas of lake-studded wilderness furnish lumber and offer facilities for hunting, fishing, and canoeing.

Aroostook, river, c.140 mi (225 km) long, rising in N Maine and winding E to the St. John River, N.B., Canada.

Ar·rah (är′rə), city (1971 pop. 92,919), Bihar state, NE India, on the Ganges Plain W of Patna. It is the administrative center for a district that produces grain, sugar cane, and oilseed.

Ar Ra·ma·di (ăr rä-mä′dē), town (1965 pop. 28,723), central Iraq, on the Euphrates River. It is the terminus of a highway across the desert from the Mediterranean Sea. The town was founded in 1869. The British won an important victory over the Turks here in 1917.

Ar·ran (ăr′ən), island (1971 est. pop. 3,564), 165 sq mi (427.4 sq km), Strathclyde region, W Scotland, in the Firth of Clyde. It is largely granitic and is wild and rocky. Its scenery and its hunting and fishing have made it a resort. Robert I hid on Arran in 1307 and launched his invasion of the mainland from here.

Ar Raq·qah (ăr räk′kä), town (1970 pop. 37,151), capital of Ar Raqqah governorate, N Syria, on the Euphrates River. Carpets are manufactured, and the town has an agricultural experimental station. Ar Raqqah was prominent during the early Abbaside caliphate and Raqqah was destroyed by the Mongols in the early 13th cent.

Ar·ras (ă-rĕs′), city (1975 pop. 79,783), capital of Pas-de-Calais dept., N France, on the canalized Scarpe River SSW of Lille. It is a communications, farm, and industrial center, with oil works and factories making machinery, metal products, and esparto goods. Of Gallo-Roman origin, it was granted (1180) a commercial charter by the crown and enjoyed international importance in banking and trade. By the 14th cent. it had become a center of wealth and culture, renowned particularly for tapestry. It was nearly destroyed during the wars between Burgundy and France (15th cent.), which ended with the Treaty of Arras (1435). Heavy bombardments in World War I destroyed much of the town, and it was further damaged in World War II. The town square, bordered by 17th cent. buildings, forms a notable ensemble of Flemish architecture.

Ar·row Lakes (ăr′ō), two expansions of the Columbia River, S British Columbia, Canada. Both lie in narrow valleys bounded by mountain ranges and are noted for their beauty. Upper Arrow Lake has an area of 88 sq mi (227.9 sq km); Lower Arrow Lake is 59 sq mi (152.8 sq km).

Ár·ta (är′tə), formerly **Am·bra·cia** (ăm-brā′shə), city (1971 pop. 19,498), capital of Árta prefecture, W Greece, in Epirus. It is a trading and shipping center for agricultural goods, including cotton, grain, citrus fruits, almonds, and olives. There is a large fishing industry, and leather goods and cotton and woolen textiles are manufactured. The city was founded (7th cent. B.C.) by Corinthian colonists. It was conquered by Rome in 189 B.C.

Ar·tem (ər-tyĕm′), city (1976 est. pop. 69,000), Primorsky Kray, Far Eastern USSR. It is a coal-mining center and has an important thermoelectric station that utilizes local coal deposits.

Ar·te·mis·i·um (är′tə-mĭsh′ē-əm), cape, N Euboea (now Évvoia), Greece, named for a great temple of Artemis. Off the cape in 480 B.C. the Greeks defeated a Persian fleet.

Ar·te·movsk (ər-tyô′mĕfsk), city (1976 est. pop. 91,000), S European USSR, in the Ukraine. An industrial center of the Donets Basin, it produces metals, mining equipment, glass, bricks, and chemicals.

Ar·te·sia (är-tē′zhə). **1.** City (1970 pop. 14,757), Los Angeles co., S Calif.; founded 1875, inc. 1959. Chiefly residential, it serves the surrounding farm area and was named for the many artesian wells in the vicinity. **2.** City (1970 pop. 10,315), Eddy co., SE N.Mex., just W of the Pecos River, in an oil, gas, farm, and livestock area; laid out 1903, inc. 1939. Artesian wells, under tremendous pressure from the nearby Sacramento Mts., irrigate a large area. The city's manufactures include petroleum products and fiberglass and plastic pipes.

Ar·thur (är′thər), county (1970 pop. 606), 706 sq mi (1,828.5 sq km), W central Nebr., in a plains region; formed 1888; co. seat Arthur.

Arthur. **1.** Village (1970 pop. 2,214), Douglas and Moultrie cos., E central Ill., ESE of Decatur, in a farm area; platted 1873, inc. 1877. It is the trade center for an agricultural and manufacturing area. **2.** Village (1970 pop. 175), seat of Arthur co., W central Nebr., NW of North Platte, in a livestock-grazing region.

Ar·tois (är-twä'), region and former province, N France, near the English Channel, between Picardy and Flanders. It is largely agricultural and occupies part of the rich Franco-Belgian coal basin. Owned in the Middle Ages by the counts of Flanders, Artois was annexed (1180) to France by Philip II through marriage. Burgundy gained (14th cent.) the territory, also through marriage. Later it was under Austrian rule, and from 1493 until its conquest (1640) by Louis XIII it was under Spanish rule.

A·ru·ba (ə-rōō'bə), island (1975 est. pop. 61,989), 69 sq mi (178.7 sq km), in the Leeward Islands group of the Netherlands Antilles; capital Oranjestad. Tourism and the refining of oil brought in from nearby Venezuela are the major industries.

A·ru Islands or **A·roe Islands** (both: ä'rōō), group of about 95 low-lying islands (1961 pop. 29,604), 3,306 sq mi (8,562.5 sq km), E Indonesia, in the Moluccas, in the Arafura Sea, SW of New Guinea. Products include sago, coconuts, tobacco, mother-of-pearl, trepang, tortoise shell, and bird of paradise plumes. The islands were discovered by the Dutch, who colonized them after 1623.

Ar·u·na·chal Pra·desh (är'ə-nä'chəl prə-dĕsh'), union territory (1971 pop. 444,744), 31,438 sq mi (81,424 sq km), NE India, bordered on the N by the Tibet region of China and on the E by Burma; capital Ziro. Formerly the North-East Frontier Agency special territory, Arunachal Pradesh became a union territory in 1972.

A·run·del (ăr'ən-dəl), municipal borough (1971 pop. 2,382), West Sussex, S England, WNW of Worthing. It is on a hillside beneath the magnificent 12th cent. Arundel Castle, seat of the dukes of Norfolk.

A·ru·sha (ə-rōō'shə), city (1967 pop. 32,452), capital of Arusha prov., NE Tanzania. It is an industrial and railroad center. Manufactures include textiles, beverages, processed foods, plastics, and electronic equipment.

Ar·vad·a (är-văd'ə), city (1978 est. pop. 83,200), Jefferson and Adams cos., N central Colo., a residential suburb of Denver; inc. 1904. Processed foods, beer, chemicals, and wood and metal products are manufactured here.

Ar·vi·da (är-vē'də), city (1971 pop. 18,448), Chicoutimi co., S Que., Canada, on the Saguenay River. It has a large aluminum smelter.

Ar·vi·ka (är'vē'kä), city (1970 pop. 15,509), Värmland co., W Sweden, on Lake Glafsfjorden. It is a commercial and industrial center.

Ar·za·mas (ər-zə-mäs'), city (1976 est. pop. 85,000), E European USSR, on the Tyosha River. A rail junction, it has food-processing plants and industries that produce leather and felt goods.

Aš or **Asch** (both: äsh), city (1975 est. pop. 12,000), W Czechoslovakia, in Bohemia, near the Bavarian border. It is a textile center and also manufactures lace, woolens, embroidery, and carpets.

A·sa·hi-da·ke (ä-sä'hē-dä'kē'), volcanic mountain, 7,513 ft (2,291.5 m) high, Daisetsu-zan National Park, central Hokkaido, Japan. The volcano has unusual alpine flora on its slopes.

A·sa·hi·ga·wa (ä-sä-hē'gä-wä), city (1975 pop. 320,526), W central Hokkaido, Japan, on the Ishikari River. Asahigawa is the commercial, industrial, and rail center of a great agricultural region. Pulp, paper, cotton yarn, lumber, wood products, and sake are among the city's industrial products.

A·sa·ka (ä-sä'kä), city (1975 pop. 81,755), Saitama prefecture, central Honshu, Japan. It is an industrial and residential suburb of Tokyo. There is an important metalworks industry in the city.

A·sa·ma, Mount (ä-sä'mä), or **A·sa·ma-ya·ma** (ä-sä'mä-yä'mä), peak, 8,340 ft (2,543.7 m) high, central Honshu, Japan, near Nagano. One of the largest and most active volcanoes in Japan, it erupted violently in 1783.

A·san·sol (ə-sən-sōl'), city (1971 pop. 157,388), West Bengal state, NE India. It is an industrial center in a coal-mining area.

As·bes·tos (ăs-bĕs'təs), town (1976 pop. 9,075), Richmond co., SE Que., Canada. Asbestos is mined in the area and asbestos products are made in the town.

As·bur·y Park (ăz'bĕr'ē, -bə-rē), city (1978 est. pop. 14,000), Monmouth co., E N.J., on the Atlantic coast; inc. 1897. It is a popular resort with a noted beach, boardwalk, convention hall, and auditorium.

As·cen·sion (ə-sĕn'shən), parish (1970 pop. 37,086), 300 sq mi (777 sq km), SE La., intersected by the Mississippi and also drained by the Amite River; formed 1807; parish seat Donaldsonville. It lies in a timber region and has oil and natural-gas wells. Its agriculture includes sugar cane, corn, rice, hay, truck, sweet potatoes, and strawberries. Farm products are processed, and textiles, chemicals, and aluminum and metal products are manufactured.

Ascension, island, 34 sq mi (88.1 sq km), in the S Atlantic, NW of St. Helena and belonging to the British St. Helena colony. Ascension is volcanic and rocky with little vegetation, but it supports considerable livestock, including sea turtles, rabbits, wild goats, and partridges. The United States maintains a missile and satellite tracking station here. Discovered by the Portuguese in 1501, Ascension was taken by the British in 1815. In 1922 it was made a dependency of St. Helena.

A·schaf·fen·burg (ä-shäf'ən-bûrg'), city (1974 est. pop. 54,535), Bavaria, S central West Germany, on the Main River. Its manufactures include clothing, machinery, precision and optical instruments, and colored paper. Once the location of a Roman garrison and later of a Frankish castle, Aschaffenburg passed to the archbishopric of Mainz in the 10th cent. It changed hands several times during the Thirty Years War (1618–48). It passed to Bavaria in 1814.

A·schers·le·ben (ä'shərs-lā'bən), city (1974 est. pop. 36,674), Halle district, W East Germany. An industrial city, it manufactures machine tools, chemicals, iron and steel, and woolen goods. There are lignite, salt, and potash mines nearby. Aschersleben was probably founded in the 11th cent. and passed to Prussia in 1813.

A·sco·li Pi·ce·no (äs'kō-lē pē-châ'nō), city (1971 pop. 55,053), capital of Ascoli Piceno prov., Marche region, central Italy, at the confluence of the Castellano and Tronto rivers. It is the market for a rich agricultural area. A Roman settlement with extensive Roman remains, the city became a free republic in the 12th cent. and passed to papal control in the 15th cent.

As·cot (ăs'kət), village, Berkshire, S central England. The famous horse races instituted by Queen Anne in 1711 are held annually in June on Ascot Heath.

As·cu·lum (ăs'kyōō-ləm), ancient town, Apulia, SE Italy, S of Foggia, on a branch of the Appian Way. Here Pyrrhus won a hard-fought battle against the Romans in 279 B.C.

A·se·nov·grad (ä-sē'nəf-gräd), city (1972 est. pop. 38,500), S central Bulgaria. It is a commercial center, with wineries and tobacco manufactures. An ancient Bulgarian stronghold, it became a trade center under Turkish rule (15th–19th cent.). Asenovgrad has several 16th cent. churches and the ruins of a 13th cent. castle.

A·shan·ti (ə-shän'tē) or **A·san·te** (ə-sän'tē), historical region, central Ghana, W Africa. The region is the source of much of Ghana's cocoa. Before the 13th cent., Akan peoples migrated into the forest belt of present-day Ghana and established small states. By the late 17th cent. the states had been welded by the Oyoko clan into the Ashanti confederation. A series of Anglo-Ashanti wars in the 19th cent. culminated in the defeat of the confederation (1896) and the annexation of Ashanti (1901) to the British Gold Coast colony. The confederation was finally restored in 1935. In 1945 the Ashanti were given representation in the executive and legislative councils of the Gold Coast. They supported an unsuccessful attempt to give Ghana a federal constitution in 1954 and resisted the centralizing measures of the Nkrumah government. The Ashanti are noted for the quality of their gold work and their colorful cloth, and were long famous for the gold-encrusted stool that was the symbol of their sovereignty.

Ash·burn (ăsh'bərn), city (1970 pop. 4,209), seat of Turner co., S Ga., ENE of Albany; founded 1889. It is a trade and processing center for a lumber, farm, and livestock area. Cottonseed products, clothing, and naval stores are made.

Ash·bur·ton (ăsh'bûr'tən), river, c.400 mi (645 km) long, Western Australia. It flows northwest to the Indian Ocean at Onslow near the entrance to Exmouth Gulf.

Ash·dod (ăsh'dŏd), city (1975 est. pop. 52,500), SW Israel, on the Mediterranean Sea. Construction is Ashdod's main industry; its manufactures include synthetic fibers, woolen yarn, and knitted goods. Nearby is the site of ancient Ashdod, which was settled as early as the Bronze Age. Conquered by the Philistines in the 12th cent. B.C., it became an important city of the Philistine Pentapolis. The city was later ruled by Judah, Egypt, Assyria, and Rome. The first modern Israeli settlement in Ashdod was made in 1955, and in 1965 the deepwater port was completed.

Ash·down (ăsh′doun), railroad city (1970 pop. 3,522), seat of Little River co., SW Ark., NNW of Texarkana near the Red River; founded 1892, inc. as a city 1937. Cotton and lumber are processed.

Ashe (ăsh), county (1970 pop. 19,571), 427 sq mi (1,105.9 sq km), NW N.C., in the Blue Ridge, bounded on the N by the Va. border, in the W by the Tenn. border, and drained by the headstreams of the New River; formed 1799; co. seat Jefferson. It has farming (tobacco, corn, white potatoes, hay, and grain), cattle and sheep raising, and sawmilling. It is included in Yadkin National Forest.

Ashe·bo·ro (ăsh′bə-rə), town (1978 est. pop. 16,300), seat of Randolph co., central N.C., in the Piedmont; inc. 1796. Its manufactures include hosiery, textiles, clothing, furniture, flashlight batteries, and electric blankets.

Ashe·ville (ăsh′vĭl′), city (1978 pop. 58,000), seat of Buncombe co., W N.C., on the French Broad River and on a plateau in the Blue Ridge Mts.; inc. 1797. Located near Great Smoky Mountains National Park and Pisgah National Forest, Asheville is a popular mountain resort. Tourism is a major business. The city is also a financial, distribution, transportation, and retail center for a tobacco and lumber area. Electronic equipment, textiles, clothing, and paper, food, and glass products are made here. Asheville's many points of interest include Biltmore, the magnificent Vanderbilt estate, and the Colburn Mineral Museum. The writer Thomas Wolfe was born and lived here; his home is a public memorial.

Ash Flat, town (1970 pop. 211), seat of Sharp co., N Ark., N of Batesville.

A·shi·ka·ga (ä-shē-kä′gä), city (1976 est. pop. 163,109), Tochigi prefecture, central Honshu, Japan. An old silk-weaving center, it is famous for its spinning and silk textile industries. The city has an ancient school (probably founded 9th cent.), which is known for its vast library of Chinese classics.

A·shi·ya (ä-shē′yä), city (1975 pop. 76,211), Hyogo prefecture, W central Honshu, Japan, on Osaka Bay. It is a residential and industrial suburb of Osaka.

Ash·kha·bad (ăsh′kə-băd′, äsh′kə-bäd′), city (1977 est. pop. 302,000), capital of the Turkmen SSR, S central Asian USSR, on the Trans-Caspian RR. The city has textile, motion-picture, and machine-building industries. Ashkhabad was founded in 1881 as a fortress. An earthquake in 1948 virtually destroyed the city, which stands in a major fault zone.

Ash·land (ăsh′lənd). **1.** County (1970 pop. 43,303), 426 sq mi (1,103.3 sq km), N central Ohio, drained by the forks of the Mohican River; formed 1846; co. seat Ashland. It has agriculture (livestock, grain, hay, and fruit), mining, and gravel pits. Paperboard products, furniture, electrical equipment, and chemicals are manufactured. Mohican State Forest, a recreation area with a flood-control dam, is here. **2.** County (1970 pop. 16,743), 1,037 sq mi (2,685.8 sq km), extreme N Wis., bounded on the N by Lake Superior and drained by the Bad River; formed 1860; co. seat Ashland. It has lumbering, iron mining, and dairying, and there is some manufacturing (wood, paper, and metal products). Part of the Gogebic Range (a source of iron ore) is here, as well as a section of Chequamegon National Forest.

Ashland. 1. Town (1970 pop. 1,921), seat of Clay co., E central Ala., S of Anniston; inc. 1866. In a timber and agricultural area, it processes broilers, eggs, and livestock. Talladega National Forest is nearby. **2.** Uninc. town (1970 pop. 14,810), Alameda co., W Calif. **3.** City (1970 pop. 1,244), seat of Clark co., SW Kansas, near the Okla. line SSE of Dodge City; laid out 1884, inc. 1886. Wheat and cattle are raised in the area. There are oil wells nearby. **4.** Industrial city (1978 est. pop. 25,900), Boyd co., E Ky., on terraces along the Ohio River near the influx of the Big Sandy; settled 1786, inc. 1854. In a region yielding coal, clay, natural gas, and timber, it is a river and railroad shipping point, with large repair yards. The city is part of a metropolitan area known for its metallurgical industries. In addition to iron and steel, Ashland's many manufactures include coke, refined oil, chemicals, leather products, clothing, and mining equipment. **5.** Town (1970 pop. 8,882), Middlesex co., E Mass., SW of Framingham; settled c.1750, inc. 1846. Thermostats are made. **6.** Village (1970 pop. 348), seat of Benton co., N Miss., ESE of Memphis, Tenn., in an agricultural area. Lumber is processed. **7.** City (1970 pop. 2,176), Saunders co., E Nebr., on the Platte River SW of Omaha, in a farm and recreation area; settled 1863, inc. 1886. The site of an Otoe Indian village is nearby. **8.** City (1978 est. pop. 18,800), seat of Ashland co., N Ohio, in a farm area; inc. 1844. Pumps, spray equipment, rubber products, adhesives, printed materials, animal medications, and machine tools are among its manufactures. **9.** City (1970 pop. 12,342), Jackson co., SW Oregon, near the Calif. border; inc. 1874. A lumbering center and a processing and shipping point for an irrigated dairy, farm, and orchard area, it is also a resort with mineral springs. It is surrounded on three sides by the Rogue River National Forest. A Shakespeare festival is held each spring and summer. **10.** Industrial borough (1970 pop. 4,737), Columbia and Schuylkill cos., E central Pa., NW of Pottsville; settled 1845, laid out 1847, inc. 1857. Anthracite coal is mined, and metal products and clothing are made. **11.** Town (1970 pop. 2,934), Hanover co., E central Va., N of Richmond; settled 1848, inc. 1858. In a dairy and truck-farming region, it was settled as a health resort. Randolph-Macon College is here. Patrick Henry and Henry Clay were born nearby. **12.** City (1970 pop. 9,615), seat of Ashland co., N Wis., on Chequamegon Bay, an arm of Lake Superior, S of the Apostle Islands; settled 1854, inc. 1887. It is a port of entry and shipping point for iron, lumber, and other goods. Manufactured products include steel, lumber products, machinery, paper, tile, and briquettes. There are granite quarries and a foundry. French explorers visited the bay shore in the 17th cent. and a French mission was founded nearby in 1665. The city grew as an iron-mining and lumbering center and as a railroad terminus.

Ashland City, town (1970 pop. 2,027), seat of Cheatham co., N central Tenn., on the Cumberland River NW of Nashville, in a timber, grain, livestock, and tobacco area; inc. 1859. Wood products and hosiery are made.

Ash·ley (ăsh′lē), county (1970 pop. 24,976), 933 sq mi (2,416.5 sq km), SE Ark., bounded on the S by the La. border, on the W by the Saline and Ouachita rivers, and drained by Bayou Bartholomew; formed 1849; co. seat Hamburg. It has agriculture (cotton, corn, hay, alfalfa, clover, potatoes, rice, peanuts, and cattle), lumber and commercial fishing operations, and manufacturing and processing industries. There is a state game refuge.

Ashley. 1. City (1970 pop. 1,236), seat of McIntosh co., S N.Dak., near the S.Dak. border SE of Bismarck, in a flour, grain, dairy, poultry, and livestock region; inc. 1920. Founded on nearby Hoskins Lake, it was moved to its present site in 1888 on the coming of the railroad. **2.** Industrial borough (1970 pop. 4,095), Luzerne co., NE Pa., S of Wilkes-Barre; settled 1810, inc. 1870. Cigars, clothing, and textiles are made.

Ash·qe·lon (ăsh′kə-lŏn), city (1975 est. pop. 47,900), SW Israel, on the Mediterranean Sea. It is a beach resort in an area of citrus groves and cotton plantations. Ashqelon's industries process agricultural products and manufacture cement, wood products, automobile parts, electronic equipment, and watches. Nearby is the site of ancient Ashqelon, or Ashkelon, whose history dates back to the 3rd millennium B.C. It was a trade center and port and a seat of worship of the goddess Astarte. Ancient Ashqelon was conquered by the Philistines in the late 12th cent. B.C., completely rebuilt, and made one of the cities of the Philistine Pentapolis. Ashqelon flourished under the Greeks and Romans; Herod, believed to have been born here, greatly enlarged the city. It was taken by the Arabs in A.D. 638, conquered by the Crusaders in 1153 and occupied by Richard I in 1191, and completely destroyed in 1270. An Israeli settlement was established here in 1948.

Ash·ta·bu·la (ăsh′tə-byo͞o′lə), county (1970 pop. 98,237), 700 sq mi (1,813 sq km), extreme NE Ohio, bounded on the N by Lake Erie and intersected by the Grand and Ashtabula rivers; formed 1811; co. seat Jefferson. Its agriculture includes cattle, corn, wheat, oats, soybeans, and alfalfa. Chemical, electrical, metallurgical, and plastics industries are here.

Ashtabula, city (1978 est. pop. 24,000), Ashtabula co., NE Ohio, on Lake Erie at the mouth of the Ashtabula River; settled c.1801 by New Englanders, inc. as a village 1831, as a city 1891. It is a port of entry; iron ore and coal are shipped. Ashtabula manufactures automobile parts, chemicals, fiberglass products, farm tools, clothing, and electric motors.

Ash·ton-un·der-Lyne (ăsh′tən-ŭn′dər-līn′), municipal borough (1971 pop. 48,865), Greater Manchester, NW England, on the Tame River.

Its industries include cotton spinning, weaving, and dyeing; coal mining; and the manufacture of diesel, gas, and oil engines.

Ash·ville (ăsh'vĭl'), city (1970 pop. 986), seat of St. Clair co., NE central Ala., on a branch of the Coosa River SW of Gadsden. Cotton ginning and lumber milling are done.

A·sia (ā'zhə, ā'shə), the world's largest continent, 17,139,000 sq mi (44,390,010 sq km), with about 2,535,333,000 people, more than half the world's total population. Asia's border with Europe lies approximately along the Urals, the Ural River, the Caspian Sea, the Caucasus, the Black Sea, the Bosporus and Dardanelles straits, and the Aegean Sea. The connection of Asia with Africa is broken only by the Suez Canal between the Mediterranean Sea and the Red Sea. In the far northeast of Asia, Siberia is separated from North America by the Bering Strait. The continent of Asia is washed on the south by the Gulf of Aden, the Arabian Sea, and the Bay of Bengal; on the east by the South China Sea, East China Sea, Yellow Sea, Sea of Japan, Sea of Okhotsk, and Bering Sea; and on the north by the Arctic Ocean.

Geologically, Asia consists essentially of ancient Precambrian rocks—the Arabian and Indian peninsulas in the south and the central Siberian plateau in the north—which enclose a central zone of folded ridges. In accordance with this underlying structure, Asia falls into the following major physiographic structures: the northern lowlands covering west central Asia and most of Siberia; the vast central highland zone of high plateaus, rising to c.15,000 ft (4,575 m) in Tibet and enclosed by some of the world's greatest mountain ranges; the southern peninsular plateaus of India and Arabia, merging, respectively, into the Ganges and Tigris-Euphrates river plains; and the lowlands of eastern Asia, especially in China, which are separated by mountain spurs of the central highland zone. Great peninsulas extend out from the mainland, dividing the oceans into seas and bays, many of them protected by Asia's numerous offshore islands. Asia's rivers, among the longest in the world, generally rise in the high plateaus and break through the great chains toward the peripheral lowlands.

Asia can be divided into five regions, each possessing distinctive physical, cultural, economic, and political characteristics. Southwest Asia (Iran and the nations of Asia Minor, the Fertile Crescent, and the Arabian Peninsula), long a strategic crossroads, is characterized by an arid climate and irrigated agriculture, great petroleum reserves, and the predominance of Islam. South Asia (Afghanistan and the nations of the Indian subcontinent) is isolated from the rest of Asia by great mountain barriers and was once entirely under British rule. Southeast Asia (the nations of the southeastern peninsula and the East Indian archipelago) is characterized by monsoon climate, maritime orientation, the fusion of Indian and Chinese cultures, and a great diversity of ethnic groups, languages, religions, and politics. East Asia (China, Mongolia, Korea, and the islands of Taiwan and Japan) is located in the mid-latitudes on the Pacific Ocean, has a strong indigenous culture, and forms the most industrialized region of Asia. Soviet Asia (in the west-central and northern third of the continent) accounts for about 75% of the area of the USSR and is the largest section of Asia controlled by one nation.

Economy. The distribution of Asia's huge population is governed by climate and topography, with the monsoons and the fertile alluvial plains determining the areas of greatest density. Less than 10% of the continent is under cultivation. Rice, by far the most important food crop, is grown for local consumption in the heavily populated countries, while countries with smaller populations are generally rice exporters. Other important crops are wheat, soybeans, peanuts, sugar cane, cotton, jute, silk, rubber, and tea. Asia's economy is predominantly agricultural, but Japan, China, Soviet Asia, India, North and South Korea, Taiwan, and Turkey are distinguished for their industrialization. In most of these countries, an iron and steel industry has grown on the basis of local coal and iron resources; Japan, the world's third-largest steel producer, is the major exception. Contributing greatly to the income of Asian countries are vital mineral exports—petroleum in southwest Asia, Soviet Asia, and Indonesia; tin in Malaysia, Thailand, and Indonesia; manganese from India; chromite from Turkey and the Philippines; and tungsten and antimony from China.

History. Asia was the site of some of the world's oldest civilizations. The empires of Sumeria, Babylonia, Assyria, Media, and Persia and the civilizations of Islam flourished in southwest Asia, while in the east the ancient civilizations of India, China, and Japan prospered. Later, nomadic tribes (Huns, Tatars, and Turks) in northern and central Asia gave rise to great westward migration. Their tribal, military-state organizations reached their highest form in the 13th-14th cent. under the Mongols, whose court was visited by early European travelers, notably the Italian Marco Polo. The Portuguese explorer Vasco da Gama reached India by sea in 1498, and in northern Asia Russian Cossacks crossed Siberia and reached the Pacific by 1640. With the formation of English, French, Dutch, and Portuguese trading companies in the 17th cent., great trade rivalry developed along the coasts of India, Southeast Asia, and China and resulted in increasing European colonial control of Asian lands. In the 19th cent. China and Japan opened their doors to foreign trade, with Japan rapidly rising to a world industrial and military power.

World War II and the conflicts of its aftermath hit Asia heavily. The center of gravity in international affairs tended to shift from Europe, the focus of both World Wars, to Asia, where the decolonization process resulted in the creation of many unstable nations. The Arab-Israeli conflict, the Korean and Vietnam wars, and the emergence of Communist governments in China, North Korea, Vietnam, Cambodia, and Laos were among the events that heightened tensions in Asia.

Asia Minor, peninsula, c.250,000 sq mi (647,500 sq km), extreme W Asia, generally coterminous with Turkey, and usually synonymous with Anatolia. It is washed by the Black Sea in the north, the Mediterranean Sea in the south, and the Aegean Sea in the west. Near the southern coast of Asia Minor are the Taurus Mts.; the rest of the peninsula is occupied by the Anatolian plateau, which is crossed by numerous mountains interspersed with lakes. The Hittites established the first major civilization in Asia Minor about 1800 B.C. Beginning in the 8th cent. B.C. Greek colonies were established on the coast lands. The conquest (6th cent. B.C.) of Asia Minor by the Persians led to the Persian Wars. Alexander the Great incorporated the region into his empire, and after his death it was divided into small states. It was reunified (2nd cent. B.C.) by the Romans, but was subject to repeated attacks by invaders. After being held by the Crusaders for a short time in the early 13th cent., Asia Minor was gradually (13th-15th cent.) conquered by the Ottoman Turks. It remained part of the Ottoman Empire until the establishment of the Republic of Turkey after World War I.

Ask·ja (äsk'yä), volcano, c.4,950 ft (1,510 m) high, E central Iceland. Its eruption of 1875 devastated a large area; it last erupted in 1961.

As·ma·ra (äs-mä'rä, äz-), city (1974 est. pop. 318,000), capital of Eritrea prov., N Ethiopia, at an altitude of c.7,300 ft (2,225 m). Textiles and clothing, processed meat, beer, shoes, and ceramics are the major industrial products. Asmara was a small village until the 1880s, when it became an administrative center. Occupied by the Italians in 1889, it became (1900) the capital of the Italian colony of Eritrea. In the 1930s Asmara was rapidly developed as a base for the Italian invasion (1935-36) of Ethiopia; later, in 1941, the city was taken by British forces.

As·nières-sur-Seine (än-yâr'sür-sĕn), industrial suburb of Paris (1975 pop. 75,431), Hauts-de-Seine dept., N central France, on the Seine River. Boats and perfumes are the major manufactures.

A·so-san (ä'sō-sän), volcanic mountain, central Kyushu, Japan. Asosan is topped by one of the world's largest calderas (circumference 75 mi/120.7 km) that contains five volcanic cones. Taka-dake (5,225 ft/1,593.6 m) is the highest cone; Naka-dake (4,340 ft/1,323.7 m) is an active volcano. Cable cars carry people over the caldera.

A·so·tin (ə-sō'tən), county (1970 pop. 13,799), 627 sq mi (1,623.9 sq km), extreme SE Wash., in a plateau and rolling hill area rising to the Blue Mts. in the SW and bordering on Oregon and Idaho; formed 1883; co. seat Asotin. It is in an agricultural area yielding livestock, dairy products, wheat, and fruit. An extensive food-processing industry is based here. The county includes part of Umatilla National Forest.

Asotin, town (1977 est. pop. 893), seat of Asotin co., SE Wash., SSE of Pullman on the Snake River.

As·pen (ăs'pən), city (1970 pop. 2,437), alt. 7,850 ft (2,394.3 m), seat of Pitkin co., S central Colo., on the Roaring Fork River W of the Sawatch Mts.; founded c.1879 by silver prospectors, inc. 1881. Once a booming silver camp, it is now a popular ski resort.

As·per·mont (ăs'pər-mŏnt'), town (1970 pop. 1,198), seat of Stonewall co., NW central Texas, NW of Abilene. It is a trade and shipping center for grain, cotton, fruit, cattle, and poultry. There is some oil in the area.

As·sab (ä-säb′), town (1970 est. pop. 15,000), Eritrea prov., E Ethiopia, a port on the Red Sea. Exports include salt, coffee, oilseeds, and hides and skins. The town has a petroleum refinery. Once the terminus of caravans from the interior of Ethiopia, Assab was acquired by a private Italian shipping company in 1869. In 1882 it was taken over by the Italian government, and in 1890 Assab was included in the colony of Eritrea.

As·sam (ə-săm′, ăs′əm), state (1971 pop. 14,630,422), c.30,000 sq mi (77,700 sq km), extreme NE India; capital Shillong. Almost completely separated from India by Bangladesh, Assam is bordered by Burma on the east and China and Bhutan on the north. Tea, grown on large plantations, is the principal crop. Rice, citrus fruit, sugar cane, sesame, cotton, and jute are also grown. The hills produce abundant timber and some coal and limestone. Assam is an important oil-producing region.

As·sa·teague Island National Seashore (ăs′ə-tēg′): *see* National Parks and Monuments Table.

As·sen (ä′sən), city (1976 est. pop. 43,783), capital of Drenthe prov., NE Netherlands. It is an administrative and industrial center. Its main growth began in 1945.

As·sin·i·boine (ə-sĭn′ə-boin′), river, 590 mi (949.3 km) long, rising in S Sask., Canada, and flowing SE into Man., then E to the Red River at Winnipeg. The Assiniboine valley is one of Canada's leading wheat-growing areas. The river was discovered in 1736, and forts were built at its mouth and near the site of Portage la Prairie. Settlement spread westward along the river from the Red River valley to the plains.

Assiniboine, Mount, 11,870 ft (3,620.4 m) high, on the British Columbia–Alta. line, Canada, on the Continental Divide in the Rocky Mts.

As·si·nie (ä-sē-nē′), town, SE Ivory Coast, on a lagoon off the Gulf of Guinea. Because of its coastal location and its contacts with the interior, Assinie became an early stopping place for European traders who sought gold and ivory. Portuguese merchants came here in the late 16th cent. French missionaries established a temporary post here in 1637, and in 1842–43 the French gained treaty rights in the town. Assinie became a center of the palm oil trade, and coffee plantations were established nearby.

As·si·si (ə-sē′zē, ə-sē′sē), town (1975 est. pop. 19,100), Umbria, central Italy. A religious and tourist center, it is situated on a hill in the Apennines with a magnificent view of the plains below. Although a well-known town in Roman times and throughout the Middle Ages, it owes its modern fame chiefly to St. Francis of Assisi, who was born here in 1182 and died here in 1226. Above the saint's tomb are two Gothic churches (both consecrated 1253).

As·suan (ăs-wăn′, äs-wän′): *see* Aswan, Egypt.

As·sump·tion (ə-sŭmp′shən), parish (1970 pop. 19,654), 357 sq mi (924.6 sq km), SE La., bounded on the W partly by the Grand River and intersected by Bayou Lafourche; formed 1807; parish seat Napoleonville. Its agriculture (sugar cane, corn, rice, and truck crops) provides the materials for its sugar, confectionery, and canning industries. The parish has oil and gas fields.

As·syr·i·a (ə-s-ĭr′ē-ə), ancient empire of W Asia. It developed around the city of Ashur, or Assur, on the upper Tigris River and south of the later capital, Nineveh. The nucleus of a Semitic state was formed by the beginning of the 3rd millennium B.C., but it was overshadowed by the greatness of Sumer and Akkad. In the 17th cent. B.C., Assyria expanded briefly, but it soon relapsed into weakness. The 13th cent. B.C. saw Assyria threatening the surrounding states, and under Tiglathpileser I Assyrian soldiers took Babylonia, and crossed northern Syria to reach the Mediterranean. This empire was, however, only ephemeral, and Assyrian greatness was to wait until the 9th cent. B.C., when Ashurnasirpal II came into power and pushed his conquests north to Urartu and west to Lebanon and the Mediterranean. Shalmaneser III attempted to continue his policies but failed to establish hegemony over the Hebrews and their Aramaic-speaking allies. In the 8th cent. B.C. conquest was pushed by Tiglathpileser III. He subdued Babylonia, defeated the king of Urartu, attacked the Medes, and established control over Syria. His successor, Shalmaneser V, besieged Samaria, the capital of Israel, in 722–721 B.C., but it was Sargon, his son, who completed the task of capturing Israel. Sargon's victory at Raphia (720 B.C.) and his invasions of Armenia, Arabia, and other lands made Assyria indisputably one of the greatest of ancient empires. His son Sennacherib devoted himself to retaining the gains his father had made. Sennacherib's successor, Esar-Haddon, defeated the Chaldaeans and carried his conquests (673–670 B.C.) to Egypt. Under Assurbanipal, Assyria reached its zenith and approached its fall. When Assurbanipal was fighting against the Chaldaeans and Elamites, an Egyptian revolt under Psamtik I was successful. The rapid decline of Assyria had begun, but the reign of Assurbanipal saw the Assyrian capital of Nineveh at its height of splendor. The library of cuneiform tablets he collected ultimately proved to be one of the most important historical sources of antiquity.

As·ti (ä′stē), city (1975 est. pop. 80,021), capital of Asti prov., in Piedmont, NW Italy, on the Tanaro River. It is a commercial and industrial center, noted for its sparkling wine.

A·stor·ga (ä-stôr′gə), town (1970 pop. 11,794), Leon prov., NW Spain, in the Cantabrian Mts. An important military center under the Romans, it regained prominence in the 8th and 9th cents. as one of the principal cities of Asturias. The cathedral, rebuilt in the 15th cent., has additions in various styles.

As·to·ri·a (ə-stôr′ē-ə). **1.** Commercial, industrial, and residential section of NW Queens borough of New York City, SE N.Y.; settled in the 17th cent., renamed for John Jacob Astor in 1839. Several 18th cent. houses remain. **2.** City (1978 est. pop. 10,000), seat of Clatsop co., NW Oregon, on the Columbia River estuary; inc. 1876. A port of entry, Astoria is the trading center for the lower Columbia basin. Its principal industries are fishing and fish processing, lumbering, and tourism; agriculture and shipbuilding are also important. The Lewis and Clark expedition spent the winter of 1805–6 at a nearby encampment, Fort Clatsop (rebuilt in 1955 and now a national memorial). Fort Astoria, a fur-trading post established in 1811 by John Jacob Astor's Pacific Fur Company, was the first permanent U.S. settlement on the Pacific coast. In the late 19th cent. Astoria grew as a coastal and river port. It later attracted Scandinavian settlers, whose descendants make up most of its present-day population.

As·tra·khan (ăs′trə-kăn′), city (1976 est. pop. 458,000), capital of Astrakhan oblast, SE European USSR. A Caspian Sea port on the Volga River's southern delta, it is a center for river transport and has shipyards, repair docks, and fish-processing plants. Astrakhan is also an important rail junction and a major transshipment center for oil, fish, grain, and wood. The capital of the khanate of Astrakhan from the 1460s, it was conquered by Ivan the Terrible in 1556.

As·tu·ri·as (ə-stŏŏr′ē-əs, ə-styŏŏr′-), region and former kingdom, NW Spain, S of the Bay of Biscay and E of Galicia. Its coal mines, exploited since Roman times, are the richest in Spain. Iron, zinc, lead, and manganese are also mined. The name Asturias is derived from an Iberian people that lived here before the Roman conquest (2nd cent. B.C.). When the Moors overran the peninsula, Christian nobles fled into the Asturian mountains. They created the first Christian kingdom of Spain and defended themselves at the battle of Covadonga. From Asturias came the Christian reconquest of Spain, as the successors of King Alfonso I extended their control over Asturias, Galicia, León, and parts of Castile, Navarre, and Vizcaya. In the 10th cent. the capital was moved from Oviedo to León, and the kingdom of Asturias became the kingdom of Asturias and León, which three centuries later was united with the kingdom of Castile. The Asturians are noted for their stubborn courage and independence—traits shown in the warfare against Napoleon, in various uprisings against the Spanish government, in the civil war of 1936–39, and in the general strike of 1962.

A·sun·ción (ä′sŏŏn-syôn′), city (1972 est. pop. 392,753), S Paraguay, capital of Paraguay, on the Paraguay River. It is the principal port and chief industrial and cultural center of Paraguay. Meat packing is the main industry. Asunción has a decidedly colonial aspect. Its outstanding structures are the government buildings, the Godoi Museum, the Church of La Encarnación, and the Panteón Nacional, a smaller version of Les Invalides in Paris. The city's botanical gardens are notable. The site of the city may have been visited by the conquistador Juan de Ayolas, but the town was founded in Aug., 1536 or 1537, by Juan de Salazar and Gonzalo de Mendoza. It became a trading post on the route to Peru and the center of the Jesuits' activities in converting the Indian population. The city developed further under the great Creole governor Hernando Arias de Saavedra (first elected 1592). In 1731 the uprising of *comuneros* under José de Antequera y Castro was one of the first major rebellions against Spanish colonial rule. The eminence of Asunción was ended by the growth of Buenos Aires, which was separated from Asunción's jurisdiction in 1617.

As·wan or **As·suan** (both: ăs-wăn′, äs-wän′), city (1970 est. pop. 206,000), capital of Aswan governorate, S Egypt, on the Nile River at the First Cataract. Long famous as a winter resort and commercial center, the city has become an important industrial center since the start nearby of hydroelectricity production in 1960. The city was called Syene or Seveneh in the Bible and is described as the southern limit of Egypt. It was a trade center, serving as the gateway to the Sudan and Ethiopia, and was the place where the annual Nile flood was first sighted in Egypt. On Elephantine Island, in the Nile opposite Aswan, and Philae Island (submerged by the Aswan High Dam complex), south of the city, are found ancient Egyptian and Roman ruins. The Aswan Dam, 3 mi (4.8 km) south of the city, was built by the British and completed in 1902. It and the barrages at Asyut in central Egypt were the chief means of storing irrigation water for the Nile valley before the completion of the Aswan High Dam. After being enlarged in 1934, the dam added c.1 million acres (405,000 hectares) of cropland along the Nile. In 1960 a hydroelectric station with an annual capacity of 2 million kilowatt hours was opened at the dam. The Aswan High Dam, about 4 mi (6.4 km) south of the Aswan Dam, was constructed from 1960 to 1970, and was dedicated in 1971. Construction was delayed by disputes with Sudan over water rights and by the withdrawal in 1956 of U.S. and British financial aid. After 1956 the Soviet Union took over much of the financing (contributing ultimately about one third of the total cost of more than $1 billion) and technical supervision of the project. Built of earth and rock fill with a core of clay and cement, the High Dam is 375 ft (114.4 m) high and 11,811 ft (3,602.4 m) long. Lake Nasser (c.2,000 sq mi/5,180 sq km), the dam's reservoir and one of the world's largest artificial lakes, has a storage capacity of c.204 billion cu yd (157 billion cu m). The creation of the lake required the relocation of 90,000 people and of many archaeological treasures.

As·yut (äs-yo͞ot′), city (1970 est. pop. 175,700), E central Egypt, on the Nile. An industrial and trading center and also the seat of a university, it is famed for its pottery, carved ivory and wood, leatherwork, and silk shawls. Asyut was the ancient Greek city of Lycopolis and later a station of the caravan trade.

A·ta·ca·ma, Desert of (ä′tə-kä′mə), arid region, c.600 mi (965 km) long, N Chile, extending S from the border of Peru. The desert itself, c.2,000 ft (610 m) above sea level, is a series of dry salt basins flanked on the west by the Pacific coastal range, averaging c.2,500 ft (765 m) high, and on the east by the Andes. There is practically no vegetation; rain has virtually never been recorded in some localities. The Atacama has been a source of great nitrate and copper wealth. The first European to cross the forbidding waste was Diego de Almagro, the Spanish conquistador, in 1537. From then until the middle of the 19th cent. it was largely ignored, but with the discovery of the use of sodium nitrate as a fertilizer and later with the invention of smokeless powder using nitroglycerin, the desert had a mining boom. When synthetic nitrates were developed after World War I, the boom collapsed. Economically, the Atacama is declining, as reserves are depleted and the desert expands into once arable land.

A·ta·mi (ä-tä′mē), city (1975 pop. 51,437), Shizuoka prefecture, central Honshu, Japan. It is a major resort, famed for its scenery and its hot springs. Atami was once the site of a geyser that, according to tradition, wrought destruction until moved by Buddhist prayers. After an earthquake in 1923 the geyser stopped erupting.

A·tas·ca·de·ro (ə-tăs′kə-dâr′ō), uninc. town (1970 pop. 10,290), San Luis Obispo co., SW Calif., on the Salinas River; founded 1913 as a model community. It is a residential and farming town.

At·a·sco·sa (ăt′ə-skō′sə), county (1970 pop. 18,696), 1,206 sq mi (3,123.5 sq km), S Texas, drained by the Atascosa River; formed 1856; co. seat Jourdanton. It has agriculture (peanuts, truck crops, strawberries, corn, grain sorghums, broomcorn, cotton, and hay), cattle ranching and dairying, oil and natural-gas wells, and clay and sand deposits.

At·ba·ra (ăt′bä-rä), river, NE Africa, rising in NW Ethiopia and flowing c.500 mi (805 km) to the Nile in Sudan. There are few permanent settlements along its banks. The Atbara's water level is low, except during the rainy season (from June to Oct.).

At·ba·rah (ăt-bä′rə), town (1973 pop. 66,116), NE Sudan, at the junction of the Atbara and Nile rivers. An important rail junction, it is also the headquarters of the Sudan railway system and has large railroad workshops.

A·tchaf·a·lay·a (ə-chăf′ə-lī′ə), navigable river, c.170 mi (275 km) long, S central La. A distributary of the Red and Mississippi rivers, the Atchafalaya flows south to the Gulf of Mexico through an extensive system of guide levees and floodways that serve as a flood control for the lower Mississippi.

Atch·i·son (ăch′ĭ-sən). **1.** County (1970 pop. 19,165), 427 sq mi (1,105.9 sq km), NE Kansas, in a gently rolling agricultural area, bounded on the E by the Missouri River and the Mo. border and drained by the Delaware River; formed 1855; co. seat Atchison. Livestock, corn, and apples provide the materials for the county's food and grain processing industry. **2.** County (1970 pop. 9,240), 549 sq mi (1,421.9 sq km), extreme NW Mo., bounded on the W by the Missouri River and drained by the Tarkio and Nishnabotna rivers; formed 1843; co. seat Rockport. It lies in a corn and livestock area. Wood products are made.

Atchison, city (1978 est. pop. 11,300), seat of Atchison co., NE Kansas, on the Missouri River; inc. as a city 1881. It is a trade and industrial center in a rich farm area. Steel castings and grain products are produced here. Atchison was founded (1854) near a military post established (1818–19) on an island in the Missouri River. The Atchison, Topeka & Santa Fe RR was chartered here in 1859, and the city boomed as an important wagon-train, river, and railroad terminal, one of the outfitting points for westward travel.

Ath·a·bas·ca (ăth′ə-băs′kə), river, 765 mi (1,230.9 km) long, rising in the Columbia snowfield of the Canadian Rockies near the Alta.–British Columbia line and flowing N through Jasper National Park, then NE and N across central Alta. to Lake Athabasca. It is the southernmost headstream of the Mackenzie River and has long been the main route to the Mackenzie valley. There are extensive deposits of oil-bearing sand along the river near McMurray.

Athabasca, Lake, fourth-largest lake of Canada, c.3,120 sq mi (8,080 sq km), c.200 mi (320 km) long and from 5 to 35 mi (8-56.3 km) wide, NE Alta. and SW Sask., at the edge of the Canadian Shield. A part of the Mackenzie River system, the lake receives the Athabasca River from the south and drains north into Great Slave Lake by way of the Slave River. Gold and uranium are found nearby.

Athabasca, Mount, 11,452 ft (3,492.9 m) high, W Alta., Canada, in the Canadian Rockies at the headwaters of the Athabasca River. It is on the edge of the Columbia snowfield, and the Saskatchewan and Athabasca glaciers flow around it.

Athabasca Pass, 5,736 ft (1,749.5 m) high, W Alta. and E British Columbia, Canada, leading from the headwaters of the Athabasca River across the Continental Divide to the Columbia River. It was discovered c.1811, and for the next 50 years it was the chief route of the Hudson's Bay men on their journeys to and from the Columbia River country.

Ath·el·ney, Isle of (ăth′əl-nē), small area formerly surrounded by marshland, Somerset, SW England. King Alfred took refuge from the Danes here in 878 and founded a Benedictine abbey in 888.

Ath·ens (ăth′ənz), city (1971 pop. 867,023), capital of Greece, E central Greece, on the plain of Attica, between the Kifisós and Ilissus rivers, near the Saronic Gulf. The capital of Attica prefecture, Athens is Greece's largest city and its administrative, economic, and cultural center. Greater Athens, which includes the port of Piraiévs and numerous suburbs, has a population of more than 2.5 million and accounts for most of Greece's industrial output. Manufactures include silk, wool, and cotton textiles, machine tools, steel, ships, food products, beverages, chemicals, pottery, printed materials, and carpets. There is a large tourist industry.

The main landmark of Athens is the Acropolis (412 ft/125.7 m), which dominates the city and on which stand the remains of the Parthenon, the Propylaea, and the Erechtheum. Occupying the southern part of Athens, the Acropolis is ringed by the other chief landmarks of the ancient city—the Pnyx, where the citizens' assemblies were held; the Areopagus; the Theseum of Hephaesteum, a well-preserved Doric temple of the 5th cent. B.C.; the old Agora and the Roman forum; the temple of Zeus or Olympieum (begun under Pisistratus in the 6th cent. B.C.); the theater of Dionysius (oldest in Greece); and the Odeum of Herodes Atticus. There are many Roman remains in the "new" quarter, built east of the original city walls by Emperor Hadrian (1st cent. A.D.). The most noteworthy Byzantine structures are the churches of St. Theodora and of the Holy Apostles, both built in the 12th cent. The city is the seat of the National and Capodistrian University (1837), a polytechnic institute, an academy of sciences, several schools of archaeology, and many museums and libraries.

History. Athens, named after its patron goddess Athena, was inhabited in the Bronze Age. According to tradition, it was governed until c.1000 B.C. by Ionian kings and then by its aristocrats until Solon began to enact liberal reforms in 594 B.C. Solon abolished serfdom and gave power to all the propertied classes, thus establishing a limited democracy. His economic reforms were largely retained when Athens came under (560–511 B.C.) the rule of the tyrant Pisistratus and his sons Hippias and Hipparchus. Cleisthenes established (c.506 B.C.) a democracy for the freemen of Athens, and the city remained a democracy during most of the years of its greatness. The Persian Wars (500–449 B.C.) made Athens the strongest Greek city-state. The powerful Athenian fleet enabled Athens to gain hegemony in the Delian League, which was created in 478–477 B.C. through the confederation of many city-states.

During the time of Pericles (443–429 B.C.) Athens reached the height of its cultural and imperial achievement. Philosophy, poetry, drama, and sculpture flourished. However, in 431 B.C. the Peloponnesian War between Sparta and Athens began. It finally ended in 404 B.C. with Athens completely humbled, its population cut in half, and its fleet reduced to a dozen ships. Sparta imposed a government of oligarchy called the Thirty Tyrants. But the city recovered rapidly, overthrew the tyrants, and rebuilt its walls and fleet. Although Athens did not again achieve hegemony over Greece, it did have a period of great prosperity before its defeat by Philip of Macedon in 338 B.C. Revolts against Macedonian rule were crushed in 322 B.C. and 262 B.C.

Through the troubled times of the Peloponnesian War and the wars against Philip, Athenian achievements in philosophy, drama, and art had continued. Aristophanes wrote comedies, Plato taught at the Academy, Aristotle compiled an incredible store of information, and Thucydides wrote a great history of the Peloponnesian War.

As the city's glory waned in the 3rd cent. B.C., its earlier contributions were spread over the world in Hellenistic culture. Athens became a minor ally of Rome. It was captured in A.D. 395 by the Visigoths, became a provincial capital of the Byzantine Empire, and passed (1205) to Othon de la Roche, a French nobleman. His successor, Guy I, made it a prosperous duchy. King Frederick II of Sicily assumed the ducal title in 1312, but Athens sank into insignificance. Ottoman Turks seized the city in 1458, marking the beginning of nearly four centuries of Ottoman rule. Modern Athens was constructed only after 1834, when it became the capital of a newly independent Greece.

Athens, county (1970 pop. 55,747), 504 sq mi (1,305.4 sq km), SE Ohio, bounded on the SE by the Ohio River, here forming the W.Va. border, and intersected by Hocking River, Shade River, and Sunday and Federal creeks; formed 1805; co. seat Athens. It has coal mines, stone quarries, agriculture (livestock, truck crops, fruit, and grain), and a printing and publishing industry.

Athens. 1. City (1975 est. pop. 17,500), seat of Limestone co., N Ala., near the Tenn. line, in a farm area; inc. 1818. It has food-processing industries and plants that make textiles, thermostats, stoves, and chemicals. Sacked and occupied by Federal troops in 1862, it was recaptured by Gen. N. B. Forrest in 1864. Fine ante-bellum buildings remain. **2.** City (1978 pop. 50,500), seat of Clarke co., NE Ga., on the Oconee River, in a piedmont area; inc. 1806. The city was founded as the site of the Univ. of Georgia. Its industries include poultry processing and the manufacture of clocks, watches, radios, and textiles. **3.** City (1978 est. pop. 18,400), seat of Athens co., SE Ohio, on bluffs overlooking the Hocking River, in a coal-mining area of the Appalachian foothills; inc. 1811. There are diverse industries in the city. Athens was surveyed in 1795–96 as the site of a university and was settled shortly thereafter. It is the seat of Ohio Univ. **4.** Industrial borough (1970 pop. 4,173), Bradford co., NE Pa., NW of Towanda at the confluence of the Chemung and Susquehanna rivers; settled c.1778, laid out 1786, inc. 1831. Its manufactures include metal products and machinery. Tioga Point Museum has Indian implements, geological specimens, and skeletal remains. **5.** City (1978 est. pop. 13,300), seat of McMinn co., E Tenn., in a farm and resort area; inc. 1829. Furniture, plastics, farm implements, dairy products, and insecticides are made. Tennessee Wesleyan College is here. **6.** City (1970 pop. 9,582), seat of Henderson co., E Texas, SE of Dallas and WSW of Tyler; settled 1848, inc. 1901. It is a shipping center for a rich oil and agricultural area. Clay is used to produce pottery, bricks, and tile.

Ath·er·ton (ăth'ər-tən), residential town (1975 est. pop. 7,905), San Mateo co., W Calif., NW of San Jose; inc. 1923.

Atherton Plateau, c.15,000 sq mi (38,850 sq km), NE Queensland, Australia, meeting the edge of the Great Dividing Range. It stretches 230 mi (370 km) southeast from Laura to Ingham with its peak at Mt. Bartle Frere, (5,287 ft/1,612.5 m). The tableland is noted for sugar cane, dairy products, rice, and timber.

A·thi (ä'tē), river, c.200 mi (320 km) long, S Kenya. Rising near Nairobi in the Kikuya Escarpment, it flows southeast through Athi Plain to the Indian Ocean. It is famous for its wild game.

Ath·lone (ăth-lōn'), urban district (1971 pop. 9,821), Co. Westmeath, central Republic of Ireland, on the Shannon River. It is an important road and rail junction and a busy inland port. Industries include the production of cotton textiles, woolens, mineral water, and furniture. The English occupied the town in the 13th cent. and built Athlone Castle. Possession of the town was disputed during succeeding centuries, and the castle was often besieged. Athlone fell to the forces of William III of Great Britain in 1691.

Ath·ol (ăth'əl), town (1970 pop. 11,185), Worcester co., N Mass.; inc. 1762. Its manufactures include tools, drills, shoes, and toys. The area was settled in 1735.

Ath·os (ăth'ŏs, ā'thŏs) or **Ak·te** (ăk'tā), easternmost of the three peninsulas of Khalkidhikí, c.130 sq mi (335 sq km), NE Greece, in Macedonia. At the southern tip of the peninsula is the virtually independent state of the monks of Mount Athos, which rises to c.6,670 ft (2,035 m). Mount Athos is a community of about 20 monasteries of the Order of St. Basil of the Orthodox Eastern Church and includes c.30 sq mi (80 sq km) of territory. The first monastery was founded c.963. The community of monks enjoyed administrative independence under the Byzantine and Ottoman empires and under the modern Greek government. Karyai, the chief town of Athos, is the seat of the Holy Community, a committee made up of one representative from each monastery, which governs the monks of Mount Athos. No woman or female animal is allowed in the religious community. The icons from Mount Athos are celebrated; the libraries contain a great wealth of Byzantine manuscripts.

A·ti·tlán (ä'tē-tlän'), volcanic lake, 53 sq mi (137.3 sq km), 17 mi (27.4 km) long and 11 mi (17.7 km) wide, SW Guatemala. One of the most magnificent lakes of Central America, it is set among lofty mountains with three inactive volcanoes nearby; Atitlán volcano (11,564 ft/3,527 m) is the tallest. The fertile lakeshore is densely populated by subsistence farmers.

At·kin·son (ăt'kĭn-sən), county (1970 pop. 5,879), 318 sq mi (823.6 sq km), S Ga., bounded on the W by the Alapaha River and drained by the Satilla River; formed 1917; co. seat Pearson. In a coastal plain agricultural area yielding cotton, corn, peanuts, tobacco, and livestock, it also has a forestry industry.

At·lan·ta (ăt-lăn'tə). **1.** City (1978 est. pop. 406,000), state capital and seat of Fulton co., NW Ga., near the Appalachian foothills; inc. as a city 1847. It is the largest city and the cultural, industrial, transportation, financial, and commercial center of the state; a port of entry; a busy air traffic hub; and one of the leading cities of the South. Manufactures include textiles, furniture, chemicals, glass, paper, lumber, steel, and leather, electrical, and aluminum products. There are flour mills, automobile and aircraft assembly plants, and printing and publishing houses. Hardy Ivy, the first settler, built a cabin here (1833) on what had been Creek Indian land. The town, founded (1837) as Terminus, the end of a railroad line, was incorporated as Marthasville in 1843 and renamed Atlanta in 1845. It became a railroad and marketing hub and in the Civil War was an important communication and supply center; it fell to Gen. W. T. Sherman on Sept. 2, 1864. Most of the city was burned on Nov. 15, before Sherman began his march to the sea. The city was rapidly rebuilt and thrived as a commercial and industrial center. It was chosen temporary state capital in 1868 and became permanent capital following a popular vote in 1877. Points of interest include the capitol (1899), housing the state library; the city hall (1929); the High Museum of Art; the state archives building, containing a historical museum and library; the building housing the huge *Cyclorama of the Battle of Atlanta;* Oakland Cemetery, containing Civil War dead; the grave of Martin Luther King, Jr.; and Grant Park, with the municipal zoo and Confederate Fort Walker (restored). The city's numerous parks are famous for their dogwood blooms, and in the area are Stone Mountain Memorial, with enormous relief carvings of Confederate figures, and Kennesaw Mountain National Battlefield Park. Atlanta is the seat of Emory Univ., Georgia Institute of Technology, Georgia State Univ., and other schools. **2.** Town (1970 pop. 600), seat of

Montmorency co., NE Mich., on a branch of the Thunder Bay River W of Alpena, in a hunting and fishing region.

At·lan·tic (ăt-lăn'tĭk), county (1970 pop. 175,043), 569 sq mi (1,473.7 sq km), SE N.J., on the coast and drained by the Great Egg Harbor and Mullica rivers; formed 1837; co. seat Mays Landing. It has farming (truck crops, dairy products, grain, fruit, and poultry), manufacturing, and fishing. There are many coast resorts along its shoreline, including Atlantic City.

Atlantic, city (1976 est. pop. 7,324), seat of Cass co., SW Iowa, on the East Nishnabotna River ENE of Council Bluffs; inc. 1869. It is a trade and processing center manufacturing feed, beverages, popcorn, wood and sheet-metal products, saws, and stoves. There are poultry-packing plants and corn and pumpkin canneries here.

Atlantic City, city (1978 est. pop. 43,100), Atlantic co., SE N.J., an Atlantic resort and convention center; settled c.1790, inc. 1854. Situated on Absecon Island, a sandbar that is 10 mi (16.1 km) long, Atlantic City was a fishing village until the construction of a railroad in 1854, when it became a fashionable resort. The first boardwalk was built in 1870. The present boardwalk, lined with hotels, shops, and amusements, is 6 mi (9.7 km) long and from 40 to 60 ft (12.2–18.3 m) wide. Five amusement piers, including the famous Steel Pier (1898), run out to sea from the boardwalk. The Miss America Pageant is held in Atlantic City every September. Legalized gambling was introduced to the city in 1978. The first Ferris wheel was built here in 1869. The board game Monopoly, which makes use of the city's street names, was invented here in 1930. Saltwater taffy, developed in Atlantic City, is the chief manufacture.

Atlantic Ocean, second-largest ocean (c.31,800,000 sq mi/82,362,000 sq km; c.36,000,000 sq mi/93,240,000 sq km with marginal seas), extending in an "S" shape from the arctic to the antarctic regions between North and South America on the W and Europe and Africa on the E. It is connected with the Arctic Ocean by the Greenland Sea and Smith Sound; with the Pacific Ocean by Drake Passage, the Straits of Magellan, and the Panama Canal; and with the Indian Ocean by the Suez Canal and the expanse between Africa and Antarctica. The shortest distance across the Atlantic Ocean (c.1,600 mi/2,575 km) is between southwest Senegal and eastern Brazil. The principal arms of the Atlantic Ocean are (in the west) Hudson and Baffin bays, the Gulf of Mexico, and the Caribbean Sea; (in the east) the Baltic, North, Mediterranean, and Black seas, the Bay of Biscay, and the Gulf of Guinea; and (in the south) Weddell Sea. The continental shelf of the Atlantic Ocean is generally narrow, with the widest sections found off northeast North America, southeast South America, and northwest Europe.

The floor of the Atlantic has an average depth of c.12,000 ft (3,660 m). It is separated from that of the Arctic Ocean by a submarine ridge extending from southeast Greenland to northern Scotland. A shallow submarine ridge also lies across the Strait of Gibraltar, separating the Mediterranean basin from the Atlantic and limiting the exchange of water between the two bodies. The Mid-Atlantic Ridge (c.300–600 mi/485–965 km wide), a submarine mountain range extending c.10,000 mi (16,100 km) from Iceland to near the Antarctic Circle, generally follows the trend of the coastlines of the continents. It divides the floor of the Atlantic Ocean into eastern and western sections that are composed of a series of deep-sea basins (abyssal plains). Its average height is c.10,000 ft (3,050 m), and a few peaks emerge as islands. The ridge, which is the center of volcanic activity and earthquakes, has a great rift that is constantly widening and filling with molten rock from the earth's interior. As a result the Western Hemisphere and Europe and Africa are moving away from each other. The greatest depth (c.28,000 ft/8,540 m) is the Milwaukee Deep, in the Puerto Rico Trench, north of Puerto Rico.

Commerce between the Mediterranean Sea and the northeast Atlantic Ocean was initiated by the Carthaginians. From the 7th cent. A.D., Scandinavians navigated the Atlantic; they probably reached North America c.1000. Trade routes along the coast of Africa were opened by Portugal in the 15th cent. and to the Western Hemisphere by Spain after the voyages of Columbus. Scientific knowledge of the ocean floor dates from the Challenger expedition (1872-76).

Atlantic Provinces, term used since 1949 to designate the Canadian provinces of Newfoundland, Nova Scotia, New Brunswick, and Prince Edward Island.

At·las Mountains (ăt'ləs), system of ranges and plateaus in NW Africa, extending c.1,500 mi (2,415 km) from SW Morocco, through N Algeria, to N Tunisia. Jebel Toubkal (13,665 ft/4,167.8 m), in southwest Morocco, is the highest peak. The Atlas Mts., predominantly folded mountains of sedimentary rock, were uplifted during the late Jurassic period. Geologically related to the Alpine system of Europe, they are separated from the Sierra Nevada of Spain by the Strait of Gibraltar and from Sicily and the Apennines of Italy by the Mediterranean Sea; the Canary Islands are a westward extension. The Atlas system is most rugged in Morocco, where, from north to south, the Rif Atlas, Middle Atlas, High or Grand Atlas (the highest part of the system), and Anti-Atlas are found; fertile lowlands separate the ranges. In Algeria the system becomes a series of plateaus, with the Tell Atlas and the Saharan Atlas rimming the extensive Plateau of the Chotts before converging in Tunisia. The Atlas Mts. are a climatic barrier between the Mediterranean basin and the Sahara Desert. The slopes facing north are generally well watered and have important farmland and forests; on these slopes are the headwaters of many streams used for irrigation. The slopes facing south and the drier areas of the system are generally covered with shrub and grasses and have salt lakes and salt flats; sheep grazing is important here. The Atlas Mts. are rich in minerals, especially phosphates, coal, iron, and oil.

At·lin Lake (ăt'lĭn), long, irregular mountain lake, c.300 sq mi (775 sq km), NW British Columbia, Canada, touching the Yukon Territory boundary. It is the source of the Yukon River.

At·more (ăt'mōr, -môr), shipping city (1975 est. pop. 8,429), Escambia co., SW Ala., NE of Mobile near the Fla. border, in a truck farm and timber area; settled 1879, inc. 1907. Textile and lumber products are made.

A·to·ka (ə-tō'kə), county (1970 pop. 10,972), 992 sq mi (2,569.3 sq km), SE Okla., drained by the Muddy Boggy and Clear Boggy creeks; formed 1907; co. seat Atoka. It has agriculture (peanuts, soybeans, and cattle) and a furniture industry.

Atoka, city (1970 pop. 3,346), seat of Atoka co., SE Okla., NE of Ardmore on Muddy Boggy Creek, in a flour, lumber, and agricultural (cotton, corn, and oats) region; founded 1867. An agreement ending Choctaw and Chickasaw Indian tribal government was signed here in 1897.

A·tra·to (ä-trä'tō), river, c.375 mi (605 km) long, rising in the Cordillera Occidental, W Colombia. It meanders north, across the base of the Isthmus of Panama, to the Gulf of Urabá. The Atrato drains a region of rain forests. Its headwaters are in Colombia's chief platinum-producing area.

At·su·gi (ät'sōō'gē), city (1976 est. pop. 113,379), Kanagawa prefecture, E central Honshu, Japan, on the Sagami River. It is an industrial and agricultural center.

At·ta·la (ăt'ə-lə), county (1970 pop. 19,570), 724 sq mi (1,875.2 sq km), central Miss., bounded on the W by the Big Black River and intersected by the Yockanookany River and Lobutcha Creek; formed 1833; co. seat Kosciusko. It has farming (corn, cotton, and hay), dairy processing, stock raising, lumbering (logging and sawmills), and concrete manufacturing.

At·tal·la (ə-tăl'ə), city (1975 est. pop. 7,212), Etowah co., NE Ala., just W of Gadsden, in a mining and cotton area; founded 1870 on the site of an Indian village; inc. 1872. Metal and textile products are made.

At·ta·wa·pis·kat (ăt'ə-wə-pĭs'kăt), river, c.465 mi (750 km) long, flowing E from Attawapiskat Lake, N Ont., Canada, then N and E into James Bay.

At·ti·ca (ăt'ĭ-kə), region of ancient Greece, a triangular area at the E end of central Greece, around Athens. According to Greek legend, the four Attic tribes were founded by Ion; in later legend Theseus combined 12 townships into a single state. This process of unification, which probably occurred over a period of time, was in all likelihood completed c.700 B.C. By the 5th cent. B.C. Athens dominated the region.

At·tle·bor·o (ăt'əl-bûr'ō), industrial city (1978 est. pop. 32,000), Bristol co., SE Mass., near the R.I. line; settled 1634, inc. as a city 1914. Its jewelry industry began in 1780; silverware, scientific instruments, and fabricated metal products are also made.

A·tuo·na (ä-twō'nä) or **A·tua·na** (ä-twä'nä), town, in the Marquesas Islands, South Pacific, in French Polynesia. Situated on the southern coast of the island of Hiva Oa, Atuona overlooks the Bay of Traitors. Gauguin lived in Atuona Valley and is buried there.

At·wa·ter (ăt'wô'tər, -wŏt'ər), city (1978 est. pop. 15,000), Merced co., central Calif., in the San Joaquin Valley; inc. 1922. It is the processing and commercial center of an irrigated farming area.

At·wood (ăt′wo͝od′), city (1970 pop. 1,658), seat of Rawlins co., NW Kansas, W of Oberline on Beaver Creek near the Nebr. border, in a grain and livestock area; founded 1878, inc. 1885.

Aube (ōb), department (1975 pop. 284,823), 2,317 sq mi (6,001 sq km), NE France, in Champagne; capital Troyes.

Au·ber·vil·liers (ō-bĕr-vē-lyā′), town (1975 pop. 72,976), Seine-Saint Denis dept., N central France, NE of Paris. It is an important industrial center where chemicals, pharmaceuticals, metals, and leather goods are produced.

Auburn. **1.** City (1978 est. pop. 26,500), Lee co., E Ala.; inc. 1839. The city's economy centers around Auburn Univ. Lumber products are also made. **2.** City (1970 pop. 6,570), seat of Placer co., N central Calif., NE of Sacramento in the gorge of the American River; inc. 1860. It is a resort and shipping center for an orchard, poultry, and livestock region. Gold was discovered here in 1848. **3.** City (1970 pop. 2,594), Sangamon co., central Ill., SSW of Springfield, in a farm and coal area; inc. 1865. **4.** City (1970 pop. 7,388), seat of De Kalb co., NE Ind., NNE of Fort Wayne; settled 1836. It is a farm trade center for an area yielding livestock, dairy products, soybeans, and grain. Automotive parts are manufactured. **5.** City (1978 est. pop. 22,700), seat of Androscoggin co., SW Maine, on the Androscoggin River opposite Lewiston; settled 1765 on the site of an Indian village, inc. 1842. Its huge shoe industry dates from c.1835. With Lewiston, Auburn forms one of the most important industrial complexes in Maine; abundant water power has spurred a great variety of manufactures. **6.** Town (1970 pop. 15,347), Worcester co., S central Mass.; inc. 1778. Its industries include warehousing and the manufacture of electronic equipment, motors, cement products, plastics, and musical instruments. **7.** City (1970 pop. 3,650), seat of Nemaha co., SE Nebr., on the Little Nemaha River S of Nebraska City, in a fruit, wheat, and dairy area. Machinery and clothing are manufactured. **8.** City (1978 est. pop. 32,000), seat of Cayuga co., W central N.Y., in the Finger Lakes region; settled 1793, inc. 1848. Its manufactures include diesel engines, rope, shoes, rugs, electronic parts, and air conditioners. The city's museum has collections of historical documents and Indian relics. The houses of William H. Seward and Harriet Tubman are preserved. **9.** City (1978 est. pop. 23,400), King co., W Wash., on the Green and White (Stuck) rivers, between Seattle and Tacoma; settled 1855, inc. 1914. It is a railroad junction and farm trade center, with large aircraft industries. Wood products are also made.

Au·burn·dale (ô′bərn-dāl′), city (1975 est. pop. 6,346), Polk co., central Fla., E of Lakeland; founded 1884, inc. 1911. A processing center for a citrus-fruit area, it also manufactures tin cans.

Au·bus·son (ō-bü-sôɴ′), town (1968 pop. 6,761), Creuse dept., central France, in the former province of Marche, on the Creuse River SE of Guéret. Its famous tapestry and carpet manufactures date from the 15th cent. Aluminum and rubber goods are also made.

Auch (ōsh), town (1975 pop. 19,388), capital of Gers dept., SW France, in Gascony, on the Gers River W of Toulouse. It is a farm market and commercial center with a variety of manufactures. One of the chief towns of Roman Gaul, it was the capital of Armagnac (10th cent.) and the capital of Gascony (17th cent.). The old part of town, steep and hilly, is topped by a flamboyant-style Gothic cathedral (15th–16th cent.).

Auck·land (ôk′lənd), city (1976 pop. 150,708; urban agglomeration pop. 698,400), NW North Island, New Zealand. It is situated on an isthmus and is the largest city and chief port of the country. The chief exports are frozen meats, dairy products, wool, hides, and wood pulp. Petroleum, iron and steel, wheat, sugar, and fertilizers are the leading imports. The major industries are engineering (including shipbuilding and boilermaking), automobile and chemical manufacturing, and food processing. It is also a fishing port and the chief base of the New Zealand navy. Auckland was founded in 1840 and was formerly (1841–65) the capital of New Zealand.

Auckland Islands, small, uninhabited group (234 sq mi/606.1 sq km), S Pacific, c.300 mi (485 km) S of South Island, New Zealand, to which they belong. There is some sealing. The islands were discovered in 1805.

Aude (ōd), department (1975 pop. 272,366), 2,406 sq mi (6,231.5 sq km), S central France, in Languedoc; capital Carcassonne.

Aude, river, 130 mi (209.2 km) long, S France. Draining the Aude dept., it rises in the Pyrenees and flows north and east into the Mediterranean Sea near Narbonne.

Au·drain (ô-drān′, ô′drān′), county (1970 pop. 25,362), 692 sq mi (1,792.3 sq km), NE central Mo., drained by the South Fork of the Salt River and the West Fork of the Cuivre River; formed 1831; co. seat Mexico. It has agriculture (corn, oats, barley, and soybeans), and cattle and poultry are raised. Coal mining and lumbering are done. Its manufactures include food products, textiles, and shoes.

Au·du·bon (ô′də-bŏn′, -bən), county (1970 pop. 9,595), 448 sq mi (1,160.3 sq km), W Iowa, drained by the East Nishnabotna River; formed 1851; co. seat Audubon. It lies in a prairie agricultural area yielding cattle, hogs, poultry, corn, hay, and oats and has coal deposits. There is a food-processing industry here, and farm machinery and equipment are manufactured.

Audubon. **1.** City (1970 pop. 2,907), seat of Audubon co., SW Iowa, on the East Nishnabotna River NE of Council Bluffs; platted 1878, inc. 1880. **2.** Borough (1970 pop. 10,802), Camden co., SW N.J., a suburb of Camden; inc. 1905. Audubon is mostly residential. It was named after John James Audubon, the ornithologist, who studied the birds of the area in 1829.

Augh·rim or **Agh·rim** (both: ŏg′rĭm, ôᴋʜ-), village, Co. Wicklow, SW Republic of Ireland. It was the scene of a battle (July 12, 1691) in which the forces of William III of Great Britain won a decisive victory over those of James II.

Au·glaize (ô-glāz′), county (1970 pop. 38,602), 400 sq mi (1,036 sq km), W Ohio, drained by the Auglaize and St. Marys rivers; formed 1848; co. seat Wapakoneta. It is basically agricultural (livestock, grain, poultry, and dairying). Textiles, rubber and plastic products, and automobile parts are manufactured. Part of Grand Lake is in the western part of the county.

Augs·burg (ŏgz′bûrg′), city (1974 est. pop. 254,053), capital of Swabia, Bavaria, S West Germany, an industrial center on the Lech River. The major industries include the manufacture of textiles, clothing, machinery, motor vehicles, and airplanes. The city is an important rail junction. Augsburg was founded (c.14 B.C.) by Augustus as a Roman garrison. In early medieval times it was controlled by the Frankish kings. It was made a free imperial city in 1276 and was later a powerful member of the Swabian League of 1488–1534. Augsburg was one of Europe's most important commercial and banking centers in the 15th and 16th cent. and was a rallying point of German science and art. Several important agreements, including the Augsburg Confession (1530), were concluded here during the Reformation. Augsburg suffered greatly in the Thirty Years War (1618–48). In 1806 it became part of Bavaria. Augsburg's many noteworthy structures include the cathedral (begun in the 9th cent.) and the 17th cent. town hall.

Au·gus·ta (ou-go͞o′stä), city (1975 est. pop. 37,367), E Sicily, Italy, on an island (formerly a peninsula) in the Ionian Sea, connected by bridge with the Sicilian mainland. It is an important port and a fishing and industrial center. Manufactures include refined petroleum, chemicals, textiles, and fertilizer. The city was a Greek settlement and then a Roman military base. It was refounded by Emperor Frederick II in 1232 and later (15th to early 16th cent.) was a thriving banking town.

Au·gus·ta (ô-gŭs′tə), county (1970 pop. 44,220), 986 sq mi (2,553.7 sq km), NW Va., mainly in the Shenandoah Valley, with the Shenandoah Mts. in the W and NW and the Blue Ridge in the SE and E; formed 1745; co. seat Staunton (independent city). Drained by headstreams of the Shenandoah River, it has agriculture, livestock raising, dairying, lumbering, manganese mining, rock quarrying, and some manufacturing. It includes parts of Shenandoah National Park, George Washington National Forest, the Appalachian Trail, and the Blue Ridge Parkway.

Au·gus·ta (ô-gŭs′tə). **1.** City (1970 pop. 2,777), seat of Woodruff co., E central Ark., on the White River NE of Little Rock, in an agricultural area; settled 1846, inc. 1861. It has cotton, lumber, and grist mills and a commercial fishing industry. **2.** City (1978 est. pop. 51,200), seat of Richmond co., E Ga.; inc. 1798. At the head of navigation on the Savannah River and protected by levees, Augusta is the trade center for a broad section known as the Central Savannah River Area. It is also an important industrial center manufacturing textiles, chemicals, bricks and tiles, fertilizers, cleansers, hospital supplies, tools, and wood, paper, metal, and plastic products. The city is a popular resort, noted especially for its golf tournaments. Augusta grew from an old river trading post existing as early as 1717 and was named by James Oglethorpe in 1735 after the mother of George III. During the American Revolution, Augusta changed hands several times. It was the capital of Georgia from 1785 to 1795. Augusta boomed after the Revolution, during the rapid expansion of the tobacco industry, followed by the growth of the cotton industry. By 1820 the city was the terminus for river boats, wagon trains, and traders. Manufacturing began in 1828, when Augusta's first textile plant began operation. During the Civil War, Augusta housed the largest Confederate powder works. The city's historical attractions include a boyhood home of Woodrow Wilson, a U.S. arsenal (1815–1955), and old homes of Georgian and classic-revival styles. Nearby is Fort Gordon, with training schools for military police, the signal corps, and the corps of engineers. **3.** City (1975 est. pop. 6,254), Butler co., S Kansas, E of Wichita on the Walnut River; settled c. 1868, inc. 1871. A trading center for a rich farm region, it is also a distributing and refining point for oil fields. **4.** City (1978 est. pop. 20,000), state capital and seat of Kennebec co., SW Maine, on the Kennebec River; inc. as a town 1797, as a city 1849. Shoes, fabrics, and paper products are manufactured. Traders visited the site even before 1628, when the Plymouth Company established a trading post. Fort Western was built in 1754, and Benedict Arnold's expedition to Quebec gathered at the fort in 1775. The settlement around the fort developed with the shipping and shipbuilding on the Kennebec, and manufacturing began in 1837, when a dam was built across the river. The capitol building (1829) was designed by Charles Bulfinch but has been considerably enlarged and remodeled.

Au·lis (ô'lĭs), small port of ancient Greece, in Boeotia, E central Greece. From here, according to tradition, the Greek fleet sailed against Troy after the sacrifice of Iphigenia.

Au·nis (ō-nēs'), small region and former province, W France, on the Atlantic coast. It includes the islands of Ré and Oléron. A part of Aquitaine, it was recovered from England in 1373 and incorporated into the French crown lands.

Au·rang·a·bad (ou-rŭng'gä-bäd'), town (1971 pop. 150,483), Maharashtra state, W India. A district administrative center, it also carries on trade in cotton and wheat. Silverware is produced. Aurangabad was founded in 1610. Nearby is the great mausoleum (1711) of Aurangzeb's empress.

Au·ri·gnac (ô-rē-nyäk'), village (1968 pop. 1,149), Haute-Garonne dept., S France, at the foot of the Pyrenees. Its caves, excavated in 1860, contain relics of prehistoric man of the Aurignacian period.

Au·ril·lac (ô-rē-yäk'), town (1975 pop. 30,863), capital of Cantal dept., S central France, in Auvergne. An industrial, communications, and market center, it is noted for its furniture, footwear, umbrellas, and Cantal cheese.

Au·ror·a (ô-rôr'ə, ə-rôr'ə), county (1970 pop. 4,183), 711 sq mi (1,841.5 sq km), SE central S.Dak., in a level farming area watered by intermittent creeks; formed 1879; co. seat Plankinton. It produces livestock, dairy products, poultry, wheat, and corn.

Aurora. 1. City (1978 est. pop. 128,000), Adams and Arapahoe cos., N central Colo., a residential suburb of Denver; inc. 1903. It is the trade center for a large farming and livestock-raising area. Electrical products, aircraft parts, and oil field equipment are manufactured. Tourism and construction are also important. Lowry Air Force Base is nearby. **2.** City (1978 est. pop. 79,600), Kane co., NE Ill., on the Fox River W of Chicago; inc. 1837. It has large railroad yards and a great variety of manufactures, including construction and highway equipment, electric tools, pumps, and heavy steel products. It was one of the first cities to use electricity for street lighting (1881). **3.** City (1970 pop. 4,293), Dearborn co., SE Ind., on the Ohio River WSW of Cincinnati; laid out 1819. Wood products are made. **4.** Village (1970 pop. 2,531), St. Louis co., NE Minn., E of the town of Virginia, at the E edge of the Mesabi iron range; inc. 1903. It is a trade center for the iron-mining region. **5.** City (1970 pop. 5,359), Lawrence co., SW Mo., in the Ozarks SW of Springfield; laid out 1870; inc. 1886. It is a trade center, with flour and feed mills and a shoe factory. Lead and zinc mines are in the region. **6.** City (1970 pop. 3,180), seat of Hamilton co., SE Nebr., W of Lincoln, in an irrigated grain and livestock area; founded 1872, inc. 1877. It has flour mills and dairies. **7.** Village (1970 pop. 6,549), Portage co., NE Ohio, SE of Cleveland on a branch of the Chagrin River; inc. 1928.

Au·sa·ble Chasm (ô-sā'bəl), gorge, 2 mi (3.2 km) long, from 20 to 50 ft (6.1–15.3 m) wide, from 100 to 200 ft (30–61 m) deep, NE N.Y. The chasm, with its rapids, waterfalls, and curious rock formations, is a popular tourist attraction; Rainbow Falls, 75 ft (23.9 m) high, is at the southern end of the gorge. The Ausable, a river rising in the Adirondack Mts. and flowing northeast to Lake Champlain, continues to carve out the gorge as it passes over the sandstone bedrock.

Ausch·witz (oush'vĭtz): *see* Oświęcim, Poland.

Aus·ter·litz (ô'stər-lĭts), town, S Czechoslovakia, in Moravia. An agricultural center, the town has sugar refineries and cotton mills. It became a seat of the Anabaptists in 1528. At Austerlitz Napoleon won (Dec. 2, 1805) his most brilliant victory by defeating the Russian and Austrian armies under Czar Alexander I and Emperor Francis II. There is a famous description of the battle in Tolstoy's *War and Peace.* The town has an 18th cent. castle, a 13th cent. church, the Renaissance Church of the Resurrection, and the Monument of Peace (built 1910–11).

Aus·tin (ô'stən), county (1970 pop. 13,831), 662 sq mi (1,714.6 sq km), SE central Texas, bounded on the E by the Brazos River and also drained by the San Bernard River; formed 1836; co. seat Bellville. It has agriculture (cotton, peanuts, corn, hay, pecans, rice, and truck farming), livestock (cattle, hogs, and poultry), dairying, and lumbering, and oil and natural-gas wells.

Austin. 1. Uninc. town (1970 pop. 4,902), Scott co., SE Ind., N of Louisville, Ky., in a farm area. It has a large cannery. **2.** City (1978 est. pop. 23,000), seat of Mower co., SE Minn., on the Cedar River near the Iowa border; inc. 1868. The industrial and commercial center of a rich farm region, it has a large meat-packing industry. Shipping and metal containers are also made. **3.** Town (1970 pop. 300), seat of Lander co., central Nev., in the W foothills of the Toiyabe Range, W of Eureka; founded 1862. Sheep are raised and silver and gold are mined. It was an important mining and trading center and post station during the early gold-rush period in Nevada. **4.** City (1978 est. pop. 326,000), state capital and seat of Travis co., S central Texas, on the Colorado River and two of the Highland Lakes; inc. 1839. It is the commercial heart of a large ranching, poultry, dairy, cotton, and grain area, with a great variety of manufactures. It is also a major convention city and an educational center—including the main campus of the Univ. of Texas. Numerous electronic and scientific research firms are located here. The site was selected in 1839 for the capital of the independent Texas republic and named in honor of Stephen F. Austin. Fear of the Mexicans and the Indians drove government officials to Houston in 1842; they returned in 1845 when Texas was admitted to the Union, and in 1870, following a referendum, Austin was made the permanent capital. Its industrial growth was spurred by the development of power and flood control projects on the Colorado River (beginning in the 1930s) and by the urgencies of World War II. The massive capitol (completed 1888), set on a hill, is the most prominent of the many state buildings; on its grounds are the state library, the old land office (1857), and two state historical museums.

Austin, Lake, 320 sq mi (828.8 sq km), in the W section of Western Australia, NNE of Perth. It is often dry.

Aus·tral·a·sia (ô'strə-lā'zhə, -shə), islands of the South Pacific, including Australia, New Zealand, New Guinea, and adjacent islands. The term is sometimes used to include all of Oceania.

Aus·tral·ia (ô-strāl'yə), smallest continent, between the Indian and Pacific oceans. It extends from east to west some 2,400 mi (3,860 km) and from north to south nearly 2,000 mi (3,220 km). With the island state of Tasmania to the south, the continent makes up the Commonwealth of Australia (1976 est. pop. 13,684,900), 2,967,877 sq mi (7,686,801 sq km); federal capital Canberra.

The Australian continent is on the whole exceedingly flat and dry. Less than 20 in. (50.8 cm) of precipitation falls annually over 70% of the land area. From the narrow coastal plain in the west the land rises abruptly in what, from the sea, appear to be mountain ranges but are actually the escarpments of a rough plateau that occupies the western half of the continent. It is generally from 1,000 to 2,000 ft (305–610 m) high but several mountain ranges rise to nearly 5,000 ft (1,525 m). In the southwest corner of the continent there is a small moist and fertile area, but the rest of Western Australia is arid, with large desert areas. The northern region also belongs to the plateau, with tropical temperatures and a winter dry season. Its northernmost section, Arnhem Land (principally given over to reservations for aborigines), faces the Arafura Sea in the north and the huge Gulf

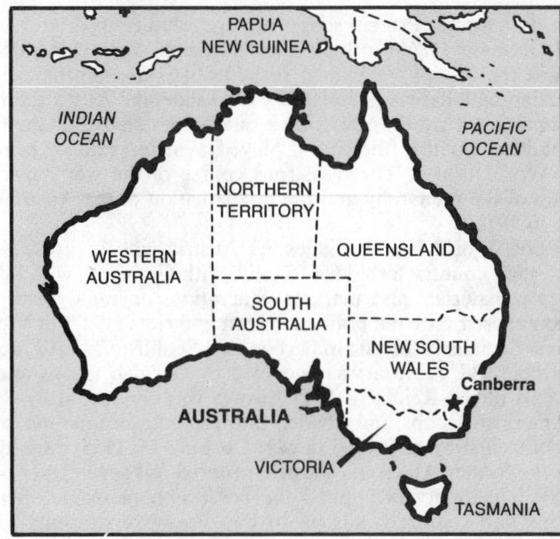

of Carpentaria on the east. On the eastern side of the gulf is the Cape York Peninsula, which is largely covered by rainforest. Off the coast of northeast Queensland is the Great Barrier Reef, the world's largest coral reef. The longest of all Australian river systems, the Murray River and its tributaries, drains the southern part of the interior basin that lies between the mountains and the great plateau. The rivers of this area are used extensively for irrigation and hydroelectric power. Australia, remote from any other continent, has many distinctive forms of plant life—notably species of giant eucalyptus—and of animal life, including the kangaroo, the koala bear, the flying opossum, the wallaby, the wombat, the platypus, and the spiny anteater; it also has many unusual birds.

Australia is divided into six states and two territories:

NAME	CAPITAL	NAME	CAPITAL
Australian Capital Territory	Canberra	South Australia	Adelaide
New South Wales	Sydney	Tasmania	Hobart
Northern Territory	Darwin	Victoria	Melbourne
Queensland	Brisbane	Western Australia	Perth

Economy. Australia is highly industrialized, and manufactured goods account for about two thirds of the total value of production. The leading manufactures are iron and steel products, transportation equipment, and machinery. Australia is one of the world's great trading nations, with one quarter to one third of its export income derived from the sale of wool, meat, and wheat. The country is self-sufficient in food, and the raising of sheep and cattle and the production of grain have long been staple occupations. Tropical and subtropical produce—citrus fruits, sugar cane, and tropical fruits—are also important, and there are numerous vineyards and dairy and tobacco farms. Some lumbering is done in the east and southeast. Australia has valuable mineral resources, including coal, iron, bauxite, uranium, and gold.

History. The groups comprising the aborigines are thought to have migrated from Southeast Asia c.20,000 years ago. They spread throughout Australia and remained isolated until the arrival of the Europeans. It seems probable that Australia was first sighted by a Portuguese, Manuel Godhino de Eredia, in 1601. Little interest was aroused, however, until the fertile east coast was observed when Capt. James Cook reached Botany Bay in 1770 and sailed north to Cape York, claiming the coast for Great Britain. In 1788 the first British settlement was made—a penal colony on the shores of Port Jackson, where Sydney now stands. By 1829 the whole continent was a British dependency. Exploration, begun before the first settlement was founded, was continued by such men as Matthew Flinders (1798), Ludwig Leichhardt (1848), and John McDouall Stuart (first to cross the continent, 1862). Sheep raising was introduced early, and before the middle of the 19th cent. wheat was being exported in large quantities to England. A gold strike in Victoria in 1851 brought a rush to that region. Other strikes were made later in the century in Western Australia. With minerals, sheep, and grain forming the base of the economy, Australia developed rapidly.

Australia was long used as a dumping ground for criminals, bankrupts, and other undesirables from the British Isles, but by the mid-19th cent. systematic, permanent colonization had completely replaced the old penal settlements. Confederation of the separate Australian colonies did not come until a constitution, drafted in 1897–98, was approved by the British Parliament and was put into operation in 1901. The Northern Territory was added to the federation in 1911.

Government. The executive power of the Commonwealth is vested in a governor general (representing the British sovereign) and a cabinet, presided over by the prime minister, which represents the party, or a coalition, holding a majority in the lower house of Parliament. The Parliament consists of two houses. The distribution of federal and state powers is roughly like that in the United States. From its early years the federal government has been noted for its liberal legislation, such as woman suffrage (1902), old-age pensions (1909), and maternity allowances (1912).

Educational Institutions. Although education is not a federal concern, government grants have aided in the establishment of state universities, including the Univ. of Sydney (1852), the Univ. of Melbourne (1854), the Univ. of Adelaide (1874), and the Univ. of Queensland (in Brisbane, 1909).

Aus·tra·lian Alps (ô-strāl′yən), chain of mountain ranges, SE Australia, making up the S part of the Eastern Highlands and forming the watershed between the Murray River system and streams flowing into the Tasman Sea. Mt. Kosciusko (7,316 ft/2,231.4 m) in the Australian Alps is the highest peak in Australia.

Australian Ant·arc·tic Territory (ănt-ärk′tĭk, -är′-), a huge section of Antarctica, c.2,360,000 sq mi (6,112,400 sq km), an external territory of Australia, situated S of lat. 60° S and between long. 45° and 160° E, excluding Adelie Coast. The territory covers almost half the continent. Australia claimed the land in 1933, but the United States does not recognize Antarctic claims south of lat. 60°. Long-term scientific research began here in 1947.

Australian Capital Territory (1976 pop. 197,578), 939 sq mi (2,432 sq km), SE Australia, an enclave within New South Wales, containing Canberra, capital of Australia. It was called the Federal Capital Territory until 1938. Most of the territory consists of an area formerly known as Yass-Canberra, which was ceded to the commonwealth by New South Wales in 1911. The remainder was added in 1915, when New South Wales ceded a part of the Jervis Bay area, providing a potential port for Canberra. The federal government is the largest employer in the territory.

Aus·tral Islands (ô′strəl), volcanic island group, South Pacific, part of French Polynesia. The group comprises seven islands, with a total land area of c.115 sq mi (300 sq km). Tubuai, the largest island (c.17 sq mi/44 sq km), was visited by Capt. James Cook in 1777 and was annexed by France in 1880. Coffee, arrowroot, tobacco, and copra are produced on the islands.

Aus·tra·sia (ô-strā′zhə, -shə), E portion of the Merovingian kingdom of the Franks in the 6th, 7th, and 8th cent., comprising, in general, parts of E France, W Germany, and the Netherlands, with its capital variously at Metz, Rheims, and Soissons. It originated in the partition (511) of the realm of the Frankish King Clovis I among his sons. Austrasia was constantly troubled by dynastic rivalries and finally became part of the Carolingian empire.

Aus·tri·a (ô′strē-ə), federal republic (1973 est. pop. 7,550,000), 32,374 sq mi (83,849 sq km), central Europe. It is bounded by Yugoslavia and Italy in the south, Switzerland and Liechtenstein in the west, West Germany and Czechoslovakia in the north, and Hungary in the east. Vienna is the capital and largest city. The Alps traverse Austria from west to east and occupy three fourths of the country. The country is drained by the Danube River and its tributaries, the Inn, the Enns, the Mürz, and the Mur.

Austria is divided into 9 provinces:

NAME	CAPITAL	NAME	CAPITAL
Burgenland	Eisenstadt	Tirol	Innsbruck
Carinthia	Klagenfurt	Upper Austria	Linz
Lower Austria	Vienna	Vienna	Vienna
Salzburg	Salzburg	Vorarlberg	Bregenz
Styria	Graz		

Economy. Forestry, cattle raising, and dairying are the main sources of livelihood in the alpine provinces. In Upper and Lower Austria and in Burgenland, tillage agriculture predominates: the chief crops are potatoes, sugar beets, barley, wheat, rye, and oats. Manufacturing and mining employ nearly half of the labor force. More than half of the industries are concentrated in the Vienna basin. Many of the country's industries were nationalized after World War II, together with the largest commercial banks. The chief manufactures are chemicals, foodstuffs, textiles, machinery, iron and steel, and metal goods. Many minerals necessary for industry (graphite, iron, magnesium, copper, zinc, and lignite) are found in Austria. The country also has deposits of natural gas, salt, and uranium, and is rich in hydroelectric power.

History. From earliest times Austrian territory has been a thoroughfare, a battleground, and a border area. It was occupied by Celts and Suebi when the Romans conquered it (15 B.C.–A.D. 10). After the 5th cent. A.D., Huns, Ostrogoths, Lombards, Bavarians, and Slavs overran the area. In 788 Charlemagne conquered it and set up the first Austrian (i.e., Eastern) March in the present Upper and Lower Austria. Colonization was encouraged and Christianity was spread energetically. The march soon fell to the Moravians and later to the Magyars, from whom it was taken (955) by Holy Roman Emperor Otto I. In 976 Otto II bestowed it on Leopold of Babenberg, founder of the first Austrian dynasty. After the death (1246) of the last Babenberg, King Ottocar II of Bohemia acquired (1251–69) Austria, Styria, Carinthia, and Carniola. The German king Rudolf I of Hapsburg then asserted (1282) his royal prerogative to reclaim the four duchies and incorporate them in his domains. Beginning with Albert II (reigned 1438–40) the rulers of the Holy Roman Empire were all Hapsburgs who considered German Austria the prized core of their dominions. In 1526 Austria, Bohemia, and Hungary were united under one crown. The Hapsburg rulers attempted to counter the rise of Protestantism by nurturing the Catholic Reformation. Under Ferdinand II, anti-Protestant vigor helped to precipitate the Thirty Years War (1618–48). Protestant Bohemia and Moravia, defeated by the Austrians (1620), became virtual Austrian provinces.

Emperor Charles VI (1711–40) secured the succession to the Hapsburg lands for his daughter, Maria Theresa, who struggled with Frederick II of Prussia in the War of the Austrian Succession and the Seven Years War. Except for the loss of Silesia, Maria Theresa held her own and later acquired parts of Poland. Her son and successor, Joseph II (reigned 1765–90), decreed a series of agrarian, fiscal, religious, and judicial reforms. In the reign of Francis II, Austria was drawn (1792) into largely unsuccessful wars with revolutionary France and with Napoleon I. Francis declared himself the Emperor of Austria in 1804 and acquiesced to the dissolution of the Holy Roman Empire in 1806. The Congress of Vienna (1814–15) did not restore to Austria its former possessions in the Netherlands and in Baden but awarded it Lombardy, Venetia, Istria, and Dalmatia. As the leading power of both the German Confederation and the Holy Alliance, Austria under the ministry (1809–48) of Prince Klemens von Metternich dominated European politics.

The revolutions of 1848 shook the Hapsburg empire. In Vienna, Emperor Ferdinand (reigned 1835–48) granted a short-lived liberal constitution and was forced to abdicate in favor of Francis Joseph (reigned 1848–1916). National tensions and political weakness grew within the empire. Prussia seized the opportunity to drive Austria out of Germany (1866) in the Austro-Prussian War, which also resulted in the loss of Venetia to Italy. In 1867 a compromise with Hungarian nationalists established a dual state, the Austro-Hungarian monarchy. Failure to provide a satisfactory status for the other nationalities, notably the Slavs, played a major role in bringing about World War I. The disastrous course of the war led to the breakup of the monarchy and the proclamation of the Austrian republic in 1918.

The post-World War I treaties left Austria with its present borders, a small country forbidden to unite with Germany, with insufficient raw materials and markets. Starvation, unemployment, and inflation caused growing political unrest and riots (1927) in Vienna. National Socialism, feeding in part on anti-Semitism, gained rapidly. In March, 1938, Austria was occupied by German troops and became part of the Reich. In 1945 Austria was conquered by Soviet and American troops and divided into separate occupation zones, each controlled by an Allied power. On May 15, 1955, a treaty between the former Allies and Austria restored full sovereignty to the country. The treaty prohibited the possession of major offensive weapons and required Austria to pay heavy reparations to the USSR. Austria, proclaiming its neutrality, joined the United Nations in 1955 and the European Free Trade Association in 1959. Under the alternating rule of the conservative People's party and the Socialist party, Austria enjoyed unprecedented prosperity in the 1960s and 1970s.

Government. Austria is governed under the revised 1929 constitution. It has a mixed presidential-parliamentary form of government. The president, elected by popular vote for a six-year term, may issue decrees. The cabinet, headed by the prime minister, is responsible to the lower house of parliament, which is popularly elected according to proportional representation. The upper house is chosen by the provincial assemblies.

Au·tau·ga (ô-tô′gə), county (1970 pop. 24,460), 599 sq mi (1,551.4 sq km), central Ala., in the Black Belt, bounded on the S by the Alabama River; formed 1818; co. seat Prattville. It is in a farming area (cotton, vegetables, and potatoes) with gently rolling land and forests. Textile manufacturing is among its various industries.

Au·teuil (ō-tö′yə), old town between the Seine and the Bois de Boulogne, absorbed (1860) into Paris, France. A favorite resort for writers (Molière, La Fontaine, and Boileau) in the 17th cent., it is now the site of a popular steeplechase track.

Au·tun (ō-tûɴ′), town (1975 pop. 21,556), Saône-et-Loire dept., E central France, on the Arroux River WNW of Chalon-sur-Saône. It is an industrial center producing metals, machinery, leather, cloth, carpets, and timber. Between the 5th and 9th cent. Autun was often attacked by barbarians. Among its Roman ruins are the remains of the town wall, an amphitheater, and the 3rd cent. gates. The Hotel Rolen (15th cent.) is now a museum. The Cathedral of St. Lazare (12th cent.) is famous for its medieval sculpture.

Au·vergne (ō-vûrn′), region and former province, S central France. The Auvergne Mts., a chain of extinct volcanoes, run north to south forming unusual and beautiful scenery. Auvergne is largely agricultural (cattle, wheat, and grapes), with cheese and many wine manufactures. The Arvennis, an ancient people, occupied Auvergne when the Romans arrived. Their chieftain, Vercingetorix, led the resistance to Caesar. Auvergne passed to the English in 1154. In the 14th cent. it was divided into the countship, dauphiny, and duchy of Auvergne. The duchy and dauphiny, which were united under the dukes of Bourbon, were confiscated (1527) by Francis I. The countship came into the royal domain in 1615.

Aux Cayes (ō kā′), **Cayes,** or **Les Cayes** (lā kā′), town (1971 pop. 22,065), SW Haiti, on the Caribbean Sea. Haiti's chief southern port, it exports mainly sugar and coffee. There are liquor distilleries in the town.

Au·xerre (ō-sâr′), town (1975 pop. 38,342), capital of Yonne dept., N central France, in Burgundy, on the Yonne River. A commercial and industrial center, it has a great variety of manufactures and an important trade in Chablis wines. Auxerre became part of Burgundy with the Treaty of Arras (1435). The Church of St. Germain (13th cent.) is built on crypts dating back to the 6th cent.

A·va (ā′və), city (1970 pop. 2,504), seat of Douglas co., S Mo., in the Ozarks SE of Springfield, in a resort, timber, and agricultural region; founded 1864. There is a cheese factory, and lumber and grain products are made.

A·va·cha (ə-vä′chə), active volcano, 8,965 ft (2,734.3 m) high, Far Eastern USSR, on S Kamchatka peninsula. It has a permanent snow cap.

Av·a·lon (ăv′ə-lŏn′). **1.** City (1970 pop. 1,520), Los Angeles co., S Calif., on Santa Catalina Island; founded 1888, inc. 1913. It has an Indian museum and is the center of the island's resort activities. **2.** Residential borough (1970 pop. 7,010), Allegheny co., SW Pa., on the Ohio River NW of Pittsburgh; settled 1802, inc. 1874.

Avalon Peninsula, 3,579 sq mi (9,269.6 sq km), SE N.F., Canada. It is nearly divided at its center by Conception Bay and St. Mary's Bay. The peninsula is the most densely populated part of the province.

Av·dir·a (äv-dîr′ə): *see* Abdera, Greece.

Ave·bur·y (āv′bər-ē), village, Wiltshire, S central England. The village, with a medieval church and Elizabethan manor house, lies within Avebury Circle, a Neolithic circular group of upright stones that are older and larger than Stonehenge but not so well preserved.

A·vei·ro (ä-vā′rō), town (1975 est. pop. 19,905), capital of Aveiro dist., NW Portugal, on the lagoon of Aveiro and at the mouth of the Vouga River, in Beira Litoral. Intersected by numerous canals, one of which connects with the Atlantic, the town is a fishing port and salt-producing center.

A·ve·lla·ne·da (ä′vā-yä-nā′dä), city (1970 pop. 337,538), Buenos Aires prov., E central Argentina, across the Riachuelo River from the Buenos Aires federal district. It is one of the most important industrial, commercial, and transportation centers in the country.

A·vel·li·no (ä′və-lē′nō), city (1975 est. pop. 58,574), capital of Avellino prov., Campania, S Italy. It is an agricultural and manufacturing center. Although damaged by an earthquake in 1930, the city has retained much of its medieval aspect. Of note are the 12th cent. cathedral and the ruins of a castle (9th-10th cent.).

Av·en·tine (ăv′ən-tīn′, -tēn′), one of the seven hills of Rome.

A·ver·no (ä-vĕr′nō), or *ancient* **A·ver·nus** (ə-vûr′nəs), small crater lake, .6 mi (.9 km) wide, Campania, S Italy, between Cumae and Puteoli, near the Tyrrhenian Sea. Its intense sulfuric vapors, caused by volcanic activity (now extinguished), supposedly killed the birds flying over it. The ancient Romans, impressed by its vapors and its gloomy aspect, regarded it as the entrance to hell; later the name was used for hell itself.

A·ver·sa (ä-vĕr′sä), city (1975 est. pop. 50,374), Campania, S Italy. It is an agricultural and transportation center, noted for its sparkling white wine. In the early 11th cent. the county of Aversa became the first possession of the Normans in Italy; it later was made part of the kingdom of Naples.

A·ve·ry (ā′və-rē), county (1970 pop. 12,655), 245 sq mi (634.6 sq km), NW N.C., in the Blue Ridge bounded on the W by the Tenn. border and drained by the North Toe River; formed 1911; co. seat Newland. It has farming (corn, potatoes, hay, vegetables, and apples), cattle raising, lumbering, and mining (mica, feldspar, and kaolin). It is in Pisgah National Forest.

Avery Island, salt dome, c.200 ft (60 m) high and 2 mi (3.2 km) in diameter, S La., in an area of sea marshes and swamps. All the cayenne peppers grown in the United States are produced on Avery Island. Rock salt has been mined here since 1791.

A·ves·ta (ä′və-stä′), city (1975 est. pop. 21,100), Kopparberg co., S central Sweden, on the Dalälven River. Aluminum and high-quality steel are manufactured here.

A·vey·ron (ä-vā-rôN′), department (1975 pop. 278,306), 3,372 sq mi (8,733.5 sq km), S central France, in Midi-Pyrénées; capital Rodez.

Aveyron, river, c.150 mi (240 km) long, Aveyron and Tarn-et-Garonne depts., S France. It rises near Sévérac-le-Château and flows west to the Tarn River northwest of Montauban.

A·vi·gnon (ä-vē-nyôN′), city (1975 pop. 90,786), capital of Vaucluse dept., SE France, on the Rhône River. It is a farm market with a wine trade and a great variety of manufactures. It was the papal see during the Babylonian captivity, from 1309 to 1378, and the residence of several antipopes from 1378 to 1408. After the Great Schism, Avignon was nominally ruled by papal legates, but the citizens actually governed themselves. In 1791, after a plebiscite, it was incorporated into France. One of the loveliest of French cities, Avignon is surrounded by ramparts (12th and 14th cent.) and has many old churches. The beautiful Gothic papal palace was built (14th cent.) atop a hill to serve as residence, fortress, and church.

A·vi·la (ä′vē-lä), town (1975 est. pop. 33,495), capital of Avila prov., central Spain, in Old Castile, on the upper Adaja River WNW of Madrid. It attracts many tourists. One of the great religious centers of Spain, Ávila has preserved much medieval architecture. Up against its turreted wall (built 11th cent.) is the imposing Cathedral of San Salvador. The Basilica of San Vicente is one of the finest Romanesque buildings in Spain.

A·vi·lés (ä-vē-lās′), town (1970 pop. 81,710), Oviedo prov., NW Spain, in Asturias, on the Bay of Biscay. Coal is exported, and there are metalworks and textile mills.

A·viz (ä-vēsh′), village, Portalegre dist., central Portugal, in Alto Alentejo. The Castilian order of the Knights of Calatrava assisted in driving the Moors from Portugal and in 1166 settled at Évora. Alfonso II granted (1211) them Aviz, and this branch of the order became known as the Order of Aviz. Later the house of Aviz was established on the Portuguese throne and ruled until 1580.

A·von (ā′vən, ăv′ən), nonmetropolitan county (1976 est. pop. 920,200) SW England; administrative center Bristol. Created under the Local Government Act of 1972 (effective 1974), it is composed of the county boroughs of Bath and Bristol and parts of the former counties of Gloucestershire and Somerset.

A·von (ā′vən, ăv′ən), name of several rivers in England. **1.** Also called Bristol Avon or Lower Avon, rising in SW England at Tetbury, Gloucestershire, and flowing 75 mi (120.7 km) E, S, and then NW through Bath and Bristol to the Severn River. It is navigable for large vessels to Bristol, an important port. **2.** Also called East Avon, rising at Devizes, Wiltshire, S England, and flowing 48 mi (77.2 km) S past Salisbury to the English Channel at Christchurch. It is navigable for small craft below Salisbury. **3.** Also called Upper Avon, the most famous of the Avon rivers, rising near Naseby, Northamptonshire, S central England, and flowing 96 mi (154.5 km) SW to the Severn River near Tewkesbury, passing Rugby, Warwick, and Stratford-upon-Avon.

A·von. **1.** (ā′vŏn) Town (1970 pop. 8,352), Hartford co., central Conn., on the Farmington River W of Hartford; inc. 1830. **2.** (ā′vŏn) Town (1970 pop. 5,295), Norfolk co., E Mass., N of Brockton; settled c.1700, inc. 1888. It is a residential community with some industrial activity.

Av·on·dale (ăv′ən-dāl′), city (1970 pop. 6,626), Maricopa co., S Ariz., in the Gila Valley W of Phoenix, in a farming and livestock-raising region; inc. 1946.

A·von Lake (ā′vŏn), city (1970 pop. 12,261), Lorain co., NE Ohio, on Lake Erie; inc. 1917. Chiefly a residential suburb, the city has an electric power plant and factories that make plastics and aluminum castings.

A·von Park (ā′vŏn), city (1970 pop. 6,712), Highlands co., S central Fla., SE of Lakeland, in a lake region; founded 1886, inc. 1913. The city processes citrus fruit.

A·voyelles (ə-voilz′), parish (1970 pop. 37,751), 832 sq mi (2,154.9 sq km), central La., partly bounded on the N and E by the Red River, and partly on the E by the Atchafalaya River; formed 1807; parish seat Marksville. Its agriculture includes cotton, corn, hay, rice, sugar cane, sweet potatoes, and livestock. It has a food-processing plant, moss and cotton gins, sugar and lumber mills, commercial fisheries, and oil and natural-gas fields. Textiles are manufactured.

A·vranches (ä-vräNsh′), town (1968 pop. 10,036), Manche dept., NW France, in Normandy, on the English Channel. Because of its proximity to the rocky island of Mont-Saint-Michel, Avranches has a large tourist trade. A Roman town, it was devastated in the Hundred Years War, the Wars of Religion, and World War II.

A·wa·ji-shi·ma (ä-wä′jē-shē′mä), island, 32 mi (51.5 km) long and from 3 to 17 mi (4.8-27.4 km) wide, Hyogo prefecture, Japan, in the Inland Sea. A relatively flat, fertile island, it produces grain and flowers and has commercial fisheries.

A·wash (ä′wäsh), river, c.500 mi (805 km) long, E Ethiopia. Originating just west of Addis Alam, it flows generally northeast into the Danakil Desert at Lake Abbé, on the border of Djibouti. It was once also known as the Hawash River.

Awe, Loch (ô), lake, 25 mi (40.2 km) long, Argyllshire, W Scotland. It is 118 ft (36 m) above sea level. The hydroelectric power facility at Cruachan (completed 1967) has a 400,000-kw capacity.

Ax·el Hei·berg Island (ăk′səl hī′bərg), 13,583 sq mi (35,180 sq km), in the Arctic Ocean, N Northwest Territories, Canada, W of Ellesmere Island. It was named by the Norwegian explorer Otto Sverdrup (who explored it 1898–1902) for one of his patrons. The island's plateau surface (3,000–6,000 ft/915–1,830 m high) is deeply indented by fjords.

Ax·holme Isle (ăks'hōm), flat lowland area, in the Parts of Lindsey, Lincolnshire, E England, c.50,000 acres (20,250 hectares), W of the Trent. Formerly marshy, it became very productive after it was drained in the early 17th cent. Its principal crops are grain and vegetables.

Ax·min·ster (ăks'mĭn-stər), rural district (1973 est. pop. 2,656), Devonshire, SW England. An early Saxon settlement (c.660), it was famous in the 18th cent. for its fine carpets. There are remains of a 13th cent. abbey.

Ax·um (äk-sōōm'): *see* Aksum, Ethiopia.

A·ya·be (ä-yä'bā), city (1975 pop. 29,000), Kyoto prefecture, W central Honshu, Japan, NW of Kyoto. It is an agricultural and communications center where raw silk and silk fabrics are manufactured.

A·ya·cu·cho (ä'yä-kōō'chō), city (1972 est. pop. 34,593), capital of Ayacucho dept., S central Peru. It is a commercial center in a rich mining region that produces gold, silver, and nickel. Tourism is also important, and there is some agricultural production. On the plains of Ayacucho, near the city, Antonio José de Sucre crushingly defeated (Dec. 9, 1824) Spanish forces under Viceroy José de la Serna. The battle secured Peruvian independence and marked the triumph of the revolutionary forces in all South America.

Ay·dın (ī-dĕn'), city (1975 pop. 59,228), capital of Aydın prov., W Turkey, on the Büyük Menderes River. It is the trade center for a farm region where olives, figs, cotton, and tobacco are grown. The city was destroyed by fire in 1922 and has been completely rebuilt.

Ayles·bur·y (ālz'bə-rē), city (1973 est. pop. 41,420), Buckinghamshire, central England. It is an agricultural market for the upper Thames valley and is famous for its ducks. There are printing works and other light industries.

Ayl·mer Lake (āl'mər), 340 sq mi (880.6 sq km), E Mackenzie dist., Northwest Territories, Canada, NE of Great Slave Lake. It is linked to the Clinton-Colden Lake in the east.

Ayr (âr), burgh (1974 est. pop. 47,991), Strathclyde region, SW Scotland, at the mouth of the Ayr River on the Firth of Clyde. Ayr is a sea resort and a port for fishing and the export of coal. It manufactures farm and mining machinery, carpets, asphalt, and shoes. In the heart of the Robert Burns country, Ayr has various Burns memorials. Until 1975 it was the county town of Ayrshire.

Ayr·shire (âr'shĭr', -shər) or **Ayr** (âr), former county, SW Scotland, on the Firth of Clyde. Its county town was Ayr. Under the Local Government Act of 1973, Ayrshire became (1975) part of the Strathclyde region.

A·yut·la (ä-yōōt'lä), town (1970 pop. 23,668), Guerrero state, S Mexico. It is the commercial center for an agricultural, cattle-raising, and lumbering area. The Plan of Ayutla, drawn up in 1854, was a reform program directed toward removing the dictator Santa Anna and convening a constituent assembly to frame a federal constitution.

A·yut·thay·a (ä'yōō-tī'ə), city (1972 est. pop. 46,664), capital of Ayutthaya prov., S central Thailand, on the Chao Phraya River. It is the trade center for a prosperous rice-growing region. Ayutthaya was the capital of a Thai kingdom founded c.1350 and was located on the site of a Khmer settlement. Destroyed by the Burmese in 1559, it was rebuilt by the Siamese in the late 16th cent. but was again devastated by the Burmese in 1767. Ayutthaya has some of the few monuments of early Siamese civilization, notably the royal palace (16th cent.) and numerous temples and pagodas.

Az·ca·po·tzal·co (äs'kä-pōt-säl'kō), city (1970 pop. 545,513), S Mexico, in the Federal District. An important rail center, with railroad yards, it is the terminus of mail and cargo traffic. Its cattle industry supplies the bulk of Mexico City's dairy products. Other industries include auto assembling, oil refining, and the manufacture of textiles, paper, and records. During Mexico's War of Independence, the city was the site (1821) of a major battle in which loyalist troops were forced to retreat by the revolutionary soldiers.

A·zer·bai·jan (ä'zĕr-bī-jän', ăz'ər-), region, c.41,160 sq mi (106,605 sq km), NW Iran. The region is bounded in the north by the Armenian and Azerbaijan SSRs and in the west by Turkey and Iraq. Azerbaijan, which includes Lake Rezaiyeh, is mountainous, with deep valleys and fertile lowlands. Grains, fruits, cotton, and tobacco are grown, and wool is produced. The region has deposits of copper, lead, and iron. There is little modern industry.

By the 8th cent. B.C. Azerbaijan had been settled by the Medes. Azerbaijan is the traditional birthplace (7th cent. B.C.) of Zoroaster, the religious teacher and prophet. After Alexander the Great conquered Persia, he appointed (328 B.C.) as governor the Persian general Atropates, who eventually established an independent dynasty. The region came to be called Atropatene or Media Atropatene and was much disputed. In the 2nd cent. B.C. it was taken by the Parthian Mithradates I, and c.226 A.D. it was captured by the Sassanian Ardashir I. Heraclius, the Byzantine emperor, briefly held the region in the 7th cent., just before the Arabs conquered it, and he converted most of its people to Islam. The Seljuk Turks dominated the region in the 11th and 12th cent., and the Mongols in the 13th cent. After being conquered by Tamerlane in the 14th cent., Tabriz became an important provincial capital of the Timurid empire. The Safavid dynasty arose (c.1500) to renew the state of Persia, and there was fierce fighting between the Ottoman Empire and Persia for Azerbaijan. After brief Ottoman control, Abbas I, shah of Persia, regained control of the region in 1603; the northern part was eventually ceded to Russia in the treaties of Gulistan (1813) and Turkmanchai (1828). In 1938 the remaining province was divided into two parts. In 1941 Soviet troops occupied Iranian Azerbaijan; they were withdrawn (May, 1946) after a Soviet-supported autonomous local government had been created. Iranian troops occupied the region in Nov., 1946, and the autonomous movement was suppressed.

Azerbaijan Soviet Socialist Republic or **A·zer·bai·dzhan** (ä'zĕr-bī-jän', ăz'ər-), constituent republic (1970 pop. 5,111,000), 33,428 sq mi (86,579 sq km), SE European USSR, in Transcaucasia; capital Baku. Strategically situated at the USSR's gateway to southwest Asia, Azerbaijan is bounded by Iran on the south, by the Caspian Sea on the east, by the Dagestan Autonomous Republic on the north, and by the Armenian Republic on the west. Azerbaijan occupies the western ranges of the Greater and Lesser Caucasus and the Kura River valley, which is the region's chief agricultural zone. Wheat, barley, corn, fruits, wine grapes, and potatoes are the leading food crops, and cotton, silk, and tobacco the foremost industrial crops. The republic's mineral resources include oil, natural gas, iron, copper, lead, zinc, limestone, pyrites, cobalt, and alunite. Among the republic's chief manufactures are machinery, electrical equipment, building materials (especially cement), steel, aluminum, chemicals, and textiles. The old craft of carpet weaving is still practiced.

The Azerbaijan SSR comprises the Transcaucasian or northern part of the historic region called Azerbaijan. The territory of the present Azerbaijan SSR was acquired by Russia from Persia through the treaties of Gulistan (1813) and Turkamanchai (1828). Soon after the Bolshevik Revolution of 1917, Russian Azerbaijan joined Armenia and Georgia to form the anti-Bolshevik Transcaucasian Federation. After its dissolution (May, 1918), Azerbaijan proclaimed itself independent but was conquered by the Red Army in 1920 and made into a Soviet republic. In 1922 Azerbaijan joined the USSR as a member of the Transcaucasian SFR. With the administrative reorganization of 1936, it became a separate republic.

Az·i·zi·a (ăz'ĭ-zē'ə): *see* Al Aziziyah, Libya.

A·zores (ə-zôrz', ā'zôrz) or *Portuguese* **A·ço·res** (ä-sō'rĭs), islands (1975 est. pop. 292,200), 905 sq mi (2,344 sq km), in the Atlantic Ocean, c.900 mi (1,448 km) W of mainland Portugal. Administratively a part of Portugal, they are divided into three districts named after their capitals: Ponta Delgada (on São Miguel), Angra do Heroísmo (on Terceira), and Horta (on Fayal). The nine main islands are São Miguel (the largest) and Santa Maria in the southeast; Terceira, Pico, Fayal, São Jorge, and Graciosa in the center; and Flores and Corvo in the northwest. The fertile soil yields many crops and supports vineyards. The islands are also a resort area. The Azores may have been known to the ancients and were included on a map in 1351. Portuguese sailors reached them in 1427 or 1431, but colonization did not begin until 1445. The islands were used as a place of exile and were also the site of naval battles between the English and the Spanish. The United States maintains air bases on the islands.

A·zov (ä-zôf'), city (1976 est. pop. 73,000), SE European USSR, a port on the Don River delta near the Sea of Azov. It is a rail junction and a fishing center and has fish-processing plants. Founded as a Greek colony in the 3rd cent. B.C., it was a trading center and fortress. It came under Kievan Russia in the 10th cent., became a Genoese colony in the 13th cent., and passed to the Turks in 1471. The Don Cossacks held the city (1637–42) but were driven out by the Turks. Peter the Great won the city in 1696 and thus opened southern routes for Russia; he was forced to cede it back to Turkey in 1711. Russia secured Azov definitively in 1774.

Azov, Sea of, N arm of the Black Sea, c.14,000 sq mi (36,260 sq km), S European USSR, in SE Ukraine. The shallow sea (maximum depth 50 ft/15.3 m) is connected with the Black Sea by the Kerch Strait. Its chief arms are the Gulf of Taganrog (in the northeast) and the Sivash Sea (in the west), which is nearly isolated from the Sea of Azov by Arabat Tongue, a narrow sandspit. The Don and Kuban rivers flow into the sea, supplying it with an abundance of fresh water but also depositing much silt that tends to make the sea more shallow. The Sea of Azov has important fisheries and accounts for a large portion of the Soviet freshwater catch.

Az·tec (ăz′tĕk), city (1970 pop. 3,354), seat of San Juan co., NW N.Mex., on the Animas River, in the foothills of the Rocky Mts. near the Colo. border. It is a trade center in a fruit-growing region.

Aztec Ruins National Monument, 27 acres (11 hectares), NW N.Mex., near Farmington; est. 1923. The ruins of a 12th cent. Pueblo Indian town contain interesting kivas, one of which has been completely restored. The ruins were named by early settlers who mistakenly believed that they were built by the Aztec Indians.

A·zu·sa (ə-zo͞o′sə), city (1978 est. pop. 26,100), Los Angeles co., S Calif., in the San Gabriel River valley; inc. 1898. It is a residential and industrial city in a citrus-fruit growing area. Its manufactures include aircraft components, electronic equipment, chemicals, lawn mowers, bicycles, and beer.

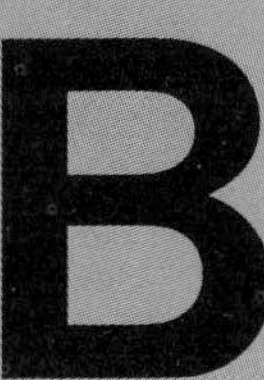

Baal·bek (bäl′bĕk, bā′əl-), ancient city, now in Lebanon, NW of Damascus. Originally it was probably devoted to the worship of Baal, the Phoenician sun god, although no traces of an early Phoenician settlement have survived. The Greeks called the city Heliopolis. It became prominent in Roman days and was made a separate colony by Augustus. The city was sacked by invaders and was destroyed by an earthquake in 1759.

Bab el Man·deb (băb ĕl măn′dĕb), strait, 17 mi (27.4 km) wide, linking the Red Sea with the Gulf of Aden and separating the Arabian Peninsula from E Africa. Control of the strategically located strait was long contested by Britain and France.

Ba·bol (bä-bōl′), town (1971 est. pop. 52,000), N Iran, near the Caspian Sea, NE of Tehran. It is the region's chief commercial center and was once the major trading center of north Iran. Processed food and textiles are produced, and fruits, tobacco, and cotton are raised nearby. Founded in the 16th cent., it was built on the site of an ancient city known as Mamter.

Bab·y·lon (băb′ə-lən, -lŏn′), ancient city of Mesopotamia. One of the most important cities of the ancient Near East, it was on the Euphrates River and was north of the cities that flourished in southern Mesopotamia in the 3rd millennium B.C. It became important when Hammurabi made it the capital of his kingdom of Babylonia. The city was destroyed (c.689 B.C.) by the Assyrians under Sennacherib, and its real splendor began after the city was rebuilt. Babylon became legendary during the time of Nebuchadnezzar (d. 562 B.C.). The Hanging Gardens were one of the Seven Wonders of the World. Among the Hebrews and the later Greeks the city was famed for its sensual living. It was captured (538 B.C.) by Cyrus the Great and was used as one of the administrative capitals of the Persian Empire. In 275 B.C. its inhabitants were removed to Seleucia, which replaced Babylon as a commercial center.

Babylon, residential and resort village (1978 est. pop. 14,200), Suffolk co., SE N.Y., on Long Island, on Great South Bay; settled 1689, inc. as a village 1893.

Bab·y·lo·ni·a (băb′ə-lō′nē-ə), ancient empire of Mesopotamia. The name is sometimes given to the whole civilization of southern Mesopotamia. Historically it is limited to the first dynasty of Babylon established by Hammurabi (c.1750 B.C.), and to the Neo-Babylonian period after the fall of the Assyrian Empire.

Ba·ca (bā′kə, bä′-), county (1970 pop. 5,674), 2,565 sq mi (6,643.4 sq km), SE Colo., bordering on Okla. and Kansas and drained by branches of the Cimarron River; formed 1889; co. seat Springfield. It is in a grain and livestock region.

Ba·cău (bə-kou′), city (1977 est. pop. 126,654), E Rumania, in Moldavia, on the Bistrița River. The administrative and industrial center of an oil-producing region, Bacău has industries that manufacture oil-field equipment. Other important products include textiles, leather and wood items, and light machinery.

Bache Peninsula (bāch), on E Ellesmere Island, in N Northwest Territories, Canada. U.S. explorer Robert Peary proved this area to be a peninsula when he explored (1898) the region. From 1926 to 1933 the Royal Canadian Mounted Police had a post here, c.800 mi (1,285 km) from the North Pole, that was the most northerly habitation in the world.

Back (băk), river, c.600 mi (965 km) long, rising in lakes, E Mackenzie dist., Northwest Territories, Canada, and flowing NE across the tundra to Chantry Inlet, S of the Boothia Peninsula.

Ba·co·lod (bä-kō′lôd), city (1975 est. pop. 196,500), capital of Negros Occidental prov., NW Negros Island, Philippines. It is an important seaport and the shipping and processing center of a major sugarcane area.

Ba·con (bā′kən), county (1970 pop. 8,233), 293 sq mi (758.9 sq km), SE central Ga., bounded on the NE by the Little Satilla River and drained by affluents of the Satilla River; formed 1914; co. seat Alma. Coastal-plain agriculture (tobacco, cotton, corn, and peanuts) and forestry are important to its economy.

Bac·tri·a (băk′trē-ə), ancient Greek kingdom in central Asia. Its capital was Bactra, present-day Balkh in north Afghanistan. Before the Greek conquest (328 B.C.), the region was an eastern province of the Persian Empire. In c.130 B.C. it fell to nomadic tribes.

Ba·cup (bă′kəp), municipal borough (1971 est. pop. 15,102), Lancashire, N England, SE of Blackburn. It has cotton mills, shoe factories, coal mines, and iron and brass foundries. The town is known for its cooperative enterprises.

Ba·da·joz (bä′də-hōz′), city (1975 est. pop. 81,710), capital of Badajoz prov., SW Spain, in Estremadura, on the Guadiana River. It is situated in a fertile agricultural region where food processing is the main industry. Badajoz was an ancient fortress city that rose to prominence under the Moors as the seat (1022–94) of a vast independent emirate. Alfonso IX of León liberated it in 1228. Thereafter Badajoz was repeatedly attacked by the Portuguese and was consequently strongly fortified. In the Peninsular War the French failed to take it in a long siege (1808–9) and succeeded in 1811 only to be driven out by Wellington in 1812 after bitter fighting. In the civil war of 1936–39 the capture (1936) of Badajoz by the Insurgents after a bloody battle was followed by hundreds of executions.

Ba·da·lo·na (bä'də-lō'nə), city (1971 pop. 162,888), Barcelona prov., NE Spain, in Catalonia. It is a Mediterranean port and an important industrial suburb of Barcelona, with textile, chemical, and glass manufactures.

Bad Axe (băd ăks), city (1970 pop. 2,999), seat of Huron co., N Mich., between Saginaw Bay and Lake Huron; settled c.1860, inc. as a village 1885, as a city 1905. It is a farm trade center. Numerous Indian rock drawings are found nearby.

Bad Ems (bät ĕmz'): *see* Ems, West Germany.

Ba·den (bäd'n), former state, SW West Germany. Stretching from the Main River in the northeast across the lower Neckar valley and along the right bank of the Rhine to the Lake of Constance, Baden bordered on France and the Rhenish Palatinate in the west, Switzerland in the south, Hesse in the north, and Bavaria and Württemberg in the east. After World War II, Baden was divided into two parts—in the south, the state of Baden (3,842 sq mi/9,950.8 sq km), occupied by France, and in the north, the state of Württemberg-Baden (1,984 sq mi/5,138.6 sq km), including part of Württemberg, occupied by U.S. armed forces. In 1952 the two states were merged with Württemberg-Hohenzollern to form Baden-Württemberg.

Ba·den (bäd'n) or **Baden-bei-Wien** (-bī-vēn'), city (1971 pop. 22,631), Lower Austria prov., E Austria, on the Schwechat River, near Vienna. The hot sulfur springs of this picturesque city have been frequented since Roman times.

Ba·den (bäd'n), town (1977 est. pop. 13,700), Aargau canton, N Switzerland, on the Limmat River. A noted spa since ancient times, the town has hot sulfur springs. It is also a manufacturing center known for aluminum ware and electrical machinery. The castle of Stein, now in ruins, was once a Hapsburg residence.

Ba·den-Ba·den (bäd'n-bäd'n), city (1975 pop. 50,201), Baden-Württemberg, SW West Germany, in the Black Forest. It is one of Europe's most fashionable spas. The city has many parks and a large casino (built 1821–24). Baden-Baden was founded as a Roman garrison in the 3rd cent.

Ba·den-Würt·tem·berg (bäd'n-wûr'təm-bûrg', vür'təm-bĕrk'), state (1974 est. pop. 9,226,100), 13,803 sq mi (35,750 sq km), SW West Germany; capital Stuttgart. It was formed in 1952 by the merger of Württemberg-Baden, Württemberg-Hohenzollern, and postwar Baden. The state borders on Switzerland in the south, France and the Rhineland-Palatinate in the west, Hesse in the north, and Bavaria in the east.

Bad Go·des·berg (bät gō'dəs-bĕrk'), part of Bonn (since 1969), North Rhine-Westphalia, W West Germany, on the Rhine River. It is the site of numerous foreign embassies and government agencies as well as residences of diplomats and government officials. It is also a resort noted for its radioactive mineral springs.

Bad Hom·burg vor der Hö·he (bät hôm'bŏŏrk fôr dĕr hö'ə), **Bad Homburg,** or **Homburg,** city (1974 est. pop. 51,465), Hesse, central West Germany, at the foot of the Taunus Mts. It is a famous spa and resort. Manufactures include foodstuffs and machinery.

Bad Ischl (bät ĭsh'əl) or **Ischl,** city (1971 pop. 12,740), in Upper Austria prov., W Austria, in the center of the Salzkammergut. A famous spa, it was the summer residence of the Austrian imperial family after 1822.

Bad Kis·sin·gen (bät kĭs'ĭng-ən), town (1970 pop. 12,672), NW Bavaria, West Germany, on the Franconian Saale. It is noted for its mineral springs.

Bad Kreuz·nach (bät kroits'näKH'), city (1974 est. pop. 43,047), Rhineland-Palatinate, W West Germany, on the Nahe River. Its manufactures include precision instruments, tires, glass, and leather. Bad Kreuznach was probably settled in the Stone Age. Its radioactive salt baths have been frequented since Roman times.

bad·lands (băd'lăndz'), area of severe erosion, usually found in semiarid climates and characterized by countless gullies, steep ridges, and sparse vegetation. Badland topography is formed on poorly cemented sediments that have few deep-rooted plants because short, heavy showers sweep away surface soil and small plants. Depressions gradually deepen into gullies. South Dakota's Big Badlands, also known as the Badlands of the White River, are the world's best and most extensive (c.2,000 sq mi/5,180 sq km) example of this topography. Gullies have cut as deep as 500 ft (152.5 m) below the plateau's surface, and differences in rock type have created colorful and spectacular formations. The Big Badlands are famous for fossils of prehistoric animals. Badlands National Monument occupies most of the region.

Bad Nau·heim (bät nou'hīm'), town (1974 est. pop. 25,929), Hesse, central West Germany, in the Taunus Mts. It is a world-famous resort, noted for its salt springs, which are used to treat heart and nerve diseases.

Bad Rei·chen·hall (bät rī'KHən-häl') or **Reichenhall,** town (1974 est. pop. 13,197), Bavaria, SE West Germany, on the Saalach River, near the Austrian border. It is a year-round health resort. Salt has been mined here since Roman times.

Ba·e·za (bä-ā'sä), city (1970 pop. 14,834), Jaén prov., S Spain, in Andalusia. It has varied manufactures. A prosperous Moorish town, it was the scene of several battles between Moors and Spaniards. It was captured definitively (c.1237) by Ferdinand III.

Baf·fin Bay (băf'ĭn), ice-clogged body of water, c.700 mi (1,125 km) long, between Greenland and NE Canada. It connects with the Arctic Ocean to the north and west and with the Atlantic Ocean to the south by way of Davis Strait. Although more than 9,000 ft (2,745 m) deep, navigation in the bay is made hazardous by many icebergs brought there by the Labrador Current. In the 1800s the bay was an important whaling station. The British explorer John Davis was first (1585) to enter the bay, which is named for William Baffin, who explored the area in 1616.

Baffin Island, 183,810 sq mi (476,068 sq km), c.1,000 mi (1,610 km) long and from 130 to 450 mi (209.7–724 km) wide, in the Arctic Ocean, E Northwest Territories, Canada. It is the fifth-largest island in the world and the easternmost member of the Arctic Archipelago. Baffin Island is geographically and geologically a continuation of Labrador, from which it is separated by Hudson Strait. Baffin Island has a deeply indented coastline with many fjords. The western side of the island is covered largely by tundra. In the east, snow-covered mountain ranges rise more than 8,000 ft (2,440 m). Most of the island's inhabitants are Eskimos who live mainly at coastal trading posts. Whaling, fur trading, and fishing are the chief occupations. Martin Frobisher visited the island between 1576 and 1578.

Bagh·dad or **Bag·dad** (both: băg'dăd, bäg-däd'), city (1970 pop. 1,300,000), capital of Iraq, central Iraq, on both banks of the Tigris River. Most of Iraq's industries are in Baghdad; they include the making of carpets, leather, textiles, cement, and tobacco products and the distilling of arrack. The present city was founded (762) on the west bank of the Tigris. Its commercial position became generally unrivaled, and under the caliph Harun ar-Rashid it rose to become one of the greatest cities of Islam. This period of its greatest glory is reflected in the *Thousand and One Nights.* After the death (809) of Harun the seat of the caliph was moved to Samarra; when the caliphate was returned later in the century, Baghdad had already been weakened by internal struggles. In 1258 the Mongols sacked the city and destroyed nearly all of its splendor. It revived but was captured again by Tamerlane (1400) and by the Persians (1524). It was repeatedly contested by Persians and Turks until 1638, when it definitively became part of the Ottoman Empire. By that time the city's population had dwindled from a peak of about 2,000,000 to only a few thousand. Baghdad was captured by the British in 1917. In 1920 it became the capital of the newly constituted kingdom of Iraq. The city was the scene of a coup in 1958 that overthrew the monarchy and established the Iraqi republic. Baghdad is rich in archaeological remains and has several museums.

Bag·ley (băg'lē), village (1970 pop. 1,314), seat of Clearwater co., NW Minn., WNW of Bemidji, in a forest and lake region; settled 1894, inc. 1899. White Earth Indian Reservation is nearby.

Ba·gui·o (bă'gē-ō, bä-gyō'), city (1975 est. pop. 100,200), Mountain prov., NW Luzon, the Philippines. Baguio is the summer capital of the country, with many government buildings. It is also a noted mountain resort situated in beautiful pine forests and is the center of a major gold and copper area. The city is noted for the wood carvings of its Igorot aborigines. Originally settled by the Spanish, Baguio developed only after the American occupation, when a modern city was laid out (1909) and roads were built (the first in 1913) to connect it with the main highways. The city was captured early (Dec., 1941) in World War II by Japanese land forces.

Ba·ha·ma Islands (bə-hä'mə), officially **Commonwealth of the Bahamas,** country (1974 est. pop. 197,000), 4,403 sq mi (11,404 sq km), in the Atlantic Ocean, consisting of 700 islands and islets and about 2,400 cays, beginning c.50 mi (80 km) off SE Florida and extending c.600 mi (965 km) SE almost to Haiti. The country does not include the Turks and Caicos Islands to the southeast, which, although geographically part of the archipelago, have been separately administered by Great Britain since 1848. The capital and principal city is Nassau, on New Providence island. The islands, composed mainly of limestone and coral, rise from a vast submarine plateau. Most of them are generally low and flat, riverless, with many mangrove swamps, brackish lakes, and coral reefs and shoals. Fresh water is obtained from rainfall and desalinization. Navigation is hazardous, and many of the outer islands are uninhabited and undeveloped. Hurricanes occasionally cause severe damage, but the climate is generally excellent. The islands' vivid subtropical atmosphere—brilliant sky and sea, lush vegetation, flocks of bright-feathered birds, and submarine gardens where multicolored fish swim among white, rose, yellow, and purple coral—as well as rich local folklore, has made the Bahamas a popular winter resort.

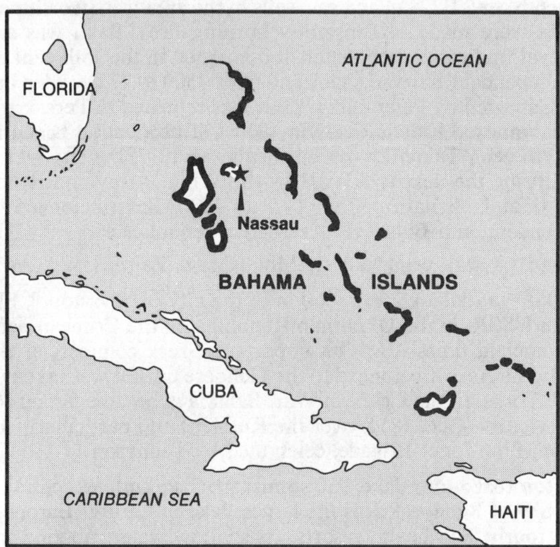

Economy. Tourism, which has grown rapidly since World War II, is by far the country's most important industry, employing a large portion of the population and accounting for most of the foreign exchange. Crawfish, lumber, cement, salt, agricultural products, and handicraft curios are exported. Sugar and oil refining industries have been introduced to diversify the economy and to increase the Bahamas' export trade.

History. Christopher Columbus first set foot in the New World in the Bahamas (1492), presumably at San Salvador, and claimed the islands for Spain. Although the aborigines were soon exterminated, the Spanish did not colonize the islands. The first settlements were made in the mid-17th cent. by the English, who later imported blacks to work cotton plantations. Woodes Rogers, the first royal governor, was appointed in 1717. Under Rogers the pirates and buccaneers, notably Blackbeard, who haunted the Bahama waters, were driven off. In 1781 the Spanish captured Nassau and took possession of the whole colony, but under the terms of the Treaty of Paris (1783) the islands were ceded to Great Britain. Plantation life gradually died out after the emancipation of slaves in 1838. Blockade-running into Southern ports in the U.S. Civil War enriched some islanders, and during the prohibition era in the United States the Bahamas became a base for rum-running. In 1950 the United States signed an agreement with Great Britain for the establishment of a proving ground and a tracking station for guided missiles.

In the 1950s black Bahamians, through the Progressive Liberal party (PLP), began to oppose successfully the ruling white-controlled United Bahamian party; but not until the 1967 elections did they win control of the government. The Bahamas were granted limited self-government in 1964. The PLP, campaigning for immediate independence, won an overwhelming victory in the 1972 elections. On July 10, 1973, the Bahamas became a sovereign state within the Commonwealth of Nations.

Government. The Bahamas are governed by the constitution of 1973 and have a parliamentary form of government. There is a bicameral legislature consisting of a 16-seat Senate and a 38-seat House of Assembly. The prime minister is the head of government, and the monarch of the United Kingdom, represented by an appointed governor-general, is the titular head of state.

Ba·ha·ram·pur (bə-hä'rəm-pôr') or **Ber·ham·pore** (bûr'əm-pôr'), town (1971 pop. 73,380), West Bengal state, E central India. Jute and rice are traded. Its industries include silk weaving, ivory carving, and the production of bell metal. An early uprising in the Indian Mutiny occurred here.

Ba·ha·wal·pur (bə-hä'wəl-po͝or'), city (1972 pop. 133,956), Punjab prov., E central Pakistan, on the Sutlej River. It is a commercial center, trading in wheat, rice, dates, and cotton. Major manufactures are textiles, machinery, and pharmaceuticals. Formerly the capital of the princely state of Bahawalpur, which was founded in the late 18th cent. and acceded to Pakistan in 1947, the city has several palaces and fine buildings.

Ba·hi·a (bä-ē'yə), state (1975 est. pop. 8,438,900), 216,612 sq mi (561,025 sq km), E Brazil, on the Atlantic Ocean; capital Salvador.

Bahia: *see* Salvador, Brazil.

Ba·hí·a Blan·ca (bä-ē'ä bläng'kä), city (1970 pop. 182,158), Buenos Aires prov., SE Argentina, a port near the head of the Bahía Blanca, an inlet of the Atlantic Ocean. It is the main commercial center and principal shipping point of the southern Pampa.

Bah·rain or **Bah·rein** (both: bä-rān', bə-), sheikdom and archipelago (1976 est. pop. 300,000), 231 sq mi (598.3 sq km), in the Persian Gulf. The two main islands, connected by a causeway, are Bahrain, or Aval, and Al Muharraq. The capital and chief port is Al Manamah, on Bahrain. The islands are flat and sandy, with a few low hills. The climate is hot and humid. The population is predominantly Arabic.

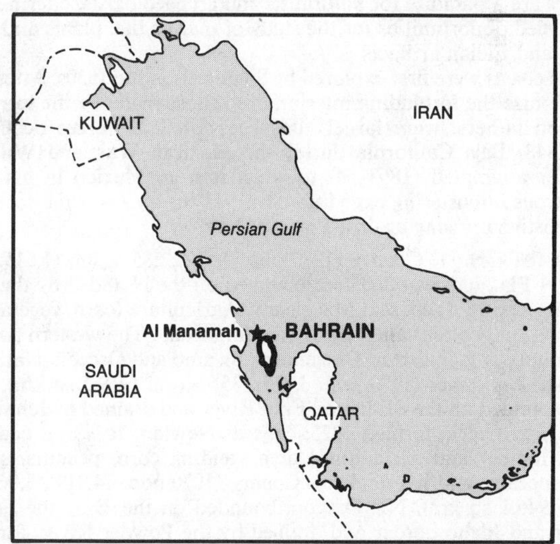

Economy. There is intensive cultivation of dates and alfalfa; cereals, fruits, and vegetables are also grown, and there are poultry and dairy industries. Oil was found in 1931, and oil revenues have financed extensive modernization projects, particularly in health and education. However, Bahrain is expected to be the first Persian Gulf nation to run dry of oil. Ship-repair, aluminum, and turbine-manufacturing industries have been started or planned to diversify the nonagricultural sector of the economy.

History. Bahrain was ruled in the 16th century by Portugal and intermittently from 1602 to 1783 by Persia. The Persians were expelled by an Arabian family that established the presently ruling dynasty of sheiks. In 1861 Bahrain became a British protectorate. In 1971, after Britain withdrew from the Persian Gulf area, Bahrain became independent. In June, 1973, a constitution was adopted limiting the powers of the sheik and granting women the right to vote.

Bahr el Gha·zal (bär ĕl gä-zäl'), region and province, SW Sudan. The region takes its name from a river that flows east to the Bahr el Jebel to form the White Nile. It is an area of swamps and ironstone plateaus. Subsistence agriculture, cattle raising, and game hunting are carried on.

Bahr el Je·bel (bä'hər ĕl jĕb'ĕl), river, 594 mi (955.8 km) long, section of the White Nile, S Sudan, Africa. The name is usually used for the White Nile between Nimule, where it enters the Sudan (as the Albert Nile), and Lake No, where it joins the Bahr el Ghazal.

Bai·kal (bī-käl'), lake: *see* Baykal.

Bai·ley (bā'lē), county (1970 pop. 8,487), 835 sq mi (2,162.7 sq km), NW Texas, on the Llano Estacado bordered on the W by N.Mex. and crossed by the Double Mountain Fork of the Brazos River; formed 1876; co. seat Muleshoe. Its diversified agriculture includes grain sorghums, cotton, wheat, alfalfa, corn, potatoes, truck crops, dairy products, poultry, beef cattle, and hogs.

Bain·bridge (bān'brĭj), city (1970 pop. 10,887), seat of Decatur co., SW Ga., on the Flint River; inc. 1829. It grew up around the site of an Indian trading post and is now a trade and industrial center as well as an inland port and barge terminal. Its manufactures include machinery, clothing, automotive parts, mobile homes, aluminum windows, and molded plastic items. Fertilizers are also produced, and salt is processed.

Baird (bârd), city (1970 pop. 1,538), seat of Callahan co., W central Texas, ESE of Abilene; settled 1880, inc. 1891. Cotton, wheat, and peanuts are processed and shipped, and oil is refined.

Ba·ja Cal·i·for·nia (bä'hə kăl'ə-fôr'nyə, -fôr'nē-ə; *Spanish* bä'hä kä'-lē-fôr'nyə) or **Lower California,** peninsula, c.760 mi (1,225 km) long and from 30 to 150 mi (48.2–241.4 km) wide, NW Mexico, separating the Gulf of California from the Pacific Ocean. The peninsula is divided into the state of Baja California (1970 pop. 856,773), 27,655 sq mi (71,626 sq km), in the north, and the state of Baja California Sur (1970 pop. 123,786), 27,979 sq mi (72,465.6 km), in the south. Except for two large coastal plains on the Pacific side, the peninsula consists largely of rugged mountain ranges averaging 5,000 ft (1,525 m). The land is generally desolate and arid. The only naturally cultivable areas are isolated mountain valleys. The peninsula and surrounding waters are a paradise for naturalists and archaeologists, offering unparalleled opportunities for the study of marine life, plants and animals, and Indian artifacts.

The coasts were first explored by Spaniards in the 1530s. Attempts to colonize the forbidding interior, even those made by the intrepid mission fathers, were largely unsuccessful. U.S. forces occupied (1847–48) Baja California during the Mexican War, and William Walker attempted (1853–54) to wrest it from Mexico in his first disastrous filibustering expedition. In 1911 the area was the scene of an abortive uprising against Porfirio Díaz.

Bak·er (bā'kər). **1.** County (1970 pop. 9,242), 585 sq mi (1,515.2 sq km), N Fla., in a wooded area bounded on the N and E by the Ga. line; formed 1861; co. seat Macclenny. Agriculture (corn, vegetables, peanuts, and cotton) and forestry are important. The western part of the county is included in Okefenokee Swamp and Osceola National Forest. **2.** County (1970 pop. 3,875), 355 sq mi (919.5 sq km), SW Ga., bounded on the SE by the Flint River and drained by Ichawaynochaway Creek; formed 1825; co. seat Newton. It is in a coastal-plain lumber and agricultural area yielding corn, peanuts, sugar cane, pecans, and livestock. **3.** County (1970 pop. 14,919), 3,068 sq mi (7,946.1 sq km), NE Oregon, bounded on the E by the Snake River and Idaho border and drained by the Powder River; formed 1862; co. seat Baker. Its economy depends on lumber, wheat, and deposits of gold, silver, and copper.

Baker. 1. City (1970 pop. 2,584), seat of Fallon co., SE Mont., E of Miles City near the N.Dak. border; inc. 1914. A natural-gas field surrounds the city, which is a trading center for a grain and livestock area. **2.** City (1970 pop. 9,354), seat of Baker co., NE Oregon, SSE of Pendleton in the Powder River valley; laid out 1865 after gold was discovered in the area, inc. 1874. It is the distributing center of a farming, dairying, stock-raising, mining, and lumbering region.

Baker, Mount, peak, 10,778 ft (3,287.3 m) high, NW Wash., in the Cascade Range E of Bellingham.

Baker Lake, c.1,000 sq mi (2,590 sq km), Keewatin dist., Northwest Territories, Canada, W of Chesterfield Inlet of Hudson Bay. It has a post of the Royal Canadian Mounted Police at its western end.

Bak·ers·field (bā'kərz-fēld'). **1.** City (1978 est. pop. 85,500), seat of Kern co., S central Calif., at the S end of the San Joaquin Valley; inc. 1898. It is an oil, mining, and agricultural center. Almost all of the major oil companies have refineries in Bakersfield. Cotton, citrus fruits, potatoes, and roses are grown in the area. Among the city's manufactures are plastics, pharmaceuticals, and processed foods. Gold was discovered in the region in 1855 and petroleum in 1899. Silver, borax, and tungsten mines are also in the vicinity. A branch of California State College is here. **2.** Town (1970 pop. 409), seat of Mitchell co., NW N.C., NW of Spruce Pine. Mica mining and lumbering are done.

Bakh·chi·sa·ray (bäкн'chē-sə-rī'), city (1970 pop. 12,000), SE European USSR, in the Ukraine. From the early 15th cent. until 1783 it was the capital of the khanate of Crimea. The palace of the khans, celebrated for its white marble fountains, was built in the 16th cent. and is now a museum.

Ba·ku (bä-ko͞o'), city (1977 est. pop. 963,000), capital of the Azerbaijan SSR, SE Asian USSR, on the Caspian Sea. Greater Baku includes almost the whole Apsheron peninsula, on which Baku proper is situated. The city is a leading Soviet industrial and cultural center and until World War II was the country's chief petroleum center. It handles one of the greatest volumes of freight (mainly oil and oil products) of any Soviet port. Oil drilling is the major economic activity, and Baku has many oil refineries and factories that produce oil-field equipment. Other important industries include shipbuilding and the processing of food and tobacco.

The city was first mentioned in a 9th cent. chronicle; but as early as the 6th cent. B.C. oil and gas wells in the area were worshiped and shrines were made of constantly burning fires. Baku was a great medieval trade and craft center. It flourished in the 15th cent. under the independent Shirvan shahs and from 1509 to 1723 under Persian rule. Captured by Peter I in 1723, it was returned to Persia in 1735. Russia annexed it definitively in 1806. Oil production began in the late 19th cent. Taken by the Bolsheviks in 1917, the city was occupied during the next two years by the White Army and its foreign allies (mainly Britain). From 1918 to 1920, Baku belonged to the independent, anti-Bolshevik Azerbaijan republic.

Bak·wan·ga (băk-wäng'gä): *see* Mbuji-Mayi, Zaire.

Bal·a·kla·va (băl'ə-klä'və), section of the city of Sevastopol, SE European USSR, in the Ukrainian Republic, on the Crimean peninsula. In ancient times it was an important Greek commercial city. In the Middle Ages it belonged to the Genoese until it was taken (1475) by the Turks. In the Crimean War, Balaklava became famous for an allied victory (Oct., 1854) over the Russians and particularly for the charge of the Light Brigade, celebrated by Tennyson.

Bal·a·ton (băl'ə-tŏn), lake, 230 sq mi (595.7 sq km), W central Hungary, SW of Budapest. It is the largest lake in Central Europe, with many tourist and health resorts. Its shallow waters abound in fish, and along the shores are fine vineyards.

Balch Springs (bôlch), town (1970 pop. 10,464), Dallas co., NE Texas, a residential suburb of Dallas; inc. 1953.

Bald·win (bôld'wĭn). **1.** County (1970 pop. 59,382), 1,578 sq mi (4,087 sq km), SW Ala., on Mobile Bay and the Gulf of Mexico; formed 1809; co. seat Bay Minette. It is drained by the Alabama, Little, Tensaw, and Perdido rivers. In a rich, level farm area, it produces fruit, potatoes, sugar cane, pecans, vegetables, gladiolus bulbs, cattle, hogs, poultry, and timber. Wood products and textiles are manufactured, and food and shellfish are processed. The delta region affords good hunting and fishing. **2.** County (1970 pop. 34,240), 255 sq mi (660.5 sq km), central Ga., drained by the Oconee River; formed 1803; co. seat Milledgeville. It is in a piedmont agricultural area yielding cotton, corn, truck crops, pecans, fruit, and livestock.

Baldwin. 1. Village (1970 pop. 612), seat of Lake co., W central Mich., SW of Cadillac, in a resort region with many lakes and streams. There are Indian village sites and mounds nearby. **2.** Uninc. city (1978 est. pop. 35,100), Nassau co., SE N.Y., on the S shore of Long Island, on Baldwin Bay; settled 1640s. A fishing center and summer resort, it has varied manufactures. **3.** Borough (1978 est. pop. 25,900), Allegheny co., SW Pa., a suburb just S of Pittsburgh, on the Monongahela River, in a bituminous-coal region; inc. 1952. Tools, wood products, and metal goods are manufactured.

Baldwin Park, city (1978 est. pop. 44,500), Los Angeles co., S Calif., a residential suburb of Los Angeles, in the fertile San Gabriel valley; settled 1870, inc. 1956. It has varied manufactures.

Bal·e·ar·ic Islands (băl'ē-ăr'ĭk), archipelago, off Spain, in the W Mediterranean, forming Baleares prov. (1974 est. pop. 564,741) of Spain. Palma is the capital. The chief islands are Majorca, Minorca, and Ibiza. Noted for their scenery and their mild climate, the Balearics are a major tourist center. After tourism, agriculture and fishing are the chief economic activities; fruit, wine, olive oil, majolica ware, and silver filigree are exported.

Inhabited since prehistoric times, the islands were occupied by Iberians, Phoenicians, Greeks, Carthaginians, Romans, and Byzantines. The Moors, who first came in the 8th cent., established (11th cent.) an independent kingdom, which became the seat of powerful pirates, harassing Mediterranean coastal cities and trade. James I of Aragón conquered (1229-35) the islands. They were included (1276-1343) in the independent kingdom of Majorca and reverted to the Aragonese crown under Peter IV.

Ba·ler (bä-lĕr′), municipality (1969 est. pop. 14,000), near the head of Baler Bay, E Luzon, Philippines. Among the finest ports on Luzon's Pacific coast, Baler fell to the Japanese in World War II and was recaptured by American forces on Feb. 12, 1945.

Ba·li (bä′lē), island and (with two offshore islets) province (1971 pop. 2,120,338), c.2,200 sq mi (5,700 sq km), E Indonesia, just E of Java across the narrow Bali Strait. Although Bali is relatively small, it is densely populated and culturally and economically one of the most important islands of Indonesia. Largely mountainous, with active volcanoes, it has a great fertile plain to the south. The Balinese are skillful farmers; rice, the chief crop, is grown with the aid of elaborate irrigation systems. Vegetables, fruits, coffee, and coconuts are also produced. Livestock is important; pigs and cattle are major export items. Industries include food processing, tourism, and handicrafts. The people are noted for their artistic skill (especially wood carving), their physical beauty, and their high level of culture.

Bali was converted to Hinduism in the 7th cent., and was under Javanese rule from the 10th to the late 15th cent. It was a refuge (1513-28) for the Hindus of Java fleeing the advance of Islam. The Dutch first landed in 1597 and the Dutch East India Company began its trade with the island in the early 17th cent. Dutch sovereignty was not firmly established until after a series of colonial wars (1846-49), and the entire island was not occupied until 1908, after the quelling of two rebellions.

Ba·lı·ke·sir (bä′lĭ-kĕ-sĭr′), city (1975 pop. 99,334), capital of Balıkesir prov., NW Turkey. It is a rail junction and the center of a fertile agricultural region.

Ba·lik·pa·pan (bä′lĭk-pä′pän), city (1971 pop. 137,340), E Borneo (Kalimantan), Indonesia, on an inlet of Makasar Strait. An important seaport and oil center with refineries, it is connected by pipeline with the oil fields of Samarinda. Timber is also exported.

Bal·kan Peninsula (bôl′kən), southeasternmost peninsula of Europe, c.200,000 sq mi (518,000 sq km), bounded by the Black Sea, Sea of Marmara, Aegean Sea, Mediterranean Sea, Ionian Sea, and Adriatic Sea. Although there is no sharp physiographic separation between the peninsula and central Europe, it is commonly considered to include Albania, continental Greece, Bulgaria, European Turkey, most of Yugoslavia, and southeastern Rumania. These six countries, successors to the Ottoman Empire, are called the Balkan States. Historically and politically the region includes all of Yugoslavia and Rumania. The peninsula is very mountainous. The mild Mediterranean-type climate, with its dry summer period, is limited to the southern and coastal areas. Covering a greater area are the humid subtropical climate in the northwest and the harsher humid continental climate in the northeast. The region as a whole is largely agricultural; fruits, grains, and grazing are important. A variety of mineral deposits are found here, including iron ore, coal, manganese, copper, lead, and zinc.

Balkans, major mountain range of the Balkan Peninsula and Bulgaria, extending c.350 mi (565 km) from E Yugoslavia through central Bulgaria to the Black Sea. It rises to 7,794 ft (2,377.2 m) at Botev, the highest peak. The Balkans are a continuation of the Carpathian Mts. The forested range is sparsely populated and rich in a variety of minerals. It acts as a climatic barrier, preventing the inland penetration of Mediterranean influences.

Balkh (bälkh), town (1967 pop. 15,000), N Afghanistan, on a dried-up tributary of the Amu Darya River. One of the world's oldest cities, it is the legendary birthplace of the prophet Zoroaster. Alexander the Great reputedly founded a Greek colony at the site c.328 B.C. The city later attained great wealth and importance as Bactra, capital of the independent kingdom of Bactria. In the early centuries A.D., Balkh, a prominent center of Buddhism, was renowned for its Buddhist monasteries and stupas. Conquered by the Arabs in 653, the city was sacked in 1221 by Genghis Khan and lay in ruins until Tamerlane rebuilt it (early 16th cent.). In 1850, Balkh became part of the unified kingdom of Afghanistan. The old city, sections of whose walls remain, is now mostly in ruins; the new city, some distance away, is an agricultural and commercial center.

Bal·khash (bäl-käsh′), city (1976 est. pop. 78,000), W Central Asian USSR, in Kazakhstan, on the N shore of Lake Balkhash. A railroad terminus, port, and copper-smelting center, it was founded as Bertys in 1929 and was renamed in 1936.

Balkhash, Lake, 6,562 sq mi (16,996 sq km), c.350 mi (565 km) long, maximum width c.45 mi (70 km), Kazakhstan, Central Asian USSR. The lake, which has an average depth of 20 ft (6 m), stretches from the Kazakh Hills in the northeast to desert steppes in the southwest. The eastern half of the lake is saline; the western half, separated from the eastern section by a sandbar and fed by the Ili River, is fresh. Lake Balkhash, which has no outlet, is slowly shrinking from evaporation.

Bal·la·rat (băl′ə-răt′), city (1976 pop. 37,863; urban agglomeration pop. 58,500), Victoria, SE Australia. It is an industrial center; clothing, food products, paper, brick and tile, and other goods are made. The city flourished during the gold rush (1860s), then declined.

Bal·lard (băl′ərd), county (1970 pop. 8,276), 259 sq mi (670.8 sq km), SW Ky., bounded on the SW by the Mississippi (at the Mo. border), on the W, NW, and N by the Ohio River (at the Ill. border), and on the S by Mayfield Creek; formed 1842; co. seat Wickliffe. In a gently rolling upland region, it has agriculture (tobacco, corn, and potatoes), clay pits, and timber.

Bal·lin·ger (băl′ən-jər), city (1970 pop. 4,203), seat of Runnels co., W central Texas, on the Colorado River SSW of Abilene; laid out 1886 with the coming of the railroad, inc. 1892. Processing plants handle diversified farm products, and there are oil wells and petrochemical plants in the area.

Ball·ston Spa (bôl′stən spä′), village (1970 pop. 4,968), seat of Saratoga co., E N.Y., SSW of Saratoga Springs, in a farm area; settled 1771, inc. 1807. There are mineral springs nearby.

Ball·win (bôl′wĭn), city (1970 pop. 10,656), St. Louis co., E Mo., a suburb of St. Louis; settled 1803 as Ballshow, renamed 1837, inc. 1950. It is mainly residential with some light industry.

Bal·ly·me·na (băl′ē-mē′nə), municipal borough (1971 pop. 16,487), Co. Antrim, NE Northern Ireland, on the Braid River. Linen, woolen goods, carpets, and tobacco products are produced here.

Bal·sam Lake (bôl′səm), village (1970 pop. 648), seat of Polk co., NW Wis., on Balsam Lake W of Rice Lake, in a dairying, farming, and resort area.

Bal·sas, Rí·o (rē′ō bäl′säs), river, c.450 mi (725 km) long, rising in the state of Puebla, E central Mexico. One of Mexico's longest rivers, it flows in a curve from south to northwest through Puebla and Guerrero states, where it waters a fertile valley, to Michoacán state, then turns southwest, passing through a hot, dry region, before emptying into the Pacific Ocean.

Bal·tic Sea (bôl′tĭk), arm of the Atlantic Ocean, c.163,000 sq mi (422,170 sq km), including the Kattegat Strait, its NW extension. The Øresund, Store Baelt, and Lille Baelt connect the Baltic Sea with the Kattegat and Skagerrak straits, which lead to the North Sea; the Kiel Canal, across the Jutland peninsula, is a more direct connection with the North Sea. The Gulf of Bothnia, the Gulf of Finland, and the Gulf of Riga are the chief arms of the Baltic Sea. Most of the Baltic is shallow, and its tides are less pronounced than those of the North Sea. The salinity of the sea is reduced by the many rivers that enter it, and parts of the sea freeze over in winter. The Baltic was frequented from ancient times, especially because of the amber found along the coast. In the late Middle Ages commerce on the Baltic was dominated by the Hanseatic League.

Baltic Shield, the continental core of Europe, composed of Precambrian crystalline rock, the oldest of Europe. The exposed portion of the Baltic Shield is found in Finland, Sweden, and Norway. During the Pleistocene epoch, great continental ice sheets scoured and depressed the shield's surface, leaving a thin covering of glacial material and innumerable lakes and streams. The ancient rocks have yielded a rich variety of minerals, especially iron and copper.

Baltic states, the countries of Estonia, Latvia, and Lithuania, bordering on the E coast of the Baltic Sea. Formed in 1918, they remained independent republics until their incorporation in 1940 into the USSR.

Bal·ti·more (bôl′tə-môr′), county (1970 pop. 620,409), 598 sq mi (1,548.8 sq km), N Md., bounded on the N by the Pa. border, on the SE by the Gunpowder River and Chesapeake Bay, and on the S and SW by the Patapsco River; formed 1659; co. seat Towson. It almost surrounds independent Baltimore city, which was separated from Baltimore co. in 1851. It is mainly a suburban area, with truck, fruit, and dairy farming, heavy and light industry, summer resorts, and large estates.

Baltimore, city (1978 est. pop. 814,000), N Md., surrounded by but politically independent of Baltimore co., on the Patapsco River estuary, an arm of Chesapeake Bay; inc. 1745. It is a port of entry, a commercial and industrial center, an important railroad point, and a great seaport with extensive anchorages and dock and storage facilities. Large amounts of coal and grain, and iron, steel, and copper products are exported. Among Baltimore's leading industries are shipbuilding, sugar and food processing, copper and oil refining, and the manufacture of chemicals, steel, clothing, aerospace equipment, fertilizer, and tin cans.

The site was first settled in the early 17th cent., but the city was not founded until 1729, when the provincial assembly authorized the building of a town. The excellent harbor soon made Baltimore an important center for the shipping of tobacco and grain. Shipbuilding, an early industry, flourished during the Revolution and the War of 1812, and in the early 1800s the famous Baltimore clippers were built. When the British occupied (1777) Philadelphia, Baltimore became the meeting place of the Continental Congress. In the War of 1812 the gallant defense of Fort McHenry inspired Francis Scott Key to write The Star-Spangled Banner. After the War of 1812, Baltimore experienced a phenomenal growth, largely because of the National Road. When the Erie Canal (completed in 1825) endangered the city's hold on the trans-Allegheny traffic, Baltimore businessmen chartered (1827) the Baltimore & Ohio Railroad to meet the competition of New York as a new ocean outlet for the West. During the Civil War, Baltimore was strongly pro-Southern in sentiment. In World Wars I and II, Baltimore was an important shipbuilding and supply-shipping center.

An important cultural and educational center, Baltimore is the seat of The Johns Hopkins Univ. and the Univ. of Baltimore. The city's many historical attractions include Flag House; the first Roman Catholic cathedral in the United States (1806–21; designed by B. H. Latrobe); the Edgar Allan Poe House (c.1830); Westminster Churchyard, where Poe is buried; Fort McHenry National Monument and Historic Shrine; the Baltimore and Ohio Transportation Museum; and numerous colonial homes. The U.S.S. *Constellation,* a national historic shrine, is docked in Baltimore.

Bal·ti·more-Wash·ing·ton Park·way (bôl'tə-môr'wŏsh'ĭng-tən pärk'wā'): *see* National Parks and Monuments Table.

Ba·lu·chi·stan (bə-lo͞o'chĭ-stăn'), region and province (1972 pop. 2,409,000), c.134,000 sq mi (347,060 sq km), Pakistan, bounded by Iran on the W, by Afghanistan on the N, and by the Arabian Sea on the S; capital Quetta. Baluchistan is largely desert land with inarable hills and mountains. Some cotton is raised and processed, and natural gas is exploited. On the coast there is trade in fish and salt.

Ba·ma·ko (bä'mä-kō'), city (1972 est. pop. 237,000), capital of Mali and of its Bamako region, SW Mali, on the Niger River. It is a river port, a railroad junction, and a major regional trade center. Manufactures include textiles, processed meat, and metal goods. Bamako was a leading center of Moslem learning under the Mali empire (c.11th–15th cent.) but by the 19th cent. had declined into a small village. In 1883 it was occupied by French troops. In 1908 Bamako became the capital of the French Sudan.

Bam·berg (băm'bûrg', bäm'bo͝orkʜ), city (1974 est. pop. 75,358), Bavaria, S West Germany, a port on the Regnitz River. It is an industrial and commercial center; its manufactures include textiles, clothing, electrical equipment, machinery, and beer. Bamberg was the capital of a powerful ecclesiastical state from 1007 to 1802. In 1803 it passed to Bavaria. Noteworthy buildings in the picturesque city include the cathedral (built mostly in the 13th cent.) and a Gothic church (14th cent.).

Bam·berg (băm'bûrg'), county (1970 pop. 15,950), 395 sq mi (1,023 sq km), S central S.C., bounded on the N by the South Fork of the Edisto River; formed 1897; co. seat Bamberg. It has agriculture (cotton, corn, soybeans, tobacco, hay, livestock, dairy products, and poultry), timber, and textile mills.

Bam·berg (băm'bûrg'), town (1970 pop. 3,406), seat of Bamberg co., S central S.C., SW of Orangeburg, in a farm and timber area; inc. 1855. Textile products are manufactured.

Ba·mi·an (bä'mē-än', bəm-yän'), town (1969 est. pop. 48,000), N central Afghanistan, in the Bamian valley on the Kunduz River. It was long a major caravan center on the route between India and central Asia. By the 7th cent. the town was a prominent center of Buddhism; the Bamian valley is lined with cave dwellings cut out of the cliffs by Buddhist monks. Particularly interesting are two great Buddha figures (probably 6th or 7th cent.) carved from rock and finished in fine plaster. A Moslem fortress town from the 9th to the 12th cent., Bamian was sacked by Genghis Khan in 1221 and never regained its former prominence.

Ba·na·ba (bə-nä'bə): *see* Ocean Island.

Ba·nat (bä-nät', bä'nät), region extending across W Rumania, NE Yugoslavia, and S Hungary, bordered on the E by Transylvania and Walachia, on the W by the Tisza River, on the N by the Mureşul River, and on the S by the Danube. Except for some eastern mountains, it is primarily an agricultural area of fertile, rolling plains. Inhabited since prehistoric times, the Banat was occupied successively by Romans, Goths, Huns, and Avars. Slavs began to settle there in the 5th cent. and Magyars in the 9th cent. After the Turkish occupation of Serbia (1459), many Serbs emigrated to the Banat, which itself became a Turkish sanjak (province) around 1552. By the Treaty of Passarowitz (1718), the Banat was made an Austrian military frontier zone. In 1779 the Banat passed to Hungary, to which it belonged until 1918, except for a brief period as an Austrian crownland. Although the Allies in World War I had promised through a secret agreement to give the Banat to Rumania, it was divided by the Treaty of Trianon (1920) between Rumania and newly independent Yugoslavia.

Ban·bur·y (băn'bĕr'ē, -bə-rē, băm'-), municipal borough (1973 est. pop. 31,060), Oxfordshire, central England, on the Cherwell River. It is an agricultural market and manufactures aluminum, fabricated steel, farm machinery, electrical apparatus, and furniture. The town still produces the spiced currant cakes for which it has been famous since the 17th cent.

Ban·da Islands (băn'də, bän'dä), group of 10 volcanic islands, c.70 sq mi (180 sq km), E Indonesia, in the Banda Sea, in the Moluccas. Nutmeg and mace are the chief products. The islands were discovered and claimed by the Portuguese in 1512. The Dutch ousted the Portuguese in the early 1600s, and the Dutch East India Company assumed control in 1619.

Ban·da Ori·en·tal (bän'dä ōr'yān-täl'), region, S Uruguay. An alluvial plain, it is Uruguay's principal livestock-raising and wheat-growing region. In the Spanish colonial period Banda Oriental was the term applied to Uruguay.

Ban·dar (bŭn'dər): *see* Masulipatam, India.

Ban·dar Ab·bas (bän-där' ä-bäs'), town (1971 est. pop. 38,000), S Iran, on the Strait of Hormoz at the mouth of the Persian Gulf. A port of strategic and commercial importance, it is the focal point of the trade routes of southern Iran. The town has food-processing and textile industries; cotton, rugs, nuts, and dates are exported. Early in the 16th cent. the Portuguese established themselves in the region, seizing the islands in the strait and using the town as a fortified mainland port. Shah Abbas I recaptured (c.1615) the town and later the islands. The Dutch (without the shah's consent) and the English (with the shah's approval) subsequently set up trading stations here.

Ban·dar-e Pah·la·vi (bän-där'ā pä-lə-vē'), city (1966 pop. 41,785), Gilan prov., NW Iran, a port on the Caspian Sea. It has fisheries and exports food products, cotton, fish, and caviar.

Ban·dar Se·ri Be·ga·wan (bän'där sĕr'ē bə-gä'wən), city (1971 pop. 36,574), capital of the sultanate of Brunei.

Ban·da Sea (băn'də, bän'dä), section of the Pacific Ocean, c.600 mi (965 km) long and c.300 mi (485 km) wide, E Indonesia, outlined by the South Molucca Islands. The deepest point is c.21,000 ft (6,405 m). The sea's reefs and currents are a hazard to shipping.

Ban·dei·ra (bän-dā'rä), highest peak of Brazil, 9,482 ft (2,892 m) high, situated on the border between Minas Gerais and Espírito Santo states, SE Brazil.

Ban·de·lier National Monument (băn-də-lîr'): *see* National Parks and Monuments Table.

Ban·de·ra (băn-dâr'ə), county (1970 pop. 4,747), 765 sq mi (1,981.4 sq km), SW central Texas, on the Edwards Plateau and drained by the Sabinal and Medina rivers; formed 1856; co. seat Bandera. It is in a hilly ranching area, with dude ranches, hunting, fishing, and some agriculture (grain sorghums, hay, corn, pecans, and poultry).

Bandera, resort city (1970 pop. 891), seat of Bandera co., SW Texas, on the Medina River NW of San Antonio. Founded by Mormons in 1854 and later a Polish settlement, this ranch and farm center has a notable museum containing pioneer relics.

Ban·djar·ma·sin (bän'jər-mä'sĭn), city (1971 pop. 281,673), capital of South Kalimantan prov., S Borneo (Kalimantan), Indonesia, on a delta island near the junction of the Barito and Martapura rivers. An important deep-water port, it is the trade center of the rich Barito basin; exports include rubber, pepper, timber, oil, coal, gold, and diamonds. There is a large oil refinery, and coal mines and sawmills are in the vicinity. In the 14th cent. Bandjarmasin was part of a Hindu kingdom, but it passed to Moslem rulers in the late 15th cent. The Dutch opened trade here in 1606. The British controlled the city for several brief periods, and in 1787 it became a Dutch protectorate.

Ban·dung or **Ban·doeng** (both: bän'do͝ong), city (1971 est. pop. 1,201,730), capital of West Java prov., W Java, Indonesia. Formerly the administrative and military headquarters of the Netherlands East Indies, it is the third-largest city in Indonesia, an industrial hub, a famous educational and cultural center, and a tourist resort known for its cool, healthful climate. Bandung is a textile center and the site of the country's quinine industry. Other manufactures include ceramics, chemicals, rubber products, and machinery. Founded by the Dutch in 1810, Bandung became important with the arrival of the railroad in the late 19th cent.

Banff (bămf), town (1971 est. pop. 3,500), SW Alta., Canada, on the Bow River in the Rocky Mts. It is a famous tourist center and a winter resort.

Banff·shire (bămf'shîr', -shər) or **Banff**, former county, NE Scotland. Its county town was Banff. Banffshire was torn by religious strife after the Reformation, and troubles continued throughout the English Civil War. After the Glorious Revolution (1688–89), the county was strongly Jacobite. In 1975 Banffshire became part of the new Grampian region.

Ban·ga·lore (băng'gə-lôr'), city (1971 pop. 1,648,232), Karnataka state, S central India, 3,000 ft (915 m) above sea level. A major industrial center and transportation hub of southern India, Bangalore has electronics and aircraft industries, textile mills, and varied manufactures. Coffee is traded. A well-planned city with numerous parks and wide streets, it is famous as a place of retirement. It was founded in 1537. Bangalore became the administrative seat of Mysore in 1831.

Bang·gai (bäng'gī), archipelago, 1,222 sq mi (3,165 sq km), off E Celebes Island, Indonesia, in the Molucca Sea. Resources include rice, sago, trepang, resin, and rattan.

Bang·ka or **Ban·ka** (both: bäng'kä, băng'kə), island (1971 pop. 384,000), c.4,600 sq mi (11,915 sq km), Indonesia, in the Java Sea, SE of Sumatra, from which it is separated by the narrow Strait of Bangka. Since c.1710, when tin was discovered, Bangka has been one of the world's principal tin-producing centers. Pepper is also produced on the island. Bangka was ceded to Britain in 1812, but in 1814 it was exchanged with the Dutch for Cochin in India.

Bang·kok (băng'kŏk'), city (1972 est. pop. 3,133,834), capital of Thailand and of Phra Nakhon prov., SW Thailand, on the E bank of the Chao Phraya River, near the Gulf of Siam. Thailand's largest city and one of the leading cities of Southeast Asia, Bangkok lies in the heart of the country's major commercial rice-growing region. Rice, tin, teak, rubber, gold, silver, hides, and processed fish are the leading exports of the city's port. Industrial plants include rice mills, cement factories, sawmills, oil refineries, and shipyards. Textiles, motor vehicles, electrical goods, and food products are also manufactured. The city is a famous jewelry trading center, dealing in silver and bronze ware and precious stones.

The city began as a small trading center and port community serving Ayutthaya, the capital of Siam, until its destruction by Burmese invaders in 1767. Thon Buri became the capital in 1769, but in 1782, King Rama I, founder of the Chakkri dynasty, built his royal palace on the east bank of the river and made Bangkok his capital. During World War II the city was occupied by the Japanese and was a target of Allied bombing raids.

Bang·la·desh (bäng-lä-dĕsh', băng-), republic (1976 est. pop. 82,900,000), 55,126 sq mi (142,776 sq km), S Asia. Dacca is the capital. Bangladesh was called East Pakistan until Dec., 1971. It borders on the Bay of Bengal in the south, the Indian states of West Bengal in the west and Assam in the north, and Burma in the southeast. A humid, low-lying, alluvial region, Bangladesh is composed mainly of the great combined delta of the Ganges, Brahmaputra, and Meghna rivers and is laced with numerous streams, distributaries, and tidal creeks, forming an intricate network of waterways that constitutes the country's chief transportation system. Bangladesh has a tropical monsoon climate with a short dry season in the winter. The low-lying delta region is subject to severe flooding from monsoon rains, cyclones, and tidal waves and usually suffers major crop damage and high loss of life.

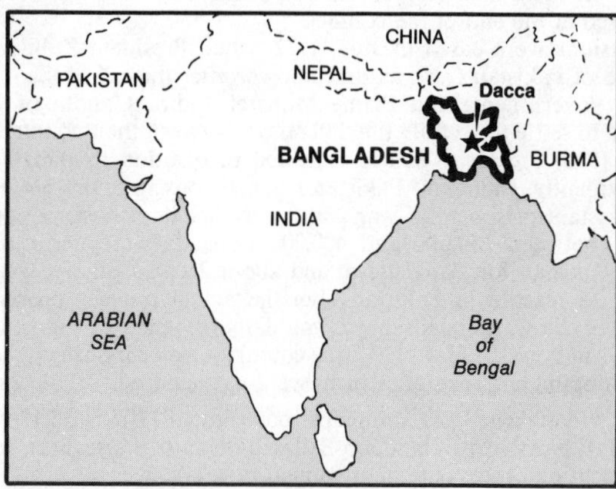

Economy. Except for natural gas (found along its eastern border) and oil (in the Bay of Bengal), Bangladesh is lacking in minerals. The country's economy is based on agriculture, with about 65% of the territory under cultivation. Jute, rice, sugar cane, tea, tobacco, and wheat are the chief crops. Bangladesh produces more than half of the world's raw jute. Fishing is also an important economic activity. Jute products, textiles, paper, processed food, and leather goods are manufactured. Raw jute and jute products account for about 90% of the country's exports, which also include tea, leather, and fish. Since the country is unable to feed itself, the most important of Bangladesh's imports is food. Raw cotton, transportation equipment, and consumer goods are other major imports.

History. Until 1757, when Robert Clive, the British statesman who laid the basis of the British Empire in India, defeated the Nawab Suraj-ud-daulah at Plassey, the area was ruled by Afghan or Mogul dynasties or by independent Moslem kings. Baber, who took Kabul in 1504 and then advanced through the northwest, established the Mogul empire in India in 1526. Thereafter, with interruptions, the Mogul empire united India until 1857. Like India and Pakistan, the area was part of imperial British India from 1857 until 1947, when India and Pakistan achieved independence; for nearly 25 years afterward, Bangladesh existed as East Pakistan, the eastern province of Pakistan.

The two provinces of Pakistan, which differed considerably in natural setting, economy, and historical background, were separated from each other by more than 1,000 mi (1,610 km) of India. The East Pakistanis, 56% of the total population, were discontented under a government centered in West Pakistan; the disparity in government investments and development funds given to each province also were troubling, especially since the eastern province's jute and tea sales supplied two-thirds of the country's foreign earnings. Efforts over the years to secure increased economic benefits and political reforms proved unsuccessful, and serious riots broke out in 1968 and 1969. The movement for greater autonomy gained momentum when, in the Dec., 1970, general elections, the Awami League under the leadership of Sheik Mujibur Rahman achieved a majority in the Pakistan National Assembly. President Mohammed Agha Yahya Khan, hoping to avert a political confrontation between East and West Pakistan, twice postponed (Mar., 1971) the opening session of the national assembly. The government's attempts to forestall the autonomy bid finally led to civil war on Mar. 25. On the following day the Awami League's leaders proclaimed the independence of Bangladesh. Yahya Khan's government outlawed the Awami League, imprisoned Sheik Mujibur Rahman on treason charges, and imposed strict press censorship. During the months of conflict an estimated one million Bengalis were killed in East Pakistan and another ten million fled into exile in India.

Finally India allied itself with Bangladesh, which it had recognized on Dec. 6, and during a two-week war (Dec. 3–16) defeated the Pakistani forces in the east. The Awami League leaders of Bangladesh's provisional government returned from exile in Calcutta, India. Sheik Mujibur Rahman, chosen president while in prison in West Pakistan, was released and allowed to return to Bangladesh in early Jan., 1972. He set up a government and assumed the premiership. Rejecting Pakistan's call for a reunited country, the sheik embarked upon the tremendous job of rehabilitating an economy devastated by months of warfare. Relations with Pakistan were hostile: Pakistan withheld recognition, and Bangladesh and India refused to repatriate more than 90,000 Pakistani prisoners of war who had surrendered at the end of the conflict.

Tensions were eased in July, 1972, when President Zulfikar Ali Bhutto of Pakistan (who assumed power after the fall of the Yahya Khan government) and Prime Minister Indira Gandhi of India agreed to settle peacefully the differences between their countries. In Feb., 1974, President Bhutto extended recognition to Bangladesh. Subsequently, India and Pakistan reached consensus on the release of Pakistani prisoners of war and the exchange of hostage populations—between 150,000 and 400,000 Bengalis were permitted to leave Pakistan for Bangladesh, and about 260,000 Biharis were allowed to resettle in Pakistan—but the actual transfer procedures were very slow. Bangladesh was gradually recognized by most of the world's nations. In Mar., 1972, the country's major industries, banks, and shipping and insurance firms were nationalized.

Ban·gor (băng'gôr, -gər), municipal borough (1971 pop. 35,178), Co. Down, E Northern Ireland, on Belfast Lough. It is a seaport, resort, and yachting center (site of an annual regatta).

Bangor 1. City (1978 est. pop. 31,900), seat of Penobscot co., S Maine, at the confluence of the Penobscot and Kenduskeag rivers; settled 1769, inc. as a town 1791, as a city 1834. It is a port of entry, commercial center, and gateway to an extensive resort and lumber region. Major industries include the production of shoes and paper, food and lumber processing, and printing. During the War of 1812 it was occupied by the British. In the 19th cent. Bangor was a shipbuilding center that carried on an extensive coastal and overseas trade in lumber, stone, and ice. **2.** Industrial borough (1970 pop. 5,425), Northampton co., E Pa., between Easton and Stroudsburg; founded 1773, inc. 1875. There are slate quarries nearby.

Bangor, municipal borough (1973 est. pop. 16,030), Gwynedd, NW Wales, at the N end of Menai Strait. Slate is shipped from adjacent Port Penrhyn. The cathedral, on the site of a 6th cent. church, dates from the 11th cent. and has been rebuilt several times.

Ban·gui (bäng-gē'), city (1971 est. pop., with suburbs, 187,000), capital of the Central African Empire, a port on the Ubangi River, near the Zaire border. Bangui is an administrative, trade, and communications center. Its manufactures include textiles, food products, beer, shoes, and soap. Bangui's port handles most of the country's international trade; the chief exports are cotton, timber, coffee, and sisal. The city was founded in 1889.

Bang·we·u·lu (băng'wē-ōō'lōō) or **Bang·we·o·lo** (băng'wē-ō'lō), lake and swamps, c.3,800 sq mi (9,840 sq km), NE Zambia. The lake is c.50 mi (80 km) long and 25 mi (40 km) wide. Commercial fishing is pursued in the lagoons of the swamps. The swamps are formed largely by the flooding of the lower Chambezi River, which enters Lake Bangweulu from the east. The lake is drained in the south by the Luapula River, a tributary of the Congo.

Ba·ni Ha·san (bä'nē hä-sän'), or **Be·ni Hasan** (bĕ'nē), village, E central Egypt, on the Nile near Al Minya. There are 39 tombs, carved out of solid rock in the XII dynasty of ancient Egypt.

Ba·ni Su·wayf (bä'nē sōō-wāf') or **Be·ni Su·ef** (bĕ'nē sōō-wāf'), city (1970 est. pop. 99,400), N central Egypt, on the Nile River. Situated in an intensely cultivated farming region, Bani Suwayf has cotton mills and sugar refineries. Alabaster is quarried near the city.

Ban·ja Lu·ka (bän'yä lōō'kä), city (1971 pop. 89,866), W Yugoslavia, in Bosnia, on the Vrbas River. It has varied manufactures, including iron goods and electrical equipment. Banja Luka was captured by the Turks in 1528. Later (1878-1918) a part of Austria-Hungary, it passed to Yugoslavia after World War I.

Ban·jul (bän'jōōl), formerly **Bath·urst** (băth'ərst), port city (1974 est. pop. 41,047), W Gambia, situated on St. Mary's Island where the Gambia River enters the Atlantic Ocean. It is the country's economic and administrative center. Its port handles oceangoing ships. Banjul's chief export is peanuts; beeswax, palm kernels and oil, and skins and hides are also shipped. Peanut processing is the chief industry. The city was founded by the British in 1816 as a trading post and a base for suppressing the slave trade.

Ban·ka (bäng'kä, băng'kə): *see* Bangka, Indonesia.

Banks (băngks), county (1970 pop. 6,833), 231 sq mi (598.3 sq km), NE Ga., drained by the headstreams of the Broad River; formed 1858; co. seat Homer. It is in a piedmont agricultural area yielding cotton, corn, hay, sweet potatoes, and fruit.

Banks Island, c.26,000 sq mi (67,340 sq km), NW Northwest Territories, Canada, in the Arctic Ocean and the Arctic Archipelago. It is the westernmost of the group and is separated from the mainland by Amundsen Gulf. Banks Island, which has many lakes, is a hilly plateau rising to c.2,000 ft (610 m) in the south. British explorer Sir Robert McClure discovered that it was an island in 1851.

Banks·town (băngks'toun), city (1976 pop. 155,830), New South Wales, SE Australia. It is a suburb of Sydney.

Bann (băn), longest river of Northern Ireland, rising as the Upper Bann in the Mourne Mts. and flowing 40 mi (64.4 km) NW through Counties Down and Armagh to the S end of Lough Neagh. It leaves the lake at its north shore as the Lower Bann and flows 40 mi (64.4 km) north to the Atlantic Ocean.

Ban·ner (băn'ər), county (1970 pop. 1,034), 738 sq mi (1,911.4 sq km), W Nebr., in an agricultural area bordering on Wyo. and drained by branches of the North Platte River; formed 1888; co. seat Harrisburg. It has oil and natural-gas wells.

Ban·ning (băn'ĭng), resort city (1970 pop. 12,034), Riverside co., S Calif., in a fruit-growing area between Mt. San Jacinto and Mt. San Gorgonio; inc. 1913. Electronic equipment, wearing apparel, plastics, and metal products are manufactured. An annual stagecoach day festival is held, and the city has a stagecoach museum.

Ban·nock (băn'ək), county (1970 pop. 52,200), 1,122 sq mi (2,906 sq km), SE Idaho, in a mountain area drained by the Bear and Portneuf rivers; formed 1893; co. seat Pocatello. Its irrigated fields produce potatoes, wheat, alfalfa, sugar beets, and dairy goods.

Ban·nock·burn (băn'ək-bûrn', băn'ək-bûrn'), moor and parish, central Scotland, on the Bannock River. Textiles are manufactured in the parish. In 1314 on the moor, a 10,000-man Scots army led by Robert Bruce routed 23,000 Englishmen under Edward II, thus climaxing Robert's struggle for Scottish independence and establishing him as king of the Scots.

Ban·ská Bys·tri·ca (băn'skä bĭs'trĭ-tsä'), city (1975 est. pop. 55,832), E central Czechoslovakia, in Slovakia, at the junction of the Bystrica and Hron rivers. It is an industrial center noted for the large plywood, pulp, and veneer factories nearby. An ancient town, Banská Bystrica was the heart of the Slovak national uprising against German occupation in 1944.

Ban·stead (băn'stĕd'), urban district (1971 pop. 44,986), Surrey, SE England, on the North Downs. The district is mainly residential and contains some highly regarded landscapes. There is a church from the Norman period and an excavated Roman villa. The area is mentioned in the Domesday Book.

Ban·try Bay (băn'trē), inlet of the Atlantic Ocean, 21 mi (33.8 km) long and 4 mi (6.4 km) wide, Co. Cork, SW Republic of Ireland.

Bar·a·boo (băr'ə-bōō'), city (1970 pop. 7,931), seat of Sauk co., central Wis., on the Baraboo River NW of Madison; founded before 1850, inc. as a village 1867, as a city 1882. It is a manufacturing and trade center in an agricultural, timber, and resort area. The Ringling Brothers' Circus began here, and the Circus World Museum was opened in 1959. Devils Lake State Park and the Dells of the Wisconsin are nearby.

Ba·ra·co·a (bä'rä-kō'ä), city (1970 pop. 20,926), Oriente prov., SE Cuba, a port near the E extremity of the island. Bananas and coffee are exported. Founded c.1512 by the Spanish explorer Diego de Velázquez, Baracoa is the oldest settlement in Cuba.

Bar·a·ga (băr'ə-gə), county (1970 pop. 7,789), 904 sq mi (2,341.4 sq km), NW Mich., in the Upper Peninsula partly bounded on the N by Keweenaw and Huron bays and drained by the Sturgeon and Silver rivers; formed 1875; co. seat L'Anse. In a dairy, orchard, and farm area, it has extensive lumbering, commercial fishing, and resorts. L'Anse and Vieux Desert Indian Reservation, a state park, and many small lakes are here.

Ba·ra·ho·na (bä'rä-ō'nä), city (1970 pop. 37,260), SW Dominican Republic, on Neiba Bay, an arm of the Caribbean Sea. Barahona has a lumber industry and is a commercial and processing center for an agricultural region.

Bar·a·nof (băr'ə-nôf'), island off SE Alaska in the Alexander Archipelago. It is more than 100 mi (160 km) long and has an area of 1,607 sq mi (4,162.1 sq km). On it is Sitka, founded by Aleksander Baranov, for whom the island is named.

Ba·ra·no·vi·chi (bə-rä'nə-vyĕ'chē), city (1976 est. pop. 123,000), Belorussia, W European USSR. It is a major railway junction and has industries that manufacture machinery, metalware, and textiles. Founded as a railway station in 1870, Baranovichi passed from the Soviet Union to Poland in 1920. In 1939 Baranovichi was again incorporated into the USSR.

Bar·ba·dos (bär-bā'dōz), island state (1970 pop. 238,141), 166 sq mi (430 sq km), in the West Indies; capital Bridgetown. The island, east of St. Vincent, in the Windward Islands, is low and rises gradually toward its highest point at Mt. Hillaby (1,104 ft/336.7 m). Although there is ample rainfall from June to December, there are no rivers, and water must be pumped from subterranean caverns. The porous soil and moderate warmth are excellent for the cultivation of sugar cane, long the island's major occupation. Other exports include mo-

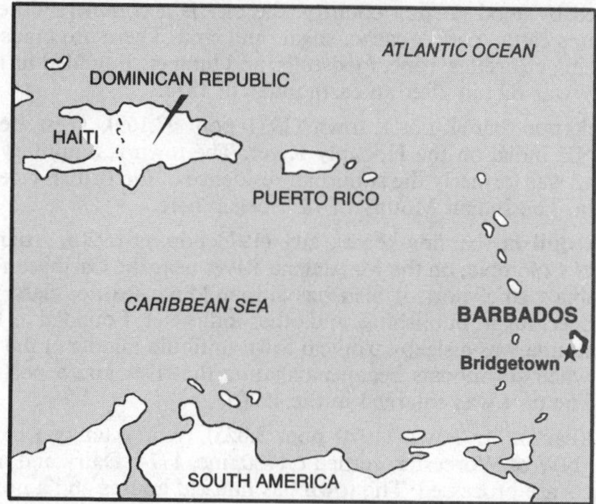

lasses and rum. Commercial fishing is also important. Tourism is the country's largest source of foreign exchange.

Although it was probably discovered by the Portuguese and named Los Barbados for the bearded fig trees they found, the first definite settlement was made by English expeditionaries in 1627 (1605, according to local tradition). Barbados remained a British colony until independence was granted in 1966. During the 19th cent. it was the administrative headquarters of the Windward Islands, but in 1885 it became a separate colony.

Bar·ba·ry Coast (bär'bə-rē), waterfront area of San Francisco, Calif., in the years after the 1849 gold rush. Gamblers, gangsters, prostitutes, and confidence men flourished, and the brothels, saloons, and disreputable boardinghouses made the Barbary Coast—named after the pirate coast of North Africa—notorious throughout the world.

Barbary States, term used for the North African states of Tripolitania, Tunisia, Algeria, and Morocco. The corsair Barbarossa led the Turkish conquest to prevent the region from falling to Spain. A last attempt by Holy Roman Emperor Charles V to drive out the Turks failed in 1541. The piracy carried on thereafter by the Moslems of North Africa began as part of the war against Spain. In the 17th and 18th cent., when the Turkish hold on the area grew weaker, the raids became less military and more commercial in character. The booty, ransom, and slaves that resulted from attacks on Mediterranean towns and shipping became the main source of revenue for local rulers. All the major European naval powers made attempts to destroy the corsairs, though, on the whole, countries trading in the Mediterranean found it more convenient to pay tribute than to undertake the expensive task of eliminating piracy. Toward the end of the 18th cent. the power of the piratical states diminished. The United States and the European powers took advantage of this decline to launch more attacks. American opposition resulted in the Tripolitan War. In 1830 France, after a three-year blockade of Algiers, began the conquest of Algeria. The Ottoman Turks were able to reassert (1835) direct control over Tripolitania and end piracy there. About the same time the sultans of Morocco were forced by France, Great Britain, and Austria to give up plans to rebuild the Moroccan fleet, and North African piracy was at an end.

Bar·ber (bär'bər), county (1970 pop. 7,016), 1,146 sq mi (2,968.1 sq km), S Kansas, in a rolling plain bordered on the S by the Okla. border and drained by the Medicine Lodge River; formed 1873; co. seat Medicine Lodge. It has gas and oil fields, gypsum mines, and agriculture (cattle and wheat).

Bar·ber·ton (bär'bər-tən), city (1978 est. pop. 26,700), Summit co., NE Ohio, an industrial suburb of Akron, on the Tuscarawas River; inc. 1892. Automobile tires and other rubber products are among its manufactures.

Bar·bour (bär'bər). **1.** County (1970 pop. 22,543), 899 sq mi (2,328.4 sq km), SE Ala., in the Black Belt bounded on the E by Lake Eufaula and the Ga. border, on the W by the Pea River; formed 1832; co. seat Clayton. Cotton, peanuts, livestock, and peanuts are produced, and textiles are manufactured. **2.** County (1970 pop. 14,030), 341 sq mi (883.2 sq km), N W.Va., on the Allegheny Plateau, drained by the Tygart River; formed 1843; co. seat Philippi. Its agricultural products include livestock, dairy, fruit, truck crops, poultry, and grain. Bituminous coal is mined, and lumber processed.

Bar·bour·ville (bär'bər-vĭl'), city (1970 pop. 3,549), seat of Knox co., SE Ky., in a valley of the Cumberland River NW of Middlesboro, in a coal, timber, farming, and horse-breeding area; founded 1800.

Bar·ce·lo·na (bär'sə-lō'nə), city (1974 est. pop. 1,816,623), capital of Barcelona prov. and chief city of Catalonia, NE Spain, on the Mediterranean Sea. Situated on a plain between the Llobregat and Besós rivers and lying between mountains and the sea, Barcelona is the second-largest city of Spain, its largest port, and its chief commercial and industrial center. Textiles, machinery, automobiles, locomotives, airplanes, and electrical equipment are the chief manufactures.

It was founded by the Carthaginians and, according to tradition, derives its name from the Barca family of Carthage. The city flourished under the Romans and Visigoths, fell to the Moors (8th cent.), and was taken (801) by Charlemagne, who included it in the Spanish March. In the 9th to 10th cent. the March became independent under the leadership of the powerful counts of Barcelona, who wrested lands to the south from the Moors, thus acquiring all Catalonia. The counts also won suzerainty over several fiefs in south France. The marriage of Count Raymond Berengar IV to the heiress of Aragón united (1137) the two lands under one dynasty. Reaching its peak around 1400, the city later declined, but enjoyed a period of prosperity as the embarkation point of the armies of Emperor Charles V. It was repeatedly (1640–52, 1715, 1808–14) occupied by the French. Barcelona was always the stronghold of Catalan separatism and was the scene of many insurrections. Later it also became the Spanish center of socialism, anarchism, syndicalism, and other radical political beliefs. It was the capital of the Catalan autonomous government (1932–39) and the seat of the Spanish Loyalist government from Oct., 1938, until its fall to Franco on Jan. 26, 1939.

Present-day Barcelona is the cultural center of Spain. A handsome modern city, it has broad avenues, bustling traffic, and striking new buildings. Its old city, with winding, narrow streets (Roman walls are still visible), has many historic structures, including the imposing Cathedral of Santa Eulalia (14th–15th cent.) with its fine cloisters.

Bards·town (bärdz'toun'), city (1970 pop. 5,816), seat of Nelson co., central Ky., SE of Louisville, in a rich farm area; settled 1775, inc. 1788. The city has distilleries, flour and lumber mills, and clothing factories.

Bard·well (bärd'wĕl'), city (1970 pop. 1,049), seat of Carlisle co., SW Ky., near the Mississippi River SW of Paducah. It is the trade center for a fertile tobacco and cotton area.

Bar·ents Island (băr'ənts, bär'-), island of Svalbard, 513 sq mi (1,328.7 sq km), in Barents Sea between Spitsbergen and Edgeøya. The island rises to 1,302 ft (397.1 m).

Barents Sea, arm of the Arctic Ocean, N of Norway and E USSR, partially enclosed by Franz Josef Land on the N, Novaya Zemlya on the E, and Svalbard on the W. Its waters are warmed by the North Atlantic Drift, so that its ports are ice-free all year.

Bar Harbor, town (1970 pop. 3,716), Hancock co., SE Maine, on Mount Desert Island and on Frenchman Bay; settled 1763, inc. 1796. It was one of the most famous resorts in New England during the 19th cent. Bar Harbor is a port of entry, with ferry connections to Yarmouth, N.S., Canada, during the summer.

Ba·ri (bä'rē), city (1976 est. pop. 384,374), capital of Bari prov. and of Apulia, S Italy, on the Adriatic Sea. It is a major seaport and an industrial and commercial center. Manufactures include chemicals, textiles, printed materials, and petroleum. Probably of Illyrian origin, Bari became a Greek and then a Roman colony. It later was controlled by the Goths, the Lombards, and the Byzantines. The Normans conquered Bari in 1071. The city was badly damaged in World War II. Noteworthy buildings include the Romanesque basilica (1087-1197), a major place of pilgrimage, with relics of St. Nicholas of Bari; the Romanesque cathedral (12th cent.); and the Hohenstaufen castle (1233).

Ba·ri·sal (bə-rĭ-säl'), city (1974 pop. 98,127), S Bangladesh, on the Ganges River delta. It is an important river port, a transshipment point for jute and rice, and a market for betel nuts and fish. There are also flour, rice, oilseed, and jute mills. The "Barisal guns," unexplained sounds resembling distant thunder or cannon, are a curious local phenomenon; they may have a seismic origin.

Ba·ri·san Mountains (bä'rē-sän'), volcanic range, c.1,000 mi (1,610 km) long, paralleling the W coast of Sumatra island, Indonesia. It rises to Mt. Kerintji (12,467 ft/3,802.4 m high).

Bar·king (bär'kĭng), borough (1971 pop. 160,499), Greater London, SE England. Barking was created in 1965 by the merger of portions of the municipal boroughs of Barking and Dagenham. The borough has automobile, engineering, chemical, and wood industries.

Bar-le-Duc (bär-lə-dük'), town (1975 pop. 19,288), capital of Meuse dept., NE France, in Lorraine. It has textile mills, iron foundries, printing plants, and metallurgical and food-processing industries. Situated in the picturesque Ornain valley, Bar-le-Duc has many old houses (16th, 17th, and 18th cent.). Until the 15th cent., it was the capital of the county (later duchy) of Bar, an irregularly shaped area stretching from the Marne to the Luxembourg frontier.

Bar·let·ta (bär-lĕt'tä), city (1976 est. pop. 78,991), Apulia, S Italy, on the Adriatic Sea. It is a seaport and a commercial and industrial center. Salt is mined nearby, and wine is produced. Barletta passed to the Goths after the fall of the Roman Empire. Later controlled by the Byzantines and the Lombards, it became a Norman city in the 12th cent.

Bar·na·ul (bär'nä-o͞ol'), city (1977 est. pop. 522,000), Altai Kray, SW Siberian USSR, on the Ob River. A port and major railway junction, Barnaul is in the heart of an agricultural area where wheat, corn, and sugar beets are grown. The city's chief industries produce cotton textiles, artificial fibers, and machinery. Barnaul was founded in 1771 as a silver-smelting center.

Barnes (bärnz), county (1970 pop. 14,669), 1,479 sq mi (3,830.6 sq km), SE N.Dak., in an agricultural area drained by the Sheyenne River; formed 1871; co. seat Valley City. Wheat, oats, barley, and potatoes are grown. It has some manufacturing.

Barnes·ville (bärnz'vĭl'), city (1970 pop. 4,935), seat of Lamar co., W central Ga., NW of Macon; settled c.1825, inc. 1854. It is a processing center in a cotton, pecan, and lumber area. Textiles, hosiery, and furniture are manufactured.

Bar·net (bär'nĕt, -nĭt), borough (1971 pop. 303,578) of Greater London, SE England. The borough was created in 1965 by the merger of the urban districts of Barnet, East Barnet, and Friern Barnet, and the municipal boroughs of Finchley and Hendon. Although mainly residential, the borough manufactures automobile and aircraft parts, electrical components, and beverages. At the Battle of Barnet (1471) during the Wars of the Roses, Edward IV of the House of York defeated the Lancastrian Richard Neville, earl of Warwick. Warwick died in the fighting.

Barns·ley (bärnz'lē), county borough (1976 est. pop. 224,400), South Yorkshire, N England. It is the railroad center of a coal region and has ironworks, linen mills, and other industries.

Barn·sta·ble (bärn'stə-bəl), county (1970 pop. 96,656), 393 sq mi (1,017.9 sq km), SE Mass., coextensive with Cape Cod; formed 1685; co. seat Barnstable. It is a summer resort area with many fine beaches. There is some agriculture and small industry. In the early 19th cent. it was a shipbuilding, shipping, and whaling center.

Barnstable, town (1970 pop. 19,842), seat of Barnstable co., SE Mass.; inc. 1639. It is a resort town on Cape Cod. Candles are produced here. Barnstable is made up of seven villages, including Hyannis. Points of interest include the home of the Revolutionary War patriot James Otis, in West Barnstable; the John F. Kennedy Memorial, in Hyannis; and several 18th cent. buildings.

Barn·sta·ple (bärn'stə-pəl), municipal borough (1973 est. pop. 17,820), Devonshire, SW England, on the Taw River estuary. The river is spanned here by a 16-arch stone bridge dating from the 13th cent. Barnstaple is a marketing town and tourist center. Gloves, pottery, bricks, tiles, furniture, and lace are manufactured.

Barn·well (bärn'wĕl), county (1970 pop. 17,176), 553 sq mi (1,432.3 sq km), W S.C., bounded on the W by the Savannah River, on the NE by the South Fork of the Edisto River; formed 1785; co. seat Barnwell. It has agriculture (asparagus, melons, cucumbers, and cotton) and a lumbering industry. Naval stores are produced.

Barnwell, city (1970 pop. 4,439), seat of Barnwell co., SW S.C., SSW of Columbia, in an agricultural area; settled 1798, inc. 1829. In the Civil War Gen. W. T. Sherman's army burned the town (1865).

Ba·ro·da (bə-rō'də), former native state, now incorporated in Gujarat state, W central India. Its chief city, Baroda (1971 pop. 467,422), on the Vishvamitri River, has cotton-textile, chemical, and metal industries, an oil refinery, and a fertilizer plant.

Bar·qui·si·me·to (bär'kē-sē-mā'tō), city (1971 pop. 330,815), capital of Lara state, NW Venezuela, on the Pan American Highway. Surrounded by good grazing country, the city is a commercial center that ships cattle, coffee, cacao, sugar, and sisal. There are industries producing cigarettes, rope, foodstuffs, and lumber. Founded in 1552, the city was rebuilt after an earthquake in 1812.

Bar·rack·pur (băr'ək-pôr'), town (1971 pop. 97,169), West Bengal state, NE India, on the Hooghly River. The town is a military station and was formerly the suburban residence of the British viceroys of India. The Indian Mutiny of 1857 began here.

Bar·ran·quil·la (bär'äng-kē'yä), city (1973 pop. 661,920), Atlántico dept., N Colombia, on the Magdalena River near the Caribbean Sea. Colombia's chief port, it also has shipbuilding, textile, glass, perfume, beer, sugar, publishing, and other industries. Founded in 1629, Barranquilla was a sleepy tropical town until the middle of the 19th cent., when steamboats began navigating the river and a port was built. The port was enlarged in the 1920s.

Bar·re (băr'ē). **1.** Town (1970 pop. 3,825), Worcester co., central Mass., NW of Worcester; settled c.1720, inc. 1774. Dairy and metal products are processed. The town has fine old houses and a notable village square. **2.** City (1978 est. pop. 9,500), Washington co., central Vt., SE of Montpelier; settled late 18th cent., inc. 1894. Granite quarrying, which began in the region in the early 19th cent., is still the largest industry.

Barre des E·crins (bär' dā-zā-krăɴ'), peak, 13,461 ft (4,105.6 m) high, in the Pelvoux group, SE France, tallest of the Dauphiné Alps.

Bar·ren (băr'ən), county (1970 pop. 28,677), 486 sq mi (1,258.7 sq km), S Ky., bounded on the SW by the Barren River and drained by several creeks; formed 1798; co. seat Glasgow. It has agriculture (livestock, tobacco, grain, dairy products, and fruit), oil and gas wells, hardwood timber, and stone quarries. There is some manufacturing. Part of Mammoth Cave National Park and other limestone caves are in the northwest part of the county.

Barren Grounds or **Barren Lands**, arctic prairie region, N Canada, NW of Hudson Bay and E of the Mackenzie River basin. Its altitude above sea level seldom reaches above 600 ft (183 m), and its geology is largely that of the Laurentian Plateau. The soil is thin, and large areas of bare rock are exposed. Drainage is inadequate, and depressions are filled with swamps, ponds, and lakes. Vast herds of caribou cross the area seasonally, and musk oxen are common in the north. The Barren Grounds were crossed by Samuel Hearne in 1770–71.

Bar·rie (băr'ē), city (1976 pop. 34,389), S Ont., Canada, on the W shore of Lake Simcoe. The city is in a mixed farming and dairying district. Leather goods, packaged meats, electrical appliances, and other goods are made.

Bar·ring·ton (băr' ĭng-tən). **1.** Village (1976 est. pop. 9,410), Cook and Lake cos., NE Ill., a suburb NW of Chicago; founded 1850, inc. 1865. It has research industries and manufactures food products and missile and aircraft parts. **2.** Suburban borough (1970 pop. 8,409), Camden co., SW N.J., SE of Camden; laid out c.1890, inc. 1917. Fiberglass products and scientific and educational materials are produced. **3.** Town (1970 pop. 17,554), Bristol co., E R.I., on the Barrington River; settled c.1670, included in Mass. until 1746, inc. 1770. It is a residential and resort area. Barrington College is here.

Bar·ron (băr'ən), county (1970 pop. 33,955), 866 sq mi (2,243 sq km), NW Wis., in an area containing many lakes and drained by the Red Cedar and Hay rivers; formed 1859 as Dallas co., renamed 1869; co.

seat Barron. It has agriculture, extensive dairying, food processing, and some industry.

Barron, city (1970 pop. 2,337), seat of Barron co., NW Wis., at the confluence of the Vermillion and Yellow rivers NW of Eau Claire; founded 1860, inc. 1887. It is a commercial and processing center for a farming and dairying area.

Bar·row (băr'ō), county (1970 pop. 16,859), 171 sq mi (442.9 sq km), NE central Ga., bounded on the S by the Apalachee River; formed 1914; co. seat Winder. It is in a piedmont agricultural area yielding cotton, corn, hay, sweet potatoes, and fruit.

Barrow, city (1970 pop. 2,104), N Alaska; inc. 1958. It is the main trade center of northern Alaska, and its population is predominantly Eskimo. Hunting of whales and polar bears, basketry, carving in ivory and bone, and the production of mukluks (sealskin or reindeer-skin boots worn by Eskimos) are important. A U.S. naval arctic research laboratory is in the city, and an air force installation, part of the Distant Early Warning Line, is nearby.

Bar·row-in-Fur·ness (băr'ō-in-fûr'nĭs), county borough (1976 est. pop. 73,400), Cumbria, NW England, on the tip of the Furness peninsula. Shipbuilding is the largest single industry. There are also immense steelworks, diesel-engine factories, armaments plants, smelting works, sawmills, and flour and paper mills. Deposits of iron ore, discovered in the late 19th cent. and responsible for the city's growth, are now much depleted.

Barrow-North Slope, borough (1970 pop. 3,451), 57,587 sq mi (149,150 sq km), N Alaska; formed after 1961; borough seat Barrow. It lies in a hunting, trapping, and oil region. In 1972 North Slope became coextensive with Barrow-North Slope.

Bar·ry (băr'ē). **1.** County (1970 pop. 38,166), 549 sq mi (1,421.9 sq km), SW Mich., in a farm area drained by the Thornapple River; organized 1839; co. seat Hastings. It produces livestock, dairy products, poultry, grain, potatoes, beans, and corn and has some light industry. There are lake resorts and a state game area here. **2.** County (1970 pop. 19,597), 783 sq mi (2,028 sq km), SW Mo., in the Ozarks drained by the White River; formed 1835; co. seat Cassville. It has agriculture (berries, apples, tomatoes, cattle, poultry, and dairy products), hardwood timber, and coal, zinc, and lead deposits. Part of Mark Twain National Forest is here.

Barry, municipal borough (1976 est. pop. 42,780), South Glamorgan, S Wales, on the Bristol Channel. It is one of Great Britain's great coal-exporting ports; cement, flour, and steel products are also exported. The leading imports are timber, grain, sand, and oil. Barry has large storage and ship-repair facilities. Other industries are flour milling, mechanical engineering, and the manufacture of clothing, plastics, and silicone. Barry is also a seaside resort.

Bar·stow (bär'stō), city (1978 est. pop. 17,300), San Bernardino co., SE Calif., on the dry Mojave River; founded in the 1880s as a silver-mining town, inc. 1947. Railroad shops, the Goldstone interplanetary tracking station, and nearby U.S. Marine Corps supply centers are major employers. Barstow is an outfitting point for expeditions into Death Valley.

Bar·thol·o·mew (bär-thŏl'ə-myōō'), county (1970 pop. 57,022), 402 sq mi (1,041.2 sq km), S central Ind., drained by the East Fork of the White River and its tributaries; formed 1821; co. seat Columbus. It has agriculture (grain, corn, tomatoes, livestock, and dairying), timber, and diversified manufacturing.

Bar·tles·ville (bär'tlz-vĭl'), city (1978 est. pop. 30,100), seat of Washington co., NE Okla., on the Caney River; inc. 1897. It is a distribution center for a ranching and rich oil-producing area. Petroleum production, marketing, and research have been major enterprises since the first well was tapped in 1897. Of interest are the Price Tower, a concrete and glass building with cantilevered floors, designed by Frank Lloyd Wright, and the Nellie Johnstone oil well, a replica of the first commercial oil well in the state.

Bart·lett (bärt'lĭt). **1.** Village (1976 est. pop. 6,510), Cook and Du Page cos., NE Ill., a suburb NW of Chicago; inc. 1891. It has diversified light industry. **2.** Village (1970 pop. 140), seat of Wheeler co., NE central Nebr., WSW of Norfolk, in a livestock and grain area.

Bar·ton (bär'tn). **1.** County (1970 pop. 30,663), 892 sq mi (2,310.3 sq km), central Kansas, in a sloping plain drained by the Arkansas River; formed 1872; co. seat Great Bend. It has extensive oil fields. Wheat and livestock are raised here. **2.** County (1970 pop. 10,431), 594 sq mi (1,538.5 sq km), SW Mo., drained by a branch of the Spring River and affluents of the Osage River; formed 1855; co. seat Lamar. It has agriculture (wheat, corn, hay, and livestock) and coal deposits.

Bar·tow (bär'tō), county (1970 pop. 32,911), 461 sq mi (1,194 sq km), NW Ga., in a valley and ridge area drained by the Etowah River; formed 1832; co. seat Cartersville. Farming (cotton, hay, sweet potatoes, corn, fruit, and livestock), mining, and manufacturing (textiles and lumber) are important to its economy.

Bartow, city (1970 pop. 12,891), seat of Polk co., central Fla.; inc. 1882. The economy is based on the production of phosphate and the raising of citrus fruit and cattle. Bartow was established in 1853 on the site of a fort built in the Seminole War.

Ba·sel (bä'zəl) or **Basle** (bäl), canton, N Switzerland, bordering on France and West Germany. The canton has been divided since 1833 into two independent half cantons—Basel-Land (1977 est. pop. 219,200), 165 sq mi (427.4 sq km), with its capital at Liestal, and Basel-Stadt (1977 est. pop. 210,700), 14 sq mi (36.3 sq km), virtually coextensive with the city of Basel (1977 est. pop. 188,800) and its suburbs. Basel is the chief rail junction and river port of Switzerland. It is also a financial center. The city is the seat of the Swiss chemical and pharmaceutical industry and of the Swiss Industries Fair; it also has an important publishing industry. Other products are metal goods, foodstuffs, and silk textiles.

Founded by the Romans as Basilia, Basel passed successively to the Alemanni, the Franks, and to Transjurane Burgundy. In the 11th cent. it became a free imperial city and the residence of prince-bishops. The celebrated Council of Basel met here in the mid-15th cent. Basel joined the Swiss Confederation in 1501. The oppressive rule of the city's patriciate over the rest of the canton led to revolts (1831–33) and the eventual split into two cantons. One of the oldest intellectual centers of Europe, Basel was the residence of Froben, Erasmus, Holbein the Younger, Calvin, Nietzsche, and the Bernoulli family. Among the city's noted structures are the cathedral (consecrated 1019), in which Erasmus is buried; the medieval gates; several guild houses; the 16th cent. town hall; and an art gallery with a valuable collection of Holbein's works.

Bash·kir Autonomous Soviet Socialist Republic (băsh-kîr') or **Bash·kir·i·a** (băsh-kîr'ē-ə), autonomous region (1970 pop. 3,819,-000), 55,444 sq mi (143,600 sq km), E European USSR, in the S Urals, occupying the Belaya River basin; capital Ufa. Bashkiria has petroleum, natural gas, coal, salt, iron, gold, copper, zinc, bauxite, and manganese deposits. Sawmilling and the production of plywood and paper are important. Grains (especially wheat, rye, and oats) are the chief agricultural products. In 1919 Bashkiria was made the first autonomous Soviet republic.

Ba·sil·don (bā'zəl-dən), urban district (1976 est. pop. 141,700), Essex, E England. The southern portion is Basildon New Town, a planned community with many factories. There are light engineering, chemical, and joinery works, milk-bottling and printing plants, and clothing and carbon-black factories.

Ba·si·li·ca·ta (bä-zē-lē-kä'tä), region (1971 pop. 602,389), 3,856 sq mi (9,987 sq km), S Italy, bordering on the Tyrrhenian Sea in the SW and on the Gulf of Taranto in the SE. It forms the instep of the Italian "boot." Potenza is the capital. The region is crossed by the Lucanian Apennines. Because of a dry climate and a scarcity of ground water, farming is difficult, although it is the occupation of most inhabitants of the generally poor region. Olives, plums, and cereals are grown, and sheep and goats are raised. There is also some fishing. The transportation network is very limited, and commerce and industry are minimal. Rome took the region in 272 B.C.; it later passed in turn to the Lombards, to the Byzantines, and (11th cent.) to the Norman duchy of Apulia. Although later a part of the kingdom of Naples, Basilicata was controlled by virtually independent feudal lords.

Ba·sin (bā'sən), town (1970 pop. 1,145), seat of Big Horn co., N Wyo., on the Bighorn River SW of Sheridan, in an oil, farm, and livestock area.

Ba·sing·stoke (bā'zĭng-stōk'), municipal borough (1973 est. pop. 60,910), Hampshire, S central England, on the North Downs. Formerly a market town trading in silk and woolens, it now has several industries, including the manufacture of agricultural machinery, precision tools and instruments, leather, clothing, and drugs. Basingstoke is mentioned as a royal manor in the Domesday Book.

Basque Provinces (băsk), including the provinces of Álava, Guipúzcoa, and Vizcaya, N Spain, S of the Bay of Biscay and bordering on France in the NE. The region includes the western Pyrenees and is bounded in the southwest by the Ebro River. It is crossed by the Cantabrian Mts. In a wider sense the name also applies to other territories largely inhabited by Basques—Spanish Navarre and Pyrénées-Atlantique dept. in France. In the densely populated coastal provinces of Vizcaya and Guipúzcoa the chief occupations are the mining of iron, lead, copper, and zinc, and metalworking, shipbuilding, and fishing. Álava is primarily agricultural; corn and sugar beets are grown, and wine and apple cider are made.

Bas·ra (bäs′rə, bŭs′-), city (1965 pop. 313,327), SE Iraq, on the Shatt al Arab. Basra is the only port in Iraq. Its commercially advantageous location, near oil fields and the Persian Gulf, has made it prosperous. Since 1948 many oil refineries have been built in the city. Petroleum products, grains, wool, and dates are exported. Basra was founded A.D. 636. Its possession was long contested by the Persians and the Turks.

Bas-Rhin (bä′răN′), department (1975 pop. 882,121), 1,848 sq mi (4,786.3 sq km), E France, in N Alsace; capital Strasbourg.

Bas·sa·no del Grap·pa (bä-sä′nō dĕl gräp′pä), city (1976 est. pop. 37,162), Venetia, NE Italy, on the Brenta River. It is an agricultural, commercial, and industrial center. First mentioned c.998, the city came under Venetian rule in 1404. In Sept., 1796, Napoleon I defeated the Austrians here. The Da Ponte family of painters, called the Bassano family after the city, had a flourishing school here in the 16th cent., and many of their works remain in the city.

Bas·sein (bə-sēn′, -sān′), town (1970 est. pop. 136,000), S Burma, on the Bassein River. Lying at the western edge of the Irrawaddy delta, Bassein is a rice-milling and export center; teak and bamboo are also handled. The British established a fort at Bassein in 1852.

Basse·terre (bäs-târ′), town (1970 est. pop. 13,055), capital of Saint Kitts-Nevis, on St. Kitts Island, British West Indies. It is one of the chief commercial depots of the Leeward Islands. Sugar refining is the leading industry. Basseterre was founded by the French in 1627.

Basse-Terre, town (1969 est. pop. 16,000), capital of Guadeloupe dept., French West Indies. It is a port that ships the products of the surrounding agricultural area. Founded by the French in 1643, it retains its French colonial atmosphere.

Bas·sett (băs′ĭt), village (1970 pop. 983), seat of Rock co., N central Nebr., W of Ainsworth. It is a trading and shipping center of a grain and dairying area.

Bass Strait (băs), channel, 80 to 150 mi (128.7-241.4 km) wide, between Tasmania and Victoria, SE Australia, connecting the Indian Ocean and Tasman Sea. It is an important fishing area. The discovery of the strait by English explorer George Bass in 1798 proved that Tasmania was not a part of the Australian continent.

Bas·ti·a (bäs-tē′ä), city (1975 pop. 50,718), NE Corsica, France, on the Tyrrhenian Sea. It is the island's largest city and chief commercial center. It has a thriving export industry, sawmills, and cigarette and food-processing plants. Founded (14th cent.) as a fort by the Genoese, it was the capital of Corsica until 1791. Its citadel (16th-17th cent.) and its many 18th cent. buildings are tourist attractions.

Bas·togne (bäs-tô′nyə), town (1970 pop. 6,816), Luxembourg prov., SE Belgium, in the Ardennes and near the border of the duchy of Luxembourg. It is a market town noted for its hams. In World War II during the Battle of the Bulge (Dec., 1944-Jan., 1945) it was held against overwhelming odds until relieved by the U.S. 3rd Army.

Bas·trop (băs′trəp), county (1970 pop. 17,297), 890 sq mi (2,305.1 sq km), S central Texas, drained by the Colorado River; formed 1836; co. seat Bastrop. Its agriculture includes cotton, corn, truck crops, peanuts, grain sorghums, dairy products, and livestock. It also has clay and lignite mines and oil and natural-gas wells.

Bastrop. 1. City (1970 pop. 14,713), seat of Morehouse parish, NE La.; founded c.1845. An industrial city in a cattle, farm, and timber area, Bastrop is the center of the huge Monroe natural-gas field (discovered 1916). Its principal manufactures are paper, wood pulp, wood products, and chemicals. **2.** City (1970 pop. 3,112), seat of Bastrop co., S central Texas, on the Colorado River SE of Austin; settled 1827, inc. 1837. It ships cotton, pecans, and poultry products and manufactures furniture and metal tools. Lignite and oil are found in the area. There are several buildings of historic interest, and two state parks are nearby.

Ba·su·to·land (bə-sōō′tō-lănd′): *see* Lesotho.

Ba·taan (bă-tăn′, -tän′), peninsula and province (1970 pop. 216,210), W Luzon, the Philippines, between Manila Bay and the South China Sea. A mountainous, thickly jungled region, it has some of the best bamboo forests in the Philippines. Early in World War II (Dec., 1941-Jan., 1942), the U.S.-Filipino army withdrew to Bataan, where it entrenched and, despite the lack of naval and air support, fought a gallant holding action that upset the Japanese timetable for conquest. The army was crippled by starvation and disease when it was finally overwhelmed on Apr. 9, 1942. The U.S. and Filipino troops captured here were subjected to the long, infamous Death March to the prison camp near Cabanatuan, in the interior; thousands perished. The battleground of Bataan is now a national shrine.

Ba·tan·gas (bə-tăng′gəs, bä-täng′gäs), city (1975 est. pop. 125,304), capital of Batangas prov., SW Luzon, the Philippines. An important port on Batangas Bay, it has a large oil refinery and serves a fertile farm area noted for its fruits, cacao, and coffee.

Ba·tan Islands (bä-tän′), island group (1970 pop. 11,398), 76 sq mi (196.8 sq km), northernmost of the Philippine islands. The Batan Islands are separated from Taiwan by the Bashi Channel (50 mi/80.5 km wide). Coal is mined, and fishing is an important industry. In World War II, Batan Island was the site of the first Japanese landing in the Philippines (Dec. 8, 1941).

Ba·ta·vi·a (bə-tā′vē-ə): *see* Djakarta, Indonesia.

Batavia. 1. City (1975 est. pop. 11,600), Kane co., NE Ill., on the Fox River W of Chicago; founded c.1834, inc. as a village 1856, as a city 1891. A wide range of metal products and other goods are manufactured here. **2.** City (1978 est. pop. 17,400), seat of Genesee co., W N.Y.; laid out 1801, inc. 1915. Situated in a farm area, Batavia has industries producing television sets, die castings, shoes, road equipment, paper boxes, and heating equipment. The city was a center of the Anti-Masonic movement in the 19th cent. **3.** Village (1970 pop. 1,894), seat of Clermont co., SW Ohio, E of Cincinnati, in a farm area; settled c.1797, inc. 1836. Tobacco, corn, and wheat are grown.

Bates (bāts), county (1970 pop. 15,468), 841 sq mi (2,178.2 sq km), W Mo., drained by the Marais des Cygnes and South Grand rivers; formed 1841; co. seat Butler. It has agriculture (corn, wheat, oats, and livestock) and coal and oil deposits.

Bates·ville (bāts′vĭl′). **1.** City (1974 est. pop. 7,085), seat of Independence co., N central Ark., in the Ozarks on the White River W of Jonesboro; settled c.1810, inc. 1841. It is the center of an area producing cotton, timber, fruit, poultry, milk, manganese, bauxite, and marble. Arkansas College is here. **2.** Town (1970 pop. 3,796), a seat of Panola co., NW Miss., near the Tallahatchie River S of Memphis, in a cotton and lumber area; founded 1855.

Bath (băth, bäth), city (1976 est. pop. 83,100), Avon, SW England, in the Avon River valley. Britain's leading winter resort, Bath has the only natural hot springs in the country. There are also engineering, printing, bookbinding, wool-weaving, and clothing industries. In the 1st cent. A.D., the Romans discovered the natural springs and built elaborate lead-lined baths with heating and cooling systems (first excavated in 1755). In Saxon times the city was destroyed and the baths buried. In the 18th cent. Richard (Beau) Nash and the architect John Wood and his son transformed Bath into England's most fashionable spa. The Woods, using Bath stone from nearby quarries, built Queen Square, the Circus, and the Royal Crescent, all excellent examples of Georgian architecture.

Bath. 1. County (1970 pop. 9,235), 287 sq mi (743.3 sq km), NE Ky., bounded on the NE by the Licking River; formed 1811; co. seat Owingsville. In a rolling upland agricultural area partly in the outer Bluegrass region, it mainly grows tobacco. Part of Cumberland National Forest is here. **2.** County (1970 pop. 5,192), 540 sq mi (1,398.6 sq km), W Va., in the Alleghenies bounded on the W by the W.Va. border and drained by the Jackson and Cowpasture rivers; formed 1791; co. seat Warm Springs. Agriculture and forestry are important. It is also a resort area, with medicinal springs at Hot Springs, Warm Springs, and Healing Springs.

Bath. 1. City (1970 pop. 9,679), seat of Sagadahoc co., SW Maine, on the Kennebec River near its mouth on the Atlantic; settled c.1670, inc. as a city 1847. It is a port of entry, with a fine harbor. Champlain and others visited or passed near this site when exploring the Kennebec River. Shipbuilding began early; many clipper ships were constructed in the 19th cent., and the Bath Iron Works began producing steel warships and commercial vessels in the 1880s. The city flourished, particularly during World Wars I and II, when a large number of destroyers were built. There is a marine museum, and many fine old mansions remain. **2.** Village (1970 pop. 6,053), seat of Steuben co., S N.Y., NW of Elmira, in a farming and manufacturing area; settled 1793, inc. 1816.

Bath·urst (băth'ərst), city (1971 pop. 17,169), New South Wales, SE Australia, on the Macquarie River; founded 1815. It is an agricultural market with food processing and other light industries and railroad workshops.

Bathurst, city (1976 pop. 16,301), N N.B., Canada, on Chaleur Bay at the mouth of the Nipisiguit River. A popular beach resort, it is also the center of an area of lead, zinc, and copper mines and has large pulp, paper, and lumber mills.

Bathurst: *see* Banjul, Gambia.

Bathurst Island, 7,609 sq mi (19,707 sq km), in the Arctic Archipelago, Franklin dist., Northwest Territories, N Canada. It is the present site of the north magnetic pole.

Bat·ley (băt'lē), municipal borough (1973 est. pop. 41,630), West Yorkshire, N central England. Heavy woolens and other textiles are the chief manufactures; tiles, carpets, mattresses, felt, biscuits, and machinery are also produced. The Bagshaw Museum in Batley illustrates the history of clothmaking.

Bat·on Rouge (băt'n ro͞ozh), city (1978 est. pop. 198,000), state capital and seat of East Baton Rouge parish, SE La., on a bluff along the E bank of the Mississippi River; inc. 1817. It is a busy deepwater port of entry; an important transportation, distributing, and commercial center for a large oil, natural gas, and farm area; and a major oil-refining hub. There are large petrochemical industries, food-processing plants, machine shops, foundries, and ironworks. Baton Rouge was founded in 1719 when the French built a fort on that strategic spot along the river. The settlement was ceded to Great Britain in 1762, captured by the Spanish in 1779, and acquired by the United States in 1815 (following a brief period when it was a part of Spanish Florida). It became state capital in 1849. In the Civil War it was captured by Adm. David Farragut after the fall of New Orleans (May, 1862). It has notable ante-bellum houses. The old capitol (1882), built in the Gothic style of the original that was burned in the Civil War, still stands. Also of interest are the governor's mansion, the old arsenal museum, and the Huey Long grave and memorial.

Bat·tam·bang (băt'əm-băng), city (1967 est. pop. 43,000), capital of Battambang prov., W Cambodia, in a great rice-growing area. The second-largest city in Cambodia, it is a market center with numerous rice mills. Textiles are also made. After the outbreak (1970) of civil war in Cambodia, the Battambang-Phnom Penh road was a prime target of the Khmer Rouge insurgents, who, by capturing it, severed Phnom Penh from its major source of rice.

Bat·ter·y, the (băt'ə-rē), park, 21 acres (8.5 hectares), S tip of Manhattan Island, New York City, site of Dutch and English fortifications. Castle Clinton, a fort built in 1808 for the defense of New York harbor, was ceded to the city in 1823 and renamed Castle Garden. It was remodeled and served as a noted amusement hall and opera house. From 1855 to 1892 it served as an immigration station, and from 1896 to 1914 it housed an aquarium. After World War II the park was remodeled, and Castle Clinton became a national monument. The park also contains a war memorial and a statue of Giovanni da Verrazano, who discovered New York harbor.

Bat·tle (băt'l), rural district (1971 pop. 33,563), East Sussex, SE England. The town grew up on the site (then a moorland) of the Battle of Hastings (1066). The victorious William the Conqueror built Battle Abbey to commemorate the event.

Battle Creek, city (1978 est. pop. 37,900), Calhoun co., S Mich., at the confluence of the Kalamazoo and Battle Creek rivers; settled 1831, inc. as a city 1859. It is an agricultural trade center and is known for its cereals, pet foods, and biscuits. Battle Creek Sanitarium (founded by Dr. J. H. Kellogg in 1866 as the Health Reform Institute) is here.

Ba·tu·mi (bä-to͞o'mī) or **Ba·tum** (bə-to͞om'), city (1976 est. pop. 117,000), capital of Adzhar Autonomous Republic, SW Asian USSR, in Georgia, on the Black Sea near the Turkish border. A major port and trade center, Batumi is an important petroleum-shipping port and has oil refineries, shipyards, and food-processing plants. It is the site of an ancient Greek colony, belonged to Georgia in the Middle Ages, fell to the Turks in the late 16th cent., and passed to Russia in 1878.

Bat Yam (bät' yäm'), city (1975 est. pop. 118,100), W central Israel, on the Mediterranean Sea, near Tel Aviv-Jaffa. It is a seaside resort and an industrial center. The city was founded in 1926.

Bau·dette (bô-dĕt'), village (1974 est. pop. 1,440), seat of Lake of the Woods co., N Minn., on Rainy River at the Canadian boundary; settled c.1910. It is a port of entry, a resort, and a lumber and farm trade center.

Baut·zen (bou'tsən), city (1974 est. pop. 45,851), Dresden dist., SE East Germany, on the Spree River. It is an industrial city, a rail junction, and the center of a kaolin-quarrying region. Manufactures include machinery, textiles, chemicals, leather and paper goods, and railroad cars. Bautzen was founded in the 10th cent.

Ba·var·i·a (bə-vâr'ē-ə), state (1970 pop. 10,479,000), 27,239 sq mi (70,549 sq km), S West Germany; capital Munich. The largest state of West Germany, Bavaria is bordered by Czechoslovakia on the east, by Austria on the southeast and south, by Baden-Württemberg on the west, by Hesse on the northwest, and by East Germany on the north. A region of rich, softly rolling hills, it is drained by several rivers and is bounded by mountain ranges.

Forestry and agriculture are important occupations; wheat, barley, sugar beets, and dairy goods are the leading products. Major industrial products include glass and ceramics, iron and steel, paper, chemicals, machinery, textiles, clothing, optical instruments, petroleum, and motor vehicles. Bavarian beer is world famous. Toys and musical instruments are made by craftsmen. Salt, graphite, iron ore, and lignite are the chief mineral resources. The scenic beauties and the regional customs of the Bavarian Alps attract tourists.

The region was inhabited by Celts when Drusus conquered it (15 B.C.) for Rome. The Baiuoarii invaded it (6th cent. A.D.) and set up the duchy to which they gave their name. It was one of the five basic duchies of medieval Germany. Christianization was completed (8th cent.) by St. Boniface. From 788 to 911, Bavaria was ruled by the Carolingians, after which the duchy (then roughly comprising Bavaria proper, present-day Austria, and part of the Upper Palatinate) came under indigenous rulers. To reduce the power of the rebellious Bavarian dukes, Emperor Otto II in 976 stripped the duchy of all but present-day Upper and Lower Bavaria and the Tyrol. After periods of rule by the Guelphs and by the Austrian Babenbergs, Bavaria passed (1180) to Otto of Wittelsbach.

Under the Wittelsbachs, who ruled until 1918, Bavaria was almost always divided among the numerous branches of the dynasty. Duke Maximilian I (1597–1651) headed the Catholic League in the Thirty Years' War and was rewarded with the Upper Palatinate and the rank of elector. Bavaria was overrun by foreign armies, notably in the War of the Spanish Succession, the War of the Austrian Succession, the War of the Bavarian Succession, and the French Revolutionary Wars. Elector Maximilian IV Joseph was proclaimed king of Bavaria as Maximilian I in 1806; he later abandoned Napoleon and joined the allies, who at the Congress of Vienna (1814–15) left him in possession of virtually all of present-day Bavaria, including the Rhenish Palatinate. King Louis I (1825–48), dethroned by the mild revolution of 1848, was succeeded by the able Maximilian II (1848–64) and the brilliant but insane Louis II (1864–86). The rural prosperity of Bavaria and the strong influence of the Catholic Church accented the hostility of Bavaria toward the rising power of Prussia. Bavaria sided with Austria in the Austro-Prussian War (1866). Defeated in that war, it acknowledged Prussian leadership and joined (1871) the German Empire.

King Louis III was dethroned in Nov., 1918, by socialists who were bloodily suppressed (1919) by the German army. Bavaria then joined the Weimar Republic. In the early 1920s, Munich became the center of the National Socialist (Nazi) movement. After World War II the Rhenish Palatinate was separated from Bavaria and was later made part of the state of Rhineland-Palatinate.

Bax·ar (bək-sär'): *see* Buxar, India.

Bax·ley (băks'lē), city (1970 pop. 3,503), seat of Appling co., SE Ga., WSW of Savannah; inc. 1875. It is a market center for an area producing tobacco, pecans, cotton, livestock, and timber.

Bax·ter (băk'stər), county (1970 pop. 15,319), 537 sq mi (1,390.8 sq km), N Ark., in a high plateau of the Ozarks, bounded on the N by the Mo. border and drained by the North Fork and White rivers; formed 1873; co. seat Mountain Home. It has agriculture (cotton, hay, corn, truck crops, and livestock) and fishing and boating resorts.

Bay (bā). **1.** County (1970 pop. 75,283), 747 sq mi (1,934.7 sq km), NW Fla., bounded on the S by the Gulf of Mexico; formed 1913; co. seat Panama City. In a flatwoods area, with a swampy coast, it has cattle raising, forestry, fishing, and some agriculture (cotton, peanuts, and sugar cane). **2.** County (1970 pop. 117,339), 446 sq mi (1,155.1 sq km), E Mich., bounded on the E by Saginaw Bay and drained by the Saginaw and Kawkawlin rivers; organized 1858; co. seat Bay City. In an agricultural area (sugar beets, beans, chicory, and dairy products), it has commercial fishing, some manufacturing, and soft-coal mining. There are resorts, hunting and fishing areas, and a state park here.

Ba·ya·món (bä'yä-mōn'), town (1970 pop. 147,552), NE Puerto Rico, a residential and industrial suburb of San Juan. Founded in 1772, it is one of the oldest settlements on the island.

Bay·bor·o (bā'bûr-ō), town (1970 pop. 665), seat of Pamlico co., E. N.C., E of New Bern, at the head of Bay River (an inlet of Pamlico Sound). Fishing and sawmilling are the principal industries.

Bay City. 1. City (1978 est. pop. 46,100), seat of Bay co., S Mich., a port of entry on the Saginaw River at its mouth on Saginaw Bay (an inlet of Lake Huron); inc. 1859 with the consolidation of several settlements along the river. Its harbor handles considerable Great Lakes and ocean shipping. Bay City is the industrial, marketing, and shipping center of a rich farm area that yields sugar beets, potatoes, and dairy products. It grew as a great lumbering center, and when the forests were depleted (after 1890) it turned to diversified manufacturing. The vicinity is rich in Indian relics. **2.** City (1970 pop. 13,445), seat of Matagorda co., S Texas, near the Colorado River and the Gulf of Mexico; inc. 1894. It is a shipping and industrial center for a region that produces oil, gas, sulfur, beef cattle, rice, cotton, soybeans, and grain sorghums. There are petrochemical plants and grass and turf farms in the city.

Ba·yeux (bä-yōō', -yœ'), town (1975 pop. 13,457), Calvados dept., N France, in Normandy, near the English Channel. It is a farm and communications center, noted for its lace industry. A Roman town, it was burned (1105) by Henry I of England. Sections of its Romanesque church withstood the fire and form a part of the remarkable Gothic cathedral built for the most part in the 13th cent. The town is particularly famous for its museum containing the Bayeux tapestry. In World War II, Bayeux was the first French city liberated by the Allies (June 8, 1944).

Bay·field (bā'fēld'), county (1970 pop. 11,683), 1,460 sq mi (3,781.4 sq km), extreme N Wis., bounded on the N by Lake Superior, on the E by Chequamegon Bay; formed 1845 as La Pointe co., renamed 1866; co. seat Washburn. A large portion of the county is on a wide peninsula. Dairying, lumbering, fishing, and agriculture are important.

Bay Islands, archipelago, 144 sq mi (373 sq km), off the N coast of Honduras, in the Caribbean Sea. Roatán is the largest island and the port of entry. The chief products are fruits and logwood, which English logcutters exploited as early as the 17th cent. British garrisoning of the islands in 1848 led to unrest, which was partially settled by the Clayton-Bulwer Treaty (1850) and relinquishment of British rights (1859) to Honduras.

Bay·kal or **Bai·kal** (both: bī-käl'), lake, 12,160 sq mi (31,494 sq km), SE Siberian USSR. It is the largest freshwater lake of Eurasia, with a width up to 50 mi (80 km) and a length of c.395 mi (635 km). Its maximum depth is 5,714 ft (1,742.8 m), making Baykal the world's deepest lake. Its only outlet is the Angara River, whose great volume is harnessed by a hydroelectric station at nearby Irkutsk. Lake Baykal is navigable and is used to float timber. Surrounded by beautiful mountain scenery, it is rich in fish, including many unusual species.

Bay·lor (bā'lər), county (1970 pop. 5,221), 875 sq mi (2,266.3 sq km), N Texas, in a prairie region drained by the Wichita River and the Salt Fork of the Brazos River; formed 1858; co. seat Seymour. Its agriculture includes cotton, wheat, oats, poultry, beef and dairy cattle, and some hogs. There are oil and natural-gas wells here.

Bay Mi·nette (mə-nět'), city (1970 pop. 6,727), seat of Baldwin co., SW Ala., NE of Mobile, in a farm and timber area; founded c.1861, inc. 1908.

Ba·yonne (bä-yôn'), town (1975 pop. 42,938), Pyrénées-Atlantiques dept., SW France, in Gascony, on the Adour River near its entrance into the Bay of Biscay. Despite a shifting sandbar at the mouth of the Adour, it is a seaport, exporting sulfur. The town also has metallurgical, chemical, aeronautical, leather, and wood industries. French and Spanish, as well as Basque, are spoken here. At Bayonne, Napoleon I forced Charles IV and Ferdinand VII of Spain to abdicate (1808). The city gives its name to the bayonet, invented here in the 17th cent. There is a Basque museum and a fine arts museum, left to the city by the painter Léon Joseph Florentin Bonnat, who was born here. Parts of the town's Roman and medieval walls are preserved.

Bay·onne (bā-ōn', -yōn'), city (1978 est. pop. 74,100), Hudson co., NE N.J.; inc. 1869. It has oil and chemical industries. Dutch traders came to this site c.1650; the British gained possession in 1664.

bay·ou (bī'ōō, bī'ō), term used mainly in U.S. Gulf states, especially La. and Miss., to describe a stationary or sluggishly moving body of water that was once part of a lake, river, or gulf and is swampy or marshy in nature.

Bay·reuth (bī-roit'), city (1974 est. pop. 66,936), capital of Upper Franconia, Bavaria, S West Germany, on the Roter Main River. It is an industrial center; its manufactures include textiles, metals, and machinery. Founded in the mid-12th cent., Bayreuth belonged to a branch of the Hohenzollern family from 1248 to 1791, when it was annexed by Prussia. It was taken by France in 1807 and passed to Bavaria in 1810. Richard Wagner lived in Bayreuth from 1872 to 1883, and annual music festivals of international importance are held in the Festspielhaus, an opera house designed by Wagner and built in 1872–76. Wagner and Franz Liszt are buried in Bayreuth.

Bay Saint Lou·is (sānt lōō'ĭs), resort city (1970 pop. 6,752), seat of Hancock co., S Miss., on St. Louis Bay and Mississippi Sound W of Gulfport; inc. 1854. Fresh- and salt-water fish are processed.

Bay Shore, uninc. city (1978 est. pop. 33,400), Islip township, Suffolk co., SE N.Y., on the S shore of Long Island, at the widest point of Great South Bay; founded 1708. It is noted as a fishing and duck-hunting center and has some light industry.

Bay Springs, town (1970 pop. 1,801), a seat of Jasper co., E central Miss., NW of Laurel, in an agricultural and timber area; settled 1896, inc. 1904.

Bay·town (bā'toun'), city (1978 est. pop. 48,900), Harris co., S Texas, at the head of Galveston Bay, on the Houston ship channel; inc. 1948 after the consolidation of Goose Creek, Pelly, and Baytown. Large volumes of oil are produced in the area, refined in Baytown, and shipped throughout the world. The city also has chemical, synthetic-rubber, and steel industries.

Ba·za (bä'zə, -thä), town (1974 est. pop. 14,290), Granada prov., S Spain, in Andalusia. It is a food-processing center for a fertile farm area noted especially for its cattle. Baza has flour mills, tanneries, and textile industries. An important city of the Moorish kingdom of Granada, it fell to the Spaniards in 1489 after a year-long siege.

Beach (bēch), city (1970 pop. 1,408), seat of Golden Valley co., SW N.Dak., near the Mont. line W of Dickinson; inc. 1910. It is a shipping center for wheat, barley, and oats.

Beach·y Head (bē'chē), high chalk cliffs (575 ft/175.4 m), on the S coast of East Sussex, S England. The Battle of Beachy Head, in the War of the Grand Alliance, was fought (1690) between an Anglo-Dutch fleet and a French fleet. Although the French won, they failed to exploit their victory.

Bea·con (bē'kən), city (1970 pop. 13,255), Dutchess co., SE N.Y., on the E bank of the Hudson River opposite Newburgh; settled 1663, inc. as a city in 1913. Beacon has textile and related industries, other varied manufactures, and a large industrial research firm. An incline railway ascends Mt. Beacon, site of a towering monument to the Revolutionary soldiers who built signal fires there to warn of the coming of the British.

Bea·cons·field (běk'ənz-fēld', bē'kənz-), urban district (1971 pop. 11,861), Buckinghamshire, central England, near High Wycombe.

Bea·dle (bēd'l), county (1970 pop. 20,877), 1,261 sq mi (3,266 sq km), E central S.Dak., in an agricultural area drained by the James River; formed 1873; co. seat Huron. It produces dairy products, poultry, grain, and livestock and has some manufacturing.

Bear (bâr), river, 350 mi (563.2 km) long, rising in the Uinta Mts., NE Utah, and flowing in a U-shaped course NW through Wyo. and Idaho, then S into Utah to enter Great Salt Lake. A perennial stream, the Bear irrigates c.50,000 acres (20,250 hectares).

Beard·more Glacier (bîrd'môr'), valley glacier, largest in the world, 260 mi (418.3 km) long, Queen Maud Mts., Antarctica. Sir Ernest Shackleton discovered it in 1908.

Beards·town (bîrdz'toun'), city (1970 pop. 6,222), Cass co., W central Ill., WNW of Springfield on the Illinois River, from which it is protected by a high levee; settled 1819, platted 1827, inc. 1837. It is a rail, trade, and shipping center for a fruit and farm area and has railroad shops, commercial fisheries, and a meat-processing plant. Gloves, flour, and steel tanks are made.

Bear Lake, county (1970 pop. 5,801), 988 sq mi (2,558.9 sq km), SE Idaho, in a mountain area bordering on Wyo. and Utah; formed 1875; co. seat Paris. In a recreation area of the Bear River valley, it produces hay, grain, livestock, and turkeys. There are oil and gas deposits in the region.

Bear Mountain. 1. Peak, 2,355 ft (718.3 m) high, NW Conn., in the Taconic Mts. NW of Salisbury and near the Mass. and N.Y. borders. It is one of the highest points in the state. **2.** Peak, 1,284 ft (391.6 m) high, SE N.Y., overlooking the Hudson River. The Bear Mt. section of the Palisades Interstate Park has facilities for both summer and winter sports. The remains of Fort Clinton, dating from the Revolutionary War, are here.

Bé·arn (bā-ärn′), former province, SW France, in the Pyrenees. Béarn was part of Roman Aquitania. It came (6th cent.) under the control of Gascony, and was made (9th cent.) a county. In 1290 it passed to the counts of Foix and in 1484 to the house of Albret. Protestantism was imposed by Jeanne d'Albret. When her son became Henry IV of France, Béarn passed to the crown. However, it remained autonomous until 1620.

Be·as (bē′äs), river, 250 mi (402.3 km) long, rising in the Himalayas and flowing generally SW through the Kulu Valley and the Siwalik Hills to join the Sutlej River, S of Amritsar, N India. It is the easternmost of the five rivers of the Punjab. The Beas marked the eastern limit of Alexander the Great's invasion of India in 326 B.C.

Be·at·rice (bē-ăt′rĭs), city (1978 est. pop. 11,200), seat of Gage co., SE Nebr., on the Big Blue River; inc. as a city 1873. It is on the old Oregon Trail and is the trading and industrial center for a grain, dairy, and livestock area. Its manufactures include metal goods, farm and garden equipment, fertilizers, hardware and electric products, store fixtures, and dairy products. Homestead National Monument is nearby.

Beat·ty·ville (bā′tē-vĭl′), city (1970 pop. 923), seat of Lee co., central Ky., on the Kentucky River SE of Lexington, in a coal, oil, timber, and farm region. Part of Daniel Boone National Forest is nearby.

Beau·fort. 1. (bō′fərt) County (1970 pop. 35,980), 831 sq mi (2,152.3 sq km), E N.C., on the Atlantic coast in a Tidewater area indented by Pamlico and Pungo river estuaries; formed 1705; co. seat Washington. It is in a fishing, lumbering, and farming (tobacco and corn) region, with many resorts along its coastline. **2.** (byo͞o′fərt) County (1970 pop. 51,136), 578 sq mi (1,497 sq km), extreme S S.C., extending along the Atlantic coast from the Savannah River to St. Helena Sound and the Combahee River; formed 1785; co. seat Beaufort. It includes several of the Sea Islands. Formerly a region of great plantations, it is now a winter-resort area, with fishing, lumbering, agriculture (truck farms, corn, and hogs), and food processing.

Beau·fort. 1. (bō′fərt) Resort and fishing town (1970 pop. 3,368), seat of Carteret co., E N.C., SE of New Bern, on the Atlantic coast at the terminus of an inland waterway; inc. c.1722. It is a port of entry. Many early 18th cent. buildings are preserved. **2.** (byo͞o′fərt) City (1970 pop. 9,434), seat of Beaufort co., S S.C., on Port Royal Island SW of Charleston. On the Inland Waterway, it is a tourist center and a canning and shipping point for a farming and fishing region. The second-oldest town in the state, it was visited early by the Spanish and French, founded in 1711, and incorporated in 1803.

Beau·fort Sea (bō′fərt), part of the Arctic Ocean, N of Alaska and Canada, between Point Barrow, Alaska, and the Canadian Arctic Archipelago. The Mackenzie River flows into the sea, which is always covered with pack ice. It was first explored in 1914.

Beau·jo·lais (bō-zhô-lā′), hilly region, Rhône dept., E central France, W of the Saône between Mâcon and Lyons. It is famous for its red wine.

Beau·mont (bō′mŏnt), city (1978 est. pop. 112,700), seat of Jefferson co., Texas, a port of entry on the Sabine-Neches Canal; inc. 1838. A ship channel provides the facilities of a modern deepwater port, with shipyards and large storage tanks. Beaumont is an important industrial and shipping center and a great oil city, with giant refineries and petrochemical complexes. Other industries are based on the forests and vast farmlands of the area.

Beaune (bōn), town (1975 pop. 16,412), Côte-d'Or dept., E France, in Burgundy. It is a noted center for Burgundy wines, with a wine school and wine research facilities.

Beau·port (bō-pôr′), city (1976 pop. 55,339), S Que., Canada, on the St. Lawrence River. It is a suburb of Quebec city. Settled in 1634, it is one of the oldest communities in Canada.

Beau·re·gard (bō′rə-gärd′), parish (1970 pop. 22,888), 1,184 sq mi (3,066.6 sq km), W La., bounded on the W by the Sabine River, here forming the Texas border; formed 1906; parish seat De Ridder. In a diversified lumber and agricultural region yielding cotton, corn, cattle, sheep, peanuts, hay, sweet potatoes, and citrus fruit, it has petroleum and natural-gas deposits. Food products are made.

Beau·vais (bō-vā′), town (1975 pop. 54,089), capital of Oise dept., N France. Tractors, ceramic tiles, textiles, and musical instruments are among its many manufactures. A Roman town, it flourished in the Middle Ages and again after the 17th cent., when Colbert established the state tapestry industry here. It was the center of the Jacquerie revolt in 1358, and in 1472 its citizens resisted Charles the Bold of Burgundy. Beauvais was severely damaged in both World Wars; in June, 1940, its tapestry factory was destroyed, and the industry was moved to Paris. The Cathedral of St. Pierre, begun in 1227 as the highest building in Christendom, was never completed.

Bea·ver (bē′vər). **1.** County (1970 pop. 6,282), 1,790 sq mi (4,636.1 sq km), extreme NW Okla., in the high plains at the base of the Panhandle, bounded on the N by the Kansas border, on the S by the Texas border, and intersected by the North Canadian and Cimarron rivers; formed 1890; co. seat Beaver. It has natural-gas wells and silica deposits. Its agriculture includes winter wheat, barley, grain sorghums, and livestock. **2.** County (1970 pop. 208,418), 440 sq mi (1,139.6 sq km), W Pa., in a mining and manufacturing area drained by the Ohio and Beaver rivers and bounded on the W by the Ohio border; formed 1800; co. seat Beaver. Metal products, glass, and chemicals are made. Its mineral resources include bituminous coal, sandstone, clay, oil, and gas. Anthony Wayne established (1792-93) one of the first army training camps here. **3.** County (1970 pop. 3,800), 2,587 sq mi (6,700.3 sq km), SW Utah, in a mountainous area drained by the Beaver River and bordering on Nev.; formed 1856; co. seat Beaver. Mining, ranching, and dry farming are important to its economy. Part of Fishlake National Forest is in the county.

Beaver. 1. Or **Beaver City,** farming and wheat-shipping town (1970 pop. 1,853), seat of Beaver co., extreme NW Okla., in the Panhandle on the North Canadian River; first settled by squatters c.1880. Beaver was the capital (1887) of the Territory of Cimarron. **2.** Residential borough (1975 est. pop. 5,600), seat of Beaver co., W Pa., NW of Pittsburgh near the junction of the Beaver and Ohio rivers, in a mineral region; laid out 1791, inc. 1802. A flagstaff marks the site of Fort McIntosh (1778), the first U.S. military post north of the Ohio River. **3.** City (1970 pop. 1,453), seat of Beaver co., SW Utah, NE of Cedar City, in an irrigated farming region; settled by Mormons 1856. There was a mining boom here in the 19th cent.

Beaver, river, rising in central Alta., Canada, and flowing 305 mi (490.7 km) E into Sask., then N to the headwaters of the Churchill River.

Beaver City, city (1970 pop. 802), seat of Furnas co., S Nebr., WSW of Hastings on Beaver Creek near the Kansas border, in a grain and dairying area.

Bea·ver·head (bē′vər-hĕd′), county (1970 pop. 8,187), 5,556 sq mi (14,390 sq km), SW Mont., in a mountainous region bordering on Idaho and bounded on the N by the Big Hole River and the Continental Divide; formed 1865; co. seat Dillon. It is crossed by the Big Hole and Beaverhead rivers. Livestock is raised. Sections of Beaverhead National Forest are in the county.

Beb·ing·ton (bĕb′ĭng-tən), municipal borough, (1973 est. pop. 62,500), Merseyside, W central England. Its frontage on the Mersey River is part of the Port of Liverpool. The borough has soap factories and freestone quarries and manufactures chemicals and margarine. The Church of St. Andrew, on a site occupied since Saxon times, dates from the 14th and 16th cent.

Bé·char (bā-shär′), formerly **Co·lomb-Bé·char** (kô-lôɴ′-), town (1974 est. pop. 71,081), capital of La Saoura dept., W Algeria. It is in a mining (coal, copper, magnesium, iron) and industrial region. Béchar also serves as a major shipping point for coal. The town was established in 1905 as a French military post.

Bech·u·a·na·land (bĕch′o͞o-ä′nə-lănd′): *see* Botswana.

Beck·er (bĕk′ər), county (1970 pop. 24,372), 1,297 sq mi (3,359.2 sq km), W Minn., in an agricultural area watered by numerous lakes; formed 1858; co. seat Detroit Lakes. It has agriculture (dairy products, livestock, and grain) and marl and peat deposits. In the north are a wildlife refuge and part of the White Earth Indian Reservation.

Beck·ham (běk'əm), county (1970 pop. 15,754), 907 sq mi (2,349.1 sq km), W Okla., in rolling plains bounded on the W by the Texas border and intersected by the North Fork of the Red River; formed 1907; co. seat Sayre. It has agriculture (cotton, wheat, and sorghums), oil and natural-gas wells, and some manufacturing.

Beck·ley (běk'lē), city (1978 est. pop. 21,200), seat of Raleigh co., S W.Va.; inc. 1927. Its major industries are coal mining, agriculture, tourism, and the production of electronic equipment.

Bed·ford (běd'fərd), municipal borough (1973 est. pop. 74,390), administrative center of Bedfordshire, central England, on the Ouse River. It is an important industrial center; diesel engines, pumps, turbines, agricultural machinery, electrical equipment, and transistors are the chief manufactures. Bedford, a battlefield for Britons and Saxons in the 6th cent., was the scene of an important Saxon defeat in 571. St. Peter's Church contains examples of Saxon stone carvings. John Bunyan is commemorated by a chapel on the site of a building where he preached in the 17th cent.

Bedford. 1. County (1970 pop. 42,353), 1,018 sq mi (2,636.6 sq km), S Pa., in a mountainous agricultural and mineral-producing area drained by the Raystown Branch of the Juniata River and bounded on the S by the Md. border, on the E by Sideling Hill, and on the NW by the Allegheny Mts.; formed 1771; co. seat Bedford. It has bituminous-coal deposits, limestone quarries, sand pits, lumbering, grain and livestock, and a fertilizer industry. **2.** County (1970 pop. 25,039), 482 sq mi (1,248.4 sq km), central Tenn., in a central basin drained by the Duck River; formed 1807; co. seat Shelbyville. Livestock is raised (especially the noted Tennessee walking horse). It also has dairying, agriculture, and some manufacturing. **3.** County (1970 pop. 26,728), 770 sq mi (1,994.3 sq km), SW central Va., partly in the Piedmont with the Blue Ridge in the NW and bounded on the SW by the Roanoke River, on the NE by the James River; formed 1754; co. seat Bedford. It has agriculture (tobacco, wheat, tomatoes, and dairying), lumbering, feldspar mining, and manufacturing (especially textiles and pulp and paper products). It includes part of Jefferson National Forest and is crossed by the Appalachian Trail.

Bedford. 1. City (1978 est. pop. 14,000), seat of Lawrence co., S Ind.; inc. 1889. Bedford limestone, which is shipped all over the world, is quarried here. The city also has several small industrial plants and a foundry. Points of interest include old stone buildings and houses and many carvings. **2.** City (1970 pop. 1,733), seat of Taylor co., SW Iowa, near the Mo. border SE of Council Bluffs; inc. 1885. Lime is produced. A state park, with a fish hatchery, and Indian mounds are nearby. **3.** City (1970 pop. 780), seat of Trimble co., N Ky., in the outer Bluegrass NE of Louisville, in a lumber, tobacco, and grain region. **4.** Town (1970 pop. 13,513), Middlesex co., E Mass., a residential suburb of Boston; settled c.1637, inc. 1729. Several pre-Revolutionary houses remain. **5.** Town (1970 pop. 5,859), Hillsborough co., S N.H., SW of Manchester; inc. 1750. It was settled in 1737 on land granted in 1733–34 to soldiers of King Philip's War and their heirs. Clothing is made here. **6.** City (1978 est. pop. 15,800), Cuyahoga co., NE Ohio, a suburb of Cleveland; settled c.1813 on the site of a Moravian settlement (1786), inc. as a city 1931. Although chiefly residential, it also has plants manufacturing office furniture, china, rubber goods, auto parts, processed foods, tools, and fixtures. **7.** Agricultural borough (1973 est. pop. 3,433), seat of Bedford co., S Pa., on a branch of the Juanita River SE of Johnstown; settled c.1750 as Raystown, laid out 1766, inc. 1795. Wood products and mining equipment are manufactured. The borough is on the site of a fort (built c.1757) used as Washington's headquarters during the Whiskey Rebellion (1794). Several colonial buildings are preserved. **8.** City (1970 pop. 10,049), Tarrant co., N Texas; settled c.1843, inc. 1954. **9.** Independent city (1975 est. pop. 6,000), S central Va., between Roanoke and Lynchburg, in a farm area; founded 1781, inc. 1890, reinc. 1912. It has tobacco warehouses and feldspar mines and manufactures textiles and rubber goods. The National Home of Elks is here. Poplar Forest, Thomas Jefferson's country home (built 1806–9), is nearby.

Bed·ford·shire (běd'fərd-shîr', -shər) or **Bedford** or **Beds** (bědz), county (1976 est. pop. 491,700), 473 sq mi (1,225.1 sq km), central England; administrative center Bedford. The terrain is generally flat, with low chalk hills in the south. Agriculture is the chief occupation. The county was a refuge for Protestants from the European continent during the English Civil War.

Bed·ling·ton·shire (běd'lĭng-tən-shîr', -shər), urban district (1971 pop. 28,167), Northumberland, NE England. Coal mining, brickmaking, and the manufacture of concrete products, shirts, and gloves are the chief industries. There is also some agriculture. The Bedlington terrier is bred in the district.

Bed·worth (běd'wûrth), urban district (1973 est. pop. 41,660), Warwickshire, central England. It is a residential and industrial area. Coal mining and brickmaking are the major economic activities. George Eliot was born nearby.

Bę·dzin (běɴ'jēn), town (1972 pop. 59,400), SE Poland, on the Czarna Przemsza River, a tributary of the Vistula. It is a coal-mining center and has industries producing metal products, machinery, chemicals, and electrical equipment. Founded in the 14th cent., Będzin passed to Russia in 1815; it was returned to Poland in 1919.

Bee (bē), county (1970 pop. 22,737), 842 sq mi (2,180.8 sq km), S Texas, drained by the Arkansas River; formed 1857; co. seat Beeville. Its agriculture includes corn, grain sorghums, cotton, broomcorn, peanuts, hogs, poultry, cattle, and sheep. Oil and natural gas are found here.

Beer·she·ba (bîr-shē'bə), city (1975 est. pop. 96,500), S Israel, principal trading city of the Negev Desert. Construction is the city's main industry. Manufactures include chemicals, textiles, ceramics, glass, plastics, and food products. The city was one of the southernmost towns of biblical Palestine; hence the expression "from Dan to Beersheba," meaning the whole of Palestine. A well believed to have been dug by Abraham is in the city. Beersheba flourished during the late Roman and Byzantine eras but was deserted soon thereafter. The Ottoman Turks re-established it c.1900 as an administrative center. Beersheba was the first city taken by the British in the Palestine campaign (1917) of World War I. Given to the Arabs in the partition of Palestine (1948), it was retaken by Israel in the Arab-Israeli War of 1948.

Bees·ton and Sta·ple·ford (bēs'tən, bē'sən; stā'pəl-fərd), urban district (1973 est. pop. 65,360), Nottinghamshire, central England. There are large pharmaceutical plants and factories that produce boilers, telecommunication equipment, fluorescent lights, textiles, pencils, cardboard boxes, and clothing. The Stapleford churchyard has an ancient Saxon cross, thought to be the oldest Christian memorial in the country.

Bee·ville (bē'vĭl'), city (1970 pop. 13,506), seat of Bee co., S Texas; settled in the 1830s, inc. 1908. Long a cow town, Beeville is the trade center of an agricultural area.

Be·har (bē-här'), state, India: *see* Bihar.

Bei·da (bā'də): *see* Al Bayda, Libya.

Bei·ra (bā'rə), region and former province, N central Portugal, S of the Douro River. The region is traversed by the Serra da Estrela, Portugal's highest mountain range. Grains, fruits, and olives are grown. Industries include fishing and the manufacture of textiles and forest products. The area had been recovered from the Moors even before Portugal was formed, but Moorish attacks continued into the 13th cent.

Bei·ra (bā'rä), city (1970 pop. 110,752), capital of Manica e Sofala district, E central Mozambique, a seaport on the Mozambique Channel (an arm of the Indian Ocean), at the mouth of the Púngoè River. A commercial center, the city grew (beginning in 1891) as the terminus of a railroad into the interior. It is a popular beach resort.

Bei·rut (bā-rōōt'), city (1972 est. pop. 720,000), W Lebanon, capital of Lebanon, on the Mediterranean Sea, at the foot of the Lebanon Mts. Beirut is an important port and commercial and financial center with food-processing industries. It was a Phoenician city and became known after 1500 B.C. as a trade center. Beirut was prominent under the Romans, when it was not only a a commercial town—with a large trade in wine and linens—but also a colony with some territory. In the 3rd cent. A.D. Beirut had a famous school of Roman law. It declined after an earthquake in 551. Beirut was captured by the Arabs in 635. The Crusaders took the city in 1110, and it was part of the Latin Kingdom of Jerusalem until 1291. After 1517 the Druses controlled the city under the Ottoman Empire. Ibrahim Pasha took it for the Egyptians (1830), but in 1840 the French and British bombarded and captured the city, enabling the Turks to return. It was taken (1918) by French troops in World War I. Beirut became the capital of Lebanon in 1920 under the French mandate. In the middle 1970s, it was decimated by the Lebanese Civil War.

Be·ja (bā'zhə), town (1970 pop. 14,760), S Portugal, capital of Beja dist. and Baixa Alentejo. It is an important trade and manufacturing center. Beja was important under the Romans. The Moors used it as a fortress city, until the Portuguese recovered it in 1162.

Be·jaï·a (bĕ-jī'ə), formerly **Bou·gie** (bōō-zhē'), city (1974 est. pop. 80,000), N Algeria, a port on the Gulf of Bejaïa (an arm of the Mediterranean Sea). Bejaïa is the principal oil port of the western Mediterranean. Exports include iron, phosphates, wines, dried figs, and plums. The city also has textile and cork industries. A minor port in Carthaginian and Roman times, Bejaïa became the capital of the Vandals in the 5th cent. It later disappeared but was refounded by the Berbers in the 11th cent. and became an important port and cultural center. After Spanish occupation (1510–55), the city was taken by the Ottoman Turks. Until it was captured by the French in 1833, Bejaïa was a stronghold of the Barbary pirates.

Be·ka·bad (byĕ-kä-bäd'), formerly **Be·go·vat** (byĕ-gō-vät'), city (1976 est. pop. 61,000), Tashkent oblast, Uzbekistan, Central Asian USSR, on the Syr Darya River. It is an important industrial center, with large iron and steel mills and cement works.

Bé·kés·csa·ba (bā'kāsh-chŏ'bŏ), city (1977 est. pop. 55,700), SE Hungary. The commercial center for a silk-raising, tobacco-growing, and hog-breeding region, Békéscsaba has meat-packing plants and flour and hemp mills. Other industries produce textiles, farm implements, and cement. The city was founded in the 13th cent. but later destroyed by the Turks. In the 18th cent. Slovak settlers helped restore Békéscsaba. Landmarks include a 13th cent. Roman Catholic church and a Lutheran cathedral.

Bel Air (bĕl âr'), town (1970 pop. 6,307), seat of Harford co., NE Md., NE of Baltimore; settled c.1782, inc. 1874. It is the trade center of an agricultural and horse-breeding area. Edwin and John Wilkes Booth were born nearby at Tudor Hall.

Be·la·ya (byĕ'lə-yə), river, c.880 mi (1,415 km) long, Bashkir Autonomous Republic, E European USSR. It rises in the Ural Mts. and winds generally northwest past Beloretsk, Sterlitamak (where it becomes navigable), and Ufa to join the Kama River.

Belaya Tser·kov (tsĕr'kəf), city (1976 est. pop. 137,000), W central European USSR, in the Ukraine, on the Ros River. It is a rail junction and an industrial and commercial center. Industries include food processing and the manufacture of machinery, shoes, and building materials. The city was founded in 1032.

Be·lém (bə-lĕm', -lāN') or **Pa·rá** (pə-rä'), city (1970 pop. 565,097), capital of Pará state, N Brazil, on the Pará River. Belém, the chief commercial center and port of the vast Amazon River basin, handles the Amazonian produce (chiefly rubber, Brazil nuts, cacao, and timber) and has processing plants. Belém was founded by Portuguese in 1616 as a military post for the defense of northern Brazil against French, English, and Dutch pirates. It reached a peak of feverish prosperity during the wild-rubber boom in the late 19th and early 20th cent., then suffered a depression that was alleviated by diversification and planned development in the 1930s. Prosperity increased also after World War II with the improvement of communications within the Amazon region. The city is known for its Goeldi museum, with ethnological and zoological collections of the Amazon basin.

Bel·fast (bĕl'făst, bĕl-făst'), county borough (1976 est. pop. 363,000), capital of Northern Ireland, county town of Co. Antrim, mainly in Co. Antrim but partly in Co. Down. It is on Belfast Lough, an inlet of the North Channel of the Irish Sea, and at the mouth of the Lagan River. The harbor is navigable to the largest ships. The great shipyards of the Harland and Wolff Company in Belfast have built some of the world's largest ocean liners. The city is also the center of the Irish linen industry; other industries include tobacco and food processing, packaging, and the manufacture of rayon, aircraft, tools and machinery, yarn, clothing, carpets, and rope. Agricultural and livestock products are the chief exports. Belfast was founded in 1177, but the present city is a product of the Industrial Revolution. French Huguenots, coming here after the revocation of the Edict of Nantes (1685), stimulated the growth of the town's linen industry. Serious rioting between Catholics and Protestants has scarred the city many times since the 19th cent.

Belfast, city (1970 pop. 5,957), seat of Waldo co., S Maine, on Penobscot Bay opposite Castine; settled 1770, inc. as a town 1773, as a city 1853. It is a port of entry in a poultry-raising and fishing area. Belfast was sacked by the British in 1779 and again in 1814. It flourished as a shipping and shipbuilding port in the 19th cent.

Bel·fort (bĕl-fôr', bā-), city (1975 pop. 54,615), capital of the Territory of Belfort (a department), E France, in Alsace. An important industrial and transportation center, it has large cotton mills and metalworks. A major fortress town since the 17th cent., it commands the Belfort Gap, or Burgundy Gate, between the Vosges and the Jura Mts., thus dominating the roads from France, Switzerland, and Germany. An Austrian possession, Belfort passed to France by the Peace of Westphalia (1648). During the Franco-Prussian War (1870–71) the garrison withstood a siege of 108 days. Partly in acknowledgment of this heroism, the Germans left Belfort and the surrounding territory to France when they annexed the rest of Alsace.

Belfort, Territory of, department (1968 pop. 118,450), 235 sq mi (608.7 sq km), E France, in Alsace, on the Swiss border; capital Belfort.

Bel·gaum (bĕl-goum'), town (1971 pop. 213,830), Karnataka state, SE India. It is an agricultural market that trades in food grains, sugar cane, cotton, tobacco, oilseed, and milk products.

Bel·gian Con·go (bĕl'jən kŏng'gō): *see* Zaire.

Bel·gium (bĕl'jəm), constitutional kingdom (1970 pop. 9,694,991), 11,781 sq mi (30,513 sq km), NW Europe. Brussels is the capital. Antwerp is the chief commercial center and one of the world's great ports. Belgium is bordered on the north by the Netherlands and the North Sea, on the east by West Germany and Luxembourg, and on the west and southwest by France. The terrain is low lying except in the Ardennes Mts. in the south. Belgium comprises two ethnic and

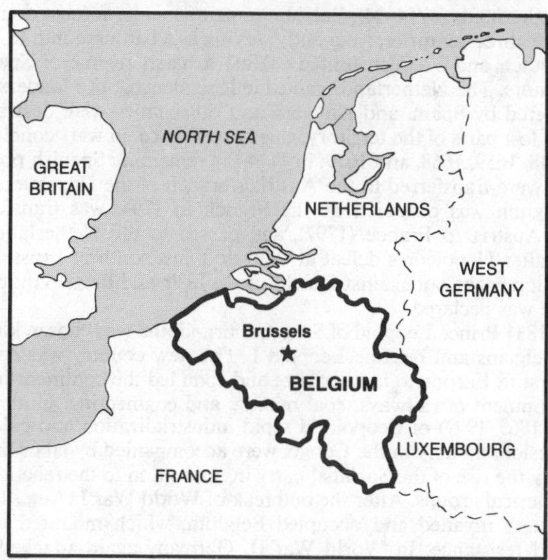

cultural regions, Flanders in the north and Wallony in the south. The dividing line runs roughly east-west just south of Brussels. Flemish (a Dutch dialect) is the official language in Flanders, French in the south. The French-speaking people are now commonly called Walloons. Brussels is bilingual.

Belgium is divided into 9 provinces:

NAME	CAPITAL	NAME	CAPITAL
Antwerp	Antwerp	Limburg	Hasselt
Brabant	Brussels	Luxembourg	Arlon
East Flanders	Gent	Namur	Namur
Hainaut	Mons	West Flanders	Brugge
Liège	Liège		

Economy. Belgium is one of the most densely populated and highly industrialized areas in Europe; emphasis is on heavy industry. Coal mining and the production of steel, chemicals, and cement are concentrated in the Sambre and Meuse valleys, in the Borinage, and in the Campine coal basin. Liège is a great steel center. Iron and steel constitute Belgium's largest single export item. Belgium also has an old, established metal-products industry; manufactures include bridges, heavy machinery, industrial and surgical equipment, motor vehicles, rolling stock, machine tools, and munitions. Shipbuilding is centered in Antwerp. Chemical products include fertilizers, dyes, pharmaceuticals, and plastics; the petrochemical industry, concentrated near the oil refineries of Antwerp, has mushroomed since World War II. Other important industries include textile production, diamond cutting, lace making, glass production, and the processing of leather and wood.

Agriculture, while engaging only a small percent of the working force, is important. Except in the marshy Campine and in the heavily forested Ardennes there is much fertile and well-watered soil. The chief crops are cereals (oats, rye, wheat, barley). Sugar beets, potatoes, and flax are also grown, and there is truck farming near the large cities. Cattle raising and dairying (especially in Flanders) are important. Flowers and chicory, grown as a winter vegetable, are valuable crops. Processed foods include beet sugar, cheese and other dairy items, and canned vegetables. Beer is made from rich hops.

History. Belgium takes its name from the Belgae, a people of ancient Gaul. The Roman province of Belgica was much larger than modern Belgium. There the Franks first appeared in the 3rd cent. A.D. The Carolingian dynasty had its roots at Herstal, in Belgium. After the divisions (9th cent.) of Charlemagne's empire Belgium became part of Lotharingia and later of the duchy of Lower Lorraine. In the 12th cent. Lower Lorraine disintegrated; the duchies of Brabant and Luxembourg and the bishopric of Liège took its place. The histories of these feudal states and of Flanders and Hainaut constitute the medieval history of Belgium. In the 15th cent. all of present Belgium passed to the dukes of Burgundy. With the marriage (1477) of Mary of Burgundy to Archduke Maximilian (later Emperor Maximilian I), the Low Countries passed to the house of Hapsburg. As part of the Spanish Netherlands (1482–1714) and then the Austrian Netherlands (1714–94), Belgium remained under foreign domination for three centuries, frequently serving as a battleground. In 1576 the Dutch and Flemish united against a harsh regime of Spanish governors. The Netherlands gained independence, but Flanders was recovered by Spain, and Antwerp and other cities were destroyed. Spain lost parts of the territory, chiefly to France, in wars concluded in 1648, 1659, 1668, and 1679; in 1714 the remaining Spanish possessions were transferred to the Austrian branch of the Hapsburgs.

Belgium was occupied by the French in 1794, was transferred from Austria to France (1797), and passed to the Netherlands in 1815 after Napoleon's defeat at Waterloo, just south of Brussels. A rebellion broke out against Dutch rule in 1830, and Belgian independence was declared.

In 1831 Prince Leopold of Saxe-Coburg-Gotha was chosen king of the Belgians and became Leopold I. The new country was among the first in Europe to industrialize and soon led the continent in the development of railways, coal mining, and engineering. Under the rule (1865–1909) of Leopold II rapid industrialization and colonial expansion, notably in the Congo, were accompanied by labor unrest and by the rise of the Socialist party in opposition to the reactionary and clerical groups. After the outbreak of World War I (Aug., 1914), Germany invaded and occupied Belgium, which mounted unexpected resistance. In World War II, Germany again attacked and occupied Belgium in May, 1940. King Leopold III (reigned 1934–51) surrendered unconditionally, but the Belgian cabinet, in exile at London, continued to oppose Germany. Liberation by British and American troops, aided by a Belgian underground army, came in Sept., 1944.

Despite much destruction during the war, the industrial plant remained relatively intact, enabling the Belgian economy to recover rapidly. Leopold III, who was barred from Belgium until July, 1950, abdicated in 1951 in favor of his eldest son, Baudouin. In 1960 the Belgian Congo (now Zaire) was given its independence. Sweeping constitutional reform in 1971 federalized the country by creating three regions—Flanders, Wallony, Brussels—with a degree of autonomy and provisions for equal political power in each. An early proponent of a united Europe and a firm advocate of collective security, Belgium is headquarters for the European Common Market and for the North Atlantic Treaty Organization (NATO).

Bel·go·rod (byĕl′gə-rət), city (1976 est. pop. 219,000), capital of Belgorod Oblast, Ukraine, S central European USSR, on the N Donets River. It is a railway junction and one of the chief centers in the USSR for the manufacture of cement and construction materials. These industries are based on nearby limestone deposits; one of the world's largest iron ore deposits is also located in the area.

Bel·go·rod-Dnes·trov·sky (byĕl′gə-rət-də-nyĕs-trôf′skē), city (1974 est. pop. 37,000), SW European USSR, in the Ukraine, a port at the mouth of the Dnestr River. It is also a rail junction and a trade center for wine. Industries include fishing and fish processing, winemaking, and meat and dairy processing. Founded by Greek colonists in the 6th cent. B.C., it later passed to Rome and Byzantium. In the 9th cent. it was a Slavic trade and political center. The city belonged to Genoa in the 14th cent. and to Moldavia in the 15th cent. The Turks acquired it in 1484. It was ceded to Russia in the early 19th cent., but was held by Rumania from 1918 to 1940 and by the Germans during World War II.

Bel·grade (bĕl′grād), city (1971 pop. 793,072), capital of Yugoslavia and of its republic of Serbia, at the confluence of the Danube and Sava rivers. It is the commercial, industrial, political, and cultural center of Yugoslavia, as well as a transportation and communications hub. Belgrade's industries include the manufacture of metals, textiles, chemicals, machine tools, and food products.

Strategically situated on land and river routes between Central Europe and the Balkans, the city grew around fortresses built by the Celts (3rd cent. B.C.), Illyrians, and Romans. Under the name of Singidinum it served as the harbor for much of Rome's Danubian fleet. Captured by the Huns, Goths, Sarmathians, and Gepids, who destroyed its forts, the city was retaken by Justinian in the 6th cent. A.D. It was held in the late 8th cent. by the Franks and from the 9th to 11th cent. by the Bulgars. It was then ruled again by Byzantium before becoming the capital of Serbia in the 12th cent. Before it fell to the Ottoman sultan Sulayman I in 1521, it was under Hungarian control. The Ottoman Turks made Belgrade their chief strategic fortress in Europe. Although the Austrians stormed it in 1688, 1717, and 1789, they were able to hold onto it only from 1718 until the Treaty of Belgrade (1739). Liberated during the Serbian uprising of 1806, Belgrade was recaptured by the Turks in 1813. The Turks finally left in 1815 but kept their garrison in the fortress until 1867. Belgrade became the capital of the kingdom of Serbia in 1882. Occupied by Austrian troops during World War I, the city was made the capital of the new kingdom of the Serbs, Croats, and Slovenes (Yugoslavia from 1929) after the war. During World War II, Belgrade suffered much damage. It was liberated by Yugoslav partisans, with Soviet aid, in 1944. The city is noted for its fine parks, palaces, museums, and churches.

Be·li·tung (bə-lē′tŏng), island (1970 est. pop. 126,000), 1,866 sq mi (4,833 sq km), Indonesia, in the Java Sea midway between Sumatra and Borneo. It has valuable tin mines (government-owned), worked chiefly by Chinese labor. Belitung is also known for its pepper.

Be·lize (bə-lēz′), British crown colony (1975 pop. 128,130), 8,867 sq mi (22,965.5 sq km), Central America, on the Caribbean Sea. Formerly named British Honduras, the name was changed to Belize in 1973. The capital is Belmopan. Belize is bounded on the north by Mexico, on the south and west by Guatemala, and on the east by the Caribbean. The land is generally low, with mangrove swamps and cays along the coast. Most of the area is heavily forested. The chief products are sugar cane, chicle, citrus fruits, and timber.

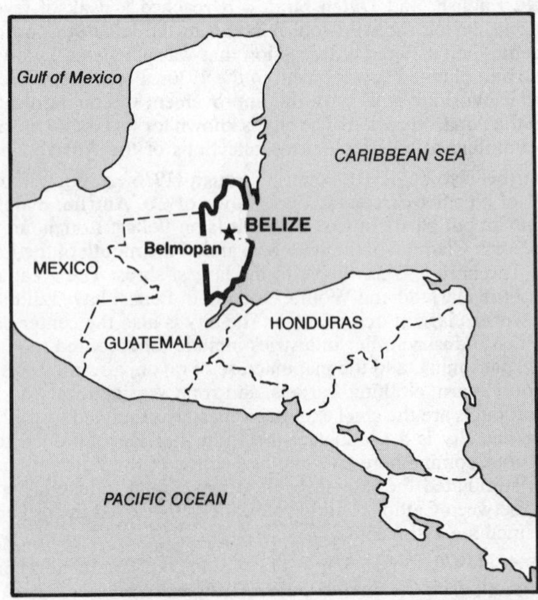

History. Once part of the Mayan civilization, the region was probably traversed by Cortés on his way to Honduras, but the Spanish made no attempt to colonization. British buccaneers, who used the cays to prey on Spanish shipping, founded the city of Belize (early 17th cent.). British settlers from Jamaica began the exploitation of timber. Spain contested British possession several times until defeated at the last Battle of St. George's Cay (1798). From 1862 to 1884 the colony was administered by the governor of Jamaica. Since 1821 Guatemala has claimed the territory as part of its inheritance from Spain. As Belize progressed toward independence, the tension between Britain and Guatemala over the issue increased. In 1964 the colony gained complete internal self-government.

Belize, city (1970 pop. 39,257), E Belize, at the mouth of the Belize River, on the Caribbean Sea. The river flows c.180 mi (290 km) generally west and is navigable almost to Guatemala; outlying cays exclude deep-draft vessels from its good harbor. Fish packing is the main industry. It was the capital of British Honduras, and later of Belize, until 1970, when it was devastated by a hurricane.

Bel·knap (bĕl'năp'), county (1970 pop. 32,367), 400 sq mi (1,036 sq km), central N.H., in a recreational and agricultural area drained by the Pemigewasset, Winnipesaukee, and Suncook rivers; formed 1840; co. seat Laconia. It has summer and winter resorts on Lake Winnipesaukee, agriculture, dairy farming, and manufacturing (textiles, hosiery, wood products, and mill machinery).

Bell (bĕl). **1.** County (1970 pop. 31,121), 370 sq mi (958.3 sq km), extreme SE Ky., in the Cumberlands bounded on the S by the Tenn. border, on the SE by the Va. border, and drained by the Cumberland River and its tributaries; formed 1867 as Josh Bell co., renamed between 1870 and 1880; co. seat Pineville. It has extensive bituminous-coal mining, with limestone quarries, iron deposits, some farming (dairy products, livestock, poultry, tobacco, and fruit), and some manufacturing. It includes Pine Mountain State Park. **2.** County (1970 pop. 124,483), 1,066 sq mi (2,760.9 sq km), central Texas, drained by the Little Leon, and Lampasas rivers; formed 1850; co. seat Belton. It is in a rich blacklands agricultural area yielding cotton, corn, oats, grain sorghums, hay, broomcorn, pecans, fruit, truck crops, livestock, and dairy products. Manufacturing, food processing, and chalk-rock quarrying are done. It includes Stillhouse Hollow Reservoir and part of Fort Hood.

Bel·laire (bə-lâr', bĕl'âr'). **1.** Village (1970 pop. 897), seat of Antrim co., NW Mich., NE of Traverse City near Bellaire Lake, in a farm and resort area. Fruit, potatoes, and vegetables are grown. **2.** City (1970 pop. 9,655), Belmont co., E Ohio, on the Ohio River below Wheeling, W.Va.; settled c.1802. Coal mining and the manufacture of enamelware and glass are done here. There are limestone and clay deposits in the area. **3.** City (1978 est. pop. 16,000), Harris co., SE Texas; inc. 1918. It is a suburb of Houston.

Bel·la·ry (bə-lä'rē), town (1971 pop. 125,127), Karnataka state, SE India. It is a district administrative center. Iron and manganese deposits are nearby.

Bel·leau Wood (bĕl'ō, bĕ-lō'), forested area in Aisne dept., N France, E of Château-Thierry. The scene of a victory over the Germans after hard fighting (June, 1918), involving chiefly U.S. troops, it was dedicated in 1923 as a permanent memorial to the American war dead.

Belle·fon·taine (bĕl-foun'tən, -fŏn'-), city (1978 est. pop. 10,700), seat of Logan co., W central Ohio; settled 1818, inc. 1835. It is a trade and rail center for a farm area. Its industries include printing and the manufacture of automobile bearings and electrical equipment.

Belle·fonte (bĕl'fŏnt), industrial borough (1970 pop. 6,828), seat of Centre co., central Pa., NE of Altoona; laid out 1795, inc. 1806. According to legend, the town was named by Talleyrand when he visited the site in 1794-95. It is a mountain resort and farm trading center. There are limestone quarries in the vicinity.

Belle Fourche (bĕl fo͝osh'), city (1970 pop. 4,236), seat of Butte co., W S.Dak., near the Wyo. border NW of Rapid City; settled 1878, platted 1890, inc. 1903. A shipping center for a cattle and sheep region, it has a beet-sugar factory, flour mills, and bentonite-processing plants.

Belle Fourche, river, c.290 mi (465 km) long, rising in NE Wyo., flowing NE and then E to the Cheyenne River in W S.Dak. The Belle Fourche project provides flood control and recreation facilities as well as irrigating c.57,000 acres (23,085 hectares) in S.Dak.

Belle Isle, Strait of, c.35 mi (55 km) long and from 10 to 15 mi (16–24 km) wide, between the island of Newfoundland and Labrador, Canada. The northern entrance to the Gulf of St. Lawrence, it is deep and free of rocks and shoals; ice blocks it from November to June. There is a strong tidal current. The tiny rock island Belle Isle (700 ft/213.5 m high), at the Atlantic entrance, has a lighthouse and is the first land sighted by ships from Europe.

Belle·ville (bĕl'vĭl'), city (1976 pop. 35,311), SE Ont., Canada, on Lake Ontario. Machinery, automotive accessories, optical lenses, and cheddar cheese are made here. Belleville is the seat of Albert College.

Belleville. **1.** City (1978 est. pop. 45,000), seat of St. Clair co., SW Ill.; inc. 1819. Coal mines here produce more than 5 million tons a year. Belleville also has farm-related industries and a great variety of manufactures, including mining equipment, industrial furnaces, machinery, dies and castings, beer, stoves, and clothing. **2.** City (1970 pop. 3,063), seat of Republic co., N central Kansas, N of Salina near the Nebr. line; laid out 1869, inc. 1878. It is the trade center for an agricultural region, and dairy products are made. A large regional fair is held in September. Nearby is a pool of brine (covering 4,000 acres/1,620 hectares) that supplied early settlers with salt. **3.** Town (1978 est. pop. 36,800), Essex co., NE N.J., on the Passaic River; settled c.1680, set off from Newark 1839, inc. 1910. Electrical equipment, fire extinguishers, water pumps, and precision instruments are among its manufactures.

Belle·vue (bĕl'vyo͞o). **1.** City (1970 pop. 8,847), Campbell co., N Ky., a suburb NNE of Covington near the Ohio River. Metal and plastic products are manufactured. **2.** City (1978 est. pop. 21,200), Sarpy co., E Nebr., a suburb of Omaha, on the Missouri River; inc. 1855. It has a meat-packing plant. The oldest city in the state, Bellevue was a trading post in the early 1800s and the site of a Presbyterian Indian mission in the 1840s and 1850s. The Strategic Aerospace Museum is in the city. **3.** City (1970 pop. 8,604), Huron and Sandusky cos., N Ohio, SSW of Sandusky, in a farm area; settled 1815, inc. 1851. It is a railroad center, and vegetable oils and metal products are made. Nearby are the Seneca Caverns. **4.** Borough (1970 pop. 11,586), Allegheny co., SW Pa., a residential suburb of Pittsburgh, on the Ohio River; settled 1802, inc. 1867. **5.** City (1978 est. pop. 68,600), King co., W Wash., opposite Seattle on Lake Washington; inc. 1953. Concrete and gravel, control systems, food products, and electronics parts are manufactured here.

Bel·ling·ham (bĕl'ĭng-hăm'). **1.** Town (1970 pop. 13,967), Norfolk co., S Mass., in a farm region; inc. 1719. **2.** City (1978 est. pop. 43,800), seat of Whatcom co., NW Wash., a port of entry on Bellingham Bay, one of the best landlocked harbors on the Pacific coast, near Canada; inc. 1904. It is an important shipping point for lumber, pulp, paper, and canned and frozen fruit.

Bel·lings·hau·sen Sea (bĕl'ĭngz-hou'zən), arm of the S Pacific Ocean, off the coast of Antarctica, stretching from Thurston Island to Alexander Island. It is named for Fabian Gottlieb von Bellingshausen, the Russian explorer, who visited the area in 1819-21.

Bel·lin·zo·na (bĕl'ĭn-zō'nə, bĕl'lĕn-tsô'nä), town (1970 pop. 16,979), capital of Ticino canton, S Switzerland, on the Ticino River, near the Italian border. It is a picturesque old town and a hub of transalpine traffic. Beverages and linoleum are produced. Possibly a Roman settlement, Bellinzona belonged at times to Lombardy, Como, Milan, France, and the Four Forest Cantons. In 1798 it became the capital of the Bellinzona canton under the Helvetic Republic and the capital of Ticino in 1803. The town is dominated by three castles (13th–15th cent.) of the dukes of Milan.

Bell·more (bĕl'môr'), uninc. residential town (1970 pop. 18,431), Nassau co., SE N.Y., on SW Long Island.

Bel·lu·no (bəl-lo͞o'nō), city (1976 est. pop. 35,519), capital of Belluno prov., Venetia, NE Italy, on the Piave River at the foot of the Dolomites. It is an agricultural and manufacturing center. A Roman town, it later belonged to various lords and was a free commune before voluntarily submitting to Venetian rule (1404-1797).

Bell·ville (bĕl'vĭl'), town (1970 pop. 49,026), Cape Prov., S South Africa, a suburb of Cape Town. Situated in a major wheat-growing region, the city ships wheat and manufactures processed lumber and synthetic textiles. Bellville was founded in 1861.

Bellville, city (1970 pop. 2,371), seat of Austin co., S Texas, WNW of Houston near the Brazos River; settled 1847, inc. 1927. It is the trade center of a farm and dairy region.

Bell·wood (bĕl'wo͝od'), village (1978 est. pop. 20,700), Cook co., NE Ill.; inc. 1900. Among its manufactures are electrical equipment and metal and asphalt products.

Bel·mont (bĕl'mŏnt'), county (1970 pop. 80,917), 539 sq mi (1,396 sq km), E Ohio, bounded on the E by the Ohio River, here forming the W.Va. border, and drained by Captina, Wheeling, and McMahon creeks; formed 1801; co. seat St. Clairsville. It has coal mines, limestone quarries, agriculture (truck crops, fruit, tobacco, grain, and dairy products), and some manufacturing.

Belmont. **1.** City (1978 est. pop. 25,800), San Mateo co., W Calif., a residential suburb midway between San Francisco and San Jose; laid out 1851, inc. 1926. The College of Notre Dame (est. 1851) is here. **2.** Town (1978 est. pop. 27,000), Middlesex co., E Mass., a residential suburb of Boston; settled 1636, inc. 1859. **3.** Village (1970 pop. 1,102), seat of Allegany co., W N.Y., NE of Olean, in a farm region; inc. 1871. Dairy products are made.

Bel·mo·pan (bĕl'mō-păn'), city (1970 est. pop. 1,000), capital of Belize, SW of Belize city. It became capital in 1970 after a hurricane devastated Belize, the former capital.

Be·lo Ho·ri·zon·te (bā'lō hôr'ə-zän'tē), city (1975 est. pop. 1,557,-464), capital of Minas Gerais state, E Brazil. The distributing and processing center of a rich agricultural and mining region, Belo Horizonte is the nucleus of a growing industrial complex; its chief manufactures are steel, steel products, and textiles. Gold, manganese, and precious stones (including diamonds) of the surrounding region are processed in the city. One of the most important inland cities of the republic, it was Brazil's first planned metropolis and was built (1895–97) to replace Ouro Prêto as the state capital. With its wide, tree-lined avenues, skyscrapers, and spacious parks, and with its beautiful surroundings and bracing climate, Belo Horizonte is a fashionable resort.

Be·loit (bə-loit'). **1.** City (1970 pop. 4,121), seat of Mitchell co., N central Kansas, on the Solomon River NW of Salina; settled 1868, inc. 1872. It is a trading center in a grain and livestock area. **2.** City (1978 est. pop. 34,600), Rock co., S Wis., on the Rock River; inc. 1846. It lies in an agricultural area. Beloit's manufactures include shoes, papermaking machinery, diesel engines, desalinization equipment, electrical equipment, and pumps. A trading post was established on the site in 1824 for trade with the Winnebago Indians, and in 1837 the first permanent settlers arrived from New England. Beloit College, founded in 1846, is in the city.

Be·lo·retsk (byĕ-lə-rĕtsk'), city (1974 est. pop. 72,000), W Siberian USSR, in the Urals on the Belaya River. One of the oldest industrial cities of the Urals region, Beloretsk is a metallurgical center, with industries that produce steel wire and cables.

Be·lo·rus·sia (bĕl'ō-rŭsh'ə, byĕ'lə-rōō'sē-ə) or **Be·lo·rus·sian Soviet Socialist Republic** (bĕl'ō-rŭsh'ən, byĕ'lə-rōō'sē-ən), constituent republic (1976 est. pop. 9,371,000), c.80,150 sq mi (207,600 sq km), W central European USSR; capital Minsk. Belorussia borders on Poland in the west, on the Lithuanian and Latvian republics in the north, on the Russian SFSR in the east, and on the Ukraine in the south. It has deposits of peat, limestone, clay, sand, chalk, dolomite, phosphorite, and rock and potassium salt. Lumbering is an important occupation. Potatoes, flax, hemp, sugar beets, rye, oats, and wheat are the chief agricultural products.

The region was colonized by East Slavic tribes from the 5th to the 8th cent. It fell (9th cent.) under the sway of Kiev and was later (12th cent.) subdivided into several Belorussian principalities forming part of the Kievan state. The dukes of Lithuania conquered Belorussia in the early 14th cent., and the region became part of the grand duchy of Lithuania, which in 1569 was merged with Poland. Through the Polish partitions of 1772, 1793, and 1795, all Belorussia passed to the Russian Empire. A battlefield in World War I and in the Soviet-Polish War of 1919–20, Belorussia experienced great devastation. In 1921 the Treaty of Riga awarded western Belorussia to Poland. The eastern and larger part formed the Belorussian SSR, which joined the USSR in 1922. Occupied by the Germans during World War II, Belorussia was one of the most devastated areas of the USSR. The republic has a separate seat in the United Nations.

Bel·sen (bĕl'zən), village, Lower Saxony, NW West Germany, NW of Celle. With nearby Bergen it was the site of an infamous Nazi concentration camp during World War II.

Bel·ton (bĕl'tən). **1.** City (1970 pop. 12,179), Cass co., W central Mo., S of Kansas City, in an agricultural area; founded 1870, platted 1871. Carry Nation lived and is buried here. An air force base is nearby. **2.** City (1970 pop. 8,696), seat of Bell co., central Texas, NNE of Austin; founded 1850, inc. 1884. It is a market center in a rich farm region. Furniture and machinery are made.

Bel·tram·i (bĕl-trăm'ē), county (1970 pop. 26,373), 2,507 sq mi (6,493.1 sq km), NW Minn., in an agricultural and resort area drained by the Upper and Lower Red lakes and the headwaters of the Mississippi; formed 1866; co. seat Bemidji. Its agriculture includes dairy products, potatoes, hay, grain, and livestock. Red Lake Indian Reservation is here.

Bel·vi·dere (bĕl'vĭ-dîr'). **1.** City (1970 pop. 14,061), seat of Boone co., N Ill., on the Kishwaukee River; inc. 1847. It is a farm trade center with food-processing industries, machine shops, and a huge automobile assembly plant. **2.** Town (1970 pop 2,722), seat of Warren co., NW N.J., on the Delaware River NNE of Phillipsburg; settled 1759, laid out 1799, inc. 1845. Building blocks, felt, hosiery, and dairy products are manufactured.

Bel·zo·ni (bĕl-zō'nē), city (1970 pop. 3,394), seat of Humphreys co., W central Miss., on the Yazoo River NNW of Yazoo City, in a rich cotton area; founded c.1827. Hardwood is among its products. Prehistoric Indian artifacts were found near here in 1951.

Be·mid·ji (bə-mĭj'ē), city (1978 est. pop. 11,400), seat of Beltrami co., N central Minn., on Bemidji and Irving lakes, through which flows the Mississippi River; inc. 1896. It is in a summer and winter resort and sport fishing area; tourism is the major industry. The city is also a trade and marketing center for the dairy farms of the region, and has lumber, wood-product, and boat manufactures. On the lakeshore stands an 18-ft (5.5-m) figure of Paul Bunyan and his ox.

Be·na·res (bə-nä'rəs, -rēz): *see* Varanasi, India.

Bend (bĕnd), city (1978 est. pop. 16,600), seat of Deschutes co., W central Oregon, on the Deschutes River, at the E foot of the Cascade Range; inc. 1904. Lumbering is the primary industry, and tourism is also important. Nearby pumice fields offer moonlike terrain for a lunar base research facility.

Ben·de·ry (bĭn-dyĕ'rē), city (1976 est. pop. 97,000), SW European USSR, in Moldavia, a port on the Dnestr River. It is a rail hub and a trade center for timber, fruits, and tobacco. Industries include the production of foodstuffs, electrical apparatus, footwear, and textiles. Historically important as the gateway of Bessarabia, the city was founded on the site of a 14th cent. Genoese colony. Captured from Moldavia by the Turks in 1538, it became a fortress on the Dnestr. It was captured by Russia in 1812. Between the world wars, Bendery belonged to Rumania; it was transferred to the USSR in 1940 but was occupied by Rumanian troops from 1941 to 1944.

Ben·di·go (bĕn'dĭ-gō), city (1976 pop. 32,573), Victoria, SE Australia. Founded in 1851 during the gold rush, Bendigo was the center for the greatest goldfield in Victoria. Mining continues, but the city is now an industrial, railroad, and commercial center in a livestock and dairy-farming region.

Be·ne·ven·to (bĕn'ə-vĕn'tō, bā'nā-vān'tō), city (1976 est. pop. 52,300), capital of Benevento prov., in Campania, S Italy. It is a trade center for wine and tobacco. Farm machinery, optical instruments, liqueur, and nougat are manufactured. A leading town of Samnium, Benevento became under the Romans an important trade center on the Appian Way. It was the capital of a powerful Lombard duchy (6th–11th cent.) that extended over much of southern Italy. Except for short periods of foreign occupation, the city was under papal rule from the 11th cent. to 1860. Noteworthy structures include the cathedral (11th–13th cent., restored after being severely damaged in World War II) and a triumphal arch erected (114 A.D.) for Trajan.

Ben·e·wah (bĕn'ə-wä'), county (1970 pop. 6,230), 791 sq mi (2,048.7 sq km), N Idaho, in a rolling, hilly area bordering on Wash. and drained by the St. Joe and St. Maries rivers; formed 1915; co. seat St. Maries. It has a lumbering industry and recreational facilities. Hay, flax, and wheat are grown, and livestock is raised. It includes part of Coeur d'Alene Indian Reservation.

Ben·gal (bĕn-gôl', bĕng-), region, 77,442 sq mi (200,575 sq km), E India and Bangladesh, on the Bay of Bengal. The inland sections are mountainous, with peaks up to 12,000 ft (3,660 m) high in the northwest, but most of Bengal is the fertile land of the Ganges-Brahmaputra alluvial plains and delta. Along the coast are richly timbered jungles, swamps, and islands.

In the 3rd cent. B.C., Bengal belonged to the empire of Asoka. It became a political entity in the 8th cent. A.D. under the Buddhist Pala kings. In the 11th cent. the Hindu Sena dynasty arose from the remnants of the Pala empire. Bengal was conquered (c.1200) by Moslems of Turki descent. When the Portuguese began their trading activities (late 15th cent.), Bengal was a part of the Mogul empire.

The British East India Company made its first settlement in 1642 and extended its occupation by conquering the native princes and expelling the Dutch and French. Moslem control of Bengal ended after the Battle of Plassey in 1757. Under British control, Bengal was a presidency of India. The population is almost equally divided between Moslems and Hindus. When India was partitioned in 1947, the presidency was divided along the line approximately separating the two religious communities.

West Bengal (1971 pop. 44,440,095), 33,928 sq mi (87,874 sq km), with its capital at Calcutta, became a state of India. A highly industrialized region, it has jute mills, steel plants, and chemical industries, all mainly centered in the Hooghlyside industrial complex. Coal is mined and petroleum is exploited. In 1950, West Bengal absorbed the state of Cooch Behar. East Bengal, overwhelmingly Moslem in population, became East Pakistan in 1947 and the independent nation of Bangladesh in 1971.

Bengal, Bay of, arm of the Indian Ocean, c.1,300 mi (2,090 km) long and 1,000 mi (1,610 km) wide, bordered on the W by Sri Lanka and India, on the N by Bangladesh, and on the E by Burma and Thailand. The Andaman and Nicobar Islands separate it from the Andaman Sea, its eastern arm.

Ben·ga·si or **Ben·gha·zi** (both: bĕn-gä'zē), city (1970 est. pop. 170,000), capital of Bengasi district, NE Libya, the main city of Cyrenaica and a port on the Mediterranean Sea. It is primarily an administrative and commercial center. Manufactures include processed food, beverages, textiles, and cement. On the site of Bengasi the Greeks founded (7th cent. B.C.) the colony of Hesperides, which was later (3rd cent. B.C.) renamed Berenice after the wife of Ptolemy III of Egypt. Under the Romans, who conquered it in the mid-1st cent. B.C., Bengasi had a large Jewish colony. In the 5th cent. A.D. the Vandals severely damaged the city, and in the 7th cent. it was captured by the Arabs. The Ottoman Turks took the city in the mid-16th cent., and they held it until it was captured by Italy in 1911. The Italians modernized the city and enlarged its port. At the start of World War II, Bengasi had about 22,000 Italian inhabitants, but they were evacuated before the city fell to the British in late 1942. From 1951 to 1972 Bengasi was the co-capital (with Tripoli) of Libya. The city is the site of the Univ. of Libya, founded in 1955.

Ben·guel·a (bĕn-gĕl'ə, bĕng-gĕl'ə), city (1970 est. pop. 40,996), W Angola, on the Atlantic. It is a rail terminus, an export point, and a commercial and fishing center. A fort was built on the site in the late 16th cent., and the city was founded in 1617. Benguela's port played an important role in slave trading.

Ben Hill, county (1970 pop. 13,171), 255 sq mi (660.5 sq km), S central Ga., bounded on the NE by the Ocmulgee River and on the W by the Alapaha River; formed 1906; co. seat Fitzgerald. It is in a coastal-plain timber and agricultural area yielding corn, tobacco, peanuts, and livestock.

Be·ni (bā'nē), river, 994 mi (1,599.3 km) long, NW and central Bolivia, rising in the E Andes and flowing N and NE to join the Mamore River. One of the major Bolivian waterways, it is particularly noted for the abundant quantities of quinine bark and rubber found along its lower course.

Be·ni·cia (bə-nē'shə), city (1977 est. pop. 8,783), Solano co., W Calif., a port on Carquinez Strait of San Francisco Bay, NE of Oakland; founded 1847, inc. 1850. From 1853 to 1854 it was the state capital. A U.S. arsenal here was established as an army post in 1849.

Be·ni Ha·san (bĕ'nē hä-sän'): *see* Bani Hasan, Egypt.

Be·nin (bə-nēn', bĕn'ĭn), republic (1977 est. pop. 3,230,000), 43,483 sq mi (112,621 sq km), W Africa, bordering on Togo in the W, on Upper Volta and Niger in the N, on Nigeria in the E, and on the Bight of Benin in the S. Porto-Novo is the capital. Benin (formerly Dahomey) falls into four main geographic regions. In the south is a narrow coastal zone fringed on the north by a series of interconnected lagoons and lakes with only two outlets to the sea. Behind the coastal region is a generally flat area of fertile clay soils; this is crossed by the wide Lama marsh, through which flows the Ouémé River. In the northwest is a region of forested mountains, from which the Mekrou and Alibori rivers flow northeast to the Niger River (which forms part of the country's northern border). In the northeast is a highland region covered mostly with savanna and containing little fertile soil.

Economy. Benin is overwhelmingly agricultural, with the majority of workers engaged in subsistence farming. The chief crops are manioc, maize, cassava, millet, sorghum, groundnuts, pulses, cacao, cotton, and palm nuts and kernels. Large numbers of goats, sheep, and pigs are raised. There is a sizable freshwater fishing industry, and some sea fish are also caught. Most of Benin's few manufactures are either processed agricultural goods or basic consumer items; the main products include palm oil, palm-kernel oil, palmetto, soap, textiles, footwear, jute sacks, cement, and ginned cotton. Mineral resources, including petroleum, chromite, low-quality iron ore, ilmenite, and titanium, have not as yet been exploited on a large scale.

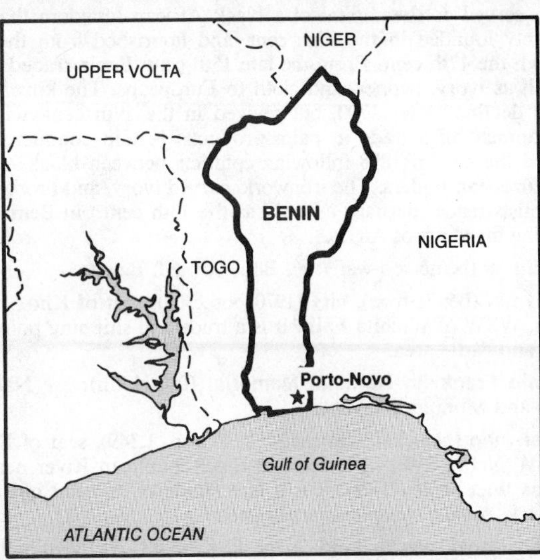

History. Little is known about the history of northern Benin. According to oral tradition, a group of Aja migrated (12th or 13th cent.) eastward from Tado on the Mono River and founded the village of Allada. Allada became the capital of Great Ardra, reaching the peak of its power in the 16th and early 17th cent. Other Aja living at Abomey organized into a strongly centralized kingdom with a standing army and gradually mixed with the local people, thus forming the Fon, or Dahomey, ethnic group. By the late 17th cent. the Dahomey were raiding their neighbors for slaves, who were then sold (through coastal middlemen) to European traders.

In order to establish direct contact with the European traders, King Agaja of Dahomey (reigned 1708-32) conquered most of the south. This expansion brought Benin into conflict with the powerful Yoruba kingdom of Oyo, which forced Benin to pay an annual tribute until 1818. Benin's involvement in the slave trade declined temporarily in the late 18th cent. The economy revived under King Gezo (reigned 1818-58), who built up the army and raided Oyo, which had been weakened by civil wars, for slaves. In 1851 Gezo signed a commercial treaty with France. His successor, King Gelele (reigned 1858-89), pursued an aggressive policy toward his neighbors and vigorously engaged in the slave trade.

During the 1880s, as the scramble for Africa among the European powers accelerated, France tried to secure its hold on the Benin coast in order to keep it out of German or British hands. Between 1895 and 1898, the French added the northern part of present-day Benin, and in 1899 the whole colony was made part of French West Africa. Under the French a port was constructed at Cotonou, railroads were built, and the output of palm products increased. In 1946 Benin became an overseas territory with its own parliament and representation in the French national assembly.

On Aug. 1, 1960, Benin became fully independent as the Dahomey Republic. The new country was plagued by governmental instability, caused by economic troubles, ethnic rivalries, and social unrest. In 1963, following demonstrations by workers and students, the armed forces staged a successful coup d'état, putting Justin Ahomadegbé into power. In 1965 the military replaced Ahomadegbé with Col. Christophe Soglo. Soglo was ousted in late 1967, and a younger army officer, Lt. Col. Alphonse Alley, came to power with the goal of re-establishing civilian rule. Elections in May, 1968, were held under a cloud of suspicion, and the results were subsequently disallowed. Later in 1968, Dr. Emile Zinsou was made president, and he gave way in 1969 to Lt. Col. Paul-Emile de Souza. Benin tried to hold elections in 1970, but severe disagreement between northern and southern politicians led to their cancellation. Instead, a three-man presidential council was formed; each member was to lead the country for two years. The first leader was Hubert Maga, who in May, 1972, was replaced without incident by Ahomadegbé. However, in Oct., 1972, the military again intervened, installing an 11-man government headed by Maj. Mathieu Kerekou. In 1975 the country's name was officially changed to the Benin Republic.

Benin, city (1971 est. pop. 122,000), S Nigeria, a port on the Benin River. Rubber, palm nuts, and timber are produced nearby and processed in Benin. Furniture and carpets are also made in the city. Benin served as the capital of a black African kingdom that was probably founded in the 13th cent. and flourished from the 14th through the 17th cent. From the late 15th cent. Benin traded slaves as well as ivory, pepper, and cloth to Europeans. The kingdom of Benin declined after 1700, but revived in the 19th cent. with the development of a trade in palm products. Britain conquered and burned the city in 1898 following conflicts between black African and European traders. The ironwork, carved ivory, and bronze portrait busts made (perhaps as early as the 13th cent.) in Benin rank with the finest art of Africa.

Be·ni Su·ef (bĕ'nē sōō-wăf'): *see* Bani Suwayf, Egypt.

Ben·ja·min (bĕn'jə-mən), city (1970 pop. 308), seat of Knox co., N Texas, WSW of Wichita Falls. It is a trade and shipping point in a grain and cattle area.

Benjamin Frank·lin National Memorial (frăngk'lĭn): *see* National Parks and Monuments Table.

Ben·kel·man (bĕng'kəl-mən), city (1970 pop. 1,349), seat of Dundy co., SW Nebr., SW of McCook on the Republican River near the Kansas line; settled 1880. It is a farm trade center and has a fish hatchery. A state recreation area is nearby.

Ben Mac·dhu·i (bĕn' măk-dōō'ē), peak, 4,296 ft (1,310.3 m) high, SW Aberdeenshire, Scotland, in the Cairngorm Mts. It is the second-highest peak in Scotland.

Ben·nett (bĕn'ĭt), county (1970 pop. 3,088), 1,181 sq mi (3,058.8 sq km), S S.Dak., on the Nebr. border in a farming area drained by the South Fork White River and several creeks; formed 1909; co. seat Martin. Corn and wheat are grown, and cattle are raised.

Ben·netts·ville (bĕn'ĭts-vĭl'), city (1970 pop. 7,468), seat of Marlboro co., NE S.C., SW of Fayetteville, N.C.; laid out 1818, inc. 1866. It is the trade and processing center for a cotton, tobacco, corn, dairy, and peach area. Textiles are made.

Ben Ne·vis (bĕn nē'vĭs, nĕv'ĭs), peak, 4,406 ft (1,343.8 m) high, Inverness-shire, W Scotland. It is the highest peak of Great Britain. Ruins of an observatory are on the summit, from which there is an impressive view, especially on the northeastern side with its precipice of more than 1,450 ft (442.3 m).

Ben·ning·ton (bĕn'ĭng-tən), county (1970 pop. 29,282), 672 sq mi (1,740.5 sq km), SW Vt., in the Green Mts. on the Mass. and N.Y. borders and drained by Batten Kill and the Deerfield, Hoosic, Walloomsac, and Mettawee rivers; organized 1779; co. seats Bennington and Manchester. It has agriculture, light industry, and ski resorts, and includes part of Green Mountain National Forest.

Bennington, town (1970 pop. 14,586), a seat of Bennington co., SW Vt.; chartered 1749, settled 1761. Major manufactures of the town are automotive batteries, paper products, electronic components, air-conditioning equipment, lubricating equipment, furniture, and lithographic products. The surrounding area has dairy farms and several ski resorts. Points of interest in Bennington include a monument that is 300 ft (91.5 m) high commemorating the Revolutionary War Battle of Bennington; the site of the first schoolhouse in Vermont; Catamount Tavern, meeting place of the Green Mountain Boys; the site of abolitionist William Lloyd Garrison's printing shop; the Old First Church (1805); and the Walloomsac Inn, opened in 1763. Bennington College is in the town.

Be·no·ni (bə-nō'nē), town (1970 pop. 151,294), Transvaal, NE South Africa, on the Witwatersrand. It is the distribution center for a gold-mining district. The chief manufacture is electrical equipment.

Ben·son (bĕn'sən), county (1970 pop. 8,245), 1,403 sq mi (3,633.8 sq km), N central N.Dak., in an agricultural area watered by the Sheyenne River; formed 1883; co. seat Minnewaukan. Devils Lake is on its eastern border. Wheat is the primary crop, with some barley and sunflowers. Machinery is manufactured. It includes part of Fort Totten Indian Reservation.

Benson, city (1970 pop. 3,484), seat of Swift co., W Minn. E of Ortonville; platted 1870, inc. as a village 1877, as a city 1908. It is a shipping center for wheat, dairy products, and livestock.

Bent (bĕnt), county (1970 pop. 6,493), 1,519 sq mi (3,934.2 sq km), SE Colo., in an irrigated agricultural area drained by the Purgatoire and Arkansas rivers; formed 1870; co. seat Las Animas. Livestock is raised, and sugar beets and cattle feed are grown. John Martin Reservoir is here.

Ben·ton (bĕn'tən). **1.** County (1970 pop. 50,476), 886 sq mi (2,294.7 sq km), extreme NW Ark., in the Ozarks bordered on the W by Okla., on the N by Mo., and drained by the White and Illinois rivers; formed 1836; co. seat Bentonville. Its agriculture includes fruit, grain, truck crops, poultry, livestock, and dairy products. It also has nurseries, stands of timber, and mineral springs. Pea Ridge National Military Park is in the northeast. **2.** County (1970 pop. 11,262), 409 sq mi (1,059.3 sq km), W Ind., bordered on the W by Ill. and drained by Sugar and Big Pine creeks; formed 1840; co. seat Fowler. It has agriculture (corn, wheat, hay, soybeans, poultry, and livestock), food-processing plants, and diversified manufacturing. **3.** County (1970 pop. 22,885), 718 sq mi (1,859.6 sq km), E central Iowa, in a prairie agricultural area drained by the Cedar River; formed 1837; co. seat Vinton. Hogs, cattle, poultry, corn, oats, and wheat are produced. There are limestone quarries in the area. **4.** County (1970 pop. 20,841), 404 sq mi (1,046.4 sq km), central Minn., in an agricultural area bounded on the W by the Mississippi; formed 1849; co. seat Foley. It processes dairy products, livestock, and grain, has marl deposits and does some light manufacturing. **5.** County (1970 pop. 7,505), 412 sq mi (1,067.1 sq km), N Miss., bounded on the N by the Tenn. border and drained by the Wolf River and Tippah Creek; formed 1870; co. seat Ashland. It has agriculture (cotton, corn, hay, and sweet potatoes) and lumbering. It includes part of Holly Springs National Forest. **6.** County (1970 pop. 9,695), 735 sq mi (1,903.7 sq km), central Mo., in the Ozarks, drained by the Osage, South Grand, and Pomme de Terre rivers and crossed by the Lake of the Ozarks; formed 1835; co. seat Warsaw. There are barite deposits and lumber, and hunting and fishing are done, as well as general farming (dairy products, livestock, and poultry). **7.** County (1970 pop. 53,776), 668 sq mi (1,730.1 sq km)., W Oregon, with the Coast Range in the W and the Willamette River valley in the E; formed 1847, small portion of Lincoln co. added 1949; co. seat Corvallis. It has agriculture, a lumber industry, dairying, and food-processing plants. Part of Siuslaw National Forest is here. **8.** County (1970 pop. 12,126), 392 sq mi (1,015.3 sq km), W Tenn., bounded on the E and NW by arms of the Kentucky Reservoir of the Tennessee River and traversed by the Big Sandy River; formed 1835; co. seat Camden. It has gravel pits, livestock and poultry raising, dairying, and diversified agriculture (corn, cotton, soybeans, peanuts, and sorghums). **9.** County (1970 pop. 67,540), 1,722 sq mi (4,460 sq km), S Wash., on the Oregon border and bisected by the Yakima River; formed 1905; co. seat Prosser. Potatoes, grapes, hops, mint, and wheat are grown. Livestock, dairy products, and poultry are also important. The Hanford Works, a U.S. Atomic Energy Commission reservation, is here.

Benton. **1.** City (1970 pop. 16,499), seat of Saline co., central Ark.; founded 1836. Its chief industry, aluminum mining and refining, is based on the extensive high-grade bauxite deposits found in the area. **2.** City (1970 pop. 6,833), seat of Franklin co., S Ill., S of Mount Vernon, in a farm and coal area; inc. 1841. Livestock, poultry, and grain are produced. Flags are made here. **3.** City (1970 pop. 3,652), seat of Marshall co., SW Ky., SE of Paducah; inc. 1845. It is in a rich farm area yielding strawberries, corn, potatoes, tobacco, and grain. An annual singing festival has been held here since 1884. **4.** Village (1970 pop. 1,493), seat of Bossier parish, NW La., N of Shreveport, in a cotton and timber region. **5.** Town (1970 pop. 640), seat of Scott co., SE Mo., near the Mississippi S of Cape Giradeau. Cotton, corn, and wheat are grown. **6.** Town (1977 est. pop. 999), seat of Polk co., SE Tenn., near the Hiwassee River ENE of Chattanooga, in a timber and farm area. Woodworking is done here.

Ben·ton·ville (bĕn'tən-vĭl'), city (1975 est. pop. 6,707), seat of Benton co., extreme NW Ark., on the Ozark plateau N of Fayetteville; settled 1837. It ships poultry, dairy products, and apples and has hatcheries, feed mills, and a cheese factory. Iron and aluminum casting and clothing are made. There are mineral springs in the area.

Bent's Old Fort National Historic Site (bĕnts): *see* National Parks and Monuments Table.

Be·nue (bā-nwā'), river, W Africa, chief tributary of the Niger, flowing c.670 mi (1,080 km) W from the United Republic of Cameroon into the Niger River at Lokoja, Nigeria.

Ben·zie (bĕn′zē), county (1970 pop. 8,593), 316 sq mi (818.4 sq km), NW Mich., bounded on the W by Lake Michingan and drained by the Betsie and Platte rivers; organized 1869; co. seat Beulah. It has agriculture (fruit, grain, truck crops, potatoes, beans, and dairy products), fisheries, and resorts. Many small lakes and part of Sleeping Bear Dunes National Seashore are here.

Bep·pu (bĕp′o͞o), city (1976 est. pop. 135,402), Oita prefecture, NE Kyushu, Japan, on Beppu Bay. It is a major fishing port and a tourist resort noted for its numerous hot springs.

Be·rat (bĕ-rät′) or **Be·ra·ti** (bĕ-rä′tē), town (1971 est. pop. 26,700), capital of Berat prov., S central Albania. It is a commercial center producing foodstuffs, textiles, and leather products. There is an oil field nearby. Built probably on the site of ancient Antipatrea, Berat fell to the Serbs in 1345 and to the Turks in 1440. A citadel, rebuilt by the Byzantines in the 13th cent., overlooks the town.

Ber·be·ra (bûr′bə-rə), city (1966 est. pop. 14,000), N Somalia, a port on the Gulf of Aden. The city, which was first described in the 13th cent. by Arab geographers, was taken in 1875 by the rulers of Egypt; when they withdrew in 1884 to fight the Mahdi in Sudan, Britain took Berbera. It served until 1941 as the winter capital of British Somaliland.

Berch·tes·ga·den (bĕrkʜ′təs-gä′dən), town (1970 pop. 4,343), Bavaria, SE West Germany, in the Bavarian Alps. It is a popular winter and summer resort. Salt has been mined there since the 12th cent. The site of Hitler's wartime villa, the Berghof, is nearby.

Ber·di·chev (byĕr-dyē′chĕf), city (1976 est. pop. 80,000), SW European USSR, in the Ukraine. It is a rail junction and the industrial and trade center of an area where sugar beets are raised. Engineering, sugar refining, tanning, and the manufacture of foodstuffs are the major industries. Founded in the 14th cent., Berdichev passed to Lithuania in 1546 and to Poland in 1569; Russia acquired it in 1793.

Ber·dyansk (bĕr′dyänsk′), formerly **O·si·pen·ko** (ŏs′ə-pĕng′kō), city (1976 pop. 117,000), S European USSR, in the Ukraine on the Berdyansk Gulf of the Sea of Azov. It is a port and a rail terminus. Industries include fishing and fish processing, flour milling, oil refining, and the production of machinery, cables, and clothing. Berdyansk is also a health and seaside resort.

Be·re·a (bə-rē′ə). **1.** City (1970 pop. 6,956), Madison co., central Ky., SSE of Lexington, in a coal and farm area. Located near the Cumberlands, it is a summer resort and the seat of Berea College. There are ancient Indian fortifications nearby. **2.** City (1978 est. pop. 20,900), Cuyahoga co., NE Ohio, a suburb of Cleveland; settled 1809, inc. as a city 1930. Berea was once famous for its sandstone quarries.

Be·re·zi·na (byĕ-rā-zē-nä′), river, c.380 mi (610 km) long, rising in NW Belorussia, E central European USSR. It flows generally south past Borisov and Bobruysk into the Dnepr River. It is navigable for most of its length. The heroic retreat across the Berezina of the remnants of Napoleon's Grand Army took place near Borisov from Nov. 26 to Nov. 29, 1812. Despite the loss of more than 20,000 men, the crossing saved Napoleon and his forces from capture.

Be·rez·ni·ki (bĭ-ryĕz-nyĭ-kē′), city (1976 est. pop. 172,000), E European USSR, a port on the Kama River. Situated in an area rich in potassium salts, Berezniki is one of the main industrial centers of the Urals and contains a huge chemical combine.

Berg (bûrg, bĕrk), former duchy, W West Germany, along the right bank of the Rhine River between the Ruhr and Sieg rivers. A county in the 12th cent., Berg passed (1348) to the dukes of Jülich and in 1380 was made a duchy. Berg passed in 1511 to Duke John III of Cleves, whose line died out in 1609, setting off a virulent struggle over succession that contributed to the outbreak of the Thirty Years War (1618–48). Ceded to France in 1806, Berg was raised to a grand duchy by Napoleon I. The Congress of Vienna assigned (1815) the duchy to Prussia.

Ber·ga·mo (bĕr′gä-mō), city (1976 est. pop. 127,816), capital of Bergamo prov., in Lombardy, N Italy, in the foothills of the Alps. It is an industrial center and an agricultural market. Manufactures include machinery, textiles, and cement. Originally a Gallic town, Bergamo became an independent commune in the 12th cent. In 1797 it was included in the Cisalpine Republic. Bergamo is divided into two sections: the old, hilltop town and the modern, lower sector.

Ber·gen (bĕr′gən), city (1970 pop. 11,046), Rostock dist., N East Germany. It is on Rügen Island and has a fishing industry.

Bergen, city (1977 est. pop. 212,692), capital of Hordaland co., SW Norway, situated on inlets of the North Sea. It is Norway's third-largest city and a major shipping and shipbuilding center. Other manufactures include processed food, textiles, steel, machinery, and electrical equipment. Founded c.1070, Bergen soon became the largest city of medieval Norway. It was often the royal seat. The city became an establishment of the Hanseatic League in the mid-14th cent. The Hansa merchants continued to have influence until the late 18th cent. During the disturbances accompanying the Reformation (16th cent.), most of the city's old churches and monasteries were destroyed. However, Bergen remained Norway's leading city until the rise of Oslo in the 19th cent. The center of Bergen was rebuilt after a severe fire in 1916. Nevertheless, the city retains many impressive monuments of its medieval past. One of its most famous buildings is Bergenhus fortress, which contains Haakon's Hall (1261); it was rebuilt after being heavily damaged in World War II.

Ber·gen (bûr′gən), county (1970 pop. 897,148), 233 sq mi (603.5 sq km), extreme NE N.J., bounded on the E by the Palisades and on the N by the N.Y. border; formed 1675; co. seat Hackensack. It is in an industrial and residential area with diversified manufactures. It includes Oradell Reservoir and part of Palisades Interstate Park.

Ber·gen op Zoom (bĕr′gən ôp zōm′), town (1977 est. pop. 42,133), North Brabant prov., SW Netherlands, on the Zoom River near its confluence with the Eastern Scheldt. It is a commercial and fishing port and its industries manufacture chemicals, machinery, and refined sugar. Bergen op Zoom was chartered c.1260 and was a major commercial rival of Antwerp until the 16th cent. It was repeatedly besieged by the Spanish and French from the 16th to the 18th cent. and by the English in 1814.

Ber·ge·rac (bĕr-zhə-räk′), town (1968 pop. 28,015), Dordogne dept., SW France, in Périgord, on the Dordogne River. It is a farm-trade and processing center. It also has boiler works, foundries, and shoe and clothing plants. Possessed by the English in the 14th cent., it was recovered in 1450 by the French. It became a Protestant stronghold and was taken (1621) by Louis XIII. A tobacco museum and an experimental tobacco institute are here.

Ber·gisch-Glad·bach (bĕr′gĭsh-glät′bäkʜ′), city (1974 est. pop. 98,679), North Rhine–Westphalia, W West Germany; chartered 1856. Manufactures include paper and metal goods, wool, pharmaceuticals, and electrical equipment.

Ber·ham·pore (bûr′əm-pôr′): *see* Baharampur, India.

Ber·ing Island (bîr′ĭng, bâr′-), largest of the Komandorski Islands, c.55 mi (90 km) long and up to c.15 mi (24 km) wide, off Kamchatka peninsula, E Far Eastern USSR, in the Bering Sea. It is low and treeless and is subject to severe windstorms. Vitus Bering was shipwrecked and died here.

Bering Sea, c.878,000 sq mi (2,274,020 sq km), northward extension of the Pacific Ocean between Siberia and Alaska. It is screened from the Pacific proper by the Aleutian Islands. The Bering Strait connects it with the Arctic Ocean. The sea's largest embayments are the Gulf of Anadyr, Norton Sound, and Bristol Bay. The warm Japan Current has little influence on the Bering Sea, which has much ice; it can usually be traversed only from June to October. The sea was first explored in the 17th cent., but not until after the voyages of Vitus Bering (1728, 1741) was the fur-seal wealth of the Bering Sea made widely known. Over the years pelagic (open-sea) sealing, practiced by Canadian and other sealing vessels, greatly reduced the herd and threatened its extinction. In 1911 Great Britain, Russia, Japan, and the United States finally agreed to prohibit pelagic sealing. For several years sealing was stopped completely, and then it was resumed but only under careful restrictions. Gradually the herd has been built up again. The 1911 agreement also prohibited the killing of sea otters, which are, however, almost extinct today.

Bering Strait, c.55 mi (90 km) wide, between extreme NE Asia and extreme NW North America, connecting the Arctic Ocean and the Bering Sea. It is usually completely frozen over from October to June. The narrowness of the strait makes it possible for small boats to cross from Chukchi Peninsula in the USSR to Seward Peninsula in Alaska. Since Alaska and Siberia were connected in the distant past, the usual theory is that the ancestors of the American Indians crossed the land bridge to North America.

Berke·ley (bûrk′lē). **1.** County (1970 pop. 56,199), 1,110 sq mi (2,874.9 sq km), SE S.C., bounded on the N by the Santee River; formed 1882; co. seat Moncks Corner. The Santee-Cooper irrigation and power development is in the north section, with dams on the

Cooper and Santee rivers forming lakes Marion and Moultrie. Cotton, dairy products, tobacco, poultry, and timber are produced. It includes part of Francis Marion National Forest. **2.** County (1970 pop. 36,356), 316 sq mi (818.4 sq km), NE W.Va., in the Eastern Panhandle, bounded on the NE by the Potomac River, on the SW by the Va. border; formed 1772; co. seat Martinsburg. Lying partly in the Great Appalachian Valley, it is drained by Opequon and Black creeks, short tributaries of the Potomac. Its agriculture includes fruit, vegetables, livestock, and dairy products. It also has limestone quarries and diversified manufacturing.

Berkeley. 1. City (1978 est. pop. 108,500), Alameda co., W Calif., on the E shore of San Francisco Bay; inc. 1878. Originally part of the Rancho San Antonio granted to the Peralta family in 1820 by the Spanish crown, the site was purchased by Americans in 1853. The settlement, at first called Oceanview, was named Berkeley in 1866. The city's industries include food processing and the manufacture of chemicals, pharmaceuticals, and metal products. A campus of the Univ. of California and several divinity schools are in Berkeley. Lawrence Radiation Laboratory, an atomic research center, is nearby. **2.** Village (1970 pop. 6,152), Cook co., NE Ill., a suburb W of Chicago; inc. 1924. **3.** City (1970 pop. 19,743), St. Louis co., E Mo.; inc. 1937. Its manufactures include aircraft, truck bodies, and brake fluid. The first International Air Meet in the United States was held in Berkeley in 1910.

Berkeley Springs, town (1970 pop. 944), seat of Morgan co., W.Va., in the Eastern Panhandle NW of Martinsburg; chartered 1776 as Bath (its official name). It is a health resort, with mineral springs. Glass-sand pits are here, and lumbering is done. Beverages, canned foods, and hosiery are manufactured.

Berk·ham·stead, formerly also **Great Berk·hamp·stead** (both: bûr′kəm-stĕd, bär′-), urban district (1973 est. pop. 15,920), Hertfordshire, central England. Berkhamstead is mainly residential but has clothing, timber, and chemical industries. It is the site of an 11th cent. royal castle in which Edgar Atheling, a claimant to the throne, submitted to William the Conqueror; Thomas à Becket lived in the castle, and Henry II held court there. John II of France was briefly imprisoned in the castle after the Battle of Poitiers (1356).

Berks (bûrks), county (1970 pop. 296,382), 864 sq mi (2,237.8 sq km), SE central Pa., in an agricultural and industrial area drained by the Schuylkill River and bounded on the NW by the Blue Mts.; formed 1752; co. seat Reading. It was first settled by Swedes. Agriculture, steel manufacturing, and retail shops are important.

Berk·shire (bärk′shîr′, -shər, bûrk′-) or **Berks** (bärks, bûrks), nonmetropolitan county (1976 est. pop. 659,000), S central England; administrative center Reading. Berkshire lies almost entirely in the basin of the Thames River, which forms its northern border. It is largely agricultural. Part of the ancient kingdom of Wessex, Berkshire was the birthplace of King Alfred. Windsor Castle, chief residence of English monarchs for centuries, is here.

Berk·shire (bûrk′shîr′, -shər), county (1970 pop. 149,402), 942 sq mi (2,439.8 sq km), W Mass., bordering on Vt., N.Y., and Conn.; formed 1761; co. seat Pittsfield. It is in a popular summer and winter resort area in the Berkshire Hills, with some manufacturing. The Appalachian Trail traverses the county from north to south.

Berk·shire Hills (bûrk′shîr′, -shər), region of wooded hills with many small lakes and streams, W Mass. The Berkshires are a southern extension of the Green Mts., but the name is generally applied to all highlands in western Mass. Mt. Greylock, 3,491 ft (1,064.8 m), is the highest point in the hills and in the state. The Berkshire Hills have numerous resorts, state parks, and forests.

Ber·lin (bûr-lĭn′), city, former capital of Germany and of Prussia, NE Germany, on the Spree and Havel rivers. It is located within the German Democratic Republic (East Germany). In 1945 it was divided into four occupation zones. The Soviet sector, known as East Berlin, is now the capital of the German Democratic Republic. The zones assigned to the British, American, and French occupation forces now constitute West Berlin.

A state of West Germany, **West Berlin** (1971 est. pop. 2,130,000; 185 sq mi/479.2 sq km) is situated more than 100 mi (161 km) inside East Germany. Although it is theoretically the West German capital, all the institutions of government are in Bonn and its representatives in the federal parliament have no vote. The chief manufactures are electrical equipment, foodstuffs, clothing, and machinery. There is a large tourist industry. At the center of the city, on the elegant street the Kurfürstendamm, is the gutted tower of the Kaiser Wilhelm Memorial Church, left unrestored as a reminder of the war. To the northeast, the large Tiergarten park contains the famous Reichstag building and the Berlin zoo and the American-designed Kongress Halle. Among West Berlin's many museums is the Dahlem Gallery in the Charlottenberg Palace.

The capital of East Germany, **East Berlin** (1970 est. pop. 1,085,441; 156 sq mi/404 sq km) has been far slower than West Berlin in recovering from wartime damage and achieving prosperity. Electrical goods are the leading products, and chemicals, machinery, and clothing are also produced. At the border with West Berlin is the imposing Brandenburg Gate. It is the western terminus of the famous tree-lined avenue, Unter den Linden. To the east along Unter den Linden are the state opera, Humboldt Univ. (the old Frederick William Univ.), and St. Hedwig's Cathedral. At its eastern end is the immense Marx-Engels Square, where formerly stood the Royal Palace. East Berlin also has a fine zoo and many museums.

History. Berlin had its beginning in two Wendish villages, Berlin and Kölln, which were chartered in the 13th cent. and merged in 1307. It assumed importance as a Hanseatic town in the 14th cent. and became the seat of the electors of Brandenburg (after 1701, kings of Prussia) in 1486. Occupied in the Seven Years War by Austrian (1757) and Russian (1760) troops and in the Napoleonic Wars by the French (1806-8), Berlin emerged from the conflicts as a center of German national feeling and an increasingly serious rival of Vienna. It was the scene of the Revolution of 1848 against King Frederick William IV. In 1866 it became the seat of the North German Confederation. After Berlin was made the capital of the German Empire in 1871, it prospered and expanded rapidly and became one of the great cities of the world.

The German military defeat of 1918 brought on a period of social and political unrest. As the capital of the Weimar Republic, Berlin suffered severe economic crises in the 1920s, but it was also a brilliant cultural capital. After the Nazis came to power in 1933, Berlin remained a notable economic, political, and educational center, and a huge inland port with a flourishing world trade. During World War II, it was repeatedly bombed by the Allies, but the heaviest destruction was caused by a Soviet artillery barrage that preceded its capture (May 2, 1945).

The division of the city into sectors by the Potsdam Conference resulted in severe tension between the Soviet Union and the Western powers. In 1948 Soviet authorities established a blockade on all land and water communications between West Berlin and West Germany. The Western powers successfully undertook to supply West Berlin by a large-scale airlift. The blockade was withdrawn in May, 1949, and the airlift ended in Sept., 1949. In that year East Berlin was proclaimed the capital of the new German Democratic Republic, and in 1950 West Berlin was established as one of the states of the Federal Republic of Germany. In the following years there were several Berlin crises, as the USSR contested the legal basis for the Western powers' presence in and access to West Berlin. To stop the flow of refugees from East to West Berlin, the Communists in Aug., 1961, erected a 29-mi (46.7-km) fortified wall along the partition line. War seemed near as Soviet and American tanks faced each other at the border crossings, but after 1962 the crisis eased. Visits across the wall and access to West Berlin from West Germany were finally regularized in the Berlin accords of 1972.

Ber·lin (bûr′lĭn). **1.** Town (1970 pop. 14,149), Hartford co., central Conn., an industrial suburb of Hartford; settled 1686, inc. 1785. Tools, metal products, and lacquers are among its manufactures. The first tinware in the United States was made here in 1740. **2.** City (1970 pop. 15,256), Coos co., NE N.H., in the White Mts. at falls of the Androscoggin; inc. 1829. In a heavily forested region, it early became the site of pulp and paper mills. Rubber products are also made. Berlin, a winter sports center, has the first ski club organized (1872) in the United States. **3.** City (1970 pop. 5,338), Green Lake and Waushara cos., central Wis., W of Oshkosh on the Fox River, in a farm area; inc. 1857. Diversified manufacturing is done.

Ber·me·jo (bĕr-mē′hô), river, c.650 mi (1,045 km) long, N Argentina, rising close to the Bolivian border and flowing SE into the Paraguay River at the Paraguay border.

Ber·mu·da (bûr-myo͞o′də), British crown colony (1970 pop. 52,330), 20 sq mi (51.8 sq km), comprising some 300 coral rocks, islets, and islands (of which some 20 are inhabited), in the Atlantic Ocean, SE of Cape Hatteras. The capital is Hamilton, on Bermuda (or Great Bermuda), the largest island. With its fine beaches, excellent climate, and picturesque sites, Bermuda is a popular year-round resort. Its coral reefs are the northernmost in the world. Although tourism is the economic mainstay, ship repairing and light industries are also important. Perfume concentrates, pharmaceuticals, textiles, and cut flowers are the chief exports.

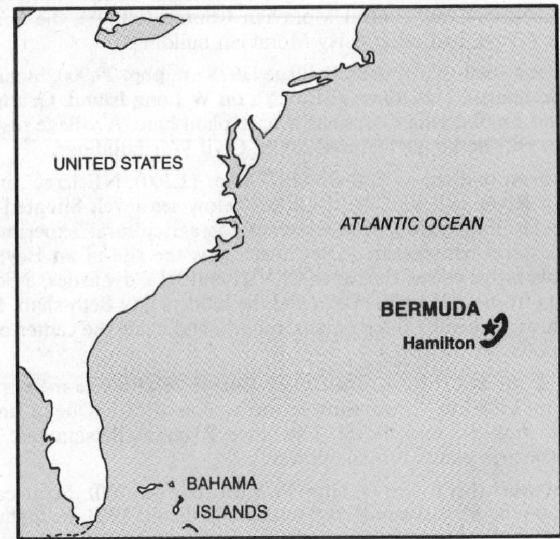

Reputedly the first person to set foot on the islands was the Spanish navigator Juan de Bermúdez (1515), but they remained uninhabited, despite visits by Spaniards and Englishmen, until Sir George Somers and a group of colonists on their way to Virginia were shipwrecked here in 1609. This incident was known to Shakespeare when he wrote *The Tempest.* Long called Somers Islands, the Bermudas were first governed by chartered companies but were acquired by the crown in 1684. The harbor of St. George was a base for privateers during the War of 1812, and the island was a center for Confederate blockade runners during the American Civil War. During World War II the islands played an important strategic role. The United States, under a 99-year lease, operates a naval and air force base. Internal self-government was granted in 1968.

Bern or **Berne** (both: bûrn, bĕrn), canton (1977 est. pop. 985,400), 2,658 sq mi (6,884.2 sq km), W central Switzerland. Its capital, Bern or Berne (1970 pop. 162,405), is also the capital of Switzerland. Situated within a loop of the Aare River, the city is a university, administrative, transportation, and industrial center. Its manufactures include precision instruments, textiles, machinery, chemicals and pharmaceuticals, and chocolate. It is also the seat of numerous international agencies, notably the Universal Postal Union (since 1875) and the International Copyright Union (since 1886). The city is largely medieval in its architecture. It has a splendid 15th cent. town hall, a noted minster (begun 15th cent.), and many picturesque patrician houses and old guild halls. An elaborate medieval clock tower and a pit in which bears (Bern's heraldic animal for seven centuries) are kept are well known to tourists. Bern was founded, according to tradition, in 1191 by Berchtold V of Zähringen as a military post. It was made (1218) a free imperial city by Emperor Frederick II when Berchtold died without an heir. Bern grew in power and population and in 1353 joined the Swiss Confederation, of which it became the leading member. Its conquests included Aargau (1415) and Vaud (1536), besides numerous smaller territories. The area was governed until 1798 by an autocratic urban aristocracy. Bern accepted the Reformation in 1528. When Switzerland was invaded (1798) by the French during the French Revolutionary Wars, Bern was occupied, its treasury pillaged, and its territories dismembered. At the Congress of Vienna (1815), Bern failed to recover Vaud and Aargau, but received the Bernese Jura. In 1848 Bern became the capital of the Swiss Confederation.

Ber·na·lil·lo (bûr'nə-lē'yō), county (1970 pop. 315,774), 1,169 sq mi (3,027.7 sq km), central N.Mex., in an irrigated livestock and farming area drained by the Rio Grande; formed 1852; co. seat Albuquerque. The center of a large retirement community, it has coal deposits, a number of military bases, and research and manufacturing facilities for atomic power. It includes part of Cibola National Forest and portions of Canoncito and Isleta Indian reservations.

Bernalillo, town (1970 pop. 2,016), seat of Sandoval co., NW N.Mex., N of Albuquerque; settled 1698 by the Spanish. It is the center of a lumbering and farming region. A nearby state monument marks the site of Coronado's headquarters (1540–42).

Bern·burg (bĕrn'bûrg', -bo͝orKH'), city (1974 est. pop. 44,428), Halle dist., central East Germany, on the Saale River. Located in a salt-mining region, it has industries that produce food products and farm machinery.

Ber·ni·na (bər-nē'nə, bĕr-nē'nä), mountain group, part of the Rhaetian Alps on the Swiss-Italian border, SE Switzerland. Piz Bernina is the highest (13,287 ft/4,052.5 m) peak. The group has many glaciers; Morteratsch Glacier is the largest. The Bernina Pass, 7,645 ft (2,331.7 m) high, from the Upper Engadine Valley, Switzerland, to the Valtellina, Italy, is crossed by a road (built 1842–65) and a railroad (built 1907–10).

Ber·ri·en (bĕr'ē-ən). **1.** County (1970 pop. 11,556), 466 sq mi (1,206.9 sq km), S Ga., drained by the Alapaha and Withlacoochee rivers; formed 1865; co. seat Nashville. It is in a coastal-plain timber and agricultural area yielding cotton, corn, tobacco, peanuts, and livestock. **2.** County (1970 pop. 163,940), 580 sq mi (1,502.2 sq km), extreme SW Mich., bounded on the S by the Ind. border, on the W by Lake Michigan, and drained by the St. Joseph, Paw Paw, and Galien rivers; organized 1831; co. seat St. Joseph. In a fruit-growing region, it also produces grain, truck crops, livestock, and dairy products. It has commercial fisheries, some manufacturing, and lake and health resorts.

Ber·ry (bĕ-rē', bĕr'ē), former province, central France. A part of Roman Aquitaine, Berry was made a county in the 8th cent., and was purchased (1101) by the French crown. In 1360 it was made a duchy. It was held as an appanage by various royal princes until 1601, when it reverted to the crown.

Ber·ry·ville (bĕr'ē-vĭl'). **1.** Resort city (1970 pop. 2,271), a seat of Carroll co., NW Ark., in the Ozarks NE of Fayetteville near the Mo. border; laid out 1850. It is a trade center for a fruit-growing and turkey-raising area. **2.** Town (1970 pop. 1,569), seat of Clarke co., N Va., in the Shenandoah Valley ESE of Winchester; laid out 1798, inc. 1870. It is in an agricultural and horse-breeding area.

Ber·si·mis (bĕr'sə-mē'), river: *see* Betsiamites.

Ber·tie (bûr'tē), county (1970 pop. 20,528), 693 sq mi (1,794.9 sq km), NE N.C., on the coastal plain, bounded on the S and SW by the Roanoke River, on the E by the Chowan River and Albemarle Sound, and drained by the Cashie River; formed 1722; co. seat Windsor. It is mainly agricultural (tobacco, corn, and peanuts), with some industry and fishing.

Ber·wick (bĕr'ĭk) or **Ber·wick·shire** (bĕr'ĭk-shîr', -shər), former county, SE Scotland. Its county town was Duns. Berwick was part of the ancient Saxon kingdom of Northumbria. For many centuries it was the scene of border strife between England and Scotland. In 1975 Berwick became part of the Borders region.

Ber·wick (bûr'wĭk), industrial borough (1978 est. pop. 12,300), Columbia co., E Pa., on the Susquehanna River, in a forest and farm area; inc. 1818. Clothing and mobile homes are produced in the city. The region abounds in fish and game.

Ber·wick upon Tweed (bĕr' ĭk; twēd), municipal borough (1973 est. pop. 11,610), Northumberland, NE England, at the mouth of the Tweed River. It is a market town and seaport and is famous for its salmon fishing. Grain is the chief export; oil and timber are imported. Other industries are shipbuilding, engineering, sawmilling, fertilizer production, and the manufacture of tweed and hosiery. The principal border town between Scotland and England, Berwick changed hands more than 13 times between 1147 and 1482, when Edward IV finally claimed it for England. It did not become officially English until 1885.

Ber·wyn (bûr'wĭn), city (1978 est. pop. 47,800), Cook co., NE Ill., a residential suburb of Chicago, on the Chicago Sanitary and Ship Canal; inc. 1891. It has varied light manufactures.

Be·san·çon (bə-zäN-sôN'), city (1975 pop. 120,315), capital of Doubs dept., E France, in Franche-Comté, on the Doubs. An industrial town with metallurgical, textile, and food-processing industries, it is especially famous for its clock and watch manufactures; its watch school is world renowned. Of Gallo-Roman origin, Besançon was made (by Emperor Frederick I) a free city. In 1648 it passed under

Spanish rule through its incorporation with Franche-Comté. After Louis XIV's second conquest of Franche-Comté (1674), Besançon became (1676) the capital of his new province.

Bes·kids (bĕs′kĭdz), mountain range of the Carpathians, extending c.200 mi (320 km) along the Polish-Czechoslovakian border and rising to 5,658 ft (1,725.7 m). The Dunajec River divides the range into eastern and western sections. Rich in coal and once having large deposits of iron ore, the Beskids became an iron and steel center in the 18th cent.

Bes·sa·ra·bi·a (bĕs′ə-rā′bē-ə), historic region, c.17,600 sq mi (45,585 sq km), SW European USSR, largely in the Moldavian SSR and in the Ukraine. Consisting mainly of a hilly plain with flat steppes, it is bounded by the Dnestr River on the north and east, the Prut on the west, and the Danube and the Black Sea on the south. As the gateway from Russia into the Danube valley, Bessarabia has been an invasion route from Asia to Europe. Greek colonies were planted on the Black Sea coast of Bessarabia as early as the 7th cent. B.C. The region was later part of Roman Dacia, but after the 4th cent. A.D. it was subject to incursions by Goths, Huns, Avars, and Magyars. Slavs first settled in Bessarabia in the 7th cent. From the 9th to the 11th cent., the area was part of Kievan Russia. Cumans and later Mongols overran Bessarabia; after the latter withdrew it was included (1367) in the newly established principality of Moldavia. In 1513 the Turks conquered Bessarabia. After the Russo-Turkish wars, the region was ceded to Russia by the Treaty of Bucharest (1812). The Crimean War resulted (1856) in Russia's cession of southern Bessarabia to Moldavia, but the Congress of Berlin (1878) returned the district to Russia. After the Bolshevik Revolution (1917) the anti-Soviet national council of Bessarabia proclaimed the region an autonomous republic; however, in 1918, Bessarabia renounced all ties with Soviet Russia and declared itself an independent Moldavian republic, later voting for union with Rumania. Although the Treaty of Paris (1920) recognized the union, Russia never accepted it, and in 1940 Rumania was forced to cede Bessarabia to the USSR.

Bes·se·mer (bĕs′ə-mər). **1.** City (1978 est. pop. 31,100), Jefferson co., N central Ala.; inc. 1887. Founded as a mining town, it was named after Sir Henry Bessemer, inventor of the Bessemer process. The surrounding area is rich in minerals, and the manufacture of iron and steel is still the city's major industry. **2.** City (1970 pop. 2,805), seat of Gogebic co., N Mich., W Upper Peninsula, near the Black River ENE of Ironwood, in the Gogebic iron-mining region; inc. as a village 1887, as a city 1889. Explosives are made, and lumber and flour are milled.

Beth·a·ny (bĕth′ə-nē). **1.** City (1970 pop. 2,914), seat of Harrison co., NW Mo., NE of St. Joseph, in a farm region; founded 1845, inc. 1858. **2.** City (1978 pop. 21,800), Oklahoma co., central Okla.; inc. 1910. Its manufactures include small airplanes and tires. Bethany was settled in 1906 by members of the Nazarene church. Bethany Nazarene College is in the city.

Beth·el (bĕth′əl), town (1970 pop. 10,945), Fairfield co., SW Conn.; inc. 1855. Bethel is noted for its hat industry, which was founded c.1800. Other manufactures include clothing, chemicals, rubber goods, metal products, and game equipment.

Bethel Park, borough (1970 pop. 34,758), Allegheny co., SW Pa., S of Pittsburgh, in an industrial area; inc. 1949 as Bethel, renamed 1960.

Be·thes·da (bə-thĕz′də), uninc. city (1978 est. pop. 77,700), Montgomery co., W central Md., a residential suburb of Washington, D.C. The area was settled in the late 17th cent. by the Scottish, English, and Irish. In 1820 they built Bethesda Presbyterian Church, from which the district takes its name. The National Institutes of Health, the National Cancer Institute, and the Naval Medical Center are here.

Beth·le·hem (bĕth′lĭ-hĕm, -lē-əm), town (1967 pop. 16,313), W Jordan. It is traditionally considered the birthplace of Jesus and is one of the world's great shrines. Situated on a hill in green, fertile country, Bethlehem is also the trade center for surrounding farming villages and for the pastoral nomads who inhabit the area. From 1099 to 1187 Crusaders controlled Bethlehem, and in 1571 the city was annexed by the Ottoman Empire. It was part of the British-administered Palestine mandate from 1922 until 1948, when it joined Jordan.

Bethlehem, town (1970 pop. 29,918), Orange Free State, E central South Africa. It is situated in a farming and livestock area and has industries producing furniture and food products. Bethlehem was founded in 1860, and its main growth began after the railroad from Natal reached here in 1905.

Bethlehem, city (1978 est. pop. 72,900), Northampton and Lehigh cos., E Pa., on the Lehigh River; inc. as a city 1917. It is one of the most important centers of steel production in the United States. Bethlehem was settled in 1740–41 by Moravians. Points of interest in Bethlehem are the Central Moravian Church (c.1803), the Schnitz House (1749), and other early Moravian buildings.

Beth·page (bĕth-pāj′), uninc. village (1978 est. pop. 29,900, including Old Bethpage), Nassau co., SE N.Y., on W Long Island. Grumman Aircraft Engineering Corp. has a large plant here. A village restoration in Old Bethpage features 20 pre-Civil War buildings.

Bet She·an (bāt shĭ-än′), town (1972 pop. 11,300), NE Israel, in the Jordan River valley, c.300 ft (90 m) below sea level. Situated in a fertile farming region, it is a center for agricultural experiments. Textiles are manufactured. Bet Shean was the site of an Egyptian administrative center during the XVIII and XIX dynasties, a Scythian city from c.625 to 300 B.C., and the biblical city Beth-shan. In 64 B.C. it was taken by the Romans, rebuilt, and made the center of the Decapolis.

Bet·si·a·mi·tes (bĕt′sē-ə-mē′tēz) or **Ber·si·mis** (bĕr′sə-mē′), river, c.240 mi (385 km) long, rising in the highlands of E Que., Canada, and flowing SE into the St. Lawrence River at Betsiamites. Two hydroelectric plants provide power.

Bet·ten·dorf (bĕt′n-dôrf′), city (1978 est. pop. 25,700), Scott co., E Iowa, on the Mississippi River; settled c.1840, inc. 1903. Its manufactures include aluminum products and farm equipment.

Beu·lah (byōō′lə), village (1970 pop. 461), seat of Benzie co., NW Mich., on the E end of Crystal Lake SW of Traverse City. Cherries, apples, and poultry are processed.

Bev·er·ley (bĕv′ər-lē), municipal borough (1971 pop. 17,124), Humberside, NE England. It is primarily a market town with some shipbuilding and such light industries as the manufacture of railroad and automobile accessories and leather. The famous large minster (13th cent.) contains the ancient "chair of peace," which gave sanctuary from the laws of man. The sanctuary, a privilege granted by Athelstan, applied in a 1 mi (1.6 km) radius around the minster. It was abolished during the Reformation. Until 1974 Beverley was the county town of the East Riding of Yorkshire.

Bev·er·ly (bĕv′ər-lē), city (1978 est. pop. 36,700), Essex co., NE Mass., on Massachusetts Bay; inc. as a city 1894. Its chief manufactures are shoe machinery and electronic equipment. Beverly was settled in 1626 by Roger Conant, one of the founders of Massachusetts. In 1775 the schooner *Hannah*, the first ship of the U.S. navy, was outfitted and commissioned by Gen. George Washington at Glover's Wharf in Beverly. In 1787 Beverly became the site of the first cotton mill in the United States. Points of interest include Balch house (1636), believed to be the oldest house in the United States; the John Cabot house (1781), which is preserved as a museum; and several other colonial buildings.

Beverly Hills. 1. City (1978 est. pop. 33,200), Los Angeles co., S Calif., completely surrounded by the city of Los Angeles; inc. 1914. Mainly residential, it is the home of many film and television personalities. **2.** Village (1970 pop. 13,598), Oakland co., SE Mich., a residential suburb of Detroit, on the Rouge River; inc. 1958.

Bexar (bâr), county (1970 pop. 830,460), 1,247 sq mi (3,229.7 sq km), S central Texas, drained by the Medina and San Antonio rivers; formed 1836; co. seat San Antonio. The Balcones Escarpment crosses the county, dividing the hills from the prairies. Its agriculture includes corn, grain sorghums, peanuts, potatoes, pecans, truck crops, cattle, sheep, goats, and poultry.

Bex·hill (bĕks′hĭl′), municipal borough (1973 est. pop. 34,680), East Sussex, SE England. It is a summer resort and has a 14th cent. manor house and an 11th cent. church.

Bex·ley (bĕks′lē), borough (1971 pop. 216,172) of Greater London, SE England. It was created in 1965 by the merger of the municipal boroughs of Bexley and Erith, the urban district of Crayford, and part of the urban district of Chislehurst and Sidcup. The borough has many parks and open areas. Erith and Crayford are industrial centers. There are engineering and chemical works, oil and resin refineries, flour and seed-crushing mills, cloth printshops, and factories that produce electrical equipment, building materials, cable, paper products, plywood, and plastics. Erith is also a yachting resort.

Bey·se·hir (bā-shĕ-hîr′), lake, 250 sq mi (647.5 sq km), SW central Turkey, W of Konya.

Bé·ziers (bā-zyā′), city (1975 pop. 84,029), Hérault dept., S France, in Languedoc. A communications and industrial center with an important trade in wines and liqueurs, it has ironworks, breweries, and factories making a great variety of products. During the Albigensian Crusade Béziers was taken (1209) by Simon de Montfort, and 20,000 people were massacred.

Bez·wa·da (bāz-wä′də): *see* Vijayawada, India.

Bha·dra·va·ti (bə-drä′və-tē), city (1971 metropolitan area pop. 101,315), Karnataka state, S India, on the Bhadra River. The city contains iron and steel plants and paper mills.

Bha·gal·pur (bä′gəl-po͝or′), city (1971 pop. 172,700), Bihar state, NE India, on the Ganges River. It is a district administrative center and the market for an agricultural region. The remains of Buddhist monasteries are in the city.

Bha·mo (bə-mō′, bä′mō), town (1960 est. pop. 16,000), NE Burma, the head of navigation on the upper Irrawaddy River. Bhamo is the market town for the surrounding hill region and is also important for its ruby mines. Formerly significant as a center of overland trade with China, it was linked in World War II by the building of the Stilwell Road to Ledo in India.

Bha·rat·pur (bə-rŭt′po͝or), city (1971 pop. 69,902), Rajasthan state, N central India. It is a district administrative center and has railroad-car and glass factories. Fans and fly whisks are fashioned from ivory and sandalwood. The city is thought to have been founded in 1733 and named after Bharat, a figure in Hindu mythology. The British captured Bharatpur in 1826.

Bhat·gaon (bät′goun) or **Bhad·gaon** (bäd′goun), city (1971 pop. 104,703), E Nepal, in a valley c.4,000 ft (1,220 m) above sea level, surrounded by high Himalayan peaks. It is a processing center for the grains, vegetables, and other crops of the surrounding area. Grazing is also important. A religious center, Bhatgaon was founded in 865.

Bhat·pa·ra (bət-pä′rə), city (1971 pop. 204,750), West Bengal state, NE India, on the Hooghly River. Once a center of Sanskrit learning, it is now part of the vast Hooghlyside industrial complex. Jute products are the chief manufactures.

Bhav·na·gar (bou-nŭg′ər), city (1971 pop. 226,072), Gujarat state, W India, on the Gulf of Cambay; the chief port on the Kathiawar peninsula. Cotton is exported. The city manufactures bricks, tiles, and metal products.

Bhi·lai·na·gar (bē′lī-nə-gər) or **Bhi·lai** (bē′lī), city, (1971 pop. 245,124), Madhya Pradesh state, central India. It is the site of a large state-owned steel industry, built with Soviet assistance.

Bhil·wa·ra (bēl-vä′rə), town (1971 pop. 82,155), Rajasthan state, NW India. The town is a market for mica, wheat, maize, cotton, and wool. Stone dressing is an important occupation.

Bhi·ma (bē′mä), river, c.400 mi (645 km) long, S India, rising in Maharashtra just E of Bombay and flowing generally SE to the Krishna River NW of Raichur.

Bho·lan Pass (bō-län′), Pakistan: *see* Bolan Pass.

Bho·pal (bō-päl′), former principality, Madhya Pradesh state, central India. Founded in the early 18th cent., Bhopal was ruled from 1844 to 1926 by the begums of Bhopal, famous women leaders, and became part of the state of Madhya Pradesh in 1956. The city of Bhopal (1971 pop. 384,859), the former capital of the principality and now the capital of Madhya Pradesh, was founded in 1728. It is a trade center with manufactures of cotton cloth, jewelry, and electrical goods.

Bhu·ba·nes·war (bo͞o′bä-nĕsh′wär), city (1971 pop. 105,514), capital of Orissa state, E central India, on a tributary of the Mahanadi River. A small village before it became the capital in 1947, it is now a model administrative center and the seat of Orissa Univ. of Agriculture and Technology. Settlements on this site date back to the reign of Asoka (3rd cent. B.C.). Bhubaneswar, a religious center, once had c.7,000 shrines around its sacred lake; the remains of c.500 still stand, displaying many styles of Hindu and Buddhist art and architecture.

Bhu·sa·wal (bo͞o-sä′vəl), town (1971 pop. 104,708), Maharashtra state, W central India. The town has large railroad workshops and several cotton factories.

Bhu·tan (bo͞o-tän′, -tăn′), kingdom (1976 est. pop. 1,200,000), 18,147 sq mi (47,000 sq km), in the E Himalayas, bordered on the S and E by India, on the N by the Tibet region of China, and on the W by Sikkim. Punaka is the traditional capital; Thimbu is the official capital. Great mountain ranges, rising in the north to Kula Kangri

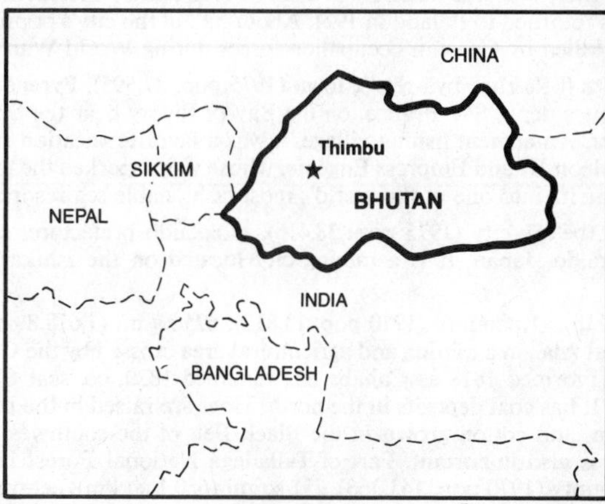

(24,784 ft/7,559.1 m), Bhutan's tallest peak, run north and south, dividing the country into forested valleys with some pastureland. Bhutan is drained by several rivers rising in the Himalayas and flowing into India. Thunderstorms and torrential rains are common; rainfall averages from 200 to 250 in. (508–635 cm) on the southern plains. The valleys, especially the Paro, are intensively cultivated.

Economy. The chief occupations are small-scale subsistence farming (with rice the chief crop) and the raising of yaks, cattle, sheep, pigs, and tanguns, a sturdy breed of pony valued in mountain transportation. Metal, wood, and leather working, papermaking, and the weaving of cloth, baskets, and mats are also important activities.

History. Although its early history is largely unknown, Bhutan seems to have existed as a political entity for centuries. In the 16th cent. the Tibetans conquered and assimilated Bhutan's native tribes, and around 1630 a refugee lama from Tibet made himself the first Dharma Raja, or spiritual ruler, of Bhutan. In 1720 the Chinese invaded Tibet and established suzerainty over Bhutan. Friction between Bhutan and Indian Bengal was followed by a British incursion into Bhutan, but the Tibetan lama's intercession with the governor-general of British India improved relations. In 1774 a British mission arrived in Bhutan to promote trade with India. British occupation of Assam in 1826, however, led to renewed border raids from Bhutan. In 1864 the British occupied part of southern Bhutan, which was formally annexed after a war in 1865; the Treaty of Sinchula provided for an annual subsidy to Bhutan as compensation. After India won independence, a treaty (1949) returned the part of Bhutan annexed by the British and allowed India to assume the former British role of subsidizing Bhutan and directing its defense and foreign relations. After Chinese Communist forces occupied Tibet in 1950, they claimed Bhutan, and the persecution of Tibetan Buddhists led India to close the Bhutanese-Tibetan border. In the 1960s Bhutan also formed a small army, trained and equipped by India. The kingdom's admission to the United Nations in 1971 was seen as strengthening its sovereignty.

Bi·a·fra, Bight of (bē-ä′frə), E bay of the Gulf of Guinea, W Africa, extending approximately from the Niger River delta, in S Nigeria, to N Gabon.

Biafra, Republic of, secessionist state of W Africa, in existence from May 30, 1967, to Jan. 15, 1970. It comprised, roughly, the East-Central, South-Eastern, and Rivers states of the Federation of Nigeria.

Bi·ak (bē-yäk′), island, 948 sq mi (2,455.3 sq km), largest of the Schouten Islands, off the NW coast of New Guinea. Taken by Japanese invaders in 1942, it was recaptured by Allied forces after heavy fighting from May 27 to June 20, 1944.

Bia·ło·wie·za (byä-lôv-yĕ′zhä), large forest, c.450 sq mi (1,165 sq km), E Poland and W USSR. It was a favorite hunting ground of Polish kings, passed to Prussia in 1795, was annexed by Russia in 1807, but was restored to Poland in 1921. In 1939, however, the forest was incorporated into the USSR. After World War II nearly half of the region was returned to Poland. Today both sections of the forest have animal preserves.

Bia·ły·stok (byä-lĭ'stôk), city (1975 pop. 195,900), capital of Białystok prov., NE Poland. It is a leading regional manufacturing center and a railway transportation point. Noted especially for its textile industry, the city also has factories producing machinery, metal goods, ceramics, food products, and precision instruments. Founded in 1310, Białystok was taken by Prussia in 1795 and by Russia in 1807; it was returned to Poland in 1921. About half of the city's population were killed by German occupation forces during World War II.

Biar·ritz (bē'ə-rĭts', byä-rēts'), town (1975 pop. 27,595), Pyrénées-Atlantiques dept., SW France, on the Bay of Biscay near the Spanish border. An ancient fishing village, it was a favorite vacation spot of Napoleon III and Empress Eugénie, whose visits sparked the growth of Biarritz into one of the world's most fashionable sea resorts.

Bi·bai (bē'bī), city (1975 pop. 38,416), Hokkaido prefecture, central Hokkaido, Japan. It is a mining city located on the Ishikari coal field.

Bibb (bĭb). **1.** County (1970 pop. 13,812), 625 sq mi (1,618.8 sq km), central Ala., in a mining and agricultural area crossed by the Cahaba River; formed 1818 as Cahaba co., renamed 1820; co. seat Centerville. It has coal deposits in the north. Hogs are raised in the plateau region, and cotton grown in the Black Belt of the southwest. Forestry is also important. Part of Talladega National Forest is here. **2.** County (1970 pop. 143,366), 251 sq mi (650.1 sq km), central Ga., drained by the Ocmulgee River; formed 1822; co. seat Macon. Textiles and wood and clay products are manufactured.

Bi·da (bē-dä'), town (1969 est. pop. 64,000), W central Nigeria. It is the trade center for a rice-growing region and is noted for its fiber, glass, and metal handicrafts. In the 19th cent. Bida was the capital of an emirate of the Moslem Fulani empire. The town was captured in 1897 by British forces.

Bid·de·ford (bĭd' ĭ-fərd), city (1970 pop. 19,983), York co., SW Maine, on the Saco River; inc. as a town 1718, as a city 1855. Samuel de Champlain, a French explorer, visited the area in 1605. The first permanent settlement was established in 1630. During the 17th cent. the town exported lumber and fish, and in 1840 the first cotton mill was built.

Bid·e·ford (bĭd'ə-fərd), municipal borough (1971 pop. 11,766), Devonshire, SW England, on the Torridge estuary. Formerly a major seaport, it still maintains some foreign trade (timber is imported) and has a boatbuilding industry. Tourism and the manufacture of gloves and concrete products are other important industries. Bideford supplied ships used in the defeat of the Spanish Armada (1588) and was a port of embarkation for colonists going to America.

Biel (bēl) or **Bienne** (byĕn), city (1977 est. pop. 88,600), Bern canton, NW Switzerland, at the NE end of the Lake of Biel. A watchmaking center, Biel also has manufactures of machinery, automobiles, and pianos. There is a 16th cent. Gothic town hall and a late Gothic church. Both French and German are spoken. The Schwab museum has archaeological relics of lake dwellings found in the Lake of Biel, or Lake of Bienne (15 sq mi/39 sq km), at the foot of the Jura Mts. The lake is connected with the Lake of Neuchâtel by the Zihl Canal.

Biel-. For some Russian names beginning thus, *see* Bel-.

Bie·le·feld (bē'lə-fĕlt), city (1974 pop. 319,611), North Rhine-Westphalia, N central West Germany. It has been noted since the 13th cent. for its handmade linens. Chartered in 1214, Bielefeld became a member of the Hanseatic League in 1270.

Biel·la (byĕl'lä), city (1976 est. pop. 56,148), Piedmont, NW Italy. It is a major cotton and wool textile manufacturing center. Biella came under the Visconti of Milan in 1353 and under the house of Savoy in 1379. Of note are several palaces (15th-16th cent.), an early Romanesque baptistery (10th cent.), and a Renaissance cathedral.

Biel·sko-Bia·ła (byĕl'skô-byä'lä), city (1975 est. pop. 120,900), S Poland, on the Biała River, a tributary of the Vistula. The city is a railway junction and has a noted woolen textile industry. Other manufactures include textile machinery, electrical equipment, and machine tools. It is also a tourist and winter sports center. Founded in the 13th cent., the city passed to Austria in 1772 and was returned to Poland in 1919.

Bi·en Hoa (bē-ĕn' wä), city (1968 est. pop. 83,000), S Vietnam, NE of Saigon. It is famous for its handmade pottery. In the city are sawmills and a rice-bag factory. A large U.S. air base was located here during the Vietnam War.

Bi·en·ville (bē-ĕn'vĭl'), parish (1970 pop. 16,024), 832 sq mi (2,154.9 sq km)., NW La., bounded partly on the W by Lake Bistineau and drained by several tributaries of the Red River; formed 1848; parish seat Arcadia. Its agriculture includes cotton, corn, sweet potatoes, and cattle. There are petroleum, natural gas, and clay deposits.

Bien·ville Lake (byăɴ-vēl', bē'ĕn-vĭl'), 392 sq mi (1,015.3 sq km), N Que., Canada. It drains into Hudson Bay via the Great Whale River.

Bié Plateau (byĕ), highland region, W section of the central plateau of Angola, SW Africa; alt. 5,000 to 6,000 ft (1,525-1,830 m). Its cool climate and ample rainfall made it a favored area for European settlement. Corn, sisal, peanuts, and coffee are raised here.

Big Bend National Park, 708,221 acres (286,830 hectares), W Texas; est. 1944. It is a triangle formed by the Rio Grande, which runs south, then north in a big bend and flows through deep canyons. The river, the desert plain, and the Chisos Mts. offer sharp contrasts in wilderness scenery, and the park has archaeological treasures, some petrified trees, vestiges of prehistoric Indian cultures, and rare forms of animal and plant life.

Big Black, river, c.330 mi (530 km), rising in N central Miss. and flowing SW to the Mississippi River below Vicksburg.

Big Blue, river, c.300 mi (485 km) long, rising in SE Nebr. near Aurora and flowing E and SE to join the Kansas River near Manhattan, Kansas. A few miles from the mouth of the river is Tuttle Creek Dam (157 ft/47.9 m high, 6,700 ft/2,043.5 m long, completed 1962), impounding Tuttle Creek Reservoir.

Big Cy·press Swamp (sī'prəs), wooded swamp region, c.2,400 sq mi (6,215 sq km), S Fla., in the W Everglades.

Big Hole National Battlefield: *see* National Parks and Monuments Table.

Big·horn (bĭg'hôrn'), river, 461 mi (741.7 km) long, formed in W central Wyo. by the confluence of the Wind and Pop Agie rivers and flowing N to join the Yellowstone River in S Mont. The Bighorn basin, part of the Missouri River basin project, has several dams that provide for flood control, irrigation, hydroelectricity, and recreation. The lake behind Yellowstone Dam is the nucleus of Bighorn Canyon National Recreation Area.

Big Horn. 1. County (1970 pop. 10,057), 5,028 sq mi (13,022.5 sq km), S Mont., in an irrigated region bordering on Wyo. and drained by the Bighorn and Little Bighorn rivers; formed 1913; co. seat Hardin. Sugar beets and beans are grown, and livestock is raised. Most of the county is within the Crow Indian Reservation. **2.** County (1970 pop. 10,202), 3,176 sq mi (8,225.8 sq km), N Wyo., in an irrigated agricultural, oil, and coal region bordering on Mont. and watered by the Bighorn River; formed 1890; co. seat Basin. Its agriculture includes sugar beets, grain, beans, and livestock.

Bighorn Mountains, range of the Rocky Mts., N central Wyo., extending c.120 mi (195 km) N into S Mont., E of the Bighorn River. Cloud Peak, 13,175 ft (4,018.4 m), is the highest point. The glaciated mountain range contains Bighorn National Forest.

bight (bīt), broad bend or curve in a coastline, forming a large open bay. The New York bight, for example, is the curve in the coast described by the southern shore of Long Island and the eastern shore of New Jersey. The term "bight" may also refer to the bay so formed.

Big Lake, town (1970 pop. 2,489), seat of Reagan co., W central Texas, WSW of San Angelo, in rough prairie country; inc. 1923. It is a marketing and shipping point for an oil and cattle area.

Big Mud·dy (mŭd'ē), river, c.135 mi (217 km) long, SW Ill., rising in Jefferson co. WNW of Mount Vernon and flowing S through Rend Lake, then SW to join the Mississippi S of Murpheysboro.

Big Rap·ids (răp'idz), city (1970 pop. 11,995), seat of Mecosta co., W central Mich., at the falls of the Muskegon River; inc. 1869. The region has extensive natural-gas wells. The city's major manufactures include shoes, machine tools, and wood products. Ferris State College here.

Big Sand·y Creek (săn'dē), river, c.200 mi (320 km) long, central and E Colo., rising in N El Paso co. and flowing ENE, then SE into the Arkansas River E of Lamar.

Big Sioux (sōō), river, 420 mi (675.8 km) long, rising in NE S.Dak. and flowing S into the Missouri River. It passes through an agricultural region that produces corn, oats, hogs, and beef cattle.

Big Spring, city (1970 pop. 28,735), seat of Howard co., W central Texas; inc. 1907. The spring for which it was named once fed a branch of the Colorado River but is now dry. The city is the trade center for a farm and livestock region. A variety of oil-related industries have been developed since the discovery of oil in 1928. Webb Air Force Base is nearby.

Big Stone, county (1970 pop. 7,941), 490 sq mi (1,269.1 sq km), W Minn., in an agricultural area bordering on S.Dak. and bounded on the W by Big Stone Lake; formed 1862; co. seat Ortonville. Grain and livestock predominate, and granite has been found in the area.

Big Tim·ber (tĭm′bər), city (1970 pop. 1,562), seat of Sweet Grass co., S Mont., on the Yellowstone River NE of Livingston; inc. 1902. In a cattle and sheep area, it has dude ranches nearby.

Bi·har or **Be·har** (both: bē-här′), state (1971 pop. 56,353,369), 67,198 sq mi (174,043 sq km), E central India; capital Patna. Bihar is bounded on the north by Nepal, on the east by West Bengal state, on the south by Orissa state, and on the west by Uttar Pradesh and Madhya Pradesh states. The predominantly agricultural northern area, crossed by the Ganges River, supports the bulk of the population. The southeastern section is one of the greatest sources of India's mineral wealth. Bihar was the scene of Buddha's early life, and Bodh Gaya is an ancient Buddhist center. Moslems occupied Bihar in 1193 and the Delhi sultans in 1497. In 1765 the British took over Bihar and merged it with Bengal. The province of Bihar and Orissa was formed in 1912, and Bihar became a separate province in 1936. About 3,150 sq mi (8,160 sq km) situated along Bihar's eastern boundary were transferred to West Bengal state in 1956. Bihar city (1971 pop. 100,052), on a tributary of the Ganges River, is an agricultural market.

Bi·ja·pur (bĭ-jä′po͝or), town (1971 pop. 103,931), Karnataka state, SE India. It is a trade and district administrative center. Cotton ginning is an important activity. Bijapur was the capital (15th–17th cent.) of the Deccan kingdom of Bijapur.

Bi·ka·ner (bē′kə-nîr′), former native state, NW India, now part of Rajasthan state. The city of Bikaner (1971 pop. 208,894), capital of the former state, was founded in 1488. There are several beautiful 16th cent. Rajput palaces.

Bi·ki·ni (bē-kē′nē), atoll, c.2 sq mi (5.2 sq km), W central Pacific, one of the Ralik Chain, Marshall Islands. It comprises 36 islets on a reef 25 mi (40.2 km) long. After its inhabitants were removed (1946) to Rongerik, Bikini was the scene of 23 U.S. atomic and hydrogen bomb tests between 1946 and 1958. The Bikini natives were transferred from Rongerik to Ujelang in 1947 and in 1949 were resettled on Kili. The atoll was declared safe for habitation in 1969.

Bi·las·pur (bē-läs′po͝or′). **1.** Former principality, Himachal Pradesh state, NW India, in the W Himalayas. The town of Bilaspur (1971 pop. 7,024), formerly the capital, trades in agricultural products. **2.** Town (1971 pop. 130,804), Madhya Pradesh state, central India. Founded in the 17th cent., the city is a district administrative center and an agricultural market.

Bi·lauk·taung Range (bē-louk′toun), mountain range, extending c.250 mi (400 km) along the Thailand-Burma border from the Dawna Range SE to the Isthmus of Kra. The western slopes of the range, which receive the heavy rains of the monsoon, have a dense covering of tropical rain forest.

Bil·ba·o (bĭl-bä′ō), city (1975 pop. 431,347), capital of Vizcaya prov., N Spain, in the Basque Provinces, on both banks of the Nervión River, near the Bay of Biscay. A leading Spanish port and commercial center since the 19th cent., it is the center of an important industrial area, with rich iron mines nearby. The production of steel and chemicals and shipbuilding are the chief industries. Founded c.1300 on the site of an ancient settlement, Bilbao flourished from a wool export trade in the 15th and 16th cent. In the 19th cent. it was three times unsuccessfully besieged by the Carlists. In the Spanish Civil War, Bilbao was the seat of the short-lived Basque autonomous government from 1936 until its capture (1937) by the Insurgents.

Bille·ric·a (bĭl-rĭk′ə), town (1975 est. pop. 34,300), Middlesex co., NE Mass., on the Concord River; settled 1637, inc. 1655. It is mainly residential. Billerica was one of the "praying Indian" towns of John Eliot. The town's historical attractions include several 17th, 18th, and 19th cent. homes and an Indian site and burial ground dating back to 1,000 B.C.

Bil·lings (bĭl′ĭngz), county (1970 pop. 1,198), 1,139 sq mi (2,950 sq km), W N.Dak., in an agricultural area watered by the Little Missouri River; formed 1879; co. seat Medora. In a lignite-rich region, it is agricultural (grain and livestock). Part of Theodore Roosevelt National Memorial Park is here.

Billings, city (1975 est. pop. 65,800), seat of Yellowstone co., S Mont., on the Yellowstone River, in a valley surrounded by seven mountain ranges; inc. as a city 1885. Founded in 1882 by the Northern Pacific RR, Billings quickly became an important shipping point and fur-trading center. Today it is a trade and manufacturing center for the surrounding area. Oil refining, sugar refining, meat packing, and flour milling are the city's major industries. Wheat, sugar, beets, livestock, and wool are traded. Custer National Forest and Yellowstone National Park are nearby.

Bi·lox·i (bĭ-lŏk′sē, -lŭk′-), city (1975 est. pop. 51,000), Harrison co., SE Miss., on a small peninsula between Biloxi Bay and Mississippi Sound, on the Gulf of Mexico; inc. as a town 1838, as a city 1896. The warm, almost tropical climate has made Biloxi a popular resort. In addition to tourism, major industries include fishing and boatbuilding, the packing and shipping of shrimps and oysters, and the manufacture of small appliances and fishing nets. The first white settlement in the lower Mississippi valley was established in 1699 across the bay at Old Biloxi (now Ocean Springs) by the French under Pierre Iberville. New Biloxi was founded in 1719 and was the capital of the French colony of Louisiana until 1722. In the city are Keesler Air Force Base and a U.S. Coast Guard station. Nearby are Beauvoir (built 1852–54), the last home of Jefferson Davis; the Biloxi Light House (built 1848); and, off the coast, Ship Island, a Union fort in the Civil War.

Bim·i·nis (bĭm′ə-nēz), island group in the Straits of Florida, forming the NW section of the Bahamas. Exceptionally good fishing attracts many tourists. According to legend, the Biminis are the location of the fountain of youth for which Juan Ponce de León searched.

Bing·en (bĭng′ən), city (1974 pop. 25,034), Rhineland-Palatinate, W West Germany, where the Nahe River enters the Rhine. A busy river port, railroad junction, and tourist center, Bingen is also noted for its wine and tobacco manufactures. Dating from pre-Roman times, Bingen was fortified (1st cent. B.C.) by Drusus. In 983 it came under the rule of the archbishops of Mainz. Near Bingen, on a rock in the Rhine, is the famous Mäuseturm, or Mouse Tower, where, according to legend, Archbishop Hatto I of Mainz was devoured (913) by mice for wronging his subjects.

Bing·ham (bĭng′əm), county (1970 pop. 29,167), 2,084 sq mi (5,397.6 sq km), SE Idaho, in an irrigated agricultural area drained by the Snake and Blackfoot rivers and the American Falls Reservoir; formed 1885; co. seat Blackfoot. Its agriculture includes potatoes, dry beans, sugar beets, and livestock.

Bing·ham·ton (bĭng′əm-tən), industrial city (1975 est. pop. 97,509), seat of Broome co., S central N.Y., at the confluence of the Chenango and Susquehanna rivers; settled 1787, inc. as a city 1867. It is the largest of the Triple Cities (Binghamton, Endicott, and Johnson City), which are famous for shoes. Many electronic products are also manufactured in the city. Binghamton grew mainly after the Chenango Canal connected it with Utica in 1837.

Bing·ley (bĭng′lē), urban district (1971 est. pop. 26,540), West Riding of Yorkshire, N England, near Bradford. It has woolen, silk, and paper mills and manufactures leather goods, machine tools, and paint.

Bin·tan (bĭn′tän) or **Bin·tang** (bĭn-täng′), biggest island, 415 sq mi (1,074.9 sq km), of Riau Archipelago, Indonesia, off the S extreme of Malay Peninsula in the S China Sea. The low mountainous island yields tin and bauxite. Other products include rubber, pepper, timber, fish, and copra.

Bí·o-Bí·o (bē′ō-bē′ō), river, c.240 mi (385 km) long, rising in the Andes of central Chile and flowing NW to the Pacific Ocean near Concepción. It forms a natural division between middle and southern Chile and is navigable for much of its length by flat-bottomed boats.

Bi·qa, Al (äl bē-kä′) or **El Bi·ka** (ĕl bē-kä′), upland valley of Lebanon and Syria, 75 mi (120.7 km) long and 5 to 9 mi (8–14.5 km) wide, between the Lebanon and Anti-Lebanon ranges. It is the highest part of the Rift Valley complex. The Biqa valley, once the heart of ancient Coele-Syria, has been the scene of warfare since the dawn of history. Al Biqa was included in a province of the Persian Empire and was later bitterly contested by the Seleucids and the Ptolemaic kings of Egypt. The city of Antioch, Turkey, was founded by King Seleucus I of Syria to dominate the region.

Bir·ken·head (bûr′kən-hĕd), county borough (1976 est. pop. 348,200),

Merseyside, W central England, at the mouth of the Mersey River and connected with Liverpool by the Mersey tunnel. Flour milling, shipbuilding, and commerce are the key industries. There are also engineering, food-processing, and clothing plants and a cattle market. It has extensive docks. The chief imports are grain and cattle; coal, flour, the byproducts of milling, and machinery are exported. Milling and shipbuilding were responsible for Birkenhead's rapid growth in the 19th cent.

Bir·ming·ham (bûr'mĭng-əm, -hăm), city (1976 est. pop. 1,058,800), West Midlands, central England. Birmingham is Britain's second-largest city (in both area and population) and is the center of a rich mining district and the hub of water, road, and rail transportation in the Midlands. The chief industries are the manufacture of automobiles, motorcycles, and bicycles and their components and accessories. Other products include electrical equipment, paint, guns, and a wide variety of metal products.

By the 15th cent., Birmingham was a market town with a large leather and wool trade; by the 16th cent. it was also known for its many metalworks. Birmingham's industrial development and population growth accelerated in the 17th and 18th cent. In 1762 Matthew Boulton and James Watt founded the Soho metalworks, where they designed and built steam engines. Joseph Priestley, the discoverer of oxygen, lived for a time in Birmingham. The town was enfranchised by the 1832 Reform Bill and was incorporated in 1838. During the 1870s, Birmingham underwent a large program of municipal improvements, including slum clearance and the development of gas and water works. Birmingham was among the first English localities to have a municipal bank, a comprehensive water-supply system, and development planning. The area of the city was enlarged in 1891 and again in 1911 under the Greater Birmingham scheme.

Birmingham was severely damaged in World War II and has been considerably rebuilt since then. Notable surviving buildings include the town hall, built in 1834 and modeled after the temple of Castor and Pollux in Rome; the 18th cent. baroque-style Cathedral of St. Philip; and the 19th cent. Cathedral of St. Chad, the first Roman Catholic cathedral to be built in England after the Reformation.

Bir·ming·ham (bûr'mĭng-hăm'). **1.** City (1975 est. pop. 463,454), seat of Jefferson co., N central Ala., in the Jones Valley near the S end of the Appalachian system; inc. 1871. It is the largest city in the state and the leading iron and steel center in the South. Iron, coal, limestone, and other natural resources from the area supply the city's great iron and steel plants and its metalworking factories. In addition, transportation equipment, construction materials, chemicals, and fabricated metals are produced. Commerce, banking, insurance, research, and government are also economically important. Founded and incorporated in 1871, Birmingham developed rapidly with the expansion of the railroads. An important trade and communications center, the city is connected with the Gulf of Mexico by canal and is a port of entry. Birmingham has a number of colleges and universities, botanical and Japanese gardens, a symphony, a ballet group, a theater, and an art museum. A Festival of Arts is held annually. Overlooking the city, on nearby Red Mt., is a huge iron statue of Vulcan, the mythical god of the forge. **2.** City (1970 pop. 26,170), Oakland co., SE Mich., on the River Rouge; settled 1819, inc. as a village 1864, as a city 1933. The city is largely residential.

Bir·nam (bûr'nəm), village (1975 est. pop. 659), Tayside, central Scotland, on the Tay opposite Dunkeld. Nearby is Birnam Hill, once forested with the Birnam Wood referred to by the three witches in *Macbeth*.

Bir·o·bi·dzhan (bîr'ō-bē-jän'): *see* Jewish Autonomous Oblast.

Bis·bee (bĭz'bē), city (1975 est. pop. 9,914), seat of Cochise co., SE Ariz., near the Mexican border; inc. 1900. It is the center of one of the greatest copper-producing areas in the country. Gold, silver, and lead are also mined. After the rich copper deposits were discovered (c.1876), the city was built in two steep-sided canyons, Mule Pass Gulch and Brewery Gulch.

Bis·cay, Bay of (bĭs'kā, -kē), arm of the Atlantic Ocean, indenting the coast of W Europe from Ushant island (Île d'Ouessant) off Brittany, NW France, to Cape Ortegal, NW Spain. The bay is noted for its sudden, severe storms and its strong currents. The rocky northeastern and southern coasts of Biscay are irregular with many good harbors. The southeastern shore is straight and sandy. There are several resorts along the French coast, notably Biarritz.

Bis·cayne Bay (bĭs'kān, bĭs-kān'), shallow, narrow inlet of the Atlantic Ocean, c.40 mi (65 km) long, SE Fla. Famous resort areas, including Miami and Miami Beach, are on the northern shore. Biscayne National Monument is at the southern end of the bay.

Bi·sce·glie (bē-shāl'yä), city (1976 est. pop. 46,296), Apulia, S Italy, on the Adriatic Sea. It is a seaport and commercial center. Conquered by the Normans in the late 11th cent., the city later developed a prosperous merchant and military fleet.

Bishop Auck·land (ôk'lənd), urban district (1973 pop. 32,940), Durham, NE England, on the Wear River. Located near the site of a Roman fort, it is a busy market area, as well as a mining town producing coal that is highly suitable for coking.

Bish·op·ville (bĭsh'əp-vĭl'), town (1970 pop. 3,404), seat of Lee co., NE S.C., NE of Columbia; inc. 1888. Clothing, farm equipment, and textiles are made, and grain and cotton seed are processed. Lee State Park is nearby.

Bis·kra (bĭs-krä'), city (1974 pop. 84,971), NE Algeria, at the foot of the Aures Mts. It is a commercial center for the nomads of the surrounding region. It was the Roman military base of Vescera; later it was an important Moslem town. After 1844 it served as a French base for operations in southern Algeria.

Bis·marck (bĭz'märk), city (1975 est. pop. 37,500), state capital and seat of Burleigh co., S central N.Dak., on hills overlooking the Missouri River; inc. 1873. A trade and distributing point for a large spring wheat, livestock, and dairy region, it is also the center for development of the rich oil reserves in nearby Williston Basin. Food items, farm machinery, woodwork, and concrete products are made. Lewis and Clark camped nearby in 1804–5. In 1872 Camp Greeley (later Camp Hancock) was erected to protect the men who were building the Northern Pacific RR. When the railroad reached the fort the next year, a town was laid out; it was subsequently named Bismarck (for Germany's chancellor) in the hope of attracting German investment in the railroad. Bismarck boomed as a river port and railroad center and as a supply point for the Black Hills gold mines (1874). It became the territorial capital in 1883. Of interest are the state capitol (1932), a skyscraper, the state historical museum, and Camp Hancock museum.

Bismarck Archipelago, volcanic group (1969 est. pop. 213,000), 19,200 sq mi (49,728 sq km), SW Pacific, a part of Papua New Guinea. The group includes New Britain (the largest island) and some 200 other islands and islets. The islands are generally mountainous and have several active volcanoes. The chief agricultural products are copra, cacao, coffee, tea, and rubber. Some copper and gold are mined. Discovered in 1616 by the Dutch explorer Willem Schouten, the group became a German protectorate in 1884. Seized by Australian forces in World War I, the islands were mandated to Australia by the League of Nations in 1920. Japan operated several naval and air bases in the islands during World War II. In 1947 Australia received trusteeship over the group from the United Nations. The archipelago was included in Papua New Guinea when it became self-governing in 1973.

Bismarck Range, mountains, Papua New Guinea, on NE New Guinea Island. The range reaches a height of 15,400 ft (4,697 m) at Mt. Wilhelm.

Bismarck Sea, sea in the SW Pacific Ocean, c.500 mi (805 km) across from E to W, NE of New Guinea and NW of New Britain. The Battle of Bismarck Sea was fought March 2–3, 1943, resulting in the destruction of the Japanese naval force by U.S. aircraft and naval units.

Bi·son (bī'sən, -zən), town (1970 pop. 406), seat of Perkins co., NW S.Dak., NNE of Rapid city. It is a trade center for a cattle and grain area. Lignite mines are in the area.

Bis·sau (bĭ-sou'), town (1970 pop. 71,169), former capital of Guinea-Bissau, a port in the Geba estuary, off the Atlantic Ocean. It is the country's largest city, major port, and administrative and military center. Bissau has been a free port since 1869 and handles some transit trade. The city was founded in 1687 by the Portuguese as a fortified port and trading center.

Bi·thyn·ia (bĭ-thĭn'ē-ə), ancient country of NW Asia Minor, in present-day Turkey. The original inhabitants were Thracians who were given some autonomy after Cyrus the Great incorporated Bithynia into the Persian Empire. Later, it was an important province of Rome. Pliny the Younger was governor of the province (A.D. c.110) under the emperor Trajan. The reign of Hadrian soon after seems to have marked the end of Bithynian prosperity. It was invaded briefly by the Goths (A.D. 298).

Bit·lis (bĭt-lĭs'), town (1975 pop. 25,085), capital of Bitlis prov., E Turkey, on a tributary of the Tigris River, at c.4,500 ft (1,375 m). Grains, fruit, and tobacco are grown nearby. Located on a passage through the Taurus Mts., it was an important caravan center for centuries and was captured by Persians, Arabs, Seljuk Turks, Byzantines, and Ottoman Turks.

Bi·to·la (bē'tô-lə), city (1971 pop. 124,648), extreme S Yugoslavia, in Macedonia. It is a commercial and industrial center for the surrounding agricultural area. Bitola was a major agricultural center in Roman times. In 1395 the Turks conquered Bitola, which became an important military and commercial center in the 15th and 16th cent. and a Balkan administrative center in the 19th cent. The city suffered much damage during the Balkan Wars (during which the Serbs took it from the Turks) and in World War I.

Bi·ton·to (bē-tôn'tō), city (1976 est. pop. 46,197), Apulia, S Italy. It is an agricultural market and is noted for its olive oil. The Spanish under Charles Bourbon defeated the Austrians here in 1734 during the War of the Polish Succession. The Apulian Romanesque cathedral (12th-13th cent.) is especially remarkable for its fine sculptures.

Bit·ter·feld (bĭt'ər-fĕlt'), city (1970 pop. 28,964), Halle dist., central East Germany, on the Mulde River. It is an industrial center and rail junction. Manufactures include chemicals, aluminum, machinery, and plastics. Lignite is mined in the region. Bitterfeld was founded in the mid-12th cent. and passed to Prussia in 1815.

Bit·ter·root (bĭt'ər-ro͞ot', -ro͝ot'), river, c.120 mi (195 km) long, rising in SW Mont. and flowing N to join the Clark Fork River near Missoula.

Bitterroot Range, part of the Rocky Mts., on the Idaho-Mont. line. The main range, running northwest-southeast, includes Trapper Peak (10,175 ft/3,103.4 m high); Mt. Garfield (10,961 ft/3,343.1 m), in an east-running spur to the south, is the highest peak. Discovered in the 1804-5 expedition of Lewis and Clark, this rugged mountain range has long been one of the most impenetrable in the United States; except for its foothills, it remains almost completely unexploited today.

Bi·wa (bē'wä), lake, c.40 mi (65 km) long and from 2-12 mi (3.2-19.3 km) wide, Shiga prefecture, S Honshu, Japan. The lake is the largest in Japan and is a popular scenic resort.

Bi·ysk (bē' ĭsk), city (1976 est. pop. 209,000), S central Siberian USSR, on the Biya River. A port and the terminus of a branch of the Turkistan-Siberia RR, Biysk manufactures food-processing equipment. The city was founded as a fortress in 1709.

Bi·zerte (bē-zĕrt'), city (1975 est. pop. 62,856), N Tunisia, on the Mediterranean Sea. It is an important port, strategically situated near the narrowest part of the Mediterranean. The city also has processing industries. Bizerte was founded by Phoenicians. In 1881 it was seized by the French, who improved and fortified the harbor. It was a German base in World War II and was heavily bombed (1943) by the Allies. Tunisian insistence that France evacuate its naval installations at Bizerte led to violent confrontations in 1961; the base was turned over to Tunisia in 1963.

Black (blăk). **1.** River rising in SE Mo. and flowing c.300 mi (485 km) SE, then SW to the White River near Newport, Ark. It is partly navigable. Clearwater Dam is on the river near Piedmont, Mo. **2.** River of N N.Y., c.120 mi (195 km) long, rising in the Adirondack Mts. and flowing mainly N and W to Black River Bay, an inlet of Lake Ontario. Its falls provide power for many factories, especially paper mills. **3.** River, c.160 mi (260 km) long, rising in central Wis. and winding SW to the Mississippi River at La Crosse, Wis. It was important in the lumbering industry and is now used to transport coal and petroleum products.

Black, river, c.500 mi (805 km) long, N Vietnam, SE Asia. It rises in Yunnan prov., central China, and flows southeast parallel to the Red River, which it joins near Sontag, Vietnam.

Black Belt, term loosely applied to several areas of the U.S. South that are characterized by black soil and excellent cotton-growing conditions.

Black·burn (blăk'bərn), county borough (1976 est. pop. 142,500), Lancashire, NW England. It was formerly a great cotton-weaving center, noted especially for calicoes. Textiles are still important, but now there are other large industries that make engineering equipment, radio parts, beer, felt, and carpets. Blackburn is also an agricultural market. The textile industry started early in the 17th cent. When James Hargreaves invented (c.1765) the spinning jenny nearby, the manufacture of cotton goods received a new impetus. The completion of the Leeds-Blackburn-Liverpool Canal in 1816 substantially aided Blackburn's economic growth.

Black Canyon of the Gun·ni·son National Monument (gŭn'ĭ-sən): *see* National Parks and Monuments Table.

Black Country, highly industrialized region, mostly in Staffordshire but partly in Worcestershire and Warwickshire, W central England.

Black·foot (blăk'fo͝ot'), city (1970 pop. 8,716), seat of Bingham co., SE Idaho, SSW of Idaho Falls, between the Blackfoot and Snake rivers near their confluence; founded 1878, inc. 1901. This livestock, dairy, and farm (potatoes, sugar beets, and grains) area is irrigated by the Snake through the Minidoka project and by the Blackfoot through the Fort Hall project. Blackfoot has many food-processing plants and is the scene of the annual Eastern Idaho Fair. Fort Hall Indian Reservation is nearby.

Black·ford (blăk'fərd), county (1970 pop. 15,888), 167 sq mi (432.5 sq km), central E Ind., drained by the Salamonie River and Lick Creek; formed 1838; co. seat Hartford City. It has agriculture (grain, soybeans, livestock, and dairy products), gas and oil wells, stone quarries, and timber. Glass is manufactured.

Black Forest or *German* **Schwarz·wald** (shvärts'vält'), mountain range, SW West Germany, extending 90 mi (145 km) between the Rhine and Neckar rivers. Feldberg is the highest (4,898 ft/1,493.9 m) peak. The range is covered by dark pine forests and cut by deep valleys and small lakes. Lumbering is an important economic activity. Orchards and cattle are found in the valleys; grains are grown in the highlands. The Black Forest is famous for its clock and toy industries. It is a year-round resort area.

Black Hawk (hôk), county (1970 pop. 132, 916), 567 sq mi (1,468.5 sq km), E central Iowa, in a prairie agricultural area drained by the Cedar River; formed 1853; co. seat Waterloo. Agriculture (hogs, cattle, poultry, corn, oats, and soybeans) and agricultural products are important. It has limestone quarries and sand pits. Farm machinery is manufactured.

Black·heath (blăk'hēth'), common, 267 acres (108 hectares) in Lewisham and Greenwich boroughs, London, England. It was the gathering place of highwaymen and of several martial groups, including the followers of Wat Tyler in 1381 and of Jack Cade in 1450.

Black Hills, rugged mountains, c.6,000 sq mi (15,540 sq km), enclosed by the Belle Fourche and Cheyenne rivers, SW S.Dak. and NE Wyo., and rising c.2,500 ft (760 m) above the surrounding Great Plains. Harney Peak, 7,242 ft (2,208.8 m) above sea level, is the highest point. The mountains received their name from the heavily forested slopes that appear black from afar. Indians, settlers, and railroad companies depended on wood from the Black Hills for fuel and building material. Gold was discovered in the hills in 1874 by an expedition led by Gen. George Custer, and the resulting gold rush drove out the Indians. Gold is still mined in the area; Homestake Mine is the largest gold mine in the United States. Other important minerals found in the hills are uranium, feldspar, mica, and silver. The Black Hills are a major recreational area.

Black·pool (blăk'po͞ol'), county borough (1971 pop. 151,311), Lancashire, NW England, on the Irish Sea. One of England's most popular seaside resorts, Blackpool has many sport and amusement facilities and a tower 520 ft (158.6 m) high, modeled on the Eiffel Tower in Paris. Blackpool's manufactures include aircraft, biscuits, and candy.

Black River Falls, city (1970 pop. 3,273), seat of Jackson co., W central Wis., on the Black River NE of La Crosse; settled before 1840, inc. 1883. It is a farm trade center. There are Indian mounds nearby.

Blacks·burg (blăks'bûrg), town (1970 pop. 9,384), Montgomery co., SW Va., in the Alleghenies W of Roanoke; settled 1745, inc. 1871. Virginia Polytechnic Institute is here, and nearby is scenic Mountain Lake, a resort.

Black Sea, inland sea, c.159,600 sq mi (413,365 sq km), between Europe and Asia, connected with the Mediterranean Sea by the Bosporus, the Sea of Marmara, and the Dardanelles. It is c.750 mi (1,210 km) long, from 75 to 350 mi (120-560 km) wide, and has a maximum depth of 7,364 ft (2,246 m). The largest arm of the Black Sea is the Sea of Azov, which joins it through the Kerch Strait. The Black Sea is enclosed by the USSR on the north and east, by Turkey on the south, and by Bulgaria and Rumania on the west. The Black Sea was once part of a large body of water that included the Caspian and

Aral seas. In the Tertiary period, it was separated from the Caspian Sea and was linked to the Mediterranean Sea. The rivers flowing into the northern part of the Black Sea carry much silt and form deltas, sandbars and lagoons along the generally low and sandy northern coast. The southern coast is steep and rocky. The Black Sea is an important navigation route and remains ice-free in winter despite severe storms. The region, especially in the Crimea and Caucasus, is a popular resort area.

The Pontus Euxinus of the ancients, the Black Sea has been navigated since prehistoric times. Its shores were colonized by the Greeks (8th–6th cent. B.C.) and later by the Romans (3rd–1st cent. B.C.). Its importance increased with the founding of Constantinople (330 A.D.). In the 13th cent. the Genoese established colonies on the Black Sea, and from the 15th to the 18th cent. it was a Turkish lake. The rise and expansion of Russia and its ambition to gain control of the Bosporus and the Dardanelles led it into protracted dispute with the Ottoman empire. In 1783 Russia annexed the Tatar Khanate of Crimea, which blocked its access to the sea, but suffered a setback as a Black Sea power as a result of the Treaty of Paris, which ended the Crimean War of 1856.

Black·shear (blăk'shîr'), city (1970 pop. 2,624), seat of Pierce co., SE Ga., near the Satilla River NE of Waycross, in a coastal-plain farm and timber area; inc. 1859.

Black·town (blăk'toun'), city (1976 est. pop. 159,724), New South Wales, SE Australia. It is a suburb of Sydney.

Black War·ri·or (wôr'ĭ-ər), navigable river, 178 mi (286.4 km) long, rising in N central Ala. and flowing generally SW to the Tombigbee River. The Black Warrior drains a rich coal- and cotton-producing area and is an important outlet for the manufactured products of Birmingham, Ala.

Black·wa·ter (blăk'wô'tər), river, c.100 mi (161 km) long, rising in Co. Kerry, SW Republic of Ireland. It flows east through the dairy region of Co. Cork and Co. Waterford before turning abruptly south and entering the Atlantic Ocean at Youghal Bay. Salmon and trout are caught in the river.

Black·well (blăk'wĕl'), city (1970 pop. 8,645), Kay co., N cental Okla., NW of Ponca City, in a wheat area; founded and inc. 1893. The city has a zinc smeltery, oil refinery, glass factory, and foundry. It is a market and shipping center for an agricultural area.

Bla·den (blăd'n), county (1970 pop. 26,477), 883 sq mi (2,287 sq km), SE N.C., on the coastal plain, bounded on the E by the South River and crossed by the Cape Fear River; formed 1734; co. seat Elizabethtown. It has agriculture (tobacco, corn, peanuts, soybeans, and blueberries), livestock raising, dairying, and some manufacturing (hardware and textiles).

Bla·dens·burg (blā'dənz-bûrg), town (1975 est. pop. 6,789), Prince Georges co., S central Md., a residential suburb of Washington, D.C.; chartered 1742, inc. 1854. The defeat (Aug. 24, 1814) at Bladensburg of American troops permitted the British to march on Washington, D.C., and burn many of the public buildings.

Bla·go·vesh·chensk (blə-gə-vyĕsh'chĭnsk), city (1976 est. pop. 171,000), capital of Amur oblast, Far Eastern USSR, at the confluence of the Amur and Zeya rivers. A river port and railroad hub, Blagoveshchensk is also an agricultural center and a supply point for the Zeya gold-mining basin. Russian pioneers settled Blagoveshchensk in 1644, but the area was returned to China in 1689. The city became a Russian army post in 1856.

Blaine (blān). **1.** County (1970 pop. 5,749), 2,649 sq mi (6,860.9 sq km), S central Idaho, in a winter resort area; formed 1895; co. seat Hailey. Mining (silver, lead, gold, zinc, and copper) and livestock raising are done. It includes parts of Sawtooth National Forest, Sawtooth National Recreation Area, Challis National Forest, and Craters of the Moon National Monument. Sun Valley ski resort is here. **2.** County (1970 pop. 6,727), 4,275 sq mi (11,072.3 sq km), N Mont., in an agricultural area bordering on Sask. and drained by the Milk River; formed 1912; co. seat Chinook. Fort Belknap Indian Reservation, Black Coulee National Wildlife Refuge, and Chief Joseph Battleground of the Bears Paw Mountain are here. **3.** County (1970 pop. 847), 710 sq mi (1,838.9 sq km), central Nebr., in an agricultural region drained by the North Loup and Middle Loup rivers; formed 1885, co. seat Brewster. It produces livestock and grain. Part of Nebraska National Forest is in the southwest section. **4.** County (1970 pop. 11,794), 917 sq mi (2,375 sq km), W central Okla., intersected by the North Canadian and Canadian rivers; formed 1891; co. seat Watonga. It has agriculture (cotton, wheat, corn, oats, livestock, and dairy products). Farm products are shipped and processed, and gypsum is mined. It includes Roman Nose State Park.

Blaine, city (1975 est. pop. 26,400), Anoka co., SE Minn., a suburb N of Minneapolis; settled 1862, inc. 1964. Construction is the major industry. The area was organized as a township in 1877.

Blair (blâr), county (1970 pop. 135,356), 530 sq mi (1,372.7 sq km), central Pa., in a mountainous agricultural and industrial area bounded on the W by the Allegheny Mts., on the E by Tussey Mt., and on the N by part of Bald Eagle Mt.; formed 1846; co. seat Hollidaysburg. It is drained by the Frankstown Branch of the Juniata River and the Little Juniata River. It has agriculture (dairy products, fruit, and grain), and limestone, sandstone, clay, shale, and bituminous-coal deposits. Metal products, railroad rolling stock, paper, and textiles are manufactured. Most of the lead used by Washington's armies in the Revolution was mined here.

Blair, city (1970 pop. 6,106), seat of Washington co., E Nebr., on bluffs overlooking the Missouri River NNW of Omaha; founded 1869. It is a farm trade center producing feed, farm equipment, and lumber. Dana College is here.

Blairs·ville (blârz'vĭl'), town (1970 pop. 491), seat of Union co., N Ga., N of Dahlonega in the Blue Ridge near Nottely Reservoir. A state agricultural experiment station is nearby.

Blake·ly (blāk'lē). **1.** City (1970 pop. 5,267), seat of Early co., SW Ga., SW of Albany near the Chattahoochee River; founded 1821. It is a trade and processing center in a timber, pecan, and peanut area. There are Indian mounds nearby. **2.** Borough (1970 pop. 6,391), Lackawanna co., NE Pa., NE of Scranton; inc. 1867. Coal is mined.

Blanc, Mont (môɴ' bläɴ), Alpine massif, on the French-Italian border, SE of Geneva. One of its several peaks, also called Mont Blanc (15,771 ft/4,810.2 m), is the highest peak in France and the second highest in Europe. The southeastern (Italian) face is a massive wall; on the northwestern slopes are numerous glaciers, the largest of which (the Mer de Glace) flows into the valley of Chamonix, a famous French resort region and starting point for mountain climbers. The first successful ascent of Mont Blanc was made in 1786. In 1965 a highway tunnel (7 mi/11.3 km long) under Mont Blanc, linking Chamonix with Courmayeur, Italy, was opened to traffic.

Blan·ca Peak (blăng'kə), 14,317 ft (4,366.7 m) high, S Colo., ENE of Alamosa. It is the highest point in the Sierra Blanca of the Sangre de Cristo Mts., a range of the Rockies.

Blan·co (blăng'kō), county (1970 pop. 3,567), 719 sq mi (1,862.2 sq km), S central Texas, in the hill country of Edwards Plateau and drained by the Pedernales and Blanco rivers; formed 1858; co. seat Johnson City. It has ranching (cattle, sheep, and goats) and agriculture (cotton, corn, oats, grain sorghums, truck crops, fruit, pecans, and turkeys). Mohair and wool are marketed. The county includes the Lyndon B. Johnson National Historic Site.

Bland (blănd), county (1970 pop. 5,423), 369 sq mi (955.7 sq km), SW Va., in a scenic region of the Allegheny Mts., bounded on the N by the W.Va. border and drained by tributaries of the New River; formed 1861; co. seat Bland. It has agriculture (livestock, corn, wheat, and hay), timber and lumber milling, and manganese deposits. It lies within Jefferson National Forest.

Bland, agricultural village (1970 pop. 300), seat of Bland co., SW Va., SSE of Bluefield. Hosiery is made, and lumber processed.

Blan·ken·burg (blăng'kən-bŏŏrk') or **Blankenburg am Harz** (äm härts), city (1974 est. pop. 18,784), Magdeburg dist., W East Germany. It is a spa located at the northern foot of the Harz Mts. and also has industries that manufacture woolens and paper. The first kindergarten was founded here (c.1840).

Blan·tyre (blăn-tīr'), city (1972 est. pop. 163,921), S Malawi, in the Shire Highlands. It is the chief commercial and industrial center of Malawi with cement, food processing, and textile industries. Blantyre was founded in 1876 as a Church of Scotland mission station.

Blantyre, town (1971 pop. 13,992), Strathclyde, S central Scotland, near Glasgow. In a coal-mining district, it has chemical, textile, and engineering works. There is a museum dedicated to David Livingstone, who was born here.

Blar·ney (blär'nē), village, Co. Cork, SE Republic of Ireland. Whoever kisses the Blarney Stone, placed in an almost inaccessible position near the top of the thick stone wall of the 15th cent. castle, is supposed to gain marvelous powers of persuasion.

Blay·don (blād'n), urban district (1971 pop. 32,018), Tyne and Wear, NE England, on the Tyne River. It manufactures iron and steel goods, bricks, and the by-products of coal from nearby mines. There are also engineering works.

Bleck·ley (blĕk'lē), county (1970 pop. 10,291), 219 sq mi (567.2 sq km), central Ga., bounded on the W by the Ocmulgee River and drained by the Little Ocmulgee River; formed 1912; co. seat Cochran. It is in a coastal-plain timber and agricultural region yielding corn, peanuts, truck crops, fruit, and livestock.

Bled·soe (blĕd'sō), county (1970 pop. 7,643), 404 sq mi (1,046.4 sq km), central Tenn., on the Cumberland Plateau and crossed by the fertile Sequatchie River valley; formed 1807; co. seat Pikeville. It has timber, dairying, livestock raising, and fruit growing.

Blen·heim (blĕn'əm), village, Bavaria, S West Germany, on the Danube River. Between Blenheim and nearby Höchstädt, John Churchill, 1st Duke of Marlborough, and Prince Eugene of Savoy defeated (Aug. 13, 1704) the French and Bavarians in one of the most important battles of the War of the Spanish Succession.

Bli·da (blē'dä), town (1974 pop. 158,947), N Algeria, at the foot of the Atlas Mts. It is an administrative center and an agricultural trading town. Blida is surrounded by gardens and by orange, olive, and almond tree plantations. The city is noted for its fruit and flower essences. Built on the site of a Roman military base, Blida was founded in 1553 by Andalusians, who developed irrigation works and orange cultivation.

Block Island (blŏk), 7 mi (11.3 km) long and 3.5 mi (5.6 km) wide, off S R.I. at the E entrance to Long Island Sound. Visited by the Dutch navigator Adriaen Block in 1614, it was settled in 1661. The murder (1637) of John Oldham, an English trader, was the direct cause of the Pequot War. Characterized by numerous small ponds, low hills, and a mild climate, the island has long been a favorite fishing and resort area. There are two lighthouses.

Bloem·fon·tein (blo͞om'fŏn-tān'), city (1970 pop. 149,836), capital of the Orange Free State and the judicial center of the Republic of South Africa. It is a transportation hub and industrial center, containing railroad workshops, food-processing plants, and factories that produce furniture, plastics, and glassware. Bloemfontein was founded in 1846 and served as the capital of the Orange Free State Republic until its capture (1900) by British forces during the South African War. Afterward, it was the site of the final negotiations (1909) that led to the establishment (1910) of the Union of South Africa.

Blois (blwä), town (1975 est. pop. 49,778), capital of Loir-et-Cher dept., central France, in Orléanais, on the Loire River. A commercial and industrial center with an outstanding trade in wines and brandies, it is also one of the most historic towns of France. The counts of Blois emerged in the 10th cent. as the most powerful feudal lords of France. The last count, childless and heavily in debt, sold his fief to Louis, duc d'Orléans, who took possession in 1397. With the accession (1498) of Louis's grandson, Louis XII, as king of France, the countship passed to the crown as part of Orléanais. The town was a favorite royal residence. Louis XII was born in the Renaissance château here. Several States-General of France were held in the château, notably in 1576–77 and in 1588.

Bloom·field (blo͞om'fēld'). **1.** Town (1975 est. pop. 20,200), Hartford co., N Conn., a suburb of Hartford, in a tobacco and dairy region; settled c.1642, inc. 1835. Aircraft parts are manufactured, and the home office of a large insurance company is here. **2.** Town (1970 pop. 2,565), seat of Greene co., SW Ind., SW of Bloomington, in a grain and livestock area. Furniture is manufactured. **3.** City (1970 pop. 2,718), seat of Davis co., SE Iowa, S of Ottumwa near the Mo. border; laid out 1844, inc. 1852. It is the center of a fine grazing area, and much wool is produced. Lake Wapello State Park is nearby. **4.** City (1970 pop. 1,584), seat of Stoddard co., SE Mo., SW of Cape Girardeau, in a farm and timber region; settled on the site of an Indian village, platted 1835, inc. 1894. **5.** City (1970 pop. 52,029), Essex co., NE N.J., an industrial and residential suburb of Newark; settled c.1660, inc. as a town 1812, as a city 1900. Electrical equipment and pharmaceuticals are made in the city, which is also the seat of Bloomfield College. Bloomfield was a supply point for both sides during the Revolutionary War.

Bloom·ing·dale (blo͞o'mĭng-dāl'), borough (1970 pop. 7,797), Passaic co., NE N.J., NW of Paterson; inc. 1918. Carbonated beverages and clothing are manufactured.

Bloom·ing·ton (blo͞o'mĭng-tən). **1.** City (1975 est. pop. 43,193), seat of McLean co., central Ill.; inc. 1839. It is an important rail, commercial, and industrial center in a rich farm and coal area. In 1856 the state Republican party was organized in Bloomington, at which time Lincoln delivered his famous "lost speech" (no copy of which is known to exist). The city is the seat of Illinois Wesleyan Univ. Adlai E. Stevenson is buried here. **2.** City (1970 pop. 43,262), seat of Monroe co., S central Ind., in a densely forested region; settled 1816, inc. 1878. Electronic machinery, electrical appliances, and elevators are manufactured. Limestone is abundant in the area. It is the seat of Indiana Univ. In the area are three state parks, Hoosier National Forest, and Lakes Monroe (Indiana's largest) and Lemon. **3.** City (1970 pop. 81,970), Hennepin co., SE Minn., a suburb adjacent to Minneapolis; inc. 1953. Its many manufactures include lawn mowers, electronic equipment, and metal products.

Blooms·burg (blo͞omz'bûrg'), industrial town (1970 pop. 11,652), seat of Columbia co., E Pa., on the Susquehanna River; settled 1772, inc. 1870. Carpets and silk are among its manufactures.

Blooms·bur·y (blo͞omz'bə-rē, -brē), residential district, Camden metropolitan borough, London, England. The British Museum and the Univ. of London are here. Bloomsbury contains numerous squares and gardens (including Bedford Square, Russell Square, and Bloomsbury Square, laid out on the Bedford estate in the 18th cent.). Many artists, writers, and students live in the district.

Blount (blŭnt). **1.** County (1970 pop. 26,853), 640 sq mi (1,657.6 sq km), N central Ala., in a hilly region drained by the Mulberry and Locust forks of the Black Warrior River; formed 1818; co. seat Oneonta. Cotton, corn, and poultry are produced. There are coal, iron, and limestone deposits in the vicinity. An iron and steel industry developed here after World War II. Wood products and textiles are manufactured. **2.** County (1970 pop. 63,744), 575 sq mi (1,489.3 sq km), E Tenn., bounded on the SE by the N.C. border, on the SW by the Little Tennessee River, and on the NW by Fort Loudoun Reservoir on the Holston River; formed 1795; co. seat Maryville. The Great Smoky Mts. are in the east and southeast. It has agriculture (corn, tobacco, hay, livestock, and dairying), lumbering, marble quarrying, and some industry.

Blounts·town (blŭnts'toun'), city (1970 pop. 2,384), seat of Calhoun co., NW Fla., near the Apalachicola River NE of Panama City, in a timber area; settled 1838, inc. 1917. Naval stores are produced.

Blount·ville (blŭnt'vĭl'), village (1970 pop. 900), seat of Sullivan co., NE Tenn., in the Great Appalachian Valley N of Johnson City.

Blue Earth (ûrth), county (1970 pop. 52,322), 740 sq mi (1,916.6 sq km), S Minn., in an agricultural area bounded on the N by the Minnesota River and drained by the Blue Earth River; formed 1853; co. seat Mankato. Its agriculture includes corn, oats, barley, livestock, and dairy products. Some food processing and manufacturing is done. Limestone deposits have been found here.

Blue Earth, city (1970 pop. 3,965), seat of Faribault co., S Minn., on the Blue Earth River W of Albert Lea and near the Iowa line; platted 1856, inc. 1874. It is a trade center for a diversified farming and food-processing area.

Blue·field (blo͞o'fēld'), city (1970 pop. 17,484), Mercer co., extreme SW W. Va., in the Allegheny Mts. adjacent to Bluefield, Va.; settled 1777, inc. 1889. It is a trade center and a shipping point for the Pocahontas coal field.

Blue·fields (blo͞o'fēldz'), town (1970 est. pop. 22,910), capital of Zelaya dept., SE Nicaragua, on Bluefields Bay at the mouth of the Escondido River. It is Nicaragua's chief Caribbean port. Bananas, hardwoods, and coconuts are exported.

Blue Island, city (1970 pop. 22,958), Cook co., NE Ill., a residential and industrial suburb of Chicago, on the Little Calumet River; inc. 1843. It has oil refineries, railroad yards and shops, canneries, and plants manufacturing electric signals, plastic products, steel forgings, glass, chemicals, and medical and dental supplies.

Blue Mountains, uplifted, eroded part of the Columbia Plateau, c.6,500 ft (1,985 m) high, NE Oregon and SE Wash. Lava flows cover much of the surface. The upper, wooded slopes are used for lumbering. Irrigated farming and cattle raising are carried on in the surrounding lowlands.

Blue Nile, river, c.1,000 mi (1,610 km) long, the chief headstream of the Nile, rising in Lake Tana, NW Ethiopia, at an altitude of c.6,000 ft (1,830 m). It flows generally south from the Lake Tana region, then west across Ethiopia, and finally northwest into the Sudan. At Khartoum the Blue Nile merges with the White Nile to form the Nile proper. The flow of the Blue Nile reaches maximum volume in the rainy season (from June to September), when it supplies about two thirds of the water of the Nile proper. The Blue Nile used to cause the annual Nile flood before the completion in 1970 of the Aswan High Dam in Egypt.

Blue Ridge, city (1970 pop. 1,602), seat of Fannin co., N Ga., in Chattahoochee National Forest N of Atlanta near the Tenn.-N.C. line; inc. 1887. It is a trade and shipping center for a farm, timber, and mine area.

Blue Ridge, range of the Appalachian Mts., extending S from S Pa. to N Ga., highest mountains in the E United States. Mt. Mitchell, 6,684 ft (2,038.6 m) high, is the tallest peak. Beginning with a narrow ridge in the north, c.10 mi (16 km) wide, the range broadens toward the south, reaching a maximum width of 70 mi (112.6 km). Receiving much rain, the region is heavily forested; wood is the area's chief resource. The Blue Ridge is a major East Coast recreation area noted for its resorts and scenery. The Appalachian Trail winds atop the range. The Blue Ridge Parkway, designed especially for motor recreation, links Shenandoah and Great Smoky Mts. national parks.

Bluff·ton (blŭf'tən), city (1970 pop. 8,297), seat of Wells co., NE Ind., on the Wabash River S of Fort Wayne; settled 1829, inc. 1858. It has limestone quarries. Dairy products are processed, and chemicals and metal and wood products are manufactured.

Blyth (blīth), municipal borough (1976 pop. 69,300), Northumberland, NE England, at the mouth of the Blyth River. It is an industrial center and seaport, with shipbuilding and ship repair and a large trade in coal and timber.

Blythe (blīth), city (1970 pop. 7,047), Riverside co., SE Calif., near the Colorado River NE of Brawley, in an irrigated farm area; laid out 1910, inc. 1916. Palo Verde College is here.

Blythe·ville (blīth'vĭl'), city (1970 pop. 24,752), a seat of Mississippi co., NE Ark., near the Mississippi River; inc. 1891. It is the trading center of the state's richest cotton area; soybeans and feed crops are also grown in the region. The city is an industrial center as well. Blytheville Air Force Base is here.

Bo·a Vis·ta (bō'ə vēsh'tə), city (1970 pop. 16,720), capital of Roraima Federal Dist., NW Brazil, on the Rio Branco. Its economy is based on the processing and shipment of minerals (gold, bauxite, quartz, and oil) found in the surrounding region.

Bo·bi·gny (bô-bē-nyē'), city (1975 est. pop. 43,125), capital of Seine-Saint Denis dept., N central France, an industrial suburb of Paris. Metals, food products, and toys are among the major manufactures.

Bo·bruysk (bə-broo'ēsk), city (1976 est. pop. 185,000), Belorussia, W central European USSR, a port on the Berezina River. It is also a railway junction and tire-manufacturing center. Bobruysk was founded in the 16th cent.

Bo·ca Ra·ton (bō'kə rə-tōn'), city (1975 est. pop. 42,363), Palm Beach co., SE Fla., on the Atlantic; inc. 1925. Boca Raton is a resort city and manufactures computers, plastic products, electrical equipment, furniture, and many other products. Florida Atlantic Univ. is here.

Bo·chum (bō'KHoom), city (1975 est. pop. 417,336), North Rhine-Westphalia, W West Germany. Mentioned in the 9th cent. and chartered in 1321, it remained a small farming community until the development of nearby coal mines in the mid-19th cent. By the late 19th cent. it was a leading center of the Ruhr iron and steel industry; since the early 1960s its importance in coal and steel production has declined. Bochum today is an industrial and commercial center, a rail and road junction, and a growing vacation spot.

Bo·den (bood'n), city (1975 est. pop. 24,428), Norrbotten co., NE Sweden, on the Luleålv River; chartered 1919. It is an important rail junction and a winter sports center.

Bo·den·see (bōd'n-zā'): *see* Constance, Lake of.

Bodh Ga·ya or **Buddh Ga·ya** (both: bood gä'yä), village (1971 pop. 6,993), Bihar state, E central India. According to tradition, Buddha received enlightenment under a pipal tree (bo tree) in Bodh Gaya.

Bo·di·na·yak·a·nur (bō'dĭ-nä-yək'ə-noor'), town (1971 pop. 54,176), Tamil Nadu state, at the foot of the Western Ghats, SE India. A Bodinayakanur state is said to have been established in 1336. The town is a market for cardamom, coffee, tea, silk, and cotton.

Bod·min (bŏd'mĭn), municipal borough (1973 est. pop. 10,430), administrative center of Cornwall, SW England. The county offices are now in Truro. Bodmin was formerly a busy market for tin and wool.

Bo·do (bō'dö), city (1977 est. pop. 31,467), capital of Nordland co., W Norway, at the mouth of the Saltfjord, N of the Arctic Circle. It is a center for coastal shipping, tourism, and fishing and serves as the port of the Sulitjelma copper and pyrite mines. Of note is Bodin Church, a medieval stone structure.

Boe·o·tia (bē-ō'shə), region of ancient Greece, N of Attica, Megaris, and the Gulf of Corinth. The early inhabitants were from Thessaly. A number of small cities scattered over the rough country—mountainous in the south, hilly in the north—may have had a sort of confederacy before the Boeotian League was formed (c.7th cent. B.C.). Thebes dominated the region and the league. After the defeat of the Persians at Plataea (479), the Greeks besieged Thebes for aiding the Persians, and the Boeotian League was disbanded. The league was temporarily revived in 457 B.C. before being defeated in the same year by Athens, which briefly attached the Boeotian cities to the Athenian empire. Thebes returned to power at the head of the league in 446, but declined after the victory of Epaminondas over the Spartans (371).

Boer·ne (bûr'nē), city (1970 pop. 2,432), seat of Kendall co., S central Texas, on Cibolo Creek NW of San Antonio; founded c.1850 by Germans, inc. 1909. It is a trade center for sheep, goat, and cattle ranches, and its clear air and rugged hills attract tourists.

Boe·roe (boo'roo): *see* Buru, Indonesia.

Bo·ga·lu·sa (bō'gə-loo'sə), city (1975 est. pop. 18,058), Washington parish, SE La.; inc. 1914. It is a manufacturing and trading center of the Pearl River valley. Bogalusa was founded in 1906 when the lumber industry established operations in this extensive pine area. Its manufactures include paper and paper products, furniture, tung oil, machine parts, and food products.

Bo·ğaz·köy or **Bo·ghaz·keui** (both: bō-äz'kô-ē), village, N central Turkey. Boğazköy (or Hattusas, as it was called) was the chief center of the Hittite empire (1400–1200 B.C.). Hugo Winckler found here (1906–7) the principal Hittite inscriptions on 10,000 tablets; this discovery greatly added to the knowledge of Hittite civilization. Among the impressive remains are huge fortifications, gates, and temples.

Bog·nor Re·gis (bŏg'nər rē'jĭs), urban district (1973 est. pop. 34,620), West Sussex, S central England. It is a seaside resort. At nearby Felpham is the cottage where the poet William Blake lived from 1801 to 1804. The title Regis was granted to the town after George V convalesced here in 1929.

Bo·gong, Mount (bō'gông), mountain, 6,508 ft (1,985 m) high, SE Victoria, SE Australia, NE of Melbourne. Part of the Australian Alps and at the southern end of the Great Dividing Range, Mt. Bogong is the highest peak in Victoria.

Bo·gor (bō'gôr), formerly **Bui·ten·zorg** (boi'tən-zôrKH), city (1971 pop. 195,882), W Java, Indonesia. At the foot of two volcanoes, it is a highland resort and an agricultural research center, known chiefly for its magnificent botanical gardens (laid out 1817). Tea is grown on the surrounding highlands, and coffee, rice, and rubber are also important crops. Automobile tires are among its manufactures. The site was selected as the resort residence of the Dutch governor-general in 1745, and the town grew around the palace.

Bo·go·tá (bō-gō-tä'), city (1973 est. pop. 2,855,065; pop. of Bogotá Special District 2,925,000), central Colombia, capital and largest city of Colombia, and capital of Cundinamarca dept. A picturesque, spacious city, Bogotá is on a high, fertile plateau (c.8,560 ft/2,610 m) in the east Andes. It is the political, social, and financial center of the republic and the marketing and processing center for a region of coffee, cocoa, and tobacco. The city is rich in splendid colonial architecture, notably the cathedral and the churches of San Ignacio and San Francisco. It has several universities and a museum with an internationally famous collection of pre-Columbian gold art.

The region was a Chibcha Indian center before the city was founded in 1538 by Jiménez de Quesada and named Santa Fé de Bogotá. As capital and archiepiscopal see of the colonial viceroyalty of New Granada, the city became an early religious and intellectual center. Alexander von Humboldt called it (c.1800) the Athens of America in honor of its cultural and scientific institutions. The intellectual impact of the French Revolution inspired Antonio Nariño and others to agitate against Spanish rule. José Acevedo y Gómez led the first successful revolt in the city against Spain in 1810. After Bolívar's decisive victory at Boyacá (1819), Bogotá became the capital of Greater Colombia; when the country was divided in 1830, Bogotá became the capital of what was later called Colombia. Much of the city was damaged during rioting in 1948 following the assassination of the radical leader Jorge Eliécer Gaitán.

Bo·he·mi·a (bō-hē'mē-ə, -hēm'yə), historic region (20,368 sq mi/ 52,753 sq km) and former kingdom, W Czechoslovakia. Bohemia is bounded by Austria in the southeast, by West and East Germany in the west and northwest, by Poland in the north and northeast, and by Moravia in the east. Its natural boundaries are the Bohemian Forest, the Erzgebirge chain, the Sudetes, and the Bohemian-Moravian heights. With Moravia and Czech Silesia, Bohemia constitutes the traditional Czech lands of Czechoslovakia, and in its broader meaning Bohemia is often understood to include this entire area, which until 1918 was a Hapsburg crown land.

Although Bohemia, with about 40% of Czechoslovakia's area and 45% of its people, is the country's most urbanized and densely inhabited region, agriculture and rural life and customs retain their importance. Central Bohemia consists of fertile lowlands and plateaus, drained by the Elbe and Vltava (Moldau) rivers. Grain, sugar beets, grapes and other fruit, flax, and the famous hops used in the breweries of Plzeň (Pilsen) are the principal crops. Mining (coal, silver, copper, lead, iron, and, at Jáchymov, radium and uranium) and textile and glass manufactures are important in the mountain districts. Prague is the center of a heavy industrial region, and Plzeň is also known for the huge Skoda works, producing machinery and munitions. Bohemia is celebrated for its spas and beautiful resorts.

History. The Romans called the area Boiohaemia after the Boii tribe, probably Celtic, which was displaced (1st-5th cent. A.D.) by Slavic settlers, the Czechs. Christianity was introduced while Bohemia was part of the great Moravian empire, from which it withdrew at the end of the century to become an independent principality. St. Wenceslaus, the first Bohemian ruler (920–29), successfully defended his land from Germanic invasion, but Bohemia became a part of the Holy Roman Empire in 950. In 1198 Ottocar I was crowned king of Bohemia, which became an independent kingdom within the empire. The conquests and acquisitions of Ottocar II (1253–78) brought Bohemia to the height of its power and its greatest extent (from the Oder to the Adriatic), but his defeat by Rudolf I of Hapsburg cost Bohemia all his conquests. The reign of Charles IV (1346–78), who was crowned Holy Roman Emperor in 1355, was the golden age of Bohemia, and Prague became the seat of the empire. In the 15th cent., religious disputes rent the kingdom, the nobles gained power, and the crown passed to the kings of Hungary.

The accession (1526) of Archduke Ferdinand (later Emperor Ferdinand I) began the long Hapsburg domination of Bohemia. Religious strife grew: Protestants forced Rudolf II to grant freedom of religion by the so-called Letter of Majesty (*Majestätsbrief*) of 1609. When, in 1618, Emperor Matthias disregarded the *Majestätsbrief*, members of the Bohemian diet protested by throwing two imperial councilors out of the windows of Hradcin Castle. The so-called Defenestration of Prague precipitated the Thirty Years War, which came to involve most of Europe. The Protestants were suppressed, and in 1627 war-devastated Bohemia was demoted from a constituent Hapsburg kingdom to an imperial crown land.

Germanization, oppressive taxation, and absentee landownership reduced the Czechs, except a few favored magnates, to misery. Leopold II tried to conciliate the population; he was the last ruler to be crowned king of Bohemia (1791). The establishment (1867) of the Austro-Hungarian monarchy thoroughly disappointed the Czech aspirations for political autonomy. Czechs entered the parliament at Vienna in 1879, but full independence was reached only at the end of World War I under the guidance of T. G. Masaryk. In 1918 Bohemia became the core of the new state of Czechoslovakia.

Bo·he·mi·an Forest (bō-hē'mē-ən, -hēm'yən), mountain range, extending c.150 mi (240 km) along the N Czechoslovakian–West German border and extending into Austria. A thickly wooded area, it rises to 4,780 ft (1,457.9 m) in the Grosser Arber. There are many marshes, swamps, and peat bogs in the Bohemian Forest. Agriculture is limited because of the harsh climate. Coal, lignite, graphite, kaolin, and granite are extracted. The region is known for its glassmaking and woodworking.

Bo·hol (bō-hôl'), island (1970 pop. 674,806), 1,491 sq mi (3,861.7 sq km), Philippines, one of the Visayan Islands, SW of Leyte. It is a major corn-producing area. Rice, cacao, and hemp are also grown, and manganese and copper are mined.

Bois de Bou·logne (bwä' də bōō-lōn', -lôn'yə), park in Paris, France, bordering on the suburb of Neuilly-sur-Seine. A favorite pleasure ground since the 17th cent., the park contains the race courses of Auteuil and Longchamps and many delightful promenades and bridle paths.

Boi·se (boi'zē, -sē), county (1970 pop. 1,763), 1,913 sq mi (4,954.7 sq km), W Idaho, in a mountain area cut by canyons of the Payette River and its branches; formed 1864; co. seat Idaho City. It has lumber, cattle, and silver, lead, and gold deposits. Arrowrock Dam and Reservoir, on the Boise River, are to the south. The Boise National Forest extends throughout the county.

Boise, city (1975 est. pop. 87,500), state capital and seat of Ada co., SW Idaho, on the Boise River; inc. 1864. The largest city in Idaho, Boise is an important trade and transportation center. Food processing and light manufacturing are the major activities, and there are many state and Federal government offices. A gold rush in the Boise valley and the establishment of a military post in 1863 led to the founding of Boise City, which grew as a distributing center for miners and became the capital of Idaho Territory in 1864. Later, particularly with the building of Arrowrock Dam (1911–15), the region was developed for farming, and Boise drew wealth from orchards and fields rather than mines.

Boise, river, c.160 mi (260 km) long, rising in SW Idaho and flowing W to join the Snake River at the Oregon line. In 1811 the Boise River, originally called Reed's River, was explored by an expedition financed by John Jacob Astor (1763–1848), an American merchant.

Boise City, town (1970 pop. 1,993), seat of Cimarron co., NW Okla., in the Panhandle WNW of Guymon; inc. 1925. It is the center of a wheat, livestock, and oil region.

Boks·burg (bŏks'bûrg'), city (1970 pop. 106,126), Transvaal prov., NE South Africa. It is an important gold- and coal-mining center. Manufactures include railroad equipment, electrical and metal goods, clay products, canned foods, and refined petroleum. Boksburg was founded in 1887.

Bo·lan Pass or **Bho·lan Pass** (both: bō-län'), gap in the central Brahui Range, W Pakistan, c.60 mi (95 km) long, alt. 5,880 ft (1,793.4 m). The pass, which is strategically located, was long used by traders, invaders, and nomadic tribes as a gateway to India.

Bol·i·var (bŏl'ə-vər), county (1970 pop. 49,409), 923 sq mi (2,390.6 sq km), W Miss., bounded on the W by the Mississippi, here forming the Ark. border, and drained by the Sunflower River; formed 1836; co. seat Cleveland. In a rich agricultural region yielding cotton, corn, and alfalfa, it does lumbering and cotton and lumber processing and has an expanding industrial base.

Bolivar. 1. City (1970 pop. 4,769), seat of Polk co., SW Mo., NNW of Springfield, in a farm area of the Ozarks; inc. 1867. Southwest Baptist College is here. In July, 1948, President Gallegos of Venezuela dedicated a statue of Simón Bolívar, for whom the city was named. **2.** Town (1970 pop. 6,674), seat of Hardeman co., SW Tenn., near the Hatchie River SSW of Jackson, in a farm, dairy, and timber area; settled 1825, inc. 1827. Shoes are manufactured.

Bo·liv·ia (bō-lĭv'ē-ə, bō-lē'vyä), republic (1973 est. pop. 5,250,000), 424,162 sq mi (1,098,580 sq km), W South America. Sucre is the legal capital and seat of the judiciary, but La Paz is the political and commercial focus of the nation. One of the two inland countries of South America, Bolivia is shut in from the Pacific in the west by Chile and Peru; in the east and north it borders on Brazil, in the southeast on Paraguay, and in the south on Argentina.

Bolivia presents a sharp contrast between high, bleak mountains and plateaus in the west and lush, tropical rain forests in the east. In the southeast it merges into the semiarid plains of the Chaco. The Andes mountain system reaches its greatest width in Bolivia. Two cordilleras, the western one tracing the border with Chile and the eastern running north and south across the center of the country, are divided by a high plateau (altiplano), most of it 12,000 ft (3,660 m) above sea level—barren, windswept, and segmented by mountain spurs. Despite the harsh conditions, the altiplano is the population center of Bolivia. In the north are Lake Titicaca, which Bolivia shares with Peru, and Lake Poopó. This region, world famous for its breathtaking scenery, was the home of one of the great pre-Columbian civilizations. The eastern mountains, consisting of three major ranges, rise to the cold, forbidding heights of the Puna plateau (as

high as 16,000 ft/4,880 m) and in the north to the snow-capped peaks of Illimani (21,184 ft/6,461.1 m) and Illampú (21,276 ft/6,489.2 m).

Bolivia is divided into nine departments:

NAME	CAPITAL	NAME	CAPITAL
Chuquisaca	Sucre	Pando	Cobija
Cochabamba	Cochabamba	Potosí	Potosí
El Beni	Trinidad	Santa Cruz	Santa Cruz
La Paz	La Paz	Tarija	Tarija
Oruro	Oruro		

Economy. Bolivia's main wealth lies in minerals. Tin is by far the most important product, but silver was once the chief metal, and copper, wolframite, bismuth, antimony, zinc, lead, and gold were also mined. Despite the importance of its mines, Bolivia still lives by a subsistence economy. More than half the people eke out a living from agriculture. Sugar cane, potatoes, corn, wheat, and rice are the leading crops. Industry is limited to processing and light manufacturing. Bolivia's mineral wealth furnishes the bulk of its exports; foodstuffs, manufactured goods, and chemicals are imported.

History. The altiplano was a center of Indian life even before the days of the Inca, but the Aymará had been absorbed into the Inca empire long before Gonzalo and Hernando Pizarro began the Spanish conquest of the Inca in 1532. In 1538 the Indians in Bolivia were defeated. Although the high, cold country was uninviting, it attracted the Spanish because of its rich silver mines, discovered as early as 1545. Exploiters poured in, forcing the Indians to work the mines. Indian laborers were also used on great landholdings. The region was made (1559) into the audiencia of Charcas, which was attached until 1776 to the viceroyalty of Peru and later to the viceroyalty of La Plata.

The revolution against Spanish control came early, with an uprising in Chuquisaca in 1809, but Bolivia remained Spanish until the campaigns of José de San Martin and Simón Bolívar; independence was won only with the victory (1824) at Ayacucho of Antonio José de Sucre. After the formal proclamation of independence in 1825, Bolívar drew up (1826) a constitution for the new republic. At the time of independence it had a seacoast, a portion of the Amazon basin, and claims to most of the Chaco; in little more than a century all these were lost. The strife-ridden internal history of Bolivia began when the first president, Sucre, was forced to resign in 1828. Territorial disputes were frequent. The nitrate deposits of Atacama proved valuable, but the mining concessions were given to Chileans. Trouble over the concessions led (1879) to the War of the Pacific. As a result Bolivia lost Atacama to Chile. After a bitter conflict, Bolivia, under President José Manuel Pando, yielded the Acre River, valuable because of its wild rubber, to Brazil in 1903 for an indemnity.

Attempts at reorganization and reform, especially by Ismael Montes, were overshadowed in the 20th cent. by military coups, rule of dictators, and bankruptcy. Foreign loans led in turn to an increase of foreign influence, strengthened by foreign interests in mines and oil fields. Conflicting claims to the Chaco, which was thought to be oil-rich, brought on yet another disastrous territorial war, this time with Paraguay (1932-35). The war and Bolivia's defeat aggravated internal discontent, and radical, conservative, and moderate programs for curing the ills of the nation were hampered by military coups and countercoups.

World War II proved a boon to the Bolivian economy by increasing demands for tin and wolframite. International pressure over pro-German elements in the government eventually forced Bolivia to break relations with the Axis and declare war (1943). Meanwhile, rising prices had aggravated the restiveness of the miners over miserable working conditions; strikes were brutally suppressed. The crisis reached a peak in Dec., 1943, when the nationalistic, pro-miner MNR (*Movimiento Nacional Revolucionario*) engineered a successful revolt. The regime, however, was not recognized by other American nations (except Argentina) until 1944, when pro-Axis elements in the MNR were officially removed. Bolivia then became a member of the United Nations.

Since then, the MNR has vied with other powers for control of the country. In 1951 civil rights and suffrage were extended to the Indians. Education, health, and construction projects were begun. In 1960 the United States, in spite of losses incurred by American investors, stepped up its program of technical and financial assistance. But economic and political factors weakened the government. Income from tin exports sank to a postwar low, thus crippling attempts at industrial diversification; technical and administrative incompetence was rife; the fiscal system, never sound, became chaotic; and an eruption of dissident groups, some fostering acts of political terror, brought all attempts at further reform to a virtual halt. A radical guerrilla movement, led by the Cuban Ernesto "Che" Guevara, was set back seriously when government troops killed Guevara in Oct., 1967. In Aug., 1971, Col. Hugo Banzer Suárez, supported by both the MNR and its traditional rightist opponent, the Bolivian Socialist Falange took control. Banzer closed the universities and returned Bolivia to a pro-U.S. foreign policy. With his power insecure, Banzer frequently arrested politicians, alleging anti-government plots. Churchmen were accused of aiding the guerrilla National Liberation Army. In June, 1974, an unsuccessful attempt to depose Banzer followed months of protests from peasants, miners, students, and opposition politicians. The government was reorganized and an all-military cabinet was installed in July.

Government. Bolivia has had more than 185 revolutions since it became independent in 1825. The latest constitution, adopted in 1967, provides for a president elected for a four-year term and a bicameral congress. However, the congress has been suspended since Sept., 1969, and no presidential election has been held since 1966.

Bol·li·gen (bôl′ ĭ-gən), town (1977 est. pop. 30,400), Bern canton, W central Switzerland. It is a dairy and industrial center.

Bol·ling·er (bŏl′ ĭn-jər, bo͝ol′ ĭng-ər, bō′lĭn-jər, -lĭng-ər), county (1970 pop. 8,820), 621 sq mi (1,608.4 sq km), SE Mo., crossed by the Castor and Whitewater rivers; formed 1851; co. seat Marble Hill. There are drainage canals in the southeast section. It has agriculture (corn, wheat, dairying, and livestock) and manufactures lumber products.

Bo·lo·gna (bə-lō′nyə, bō-lô′nyä), city (1976 est. pop. 485,643), capital of Emilia-Romagna and of Bologna prov., N central Italy, at the foot of the Apennines and on the Aemilian Way. It is a commercial and industrial center and a railroad junction. Manufactures include farm machinery, motor vehicles, metal goods, and chemicals.

Originally an Etruscan town called Felsina, it became a Roman colony in 189 B.C. The city came under Byzantine rule in the 6th cent. A.D. and later passed to the papacy. In the early 12th cent. a strong free commune was established. The victory of Bologna over Emperor Frederick II at Fossalta (1249) added political power to the city. In politics the rivalry between the Guelphs and the Ghibellines enabled several ambitious families to seize power (13th-15th cent.). In 1506, Pope Julius II re-established papal rule, which was interrupted in 1797, when Bologna was made the capital of the Cispadane Republic, but resumed in 1815 after the Congress of Vienna. The coronation of Charles V at Bologna (1530) was the last imperial crowning by a pope. There were unsuccessful revolts against papal rule in 1831, 1843, and 1848, and in 1860 Bologna voted to unite with the kingdom of Sardinia. The city was heavily bombed by the Allies in World War II. Bologna's famous university originated (c.1088) with its Roman law school (founded A.D. 425); medical and theological faculties and courses in the liberal arts were added in the 14th cent. The city has long been a center of printing, and its observatory (founded 1712) is the oldest in Italy. Bologna has retained a marked medieval aspect; many streets are arcaded. Noteworthy structures include the Palazzo Comunale (13th and 15th-16th cent.); the Renaissance-style Palazzo del Podesta; the palace of King Enzio (13th cent.); the Basilica of San Petronio (begun in 1390); the Church of Santo Stefano; the Church of San Giacomo Maggiore (founded 1267, major alterations in the 15th cent.); the Church of San Domenico (early 13th cent.); and the Archiginnasio (once the seat of the university and now a library).

Bol·ton (bōl′tən) or **Bol·ton-le-Moors** (bōl′tən-lə-mŏŏrz), borough (1976 est. pop. 261,000), Greater Manchester, NW England. Since the late 18th cent., when spinning factories were built and a canal (1791) was constructed to Manchester, Bolton has been a cotton-textile center. Prior to that time, woolen weaving was important. It also has factories that pack poultry and produce textile and other machinery, chemicals, leather goods, furniture, carpets, and paper. Samuel Crompton, inventor of the spinning mule (1779), was born nearby and is buried in Bolton. Sir Richard Arkwright invented the "water frame" here c.1768.

Bol·za·no (bōl-tsä′nō), city (1971 pop. 103,267), capital of Bolzano prov., in Trentino-Alto Adige, N Italy, on the Isarco River near its confluence with the Adige. It is the center of the German-speaking part of south Tyrol and is a tourist and health resort noted for its Alpine scenery and mild climate. Its position on the Brenner road has made it the chief commercial center of the area since the Middle Ages, when important fairs were held here. The city's manufactures today include steel, plastics, aluminum products, and woolen goods. Bolzano was part of the bishopric of Trent from the 11th cent. until the 16th cent., when it was ceded to the Hapsburgs. It was awarded to Italy in 1919. The city was severely damaged in World War II.

Bo·ma (bō′mə), city (1970 est. pop. 33,143), Bas-Zaire region, W Zaire, on the Congo estuary. A port and railhead, it exports tropical timber, bananas, cacao, and palm products.

Bom·bay (bŏm-bā′), former state, W central India, on the Arabian Sea. The state contained within its borders the former Portuguese colonies of Goa, Daman, and Diu. In 1960 it was divided into the new states of Gujarat and Maharashtra.

Bombay, city (1971 pop. 5,970,975), capital of Maharashtra state, W central India, occupying c.25 sq mi (65 sq km) on Bombay and Salsette islands just off the coast. Bombay has the only natural deep-water harbor in western India. It is a transportation hub and industrial center. Industries include textile and chemical manufacturing and petroleum refining. There is an extensive system of hydroelectric stations, and nearby at Trombay is a nuclear reactor. The area of the city was ceded (1534) to Portugal by the Sultan of Gujarat. After it passed to Great Britain in 1661, Bombay was the headquarters (1668–1858) of the East India Company in India, and during the American Civil War it expanded to meet the world demand for cotton and became a leading cotton-spinning and weaving center.

Bo·mo·seen, Lake (bō-mə-sēn′), 7.5 mi (12 km) long, 1.5 mi (2.4 km) wide, W Vt., largest lake wholly within the state. Surrounded by wooded hills, it is a popular summer resort.

Bo·mu (bō′mŏŏ), river, c.500 mi (805 km) long, rising in NE Zaire and flowing generally W. The Bomu merges with the Uele to form the Ubangi, a tributary of the Congo.

Bon·aire (bô-nâr′), island (1975 est. pop. 8,785), 112 sq mi (290.1 sq km), in the Leeward Islands group of the Netherlands Antilles. Its good harbor has made Bonaire an export point. Sisal and salt are produced on the island, and goats and sheep are raised. Tourism is increasingly important.

Bo·nam·pak (bō-näm-päk′), ruined city of the Late Classic period of the Maya, close to Tuxtla, in Chiapas, S Mexico. Discovered in 1946, it consists of a group of temples, one of which is remarkable for its well-preserved frescoes depicting scenes of Mayan life in considerable detail.

Bo·nan·za Creek (bə-năn′zə, bō-), stream, c.20 mi (30 km) long, W Yukon Territory, Canada, flowing NW to the Klondike River near Dawson. The first gold strike in the Yukon occurred here in 1896.

Bon·a·vis·ta Bay (bŏn′ə-vĭs′tə), arm of the Atlantic Ocean, c.40 mi (65 km) long and 40 mi (65 km) wide, E N.F., Canada. The bay is irregular and filled with islands. Cape Bonavista, the headland of the Bonavista Peninsula, marks the southern entrance to the bay and is the reputed landfall (1497) of John Cabot.

Bond (bŏnd), county (1970 pop. 14,012), 378 sq mi (979 sq km), SW central Ill., drained by the Kaskaskia River and Shoal Creek; formed 1817; co. seat Greenville. Agriculture (corn, wheat, poultry, and livestock) is of the greatest importance, with some manufacturing and retail trade. It has bituminous-coal mines and natural-gas wells.

Bône (bōn): *see* Annaba, Algeria.

Bon·ham (bŏn′əm), city (1970 pop. 7,698), seat of Fannin co., NE Texas, NE of Dallas near the Red River; platted 1837. It processes and ships cheese, cotton, and truck crops. Manufactures include trailers, pumps, and cables. A state park is nearby.

Bon Homme (bŏn′əm), county (1970 pop. 8,577), 560 sq mi (1,450.4 sq km), SE S.Dak., on the Nebr. border and bounded on the S by the Missouri River; formed 1862; co. seat Tyndall. It has diversified farming, dairying, and grain raising. There are a number of state parks and recreation areas here.

Bon·i·fay (bŏn′ĭ-fā), town (1970 pop. 2,068), seat of Holmes co., NW Fla., W of Marianna, in a farm and livestock area. Plywood is made.

Bo·nin Islands (bō′nĭn), volcanic island group (1967 est. pop. 200), c.40 sq mi (100 sq km), Tokyo prefecture, Japan, in the W Pacific Ocean S of Tokyo. The principal products are sugar cane, cocoa, bananas, and pineapples. Discovered by the Japanese in the 16th cent. and later by the Spanish, the islands were claimed by the British in 1827 and by Japan in 1875. In World War II the islands formed a major Japanese military stronghold and were the scene of land, sea, and air battles. The U.S. navy occupied the islands in 1945. Japan regained technical sovereignty over them in 1951, but they continued to be under U.S. military administration until 1968, when they were returned to Japan.

Bonn (bŏn, bôn), city (1974 pop. 283,891), capital of the Federal Republic of Germany, North Rhine-Westphalia, W West Germany, on the Rhine River. Manufactures include light-metal products, ceramics, office equipment, chemicals, and pharmaceuticals. Bonn was founded in the 1st cent. A.D. as a Roman garrison. It was devastated by the Normans in the 9th cent. and later became the residence (1238–1794) of the electors of Cologne and the scene of the coronations of Frederick the Handsome (1314) and Charles IV (1346) as kings of the Romans. During the Palatinate Succession War (1689), Bonn was destroyed by Elector Frederick III of Brandenburg. The city was later rebuilt in the baroque style. Bonn was occupied (1794) and annexed (1798–1814) by France. In 1815 it passed to Prussia. In 1948–49 delegates from the parts of Germany occupied by France, Great Britain, and the United States met in Bonn and drafted a constitution for the Federal Republic of Germany. In 1949 Bonn was also made the capital of West Germany. The house where Ludwig van Beethoven was born (1770) has been preserved and is a museum. Bonn is the seat of a famous university (founded 1784), whose main building formerly was the electoral palace (built 1697–1725). The city has a noteworthy church (11th–13th cent.).

Bon·ner (bŏn′ər), county (1970 pop. 15,560), 1,736 sq mi (4,496.2 sq km), extreme N Idaho, in a mountainous area bordering on Mont. and Wash., and drained by the Priest and Pend Oreille rivers and Clark Fork; formed 1907; co. seat Sandpoint. There are silver, lead, and copper deposits, and dairy products and lumber are processed.

Bon·ners Ferry (bŏn′ərz), city (1970 pop. 1,909), seat of Boundary co., N Idaho, NNE of Sandpoint, on the Kootenai River near the Canadian boundary; inc. 1899. Established (1863) as a ferry and trading post on the Kootenai River trail to the British Columbia gold fields, it grew as a lumbering town and today is a farm center in an irrigated area. Wheat is the main crop.

Bon·ne·ville (bŏn′ə-vĭl′), county (1970 pop. 52,457), 1,836 sq mi (4,755.2 sq km), SE Idaho, in a mountainous area bordering on Wyo.; formed 1911; co. seat Idaho Falls. The irrigated Snake River valley extends through the northwest part of the county. Sugar beets, seed peas, beans, potatoes, and rye are grown. Livestock is raised. Parts of Targhee and Caribou national forests are here.

Bonneville, Lake, ancient lake, once covering c.19,500 sq mi (50,505 sq km), NW Utah. The lake expanded during the period of heavy precipitation brought on by the advancing glaciers of the Pleistocene epoch. At the end of the Pleistocene epoch the lake's area rapidly shrank. Its six terraces still exist and locate the different lake levels. Great Salt Lake, Lake Sevier, and Utah Lake are remnants of Lake Bonneville.

Book·er T. Washington National Monument (bŏŏk′ər; wôsh′ĭng-tən): *see* National Parks and Monuments Table.

Boone (bōōn). **1.** County (1970 pop. 19,073), 593 sq mi (1,535.9 sq km), N. Ark., in the Ozarks bordered on the N by Mo. and drained by small tributaries of the White River; formed 1869; co. seat Harrison. It has agriculture (fruit, hops, cotton, truck crops, and livestock), timber, lead and zinc mines, and deposits of marble. Mystic Cavern and part of Bull Shoals Lake are here. **2.** County (1970 pop. 25,440), 283 sq mi (733 sq km), N Ill., on the Wis. border and drained by the Kishwaukee River; formed 1837; co. seat Belvidere. Its agriculture includes livestock, corn, wheat, oats, truck, and poultry. Dairy and grain products, canned foods, sewing machine, scales, machine parts, clothing, and hardware are manufactured. **3.** County (1970 pop. 30,870), 427 sq mi (1,105.9 sq km), central Ind., drained by Sugar Creek and the Eel River; formed 1830; co. seat Lebanon. In an agricultural area yielding grain, truck crops, livestock, and dairy products, it also has diversified manufacturing. **4.** County (1970 pop. 26,470), 573 sq mi (1,484.1 sq km), central Iowa, in a prairie area drained by the Des Moines River; formed 1846; co. seat Boone. Its agriculture includes hogs, cattle, poultry, corn, oats, and soybeans. Bituminous coal is mined in the central region. There are a number of state parks and a wildlife research station here. **5.** County (1970 pop. 32,812), 249 sq mi (644.9 sq km), N Ky., in the outer Blue Grass bounded on the N and W by the Ohio River, here forming the Ind. and Ohio borders, and drained by several creeks; formed 1798; co. seat Burlington. In a gently rolling agricultural area yielding burley tobacco, corn, fruit, livestock, poultry, and dairy products, it includes Big Bone Lick. **6.** County (1970 pop. 80,935), 685 sq mi (1,774.2 sq km), central Mo., bounded on the E by the Cedar River, on the SW by the Missouri River; formed 1820; co. seat Columbia. Corn, wheat, oats, and hay are grown. It also has livestock, lumber, coal mines, limestone deposits, and some manufacturing. **7.** County (1970 pop. 8,190), 683 sq mi (1,769 sq km), E central Nebr., drained by the Cedar River; formed 1875; co. seat Albion. It is mainly agricultural (livestock, grain, dairy products, and poultry. **8.** County (1970 pop. 25,118), 501 sq mi (1,297.6 sq km), SW W.Va., in a bituminous-coal region on the Allegheny Plateau, drained by the Coal and Mud rivers; formed 1847; co. seat Madison. In an agricultural area yielding fruit, tobacco, and livestock, it also has stands of timber and oil and natural-gas fields.

Boone. 1. City (1975 est. pop. 12,394), seat of Boone co., central Iowa, on the Des Moines River; inc. 1865. It is a railroad and industrial center with plants making machinery, steel fabrications, and plastic signs. It was laid out (1865) by the railroad, which built a long, high double-track bridge there. **2.** Town (1970 pop. 8,754), seat of Watauga co., NW N.C., in the Blue Ridge SE of Bristol, Tenn.; settled in the 18th cent. It is a resort in a farm, livestock, and timber area.

Boones·bor·o (bo͞onz'bûr'ō, -bŭr'ō), former settlement, central Ky., on the Kentucky River. It was named for Daniel Boone, who in 1775 built a small fort there under orders from the Transylvania Company. The seat of the government of Transylvania for several years, Boonesboro was later abandoned because of repeated Indian attacks.

Boone·ville (bo͞on'vĭl'). **1.** City (1970 pop. 3,239), a seat of Logan co., W Ark., SE of Fort Smith, in a farm, timber, and natural-gas area of the Ouachita foothills; settled c.1821. Ozark National Forest and Mt. Magazine are nearby. **2.** City (1970 pop. 126), seat of Owsley co., E central Ky., in the Cumberlands on the South Fork of the Kentucky River WNW of Hazard, in an agricultural region. **3.** Town (1970 pop. 5,895), seat of Prentiss co., NE Miss., S of Corinth, in a lumbering area; settled 1859, inc. 1869. Northeast Mississippi Junior College is here.

Boon·ton (bo͞on'tən), town (1970 pop. 9,261), Morris co., N.J., NNE of Morristown; settled 1760, inc. 1867. An important producer of iron in the 19th cent., Boonton now produces plastics and makes food and dairy products.

Boon·ville (bo͞on'vĭl'). **1.** City (1970 pop. 5,736), seat of Warrick co., SW Ind., NE of Evansville; platted 1818. It has bituminous coal mines and manufactures bricks and food and dairy products. **2.** City (1970 pop. 7,514), seat of Cooper co., central Mo., on the Missouri River W of Columbia, in a grain and livestock area; laid out 1817. A Civil War battle, resulting in a Union victory, was fought nearby on June 17, 1861.

Boo·thi·a Peninsula (bo͞o'thē-ə), 12,483 sq mi (32,331 sq km), S central Franklin dist., Northwest Territories, Canada. It is the northernmost (71°58′ N) tip of the North American mainland. It is almost an island, being connected with the mainland only by the narrow Isthmus of Boothia. Topographically and in climate it is like the islands of the Arctic Archipelago. A narrow strait separates it in the north from Somerset Island. To the east the Gulf of Boothia separates it from Baffin Island. It is virtually uninhabited except for a few hundred settlers at Spence Bay and Thom Bay. The peninsula was discovered and explored (1829-33) by John Ross, the British explorer, and named for a patron of the expedition, Sir Felix Booth. Near the southwest end the expedition of Sir John Franklin, the British explorer, ended in tragedy. Roald Amundsen, a Norwegian, explored the peninsula in 1903-5.

Boo·tle (bo͞ot'l), county borough (1973 est. pop. 71,160), Merseyside, NW England, at the mouth of the Mersey River. It has extensive docks adjacent to those of Liverpool. Besides shipping, Bootle's industries include tanning, tin smelting, engineering, and flour milling.

Bo·ra-Bo·ra (bō'rä-bō'rä), volcanic island, 15 sq mi (38.9 sq km), South Pacific, in the Leeward group of the Society Islands, French Polynesia. It is a mountainous island, with a good harbor, which is a large lagoon surrounded by coral islets. Copra, oranges, and vanilla are produced.

Bo·rah, Mount (bôr'ə), peak, 12,662 ft (3,861.9 m) high, central Idaho, in the Lost River Mts. It is the highest point in the state.

Bo·rås (bo͞o-rōs'), city (1975 est. pop. 105,177), Älvsborg co., SW Sweden, on the Viskan River. It is a transportation and commercial center and has numerous cotton and woolen textile factories. Borås was founded in 1632 by Gustavus II.

Bor·deaux (bôr-dō'), city (1975 est. pop. 223,131), capital of Gironde dept., SW France, on the Garonne River. Bordeaux is a major economic and cultural center, and a busy port accessible to oceangoing ships from the Atlantic through the Gironde River. Although Bordeaux has important shipyards and industries, its principal source of wealth is its wine trade. Known as Burdigala by the Romans, Bordeaux was the capital of the province of Aquitania and a prosperous commercial city. Its importance declined under Visigothic and Frankish rule (c.5th cent.), but was revived when the city became (11th cent.) the seat of the dukes of Aquitaine. Eleanor of Aquitaine, who was born here, precipitated through her successive marriages to Louis VII of France and Henry II of England the long struggle between the two nations. As a result of these wars Bordeaux came under English rule, which lasted from 1154 to 1453. The city's commercial importance dates from this period. Reconquered by France, Bordeaux became capital of the province of Guienne. It reached the height of its prosperity in the 18th cent. Bordeaux was the center of the Girondists in the French Revolution and the site of the National Assembly of 1871 that established the Third Republic. In 1914 and again in 1940, the city was the temporary seat of the French government. The Place des Quinconces, with its statues of Montaigne and Montesquieu, who were both born nearby, dominates the center of the city. Other points of interest are the Gothic cathedral of St. André, several art museums, and some elegant 18th cent. buildings.

Bor·den (bôr'dn), county (1970 pop. 888), 907 sq mi (2,349.1 sq km), W Texas, drained by the Colorado River; formed 1876; co. seat Gail. The Cap Rock escarpment of the Llano Estacado runs from north to south through the center of the county. It has cattle and sheep ranches, some agriculture (cotton, grain sorghums, fruit, and truck crops), and oil wells.

Border, the (bôr'dər), region surrounding the boundary between England and Scotland. From the coast near Berwick along the Tweed River through the Cheviot Hills and on to Solway Firth, the narrow, rugged country is dotted with sites of battles between the Scots and the English. The wild country figures in legend, in folklore, and particularly in the Border ballads.

bore (bôr), inrush of water that advances upstream with a wavelike front caused by the progress of incoming tide from a wide-mouthed bay into its narrower portion. The tidal movement tends to be retarded by friction as it reaches the shallower water and meets the river current; it therefore piles up and forms a low wall of water that moves upstream with considerable force and velocity as the tide continues to rise.

Bor·gå (bôr'gō) or **Por·voo** (pôr'vō), city (1975 est. pop. 18,740), Uusimaa prov., S central Finland, on the Gulf of Finland at the mouth of the Porvoonjoki River. It is an export center for forest products and has plywood and cellulose mills, breweries, and a publishing industry. A trade center in the early Middle Ages, it was chartered in 1350. In 1809 Alexander I of Russia granted Finland a constitution at Borgå.

Bor·ger (bôr'gər), city (1975 est. pop. 14,198), Hutchinson co., extreme N Texas, in the Panhandle; inc. 1930. After the discovery of oil in 1925, Borger grew as the industrial center of a vast natural-gas and oil field. In the area are refineries, carbon-black plants, synthetic-rubber factories, and related enterprises.

Bor·ger·hout (bôr'KHər-hout), city (1977 est. pop. 46,079), Antwerp prov., N Belgium, on the Albert Canal, a suburb of Antwerp.

Bor·läng·e (bôr'lĕng'ə), city (1975 est. pop. 46,208), Kopparberg co., S central Sweden, on the Dalälven River; chartered 1944. It has sawmills, machine shops, and factories manufacturing iron, steel, and paper.

Bor·ne·o (bôr'nē-ō), island (1970 est. pop. 6,800,000), c.287,000 sq mi (743,330 sq km), largest of the Malay Archipelago and third-largest island in the world, SW of the Philippines and N of Java. Indonesian Borneo (called Kalimantan by the Indonesians) covers over 70% of the total area; the Malaysian states of Sabah and Sarawak and the British-protected sultanate of Brunei stretch across the north coast. The island largely consists of dense jungle and mountains. Much of the terrain is virtually impassable, and large areas are unexplored. The island is one of the most sparsely populated regions in the world. Kalimantan contains Indonesia's greatest expanse of tropical rain forests, including valuable stands of camphor, sandalwood, and ironwood, and many palms. The thick jungle and myriad insects discourage large-scale agriculture, but rice, sago, tobacco, millet, coconuts, pepper, sweet potatoes, sugar cane, coffee, and rubber are grown. Kalimantan also contains some of Indonesia's most productive oil fields (discovered in 1888). Coal has been mined for more than a century, and gold since earliest times. Other mineral resources include industrial diamonds, bauxite, and extensive reserves of low-grade iron ore. Portuguese traders visited Borneo in 1521, and shortly thereafter the Spanish established trade relations with the island. The Dutch arrived in the early 1600s, and the English c.1665. Dutch influence was established on the west coast in the early 1800s and was gradually extended to the south and east. The British adventurer James Brook took the north edge of the island in the 1840s, and present-day Sabah, Sarawak, and Brunei were declared British protectorates in 1880. The final boundaries were defined in 1905. In World War II the Japanese held the island from 1942 to 1945. Dutch Borneo became part of the republic of Indonesia in 1950. The union of Sabah and Sarawak with the federation of Malaysia in 1963 was resented by Indonesians; Indonesian guerrilla raids against both areas continued sporadically until Aug., 1966.

Born·holm (bôrn'hôlm'), island group (1976 est. pop. 47,126), 227 sq mi (588 sq km), extreme E Denmark, in the Baltic Sea, near Sweden. Farming, fishing, handicrafts, and tourism are the chief occupations; granite and kaolin are the main exports. Bornholm was divided (1149) between Denmark and Sweden, ruled (1327–1522) by the Danish archbishops, governed (1525–76) by Lübeck merchants, and ceded (1658) to Denmark. After Germany's surrender (May, 1945) in World War II, German forces made a desperate stand on Bornholm before Soviet troops forced them to surrender.

Bo·ro·bu·dur (bōr'ə-bə-dŏŏr'), ruins of one of the finest Buddhist monuments, in central Java, Indonesia. Built by the Sailendras of Sumatra, this manificent shrine dates from about the 9th cent. It is a huge, truncated pyramid, covered with intricately carved blocks of stone that illustrate episodes in the life of the Buddha. A seated Buddha within may be seen from three platforms above the seven stone terraces that encircle the pyramid.

Bo·ro·di·no (bə-rə-dyĭ-nô'), village, central European USSR, W of Moscow. It was the site, on Sept. 7, 1812, of a battle between Napoleon's Grande Armée and Gen. Mikhail Kutuzov's Russian forces defending Moscow. The battle, which cost some 108,000 casualties, is described in Tolstoy's *War and Peace.*

Bos·co·re·a·le (bôs'kō-rā-ä'lā), town (1971 pop. 18,674), in Campania, S Italy, at the foot of Vesuvius. Roman villas have been excavated in the town. A celebrated collection of gold coins, jewelry, and silverwork dating from the 1st and 2nd cent. A.D. was unearthed here in the late 1800s.

Bos·ni·a and Her·ce·go·vi·na (bŏz'nē-ə; hĕr'tsə-gō-vē'nə), constituent republic of Yugoslavia (1971 pop. 3,742,852), 19,741 sq mi (51,129 sq km), W central Yugoslavia. It consists of two regions—Bosnia in the north, and Hercegovina in the south. Sarajevo, in Bosnia, is the capital. Half of the area is forested, and timber is an important product of Bosnia. About one fourth of the republic's land is cultivated; corn, wheat, and flax are the principal products of Bosnia, and tobacco, cotton, fruits, and grapes of Hercegovina. There are large deposits of lignite, iron ore, and bauxite, as well as smaller quantities of copper and manganese. Despite some industrialization, it remains one of the poorer areas of Yugoslavia.

Bosnia was settled by Serbs in the 7th cent.; it appeared as an independent country by the 12th cent. Medieval Bosnia reached the height of its power in the second half of the 14th cent., when it controlled many surrounding territories. Bosnia also annexed the duchy of Hum, which, however, regained autonomy in 1448 and became known as Hercegovina. Bosnia fell to the Turks in 1463. Hercegovina held out until 1482, when it too was occupied and joined administratively to Bosnia. After the Russo-Turkish War of 1877–78, Bosnia and Hercegovina were placed under Austro-Hungarian administration. Serbian nationalism mounted when the area was completely annexed in 1908. The assassination (1914), by a Serbian nationalist, of Archduke Francis Ferdinand in Sarajevo precipitated World War I. In 1918 Bosnia and Hercegovina were annexed to Serbia. The dismemberment of Yugoslavia during World War II led to Bosnia and Hercegovina's incorporation into the German puppet state of Croatia. In 1946 Bosnia and Hercegovina became one of the six constituent republics of Yugoslavia.

Bos·po·rus (bŏs'pər-əs), strait, c.20 mi (32 km) long and c.2,100 ft (640 m) wide at its narrowest, separating European from Asian Turkey and joining the Black Sea with the Sea of Marmara. At its narrowest point are two famous castles: Anadolu Hisar, built in 1390, on the Asian side and Rumeli Hisar, completed in 1452, on the European side. The Bosporus Bridge, one of the world's longest suspension bridges (3,524 ft/1,074.8 m long; opened 1973) spans the strait at İstanbul.

Bos·que (bŏs'kē), county (1970 pop. 10,966), 990 sq mi (2,564.1 sq km), central Texas, bounded on the NE and E by the Brazos River and drained by the Bosque River; formed 1854; co. seat Meridian. It is mainly agricultural, yielding oats, corn, grain sorghums, wheat, cotton, peanuts and pecans, livestock, poultry, and dairy products. Meridian State Park is here.

Bos·sier (bō'zhâr), parish (1970 pop. 63,703), 849 sq mi (2,198.9 sq km), NW La., drained by the Bodcau Bayou and bounded on the N by the Ark. border, on the W by the Red River, and on the SE by Lake Bistineau; formed 1843; parish seat Benton. In an agricultural area (cotton, corn, hay, fruit, sweet potatoes, and peanuts), it has oil and natural-gas fields, refineries, and pipelines, timber and lumber milling, cotton ginning, and some industry.

Bossier City, city (1970 pop. 41,595), Bossier parish, NW La., on the Red River, across from Shreveport, with which it is connected by several bridges; inc. 1907. Barksdale Air Force Base is the major employer.

Bos·ton (bô'stən, bŏs'tən), municipal borough (1973 est. pop. 26,700), Lincolnshire, E central England, on the Witham River. Boston's fame as a port dates from the 13th cent., when it was a Hanseatic port trading wool and wine. Boston now exports coal, grain, agricultural machinery, potatoes, and cattle. It is also a shellfish center and a market for a rich lowland farm area. Puritans under John Cotton sailed in 1633 from Boston to Massachusetts Bay. St. Botolph's Church is on the site of a 7th cent. monastery, founded by St. Botolph, for whom the town is named. The guildhall, begun in 1545, was restored in 1911 and is now a museum.

Boston. 1. City (1975 est. pop. 601,000), state capital and seat of Suffolk co., E Mass., at the head of Boston Bay; inc. 1822. The largest city in New England, Boston is a major financial center, a leading port, and an important market for fish and wool. Its industries include publishing, food processing, and the manufacture of shoes, textiles, machinery, and electronic equipment. Established by the elder John Winthrop in 1630 as the main colony of the Massachusetts Bay Company, Boston was an early center of American Puritanism, with vigorous intellectual life. The Boston Public Latin School was opened in 1635; Harvard Univ. was founded at nearby Cambridge in 1636; a public library was started in 1653; and the first newspaper in the Colonies, the *Newsletter,* appeared in 1704. As the American Revolution approached, Boston became a center of opposition to the British. The Battle of Bunker Hill (June 17, 1775) was one of the first battles of the Revolution, and Boston was under siege until the British withdrew in Mar., 1776. Shortly thereafter, the city entered a period of prosperity that lasted until the middle of the 19th cent. Prominent Boston families made fortunes from shipping and from textile mills and shoe factories. These families patronized the arts and letters, making Boston "the Athens of America," and backed reformers, notably the abolitionists. The growth of industry brought many immigrants (at first mostly Irish), and Boston changed from a commercial city surrounded by farms to an industrial metropolis. The city of today, with its broad avenues running into the crooked narrow streets of colonial Boston, cherishes the landmarks of the past: the 17th cent. house in which Paul Revere lived; Old North Church; Old South Meetinghouse, a rallying place for patriots during the Revolution; the old statehouse (1713), now a museum; the Boston Common, one of the oldest public parks in the country; Faneuil Hall; the golden-domed statehouse, with its façade designed by Charles Bulfinch; and the red-brick houses of Louisburg Square. Among notable Boston churches are King's Chapel, the birthplace of American Unitarianism (1785); the Mother Church of Christian Science; and Trinity Church (1872-77). Boston is one of the great cultural centers of the nation. In the city are the Massachusetts Historical Society (founded 1791); the Boston Athenaeum (1807); the Boston Public Library; the New England Conservatory of Music; the Boston Symphony Orchestra; the Museum of Fine Arts; and the Isabella Stewart Gardner Museum. Harvard Medical School is in Boston proper, as are the New England Medical Center and Massachusetts General Hospital. Other educational institutions include Boston College, Boston Univ., Simmons College, Emerson College, Emmanuel College, and Northeastern Univ. The restored U.S.S. *Constitution* (launched in 1797) is berthed here. **2.** Village (1970 pop. c.100), seat of Bowie co., NE Texas, near the Red River W of Texarkana just S of New Boston, in an agricultural area.

Bos·worth Field (bŏz'wərth), Leicestershire, central England. It was the scene of the battle (1485) at which Richard III was killed and the crown was passed to his opponent, Henry VII, first of the Tudors.

Bot·a·ny Bay (bŏt'ə-nē, bŏt'nē), inlet, New South Wales, SE Australia, just S of Sydney. It was visited in 1770 by James Cook, who proclaimed British sovereignty over the east coast of Australia. The site of the landing is marked by a monument on Inscription Point. The bay was named by Cook and Sir Joseph Banks because of the interesting flora on its shores.

Bot·e·tourt (bŏt'ə-tŏt), county (1970 pop. 18,193), 549 sq mi (1,421.9 sq km), W Va., mainly in the Great Appalachian Valley, bounded on the SE by the Blue Ridge and drained by the James River and Craig Creek; formed 1770; co. seat Fincastle. It has agriculture (dairying, beef cattle, apples, and peaches), extensive mining, and many small industries. It includes part of Jefferson National Forest.

Both·ni·a, Gulf of (bŏth'nē-ə), N arm of the Baltic Sea, 400 mi (643.6 km) long and 50-150 mi (80.5-241.4 km) wide, between Sweden and Finland.

Bot·swa·na (bŏt-swä'nə), formerly **Bech·u·a·na·land** (bĕch'ōō-ä'nə-lănd'), republic (1971 pop. 630,379), 231,804 sq mi (600,372 sq km), S central Africa; capital Gaborone. Botswana is bordered by Namibia on the west and north, by Zambia at a narrow strip in the north, by Rhodesia on the east, and by the Republic of South Africa

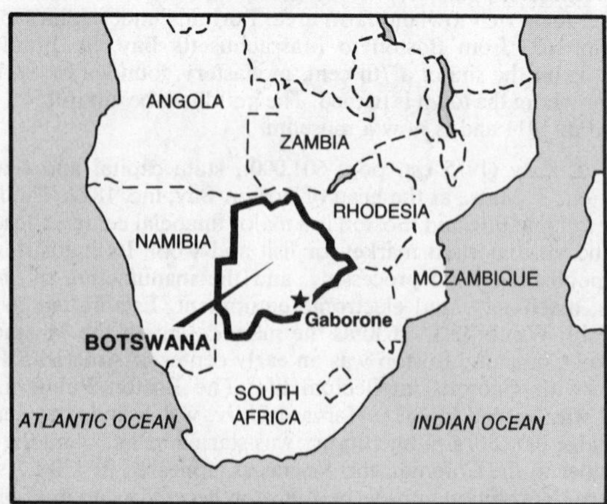

on the east and south. The terrain is mostly an arid plateau (c.3,000 ft/915 m high) of rolling land. In the east are hills. The Kalahari Desert lies in the south and west. In the northwest the Okavango River drains into the vast region of the Okavango swamp and Lake Ngami, thus forming a huge marshland.

Economy. Most of Botswana's people are pastoralists, and cattle raising and the export of beef and other cattle products are the chief economic activities. The country's water shortage and consequent lack of sufficient irrigation facilities have hampered agriculture; only a small percentage of the potentially arable land is under cultivation. Sorghum, maize, millet, and beans are the principal subsistence crops, and cotton, peanuts, and sunflowers are the main cash crops. Many citizens of Botswana work in the mines of South Africa, and lesser numbers are employed in Rhodesia. The only known minerals in the country at the time of independence were manganese and some gold and asbestos. Large nickel, copper, and diamond deposits have since been found, as well as salt and soda ash; antimony and sulfur are known to exist, and the discovery of oil is a serious possibility. Vast coal deposits are also being worked. Development of a tourist industry has been based partly on the attraction of one of Africa's few remaining large natural game reserves.

History. San (Bushmen) were the original inhabitants of what is now Botswana. In the 18th cent. the Tswana supplanted the San, who remained as serfs. Beginning in the 1820s, the region was disrupted by the expansion of the Zulu and their offshoot, the Ndebele. A new threat arose in the late 19th cent. with the incursion of Boers from neighboring Transvaal. After gold was discovered in the region in 1867, the Transvaal government sought to annex parts of Botswana. Although the British forbade annexation, the Boers continued to encroach on tribal lands during the 1870s and 80s. German colonial expansion in South West Africa caused the British to establish (1884-85) a protectorate called Bechuanaland. The southern part of the area was incorporated into Cape Colony in 1895. Until 1961, Bechuanaland was administered by a resident commissioner in South Africa. Britain provided for the eventual transfer of Bechuanaland to the Union of South Africa, which was established in 1910. However, the rise of the National party in South Africa in 1948 and its pursuit of apartheid turned British opinion against the incorporation of Bechuanaland into South Africa. Britain granted internal self-government in 1965 and full independence on Sept. 30, 1966. Despite the country's close ties with its white-ruled neighbors, Botswana has increasingly lent support to liberation groups fighting in Rhodesia and South Africa.

Bot·ti·neau (bŏt'ĭ-nō'), county (1970 pop. 9,496), 1,677 sq mi (4,343.4 sq km), N N.Dak., on the Canadian border in an agricultural region watered by the Souris River; formed 1873; co. seat Bottineau. The Turtle Mts. are in the northeast. It has timber and diversified agriculture (wheat, oats, barley, flax, sunflowers, and beef cattle). Oil has been found here.

Bottineau, city (1970 pop. 2,760), seat of Bottineau co., N N.Dak., near the Canadian boundary W of the Turtle Mts., in a farm and timber region; settled 1883, inc. 1904. It is a resort center for a hunting and fishing area.

Bot·trop (bôt'rôp), city (1975 est. pop. 101,495), North Rhine-Westphalia, W West Germany, in the Ruhr district. It was a small town until 1863, when it began to develop as a coal-mining center. The city is today also an industrial center; its manufactures include chemicals, electrical equipment, and textiles.

Boua·ké (bwä-kā'), town (1969 est. pop. 161,300), central Ivory Coast. It is a transportation hub and a commercial center and was once the crossroads for the caravan trade. Tobacco products are produced in the town, and gold and manganese are found nearby.

Bouches-du-Rhône (bōōsh'dü-rōn'), department (1975 est. pop. 1,632,974), 1,974 sq mi (5,112.7 sq km), in Provence, SE France; capital Marseilles.

Bou·gain·ville (bōō'gən-vĭl', bōō'găN-vēl'), volcanic island (1964 pop. 64,100), c.3,880 sq mi (10,050 sq km), SW Pacific, largest of the Solomon Islands. With the neighboring island of Buka, it forms a part of Papua New Guinea. Bougainville is rugged and densely forested. There are several good harbors. The economy is mainly agricultural; major exports are copra, ivory nuts, green snails, cocoa, tortoise shells, and trepang. Copper and gold are mined. The island was discovered in 1768 by the French navigator Louis de Bougainville. Unlike the rest of the Solomon Islands, which became a British territory, Bougainville and Buka became part of German New Guinea in 1884. Occupied by Australian forces during World War I, Bougainville was mandated to Australia by the League of Nations in 1920. During World War II the island was the last Japanese stronghold in the Solomons.

Bou·gie (bōō-zhē'): *see* Bejaïa, Algeria.

Boul·der (bōl'dər), county (1970 pop. 131,889), 753 sq mi (1,950.3 sq km), N central Colo., in an irrigated agricultural and livestock region; formed 1861; co. seat Boulder. Sugar beets, beans, fruit, and grain are grown. It also has coal, gold, tungsten, and fluorspar mines and diversified industries. Part of Rocky Mountain National Park is in the county.

Boulder. **1.** City (1975 pop. 77,000), seat of Boulder co., N central Colo.; inc. 1871. Situated c.5,350 ft (1,631.8 m) above sea level, it is a major resort of the Rocky Mts. and has mineral springs. Its manufactures include aircraft, computers, electronic equipment, chemicals, and sporting goods. The Univ. of Colorado and the National Center for Atmospheric Research are in the city. **2.** Town (1970 pop. 1,342), seat of Jefferson co., SW Mont., on the Boulder River S of Helena; inc. 1911. The trade center of a ranching and mining area, it is also a health resort with mineral springs nearby.

Boulder City, residential city (1975 est. pop. 6,126), Clark co., S Nev., just W of Hoover Dam near Lake Mead; inc. 1959. Built (1932) by the Federal government as headquarters during the dam's construction, it became a self-governing municipality by act of Congress in 1958. It is a year-round tourist center.

Bou·logne-Bil·lan·court (bōō-lô'nyə-bē'yäɴ-kŏŏr'), city (1968 pop. 109,380), Hauts-de-Seine dept., N central France, a suburb SW of Paris. One of the largest automobile factories in France is in the city. Other manufactures include airplanes, electrical goods, chemicals, bicycles, and processed foods.

Bou·logne-sur-Mer (bōō-lô'nyə-sür-měr'), city (1975 est. pop. 48,440), Pas-de-Calais dept., N France, in Picardy, on the English Channel. It is a great commercial seaport and the leading fishing port of France. It has canning and shipbuilding industries. From here the Romans sailed (A.D. 43) to conquer Britain, and Napoleon assembled an invasion fleet (which never sailed) in 1803–5. The port was a main base for British armies in World War I and a German submarine base in World War II. Most of the city was destroyed during the latter conflict. The Cathedral of Notre Dame (built 19th cent.; damaged 1941 and since restored) is a pilgrimage shrine.

Bound·a·ry (boun'də-rē), county (1970 pop. 5,484), 1,275 sq mi (3,302.3 sq km), extreme N Idaho, in a mountainous area drained by the Kootenai River and bordering on British Columbia, Wash., and Mont.; formed 1884; co. seat Bonners Ferry. It has lumbering, diversified farming and cattle raising, dairying, and some mining (lead, zinc, silver, and molybdenum). Kaniksu National Forest extends throughout much of the county.

Boundary Peak, 13,145 ft (4,009.2 m) high, SW Nev., in the White Mts. near the Calif. line. It is the highest point in the state.

Bound Brook (bound' brŏŏk'), borough (1975 est. pop. 9,468), Somerset co., N central N.J., on the Raritan River; settled 1681, inc. 1891. It has large orchid and gardenia nurseries and chemical manufactures. In the Revolution, George Washington maintained an outpost here, and American forces were defeated (Apr., 1777) by Cornwallis.

Boun·ti·ful (boun'tĭ-fəl), city (1975 est. pop. 29,900), Davis co., N central Utah, a residential suburb N of Salt Lake City; settled by Mormons, inc. 1892.

Boun·ty Islands (boun'tē), outlying island group, .5 sq mi (1.3 sq km), New Zealand, in the South Pacific ESE of Dunedin. Capt. William Bligh discovered the 13 islets in 1788.

Bour·bon (bûr'bən). **1.** County (1970 pop. 15,215), 639 sq mi (1,655 sq km), SE Kansas, in a rolling plains area bordered on the E by Mo. and watered by the Marmaton and Little Osage rivers; formed 1855; co. seat Fort Scott. It has stock raising, dairying, and diversified agriculture. There are coal deposits and scattered oil and gas fields. It includes Fort Scott Historic Area. **2.** County (1970 pop. 18,476), 300 sq mi (777 sq km), N central Ky., drained by the South Fork of the Licking River; formed 1785; co. seat Paris. It is in a gently rolling agricultural area (burley tobacco, livestock, dairy products, and poultry) of the Bluegrass region. Limestone is quarried, and some manufacturing is done. Paris had one of the earliest (1790) distilleries in the state; whiskey from the region was supposedly first called bourbon after the county.

Bour·bon·nais (bŏŏr-bô-nā'), former province, central France, in the part of the Massif Central. It had approximately the same area as today's Allier dept. The counts (later dukes) of Bourbon held the Bourbonnais as an appanage until 1527, when Francis I of France confiscated it.

Bourg-en-Bresse (bŏŏr'käɴ-brĕs') or **Bourg** (bŏŏr), town (1975 est. pop. 42,181), capital of Ain dept., in Burgundy, E central France. It is a major transportation hub, farm market, and gastronomic center. Machinery, morocco leather, shoes, and ceramics are made.

Bourges (bŏŏrzh, bŏŏrzh), city (1975 est. pop. 77,300), capital of Cher dept., central France. It is a transportation center with foundries, arsenals, breweries, printing plants, and aeronautical and food industries. Known as Avaricum, Bourges was the Roman capital of Aquitania. Louis XI founded (1463) the Univ. of Bourges, which was abolished in the French Revolution. The Cathedral of St. Etienne (13th cent.), one of the glories of French Gothic, is remarkable in that it has no transept. Jacques Cœur, whose splendid house still stands, was born in Bourges.

Bour·get, Le (lə bŏŏr-zhā'), town (1968 pop. 49,302), Seine-Saint-Denis dept., N central France. One of the major airports of Paris is here. Charles Lindbergh landed at Le Bourget after his transatlantic flight of 1927.

Bour·gogne (bŏŏr-gô'nya), region: *see* Burgundy.

Bourne·mouth (bôrn'məth), county borough (1976 est. pop. 144,100), Dorset, S central England, on Poole Bay. It has grown since the middle of the 19th cent. from a small fishing village in the sheltered, pine-wooded valley of the Bourne into a popular resort and fine-arts center. It has an excellent sandy beach, a fine climate, and numerous parks. Mary Shelley is buried in the parish churchyard.

Bou·vines (bōō-vēn'), village (1968 pop. 560), Nord dept., N France, in Flanders. In an epochal battle in 1214, Philip II of France defeated the joint forces of King John of England, Emperor Otto IV, and the count of Flanders, establishing the power of the French monarchy.

Bow (bō), river, 315 mi (506.8 km) long, rising in the Rocky Mts., S Alta., Canada, and flowing SE through Banff National Park. It emerges from the mountains in the Bow River Pass and continues past Calgary southeastward across the plains to its junction with the Belly River to form the South Saskatchewan River.

Bow·bells (bō'bĕlz) city (1970 pop. 584), seat of Burke co., NW N.Dak., NW of Minot, in a livestock, poultry, wheat, grain, and flax area. There are coal mines and dairy farms here.

Bow·er·y, the (bou'ə-rē, bou'rē) section of lower Manhattan, New York City. The Bowery, the street that gives the area its name, was once a road to the farm, or *bouwerie*, of Peter Stuyvesant. The mail route (est. 1673) to Boston traveled this road. By the 1860s and 1870s it had many fine theaters. Later the section became notorious for its saloons, dance halls, swindlers, petty criminals, and derelicts.

Bow·ie (bōō'ē), county (1970 pop. 68,909), 891 sq mi (2,307.7 sq km), NE corner of Texas, bounded on the N by the Red River, here forming the Okla. and Ark. borders, on the E by Ark., on the S by the Sulphur River and Lake Texarkana; formed 1840; co. seat Boston. Partly forested with pine, oak, and gum trees, it has lumbering, agriculture (cotton, corn, grain, fruit, pecans and peanuts, truck crops, livestock, and dairy products), and some manufacturing.

Bowie, city (1975 est. pop. 37,323), Prince Georges co., W central Md.; inc. 1916. It is mainly a residential community. Points of interest include the Woodward Mansion (c.1743), which now serves as the city hall, and Belair Stables, now a historical museum.

Bowl·ing Green. (bō'lĭng grēn'). **1.** City (1970 pop. 36,705), seat of Warren co., S Ky., on the Barren River; inc. 1812. It is a shipping and marketing center for an area producing tobacco, corn, livestock, and dairy items. Textiles, apparel, automobile parts, woodwork, and heavy equipment are manufactured in the city. Bowling Green was occupied by the Confederates at the beginning of the Civil War until the Federal advance forced them to retreat in 1862. The city is the seat of Western Kentucky Univ. **2.** City (1970 pop. 2,936), seat of Pike co., NE Mo., SSE of Hannibal, in a farming region; settled 1819, inc. 1823. Rock and limestone quarries are in the area. The city has a statue of Champ Clark, who lived here. **3.** City (1970 pop. 21,760), seat of Wood co., NW Ohio, in a farm area; inc. 1855. Tomato products, hydraulic hoists, and plastics are the chief manufactures. **4.** Town (1970 pop. 528), seat of Caroline co., E Va., SSE of Fredericksburg, in an agricultural area. Baskets and excelsior are made. Camp Hill Military Reservation is just to the east.

Bow·man (bō'mən), county (1970 pop. 3,901), 1,170 sq mi (3,030.3 sq km), extreme SW N.Dak., in an agricultural area watered by the Little Missouri and Grand rivers; formed 1883; co. seat Bowman. It is mostly agricultural (livestock, dairying, poultry, grain, and vegetables), with an extensive coal-mining operation and several oil and natural-gas wells.

Bowman, city (1970 pop. 1,762), seat of Bowman co., extreme SW N.Dak., SSW of Dickinson; settled 1907, inc. 1908. It is a trade center for a livestock, dairy, and wheat region. Lignite is mined nearby.

Box Butte (bŏks' byo͝ot'), county (1970 pop. 10,094), 1,065 sq mi (2,758.4 sq km), NW Nebr.; formed 1886; co. seat Alliance. In an agricultural area yielding wheat, sugar beets, corn, and beans, it has some ranching and light industry.

Box El·der (ĕl'dər), county (1970 pop. 28,129), 5,627 sq mi (14,573.9 sq km), NW Utah, in an agricultural area bordering on Nev. and Idaho; formed 1856; co. seat Brigham City. Its agriculture includes wheat, sugar beets, fruit, alfalfa, corn, tomatoes, potatoes, onions, cattle, and sheep. Great Salt Lake occupies much of the southeast portion of the county, and part of the Great Salt Lake Desert is in the southwest.

Bo·ya·cá (bō'yä-kä'), town (1968 est. pop. 7,700), N central Colombia, near Tunja. At Boyacá on Aug. 7, 1819, revolutionary forces under Simón Bolívar won the decisive engagement that assured the independence of present-day Colombia and Venezuela from Spain.

Boyd (boid). **1.** County (1970 pop. 52,376), 159 sq mi (411.8 sq km), NE Ky., bounded on the NE by the Ohio River, forming the Ohio border, and on the E by the Big Sandy River, forming the W.Va. border; formed 1860; co. seat Catlettsburg. In a hilly industrial and mineral region, it includes part of the Huntington, W.Va., and Ashland, Ky., metropolitan district, with extensive manufacturing. It has bituminous-coal mines, oil and gas wells, clay pits, limestone and sandstone quarries, and iron deposits. Some farming is done (livestock, fruit, and tobacco). **2.** County (1970 pop. 3,752), 538 sq mi (1,393.4 sq km), N Nebr., in an agricultural area bounded on the N by the S.Dak. border, on the S by the Niobrara River, on the NE by the Missouri River; formed 1891; co. seat Butte. It is drained by the Keya Paha River and Ponca Creek. It has general farming and ranching and a small amount of manufacturing.

Boyd·ton (boid'tən), town (1970 pop. 541), seat of Mecklenburg co., S Va., SW of Petersburg, in an agricultural area. Tobacco is grown, and sawmilling is done.

Boyle (boil), county (1970 pop. 21,861), 181 sq mi (468.8 sq km), central Ky., bounded on the E by the Dix River and drained by the Salt River and Beech Fork; formed 1842; co. seat Danville. It is in a gently rolling and partly hilly upland area, partly in the outer Bluegrass region. Agriculture, light industry, and government services are important. It includes Perryville National Cemetery.

Boyne (boin), river, c.70 mi (115 km) long, rising in the Bog of Allen, Co. Kildare, E Republic of Ireland, and flowing NE through Co. Meath, past Trim, to the Irish Sea near Drogheda. Salmon is caught in the river. In the Battle of the Boyne (July, 1690) near Drogheda, the armies of King William III defeated the Catholic James II, who fled to France.

Boyn·ton Beach (boin'tən), city (1975 est. pop. 30,944), Palm Beach co., SE Fla., on the Atlantic coast; inc. 1920. It is a beach resort.

Boys Town (boiz), village (1970 pop. 989), Douglas co., E Nebr.; inc. 1936. The noted community was founded in 1917 by Father Edward J. Flanagan (1886–1948) for homeless or abandoned boys. The village is governed by the boys themselves and maintained by voluntary contributions.

Boz·ca·a·da (bōz'jä-ä-dä') or **Ten·e·dos** (tĕn'ə-dŏs), island (1970 pop. 2,030), 15 sq mi (38.9 sq km), NW Turkey, in the Aegean Sea. The strategically located island was a station of the Greek fleet during the Trojan War. Xerxes used it (5th cent. B.C.) as a base for the Persian fleet. The Ottoman Turks captured it in 1657.

Boze·man (bōz'mən), city (1975 est. pop. 19,847), seat of Gallatin co., SW Mont.; inc. 1883. The city is named after John M. Bozeman, a pioneer who led the first settlers here in 1864. Bozeman is the center of a farming and stock-raising area. Tourism is an important source of revenue; the city is the headquarters of Gallatin National Forest, and Yellowstone National Park is nearby.

Bra·bant (brə-bănt', brä'bənt, brä'bänt, brä-bäN'), province (1976 est. pop. 2,217,934), 1,268 sq mi (3,284.1 sq km), central Belgium; capital Brussels.

Brabant, duchy of, former duchy, now divided between Belgium (Brabant and Antwerp provs.) and the Netherlands (North Brabant prov.). The duchy of Brabant emerged (1190) from the duchy of Lower Lorraine. In 1430 it passed to Philip the Good of Burgundy, and in 1477 it was taken by the Hapsburgs. Brabant owed its extraordinary prosperity during the Middle Ages to its wool and other textile industries and to the commercial enterprise of the inhabitants of its cities and towns. In 1830 southern Brabant led the revolt against Dutch rule that resulted in independence for Belgium.

Brack·en (brăk'ən), county (1970 pop. 7,227), 204 sq mi (528.4 sq km), N Ky., bounded on the N by the Ohio River, forming the Ohio border, and on the S by the North Fork of the Licking River; formed 1796; co. seat Brooksville. In a gently rolling upland agricultural area in the outer Bluegrass region, its most important crop is tobacco. There is some shipping and manufacturing.

Brack·ett·ville (brăk'ĭt-vĭl'), town (1970 pop. 1,539), seat of Kinney co., W Texas, ESE of Del Rio; inc. 1928. Cattle, sheep, and goats are raised. Laughlin Air Force Base is nearby.

Brack·nell (brăk'nəl), new town and civil parish (1971 pop. 37,279), Easthampstead rural dist., Berkshire, S England. Bracknell was designated one of the New Towns in 1949 to alleviate overpopulation in London. Its industries include the manufacture of boilers, gasoline pumps, tools, clothing, and sealing compounds.

Bra·den·ton (brăd'n-tən), city (1975 est. pop. 25,004), seat of Manatee co., SW Fla., on Tampa Bay at the mouths of the Braden and Manatee rivers; inc. 1903. A popular winter resort with excellent fishing, it is also a shipping center for the citrus fruit and truck crops of the area. Travertine is quarried and refined here. Hernando DeSoto is believed to have landed near here in 1539; the DeSoto National Memorial is to the west. The area was settled (1850s) by Joseph Braden, whose castlelike home is a local landmark.

Brad·ford (brăd'fərd), county borough (1971 pop. 293,756), West Yorkshire, N central England, on a small tributary of the Aire River. It is a center of the worsted industry, which dates from the Middle Ages. Besides woolens, other fabrics (including synthetics) are made. Electroplating, electrical engineering, and the manufacture of machinery and automobiles are also important industries. There are stone quarries nearby. Bradford's landmarks include the memorial hall, dedicated to Edmund Cartwright, inventor of the power loom, and the Conditioning House, a unique textile-testing establishment.

Bradford 1. County (1970 pop. 14,625), 293 sq mi (758.9 sq km), N Fla., bounded on the S by the Santa Fe River; formed 1858; co. seat Starke. In a wooded area dotted with lakes, it has agriculture (corn, strawberries, vegetables, tobacco, pecans, and livestock) and lumbering. **2.** County (1970 pop. 57,962), 1,148 sq mi (2,973.3 sq km), NE Pa., in a hilly agricultural and industrial region drained by the Susquehanna River; formed 1812; co. seat Towanda. It was settled in 1763 by Moravian missionaries. Textiles, clothing, and metal products are manufactured.

Bradford, city (1975 est. pop. 11,941), McKean co., NW Pa., in the Alleghenies, near the N.Y. line; settled c.1823, inc. as a city 1879. The growth of the city was initiated by the discovery of oil (c.1871), and oil refining is still a major industry. Other products include electronic components, steel couplings, cutlery, chemicals, and explosives. The area is popular for hunting and fishing.

Brad·ley (brăd'lē). **1.** County (1970 pop. 12,778), 649 sq mi (1,680.9 sq km), S Ark., bounded on the SW by the Ouachita River and on the E and SE by the Saline River; formed 1840; co. seat Warren. It has agriculture (cotton, potatoes, and truck crops) and timber. Wood products are manufactured. **2.** County (1970 pop. 50,686), 334 sq mi (865.1 sq km), SE Tenn., bounded on the S by the Ga. border and on the N by the Hiwassee River; formed 1835; co. seat Cleveland. In an agricultural area yielding corn, cotton, soybeans, dairy products, and beef cattle, it also has diversified industry.

Bra·dy (brā'dē), city (1970 pop. 5,557) seat of McCulloch co., central Texas, on Brady Creek ESE of San Angelo; settled c.1876, inc. 1906. It processes peanuts, turkeys, milk, cotton, and wool from farms and ranches in surrounding hills.

Bra·ga (brä'gə), city (1975 est. pop. 48,735), capital of Braga dist., NW Portugal, in Minho. It is an agricultural trade center with minor industry. The ancient Bracara Augusta, it had considerable importance in Roman days. As the seat of Portugal's titular primate, the city is still a religious center. In the old cathedral is the tomb of Henry of Burgundy. Nearby is a summer resort with the well-known Church of Bom Jesus do Monte.

Bra·gan·ça (brə-gäN'sə) or **Bra·gan·za** (-zə), town (1975 est. pop.

9,310), capital of Bragança dist., NE Portugal, in Trás-os-Montes. Textiles are produced. Its castle was seat of the Braganza family, long the royal family of Portugal.

Brah·ma·pu·tra (brä'mə-pōō'trə), river, c.1,800 mi (2,895 km) long, rising in the Kailas range of the Himalayas, SW Tibet, and flowing through NE India to join with the Ganges River in central Bangladesh to form a vast delta. It is navigable for large craft c.800 mi (1,290 km) upstream. The river's lower course is sacred to Hindus.

Bră·i·la (brə-ē'lä), city (1977 est. pop. 194,633), SE Rumania, in Walachia, on the Danube River. The chief grain-shipping port of Rumania, it is also a major industrial and commercial city. Machinery, metals, foodstuffs, and textiles are the principal products. Brăila probably dates from Greek times. It was taken by the Turks c.1550 and played an important role in the Russo-Turkish Wars. The Treaty of Adrianople (1829) awarded the city to Rumania.

Brai·nerd (brā'nərd), city (1975 est. pop. 11,672), seat of Crow Wing co., central Minn., on the Mississippi River, in a pine-forested and lake region; inc. 1881. Founded (1870) by the Northern Pacific RR, it is still a railroad center with repair shops. Lumbering and related enterprises are its economic mainstays.

Brain·tree (brān'trē'), town (1975 est. pop. 36,804), Norfolk co., E Mass., a suburb of Boston; inc. 1640. Abrasives and rubber goods are among its manufactures. John Hancock was born here.

Braintree and Bock·ing (bŏk'ĭng), urban district (1973 est. pop. 26,300), Essex, E England, between the Pant (Blackwater) and Brain river valleys. There are textile, plastic, and metal-product industries.

Brak·pan (brăk'păn), city (1970 pop. 113,115), Transvaal prov., NE South Africa. It is a gold- and coal-mining center.

Bramp·ton (brăm'tən, brămp'-), town (1976 est. pop. 103,459), S Ont., Canada, NW of Toronto. It is noted for its greenhouses. Automobiles, optical goods, and other products are made.

Branch (brănch), county (1970 pop. 37,906), 506 sq mi (1,310.5 sq km), S Mich., bounded on the S by the Ind. border and drained by the St. Joseph and Coldwater rivers; organized 1833; co. seat Coldwater. It has agriculture (fruit, grain, corn, livestock, and dairy products), some manufacturing, and marl and clay deposits. There are resorts on its many small lakes.

Bran·den·burg (brăn'dən-bûrg', brän'dən-bŏŏrkh'), former state, c.10,400 sq mi (26,935 sq km), central East Germany. As constituted in 1947 under Soviet military occupation, Brandenburg consisted of the former Prussian province of Brandenburg minus those parts of the province lying east of the Oder and Neisse rivers. It became (1949) one of the states of the German Democratic Republic, but was abolished as an administrative unit in 1952.

Slavic-speaking Wends inhabited Brandenburg at the time of its acquisition (12th cent.) by Albert the Bear. Albert's descendants, the Ascanians, ruled Brandenburg until their extinction in 1320. Emperor Charles IV conferred (1373) it on his son Wenceslaus. When Wenceslaus became (1378) German king, Brandenburg went to his brother, later Emperor Sigismund, who in 1417 formally transferred it to Frederick I of the house of Hohenzollern. Although it suffered heavily in the Thirty Years War (1618–48), Brandenburg emerged as a military power under Frederick William, the Great Elector (reigned 1640–88). Thereafter, it became an integral part of Prussia.

Brandenburg, city (1974 est. pop. 94,071), Potsdam district, central East Germany, a port on the Havel River. It is an industrial center and rail junction. Manufactures include steel, textiles, machinery, and motor vehicles.

Bran·den·burg (brăn'dən-bûrg'), city (1970 pop. 1,637), seat of Meade co., N central Ky., on the Ohio River SW of Louisville, in a farm area; inc. 1825. Wheat and tobacco are grown.

Bran·don (brăn'dən), city (1976 est. pop. 34,901), SW Man., Canada, on the Assiniboine River. The center of a wheat-raising area, Brandon has an extensive trade in farm products and machinery.

Brandon. 1. Uninc. village (1970 pop. 12,749), Hillsborough co., W Fla., a suburb just E of Tampa. Chiefly residential, it is also a retail and service center. Citrus fruits and vegetables are grown in the area, and there are many cattle and dairy farms. **2.** Town (1970 pop. 2,685), seat of Rankin co., S central Miss., E of Jackson, in a truck-farm, cotton, and natural-gas area; inc. 1831.

Bran·ford (brăn'fərd), town (1975 est. pop. 22,004), New Haven co., S Conn., on Long Island Sound; settled 1644, inc. as a town 1930. Formerly a shipping and fishing center, the town is now mainly residential and manufactures prestressed concrete forms, automotive parts, wire, and other products.

Brant·ford (brănt'fərd), city (1976 est. pop. 66,950), S Ont., Canada, on the Grand River. It is a leading manufacturing city, noted particularly for its large farm implement factories. The city was named for the Mohawk chieftain Joseph Brant, who led the Six Nations of the Iroquois to the region after the American Revolution and who is buried in the old Mohawk Church near the city. Alexander Graham Bell was living in Brantford in 1876 when he made his first successful experiment in the transmission of sound by electric wire. A museum, formerly his home, exhibits the first telephone.

Brant·ley (brănt'lē), county (1970 pop. 5,940), 447 sq mi (1,157.7 sq km), SE Ga., bounded on the N and E by the Satilla River; formed 1920; co. seat Nahunta. It has coastal-plain agriculture (tobacco, corn, melons, and sweet potatoes) and timber.

Bras d'Or Lake (brä dôr), arm of the Atlantic Ocean, c.360 sq mi (930 sq km), indenting deeply into Cape Breton Island, N.S., SE Canada, and occupying much of the interior. A narrow channel links it with the sea. In 1907 Alexander Graham Bell founded at Baddeck the Aerial Experiment Association, and on Feb. 23, 1909, J. A. D. McCurdy piloted his airplane, the *Silver Dart,* a distance of half a mile.

Bra·sí·lia (brä-zēl'yä), capital city and federal district (1975 est. pop. 750,000) of Brazil, 2,264 sq mi (5,863.8 sq km), an enclave in the SW portion of Goiás state. One of the newest cities of the world, it was inaugurated in 1960. It is situated in the highlands of central Brazil, and its ultramodern public buildings (designed by Oscar Niemeyer) dominate the sparsely settled countryside. The removal of the capital from Rio de Janeiro to the interior, to encourage the development of central Brazil, had long been advocated. The city was laid out (1957) by the Brazilian architect Lúcio Costa.

Bra·şov (brä-shôv'), city (1977 est. pop. 257,150), central Rumania, in Transylvania, at the foot of the Transylvanian Alps. The city is a road and rail junction and a major industrial center. Tractors, trucks, machinery, chemicals, and textiles are among the chief manufactures. The city is also a noted resort and winter sports center. Founded in the 13th cent. by the Teutonic Knights, Braşov was a major center of trade and industry in the Middle Ages. It enjoyed considerable autonomy under the Hapsburg empire. After World War I the city, along with Transylvania, was ceded by Hungary to Rumania. Parts of the medieval town wall and the 17th cent. citadel remain intact.

Bra·ti·sla·va (brăt'i-slä'və), city (1975 est. pop. 340,902), S Czechoslovakia, on the Danube River near the Austrian and Hungarian borders. It is Czechoslovakia's third-largest city and the traditional capital of Slovakia. Bratislava is also an important road and rail center and a leading Danubian port. Industries include mechanical engineering, machine building, oil refining, food processing, and the manufacture of chemicals, textiles, electrical equipment, paper, wood products, and beer. Forests, vineyards, and large farms surround the city, which has an active trade in agricultural products. It is also a popular tourist center.

A Roman outpost called Posonium by the 1st cent. A.D., Bratislava became a stronghold of the Great Moravian Empire in the 9th cent. After the death of Ottocar II (1278), Bratislava and much of southern and eastern Slovakia fell under Hungarian rule. From 1541, when the Turks captured Buda, until 1784, Bratislava served as Hungary's capital. Hungarian kings were crowned here until 1835, and Bratislava was the meeting place of the Hungarian diet until 1848. Inhabited largely by German traders before the 19th cent., the city then became predominantly Magyar. In the 19th cent. it was the center of the emerging Slovak national revival, and after the union of the Czech and Slovak territories in 1918 it was incorporated into Czechoslovakia. From 1939 until 1945, Bratislava was the capital of a nominally independent Slovak republic that was governed by a pro-German regime.

Brat·tle·bor·o (brăt'l-bûr'ō, -bŭr'ō), town (1975 est. pop. 11,977), Windham co., SE Vt., on the Connecticut River; chartered 1753. The town grew near Fort Dummer, which was established in 1724 to protect the settlers from Indians. Once an artists' colony, Brattleboro is now a center for winter sports. Its manufactures include optical goods, paper and wood products, books, and purses.

Braun·schweig (broun'shvīk): *see* Brunswick, West Germany.

Braw·ley (brô'lē), city (1975 est. pop. 14,012), Imperial co., SE Calif., in an agricultural area of the Imperial Valley SE of the Salton Sea; inc. 1908. Cattle feeding and the production of beet sugar are the major industries.

Brax·ton (brăk′stən), county (1970 pop. 12,666), 517 sq mi (1,339 sq km), central W.Va., on the Allegheny Plateau, drained by the Elk and Little Kanawha rivers; formed 1836; co. seat Sutton. The county raises livestock and produces fruit and tobacco. It lies in a timber, coal, oil, and natural-gas area and has granite quarries.

Bra·zil (brə-zĭl′), republic (1973 est. pop. 99,000,000), 3,286,470 sq mi (8,511,957 sq km), E South America; capital Brasília. By far the largest of the Latin American countries, Brazil occupies nearly half the continent of South America, stretching from the Guiana Highlands in the north to the plains of Uruguay and Paraguay in the south. In the west it spreads to the equatorial rain forest, bordering on Bolivia, Peru, and Colombia; in the east it juts far out into the Atlantic toward Africa.

Its vast extent covers a great variety of land and climate, for although Brazil is mainly in the tropics (the equator crosses it in the north and the Tropic of Capricorn crosses it in the south), the southern part of the great central upland is cool and yields the produce of temperate lands. The people are also diverse in origin. Portuguese is the official language, and a large part of the population is at least nominally Roman Catholic. Most of the estimated 150,000 Indians (chiefly of Tupí or Guaraní linguistic stock) are found in the rain forests of the Amazon River basin, which occupies all the north and north-central portions of Brazil. Wild rubber, once of great economic importance, and other forest products are gathered in the Amazon region, but the area is still largely of potential rather than actual economic value. Southeast of the Amazon mouth is the great

seaward outthrust of Brazil, the region known as the Northeast, which was the center of the great sugar culture that for centuries dominated Brazil. In this region the general pattern is a narrow coastal plain (formerly supporting the sugar-cane plantations and now given over to diversified subtropical crops) and a semiarid interior subject to recurrent droughts. The "bulge" of Brazil reaches its turning point at the Cape of São Roque.

South of the "corner" of Brazil, the characteristic pattern of Brazilian geography emerges: the narrow and interrupted coastal lowlands are bordered on the west by an escarpment; in some places, however, the escarpment actually reaches the sea. Above the escarpment is the great Brazilian plateau, which tapers off in the south, where it is succeeded by the plains of the Rio de la Plata country. The escarpment itself appears from the sea as a mountain range, generally called the Serra do Mar, and the plateau is interrupted by mountainous regions, such as that in Bahia. There are a number of excellent harbors farther south, including Rio de Janeiro, the former capital, one of the most beautiful harbors in the world.

In the east and southeast is the heavily populated region of Brazil—the states that in the 19th and 20th cent. received the bulk of European immigrants and took hegemony away from the old Northeast. São Paulo state has 50% of all of Brazil's industry and a well-developed agriculture. The city of São Paulo on the plateau has continued the vigorous and aggressive development that marked the region in the 17th and 18th cent., when the *paulistas* went out in the famed *bandeiras* (raids), searching for Indian slaves and gold and opening the rugged interior. They were largely responsible for the development of the gold and diamond mines of Minas Gerais state. Minas has some of the finest iron reserves in the world, as well as other mineral wealth, and is becoming industrialized. Settlement also spread from São Paulo southward, particularly in the 19th and early 20th cent. when coffee had become the basis of Brazilian wealth. The southernmost states, developed to a large extent by German and Slavic immigrants, are primarily cattle-growing areas with increasing industrial importance. Frontier development is continuing in central Brazil. The federal district of Brasília was carved out of the neighboring plateau, to the east. The national capital was transferred to the planned city of Brasília in 1960.

Brazil is divided into a federal district, 22 states, and 4 (federal) territories:

NAME	CAPITAL	NAME	CAPITAL
Acre	Rio Branco	Minas Gerais	Belo Horizonte
Alagoas	Maceió	Pará	Belém
Amapá Territory	Macapá	Paraíba	João Pessoa
Amazonas	Manaus	Paraná	Curitiba
Bahia	Salvador	Pernambuco	Recife
Ceará	Fortaleza	Piauí	Teresina
Espírito Santo	Vitória	Rio de Janeiro	Niterói
Federal District		Rio Grande do Norte	Natal
Fernando de Noronha Territory		Rio Grande do Sul	Pôrto Alegre
	Brasília	Rondônia Territory	Pôrto Velho
Goiás	Goiânia	Roraima Territory	Boa Vista
Guanabara	Rio de Janeiro	Santa Catarina	Florianópolis
Maranhão	São Luís	Saõ Paulo	São Paulo
Mato Grosso	Cuiabá	Sergipe	Aracaju

Economy. Despite a high annual growth rate and recent industrialization, Brazil is still an agricultural country. Agriculture employs about 60% of the labor force and accounts for 70% of the exports. The major commercial crops are coffee, cocoa, cotton, sugar cane, oranges, bananas, and beans. Cattle, pigs, and sheep are the most numerous livestock. Besides iron, Brazil is an important producer of coal, manganese, chrome, industrial diamonds, quartz crystal, and many other minerals. The leading manufacturing industries produce cotton textiles, paper, fertilizer, and asphalt. Motor vehicle production is increasing. Brazil is the world's leading coffee exporter. Other exports are iron, cotton, and sugar.

History. Whether or not Brazil was known to Portuguese navigators in the 15th cent. is still an unsolved problem, but the coast was visited by the Spanish mariner Vicente Yáñez Pinzón before the Portuguese under Pedro Alvares Cabral in 1500 claimed the land, which came within the Portuguese sphere as defined in the Treaty of Tordesillas (1494). Little was done to support the claim, but the name Brazil is thought to derive from the Portuguese word for the red color of brazilwood, which the early visitors gathered. The first permanent settlement was not made until 1532, and that was at São Vicente in São Paulo. Development of the Northeast was begun about the same time. Salvador was founded in 1539, and 12 captaincies were established, stretching inland from the Brazilian coast. French Huguenots established themselves (1555) on an island in Rio de Janeiro harbor and were routed in 1567 by a force under Mem de Sá, who then founded the city of Rio de Janeiro. The Dutch made their first attack on Salvador (Bahia) in 1624, and in 1633 the vigorous Dutch West India Company was able to capture and hold not only Salvador and Recife but the whole of the Northeast; the region was ably ruled by John Maurice of Nassau. No aid was forthcoming from Portugal, which had been united with Spain in 1580 and did not regain its independence until 1640. It was a naval expedition from Rio itself that drove out the Dutch in 1654.

Farther south, the *bandeirantes* from São Paulo had been trekking westward since the beginning of the 17th cent., thrusting far into Spanish territory and extending the western boundaries of Brazil. The Portuguese also had ambitions to control the Banda Oriental (present Uruguay) and in the 18th cent. came into conflict with the Spanish there; the matter was not completely settled even by the independence of Uruguay in 1828. Meanwhile in the Northeast the plantations were furnishing most of the sugar demanded by Europe.

The native Indians were not adaptable to the backbreaking labor of the cane fields, and Negro slaves were imported in large numbers. Dependence on a one-crop economy was lessened by the development of the mines in the interior, particularly those of Minas Gerais, where gold was discovered late in the 17th cent. Mining towns sprang up, and Ouro Prêto became in the 18th cent. a major intellectual and artistic center. The center of development began to swing south, and Rio de Janeiro, increasingly important as an export center, supplanted Salvador as the capital of Brazil in 1763. Ripples from intellectual stirrings in Europe that preceded the French Revolution and the successful American Revolution brought on an abortive plot for independence among a small group of intellectuals in Minas.

When Napoleon's forces invaded Portugal, the king of Portugal, John VI, fled (1807) to Brazil, and on his arrival (1808) in Rio de Janeiro that city became the capital of the Portuguese Empire. The ports of the colony were freed of mercantilist restrictions, and Brazil became a kingdom, of equal status with Portugal. In 1821 the king returned to Portugal, leaving his son behind as regent of Brazil. New policies by Portugal toward Brazil, tightening colonial restrictions, stirred up wide unrest. The young prince eventually acceded to popular sentiment, and advised by the Brazilian José Bonifácio, uttered the fateful cry of independence on Sept. 7, 1822, on the banks of the little Ipiranga River. He became Pedro I, emperor of Brazil. Pedro's rule, however, gradually kindled increasing discontent in Brazil, and in 1831 he had to abdicate in favor of his son, Pedro II. The reign of this popular emperor saw the foundation of modern Brazil. Ambitions directed toward the south were responsible for involving the country in the war (1851-52) against the Argentine dictator, Juan Manuel de Rosas, and again in the War of the Triple Alliance (1865-70) against Paraguay. Brazil drew little benefit from either; far more important were the beginnings of the large-scale European immigration that was to make southeast Brazil the economic heart of the nation. Railroads and roads were constructed, and today the region has an excellent transportation system.

The plantation culture of the Northeast was already crumbling by the 1870s, and the growth of the movement to abolish slavery threatened it even more. The slave trade had been abolished in 1850, and a law for gradual emancipation was passed in 1871. In 1888, while Pedro II was in Europe and his daughter Isabel was governing Brazil, slavery was completely abolished. The planters thereupon withdrew their support of the empire, enabling republican forces, aided by a military at odds with the emperor, to triumph. By a bloodless revolution in 1889 the republic was established. The rivalry of the states and the power of the army in government caused the political situation to remain uneasy. The expanding market for Brazilian coffee and more particularly the wild-rubber boom brought considerable wealth as the 19th cent. ended, but the creation of rubber plantations in the Far East brought the wild-rubber boom to a halt and hurt the economy of the Amazon region after 1912. Brazil sided with the Allies in World War I, declaring war in Oct., 1917, and shared in the peace settlement, but later (1926) it withdrew from the League of Nations.

Measures to reverse the country's growing economic dependence on coffee were taken by Getúlio Vargas, who came into power through a revolution in 1930. By changing the constitution (notably in 1937) and establishing a type of corporative state he centralized government (the *Estado Novo*—new state) and began the forced development of basic industries and diversification of agriculture. His dictatorial rule, although it aroused much opposition, reflected a new consciousness of nationality. World War II brought a new boom (chiefly in rubber and minerals) to Brazil, which joined the Allies on Aug. 22, 1942, and took a large part in inter-American affairs. In 1945 the army forced Vargas to resign, but Brazil's economic growth was plagued by inflation, and this issue enabled Vargas to be elected in 1950. His second administration was marred by economic problems and corruption, and in 1954 he resigned and committed suicide. Juscelino Kubitschek was elected president in 1955. Under Kubitschek the building of Brasília and an ambitious program of highway and dam construction were undertaken. The inflation problem persisted. In 1960 Jânio Quadros was elected by the greatest popular margin in Brazilian history. But his autocratic manner and reform program aroused great opposition, and he resigned within seven months. Vice President João Goulart was the legal successor. Military leaders and conservatives opposed to him forced constitutional changes creating a parliamentary government and weakening the presidency (1961). In 1963, however, full presidential powers were restored by plebiscite.

Weakened by political strife and seemingly insurmountable economic chaos, the leftist administration of Goulart demanded radical constitutional changes. In 1964 a military insurrection deposed Goulart. Goulart's supporters and other leftists were removed from power and influence throughout Brazil, and the new president was given far-reaching powers. In 1965, after anti-military forces won elections in two states, the president's extraordinary powers were extended, and all political parties were dissolved. A new constitution was adopted in 1967. In 1968, in the face of student protests and criticism from the church against the military regime, President Costa e Silva recessed Congress and assumed one-man rule. Terrorism of the right and left (several diplomats were captured by leftist guerrillas) became a feature of Brazilian life. Government security forces conducted an intensive campaign that resulted in the death or exile of a number of the more prominent guerrilla leaders. The policies of the military were continued under the successor regime of Emilio Garrastazú Medici (1969-74). Despite minor periodic disturbances the power was passed to president Ernesto Geisel in early 1974.

Several political factions had hoped that the nationwide 1976 municipal elections might lead to a return to full democracy. In early 1977, after some squabbles with Congress over a constitutional reform of the judiciary, president Geisel suspended Congress on Apr. 1 and enacted the reform by decree on Apr. 13. Additional constitutional changes were made subsequently, including indirect elections for state governors (elected by each state's electoral college) and for one third of the senate. These measures evoked widespread criticism from the press and the opposition party. In early 1978 president Geisel announced the selection of Gen. Joao Batista de Oliveira Figueiredo as his designated successor. Gen. Figueiredo was elected in Nov., 1978, and started his term of office in Mar., 1979. As an exception his term will run for six years; after that the presidential term reverts to a five-year period.

Government. Brazil is governed by the 1967 constitution, which has been amended frequently. Authority is vested in the president, who is elected for five years by an electoral college consisting of members of congress and the state legislatures. The bicameral congress is popularly elected, except for one third of the senate which is elected by electoral colleges. All 66 senators serve for eight years and are elected in rotation. The 310 deputies serve for four years. The president may unilaterally intervene in state affairs, although each state has its own governor and legislature. ◊ *See Appendix.*

Brazil, city (1970 pop. 8,163), seat of Clay co., W Ind., ENE of Terre Haute. A railroad center in a farm and mining area, it produces coal and makes clay products.

Bra·zo·ri·a (brə-zôr'ē-ə), county (1970 pop. 108,312), 1,423 sq mi (3,685.6 sq km), SE Texas, on the Gulf of Mexico and drained by the Brazos and San Bernard rivers; formed 1836; co. seat Angleton. It has agriculture (cotton, corn, rice, wheat, soybeans, fruit, truck crops, livestock, and dairy products), oil, sulfur, and natural-gas deposits, and an extensive chemical industry.

Braz·os (brăz'əs), county (1970 pop. 57,978), 583 sq mi (1,510 sq km), SE Texas, bounded on the E by the Navasota River, on the SW by the Brazos River; formed 1841; co. seat Bryan. It is in a rich agricultural area yielding cotton, corn, alfalfa, peanuts, fruit, truck crops, dairy products, and livestock. Fuller's earth is mined.

Brazos, river, 870 mi (1,400 km) long (1,210 mi/1,946.9 km long with its main tributary), rising in E N.Mex. From its source it flows southeast across Texas to enter the Gulf of Mexico at Freeport.

Braz·za·ville (brăz'ə-vĭl', brä-zä-vēl'), city (1974 est. pop. 289,700), capital of the People's Republic of the Congo, on Stanley (Malebo) Pool of the Congo River. It is the nation's largest city and its administrative, communications, and economic center. The chief industries are beverage processing, tanning, and the manufacture of construction materials, matches, and textiles. There are also machine shops. The city was founded in 1880 by Savorgnan de Brazza, the French explorer. It was the capital of French Equatorial Africa from 1910 to 1958 and was the center of Free French forces in Africa during World War II. The city's main growth began after 1945.

Bre·a (brā'ə), city (1975 est. pop. 21,599), Orange co., S Calif.; inc. 1917. It is an industrial, commercial, and residential community in an oil and citrus-fruit area.

Breath·itt (brĕth'ĭt), county (1970 pop. 14,221), 494 sq mi (1,279.5 sq km), E central Ky., in the Cumberlands, drained by the North and

Middle forks of the Kentucky River and several creeks; formed 1839; co. seat Jackson. In a mountainous agricultural area yielding poultry, cattle, apples, potatoes, tobacco, soybeans, corn, and truck crops, it has bituminous-coal mines and timber. It has been called "Bloody Breathitt" because of the many mountain feuds that occurred here.

Brèche de Ro·land (brĕsh' də rô-län'), narrow gorge (alt. 9,200 ft/2,806 m), Hautes-Pyrénées dept., SW France, in the Pyrenees. It leads into the Cirque de Gavarnie, a natural amphitheater. According to legend Roland, one of Charlemagne's knights, created the breach with his sword.

Breck·en·ridge (brĕk'ən-rĭj'). **1.** Town (1970 pop. 548), seat of Summit co., central Colo. on the Blue River just S of Gore Range and WSW of Denver. Once a trade center (pop. c.8,000) for a group of gold-mining camps, it still mines gold and silver. Livestock is raised. **2.** City (1970 pop. 4,200), seat of Wilkin co., W Minn., on the Red River opposite Wahpeton, N.Dak.; laid out 1857, inc. 1908. It is a railroad center in a farm area. **3.** City (1970 pop. 5,944), seat of Stephens co., N Texas, on Gonzales Creek W of Fort Worth, in a cattle and farm area; settled 1876, inc. 1919. Its industries grew largely after oil and natural gas were discovered.

Breck·in·ridge (brĕk'ĭn-rĭj'), county (1970 pop. 14,789), 554 sq mi (1,434.9 sq km), NW Ky., bounded on the NW by the Ohio River, here forming the Ind. border, and on the S by the Rough River; formed 1799; co. seat Hardinsburg. It is drained by South Fork Panther and Sinking creeks. In a rolling agricultural area yielding burley tobacco, corn, wheat, hay, livestock, and cotton, it has timber tracts, limestone quarries, asphalt deposits, and some manufacturing (especially textiles).

Brecks·ville (brĕks'vĭl'), city (1970 pop. 9,137), Cuyahoga co., NE Ohio, a suburb SSE of Cleveland, in a farm area; settled c.1811, inc. 1921.

Brec·on·shire (brĕk'ən-shîr', -shər), or **Brecon,** former county, S Wales. The county town was Brecknock. In a region that may have been inhabited during the Stone and Bronze ages, Breconshire was seized from the Welsh princes by the Normans in 1092. In 1974 Breconshire was divided among of the new nonmetropolitan counties of Gwent, Mid Glamorgan, and Powys.

Bre·da (brā-dä'), city (1977 est. pop. 152,000), North Brabant prov., S Netherlands, at the confluence of the Mark and Aa rivers. It is an industrial and transportation center; its manufactures include machinery, textiles, and canned foods. Breda was founded in the 11th cent. The city was besieged (1624–25) by the Spaniards under Ambrogio Spinola; the surrender of its heroic garrison is the subject of a famous painting by Velázquez.

Bre·genz (brā'gĕnts), city (1971 pop. 22,839), capital of Vorarlberg prov., extreme W Austria, on the Lake of Constance. It is a lake port and a winter sports center. Textiles, food products, and machinery are manufactured. Located on a site settled in the Bronze Age, Bregenz was chartered c.1200. Nearby is the Bregenz Forest, a densely wooded highland noted for its scenic beauty.

Brei·sach (brī'zäкн), town (1970 est. pop. 5,000), Baden-Württemberg, SW West Germany, on the Rhine River. Its manufactures include wine and paper. An old town, it has long been coveted because of its strategic location. Fortified by the Romans, who called it Mons Brisiacus, it became an imperial town in 1275. Bernhard of Saxe-Weimar took the town in 1638. Louis XIV secured it for France in the Peace of Westphalia (1648) and ceded it back to the emperor in the Treaty of Ryswick (1697). The French repeatedly captured Breisach during the 18th cent. but gave it to Baden in 1805.

Bre·men (brĕm'ən, brā'mən), city (1974 est. pop. 579,430), capital of the state of Bremen, N West Germany, on the Weser River. Known as the Free Hanse City of Bremen, it is West Germany's second-largest port and is a commercial and industrial center trading in cotton, wool, tobacco, and copper. Manufactures include ships, steel, machinery, electrical equipment, textiles, beer, and foodstuffs, including roasted coffee. Bremen is Germany's oldest port city. It was made an archbishopric in 845, and under Archbishop Adalbert (1043–72) it included all of Scandinavia, Iceland, and Greenland. The archbishops held temporal sway over a large area between the Weser and Elbe rivers, but the city of Bremen itself remained virtually independent as its importance grew. In 1358 it became one of the leading members of the Hanseatic League. It accepted the Reformation in 1522, and in 1646 it was made a free imperial city. It stubbornly fought to preserve this status after the archbishopric had been assigned to Sweden by the Peace of Westphalia and later was ceded (1719) by Sweden to the elector of Hanover (George I of England). Bremen was occupied by France from 1810 to 1813. The city joined the German Empire in 1871. After World War I there was a short-lived (1918–19) socialist republic of Bremen. The city was badly damaged by bombs during World War II, but numerous historic monuments remain, including the Gothic city hall (1405–9); the statue of Roland, the medieval hero, which was erected in 1404 as a symbol of the city's freedom; and the cathedral (begun 1043), a blend of Romanesque and Gothic styles. The state of Bremen (1974 est. pop. 805,000), 156 sq mi (404 sq km), was formed in 1947 by combining Bremen and Bremerhaven.

Bre·mer (brē'mər), county (1970 pop. 22,737), 439 sq mi (1,137 sq km), NE Iowa, in a rolling prairie agricultural area drained by the Cedar, Wapsipinicon, and Shellrock rivers; formed 1851; co. seat Waverly. Corn, oats, soybeans, dairy products, and livestock are important. It also has limestone quarries, sand and gravel pits, and some industry.

Brem·er·ha·ven (brĕm'ər-hā'vən, brā'mər-hä'fən), city (1974 est. pop. 144,529), in the state of Bremen, N West Germany, at the mouth of the Weser River, near the North Sea. It is one of the largest fishing ports in Europe and is a major passenger and freight port. Founded in 1827, Bremerhaven in 1939 was absorbed by Wesermünde. In 1947 the combined municipality was renamed Bremerhaven and returned to the state of Bremen. The first regular ship service between continental Europe and the United States was started in Bremerhaven in 1847.

Brem·ers·dorp (brĕm'ərz-dôrp): *see* Manzini, Swaziland.

Brem·er·ton (brĕm'ər-tən), city (1970 pop. 35,307), Kitsap co., NW Wash., an excellent harbor on an arm of Puget Sound; inc. 1901. The city was platted (1891) when the area was selected as the site for the U.S. Puget Sound Naval Shipyard. There are also some logging and wood-product enterprises, and tourism is important. Bremerton is the gateway to the Olympic Peninsula, with easy access to the Cascade and Olympic mts. It is surrounded on three sides by water, and numerous ferries ply the inland seas of Puget Sound, linking the city to nearby resort islands. The U.S.S. *Missouri*, docked here, is a national shrine; it was the scene of the official Japanese surrender at the end of World War II.

Bren·ham (brĕn'əm), city (1970 pop. 8,922), seat of Washington co., S central Texas, NW of Houston, in an oil area; founded 1844. A center of rich Brazos River valley plantations before the Civil War, it later became a farm, dairy, and egg market with some manufacturing. Blinn College is here.

Bren·ner Pass (brĕn'ər), Alpine pass, 4,495 ft (1,371 m) high, connecting Innsbruck, Austria, with Bolzano, Italy. The lowest of the principal Alpine passes, it was an important Roman route through which many invasions of Italy were made. A long carriage road was built c.1772, and the railroad was completed in 1867. The pass became the border between Italy and Austria after World War I.

Brent (brĕnt), borough (1971 pop. 278,541) of Greater London, SE England. Brent was created in 1965 by the merger of the municipal boroughs of Wembly and Willesden. The area is a rail and industrial center. At Wembly is a large sports stadium that was originally built for the British Empire Exposition of 1924–25.

Brent·wood (brĕnt'wŏŏd'), urban district (1971 pop. 57,976), Essex, SE England. It is mainly residential but produces some agricultural equipment, film, and prefabricated concrete. Brentwood was on an important coach road from London to Colchester; the 15th cent. White Hart Inn remains standing.

Brentwood. 1. City (1970 pop. 11,248), St. Louis co., E Mo., a residential suburb W of St. Louis; inc. 1919. Its manufactures include pencils, leather goods, women's apparel, hospital and pharmaceutical supplies, and plastic products. **2.** Uninc. town (1970 pop. 47,100), Suffolk co., SE N.Y., on central Long Island, in the town of Islip. It is mainly residential, with some light industry. **3.** Borough (1970 pop. 13,732), Allegheny co., W Pa., a residential suburb of Pittsburgh; inc. 1915. There is some light industry.

Bre·scia (brā'shä), city (1971 pop. 210,067), capital of Brescia prov., Lombardy, N Italy. It is a commercial and industrial center and a railroad junction. Manufactures include machinery, firearms, textiles, and processed food. A Gallic town, it later became a Roman stronghold (1st cent. B.C.) and then the seat of a Lombard duchy. In the 12th cent. it was made an independent commune. It subsequently fell under the domination of a long series of outside powers (including Verona, Milan, Venice, and Austria), until it united with Italy in 1860. Of note in Brescia are Roman remains; the Romanesque Old Cathedral (11th cent.); the baroque New Cathedral (17th cent.); the Lombard-Romanesque Church of San Francesco; and a Renaissance-style city hall.

Bres·lau (brĕs'lou, brĕz'-): *see* Wrocław, Poland.

Bres·sa·no·ne (brä'-sä-nô'nä) or *German* **Bri·xen** (brĭk'sən), town (1971 pop. 16,025), Trentino-Alto Adige, N Italy, on the Brenner Road, and at the confluence of the Isarco and Rienza rivers. Bressanone was ruled by prince-bishops from the 11th cent. In 1803 it passed to Austria as a part of the Tyrol. In 1919 it became part of Italy. It retains a mixed German and Italian population.

Bresse (brĕs), region, in Burgundy, E France, between the Ain and Saône rivers. A fertile farm area, it is famous for its chickens and wines. Bresse was part of the duchy of Savoy until 1601, when it was ceded to France and added to Burgundy prov.

Brest (brĕst), city (1968 pop. 159,857), Finistère dept., NW France, on an inlet of the Atlantic Ocean. It is a commercial port and an important naval station. Electronics equipment and clothing are the chief manufactures. The city dates from Gallo-Roman times. The spacious, landlocked harbor was created in 1631 by Cardinal Richelieu as a military base and arsenal. In 1683, during the reign of Louis XIV, Marshal Vauban built the ramparts and a castle. The French repulsed the English in 1694 off Brest; in 1794 the English, under Lord Howe, defeated the French fleet. During World War II the Germans had a huge submarine base at Brest. Their heavily fortified submarine pens showed few cracks under Allied air raids; but the city itself was almost completely destroyed. The German garrison capitulated to U.S. troops in 1944.

Brest formerly **Brest-Li·tovsk** (brĕst'lĭ-tôfsk'), city (1970 pop. 122,000), capital of Brest oblast, W European USSR, in Belorussia, at the confluence of the Western Bug and Mukhavets rivers near the Polish border. It is a major industrial, commercial, and transportation center. Industries include shipbuilding, food processing, and the production of metals, textiles, and electrical machinery. Founded by Slavs in 1017, the city was conquered by the Mongols in 1241 and by Lithuania in 1319. In 1569 it became capital of the newly merged Polish and Lithuanian state. Brest passed to Russia in the third partition of Poland (1795). German forces took the city in 1915 and three years later signed the Treaty of Brest-Litovsk with Soviet Russia. Held by Poland between the world wars, Brest was regained by the USSR in 1939, occupied by Germany from 1941–44, and finally liberated by the Soviet army.

Bre·vard (brə-värd'), county (1970 pop. 230,006), 1,011 sq mi (2,618.5 sq km), central Fla., bounded on the E by the Atlantic and drained by the St. Johns River; formed 1844; co. seat Titusville. It is in a lowland region bordered by the barrier beaches enclosing Merritt Island and the Indian River and Banana River lagoons. Citrus fruit and truck crops are grown. Tourism, fishing, and the aerospace industry are also important. It includes John F. Kennedy Space Center, Cape Canaveral, site of NASA rocket launchings, Patrick Air Force Bace, and sections of Canaveral National Seashore and the Astronauts Trail.

Brevard, resort town (1970 pop. 5,243), seat of Transylvania co., SW N.C., in the Blue Ridge SSW of Asheville and near the S.C. line; inc. 1867. Leather, plastic, and paper products are made, and silicon is processed. Brevard has a junior college and an annual music festival.

Brew·er (bro͞o'ər), industrial city (1970 pop. 9,300), Penobscot co., S Maine, on the Penobscot River opposite Bangor; settled 1770, inc. as a town 1812, as a city 1889. It flourished as a shipbuilding center and a port that exported lumber and ice during the 19th cent. Brewer now has pulp and paper mills.

Brew·ster (bro͞o'stər), county (1970 pop. 7,780), 6,208 sq mi (16,078.7 sq km), extreme W Texas, bounded on the S by the Big Bend of the Rio Grande, here forming the Mexican border, and drained by numerous rivers; formed 1887; co. seat Alpine. In a rugged mountainous area, it has cattle, sheep, goat, and horse ranches, tourist facilities, some irrigated agriculture, and copper, sulfur, sand and gravel, and fluorite deposits. Big Bend National Park is here.

Brewster, village (1970 pop. 54), seat of Blaine co., central Nebr., NW of Grand Island on the North Loup River, in a livestock and grain area.

Brew·ton (bro͞ot'n), city (1970 pop. 6,747), seat of Escambia co., S Ala., SW of Andalusia near the Fla. border; settled 1861, inc. 1885. In a cotton, timber, and livestock area, it makes textile and lumber products.

Bri·ar·cliff Manor (brī'ər-klĭf'), residential village (1970 pop. 6,521), Westchester co., SE N.Y., SE of Ossining; settled 1896, inc. 1902.

Bric·es Cross Roads National Battlefield Site (brī'sĕz, -sĭz): *see* National Parks and Monuments Table.

Bridg·end (brĭj'ĕnd'), urban district (1971 est. pop. 14,531), Glamorganshire, S Wales, on the Ogwr River. It is a cattle market and manufactures electric transformers and pharmaceuticals. There are remains of an ancient Norman castle and a 12th cent. fortified church.

Bridge·port (brĭj'pôrt'). **1.** Village (1970 pop. 500), seat of Mono co., E Calif., NW of Bishop near the Nev. border, in a mining, stock-raising, and resort area. Bridgeport Reservoir and Bodie Historical State Park are nearby. **2.** City (1970 pop. 156,542), Fairfield co., SW Conn., on Long Island Sound; inc. 1836. It is a port of entry and the chief industrial city in the state. Its manufactures include electrical appliances and equipment, firearms, ammunition, helicopters, gas turbine engines, metal products, trucks, building materials, and aerosol products. Bridgeport was settled in 1639 and grew as a fishing community. The Barnum Institute of Science and History commemorates the showman P. T. Barnum, who lived in Bridgeport and whose circus wintered here. **3.** City (1970 pop. 1,490), seat of Morrill co., W Nebr., on the North Platte River SE of Scottsbluff. It is a trade center for an irrigated grazing, alfalfa, and sugar-beet area. Chimney Rock National Historic Site is nearby. **4.** Borough (1970 pop. 5,630), Montgomery co., SE Pa., NW of Philadelphia; settled 1829, inc. 1851. Dolomite is mined here, and textiles and metal products are manufactured.

Bridge·ton (brĭj'tən). **1.** City (1970 pop. 19,992), St. Louis co., E Mo., on the Missouri River; settled c.1765, inc. 1843. Refrigerators are among its manufactures. **2.** City (1970 pop. 20,435), seat of Cumberland co., S N.J., on the Cohansey River; settled 1686, inc. 1865. Once a rural farm center, it is now highly industrialized, with glassworks, fertilizer plants, and food-processing, textile, and garment industries. Bridgeton's downtown is Victorian in appearance, but the city has several 18th cent. buildings, including Potter's Tavern (recently restored) and a Presbyterian church (1792). The city's liberty bell, now in the county courthouse lobby, rang on July 7, 1776, for the reading of the Declaration of Independence.

Bridge·town (brĭj'toun'), city (1970 pop. 8,868), capital, commercial center, and chief port of Barbados, West Indies. It is, in addition, a tourist and health resort. Sugar, rum, and molasses are the leading exports. The city was founded by the British in 1628.

Bridge·ville (brĭj'vĭl'), industrial borough (1970 pop. 6,717), Allegheny co., SW Pa., SW of Pittsburgh; inc. 1901. Metal products, chemicals, beverages, and paint are made.

Bridge·wa·ter (brĭj'wô'tər), town (1970 pop. 11,829), Plymouth co., E Mass.; inc. 1656. Its iron industry dates from colonial times.

Bridg·wa·ter (brĭj'wô'tər), municipal borough (1971 pop. 26,598), Somerset, SW England, on the Parrett River estuary. It is a port for seaborne traffic and a market town. Bridgwater is the only place in England that produces bathbricks, which are made from clay and sand deposited by the river and are used for scouring metals.

Brid·ling·ton (brĭd'lĭng-tən, bûr'-), municipal borough (1971 pop. 26,729), Humberside, NE England. It has a well-protected harbor on Bridlington Bay, and its beaches and pavilions make it a popular holiday resort. The Royal Yorkshire Yacht Club has its headquarters here. Bridlington is an ancient market town and port. An Augustinian priory founded during the reign of Henry I has been restored. Of interest are Roman and early British remains.

Brie (brē), region, Marne and Seine-et-Marne depts., N France, E of Paris. Rich in wheat and cattle, it is famous for Brie cheese.

Bri·er·ley Hill (brī'ər-lē), urban district (1971 pop. 56,377), Staffordshire, W central England, near Birmingham. Its industries include food processing and the manufacture of steel, glass, ceramics, and metal products. The parish of Kingswinford, included in the district, dates from the 11th cent.

Brig·an·tine (brĭg'ən-tēn'), resort city (1970 pop. 6,741), Atlantic co., SE N.J., on the Atlantic coast NE of Atlantic City, on an island; inc. 1924.

Brig·ham City (brĭg'əm), city (1970 pop. 14,007), seat of Box Elder co., N Utah; inc. 1869. It is the center of a large farm area served by the Ogden River project. Sheep, cattle, wheat, sugar beets, garden crops, and orchard fruit are raised. The city has woolen mills, granaries, and food-processing plants, and a sugar refinery is nearby. It was founded as Box Elder in 1851, and its name was changed to honor Brigham Young in 1856. Golden Spike National Historic Site, which marks the spot in which the last railroad spike was driven in 1869, is nearby.

Brig·house (brĭg'hous'), municipal borough (1971 pop. 34,111), West Yorkshire, N central England, on the Calder River. It is a center of woolen, cotton, and silk milling and produces carpets, leather goods, machinery, radio and television equipment, dyes, and soap. Stone quarries are nearby. Also in the vicinity is the traditional grave of Robin Hood.

Brigh·ton (brīt'n), county borough (1971 pop. 166,081), East Sussex, SE England. The largest and most popular resort in southern England, Brighton also has engineering works and factories that manufacture office machinery, machine tools, electrical apparatus, vacuum cleaners, shoes, and paint. Formerly a small fishing village, it became a fashionable resort and was patronized, starting in 1783, by the Prince of Wales (later George IV), who built the Royal Pavilion. The Univ. of Sussex is in Brighton.

Brighton, city (1970 pop. 8,309), seat of Adams co., N central Colo., N of Denver; inc. 1887. It is a shipping and processing center in a rich farm area and has a beet-sugar plant.

Brigue and Tende (brēg; täɴd) or *Italian* **Bri·ga-Ten·da** (brē'gä-tĕn'-dä), two small districts (1968 pop. 2,726), Alpes-Maritimes dept., SE France, on the French-Italian border. With several smaller frontier areas in the Mont Cenis and Mont Blanc regions, they were ceded to France by Italy in 1947 after a referendum.

Brin·da·ban (brĭn'də-bŭn'): *see* Vrindaban, India.

Brin·di·si (brĭn'də-zē, brēn'dē-zē), city (1971 pop. 79,784), capital of Brindisi prov., in Apulia, S Italy. A modern port on the Adriatic Sea, it has been noted since ancient times for its traffic with the eastern Mediterranean. Manufactures include petrochemicals, plastics, and food products. Its excellent harbor was a Roman naval station (known as Brundisium), a chief embarkation point for the Crusaders (12th-13th cent.), and an important Italian naval base in World War I. One of the two columns marking the terminus of the Appian Way still stands.

Bris·bane (brĭz'băn, -bən), city (1977 est. pop. 722,700; urban agglomeration pop. 940,800), capital of Queensland, E Australia, on the Brisbane River above its mouth on Moreton Bay. It has shipyards, oil refineries, food-processing plants, textile mills, automobile plants, and railroad workshops. Principal exports are wool, meat, fruit, sugar, and coal and other minerals. The area was settled in 1824 as a penal colony. In 1925 the Greater Brisbane Act unified the administration of 19 formerly separate localities.

Bris·coe (brĭs'kō), county (1970 pop. 2,794), 874 sq mi (2,263.7 sq km), NW Texas, bounded on the E by the Cap Rock escarpment of the Llano Estacado and drained by Tule Creek and the Prairie Dog Town Fork of the Red River; formed 1876; co. seat Silverton. It has cattle ranches, agriculture (wheat, grain sorghums, barley, oats, alfalfa, cotton, peas, some fruit, truck crops, dairy products, sheep, and poultry), and deposits of clay and fuller's earth.

Bris·tol (brĭs'təl), county borough (1976 est. pop. 416,300), Avon, SW England, at the confluence of the Avon and Frome rivers. Bristol is a leading international port. Automobiles, tractors, machinery, clay, chemicals, coke, and tea are exported from Bristol; wine, grain, petroleum, tobacco, dairy products, fruit, and lumber are imported. General and nuclear engineering and the design and manufacture of aircraft are the largest industries; others are flour milling, printing, and the manufacture of paper, footwear, and tobacco products.

Bristol has been a trading center since the 12th cent. First chartered as a city in 1155, it became a separate county by order of Edward III in 1373, the first provincial town to receive this honor. During the reign of Edward III the manufacture of woolen cloth was developed. In the 18th cent. Bristol was active in the colonial triangular trade: English goods went to Africa; African slaves to the West Indies; and West Indian sugar, rum, and tobacco to Bristol. In 1838 the *Great Western,* one of the first transatlantic steamships, was launched from Bristol. The port declined during the late 18th and early 19th cent. because of competition from Liverpool, the end of slave trading, and the decline of the West Indian trade. It revived in the mid-19th cent. Points of interest include the 14th cent. church of St. Mary Redcliffe, known for its fine architecture; a 14th cent. cathedral (rebuilt 1868–88) with a Norman chapter house and gateway; the Merchant Venturers' Almshouses; University Tower; and some notable examples of Regency architecture.

Bristol. 1. County (1970 pop. 444,301), 556 sq mi (1,440 sq km), SE Mass., on Buzzards Bay and the Atlantic, bounded on the W by R.I. and intersected by the Taunton River; formed 1685; co. seat Taunton. Its early activities of whaling and shipping gave way to textile milling after the mid-19th cent. There are many resorts along the coastline. **2.** County (1970 pop. 45,937), 25 sq mi (64.8 sq km), E R.I., bounded on the SW by Narragansett Bay, on the NE by the Mass. border, and on the SE by Mt. Hope Bay; inc. 1747; co. seat Bristol. It has resorts, some agriculture (dairy products, poultry, and truck crops), fishing and shipbuilding industries, and diversified light manufacturing.

Bristol. 1. Industrial city (1970 pop. 55,487), Hartford co., central Conn., on the Pequabuck River; settled 1727, inc. 1785. Its clock-making industry dates from 1790. It also has a steel mill and plants that make ball bearings, mechanical springs, electric and electronic equipment, paper boxes, and a great variety of metal parts. The American Clock and Watch Museum is in the city. **2.** City (1970 pop. 626), seat of Liberty co., NW Fla., on the Apalachicola River W of Tallahassee. Lumber is processed. **3.** Industrial borough (1970 pop. 12,085), Bucks co., SE Pa., on the Delaware River opposite Burlington, N.J.; settled 1697, inc. 1720. Its many manufactures include paper, chemicals, textiles, aircraft parts, and metal products. It was once a busy river port with important shipbuilding activities. Among its historic structures is the Friends Meetinghouse, built c.1710. A restoration of 17th and 18th cent. buildings is to the north, and a replica of William Penn's country manor is to the northeast. **4.** Town (1970 pop. 17,860), seat of Bristol co., E R.I., a port of entry on Narragansett Bay; inc. as a Plymouth Colony town 1681, ceded to Rhode Island 1746. An early center of commercial trade, the port was (18th–19th cent.) a base for slave trading, privateering, whaling, and shipbuilding. Bristol is still a yachting and yacht-building center and has a large rubber industry. King Philip's War (1675–76) began and ended on the site of the town, and a monument on Mt. Hope marks the spot where King Philip fell. The Haffenreffer Museum of Anthropology has notable collections of Indian relics. On Hope St. is a row of preserved colonial homes. In 1938 and 1954 hurricanes caused heavy damage to the shore and harbor. **5.** Industrial city on the Tenn.-Va. line, Sullivan co., Tenn. (1970 pop. 20,064), independent and in no county in Va. (1970 pop. 14,857); settled 1749 as Sapling Grove, inc. as separate towns 1856, as Bristol city 1890. The two cities, although separate municipalities, are economically a unit that is the transportation and processing center of a mountainous region producing tobacco, coal, and livestock. Shelby's Fort, built in 1771, was frequented by Daniel Boone and other early pioneers. Two hundred years of controversy preceded the location of the state line down the middle of State Street. King College is in the Tenn. section of the city, and Virginia Intermont College is on the Va. side. Bristol Caverns, largest in the Smoky Mts., are nearby.

Bristol A·von (ā'vən, ăv'ən), river, England: *see* Avon.

Bristol Bay, borough (1970 pop. 1,147), 531 sq mi (1,375.3 sq km), SW Alaska, on Bristol Bay, an arm of the Bering Sea, between the Alaskan mainland and Alaska Peninsula; formed after 1961; borough seat Naknek. It is in a salmon-fishing region and has canneries.

Bristol Channel, inlet of the Atlantic Ocean, c.85 mi (135 km) long and from 5 to 50 mi (8–80.5 km) wide, stretching W from the mouth of the River Severn and separating Wales from SW England. Along the coast of southern Wales there is a great concentration of economic activity, and the channel serves as a major shipping corridor.

Brit·ain (brĭt'n), alternate name for the United Kingdom of Great Britain and Northern Ireland. It is derived from *Britannia,* the name given by the Romans to the portion of the island of Great Britain that they occupied.

Brit·ish Co·lum·bi·a (brĭt'ĭsh kə-lŭm'bē-ə), province (1975 pop. 2,457,000), 366,255 sq mi (948,600 sq km), including 6,976 sq mi (18,068 sq km) of water surface, W Canada; capital Victoria. British Columbia, the westernmost province of Canada, is bounded on the east by Alta., on the south by Mont., Idaho, and Wash., on the west by the Pacific Ocean, on the northwest by Alaska, and on the north by the Yukon and by the Mackenzie dist. of the Northwest Territories. Off its deeply indented Pacific coast lie many islands, notably Vancouver Island and the sparsely inhabited Queen Charlotte Islands.

The province is almost wholly mountainous, with the Rocky Mts. in the southeast, the Coast Mts. along the Pacific, and the Stikine Mts. in the northwest. Chief of the many rivers is the Fraser; others include the upper Columbia, the Kootenay, the Peace, the Stikine, the Nass, and the Skeena. Hydroelectric power in British Columbia is highly developed; the largest station, at Kemano on the Nechako River, serves one of the biggest aluminum plants in the world, at Kitimat. Large areas of central and northern British Columbia are sparsely settled, almost three fourths of the population crowding the southwest coastal tip. Less than 10% of the province can be used for grazing or cultivation, since nearly three fourths of the land is covered with forests. British Columbia attracts millions of visitors annually, and the land is a hunting and fishing paradise. There are four national parks—Glacier, Mt. Reveistoke, Yoho, and Kootenay—and hundreds of provincial parks and camping grounds.

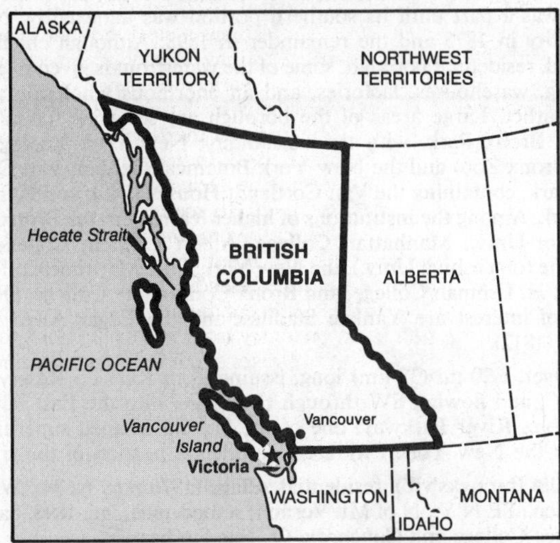

Economy. Lumbering and related enterprises (such as pulp and paper manufacturing) are the province's major industries. Next in importance is mining. The silver mine at Kimberley is the largest in the world; copper, gold, iron ore, lead, and zinc are also found in large quantities. British Columbia ranks first among the provinces in fishing; the most important catches are salmon, halibut, and herring. Beef is also an important product. Cattle are raised along the Fraser River on sprawling ranches. Other industries include food processing and the manufacture of chemicals, furniture, transportation equipment, and electrical items.

History. The area was originally inhabited by Indians of the Pacific Northwest. In 1778 Capt. James Cook, on his last voyage, explored the coast and claimed the area for Great Britain. Rival British and Spanish claims for the area were resolved by the Nootka Convention in 1790. The British sent George Vancouver to take possession of the land, and from 1792 to 1794 he explored and mapped the coast. Early in the 19th cent. fur traders and explorers of the North West Company crossed the mountains to establish posts in New Caledonia, as the region was then called. After the Hudson's Bay Company absorbed the North West Company in 1821, the region became a preserve of the new company. Rival British and American claims to the area were settled in 1846 when the boundary was set at the 49th parallel. Partly as protection against further American expansion, Vancouver Island was ceded (1849) by the Hudson's Bay Company and became a crown colony.

In 1858 gold was discovered in the sand bars of the Fraser River. The great gold rushes that resulted brought profound changes. Fort Victoria (est. 1843) boomed as a supply base for the miners, and a town quickly sprang up around it. Some 30,000 miners moved into what was then unorganized territory; this led to the creation (1858) of a new colony on the mainland, called British Columbia, and the end of the Hudson's Bay Company's supremacy. In 1863 the newly settled territory about the Stikine River was added. In 1866 Vancouver Island and British Columbia were combined, and in 1871 this united British Columbia voted to join the new Canadian confederation. The Canadian Pacific Railway finally reached Vancouver in 1885, and a new era began. By providing access to new markets, the railroads furthered agriculture, mining, and lumbering. Vancouver grew as a busy port, serving many provinces. A long dispute with the United States over the Alaska boundary was finally settled by the Alaska Boundary Commission in 1903.

British Commonwealth: *see* Commonwealth of Nations.

British East Af·ri·ca (ăf'rĭ-kə), inclusive term for several former British dependencies, especially Kenya, Uganda, Tanganyika, and Zanzibar.

British Empire, overseas territories linked to Great Britain in a variety of constitutional relationships, established over a period of three centuries. At its height in the late 19th and early 20th cent., the empire included territories on all continents, comprising about one quarter of the world's population and area.

In 1931 the Statute of Westminister officially recognized the independent and equal status under the crown of the former dominions within a British Commonwealth of Nations. After World War II self-government advanced rapidly. In 1947 independence was granted to the new states of India and Pakistan. In 1948 the mandate over Palestine was relinquished, and Burma gained independence as a republic. Other parts of the empire, notably in Africa, gained independence and subsequently joined the British Commonwealth. In 1978 Great Britain still administered, as colonies, protectorates, or trust territories, many dependencies throughout the world. They included Brunei and Hong Kong in eastern Asia; the Gibraltar in the Mediterranean; the Falkland Islands, Bermuda, and St. Helena in the Atlantic; the Cayman Islands, British Virgin Islands, Turks and Caicos Islands, and several of the Leeward and Windward Islands in the West Indies; Belize in Central America; and Pitcairn Island, the Solomon Islands, the New Hebrides Islands (jointly with France), and the Gilbert and Ellice Islands in the Pacific.

British Gui·a·na (gē-ä'nə, -ăn'ə): *see* Guyana.

British Hon·du·ras (hŏn-do͝or'əs,-dyo͞o'-): *see* Belize.

British West Af·ri·ca (ăf'rĭ-kə), former inclusive term for the colonies of Cameroons, Gambia, Gold Coast, Nigeria, Sierra Leone, and Togoland.

British West In·dies (ĭn'dēz): *see* West Indies.

Brit·ta·ny (brĭt'ə-nē), region and former province, NW France. It is a peninsula between the English Channel on the north and the Bay of Biscay on the south. The coast, particularly at the western tip, is irregular and rocky, with natural harbors and numerous islands. The economy of the region is based on agriculture, fishing, and tourism. Apples are grown extensively inland. Industry includes shipbuilding, food processing, and automobile manufacturing. There is a nuclear power plant in the Arrée Mts. A part of ancient Armorica, the area was conquered by Julius Caesar and became part of the Roman province of Lugdunensis. It received its modern name when it was settled (c.500) by Britons whom the Anglo-Saxons had driven from Britain. Breton history is a long struggle for independence—first from the Franks (5th–9th cent.), then from the dukes of Normandy and the counts of Anjou (10th–12th cent.), and finally from England and France. In 1196 Arthur I, an Angevin, was acknowledged as duke. His death (1203) without a direct heir led to the War of the Breton Succession (1341–65), a part of the Hundred Years War (1337–1453). With the end of the Breton war, the dukedom was won by the house of Montfort. King Francis I formally incorporated the duchy into France in 1532. Breton nationalism grew in the 19th cent., but was successfully repressed by the French government. In recent years emigration has resulted in a serious decline in the region's population.

Brit·ton (brĭt'n), city (1970 pop. 1,465), seat of Marshall co., NE S.Dak., near the N.Dak border NNE of Aberdeen; settled 1883. It is the trade center of a grain, livestock, and dairy region.

Bri·xen (brĭk'sən): *see* Bressanone, Italy.

Br·no (bûr'nô), city (1970 pop. 335,918), central Czechoslovakia, at the confluence of the Svratka and Svitava rivers. It is the second-largest city of Czechoslovakia. Brno is an industrial center, known particularly for its woolen industry and for its manufacture of textiles, machinery (notably tractors), machine tools, and armaments. The famous Bren gun, later made in Enfield, England, was developed in Brno. Tourism is also important. Originally the site of a Celtic settlement, Brno became part of the kingdom of Bohemia in 1229. King Wenceslaus I made it a free city by royal decree in 1243, and Brno flourished in the 13th and 14th cent. The city was besieged in 1645 by the Swedes and served as headquarters for Napoleon I during the Battle of Austerlitz in 1805. In the 19th cent. Brno became one of the foremost manufacturing towns of the Austrian empire. Masaryk Univ. (founded 1919), Beneš Technical College, a music conservatory, and several fine museums are located in the city.

Broach (brōch), town (1971 pop. 92,263), Gujarat state, W India, on the Gulf of Cambay. A port at the mouth of the Namada River, Broach ships cotton and timber and manufactures textiles. Broach was an important Buddhist center in the 7th cent.

Broad (brôd), river, c.150 mi (240 km), rising on the E slope of the Blue Ridge SE of Asheville, N.C., and flowing SE into S.C., then S to Columbia, where it joins the Saluda River to form the Congaree River. It is a source of hydroelectric power.

Broads, the (brôdz), region, c.5,000 acres (2,025 hectares), mainly in Norfolk, E England, extending inland to Norwich from the coast. It is composed of wide, interlocking shallow lakes. The Broads is a vacation center and wildlife sanctuary.

Broad·stairs and Saint Pe·ter's (brôd'stârz'; sānt pē'tərz), urban district (1971 pop. 19,996), Kent, SE England. It is a residential area and resort and was once a retreat of Charles Dickens, whose residence here is now called Bleak House.

Broa·dus (brô'dəs), town (1970 pop. 799), seat of Powder River co., SE Mont., on Powder River SSE of Miles City. Coal is mined. It is also a trading point for an agricultural region producing livestock, grain, and wool.

Broad·view (brôd'vyōō'), residential village (1970 pop. 9,623), Cook co., NE Ill., a suburb W of Chicago; settled 1835, inc. 1913. Electrical equipment and metal products are manufactured.

Broadview Heights, village (1970 pop. 11,463), Cuyahoga co., NE Ohio, a suburb of Cleveland; inc. 1926.

Broad·wa·ter (brôd'wô'tər), county (1970 pop. 2,526), 1,193 sq mi (3,089.9 sq km), W central Mont., drained by the Missouri River; formed 1897; co. seat Townsend. In a livestock and grain area, it has gold mines.

Broad·way (brôd'wā'), thoroughfare, New York City and New York State. The longest street in the world, it begins at Bowling Green near the tip of Manhattan Island and extends 150 mi (241.4 km) north to Albany. Within New York City, it passes through the Wall Street financial center, the garment center, the theater district, Lincoln center, and the residential Upper West Side. The street was laid out by the Dutch and was the principal street of New Amsterdam. It was extended northward as the colony grew.

Brock·en (brôk'ən), granite peak, 3,747 ft (1,142.8 m) high, W East Germany. It is the highest peak of the Harz Mts. Popular legend makes it the meeting place of the Walpurgis Night or Witches' Sabbath.

Brock·port (brŏk'pôrt'), village (1970 pop. 7,878), Monroe co., NW N.Y., on the Barge Canal WNW of Rochester, in a farm and food-processing region; inc. 1829. A campus of the State Univ. of New York is here.

Brock·ton (brŏk'tən), industrial city (1970 pop. 89,040), Plymouth co., E Mass.; settled c.1700, set off from Bridgewater 1821, inc. as a city 1881. It has a large shoe and leather products industry. Textiles and clothing, machinery and machine tools, plastics, and electrical and electronic equipment are also produced.

Brock·ville (brŏk'vĭl'), city (1971 pop. 19,765), SE Ont., Canada, on the St. Lawrence River. It is in a rich dairy region. The city's manufactures include telecommunications equipment, power tools, and baby foods. In summer it is a tourist resort.

Bro·ken Ar·row (brō'kən ăr'ō), city (1970 pop. 11,787), Tulsa co., NE Okla., a suburb of Tulsa.

Broken Bow (bō), city (1970 pop. 3,734), seat of Custer co., central Nebr., NW of Grand Island; settled 1880, inc. as a town 1882, as a city 1888. It is a shipping center for a livestock and grain area.

Broken Hill, city (1971 pop. 29,743), New South Wales, SE Australia, near the South Australia border. Since 1884 it has been a principal center of zinc and silver mining in Australia.

Brom·ley (brŭm'lē, brŏm'-), borough (1971 pop. 304,357) of Greater London, SE England. The borough was created in 1965 by the merger of the former municipal boroughs of Bromley and Beckenham, the urban districts of Orpington and Penge, and part of the urban district of Chislehurst and Sidcup. Bromley is mainly residential. The Crystal Palace, site of the 1851 Great Exhibition, was destroyed by fire in 1936.

Broms·grove (brŏmz'grōv'), urban district (1971 pop. 40,669), Hereford and Worcester, central England. Bromsgrove is an ancient market town and road junction. It is predominantly residential but has some industry.

Bron·son (brŏn'sən), town (1970 pop. 698), seat of Levy co., N Fla., SW of Gainesville. Lumber and naval stores are produced.

Bronx, the (brŏngks), borough of New York City, coextensive with Bronx co. (1970 pop. 1,471,701; 41 sq mi (106.2 sq km), SE N.Y.; settled 1641 by Jonas Bronck (a Dane acting for the Dutch West India Company), chartered as a part of Greater New York City 1898. The only mainland borough of New York City, it comprises the southern part of a peninsula bordered on the west by the Hudson River, on the southwest by the Harlem River (which separates it from Manhattan), on the south by the East River, and on the east by Long Island Sound. To the north is Westchester co., of which the Bronx was a part until its southern portion was annexed by New York City in 1875 and the remainder in 1898. Although chiefly a crowded, residential borough, some of the waterfront is given over to shipping, warehouses, factories, and an enormous wholesale produce market. Large areas of the borough are set aside for parks, notably Bronx Park, with the outstanding New York Zoological Park (Bronx Zoo) and the New York Botanical Garden; Van Cortlandt Park, containing the Van Cortlandt House (1748); and Pelham Bay Park. Among the institutions of higher learning in the Bronx are Fordham Univ., Manhattan College, Albert Einstein College of Medicine (of Yeshiva Univ.), the New York State Maritime College, Herbert H. Lehman College, and Bronx Community College. Other points of interest are Yankee Stadium and the Edgar Allan Poe cottage (1812).

Bronx, river, c.20 mi (32 km) long, issuing from Kensico Reservoir, SE N.Y., and flowing SW through the Bronx into the East River. The Bronx River Parkway, one of the first landscaped superhighways in the New York City area, parallels a portion of the river.

Bronx·ville (brŏngks'vĭl'), residential village (1970 pop. 6,674), Westchester co., SE N.Y., N of Mt. Vernon; settled 1664, inc. 1898. Sarah Lawrence College and Concordia College are here.

Brooke (brŏŏk), county (1970 pop. 29,685), 89 sq mi (230.5 sq km), N W.Va., in the Northern Panhandle bounded on the W by the Ohio River, here forming the Ohio border, on the E by the Pa. border; formed 1797; co. seat Wellsburg. It has bituminous-coal mines, some agriculture (fruit, grain, sheep, and poultry), and diversified manufactures of iron, steel, glass, paper, and chemical products.

Brook·field (brŏŏk'fēld'). **1.** Town (1970 pop. 9,688), Fairfield co., SW Conn., near the Housatonic River NNE of Danbury; inc. 1788. It is a resort in a truck-farming area. **2.** Village (1970 pop. 20,284), Cook co., NE Ill., a residential suburb of Chicago; inc. 1893. The noted Chicago Zoological Park is here. **3.** City (1970 pop. 5,491), Linn co., N central Mo., SW of Kirksville, in a farm and coal region; founded 1859. It is a railroad shipping point. Gen. John J. Pershing was born in nearby Laclede; his home is now a shrine. Pershing State Park is nearby. **4.** City (1970 pop. 32,140), Waukesha co., SE Wis., a suburb of Milwaukee; inc. 1954. It has iron foundries and light manufacturing.

Brook·ha·ven (brŏŏk'hā'vən). **1.** City (1970 pop. 10,700), seat of Lincoln co., SW Miss.; inc. 1859. It is situated in a dairy, timber, and farm area; nearby are oil and gas fields. The city's manufactures include mobile homes, electronic equipment, lawn mowers, and thermometers. **2.** Borough (1973 est. pop. 7,262), Delaware co., SE Pa., SW of Philadelphia; inc. 1945. Metal products are made.

Brook·ings (brŏŏk'ĭngz), county (1970 pop. 22,158), 800 sq mi (2,072 sq km), E S.Dak., on the Minn. border in an agricultural area drained by the Big Sioux River; formed 1862; co. seat Brookings. Dairy products, livestock, poultry, and corn are produced. It has a number of small lakes.

Brookings, city (1970 pop. 13,717), seat of Brookings co., E S.Dak., on the Big Sioux River; inc. 1883. A trade center in a livestock and grain region, the city is an important seed-processing point. Other industries produce medical and dental equipment; aluminum windows, doors, and awnings; concrete products; and fabricated structural steel.

Brook·line (brŏŏk'līn'), town (1970 pop. 58,689), Norfolk co., E

Mass., a residential suburb adjacent to Boston; settled 1630s, set off from Boston and inc. 1705. It was known as "Muddy River" when part of Boston. The birthplace of President John F. Kennedy in Brookline is a national historic site.

Brook·lyn (brŏŏk'lĭn). **1.** Town (1970 pop. 4,965), Windham co., E. Conn., ENE of Willimantic; settled c.1703, inc. 1786. It has a number of 18th cent. houses and a monument to Gen. Israel Putnam, whose farm and tavern are preserved. **2.** Uninc. city (1970 pop. 13,896), Anne Arundel co., central Md. **3.** Borough of New York City (1970 pop. 2,601,852), 71 sq mi (183.9 sq km), coextensive with Kings co., SE N.Y., at the SW extremity of Long Island; settled 1636, chartered as a part of Greater New York 1898. Brooklyn is a residential and industrial region, with the largest population of the city's five boroughs. Among its manufactures are machinery, textiles, paper products, and chemicals. The borough is the center of an important foreign and domestic commerce and has extensive waterfront facilities. Hollanders and Walloons settled about Gowanus and Wallabout bays in 1636 and 1637; about nine years later Dutch farmers established the hamlet of Breuckelen, near the present borough hall. Becoming Brooklyn under the English, it was incorporated as a village (Brooklyn Ferry) in 1816 and was chartered as a city in 1834. As it grew, Brooklyn absorbed many settlements and villages. In 1855 it became the third-largest city in the United States. In 1898, when it became a borough of New York City, its population was about one million. Among the numerous educational institutions in the borough are Brooklyn College, Polytechnic Institute of New York, Pratt Institute, and Long Island Univ. The New York Naval Shipyard was located on the East River from 1801 until its closing in the late 1960s, at which time the installation was turned over to private enterprise. Fort Hamilton (built 1831 as a harbor defense) overlooks the Narrows of New York Bay. Near beautiful Prospect Park, the scene of fierce fighting in the Revolution, is the main building of the Brooklyn Public Library. Also in that area are the Brooklyn Museum, with noted collections of Egyptian, Oriental, and primitive art; the Brooklyn Botanic Garden; and the Brooklyn Children's Museum. Among the many structures that give the borough its appellation "City of Churches" are the Reformed Protestant Dutch Church of Flatbush (first built 1654; rebuilt 1796), St. Ann's Episcopal Church (est. 1784), and Plymouth Church of the Pilgrims, where Henry Ward Beecher preached. Other points of interest in the borough include Coney Island, with its beach and amusement park; Sheepshead Bay, a fishing and boating center; the invaluable historical library of the Long Island Historical Society; the New York Aquarium (at Coney Island); Brooklyn Heights Historic District; and the Lefferts Homestead (1777). **4.** City (1970 pop. 13,142), Cuyahoga co., NE Ohio, a residential suburb of Cleveland; inc. 1867.

Brooklyn Center, city (1970 pop. 35,173), Hennepin co., SE Minn., a residential suburb of Minneapolis; inc. 1911. It has some diversified light industry.

Brooklyn Park, city (1970 pop. 26,230), Hennepin co., SE Minn., a suburb of Minneapolis; chartered as a city 1969. Potatoes are grown and wood products are made

Brook Park (brŏŏk'), city (1970 pop. 30,774), Cuyahoga co., NE Ohio, a suburb of Cleveland; inc. 1914.

Brooks (brŏŏks) **1.** County (1970 pop. 13, 743), 492 sq mi (1,274.3 sq km), S. Ga., bounded on the S by the Fla. line and on the E by the Little and Withlacoochee rivers; formed 1858; co. seat Quitman. It sustains coastal-plain agriculture (tobacco, cotton, corn, melons, peanuts, and livestock) and a lumber industry. **2.** County (1970 pop. 8,005), 908 sq mi (2,351.7 sq km), S Texas, in a diversified irrigated agricultural area; formed 1911; co. seat Falfurrias. It has oil and natural-gas fields and gypsum, clay, and slate deposits. Its agriculture includes livestock, hay, grain sorghums, vegetables, watermelons, and poultry.

Brooks Range, mountain chain, northernmost part of the Rocky Mts., extending c.600 mi (965 km) from E to W across N Alaska. Rugged, barren, snow-covered, and uninhabited, Brooks Range separates the oil-rich Arctic Ocean coastal plain from the Yukon River basin.

Brooks·ville (brŏŏks'vĭl'). **1.** City (1970 pop. 4,060), seat of Hernando co., W Fla., N of Tampa, in a timber, citrus-fruit, and limestone area; inc. 1925. **2.** City (1970 pop. 609), seat of Bracken co., N Ky., E of Maysville, in a farm and dairy area. Burley tobacco, corn, and wheat are grown.

Brook·ville (brŏŏk'vĭl'). **1.** Town (1970 pop. 2,864), seat of Franklin co., SE Ind., on the Whitewater River SSW of Richmond; settled 1804, platted 1808. It is a farm trade center. Mounds State Recreational Area is nearby. **2.** Industrial borough (1970 pop. 4,314), seat of Jefferson co., W Pa., NE of Pittsburgh; settled 1801, laid out 1730, inc. 1843. It is in an area yielding bituminous coal, natural gas, and oil. Steel products and radio tubes are manufactured.

Broome (brōōm, brŏŏm), county (1970 pop. 221,815), 710 sq mi (1,838.9 sq km), S N.Y., bounded on the S by the Pa. border and drained by the Susquehanna, Chenango, and Otselic rivers; formed 1806; co. seat Binghamton. It has extensive manufacturing and some agriculture. Chenango Valley State Park is here.

Broom·field (brōōm'fēld', brŏŏm'-), city (1970 pop. 7,261), Boulder and Jefferson cos., N central Colo., a suburb NNW of Denver, in a diversified agricultural area.

Brow·ard (brou'ərd), county (1970 pop. 620,100), 1,218 (3,154.6 sq km), S Fla., in the Everglades, bounded on the E by the Atlantic Ocean and intersected by drainage canals running from Lake Okeechobee to the coast; formed 1915; co. seat Fort Lauderdale. The coastal fringe is a tourist, truck-farming, and dairying area, with some citrus groves and limestone quarries. It has peat deposits. Part of a Seminole Indian reservation is here.

Brown (broun). **1.** County (1970 pop. 5,586), 307 sq mi (795.1 sq km), W Ill., bounded on the SE by the Illinois River, on the NE by the La Moine River; formed 1839; co. seat Mount Sterling. It has agriculture (livestock, corn, wheat, oats, and poultry) and bituminous-coal mines. Butter, cheese, and bricks are manufactured. **2.** County (1970 pop. 9,057), 324 sq mi (839.2 sq km), S central Ind., drained by Monroe Lake and Bean Blossom and Salt creeks; formed 1836; co. seat Nashville. Timber and agriculture (grain, tobacco, fruit, and livestock) are important. Brown County State Park and part of Hoosier National Forest are here. **3.** County (1970 pop. 11,685), 578 sq mi (1,497 sq km), NE Kansas, in a gently rolling corn-belt region, bordered on the N by Nebr. and watered by headstreams of the Delaware River; formed 1855; co. seat Hiawatha. It is agricultural, with grain, livestock, and poultry. **4.** County (1970 pop. 28,887), 610 sq mi (1,579.9 sq km), S Minn., in an agricultural area bounded on the N by the Minnesota River and drained by the Cottonwood and Little Cottonwood rivers; formed 1855; co. seat New Ulm. Corn, oats, and soybeans are grown, and cattle and hogs are raised. It has some manufacturing. **5.** County (1970 pop. 4,021), 1,218 sq mi (3,154.6 sq km), N Nebr., in an agricultural region bounded on the N by the Niobrara River and drained in the S by the Calamus River; formed 1883; co. seat Ainsworth. Livestock is raised, and corn is grown. It has a number of small lakes. **6.** County (1970 pop. 26,635), 490 sq mi (1,269.1 sq km), SW Ohio, bounded on the S by the Ohio River, here forming the Ky. border, and drained by the East Fork of the Miami, River; formed 1817; co. seat Georgetown. It has agriculture (livestock, tobacco, grain, poultry, and dairy products) and some manufacturing. Grant Schoolhouse and John Rankin House state memorials are here. **7.** County (1970 pop. 36,920), 1,677 sq mi (4,343.4 sq km), NE S.Dak., on the N.Dak. border in an agricultural region drained by the James River and numerous creeks; formed 1879; co. seat Aberdeen. Wheat, corn, barley, oats, and sunflowers are grown. There is some manufacturing. **8.** County (1970 pop. 25,877), 938 sq mi (2,429.4 sq km), central Texas, bounded on the S by the Colorado River and drained by Pecan Bayou; formed 1856; co. seat Brownwood. Ranching and agriculture (peanuts, oats, grain sorghums, cotton, wheat, pecans, fruit, and vegetables) are important. It also has oil and gas wells and some manufacturing and industry. Lake Brownwood, with a state park, is here. **9.** County (1970 pop. 158,244), 525 sq mi (1,359.8 sq km), E Wis., bounded on the N by Green Bay and drained by the Fox River; formed 1818; co. seat Green Bay. It is a processing and shipping center of a dairying and lumbering area with some manufacturing.

Brown Deer, village (1970 pop. 12,582), Milwaukee co., SE Wis., on the Milwaukee River; inc. 1955. It is a residential suburb north of Milwaukee.

Brown·field (broun'fēld'), town (1970 pop 9,647), seat of Terry co., NW Texas, in the Llano Estacado SW of Lubbock; inc. 1926. The town's growth has been spurred by an oil industry, cattle raising, and plants processing chemicals and food.

Browns·town (brounz'town'), town (1970 pop. 2,376), seat of Jackson co., S Ind., SE of Bloomington; laid out 1816. It has canneries and diverse manufacturing. Hoosier National Forest is nearby.

Browns·ville (brounz'vĭl'). **1.** City (1970 pop. 542), seat of Edmonson co., central Ky., near Green River just W of Mammoth Cave National Park and NE of Bowling Green, in an agricultural area. A game refuge is nearby. Many Indian artifacts have been found in the vicinity. **2.** Borough (1975 est. pop. 5,213), Fayette co., SW Pa., on the Monongahela River SSE of Pittsburgh; laid out 1785, inc. 1815. **3.** City (1970 pop. 7,011), seat of Haywood co., W Tenn., near the Hatchie River W of Jackson, in a farm and timber area; settled c.1810, inc. 1870. The city has cotton gins and sawmills. **4.** City (1970 pop. 52,522), seat of Cameron co., extreme S Texas, on the Rio Grande near its mouth at the Gulf of Mexico; inc. 1850. It is an important port of entry across the river from Matamoros, Mexico; a deepwater channel (completed 1936) accommodates ocean vessels. Brownsville is a trade, processing, and distributing point for the rich, irrigated lower Rio Grande valley, and has many industries, especially those connected with oil and natural gas. Other products include shrimp, electronic equipment, and aircraft parts. The establishment of Fort Texas here by Gen. Zachary Taylor in 1846 invited a Mexican attack that precipitated the Mexican War. The fort was renamed (1846) for Major Jacob Brown, killed while commanding its defense. Active until 1944, Fort Brown was held briefly by Union forces in the Civil War. The town of Brownsville grew around the fort and was a cattle-shipping point in the late 19th cent.

Brown·wood (broun'wŏŏd'), city (1970 pop. 17,368), seat of Brown co., central Texas; inc. 1876. It is an industrial community; its products include brick, clothing, glass, furniture, feather products, mobile homes, plastic pipe, food products, beverage cartons, concrete mixers, reflective products, sportswear, cable, and wire. Brownwood processes and ships pecans, peanuts, cattle, wool, poultry, and meat from the surrounding agricultural area.

Bru·ay-en-Ar·tois (brü-ā'äN-är-twä'), town (1968 pop. 28,628), Pas-de-Calais dept., NE France, on the Loire River. Primarily a coal-mining center, the town also produces fuels, boilers, clothing, beer, and candy.

Bruck an der Mur (brŏŏk' än dĕr mŏŏr'), city (1971 pop. 16,400), Styria prov., E central Austria, at the confluence of the Mur and the Mürz rivers. Manufactures include metal products and paper. Bruck was founded in 1263.

Bruges (brōōzh), city (1970 pop. 51,300), capital of West Flanders prov., NW Belgium, connected by canal with Zeebrugge (on the North Sea), its outer port. It is a commercial, industrial, and tourist center and a rail junction. Manufactures include lace, textiles, ships, railroad cars, communications equipment, chemicals, and processed food.

Bruges was founded on an inlet of the North Sea in the 9th cent. and became (11th cent.) a center of trade with England. In the 13th and 14th cent. it flourished as the major entrepôt port of the Hanseatic League and as one of the chief wool-processing centers of Flanders. Despite frequent political disturbances, Bruges continued to prosper until the Flemish wool industry declined (early 15th cent.) as a result of foreign competition. In addition, the North Sea inlet on which Bruges was located silted up completely by 1490, and the city lost its access to the sea and to its outer ports. The commercial and industrial revival of Bruges began in 1895, with the start of extensive repairs to its port; in 1907 the Zeebrugge canal was opened. The city was occupied by the Germans in World Wars I and II. Among its noted structures are the Hospital of St. John (12th cent.), containing several masterpieces by Hans Memling; the 13th cent. market hall, with its famous carillon; the city hall (14th cent.); and the Church of Notre Dame (13th-15th cent.), with the tombs of Charles the Bold and Mary of Burgundy and with Michelangelo's *Virgin*.

Brule (brōōl, brōō'lə), county (1970 pop. 5,870), 818 sq mi (2,118.6 sq km), S central S. Dak., in an agricultural area bounded on the W by the Missouri River; formed 1875; co. seat Chamberlain. Dairying and diversified farming are done.

Bru·nei (brōō-nī'), sultanate (1971 pop. 135,665), 2,226 sq mi (5,765.3 sq km), NW Borneo, on the South China Sea; a British protectorate since 1888. Its two sections are surrounded by Sarawak, Malaysia. Oil is Brunei's main export. Rubber is also produced, and cassava, pineapples, bananas, rice, and other crops are raised. A native sultanate was established on Brunei in the 15th cent. At one time the sultan controlled nearly all of Borneo, but by the 19th cent. his power had declined and Brunei had become a haven for pirates. In 1888 the British established a protectorate over Brunei, administered by a British resident, although the sultan retained formal authority. The Japanese overran the area during World War II. In 1959 a

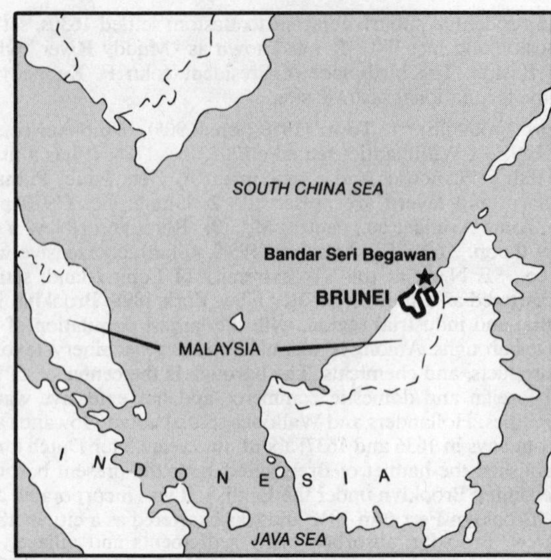

written constitution went into effect. Under it, as amended in 1965, the sultanate remains and the protectorate is governed by a chief minister, council of ministers, and elected legislative council. There was a leftist revolt in 1962. The Federation of Malaysia was planned to include Brunei, but at the last moment the sultan refused to join. The capital and major port of Brunei is Bandar Seri Begawan (formerly Brunei).

Brü·nig Pass (brü'nĭkн'), 3,396 ft (1,035.8 m) high, ancient route between the Forest Cantons and the Bernese Alps, central Switzerland. It is crossed by a highway and a railroad.

Bruns·wick (brŭnz'wĭk) or **Braun·shweig** (broun'shvīk), former state, E West Germany and W East Germany, surrounded by the former Prussian provinces of Saxony, Hanover, and Westphalia. In 1946 it was included (except for several small territories placed in East Germany) in the West German state of Lower Saxony. The duchy of Braunschweig emerged (13th cent.) from the remnants of the domains of Henry the Lion, the duke of Saxony. The Guelphic house repeatedly divided into several branches, the main ones being Braunschweig-Wolfenbüttel and Braunschweig-Lüneburg. In 1692 the duke of Braunschweig-Lüneburg became elector of Hanover. The Braunschweig-Wolfenbüttel line (itself a cadet branch of the Lüneburg line since 1634) ruled over Braunschweig and had, among its dukes, the famous generals Charles William Ferdinand (1735-1806) and Frederick William (1771-1815). Frederick William recovered (1813) the duchy, which Napoleon I had incorporated (1807) into the kingdom of Westphalia. The line became extinct in 1884, and Braunschweig was ruled by regents until 1913, when Ernest Augustus of Cumberland, grandson of King George V of Hanover, was made duke. A member of the North German Confederation from 1866 and of the German Empire from 1871, Braunschweig became a republic in 1918 and then joined the Weimar Republic.

Brunswick or **Braunschweig**, city (1970 pop. 223,700), Lower Saxony, E West Germany, on the Oker River. It is an industrial and commercial center; its manufactures include pianos, optical equipment, food products, and printed materials. Motor vehicles are assembled here. Reputedly founded c.861 and chartered in the 12th cent., Braunschweig became (13th cent.) a prominent member of the Hanseatic League. In 1830 the city became a self-governing municipality. The city has a 12th cent. Romanesque cathedral, which contains the tombs of Henry the Lion (d. 1195) and Emperor Otto IV (d. 1218); several Gothic churches; and a famous fountain representing Till Eulenspiegel, the legendary prankster. The philosopher and dramatist Gotthold Lessing (1729-81) is buried in Brunswick.

Brunswick. 1. County (1970 pop. 24,223), 855 sq mi (2,214.5 sq km), SE N.C., bounded on the E by Cape Fear River, on the S by the Atlantic, on the SW by the S.C. border, and on the W by the Waccamaw River; formed 1764; co. seat Southport. It is in a forested and swampy tidewater area, with farming (tobacco, corn, fruit, sweet potatoes, and livestock), fishing, sawmilling, and resorts along the coastline. **2.** County (1970 pop. 16,172), 579 sq mi (1,499.6 sq km), S Va., bounded on the S by the N.C. border, on the N by the Nottoway River, and drained by the Meherrin River; formed 1732; co. seat

Lawrenceville. It has agriculture (tobacco, grain, cotton, hay, sweet potatoes, peanuts, dairying, and livestock), lumber milling, manufacturing, and some processing of farm products.

Brunswick. **1.** City (1970 pop. 19,585), seat of Glynn co., SE Ga., on St. Simon's Sound near the Atlantic coast; laid out 1771-72, inc. 1856. It is a port of entry, and its sheltered harbor is used by coastal freighters and fishing and shrimping fleets. The gateway to offshore resort islands, Brunswick has a large seafood-processing industry and a great variety of manufactures, based principally upon forest products. The city was named for George III of the house of Brunswick (Hanover). **2.** Town (1970 pop. 16,195), Cumberland co., S Maine, on the Androscoggin River and Casco Bay, in a resort area; settled as a trading post in 1628, inc. 1738. It is a growing commercial center for southern Maine, with plants that make footwear, clothing, and paint brushes. Bowdoin College (1794) and a naval air station are in Brunswick. Henry Wadsworth Longfellow taught at Bowdoin College. A house dating from 1808 was once his home. Nathaniel Hawthorne's first novel, *Fanshawe* (1828), was printed in the town. In 1851 Harriet Beecher Stowe, then a Bowdoin faculty wife, wrote *Uncle Tom's Cabin* here; her house is a national landmark. In the first half of the 19th cent. Brunswick enjoyed prosperity based on shipbuilding. After the Civil War, textiles became the chief industry. The town's textile mill closed in 1955. **3.** City (1970 pop. 15,832), Medina co., N Ohio, a suburb of Cleveland; settled 1815 as part of the Connecticut Western Reserve, inc. 1960. A small farm community for many years, its population grew with the housing boom after World War II. It has a tire retread plant and a factory that makes powdered metals for roof coatings.

Bru·ny (bro͞o'nē), island, 149 sq mi (358.9 sq km), off the SE coast of Tasmania, Australia. Sheep and dairy products are produced.

Brussels (brŭs'əlz), city (1970 pop. 161,080), capital of Belgium and of Brabant prov., central Belgium, on the Senne River and at the junction of the Charleroi-Brussels and Willebroek canals. The city is officially bilingual (French and Flemish). Brussels is an important commercial, financial, industrial, administrative, and cultural center and a major rail junction. Among its varied manufactures are pharmaceuticals, electronics equipment, machine tools, rubber, processed food, and lace. It is the seat of the Council of Ministers and of the Commission of the European Communities; of the Economic and Social Committee of the European Economic Community; and of the North Atlantic Treaty Organization.

Brussels was inhabited by the Romans and later (7th cent. A.D.) by the Franks. The city developed into a center of the wool industry in the 13th cent. It became (1430) the seat of the dukes of Burgundy and later (1477) of the governors of the Spanish (after 1714, Austrian) Netherlands. The city suffered heavily in the wars fought in the Low Countries in the 16th to 18th cent. From 1815 to 1830 it was, with The Hague, the alternate meeting place of the Netherlands parliament; in 1830 it became the capital of Belgium.

The historical nucleus of the city, the medieval and Renaissance Grand' Place, a large square, is the site of the Gothic city hall (15th cent.); the Renaissance-style Maison du Roi or Broodhuis (13th cent.), meeting place of the old States-General of the Netherlands; and a number of rebuilt Gothic guildhalls. Other noteworthy buildings include the Collegiate Church of St. Michael and St. Gudule (founded in the 11th cent. and rebuilt in the 13th-15th cent.), which contains many noted Flemish paintings; the late-18th cent. Palais de la Nation (parliament building); and the Palais du Roi (royal palace). Brussels has excellent art museums and a botanical garden.

Brut·ti·um (brŭt'ē-əm), ancient region, S Italy, roughly occupying the present Calabria, the "toe" of the Italian peninsula. Bruttium faced Sicily across the Strait of Messina. It was settled (8th cent. B.C.) along the coast by Greek colonists. The Romans conquered Bruttium in the 3rd cent. B.C. The region passed to Byzantium after the fall of Rome and became known as Calabria.

Bry·an (brī'ən). **1.** County (1970 pop. 6,539), 439 sq mi (1,137 sq·km), SE Ga., bounded on the SE by the Atlantic, on the NE by the Ogeechee River, on the W by the Canoochee River; formed 1793; co. seat Pembroke. It includes Ossabaw Island and has coastal-plain agriculture (corn, truck crops, sugar cane, and livestock) and forestry and fishing industries. Part of Fort Stewart is within the county. **2.** County (1970 pop. 25,552), 889 sq mi (2,302.5 sq km), S central Okla., bounded on the S by the Red River, here forming the Texas border, and on the W by Lake Texoma; formed 1907; co. seat Durant. It has agriculture (cotton, corn, oats, peanuts, livestock, and dairy products) and some manufacturing. Stone quarrying is done. Dennison Dam, a waterpower source on the Red River, is in the southwest.

Bryan. **1.** City (1970 pop. 7,008), seat of Williams co., NW Ohio, NNW of Defiance, in a trade and industrial area; laid out 1840, inc. as a village 1849, as a city 1941. Furnaces, automobile and truck parts, and electrical applicances are among its manufactures. **2.** City (1970 pop. 33,719), seat of Brazos co., E central Texas; inc. 1872. Settled in the early 19th cent. in an area of large plantations, Bryan was long a cotton center. Farms producing alfalfa, truck crops, dairy goods, and poultry now occupy much of the land. Bryan's manufactures include aluminum products, furniture, building materials, agricultural chemicals, electronic components, and laboratory research equipment. The Research and Development Center of Texas A & M Univ. is in Bryan.

Bry·ansk (brē-änsk'), city (1970 pop. 1,582,000), capital of Bryansk oblast, central European USSR, on the Desna River. The city is a transportation hub. There are ironworks and locomotive, machine, and cement plants. Bryansk is also a major distributing center for natural gas. Originally called Brinyu and later Debryansk, the city was first known in 1146. Bryansk later passed to Lithuania and in the 16th cent. was annexed by Muscovy.

Bryce Canyon National Park (brīs), 36,010 acres (14,584 hectares), SW Utah; est. 1924. The Pink Cliffs of the Paunsaugunt Plateau, c.2,000 ft (610 m) high, were formed by water, frost, and wind action on alternate strata of softer and harder limestone; the result is colorful and unique erosional forms, including miniature cities, cathedrals, and spires.

Bryn Mawr (brĭn' mär'), uninc. village (1970 pop. 9,500), Montgomery co., SE Pa., a suburb of Philadelphia. It is the seat of Bryn Mawr College, opened in 1885 by the Society of Friends.

Bry·son City (brī'sən), town (1970 pop. 1,290), seat of Swain co., SW N.C., in the Great Smoky Mts. WSW of Asheville. It is a tourist center in a timber, mining, and farming area.

Bu·bas·tis (byo͞o-băs'tĭs), ancient city, NE Egypt, in the Nile delta. Capital of Egypt in the XXII and XXIII dynasties, it began to decline after the second Persian conquest (343 B.C.). Bubastis was the center of the worship of the cat-headed goddess Bast. Excavations were made in 1886, 1887, and 1906. Among the finds were a chapel of the VI dynasty (proving that the site dates back to the Old Kingdom) and a great temple built in the 8th cent. B.C.

Bu·ca·ra·man·ga (bo͞o'kä-rä-mäng'gä), city (1968 est. pop. 250,000), capital of Santander dept., N central Colombia, in the E highlands of the Andes. A leading commercial city, Bucaramanga is in the center of a rich coffee and tobacco area. Founded in 1622, the city preserves many monuments from the colonial period.

Bu·chan·an (byo͞o'kăn'ən, bə-). **1.** County (1970 pop. 21,762), 569 sq mi (1,473.7 sq km), E Iowa, in a prairie agricultural area drained by the Wapsipinicon River; formed 1837; co. seat Independence. Its agriculture includes cattle, hogs, poultry, corn, and oats. There are many sand and gravel pits and some limestone quarries. **2.** County (1970 pop. 86,915), 404 sq mi (1,046.4 sq km), NW Mo., bounded on the W by the Missouri River, here forming the Kansas border, and drained by the Little Platte River; formed 1838; co. seat St. Joseph. It has agriculture (corn, oats, wheat, hay, apples, and livestock) and some manufacturing. **3.** County (1970 pop. 32,071), 508 sq mi (1,315.7 sq km), SW Va., in the Alleghenies, bounded on NW by the Ky. border, on the NE by the W.Va. border, and drained by the Levisa Fork of the Big Sandy River; formed 1858; co. seat Grundy. Coal mining, lumbering, and manufacturing are important. Its agricultural products include soybeans, hay, potatoes, corn, and oats.

Buchanan, city (1970 pop. 800), seat of Haralson co., NW Ga., S of Cedartown near the Ala. border. Clothing is manufactured.

Bu·cha·rest (bo͞o'kə-rĕst, byo͞o'-), city (1977 est. pop. 1,565,872), capital and largest city of Rumania, SE Rumania, in Walachia, on the Dîmboviţa River, a tributary of the Danube. It is Rumania's chief industrial and communications center. Machine-building, metalworking, engineering, oil refining, food processing, and the manufacture of textiles, chemicals, automobiles, and footwear are the chief industries. The city, probably founded in the late 14th cent., was a military fortress and commercial center astride the trade routes to Constantinople. In 1698 the city became the capital of Walachia; after the union (1859) of Walachia and Moldavia it was made (1861) the capital of Rumania. During World War I, Bucharest was occupied (1916-18) by the Central Powers. After Rumania's surrender to the Allies (Aug., 1944) in World War II, German planes severely bombed the city; Soviet troops entered on Aug. 31, by which time a coalition of leftist parties had seized power. Bucharest served as headquarters of the Cominform from 1948 to 1956. Today it is a modern city, with fine parks, libraries, museums, and theaters. Landmarks include the Metropolitan Church (1649), the Radu Voda (1649) and Stavropoleos (1724–30) churches, and the Athenaeum, devoted to art and music.

Buch•en•wald (bo͝o'kən-wôld', bo͝o'KHən-vält'), village, Erfurt dist., SW East Germany, in the Buchenwald forest, near Weimar. It was the site of a Nazi concentration camp.

Buck•han•non (bŭk-hăn'ən), city (1970 pop. 7,261), seat of Upshur co., N W.Va., on the Buckhannon River S of Clarksburg; settled 1770, inc. 1852. Its principal industries are coal mining and lumbering. West Virginia Wesleyan College is here.

Buck•ha•ven and Meth•il (bŭk'hā'vən; mĕth'ĭl), burgh (1971 pop. 21,318), Fife, E Scotland, on the Firth of Forth. Methil is a leading port. In the burgh is Wemyss Castle (13th cent.), where Mary Queen of Scots met Lord Darnley in 1565.

Buck•ie (bŭk'ē), burgh (1971 est. pop. 21,318), Grampian, NE Scotland, on the Moray Firth. A leading herring port, it has an engineering plant and manufactures whiskey and barrels.

Buck•ing•ham (bŭk'ĭng-hăm'), county (1970 pop. 10,597), 576 sq mi (1,491.8 sq km), central Va., bounded on the NW and N by the James River, on the S by the Appomatox River; formed 1761; co. seat Buckingham. It has extensive lumbering and slate quarrying. Its agriculture includes grain, fruit, livestock, and dairy products.

Buckingham, town (1970 pop. 218), seat of Buckingham co., central Va., ENE of Lynchburg. It is also called Buckingham Court House.

Buck•ing•ham•shire (bŭk'ĭng-əm-shîr', -shər), **Buckingham,** or **Bucks** (bŭks), nonmetropolitan county (1976 est. pop. 512,000), central England; administrative center Aylesbury. The Thames River forms the southern boundary of the county. In south Buckinghamshire are the chalky Chiltern Hills with their beech forests. The region is mostly agricultural; barley, wheat, oats, and beans are the chief crops of the fertile Vale of Aylesbury in the north. Cattle, pigs, sheep, and poultry are raised farther south. The county has extensive Roman and pre-Roman remains.

Buck Island Reef National Monument (bŭk): *see* National Parks and Monuments Table.

Bucks (bŭks), county (1970 pop. 416,728), 617 sq mi (1,598 sq km), SE Pa., in an agricultural and residential area bounded on the E and SE by the Delaware River, here forming the N.J. border; formed 1682; co. seat Doylestown. Textiles and metal products are manufactured, and limestone, sandstone, and granite are quarried. George Washington crossed the Delaware at Washington Crossing here. The county includes a number of state parks and old stone buildings.

Bu•cy•rus (byo͞o-sī'rəs), city (1970 pop. 13,111), seat of Crawford co., N central Ohio, on the Sandusky River, in a farm area; settled 1818, inc. 1886. It is a trade center and has varied manufactures.

Bu•da•pest (bo͞o'də-pĕst'), city (1977 est. pop. 2,055,646), capital of Hungary, N central Hungary, on both banks of the Danube. The largest city of Hungary and its industrial, cultural, and transportation center, Budapest has varied manufactures, notably machinery, iron and steel, chemicals, pharmaceuticals, and textiles. Together with its industrial suburbs, the city accounts for about half of Hungary's total industrial production.

Budapest was formed in 1873 by the union of Buda and Óbuda on the right bank of the Danube River with Pest on the left bank. Buda, situated among a series of hills, was traditionally the center of government buildings, palaces, and villas belonging to the landed gentry. Pest, a flat area, has long been a commercial and industrial center. Educational and cultural institutions in the city include Roland Eötvös Univ. (1635), the Hungarian Academy of Sciences, the National Széchenyi Library, the National Museum, the National Theater, and the State Opera House.

Aquincum, the Roman capital of Lower Pannonia, was near the modern Óbuda, and Pest developed around another Roman town. Both cities were destroyed by Mongols in 1241, but in the 13th cent. King Béla IV built a fortress (Buda) on a hill. Buda became the capital of Hungary in 1361, reaching its height as a cultural center under Matthias Corvinus. Pest fell to the Turks in 1526, Buda in 1541. When Charles V of Lorraine conquered them for the Hapsburgs in 1686, both Buda and Pest were in ruins. They were resettled, Buda with Germans, Pest with Serbs and Hungarians. In the 19th cent. Pest flourished as an intellectual and commercial center; after the flood of 1838, it was rebuilt on modern lines. Buda became largely a residential sector. After the union of Buda and Pest in 1873, the united city grew rapidly as one of the two capitals of the Austro-Hungarian monarchy. With the collapse of the monarchy (Oct., 1918), Budapest became the capital of an independent Hungary.

Bu•daun (bə-doun'), town (1971 pop. 72,109), Uttar Pradesh state, N India, on the Sot River. An administrative center, it trades in grain, cotton, sugar cane, and oilseed.

Buddh Ga•ya (bo͝od gä'yä): *see* Bodh Gaya, India.

Bue•na Park (bwā'nə), city (1970 pop. 63,646), Orange co., S Calif.; inc. 1953. Food is processed, and tourism is an important industry. Knott's Berry Farm, a re-created gold rush town, is here.

Bue•na•ven•tu•ra (bwā'nə-vĕn-to͝or'ə, -tyo͝or'ə, bwā'nä-vän-to͞o'rä), city (1968 est. pop. 78,700), W Colombia, a port on the Pacific Ocean. The city, located on Cascajal Island in Buenaventura Bay, is the shipping point for the tobacco and sugar of the Cauca valley. Coffee, platinum, gold, and hides are also exported. The original settlement was founded in 1545.

Bue•na Vis•ta (bwā'nä vēs'tä), locality just S of Saltillo, Coahuila, Mexico, where a battle of the Mexican War was fought on Feb. 22–23, 1847. U.S. forces led by Zachary Taylor defeated Mexican troops commanded by Santa Anna.

Bue•na Vis•ta (byo͞o'nə vĭs'tə), county (1970 pop. 20,693), 573 sq mi (1,484.1 sq km), NW Iowa, in a prairie agricultural area drained by the Little Sioux River and headstreams of the Raccoon, Boyer, and Maple rivers; formed 1851; co. seat Storm Lake. Hogs, cattle, poultry, corn, oats, and soybeans are produced. It has sand and gravel pits and includes Storm Lake and Wanata State Park.

Bue•na Vis•ta (byo͞o'nə vĭs'tə). **1.** Resort town (1970 pop. 1,962), alt. 7,800 ft (2,379 m), Chaffee co., central Colo., on the Arkansas River W of Colorado Springs; founded and inc. 1879. It was once the center of a rich silver-mining region. Today livestock and grain are raised. **2.** City (1970 pop. 1,486), seat of Marion co., W Ga., SE of Columbus, in a farm and timber area; settled 1830. Cotton cloth is made. **3.** Independent city (1975 est. pop. 6,683), W Va., in the Blue Ridge ESE of Lexington; founded 1889, inc. 1891. Its manufactures include paper, bricks, and plastic and rubber products. Southern Seminary Junior College is here, and George Washington National Forest is nearby.

Bue•nos Ai•res (bwā'nəs ī'rēz, âr'ēz, bwā'nōs ī'rās), city and federal district (1977 est. pop. 2,972,453; metropolitan area 8,625,000), the capital of Argentina, E Argentina, on the Río de la Plata. One of the largest cities of Latin America, Buenos Aires is Argentina's chief port and its financial, industrial, commercial, and social center. Located on the eastern edge of the Pampa, Argentina's most productive agricultural region, and linked with Uruguay, Paraguay, and Brazil by a great inland river system, the city is the distribution hub and trade outlet for a vast area. Meat, meat products, grain, dairy products, hides, wool, flax, and linseed oil are the chief exports. Buenos Aires, the most heavily industrialized city of Argentina, is a major food-processing center, with huge meat-packing and refrigeration plants and flour mills. Other leading industries are metalworking, automobile manufacturing, oil refining, printing and publishing, machine building, and the production of textiles, chemicals, paper, clothing, beverages, and tobacco products.

Among the numerous educational, scientific, and cultural institutions are the Univ. of Buenos Aires (est. 1821), several private universities, the National Library, and the Teatro Colon, one of the world's most famous opera houses. *La Prensa* and *La Nacion* are daily newspapers famous throughout the Spanish-speaking world. The city has a modern subway system and is a railroad hub.

The city was first founded in 1536 by a Spanish royal gold-seeking expedition under Pedro de Mendoza. However, Indian attacks forced the settlers in 1539 to move Asunción (now the capital of Paraguay). A second and permanent settlement was planted in 1580 by Juan de Garay, who set out from Asunción. Although Spain long neglected Buenos Aires in favor of the riches of Mexico and Peru, the settlement's growth was enhanced by the development of trade, much of it contraband. During the 17th cent. the city ceased to be endangered by Indians, but French, Portuguese, and Danish raids were frequent. Buenos Aires remained subordinate to the Spanish viceroy in Peru until 1776, when it became the capital of a newly created viceroyalty of the Río de la Plata, including much of present-day Argentina, Uruguay, Paraguay, and Bolivia. In 1806, when Spain was allied with France during the Napoleonic Wars, the colonial militia defeated British troops without Spanish help, stimulating the drive for independence from Spain. Another British attack was repelled the following year.

On May 25, 1810, armed citizens of the cabildo, or town council, successfully demanded the resignation of the Spanish viceroy and established a provisional representative government. A long conflict ensued between the unitarians, strongest in Buenos Aires prov., who advocated a centralized government dominated by the city of Buenos Aires, and the federalists, mostly from the interior provinces, who supported provincial autonomy and equality. In 1853 the city and province of Buenos Aires refused to participate in a constituent congress and seceded from Argentina. National political unity was finally achieved when Bartolomé Mitre became Argentina's president in 1862 and made Buenos Aires his capital. Bitterness continued, however, until 1880, when the city was detached from the province and federalized. A new city, La Plata, was built as the provincial capital. Argentine railroad construction in the second half of the 19th cent. stimulated settlement and cultivation of the pampas. The city's spectacular economic development has attracted immigration from all over the world.

Buf·fa·lo (bŭf'ə-lō'). **1.** County (1970 pop. 31,222), 952 sq mi (2,465.7 sq km), S central Nebr., in an agricultural area drained by the South Loup and Wood Rivers and bounded on the S by the Platte River; formed 1870; co. seat Kearney. It produces livestock, grain, and potatoes. Wild game is hunted here. **2.** County (1970 pop. 1,739), 482 sq mi (1,248.4 sq km), central S.Dak., in an agricultural area bounded on the W by the Missouri River; formed 1864; co. seat Gannvalley. Wheat is a major crop. Part of the Crow Creek Indian Reservation is in the west. **3.** County (1970 pop. 13,743), 712 sq mi (1,844.1 sq km), W Wis., bounded on the W by the Chippewa River, on the SW by the Mississippi, here forming the Minn. line, and on the SE by the Trempealeau River; formed 1853; co. seat Alma. It is drained by the Buffalo River. In a hilly dairy and livestock area, it also produces poultry, clover, corn, oats, and timber.

Buffalo. **1.** Village (1970 pop. 3,275) seat of Wright co., E Minn., NW of Minneapolis, in a lake region; settled c.1855, inc. 1887. Clothing and wood products are made. **2.** City (1970 pop. 1,915), seat of Dallas co., SW Mo., NE of Springfield, in a farm region; inc. 1845. **3.** City (1977 est. pop. 420,000), seat of Erie co., W N.Y., on Lake Erie and the Niagara and Buffalo rivers; inc. 1832. With more than 37 mi (60 km) of waterfront, it is an important port of entry and one of the largest grain-distributing ports in the United States. It is also a major railroad hub. Buffalo is a great flour-milling center and has an enormous steel mill, many automobile plants, some of the world's largest electrochemical and electrometallurgical industries, and numerous other diversified manufactures. In 1803 a village was laid out on the site of modern Buffalo by Joseph Ellicott for the Holland Land Company. The village was almost destroyed by fire (1813) in the War of 1812 and recovered slowly until the opening of the Erie Canal in 1825. Transportation was a primary factor in the city's growth, and Buffalo became a major Great Lakes port. Its educational institutions today include the State Univ. of New York at Buffalo, State Univ. College of Arts and Science at Buffalo, Canisius College, and D'Youville College. Of interest are the Albright-Knox Art Gallery, the Buffalo Museum of Science, the county historical museum, and the Buffalo Zoological Gardens. The Peace Bridge (1927) connects Buffalo with Fort Erie, Canada. Grover Cleveland became mayor of Buffalo in 1882. In 1901, at the Pan-American Exposition, President William McKinley was assassinated; Theodore Roosevelt took the presidential oath in Buffalo. The McKinley monument and the Theodore Roosevelt Inaugural National Historic Site commemorate the two events. Millard Fillmore's home was in Buffalo. **4.** Town (1970 pop. 1,579), seat of Harper co., NW Okla., NNW of Woodward, in a farm and livestock area; founded 1907, inc. 1908. There are oil and gas wells in the vicinity. **5.** Town (1970 pop. 393), seat of Harding co., NW S.Dak., N of Rapid City. It is a trade center for a sheep and cattle region. **6.** City (1970 pop. 3,394), seat of Johnson co., N Wyo., SSE of Sheridan on Clear Creek; founded c.1880, inc. 1884. It is the trade center for a ranch, farm, and resort area and the eastern gateway to the Bighorn National Forest. Lake De Smet is to the north.

Buffalo Grove, village (1970 pop. 11,799), Cook and Lake cos., NE Ill.; inc. 1958.

Buffalo National River: *see* National Parks and Monuments Table.

Bu·ford (byōō'fərd), city (1975 est. pop. 8,693), Gwinnett and Hall cos., N Ga., NE of Atlanta, in a farm area; inc. 1872.

Bug (bōōg, bŭg, bŏŏk) or **Western Bug,** river, c.480 mi (770 km) long, rising in the Volhynian-Podolian hills, the Ukraine, W European USSR, and flowing N along the Polish-Ukrainian and Polish-Belorussian borders past Brest and then NW through Poland to join the Vistula River near Warsaw.

Bug or **Southern Bug,** river, c.490 mi (790 km) long, rising in the Volhynian-Podolian hills, the Ukraine, W European USSR. The Bug, flowing generally southeast into the Black Sea, is navigable for c.100 mi (160 km) from Voznesensk to its mouth.

Bui·ten·zorg (boi'tən-zôrKH): *see* Bogor, Indonesia.

Bu·jum·bu·ra (bōō'jəm-bŏŏr'ə), city (1971 est. pop. 57,200), capital of Burundi and of Bujumbura prov., W Burundi, a port on Lake Tanganyika. It is Burundi's largest city and its administrative, communications, and economic center. Manufactures include food products, cement and other building materials, textiles, soap, shoes, and metal goods. Livestock and agricultural produce from the surrounding region are traded in the city. Bujumbura is Burundi's main port and ships most of the country's chief export, coffee, as well as cotton, skins, and tin ore. The city attracts many tourists. A small village in the 19th cent., Bujumbura grew after it became (1899) a military post in German East Africa. After World War I it was made the administrative center of the Belgian Ruanda-Urundi League of Nations mandate. Its name was changed from Usumbura to Bujumbura when Burundi became independent in 1962.

Bu·ka·vu (bōō-kä'vōō), city (1970 pop. 135,000), capital of Kivu region, E Zaire, a port on Lake Kivu. It is an administrative, commercial, and transportation center. Hides and coffee are processed. The city was founded in 1901.

Bu·kha·ra (bə-kä'rə), city (1970 pop. 112,000), capital of Bukhara oblast, S Central Asian USSR, in Uzbekistan, in the Zeravshan River valley. On the Shkhrud irrigation canal system, it is the center of a large cotton district and has textile mills as well as cotton-ginning industries and the largest karakul skin processing plant in the USSR. First mentioned in Chinese chronicles in the 5th cent. A.D., Bukhara is one of the oldest trade and cultural centers in central Asia. It came under the Arab caliphate in the 8th cent. and became a major center of Islamic learning. From the 16th cent. to 1920 it was the capital of the khanate of Bukhara, which was ceded to Russia in 1868. From 1920 to 1924 it was the capital of the Bukhara People's Republic.

Bukhara, emirate of, former state, central Asia, in Turkistan, in the Amu Darya River basin. It was a trade, transport, and cultural center of the Islamic world. The Seljuk Turks ruled from 1004 to 1133; later, the realm was conquered by Genghis Khan (1220) and in the 14th cent. by Tamerlane. The Bukhara emirate was founded by the Uzbek Khan Sheybani early in the 16th cent. Defeated by Russia in 1866, the emirate became a Russian protectorate in 1868. In 1920 the Bukhara People's Soviet Republic was established and lasted until 1924. In the same year it was proclaimed a socialist republic and was included in the USSR; a few months later, however, it was divided between Uzbekistan, Tadzhikistan, and Turkmenistan.

Bu·ko·vi·na (bōō-kə-vē'nə), historic region of E Europe, in W Ukraine and NE Rumania. Traversed by the Carpathian Mts. and the upper Prut and Siretul rivers, it is heavily forested and produces timber, textiles, grain, and livestock. Petroleum and salt are produced in quantity; other mineral resources include manganese, iron, and copper. A part of the Roman province of Dacia, Bukovina was overrun after the 3rd cent. A.D. by the Huns and other nomads. It later (10th–13th cent.) belonged to the Kievan state. In 1514 Bukovina, then part of Moldavia, became tributary to the Turkish sultans. Ceded by the Ottoman Empire to Austria in 1775, it was at first a district of Galicia but in 1848 was made a separate Austrian crownland. The Treaty of Saint-Germain (1919) gave only the southern part of Bukovina to Rumania, but the subsequent Treaty of Sèvres awarded Rumania the entire region. In a treaty of June, 1940, Rumania ceded the northern part of Bukovina (c.2,140 sq mi/5,545 sq km) to the USSR, which incorporated it into the Ukrainian SSR. The remainder of the area (c.1,890 sq mi/4,895 sq km) forms one of the historical provinces of Rumania.

Bu·la·wa·yo (bōō-lə-wä'yō), city (1970 est. pop. 70,000), SW Rhodesia. It is the second-largest city of Rhodesia and an important industrial, commercial, and railroad center. Among its manufactures are textiles, motor vehicles, metal products, and cement. It was founded by the British in 1893.

Bul·gar·ia (bŭl-gâr′ē-ə), republic (1973 est. pop. 8,620,000), 42,823 sq mi (110,912 sq km), SE Europe, on the E Balkan Peninsula. It is bounded by the Black Sea on the east, by Rumania on the north, by Yugoslavia on the west, by Greece on the south, and by European Turkey on the southeast. Sofia is the capital. Central Bulgaria is

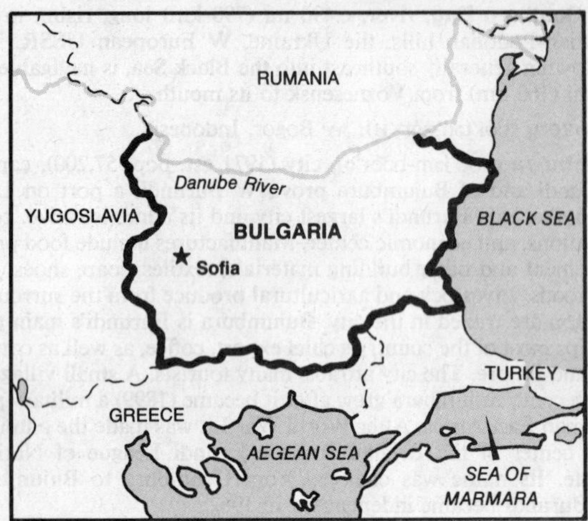

traversed from east to west by ranges of the Balkan Mts. A fertile plateau runs north of the Balkans to the Danube River, which forms most of the northern border. In the southwest is the Rhodope Range. The Thracian plain lies south of the Balkans and east of the Rhodope.

Bulgaria is divided into 27 provinces and the city commune of Sofia:

NAME	CAPITAL	NAME	CAPITAL
Blagoevgrad	Blagoevgrad	Shumen	Shumen
Burgas	Burgas	Silistra	Silistra
Gabrovo	Gabrovo	Sliven	Sliven
Khaskovo	Khaskovo	Smolyan	Smolyan
Kŭrdzhali	Kŭrdzhali	Sofia (city commune)	
Kyustendil	Kyustendil	Sofia	Sofia
Lovech	Lovech	Stara Zagora	Stara Zagora
Mikhaylovgrad	Mikhaylovgrad	Tolbukhin	Tolbukhin
Pazardzhik	Pazardzhik	Tŭrgovishte	Tŭrgovishte
Pernik	Pernik	Varna	Varna
Pleven	Pleven	Veliko Tŭrnovo	Veliko Tŭrnovo
Plovdiv	Plodiv	Vidin	Vidin
Razgrad	Razgrad	Vratsa	Vratsa
Ruse	Ruse	Yambol	Yambol

Economy. Traditionally an agricultural country, Bulgaria has been considerably industrialized since World War II. The leading industries are engineering, metallurgy, and the production of chemicals and fertilizers. Agriculture, however, remains the chief occupation; the principal crops are wheat, corn, barley, and sugar beets. Grapes and other fruit, as well as roses, are grown, and much stock is raised. Most of the land was collectivized by 1958. Bulgaria's mineral resources include lignite, bauxite, iron ore, lead, zinc, and oil and natural gas. The chief exports are foodstuffs and attar of roses; manufactured goods and fuels are the leading imports.

History. Ancient Thrace and Moesia, which modern Bulgaria occupies, were settled (6th cent. A.D.) by Slavic tribes. Beginning in 679–80, Bulgar tribes from the banks of the Volga crossed the Danube and merged with the Slavs, The first Bulgarian empire (681–1018), emerged as a significant Balkan power, and in 809 the khan Krum (ruled 803–814) captured Sofia from the Byzantines. In the 10th cent. Bulgaria crumbled under the attacks of a reinvigorated Byzantium, and in 1018 it was annexed. The second Bulgarian empire (1186–1396) reached its height under Ivan II, whose rule (1218–1241) extended over nearly the whole Balkan Peninsula except Greece. After the battles of Kossovo (1389) and Nikopol (1396) Bulgaria was absorbed into the Ottoman Empire.

Turkish rule was often oppressive, and rebellions were frequent. A determined effort was made to destroy Bulgarian Christianity and the Bulgarian language. In 1876 a rebellion, led by Stefan Stambulov, broke out. The subsequent Turkish reprisals provided a reason for the Russians to liberate (1877–78) their brother Slavs and create an enlarged, autonomous Bulgaria. In order to avert the expansion of Russian influence in the Balkans, Austria-Hungary and Great Britain insisted that Russia accept a treaty revision (1878) that reduced the country to the territory between the Danube and the Balkans. However, Alexander, first prince of Bulgaria, regained most of the territory south of the Balkans by annexation (1885). His successor, Prince Ferdinand (reigned 1887–1918), proclaimed Bulgaria independent in 1908 with himself as czar.

The kingdom emerged from the Balkan Wars (1911–12) and World War I, in which it sided with the Axis, with again diminished borders. Almost all of Macedonia was lost. An era of political confusion ensued, dominated by the violent activities of an irredentist Macedonian terrorist group. Bulgaria saw in an alliance with Germany in World War II an opportunity to satisfy its territorial claims. Boris III (reigned 1918–48) died and was succeeded by his young son Simeon II.

Soviet troops entered the country in Sept., 1944. After a short period of coalition rule, the Communists succeeded in taking over the government. The monarchy was abolished, and in 1946 Bulgaria was proclaimed a republic with Georgi Dimitrov as premier. Industry was nationalized and farms collectivized. Bulgaria has closely followed the Soviet Union in its domestic and foreign policies. The nation joined the Council for Economic Mutual Assistance (1949), the Warsaw Treaty Organization (1955), and the United Nations (1955). Communist Party leader Todor Zhiukov became chairman of the Council of State in 1976.

Government. A new constitution, adopted in 1971, provided for a unicameral national assembly to be elected every five years. The assembly elects a council of state and the cabinet of ministers. But actual power resides in the Communist party, which heads the Fatherland Front, a grouping of organizations that support the regime.

Bul·litt (bo͝ol′ ĭt), county (1970 pop. 26,090), 300 sq mi (777 sq km), NW Ky., drained by the Salt River and bounded on the W by the Ohio River, on the SW by the Rolling Fork River; formed 1796; co. seat Shepherdsville. In a rolling agricultural area yielding livestock, grain, burley tobacco, and dairy products, it has some manufacturing, especially whiskey and food products.

Bul·loch (bo͝ol′ək), county (1970 pop. 31,585), 685 sq mi (1,774.2 sq km), E Ga., bounded on the NE by the Ogeechee River; formed 1796; co. seat Statesboro. It is in a coastal-plain timber and agricultural area yielding cotton, tobacco, corn, peanuts, and livestock.

Bul·lock (bo͝ol′ək), county (1970 pop. 11,824), 615 sq mi (1,592.9 sq km), SE Ala., in a rolling agricultural area watered by headstreams of the Conecuh and Pea rivers; formed 1866; co. seat Union Springs. Its agriculture includes cattle, soybeans, cotton, and peanuts. It also has a lumbering industry and some manufacturing.

Bull Run (bo͝ol′ rŭn), small stream, NE Va., SW of Washington, D.C. Two important battles of the Civil War were fought here. In the first Battle of Bull Run (or First Battle of Manassas), on July 21, 1861, unseasoned Union volunteers retreated, but the equally inexperienced Confederates were in no condition to make an effective pursuit. The Second Battle of Bull Run (or Second Battle of Manassas), Aug. 29–30, 1862, was also a victory for the Confederates. Both battlefields are included in Manassas National Battlefield Park.

Bun·combe (bŭng′kəm), county (1970 pop. 145,056), 657 sq mi (1,701.6 sq km), W N.C., in the Blue Ridge, crossed by the French Broad River; formed 1791; co. seat Asheville. It has agriculture (tobacco, corn, dairy products, cattle), lumbering, textile manufacturing, and a tourist industry. The Black Mts. are in the east, and Pisgah National Forest is in the east and southwest.

Bun·da·berg (bŭn′də-bûrg′), city (1971 pop. 27,394), Queensland, E Australia, on the Burnett River. It is a sugar-refining center and a port.

Bun·ker Hill (bŭng′kər), height (107 ft/32.6 m) in Charlestown, Boston, Mass., near which the first major battle of the American Revolution was fought (June 17, 1775). When Continental forces learned of a British plan to take the heights of Dorchester and Charlestown, William Prescott was sent to occupy Bunker Hill. Prescott instead chose the neighboring Breed's Hill to the southeast, but the engagement that ensued has become known as the Battle of Bunker Hill. The British commander William Howe was ordered to attack the American position, and after two slaughterous failures a third charge dislodged the Americans, who had run out of powder. The British victory failed to break the Continental siege of Boston, and the American defense heightened colonial morale and resistance.

Bun·nell (bə-nĕl'), city (1970 pop. 1,687), seat of Flagler co., NE Fla., S of St. Augustine, in a timber area. Citrus fruit and potatoes are grown.

Bu·rai·mi (bo͞o-rī'mē), group of small oases, SE Arabia, on the border between Abu Dhabi and Oman. In the 1950s the area, rich in oil, was claimed by Saudi Arabia, causing a dispute with Great Britain, which at the time was the protector of Oman and Abu Dhabi.

Bu·ra·no (bo͞o-rä'nō), former town, now part of Venice, in Venetia, NE Italy, built on four islets in the Lagoon of Venice. It is a fishing center and has been famous for its lace since the 15th cent.

Bur·bank (bûr'băngk), city (1970 pop. 88,871), Los Angeles co., S Calif.; inc. 1911. Aircraft manufacturing is the major industry. Several motion-picture and television studios are in Burbank.

Bur·dwan (bûr-dwän'), town (1971 pop. 144,970), West Bengal state, E central India. It has cutlery and tool industries but is chiefly known for its 108 linga temples. Rice is the chief product of the surrounding area.

Bur·eau (byo͞o'rō), county (1970 pop. 38,541), 868 sq mi (2,248.1 sq km), N Ill., bounded on the SE by the Illinois River and drained by the Green and Spoon rivers; formed 1837; co. seat Princeton. It is crossed by the old Illinois and Mississippi Canal. It has agriculture (corn, oats, wheat, fruit, livestock, and poultry), bitumous-coal mines, and sand and gravel pits. Its manufactures include dairy products, furniture, brick and tile, vinegar, sealing wax, feed, clothing, and cigars.

Bu·re·ya (bo͞o-rā'ä), mountain range, Khabarovsk Kray, SE Far Eastern USSR, extending into NE China as the Lesser Khingan range. The site of the Bureya coal basin, it rises to c.7,150 ft (2,180 m) and yields iron and coal. The Bureya River, c.445 mi (715 km) long, rises in the northern part of the range and flows southwest to join the Amur River.

Bur·gas (bo͝or-gäs'), city (1968 est. pop. 126,500), SE Bulgaria, on the Black Sea. It is an important port and commercial center. Fishing and fish canning, flour milling, sugar refining, copper mining, and soap making are carried on in Burgas. The city was founded (18th cent.) on the site of a 14th cent. fortified town.

Bur·gaw (bûr'gô), town (1970 pop. 1,744), seat of Pender co., SE N.C., N of Wilmington, in a farm area. It has sawmills.

Burg·dorf (bo͝orkʜ'dôrf'), town (1970 pop. 15,888), NW Switzerland, on the Emme River. It is a manufacturing and cheese-trading town. There is a 12th cent. castle in which J. H. Pestalozzi, the educational reformer, held (1799–1804) his first school.

Bur·gen·land (bo͝or'gən-länt'), province (1971 pop. 272,000), 1,530 sq mi (3,962.7 sq km), E Austria; capital Eisenstadt. Its territory was transferred from Hungary by the treaties of Saint-Germain (1919) and Trianon (1920).

Bur·gos (bo͝or'gōs), city (1970 pop. 119,915), capital of Burgos prov., N Spain, in Old Castile, on a mountainous plateau c.2,800 ft (855 m) above sea level, near the Arlanzón River. It is an important trade center with a large tourist industry. It was one of the ancient capitals of Castile but is chiefly known for its outstanding architecture and great historic tradition. Founded c.855, it became the capital of the kingdom of Castile under Ferdinand I (1035). The royal residence was moved (1087) to Toledo, and Burgos lost some of its cultural importance. In the civil war of 1936–39, Burgos was the capital of Franco's regime. Its most notable building is the cathedral of white limestone, begun in 1221, one of the finest examples of Gothic architecture in Europe.

Bur·gun·dy (bûr'gən-dē), or *French* **Bour·gogne** (bo͞or-gô'nyə), historic region, E France. The name once applied to a large area embracing several kingdoms, a free county, and a duchy. The present region is identical with the province of Burgundy of the 17th and 18th cent. Burgundy west of the Saône River is generally hilly; the southeast includes the southern spurs of the Jura Mts.; the center is a lowland, extending south almost to the junction of the Saône and Rhône rivers. A rich agricultural country, Burgundy is especially famous for the wine produced in the Chablis region, the mountains of the Côte d'Or, and the Saône and Rhône valleys.

The territory, conquered by Caesar in the Gallic Wars, was divided into the Roman provinces of Lugdunensis and Belgic Gaul (later Upper Germany). In c.480 A.D. the country was conquered by the Burgundii, a tribe from Savoy. The Burgundii accepted Christianity and formed the First Kingdom of Burgundy, which at its height covered southeast France. Conquered (534) by the Franks, it was throughout the Merovingian period subjected to numerous partitions. The kingdoms of Cisjurane Burgundy in the south and Transjurane Burgundy in the north were united (933) in the Second Kingdom of Burgundy. A smaller area, corresponding roughly to present Burgundy, was created as the duchy of Burgundy by Emperor Charles II in 877.

The golden age of Burgundy began (1364) when John II of France bestowed the fief on his son, Philip the Bold. Philip and his successors acquired vast territories, including most of the present Netherlands, Belgium, and northeast France. In the early 15th cent. Burgundy had the most important trade, industry, and agriculture of Europe. The wars of ambitious Duke Charles the Bold, however, proved ruinous; he was defeated and killed by the Swiss at Nancy (1477). His daughter, Mary of Burgundy, by marrying Archduke Maximilian of Austria (1477), brought most of the Burgundian possessions to the house of Hapsburg. The duchy itself was seized (1477) by Louis XI, who incorporated it into France as a province.

Bur·han·pur (bo͝or'hän-po͝or'), town (1971 pop. 105,349), Madhya Pradesh state, W central India, on the Tapti River. It trades in cotton and oilseed, and is known for its gold and silver embroidery.

Bu·rias (bo͝or'yäs), island (1960 pop. 15,918), 164 sq mi (424.8 sq km), Masbate prov., SE of Luzon, Philippines, across the Burias Pass. Generally mountainous, it is noted for coconuts and rice.

Burk·bur·nett (bûrk'bər-nĕt'), city (1970 pop. 9,230), Wichita co., N Texas, near the Okla. line; inc. 1913. A shipping center for livestock, cotton, and wheat, it also has many oil wells and refineries. The area's first big gusher (1918) brought a boom that transformed the quiet little community into one of the wildest and roughest of all the oil towns; at one time its population approached 30,000.

Burke (bûrk). **1.** County (1970 pop. 18,255), 832 sq mi (2,154.9 sq km), E Ga., bounded on the NE by the Savannah River, here forming the S.C. border, and drained by Brier Creek; formed 1777; co. seat Waynesboro. It is in a coastal-plain cotton region that also produces truck crops, melons, and pecans. **2.** County (1970 pop. 60,364), 506 sq mi (1,310.5 sq km), W central N.C., in the Piedmont, drained by the Catawba River; formed 1777; co. seat Morganton. It has hydroelectric plants, dairying, poultry raising, farming (sweet potatoes, hay, corn, and wheat), and diversified manufacturing (furniture, hosiery, and textiles). Part of Pisgah National Forest is here. **3.** County (1970 pop. 4,739), 1,121 sq mi (2,903.4 sq km), NW N.Dak., in a rich agricultural area bordered on the N by Sask., Canada, and drained by the Upper Des Lacs Lake; formed 1910; co. seat Bowbells. Wheat, barley, oats, flax, rye, alfalfa, and sweet clover are grown. It also has lignite deposits, coal mines, and oil wells.

Burke, city (1970 pop. 892), seat of Gregory co., S S.Dak., SE of Pierre. It is a trade center for a farming and ranching region. Livestock, dairy products, and grain are its major products.

Burkes·ville (bûrks'vĭl'), city (1970 pop, 1,717), seat of Cumberland co., S central Ky., on the Cumberland River near the Tenn. line SE of Bowling Green; inc. as a town 1810, as a city 1926. It is a resort in the Cumberland foothills.

Bur·leigh (bûr'lē), county (1970 pop. 40,714), 1,625 sq mi (4.208.8 sq km), S central N.Dak., in an agricultural area drained by Apple Creek and bounded on the W by the Missouri River; formed 1873; co. seat Bismarck. It is mainly agricultural (wheat, oats, corn, barley, alfalfa, dairy products, and cattle and hogs). Farm machinery is manufactured.

Bur·le·son ((bûr'lə-sən), county (1970 pop. 9,999), 679 sq mi (1,758.6 sq km), S central Texas, bounded on the NE and E by the Brazos River, on the S by Lake Somerville; formed 1846; co. seat Caldwell. It has diversified agriculture (cotton, corn, grain sorghums, peanuts, and pecans), oil and natural-gas wells, coal deposits, and some manufacturing.

Burleson, city (1975 est. pop. 10,480), Johnson and Tarrant cos., N central Texas, S or Fort Worth, in a farm region; est. 1881.

Bur·ley (bûr'lē), city (1976 est. pop 8,773), seat of Cassia co., S Idaho, on the Snake River ESE of Twin Falls; founded 1905, inc. 1906. In a farm area irrigated by the Minidoka project, it has mills processing sugar beets, alfalfa, and potatoes. It is the headquarters of Minidoka National Forest.

Bur·lin·game (bûr'lĭn-gām', -lĭng-gām'), city (1970 pop. 27,320), San Mateo co., W Calif., on San Francisco Bay; founded 1868, inc. 1908. Burlingame is mainly residential, with some light industries.

Bur·ling·ton (bûr'lĭng-tən), town (1971 pop. 87,023), SE Ont., Canada, on Lake Ontario. It is a suburb of Hamilton.

Burlington, county (1970 pop. 323,132), 819 sq mi (2,121.2 sq km), central N.J., bounded on the NW by the Delaware River, on the S by the Mullica River, and drained by Rancocas Creek, branches of the Wading River, and the Haynes, Friendship, and Bass rivers; formed 1681; co. seat Mount Holly. It is agricultural (blueberries and cranberries, grain, and dairy products), with horse-breeding farms and industrial parks. McGuire Air Force Base, part of Fort Dix, and several state parks are here.

Burlington. 1. Town (1970 pop. 2,828), seat of Kit Carson co., E Colo., near the Kansas border ESE of Denver, in a grain and livestock area; inc. 1888. Sugar beets are grown. **2.** Town (1975 est. pop 5,246), Hartford co., central Conn., W of Hartford; settled 1740, inc. 1806. It has a state trout hatchery. **3.** City (1970 pop. 32,366), seat of Des Moines co., SE Iowa, on four hills overlooking the Mississippi (spanned here by rail and highway bridges); inc. 1836. It is a farm, shipping, and manufacturing center with railroad shops and docks. Zebulon Pike selected this spot for a fort in 1805. Burlington was the temporary capital of Wisconsin Territory (1837) and of Iowa Territory (1838–40). One of the oldest newspapers in the state, the Burlington *Hawk-Eye*, is still published. **4.** City (1970 pop. 2,099), seat of Coffey co., E Kansas, S of Topeka; inc. 1870. It is the trade center of a farm and timber area. **5.** Town (1970 pop. 550), seat of Boone co., NE Ky., WSW of Covington in the outer Bluegrass region. It is a farm trade center. Big Bone Lick, where many bones of prehistoric mammals have been discovered, is to the south. **6.** Town (1970 pop. 21,980), Middlesex co., E Mass., a residential suburb of Boston, in a farm area; settled 1641, inc. 1799. Its pre-Revolutionary meetinghouse still stands. **7.** City (1970 pop. 12,010), Burlington co., W N.J., on the Delaware River between Trenton and Camden, in a rich farm area; settled 1677 by Friends, inc. 1733. A shipping point for farm and dairy products, it also has varied manufactures. Burlington grew mainly as a port; it was capital of West Jersey from 1681 until the union of East and West Jersey (1702), and thereafter until 1790 was alternate capital with Perth Amboy. The first colonial money was printed here by Benjamin Franklin in 1726, and the first newspaper in New Jersey appeared in 1777. The Friends' school (1792; now the Y.W.C.A.) and meetinghouse (1784) still stand. The birthplaces of James Fenimore Cooper and James Lawrence are preserved. **8.** City (1970 pop. 35,930), Alamance co., N N.C., on the Haw River; settled c.1700, inc. 1866. It is a great textile center in a heavily industrialized area, with plants manufacturing textiles, hosiery, and yarn. In May, 1771, 2,000 colonial "Regulators" clashed with British troops south of Burlington; the site is in Alamance Battleground State Park. **9.** City (1970 pop. 38,633), seat of Chittenden co., NW Vt., on Lake Champlain; settled 1773, inc. 1865. The largest city in the state, it is a port of entry and a major industrial center. Missile and ordnance parts, data-processing machinery, textiles, canned goods, and wood and steel products are its chief manufactures. Battery Park, famous for sunset views, was the scene of an abortive British naval attack (Aug. 3, 1813) during the War of 1812. The city is the seat of the Univ. of Vermont. American Revolutionary hero Ethan Allen spent his last years near Burlington village (part of his farm is included in Ethan Allen Park) and is buried nearby. The Burlington *Free Press* (founded 1827) became the state's first daily newspaper in 1848. The philosopher John Dewey was born in the city. **10.** City (1975 est. pop. 8,548), Racine co., SE Wis., on the Fox River WSW of Racine; settled 1835, inc. as a village 1896, as a city 1900. Floor covering is made. A Mormon colony was established (1844–49) near here.

Bur·ma, Union of (bûr'mə) republic (1975 est. pop. 31,240,000), 261,789 sq mi (678,034 sq km), SE Asia; capital Rangoon. Burma is bounded on the west by Bangladesh, India, and the Bay of Bengal, on the north and northeast by China, on the east by Laos and Thailand, and on the south by the Andaman Sea. The most densely populated part of the country is the valley of the Irrawaddy River, which, with its vast delta, is one of the main rice-growing regions of the world. The valley is surrounded by a chain of mountains that stem from the eastern Himalayas and spread out roughly in the shape of a giant horseshoe; the ranges and river valleys of the Chindwin (a tributary of the Irrawaddy) and of the Sittang and the Salween (both to the east of the Irrawaddy) run from north to south. In the mountains of northern Burma (rising to more than 19,000 ft/5,795 m) and along the India-Burma frontier live various Mongoloid peoples; the most important are the Kachins and the Chins. Between the Bay of Bengal and the hills of the Arakan Yoma is the Arakan, a narrow coastal plain. In eastern Burma on the Shan Plateau is the Shan State, home of the Shans, a Tai race closely related to the Siamese. South of the Shan State are the mountainous Kayah State and the Kawthule State; the Karens, who inhabit this region, are of Tai-Chinese origin, and many are Christians. South of the Kawthule State is the Tenasserim region, a long, narrow strip of coast extending to the Isthmus of Kra. Most of Burma has a tropical, monsoon climate; however, north of the Pegu Hills around Mandalay is the so-called Dry Zone with a rainfall of 20 to 40 in. (51–102 cm). On the Shan Plateau temperatures are moderate.

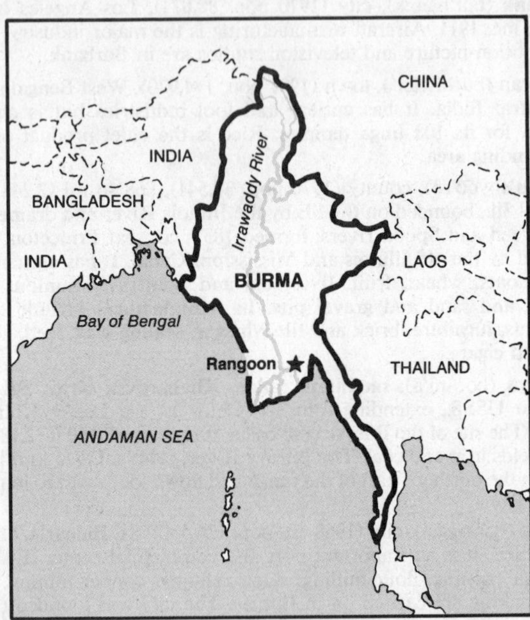

Economy. Most of the population work in agriculture and forestry, and rice accounts for about half of the agricultural output. (Until 1964, Burma was the world's largest rice exporter.) Other important crops are sugar cane, groundnuts, and pulses. Burma's forests, which are government-owned, are the source of teak and other hardwoods. The country is rich in minerals including petroleum, tin, tungsten, lead, silver, and zinc. Coal and iron deposits have also been found. Aside from food-processing establishments, there are few manufacturing industries in Burma. Rice and teak are the leading exports.

History. Burma's early history is mainly the story of the struggle of the Burmans against the Mons, or Talaings (of Mon-Khmer origin, now assimilated). In 1044 King Anawratha established Burman supremacy over the Irrawaddy delta and adopted Hinayana Buddhism from the Mons. His capital, Paga, "the city of a thousand temples," was the seat of his dynasty until it was conquered by Kublai Khan in 1287. In the 16th cent. the Burman Toungoo dynasty unified the country and initiated the permanent subjugation of the Shans to the Burmans. In the 18th cent. the Mons of the Irrawaddy delta overran the Dry Zone. In 1758 Alaungapaya rallied the Burmans, crushed the Mons, and established his capital at Rangoon. He extended Burman influence to areas in present-day India (Assam and Manipur) and Thailand.

Burma was ruled by his successors (the Konbaung dynasty) when friction with the British over border areas in India led to war in 1824. The Treaty of Yandabo (1826) forced Burma to cede to British India the Arakan and Tenasserim coasts. In a second war (1852) the British occupied the Irrawaddy delta. Fear of growing French strength in the region, in addition to economic considerations, caused the British to instigate the third Anglo-Burman War (1885). The Burman king was captured, and the remainder of the country was annexed to India. Under British rule rice cultivation in the delta was expanded, an extensive railroad network was built, and the natural resources of Burma were developed.

In 1923 a system of "dyarchy," already in effect in the rest of

British India, was introduced, whereby a partially elected legislature was established and some ministers were made responsible to it. In 1935 the British gave Burma a new constitution (effective 1937), which separated the country from British India and provided for a fully elected assembly and a responsible cabinet. During World War II, Burma was invaded and quickly occupied by the Japanese. Allied forces drove the Japanese out of Burma in Apr., 1945.

In Jan., 1948, Burma became an independent republic outside the British Commonwealth of Nations. The new constitution provided for a bicameral legislature with a responsible prime minister and cabinet. Non-Burman areas were organized as the Shan, Kachin, Kawthule, and Kayah states and the Chin Special Division; each possessed a degree of autonomy. The government, controlled by the socialist Anti-Fascist People's Freedom League (AFPFL), was soon faced with armed risings of Communist rebels and of Karen tribesmen, who wanted a separate Karen nation.

In foreign affairs Burma has followed a generally neutralist course. It refused to join the Southeast Asia Treaty Organization and was one of the first countries to recognize the Communist government in China. The AFPFL leaders intended to socialize the country rapidly, but lower rice prices after the Korean War and a shortage of trained personnel forced the abandonment of most of the plans. With a breakdown of order threatening, Premier U Nu invited General Ne Win, head of the army, to take over the government (Oct., 1958). After the 1960 elections, which were won by U Nu's faction, civilian government was restored. However, as rebellions among the minorities flared and opposition to U Nu's plan to make Buddhism the state religion mounted, conditions deteriorated rapidly. In Mar., 1962, Ne Win staged a military coup, discarded the constitution, and established a Revolutionary Council, made up of military leaders who ruled by decree.

The Revolutionary Council fully nationalized the industrial and commercial sectors of the economy. Insurgency became a major problem of the Ne Win regime. Pro-Chinese Communist rebels—the "White Flag" Communists—were active in the northern part of the country, where, from 1967 on, they received aid from Communist China; the Chinese established links with the Shan and Kachin insurgents as well. The deposed U Nu, who managed to leave Burma in 1969, also organized an anti-Ne Win movement among the Shans, Karens, and others in the east. However, in 1972 U Nu split with minority leaders over their assertion of the right to secede from Burma. By the early 1970s the various insurgent groups controlled about one third of Burma. A new constitution, providing for a unicameral legislature and one legal political party, took effect in Mar., 1974, and the Revolutionary Council was disbanded.

Burma Road, in China and Burma, extending from the Burmese railhead of Lashio to K'un-ming, Yünnan prov., China. About 700 mi (1,125 km) long and constructed through rough mountain country, it was undertaken by the Chinese after the start of the Sino-Japanese war in 1937 and completed in 1938. The road is now in a state of disrepair.

Bur·net (bûr'nĭt), county (1970 pop. 11,420), 996 sq mi (2,579.6 sq km), central Texas, bounded on the W by the Colorado River; formed 1852; co. seat Burnet. It has agriculture (cotton, corn, grain sorghums, legumes, hogs, poultry, and dairy products), cattle ranches, and granite and limestone quarries. Tourism, hunting, and fishing are also important. Longhorn Cavern State Park is here.

Burnet, town (1970 pop. 2,864), seat of Burnet co., central Texas, NW of Austin; inc. 1885. It grew about Fort Croghan, a frontier post established in 1849, and is now a resort and trade and shipping center for an agricultural region.

Bur·nett (bûr-nĕt'), county (1970 pop. 9,276), 840 sq mi (2,175.6 sq km), NW Wis., bounded on the W by the St. Croix River, here forming the Minn. border, and watered by numerous lakes and small rivers; formed 1856; co. seat Grantsburg. Dairy products, lumber, and tourism are important to its economy.

Burn·ley (bûrn'lē)), county borough (1971 pop. 76,483), Lancashire, NW England. Coal mining and cotton weaving, the keys to Burnley's growth, are still important industries. Electrical heating appliances, kitchen equipment, and gas turbines are manufactured.

Burns (bûrnz), city (1970 pop. 3,293), seat of Harney co., SE central Oregon, on the Silvies River N of Malheur; inc. 1899. The livestock center of Oregon, it also has lumber mills.

Burns·ville (bûrnz'vĭl'), resort town (1970 pop. 1,348), seat of Yancey co., W central N.C., in the Blue Ridge NE of Asheville.

Bur·rill·ville (bûr'əl-vĭl'), town (1970 pop. 10,087), Providence co., NW R.I.; inc. 1806. Its manufactures include textiles and plastics.

Bur·sa (bo͝or-sä'), city (1970 pop. 275,917), capital of Bursa prov., NW Turkey. The market center of a rich agricultural region, Bursa is a commercial and industrial center, noted for its silk textiles. Founded at the end of the 3rd cent. B.C., it was captured by the Seljuk Turks in 1075, taken by the Crusaders in 1096, and in 1204 passed to the Byzantines. In 1326 it became the Ottoman capital and was embellished with mosques, baths, and a caravansary. It was sacked by Tamerlane in 1402. There are many fine old mosques, notably the Green Mosque (1421).

Burt (bûrt), county (1970 pop. 9,247), 483 sq mi (1,251 sq km), E Nebr., in an agricultural area bounded on the E by the Missouri River and the Iowa border and drained by Logan Creek; formed 1854; co. seat Tekamah. Grain, livestock, and dairy products are important.

Bur·ton upon Trent (bûr'tən; trĕnt), county borough (1971 pop. 50,175), Staffordshire, W central England, on the Trent River and the Grand Trunk Canal. Brewing, begun here by Benedictine monks, is the most famous industry. Other manufactures include foundry products, tires, footwear, chemicals, and locomotives. Remains of a Benedictive abbey founded in 1002 are in the borough.

Bu·ru or **Boe·roe** (both: bo͞o'ro͞o), island c.3,500 sq mi (9,065 sq km), E Indonesia, in the Moluccas, W of Ceram. Forest products, including cajeput oil, gums and resins, and timber, are exported.

Bu·run·di (bə-ro͝on'dē), republic (1973 est. pop. 3,725,000), 10,747 sq mi (27,835 sq km), E central Africa, bordering on Rwanda in the N, on Tanzania in the E, on Lake Tanganyika in the SW, and on Zaire in the W; capital Bujumbura. A narrow area in the west is part of the western branch of the Great Rift Valley. To the east of this region are mountains, which run north and south and reach an altitude of c.8,800 ft (2,685 m). Further east is a region of broken plateaus with somewhat lower elevations (c.4,500–6,000 ft/1,375–1,830 m), where most of the population lives.

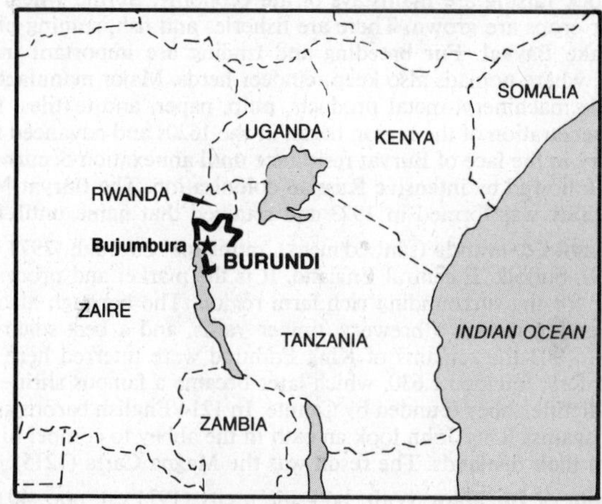

Economy. Burundi is almost entirely agricultural. Beans, cassava, maize, and plantains are grown. Coffee (Burundi's chief export), cotton, and tea are also cultivated. Large numbers of cattle, goats, and sheep are raised; however, the animals play a small role in the economy. Manufactures include basic consumer goods, such as processed food, beverages, clothing, and footwear. Bastnaesite, cassiterite, kaolin, and gold are mined in small quantities. Burundi's imports usually considerably exceed the value of its exports.

History. The Twa were the original inhabitants of Burundi and were followed (c.1200), and then outnumbered, by the Hutu. Probably in the 15th cent., the Tutsi migrated into the area from the northeast, gained dominance over the Hutu, and established several states. By the 19th cent., the country was ruled by the mwami (king)—a Tutsi who controlled the other Tutsi of the region in a vassal relationship. In 1890 Burundi (along with Rwanda) became part of German East Africa. During World War I, Belgian forces occupied Burundi (1916), and in 1919 it became part of the Belgian League of Nations mandate of Ruanda-Urundi (which in 1946 became a UN trust territory). Under the Germans and Belgians, Christianity spread, but the traditional social structure of Burundi was not altered, and there was little economic development.

On July 1, 1962, the country became an independent kingdom. The mid-1960s were marked by fighting and power struggles between the Tutsi and Hutu. In 1965 a coup attempted by the Hutu failed, and the Tutsi retaliated by executing most Hutu political leaders and many other Hutu. After two successful coups in 1966, a republic was established and Michel Micombero, a Tutsi, became president. Following an attempted coup in 1969, Micombero concentrated power in his hands and headed the country's only legal political party, the Unity and National Progress party; a new constitution was adopted in 1970. Renewed fighting between the Tutsi and Hutu in the early 1970s resulted in the deaths of many thousands of Hutu. On Nov. 1, 1976, Micombero was ousted in a military coup led by Lt. Col. Jean-Baptiste Bagaza, who suspended the constitution and assumed the presidency.

Bur·well (bûr'wĕl), city (1970 pop. 1,341), seat of Garfield co., central Nebr., NNW of Grand Island; platted 1883. It is a trade center for a cattle area.

Bur·y (bĕr'ē), county borough (1971 pop. 67,776), Greater Manchester, NE England, on the Irwell River and linked by canal with Bolton and Manchester. Bury has factories for the spinning, weaving, and bleaching of cotton. Hats, paper, machines, and boilers are among its other manufactures.

Bur·yat Autonomous Soviet Socialist Republic (bo͝or-yät'), autonomous republic (1970 pop. 812,000), c.135,600 sq mi (351,205 sq km), SE Siberian USSR, N of Mongolia, extending between Lake Baykal and the Yablonovy Mts.; capital Ulan-Ude. The republic is mountainous and heavily forested and has rivers and lakes that are rich in fish and that provide hydroelectric power. In the mountains are valuable deposits of coal, iron ore, tungsten, molybdenum, gold, wolfram, nickel, bauxite, and manganese. Mining, lumbering, and livestock raising are mainstays of the economy. Spring wheat and fodder crops are grown. There are fisheries and fish-canning plants on Lake Baykal. Fur breeding and trading are important in the north, where nomads also keep reindeer herds. Major manufactures include machinery, metal products, pulp, paper, and textiles. Russian penetration of the region began in the 1620s and advanced for a century in the face of Buryat resistance until annexation occurred in 1727, followed by intensive Russian colonization. The Buryat-Mongol ASSR was formed in 1923 and retained that name until 1958.

Bury Saint Ed·munds (sānt ĕd'mənz), municipal borough (1971 pop. 25,629), Suffolk, E central England. It is the market and processing center for the surrounding rich farm region. The borough also has engineering works, a brewery, timber yards, and a beet-sugar factory. In 903 the remains of King Edmund were interred here in a monastery, founded c.630, which later became a famous shrine and Benedictine abbey founded by Canute. In 1214 English barons struggling against King John took an oath in the abbey to compel him to accept their demands. The result was the Magna Carta (1215).

Bu·shehr or **Bu·shire** (both: bo͞o-shîr'), city (1971 est. pop. 40,000), SW Iran, on the Persian Gulf. It is one of the chief ports of Iran. Carpets, agricultural products, cotton, and wool are exported. Bushehr was founded in 1736. It was used by the British as a base for their Persian Gulf fleet in the 18th cent.

Bush·ey (bo͝osh'ē), urban district (1971 pop. 23,729), Hertfordshire, SE England. Bushey is a residential district just north of Greater London. The local church contains windows by William Morris.

Bush·nell (bo͝osh'nəl), city (1970 pop. 700), seat of Sumter co., central Fla., NNE of Tampa. It is a shipping point in an agricultural area. Dade Battlefield Memorial Park, commemorating the massacre of Maj. Francis L. Dade and his men by Indians (1835), is nearby.

Bush·y Run (bo͝osh'ē rŭn), locality near Greensburg, SW Pa., site of a Pontiac's War battle (Aug. 5–6, 1763) in which Col. Henry Bouquet routed the Indians. The British victory opened the way for the relief of Fort Pitt.

Bus·sa·co or **Bu·sa·co** (both: bo͞o-sä'kō), locality, W central Portugal, in Beira, near Coimbra and around Mt. Bussaco. Now a summer resort, it was formerly a place of seclusion and penitence for monks. At Bussaco in 1810, British and Portuguese troops under Wellington decisively defeated the French in the Peninsular War.

Bu·sto Ar·si·zio (bo͞o'stō är-sē'tsyō), city (1971 pop. 78,632), Lombardy, N Italy. It is a leading center of the Italian cotton industry; metal goods and shoes are also manufactured.

Bu·ta·ri·ta·ri (bə-tär'ē-tär'ē) or **Ma·kin** (mä'kĭn, mä'-), triangular atoll (4.5 sq mi/11.7 sq km), central Pacific, in the Gilbert Islands. Butaritari became a part of the British colony of the Gilbert and Ellice Islands in 1915. During World War II it was the first central Pacific island to be regained by the Allies (Nov., 1943). Butaritari was formerly called Pitt Island.

Bute·shire (byo͞ot'shîr', -shər) or **Bute** (byo͞ot), former county, W Scotland. It consisted primarily of the islands of Bute, Aran, and the Cumbraes. The county town was Rothesay. In 1975 Buteshire became part of the Strathclyde region.

But·ler (bŭt'lər). **1.** County (1970 pop. 22,007), 773 sq mi (2,002.1 sq km), S Ala., in the coastal plain drained by branches of the Sepulga River; formed 1819; co. seat Greenville. Agriculture (cattle, hogs, cotton, peanuts, soybeans, corn, and tobacco), lumbering, and industry are important. **2.** County (1970 pop. 16,953), 582 sq mi (1,507.4 sq km), N central Iowa, in a rolling prairie agricultural area drained by the Shellrock River; formed 1851; co. seat Allison. Its agriculture includes hogs, cattle, poultry, corn, oats, and soybeans. Limestone quarries and sand and gravel pits are in the area. **3.** County (1970 pop. 38,658), 1,445 sq mi (3,742.6 sq km), SE Kansas, in a rolling plain drained by the Walnut River; formed 1855; co. seat El Dorado. It has livestock, grain fields, and highly productive oil wells. **4.** County (1970 pop. 9,723), 443 sq mi (1,147.4 sq km), W central Ky., drained by the Green and Barren rivers; formed 1810; co. seat Morgantown. In a rolling agricultural area yielding livestock, grain, and tobacco, it has coal mines, stone quarries, and tracts of timber. **5.** County (1970 pop. 33,529), 715 sq mi (1,851.9 sq km), SE Mo., in the Ozarks, bounded on the E by the St. Francis River, on the S by the Ark. border, and drained by the Black River; formed 1849; co. seat Poplar Bluff. It has agriculture (cotton, rice, corn, and livestock), lumber, and fire-clay and iron mines. Parts of Lake Wappapello and Mark Twain National Forest are here. **6.** County (1970 pop. 9,461), 582 sq mi (1,507.4 sq km), E Nebr., in an agricultural area bounded on the N by the Platte River and drained in the SW by the Big Blue River; formed 1868; co. seat David City. Livestock, dairy products, and grain are important. **7.** County (1970 pop. 226,207), 471 sq mi (1,219.9 sq km)., extreme SW Ohio, bounded on the W by the Ind. border and intersected by the Miami River and its tributaries; formed 1803; co. seat Hamilton. It has limestone quarries, agriculture (livestock, dairy products, grain, poultry, and tobacco), and some manufacturing. **8.** County (1970 pop. 127,941), 794 sq mi (2,056.5 sq km), W Pa., in an industrial area drained by tributaries of the Beaver River; formed 1800; co. seat Butler. There are bituminous coal mines, oil and gas wells, and limestone quarries. Iron and steel products, chemicals, plate glass, clay products, butter, and grain are produced.

Butler. 1. Town (1970 pop. 2,064), seat of Choctaw co., W Ala., WSW of Selma near the Miss. border; formed 1819, inc. 1848. Lumber, peanut, and textile products are made here. There are oil and minerals in the vicinity. **2.** Town (1970 pop. 1,589), seat of Taylor co., W central Ga., SW of Macon, in a lumber and farm area; inc. 1854. **3.** City (1970 pop. 2,394), De Kalb co., NE Ind., NE of Fort Wayne, in an agricultural area; settled 1841, inc. 1866. Automotive equipment is manufactured. **4.** City (1970 pop. 3.984), seat of Bates co., W Mo., NE of Fort Scott, Kansas; laid out 1854. It is known for its livestock auctions and annual horse show. Optical lenses are made. **5.** Borough (1975 est. pop. 7,730), Morris co., N N.J., NW of Paterson; settled 1695, inc. 1901. Trucks crops are grown, and rubber products manufactured. **6.** City (1970 pop. 18,691), seat of Butler co., W Pa.; inc. as a borough 1817, as a city 1917. It is located in an area rich in coal, natural gas, oil, and limestone. Among its manufactures are steel, railroad-car parts, copper tubing, machinery, and petroleum products.

Butte (byo͞ot). **1.** County (1970 pop. 101,969), 1,665 sq mi (4,312.4 sq km), N central Calif., in the flatlands of the Sacramento Valley rising to c.6,600 ft (2,013 m) in the Sierra Nevada to the E; formed 1850; co. seat Oroville. It is drained by the Feather, Butte, and Sacramento rivers. It has extensive agriculture and ranching, lumbering, and gold dredging. Lumber and rice milling, olive-oil extracting, and fruit canning are done. It includes Oroville Reservoir and parts of Lassen and Plumas national forests. **2.** County (1970 pop. 2,925), 2,240 sq mi (5,801.6 sq km), SE central Idaho, in a livestock-grazing area watered by the Big and Little Lost rivers; formed 1917; co. seat Arco. It has mineral resources, especially silver, zinc, and tungsten. Craters of the Moon National Monument and parts of Chehalis and Salmon national forests are here. A National Reactor Testing Station occupies much of the eastern portion of the county. **3.** County (1970 pop. 7,825), 2,250 sq mi (5,827.5 sq km), W S.Dak., in an agricultural area bounded on the W by the Wyo. and Mont. borders and drained by the Belle Fourche River and Reservoir; formed 1881; co. seat Belle Fourche. Dairy products, sugar beets, grain, wool, and cattle are important. There is some manufacturing. The geographic center of the United States is here.

Butte. **1.** City (1970 pop. 23,368), seat of Silver Bow co., SW Mont.; inc. 1879. It is a trade, distribution, and industrial center. The mining industry has dominated the city's economy since its establishment in 1862. Copper is the major product, and zinc, silver, manganese, gold, lead, and arsenic are also extracted from the numerous mines in the region. First a gold-mining camp, then a silver center, Butte gained importance when copper was discovered (c.1880) and Marcus Daly with his Anaconda Copper Mining Company began to exploit the "richest hill on earth." The Montana College of Mineral Science and Technology is in the city. Local attractions include tours of the mines, a mining museum, and the Columbia Gardens recreational area, maintained for the public by the Anaconda Company. **2.** Village (1970 pop. 575), seat of Boyd co., N Nebr., NW of Norfolk near the S.Dak. border between the Niobrara River and Ponca Creek.

Butts (bŭts), county (1970 pop. 10,560) 185 sq mi (479.2 sq km), central Ga., bounded on the E by the Ocmulgee River; formed 1825; co. seat Jackson. It is in a piedmont agricultural area, growing cotton, corn, truck crops, and fruit. Indian Springs State Park is here.

Bu·tu·an (bo͞o-to͞o'än), city (1970 est. pop. 116,900), capital of Agusan del Norte prov., NE Mindanao, Philippines. It is a port on the Agusan River near its mouth at Butuan Bay.

Bux·ar or **Bax·ar** (both: bək-sär'), village (1971 pop. 31,694), West Bengal state, E central India. A British victory over the Nawab of Oudh at Buxar in 1764 assured British control of the Bengal area.

Bux·ton (bŭks'tən), municipal borough (1971 pop. 20,316), Derbyshire, central England, on the Wye River in Peak District National Park. It is c.1,000 ft (305 m) high; the "old town" is on a hill above it. Buxton is a year-round resort, with mineral springs and baths.

Bü·yük Men·de·res (bo͞o-yŭk' mĕn'də-rēs'), river, Turkey: *see* Maeander.

Bu·zău (bo͞o-zû'o͞o), city (1968 est. pop. 55,000), SE Rumania, in Walachia, on the Buzău River. It is an important railroad junction and a market for petroleum, timber, and grain. Buzău also has oil refineries, foundries, distilleries, and a textile industry.

Buz·zards Bay (bŭz'ərdz), inlet of the Atlantic Ocean, 30 mi (48.3 km) long, from 5 to 10 mi (8–16 m) wide, SE Mass., connected with Cape Cod Bay by the Cape Cod Canal and bounded on the SE by the Elizabeth Islands. Its shores are very irregular.

Byb·los (bĭb'ləs), ancient city, Phoenicia, a port NNE of modern Beirut, Lebanon. The principal city of Phoenicia during the 2nd millennium B.C., it long retained importance as an active port under the Persians. Byblos was the chief center of the worship of Adonis. Because of its papyruses, it was also the source of the Greek word for *book* and, hence, of the name of the Bible. Excavations of Byblos, especially since 1922, have shown that trade existed between Byblos and Egypt as early as c.2800 B.C.

Byd·goszcz (bĭd'gôshch), city (1970 pop. 280,460), capital of Bydgoszcz prov., N central Poland, on the Brda River, a tributary of the Vistula. One of Poland's major inland ports, it stands on the Bydgoszcz Canal (built 1773–74), which links the Brda and Noteć rivers and is part of the Vistula-Oder waterway. Its chief industries produce machinery and machine tools, electrical equipment, metal goods, precision instruments, and chemicals. Chartered in 1346, the city developed during the Middle Ages around the site of a prehistoric fort. In the 15th and 16th cent. it became an important commercial center. It passed to Prussia in 1772 and was returned to Poland in 1919. Occupied by German forces from 1939 to 1945, the city suffered heavy damage in World War II.

Byelo-. For some names beginning thus, *see* Belo-; e.g., for Byelorussia, *see* Belorussia.

Byrds·town (bûrdz'toun'), town (1970 pop. 582), seat of Pickett co., N Tenn., near the Ky. border NW of Knoxville. Dale Hollow Reservoir is nearby.

By·tom (bĭ'tôm), city (1970 pop. 186,993), SW Poland, in the Katowice mining region. An important industrial center, it has factories producing metal products and furniture. A Polish king built a fortress on the site in the 11th cent., and by the 12th cent. the lead and zinc mines of the region were being exploited. The city was chartered in 1254. The Hapsburgs held the city from 1526 until 1742, when it passed to Prussia. In a plebescite after World War I a majority of the population voted to join Poland, but Germany held onto the city. It was finally incorporated into Poland in 1945.

By·zan·ti·um (bĭ-zăn'shē-əm, -tē-əm, bĭ-), ancient city of Thrace, on the site of the present-day Istanbul, Turkey. Founded by the Greeks in 667 B.C., it early rose to importance because of its position on the Bosporus. In the Peloponnesian War it was captured and recaptured by the contending forces. It was taken (A.D. 196) by Roman Emperor Septimius Severus. Constantine I ordered (A.D. 330) a new city built here; this was Constantinople, later the capital of the Byzantine Empire.

Ca·bar·rus (kə-bâr'əs), county (1970 pop. 74,629), 360 sq mi (932.4 sq km), S central N.C., in a piedmont region drained by the Rocky River; formed 1792; co. seat Concord. It has agriculture (cotton, corn, wheat, hay, dairying, and poultry) and timber (pine and oak). Sawmilling and textile manufacturing are done.

Cab·ell (kăb'əl), county (1970 pop. 106,918), 279 sq mi (722.6 sq km), W W.Va., bounded on the NW by the Ohio River and the Ohio border and drained by the Guyandot and Mud rivers; formed 1809; co. seat Huntington. Its agriculture includes dairy products, livestock, poultry, corn, tobacco, fruit, truck crops, and grain. There is bituminous-coal mining and oil and natural-gas extraction in the county. Food products, furniture, glass products, brick and clay tile, metal products, and machinery are manufactured.

Ca·bin·da (kə-bĭn'də), Angolan enclave (1968 est. pop. 55,919), c.2,800 sq mi (7,250 sq km), W Africa. The territory is bounded on the north by the Congo Republic, on the east and south by Zaire, and on the west by the Atlantic Ocean. Cabinda was once geographically part of Angola but was separated from it in 1885 when the Belgian Congo (now Zaire) acquired a corridor to the sea along the lower Congo River. Largely tropical forest, the region produces hardwoods, coffee, cacao, crude rubber, and palm-oil products. Petroleum production began in 1968 and increased in the 1970s.

Cab·i·net Mountains (kăb'ə-nĭt), range of the Rocky Mts. in NW Mont. and N Idaho, extending c.65 mi (105 km). Its highest point is Snowshoe Peak (c.9,000 ft/2,745 m).

Cab·ot Strait (kăb'ət), channel, c.60 mi (96.5 km) wide, between SW N.F. and N Cape Breton Island, Canada, connecting the St. Lawrence River with the Atlantic Ocean.

Ca·bril·lo National Monument (kə-brĭl'ō, -brē'ō, -yō): *see* National Parks and Monuments Table.

Cá·ce·res (kä'sā-rās, *-thā-*), city (1975 est. pop. 58,870), capital of Cáceres prov., W central Spain, in Estremadura. Products of cork, leather, pottery, and cloth are made here. Cáceres was an important Roman colony. It fell to the Moors in the 8th cent. but was recaptured (1229) by Alfonso IX.

Cache (kăsh), county (1970 pop. 42,331), 1,175 sq mi (3,043.3 sq km), N Utah, in an irrigated agricultural area bordering on Idaho and drained by the Bear River; formed 1856; co. seat Logan. Hay, sugar beets, fruit, and truck crops are grown. Part of Cache National Forest is in the county.

Cache la Pou·dre (lə po͞o'drə), river, N Colo., rising in the NW corner of Rocky Mountain National Park and flowing c.125 mi (200 km) N, E, and SE to the South Platte River near Greeley. It waters a large sugar-beet and livestock region.

Cad·do (kăd'ō). **1.** Parish (1970 pop. 230,184), 900 sq mi (2,331 sq km), extreme NW La., on the Texas-Ark. border and bounded on the E by the Red River; formed 1838; parish seat Shreveport. It has natural-gas wells, and oil and gas refineries and pipelines. Its agriculture includes cotton, corn, hay, and peanuts. **2.** County (1970 pop. 28,931), 1,275 sq mi (3,302.3 sq km), W central Okla., in a hilly agricultural area intersected by the Washita River; formed 1907; co. seat Anadarko. Peanuts, wheat, cotton, and alfalfa are grown. There are oil and natural-gas wells, an oil refinery, and some manufacturing plants.

Cad·il·lac (kăd'ə-lăk'), city (1975 est. pop. 9,300), seat of Wexford co., NW Mich., SSE of Traverse City; settled c.1871 on the site of a French trading post, inc. as a village 1875, as a city 1877. It is a trade center for a farming, lumbering, fishing, and resort area and has lumber mills, furniture factories, and other industrial plants.

Ca·diz (kā'dĭz). **1.** City (1970 pop. 1,987), seat of Trigg co., SW Ky., W of Hopkinsville near the Cumberland River. It is a trade and shipping center for a tobacco, grain, livestock, and timber area. The Brelsford Caves and a U.S. wildlife refuge are nearby. **2.** Village (1970 pop. 3,060), seat of Harrison co., E Ohio, SW of Steubenville; laid out 1803–4, inc. 1818. It is the trading and marketing center of a farm area. Coal mining is done, and sheet metal is made.

Cá·diz (kä'dēth), city (1975 est. pop. 140,862), capital of Cádiz prov., SW Spain, in Andalusia, on the Bay of Cádiz. Picturesquely situated on a promontory, it is today chiefly a port exporting wines and other agricultural items and importing coal, iron, and foodstuffs. Shipbuilding and fishing are other industries. The clean, white city has palm-lined promenades and parks. Its 13th cent. cathedral, originally Gothic, was rebuilt in Renaissance style; the new cathedral was begun in 1722. The Phoenicians founded here (c.1100 B.C.) the port of Gadir, which became a market for tin and silver. It was taken (c.500 B.C.) by the Carthaginians and passed late in the 3rd cent. B.C. to the Romans, who called it Gades. It flourished until the fall of Rome, but suffered from the barbarian invasions and declined further under the Moors. After its reconquest (1262) by Alfonso X of Castile, its fortifications were rebuilt. The discovery of America revived its prosperity. Columbus sailed from Cádiz on his second voyage (1495). In 1587 Sir Francis Drake burned a Spanish fleet in its harbor, and in 1596 the Earl of Essex attacked and partly destroyed the city. In 1718 Cádiz became the official center for New World trade. After Spain lost its American colonies, the city declined. During the siege by the French—which Cádiz resisted for two years (1810–12) until relieved by Wellington—the Cortes assembled in the city and issued a liberal constitution for Spain (Mar., 1812).

Cae·li·an (sē'lē-ən), one of the seven hills of Rome.

Caen (käɴ), city (1975 pop. 119,474), capital of Calvados dept., N France, in Normandy, on the Orne River. A busy port and the commercial center of the rich Calvados region, it is highly industrialized; the nearby iron-ore mines are the second largest in France. The city's manufactures include automobiles, heavy equipment, electronic gear, and textiles (especially lace). Caen's importance dates from the 11th cent., when it was a favorite residence of William the Conqueror. During the French Revolution it was a rallying place for the federalists. The town, an architectural gem, was largely destroyed during the Normandy campaign of World War II. However, three outstanding examples of 11th cent. Norman architecture survived: the Abbaye aux Hommes, founded by William the Conqueror, who is buried here; the Abbaye aux Dames, founded by Queen Matilda; and the Church of St. Nicholas.

Caer·nar·von (kär-när'vən), municipal borough (1973 est. pop. 8,840), administrative center of Gwynedd, NW Wales, on Menai Strait. Petroleum is imported and slate exported. Tourism is important. The castle, begun by Edward I c.1284, is a fine example of a medieval fortress. The Prince of Wales is invested at Caernarvon. Until 1974 the borough was county town of Caernarvonshire.

Caer·nar·von·shire (kär-när'vən-shîr', -shər), former county, NW Wales. The county town was Caernarvon. Historical remains include evidence of Roman settlement. In 1974 Caernarvonshire became part of the new nonmetropolitan county of Gwynedd.

Caer·phil·ly (kär-fĭl'ē), urban district (1973 est. pop. 42,190), Mid Glamorgan, S Wales. In a coal area, it is also a market center and is noted for its cheese. Its 13th cent. castle is the largest in Wales.

Cae·sa·re·a Pal·e·sti·nae (sĕs-ə-rē'ə păl-ĭ-stī'nē, sēz-, sĕz-), old city, NW Palestine, S of Mt. Carmel. It was taken (104 B.C.) by Alexander Jannaeus, leader of the Maccabees, and was made (30 B.C.) the capital of Herod the Great. The Jewish citizens were massacred by the Romans in A.D. 66.

Caesarea Phil·ip·pi (fĭl-ĭp'ī), ancient city, N Palestine, at the foot of Mt. Hermon. It was built by Philip the Tetrarch in the 1st cent. A.D. Its site (Paneas) had long been a center for the worship of Pan.

Ca·glia·ri (kä'lyä-rē), city (1976 est. pop. 240,256), capital of Sardinia and of Cagliari prov., S Sardinia, Italy, on the Gulf of Cagliari (an arm of the Mediterranean Sea) and at the mouth of the Mannu River. It is the largest city in Sardinia and is a modern port and an industrial center. A flourishing Carthaginian city, it was taken by Rome in 238 B.C. Cagliari endured Arab invasions in the 8th and 9th cent. A.D. The city was the site of a submarine base in World War II and was heavily bombed by the Allies. Noteworthy structures include the Romanesque-Gothic cathedral (13th cent.) and the Basilica of San Saturnino (5th cent.).

Ca·guas (kä'gwäs), city (1970 pop. 63,215), E central Puerto Rico. Largest of Puerto Rico's inland cities, Caguas is an industrial center. Sugar refining is done.

Ca·ha·ba (kə-hô'bə), river, central Ala., rising in the mountains NE of Birmingham and flowing c.200 mi (320 km) SW and S into the Alabama River near Selma.

Ca·ho·ki·a (kə-hō'kē-ə), village (1970 pop. 20,649), St. Clair co., SW Ill., a residential suburb of East St. Louis, on the Mississippi River; inc. 1927. The first permanent settlement in Illinois, it was named for a tribe of the Illinois Indians. The French established a mission in 1699 and a fur-trading post later. Cahokia was occupied by the British in 1765 and captured by the Americans under George Rogers Clark in 1778.

Cahokia Mounds, approximately 85 Indian earthworks in Cahokia State Park, SW Ill., near East St. Louis. It is the largest group of mounds north of Mexico. Monks' Mound, a rectangular, flat-topped earthwork, 100 ft (30.5 m) high with a 17-acre (6.9-hectare) base, is the largest mound; it is named for Trappist monks who settled on top of it in the early 19th cent. The people who constructed the mounds were village dwellers living in a fertile river-bottom area; their culture flourished from c.1300 to c.1700.

Ca·hors (kä-ôr'), town (1975 pop. 20,311), capital of Lot dept., S central France, in Quercy, on the Lot River. A commercial center, it has canneries, distilleries, and factories making a great variety of products. It was an important Roman town. Ruled by its bishops until the 14th cent., it was one of the major banking centers of medieval Europe. The old part of Cahors is of great architectural interest. Part of the medieval fortifications, including a fortified bridge, still stand. The Cathedral of St. Étienne (12th–15th cent.), with Byzantine cupolas, and the palace of John XXII (begun 14th cent.; never completed) are among its many edifices.

Cai·cos Islands (kā'kəs): *see* Turks and Caicos Islands.

Cairn·gorms (kârn'gôrmz), group of mountains forming part of the Grampian system, central Scotland, between the Dee and the upper Spey rivers. They rise to c.4,300 ft (1,310 m). The name "cairngorm" is given to an ornamental yellow or brown quartz found in the mountains. The region is being developed for winter sports.

Cairns (kârnz), city (1975 est. pop. 35,200), Queensland, NE Australia, on Trinity Bay. It is a principal sugar port of Australia; lumber and other agricultural products are also exported. The city's proximity to the Great Barrier Reef has made it a tourist center.

Cai·ro (kī'rō), city (1976 pop. 8,000,133), capital of Egypt and its Cairo governorate, N Egypt, a port on the Nile River near the head of its delta. The city includes two islands in the Nile, Zamalik (Gezira) and Rawdah (Roda), which are linked to the mainland by bridges. Cairo's manufactures include textiles, food products, pharmaceuticals, chemicals, plastics, and metals.

Babylon, a Roman fortress city, occupied a part of southeast Cairo now known as Old Cairo. Cairo was founded in 969 by the Fatimids to replace nearby Al Qatai as the capital of Egypt. In the 12th cent. Saladin ended Fatimid rule. To defend the city against an attack by Crusaders, he erected (c.1179) the citadel, which still stands, and extended the walls of the city, parts of which remain. Cairo prospered under the rule of the Mamelukes, who added many buildings of high artistic merit, but the city declined after it was conquered (1517) by the Ottoman Empire. At the time of its capture (1798) by French forces led by Napoleon I, the city had about 250,000 inhabitants. British and Turkish forces ousted the French in 1801, and Cairo was returned to Ottoman control. During World War II, Cairo was the Allied headquarters and supply center for the Middle East and was the site (1943) of the Cairo Conference. From 1958 to 1961 the city was the capital of the United Arab Republic, which joined Egypt and Syria.

Today much of Cairo is modern, with wide streets; its famed mosques, palaces, and city gates are found mostly in the older sections. The mosques of Amur (7th cent.), Ibn Tulun (876-79), Hasan (c.1356), and Qait Bay (1475) are especially noted for their bold design. The Mosque of Al Azhar (970) and adjoining buildings house Al Azhar Univ., considered the world's leading center of Koranic studies. The Univ. of Cairo is nearby, in Al Jizah. Cairo has many museums; the Egyptian National Museum is especially noted for its holdings of ancient Egyptian art. The Nilometer, a graduated column first built in 716 and used to measure the Nile water level, is on Rawdah island, where the infant Moses is believed to have been found in the bulrushes.

Cai·ro (kâr'ō, kä'rō). **1.** City (1970 pop. 8,061), seat of Grady co., SW Ga., near the Fla. line WNW of Thomasville, in a fertile farm area; settled 1866, inc. 1870. It is a processing and shipping center for cane sugar and syrup, tobacco, pecans, peanuts, and cucumbers. There are lumber, veneer, and pulp mills and furniture and box factories here. **2.** City (1978 est. pop. 5,800), seat of Alexander co., extreme S Ill., on a levee-protected tongue of land between the Mississippi and Ohio rivers; inc. 1857. It is a center for shipping by river, rail, and highway and the processing and distributing point for a large and fertile farm area. Manufactures include flour, lumber, cottonseed oil, textiles, woodwork, and silica. The city and surrounding area are popularly called "Egypt" because of the deltalike geographic similarity. Settlement was attempted here in 1818, but permanent settlement did not begin until 1837. Cairo was a strategic point in the Civil War; it was Gen. Grant's headquarters during much of his Western campaign. The city has often been endangered by floods, but Federal flood control projects have decreased the danger. Fort Defiance State Park, the site of a Civil War fort, on the southern edge of town, offers a magnificent view of the convergence of the Ohio and Mississippi rivers.

Caith·ness (kāth'nĕs, kāth-nĕs'), former county, NE Scotland. Wick was the county town. Originally part of the Pictish nation, Caithness was absorbed into the Viking earldom in the 9th cent. and reverted to Scottish rule only in 1202. It was the scene of frequent clan warfare until the end of the 17th cent. In 1975 Caithness became part of the Highland region.

Ca·ja·mar·ca (kä'hä-mär'kä), city (1972 pop. 37,608), N Peru. An important commercial center, Cajamarca is situated at an altitude of c.9,000 ft (2,745 m) and has a cool, dry climate. Grains and alfalfa are raised in the region, and gold, silver, and copper come from nearby mines. Inca ruins and nearby thermal springs attract many tourists.

Ca·la·bri·a (kə-lā'brē-ə), region (1975 est. pop. 2,034,425), 5,822 sq mi (15,079 sq km), S Italy, a peninsula projecting between the Tyrrhenian Sea and the Ionian Sea, separated from Sicily by the narrow Strait of Messina. It forms the toe of the Italian "boot." Catanzaro is the capital. The region is generally mountainous, with narrow coastal strips. Farming is the main occupation; olives, plums, grapes, citrus fruit, and wheat are grown, and sheep and goats are raised. Fishing is well developed along the Strait of Messina. The ancient Bruttium, the region was named Calabria in the 8th cent.; before then Calabria referred to the present southern Apulia. In the 11th cent. Calabria was part of the Norman kingdom of Sicily and after 1822 became part of the kingdom of Naples. The region was conquered by Garibaldi in 1860. Feudal landholding patterns, malaria, destructive earthquakes (particularly in 1905 and 1908), droughts, and poor transportation facilities have hindered the economic development of the region and resulted in large-scale emigration.

Ca·lah (kā'lə) or **Ka·lakh** (kä'läKH), ancient city of Assyria, S of Nineveh and therefore S of present Mosul, Iraq. Known as Calah in the Bible, it is the same as the ancient Nimrud, named after a legendary Assyrian hunting hero. Calah emerged as a famous city when Ashurnasirpal II chose (c.880 B.C.) the site for his capital. Excavations carried on since the mid-19th cent. have revealed remarkable bas-reliefs, ivories, and sculptures.

Ca·la·hor·ra (kä'lä-ôr'rä), town (1970 pop. 16,340), Logroño prov., NE Spain, in Old Castile, on the Cidacos River near its confluence with the Ebro. Calahorra is a farm (cereals and grapes) and manufacturing center. Known in ancient times as Calagurris, it is the place where Pompey unsuccessfully besieged (76–72 B.C.) the rebel Sertorius. An old cathedral (c.5th cent.; restored 15th cent.) and some Roman ruins survive today.

Ca·lais (kă-lā', kăl'ā), city (1975 pop. 78,820), Pas-de-Calais dept., N France, in Picardy, on the Strait of Dover. An industrial center with a great variety of manufactures, it has been a major commercial seaport and a communications center with England since the Middle Ages. In 1347, after a siege of 11 months, Calais fell to Edward III of England. A bronze monument by Rodin commemorates the famous episode of the six burghers who offered their lives to save the town. The city was recovered (1558) by the French. It was the scene of much fighting in World War II.

Cal·a·ver·as (kăl'ə-vâr'əs), county (1970 pop. 13,585), 1,032 sq mi (2,672.9 sq km), central Calif., in the Sierra Nevada bounded on the N by the Mokelumne River and on the S by the Stanislaus River; formed 1850; co. seat San Andreas. It is a winter sports area, with gold mining, lumbering, stock grazing, and general farming. Stanislaus National Forest is here.

Cal·ca·sieu (kăl'kə-sōō), parish (1970 pop. 145,415), 1,105 sq mi (2,862 sq km), extreme SW La., bounded on the W by the Sabine River, here forming the Texas border; formed 1840; parish seat Lake Charles. It has agriculture (rice, corn, cotton, soybeans, and sweet potatoes), stock raising, oil and natural-gas wells, sulfur mines, sand and gravel pits, lumbering, and diversified manufacturing.

Calcasieu, river c.200 mi (320 km) long, rising in W central La. and flowing S through Lake Charles and Calcasieu Lake to the Gulf of Mexico. The river, which is partly navigable, connects the port of Lake Charles city with the Intracoastal Waterway and the Gulf of Mexico.

Cal·cut·ta (kăl-kŭt'ə), city (1971 pop. 3,141,180), capital of West Bengal state, E India, on the Hooghly River. It is the second-largest city in India and one of the largest in the world. The population of Greater Calcutta in 1971 was 7,005,362. Its area is 228.5 sq mi (591.8 sq km). Calcutta is the chief port and major industrial center of eastern India; jute is milled, and textiles, chemicals, paper, and metal products are manufactured. Calcutta was founded c.1690 by the British East India Company. In 1756 the nawab of Bengal captured Calcutta and killed most of its garrison by imprisoning it overnight in a small, stifling room, known as the notorious "black hole." Robert Clive retook the city in 1757. From 1833 to 1912, Calcutta was the capital of India. The Univ. of Calcutta (founded 1857) and the Indian Museum, which houses one of the world's outstanding natural history collections, are in the city.

Cald·well (kôld'wĕl). **1.** County (1970 pop. 13,179), 357 sq mi (924.6 sq km), W Ky., in a rolling agricultural area bounded on the NE by the Tradewater River; formed 1809; co. seat Princeton. It has fluorspar and coal mines, timber tracts, and stone quarries. Some manufacturing is done. Part of Pennyrile State Forest is here. **2.** Parish (1970 pop. 9,354), 550 sq mi (1,424.5 sq km), NE central La., bounded on the E by the Boeuf River and intersected by the Ouachita River and Bayou Castor; formed 1838; parish seat Columbia. Farming (cotton, corn, sweet potatoes, livestock, and hay) and forestry are important. It has natural-gas wells, and cotton ginning and lumber milling are done. **3.** County (1970 pop. 8,351), 430 sq mi (1,113.7 sq km), NW Mo.; formed 1836; co. seat Kingston. It has agriculture (corn, wheat, oats) and some coal, and its industry includes shoe manufacturing. **4.** County (1970 pop. 56,699), 476 sq mi (1,232.8 sq km), W central N.C., bounded on the NW by the Blue Ridge and drained by the Catawba and Yadkin rivers; formed 1841; co. seat Lenoir. Tobacco is its main crop, and textiles and furniture are manufactured. **5.** County (1970 pop. 21,178), 544 sq mi (1,409 sq km), S central Texas, bounded on the SW by the San Marcos River; formed 1848; co. seat Lockhart. It has agriculture (cotton, corn, grain sorghums, vegetables, watermelons, pecans, peanuts, and livestock), oil and natural-gas fields, and some processing and manufacturing plants.

Caldwell. **1.** City (1978 est. pop. 15,600), seat of Canyon co., SW Idaho, on the Boise River; inc. 1890. On the site of an Oregon Trail camping ground, the city is now a major processing and distribution center for an agricultural and livestock area. Mobile homes and recreational vehicles are manufactured here. **2.** City (1970 pop. 1,540), Sumner co., S central Kansas, SSW of Wichita near the Okla. border, in a wheat and grazing area; laid out 1871, inc. 1879. An important trading point on the old Chisholm Trail, it had a boom period in the 1880s. **3.** Industrial borough (1970 pop. 8,677), Essex co., NE N.J., NW of Newark; settled before 1785, inc. 1892. The Presbyterian parsonage in which Grover Cleveland was born is now a museum. **4.** Village (1970 pop. 2,082), seat of Noble co., SE Ohio, on Duck Creek S of Cambridge; laid out 1857, inc. 1870. There are coal and salt mines and oil wells in the vicinity. **5.** City (1970 pop. 2,308), seat of Burleson co., E central Texas, ENE of Austin, in a cotton and corn area. Cottonseed oil is manufactured.

Cal·e·do·ni·a (kăl'ĭ-dō'nyə, -nē-ə), Roman name for that part of the island of Great Britain that lies N of the firths of Clyde and Forth. The name first occurs in the works of Lucan (1st cent. A.D.) and has been used in modern times rhetorically and poetically to mean all of Scotland or the Scottish Highlands.

Caledonia, county (1970 pop. 22,789), 612 sq mi (1,585 sq km), NE Vt., partly bounded on the E by the Connecticut River; organized 1792; co. seat St. Johnsbury. It is drained by the Passumpsic, Moose, Lamoille, and Wells rivers. In a dairying and lumbering area, it has granite quarries and manufactures scales, machinery, and paper.

Caledonia, village (1970 pop. 2,619), seat of Houston co., extreme SE Minn., near the Iowa-Wis. border SW of La Crosse; settled 1855. It is a trade and manufacturing center in a farm area.

Cal·e·do·ni·an Canal (kăl'ĭ-dōn'yən, -dō'nē-ən), waterway, c.60 mi (95 km) long, cutting across Highland region, N Scotland, from Moray Firth to Loch Linnhe by way of the Great Glen. Built in two phases (1803-22 and 1843-47; opened 1822) to save shallow-draft vessels the circuitous route around northern Scotland, it is of little use today except for pleasure craft.

Ca·lex·i·co (kə-lĕk'sĭ-kō), city (1978 est. pop. 13,400), Imperial co., S Calif., at the Mexican border; inc. 1908. A port of entry from its adjacent sister city of Mexicali, Mexico, it is also a trade center in the southern part of the fertile Imperial Valley.

Cal·ga·ry (kăl'gə-rē), city (1976 pop. 469,917), S Alta., Canada, at the confluence of the Bow and Elbow rivers. Calgary is a wholesale and processing center for a large agricultural and stock-raising area. The city began (1875) as a fort of the Northwest Mounted Police. The Calgary Stampede, inaugurated 1912, is an annual rodeo.

Cal·houn (kăl-ho͞on'). **1.** County (1970 pop. 103,092), 610 sq mi (1,580 sq km), E Ala., in an agricultural area bounded on the W by the Coosa River; formed 1832; co. seat Anniston. Textiles are manufactured, and there are deposits of iron ore, limestone, bauxite, and barites. It includes part of Talladega National Forest. **2.** County (1970 pop. 5,573), 629 sq mi (1,629.1 sq km), bounded on the E by Moro Creek and on the S by the Ouachita River; formed 1850; co. seat Hampton. It has agriculture (cotton, corn, and truck crops) and timber. Lumber milling and cotton ginning are done. **3.** County (1970 pop. 7,624), 561 sq mi (1,453 sq km), NW Fla., in a lowland area bounded on the E by the Apalachicola River and drained by the Chipola River; formed 1838; co. seat Blountstown. Agriculture (corn, peanuts, sugar cane, vegetables, and livestock) and forestry (lumber and naval stores) are important to its economy. **4.** County (1970 pop. 6,606), 289 sq mi (748.5 sq km), SW Ga., intersected by Ichawaynochaway Creek; formed 1854; co. seat Morgan. It is in a coastal-plain area yielding cotton, corn, truck crops, peanuts, and pecans. **5.** County (1970 pop. 5,675), 247 sq mi (639.7 sq km), W Ill., bounded by the Mississippi and Illinois rivers, which join at the county's SE tip; formed 1835; co. seat Hardin. In an apple-growing region, it also produces corn, wheat, livestock, dairy products, and vinegar. **6.** County (1970 pop. 14,292), 572 sq mi (1,481.5 sq km), central Iowa, in a prairie agricultural area drained by the Raccoon River; formed 1851; co. seat Rockwell City. It has bituminous-coal deposits and sand and gravel pits. Corn, oats, soybeans, hogs, cattle, and poultry are produced. **7.** County (1970 pop. 141,963), 709 sq mi (1,836.3 sq km), S Mich., drained by Battle Creek and the Kalamazoo and St. Joseph rivers; formed 1833; co. seat Marshall. In an agricultural area, its major crop is corn. Livestock is raised, and it has both heavy and light industry. Kellogg Bird Sanctuary is here. **8.** County (1970 pop. 14,625), 575 sq mi (1,489.3 sq km), N central Miss., drained by the Skuna and Yalobusha rivers; formed 1852; co. seat Pittsboro. Its major crops are cotton, soybeans, and sweet potatoes. It also has stock raising, a lumber industry, and bauxite, lignite, and clay deposits. **9.** County (1970 pop. 10,780), 377 sq mi (976.4 sq km), central S.C., bounded on the N by the Congaree River, on the NE by Lake Marion; formed 1908; co. seat St. Matthews. Cotton is its major crop; soybeans, feed grains, and timber are also important. Cattle and hogs are raised, and there is some industry. **10.** County (1970 pop. 17,831), 527 sq mi (1,364.9 sq km), S Texas, on the Gulf Coast, indented by San Antonio, Lavaca, and Matagorda bays; formed 1846; co. seat Port Lavaca. Fisheries, seafood packing, agriculture (cotton, corn, sorghums, rice, fruit, vegetables, and flax), livestock raising, and dairying are important. It also has oil and natural-gas wells and refineries and resorts along the coast. **11.** County (1970 pop. 7,046), 281 sq mi (727.8 sq km), central W.Va., on the Allegheny Plateau and drained by the Little Kanawha River; formed 1856; co. seat Grantsville. The county raises livestock and produces fruit and tobacco. It has oil and natural-gas deposits and some coal and timber. Rubber and plastic products are manufactured.

Calhoun. **1.** City (1970 pop. 4,748), seat of Gordon co., NW Ga., on the Oostanaula River NNW of Atlanta; inc. 1852. It is a textile-manufacturing center and also has a lumber industry. During the Civil War the city was destroyed (1864) by Gen. William Tecumseh Sherman, and was later rebuilt. **2.** City (1970 pop. 901), seat of McLean co., W Ky., on the Green River NE of Madisonville, in an agricultural, coal, and timber area. Canned goods are produced.

Ca·li (kä'lē), city (1973 est. pop. 923,446), capital of Valle del Cauca dept., W Colombia, on the Cali River. It is an industrial and commercial center of the upper Cauca River valley. Livestock, minerals, lumber, and farm products are shipped through the city; and tires, tobacco products, textiles, paper, chemicals, and building materials are manufactured. Cali is also a tourist center. The city was founded in 1536, but its growth is relatively recent, with the population more than doubling in the 1950s.

Cal·i·cut (kăl'ĭ-kət) or **Ko·zhi·kode** (kō'zhə-kōd'), city (1971 pop. 333,980), Kerala state, SW India, on the Malabar coast of the Arabian Sea. Timber, cashew nuts, spices, tea, and coffee are exported. Calicut was (1498) Vasco da Gama's first Indian port of call, and the city soon became a center for European traders. The term "calico" was first applied to Calicut cotton cloth.

Cal·i·for·nia (kăl'ə-fôrn'yə, -fôr'nē-ə), state (1975 est. pop. 21,165,-000), 158,693 sq mi (411,015 sq km), W United States, admitted as the 31st state of the Union in 1850. The capital is Sacramento. California is bounded on the north by Oregon, on the east by Nev. and Ariz. (from which it is separated by the Colorado River), on the south by Mexico, and on the west by the Pacific Ocean.

Ranking first among the U.S. states in population and third in area, California has a diverse topography and climate. A series of low mountains known as the Coast Ranges extends along the 1,200-mi (1,930.8-km) coast. The region from Point Arena, north of San Francisco, to the southern part of the state is subject to earthquakes caused by the San Andreas fault. The Coast Ranges receive heavy rainfall in the north, where the giant redwood forests prevail, but the climate of these mountains is considerably drier in southern California, and south of the Golden Gate no major rivers reach the ocean. Behind the coastal ranges lies the great Central Valley, drained by the Sacramento and San Joaquin rivers. In the southeast are vast wastelands, notably the Mojave Desert. Rising east of the Central Valley is the Sierra Nevada range. Death Valley is east of the southern Sierra Nevada.

California is divided into 58 counties:

NAME	COUNTY SEAT	NAME	COUNTY SEAT
Alameda	Oakland	Colusa	Colusa
Alpine	Markleeville	Contra Costa	Martinez
Amador	Jackson	Del Norte	Crescent City
Butte	Oroville	El Dorado	Placerville
Calaveras	San Andreas	Fresno	Fresno

NAME	COUNTY SEAT	NAME	COUNTY SEAT
Glenn	Willows	San Benito	Hollister
Humboldt	Eureka	San Bernardino	San Bernardino
Imperial	El Centro	San Diego	San Diego
Inyo	Independence	San Francisco	San Francisco
Kern	Bakersfield	San Joaquin	Stockton
Kings	Hanford	San Luis Obispo	San Luis Obispo
Lake	Lakeport	San Mateo	Redwood City
Lassen	Susanville	Santa Barbara	Santa Barbara
Los Angeles	Los Angeles	Santa Clara	San Jose
Madera	Madera	Santa Cruz	Santa Cruz
Marin	San Rafael	Shasta	Redding
Mariposa	Mariposa	Sierra	Downieville
Mendocino	Ukiah	Siskiyou	Yreka
Merced	Merced	Solano	Fairfield
Modoc	Alturas	Sonoma	Santa Rosa
Mono	Bridgeport	Stanislaus	Modesto
Monterey	Salinas	Sutter	Yuba City
Napa	Napa	Tehama	Red Bluff
Nevada	Nevada City	Trinity	Weaverville
Orange	Santa Ana	Tulare	Visalia
Placer	Auburn	Tuolumne	Sonora
Plumas	Quincy	Ventura	Ventura
Riverside	Riverside	Yolo	Woodland
Sacramento	Sacramento	Yuba	Marysville

Economy. Although agriculture is second to industry as the basis of the state's economy, California is a leading state in the production of fruits and vegetables and is the largest producer in the United States of many crops, including tomatoes, carrots, lettuce, asparagus, broccoli, spinach, and strawberries. The state's most valuable crops are hay, grapes, tomatoes, and cotton. Cattle and dairy products also contribute a great deal of farm income. The state produces the major share of American wine. The gathering and packing of crops is done largely by seasonal migrant labor. Fishing is another important industry. Much of the state's manufacturing depends on the processing of farm produce and upon such local natural resources as mineral deposits and forests. Petroleum is the state's most valuable mineral; in the late 1960s California ranked third in the country in oil production. Other important products are natural gas, cement, and sand and gravel. Since World War II heavy industry in the state has increased enormously, notably in the manufacture of transportation equipment, electronic equipment, machinery, and metal products. Defense-contract industries, particularly in southern California, represent a major base of the region's economy and have contributed to the growing wealth and population of the area. California has long been a major U.S. center for motion-picture and television film production. Tourism is an important source of income. Disneyland, San Francisco and the Golden Gate Bridge, the giant sequoia, many national parks and forests, and beautiful beaches are among California's numerous attractions.

History. The first voyage (1542) to Alta California (Upper California), as the region north of Baja California (Lower California) came to be known, was commanded by the Spanish explorer Juan Rodriguez Cabrillo. In 1579 Sir Francis Drake landed near Point Reyes, north of San Francisco, and claimed the region for Queen Elizabeth I. Colonization was slow, but finally in 1769 Gaspar de Portolá, governor of the Californias, led an expedition up the Pacific coast and established a colony on San Diego Bay. The following year he explored the area around Monterey Bay. Soon afterward Monterey became the capital of Alta California. Franciscans later founded several missions that extended as far north as Sonoma, north of San Francisco. The missionaries sought to Christianize the Indians but also forced them to work as manual laborers. Cattle raising was of primary importance, and hides and tallow were exported. In 1776 Juan Bautista de Anza founded San Francisco, where he established a military outpost.

The early colonists, called the Californios, for the most part were not interfered with by the central government of New Spain or later (1820s) by that of Mexico. Under Mexican rule the missions were secularized (1833–34) and the Indians released from their servitude. Colonization of California remained largely Mexican until the 1840s. Russian fur traders had penetrated south to the California coast and established Fort Ross, north of San Francisco, in 1812. Jedediah Strong Smith and other trappers made the first U.S. overland trip to the area in 1826, but U.S. settlement did not become significant until the 1840s. After having briefly asserted the independence of California in 1836, the Californios drove out the last Mexican governor in 1845. Under the influence of the American explorer John C. Frémont, U.S. settlers set up (1846) a republic at Sonoma. The news of war between the United States and Mexico (1846–48) reached California soon afterward. On July 7, 1846, Commodore John D. Sloat captured Monterey, the capital, and claimed California for the United States.

By the Treaty of Guadalupe Hidalgo (1848), Mexico formally ceded the territory to the United States. In the same year, while establishing a sawmill for John Sutter near Coloma, James W. Marshall discovered gold and touched off the California gold rush. The forty-niners, as the miners were called, came in droves. San Francisco rapidly became a boom city. With the gold rush came a huge increase in population and a pressing need for civil government. In 1849 Californians sought statehood, and California entered the Union as a nonslavery state by the Compromise of 1850. San Jose, Monterey, Vallejo, and Benicia each served as the capital before it was finally moved to Sacramento in 1854. Communication and transportation depended upon ships, the stagecoach, the pony express, and the telegraph until the transcontinental railroad was completed, with imported Chinese labor, in 1869. A railroad-rate war (1884) and a boom in real estate (1885) fostered a new wave of overland immigration.

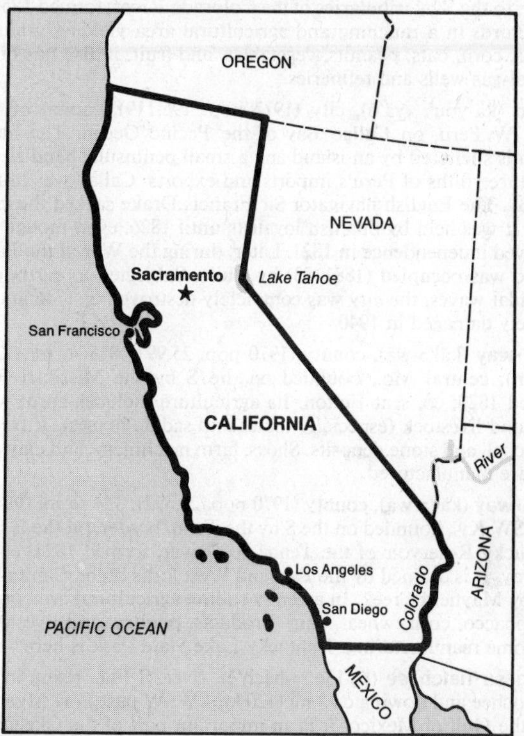

By the turn of the century the discovery of oil, industrialization resulting from the increase of hydroelectric power, and expanding agricultural development attracted more settlers. Los Angeles grew rapidly in this period and in population soon surpassed San Francisco, which suffered greatly after the great earthquake and fire of 1906. Successive waves of settlers arrived in California, attracted by a new real-estate boom in the 1920s, by the promise of work in the 1930s, and by ship and aircraft production during World War II. Prosperity and rapid population growth continued after the war. The turbulent 1960s brought riots in Watts, a black section of Los Angeles (1965); student demonstrations and protests over the issue of free speech in the state universities (1964); and a long struggle by migrant farm workers for unionization and better conditions.

Government. The state's present constitution dates from 1879 and provides for initiative, referendum, and recall of public officials. The executive branch is headed by a governor elected for a four-year term. California's legislature has a senate with 40 members elected for four-year terms and an assembly with 80 members elected for two years.

Educational Institutions. Among the state's more prominent institutions of learning are the Univ. of California, with eight campuses; Occidental College and the Univ. of Southern California, at Los Angeles; Stanford Univ., at Stanford; the California Institute of Technology, at Pasadena; Mills College, at Oakland; and the Claremont Colleges, at Claremont.

California. **1.** City (1970 pop. 3,105), seat of Moniteau co., central Mo., WNW of Jefferson City, in a livestock, dairy, and grain area; founded 1845, inc. 1857. Harnesses, saddles, and pork products are made. **2.** Borough (1970 pop. 6,635), Washington co., SW Pa., on the Monongahela River S of Pittsburgh, in a coal-mining region; laid out c.1850, inc. 1853. California State College is here.

California, Gulf of, arm of the Pacific Ocean, c.700 mi (1,125 km) long and 50 to 130 mi (80–210 km) wide, NW Mexico, separating Baja California from the Mexican mainland. The gulf is part of a depression in the earth's surface that extends inland to the Coachella Valley in southern Calif. The Imperial Valley and the Salton Sea, once part of the gulf, have been cut off from it by the growth of the Colorado River delta. The gulf deepens from north to south; its greatest depth is c.8,500 ft (2,595 m). Storms and tidal currents hinder navigation in the gulf. Commercial and sport fishing thrive; pearl, sponge, and oyster beds are harvested. The region, first explored in 1538, is a developing tourist center.

Cal·la·han (kăl'ə-hăn), county (1970 pop. 8,205), 857 sq mi (2,219.6 sq km), central Texas, drained to the N by tributaries of the Brazos River, to the S by tributaries of the Colorado River; formed 1858; co. seat Baird. In a ranching and agricultural area yielding grain sorghums, corn, oats, peanuts, vegetables, and fruit, it also has oil and natural-gas wells and refineries.

Cal·lao (kä-you', -yä'ō), city (1972 pop. 196,119), capital of Lima dept., W Peru, on Callao Bay of the Pacific Ocean. The harbor, which is sheltered by an island and a small peninsula, handles more than three fifths of Peru's imports and exports. Callao was founded in 1537. The English navigator Sir Francis Drake sacked the city in 1578. It was held by Spanish loyalists until 1826, even though Peru achieved independence in 1821. Later, during the War of the Pacific, Callao was occupied (1881–83) by Chile. Subjected to earthquakes and tidal waves, the city was completely destroyed in 1746 and was severely damaged in 1940.

Cal·la·way (kăl'ə-wā), county (1970 pop. 25,991) 835 sq mi (2,162.7 sq km), central Mo., bounded on the S by the Missouri River; formed 1820; co. seat Fulton. Its agriculture includes corn, wheat, oats, and livestock (especially mules and saddle horses). It has fireclay, coal, and stone deposits. Shoes, farm machinery, and clay products are manufactured.

Cal·lo·way (kăl'ə-wā), county (1970 pop. 27,692), 384 sq mi (994.6 sq km), SW Ky., bounded on the S by the Tenn. border, on the E by the Kentucky Reservoir of the Tennessee River; formed 1821; co. seat Murray. It is drained by the East and West forks of the Clarks River and by Mayfield Creek. In a gently rolling agricultural area producing tobacco, corn, wheat, dairy products, poultry, and livestock, it has some manufacturing. Kentucky Lake State Park is here.

Ca·loo·sa·hatch·ee (kə-lo͞o'sə-hăch'ē), river, S Fla., rising in Lake Hicpochee and flowing c.75 mi (120 km) WSW past Fort Myers and into the Gulf of Mexico. It is an important part of the Okeechobee Waterway.

Cal·pe (kăl'pē), ancient name, possibly Phoenician in origin, of Gibraltar. It is one of the Pillars of Hercules at the eastern end of the Strait of Gibraltar.

Cal·ta·gi·ro·ne (käl'tä-jîr-ō'nā), city (1976 est. pop. 37,860), SE Sicily, Italy. An agricultural and sulfur-mining center, it has been famous for its majolica ware since the 9th cent.

Cal·ta·nis·set·ta (käl'tə-nĭ-sĕt'ə), city (1976 est. pop. 54,100), capital of Caltanissetta prov., central Sicily, Italy. It is an agricultural center and an important sulfur-producing center.

Cal·u·met (kăl'yo͞o-mĕt'), industrial region of NW Ind. and NE Ill., along the S shore of Lake Michigan. It has one of the world's greatest concentrations of heavy industry, especially steel manufacturing.

Calumet, county (1970 pop. 27,604), 322 sq mi (834 sq km), E Wis., bounded on the W by Lake Winnebago and drained by the Manitowoc River; formed 1836; co. seat Chilton. It has dairying, farming (grain, peas, and clover), and some varied manufacturing.

Calumet City, city (1978 est. pop. 39,900), Cook co., NE Ill., an industrial suburb in the greater Chicago metropolitan area, near the Ind. line; settled 1868, inc. 1911. It has steelworks and chemical and meat-packing industries.

Cal·va·dos (käl-vä-dôs'), department (1975 est. pop. 560,967), 2,137 sq mi (5,534.8 sq km), in Normandy, N France, on the English Channel; capital Caen.

Cal·vert (kăl'vərt), county (1970 pop. 20,682), 217 sq mi (562 sq km), S Md., on a narrow tidewater peninsula bounded on the E by Chesapeake Bay, on the S and W by the Patuxent River; formed 1654; co. seat Prince Frederick. In a resort and agricultural area yielding tobacco, grain, vegetables, and livestock, it has fisheries, a lumber industry, and some boatbuilding. Many historic houses are here.

Cam (kăm), river, c.40 mi (65 km) long, E central England, flowing NE past Cambridge to join the Ouse River S of Ely.

Ca·ma·güey (kä'mä-gwā'), city (1970 pop. 196,854), capital of Camagüey prov., E Cuba. The island's third-most populous city, Camagüey is an important commercial and transportation center. Its economy is based on agriculture and cattle raising and related industries. Founded in 1514 as Santa Maria del Puerto Principe, the city was moved to its present site in 1528 and renamed for the Indian village that previously occupied site. The city, which has retained much of its Spanish colonial atmosphere, is noted for its churches, mansions, and narrow twisting streets.

Ca·margue (kä-märg'), island, c.215 sq mi (555 sq km), Bouches-du-Rhône dept., SE France, in the Rhône delta. Formed by sedimentation, the marshy island has numerous shallow lagoons.

Ca·ma·ril·lo (kăm'ə-rĭl'ō), city (1978 est. pop. 27,700), Ventura co., S Calif.; inc. 1964. It is the center of a fertile farm area where citrus fruits and flowers are grown. Camarillo also has electronic and aerospace industries and manufactures magnetic tape and containers.

Cam·as (kăm'əs), county (1970 pop. 728), 1,054 sq mi (2,729.9 sq km), S central Idaho, in a livestock-grazing and dry-farming area; formed 1917; co. seat Fairfield. It produces grain and timber and has deposits of lead, silver, zinc, gold, and copper. There are some summer and winter resorts here.

Camas, city (1970 pop. 5,790), Clark co., SW Wash., on the Columbia River E of Vancouver; settled 1846, platted 1883, inc. 1906. Large lumber and pulp and paper mills are here, as well as plants processing poultry and fruit.

Cam·bay (kăm-bā'), town (1971 pop. 62,133), Gujarat state, W India, on the Mahi River estuary. The industries of Cambay include textile weaving and carpet making. Oil and natural gas are found nearby. Once a great port under the Moslem rulers of Gujarat (14th–15th cent.), Cambay lost its importance when the harbor silted up.

Cam·bo·di·a (kăm-bō'dē-ə), officially **Khmer Republic** (kmĕr), country (1975 est. pop. 8,110,000), 69,898 sq mi (181,036 sq km), SE Asia. Phnom Penh is the capital. Cambodia is bordered by Laos on the north, by Vietnam on the east, by the Gulf of Siam on the south, and by Thailand on the west and north. The heart of the country is a saucer-shaped, gently rolling alluvial plain drained by the Mekong River and shut off by mountain ranges; the Dangrek Mts. form the frontier with Thailand in the northwest and the Cardamom Mts. are in the southwest. About half the land is tropical forest. In general, Cambodia has a tropical monsoon climate. During the rainy season the Mekong swells and backs into the Tônlé Sap (Great Lake), increasing the size of the lake almost threefold. The seasonal rise of the Mekong floods almost 400,000 acres (162,000 hectares) around the lake, leaving rich silt when the waters recede.

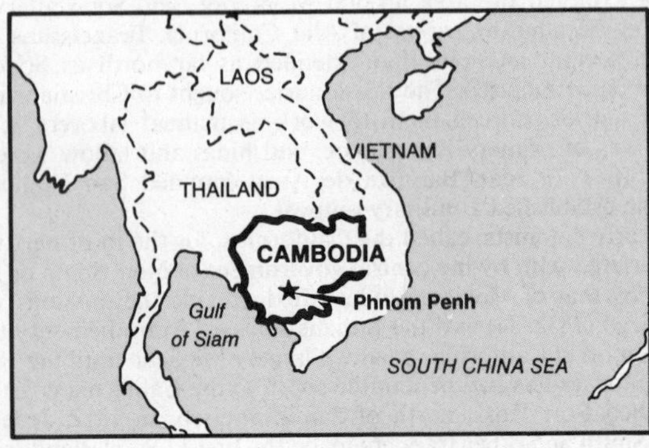

Economy. Rice is the country's chief crop. Livestock raising (cattle, buffalo, poultry, and hogs) and extensive fishing supplement the diet. Corn, vegetables, fruits, peanuts, tobacco, cotton, and sugar palms are also raised. Pepper is grown in the south, and great amounts of rubber are produced on large plantations. In the early

1970s, however, heavy fighting in the countryside put almost all of the rubber plantations out of operation. Rice and rubber are traditionally the principal exports of Cambodia. Inadequate transportation hampers exploitation of the country's vast forests. Mineral resources are limited; phosphate rock, limestone, semiprecious stones, and salt are extracted. The country's industries are based primarily on the processing of agricultural, fish, and timber products.

History. The Funan empire was established in what is now Cambodia in the 1st cent. A.D. By the 3rd cent. the Funanese had conquered their neighbors and extended their sway to the lower Mekong River. In the 4th cent., according to Chinese records, an Indian Brahman extended his rule over Funan, introducing Hindu customs, the Indian legal code, and the alphabet of central India. In the 6th cent. Khmers from the rival Chen-la state to the north overran Funan. With the rise of the Khmer Empire, Cambodia became dominant in southeast Asia.

After the fall of the empire (15th cent.), however, Cambodia was the prey of stronger neighbors. The kings of Siam and the lords of Hue alike asserted overlordship and claims to tribute. Intrigue and wars on Cambodian soil continued into the 19th cent., and in 1854 the king of Cambodia appealed for French intervention. A French protectorate was formally established in 1863, and French influence was consolidated by a treaty in 1884. Cambodia became part of the Union of Indochina in 1887. In World War II the country was under Japanese occupation.

In Jan., 1946, France granted Cambodia self-government within the French Union; a constitution was promulgated in May, 1947. Complete independence was finally granted in 1953. Early in 1954 Communist Vietminh troops from Vietnam invaded Cambodia. The Geneva Conference of 1954 led to an armistice providing for the withdrawal of all foreign forces from Cambodia. Cambodia was admitted into the United Nations in 1955. King Norodom Sihanouk abdicated in Mar., 1955, in order to enter politics; his father succeeded him as monarch. Sihanouk subsequently formed the Popular Socialist party and served as premier. He permitted the use of Cambodian territory as a supply base and refuge by North Vietnamese and Vietcong troops while accepting military aid from the United States to strengthen his forces against Communist infiltration. Following a series of border incidents involving South Vietnamese troops, Cambodia in 1965 severed diplomatic relations with the United States. Sihanouk remained on friendly terms with the Communist countries, especially Communist China, and established close relations with France. Economic conditions deteriorated, and North Vietnamese and Vietcong troops continued to infiltrate. In the spring of 1969 the United States instituted aerial attacks against Communist strongholds in Cambodia.

As Communist infiltration increased, Sihanouk began to turn more toward the West, and in July, 1969, diplomatic ties with the United States were restored. Relations with South Vietnam and Thailand, after years of border disputes and incidents, began to improve. In Aug., 1969, Lt. Gen. Lon Nol, the defense minister and supreme commander of the army, became premier, with Sihanouk delegating considerable power to him. Sihanouk began negotiating for the removal of Vietcong and North Vietnamese troops, who now numbered over 50,000 and occupied large areas of Cambodia. His actions, however, did not ease the growing concern of many army leaders. Discontent was further heightened by rising inflation, ruinous financial policies, and governmental corruption and mismanagement. On Mar. 18, 1970, while Sihanouk was in Moscow, Lon Nol led a right-wing coup deposing Sihanouk as chief of state. Sihanouk subsequently set up a government-in-exile in Peking. Soon after the coup, Cambodian troops began engaging Communist forces on Cambodian soil. In Apr., 1970, U.S. and South Vietnamese troops entered Cambodia to attack Communist bases and supply lines. U.S. ground forces were withdrawn by June 30, but South Vietnamese troops remained, occupying heavily populated areas.

The inevitable destruction of villages and killing of civilians alienated many Cambodians and may have created considerable sympathy for the Communists. The number of Cambodian Communists (known as the Khmer Rouge) increased from about 3,000 in Mar., 1970, to over 30,000 within a few years. A raging civil war emerged, fought by Cambodians but financed by the United States, North Vietnam, and Communist China. On Oct. 9, 1970, the national assembly declared Cambodia a republic and changed the country's name to the Khmer Republic. By that time, however, the national government controlled less than one third of Cambodia's total land area. In Feb., 1971, Lon Nol suffered a paralytic stroke and vice premier Sisowath Sirik Matak assumed power (although Lon No technically remained premier). After a major government defeat (Dec., 1971) on a highway north of Phnom Penh, most of Cambodian territory east of the Mekong River fell to the insurgents. In 1972 student agitation in Phnom Penh led to the resignation (Mar. 10) of chief of state Cheng Heng, who transferred his post to the ailing Lon Nol. Two days later Lon Nol dissolved the government and declared himself president as well as chief of state and commander in chief. A new constitution providing for a presidency was approved by popular referendum in April, and Lon Nol was formally elected president in June, 1972; the defeated candidates charged irregularities in the election.

Meanwhile, more and more territory fell into Communist hands. The government's military position became desperate. In Sept., 1972, severe food shortages in Phnom Penh sparked two days of rioting and large-scale looting in which government troops participated. Lon Nol, aided by his brother Lon Non, exerted an increasingly oppressive rule, with massive political arrests and newspaper seizures. U.S. pressure for a more representative government finally resulted (Apr., 1973) in the appointment of a member of the opposition party, In Tam, to the premiership, but In Tam resigned in Dec., 1973 and was succeeded by Long Boret of the ruling party. The Khmer Rouge insurgents launched a large-scale attack against Cambodia's third-largest city, Kompong Cham, in Sept., 1973, and shelled Phnom Penh in 1974 and early 1975.

Cam·borne-Red·ruth (kăm'bôrn-rĕd'ro͞oth, -bərn-), urban district (1971 pop. 24,029), Cornwall, SW England. The neighboring urban districts of Camborne and Redruth were combined in 1934. Tin and copper mines in the area have been greatly depleted.

Cam·brai (käN-brā'), city (1975 est. pop. 39,049), Nord dept., N France, a port on the Escaut (Scheldt) River. It has long been known for its fine textiles and gave its name to cambric, first manufactured here. Clay, metal, and wood products are also manufactured. An episcopal see since the 4th cent., Cambrai was ruled by bishops until it was seized by Spain (1595) and by France (1677).

Cam·bria (kăm'brē-ə), ancient name of Wales.

Cambria, county (1970 pop. 186,785), 695 sq mi (1,800 sq km), central Pa., in a mountainous manufacturing and coal-mining area bounded on the E by the Allegheny Mts.; formed 1804; co. seat Ebensburg. Drained by the Conemaugh River and the West Branch of the Susquehannah River, it has some agriculture (pears, potatoes, and honey), manufacturing (metals and metal products), and deposits of bituminous coal, clay, and limestone.

Cam·bri·an Mountains (kăm'brē-ən), rugged upland plateau occupying most of Wales and rising to 2,970 ft (905.9 m). The area has deep lakes and is cut by numerous river valleys.

Cam·bridge (kām'brĭj), municipal borough (1976 est. pop. 106,400), administrative center of Cambridgeshire, E central England, on the Cam River. It is an ancient market town and the former county town of Cambridgeshire and Isle of Ely. Although light industries such as the manufacture of agricultural tools, precision instruments, radios, and cement have developed on the outskirts, the town is most famous as the site of Cambridge Univ. Originally the site of a Roman fort, the town was an administrative and trading center in Anglo-Saxon times. William I built a fort and mint here. Two monastic establishments were built in early medieval times. The university was founded in the 13th cent. The present town still maintains much of its medieval atmosphere and appearance. There are many old inns, hostels, houses, winding streets, and narrow passages that have not altered greatly with time. Cambridge abounds in medieval churches, including St. Benet's or Bene't's, the oldest, dating back to the late Saxon period, and the Church of the Holy Sepulchre, one of the four Norman round churches in England.

Cambridge. 1. Village (1970 pop. 2,095), seat of Henry co., NW Ill., SE of Rock Island, in a farm area; inc. 1861. **2.** City (1978 est. pop. 11,700), seat of Dorchester co., E Md., Eastern Shore, a port of entry on the Choptank River at its mouth on Chesapeake Bay; founded 1684, inc. as a city 1884. It is a fishing and yachting center and has shipyards, seafood and vegetable canneries, and electronic, clothing, and printing industries. The Old Trinity Church (c.1675; restored 1960) is said to be the oldest church in the United States still in use. **3.** City (1978 est. pop. 103,400), seat of Middlesex co., E Mass., across the Charles River from Boston; settled 1630 as New Towne, inc. as a city 1846. A famous educational and research center, it is the seat of Harvard Univ. (founded 1636), Massachusetts Institute of

Technology, Lesley College, and several theological seminaries. It is also an industrial city; its manufactures include electrical machinery, scientific instruments, rubber goods, glass, wire cables, and machine shop products. Its printing and publishing industry dates from about 1639, when Stephen Daye established the first printing press in America. Cambridge was a gathering place for colonial troops. Craigie House (1759), which served as Washington's headquarters (1775–76), was the home of Longfellow from 1837 until his death in 1882. Other historic structures are Elmwood (1767), the birthplace and home of James Russell Lowell, the Cooper-Frost-Austin house (c.1657), and the Episcopal church (1761). Lowell, Longfellow, Mary Baker Eddy, and many other notable people are buried in Mt. Auburn Cemetery. **4.** Village (1970 pop. 2,720), seat of Isanti co., E Minn., on the Rum River N of Minneapolis, in a farm area; settled 1856, inc. 1876. Dairy products and wool are processed. **5.** Industrial city (1978 est. pop. 12,800), seat of Guernsey co., E central Ohio, in a farm, coal, natural-gas, and clay area; settled 1798 by immigrants from the isle of Guernsey, inc. 1837.

Cam·bridge·shire (kăm'brĭj-shĭr', -shər), nonmetropolitan county (1976 est. pop. 563,000), E central England; administrative center Cambridge. It was formed in 1974 and includes the former county of Cambridgeshire and Isle of Ely.

Cambridgeshire and Isle of E·ly (ē'lē), former county, E central England. The county town was Cambridge. Most of the area is alluvial fenland. Efforts to reclaim the fens date back to the days of Roman occupation; they were finally drained in the 17th cent. In 1974 Cambridgeshire and Isle of Ely became part of the new nonmetropolitan county of Cambridgeshire.

Cam·den (kăm'dən), borough (1971 pop. 200,784) of Greater London, SE England. Camden was created in 1965 by the merger of the metropolitan London boroughs of Hampstead, Holborn, and St. Pancras. Hampstead is a residential district popular with writers and artists. John Keats, John Constable, George Du Maurier, and Kate Greenaway, as well as Karl Marx, lived here. It is also known as a piano-making center. Holborn also houses the British Museum, the Univ. of London, Gray's Inn and Lincoln's Inn, law courts, the Royal College of Surgeons, and Hatton Garden, known for its trade.

Camden. 1. County (1970 pop. 11,334), 653 sq mi (1,691.3 sq km), extreme SE Ga., intersected by the Saltilla River and bounded on the S by the St. Marys River at the Fla. border, on the E by the Atlantic, on the N by the Little Saltilla River; formed 1772; co. seat Woodbine. Lumbering, fishing, and farming (corn, sugar cane, truck crops, and livestock) are important to its economy. **2.** County (1970 pop. 13,315), 640 sq mi (1,657.6 sq km), S central Mo., in the Ozarks, crossed by Lake of the Ozarks and drained by the Niangua River; formed 1841 as Kinderhook, renamed 1843; co. seat Camdenton. In a resort and agricultural region raising corn, wheat, potatoes, and livestock, it has oak and pine timber and lead, barite, and calcite mines. **3.** County (1970 pop. 456,291), 221 sq mi (572.4 sq km), SW N.J., bounded on the NW by the Delaware River and drained by the Great Egg Harbor and Mullica rivers and Big Timber Creek; formed 1844; co. seat Camden. It has agriculture (fruit, vegetables, poultry, and dairy products), marine terminals, railroad shops, shipbuilding and food-processing industries, and varied manufactures. **4.** County (1970 pop. 5,453), 239 sq mi (619 sq km), NE N.C., bounded on the N by the Va. border, on the SE by the North River estuary, on the S by Albemarle Sound, and on the SW by the Pasquotank River; formed 1777; co seat Camden. In a tidewater area, partly in the Dismal Swamp, it has a lumbering industry, agriculture (corn, soybeans, cotton, and vegetables), duck hunting, and fishing.

Camden. 1. Town (1970 pop. 1,742), seat of Wilcox co., SW central Ala., SSW of Selma, in a lumbering and corn-growing area; inc. 1841. **2.** City (1978 est. pop. 15,800), seat of Ouachita co., S Ark., on the Ouachita River; inc. 1847. It is a railroad and river shipping point. Its manufactures include paper, pottery, furniture, air conditioners, and house trailers. **3.** Resort town (1970 pop. 3,492), Knox co., S Maine, on Penobscot Bay N of Rockland; settled 1769, inc. 1791. Textile goods and electrical equipment are made here. A state park is nearby. **4.** Industrial city (1978 est. pop. 87,600), seat of Camden co., W N.J., a port of entry on the Delaware River opposite Philadelphia, settled 1681, inc. 1828. The arrival of the Camden and Amboy RR in 1834 spurred the city's growth as a commercial, shipbuilding, and manufacturing center. Manufactures include canned foods, electric and electronic goods, and paper and wood products. Walt Whitman's home is preserved, and the poet is buried in the city, where he lived from 1873. **5.** Village (1970 pop. 2,936), Oneida co., central N.Y., NW of Rome, in a farm and dairy area; inc. 1834. Furniture, copper wire, glass, and canned goods are manufactured. **6.** Village (1970 pop. 300), seat of Camden co., NE N.C., near the Pasquotank River ENE of Elizabeth City. **7.** City (1970 pop. 8,532), seat of Kershaw co., N central S.C., near the Wateree River and NE of Columbia; settled c.1735, inc. 1791. In a longleaf pine and farm area, it is a trade and processing center. Textiles, clothing, candy, and wood products are made. In the Carolina campaign of the American Revolution, the battles of Camden (Aug. 16, 1780) and Hobkirks Hill (Apr. 25, 1781) were fought in the neighborhood. The city was practically destroyed by fire when the British evacuated it on May 8, 1781. In the Civil War it was taken (Feb. 24, 1865) by a part of Sherman's army and again partially burned. Since the 1880s Camden has become a winter resort; it has beautiful estates and a number of ante-bellum houses. **8.** Town (1970 pop. 3,052), seat of Benton co., NW Tenn., near the Tennessee River W of Nashville; laid out 1836, inc. 1899. Corn, cotton, and peanuts are grown here.

Cam·den·ton (kăm'dən-tən), city (1970 pop. 1,636), seat of Camden co., central Mo., in the Ozarks SW of Jefferson City; founded 1929, inc. 1930. It is a resort near Lake of the Ozarks. Lead and calcite mines are in the area.

Cam·e·lot (kăm'ə-lŏt), in Arthurian legend, the seat of King Arthur's court. The origin of the name is unknown. It has been variously located at Cadbury Camp, Somerset, and at Winchester, Camelford, and Caerleon—all in Great Britain.

Cam·er·on (kăm'ər-ən). **1.** Parish (1970 pop. 8,194), 1,444 sq mi (3,740 sq km), extreme SW La., in a marshy coastal area bounded on the W by Sabine Lake and the Sabine River, on the S by the Gulf of Mexico; formed 1820; parish seat Cameron. In a stock-raising, fishing, hunting, and trapping region, it also produces poultry, vegetables, and fruit and has oil and natural-gas wells. **2.** County (1970 pop. 7,096), 401 sq mi (1,038.6 sq km), N central Pa., in a mountainous area drained by Sinnemahoning Creek; formed 1860; co. seat Emporium. Lumbering was once important here; today the county is in a recreational area producing dairy products, leather, electronic and surgical equipment, and explosives. It has bituminous coal, sandstone, shale, and natural-gas deposits. **3.** County (1970 pop. 140,368), 896 sq mi (2,320.6 sq km), extreme S Texas, in a rich irrigated agricultural area bounded on the S by the Rio Grande, on the E by Laguna Madre, an inlet of the Gulf of Mexico; formed 1848; co. seat Brownsville. It is a shipping and processing area for citrus fruit, vegetables, cotton, dairy products, and some livestock.

Cameron. 1. Town (1970 pop. 950), seat of Cameron parish, extreme SW La., on the Calcasieu River between Lake Calcasieu and the Gulf Coast. Shrimp and cotton are processed. **2.** City (1970 pop. 5,546), seat of Milam co., E central Texas, SSE of Waco, in a farm and cattle area; founded 1846, inc. 1888. There are fertilizer plants, cotton gins, grain and cottonseed mills, and a poultry hatchery here.

Cam·e·roon, United Republic of (kăm'ə-ro͞on'), republic (1977 est. pop. 6,575,000), 183,568 sq mi (475,441 sq km), W central Africa. It is bordered on the west by the Gulf of Guinea, on the northwest by Nigeria, on the northeast by Chad, on the east by the Central African Empire, on the south by the Congo Republic, Gabon, and Equatorial Guinea, and on the southwest by the Bight of Biafra. Yaoundé is the capital. Cameroon is triangular in shape. A coastal strip 10 to

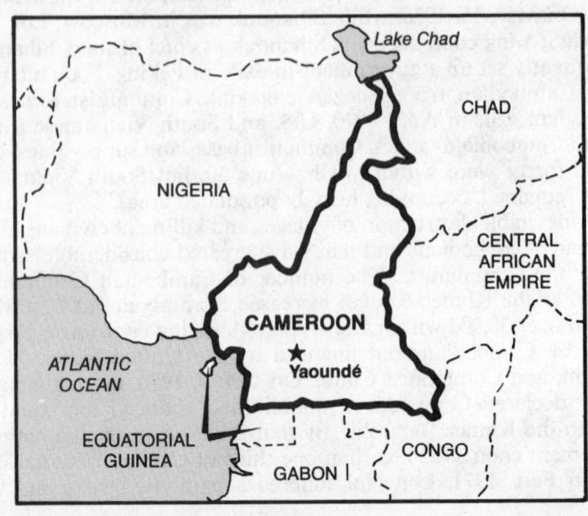

50 mi (16-80.5 km) wide in the southwest is covered with swamps and dense, tropical rain forests; it has one of the wettest climates in the world, with an average annual rainfall of 152 in. (386 cm). Near the coast are volcanic peaks. Beyond the coastal marshes and plains, the land rises to a densely forested plateau c.1,000 ft (305 m) above sea level. The interior of the country is a plateau c.2,500 to 4,000 ft (765-1,220 m) high, where forests give way to savanna. This plateau forms a barrier between the agricultural south and the pastoral north. The extreme northern regions, near Lake Chad, are dry, thornbush lands.

The country consists of the former French Cameroons and the southern portion of the former British Cameroons. The French, or eastern, section constitutes four fifths of the country and supports the bulk of the population. With more than 150 ethnic groups, Cameroon has one of the most diverse populations in Africa. Bantu-speaking peoples, such as the Douala, predominate along the southern coast and in the forested areas. In the highlands are the Bamiléké. Important northern groups include the Fulani and the Kirdi.

Economy. Agriculture is the mainstay of the country's economy. Cameroon is one of the world's leading cocoa producers; coffee, bananas, palm products, tobacco, peanuts, and rubber, all grown mainly on plantations, are also important. Cotton production is centered in the Benue River valley. Only about 10% of the country's land is cultivated. The principal subsistence crops are bananas, cassava, plantains, peanuts, millet, sorghum, and manioc. Fishing and forestry follow agriculture as leading occupations, but the vast timber reserves remain largely untapped. Cameroon's mineral resources include gold, diamonds, bauxite, tin, and mica. Prospecting for oil and natural gas is under way.

History. Throughout history the region witnessed numerous invasions and migrations, especially by the Fulani, Hausa, Fang, and Kanuri. Contact with Europeans began in 1472, when the Portuguese reached the Wuori River estuary, and a large-scale slave trade ensued, carried on by European traders. In the 19th cent., palm oil and ivory became the main items of commerce. The British established commercial hegemony over the coast in the early 19th cent., but were supplanted by the Germans in 1884. The Germans began constructing the port of Douala and then advanced into the interior. An additional area was acquired from France in 1911 as compensation for the surrender of German rights in Morocco. After World War I, the area ceded in 1911 was rejoined to French Equatorial Africa, and in 1919 the remainder of Cameroon was divided between France and Britain. Little social or political progress was made in either area.

In the 1950s guerrilla warfare raged in the French Cameroons instigated by the radical nationalist Union of the Peoples of the Cameroons, which demanded immediate independence and union with the British Cameroons. On Jan. 1, 1960, the territory became independent, with Ahmadou Ahidjo as its first president. The British-administered territory was divided into two zones, both administratively linked with Nigeria. In a UN-sponsored plebiscite in early 1961, the north voted for union with Nigeria, and the south for incorporation into Cameroon. National integration proceeded gradually. In 1966 the dominant political parties in the east and west merged into the Cameroon National Union (CNU). In 1972 the population voted favorably on a national referendum to adopt a new constitution setting up a unitary state.

Cameroon Mountain, active volcano, 13,353 ft (4,072.7 m) high, in the Cameroon Highlands, W Cameroon. It is the highest point in western Africa.

Cam·e·roons (kăm'ə-ro͞onz'), former German colony, W Africa, on the Gulf of Guinea and extending N to Lake Chad. Germany's penetration of the area began in 1884. A portion of French Equatorial Africa was added in 1911 in return for the surrender of German rights in Morocco. In World War I French and British troops occupied the Cameroons. After the war the territory ceded in 1911 was rejoined to French Equatorial Africa, and in 1919 the remainder of the Cameroons was divided into French and British zones. British Cameroons consisted of two noncontiguous sections lying on the eastern border of Nigeria; the more southerly extended to the coast. French Cameroons was administered as a separate territory with the capital at Yaoundé. In 1960 French Cameroons became the Cameroon Republic; in 1961 the southern section of British Cameroons was joined to the Cameroon Republic to form the Federal Republic of Cameroon (United Republic of Cameroon after 1972), while the northern section passed to Nigeria.

Ca·mil·la (kə-mĭl'ə), city (1970 pop. 4,987), seat of Mitchell co., SW Ga., S of Albany, in a farm and timber area; inc. 1858. Cotton, corn, and livestock are raised.

Camp (kămp), county (1970 pop. 8,005), 190 sq mi (492.1 sq km), NE Texas, in an agricultural area; founded 1874; co. seat Pittsburg. It is known for its sweet potatoes and black-eyed peas and also produces cotton, peanuts, fruit, vegetables, cattle, hogs, and poultry. Lumbering, lumber milling, and some manufacturing are done.

Cam·pa·gna di Ro·ma (käm-pä'nyä dē rô'mä), low-lying region surrounding the city of Rome, c.800 sq mi (2,070 sq km), Campania, central Italy. A favorite residential area in Roman times, it was later largely abandoned for centuries because of the prevalence of malaria and the lack of sufficient water for cultivation. Much of the region was reclaimed in the 19th and 20th cent.

Cam·pa·nia (käm-pä'nyä), region (1975 est. pop. 5,280,268), 5,249 sq mi (13,595 sq km), central Italy, extending from the Apennines W to the Tyrrhenian Sea and from the Garigliano River S to the Gulf of Policastro. It includes the islands of Capri, Ischia, and Procida. Naples is the capital of Campania. The central coast of the region is mostly high and rocky, with volcanic ridges and the crater of Vesuvius. However, the northern and southern coastal areas are fertile plains, famous since ancient times for their agricultural output. The interior of Campania is mountainous. Farm products of the region include grapes, citrus fruit, olives, apricots, grain, and vegetables. Manufactures include textiles, shoes, chemicals, pharmaceuticals, refined petroleum, metal goods, wine, and motor vehicles. There is also a thriving tourist industry. Various Italic tribes, Greek colonists, Etruscans, and Samnites lived in the region before it was conquered (4th-2nd cent. B.C.) by Rome. After the fall of Rome the Goths and the Byzantines occupied the region; it later became part of the Lombard duchy of Benevento. In the 11th cent. the Normans conquered Campania, and in the 12th cent. it became part of the kingdom of Sicily. Naples soon rose to prominence, and after the Sicilian Vespers revolt (1282) it was made the capital of a separate kingdom. In World War II there was heavy fighting around Naples after the Allied landing (Sept., 1943) at Salerno.

Camp·bell (kăm'bəl). **1.** County (1970 pop. 88,704), 151 sq mi (391.1 sq km), N Ky., in a gently rolling upland area in the outer Bluegrass region, bounded on the N and E by the Ohio River, here forming the Ohio border, on the W by the Licking River; formed 1794; co. seats Alexandria and Newport. It is chiefly industrial and residential, with some agriculture (vegetables, corn, alfalfa, tobacco, dairy products, poultry, and livestock). **2.** County (1970 pop. 2,866), 732 sq mi (1,895.9 sq km), N S.Dak., in an agricultural area bounded on the W by the Missouri River; formed 1873; co. seat Mound City. Dairy products, livestock, wheat, corn, and barley are produced. It includes a state game refuge. **3.** County (1970 pop. 26,045), 450 sq mi (1,165.5 sq km), NE Tenn., with the Cumberlands in the NW and bounded on the N by the Ky. border, on the SE by the Clinch River; formed 1806; co. seat Jacksboro. It has bituminous-coal mining, lumbering, and some agriculture (livestock, fruit, tobacco, corn, and hay). **4.** County (1970 pop. 43,319), 524 sq mi (1,357.2 sq km), SW central Va., bounded on the N by the James River, on the S by the Roanoke River; formed 1782; co. seat Rustburg. Agriculture (tobacco, corn, hay, and wheat), dairying, and poultry raising are important. **5.** County (1970 pop. 12,957), 4,755 sq mi (12,315.5 sq km), NE Wyo., in a grain and livestock area bordering on Mont. and watered by the Little Powder and Belle Fourche rivers; formed 1911; co. seat Gillette. It has coal deposits.

Campbell. 1. City (1978 est. pop. 25,400), Santa Clara co., W Calif., in the fertile Santa Clara valley; founded 1885, inc. 1952. A processing center for fruits and vegetables, it has a huge fruit-drying facility. **2.** City (1970 pop. 12,577), Mahoning co., NE Ohio, on the Mahoning River, adjacent to Youngstown; inc. 1908. It has extensive iron and steel works.

Camp·bells·ville (kăm'bəlz-vĭl'), city (1970 pop. 7,598), seat of Taylor co., central Ky., SSW of Lexington, in an area of farms, timber, and limestone; inc. 1817. Tobacco is a major crop. The city has a bottling works and produces furniture, concrete, clothing, dairy products, and feed.

Camp Da·vid (dā'vĭd), U.S. presidential retreat, Md.: *see* Catoctin Mountain Park under National Parks and Monuments Table.

Cam·pe·che (kăm-pā'chā), city (1970 pop. 69,506), capital of Campeche state, SE Mexico, on the Yucatán peninsula. It is fortified and surrounded by 18th cent. walls. Fish canning is the chief industry. The harbor is shallow, and vessels must anchor far from shore. Campeche, once the site of the pre-Columbian town called Kimpech (whose remains are still observable), was founded in 1540. It was sacked frequently by English buccaneers. From 1862 to 1864 French forces blockaded the city.

Cam·pi·na Gran·de (kəm-pē'nä grän'dĭ), city (1970 pop. 163,206), Paraíba state, NE Brazil, on the Borborema plateau. It is an important commercial and financial center and a shipping point for products from the Brazilian interior (hides and skins, cotton, and agave). Textiles, leather goods, cheese, and butter are the principal products.

Cam·pi·nas (kəm-pē'nəs), city (1970 pop. 328,629), São Paulo state, S Brazil. It is a growing industrial and financial city, the processing and distributing center for a diversified agricultural region, and a major transportation hub. Consumer products, agricultural tools, and railroad equipment are among its manufactures. The city was founded in the 18th cent. Coffee cultivation in the region brought prosperity by the late 19th cent.

Cam·po·bas·so (käm'pō-bäs'sō), city (1976 est. pop. 45,343), capital of Molise and of Campobasso prov., S central Italy. It is an agricultural and industrial center.

Cam·po·bel·lo (kăm'pō-bĕl'ō), island, 9 mi (14.5 km) long and 3 mi (4.8 km) wide, in Passamaquoddy Bay, N.B., Canada, just off the coast of Maine. The island passed to Canada by the Convention of 1817. President Franklin Delano Roosevelt had a summer home here for many years. It is now preserved in Roosevelt-Campobello International Park.

Cam·pos (käm'po͞os), city (1970 pop. 153,310), Rio de Janeiro state, SE Brazil, on the Paraíba River near its mouth. It is the commercial hub of a rich agricultural region and a transportation center. More than half of the state's sugar output is produced in Campos. There are also distilleries in the city. Campos was founded in the early 17th cent. and under the empire was an important slave center.

Camp·ton (kămp'tən), city (1970 pop. 419), seat of Wolfe co., E central Ky., in the Cumberlands ESE of Winchester. Cumberland National Forest is nearby.

Cam Ranh Bay (käm rän), inlet of the South China Sea, 10 mi (16 km) long and 20 mi (32 km) wide, SE Vietnam. It is an excellent harbor linked to the sea by a strait (1 mi/1.6 km wide). The bay was the site of one of the largest U.S. military facilities (est. 1965) during the Vietnam War.

Cam·rose (kăm'rōz), city (1976 pop. 10,104), central Alta., Canada. It is a railroad center in a mixed farming area.

Can·a·da (kăn'ə-də), country (1975 pop. 22,800,000), 3,851,787 sq mi (9,976,128 sq km), N North America. The capital is Ottawa. Canada occupies all of North America north of the United States (and east of Alaska) except for the French islands of St. Pierre and Miquelon. It is bounded on the east by the Atlantic Ocean, on the north by the Arctic Ocean, and on the west by the Pacific Ocean and Alaska. A transcontinental border, formed in part by the Great Lakes, divides Canada from the United States; Nares and Davis straits separate Canada from Greenland. The Arctic Archipelago extends far into the Arctic Ocean. Hudson Bay and the Gulf of St. Lawrence indent the east coast and the Inside Passage extends along the west coast.

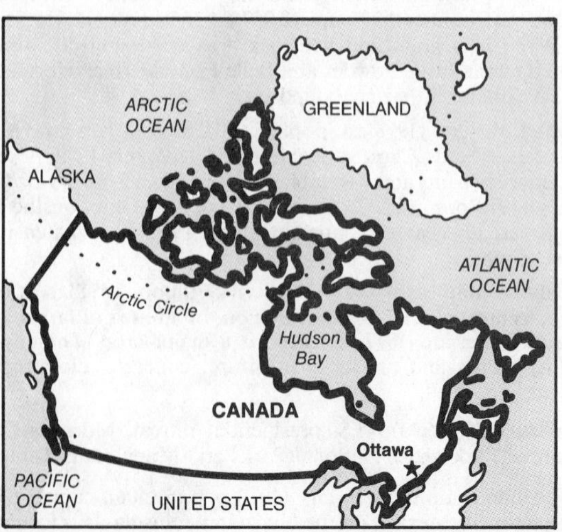

Canada has a bowl-shaped geologic structure rimmed by highlands, with Hudson Bay at the lowest point. The country has eight major physiographic regions—the Canadian Shield, the Hudson Bay Lowlands, the Western Cordillera, the Interior Lowlands, the Great Lakes-St. Lawrence Lowlands, the Appalachians, the Arctic Lowlands, and the Innuitians. The exposed portions of the Canadian Shield cover more than half of Canada. This once-mountainous region, which contains the continent's oldest rocks, has been worn low by erosion over the millenniums. Its upturned eastern edge is indented by fjords. In the center of the Shield are the Hudson Bay Lowlands, encompassing Hudson Bay and the surrounding marshy land. The Western Cordillera, a geologically young mountain system parallel to the Pacific coast, is composed of a series of north-south trending ranges and valleys that form the highest and most rugged section of the country; Mt. Logan (19,850 ft/6,054.3 m) is the highest point in Canada. Part of this region is made up of the Rocky Mts. and the Coast Mts., which are separated by plateaus and basins. The islands off western Canada are partially submerged portions of the Coast Mts. The Western Cordillera is also rich in minerals and timber and potential sources of hydroelectric power.

Between the Rocky Mts. and the Canadian Shield are the Interior Lowlands, a vast region divided into the prairies, the plains, and the Mackenzie Lowlands. The prairies are Canada's granary, while grazing is important on the plains. The smallest and southernmost region is the Great Lakes-St. Lawrence Lowlands, Canada's heartland. Dominated by the St. Lawrence River and the Great Lakes, the region provides a natural corridor into central Canada, and the St. Lawrence Seaway gives the interior cities access to the Atlantic. This section, which is composed of gently rolling surface on sedimentary rocks, is the location of extensive farmlands, large industrial centers, and most of Canada's population. In southeast Canada and on Newfoundland is the northern end of the Appalachian Mt. system, an old and geologically complex region with a generally low and rounded relief. The Arctic Lowlands and the Innuitians are the most isolated areas of Canada and are barren and snow covered for most of the year. The Arctic Lowlands comprise much of the Arctic Archipelago and contain sedimentary rocks that may have oil-bearing strata. In the extreme north, mainly on Ellesmere Island, is the Innuitian Mt. system, which rises to c.10,000 ft (3,050 m).

Canada is divided into 10 provinces and 2 territories:

NAME	CAPITAL	NAME	CAPITAL
Alberta	Edmonton	Nova Scotia	Halifax
British Columbia	Victoria	Ontario	Toronto
Manitoba	Winnipeg	Prince Edward Island	Charlottetown
New Brunswick	Fredericton	Quebec	Quebec
Newfoundland	St. John's	Saskatchewan	Regina
Northwest Territories	Yellowknife	Yukon Territory	Whitehorse

Economy. Manufacturing is Canada's most important economic activity, engaging 22% of the work force and accounting for more than half the value of all Canadian production. The leading products are motor vehicles, pulp and paper, processed meat, petroleum, iron and steel, dairy products, and processed metals. Industries are centered in Ontario, Quebec, and to a lesser extent, British Columbia. Agriculture contributes about one tenth of the value of production. The sources of the greatest farm income are livestock and dairy products. Among the biggest income-earning crops are wheat, oats, barley, and corn. Apples and peaches are the principal fruits.

Canada is the world's largest source of asbestos, nickel, zinc, and silver, and the second-largest source of potash, molybdenum, gypsum, uranium, and sulfur. Other mineral resources include copper, lead, and petroleum. Fishing is also an important economic activity; cod and lobster from the Atlantic and salmon from the Pacific are the principal catches.

History. An unknown number of Indians and Eskimos inhabited Canada before the white man arrived. The Vikings landed in Canada A.D. c.1000. John Cabot, sailing under English auspices, touched the east coast in 1497. In 1534 the Frenchman Jacques Cartier planted a cross on the Gaspé Peninsula. The first permanent white settlement in Canada was founded in 1605 by the sieur de Monts and Samuel de Champlain at Port Royal (now Annapolis Royal, N.S.) in Acadia. A trading post was established in Quebec in 1608. Meanwhile the English, moving to support their claims under Cabot's discoveries, attacked Port Royal (1614) and captured Quebec (1629). However, the French regained Quebec (1632) and began to establish new settlements. The French were primarily interested in fur trading. Between 1608 and 1640, fewer than 300 settlers arrived. The sparse French settlements sharply contrasted with the relatively dense English settlements along the Atlantic coast to the south. Under a policy initiated by Champlain, the French supported the Huron Indians in their warfare against the Iroquois; later, when the Iroquois crushed the Huron, the French colony came near extinction.

Exploration, however, continued. Both missionaries and traders extended French knowledge and influence. The greatest of all the empire builders in the west was Robert Cavelier, sieur de La Salle, who descended the Mississippi to its mouth. Meanwhile, the English had claims on Acadia, and the Hudson's Bay Company in 1670 began to vie for the lucrative fur trade of the West. When the long series of wars between Britain and France broke out in Europe, they were paralleled in North America by the French and Indian Wars. The Peace of Utrecht (1713) gave Britain Acadia, the Hudson Bay area, and Newfoundland. In 1759 Wolfe defeated Montcalm on the Plains of Abraham, bringing about the fall of Quebec to the British. By the Treaty of Paris in 1763, France ceded all its North American possessions to Britain (except Louisiana, which went to Spain).

The French residents of Quebec strongly resented the Royal Proclamation of 1763, which imposed British institutions on them. Many of its provisions, however, were reversed by the Quebec Act (1774), which granted important concessions to the French and extended Quebec's borders westward and southward. The Constitutional Act of 1791 divided Quebec into Upper Canada (present-day Ontario), predominantly British and Protestant, and Lower Canada (present-day Quebec), predominantly French and Catholic. Each new province had its own legislature and institutions.

After 1815 thousands of immigrants came to Canada from Scotland and Ireland. Movements for political reform arose. The British emissary Lord Durham recommended (1839) the union of Upper and Lower Canada under responsible government. The two Canadas were made one province by the Act of Union (1841) and responsible government was achieved in 1849 (it had been granted to the Maritime Provinces in 1847). Federation of all the Canadian provinces was achieved with the British North America Act in 1867. The four original provinces were Ontario (Canada West), Quebec (Canada East), Nova Scotia, and New Brunswick.

The new federation acquired the vast possessions of the Hudson's Bay Company in 1869. The Red River Settlement became the province of Manitoba in 1870. In 1873 Prince Edward Island joined the federation, and Alberta and Saskatchewan were admitted in 1905. Newfoundland joined in 1949. Canada's first prime minister was John A. Macdonald (served 1867–73 and 1878–91), who sponsored the Canadian Pacific Railway. Between 1891 and 1914 more than three million people came to Canada, largely from continental Europe. In the same period, mining operations were begun in the Klondike and the Canadian Shield.

Since World War II uranium, iron, and petroleum resources have been exploited; uses of atomic energy have been developed; and hydroelectric and thermal plants have been built to produce electricity for new and expanded industries. A major problem for Canada in recent decades has been to prevent economic domination by the United States. The St. Lawrence Seaway was opened in 1959. The government of Prime Minister Pierre Elliot Trudeau (1968–) was faced with the increasingly violent separatist movement active in Quebec in the late 1960s and 70s. ◊ *See Appendix.*

Government. Canada is an independent constitutional monarchy and a member of the Commonwealth of Nations. The British monarch is also the monarch of Canada and is represented in the country by the office of governor general. The basic constitutional document is the British North America Act of 1867. The Canadian federal government has authority in all matters not specifically reserved to the provincial governments. The provincial governments have power in the fields of property, civil rights, education, and local government. They may levy only direct taxes. The federal government may veto any provincial law. Power on the federal level is exercised by the Canadian Parliament and the cabinet of ministers, headed by the prime minister. The Parliament has two houses: the Senate and the House of Commons. There are a maximum of 110 senators, apportioned among the provinces and appointed by the governor general upon the advice of the prime minister. Members of the House of Commons are elected, largely from single-member constituencies. After the 1971 census there were 264 members. Elections must be held at least every five years. The Commons may be dissolved and new elections held at the request of the prime minister.

Ca·na·di·an (kə-nā'dē-ən), county (1970 pop. 32,245), 897 sq mi (2,323.2 sq km), central Okla., intersected by the North Canadian and Canadian rivers; formed 1890; co. seat El Reno. In a stock-raising, dairying, and agricultural area, it has oil fields and some industry. Wheat, oats, barley, alfalfa, corn, and sorghums are grown.

Canadian, town (1970 pop. 2,292), seat of Hemphill co., extreme N Texas, in the Panhandle NE of Amarillo on the Canadian River; inc. 1908. Founded in 1887 with the coming of the railroad, it is a cattle and farm trade center.

Canadian, river, 906 mi (1,457.8 km) long, rising in NE N.Mex. and flowing E across N Texas and central Okla. into the Arkansas River in E Okla.

Canadian Shield or **Lau·ren·tian Plateau** (lôr-ĕn'shən), U-shaped region of ancient rock, the nucleus of North America, stretching N from the Great Lakes to the Arctic Ocean. Covering more than half of Canada, it also includes most of Greenland and extends into the United States as the Adirondack Mts. The first part of North America to be permanently elevated above sea level, it has remained almost wholly untouched by successive encroachments of the sea upon the continent. It is the earth's greatest area of exposed Archaean-age rock; the metamorphic rocks of which it is largely composed were probably formed in the Precambrian era. During the Pleistocene epoch, continental ice sheets depressed the land surface (creating such features as Hudson Bay), scooped out thousands of lake basins, and carried away much of the region's soil. The southern part of the shield has thick forests while the north is covered with tundra.

Ça·nak·ka·le (chä'näk-kä-lĕ'), city (1975 pop. 30,760), capital of Çanakkale prov., NW Turkey, on the Asian shore of the Dardanelles.

Çanakkale Bo·ğa·zı (bō'ä-zĭ'): *see* Dardanelles.

ca·nal (kə-năl'), an artificial waterway constructed for navigation or for the movement of water. The digging of canals for irrigation probably dates back to the beginning of agriculture, and traces of canals have been found in the regions of ancient civilizations. Canals are also used to control flood waters, drain wetlands, supply water for municipal and industrial purposes, or generate electricity.

Ca·nal du Mi·di (kä-näl' dü mē-dē'), canal, c.150 mi (240 km) long, linking Sète and Toulouse, S France. It carries barge traffic between the Atlantic Ocean and the Mediterranean Sea.

Can·an·dai·gua (kăn'ən-dā'gwə), city (1970 pop. 10,488), seat of Ontario co., W central N.Y., in the Finger Lakes region, at the N end of Canandaigua Lake; settled 1789, inc. 1913. It is a resort and farm trade center. The county historical-society museum contains a copy of a treaty with the Iroquois Confederacy signed here in 1794 by Timothy Pickering. The courthouse was the scene of Susan B. Anthony's trial (1873) for voting.

Ca·nar·y Islands (kə-nâr'ē), group of seven islands (1974 est. pop. 1,244,073), 2,808 sq mi (7,272.7 sq km), off Spanish Sahara, in the Atlantic Ocean. They constitute two provinces of Spain, Santa Cruz de Tenerife and Las Palmas. The islands, of volcanic origin, are rugged; Mt. Teide (12,162 ft/3,709.4 m), on Tenerife, is the highest point. Pliny mentions an expedition to the Canaries c.40 B.C., and they may have been the Fortunate Islands of later classical writers. They were occasionally visited by Arabs and by European travelers in the Middle Ages. The Treaty of Alcácovas (1479) between Portugal and Spain recognized Spanish sovereignty over the Canaries; conquest of the Guanches, the indigenous inhabitants of the islands, was completed in 1496. The islands became an important base for voyages to the Americas and were frequently raided by pirates and privateers. Wine was the main export of the Canaries until the grape blight of 1853; its place was taken by cochineal until aniline dyes came into general use; sugar cane then became the chief commercial crop. Today the leading exports are bananas, tomatoes, potatoes, and tobacco, which are grown where irrigation is possible. There is fishing on the open seas, and the Canaries, with their warm climate and fine beaches, have become a major tourist center.

Ca·nav·er·al, Cape (kə-năv'ər-əl), low, sandy promontory extending E into the Atlantic Ocean from a barrier island, E Fla., separated from Merritt Island by the Banana River, a lagoon; named (1963) Cape Kennedy in memory of President John F. Kennedy, it reverted to its original name in 1973. The John F. Kennedy Manned Space Flight Center of the National Aeronautics and Space Administration is located here.

Can·ber·ra (kăn'bĕr'ə, -bə-rə), city (1976 pop. 193,000), capital of Australia, in the Australian Capital Territory, SE Australia. The federal government is the largest employer in Canberra. The site chosen (1908) for the capital city was first settled in 1824. In 1913 Canberra officially became the capital of the commonwealth (succeeding Melbourne); the transfer of federal functions was not completed until after World War II.

Can·di·a (kăn'dē-ə): *see* Iráklion, Crete.

Cand·ler (kănd'lər), county (1970 pop. 6,412), 250 sq mi (647.5 sq km), E central Ga., drained by the Canoochee River; formed 1914; co. seat Metter. Coastal-plain agriculture (cotton, corn, truck crops, tobacco, and livestock) and timber are its major resources.

Can·dle·wood Lake (kăn'dl-wŏŏd'), 8.4 sq mi (21.8 sq km), W Conn. It is formed behind a power dam south of the Rocky River's junction with the Housatonic River. Along its shoreline are summer resorts and recreational facilities.

Can·do (kăn'dōō), city (1970 pop. 1,512), seat of Towner co., NE N.Dak., NNW of Devils Lake; named 1884, inc. 1901. It is a trade center for a grain, livestock, and dairy region.

Ca·ne·a (kə-nē'ə): *see* Khaniá, Crete.

Ca·ney Fork (kā'nē), river, 144 mi (231.7 km) long, rising in central Tenn. and flowing NW to the Cumberland River. On Caney Fork are Great Falls Dam and Center Hill Dam, which provide flood control and power for the surrounding area.

Can·field (kăn'fēld'), village (1975 est. pop. 6,891), Mahoning co., NE Ohio, SW of Youngstown, surveyed 1798, inc. 1849. It is a trade and manufacturing center for a mine, farm, and timber area.

Can·ia·pis·cau (kăn'yə-pĭs'kou), river: *see* Kaniapiskau.

Can·nae (kăn'ē), ancient village, Apulia, SE Italy, scene in 216 B.C. of Hannibal's crushing defeat of the Romans.

Can·nel·ton (kăn'əl-tən), city (1970 pop. 2,280), seat of Perry co., S central Ind., on the Ohio River NE of Owensboro, Ky.; laid out 1835. It is a shipping and manufacturing center in an agricultural, coal, clay, and timber area.

Cannes (kăn, kän), town (1975 pop. 70,527), Alpes-Maritimes dept., SE France. An important and fashionable resort on the French Riviera, Cannes also has shipbuilding and textile industries. Churches from the 16th and 17th cent. are in the old part of town. An international film festival is held in Cannes each spring.

Can·nock (kăn'ək), urban district (1976 est. pop. 84,800), Staffordshire, W central England. It is a mining town dependent upon the rich coal deposits of Cannock Chase, a nearby moorland.

Can·non (kăn'ən), county (1970 pop. 8,467), 270 sq mi (699.3 sq km), central Tenn., drained by small affluents of the Cumberland and Stones rivers; formed 1836; co. seat Woodbury. Livestock raising, dairying, lumbering, and truck farming are done.

Can·on City (kăn'yən), city (1970 pop. 9,206), seat of Fremont co., S central Colo., at the mouth of the Grand Canyon of the Arkansas River; laid out 1859, inc. 1872. It is a health and tourist resort in a scenic area with mineral springs and marble and limestone quarries.

Can·ons·burg (kăn'ənz-bûrg'), borough (1970 pop. 11,439), Washington co., SW Pa.; inc. 1802. It is an industrial center in a coal-mining area. Its varied manufactures include steel and metal products and pottery. The Log Cabin School (est. 1777; the first school west of the Alleghenies) is preserved. The Black Horse Tavern was a famous gathering place for leaders of the Whiskey Rebellion (1794).

Ca·no·pus (kə-nō'pəs), ancient city of N Egypt, E of Alexandria near the village of Abu Qir. In Hellenistic times Canopus was known as a pleasure city for the rich. The Decree of Canopus, issued there in 238 B.C. and found at Tanis, has been of value in studying the ancient Egyptian language.

Ca·no·sa di Pu·glia (kä-nô'sä dē pōō'lyä), city (1976 est. pop. 30,855), Apulia, S Italy, on the Ofanto River. It is a commercial and agricultural center. The city flourished under the Romans and was noted for its wool and its fine vases, many of which have been unearthed in nearby tombs (3rd and 4th cent. B.C.).

Ca·nos·sa (kä-nôs'sä), village, in Emilia-Romagna, N central Italy, in the Apennines. There are ruins of a 10th cent. castle. In Jan., 1077, the castle was the scene of penance done by Emperor Henry IV to obtain from Pope Gregory VII the withdrawal of the excommunication against him.

Can·ta·bri·an Mountains (kăn-tā'brē-ən), N Spain, extending c.300 mi (480 km) along the Bay of Biscay from the Pyrenees to Cape Finisterre. Torre de Cerredo (8,688 ft/2,649.8 m) in the Europa group in the central section is the highest peak. The mountains are rich in minerals, especially coal and iron.

Can·tal (käɴ-täl'), department (1975 pop. 166,549), 2,217 sq mi (5,742 sq km), S central France, in Auvergne; capital Aurillac.

Can·ter·bur·y (kăn'tər-bĕr'ē), city (1976 pop. 128,669), New South Wales, SE Australia. It is a suburb of Sydney.

Canterbury, county borough (1973 est. pop. 34,510), Kent, SE England, on the Stour River. Economically unimportant except for tourism, Canterbury is famous as the long-time spiritual center of England. In 597 St. Augustine went to England from Rome and founded an abbey at Canterbury. The early cathedral was burned and rebuilt several times. After the murder (1170) of Thomas à Becket and the penance of Henry II, Canterbury became famous throughout Europe as the object of pilgrimage, and the *Canterbury Tales* of Chaucer relate the stories told by a fictional group of pilgrims. The present cathedral, constructed from 1070 to 1180 and from 1379 to 1503, is a magnificent structure, its architecture embodying the styles of several periods and various architects. During World War II the cathedral was the object of severe German bombing raids (June, 1942), which destroyed the library and many other surrounding buildings, but the cathedral itself received no direct hits. The city of Canterbury has a 14th cent. gate and remains of the old city walls. St. Martin's Church (established before St. Augustine's arrival and known as the Mother Church of England), the old pilgrims' hostel called the Hospital of St. Thomas, and several fine old inns are here.

Can·ti·gny (käɴ-tē-nyē'), village, Somme dept., N France, S of Amiens. It is the site of a monument commemorating the first offensive operation (May, 1918) of American forces in World War I.

Can·ton (kăn'tŏn', kăn'tŏn'), city (1970 est. pop. of 2,300,000), capital of Kwangtung prov., S China, a major deepwater port on the Canton River delta. Canton is the transportation, industrial, financial, and trade center of southern China. It has shipyards, a steel complex, paper and textile mills, and diversified manufacturing plants. Canton became a part of China in the 3rd cent. B.C. Hindu and Arab merchants reached Canton in the 10th cent., and the city became the first Chinese port regularly visited by European traders. In 1511 Portugal secured a trade monopoly, but it was broken by the British in the late 17th cent.; in the 18th cent. the French and Dutch were also admitted. Trading, however, was restricted until the Treaty of Nanking (1842) following the Opium War, which opened the city to foreign trade. Canton was the seat of the revolutionary movement under Sun Yat-sen in 1911; the Republic of China was proclaimed here. In 1927 Canton was briefly the seat of one of the earliest Communist communes in China. The fall of Canton in late Oct., 1949, signaled the Communist takeover of all China.

Can·ton (kăn'tən). **1.** Town (1975 est. pop. 7,463), Hartford co., N central Conn., NW of Hartford; settled 1737, inc. 1806. The town includes the village of Collinsville (1970 pop. 2,897), where edged tools have been produced since 1826. **2.** City (1970 pop. 3,654), seat of Cherokee co., N Ga., on the Etowah River N of Atlanta; inc. 1833. There are cotton mills and poultry farms here. **3.** City (1978 est. pop. 14,300), Fulton co., W central Ill., in the corn belt; inc. 1849. It is a trade and industrial center for a coal and farm area. Its industries include coal mining and the manufacture of farm equipment and clothing. **4.** Town (1970 pop. 17,100), Norfolk co., E Mass., a residential and industrial suburb of Boston; settled 1630, inc. 1797. Rubber goods, textiles, plastics, and paper products are manufactured. Paul Revere operated a copper-rolling mill here. **5.** City (1970 pop. 10,503), seat of Madison co., W central Miss.; inc. 1836. It is a trade and processing center in a cotton, truck farm, and timber area. There are a number of fine old ante-bellum houses here. **6.** Village (1970 pop. 6,398), seat of St. Lawrence co., N N.Y., on the Grass River ESE of Ogdensburg; settled 1799, inc. 1845. has a campus of the State Univ. of New York Agricultural and Technical College. **7.** City (1978 est. pop. 97,600), seat of Stark co., NE Ohio, at the junction of three branches of Nimishillen Creek; inc. 1822. It is a steel-processing center in an iron and steel area. Other manufactures include roller bearings, heavy office equipment, water softeners, and forgings. William McKinley lived in Canton; his grave and monument are in the McKinley State Memorial. **8.** City (1970 pop. 2,665), seat of Lincoln co., SE S.Dak., on the Big Sioux River S of

Sioux Falls; settled 1868, inc. 1881. It is a trade and shipping point for a livestock and grain region. **9.** Town (1970 pop. 2,283), seat of Van Zandt co., NE Texas, WNW of Tyler; settled 1850. It is a commercial center in a ranch and farm area producing peas, peaches, and poultry.

Can·ton (kăn'tŏn', kăn'tŏn') or **Pearl** (pûrl), river, 110 mi (177 km) long, S Kwangtung prov., S China. Formed at Canton by the confluence of the Si and Pei rivers, it flows east then south past Canton and Huang-pu island to form a large estuary between Hong Kong and Macao. The river links Canton to Hong Kong and the South China Sea and is one of China's most important waterways. The estuary, called Boca Tigris, is kept open for ocean vessels by dredging.

Can·ton Island (kăn'tən), coral atoll (1970 est. pop. 421), 3.5 sq mi (9.1 sq km), central Pacific, largest of the Phoenix Islands, SE of Honolulu, Hawaii. Annexed by the British at the end of the 19th cent., the island was also claimed by American guano companies. In 1939 Great Britain and the United States agreed on joint control of Canton and nearby Enderbury Island for 50 years.

Can·yon (kăn'yən), county (1970 pop. 61,288), 580 sq mi (1,502.2 sq km), SW Idaho, in an agricultural area bordering on Oregon, bounded on the S by the Snake River, and drained by the Boise River; formed 1891; co. seat Caldwell. Dairy products, livestock, hay, sugar beets, fruit, and vegetables are produced.

Canyon, city (1970 pop. 8,333), seat of Randall co., extreme N Texas, in the Panhandle S of Amarillo; settled 1892, inc. 1906. A trade center in a farming and stock-raising area, it is the seat of West Texas State Univ.

Canyon City, town (1970 pop. 600), seat of Grant co., NE central Oregon, N of Burns near the John Day River, in a livestock area at an altitude of 3,194 ft (974.2 m).

Canyon de Chel·ly National Monument (də shā'), 83,840 acres (33,955.2 hectares), NE Ariz.; est. 1931. The area contains the ruins of several hundred prehistoric Indian villages, most of them built A.D. 350–1300. The spectacular cliff dwellings include Mummy Cave, with a three-story tower house, and have numerous pictographs in rock shelters and on cliff faces.

Can·yon·lands National Park (kăn'yən-lăndz'), 257,640 acres (104,344 hectares), SE Utah; est. 1964. Located in a desert region, the park contains a maze of deep canyons and many unusual features carved by wind and water, including spires, pinnacles, and arches; surrounding mesas rise more than 7,800 ft (2,379 m). Cataract Canyon contains one of the world's largest exposures of red sandstone. Island in the Sky, a plateau overlooking the junction of the Green and Colorado rivers, has walls that drop in giant steps 2,200 ft (671 m) to the canyon floor. Upheaval Dome, pushed upward by the pressure of surrounding rock on underground salt deposits, contains a crater 1 mi (1.6 km) wide and 1,500 ft (457.5 m) deep.

Cap de la Ma·de·leine (käp də lä mäd-lĕn'), city (1976 pop. 32,126), S Que., Canada, at the confluence of the St. Maurice and St. Lawrence rivers. Newsprint and paper products, plywood, aluminum products, and clothing are manufactured here.

Cape (kāp) or **Cape of.** For names of actual capes, see the specific element of the name, e.g., Hatteras, Cape; Good Hope, Cape of. Other entries beginning with Cape, such as Cape May, N.J., and Cape Sable Island, are entered under Cape.

Cape Bre·ton Island (brĕt'n), island (1976 pop. 128,229), 3,970 sq mi (10,282.3 sq km), forming the NE part of N.S., Canada, and separated from the mainland by the narrow Gut of Canso. Gently sloping in the south, the island rises to rugged hills in the wilder northern part. There are many summer resorts on the lakes and fishing villages on the coast. In the northeast are steelworks. The Cabot Trail, a scenic road through Cape Breton Highlands National Park, commemorates the discovery of Cape Breton Island in 1497 by John Cabot. The island was a French possession from 1632 to 1763. After the Peace of Utrecht (1713) many Acadians migrated here from mainland Nova Scotia. They renamed the island Ile Royale and established the fortress at Louisburg. With the final cession of Canada to the British (1763), Cape Breton was attached to Nova Scotia. It was made a separate colony in 1784, with Sydney as its capital, but was rejoined to Nova Scotia in 1820.

Cape Coast, town (1970 pop. 71,594), capital of Central Region, S Ghana, on the Gulf of Guinea. The town is an export port and fishing center. It grew up around European forts built in the 17th cent. The British made it their headquarters in 1664. It was capital of the Gold Coast until 1877.

Cape Fear River (fîr), 202 mi (325 km) long, formed in E central N.C. by the junction of the Deep and Haw rivers, and flowing SE to enter the Atlantic Ocean N of Cape Fear. Dams and locks make the river navigable to Fayetteville; its estuary forms part of the Intracoastal Waterway.

Cape Gi·rar·deau (jĭ-rär'dō, jĭr'ər-dō'), county (1970 pop. 49,350), 576 sq mi (1,491.8 sq km), SE Mo., bounded on the E by the Mississippi and crossed by the White River; formed 1812; co. seat Jackson. In an agricultural area yielding corn, wheat, and hay, it has limestone quarries and stands of timber. Food products, textiles, furniture, rubber and plastic products, shoes, cement, pottery, and metal products are manufactured.

Cape Girardeau, city (1978 est. pop. 32,400), Cape Girardeau co., SE Mo., overlooking the Mississippi River; founded 1793, inc. as a city 1843. It is a transportation, trade, and distribution center with factories that manufacture a variety of products. Its position on the river, near the confluence with the Ohio River, spurred its early growth. During the Civil War it was occupied by Union forces, and four forts were built here. Fort D (1861) and other old buildings are among today's points of interest.

Cape Look·out National Seashore (lo͝ok'out'): *see* National Parks and Monuments Table.

Cape May (mā), county (1970 pop. 59,554), 267 sq mi (691.5 sq km), S extremity of N.J., occupying Cape May Peninsula between the Atlantic Ocean and Delaware Bay; formed 1692; co. seat Cape May Court House. Its coastal resorts have been popular since the mid-19th cent. Fisheries, agriculture (especially truck-garden crops and lima beans), and light industry are also important to its economy.

Cape May, city (1970 pop. 4,392), Cape May co., S N.J., at the end of Cape May peninsula, on the Atlantic Ocean; settled in the 1600s, inc. 1857. One of the nation's oldest beach resorts, it became popular in the mid-19th cent., when it was known as the "President's Playground"; Lincoln, Grant, Arthur, Buchanan, Hayes, and Benjamin Harrison vacationed here. The city's various mansions and Victorian hotels are notable examples of 19th cent. architecture.

Cape May Court House, uninc. village (1970 pop. 2,062), seat of Cape May co., S N.J., NNE of Cape May city; laid out 1703. A historical museum is here.

Cape Province, formerly **Cape of Good Hope Colony,** province (1970 pop. 6,827,756), 278,465 sq mi (721,224 sq km), S Republic of South Africa. The capital and largest city is Cape Town, which is also the country's legislative capital. Grain, fruit, tobacco, and chicory are cultivated, chiefly in the fertile coastal regions; cattle, sheep, and goats are raised in the interior. Marine fishing is pursued, especially in the southwest, and diamonds, iron ore, manganese, asbestos, and copper are mined. Manufactures include textiles, clothing, processed foods, wine and liquor, motor vehicles, refined petroleum, and footwear.

Although the Cape of Good Hope was first circumnavigated in 1488 by Bartolomeu Dias and later (1497) by Vasco da Gama, the first European settlement of the region was not made until 1652, when Jan van Riebeeck founded a resupply station for the Dutch East India Company on Table Bay. The company brought Dutch settlers to Cape Town, who were called Boers. In 1689 French Huguenots began to arrive; they developed the wine industry. The company ruled the Cape until 1795, except for a brief period (1781–84) of French occupation. In 1779 the first of numerous frontier wars (continuing until 1877) between Europeans and the Xhosa (a Bantu-speaking people) erupted. These so-called Kaffir Wars were mainly over land and cattle.

During the French Revolutionary and Napoleonic Wars (1792–1815), Britain occupied the Cape from 1795 to 1803, when the Dutch regained control; Holland formally ceded it to Great Britain in 1806. The British named the territory Cape of Good Hope Colony and encouraged immigration from England. The new British settlers soon conflicted with the Boers over anglicization of the courts, control of farm- and pastureland, and slaveholding. Beginning in 1835 many Boers left Cape Colony. They founded a temporary republic in Natal and longer lasting republics in the Transvaal and Orange Free State. In 1872 Cape Colony received internal self-government. In 1867 diamonds were discovered in the Kimberley region, which in 1880 was annexed by the Cape. The British and the remaining

Boers generally cooperated until the 1890s, when the British sought to unite the Transvaal and the Orange Free State with the Cape and Natal. In 1895-96 L. S. Jameson staged an unsuccessful raid from Cape Colony into the Transvaal, which greatly increased tension between Britons and Boers. The South African War (1899-1902) followed soon thereafter. In 1910 the Cape Colony joined with Natal, the Transvaal, and the Orange Free State to become a founding province of the Union of South Africa.

Cape Town or **Cape·town** (kāp'toun'), city (1970 pop. 697,514), legislative capital of the Republic of South Africa and capital of Cape Province, a port on the Atlantic Ocean. The city lies at the foot of Table Mt. (3,567 ft/1,088 m) and on the shore of Table Bay. Cape Town is a commercial and industrial center; food processing, winemaking, printing, and the manufacture of clothing and plastic and leather goods are the chief industries. An important port, Cape Town exports mainly gold, diamonds, and fruits. Tourism is of growing economic importance for the city, with its beaches and pleasant climate. Cape Town was founded in 1652 by Governor Jan van Riebeeck as a supply station for the Dutch East India Company. In 1795 the British occupied the city. It was returned to the Dutch in 1803 but recaptured in 1806 by the British, who established Cape of Good Hope Colony with Cape Town as capital. Cape Town's attractions include the Castle, a fortress dating from 1666, and the Dutch Reformed church (begun 1699).

Cape Verde (vûrd), republic (1970 pop. 272,017), c.1,560 sq mi (4,040 sq km), W Africa, in the Atlantic Ocean about 300 mi (480 km) W of Dakar, Senegal. The capital is Praia. It is an archipelago made up of 10 islands and 5 islets, which fall into two main groups—the Barlavento, or Windward, in the north and the Sotavento, or Leeward, in the south. The islands are mountainous and of volcanic origin; the only active volcano is at the archipelago's highest point, Cano (c.9,300 ft/2,835 m). About 60% of the population is of mixed black African and European descent, and most of the rest are black Africans; there are also a few Portuguese settlers.

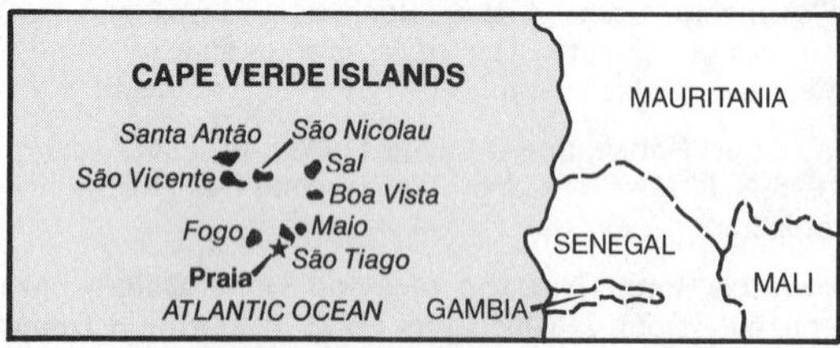

Economy. Farming, the main economic activity, is severely limited by the small annual rainfall. Occasionally, as in the early 1970s, there are severe droughts. The main crops are maize, bananas, potatoes, tomatoes, pulses, arabica coffee, groundnuts, and sugar cane. Goats, hogs, cattle, and sheep are raised. Tuna and lobster are the main catches of a small, but growing, fishing industry. Puzzolana and salt are the only minerals extracted. The islands' manufactures are limited to processed food, beverages, and tobacco products.

History. The Cape Verde Islands probably were discovered (1456) by Luigi da Cadamosto, a navigator in the service of Prince Henry of Portugal; at that time they were uninhabited. Colonists from Portugal began to settle here in 1462. Soon thereafter, black Africans from western Africa were brought to the islands as slaves. Later a Portuguese penal colony was established, and some of the convicts remained after their terms had been completed. Slavery was abolished on the islands in 1876. Portuguese Guinea (now Guinea-Bissau) was administered as part of the Cape Verde Islands until 1879. In 1951 the status of the islands was changed from colony to overseas province. The Cape Verde Islands became independent on July 5, 1975. At that time a national assembly was chosen, and all 56 members selected were from the African Party for the Independence of Guinea-Bissau and Cape Verde (PAIGCC), the principal independence movement of the African republic of Guinea-Bissau; the aim of the Cape Verde branch of the party is the eventual union of the two countries. In 1976 a protocol was signed with Guinea-Bissau establishing a common judicial system. By the late 1970s, the Cape Verde Islands continued to be plagued by economic difficulties caused by the decade-long drought. More than 300,000 people emigrated after 1966 and the unemployment rate reached 80% by 1976.

Cape York Peninsula (yôrk), 280 mi (450.5 km) long, N Queensland, Australia, between the Gulf of Carpentaria and the Coral Sea. It is largely tropical jungle and sparsely populated.

Cap-Ha·ï·tien (kăp'ä-ē-syăɴ'), city (1971 pop. 46,217), N Haiti, on the Atlantic Ocean. Haiti's second-largest city, it is a major seaport, commercial center, and tourist attraction. Coffee, cacao, and sugar are exported. Founded by the French c.1670, the city was the capital of colonial Haiti for a century. In 1791 Cap-Haïtien was captured by Toussaint L'Ouverture, leader of a slave rebellion. From 1811 to 1820 it served as capital of the kingdom of Henri Christophe.

Cap·i·tol (kăp'ĭt-l), seat of the U.S. government at Washington, D.C. It is the city's dominating monument, built on an elevated site that was chosen by George Washington in consultation with Maj. Pierre L'Enfant. The building as it now stands was begun in 1793, took many years to build, and is the result of the work of several architects, including William Thornton, James Hoban, B. H. Latrobe, and Charles Bulfinch.

Cap·i·to·la (kăp'ĭ-tō'lə), city (1975 est. pop. 7,405), Santa Cruz co., W Calif., on Monterey Bay E of Santa Cruz; inc. 1949. It is a seaside resort. There are nurseries here, and the city is noted for its begonias. A state park is nearby.

Cap·i·to·line Hill (kăp'ĭ-tō-līn) or **Capitol**, highest of the seven hills of ancient Rome and historic and religious center of the city.

Capitol Reef National Park: *see* National Parks and Monuments Table.

Cap·pa·do·cia (kăp-ə-dō'shə), ancient region of Asia Minor, watered by the Halys River (the modern Kizil Irmak), in present E central Turkey. The name was applied at different times to territories of varying size. At its greatest extent Cappadocia stretched from the Halys valley east to the Euphrates River, from the Black Sea south to the heights of the Taurus and Anti-Taurus ranges. Mostly a high plateau, it was famous for its mineral resources, particularly its copper and iron. In c.1800 B.C. Cappadocia was the heart of an old Hittite state. Later the Persians controlled Cappadocia. It did not yield fully to the conquest of Alexander the Great, and during the 3rd cent. B.C. it gradually developed as an independent kingdom. In the 2nd and 1st cent. B.C. the Cappadocian dynasty maintained itself largely by siding with Rome. In A.D. 17 Rome annexed the region as a province.

Ca·pri (kä'prē, kə-prē'), island (1971 pop. 11,962), 4 sq mi (10.4 sq km), Campania, S Italy, in the Bay of Naples off the tip of the Sorrento Peninsula. It is an international tourist center, celebrated for its striking scenery, delightful climate, and luxurious vegetation. The Blue Grotto is the most famous of the many caves along the island's high, precipitous coast. On the island are remains of the 12 fine villas built by the Roman emperors Augustus and Tiberius.

Cap·u·a (kăp'o͞o-ə, kä'pwä), town (1976 est. pop. 18,333), Campania, S Italy, on the Volturno River. It is an agricultural center and occupies the site of ancient Casilinum. Ancient Capua, situated 3 mi (4.8 km) to the southeast, was a Roman town strategically located on the Appian Way. During the second of the Punic Wars it went over (216 B.C.) to the side of Hannibal, but was retaken by Rome in 211 B.C. Later it was an important colony under the Roman Empire. After Capua was destroyed (A.D. 841) by the Arabs, its inhabitants moved to Casilinum and founded modern Capua.

Ca·pu·lin Mountain National Monument (kə-po͞o'lĭn, -pyo͞o'-, kăp'ə-lən, -yə-): *see* National Parks and Monuments Table.

Ca·rac·as (kə-răk'əs, -rä'kəs, kä-rä'käs), city (1971 pop. 1,658,500), N Venezuela, the capital and largest city of the country, near the Caribbean Sea. Its port is La Guaira. At an elevation of c.3,100 ft (945 m), Caracas is the commercial, industrial, and cultural hub of the nation. As a result of the oil boom of the 1950s the city expanded prodigiously. Enormous sums were spent on public works, notably the futuristic University City and an impressive highway cloverleaf, known to Caracans as "the octopus." The symbol of the new Caracas is the twin-towered complex housing government offices known as Centro Bolívar. In addition to oil refining, industries include textile milling, sugar refining, and meat-packing. Caracas was founded in 1567 as Santiago de León de Caracas. The city was sacked by the English in 1595 and by the French in 1766. Two of South America's great revolutionary leaders, Francisco de Miranda (1750) and Simón Bolívar (1783) were born in the city. Independence from Spain was declared in Caracas in July, 1811. However, the city was almost completely destroyed by an earthquake on Mar. 26, 1812. Bolívar captured the city in Aug., 1813, but abandoned it after a crushing defeat in June, 1814. Finally, after his victory at Carabobo, he made a triumphal entry in June, 1821.

Car·bon (kär'bən). **1.** County (1970 pop. 7,080), 2,070 sq mi (5,361.3 sq km), S Mont., in an irrigated agricultural and mining region bordering on Wyo. and drained by the Clarks Fork of the Yellowstone River; formed 1895; co. seat Red Lodge. It has deposits of coal and natural gas. **2.** County (1970 pop. 50,573), 405 sq mi (1,049 sq km), E Pa., in a mountainous region drained by the Lehigh River and bounded on the S by the Blue Mts., on the E by the Pocono plateau; formed 1843; co. seat Mauch Chunk (Jim Thorpe). It was settled (1746) by Moravian missionaries. Light manufacturing (metal products and textiles), coal mining, agriculture, and tourism are important. **3.** County (1970 pop. 15,647), 1,474 sq mi (3,817.7 sq km), central Utah, in a coal-mining region drained by the Green and Price rivers; formed 1894; co. seat Price. It has industry and some agriculture (sugar beets, hay, and grain). **4.** County (1970 pop. 13,354), 7,905 sq mi (20,474 sq km), S Wyo., in a grain and livestock area bordering on Colo. and watered by the North Platte and Medicine Bow rivers; formed 1868; co. seat Rawlins. It has coal, oil and uranium deposits. Cattle and sheep are raised.

Car·bon·dale (kär'bən-dāl'). **1.** City (1978 est. pop. 21,800), Jackson co., S Ill.; inc. 1869. It is a railroad division point and the retail center of a coal-mining and farming area. Southern Illinois Univ. is the major employer. **2.** Industrial city (1970 pop. 12,478), Lackawanna co., NE Pa., on the Lackawanna River; inc. 1851. Its important activities are anthracite-coal mining and the manufacture of mining machinery, machine shop products, chemicals, and clothing.

Car·cas·sonne (kär-kä-sôn'), city (1975 pop. 42,154), capital of Aude dept., S France, in Languedoc. The old city, a medieval fortress atop a hill, is one of the architectural marvels of Europe. The new city, across the Aude River, is a farm trade center with rubber, shoe, and textile manufactures. The Romans fortified the hilltop site in the 1st cent. B.C.; towers built (c.6th cent.) by the Visigoths are still intact. A stronghold of the Albigenses, the fortess was taken by Simon de Montfort in 1209. It yielded to the king in 1247, at which time Louis IX (St. Louis) founded the new city across the river. The outer ramparts of the fortress were constructed during St. Louis's reign, and the work was continued, with intricate defense devices, under Philip III. When completed, the fortress was widely considered impregnable; Edward the Black Prince was stopped at its walls in 1355. The ramparts were gradually abandoned and fell into disrepair; they were restored by Viollet-le-Duc in the 19th cent.

Car·che·mish (kär'kə-mĭsh, kär-kē'-), ancient city, Turkey, on the Euphrates River, at the Syrian border SE of Gaziantep. It was an important Neo-Hittite city and was prosperous in the 9th cent. B.C. before it was destroyed by the Assyrians. Among the excavated remains are sculptured neo-Hittite reliefs with hieroglyphic Hittite inscriptions.

Cardamom Mountains, mountain group extending c.100 mi (160 km) along the Thai-Cambodian border, E of Chanthaburi, SE Thailand. Ta Det (3,667 ft/1,118.4 m) is the highest peak. The mountains receive monsoon rains and have a dense vegetation cover.

Cár·de·nas (kär'dā-näs), city (1970 pop. 55,209), N central Cuba, a port on Cárdenas Bay. It processes and exports sugar and sisal and has industries producing tobacco, beer, and soap. A fishing fleet is based here. The city was founded in 1828 as a shipping point for sugar.

Car·diff (kär'dĭf), county borough (1973 est. pop. 281,500), administrative center of Mid Glamorgan, S Wales, on the Taff River near its mouth on the Bristol Channel. Until the early 20th cent. Cardiff was one of the greatest coal-shipping ports in the world. Present industries include shipbuilding and repairing, metal casting, engineering, oil and gasoline distribution, and food processing. Cardiff Castle was first built in 1090 on the site of a Roman fort. Until 1974 Cardiff was the county town of Glamorganshire.

Car·di·gan (kär'dĭ-gən), municipal borough (1971 est. pop. 3,800), Dyfed, W Wales, on the Teifi River near its mouth on Cardigan Bay. It is a cattle and dairy center. Until 1974 Cardigan was the county town of Cardiganshire.

Car·di·gan·shire (kär'dĭ-gən-shîr', -shər), former county, W Wales, on Cardigan Bay. The county town was Cardigan. The county long resisted English influence, and the Welsh language and Welsh customs are well preserved. In 1974 Cardiganshire became part of the new nonmetropolitan county of Dyfed.

Car·i·a (kâr'ē-ə), ancient region of SW Asia Minor, S of the Maeander River, which separated it from Lydia. The region was settled by both Dorian and Ionian colonists. Caria was a center of the Ionian revolt (c.499 B.C.) that was a prelude to the Persian Wars. Alexander the Great conquered Caria, and it changed hands often in the wars after his death. In 125 B.C. it was made a Roman province.

Car·ib·be·an Sea (kâr'ĭ-bē'ən, kə-rĭb'ē-ən), tropical sea, c.750,000 sq mi (1,942,500 sq km), arm of the Atlantic Ocean, Central America. It is bordered on the north and east by the West Indies, on the south by South America, and on the west by the Central American isthmus. The Caribbean is linked to the Gulf of Mexico by the Yucatán Channel; to the Atlantic by many straits, of which the Windward Channel and Mona Passage are the most important; and to the Pacific Ocean by the Panama Canal. Geologically, the Caribbean Sea consists of two main basins separated by a broad, submarine plateau; Bartlett Deep, a trench between Cuba and Jamaica, contains the Caribbean's deepest point (22,788 ft/6,950.3 m below sea level). The Caribbean's water is clear, warm, and less salty than the Atlantic. It has a counterclockwise current; water enters through the Lesser Antilles, is warmed, and exits via the Yucatán Channel, where it forms the Gulf Stream. Volcanic activity and earthquakes are common in the Caribbean, as are destructive hurricanes that originate over the sea or in the Atlantic.

The Caribbean was discovered by Christopher Columbus in 1493 and was named for the Carib Indians. Spain claimed the area, and its ships searched for treasure. With the discovery of the Pacific Ocean in 1513 the Caribbean became the main route of Spanish expeditions and, later, of convoys. Pirates and warships of rival powers preyed on Spanish ships in the Caribbean. Although Spain controlled most of the sea, Britain, France, Holland, and Denmark established colonies on the islands along the eastern fringe. The 1800s brought U.S. ships into the Caribbean, especially after 1848, when many goldseekers crossed the sea to reach California via Panama. The 1914 opening of the Panama Canal paved the way for increased U.S. interest and involvement in this strategic sea. U.S. policy since the Monroe Doctrine of 1823 has been to exclude foreign powers from the Caribbean; however, in 1959, Cuba became the first country to come under strong foreign (Soviet) influence.

Car·ib·bees (kâr'ĭ-bēz), name sometimes applied to the islands of the Caribbean or even to all the West Indies. More specifically, the Caribbees are the Lesser Antilles and include the Leeward Islands, the Windward Islands, and the Virgin Islands.

Car·i·boo Mountains (kăr'ə-bo͞o), range, c.200 mi (320 km) long, E British Columbia, Canada, rising to 11,750 ft (3,583.8 m) at Mt. Sir Wilfrid Laurier. It runs roughly parallel with the main Rocky Mt. range to the northeast, from which it is separated by the Rocky Mt. Trench. In the foothills to the west is the Cariboo dist., scene of the famous Cariboo gold rush of 1860.

Car·i·bou (kăr'ə-bo͞o'), county (1970 pop. 6,534), 1,746 sq mi (4,522.1 sq km), SE Idaho, in a mountainous area bordering on Wyo. and crossed by the Blackfoot River; formed 1919; co. seat Soda Springs. Wheat, livestock, and dairy products are produced. It also has phosphate mines.

Caribou, town (1970 pop. 10,419), Aroostook co., NE Maine, on the Aroostook River; inc. 1859. A processing and shipping hub for a great potato-growing region, it is also a winter sports center.

Ca·rin·thi·a (kə-rĭn'thē-ə), province (1971 pop. 525,728), c.3,680 sq mi (9,531 sq km), S Austria; capital Klagenfurt. Predominantly mountainous, it borders on Italy and Yugoslavia in the south.

Car·lin·ville (kär'lĭn-vĭl'), city (1970 pop. 5,675), seat of Macoupin co., W central Ill., SW of Springfield, in a coal, natural-gas, and farm area; settled c.1829, inc. 1837. Blackburn College is here.

Car·lisle (kär-līl', kär'līl), county borough (1973 est. pop. 70,930), administrative center of Cumbria, NW England, near the junction of the Caldew, Eden, and Petteril rivers. It is an important rail center and manufactures textiles, biscuits, and metal products. There is also an important livestock auction. The Roman camp Luguvallium, near Hadrian's Wall, was here. The site figured prominently in the border warfare between the English and the Scots during the Middle Ages. Mary Queen of Scots was imprisoned here in 1568. Until 1974 Carlisle was the county town of Cumberland.

Carlisle, county (1970 pop. 5,354), 196 sq mi (507.6 sq km), SW Ky., bounded on the W by the Mississippi at the Mo. border, on the N by Mayfield Creek; formed 1886; co. seat Bardwell. In a gently rolling agricultural area, partly in the Mississippi flood plain, it produces tobacco, livestock, corn, cotton, and dairy products.

Carlisle. **1.** City (1970 pop. 1,579), seat of Nicholas co., N Ky., NE of Lexington, in a tobacco region. **2.** Industrial borough (1978 est. pop. 19,600), seat of Cumberland co., S Pa.; inc. 1782. Its manufactures include shoes, rugs, and quartz crystals. A munitions depot during the Revolution, Carlisle was a headquarters for Washington during the Whiskey Rebellion in 1794. Molly Pitcher is buried in the Old Graveyard here.

Car·low (kär'lō), county (1971 pop. 34,237), 346 sq mi (896.1 sq km), SE Republic of Ireland; co. town Carlow. The granitic uplands of the Blackstairs Mts. in the southeast are a conspicuous feature in an otherwise fertile lowland region. Wheat, barley, and sugar-beet farming, cattle raising, and dairying are done. There are also flour-milling, malting, and sugar-refining industries. Organized as a county in the early 13th cent., Carlow was strategically situated on the southern edge of the English Pale.

Carlow, urban district (1971 pop. 9,588), county town of Co. Carlow, SE Republic of Ireland, on the Barrow River. It is an agricultural market in a dairy region, with sugar refining, flour milling, brewing, and shoe manufacturing. There are ruins of a 12th cent. castle. Of strategic importance, Carlow was burned in 1405 and in 1577.

Carl Sand·burg Home National Historic Site (sănd'bûrg'): *see* National Parks and Monuments Table.

Carls·bad (kärlz'băd). **1.** Resort city (1978 est. pop. 25,700), San Diego co., S Calif., on the Pacific coast; settled in the 1880s, inc. 1952. It has an electronic industry, machine shops, and a crystal silica quarry. Major agricultural products are tomatoes and flowers. The discovery of mineral springs with waters identical to those at Karlavy Vary (Carlsbad), now in Czechoslovakia, led to the settlement and naming of the town. **2.** City (1978 est. pop. 24,900), seat of Eddy co., SE N.Mex., on the Pecos River, in a grazing and irrigated farm area; settled 1888, inc. 1918. Great quantities of potash are mined and refined here. Other industries include agriculture, ranching, and tourism. The Carlsbad reclamation project, begun in 1906, serves more than 20,000 acres (8,100 hectares).

Carlsbad Caverns National Park, 46,753 acres (18,935 hectares), SE N.Mex., in the Guadalupe Mts.; est. as a national park 1930. These limestone caves, with remarkable stalactite and stalagmite formations and huge chambers, began forming 60 million years ago as groundwater started dissolving the rock. The caverns, among the largest in the world, were discovered c.1900 and still have not been completely explored. The Big Room, 754 ft (230 m) below the surface, is the most majestic of the many chambers.

Carl·ton (kärl'tən), county (1970 pop. 28,072), 860 sq mi (2,227.4 sq km), E Minn., in an agricultural area bordering on Wis. and drained by the St. Louis River; formed 1857; co. seat Carlton. Wood and dairy products, potatoes, livestock, poultry, and paper are produced.

Carlton, village (1970 pop. 884), seat of Carlton co., E Minn., on the St. Louis River near the Wis. border WSW of Duluth, in a diversified farming region. Dairy products and flour are made.

Car·lyle (kär-līl', kär'līl), city (1975 est. pop. 3,305), seat of Clinton co., S Ill., on the Kaskaskia River E of East St. Louis; laid out 1818, inc. 1837. It is a manufacturing and trade center for a grain-producing area.

Car·mar·then (kər-mär'*th*ən, kär-), municipal borough (1973 est. pop. 12,860), administrative center of Dyfed, S Wales, on the Towy River. It is a a cattle market and a dairy center. In the Middle Ages Carmarthen was an important wool port. Its old castle (now in ruins) was the headquarters of Welsh chieftains. Until 1974 Carmarthen was the county town of Carmarthenshire.

Car·mar·then·shire (kər-mär'*th*ən-shîr', -shər, kär-), former county, S Wales. The county town was Carmarthen. There are remains of prehistoric and Roman settlements. In 1974 Carmarthenshire became part of the new nonmetropolitan county of Dyfed.

Car·mel **1.** (kär-mĕl') or **Car·mel-by-the-sea,** (kär-mĕl'bī-thə-sē'). Village (1970 pop. 4,525), Monterey co., S Calif., at the neck of Monterey peninsula on Carmel Bay; inc. 1916. It is a tourist spot as well as an artists' and writers' community; art shows and an annual Bach festival are held in the village. Mission San Carlos Borromeo, the burial place of Father Junípero Serra, is nearby. **2.** (kär'məl) Village (1970 pop. 3,395), seat of Putnam co., SE N.Y., W of Danbury, Conn., on Lake Glenida. It is a trade center for a dairying and summer-resort area. Lake Carmel and many reservoirs are nearby.

Car·mel, Mount (kär'məl), mountain ridge, NW Israel, extending 13 mi (21 km) NW from the plain of Esdraelon to the Mediterranean Sea, where it ends in a promontory marking the S limit of the Bay of Haifa. Its highest point is 1,792 ft (546.6 m), and it is one of the most striking physical features of Israel.

Car·mi (kär'mī), city (1970 pop. 6,033), seat of White co., SE Ill., NE of Harrisburg on the Little Wabash River; platted 1816, inc. 1819. It is a rail and trade center in a farm and oil-producing region.

Car·mo·na (kär-mō'nä), town (1970 pop. 24,378), Seville prov., SW Spain, in Andalusia. It is a farm center for an area raising cattle, cereals, fruits, and olives. Ferdinand III of Castile took Carmona from the Moors in 1247 after a year-long siege. It has numerous examples of Gothic, Moorish, and baroque architecture.

Car·nac (kär'năk, kär-näk'), town (1968 pop. 3,681), Morbihan dept., NW France, in Brittany, at the foot of the Quiberon peninsula. It is the site of remarkable megalithic monuments, particularly menhirs, which extend along the coast in 11 parallel rows.

Car·na·tic (kär-năt'ĭk), region, SW India, on the Arabian Sea. The early European settlers sometimes applied the term Carnatic to all of southern India. The region was the site of the earliest European settlements in India. During the 18th cent. the Carnatic plains became the arena for the struggle between Great Britain and France for supremacy in India.

Car·ne·gie (kär-nā'gē), borough (1970 pop. 10,864), Allegheny co., SW Pa., an industrial suburb of Pittsburgh; inc. 1894. A steel town, it also has coal mines and plants making chemicals and electrical equipment.

Carnes·ville (kärnz'vĭl'), city (1970 pop. 510), seat of Franklin co., NE Ga., SSE of Toccoa.

Car·nic Alps (kär'nĭk), range of the E Alps in S Austria and NE Italy, extending c.60 mi (96.5 km). The highest elevation is Kellerwand (c.9,220 ft/2,810 m).

Car·ni·o·la (kär-nē-ō'lə), historic region, NW Yugoslavia, in Slovenia. The first known inhabitants, a Celtic tribe, were displaced by the Romans, who made Carniola part of their province of Pannonia. Slovenes settled Carniola in the 6th cent. Charlemagne later incorporated it into his empire. In 1269 the region was acquired by Bohemia. It passed to the Austrian Hapsburgs in 1282. After World War I, Carniola was divided between Italy and Yugoslavia, but the Italian part passed to Yugoslavia in 1947.

Car·o (kâr'ō), village (1970 pop. 3,701), seat of Tuscola co., S Mich., SE of Bay City; settled 1867, inc. 1871. It is a farm trade center, and beet sugar and canned goods are produced.

Car·o·line (kăr'ə-līn). **1.** County (1970 pop. 19,781), 320 sq mi (828.8 sq km), E Md., on the Eastern Shore bounded on the E by the Del. border, on the W by the Choptank River; formed 1773; co. seat Denton. In an agricultural area (vegetables, fruit, dairy products, corn, wheat, poultry, and livestock), it also has vegetable-canning and poultry-dressing plants and some lumbering. **2.** County (1970 pop. 13,925), 544 sq mi (1,409 sq km), E Va., bounded on the NE by the Rappahannock River, on the SW and S by the North Anna and Pamunkey rivers; formed 1728; co. seat Bowling Green. It has agriculture (vegetables, tobacco, and potatoes), livestock raising, and processing industries (canned foods and lumber).

Caroline Islands, archipelago (1973 est. pop. 75,394), c.830 sq mi (2,150 sq km), W Pacific, just N of the equator. It was included in 1947 in the U.S. Trust Territory of the Pacific Islands under United Nations trusteeship. The islands are fertile and rich in minerals. There are deposits of phosphate, guano, bauxite, and iron; coconuts, sugar cane, and tapioca are produced. The chief exports are dried bonito, copra, and tapioca. There is evidence of Chinese contact with the western islands in the 7th cent. A.D. The first Europeans to visit the Carolines were the Spanish in 1526, but the islands did not come under Spain's control until 1886. After the Spanish-American War the islands were sold (1899) to Germany. They were occupied in 1914 by the Japanese, who in 1920 were given a League of Nations mandate over them. Annexed to Japan in 1935, the islands were heavily bombed prior to American occupation during World War II.

Ca·ro·ní (kä-rō-nē'), river, E Venezuela, rising in the SE near the Guyana border and flowing N 550 mi (885 km) to join the Orinoco River.

Car·pa·thi·an Mountains (kär-pā'thē-ən) or **Car·pa·thi·ans** (-ənz), major mountain system of central and E Europe, extending c.930 mi (1,495 km) along the N and E sides of the Danubian plain. The geologically young mountains, part of the main European chain, link the Alps with the Balkans. The Carpathians begin in southeast Czechoslovakia and extend northeast to the Polish-Czechoslovak border. There the Northern Carpathians, comprising the Beskids and the Tatra, run east along the border, then southeast through the west Ukraine; in Rumania they are continued by the Transylvanian Alps (or Southern Carpathians), which extend southwest to the Danube River. The Carpathians are rich in minerals and timber. The region's cold winters and hot summers make it a year-round resort.

Car·pen·tar·i·a, Gulf of (kär'pən-târ'ē-ə), arm of the Arafara Sea, 305 mi (490.7 km) wide and 370 mi (595.3 km) long, indenting the N coast of Australia. On its eastern shore are bauxite deposits.

Car·pen·ters·ville (kär'pən-tərz-vĭl'), village (1978 est. pop. 26,700), Kane co., NE Ill., on the Fox River; inc. 1887.

Car·ran·tuo·hill (kă'rən-tōō'əl), mountain, 3,414 ft (1,041.3 m) high, Co. Kerry, SW Republic of Ireland, in Macgillicuddy's Reeks. It is the highest peak in Ireland.

Car·ra·ra (kə-rär'ə, kär-rä'rä), city (1976 est. pop. 70,181), Tuscany, N central Italy, near the Ligurian Sea. It is the center of the Italian marble industry. Chemicals and metal goods are manufactured.

Carr·hae (kâr'ē), Roman name for the ancient Mesopotamian city of Haran. The name Carrhae is best known because of the Battle of Carrhae in 53 B.C., in which M. Licinius Crassus was defeated by the Parthians.

Car·rick-on-Shan·non (kăr'ĭk-ŏn-shăn'ən), town (1971 pop. 6,411), county town of Co. Leitrim, N Republic of Ireland. It is a farm market and a center for trout fishing.

Car·ring·ton (kâr'ĭng-tən), city (1976 est. pop. 2,637), seat of Foster co., E central N.Dak., NNW of Jamestown, in a dairy, wheat, and livestock area; laid out 1882, inc. 1900.

Car·ri·zo Springs (kə-rē'zō), city (1970 pop. 5,374), seat of Dimmit co., SW Texas, SW of San Antonio; settled 1862, inc. 1910. Formerly a cow town, it is now the center of an irrigated farm and oil-processing area.

Car·ri·zo·zo (kär-ĭ-zō'zō), town (1970 pop. 1,123), seat of Lincoln co., S central N.Mex., WNW of Roswell at an altitude of 5,425 ft (1,654.6 m); laid out 1899. It is a health resort and a trade and shipping center of a livestock and farm area.

Car·roll (kăr'əl). **1.** County (1970 pop. 12,301), 634 sq mi (1,642.1 sq km), NW Ark., in the Ozarks bounded on the N by the Mo. border and drained by the White and King rivers; formed 1833; co. seats Berryville and Eureka Springs. Its agriculture includes fruit, grain, hay, truck crops, livestock, and dairy products. Some manufacturing and lumbering (pine and oak) is done. The county also has mineral springs and health resorts. **2.** County (1970 pop. 45,404), 495 sq mi (1,282.1 sq km), W Ga., bounded on the W by the Ala. line and on the SE by the Chattahoochee River; formed 1826; co. seat Carrollton. In a cotton-growing and textile-manufacturing area drained by the Little Tallapoosa River, it has livestock and a lumbering industry. **3.** County (1970 pop. 19,276), 456 sq mi (1,181 sq km), NW Ill., bounded on the W by the Mississippi and drained by the Plum River and Elkhorn Creek; formed 1839; co. seat Mount Carroll. Dairy products, livestock, corn, hay, wheat, oats, and poultry are produced. **4.** County (1970 pop. 17,734), 374 sq mi (968.7 sq km), NW central Ind., intersected by the Wabash River and drained by the Tippecanoe River and Wildcat and Deer creeks; formed 1828; co. seat Delphi. It has agriculture (grain, dairy products, and livestock), meat and poultry processing, lumber milling, and other manufacturing. **5.** County (1970 pop. 22,912), 574 sq mi (1,486.7 sq km), W central Iowa, in a prairie agricultural area drained by the Raccoon, Middle Raccoon, and East and West Nishnabotna rivers; formed 1851; co. seat Carroll. Cattle, hogs, poultry, corn, oats, and wheat are produced. A state park is here. **6.** County (1970 pop. 8,523), 130 sq mi (336.7 sq km), N Ky., in a gently rolling upland agricultural area in the outer Bluegrass region, drained by the Kentucky River and bounded on the N by the Ohio River, here forming the Ind. border, on the SE by Eagle Creek; formed 1838; co. seat Carrollton. Its agriculture includes grain, tobacco, livestock, poultry, and dairy products. It also has sand and gravel pits and some manufacturing. **7.** County (1970 pop. 69,006), 456 sq mi (1,181 sq km), N Md., in a piedmont region bounded on the NW by the Monocacy River, on the N by the Pa. border; formed 1836; co. seat Westminster. Dairying, stock and poultry raising, and agriculture (corn, wheat, potatoes, vegetables, and apples) are important. It has soapstone quarries, some manufacturing, and vegetable-canning plants. **8.** County (1970 pop. 9,397), 638 sq mi (1,652.4 sq km), central Miss., drained by the Big Black and Yalobusha rivers; formed 1833; co. seats Carrollton and Vaiden. It is in an agricultural (cotton, corn, and hay) and timber region. **9.** County (1970 pop. 12,565), 697 sq mi (1,805.2 sq km), NW central Mo., bounded on the S by the Missouri, on the E by the Grand River; formed 1833; co. seat Carrollton. In an agricultural area yielding wheat, corn, oats, and livestock, it has a food-processing industry and some coal deposits. **10.** County (1970 pop. 18,548), 938 sq mi (2,429.4 sq km), E N.H., in an agricultural and recreational area bordering on Maine and drained by the Saco, Ellis, and Ossipee rivers; formed 1840; co. seat Ossipee. Quarrying and some manufacturing (wood products and machinery) are done, and dairy products and poultry are produced. White Mountain National Forest is in the north. **11.** County (1970 pop. 21,579), 390 sq mi (1,010.1 sq km), E Ohio, in an agricultural area drained by a number of small creeks; formed 1832; co. seat Carrollton. It has coal mines and fire-clay quarries, and rubber, plastics, and metal products are manufactured. **12.** County (1970 pop. 25,741), 596 sq mi (1,543.6 sq km), NW Tenn., drained by the Big Sandy River and headstreams of the Obion River; formed 1821; co. seat Huntingdon. It is in an agricultural area producing cotton, corn, livestock, potatoes, and truck crops. **13.** County (1970 pop. 23,092), 496 sq mi (1,284.6 sq km), SW Va., in the Blue Ridge, bounded on the S by the N.C. border and drained by the New River; formed 1842; co. seat Hillsville. Agriculture (fruit, clover, grain, and cabbage), livestock raising, and dairying are important. It has some manufacturing (especially textiles and wood products) and pyrrhotite mining. It includes part of Jefferson National Forest and is traversed by the Appalachian Trail and the Blue Ridge Parkway.

Carroll, city (1976 est. pop. 9,218), seat of Carroll co., W central Iowa, W of Ames; inc. 1869. It is a trade and manufacturing center, with a poultry-processing industry. Swan Lake State Park is nearby.

Car·roll·ton (kăr'əl-tən). **1.** Town (1970 pop. 923), seat of Pickens co., W Ala., W of Tuscaloosa. Lumber milling and cotton ginning are done. **2.** City (1970 pop. 13,520), seat of Carroll co., W Ga., on the Little Tallapoosa River; inc. 1897. A trade center for a fertile farm area, it has textile-dyeing plants and factories making wires and chrome plating. **3.** City (1970 pop. 2,866), seat of Greene co., W Ill., NW of Alton; settled 1818, laid out 1821, inc. 1861. It is a farm trade center. **4.** City (1970 pop. 3,884), seat of Carroll co., N Ky., on the Ohio River at the mouth of the Kentucky River NE of Louisville, in a farm area; inc. 1794 as Port William, renamed 1838. The city has tobacco storehouses. Butler Memorial State Park is nearby. **5.** Town (1970 pop. 295), a seat of Carroll co., central Miss., E of Greenwood. Lumber and cotton are processed. **6.** City (1970 pop. 4,847), seat of Carroll co., N central Mo., ENE of Kansas City, in a dairy and farm area; settled 1818, inc. 1833. **7.** Village (1970 pop. 2,817), seat of Carroll co., NE Ohio, SE of Canton, in an agricultural and manufacturing area; laid out 1815. There are coal fields nearby. **8.** City (1978 est. pop. 31,800), Dallas and Denton cos., N Texas, a suburb of Dallas, in a rapidly growing area. Metal products, aircraft parts, and electronic equipment are the major products.

Car·son (kär'sən), county (1970 pop. 6,358), 900 sq mi (2,331 sq km), extreme N Texas, in the high plains of the Panhandle, drained by McClellan Creek, forks of the Red River, and small tributaries of the Canadian River; formed 1876; co. seat Panhandle. It is in a cattle-ranching and grain area underlaid by the Panhandle natural-gas and oil field, one of the largest in the world.

Carson. 1. City (1978 est. pop. 78,700), Los Angeles co., S Calif., an industrial and residential suburb of Los Angeles; inc. 1968. Oil refining is the major industry, but fabricated metals, paper, and many other products are manufactured. **2.** Village (1970 pop. 466), seat of Grant co., S N.Dak., SW of Bismarck. There are flour mills, coal mines, and dairy farms in the area.

Carson City, city (1978 est. pop. 28,700), state capital, W Nev., in the Carson valley; inc. 1875. It is a trade center for a mining and agricultural area. The state government is a major employer, and tourism is important. The city was laid out in 1858 on the site of Eagle Station, a trading post established (1851) on the immigrant trail from Salt Lake City to California. It became important after the discovery (1859) of the Comstock Lode. In 1861, when the Territory of Nevada was created, the city was made the capital, and in 1864 it became the state capital.

Car·ta·ge·na (kär′tə-jē′nə, kär′tä-hā′nä), city (1973 pop. 313,305), capital of Bolívar dept., NW Colombia, a port on the Bay of Cartagena in the Caribbean Sea. Oil refining and the manufacture of sugar, tobacco, hides, textiles, and cosmetics are the principal industries. Tourism is also important. Cartagena was founded in 1533 and became the treasure city of the Spanish Main, where precious stones and minerals from the New World awaited transshipment to Spain. Although the harbor was guarded by 29 stone forts and the city was encircled by a high wall of coral, Cartagena suffered sackings and invasions—in 1544, 1560, and 1586 (by Sir Francis Drake). In 1741 it withstood a three-month British siege. The city was one of the first to declare (1811) absolute independence from Spain. Known as the Republic of Cartagena, it was one of the bases used by Simón Bolívar to launch his campaign to liberate Venezuela. In 1815 the city was besieged and captured by the Spanish general Pablo Morillo, who inflicted savage reprisals on the population. Captured by rebel forces in 1821, Cartagena was incorporated into Colombia. Shady plazas and narrow cobblestone streets make Cartagena one of the most picturesque cities in Latin America.

Cartagena, city (1970 pop. 120,000), Murcia prov., SE Spain, on the Mediterranean Sea. A major seaport and naval base, it has a fine natural harbor, protected by forts, with a naval arsenal and important shipbuilding and metallurgical industries. Lead, iron, and zinc are mined and processed nearby. The city was founded c.225 B.C. and soon became a flourishing port, the chief Carthaginian base in Spain. Captured (209 B.C.) by Scipio Africanus Major, it continued to flourish under the Romans. The Moors, who took it in the 8th cent., later included it in Murcia. The Spaniards recovered it definitively in the 13th cent. Cartagena served as the Loyalist naval base during the civil war (1936–39).

Car·ta·go (kär-tä′gō), city (1976 est. pop. 23,100), central Costa Rica. The raising of livestock and the production of coffee are its main industries. Cartago was founded in 1563. It was the political center of Costa Rica until independence was won from Spain in 1821. It was destroyed by an eruption (1723) of Irazú volcano and was severely damaged by earthquakes in 1822, 1841, and 1910.

Car·ter (kär′tər). **1.** County (1970 pop. 19,850), 402 sq mi (1,041.2 sq km), NE Ky., in a hilly agricultural area drained by the Little Sandy River; formed 1838; co. seat Grayson. Tobacco, corn, poultry, livestock, dairy products, lespedeza, clover, fruit, and wheat are produced. It has clay, sand, and gravel pits, stone quarries, coal mines, iron and asphalt deposits, and timber tracts. Stone, clay, and glass products are manufactured. Limestone caverns and natural bridges attract many tourists. **2.** County (1970 pop. 3,878), 506 sq mi (1,310.5 sq km), SE Mo., in the Ozarks, drained by the Current River; formed 1859; co. seat Van Buren. It lies in an agricultural, livestock, and timber area. Wood products are manufactured. Part of Clark National Forest is here. **3.** County (1970 pop. 1,956), 3,313 sq mi (8,580.7 sq km), SE Mont., in a plains grazing area bordering on S.Dak. and Wyo. and drained by the Little Missouri River; formed 1917; co. seat Ekalaka. Sections of Custer National Forest are in the north and east. **4.** County (1970 pop. 37, 349), 830 sq mi (2,149.7 sq km), S Okla., drained by the Washita River and Walnut and Caddo creeks; formed 1907; co. seat Ardmore. Its oil and natural-gas wells and refineries are important. It also has stock raising, agriculture (grasses, pecans, peanuts, cotton, corn, and dairy products), manufacturing, and a tourist industry. **5.** County (1970 pop. 43,259), 348 sq mi (901.3 sq km), NE Tenn., bordering on N.C. in the S and SE and drained by the Watauga and Doe rivers; formed 1796; co. seat Elizabethton. It has timber, agriculture (tobacco, grain, livestock, and fruit), and iron-ore deposits. There is some manufacturing.

Car·ter·et (kär′tə-rĕt′), county (1970 pop. 31,603), 532 sq mi (1,377.9 sq km), E N.C., in a tidewater area bounded on the S by Onslow Bay, on the E by Raleigh Bay, and on the NE by the Neuse River and Pamlico Sound; formed 1722; co. seat Beaufort. It is bordered by Bogue and Portsmouth islands and other barrier beaches. Shipbuilding, fishing, sawmilling, and farming are important. There are resorts along the coast. The west part of the county is included in Croatan National Forest.

Carteret, borough (1978 est. pop. 22,000), Middlesex co., NE N.J., on Arthur Kill, opposite Staten Island; inc. 1906. It has oil and copper refineries and industries producing steel, chemicals, and cigars.

Car·ters·ville (kär′tərz-vĭl′), city (1970 pop. 10,138), seat of Bartow co., NW Ga., on the Etowah River NW of Atlanta, in a piedmont mining area; inc. as a town 1850, as a city 1872. Rubber products and textiles are made here, and lime, manganese, iron ore, barium, and ocher are mined in the area. There are Indian mounds nearby.

Car·thage (kär′thĭj), ancient city, on the N shore of Africa, on a peninsula in the Bay of Tunis and near modern Tunis. The Latin name, Carthago or Cartago, was derived from the Phoenician name, which meant "new city" (the old city being Utica). It was founded from Tyre in the 9th cent. B.C. The city-state built up trade and in the 6th and 5th cent. B.C. began to acquire dominance in the western Mediterranean. In the 3rd cent. B.C., Rome challenged that dominance in the Punic Wars (so called after the Roman name for the Carthaginians, Poeni, i.e., Phoenicians). The first of these wars (264-241) cost Carthage its holdings on Sicily, but Hamilcar Barca compensated for the loss by undertaking conquest in Spain. The Second Punic War (218–201) again ended in defeat, although the Carthaginian general was the formidable Hannibal. All its warships and its possessions outside Africa were lost, but Carthage recovered commercially and remained prosperous. The Third Punic War (149-146 B.C.) ended with the total destruction of Carthaginian power and the razing of the city. A new city was founded in 44 B.C. and under Augustus became an important center of Roman administration. Carthage was later (A.D. 439-533) the capital of the Vandals. Although practically destroyed by Arabs in 698, the site was populated for many centuries afterward.

Carthage. 1. City (1970 pop. 3,350), seat of Hancock co., W Ill., near the Mississippi E of Keokuk, Iowa; laid out 1833, inc. 1837. It is a farm trade center. The jail where Joseph Smith was killed (1844) has been preserved and is regarded as a shrine by many Mormons. **2.** Town (1970 pop. 3,031), seat of Leake co., central Miss., near the Pearl River NE of Jackson, in a farm and timber area. **3.** City (1970 pop. 11,035), seat of Jasper co., SW Mo., on the Spring River, in a rich farm area; inc. 1873. Its gray marble quarries are the largest of their kind in the world. A Civil War battle was fought here July 5, 1861; the city was burned and was rebuilt after the war. **4.** Village (1970 pop. 3,889), Jefferson co., N N.Y., on the Black River E of Watertown; settled before 1801, inc. 1841. It has diversified manufactures. **5.** Town (1970 pop. 1,034), seat of Moore co., central N.C., SW of Sanford, in a fruit and tobacco area; inc. 1796. **6.** Town (1970 pop. 2,491), seat of Smith co., central Tenn., on the Cumberland River ENE of Nashville; founded 1804, inc. 1817. A tobacco center, it also produces cheese and textiles. **7.** City (1970 pop. 5,392), seat of Panola co., E Texas, near the Sabine River SE of Longview; founded 1848. It is the site of a large natural-gas field and the source of many interstate pipelines. The region also yields oil and pinewood.

Ca·ruth·ers·ville (kə-rŭ*th*′ərz-vĭl′), city (1970 pop. 7,350), seat of Pemiscot co., extreme SE Mo., on the Mississippi River NW of Dyersburg, Tenn., in a cotton, farm, and timber region; founded 1794, platted and inc. 1857.

Car·ver (kär′vər), county (1970 pop. 28,331), 358 sq mi (927.2 sq km), S central Minn., in an agricultural area bounded on the SE by the Minnesota River and watered by small lakes; formed 1855; co. seat Chaska.

Cas·a·blan·ca (kăs′ə-blăng′kə, kăz′-, kä′sä-bläng′kä), city (1971 pop. 1,506,373), W Morocco, on the Atlantic Ocean. It is the largest city of Morocco and handles over two thirds of the country's commerce. Phosphates comprise 75% of the total export traffic, and petroleum products are the major imports. The city's leading industries produce textiles, glass, and bricks. Casablanca is on the site of Anfa, a prosperous town that the Portuguese destroyed in 1468; they resettled it in 1515 under its present name. During World War II, Casablanca was the scene of one of the three major Allied landings in North Africa (Nov., 1942) and of a conference between Franklin Delano Roosevelt and Winston Churchill (Nov., 1943).

Ca·sa·le Mon·fer·ra·to (kä-sä′lā mōn′fə-rä′tō) or **Casale,** city (1976 est. pop. 43,444), Piedmont, NW Italy, on the Po River. Manufactures include cement and electrical appliances, and much wine is produced in the region. Of note are the Romanesque cathedral (12th cent.) and the citadel (15th cent., now a barracks).

Cas·cade (kăs-kād′), county (1970 pop. 81,804), 2,660 sq mi (6,889.4 sq km), W central Mont., drained by the Missouri, Sun, and Smith rivers; formed 1887; co. seat Great Falls. In an agricultural region yielding livestock, grain, and diary products, it also has coal, silver, zinc, lead, and gold mines. Part of Lewis and Clark National Forest is in the southwest.

Cascade, city (1970 pop. 833), alt. 4,800 ft (1,464 m), seat of Valley co., W central Idaho, N of Boise on the North Fork of the Payette River; founded 1912–13, inc. 1917. It is the trade center for a lumber, agriculture, dairying, and livestock area. Gold, silver, and lead mines are nearby.

Cascade Range, mountain chain, c.700 mi (1,125 km) long, extending S from British Columbia to N Calif., where it joins the Sierra Nevada; it parallels the Coast Ranges, 100–150 mi (161–241.4 km) inland from the Pacific Ocean. Many of the range's highest peaks are volcanic cones, covered with snowfields and glaciers. Mt. Rainier (14,410 ft/4,395.1 m) is the highest point in the range. The Klamath, Columbia, and Fraser rivers flow from east to west across the range. Of the many lakes in the Cascades, Crater Lake and Lake Chelan are the most famous. National forests cover an extensive area. Receiving more than 100 in. (254 cm) of precipitation annually, the Cascades are a major source of water in the U.S. Northwest. Hydroelectricity is generated on the western slope; irrigation is used in the fertile eastern side valleys. The Cascade Tunnel, 8 mi (12.9 km), is the longest railroad tunnel in North America.

Cas·co Bay (kăs′kō), deep inlet of the Atlantic Ocean, 200 sq mi (518 sq km), SW Maine. The bay, with its more than 200 wooded, hilly islands, has many summer estates and resorts.

Ca·ser·ta (kä-zĕr′tä), city (1976 est. pop. 65,074), capital of Caserta prov., Campania, S central Italy. It is an agricultural and commercial center and a transportation junction. The surrender of the German forces in Italy to the Allies took place here on Apr. 29, 1945.

Ca·sey (kā′sē), county (1970 pop. 12,930), 435 sq mi (1,126.7 sq km), central Ky., bounded on the E by Fishing Creek and drained by the Green River; formed 1806; co. seat Liberty. In a hilly agricultural area producing livestock, corn, hay, and tobacco, it has stone quarries, timber, and lumber mills.

Cas·i·li·num (kăs′ĭ-lī′nəm), ancient town, Campania, S Italy, N of present-day Naples. Founded (c.600 B.C.) probably by the Etruscans, it became (5th cent. B.C.) the capital of the Samnites.

Ca·si·quia·re (kä-sē-kyä′rā), river, c.100 mi (160 km) long, S Venezuela. Also called the Canal Casiquiare, it is a branch of the Orinoco and flows southwest to the Río Negro, thus linking the Orinoco and Amazon basins.

Cas·per (kăs′pər), city (1978 est. pop. 43,700), alt. 5,123 ft (1,562.5 m), seat of Natrona co., E central Wyo., on the North Platte River; inc. 1889. It is a rail, distributing, processing, and trade center in a farming, ranching, and mineral-rich area. An oil boom town since the first well was tapped in 1890, it has large oil refineries and many oil-affiliated industries. Open-pit uranium mining nearby is important, and gas, coal, and bentonite deposits are also exploited. The city has wool and livestock markets, meat-packing plants, and a growing tourist industry. At this fording place on the Oregon Trail the Mormons in 1847 established a ferry, which was in the 1850s superseded by Platte Bridge. The city was founded (1888) with the coming of the railroad and grew with the discovery of oil at Salt Creek, followed by the Teapot Dome and Big Muddy finds. Old Fort Caspar Museum (the fort has been restored; a clerk's error accounts for the later spelling of the name) and Casper Mt. (c.8,000 ft/2,440 m high) are nearby.

Cas·pi·an Sea (kăs′pē-ən), salt lake, c.144,000 sq mi (372,960 sq km), USSR and Iran, between Europe and Asia; the largest inland body of water in the world. The largest part lies in Soviet territory; only the extreme southern shore belongs to Iran. The Caspian is 92 ft (28 m) below sea level. It reaches its maximum depth, c.3,200 ft (975 m), in the south; the shallow northern half averages only about 17 ft (5.2 m). The Caspian receives the Volga (which supplies more than 75% of its inflow), Ural, Emba, Kura, and Terek rivers, but has no outlet. Variations in evaporation account for the great changes in the size of the sea during the course of history. The construction of large dams and lakes on the Volga is the major reason for the recent lowering of the Caspian's water level.

Cass (kăs). **1.** County (1970 pop. 14,219), 370 sq mi (958.3 sq km), W central Ill., bounded on the N by the Sangamon River, on the W by the Illinois River; formed 1837; co. seat Virginia. It has agriculture (corn, wheat, soybeans, fruit, sweet potatoes, livestock, poultry, and dairy products), commercial fisheries, river and rail shipping, and some light industry. **2.** County (1970 pop. 40,456), 415 sq mi (1,074.9 sq km), N central Ind., intersected by the Wabash River and drained by the Eel River and Deer Creek; formed 1828; co. seat Logansport. Its agriculture includes grain, fruit, truck crops, livestock, poultry, and dairy products. Some manufacturing and shipping is done. **3.** County (1970 pop. 17,007), 559 sq mi (1,447.8 sq km), SW Iowa, in a prairie agricultural area drained by the East Nishnabotna and West Nodaway rivers; formed 1851; co. seat Atlantic. Cattle, hogs, poultry, and corn are produced. It also has coal deposits. **4.** County (1970 pop. 43,312), 488 sq mi (1,263.9 sq km), SW Mich., in a lake and farm region bounded on the S by the Ind. border and drained by the St. Joseph River; formed 1829; co. seat Cassopolis. It has agriculture (grain, truck crops, peppermint, fruit, livestock, and dairy products), resorts, and some manufacturing. **5.** County (1970 pop. 17,323), 1,998 sq mi (5,174.8 sq km), N central Minn., in an agricultural area bounded on the S by the Crow Wing River, on the N by the Mississippi; formed 1851; co. seat Walker. Dairy products, livestock, and potatoes are produced. It also has peat deposits. Parts of the Greater Leech Lake Indian Reservation and Chippewa National Forest are here. **6.** County (1970 pop. 39,448), 698 sq mi (1,807.8 sq km), W Mo., drained by the South Grand River; formed as Van Buren co. 1835, renamed 1849; co. seat Harrisonville. Its agriculture includes corn, wheat, oats, cattle, and poultry. It also has gas wells and some industry. **7.** County (1970 pop. 18,076), 552 sq mi (1,429.7 sq km), SE Nebr., in an agricultural region bounded on the E by the Missouri River and the Iowa border, on the N by the Platte River; formed 1868; co. seat Plattsmouth. Grain, livestock, and dairy products are produced. **8.** County (1970 pop. 73,653), 1,749 sq mi (4,529.9 sq km), E N.Dak., in a rich agricultural area bounded on the E by the Red River of the North and drained by the Maple and Sheyenne rivers; formed 1873; co. seat Fargo. Among its crops are wheat, sunflowers, and sugar beets. Tractors are manufactured. **9.** County (1970 pop. 24,133), 941 sq mi (2,437.2 sq km), NE Texas, in a rolling, partly forested area bounded on the E by the Ark. and La. borders, on the N by the Sulphur River; formed 1846; co. seat Linden. Lumbering, agriculture (cotton, corn, peanuts, peas, sweet potatoes, fruit, and truck crops), livestock raising, and dairying are important. It has oil and natural-gas wells, lignite, clay, and iron deposits, and some processing and manufacturing plants.

Cas·sia (kăsh′ə), county (1970 pop. 17,017), 2,544 sq mi (6,589 sq km), S Idaho, in an irrigated agricultural area bordering on Utah and Nev. and bounded on the N by the Snake River; formed 1879; co. seat Burley. Potatoes, sugar beets, dry beans, and livestock are produced.

Cas·si·no (kə-sē′nō, käs-sē′nō), town (1976 est. pop. 25,300), in Latium, central Italy, in the Apennines, on the Rapido River. It is a commercial and agricultural center. The peace between Emperor Frederick II and Pope Gregory IX was signed here in 1230. During World War II (late 1943) the town and the nearby Benedictine abbey of Monte Cassino were strongly defended by Germans blocking the Allied advance on Rome. The Allies finally captured the German positions in May, 1944. Cassino was reduced to rubble but was largely rebuilt.

Cas·sop·o·lis (kə-sŏp′ə-lĭs), resort village (1970 pop. 2,108), seat of Cass co., SW Mich., NE of South Bend, Ind., in a farm area; settled 1831, inc. 1863. There are Indian mounds in the vicinity.

Cass·ville (kăs′vĭl′), city (1970 pop. 1,910), seat of Barry co., SW Mo., SE of Joplin; platted 1845, inc. 1847. A resort and fishing center in the Ozarks, it is also a shipping point for a fruit, dairy, and poultry region. Roaring River State Park is nearby.

Cas·tel Gan·dol·fo (käs-tĕl′ gän-dôl′fō), town (1976 est. pop. 3,200), in Latium, central Italy, in the Alban Hills, overlooking Lake Albano. Possibly occupying the site of ancient Alba Longa, it is the papal summer residence. The papal palace (17th cent.), its magnificent gardens, the Vatican observatory (founded 1936), and the Villa Barbarini enjoy extraterritorial rights.

Cas·tel·lam·ma·re di Sta·bia (käs-tĕl′läm-mä′rā dē stä′byä), city (1976 est. pop. 73,049), in Campania, S Italy, on the Bay of Naples. A summer resort and spa, it has thermal mineral springs that have been used since Roman times. It is also a commercial and industrial center, with navy yards founded in 1783. Manufactures include food products, paper, and cement. The city was built on the site of Stabiae, a favorite Roman resort, which was buried in the eruption of Mt. Vesuvius in A.D. 79.

Cas·tel·lón de la Pla·na (käs-tĕl-yōn′ dā lä plä′nä), city (1975 est. pop. 108,650), capital of Castellón de la Plana prov., E Spain, in Valencia. It is a farm center with fishing, mining, and handicraft industries.

Cas·tile (kăs-tēl'), region and former kingdom, central and N Spain, traditionally divided into Old Castile (1974 est. pop. 2,228,542) in the north and New Castile (1974 est. pop. 5,456,846) in the south. Castile is generally a vast underdeveloped region surrounding the highly industrialized Madrid area. It includes most of the high plateau of central Spain, across which rise the rugged Sierra de Guadarrama and the Sierra de Gredos, forming a natural boundary between Old and New Castile. Old Castile has grain growing and sheep raising; in more fertile areas, especially in New Castile, olive oil and grapes are produced. Scattered forests yield timber and naval stores. Agricultural methods are largely primitive, but irrigation, introduced by the Romans and the Moors, has progressed significantly in recent decades. Mineral resources, except for the rich mercury mines of Almadén, are of minor economic importance. Old Castile at first was a county of the kingdom of León, with Burgos as its capital. Its nobles secured virtual autonomy by the 10th cent. León was first united with Castile in 1037, but complex dynastic rivalries delayed the permanent union of the two realms, which was achieved under Ferdinand III in 1230. The Castilian kings played a leading role in the fight against the Moors, from whom they wrested New Castile. In 1479, a personal union of Castile and Aragón was established under Isabella I and her husband, Ferdinand II of Aragón. The union was confirmed with the accession (1516) of their grandson, Charles I (later Emperor Charles V), to the Spanish kingdoms.

Cas·til·lo de San Mar·cos National Monument (kăs-tē'yō də săn mär'kəs): *see* National Parks and Monuments Table.

Cas·tle·bar (kăs'əl-bär'), urban district (1971 pop. 5,979), county town of Co. Mayo, W Republic of Ireland. It is a market for a farm area. Cured bacon and manufactured hats are products of the town. Castlebar was occupied by the French in 1798.

Cas·tle Clin·ton National Monument (kăs'əl klĭn'tən): *see* Battery, the.

Castle Dale, city (1970 pop. 541), alt. 5,771 ft (1,760.2 m), seat of Emery co., central Utah, on the headstream of the San Rafael River SSW of Price. There are coal and radioactive ore deposits in the vicinity.

Cas·tle·ford (kăs'əl-fərd), municipal borough (1973 est. pop. 37,650), West Yorkshire, central England, at the junction of the Aire and Calder rivers. Chartered as a municipal borough in 1955, it has bottleworks, chemical works, and collieries. The site of an ancient Roman town lies within its borders.

Castle Peak, mountain, 14,259 ft (4,349 m) high, W central Colo., NNE of Crested Butte. It is the highest elevation in the Elk Mts. range of the Rockies.

Cas·tres (käs'trə), city (1975 pop. 45,978), Tarn dept., SW France, on the Agout River. It has been a textile center since the 13th cent., and its machine tools are known worldwide. Wood products, especially furniture, are also manufactured. Once the site of a Roman encampment, Castres grew around a Benedictine monastery founded in 647 A.D. Protestantism took hold in the 16th cent. but was suppressed by Louis XIII.

Cas·tries (kăs'trē', käs'trēs'), town (1970 pop. 39,132), capital and commercial center of St. Lucia, British West Indies. Its excellent landlocked harbor is one of the best in the West Indies. Castries was founded by the French in 1650.

Cas·tro (kăs'trō), county (1970 pop. 10,394), 876 sq mi (2,268.8 sq km), NW Texas, on the Llano Estacado in an irrigated agricultural area; organized 1891; co. seat Dimmitt. Cattle, wheat, grain, sorghums, cotton, potatoes, sugar beets, and truck crops are produced.

Cas·trop-Rau·xel (käs'trôp-rouk'səl), city (1975 pop. 83,421), North Rhine-Westphalia, W West Germany, on the Rhine-Herne Canal, an industrial city of the Ruhr district. Chemicals and other light industrial goods are produced here.

Cas·well (kăz'wəl, -wĕl), county (1970 pop. 19,055), 435 sq mi (1,126.7 sq km), N N.C., in a piedmont region bounded on the N by the Va. border and drained by the Dan River; formed 1777; co. seat Yanceyville. It has agriculture (tobacco, corn, wheat, and hay), lumbering, and some manufacturing.

Cat·a·hou·la (kăt'ə-hoo'lə), parish (1970 pop. 11,769), 732 sq mi (1,895.9 sq km), E La., bounded on the E by the Black and Tensas rivers, on the S by the Red River and Big Saline Bayou, and intersected by the Ouachita and Little rivers; formed 1808; parish seat Harrisonburg. In an agricultural area yielding cotton, corn, hay, sugar cane, livestock, and poultry, it also has a lumbering industry, fisheries, and sand and gravel pits. Cotton is ginned.

Cat·a·li·na Island (kăt'ə-lē'nə): *see* Santa Catalina.

Cat·a·lo·nia (kăt-ə-lō'nyə, -lō'nē-ə), region (1974 est. pop. 5,534,770), NE Spain, stretching from the Pyrenees at the French border S along the Mediterranean Sea. Barcelona is the historic capital. Mostly hilly, with pine-covered mountains, it also has some highly fertile plains. Cereals, olives, and grapes are grown, and one third of the wines of Spain are produced here. The seacoast has fine harbors and an active tourist industry. Trade has been active along the coast since Greek and Roman times. The history of medieval Catalonia is that of the counts of Barcelona, who emerged (9th cent.) as the chief lords in the Spanish March founded by Charlemagne. United (1137) with Aragón through marriage, Catalonia preserved its own laws, its cortes, and its own language (akin to Provençal). In the cities, notably Barcelona, the burgher and merchant classes grew very powerful. Catalan traders rivaled those of Genoa and Venice, and their maritime code was widely used in the 14th cent. Catalonia failed in its rebellion (1461–72) against John II of Aragón, and after the union (1479) of Aragón and Castile, Catalonia declined. In the Thirty Years War (1618–48), Catalonia rose against Philip IV, and in the War of the Spanish Succession it sided with Archduke Charles against Philip V, who in reprisal deprived it of its privileges. In the late 19th and 20th cent. it was a center of socialist and anarchist strength. In 1932 the Catalans established a separate government, which in 1932 won autonomy from the Spanish Cortes. In the civil war of 1936–39, Catalonia sided with the Loyalists and suffered heavily. It fell to Franco in Feb., 1939.

Ca·ta·nia (kə-tā'nyə, kä-tä'nyä), city (1976 est. pop. 399,773), capital of Catania prov., E Sicily, Italy, on the Gulf of Catania, an arm of the Ionian Sea, and at the foot of Mt. Etna. It is a busy port and a major commercial and industrial center. Manufactures include chemicals, silk and cotton textiles, and asphalt. The city also has a fishing industry. Founded in the 8th cent. B.C., Catania was a flourishing Greek town and was later a Roman colony. It was rebuilt after earthquakes in 1169 and 1693 and after a severe volcanic eruption in 1669. The city was heavily damaged in World War II. Points of interest include the extensive Bellini Gardens (named for the 19th cent. composer, who was born in Catania) and Ursino castle, built (13th cent.) by Emperor Frederick II.

Ca·tan·za·ro (kä-tän-dzä'rō), city (1971 pop. 85,316), capital of Catanzaro prov. and of Calabria, S Italy, on a hill above the Ionian Sea. It is a commercial center, with flour mills and distilleries.

Ca·taw·ba (kə-tô'bə), county (1970 pop. 90,873), 406 sq mi (1,051.5 sq km), W central N.C., in a piedmont area bounded on the E and N by the Catawba River; formed 1842; co. seat Newton. It has agriculture (tobacco, cotton, corn, wheat, hay, livestock, dairy products, and poultry) and timber (pine and oak). Textiles and furniture are manufactured. There are hydroelectric plants in the area.

Catawba, river, rising in the Blue Ridge, W N.C., and flowing 250 mi (402.3 km) generally S into S.C., where it becomes the Wateree River.

Ca·teau, Le (lə kä-tō'), town (1968 pop. 9,314), Nord dept., N France, in French Flanders. It has textile, metallurgical, and ceramic industries. In a treaty signed here in 1559, the last English foothold on the continent was returned to France. A museum contains many works by Matisse, who was born here.

Ca·ter·ham and War·ling·ham (kā'tər-əm; wôr'lĭng-əm), urban district (1973 est. pop. 35,840), Surrey, SE England. A residential suburb of London, it has engineering, chemical, perfume, and printing industries.

Ca·thay (kə-thā'), medieval name for China. It was popularized by Marco Polo (c.1254–c.1324) and usually applied only to China north of the Yangtze River.

Cath·lam·et (kăth-lăm'ĭt), town (1970 pop. 647), seat of Wahkiakum co., SW Wash., W of Longview on the Columbia River, in a fishing, agricultural, and lumbering region.

Cat·letts·burg (kăt'lĭts-bûrg'), city (1970 pop. 3,420), seat of Boyd co., E Ky., on the bank of the Ohio River at the mouth of the Big Sandy River near Ashland, in an manufacturing and oil-refining area; settled 1808.

Ca·to·che, Cape (kä-tō'chā), extremity of Yucatán peninsula, SE Mexico. It was the first Mexican land seen by the Spanish (1517).

Ca·toc·tin Mountain Park (kə-tŏk'tĭn): *see* National Parks and Monuments Table.

Ca·too·sa (kə-to͞o'sə), county (1970 pop. 28,271), 167 sq mi (432.5 sq km), NW Ga., bordered on the N by Tenn.; formed 1853; co. seat Ringgold. Its agriculture includes cotton, potatoes, corn, fruit, and livestock. Textiles are manufactured. The county includes parts of Chickamauga and Chattanooga National Military Park and Chattahoochee National Forest.

Cat·ron (kə-trŏn'), county (1970 pop. 2,198), 6,898 sq mi (17,865.8 sq km), W N.Mex., in a stock-grazing, mining, and quarrying region watered by the San Francisco and Tularosa rivers and bounded on the W by the Ariz. border; formed 1921; co. seat Reserve. Ranching and lumber milling are done, and grain and alfalfa are grown. It includes parts of the Gila and Apache national forests. Gila Cliff Dwellings National Monument is in the southeast.

Cats·kill (kăt'skĭl'), village (1970 pop. 5,317), seat of Greene co., SE N.Y., on the Hudson River; settled 17th cent. by Dutch, inc. 1806. It is a gateway to resorts in the Catskill Mts. Thomas Cole lived and painted in the village.

Catskill Mountains, dissected plateau of the Appalachian Mt. system, SE N.Y., just W of the Hudson River, to which it descends abruptly in places. This glaciated region is well wooded and rolling with deep gorges and many beautiful waterfalls. Most of the summits are c.3,000 ft (915 m) above sea level. Close to New York City, the area is a popular summer and winter resort.

Cat·ta·rau·gus (kăt'ə-rô'gəs), county (1970 pop. 81,666), 1,335 sq mi (3,457.7 sq km), W N.Y., bounded on the S by the Pa. border, intersected by the Allegheny River, and drained by Cattaraugus, Conewango, and Ischua creeks; formed 1808; co. seat Little Valley. In a dairying, oil producing, farming, and stock and poultry raising area, it also has diversified manufacturing, natural-gas wells, and sand and gravel pits. It includes Allegany State Park, Allegany Indian Reservation, and part of Cattaraugus Indian Reservation.

Cau·ca (kou'kä), river, c.600 mi (965 km) long, rising in the Cordillera Central, near Popayán, W Colombia. It flows north in a rift valley between the Cordillera Central and Cordillera Occidental to the Magdalena River. It is navigable in its lower course and drains a fertile valley that has many minerals, including gold.

Cau·ca·sus (kô'kə-səs), region and mountain system, SE European USSR. The mountain system extends c.750 mi (1,205 km) from the mouth of the Kuban River on the Black Sea southeast to the Apsheron peninsula on the Caspian Sea. As a divide between Europe and Asia, the Caucasus has two major regions—North Caucasia and Transcaucasia. North Caucasia, composed mainly of plain (steppe) areas, begins at the Manych Depression and rises to the south, where it runs into the main mountain range, a series of chains running northwest-southeast. The Caucasus Mts. are crossed by several passes, which connect North Caucasia with the second major section, Transcaucasia. This region includes the southern slopes of the Caucasus Mts. and the depressions that link them with the Armenian plateau. Oil is the major product in the Caucasus. Iron, steel, and manganese are also important. Power for these industries is produced at several large hydroelectric stations. On the mountain slopes, which are densely covered by pine and deciduous trees, there is stock raising. In the valleys, citrus fruits, tea, cotton, grain, and livestock are raised. Along the Black Sea coast there are many resorts and summer homes.

The Caucasus figured greatly in the legends of ancient Greece. Persians, Khazars, Arabs, Huns, Turko-Mongols, and Russians have invaded and migrated into the Caucasus and have given the region its ethnic and linguistic complexity. The Russians assumed control in the 19th cent. after a series of wars with Persia and Turkey. In World War II the invading German forces launched (July, 1942) a major drive to seize or neutralize the vast oil resources of the Caucasus. In Jan., 1943, the Soviets launched a winter offensive and by Oct. had driven the Germans from the region.

Cau·to (kou'tō), longest river in Cuba, c.150 mi (240 km) long, rising in the Sierra Maestra and flowing NW and W to the Caribbean Sea just N of Manzanillo.

Cau·ve·ry (kô'və-rē), river, c.475 mi (765 km) long, rising in the Western Ghats and flowing SE across a plateau to the Bay of Bengal, S India. At its mouth is a great, fertile delta that is irrigated by an extensive canal system, one of the oldest in India. Before entering the delta, the river is divided by Sivasamudram Island and drops 320 ft (98 m), forming Cauvery Falls. On the left falls is India's first hydroelectric plant (built 1902), which supplies most of southern India with power.

Cav·a·lier (kăv'ə-lîr'), county (1970 pop. 8,213), 1,513 sq mi (3,918.7 sq km), NE N.Dak., in a prairie area bordering on Man., Canada; formed 1873; co. seat Langdon. Wheat, barley, oats, and potatoes are grown. Dairy products and livestock are also important.

Cavalier, city (1970 pop. 1,381), seat of Pembina co., extreme NE N.Dak., on the Tongue River NNW of Grand Forks; founded 1875, inc. 1885. It is a processing center for a livestock, dairy, and farm region.

Cav·an (kăv'ən), county (1971 pop. 52,618), 730 sq mi (1,890.7 sq km), N Republic of Ireland; co. town Cavan. It is a hilly region of lakes and bogs. Pastoral agriculture is the chief occupation. Cavan was organized as a shire of Ulster prov. in 1584.

Cavan, urban district (1971 pop. 3,268), county town of Co. Cavan, N Republic of Ireland.

Ca·vi·te (kä-vē'tā), city (1970 pop. 75,739), Cavite prov., SW Luzon, the Philippines. The city, situated on a small peninsula in Manila Bay, has been important as a naval base and trade center since the days of the Spanish. In the Spanish-American War it was captured by Adm. George Dewey on May 1, 1898. The United States established a major naval base at Sangley Point just opposite the city proper. In World War II this base was bombed (Dec. 10, 1941) by the Japanese and virtually destroyed.

Caw·dor (kô'dər), village, Highland region, NE Scotland, SW of Nairn. Cawdor Castle, whose earliest construction dates from 1454, was represented by Shakespeare, following tradition, as the scene of the slaying (1040) of Duncan by Macbeth.

Ca·xi·as do Sul (kə-shē'əs də so͞ol), city (1975 est. pop. 107,487), Rio Grande do Sul state, S Brazil. It is an important metallurgical center and has the most extensive vineyards in Brazil. The city was founded in 1875.

Cay·enne (kī-ĕn', kā-ĕn'), city (1967 pop. 24,518), capital of French Guiana, on Cayenne Island at the mouth of the Cayenne River. The city has a shallow harbor. Timber, rum, essence of rosewood, and gold are exported. Cayenne was founded by the French in 1643, but it was wiped out by an Indian massacre and was not resettled until 1664. Throughout the 17th cent. the city and its surrounding region were sharply contested by Great Britain, France, and the Netherlands. It was occupied (1808–16) by both the British and the Portuguese. From 1851 to 1946 the city was the center of French penal settlements in Guiana. In the city are the Pasteur Institute, which specializes in the study of tropical diseases, and several buildings from the colonial period. The city gives its name to cayenne pepper, a very sharp condiment found on the island in abundance.

Cay·man Islands (kā-măn', kā'mən), archipelago (1970 pop. 10,652), 100 sq mi (259 sq km), British West Indies. Georgetown, the capital and chief port, is on Grand Cayman. The inhabitants engage in shipbuilding, turtle and shark fishing, coconut raising, and lumbering; exports include green turtles, turtle shells, shark skins, coconuts, and dyewood. Tourism is also a major industry. The islands were discovered by Christopher Columbus in 1503.

Cay·u·ga (kī-yo͞o'gə, kā-, kə-), county (1970 pop. 77,439), 699 sq mi (1,810.4 sq km), W central N.Y., bounded on the N by Lake Ontario and extending into the Finger Lakes region on the S; formed 1799; co. seat Auburn. In a dairying and farming area yielding hay, grain, potatoes, and truck crops, it has diversified manufacturing and resorts on scenic Cayuga and Owasco lakes.

Cayuga Lake, 38 mi (61.1 km) long and 1 to 3.5 mi (1.6–5.6 km) wide, W central N.Y. It is the longest of the Finger Lakes and is connected by canal and by the Seneca River with the Barge Canal.

Ce·a·nan·nus Mor (sē'ə-năn'əs môr) or **Kells** (kĕlz), urban district (1971 pop. 2,395), Co. Meath, NE Republic of Ireland, on the Blackwater River. It is a market town and was once a royal residence for Irish kings. Noteworthy are the relic of an ancient monastery founded in the 6th cent. by St. Columba, the round tower, and several ancient crosses. The Book of Kells, found in the ancient monastery and believed to have been written in the 8th cent., is generally regarded as the finest example of Celtic illumination.

Ce·a·rá (sā'ə-rä'), state (1975 est. pop. 5,111,600), 57,149 sq mi (148,016 sq km), NE Brazil, on the Atlantic; capital Fortaleza.

Ce·bu (sā-bo͞o'), island (1970 pop. 1,632,642), 1,702 sq mi (4,408.2 sq km), one of the Visayan Islands, Philippines, between Leyte and Negros. The island is a leading peanut and corn producer; rice, sugar cane, coconuts, and hemp are also grown. There are major coal and copper deposits. Fertilizer is made from local pyrite. Magellan landed on the island in 1521; the wooden cross he planted is a major tourist attraction. The city of Cebu (1970 pop. 342,116) is an important harbor and the trade and manufacturing center of the Visayan Islands.

Cec·il (sē'səl), county (1970 pop. 53,291), 352 sq mi (911.7 sq km), extreme NE Md., at the head of Chesapeake Bay at the base of the Eastern Shore, bounded on the S by the Sassafras River, on the E by the Del. border, on the N by the Pa. border, and on the W by the Susquehanna River; formed 1674; co. seat Elkton. Its agriculture includes wheat, corn, hay, truck crops, fruit, livestock, and dairy products. It also has granite quarries, kaolin and sand pits, commercial fisheries, and some manufacturing.

Ce·dar (sē'dər). **1.** County (1970 pop. 17,655), 585 sq mi (1,515.2 sq km), E Iowa, in a prairie agricultural area drained by the Cedar River; formed 1837; co. seat Tipton. Hogs, cattle, corn, oats, hay, and wheat are produced. It has limestone quarries and many small industries. **2.** County (1970 pop. 9,424), 496 sq mi (1,284.6 sq km), W Mo., in an Ozark region drained by the Sac River; formed 1845; co. seat Stockton. It lies in a timber, coal, and agricultural region yielding corn, wheat, oats, and livestock. Food processing is done, and textiles and shoes are manufactured. **3.** County (1970 pop. 12,192), 743 sq mi (1,924.4 sq km), NE Nebr., in an agricultural region bounded on the N by the Missouri River and the S.Dak. border; formed 1857; co. seat Hartington. It has agriculture (hogs, corn, soybeans, and dairy products) and some light industry.

Cedar Breaks National Monument: *see* National Parks and Monuments Table.

Cedar City, town (1970 pop. 8,946), Iron co., SW Utah, at the base of the Wasatch Mts. E of the Escalante Desert; inc. 1868. With nearby Parowan, it was founded in 1851 by the Mormon "iron mission," sent to develop coal and iron deposits. Today it is a tourist center in an iron ore and ranching area. Zion National Park, Bryce Canyon National Park, and Cedar Breaks National Monument are nearby.

Cedar Falls, city (1978 est. pop. 34,000), Black Hawk co., N Iowa, on the Cedar River; inc. 1854. It developed as a milling center in the late 19th cent. after the coming of the railroad. Its manufactures include pumps, farm machinery, tools and dies, golfing equipment, and refuse disposal equipment.

Cedar Rap·ids (răp'ĭdz), city (1978 est. pop. 110,400), seat of Linn co., E central Iowa, on the Cedar River; inc. as a city 1856. It is named for the surging rapids in the river. One of Iowa's principal commercial and industrial cities, Cedar Rapids is a distribution and rail center for an extensive agricultural area. The city's major manufactures are cereals, communications equipment, farm and road machinery, syrup, plastic products, and gymnastic equipment. Points of interest include a large Masonic library (1884) and an art museum with a collection by the American artist Grant Wood.

Ce·dar·town (sē'dər-toun'), industrial city (1970 pop. 9,253), seat of Polk co., NW Ga., near the Ala. border WNW of Atlanta; settled on the site of a Cherokee village, inc. 1854. Chemicals and textiles are manufactured.

Ce·dros (sā'drōs), island, 134 sq mi (347.1 sq km), NW Mexico, in the Pacific off the coast of Baja California. It is sparsely inhabited and has varied wildlife, such as deer, reptiles, and sea lions.

Ce·fa·lù (chā-fä-lo͞o'), town (1976 est. pop. 11,300), N Sicily, Italy, a port on the Tyrrhenian Sea. It is a commercial and fishing center and a seaside resort. Its famous cathedral, started in 1131, is one of the finest examples of Norman architecture in Sicily.

Ce·la·ya (sā-lä'yä), city (1970 pop. 79,977), Guanajuato state, W central Mexico. In a region watered by the Lerma irrigation works, Celaya is the center of a prosperous agricultural area. Founded in 1571, Celaya was frequently involved in Mexican wars.

Cel·e·bes (sĕl'ə-bēz, sə-lē'-) or **Su·la·we·si** (so͞o'lä-wä'sē), island (1970 est. pop., including offshore islands, 8,925,000), c.73,000 sq mi (189,070 sq km), largest island in E Indonesia, E of Borneo, from which it is separated by the Makasar Strait. Extremely irregular in shape, it comprises four large peninsulas separated by three gulfs. The terrain is almost wholly mountainous, with many active volcanoes. Asian and Australian elements are commingled in the fauna, which includes the babirusa (resembling swine), the small wild ox called anoa (found only in the Celebes), the baboon, some rare species of parrot, and a large number of crocodiles. Valuable stands of timber cover much of the island; many forest products are exported. Mineral resources include nickel, gold, diamonds, sulfur, and low-grade iron ore. Celebes is a major source of copra, and corn, rice, cassava, yams, tobacco, and spices are grown. The Portuguese first visited the island in 1512. The Dutch expelled the Portuguese in the 1600s and conquered the natives in the Makasar War (1666–69). In 1950 Celebes became one of 10 provinces of the newly created republic of Indonesia; it has since been divided into 4 provinces. The Celebes Sea is north of the island, between it and the Philippines.

Ce·li·na. 1. (sə-lī'nə) City (1975 est. pop. 8,000), seat of Mercer co., W Ohio, on Lake St. Marys SW of Lima; settled 1834, inc. as a village 1861. It is a summer resort and has some industry. **2.** (sə-lē'nə) Town (1970 pop. 1,370), seat of Clay co., N Tenn., on the Cumberland River at the mouth of the Obey River NE of Nashville; inc. 1909.

Cel·le (tsĕl'ə), city (1974 est. pop. 74,845), Lower Saxony, N West Germany, on the Aller River. Its manufactures include food products, machinery, chemicals, and textiles. Celle was chartered in 1294. Its castle houses a famous 17th cent. baroque theater.

Ce·nis, Mont (môɴ sə-nē'), Alpine pass, 6,831 ft (2,083.5 m) high, on the French-Italian border. It is one of the great invasion routes in Italian history. The Mont Cenis railroad tunnel, c.8 mi (13 km) long, was built in 1871.

Cen·ter (sĕn'tər). **1.** Village (1970 pop. 111), seat of Knox co., NE Nebr., NNW of Norfolk on Bazile Creek. **2.** Village (1970 pop. 619), seat of Oliver co., central N.Dak. NW of Bismarck on Square Butte Creek, in a coal-mining, livestock, and farming region. Poultry, wheat, corn, and potatoes are grown. **3.** City (1970 pop. 4,989), seat of Shelby co., E Texas, near the Sabine River NE of Nacogdoches, in a pinewoods area; founded 1866, inc. 1903.

Center City, village (1970 pop. 324), seat of Chisago co., E Minn., near the St. Croix River NNE of St. Paul, in a resort and lake region. Grain and potatoes are grown, and livestock is raised.

Cen·ter·ville (sĕn'tər-vĭl'). **1.** City (1970 pop. 6,531), seat of Appanoose co., S central Iowa, SW of Ottumwa, in a coal area; platted 1846, inc. 1855. It has varied manufactures. Sharon Bluffs State Park is nearby. **2.** Village (1970 pop. 209), seat of Reynolds co., SE Mo., in the Ozarks SE of Salem. It is a resort. **3.** City (1970 pop. 10,333), Montgomery co., SW Ohio, a residential suburb of Dayton; inc. 1879. It has a small industrial park. **4.** Town (1970 pop. 2,592), seat of Hickman co., central Tenn., on the Duck River SW of Nashville, in a farm and phosphate-mining area; inc. 1911. **5.** City (1970 pop. 831), seat of Leon co., E central Texas ESE of Waco. It is a trade center in an agricultural and timber area.

Cen·tral Af·ri·can Empire (sĕn'trəl ăf'rĭ-kən), formerly **Central African Republic** (1975 est. pop. 1,800,000), 240,534 sq mi (622,983 sq km), central Africa. Bangui is the capital. The landlocked nation is bordered by Chad in the north, Sudan in the east, Zaire and the Congo Republic in the south, and Cameroon in the west. The terrain consists of a 2,000–3,000 ft (610–915 m) undulating plateau, with dense tropical forests in the south and a semidesert area in the east. The country is drained by numerous rivers, but only the Ubangi is commercially navigable. Rainfall is heavy in the south. There are no railroads, and the network of all-weather roads is inadequate; rivers are the chief means of transportation. Population density is only about six persons per square mile.

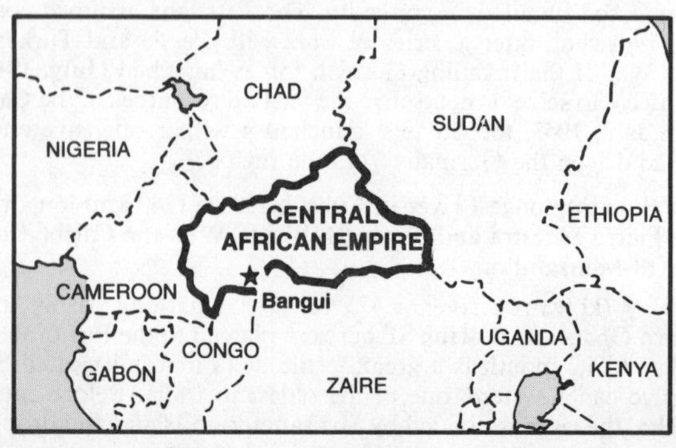

Economy. The overwhelming majority of the people are engaged in agriculture. Cassava, millet, rice, and peanuts are grown for subsistence. The principal cash crops and exports are cotton and coffee; cocoa, rubber, and palm products are raised in the southwest. Timber is also an important product and export. There have been recent attempts to develop a livestock (mainly cattle) industry, despite unfavorable climate and the prevalence of the tsetse fly. Mining, formerly limited to diamonds (another leading export), has become increasingly important with the extraction of uranium, begun in 1972.

History. Between the 16th and the 19th cent., much of the region was subject to devastating slave raids. The Baya people, seeking refuge from the Fulani of Cameroon, arrived in what is now the Central African Empire in the early 19th cent.; the Banda, fleeing the Arab slave raiders of Sudan, came later in the century. French expeditions, pushing out from the Congo and making treaties with local tribal chiefs, occupied the area in 1887. In 1946 the colony gained representation in the French parliament and in 1958 opted for membership in the French Community. Full independence was attained on Aug. 13, 1960, under President David Dacko. The Central African Republic had a parliamentary government until Dec., 1965, when a military coup led by Col. Jean-Bédel Bokassa overthrew the Dacko regime, dissolved the national assembly, and abrogated the constitution. In Dec., 1977, Bokassa crowned himself emperor in a pageant fashioned after the coronation of Napoleon and renamed the country the Central African Empire.

Central A·mer·i·ca (ə-mĕr'ə-kə), narrow, southernmost portion of the continent of North America, linked to South America by the Isthmus of Panama. It separates the Caribbean Sea from the Pacific Ocean. From a geological standpoint, Central America includes the land (c.276,400 sq mi/715,875 sq km) between the Isthmus of Tehuantepec and the Isthmus of Panama. Although it includes four states and one territory of Mexico and excludes Panama (which occupies an arm of South America), the term is generally applied to Belize, Guatemala, Honduras, El Salvador, Nicaragua, Costa Rica, and Panama. The mountains of northern Central America are an extension of the mountain system of western North America and are related to the islands of the West Indies. The middle portion of Central America is an active zone of volcanoes and earthquakes. The ranges of south Central America are outliers of the Andes Mts. of South America. Central America's climate varies with altitude from tropical to cool. The eastern side of the region receives heavy rainfall. Bananas, coffee, and cacao are the chief crops of Central America, and gold and silver are mined here.

Central City. 1. Town (1970 pop. 228), alt. 8,560 ft (2,610.8 m), seat of Gilpin co., N central Colo., in Clear Creek Canyon W of Denver; inc. 1886. It boomed after the discovery of gold in 1859. Although some mining is still done in the area, the city is a ghost town and a tourist center. There is an annual summer festival in the opera house (opened 1878). The Teller House, a famous frontier hotel built in 1872, is next door. **2.** Town (1970 pop. 5,450), Muhlenberg co., W Ky., S of Owensboro; settled as Morehead's Horse Mill, renamed after 1870 when the railroad reached here. It is the trade center of a coal, oil, timber, and farm area. **3.** City (1970 pop. 2,803), seat of Merrick co., E central Nebr., on the Platte River WNW of Lincoln; platted 1864. Meat products are processed. The site of Lone Tree Ranch, a stopping place on the Overland Trail, is nearby.

Cen·tra·lia (sĕn-trāl'yə, -trā'lē-ə). **1.** City (1978 est. pop. 15,200), Clinton and Marion cos., S Ill., in an oil, coal, farm, and fruit region; inc. 1859. Founded in 1853 by the Illinois Central RR, it is the shipping center for the products of the area. The city has varied manufactures, including clothing, candy, and stoves and heaters. **2.** City (1978 pop. 10,600), Lewis co., SW Wash., at the confluence of the Chehalis and Skookumchuck rivers; inc. 1889. It is a railroad junction and a farm trade center, with a great lumbering industry. A massive electric steam plant and two nearby dams make the city a major power center.

Central Park, largest park (840 acres/340.2 hectares) in Manhattan, New York City. The land, acquired by the city in 1856, was improved according to the plans of U.S. landscape architects Frederick L. Olmsted and Calvert Vaux. The park has rolling terrain with lakes and ponds, greeneries, bridle paths, walks, and park drives. There are many playgrounds and other recreational facilities. The Metropolitan Museum of Art stands in the park on Fifth Ave.; other points of interest include a formal garden, a zoo, an Egyptian obelisk called Cleopatra's Needle, the Mall, and an open-air theater.

Central U·tah project (yo͞o'tô', -tä'), N central Utah; begun 1959 near Vernal, Utah, by the U.S. Bureau of Reclamation in conjunction with the Colorado River storage project. Water, collected from streams in the Uinta Mts., is carried across the Wasatch Range to the densely populated Salt Lake City region by a system of dams, reservoirs, tunnels, aqueducts, and canals.

Central Valley, great trough of central Calif., c.450 mi (725 km) long and c.50 mi (80 km) wide, between the Sierra Nevada and the Coast Ranges. The Sacramento and San Joaquin rivers drain most of the valley before converging in a huge delta and flowing into San Francisco Bay. With its long growing season and rich soil, the valley has the largest single concentration of fruit farms and vineyards in the United States; cotton, grain, and vegetables are also grown. Precipitation ranges from 30 in. (76.2 cm) in the north to 6 in. (15.2 cm) in the south. The Central Valley was first seen by Spanish explorers in the 1500s but remained virtually uninhabited until 1848, when gold was discovered nearby. In the late 1800s the valley became a rich agricultural region, with wheat as the main crop. Irrigation was introduced in the 1880s.

Cen·tre (sĕn'tər), county (1970 pop. 99,267), 1,115 sq mi (2,887.9 sq km), central Pa., in a mountainous region bounded on the NW by the West Branch of the Susquehanna River; formed 1800; co. seat Bellefonte. It has limestone, bituminous-coal, clay, and sandstone deposits, dairy farms, and manufacturing.

Centre, town (1970 pop. 2,418), seat of Cherokee co., NE Ala., NE of Gadsen, in a farm and timber region; settled c.1840, inc. 1937.

Cen·tre·ville (sĕn'tər-vĭl'). **1.** Or **Centerville,** city (1970 pop. 2,233), seat of Bibb co., central Ala., SSW of Birmingham near the Cahaba River, in a timber and cotton area. Talladega National Forest is nearby. **2.** City (1970 pop. 11,378), St. Clair co., SW Ill., a suburb of East St. Louis. **3.** Town (1970 pop. 1,853), seat of Queen Annes co., E Md., on the Eastern Shore S of Chestertown, in an agricultural area; founded c.1788, laid out 1792, inc. 1794. Historic houses in the area include Poplar Grove (1700), with the oldest boxwood garden in the state, and Walnut Grove (1681–85). **4.** Or **Centerville,** village (1970 pop. 1,044), seat of St. Joseph co., SW Mich., on the Prairie River S of Kalamazoo, in an agricultural area; settled 1826.

Ceph·a·lo·ni·a (sĕf'ə-lōn'yə, -lō'nē-ə): *see* Kefallinía, Greece.

Ce·ram (sā'räm), island (1970 est. pop., including offshore islands, 100,000), c.6,600 sq mi (17,095 sq km), E Indonesia, W of New Guinea, second largest of the Moluccas. The interior has dense rain forests and is largely unexplored. Copra, resin, sago, and fish are important commercial products. Oil is exploited in the northeast. Portuguese missionaries were active here in the 16th cent. Dutch trading posts were opened in the early 17th cent., and the island came under nominal Dutch control c.1650.

Cer·ro de Pas·co (sĕr'rō dā päs'kō), city (1972 pop. 35,975), capital of Pasco dept., central Peru. At an altitude of 13,973 ft (4,261.8 m), it is one of the highest cities in the world. Cerro de Pasco is noted for its silver mines, which, according to tradition, were discovered in 1630. From the nearby Minasraga mines comes about 80% of the world's supply of vanadium.

Cer·ro Gor·do (sĕr'ə gôr'də), county (1970 pop. 49,223), 576 sq mi (1,491.8 sq km), N Iowa, in a rolling prairie agricultural area drained by the Shellrock River and Lime Creek; formed 1851; co. seat Mason City. Cattle, hogs, poultry, and grain are produced. There are limestone quarries and sand and gravel pits here.

Ce·se·na (chā-zā'nä), city (1976 est. pop. 67,500), in Emilia-Romagna, N central Italy, on the Sávio River. It is an agricultural market and a food-processing center. Cesena flourished (1379–1465) under the Malatesta family, who built (15th cent.) a castle on a hill overlooking the city. The castle includes the splendid Renaissance-style Malatestiana Library.

Čes·ké Bu·dě·jo·vi·ce (chĕs'kā bo͝o'dyĕ-yô-vĭ-tsĕ), city (1975 est. pop. 82,528), SW Czechoslovakia, in Bohemia, on the Vltava River. An important road and rail hub and river port, České Budějovice is famous for its breweries. Other industries produce machinery, enamelware, food products, and pencils. The city was founded in the 13th cent. and is noted for its inner town, with an arcaded square.

Ce·ti·nje (tsĕ'tĭ-nyĕ), town (1971 pop. 11,892), SW Yugoslavia, in Montenegro. It grew around a monastery founded in 1485 and was the capital of Montenegro until 1945. The monastery, the burial place of the Montenegrin princes, and the former royal palace (now a museum) remain.

Ce·u·ta (sä-ōō'tä, *th*ä-), city (1974 est. pop. 65,235), NW Africa, a possession of Spain, on the Strait of Gibraltar. An enclave in Morocco, Ceuta is administered as an integral part of Cádiz prov., Spain. It is located on a peninsula whose promontory forms one of the Pillars of Hercules. The city is a free port, with a large harbor and ample wharves; it is also a refueling and fishing port. Food processing is an important activity. The city was held by Carthaginians, Romans, Vandals, Byzantines, and Arabs (711). Taken by Portugal in 1415, it then passed (1580) to Spain. It has remained Spanish despite several attacks, notably a prolonged siege (1694–1720).

Cé·vennes (sā-vĕn'), mountain range, S France, bordering the Massif Central on the SE. The Cévennes proper occupy the central section of a mountainous arc (average height 3,000 ft/915 m), swinging generally northeast from the Montagne Noire (northeast of Toulouse) to Mont Pilat (southwest of Lyons). Between the Cévennes proper and the Montagne Noire are the Causses—barren limestone plateaus intersected by deep chasms and ravines.

Cey·lon (sĭ-lŏn'): *see* Sri Lanka.

Cha·blis (shä-blē'), village (1968 pop. 1,982), Yonne dept., central France, in Burgundy. It is famous for the white wine named for it.

Cha·co (chä'kō), plain: *see* Gran Chaco.

Chaco Canyon National Monument: *see* National Parks and Monuments Table.

Chad (chăd, chäd), republic (1977 est. pop. 4,150,000), 495,752 sq mi (1,284,000 sq km), N central Africa. Ndjamena is the capital. Chad is bordered by the Central African Empire on the south, Sudan on the east, Libya on the north, and Cameroon, Niger, and Nigeria on the west. The terrain in the south is wooded savanna; it becomes brush country near Lake Chad. The only important rivers are the Chari and the Logone, both of which flow into Lake Chad and are used for irrigation and seasonal navigation. Northern Chad is a desert that merges with the southern Sahara; areas of the mountainous Tibesti region are 11,000 ft (3,355 m) high.

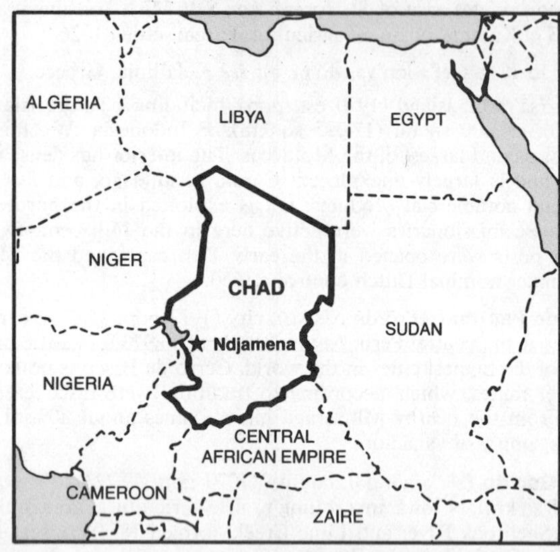

Economy. Chad's economy is based primarily on sedentary agriculture and nomadic pastoralism. The best farming zone is in the south, where rainfall is sufficient for the cultivation of cotton and peanuts (the country's leading cash crops) for export and some subsistence crops. Natron is the country's chief mineral; tungsten has been found in the arid Tibesti region. Industry is limited to food processing and the production of textiles and light consumer goods.

History. Traditionally, the region around Chad was a focal point for trans-Saharan trade routes. Arab traders penetrated the area in the 7th cent., followed by nomads from North Africa. They eventually established the state of Kanem, which reached its zenith in the 13th cent. Its kings converted to Islam, the religion also practiced by the successor state of Bornu. The Wadai and Bagirmi empires arose in the 16th cent.; they warred with Bornu and in the 18th cent. surpassed it in power. By the early 1890s all of these states fell under the control of Sudan. French expeditions advanced into the region in 1890, and by 1913 the conquest of Chad was completed. Full independence from France was attained on Aug. 11, 1960, with Ngarta Tombalbaye as the first president. Tombalbaye steadily strengthened his control over the country, and by 1965 it had become a one-party state. Discontent among northern Moslem tribes evolved into full-scale guerrilla war in 1966. Chad requested French troops to help battle the guerrillas. These troops were withdrawn in 1971, and the revolt was over by 1973. Chad suffered severely from the drought of the late 1960s and early 1970s. In 1975, during a military coup, Tombalbaye was killed, and Brig. Gen. Félix Malloum took control of the country. Fighting in the north resumed in 1977.

Chad, Lake, N central Africa. It lies mainly in the Republic of Chad and partly in Nigeria, Cameroon, and Niger. The size of the lake varies seasonally from c.4,000 to c.10,000 sq mi (10,360–25,900 sq km). It is divided into north and south basins, neither of which is generally more than 25 ft (7.6 m) deep.

Chad·ron (shăd'rən), city (1970 pop. 5,921), seat of Dawes co., NW Nebr., near the S.Dak. line N of Alliance, in a livestock and grain area of the Great Plains region; founded 1885. It is a railroad division point and manufactures flour, feed, and milk products. Chadron State College is here. The Museum of the Fur Trade is nearby.

Chaf·fee (chă'fē), county (1970 pop. 10,162), 1,039 sq mi (2,691 sq km), central Colo., in a cattle area drained by the Arkansas River; formed 1879; co. seat Salida. Grain and truck-garden crops are grown. It includes part of San Isabel National Forest.

Cha·har (chä'här'), former province, N China. Chang-chia-k'ou was the capital. It was abolished as a province in 1952; most of it was incorporated into the Inner Mongolian Autonomous Region, and the rest was divided between Shansi and Hopeh provs.

Chal·ce·don (kăl'sĭ-dŏn, -dən, kăl-sē'dən), ancient Greek city of Asia Minor, on the Bosporus in the suburbs of present-day Istanbul. It was founded in 685 B.C. and passed to Rome in A.D. 74.

Chal·cid·i·ce (kăl-sĭd'ĭ-sē): *see* Khalkidikí, Greece.

Chal·cis (kăl'sĭs): *see* Khalkís, Greece.

Chal·dae·a or **Chal·de·a** (both: kăl-dē'ə), ancient region, properly the southernmost portion of the valley of the Tigris and Euphrates rivers. Sometimes it is extended to include Babylonia and thus comprises all of southern Mesopotamia. The Chaldaeans were a Semitic people who first came into south Babylonia c.1000 B.C. Their empire flourished under Nebuchadnezzar II, but declined rapidly thereafter and came to an end when Babylon fell to Cyrus the Great in 539 B.C.

Cha·leur Bay (shə-lōōr'), inlet of the Gulf of St. Lawrence, c.85 mi (135 km) long and from 15 to 25 mi (24–40 km) wide, between N N.B. and the Gaspé Peninsula, E Que., Canada. It is a famous fishing ground for cod, herring, mackerel, and salmon. The bay was discovered and named by Jacques Cartier in 1534.

Chal·lis (chăl'ĭs), city (1970 pop. 784), alt. 5,400 ft (1,647 m), seat of Custer co., central Idaho, SSW of Salmon, in a grain and livestock area.

Chal·mette (shăl-mĕt'), village (1970 est. pop. 23,400), seat of St. Bernard parish, extreme SE La., on the E bank of the Mississippi just below New Orleans. There is a large sugar refinery here. Petroleum products are manufactured.

Chalmette National Historical Park: *see* National Parks and Monuments Table.

Châ·lons-sur-Marne (shä-lôN'sür-märn'), city (1975 pop. 52,275), capital of Marne dept., NE France, in Champagne, on the Marne River. It is a commercial and industrial center. Among its manufactures are electrodes, paper, hosiery, foundry products, and musical and precision instruments. In 451 the Huns under Attila were defeated here by Aetius.

Cha·lon-sur-Saône (shä-lôN'sür-sōn'), town (1975 pop. 58,187), Saône-et-Loire dept., E central France, in Burgundy, on the Saône River and the Canal Central. It is an inland port with a large wine and grain trade. Its many manufactures include metal products, electrical equipment, barges, textiles, chemicals, and glass. Of pre-Roman origin, it was the scene of 10 church councils, most notably the one convoked by Charlemagne in 813.

Cham·ber·lain (chăm'bər-lăn), city (1970 pop. 2,626), seat of Brule co., S S.Dak., on the Missouri River SE of Pierre, in a livestock and grain area; platted 1880, inc. 1882. An early ferrying point for passengers and freight on the river, the city has become a trade and recreation center. Crow Creek and Lower Brule Indian reservations are nearby.

Cham·bers (chām′bərz). **1.** County (1970 pop. 36,356), 598 sq mi (1,548.8 sq km), E Ala., drained on the NW by the Tallapoosa River and bounded on the E by the Chattahoochee River and the Ga. border; formed 1832; co. seat Lafayette. Lumbering, cattle raising, and agriculture are important. It has a large amount of industry. **2.** County (1970 pop. 12,187), 618 sq mi (1,600.6 sq km), SE Texas, on the Gulf coastal plains bounded on the S by East Bay and indented by Galveston Bay; formed 1858; co. seat Anahuac. In an oil-producing area, it has natural-gas and sulfur deposits and diversified agriculture (rice, corn, wheat, citrus fruits, truck crops, fruit, cattle, hogs, and dairy products).

Cham·bers·burg (chām′bərz-bûrg′), borough (1978 est. pop. 16,700), seat of Franklin co., S Pa., in a fertile farm area; settled 1730, inc. 1803. Food products, steam and pneumatic hammers, sheet-metal goods, clothing, and concrete and lumber products are manufactured. Chambersburg was the headquarters of abolitionist John Brown in 1859 and of Gen. Robert E. Lee before the Battle of Gettysburg. The town was burned by Confederate cavalry in July, 1864, after refusing to pay an indemnity of $100,000 in gold.

Cham·bé·ry (shäɴ-bā-rē′), town (1975 pop. 54,415), capital of Savoie dept., E France, in the Alpine trough. It is a communications center with many manufactures. It was the capital of Savoy from 1232 to 1562.

Cham·bly (shäɴ-blē′), city (1976 pop. 11,815), S Que., Canada, on the Richelieu River, E of Montreal. Chambly Fort was built in 1665 and was a strategic point in the defense of New France against the British and the Iroquois. The British captured it in 1760. It was seized by the invading Americans in 1775 and burned when they withdrew in 1776. The partially restored fort is a national historic site.

Cham·bord (shäɴ-bôr′), château, park, and village (1968 pop. 267), all owned by the state, in Loir-et-Cher dept., N central France. The huge Renaissance château, built by Francis I, is set in an immense park and forest (c.13,600 acres/5,510 hectares). Louis XV gave Chambord to Maurice de Saxe, who died here in 1750. Napoleon I later presented it to Marshal Berthier, and in 1821 it went by national subscription to the duke of Bordeaux. Repurchased by the state in 1932, Chambord is now open to the public.

Cham·i·zal National Memorial (shăm′ĭ-zăl′, chä′mē-säl′): *see* National Parks and Monuments Table.

Cha·mo·nix (shä-mô-nē′), town (1975 pop. 6,285), Haute-Savoie dept., E France, at the foot of Mont Blanc. The principal base for climbing Mont Blanc and for visiting the Mer de Glace, it is a popular summer and winter resort. It has the world's highest (12,605 ft/3,844.5 m) aerial cable car.

Cham·pagne (shăm-pān′, shäɴ-pä′nyə), region and former province, NE France. Abutting in the west on the Paris basin, Champagne is a generally arid, chalky plateau. Agriculture, except in the Ardennes dept., is mostly confined to the valleys. Crests divide the plateau from northwest to southeast into several areas. In the east, bordering on Lorraine, is the so-called Champagne Humide, largely agricultural, and the Langres Plateau. In the center is the Champagne Pouilleuse, a bleak and eroded plain, traditionally used for sheep grazing. A narrow strip along the westernmost crest of Champagne is extremely fertile, and the small area around Rheims and Epernay furnishes virtually all of the champagne wine exported by France.

Cham·paign (shăm-pān′). **1.** County (1970 pop. 163,281), 1,000 sq mi (2,590 sq km), E Ill., in a prairie agricultural area drained by the Sangamon, Kaskaskia, and Embarrass rivers and by the South Fork of the Vermilion River; formed 1833; co. seat Urbana. Its agriculture includes corn, wheat, soybeans, alfalfa, oats, livestock, and poultry. It also has diversified manufacturing. Chanute Air Force Base and Lake of the Woods Park are here. **2.** County (1970 pop. 30,491), 433 sq mi (1,121.5 sq km), W central Ohio, in an agricultural area intersected by the Mad River, Darby Creek, and Buck and Little Darby creeks; formed 1805; co. seat Urbana. Livestock, corn, wheat, soybeans, poultry, honey, and dairy products are produced. It has sand and gravel pits and some manufacturing and includes a state park and the Ohio Caverns.

Champaign, city (1978 est. pop. 60,100), Champaign co., E central Ill.; inc. 1860. It is a commercial and industrial center in a fertile farm area. Its manufactures include metal products, academic apparel, and electrical equipment. Champaign, founded in 1855 with the arrival of the Illinois Central RR, was first called West Urbana.

Champ-de-Mars (shäɴ-də-märs′), former parade ground of Paris, France, between the École militaire and the Seine River. Here, on July 14, 1790, Louis XVI took an oath to uphold the new constitution. On its vast grounds several expositions were held, notably that of 1889, when the Eiffel Tower was erected.

Cham·plain, Lake (shăm-plān′), 125 mi (201 km) long and from 0.5 to 14 mi (0.8–22.5 km) wide, forming part of the N.Y.-Vt. border and extending into Que., Canada. Lake Champlain lies in a broad valley between the Adirondacks and the Green Mts. A link in the Hudson-St. Lawrence waterway, the lake is connected with the Hudson by the Champlain division of the Barge Canal; the Richelieu River connects the lake with the St. Lawrence. The region is noted for its beautiful scenery and has many resorts. The lake, discovered by Samuel de Champlain in 1609, was the scene of battles in the French and Indian Wars and the American Revolution at Crown Point and Ticonderoga, of a naval engagement in 1776, and of the important American victory of Thomas Macdonough in the War of 1812.

Champs É-ly-sées (shäɴ zā-lē-zā′), avenue of Paris, France, leading from the Place de la Concorde to the Arc de Triomphe. It is celebrated for its tree-lined beauty and the elegance of its cafés, theaters, and shops. Begun by Louis XIV and completed by Louis XV, it led through open country until the early 19th cent.

Chan-chiang or **Chan·kiang** (both: chän′jē-äng′), official Chinese name for the former French territory of Kwangchowan (325 sq mi/841.8 sq km) on Kuang-chou Bay, S Kwangtung prov., China. It was leased from China in 1898 for 99 years but was returned in 1945. Its chief city, Fort-Bayard, was renamed Chan-chiang (1970 est. pop. 220,000) and since 1955 has been developed as a major seaport and regional trade center. The city has textile, chemical, shipbuilding, and electric power industries.

Chan·der·na·gor (chŭn′dər-nə-gôr′), town (1971 pop. 75,960), West Bengal state, E India, on the Hooghly River, a suburb of Calcutta. Founded by the French in 1686, it was of great commercial importance until the 19th cent. It became part of India in 1951.

Chan·di·garh (chŭn′dē-gər), union territory (1971 pop. 256,972), 44 sq mi (114 sq km) and city (1971 pop. 218,807), NW India. The city was designed by the architect Le Corbusier and built on a site chosen for its healthy climate and plentiful water supply.

Chan·dler (chăn′dlər). **1.** City (1970 pop. 13,763), Maricopa co., S central Ariz., in the Salt River valley; inc. 1920. It is a residential community in an area that produces cotton, alfalfa, and citrus fruit. Sugar is processed, and computer components, mobile homes, and containers are manufactured. Williams Air Force Base is nearby. **2.** City (1970 pop. 2,529), seat of Lincoln co., central Okla., NE of Oklahoma City; settled 1891. It is a trade center in a farm area noted especially for pecans.

Chang-chia-k'ou or **Chang·kia·kow** (both: chäng′jē-ä′kou′), city (1970 est. pop. 1,000,000), NW Hopeh prov., China, near a gateway of the Great Wall and on the Peking-USSR RR. A major trade center for northern China and Mongolia, it has food-processing plants, machine shops, and tanneries. The meeting place of caravans traveling from Peking to Ulan Bator, it was an important military center under the Manchu dynasty but declined somewhat after the opening (1905) of the Trans-Siberian RR.

Ch'ang-ch'un or **Chang·chun** (both: chäng′cho͞on′), city (1970 est. pop. 1,500,000), capital of Kirin prov., NE China, on the railroad between Harbin and Lü-ta. An industrial city, it is the country's major center of motor vehicle production, with enormous truck and tractor works. Railroad cars, tires, pharmaceuticals, and textiles are also manufactured. Ch'ang-ch'un has government-owned motion picture studios. As Hsinking, it was the capital of the former state of Manchukuo (1932–45).

Ch'ang-hua or **Chang-hwa** (both: chäng′hwä′), city (1973 est. pop. 111,000), central Taiwan. It is a transportation center as well as a market for rice, oranges, and pineapples. The city's industries produce wood and paper products, textiles, canned food, refined sugar, and machinery. Settled in the 17th cent., Ch'ang-hua was once an important fort.

Ch'ang-sha or **Chang·sha** (both: chäng′shä′), city (1970 est. pop. 850,000), capital of Hunan prov., S China, on the Hsiang River. The name, which means "long sandbank," is derived from an island in the river. Ch'ang-sha is an agricultural distribution and market center, an important stop on the Peking-Canton RR, and a river port. Rice is processed, meats are canned, and paper products, fertilizer, trucks, ceramics, and a wide variety of handicrafts are made. The city was founded in the early 3rd cent. B.C. and has long been noted as a literary and educational center. It became a treaty port in the early 1900s. Mao Tse-tung was educated in Ch'ang-sha, and in 1927 he led a Communist uprising here.

Ch'ang-te or **Chang·teh** (both: chäng'dŭ'), city (1970 est. pop. 225,000), N Hunan prov., China, on the Yüan River. Formerly a treaty port, it is now a storage and shipping point for tung oil, medicinal herbs, and wood. Manufactures include ceramics, machine tools, textiles, leather, and processed foods.

Chan·nel Islands (chăn'əl), archipelago (1971 pop. 125,243), 75 sq mi (194.3 sq km), 10 mi (16 km) off the coast of Normandy, France, in the English Channel. The main islands are Jersey, Guernsey, Alderney, and Sark; all the islands are dependencies of the British crown. The mild and sunny climate and the fertile soil have made the islands chiefly agricultural. Large quantities of vegetables, fruits, and flowers are shipped to English markets. Dairying is the chief occupation of the islanders. The famous Jersey and Guernsey breeds of cattle are kept pure by local laws. The islands are a favorite resort of tourists and vacationers. The archipelago is divided into two administrative bailiwicks, each with its own lieutenant governor appointed by the crown, its own chief magistrate and legislature, and its own judicature. The inhabitants are mostly of Norman descent, but on Alderney the stock is mainly English. The English language is spoken everywhere, although French is the official language of Jersey. In the 10th cent. the isles became possessions of the duke of Normandy. At the Norman conquest they were joined to the English crown; they remained under the control of King John and England in 1204 when Philip II of France confiscated the duchy of Normandy. The French attempted unsuccessfully to re-establish control in the 14th cent. In World War II, after the evacuation of some 10,000 military and civilian personnel, the islands were occupied (1940) by German forces.

Channel Islands National Monument: *see* National Parks and Monuments Table.

Chan·ning (chăn'ĭng), town (1970 pop. 336), seat of Hartley co., extreme N Texas, in the high plains of the Panhandle NW of Amarillo. It is a market and shipping point for a cattle and grain area.

Chan·til·ly (shăn-tĭl'ē, shäN-tē-yē'), town (1968 pop. 10,246), Oise dept., N France, N of Paris. In the 18th cent., Chantilly gained renown for its fine lacework and the beauty of its porcelain. It is now a fashionable horse-racing center. The château, rebuilt in the 19th cent., contains the famous art museum of Condé.

Chan·til·ly (shăn-tĭl'ē), village (1970 pop. 200), Fairfax co., NE Va., W of Washington, D.C. It was the scene of a Civil War battle (Sept. 1, 1862), in which Union troops, retreating after the Second Battle of Bull Run, checked Stonewall Jackson's troops.

Cha·nute (shə-no͞ot'), city (1970 pop. 10,341), Neosho co., SE Kansas, on the Neosho River; inc. 1873 following the consolidation of four contiguous towns. It is a processing and trade center for a rich agricultural region, with a great variety of manufactures. Nearby is the site of a mission (1824–29), the first in Kansas.

Chao Phra·ya (chou prä-yä'), **Mae Nam Chao Phraya,** or **Me·nam Chao Phraya** (both: mä-näm'), chief river of Thailand, c.140 mi (225 km) long, formed by the confluence of the Ping (c.300 mi/480 km long) and the Nan (c.500 mi/805 km) rivers at Nakhon Sawan, W central Thailand. It flows south past Bangkok to the Gulf of Siam and is navigable for its entire length. With its tributaries, the Chao Phraya drains most of west Thailand; its valley is the country's main rice-producing region.

Cha·pa·la, Lake (chä-pä'lä), c.50 mi (80 km) long and 8 mi (12.9 km) wide, W Mexico, in Jalisco and Michoacán states. It is the largest lake in Mexico. Set in a depression on the central plateau, Lake Chapala is fed by the Lerma River and drained by the Río Grande de Santiago. It is a popular scenic resort. Fishing is an important native occupation. Since the early 1950s the waters have been receding at an alarming rate, and the lake is rapidly becoming choked with water hyacinths.

Chap·el Hill (chăp'əl), town (1978 est. pop. 31,900), Orange and Durham cos., central N.C., at the edge of the Piedmont; founded 1792, inc. 1851. It is the seat of the Univ. of North Carolina.

Chap·pell (chăp'əl), city (1970 pop. 1,204), seat of Deuel co., W Nebr., on a tributary of the South Platte River W of North Platte; platted 1884. It is in an agricultural area yielding wheat and sugar beets.

Cha·pul·te·pec (chä-po͝ol'tā-pĕk'), rocky hill S of Mexico City. It was originally developed as a playground for Aztec emperors. A castle built in the late 18th cent. as a summer home for the Spanish viceroys later became the traditional home of the rulers of Mexico. Chapultepec, heavily fortified, was the scene of spectacular fighting during the Mexican War; U.S. Gen. Winfield Scott ordered the storming of Chapultepec on Sept. 12, 1847, and it fell the next day. Both Emperor Maximilian and, later, Porfirio Díaz beautified the grounds and embellished the castle. The castle is now a museum of colonial history and ethnography.

Char·don (shär'dən), village (1970 pop. 3,991), seat of Geauga co., NE Ohio, ENE of Cleveland; settled 1812, inc. 1872. Maple sugar and syrup are produced.

Char·dzhou (chär-jō'), city (1976 est. pop. 110,000), capital of Chardzhou oblast, SW Central Asian USSR, in the Turkmen Republic, on the Amu Darya River. An inland port, it has shipyards and is a cotton and silk manufacturing center.

Cha·rente (shä-räNt'), department (1975 pop. 337,064), 2,298 sq mi (5,951.8 sq km), W France; capital Angoulême.

Charente, river, 220 mi (354 km) long, rising near Limoges, W France, and flowing W to the Bay of Biscay. The river flows through an important cattle-raising region. Along its western course are the celebrated vineyards from which cognac brandy is made.

Cha·rente-Ma·ri·time (shä-räNt'mä-rē-tēm'), department (1975 pop. 497,859), 2,644 sq mi (6,848 sq km), W France, on the Atlantic coast, formerly Charente-Inférieure; capital La Rochelle.

Cha·ri or **Sha·ri** (both: shä'rē), longest river of interior drainage in Africa, c.650 mi (1,045 km) long, rising in the uplands of the Central African Republic, N central Africa. It flows northwest across south Chad and enters Lake Chad through a wide delta. During the summer rainy season, the river floods much of the surrounding area.

Char·ing Cross (chăr'ĭng), square and district, Westminster metropolitan borough, London, England, at the W end of the Strand near Trafalgar Square. Charing Cross Station is one of London's busiest terminals. Road distances from London are traditionally measured from here.

Char·i·ton (shăr'ĭ-tən), county (1970 pop. 11,084), 759 sq mi (1,965.8 sq km), N central Mo., bounded on the S by the Missouri River, on the W by the Grand River, and drained by the Chariton River; formed 1820; co. seat Keytesville. It lies in a corn, wheat, oats, and livestock area and has bituminous-coal deposits. Textiles and metal products are manufactured.

Chariton, city (1970 pop. 5,009), seat of Lucas co., S central Iowa, W of Ottuma; inc. 1857. It is a farm trade and coal-mining center, with railroad shops and factories producing clothing and industrial machinery. Red Haw Lake State Park is nearby.

Chariton, river, c.280 mi (450 km) long, S central Iowa and N central Mo., rising S of Osceola, Iowa, and flowing E then S across the Mo. border to join the Missouri River above Glasgow.

Char·le·roi (shär'lə-rwä'), town (1976 est. pop. 229,474), Hainaut prov., S Belgium, on the Sambre River and the Charleroi-Brussels Canal. It is a commercial and industrial center and a rail junction. Manufactures include steel, glass, machinery, processed food, and chemicals. Coal and iron are mined in the region. Charleroi was founded in 1666 and named for Charles II of Spain. It was of strategic importance in the wars of the 17th and 18th cent. The Germans won a battle here (1914) in World War I.

Charles (chärlz), county (1970 pop. 47,678), 458 sq mi (1,186.2 sq km), S Md., on the SW shore of the Md. peninsula in a tidewater agricultural area bounded on the W and S by the Potomac River, on the E by the Patuxent River; formed 1658; co. seat La Plata. It produces tobacco and has some lumbering, commercial fishing, and a naval ordnance installation. The region, one of the earliest settled in Md., has many beautiful old buildings.

Charles, Cape, E Va., the S point of the Eastern Shore, separating Chesapeake Bay from the Atlantic Ocean. Cape Charles is opposite Cape Henry at the entrance to the bay and has ferries to the Va. mainland.

Charles·bourg (shärl'bo͝or', shärl'bo͝org'), city (1976 pop. 63,147), S Que., Canada, a suburb of Quebec city. One of the oldest parishes in the province, it includes part of the seigniory first granted to the Jesuits in 1626 and was settled in 1659.

Charles City, county (1970 pop. 6,158), 184 sq mi (476.6 sq km), E Va., in a tidewater region on a peninsula between the James and Chickahominy rivers; formed 1634; co. seat Charles City. It has agriculture (truck crops, poultry, and dairying), river fisheries, and lumbering. One of the oldest counties in the state, it has many beautiful old estates, including Westover, Berkeley, and Shirley.

Charles City. 1. City (1974 est. pop. 9,119), seat of Floyd co., N Iowa, on the Cedar River NNW of Waterloo; settled c.1850, platted 1854, inc. 1869. Tractors and feed supplements are produced. Indian mounds have been excavated nearby. **2.** Village, seat of Charles City co., E Va., SE of Richmond, near the James River. Many fine old estates are nearby.

Charles Mix (mĭks), county (1970 pop. 9,994), 1,097 sq mi (2,841.2 sq km), S S.Dak., in an agricultural area watered by Lake Andes and bounded on the SW by the Missouri River; formed 1862; co. seat Lake Andes. Corn, grain sorghums, livestock, and dairy products are produced. It includes Fort Randall Dam and Fort Randall Historic Restoration.

Charles Mound, hill, 1,241 ft (378.5 m) high, NW Ill., near the Wis. line. It is the highest point in the state.

Charles·ton (chärl'stən), county (1970 pop. 247,650), 945 sq mi (2,447.6 sq km), SE S.C., extending along the Atlantic coast between the mouth of the South Edisto River and the mouth of the South Santee River; formed 1785; co. seat Charleston. In ante-bellum days it was a rich plantation area. Today it is the manufacturing and shipping center of the state, with some agriculture (sweet potatoes, corn, and vegetables), fisheries, and hunting. There are many resorts along the coast. Part of Francis Marion National Forest is here.

Charleston. 1. Town (1976 est. pop. 1,761), a seat of Franklin co., NW Ark., ESE of Fort Smith, in a dairy, poultry, cotton, grain, and coal area. **2.** City (1978 est. pop. 19,200), seat of Coles co., E Ill.; inc. 1835. Shoes, electronic equipment, farm buildings, and tools are manufactured in this industrial, rail, and trade center located in an agricultural area. A Lincoln-Douglas debate was held in Charleston on Sept. 8, 1858. Local attractions include nearby Lincoln Log Cabin State Park (the site of Thomas Lincoln's reconstructed farmhouse) and Fox Ridge State Park. Eastern Illinois Univ. is here. **3.** City (1970 pop. 2,821), a seat of Tallahatchie co., NW Miss., NNE of Greenwood, in a cotton, timber, and livestock area. **4.** City (1970 pop. 5,131), seat of Mississippi co., SE Mo., SW of Cairo, Ill., in a farm, timber, and cotton region; laid out and inc. 1837. Shoes and wood products are made. **5.** City (1978 est. pop. 56,300), seat of Charleston co., SE S.C.; founded 1680, inc. 1783. The oldest city in the state and one of the chief ports of entry in the southeastern United States, Charleston lies on a low, narrow peninsula between the Ashley and Cooper rivers at the head of the bay formed by their confluence. In the bay, or bordering on it, are several islands, including Fort Sumter and Castle Pinckney. Among the city's varied manufactures are fertilizers, chemicals, steel, asbestos, cigars, pulp and paper, and textiles and clothing. Charleston is the headquarters for the 6th U.S. naval district and for the U.S. air force defense command. The extensive military facilities include a Polaris submarine base and a huge navy yard (est. 1901). The English under William Sayle settled (1670) at Albemarle Point, on the western bank of the Ashley River. They later moved (1680) to Oyster Point, where their capital, Charles Town (as it was first called), had been laid out. The city, surviving Spanish and Indian threats, became the most important seaport in the Southern colonies (exporting indigo, rice, and deerskins) and the leading center of wealth and culture in the South. Non-English immigrants, among whom the French Huguenots were most prominent, added a cosmopolitan touch. In the American Revolution, after being successfully defended (1776, 1779), Charleston was surrendered (May 12, 1780) to the British, who held it until Dec. 14, 1782. The capital was moved to Columbia in 1790, but Charleston remained the social and economic center of the region. The South Carolina ordinance of secession (Dec., 1860) was passed in Charleston, and the city was the scene of the precipitating act of the Civil War, the firing on Fort Sumter (Apr. 12, 1861). With its harbor blockaded and the city itself under virtual siege by Union forces (1863–65), Charleston suffered partial destruction but did not fall until Feb., 1865, after it had been isolated by Sherman's army. A violent earthquake on Aug. 31, 1886, took many lives and made thousands homeless, and periodic hurricanes and tornadoes (one in 1938 was particularly severe) have also caused great damage. Despite these repeated devastations, many charming colonial buildings survive; outstanding among them are St. Michael's Episcopal Church (begun 1752), noted for its chimes, and the Miles Brewton house (1765–69). Among the many other points of interest are the Old Powder Magazine (1719); the Old Slave Mart Museum and Gallery; the Gibbes Art Gallery; the Charleston Museum (1773), one of the oldest museums in the country; and Fort Sumter National Monument. The waterfront, called the Battery, and the Grace Memorial Bridge over the Cooper River are famous Charleston landmarks. Cabbage Row surrounds a court that was the original Catfish Row of DuBose Heyward's novel *Porgy*. The city's picturesque old homes and winding streets, historic attractions, and unique charm, with its pleasant climate, nearby beaches, and beautiful gardens, attract thousands of visitors each year. The annual azalea festival is an important event. The city is the seat of the College of Charleston (1790), which in 1837 became the first municipal college in the United States. **6.** City (1978 est. pop. 64,800), state capital and seat of Kanawha co., W central W.Va., on the Kanawha River where it is joined by the Elk River; inc. 1794. Charleston is an important transportation and trading center for the highly industrialized Kanawha valley and a major chemical, glass, and metal producing area. Additional manufactures are based on the salt, coal, natural gas, clay, sand, timber, and oil of the region. The city grew around the site of Fort Lee (1788). Daniel Boone lived here from 1788 to 1795. The capital was transferred here from Wheeling in 1870, moved back to Wheeling in 1875, then returned to Charleston in 1885 after an election to determine the permanent site. The capitol building (completed 1932) was designed by Cass Gilbert.

Charles·town (chärlz'toun'), town (1970 est. pop. 2,800) on the island of Nevis, British West Indies. It is a port that ships goods to St. Kitts. Cotton, sugar cane, livestock, and some food crops are raised. Alexander Hamilton was born in Charlestown.

Charlestown. 1. City (1970 pop. 5,933), Clark co., S Ind., near the Ohio River NE of New Albany; laid out 1808. Chemical products are made. **2.** Former city, now part of Boston, Middlesex co., E Mass., on Boston Harbor, between the Mystic and the Charles rivers; settled 1629, included in Boston 1874. It is the oldest part of Boston. The Battle of Bunker Hill was fought here on June 17, 1775.

Charles Town, city (1970 pop. 3,023), seat of Jefferson co., W.Va., in the Eastern Panhandle SW of Harpers Ferry; laid out 1786. It is a trading center for a farming and horse-breeding area. John Brown was tried (1859) and hanged here.

Char·le·ville-Mé·zières (shär'lə-vēl'mā-zyâr'), town (1975 pop. 60,176), capital of Ardennes dept., NE France, on the Meuse River, in Champagne. It was formed in 1966 when the twin cities of Charleville and Mézières were merged. It is a commercial and metalworking center. Mézières was an old fortified town, founded in the 9th cent.; Charleville was founded in 1606. The area has often been captured by the Germans (1815, 1870, 1914, 1940), and Mézières in particular suffered heavy damage in both World Wars. Its recovery (1918) by the Allies marked the last major battle of World War I.

Char·le·voix (shär'lə-voi'), county (1970 pop. 16,541), 414 sq mi (1,072.3 sq km), NW Mich., bounded on the NW by Lake Michigan and watered by Walloon Lake and Lake Charlevoix; organized 1869; co. seat Charlevoix. It has dairying, agriculture (livestock, potatoes, cherries, apples, grain, corn, and seed), flour and lumber mills, fisheries, many recreational facilities, and some manufacturing.

Charlevoix, city (1970 pop. 3,519), seat of Charlevoix co., NW Mich., on Lake Michigan NE of Traverse City; settled 1852, inc. as a village 1879, as a city 1905. It is a port of entry, a summer resort, with fishing on Lake Charlevoix and other small lakes, and a winter sports center. The vicinity is rich in Indian remains.

Char·lotte (shär'lət). **1.** County (1970 pop. 27,559), 705 sq mi (1,826 sq km), SW Fla., bounded on the W by the Gulf of Mexico; formed 1921; co. seat Punta Gorda. In a lowland cattle and vegetable area, it also has a fishing industry. **2.** County (1970 pop. 12,366), 468 sq mi (1,212.1 sq km), S Va., in a rolling agricultural region bounded on the W and S by the Roanoke River; formed 1765; co. seat Charlotte Court House. Tobacco, corn, clover, beef cattle, and dairy products are produced. Lumber milling is done.

Char·lotte. 1. (shär-lŏt', shər-) City (1970 pop. 8,244), seat of Eaton co., S Mich., SW of Lansing, in a dairy and farming region; settled before 1840, inc. as a village 1863, as a city 1871. Aluminum products and furniture are made. **2.** (shär'lət) City (1978 est. pop. 286,400), seat of Mecklenburg co., S N.C.; inc. 1768. The largest city of the state and the foremost commercial and industrial center of the Piedmont region, Charlotte is a transportation hub and distribution point for the Carolina manufacturing belt, now the nation's leading textile area. The bountiful hydroelectric power from the Catawba River serves the city's industries. Its products include textiles, chemicals, apparel, machinery, food, and printed materials. Charlotte, named for Queen Charlotte, wife of King George III of England, was settled c.1750. The Mecklenburg Declaration of Independence was signed here in May, 1775. In his brief occupation of the city (Sept.-Oct., 1780), British Gen. Charles Cornwallis called it a "hornet's nest of rebellion." The Univ. of North Carolina at Charlotte, Queens College, and Johnson C. Smith Univ. are in the city. The Mint Museum of Art is a reproduction of the U.S. Mint, located here from 1837 until 1913. President James K. Polk was born in Charlotte. **3.** (shär'lət) Town (1970 pop. 610), seat of Dickson co., N central Tenn., W of Nashville, in a lumbering area.

Char·lotte A·ma·lie (shär'lət ə-mäl'yē), town (1970 pop. 12,372), capital of the Virgin Islands of the United States, on St. Thomas Island. It is the commercial center of the islands, a free port, and a popular tourist resort. Founded in the late 17th cent., Charlotte Amalie was a center of Danish colonial life. It became important as a trading center during the American Civil War. It was renamed St. Thomas in 1921, but the former Danish name was restored in 1937.

Char·lotte Court House (shär'lət), town (1970 pop. 539), seat of Charlotte co., S Va., SE of Lynchburg. It is a trade center in a tobacco and timber area.

Char·lottes·ville (shär'ləts-vĭl'), city (1978 est. pop. 41,300), independent city and seat of Albemarle co., central Va., on the Rivanna River, in a piedmont farm region known for its apples; founded 1762, chartered as a city 1888. Textiles are made here. Charlottesville is the seat of the Univ. of Virginia. Nearby are Monticello, home of Thomas Jefferson, and Ash Lawn, home of James Monroe.

Char·lotte·town (shär'lət-toun'), city (1976 pop. 17,063), capital and chief port, P.E.I., E Canada, on the S coast. Food processing and tourism are the main industries. The French established (c.1720) a fort and settlement across the harbor. Charlottetown itself was laid out by the British in 1768. Its growth was slow until the middle of the 19th cent., when it became noted for the sailing vessels it built for fishing and lumber transport. The Charlottetown Conference of the Maritime Provinces (1864) was the first step toward Canadian confederation.

Charl·ton (chärl'tən), county (1970 pop. 5,680), 799 sq mi (2,069.4 sq km), SE Ga., in a flatwoods area drained by the Suwannee River and bounded on the S and SE by the St. Marys River, here forming the Fla. border, and on the NE by the Saltilla River; formed 1854; co. seat Folkston. Corn, tobacco, and livestock are raised, and sawmilling is done.

Char·tres (shär'trə), city (1975 pop. 38,928), capital of Eure-et-Loir dept., NW France, in Orléanais, on the Eure River. Chartres is of great historic and artistic interest; it is also a regional market with many industries, including metallurgy, and the production of chemicals and electronic equipment. An ancient town, it was the probable site of the great assemblies of the druids. Chartres became a possession of the French crown in 1286. Its fame today stems largely from its magnificent Gothic Cathedral of Notre Dame (12th to 13th cent.), remarkable for its two spires (375 ft/114.4 m and 350 ft/106.8 m), its stained glass windows, and its superb sculpture. Inside the cathedral Henry IV was crowned king of France (1594).

Char·treuse, Grande (gräNd' shär-trōōz'), mountainous massif, Isère dept., SE France, in the Dauphiné Alps. Chamechaude Peak (6,847 ft/2,088.3 m) is the highest point. St. Bruno founded (1084) the famous monastery La Grande Chartreuse, the principal seat of the Carthusians until 1903, when the order was expelled from France. The Carthusians returned to their monastery in 1941. Chartreuse liqueur originated here.

Chase (chās). **1.** County (1970 pop. 3,408), 774 sq mi (2,004.7 sq km), E central Kansas, in a hilly agricultural area drained by the Cottonwood River; formed 1859; co. seat Cottonwood Falls. Livestock and grain are produced. It also has scattered gas fields. A monument to Knute Rockne is here. **2.** County (1970 pop. 4,129), 894 sq mi (2,315.5 sq km), S Nebr., in an agricultural area bounded on the W by the Colo. border and drained by Frenchman Creek; formed 1886; co. seat Imperial. Livestock and grain are produced. It includes Champion Lake and Enders Reservoir state recreational areas.

Chas·ka (chăs'kə), city (1972 est. pop. 5,398), seat of Carver co., E Minn., on the Minnesota River SW of Minneapolis; settled 1853, inc. as a village 1871, as a city 1891. It is in a dairy-farming area and has a beet-sugar refinery. The city park contains Indian mounds.

Cha·teau·gay (shăt'ə-gā'), river, c.50 mi (80 km) long, rising in Chateaugay Lake in the Adirondacks, NE N.Y., and flowing through Que., Canada, to empty into the St. Lawrence below Montreal. In the War of 1812 the Battle of Chateaugay was fought (1813) between an American invading force of 7,000 under Gen. Wade Hampton and some 750 Canadians and Indians. The Americans were defeated and had to abandon their plan to attack Montreal.

Châ·teau·roux (shä-tō-rōō'), city (1975 pop. 53,429), capital of Indre dept., central France, on the Indre River. It has textile, metal, and food-processing industries.

Châ·teau-Thier·ry (shä-tō' tyĕ-rē'), town (1968 pop. 11,629), Aisne dept., N France, on the Marne River. The town was the focal point of the second Battle of the Marne (1918), which ended the last German offensive of World War I. An imposing monument to the U.S. soldiers who fought in the battle is just outside the town. The birthplace of Jean de La Fontaine is preserved as a museum.

Châ·tel·le·rault (shä-tĕl-rō'), town (1975 pop. 37,080), Vienne dept., W central France. It is an industrial center where armaments, cutlery, camping equipment, plywood, and clothing are produced. The house where René Descartes spent his childhood is now a museum.

Chat·ham (chăt'əm). **1.** Town (1971 pop. 7,833), Northumberland co., E N.B., Canada, on the estuary of the Miramichi River NNW of Moncton. It has an excellent harbor with steamer communication to other river and coastal points, and has shipyards, pulp mills, foundries, and a large lumber trade. **2.** City (1976 pop. 38,685), S Ont., Canada, E of Detroit, Mich., on the Thames River. It is an industrial center in a rich mixed farming and fruit-raising region.

Chatham, municipal borough (1971 pop. 56,921), Kent, SE England, on the Medway River. Chatham is a great naval station, with well-equipped dockyards, dry docks, and shipbuilding and repairing equipment. The first dockyard was established by Elizabeth I in 1588.

Chatham. 1. County (1970 pop. 187,816), 445 sq mi (1,152.6 sq km), E Ga., bounded on the N by the Savannah River, here forming the S.C. border, on the E by the Atlantic Ocean, on the S by the Ogeechee River; formed 1777; co. seat Savannah. It has agriculture (truck crops, poultry, and dairy products), manufacturing and shipping, and a fishing industry. Two state parks, Savannah and Wassaw Island national wildlife refuges, and Fort Pulaski National Monument are here. **2.** County (1970 pop. 29,554), 707 sq mi (1,831.1 sq km), central N.C., in a piedmont area drained by the Deep and Haw rivers, joining at the SE border to form the Cape Fear River; formed 1770; co. seat Pittsboro. Poultry raising, farming (tobacco and corn), textile manufacturing, and sawmilling are done.

Chatham. 1. Surburban borough (1970 pop. 9,566), Morris co., NE N.J., on the Passaic River W of Newark; settled 1749, inc. 1897. **2.** Town (1970 pop. 1,801), seat of Pittsylvania co., S Va., in the Blue Ridge foothills N of Danville; inc. 1852.

Chatham Islands, island group (1968 est. pop. 520), 373 sq mi (966.1 sq km), South Pacific, c.425 mi (685 km) E of New Zealand, to which it belongs. The two largest islands are Chatham Island, which has a large central lagoon, and Pitt Island. The inhabitants engage mainly in sheep raising, sealing, and fishing. The islands were discovered by Britons in 1791.

Châ·til·lon-sur-Seine (shä-tē-yôN'sür-sĕn'), town (1968 pop. 6,746), Côte d'Or dept., N central France, in Burgundy, on the Seine River. It was the site of unsuccessful peace negotiations (1814) between Napoleon I and his opponents.

Chat·om (chăt'əm), town (1970 pop. 1,059), seat of Washington co., SW Ala., NNW of Mobile, in a timber and agricultural area; inc. 1949. Chemicals are manufactured.

Chats·worth (chăts'wôrth'), city (1970 pop. 2,706), seat of Murray co., NW Ga., near the Tenn. border NE of Rome, in a talc-mining area. Corn and cotton are grown.

Chat·ta·hoo·chee (chăt'ə-hōō'chē), county (1970 pop. 25,813), 253 sq mi (655.3 sq km), W Ga., bounded on the W by the Chattahoochee River, here forming the Ala. border; formed 1854; co. seat Cusseta. Truck crops, poultry, and dairy products are processed. The major part of the county is included in Fort Benning.

Chattahoochee, town (1970 pop. 7,944), Gadsden co., NW Fla., at the junction of the Chattahoochee, Flint, and Apalachicola rivers, at the Ga. border NW of Tallahassee; inc. 1921. Furniture and wood products are made. Jim Woodruff Dam is nearby.

Chattahoochee, river, 436 mi (701.5 km) long, rising in N Ga., and flowing generally S to join the Flint River in Lake Seminole on the Ga.-Fla. line.

Chat·ta·noo·ga (chă'tə-nōō'gə), city (1978 est. pop. 154,600), seat of Hamilton co., E Tenn., on both sides of the Tennessee River near the Ga. line; inc. 1839. It is a port of entry. Foremost among its many manufactures are textile and metal products, chemicals, and primary metals. It is also a resort center, almost entirely surrounded by mountains, with many historical and tourist attractions on or near Lookout Mt., Missionary Ridge, and Signal Mt. The Cherokees were defeated here in 1794, and a trading post was established in 1810, followed by the Brainerd mission in 1817. A center first of salt shipping and then of cotton shipping, the city expanded with the arrival of the railroads in the 1840s. Chattanooga was of great strategic importance in the Civil War and was finally taken by Union forces in Nov., 1863. Northern industrialists developed the iron industry here during the 1870s. Electric power, augmented by the Tennessee Valley Authority project after 1933, has played an important role in the city's development. Points of interest include the Rock City Gardens, with unusual lichen-covered sandstone formations, a wildlife sanctuary, historic cemeteries, and numerous old buildings.

Chat·too·ga (chə-tōō'gə), county (1970 pop. 20,541), 317 sq mi (821 sq km), extreme NW Ga., in a hilly region bordered on the W by Ala.; formed 1838; co. seat Summerville. Its agriculture includes cotton, corn, hay, sweet potatoes, fruit, and livestock. Textile manufacturing and sawmilling are also done.

Chaudière Falls, in the Ottawa River in the heart of the city of Ottawa, Ont., Canada. The river is narrowed by rocky cliffs to a width of c.200 ft (60 m) and drops 50 ft (15.3 m) in a series of cascades.

Chau·mont (shō-mōɴ'), town (1975 pop. 27,226), capital of Haute-Marne dept., NE France, in Champagne, at the confluence of the Marne and Saize rivers. It is a railroad and industrial center. Iron is mined nearby. The Treaty of Chaumont, signed on Mar. 1, 1814, by England, Russia, Prussia, and Austria, laid the foundation for the Holy Alliance.

Chau·tau·qua (shə-tô'kwə). **1.** County (1970 pop. 4,642), 647 sq mi (1,675.7 sq km), SE Kansas, in a hilly agricultural area bordered on the S by Okla. and drained by the Caney River; formed 1875; co. seat Sedan. It produces livestock and grain and has oil and natural-gas fields. **2.** County (1970 pop. 147,305), 1,080 sq mi (2,797.2 sq km), extreme W N.Y., in a dairying and grape-growing area bounded on the NW by Lake Erie, on the W and S by Penn.; formed 1808; co. seat Mayville. General agriculture (livestock, seed, and truck crops) and fishing are also important. Metal and wood products are manufactured. It has many lake resorts.

Chautauqua, resort village (1970 pop. 500), Chautauqua co., W N.Y., on the W shore of Chautauqua Lake. The Chautauqua movement of adult education was founded here in 1874. Today the Chautauqua Institution sponsors summer concerts and lectures.

Chaux-de-Fonds, La (lä shō'də-fôɴ'), city (1971 pop. 42,347), Neuchâtel canton, NW Switzerland, in the Jura Mts., near the French border. It is a watch-manufacturing centers.

Chav·es (chăv'ĭs), county (1970 pop. 43,335), 6,094 sq mi (15,783.5 sq km), SE N.Mex., in a livestock and irrigated agricultural area watered by the Pecos and Salt rivers and Rio Hondo; formed 1889; co. seat Roswell. Cotton, alfalfa, and dairy products are processed. It includes Bottomless Lakes State Park.

Cha·vín de Huan·tar (chä-vēn' dā wän'tär), archaeological site in the NE highlands of Peru. It was probably the chief ceremonial and urban center of the earliest civilization (fl. c.700 B.C.–c.200 B.C.) of the Andes, now called the Chavín. Highly developed and sophisticated, the Chavín built large temples with painted relief sculpture of mythical beasts and produced boldly designed ceramics, gold objects, and textiles.

Chea·dle and Gat·ley (chēd'l; găt'lē), urban district (1973 est. pop. 62,460), Greater Manchester, NW England. The district is both residential and industrial. Industries include engineering works and the manufacture of chemicals, drugs, and bricks.

Chea·ha (chē'hô), peak, 2,407 ft (734.1 m) high, E Ala., in the Talladega Mts. It is the highest point in the state.

Cheap·side (chēp'sīd'), street and district, in the City of London, England, the most important market center of medieval London. It runs from St. Paul's churchyard to the Bank of England. Tournaments and some executions took place here. It was also the site of the Mermaid Tavern, a meeting place for Elizabethan poets and playwrights. A number of ancient guildhalls were destroyed or damaged during World War II.

Chea·tham (chē'təm), county (1970 pop. 13,199), 305 sq mi (790 sq km), N central Tenn., drained by the Cumberland and Harpeth rivers; formed 1856; co. seat Ashland City. Its agriculture includes grain, tobacco, and livestock.

Che·bok·sa·ry (chĕ'bŏk-sä'rē), city (1976 est. pop. 278,000), capital of Chuvash ASSR, NW European USSR, a port on the Volga River. It is the center of an agricultural region.

Che·boy·gan (shə-boi'gən), county (1970 pop. 16,573), 725 sq mi (1,877.8 sq km), N Mich., bounded on the N by the Straits of Mackinac and drained by the Pigeon, Black, and Sturgeon rivers; organized 1853; co. seat Cheboygan. In a livestock, dairying, and farming (potatoes and fruit) area, it also has commercial fishing, limestone quarrying, sawmilling, and some manufacturing. It includes a number of lakes and state parks.

Cheboygan, city (1970 pop. 5,553), seat of Cheboygan co., N Mich., on the S channel of the Straits of Mackinac at the mouth of the Cheboygan River; settled c.1845, inc. as a village 1887, as a city 1889. Once an important lumber town, it is now a resort, farm trade center, and commercial fishing port. Paper is made here.

Che·chen-In·gush Autonomous Soviet Socialist Republic (chĭ-chĕn'ĭn-gōōsh'), autonomous republic (1970 pop. 1,065,000), 7,452 sq mi (19,300.7 sq km), SE European USSR, in the N Caucasus; capital Grozny. The Grozny fields represent a major source of Soviet oil; the republic also has sizable deposits of natural gas, limestone, marl, gypsum, alabaster, and sulfur. Mineral waters make the region an important health center. Industries include oil refining, food processing, wine and cognac making, fruit canning, and the manufacture of chemicals and oil field equipment. Known since the 17th cent., the Chechen people became the most active opponents of czarist Russia's conquest and occupation (1818–1917) of the Caucasus. The Ingush first settled in the lowlands in the 17th cent. The Bolsheviks seized the region in 1918 but were dislodged the following year. With Soviet power re-established, the area was included in 1921 in the Mountain People's Republic. The Chechen Autonomous Oblast was created in 1922 and the Ingush Autonomous Oblast in 1924; the two were joined in 1934 to form the Chechen-Ingush Autonomous Oblast, which became an autonomous republic in 1936.

Ched·dar (chĕd'ər), village, Somerset, SW England. It is chiefly a tourist center. Limestone is quarried, and strawberries are grown. The town gives its name to the famous cheese, which has been made here since at least the 16th cent.

Cheek·to·wa·ga (chēk'tə-wä'gə), uninc. town (1970 pop. 113,844), Erie co., W N.Y., E of Buffalo.

Che·foo (chē'fōō'): *see* Yen-t'ai, China.

Che·ha·lis (chə-hā'lĭs), city (1975 est. pop. 5,800), seat of Lewis co., SW Wash., S of Seattle on the Chehalis River, in a fruit, dairy, and poultry region; inc. 1883. Lumbering and food processing are done.

Che·ju (chā'jōō'), island (1975 pop. 412,021), c.700 sq mi (1,815 sq km), c.60 mi (95 km), South Korea, SW of the Korean peninsula. Cheju is of volcanic origin and rises to c.6,400 ft (1,950 m). Fishing, dairy farming, and livestock breeding are the chief occupations on the mountainous, heavily wooded island.

Che·kiang (jŭ'jē-äng'), province (1968 est. pop. 31,000,000), c.40,000 sq mi (103,600 sq km), SE China, on the East China Sea; capital Hangchow. Except for the level area in the north, Chekiang is mountainous. Over one third of the area is forested; pine and bamboo predominate. Most of Chekiang has a wet climate, with a long frost-free period and high summer temperatures. Rice is the leading food crop and tea the major industrial crop. Cotton, wheat, hemp, corn, and sweet potatoes are also grown. There are tung and mulberry trees, and some silk is produced. Fishing is extensive. Iron, aluminum, coal, and fluorspar are mined. Chekiang, part of the kingdom of Wu, passed into the Chinese orbit in the 3rd cent. B.C. It flourished in the 12th and 13th cent. as the center of the Southern Sung dynasty. Chekiang passed to Manchu control in 1645. It was devastated in the Taiping Rebellion (1850–65), was partly occupied by the Japanese in the Second Sino-Japanese War, and fell to the Communists in 1949.

Che·lan (shə-lăn′), county (1970 pop. 41,103), 2,918 sq mi (7,557.6 sq km), N central Wash., in a mountainous area of the Cascade Range bounded on the E by the Columbia River; formed 1899; co. seat Wenatchee. In an irrigated fruit and livestock region, it is watered by the Entiat, White, and Wenatchee rivers. Lumber, minerals, manufactures, and winter sports are also important to its economy. The county includes Wenatchee National Forest, Lake Chelan National Recreation Area, and part of North Cascades National Park.

Chelan, Lake, 55 mi (88.5 km) long and from 1 to 2 mi (1.6-3.2 km) wide, located in a deep narrow gorge in the Cascade Range, NW Wash. It is the third-deepest freshwater lake in the United States. Fed by streams from the Cascade Range, the lake flows into the Columbia River via the Chelan River. Lake Chelan Dam, built at the lake's outlet, generates electricity. The northern part of the lake is part of the Lake Chelan National Recreation Area.

Ché·liff (shā-lēf′), river, c.420 mi (675 km) long, N Algeria. It rises in the Amour Mts. of the Saharan Atlas and empties into the Mediterranean Sea near Mostaganem. The Chéliff, the longest river in Algeria, is not navigable, but its waters are used for irrigation and hydroelectric power.

Chełm (KHĕlm), city (1975 est. pop. 45,500), E Poland. It is a railway junction and has industries manufacturing metals, farm tools, machinery, furniture, and liquors. An old Slavic settlement, Chełm was chartered in 1233. It passed to Poland in 1377, to Austria in 1795, and to Russia in 1815. The Treaty of Brest-Litovsk (1918) transferred the city to Ukraine, but it passed to Poland in 1921.

Chełm·no (KHĕlm′nô), city (1970 pop. 38,800), N central Poland. Its industries manufacture metals, bricks, and farm tools. Chartered in 1223, it was transferred to the Teutonic Knights in 1228, passed to Poland in 1466, and was included in Prussia in 1772. It reverted to Poland in 1919.

Chelms·ford (chĕms′fərd, chĕmz′-), municipal borough (1973 est. pop. 58,320), administrative center of Essex, SE England. It is a market center, especially for cattle. Manufactures include electrical equipment, radios, ball bearings, rope, and agricultural equipment. A Roman town on this site was excavated in 1849.

Chelms·ford (chĕmz′fərd, chĕlmz′-), town (1970 pop. 31,432), Middlesex co., NE Mass.; inc. 1655. It is chiefly a residential town with wool and nylon industries and granite quarries.

Chel·sea (chĕl′sē). **1.** City (1978 est. pop. 22,200), Suffolk co., E Mass., a suburb of Boston; settled 1624, inc. as a town 1739, as a city 1857. Its industries include printing and the manufacture of rubber and plastic products, electrical machines, shoes and shoe accessories, and paint. At the Battle of Chelsea Creek (1775) Revolutionary forces made one of their first captures of a British ship. **2.** Town (1970 pop. 983), seat of Orange co., E central Vt., on the First Branch of the White River SSE of Montpelier; granted by N.Y. c.1770, chartered by Vt. 1781, settled 1784. It is in a farm, dairy, and maple-sugar area.

Chel·ten·ham (chĕlt′nəm), municipal borough (1976 est. pop. 86,500), Gloucestershire, W central England. It has been a health and holiday resort since the discovery of mineral springs in 1716. Products include bricks, beer, rubber goods, and anesthetics. There are numerous Regency houses and Georgian squares.

Che·lya·binsk (chĭ-lyä′bĭnsk), city (1977 est. pop. 1,007,000), capital of Chelyabinsk oblast, W Siberian USSR, in the S foothills of the Urals. One of the major metallurgical and industrial centers of the USSR, Chelyabinsk produces steel and agricultural machinery and processes ore. Founded in 1736 as a Russian frontier outpost, it was chartered in the 1740s.

Che·lyus·kin, Cape (chĭ-lyo͞os′kĭn), northernmost point (lat. 77°43′ N) of Asia, Krasnoyarsk Kray, N central Siberian USSR. It is named after the Russian navigator who discovered it in 1742.

Chem·nitz (kĕm′nĭts): *see* Karl-Marx-Stadt, East Germany.

Che·mung (shə-mŭng′), county (1970 pop. 101,537), 415 sq mi (1,074.9 sq km), S N.Y., in a rolling hilly area bounded on the S by the Pa. border and drained by Cayuta Creek; formed 1836; co. seat Elmira. It has agriculture (dairy products, poultry, and apples), sand and gravel pits, and some manufacturing. Woodlawn National Cemetery and Newtown Battlefield State Park are here.

Che·nab (chē-näb′), one of the five rivers of the Punjab, 675 mi (1,086.1 km) long, rising in the Punjab Himalayas, W Kashmir, and flowing NW then SW through Pakistani Punjab to join the Sutlej River.

Che·nang·o (shə-năng′gō), county (1970 pop. 46,368), 908 sq mi (2,351.7 sq km), central N.Y., in a dairying area bounded on the E by the Unadilla River and drained by the Susquehanna, Otselic, and Chenango rivers; formed 1798; co. seat Norwich. Fruit, maple sugar, general farm crops, and poultry are produced. It has some lumbering and diversified manufacturing. Gladding International Fishing Museum is here.

Chen-chiang (jŭn′jē-äng′) or **Chin·kiang** (jĭn′jē-äng′, chĭn′kyăng′), city (1970 est. pop. 250,000), S Kiangsu prov., China, a port at the junction of the Grand Canal with the Yangtze River. An important commercial and industrial center, it is known for its silk, vinegar, and pickled vegetables. Other processed foods, pharmaceuticals, machine tools, and paper products are also made. Chen-chiang was known in the Sung dynasty (12th cent.), flourished under the Ming and Manchu dynasties, was held by the Taipings and ravaged (1857), and was opened to foreign trade in 1859. It was a British concession until 1927, when it was returned to China.

Cheng-chou or **Cheng·chow** (both: jŭng′jō′), city (1970 est. pop. 1,500,000), capital of Honan prov., E central China. An important railroad center, it is a textile center and a flourishing industrial city. Manufactures include chemicals, aluminum, fertilizer, processed meats, agricultural machinery, and electrical equipment.

Ch'eng-te or **Cheng·teh** (both: chŭng′dŭ′), city (1970 est. pop. 200,000), N Hopeh prov., China, near the Luan River. It is a distribution center for lumber products, fruits, and pharmaceuticals, and has an iron mine. The former summer capital of the Ch'ing dynasty (1644-1911), Ch'eng-te is surrounded by large parks with lakes, palaces, and pavilions.

Ch'eng-tu or **Cheng·tu,** (both: chŭng′do͞o′), city (1970 est. pop. 2,000,000), capital of Szechwan prov., SW China, on the Min River. It is a port and the commercial center of the Ch'eng-tu plain, the main farming area of Szechwan. Products include textiles, processed foods, chemicals, machinery, and paper. High-grade iron ore is mined nearby. Ch'eng-tu, an old walled city, was the capital of the Shu Han dynasty (3rd cent. A.D.) and one of the earliest (9th cent. A.D.) printing centers in China. It has been a cultural center since ancient times.

Che·non·ceaux (shə-nôN-sō′), village, Indre-et-Loire dept., W central France, in Touraine, on the Cher River. It is famous for its château (built 1515-22), the residence of both Diane de Poitiers and Catherine de' Medici.

Cher (shâr), department (1975 pop. 316,350), 2,791 sq mi (7,228.7 sq km), central France, in Berry; capital Bourges.

Che·raw (chē′rô), town (1970 pop. 5,627), Chesterfield co., NE S.C., on the Pee Dee River NE of Columbia, in a fertile farm region; settled in the mid-18th cent. by Welsh immigrants; inc. 1820. Textiles, clothing, bricks, and lumber products are among its manufactures. Notable among the old buildings is St. David's Episcopal Church (1768). In its graveyard are buried some 50 British soldiers who died of smallpox during the American Revolution.

Cher·bourg (shâr′bo͝org, shĕr-bo͞or′), city (1975 pop. 32,536), Manche dept., NW France, in Normandy, on the English Channel, at the tip of the Cotentin peninsula. It is a naval base and seaport with related industries. The site has been settled since ancient times and was frequently fought over by the French and English because of its strategic value. In World War II it was attacked by Allied forces (June 21-22, 1944), and on June 27 it was captured along with 35,000 prisoners.

Che·rem·kho·vo (chə-rĭm-kô′və), city (1976 est. pop. 88,000), SE Siberian USSR, on the Trans-Siberian RR. The center of the Cheremkhovo coal basin, the city forms part of an industrial complex based mainly on coal, oil refining, and chemical production.

Che·re·po·vets (chə-rĭ-pə-vyĕts′, -pô′vyĭts), city (1976 est. pop. 238,000), NE European USSR, on the Rybinsk Reservoir. A rail and water transportation center of the Volga-Baltic Waterway, it has an iron and steel complex. Cherepovets arose (14th cent.) as a settlement around a monastery.

Cher·kas·sy (chĕr-kä′sē), city (1976 est. pop. 221,000), capital of Cherkassy oblast, in Ukraine, S European USSR, a port on the Dnepr River. Cherkassy has important chemical-fiber and fertilizer industries. Founded at the end of the 13th cent., it was a fortress in the 14th cent. and passed to Russia in 1793.

Cher·kess Autonomous Oblast (chĕr-kĕs'): *see* Karachay-Cherkess Autonomous Oblast.

Cher·kessk (chĕr-kĕsk'), city (1976 est. pop. 82,000), capital of Karachay-Cherkess Autonomous Oblast, Stavropol Kray, SE European USSR, on the Kuban River. Founded in 1825, it manufactures electrical equipment and shoes and has food-processing plants.

Cher·ni·gov (chĕr-nyē'gəf), city (1976 est. pop. 225,000), capital of Chernigov oblast, W central European USSR, in the Ukraine, on the Desna River. It is a rail junction, a river port, and an air and highway transport hub. Industries include ship repairing, woodworking, food and wool processing, and the manufacture of metal goods and machinery. First mentioned in 907, Chernigov is one of the oldest cities of Kievan Russia. The city declined after the Mongol invasion of 1239. It passed to Lithuania in the 14th cent. and to Russia in the 16th cent. It was under Polish control during part of the 17th cent.

Cher·nov·tsy (chĕr-nôf'tsē), city (1970 pop. 187,000), capital of Chernovtsy oblast, SW European USSR, in the Ukraine, on the Prut River and in the Carpathian foothills. It is the economic, cultural, and scientific center of the region of Bukovina. Industries include woodworking, food processing, and the manufacture of machinery, textiles, chemicals, footwear, and hosiery. One of Russia's oldest towns, Chernovtsy was part of Kievan Russia. It passed to Austria in 1775 and in 1849 became the capital of Bukovina. During the 19th and early 20th cent., the city was a center of the Ukrainian national movement. With the dissolution of Austria-Hungary in 1918, Chernovtsy was transferred to Rumania, which held it until the USSR seized northern Bukovina in 1940.

Cher·o·kee (chĕr'ə-kē', chĕr'ə-kē'). **1.** County (1970 pop. 15,606), 556 sq mi (1,440 sq km), NE Ala., in an agricultural area bordering in the E on Ga. and watered by Weiss Lake and the Coosa River; formed 1836; co. seat Centre. Cotton, corn, and livestock are raised. It has deposits of coal and limestone, and lumber milling and iron mining are done. **2.** County (1970 pop. 31,059), 415 sq mi (1,074.9 sq km), NW Ga., drained by the Etowah River and Allatoona Lake; formed 1831; co. seat Canton. It is in a piedmont agricultural area yielding cotton, corn, sweet potatoes, fruit, poultry, and livestock. It also has textile mills and marble quarries. **3.** County (1970 pop. 17,269), 573 sq mi (1,484.1 sq km), NW Iowa, in a prairie agricultural area drained by the Little Sioux River; formed 1851; co. seat Cherokee. Hogs, cattle, and poultry are raised, and corn, oats, and soybeans grown. **4.** County (1970 pop. 21,549), 587 sq mi (1,520.3 sq km), extreme SE Kansas, in a gently rolling agricultural area bordered on the S by Okla. and on the E by Mo.; formed 1866; co. seat Columbus. It is drained by the headwaters of the Spring River. Livestock, grain, poultry, and dairying are important. It also has timber and deposits of coal, lead, and zinc. **5.** County (1970 pop. 16,330), 452 sq mi (1,170.7 sq km), extreme W N.C., partly in the Blue Ridge and bounded on the S by the Ga. border, on the W and NW by the Tenn. border; formed 1839; co. seat Murphy. It is drained by the Hiwassee and Nottely rivers. Its agriculture includes tomatoes, cucumbers, green beans, livestock, and poultry. In a forested area, it also has marble quarries, sawmills, and some manufacturing. Part of Nantahala National Forest is in the county. **6.** County (1970 pop. 23,174), 756 sq mi (1,958 sq km), E Okla., bounded on the W by the Neosho River and intersected by the Illinois River; formed 1907; co. seat Tahlequah. In a stock-raising, agricultural (fruit, truck crops, corn, cotton, and grain), and dairying area, it also has food-processing and lumbering industries and some manufacturing. It includes a number of state parks and recreational areas. **7.** County (1970 pop. 36,791), 394 sq mi (1,020.5 sq km), N S.C., drained by the Broad River and bounded on the N by the N.C. border, on the S by the Pacolet River; formed 1798; co. seat Gaffney. It is primarily agricultural (cotton, grain, livestock, and dairy products), with textile mills and some other manufacturing. There are limestone and feldspar deposits here. It includes Cowpens National Battlefield and Kings Mountain National Military Park. **8.** County (1970 pop. 32,008), 1,049 sq mi (2,716.9 sq km), E Texas, bounded on the W by the Neches River and partly on the E by the Angelina River; formed 1846; co. seat Rusk. In a partly wooded area with extensive lumbering, it has agriculture (tomatoes and other vegetables, fruit, potatoes, and peanuts), minerals (iron, lignite, clay, silica, and salt), and oil and natural-gas wells. Roses are grown, and there is some manufacturing. It includes Jim Hogg State Park.

Cherokee. 1. City (1970 pop. 7,272), seat of Cherokee co., NW Iowa, on the Little Sioux River NE of Sioux City; settled 1856, inc. 1873. It is a rail, trade, and industrial center for a farm, dairy, and livestock area. Pilot Rock, a large glacial boulder that served as a landmark for early settlers, is nearby. **2.** City (1970 pop. 2,119), seat of Alfalfa co., N Okla., NW of Enid. It is the trading and processing center for an area growing wheat, corn, alfalfa, and sorghums. Great Salt Plains Lake and State Park are nearby.

Cher·ry (chĕr'ē), county (1970 pop. 6,846), 5,966 sq mi (15,451.9 sq km), N Nebr., in an agricultural region drained by the North Loup and Niobrara rivers and bordered on the N by S.Dak.; formed 1877; co. seat Valentine. Livestock and grain are produced. It includes Samuel R. McKelvie National Forest, Fort Niobrara and Valentine national wildlife refuges, and a number of state recreational areas.

Cher·so·nese (kûr'sō-nēs') or **Cher·so·ne·sus** (-nē'səs), name applied in ancient geography to several regions, such as Crimea (Chersonesus Taurica or Scythia); Gallipoli Peninsula (Chersonesus Thracica); Malay Peninsula (Chersonesus Aurea); and Jutland (Chersonesus Cimbrica).

Chert·sey (chûrt'sē), urban district (1976 est. pop. 74,800), Surrey, SE England. Its market gardens serve London. There are varied engineering works.

Ches·a·peake (chĕs'ə-pēk), city (1978 est. pop. 112,000), independent and in no county, SE Va.; inc. 1963. Chesapeake was created (1963) by merging the former city of South Norfolk with all of Norfolk co. Within its vast area are residential sections, much farmland, with related agricultural industries, and a large part of the Dismal Swamp. There are also industries manufacturing a great variety of products, including fertilizer, chemicals, lumber and wood items, steel equipment, and cement. The Battle of Great Bridge was fought (1775) in Chesapeake.

Chesapeake and Del·a·ware Canal (dĕl'ə-wâr), sea-level canal, 19 mi (30.6 km) long, 250 ft (76.3 m) wide, and 27 ft (8.2 m) deep, connecting the head of Chesapeake Bay with the Delaware River. Built in 1824–29, the canal was bought by the Federal government in 1919 and later was enlarged and modernized. It is part of the Intracoastal Waterway.

Chesapeake and O·hi·o Canal (ō-hī'ō), former waterway, c.185 mi (300 km) long, from Washington, D.C., to Cumberland, Md., running along the N bank of the Potomac River. A successor to the Potomac Company's (1784–1828) navigation improvement project, the Chesapeake and Ohio Canal was planned to extend west to Pittsburgh. Financial and labor problems, as well as opposition from the rival Baltimore and Ohio RR, delayed completion to Cumberland until 1850. Although extension to Pittsburgh proved impractical, the canal experienced a busy period in the 1870s carrying coal from the Cumberland mines. The canal was used until it was damaged by floods in 1924. It was sold in 1938 to the U.S. government. The canal, partially restored, was made a national monument in 1961. In 1971 it became a national historic park.

Chesapeake Bay, inlet of the Atlantic Ocean, c.200 mi (320 km) long, from 3 to 30 mi (4.8–48.3 km) wide, and 3,237 sq mi (8,383.8 sq km), separating the Delmarva Peninsula from the mainland, E Md. and E Va. The bay is the drowned mouth of the Susquehanna River and also is fed by many other rivers including the Potomac, Rappahannock, and James. Chesapeake Bay is entered from the Atlantic Ocean through a 12-mi (19.3-km) wide gap between Capes Henry and Charles, Va. It is an important part of the Intracoastal Waterway. The English colonist John Smith explored and charted Chesapeake Bay in 1608.

Chesh·ire (chĕsh'ər) or **Ches·ter** (chĕs'tər), county (1976 est. pop. 916,400) W central England; administrative center Chester. The terrain is generally low, flat, and fertile. The county is important agriculturally and industrially. Cheshire was made a palatinate by William I and maintained some of its privileges as such until 1830. The numerous black-and-white-timbered manor houses attest to the county's prosperity in the 16th and 17th cent. In 1974 most of Cheshire became part of the new nonmetropolitan county of Cheshire; northwest Cheshire became part of the new metropolitan county of Merseyside, and northeast Cheshire became part of the new metropolitan county of Greater Manchester.

Cheshire, county (1970 pop. 52,364), 715 sq mi (1,851.9 sq km), SW N.H., drained by the Connecticut and Ashuelot rivers; formed 1769; co. seat Keene. It has some agriculture, granite quarries, and feldspar mines. Textiles, wood products, and shoes are manufactured. There are a number of lake resorts here.

Cheshire, town (1970 pop. 19,051), New Haven co., S central Conn., in a farm area; settled 1695, inc. 1780. It is chiefly residential, with some machine shop manufactures.

Ches·hunt (chĕs'ənt), urban district (1976 est. pop. 76,800), Hertfordshire, SE England. A suburb of London, it is a prominent market-gardening district.

Ches·ter (chĕs'tər), county borough (1973 est. pop. 61,370), administrative center of Cheshire, W central England, on a sandstone height above the Dee River. Its manufactures include electrical switches, paint, and window panes. Tourism is also economically important. Formerly Chester had great military importance, and it was a significant port for centuries. Under the name Castra Devana or Deva it was the headquarters of the Roman 20th legion. It was ravaged by Æthelfrith of Northumbria in the 7th cent. and the Danes in the 9th cent. William I took it in 1070 and the following year granted it to his nephew, Hugh Lupus, as a palatine earldom. Chester served the English crown as a defensive bastion and was used as a base for operations against Wales from 1275 to 1284. Its greatest prosperity as a port was from c.1350 to 1450. Modern Chester is medieval in appearance. It is the only city in England that still possesses its entire wall. Interesting features are the red sandstone wall with a walk along the top; Agricola's Tower; 15th and 16th cent. timbered houses; the Roodee, on which races have been held since 1540; and The King's School, a public school founded by Henry VIII in 1541. Characteristic of Chester are the Rows, a double tier of shops formed by recessing the second stories of the buildings along the main streets. This creates a sheltered walk upon the roofs of the street-level stores.

Chester. 1. County (1970 pop. 277,746), 760 sq mi (1,968.4 sq km), SE Pa., in an industrial and agricultural area drained by the Schuylkill River and Brandywine Creek; formed 1682; co. seat West Chester. Iron and steel, canned goods, and clothing are produced. There are granite and limestone quarries. **2.** County (1970 pop. 29, 811), 585 sq mi (1,515.2 sq km), N S.C., in a mainly agricultural area bounded on the W by the Broad River, on the E by the Catawba River; formed 1785; co. seat Chester. Its agriculture includes cotton, livestock, corn, oats, dairy products, and hay. Some manufacturing is done. Part of Sumter National Forest is here. **3.** County (1970 pop. 9,927), 285 sq mi (738.2 sq km), SW Tenn., drained by the South Fork of the Forked Deer River; formed 1879; co. seat Henderson.

Chester. 1. Town (1970 pop. 2,982), Middlesex co., S Conn., on the Connecticut River SE of Middletown; settled 1692, inc. 1836. Shipbuilding was important here in the early 19th cent. Today the town is a boating center. **2.** City (1970 pop. 5,310), seat of Randolph co., SW Ill., on the Mississippi River SSE of St. Louis; founded 1819, inc. 1835. It is a shipping center for a mine and quarry area. Feed, flour, and hosiery are made. Fort Kaskaskia State Park is nearby. **3.** Town (1970 pop. 936), seat of Liberty co., N Mont., NNE of Great Falls, in a wheat area; settled in the 1880s, inc. 1910. There are oil and natural-gas wells in the vicinity. **4.** City (1978 est. pop. 45,500), Delaware co., SE Pa., on the Delaware River, an industrial suburb of Philadelphia; settled c.1644 by Swedes, inc. as a city 1866. It is a port of entry and has an important shipbuilding industry that dates from before the Civil War. There are also steel mills, oil refineries, automobile assembly plants, and factories making a huge variety of products, including aircraft parts, chemicals, and electrical equipment. The oldest city in the state, Chester (established as Upland) was the site of William Penn's first landing (1682) in America. Penn renamed the settlement and convened (1682) the first assembly of the province here. Historic attractions include the foundations of the original settlement, in Governor Printz Park; the Morton Homestead (1654); the Caleb Pusey House, at Landingford Plantation (1683); the old courthouse (1724); and the Washington House (1747). **5.** City (1970 pop. 7,045), seat of Chester co., N S.C., NNW of Columbia; settled in the late 18th cent., inc. as a town 1849, as a city 1893. Textiles, clothing, fertilizer, and lumber and dairy products are made. **6.** Village (1970 pop. 5,556), Chesterfield co., E central Va., near the James River S of Richmond. Sand and gravel are produced. Pocahontas State Park and State Forest are nearby.

Ches·ter·field (chĕs'tər-fēld'), municipal borough (1976 est. pop. 93,900), Derbyshire, central England. An important industrial center, the borough produces mining equipment, railroad cars, metal products, and many other goods.

Chesterfield. 1. County (1970 pop. 33,667), 793 sq mi (2,053.9 sq km), N S.C., bounded on the E by the Pee Dee River, on the W by the Lynches River, and on the N by the N.C. border; formed 1798; co. seat Chesterfield. In a mainly agricultural (cotton, corn, and tobacco) and timber area, it has some manufacturing. The Sand Hills Development Project for reforestation and game preservation is in the south. **2.** County (1970 pop. 77,045), 460 sq mi (1,191.4 sq km), E central Va., bounded on the SW, S, and SE by the Appomatox River and on the N and NE by the James River; formed 1749; co. seat Chesterfield. In an agricultural (corn, tobacco, peanuts, and cotton) and dairying area, it also has fisheries and large tobacco-processing and rayon plants. Livestock and poultry are raised.

Chesterfield. 1. Town (1970 pop. 1,667), seat of Chesterfield co., NE S.C., near the N.C. border NE of Columbia; settled c.1798, inc. 1872. Metal and lumber products, textiles, and clothing are made. **2.** Village (1970 pop. 400), seat of Chesterfield co., E central Va., S of Richmond. It is sometimes called Chesterfield Court House.

Ches·ter·town (chĕs'tər-toun'), town (1970 pop. 3,476), seat of Kent co., NE Md., on the Chester River NE of Annapolis; laid out 1706, inc. 1806. It is a port of entry and a trading center for a resort and farming region. It has food-processing plants, fisheries, and fertilizer factories. Washington College is here.

Che·sun·cook Lake (chĭ-sŭn'ko͝ok), 22 mi (35.4 km) long and from 1 to 4 mi (1.6–6.4 km) wide, N central Maine. The lake is in a noted hunting and fishing region. Baxter State Park is nearby.

Chev·i·ot Hills (chĕv'ē-ət, chē'vē-), range, c.35 mi (56 km) long, extending along part of the border between Scotland and England. The highest point is The Cheviot (2,676 ft/816.2 m). Since World War II the hills have been reforested.

Chev·y Chase (chĕv'ē chās'), village (1978 est. pop. 23,000), Montgomery co., W central Md., a residential suburb of Washington, D.C.; inc. 1914.

Chey·enne (shī-ăn', -ĕn'). **1.** County (1970 pop. 2,396), 1,772 sq mi (4,589.5 sq km), E Colo., in an agricultural area bordering on Kansas and drained by the Big Sandy Creek; formed 1889; co. seat Cheyenne Wells. Wheat and livestock are important. It also has oil wells. **2.** County (1970 pop. 4,256), 1,027 sq mi (2,659.9 sq km), extreme NW Kansas, in a farming and cattle region drained by the South Fork of the Republican River and bordered on the N by Nebr., on the W by Colo.; formed 1886; co. seat St. Francis. Wheat and barley are its major crops. **3.** County (1970 pop. 10,778), 1,186 sq mi (3,071.7 sq km), W Nebr., in an agricultural area bordered on the S by Colo. and drained by Lodgepole Creek; formed 1870; co. seat Sidney. Dairy products, grain, and livestock are produced.

Cheyenne. 1. Town (1970 pop. 892), seat of Roger Mills co., W Okla., on the Washita River near the Texas border NW of Elk City, in a livestock, grain, and cotton area; founded c.1892. Cement blocks are manufactured, and cotton is ginned. The Battle of Washita (Nov. 23, 1868), between troops led by Gen. George Custer and Cheyenne Indians, took place nearby. **2.** City (1978 est. pop. 47,000), alt. 6,062 ft (1,848.9 m), state capital and seat of Laramie co., SE Wyo., near the Colo. and Nebr. borders; inc. 1867. It is a market for sheep and cattle ranches and a shipping center with good transportation facilities. The city sprang up after the Union Pacific RR selected this site for a division point in 1867. It was made territorial capital in 1869. In the 1870s the development of the area as a cattle-ranching section and the opening of the Black Hills gold fields stimulated the city's growth. Cheyenne revives its past annually with a Frontier Days celebration, first held in 1897.

Cheyenne, river, 527 mi (848 km) long, rising in E Wyo. and flowing NE to the Missouri River near Pierre, S.Dak. The Cheyenne basin is part of the Missouri River basin project.

Cheyenne Wells, town (1970 pop. 982), seat of Cheyenne co., E Colo., E of Colorado Springs near the Kansas border; settled in the 1860s, inc. 1890. Wheat, corn, and feed are produced.

Chia-i or **Chia·yi** (both: jē-ä'ē'), city (1973 est. pop. 246,513), S Taiwan. It is an agricultural market and headquarters for the Chia-i irrigation system.

Chia-ling (jē-ä'lĭng') or **Kia·ling** (kyä'lĭng), river, c.450 mi (725 km) long, rising in S Kansu prov., central China, and flowing S to join the Yangtze River at Chungking.

Chia-mu-ssu (jē-ä'mo͞o'so͞o') or **Kia·mu·sze** (kyä'mo͞o'so͞o', jē-ä'-), city (1970 est. pop. 275,000), E Heilungkiang prov., China. It is the chief port on the lower reaches of the Sungari River, with aluminum, lumber, paper, textile, and beet-sugar-processing industries. The city was formerly the capital of Hokiang prov.

Chiang·mai (jē-äng'mī') or **Chieng·mai** (jē-ĕng'-), city (1972 pop. 93,353), capital of Chiangmai prov., N Thailand, on the Ping River, near the Burmese border. It is the economic, cultural, and religious center of the northern provinces. The city is a shipping point for the agricultural products of the surrounding region. Long the center of Thailand's teak industry, Chiangmai also produces silver and wood articles, pottery, and silk and cotton goods.

Chi·an·ti, Mon·ti (mōn'tē kyän'tē, kē-än'tē), small range of the Apennines, c.15 mi (25 km) long, in Tuscany, central Italy, W of the Arno River, rising to c.3,000 ft (915 m). The celebrated Chianti wines are produced on its slopes.

Chi·a·tu·ra (chē-ə-tōōr'ə), city (1970 pop. 25,000), SE European USSR, in Georgia, on the Kvirila River. One of the world's largest manganese producers, Chiatura alone accounted for half of the world's manganese trade before World War I.

Chi·ba (chē'bä), city (1976 est. pop. 683,514), capital of Chiba prefecture, central Honshu, Japan, on Tokyo Bay. It is a manufacturing center noted for textiles and paper products. The city has an 8th cent. Buddhist temple.

Chi·ca·go (shĭ-kä'gō), city (1970 pop. 3,369,359), seat of Cook co., NE Ill., on Lake Michigan; inc. 1837. With a metropolitan population of almost 7 million people, it is the commercial, financial, industrial, and cultural center for a vast region and a great midcontinental shipping point. It is a port of entry; a major Great Lakes port, located at the junction of the St. Lawrence Seaway with the Mississippi River system; the busiest air center in the country; and an important rail and highway hub. Chicago has large grain mills and elevators, iron and steel works, steel-fabrication plants, stockyards, meat-packing establishments, and printing and publishing houses. Among its many other products are machinery, musical instruments, electronic equipment, furniture, chemicals, household appliances, foods, and clothing.

Chicago covers more than 200 sq mi (520 sq km); it extends more than 20 mi (32 km) along the lakefront, then sprawls inland to the west. The city's arteries are its boulevards, expressways, and a system of elevated railways (part of it a subway). The elevated lines extend into the heart of the city, making a huge rectangle for passenger convenience in transferring from one to another. This is the celebrated Loop, which gives its name to the downtown section. In or near the center of the city are the Merchandise Mart, the world's largest commercial building; the Chicago Board of Trade building; and the Chicago Civic Opera. On the lakefront, which has many beaches, are Grant Park, with the Art Institute of Chicago, the Chicago Natural History Museum, the Adler Planetarium, the Buckingham Memorial Fountain, and the John G. Shedd Aquarium.

To the north along the lakefront is Michigan Boulevard, which proceeds past the rich hotels of the "gold coast" and enters the residential district. In this section lies Lincoln Park, with the Chicago Historical Society building, the Chicago Academy of Sciences, a zoological garden, and a conservatory. The south side of Chicago is the seat of the Univ. of Chicago, with its imposing Gothic buildings and attractive spaciousness. Nearby is Jackson Park, with the Museum of Science and Industry. Much of the south side is, however, given over to industry and to poor residential areas. The west side extends over a vast area. It is a region of nationalities, which, though crowded next to each other physically, are more or less separate culturally. In the west, too, are large industrial areas and two well-known parks—Garfield Park, with its noted conservatory, and Humboldt Park.

Notable as dividing lines in the city are the two branches of the Chicago River. In early days the narrow watershed between it and the Des Plaines River (draining into the Mississippi through the Illinois River) offered an easy portage that led explorers, fur traders, and missionaries to the great central plains. Father Marquette and Louis Jolliet arrived here in 1673, and the spot was well known for a century before Jean Baptiste Point Sable (or Point du Sable) set up a trading post at the mouth of the river. John Kinzie, who succeeded him as a trader, is usually called the father of Chicago. The military post, Fort Dearborn, was established in 1803. In the War of 1812 its garrison perished in one of the most famous tragedies of Western history. Fort Dearborn was rebuilt in 1816, and the construction of the Erie Canal in the next decade speeded the settling of the Middle West and the growth of Chicago. By 1860 a number of railroad lines connected Chicago with the rest of the nation, and the city was launched on its career as the great midcontinental shipping center. Gurdon S. Hubbard had already helped establish the meat-packing industry, with its large stockyards. In 1871 the shambling city built of wood was almost entirely destroyed by a great fire (which legend says was started when Mrs. O'Leary's cow kicked over a lantern). The fire, one of the most famous disasters of U.S. history, killed several hundred people, rendered 90,000 homeless, and destroyed some $200 million worth of property. Chicago was rebuilt as a city of stone and steel. With industry came labor troubles, highlighted by the Haymarket Square riot of 1886 and the great strikes at Pullman in 1894.

The city, although proud of its reputation for brawling lustiness, also has been the center of Middle Western culture. Most notable in the development of American thought and art was the World's Columbian Exposition of 1893. One of the architects at the fair was Louis H. Sullivan who, with D. H. Burnham, John W. Root, Frank Lloyd Wright, and others, made Chicago a leading architectural center; it was here that a distinctive U.S. contribution to architecture, the skyscraper, came into being. Chicago's continuing interest in this type of structure is seen in the John Hancock Center (1968), the Standard Oil building (1973), and the Sears Tower (1974).

The first decade of the 20th cent. saw the development of many agencies concerned with civic improvement. However, between World War I and 1933, Chicago earned unenviable renown as the home ground of gangsters—Al Capone being perhaps the most notorious—and its reputation for gangster warfare persisted long after that violent era had passed. Despite the worldwide depression of the 1930s, Chicago's world's fair, the Century of Progress Exposition (1933–34), proved how greatly the city had prospered and advanced. On Dec. 2, 1942, under the west stand of the Univ. of Chicago's Stagg Field, a group of scientists working on the government's atomic bomb project achieved the world's first nuclear chain reaction. With the war came a considerable growth of the Chicago metropolitan area. Chicago's many cultural attractions and points of interest help make it a popular convention city. Among the many political conventions held here were the Republican national conventions of 1952 and 1960, and the Democratic national conventions of 1952, 1956, and 1968.

Chicago, river, formed in Chicago by the junction of its North Branch (24 mi/38.6 km long) and South Branch (10 mi/16.1 km long), and flowing SE via a canal into the Des Plaines River at Lockport, Ill. The river formerly flowed east, then northeast via a channel, into Lake Michigan. Its course was reversed by the Chicago Sanitary and Ship Canal, built (1892–1900) on the South Branch to prevent the pollution of Lake Michigan by Chicago's sewage; locks prevent the river from entering the lake.

Chicago Heights, city (1978 est. pop. 39,300), Cook co., NE Ill., S of Chicago; settled in the 1830s, inc. as a city 1901. It is an industrial community where steel, automobile bodies, castings, railroad cars, and chemicals are manufactured.

Chicago Portage Railroad National Historic Site: *see* National Parks and Monuments Table.

Chi·chén It·zá (chē-chān' ēt'sä), city of the ancient Maya, central Yucatán, Mexico. It was founded c.514 around two large cenotes, or natural wells. After being defeated by Mayapán in 1194, the Itzá abandoned the city. Spanning two great periods of Maya civilization, Chichén Itzá shows both Classic and Post-Classic architectural styles. The Classic style is massive, with heavy decorative sculpture and cramped interiors. The later buildings have plainer, more austere lines, with the sculpture based on the Mexican feathered-serpent motif and columns. Toltec influence is strong.

Chich·es·ter (chĭch'ĭ-stər), municipal borough (1973 est. pop. 20,940), administrative center of West Sussex, S England. Chichester is an agricultural and yachting center and has some light industry. Called Regnum by the Romans, it was conquered by Ælla and his sons, who landed near Selsey in 477 and later (c.491) founded the kingdom of the South Saxons. In the Middle Ages Chichester was an important port, trading in wheat and wool. A portion of the medieval walls still stands.

Chi·chi·bu (chē'chē-bōō), city (1975 pop. 61,798), Saitama prefecture, central Japan, on the Ara River. It is a center for agricultural products and for the manufacture of silk fabrics. The city's Shinto shrine is a major tourist attraction.

Ch'i-ch'i-ha-erh (chē'chē'här') or **Tsi·tsi·har** (tsē'tsē'här'), city (1970 est. pop. 1,500,000), S central Heilungkiang prov., China, a port on the Nen River near the Great Khingan Mts. It is a processing center for soybeans, grain, and sugar beets. Manufactures include locomotives, machine tools, paper products, and cement. Ch'i-ch'i-ha-erh was founded in 1691 as a Chinese fortress and was formerly the capital of Hokiang and Heilungkiang provs.

Chick·a·hom·i·ny (chĭk-ə-hŏm'ī-nē), river, c.90 mi (145 km) long, rising NW of Richmond, Va., and flowing SE to the James River. In the Civil War there was heavy fighting (1862) along its banks.

Chick·a·mau·ga (chĭk'ə-mô'gə), city (1970 pop. 1,842), Walker co., NW Ga., S of Chattanooga, Tenn., in an agricultural area; inc. 1891. Nearby, at Chickamauga Creek, Confederate troops led by Gen. Braxton Bragg defeated a Union army (Sept. 19–20, 1863). Fort Oglethorpe is in the vicinity.

Chick·a·saw (chĭk'ə-sô'). **1.** County (1970 pop. 14,969), 505 sq mi (1,308 sq km), NE Iowa, in a rolling prairie agricultural area drained by the Cedar, Wapsipinicon, and Little Cedar rivers; formed 1851; co. seat New Hampton. Corn, hogs, cattle, grain, and soybeans are produced. It has limestone quarries and sand and gravel pits. **2.** County (1970 pop. 16,805), 506 sq mi (1,310.5 sq km), NE central Miss., drained by the Yalobusha River and tributaries of the Tombigbee River; formed 1836; co. seats Houston and Okolona. In a timber and stock-raising area, its agriculture includes cotton, corn, and dairy products. There are clay and iron deposits here. It includes part of Tombigbee National Forest.

Chickasaw Bayou, arm of the Mississippi River, W Miss., N of Vicksburg. In the Battle of Chickasaw Bluffs on its heights (Dec. 29, 1862), Gen. Sherman's attempt to take Vicksburg was thwarted.

Chick·a·sa·whay (chĭk'ə-sô'wā), river, c.210 mi (340 km) long, SE Miss., rising as Okatibbee Creek N of Meridian and flowing S to join the Leaf River and form the Pascagoula River.

Chick·a·sha (chĭk'ə-shā), city (1978 est. pop. 14,800), seat of Grady co., S central Okla., on the Washita River; inc. 1898. It lies in an agricultural and oil-producing area. Manufactures include mobile homes, transistor components, lenses, and shock absorbers.

Chi·cla·yo (chē-klä'yō), city (1972 pop. 148,932), capital of Lambayeque dept., NW Peru. On the coastal desert between the Andes and the Pacific, Chiclayo may go years at a time with no rainfall. However, by utilizing short Andean streams for irrigation, Chiclayo raises considerable sugar cane and a major part of the country's rice.

Chi·co (chē'kō), city (1978 est. pop. 25,200), Butte co., N Calif., in a region noted for its almond production; inc. 1872. Principal manufactures are processed almonds, matches, and wood products. California State Univ. at Chico and a U.S. botanical experiment station are in the city. Lassen Volcanic National Park is to the northeast.

Chic·o·pee (chĭk'ə-pē), industrial city (1978 est. pop. 55,000), Hampden co., SW Mass., at the confluence of the Chicopee and Connecticut rivers; settled c.1641, set off from Springfield 1848, inc. as a city 1890. Rubber and rubber products, sporting goods, machinery, and firearms are among the city's manufactures.

Chi·cot (shē'kō), county (1970 pop. 18,164), 647 sq mi (1,675.7 sq km), SE Ark., bounded on the S by the La. border, on the E by the Mississippi River; formed 1823; co. seat Lake Village. Its agriculture includes cotton, corn, oats, hay, pecans, and truck crops. Lumber milling and cotton ginning are done. An abandoned channel of the Mississippi has been converted into Lake Chicot, a hunting and fishing resort area.

Chi·cou·ti·mi (shĭ-kōō'tĭ-mē'), city (1976 pop. 57,737), S Que., Canada, at the confluence of the Chicoutimi and Saguenay rivers. The city is the cultural and economic center of the Saguenay area. It has aluminum plants and pulp and paper mills. A Jesuit mission was established here in 1676.

Chiem·see (kēm'zā'), lake, 31 sq mi (80.3 sq km), SE West Germany, SE of Munich. The largest lake entirely within West Germany, it has many resorts along its shores. On the largest of three islands is a palace built by Louis II of Bavaria in imitation of Versailles.

Chieng·mai (jē-ĕng'mī'): *see* Chiangmai, Thailand.

Chie·ti (kyĕt'ē), city (1976 est. pop. 55,530), capital of Chieti prov., Abruzzi region, central Italy, on the Pescara River, near the Adriatic Sea. It is a commercial and industrial center. Manufactures include textiles, iron goods, and construction materials. The city occupies the site of the Roman Teate Marrucinorum, of which ruins remain. Chieti was in the duchy of Benevento (7th cent.), fell to the Normans (1078), and thereafter was in the kingdom of Naples.

Chi·ga·sa·ki (chē'gə-sä'kē), city (1976 est. pop. 156,630), Kanagawa prefecture, central Honshu, Japan, on Sagami Bay. It is a fashionable resort with a large electronics industry.

Chig·nec·to (shĭg-nĕk'tō), isthmus connecting N.S., Canada, with the Canadian mainland, between Chignecto Bay and Northumberland Strait. It is c.17 mi (27 km) across at its narrowest point.

Chig·well (chĭg'wĕl), urban district (1973 est. pop. 54,220), Essex, SE England. It is a residential suburb of London. Portions of Epping and Hainault forests are in the district. The Chigwell public school was founded in 1629. Part of the urban district was included in Redbridge, a borough of Greater London, in 1965.

Chih·li, Gulf of (chē'lē'), China: *see* Po Hai.

Chi·hua·hua (chē-wä'wä), city (1974 est. pop. 327,300), capital of Chihuahua state, N Mexico. It lies in a valley almost encircled by hills. Chihuahua is the only large rail and commercial center of a vast northern area. Although agriculture is important, the city's economy depends chiefly on nearby mines; smelting and other mining processes constitute the main industries. Founded in the early 18th cent., Chihuahua prospered despite Indian raids. The city was occupied briefly by U.S. forces in 1846 and served as the headquarters of Benito Juárez until French troops took it in 1865; it now has many American residents. There are several good examples of 18th cent. colonial architecture.

Chil·dress (chĭl'drĭs), county (1970 pop. 6,605), 701 sq mi (1,815.6 sq km), N Texas, in the rolling prairies of the SE Panhandle and drained by the Prairie Dog Town Fork of the Red River; formed 1876; co. seat Childress. In a livestock, dairying, and agricultural area yielding cotton, wheat, grain, sorghums, alfalfa, fruit, and truck crops, it has some light industry. It acquired part of Harmon co., Okla., when the 100th meridian was relocated in 1930.

Childress, city (1970 pop. 5,408), seat of Childress co., extreme N Texas, in the Panhandle SE of Amarillo; inc. 1888. It is a railroad division point and a market and processing center for a region producing wheat, cotton, poultry, and milk.

Chil·e (chĭl'ē, chē'lā), republic (1972 pop. 10,044,940), 292,256 sq mi (756,943 sq km), S South America, W of the continental divide of the Andes. Santiago is the capital and the largest city. A long narrow strip of land (no more than c.265 mi/425 km wide) between the Andes and the Pacific Ocean, Chile stretches c.2,880 mi (4,635 km) from near lat. 18° S to Cape Horn (lat. 56° S), including at its southern end the Strait of Magellan and Tierra del Fuego, an island shared with Argentina. Chile is bordered by Peru on the north, Bolivia on the northeast, and Argentina on the east. In the Pacific Ocean, which forms the nation's western and southern borders, are Chile's several island possessions, including Easter Island, the Juan Fernández islands, and the Diego Ramírez islands. Chile also claims a sector of Antarctica.

The country is composed of three distinct and parallel natural

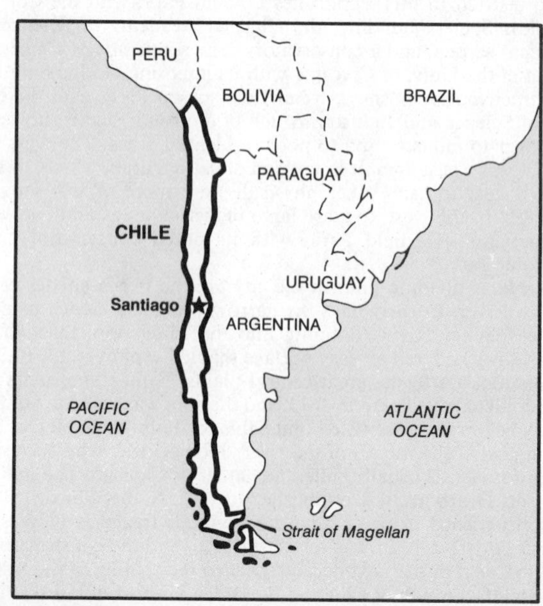

regions—from east to west, the Andes, the central lowlands, and the Coast Ranges. Chile is located along an active zone in the earth's crust and experiences numerous earthquakes, some of great magnitude. The climate varies from hot desert in the north through Mediterranean-type in the central portion to the cool and humid marine west coast type in the south. The rivers of Chile are generally short and swift flowing, rising in the well-watered Andean highlands and flowing generally west to the Pacific Ocean.

In northern Chile is the sunbaked Desert of Atacama, which, toward the south, gradually becomes a semiarid steppe with limited vegetation. The barren landscape of the north extends from the coast to the Andes, where snow-capped peaks tower above the desert. The middle portion of the country has fertile soils and is the nation's most populous and productive region. Between the Andes and the Coast Ranges is the Vale of Chile, a long fertile valley divided into basins by Andean spurs. Southern Chile is cold and humid, with dense forests, heavy rainfall, snow-covered peaks, glaciers, and islands. Because of subsidence of the earth's crust, the Coast Ranges and the central lowlands have been partially submerged, forming the extensive archipelago of southern Chile, an area of craggy islands (notably Chiloé), numerous channels, and deep fjords.

Chile is divided into 25 provinces:

NAME	CAPITAL	NAME	CAPITAL
Aconcagua	San Felipe	Llanquihue	Puerto Montt
Antofagasta	Antofagasta	Magallanes	Punta Arenas
Arauco	Lebu	Malleco	Angol
Atacama	Copiapó	Maule	Cauquenes
Aysén	Puerto Aysén	Ñuble	Chillán
Bío-Bío	Los Ángeles	O'Higgins	Rancagua
Cautín	Temuco	Osorno	Osorno
Chiloé	Ancud	Santiago	Santiago
Colchagua	San Fernando	Talca	Talca
Concepción	Concepción	Tarapáca	Iquique
Coquimbo	La Serena	Valdivia	Baldivia
Curicó	Curicó	Valparaíso	Valparaíso
Linares	Linares		

Economy. Chile's economy is based on the export of minerals, which accounts for more than 85% of the total value of exports. The country has great potential for the development of hydroelectric power, which already accounts for more than half of its electrical output. Although agriculture is the main occupation of about a third of the population, it only accounts for about 10% of the national wealth and produces less than half of the domestic needs; the production of an adequate food supply remains Chile's major economic problem. The country has a growing wine industry. Sheep raising is the chief pastoral occupation, providing wool and meat for domestic use and for export. Fishing is an important economic activity. Since World War I, Chile has developed an industrial capacity to process its raw materials and to manufacture various consumer goods. The major industrial products are processed food, fish meal, textiles, iron and steel, paper, lumber, chemicals, and leather goods.

History. Before the arrival of the Spanish in the 16th cent., the Araucanian Indians had long been in control of the land. Diego de Almagro, who was sent by Francisco Pizarro from Peru to explore the southern region, led a party of men through the Andes into the central lowlands of Chile but was unsuccessful (1536) in establishing a foothold there. In 1540 Pedro de Valdivia marched into Chile and, despite stout resistance from the Araucanians, founded Santiago (1541). Although Chile was unattractive to the Spanish because of its isolation from Peru and its lack of precious metals (copper was discovered much later), the Spanish developed a pastoral society here. During the long colonial era, the mestizos became a tenant farmer class; although technically free, most were in practice bound to the soil. During most of the colonial period Chile was a captaincy general dependent upon the viceroyalty of Peru, but in 1778 it became a separate division virtually independent of Peru.

The movement toward independence began in 1810 under the leadership of Juan Martínez de Rozas and Bernardo O'Higgins. O'Higgins formally proclaimed Chile's independence Feb. 12, 1818, at Talca and established a military autocracy that characterized the republic's politics until 1833. During this time the British expatriot Lord Cochrane, commanding the Chilean navy, cleared (1819–20) the coast of Spanish shipping. Since the colonial aristocracy and the clergy had been discredited because of royalist leanings, the army, plus a few intellectuals, established a government devoid of democratic forms. Yet with the centralistic constitution of 1833, fashioned largely by Diego Portales on Chile's particular needs, a foundation was laid for the gradual emergence of parliamentary government and a long period of stability. Chileans obtained the right to work the nitrate fields in the Atacama, which then belonged to Bolivia. Trouble over the concessions led in 1879 to open war. Chile was the victor and added valuable territories taken from Bolivia and Peru; a long-standing quarrel also ensued, the Tacna-Arica Controversy, which was finally settled in 1929. Chile's serious border troubles with Argentina ended in 1904 and was symbolized by the dedication of the Christ of the Andes.

With the exploitation of nitrate and copper by foreign interests, chiefly the United States, prosperity continued. The Transandine Railway was completed (1910), and many more railroads were built. Industrialization, which soon raised Chile to a leading position among South American nations, was begun. Meanwhile, internal struggles between the executive and legislative branches of the government intensified. A congressional dictatorship (with a figurehead president and cabinet ministers appointed by the congress) controlled the government until the constitution of 1925, which provided for a strong president. Although Chile enjoyed economic prosperity between 1926 and 1931, it was hard hit by the world economic depression, largely because of its dependence on mineral exports and fluctuating world markets. The rise of the laboring classes was marked by unionization, and there were many Marxists who advocated complete social reform. The struggle between radicals and conservatives led to a series of social experiments and to counterattempts to suppress the radicals (especially the Communists) by force. Chile broke relations with the Axis (1943) and declared war on Japan in 1945.

Chronic inflation and repeated labor crises characterized the postwar period. In the 1964 presidential election and in the 1965 congressional elections, the Christian Democratic party won overwhelming victories over the Socialist-Communist coalition. Advances were made in land reform, education, housing, and labor. The government assumed a controlling interest in U.S.-owned copper mines while cooperating with U.S. companies in their management and development. In 1970 Salvador Allende Gossens, head of the Popular Unity party, a coalition of leftist political parties, became the first Marxist to be elected president by popular vote in Latin America. Allende nationalized many private companies, instituted programs of land reform, and sought closer ties with Communist countries. Continuing widespread domestic problems, including spiraling inflation, lack of food and consumer goods, and stringent government controls, led to a series of violent strikes and demonstrations. In Sept., 1973, a military coup resulted in Allende's death (by suicide, according to the military junta that succeeded him) and in the execution, detention, or expulsion from Chile of thousands of people. Gen. Augusto Pinochet Ugarte took control of the country, promising a more moderate economic policy and the restoration of a pro-Western foreign policy. However, the economy continued to deteriorate, even though the government sought to return private enterprise to Chile. Work proceeded on the drafting of Chile's third constitution, which was to include articles preventing the election of a minority government. In June, 1974, Pinochet assumed the position of head of state.

Chi-lin (jē'lĭn') or **Ki·rin** (kē'rĭn'), city (1970 est. pop. 1,200,000), central Kirin prov., China, on the Sungari River. It is a shipping port and a commercial and industrial center, with large chemical plants. Oil is refined, and fertilizer, cement, lumber, and sugar are also produced. Chi-lin was the capital of Kirin prov. until 1954.

Chil·ko Lake (chĭl'kō), 40 mi (64.4 km) wide and 1 to 4 mi (1.6–6.4 km) wide, SW British Columbia, Canada, NNW of Vancouver in the Coast Range at the foot of Good Hope Mt. and Mt. Queen Bess. It is drained by the Chilko River, c.60 mi (95 km) long.

Chil·koot Pass (chĭl'kōōt'), alt. c.3,500 ft (1,070 m), in the Coast Mts., on the British Columbia–Alaska line. It was long used by the Chilkoot Indians as a link between the Pacific coast and the Yukon River valley. After the Klondike gold strike (1896), the pass became an important route to the interior.

Chil·lán (chē-yän'), city (1970 pop. 87,555), capital of Ñuble prov., S central Chile. Located in Chile's central valley, the city is a leading agricultural and commercial center. Founded in the 16th cent., it was destroyed twice (1751 and 1939) by earthquakes.

Chil·li·coth·e (chĭl'ĭ-kŏth'ē). **1.** City (1970 pop. 6,052), Peoria co., N central Ill., on the Illinois River NNE of Peoria, in a farm and bituminous-coal area; founded 1836, inc. 1861. **2.** City (1975 est.

pop. 9,400), seat of Livingston co., N central Mo., near the Grand River NE of Kansas City; laid out 1837. It is the trade center of a farm area, with coal and limestone deposits nearby. **3.** City (1978 est. pop. 23,500), seat of Ross co., S central Ohio, on the Scioto River; inc. 1802. It is the trade and distribution center of a farming area that specializes in raising cattle and hogs and growing corn. Long noted for its large paper mills, Chillicothe also manufactures aluminum cooking utensils, shoes, floor tiles, and railroad-car springs. Founded in 1796 by settlers from Virginia, in 1800 it became the capital of the Northwest Territory; from 1803 to 1810 and from 1812 to 1816 it was the capital of Ohio. Chillicothe grew in the 19th cent. as an inland port on the Ohio and Erie Canal.

Chi·lo·é (chē-lō-ā'), island, 3,241 sq mi (8,394.2 sq km), a part of Chiloé prov., off S Chile. It is separated from the mainland by the Corcovado and Ancud gulfs and the Chacao channel. It is the largest of the Chilean islands and the only one that has been successfully settled. A rainy climate favoring the growth of wet and dense evergreen forests makes it one of the world's last virgin frontiers. The settlers raise wheat and potatoes and export timber. Wrested from the Indians by the Spanish in 1567, Chiloé was the last stronghold of Spanish royalists, who were not driven out until 1826.

Chiltern Hundreds, the obsolete (since the 19th cent.) administrative districts of Stoke, Burnham, and Desborough in Buckinghamshire, S central England. The stewardship of the Chiltern Hundreds is an obsolete office with only a nominal salary. It is, however, legally an office of profit under the crown and, as such, may not be held by a member of Parliament. Since members of Parliament may not resign, "applying for the Chiltern Hundreds" or for the similarly obsolete stewardship of the Manor of Northstead is the method by which a member gives up his seat.

Chil·ton (chĭl'tən), county (1970 pop. 25,180), 699 sq mi (1,810.4 sq km), central Ala., in an agricultural area bounded on the E by the Coosa River and drained by the Mulberry River; formed 1868 as Baker co., renamed 1874; co. seat Clanton. It has cotton and truck farms and lumber mills. Part of Talladega National Forest is in the southwest.

Chilton, city (1970 pop. 3,030), seat of Calumet co., E Wis., E of Lake Winnebago on a branch of the Manitowoc River; settled 1847, inc. 1877. It is a trade and processing center for a dairy and farm area. Metal and wood products are made.

Chi-lung (jē'lŏŏng'), **Ki·lung,** or **Kee·lung** (both: kē'-), city (1975 est. pop. 341,400), N Taiwan, on the East China Sea. It is the principal port and naval base of Taiwan. Shipbuilding is an important industry. Chemicals, machinery, fertilizers, and marine products are also produced. Coal and gold are mined nearby. Occupied by the Spanish in 1626, it passed (1641) to the Dutch, who lost it to invading Chinese in 1662. The port was opened to Western trade in 1860. Captured by the Japanese in 1895, it remained under their rule until 1945.

Chim·bo·razo (chĭm'bō-rä'zō, chĕm'bō-rä'sō), inactive volcano, 20,561 ft (6,271.1 m) high, central Ecuador. Its summit is always snow-capped. Explored by Alexander von Humboldt in 1802, it was first scaled in 1880 by Edward Whymper.

Chim·kent (chĭm-kyĕnt'), city (1976 est. pop. 296,000), capital of Chimkent Oblast, Central Asian USSR, in Kazakhstan, on the Turkistan-Siberia RR. It has large zinc and lead smelters and machine, chemical, and textile industries. Founded in the 12th cent., Chimkent was taken by Russia in 1864.

Chim·ney Rock National Historic Site (chĭm'nē): *see* National Parks and Monuments Table.

Chi·na, Great Wall of (chī'nə), fortifications, c.1,500 mi (2,415 km) long, winding across N China from Kansu prov. to Hopeh prov. on the Yellow Sea. The wall, running mostly along the southern edge of the Mongolian plain, was erected to protect China from northern nomands. It is an amalgamation of many walls built in ancient times; the first unified wall was built in the 3rd cent. B.C. by the Ch'in dynasty. Laborers were conscripted from all over China to build it, and many of them died during the project. The wall's present form dates substantially from the Ming dynasty (1368–1644). It averages 25 ft (7.6 m) in height and is 15 to 30 ft (4.6–9.2 m) thick at the base, sloping to 12 ft (3.7 m) at the top.

China, People's Republic of, country (1976 est. pop. 853,000,000), 3,691,502 sq mi (9,560,990 sq km), E Asia; capital Peking. The most populous country in the world and the second largest (after the USSR), China has a 4,000-mi (6,436-km) coast that fronts on the Yellow Sea, the East China Sea, and the South China Sea. It is bounded on the east by the USSR and North Korea, on the north by the USSR and the Mongolian People's Republic, on the west by the USSR and Afghanistan, and on the south by Pakistan, India, Nepal, Sikkim, Bhutan, Burma, Laos, and Vietnam.

China is divided into the following geographic regions: the 12,000-ft-high (3,660-m) Tibetan plateau, bounded in the north by the Kunlun mountain system; the Tarim and Dzungarian basins of Sinkiang, separated by the Tien Shan Mts.; the vast Inner Mongolian tableland; the eastern highlands and central plain of Manchuria; and what has been traditionally called China proper. This last region, which contains some four fifths of the country's population, falls into three divisions. North China, which coincides with the Huang Ho (Yellow River) basin and is bounded in the south by the Tsingling Mts., includes the loess plateau of the northwest, the northern China plain, and the mountains of the Shantung peninsula. Central China, watered by the Yangtze River, includes the basin of Szechwan, the central Yangtze lowlands, and the Yangtze delta. South China includes the plateau of Yünnan and Kweichow and the valleys of the Si and Canton rivers. The Tsingling Mts. are the major dividing range not only between semiarid northern China and the more humid central and southern China but also between the grain-growing economy of the north and the rice economy of the south.

China is divided into 21 provinces, 5 autonomous regions, and 2 independent cities:

NAME	CAPITAL	NAME	CAPITAL
Anhwei	Ho-fei	Liaoning	Shen-yang
Chekiang	Hangchow	Ninghsia Hui Autonomous Region	Ying-ch'uan
Fukien	Fu-chou		
Heilungkiang	Harbin		
Honan	Cheng-chou		Peking (independent city)
Hopeh	Shih-chia-chuang		
Hunan	Ch'ang-sha		Shanghai (independent city)
Hupeh	Wu-han		
Inner Mongolian Autonomous Region	Hu-ho-hao-t'e	Shansi	T'ai-yüan
		Shantung	Chi-nan
		Shensi	Hsi-an
Kansu	Lan-chou	Sinkiang Uigur Autonomous Region	Wu-lu-mu-ch'i
Kiangsi	Nan-ch'ang		
Kiangsu	Nanking		
Kirin	Ch'an-ch'un	Szechwan	Ch'eng-tu
Kwangsi Chuang Autonomous Region	Nan-ning	Tibet Autonomous Region	Lhasa
Kwangtung	Canton	Tsinghai	Hsi-ning
Kweichow	Kuei-yang	Yünnan	K'un-ming

Economy. Agriculture is by far the leading occupation in China, involving about 80% of the population, although extensive rough, high terrain and large arid areas—especially in the west and north—limit cultivation to only about 11% of the land surface. Agricultural production is largely restricted to the east. China is the world's largest producer of rice, sweet potatoes, kaoliang, millet, barley, peanuts, and tea. Those, together with wheat (in which China ranks third in world production), other grains, corn, soybeans, and potatoes, are the most important crops. Cotton is the most valuable cash crop, followed by oilseeds, silk, tea, tobacco, ramie, jute, hemp, sugar cane, and sugar beets. Livestock raising on a large scale is confined to the border regions and provinces in the north and west; it is mainly of the nomadic pastoral type. China ranks third in world production of sheep and fifth in cattle production. Hogs and poultry are widely raised in China proper, furnishing important export staples, such as hog bristles and egg products. Fish supply most of the animal protein in the diet, and both inland and marine fishing are important.

China is one of the world's major mineral-producing countries; there has been extensive exploration since 1950 and significant new deposits have been found. Coal is the most abundant mineral (China ranks with the United States and the USSR in production and reserves); high-quality, easily-mined coal is found throughout the country, but especially in the north and northeast. China also has extensive iron-ore deposits. China used to import about 90% of its petroleum, but new fields were discovered in the 1960s, and the country is approaching self-sufficiency in crude oil. Refining operations are being improved. Massive deposits off the coasts are believed to exceed all the world's known oil reserves. China's leading export minerals are tungsten (China has the world's largest supply), antimony, tin, molybdenum, bismuth, mercury, magnesite, and salt. China is among the world's three top producers of tin, tungsten, antimony, and magnesite, and ranks second (after the United States) in the production of salt, seventh in manganese, and eighth in lead ore. There are large deposits of uranium in the northwest, especially in Sinkiang; new mines have also opened in Kiangsi and Kwangtung provs. Aluminum is found in many parts of the country. China also has deposits of gold, zinc, copper, fluorite, asbestos, phosphate rock, pyrite, and sulfur.

Important industrial products are manufactures that serve agriculture, such as farm machinery and fertilizers, as well as machine tools, iron and steel, textiles, processed foods, and building materials. Before 1945 heavy industry was concentrated in the northeast, but important centers have now been established in other parts of the country. Brick, tile, cement, and food-processing plants are found in almost every province. Shanghai and Canton are the traditionally great textile centers, but many new mills have been built, concentrated mostly in the cotton-growing provinces of northern China and along the Yangtze River. The domestic handicraft industry produces most of the consumer goods and such export products as porcelain and lacquer articles. Rivers and canals (notably the Grand Canal, which connects the Huang Ho and the Yangtze rivers) remain important transportation arteries. The east and northeast are well served by railroads and highways, and there are now major rail and road links with the interior.

History. The fossils of *Sinanthropus pekingensis* found in northern China are the earliest discovered protohuman remains in northeast Asia. About 20,000 years ago, after the last glacial period, modern man appeared in the Ordos desert region. The subsequent culture shows marked similarity to that of the higher civilizations of Mesopotamia, and some scholars argue a Western origin for Chinese civilization. However, since the 2nd millennium B.C. a unique and fairly uniform culture has spread over almost all of China. The substantial linguistic and ethnological diversity of the south and the far west result from their having been infrequently under the control of central government. China's history is traditionally viewed as a continuous development with certain repetitive tendencies. The area under political control tends to expand from the east Huang Ho and Yangtze basins, the heart of Chinese culture, and then, under outside military pressure, to shrink back. Conquering barbarians from the north and the west supplant native dynasties, take over Chinese culture, lose their vigor, and are expelled in a surge of national feeling. Following a disordered and anarchic period a new dynasty may arise. Its predecessor, by engaging in excessive warfare, tolerating corruption, and failing to keep up public works, has forfeited the right to rule—in the traditional view, he has lost "the mandate of Heaven." The administrators change, central authority is re-established, public works constructed, taxation modified and equalized, and land redistributed. After a prosperous period disintegration reappears, inviting barbarian intervention or native revolt.

Although traditionally supposed to have been preceded by the semilegendary Hsia dynasty, the Shang dynasty (c.1523-1027 B.C.) is the first in documented Chinese history. During the succeeding, often turbulent, Chou dynasty (c.1027-256 B.C.), Confucius, Lao-tze, and Mencius lived, and the literature that until recently formed the basis of Chinese education was written. The use of iron was the main material advance. The semibarbarous Ch'in dynasty (221-207 B.C.) first established the centralized imperial system that was to govern China during stable periods. The Great Wall was begun in this period. The native Han dynasty period (202 B.C.-A.D. 220), traditionally deemed China's imperial age, is notable for long peaceable rule, expansionist policies, and great artistic achievement. The Three Kingdoms period (A.D. 220-65) opened four centuries of warfare among petty states and of invasions of the north by the barbarian Hsiung-nu (Huns). In this inauspicious time Buddhism, which had earlier entered from India, and Taoism, a native cult, grew and seriously endangered Confucianism. Indian advances in medicine, mathematics, astronomy, and architecture were adopted. Art, particularly figure painting and decoration of Buddhist grottoes, flourished. Feudalism partly revived under the Tsin dynasty (A.D. 265-420) with the decay of central authority.

Under the Sui (581-618) and the T'ang (618-906) a vast domain, much of which had first been assimilated to Chinese culture in the preceding period, was unified. The civil service examination system based on the Chinese classics and a renaissance of Confucianism were important developments of this brilliant era. Its fresh and vigorous poetry is especially noted. The end of the T'ang was marked by a withdrawal from conquered border regions to the center of Chinese culture. The period of the Five Dynasties and the Ten Independent States (906-60), chaotic and depraved, was followed by the Sung dynasty (960-1279), a time of scholarly studies and artistic progress, marked by authentication of the Confucian literary canon and the improvement of printing techniques through the invention of movable type. The poetry of the Sung period was derivative, but a new popular literary form, the novel, appeared at that time. Neo-Confucianism developed systematically. Gunpowder was first used for military purposes in this period.

While the Sung ruled central China, barbarians created northern empires that were swept away by the Mongols under Genghis Khan. His grandson Kublai Khan, founder of the Yüan dynasty (1260-1368), retained Chinese institutions. The great realm of Kublai was described in all its richness by one of the most celebrated of all travelers, Marco Polo. Improved roads and canals were the dynasty's main contributions to China. The Ming dynasty (1368-1644) set out to restore Chinese culture by a study of Sung life. Its initial territorial expansion was largely lost by the early 15th cent. European trade and European infiltration began with Portuguese settlement of Macao in 1557 but immediately ran into official Chinese antiforeign policy. Meanwhile the Manchu peoples advanced steadily south in the 16th and the 17th cent. and ended with complete conquest of China by 1644 and with establishment of the Ch'ing (Manchu) dynasty (1644-1912). Under emperors K'ang Hsi (reigned 1662-1722) and Ch'ien Lung (reigned 1735-96), China was perhaps at its greatest territorial extent.

The Ch'ing opposition to foreign trade, at first even more severe than that of the Ming, relaxed ultimately, and in 1834, Canton was opened to limited overseas trade. Great Britain, dissatisfied with trade arrangements, provoked the Opium War (1839-42), obtained commercial concessions, and established extraterritoriality. Soon France, Germany, and Russia successfully put forward similar demands. The Ch'ing regime, already weakened by internal problems, was further enfeebled by the devastating Taiping Rebellion (1848-65) and Japan's military success in 1894-95. Great Britain and the United States promoted the Open Door Policy—that all nations enjoy equal access to China's trade; nonetheless, China was divided into separate zones of influence. Chinese resentment of foreigners grew, and the Boxer Uprising (1900), encouraged by Empress Tz'u Hsi, was a last desperate effort to suppress foreign influence. Belated domestic reforms failed to stem a revolution long-plotted, chiefly by Sun Yat-sen, and in 1911 the Ch'ing dynasty was overthrown.

Sun, the first president, resigned early in 1912 in favor of Yüan Shih-K'ai, who commanded the military power; Yüan established a repressive rule, which led Sun's followers to revolt sporadically. Early in World War I Japan seized the German leasehold in Shantung prov. China entered World War I on the Allied side in 1917, but at the Versailles peace conference was unable to prevent Japan from being awarded the Shantung territory. At the Washington Conference (1921-22) Japan finally agreed to withdraw its troops from Shantung and restore full sovereignty to China. The Nine-Power Treaty, signed at the Conference, guaranteed China's territorial integrity and the Open Door Policy.

Meanwhile, China was disintegrating into rival warlord states. Civil war raged between Sun's new revolutionary party, the Kuomintang, which established a government in Canton and received the support of the southern provinces, and the national government in Peking, supported by warlords in the north. Labor agitation, especially against foreign-owned companies, became more common, and resentment against Western religious ideas grew. In 1921 the Chinese Communist party was founded. Failing to get assistance from the Western countries, Sun made an alliance with the Communists and sought aid from the USSR. In 1926 Chiang Kai-shek led the army of the Kuomintang northward to victory. Chiang reversed Sun's policy of cooperation with the Communists and executed many of their leaders. Chiang established (1928) a government in Nanking and obtained foreign recognition.

A Communist government was set up in the early 1930s in Kiangsi, but Chiang's continued military campaigns forced (1934) them on the long march to the northwest, where they settled in Shensi. Japan, taking advantage of China's dissension, occupied Manchuria in 1931 and established (1932) the puppet state of Manchukuo. While Japan moved southward from Manchuria, Chiang chose to campaign against the Communists. In July, 1937, the Japanese attacked and invaded China proper. By 1940 northern China, the coastal areas, and the Yangtze valley were all under Japanese occupation. With Japan's attack (1941) on U.S. and British bases and the onset of World War II in Asia, China received U.S. and British aid.

The country was much weakened at the war's close. The end of the Japanese threat and the abolition of extraterritoriality did not bring peace to the country. The hostility between the Chinese Nationalists and the Communists flared into full-scale war as both raced to occupy the territories evacuated by the Japanese. When the Russians withdrew from Manchuria, which they had occupied in accordance with agreements reached at the Yalta Conference, they turned the Japanese military equipment in that area over to the Chinese Communists, giving them a strong foothold in what was then the industrial core of China. Elsewhere in the country, Chiang's Nationalists, supplied by U.S. arms, were generally successful until 1947, when the Communists, led by Mao Tse-tung, gained the upper hand. Peking fell to the Communists without a fight in Jan., 1949. The Communists proclaimed a central people's government on Oct. 1, 1949. The seat of the Nationalist government was moved to Taiwan in Dec., 1949. The new Communist government was immediately recognized by the USSR, and shortly thereafter by Great Britain, India, and other nations. Recognition was, however, refused by the United States.

The Communists brought the soaring inflation under control and effected a more equitable distribution of food. A land-reform program was launched, and police control was tightened. During the first five-year plan (1953–57), agriculture was collectivized and industry was nationalized. With USSR assistance, construction of many modern large-scale plants was begun, and railroads were built to link the new industrial complexes of the north and northwest. On the international scene, Chinese Communist troops took possession of Tibet in Oct., 1950. That same month Chinese forces intervened in the Korean War to meet a drive by United Nations forces toward the Manchurian border. Large-scale Chinese participation in the war persisted until the armistice of July, 1953, after which China emerged as a diplomatic power in Asia.

The Great Leap Forward, an economic program aimed at making China a major industrial power overnight, was underway by 1958. It featured the expansion of cooperatives into communes, which disrupted family life but offered a maximum use of the labor force. The program was not successful. The worst weather conditions in a century brought three successive crop failures (1959–61), with the ensuing food shortages dramatizing the dangers of neglecting agricultural development while emphasizing industrial expansion. The industrialization program, pushed too fast, resulted in the overproduction of inferior goods and the deterioration of the industrial plant. A severe blow was the termination of Soviet aid in 1960 and the withdrawal of Soviet technicians and advisers—events that revealed a growing ideological rift between China and the USSR. The rift, which began with the institution of a de-Stalinization policy by the Soviets in 1956, widened considerably after the USSR adopted a more conciliatory approach toward the West in the cold war. There were massive military buildups along the USSR-Chinese border, and border clashes erupted in Manchuria and Sinkiang. Meanwhile, the Communist government insisted upon its right to Taiwan, but the United States made clear its intention to defend that island against direct attack, having even given (1955) a qualified promise to defend the Nationalist-held offshore islands of Quemoy and Matsu as well. China's relations with other Asian nations, at first cordial, were affected by China's encouragement of Communist activity within their borders, the suppression of a revolt in Tibet (1959–60), and an undeclared border war with India in late 1962 over disputed territory. In the Vietnam War, China provided supplies, armaments, and technical assistance to North Vietnam.

In the late 1960s and early 1970s the emphasis of China's foreign policy changed from revolutionary to diplomatic; new contacts were established, and efforts were made to improve relations with many governments. China continued to strengthen its influence with other underdeveloped nations, extending considerable economic aid to countries in South America, Africa, and Asia. Important steps in Chinese progression toward recognition as a world power were the successful explosions of China's first atomic bomb (1964) and of its first hydrogen bomb (1967), and the launching of its first satellite (1970). Internal dissension and power struggles were revealed in such domestic crises as the momentous Cultural Revolution (1966–69); the death (1971) in an airplane crash of defense minister Lin Piao while he was allegedly fleeing to the Soviet Union after an abortive attempt to assassinate Mao and establish a military dictatorship; and a major propaganda campaign launched in 1973, which mobilized the masses against such widely ranging objects of attack as Lin Piao, the teachings of Confucius, and cultural exchanges with the West. Economically, the emphasis in the 1960s and early 1970s was on agriculture. After the Cultural Revolution, economic programs were initiated featuring the establishment of many small factories in the countryside and stressing local self-sufficiency. Both industrial and agricultural production records were set in 1970, and, despite serious droughts in some areas in 1972, output continued to increase steadily.

Long-standing objections to the admission of Communist China to the United Nations were set aside by the United States in 1971; that Oct., Communist delegates were seated and, despite the opposition of the United States, which favored a "two-China" membership, the Nationalist delegation was expelled. A breakthrough in the hostile relations between the United States and Communist China came with the dramatic visit of President Richard M. Nixon to Peking in Feb., 1972. Although U.S. support of Taiwan remained a sensitive issue, the visit resulted in a joint agreement to work toward peace in Asia and to develop closer economic, cultural, and diplomatic ties.

China, Republic of, or **Na·tion·al·ist China** (năsh'ən-əl-ĭst): *see* Taiwan.

Chi-nan (jē'nän') or **Tsi·nan** (tsĭ'nän'), city (1970 est. pop. 1,500,000), capital of Shantung prov., E China. It is a railroad junction and an industrial center with textile mills, food-processing establishments, machine shops, paper mills, agricultural machinery, chemicals, and fertilizer. An ancient walled city, Chi-nan was a provincial center as early as the 12th cent. It fell to the Communists in Sept., 1948, with the loss of some 75,000 Nationalist troops.

Chin·dwin (chĭn'dwĭn'), river, c.550 mi (885 km) long, rising in the hills of N Burma and flowing generally S into the Irrawaddy (of which it is the chief tributary).

Ch'ing-tao (chĭng'dou') or **Tsing·tao** (tsĭng'tou', chĭng'dou'), city (1970 est. pop. 1,900,000), SE Shantung prov., E China, on the Yellow Sea. With an excellent ice-free harbor, it is a major port and has textile mills, food-processing and tobacco-processing establishments, machine shops, paper mills, and plants making diesel locomotives and railroad cars, tires, fertilizers, rubber products, chemicals, and metal items. Leased to Germany in 1898 as part of the Kiaochow territory, Ch'ing-tao was held by the Japanese from 1914 to 1922 and was a marine and naval base for the United States from 1945 to 1949, when it was abandoned and fell to the Communists.

Ching-te-chen (jĭng'dŭ'jŭn') or **Fow·li·ang** (fo͞o'lē-äng'), city (1970 est. pop. 300,000), NE Kiangsi prov., China, on the Chang River. It is famous for its fine porcelain, made since the Han dynasty (202 B.C.–A.D. 220) from the kaolin found nearby.

Chin·hae (chĕn'hă'), city (1975 pop. 103,657), SE South Korea, on the Korea Strait. It is an important fishing port and naval base.

Chin Hills (chĭn), mountain range, W Burma, along the boundary between Burma and Assam, India. Mt. Victoria rises to 10,018 ft (3,055.5 m). The range is covered with pine and teak forests.

Chin·ju (jĭn'jo͞o'), city (1975 pop. 121,400), capital of South Kyongsang prov., S South Korea. It is a transportation and agricultural center, with industries producing food products and textiles.

Chin·kiang (jĭn′jē-äng′, chĭn′kyăng′): *see* Chen-chiang, China.

Ch'in Ling (chĭn′ lĭng′), mountain range, China: *see* Tsinling.

Chin·nam·po (chē′näm′pō′): *see* Nampo, North Korea.

Chi·no (chē′nō), city (1978 est. pop. 33,600), San Bernardino co., S Calif.; founded 1887, inc. 1910. It is the business and processing center of a diversified farming (notably dairying) area.

Chi·non (shē-nôɴ′), town (1975 pop. 5,391), Indre-et-Loire dept., W central France, in Touraine, on the Vienne River. Chinon was an important medieval town and many buildings (notably three churches) from that period are preserved. Its castle, overlooking the river, consists of three distinct fortresses built from the 11th to the 15th cent. by Philip II of France, Richard I of England, and Henry II of France (who died here in 1559).

Chi·nook (shĭ-no͝ok′, chĭ-), town (1970 pop. 1,813), seat of Blaine co., N Mont., on the Milk River NE of Great Falls, in an irrigated agricultural and livestock area; settled 1889, inc. 1901. There are natural-gas wells in the vicinity.

Chiog·gia (kyôd′jä), city (1976 est. pop. 37,800), Venetia, NE Italy, on a small island at the S end of the Lagoon of Venice (an arm of the Gulf of Venice), connected to the mainland by a bridge. In 1379–80 several naval battles were fought off Chioggia in the war between Venice and Genoa.

Chi·os (kī′ŏs), island, Greece: *see* Khíos.

Chip·ley (chĭp′lē), city (1970 pop. 3,347), seat of Washington co., NW Fla., WNW of Tallahassee, in a fruit and timber area; founded 1882. Falling Waters is nearby.

Chip·pe·wa (chĭp′ə-wä′, -wə). **1.** County (1970 pop. 32,412), 1,590 sq mi (4,118.1 sq km), E Upper Peninsula, Mich., bounded on the E by the Canadian border, on the N by Whitefish Bay, and on the S by Lake Huron; formed 1826; co. seat Sault Ste. Marie. It is drained by the Pine, Tahquamenon, and Munuscong rivers. In an agricultural (flax and potatoes), dairying, and lumbering area, it has commercial fisheries, some manufacturing, and many resorts. Part of Hiawatha National Forest and a number of state parks and recreational areas are here. **2.** County (1970 pop. 15,109), 582 sq mi (1,507.4 sq km), SW central Minn., bounded on the W by the Minnesota River and Lac qui Parle, and drained by the Chippewa River; formed 1862; co. seat Montevideo. It lies in an agricultural area that produces corn, oats, barley, livestock, and poultry. It has some industry (wood products, electronic equipment, and transportation equipment). **3.** County (1970 pop. 47,717), 1,018 sq mi (2,636.6 sq km), W central Wis., in a dairying and stock-raising area drained by the Chippewa River; formed 1845; co. seat Chippewa Falls. Shoes and plastics are manufactured. A state park is here.

Chippewa, river, c.200 mi (320 km) long, rising in several forks in the lake region of N Wis., and flowing SW to the Mississippi, which it enters at the foot of Lake Pepin.

Chippewa Falls, city (1978 est. pop. 12,400), seat of Chippewa co., W central Wis., on the Chippewa River; settled 1837, inc. as a city 1869. Originally a lumbering town, Chippewa Falls once had the world's largest sawmill. Today it is a trade and transportation center in a region of beef-cattle and dairy farms. Its industries include meat packing and the manufacture of shoes, plastics, tools, and dies.

Chip·ut·net·i·cook Lakes (chĭp′o͞ot-nĕt′ē-ko͝ok′), chain on the Maine-N.B. border, forming the international boundary for 28 mi (45.1 km). It is the source of the St. Croix River.

Chir·i·ca·hua National Monument (chĭr′ĭ-kä′wə): *see* National Parks and Monuments Table.

Chi·sa·go (shĭ-sô′gō), county (1970 pop. 17,492), 419 sq mi (1,085.2 sq km), E Minn., bounded on the E by the St. Croix River, here forming the Wis. border; formed 1851; co. seat Center City. In an agricultural region yielding dairy products, livestock, poultry, and grain, it has some industry. Plastics, tools, and transportation equipment are manufactured.

Chisholm Trail, route over which vast herds of cattle were driven from San Antonio, Texas, N to Abilene, Kansas, after the Civil War. With the development of railroads and the introduction of barbed-wire fencing, the trail fell into disuse, although traces of it can still be seen.

Chi·ta (chē-tä′), city (1976 est. pop. 290,000), capital of Chita oblast, SE Siberian USSR, at the confluence of the Chita and Ingoda rivers. Machines and food-processing equipment are manufactured. It was founded in 1653.

Chit·ta·gong (chĭt′ə-gŏng), city (1974 est. pop. 416,733), capital of Chittagong division, SE Bangladesh, on the Karnafuli River near the Bay of Bengal. An important rail terminus, it is the chief port of Bangladesh, with modern facilities for oceangoing vessels. Jute, tea, and skins and hides are the major exports. Offshore oil installations were set up during the 1960s. The city has large cotton- and jute-processing mills, tea and match factories, chemical and engineering works, an iron and steel mill, and fruit-canning, leather-processing, and shipbuilding industries. The port was known to the civilized world by the early centuries A.D. and was used by Arakan, Arab, Persian, Portuguese, and Mogul sailors. British control began in 1760. In the city are notable Hindu temples and Buddhist ruins.

Chit·ten·den (chĭt′n-dən), county (1970 pop. 99,131), 532 sq mi (1,377.9 sq km), NW Vt., bounded on the W by Lake Champlain; organized 1782; co. seat Burlington. It is drained by the Winooski and Lamoille rivers. Dairy and maple products, fruit, granite, and talc are important to its economy. Textiles, wood and metal products, building materials, and clothing are manufactured.

Chit·toor (chĭ-to͝or′), city (1971 pop. 63,041), Andhra Pradesh state, SE India, in the Poini River valley. Chittoor is a market for grain, sugar cane, and peanuts. The city is surrounded by mango and tamarind groves, and cattle are bred in the area. Chittoor was a British military post until 1884.

Chiu-chiang or **Kiu·kiang,** (both: jē-ō′jē-äng′) city (1970 est. pop. 120,000), N Kiangsi prov., China, on the Yangtze River. In a major tea-growing area, it is a large processing, marketing, and shipping point.

Chiu·si (kyo͞o′sē), town (1971 pop. 8,756), in Tuscany, central Italy, in the Apennines. Chiusi was one of the 12 sovereign towns of ancient Etruria; its semilegendary king Lars Porsena is said to have marched from there against Rome (c.500 B.C.). The town was taken by Rome (c.225 B.C.). Many Etruscan ruins have been found, including tombs dating from the 5th cent. B.C., and there is an excellent Etruscan museum. There are also Christian catacombs.

Chka·lov (chkä′ləf): *see* Orenburg, USSR.

Choc·taw (chŏk′tô). **1.** County (1970 pop. 16,589), 911 sq mi (2,359.5 sq km), W Ala., in the Black Belt bordering on Miss. and bounded on the E by the Tombigbee River; formed 1847; co. seat Butler. Cotton, corn, and peanuts are grown; livestock is raised; and lumber milling is done. **2.** County (1970 pop. 8,440), 417 sq mi (1,080 sq km), central Miss., drained by the Big Black, Noxubee, and Yockanookany rivers; formed 1833; co. seat Ackerman. In a stock-raising and lumbering area, its agriculture includes cotton, corn, lespedeza, and sweet potatoes. **3.** County (1970 pop. 15,141), 781 sq mi (2,022.8 sq km), SE Okla., bounded on the S by the Red River, here forming the Texas border, and drained by Muddy Boggy and Clear Boggy creeks and the Kiamichi River; formed 1907; co. seat Hugo. In an agricultural area yielding cotton, grain, livestock, peanuts, and corn, it has food-processing plants and some manufacturing.

Choc·taw·hatch·ee (chŏk′tə-hăch′ē), river, rising in S Ala. and flowing 140 mi (225.3 km) S into NW Fla. It empties into Choctawhatchee Bay, an inlet of the Gulf of Mexico, east of Pensacola, Fla. The bay forms part of the Gulf Intracoastal Waterway.

Choi·seul (shwä-zœl′), one of the Solomon Islands (1970 pop. 8,021), 1,170 sq mi (3,030.3 sq km), SW Pacific Ocean, SE of Bougainville Island. It is volcanic, with a peak of 2,470 ft (753.4 m) and is almost completely surrounded by a barrier reef. During World War II it was occupied in 1942 by Japan and taken by the United States in 1943.

Cho·let (shô-lā′), city (1975 pop. 52,976), Maine-et-Loire dept., W France, in Poitou, on the Maine River. Cholet, a livestock market, has textile, metallurgical, and other industries.

Cho·lu·la (chō-lo͞o′lä), city (1970 pop. 15,399), Puebla state, E central Mexico. The site of a pre-Columbian pyramid of great antiquity, the city was a Toltec center and, when the Spanish came, was an Aztec sacred city devoted to the worship of Quetzalcoatl. Hernán Cortés destroyed the city in 1519; from 5,000 to 10,000 Indians were killed. The picturesque city attracts many tourists.

Cho·mo Lha·ri (chō′mō lär′ē), peak, 23,997 ft (7,319 m) high, on the Bhutan-China border, in the Himalayas. It is sacred to the Tibetans.

Chong·ju (chŭng'jo͞o'), city (1975 pop. 192,734), capital of North Chungchong prov., W central South Korea. It is a marketing and processing center for the surrounding agricultural region.

Chon·ju (chŭn'jo͞o'), city (1975 est. pop. 311,432), capital of North Cholla prov., SW South Korea. It is a transportation and agricultural center in the heart of the country's most densely populated and richest rice-growing area. Food processing and textile manufacturing are the chief industries.

Chor·ley (chôr'lē), municipal borough (1973 est. pop. 31,800), Lancashire, NW England. Manufactures include cotton goods and cotton-mill machinery, rayon goods, rubber products, and footwear.

Chor·zów (KHô'zho͝of), city (1975 est. pop. 156,300), S Poland. A center of the Katowice mining and industrial region, it has iron, steel, and nitrogen plants, zinc foundries, and factories producing heavy machinery. It passed from Germany to Poland in 1921.

Cho·shi (chō'shē), city (1975 pop. 90,374), Chiba prefecture, central Honshu, Japan, on the Kashimada Sea at the mouth of the Tone River. It is a fishing center. Great quantities of soy are produced.

Cho·teau (shō'tō), city (1970 pop. 1,586), seat of Teton co., N Mont., on the Teton River NW of Great Falls, in a livestock and dairying area; settled 1893, inc. 1913.

Chotts, Plateau of the (shŏts), plateau region of the Atlas Mts., c.3,500 ft (1,070 m) high, N Algeria, N Africa. The plateau is c.125 mi (200 km) wide in the west, narrowing in the east to become a series of valleys. Enclosed by the Tell Atlas in the north and the Saharan Atlas in the south, the region has a semiarid climate and is dotted with salt lakes and salt flats.

Chou-shan Archipelago (jō'shän'), NE Chekiang prov., China, in the East China Sea, at the entrance to Hangchow Bay. It includes the main island of Chou-shan and about 100 lesser islands. The archipelago forms the richest fishing grounds off the China coast.

Chou·teau (shō'tō), county (1970 pop. 6,473), 3,920 sq mi (10,152.8 sq km), N central Mont., in an agricultural area drained by the Missouri, Teton, and Marias rivers; formed 1865; co. seat Fort Benton. Livestock and grain are the mainstays of its economy.

Cho·wan (chō-wŏn'), county (1970 pop. 10,764), 180 sq mi (466.2 sq km), NE N.C., in a coastal-plain area bounded on the W by the Chowan River, on the S by Albemarle Sound; formed 1672; co. seat Edenton. Corn, peanuts, soybeans, tobacco, and hogs are produced. It has fisheries, a lumbering industry, and some manufacturing.

Christ·church (krīst'chûrch', krĭs'-), municipal borough (1976 est. pop. 36,700), S central England, on Christchurch Bay at the confluence of the Avon and Stour rivers. Its industries include aircraft manufacturing and salmon fishing. Christchurch is also a resort. The town's history dates back to Anglo-Saxon times.

Christchurch, city (1976 pop. 171,987), E South Island, New Zealand, at the base of Banks Peninsula. Industries include tanning, meat-packing, and woolens manufacturing. The Univ. of Canterbury was founded in the city in 1873.

Chris·tian (krĭs'chən). **1.** County (1970 pop. 35,948), 709 sq mi (1,836.3 sq km), central Ill., bounded on the N by the Sangamon River; formed 1839 as Dane co., renamed 1840; co. seat Taylorville. It has agriculture (corn, wheat, oats, soybeans, dairy products, livestock, and poultry), commercial rose growing, and bituminous-coal mining. Clothing, soybean and paper products, tools, and feed are manufactured. **2.** County (1970 pop. 56,224), 725 sq mi (1,877.8 sq km), SW Ky., bounded on by S by the Tenn. border, on the NE by a headstream of the Pond River; formed 1796; co. seat Hopkinsville. It is drained by the Tradewater and Little rivers and a branch of the Red River. In a gently rolling agricultural area (tobacco, corn, wheat, and livestock), it has coal mines, gas wells, hardwood timber, and some manufacturing. It includes parts of Fort Campbell and Pennyrile State Forest. **3.** County (1970 pop. 15,124), 567 sq mi (1,468.5 sq km), SW Mo., in a resort area of the Ozarks drained by the James River; formed 1859; co. seat Ozark. Its agriculture includes corn, wheat, oats, strawberries, livestock, and dairy products. Lumbering is done, and textiles are manufactured. Part of Mark Twain National Forest is here.

Chris·tians·burg (krĭs'chənz-bûrg'), town (1970 pop. 7,857), seat of Montgomery co , SW Va., WSW of Roanoke; founded 1792, inc. 1916. It has stockyards and textile factories and processes vegetables and lumber.

Chris·tians·håb (krĭs'tyäns-hôp), town (1969 pop. 1,588) in Christianshåb dist. (1969 pop. 1,841), W Greenland, on Disko Bay. The town was founded in 1734. It has a shrimp-canning factory.

Chris·tian·sted (krĭs'chən-stĕd'), town (1970 pop. 3,020), chief city of St. Croix, one of the U.S. Virgin Islands. It is a shipping port for sugar and rum; tourism is the leading industry. Founded in 1733, Christiansted served briefly as capital of the Danish West Indies.

Christ·mas Island (krĭs'məs), island (1969 pop. 3,500), 60 sq mi (155.4 sq km), in the Indian Ocean S of Java. It has extensive deposits of phosphate of lime. The island was annexed by Great Britain in 1888 and became part of the former Straits Settlements in 1889. In 1958 it passed under Australian administration.

Christmas Island, largest atoll in the Pacific (1968 pop. 367), 222 sq mi (575 sq km), in the Line Islands, a part of the British colony of the Gilbert Islands. The island is worked as a copra plantation by the British government, and most of the inhabitants work in the industry. The atoll was discovered by Capt. James Cook in 1777, annexed by Great Britain in 1888, and included in the former Gilbert and Ellice Islands colony in 1919. British nuclear tests were conducted on the atoll in 1957 and 1958 and U.S. tests in 1962. The United States claims sovereignty over Christmas Island.

Ch'üan-chou (chə-wän'jō') or **Tsin·kiang** (jĭn'jē-äng'), town (1970 est. pop. 130,000), SE Fukien prov., China, on an inlet of Formosa Strait. Local handicrafts, machine tools, and fertilizer are produced.

Chu·but (cho͞o-bo͞ot'), river, c.500 mi (805 km) long, rising in the Andes of SW Argentina and flowing E across Chubut prov. to the Atlantic Ocean at Rawson. Sheep raising and fruit growing are important along the river's lower course.

Chu-chou or **Chu·chow** (both: jo͞o'jō'), town (1970 est. pop. 350,000), E central Hunan prov., China, on the Hsiang River. It is a railroad center. Trucks and fertilizers are manufactured, uranium is processed, and lead and zinc are mined nearby.

Chud·sko·ye, Lake (cho͞ot'-skə yə), or **Lake Pei·pus** (pī'pəs), c.1,390 sq mi (3,600 sq km), dividing the Estonian Republic from W Pskov oblast, NW European USSR.

Chu·gach Mountains (cho͞o'găch), one of the Pacific coastal ranges, S Alaska, extending from the St. Elias Mts., on the Alaska-Yukon border, NW to the Manuska River. Mt. Marcus Baker, 13,176 ft (4,018.7 m), is the highest peak. Rugged, with forested lower slopes and glacier-covered summits, the Chugach Mts. are a barrier for movement inland from the coast.

Chu·gu·chak (cho͞o'go͞o'chäk'): *see* T'a-ch'eng, China.

Chuk·chi Peninsula (cho͝ok'chē), NE extremity of Asia, terminating in Cape Dezhnev, Far Eastern USSR. Washed by the East Siberian and Chukchi seas in the northeast, the peninsula is the eastern extension of the Anadyr mountain range. It is included in the Chukchi National Okrug (1970 pop. 101,000); capital Anadyr. The okrug has mining (tin, lead, zinc, gold, and coal), hunting and trapping, reindeer raising, and fishing. Formed in 1930, the okrug is now part of the Magadan oblast.

Chu·la Vis·ta (cho͞o'lə vĭs'tə), city (1978 est. pop. 79,900), San Diego co., S Calif., on San Diego Bay; inc. 1911. Citrus fruits and vegetables are grown in the area, and aircraft engines and men's slacks are manufactured.

Chu·lym (cho͞o-lĭm'), river, c.1,075 mi (1,730 km) long, Krasnoyarsk Kray, S central Siberian USSR. It rises in the eastern slopes of the Kuznetsk Ala-Tau and flows north and west through Krasnoyarsk Kray and Tomsk oblast into the Ob. Its lower course is navigable.

Chun·chon (cho͞on'chŭn'), city (1975 pop. 140,521), capital of Kangwon prov., N South Korea. It is an important market town and rice-processing center. Textiles, silk yarn, and raw silk are also produced. Tungsten, mica, and fluorspar are mined nearby.

Chung·ju (cho͝ong'jo͞o'), city (1975 pop. 73,100), central South Korea. Chungju is an important agricultural center.

Chung·king (cho͝ong'kĭng') or **Ch'ung-ch'ing** (-chĭng'), city (1970 est. pop. 3,500,000), SE Szechwan prov., China, at the junction of the Yangtze and Chia-ling rivers. The commercial center of western China, it commands a large river trade. Surrounded on three sides by water, it is situated on a rock promontory; all supplies from the river front must be carried by stairway or inclined railway. A flourishing industrial city, it has railroad shops, shipyards, a large-scale integrated steel complex, cotton and silk mills, chemical and cement plants, food-processing establishments, machine shops, paper mills, and a developing motor vehicle industry. Large coal and iron mines are nearby. Chungking was opened as a treaty port in 1891. In Nov., 1937, just before the Japanese capture of Nanking in the Second Sino-Japanese War, the capital of China was transferred to Chungking, where it remained until the end of hostilities. The city was taken by the Communists on Nov. 30, 1949.

Chu·qui·ca·ma·ta (chōō'kē-kä-mä'tä), town, N Chile, on the W slopes of the Andes. At an elevation of 10,435 ft (3,182.7 m), Chuquicamata has one of the world's largest copper mines. The extensive open-pit mining of the region dates to 1915.

Chur (kōōr), city (1977 pop. 32,800), capital of Grisons canton, E Switzerland, on the Plessur River. Manufactures include foodstuffs (especially chocolate), textiles, and metal products. Chur was capital of the Roman province of Rhaetia. In the 5th cent. it became an episcopal see; the bishops were later made princes of the Holy Roman Empire.

Church·ill (chûr'chĭl), county (1970 pop. 10,513), 4,885 sq mi (12,652.2 sq km), W central Nev., in the Great Basin and watered by Lahontan Reservoir; formed 1861; co. seat Fallon. In an agricultural area (alfalfa and livestock), it has gold and silver deposits.

Churchill. 1. River, c.600 mi (965 km) long, issuing as the Ashuanipi River from Ashuanipi Lake, SW Labrador, Canada, and flowing in an arc N, then SE through a series of lakes to Churchill Falls and McLean Canyon, then NE past Goose Bay and through Melville Lake and Hamilton Inlet to the Atlantic Ocean near Rigolet. The river has the greatest hydroelectric power potential of any river in North America, and Churchill Falls is the site of one of the world's largest hydroelectric power plants. Formerly known as the Hamilton River, it was renamed (1965) in honor of Sir Winston Churchill. **2.** River, c.1,000 mi (1,610 km) long, issuing from Methy Lake, NW Sask., Canada, and flowing SE, E, and NE across the lowlands of N Sask. and N Man. to Hudson Bay at Churchill. Once a famous furtrade route, it was discovered (1619) by Jens Munck, a Scandinavian searching for the Northwest Passage. In 1717 the Hudson's Bay Company established a trading post, later called Fort Prince of Wales, near the mouth of the river. A massive stone fort replaced the post in 1732. Captured (1782) by the French under Jean La Pérouse, the fort was regained by the British and renamed Fort Churchill.

Churchill Falls, once spectacular waterfalls of the upper Churchill River, 245 ft (74.7 m) high, SW Labrador, Canada; known as Grand Falls until renamed (1965) in honor of Sir Winston Churchill. The falls were discovered (1839) by John McLean, a trader of the Hudson Bay Company, and were rediscovered in 1891, after having been generally forgotten because of their remote location. Churchill Falls is the site of the world's largest underground power plant (put into operation in June, 1972). The falls are expected to dry up as the power plant diverts greater amounts of water to drive the nearby underground turbines.

Chu·so·va·ya (chōō-sə-vī'ə), river, c.460 mi (740 km) long, E European USSR. It rises in the central Urals and flows northwest through a major industrial region to join the Kama River at Perm. The Chusovaya is navigable c.250 mi (400 km).

Chu·vash Autonomous Soviet Socialist Republic (chōō-väsh'), autonomous republic (1970 pop. 1,224,000), 7,066 sq mi (18,301 sq km), E central European USSR, in the middle Volga river valley; capital Cheboksary. The region has peat bogs and deposits of limestone, dolomite, clays, sands, and phosphorites. Grain, potatoes, flax, hemp, fruit, and sugar beets are grown, and livestock is raised. Lumbering and woodworking are important occupations. Among the republic's other industries are oil and natural gas refining, metalworking, and food and flax processing. Conquered by the Mongols in the 13th and 14th cent., the Chuvash came under Russian rule in 1552. The Chuvash Autonomous Oblast was established in 1920; it became an autonomous republic in 1925.

Chu·zen·ji (chōō-zĕn'jē), mountain lake, c.5 sq mi (13 sq km), Tochigi prefecture, central Honshu, Japan, in Nikko National Park. The lake is famed for its beauty. The Kegon waterfall (350 ft/106.8 m high) spills from the lake's outlet.

Cic·e·ro (sĭs'ə-rō), town (1978 est. pop. 61,600), Cook co., NE Ill., an industrial and residential suburb adjoining Chicago; inc. 1867.

Cie·go de Á·vi·la (syā'gō dā ä'vē-lä), city (1970 pop. 60,910), Camagüey prov., central Cuba. An important processing center in a sugarcane region, it is also Cuba's leading producer of pineapples and oranges. Cattle raising is another major industry.

Cien·fue·gos (syān-fwā'gōs), city (1970 pop. 85,248), Las Villas prov., central Cuba, a port on the Caribbean Sea. It is the marketing and processing center of a region producing sugar cane, tobacco, coffee, and rice. Established in 1819 by French emigrants from Louisiana, Cienfuegos was destroyed by a tropical storm in 1825 and later rebuilt. In 1957 members of its naval academy staged an unsuccessful revolt against Cuban dictator Fulgencio Batista. Reported Soviet efforts to build a submarine base at the Cienfuegos harbor in 1970 ceased after the U.S. government expressed strong opposition.

Ci·li·cia (sə-lĭsh'ə, -ē-ə), ancient region of SE Asia Minor, in present S Turkey, between the Mediterranean and the Taurus Mts. The area was under the domination of the Assyrian Empire before it became part of the Persian Empire. Greeks early settled on the coast, and Cilicia was hellenized to a great extent. The region became part of the Roman Empire in 67 B.C. Later Cilicia was included in the Byzantine Empire and in the 8th cent. was invaded by the Arabs. In 1080 Prince Reuben set up an Armenian state here, which became a kingdom in 1098 and is generally called Little Armenia. The Armenians maintained their independence against the Turks until 1375, when the Mamelukes conquered them.

Ci·li·cian Gates (sə-lĭsh'ən), mountain pass, S Turkey, leading across the Taurus Mts. Known to the ancients as the Pylae Ciliciae, it follows the gorge of the Gökoluk River. The gates have served for centuries as a natural highway linking Anatolia with the Mediterranean coast.

Cim·ar·ron (sĭm'ə-rŏn', -rōn'), county (1970 pop. 4,145), 1,843 sq mi (4,773.4 sq km), extreme NW Okla., in the high plains of the Panhandle, bounded on the N by Colo., on the W by N.Mex., on the S by Texas; formed 1907; co. seat Boise City. It is drained by the Cimarron and North Canadian rivers. Farming (broomcorn, wheat, barley, and corn) and ranching are important.

Cimarron, city (1970 pop. 1,373), seat of Gray co., SW Kansas, on the Arkansas River WNW of Dodge City, in a farm area; inc. 1885.

Cimarron, river, 698 mi (1,123 km) long, rising in NE N.Mex., and flowing generally E to the Arkansas River, W of Tulsa, Okla. The river winds through a thinly populated area where cattle and wheat are raised. Sections of its bed are dry during most of the year.

Cimarron, Territory of, now the Panhandle of Okla. It was settled in the early 1800s by cattle ranchers, many of them squatters. To protect their claims they attempted, in 1887, to create a separate territorial government at Beaver, Okla. After subsequent efforts toward this end failed in the U.S. Congress, Cimarron became part of the Oklahoma Territory in 1890.

Cîm·pu·lung (kəm-pōō-lōōng'), town (1974 est. pop. 27,972), S central Rumania, in Walachia, on the S slope of the Transylvanian Alps. A commercial center, it has industries producing textiles and paper. It is also a summer resort. Founded in the 12th cent. by German colonists, Cîmpulung became the capital of Walachia in the 13th cent.

Cin·cin·na·ti (sĭn'sə-năt'ē, -năt'ə), city (1978 est. pop. 390,000), seat of Hamilton co., extreme SW Ohio, on the Ohio River opposite Covington, Ky.; inc. as a city 1819. Cincinnati is the industrial, commercial, and cultural center for an extensive area including numerous suburbs in Ohio and Kentucky. It is also a port of entry with a large river front and good transportation facilities. Machine tools, transportation equipment (automobiles and parts, truck bodies, aircraft engines), radar equipment, electrical machinery, metal goods, and cosmetics are the chief manufactures. Cincinnati was founded in 1788 as Losantiville; in 1790 Arthur St. Clair, the first governor of the Northwest Territory, renamed it Cincinnati for the Society of Cincinnati, a group of Revolutionary War officers. It was the first seat of the legislature of the Northwest Territory and a busy transshipping center for early settlers. After the opening of the Ohio and Erie Canal (c.1832), the city developed as a shipping point for farm products and meat. Cincinnati suffered disastrous floods in 1884 and 1937, but Federal and state flood-control projects have now greatly reduced the danger. The Univ. of Cincinnati and several other educational institutions are in the city. William Howard Taft and his son Robert A. Taft were born in Cincinnati, where the Taft family has long been prominent. Cincinnati's landmarks are the Taft Museum; Eden Park, with the Cincinnati Art Museum; a museum of natural history; and zoological gardens.

Cinque Ports (sĭngk), name applied to an association of maritime towns in Sussex and Kent, SE England. They originally numbered five: Hastings, Romney (now New Romney), Hythe, Dover, and Sandwich. The association was informally organized in the 11th cent., and a formal charter was drawn up in the 13th cent. In the 12th cent., Winchelsea and Rye were added with privileges and duties similar to those of the founding members. Later, neighboring places were added as "limbs" or "members." The Cinque Ports reached the peak of their significance during the Anglo-French struggle in the 14th cent.

Cin·to, Mon·te (môn'tā chēn'tō), peak, 8,891 ft (2,711.8 m) high, NW Corsica, France, NW of Corte. It is the highest point on Corsica.

Cin·tra or **Sin·tra** (both: sēn'trə), town (1974 est. pop. 15,994), Lisboa dist., W Portugal, in Estremadura. The region has orange groves and vineyards as well as marble quarries, but Cintra is known primarily for its beautiful mountain location. It flourished as a Moorish city, and there are still ruins of a Moorish castle. It was permanently retaken from the Moors by Alfonso I in 1147 and thereafter was a favorite residence of the Portuguese monarchs.

Ci·pan·go (sĭ-păng'gō), poetic name for Japan. It was used by Marco Polo.

Cir·cas·si·a (sər-kăsh'ē-ə), historic region, encompassing roughly the area between the Black Sea, the Kuban River, and the Caucasus, now largely the Krasnodar kray of SE European USSR. The Circassians are a Moslem people, whose Russian name is Cherkess and whose native name is Adyge. Known in antiquity, they were Christianized in the 6th cent. A.D. but adopted Islam in the 17th cent. after coming under the rule of the Ottoman Empire. In 1829 the Ottoman Turks were forced to cede Circassia to Russia. In the many Russo-Turkish wars in the first half of the 19th cent., the Circassians bitterly fought the Russians. After the Russian conquest of the area, many Circassians migrated to Turkey (1861–64).

Cir·cle (sûr'kəl), shipping town (1970 pop. 964), seat of McCone co., E central Mont., NW of Glendive, in a grain and livestock area; inc. 1929.

Cir·cle·ville (sûr'kəl-vĭl'), city (1970 pop. 11,687), seat of Pickaway co., S central Ohio, on the Scioto River, in a farm area; inc. 1853. Corn, hogs, and poultry are processed in the city. Circleville was laid out in 1810 within the remains of a circular fort allegedly erected by Mound Builders. Its growth was spurred by the building of the Ohio and Erie Canal.

Cis·al·pine Republic (sĭs-ăl'pīn'), Italian state created by Napoleon Bonaparte in 1797 by uniting the Transpadane and Cispadane republics, which he had established (1796) N and S of the Po River. In 1802 it became the Italian Republic and in 1805, with the addition of Venetia, the Napoleonic kingdom of Italy. It was broken up by the Congress of Vienna in 1815.

Ci·thae·ron (sĭ-thē'rən), mountain range, c.10 mi (16 km) long, central Greece, between Boeotia in the N and Attica in the S. It rises to 4,623 ft (1,410 m). The range was the scene of many events in Greek mythology.

Cit·rus (sĭt'rəs), county (1970 pop. 19,196), 560 sq mi (1,450.4 sq km), central Fla., bounded on the W by the Gulf of Mexico, on the N and E by the Withlacoochee River; formed 1887; co. seat Inverness. Agriculture (corn, peanuts, citrus fruit, cattle, and hogs), fishing, and lumbering are important. It also has deposits of phosphate, limestone, and clay.

Cit·tà Vec·chia (chēt-tä' vĕk'kyä) or **Città No·ta·bi·le** (nō-tä'bē-lā), town, central Malta. It was the capital of Malta until supplanted by Valletta (1570). The town has a large 17th cent. cathedral, the old palace of the grand masters of the Knights of Malta (Knights Hospitalers), and catacombs, some of which are pre-Christian.

cit·y (sĭt'ē), densely populated urban center, larger than a village or a town, whose inhabitants are engaged primarily in commerce and industry. In the legal sense, in the United States a city is an incorporated municipality.

City of Da·vid (dā'vĭd), epithet of Bethlehem, the birthplace of David, and of Jerusalem, his capital.

City of Ref·uge National Historical Park (rĕf'yo͞oj): *see* National Parks and Monuments Table.

Ciu·dad (syo͞o-*thäth*', -däd'), for cities whose names begin thus but are not so listed, see under the following name; e.g., for Ciudad Juarez, *see* Juarez.

Ciudad Bolívar (bō-lē'vär), city (1971 pop. 103,728), capital of Bolívar state, E Venezuela, an inland port on the Orinoco River. It is the commercial center of the eastern llanos, the Orinoco basin, and the Guiana Highlands. Wood products and leather are produced, and hides, cattle, and gold are exported. The city was founded in 1764 and called Angostura. The Congress of Angostura (1819) made Simon Bolívar president of Venezuela and later in the same year decreed the formation of the republic of Greater Colombia, with Bolívar as president.

Ciudad Re·al (rā-äl'), city (1975 est. pop. 45,025), capital of Ciudad Real prov., central Spain, in New Castile, on a fertile plain between the Jabalón and Guadiana rivers. It is an agricultural market place, with farm-related industries. Ciudad Real was founded by Alfonso X of Castile in the 13th cent. and preserves some of its medieval flavor.

Ciudad Ro·dri·go (rō-*th*rē'gō, rō-drē'-), town (1970 pop. 13,320), Salamanca prov., central Spain, in León, on the Agueda River near the Portuguese border. It is a trade center for a cattle-raising area. Originally a Roman settlement, the town was abandoned and reestablished in the 12th cent. as a fortress. It has been declared a historic monument.

Ci·vi·ta·vec·chia (chē'vē-tä-vĕk'kyä, -vĕk-kē-ä), city (1976 est. pop. 47,172), in Latium, W central Italy, on the Tyrrhenian Sea. The harbor, favored by Trajan, is still the chief port of Rome. Industries include fishing and petroleum refining. The arsenal in Civitavecchia was built by Bernini, and Michelangelo directed the final stages of the construction of the powerful citadel (begun 1508, nearly destroyed in World War II).

Clack·a·mas (klăk'ə-məs), county (1970 pop. 166,088), 1,884 sq mi (4,879.6 sq km), NW Oregon, drained by the Willamette and Clackamas rivers; formed 1843; co. seat Oregon City. In a dairying and agricultural area yielding poultry, fruit, truck crops, grain, and hay, it has a lumbering industry and recreational facilities. Mount Hood National Forest is in the east.

Clack·man·nan·shire (klăk-măn'ən-shîr', -shər), former county, central Scotland, at the head of the Firth of Forth. Alloa was the county town. In 1975 Clackmannanshire became part of the new Central region.

Clac·ton (klăk'tən), urban district (1973 est. pop. 39,380), Essex, E central England. It is a seaside resort situated on high cliffs.

Clai·borne (klā'bərn). **1.** Parish (1970 pop. 17,024), 766 sq mi (1,983.9 sq km), N La., bordered on the N by Ark. and drained by Bayou D'Arbonne; formed 1828; parish seat Homer. It has oil wells, natural-gas plants, lumbering, and agriculture (cotton, corn, hay, and peanuts). **2.** County (1970 pop. 10,086), 486 sq mi (1,258.7 sq km), SW Miss., bounded on the W by the Mississippi River, here forming the La. border, and on the N by the Big Black River; formed 1802; co. seat Port Gibson. In a cotton and corn area, it also has cattle and a lumbering industry. **3.** County (1970 pop. 19,420), 445 sq mi (1,152.6 sq km), NE Tenn., drained by the Powell River and bounded on the N by the Ky. and Va. borders, on the S by the Clinch River; formed 1801; co. seat Tazewell. The Cumberland Mts. traverse the northwest section of the county. Its economy is based on coal mining, lumbering, woodworking, and farming (corn, livestock, and some tobacco). Part of Norris Reservoir is here. Cumberland Gap National Historical Park is in the north.

Claire, Lake (klâr), 545 sq mi (1, 411.6 sq km), NE Alta., Canada, W of Lake Athabasca.

Clair·ton (klâr'tən), city (1970 pop. 15,051), Allegheny co., SW Pa., an industrial suburb of Pittsburgh, on the Monongahela River; settled 1770, inc. 1903. Its extensive steelworks turn out a great variety of products. Coal mines and oil wells are also found in the area, and coke, coke by-products, and chemicals are important manufactures.

Clal·lam (klăl'əm), county (1970 pop. 34,770), 1,753 sq mi (4,540.3 sq km), NW Wash., on the Olympic Peninsula bounded on the W by the Pacific Ocean, on the N by the Strait of Juan de Fuca; formed 1854; co. seat Port Angeles. It depends mainly on fishing and lumbering. Tourism is also important. The county includes parts of Olympic National Forest and Olympic National Park.

Clan·ton (klăn'tən), city (1970 pop. 5,868), seat of Chilton co., central Ala., NW of Montgomery, in a cotton, timber, and fruit area; inc. 1873. It has cotton and planing mills.

Clare (klâr), county (1971 pop. 75,008), 1,231 sq mi (3,188.3 sq km), W Republic of Ireland, between Galway Bay and the Shannon River; co. town Ennis. The terrain is broken and hilly, with many bogs and lakes, and the coastline is especially rugged; much of the land is completely barren. The region came under the control of the Anglo-Norman Clare family in the 13th cent.

Clare, county (1970 pop. 16,695), 572 sq mi (1,481.5 sq km), central Mich., drained by the Muskegon, Tobacco, and Cedar rivers; organized 1871; co. seat Harrison. In a hunting, fishing, and resort area, it has oil and natural-gas wells, some manufacturing, and diversified agriculture (livestock, dairy products, poultry, grain, potatoes, corn, beans, and sugar beets). A state park is here.

Clare·mont (klâr'mŏnt'). **1.** City (1978 est. pop. 26,300), Los Angeles co., S Calif., in a citrus farm area at the foot of the San Gabriel Mts.; inc. 1907. It is mainly residential. The Claremont Colleges, a theological school, and a large botanical garden are there. **2.** City (1978 est. pop. 13,700), Sullivan co., SW N.H., in a farm and dairy area, on the Sugar River near its junction with the Connecticut; inc. 1764. It is a summer resort and has plants manufacturing shoes, textiles, machinery, and paper.

Clare·more (klâr'môr), city (1970 pop. 9,084), seat of Rogers co., NE Okla., NE of Tulsa; settled in the late 19th cent., inc. as a town 1896, as a city 1908. It is a health resort, with mineral springs. Will Rogers was born nearby.

Clar·en·don (klăr'ən-dən), county (1970 pop. 25,604), 599 sq mi (1,551.4 sq km), E central S.C., in an agricultural area bounded on the S by Lake Marion and drained by the Black River; formed 1856; co. seat Manning. Cotton, tobacco, and truck crops are grown. It also produces dairy products and timber.

Clarendon. 1. City (1970 pop. 2,563), seat of Monroe co., E central Ark., at the junction of the White and Cache rivers ESE of Little Rock, in a rich farm and timber area; settled c.1819. **2.** City (1970 pop. 1,974), seat of Donley co., extreme N Texas, in the Panhandle ESE of Amarillo; inc. 1901. Settled in the 1870s, it was moved to its present site in 1887 and is now a shipping and processing center for a cattle, grain, and cotton area.

Cla·rin·da (klə-rĭn'də), city (1977 est. pop. 5,312), seat of Page co., SW Iowa, on the West Nodaway River near the Mo. border ESE of Shenandoah; inc. 1866. It has some industry. Iowa Western Community College is here.

Clar·i·on (klăr'ē-ən), county (1970 pop. 38,414), 599 sq mi (1,551.4 sq km), W central Pa., in a coal-mining plateau area drained by the Clarion River and bounded on the SW by the Allegheny River, on the S by Redbank Creek; formed 1839; co. seat Clarion. Scotch-Irish settlers came here c.1800. It has iron, lumber, and oil industries, bituminous-coal mines, and some agriculture. Glass products and rubber goods are manufactured.

Clarion. 1. City (1970 pop. 2,972), seat of Wright co., N central Iowa, SW of Mason City, in a corn and poultry region; inc. 1881. **2.** Borough (1970 pop. 6,095), seat of Clarion co., W central Pa., on the Clarion River, a tributary of the Allegheny River, SE of Franklin, in a bituminous-coal region; laid out 1840, inc. 1841. Glass, lumber, and beverages are produced. Clarion State College is here.

Clark (klärk). **1.** County (1970 pop. 21,537), 878 sq mi (2,274 sq km), S central Ark., bounded on the N by De Gray Lake, on the S by the Little Missouri River, on the SE by the Ouachita River; formed 1818; co. seat Arkadelphia. It has agriculture (cotton, corn, fruit, poultry, and livestock) and stands of timber. Clothing, metal products, and boats are manufactured. **2.** County (1970 pop. 741), 1,751 sq mi (4,535.1 sq km), E Idaho, in a mountainous area bordering in the N on Mont. and drained by Medicine Lodge Creek and the Camas River; formed 1919; co. seat Dubois. Livestock (sheep, beef cattle, and horses), lumbering, and tourism are important. Part of Targhee National Forest is here. **3.** County (1970 pop. 16,216), 505 sq mi (1,308 sq km), E Ill., bounded on the E by the Ind. border, on the SE by the Wabash River, and drained by the North Fork of the Embarrass River; formed 1819; co. seat Marshall. In an agricultural area yielding corn, wheat, oats, livestock, and poultry, it has oil and natural-gas wells. Shoes, limestone products, hardware, and paint are manufactured. **4.** County (1970 pop. 75,876), 384 sq mi (994.6 sq km), SE Ind., bounded on the SE by the Ohio River, here forming the Ky. line; formed 1801; co. seat Jeffersonville. It is mainly agricultural, yielding grain, tobacco, and livestock, and has some manufacturing. **5.** County (1970 pop. 2,896), 984 sq mi (2,548.6 sq km), SW Kansas, in a gently rolling prairie area bordered on the S by Okla. and drained by the Cimarron River; formed 1885; co. seat Ashland. It has oil and gas fields, and livestock and grain are produced. **6.** County (1970 pop. 24,090), 259 sq mi (670.8 sq km), central Ky., in a gently rolling agricultural area bounded on the S by the Kentucky River and drained by several creeks; formed 1792; co. seat Winchester. In the Bluegrass, it has limestone quarries and some manufacturing. **7.** County (1970 pop. 8,260), 509 sq mi (1,318.3 sq km), extreme NE Mo., bounded on the E by the Mississippi and Des Moines rivers and drained by the Fox and Wyaconda rivers; formed 1836; co. seat Kahoka. Wheat, corn, oats, and soybeans are grown. Beef cattle and dairy products are also important. It has a mineral industry and stone, sand, and gravel pits. **8.** County (1970 pop. 273,288), 7,874 sq mi (20,393.7 sq km), SE Nev., in a mining and livestock-grazing area bordering on Ariz. and Calif. and drained by the Colorado and Virgin rivers; formed 1909; co. seat Las Vegas. It has magnesium deposits and a number of military and research installations. Tourism is of major importance. **9.** County (1970 pop. 157,115), 402 sq mi (1,041.2 sq km), W central Ohio, intersected by the Mad and Little Miami rivers; formed 1818; co. seat Springfield. It has agriculture (grain, dairy products, and livestock), limestone quarries, sand and gravel pits, and some manufacturing. George Rogers Clark Memorial Park is here. **10.** County (1970 pop. 5,515), 964 sq mi (2,496.8 sq km), E central S. Dak., in an agricultural area watered by numerous intermittent creeks; formed 1873; co. seat Clark. Potatoes, corn, wheat, oats, flax, soybeans, dairy products, and livestock are produced. **11.** County (1970 pop. 128,454), 627 sq mi (1,623.9 sq km), SW Wash., in rolling hills sloping S to the valley of the Columbia River; formed 1844; co. seat Vancouver. In an ocean-shipping and agricultural area, it has highly diversified industries. Pulp and paper products, metals, machinery, chemicals, and textiles are made. **12.** County (1970 pop. 30,361), 1,222 sq mi (3,165 sq km), central Wis., in a dairying area drained by the Black and Eau Claire rivers; formed 1853; co. seat Neillsville. Cheese is its main product. Lumbering, stock raising, and vegetable canning are also important.

Clark, city (1970 pop. 1,356), seat of Clark co., E S.Dak., W of Watertown; settled 1879, inc. 1904. It is the trade center of a potato, grain and livestock region.

Clarke (klärk). **1.** County (1970 pop. 26,724), 1,232 sq mi (3,190.9 sq km), SW Ala., in a heavily forested area in the Black Belt and bounded on the W by the Tombigee River, on the SE by the Alabama River; formed 1812; co. seat Grove Hill. Cotton, corn, and livestock are produced. Lumber milling is also important. **2.** County (1970 pop. 65,177), 125 sq mi (323.8 sq km), NE central Ga., drained by the Oconee River; formed 1801; co. seat Athens. In a piedmont agricultural area yielding cotton, corn, truck crops, poultry, and dairy products, it also has a textile industry. **3.** County (1970 pop. 7,581), 429 sq mi (1,111.1 sq km), S Iowa, in a rolling prairie agricultural area; formed 1846; co. seat Osceola. Its agriculture includes hogs, cattle, poultry, corn, and wheat. There are bituminous-coal deposits. **4.** County (1970 pop. 15,049), 697 sq mi (1,805.2 sq km), E Miss., bordering in the E on Ala., and drained by the Chickasawhay River; formed 1812; co. seat Quitman. Cotton and corn are grown. It also has oil and natural-gas fields and stands of timber. Clothing, chemicals, and plastics are manufactured. **5.** County (1970 pop. 8,102), 174 sq mi (450.7 sq km), N Va., in the N Shenandoah Valley bordered on the N by W.Va. and drained by Opequon Creek; formed 1836; co. seat Berryville. In a rich agricultural area (apples, corn, and wheat), it also has dairying and horse breeding. The Blue Ridge is in the southeast.

Clarkes·ville (klärks'vĭl'), city (1970 pop. 1,294), seat of Habersham co., NE Ga., NE of Atlanta, in a truck-farming area. It has lumber and grist mills and food-processing plants.

Clark Fork, river c.360 mi (579.2 km) long, part of the Columbia River system, rising in SW Mont. near Butte and flowing N to Deer Lodge and then NW past Missoula and on to Pend Oreille Lake in the Idaho Panhandle.

Clarks·burg (klärks'bûrg'), city (1978 est. pop. 22,800), seat of Harrison co., N central W.Va., at the confluence of Elk Creek and the West Fork of the Monongahela River; inc. 1795. It is an industrial and shipping center for an area of coal mines, oil and natural-gas fields, and grazing lands. Glass and glass products are the chief manufactures. The city was an important Union supply base in the Civil War.

Clarks·dale (klärks'dāl'), city (1970 pop. 21,673), seat of Coahoma co., NW Miss., on the Sunflower River; inc. 1882. It is a processing and distributing center for a cotton area. Its manufactures include paper, conveyor belts, house trailers, locks, and rubber products.

Clarks·ton (klärks'tən), city (1970 pop. 6,312), Asotin co., SE Wash., on the Snake River opposite Lewiston, Idaho; platted 1896, inc. 1902. It is a shipping and processing center in a grain, fruit, timber, and livestock area.

Clarks·ville (klärks'vĭl'). **1.** City (1974 est. pop. 4,905), seat of Johnson co., NW Ark., near the Arkansas River ENE of Fort Smith, in a region producing peaches, coal, natural gas; settled c.1820. Lumber, metal products, and bricks and tile are made. College of the Ozarks is here. **2.** Town (1970 pop. 13,806), Clark co., S central Ind., on the Ohio River opposite Louisville, Ky.; founded 1784 by George Rogers Clark. Soap is the chief manufacture. **3.** City (1978 est. pop. 57,400), seat of Montgomery co., NW Tenn., on the Cumberland and Red rivers, in a farm, livestock, and tobacco region; platted 1784, inc. as a city 1855. It is an important market and processing center. Its industries include meat packing and the manufacture of snuff, footwear, tires, and air-conditioning equipment. **4.** City (1970 pop. 3,346), seat of Red River co., extreme NE Texas, NE of Dallas, in a blackland farm, livestock, and timber area; settled 1828, inc. 1837.

Clat·sop (klăt'səp), county (1970 pop. 28,473), 805 sq mi (2,085 sq km), NW Oregon, in the NW extremity of the state, drained by the Nehalem River and bounded on the N by the Columbia River, on the W by the Pacific Ocean; formed 1844; co. seat Astoria. Fishing, lumbering, agriculture, and tourism are important.

Claude (klôd), city (1970 pop. 992), seat of Armstrong co., extreme N Texas, in the Panhandle ESE of Amarillo. It is the trade and shipping center for a livestock and grain area.

Claus·thal-Zel·ler·feld (klous'täl-tsĕl'ər-fĕlt'), town (1970 pop. 14,821), Lower Saxony, E West Germany, a resort in the Harz Mts. Its manufactures include textiles and wood products. The town was once a center for the mining of copper, zinc, and lead ores.

Clav·er·ack (klăv'ə-rĭk,-răk), town (1970 pop. 5,711), Columbia co., E N.Y., SE of Hudson, in a dairy and farm region. Among its old buildings is one of the Van Rensselaer manor houses.

Clax·ton (klăks'tən), city (1970 pop. 2,669), seat of Evans co., SE Ga., on the Canoochee River WNW of Savannah, in the coastal plain.

Clay (klā). **1.** County (1970 pop. 12,636), 603 sq mi (1,561.8 sq km), E Ala., in an agriculture and lumbering area; formed 1866; co. seat Ashland. Poultry and livestock are produced. Part of Talladega National Forest is in the west and north. **2.** County (1970 pop. 18,771), 639 sq mi (1,655 sq km), extreme NE Ark., bounded on the N and E by the Mo. border and drained by the St. Francis, Black, and Current rivers; formed 1873 as Clayton co., renamed 1875; co. seats Piggott and Corning. It is in a hardwood-timber and agricultural area yielding fruit, cotton, grain, livestock, and dairy products. Shoes, clothing, and automobile parts are manufactured. **3.** County (1970 pop. 32,059), 598 sq mi (1,548.8 sq km), NE Fla., bounded on the E by the St. Johns River; formed 1858; co. seat Green Cove Springs. Stock raising, dairying, farming (particularly corn, vegetables, and peanuts), and lumbering are important to its economy. The county also has clay pits and a number of small lakes in the west. **4.** County (1970 pop. 3,636), 224 sq mi (580.2 sq km), SW Ga., bounded on the W by the Chattahoochee River, here forming the Ala. border; formed 1854; co. seat Fort Gaines. In a coastal-plain agricultural area producing cotton, corn, peanuts, truck crops, and livestock, it also has stands of timber. **5.** County (1970 pop. 14,735), 464 sq mi (1,201.8 sq km), S central Ill., drained by the Little Wabash River; formed 1824; co. seat Louisville. It has agriculture (corn, wheat, fruit, dairy products, and livestock) and oil and natural-gas wells. Shoes, food products, clothing, and furniture are manufactured. **6.** County (1970 pop. 23,933), 364 sq mi (942.8 sq km), W central Ind., drained by the Eel River and Birth Creek; formed 1825; co. seat Brazil. Its agriculture includes grain, soybeans, livestock, and dairy products. It also has clay pits and bituminous-coal mines. Clay products are manufactured. **7.** County (1970 pop. 18,464), 570 sq mi (1,476.3 sq km), NW Iowa, in a rolling prairie agricultural area drained by the Little Sioux and Ocheyedan rivers; formed 1851; co. seat Spencer. Hogs, cattle, poultry, corn, oats, and soybeans are produced. **8.** County (1970 pop. 9,890), 635 sq mi (1,644.7 sq km), N Kansas, drained by the Republican River; formed 1866; co. seat Clay Center. In a rolling to hilly agricultural region, it depends mainly on grain and livestock. **9.** County (1970 pop. 18,481), 474 sq mi (1,227.7 sq km), SE Ky., in the Cumberland foothills, drained by the South Fork of the Kentucky River and its headstreams; formed 1806; co. seat Manchester. It has bituminous-coal mines, hardwood timber, and agriculture (tobacco, corn, livestock, apples, and hay). Most of the county is included in Daniel Boone National Forest. **10.** County (1970 pop. 46,608) 1,050 sq mi (2,719.5 sq km), W Minn., bounded on the W by N.Dak. and the Red River of the North and drained by the Buffalo River; formed 1858 as Breckinridge co., renamed 1862; co. seat Moorhead. In an agricultural area yielding wheat, potatoes, sugar beets, livestock, and dairy products, it has food-processing and boatbuilding industries. Plastic and metal products are made. **11.** County (1970 pop. 18,840), 414 sq mi (1,072.3 sq km), E Miss., drained by the Tombigbee River and its small affluents; formed 1871 as Colfax co., renamed 1876; co. seat West Point. It has agriculture (cotton, corn, hay, livestock, and dairy products), sand and gravel pits, and stands of timber. Meat is packed, and textiles, metal products, and sporting goods manufactured. **12.** County (1970 pop. 123,702), 413 sq mi (1,069.7 sq km), NW Mo., bounded on the S by the Missouri River; formed 1822; co. seat Liberty. In an agricultural and coal-mining area, it has highly diversified industries. Its manufactures include flour, textiles, furniture, chemicals, plastics, concrete, and petroleum, paper, wood, and metal products. **13.** County (1970 pop. 8,266), 570 sq mi (1,476.3 sq km), S Nebr., in an agricultural area drained in the SW by the Little Blue River; formed 1871; co. seat Clay Center. Cattle, corn, wheat, oats, barley, and soybeans are important. **14.** County (1970 pop. 5,180), 209 sq mi (541.3 sq km), W N.C., in a mountainous region drained by the Hiwassee River and bounded on the S by the Ga. border, on the NE by the Nantahala River; formed 1861; co. seat Hayesville. It has agriculture (corn, hay, apples, livestock, dairy products, and poultry), lumbering, many resorts, and some light industry. Nantahala National Forest is in the northeast. **15.** County (1970 pop. 12,923), 405 sq mi (1,049 sq km), SE S.Dak., in an agricultural area bordering on Nebr., drained by the Vermillion River and bounded on the S by the Missouri River; formed 1862; co. seat Vermillion. Grain, livestock, and dairy products are produced. **16.** County (1970 pop. 6,624), 233 sq mi (603.5 sq km), N Tenn., bounded on the N by the Ky. border and drained by the Obey and Cumberland rivers; formed 1870; co. seat Celina. Coal mining, lumbering, and agriculture (tobacco, grain, and livestock) are important. Clothing, automobile parts, and electrical equipment are manufactured. It includes part of Dale Hollow Reservoir. **17.** County (1970 pop. 8,079), 1,101 sq mi (2,851.6 sq km), N Texas, bounded on the N by the Red River, here forming the Okla. border, and drained by the Wichita and Little Wichita rivers; formed 1857; co. seat Henrietta. In a livestock, agricultural (grain, cotton, corn, dairy products, and poultry), and oil-producing area, it also has some manufacturing. **18.** County (1970 pop. 9,330) 342 sq mi (885.8 sq km), central W.Va., on the Allegheny Plateau, drained by Elk River; formed 1858; co. seat Clay. It has oil and gas wells, some coal mines, and timber. Its agriculture includes livestock, fruit, and tobacco.

Clay, town (1970 pop. 479), seat of Clay co., central W.Va., on the Elk River ENE of Charleston, in a timber and agricultural region. There are coal mines and gas and oil wells in the area.

Clay Center. 1. City (1970 pop. 4,963), seat of Clay co., NE Kansas, on the Republican River NW of Manhattan; laid out 1862, inc. 1875. It is a shipping and processing center for a farm and dairy region. **2.** City (1970 pop. 952), seat of Clay co., S Nebr., ESE of Hastings, in a grain and dairying area. Flour is produced.

Clay·ton (klā'tən). **1.** County (1970 pop. 98,126) 149 sq mi (385.9 sq km), NW central Ga., drained by the Flint River; formed 1858; co. seat Jonesboro. It supports piedmont agriculture (cotton, corn, potatoes, truck crops, fruit, and livestock) and lumbering. **2.** County (1970 pop. 20,606), 778 sq mi (2,015 sq km), NE Iowa, bounded on the E by the Mississippi River, here forming the Wis. border; formed 1837; co. seat Elkader. It is drained by the Turkey and Volga rivers. In a dairying and agricultural (hogs, cattle, corn, livestock, poultry, and hay), area, it has limestone quarries, sand and gravel pits, and lead and zinc deposits. Recreational facilities are plentiful here.

Clayton. 1. Town (1970 pop. 1,626), seat of Barbour co., SE Ala., W of Eufaula, in a cotton and timber area. **2.** City (1970 pop. 1,569), seat of Rabun co., extreme NE Ga., in the Blue Ridge in Chattahoochee National Forest NNW of Toccoa; inc. 1821. It is a resort. **3.** City (1978 est. pop. 15,200), seat of St. Louis co., E central Mo., a suburb of St. Louis; inc. 1919. **4.** Residential borough (1970 pop. 5,193), Gloucester co., SW N.J., SSE of Glassboro; settled c.1775, inc. 1924. **5.** Town (1970 pop. 2,931), seat of Union co., NE N.Mex., near the Texas and Okla. borders; laid out 1887, inc. 1908. It is a trade and shipping center for a ranch and farm region.

Cla·zom·e·nae (klə-zŏm'ə-nē), ancient city of W Asia Minor, W of present-day Izmir, Turkey. It was one of the 12 Ionian cities of Asia Minor. The city was founded on the mainland but was later moved to a small island, and Alexander the Great built a causeway to it. The town continued to flourish through the Hellenistic and Roman periods.

Clear Creek (klîr), county (1970 pop. 4,819), 394 sq mi (1,020.5 sq km), N central Colo., in a mining and livestock-grazing area drained by Clear Creek; formed 1861; co. seat Georgetown. It has gold, silver, lead, copper, and zinc mines. Parts of Pike National Forest, Arapaho National Forest, and the Front Range are here.

Clear·field (klîr'fēld'), county (1970 pop. 74,619), 1,139 sq mi (2,950 sq km), central Pa., in a hilly upland area drained by the West Branch of the Susquehanna River; formed 1804; co. seat Clearfield. Bituminous-coal mining, manufacturing (metal and clay products and leather), and tourism are important.

Clearfield. **1.** Borough (1970 pop. 8,176), seat of Clearfield co., W central Pa., on the West Branch of the Susquehanna River in the Alleghenies N of Altoona; laid out 1805, inc. 1840. In a coal and clay area, it has diversified manufactures. **2.** City (1970 pop. 13,316), Davis co., N Utah; inc. 1922. Hill Air Force Base and a naval supply depot are the major employers.

Clear Lake. **1.** City (1973 est. pop. 6,873), Cerro Gordo co., N central Iowa, on Clear Lake just W of Mason City; inc. 1871. It is in a popular resort area near a state park. **2.** City (1970 pop. 1,157), seat of Deuel co., E S.Dak., ESE of Watertown, in a dairy, livestock, and grain region.

Clear Lake, 65 sq mi (168.4 sq km), W Calif., in wooded hills NW of San Francisco. It is the largest freshwater lake entirely within the state and is a fishing resort.

Clear·wa·ter (klîr'wô'tər). **1.** County (1970 pop. 10,871), 2,522 sq mi (6,532 sq km), N Idaho, in a mountainous area bounded on the E by the Bitterroot Range and the Mont. border and crossed by the North Fork of the Clearwater River; formed 1911; co. seat Orofino. Lumbering, stock raising, and dairying are important. Clearwater National Forest and the Clearwater Mts. extend throughout much of the county. Dworshak Reservoir is in the west. **2.** County (1970 pop. 8,013), 1,000 sq mi (2,590 sq km), NW Minn., drained by the Clearwater River and by headwaters of the Mississippi River in the SE; formed 1902; co. seat Bagley. In a timber and agricultural region yielding potatoes, livestock, and dairy products, it also has peat deposits. Part of Red Lake Indian Reservation is in the north.

Clearwater, residential and resort city (1978 est. pop. 72,300), seat of Pinellas co., W Fla., on the Pinellas peninsula, on Clearwater Bay and the Gulf of Mexico; inc. 1891. Its thriving tourist industry dates from 1896. A landscaped causeway connects the city proper with a 4-mi (6.4-km) long island of white sand beaches fronting on the Gulf.

Clearwater. **1.** River, SW Alta., Canada, rising in the Rocky Mts. in Banff National Park and flowing c.100 mi (160 km) generally NE in a winding course. It joins the North Saskatchewan River at Rocky Mountain House. **2.** River, central Canada, rising in NW Sask. S of Lake Athabasca and flowing c.130 mi (210 km) W, crossing into NE Alta. where it joins the Athabasca River at Fort McMurray.

Clearwater, river, c.190 mi (305 km) long, rising in several branches in the Bitterroot Range, N Idaho, and flowing W to join the Snake River at Lewiston, Idaho. The gold-mining era in Idaho began in 1860, when gold was discovered and mining camps were set up on the river's southern fork.

Clearwater Mountains, range, N central Idaho, bounded on the S and W by the Salmon River, on the E by the Bitterroot Range. The highest elevations (8,000–9,000 ft/2,440–2,745 m) are in the southern part of the range.

Cle·burne (klē'bərn). **1.** County (1970 pop. 10,996), 574 sq mi (1,486.7 sq km), E Ala., in a piedmont region crossed by the Tallapoosa River and bordering on Ga.; formed 1866; co. seat Heflin. Lumber milling is done, and cotton, corn, and livestock are produced. It includes part of Talladega National Forest in the west and north. **2.** County (1970 pop. 10,349), 554 sq mi (1,434.9 sq km), N central Ark., in the Ozarks and drained by the Little Red River and its tributaries; formed 1883; co. seat Heber Springs. Its agriculture includes cotton, livestock, poultry, and dairy products. Cotton ginning, lumber milling, and stone quarrying are also done.

Cleburne, city (1978 pop. 15,500), seat of Johnson co., N Texas; inc. 1907. It is a rail, processing, and medical center in a farming area. The city has railroad shops, cotton mills, limestone-processing plants, and factories producing a variety of products. Two rodeos are held here annually.

Clee·thorpes (klē'thôrps'), municipal borough (1973 est. pop. 37,200), Humberside, E central England, on the Humber River estuary. It is a popular resort, with many recreational facilities. The nearby Church of Old Clee was dedicated in 1192.

Clem·son (klĕm'sən), town (1970 pop. 5,578), Anderson and Pickens cos., NW S.C., SW of Spartanburg; inc. 1892. It is the seat of Clemson Univ. On the campus is Fort Hill, the former home of John C. Calhoun.

Clerk·en·well (klärk'ən-wəl, -wĕl), district of Islington borough, London, England, named from a well near which the London parish clerks held performances of miracle plays. It has many shops of jewelers, watchmakers, and opticians.

Cler·mont (klĕr'mŏnt), county (1970 pop. 95,372), 458 sq mi (1,186.2 sq km), SW Ohio, intersected by the East Fork of the Little Miami River and bounded on the SW by the Ohio River, here forming the Ky. border, and on the NW by the Little Miami River; formed 1800; co. seat Batavia. In an increasingly industrial area, its agriculture includes livestock, corn, wheat, tobacco, fruit, and dairy products. It also has nurseries.

Cler·mont-Fer·rand (klĕr-môɴ'fĕ-räɴ'), city (1975 pop. 156,900), capital of Puy-de-Dôme dept., central France, in Auvergne, on the Tiretaine River. It is an industrial center and has tire factories and metallurgical works. The capital of the former province of Auvergne, it was formed in 1731 by the merger of Clermont and Montferrand. Clermont was built in Roman times. An episcopal see since the 3rd cent., it was the site of several church councils, notably that of 1095, where Pope Urban II preached the First Crusade. The city is built largely of the dark volcanic rock of the region. The Gothic Cathedral of Notre-Dame (13th-14th cent.) and the Romanesque Church of Notre-Dame du Port (12th cent.) are among the notable buildings. Blaise Pascal was born in Clermont-Ferrand.

Cleve·land (klēv'lənd), nonmetropolitan county (1976 est. pop. 567,900), NE England, created under the Local Government Act of 1972 (effective 1974); administrative center Middlesborough. It is composed of the county boroughs of Harlepool and Teesside and parts of the former counties of Durham and Yorkshire.

Cleveland. **1.** County (1970 pop. 6,605), 601 sq mi (1,556.6 sq km), S Ark., bounded on the W by Moro Creek and intersected by the Saline River; formed 1873 as Dorsey co., renamed 1885; co. seat Rison. Its agriculture includes cotton, corn, hay, and livestock. It also has red clay and gravel pits and a lumbering industry. Clothing is manufactured. **2.** County (1970 pop. 72,556), 468 sq mi (1,212.1 sq km), SW N.C., in a piedmont agricultural region bounded on the S by the S.C. border and drained by the Broad River and its affluents; formed 1841; co. seat Shelby. Cotton and corn are grown; textiles and lumber products are manufactured. **3.** County (1970 pop. 81,839), 527 sq mi (1,364.9 sq km), central Okla. bounded on the SW by the Canadian River and drained by the Little River; formed 1890; co. seat Norman. It has diversified agriculture (cotton, cattle, hogs, poultry, grain, sorghums, corn, and dairy products), oil and natural-gas fields, and some manufacturing and food processing.

Cleveland. **1.** City (1970 pop. 1,353), seat of White co., NE Ga., N of Gainesville in the foothills of the Blue Ridge. It is a resort center in a lumber area. Pottery is manufactured. Truett McConnell Junior College is here. **2.** City (1970 pop. 13,327), seat of Bolivar co., NW Miss., in the rich delta cotton country; inc. 1886. It is a farm market center (rice and soybeans are also grown in the area), and its manufactures include pharmaceuticals, aluminum doors, tiles, and pens and pencils. **3.** City (1978 est. pop. 595,000), seat of Cuyahoga co., NE Ohio, a port of entry on Lake Erie at the mouth of the Cuyahoga River; laid out (1796) by Moses Cleaveland, chartered as a city 1836. It is a Great Lakes shipping point and one of the nation's leading iron and steel centers. In addition to many metallurgical manufactures, it has chemical, oil-refining, electrical, automobile, garment, and food-processing industries. Cleveland grew rapidly after the opening of the first section of the Ohio and Erie Canal in 1827 and the arrival of the railroad in 1851. Its central location midway between the coal and oil fields of Pennsylvania and (via the Great Lakes) the Minnesota iron mines spurred its industrialization; it was here that John D. Rockefeller began his oil dynasty. In the 1960s and 1970s Cleveland was plagued by racial and financial problems. In

1978 it was the first U.S. city to default on its bank loans. **4.** City 1978 est. pop. 26,100), seat of Bradley co., SE Tenn., in a farm and timber area; inc. 1838.

Cleveland Heights, city (1978 pop. 47,400), Cuyahoga co., NE Ohio, a residential suburb of Cleveland; inc. 1903. It is known for its beautiful homes. Forest Hills Park, once part of an estate owned by John D. Rockefeller, has recreational facilities.

Cleves (klēvz), city (1970 pop. 43,447), North Rhine-Westphalia, W West Germany, near the Dutch border. Its manufactures include shoes and food and tobacco products. It is a rail junction and popular resort.

Cleves, duchy of, former state, W West Germany, on both sides of the lower Rhine, bordering on the Netherlands. Cleves was held by France during the French Revolutionary Wars and in 1815 was returned to Prussia.

Clew Bay (klo͞o), inlet of the Atlantic Ocean, c.15 mi (25 km) long and 10 mi (16.1 km) wide, Co. Mayo, W Republic of Ireland. There are about 300 islands in the eastern part of the bay, some of which are cultivated.

Cli·chy (klē-shē'), suburb N of Paris (1975 pop. 47,764), Hauts-de-Seine dept., N central France. It is a modern industrial city with iron works; automobile parts, metal products, machinery, and plastics are also manufactured. Clichy was once a residence of Merovingian kings. The Church of St. Vincent de Paul, named for the saint who was parish priest to Clichy, is a major landmark.

Cliff·side Park (klĭf'sīd'), borough (1970 pop. 18,891), Bergen co., NE N.J., on the Palisades above the Hudson River, opposite New York City; inc. 1895. A residential suburb, it has some light industry.

Clif·ton (klĭf'tən). **1.** Town (1970 pop. 5,087), seat of Greenlee co., SE Ariz., on the San Francisco River near the N.Mex. border; settled 1872. It is a trade center in a mining and livestock region and has a copper smelter. Hot springs bubble up near the town's center. **2.** Industrial city (1978 est. pop. 77,500), Passaic co., NE N.J., on the Passaic River; settled 1685, set off from Passaic and inc. 1917. It has steel, textile, chemical, and electronic industries.

Cli·max (klī'măks'), uninc. village (1970 pop. 150), alt. 11,320 ft (3,452.6 m), Lake co., central Colo., SW of Denver. It has one of the world's largest molybdenum mines. First discovered here by gold seekers, this rare metal was mistaken for lead and ignored until 1900, when it was identified by the Colorado School of Mines. The Univ. of Colorado has an observatory here.

Clinch (klĭnch), county (1970 pop. 6,405), 796 sq mi (2,061.6 sq km), S Ga., on the Fla. border in a flatwoods area drained by the Suwannee River; formed 1850; co. seat Homerville. Part of the Okefenokee Swamp is in the east. Forestry and farming (corn, sweet potatoes, and livestock) are important to its economy.

Clinch, river, c.300 mi (480 km) long, formed by the junction of two forks in SW Va., and flowing generally SW across E Tenn. to the Tennessee River at Kingston. The river is an important part of the Tennessee Valley Authority.

Clin·ton (klĭn'tən). **1.** County (1970 pop. 28,315), 434 sq mi (1,124.1 sq km), SW Ill., in an agricultural area bounded on the S by the Kaskaskia River; formed 1824; co. seat Carlyle. Corn, wheat, fruit, livestock, poultry, and dairy products are produced. It has bituminous-coal mines, oil wells and refineries, and diversified manufactures (flour, shoes, metal articles, and paper, glass, and food products). **2.** County (1970 pop. 30,547), 407 sq mi (1,054.1 sq km), central Ind., drained by Sugar Creek and forks of Wildcat Creek; formed 1830; co. seat Frankfort. It has agriculture (grain, livestock, soybeans, and apples), bituminous-coal mines, oil refining at Frankfort, and diversified manufacturing. **3.** County (1970 pop. 56,749), 695 sq mi (1,800 sq km), E Iowa, in a prairie agricultural area bounded on the E by the Mississippi, here forming the Ill. border, and on the SW and S by the Wapsipinicon River; formed 1840; co. seat Clinton. Hogs, cattle, corn, and oats are produced. It has limestone quarries and some industry. **4.** County (1970 pop. 8,174), 190 sq mi (492.1 sq km), S Ky., in a hilly agricultural region in the Cumberland foothills, drained by several creeks and bounded on the S by the Tenn. border, on the NW by the Cumberland River; formed 1836; co. seat Albany. Tobacco, livestock, grain, poultry, and dairy products are produced. It also has coal mines and stands of timber. Part of Dale Hollow Reservoir is here. **5.** County (1970 pop. 48,492), 571 sq mi (1,478.9 sq km), S central Mich., drained by the Maple, Looking Glass, and Grand rivers and Stony Creek; formed 1839; co. seat St. Johns. In a dairying and agricultural area yielding grain, sugar beets, peppermint, beans, fruit, and livestock, it also has oil refineries and some manufacturing. **6.** County (1970 pop. 12,462), 420 sq mi (1,087.8 sq km), NW Mo., drained by the Grand River; formed 1833; co. seat Plattsburg. It is primarily an agricultural area with corn, wheat, and oat farming and mule raising. There is some mining, and textiles and metals are manufactured. **7.** County (1970 pop. 72,934), 1,059 sq mi (2,742.8 sq km), extreme NE N.Y., bounded on the N by the Que. border, on the E by Lake Champlain; formed 1788; co. seat Plattsburgh. In a mountain and lake resort area, it is drained by the Saranac, Salmon, and Great Chazy rivers. Tourism, dairying, farming, and lumbering are important. It also has iron mines, paper mills, and varied manufacturing plants. **8.** County (1970 pop. 31,464), 410 sq mi (1,061.9 sq km), SW Ohio, drained by forks of the Little Miami River; formed 1810; co. seat Wilmington. In a stock-raising, grain-farming, and dairying area, it has some manufacturing. **9.** County (1970 pop. 37,721), 899 sq mi (2,328.4 sq km), N central Pa., in a forested mountain area drained by the West Branch of the Susquehanna River and Bald Eagle Creek; formed 1839; co. seat Lock Haven. It has clay and bituminous-coal deposits, limestone quarries, stands of timber, and some agriculture. Meat products, paper, leather, and textiles are manufactured.

Clinton. 1. City (1975 est. pop. 1,162), seat of Van Buren co., N central Ark., N of Little Rock, in a diversified agricultural area; inc. as a city 1938. **2.** Resort town (1970 pop. 10,267), Middlesex co., S Conn., on Long Island Sound; settled 1663, set off from Killingworth and inc. 1838. A monument commemorates the early years of the school that later became Yale Univ. **3.** City (1977 est. pop. 7,604), seat of De Witt co., central Ill., N of Decatur, in a farm area; settled 1836, inc. 1885. Abraham Lincoln frequently argued law cases here. **4.** City (1970 pop. 5,340), Vermillion co., W central Ind., on the Wabash River N of Terre Haute; settled 1818, laid out 1829. It has coal mines and small factories making clothing and other products. **5.** City (1978 est. pop. 33,200), seat of Clinton co., E central Iowa, on the Mississippi, in a rich corn and livestock area; inc. 1859. An industrial and rail center, it has food-processing and diverse manufacturing industries. Clinton grew as a lumbering town. **6.** City (1970 pop. 1,618), seat of Hickman co., SW Ky., near the Mississippi River SW of Paducah; platted 1826, inc. 1831. It is a shipping point in a corn, cotton, and tobacco region. Columbus-Belmont Battlefield State Park is nearby; it is on the old site of the city of Columbus (now moved to higher ground), scene of early settlement and a Confederate stronghold at the start of the Civil War. **7.** Town (1970 pop. 1,884), seat of East Feliciana parish, SE La., NNE of Baton Rouge, in a farm and timber area; inc. 1852. **8.** Uninc. village (1970 pop. 4,700), Prince Georges co., W central Md., SE of Washington, D.C. Andrews Air Force Base is nearby. **9.** Industrial town (1970 pop. 13,383), Worcester co., E central Mass., on the Nashua River, near Wachusett Reservoir, in a farm and wooded area; settled c.1654, set off from Lancaster and inc. 1850. Once an important textile center, it now has chemical and metallurgical industries. **10.** Town (1976 est. pop. 12,100), Hinds co., W central Miss., just W of Jackson; settled c.1823 on the site of an early Indian agency; inc. 1830. Mississippi College is here. **11.** City (1970 pop. 7,504), seat of Henry co., W central Mo., SE of Kansas City, in an agricultural and coal region; inc. 1836. **12.** Town (1976 est. pop. 7,893), seat of Sampson co., SE N.C., E of Fayetteville; laid out 1818, inc. 1852. It is the trade and shipping center of a large agricultural area. Food is processed, and electronic equipment, furniture, and clothing are made. **13.** City (1970 pop. 8,513), Custer co., W central Okla., on the Washita River W of Oklahoma City; founded 1903. It is a shipping center for a wheat and livestock area, and its industries include brick manufacturing and poultry processing. **14.** Town (1970 pop. 8,138), Laurens co., NW S.C., S of Spartanburg; settled c.1809, inc. 1890. Apparel and paper, wood, and metal products are manufactured. Presbyterian College is here. **15.** Town (1976 est. pop. 4,860), seat of Anderson co., E central Tenn., on the Clinch River NW of Knoxville, in a farm and timber area; inc. 1835.

Clin·ton-Col·den Lake (klĭn'tən-kōl'dən), 253 sq mi (655.3 sq km), E central Mackenzie dist., Northwest Territories, Canada, NE of Great Slave Lake.

Clint·wood (klĭnt'wo͝od'), town (1970 pop. 1,320), seat of Dickenson co., extreme SW Va., E of Jenkins, Ky., in a timber, coal, and farm area; inc. 1894.

Clip·per·ton Island (klĭp'ər-tən), uninhabited atoll, c.2 sq mi (5.2 sq km), in the Pacific Ocean, c.800 mi (1,290 km) SW of Mexico. It was used as a base by John Clipperton, an English pirate. The French claimed it in 1858, the Americans held it for a time in the Spanish-American War, and Mexican troops occupied it in 1897. The conflict between France and Mexico was referred to the king of Italy for arbitration in 1908. The award was made (1931) in favor of France, and Mexico surrendered the island in 1932.

Clith·er·oe (klĭth'ə-rō), municipal borough (1971 est. pop. 13,191), Lancashire, N England, N of Accrington. A market town for a rural area in the Ribble Valley, it also has mills for weaving, bleaching, and printing cotton, paper mills, and nearby quarries. Its Norman keep, said to be the smallest in England, is now a war memorial.

Clon·mel (klŏn-mĕl'), municipal borough (1971 pop. 11,622), administrative center of South Riding, Co. Tipperary, S Republic of Ireland, on the Suir River. Footwear, cider, enamelware, tubular steel furniture, perambulators, and canned meat are produced here. It is also a tourist center with good hunting and salmon fishing. Laurence Sterne was born here.

Clon·tarf (klŏn-tärf'), suburb of Dublin, Co. Dublin, E Republic of Ireland. It was the scene of a decisive defeat (1014) of the Danes by the Irish under Brian Boru, who was killed in the fighting.

Clo·quet (klō-kā'), city (1974 est. pop. 11,439), Carlton co., NE Minn., on the St. Louis River WSW of Duluth; settled 1879, inc. as a village 1880, as a city 1904. Paper and wood products are made.

Cloud (kloud), county (1970 pop. 13,466), 711 sq mi (1,841.5 sq km), N Kansas, in a plains region drained by the Republican and Solomon rivers; formed 1867; co. seat Concordia. Wheat growing and stock raising are important.

Cloud Peak, 13,175 ft (4,018.4 m) high, N Wyo., W of Buffalo. It is the highest peak in the Bighorn Mts. and has a notable glacier.

Clo·vel·ly (klō-vĕl'ē), village, Devonshire, SW England. The main street is a cobbled staircase rising 400 ft (122 m) up from the quay, where there is a massive stone pier. Clovelly Dykes, an ancient British encampment, is nearby.

Clo·vis (klō'vĭs). **1.** City (1978 est. pop. 28,300), Fresno co., S central Calif., near the foothills of the Sierra Nevada range; inc. 1912. It is a trade center in a farm and vineyard area. **2.** City (1978 est. pop. 32,100), seat of Curry co., E N.Mex., near the Texas line; inc. 1909. It is a railroad division point, the trade center of a cattle and irrigated farm area (with large stockyards), and the home of Cannon Air Force Base. A county fair and a rodeo are annual events here.

Cluj (klo͞ozh), city (1977 est. pop. 262,421), W central Rumania, in Transylvania, on the Someşul River. It is the administrative center of an agricultural and mineral-rich area. Its diverse manufactures include machinery, metal products, electrical equipment, chemicals, textiles, and footwear. The city is also a noted educational center. Cluj was founded by German colonists in the 12th cent. and became a thriving commercial and cultural center in the Middle Ages. It was made a free city in 1405 by the king of Hungary and became (16th cent.) the chief cultural and religious center of Transylvania. It was incorporated into Austria-Hungary in 1867 and was transferred to Rumania in 1920. Hungarian forces occupied the city during World War II.

Clu·ny (klo͞o'nē, klü-nē'), former abbey, Saône-et-Loire dept., E France, founded (910) by St. Berno, a Burgundian monk. He and his successors, all vigorous reformers, made their abbey the center of the Cluniac Order. Cluny became one of the chief religious and cultural centers of Europe.

Clu·tha (klo͞o'thə), longest river, c.200 mi (320 km), on South Island, New Zealand. Near its source at Luggate there is a hydroelectric plant.

Clw·yd (klo͞o'ĭd), nonmetropolitan county (1976 est. pop. 376,000), N Wales, created under the Local Government Act of 1972 (effective 1974); administrative center Mold. It comprises the former county of Flintshire and portions of the former counties of Denbigh and Merioneth.

Clyde (klīd), principal river of SW Scotland, 106 mi (170.6 km) long, rising in the Southern Uplands and flowing generally NW through Glasgow to the Firth of Clyde. The river has been deepened and widened and is navigable for oceangoing vessels to Glasgow. The Firth of Clyde, c.50 mi (80 km) wide and 2 to 25 mi (3.2–40 km) wide, an arm of the North Channel, extends southwest from Dunoon to Ailsa Craig.

Clyde·bank (klīd'băngk'), burgh (1976 est. pop. 55,902), Strathclyde region, W central Scotland, on the N bank of the Clyde River. The chief industry is shipbuilding.

Cni·dus or **Cni·dos** (both: nī'dəs), ancient Greek city of Caria, SW Asia Minor, on Cape Krio, in present SW Asiatic Turkey. It was partly on the peninsula and partly on an island that had been created by cutting through the peninsula. One of the cities of the Dorian Hexapolis, it sought to maintain its independence but fell (540 B.C.) under Persian rule. One of the most famous statues of the ancient world, Aphrodite by Praxiteles, was here. Cnidus retained its importance in Roman times.

Cnos·sus or **Knos·sos** (both: nŏs'əs), ancient city of Crete, on the N coast, near modern Iráklion. The site was occupied long before 3000 B.C. and was the center of an important Bronze Age culture. It is from a study of the great palace, as well as other sites in Crete, that knowledge of the Minoan civilization has been drawn. The city was destroyed before 1500 B.C. (possibly by earthquake) and was splendidly rebuilt, only to be destroyed again c.1400 B.C., probably at the hands of invaders from the Greek mainland. Cnossus later became a flourishing Greek city, and it continued to exist through the Roman period until the 4th cent. A.D. In Greek legend it was the capital of King Minos and the site of the labyrinth.

Coachella Valley, arid region, SE Calif., N of the Salton Sea. Water is brought into the region by artesian wells and by the Coachella Canal (123 mi/198 km long), a branch of the All-American Canal built between 1938 and 1948; more than 100,000 acres (40,500 hectares) have been irrigated.

Coa·ho·ma (kō-hō'mə), county (1970 pop. 40,447), 570 sq mi (1,467.3 sq km), NW Miss., bounded on the NW and W by the Mississippi River, here forming the Ark. border, and drained by the Sunflower River; formed 1836; co. seat Clarksdale. It lies in a rich timber and agricultural area yielding cotton, corn, cattle, and hogs. Its varied manufactures include meat products, soybean and cottonseed oil, beverages, wood products, furniture, chemicals, tires, hardware, and farm machinery.

Coal (kōl), county (1970 pop. 5,525), 526 sq mi (1,362.3 sq km), S central Okla., drained by Clear Boggy and Muddy Boggy creeks; formed 1907; co. seat Coalgate. In an agricultural area (corn, cotton, sorghums, cattle, hogs, and poultry), it has oil and natural-gas wells, coal mines, and some manufacturing.

Coal·gate (kōl'gāt'), city (1970 pop. 1,859), seat of Coal co., SE Okla., NE of Ardmore, in a farm region. There are oil and natural-gas wells in the area.

Coal·ville (kōl'vĭl'), urban district (1973 est. pop. 28,740), Leicestershire, central England. Besides coal mining, it has hosiery, footwear, and plastics industries.

Coalville, city (1970 pop. 864), seat of Summit co., N Utah, SE of Ogden, in a livestock and poultry area at an altitude of 5,571 ft (1,699.2 m). There are coal mines in the area.

Co·a·mo (kō-ä'mō, kwä'-), town (1970 pop. 12,077), S central Puerto Rico, on the Coamo River. It is the trade center of a sugar and tobacco region and has garment factories. The town was founded in the 16th cent.

Coast Mountains (kōst), range, W British Columbia and SE Alaska, extending c.1,000 mi (1,610 km) parallel to the Pacific coast, from the mountains of Alaska near the Yukon border to the Cascade Range near the Fraser River. Mt. Waddington (13,260 ft/4,044.3 m) is the highest peak. The geologically complex range, composed mainly of metamorphic rocks, slopes steeply to the Pacific Ocean, where the shoreline is deeply indented by fjords. The Coast Mts. have been heavily eroded by mountain glaciers; numerous rivers have cut deep gorges across the range.

Coast Ranges, series of mountain ranges along the Pacific coast of North America, extending from SE Alaska to Baja California, and from 2,000 to 20,000 ft (610–6,100 m) high. The Coast Ranges are rugged, geologically young mountains formed by faulting and folding and are composed mainly of granitic rock; the northern third is glaciated. North of San Francisco the ranges are humid and thickly forested; the southern parts are dry and covered with brush and grass. Lumbering, mining, and tourism are important industries.

Coat·bridge (kōt'brĭj'), burgh (1971 pop. 52,131), Strathclyde region, S central Scotland. In Coatbridge a variety of iron and steel products are manufactured.

Coates·ville (kōts'vĭl'), city (1978 est. pop. 12,400), Chester co., SE Pa., on Brandywine Creek, in a farm area; settled c.1717, inc. as a city 1916. It is a steel center. The Revolutionary Battle of Brandywine (Sept. 11, 1777) was fought to the south of the city.

Coats Island (kōts) 2,210 sq mi (5,723.9 sq km), N Hudson Bay, E Keewatin dist., Northwest Territories, Canada, S of Southampton Island, from which it is separated by Fisher Strait.

Coats Land, ice-covered region forming the SE shore of Weddell Sea, Antarctica, W of Queen Maud Land. It was discovered by a Scottish explorer, W. S. Bruce, in 1904.

Co·at·za·co·al·cos (kō-ät'sə-kō-äl'kəs, kwät'sä-kwäl'kōs), city (1970 pop. 69,753), Veracruz state, E central Mexico, at the mouth of the Coatzacoalcos River on the Gulf of Campeche. The city is an important commercial center. Oil, sulfur, and timber are exported, and the port facilities have been enlarged to enable Coatzacoalcos to handle foreign trade.

Cobb (kŏb), county (1970 pop. 196,793), 348 sq mi (901.3 sq km), NW central Ga., bounded on the SE by the Chattahoochee River; formed 1832; co. seat Marietta. It is in a piedmont timber and agricultural area yielding cotton, corn, poultry, and dairy products.

Cóbh (kōv), urban district (1971 pop. 6,076), Co. Cork, S Republic of Ireland, on the S shore of Great Island in Cork Harbour. There are large docks and stations of naval stores. Situated on slopes above the harbor and having a fine climate, Cóbh has become a seaside resort. It is the headquarters of the Royal Cork Yacht Club, the oldest yacht club in the world (founded in the early 18th cent.).

Co·bourg (kō'bûrg'), town (1976 pop. 11,082), S Ont., Canada, on Lake Ontario E of Toronto. It is a popular summer resort and a port serving a rich farming, dairying, and timber area. Plastics, machinery, chemicals, and furniture are manufactured.

Cobourg Peninsula, c.50 mi (80 km) long and 25 mi (40 km) wide, N Northern Territory, Australia, E of Melville Island. It is a reserve for native flora and fauna.

Co·burg (kō'bûrg, -bŏŏrкн), city (1974 est. pop. 46,646), Bavaria, E central West Germany, on the Itz River. It has metal, glass, and ceramics industries and is known for its toys and Christmas ornaments. Mentioned in the 11th cent., Coburg was the alternate capital (with Gotha) of Saxe-Coburg-Gotha from 1826 to 1918 and joined Bavaria in 1920. The large ducal castle (16th cent.) was the residence of Martin Luther in 1530.

Co·cha·bam·ba (kō'chä-bäm'bä), city (1976 pop. 204,414), alt. c.8,400 ft (2,560 m), capital of Cochabamba dept., W central Bolivia. It is a commercial center in an agricultural region that ships grains, fruits, and cattle. Founded in 1574, the city was called Villa de Oropeza and was renamed in 1786. Cochabamba has many historic buildings, including a convent, with five paintings by the Spanish artist Goya, and a monument to the women of the city who fought and died in the Bolivian war of independence (1815).

Co·chin (kō'chĭn, kŏch'ĭn), former princely state, SW India, on the Arabian Sea. It is now part of Kerala state. The city of Cochin (1971 pop. 438,420) is the finest port south of Bombay. Cochin has a naval base and shipbuilding industry. Tires, paper, chemicals, and tiles are manufactured. After Vasco da Gama visited Cochin (1502), the Portuguese established a settlement here. The Dutch captured it in 1663 and the British in 1795.

Cochin China, historic region (c.26,500 sq mi/68,635 sq km) of Vietnam, SE Asia. Cochin China was bounded by Cambodia on the northwest and north, by the historic region of Annam on the northeast, by the South China Sea on the east and south, and by the Gulf of Siam on the west. It included the rich Mekong delta. Cochin China was originally part of the Khmer Empire. In the 17th cent. the Annamese (later called Vietnamese) gradually infiltrated through the mouths of the Mekong, increasing their commercial influence until in the middle of the 18th cent. they became masters of the region. After the French occupied Saigon (1859), Annam ceded to France both eastern Cochin China (1862) and western Cochin China (1867). After World War II the status of Cochin China became a major issue in the relations between France and Vietnam. Constituted (1946) as an independent republic within the Federation of Indochina, Cochin China was later (1949) permitted by the French to join with Annam and Tonkin in Vietnam. After 1954, when Vietnam was partitioned, Cochin China became the heartland of South Vietnam. It is now part of the reunited Vietnam.

Co·chi·nos Bay (kə-chē'nəs) or **Bay of Pigs** (pĭgz), inlet, SW central Cuba. It was the scene of an unsuccessful invasion (Apr. 17, 1961) by Cuban exiles supported by the United States.

Co·chise (kō-chēs', -chēz'), county (1970 pop. 61,918), 6,256 sq mi (16,203.4 sq km), SE Ariz., in a mountainous and irrigated farming area bordering in the E on N.Mex., in the S on Mexico; formed 1881; co. seat Bisbee. Livestock and alfalfa are produced. It has copper, silver, gold, and gypsum deposits, dude ranches, national forests, and many other recreational facilities.

Coch·ran (kŏk'rən), county (1970 pop. 5,326), 782 sq mi (2,025.4 sq km), NW Texas, on the Llano Estacado and bounded on the W by the N.Mex. border; formed 1876; co. seat Morton. It has oil wells, cattle ranches, and agriculture (cotton, grain sorghums, wheat, some fruit, truck crops, hogs, poultry, and some dairy products).

Cochran, city (1970 pop. 5,161), seat of Bleckley co., central Ga., SSE of Macon, in a farm area; inc. 1870. Lumber and textiles are made, and food is processed.

Cocke (kŏk), county (1970 pop. 25,283), 424 sq mi (1,098.2 sq km), E Tenn., bordered in the SE by N.C. and drained by the French Broad, Pigeon, and Nolichucky rivers; formed 1797; co. seat Newport. In a forest and farm region, it has iron-ore and granite deposits. Corn, tobacco, fruit, dairy products, and livestock are produced. It includes parts of Great Smoky Mountains National Park, Cherokee National Forest, and Douglas Reservoir.

Co·coa (kō'kō), city (1978 est. pop. 14,900), Brevard co., E Fla., on the Indian River (a lagoon), a segment of the Intracoastal Waterway; inc. 1895. It is a tourist center in a region where citrus fruits are grown. An 8-mi (12.9-km) causeway leads from the city over Indian River to Merritt Island, Cocoa Beach, and Cape Canaveral.

Cocoa Beach, resort town (1975 est. pop. 12,800), Brevard co., E Fla., ESE of Cocoa, on a barrier beach between Banana River (a lagoon) and the Atlantic Ocean. Nearby is Patrick Air Force Base.

Co·co·ni·no (kō'kə-nē'nō), county (1970 pop. 48,326), 18,540 sq mi (48,018.6 sq km), N Ariz., in a plateau and mesa area; formed 1891; co. seat Flagstaff. It is the second-largest county in the United States. Tourism, lumbering, ranching, and agriculture (beans and potatoes) are important. Part of Grand Canyon National Park is in the north, and Wupatki National Monument is on the Little Colorado River in the east. Part of Hualapai Indian Reservation is in the west.

Co·cos Islands (kō'kŏs) or **Kee·ling Islands** (kē'lĭng), two separate atolls comprising 27 coral islets (1970 pop. 611), 5.5 sq mi (14.2 sq km), in the Indian Ocean, c.1,400 mi (2,255 km) SE of Sri Lanka. They are under Australian administration. Discovered in 1609 by Capt. William Keeling of the East India Company, the Cocos were settled in 1826 by Alexander Hare, an Englishman. A second settlement was founded in 1827 by John Clunies-Ross, a Scottish seaman. In 1857 the islands were annexed to the British crown. Queen Victoria granted the lands to the Clunies-Ross family in 1886 in return for the right to use any land on the islands for public purposes. In 1903 the islands were included in the Straits Settlements, and in 1955 they were placed under Australian administration. Only three of the islands are inhabited. The economy is based on the production of copra and on aviation and government facilities maintained by the Australian government.

Cod, Cape (kŏd), narrow peninsula of glacial origin, 399 sq mi (1,033.4 sq km), SE Mass., extending 65 mi (104.6 km) E and N into the Atlantic Ocean. It is generally flat, with sand dunes, low hills, and numerous lakes. The cape's familiar hook shape is a result of the action of winds and ocean currents on the sand and gravel. Bartholomew Gosnold, an English explorer, visited the cape in 1602 and named it for the abundant codfish found in surrounding waters. Fishing, whaling, shipping, and salt making were important until the late 1800s; tourism and cranberry growing are now the main industries. Candle making and boat building are also carried on. The Cape Cod Canal, a lockless canal 17.5 mi (28.2 km) long, 32 ft (9.8 m) deep was built (1910–14) with private funds. It was purchased by the U.S. government in 1927. The canal accommodates oceangoing vessels and cuts the distance between New York City and Boston by 75 mi (121 km). Parts of Cape Cod constitute Cape Cod National Seashore (44,600 acres/18,063 hectares; est. 1961). It contains beaches, sand dunes, heathlands, marshes, freshwater ponds, and historic sites, including the first Marconi Wireless Station in the United States.

Cod·ing·ton (kŏd'ĭng-tən), county (1970 pop. 19,140), 691 sq mi

(1,789.7 sq km), E S.Dak., in an agricultural area drained by the Big Sioux River; formed 1877; co. seat Watertown. Dairy products, poultry, corn, potatoes, soybeans, and artichokes are produced. It has some manufacturing.

Co·dy (kō'dē), city (1970 pop. 5,161), seat of Park co., NW Wyo., on the Shoshone River in a sheep, cattle, and irrigated farm area; founded and inc. 1901 by William F. Cody (Buffalo Bill). It is a tourist resort at the eastern entrance to Yellowstone National Park, with dude ranches and a colorful old frontier town flavor. Of interest are the Buffalo Bill Historical Center, containing Cody memorabilia; the Whitney Gallery of Western Art, housing a notable collection of art of the Old West; and an annual rodeo.

Coeur d'A·lene (kûr də-lān'), city (1970 pop. 16,228), seat of Kootenai co., N Idaho, near the Wash. line; inc. 1907. It is a tourist and lumbering center situated on Coeur d'Alene Lake west of the Coeur d'Alene Mts., and the gateway to a beautiful summer and winter resort area. The city has numerous lumber mills, grass-seed farms, and plants making electronic items and prefabricated homes. Fort Coeur d'Alene (later Fort Sherman) was established here in 1876. The city grew around the fort after the discovery (1883) of the fabulously rich silver, lead, and zinc lodes and after the mining boom of 1884.

Cof·fee (kô' fē, kŏf'ē). **1.** County (1970 pop. 34,872), 677 sq mi (1,753.4 sq km), SE Ala., in a coastal-plain region drained by the Pea River; formed 1841; co. seat Elba. It is mainly agricultural, producing hogs, peanuts, cattle, poultry, and cotton. It also has a lumbering industry and some manufacturing. **2.** County (1970 pop. 22,828), 613 sq mi (1,587.7 sq km), S central Ga., bounded on the N by the Ocmulgee River and drained by the Saltilla River; formed 1854; co. seat Douglas. It is in a coastal-plain lumbering and farming area yielding tobacco, cotton, corn, peanuts, and livestock. **3.** County (1970 pop. 32,572), 435 sq mi (1,126.7 sq km), central Tenn., partly in the Cumberlands, bounded on the SE by the Elk River and drained by the Duck River; formed 1846; co. seat Manchester. It has tracts of timber and coal deposits. Its agriculture includes corn, soybeans, tobacco, potatoes, livestock, and dairy products.

Cof·fee·ville (kô' fē-vĭl', kŏf 'ē-), town (1970 pop. 1,024), a seat of Yalobusha co., N central Miss., NNE of Grenada.

Cof·fey (kô' fē, kŏf 'ē), county (1970 pop. 7,397), 617 sq mi (1,598 sq km), E Kansas, in a rolling plains area drained by the Neosho River; formed 1859; co. seat Burlington. Livestock, grain, and timber are important. It has scattered oil fields in the south.

Cof·fey·ville (kô' fē-vĭl', kŏf 'ē-), city (1978 est. pop. 14,000), Montgomery co., SE Kansas, on the Verdigris River near the Okla. border, in a farm and oil area; inc. 1872. It is a trading and distributing center, with oil refineries and plants producing foundry and machine-shop products, inorganic chemicals, power-transmission equipment, and milk and dairy items. With the coming of the railroad (1870), Coffeyville grew as a cattle-shipping point. Oil and natural gas were discovered in the area in 1902.

Co·gnac (kô-nyăk'), city (1975 pop. 22,237), Charente dept., W France, in Angoumois, on the Charente River. The French brandy to which Cognac gives its name has been manufactured and exported from the city since the 18th cent.

Co·has·set (kō-hăs' ĭt), resort town (1970 pop. 6,954), Norfolk co., E Mass., on the South Shore of Massachusetts Bay E of Hingham; settled c.1647, inc. 1770. It has a summer theater. A lighthouse has been maintained off the shore here since 1850, and Cohasset had the first U.S. lifeboat service.

Co·hoes (kə-hōz'), city (1978 est. pop. 16,800), Albany co., E N.Y., near Albany, at the confluence of the Mohawk and Hudson rivers; settled by the Dutch 1665, inc. 1869. Its manufactures include textiles, knitted goods, paper products, boats, and electrical appliances. The first power-operated knitting mill was opened here in 1832.

Coim·ba·tore (kwĭm-bə-tôr'), town (1971 pop. 353,469), Tamil Nadu state, SE India. It commands the approach to the Palghat Gap, the major pass through the Western Ghats. The British obtained undisputed possession of Coimbatore in 1799. Glassware, fertilizer, electrical goods, cement, and synthetic gems are produced. Coimbatore is also a market for tea, cotton, cardamom, cinchona, and teak.

Co·im·bra (kō-ēm'brə), city (1970 pop. 55,985), capital of Coimbra dist., W central Portugal, on the Mondego River, in Beira Litoral. The old capital of Beira, it is a market center with small industries but is known chiefly for its famous university, which was founded (1292) by King Diniz in Lisbon but was moved temporarily to Coimbra in 1308 and permanently in 1540.

Coke (kōk), county (1970 pop. 3,087), 915 sq mi (2,369.9 sq km), W Texas, in a ranching area crossed by the Colorado River; formed 1889; co. seat Robert Lee. Its agriculture includes poultry, dairy products, grain sorghums, peanuts, fruit, truck crops, and cotton. There are oil, natural-gas, limestone, and clay deposits.

Col·bert (kŏl'bərt), county (1970 pop. 49,632), 596 sq mi (1,543.6 sq km), NW Ala., bounded on the W by the Miss. border; formed 1867; co. seat Tuscumbia. There are deposits of asphalt, bauxite, and limestone. TVA developments have stimulated industry in the region. Aluminum and other metal products are manufactured.

Col·by (kōl'be), city (1970 pop. 4,658), seat of Thomas co., NW Kansas, NE of Sharon Springs; inc. 1886. It is a shipping trade center for a wheat-growing area. Colby Community College is here, and an agricultural experiment station is nearby.

Col·ches·ter (kōl'chĭs-tər, -chĕs'-), municipal borough (1973 est. pop. 79,600), Essex, SE England, on the Colne River. It is a grain and cattle market. The oyster fisheries of the Colne are important; an annual event is the October oyster feast. Other industries are flour milling, malting, and the making of boilers, gas engines, shoes, and farm machinery. Colchester was one of the great cities of pre-Roman Britain, the capital of the ruler Cunobelin (Shakespeare's Cymbeline). It became an important Roman colony and was the particular object of attack (A.D. 61) by Boadicea. During the English Civil War the town was taken (1648) after a long siege by Parliamentarians. Of interest are the Roman walls (more completely preserved than elsewhere in England) and the massive Norman castle, part of which houses a museum of Roman antiquities.

Col·ches·ter (kōl'chĕs'tər). **1.** Town (1970 pop. 6,603), New London co., SE central Conn., E of Middletown on the Salmon River; settled and inc. 1699. Leather goods and clothing are made. The old Comstock covered bridge became state property c.1934. **2.** Resort town (1970 pop. 8,776), Chittenden co., NW Vt., on Lake Champlain near Burlington and Winooski, in an agricultural area; chartered 1763.

Col·chis (kŏl'kĭs), ancient country on the E shore of the Black Sea in the Caucasus region. Centered about the fertile valley of the Phasis River (the modern Rion), Colchis corresponds to the present-day region of Mingrelia in the Georgian SSR. In Greek legend it was the land where the Golden Fleece was sought by Jason and the Argonauts. Greek trading posts were estabilished in Colchis, but the land remained independent until conquered (c.100 B.C.) by Mithradates VI of Pontus. After the time of Trajan to the end of the Roman Empire, Rome exerted considerable influence on the region.

Cold Harbor (kōld), locality, Hanover co., E central Va., N of the Chickahominy River ENE of Richmond. Two Civil War battles were fought here: Gaines Mill (1862) and the Battle of Cold Harbor (June 1-3, 1864), with heavy losses on both sides.

Cold·spring (kōld'sprĭng'), village (1970 pop. 488), seat of San Jacinto co., E Texas, near the San Jacinto River NNE of Houston, at the edge of Sam Houston National Forest in an agricultural, cattle, and timber area.

Cold·stream (kōld'strēm'), burgh (1971 pop. 1,270), Borders region, SE Scotland, on the English border. George Monck raised troops here in 1660 for his march into England that resulted in the restoration of Charles II to the throne. The regiment became known as the Coldstream Guards, one of the regiments of guards of the royal household.

Cold·wa·ter (kōld'wô'tər). **1.** City (1970 pop. 1,016), seat of Comanche co., S central Kansas, near the Okla. line SE of Dodge City; inc. 1884. It is a trade center for a wheat area. **2.** City (1975 est. pop. 8,700), seat of Branch co., S central Mich., SSE of Battle Creek, in a lake region; settled 1830, inc. as a village 1837, as a city 1861. It is a farm trade center and has diversified manufacturing.

Coldwater, river, NW Miss., rising near the Mississippi River and the Ark. and Tenn. borders and flowing 220 mi (354 km) generally S to join the Tallahatchie River near Lambert.

Cole (kōl), county (1970 pop. 46,228), 385 sq mi (997.2 sq km), central Mo., in the Ozarks bounded on the N by the Missouri River, on the E by the Osage River; formed 1820; co. seat Jefferson. It has agriculture (wheat, corn, and livestock), dairy-processing plants, and zinc, lead, barite, copper, and clay mines. Chemicals and primary metals are manufactured.

Cole·man (kōl'mən), county (1970 pop. 10,288), 1,280 sq mi (3,315.2 sq km), central Texas, bounded on the S by the Colorado River and drained by its tributaries; formed 1858; co. seat Coleman. Agriculture (oats, wheat, grain sorghums, cotton, vegetables, fruit, and pecans), livestock, and dairying are important. It also has oil and natural-gas wells, clay mines, and coal deposits.

Coleman, city (1970 pop. 5,608), seat of Coleman co., central Texas, WNW of Brownwood; founded 1876, inc. 1877. It is the market and processing center of a region yielding oats and other grains, cotton, oil, natural gas, cattle, sheep, and turkeys.

Cole·raine (kol-rān'), municipal borough (1971 pop. 14,871), Co. Londonderry, N Northern Ireland, near the mouth of the Bann River. Coleraine is a port. Its industries include distilling, linen milling, the curing of ham and bacon, bog-iron mining, and salmon fishing. In 1613 James I gave the site of the town to the corporations of the City of London for development.

Coles (kōlz), county (1970 pop. 47,815), 507 sq mi (1,313.1 sq km), E central Ill., drained by the Kaskaskia, Embarrass, and Little Wabash rivers; formed 1830; co. seat Charleston. In an agricultural area (corn, wheat, broomcorn, soybeans, livestock, poultry, and dairy products), it also has diversified manufacturing.

Col·fax (kōl'făks). **1.** County (1970 pop. 9,498), 405 sq mi (1,049 sq km), E Nebr., in an agricultural area bounded on the S by the Platte River; formed 1869; co. seat Schuyler. Flour, grain, and livestock are produced. **2.** County (1970 pop. 12,170), 3,765 sq mi (9,751.4 sq km), in a livestock and coal-mining area bordering on Colo. in the N and drained by the Canadian River; formed 1869; co. seat Raton. Philmont National Boy Scout Ranch is in the county.

Colfax. 1. Town (1970 pop. 1,892), seat of Grant parish, central La., on the Red River, in a timber and farm area; founded c.1870, inc. 1878. **2.** City (1970 pop. 2,664), seat of Whitman co., SE Wash., on the Palouse River S of Spokane; settled 1870 as a cow town, inc. 1878. It is a marketing center for a rich farm area.

Co·li·ma (kō-lē'mä), city (1970 pop. 58,450), capital of Colima state (2,010 sq mi/5,205.9 sq km), SW Mexico. It is a marketing and processing center for the surrounding agricultural region. The city was founded in 1523 by the Spanish explorer Gonzalo de Sandoval.

Coll (kŏl), island 12 mi (19.3 km) long and 4 mi (6.4 km) wide, Argyllshire, W Scotland, one of Inner Hebrides. Coll has dairy and cattle farms and exports potatoes and cheese.

Col·lege (kŏl'ĭj), uninc. village (1970 pop. 3,000), central Alaska, just NW of Fairbanks. It is the seat of the Univ. of Alaska.

College Park. 1. City (1978 est. pop. 26,000), Clayton and Fulton cos., NW Ga., a residential suburb of Atlanta; inc. 1891. **2.** City (1978 est. pop. 27,900), Prince Georges co., W central Md., a residential suburb of Washington, D.C.; settled 1745, inc. 1945. It is the seat of the Univ. of Maryland, and its economy is centered on the university, research institutions, and electronics plants.

College Station, city (1978 est. pop. 30,100), Brazos co., E central Texas, in a livestock and cotton region; inc. 1938. Texas Agricultural and Mechanical Univ. is here.

Col·le·ton (kŏl'ə-tən), county (1970 pop. 27,622), 1,048 sq mi (2,714.3 sq km), S S.C., in a rural area bounded on the NE and E by the Edisto River, on the SW by the Combahee River, and on the S by St. Helena Sound; formed 1798; co. seat Walterboro. Timber, cotton, corn, and livestock are processed. It is also a tourist area, with good hunting and fishing.

Col·lier (kŏl'yər), county (1970 pop. 38,040), 2,006 sq mi (5,195.5 sq km), S Fla., bounded on the W by the Gulf of Mexico; formed 1923; co. seat East Naples. Big Cypress Swamp and the Everglades cover most of the county. The Ten Thousand Islands run along the coastline. Truck farming, fishing, and lumbering are important. It also has some oil wells.

Col·lin (kŏl' ĭn), county (1970 pop. 66,920), 867 sq mi (2,245.5 sq km), N Texas, in a rich blackland-prairie agricultural area drained by the East Fork of the Trinity River; formed 1846; co. seat McKinney. A leading corn-growing area, it also produces cotton, grain, alfalfa, pecans, fruit, truck crops, dairy products, and livestock (cattle, hogs, sheep, and poultry).

Col·lings·wood (kŏl' ĭngz-wo͝od'), borough (1970 pop. 17,422), Camden co., SW N.J.; settled 1682 by Quakers, inc. 1888. It has some light industry.

Col·lings·worth (kŏl'ĭngz-wûrth), county (1970 pop. 4,755), 899 sq mi (2,328.4 sq km), extreme N Texas, in the Panhandle on the Okla. border and drained by the Salt Fork of the Red River; formed 1876; co. seat Wellington. It acquired parts of Beckham and Harmon cos., Okla., when the 100th meridian was relocated in 1930. It is mainly agricultural, yielding cotton, sorghum, peanuts, fruit, truck crops, beef cattle, poultry, hogs, and dairy products.

Collingwood, town (1976 pop. 11,114), S Ont., Canada at the S end of Georgian Bay, an arm of Lake Huron. Collingwood has one of the largest dry docks on the Great Lakes.

Col·lins (kŏl' ĭnz), town (1974 est. pop. 2,245), seat of Covington co., S central Miss., NW of Hattiesburg, in a farm area.

Col·lins·ville (kŏl' ĭnz-vĭl'). **1.** City (1978 est. pop. 20,000), Madison and St. Clair cos., SW Ill.; settled 1817, inc. 1872. It is a former coal-mining center where food products and garments are now manufactured. Cahokia Mounds State Park is nearby. **2.** Uninc. town (1970 pop. 6,015), Henry co., S central Va., just NW of Martinsville.

Col·mar or **Kol·mar** (both: kōl'mär, kôl-mär'), city (1975 pop. 64,771), capital of Haut-Rhin dept., E France, in Alsace, on the Lauch River and the Logelbach Canal. Colmar has textile and other industries. It became a free city of the Holy Roman Empire in 1226, and Louis XIV made it the capital of Alsace in 1673. The old section of Colmar retains its medieval architecture.

Co·logne (kə-lōn'), or *German* **Köln** (kœln), city (1974 pop. 1,022,-075), North Rhine-Westphalia, W West Germany, on the Rhine River. It is a commercial center, a rail and road junction, and a river port. Its manufactures include iron, steel, heavy machinery, chemicals, textiles, printed materials, and eau de cologne.

A Roman garrison in the 1st cent. B.C., Cologne was made a Roman colony in A.D. 50 by Emperor Claudius. The city passed under Frankish control in the 5th cent. The episcopal see, established here in the 4th cent., was made an archdiocese under Charlemagne. Its archbishops, who later ruled a strip of land on the west bank of the Rhine as princes of the Holy Roman Empire, acquired great power and ranked third among the electors. Cologne was self-governing after 1288, became a free imperial city in 1475, and as a member of the Hanseatic League, flourished as a commercial center until the 16th cent. Its decline was hastened by the expulsion of the Jews (15th cent.) and the restrictions imposed on Protestants (16th cent.). Cologne was seized by the French in 1794. The city passed to Prussia in 1815. In the 19th cent. Cologne prospered again as an industrial center and as the main port of northwest Germany.

Old Cologne, with its numerous historic buildings, was severely damaged by aerial bombardment in World War II. The famous Gothic cathedral, begun in 1248 and the largest in northern Europe, was closed from the end of the war until 1956. Other historic buildings in the city include the Romanesque churches of St. Maria im Kapitol, of St. Gereon, of the Holy Apostles, and of St. Andreas (where Albertus Magnus, the 13th cent. scholastic, is buried).

Co·lo·ma (kə-lō'mə, kō-), village (1970 pop. 280), El Dorado co., N central Calif., on the American River NE of Sacramento. Gold was discovered at Sutter's Mill here by James W. Marshall in 1848. A state historic monument and the Marshall Gold Discovery State Park commemorate the event.

Co·lomb-Bé·char (kô-lôn'bā-shär'): *see* Béchar, Algeria.

Co·lombes (kô-lôɴb'), city (1975 pop. 83,390), Hauts-de-Seine dept., N central France, on the Seine River. An industrial suburb of Paris, Colombes has fuel refineries, foundries, and publishing houses.

Co·lom·bi·a (kə-lŭm'bē-ə, kō-lōm'byä), republic (1973 est. pop. 22,750,000), 439,735 sq mi (1,138,914 sq km), NW South America; capital Bogotá. The only South American country with both a Caribbean and a Pacific coastline, Colombia is bounded on the northwest by Panama, on the northeast by Venezuela, on the south by Ecuador and Peru, and on the southeast by Brazil.

By far the most prominent physical features are the three great Andean chains that fan north from Ecuador. The Andean interior is the heart of the country, where in pre-Columbian days the highly advanced Chibcha lived. It has the largest concentration of population and is the area of large-scale cultivation of coffee, Colombia's major crop. Of the three principal Andean ranges, the Western Cordillera is of the least economic importance. The Central Cordillera has a towering chain of volcanoes and is the divide between the valleys of the Magdalena and the Cauca rivers. With improved transportation, the introduction of coffee culture, the exploitation of high-grade coal reserves, and an enormous increase of the white population, it is the economic and industrial core of the republic. The Eastern Cordillera is the longest chain. Its western slopes yield coffee, and in its intermontane basins grains and cattle are raised. The area is rich in iron, coal, and emeralds.

To the east of the Andes lies more than half of Colombia's territory, a vast undeveloped lowland. The plains are crossed by navigable rivers, tributaries of the Orinoco and Amazon systems. The northern section consists of savannas (the llanos), which are devoted to a large extent to cattle and sheep grazing. The dense jungles of the extreme southeast are of negligible economic importance. A fourth mountain chain, the Cordillera del Chocó, runs parallel to the Pacific. The range's slopes yield dyewoods and hardwoods, rubber, tagua nuts (vegetable ivory), and gold and platinum. In the north, separating the La Guajira peninsula from the rest of the country, is the magnificent Sierra Nevada de Santa Marta. The difficult terrain in Colombia limits road and rail transportation, making air and water travel especially important.

Colombia is divided into 22 departments, 4 intendencies, 5 commissaries, and a special district:

NAME	CAPITAL	NAME	CAPITAL
Amazonas (intendency)	Leticia	Juila	Neiva
Antioquia	Medellin	La Guajira	Riohacha
Arauca (commissary)	Arauca	Magdalena	Santa Marta
Atlántico	Barranquilla	Meta	Villavicencio
Bolivar	Cartagena	Nariño	Pasto
Boyacá	Tunja	Norte de Santander	Cúcuta
Caldas	Manizales	Putumayo (commissary)	Mocoa
Caquetá (intendency)	Florencia	Quindío	Armenia
Casanare (intendency)	Yopal	Risaralda	Pereira
Cauca	Popayán	San Andrés y Providencia (intendency)	San Andres
Cesar	Valledupar	Santander	Bucaramanga
Chocó	Quibdo	Sucre	Sincelejo
Córdoba	Monteria	Tolima	Ibagué
Cundinamarca	Bogotá	Valle del Cauca	Cali
Distrito Especial	Bogotá	Vaupés (commissary)	Mitú
Guainía (commissary)	San Felipe	Vichada (commissary)	Puerto Carreño

Economy. Agriculture is the chief source of income in Colombia. Coffee is by far the major crop and its price on the world market has affected Colombia's economic health. Among the commercial crops, coffee is grown between elevations of 3,000 and 6,000 ft (915 and 1,830 m); bananas, cotton, sugar cane, oil palm, and tobacco are grown at lower elevations. Between 6,000 and 10,000 ft (1,830 and 3,050 m) potatoes, beans, grains, and temperate zone fruit and vegetables are grown. Colombia is rich in minerals, including petroleum, iron, coal, gold, silver, platinum, and emeralds. The saltworks at Zipaquira, near Bogotá, are world famous. Manufacturing has expanded greatly in recent decades, although it is heavily dependent on imported materials. Beverages and processed foods, textiles, metal products, and chemicals are the chief manufactures.

History. After the Spanish conquest the area of present-day Colombia formed the nucleus of New Granada. The struggle for independence was, as in all Spanish-American possessions, precipitated by the Napoleonic invasion of Spain. The revolution lasted nine years before the victory of Simón Bolívar at Boyacá (1819) secured the independence of Greater Colombia. The new state Bolívar created included what is now Venezuela, Panama, and (after 1822) Ecuador, as well as Colombia. While Bolívar, who had been named president, headed campaigns in Ecuador and Peru, the vice president, Francisco de Paula Santander, administered the new nation. Santander advocated a union of federal sovereign states, while Bolívar championed a centralized republic. Although Bolívar's authority prevailed (1828), Greater Colombia soon fell apart. In 1830 Venezuela and Ecuador became separate nations. The remaining territory emerged as the republic of New Granada.

Through the 19th cent. and into the 20th cent. political unrest and civil strife reappeared constantly. Strong parties developed along conservative and liberal lines; the conservatives favored centralism and participation by the church in government and education, and the liberals supported federalism, anticlericalism, and some measure of social legislation and fiscal reforms. While Tomás Cipriano de Mosquera was president, a treaty was concluded (1846) granting the United States transit rights across the Isthmus of Panama. A new constitution in 1858 created a confederation of nine states called Granadina, eventually (1863) called the United States of Colombia. The antifederalist revolution of 1885 led one year later, during the presidency of Rafael Núñez, to the formation of the republic of Colombia and enactment of a conservative constitution. In 1899, five years after Núñez's death, civil war of unprecedented violence broke out and raged for three years. As many as 100,000 people were killed before the Conservatives emerged victorious. When the United States acquired the right to complete the Panama Canal (although the agreement was later rejected by the Colombian congress), the republic of Panama, aided by the United States, achieved its independence from Colombia (1903). Colombia recognized (1914) Panama's independence in exchange for rights in the Canal Zone and the payment of an indemnity from the United States. For the next four decades political life remained fairly peaceful, although there was economic and social unrest in the 1920s and 1930s.

Colombia participated in World War II on the Allied side. During and after the war years, internal divisions worsened. In 1948, while an Inter-American Conference was being held in Bogotá, the leftist Liberal leader Jorge Eliécer Gaitán, under whom the party had reunited, was assassinated, precipitating violent riots and acts of vandalism. The country was plunged into a decade of civil strife, martial law, and violent rule that cost hundreds of thousands of lives. An archconservative dictator, Laureano Gómez, took power in 1950 but in 1953 was ousted by a coup led by Gustavo Rojas Pinilla, the head of the armed forces. Repressive measures continued, fiscal reforms failed, the country was plunged into debt, and Rojas Pinilla became implicated in scandalously corrupt schemes. A military junta, backed by liberals and conservatives alike, ousted Rojas Pinilla in 1957. He returned to politics (1970) as the champion of the underprivileged. Colombia's economy began to recover from the setbacks of the early 1970s as economic diversification and incentives to lure foreign capital into the country were initiated. However, a high inflation rate continued to impede economic growth.

Government. Colombia is governed under an 1886 constitution. The president serves a four-year term. The legislature, subservient to the president, consists of a senate and chamber of deputies. The supreme court is chosen by the president and the legislature.

Co·lom·bo (kə-lŭm'bō), largest city (1973 est. pop. 618,000) and capital of Sri Lanka (Ceylon), a port on the Indian Ocean near the mouth of the Kelani River. The city's major sections are the old area of narrow streets and colorful market stalls; the modern commercial, business, and government area around a 16th cent. Portuguese fort; and Cinnamon Gardens, a wealthy residential and recreational area. Colombo has one of the world's largest man-made harbors. Gem cutting and ivory carving are among Colombo's specialties; other industries include food and tobacco processing, metal fabrication, engineering, and the manufacture of chemicals, textiles, glass, cement, leather goods, clothing, furniture, and jewelry.

Colombo was probably known to Greco-Roman, Arab, and Chinese traders more than 2,000 years ago as an open anchorage for oceangoing ships. Moslems settled here in the 8th cent. The Portuguese arrived in the 16th cent. and built a fort to protect their spice trade. The Dutch, also coveting this trade, gained control in the 17th cent. In 1796 Colombo passed to the British, who made it the capital of their crown colony of Ceylon in 1802. In the 1880s Colombo replaced Galle as Ceylon's chief port and became a major refueling and supply center for merchant ships on the Europe-Far East route. Colombo served as an Allied naval base in World War II and was made the capital of independent Ceylon in 1948.

Co·lón (kō-lōn'), city (1970 pop. 25,986), Matanzas prov., W central Cuba. Colón's sugar industry reached its peak in the mid-19th cent. and has since declined. The city was founded in 1818.

Colón, city (1970 pop. 69,650), Panama, at the Caribbean end of the Panama Canal. Colón is an important port and commercial center. It was made a free trade zone in 1953. The city was founded in 1850 by Americans working on the trans-Panama railroad and was named Aspinwall until 1890. The city was often scourged by yellow fever until the sanitary work associated with the construction of the canal was completed under W. C. Gorgas.

Co·lo·nia (kō-lō'nyä), city (1975 pop. 110,860), capital of Colonia dept., S Uruguay, on the Río de la Plata. It is a resort city, a port, and the trade center for a rich agricultural region. The city was founded by the Portuguese in 1680.

Co·lo·ni·al Heights. (kə-lō'nē-əl), independent city (1970 pop. 15,097), SE Va.; inc. as a city 1948. Metal awnings and paint are manufactured, tires are retreaded, and whiskey is bottled in the city. Of particular interest is the Violet Bank Library and Museum and the giant cucumber tree in front of it. In 1864, during the Civil War, Gen. Robert E. Lee made his headquarters under the tree while directing the defense of Petersburg.

Colonial National Historical Park, 9,430 acres (3,819.2 hectares), SE Va., mainly on the peninsula between the York and James rivers; created 1930 as Colonial National Monument, renamed 1936. The park embraces a historic region that includes Cape Henry and Yorktown, Jamestown, and Williamsburg. The Colonial Parkway, part of the park, links the three old towns. Archaeological studies and reconstruction of old places of interest have been carried on.

Col·on·say (kŏl'ən-zā, -sā), island, 17 sq mi (44 sq km), Argyll, NW Scotland, one of the Inner Hebrides. Crofting and cheese making are the main occupations.

Col·o·ra·do (kŏl'ə-rä'dō, -răd'ə), state (1975 est. pop. 2,524,000) 104,247 sq mi (270,000 sq km), W central United States, one of the Rocky Mt. states, admitted as the 38th state of the Union in 1876 (and therefore known as the "Centennial State"). Denver is the capital. Colorado is bounded on the north by Wyo. and Neb., on the east by Neb. and Kansas, on the south by Okla. and N.Mex., and on the west by Utah.

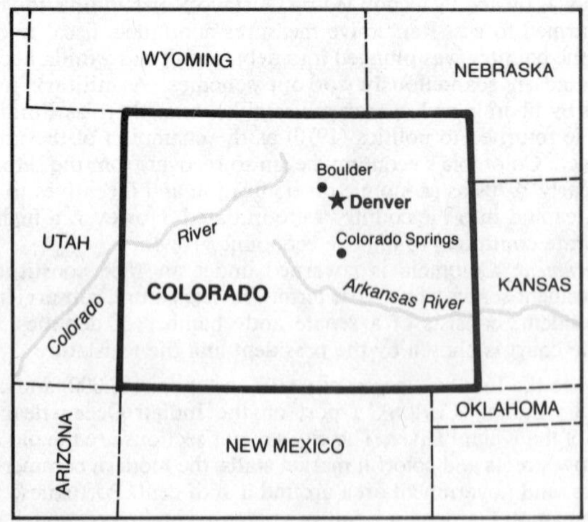

The plains of Colorado's eastern section are part of the High Plains section of the Great Plains. On their western edge the plains give way to the foothills of the Rocky Mts., which run north-south through central Colorado. The mountains are divided into several ranges that make up two generally parallel belts, with the Front Range and a portion of the Sangre de Cristo Mts. on the east and the Park Range, Sawatch Mts., and San Juan Mts. on the west. The mountain ranges are separated by high valleys and basins called parks. These include North Park, Middle Park, South Park, and San Luis Park. The Continental Divide runs north-south along the Rocky Mts. in Colorado. One of the most scenic states in the country, Colorado's parks include Rocky Mountain National Park, Black Canyon of the Gunnison National Monument with its narrow gorge cut by the Gunnison River, Dinosaur National Monument in northwest Colorado, and Great Sand Dunes National Monument in south central Colorado. Mesa Verde National Park, once the home of Indian cliff dwellers, is located in the southwestern corner of the state. Most of western Colorado is occupied by the Colorado Plateau, where many canyons have been formed by the action of the Colorado and Gunnison rivers. Colorado has a mean elevation of c.6,800 ft (2,075 m) and numerous peaks over 14,000 ft (4,270 m) high.

Colorado is divided into 63 counties:

NAME	COUNTY SEAT	NAME	COUNTY SEAT
Adams	Brighton	Lake	Leadville
Alamosa	Alamosa	La Plata	Durango
Arapahoe	Littleton	Larimer	Fort Collins
Archuleta	Pagosa Springs	Las Animas	Trinidad
Baca	Springfield	Lincoln	Hugo
Bent	Las Animas	Logan	Sterling
Boulder	Boulder	Mesa	Grand Junction
Chaffee	Salida	Mineral	Creede
Cheyenne	Cheyenne Wells	Moffat	Craig
Clear Creek	Georgetown	Montezuma	Cortez
Conejos	Conejos	Montrose	Montrose
Costilla	San Luis	Morgan	Fort Morgan
Crowley	Ordway	Otero	La Junta
Custer	Westcliffe	Ouray	Ouray
Delta	Delta	Park	Fairplay
Denver	Denver	Phillips	Holyoke
Dolores	Dove Creek	Pitkin	Aspen
Douglas	Castle Rock	Prowers	Lamar
Eagle	Eagle	Pueblo	Pueblo
Elbert	Kiowa	Rio Blanco	Meeker
El Paso	Colorado Springs	Rio Grande	Del Norte
Fremont	Canon City	Routt	Steamboat Springs
Garfield	Glenwood Springs	Saguache	Saguache
Gilpin	Central City	San Juan	Silverton
Grand	Hot Sulphur Springs	San Miguel	Telluride
Gunnison	Gunnison	Sedgwick	Julesburg
Hinsdale	Lake City	Summit	Breckenridge
Huerfano	Walsenburg	Teller	Cripple Creek
Jackson	Walden	Washington	Akron
Jefferson	Golden	Weld	Greeley
Kiowa	Eads	Yuma	Wray
Kit Carson	Burlington		

Economy. Agriculture, especially the raising of cattle and sheep, is economically important in the state. Crops, which include wheat, hay, corn, and sugar beets, accounted for less than a quarter of all farm income in 1970. In the 1950s manufacturing displaced agriculture as the major source of income in the state. Food processing is the main industry. Other important industries include the manufacture of nonelectrical equipment, transportation equipment, and electrical equipment, printing and publishing, and the production of stone, clay, and glass products, fabricated metals, chemicals, and lumber. Tourism also plays a vital role in the economy. Gold, the lure to exploration and settlement of Colorado, was the first of many useful minerals to be discovered there. In 1970 molybdenum was the most valuable mineral produced in the state; Colorado has the world's largest known deposit of that mineral. Other leading minerals are petroleum, coal, sand and gravel, and uranium. Large coal and oil deposits provide considerable resources for the generation of electricity. Hydroelectric power is also used.

History. Colorado's earliest known inhabitants were the Basket Makers and the Cliff Dwellers. The first white man to enter the region was probably the Spanish conquistador Francisco Vásquez de Coronado in the 16th cent. Spain subsequently claimed (1706) the territory, although no Spanish settlements were established there. Part of the area was also claimed for France as part of the Louisiana Territory. At the end of the French and Indian Wars (1763), France secretly ceded the Louisiana Territory, including much of Colorado, to Spain. The French regained the area in 1800 and sold it to the United States in the Louisiana Purchase of 1803. The Federal government sent expeditions to Colorado under Zebulon M. Pike (1806), Stephen H. Long (1819-20), and John C. Frémont (1842-43 and 1845). Settlement in the area did not begin, however, until the United States acquired the remainder of present-day Colorado from Mexico by the Treaty of Guadalupe Hidalgo in 1848.

It was the discovery of gold that first brought large numbers of white men to Colorado. Prospectors led by Green Russell discovered gold in 1858 at Cherry Creek, where the city of Denver now stands. The next year John Gregory made a great strike on the site of present-day Central City, and the lusty, lawless days of the mining boom began. The area in which the gold fields were located was then part of the U.S. Kansas Territory. The miners sought separate territorial status in 1859 and formed the illegal Territory of Jefferson, which operated until the bill for territorial status was passed by Congress in 1861. A bill granting Colorado's statehood was finally passed by Congress in 1876. Meanwhile, warfare culminated in the defeat of the Plains Indians after the Indian Wars (1861-69) and the Buffalo War (1873-74). Today Colorado's Indians live mainly on the Southern Ute Reservation and in the Denver area.

The completion (1870) of a railroad link from Denver to the Union Pacific in Cheyenne, Wyo., and later railroad construction helped to stimulate the extension of farming and ranching as mining communities became ghost towns. Between 1870 and 1880 population increased almost fivefold. Denver briefly became the largest receiving market for sheep, and a smelting industry was established. In the 1890s, despite a rich silver strike at Creede and the discovery of the state's richest gold field at Cripple Creek, Colorado suffered a depression. Labor conflicts, disputes over railway franchises, and warfare between sheep and cattle interests also plagued the state at the turn of the century.

By 1910, with the improvement of national economic conditions, Colorado settled down to a predominantly agricultural economy. During World War I the price of silver soared again and the economy prospered. The droughts of 1935 and 1937 and the stock-market crash of 1929 brought hardship to many. Since the mid-1960s Colorado has experienced a large influx of new residents and rapid urban growth and development, especially along a strip (c.150 mi/240 km long) centered on Denver and stretching from Fort Collins and Greeley in the north to Pueblo in the south.

Government. Colorado's constitution was drawn up in 1876. The governor of the state is popularly elected and serves for a term of four years. The legislature is made up of a senate with 35 members elected for four-year terms and a house of representatives with 65 members elected for two-year terms.

Educational Institutions. Among Colorado's institutions of higher learning are the Univ. of Colorado, at Boulder; the Univ. of Denver, at Denver; Colorado State Univ., at Fort Collins; and the United States Air Force Academy, at Colorado Springs.

Colorado, county (1970 pop. 17,638), 950 sq mi (2,460.5 sq km), S Texas, drained by the Colorado and San Bernard rivers; formed 1836; co. seat Columbus. In an agricultural area (rice, potatoes, cotton, wheat, corn, grain sorghums, truck crops, fruit, livestock, and dairy products), it also has gravel and sand pits, natural-gas and oil wells, and a lumber industry.

Co·lo·ra·do (kō-lō-rä'dō, *-thō*), river, c.550 mi (885 km) long, rising from tributaries in the Andes and flowing SE across S central Argentina to the Atlantic Ocean. It marks the northern limit of Patagonia.

Col·o·ra·do (kŏl'ə-rä'dō, -răd'ə). **1.** River, 1,450 mi (2,333 km) long, SW United States, rising in the Rocky Mts. of N Colo., and flowing generally SW through Colo., Utah, Ariz., between Nev. and Ariz., and Ariz. and Calif., and then into Mexico, emptying into the Gulf of California. It drains c.244,000 sq mi (631,960 sq km). The Gunnison, Green, San Juan, and Little Colorado are the main tributaries of the upper basin of the Colorado; the Gila is the chief tributary of the lower basin. Silt deposited by the Colorado has formed a great delta across the northern part of the Gulf of California, cutting off the head of the gulf; Salton Sea is a remnant of the severed part. The river flows through c.1,000 mi (1,610 km) of canyons, of which the most spectacular is the Grand Canyon. Many national parks, monuments, and recreational areas are located along the river banks. The Colorado's waters are used for power and irrigation. The mouth of the river was seen by Francisco de Ulloa in 1539; the lower part was explored by Hernando de Alarcón in 1540. Controversies over water rights on the Colorado have long raged between the United States and Mexico and among the bordering states; treaties and compacts now regulate the river's use. **2.** River, 894 mi (1,438.5 km) long, rising in the Llano Estacado, NW Texas, and flowing SE to Matagorda Bay, an inlet of the Gulf of Mexico. It drains c.41,500 sq mi (107,485 sq km). Destructive floods, which prevented private development of the river for power, led the Texas legislature to set up the Lower, Central, and Upper Colorado River authorities to undertake projects for flood control, power plants, and irrigation.

Colorado City, city (1970 pop. 5,227), seat of Mitchell co., W Texas, on the Colorado River WSW of Sweetwater. It prospered from shipping cattle in the late 19th cent. and later from processing cotton and refining oil. It is now the center of a farm area growing cotton, grains, peanuts, and other crops.

Colorado National Monument: *see* National Parks and Monuments Table.

Colorado Plateau, physiographic region of North America, c.150,000 sq mi (388,500 sq km), SW United States, in Ariz., Utah, Colo., and N.Mex. It is characterized by broad plateaus, ancient volcanic mountains at altitudes of c.5,000 to 13,000 ft (1,525-3,965 m), and deeply dissected great canyons carved into nearly horizontal and often brightly colored sedimentary and volcanic rocks; the Grand Canyon of the Colorado River is part of the region, as are a number of U.S. national parks and monuments. Indian reservations occupy about one third of the mostly semiarid and sparsely vegetated area; about one half of the public land is used for grazing. Ancient cliff dwellings at Mesa Verde National Park and Canyon de Chelly National Monument are of archaeological interest.

Colorado Springs, city (1970 pop. 135,060), seat of El Paso co., central Colo., on Monument and Fountain creeks, at the foot of Pikes Peak; inc. 1886. It is a beautiful residential and year-round vacation and health resort city, with thriving industries producing a wide variety of products. The town of El Dorado (later Colorado City) was founded on Fountain Creek by gold miners in 1859. In 1871 Gen. William Palmer and the Denver and Rio Grande RR established the modern city of Fountain Colony nearby; the name was changed to Colorado Springs because of the many mineral springs in the area. The city grew as a summer and health resort, absorbing the earlier community of Colorado City in 1917. The United States Air Force Academy is nearby.

Col·quitt (kŏl'kwĭt), county (1970 pop. 32,298), 563 sq mi (1,458.2 sq km), S Ga., bounded on the E by the Little River and drained by the Ochlockonee River and its affluents; formed 1856; co. seat Moultrie. It is in a coastal-plain timber and agricultural area yielding cotton, tobacco, corn, peanuts, melons, truck crops, and livestock.

Colquitt, city (1970 pop. 2,026), seat of Miller co., SW Ga., on Spring Creek WNW of Thomasville; inc. 1860. It is a farm trade center.

Col·ton (kōl'tən), city (1970 pop. 20,016), San Bernardino co., S Calif., a suburb of San Bernardino, in a rich citrus and farm area; inc. 1887.

Co·lum·bi·a (kə-lŭm'bē-ə). **1.** County (1970 pop. 25,952), 768 sq mi (1,989.1 sq km), SW Ark., bounded on the S by the La. border and drained by Bayou Dorcheat; formed 1852; co. seat Magnolia. It has agriculture (cotton, corn, and hogs), oil and natural-gas wells, and a lumber industry. Food products, textiles, chemicals, and rubber, plastic, and metal goods are manufactured. **2.** County (1970 pop. 25,250), 786 sq mi (2,035.7 sq km), N Fla., bounded on the N by the Ga. border, on the NW by the Suwannee River, on the S by the Santa Fe River; formed 1832; co. seat Lake City. In a flatwoods and agricultural area yielding corn, peanuts, cotton, tobacco, and livestock, it also has some phosphate deposits in the south. It includes parts of Okefenokee Swamp and Osceola National Forest. **3.** County (1970 pop. 22,327), 290 sq mi (751.1 sq km), E Ga., bounded on the E by the Savannah River, here forming the S.C. line, and on the N by Clark Hill Reservoir; formed 1790; co. seat Appling. Farming (cotton, corn, and potatoes), livestock ranching, and sawmilling are done. **4.** County (1970 pop. 51,519), 643 sq mi (1,665.4 sq km), SE N.Y., drained by Kinderhook Creek and bordered on the E by Mass., on the W by the Hudson River; formed 1786; co. seat Hudson. Dairying, poultry raising, and farming (apples, corn, grain, and potatoes) are important. It also has limestone quarries. Part of Taconic State Park, several small lakes, and summer and winter resorts are here. **5.** County (1970 pop. 28,790), 640 sq mi (1,657.6 sq km), NW Oregon, bounded on the N and E by the Columbia River; formed 1854; co. seat St. Helens. Lumber milling, dairying, and salmon fishing are done. It has iron-ore deposits. **6.** County (1970 pop. 55,114), 484 sq mi (1,253.6 sq km), E central Pa., in a hilly region extending from the Pocono plateau in the N to anthracite-coal fields in the S, and drained by the Susquehanna River; formed 1813; co. seat Bloomsburg. Metals, railroad parts, motor vehicles, textiles, and carpets are made. **7.** County (1970 pop. 4,439), 860 sq mi (2,227.4 sq km), SE Wash., on the Oregon border in a plateau region drained by the Tucannon and Touchet rivers and bounded on the N by the Snake River; formed 1875; co. seat Dayton. It is in an agricultural area yielding wheat, barley, oats, apples, timber, and livestock. Food processing and canning is done. The Blue Mts. are in the south, and part of Umatilla National Forest is here. **8.** County (1970 pop. 40,150), 778 sq mi (2,015 sq km), S central Wis., drained by the Wisconsin, Fox, Crawfish, and Baraboo rivers; formed 1846; co. seat Portage. In a dairying, stock-raising, and farming (grain, potatoes, and tobacco) area, it processes vegetables and dairy products and has some manufacturing.

Columbia. **1.** Rural town (1970 pop. 3,129), Tolland co., E central Conn., W of Willimantic; settled c.1695, set off from Lebanon 1804. Here Eleazar Wheelock established the Indian charity school that became Dartmouth College. **2.** City (1970 pop. 4,188), Monroe co., SW Ill., WSW of Belleville; inc. 1859. Limestone quarries are in the area. **3.** Town (1970 pop. 3,234), seat of Adair co., S central Ky., ENE of Glasgow, in a tobacco and livestock area; settled c.1793. Lindsey Wilson College is here. **4.** City (1970 pop. 1,000), seat of Caldwell parish, NE La., on the Ouachita River S of Monroe, in a farm and timber region. **5.** City (1970 pop. 7,587), seat of Marion co., S Miss., on the Pearl River WSW of Hattiesburg, in an agricultural area. It was the state capital for part of 1821. **6.** City (1970 pop. 58,812), seat of Boone co., central Mo.; inc. 1826. The trade center of a farm and coal area, it is best known as the seat of the Univ. of Missouri and Stephens College. The city is a medical center, with the university hospital, a state cancer hospital, and a state regional mental health clinic. **7.** Town (1970 pop. 902), seat of Tyrrell co., NE N.C., in fertile tidewater area on the Scuppernong River just S of Albemarle Sound. The town was a trading post before 1700; its name was changed several times before it became Columbia in 1810. **8.** Industrial borough (1970 pop. 11,237), Lancaster co., SE Pa., on the Susquehanna River; settled by Quakers c.1730, inc. 1814. The borough was originally called Wright's Ferry; its name was changed in 1789 when it narrowly missed Congressional selection as the permanent U.S. capital. **9.** City (1970 pop. 113,542), state capital, and seat of Richland co., central S.C., at the head of navigation on the Congaree River; inc. 1805. It is an important trade and commercial point in the heart of a rich farm region. Its industries include printing and the manufacture of textiles, clothing, plastics, electronic equipment, office machinery, and glass and stone products. A trading post flourished nearby in the early 18th cent. In 1786 the site was chosen for the new state capital because of its central location; the legislature first met in its new quarters in 1790. During the Civil War, Gen. Sherman's army entered Columbia on Feb. 17, 1865. That night most of the city was burned by drunken Union soldiers and was almost totally destroyed. An educational center, Columbia is the seat of the Univ. of South Carolina, Benedict College, and Columbia College. Notable buildings include the statehouse (begun 1855, damaged in 1865, completed 1901), Woodrow Wilson's boyhood home (1870), and several ante-bellum houses. **10.** City (1970 pop. 21,471), seat of Maury co., central Tenn., on the Duck River; inc. 1817. Once a noted mule market and racing-horse center, it is now the trade and processing hub of a fertile area producing beef cattle and burley tobacco and a shipping point for the region's limestone and phosphate deposits. Columbia's many fine ante-bellum homes include the James K. Polk House (1816).

Columbia, river, c.1,210 mi (1,945 km) long, rising in Columbia Lake, SE British Columbia, Canada. It flows first northwest in the Rocky Mt. Trench, then hooks sharply about the Selkirk Mts. to flow south through Upper Arrow Lake and Lower Arrow Lake and receive the Kootenai River (spelled Kootenay in Canada) before entering the United States after a course of 465 mi (748.2 km). It continues south through Wash. and just below the mouth of the Spokane River is forced by lava beds to make a great bend westward before veering south again, running the while entrenched in a narrow valley through the Columbia Plateau. Its chief tributary, the Snake River, joins it just before it turns west again. The Columbia then forms part of the Wash.-Oregon border before entering the Pacific Ocean through a wide estuary west of Portland, Oregon. The Columbia River has created regal gorges by cutting through the Cascades and the Coast Ranges; it is fed by the Cowlitz and Willamette rivers, which drain the Puget trough between those ranges. Grand Coulee, now a reservoir in the Columbia basin project, was a former stream channel of the Columbia River. It was created during the Ice Age when the Columbia's course was blocked by ice, forcing it to cut a new channel through the Columbia Plateau. When the ice receded the river resumed its former channel.

The Columbia River, commanding one of the great drainage basins of North America (c.259,000 sq mi/670,810 sq km), was discovered by Robert Gray, an American explorer, in 1792 and is named for his vessel, the *Columbia.* It was first actually entered by a British naval officer, William R. Broughton, later the same year. The first whites to arrive overland were the members of the Lewis and Clark expedition and the fur traders. The river was the focus of the American settlement that created Oregon, and the river was itself sometimes called the Oregon River or the River of the West. Irrigation was begun early, and some tributaries were used to water cropland and orchards, as in, for example, the valleys of the Wenatchee and Yakima rivers. After 1932 plans gradually developed to use the Columbia River to its ultimate possibility and the Columbia basin project was established.

Columbia City, city (1977 est. pop. 5,022), seat of Whitley co., NE Ind., WNW of Fort Wayne. It is a resort center in a lake region and has some manufacturing.

Columbia Heights, city (1970 pop. 23,997), Anoka co., SE Minn., a residential suburb adjoining Minneapolis, on the Mississippi River; inc. 1921. It has many varied manufactures.

Co·lum·bi·an·a (kə-lŭm'bē-ăn'ə), county (1970 pop. 108,310), 535 sq mi (1,385.7 sq km), E Ohio, drained by the Little Beaver River and Sandy and Yellow creeks and bounded on the E by the Penn. border, on the SE by the Ohio River; formed 1803; co. seat Lisbon. In an agricultural area yielding livestock, dairy products, fruit, grain, and truck crops, it also has coal mines and clay pits.

Columbiana, town (1970 pop. 2,248), seat of Shelby co., central Ala., SSE of Birmingham; founded c.1825. It is a trade center for a corn, cotton, and timber area. During the Civil War Confederate munitions were made here.

Columbia Plateau, physiographic region of North America, c.100,000 sq mi (259,000 sq km), NW United States, between the Rocky Mts. and the Cascade Range in Wash., Oregon, and Idaho. Most of the plateau is underlaid by deposits, more than 10,000 ft (3,050 m) thick in places, of lava (mainly basalt) interbedded with sedimentary rock; older rocks outcrop in the Blue and Wallowa mts. Young lavas, scattered cinder cones, volcanic ash, and barren landscapes are features of the Snake River plain in the south. Older, decayed lavas, much modified by accumulations of loess, occur in the north in the Columbia basin section; coulees (dry river canyons) and scablands (extensively eroded basalt surfaces), both carved by glacial meltwaters, are features of the region. The Columbia Plateau is an important agricultural and grazing area and is a major source of hydroelectric power.

Co·lum·bus (kə-lŭm'bəs), county (1970 pop. 46,937), 939 sq mi (2,432 sq km), SE N.C., in a forested and partly swampy tidewater area drained by the Waccamaw River and bounded on the SW by the S.C. border, on the NW by the Lumber River; formed 1808; co. seat Whiteville. Tobacco, corn, and sweet potatoes are grown, and sawmilling is done.

Columbus. **1.** City (1970 pop. 166,565), consolidated government area, W Ga., at the head of navigation on the Chattahoochee River; settled and inc. 1828 on the site of a Creek Indian village. Columbus is a port of entry situated at the foot of a series of falls that extend more than 30 mi (48 km) and provide extensive water power. An important industrial and shipping center with many giant textile mills (the first was built in 1838), it also has ironworks, food-processing plants, and factories producing lumber, chemicals, crushed granite, furniture, hospital equipment, concrete, wood and rubber products, and beverages. Columbus, carved out of the wilderness, was built according to plan and remained a busy river port until the arrival of the railroads in the 1850s. Its river traffic has been revitalized with the completion of a series of locks and dams providing access to the Gulf of Mexico. During the Civil War, Columbus was an important Confederate industrial center. It was captured by Federal troops one week after Lee's surrender at Appomattox. Its industrial growth received added impetus in the early 20th cent. with the development of hydroelectric power plants. There are many ante-bellum homes in the city, and its oldest section has been marked for restoration and preservation. **2.** City (1970 pop. 26,457), seat of Bartholomew co., S central Ind., on the East Fork of the White River; inc. 1821. Its many manufactures include automotive parts, diesel engines, castings, metal furniture, electric controls, and plastic components. In the Civil War, Columbus served as a depot for Union armies. Both the railroads and the war brought industries, which remain to this day. The city is known for its outstanding architecture. **3.** City (1970 pop. 3,356), seat of Cherokee co., extreme SE Kansas, SSW of Pittsburg; laid out 1868, inc. 1871. It is a trade center for a farming area. There are coal and zinc deposits in the vicinity. **4.** City (1970 pop. 25,795), seat of Lowndes co., NE Miss., on the Tombigbee River; inc. 1821. The trade, processing, and shipping center of a large cotton, livestock, dairy, and timber area, it also has marble works and garment factories. Franklin Academy, the first free school in the state, now part of the public school system, was opened in 1821. Mississippi Univ. for Women and Columbus Air Force Base are here. A pilgrimage for tourists to the city's many beautiful ante-bellum homes is conducted each year. **5.** Town (1970 pop. 1,173), seat of Stillwater co., S central Mont., at the junction of the Stillwater and Yellowstone rivers WSW of Billings, in a coal, grain, livestock, and sugar-beet area; inc. 1907. **6.** City (1970 pop. 15,471), seat of Platte co., E central Nebr., in a prairie region at the confluence of the Loup and Platte rivers; inc. 1857. It is a railroad, manufacturing, and trade center for a livestock, dairy, and grain area. **7.** Village (1970 pop. 241), Luna co., SW N.Mex., just N of the Mexican border, in a mining area. On Mar. 9, 1916, the village was raided by some 800 Mexican irregulars under Pancho Villa. Several American citizens were killed and much property was damaged. President Woodrow Wilson thereupon ordered Gen. John J. Pershing's punitive expedition into Mexico against Villa. **8.** Town (1970 pop. 731), seat of Polk co., W N.C., SE of Hendersonville near the S.C. border. It has a large woolen mill, and drugs are manufactured. **9.** City (1970 pop. 540,025), state capital and seat of Franklin co., central Ohio, on the Scioto River; inc. as a city 1834. It is a port of entry and a major industrial and trade center in a rich farm region. Its many manufactures include household appliances, aircraft and missiles, automatic controls, foundry and machine-shop products, glass items, processing equipment, and coated fabrics. Columbus was laid out as state capital in 1812, but did not take over the government from Chillicothe until 1816. Its growth was stimulated by the development of transportation facilities—a feeder canal to the Ohio and Erie Canal, which was opened in 1831; the National Road, which reached the city in 1833; and the railroad, which arrived in 1850. Today the city is the seat of Ohio State Univ. and many other educational institutions. Landmarks include the state capitol; the Columbus Gallery of Fine Arts; the library and museum of the state archaeological and historical society; the headquarters of the American Rose Society, with one of the world's largest rose gardens; Camp Chase Confederate cemetery, with the graves of soldiers who died in the Civil War prison camp here; and the vast state fair grounds. **10.** City (1970 pop. 3,342), seat of Colorado co., S Texas, on the Colorado River W of Houston; founded 1823; inc. 1928. The old city has wide, oak-bordered streets and yards. It is the trade center of a diversified farm area, with oil and natural-gas wells and gravel pits.

Co·lu·sa (kə-lo͞o'sə), county (1970 pop. 12,430), 1,153 sq mi (2,986.3 sq km), N central Calif., in a farming, stock-raising, dairying, and bee-keeping area bounded on the E by the Sacramento River; formed 1850; co. seat Colusa. It rises from the lowlands along the Sacramento River in the east to the Coast Ranges in the west. Rice, barley, sugar beets, alfalfa, almonds, prunes, and vegetables are grown. Gold mining, sand and gravel quarrying, and hunting are also important. Part of Mendocino National Forest and a wildlife refuge are here.

Colusa, city (1970 pop. 3,842), seat of Colusa co., N central Calif., on the Sacramento River NNW of Sacramento; founded 1850, inc. 1870. It is the center of an irrigated rice-producing area.

Col·ville (kŏl'vĭl), city (1970 pop. 3,742), seat of Stevens co., NE Wash., on the Colville River NNW of Spokane; founded 1825 as a Hudson's Bay Company post, platted 1883, inc. 1890. It is in a mining, timber, and farm area.

Colville, river, c.375 mi (605 km) long, rising in the De Long Mts. of the Brooks Range, NW Alaska, and flowing across the tundra, E then N, to the Arctic Ocean. The river, frozen for most of the year, floods each spring as ice on its upper course melts. Coal, oil, and natural gas are found in the valley.

Col·wyn (kŏl'wĭn), borough (1970 pop. 3,169), Delaware co., SE Pa., SW of Philadelphia; inc. 1894. It has machine shops.

Colwyn Bay, municipal borough (1971 pop. 25,535), Clwyd, N Wales. It is a popular seaside resort.

Co·mal (kō-măl'), county (1970 pop. 24,165), 567 sq mi (1,468.5 sq km), S central Texas, drained by the Guadalupe River; formed 1846; co. seat New Braunfels. In a ranching, dairying, timber, and agricultural (corn, sorghum, oats, cotton, and pecans) area, it also has limestone and clay deposits and some manufacturing.

Co·man·che (kə-măn'chē). **1.** County (1970 pop. 2,702), 800 sq mi (2,072 sq km), S Kansas, in a level to rolling prairie region bordered on the S by Okla. and drained by the Salt Fork of the Arkansas River; formed 1885; co. seat Coldwater. It is in a cattle-raising and grain-growing area. **2.** County (1970 pop. 108,144), 1,088 sq mi (2,817.9 sq km), SW Okla., drained by Cache, West Cache, and Beaver creeks; formed 1907; co. seat Lawton. It has granite and limestone quarries, oil wells, sand and gravel pits, and some agriculture (livestock, cotton, wheat, hay, oats, and sorghums). The Wichita Mts., Fort Sill, and a wildlife refuge are here. **3.** County (1970 pop. 11,898), 966 sq mi (2,501.9 sq km), central Texas, drained by the Leon and South Leon rivers; formed 1856; co. seat Comanche. In a leading peanut-producing area, it also markets fruit, pecans, grain, truck crops, and livestock. Wool and mohair are processed. There are oil and natural-gas wells and clay deposits here.

Comanche, town (1970 pop. 3,933), seat of Comanche co., central Texas, SW of Fort Worth; founded 1858, inc. 1873. Long a cow town, Comanche retains an Old West air and still ships cattle and sheep. It also processes peanuts and pecans, dairy products, and poultry.

Co·ma·ya·gua (kō'mä-yä'gwä), town (1974 pop. 15,941), W central Honduras. Founded in 1537, Comayagua was the most important city of colonial Honduras. In the political struggle following independence from Spain (1821), Comayagua, the Conservative stronghold, rivaled Tegucigalpa, seat of the Liberal faction. The cities alternated as capital of the republic, but in 1880 Tegucigalpa became the permanent capital. Today Comayagua is the center of an agricultural and mining region. It has a fine colonial cathedral and other colonial landmarks.

Com·ba·hee (kŭm'bē), river, c.140 mi (225 km) long, S S.C., formed by the junction of the Salkehatchie and Little Salkehatchie rivers and flowing SE to St. Helena Sound, an inlet of the Atlantic Ocean.

Co·mil·la (kō-mĭl'ə), town (1974 pop. 86,446), E Bangladesh, on the Gumti River. It is a collection point for hides and skins and has a noted cottage industry in cane and bamboo basketry.

Com·mack (kŏm'ăk), uninc. town (1970 pop. 24,200), Suffolk co., SE N.Y., on central Long Island. It is chiefly residential.

Com·ma·ge·ne (kŏm'ə-jē'nē), ancient district of N Syria, on the Euphrates River and S of the Taurus Mts., now in SE Asiatic Turkey. The fertile agricultural district was made part of the Assyrian Empire and later of the Persian Empire. In the period after Alexander the Great, it became independent under the Seleucid kings of Syria. Vespasian permanently annexed Commagene (A.D. 72).

Com·mand·er Islands (kə-măn'dər): *see* Komandorski Islands.

Com·merce (kŏm'ərs). **1.** City (1970 pop. 10,635), Los Angeles co., S Calif., a suburb of Los Angeles; inc. 1960. An important transportation hub, Commerce is the home of several large corporations; manufactures range from telephones to chemicals. **2.** City (1970 pop. 9,534), Hunt co., NE Texas, NE of Dallas; settled 1874. Primarily a cotton center, it makes clothing as well as dairy and wood products. East Texas State Univ. is here.

Commerce City, city (1970 pop. 17,407), Adams co., N central Colo., an industrial suburb of Denver; inc. 1952.

Com·mon·wealth of Nations (kŏm'ən-wĕlth'), voluntary association of Great Britain and its dependencies, certain former British dependencies that are now sovereign states and their dependencies, and the associated states (states with full internal government but whose external relations are governed by Britain). At its foundation under the Statute of Westminster in 1931, the Commonwealth was composed of Great Britain, the Irish Free State (now the Republic of Ireland), Canada, Newfoundland (since 1949 part of Canada), Australia, New Zealand, and South Africa. As of 1977 the other sovereign members (with date of entry) were: India (1947), Sri Lanka (as Ceylon, 1948), Ghana (1957), Malaysia (as Federation of Malaya, 1957), Nigeria (1960), Cyprus (1961), Sierra Leone (1961), Tanzania (as Tanganyika, 1961), Jamaica (1962), Trinidad and Tobago (1962), Uganda (1962), Kenya (1963), Malawi (1964), Zambia (1964), Malta (1964), The Gambia (1965), Singapore (1965), Guyana (1966), Botswana (1966), Lesotho (1966), Barbados (1966), Mauritius (1968), Swaziland (1968), Western Samoa (1970), Tonga (1970), Fiji (1970), Bangladesh (1972), the Bahamas (1973), Grenada (1974), Papua New

Guinea (1975), and Seychelles (1976). Ireland, South Africa, and Pakistan withdrew in 1949, 1961, and 1972, respectively. Nauru became a special member in 1968. The associated states in 1977 were: Antigua (1967); St. Kitts-Nevis (1967); Dominica (1967); St. Lucia (1967); and St. Vincent (1969).

Co·mo (kō'mō), city (1971 pop. 97,395), capital of Como prov., Lombardy, N Italy, at the SW end of Lake Como, near the Swiss border. It is primarily a tourist center. Originally a Roman colony, Como became an independent commune in the 11th cent. and was frequently at war with, and ruled by, Milan. It later came under Spanish and Austrian control and was liberated by Garibaldi in 1859.

Como, Lake, c.56 sq mi (145 sq km), 30 mi (48 km) long and from .5 to 2.5 mi (0.8–4 km) wide, in Lombardy, N Italy. Lake Como is a natural widening of the Adda River, which feeds and drains the lake. Situated in the foothills of the Alps, the lake is one of the most beautiful of Europe.

Co·mo·do·ro Ri·va·da·via (kō'mō-*thō*'rō rē'vä-*thа*'vyä), town (1970 pop. 78,479), Chubut prov., S Argentina, on the Gulf of San Jorge, an inlet of the Atlantic Ocean. The major center of oil production in Argentina, it is connected by a 1,100-mi (1,770-km) pipeline with Buenos Aires.

Com·o·ro Islands (kŏm'ə-rō), republic (1974 est. pop. 292,000), 838 sq mi (2,170.4 sq km), an archipelago in the Indian Ocean, at the N end of the Mozambique Channel, between the Malagasy Republic and Mozambique. The capital and largest city is Moroni. The Comoro Islands comprise the four main islands of Grande Comore—on which Moroni is located—Anjouan, Mayotte, and Mohéli, and numerous coral reefs and islets. They are volcanic in origin and have a tropical climate. The islands' economy is largely agricultural; the main farming areas are held by foreign companies and feudalistic local landowners. Vanilla, copra, cocoa, sisal, cloves, and essential

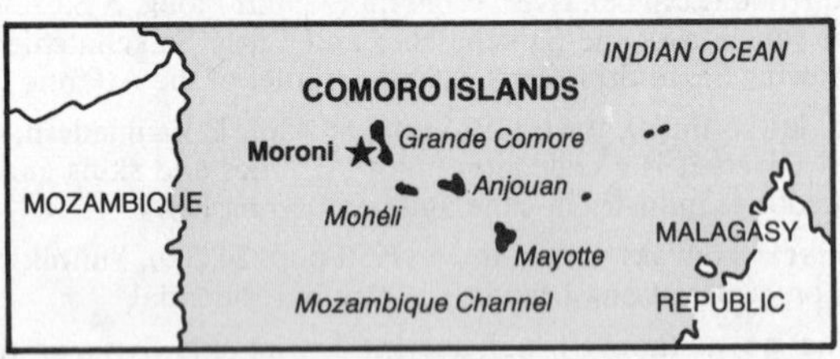

oils are the major crops. The islands were populated by successive waves of immigrants from Africa, Indonesia, Madagascar, and Arabia. In 1841 the French persuaded the king of Mayotte to cede Grande Comore. The other islands were ceded between 1866 and 1909. All were occupied by the British during World War II. In 1946 the islands were granted administrative autonomy within the French Union. The territorial assembly voted in Dec., 1958, to remain in the French Republic as an overseas territory. By 1968 internal self-government was achieved. In 1975 the islands declared themselves independent, although Mayotte voted to remain allied with France.

Com·piègne (kôN-pyĕn'yə), city (1975 pop. 37,699), Oise dept., N France, in Île-de-France, on the Oise River. It is an industrial center with varied manufactures; a large glassworks is located in the suburbs. As far back as the Merovingian period (7th cent.), Compiègne had been the site of royal gatherings; from the 17th to 19th cent. French monarchs used it as a summer residence. The forest of Compiègne was a royal hunting ground. Joan of Arc was captured (1430) by the Burgundians at Compiègne. In a railroad car in the forest the armistice ending World War I was signed; in 1940 Hitler forced the French to surrender in the same car (which was later taken to Germany and destroyed).

Comp·ton (kŏmp'tən), city (1970 pop. 78,547), Los Angeles co., S Calif., a residential and industrial suburb between Los Angeles and Long Beach; inc. 1888. It has aircraft, electronic, oil, chemical, and steel industries.

Com·tat Ve·nais·sin (kôN-tä' vĕ-nĕ-săN') or **Comtat,** region of SE France, Vaucluse dept., comprising the territory around Avignon. Well-irrigated, it is a truck-farming and fruit-growing area. Comtat Venaissin was given by King Philip III to Pope Gregory X in 1274. Succeeding French kings sought to regain the region, but it remained in papal hands until 1791, when a plebiscite was held and the inhabitants voted to reunite with France.

Con·a·kry (kŏn'ə-krē), city (1972 pop. est., with suburbs, 290,000), capital of Guinea and its Conakry region, SW Guinea, a port on the Atlantic Ocean. Located on Tombo Island and connected with the mainland by a causeway, Conakry is Guinea's largest city and its administrative, communications, and economic center. Its economy revolves largely around the port. The few local manufactures include food products and beverages; iron ore and bauxite were mined nearby until the late 1960s. In 1887 Conakry was occupied by French forces. Its main growth dates from World War II, and today it is a modern city with wide boulevards and fine botanical gardens.

Con·cep·ción (kôn'sĕp-syôn'), city (1970 pop. 189,929), capital of Concepción prov., S central Chile, near the mouth of the Bío-Bío River. It is an industrial and commercial center. Its port, Talcahuano, just north of the city, ships the products of the surrounding rich agricultural region. Concepción's industries produce glass, textiles, sugar, hides, and steel. Founded in 1550, the city was besieged and destroyed by the Araucanians in 1554–55. It was completely destroyed by earthquakes in 1570, 1730, 1751, 1835, and 1939, and was severely damaged in 1960.

Concepción del Ur·u·guay (dĕl yŏŏr'ə-gwī', -gwā', ōō'rōō-gwī'), city (1970 pop. 73,720), Entre Ríos prov., NE Argentina, a port on the Uruguay River. It ships the grain and beef of the surrounding region. The city was founded in 1778 and was twice the capital of Argentina in the 19th cent.

Con·cho (kŏn'chō), county (1970 pop. 2,937), 1,004 sq mi (2,600.4 sq km), W central Texas, on the Edwards Plateau at an altitude of c.1,600 to 2,100 ft (490–640 m), bounded on the NE by the Colorado River and drained by the Concho River and Brady Creek; formed 1858; co. seat Paint Rock. It is in a stock-raising (sheep, goats, and cattle) area, with dairying, poultry raising, and some agriculture (grain, sorghums, and cotton). Wool and mohair are marketed.

Con·chos (kōn'chōs), river, c.350 mi (565 km) long, rising in S Chihuahua state, N Mexico, and flowing N and NE to the Rio Grande. Dams along its middle course provide water for extensive cotton oases just south of the city of Chihuahua.

Con·cord. 1. (kŏng'kərd) Residential city (1970 pop. 85,164), Contra Costa co., W central Calif., in an oil and farm region; settled c.1852, inc. 1905. Electronic equipment is made. A U.S. naval ammunition depot is nearby. **2.** (kŏng'kərd) Town (1970 pop. 16,148), Middlesex co., E Mass., on the Concord River; inc. 1635. Electronic and wood products are made. The site of the Revolutionary Battle of Concord on Apr. 19, 1775 is marked by Daniel Chester French's bronze *Minuteman.* Concord has many fine old houses, some opened as memorials to noted occupants—Ralph Waldo Emerson, Louisa May and Bronson Alcott, Nathaniel Hawthorne, and Henry David Thoreau. An antiquarian museum and the Old Manse, built in 1769 by Emerson's grandfather and made famous by Thoreau, are here. **3.** (kŏng'-kərd) City (1970 pop. 30,022), state capital and seat of Merrimack co., S central N.H., on the Merrimack River; settled 1725–27, inc. as Rumford, Mass., in 1733 and as Concord, N.H., in 1765. Famous for its granite, the city also has a printing industry and plants making leather goods, electrical equipment, furniture, stone and clay products, textiles and apparel, metalware, and food. It became the state capital in 1808, and its growth was further aided by the building of the Middlesex Canal in 1815. St. Paul's school (preparatory) and the house of Franklin Pierce (now a museum) are in Concord. **4.** (kôn'-kôrd') City (1970 pop. 18,464), seat of Cabarrus co., central N.C., near the edge of the Piedmont; settled 1796, inc. 1837. Located in a livestock and grain area, it is also a thriving cotton textile center. In addition to a great variety of cotton goods, its manufactures include foods and metal products. Gold discovered nearby in 1799 started a gold rush. Concord is the seat of Barber-Scotia College.

Con·corde, Place de la (pläs də lä kôN-kôrd'), large square, Paris, France. It is bounded by the Tuilleries gardens; the Champs Élysées; the Seine River; and a façade of buildings divided by a vista of the Madeleine Church. The Pont de la Concorde, a monumental bridge, leads from the Place to the other side of the Seine.

Con·cor·dia (kông-kôr'dyä), city (1970 pop. 110,401), Entre Ríos prov., NE Argentina, a port on the Uruguay River. It is the distributing center of a farm and stock-raising district.

Con·cor·di·a (kən-kôr'dē-ə, kəng-), parish (1970 pop. 22,578), 718 sq mi (1,859.6 sq km), E central La., bounded on the E by the Mississippi, on the W by the Black and Tensas rivers, on the S by the Red River; formed 1805; parish seat Vidalia. Its agriculture includes cotton, corn, dairy products, and livestock. It also has oil and natural-gas fields, lumber, and sand and gravel pits. Cotton ginning and lumber milling are done. Oxbow lakes along the Mississippi afford good fishing.

Concordia, city (1970 pop. 7,221), seat of Cloud co., N central Kansas, on the Republican River N of Salina; founded 1870, inc. 1872. It is a farm trade center, with processing plants.

Con·don (kŏn'dən), city (1975 est. pop. 905), seat of Gilliam co., N Oregon, SW of Pendleton; inc. 1901. In a livestock and timber region, it also grows wheat.

Co·ne·cuh (kə-nā'kə), county (1970 pop. 15,645), 850 sq mi (2,201.5 sq km), S Ala., in a coastal plain region drained by the Sepulga River; formed 1818; co. seat Evergreen. Cotton and peanuts are grown, and lumber is milled.

Co·ne·jos (kə-nā'əs, -həs), county (1970 pop. 7,846), 1,271 sq mi (3,291.9 sq km), S Colo., in an irrigated agricultural area on the N.Mex. border, bounded on the E by the Rio Grande and drained by the Conejos River; formed 1891; co. seat Conejos. Livestock, hay, and potatoes are important.

Conejos, village (1970 pop. 100), seat of Conejos co., S Colo., in the SE foothills of the San Juan Mts. on Conejos River SSW of Alamosa; founded 1854. In the San Luis Valley at an altitude of 7,880 ft (2,403.4 m), it is one of the oldest towns in the state.

Co·ney Island (kō'nē), beach resort and amusement center of S Brooklyn borough of New York City, SE N.Y., on the Atlantic Ocean. The tidal creek that once separated the island from the mainland has been filled in, making the area a peninsula. More than a million persons throng to Coney Island on hot weekends and holidays, attracted by the beach, the 2-mi (3.2-km) boardwalk, the New York Aquarium, and the many other entertainment devices, eating places, and souvenir stands. High-rise apartments have replaced much of the amusement area since the 1950s.

Con·go (kŏng'gō) or **Zaire** (zâr), river of equatorial Africa, c.2,720 mi (4,375 km) long, formed by the waters of the Lualaba River and its tributary, the Luvua River, and flowing generally N and W through Zaire to the Atlantic Ocean. One of the longest rivers in the world, the Congo River drains c.1,425,000 sq mi (3,690,750 sq km).

The Lualaba River, considered to be the upper Congo River, rises in southeast Zaire, flows north over rapids and falls to Bukama, and thence across a vast plain and through a series of marshy lakes to receive the Luvua River at Ankoro. The Luvua River has its most remote source in the Chambeshi River, which rises in north Zambia and flows southwest into swamps around Lake Bangweulu; it emerges from the swamps as the Luapula River, continues north along the Zaire-Zambia border into Lake Mweru, exits from there as the Luvua River, and continues northwest to the Lualaba River. A third major headstream is the Lukuga River, which drains from Lake Tanganyika and joins the Lualaba River near Kabalo. From Kabalo, the Lualaba River flows north to Kisangani in a varied course marked by a deep and narrow gorge (the Gates of Hell) and a section of seven cataracts—known as Stanley Falls—between Ubundi and Kisangani that marks the end of the Lualaba and the beginning of the Congo River proper.

Below Kisangani, the Congo flows west and southwest, in a great curve unbroken by falls or rapids for about 1,090 mi (1,755 km) to Kinshasa. For most of its middle section the Congo is from 4 to 10 mi (6.4–16.1 km) wide, with many islands and sandbars. Because its many large tributaries drain areas with alternating rainy seasons on either side of the equator, the Congo has a fairly constant flow throughout the year. Between Bolobo and Kwamouth the Congo narrows in width to between 1 mi and 1.5 mi (1.6–2.4 km) but then widens to form lakelike Stanley Pool. From the western end of Stanley Pool, the Congo descends 876 ft (267.2 m) in a series of 32 rapids, known as Livingstone Falls, to the port of Matadi. Below Matadi the Congo is navigable by oceangoing vessels and, despite such hazards as the whirlpools of the Devil's Cauldron, shifting sandbars, and sharp bends in the river, forms one of the largest natural harbors in Africa. The Congo River enters the Atlantic Ocean between Banana Point, Zaire, and Sharks Point, Angola, and dredging is required to keep a navigable channel open. With railroads to bypass major falls, the Congo River and its tributaries form a system of navigable waterways c.9,000 mi (14,480 km) long. The river is Africa's largest potential source of hydroelectric power.

The mouth of the Congo River was visited (1482) by Diogo Cão, the Portuguese navigator. It became known as the Zaire River (a corruption of the local name Mzadi meaning "great water") and was later referred to as the Congo River (for the Kongo kingdom located near its mouth); it was renamed Zaire River by the government of Zaire in 1971. The Congo's lower course was traced upstream as far as Isangila by a British force under Capt. J. K. Tuckey in 1816, and its upper headwaters by the missionary David Livingstone, who followed the Lualaba River to Nyangwe in 1871. The journalist Henry Stanley traveled from Nyangwe to Isangila and on to Boma during his great transcontinental journey (1874–77), thus proving the headwaters to be tributaries of the Congo River and not sources of the Nile as hypothesized by Livingstone.

Congo, People's Republic of the, republic (1977 est. pop. 1,405,000), 132,046 sq mi (342,000 sq km), W central Africa, bordered on the W by Gabon, on the N by Cameroon and the Central African Empire, on the E and SE by Zaire, and on the SW by Cabinda and the Atlantic Ocean. The capital is Brazzaville. The terrain is covered mainly by dense tropical rain forest, with stretches of wooded savanna. Tributaries of the Congo and Ubangi rivers, which separate the Congo from Zaire, flow through the country.

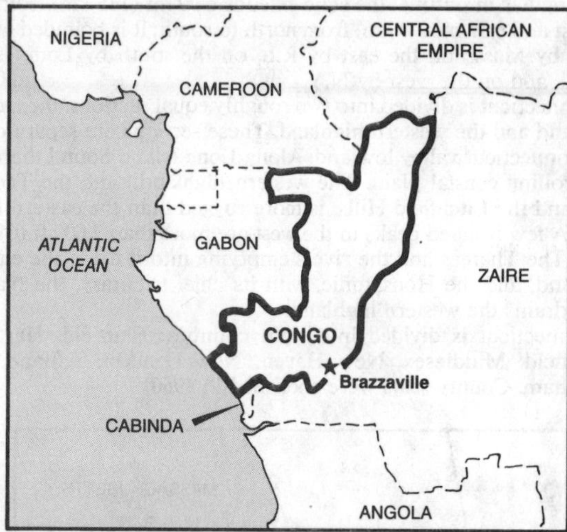

Economy. Agriculture and forestry are the chief economic activities in the Congo. The major subsistence crops are cassava and yams. Sugar cane and tobacco, raised primarily on plantations, are the leading export crops, followed by coffee, cocoa, palm products, and groundnuts. Timber is also a major export. Diseases restrict cattle raising, and fishing is not well developed. Industry is limited mainly to the processing of agricultural and forest products. Mining is increasingly important, with potash and oil the principal exports; petroleum resources are being rapidly depleted, however.

History. Pygmies, migrating from the Zaire region, were probably the first inhabitants of what is now the Congo. They were followed by the Bakongo, the Bateke, and the Sanga, who arrived in the 15th cent. After the coastal areas were explored by the Portuguese, commerce developed between Europeans and the coastal Africans. Europeans penetrated inland in the late 19th cent. Between 1889 and 1910, the Congo (called the French Congo and later the Middle Congo) was administered primarily by French companies exploiting the area's rubber and ivory resources. Scandals over the practice of forced labor and porterage broke out and France restricted the role of the concessionaires. In 1910 the Congo became a colony in French Equatorial Africa. Renewed forced labor and other abuses sparked an African revolt in 1928.

The Free French forces made the Congo a bastion of their struggle against the Germans and the Vichy regime during World War II. In 1946 the region was granted representation in the French parliament. Independence was achieved on Aug. 15, 1960, with Fulbert Youlou as the first president. He was succeeded by Alphonse Massamba-Débat, who in 1964 founded a Marxist-Leninist party and proclaimed a noncapitalist path of economic development. Tensions between the government and the army grew, and in 1968 Marien Ngouabi, an army commander, seized power. He followed his predecessor's socialist policies, but created his own Marxist-Leninist type of party, the Congolese Workers party. After an attempted coup in 1972, Ngouabi purged opponents. In June, 1973, a new constitution was approved, providing for popularly elected national, regional, and local assemblies. In Jan., 1975, Ngouabi was sworn in for a five-year term as president. Later that year an economic and technical aid pact was signed by the USSR and the Congo. The assassination of Ngouabi, on Mar. 18, 1977, led to a series of executions. Finally, in Apr., Col. Joachima Yombi Opango, chief of staff of the Congolese army, assumed the presidency.

Congo, Republic of the: *see* Zaire.

Con·jee·ve·ram (kən-jē'və-rəm): *see* Kanchipuram, India.

Con·naught or **Con·nacht** (both: kŏn'ôt, -əKHt), province (1971 pop. 390,902), 6,611 sq mi (17,122 km), W Republic of Ireland. It was one of the ancient kingdoms of Ireland, whose rulers, the O'Connors, were supplanted by the Anglo-Norman De Burghs in the 13th cent.

Con·ne·aut (kŏn'ē-ŏt'), city (1970 pop. 14,552), Ashtabula co., extreme NE Ohio, on Lake Erie, near the Pa. line; settled 1799, inc. 1834. It is a port of entry and a vacation resort.

Con·nec·ti·cut (kə-nĕt'ĭ-kət), state (1975 pop. 3,096,000) 5,009 sq mi (12,973 sq km), NE United States, southernmost of the New England states, one of the Thirteen Colonies. Hartford is the capital. Rectangular in outline, the state extends c.90 mi (145 km) from east to west and c.55 mi (90 km) from north to south; it is bounded on the north by Mass., on the east by R.I., on the south by Long Island Sound, and on the west by N.Y.

Connecticut is divided into two roughly equal sections, the eastern highland and the western highland. These sections are separated by the Connecticut valley lowland. Along Long Island Sound there is a low, rolling coastal plain. The western highland, with the Taconic Mts. and the Litchfield Hills, is more rugged than the eastern highland. A few isolated peaks in the west are more than 2,000 ft (610 m) high. The Thames and the rivers emptying into it drain the eastern highland, and the Housatonic, with its chief tributary, the Naugatuck, drains the western highland.

Connecticut is divided into eight counties: Fairfield, Hartford, Litchfield, Middlesex, New Haven, New London, Tolland, and Windham. County seats were abolished in 1960.

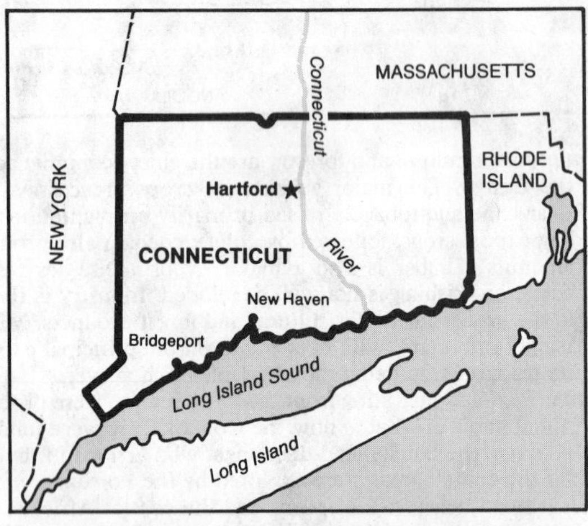

Economy. Though famed for its rural loveliness, Connecticut is heavily industrialized: textiles, typewriters, silverware, sewing machines, clocks, and watches are among its many products. The state's principal industries produce transportation equipment, nonelectrical and electrical machinery, fabricated metals, primary metals, and chemicals. Groton is an important center for submarine building. Agriculture accounts for only a small share of income in the state; dairy products, eggs, and tobacco are the leading farm items. High-grade broadleaf tobacco, used in making cigar wrappers, has been a speciality since the 1830s. Few minerals are produced; stone, sand, and gravel account for most income derived from mining. Insurance is an important industry in Connecticut, and Hartford is the home offices of many insurance companies.

History. In 1614 Adriaen Block, a Dutchman, sailed through Long Island Sound and discovered the Connecticut River. The Dutch built a small fort in 1633 on the site of present-day Hartford, but they abandoned it in 1654. Edward Winslow of Plymouth Colony was apparently the first Englishman to visit (1632) Connecticut, and in 1633 members of the Plymouth Colony established a trading post on the site of Windsor. By 1635 Puritan settlers from the Massachusetts Bay Company were flocking to the Connecticut valley. Land was purchased from the natives, who were on the whole friendly. The Pequot Indians resisted white settlement, but they were defeated by the English in the short Pequot War of 1637. Not until King Philip's War in 1675–76 was there further serious trouble with the Indians. In 1638–39 representatives of the three Connecticut River towns—Hartford, Windsor, and Wethersfield—met at Hartford and formed the colony of Connecticut. They also adopted the Fundamental Orders, which established a government for the colony. A second colony, Saybrook, had been established at the mouth of the Connecticut River in 1635 by a group of Englishmen who sold it to Connecticut colony in 1644. A Puritan settlement, New Haven, was established in 1638 and remained separate until its acquisition by Connecticut in 1662, when Gov. John Winthrop secured the first royal charter for Connecticut from King Charles II. The new charter confirmed the Fundamental Orders and subsequent laws so that government went on much as before, except for a brief interruption from 1687 to 1689 when the English tried to assert control over the colony and dispatched an administrator, Sir Edmund Andros, to Connecticut. Andros sought to recover the charter from the colonists, who hid it in an oak tree that came to be known as the Charter Oak.

In 1708 Congregationalism was established as the official religion of the colony; later the Anglicans (1727), the Baptists, and the Quakers (1729) were exempted from contributing to the support of the Established Church. Other dissenting groups were treated harshly. Connecticut thus occupied a position midway between the more autocratic ecclesiastical system of Massachusetts and the liberal one of Rhode Island.

Connecticut's agrarian economy was gradually being transformed, as a small but vigorous merchant class arose. The colony came to resent England's increasingly burdensome commercial and colonial policy. In 1776 the patriot governor, Jonathan Trumbull, was re-elected almost unanimously (Connecticut and Rhode Island were the only colonies privileged to elect their chief executives), and he was the only governor of any colony to be retained in office after the outbreak of the American Revolution. Connecticut was one of the first states to approve the Federal Constitution.

The Embargo Act of 1807 was vehemently denounced throughout New England; the ports on Long Island Sound and on the Connecticut River had developed a lively carrying trade with which the embargo interfered. The War of 1812 was also so unpopular that New England Federalists, meeting at the Hartford Convention in late 1814, considered secession. In 1818 the Jeffersonians came into power in the state, and a new constitution, replacing the old charter of 1662, was adopted. It disestablished the Congregational Church and greatly extended the franchise, although universal manhood suffrage was not proclaimed until 1845.

Artisans and craftsmen had become increasingly numerous in late colonial days. Modern mass production had its beginning in the state when Eli Whitney established (1798) at New Haven a firearms factory that began making guns with standardized, interchangeable parts. Earlier, in 1793, he had invented and manufactured the cotton gin at New Haven. The manufacture of notions (buttons, pins, needles, metal goods, and clocks) gave rise to the enterprising "Yankee peddler," who, with horse and team, covered the nation hawking his wares. Connecticut's insurance industry also developed during this period, and in 1810 the Hartford Fire Insurance Company was established.

Connecticut, which had placed limitations on slavery in 1784 and abolished it in 1848, supported the Union during the Civil War with nearly 60,000 troops and an able Secretary of the Navy, Gideon Welles. During and after the war, industry expanded greatly. Immigration provided a cheap labor supply as English, Scottish, and many Irish immigrants were followed by French Canadians, Italians, Poles, and others. Connecticut's industries have grown and developed steadily since World War I, except for the period of the Great Depression. In 1954 the world's first nuclear-powered submarine was launched at Groton.

Government. The 1965 state constitution provides for the election of both houses of the general assembly, as the legislature is called, on the basis of election districts apportioned according to population. Connecticut's state senate has 36 members and its house of representatives has 177; both houses are elected for two-year terms.

Educational Institutions. Institutions of higher learning in Connecticut include Yale Univ., at New Haven; Trinity College, at Hartford; Wesleyan Univ., at Middletown; the Univ. of Connecticut, at Storrs; and the United States Coast Guard Academy and Connecticut College, at New London.

Connecticut river, longest river in New England, 407 mi (654.9 km) long, rising in Connecticut Lakes, N N.H., and flowing S along the Vt.-N.H. line, then across Mass. and Conn. to enter Long Island Sound at Old Saybrook, Conn. It drains c.11,000 sq mi (28,490 sq km). The river is navigable to Hartford, Conn. The Connecticut Valley is one of the best agricultural regions in New England.

Connellsville, city (1975 est. pop. 10,700), Fayette co., SW Pa., on the Youghiogheny River in the Allegheny Mts.; settled c.1770, inc. as a borough 1806, as a city 1911. A major producer of coal and coke, the city also has railroad shops; its manufactures include glass, iron, and steel products.

Con·ne·ma·ra (kŏn'ə-mär'ə), wild, mountainous region, Co. Galway, W Republic of Ireland, lying between the Atlantic Ocean and Loughs Corrib and Mask. It is a well-known vacation area, with many mountains, lakes, streams, and glens. The peat bogs of southern Connemara are major fuel sources.

Con·ners·ville (kŏn'ərz-vĭl'), city (1975 est. pop. 17,800), seat of Fayette co., E central Ind., on the Whitewater River, in a farm area; founded 1813, inc. as a city 1870. Whitewater State Park and several historic covered bridges are nearby.

Con·rad (kŏn'răd), city (1970 pop. 2,770), seat of Pondera co.; NW Mont.; NNW of Great Falls; in an oil and farm area; inc. 1911.

Con·roe (kŏn'rō), city (1975 est. pop. 14,900), seat of Montgomery co., SE Texas; inc. 1885. Long a pine-lumbering town, it prospered after oil was discovered in 1932. Other natural resources in the area are timber, clays, and gas. Farm products include beef and dairy cattle and feed for livestock and poultry.

Con·sett (kŏn'sĕt, -sit), urban district (1973 est. pop. 35,080), Durham, NE England. There are coal mines, iron and steel plants, and nurseries in Consett. The district has associations with the Roman, Saxon, and Norman conquests. A German colony of swordmakers settled in Consett in the 17th cent.

Con·sho·hock·en (kŏn'shə-hŏk'ən), industrial borough (1970 pop. 10,195), Montgomery co., SE Pa., on the Schuylkill River, in a fertile farm area that also has clay pits; inc. 1850.

Con·stance (kŏn'stəns) or *German* **Kon·stanz** (kŏn-stänts), city (1975 pop. 65,963), Baden-Württemberg, S West Germany, on the Rhine River at the W end of the Lake of Constance and near the Swiss border. Its manufactures include textiles, chemicals, and electrical equipment. The city is also a tourist center. Constance was founded as a Roman fort in the 4th cent. A.D. and became an episcopal see at the end of the 6th cent. The bishops became powerful and held large territories, including much of Baden-Württemberg and Switzerland, as princes of the Holy Roman Empire. Located on a trade route between Germany and Italy, Constance became a free imperial city in 1192. During the Council of Constance (1414–18), John Huss was burned at the stake. In 1531 the city, which had accepted the Reformation, joined the Schmalkaldic League. Emperor Charles V, after defeating the League, deprived Constance of its free imperial status and gave it to his brother, later Emperor Ferdinand I. Constance was in Austrian hands from 1548 until it was ceded (1805) to Baden. The bishopric was suppressed in 1821, and the diocese was abolished in 1827. Among the numerous historic buildings in Constance are the cathedral (11th cent.; additions 15th and 17th cent.); the Council building (1388); and a former Dominican convent (now a hotel).

Constance, Lake of, or *German* **Bo·den·see** (bōd'n-zā'), lake, 208 sq mi (538.7 sq km), bordering on Switzerland, West Germany, and Austria. It is 42 mi (67.6 km) long and has a maximum depth of 827 ft (252.2 m). The lake is fed and drained by the Rhine River and divides near the city of Constance into two arms. Fruit is grown on the lake's fertile shores, and wine making and fishing are major industries. Remains of lake dwellings have been found.

Con·stan·ţa (kôn-stän'tsä), city (1974 est. pop. 256,875), SE Rumania, on the Black Sea. It is a major railroad junction and industrial city, Rumania's main seaport, a naval and air base, and a seaside resort. The city was founded in the 7th cent. B.C. as a Greek colony and came under Roman rule in 72 B.C. Ovid lived in exile here. Constantine I (4th cent. A.D.) named the city Constantiniana and made it an episcopal see. It was captured by the Turks in 1413. Rumania acquired it in 1878.

Con·stan·tine (kŏn'stən-tēn), city (1974 est. pop. 350,183), capital of Constantine dept., NE Algeria, on the gorge of the Rhumel River. A major inland city, it is the railhead of a prosperous and diverse agricultural area. Constantine is also a center of the grain trade and has flour mills, a tractor factory, and industries producing textiles and leather goods. Founded by Carthaginians, Constantine became the capital and commercial center of Numidia. Under Roman rule it was a major grain-shipping point and one of the wealthiest cities of Africa. Destroyed (A.D. 311) during the war preceding the accession of Constantine I, it was rebuilt by Constantine himself and named in his honor. The city was pillaged by the Vandals in the 5th cent. The Turks captured it in the 16th cent. By the time of the French conquest in 1837 the governor of Constantine had become virtually independent of the Ottoman Empire.

Con·stan·ti·no·ple (kŏn'stăn-tə-nō'pəl), former capital of the Byzantine Empire and of the Ottoman Empire, since 1930 officially called İstanbul. It was founded (A.D. 330) at ancient Byzantium as the new capital of the Roman Empire by Constantine I, after whom it was named. The largest and most splendid European city of the Middle Ages, Constantinople shared the glories and vicissitudes of the Byzantine Empire, which in the end was reduced to the city and its environs. Although besieged innumerable times by various peoples, it was taken only three times—in 1204 by the army of the Fourth Crusade, in 1261 by Michael VIII, and in 1453 by the Ottoman Sultan Mohammed II.

Built on seven hills, the city on the Bosporus presented the appearance of an impregnable fortress enclosing a sea of magnificent palaces and gilded domes and towers. In the 10th cent., it had a cosmopolitan population of about 1 million. The Church of Hagia Sophia, the sacred palace of the emperors (a city in itself); the huge hippodrome, center of the popular life; and the Golden Gate, the chief entrance into the city; were among the largest of the scores of churches, public edifices, and monuments that lined the broad arcaded avenues and squares. Constantinople had a great wealth of artistic and literary treasures before it was sacked in 1204 and 1453. Virtually depopulated when it fell to the Ottoman Turks, the city recovered rapidly. The Ottoman sultans, whose court was called the Sublime Porte, embellished Constantinople with many beautiful mosques, palaces, monuments, fountains, baths, aqueducts, and other public buildings. After World War I the city was occupied (1918–23) by the Allies. In 1922 the last Ottoman sultan was deposed and Ankara became (1923) the new capital of Turkey.

Con·sti·tu·tion Island (kŏn'stĭ-to͞o'shən, -tyo͞o'-), in the Hudson River opposite West Point, SE N.Y.; part of the U.S. Military Academy. The ruins of Fort Constitution, built in 1775, are here. During the American Revolution, a chain was stretched across the Hudson at Constitution Island to prevent the ascent of British ships.

Con·ti·nen·tal Di·vide (kŏn'tə-nĕn'tl dĭ-vīd'), the "backbone" of a continent. In North America, from north Alaska to N.Mex,. it is the great ridge of the Rocky Mts., which separates westward-flowing streams from eastward-flowing waters. In southwest N.Mex. the divide crosses an area of low relief; it becomes more distinct in northern Mexico, where it follows the Sierra Madre Occidental. In the United States it has sometimes been called the Great Divide, a name also occasionally used to designate the whole Rocky Mt. system.

Con·tra Cos·ta (kŏn'trə kŏs'tə), county (1970 pop. 555,805), 734 sq mi (1,901.1 sq km), W Calif., in an industrial and residential area along a 70-mi (112.6-km) waterfront along San Francisco, San Pablo, and Suisun bays; formed 1850; co. seat Martinez. In a major shipping area, it has oil refineries, steelworks, shipyards, and other diversified manufacturing plants. Its agriculture, especially in the fertile delta of the San Joaquin River to the northeast, is slowly declining as new suburbs are built.

Con·vent (kŏn'vĕnt), village (1970 pop. 400), seat of St. James parish; SE central La.; on the E bank of the Mississippi W of New Orleans; in a tobacco-growing area. Gas and oil fields are nearby.

Con·verse (kŏn'vûrs), county (1970 pop. 5,938), 4,282 sq mi (11,090.4 sq km), E Wyo., in a grain, sugar-beet, and livestock area drained by the North Platte River; formed 1888; co. seat Douglas. It has coal and oil deposits. Part of Medicine Bow National Forest is here.

Con·way (kŏn'wā'), county (1970 pop. 16,805), 560 sq mi (1,450.4 sq km), central Ark., bounded on the S by the Arkansas River and drained by Cadron Creek; formed 1825; co. seat Morrilton. It has agriculture (cotton, corn, hay, and livestock) and a lumbering industry. Food products, textiles, paper, plastics, and transportation equipment are manufactured.

Conway. 1. City (1973 est. pop. 16,772), seat of Faulkner co., central Ark., in a farm and cotton area; inc. 1873. It is a trade and industrial center. Conway was settled (c.1865) near the site of a French trading post (c.1770). It is the seat of Hendrix College. **2.** Town (1975 est. pop. 8,100), seat of Horry co.; E S.C.; on the Waccamaw River SE of Florence; settled c.1768, inc. 1855. Its manufactures include curtains, wood and metal products, and processed meat and poultry.

Conway, municipal borough (1971 pop. 12,158), Gwynedd, N Wales, at the mouth of the Conway River. Conway is a picturesque town with several notable old structures. A high wall (13th cent.) encloses the old town.

Con·yers (kŏn′yərz), city (1970 pop. 4,890), seat of Rockdale co., N central Ga., ESE of Atlanta, in a farm area; inc. 1854. Steel products are made.

Cooch Be·har (ko͝och′ bə-här′), former princely state, now part of West Bengal state, E India. It lies in a low, poorly drained plain. Rice, tobacco, and jute are grown. The chief town, Cooch Behar (1971 pop. 53,734), is a district administrative center and market town.

Cook (ko͝ok). **1.** County (1970 pop. 12,129), 233 sq mi (603.5 sq km), S Ga., bounded on the W by the Little River; formed 1918; co. seat Adel. It is in a coastal-plain timber and agricultural area yielding tobacco, cotton, corn, peanuts, fruit, truck crops, cattle, hogs, and poultry. **2.** County (1970 pop. 5,493,766), 954 sq mi (2,470.9 sq km), NE Ill., in a highly industrial urban area bounded on the E by Lake Michigan and the Ind. border and traversed by the Chicago and Des Plaines rivers; formed 1831; co. seat Chicago. It has many residential communities, recreational areas, and truck-farming and dairying districts. **3.** County (1970 pop. 3,423), 1,346 sq mi (3,486.1 sq km), extreme NE Minn., bounded on the S by Lake Superior, on the N by a chain of lakes forming the Ont., Canada, border; formed 1874; co. seat Grand Marais. Tourism, agriculture (potatoes, poultry, and dairy products), lumbering, and fishing are important. Most of the county lies within Superior National Forest. Grand Portage Indian Reservation is in the east.

Cook, Mount, 12,349 ft (3,766.4 m) high, on South Island, New Zealand, in the Southern Alps. It is the highest peak of New Zealand.

Cooke (ko͝ok), county (1970 pop. 23,471), 909 sq mi (2,354.3 sq km), N Texas, bounded on the N by the Red River, here forming the Okla. border, and drained by the Elm Fork of the Trinity River; formed 1848; co. seat Gainesville. Its agriculture includes grain, cotton, corn, peanuts, fruit, truck crops, dairy products, and livestock. It also has timber, oil and natural-gas wells, and clay and sand deposits. There is some manufacturing.

Cooke·ville (ko͝ok′vĭl′), city (1975 est. pop. 17,070), seat of Putnam co., N central Tenn.; inc. 1854. It is a farm trade center with plants making filters, automobile accessories, brushes, clothing, and heating elements.

Cook Inlet, c.150 mi (240 km) long, inlet of the Gulf of Alaska, S Alaska, on the W side of Kenai Peninsula. Noted for its salmon and herring fisheries, Cook Inlet has the largest tidal bore in the United States. Capt. James Cook sailed through it in 1778.

Cook Islands, group (1970 est. pop. 22,000), 90 sq mi (233.1 sq km), South Pacific, SE of Samoa. Avarua on Rarotonga Island is the administrative center of the group. Fruit juices, citrus fruits, clothing, copra, tomatoes, pearl shell, handicrafts, and jewelry are produced. The southern islands were probably occupied by the Polynesians c.1,500 years ago. Spaniards visited the islands in the late 16th and early 17th cent. Capt. James Cook sighted some of the islands in 1773; others were not discovered until the 1920s. The London Missionary Society was a powerful influence in the southern islands during the 19th cent. The group was proclaimed a British protectorate in 1888 and was annexed by New Zealand in 1901. Although under New Zealand sovereignty, the Cook Islands achieved internal self-government in 1965.

Cook Strait, channel, c.15 mi (24 km) wide, between North Island and South Island, New Zealand. It was discovered in 1770 by Capt. James Cook.

Coo·lidge (ko͞o′lĭj), city (1970 pop. 5,314), Pinal co., S central Ariz., SE of Phoenix in the Casa Grande valley; laid out 1925, inc. 1945. Irrigated by Coolidge Dam, the area has diversified farming. Ruins of an ancient Indian civilization may be seen at nearby Casa Grande National Monument.

Coon Rapids (ko͞on), city (1975 est. pop. 34,900), Anoka co., SE Minn., on the Mississippi River; inc. 1952. It is a suburb of Minneapolis-St. Paul. It has an aerospace research facility and plastic and metallurgical industries.

Coo·per (ko͞o′pər), county (1970 pop. 14,732), 563 sq mi (1,458.2 sq km), central Mo., bounded on the N by the Missouri River and drained by the Lamine River; formed 1818; co. seat Boonville. Its agriculture includes wheat, corn, oats, cattle, and poultry. It also has barite mines and some industry (wood and plastic products).

Cooper, town (1970 pop. 2,258), seat of Delta co., NE Texas, NE of Dallas, in a blackland cotton area; settled in the 1870s.

Coo·pers·town (ko͞o′pərz-toun′). **1.** Residential village (1970 pop. 2,403), seat of Otsego co., E central N.Y., on the Susquehanna River and Otsego Lake; inc. 1807. It was founded by William Cooper, who brought his family here in 1787. His son, James Fenimore Cooper, lived here after 1833, and the region is described in his *Leatherstocking Tales.* Fenimore House is the headquarters of the New York State Historical Association. Other museums include Cooperstown Indian Museum, Farmers' Museum, and the National Baseball Hall of Fame and Museum. **2.** City (1970 pop. 1,485), seat of Griggs co., E N.Dak., NW of Fargo, in a livestock, dairy, and grain region; founded 1882, inc. 1907.

Coorg (ko͝org), former state, Karnataka state, SW India. An independent Hindu dynasty ruled Coorg from the late 16th cent. until it was annexed by the British in 1834. It was administered by a British chief commissioner until India became independent in 1947.

Coos (ko͞os). **1.** County (1970 pop. 34,291), 1,825 sq mi (4,726.8 sq km), N N.H., in a recreational area bordering on Vt., Maine, and Que., Canada; formed 1803; co. seat Lancaster. It is drained by the Androscoggin and Connecticut rivers. Lumbering is important, and pulp and paper and wood products are manufactured. **2.** County (1970 pop. 56,515), 1,604 sq mi (4,154.4 sq km), SW Oregon, bounded on the W by the Pacific Ocean and on the E by the Coast Range; formed 1853; co. seat Coquille. In a lumbering area, it produces livestock and dairy products.

Coo·sa (ko͞o′sə), county (1970 pop. 10,662), 648 sq mi (1,678.3 sq km), E central Ala., bounded on the W by the Coosa River; formed 1832; co. seat Rockford. Its agriculture includes cotton, corn, and livestock. It has marble and granite quarries and a large tin mine.

Coosa, river, 286 mi (460.2 km) long, rising in NW Ga. and flowing SW through E Ala., joining the Tallapoosa near Montgomery, Ala., to form the Alabama River. Locks and dams make the river navigable for barges to Rome, Ga.

Coos Bay (ko͞os), city (1975 est. pop. 12,900), Coos co., SW Oregon, a port of entry on Coos Bay; founded 1854 as Marshfield, inc. 1874, renamed 1944. Lumbering, shipping, tourism, fishing, and canning are important industries.

Co·pán (kō-pän′), ruined city of the Maya, W Honduras, near the village of Copán. Noted for fine sculptured stelae and in particular for the Hieroglyphic Stairway (containing nearly 2,000 glyphs), Copán was, perhaps, the center of knowledge where Mayan astronomical learning, as applied to chronology, achieved its most accurate expression.

Co·pen·ha·gen (kō′pən-hā′gən, -hä′-), city (1971 pop. 625,678), capital of Denmark and of Copenhagen co., E Denmark, on E Sjaelland and N Amager islands and on the Øresund. It is a major commercial, fishing, and naval port and is Denmark's chief commercial, industrial, and cultural center. Manufactures include pharmaceuticals, processed food, beer, textiles, plastics, marine engines, furniture, and the celebrated Copenhagen Ware. There are also iron foundries and large shipyards.

Copenhagen was a trading and fishing center by the early 11th cent. The city was twice destroyed by the Hanseatic League but successfully resisted (1428) a third attack. Copenhagen replaced Roskilde as the Danish capital in 1443. The city exacted tolls from all ships passing through the Øresund until 1857. Having resisted (1658–59) a Swedish siege, Copenhagen was relieved by the Dutch. In 1660 peace between Denmark and Sweden was negotiated here. Copenhagen became involved in the war between Napoleonic France and England in the early 19th cent. The news that Denmark, by a secret convention, was about to join Napoleon's Continental System and to join in the war on England led the British government to send an expeditionary force to seize the Danish fleet. When the Danes refused to surrender, the British landed troops in 1807 and bombarbed Copenhagen. However, the city recovered quickly after the Napoleonic Wars, and its industrial base grew rapidly in the

19th cent. In World War II, Copenhagen was occupied (1940–45) by the Germans, and its shipyards were bombed by the Allies. The city itself was only slightly damaged, and it retained the charm and design that had resulted in its being called "the Paris of the North."

Famous landmarks of the city include the Charlottenborg Palace (17th cent.), Amalienborg Square, enclosed by four 18th cent. palaces, one of which has been the royal residence since 1794; the citadel (c.1662); the city hall (1894–1905); the famous round tower, which the astronomer Tycho Brahe (1546–1601) used as an observatory; and the Cathedral of Our Lady (c.1209; rebuilt in the early 19th cent.). The island of Slotsholmen, surrounded by a moat on three sides and by the harbor on the fourth, supports an impressive complex of buildings, notably Christiansborg Palace (18th cent.; restored 1916), erected on the site of Archbishop Absalon's original castle and now housing the Danish parliament, supreme court, and foreign office; the Thorvaldsen Museum (opened 1848); and the stock exchange (17th cent.). Favorite spots in the city include the Tivoli amusement park (opened 1843) and the waterfront Langelinie Promenade, near which is the famous statue of Hans Christian Andersen's *Little Mermaid.*

Co·piague (kō'pāg'), uninc. residential town (1975 est. pop. 21,000), Suffolk co., SE N.Y., on the S shore of Long Island.

Co·pi·ah (kə-pī'ə, kō-), county (1970 pop. 24,764), 780 sq mi (2,020.2 sq km), SW Miss., bounded on the E by the Pearl River and drained by the Homochitto River; formed 1823; co. seat Hazlehurst. It has agriculture (cotton, fruit, truck crops, and cattle), a lumbering industry, and sand and gravel pits. Textiles and wire are manufactured.

Co·pia·pó (kō'pyä'pō'), city (1970 pop. 51,809), capital of Atacama prov., N central Chile, on the Copiapó River. An industrial city at the southern edge of the Desert of Atacama, Copiapó has industries that ship and process the copper, gold, and silver of the surrounding region. The city was founded in 1540.

Cop·per (kŏp'ər), river, c.300 mi (480 km) long, rising in the Wrangell Mts., SE Alaska, and flowing S through the Chugach Mts. to the Gulf of Alaska. The Indians obtained copper from the deposits near the upper river; these deposits attracted the attention of the Russians and later the Americans, but exploration was difficult because of the river's currents and the glaciers near its mouth. The Kennecott mine (discovered 1898) was finally developed and was reached by the building (1908–11) of the Copper River and Northwestern RR from Cordova, following the river along part of its lower valley. The mine was abandoned in 1938.

Cop·per·as Cove (kŏp'ə-rəs), town (1975 est. pop. 17,500), Coryell co., central Texas. A farm and ranch center, it grew with the establishment of nearby U.S. Fort Hood.

Cop·per·belt (kŏp'ər-bĕlt'), mining region, N central Zambia, central Africa. The Copperbelt is one of the richest sources of copper in the world. Cobalt, selenium, silver, and gold are also produced.

Cop·per·mine (kŏp'ər-mīn'), river, 525 mi (844.7 km) long, rising in Lac de Gras, central Mackenzie dist., Northwest Territories, Canada, and winding NW to enter the Arctic Ocean at Coronation Gulf. Its many falls gives it great hydroelectric power potential.

Cop·tos or **Cop·tus** (both: kŏp'təs), ancient city of Egypt, on the right bank of the Nile, N of modern Luxor. Remains of the Temple of Min, patron god of Coptos, have been found here as well as relics from the time of Ramses II and Thutmose III.

Co·quil·hat·ville (kô-kē-yä-vēl'): *see* Mbandaka, Zaire.

Co·quille (kō-kēl'), city (1970 pop. 4,437), seat of Coos co., SW Oregon, SSE of Coos Bay on the Coquille River near the Pacific coast; settled in the 1870s, inc. 1901. It is a lumbering and dairying center.

Co·quim·bo (kō-kēm'bō), city (1970 pop. 55,360), N central Chile, on a sheltered bay of the Pacific. In 1922 Coquimbo was severely damaged by a tidal wave following an earthquake.

Cor·al Ga·bles (kôr'əl gā'bəlz), city (1975 est. pop. 39,900), Dade co., SE Fla., on Biscayne Bay; inc. 1925. Founded at the height of the Florida land boom, Coral Gables is mainly residential. Electronic equipment, processed meat, and furniture are among its products. The Univ. of Miami is in the city.

Coral Sea, SW arm of the Pacific Ocean, between Australia, New Guinea, and the New Hebrides. The Great Barrier Reef lies along its western edge. During World War II it was the scene of a major U.S. victory against the Japanese in 1942.

Cor·al·ville (kôr'əl-vĭl'), city (1970 pop. 6,130), Johnson co., E central Iowa, a suburb NW of Iowa City; named 1866. A large dam and reservoir are nearby.

Co·ra·op·o·lis (kôr'ē-ŏp'ə-lĭs), borough (1970 pop. 8,435), Allegheny co., SW Pa., on the Ohio River NW of Pittsburgh, in a farm region; settled c.1760, inc. 1886. Varied products are manufactured.

Cor·bin (kôr'bĭn), city (1970 pop. 7,317), Knox and Whitley cos., SE Ky., NW of Middlesboro, in a timber area. A railroad center and a shipping point for the region's coal, the city lies on the old Wilderness Road in the Cumberland Plateau. It developed after the advent of the railroad (1883) and was incorporated in 1894.

Cor·by (kôr'bē), urban district (1976 pop. 55,600), Northamptonshire, central England. Situated over one of the world's largest ironstone fields, Corby has grown rapidly since the 1930s, when new techniques of steel production were developed.

Cor·co·va·do (kôr'kə-vä'dō), peak, 3,309 ft (1,009.2 m) high, overlooking Rio de Janeiro, Brazil. It is served by a funicular railroad and has a 125-ft (38.1-m) high concrete figure of Christ the Redeemer on the top.

Cor·dele (kôr-dēl'), city (1975 est. pop. 12,100), seat of Crisp co., S central Ga., on a branch of the Flint River; founded and inc. 1888. It is a shipping, commercial, and processing center in a timber and farm area.

Cor·dell (kôr-dĕl'), city (1970 pop. 3,261), seat of Washita co., W central Okla., WSW of Oklahoma City near the Washita River, in a cotton and farm area.

Cor·dil·le·ras (kôr'dĭl-yâr'əz, kôr-dĭl'ər-əz, kōr'dē-yä'räs), general name for the entire chain of mountain systems of W North America, extending from N Alaska to Nicaragua, and of W South America, where the mountains stretching from Panama to Cape Horn are known locally as the Cordillera de Los Andes (Andes Mts.). Some geographers use the term "cordillera" for any extensive group of mountain systems.

Cór·do·ba (kôr'dō-vä), city (1970 pop. 798,663), capital of Córdoba prov., central Argentina, on the Río Primero. It is a cultural and commercial center. Irrigation has transformed the surrounding countryside, formerly devoted to cattle ranches, into orchards, grain fields, and vineyards. The city is also a popular tourist and health resort. Córdoba was founded in 1573. The advent of the railroad in the 19th cent. increased prosperity. The university (founded 1613) made Córdoba an early intellectual center of South America.

Córdoba, city (1970 pop. 78,495), Veracruz state, E central Mexico. It is the commercial and processing center of a fertile coffee, sugarcane, and tropical-fruit region. Sugar milling is the chief industry. Córdoba was founded in 1617. The Spanish viceroy O'Donojú and the Mexican revolutionary Agustín de Iturbide signed a treaty here in 1821 that established Mexico's independence.

Córdoba or **Cor·do·va** (kôr'dō-vä), city (1975 pop. 250,903), capital of Córdoba prov., S Spain, in Andalusia, on the Guadalquivir River. Modern industries in the city include brewing, distilling, textile manufacturing, and metallurgy. Of Iberian origin, Córdoba flourished under the Romans, then passed to the Visigoths (572) and the Moors (711). Under the Umayyad dynasty it became the seat (756–1031) of an independent emirate, later called caliphate, which included most of Moslem Spain. The city was then one of the greatest and wealthiest in Europe, renowned for its architectural glories and its gold, silver, silk, and leather work. The city reached its zenith under Abd ar-Rahman III but declined after the fall of the Umayyads and became subject to Seville in 1078. Ferdinand III of Castile conquered it in 1236; in 1238 the great mosque became a cathedral.

Cor·fu (kôr'fōō): *see* Kérkira, Greece.

Cor·inth (kôr'ĭnth) or **Kó·rin·thos** (kôr'ĭn-thôs), city (1971 pop. 20,773), capital of Corinth prefecture, S Greece, in the NE Peloponnesus, on the Gulf of Corinth. It is a port and major transportation center trading in olives, tobacco, raisins, and wine. Founded in 1858 after the destruction of Old Corinth by an earthquake, it was rebuilt after another earthquake in 1928.

Strategically situated on the Isthmus of Corinth and protected by the fortifications on the Acrocorinthus, Corinth was one of the largest, wealthiest, most powerful, and oldest cities of ancient Greece. Dating from Homeric times, it was conquered by the Dorians. In the 7th and 6th cent. B.C., it became a flourishing maritime power. The natural rival of Athens, Corinth was traditionally allied with Sparta. Athenian assistance to the rebellious Corinthian colonies was a direct cause of the Peloponnesian War (431-404 B.C.). During the Corinthian War (395-387 B.C.), however, Corinth joined with Athens against the tyrannical rule of Sparta. After the battle of Chaeronea (338 B.C.) Corinth was garrisoned by Macedonian troops. It became (224 B.C.) a leading member of the Achaean League and in 146 B.C. was destroyed by the victorious Romans. Julius Caesar restored it (46 B.C.). Corinth was again laid waste by the invading Goths (A.D. 395) and by an earthquake in 521. Early in the 13th cent., Corinth was conquered by Geoffroi I de Villehardouin as a sequel to the Fourth Crusade. It was taken by the Ottoman Turks in 1458, and in 1687 was seized by Venice, which lost it to the Turks in 1715. In 1822 it was captured by Greek insurgents. Ancient ruins at Old Corinth include the market place, fountains, the temple of Apollo, and a Roman amphitheater. Paul preached here, and wrote two epistles to the infant Corinthian church.

Corinth, city (1975 est. pop. 10,900), seat of Alcorn co., extreme NE Miss., near the Tenn. line, in a livestock and farm area; founded c.1855. Manufactures include telephone equipment, textiles, clothing, and dairy products. During the Civil War, Corinth was a strategic railroad center. Gen. William S. Rosecrans repulsed the Confederates in heavy fighting here on Oct. 3-4, 1862.

Corinth, Gulf of, inlet of the Ionian Sea, c.80 mi (130 km) long and from 3 to 20 mi (4.8-32 km) wide, indenting central Greece and separating the Peloponnesus from the Greek mainland.

Corinth, Isthmus of, c.20 mi (32 km) long and 4-8 mi (6.4-12.9 km) wide, connecting central Greece with the Peloponnesus, between the Gulf of Corinth and the Saronic Gulf. It is crossed by the 4-mi (6.4-km) Corinth Canal, built between 1881 and 1893, which connects the Aegean and the Adriatic seas. Parallel to the canal are ruins of the ancient Isthmian Wall, which was restored (3rd-6th cent. A.D.) by Byzantine emperors to defend the Peloponnesus.

Co·rin·to (kō-rēn'tō), town (1970 est. pop. 12,985), NW Nicaragua, on the Pacific Ocean. It is Nicaragua's leading port. Coffee, sugar, hides, and woods are exported.

Cork (kôrk), county (1971 pop. 352,883), 2,881 sq mi (7,461.8 sq km), SW Republic of Ireland; co. town Cork. Largest of the Irish counties, it has a rocky and much-indented coast line, wild rugged mountains, and fertile valleys. The main occupations are farming and fishing. There are prehistoric remains (dolmens and stone circles) and ruins of medieval abbeys and churches.

Cork, county borough (1971 pop. 128,645), county town of Co. Cork, S Republic of Ireland, on the Lee River near its mouth on Cork Harbour. The oldest part of the town is on an island between the north and south branches of the Lee, now crossed by numerous bridges. Automobiles, rubber, leather, cotton, and woolen goods, paint, processed foods, flour, and whiskey are manufactured. St. Finbarr is supposed to have founded an abbey on the site early in the 7th cent. In the 9th cent. the Danes occupied Cork and walled it. Dermot MacCarthy ousted the Danes and in 1172 swore allegiance to Henry II of England. Oliver Cromwell occupied Cork in 1649, and the Duke of Marlborough in 1690. Many public buildings were destroyed in the nationalist disturbances of 1920, and the Sinn Fein lord mayor was murdered by the constabulary. The Protestant St. Finbarr's Cathedral, the Roman Catholic cathedral, the Church of St. Ann, and the Carnegie library are noteworthy.

Corn Belt (kôrn' bĕlt'), major agricultural region of the U.S. Midwest, in the N central plains, centered in Iowa and Ill. and extending into S Minn., SE S.Dak., E Nebr., NE Kansas, N Mo., Ind., and W Ohio. Large-scale commercial and mechanized farming prevails in this region of deep, fertile, well-drained soils and long, hot, humid summers. The belt produces more than half of the U.S. corn crop, mainly as feed for livestock, especially hogs, which are the main source of cash income.

Cor·nel·ia (kôr-nēl'yə), city (1970 pop. 3,014), Habersham co., NE Ga., NE of Atlanta, in the Blue Ridge foothills; inc. 1887. It is a trade center of an apple-growing region.

Cor·ner Brook (kŏr'nər), city (1971 pop. 26,309), W central N.F., Canada, on the Humber River. It has a large pulp and paper mill. Nearby is Gros Morne National Park.

Cor·ning (kôr'nĭng). **1.** City (1972 est. pop. 3,136), a seat of Clay co., NE Ark., near the Mo. border NNW of Paragould, in a cotton, rice, and timber area. Furniture is made here. **2.** City (1970 pop. 3,573), Tehama co., N Calif., NNW of Sacramento; laid out 1882, inc. 1907. It is the processing and shipping center of an extensive olive-growing region. Woodson Bridge State Park is nearby. **3.** City (1970 pop. 2,095), seat of Adams co., SW Iowa, WSW of Creston, in a farm, livestock, and dairy area; platted 1855, inc. 1871. An Icarian communist settlement was here from 1858 to 1898. **4.** City (1975 est. pop. 15,300), Steuben co., S N.Y., on the Chemung River, in a dairy and vineyard region; settled 1788, inc. as a city 1890. The glass industry, for which the city is famous, began here in 1868, and the Corning glass museum is a major tourist attraction today. In 1972, in the wake of Hurricane Agnes, the city was heavily damaged by floodwaters from the Chemung River.

Cor·no, Mon·te (mōn'tā kôr'nō), highest peak of the Apennines, c.9,560 ft (2,915.8 m) high, in the Gran Sasso d'Italia range, Abruzzi, central Italy. It is snow-capped for most of the year.

Corn·wall (kôrn'wôl, -wəl), nonmetropolitan county (1976 pop. including Isles of Scilly 407,100), 1,357 sq mi (3,514.6 sq km), SW England; administrative center Bodmin. Cornwall is a peninsula bounded by the English Channel and the Atlantic Ocean. It terminates in the west with the rugged promontory of Lands End. Farming and fishing are important. Engineering, ship repairing, and rock quarrying are the only industries. Cornish tin and copper mines were known to ancient Greek traders, and during World War II the old mines were reworked. Cornwall's climate, the picturesque coastal towns, and the romance of its past, interwoven with Arthurian legend and tales of piracy, have made the region popular with tourists. The Cornish language, related to the Welsh and Breton tongues, did not die out until the 18th cent.

Cornwall, manufacturing city (1971 pop. 47,116), SE Ont., Canada, on the St. Lawrence River. Its principal manufactures are cotton and rayon textiles, paper, chemicals, and electronic equipment.

Co·ro (kōr'ō), city (1971 pop. 68,701), capital of Falcón state, NW Venezuela, near the Caribbean Sea at the base of the Paraguaná peninsula. The development of the oil industry on the peninsula stimulated rapid expansion of the city. Founded in 1527, Coro became the base for Spanish explorations into the interior. From 1528 to 1546 it was mortgaged by the Spanish to a German banking house, and German adventurers explored the region.

Cor·o·man·del Coast (kôr'ə-măn'dĕl), E coast of Tamil Nadu and Andhra Pradesh states, SE India, stretching more than 400 mi (644 km) from Point Calimere, opposite the N tip of Sri Lanka, to the delta of the Krishna River. The inland coastal plain is bounded by the Eastern Ghats. The name probably stems from Cholomandalam, i.e., land of the Cholas, an empire that ruled the region from the 9th to the 12th cent.

Co·ro·na (kə-rō'nə), city (1975 est. pop. 29,800), Riverside co., S Calif.; inc. 1896. Citrus fruits are processed and castings, plywood paneling, fiberglass insulation, pipes, valves, and mobile homes are manufactured in the city. Cleveland National Forest is nearby.

Co·ro·na·do (kŏr'ə-nä'dō), city (1975 est. pop. 29,800), San Diego co., S Calif., on a peninsula on the W side of San Diego Bay; inc. 1890. It is a well-known beach resort. Adjacent to the city are a large U.S. naval air station and a naval amphibious base. Points of interest include the Hotel del Coronado, a state historical monument.

Coronado National Memorial: *see* National Parks and Monuments Table.

Co·ro·nel (kō-rō-nĕl'), city (1970 pop. 37,312), S central Chile, a port on the Pacific Ocean. It is a shipping point for coal from nearby mines. In a naval engagement off Coronel on Nov. 1, 1914, German Adm. Graf von Spee defeated a British squadron under Sir Christopher Cradock.

Cor·pus Chris·ti (kôr'pəs krĭs'tē), city (1975 est. pop. 210,000), seat of Nueces co., S Texas; inc. 1852. It is a busy port of entry on Corpus Christi Bay at the entrance to Nueces Bay (an inlet at the mouth of the Nueces River). The city is a petroleum and natural-gas center, with much heavy industry. It has oil refineries, smelting plants, chemical works, and food-processing establishments, as well as a large shrimp fleet and an important fishing industry.

Tradition holds that the bay was named by the Spanish explorer Alonzo Alvarez de Pineda who discovered it on Corpus Christi Day in 1519, but there is evidence that it was named instead by the first settlers, who arrived from the lower Rio Grande valley in the 1760s. In 1839 Col. H. L. Kinney founded a trading post here, and traders, adventurers, and ne'er-do-wells collected in a raffish colony on land claimed by both Texas and Mexico. It was briefly captured by the U.S. navy in the Civil War and later served as a supply and shipping point for sheep and cattle. It developed industrially after the discovery of oil in the area and the completion (1926) of a deepwater channel past Mustang Island.

Cor·reg·gio (kə-rĕj'ō, -rĕj'ē-ō, kōr-răd'jō), town (1971 pop. 20,301), in Emilia-Romagna, N central Italy. It is an agricultural market and a cheese-manufacturing center. The painter Antonio Allegri was born here (1494) and was called Correggio after the town.

Cor·reg·i·dor (kə-rĕg'ĭ-dôr'), historic fortified island (c.2 sq mi/5 sq km), at the entrance to Manila Bay, just off Bataan peninsula of Luzon Island, Philippines. From the days of the Spanish, Corregidor and its tiny neighboring islets—El Fraile, Caballo, and Carabao—guarded the entrance to Manila Bay, serving as an outpost for the defense of Manila. The Spanish also maintained a penal colony on Corregidor. When the Americans acquired the Philippine Islands after the Spanish-American War of 1898, they elaborately strengthened those defenses. Corregidor was honeycombed with tunnels to serve as ammunition depots, and Fort Mills and Kindley Field were established. Fort Drum was built on El Fraile, Fort Hughes on Caballo, and Fort Frank on Carabao. The new fortifications were deemed so formidable that Corregidor became known as the Gibraltar of the East. In the early phase of World War II, Corregidor's batteries guarded the entrance to Manila Bay and protected the flank of the large U.S.-Filipino army concentrated on Bataan. For five months Corregidor was subjected to one of the most intense continuous bombardments of the entire war. After the fall of Bataan, about 10,000 U.S. and Filipino troops under Lt. Gen. Jonathan M. Wainwright fought gallantly on for a month. They were hopelessly cut off from all supplies and aid. Corregidor was finally invaded early in May, 1942, and the garrison was forced to surrender. The island was recaptured in Mar., 1945, by U.S. paratroopers and shore landing parties. It is now a national shrine.

Cor·rèze (kô-rĕz'), department (1975 pop. 240,363), 2,263 sq mi (5,861.2 sq km), S central France, in Limousin; capital Tulle.

Cor·rib, Lough (lŏKH kŏr'ĭb), lake, 68 sq mi (176.1 sq km), Cos. Galway and Mayo, W Republic of Ireland. The irregularly shaped lake, which is 27 mi (43.4 km) long, drains into Galway Bay through the Corrib River.

Cor·rien·tes (kôr-ryān'tās), city (1970 pop. 137,823), capital of Corrientes prov., NE Argentina, a port on the Paraná River. It is the commercial center of a rich pastoral and agricultural region. An important cultural center, it has several institutions of higher education, museums, and historical monuments. Corrientes was founded in 1588 and survived fierce Indian attacks during the late 16th and early 17th cent. In 1762 an uprising against the colonial governor foreshadowed the wars of independence from Spain.

Cor·ry (kôr'ē), city (1970 pop. 7,435), Eric co., NW Pa., SE of Erie; settled 1795, inc. as a borough 1863, as a city 1866. Steel products and furniture are made here.

Cor·si·ca (kôr'sĭ-kə), island (1975 pop. 289,842), 3,367 sq mi (8,720.5 sq km), a department of metropolitan France, SE of France and N of Sardinia, in the Mediterranean Sea. Ajaccio is the capital. Olive oil, wine, timber, wheat, and cheese are produced. The island has developed a tourist industry. Much of the island is wild, covered by undergrowth, or maquis; the flowers of the maquis produce a fragrance that carries far out to sea and has earned for Corsica the name "the scented isle." The maquis long provided ideal hideouts for bandits, and banditry was not fully suppressed until the 1930s. Blood feuds between clans also persisted into modern times.

After having belonged to the Romans (3rd cent. B.C.-5th cent. A.D.), the Vandals, the Byzantines, and the Lombards, the island was granted (late 8th cent.) by the Franks to the papacy. It was threatened by the Arabs from c.800 to 1100. In 1077 Pope Gregory VII ceded Corsica to Pisa. Pisa and Genoa and later Genoa and Aragón battled for Corsica. In the mid-15th cent. the actual administration of the island was taken up by the Bank of San Giorgio in Genoa; Genoese rule was harsh and unpopular. In 1755 Pasquale Paoli headed a rebellion against Genoa, but its success resulted only in the cession (1768) of Corsica to France. One consequence of the transfer was the French citizenship of Napoleon I, who was born in 1769 at Ajaccio. With British support Paoli expelled the French in 1793, and in 1794 Corsica voted its union with the British crown. The French (under Napoleon) recovered it, however, in 1796, and French possession was guaranteed at the Congress of Vienna (1815). In World War II, Corsica was occupied by Italian and German troops. Late in 1943 the population revolted, and, with the assistance of a Free French task force, the Axis forces were driven out.

Cor·si·ca·na (kôr-sĭ-kăn'ə), city (1975 est. pop. 19,600), seat of Navarro co., E central Texas; inc. 1848. It is an oil center with wells and refineries and additional industries that depend on the cotton, small grains, and Hereford cattle produced in the surrounding blackland farm area. The discovery of oil when a city water well was being dug (1894) caused dismay at first but led to the drilling (1895) of the first commercial oil well west of the Mississippi and the building (1898) of the first refinery in Texas.

Cor·son (kôr'sən), county (1970 pop. 4,994), 2,470 sq mi (6,397.3 sq km), N S.Dak., in an agricultural and cattle-raising area bordered on the N by N.Dak., on the E by the Missouri River, and drained by the Grand River; formed 1909; co. seat McIntosh. Wheat and flax are grown, and lignite is mined. The county is included in Standing Rock Indian Reservation.

Cor·te Ma·der·a (kôr'tā mə-dâr'ə), town (1970 pop. 8,464), Marin co., W Calif., N of San Francisco; inc. 1916.

Cor·tez (kôr'tĕz, kôr-tĕz'), city (1970 pop. 6,032), seat of Montezuma co., SW Colo., in a fruit, grain, and livestock area at an altitude of 6,198 ft (1,890.4 m); founded 1887, inc. 1902. Food is processed, and oil-field machinery, sheet metal, and concrete are produced. Mesa Verde National Park and Yucca House National Monument are nearby.

Cort·land (kôrt'lənd), county (1970 pop. 45,894), 502 sq mi (1,300.2 sq km), central N.Y., in a dairying and farming area drained by the Tioughnioga River; formed 1808; co. seat Cortland. Beans, potatoes, truck crops, and clover are grown. Its diversified manufactures include typewriters, paint, and clothing.

Cortland, city (1970 pop. 19,621), seat of Cortland co., central N.Y., in a fertile farm area; settled 1791, inc. as a city 1900. Metal products and automotive and aircraft parts are among the manufactures.

Cor·to·na (kôr-tō'nə, kōr-tō'nä), town (1976 pop. 22,597), Tuscany, central Italy. It is an agricultural and tourist center. One of the 12 important Etruscan cities, Cortona later (310 B.C.) united with Rome. The town passed to Florence in the early 15th cent. Landmarks include the Romanesque cathedral (remodeled during the Renaissance), the Palazzo Pretorio (13th cent.), and the Church of San Francesco (begun (1245).

Ço·rum (chō-rōōm'), city (1975 pop. 64,839), capital of Çorum prov., N central Turkey. It is the trade center for a farm region where grains, fruits, sheep, and goats are raised. The city's manufactures include copper and leather goods. Important Hittite remains have been found here.

Co·rum·bá (kōō-rōōm-bä'), city (1970 pop. 81,838), Mato Grosso state, SW Brazil, on the Paraguay River. A river port, it is a trade center for a large pastoral region and has varied light industries. Founded as a military outpost and colony in 1778, it became strategically important with the opening of the Paraguay River to international trade after the Paraguayan War (1865–70).

Co·ru·ña, La (lä kō-rōō'nyä), city (1975 est. pop. 206,776), capital of La Coruña prov., NW Spain, in Galicia. It is a busy Atlantic port, a distribution center for the surrounding farm area, and a summer resort spot. It has shipyards, metalworks, and an important fishing industry. La Coruña reached its height as a port and a textile center in the late Middle Ages. The Armada sailed from its harbor in 1588. The city was sacked by Sir Francis Drake in 1598.

Co·run·na (kə-rŭn'ə), city (1970 pop. 2,829), seat of Shiawassee co., S central Mich., on the Shiawassee River W of Flint; inc. as a village 1858, as a city 1869.

Cor·val·lis (kôr-văl'ĭs), city (1975 est. pop. 37,100), seat of Benton co., NW Oregon, on the Willamette River; inc. 1857. It is a food-processing hub in the heart of the fertile Willamette River valley. Corvallis is the seat of Oregon State Univ.

Cor·y·don (kŏr'ĭ-dən, -dŏn'). **1.** Town (1970 pop. 2,719), seat of Harrison co., S central Ind., WSW of Louisville, Ky.; laid out 1808. It is a processing center in a farm area. Furniture is manufactured. Corydon was the territorial capital (1813–16) and state capital (1816–25). **2.** Town (1970 pop. 1,745), seat of Wayne co., S central Iowa, S of Chariton; settled 1849, inc. 1867.

Cor·yell (kôr-yĕl'), county (1970 pop. 35,311), 1,043 sq mi (2,701.4 sq km), central Texas, drained by the Leon River; formed 1854; co. seat Gatesville. In a primarily agricultural area yielding oats, corn, sorghums, cotton, hay, peanuts, livestock, dairy products, and poultry, it has good hunting and fishing. It includes Mother Neff State Park and part of Fort Hood.

Co·sen·za (kō-zān'tsä), city (1976 est. pop. 102,375), capital of Cosenza prov., Calabria, S Italy. It is an agricultural and industrial center. Manufactures include textiles, furniture, and lumber. The chief city of the ancient Brutii, it was taken by the Romans in 204 B.C. According to tradition, Alaric I (c.370–410 A.D.), the Visigothic king, was buried in the bed of the Busento River at Cosenza.

Co·shoc·ton (kə-shŏk'tən), county (1970 pop. 33,486), 562 sq mi (1,455.6 sq km), central Ohio, drained by the Muskingum, Tuscarawas, and Walhonding rivers; formed 1811; co. seat Coshocton. It has agriculture (livestock, dairy products, and grain), coal mines, sand and gravel pits, and some manufacturing.

Coshocton, city (1975 est. pop. 13,500), seat of Coshocton co., central Ohio, where the Tuscarawas and Walhonding rivers meet to form the Muskingum; inc. 1833. A warlike tribe of Delawares had a village here of the same name; in 1764 the expedition of Col. Henry Bouquet freed a number of white prisoners and established a peace treaty.

Cos·ta Bra·va (kōs'tä brä'vä), coastal strip, NE Spain, in Catalonia, near the French border on the Mediterranean. The area has enjoyed a booming tourist industry since the end of World War II.

Cos·ta Me·sa (kōs'tə mā'sə), city (1975 est. pop. 78,200), Orange co., S Calif.; inc. 1953. Boatbuilding and the manufacture of electronic equipment and tools are the major industries. Orange Coast College and Southern California College are in the city.

Cos·ta Ri·ca (kōs'tə rē'kə), republic (1970 pop. 1,800,000), 19,575 sq mi (50,700 sq km), Central America, bounded on the N by Nicaragua, on the E by the Caribbean Sea, on the SE by Panama, and on the S and W by the Pacific Ocean; capital San José.

Costa Rica is divided into seven provinces:

NAME	CAPITAL	NAME	CAPITAL
Alajuela	Alajuela	Limón	Limón
Cartago	Cartago	Puntarenas	Puntarenas
Guanacaste	Liberia	San José	San José
Heredia	Heredia		

Economy. Costa Rica is an agricultural country. Coffee, bananas, cocoa, and sugar cane are grown. Cattle and grain are raised.

History. Although Columbus skirted the Costa Rican coast in 1502, the Spanish conquest did not begin until 1563. The region was administered as part of the captaincy general of Guatemala. Few of the native Indians survived, and the colonists, unable to establish a hacienda system based on Indian labor, generally became small landowners. Westward expansion into the plateau began in the 18th cent. Costa Rica became independent from Spain in 1821. From 1822 to 1823 it was part of the Mexican Empire of Augustín de Iturbide. It then became part of the Central American Federation until 1838, when the sovereign republic of Costa Rica was proclaimed.

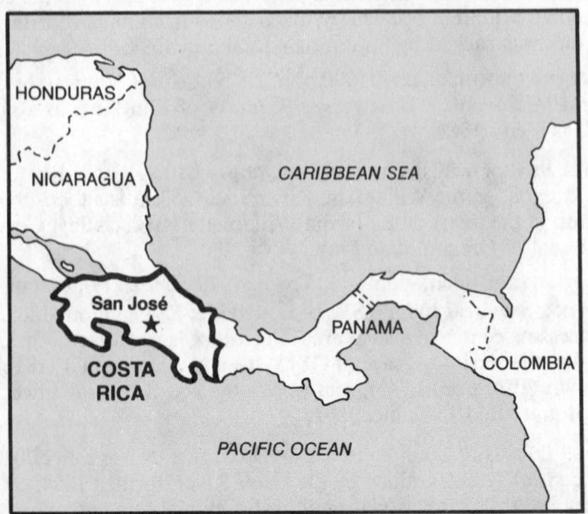

The cultivation of coffee, introduced in the 19th cent., helped create a landed oligarchy that dominated the country until 1870. In 1874 Minor Cooper Keith founded Limón and introduced banana cultivation. Keith also started the United Fruit Company.

Costa Rica's pattern of orderly, democratic government was broken in 1917, when Federico Tinoco overthrew the elected president, Alfredo González. The United States opposed Tinoco, and he was deposed in 1919. In 1948, in a close presidential election, Otilio Ulate appeared to have won, but the incumbent, Teodoro Picado, accused Ulate's supporters of fraud and obtained a congressional invalidation of the election. A six-week civil war ensued, at the conclusion of which a junta led by José Figueres Ferrer assumed power. In 1949 a new constitution was adopted, and the junta transferred power to Ulate as the elected president.

Government. The country is governed under the 1949 constitution. The president serves a four-year term and may not be immediately re-elected. The unicameral legislature is also elected for four years. There is universal adult suffrage, and voting is compulsory.

Cos·til·la (kōs-tē'ə, -yə), county (1970 pop. 3,091), 1,215 sq mi (3,146.9 sq km), S Colo., in an irrigated agricultural area on the N.Mex. border and bounded on the W by the Rio Grande; formed 1861; co. seat San Luis. Livestock, hay, and potatoes are produced. Tourism is important here, and some mining is done.

Co·ta·ba·to (kō'tä-bä'tō), city (1970 est. pop. 51,900), Cotabato prov., W Mindanao, Philippines, near the mouth of the Mindanao River on Moro Gulf. Its port serves a vast, fertile farm area.

Côte-d'Or (kōt-dôr'), department (1975 pop. 456,070), 3,384 sq mi (8,764.6 sq km), E France, largely in Burgundy, partly in Champagne; capital Dijon.

Co·ten·tin (kô-tän-tăn'), region of N France, in Normandy, roughly coinciding with the peninsula formed by Manche dept. and extending into the English Channel. The lambs of the Cotentin breed of sheep are highly esteemed for their meat.

Côtes-du-Nord (kōt'dü-nôr'), department (1975 pop. 525,556), 2,655 sq mi (6,876.5 sq km), NW France, in Brittany, on the English Channel; capital Saint-Brieuc.

Co·to·nou (kō-tō-nōō'), city (1975 est. pop. 178,000), capital of Atlantique dept., S Benin, on the Gulf of Guinea. It is a seaport and commercial center. The city has small-scale industries; manufactures include palm oil and cake, peanut oil, textiles, cement and other construction materials, aluminum sheet, beverages, and processed seafood. Motor vehicles and bicycles are assembled, and there are sawmills in the city. Drilling for offshore oil is carried on nearby. Cotonou was originally a small state that was dominated by the kingdom of Dahomey from the 18th cent. In 1851 the French made a treaty with the Dahomean king Gezo that allowed them to establish a trading post at Cotonou. In 1883 the French navy forcibly occupied the city to forestall British ambitions in the area. Britain confirmed France's right to Cotonou in 1885.

Co·to·pax·i (kō-tō-păk'sē), active volcano, 19,347 ft (5,900.8 m) high, N central Ecuador. A symmetrical snowcapped cone, it is one of the most beautiful peaks of the Andes and one of the highest volcanoes in the world. It is continuously active, and frequent eruptions have caused severe damage. Cotopaxi was first scaled in 1872.

Cots·wold Hills (kŏts'wōld', -wəld), range, mainly in Gloucestershire, W England, extending c.50 mi (80 km) NE from Bath. Cleeve Cloud (c.1,080 ft/330 m) is the highest point. Its crest line forms the Thames-Severn watershed. The region is famous for Cotswold sheep and for its picturesque stone houses. Noteworthy are the many megalithic monuments and long barrows.

Cot·tage Grove (kŏt'ĭj grōv'), village (1975 est. pop. 16,900), Washington co., SE Minn., near the St. Croix River; inc. 1965. Machined-metal products are among the manufactures.

Cott·bus or **Kot·tbus** (both: kŏt'bəs, kôt'bōōs), city (1974 est. pop. 94,293), capital of Cottbus dist., E East Germany, on the Spree River. It is an industrial center and rail junction. Manufactures include textiles, metal products, and processed food. Cottbus developed as a market center in the late 12th cent. and passed to Brandenburg in the mid-15th cent. It was annexed by Saxony in 1635 and was taken by Prussia in 1815.

Cot·ti·an Alps (kŏt'ē-ən), section of the W Alps between SE France and N Italy; highest elevation Mount Viso, 12,602 ft (3,843.6 m).

Cot·tle (kŏt'l), county (1970 pop. 3,204), 900 sq mi (2,331 sq km), NW

Texas, in rolling country drained by the Pease and Wichita rivers; formed 1876; co. seat Paducah. It is primarily agricultural, yielding cotton, grain sorghums, wheat, beef cattle, horses, hogs, sheep, poultry, and dairy products.

Cot·ton (kŏt'n), county (1970 pop. 6,832), 651 sq mi (1,686.1 sq km), S Okla., bounded on the S by the Red River, here forming the Texas border, and drained by the Deep Red Run and by Beaver and Cache creeks; formed 1912; co. seat Walters. In an agricultural area (livestock, cotton, wheat, oats, and poultry), it has oil and gas wells and some manufacturing.

Cotton Belt, major agricultural region of the SE United States, on the Atlantic and Gulf coastal plains and the Piedmont upland, extending through N.C., S.C., Ga., Ala., Miss., W Tenn., E Ark., La., E Texas, and S Okla., and also into small areas of SE Mo., SW Ky., N Fla., and SE Va. The belt has the climatic conditions necessary for cotton to thrive—high temperatures, from 30 to 55 in. (76.2–139.7 cm) annual rainfall, and a 200-day growing season. The Cotton Belt, no longer continuous, is made up of many separate intensive production areas; corn, wheat, soybeans, peanuts, beans, and livestock are important in the intervening areas.

Cot·ton·wood (kŏt'n-wŏŏd'), county (1970 pop. 14,887), 640 sq mi (1,657.6 sq km), SW Minn., watered by the Des Moines River; formed 1857; co. seat Windom. It lies in an agricultural area yielding corn, oats, barley, and livestock, and has a food-processing industry (meat-packing and dairy products). It manufactures textiles and farm, lawn, and garden machinery.

Cottonwood, river, c.140 mi (225 km) long, SW Minn., rising near the S.Dak. border and flowing E to join the Minnesota River below New Ulm.

Cottonwood Falls, city (1970 pop. 987), seat of Chase co., E central Kansas, SW of Topeka; founded 1858 by Free State settlers, inc. 1872. It is a trade and shipping center in a farm area.

Co·tul·la (kə-tōō'lə), city (1970 pop. 3,415), seat of La Salle co., SW Texas, on the Nueces River SW of San Antonio; inc. 1910. It has developed from a cow town in mesquite-brush country to a shipping center for truck crops and livestock and a tourist stop on the Pan American Highway.

Cou·ders·port (kou'dərz-pôrt'), industrial borough (1970 pop. 2,831), seat of Potter co., N central Pa., on the W bank of the Allegheny River W of Wellsboro; laid out 1807, inc. 1848. Its manufactures include textiles, leather goods, and electrical equipment.

Cou·lee Dam National Recreation Area (kōō'lē): *see* National Parks and Monuments Table.

Coun·cil (koun'səl), village (1970 pop. 899), seat of Adams co., W Idaho, NW of Cascade on the Weiser River. Lumber, apples, and sugar beets are processed.

Council Bluffs (blŭfs), city (1975 est. pop. 60,500), seat of Pottawattamie co., SW Iowa, on and below bluffs overlooking the Missouri River, opposite Omaha, Nebr.; inc. 1853. It was the site of an Indian trading post and of a Pottawattamie Indian mission before 1846, when it was settled by Mormons and named Kanesville. When the Mormons departed in 1852, the settlement was renamed Council Bluffs. The city is an important trade and industrial center for a large agricultural area. It has grain elevators, and its manufactures include processed foods, cast-iron pipes, farm equipment, electronic equipment, and fabricated metals.

Council Grove, city (1970 pop. 2,403), seat of Morris co., E central Kansas, on the Neosho River SW of Topeka, in a farm and cattle area; settled as a trading post 1857, inc. 1858. The site was an important stop on the old Santa Fe Trail.

coun·ty (koun'tē), division of local government in the United States, Great Britain, and many Commonwealth countries. The county developed in England from the shire, a unit of local government that originated in the Saxon settlements of the 5th cent. In the United States there are approximately 3,000 counties, most of which are either rural or suburban. Louisiana, influenced by the French, has parishes, which are essentially similar to counties.

Coupe·ville (kōōp'vĭl'), town (1977 est. pop. 924), seat of Island co., NW Wash., on Whidbey Island NW of Everett. It is a shore resort.

Cour·an·tyne (kôr'ən-tīn), river, rising in S Guyana and flowing c.450 mi (725 km) N to the Atlantic Ocean. It forms the boundary between Guyana and Surinam.

Cour·be·voie (kōōr'bə-vwä'), city (1968 pop. 58,283), Hauts-de-Seine dept., N central France, on the Seine River. An industrial suburb of Paris, Courbevoie manufactures automobiles, bicycles, perfumes, and pharmaceuticals. There are also electrical industries, foundries, and copper works.

Cour·land or **Kur·land** (both: kōōr'lənd, -länt'), historic region and former duchy, W European USSR, in Latvia, between the Baltic Sea and the Western Dvina River. It is an agricultural and wooded lowland. The early Baltic tribes—Letts and Kurs—who inhabited the region were subjected in the 13th cent. by the Livonian Brothers of the Sword. In the Northern War (1700–21), Courland was taken (1701) by Charles XII of Sweden. Empress Anna, who was, by marriage, duchess of Courland before her accession in Russia, forced (1737), the nobles of Courland to elect her favorite, Ernst Johann von Biron, their duke. Russian influence became paramount, and with the third partition of Poland (1795) the duchy passed to Russia. In 1918, Courland was incorporated into Latvia, except for a strip of the southern coast that went to Lithuania.

Cour·land Lagoon (kōōr'lănd): *see* Kursky Zaliv.

Court·land (kôrt'lənd), town (1970 pop. 899), seat of Southampton co., SE Va., on the Nottoway River W of Suffolk.

Cou·shat·ta (kōō-shăt'ə), town (1970 pop. 1,492), seat of Red River parish, NW La., near the Red River SE of Shreveport, in a cotton and cattle area; settled c.1870.

Cov·en·try (kŏv'ən-trē), county borough (1976 est. pop. 336,800), West Midlands, central England. It is an industrial center noted for automobile production; tractors, airplanes, machine tools, synthetic textiles, electrical equipment, and engineering products are also made. Lady Godiva and her husband founded a Benedictine abbey in the town in 1043. By the 14th cent. Coventry, a flourishing market and textile-weaving town, was one of the five largest towns in England. The entire central portion of Coventry, including the 14th cent. Cathedral of St. Michael, was destroyed in an 11-hour air raid in Nov., 1940. A new cathedral, alongside the ruins of the old one, was completed in 1962.

Coventry. 1. Town (1970 pop. 8,140), Tolland co., NE Conn., bordered by the Willimantic river; settled c.1700, inc. 1712. Silk goods are produced. Nathan Hale was born here, and his house has been restored. **2.** Town (1970 pop. 22,947), Kent co., W R.I.; settled 1643, set off from Warwick and inc. 1741. Formerly a noted lace center, it still has textile industries, but today glass, chemicals, and pharmaceuticals are also important. Coventry's many historic structures include the Payne House (1668) and Nathanael Greene's homestead (1770).

Co·vi·lhã (kōō-vē-lyăɴ'), town (1970 municipal pop. 60,768), E central Portugal, in Beira Baixa. It had a famous fair in medieval times and is still a trade center as well as a textile milling town.

Co·vi·na (kō-vē'nə), city (1975 est. pop. 32,400), Los Angeles co., S Calif.; inc. 1901. Citrus fruits are processed, and medical supplies and fabricated-metal products are made. The area was settled in 1842, citrus crops were introduced in 1886, and the citrus industry reached its peak in the 1930s when Covina was one of the world's largest producers.

Cov·ing·ton (kŭv'ĭng-tən). **1.** County (1970 pop. 34,079), 984 sq mi (2,548.6 sq km), S Ala., in a coastal plain area bordering on Fla. and drained by the Conecuh and Yellow rivers; formed 1821; co. seat Andalusia. Cotton, corn, and hogs are produced. It has some manufacturing. Part of Conecuh National Forest is here. **2.** County (1970 pop. 14,002), 416 sq mi (1,077.4 sq km), S Miss., drained by tributaries of the Leaf River; formed 1819; co. seat Collins. Its agriculture includes cotton, corn, truck crops, poultry, and dairy products. Lumbering is done, and appliances and clothing are manufactured.

Covington. 1. City (1970 pop. 10,267), seat of Newton co., N central Ga.; inc. 1854. It is a processing and market center in a cotton area. Natural and synthetic textiles are manufactured in the city. Points of interest include ante-bellum homes spared by Gen. W. T. Sherman in his march (1864) to the sea. **2.** City (1970 pop. 2,641), seat of Fountain co., W central Ind., on the Wabash River E of Danville, Ill.; laid out 1826. It ships fruit and farm produce. **3.** City (1975 est. pop. 50,500), Kenton co., N central Ky., at the confluence of the Ohio and Licking rivers; inc. 1815. It is an industrial center, connected by bridges with Cincinnati across the Ohio and Newton across the Licking. There are tobacco and meat-packing establishments and plants making a great variety of products, including paper, sheet metal, metal fabricators, machine tools, and electrical equipment. A ferry and a tavern were established here c.1801, and the city was first settled in 1812. Among its points of interest are the suspension bridge to Cincinnati (designed by J. A. Roebling), Devou Park, with a museum of natural history, Cathedral Basilica of the Assumption, the tiny Monte Casino chapel, the Garden of Hope; and the Carneal House (1815). **4.** City (1970 pop. 7,170), seat of St. Tammany parish, SE La., near Lake Pontchartrain N of New Orleans, in a timber, farm, and fishing area; settled 1769, inc. 1813. Two state parks are nearby. **5.** Village (1970 pop. 2,575), Miami co., W central Ohio, NNW of Dayton; settled 1807, inc. 1835. It is a farm trade center, with creameries. **6.** Town (1975 est. pop. 5,898), seat of Tipton co., W central Tenn., near the Mississippi and Hatchie rivers NNE of Memphis, in a farm and cotton area; inc. 1826. Textiles, chemicals, and ceramics are produced. **7.** City (1975 est. pop. 9,100), seat of Alleghany co. but politically independent, W central Va., near the W.Va. line, on the Jackson River in a valley surrounded by mountains; laid out 1819, inc. as a city 1952. Paper, furniture, chemical fibers, and film are manufactured in Covington. There is excellent hunting and fishing in the area.

Cow·ans·ville (kou'ənz-vĭl'), town (1971 pop. 11,902), S Que., Canada, on the Yamaska River, SE of Montreal. Its manufactures include textiles, furniture, electronic equipment, and chemicals.

Cowes (kouz), urban district (1971 pop. 18,895), Isle of Wight, S England. A resort town with lovely promenades, it is also the main port of the island and the center for yachting in the British Isles. Cowes became the headquarters of the Royal Yacht Club in 1838, and fashionable regattas are held annually. Queen Victoria died in Osborne House in East Cowes.

Co·we·ta (kə-wē'tə), county (1970 pop. 32,310), 443 sq mi (1,147.4 sq km), W Ga., bounded on the NW by the Chattahoochee River; formed 1826; co. seat Newman. In a timber and agricultural region yielding cotton, corn, melons, peaches, pecans, and livestock, it also has a textile industry.

Cow·ley (kou'lē), county (1970 pop. 35,012), 1,136 sq mi (2,942.2 sq km), S Kansas, in a gently rolling to hilly area bordering in the S on Okla. and drained in the W by the Arkansas and Walnut rivers; formed 1870; co. seat Winfield. Its agriculture includes wheat, cattle, and hogs. It has extensive oil and natural-gas fields, refineries, flour mills, and meat-packing plants.

Cow·litz (kou'lĭts), county (1970 pop. 68,616), 1,146 sq mi (2,968.1 sq km), SW Wash., in a rolling hill area rising to the foothills of the Cascade Range in the E and watered by the Cowlitz, Lewis, and Columbia rivers; formed 1854; co. seat Kelso. Logging and fishing are important. Its agriculture includes livestock, dairy products, fruit, hay, and truck crops.

Cowlitz, river, 130 mi (209.2 km) long, SW Wash., rising in the Cascade Mts. SE of Mt. Rainier and flowing W and S to the Columbia River near Longview.

Cow·pens (kou'pĕnz'), town (1970 pop. 2,109), Spartanburg co., NW S.C., ENE of Spartanburg; inc. 1880. The Revolutionary War Battle of Cowpens was fought on Jan. 17, 1781, between American militia under Daniel Morgan and a British force under Sir Banastre Tarleton. The resulting defeat and heavy losses of the British were serious blows to Cornwallis's plans for the subjugation of the Carolinas. The battle is commemorated by Cowpens National Battlefield Site.

Crad·ock (krăd'ək), town (1970 pop. 22,329), Cape Province, SE South Africa, on the Great Fish River; founded as a frontier outpost in 1811. It is a trade and distribution center.

Craf·ton (krăf'tən), residential borough (1970 pop. 8,233), Allegheny co., SW Pa., W of Pittsburgh, in a coal-mining area; inc. c.1893. Paper products are made.

Craig (krāg). **1.** County (1970 pop. 14,722), 764 sq mi (1,978.8 sq km), NE Okla., bounded on the N by the Kansas border and drained by the Neosho River; formed 1907; co. seat Vinita. In a stock-raising and agricultural (grain, poultry, corn, and dairy products) area, it has oil and natural-gas wells, coal mines, and stands of hardwood timber. Some manufacturing is done. **2.** County (1970 pop. 3,524), 336 sq mi (870.2 sq km), W Va., in the Alleghenies bounded on the W by the W.Va. border and drained by Craig Creek; formed 1851; co. seat New Castle. It has timber, iron-ore deposits, and varied agriculture (grain, livestock, and poultry). Entirely within Jefferson National Forest, it also has mountain resorts, mineral springs, and good hunting and fishing areas.

Craig, city (1975 est. pop. 4,629), seat of Moffat co., NW Colo., on the Yampa River at an altitude of c.6,200 ft (1,890 m); inc. 1908. It is the shipping point for an oil, livestock, and grain area.

Craig·head (krāg'hĕd'), county (1970 pop. 52,068), 717 sq mi (1,857 sq km), NE Ark., intersected by Crowley's Ridge, drained by the St. Francis and Cache rivers; formed 1859; co. seats Jonesboro and Lake City. Its agriculture includes cotton, corn, rice, fruit, poultry, and livestock. It has gravel pits and manufacturing (meat-packing, rice-milling, wood, paper, and aluminum products, and hardware).

Cra·io·va (krä-yô'vä), city (1977 est. pop. 222,399), SW Rumania, in Walachia, on the Jiu River, a tributary of the Danube. It is the industrial center of the agricultural and mineral-rich Oltenia region and is an important market for grain. Machine building, food processing, and the manufacture of electrical equipment are the chief industries. Built on the site of a Roman settlement, Craiova became the capital of Oltenia in 1492. It was destroyed by an earthquake in 1790 and burned by the Turks in 1802. Craiova has several museums containing prehistoric and Roman relics.

Cran·brook (krăn'brŏŏk'), city (1976 pop. 13,510), SE British Columbia, Canada. It is a lumbering center.

Cran·don (krăn'dən), city (1970 pop. 1,582), seat of Forest co., NE Wis., ESE of Rhinelander, in a wooded lake region; inc. 1898. It is a resort.

Crane (krān), county (1970 pop. 4,172), 796 sq mi (2,061.6 sq km), W Texas, on the W part of Edwards Plateau at an altitude of c.2,000 ft (610 m), and bounded on the S by the Pecos River; formed 1887; co. seat Crane. In an oil-producing region, it does some ranching and farming (fruit and truck crops).

Crane, city (1970 pop. 3,427), seat of Crane co., W Texas, S of Odessa, in a dairy and oil region; founded 1926, inc. 1933.

Cran·ford (krăn'fərd), residential township (1970 pop. 27,391), Union co., NE N.J., W of Elizabeth; inc. 1871. Metal products are manufactured. Union College is here.

Cran·ston (krăn'stən), industrial city (1975 est. pop. 73,300), Providence co., central R.I., a residential suburb of Providence; inc. as a town 1754, as a city 1910. Its manufactures include machinery, chemicals, textiles, and beer. In the 19th cent. Cranston was an important textile center. The Friends Meeting House (1729) and several pre-Revolutionary buildings remain standing.

Cra·ter Lake National Park (krā'tər), 160,290 acres (64,917.5 hectares), SW Oregon, in the Cascade Range; est. 1902. Crater Lake, 20 sq mi (51.8 sq km), lies in a huge pit that was created when the top of a prehistoric volcano was blown off by a violent eruption. The second-deepest lake (1,932 ft/589.3 m) in North America, Crater Lake is 6 mi (9.7 km) wide, lies 6,164 ft (1,880 m) above sea level, and is surrounded by cliffs that are from 500 to 2,000 ft (152.5–610 m) high. Having no inlet or outlet, the lake was formed by rain and snowfall, and its waters are maintained by precipitation. The lake was discovered in 1853 by prospectors, who called it Deep Blue Lake because of the intense blue of the water; it was renamed Crater Lake in 1869. A scenic highway follows the rim of the crater. Wizard Island, a cinder cone 776 ft (236.7 m) high, near the lake's western shore, was also formed by volcanic activity.

Craters of the Moon National Monument, 53,545 acres (21,685.7 hectares), S central Idaho; est. 1924. This region, composed of several closely grouped volcanoes, is suggestive of a telescopic view of the moon. Volcanic activity dating back c.20,000 years has left behind cinder cones, tree molds, craters, and other interesting formations.

Cra·ven (krā'vən), county (1970 pop. 62,554), 725 sq mi (1,877.8 sq km), E N.C., near the Atlantic coast, in a tidewater area drained by the Neuse and Trent rivers; formed 1712; co. seat New Bern. Farming (tobacco and corn), lumbering, and fishing are important.

Craw·ford (krô'fərd). **1.** County (1970 pop. 25,677), 598 sq mi (1,548.8 sq km), NW Ark., bounded on the W by the Okla. border, on the S by the Arkansas River; formed 1820; co. seat Van Buren. It has agriculture (fruit, cotton, truck crops, and livestock), zinc deposits, hardwood timber, and natural-gas wells. Food is processed, and furniture and wood and rubber products are manufactured. The county has mountain resorts and includes part of Ozark National Forest. **2.** County (1970 pop. 5,748), 313 sq mi (810.7 sq km), central Ga., bounded on the SW by the Flint River and intersected by the Fall Line; formed 1812; co. seat Knoxville. It is in a timber and agricultural area yielding cotton, corn, vegetables, peaches, pecans, and livestock. **3.** County (1970 pop. 19,824), 442 sq mi (1,144.8 sq km), SE Ill., bounded on the E by the Wabash River and drained by the Embarrass River; formed 1816; co. seat Robinson. Its agriculture includes livestock, corn, wheat, oats, hay, alfalfa, poultry, and dairy products. Oil drilling and refining and lumber milling are done. China, pottery, oil-well supplies, food products, glycerin, and caskets are manufactured. Ports along the Wabash River were important trade centers in the 19th cent. **4.** County (1970 pop. 8,033), 312 sq mi (808.1 sq km), S Ind., bounded on the S by the Ohio River and the Ky. border, on the E by the Blue River; formed 1818; co. seat English. In an agricultural area yielding grain, tobacco, truck crops, poultry, livestock, and dairy products, it has limestone quarries and a lumbering industry. Canned food, wood products, and cement are produced. **5.** County (1970 pop. (19,116), 716 sq mi (1,854.4 sq km), W Iowa, in a prairie agricultural area drained by the Boyer River; formed 1851; co. seat Denison. Hogs, cattle, poultry, corn, oats, and wheat are produced. It has bituminous-coal deposits. **6.** County (1970 pop. 37,850), 598 sq mi (1,548.8 sq km), SE Kan., in an agricultural area bordering in the E on Mo.; formed 1867; co. seat Girard. Livestock, grain, poultry, and dairy products are important. It has extensive coal deposits, zinc and lead mines, and scattered gas and oil fields in the west. Diversified manufacturing is done. **7.** County (1970 pop. 6,482), 563 sq mi (1,458.2 sq km), N central Mich., drained by the Manistee River and branches of the Au Sable River; organized 1879; co. seat Grayling. In an agricultural area yielding livestock, potatoes, grain, hay, and dairy products, it has lumber mills and some small industry. Hunting, fishing, and winter sports are popular here. The county includes part of Huron National Forest and Lake Margrethe. **8.** County (1970 pop. 14,828), 760 sq mi (1,968.4 sq km), SE central Mo., in the Ozarks and drained by the Meramac River; formed 1829; co. seat Steelville. It has agriculture (wheat, corn, hay, and livestock), timber, and fire-clay, sulfur, and iron deposits. Hats, shoes, and metal products are manufactured. Part of Mark Twain National Forest is in the county. **9.** County (1970 pop. 50,364), 404 sq mi (1,046.4 sq km), N central Ohio, drained by the Sandusky; formed 1815; co. seat Bucyrus. In an agricultural area (livestock, dairy products, grain, and poultry), it has stands of hardwood timber and some manufacturing. **10.** County (1970 pop. 81,342), 1,016 sq mi (2,631.4 sq km), NW Pa., in a manufacturing and agricultural area drained by French Creek; formed 1800; co. seat Meadville. It has oil and gas deposits and sand pits. Metal products, textiles, food products, and chemicals are manufactured. **11.** County (1970 pop. 15,252), 586 sq mi (1,517.7 sq km), SW Wis., bounded on the W by the Mississippi, here forming the Iowa border, on the S by the Wisconsin River; formed 1818; co. seat Prairie du Chien. It is drained by the Kickapoo River. In a dairying and stock-raising area, it has grain and tobacco farms. It is an important cheese-making center.

Craw·fords·ville (krô'fərdz-vĭl'), city (1975 est. pop. 13,300), seat of Montgomery co., W central Ind.; inc. 1866. It is the trading center of an agricultural and dairy region. Major industries include printing and binding and the manufacture of nails and wire, plastic, and metal products. Wabash College is here.

Craw·ford·ville (krô'fərd-vĭl'). **1.** Village (1970 pop. 750), seat of Wakulla co., NW Fla., SSW of Tallahassee, in a lumbering and agricultural area. It is at the edge of Apalachicola National Forest. **2.** City (1970 pop. 735), seat of Taliaferro co., NE Ga., WNW of Augusta; inc. 1826. Liberty Hall, the home of Alexander H. Stephens, is included in a memorial state park here.

Craw·ley (krô'lē), new town and urban district (1976 est. pop. 72,600), West Sussex, SE England. Crawley was designated one of the New Towns in 1946 to alleviate overpopulation in London. There are many industries, including precision engineering.

Cré·cy (krā-sē'), officially **Crécy-en-Pon·thieu** (-äɴ-pôɴ-tyœ'), village, Somme dept., N France. A nearby forest is popular for camping. At Crécy, on Aug. 26, 1346, Edward III of England defeated Philip VI of France in the Hundred Years' War.

Creede (krēd), town (1970 pop. 653), seat of Mineral co., SW Colo., on a headstream of the Rio Grande, in the San Juan Mts. NW of Alamosa; founded 1890 when silver was discovered. The mines were closed down after 1893 when the price of silver declined.

Creek (krēk), county (1970 pop. 45,532), 936 sq mi (2,424.2 sq km), central Okla., drained by the Cimarron and Arkansas rivers and the Deep Fork; formed 1907; co. seat Sapulpa. In an agricultural area yielding grain, cotton, corn, sorghums, livestock, and dairy products, it has extensive oil and natural-gas fields and refineries. Some manufacturing is done.

Cre·mo·na (krĭ-mō'nə, krā-mô'nä), city (1976 est. pop. 82,489), capital of Cremona prov., Lombardy, N Italy, on the Po River. It is an agricultural market and an industrial center. Originally (3rd cent. B.C.) a Roman colony, Cremona was in the Middle Ages an independent commune frequently at war with Milan until its surrender to that city in 1344. It was known in the Middle Ages as a center of learning, in the late Renaissance for a school of painting founded (16th cent.) by Giulio Campi, and later (17th–18th cent.) for the violins made by the Amati, the Guarneri, the Stradivari, and their successors.

Cren·shaw (krĕn'shô), county (1970 pop. 13,188), 611 sq mi (1,582.5 sq km), S central Ala., in a level farm region drained by the Conecuh River and Patsaliga Creek; formed 1866; co. seat Luverne. Cotton, peanuts, corn, livestock, and lumber are produced.

Crĕs (tsrĕs), island (1961 pop. 4,113), 158 sq mi (409.2 sq km), in the Adriatic Sea, off Croatia, NW Yugoslavia. Formerly in Austria-Hungary, it passed to Italy in 1918 and to Yugoslavia in 1947. Fruit growing, fishing, and sheep raising are the chief occupations.

Cres·cent City (krĕs'ənt), city (1970 pop. 2,586), seat of Del Norte co., N Calif., on the Pacific coast near the Oregon border NNW of Eureka, in a fishing and timber region; laid out 1852, inc. 1854.

Cres·co (krĕs'kō), city (1970 pop. 3,927), seat of Howard co., NE Iowa, near the Minn. border WNW of Decorah; inc. 1868. It is a trade and processing center for a farm, dairy, and livestock area. Some manufacturing is done.

Cress·kill (krĕs'kĭl), suburban borough (1970 pop. 8,298), Bergen co., NE N.J., near the Hudson River opposite Yonkers, N.Y.; inc. 1894.

Crest Hill (krĕst), city (1970 pop. 7,460), Will co., NE Ill., just N of Joliet; inc. 1960.

Cres·ton (krĕs'tən), city (1970 pop. 8,234), seat of Union co., SW Iowa, SW of Des Moines; inc. 1871. It is a trade, rail, and manufacturing center for a bluegrass and corn area.

Crest·view (krĕst'vyōō'), city (1970 pop. 7,952), seat of Okaloosa co., NW Fla., NE of Pensacola near the Ala. line, in a lumber area; inc. 1917.

Crest·wood (krĕst'wŏŏd'). **1.** Village (1973 est. pop. 7,557), Cook co., NE Ill., a suburb SSW of Chicago; inc. 1928. **2.** City (1975 est. pop. 14,500), St. Louis co., E central Mo., a suburb of St. Louis; inc. as a city 1949. Located in a truck-farming area, it is mostly residential with some light industry. The Thomas Sappington House (1808; restored 1965) is a good example of Federal architecture.

Crete (krēt), island (1971 pop. 456,642), c.3,235 sq mi (8,380 sq km), SE Greece, in the E Mediterranean Sea, c.60 mi (100 km) from the Greek mainland. The largest of the Greek islands, it extends c.160 mi (260 km) from east to west and marks the southern limit of the Aegean Sea, the southern part of which is also called the Sea of Crete. The rocky northern coast of Crete is deeply indented, and the interior is largely mountainous, culminating in Mt. Ida (8,058 ft/2,457.7 m). Crete has many small farms, whose chief crops are grains, olives, and oranges, and food processing is its main industry. Sheep, goats, and dairy cattle are also raised. Crete had one of the world's earliest civilizations, the Minoan civilization, which reached its greatest power and prosperity c.1600 B.C. Later, for reasons still obscure, its power suddenly collapsed; but Crete flourished again after the Dorian Greeks settled on the island and established city-states. Although important as a trade center, Crete played no significant part in the political history of ancient Greece. It was conquered (68 B.C.–67 B.C.) by the Romans, passed (A.D. 395) to the Byzantines, fell (824) to the Arabs, was reconquered by the Byzantines (961), and passed to Venice in 1204.

The Cretans were not displeased at changing masters when the Ottoman Turks conquered (1669) virtually the whole island after a 24-year war. A series of revolts against the Turks in the 19th cent. reached a climax in the insurrection of 1896–97 that led to war (1897) between Greece and Turkey. An autonomous Cretan state was formed under nominal Turkish rule, but the population preferred union with Greece, which was proclaimed in 1908 and internationally recognized in 1913.

Cré·teil (krā-tā'), city (1975 est. pop. 59,023), capital of Val-de-Marne dept., N central France, on the Marne River. Gold and silver items, pencils, and varnish are produced.

Creuse (krœz), department (1968 pop. 156,876), 2,146 sq mi (5,558.1 sq km), central France, in the Massif Central; capital Guéret.

Creu·sot, Le (lə krœ-zō'), city (1968 pop. 34,102), Saône-et-Loire dept., E central France, in Burgundy. Situated in a coal-mining region, it is the site of the large Schneider iron and steel mills and munitions factories (founded 1837).

Creve Coeur (krēv' ko͝or'). **1.** Village (1973 est. pop. 6,594), Tazewell co., N central Ill., S of Peoria; inc. 1921. **2.** City (1976 est. pop. 10,660), St. Louis co., E Mo., a suburb WNW of St. Louis; inc. 1949.

Crewe (kro͞o), municipal borough (1971 pop. 51,302), Cheshire, W central England. It is an important railroad junction with large locomotive and car works.

Cri·me·a (krī-mē'ə), peninsula (1970 pop. 1,814,000), c.10,000 sq mi (25,900 sq km), extreme S European USSR, linked with the mainland by the Perekop Isthmus. The peninsula, administratively part of the Ukraine, is coterminous with the Crimea oblast, of which Simferopol is the capital. The peninsula is bounded on the south and west by the Black Sea.

Known in ancient times as Tauris, the peninsula became (5th cent. B.C.) the kingdom of Cimmerian Bosporus, which later came under Greek influence. Ionian and Dorian Greeks began to colonize the coast in the 6th cent., and the peninsula became the major source of wheat for ancient Greece. In the 1st cent. A.D., the Greek part of the peninsula became a Roman protectorate. Its Greek name was then Latinized into Chersonesus Taurica. During the next millennium the area was overrun by Ostrogoths, Huns, Khazars, Cumans, and in 1239, by the Mongols. The southern shore was mostly under Byzantine control from the 6th to the 12th cent. Trade relations were established (11th–13th cent.) with Kievan Russia. In the 13th cent. Genoa founded prosperous coastal commercial settlements. After Tamerlane's destruction of the Golden Horde, the Tatars established (1475) an independent khanate in north and central Crimea. In the late 15th cent. both the khanate and the southern coastal towns were conquered by the Ottoman Empire. Russian armies first invaded the Crimea in 1736. Catherine II forced Turkey to recognize the khanate's independence in 1774, and in 1783 she annexed it outright.

During the Crimean War (1853–56), parts of the remaining Tatar population were resettled in the interior of Russia. After the Bolshevik Revolution (1917) an independent Crimean republic was proclaimed; but the region was soon occupied by German forces and then became a refuge for the White Army. In 1921 a Tatar Autonomous Soviet Socialist Republic was created. During World War II, German invaders took the Crimea after an eight-month siege. Accused by the Soviet government of collaborating with the Germans, the Crimean Tatars were forcibly removed from their homeland after the war and resettled in distant parts of the Asian USSR. The republic itself was dissolved (1945) and made into an oblast of the Russian SFSR; in 1954 it was transferred to the Ukraine.

Cri·şa·na-Ma·ra·mu·reş (krĭ-shä'nä-mä'rä-mo͞o'rĭsh), historic province, NW Rumania, between Transylvania and Hungary. The region occupies the easternmost part of the Hungarian plain and the western foothills of the Transylvanian Alps. It is largely agricultural. Crişana-Maramureş was part of Hungary until 1919.

Crisp (krĭsp), county (1970 pop. 18,087), 296 sq mi (766.6 sq km), S central Ga., bounded on the W by the Flint River and Lake Blackshear, formed 1905; co. seat Cordele. It is in a coastal-plain timber and agricultural area yielding cotton, corn, peanuts, pecans, fruit, and truck crops.

Crit·ten·den (krĭt'n-dən). **1.** County (1970 pop. 48,106), 608 sq mi 1,574.7 sq km), E Ark., in a delta region bounded on the E by the Mississippi River, here forming the Tenn. border; formed 1825; co. seat Marion. It has agriculture (cotton, corn, soybeans, and hay), and hardwood timber. Paper, iron and steel, and metal products are produced. **2.** County (1970 pop. 8,493), 365 sq mi (945.4 sq km), W Ky., bounded on the NW by the Ohio River, here forming the Ill. border, on the NE by the Tradewater River, and on the SW by the Cumberland River; formed 1824; co. seat Marion. In a rolling agricultural area yielding livestock, poultry, grain, dairy products, and burley tobacco, it has limestone quarries.

Croagh·pat·rick (krō-păt'rĭk) mountain, 2,510 ft (765.6 m) high, Co. Mayo, W Republic of Ireland, near Clew Bay. Legend connects it with St. Patrick, and its summit has long been a place of pilgrimage.

Cro·a·tia (krō-ā'shə), constituent republic of Yugoslavia (1971 pop. 4,422,564), 21,824 sq mi (56,524 sq km), NW Yugoslavia. Zagreb is the capital. It includes Croatia proper, Slavonia, Dalmatia, and most of Istria. Western Croatia lies in the Dinaric Alps; the eastern part, drained by the Sava and Drava rivers, is mostly low lying and agricultural. The Pannonian plain is the chief farming region. More than one third of Croatia is forested, and lumber is a major export. The region is the leading coal producer of Yugoslavia, and also has deposits of bauxite, copper, petroleum, and iron ore. The republic is the most industrialized and prosperous area of Yugoslavia. Tourism, especially along the Adriatic coast, is important to the economy.

A part of the Roman province of Pannonia, Croatia was settled in the 7th cent. by Croats, who accepted Christianity in the 9th cent. A kingdom from the 10th cent., Croatia reached its peak of power in the 11th cent., but internecine strife facilitated its conquest in 1091 by King Ladislaus I of Hungary. Although Croatia remained linked with Hungary for eight centuries, the Croats retained their own diet and were governed by a ban, or viceroy. In 1527 the Croatian feudal lords agreed to accept the Hapsburgs as their kings in return for common defense against the Ottoman Turks. During the 19th cent. Hungary imposed Magyarization on Croatia and promulgated (1848) laws that seriously jeopardized Croatian autonomy. But after the dual Austro-Hungarian monarchy was established in 1867, Croatia, united with Slavonia, became an autonomous Hungarian crown land.

With the collapse of Austria-Hungary (1918), the kingdom of Serbs, Croats, and Slovenes was formed. Croatian separatist agitation persisted, however, and during World War II an independent Croatian state was established (1941) under Italian and German military control. A large part of the population joined the anti-fascist Yugoslav partisan forces under Josip Broz Tito, himself a native of Croatia. In 1945 Croatia became one of the six republics of reconstituted Yugoslavia.

Crock·ett (krŏk'ĭt). **1.** County (1970 pop. 14,402), 269 sq mi (696.7 sq km), W Tenn., bounded on the SW by the South Fork of the Forked Deer River; formed 1871; co. seat Alamo. Its agriculture includes cotton, corn, livestock, and truck crops. Strawberries and tomatoes are shipped. **2.** County (1970 pop. 3,885), 2,794 sq mi (7.236.5 sq km), W Texas, in the rough prairies and woodlands of the Edwards Plateau and bounded on the W by the Pecos River; founded 1875; co. seat Ozona. It is at an altitude of c.1,500–2,500 ft (460–765 m) in an oil, natural-gas, and livestock (sheep, goats, and cattle) region.

Crockett, city (1970 pop. 6,616), seat of Houston co., E Texas, SSE of Palestine; founded in the 1830s. In a lumbering and farming region, it manufactures wood products.

Crom·ar·ty Firth, (krŏm'ər-tē), deep narrow inlet of Moray Firth, c.15 mi (25 km) long, N Scotland. Its narrow entrance is protected by the headlands of the Sutor rocks, more than 400 ft (122 m) high.

Crom·well (krŏm'wĕl, -wəl), town (1970 pop. 7,400), Middlesex co., central Conn., on the Connecticut River below Hartford; settled c.1650, set off from Middletown 1851. Tools and toys are made.

Crook (kro͝ok). **1.** County (1970 pop. 9,985), 2,980 sq mi (7,718.2 sq km), central Oregon, in a timber, livestock, and dairy area drained by the Crooked River; formed 1882; co. seat Prineville. Quicksilver is mined. It includes part of Ochoco National Forest. **2.** County (1970 pop. 4,535), 2,882 sq mi (7,464.4 sq km), NE Wyo., in a grain and livestock region bordered on S. Dak. and Mont. and watered by the Belle Fourche and Little Missouri rivers; formed 1875; co. seat Sundance. Oil, lumber, and coal are important. It includes Devils Tower National Monument, a section of Black Hills National Forest, and part of the Black Hills in the northeast.

Crooks·ton (kro͝oks'tən), city (1975 est. pop. 8,499), seat of Polk co., NW Minn., on the Red Lake River SE of Grand Forks, N. Dak.; settled 1872, inc. 1879. It is a trade and shipping center for livestock and farm products.

Cros·by (krôz'bē), municipal borough (1973 est. pop. 56,750), Merseyside, NW England, on Liverpool Bay. Formed in 1937 from the urban districts of Great Crosby and Waterloo-with-Seaforth, Crosby is primarily residential. The local history of Crosby dates back more than 1,000 years. The Merchant Taylor's School was founded in 1620.

Crosby, county (1970 pop. 9,085), 911 sq mi (2,359.5 sq km), NW Texas, on the edge of the Llano Estacado at an altitude of 2,100–3,400 ft (640.5–1,037 m); formed 1876; co. seat Crosbyton. It is drained by the White River and the Double Mountain Fork of the Brazos River. In a cattle-ranching and agricultural (cotton, sorghums, wheat, forage, dairy products, hogs, sheep, and poultry) area, it has some light industry.

Crosby, city (1970 pop. 1,545), seat of Divide co., extreme NW N.Dak., near the Canadian boundary; inc. 1911. A port of entry, it is a trade center for a livestock and grain region.

Cros·by·ton (krôz'bē-tən), city (1970 pop. 2,251), seat of Crosby co., NW Texas, ENE of Lubbock, in a farm and ranch region of the Llano Estacado; founded 1908.

Cross (krôs), county (1970 pop. 19,783), 625 sq mi (1,618.8 sq km), E Ark., drained by the St. Francis River; formed 1862; co. seat Wynne. Its agriculture includes cotton, corn, rice, wheat, soybeans, hay, peaches, vegetables, and livestock. It also has fisheries and a lumbering industry. Clothing, shoes, chemicals, and rubber products are manufactured.

Cross City, town (1970 pop. 2,268), seat of Dixie co., N Fla., W of Gainesville; inc. 1924. It is a trade, shipping, and processing center for a farm region, with lumbering and fishing industries.

Cros·sett (krô-sĕt'), city (1975 est. pop. 6,295), Ashley co., SE Ark., ESE of El Dorado near the La. border; inc. 1947. Settled in 1899 as a company lumber town, it was later sold to the inhabitants. Lumber, paper, and chemicals are now manufactured.

Cross-Flor·i·da Waterway: (krôs'flôr'ĭ-də): *see* Okeechobee Waterway.

Cross·ville (krôs'vĭl), town (1970 pop. 5,381), seat of Cumberland co., central Tenn., W of Knoxville, in a coal, timber, and farm region; founded c.1856, inc. 1901.

Cro·to·na (krō-tō'nə) or **Cro·ton** (krō'tən), ancient city, S Italy, on the E coast of Bruttium (now Calabria), a colony of Magna Graecia founded c.708 B.C. Pythagoras established a school here, which exerted a notable political and moral influence. The nearby temple of Hera Lacinia was the religious shrine of Magna Graecia. The height of the city's prosperity was reached after the army, led by the athlete Milo, destroyed the rival town of Sybaris (510 B.C.). Crotona then became involved in wars and soon declined. It was captured by the Romans in 277 B.C.

Cro·ton Aqueduct (krō'tən), 38 mi (61.1 km) long, SE N.Y., carrying water from the Croton River basin to New York City; built 1837–42. It was one of the earliest modern aqueducts in the United States.

Cro·ton-on-Hud·son (krō'tən-ŏn-hŭd'sən), residential village (1970 pop. 7,523), Westchester co., SE N.Y., on the E bank of the Hudson River NNW of Ossining; settled 1609, inc. 1898. Van Cortlandt Manor, dating from the 17th cent. and thought to be the oldest surviving Dutch manor house in America, has been restored.

Crow·ell (krō'əl), city (1970 pop. 1,399), seat of Foard co., N Texas, W of Wichita Falls, in a cotton, wheat, and cattle area; settled 1887, inc. 1908. The town was rebuilt after a tornado in 1942.

Crow·ley (krou'lē), county (1970 pop. 3,086), 803 sq mi (2,079.8 sq km), SE central Colo., in an irrigated agricultural region bounded on the S by the Arkansas River; formed 1911; co. seat Ordway. Livestock, sugar beets, and feed grains are produced.

Crowley, city (1975 est. pop. 16,600), seat of Acadia parish, SW La.; inc. 1888. It is a shipping, milling, and storing center for one of the nation's largest rice-growing areas and has a rice experiment station. Oil and natural-gas wells are located nearby.

Crown Point (kroun). **1.** City (1970 pop. 10,931), seat of Lake co., NW Ind.; inc. 1868. Film is processed, and truck conveyors, golf balls, feed grinders, and cabinets are made. **2.** Town (1970 pop. 1,857), Essex co., NE N.Y., on Lake Champlain. Crown Point is a summer resort on a historic site. The French realized the strategic importance of this point on the route from New York to Canada and in 1731 began building Fort St. Frederic. In the French and Indian Wars the fort successfully resisted (1755–56) early English attacks but was demolished (1759) before the advance of Jeffrey Amherst. The British began to build Fort Amherst (renamed Fort Crown Point) in 1759. In the American Revolution, Crown Point was finally abandoned (June 22, 1777) to Gen. John Burgoyne.

Crow Wing (krō' wĭng'), county (1970 pop. 34,826), 995 sq mi (2,577 sq km), central Minn., drained by the Mississippi River, and watered by numerous lakes; formed 1857; co. seat Brainerd. It lies in a resort and agricultural (dairy products, livestock, and potatoes) area. There is iron mining in the east. Its manufactures include lumber, textiles, paper, plastics, and metal products.

Croy·don (kroid'n), borough (1971 pop. 331,851) of Greater London, SE England. The borough was created in 1965 by the merger of the county borough of Croydon with the urban district of Coulsdon and Purley. Scientific instruments, internal-combustion engines, and electronic equipment are manufactured.

Crys·tal (krĭs'təl), city (1970 pop. 30,925), Hennepin co., SE Minn., a suburb of Minneapolis.

Crystal City. 1. City (1970 pop. 3,898), Jefferson co., E central Mo., on the Mississippi SSW of St. Louis, in an area of silica pits; settled 1900, inc. 1911. The Pittsburgh Plate Glass Company is the major source of employment here. **2.** City (1970 pop. 8,104), seat of Zavala co., SW Texas, SW of San Antonio. In an irrigated winter-garden area, it boasts of being the Spinach Capital of the World.

Crystal Falls, city (1970 pop. 2,000), seat of Iron co., N Mich., W Upper Peninsula, near the falls of the Paint River W of Mansfield; inc. as a village 1889, as a city 1899. It is a thriving mining city on the Menominee iron range. Nearby lakes attract tourists.

Crystal Lake, city (1970 pop. 14,541), McHenry co., NE Ill., in a dairy farm and lake resort area; inc. 1874. Electrical components, drills, and tools are manufactured.

Cse·pel (chĕ'pĕl), island, c.100 sq mi (260 sq km), in the Danube, N central Hungary, just S of Budapest. In the northern section are the city and harbor of the same name, and there is an international free port. An industrial suburb of Budapest, the city of Csepel has ironworks and steelworks, an oil refinery, munitions factories, and motorcycle works. The rest of the island is agricultural.

Ctes·i·phon (tĕs'ĭ-fŏn'), ruined ancient city, SE of Baghdad, Iraq, on the left bank of the Tigris opposite Seleucia and at the mouth of the Diyala River. After 129 B.C. it was the winter residence of the Parthian kings. Ctesiphon grew rapidly and was of renowned splendor. The Romans captured it in warring against Parthia. In 637 it was taken and plundered by the Arabs; it was abandoned by them when Baghdad became the capital of the Abbassides. Its site marks the farthest advance of Great Britain against Turkey in World War I. It is noted for its impressive ruins.

Cuan·za or **Kwan·za** (both: kwän'zə), river, 600 mi (965.4 km) long, rising in central Angola and flowing NW and W to the Atlantic. Its lower course, which is navigable for c.160 mi (260 km), was the original route of Portuguese penetration into northern Angola.

Cuau·tla (kwou'tlä), city (1970 pop. 67,869), Morelos state, S Mexico, in the Cuautla River valley. It is the heart of a sugar-cane and rice district. Cuautla's hot springs and lovely scenery make it a popular resort and tourist attraction.

Cu·ba (kyo͞o'bə), republic (1970 pop. 8,553,395, including the Isle of Pines), 44,218 sq mi (114,524 sq km), consisting of the island of Cuba and numerous adjacent islands. Havana is the capital. Cuba is the largest and westernmost of the West Indies and lies strategically at the entrance of the Gulf of Mexico. The south coast is washed by the Caribbean Sea, the north coast by the Atlantic Ocean, and in the

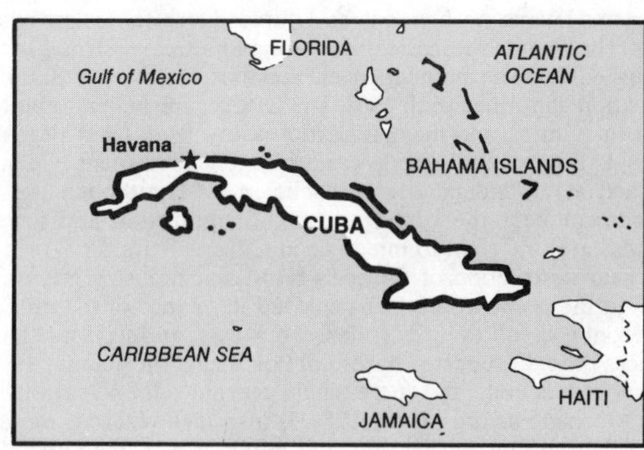

east the Windward Channel separates Cuba from Haiti. The shores are often marshy and are fringed by coral reefs and cays. Cuba has three mountain regions: the wild and rugged Sierra Maestra in the east, a lower range, the scenic Sierra de los Órganos, in the west; and the Sierra de Trinidad, a picturesque mass of hills amid the plains and rolling country of central Cuba, a region of vast sugar plantations. The rest of the island is level or rolling.

Cuba is divided into six provinces:

NAME	CAPITAL	NAME	CAPITAL
Camagüey	Camagüey	Matanzas	Matanzas
La Habana	Havana	Oriente	Santiago de Cuba
Las Villas	Santa Clara	Pinar del Rio	Pinar del Rio

Economy. The topography, the semitropical and generally uniform climate, and the soil are suitable for various crops, but sugar cane has been dominant since the late 18th cent.; it is grown on about two thirds of all crop land. Cattle raising is second in production value. An excellent tobacco is grown, and coffee, rice, corn, citrus fruits, and sweet potatoes are important. Large-scale fishing operations have been encouraged under the Castro regime, and that industry is now one of the largest in Latin America. Manufacturing is centered chiefly in the processing of agricultural products; sugar milling has long been the largest industry, and Cuba is also known for its tobacco products. Some consumer goods (textiles, fertilizer, cement, etc.) are also manufactured. Large amounts of nickel, copper, chromite, and manganese are mined, as well as lesser quantities of salt, lead, zinc, gold, silver, and oil. Limestone, clay, gypsum, and sulfur production easily meet the country's needs. There are immense iron reserves, but problems of extraction and purification are great, and iron production is still almost negligible.

History. The island was discovered in 1492 by Christopher Columbus. The Spanish conquest began in 1511 under the leadership of Diego de Velázquez. Cuba served as the staging area for Spanish explorations of the Americas. As an assembly point for treasure fleets, it offered a target for French and British buccaneers. The native Arawak Indians soon died off and were replaced as laborers by Negro slaves, who contributed much to the cultural evolution of the island. Despite pirate attacks and the trade restrictions of Spanish mercantilist policies, Cuba, the Pearl of the Antilles, prospered. In the imperial wars of the 18th cent. other nations coveted the Spanish possession, and in 1762 a British force captured and briefly held Havana. Cuba was returned to Spain by the Treaty of Paris in 1763 and remained Spanish even as most of Spain's possessions became (early 19th cent.) independent republics. The slave trade reached its peak in 1817. Sporadic slave revolts were brutally suppressed by the Spaniards. Representation at the Spanish Cortes, granted in 1810, was withdrawn, yet neither internal discontent nor filibustering expeditions (1848–51) led by Narciso López, achieved results. The desire of U.S. Southerners to acquire the island as a slave state also failed. Cuban discontent grew and finally erupted (1868) in the Ten Years' War, which ended (1878) in a truce, with Spain promising reforms and greater autonomy. Spain failed to carry out most of the reforms, although slavery was abolished (1886) as promised.

In 1895 a second war of independence was launched with the brilliant writer José Martí as its leader. Strong sentiment in the United States favored the rebels, which after the sinking of the *Maine* in Havana harbor led the United States to declare war on Spain. The Spanish forces capitulated, and a treaty, signed in 1898, established Cuba as an independent republic, although U.S. military occupation continued until 1902. The U.S. regime helped rebuild the war-torn country, and the conquest of yellow fever by Walter Reed, Carlos J. Finlay, and others was a heroic achievement. Cuba was launched as an independent republic in 1902, although the Platt Amendment kept the island under U.S. protection and gave the United States the right to intervene in Cuban affairs. In World War I the near-destruction of Europe's beet-sugar industry raised sugar prices to the point where Cuba enjoyed its "dance of the millions." The boom was followed by collapse, however, and wild fluctuations in prices brought repeated hardship. Politically, the country suffered fraudulent elections and increasingly corrupt administrations. Gerardo Machado as president (1925–33) instituted vigorous measures, forwarding mining, agriculture, and public works, then abandoned his great projects in favor of suppressing opponents.

Machado was overthrown in 1933, and from then until 1959 Fulgencio Batista y Zaldívar, a former army sergeant, dominated the political scene, either directly as president or indirectly as army chief of staff. Under Franklin Delano Roosevelt's administration the Platt Amendment was abandoned in 1934, the sugar quota was revised, and tariff rulings were changed to favor Cuba. However, economic problems continued, complicated by the difficulties of U.S. ownership of many of the sugar mills and the continuing need for diversification. In Mar., 1952, shortly before scheduled presidential elections, Batista seized power through a military coup. A revolt in 1953 by Fidel Castro was abortive, but in 1956 Fidel Castro took to the Sierra Maestra, where, aided by Ernesto "Che" Guevara, he waged a guerrilla war. The United States withdrew military aid to Batista in 1958; his army demoralized, Batista finally fled on Jan. 1, 1959. Castro, supported by young professionals, students, urban workers, and some farmers, was soon in control of the nation.

The expropriation of U.S. landholdings, banks, and industrial concerns, and an intensive program of vilification against the United States, led to the breaking (Jan., 1961) of diplomatic relations by the U.S. government. That same year Castro openly proclaimed his allegiance with the Communist camp. Meanwhile Cuban exiles were pouring into the United States by the thousands; one result of their activities was an invasion force (trained mostly in Florida and Guatemala under the supervision of the U.S. Central Intelligence Agency) which landed on Girón Beach in the Bay of Pigs, Cuba, in Apr., 1961. It was quickly crushed—a debacle especially humiliating to the United States because of its involvement. The following year the USSR began to buttress Cuba's military power and to build missile bases on the islands. In a dramatic confrontation President John F. Kennedy demanded (Oct., 1962) the dismantling of the missiles and ordered naval vessels to blockade the island, preventing further importation of offensive weapons. After a period of great world tension, during which several Soviet vessels turned away from Cuba, Soviet Premier Khrushchev (despite fiery denunciations by Castro and by Chinese Communists) agreed to withdraw the missiles.

Castro's relations with other Latin American countries were harmed by his announced intention of spreading his revolution to those countries by guerrilla warfare. In Feb., 1962, the Organization of American States formally excluded Cuba from its council, and by Sept., 1964, all Latin American nations except Mexico had broken diplomatic and economic ties with Cuba. After the death (1967) of Guevara while engaged in guerrilla activity in Bolivia, Cuban attempts to encourage revolution in other countries abated, and by the early 1970s the Castro government exhibited an interest in regaining the friendship of the Latin American nations and resumed diplomatic relations with several of them. In Cuba, Castro has remained in firm control; most of the thousands who had opposed him have fled the island.

Cú·cu·ta (kōō'kōō-tä), city (1968 est. pop. 167,400), capital of Norte de Santander dept., NE Colombia, near the Venezuelan border, on the E cordillera of the Colombian Andes. An industrial city, Cúcuta is the center of a rich coffee, oil, and mineral region. The city was founded in 1733. At Cúcuta the constituent congress of 1821 met to draft the constitution of Greater Colombia (present-day Venezuela, Ecuador, and Colombia). The city was rebuilt after an earthquake in 1875.

Cud·a·hy (kŭd'ə-hē), city (1970 pop. 22,078), Milwaukee co., SE Wis., an industrial suburb of Milwaukee, on Lake Michigan; inc. 1906. It was founded in 1892 by John and Patrick Cudahy as a site for their meat-packing enterprise, which remains a major industry. The city also produces pipe fittings, valves, drop forgings, packaging and bottling machinery, cranes, and truck seats.

Cud·da·lore (kŭd-ə-lôr'), town (1971 pop. 101,345), Tamil Nadu state, SE India. It is a port on the Bay of Bengal. Peanut products, cashew nuts, and sugar are the chief exports.

Cud·da·pah (kŭd'ə-pä'), city (1971 pop. 66,238), Andhra Pradesh state, S central India. It is a district administrative center and a market for peanuts, cotton, tumeric, and onions. Paint and varnish are manufactured, and asbestos and barite are processed. Melons from the district are famous.

Cuen·ca (kwĕng'kä), city (1974 pop. 104,667), alt. c.8,000 ft (2,440 m), S central Ecuador. Founded in 1557, Cuenca is in one of the richest agricultural basins of the Ecuadorian Andes and is the commercial center of southern Ecuador. Cuenca is known as the "marble city" because of its many fine buildings, including the cathedral, government palace, and university.

Cuenca, city (1975 est. pop. 36,806), capital of Cuenca prov., E central Spain, in New Castile, at the confluence of the Huecar and Júcar rivers, c.3,000 ft (915 m) above sea level. This historic town retains its medieval character in the narrow streets, clustered houses, and bridges. The city was taken (1177) from the Moors by Alfonso VIII of Castile. Cuenca was badly damaged in the Peninsular War and the Second Carlist War. Nearby is the Ciudad Encantada, or enchanted city, a fantastic labyrinth of eroded rocks.

Cuer·na·va·ca (kwâr'nä-vä'kä), city (1974 est. pop. 239,800), capital of Morelos state, S Mexico, in the Cuernavaca valley. The city has flour mills and beverage, textile, and cement industries. Cuernavaca is also a popular tourist and health resort. In the city are beautiful churches, monasteries, a 16th cent. Franciscan convent, a palace built by Hernán Cortés and now decorated with murals by Diego Rivera, and a formal garden that was frequented by Emperor Maximilian and Empress Carlotta.

Cue·ro (kwâr'ō), city (1970 pop. 6,956), seat of De Witt co., S Texas, on the Guadalupe River ESE of San Antonio; founded 1872, inc. 1873. An early cow town, it now raises turkeys and processes cotton, milk, and other farm produce.

Cu·ia·bá (ko͞o'yə-bä'), city (1970 pop. 100,865), capital of Mato Grosso state, W Brazil, at the head of navigation on the Cuiabá River. Founded in the gold rush of the early 18th cent., the city is a trading center for an extensive cattle-raising and agricultural area.

Cui·to (kwē'tō), river, SE Angola, rising in the central plateau and flowing c.400 mi (645 km) generally SSE to the Okavango River.

Cul·ber·son (kŭl'bər-sən), county (1970 pop. 3,429), 3,848 sq mi (9,966.3 sq km), extreme W Texas, in a scenic mountain and plateau region bounded on the N by the N.Mex. border; formed 1911; co. seat Van Horn. In a cattle-raising area, it has a tourist trade, saltworks, silver, copper, and mica mines, and some stands of timber.

Cul·go·a (kŭl-gō'ə), river, c.200 mi (320 km) long, SE Queensland and NE New South Wales, Australia, a tributary of the Darling River.

Cu·lia·cán (ko͞o'lyä-kän'), city (1974 est. pop. 228,000), capital of Sinaloa state, W Mexico, on the Culiacán River. It is situated on a hot coastal plain that produces tropical fruits, sugar cane, cotton, beans, and maize; cattle-raising is also important. Culiacán, founded in 1531, figured prominently in the early Spanish colonial period as a point of departure for northern expeditions, notably that of Francisco Coronado in 1540.

Cull·man (kŭl'mən), county (1970 pop. 52,445), 743 sq mi (1,924.4 sq km), N Ala., in a hilly agricultural area bounded on the E by Mulberry Fork; formed 1877; co. seat Cullman. Cotton, corn, and strawberries are grown. It also has lumber mills and coal mines.

Cullman, city (1975 est. pop. 13,000), seat of Cullman co., N Ala.; inc. 1875. It is a shipping and trade center for a cotton, timber, and dairy region. Cullman was settled in 1873 by German immigrants.

Cul·lod·en Moor (kə-lŏd'n, -lō'dən), moorland in Inverness-shire, NE Scotland, in the Highland region. Here, on Apr. 16, 1746, English forces defeated the Highlanders under Prince Charles Edward Stuart, thus ending the Jacobite uprising of 1745.

Cul·pep·er (kŭl'pĕp'ər), county (1970 pop. 18,218), 389 sq mi (1,007.5 sq km), N Va., in a piedmont region bounded on the S by the Rapidan River, on the N and NE by the Rappahannock River; formed 1748, Madison co. set off 1792, Rappahannock co. set off 1833; co. seat Culpeper. It is in a rich livestock, dairying, poultry-raising, and agricultural (fruit, grain, tobacco, and truck crops) area.

Culpeper, town (1970 pop. 6,056), seat of Culpeper co., N Va., WNW of Fredericksburg; founded 1759, inc. 1898. It is the trade and shipping center of a rich agricultural area. Several Civil War engagements were fought nearby.

Cul·ver City (kŭl'vər), city (1975 est. pop. 39,400), Los Angeles co., S Calif., a residential suburb of Los Angeles; inc. 1917. It is a center of the U.S. motion-picture industry, which began in the city c.1915. The city's chief industrial products are electronic and aerospace equipment.

Cu·mae (kyo͞o'mē), ancient city of Campania, Italy, near Naples. According to Strabo, it was the earliest Greek colony in Italy or Sicily, and it seems to have been founded c.750 B.C. by Chalcis. The area has yielded earlier non-Greek archaeological finds. Cumae founded a number of colonies and grew to be a great power. It repulsed Etruscan and Umbrian attacks, but fell in the late 5th cent. B.C. to the Samnites. Cumae supported Rome in the 2nd cent. B.C. and adopted Roman culture; ultimately its inhabitants became Roman citizens. As neighboring cities rose to power, Cumae declined, although it did not disappear until the 13th cent. A.D.

Cu·ma·ná (ko͞o'mä-nä'), city (1971 pop. 119,751), capital of Sucre state, NE Venezuela, on the Manzanares River near its mouth on the Gulf of Cariaco, an inlet on the Caribbean Sea. Coffee, tobacco, cacao, and sugar are exported. Founded in 1521 to exploit the pearl fisheries near Margarita island, Cumaná was often raided by the Dutch and British in the 16th and 17th cent. Frequently a victim of earthquakes, the city was severely damaged in 1929.

Cum·ber·land, (kŭm'bər-lənd), former county of N England, bordering on the Irish Sea to the W, Solway Firth to the N and W, and Scotland to the N. The county town was Carlisle. Cumberland was the scene of many centuries of border strife between England and Scotland. In 1974 Cumberland became part of the new nonmetropolitan county of Cumbria.

Cumberland. 1. County (1970 pop. 9,772), 347 sq mi (898.7 sq km), SE central Ill., drained by the Embarrass River; formed 1843; co. seat Toledo. Its agriculture includes livestock, fruit, poultry, seed, broomcorn, hay, beans, corn, and wheat. Brooms, shoes, flour, cheese and other dairy products, and gloves are manufactured. **2.** County (1970 pop. 6,850), 313 sq mi (810.7 sq km), S Ky., bordered on the S by Tenn. and drained by the Cumberland River; formed 1798; co. seat Burkesville. In a hilly agricultural area in the Cumberland foothills, it produces livestock, grain, burley tobacco, poultry, and fruit. It also has stone quarries and oil deposits. Part of Dale Hollow Reservoir is here. **3.** County (1970 pop. 192,528), 881 sq mi (2,281.8 sq km), SW Maine, drained by the Fore, Presumpscot, Nonesuch, Royal, and Stroudwater rivers; formed 1760; co. seat Portland. Truck farming, dairying, food canning, fishing, and printing are done. Paper and wood products, shoes, furniture, hardware, clay products, and textiles are manufactured. There are many resort areas in the county. **4.** County (1970 pop. 121,374), 503 sq mi (1,302.8 sq km), S N.J., bounded on the SW by Delaware Bay and drained by the Maurice River and Cohansey Creek; formed 1748; co. seat Bridgeton. In an agricultural area yielding poultry, truck crops, fruit, and dairy products, its manufactures include stone, concrete, glass, and iron products, textiles, clothing, and canned goods. Part of the pine-barrens region (with timber, cranberries, and huckleberries) is within the county. **5.** County (1970 pop. 212,042), 661 sq mi (1,712 sq km), S central N.C., in a sand-hill region drained by the Cape Fear and South Black rivers; formed 1754; co. seat Fayetteville. Its agriculture includes tobacco, corn, soybeans, cotton, and peanuts. Rubber, chemicals, metal products, and textiles are manufactured. Fort Bragg Military Reservation is in the northwest. **6.** County (1970 pop. 158,177), 555 sq mi (1,437.5 sq km), S Pa., in an agricultural and manufacturing area bounded on the E by the Susquehanna River and drained by Conodoguinet Creek; formed 1750; co. seat Carlisle. Metals and metal products, textiles, leather products, and flour are manufactured. It also has clay and limestone deposits and some agriculture (grain and fruit). **7.** County (1970 pop. 20,733), 679 sq mi (1,758.6 sq km), E central Tenn., on the Cumberland Plateau drained by the Sequatchie River; formed 1856; co. seat Crossville. Its agriculture includes corn, hay, potatoes, and livestock. There are stone quarries, coal deposits, and tracts of timber here. **8.** County (1970 pop. 6,179) 288 sq mi (745.9 sq km), central Va., drained by the Willis River and bounded on the S and SE by the Appomattox River, on the NE by the James River; formed 1749; co. seat Cumberland. In an agricultural area yielding tobacco, cattle, and dairy products, it also has a lumbering industry.

Cumberland 1. City (1975 est. pop. 27,000), seat of Allegany co., NW Md., on the North Branch of the Potomac; settled 1750, inc. 1815. It is an important railroad and shipping center for a coal-mining area. Its manufactures include textiles, synthetic fibers, tires, glass, metal products, petrochemicals, propellants, and plastics. Cumberland grew around the site of a trading post established (1750) by the Ohio Company at a natural gateway through the Appalachians to the Ohio valley. Fort Cumberland (built 1754) was the base of operations for the ill-fated Braddock expedition (1755) against the French and Indians and the site of Washington's first military headquarters (1757). **2.** Town (1975 est. pop. 26,700), Providence co., NE R.I., on the Blackstone River and the Mass. line; included in Mass. until 1746, inc. as a R.I. town 1747. Its manufactures include textiles and metal and fiberglass products. The Ballou Meetinghouse dates from c.1740. **3.** Village (1970 pop. 200), seat of Cumberland co., central Va., W of Richmond, in an agricultural area. Tobacco is grown.

Cumberland, river, 687 mi (1,105.4 km) long, rising in E Ky., and winding generally SW through Ky. and Tenn., then NW to the Ohio River near Paducah, Ky. It drains c.18,500 sq mi (47,915 sq km).

Locks and canals make the river navigable for small craft.

Cumberland Gap, natural passage through the Cumberland Plateau, near the point where Va., Ky., and Tenn. meet. The gap was formed by the erosive action of a stream that once flowed there. It was discovered and named in 1750 by Dr. Thomas Walker, leader of a land company exploration party. Daniel Boone's Wilderness Road ran through the gap. A strategic point in the Civil War, the gap was held alternately by Confederate and Union forces. Cumberland Gap National Historic Park was established in 1955.

Cumberland Island National Seashore: *see* National Parks and Monuments Table.

Cumberland Plateau or **Cumberland Mountains,** SW division of the Appalachian Mt. system, extending NE to SW throughout parts of W.Va., Va., Ky., and Tenn. into N Ala. Black Mt. is the highest point (4,145 ft/1,264.2 m). On the east the plateau rises sharply from the Great Valley of eastern Tenn.; on the west the slope is rough and broken. The plateau is the source of the Cumberland River.

Cum·ber·nauld (kŭm'bər-nôld'), new town and burgh (1974 est. pop. 38,287), Strathclyde region, Scotland. Cumbernauld's industries include food processing and the manufacture of adding machines and adhesive products.

Cum·bri·a (kŭm'brē-ə), nonmetropolitan county (1976 est. pop. 473,600), extreme NW England, created under the Local Government Act of 1972; administrative center Carlisle. It is composed of the county boroughs of Barrow-in-Furness and Carlisle, the former counties of Cumberland and Westmorland, and parts of the former counties of Lancaster and Yorkshire (West Riding).

Cum·bri·an Mountains (kŭm'brē-ən), mountains of the Lake District, NW England. Scafell Pike (3,210 ft/979 m) is the highest point. The range is studded with lakes and narrow valleys.

Cum·ing (kŭm'ĭng), county (1970 pop. 12,034), 571 sq mi (1,478.9 sq km), NE Nebr., in an agricultural area drained by the Elkhorn River; formed 1855; co. seat West Point. Corn, soybeans, alfalfa, and grains are grown. Some food processing and mining are done.

Cum·ming (kŭm'ĭng), town (1970 pop. 2,031), seat of Forsyth co., N Ga., NNE of Atlanta, in a farm area; founded 1832. Lumber and cotton are processed.

Cu·ne·ne or **Ku·ne·ne** (both: kōō-nā'nə), river, rising in W central Angola and flowing c.750 mi (1,205 km) S and W to the Atlantic. Its lower course forms part of Angola's border with Namibia.

Cu·ne·o (kōō'nā-ō), city (1976 est. pop. 56,088), capital of Cuneo prov., Piedmont, NW Italy, on the Stura River, near the Maritime Alps. It is an agricultural and industrial center and a transportation junction. Manufactures include textiles, beer, and chemical fertilizers. Silkworms are raised. A possession of the house of Savoy after 1382, Cuneo endured numerous sieges (especially in 1799, when it fell to Austrian and Russian forces).

Cu·per·ti·no (kōō'pər-tē'nō, kyōō'-), city (1975 est. pop. 23,900), Santa Clara co., W Calif., W of San Jose; settled in the 1850s, inc. 1955. In a fruit-growing region, it makes electronic components.

Cu·ra·çao (kyŏŏr'ə-sou', -sō', kŏŏr'-), island (1975 est. pop. 156,209), 178 sq mi (461 sq km), largest of the Netherlands Antilles, in the Dutch West Indies. Curaçao is semiarid; most of the plant life is of desert character. Oil refining is the principal industry, and the island has some of the world's largest refineries, receiving oil from the enormous reserves at nearby Lake Maracaibo, Venezuela. Other island industries include tourism (Curaçao is a free port), shipbuilding, and the manufacture of cement, paint, and tiles. Discovered in 1499, Curaçao was not settled by the Spanish until 1527. The Dutch captured it in 1634 and remained in possession except for a brief period of British rule during the Napoleonic Wars. In the 18th cent. Curaçao was a base for a flourishing Dutch entrepôt trade. Prosperity declined after the abolition of slavery in 1863 but revived with the introduction of the petroleum industry in the early 20th cent. Curaçao was the scene of severe racial strife and rioting in 1969.

Cu·re·can·ti National Recreation Area (kyŏŏr'ə-kăn'tē): *see* National Parks and Monuments Table.

Cu·ri·có (kōō-rē-kō'), city (1970 pop. 59,621), capital of Curicó prov., central Chile, near the Mataquito River. Founded in 1743, Curicó is the metropolis of a flourishing agricultural region noted for livestock raising. The town was rebuilt after an earthquake destroyed it in 1928.

Cu·ri·ti·ba (kōō-rē-tē'bä), city (1970 pop. 483,038), capital of Paraná state, SE Brazil. A commercial and processing center for an expanding agricultural and ranch area, it was founded in 1654 but was of little significance until the late 19th and early 20th cent., when immigrants (chiefly Germans, Italians, and Slavs) began to develop the Paraná hinterland.

Cur·ragh, the (kûr'ə), undulating plain or common, 4,885 acres (1,978.4 hectares), Co. Kildare, E Republic of Ireland. It has been a military camp since 1646.

Cur·ri·tuck (kûr'ĭ-tŭk), county (1970 pop. 6,976), 273 sq mi (707.1 sq km), extreme NE N.C., in a tidewater area bordered by a barrier beach enclosing Currituck Sound and bounded on the N by the Va. border, on the E by the Atlantic, on the S by Albemarle Sound, and on the SW by the North River; formed 1672; co. seat Currituck. In a timber and agricultural (corn, soybeans, cotton, and truck crops) area, it has sawmills, a fishing industry, and coastal resorts.

Currituck, village (1970 pop. 250), seat of Currituck co., extreme NE N.C., on Currituck Sound NE of Elizabeth City. It is a fishing and hunting resort.

Cur·ry (kûr'ē). **1.** County (1970 pop. 39,517), 1,403 sq mi (3,633.8 sq km), E N.Mex., in an agricultural area bordering on Texas; formed 1909; co. seat Clovis. Grain and livestock are raised. It also has some mining and manufacturing. **2.** County (1970 pop. 13,006), 1,622 sq mi (4.201 sq km), SW Oregon, in the SW extremity of the state bounded on the W by the Pacific Ocean, on the S by the Calif. border; formed 1855; co. seat Gold Beach. It is drained by the Rogue River. Agriculture, lumber milling, fishing, and tourism are important. Siskiyou National Forest is here.

Cur·tea-de-Ar·ges (kŏŏr'tyä-dā-är'zhĕsh), town (1974 est. pop. 20,209), S central Rumania, in Walachia, on the S slope of the Transylvanian Alps. A trade center, it has industries producing tiles and textiles. Its 16th cent. Byzantine cathedral (rebuilt 19th cent.) became the burial place of the kings of Rumania.

Cush·ing (kŏŏsh'ĭng), city (1970 pop. 7,529), Payne co., central Okla., near the Cimarron River WSW of Tulsa; settled 1892, inc. as a town 1894, as a city 1913. It is the trading and industrial center of a farm and oil-producing area.

Cus·se·ta (kə-sē'tə), town (1970 pop. 1,251), seat of Chattahoochee co., W Ga., SE of Columbus near the Ala. border. Sawmilling is done.

Cus·ter (kŭs'tər). **1.** County (1970 pop. 1,120), 737 sq mi (1,908.8 sq km), S central Colo., in a livestock-grazing area drained by branches of the Arkansas River; formed 1877; co. seat Westcliffe. Parts of San Isabel National Forest are in the east and west. **2.** County (1970 pop. 2,967), 4,933 sq mi (12,776.5 sq km), central Idaho, in a mountainous area crossed by the Salmon River; formed 1881; co. seat Challis. Stock raising and mining are done. Challis and Sawtooth national forests are here. **3.** County (1970 pop. 12,174), 3,765 sq mi (9,751.4 sq km), SE Mont., in an agricultural area drained by the Yellowstone, Tongue, and Powder rivers; formed 1877; co. seat Miles City. It has livestock ranches. **4.** County (1970 pop. 14,092), 2,562 sq mi (6,635.6 sq km), central Nebr., in an agricultural region drained by the Middle Loup and South Loup rivers and Mud Creek; formed 1877; co. seat Broken Bow. Livestock and grain are raised. **5.** County (1970 pop. 22,665), 999 sq mi (2,587.4 sq km), W Okla., intersected by the Washita and Canadian rivers; formed 1907; co. seat Arapaho. In an agricultural region yielding wheat, cotton, corn, rye, barley, oats, livestock, and poultry, it has oil and natural-gas deposits and some manufacturing. **6.** County (1970 pop. 4,698), 1,552 sq mi (4,019.7 sq km), SW S.Dak., in an agricultural and mining area on the Wyo. border; formed 1875; co. seat Custer. It is watered by the Cheyenne River. Dairy products, livestock, poultry, grain, and timber are produced. Wind Cave National Park and Jewel Cave National Monument are here. The county includes part of Black Hills National Forest.

Custer, city (1970 pop. 1,597), seat of Custer co., SW S.Dak., SSW of Rapid City, in a mining, timber, and livestock area. The oldest town in the Black Hills, it was laid out in 1875 after gold was discovered nearby. It has an active tourist trade.

Custer Battlefield National Monument: *see* National Parks and Monuments Table.

Custis-Lee Mansion (kŭs'tĭs-lē'): *see* Arlington House National Memorial.

Cut Bank (kŭt′ băngk′), town (1970 pop. 4,004), seat of Glacier co., NW Mont., NNW of Great Falls; settled 1910, inc. 1911. It is the trade center of a livestock, petroleum, and natural-gas area. Blackfeet Indian Reservation is to the west.

Cutch (kŭch), district, India: *see* Kutch.

Cuth·bert (kŭth′bərt), city (1970 pop. 3,972), seat of Randolph co., SW Ga., SSE of Columbus; inc., 1834. Farm products are processed, and plywood is manufactured.

Cut·tack (kŭt′ək), city (1971 pop. 194,036), Orissa state, E central India. Founded in the 10th cent., it was long famous for gold and silver filigree work.

Cux·ha·ven (ko͝oks-hä′fən), city (1974 est. pop. 60,669), Lower Saxony, N West Germany, at the mouth of the Elbe River. A North Sea fishing and passenger port, it is also a summer resort. Its manufactures include machinery and textiles.

Cuy·a·ho·ga (kī′ə-hō′gə), county (1970 pop. 1,720,835), 456 sq mi (1,181 sq km), N Ohio, in a heavily industrial area bounded on the N by Lake Erie and drained by the Cuyahoga and Rocky rivers; formed 1810; co. seat Cleveland. Its agriculture includes vegetables, poultry, and dairy products.

Cuyahoga, river, c.80 mi (130 km) long, flowing SW through Cuyahoga Falls, then N to Lake Erie, NE Ohio, forming part of Cleveland harbor. By the late 1960s, the Cuyahoga was one of the most polluted rivers in the United States and was a major cause of Lake Erie's deterioration.

Cuyahoga Falls, city (1975 est. pop. 48,300), Summit co., NE Ohio, on the Cuyahoga River; inc. 1836. A suburb of Akron, Cuyahoga Falls is both residential and industrial, with a milk plant and factories that make metal products, rubber goods, and machinery.

Cuz·co or **Cus·co** (both: ko͞os′kō), city (1972 pop. 67,658), alt. 11,207 ft (3,418.1 m), capital of Cuzco dept., S Peru, at the confluence of the Huatanay and Tullamayo rivers. Predominantly Indian in population, it is a trading center for agricultural produce and for woolen textiles produced in the Cuzco mills. Cuzco was founded, according to legend, by Manco Capac, first of the Inca rulers. The city had massive palaces and temples (most notably the Temple of the Sun, now the site of a Dominican convent), which were lavishly decorated with gold medallions and ornaments. When Francisco Pizarro entered the city in 1533, it was plundered; and on its ruins the conquerors and their descendants built the colonial city, using the ancient walls (many of which are still visible) as foundations for new buildings. A severe earthquake in 1950 destroyed much of the city, but most of the historic buildings have been restored.

Cwm·bran (ko͞om-brän′), urban district (1971 pop. 31,614), Gwent, SE Wales. Cwmbran was created under the New Town Act of 1946 to house employees of nearby steelworks.

Cy·cla·des (sĭk′lə-dēz), island group (1971 pop. 86,337), c.1,000 sq mi (2,590 sq km), SE Greece, a part of the Greek archipelago, in the Aegean Sea stretching SE from Attica. The name was originally used to indicate those islands forming a rough circle around Delos. The Cyclades include about 220 islands. Largely mountainous, with a dry and mild climate, the islands produce wine, fruit, wheat, olive oil, and tobacco, and attract many tourists. Iron, manganese, and sulfur are mined, and marble is quarried. In 1829 the Cyclades passed from the Ottoman Empire to Greece.

Cy·me (sī′mē), ancient Greek city of W Asia Minor, on the Ionian Sea and N of the present Smyrna in W Asiatic Turkey. It was the largest and most important of the 12 cities of Aeolis.

Cyn·thi·an·a (sĭn′*thē*-ăn′ə), city (1970 pop. 6,356), seat of Harrison co., N Ky., NNE of Lexington on the South Fork of the Licking River, in a farm area; founded 1793, inc. 1806. It has tobacco warehouses, a distillery, and food-processing plants.

Cy·press (sī′prəs), city (1970 pop. 31,569), Orange co., S Calif.; inc. 1956. Forest Lawn–Cypress, a branch of the famous cemetery in Glendale, Calif., is a major employer. Los Alamitos Naval Air Station is just outside the city limits.

Cy·prus (sī′prəs), republic (1971 est. pop. 634,000), 3,578 sq mi (9,267 sq km), an island in the E Mediterranean Sea, c.40 mi (60 km) S of Turkey and c.60 mi (100 km) W of Syria. The capital is Nicosia. Two mountain ranges traverse the island from east to west; the highest point is Mt. Olympus (6,406 ft/1,953.8 m), in the southwest. Between the ranges lies a wide plain, the chief agricultural region.

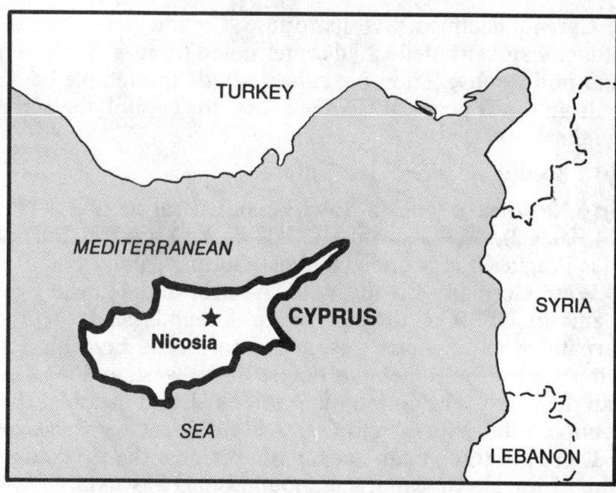

Economy. Crops include grapes (used for wine making), cereals, olives, tobacco, citrus fruits, and cotton. Sheep and goats are raised, and silkworms are cultivated. Fishing is an important livelihood. Among the mineral resources are pyrites, ocher, chrome, asbestos, gypsum, umber, and copper.

History. Excavations have proved the existence of a Neolithic culture on Cyprus in the period from 4000 B.C. to 3000 B.C. Contact with the Middle East and, after 1500 B.C., with Greece greatly influenced Cypriot civilization. Phoenicians settled on the island c.800 B.C. Cyprus fell successively to the Assyrians, Egyptians, Persians, Macedonian Greeks, Egyptians, and finally the Romans (58 B.C.). After A.D. 395 Cyprus was ruled by the Byzantines until 1191, when Richard I of England conquered it. In 1489 Cyprus was annexed by Venice. The Turks conquered it in 1571. At the Congress of Berlin (1878) the Ottoman Empire placed Cyprus under British administration, and in 1914 Britain annexed it outright.

Under British rule the movement among the Greek Cypriot population for union (*enosis*) with Greece was a constant source of tension. Agitation flared into violence in 1954. The conflict, tantamount to civil war, was aggravated by Turkish support of Turkish Cypriot demands for partition of the island. A settlement was finally reached providing for Cypriot independence in 1960, but large-scale fighting between Greek and Turkish Cypriots erupted several times in the 1960s and 1970s. In 1974 the government of Archbishop Makarios III, president of Cyprus, was overthrown by proponents of *enosis.* Turkey invaded northern Cyprus. An unsponsored cease-fire permitted Turkish forces to remain. Makarios returned from exile and again served as president until his death in 1977. An independent Turkish Cypriot state in northern Cyprus, comprising about 40% of the island, was proclaimed in 1975. The Greek Cypriots refused to accept partition, and talks between the two communities failed to produce a settlement.

Government. The constitution, in effect since 1960, provides for a president and vice president, elected by the Greek and Turkish communities, respectively. There is a bicameral legislature. In 1975 the Turkish Cyprists established a Turkish Federated State of Cypress with its own elected president and legislature.

Cyr·e·na·i·ca (sîr′ə-nā′ĭ-kə), historic region, E Libya, bordering on the Mediterranean Sea. The Greeks colonized northern Cyrenaica in the 7th cent. B.C., founding numerous settlements. In the mid-1st cent. B.C., Cyrenaica became a Roman province. In A.D. 115-16 there was a large-scale but unsuccessful revolt of Jewish settlers. When Rome was divided (4th cent.) into the Eastern and Western empires, Cyrenaica came under the Byzantines. In 642 the Arabs conquered Cyrenaica and many of them settled in the region from the 9th to 11th cent. The Ottoman Turks captured the area in the mid-16th cent.

Cy·re·ne (sī-rē′nē), ancient city near the N coast of Africa, in Cyrenaica. It was a Greek colony founded c.630 B.C. Having important commerce with Greece, the little city-state flourished. In the late 6th cent. Cyrene submitted to the Persians, but later (after 480 B.C.) became independent again. Although the city became subject to Alexander the Great in 331 and was later practically annexed by the Ptolemies of Egypt, it seems to have had nominal independence until the marriage of Berenice, daughter of Cyrene's king, to Ptolemy III. Cyrene remained part of the Ptolemaic kingdom until 96 B.C. It was later the center of a Roman province. Under the Roman emperor Trajan there were Jewish uprisings, which were severely punished; Cyrene declined. At its prime Cyrene was a large and beautiful city and an intellectual center noted for its schools of medicine and philosophy. Extensive ruins include the temple of Apollo (dating from the 7th cent. B.C.), the agora, the capitol, the acropolis, and the theater.

Cy·the·ra (sə-thîr'ə), island: *see* Kíthira.

Czech·o·slo·va·ki·a (chĕk'ə-slō-vä'kē-ə), federal republic (1973 est. pop. 14,500,000), 49,370 sq mi (127,868 sq km), central Europe. Its capital is Prague. It is bounded by East Germany and Poland on the north, West Germany on the west, Austria and Hungary on the south, and the USSR on the east. Natural boundaries are formed by the Carpathian and Sudetes mts. in the north, the Erzgebirge range in the northwest, the Bohemian Forest in the west, and the Danube River in the south. The country is landlocked, and the chief rivers—the Danube, Elbe, Vltava, and Oder—are of great economic importance. There are three main geographic regions: the Bohemian plateau, the Moravian lowland, and mountainous Slovakia.

The republic, founded in 1918 as the Austro-Hungarian Monarchy fell apart, comprises Slovakia and the traditional Czech lands of Bohemia, Moravia, and Czech Silesia. The two administrative divisions of the country are the Czech Socialist Republic (with its capital in Prague) and the Slovak Socialist Republic (with its capital in Bratislava). The population is largely Slavic, consisting chiefly of Czechs (about two thirds of the total) and Slovaks (a little less than one third).

Economy. Czechoslovakia is highly industrialized. More than one third of the total labor force is in industry. The major manufactured goods include machinery, transportation equipment, and other metal goods; iron and steel; chemicals; food products; and textiles and footwear. The country has large reserves of bituminous coal and lignite and some iron ore and other minerals, but iron must be imported to meet industrial needs. Agriculture accounts for only a small part of Czechoslovakia's national income and employs well under one fifth of the labor force, but it is highly developed and efficient. Ninety percent of the farmland is collectivized, and half of it is devoted to grains. Major crops are sugar beets, potatoes, fodder, wheat, barley, and oats. Additional grain must be imported.

History. The union of the Czech lands and Slovakia was officially proclaimed in Prague on Nov. 14, 1918; Thomas A. Masaryk was the first president of the new republic. Ruthenia was added in 1920. Czechoslovakia inherited the greater part of the industries of the Austro-Hungarian Monarchy, but the new state suffered from antagonistic ethnic elements. Slovaks, Ruthenians, and ethnic Germans and Magyars all demanded greater autonomy on border revisions. The nationality problem led to a European crisis when the German nationalist minority demanded the union of their districts with Germany. Threatening war, Hitler induced Great Britain and France to agree in the Munich Pact (1938) to the cession of the Bohemian borderlands (Sudetenland). Poland and Hungary obtained territorial cessions shortly thereafter. The whole country surrendered to German control in Mar., 1939.

At the end of World War II, Czechoslovakia was liberated by Soviet and American troops (Apr., 1944–May, 1945), who installed a provisional government under President Eduard Beneš. The country's pre-1938 territory was restored, except for Ruthenia, which was ceded to the USSR. In the elections of 1946 the Communists emerged as the strongest party (obtaining one third of the votes) and became the dominant party in a governing coalition. The Communists then began a campaign of political agitation and intrigue that gave them complete control of the government in Feb., 1948. Political and cultural liberty was curtailed, and purge trials were conducted from 1950 to 1952.

A movement toward liberalization was initiated in 1963. In Jan., 1968, Alexander Dubček replaced Antonin Novotný as party leader and undertook a startling program of democratization. Press censorship was reduced, and the restoration of a genuinely democratic political life seemed possible. Alarmed at what it construed to be a threat to Soviet security, the USSR with its Warsaw Pact allies invaded Czechoslovakia in Aug., 1968. Despite opposition by the populace, the USSR forced the repeal of most of the reforms. In 1969 Dubček was replaced as party leader by Gustáv Husák. In the 1970s there were many efforts to stamp out dissent, including mass arrests, union purges, and the persecution of the signers of a human rights manifesto (1977).

Government. Under the 1968 constitution, Czechoslovakia has a bicameral federal legislature elected every five years. The federal president, who is elected by the legislature, appoints the premier and ministers. The Czech and Slovak republics both have a council and assembly. The federal government deals with defense, foreign affairs, and certain economic matters. The Communist party, which is the actual source of power in the country, heads the National Front.

Czę·sto·cho·wa (chĕɴ'stə-kô'və), city (1975 est. pop. 200,300), S Poland, on the Warta River. It is an industrial center, known especially for its iron and steel plant and iron-smelting works. Other industries include sawmilling, papermaking, and the manufacture of metals, chemicals, and textiles. Iron ore is mined in the vicinity. Częstochowa is a celebrated religious center, and a world-famous place of pilgrimage.

Dą·bro·wa Gór·ni·cza (dôɴ-brô'vä gôr-nē'chä), city (1970 pop. 61,660), SE Poland, on the Czarna Przemsza River, a tributary of the Vistula. It is a railway junction and a center of the Katowice mining and industrial region. Coal mining in the area dates from 1796. Metals, machine tools, glass, and electric motors are manufactured in the city. Dąbrowa Górnicza passed to Prussia in 1795 and to Russia in 1815. The city reverted to Poland in 1919.

Dac·ca (dăk'ə), city (1974 pop. 1,310,976), capital of Bangladesh, on a channel of the Dhaleswari River, in the heart of the world's largest jute-growing region. It is the industrial, commercial, and administrative center of Bangladesh, with an active trade in jute, rice, oilseeds, sugar, and tea. Manufactures include textiles, cotton saris, jute products, paper, chemicals, hosiery, shoes, matches, soap, and glass. There are also printing and engineering plants.

Dacca's history dates back to A.D. c.1000, but the city achieved its glory as the 17th cent. Mogul capital of Bengal. English, French, and Dutch industrialists set up factories here in the 17th and 18th cent., and Dacca passed under British rule in 1765. It became the capital of East Pakistan in 1947. The city was surrendered by the Pakistani army to Indian troops in Dec., 1971, and a few days later became capital of the provisional government of Bangladesh.

Da·chau (dä'KHOU'), city (1974 est. pop. 33,691), Bavaria, S West Germany, on the Amper River; chartered in 1391. It has industries that manufacture machinery, electrical equipment, and textiles. Nearby was (1933–45) an infamous Nazi concentration camp, which today has a number of memorials and a museum.

Da·ci·a (dā'shē-ə, -shə), ancient name of the European region corresponding roughly to modern Rumania (including Transylvania). It was inhabited before the Christian era by a people who were called Getae by the Greeks and were called Daci by the Romans. They were a people of advanced material culture, with a tribal organization. Trajan invaded Dacia in A.D. 102 and again in 105. He established a large number of colonies, and Dacia became a Roman province. The Goths invaded (250–70) the region, and Aurelian was obliged to concede Dacia. It was the Roman colonists in Dacia who formed the Latin-speaking nucleus that established the Romance language Rumanian, which is still spoken in this region.

Dade (dād). **1.** County (1970 pop. 1,267,792), 2,042 sq mi (5,288.8 sq km), S Fla., in a lowland area at the tip of the peninsula and bordered by the Florida Keys, enclosing Biscayne Bay on the E and part of Florida Bay in the S; formed 1836; co. seat Miami. The coastal fringe is a tourist area, with agriculture (truck crops, citrus fruits, dairy products, and poultry), limestone quarries, and sand pits. The interior lies in the Everglades and includes part of Everglades National Park. Food and wood products, building materials, clothing, and chemicals are manufactured. Fishing is also important. **2.** County (1970 pop. 9,910), 165 sq mi (427.4 sq km), extreme NW Ga., bounded on the N by Tenn., on the W by Ala.; formed 1836; co. seat Trenton. In a lumbering and agricultural area yielding fruit, vegetables, corn, grain, cotton, livestock, and poultry, it also has coal deposits. **3.** County (1970 pop. 6,850), 504 sq mi (1,305.4 sq km), SW Mo., in an Ozark region drained by the Sac and Horse rivers and Stockton Lake; formed 1841; co. seat Greenfield. It has agriculture (wheat, hay, corn, and livestock), coal mines, and limestone quarries. Clothing and hats are manufactured.

Dade City, city (1973 est. pop. 4,257), seat of Pasco co., W Fla., NNE of Tampa; settled 1862, inc. 1889. It is the processing center of a citrus-fruit area, with kaolin deposits nearby.

Dade·ville (dād'vĭl'), town (1970 pop. 2,847), seat of Tallapoosa co., E central Ala., near Martin Lake SE of Alexander City, in a timber and mineral area; founded 1832, inc. 1837. Textile products are made. Horseshoe Bend National Military Park is nearby.

Da·dra and Na·gar-Ha·ve·li (dä'drä; nä'gär-hä-vā'lē), union territory (1971 pop. 74,165), 188 sq mi (486.9 sq km), E central India, on the Arabian Sea. Portugal colonized these two enclaves in the mid-16th cent. India occupied them in 1954. Despite a ruling by the International Court of Justice at The Hague that upheld Portugal's claim to the areas, they were incorporated into India as a single union territory in 1961.

Da·ges·tan Autonomous Soviet Socialist Republic (dä'gĭ-stän'), autonomous region (1970 pop. 1,429,000), c.19,400 sq mi (50,245 sq km), SE European USSR, bounded on the E by the Caspian Sea; capital Makhachkala. Except for the Caspian plain and the irrigated lowlands in the north, the terrain is one of steep mountains divided by valleys. Difficulty of access has left most of Dagestan's mineral resources untapped; however, important quantities of oil and natural gas have been extracted along the coast. The irrigated lowlands support winter wheat, corn, sunflowers, fruits, and wine grapes. The republic's major industries produce canned fruit, wine, oil, machines, chemicals, textiles, and wood products. An ancient area of human settlement, Dagestan belonged to Caucasian Albania in the 1st millennium B.C. It was later invaded by Huns, Persian Sassanids, and in the 7th cent. A.D. Arabs, who introduced Islam. Taken by the Turks in the 11th cent. and the Mongols in the 13th cent., the region became the center of a struggle between Turkey and Persia in the 15th cent. It was a Persian province when Russia annexed it by the Treaty of Gulistan in 1813. Dagestan came under Soviet rule in 1920 and in 1921 was made an autonomous republic.

Dag·gett (dăg'ĕt, -ĭt), county (1970 pop. 666), 706 sq mi (1,828.5 sq km), NE Utah, in a mountainous area bordering on Wyo. and Colo. and crossed by the Green River; formed 1918; co. seat Manila.

Dah·lon·e·ga (də-lŏn'ə-gə), city (1970 pop. 2,658), seat of Lumpkin co., N central Ga., NNE of Atlanta, in a hunting and fishing area; inc. 1833. A mountain trade center at the edge of Chattahoochee National Forest, it was settled after the opening of gold mines in 1828. A branch of the U.S. mint was here (1836–61). The Dahlonega Gold Museum allows visitors to pan for gold.

Da·ho·mey (də-hō'mē): *see* Benin.

Dain·ger·field (dān'jər-fēld'), town (1970 pop. 2,630), seat of Morris co., NE Texas, N of Longview, in a truck-farm, cattle, and dairy area. A state park is nearby.

Dai·ren (dī'rĕn'): *see* Ta-lien, China.

Dai·se·tsu-zan (dī-sā'tsōō-zäN), group of volcanic peaks, central Hokkaido, Japan, rising to 7,513 ft (2,291.5 m) at Asahi-dake. They are part of Daisetsu-zan National Park (895 sq mi/2,318 sq km; est. 1934), the largest national park in Japan.

Da·kar (də-kär', dä-), city (1973 est. pop. 667,000), capital of Senegal, W Senegal, on Cape Verde Peninsula, a port on the Atlantic Ocean. Manufactures include refined sugar, peanut oil, fertilizers, cement, and textiles. Artisans make garments, furniture, and jewelry. Dakar grew up around a French fort built in 1857 to protect the merchants and residents of nearby Gorée Island (now a Dakar suburb). Dakar's importance increased significantly after 1855, when a railroad linked it with the Senegal River. In 1887 it was made a commune, and in 1902 it replaced St. Louis as the capital of French West Africa. In 1940 Free French forces under Gen. Charles De Gaulle fought unsuccessfully to free Dakar from Vichy control, but in late 1942 U.S. forces occupied the city and stayed until the end of World War II. Dakar was the capital of the short-lived (1959–60) Mali Federation.

Da·ko·ta (də-kō'tə). **1.** County (1970 pop. 139,808), 571 sq mi (1,478.9 sq km), SE Minn., bounded on the N and NE by the Mississippi River, on the NW by the Minnesota River; formed 1849; co. seat Hastings. It lies in an agricultural area yielding grain, livestock, and dairy products and has a food-processing industry. Textiles, paper, chemicals, and wood, glass, and metal products are manufactured. **2.** County (1970 pop. 13,137), 255 sq mi (660.5 sq km), NE Nebr., in an agricultural region bounded on the E and NE by the Missouri River at the Iowa-S.Dak. border; formed 1855; co. seat Dakota City. Food processing is done.

Dakota City. **1.** Town (1974 est. pop. 867), seat of Humboldt co., N central Iowa, on the East Des Moines River just E of Humboldt, in a livestock and grain area. **2.** Village (1976 est. pop. 1,234), seat of Dakota co., NE Nebr., on the Missouri River just S of Sioux City, Iowa. Timber and grain are produced.

Da Lat (dä' lät'), city (1971 est. pop. 86,600), SE Vietnam, in the central highlands at an altitude of c.5,000 ft (1,525 m). Developed by the French as a health resort and hunting center, it is surrounded by mountains that rise to 7,380 ft (2,250.9 m).

Dale (dāl), county (1970 pop. 52,938), 560 sq mi (1,450.4 sq km), SE Ala., in the coastal plain and drained by the Choctawhatchee River; formed 1818; co. seat Ozark. Its agriculture includes cotton, peanuts, corn, and hogs. Lumber milling is done. Part of Fort Rucker is here.

Dal·hart (dăl'härt), city (1970 pop. 5,705), seat of Dallam co., extreme N Texas, in the NW corner of the Panhandle NW of Amarillo; founded 1901 with the coming of the railroad; inc. 1903. It is a shipping point for a cattle and grain area.

Dal·lam (dăl'əm), county (1970 pop. 6,012), 1,494 sq mi (3,869.5 sq km), extreme NW Texas, in a high, treeless plain of the Panhandle, bordered on the N by Okla., on the W by N.Mex.; formed 1876; co. seat Dalhart. It is in an agricultural area yielding cattle, corn, wheat, grain sorghums, alfalfa, and dairy products.

Dal·las (dăl'əs). **1.** County (1970 pop. 55,296), 976 sq mi (2,527.8 sq km), S central Ala., in the Black Belt and drained by the Alabama and Cahaba rivers; formed 1818; co. seat Selma. Cotton, dairy products, livestock, and timber are important. It has an increasing amount of light industry. **2.** County (1970 pop. 10,022), 672 sq mi (1,740.5 sq km), S central Ark., drained by the Ouachita and Saline rivers; formed 1845; co. seat Fordyce. Its agriculture includes cotton, truck crops, fruit, livestock, poultry, and dairy products. Lumbering, hunting, and fishing are also done. Clothing and wood products are manufactured. **3.** County (1970 pop. 26,085), 597 sq mi (1,546.2 sq km), central Iowa, in a prairie agricultural area drained by the Raccoon River and Beaver Creek; formed 1846; co. seat Adel. Livestock is raised, and corn, oats, and wheat are grown. It also has coal mines. **4.** County (1970 pop. 10,054), 537 sq mi (1,390.8 sq km), SW central Mo., in the Ozarks and drained by the Niangua and Little Niangua rivers; formed 1841 as Niangua co., renamed 1844; co. seat Buffalo. It lies in an agricultural area producing dairy cattle and poultry. It has stands of oak and some manufacturing. **5.** County (1970 pop. 1,327,-695), 859 sq mi (2,224.8 sq km), NE Texas, in a blackland prairie area drained by branches of the Trinity River; formed 1846; co. seat Dallas. Its agriculture includes cotton, corn, grain, clover, fruit, truck crops, pecans, livestock, poultry, and dairy products. It also has anthracite-coal mines, oil and natural-gas wells, sand and gravel pits, and lumbering and food-processing industries. Among its manufactures are wood products, furniture, paper, chemicals, plastics, concrete, metal products, machinery, and transportation equipment. The county is dominated by Dallas, a financial, commercial, and industrial center.

Dallas. **1.** City (1970 pop. 2,133), seat of Paulding co., NW Ga., WNW of Atlanta, in a farm and timber area; inc. 1854. **2.** City (1970 pop. 6,361), seat of Polk co., NW Oregon, in the Willamette River valley W of Salem; founded 1852, inc. 1901. In a dairy region, it also grows fruit and processes lumber and leather. **3.** City (1975 est. pop. 859,000), seat of Dallas co., N Texas, on the Trinity River near the junction of its three forks; inc. 1871. Its manufactures include cotton-processing machinery, textiles, leather goods, and aircraft and electronic equipment. Oil is refined, and there are meat-packing plants. Founded c.1841, Dallas was early populated by French settlers who abandoned the nearby Fourierist community, La Réunion. Dallas developed as a cotton market in the 1870s. Later it became the financial and commercial center of the Southwest. A branch of the Univ. of Texas, a theological seminary, and the schools of dentistry and nursing of Baylor Univ. are in Dallas. A noted fashion center, the city is also known for its museums, its musical activities, and its interest in literature and the drama (the Dallas Theatre Center boasts the only public theater ever designed by Frank Lloyd Wright). The city is served by an expanding reservoir system utilizing the waters of the Trinity River. The Dallas-Fort Worth Regional Airport, the largest commercial airport in the world, was opened in 1974.

Dalles, The (dălz): *see* The Dalles, Oregon.

Dal·ma·tia (dăl-mā'shə), historic region of Yugoslavia and province of Croatia, extending along the Adriatic Sea, approximately from Rijeka (Fiume) to the Gulf of Kotor. Except for a coastal lowland, Dalmatia is generally mountainous, rising to the Dinaric Alps. The coast, famed for its scenic beauty and its resorts, has many bays and excellent harbors protected by a chain of islands. Agriculture, fishing, and tourism are the principal economic activities.

Long in conflict with Rome, Dalmatia was definitively subdued by Augustus (35 B.C.-33 B.C.) and was incorporated with part of Illyria as a Roman province. It was overrun by the Ostrogoths (5th cent. A.D.), reconquered by the Byzantine Empire (6th cent.), and settled, except in the coastal cities, by the Slavs in the 7th cent. By the 10th cent. it was divided between the kingdoms of Croatia (north) and Serbia (south), while Venice held several ports and islands. After several centuries of struggle, chiefly between Venice and the crowns of Hungary and Croatia, the coastal islands and most of Dalmatia, except Dubrovnik, were under Venetian control by 1420. Hungary retained the Croatian part, which in 1526 passed to the Turks but was recovered by the Treaty of Karlowitz (1699). The Treaty of Campo Formio (1797) gave Venetian Dalmatia to Austria, and the Treaty of Pressburg (1805) gave it to France. It was first attached to Napoleon's Italian kingdom but in 1809 was incorporated into the Illyrian provinces. The Congress of Vienna restored (1815) it to Austria, where it was made (1861) a crown land, with its capital at Zara. By the secret Treaty of London (1915) the Allies promised Dalmatia to Italy in return for Italian support in World War I. In Dec., 1918, it became part of the newly established kingdom of Serbs, Croats, and Slovenes (after 1929 Yugoslavia), but Italy continued to claim Dalmatia. The Treaty of Rapallo (1920) gave Dalmatia to Yugoslavia, except for Zara and several islands, which subsequently passed to Italy. During World War II, Italy held most of Dalmatia, and after the war it was returned to Yugoslavia. The Italian peace treaty of 1947 gave Yugoslavia the islands that had been ceded to Italy after World War I.

Dal·ton (dôl'tən). **1.** City (1975 est. pop. 20,700), seat of Whitfield co., extreme NW Ga., in the Appalachian Valley; inc. 1847. It is a highly industrialized city in a farm area; its large tufted-textile industry was begun in the late 1800s. In the Civil War, Dalton (Confederate headquarters after the Chattanooga campaign) fell to Gen. W. T. Sherman in the Atlanta campaign (1864). The Chickamauga and Chattanooga National Military Park is nearby. **2.** Town (1970 pop. 7,505), Berkshire co., W Mass., in the Berkshires NE of Pittsfield; settled 1755, inc. 1784. Its paper industry (including bank-note paper) dates from 1801. Woolen goods are made here.

Da·ly City (dā'lē), city (1975 est. pop. 70,500), San Mateo co., W Calif., a suburb of San Francisco; inc. 1911. Settled in 1906 by refugees from the San Francisco earthquake, Daly City is still primarily residential. The Cow Palace, scene of the 1964 Republican national convention, is here.

Da·man (də-män'): *see* Goa, Daman, and Diu.

Da·man·hur (dä-män-ho͝or'), city (1970 est. pop. 161,400), capital of Buhayrah governorate, N Egypt, in the Nile River delta. It is a communications center and a market for cotton and rice. In Roman times it was called Hermopolis Parva.

Da·mas·cus (də-măs'kəs), city (1970 est. pop. 835,000), capital of Syria and of Damascus governorate, SW Syria, on the E edge of the Anti-Lebanon Mts. It is Syria's administrative, financial, and communications center. Manufactures include textiles, metalware, refined sugar, glass, furniture, cement, leather goods, preserves, confections, and matches. The old city lies south of the Barada River, and the new town (greatly extended since 1926) lies north of the river. Points of interest include the Great Mosque, the quadrangular citadel (originally Roman; rebuilt 1219), a 16th cent. Moslem monastery, and Azm Palace (1749).

Damascus has been inhabited since prehistoric times. There was a city on its site even before the time (c.2000 B.C.) of Abraham. It was probably held by the Egyptians before the Hittite period (2nd millennium B.C.) and was later ruled by the Israelites, Assyrians, Persians, Macedonian Greeks, Seleucids, Armenians, and finally the Romans (64 B.C.). Under Roman rule Damascus became a thriving commercial city, noted for its woolen cloth and grain, and was early converted to Christianity. After the permanent split (A.D. 395) of the Roman Empire, it became a provincial capital of the Byzantine Empire. The Arabs occupied it permanently in 635. The city was then gradually converted to Islam, and the Christian church built (379) by Roman Emperor Theodosius I was rebuilt (705) as the Great Mosque. Damascus was the seat of the caliphate under the Umayyads from 661 until 750. It thereafter fell prey to new conquerors—the Egyptians, the Karmathians, the Seljuk Turks (1076), and the Saracens. It continued to prosper under the Saracens; its bazaars sold brocades (damask), wool, and the famous swords and other ware of the Damascene metalsmiths. In 1260 the city fell to the Mongols, and it was sacked c.1400 by Tamerlane. In 1516 it passed to the Ottoman Turks, and for 400 years it remained in the Ottoman Empire.

After World War I the city became the capital of one of the French Levant States mandated under the League of Nations. In 1925-26, Damascus joined a revolt against the French, who shelled and badly damaged the city. It became the capital of independent Syria in 1941.

Da·ma·vand or **De·ma·vend** (both: dĕm'ə-vĕnd'), volcanic cone, 18,934 ft (5,774.9 m) high, in the Elburz range, N Iran.

Dam·i·et·ta (dăm'ē-ĕt'ə): *see* Dumyat, Egypt.

Da·mo·dar (dä'mō-där), river, 370 mi (595.3 km) long, rising in Bihar state, E India, and flowing SE through West Bengal state to join the Hooghly River below Calcutta. Its dams supply electricity to Calcutta and the Hooghlyside industrial district.

Dam·pi·er Archipelago (dăm'pē-ər, -pîr'), group of about 20 small islands, in the Indian Ocean off the NW coast of Australia. The archipelago is named after its discoverer, the English explorer and buccaneer William Dampier.

Dan·a·kil (dăn'ə-kĭl'), desert region, c.350 mi (565 km) long and 50-250 mi (80-400 km) wide, NE Ethiopia and N Djibouti Republic, part of the Great Rift Valley. It is bounded by the Red Sea on the north and east. The principal inhabitants are the Danakils and Afar people who tend sheep, goats, camels, and cattle for their livelihood.

Da Nang (də năng', dä näng), formerly **Tou·rane** (to͞o-răn', -rän'), city

(1971 est. pop. 437,700), central Vietnam, a port on the South China Sea. Da Nang was the scene (1535) of the first landing of Europeans in Vietnam. The city was originally ceded to France by Annam in 1787 and became (after 1858) a French concession. It was the site of a huge U.S. military base during the Vietnam War.

Dan·bur·y (dăn'bĕr'ē, -bə-rē). **1.** City (1978 est. pop. 56,500), Fairfield co., SW Conn., in a farm area; settled 1685, inc. as a city 1889. Its hat industry dates from 1780. Other manufactures include electronic equipment, plastics, machinery, and furniture. An early military depot, Danbury was the object of Gen. William Tryon's 1777 raid, which resulted in the destruction of much of the village. The David Taylor House (1750) and the Dodd House (1770)—both included in the Scott-Fanton Museum—are noteworthy. Danbury is also known for its state fairs, held annually since the early 1800s. **2.** Town (1970 pop. 152), seat of Stokes co., N N.C., on the Dan River N of Winston-Salem. Hanging Rock State Park is nearby.

Dan·da·rah (dän'də-rä') or **Den·de·ra** (dĕn'də-rä'), town, N Egypt, on the Nile River. Nearby is the site of the ancient Greek city of Tentyra. There is a large, well-preserved temple of Hathor (1st cent. B.C.) and a temple of Isis, which contains cultist inscriptions.

Dan·dridge (dăn'drĭj'), town (1970 pop. 1,280), seat of Jefferson co., E Tenn., on Douglas Lake E of Knoxville, in an agricultural and food-processing area.

Dane (dān), county (1970 pop. 290,272), 1,197 sq mi (3,100.2 sq km), S Wis., bounded on the NW by the Wisconsin River and drained by the Sugar River; formed 1836; co. seat Madison. In a dairy-farming and mining region, it has several large lakes. Dairy and grain products, textiles, and chemicals are produced.

Dan·iels (dăn'yəlz), county (1970 pop. 3,083), 1,443 sq mi (3,737.4 sq km), NE Mont., in an agricultural area bordering in the N on Sask., Canada, and drained by forks of the Poplar River; formed 1920; co. seat Scobey. Grain and livestock are important. It includes part of Fort Peck Indian Reservation in the south.

Dan·iels·ville (dăn'yəlz-vĭl'), city (1970 pop. 378), seat of Madison co., NE Ga., NE of Athens, in an agricultural area.

Dans·ville (dănz'vĭl'), village (1970 pop. 5,436), Livingston co., W central N.Y., SSE of Geneseo, in a nursery and truck-farm area; settled 1795, inc. 1845. Shoes are made. Clara Barton founded (1881) the first local chapter of the American Red Cross here.

Dan·ube (dăn'yo͞ob), river of central and SE Europe, c.1,770 mi (2,850 km) long, with a drainage basin of c.320,000 sq mi (828,800 sq km); it is second in length only to the Volga among European rivers. The Danube rises in the Black Forest in southwest West Germany and flows northeast across southern West Germany past Ulm and Regensburg, where it turns southeast to enter Austria at Passau. It continues its southeastern course through Upper and Lower Austria, past Linz and Vienna; this section is particularly famous for its scenery. It then forms the border between Czechoslovakia and Hungary from Bratislava to Szob, where it turns south and flows across the Great Alföld of central Hungary past Budapest. After entering Yugoslavia above Belgrade, it turns southeast, then east, and flows through narrow gorges, forming part of the Yugoslavia-Rumania border. After passing the Iron Gate, the Danube broadens again and forms most of the Rumania-Bulgaria border. Near Silistra it leaves the Bulgarian border and turns north, passing through eastern Rumania to Galaţi, where it divides into a delta before entering the Black Sea. Navigable by barges from Ulm, the Danube is a vital traffic artery. The Danube is linked to the Main and Rhine rivers by the Rhine-Main-Danube Canal. River navigation is impeded by ice in winter and by varying seasonal water levels.

Under the Roman Empire the Danube was the northern border against the barbarian world. As Rome declined, the Danubian plains for centuries attracted invading hordes. The Danube increased in commercial importance in the era of the Crusades, but commerce suffered (15th–16th cent.) after the Turks gained control of its course from the Hungarian plain to the Black Sea. In the 19th cent. the Danube's economic importance as an international waterway increased. By the Treaty of Versailles (1919) the Danube was internationalized, and an international commission was set up with jurisdiction over the course from Ulm to Brăila. Germany repudiated the internationalization in 1936 and forced both the navigation (1939) and international (1940) commissions to dissolve.

Dan·vers (dăn'vərz), town (1978 est. pop. 24,400), Essex co., NE Mass.; settled in the 1630s, set off from Salem 1752, inc. as a town 1757. Electrical equipment and shoes are the chief products. The Salem witchcraft incidents began here in 1692. John Greenleaf Whittier spent his later years here, and Israel Putnam was born in Danvers. Numerous buildings of historical and architectural interest are preserved, several dating from the 1600s.

Dan·ville (dăn'vĭl'). **1.** City (1975 est. pop. 1,712), a seat of Yell co., W central Ark., SW of Russellville on the Petit Jean River; founded 1841. Cotton ginning and lumber milling are done. **2.** Town (1970 pop. 9,500), Contra Costa co., W Calif., E of Oakland; settled 1858. **3.** City (1978 est. pop. 43,800), seat of Vermilion co., E Ill., on the Vermilion River at the Ind. line; inc. 1839. It is a commercial and industrial center in a dairy, farm, and coal area. The city has railroad shops, meat-packing establishments, lumber and flour mills. **4.** Residential town (1970 pop. 3,771), seat of Hendricks co., W central Ind., W of Indianapolis, in a farm area; settled 1824, inc. 1835. **5.** City (1975 est. pop. 12,038), seat of Boyle co., central Ky.; settled 1775, inc. 1836. It is a manufacturing center in an agricultural region. One of the oldest settlements in the state, Danville was a seat of government (1784–92) by act of the Va. legislature. The state constitutional conventions were held here. **6.** Borough (1970 pop. 6,176), seat of Montour co., E central Pa., on the Susquehanna River WSW of Bloomsburg; laid out 1790, inc. 1849. Iron and steel are manufactured here. The area is rich in deposits of coal, iron ore, and limestone. **7.** Independent city (1978 est. pop. 45,300), S central Va., on the Dan River, founded 1793, inc. 1870. It is a market center for tobacco and textiles. The Sutherlin Mansion, the "Last Capitol of the Confederacy," is a historical landmark.

Dan·zig (dăn'sĭg): *see* Gdańsk, Poland.

Dar·bhan·ga (dər-bŭng'gə), city (1971 pop. 132,059), Bihar state, NE India, on the Baghmati River. It is a district administrative center in a sugar-cane and tobacco area. Darbhanga passed to the Mogul empire in the 14th cent. The British assumed control in 1765.

Dar·by (där'bē), borough (1975 est. pop. 12,834), Delaware co., SE Pa., a suburb adjacent to Philadelphia; settled by Quakers 1682, inc. 1853. Although residential, it has some industry. One of the oldest settlements in the state, it has many colonial landmarks.

Dar·da·nelle (där'dn-ĕl'), city (1975 est. pop. 3,684), a seat of Yell co., W central Ark., on the Arkansas River ESE of Fort Smith, in a livestock and timber area. It was platted in 1843 on the site of an Indian post. Mt. Nebo State Park is nearby.

Dar·da·nelles (där'dn-ĕlz') or **Ça·nak·ka·le Bo·ğa·zı** (chä'näk-kä-lĕ' bō'ä-zĭ'), strait, c.40 mi (65 km) long and from 1 to 4 mi (1.6 to 6.4 km) wide, connecting the Aegean Sea with the Sea of Marmara and separating the Gallipoli peninsula of European Turkey from Asian Turkey. It was called the Hellespont in ancient times and was the scene of the legend of Hero and Leander. Controlling navigation between the Black Sea and the Mediterranean, the Dardanelles and Bosporus straits have long been of immense strategic and commercial importance. Throughout the time of the Byzantine and Ottoman empires the straits were essential to the defense of Constantinople. By 1402 the Dardanelles was under the control of the Ottoman Turks. With brief interruptions, the passage has remained in Turkish hands until the present.

Early in 1915 an Anglo-French fleet sought unsuccessfully to force the Dardanelles and take Constantinople. A second attempt was also unsuccessful, but after the final Turkish collapse an Allied fleet passed (Nov., 1918) the straits and occupied Constantinople. The Treaty of Sèvres (1920) with Turkey internationalized and demilitarized the straits zone, but it was superseded by the Treaty of Lausanne (1923). The zone was restored to Turkey, but was to remain demilitarized. Secretly, however, Turkey soon began to refortify the zone, and in 1936, by the Montreux Convention, it was formally permitted to remilitarize it. The Montreux Convention was essentially enforced by Turkey through World War II until Jan., 1945, when it was modified by the opening of unrestricted passage for Allied supplies to the Soviet Union.

Dare (dâr), county (1970 pop. 6,995), 388 sq mi (1,004.9 sq km), NE N.C., in a forested and swampy tidewater area bounded on the E by the Atlantic Ocean, on the N by Albemarle Sound, and on the W by the Alligator River; formed 1870; co. seat Manteo. Dairying, farming (grapes, sweet potatoes, and corn), fishing, and sawmilling are important. Cape Hatteras National Seashore, Fort Raleigh National Historic Site, and the Wright Brothers National Memorial are here.

Dar-es-Sa·laam (där'ĕs-sä-läm'), city (1970 est. pop. 343,900), capital of Tanzania, on an arm of the Indian Ocean. It is the country's administrative, transportation, communications, and economic center. The major industries produce foods and beverages, oil, textiles, and pharmaceuticals. Although situated on a lagoon with only limited access to the sea, Dar-es-Salaam is Tanzania's main port. Founded in 1866 by the sultan of Zanzibar, who built his summer palace here, Dar-es-Salaam was a small town when German forces occupied it in 1887. In 1891 it became the capital of German East Africa, but its main growth began during World War II. Today it is a modern city, with wide, shaded streets. It is the site of the Univ. of Dar-es-Salaam (1961) and the National Museum of Tanzania.

Dar·fur (där-fo͝or'), province and former sultanate, W Sudan. The region is mountainous, much of the terrain is dry plateau, and there are sand dunes in the extreme north. Darfur's economy is based on subsistence agriculture. Prehistoric Darfur was inhabited by peoples related to the predynastic Egyptians. The royal family of Cush, which fell A.D. c.350, may have established a dynasty in Darfur. Christian kingdoms emerged in the period between 900 and 1200, but were destroyed by Moslems from Kanem in the middle 13th cent. Fur, a major black African kingdom probably founded in the 15th cent., pushed aside the Kanem rulers in the 17th cent. Fur was conquered by the Egyptians in 1874 and by the Mahdists of Sudan in 1883. With the fall of the Mahdist state in 1898, Darfur became a semiautonomous sultanate under Anglo-Egyptian suzerainty. The sultan attempted to expel the foreign influence during World War I, but his forces were defeated by the British in 1916, and Darfur was incorporated into Sudan as a province.

Dar·i·en (dâr'ē-ĕn'), E part of Panama between the Gulf of Darien on the E and the Gulf of San Miguel on the W. In 1513 Vasco Núñez de Balboa led an expedition across Darien and became the first European to view the Pacific Ocean from the New World.

Dar·i·en (dâr'ē-ĕn', dâr'ē-ĕn'). **1.** Residential town (1975 est. pop. 20,229), Fairfield co., SW Conn., on Long Island Sound; settled c.1641, inc. 1820. Many 18th cent. houses remain. **2.** City (1970 pop. 1,826), seat of McIntosh co., SE Ga., on the Altamaha River NNE of Brunswick; inc. 1816. A fishing port, it was founded in 1736 by a group of Scottish Highlanders recruited by James Oglethorpe to supersede Spanish influence in the area.

Dar·jee·ling (där-jē'lĭng), town (1971 pop. 42,873), West Bengal state, NE India, near the Sikkim border. Its most famous product is tea. The town is a market for grains, fruit, and vegetables. Situated at an altitude of c.7,000 ft (2,135 m), Darjeeling is also a Himalayan resort commanding majestic views of Mt. Everest and Kanchenjunga.

Darke (därk), county (1970 pop. 49,141), 605 sq mi (1,567 sq km), W Ohio, bounded on the W by the Ind. border and drained by Greenville Creek and the Stillwater River; formed 1816; co. seat Greenville. In a rich agricultural area yielding livestock, grain, tobacco, tomatoes, and fruit, it has clay pits and some manufacturing.

Dar·ling (där'lĭng), river, 1,702 mi (2,738.5 km) long, rising in the Eastern Highlands, NE New South Wales and SE Queensland, Australia, and flowing SW across New South Wales into the Murray River at Wentworth. It is the longest river in Australia. The river was discovered in 1828 by Charles Sturt, an English explorer.

Darling Downs (dounz), tableland, 27,610 sq mi (71,510 sq km), SE Queensland, Australia, W of the Great Dividing Range. Settled in 1840 by sheep grazers, this grassland region has become an important farming and dairying area; it is in Australia's wheat belt.

Dar·ling·ton (där'lĭng-tən), county borough (1973 est. pop. 85,120), Durham, NE England, on the Skerne River near its junction with the Tees River. It is a railroad center, with extensive locomotive works, iron foundries, steel plants, and worsted factories.

Darlington, county (1970 pop. 53,442), 545 sq mi (1,411.6 sq km), NE S.C., bounded on the E by the Pee Dee River; formed 1785; co. seat Darlington. Tobacco, cotton, corn, and hogs are raised.

Darlington. 1. Town (1970 pop. 6,990), seat of Darlington co., NE S.C., NW of Florence, in a cotton and tobacco region; settled 1798, inc. 1835. Textiles and wood and paper products are manufactured. **2.** City (1970 pop. 2,351), seat of Lafayette co., S Wis., on the Pecatonica River SSW of Madison; inc. 1877.

Darm·stadt (därm'stăt, -shtät'), city (1974 est. pop. 138,871), Hesse, central West Germany. It is a commercial, industrial, and transportation center; its manufactures include chemicals, steel, machinery, and printed materials. Darmstadt was chartered in 1330. It passed to the landgraves of Hesse in 1479.

Dart·ford (därt'fərd), municipal borough (1973 est. pop. 44,130), Kent, SE England, near London. Its industries include flour milling and the manufacture of paper, drugs, chemicals, and cement. The rebellion led by Wat Tyler started here in 1381.

Dart·moor (därt'mo͝or'), upland region, 365 sq mi (945.4 sq km), Devonshire, SW England. A picturesque wasteland, its bare granite peaks (tors) rise 1,600 to 2,000 ft (488-610 m). During the Middle Ages it was an important tin-mining region. China clay and granite have been extracted from Dartmoor since the 12th cent. There are many remains of Bronze Age settlers.

Dart·mouth (därt'məth), city (1976 pop. 65,341), S N.S., Canada, on Halifax harbor, an inlet of the Atlantic Ocean. The city has large sugar and oil refineries and an automobile assembly plant.

Dartmouth, municipal borough (1973 est. pop. 6,720), Devonshire, SW England, on the Dart estuary. The principal feature of the town today is the Royal Naval College.

Dartmouth, residential and resort town (1975 est. pop. 22,229), Bristol co., SE Mass., on Buzzards Bay, in a dairy region; settled c.1650, inc. 1664. Farming, fishing, and summer tourism are its economic mainstays. The town was practically annihilated in King Philip's War but was rebuilt and later became a shipbuilding center.

Dar·wen (där'wĭn, dăr'ən), municipal borough (1971 pop. 28,880), Lancashire, NW England. Engineering and the manufacture of wallpaper, paint, and plastics are the major industries.

Dar·win (där'wĭn), city (1976 pop. 36,900), capital of the Northern Territory, N Australia, on Port Darwin, an inlet of the Timor Sea. Remotely situated on the sparsely settled north coast, Darwin is important largely because of its convenient position as an air stop on the Singapore-Sydney route. In World War II the city was heavily bombed by the Japanese; later a military airdrome, fuel-oil installations, and a wharf were built, and Darwin became a key Allied base. Originally called Palmerston, the town was renamed (1911) for Charles Darwin because its site had been discovered (1839) during a voyage of Darwin's ship, the *Beagle*. The city was almost completely destroyed by a hurricane in Dec., 1974.

Dar·yal (där-yăl', dər-) or **Dar·i·el** (dăr'ē-ĕl'), pass, c.3,950 ft (1,205 m) high, SE European USSR, in Georgia, in the central Greater Caucasus Mts. below Mt. Kazbek. Situated above the Terek River, it is noted for its wild grandeur. The Georgian Military Road crosses the pass, which has long been significant as an invasion route.

Dasht-e Ka·vir (däsht'ĕ kä-vîr'), salt desert, c.500 mi (805 km) long and c.200 mi (320 km) wide, SE of the Elburz Mts., N central Iran. It is a huge basin of interior drainage named after the kavirs, or salt marshes, located here. Extending southward from the Dasht-e Kavir is the Dasht-e Lut, a sand and stone desert, c.300 mi (480 km) long and c.200 mi (320 km) wide; it consists primarily of dried-out kavirs.

Dau·gav·pils (dou'găf-pēls'), city (1976 est. pop. 112,000), W European USSR, in Latvia, on the Western Dvina River. It is a rail junction and commercial center. The city's industries produce lumber, food products, iron, and textiles. It was founded (1278) by the Livonian Knights and became a strategic fortress. Passing (1561) to the combined kingdom of Lithuania and Poland, it was ceded to Russia in the first partition of Poland (1772).

Dauphin, county (1970 pop. 223,713), 520 sq mi (1,346.8 sq km), S central Pa., bounded on the W by the Susquehanna River and drained by numerous creeks; formed 1785; co. seat Harrisburg. The Blue Mts. divide the mountainous anthracite-coal region in the north from the agricultural and industrial region in the south. Chocolate, metal products, shoes, and clothing are manufactured.

Dau·phi·né (dō-fē-nā'), region and former province, SE France, bordering on Italy. In the east the Alps culminate in the Barre des Écrins; their magnificent scenery attracts many tourists. The lower districts are fertile and warm, with vineyards and mulberry shrubs. Some iron is mined, and water power is harnessed for industry. In the kingdom of Provence (879) and after 933 in that of Arles, Dauphiné was nominally part of the Holy Roman Empire. In 1349 the area was sold to France by Dauphin Humbert II, and for the next century it was governed as a separate province by the eldest son of the king of France. In 1457 it was annexed by the crown.

Da·van·ge·re (dä-vŭng'gĕ-rĕ), town (1971 pop. 121,110), Karnataka state, SW India. It is a market for blankets, textiles, vegetable oil, and cotton. There is a machine-tool factory in the suburbs.

Da·vao (dä'vou'), city (1970 pop. 190,700), Davao del Sur prov., SE Mindanao, Philippines, at the mouth of the Davao River on Davao Gulf. The chief commercial center and major port of Mindanao, Davao serves a prosperous region that produces hemp, coffee, cacao, and timber. The city and port were seized by Japanese landing parties on Dec. 20, 1941, and used as a base for operations in the Dutch East Indies. In 1945 after most of the Philippines had been liberated, Japanese forces clung stubbornly to the city, and its recovery involved heavy fighting. Davao has a land area of 748 sq mi (1,937.3 sq km), making it one of the largest cities in the world.

Dav·en·port (dăv'ən-pôrt'). **1.** City (1978 est. pop. 101,510), seat of Scott co., E central Iowa, on the Mississippi River; inc. 1836. Bridges connect it with Rock Island, Moline, and East Moline, Ill. Davenport is an important rail, commercial, and industrial center. Heavy industrial and agricultural equipment are among the city's manufactures. An early trading post was on the site, and the treaty ending the Black Hawk War was signed here in 1832. Davenport prospered with the arrival (1856) of the first railroad to bridge the Mississippi. **2.** Town (1977 est. pop. 1,471), seat of Lincoln co., E central Wash., W of Spokane, in a wheat-growing area; settled 1880, inc. 1890. It is a farm trade center and has grain elevators and mills.

Dav·en·try (dăv'ən-trē, dān'trē), municipal borough (1971 est. pop. 11,813), Northamptonshire, central England. Footwear is manufactured here. Borough Hill, just to the east, has neolithic fortifications and vestiges of a Roman camp.

Da·vid City (dā'vĭd), city (1970 pop. 2,380), seat of Butler co., E central Nebr., SSE of the Platte River, in a livestock and grain area; inc. 1878. It is a trade and processing center.

Da·vid·son (dā'vĭd-sən). **1.** County (1970 pop. 95,627), 546 sq mi (1,414.1 sq km), central N.C., in a piedmont region bounded on the W by the Yadkin River; formed 1822; co. seat Lexington. Its agriculture includes tobacco, corn, wheat, sweet potatoes, hay, and dairy products. It also has a lumbering industry and textile and furniture factories. **2.** County (1970 pop. 447,877), 532 sq mi (1,377.9 sq km), N central Tenn., intersected by the Cumberland River; formed 1783; co. seat Nashville. In an agricultural region yielding livestock, grain, dairy products, and tobacco, it has diversified manufacturing.

Da·vie (dā'vē), county (1970 pop. 18,855), 264 sq mi (683.8 sq km), central N.C., in a piedmont region drained on the E by the Yadkin River; formed 1836; co. seat Mocksville. Its agriculture includes tobacco, wheat, corn, cotton, and hay. Textiles and wood products are manufactured.

Da·viess (dā'vĭs). **1.** County (1970 pop. 26,602), 430 sq mi (1,113.7 sq km), SW Ind., bounded on the W and S by forks of the White River; formed 1816; co. seat Washington. It lies in an agricultural area yielding grain, fruit, and livestock and has bituminous-coal mines and oil and natural-gas fields. Clothing and wood and metal products are manufactured. **2.** County (1970 pop. 79,486), 466 sq mi (1,207 sq km), NW Ky., in a rolling agricultural area bounded on the N by the Ohio River, here forming the Ind. border, on the W by the Green River; formed 1815; co. seat Owensboro. It is drained by Panther Creek. Tobacco, corn, wheat, cattle, and hogs are produced. It also has oil and gas wells, coal mines, clay, sand, and gravel pits, tracts of timber, and some industry. **3.** County (1970 pop. 8,420), 563 sq mi (1,458.2 sq km), NW Mo., drained by the Grand River; formed 1836; co. seat Gallatin. Its agriculture includes corn, wheat, hay, oats, vegetables, livestock, and dairy products. Some manufacturing is done.

Da·vis (dā'vĭs). **1.** County (1970 pop. 8,207), 509 sq mi (1,318.3 sq km), SE Iowa, on the Mo. border and drained by the Des Moines, Fox, Fabius, and North Wyaconda rivers; formed 1843; co. seat Bloomfield. In a stock-raising (sheep, hogs, and cattle) and agricultural (corn, soybeans, and hay) area, it has bituminous-coal deposits. **2.** County (1970 pop. 99,028), 297 sq mi (769.2 sq km), N Utah, in an irrigated agricultural area; formed 1850; co. seat Farmington. Its agriculture includes livestock, hay, sugar beets, fruit, and truck crops. Food processing is done. Parts of Great Salt Lake and Wasatch National Forest are in the county.

Davis, city (1978 est. pop. 36,000), Yolo co., central Calif.; settled in the 1850s, inc. 1917. Canned foods and steel products are manufactured. The Univ. of California at Davis is here.

Davis Mountains, W Tex., SE of El Paso. Baldy Peak, 8,382 ft (2,556.5 m), is the highest peak. Forested slopes, great springs, and deep canyons attract tourists. Fort Davis, est. 1854 as a border outpost, is now a national historic site.

Da·vi·son (dā'vĭ-sən), county (1970 pop. 17,319), 432 sq mi (1,118.9 sq km), SE central S.Dak., in an agricultural area watered by several creeks; formed 1873; co. seat Mitchell. Livestock, dairy products, poultry, and corn are produced. It has some manufacturing.

Davison, city (1976 est. pop. 6,193), Genesee co., S Mich., E of Flint, in a farm area; settled 1836, inc. as a village 1889, as a city 1939.

Davis Strait, c.400 mi (645 km) long and c.180 mi (290 km) wide at the narrowest point, between Greenland and Baffin Island, NE Canada, connecting the Atlantic Ocean and Baffin Bay. Large amounts of ice and icebergs move south through the strait. The British explorer John Davis sailed through it in 1587.

Da·vos (dä-vôs'), town (1977 est. pop. 11,500), Grisons canton, E Switzerland, on the Landwasser River. It is a famous winter sports center and a health resort for the tubercular.

Dawes (dôz), county (1970 pop. 9,761), 1,389 sq mi (3,597.5 sq km), NW Nebr., in an irrigated agricultural area drained by the White and Niobrara rivers and bordered on the N by S.Dak.; formed 1885; co. seat Chadron. Grain, sugar beets, beans, and potatoes are grown.

Daw·son (dô'sən). **1.** County (1970 pop. 3,639), 213 sq mi (551.7 sq km), N Ga., in the Blue Ridge, drained by the Chestatee River; formed 1857; co. seat Dawsonville. It has agriculture (cotton, corn, hay, sweet potatoes, and poultry) and a lumbering industry. Part of Chattahoochee National Forest is in the north. **2.** County (1970 pop. 11,269), 2,370 sq mi (6,138.3 sq km), E Mont., in a grain and livestock region drained by the Yellowstone and Redwater rivers; formed 1869; co. seat Glendive. Oil was discovered in the county in 1951. **3.** County (1970 pop. 19,771), 975 sq mi (2,525.3 sq km), S central Nebr., in an agricultural area drained by the Platte River; formed 1871; co. seat Lexington. Grain, livestock, and dairy and poultry products are important. **4.** County (1970 pop. 16,604), 899 sq mi (2,328.4 sq km), NW Texas, in the high plains at an altitude of 2,600–3,200 ft (793–976 m) and drained by intermittent Sulphur Springs Creek and the Colorado River; formed 1876; co. seat Lamesa. In a dairying and agricultural area yielding grain sorghums, cotton, soybeans, legumes, fruit, truck crops, cattle, hogs, and poultry, it also has natural-gas and oil wells.

Dawson or **Dawson City,** city (1971 pop. 762), W Yukon Territory, Canada, at the confluence of the Yukon and Klondike rivers. It is the trade center of the Klondike mining region and a tourist center. During the gold rush of 1898 Dawson was a boom town, reported to have a population of c.20,000. The territorial capital was moved from Dawson to Whitehorse in 1952.

Dawson, city (1970 pop. 5,383), seat of Terrell co., SW Ga., NW of Albany; settled 1856; inc. 1872. It is a trade center in a coastal-plain farm and timber area. Peanut and cottonseed products are made.

Dawson Creek, city (1976 pop. 10,528), E British Columbia, Canada, near the Alta. border, on Dawson Creek and NE of Prince George. It is the southern terminus of the Alaska Highway.

Daw·son·ville (dô'sən-vĭl'), town (1970 pop. 288), seat of Dawson co., N Ga., WNW of Gainesville, Fla. It is an agricultural trade center. Sawmilling is done.

Dax (däks), town (1975 pop. 19,137), Landes dept., SW France, in Gascony, on the Adour River. It has long been famous for its hot mineral springs.

Day (dā), county (1970 pop. 8,713), 1,030 sq mi (2,667.7 sq km), NE S.Dak., in an agricultural region watered by the Mud River and numerous lakes; formed 1875; co. seat Webster. Dairy produce and grain are important to its economy.

Dayr az Zawr (dĕr' ăz zôr') or **Deir ez Zor** (ĕz), town (1970 pop. 66,164), capital of Dayr az Zawr governorate, E Syria, on the Euphrates River. It is a prosperous farming town, with a cattle-breeding center and an agricultural school. Salt rock mines are nearby. The modern town was built by the Ottoman Empire in 1867 to curb Arab tribes of the Euphrates region. France occupied Dayr az Zawr in 1921. It was taken by Britain in 1941, and in 1946 it became part of independent Syria.

Day·ton (dāt'n). **1.** City (1970 pop. 8,751), Campbell co., N central Ky., on the Ohio River NNE of Covington and across the river from Cincinnati, Ohio; settled 1848; inc. 1867. **2.** City (1978 est. pop. 187,400), seat of Montgomery co., SW Ohio, on the Great Miami

River where it is joined by the Stillwater River; inc. 1805. It is a port of entry, the industrial, trade, and distributing point for a fertile farm area, and an aviation center. Its chief products are cash registers, air conditioners, home appliances, and automobile parts and accessories. Dayton grew with the extension of canals and railroads and with the industrial demands of the Civil War. It was the home of the Wright brothers, who, after their flight near Kitty Hawk, N.C., set up a research aircraft plant in Dayton. Carillon Park, which contains a restored Wright brothers' airplane and a fine carillon tower, is here. **3.** City (1970 pop. 4,361), seat of Rhea co., SE Tenn., on the Tennessee River NNE of Chattanooga, in a farm, coal, and timber area; settled 1820, inc. 1895. It was the scene of the so-called Monkey Trial, in which John T. Scopes was found guilty of violating a state law prohibiting the teaching of evolution in public schools. **4.** City (1977 est. pop. 2,650), seat of Columbia co., SW Wash., on the Touchet River NE of Walla Walla, in a rich grain, fruit, and livestock area; platted 1871, inc. 1876. Food and lumber are processed.

Day·to·na Beach (dā-tō'nə), city (1978 est. pop. 49,000), Volusia co., NE Fla., on the Atlantic coast and Halifax River (a lagoon); inc. 1876. The center of a major urban area comprising eight cities, Daytona Beach is a popular year-round resort. The city was founded in 1870 in an area first settled by Spanish Franciscans in the late 16th and 17th cent. Noted for its hard, white beach, Daytona Beach has been the scene of automobile racing since 1902.

Dead Sea (dĕd), salt lake, c.390 sq mi (1,010 sq km), extending c.45 mi (70 km) in the Jordan trough of the Great Rift Valley between the Ghor on the N and Wadi Arabah on the S, on the Israel-Jordan border. The surface of the Dead Sea, 1,292 ft (394 m) below sea level, is the lowest point on earth. Situated between steep, rocky cliffs, 2,500 to 4,000 ft (762.5–1,220 m) high, the lake is fed by the Jordan River and a number of small streams; it has no outlet. One of the saltiest water bodies in the world, the sea yields large amounts of mineral salts; potash and bromine are commercially extracted.

Dead·wood (dĕd'wo͝od'), city (1970 pop. 2,409), seat of Lawrence co., W S.Dak., in the Black Hills; settled 1876 after the discovery of gold. It is a tourist center for the Black Hills and a trading hub for a lumbering, stock-raising, and mining region, with ore smelting and refining operations. The city of Deadwood Gulch (so called because the trees had been killed by fire) boomed and waned with the alternate discovery and abandonment of nearby gold and silver mines. Its early history is commemorated by the Adams Memorial Museum, several monuments, and an annual "Days of '76" celebration in August. The graves of such famous Deadwood citizens as Wild Bill Hickok (who was shot in the back in a saloon here during a card game) and Calamity Jane are in Deadwood.

Deaf Smith (dĕf' smĭth', dĕf'), county (1970 pop. 18,999), 1,510 sq mi (3,910.9 sq km), NW Texas, in the high plains of the Panhandle bordering in the W on N.Mex.; formed 1876; co. seat Hereford. It is a leading producer of cattle and wheat. Grain sorghums, corn, and truck crops are grown on its irrigated farmland. Food processing and some manufacturing are done.

Deal (dēl), municipal borough (1973 est. pop. 26,840), Kent, SE England, on the Downs, an important passage for Channel shipping. It is a popular holiday resort. There is some boatbuilding; other industries are brushmaking, the production of plastics, and precision engineering. Deal was the reputed landing place of Julius Caesar in 55 B.C. and, later, one of the Cinque Ports.

Dear·born (dîr'bôrn', -bərn), county (1970 pop. 29,430), 306 sq mi (792.5 sq km), SE Ind., bounded on the E by the Ohio border, on the SE by the Ohio River, here forming the Ky. border, and drained by the Whitewater and Laughery rivers; formed 1803; co. seat Lawrenceburg. In an agricultural area yielding livestock, tobacco, and truck crops, it has whiskey distilleries and iron and steel works. Wood products, paper, and glassware are manufactured.

Dearborn, city (1978 est. pop. 95,500), Wayne co., SE Mich., on the Rouge River adjoining Detroit; settled 1795, inc. as a city 1929. Dearborn is a major warehousing and distribution center. Automobiles, bricks, steel, tools and dies, and metal products are manufactured. The Edison Institute of Technology, which includes Greenfield Village, birthplace of Henry Ford, is here. Ford's large estate, Fair Lane, designated as a national historic landmark, is now part of the Univ. of Michigan's Dearborn campus.

Death Valley National Monument (dĕth), 1,907,760 acres (772,643 hectares), SE Calif. and SW Nev.; est. 1933. Death Valley, 140 mi (225.3 km) long, is a deep, arid basin, bordered on the west by the Panamint Range and on the east by the Amargosa Range. In summer the valley has some of the highest air temperatures (134°F/57°C) and ground temperatures (165°F/74°C) in the world. Less than 2 in. (5 cm) of rain falls annually. Salt and alkali flats, unique rock formations, and briny pools are found here. Badwater, in the south-central part of Death Valley, is 282 ft (86 m) below sea level, the lowest point in North America. Death Valley was named by gold seekers who mistakenly undertook to cross this desolate region in 1849 on their way to the California gold fields. The valley itself yielded gold and silver in the 1850s, and in the 1880s borax was discovered and taken out by mule-drawn wagons. The valley was much publicized by the American adventurer Walter Scott ("Death Valley Scotty"), whose ostentatious home, Scotty's Castle, is a show-place here.

Deau·ville (dō'vĭl, dō-vēl'), town (1975 pop. 5,664), Calvados dept., N France, on the English Channel. A fashionable resort, it has a famous racecourse and a gambling casino.

De Ba·ca (də bä'kə), county (1970 pop. 2,547), 2,358 sq mi (6,107.2 sq km), E N.Mex., in a grazing and grain area watered by the Pecos River; formed 1917; co. seat Fort Sumner. Lake Sumner State Park is here.

Dę·blin (dĕɴ'blēn, dĕm'blĭn), city (1966 pop. 11,700), E Poland, on the Vistula River. It is a railway junction and one of the main crossings of the Vistula. Founded as a fortress by Czar Nicholas I in 1837, it was captured by the Germans in 1915 but was returned to Poland after World War I.

De·bre·cen (dĕ'brĕt-sĕn), city (1976 est. pop. 187,103), E Hungary, the economic and cultural center of the Great Plain (Alföld) region E of the Tisza River. It is an industrial city that produces railway cars, agricultural machinery, medical instruments, and pharmaceuticals. Known in the 13th cent., the city grew as a market for cattle and grain. It became the stronghold of Hungarian Protestantism in the 16th cent., and its Calvinist college later formed the nucleus of a university. Under the Turkish occupation of Hungary (16th–17th cent.), Debrecen enjoyed semiautonomous status and often served as a refuge for peasants fleeing the Turks. The wars in the late 17th cent. between Christian Europe and the Turks ruined the city's economy. Debrecen became the center of Hungarian resistance against Austrian rule in the 19th cent. On Apr. 14, 1849, Louis Kossuth proclaimed Hungary's independence in the great church in the heart of Debrecen. Russian troops, who had helped the Hapsburgs crush the Hungarian uprising, occupied the city briefly. In 1944–45 Debrecen served as provisional capital of Hungary.

De·ca·tur (dĭ-kā'tər). **1.** County (1970 pop. 22,310), 575 sq mi (1,489.3 sq km), SW Ga., bounded on the S by the Fla. border, on the W by the Spring River and Lake Seminole, and drained by the Flint River; formed 1823; co. seat Bainbridge. In a coastal-plain agricultural area yielding corn, sugar cane, peanuts, pecans, and livestock, it also has fuller's-earth mines. **2.** County (1970 pop. 22,738), 370 sq mi (958.3 sq km), SE central Ind., drained by Flatrock and Sand creeks; formed 1821; co. seat Greensburg. It has agriculture (grain, tobacco, and livestock), limestone quarries, and some oil and natural-gas deposits. Clothing, plastics, and metal products are manufactured. **3.** County (1970 pop. 9,737), 530 sq mi (1,372.7 sq km), S Iowa, in a rolling prairie agricultural area on the Mo. border; formed 1846; co. seat Leon. It is drained by the Thompson and Weldon rivers. Hogs, cattle, poultry, corn, and alfalfa are produced. It also has bituminous-coal deposits. **4.** County (1970 pop. 4,988), 899 sq mi (2,328.4 sq km), NW Kansas, in a flat to rolling area bordered on the N by Nebr. and watered by the Prairie Dog River and Sappa and Beaver creeks; formed 1879; co. seat Oberlin. Grain and livestock are important. It has oil fields and some industry. **5.** County (1970 pop. 9,457), 337 sq mi (872.8 sq km), W Tenn., bounded on the E and S by the Tennessee River; formed 1845; co. seat Decaturville. In an agricultural area yielding livestock, dairy products, cotton, corn, and hay, it has some manufacturing.

Decatur. 1. Industrial city (1978 est. pop. 39,500), seat of Morgan co., N Ala., on the Tennessee River; inc. 1826. It is a commercial and manufacturing center, with shipyards and industries thriving on power supplied by the Tennessee Valley Authority (TVA). Textiles, plastics, chemicals, bricks, tires, and trailers are among the city's manufactures. A settlement known as Rhodes Ferry was here when President James Monroe directed (1820) that a site be selected near a great river to honor Stephen Decatur, who had been killed in a duel. The city grew as a cotton center. During the Civil War it was continually raided by Federal forces; only two houses and the imposing state bank (1832) survived. The present city was formed (1927) by the union of Decatur and Albany (formerly New Decatur). **2.** City (1978 est. pop. 18,800), seat of De Kalb co., NW Ga., a residential suburb of Atlanta; inc. 1823. On nearby Stone Mountain is a spectacular Confederate memorial featuring carved figures of Gen. Robert E. Lee, Gen. Stonewall Jackson, and Confederate President Jefferson Davis. **3.** City (1978 est. pop. 88,900), seat of Macon co., central Ill., on the Sangamon River (dammed here to form Lake Decatur); inc. 1839. A railroad and industrial center in a rich farm and livestock area, Decatur has railroad repair shops and huge plants for processing corn and soybeans. The city's manufactures include tires, tractors, machinery, and automobile equipment. Coal deposits underlie the area. Points of interest include the Lincoln Log Cabin Courthouse, where Abraham Lincoln practiced law; Lincoln Square, where he received his first endorsement for the presidential nomination; and the city library, which has a Lincoln collection. The Grand Army of the Republic was organized in Decatur in Apr., 1866. **4.** City (1970 pop. 8,445), seat of Adams co., NE Ind., SE of Fort Wayne; platted 1836. It is a processing point in a timber, farm, and livestock area. **5.** Town (1970 pop. 1,311), seat of Newton co., E central Miss., WNW of Meridian. **6.** Town (1975 est. pop. 1,021), seat of Meigs co., SE Tenn., NE of Chattanooga, in a farm area. **7.** City (1970 pop. 3,240), seat of Wise co., N Texas, near the old Chisholm Trail, NNW of Fort Worth. In an agricultural and oil region, it has processing plants. East Central Junior College is here.

De·ca·tur·ville (dĭ-kā'tər-vĭl'), town (1970 pop. 958), seat of Decatur co., W Tenn., E of Jackson, near the Kentucky Reservoir. It has sawmills and limestone quarries.

Dec·can (dĕk'ən), region of India, sometimes defined as all India S of the Narbada River; in a more limited sense the plateau of central peninsular India, including approximately all Karnataka and S Andhra Pradesh, SE Maharashtra, and NW Tamil Nadu. The rich volcanic soil is used for growing cotton. The last of the great Mogul emperors, Aurangzeb, exhausted the power of his empire in a futile attempt (1683–1707) to absorb the region. It was in the Deccan that the Hindus began to regain (early 18th cent.) political and military power in India. Here in the late 18th cent. the British decisively defeated the French in their struggle for India.

Dě·čín (dyĕ'chēn), city (1975 est. pop. 46,633), Czechoslovakia, in Bohemia, on the Elbe. It includes (since 1950) the city of Podmokly on the left bank of the Elbe. A center of a coal-mining region, it is also a river port and an industrial center. Founded in 1128, it was incorporated into Czechoslovakia in 1918.

De·cor·ah (də-kôr'ə), city (1970 pop. 7,237), seat of Winneshiek co., NE Iowa, near the Minn. border on the Upper Iowa River ENE of Charles City; inc. 1857. Luther College, with a museum commemorating the early history of the area, is here.

Ded·ham (dĕd'əm), town (1978 est. pop. 26,600), seat of Norfolk co., E Mass., on the Charles River, a suburb of Boston; inc. 1636. America's oldest frame house, the Fairbanks House (1636), is here.

Dee (dē). **1.** River, c.90 mi (145 km) long, rising in the Cairngorms, SW Aberdeenshire, E Scotland, and flowing E past Ballater to the North Sea through an artificial channel at Aberdeen. The channel was constructed (1872) to improve Aberdeen's harbor. Celebrated for its beauty, the Dee also has notable salmon fisheries. **2.** River, c.50 mi (80 km) long, rising in N Kirkcudbrightshire, SW Scotland, and flowing generally S to the Irish Sea. There are five power stations in the Dee basin.

Dee, river, c.70 mi (112 km) long, rising in the Cambrian Mts., Merionethshire, NW Wales, and flowing NE through Bala Lake, then meandering through a picturesque course NE, N, and NW past Chester to the Irish Sea. At low tide the long, broad, shallow estuary is an expanse of sand, across which the narrow stream flows.

Deep (dēp), river, c.125 mi (200 km) long, N central N.C., rising E of Winston-Salem and flowing SE and E to connect with the Haw River, thus forming the Cape Fear River.

Deer·field (dîr'fēld'). **1.** Village (1975 est. pop. 19,007), Cook and Lake cos., NE Ill., a residential suburb of Chicago; inc. 1903. Bakery products and communications and construction equipment are made. **2.** Town (1970 pop. 3,850), Franklin co., NW Mass., on the Deerfield River; inc. 1677. In the Indian massacre of 1704 nearly 50 inhabitants were killed, and most of the survivors were taken to Canada; many were killed on the way. Old Deerfield St. is lined with 18th cent. houses. Deerfield Academy is here.

Deerfield Beach, town (1978 est. pop. 28,200), Broward co., SE Fla., on the Atlantic coast; inc. 1925.

Deer Lodge (dîr' lŏj'), county (1970 pop. 15,652), 740 sq mi (1,916.6 sq km), SW Mont., in a mining and agricultural area bounded on the SW by the Continental Divide, on the SE by the Big Hole River; formed 1865; co. seat Anaconda. One of the world's largest smelters is here. Parts of Deerlodge National Forest are in the county.

Deer Lodge, city (1970 pop. 4,306), seat of Powell co., W central Mont., SW of Helena, in a mining and ranching area; inc. 1888. Founded as La Barge in 1862 after gold was discovered in Deer Lodge Valley, it was renamed in 1864. Montana State Prison is here.

De·fi·ance (dĭ-fī'əns), county (1970 pop. 36,949), 410 sq mi (1,061.9 sq km), NW Ohio, bounded on the W by the Ind. line and intersected by the Maumee, Auglaize, and Tiffin rivers; formed 1845; co. seat Defiance. Its agriculture includes livestock, grain, and truck crops, and it has diversified manufacturing.

Defiance, city (1978 est. pop. 15,500), seat of Defiance co., NW Ohio, at the confluence of the Auglaize and Maumee rivers, in a farm area; settled 1790, inc. 1836. Its manufactures include machinery, food products, and fabricated metal items. Anthony Wayne built Fort Defiance here in 1794.

De Fu·ni·ak Springs (də fyŏŏ'nē-ăk'), town (1970 pop. 4,966), seat of Walton co., NW Fla., NW of Panama City, in a diversified agricultural area. Lake De Funiak is formed by a great spring in the center of the town.

Deh·ra Dun (dā'rə dŏŏn'), city (1971 pop. 166,073), Uttar Pradesh state, N central India. It is a district administrative headquarters and a trade center for surrounding hill areas. An institute for research in forestry, with an associated museum and botanical garden, is here.

Deir ez Zor (dĕr' ĕz zôr'): *see* Dayr az Zawr, Syria.

De Kalb (dĭ kălb'). **1.** County (1970 pop. 41,981), 778 sq mi (2,015 sq km), NE Ala., in a hilly region bordering on Ga.; formed 1836; co. seat Fort Payne. It has agriculture (cotton, corn, and livestock) and deposits of coal, iron, limestone, and fuller's earth. The Sand Mts. extend throughout much of the county. **2.** County (1970 pop. 415,387), 269 sq mi (696.7 sq km), NW central Ga., including part of the Atlanta metropolitan area; formed 1822; co. seat Decatur. It is in a manufacturing, dairying, poultry, and truck-farming region. Stone Mountain Memorial is here. **3.** County (1970 pop. 71,654), 636 sq mi (1,647.2 sq km), N Ill., in an agricultural and manufacturing area drained by branches of the Kishwaukee River; formed 1837; co. seat Sycamore. Metal, brick, tile, cement, and dairy products are made. **4.** County (1970 pop. 30,837), 365 sq mi (945.4 sq km), NE Ind., bounded on the E by the Ohio border and drained by the St. Joseph River; formed 1835; co. seat Auburn. It lies in an agricultural area yielding livestock, poultry, truck crops, soybeans, corn, wheat, oats, and dairy products. Wood and metal products, paper, rubber, and iron and steel are manufactured. **5.** County (1970 pop. 7,305), 423 sq mi (1,095.6 sq km), NW Mo., drained by branches of the Grand and Platte rivers; formed 1845; co. seat Maysville. It has agriculture (corn, wheat, oats, soybeans, and livestock) and some manufacturing. **6.** County (1970 pop. 11,151), 278 sq mi (720 sq km), central Tenn., in an agricultural and manufacturing area drained by the Caney Fork of the Cumberland River; formed 1837; co. seat Smithville. Corn, hay, and tobacco are grown.

De Kalb. 1. City (1978 est. pop. 29,800), De Kalb co., N Ill., in a farm area; inc. 1861. Vegetables and fruit are canned, and motors, musical instruments, wire screening, electrical equipment, and plastics are manufactured. The growth of the city was stimulated in the 1870s by the development of workable barbed wire. De Kalb is the seat of Northern Illinois Univ. **2.** Town (1970 pop. 1,072), seat of Kemper co., E Miss., N of Meridian. It has sawmills and cotton gins.

Del·a·go·a Bay (dĕl'ə-gō'ə), inlet of the Indian Ocean, c.55 mi (90 km) long and 20 mi (32 km) wide, S Mozambique, SE Africa. The bay was discovered (1502) by António do Campo, one of Vasco da Gama's captains. The area was first explored (1544) by Lourenço Marques, the Portuguese trader. In the 1700s Dutch and Austrian trading companies tried to establish posts on the bay, but were driven out by malaria and the Portuguese. In 1787 Portugal built a fort here, around which Lourenço Marques (now Maputo) grew. In the mid-1800s, Portugal's claim to the area was challenged by Great

Britain and by the Transvaal when it was realized that the bay provided a major access route to the Kimberley diamond mines. The Transvaal recognized Portugal's sovereignty in 1869, and in 1875 France, acting as arbiter, awarded the area to Portugal.

De Land (dĭ lănd'), resort city (1978 est. pop. 13,500), seat of Volusia co., NE Fla.; inc. 1882. It has dairies, citrus packing plants, and lumber mills. Other products are electrical and electronic parts, wearing apparel, and medical supplies. Nearby are Indian burial grounds and Ponce de Leon Springs.

De·la·no (də-lā'nō, dĕl'ə-nō), city (1978 est. pop. 15,400), Kern co., S central Calif., in the fertile San Joaquin valley; inc. 1915. The city's economy is based on agriculture and related enterprises, especially vineyards and wineries.

Del·a·ware (dĕl'ə-wâr, -wər), state (1975 pop. 576,000), 2,057 sq mi (5,327.6 sq km), United States, one of the Middle Atlantic States and one of the Thirteen Colonies. The capital is Dover. With Md. and Va., the state occupies the peninsula between Chesapeake Bay and Delaware Bay. Delaware is situated on the northeast portion of the

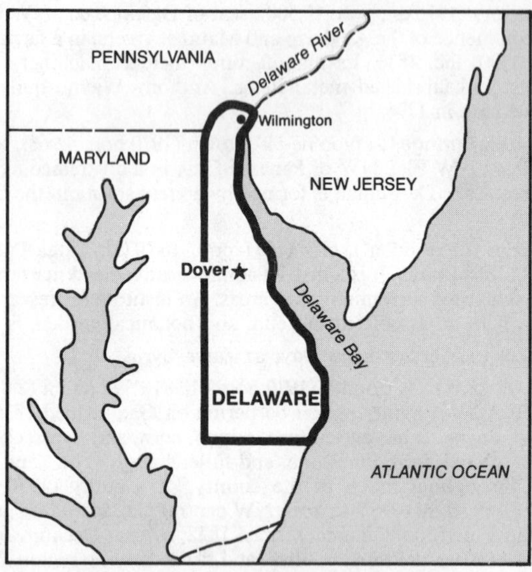

peninsula, facing the Delaware River, which broadens into Delaware Bay; the bay in turn joins the Atlantic Ocean at Cape Henlopen. N.J. lies across the river and bay from Delaware, which has water along its entire eastern edge; elsewhere Delaware is bounded by Pa. on the north and by Md. on the west and the south.

Delaware is divided into three counties:

NAME	COUNTY SEAT	NAME	COUNTY SEAT
Kent	Dover	Sussex	Georgetown
New Castle	Wilmington		

Economy. Delaware is chiefly an industrial state, although agriculture is still important. The leading agricultural products are broilers, corn, soybeans, dairy products, potatoes, and hay. Much of the state's wealth comes from industries around Wilmington, especially the chemical industry that was founded by the Du Pont family in the 19th cent. In addition to chemicals and chemical products, industries manufacture food products, rubber and plastic products, and primary and fabricated metals. Some of the nation's largest corporations have their home offices in Wilmington, because of Delaware's lenient laws governing business taxation.

History. Long before white men explored the Delaware area, it was inhabited by the Delaware Indians—notably the Nanticoke in the south and the Minqua in the north. In 1609 Henry Hudson sailed into Delaware Bay. A year later the British captain Sir Samuel Argall, bound for the colony of Virginia, also sailed into the bay. From the time of its discovery, the region was contested by the Dutch and English. The first settlement was established (1631) by the Dutch on the site of the present-day town of Lewes. However, within a year it was attacked and utterly destroyed by the Indians. This attack notwithstanding, the Indians were generally friendly and willing to trade with the newcomers. In 1637-38 the Dutchman Peter Minuit directed a colonizing expedition for Swedes in order to organize New Sweden. Fort Christina was founded in 1638 on the site of modern Wilmington and was named in honor of the queen of Sweden. The colony grew with the arrival of Swedish, Finnish, and Dutch settlers. Peter Stuyvesant, governor of New Netherland, sailed to the Delaware region in 1651 and established Fort Casimir on the Delaware shore at the site of modern New Castle. The Swedes captured the fort by surprise in 1654, but Stuyvesant returned with an expedition in 1655 and conquered all New Sweden. In 1664 the English seized the Dutch holdings on the Delaware, and after 1674 the colony remained firmly in their hands until the American Revolution.

The English duke of York (later James II) annexed the region to New York, land granted him earlier by Charles II. In 1682 the duke transferred the claim to William Penn. The three counties of Delaware thus became the Three Lower Counties of Pennsylvania. The Penn charter of 1701 provided a separate assembly for the Three Lower Counties, which thereafter maintained partial autonomy. In 1776 the colony became a state. Delaware was a leader in the movement for revision of the form of government under the Articles of Confederation and in 1787 became the first state to ratify the new Constitution of the United States. The late 18th cent. also marked the beginning of industry in Delaware with the establishment of gristmills on the Brandywine and Christina rivers. Wilmington became a center for the manufacture of cloth, paper, and flour. In 1802 Eleuthère Irénée Du Pont established a gunpowder mill on the Brandywine River, thus originating the family chemical industry that grew until a federal antitrust suit broke it up in 1912.

Delaware was a slave state, but in the early 19th cent. the number of slaves in the state declined. In the Civil War Delaware remained loyal to the Union, but pro-Southern feeling increased during the course of the war. Many European immigrants came to the state in the late 19th and early 20th cent., settling in the Wilmington area. Delaware's industries flourished during the 19th cent. as transportation facilities improved. Industry continued to expand in the 20th cent., especially during World Wars I and II.

Government. Under the provisions of the 1897 constitution, the governor is elected to a four-year term. The state legislature, called the general assembly, is made up of a senate with 21 members elected for four years and a house of representatives with 41 members elected for two years.

Educational Institutions. The most prominent in the state is the Univ. of Delaware, at Newark.

Delaware. 1. County (1970 pop. 129,219), 400 sq mi (1,036 sq km), E central Ind., drained by the Mississinewa River and the West Fork of the White River; formed 1827; co. seat Muncie. Its agriculture includes corn, grain, soybeans, tomatoes, hogs, and dairy products. Food processing and shipping is done, and wood products, paper, concrete, iron and steel, and metal products are manufactured. **2.** County (1970 pop. 18,770), 573 sq mi (1,484.1 sq km), E Iowa, in a prairie agricultural area drained by the Maquoketa River; formed 1837; co. seat Manchester. Hogs, cattle, poultry, corn, oats, and rye are produced. There are limestone quarries here. **3.** County (1970 pop. 44,718), 1,458 sq mi (3,776.2 sq km), S N.Y., in the Catskills bounded on the NW by the Susquehanna River, on the SW by the Delaware River; formed 1797; co. seat Delhi. It is drained by Beaver Kill and the Charlotte River. In a summer and winter resort area, it is a leading dairying county and also produces hay, truck crops, and poultry. Some lumbering and diversified manufacturing are done. **4.** County (1970 pop. 42,908), 450 sq mi (1,165.5 sq km), central Ohio, intersected by the Olentangy and Scioto rivers; formed 1808; co. seat Delaware. In an agricultural area yielding livestock, dairy products, grain, and fruit, it has limestone quarries and some manufacturing plants. **5.** County (1970 pop. 17,767), 707 sq mi (1,831.1 sq km), NE Okla., partly in the Ozarks and bordered on the E by Ark. and Mo.; formed 1907; co. seat Jay. Stock raising, dairying, and some farming (fruits, berries, corn, wheat, and oats) are done. Lake of the Cherokees Recreational Area is here. **6.** County (1970 pop. 603,456), 185 sq mi (479.2 sq km), SE Pa., in an industrial and residential area bounded on the E by Philadelphia, on the SE by the Delaware River, and on the S by the Del. border; formed 1789; co. seat Media. Oil refining and shipbuilding are done, and textiles and metal and paper products are made.

Delaware, city (1978 est. pop. 20,300), seat of Delaware co., central

Ohio, on the Olentangy River; inc. as a city 1903. A trade center in a rich farm area, it has some manufacturing. Ohio Wesleyan Univ. is in the city, which is also the birthplace of Rutherford B. Hayes.

Delaware, river, c.280 mi (450 km) long, rising in the Catskill Mts., SE N.Y., and flowing SE along the N.Y.-Pa. border to Port Jervis, then between Pa. and N.J. and generally S to Delaware Bay, an estuary (52 mi/83.7 km long) between N.J. and Del. Reservoirs and dams on its headstreams provide flood control and water supply. The lower Delaware River, from Trenton, N.J. (the head of navigation), past Philadelphia, Pa. (an ocean port), to Wilmington, Del., flows through a highly industrialized area. The Chesapeake and Delaware Canal links the river with Chesapeake Bay.

Delaware Water Gap, scenic gorge, 2 mi (3.2 km) long, cut by the Delaware River through Kittatinny Mt., on the N.J.-Pa. line. The gap, parts of the wooded Kittatinny Mt., several islands, and c.40 mi (64 km) of river bank are included in Delaware Water Gap National Recreation Area.

De·lé·mont (də-lā-môɴ'), town (1977 est. pop. 11,400), Bern canton, NW Switzerland. A watchmaking center of the Bernese Jura, it was once the residence (1528–1792) of the prince-bishops of Basel.

Delft (dĕlft), city (1977 est. pop. 85,118), South Holland prov., W Netherlands. It has varied industries and is noted for its ceramics (china, tiles, and pottery) known as delftware. Founded in the 11th cent. and chartered in 1246, Delft was an important commercial center until superseded (17th cent.) by Rotterdam. The aspect of old Delft has changed little since Jan Vermeer, who was born and lived here (17th cent.), painted his famous *View of Delft.*

Del·hi (dĕl'ē), union territory and city, N central India. The union territory (1971 pop. 4,065,698), 573 sq mi (1,484.1 sq km), is on the Delhi plain, which is crossed by the Jumna River and stretches between the Indus valley and the alluvial plain of the Ganges. Throughout India's history the region of Delhi, commanding roads in all directions, was the key to the empire. From the earliest times many cities rose and fell there, and within 50 sq mi (129.5 sq km) south of New Delhi are more important dynastic remains than exist in any other area of the country. The city of Delhi, or Old Delhi (1971 metropolitan area pop. 3,629,842), on the Jumna River, adjoins New Delhi in the east-central part of the territory and is enclosed by high stone walls erected in 1638 by Shah Jahan. Within the walls he built the famous Red Fort—so called for its walls and gateways of red sandstone—that contained the imperial Mogul palace. Just south of the fort, on the Jumna's bank, is Rajghat, where Gandhi's body was cremated; it is now one of the most revered shrines in India. In the northwest, beyond the old walls, there are residences, hotels and clubs, the Univ. of Delhi, and an amphitheater (built 1911) that marks the site of the ceremony in 1877 in which Queen Victoria was proclaimed empress of India. The present city of Old Delhi did not become important until Shah Jahan made it the capital of the Mogul empire in 1638. It was sacked (1739) by the Persian Nadir Shah. The city was held by the Mahrattas from 1771 until 1803, when the British took it. During the Indian Mutiny of 1857 it was held for five months by the rebels. It was (1912–31) interim capital of India until New Delhi was officially inaugurated and is today a commercial center.

Del·hi (dĕl'hī), village (1970 pop. 3,017), seat of Delaware co., S N.Y., in the Catskills, on the West Branch of the Delaware River SE of Oneonta; settled c.1785, inc. 1821. Dairy products are made.

Dells of the Wisconsin or **The Dells** (dĕlz), scenic part of the Wisconsin River, central Wis., NW of Portage. The river has cut a deep gorge through 8 mi (12.9 km) of sandstone, which is carved into caves, pinnacles, and other curious shapes.

Del·mar (dĕl'mär'), town (pop. in Wicomico co., Md., 1,191; in Sussex co., Del., 943), on the Del.-Md. border NE of Salisbury, Md. It has two town councils, a president, and a mayor. The town is a trading and shipping center in a farm area. Clothing is made.

Del·mar·va peninsula (dĕl-mär'və), c.180 mi (290 km) long, separating Chesapeake Bay on the W from Delaware Bay and the Atlantic Ocean on the E. The western coast of the peninsula is irregular and marshy; the eastern shore is straight with sandy beaches and offshore bars.

Del·men·horst (dĕl'mən-hôrst), industrial city (1974 est. pop. 70,992), Lower Saxony, N West Germany, near Bremen. Manufactures include textiles, linoleum, processed food, and machinery.

Del Norte (dĕl'nôrt'), county (1970 pop. 14,580), 1,003 sq mi (2,597.8 sq km), NW Calif., in a mountainous area bounded on the N by the Oregon border, on the W by a coastal strip and the Pacific Ocean; formed 1857; co. seat Crescent City. It is drained by the Smith and Klamath rivers. In a lumbering and sawmilling region, it also has dairying, cattle and sheep raising, mining, quarrying, ocean fishing, and processing industries. Redwood forests, game fishing, and hunting attract vacationers.

Del Norte, town (1970 pop. 1,569), seat of Rio Grande co., S Colo., NW of Monte Vista on the Rio Grande; settled 1871, inc. 1885. At an altitude of 7,800 ft (2,379 m), the area has irrigated farms and gold and silver mines. Rio Grande National Forest is nearby.

De·los (dē'lŏs, dĕl'ōs), island, c.1 sq mi (2.6 sq km), SE Greece, in the Aegean Sea, smallest of the Cyclades. In Greek mythology, Leto gave birth to Apollo and Artemis on Delos, and the island was particularly sacred to Apollo. The temple of Apollo was the seat of the treasury of the Delian League until it was removed (454 B.C.) to Athens. In the 2nd cent. B.C. Delos had a flourishing slave market which continued to thrive even after a slave rebellion c.130 B.C. In 88 B.C. the island was sacked by Mithridates VI of Pontus; it never recovered and Delos was abandoned toward the end of the 1st cent. B.C. It is virtually uninhabited today. Excavations conducted since the 1870s have revealed remains of temples, commercial buildings, theaters, private houses, and numerous inscriptions.

Del·phi (dĕl'fī), locality in Phocis, Greece, near the foot of the S slope of Mt. Parnassós, c.6 mi (10 km) NE of the port of Cirrha. It was the seat of the Delphic Oracle, the most famous of ancient Greece.

Delphi, city (1970 pop. 2,582), seat of Carroll co., NW Ind., on the Wabash River NE of Lafayette; founded 1828. It is a farm trade center and produces truck bodies and farm equipment.

Del·ray Beach (dĕl'rā'), resort city (1978 est. pop. 30,700), Palm Beach co., SE Fla., on the Atlantic coast; settled 1895, inc. 1911. Mostly residential, Delray Beach is also the trade center for a citrus-fruit and vegetable-growing region. The city's vast flower farms are noted especially for chrysanthemums and gladiolas.

Del Ri·o (dĕl rē'ō), city (1978 est. pop. 26,500), seat of Val Verde co., W Texas, a port of entry on the Rio Grande opposite Ciudad Acuña, Mexico; founded 1868, inc. 1911. It is the marketing and distributing center for a region known for its sheep, lambs, wool, and mohair. Farms irrigated from the nearby San Felipe Springs also yield alfalfa, truck crops, fruits, and especially grapes. Laughlin Air Force Base, a jet training command, is to the east.

del·ta (dĕl'tə), the alluvial plain formed at the mouth of a river where the stream loses velocity and deposits part of its load of sediment. No delta is formed if the coast is sinking or if there is an ocean or tidal current strong enough to prevent deposition of sediment. A deltaic plain is usually very fertile but subject to floods. The three main varieties of deltas are the arcuate delta (as that of the Nile), the bird's-foot delta (as that of the Mississippi), and the cuspate delta (as that of the Tiber).

Delta. 1. County (1970 pop. 15,286), 1,157 sq mi (2,996.6 sq km), W Colo., in a coal-mining and agricultural area drained by the Gunnison River; formed 1883; co. seat Delta. Fruit, beans, and hay are grown, and livestock raised. Part of Grand Mesa National Forest is in the north. **2.** County (1970 pop. 35,924), 1,180 sq mi (3056.2 sq km), S Upper Peninsula, Mich., bounded on the S by Lake Michigan and arms of Green Bay and drained by the Ford, Escanaba, Sturgeon, and Whitefish rivers; organized 1861; co. seat Escanaba. Lumbering, dairying, and agriculture (fruit, potatoes, truck crops, livestock, and poultry) are important. It also ships iron ore and grain and has heavy manufacturing and a fishing industry. There are resorts in the area, which includes Hiawatha National Forest. **3.** County (1970 pop. 4,927), 276 sq mi (714.8 sq km), NE Texas, in a rich prairie region between the North and South forks of the Sulphur River; formed 1870; co. seat Cooper. It lies in an agricultural area that produces cotton, corn, hay, cattle feed, peanuts, fruit, truck crops, dairy products, cattle, hogs, mules, and poultry. It has some lumbering and manufacturing.

Delta, city (1970 pop. 3,694), seat of Delta co., W Colo., on the Gunnison River SE of Grand Junction, in a farm area on the site of a fur trappers' fort built in 1830; inc. 1882. It has a beet-sugar refinery and is headquarters for the Uncompahgre National Forest.

Dem·a·vend (dĕm'ə-vĕnd'), peak, Iran: *see* Damavand.

Dem·e·rar·a (dĕm′ə-rär′ə), river, c.200 mi (320 km) long, rising in the Guiana Highlands, E Guyana, and flowing N to the Atlantic Ocean. The Demerara is navigable for oceangoing vessels to Mackenzie, an important exporting center for bauxite and kaolin.

Dem·ing (dĕm′ĭng), village (1970 pop. 8,343), seat of Luna co., SW N.Mex., in the valley of the Mimbres River; settled 1880, inc. 1902. It is a health resort and the rail and trade center of a farm, ranch, and copper-mine region.

De·nain (də-năN′), city (1975 pop. 26,204), Nord dept., N France. It has coal fields, ironworks, and steel mills. In 1712, during the War of the Spanish Succession, the French defeated the Austrians here.

Den·bigh·shire (dĕn′bē-shîr′, -shər) or **Den·bigh** (dĕn′bē), former county, N Wales, bounded on the N by the Irish Sea and on the W by the River Conway. The county town was Denbigh. In 1974 Denbighshire was divided between the new nonmetropolitan counties of Clwyd and Gwynedd.

Den·de·ra (dĕn′də-rä): *see* Dandarah, Egypt.

Den·der·mon·de (dĕn′dər-môn′də), town (1977 est. pop. 41,066), East Flanders prov., central Belgium, at the confluence of the Dender and Scheldt rivers. Manufactures include carpets and linen. In 1667 Dendermonde held off the French under Louis XIV by opening dikes and flooding the countryside.

Den Hel·der (dŭn hĕl′dər), city (1977 est. pop. 60,828), North Holland prov., NW Netherlands, on the North Sea. It is the main base of the country's navy.

Den·i·son (dĕn′ĭ-sən). **1.** City (1970 pop. 6,218), seat of Crawford co., W central Iowa, on the Boyer River NNE of Council Bluffs, in an area producing cattle, poultry, grain, and dairy foods; platted 1856, inc. 1876. Indian artifacts have been found nearby. **2.** City (1978 est. pop. 22,000), Grayson co., N Texas, near the Red River; inc. 1873. It is a rail center with textile mills, garment factories, and plants that manufacture a great variety of products. The town was founded by the railroad in 1872 on the site of an old stagecoach station. Wealth from industry has preserved the beauty of gracious and well-shaded streets, and local funds have converted (1968) the downtown area into a modern shopping park. Dwight D. Eisenhower was born here, and his birthplace is open to visitors.

De·niz·li (dĕ-nēz-lē′), city (1975 pop. 106,704), capital of Denizli prov., W Turkey. It is an agricultural market center and the gateway for excursions to the nearby ruins of Laodicea and Hierapolis. The city was badly damaged by earthquakes in 1710 and 1899.

Den·mark (dĕn′märk′), kingdom (1971 pop. 4,950,597), 16,629 sq mi (43,069 sq km), N Europe, bordering on West Germany in the S, on the North Sea in the W, on the Skagerrak in the N, and on the Kattegat and the Øresund in the E. The southernmost of the Scandinavian countries, Denmark proper includes most of the Jutland peninsula, several major islands, notably Sjaelland, and about 450 other islands. The Faeroe Islands and Greenland, semiautonomous parts of the Danish realm, lie to the northwest. Copenhagen is Denmark's capital.

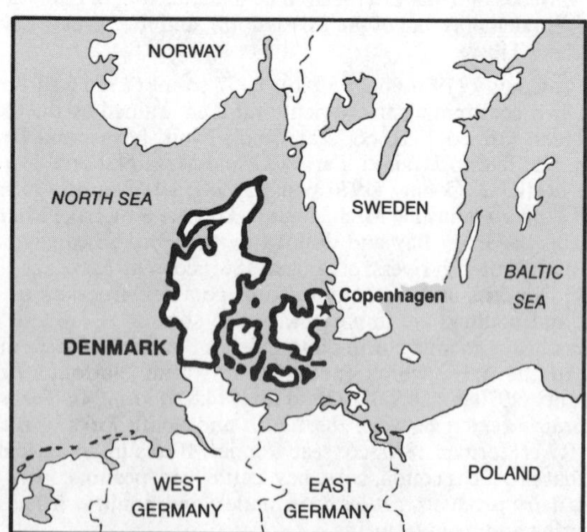

Denmark is divided into 14 counties and 2 communes:

NAME	CAPITAL	NAME	CAPITAL
Århus	Århus	Ringkøbing	Ringkøbing
Bornholm	Rønne	Roskilde	Roskilde
Copenhagen (commune)		Sønderjylland	Åbenrå
Copenhagen	Copenhagen	Storstrøm	Nykøbing Falster
Frederiksberg (commune)			
Frederiksberg	Hillerød	Vejle	Vejle
Fyn	Odense	Vestsjælland	Sorø
Nordjylland	Ålborg	Viborg	Viborg
Ribe	Ribe		

Economy. Traditionally an agricultural country, Denmark after 1945 greatly expanded its industrial base. The main commodities raised are livestock, root crops (beets, kohlrabi, and potatoes), and cereals (barley, oats, and wheat). There is a large fishing industry, and Denmark possesses a commercial shipping fleet of considerable size. The leading manufactures include food products, beer, metal goods, chemicals, and electronic equipment.

History. The Danes probably settled Jutland by c.10,000 B.C. They had an important role in the Viking (or Norse) raids on Western Europe and England (9th-11th cent. A.D.). Harold Bluetooth (d. c.985) was the first Christian king of Denmark. His son, Sweyn (reigned c.986–1014), conquered England. From 1018 to 1035 Denmark, England, and Norway were united under King Canute (Knut). Later, Waldemar I (reigned 1157–82) and Waldemar II (reigned 1202–41) were energetic rulers who established Danish hegemony over N Europe. In 1282 Eric V (reigned 1259–86) was forced to submit to the Great Charter, which established annual parliaments and a council of nobles who shared the king's power. Queen Margaret achieved (1397) the Kalmar Union of the Danish, Norwegian, and Swedish crowns; union with Sweden was dissolved in 1523, but Norway remained part of Denmark until 1814.

During the reign of Christian III (1535–59) Lutheranism became the established religion. In the late 16th and early 17th cent. Denmark had a brilliant intellectual and cultural life. Frederick III (reigned 1648–70) and Christian V (reigned 1670–99) were able to make the kingdom an absolute monarchy with the support of peasants and townspeople. The later 18th cent. was marked by important social reforms: serfdom was abolished (1788), and peasant proprietorship was encouraged. By the Treaty of Kiel (1814), Denmark lost Norway to Sweden and Helgoland to England. Schleswig-Holstein, long disputed with Prussia, was yielded in 1864. This loss of about one third of the Danish territory was, however, offset by great economic gains in the second half of the 19th cent. Prosperity was achieved largely by persuading the farmers to specialize in dairy and pork products rather than in grain.

Denmark remained neutral in World War I and recovered North Schleswig after a plebiscite in 1920. In the interwar period and after World War II Denmark adopted much social welfare legislation and a system of progressive taxation. German forces occupied the country (1940–45) and placed King Christian X (reigned 1912–47) under house arrest in 1943. After the war Denmark recovered quickly, became (1945) a charter member of the United Nations, and, breaking a long tradition of neutrality, joined the North Atlantic Treaty Organization in 1949. In 1972 Queen Margaret (Margrethe) II succeeded her father, King Frederick IX.

Government. Denmark is a constitutional monarchy, governed according to the 1953 constitution. Legislative power is vested in the monarch (who is also head of state) in conjunction with the unicameral Folketing (parliament) of 179 elected members. Executive power is exercised by the monarch through his ministers (led by the prime minister), who are responsible to the Folketing and must have the support of the majority of that body.

Denmark Strait, passage, c.300 mi (480 km) long and 180 mi (290 km) wide at the narrowest point, between Greenland and Iceland.

Den·nis (dĕn′ĭs), resort town (1975 est. pop. 9,503), Barnstable co., SE Mass., on Cape Cod NE of Yarmouth; settled 1639, set off from Yarmouth 1793. Cranberries are grown.

Dent (dĕnt), county (1970 pop. 11,457), 756 sq mi (1,958 sq km), SE central Mo., in the Ozarks, drained by the Meramec River; formed 1851; co. seat Salem. It lies in a livestock and farm (corn, wheat, and hay) region and has oak and pine timber, iron and zinc mines, and some manufacturing (clothing, shoes, and chemical products).

Dent Blanche (däɴ' bläɴsh'), peak, 14,318 ft (4,367 m) high, Valais canton, S Switzerland, in the Pennine Alps.

Dent du Midi (däɴ' dü mē-dē'), mountain group in the Alps, Vaud canton, SW Switzerland. It rises to 10,695 ft (3,262 m).

Den·ton (dĕn'tən), county (1970 pop. 75,633), 911 sq mi (2,359.5 sq km), N Texas, watered by the Elm Fork of the Trinity River and the Garza-Little Elm Reservoir; formed 1846; co. seat Denton. It is in a rich agricultural area yielding cotton, wheat, oats, corn, hay, pecans, peanuts, fruit, truck crops, dairy products, poultry, cattle, sheep, and hogs. It has some mining. Clothing, wood and metal products, plastics, concrete, and machinery are manufactured.

Denton. 1. Town (1970 pop. 1561), seat of Caroline co., Eastern Shore, Md., NNW of Salisbury, in a farm region; settled c.1765, inc. 1802. **2.** City (1978 est. pop. 43,500), seat of Denton co., N Texas; inc. 1866. It is the processing, trade, and distribution center of a large agricultural area. It has flour mills, food-processing establishments, and plants making building materials, machine tools, and clothing.

D'En·tre·cas·teaux Islands (däɴ'trə-käs-tō'), volcanic group, SW Pacific, SE of New Guinea, part of Papua New Guinea. The group, with a total land area of c.1,200 sq mi (3,110 sq km), is mountainous and has several extinct volcanoes, hot springs, and geysers. Coconuts and pearl shells are the chief products.

Den·ver (dĕn'vər), city (1978 est. pop. 473,900), alt. 5,280 ft (1,610.4 m), state capital, coextensive with Denver co. (68 sq mi/176.2 sq km), N central Colo., on the South Platte River at the mouth of Cherry Creek; inc. 1861. It is a port of entry and a processing, shipping, and distributing point for an extensive agricultural area. Denver has stockyards and meat-packing plants, railroad shops, fruit and vegetable canneries, feed and flour mills, and many electronics plants. Foremost among its manufactures are rubber goods and luggage. Tourism is also important. The city was made territorial capital in 1867. The rich gold and silver strikes of the late 1870s and the 1880s brought prosperity to the city. The Univ. of Denver is here.

De Pere (də pîr'), city (1978 est. pop. 14,800), Brown co., E central Wis., on the Fox River; inc. 1857; De Pere and West De Pere consolidated 1890. A channel 20 ft (6 m) wide allows port traffic from Green Bay as far as De Pere, the last upstream dock. Wood and paper products, industrial plating, aluminum, and metal parts are among the manufactures. A mission, founded here in 1671, was burned, rebuilt (1685), and used until 1717. De Pere grew in the 19th cent. as a lumber town, port, and commercial center.

De·pew (də-pyōō'), village (1978 est. pop. 27,200), Erie co., W central N.Y., a suburb of Buffalo; founded 1892, inc. 1894. Printing is a major industry. Depew's diverse manufactures include transportation equipment, chemicals, and prefabricated concrete.

De Queen (dĭ kwēn'), city (1974 est. pop. 4,083), seat of Sevier co., SW Ark., near the Okla. border S of Fort Smith. It is the shipping center of a farm and timber area.

De·ra Gha·zi Khan (dā'rə gä'zē kän'), town (1972 est. pop. 71,429), E central Pakistan, on the Indus canal. It is an administrative center in a wheat and millet area. The town was founded in the late 15th cent.

Dera Is·mail Khan (ĭs'mīl), town (1972 est. pop. 57,500), N central Pakistan, W of the Indus River. It is known for its lacquered woodwork, glass and ivory ware, mats, and sarongs. The old town, founded in 1469, was washed away by the Indus River; the new town was laid out in 1823. Oil has been found in the district.

Der·bent (dyĭr-byĕnt'), city (1976 est. pop. 66,000), SE European USSR, in Dagestan, on the Caspian Sea. It stands on a narrow strip of land that forms a natural pass (the Caspian or Iron Gates) between the Caucasian foothills and the sea. Orchards and vineyards are cultivated, and fishing is an important occupation. Industries include food processing and the production of textiles and bricks. There are oil and natural gas deposits in the area. Derbent was founded (5th or 6th cent. A.D.) by the Persians as a strategic fortress at the Iron Gates. There are remains of the Caucasian Wall (also called Alexander's Wall), built by the Persians in the 6th cent. as a bulwark against northern invaders. The Arabs, who took Derbent in 728, made it a commercial and cultural center. Passing (1220) to the Mongols and later recovered by Persia, Derbent was briefly held (1722) by Peter I of Russia and was annexed to Russia in 1806.

Der·by (där'bē, dûr'-), county borough (1976 est. pop. 213,700), administrative center of Derbyshire, central England, on the Derwent River. Derby is a rail center with large engineering works. Manufactures include airplane engines, pottery, synthetic textiles, machinery, and chemicals. Derby was a Roman settlement and (in the 9th cent.) one of the Five Boroughs of the Danes. England's first silk mill was built here in 1718. Noteworthy are the Cathedral of All Saints, with its Perpendicular tower (1509–27), the Roman Catholic Church of St. Mary, the arboretum, and a grammar school founded in 1160.

Der·by (dûr'bē). **1.** City (1975 est. pop. 11,983), New Haven co., SW Conn., at the confluence of the Naugatuck and Housatonic rivers, opposite Shelton; founded 1642 as a trading post, inc. as a city 1893. Its copper industry and pin manufactures date from the 1830s. **2.** City (1975 est. pop. 8,269), Sedgwick co., S central Kansas, SSE of Wichita; settled 1876 as El Paso, name changed 1957. There are aircraft plants in the area.

Der·by·shire (där'bē-shîr', -shər, dûr'-) or **Der·by** (där'bē, dûr'-), nonmetropolitan county (1976 est. pop. 887,600), 997 sq mi (2,582.2 sq km), central England; administrative center Derby. The terrain of the county is flat in the south, rising in the north to more than 2,000 ft (610 m) in the Peak district. Much of the county is used for agriculture. In the eastern part of the county there are rich coal deposits. In the Anglo-Saxon period Derbyshire was part of the kingdom of Mercia. It has pre-Roman, Roman, and Norman remains.

Derg, Lough (dûrg'). **1.** Expansion of the Shannon River, 23 mi (37 km) long and 1 to 5 mi (1.6–8 km) wide, W central Republic of Ireland. On the lake is the republic's first (1927) major hydroelectric power plant. **2.** Lake, c.20 sq mi (50 sq km), Co. Donegal, NW Republic of Ireland. Station Island, traditional scene of St. Patrick's Purgatory, is a famous place of pilgrimage.

De Rid·der (də rĭd'ər), city (1970 pop. 8,030), seat of Beauregard parish, SW La., NNW of Lake Charles; inc. 1907. It is the industrial and trade center of a timber, farm, and sheep-raising district.

Der·na (dĕr'nə): *see* Darnah, Libya.

Der·ry (dĕr'ē): *see* Londonderry, Northern Ireland.

Derry, town (1975 est. pop. 15,259), Rockingham co., SE N.H.; set off from Londonderry 1827. Shoes and wood products are made. Robert Frost farmed in Derry and taught school here.

Der·went (dûr'wənt), river, c.60 mi (95 km) long, rising in the Pennines, Derbyshire, central England, and flowing SE past Derby to the River Trent.

Des·a·gua·de·ro (dā'sä-gwä-*th*ā'rō), river, c.200 mi (320 km) long, flowing SE from Lake Titicaca to Lake Poopó, W Bolivia. It is used for irrigation in its northern course.

Des Arc (dĕz ärk'), town (1975 est. pop. 2,121), a seat of Prairie co., E central Ark., on the White River ENE of Little Rock, in a cotton and rice area; settled by the French in the 1820s, inc. 1854.

Des·chutes (dā-shōōt'), county (1970 pop. 30,442), 3,031 sq mi (7,850.3 sq km), central Oregon, drained by the Deschutes River; formed 1916; co. seat Bend. It has agriculture (potatoes, alfalfa, and cattle), lumber, and diatomite, manganese, lead, and zinc mines. Wood products, clothing, and furniture are manufactured. The county attracts hunters and fishermen and includes parts of Willamette and Deschutes national forests.

Deschutes, river, c.240 mi (385 km) long, rising in several lakes in the Cascade Range, W central Oregon, and flowing NE to the Columbia River. The U.S. Bureau of Reclamation has developed the stream and its main tributary, the Crooked River, for power and irrigation.

De·se·a·do (dā'sā-ä'dō), river, 380 mi (611.4 km) long, S Argentina, rising in the Andes S of Lake Buenos Aires and flowing generally E to the Atlantic Ocean at Puerto Deseado.

des·ert (dĕz'ərt), arid region, usually partly covered by sand, having scanty vegetation or sometimes almost none, and capable of supporting only a limited and specially adapted animal population. The largest desert regions of the world lie between 20° and 30° north and south of the equator, either where mountains intercept the paths of the trade winds or where atmospheric high-pressure areas cause descending air currents and lack of precipitation. Other factors contributing to the formation of deserts include the amount of sunshine, rate of evaporation of water, and range of temperature. An area having an annual rainfall of 10 in. (25.4 cm) or less is considered to be a desert. Europe is the only continent without deserts.

De·sha (də-shā'), county (1970 pop. 18,761), 736 sq mi (1,906.2 sq km), SE Ark., bounded on the E by the Mississippi River and drained by the Arkansas and White rivers; formed 1838; co. seat

Arkansas City. It has agriculture (cotton, corn, hay, rice, and livestock) and stands of timber. Clothing and leather products are manufactured.

De·shi·ma (dā'shĭ-mä'), artificial island, c.40 acres (16 hectares), Nagasaki prefecture, W Kyushu, Japan, in Nagasaki harbor. Dutch traders were restricted (1641–1858) to this island after Japan closed (17th cent.) its borders to foreign trade.

De Smet (dĭ smĕt'), city (1970 pop. 1,336), seat of Kingsbury co., E central S.Dak., E of Huron; settled 1878, inc. 1888. In a dairy and livestock region, it raises grain.

Des Moines (də moin', moinz'), county (1970 pop. 46,982), 409 sq mi (1,059.3 sq km), SE Iowa, in a prairie agricultural area bounded on the E by the Mississippi, here forming the Ill. border, and on the S by the Skunk River; formed 1834; co. seat Burlington. Hogs, cattle, corn, and soybeans are produced.

Des Moines. 1. City (1978 est. pop. 191,000), state capital and seat of Polk co., S central Iowa, at the junction of the Des Moines and Raccoon rivers; inc. as Fort Des Moines in 1851, chartered as Des Moines in 1857. It is an important industrial and transportation center in the heart of the Corn Belt. Printing and publishing, agricultural processing, and the manufacture of machinery are among its many industries. The city is also the home office of numerous insurance companies. Settled by homesteaders, Des Moines became the capital of Iowa in 1857. Places of interest include the capitol (1871–84), the Des Moines Art Center, the Center of Science and Industry, the State Historical, Memorial, and Art Building, and c.1,700 acres (690 hectares) of parks. The city suffered a severe flood in 1954; dams and reservoirs on the Des Moines River now provide flood control. **2.** Residential city (1975 est. pop. 5,496), King co., W central Wash., on Puget Sound S of Seattle; settled in the 1870s, inc. 1959. Saltwater State Park is nearby.

Des Moines, river, 535 mi (860.8 km) long, rising in SW Minn. and flowing SE across Iowa to the Mississippi River at Keokuk, SE Iowa. Flowing through rich farmland, the river floods in the spring and is nearly dry in late September; dams now regulate its flow.

Des·na (dĕs-nä', dyĭs-), river, c.740 mi (1,190 km) long, W European USSR, partly in the Ukraine, rising SE of Smolensk and flowing S and SW to join the Dnepr above Kiev.

De So·to (də sō'tō). **1.** County (1970 pop. 13,060), 648 sq mi (1,678.3 sq km), S central Fla., drained by the Peace River; formed 1887; co. seat Arcadia. In an area with rolling terrain, many lakes, and some swamps, it is mainly agricultural, producing citrus fruit, vegetables, corn, cattle, and poultry. **2.** Parish (1970 pop. 22,764), 899 sq mi (2,328.4 sq km), NW La., bounded on the W by the Texas border, on the SW by the Sabine River, on the NE by Bayou Pierre and several lakes; formed 1843; parish seat Mansfield. Its agriculture includes cotton, corn, hay, and peanuts. It also has oil and natural-gas wells, timber, and some industry. **3.** County (1970 pop. 35,885), 478 sq mi (1,238 sq km), extreme NW Miss., bounded on the W by the Mississippi River, here forming the Ark. border, on the N by the Tenn. border, and drained by the Coldwater River and Arkabutla Reservoir; formed 1836; co. seat Hernando. In an agricultural area yielding cotton, corn, livestock, and dairy products, it has diversified manufactures, including furniture, chemicals, plastics, concrete, and iron and steel.

De Soto National Memorial: *see* National Parks and Monuments Table.

Des Plaines (dĕs plānz'), city (1978 est. pop. 54,700), Cook co., NE Ill., a suburb of Chicago on the Des Plaines River; inc. 1925. Its manufactures include electrical and electronic equipment, chemicals, cylinders, oil products, and cosmetics. It was founded in the 1830s as the town of Rand; the name was changed in 1869; Riverview was annexed in 1925.

Des·sau (dĕs'ou), city (1974 est. pop. 100,820), Halle dist., central East Germany, at the confluence of the Elbe and Mulde rivers. It is an industrial city, river port, and rail and road transport center. Manufactures include machinery, chemicals, paper, and processed food. Dessau was first known as a German settlement in 1213. From 1925 to 1932 it was the seat of the Bauhaus art school, headed by Walter Gropius. The city was severely damaged in World War II.

Des·sye (dā'syā), town (1974 est. pop. 54,910), capital of Wallo prov., central Ethiopia, in the Great Rift Valley. It is an administrative, military, and commercial center.

Det·mold (dĕt'mōlt'), city (1974 est. pop. 65,677), North Rhine-Westphalia, N central West Germany. Once the capital of Lippe, it is now a furniture-manufacturing center and summer resort.

De·troit (dĭ-troit', dē-), city (1978 est. pop. 1,245,000), seat of Wayne co., SE Mich., on the Detroit River between lakes St. Clair and Erie; inc. as a city 1815. Detroit is a port of entry and a major Great Lakes shipping and rail center. Its early carriage industry helped Henry Ford and others to make Detroit the "automobile capital of the world." Detroit's other industries include steel mills, drug manufactures, and food-processing plants. Detroit leads the nation in the production of gray-iron foundry products, metal stampings, and machine tools. It is high among national producers of tires, paint, wire goods, and industrial inorganic chemicals. Extensive salt mines lie under the southwestern section of the city.

A French fort and fur-trading settlement founded here in 1701 by Antoine de la Mothe Cadillac and called *Ville d'etroit,* city of the strait, were captured by the British in 1760. Three years later the British withstood a long siege in Pontiac's Rebellion. American control, resulting from Jay's Treaty, was established in 1796. Detroit was first the territorial and then the state capital from 1805 to 1847. Fire in 1805 nearly destroyed all of the several hundred buildings in the town, but the settlement was rebuilt from a design by Pierre C. L'Enfant. Detroit was surrendered in 1812 by William Hull to British forces under Isaac Brock, but was recovered by William Henry Harrison in 1813. With the development of land and water transportation, the city grew rapidly during the 1830s. It assumed great importance after the mid-19th cent. as a shipping, shipbuilding, and manufacturing center. In July, 1967, race riots in Detroit caused property damage estimated at $150 million.

Detroit Lakes, city (1970 pop. 5,797), seat of Becker co., W central Minn., E of Moorhead; settled c.1858, inc. as a village 1880, as a city 1900. It is a summer resort and a farm trade center.

Det·ti·foss (dĕt' ĭ-fôs), waterfall, in the Jökulsá á Fjöllum River, NE Iceland. Iceland's most impressive fall, it drops 144 ft (44 m) into a long canyon.

Det·ting·en (dĕt'ĭng-ən), village (1967 est. pop. 3,600), NW Bavaria, West Germany, on the Main River. English, Austrian, and Hanoverian troops commanded by George II of England defeated (1743) the French in the War of the Austrian Succession.

Deu·el (do͞o'əl, dyo͞o'-). **1.** County (1970 pop. 2,717), 435 sq mi (1,126.7 sq km), W Nebr., in an agricultural area bounded on the S by the Colo. border and drained by Lodgepole Creek and the South Platte River; formed 1888; co. seat Chappell. Its agriculture includes grain, livestock, dairy produce, and sugar beets. **2.** County (1970 pop. 5,686), 636 sq mi (1,647.2 sq km), E S.Dak., in a farming area on the Minn. border; formed 1862; co. seat Clear Lake. In a region watered by numerous lakes and creeks, it produces livestock, grain, potatoes, and dairy products.

Deux-Sè·vres (dœ-sĕv'rə), department (1975 pop. 335,829), 2,318 sq mi (6,003.6 sq km), W France, largely in Poitou; capital Niort.

De Valls Bluff (dĭ vălz'), town (1970 pop. 622), a seat of Prairie co., E central Ark., on the White River E of Little Rock. It is a shipping center for fish, rice, cotton, and hay.

De·ven·ter (dā'vən-tər), city (1977 est. pop. 65,140), Overijssel prov., E central Netherlands, on the IJssel River. It is an industrial center with machine shops, foundries, textile plants, and carpet manufactures. A member of the Hanseatic League in the Middle Ages, it was a prosperous commercial city and a center of piety and learning; Thomas à Kempis and Erasmus of Rotterdam studied here. Deventer has retained many medieval and Renaissance structures, including the Groote Kerk (church, built 8th cent., and later rebuilt), the Mariakerk (15th cent.), the weighhouse (16th cent.), and the town hall (17th cent.).

Dev·ils (dĕv'əlz), river, 100 mi (160 km) long, SW Texas, flowing S into the Rio Grande NW of Del Rio. Power dams along its length create two lakes, Walk and Devils.

Devils Island, the smallest and southernmost of the Îles du Salut, in the Caribbean Sea off French Guiana. A penal colony founded in 1852, it was used primarily for political prisoners, among them Alfred Dreyfus. Although conditions were probably not as sordid as in other prison camps in French Guiana, the island's name became synonymous with the horrors of the system. The penal colonies were phased out between 1938 and 1951.

Devils Lake, city (1974 est. pop. 7,354), seat of Ramsey co., NE N.Dak., near Devils Lake W of Grand Forks; settled 1882, inc. 1887. It is a trade and processing center of a wheat, livestock, and dairy region. Nearby lakes attract tourists. Lake Region Community College is here, and Fort Totten Indian Reservation is nearby.

Devils Post·pile National Monument (pōst'pīl'): *see* National Parks and Monuments Table.

Devils Tower National Monument, 1,347 acres (545.5 hectares), overlooking the Belle Fourche River, NE Wyo.; first designated U.S. national monument (est. 1906). Devils Tower, 865 ft (263.8 m) high and narrowing in width from 1,000 ft (305 m) at its base to 250 ft (76.3 m) at its summit, is a cluster of rock columns formed by the cooling and crystallization of molten matter. Generally acknowledged by geologists to be a remnant of an ancient lava intrusion in sedimentary strata, the exact form of its origin and its relation to the Black Hills is debated. Through the ages the surrounding sedimentary material has been removed by wind and water action, especially that of the Belle Fourche River, leaving the more resistant igneous-rock tower to dominate the skyline. Devils Tower has long served as a landmark for travelers and explorers.

Dev·on (dĕv'ən) or **Dev·on·shire** (dĕv'ən-shîr', -shər), county (1976 est. pop. 942,100), 2,591 sq mi (6,710.7 sq km), SW England; administrative center Exeter. It is a land of rolling hills, dominated by Dartmoor and Exmoor, upland areas of forests and rugged stone. The county was occupied in Paleolithic times; numerous habitation sites and ceremonial centers have been excavated. Devonshire was incorporated into Wessex early in the 8th cent. In Elizabethan times the county reached its greatest maritime importance.

Devon Island, c.20,900 sq mi (54,130 sq km), E Franklin dist., Northwest Territories, Canada, between Baffin and Ellesmere islands.

Dew·ey (dōō'ē, dyōō'ē). **1.** County (1970 pop. 5,656), 1,018 sq mi (2,636.6 sq km), W Okla., intersected by the Canadian and North Canadian rivers; formed 1907; co. seat Taloga. In an agricultural area yielding wheat, broomcorn, cotton, oats, and livestock, it has oil and natural-gas wells, bentonite and fuller's-earth mines, and some manufacturing. **2.** County (1970 pop. 5,170), 2,351 sq mi (6,089.1 sq km), N central S.Dak., in an agricultural and cattle-raising area drained by the Moreau River and bounded on the E by the Missouri River; formed 1883, annexed Armstrong co. 1954; co. seat Timber Lake. It processes dairy produce, flax, and wheat. Lignite mines are here. It is entirely within the Cheyenne River Indian Reservation.

De Witt (də wĭt'). **1.** County (1970 pop. 16,975) 399 sq mi (1,033.4 sq km), central Ill., in a grain-growing area drained by Salt Creek; formed 1839; co. seat Clinton. Corn, wheat, soybeans, oats, livestock, and poultry are produced. Its manufactures include dairy products, clothing, and patent medicines. **2.** County (1970 pop. 18,660), 910 sq mi (2,356.9 sq km), S Texas, drained by the Guadalupe River; formed 1846; co. seat Cuero. It lies in a poultry and egg-producing area, with cattle ranching, hog and sheep raising, and farming (corn, maize, cotton, peanuts, grain sorghums, fruit, and truck crops). It also has oil and natural-gas wells and food-processing plants.

De Witt. 1. City (1970 pop. 3,728), a seat of Arkansas co., E central Ark., E of Pine Bluff, in a rice and cotton area. **2.** Uninc. town (1970 pop. 10,032), Onondaga co., central N.Y., a suburb of Syracuse.

Dews·bur·y (dōōz'bə-rē), county borough (1973 est. pop. 50,560), West Yorkshire, N central England, on the Calder River. It is a commercial center for heavy woolen textiles. Other industries are coal mining and the manufacture of chemicals.

Dez·ful (dĕz-fōōl'), city (1966 pop. 84,499), Khuzestan prov., W Iran, on the Dez River, near the site of ancient Susa. It is the trade center for an irrigated farm region. Petroleum is produced nearby.

Dezh·nev, Cape (dĕzh'nəf, dyăsh'nyəf), or **East Cape,** northeasternmost point of Asia, Far Eastern USSR, on Chukchi Peninsula and on the Bering Strait. It is named for the Russian navigator who discovered it in 1648.

Dhar (där), town (1971 pop. 36,164), Madhya Pradesh state, central India. It is a market for cotton, grains, and oilseed. Dhar was the capital of the kingdom of Malwa and a center of Hindu learning from the 9th to 14th cent.

Dhar·war (där-wär'): *see* Hubli-Dharwar, India.

Dhau·la·gi·ri (dou'lə-gîr'ē), peak, 26,810 ft (8,177 m) high, Nepal, in the Himalayas. It was first scaled in 1960.

Dhu·li·a (dōō'lē-ə), city (1971 pop. 137,089), Maharashtra state, W central India, on the Panjhra River. Dhulia is a district administrative center. Cotton and woolen goods are manufactured.

Di·a·blo, Mount (dē-ä'blō, dī-ăb'lō), 3,849 ft (1,174 m) high, N Diablo Range, W Calif., E of Oakland. The peak is isolated in a state park.

Di·a·man·ti·na (dī'ə-mən-tē'nə), river, 560 mi (901 km) long, SW Queensland, Australia, flowing generally SW. It is a tributary of the Warburton River.

Dia·mond Head (dī'mənd, dī'ə-), peak, 761 ft (232.1 m) high, along the rim of an extinct volcano, SE Oahu Island, Hawaii. A prominent point in the Honolulu skyline, Diamond Head was designated a national natural landmark to protect its slopes from the commercial development along world-famous Waikiki Beach. The crater was the site of an ancient Hawaiian burial ground.

Di·bon (dī'bŏn), ancient city, E of the Dead Sea, now a ruin called Dhiban. The Moabite Stone was found here, and important remains from the Moabite period have been excavated.

Di·bru·garh (dĭb'rōō-gär'), town (1971 pop. 80,344), Assam state, NE India, at the confluence of the Brahmaputra and Bibru rivers. The town is surrounded by hills and is often threatened by monsoon floods. There are four airfields, which were used by the British against Japanese forces in Burma during World War II. Dibrugarh is surrounded by tea plantations and has tea and plywood factories, oilseed mills, and railroad shops.

Dick·ens (dĭk'ənz), county (1970 pop. 3,737), 930 sq mi (2,408.7 sq km), NW Texas, in the rolling plains just below Cap Rock escarpment of the Llano Estacado; formed 1876; co. seat Dickens. It lies in a cattle-ranching area, with agriculture (cotton, grain sorghums, wheat, alfalfa, some fruit and truck crops, dairy products, hogs, and poultry) and mining (gypsum, caliche, and gravel).

Dickens, city (1970 pop. 295), seat of Dickens co., NW Texas, on the plain below Cap Rock escarpment E of Lubbock. It is a trading point for an agricultural and cattle-ranching area.

Dick·en·son (dĭk'ən-sən), county (1970 pop. 16,077), 335 sq mi (867.7 sq km), SW Va., in the Cumberlands, bounded on the N and NW by the Ky. border and drained by Russell Fork, a headstream of the Big Sandy River; formed 1880; co. seat Clintwood. It has bituminous-coal mines, agriculture (grain, potatoes, fruit, tobacco, and livestock), and some timber. Clothing is manufactured.

Dick·ey (dĭk'ē), county (1970 pop. 6,976), 1,144 sq mi (2,963 sq km), SE N.Dak., in an agricultural area drained by the James River and bordering in the S on S.Dak.; formed 1881; co. seat Ellendale. Grain and dairy products are produced. It has some light industry.

Dick·in·son (dĭk'ən-sən). **1.** County (1970 pop. 12,565), 382 sq mi (989.4 sq km), NW Iowa, in a glaciated prairie agricultural area on the Minn. border and drained by the Little Sioux River; formed 1851; co. seat Spirit Lake. The chief lake region in the state, it has many state parks and resorts. Some lumbering is done. Cattle, hogs, poultry, corn, oats, and hay are produced. **2.** County (1970 pop. 19,993), 855 sq mi (2,214.5 sq km), central Kansas, in a gently rolling area drained by the Smoky Hill River; formed 1857; co. seat Abilene. Grain and livestock are raised. **3.** County (1970 pop. 23,753), 757 sq mi (1,960.6 sq km), SW Upper Peninsula, Mich., bounded on the SW by the Wis. border and drained by the Menominee and Ford rivers; organized 1891; co. seat Iron Mountain. In a dairying and agricultural area yielding livestock, poultry, potatoes, truck crops, hay, and fruit, it also has manufacturing. Some lumbering and iron mining are done. There are hunting, fishing, and winter resorts here.

Dickinson. 1. City (1975 est. pop. 12,492), seat of Stark co., SW N.Dak., on the Heart River; inc. 1919. It is a processing and shipping center for a livestock, dairy, and wheat region. A briquette-producing plant utilizes the area's extensive lignite reserves. **2.** Uninc. town (1970 pop. 10,776), Galveston co., S Texas, on a navigable bayou that flows into Galveston Bay.

Dick·son (dĭk'sən), county (1970 pop. 21,977), 485 sq mi (1,256.2 sq km), NW central Tenn., bounded on the NE by the Cumberland River and drained by the Harpeth River; formed 1803; co. seat Charlotte. Livestock, dairy products, and timber are important. It also has iron-ore deposits and oil wells. Clothing, boats, wood and metal products, and motor vehicle parts are manufactured.

Dié·go-Sua·rez (dyā'gō-swä'rĕs) or **An·tsi·ra·ne** (än'tsə-rä'nə), town (1970 est. pop. 38,600), N Malagasy Republic, at the tip of

Madagascar, on Diégo-Suarez Bay. The area was ceded to France in 1885, at which time the town became the capital of the French colony; France has maintained a naval base here since 1901.

Dien·bien·phu (dyĕn'byĕn'fo͞o'), former French military base, NW Vietnam, near the Laos border. It was the scene in 1954 of the last battle between the French and the Viet Minh forces of Ho Chi Minh in Indochina. The French occupied the base by parachute drop in Nov., 1953; this move prevented a Viet Minh thrust into Laos and provided support for indigenous forces in that area. Although the base could be supplied only by air, the French military felt its position was tenable. By Mar., 1954, some 49,500 Viet Minh troops had encircled Dienbienphu, where 13,000 French soldiers were firmly entrenched. The first Viet Minh assault came on Mar. 13, and by the end of Apr., despite massive French air bombardment, the French defense area had been reduced to 2 sq mi (5.2 sq km). Desperate pleas for U.S. intervention were unsuccessful, and on May 7, after a 56-day siege, the French positions fell. This defeat signaled the end of French power in Indochina.

Di·eppe (dē-ĕp'), city (1975 pop. 25,822), Seine-Maritime dept., N France, in Normandy, at the mouth of the Arques River on the English Channel. It is a fishing and commercial port, a manufacturing center, and a beach resort. Channel steamers sail from Dieppe to Newhaven, England. Dieppe was frequently involved in the wars between England and France. In the late 17th cent. it suffered severely from the Dragonnades of Louis XIV and an Anglo-Dutch naval bombardment (1694). In World War II Dieppe was the object of a costly commando attack (Aug. 19, 1942) to test the strength of the German defenses. The Allied forces, mostly Canadians, lost two thirds of their men in casualties.

Die·ti·kon (dĕ'tē-kôn), town (1970 pop. 22,705), Zürich canton, N Switzerland, on the Limmat River. It is a suburb of Zürich.

Digh·ton (dīt'n), city (1970 pop. 1,540), seat of Lane co., W central Kansas, W of Great Bend, in a farm and grain area; inc. 1887.

Digne (dēn'yə), city (1975 pop. 13,645), capital of Alpes-de-Haute, Provence dept., SE France, in Provence. Points of interest include the Notre-Dame-de-Bourg Cathedral (13th–14th cent.) and a museum housing Gallo-Roman artifacts.

Di·jon (dē-zhôɴ'), city (1975 pop. 151,705), capital of Côte-d'Or dept., E France. The old capital of Burgundy, it is a transportation hub and industrial center with food, metal-products, and electronics industries. Its mustard and cassis (black currant liqueur) are famous. Dijon is also an important shipping point for Burgundy wine. Founded in ancient times, Dijon began to flourish when the rulers of Burgundy made it their residence (11th cent.). Even after Burgundy was reunited with France (late 15th cent.), Dijon remained a thriving cultural center. Dijon Univ. was founded in 1722. Among the city's valued artworks are the funeral statues of the dukes of Burgundy by Claus Sluter and his disciples, housed in a museum in the town hall, originally the ducal palace (12th cent.; largely rebuilt 17th–18th cent.). Other remarkable buildings include the Cathedral of St. Bénigne (13th–14th cent.), the Church of Notre Dame (13th cent., in Burgundian Gothic), St. Michael's Church (Renaissance), the Hôtel Aubriot (14th cent.; now containing a museum), and the palace of justice (15th–16th cent.).

Dil·lon (dĭl'ən), county (1970 pop. 28,838), 407 sq mi (1,054.1 sq km), NE S.C., bounded on the NE by the N.C. border and drained by the Little Pee Dee River; formed 1910; co. seat Dillon. It lies in an agricultural area that produces tobacco, cotton, corn, wheat, and paprika. It also has a timber industry, food-processing plants, and textile mills. Clothing and furniture are made.

Dillon. 1. City (1970 pop. 4,548), seat of Beaverhead co., SW Mont., on the Beaverhead River SSW of Butte; founded 1879, inc. 1885. It ships large quantities of wood and seed potatoes. Western Montana College of Education is here. **2.** Town (1970 pop. 6,391), seat of Dillon co., NE S.C., near the N.C. border NE of Florence, in a timber area; founded c.1886, inc. 1888. It has flour and textile mills.

Di·mi·trov·grad (dĭ-mē'trôv-grät'), city (1969 est. pop. 44,200), S Bulgaria, on the Maritsa River. The city has one of Bulgaria's largest cement works, as well as several thermoelectric power stations that provide power to coal-mining areas nearby.

Dim·mit (dĭm' ĭt), county (1970 pop. 9,039), 1,341 sq mi (3,473.2 sq km), S Texas, on the plains of the Rio Grande and drained by the Nueces River; formed 1858; co. seat Carrizo Springs. It lies in a winter truck-farming area irrigated by artesian wells and also produces cattle, dairy products, hogs, and poultry. It has large oil and gas fields and some manufacturing (clothing and electronic equipment). There are hunting and recreational areas in the county.

Dim·mitt (dĭm'ĭt), city (1970 pop. 4,327), seat of Castro co., NW Texas, on the Llano Estacado NW of Plainview. It is the center of a farm area producing vegetables, wheat, corn, and cattle.

Di·mo·na (dĭ-mō'nə), town (1975 est. pop. 27,400), S Israel, in the Negev Desert. It is the seat of the Negev Nuclear Research Center. Mining and the production of textiles, chemicals, and processed minerals are also important. The town was founded in 1955.

Di·naj·pur (dĭ-näj'po͝or'), town (1974 pop. 61,866), N Bangladesh, on the Punarbhaba River. It is a road junction and the administrative center for a district where rice and sugar cane are grown.

Di·nan (dē-näɴ'), town (1968 pop. 13,137), Côtes-du-Nord dept., NW France, in Brittany, on a bluff above the Rance River. It is an industrial town producing textiles and machine tools. Beautifully situated, with trees and gardens, it has medieval walls and towers, a 14th cent. chateau, and the remarkable Church of Saint-Sauveur (12th–16th cent.).

Di·nant (dē-näɴ'), town (1970 pop. 9,747), Namur prov., S Belgium, on the Meuse River. It is a commercial and industrial center, a tourist resort, and the gateway to the nearby limestone Han Grotto. Fortified since Merovingian times, Dinant was noted in the Middle Ages for its metalware. The town was sacked by Charles the Bold in 1466 in the revolt of Liège.

Di·nar·ic Alps (dĭ-nâr'ĭk), mountain system, extending c.400 mi (645 km) along the E coast of the Adriatic Sea from the Isonzo River, NE Italy, through Yugoslavia to the Drin River, N Albania. The highest peak is Jezerce (8,833 ft/2,694.1 m) in northern Albania. The partially submerged western part of the system forms the numerous islands and harbors along the Yugoslav coast. The rugged mountains, composed of limestone and dolomite, are a barrier to travel from the coast to the interior; there are no natural passes. Sinkholes and caverns dominate the landscape. The region is sparsely populated and forestry and mining are the chief economic activities.

Din·di·gul (dĭn'dĭ-gŭl), town (1971 pop. 128,429), Tamil Nadu state, S India. It is a trade center for hides, food grains, coffee, and spices. Cigars are the principal manufacture.

Di·no·saur National Monument (dī'nə-sôr'): *see* National Parks and Monuments Table.

Din·wid·die (dĭn-wĭd'ē, dĭn'wĭd'ē), county (1970 pop. 25,046), 507 sq mi (1,313.1 sq km), SE Va., bounded on the N by the Appomattox River, on the SW by the Nottoway River; formed 1752; co. seat Dinwiddie. In an agricultural area yielding tobacco, peanuts, hogs, cotton, hay, and truck crops, it has sawmills, granite quarries, and some manufacturing (clothing, furniture, and paper products).

Dinwiddie, village (1970 pop. 250), seat of Dinwiddie co., SE Va., SW of Petersburg. The Battle of Five Forks was fought near here (Mar.–Apr., 1865).

Di·o·mede Islands (dī'ə-mēd'), pair of rocky islands in Bering Strait between Alaska and Siberia. The boundary between the United States and the USSR as well as the International Date Line pass between them. They were discovered (1728) by the Danish explorer Vitus Bering.

Diour·bel (dyo͞or-bĕl'), town (1967 est. pop. 36,000), W Senegal, on the railroad from Dakar to the Niger River in Mali. The market for a peanut-growing region, it produces peanut oil as well as beverages and perfume. Diourbel has a beautiful mosque.

Di·re·da·wa or **Di·re Da·wa** (both: dē'rə-dä'wä), city (1974 est. pop. 72,860), Harar prov., E Ethiopia. It is a commercial and industrial center. Manufactures include processed meat, vegetable oil, textiles, and cement. Diredawa was founded in 1902 when the railroad from Djibouti reached the area.

Dirk Har·tog (dûrk här'tôg), island, 239 sq mi (619 sq km), Australia, in the Indian Ocean off the W coast of Western Australia. It has sandstone cliffs and a sheep run.

Dis·ap·point·ment, Cape (dĭs'ə-point'mənt), projection into the Pacific Ocean, SW Wash., on the N side of the mouth of the Columbia River. It was named in 1788 by English Capt. John Meares, who rounded it when searching for the fabled River of the West and was "disappointed" because he could not enter the river.

Dis·ko (dĭs'kō), island, 3,312 sq mi (8,578.1 sq km), in the Davis Strait off W Greenland. It is mountainous (rising to 6,296 ft/1,920.3 m) and partly glaciated. Telluric iron and lignite have been found here. In Disko Bay to the south are some of the world's largest shrimp beds. Disko was first reached (c.985) by Eric the Red.

Dis·mal Swamp (dĭz'məl), SE Va. and NE N.C. With dense forests and tangled undergrowth, it is a favorite site for sportsmen and naturalists. Thought to have once covered c.2,200 sq mi (5,700 sq km) the swamp has been reduced by drainage to less than 600 sq mi (1,554 sq km). The bottom is composed of organic material deposited by fallen trees and other vegetation. Visited by American colonist William Byrd in 1728, Dismal Swamp was surveyed in 1763 by George Washington, who was a member of a company organized to drain it.

District of Co·lum·bi·a (kə-lŭm'bē-ə), Federal district (1978 est. pop. 689,000), c.70 sq mi (180 sq km), on the E bank of the Potomac River, coextensive with the city of Washington, D.C. (the capital of the United States). The District was established by congressional acts of 1790 and 1791 and selected by George Washington. It was originally a 10-mi (16.1-km) square (100 sq mi/259 sq km), with Md. and Va. granting land on each side of the river, including the town of Georgetown and the county of Alexandria. Alexandria co. was returned to Va. in 1846. The city continued to grow on the east bank of the river, and in 1878, when Georgetown became a part of Washington, the city of Washington and the District of Columbia became one and the same.

Di·u (dē'o͞o): *see* Goa, Daman, and Diu.

Di·vide (dĭ-vīd'), county (1970 pop. 4,564), 1,303 sq mi (3,374.8 sq km), extreme NW N.Dak., in a hilly area watered by small creeks and lakes and bounded on the W by Mont., on the N by Sask., Canada; formed 1910; co. seat Crosby. In a lignite-mining area, its agriculture includes grain, livestock, and dairy products.

Dix·ie (dĭk'sē), county (1970 pop. 5,480), 692 sq mi (1,792.3 sq km), N Fla., bounded on the S and W by the Gulf of Mexico, on the E by the Suwannee River; formed 1921; co. seat Cross City. In a flatwoods area, with swamps in the central part and small lakes in the northeast, it is mainly agricultural (corn, peanuts, and cattle). Fishing and lumbering are also done.

Dix·on (dĭk'sən), county (1970 pop. 7,453), 480 sq mi (1,243.2 sq km), NE Nebr., in an agricultural region bounded on the N by the Missouri River and the S.Dak. border and watered by Logan Creek; formed 1888; co. seat Ponca. Livestock and grain are raised.

Dixon. 1. City (1978 est. pop. 15,000), seat of Lee co., N Ill., on the Rock River; founded 1830, inc. 1857. Electronic equipment, cement, and paper products are manufactured. **2.** City (1970 pop. 572), seat of Webster co., W Ky., SSW of Henderson, in an agricultural, timber, and coal-mining region.

Di·ya·la (dĭ-yă'lă, -yăl'ə), river, c.275 mi (440 km) long, rising as the Sirvan River in NW Iran and flowing SW through the Zagros Mts. into E Iraq, where it enters the Tigris River S of Baghdad. The Diyala is unnavigable, but its valley is an important trade route between Iran and Iraq.

Di·yar·ba·kır (dē-yär'bä-kĭr'), city (1975 pop. 169,746), capital of Diyarbakır prov., SE Turkey, on the Tigris River. It is the trade center for a region producing grains, melons, cotton, and copper ore. Manufactures of the city include flour, wine, textiles, and machinery. A Roman colony from A.D. 230, the city was taken (mid-4th cent.) by Shapur II of Persia. It was conquered by the Arabs in 638 and later was held by the Seljuk Turks and Persians. The Ottoman Turks captured Diyarbakır in 1515. The city retains the magnificent black basalt fortification walls mainly constructed by Constantine I in the 4th cent.

Dja·ja·pu·ra (jä'yə-po͝or'ə), formerly **Su·kar·na·pu·ra** (so͞o-kär'nə-po͝or'ə), town, capital of Irian Barat (Indonesian New Guinea), Indonesia. A regional trade center and seaport, it is on Humboldt Bay (an inlet of the Pacific) near the border of the Australian territory of New Guinea. Occupied by the Japanese in World War II, it was liberated by U.S. forces in Apr., 1944.

Dja·kar·ta or **Ja·kar·ta** (both: jə-kär'tə, jä-kär'tä), formerly **Ba·ta·vi·a** (bə-tā'vē-ə), city (1971 pop. 4,576,009), capital of Indonesia, NW Java, on Djakarta Bay, an inlet of the Java Sea. The city is divided into two sections—the old town in the north, with Javanese, Chinese, and Arab quarters, and a modern residential garden suburb in the south. With its many canals and drawbridges, Djakarta somewhat resembles a Dutch town. The Dutch founded (c.1619) the fort of Batavia near the Javanese settlement of Djakarta, repulsing English and native attempts to oust them. Batavia became the headquarters of the Dutch East India Company and was a major trade center in the 17th cent. It declined in the 18th cent., following rebellions against the Dutch, but prospered again with the introduction of plantation cultivation in the 19th cent. From 1811 to 1814 Djakarta was the center of British rule in Java. Batavia was renamed Djakarta in Dec., 1949, and was proclaimed the capital of newly independent Indonesia.

Djam·bi or **Jam·bi** (both: jäm'bē), city (1971 pop. 158,559), SE Sumatra, capital of Djambi prov., Indonesia, a port at the head of navigation on the Hari River. It is the shipping and commercial center of an area producing oil, rubber, and timber.

Djen·né or **Jen·né** (both: jĕ-nā'), town, S central Mali, on the Bani River. It is an agricultural market center. Founded in the 8th cent., Djenné became (13th cent.) a great market for gold, slaves, and salt and rivaled Timbuktu in prosperity and culture. Djenné resisted a series of attacks by the kings of ancient Mali but finally fell c.1473. From the 16th cent. the town declined; it came under French control in the late 19th cent.

Dji·bou·ti (jĭ-bo͞o'tē), republic (1978 est. pop. 250,000), c.8,500 sq mi (22,015 sq km), E Africa, on the Gulf of Aden bounded on the N and W by Ethiopia, on the S by the Somali Republic, and on the E by the Gulf of Aden. Its capital is Djibouti. Largely a stony desert with isolated plateaus and highlands, it has a generally dry and torrid climate. The territory is economically underdeveloped, and nomadic pastoralism is the chief occupation. Manufacturing is limited to shipbuilding and repair, building and construction, production of compressed or liquid gas, and the manufacture of foodstuffs.

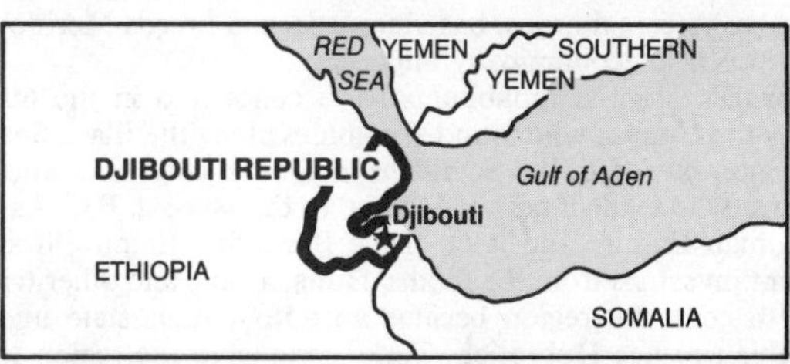

Strategically situated, the republic commands the strait between the Gulf of Aden and the Red Sea. France first obtained a foothold here in 1862. By 1896 the area was organized as a colony. It remained a colony until 1946, at which time it became a territory (the Afars and the Issas) within the French Union. Membership in the French Community followed in 1958. In June, 1977, the territory became independent.

Djibouti, town (1970 est. pop. 62,000), capital of the Djibouti Republic, NE Africa, a port on the Gulf of Tadjoura (an inlet of the Gulf of Aden). It is the republic's only sizable town. Its importance results from the large transit trade it enjoys as a terminus of the railroad from Addis Ababa. Activity at its port declined drastically after the Suez Canal was closed in 1967. The only important industry is the production of salt from the sea. There is a camel market in the town. Djibouti was founded by the French c.1888 and became the capital of French Somaliland in 1892.

Djok·ja·kar·ta (jŏk'yä-kär'tä): *see* Jogjakarta, Indonesia.

Dmi·trov (də-mē'trôf), city (1976 est. pop. 57,000), N central European USSR, on the Moscow Canal. It is a river port and industrial city. Products include machinery, iron, cellulose, reinforced concrete, clothing, and gloves. Dmitrov was founded in 1154.

Dne·pr (də-nyĕ'pər) or **Dnie·per** (nē'pər), river, c.1,430 mi (2,300 km) long, W European USSR, rising in the Valdai Hills W of Moscow and flowing generally S past Smolensk, through Belorussia, then through the Ukraine. It empties into the Black Sea. Since the construction (1932) of the Dneproges Dam, the Dnepr is navigable for virtually its entire course. The river was (9th–11th cent.) a commercial route for the Slavs and Byzantines.

Dne·pro·dzer·zhinsk (nĕp'rō-dər-zhĭnsk', də-nyĕ'prə-dyĕr-zhĭnsk'), city (1976 est. pop. 248,000), S European USSR, in the Ukraine, a port on the Dnepr River. It is a major industrial center with iron and steel, machine-tool, chemical, and cement plants. Originally called Kamenskoye, it was industrialized in the late 19th cent. and renamed Dneprodzerzhinsk in 1936.

Dne·pro·pe·trovsk (nĕp'rō-pə-trôfsk', də-nyĕ'prə-pyĕ-trôfsk'), city

(1974 est. pop. 995,000), capital of Dnepropetrovsk oblast, S European USSR, in the Ukraine, on the Dnepr River. It is a major industrial center with a huge iron and steel industry. The city also has plants producing heavy machinery, chemicals, rolling stock, and food products. Founded in 1787 by Potemkin on the site of a Cossack village, it was named Ekaterinoslav (Yekaterinoslav) for Catherine II. It was called Novorosiysk from 1791 to 1802 and Katerineslav until 1926. The city was occupied (1941–43) by German forces during World War II.

Dnes·tr (də-nyĕs'tər) or **Dnies·ter** (nēs'tər), river, c.850 mi (1,370 km) long, SW European USSR, forming part of the border between the Ukraine and the Moldavian Republic. It rises in the Carpathian Mts., flows generally southeast through the Ukraine and the Moldavian Republic, and empties through an estuary into the Black Sea southwest of Odessa. The Dnestr formed the Rumanian-Soviet border from 1918 to 1940, when the USSR recovered Bessarabia.

Do·ab (dō'ăb, dō-äb'), term applied in India to a tract of land between two converging rivers. The Doab, unqualified by the names of any rivers, designates the tract in Uttar Pradesh state between the Ganges and Jumna rivers, extending from the Siwalik Hills to the rivers' confluence at Allahabad. This well-irrigated region is the greatest wheat-growing area of the state.

Dobbs Ferry (dŏbz), village (1975 est. pop. 9,866), Westchester co., SE N.Y., on the Hudson River, a suburb of New York City; inc. 1873. It is mostly residential but has a chemical research laboratory.

Do·bru·ja (dō'brŏŏ-jə, dô'-), historic region, c.9,000 sq mi (23,310 sq km), SE Europe, in SE Rumania and NE Bulgaria, between the lower Danube River and the Black Sea. Dobruja comprises a low coastal strip and a hilly and forested inland. Largely agricultural, the region grows cereal crops, has vineyards, and breeds Merino sheep. Tourism is also economically important.

Dobruja's original inhabitants were conquered in the 6th cent. B.C. by the Greeks, who founded colonies along the Black Sea coast. The region passed to the Scythians in the 5th cent. B.C. and to the Romans (who made it part of Moesia) in the 1st cent. B.C. As part of the Roman Empire and later of the Byzantine Empire, it suffered frequent invasions from the Goths, Huns, Avars, and other tribes. In the 14th cent. the region became an autonomous state under the Walachian prince Dobrotich. Turks conquered the region in 1411, and for the next five centuries it remained a sparsely populated and barely cultivated territory of the Ottoman Empire. In 1878 the Congress of Berlin awarded northern Dobruja to Rumania and southern Dobruja to Bulgaria. As a result of the second Balkan War Bulgaria ceded (1913) southern Dobruja to Rumania. The Treaty of Neuilly, signed in 1919 between Bulgaria and the Allies of World War I, gave all of Dobruja to Rumania. In 1940, however, the German-imposed Treaty of Craiova forced Rumania to transfer southern Dobruja to Bulgaria.

Dob·son (dŏb'sən), town (1970 pop. 933), seat of Surry co., N N.C., NW of Winston-Salem near the Va. border. Surry Community College is here.

Do·ce (dō'sə), river, c.360 mi (580 km) long, E Brazil, rising in S central Minas Gerais state and flowing generally E through central Espírito Santo to the Atlantic Ocean NE of Vitória. Iron ore and semiprecious stones are found throughout the river valley.

Dod·dridge (dŏd'rĭj), county (1970 pop. 6,389), 319 sq mi (826.2 sq km), N W.Va., on the Allegheny Plateau, drained by Middle Island Creek and the South Fork of the Hughes River; formed 1845; co. seat West Union. Its agriculture includes fruit, potatoes, corn, tobacco, and livestock. It has oil and natural-gas wells and some coal mines and timber. There is also a textile industry in the county.

Do·dec·a·nese (dō-dĕk'ə-nēs', -nēz'), island group (1971 pop. 121,017), c.1,035 sq mi (2,680 sq km), SE Greece, in the Aegean Sea, between Asia Minor and Crete, comprising the greater part of the group known as the Southern Sporades. Despite its name ("twelve islands"), it consists of about 20 islands and islets. The city of Rhodes, on the largest of the islands, is the administrative seat. Agriculture, livestock raising, fruit growing, and sponge diving are the main occupations. Centers of ancient Greek culture, the Dodecanese were held by the Ottoman Turks from 1522 until 1912, when they were occupied by Italy during the Italo-Turkish War. The islands were captured by the Allies during World War II, and in 1947 they formally passed to Greece.

Dodge (dŏj). **1.** County (1970 pop. 15,658), 499 sq mi (1,292.4 sq mi), S central Ga., bounded on the SW by the Ocmulgee River and drained by the Little Ocmulgee River; formed 1870; co. seat Eastman. Coastal-plain agriculture (cotton, corn, peanuts, pecans, and livestock) and lumber are important to its economy. It also has a textile industry. **2.** County (1970 pop. 13,037), 435 sq mi (1,126.7 sq km), SE Minn., drained by branches of the Zumbro River; formed 1855; co. seat Mantorville. It is in an agricultural area yielding dairy products, livestock, and grain. There is some manufacturing. **3.** County (1970 pop. 34,782), 529 sq mi (1,370.1 sq km), E Nebr., in an agricultural area bounded on the S by the Platt River and drained by the Elkhorn River; formed 1854; co. seat Fremont. Flour, grain, livestock, dairy products, and poultry are produced. **4.** County (1970 pop. 69,004), 892 sq mi (2,310.3 sq km), S central Wis., drained by the Rock River and its tributaries; formed 1836; co. seat Juneau. Its agriculture includes dairy products, livestock, and grain. It also has iron mines and food-processing plants.

Dodge City, city (1970 pop. 14,127), seat of Ford co., SW Kansas, on the Arkansas River; inc. 1875. The distributing center for a wheat and livestock area, it also produces agricultural implements and supplies. Laid out in 1872 near Fort Dodge (1864) on the old Santa Fe Trail, it soon flourished as the Santa Fe railhead and became a wild and rowdy cow town; Wyatt Earp and Bat Masterson were among its famous residents. The city hall, formerly located on the site of Boot Hill, an early cowboy burial ground, has been removed to permit enlargement of that tourist attraction.

Dodge·ville (dŏj'vĭl'), city (1970 pop. 3,255), seat of Iowa co., SW Wis., WSW of Madison; settled 1827, inc. as a village 1858, as a city 1889. Once a lead-mining center, it is now a trade and shipping point for a farm and dairy area. Governor Dodge State Park is nearby.

Dog·ger Bank (dô'gər), extensive sandbank, c.6,800 sq mi (17,610 sq km), central North Sea, between Great Britain and Denmark. Covered by shallow water (c.55–120 ft/17–37 m deep), it is a major breeding ground for many types of fish.

Do·ha (dō'hä), city (1971 est. pop. 95,000), capital of Qatar, SE Arabia, on the Persian Gulf. Doha was a small fishing village until oil production in Qatar began in 1949.

Dôle (dōl), city (1975 pop. 29,295), Jura dept., E France, in Franche-Comté, on the Doubs River. There are metallurgical and food industries. Dôle was the capital of Franche-Comté until Louis XIV conquered the region. The university, founded (1422) by Philip the Good of Burgundy, was also transferred to Besançon at that time. Louis Pasteur was born in Dôle; his home is now a museum.

Dol·o·mites (dŏl'ə-mīts') or **Dol·o·mite Alps** (dŏl'ə-mīt'), Alpine group, N Italy, between the Isarco and Piave rivers, named for the dolomitic limestone of which it is composed. Famous for their strikingly bold outline (a stairstep effect created by erosion of alternate layers of soft and hard rock) and for their vivid colors at sunrise and sunset, the Dolomites are ideal for mountain climbing.

Do·lo·res (də-lôr'ĭs), county (1970 pop. 1,641), 1,028 sq mi (2,662.5 sq km), SW Colo., in a zinc-mining and agricultural area bounded on the W by the Utah border and drained by the Dolores River; formed 1881; co. seat Dove Creek. It includes parts of San Juan and Uncompahgre national forests.

Dolores, river, 230 mi (370 km) long, E Utah and SW Colo., rising in the NW San Juan Mts. and flowing SW and NNW across the Utah border to join the Colorado River.

Dol·ton (dôl'tən), village (1978 est. pop. 26,500), Cook co., NE Ill., on the Calumet River, S of Chicago; settled 1832, inc. 1892. It lies in a truck-farming area. Steel, aluminum products, glass, chemicals, paper bags, and chain belts are manufactured.

Dom (dōm), peak, 14,942 ft (4,557.3 m) high, Valais canton, S Switzerland, in the Mischabelhörner group. It is the highest peak entirely in Switzerland.

Dom·i·ni·ca (dŏm'ə-nē'kə, də-mĭn' ĭ-kə), island (1970 pop. 70,302), c.290 sq mi (750 sq km), the most important of the Windward Islands, British West Indies. Roseau is the capital and chief port. The island, of volcanic origin, is mountainous and forested, with fertile soil. Bananas, citrus fruits, copra, cacao, spices, vanilla beans, mangoes, and tobacco are grown. The island was discovered by Columbus in 1493. English and French attempts at settlement were thwarted by the hostile Carib Indians. An Anglo-French treaty of 1748 left Dominica in Carib hands, but both powers continued to covet it. The island definitively passed to the British in 1815. Dominica is an associated state of Britain with full internal self-government.

Do·min·i·can Republic (də-mĭn'ĭ-kən), republic (1970 pop. 4,011,589), 18,816 sq mi (48,733 sq km), West Indies, on the E two thirds of the island of Hispaniola. The capital is Santo Domingo.

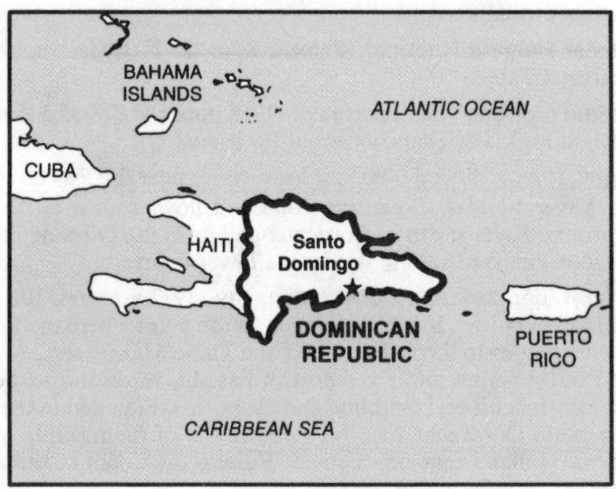

Economy. The country is predominantly agricultural. Sugar cane is the chief crop, and sugar is the chief product, accounting for 50% of exports. Other major crops are cocoa, coffee, bananas, tobacco, and rice. There are deposits of rock salt, bauxite, copper, platinum, zinc, gold, silver, and nickel; mining has gained importance in recent years. The country has various light industries. In the late 1960s and early 1970s tourism became increasingly important.

History. The history of the country has been unusually turbulent and has been closely linked with that of the neighboring republic of Haiti. After Spain ceded the colony of Santo Domingo to France by the Treaty of Basel (1795), the area now known as the Dominican Republic was conquered by Haitians under Toussaint L'Ouverture. Toussaint was defeated by the French, who invaded Haiti under General Charles Victor Leclerc. In 1808 the people revolted and in 1809, with the aid of an English squadron, ended French control of the city of Santo Domingo. Spanish rule was re-established. In 1821 the inhabitants expelled the Spanish governor, but in 1822 they were reconquered by the Haitians. A revolt broke out in 1844, the Haitians were defeated, and a republic was established under Pedro Santana. Frequent revolts as well as continued Haitian attacks led Santana to make his country a province of Spain in 1861, but opposition under Buenaventura Báez was so severe that Spain withdrew in 1865. Unable to preserve order, Báez himself negotiated a treaty of annexation with the United States, which the Dominicans approved but which the U.S. Senate failed to ratify. All semblance of order vanished. There were kaleidoscopic changes in the presidency and a long (1882–99), ruthless dictatorship under Ulíses Heureaux, ended by his assassination. The republic was hopelessly bankrupt by 1905 and faced intervention by European powers. U.S. President Theodore Roosevelt arranged a U.S. customs receivership. Although there was a marked improvement in finances, fiscal control brought virtual political domination by the United States. Disorder continued, however, and the country was occupied by U.S. marines in 1916. They were withdrawn in 1924 and the customs receivership terminated in 1941.

After the overthrow of Horacio Vásquez in 1930, Rafael Trujillo Molina became dictator. Trujillo suppressed domestic opposition, and he and his retinue gradually turned the country into a private fiefdom. In 1960, after finding Trujillo guilty of instigating an attempt on the life of President Betancourt of Venezuela, the Organization of American States (OAS) imposed diplomatic and economic sanctions on his regime. The United States broke diplomatic relations with the Dominican Republic in Aug., 1960. In 1961 Trujillo was assassinated. Joaquín Balaguer, who had been named president by Trujillo in 1960, initiated democratization measures, but was deposed (Jan., 1962). The governing council, after surviving a military coup, promulgated (Sept., 1962) a new constitution.

In Dec., 1962, in their first free election since 1924, the Dominicans elected Juan Bosch president by a substantial majority. Bosch committed himself to an ambitious program of reforms. He was inaugurated in an atmosphere of confidence and optimism, but right-wing opposition to Bosch led to his overthrow in Sept., 1963. A civilian triumvirate was installed by the military leaders. In 1965 civil war broke out in Santo Domingo. Initially announcing that U.S. civilians were endangered, U.S. President Lyndon B. Johnson sent troops to the Dominican Republic. Subsequently the United States claimed that Communists had infiltrated the pro-Bosch forces. The OAS called for a cease-fire and, at the suggestion of the United States, formed a peace-keeping force. OAS forces gradually replaced those of the United States. A compromise agreement was finally reached in Sept., 1965. After much difficulty, OAS-supervised elections were held in June, 1966, with Balaguer and Bosch the leading candidates. Balaguer won; the OAS forces were subsequently withdrawn. Authoritarianism continued under Balaguer, who enjoyed the support of the right, the military, and the church. Balaguer was re-elected in 1970 and 1974. The political climate, however, remained uneasy, with the economy stagnant and many workers emigrating to the United States and Puerto Rico.

Government. The country is governed under the 1966 constitution. The president, senate, and chamber of deputies are all directly elected for four-year terms.

Dom·ré·my-la-Pu·celle (dôɴ'rā-mē'lä-pü-sĕl'), village, Vosges dept., E France, in Lorraine, on the Meuse River. The house in which Joan of Arc was born (c.1412) is now a museum.

Don (dŏn, dôn), river, SW European USSR, rising SE of Tula and flowing c.1,200 mi (1,930 km), first SE past Voronezh, then SW into the Sea of Azov. At its eastern bend the Don is linked by a canal (c.65 mi/105 m long), with the Volga River near Volgograd. Navigable for c.850 mi (1,370 km) and accessible to seagoing vessels as far as Rostov-na-Donu, the Don is an important artery for grain, coal, and lumber shipments. Fisheries, especially in its lower course, are also valuable. Known to the ancients as the Tanaïs, the Don has been a trading channel since Scythian times.

Do·ña A·na (dōn'yə ăn'ə), county (1970 pop. 69,773), 3,804 sq mi (9,852.4 sq km), S N.Mex., bounded on the S by the Texas and Mexico borders and watered by the Rio Grande; formed 1852; co. seat Las Cruces. It is in an irrigated agricultural region yielding cotton, alfalfa, dairy products, and livestock, and has some industry (food products, concrete, and aerospace equipment). The eastern part of the county includes sections of Fort Bliss, White Sands National Monument, and White Sands Missile Range.

Don·ald·son·ville (dŏn'əld-sən-vĭl'), city (1970 pop. 7,367), seat of Ascension parish, SE La., on the Mississippi SSE of Baton Rouge. It was founded as a trading post in 1806, and plantations and farms were later established in the area. It was the temporary state capital (1830–31), and is now a trade center for an area producing sugar, rice, cotton, lumber, and oil.

Don·al·son·ville (dŏn'əl-sən-vĭl'), city (1970 pop. 2,907), seat of Seminole co., extreme SW Ga., near the Chattahoochee River NW of Bainbridge; inc. 1897. It is in a farm and timber area.

Do·nau·wörth (dō'nou-vûrt'), town (1970 pop. 11,539), Bavaria, S West Germany, a port at the confluence of the Donau and Wörnitz rivers. Its manufactures include machinery, airplanes, lace, and dolls. Historically a Swabian town, Donauwörth became (mid-13th cent.) the seat of the dukes of Upper Bavaria. It was made (14th cent.) an imperial city and adopted the Reformation in 1555. The efforts of Maximilian I of Bavaria to re-establish (1607) Catholicism in the city led to the formation of the Protestant Union and, in part, to the Thirty Years' War. Donauwörth passed to Bavaria in 1714.

Don·cas·ter (dŏng'kəs-tər), county borough (1973 est. pop. 81,530), South Yorkshire, N central England, on the Don River. Doncaster is a communications center and a market for fruits, vegetables, and livestock. There are slaughterhouses, railroad shops, steel mills, and food-processing plants. Coal mining is the main industry. Other manufactures include metal products, electrical equipment, agricultural implements, clothing, and nylon.

Don·e·gal (dŏn'ĭ-gôl', dŏn'ĭ-gôl'), county (1971 pop. 108,344), 1,865 sq mi (4,830.4 sq km), N Republic of Ireland, on the Atlantic Ocean; co. town Lifford. Although agriculture is the leading industry, only one third of the land is suitable for cultivation. Oats and potatoes are the chief crops. Fishing and tourism are also important industries. Gaelic is still spoken in the highland region. In ancient times the kingdom of Tyrconnell, Donegal was not organized as a county until the reign of Elizabeth I of England.

Donegal, rural district (1971 pop. 10,046), Co. Donegal, NW Republic of Ireland, on the River Eske at the head of Donegal Bay.

Do·nets (də-nĕts′, -nyĕts′), river, c.650 mi (1,045 km) long, SW European USSR, mainly in the Ukraine. A tributary of the Don, it rises northeast of Belgorod and flows generally southeast to join the lower Don. Its lower course is navigable.

Donets Basin, abbreviated as **Don·bas** or **Don·bass** (both: dŏn′băs, dən-bäs′), industrial region (c.10,000 sq mi/25,900 sq km), S European USSR, N of the Sea of Azov and W of the Donets River. The Donets Basin is one of the main coal-producing and steel-manufacturing areas of the USSR and forms one of the densest industrial concentrations in the world. Two types of coal predominate in the Donbas: anthracite, in the south and east (used mainly by thermal power stations), and bituminous, in the southwest and north (used mainly for coking). Other minerals besides coal are produced in the region, and there are also heavy-machinery, chemical, and power plants. The development of the Donets Basin began c.1870.

Do·netsk (də-nĕtsk′, -nyĕtsk′), city (1977 est. pop. 984,000), capital of Donetsk oblast, S European USSR, in Ukraine, on the Kalmius River. The largest industrial center of the Donets Basin and one of the largest in the Soviet Union, it has coal mines, iron and steel mills, machinery works, and chemical plants.

Dong Nai (dông′ nī′) or **Don·nai** (dô-nā′), river, c.300 mi (485 km) long, rising in the mountains of S Vietnam. It flows southwest past Bien Hoa and joins with the Saigon River below Saigon to form an extensive delta on the South China Sea. There are rapids and waterfalls on its upper reaches.

Don·i·phan (dŏn′ə-fən), county (1970 pop. 9,107), 391 sq mi (1,012.7 sq km), extreme NE Kansas, in a fertile agriculture area bounded on the E and N by the Missouri River, here forming the Mo. border; formed 1855; co. seat Troy. Corn is the most important crop.

Doniphan, resort city (1970 pop. 1,850), seat of Ripley co., SE Mo., on the Current River WSW of Poplar Bluff, in a timber and farm area; settled c.1847, inc. 1898.

Don·ley (dŏn′lē), county (1970 pop. 3,641), 909 sq mi (2,354.3 sq km), NW Texas, in the Panhandle, just below the high plains, and traversed by the Salt Fork of the Red River; formed 1876; co. seat Clarendon. Its agriculture includes cotton, grain, cattle, hogs, poultry, and dairy products. It also has some manufacturing.

Don·ny·brook (dŏn′ē-bro͝ok′), parish and suburb of Dublin, Co. Dublin, E central Republic of Ireland. It was famous for its annual fair, licensed by King John of England in 1204 and suppressed in 1855 because of its disorderliness. The disorder gave rise to the term "donnybrook," meaning an uproarious brawl.

Doo·ly (do͞o′lē), county (1970 pop. 10,404), 394 sq mi (1,020.5 sq km), central Ga., bounded on the W by the Flint River; formed 1821; co. seat Vienna. Coastal-plain agriculture (cotton, corn, peanuts, pecans, and truck crops) and forestry are important.

Doon (do͞on), river, c.30 mi (48 km) long, Ayrshire, SW Scotland, flowing NW through Loch Doon (6 mi/9.7 km long) to the Firth of Clyde S of Ayr. Robert Burns celebrated its beauty in his poetry.

Door (dôr), county (1970 pop. 20,106), 491 sq mi (1,271.7 sq km), NE Wis., on Door Peninsula, bounded on the E by Lake Michigan, on the W by Green Bay; formed 1851; co. seat Sturgeon Bay. In a resort area, it has cherry orchards, quarries, fisheries, and dairy farms. Shipbuilding is done.

Dor (dôr) or **Do·ra** (dôr′ə), Canaanite seaport, ancient Palestine. It was a Phoenician outpost and was rebuilt by the Romans; still visible are the ruins of a temple and a theater. Later it was fortified by the Crusaders.

Dor·ches·ter (dôr′chĭs-tər, -chĕs′-), municipal borough (1973 est. pop. 13,880), administrative center of Dorset, S central England. Dorchester is a busy agricultural market, especially for sheep and lambs. Printing, leatherworking, brewing, and the manufacture of agricultural machinery are important industries.

Dorchester. 1. County (1970 pop. 29,405), 594 sq mi (1,538.5 sq km), E Md., on a marsh-fringed peninsula on the Eastern Shore, bounded on the N by the Choptank River, on the SE by the Nanticoke River, on the E by the Del. border; formed 1668; co. seat Cambridge. Its agriculture includes fruit, vegetables and truck crops, corn, wheat, dairy products, and poultry. It also has a large seafood industry, canneries and packing houses, lumber and flour mills, and boatyards. Fishing, hunting, and boating attract visitors. Blackwater National Wildlife Refuge is here. **2.** County (1970 pop. 32,276), 569 sq mi (1,473.7 sq km), SE S.C., bounded on the W by the Edisto River; formed 1897; co. seat St. George. It has agriculture (cotton, tobacco, corn, and livestock) and timber and food-processing industries. Textiles and cement, clay, and metal products are manufactured. In the 18th and 19th cent. it was the site of rich rice and indigo plantations. The county includes a winter resort area and Middleton Place, with its famous formal gardens.

Dorchester Heights National Historic Site: *see* National Parks and Monuments Table.

Dor·dogne (dôr-dôn′yə), department (1968 pop. 374,073), 3,546 sq mi (9,184.1 sq km), SW France; capital Périgueux.

Dordogne, river, c.305 mi (490 km) long, rising near the Puy de Sancy in the Auvergne Mts., S central France. It flows southwest to join the Garonne River north of Bordeaux and form the Gironde. There are famous vineyards along the river's lower course.

Dor·drecht (dôr′drĕKHt) or **Dort** (dôrt), city (1977 est. pop. 102,688), South Holland prov., SW Netherlands, at the point where the Lower Merwede divides to form the Noord and Oude Maas rivers. An important rail junction and river port, it has shipyards and manufactures heavy machinery, clothing, and chemicals. Founded in the early 11th cent., Dordrecht was the scene (1572) of the meeting of the Estates of Holland that proclaimed William the Silent stadtholder. Dordrecht has a 14th cent. Gothic church.

Do·ris (dôr′ĭs), mountainous district, central Greece, inland between the Gulf of Corinth and the Malian Gulf. It was the traditional homeland of the Dorians, who may, in fact, have paused here during their invasion of Greece. Sparta gave Doris military aid during the 5th cent. B.C.

Dor·king (dôr′kĭng), urban district (1973 est. pop. 22,410), Surrey, SE England. It is a market town and residential suburb of London.

Dor·mont (dôr′mŏnt), borough (1970 pop. 12,856), Allegheny co., SW Pa., a residential suburb of Pittsburgh; settled c.1790, inc. 1909.

Dor·noch (dôr′nŏKH, -nəKH), burgh (1971 pop. 838), Highland region, N Scotland, on Dornoch Firth. It is a summer resort. Dornoch has a 13th cent. cathedral (rebuilt in the 19th cent.). The last burning for witchcraft in Scotland took place in Dornoch in 1722.

Dor·set (dôr′sĕt, -sĭt) or **Dor·set·shire** (dôr′sĕt-shîr′, -shər, -sĭt-), nonmetropolitan county (1976 est. pop. 575,800), 978 sq mi (2,533 sq km), SW England, on the English Channel; administrative center Dorchester. The rolling agricultural country is crossed by the North Dorset and South Dorset downs, chalk ranges running east and west. The county's pre-Roman antiquities include Maiden Castle. Dorset was part of the Anglo-Saxon kingdom of Wessex.

Dort·mund (dôrt′mənd, -mo͝ont), city (1970 pop. 636,954), North Rhine-Westphalia, W West Germany, a port on the Dortmund-Ems Canal. Its manufactures include iron, steel, and beer. Coal is mined nearby. First mentioned c.885, Dortmund flourished from the 13th cent. as a member of the Hanseatic League but later (17th cent.) declined. From the mid-19th cent. the city grew as an industrial center. It was badly damaged during World War II.

Dor·val (dôr-väl′), city (1976 pop. 19,131), S Que., Canada, on the S shore of Montreal Island on the St. Lawrence River. It is the site of Montreal's international airport.

Do·ry·lae·um (dŏr′ə-lē′əm), ancient city of N Phrygia, Asia Minor, now in NW Turkey. It was an important trading city of the Romans but later fell to ruins. At this site on July 1, 1097, the Christians of the First Crusade defeated the Seljuk Turks.

Do·than (dō′thən), city (1978 est. pop. 45,600), seat of Houston co., SE Ala., near the Fla. line, in a timber and peanut-growing area; inc. 1885. Manufactures include lumber products, furniture, toys, farming tools, truck and trailer bodies, and textile goods. A nuclear power plant and hot mineral springs are located in Dothan.

Dou·ai (do͞o-ā′), town (1975 pop. 45,239), Nord dept., N France, in French Flanders, on the Scarpe River. It is a major industrial and commercial center of the northern coal region. The chief manufactures are foundry products, wire, boilers, springs, and chemicals. Probably a Roman fortress built in the 4th cent., Douai was a possession of the counts of Flanders during the Middle Ages. Because of its prosperity as a center of the cloth trade, the town received a charter (1228) granting some autonomy. With the Hundred Years' War (1337–1453) and the resulting curtailment of English wool imports, the town declined and passed in 1384 to the dukes of Burgundy and in 1477 to the Spanish Hapsburgs. Louis XIV seized Douai in 1667, and after the War of the Spanish Succession (1701–14), the town was permanently restored to France by the Peace of Utrecht (1713). Under the patronage of Philip II of Spain a Roman Catholic college was established in Douai for English priests. At the college the Old Testament of the Douay Bible was prepared in 1609.

Dou·ble Springs (dŭb'əl sprĭngz'), town (1970 pop. 957), seat of Winston co., NW Ala., W of Cullman, in a lumber and cotton area.

Doubs (do͞o), department (1975 pop. 471,082), 2,019 sq mi (5,229.2 sq km), E France, in Franche-Comté, bordering on the Jura Mts. and on Switzerland; capital Besançon.

Doubs, river, c.270 mi (435 km) long, rising in the Jura Mts., E France, and flowing NE, forming part of the French-Swiss border, then looping into W Switzerland before turning back into France where it meanders SW to empty into the Saône River.

Doug·ga (do͞o'gə), village, Tunisia, SW of Tunis. It is a tourist spot noted for the ruins of the ancient city of Thugga, including a Punic mausoleum (2nd cent. B.C.); temples, arches, a theater, a circus, and an aqueduct of Roman times; and a Byzantine fortress.

Dou·gher·ty (dō'ər-tē), county (1970 pop. 89,639), 326 sq mi (844.3 sq km), SW Ga., intersected by the Flint River; formed 1853; co. seat Albany. Coastal-plain agriculture (pecans, peanuts, corn, and cattle) and timber characterize the area.

Doug·las (dŭg'ləs), municipal borough (1971 pop. 20,389), capital of the Isle of Man, Great Britain. It is a popular resort on the Irish Sea, and tourism is the chief industry. There are also light-engineering, knitting, and carpet-weaving factories. The Manx Museum has a collection of the natural history and antiquities of the Isle of Man.

Douglas. 1. County (1970 pop. 8,407), 843 sq mi (2,183.4 sq km), central Colo., in a wheat and livestock area bounded on the W by the South Platte River; formed 1861; co. seat Castle Rock. Part of Pike National Forest is in the west. **2.** County (1970 pop. 28,659), 201 sq mi (520.6 sq km), NW central Ga., in a piedmont area bounded on the SE by the Chattahoochee River; formed 1870; co. seat Douglasville. It is mainly agricultural (cotton, corn, vegetables, fruit, livestock, and poultry), with stands of timber. **3.** County (1970 pop. 18,997), 420 sq mi (1,087.8 sq km), E central Ill., drained by the Embarrass and Kaskaskia rivers; formed 1859; co. seat Tuscola. Its agriculture includes corn, wheat, soybeans, broomcorn, livestock, poultry, and dairy products. Brooms, road machinery, oil tanks, caskets, office equipment, and food and wood products are manufactured. **4.** County (1970 pop. 57,932), 468 sq mi (1,212.1 sq km), E Kansas, in a gently sloping agricultural area bounded on the N by the Kansas River; formed 1855; co. seat Lawrence. Grain, livestock, and truck crops are raised. It also has scattered oil and gas fields and some manufacturing. **5.** County (1970 pop. 22,910), 647 sq mi (1,675.7 sq km), W central Minn., watered by the Long Prairie River and numerous lakes; formed 1858; co. seat Alexandria. In an agricultural area yielding grain, poultry, livestock, and dairy products, it also has some manufacturing and a thriving tourist industry. **6.** County (1970 pop. 9,268), 809 sq mi (2,095.3 sq km), S Mo., in the Ozarks and drained by the North Fork of the White River; formed 1857; co. seat Ava. It has agriculture (corn, poultry, and dairy cattle), stands of pine and oak, and sand and gravel pits. Wood products and sporting goods are manufactured. Honey Branch Cave and sections of Mark Twain National Forest are here. **7.** County (1970 pop. 389,455), 335 sq mi (867.7 sq km), E Nebr., bounded on the W by the Platte River, on the E by the Missouri River, here forming the Iowa border, and drained by the Elkhorn River; formed 1854; co. seat Omaha. Livestock and grain are raised. Food is processed, and textiles, wood products, furniture, paper, chemicals, plastics, glass, cement, and metal products and machinery are manufactured. There are also limestone quarries and sand and gravel pits. **8.** County (1970 pop. 6,882), 724 sq mi (1,875.2 sq km), W Nev., in an irrigated region bordering on Calif. and watered by the Carson River, with foothills of the Sierra Nevada throughout the county and part of Lake Tahoe in the NW; formed 1861; co. seat Minden. In an agricultural area yielding alfalfa, grain, cattle, sheep, and dairy products, it has a lumber industry and some manufacturing. Its casinos and summer and winter resorts are especially popular. The county includes Washoe Indian Reservation. **9.** County (1970 pop. 71,743), 5,062 sq mi (13,110.6 sq km), SW Oregon, in a mountainous region bounded on the W by the Pacific Ocean and crossed by the Umpqua River and numerous creeks; formed 1852; co. seat Roseburg. Lumbering and agriculture (field crops, fruit, livestock, and poultry) are important. It also has nickel mines and stone quarries. There are many parks and recreational areas here. Part of Siuslaw National Forest is in the west; part of Umpqua National Forest is in the east. **10.** County (1970 pop. 4,569), 435 sq mi (1,126.7 sq km), SE S.Dak., in an agricultural and cattle-raising area; formed 1883; co. seat Armour. Its economy depends on livestock, cattle feed, dairy produce, poultry, and grain. **11.** County (1970 pop. 16,787), 1,841 sq mi (4,768.2 sq km), central Wash., in a plateau area bounded on the N and W by the Columbia River; formed 1883; co. seat Waterville. In an agricultural area yielding cattle, wheat, and fruit, especially apples, it has clay and limestone deposits. Paper and metal products are manufactured. **12.** County (1970 pop. 44,657), 1,310 sq mi (3,392.9 sq km), extreme NW Wis., in a generally wooded area bounded on the W by the Minn. border, on the N by the St. Louis River and Lake Superior, and drained by the St. Croix, Eau Claire, Upper Tamarack, and Nemadji rivers and numerous small lakes; formed 1854; co. seat Superior. It has some agriculture (grain, cattle, and dairy products), but depends primarily on lumbering, shipping, petroleum refining, and diversified manufacturing (boats, machinery, and food, wood, and metal products).

Douglas. 1. City (1978 est. pop. 13,600), Cochise co., SE Ariz., at the Mexican border; inc. 1905. The mining and smelting of copper have been important since 1900; the city grew around a copper smelter. Douglas is also a ranching center and a border station, with plants that manufacture wearing apparel and electronic parts. Gypsum and tungsten mines and limestone quarries are in the area. **2.** City (1970 pop. 10,195), seat of Coffee co., S central Ga.; inc. 1895. It is a livestock and tobacco market, with food-processing plants, mobile home manufacturers, and garment factories. **3.** Town (1970 pop. 2,677), seat of Converse co., E central Wyo., on the North Platte River ESE of Casper; laid out 1886 with the coming of the railroad. It is a trade center for a livestock, farm, and oil area.

Doug·las·ville (dŭg'ləs-vĭl'), city (1970 pop. 5,472), seat of Douglas co., NW Ga., W of Atlanta; settled on the site of an Indian village, inc. 1875. Textiles, wood products, and asphalt are made.

Dou·ro (dō'ro͞o) or **Due·ro** (dwā'rō), river, c.475 mi (765 km) long, rising in the Sierra de Urbión, N central Spain. It flows west across northern Spain, then southwest to form part of the Spanish-Portuguese border before flowing west across northern Portugal to the Atlantic Ocean at Oporto. Silting, rapids, and deep gorges combine to make the Douro unnavigable.

Dove (dŭv), river, c.40 mi (65 km) long, rising in the Pennines, Derbyshire, central England, and flowing S and SE to the River Trent near Burton upon Trent. Its watercourse was a haunt of Izaak Walton and still provides fishing.

Dove Creek, town (1970 pop. 619), seat of Dolores co., SW Colo., between the Dolores River and the Utah border, NW of Durango, in a lumber and agricultural area.

Do·ver (dō'vər), municipal borough (1973 est. pop. 34,160), Kent, SE England, on the Strait of Dover, beneath chalk cliffs c.375 ft (115 m) high. The small Dour River flows through the town. Dover is a resort and an important port for travel and shipping to the Continent; it was chief among the members of the Cinque Ports. The Romans fortified the place and called it Dubris. In Anglo-Saxon times there was also a fort here. In 1216 Dover was defended by Hubert de Burgh against a French attack. It was the landing place of Charles II in 1660. Dover was the center of English Channel defense and an important naval base in World War I. For four years in World War II it was a constant target of German long-range guns. In the cliffs a series of subterranean caves and tunnels once used by smugglers were put to use as shelters from 1940 to 1944. Noteworthy are Shakespeare Cliff (the first coal in Kent was discovered here in 1822); the 13th cent. Maison Dieu Hall; Dover Castle on the cliffs, of Roman or Saxon origin; the lighthouse in the castle, partly Roman; and the Church of St. Mary, also in the castle, of Saxon origin with Roman brick.

Dover. 1. City (1978 est. pop. 23,100), state capital, and seat of Kent co., central Del., on the St. Jones River; founded 1683 on orders of William Penn, laid out 1717, inc. as a city 1929. In a rich farming and fruit-growing region, it is a shipping and canning center with varied industries. Dover Air Force Base, one of the largest air cargo terminals in the world, is a major factor in the city's economy. The old statehouse on the green, built in part in 1722 as the county courthouse, has been the capitol since 1777. Numerous historic houses and sites remain. The state museum is in the Old Presbyterian

Church (1790). **2.** City (1978 est. pop. 20,600), seat of Strafford co., SE N.H., on the Bellamy, Salmon Falls, and Cocheco rivers near their confluence with the Piscataqua River; settled 1623, inc. as a city 1855. Among its many manufactures are shoes, printing presses, office machines, and farm equipment. The first permanent settlement in New Hampshire, Dover was organized in 1633. An Indian massacre occurred in 1689. In 1812 the first cotton factory was established and the town thrived as a textile center. Dover's historic attractions include the garrison house (late 1600s), the Hale house (1806), where Lafayette and James Monroe stayed, and a library that was organized in 1792. **3.** Industrial town (1978 est. pop. 14,900), Morris co., N central N.J., on the Rockaway River; settled 1722, inc. as a town 1869. In a rich iron-ore area, the town grew as an iron-manufacturing center on the old Morris Canal. It still has important iron and steel works as well as a great variety of manufactures. **4.** City (1978 est. pop. 11,100), Tuscarawas co., E Ohio, on the Tuscarawas River, in a farm, coal, and fire-clay area; inc. 1867. Electrical equipment, sheet steel, masonite, chemicals, knives, and wire are among its products. Many people of Swiss descent live here. **5.** Town (1970 pop. 1,179), seat of Stewart co., NW Tenn., on the Cumberland River W of Clarksville, in an agricultural area. Fort Donelson National Military Park is nearby.

Dover, Strait of, separating Great Britain from France and connecting the English Channel with the North Sea. It is 21 mi (33.8 km) wide between Dover and Cape Gris-Nez, near Calais, and is called Pas-de-Calais by the French. In 1588 the Spanish Armada was checked by the English in the Strait.

Do·ver-Fox·croft (dō'vər-fŏks'krôft'), town (1970 pop. 3,102), seat of Piscataquis co., central Maine, NW of Bangor. It was formed in 1922 by the union of the towns of Foxcroft (settled 1806, inc. 1812) and Dover (settled 1806, organized as a plantation 1812, inc. as a town 1822). The two parts are on opposite sides of the Piscataquis River, which furnishes power for the town's textile and lumber mills.

Dov·re·fjell (dôv'rə-fyĕl'), mountainous region of S Norway, c.100 mi (160 km) long and 40 mi (60 km) wide, culminating in Snøhetta (7,500 ft/2,287.5 m high). It is a symbol of steadfastness and independence in Norwegian folklore and literature.

Down (doun), county (1971 pop. 310,617, excluding Belfast, which lies partly within the county), 952 sq mi (2,465.7 sq km), SE Northern Ireland; co. town Downpatrick. The undulating land surface rises to the Mourne Mts. in the south. Agriculture is the most important activity. Manufactures include linen, cotton, rayon, clothing, and processed foods. Fishing is also economically important.

Down East, also **down East,** name for New England, especially Maine.

Dow·ners Grove (dou'nərz grōv'), village (1978 est. pop. 39,000), Du Page co., NE Ill.; settled 1832, inc. 1873. Its manufactures include baked goods, plastic and metal products, electrical components, and chain belts.

Dow·ney (dou'nē), city (1978 est. pop. 86,400), Los Angeles co., S Calif., a residential and industrial suburb between Los Angeles and Long Beach; inc. 1957. Its many manufactures include metal products, aircraft, missiles, and chemicals.

Dow·nie·ville (dou'nē-vĭl'), village (1970 pop. 500), seat of Sierra co., NE Calif., in the Sierra Nevada on North Yuba River SW of Portola. Gold is mined in the region.

Dow·ning Street (dou'nĭng), Westminster, London, England. On the street are the British Foreign Office and, at No. 10, the residence of the first lord of the Treasury, who is usually (although not necessarily) the prime minister of Great Britain. Since nearly all prime ministers from the time of Robert Walpole (18th cent.) have lived at No. 10, it has come to designate the British government.

Dow·ning·ton (dou'nĭng-tən), borough (1975 est. pop. 8,841), Chester co., SE Pa., on a branch of Brandywine Creek W of Philadelphia; settled c.1730, inc. 1859. Paper and paper-mill machinery are made.

Down·pat·rick (doun-păt'rĭk), urban district (1971 pop. 7,403), county town of Co. Down, E Northern Ireland, at the SW extremity of Strangford Lough. The town has linen mills and is a market for an area where oats and flax are grown and sheep are raised. Hunting is popular in the vicinity. The seat of the diocese of Down, Downpatrick has long been a religious center; St. Patrick is said to have founded a church here c.440. The present cathedral dates from 1790.

Downs, North, and **South Downs** (dounz), parallel ranges of chalk hills, SE England. They rise to 965 ft (294.3 m) at Leith Hill. The North Downs range extends c.100 mi (160 km) through Surrey and Kent. It is separated by The Weald from the South Downs (c.65 mi/105 km long) in Sussex and eastern Hampshire.

Downs, The, roadstead, c.8 mi (13 km) long and 6 mi (9.7 km) wide, between North Foreland and South Foreland, off Deal, Kent, SE England, in the English Channel. It is protected, except from strong south winds, by the Goodwin Sands and the coast. Two naval battles were fought nearby—between the Dutch and the Spanish in 1639 and between the British and the French in 1666.

Doyles·town (doilz'toun'), borough (1975 est. pop. 8,400), seat of Bucks co., SE Pa., N of Philadelphia; settled 1745, inc. 1838. The trade center of a dairy and farm area, it also produces textiles. The archaeologist Henry Mercer was born here, and his castle, inlaid with tiles of his own design, is preserved. Also here is a historical museum with a collection of early American tools and utensils.

Dra·chen·fels (drä'ᴋʜən-fĕls'), mountain, 1,053 ft (321.2 m) high, in

Dra·cut (drā'kət), town (1970 pop. 18,214), Middlesex co., NE Mass., near the N.H. line; settled 1664, inc. 1702. The commercial center of a fertile farm region, the town manufactures textiles.

Dra·gui·gnan (drä-gē-nyäɴ'), town (1968 pop. 19,465), capital of Var dept., SE France, in Provence. It is a rural town with some medieval remains.

Dra·kens·berg Range (drä'kənz-bûrg'), South Africa, extending 700 mi (1,126.3 km) NE-SW in Natal, Lesotho, Orange Free State, and Transvaal. Thabana-Ntlenyana, at 11,425 ft (3,484.6 m), is the highest point. The mountains are part of the escarpment that forms the southern edge of the central plateau of Africa.

Drake's Bay (drāks), inlet of the Pacific Ocean, formed by the San Andreas fault, W Calif., NW of San Francisco. Point Reyes forms its outer arm. The bay was visited by Sir Francis Drake in 1579.

Drake Strait or **Passage** (drāk), c.500 mi (805 km) long, Antarctica, between Cape Horn and S Shetland Islands. It connects the South Pacific Ocean with the South Atlantic Ocean.

Drá·ma (drä'mä), city (1971 pop. 29,692), capital of Dráma prefecture, NE Greece, in Macedonia. It is the trade center for a tobacco-producing region.

Dram·men (drä'mən), city (1977 est. pop. 50,834), capital of Buskerud co., SE Norway, at the mouth of the Dramselva River. It is a commercial and fishing port and a trade and industrial center. Manufactures include forest products, textiles, and metal goods.

Dran·cy (dräɴ-sē'), city (1975 pop. 64,430), Seine-Saint Denis dept., N central France. An industrial suburb northeast of Paris, Drancy produces automobile brakes, aircraft, and hardware. During World War II Drancy was the site of a Nazi concentration camp.

Dran·gi·a·na (drăn'jē-ā'nə, -ăn'ə), ancient country, part of the Persian Empire, between Aria on the north and Gedrosia on the south. It was conquered (330 B.C.) by Alexander the Great and incorporated into his empire. Drangiana is the modern Seistan region of Afghanistan and eastern Iran.

Dra·va or **Dra·ve** (both: drä'və), river, c.450 mi (725 km) long, rising in the Carnic Alps, N Italy. It flows generally east through southern Austria (where it is called the Drau) and enters northwest Yugoslavia. It forms part of the Yugoslav-Hungarian border before joining the Danube River east of Osijek.

Dren·the (drĕn'tə), province (1977 est. pop. 409,874), c.1,030 sq mi (2,670 sq km), NE Netherlands, bordering on West Germany in the E; capital Assen.

Dres·den (drĕz'dən), city (1974 est. pop. 507,692), capital of Dresden dist., SE East Germany, on the Elbe River. It is an industrial and cultural center, a rail junction, and a large inland port. Manufactures include precision and optical instruments, chemicals, clothing, processed food, ceramics, and glass.

Originally a Slavic settlement called Drezdane, Dresden was settled with Germans by the margrave of Meissen in the 13th cent. From 1485 until 1918 it was the residence of the dukes, then the electors, and later the kings, of Saxony. Prussia occupied Dresden in the Second Silesian War, but withdrew after the Treaty of Dresden (1745). In the Seven Years' War Dresden was again occupied (1756) by the Prussians. In the late 17th and 18th cent. Dresden became a center of the arts and an outstanding showplace of baroque and rococo architecture. In the late 18th and early 19th cent. it was a leading center of the romantic movement, and in the late 19th and early 20th cent. it was a center of German opera. Ranked as one of the world's most beautiful cities before World War II, Dresden was severely damaged by British and U.S. bombing during the war (Feb., 1945). Most of the fabulous art collection, acquired by the court in the 18th and 19th cent., was safely kept through the war outside Dresden, but many art objects were afterward moved to the USSR. Dresden china was originally made in Dresden, but the factory was moved to Meissen in the early 18th cent.

Dresden, town (1970 pop. 1,939), seat of Weakley co., NW Tenn., W of Paris, in a timber, cotton, and farm area; laid out 1825, inc. 1827.

Dreux (drœ), town (1975 pop. 33,101), Eure-et-Loir dept., N central France. It is an industrial center where foundry products, boilers, metal products, radio and television equipment, and automobile paints are manufactured. An old Gallo-Roman city, Dreux was finally united with the French crown under Louis XV (18th cent.).

Drew (drōō), county (1970 pop. 15,157), 836 sq mi (2,165.2 sq km), SE Ark., drained by Bayou Bartholomew and the Saline River; formed 1846; co. seat Monticello. Its agriculture includes cotton, hay, corn, truck crops, fruit, and livestock. It has timber and textile mills and manufactures transportation equipment.

Drew·ry's Bluff (drōō'rēz), high ground on the S bank of the James River, E Va., S of Richmond; scene of two engagements in the Civil War. On May 15, 1862, the Confederates, positioned on the bluff, repulsed Union gunboats that were part of Gen. George McClellan's Peninsular campaign. On May 16, 1864, Union Gen. Benjamin Butler was defeated by a greatly inferior Confederate force under Gen. Pierre Beauregard.

Drift·less Area (drĭft'lĭs), region, c.13,000 sq mi (33,670 sq km), largely in SW Wis. but extending into SE Minn., NE Iowa, and NW Ill. The continental glacier that covered most surrounding regions did not touch this area, which abounds in caves and sinkholes and has residual, well-drained soil. The Federal government prohibited farming in the Driftless Area until the 1840s because it was an important lead-mining region. In the mid-1800s the area was settled by European immigrants.

Driggs (drĭgz), village (1970 pop. 727), seat of Teton co., E Idaho, near the Wyo. border ENE of Idaho Falls in the valley of the Teton River. It is the trade center for a farm and dairy region. Cheese and flour are made. There is a large coal bed nearby.

Drin (drēn), river, c.175 mi (280 km) long, formed at Kukës, NE Albania, by the confluence of the White Drin and the Black Drin, which rise in Yugoslavia. It is the largest river of Albania. The Drin flows generally west through deep gorges and then onto the coastal plain before turning south and entering the Adriatic Sea.

Dri·na (drē'nä), river, c.285 mi (460 km) long, formed by the confluence of the Piva and Tara rivers, S central Yugoslavia. It flows generally north through central Yugoslavia to the Sava River.

Dro·ghe·da (drô'ə-də, droi'də), urban district (1971 pop. 19,744), Co. Louth, E central Republic of Ireland, on the Boyne River. The town is a port and has cement-processing works, breweries, ironworks, and linen, cotton, and lumber mills. Drogheda was a Danish stronghold in the 10th cent. Oliver Cromwell stormed the town in 1649 and massacred the inhabitants. The Battle of the Boyne was fought at Drogheda in 1690.

Dro·go·bych (drə-gô'bĭch), city (1976 est. pop. 66,000), Lvov oblast, SW European USSR, in the Ukraine, in the N Carpathian foothills. It is the major petroleum-refining center of the Borislav oil field. An old Ukrainian settlement, Drogobych belonged to Kievan Russia until the 14th cent., when it passed to Poland. It was taken by Austria in 1772 but reverted to Poland in 1919; in 1939 it was included in the Ukraine.

Droit·wich (droit'wĭch'), municipal borough (1973 est. pop. 13,950), Worcestershire, W central England. It is a spa noted for its brine baths. Droitwich had a flourishing salt industry from Roman times until 1922, when the plants were moved to nearby Stoke Prior.

Drôme (drōm), department (1968 pop. 342,891), 2,519 sq mi (6,524.2 sq km), SE France; capital Valence.

Drum·mond·ville (drŭm'ənd-vĭl'), city (1976 pop. 29,286), S Que., Canada, on the St. Francis River, NE of Montreal. Its manufactures include textiles, paper and wood products, and rubber goods.

Dry Tor·tu·gas (drī' tôr-tōō'gəz), island group, off S Fla., W of Key West. Named by the Spanish explorer Ponce de León in 1513, the islands later became a pirate base. They are famous for their bird and marine life.

Duar·te (dwär'tē), city (1970 pop. 14,981), Los Angeles co., S Calif.; settled c.1841, inc. 1957. It is residential, with light industry and warehousing. The City of Hope National Medical Center is here.

Du·bai (dōō-bī'), sheikdom (1970 est. pop. 60,000), c.1,500 sq mi (3,885 sq km), part of the federation of United Arab Emirates, E Arabia, on the Persian Gulf. Little is known of the early history of Dubai, but it appears to have been a dependency of Abu Dhabi until 1833. Along with the other sheikdoms that now compose the federation, it became a British protectorate in the 19th cent. Oil was discovered in Dubai in the early 1960s.

Du·bawnt (dōō-bônt'), river, 580 mi (933.2 km) long, rising in Wholdaia Lake, SE Mackenzie dist., Northwest Territories, Canada, and flowing NE to Dubawnt Lake (c.1,600 sq mi/4,145 sq km), then E across Keewatin dist. to Baker Lake at the head of Chesterfield Inlet of Hudson Bay.

Dü·ben·dorf (dü'bən-dôrf'), town (1970 pop. 19,639), Zürich canton, NE Switzerland. Tobacco products and chemicals are produced.

Dub·lin (dŭb'lĭn), county (1971 pop. 852,219), 327 sq mi (847 sq km), E central Republic of Ireland, on the Irish Sea; co. town Dublin. The area is low-lying in the north and center, rising to the Wicklow Mts. in the south. The rural area, upon which the city has increasingly encroached, is devoted to dairy farming and the raising of wheat, barley, and potatoes. Organized as a county by King John of England in the early 13th cent., Dublin was the heart of the English Pale and was strongly guarded by castles along its boundaries.

Dublin, county borough (1971 pop. 567,866), capital of the Republic of Ireland and county town of Co. Dublin, on Dublin Bay at the mouth of the Liffey River. Its harbor, with shipyards, docks, and quays, dates from 1714. The old Royal and Grand canals, connecting Dublin with the interior, have been superseded by railroads for most commercial traffic. Dublin's chief industries are brewing, textile manufacturing (silk making was introduced by Huguenot refugees in the 16th cent.), distilling, shipbuilding, food processing, and the manufacture of foundry products, glass, and cigarettes.

Dublin was a Danish town until 1014, when Brian Boru defeated the Danes at nearby Clontarf. The Danes established themselves again until Richard Strongbow, 2nd earl of Pembroke, captured the city for the English in 1170. In 1209 occurred the Black Monday massacre of English residents. Edward Bruce unsuccessfully assaulted the town in the early 14th cent. In the English Civil War the city surrendered (1647) to the Parliamentarians, and Oliver Cromwell landed here in 1649. From 1782 to 1800, when the Irish Parliament enjoyed temporary independence of England, Dublin experienced a prosperous and stimulating era; many of the city's buildings date from this period. After the Act of Union of 1800 many wealthy aristocrats moved to London, and the years of prosperity ended.

In the 19th cent. Dublin saw much bloodshed in connection with nationalist efforts to free Ireland from English rule—the insurrection led by Robert Emmet in 1803; the 1867 uprising of the Fenian movement; and the murder (1882) of Lord Frederick Cavendish, chief secretary for Ireland. Dublin also became the center of a Gaelic renaissance: the Gaelic League was founded here in 1893, and the Abbey Theatre began producing Irish plays. In 1913 the city was paralyzed by strikes, eventually culminating in the Easter Rebellion of 1916. The early troubles of the Irish Free State led to the worst period of bloodshed in Dublin's history.

Today Dublin is the seat of the Irish legislature, the Dáil Éireann, in Leinster House. The Univ. of Dublin or Trinity College (founded 1591) has in its library the famous Book of Kells and a copy of every book published in the British Isles. University College (Roman Catholic) was incorporated in 1909 as part of the National Univ. of Ireland; mastery of the Gaelic language is a requirement for its students. The city's earliest church, Christ Church, was founded in 1038. St. Patrick's is the national cathedral of the Protestant Church of Ireland; Jonathan Swift, buried there, was dean from 1713 to 1745.

Dublin. 1. Uninc. town (1970 pop. 13,641), Alameda co., W Calif., a suburb in the San Francisco-Oakland area. Photographic supplies, security equipment, and telephone parts are manufactured, and aircraft research is conducted. **2.** City (1978 est. pop. 14,900), seat of Laurens co., central Ga., on the Oconee River; inc. 1812. It is a commercial and industrial center. Lumbering and the manufacture of wood products are its chief industries.

Dub·na (do͞ob-nŭ'), town (1976 est. pop. 50,000), Moscow oblast, central European USSR, near the confluence of the Volga and Dubna rivers. Founded in 1956, it is the seat of the Joint Institute for Nuclear Research.

Du·bois (do͞o-bois'), county (1970 pop. 30,934), 433 sq mi (1,121.5 sq km), SW Ind., bounded partly on the N by the East Fork of the White River and drained by the Patoka River; formed 1817; co. seat Jasper. In an agricultural area yielding corn, wheat, poultry, and dairy products, it has bituminous-coal and clay deposits, sand and gravel pits, timber, and stone quarries.

Dubois, village (1970 pop. 400), seat of Clark co., E Idaho, NW of Rexburg, in a livestock and grain area.

Du·brov·nik (do͞o'brôv-nĭk), city (1971 pop. 31,106), SW Yugoslavia, in Croatia, on a promontory jutting into the Adriatic Sea. It is a seaport and a tourist center. Industries include oil refining, slate mining, and metalworking. Dubrovnik was founded in the 7th cent. by Roman refugees fleeing Slav incursions. Later, Slavic people settled in the city, which became a link between the Latin and Slavic civilizations. Dubrovnik rose to eminence as a powerful merchant republic. Though it was under the protectorate of the Byzantine Empire until 1205, of Venice until 1358, of Hungary until 1526, and of Turkey until 1806, Dubrovnik remained a virtually independent republic until it was abolished in 1808 by Napoleon I and included in the Illyrian Provinces. The Congress of Vienna assigned (1815) it to Austria, and in 1918 it was included in Yugoslavia. Medieval Dubrovnik was the center of Serbo-Croatian culture and literature. The city retains much of its medieval architecture.

Du·buque (də-byo͞ok'), county (1970 pop. 90,609), 608 sq mi (1,574.7 sq km), E Iowa, in a prairie agricultural area bounded on the E by the Mississippi, here forming the Wis. and Ill. borders, and drained by the North Fork of the Maquoketa River; formed 1834; co. seat Dubuque. Hogs, cattle, corn, and oats are produced. It has lead and zinc deposits, limestone quarries, and diversified manufacturing.

Dubuque, city (1978 est. pop. 63,200), seat of Dubuque co., NE Iowa, on the Mississippi River; chartered 1841. It is a trade, industrial, and rail center and a river port for an agricultural and dairy area. It has railroad shops, shipyards, food-processing plants, a brewery, and factories that make cast-iron and sheet-metal products, chemicals, and machinery. One of the oldest cities in the state, it was named for Julien Dubuque, who had settled nearby c.1788. Indian title to the territory ended with the Black Hawk Treaty of 1833, and white settlers began to pour in. The town developed first as a mining town, then as a lumbering and milling center.

Du·chesne (do͞o-shān'), county (1970 pop. 7,299), 3,255 sq mi (8,430.5 sq km), NE Utah, in a mountainous area drained by the Uinta and Duchesne rivers; formed 1913; co. seat Duchesne. Livestock, dairy products, wheat, hay, alfalfa, and sugar beets are produced. The county has oil and gas wells and some manufacturing. It includes Kings Peak, the highest point in the state, sections of Ashley National Forest, and part of Uintah and Ouray Indian Reservation.

Duchesne, city (1970 pop. 1,094), seat of Duchesne co., NE Utah, NNE of Price, near the junction of the Strawberry and Duchesne rivers, in a mountainous agricultural area.

Duck Lake (dŭk), small lake, central Sask., Canada, SW of Prince Albert. It was the scene of the first encounter in Riel's Rebellion in 1885, in which a large group of métis (persons of mixed French and Indian descent) under Gabriel Dumont defeated a detachment of Northwest Mounted Police.

Du·din·ka (do͞o-dĭng'kə, do͞o-dyĕn'kə), city (1975 est. pop. 23,000), capital of the Taymyr National Okrug, Krasnoyarsk Kray, N Siberian USSR, on the Yenisei River. It is the river port for the Norilsk mining area. Founded in 1616 as a winter outpost, Dudinka became a city in 1951.

Dud·ley (dŭd'lē), county borough (1976 est. pop. 300,200), West Midlands, W central England. Dudley's famed iron, coal, and limestone industries have been declining since c.1870. Today industries include engineering works and steelworks, metallurgy, glass cutting, and leatherworking. The town developed around Dudley Castle in the 13th cent.

Due·ro (dwā'rō), river: *see* Douro.

Duis·burg (dœs'bo͝orkh), city (1975 pop. 599,799), North Rhine-Westphalia, W West Germany, at the confluence of the Rhine and Ruhr rivers. Located in the Ruhr district, it is the largest inland port in Europe and a center of West German steel production. Other manufactures include textiles, chemicals, and metal and wood products. Duisburg was a port in Roman times. It passed to the duchy of Cleves in 1290 and in 1614 was acquired, with Cleves, by Brandenburg. Its growth as an industrial center dates from c.1850. As a center of the German armaments industry, the city was heavily bombed during World War II.

Duke of York Islands (do͞ok', dyo͞ok'; yôrk'), group of 13 coral islands (1969 est. pop. 5,870), 23 sq mi (59.6 sq km), SW Pacific, in the Bismarck Archipelago, part of Papua New Guinea. There are several coconut plantations.

Dukes (do͞oks, dyo͞oks), county (1970 pop. 6,117), 106 sq mi (274.5 sq km), SE Mass., comprising the islands of Martha's Vineyard and Elizabeth Islands; formed 1695; co. seat Edgartown. In an agricultural and fishing area, it has many summer resorts and a state lobster hatchery.

Du·luth (də-lo͞oth'), city (1978 est. pop. 93,000), seat of St. Louis co., NE Minn., at the W end of Lake Superior, at the head of lake navigation and opposite Superior, Wis.; inc. 1870. It is a commercial, industrial, and cultural center of northern Minn. and the gateway to a resort region. Huge amounts of grain, iron ore (especially taconite), and bulk cargo are shipped on lake freighters and ocean vessels. Manufactures include steel and cement; metal and wood products; electrical equipment; textiles; and prepared foods. Indian settlements were found here (1670s) by the early explorers and fur traders. Permanent settlement began c.1852. Built largely on rocky bluffs overlooking the lake, the city was at first a trade and shipping center for the timber country. Discovery of iron (1865) in the Mesabi range made it the chief ore-shipping point for the nation's steel mills. With the opening of the St. Lawrence Seaway (1959), it became one of the leading ports on the Great Lakes for the export of grain. Points of interest include the huge Aerial Lift Bridge, linking the city to 7 mi (11.3 km) of sand beach on Park Point; the Skyline Blvd., winding high above the city for 15 mi (24.1 km); and Leif Erikson Park, with its replica of the Viking ship that sailed from Norway in 1922.

Du·ma·gue·te (do͞o'mä-gā'tā), city (1970 est. pop. 21,811), capital of Negros Oriental prov., SE Negros, the Philippines. A busy interisland-shipping port, it is also a trade and cultural center.

Du·mas (do͞o'məs), city (1970 pop. 9,771), seat of Moore co., extreme N Texas, N of Amarillo on the plains of the Panhandle; founded 1892, inc. 1930. Originally a cattle town and later a wheat-shipping center, Dumas turned to industry with the development of extensive natural-gas fields. It has a helium plant, a zinc smelter, an oil refinery, and carbon-black plants.

Dum·bar·ton (dŭm-bär'tən), royal burgh (1974 est. pop. 25,440), Strathclyde region, W Scotland, at the confluence of the Leven and Clyde rivers. It is a shipbuilding center and has engineering works. Aircraft, bricks, and whiskey are produced. Castles that played an important role in Scottish history were built on Dumbarton Rock, a 250-ft (76.3-m) high hill of basalt, from at least the 5th cent., when Dumbarton was the capital of the kingdom of Strathclyde. It was granted a royal charter in 1222. Until 1975 Dumbarton was the county town of Dunbartonshire.

Dum·dum (dŭm'dŭm), town (1971 pop. 31,232), West Bengal state, E central India, a suburb of Calcutta. In the 19th cent. its arsenal was the first to manufacture lead-nosed bullets that spread on impact, inflicting a tearing wound.

Dum·fries (dŭm-frēs', dəm-), burgh (1974 est. pop. 29,431), SE Scotland, on the Nith River. The chief manufactures are nitrocellulose, hosiery, knitwear, rubber goods, and canned milk. Dumfries was sacked by the English in 1448, 1536, and 1570. Robert Burns lived in Dumfries (1791–96) and is buried in St. Michael's Church. Until 1975 Dumfries was the county town of Dumfriesshire.

Dumfries and Gal·lo·way (găl'ə-wā'), region (1976 est. pop. 143,585), SE Scotland. The region was formed in 1975.

Dum·fries·shire (dŭm-frēs'shîr', -shər, dəm-) or **Dumfries,** former county SE Scotland. Dumfries was the county town. In 1975 Dumfriesshire became part of the new Dumfries and Galloway region.

Du·mont (do͞o'mŏnt), borough (1970 pop. 20,155), Bergen co., NE N.J.; settled 1677 by the Dutch, inc. 1894. It is a residential suburb of Hackensack.

Dum·yat (do͞om-yät') or **Dam·i·et·ta** (dăm'ē-ĕt'ə), city (1970 est. pop. 98,000), capital of Dumyat governorate, N Egypt, on Lake

Manzala near the Mediterranean Sea. It is a manufacturing and trade center. Its products include glassware, cotton, silk, and rayon textiles, and processed rice and fish. Of commercial and strategic importance in the Middle Ages, Dumyat was pillaged by the Byzantines and by the Sicilian Normans. It was captured and held by Crusaders from 1219 to 1221 and again (under Louis IX of France) from 1249 to 1250.

Dun·bar (dŭn-bär′), burgh (1971 pop. 4,586), Lothian region, SE Scotland, on the North Sea. It is a fishing center and seaside resort. Mary Queen of Scots was abducted to Dunbar Castle by the earl of Bothwell and stayed there the night before her defeat at Carberry Hill (1567). Oliver Cromwell defeated the Scots here in 1650.

Dun·bar·ton·shire (dŭn-bär′tən-shîr′, -shər) or **Dum·bar·ton·shire** (dŭm-), former county, W central Scotland. Dumbarton was the county town. In 1975 Dunbartonshire became part of the Strathclyde region.

Dun·can (dŭng′kən), city (1978 est. pop. 20,200), seat of Stephens co., SW Okla., in an oil, farm, and cattle area; inc. 1892. Its economy is based chiefly upon the oil industry; there are oil refineries and meat-packing plants in the city. Oil-drilling equipment and clothing are manufactured here. The city is located on the old Chisholm Trail.

Dun·cans·bay Head or **Dun·cans·by Head** (both: dŭng′kənz-bē), sandstone cliff, 210 ft (64 m) high, in Caithness, NE extremity of the Scottish mainland.

Dun·can·ville (dŭng′kən-vĭl′), city (1978 est. pop. 26,200), Dallas co., N Texas, a suburb of Dallas; est. 1882, inc. 1947. It is mostly residential with some light manufacturing.

Dun·dalk (dŭn-dôk′), urban district (1971 pop. 21,672), county town of Co. Louth, NE Republic of Ireland, near the mouth of the Castletown River at Dundalk Bay. It has tobacco, clothing, and shoe factories, distilleries, breweries, flour mills, linen mills, and fisheries. Dundalk is also a port and a railroad center.

Dundalk, city (1978 est. pop. 88,800), Baltimore co., NE Md., a suburb of Baltimore, on the Patapsco River; inc. 1946. It has one of the world's largest steel plants and a busy marine terminal. There are also factories making radio and electronic parts, auto bodies, and yachts.

Dun·das (dŭn-dăs′, dŭn′dəs), town (1976 pop. 19,179), S Ont., Canada. It is a suburb of Hamilton and is at the head of the Desjardins Canal, which formerly gave it water connection with Hamilton and other ports. The canal is no longer in use.

Dun·dee (dŭn-dē′, dŭn′dē), city (1976 est. pop. 194,420), Tayside region, E central Scotland, on the Firth of Tay. It is a port and manufacturing city. Jute processing (using jute imported from Asia) is the largest industry; others include engineering, shipbuilding and repairing, textiles (especially linen), food processing (the marmalade is famous), and the manufacture of brick, plastics, metal products, linoleum, and electrical products. Dundee was a center of the Reformation and a stronghold of the Covenanters in the religious wars.

Dun·dy (dŭn′dē), county (1970 pop. 2,926), 921 sq mi (2,385.4 sq km), extreme SW Nebr., bordering in the W on Colo., in the S on Kansas, and drained by the Republican River; formed 1873; co. seat Benkelman. Its agriculture includes grain, cattle, dairy products, and poultry. It also has some mining and manufacturing.

Dun·e·din (dŭn-ēd′n), city (1976 pop. 82,546), SE South Island, New Zealand, at the head of Otago Harbor. Dunedin is an important port and industrial center. The Univ. of Otago (the first in New Zealand), Knox College, and Anglican and Roman Catholic theological schools and cathedrals are in the city. It was founded in 1848 as a Free Church of Scotland settlement.

Dunedin, resort city (1978 est. pop. 26,000), Pinellas co., W Fla., on the Gulf Coast and St. Joseph Sound (part of the Intracoastal Waterway); founded by Scots in 1870, inc. 1898. It is a processing center for citrus fruit. The city is connected to Dunedin Beach, an island, by a causeway.

Dun·ferm·line (dŭn-fûrm′lĭn, dŭm-), burgh (1974 est. pop. 53,418), Fife region, E central Scotland, on the Firth of Forth. It is a center for the manufacture of table linen and terylene, a synthetic fabric, and has silk mills, collieries, and engineering works. Dunfermline abbey, founded by Malcolm III of Scotland in the 11th cent., holds his remains and those of his wife, St. Margaret, and of Robert I. The palace was a favorite seat of Scottish kings and was the birthplace of Charles I of England. Andrew Carnegie, the industrialist, was born in Dunfermline, which is now the headquarters of the Carnegie Trusts. Carnegie gave the town its library and Pittencrieff Glen, a 60-acre (24.3-hectare) public park.

Dunge·ness (dŭnj′nĕs′), flat shingle headland, c.6,000 acres (2,430 hectares), forming the E extremity of Kent, SE England. The Dutch defeated the English in a battle here (1652). There are wildlife sanctuaries in the area. Work on a nuclear power station was begun here in 1960. The headland is slowly extending seaward by accumulation of the shingle.

Dun·keld (dŭn-kĕld′), village, Perthshire, central Scotland. It was a center of Celtic Christianity and the seat of the early Scottish kings.

Dun·kirk (dŭn′kûrk′), town (1975 pop. 83,163), Nord dept., N France, on the North Sea. It is a leading French port and one of Western Europe's major iron and steel centers. Other important industries are oil production and refining, shipbuilding, food processing, brewing, and the manufacture of textiles and electrical equipment. Probably founded c.7th cent. A.D. and often fortified, Dunkirk played a key role in the struggles in Europe that extended over centuries; it was ruled successively by Flanders, Burgundy, Austria, France, Great Britain, and Spain. Ceded briefly in the 1650s to Oliver Cromwell, it was bought back permanently from Charles II by Louis XIV in 1662. The town withstood an Anglo-Dutch bombardment in 1694 and an English siege in 1793. During World War II more than 300,000 Allied troops cut off from retreat on land by the German breakthrough to the French Channel ports were evacuated (May 26–June 4, 1940) from Dunkirk. The retreat is considered one of the epic actions of naval history.

Dunkirk, city (1978 est. pop. 15,600), Chautauqua co., SW N.Y., on Lake Erie; founded c.1800, inc. as a city 1880. It is a port of entry. Dunkirk produces wines and other grape products. The city also manufactures steel, food products, pet foods, and clothing.

Dun·klin (dŭng′klĭn), county (1970 pop. 33,742), 543 sq mi (1,406.4 sq km), SE Mo., bounded on the W and S by the Ark. border and drained by the St. Francis River and numerous drainage canals; formed 1845; co. seat Kennett. Its agriculture includes cotton, corn, wheat, melons, strawberries, and livestock. Clothing and wood products are manufactured.

Dun Laoghai·re (dŭn lâr′ə), borough (1971 pop. 53,171), Co. Dublin, E central Republic of Ireland, on the Irish Sea. It is the main passenger and mail port for Dublin and a seaside resort.

Dun·lap (dŭn′lăp), city (1970 pop. 1,672), seat of Sequatchie co., SE Tenn., on the Sequatchie River, NNW of Chattanooga, in a dairy, livestock, poultry, and fruit region; inc. 1909.

Dun·more (dûn-môr′, dŭn′môr), borough (1978 est. pop. 17,400), Lackawanna co., NE Pa., an industrial suburb of Scranton; inc. 1783. It is the center of an anthracite-coal region.

Dunn (dŭn). **1.** County (1970 pop. 4,895), 2,992 sq mi (7,749.3 sq km), W N.Dak., drained by the Little Missouri and Knife rivers; formed 1883; co. seat Manning. In an agricultural area yielding wheat, flax, oats, cattle, dairy products, and poultry, it also has coal mines and oil wells. Part of Fort Berthold Indian Reservation is in the north. **2.** County (1970 pop. 28,991), 858 sq mi (2,222.2 sq km), W Wis., in a dairying area drained by forks of the Red Cedar and Chippewa rivers; formed 1854; co. seat Menomonie. Livestock and lumber are also important. Clothing and farm equipment are manufactured.

Dun·si·nane (dŭn′sə-nān′, dŭn′sə-nān′), westernmost of the Sidlaw Hills, 1,012 ft (308.7 m) high, Perthshire, central Scotland. On its summit are ruins of a fort, called Macbeth's Castle; it is the traditional scene of Macbeth's final defeat as related by Shakespeare.

Dun·sta·ble (dŭn′stə-bəl), municipal borough (1973 est. pop. 32,090), Bedfordshire, SE England. Located at the meeting point of the ancient Icknield and Watling streets, Dunstable is a developing residential and industrial district, with printing plants and extensive automobile works. There are interesting traces of Stone and Bronze Age civilizations, including the Maiden Bower and Five Knolls earthworks; one of the knolls, excavated in 1926, contained remains and ornaments of a woman of c.2000 B.C.

Du Page (do͞o pāj′), county (1970 pop. 490,822), 331 sq mi (857.3 sq km), NE Ill., partly in the Chicago suburban area and drained by the Des Plaines River; formed 1839; co. seat Wheaton. In an agricultural and dairying region, it has many residential communities, diversified manufacturing, railroad shops, limestone quarries, and nurseries.

Du·plin (dōō′plĭn), county (1970 pop. 38,015), 815 sq mi (2,110.9 sq km), SE N.C., in a forested coastal-plain area drained by the Cape Fear River; formed 1749; co. seat Kenansville. Its agriculture includes tobacco, corn, soybeans, cotton, and poultry. It also has textile and timber industries.

Du·pree (dōō-prē′), town (1970 pop. 523), seat of Ziebach co., N central S.Dak., NW of Pierre on the Elm Creek. It is a trading center in a livestock, poultry, flax, and wheat area.

Du·que de Ca·xi·as (dōō′kē də kə-shē′əs, kăsh′yəs), city (1970 pop. 256,582), Rio de Janeiro state, SE Brazil, on Guanabara Bay. It is a commercial and industrial suburb of Rio de Janeiro.

Du·quesne (dōō-kān′, dyōō-), city (1970 pop. 11,410), Allegheny co., SW Pa., on the Monongahela River, opposite McKeesport, in a coal region; settled 1789, laid out 1885 by the Duquesne Steel Company, inc. as a city 1917. Iron and steel are produced in the city.

Du·ra (dŏŏr′ə) or **Eu·ro·pus** (yōō-rō′pəs), ancient city of Syria, on the Euphrates River and E of Palmyra, sometimes called Dura-Europus or Dura-Europos. Founded c.300 B.C., it was taken (2nd cent. A.D.) by the Parthians and (A.D. 165) by Rome. It remained a Roman city until it was seized (A.D. c.257) by Shapur I of Persia. Dura was then abandoned to the desert. Modern excavations have yielded rich finds, supplying much information on life, history, and art in Mesopotamia from Hellenistic through Roman times.

Du·rance (dü-räNs′), river, c.180 mi (290 km) long, rising in SE France at the foot of Montgenèvre Pass on the Italian border and flowing SW, then NW before entering the Rhône River at Avignon.

Du·rand (də-rănd′), city (1970 pop. 2,103), seat of Pepin co., W Wis., on the Chippewa River SW of Eau Claire; inc. 1887.

Du·ran·go (dōō-räng′gō), city (1974 est. pop. 182,600), capital of Durango state, N central Mexico, along the highway linking Mexico City with El Paso, Texas. Minerals are the chief product, but the city is also an agricultural and commercial center. Founded as a mining town in 1563, Durango served as capital of the region of Nueva Viscaya. In the 19th cent. the city suffered frequent scorpion plagues.

Durango, city (1978 est. pop. 11,800), seat of La Plata co., SW Colo., on the Animas River; inc. 1881. It is situated in a mountainous region at an altitude of c.6,500 ft (1,985 m). The economy is based on farming, mining, lumbering, and tourism. The mining of carnotite ore for uranium brought about a boom in 1948.

Du·rant (də-rănt′), city (1970 pop. 11,118), seat of Bryan co., S central Okla., in the Red River valley farm area; inc. 1873. It is the commercial and processing center for an agricultural region where peanuts, cotton, wheat, and oil are produced and cattle are raised.

Dur·ban (dûr′bən), city (1970 pop. 736,852), Natal prov., E South Africa, on Durban Bay, an arm of the Indian Ocean. Durban is an industrial center, a major seaport, and a year-round resort. Industries include shipbuilding and ship repairing, petroleum refining, sugar refining, whaling and fishing, automobile assembly, and the manufacture of food products, paint, chemicals, fertilizers, soap, footwear, and textiles. Sugar cane is grown on nearby estates. The site of Durban was discovered in 1497 by Vasco da Gama, but was not settled until 1824, when Britons arrived. The city, first called Port Natal, was renamed Durban in 1835 for Sir Benjamin D'Urban, then governor of Cape Colony. In 1842 Boers besieged British troops in the Old Fort (now a museum). Durban became the chief commercial city of Natal after 1887, when the bay was dredged. The city was the site of the national convention (1908–09) that paved the way for the creation in 1910 of the Union of South Africa.

Dü·ren (dür′ən), city (1974 est. pop. 88,579), North Rhine-Westphalia, W West Germany, on the Ruhr River. It is a transportation and industrial center; manufactures include iron, steel, glass, textiles, and chemicals. Düren was a center of Carolingian culture. In 1543 it was captured and burned by the troops of Emperor Charles V. The city was severely damaged during World War II.

Dur·ga·pur (dŏŏr′gä-pŏŏr), city (1971 pop. 207,232), West Bengal state, E central India, on the Damodar River. Located in an area of iron and coal mines, the city has an iron and steel plant, completed in 1962 with British aid.

Dur·ham (dûr′əm), nonmetropolitan county (1976 est. pop. 610,400), NE England, on the North Sea between the Tees and Tyne rivers; administrative center Durham. The region is low-lying along the coast, rising inland to the Pennines. The area was occupied by the Romans and subsequently became part of the Anglo-Saxon kingdom of Northumbria. In 1974 most of the former county of Durham plus a small area of the North Riding of Yorkshire was reorganized as this new county. Northeastern Durham became part of the new metropolitan county of Tyne and Wear, and southeastern Durham became part of the new nonmetropolitan county of Cleveland.

Durham, municipal borough (1973 est. pop. 29,490), administrative center of Durham, NE England, on the sides of a hill nearly encircled by the Wear River. There are small factories producing organs and carpets. In 995 the relics of St. Cuthbert were brought to Durham (then Dunholme), and a church was built as his shrine. The present cathedral, begun on the same site in 1093, is considered the finest example of Norman architecture in the country. It contains the tomb of the Venerable Bede (d. 735).

Durham, county (1970 pop. 132,681), 295 sq mi (764.1 sq km), NE central N.C., in a fertile piedmont region drained by headstreams of the Neuse River; formed 1881; co. seat Durham. Tobacco is grown and cured here. The county also has stone quarries and pine forests.

Durham. 1. Town (1970 pop. 8,869), Strafford co., SE N.H., on the Oyster River SW of Dover; settled 1635, inc. 1732. The early settlement suffered greatly during the Indian wars of 1694 and 1704. The town is the seat of the Univ. of New Hampshire. **2.** City (1978 est. pop. 105,400), seat of Durham co., N central N.C., in the Piedmont; inc. 1867. A prominent center for the marketing and processing of tobacco, Durham is a major cigarette manufacturer. It also has textile and hosiery mills and an important insurance industry. The area was settled c.1750. Durham is the seat of Duke Univ.

Dur·rës (dŏŏr′əs), city (1971 est. pop. 55,000), capital of Durrës prov., W Albania, on the Adriatic Sea. The chief seaport of Albania and the leading commercial and communications center, it has a dockyard, a power plant, and industries that manufacture clothing, foodstuffs, and leather and tobacco products. Founded (c.625 B.C.) as Epidamnus, a joint colony of Corinth and Corcyra, it became an important trade center. The quarrel between Corinth and Corcyra over Epidamnus helped to precipitate (431 B.C.) the Peloponnesian War. Durrës passed to Rome in 229 B.C. and became a military and naval base known as Dyrrhachium. The city passed to the Byzantine empire in the 8th cent., to the Normans of Sicily in 1185, to Naples in 1272, and to Serbia in 1336. Venice captured it in 1392 and held it until 1501 when it passed to the Turks. Under Turkish rule Durrës declined rapidly and almost disappeared. It was occupied (1912) by the Serbs in the First Balkan War, but was assigned to Albania in 1913. Italy (1915) and Austria (1916–18) also occupied the city. Durrës was the capital of Albania from 1913 to 1920, and revived thereafter as the country's chief seaport. It suffered heavy damage during World War II. The city, with its many mosques, has an Oriental character.

Du·shan·be (dōō-shăn′bə, -shän′-), city (1977 est. pop. 460,000), capital of the Tadzhik Soviet Socialist Republic, Central Asian USSR. It is a major industrial and cultural center in a rich agricultural area. Coal, lead, and arsenic are mined nearby. A leading cotton textile center, Dushanbe also produces silk, textile machinery, clothing, leather goods, tractor parts, and foodstuffs.

Düs·sel·dorf (düs′əl-dôrf′), city (1975 pop. 675,437), capital of North Rhine-Westphalia, W West Germany, at the confluence of the Rhine and Düssel rivers. It is a major industrial, financial, and commercial center, a busy inland port, and an important rail junction. Its manufactures include iron, steel, machinery, chemicals, textiles, clothing, and paper. Chartered in 1288, Düsseldorf was occupied by France in 1795 and in 1815 became part of Prussia. Its industrial growth dates from c.1870. After World War I it was occupied again by France from 1921 to 1925. The city was badly damaged during World War II. Present-day Düsseldorf is an elegant city and a cultural center, with noted theaters and museums, an art academy, and a university. Heinrich Heine, the poet, was born (1797) here.

Dust Bowl (dŭst′ bōl′), areas of the U.S. prairie states that suffer from dust storms. The storms may smother growing crops, destroy pasturage, and injure health; but their most serious effect is the removal of topsoil. During World War I the high price of wheat and the needs of Allied troops encouraged farmers to grow more wheat by plowing and seeding areas that were formerly used only for grazing. After years of good yields, livestock were returned to graze the area; their hooves pulverized the unprotected soil. In 1934 strong winds blew the soil into huge clouds, and each succeeding year, from Dec. to May, the dust storms have recurred. However, the Dust Bowl, which covered 25,000 sq mi (64,750 sq km) at its greatest extent in the late 1930s, has been shrinking because of increased rainfall, regrassing, and erosion-preventing measures such as contour farming.

Dutch·ess (dŭch′ ĭs), county (1970 pop. 222,295), 813 sq mi (2,105.7 sq km), SE N.Y., bounded on the W by the Hudson River, on the E by the Conn. border; formed 1683; co. seat Poughkeepsie. In a resort, residential, and dairy-farming area, it also has limestone quarries and diversified light industry. The Vanderbilt Mansion and Hyde Park, home of Franklin D. Roosevelt, are here—both are national historic sites.

Dutch Gui·an·a (gē-ăn′ə, -ä′nə): *see* Surinam.

Du·val (do͞o-văl′). **1.** County (1970 pop. 528,865), 766 sq mi (1,983.9 sq km), NE Fla., in a lowland area drained by the St. Johns River and bounded on the E by the Atlantic Ocean; formed 1822; co. seat Jacksonville. Most of the county is included within the city limits of Jacksonville. Fishing, forestry, and agriculture (corn, vegetables, and dairy and poultry products) are important. There are resorts along the coastline. **2.** County (1970 pop. 11,722), 1,818 sq mi (4,708.6 sq km), S Texas; formed 1858; co. seat San Diego. In a ranching area, it also produces cotton, peanuts, corn, fruit, truck crops, and grain sorghums. The county also has extensive oil deposits, sulfur and salt mines, and some manufacturing.

Dux·bur·y (dŭks′bĕr′ē, -bə-rē), resort town (1970 pop. 7,636), Plymouth co., SE Mass., on Duxbury Bay SE of Boston; settled c.1624, inc. 1637. Plymouth colonists, including Miles Standish, William Brewster, and John Alden, settled here.

Dvi·na (dvē-nä′). **1.** Or **Northern Dvina,** river, c.465 mi (750 km) long, N European USSR. It is formed near Veliki Ustyug by the union of the Sukhona and Yug rivers, flows north past Kotlas, then turns northeast, and empties into Dvina Bay, an arm of the White Sea, just below Arkhangelsk. **2.** Or **Western Dvina,** river, c.635 mi (1,020 km) long, NW European USSR. Rising in the Valdai Hills, it flows south and then generally west through Belorussia and Latvia, past Daugavpils and Riga, into the Gulf of Riga, an arm of the Baltic Sea. It is navigable in its upper course, but, because of rapids, only partly navigable in its lower course.

Dvůr Krá·lo·vé nad La·bem (dvo͞or krä′lô-vâ nät lä′bĕm), town (1970 pop. 16,220), NW Czechoslovakia, in Bohemia, on the Labe (Elbe) River. Among its manufactures are cement, cotton and linen textiles, machinery, and beer. Founded in 1139 as a ducal palace, the town was given by King Wenceslaus II of Bohemia to his queen, Elizabeth, in the late 13th cent.

Dy·er (dī′ər), county (1970 pop. 30,427), 527 sq mi (1,364.9 sq km), NW Tenn., bounded on the W by the Mississippi River, on the SW by the Forked Deer River, and drained by the Obion River; formed 1823; co. seat Dyersburg. In a fertile cotton and livestock-grazing area, it also has farms growing corn, soybeans, and wheat. Clothing and lumber and metal products are manufactured.

Dy·ers·burg (dī′ərz-bûrg′), city (1978 est. pop. 15,000), seat of Dyer co., NW Tenn., near the Mississippi River; inc. 1850. It is a processing and industrial center for a rich cotton and farm belt.

Dy·fed (dĭv′ĕd), nonmetropolitan county (1976 est. pop. 323,100), W Wales, created under the Local Government Act of 1972 (effective 1974); administrative center Carmathen. It comprises the former counties of Cardiganshire, Carmarthenshire, and Pembroke.

Dykh-Tau (dĭKH′tou′), peak, c.17,000 ft (5,185 m) high, SE European USSR, in the central Greater Caucasus.

Dzer·zhinsk (dyĕr-zhĭnsk′), city (1976 est. pop. 245,000), W European USSR, a port on the Oka River. There are chemical, textile, and cable industries. The city was called Chernorech until about 1919 and Rastyapino until 1929.

Dzham·bul (jəm-bo͞ol′), city (1976 est. pop. 246,000), capital of Dzhambul oblast, Central Asian USSR, on the Talas River and the Turkistan-Siberia RR. Industries include food processing and the manufacture of chemicals, metal products, and leather goods. Founded in the 7th cent., it was ruled by Arabs in the 8th and 9th cent. In 1864 it passed to Russia. It was called Aulie-Ata until 1938.

Dzier·żo·niów (jîr-zhô′nyo͞of), town (1975 est. pop. 35,400), SW Poland. It is a manufacturing center known for its textiles (especially woolens) and for its machine-building and electrical equipment industries. Two treaties were signed here. By the first (1790), Austria promised Prussia to renounce acquisition of Turkish territory; by the second (1813), Austria conditionally consented to join the coalition against Napoleon I.

Dzun·gar·i·a (zo͝ong-gâr′ē-ə), region, c.300,000 sq mi (777,000 sq km) of Sinkiang Uigur Autonomous Region, NW China. It is a largely steppe and semidesert basin surrounded by high mountains. Wheat, barley, oats, and sugar beets are grown, and cattle, sheep, and horses are raised. The fields are irrigated with melted snow from the permanently white-capped mountains. Dzungaria has deposits of coal, iron, and gold, as well as large oil fields. It was ruled by a confederation of Western Mongols that established (17th cent.) a large empire in central Asia. The region passed to the Chinese in the mid-18th cent. At the eastern end of the Dzungarian Ala-Tau, on the Russian-Chinese border, lies the Dzungarian Gate, which for centuries was used as an invasion route by conquerors from central Asia.

Eads (ēdz), town (1970 pop. 795), seat of Kiowa co., E Colo., NE of La Junta, in a grain, livestock, and natural-gas area. Dairy and poultry products are made.

Ea·gle (ē′gəl), county (1970 pop. 7,498), 1,685 sq mi (4,364.2 sq km), NW central Colo., in the Rocky Mts. and drained by the Colorado and Eagle rivers; formed 1883; co. seat Eagle. It has livestock, some farming, timber, and light industry. Lead and zinc ores and some silver, gold, and copper are mined.

Eagle, town (1970 pop. 790), seat of Eagle co., NW central Colo., on the Eagle River NW of Leadville. It is a trading point in a farming, mining, and lumbering area.

Eagle Pass, city (1975 est. pop. 19,098), seat of Maverick co., W Texas, a port of entry on the Rio Grande opposite Piedras Negras, Mexico; inc. 1918. Linked by highway with Mexico City, it is a tourist resort and a shipping and processing point for vegetables grown in the richly cultivated lowlands along the Rio Grande. The city has also prospered from mineral processing and international trade. The site of a U.S. army camp during the Mexican War, it was on one of the main routes to California during the gold rush.

Eagle River. 1. Village (1970 pop. 150), seat of Keweenaw co., Mich., NW Upper Peninsula, on Lake Superior NE of Houghton. It is a shipping and distributing point and a summer resort. **2.** City (1970 pop. 1,326), seat of Vilas co., N Wis., NNE of Rhinelander, in a lake and woods region; inc. as a village 1923, as a city 1937. It is a center for summer and winter sports.

Ea·ling (ē′lĭng), borough (1971 pop. 299,450) of Greater London, SE England. The borough was created in 1965 by the merger of the municipal boroughs of Acton, Ealing, and Southall. It is highly industrialized; motor vehicles, scientific instruments, glass, plastics, and engineering products are manufactured.

Ear·ly (ûr'lē), county (1970 pop. 12,682), 525 sq mi (1,359.8 sq km), SW Ga., bounded on the W by the Chattahoochee River, here forming the Ala. border, and drained by Spring Creek; formed 1818; co. seat Blakely. In a coastal-plain area yielding cotton, corn, peanuts, and pecans, it has stands of timber and limestone quarries. Wood and paper products and textiles are manufactured.

Earn, Loch (ûrn), lake, 7 mi (11.3 km) long and 1 mi (1.6 km) wide, Perthshire, central Scotland. Earn River (46 mi/74 km long), the lake's outlet, flows eastward into the Firth of Tay.

Eas·ley (ēz'lē), city (1975 est. pop. 12,161), Pickens co., NW S.C., in the foothills of the Blue Ridge Mts.; inc. 1874. Cotton is grown, and clothing, textiles, and textile machinery are manufactured.

East An·gli·a (ēst' ăng'glē-ə), kingdom of Anglo-Saxon England. It was settled in the late 5th cent. by Angles. Little is known of its early history, but its large size and the fact that it was protected by fens probably made it one of the most powerful English kingdoms in the late 6th cent. East Anglia was a dependency of the kingdom of Mercia for long periods after 650. In 825 the East Anglians rebelled against Mercia, with the help of Egbert of Wessex, but thereafter their kingdom was a dependency of Wessex. The great Danish invading army was quartered (865–66) in East Anglia and returned (869) to conquer the kingdom completely, to destroy its monasteries, and to murder its young ruler, St. Edmund. When King Alfred of Wessex first defeated the Danes in the 870s, they retired to an area that included East Anglia, and the treaty of 886 confirmed the region as part of the Danelaw. Its Danes gave aid to later Viking invaders and continued to harass Wessex until Edward the Elder finally defeated their army in 917. After that time, East Anglia was an earldom of England.

East Au·ror·a (ô-rôr'ə, ə-rôr'ə), village (1970 pop. 7,033), Erie co., W central N.Y., SE of Buffalo in a manufacturing and agricultural area; settled 1804, inc. 1874. The Roycroft Colony of craftsmen and the Roycroft Press were established here by Elbert Hubbard.

East Bat·on Rouge (băt'n ro͞ozh'), parish (1970 pop. 285,167), 462 sq mi (1,196.6 sq km), SE central La., bounded on the E by the Amite River, on the W by the Mississippi; formed 1810; parish seat Baton Rouge. It is a shipping and commercial center for a large region, with oil and gas fields, timber, and sand and gravel pits. Its agriculture includes corn, cotton, sugar cane, poultry, cattle, hay, sweet potatoes, and dairy products. Furniture, chemicals, cement, concrete, metal products, and machinery are manufactured.

East Ber·lin (bûr-lĭn', bĕr-lēn'): *see* Berlin, Germany.

East·bourne (ēst'bôrn, -bərn), county borough (1976 est. pop. 73,200), East Sussex, SE England. It is a popular resort with a 3-mi (4.8-km) terraced promenade along the sea front. There are glass, soap, brewing, and boatbuilding industries.

East Car·roll (kăr'əl), parish (1970 pop. 12,884), 432 sq mi (1,118.9 sq km), extreme NE La., bounded on the E by the Mississippi River, on the W by Bayou Macon, on the N by the Ark. border; formed 1877; parish seat Lake Providence. It has farming (cotton, corn, oats, soybeans, and livestock), sawmills, natural-gas wells, and fisheries.

East Chi·ca·go (shĭ-kä'gō, -kô'-), city (1978 est. pop. 42,700), Lake co., extreme NW Ind., on Lake Michigan, in the heavily industrialized Calumet region; inc. 1889. Its Indiana Harbor on Lake Michigan is connected with the Grand Calumet River by a ship and barge canal. The city has important steelworks. There are also oil refineries, railroad equipment shops, and chemical plants.

East Chi·na Sea (chī'nə), arm of the Pacific Ocean, c.480,000 sq mi (1,243,200 sq km), bounded on the E by the Kyushu and Ryukyu islands, on the S by Taiwan, and on the W by China. It is connected with the South China Sea by the Formosa Strait and with the Sea of Japan by the Korea Strait; it opens in the north to the Yellow Sea.

East Cool·gar·die Gold·field (ko͞ol-gär'dē gōld'fēld'), Western Australia, SW Australia. Gold was discovered in 1892.

Eas·ter Island (ē'stər), island (1970 est. pop. 1,600), 46 sq mi (119.1 sq km), South Pacific, c.2,200 mi (3,540 km) W of Chile, to which it belongs. Of volcanic origin, Easter Island is mostly covered with grasslands and is swept by strong trade winds. The inhabitants are of Polynesian stock. The land is fertile, and farming is the principal occupation; sheep are raised, and maize, sweet potatoes, figs, bananas, pineapples, melons, and vegetables are grown. The inhabitants are citizens of Chile but do not pay taxes and are not subject to military conscription.

Easter Island was named on Easter Day, 1722, by the Dutch navigator Jakob Roggeven. At that time the population was about 4,000, but with the spread of European epidemics, especially smallpox, and the marauding of Spanish slavers, the population was reduced to slightly more than 100 by 1887. Chile annexed the island in 1888. Easter Island has long been famous for the hieroglyphs and remarkable monolithic stone heads that have evoked various legends and theories as to their origins. The statues, carved from tufa, a soft volcanic stone, range in height from 10 to 40 ft (3–12 m), some weighing more than 50 tons.

East·ern Desert (ē'stərn): *see* Arabian Desert.

Eastern High·lands (hī'ləndz), c.2,400 mi (3,860 km) long, general name for the mountains and plateaus roughly paralleling the E and SE coasts of Australia (including Tasmania) and forming the Continental Divide. It rises to Mt. Kosciusko (7,316 ft/2,231.4 m). Rugged, with many gorges and few gaps, the Eastern Highlands long hindered Australia's westward expansion. Rich in minerals, the highlands contain most of Australia's coal fields; gold, copper, tin, oil, and natural gas are also extracted.

East Fe·li·ci·an·a (fə-lĭsh'ē-ăn'ə), parish (1970 pop. 17,657), 454 sq mi (1,175.9 sq km), E La., bounded on the E by the Amite River, on the N by the Miss. border, on the W by the Thompson River; formed 1824; parish seat Clinton. It has agriculture (cotton, corn, hay, and sweet potatoes), stands of timber, and sand and gravel pits. Wood products are manufactured.

East Flan·ders (flăn'dərz), province (1976 est. pop. 1,325,222), 1,147 sq mi (2,970.7 sq km), NW Belgium, bordering on the Netherlands in the N; capital Ghent.

East Fries·land (frēz'lənd), region and former duchy, c.1,100 sq mi (2,850 sq km), Lower Saxony, NW West Germany, on the North Sea. It includes the East Frisian Islands and is separated in the west from the Netherlands by an inlet of the North Sea. The extensive moors and marshlands of East Friesland have been partly reclaimed. Cattle raising, sheep raising, and farming are carried on, and there is fishing along the coastline. East Friesland became a county of the Holy Roman Empire in 1454, was raised to a duchy in 1654, passed to Prussia in 1744, and was attached to Hanover in 1815.

East Ger·ma·ny (jûr'mə-nē): *see* Germany.

East·hamp·ton (ēst-hămp'tən), town (1975 est. pop. 15,168), Hampshire co., W Mass.; inc. 1809. It is primarily a manufacturing town with diversified light industry. Easthampton was settled in 1664. In the 1820s Samuel Williston founded the town's textile industry.

East Hamp·ton (hămp'tən). **1.** Residential, manufacturing, and agricultural town (1970 pop. 7,078), Middlesex co., central Conn., E of the Connecticut River SE of Hartford; settled c.1710, inc. as Chatham in 1767, renamed 1915. Metal products are made. **2.** Resort and residential village (1970 pop. 1,753), Suffolk co., SE N.Y., on E Long Island, ENE of Southhampton in a diversified farming area; settled 1648, considered part of Conn. until 1664, inc. 1920.

East Hart·ford (härt'fərd), town (1978 est. pop. 52,200), Hartford co., central Conn., on the Connecticut River opposite Hartford; settled c.1640, inc. 1783. East Hartford is a major trucking and warehousing center. Tobacco is grown and processed and bulk oil is stored and distributed there. Fabricated steel, precision parts, aircraft engines, appliances, brushes, and candy are the chief manufactures.

East In·dies (ĭn'dēz), name used primarily for Indonesia, but also more widely to include SE Asia. It once referred chiefly to India.

East Kil·bride (kĭl'brīd), town (1974 pop. 67,834), Strathclyde region, S central Scotland. Established in 1946 under the New Towns Act to absorb the growing population of Glasgow, East Kilbride has engineering works and manufactures automobile and aircraft engines.

East·land (ēst'lənd), county (1970 pop. 18,092), 955 sq mi (2,473.5 sq km), N central Texas, drained by the Leon River; formed 1858; co. seat Eastland. Its agriculture includes peanuts, cotton, corn, oats, sorghums, and dairy products. Poultry, hogs, cattle, sheep, and goats are raised, and mohair is marketed.

Eastland, city (1970 pop. 3,178), seat of Eastland co., N central Texas, E of Abilene; laid out 1875, inc. 1897. It is the center of a farm, dairy, and oil region and has small factories. Much publicity was given in the late 1920s to a horned toad said to have remained alive 31 years in the cornerstone of the city's old courthouse.

East Lan·sing (lăn'sĭng), city (1978 est. pop. 51,600), Ingham co., S central Mich., a suburb of Lansing, on the Red Cedar River; inc. 1907. It is a residential city and the seat of Michigan State Univ.

East·leigh (ēst'lē'), municipal borough (1973 est. pop. 46,340), Hampshire, S central England; inc. 1936. Its industries include workshops for the British Railways and two large cable factories.

East Liv·er·pool (lĭv'ər-pōōl'), industrial city (1978 est. pop. 21,200), Columbiana co., NE Ohio, on the Ohio River near the Pa. and W.Va. borders; settled 1798 as St. Clair, called Fawcett's Town until inc. as East Liverpool in 1834. Extensive clay deposits are used in making pottery, brick, and tile. A ceramics center since c.1839, it has a museum housing a historical pottery collection.

East Lon·don (lŭn'dən), city (1970 pop. 119,727), Cape Prov., SE South Africa, on the Indian Ocean. The city grew around a British military post founded in 1847. Its harbor was developed after 1886. East London's manufactures include automobiles, furniture, textiles, clothing, footwear, processed food, and glass. There is a large fishing industry. The city is also a popular seaside resort.

East Los An·ge·les (lôs ăn'jə-ləs, ăng'gə-, -lēz'), uninc. city (1970 pop. 92,000), Los Angeles co., S Calif., a suburb E of Los Angeles.

East Lo·thi·an (lō'*th*ē-ən), former county, SE Scotland. The county town was Haddington. The county was part of the Anglo-Saxon kingdom of Northumbria. It suffered severely in the border warfare (13th–16th cent.) between England and Scotland. In 1975 East Lothian became part of the Lothian region.

East Lyme (līm), town (1975 est. pop. 13,575), New London co., SE Conn., on Long Island Sound; settled c.1660, inc. 1839. The town has light diversified industry. Its many colonial houses include the Thomas Lee House (c.1660), which has been restored.

East·main (ēst'mān'), river, c.510 mi (820 km) long, rising in the Otish Mts., central Que., Canada, and flowing W into James Bay. Three miles (4.8 km) from its mouth is East Main (founded 1685), one of the oldest Hudson's Bay Company posts.

East·man (ēst'mən), city (1970 pop. 5,416), seat of Dodge co., central Ga., SSE of Macon, in an agricultural and timber area; founded after the coming of the railroad (1871).

East Mo·line (mō-lēn'), city (1970 pop. 20,832), Rock Island co., NW Ill., an industrial suburb of Moline, on the Mississippi River; inc. 1907. East Moline, Moline, Rock Island, and Davenport, Iowa, are known as the Quad Cities. Farm equipment is made.

Eas·ton (ē'stən). **1.** Town (1970 pop. 6,809), seat of Talbot co., Md., on the Eastern Shore on Chesapeake Bay SE of Annapolis; settled 1710, inc. 1790. It is a trade center of a farm region. Food processing is done. The Friends Meeting House dates from c.1684. **2.** Town (1975 est. pop. 14,221), Bristol co., E. Mass., in a farm area; settled 1694, inc. 1725. It has a foundry and assorted manufactures. **3.** Industrial city (1978 est. pop. 28,600), seat of Northampton co., E. Pa., at the junction of the Delaware and Lehigh rivers; founded 1751 by Thomas Penn, inc. as a city 1886. Cement, paper products, electronic equipment, machinery, and steel are among its manufactures. During canal days Easton was a coal-receiving port in the Pennsylvania Dutch region.

East Or·ange (ôr'ĭnj), city (1978 est. pop. 72,300), Essex co., NE N.J.; settled 1678, separated from Orange and inc. 1863. A residential city adjacent to Newark, it is the seat of Upsala College.

East Pe·o·ri·a (pē-ôr'ē-ə), city (1975 est. pop. 21,822), Tazewell co., N central Ill., on the Illinois River opposite Peoria; inc. 1919. It is a rail and warehousing center for central Illinois. Tractors, earth-moving machinery, and diesel engines are made.

East Point, city (1978 est. pop. 35,700), Fulton co., NW Ga., an industrial suburb of Atlanta; inc. 1887. Textiles, machinery, fertilizer, chemicals, and paper are among the manufactures.

East·port (ēst'pôrt'), city (1970 pop. 1,989), SE Maine, on Moose Island in SE Passamaquoddy Bay; settled c.1780, inc. as a town 1798, as a city 1893. A port of entry, Eastport was occupied by the British in the Revolution and in the War of 1812. Fishing and fish processing are the chief industries; the country's first sardine cannery was built here c.1875. The tidal variation at Eastport averages 18 ft (5.5 m), the highest on the eastern coast of the United States.

East Prov·i·dence (prŏv'ĭ-dəns), city (1978 est. pop. 49,900), Providence co., E R.I., on the Providence and Seekonk rivers; inc. as a city 1958. A wholesale and distribution center for petroleum products, East Providence is also the site of factories producing metal goods and machinery. Originally part of Mass., it was organized as a R.I. town 1862.

East Prus·sia (prŭsh'ə), former province of Prussia, extreme NE Germany. From 1919 to 1939 it was separated from the rest of Germany by the Polish Corridor. East Prussia bordered on Poland and Lithuania in the south and east and stretched to Memel and the Baltic Sea in the north and northeast. In 1945, at the end of World War II, East Prussia was overrun by Soviet troops and about 600,000 of its inhabitants were killed. At the Potsdam Conference (1945) East Prussia was divided by two transfers; the transfers were made permanent by treaties between West Germany and Poland and the USSR that were signed and ratified between 1970 and 1972. The northern part was assigned at Potsdam to the USSR. The rest was incorporated into Poland as Olsztyn prov. Most Germans remaining at the end of the war were expelled by the Polish and Soviet governments shortly after its end.

East River, tidal strait, 16 mi (25.7 km) long and 600–4,000 ft (183–1,220 m) wide, connecting Upper New York Bay and Long Island Sound, New York City, and separating the boroughs of Manhattan and the Bronx from Brooklyn and Queens. The East River is linked with the Hudson River at the northern end of Manhattan island by the Harlem River. Hell Gate, at the junction of the Harlem and East rivers, was named for its treacherous currents and rocky reefs.

East Ruth·er·ford (rŭth'ər-fərd), industrial borough (1970 pop. 8,536), Bergen co., NE N.J., SE of Passaic; inc. 1894. Medical and surgical supplies and asbestos products are made.

East Saint Lou·is (sānt' lōō'ĭs), city (1978 est. pop. 53,600), St. Clair co., SW Ill., on the Mississippi River opposite St. Louis, with which it is connected by five bridges; inc. 1859. It is an important transportation hub, industrial center, and livestock market. East St. Louis has major chemical and meat-packing industries and factories that produce a great variety of products, chiefly those using the clay, stone, and silica of the area. The first settlement here was in 1765.

East Sus·sex (sŭs'ĭks), nonmetropolitan county (1976 est. pop. 655,600), extreme SE England; administrative center Lewes. Created under the Local Government Act of 1972 (effective 1974), it is composed of the county boroughs of Brighton, Eastbourne, and Hastings, and parts of the former county of East Sussex.

East·ville (ēst'vĭl'), town (1970 pop. 203), seat of Northampton co., E Va., NE of Cape Charles, in a truck-farming region. County records beginning in 1632 are preserved here.

East Wind·sor (wĭn'zər), town (1970 pop. 8,513), Hartford co., N Conn., on the Connecticut River NE of Hartford, in a farm area; set off from Windsor 1768. Propellers and fertilizer are made. Jonathan Edwards was born here.

Ea·ton (ēt'n), county (1970 pop. 68,892), 571 sq mi (1,478.9 sq km), S Mich., drained by the Grand and Thornapple rivers and by Battle Creek; formed 1829; co. seat Charlotte. It lies in an agricultural area that produces maple sugar, grain, corn, beans, sugar beets, fruit, poultry, dairy products, and livestock. It has some mining and diversified industry (food products, lumber, furniture, glassware, concrete, iron and aluminum, metal products, and machinery).

Eaton, city (1970 pop. 6,020), seat of Preble co., W Ohio, W of Dayton on Sevenmile Creek. It is a trading center for an agricultural area yielding tobacco and grain. Fertilizer, gloves, tobacco boxes, and food products are manufactured.

Ea·ton·ton (ēt'n-tən), city (1970 pop. 4,125), seat of Putnam co., N central Ga., NNE of Macon, in a cotton, lumber, and dairying area; inc. 1808. There is a monument to Joel Chandler Harris, who was born here.

Ea·ton·town (ēt'n-toun'), borough (1975 est. pop. 15,263), Monmouth co., E central N.J.; inc. 1926. A residential borough in a truck-farming region, it is named for Thomas Eaton, who built a gristmill here c.1670. The mill's site is now a landmark.

Eau Claire (ō klâr'), county (1970 pop. 67,219), 649 sq mi (1,680.9 sq km), W Wis., drained by the Chippewa and Eau Claire rivers; formed 1856; co. seat Eau Claire. In a dairying region, it has agriculture (clover, corn, oats, and livestock) and some manufacturing (concrete, tires, and metal-working machinery).

Eau Claire, city (1978 est. pop. 49,500), seat of Eau Claire co., W central Wis., on the Chippewa River at the mouth of the Eau Claire River, in a hilly lake region; inc. 1872. Manufactures include defense items, kitchen appliances, tires, paper products, and dairy goods. A trading port was here in the late 18th cent. The city grew from several sawmills established in the mid-1800s.

E·bal, Mount (ē'bəl), 3,084 ft (940.6 m) high, in the Samarian hills, NW Jordan. On Ebal, according to the Bible, Joshua built the altar and monument inscribed with the Mosaic law.

Eb·bw Vale (ĕb'o͞o vāl'), urban district (1973 est. pop. 25,670), SE Wales. It is an industrial center with steelworks, tin-plate factories, and collieries.

Eb·ens·burg (ĕb'ənz-bûrg'), agricultural borough (1970 pop. 4,318), seat of Cambria co., SW Pa., in the Alleghenies W of Altoona; settled 1800, laid out 1806, inc. 1825. A summer resort, it also has farms and coal mines.

E·be·tsu (ā-bă'tso͞o), city (1975 pop. 77,624), Hokkaido prefecture, central Hokkaido, Japan. It is an industrial suburb of Sapporo and the site of a huge electric power company.

E·bo·li (ĕb'ə-lē, ā'bō-), town (1976 est. pop. 27,815), in Campania, S Italy. It is an agricultural and industrial center. A medieval castle dominates the town. Nearby are the ruins of Eburum, which was colonized by the Greeks, flourished under the Romans, and was destroyed (5th cent. A.D.) by the Visigoths.

E·bro (ā'brō, ē'brō), longest river entirely in Spain, c.575 mi (925 km) long, rising in the Cantabrian Mts., N Spain, and flowing SE between the Pyrenees and the Iberian Mts. past Logroño and Zaragoza. It empties through a wide delta into the Mediterranean below Tortosa. The river is of little use for inland navigation because of varying volume. During the Spanish Civil War a battle ultimately lost by the Loyalists was fought along the Ebro (Aug.-Nov., 1938).

E·ca·te·pec (ā-kä'tā-pĕk'), city (1970 pop. 220,918), Mexico state, S central Mexico. It is an industrial center, with ironworks and chemical and paper factories. Ecatepec was the site of an Aztec kingdom established in the 12th cent.

Ec·bat·an·a (ĕk-băt'ən-ə, ĕk-bə-tä'nə), capital of ancient Media; later the summer residence of Achaemenid and Parthian kings, beautifully situated at the foot of Mt. Elvend NE of Behistun. In 549 B.C. it was captured by Cyrus the Great. It possessed a royal treasury and was plundered in turn by Alexander, Seleucus, and Antiochus III. The site has never been thoroughly excavated, since it is covered by the modern city of Hamadan, Iran.

Ec·cles (ĕk'əlz), municipal borough (1971 pop. 38,413), Greater Manchester, NW England, on the Manchester Ship Canal. There are light and heavy engineering, chemical, rubber, plastics, and textile industries. Eccles cakes are famous. The parish church is said to date from 1111, although most of the present building is of the 15th cent.

Ech·mi·a·dzin (ĕch'mē-ä-dzēn'), town (1967 est. pop. 26,000), SE European USSR, in Armenia, in the Araks River valley. It has winemaking and plastics industries. Known since the 6th cent. B.C., Echmiadzin (which was called Vagarshapat until 1945) was the capital of the ancient kingdom of Armenia (A.D. 184–344). The Echmiadzin monastery has been the residence since 1441 of the patriarch of the Armenian Church; inside the monastery walls is the cathedral founded in 303 by St. Gregory the Illuminator.

Ech·ols (ĕk'əlz), county (1970 pop. 1,924), 425 sq mi (1,100.8 sq km), S Ga., bordered on the S by Fla. and drained by the Alapaha and Suwannee rivers; formed 1858; co. seat Statenville. It is in a coastal-plain timber and agricultural area yielding corn, tobacco, truck crops, and livestock. Some textile manufacturing is done.

Ech·ter·nach (ĕᴋʜ'tər-näᴋʜ), town, E Grand Duchy of Luxembourg, on the Sauer River, at the West German border. It is an agricultural, industrial, and tourist center, with mineral springs that have been frequented since Roman times.

E·ci·ja (ā'*th*ē-hä, ā'sē-), city (1970 pop. 36,056), Seville prov., S Spain, in Andalusia, on a hill overlooking the Genil River. It is an agricultural center for an area that produces olives, cereal, and cotton. Of pre-Roman origin, Écija was recovered from the Moors by Ferdinand III in 1240.

E·corse (ĭ-kôrs', ē'kôrs), industrial city (1975 est. pop. 15,616), Wayne co., SE Mich., on the Detroit River; settled c.1815, inc. as a city 1941. Steel, automobile parts, furniture, and metal products are the principal manufactures.

Ec·tor (ĕk'tər), county (1970 pop. 92,660), 907 sq mi (2,349.1 sq km), W Texas, in the S Llano Estacado and Edwards Plateau; formed 1887; co. seat Odessa. It is a leading petroleum producer. Cattle ranching and some farming (grain sorghums and truck crops) are done. Carbon black and oil-field supplies are manufactured. The county also has large deposits of potash.

Ec·ua·dor (ĕk'wə-dôr), republic (1973 est. pop. 6,600,000), 109,483 sq mi (283,561 sq km), W South America bounded on the N by Colombia, on the S and E by Peru, and on the W by the Pacific Ocean. The capital is Quito. The Andes, dominating the country, cut across Ecuador in two ranges and reach their greatest altitude in the snow-capped volcanic peaks of Chimborazo (20,561 ft/6,271.1 m) and Cotopaxi (19,347 ft/5,900.8 m). Within the mountains are high, often fertile valleys. Earthquakes are frequent and often disastrous. West of the Andes is a region of tropical jungle, through which run the tributaries of the Amazon River.

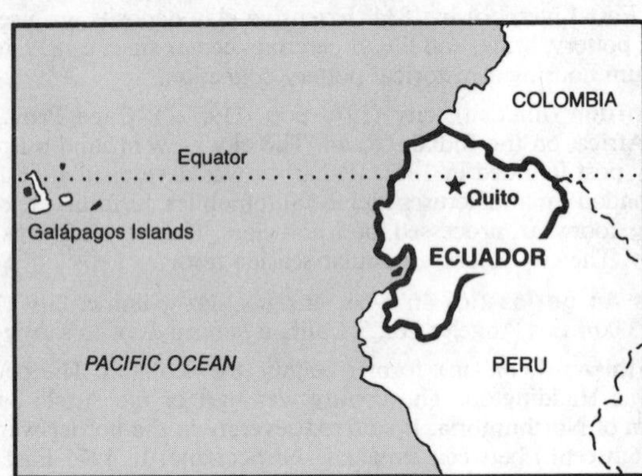

Ecuador is divided into 20 provinces:

NAME	CAPITAL	NAME	CAPITAL
Azuay	Cuenca	Imbabura	Ibarra
Bolívar	Guaranda	Loja	Loja
Cañar	Azogues	Los Ríos	Babahoyo
Carchi	Tulcán	Manabí	Portoviejo
Chimborazo	Riobamba	Morona-Santiago	Macas
Cotopaxi	Latacunga	Napo	Tena
El Oro	Machala	Pastaza	Puyo
Esmeraldas	Esmeraldas	Pichincha	Quito
Galápagos Islands	Puerto Baquerizo Moreno	Tungurahua	Ambato
Guayas	Guayaquil	Zamora-Chinchipe	Zamora

Economy. Ecuador's chief exports are oil, bananas, coffee, and cocoa. Other exports include forest products, fish, sugar, rice, copper, and panama hats. Two thirds of the work force engages in agriculture. Corn, barley, rice, and wheat are grown for subsistence. Manufacturing industries are few and small-scale. Large deposits of oil are located also in the northeast.

History. The country is still torn by the personal and factional rivalry that began with the Spanish conquest. Francisco Pizarro's subordinate, Benalcázar, entered the area in 1533. Not finding the wealth of the mythical El Dorado, he and other conquistadores moved restlessly on. The region became a colonial backwater. It was at various times subject to Peru and to New Granada. After an abortive independence movement in 1809, the region was finally liberated by Antonio José de Sucre in the Battle of Pichincha (1822) and was joined by Simón Bolívar to Greater Colombia. With the dissolution of the union in 1830, Ecuador, geographically isolated, became a separate state (four times its present size). Boundary disputes led to frequent invasions by Peruvians in the 19th and 20th cent. In 1942 Ecuador signed a treaty ceding a large area to Peru, but in 1960 it renounced the treaty.

Bitter internecine struggles between Conservatives and Liberals marked the political history of Ecuador in the 19th cent. The Conservatives supported entrenched privileges and the dominance of the Roman Catholic Church; the Liberals sought social reforms. In 1925 the army replaced the coastal banking interests, dominant since 1916, as the ultimate source of power. Military juntas supported various rival factions, and between 1931 and 1940, 12 presidents were in office. José María Velasco Ibarra and Galo Plaza Lasso alternately occupied the presidency from 1944 to 1956. The first Conservative to rule in 60 years, Camilo Ponce Enríquez, followed (1956–60), but Velasco Ibarra was elected again in 1960. He was forced to resign the following year. His legal successor was deposed by a junta in 1963. The military removed the junta in 1966. A constitutional assembly installed Otto Arosemena Gómez as provisional president and drafted the country's 17th constitution. Velasco Ibarra was elected for the fifth time in 1968. Two years later, faced with economic problems and protests by leftist students, he assumed absolute power. The military deposed him in Feb., 1972. The new regime, headed by Brig. Gen Guillermo Rodríguez Lara, was in turn deposed (Jan., 1976) by a military coup led by Vice Adm. Alfredo Poreda.

Ecuador's relations with the United States deteriorated in the early 1970s after Ecuador claimed that its territorial waters extended 200 mi (322 km) out to sea. Several U.S. fishing boats were seized by Ecuadorians, and U.S. aid to the country was suspended. In the same period Ecuador became Latin America's second-largest oil producer. Increased oil revenues and exploitation of the country's vast forests raised the prospect of rapid economic development.

E·dam (ā-däm', ē'dəm), town (1976 est. pop. 21,507), North Holland prov., N central Netherlands, on IJsselmeer lake; chartered 1357. It is a picturesque town that attracts many tourists. Edam is noted for its cheese; it also has fisheries.

Ed·dy (ĕd'ē). **1.** County (1970 pop. 41,119), 4,163 sq mi (10,782.2 sq km), SE N.Mex., bordered on the S by Texas and watered by the Pecos River; formed 1889; co. seat Carlsbad. It is in an irrigated agricultural region yielding cotton, alfalfa, livestock, and dairy products and has oil and natural-gas wells, salt mines, and large deposits of potash. Cottonseed oil, chemicals, concrete, and machinery are produced. It includes Carlsbad Caverns National Park and part of Lincoln National Forest. **2.** County (1970 pop. 4,103), 635 sq mi (1,644.6 sq km), E central N.Dak., drained by the Sheyenne and James rivers; formed 1885; co. seat New Rockford. Its agriculture includes wheat, poultry, livestock, and dairy products. Some mining and manufacturing are done. Part of Fort Totten Indian Reservation is in the north.

Ed·dy·stone (ĕd'ĭ-stən), lighthouse, 135 ft (41.2 m) high, on dangerous rocks in the English Channel, S of Plymouth, SW England. It is the fourth lighthouse on the site (the first was begun in 1696) and was built between 1878 and 1882 by Sir James Douglass.

Ed·dy·ville (ĕd'ē-vĭl'), city (1970 pop. 1,981), seat of Lyon co., W Ky., E of Paducah, in a tobacco, corn, limestone, and timber area. Nearby are the ruins of an iron furnace where William Kelly invented a method of making steel.

Eden, name of several rivers in England and Scotland. The principal one rises in Westmorland, northern England, and flows 65 mi (104.6 km) northwest through Cumberland, past Carlisle, into Solway Firth. The Vale of Eden is a rich farming region.

E·den·ton (ēd'n-tən), town (1970 pop. 4,956), seat of Chowan co., NE N.C., on Albemarle Sound SW of Elizabeth City; inc. 1722. Its industries are based on fish, peanuts, and cotton. In 1774 a group of townswomen held the Edenton Tea Party, at which they resolved not to use tea or other English goods until the repeal of acts oppressive to the Colonies.

E·der (ā'dər), river, c.110 mi (175 km) long, rising near Siegen, central West Germany, and flowing E to the Fulda River. The Eder dam, at Hemfurth, impounds one of the largest reservoirs in West Germany.

E·des·sa (ĭ-dĕs'ə), ancient city of Mesopotamia, on the site of modern Urfa, Turkey. From c.137 B.C. it was the capital of the independent kingdom of Osroene. It later became a Roman city. Edessa was a center of Christianity by the 3rd cent. A.D. and became one of the major religious centers of the Byzantine Empire. The city fell to the Arabs in 639 and remained in Moslem hands until captured by the Crusaders in 1097. Baldwin became the ruler of Edessa, and when he became king of Jerusalem he turned it over to one of his cousins. The city, however, fell to the Moslems in 1144 and passed to the Turks in 1637.

Ed·fu (ĕd'fo͞o'): *see* Idfu, Egypt.

Ed·gar (ĕd'gər), county (1970 pop. 21,591), 628 sq mi (1,626.5 sq km), E Ill., bordered on the E by Ind. and drained by small tributaries of the Wabash River; formed 1823; co. seat Paris. In an agricultural area yielding corn, wheat, broomcorn, soybeans, poultry, and livestock, it also has deposits of bituminous coal. Clothing, metal products, turbines, and electronic components are manufactured.

Ed·gard (ĕd'gərd, -gärd), village (1970 pop. 400), seat of St. John the Baptist parish, SE La., on the Mississippi WNW of New Orleans, in a sugar-cane area. Sugar refining is done.

Ed·gar·town (ĕd'gər-town'), resort town (1970 pop. 1,481), seat of Dukes co., SE Mass., on E Martha's Vineyard; settled 1642, inc. 1671. Stately old houses, built when the town was a prosperous whaling center, survive.

Edge·combe (ĕj'kəm), county (1970 pop. 52,371), 510 sq mi (1,320.9 sq km), NE N.C., in a forested and swampy area on the coastal plain, bounded on the N by Fishing Creek and crossed by the Tar River; formed 1735; co. seat Tarboro. It has agriculture (tobacco, corn, peanuts, and soybeans) and processes tobacco. Clothing, furniture, chemicals, plastics, machinery, and concrete are manufactured.

Edge·field (ĕj'fēld'), county (1970 pop. 15,692), 482 sq mi (1,248.4 sq km), W S.C., bounded on the SW by the Savannah River; formed 1785; co. seat Edgefield. In a sparsely settled region, it has agriculture (cotton, asparagus, and peaches), a logging industry, and some textile manufacturing. It includes part of Sumter National Forest.

Edgefield, town (1970 pop. 2,750), seat of Edgefield co., W S.C., WSW of Columbia, in an agricultural area; inc. 1830. Meat and poultry are processed.

Ed·ge·ø·ya (ĕd'yə-œ-yä) or **Edge Island** (ĕj), island of the Svalbard group, 1,942 sq mi (5,029.8 sq km), Norway, in the Barents Sea, E of Spitsbergen Island. An ice field covers the southeastern portion of the island.

Ed·hes·sa (ĕd'ĕ-sä) or **Vo·de·na** (vô-*th*ā-nä'), city (1971 pop. 13,967), capital of Pella prefecture, N Greece, in Macedonia. It is an agricultural trading center. Textiles and rugs are manufactured. Known as Aegae in antiquity, it was the earliest seat of the Macedonian kings. After Philip II moved (4th cent. B.C.) the Macedonian capital to Pella, it continued to be the royal burial place.

E·di·na (ĭ-dī'nə). **1.** Village (1975 est. pop. 47,989), Hennepin co., E Minn., a suburb of Minneapolis. It is chiefly residential and is in an area with more than 60 lakes and much park land. **2.** City (1970 pop. 1,574), seat of Knox co., NE Mo., ESE of Kirksville, in a farm and timber region; inc. 1879.

Ed·in·burg (ĕd'n-bûrg'). **1.** Town (1975 est. pop. 5,079), Bartholomew and Johnson cos., S central Ind., SSE of Indianapolis; settled in the 1820s, inc. 1853. Plastics and veneers are produced. Camp Atterbury is nearby. **2.** City (1978 est. pop. 22,200), seat of Hidalgo co., extreme S Texas; inc. 1919. It is a processing center in the irrigated portion of the lower Rio Grande valley. Agricultural products include packaged meats, dairy items, citrus fruits, vegetables, and cotton. Oil and gas are produced from surrounding fields, and oil-field machinery, men's slacks, and paper, plastic, and metal products are manufactured. Pan American Univ. is in Edinburg. In 1967 the city suffered extensive damage from a hurricane.

Ed·in·burgh (ĕd'n-bûr'ə), city (1976 est. pop. 467,097), Lothian region, royal burgh, capital of Scotland and formerly the county town of Midlothian, on the Firth of Forth. Leith, part of the city since 1920, is Edinburgh's port. The city is famous in Scottish legend and literature as Dunedin or "Auld Reekie." It is divided into two sections: the Old Town, on the slope of Castle Rock, dates from the 11th cent. and contains most of the city's historic sites; the New Town spread to the north in the late 18th cent. Edinburgh is a large brewing center, has a thriving publishing industry, and produces great quantities of high-grade paper. There are metalworks and rubber and engineering works. Other industries are distilling, the manufacture of glassware, drugs, and chemicals, and shipbuilding.

Edinburgh's history began when Malcolm III of Scotland erected a castle here in the late 11th cent., and his wife built the Chapel of St. Margaret, the city's oldest surviving building. A town grew up around the castle and was chartered in 1329 by Robert I. It grew steadily despite repeated sacking and burning by the English in the border wars and became the capital city of Scotland in 1437. James IV was the first monarch to make Edinburgh his regular seat. The rooms of Mary Queen of Scots are preserved in Holyrood palace. The city lost importance when James VI became king of England in 1603 and commerce and society followed the court to London. After the Act of Union with England in 1707 dissolved the Scottish Parliament, Edinburgh retained the Supreme Courts of Law, which meet in Parliament House.

Edinburgh blossomed as a cultural center in the 18th and 19th cent. around the figures of the philosophers David Hume and Adam Smith and the writers Robert Burns and Sir Walter Scott. The *Edinburgh Review*, founded in 1802, added to the city's literary reputation. The Edinburgh International Festival of Music and Drama, held every summer since 1947, is world-famous. Other features of interest are the National War Memorial; the collections of the Royal Scottish Academy, the National Gallery of Scotland, and the Royal Scottish Museum; the National Library; Princes St.; the Royal Botanic Gardens; the house of the Protestant reformer John Knox; the Church of St. Giles's, dating from the 12th cent.; and the site of the famous prison, Old Tolbooth, which figures in Scott's novel, *The Heart of Midlothian*. The Univ. of Edinburgh, founded 1583, has noted faculties of medicine, law, divinity, music, and the arts.

Ed·in·burgh·shire (ĕd'n-bûr'ə-shîr', -shər): *see* Midlothian, Scotland.

E·dir·ne (ĕ-dîr'nĕ), formerly **A·dri·a·no·ple** (ā'drē-ə-nō'-pəl), city (1975 pop. 63,290), capital of Edirne prov., NW Turkey, in Thrace. It is the commercial center for a farm region where grains, fruits, and tobacco are grown and cattle and sheep are raised. Manufactures of Edirne include cheese and textiles. The city was founded (A.D. c.125) by Hadrian, the Roman emperor. Of great strategic importance and strongly fortified, the city has had a turbulent history. The defeat (378) of Emperor Valens by the Visigoths at Adrianople left Greece open to invasion by barbarian tribes. Later conquered by the Avars, the Bulgarians, and the Crusaders, the city passed to the Ottoman Turks in 1361 and was the residence of the Ottoman sultans until the conquest of Constantinople in 1453. Russia captured the city twice (1829 and 1878) during the Russo-Turkish Wars. It fell (1913) to Bulgaria in the First Balkan War but was restored to Turkey after the Second Balkan War. It passed to Greece by the Treaty of Sèvres (1920), but was again restored to Turkey by the Treaty of Lausanne (1923).

Ed·i·son National Historic Site (ĕd'ĭ-sən): *see* National Parks and Monuments Table.

Ed·mond (ĕd'mənd), city (1970 pop. 26,300), Oaklahoma co., central Okla.; settled 1889. It is a trading center with small industries and a huge oil field.

Ed·mon·son (ĕd'mən-sən), county (1970 pop. 8,751), 304 sq mi (787.4 sq km), SW central Ky., drained by the Green and Nolin rivers; formed 1825; co. seat Brownsville. It is a major producer of asphalt. Its agriculture includes livestock, dairy products, poultry, grain, and burley tobacco. Some clothing is manufacuted. The county includes most of Mammouth Cave National Park.

Ed·mon·ton (ĕd'mən-tən), city (1976 pop. 461,361), provincial capital, central Alta., Canada, on the North Saskatchewan River. Edmonton is a major market center for farm and petrochemical products. The Univ. of Alberta is in the city.

Edmonton, city (1970 pop. 958). seat of Metcalfe co., S Ky., E of Glasgow, in an agricultural and timber area. Corn, burley tobacco, and wheat are grown.

Ed·munds (ĕd'məndz), county (1970 pop. 5,548), 1,153 sq mi (2,986.3 sq km), N S.Dak., drained in the E by intermittent creeks; formed 1873; co. seat Ipswich. In an agriculture area yielding grain, poultry, livestock, and dairy products, it also has some manufacturing.

Ed·munds·ton (ĕd'məns-tən), city (1976 pop. 12,710), NW N.B., Canada, at the confluence of the St. John and Madawaska rivers, at the U.S. border. It has a large pulp mill and is a hunting and fishing base. Settled c.1785 by Acadians, it was known as Petit Sault to the French and Little Falls to the English; it was renamed in 1850.

Ed·na (ĕd'nə), city (1970 pop. 5,332), seat of Jackson co., S Texas, SW of Houston on the Gulf Coast plain; inc. 1926. The region has oil fields, cattle ranches, and farms producing rice, cotton, and other crops. A hurricane caused extensive damage here in 1961.

Ed·ward Ny·an·za (ĕd'wərd nī-ăn'zə, nē-) or **Lake Edward,** 830 sq mi (2,150 sq km), in the Great Rift Valley, central Africa, on the Zaire-Uganda border. It lies at an altitude of c.3,000 ft (915 m), is c.47 mi (76 km) long, and has a maximum width of c.32 mi (52 km). Edward Nyanza is connected with the Nile by the Semliki River. The lake was discovered in 1889 by Henry Morton Stanley.

Ed·wards (ĕd'wərdz). **1.** County (1970 pop. 7,090), 225 sq mi (582.8 sq km), SE Ill., bounded on the E by Bonpas Creek and drained by the Little Wabash River; formed 1814; co. seat Albion. Its agriculture includes corn, wheat, soybeans, fruit, truck crops, livestock, and dairy products. It also has some mining, commercial fishing, and manufacturing (clothing and automobile parts). **2.** County (1970 pop. 4,581), 614 sq mi (1,590.3 sq km), SW central Kansas, drained by the Rattlesnake and Arkansas rivers; formed 1874; co. seat Kinsley. In a prairie agricultural area yielding wheat, livestock, and dairy products, it also has some mining. Machinery is manufactured. **3.** County (1970 pop. 2,107), 2,075 sq mi (5,374.3 sq km), SW Texas, on the Edwards Plateau, bounded on the E by the Nuceces River and drained by the South Llano, Dry Devils, and West Nueces rivers; formed 1858; co. seat Rocksprings. Ranching is the most important activity. There are also deposits of silver, iron, sulfur, coal, kaolin, natural gas, and some oil.

Ed·wards·ville (ĕd'wərdz-vĭl'). **1.** City (1975 est. pop. 11,811), seat of Madison co., SW Ill.; inc. 1819. It is mainly residential, with many citizens commuting to St. Louis. **2.** Industrial borough (1975 est. pop. 5,110), Luzerne co., NE Pa., on the Susquehanna River NW of Wilkes-Barre; inc. 1884. Its manufactures include clothing and food products. Anthracite coal is mined here.

E·fa·te (ĕ-fä'tē), volcanic island (1967 pop. 10,008), c.300 sq mi (780 sq km), South Pacific, most important island of the New Hebrides. Efate produces copra, coffee, and sandalwood.

Ef·fi·gy Mounds National Monument (ĕf'ĭ-jē moundz): *see* National Parks and Monuments Table.

Ef·fing·ham (ĕf'ĭng-hăm'). **1.** County (1970 pop. 13,632), 480 sq mi (1,243.2 sq km), E Ga., bounded on the NE and E by the Savannah River, here forming the S.C. border, and on the W by the Ogeechee River; formed 1777; co. seat Springfield. It is in a coastal-plain timber and agricultural area yielding truck crops, fruit, poultry, and livestock. Textiles are manufactured. **2.** County (1970 pop. 24,608) 483 sq mi (1,251 sq km), SE central Ill., drained by the Little Wabash River; formed 1831; co. seat Effingham. It has agriculture (corn, wheat, oats, poultry, and livestock), food-processing plants, and petroleum-related industries.

Effingham, city (1976 est. pop. 10,772), seat of Effingham co., E central Ill., SE of Decatur, in a farm, dairy, and timber area; founded 1854, inc. 1861. Household appliances are made here.

E·ga·di Islands (ĕg'ə-dē) or **Ae·ga·di·an Isles** (ē-gā'dē-ən), archipelago (1968 est. pop. 5,800), c.15 sq mi (40 sq km), W Sicily, Italy, in the Mediterranean Sea. Fishing is the main occupation, and the most important tuna fisheries of Sicilý are here. A Roman naval victory over the Carthaginians, fought near the islands in 241 B.C., ended the first of the Punic Wars.

E·ger (ĕ'gĕr), city (1977 est. pop. 59,000), NE Hungary, on the Eger River. It is the commercial center of a wine-producing region and has food- and tobacco-processing plants and a mechanical engineering industry. There are mineral springs nearby. One of the first Magyar settlements in central Europe, Eger was destroyed (13th cent.) by the Tatars, rebuilt and fortified, and captured in 1596 by the Turks, who held it for nearly 150 years.

E·ger·sund (ā'gər-so͞on'), town (1977 est. pop. 11,506), Rogaland co., S Norway, a modern fishing port on the North Sea. Often mentioned in the Norwegian sagas, Egersund was a busy port as early as the Middle Ages.

Eg·ham (ĕg'əm), urban district (1971 pop. 30,510), Surrey, SE England. Light engineering and gravel working are the main industries. Runnymede and part of Windsor Great Park, the royal estate, are in the district.

Eg·mont, Mount (ĕg'mŏnt), dormant volcanic cone, 8,260 ft (2,519.3 m) high, on North Island, New Zealand. Symmetrical and snow-capped, it dominates the island's west side.

E·gypt (ē'jĭpt), officially Arab Republic of Egypt, republic (1973 est. pop. 35,330,000), 386,659 sq mi (1,001,447 sq km), NE Africa and SW Asia, bordering on the Mediterranean Sea in the N, on Israel and the Red Sea in the E, on Sudan in the S, and on Libya in the W. Its capital is Cairo.

Egypt north of Cairo is often called Lower Egypt and south of Cairo, Upper Egypt. The principal physiographic feature of the country is the Nile River, which flows from south to north through eastern Egypt for c.900 mi (1,450 km). In the far south is Lake Nasser, a vast artificial lake impounded by the Aswan High Dam (built 1960-70), and in the north, below Cairo, is the great Nile delta (c.8,500 sq mi/22,000 sq km). Bordering the Nile between Aswan and Cairo are narrow strips of cultivated land; there are broad re-

gions of tilled land in the delta. The vast majority of Egypt's inhabitants live in the Nile valley and delta, and the rest of the country (about 96% of Egypt's total land area) is sparsely populated. West of the Nile is the extremely arid Libyan (or Western) Desert, a generally low-lying region largely covered with sand dunes or barren rocky plains. In southwest Egypt the desert rises to the Jilf al Kabir plateau. East of the Nile is the Arabian (or Eastern) Desert, a barren dissected highland area. The Sinai peninsula is a plateau broken by deep valleys.

Economy. Egypt is predominately an agricultural country. The principal crops are cotton, rice, maize, wheat, millet, onions, beans, barley, tomatoes, sugar cane, citrus fruit, and dates. Most agricultural produce is marketed, and cotton is by far the leading cash crop. Large numbers of poultry, cattle, sheep, buffalo, donkeys, and goats are raised. The leading manufactures are refined petroleum, chemicals (including fertilizer), textiles, clothing, processed food, construction materials (especially cement), iron and steel, and metal products. The principal minerals produced are petroleum, natural gas, phosphates, salt, iron ore, manganese, coal, and gold. The state owns much of the economy and plays a decisive role in economic planning. Economic growth was hindered somewhat after 1945 by the large proportion of funds and energy devoted to armaments and war.

History. The Nile valley was the seat of one of the earliest civilizations built by man. From the beginning there was a concept of the divinity of the king (pharaoh), which lasted from the time that Egypt was first united (c.3200 B.C.) until the ultimate fall of Egypt to the Romans. According to tradition, it was Menes (or Narmer) who as king of Upper Egypt conquered the rival kingdom of Lower Egypt in the Nile delta, thus forming the single kingdom of Egypt. Trade flourished; under the II dynasty (2884-2780 B.C.) there was trade with areas as far north as the Black Sea. During the III dynasty (2780-2680 B.C.) sun worship, mummification, and the building of stone monuments were begun. The kings of the IV dynasty (2680-2565 B.C.) were the builders of the great pyramids at Al Jizah. The V to the VII dynasties (2565-2258 B.C.) are remarkable for their records of trading expeditions with armed escorts. Although Egypt flourished culturally and commercially during this period, it started to become less centralized and weaker politically. In the 23rd cent. B.C. the Old Kingdom fell apart; it was not until 2134 B.C. that power was again centralized, this time at Thebes. The Middle Kingdom, founded at the end of the XI dynasty (c.2000 B.C.), reached its zenith under the XII (2000-1786 B.C.). Order was preserved, the draining of Al Fayyum was begun (adding a new and fertile province), a uniform system of writing was adopted, and civilization reached a new peak. After the XII dynasty Egypt passed for more than a century under the Hyksos, who were expelled from Egypt by Amasis I (Ahmose I), founder of the XVIII dynasty.

The XVIII dynasty (1570-c.1342 B.C.), the first of the New Kingdoms, is the most important and the best-recorded period in Egyptian history. Ancient Egypt reached its height. Its boundaries were extended into Asia, with a foreign province reaching the Euphrates. Architecture was at its zenith with the enormous buildings at and around Thebes. Egyptian civilization seems to have worn out rapidly after conflicts with the Hittites under the XIX dynasty (1342-1200 B.C.) and with sea raiders under the XX dynasty (1200-1085 B.C.). With the disappearance of the weak XXI dynasty (1085-945 B.C.), a Libyan dynasty came to power. This was succeeded by the alien rules of the Nubians from the south and the Assyrians from the east. By 650 B.C., Egypt was once more independent and orderly. Attempts to re-establish Egyptian power in Asia were turned back (605 B.C.) by the Babylonian king Nebuchadnezzar, and Egypt fell easy prey (525 B.C.) to the armies of Cambyses of Persia. The Persians maintained their hegemony until 405 B.C. and again became dominant in 341 B.C. Egypt, rich and ill-defended, fell to Alexander the Great with no resistance in 332 B.C.

When Alexander's brief empire faded, Egypt fell to his general Ptolemy, who became king as Ptolemy I. The great city of Alexandria became the intellectual center and fountainhead of the Hellenistic world. The Ptolemies maintained a formidable empire for more than two centuries, but the rising power of Rome soon overshadowed it. In 58 B.C. the Romans obtained a foothold in Egypt. Octavian (later Emperor Augustus) annexed Egypt to Rome, putting to death Cleopatra's son, Ptolemy XIV, who was the last of the Ptolemies. Egypt became a granary for Rome. In the 2nd cent. A.D., strife between Jews and Greeks in Alexandria brought massacres. Christianity was welcomed and Monophysitism became the national faith; out of this arose the Coptic Church. The hostility of the people to the Orthodox Byzantine emperors and officials probably helped Persia to gain Egypt briefly (A.D. 616-c.628).

The Arab conquest of Egypt (A.D. 639-42), only some 20 years after the rise of Islam, made the country an integral part of the Moslem world. The Umayyad caliphate (661-749) was followed by the Abbaside caliphate, which fell to the Fatimids (ruled (969-1171). The strain of the Crusades may have been responsible for the fall of the Fatimids, which led to the founding by Saladin of the Ayyubid dynasty. The later Ayyubid rulers came excessively under the control of their soldiers and advisers, the Mamelukes, who in 1250 seized the country. Until 1517, when Egypt was conquered by the Ottoman Turks, the Mamelukes maintained their turbulent rule. Under Ottoman power, the Egyptian pasha (governor) was compelled to consult the Mameluke beys (princes), who continued in control of the provinces. Napoleon Bonaparte undertook the French occupation of Egypt (1798-1801). French withdrawal was followed by the rise of Mohammed Ali, a former common soldier, who was appointed (1805) Egyptian pasha by the Ottoman emperor. He permanently destroyed (1811) the Mamelukes' power by massacring their leaders. Using Europe as a model, Mohammed Ali introduced political, social, and educational reforms.

The Suez Canal, completed in 1869, put Egypt into deep financial debt and opened the door to British intervention in Egyptian affairs. Egypt's financial problems led to subordination of the country to great power interests. A nationalist and military revolt (1881-82) prompted the British to intervene; they remained and consolidated their control. Egyptian independence was granted in 1922, and a new constitution was proclaimed that made Egypt a kingdom under Fuad I. The British protectorate was maintained until 1936, however, and British troops did not withdraw completely until 1956. During World War II Egypt remained officially neutral. Egypt bitterly opposed the UN partition of Palestine in 1948 and took a leading role in the first Arab-Israeli War. Israeli forces, however, held the Egyptians to slight gains in the southern Negev. On July 23, 1952, the military headed by Gen. Mohammed Naguib took power by coup d'état. King Farouk abdicated; in 1953 the monarchy was abolished and a republic was set up with Naguib as president. Col. Gamal Abdal Nasser was elected president in 1954.

Under Nasser, tension increased with Israel and its Western allies. When the United States and Great Britain withdrew their pledges of financial aid for the building of the Aswan High Dam in 1956, Nasser nationalized the Suez Canal. Israel, Great Britain, and France responded by invading the Sinai peninsula. They soon yielded to worldwide pressure and withdrew their troops. The canal remained in Egyptian hands. On Feb. 1, 1958, Syria and Egypt merged as the United Arab Republic, but the union was dissolved in 1961. Egypt embarked on a program of industrialization and nationalization with Soviet assistance. Nasser assumed nearly absolute control in 1967 by taking over the premiership and the leadership of the Arab Socialist Union (ASU), the country's sole political party. In June of 1967 Israel, provoked by Arab troops massed on its borders, launched air and ground attacks against Egypt and after six days achieved a decisive victory. When the UN cease-fire went into effect, Israel held the Sinai peninsula and the east bank of the Suez Canal.

After Nasser's sudden death in 1970, Vice President Anwar al-

Sadat succeeded him as president. In 1972 Sadat suddenly ousted all Soviet military personnel stationed in Egypt and placed Soviet bases and equipment under Egyptian control. Thereafter, relations improved with the West. Another war with Israel broke out on Oct. 6, 1973, when Egyptian forces crossed the Suez Canal and established footholds in the Israeli-occupied Sinai peninsula. Israeli forces counterattacked strongly. A UN-supervised cease-fire was arranged after 18 days of fighting. In 1974 an agreement was reached providing for a disengagement of military forces on the Suez front; Egypt regained both banks of the Suez Canal (closed since 1967), which it cleared of obstacles and reopened in 1975. A result of the intense U.S. effort to secure an Arab-Israeli settlement was the improvement of relations between the United States and Egypt, coupled with further deterioration in Egyptian relations with the USSR and with more militantly anti-Israeli Arab states. The 1970s brought severe inflation and social unrest, but Sadat maintained his rule, suppressing domestic critics. Diplomatic attempts to regain the Israeli-occupied Sinai failed. Sadat vainly tried to break the stalemate in negotiations by visiting Israel in 1977. Finally, in 1979, a peace treaty was signed. As a consequence, a number of Arab nations severed diplomatic relations with Egypt.

Government. Egypt is governed under the constitution of 1971. The president is the head of state and the prime minister is the head of government. There is a 360-seat national assembly whose members serve five-year terms. The Arab Socialist Union (ASU) is the only legal political party.

Ei·der (ī'dər), river, 117 mi (188.3 km) long, rising S of Kiel, N West Germany, and flowing N to the Kiel Canal before turning W and meandering to the North Sea. It is navigable for most of its length.

Eids·voll or **Eids·vold** (both: āts'vôl'), town (1970 pop. 2,906), Akershus co., SE Norway, near Lake Mjøsa. The present constitution of Norway was proclaimed (1814) at Eidsvoll Manor by an assembly of Norwegian patriots.

Eind·ho·ven (īnt'hō'vən), city (1977 est. pop. 192,566), North Brabant prov., S Netherlands, on the Dommel River. Chartered in 1232, Eindhoven rapidly expanded after the founding (1891) of the Philips electrical works. In World War II Eindhoven was taken (Sept., 1944) by Allied troops in a major airborne operation.

Ein·sie·deln (īn-zē'dəln), town (1977 est. pop. 9,800), Schwyz canton, E central Switzerland. Einsiedeln is the most famous pilgrimage center in Switzerland and one of the most noted in Europe. Its important Benedictine abbey, founded in the 10th cent., was built on the supposed site of the cell of St. Meinrad, a 9th cent. martyr. The monastery (rebuilt in the early 18th cent.) is one of the largest and finest examples of Swiss baroque architecture. Its church contains the sacred image of the "Black Virgin."

Eir·e (âr'ə): *see* Ireland; Ireland, Republic of.

Ei·se·nach (ī'zə-näKH'), city (1974 est. pop. 49,954), Erfurt dist., SW East Germany. Manufactures include machine tools, processed food, textiles, electrical supplies, agricultural machinery, and motor vehicles. There are salt mines and saline springs in the region. Eisenach was founded c.1150 and was chartered in 1283. The city passed to the house of Wettin in 1440, to Saxony in 1485, and to Saxe-Weimar in 1741. It often served as a residence of the electors of Saxony and the dukes of Saxe-Weimar. The German Social Democratic Party was founded here (1869) at the Congress of Eisenach. Johann Sebastian Bach was born (1685) in Eisenach and Martin Luther studied here (1498–1501).

Ei·sen·erz (ī'zən-ĕrts'), town (1971 pop. 11,563), in Styria prov., central Austria, at the N foot of the Erzberg. There are large ironworks based on iron-ore deposits that have been mined for more than 1,000 years.

Ei·sen·how·er National Historic Site (ī'zən-hou'ər): *see* National Parks and Monuments Table.

Ei·sen·hüt·ten·stadt (ī'zən-hüt'ən-shtät'), city (1974 est. pop. 46,455), Frankfurt dist., E East Germany. Manufactures include iron and steel. It was founded in 1951 as a residential town.

Ei·sen·stadt (ī'zən-shtät'), town (1971 pop. 10,059), capital of Burgenland, E Austria, at the foot of the Leitha Mts. The composer Joseph Haydn (1732–1809), who lived in Eisenstadt for many years, is buried in the Bergkirche, an 18th cent. church. The Esterházy palace (14th cent.; redone 17th cent. in baroque style) still stands.

Eis·le·ben (īs'lā'bən), city (1974 pop. 29,297), Halle dist., central East Germany, at the foot of the Harz Mts. It is an industrial city and has been a copper-mining center since the 14th cent. Manufactures include processed food, clothing, and wood products. Eisleben was first mentioned c.1000 as a market settlement. It passed to Prussia in 1815. In Eisleben are the house in which Martin Luther was born (1483), the church where he was baptized, and the house in which he died (1546).

E·ka·lak·a (ē'kə-lăk'ə), town (1970 pop. 663), seat of Carter co., SE Mont., SE of Miles City, in a timber, livestock, and grain area. A local museum contains animal fossils.

E·ki·bas·tuz (ĕ-kē'bäs-to͞os'), city (1976 est. pop. 54,000), Central Asian USSR, in Kazakhstan. It is the industrial center of a bituminous-coal basin, which has coal reserves estimated at 8 billion tons. Although coal mining began in the late 19th cent., the city became industrialized only in the 1950s when the railroad reached it.

El·am (ē'ləm), ancient country of Asia, N of the Persian Gulf and E of the Tigris, now in W Iran. A civilization seems to have been established very early, probably in the late 4th millennium B.C. The capital was Susa, and the country is sometimes called Susiana. In historical times the Elamites were known as a warlike people who rivaled and threatened Babylonia. The Elamites seem to have maintained their independence steadily, despite invasions and counterinvasions. At the beginning of the 2nd millennium the Elamites invaded Babylon and founded a dynasty at Larsa. Shortly thereafter they became masters of Erech, Babylon, and Isin. In the 18th cent. B.C. Hammurabi was able to keep the Elamites from expanding. The golden age of Elam came in the 13th and 12th cent. B.C. when there was a literary renaissance and great development of architecture and sculpture. In the 7th cent. B.C. Sargon of Assyria, Sennacherib, and Esar-haddon all attacked the Elamites, but Susa fell only to Assurbanipal, who sacked the city. Susa later became a favored provincial capital of Persia.

El As·nam (ĕl äs-näm'), city (1974 est. pop. 80,500), capital of El Asnam dept., N Algeria, on the Chéliff River. It is the center of an important cereal and citrus-fruit region.

E·lat or **Ei·lat** (both: ā'lät) city (1975 est. pop. 15,800), S Israel, a port on the Gulf of Aqaba, an arm of the Red Sea. It is strategically located near the Sinai peninsula, Jordan, and Saudi Arabia and is Israel's gateway to Africa and the Far East. A tourist center with small industries, the city is located near copper mines.

El·ba (ĕl'bə), city (1970 pop. 4,634), seat of Coffee co., SE Ala., NW of Dothan on the Pea River, in a farm and livestock area. Truck trailers and meat products are among its manufactures.

Elba, island, 86 sq mi (222.7 sq km), Tuscany, central Italy, in the Tyrrhenian Sea, 6 mi (9.7 km) from the Italian mainland, part of the Tuscan archipelago. Iron ore has been mined here since Etruscan and Roman times. Wine, olive oil, and fruit are produced, and there is a large tourist industry. Elba has come under numerous foreign powers, including Syracuse (mid-5th cent. B.C.), Pisa (11th cent. A.D. 1399), Spain, and Naples. It was briefly (May, 1814–Feb., 1815) a sovereign principality under the exiled Napoleon I, who improved the island's roads and agriculture. After Napoleon's dramatic escape from Elba and his subsequent exile to St. Helena Island, Elba passed to Tuscany.

El·ba·san (ĕl-bä-sän') or **El·ba·sa·ni** (-sä'nē), town (1971 est. pop. 43,200), capital of Elbasan prov., central Albania, on the Shkumbin River. It is located in a fertile agriculture region where tobacco, olives, and fruit are raised.

El Ba·yadh (ĕl bä-yäd'), town (1974 est. pop. 21,200), N Algeria. It is an important market for sheep. The town developed around a French military post established in 1852.

El·be (ĕl'bə), a major river of central Europe, c.725 mi (1,165 km) long, rising in the Krkonoše Mts., NW Czechoslovakia, and traversing NW Czechoslovakia in a wide arc. It then cuts through steep sandstone cliffs, enters East Germany, and flows generally northwest through central East Germany and onto the North German plain. The Elbe forms part of the East German–West German border before flowing across northern West Germany and into the North Sea at Cuxhaven. In Hamburg the river divides into two arms before forming a 60-mi (96.5-km) long estuary. The Elbe is navigable for c.525 mi (845 km). A canal system connects the Elbe with Berlin and the Oder River to the east, with the Ruhr region and the Weser and Rhine rivers to the west, and with the Baltic Sea to the north. Known as the Albis to the Romans, the river marked the farthest

Roman advance into Germany (9 B.C.) and was later the eastern limit of Charlemagne's conquests. The Treaty of Versailles (1919) internationalized its course from the Vltava River to the sea, but Germany repudiated its internationalization after the Munich Pact (1938). In 1945 the river was made part of the demarcation line between East and West Germany.

El·bert (ĕl'bərt). **1.** County (1970 pop. 3,903), 1,864 sq mi (4,827.8 sq km), E central Colo., drained by forks of the Bijou River and by the Big Sandy, Kiowa, and Box Elder rivers; formed 1874; co. seat Kiowa. In a wheat and cattle-ranching area, it has oil and natural-gas wells and some industry. **2.** County (1970 pop. 17,262), 362 sq mi (937.6 sq km), NE Ga., bounded on the E by the Savannah River, here forming the S.C. border, and on the W and S by the Broad River; formed 1790; co. seat Elberton. Its agriculture includes cotton, corn, oats, and fruit. It also has a textile industry.

Elbert, Mount, peak, 14,431 ft (4,401.5 m) high, central Colo. It is the highest point in the state and the tallest peak in the U.S. Rocky Mts.

El·ber·ton (ĕl'bər-tən), industrial city (1970 pop. 6,438), seat of Elbert co., NE Ga., between the Broad and Savannah rivers ENE of Athens, in a cotton area; settled in the 1780s. It is the center of the state's important granite industry; the first quarry was opened here in 1882. Textiles and tools are also produced.

El·beuf (ĕl-bœf'), town (1975 pop. 19,116), Seine-Maritime dept., NW France, in Normandy, on the Seine River. It is the center of an industrial complex and is a river port for the shipping of coal. The town has been famous as a woolen center since the 16th cent. but was heavily industrialized only after World War II. Elbeuf's history dates back to Roman times.

El·bląg (ĕl'blôNG) or **El·bing** (ĕl'bĭng), city (1975 est. pop. 97,300), N Poland. A seaport near the Vistula Lagoon, it has shipyards, machinery plants, and an important metallurgical industry. In 1237 the Teutonic Knights built a castle, around which developed a settlement. Elbląg joined the Hanseatic League in the late 13th cent. It revolted against the Teutonic Knights c.1450 and submitted to the rule of Poland. The city was ceded to Prussia in 1772. It suffered heavy damage in World War II, after which it passed to Poland.

El·bow Lake (ĕl'bō), village (1970 pop. 1,484), seat of Grant co., W central Minn., S of Fergus Falls. It is a farm trade center.

El·brus, Mount (ĕl-bro͝oz'), highest mountain of the Caucasus, SE European USSR, in Georgia, formed by two extinct volcanic cones, respectively 18,481 ft (5,636.7 m) and 18,356 ft (5,598.6 m) high.

El·burz (ĕl-bo͝orz'), mountain range, N Iran, between the Caspian Sea and the central Iranian plateau, rising to 18,934 ft (5,774.9 m) in Mt. Damavand, the highest peak in Iran. The range consists of steep, narrow, parallel ridges crossed by the Safid Rud.

El Ca·jon (ĕl kə-hōn'), city (1978 est. pop. 68,100), San Diego co., S Calif.; inc. 1912. Electronic equipment, missile parts, and metal products are among the manufactures of this rapidly growing city.

El Cen·tro (ĕl sĕn'trō), city (1978 est. pop. 23,600), seat of Imperial co., SE Calif., near the Mexican border; inc. 1908. It is a processing and shipping center for a heavily irrigated agricultural region.

El·che (ĕl'chā), city (1970 pop. 101,271), Alicante prov., SE Spain, in Valencia. It is surrounded by a grove of date palms. The city's industries produce leather, soap, oil, and palm products. Iberian, Greek, Roman, and Arabic artifacts have been found in the area.

El·do·ra (ĕl-dôr'ə), city (1970 pop. 3,223), seat of Hardin co., central Iowa, on the Iowa River NNW of Marshalltown; settled 1851, inc. 1896. Pine Lake State Park is nearby.

El·do·ra·do (ĕl'də-rā'dō), town (1970 pop. 1,446), seat of Schleicher co., W Texas, on the Edwards Plateau S of San Angelo, in a sheep-ranching area. There are oil and gas wells in the vicinity.

El Do·ra·do (ĕl' də-rā'dō), county (1970 pop. 43,833), 1,725 sq mi (4,467.8 sq km), E Calif., in the Sierra Nevadas, bounded on the N by the Rubicon River, on the NE by Lake Tahoe and the Nev. border, on the S by the Cosumnes River, on the W by Folsom Lake; formed 1850; co. seat Placerville. It has limestone quarries, a lumber industry, fruit orchards (especially Bartlett pears), and some industry (plastic and concrete products). Tourism, winter sports, and hunting and fishing are particularly important. Coloma, the site of the discovery that prompted the 1848 gold rush, and other old mining towns survive. The county includes parts of El Dorado and Toiyabe national forests.

El Dorado. 1. City (1978 est. pop. 24,900), seat of Union co., S central Ark; inc. 1845. The discovery of oil in 1921 made it the oil center of the state. The city has oil refineries, chemical plants, and poultry-packing houses. **2.** City (1975 est. 11,694), seat of Butler co., SE Kansas, on the Walnut River, on the edge of the Flint Hills, in a grain and livestock area; inc. 1871. Since the discovery (1915) of oil in the region, El Dorado has been a refining and shipping point for petroleum. Aluminum and plastic products are made.

E·lec·tric Peak (ĭ-lĕk'trĭk), mountain, 11,155 ft (3,402.3 m) high, SW Mont., in Yellowstone National Park near the Wyo. border. It is the highest point in the Gallatin Range.

El·e·phan·ta (ĕl-ə-făn'tə), island, c.2 sq mi (5.2 sq km), in Bombay harbor, Maharashtra state, W India. It is noted for six Brahmanic caves, carved (8th cent.) from solid rock some 250 ft (76 m) above sea level. The Great Cave, the largest (130 ft/40 m long), contains gigantic pillars supporting its roof and colossal statuary, especially the famous three-headed bust of the Hindu god Siva.

El·e·phan·ti·ne (ĕl'ə-făn-tī'nē'), island, SE Egypt, in the Nile below the First Cataract, near Aswan. In ancient times it was a military post guarding the southern frontier of Egypt. The Elephantine papyruses, which date from the 5th cent. B.C. and describe a colony of Jewish mercenaries, were found here.

E·leu·sis (ĭ-lo͞o'sĭs), ancient city of Attica, Greece, NW of Athens. Through ancient times it was the seat of the Eleusinian mysteries. Excavation of the cemetery here began in 1952; graves were found that date from the 7th and 8th cent. B.C. A temple to Demeter and a type of theater with rock-cut seats for about 3,000 spectators were uncovered near the modern village of Eleusis.

El·gin (ĕl'gĭn), royal burgh (1974 est. pop. 17,589), Grampian region, NE Scotland, on the Lossie River. Elgin is the market town for Morayshire's farm belt. Scotch whisky and woolen textiles are manufactured. It became the cathedral town for the see of Moray in 1224, when Elgin Cathedral was founded. Called "the Lantern of the North," the cathedral was reputedly Scotland's finest piece of early Gothic architecture. Its ruins still stand. Until 1975 Elgin was the county town of Morayshire.

El·gin. 1. (ĕl'jĭn) City (1978 est. pop. 62,800), Cook and Kane cos., NE Ill., on the Fox River; inc. 1854. Elgin is a railroad, trade, and industrial city and the home of the Elgin watch factories. Among other manufactures are household appliances, gaskets, electrical and electronic equipment, paper products, precision instruments, and machinery. **2.** (ĕl'gĭn) City (1970 pop. 3,832), Bastrop co., S central Texas, E of Austin, in a truck-farm, dairy, and oil region; settled 1867, inc. 1890. Brick is produced.

El·gin·shire (ĕl'gĭn-shîr', -shər): *see* Morayshire, Scotland.

El·gon, Mount (ĕl'gŏn), extinct volcano, central Africa, on the Kenya-Uganda border. Its highest peak is Wagagai (14,178 ft/4,324.3 m).

E·lis (ē'lĭs), region of ancient Greece, in W Peloponnesus, W of Arcadia. It was divided into three parts—Elis proper, Pisatis, and Triphylia. A plain watered by the Alpheus and the Peneus rivers, Elis was notable as a place for breeding horses and growing flax.

E·lis·a·beth·ville (ĭ-lĭz'ə-bəth-vĭl'): *see* Lubumbashi, Zaire.

E·liz·a·beth (ĭ-lĭz'ə-bəth). **1.** City (1978 est. pop. 100,600), seat of Union co., NE N.J., on Newark Bay; inc. 1855. It is an important shipping and transportation area, with some of the world's largest containerized dock facilities. Elizabeth's manufactures include sewing machines, foundry products, cord, office supplies, chemicals, biscuits, toys, cans, swimming-pool equipment, and industrial valves. The area, purchased from Indians in 1664, was the home and provincial capital of Gov. Philip Carteret and from 1668 to 1682, the meeting place of the province's assembly. Chartered as the town of Elizabeth in 1740, it was the scene of several Revolutionary clashes; many buildings were burned (1780). **2.** Town (1970 pop. 821), seat of Wirt co., W W.Va.; on the Little Kanawha River SE of Parkersburg, in an agricultural area. Lumber milling is done.

Elizabeth City, city (1978 est. pop. 14,400), seat of Pasquotank co., NE N.C., a port of entry on the Pasquotank River; settled mid-1600s, inc. 1793. It is a trade and shipping center for the diversified farm products of the Alkermarle Sound area. Cabinets, textiles, wearing apparel, and lumber are manufactured. The area was first visited (1584) and mapped by a scouting expedition from Roanoke Island. The first General Assembly of Carolina met here in 1665.

E·liz·a·beth·ton (ĭ-lĭz'ə-bəth-tən), city (1978 est. pop. 12,200), seat of

Carter co., NE Tenn., on the Watauga River; inc. 1799. It is an industrial center where rayon is produced. There are manganese deposits in the area.

E·liz·a·beth·town (ĭ-lĭz′ə-bəth-toun′). **1.** Village (1970 pop. 436), seat of Hardin co., extreme SE Ill., on the Ohio River SSE of Harrisburg, in a mining and farming area. Cavern Rock State Park is nearby. **2.** City (1975 est. pop. 14,152), seat of Hardin co., central Ky.; inc. 1797. Among its manufactures are steel magnets, automobile hoses, and men's slacks. Points of interest include the Lincoln Heritage House, built by Abraham Lincoln's father, and many antebellum homes. Fort Knox is nearby. **3.** Village (1970 pop. 607), seat of Essex co., NE N.Y., in the Adirondacks on the Bouquet River SSW of Plattsburg. Lumbering is done. **4.** Town (1970 pop. 1,418), seat of Bladen co., S N.C., on Cape Fear River SSE of Fayetteville, in a cotton, tobacco, and peanut area; settled c.1738. **5.** Borough (1970 pop. 8,072), Lancaster co., SE Pa., SE of Harrisburg; settled c.1735, laid out 1751, inc. 1827. Manufactures include shoes, chocolate, and clothing.

Elk (ĕlk). **1.** County (1970 pop. 3,858), 647 sq mi (1,675.7 sq km), SE Kansas, watered by the Elk River; formed 1875; co. seat Howard. In a gently rolling agricultural area yielding grain and cattle, it also has limestone quarries, oil and natural-gas fields, and some manufacturing. **2.** County (1970 pop. 37,770), 809 sq mi (2,095.3 sq km), NW central Pa., in a forested upland drained by the Clarion River and numerous creeks; formed 1843; co. seat Ridgway. It is primarily industrial; its manufactures include electric and electronic components, metal products, paper, inks and dyes, steel, and wood products. There are bituminous-coal mines, clay, sandstone, and limestone deposits, oil and natural-gas wells, and stands of timber here. The county includes part of Allegheny National Forest and has recreational and tourist facilities.

Elk. 1. River, rising on the W slope of the Cumberland Mts. W of Altamont, Tenn., and meandering c.200 mi (320 km) generally WSW to empty into Wheeler Lake W of Athens, Ala. **2.** River, 172 mi (276.7 km) long, rising in the Alleghenies in central W.Va. and flowing N, W, and NW to join the Kanawha River at Charleston.

El·ka·der (ēl-kā′dər, -kä′-), town (1970 pop. 1,592), seat of Clayton co., NE Iowa, on the Turkey River NW of Dubuque; inc. 1868. The ruins of Communia, a cooperative town settled c.1850, are nearby.

Elk Grove Village, village (1978 est. pop. 28,100), Cook and Du Page cos., NE Ill., a suburb of Chicago; inc. 1956. Elk Grove Village has a large industrial park. Its manufactures include electronic components, electrical equipment, and rubber products.

Elk·hart (ĕlk′härt, ĕl′kärt), county (1970 pop. 126,529), 468 sq mi (1,212.1 sq km), bordered on the N by Mich. and drained by the Elkhart River; formed 1830; co. seat Goshen. In an agricultural area yielding soybeans, corn, wheat, oats, potatoes, mint, onions, livestock, poultry, and dairy products, it also has timber and sand and gravel pits. Its manufactures include textiles and clothing, paper, chemicals, furniture, and wood, rubber, plastic, and metal products.

Elkhart. 1. City (1978 est. pop. 43,200), Elkhart co., N Ind., at the confluence of the Elkhart and St. Joseph rivers; settled 1824, inc. 1877. Its manufactures include musical instruments, electrical equipment, and pharmaceuticals. **2.** Town (1970 pop. 2,089), seat of Morton co., extreme SW Kansas, on the Okla. border SW of Garden City, in an oil, natural-gas, and farm area. It has diversified industries. Cimarron National Grassland is nearby.

Elk·horn (ĕlk′hôrn′), city (1970 pop. 3,992), seat of Walworth co., SE Wis., SE of Milwaukee, in a farm and dairy area; settled 1837, inc. as a village 1852, as a city 1897. Musical instruments and electrical equipment are made.

Elkhorn, river, rising in NE Nebr. and flowing c.333 mi (535.8 km) SE to join the Platte River near Omaha.

El·kins (ĕl′kĭnz), city (1970 pop. 8,287), seat of Randolph co., NE W.Va., on the Tygart River SE of Clarksburg, in a timber and mining area; inc. 1890. Davis and Elkins College is here, and Monongahela National Forest is nearby.

Elk Mountains, range of the Rocky Mts., W central Colo. Castle Peak, 14,259 ft (4,349 m), is the highest summit.

El·ko (ĕl′kō), county (1970 pop. 13,958), 17,163 sq mi (44,452.2 sq km), extreme NE Nev., in a mountain and plateau area crossed by the Humboldt River and bordering on Idaho and Utah; formed 1868; co. seat Elko. Cattle and sheep are raised; chemical products and tourism are also important. The county includes Te-Moak Indian Reservation, part of Duck Valley Indian Reservation, and sections of Humboldt National Forest.

Elko, city (1970 pop. 7,621), seat of Elko co., NE Nev., on the Humboldt River N of the Diamond Mts.; inc. 1917. Originally a campsite on the pioneer trail from Salt Lake City to Calif., it was founded (1868) when the Central Pacific RR arrived. It is today a trade and transportation center for a large ranching area.

Elk Point, city (1970 pop. 1,372), seat of Union co., extreme SE S.Dak., S of Sioux Falls; founded 1861. It is a shipping center for grain and livestock.

Elk River, village (1970 pop. 2,252), seat of Sherburne co., SE Minn., on the Mississippi River NNW of Minneapolis; platted 1865, inc. 1881. Concrete products are made.

Elk·ton (ĕlk′tən). **1.** City (1970 pop. 1,612), seat of Todd co., SW Ky., ESE of Hopkinsville near the Tenn. border, in a tobacco, livestock, grain, and fruit area. At Fairview, a few miles west, a memorial park and monument mark the birthplace of Jefferson Davis. **2.** Town (1970 pop. 5,362), seat of Cecil co., NE Md., NE of Baltimore; founded 1681, inc. 1787. Pulp and paper, textiles, clothing, and explosives are manufactured.

El·la·ville (ĕl′ə-vĭl′), city (1970 pop. 1,391), seat of Schley co., W central Ga., N of Americus, in a farm and timber region.

El·len·dale (ĕl′ən-dāl′), city (1973 est. pop. 1,792), seat of Dickey co., SE N.Dak., near the S.Dak. border S of Jamestown, in a livestock and grain area; inc. 1889.

El·len Mountain (ĕl′ən), mountain, 11,485 ft (3,503 m) high, NE Garfield co., S Utah. It is the highest peak of the Henry Mts.

El·lens·burg (ĕl′ənz-bûrg′), city (1975 est. pop. 12,924), seat of Kittitas co., central Wash., on the Yakima River; inc. 1886. It is the trade and processing center for a region in which cattle raising, logging, and diversified farming are carried on. Central Washington State College is here. An annual rodeo is held in Ellensburg.

Elles·mere Island (ĕlz′mîr), 82,119 sq mi (212,688 sq km), c.500 mi (805 km) long, in the Arctic Ocean, N Northwest Territories, Canada, northernmost island of the Arctic Archipelago. It is separated from Greenland by a narrow passage. The island's coast is indented by deep fjords; an ice cap covers most of the island's east side. There are scientific stations and some Eskimo settlements on the island. First sighted by the British explorer William Baffin in 1616, Ellesmere Island was explored in the latter half of the 19th cent.

Ellesmere Port, municipal borough (1973 est. pop. 63,870), Cheshire, W central England, on the Manchester Ship Canal. It is an important oil-refining center and the distribution center for imported commodities such as iron ore and grain. The town also has dye works and various light industries.

El·lice Islands (ĕl′ĭs) or **La·goon Islands** (lə-gōōn′), group of atolls (1968 pop. 6,332), 9.5 sq mi (24.6 sq km), South Pacific. The group includes nine low coral atolls with pandanus and coconut groves. Copra is the chief export. Discovered in 1764 by Capt. John Byron and made a British protectorate in 1892, the group was included in the Gilbert and Ellice Island colony in 1915.

El·li·cott City (ĕl′ĭ-kŏt′), uninc. town (1970 pop. 2,000), seat of Howard co., central Md., on the Patapsco River WSW of Baltimore; inc. 1867, reverted to uninc. status 1935. It is a trade center in an agricultural and manufacturing area.

El·li·jay (ĕl′ə-jā′), city (1970 pop. 1,326), seat of Gilmer co., N Ga., on the Coosawattee River NE of Rome; inc. 1834. Bedspreads and lumber are manufactured.

El·li·ott (ĕl′ē-ət, ĕl′yət), county (1970 pop. 5,933), 240 sq mi (621.6 sq km), NE Ky., in a hilly region drained by the Little Sandy River and several creeks; formed 1869; co. seat Sandy Hook. Its agriculture includes tobacco, corn, hay, fruit, and livestock. It also has bituminous-coal mines, a lumbering industry, and yarn mills.

El·lis (ĕl′ĭs). **1.** County (1970 pop. 24,730), 900 sq mi (2,331 sq km), central Kansas, drained in the N by the Saline River, in the S by the Big Smoky River; formed 1867; co. seat Hays. In a prairie agricultural area yielding wheat and livestock, it has oil fields in the northeast and some manufacturing. **2.** County (1970 pop. 5,129), 1,242 sq mi (3,216.8 sq km), NW Okla., bounded on the W by the Texas border, on the S by the Canadian River, and drained by Wolf Creek; formed 1907; co. seat Arnett. It has agriculture (wheat, barley, broomcorn, poultry, and livestock), oil and natural-gas wells, and some manufacturing. **3.** County (1970 pop. 46,638), 950 sq mi (2,460.5 sq km), NE central Texas, in a rich blackland agricultural area bounded on the E by the Trinity River; formed 1849; co. seat Waxahachie. Its agriculture includes cotton, corn, grains, clover seed, pecans, fruit, truck crops (especially onions), livestock, and dairy products. There are limestone quarries and clay pits here. The county has a large brick-making industry and some manufacturing and processing plants.

Ellis Island, c.27 acres (10.9 hectares), in Upper New York Bay, SW of Manhattan island. Government property since 1808, it was long the site of an arsenal and a fort. From 1892 until 1943 it served as the chief immigration station of the country. Ellis Island is part of the Statue of Liberty National Monument.

El·lis·ville (ĕl'ĭs-vĭl'), city (1970 pop. 4,643), a seat of Jones co., SE Miss., SSW of Laurel, in a farm, dairy, and timber area. In the Civil War the county was a center of anti-Confederate activity.

El·lo·ra (ĭ-lōr'ə), village, E central Maharashtra state, India. Extending more than 1 mi (1.6 km) on a hill are 34 rock and cave temples (5th–13th cent.), most of them Hindu but some Buddhist and Jain. The most remarkable building is the great Kailasa temple, dedicated to the god Shiva, who is enshrined as a giant lingam in the innermost sanctuary. The temple is a free-standing structure, carved like a statue from the surrounding hillside. Mythological and animal figures are profusely carved on nearly all the surfaces.

El·lore (ĭ-lōr'): *see* Eluru, India.

Ells·worth (ĕlz'wûrth'), county (1970 pop. 6,146), 723 sq mi (1,872.6 sq km), central Kansas, drained by the Smoky Hill River; formed 1867; co. seat Ellsworth. In a rolling-plain ranching and wheat-farming area, it has oil and natural-gas fields, salt mines, and limestone quarries. Wood products and automobile parts are manufactured.

Ellsworth. 1. City (1970 pop. 2,080), seat of Ellsworth co., central Kansas, on the Smoky Hill River W of Salina; laid out 1867, inc. 1868. It grew as a cow town in the 1870s and is now a farm trade center for a wheat-growing area. **2.** City (1970 pop. 4,603), seat of Hancock co., S Maine, N of Mt. Desert Island; settled 1763, inc. as a town 1800, as a city 1869. It was once a shipping and shipbuilding center. Today it is a summer resort. Lamoine State Park is nearby. **3.** Village (1970 pop. 1,983), seat of Pierce co., W central Wis., WSW of Eau Claire; inc. 1887.

Elm·hurst (ĕlm'hûrst), city (1978 est. pop. 44,300), Du Page co., NE Ill., a suburb of Chicago; settled 1843, inc. 1910. A residential city in a truck-farming area, it also has three industrial parks.

El·mi·ra (ĕl-mī'rə), city (1978 est. pop. 35,600), seat of Chemung co., extreme S central N.Y., on the Chemung River; settled 1788, inc. 1864. It is a distributing and manufacturing center with plants making food items, electronic equipment, fire engines, automobile parts, and iron and steel products. The Treaty of Painted Post, ending warfare between settlers and the Iroquois, was signed here in 1791. Mark Twain spent many summers in Elmira and is buried here.

El Mon·te (ĕl mŏn'tē), city (1978 est. pop. 67,500), Los Angeles co., S Calif.; inc. 1912. An industrial and commercial city in the San Gabriel Valley, the city is also known for its walnut groves. El Monte was founded in 1852 by westward-bound pioneers on the Santa Fe Trail.

El·more (ĕl'môr'). **1.** County (1970 pop. 33,661), 628 sq mi (1,626.5 sq km), E central Ala., in the coastal plain, bounded on the E and S by the Tallapoosa River and drained by the Coosa River; formed 1866; co. seat Wetumpka. In a lumbering region, it produces cotton, corn, and livestock. Textiles, clothing, and wood, brick, and clay products are manufactured. **2.** County (1970 pop. 17,479), 3,048 sq mi (7,894.3 sq km), SW central Idaho, in an irrigated livestock region bounded on the N by the Boise River, on the S by the Snake River; formed 1889; co. seat Mountain Home. It has a lumbering industry and deposits of gold, silver, copper, lead, and zinc. Part of Boise National Forest occupies the northern section.

El Mor·ro National Monument (ĕl môr'ō): *see* National Parks and Monuments Table.

El O·beid (ĕl ō-bād'): *see* Al Ubayyid, Sudan.

El Pas·o (ĕl păs'ō). **1.** County (1970 pop. 235,972), 2,158 sq mi (5,589.2 sq km), E central Colo., in a mining area drained by Fountain Creek; formed 1861; co. seat Colorado Springs. Its agriculture includes wheat, corn, barley, rye, sugar beets, potatoes, broomcorn, fruit, truck crops, and livestock. It also has food-processing and lumbering industries. The county, which includes Pikes Peak and part of Pike National Forest, has good recreational facilities. **2.** County (1970 pop. 359,291), 1,054 sq mi (2,729.9 sq km), extreme SW Texas, bounded on the N by the N.Mex. border, on the W and S by the Rio Grande, here forming the Mexico border; formed 1850; co. seat El Paso. In an irrigated high-plateau region, it has agriculture (cotton, alfalfa, fruit, truck crops, cattle, poultry, and dairy products), industry in the El Paso area, and some minerals (copper, tin, lead, zinc, and borax). It also has deposits of limestone, glass sand, and clay.

El Paso, city (1978 est. pop. 402,000), seat of El Paso co., extreme W Texas, on the Rio Grande opposite Juárez, Mexico; inc. 1873. Located in a region of cattle ranches and cotton and vegetable farms, the city is a port of entry and a commercial, industrial, and mining center. Manufactures include refined petroleum, processed copper, foodstuffs, clothing, and machinery. El Paso is in the region once known as El Paso del Norte, so called because of the route through the mountains to the north. In the 16th and 17th cent. missionaries, soldiers, and traders came to the region. After the U.S.-Mexico border was set, settlement increased, and the coming of the railroad in 1881 prefaced the growth of a rendezvous of cowboys, exiles, traders, and adventurers into a great commercial city. After the settlement in 1963 of the Chamizal border dispute, a small area of El Paso was transferred to Mexico.

El Re·no (ĕl rē'nō), city (1975 est. pop. 14,810), seat of Canadian co., central Okla.; inc. 1889. It is a rail and marketing center, with railroad shops and grain mills.

El Sal·va·dor (ĕl săl'və-dôr', säl'vä-*th*ôr'), republic (1970 est. pop. 3,533,628), 8,260 sq mi (21,393 sq km), Central America, bounded on the S by the Pacific Ocean, on the W by Guatemala, and on the N and E by Honduras. The capital is San Salvador. Two volcanic ranges, running roughly west to east, segment the country, but in between are broad, fertile valleys, such as that of the Lampa, the principal river. There are several fairly large lakes.

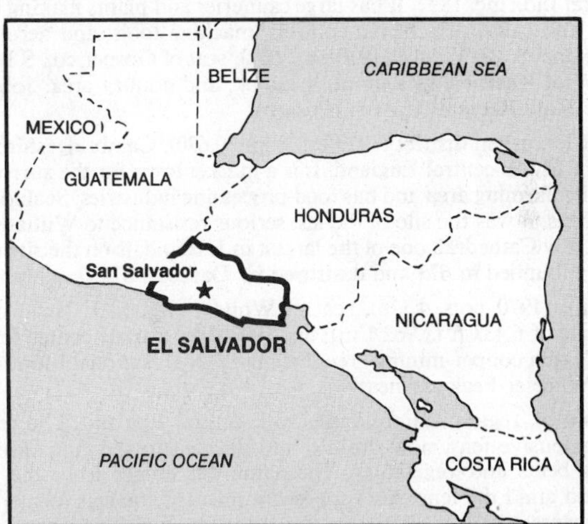

Economy. El Salvador's economy is primarily agricultural, although it is more highly industrialized than its neighbors. Two thirds of the land is used for either crops or pasturage. Rice is the chief subsistence crop; coffee is grown for export. Textiles, processed foods, and footwear are among El Salvador's leading manufactures.

History. Pedro de Alvarado led the Spanish conquest (1524) of El Salvador. The region was governed under the captaincy general of Guatemala. With independence from Spain in 1821, it became briefly a part of the Mexican Empire of Augustín de Iturbide, and after the empire collapsed (1823), El Salvador joined the Central American Federation. After the dissolution of the federation (1839), the republic was plagued by frequent interference from the dictators of neighboring countries.

The primacy of coffee cultivation began in the second half of the 19th cent. and led to the predominance of landed proprietors. The economy became vulnerable to fluctuations in the world market price for coffee. In 1931 Maximiliano Hernández Martínez, capitalizing on discontent caused by the collapse of coffee prices, led a coup d'état. His dictatorship lasted until 1944, after which there was chronic political unrest. Under the authoritarian rule of Maj. Oscar Osorio (1950–56) and Lt. Col. José María Lemus (1956–60) considerable economic progress was made. Lemus was overthrown by a coup, and after a confused period a junta composed of leaders of the National Conciliation Party came to power in June, 1961. Relations with Honduras deteriorated in the late 1960s. War broke out following a tension-filled soccer match between the two nations in July, 1969. After four days the Organization of American States imposed a cease-fire. The Salvadoran forces that had invaded Honduras were withdrawn. In the early 1970s there were several instances of leftist terrism. In 1977 Gen. Carlos Humberto Romero became president.

Government. El Salvador is governed under the 1962 constitution. The president is popularly elected for a five-year term and may not succeed himself. The members of the unicameral legislature are elected for two-year terms under a system of proportional representation.

El Se·gun·do (ĕl sē-gŭn'dō, -go͝on'-), industrial city (1975 est. pop. 15,127), Los Angeles co., S Calif., on Santa Monica Bay; inc. 1917. Its products include electronic equipment, aircraft and space vehicles, and petroleum. It was founded (1911) as an oil town.

El·si·nore (ĕl'sə-nôr'): *see* Helsingør, Denmark.

Els·mere (ĕlz'mîr'). **1.** Town (1970 pop. 8,415), New Castle co., NE Del., a suburb W of Wilmington; inc. 1909. **2.** City (1970 pop. 5,161), Kenton co., N central Ky., a suburb SW of Covington.

E·lu·ru (ĭ-lo͝or'o͞o) or **El·lore** (ĭ-lōr'), town (1971 pop. 127,047), Andhra Pradesh state, E central India. Carpet making and tanning are important. Extensive ruins north of Eluru are thought to be the site of the capital of a Buddhist kingdom (c.600–1000 A.D.).

El·vend (ĕl-vĕnd', ĕl'vĕnd'), mountain: *see* Alvend.

El·ve·rum (ĕl'və-ro͝om), town (1970 est. pop. 14,000), Hedmark co., SE Norway, on the Glåma River. An important military training center, it was badly damaged in World War II during the German invasion (1940).

El·wood (ĕl'wo͝od'). **1.** City (1975 est. pop. 10,581), Madison co., central Ind.; inc. 1872. It has large canneries and plants making cans, wire and cables, electronic controls, machine tools, and aerospace components. **2.** Village (1970 pop. 601), seat of Gosper co., S Nebr., WSW of Kearney, in a grain, livestock, and poultry area. Johnson Lake State Recreation Area is nearby.

E·ly (ē'lē), urban district (1973 est. pop. 10,630), Cambridgeshire and Isle of Ely, E central England. It is a market town for the surrounding rich farming area and has food-processing industries. Secluded in the Fens, it was the site of the last serious resistance to William I in 1071. Ely Cathedral, one of the largest in England, is on the site of an abbey founded in 673 and destroyed by Danes in 870.

Ely, city (1970 pop. 4,176), seat of White Pine co., E Nev., at an altitude of 6,433 ft (1,962.1 m); inc. 1907. It is a trade center for the state's rich copper-mining area. Lehman Caves National Monument and Wheeler Peak are nearby.

Ely, Isle of, region, Cambridgeshire, E central England. The region has extensive fens, now drained and devoted to the cultivation of sugar beets and vegetables. The name *Isle* comes from the high ground amid the fens; *Ely* supposedly refers to the eels formerly in the waters.

E·ly·ria (ĭ-lîr'ē-ə), city (1978 est. pop. 50,600), seat of Lorain co., N Ohio, on the Black River; inc. 1833. It is a farm trade and industrial center. Cascade Park, with waterfalls, caves, nature trails, and a zoo, is in the heart of the city.

E·man·u·el (ĭ-măn'yo͞o-əl), county (1970 pop. 18,357), 686 sq mi (1,776.7 sq km), E central Ga., bounded on the NE by the Ogeechee River and drained by the Little Ohoopee and Canoochee rivers; formed 1812; co. seat Swainsboro. It lies in a coastal-plain timber and agricultural area yielding cotton, corn, tobacco, sweet potatoes, peanuts, hogs, and cattle. Clothing, plastics, metal products, and machinery are made.

Em·ba (ĕm'bə), river, c.400 mi (645 km) long, SW Central Asian USSR, in Kazakhstan, rising in the Mugodzhar Mts. and flowing SW into the Caspian Sea. The lower course traverses a region of flat salt domes, characteristic of the petroleum-rich Emba fields.

Em·bar·rass or **Em·bar·ras** (both: ăm'brô'), river, 185 mi (297.7 km) long, rising in E Ill. and flowing S then SE to the Wabash River near Vincennes, Ind.

Em·den (ĕm'dən), city (1974 est. pop. 53,911), Lower Saxony, NW West Germany, at the mouth of the Ems River. A major North Sea port, it has extensive shipyards and herring fisheries. Manufactures include chemicals, machinery, and motor vehicles. It was known in the 10th cent. and passed to East Friesland in 1453. It went to Prussia in 1744, passed to Hanover in 1815, and was regained by Prussia in 1866. Its modern development dates from the late 19th cent., when the Dortmund-Ems Canal was constructed and the industrialization of the Ruhr district accelerated.

Em·e·ry (ĕm'ə-rē, ĕm'rē), county (1970 pop. 5,137), 4,442 sq mi (11,504.8 sq km), E central Utah, bounded on the E by the Green River and drained by the Price and San Rafael rivers and Muddy Creek; formed 1880; co. seat Castle Dale. In a livestock and grain area, it has bituminous-coal mines and some manufacturing. Part of Manti-La Sal National Forest is in the northwest.

E·mi Kous·si (ā'mē ko͞o'sə), volcanic mountain peak, 11,204 ft (3,417.2 m) high, in the Tibesti massif, NW Chad. The extinct crater is 4,000 ft (1,220 m) deep and 12 mi (19.3 km) wide.

E·mi·lia-Ro·ma·gna (ā-mē'lyä-rō-mä'nyä), region (1975 est. pop. 3,935,834), 8,542 sq mi (22,124 sq km), N central Italy, bordering on the Adriatic Sea in the E; capital Bologna. The region includes a fertile, low-lying plain in the north and east, and the Apennine Mts. in the south and west. Agriculture is the chief occupation. Emilia-Romagna has large deposits of petroleum and natural gas.

Emilia takes its name from the Aemilian Way, a Roman road (laid out 187 B.C.) that crossed the region from Piacenza to Rimini. After the fall of Rome, the region was conquered (5th cent. A.D.) by the Lombards. Divided into several duchies and counties, Emilia was conquered by the Franks in the 8th cent. Emilia was held by the French from 1797 to 1814. It played an important role in the Risorgimento, and there were revolts against foreign rule in 1821, 1831, and 1848–49. In 1860 all of Emilia-Romagna was joined to the kingdom of Sardinia, which in 1861 became the kingdom of Italy. In the 20th cent. Emilia (especially Bologna) has been a center of radical activity. The region suffered severe flooding in 1966.

Em·i·nence (ĕm'ə-nəns), city (1970 pop. 520), seat of Shannon co., S Mo., in the Ozarks on Jacks Fork of the Current River S of Salem. It has resorts and copper mines.

Em·men (ĕm'ən), city (1977 est. pop. 34,600), Drenthe prov., NE Netherlands. Manufactures include textiles and electronic equipment. Peat is produced in the region.

Emmen, town (1977 est. pop. 22,700), Lucerne canton, central Switzerland, on the Reuss River. Textiles, electrical and iron goods, and airplanes are made in the town.

Em·men·tal (ĕm'ən-täl'), valley of the Emme River, W central Switzerland. In a region devoted to farming, cattle raising, and dairying, it produces some of the finest cheese in Switzerland.

Em·met (ĕm'ĭt). **1.** County (1970 pop. 14,009), 395 sq mi (1,023.1 sq km), N Iowa, on the Minn. border in a glaciated, rolling prairie region dotted with small lakes and drained by forks of the Des Moines River; formed 1851; co. seat Estherville. It has agriculture (corn, oats, soybeans, cattle, hogs, and poultry), food-processing industries, and feed mills. Farm machinery is made. It includes Fort Defiance State Park. **2.** County (1970 pop. 18,331), 461 sq mi (1,194 sq km), N Mich., bounded on the W by Little Traverse Bay and Lake Michigan, on the N by the Straits of Mackinac; formed 1840 as Tonedagana co., renamed 1843; co. seat Petoskey. Its agriculture includes potatoes, grain, fruit, livestock, and dairy products. It also has limestone deposits, a lumbering industry, and some manufacturing. The county includes numerous lakes and year-round resorts for hunting, fishing, camping, and winter sports.

Em·mets·burg (ĕm'ĭts-bûrg'), city (1970 pop. 4,150), seat of Palo Alto co., N central Iowa, SSE of Estherville on the West Fork of the Des Moines River; founded 1856.

Em·mett (ĕm'ĭt), city (1970 pop. 3,945), seat of Gem co., SW Idaho, on the Payette River NW of Boise; founded 1864 as a trading post, inc. as a village 1900, as a city 1909. It is the processing center for an area producing cherries and other fruit, livestock, dairy goods, and some lumber.

Em·mons (ĕm'ənz), county (1970 pop. 7,200), 1,503 sq mi (3,892.8 sq km), S N.Dak., bounded on the W by the Missouri River, on the S by the S.Dak. border; formed 1879; co. seat Linton. It has agriculture (mainly grain and livestock) and some manufacturing.

Em·or·y (ĕm′ə-rē), village (1970 pop. 693), seat of Rains co., NE Texas, SE of Greenville. It is a trade center in a truck-farming and cotton area. There are cotton gins and feed mills here.

Em·po·li (ĕm′pō-lē), town (1976 est. pop. 45,906), Tuscany, central Italy, on the Arno River. It is a commercial and industrial center. Manufactures include textiles, glass, and chemicals.

Em·po·ri·a (ĕm-pôr′ē-ə). **1.** City (1978 est. pop. 22,000), seat of Lyon co., E central Kansas, in the Flint Hills between the Neosho and Cottonwood rivers; inc. 1857. It is a commercial and shipping (railroad and highway) center for a large cattle and farm area. It has grain elevators, stockyards, industries processing beef and soybeans, and plants making printing equipment, baked goods, and steel tanks. William Allen White made the Emporia *Gazette* nationally known. **2.** Town (1970 pop. 5,300), seat of Greensville co., but politically independent, SE Va., on the Meherrin River S of Petersburg; inc. 1887. It is a processing center in a peanut, cotton, and timber area. The courthouse dates from 1787.

Ems (ĕms) or **Bad Ems** (bät), town (1970 pop. 10,487), Rhineland-Palatinate, W West Germany, on the Lahn River. Chartered in 1324 as an important lead and silver mining center, it has been one of Europe's most famous spas since the late 17th cent.

Ems, river, 208 mi (334.7 km) long, rising in the Teutoburger Wald, NW West Germany, and flowing NW into the North Sea near Emden. Its wide mouth is called the Dollart. The Emsland is a swampy region between the lower course of the Ems and the Dutch border.

En·chant·ed Me·sa (ĕn-chănt′ĭd mā′sə), sandstone butte, 430 ft (131.2 m) high, central N.Mex., near the pueblo of Acoma. According to legend, the mesa was the home of the Pueblo Indian people until an earthquake destroyed the only approach.

En·ci·ni·tas (ĕn′sə-nē′təs), uninc. town (1970 pop. 5,375), San Diego co., S Calif., on the Pacific coast N of San Diego, in a farm area; settled 1854. Known for its flowers, the town has an annual national flower show.

En·der·by Land (ĕn′dər-bē), projection stretching from Ice Bay on Prince Olav Coast to Edward VIII Bay, Antarctica, at about the same longitude as Madagascar. It was first explored (1831-32) by John Biscoe, who claimed it for Australia.

En·di·cott (ĕn′dĭ-kət), village (1978 est. pop. 16,700), Broome co., S central N.Y., on the Susquehanna River; settled c.1795, inc. 1906. Shoes and business machines are the chief manufactures.

En·field (ĕn′fēld′), borough (1971 pop. 266,788) of Greater London, SE England. The borough was created in 1965 by the merger of the municipal boroughs of Enfield, Edmonton, and Southgate. It is residential, with important concentrations of industry. The poets John Keats and William Cowper lived in Edmonton.

Enfield. 1. Town (1970 pop. 46,189), Hartford co., N Conn., on the Connecticut River at the Mass. line, in a tobacco-growing area; settled c.1680. Among its many manufactures are carpets and plastic products. The site of a Shaker settlement (c.1780-1915) was bought for a state prison farm. **2.** Town (1970 pop. 2,345), Grafton co., W N.H., on Mascoma River and Lake, E of Lebanon. A charter was granted to settlers in 1761, and a Shaker community was established in 1793. Woolen fabrics are made.

En·ga·dine (ĕng′gə-dēn), valley of the upper Inn River, Grisons canton, E Switzerland, in the Rhaetian Alps. It extends for c.60 mi (95 km) northeast from Maloja Pass to the Austrian border and consists of the Upper and the Lower Engadine. Noted for its scenery and climate, the Engadine is famous as a tourist and health center.

En·ga·ño, Cape (ĕn-gä′nyō), N tip of Palau Island, NE Luzon, Philippines, at the entrance to Babuyan Channel. In a naval battle fought (Oct. 25, 1944) off this cape, the U.S. fleet was victorious over Japanese forces.

En·gel·berg (ĕng′əl-bĕrKH′), town (1970 pop. 2,841), Obwalden half canton, central Switzerland. It is a winter and summer resort and has an early 12th cent. Benedictine abbey.

En·gels (ĕn′gĭls), city (1976 est. pop. 159,000), E European USSR, a port on the Volga River. It has a large chemical fiber complex. It was the capital (1924-41) of the former German Volga ASSR.

Eng·land (ĭng′glənd), part of the United Kingdom of Great Britain and Northern Ireland (1971 pop. of England, 45,870,062), 50,334 sq mi (130,365 sq km). It is bounded by Wales and the Irish Sea on the west and Scotland on the north. The English Channel, the Strait of Dover, and the North Sea separate it from the continent of Europe. The Isle of Wight, off the southern mainland in the English Channel, and the Scilly Islands, in the Atlantic Ocean off the southwestern tip of the mainland, are considered part of England. The Thames and the Severn are the longest rivers.

Behind the white chalk cliffs of the southern coast lie the gently rolling downs and wide plains stretching to the Chiltern Hills and the Cotswold Hills. Along the east coast are the lowlands of Norfolk, reaching up to the Fens, formerly marshy country on both sides of The Wash, an inlet of the North Sea. In the east and southeast, river estuaries lead to some of England's great commercial and industrial centers. The north of England, above the Humber, is mountainous; the chief highlands are the Cumbrian Mts. in the northwest and the Pennines, which run north-south in north-central England. The famous Lake District, in the Cumbrians, has England's highest points. The center of England, the Midlands, is a large plain, interrupted and bordered by hills. In the Midlands are the industrial centers of Birmingham and the Black Country. On the Lancashire plain is the great city of Manchester, the center of the English textile industry. Durham and Western Yorkshire are also highly industrialized, but Eastern Yorkshire is an area of bleak moors and wolds, and the upper reaches of Northumberland are sparsely populated. In the west and southwest the border with Wales and the peninsula of Devonshire and Cornwall have a hilly, upland terrain.

In 1974 the 45 historic counties of England were abolished and replaced by 6 metropolitan counties and 39 nonmetropolitan counties. Greater London constitutes a separate administrative unit.

NAME	ADMINISTRATIVE CENTER
Metropolitan counties:	
Greater Manchester	
Merseyside	
South Yorkshire	
Tyne and Wear	
West Midlands	
West Yorkshire	
Nonmetropolitan counties:	
Avon	Bristol
Bedfordshire	Bedford
Berkshire	Reading
Buckinghamshire	Aylesbury
Cambridgeshire	Cambridge
Cheshire	Chester
Cleveland	Middlesbrough
Cornwall	Bodmin
Cumbria	Carlisle
Derbyshire	Derby
Devon	Exeter
Dorset	Dorchester
Durham	Durham
East Sussex	Lewes
Essex	Chelmsford
Gloucestershire	Gloucester
Hampshire	Winchester
Hereford and Worcester	Worcester
Hertfordshire	Hertford
Humberside	Kingston upon Hull
Isle of Wight	Newport
Kent	Maidstone
Lancashire	Lancaster
Leicestershire	Leister
Lincolnshire	Lincoln
Norfolk	Norwich
North Yorkshire	Northallerton
Northamptonshire	Northampton
Northumberland	Newcastle upon Tyne
Nottinghamshire	Nottingham
Oxfordshire	Oxford
Salop	Shrewsbury
Somerset	Taunton
Staffordshire	Stafford
Suffolk	Ipswich
Surrey	Kingston upon Thames
Warwickshire	Warwick
West Sussex	Chicester
Wiltshire	Salisbury

For the history of England as well as information on government and economy, *see* Great Britain.

En·gle·wood (ĕng′gəl-wŏŏd′). **1.** City (1978 est. pop. 35,600), Arapahoe co., N central Colo., on the South Platte River, a residential and industrial suburb of Denver; inc. 1903. It has ironworks, dairy-processing plants, and factories producing precision instruments. **2.** City (1978 est. pop. 22,100), Bergen co., NE N.J., a residential suburb of the N.Y.-N.J. metropolitan area; inc. as a city 1899. It was originally settled by the Dutch in the 17th cent. **3.** Village (1970 pop.

7,885), Montgomery co., SW Ohio, a suburb NW of Dayton; inc. 1914. Machine tools and farm implements are made.

Eng·lish (ĭng'glĭsh), town (1970 pop. 664), seat of Crawford co., S Ind., W of New Albany, in an agricultural and timber area. Wagon parts and baskets are manufactured.

English Ba·zar (bə-zär'), town (1971 pop. 61,713), West Bengal state, E central India, on the Mahananda River. The British East India Company established factories for the production of silk and cotton fabrics, for which English Bazar is still known. Jute, mulberries, and mangoes are grown.

English Channel, arm of the Atlantic Ocean, c.350 mi (565 km) long, between France and Great Britain. It is 112 mi (180 km) wide at its west entrance, between Land's End, England, and Ushant, France. Its greatest width, c.150 mi (240 km), is between Lyme Bay and the Gulf of St.-Malo; between Dover and Cape Gris-Nez it is 21 mi (33.8 km) wide. At the east end the Strait of Dover connects the Channel with the North Sea. A train-ferry service to carry passengers and freight without change between Paris and London was opened between Dover and Dunkirk in 1936. In 1785 J. P. Blanchard and Dr. John Jeffries crossed the Channel by balloon; the first to swim across was Matthew Webb (1875); and the first airplane crossing was made in 1909. The construction of a Channel railroad tunnel, long under discussion, was begun in 1974; however, shortly thereafter the project was postponed indefinitely because of a shortage of funds.

E·nid (ē'nĭd), city (1978 est. pop. 51,400), seat of Garfield co., N central Okla.; inc. 1893. It is an important trade and processing center for an area rich in wheat, dairy cattle, and poultry. Oil fields are nearby. Phillips Univ. and Vance Air Force Base are here.

E·ni·we·tok (ĕ-nē'wə-tŏk, ĕn'ĭ-wē'tŏk), uninhabited circular atoll, central Pacific, one of the Ralik Chain in the Marshall Islands. Eniwetok is c.50 mi (80 km) in circumference and comprises about 40 islets surrounding a large lagoon. Mandated to Japan by the League of Nations in 1920, Eniwetok was captured in World War II by U.S. forces. Designated an atomic proving station, it was the site of atomic tests in 1948, 1951, 1952, and 1954.

Enk·hui·zen (ĕngk'hoi'zən), town (1976 est. pop. 13,430), North Holland prov., N central Netherlands, a port on IJsselmeer. It is a commercial and industrial center and has shipyards. The town played an important part in the Dutch struggle for independence, was a prosperous fishing and trading center in the 17th cent., and has retained numerous buildings of the 16th and 17th cent.

En·kö·ping (ān'chœ'pĭng), city (1975 est. pop. 23,400), Uppsala co., E Sweden, near Stockholm. It is an industrial center whose manufactures include machinery and vehicles. Enköping was founded c.1300 and is one of Sweden's oldest cities.

En·na (ĕn'ə, ĕn'nä), town (1976 est. pop. 29,514), capital of Enna prov., central Sicily, Italy. It is an agricultural market, resort, and sulfur-mining center. Enna was taken by Syracuse (396 B.C.) and by Rome (258 B.C.) and played a major part in the Sicilian slave rebellion that occurred from 135 B.C. to 132 B.C. The town was later captured by the Arabs (9th cent.) and the Normans (11th cent.).

En·nis (ĕn'ĭs), city (1975 est. pop. 11,855), Ellis co., N Texas; inc. 1872. It is a trading, financial, rail, and processing center in a rich blackland area that produces cotton and grain. Ennis also has plants that manufacture clothing, office supplies, and beds. The city, which was settled by Czechs, sponsors a national polka festival each May.

En·nis·kil·len (ĕn'ĭs-kĭl'ən), municipal borough (1971 pop. 6,553), county town of Co. Fermanagh, SW Northern Ireland, on an island in the Erne River between Upper and Lower Loughs Erne. Farm produce is traded and hosiery is manufactured here. In 1689 the forces of William III defeated those of James II at Enniskillen.

Enns (ĕns), town (1971 pop. 9,622), Upper Austria prov., N central Austria, on the Enns River near its confluence with the Danube. One of Austria's oldest towns, Enns was established as a fortress in the 10th cent. and was chartered in 1212. The picturesque town retains part of its medieval walls, a 16th cent. fortress, and a Gothic parish church (13th–15th cent.). Lorch, incorporated into Enns in 1938, is on the site of a Roman camp established (A.D. c.170) by Marcus Aurelius.

En·sche·de (ĕn'-sKHə-dā'), city (1977 est. pop. 141,423), Overijssel prov., E Netherlands, near the West German border. It is a textile and machinery manufacturing center. Enschede was largely destroyed by fire in 1862, but was later rebuilt.

En·se·na·da (ĕ'sə-nä'də, än'sā-nä'dä), city (1970 pop. 77,687), Baja California state, NW Mexico. Cereal growing, cattle raising, and fishing are the chief occupations. Tourism is also important.

En·teb·be (ĕn-tĕb'ə), town (1969 pop. 21,096), S Uganda, on Victoria Nyanza, near Kampala. Located in a region producing cotton, coffee, and plantains, it was founded in 1893. From 1894 to 1962 it was the administrative capital of the British Uganda Protectorate.

En·ter·prise (ĕn'tər-prīz'). **1.** City (1970 pop. 15,591), Coffee co., SE Ala.; inc. 1896. It is a peanut-shipping center with lumber and textile mills and plants making concrete. The region's diversified farming began after the boll weevil destroyed (1910–15) the cotton; in gratitude for the resulting prosperity, the city erected (1919) a monument to the boll weevil. **2.** Uninc. town (1970 pop. 11,486), Shasta co., N central Calif. A state park is nearby. **3.** City (1970 pop. 1,680), seat of Wallowa co., NE Oregon, on the Wallowa River ENE of La Grande; platted 1885, replatted 1887, inc. 1889. It is a trade center for a ranching and lumbering area.

E·nu·gu (ə-nōō'gōō), city (1971 est. pop. 167,100), SE Nigeria. It is an industrial center. Furniture, ceramics, textiles, shoes, asbestos, cement, and steel are the chief products. Enugu developed as a mining town after the discovery of coal in 1909. The city served as capital of Nigeria's Southern Region (1929–39) and Eastern Region (1939–67) and of the secessionist state of Biafra (1967–70).

É·per·nay (ā'-pĕr-nā'), town (1975 pop. 29,677), Marne dept., NE France, on the Marne River. It is, next to Rheims, the largest manufacturing center for champagne wine and the headquarters of some of the oldest firms, notably Moët. The wine is stored in caves (open to tourists), which form a labyrinth some 30 mi (48 km) long in the surrounding hills. Heavy fighting during World War I destroyed many old buildings.

Eph·e·sus (ĕf'ə-səs), ancient Greek city of Asia Minor, near the mouth of the Caÿster River (modern Küçük Menderes), in what is today W Turkey, S of Smyrna (now Izmir). One of the greatest of the Ionian cities, it became the leading seaport of the region. The Greek city was near an old center of worship of a native nature goddess, who was equated with the Greek Artemis, and c.550 B.C. a large temple was built. The temple was burned down in the 4th cent. B.C., but rebuilding was begun before Alexander the Great took Ephesus in 334. After it passed (133) to the Romans it retained its hegemony and was the leading city of the province of Asia. Its great temple of Artemis, called by the Romans the temple of Diana, was considered one of the Seven Wonders of the World. Ephesus became a center of Christianity and was visited by St. Paul. The city was sacked by the Goths in A.D. 262, and the temple was destroyed. The seat of a church council in 431, Ephesus was abandoned after the harbor silted up. Excavations (1869–74) of the ruins of the temple brought to light many artifacts. Later excavations uncovered important Roman and Byzantine remains.

Eph·ra·ta. **1.** (ĕf'rə-tə), Industrial borough (1975 est. pop. 11,118), Lancaster co., SE Pa., in a prosperous farm area; inc. 1891. A noted semimonastic religious community was founded (c.1732) here by Seventh-Day Baptists under the leadership of Johann Conrad Beissel. This austere colony, the Ephrata Cloisters, was famous for its music and established (1745) one of the earliest printing presses in the country. The well-preserved buildings are now maintained as a monument by the state. **2.** (ē-frā'tə) City (1970 pop. 5,255), seat of Grant co., central Wash., ESE of Wenatchee, in an irrigated farm region; settled 1882, platted 1901, inc. 1909.

Ep·i·dau·rus (ĕp'ĭ-dôr'əs), ancient city of Greece, on an inlet of the Saronic Gulf, NE Peloponnesus. It was celebrated as the site of the temple of Asclepius, which dates from the 4th cent. B.C. and is renowned for its beautiful sculpture.

É·pi·nal (ā-pē-näl'), town (1968 pop. 39,525), capital of Vosges dept., E France, in Lorraine, on the Moselle. Although considerably damaged during World War II, the city today is an active industrial center, with textile and metal industries and plants making morocco leather, precision instruments, and bicycles.

E·pi·rus (ĭ-pī'rəs), ancient country of Greece, on the Ionian Sea and W of Macedon and Thessaly, a region now occupied by NW Greece and S Albania. It was inhabited from very early times by Epirote tribes, barely known to the Greeks. The tribes were molded into a state in the 4th cent. B.C. A subsequent kingdom reached its height in the 3rd cent. B.C. The Epirotes sided with Macedon in the wars against Rome, and Epirus was sacked (167 B.C.) and put under Roman dominion. It was a more or less neglected portion of the Byzantine Empire. After the Crusaders conquered Constantinople, the despotate of Epirus, larger than ancient Epirus, was set up. At the end of the 18th cent. Ali Pasha, the pasha of Yannina, set up an independent state in Epirus and Albania.

Ep·ping (ĕp'ĭng), urban district (1971 est. pop. 11,681), Essex, E England, near London, on the N edge of Epping Forest, an ancient royal forest that is now a public park of 5,600 acres (2,268 hectares).

Ep·som and Ew·ell (ĕp'səm; yo͞o'əl), municipal borough (1976 est. pop. 70,700), Surrey, SE England. Epsom salts were first prepared from the town's mineral waters in 1618. Epsom was a popular spa in the 17th and 18th cent. The town is now famous for horse racing at Epsom Downs.

e·qua·tor (ĭ-kwā'tər), imaginary great circle around the earth, everywhere equidistant from the two geographic poles and forming the base line from which latitude is reckoned. The equator, which measures c.24,000 mi (38,615 km), is designated as lat. 0°. It intersects northern South America, central Africa, and Indonesia.

E·qua·to·ri·al Guin·ea (ē'kwə-tôr'ē-əl gĭn'ē, ĕk'wə-), republic (1977 est. pop. 320,000), 10,830 sq mi (28,050 sq km), W central Africa, including the islands of Macías Nguema (formerly Fernando Po), Pigalu (formerly Annobón), Corisco, Elobey Grande, and Elobey Chico in the Gulf of Guinea, and Río Muni on the African mainland. Río Muni (93% of the nation's land area) is bordered on the north by Cameroon, on the east and south by Gabon, and on the west by the Gulf of Guinea. Malabo is the capital.

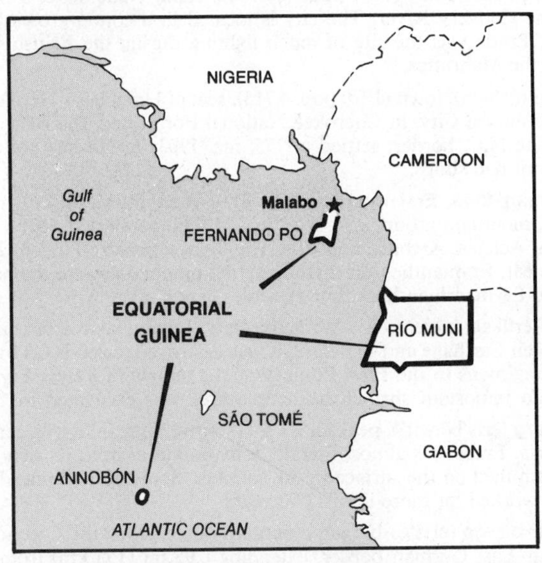

Macías Nguema is made up of three extinct volcanoes, the loftiest of which is c.9,870 ft (3,010 m) high. The island has abundant fertile volcanic soil. Río Muni, located just north of the equator, is made up of lowland along the coast, which gradually rises in the interior to a maximum height of c.3,600 ft (1,100 m). Río Muni includes three major rivers—the Campo, which forms part of the northern boundary; the Benito, located in the center; and Río Muni, which forms part of the southern boundary.

Economy. Equatorial Guinea is almost exclusively agricultural. The money economy is based on the production of cacao (mostly on Macías Nguema) and coffee and timber (in Río Muni). Other agricultural products include plantains, cassava, and palm oil. Manufactures are limited to basic consumer items and processed cacao and timber. There is a small fishing industry.

History. Macías Nguema was discovered and claimed by Fernão do Po, a Portuguese navigator, in 1472, and Pigalu was also claimed. In 1778 Portugal ceded the islands Spain. The Spanish sent settlers to the islands, but they died of yellow fever, and by 1781 the region was abandoned by the Europeans. From 1827 to 1843 the British leased bases at Malabo and San Carlos from Spain for use by their antislavery patrols. In 1844 the Spanish reacquired Macías Nguema and began to occupy it. In 1879 a Cuban penal settlement was established here. The general region of Río Muni was awarded to Spain at the Conference of Berlin in 1885; the islands and Río Muni were grouped together as the colony of Spanish Guinea. Under the Spanish economic development was largely confined to Macías Nguema. By 1960 about 6,000 Europeans (mostly Spanish) were living in the colony; they controlled the production of cacao and timber. In 1959 the colony was reorganized into two overseas provinces of Spain. However, nationalists demanded independence.

In 1963 Spain granted the country (renamed Equatorial Guinea) limited autonomy and on Oct. 12, 1968, granted it complete independence. The first president was Francisco Macías Nguema; in 1972 he was appointed president for life. In 1969 there were violent anti-European demonstrations in Río Muni and most whites left the country.

Equatorial Islands: *see* Line Islands.

E·rath (ē'răth'), county (1970 pop. 18,141), 1,085 sq mi (2,810.2 sq km), N central Texas, drained by the Bosque River; formed 1856; co. seat Stephenville. Its agriculture includes peanuts, cotton, wheat, corn, hay, pecans, fruit, truck crops, livestock, and dairy products. It also has oil and natural-gas wells and some remaining deposits of coal. Textiles are manufactured.

Er·e·bus, Mount (ĕr'ə-bəs), volcanic peak, 12,280 ft (3,745.4 m) high, on Ross Island, in the Ross Sea, E Antarctica. One of the loftiest volcanoes of the world, it was discovered in 1841 by the British explorer James C. Ross and first climbed in 1908.

E·rech (ē'rĕk) or **U·ruk** (o͞o'ro͝ok), ancient Sumerian city of Mesopotamia, on the Euphrates and NW of Ur (in present-day S Iraq). Erech, dating from the 5th millennium B.C., was the largest city in southern Mesopotamia and an important religious center. There have been excavations at the site since 1912.

E·re·tri·a (ĭ-rē'trē-ə), ancient city of Greece, in Euboea (now Évvoia), SE of Chalcis (now Khalkis), its rival. Eretria supported (499 B.C.) the Ionian cities in the revolt against Persian control. In retaliation Darius I destroyed (490 B.C.) the city. Athens planted (c.445 B.C.) a colony here, which revolted in 411 B.C. with the rest of Euboea. Excavations have uncovered the city walls, temple of Apollo, theater, and some public buildings.

Er·furt (ĕr'fo͝ort), city (1974 est. pop. 202,979), capital of Erfurt dist., SW East Germany, on the Gera River. It is an industrial and horticultural center. Manufactures include machinery, chemicals, precision and optical instruments, clothing, processed food, and electrical equipment. Erfurt was mentioned by St. Boniface in the 8th cent., and Charlemagne later made it a center for trade with the Slavs. Martin Luther studied (1501–5) at its university (opened 1392, closed 1816), and he took his vows as a friar at its monastery. Erfurt was a free imperial city and a member of the Hanseatic League. It passed (1664) to the electors of Mainz and (1802) to Prussia.

Er·icht, Loch (ĕr'ĭKHt), lake, 15 mi (24.1 km) long and 1 mi (1.6 km) wide, Inverness-shire and Perthshire, central Scotland. It is drained by the River Ericht. The lake is famous for trout.

E·ri·du (ā'rĭ-do͞o), ancient city of Sumer, Mesopotamia, near the Euphrates, S of Ur (in present-day S Iraq). Excavations conducted from 1946 to 1949 revealed that Eridu was the earliest known settlement in southern Mesopotamia and dated from c.5000 B.C.

E·rie (ĭr'ē). **1.** County (1970 pop. 1,113,491), 1,054 sq mi (2,729.9 sq km), W N.Y., bounded on the W by Lake Erie, on the N by Tonawanda Creek, on the S by Cattaraugus Creek; formed 1821; co. seat Buffalo. Its agriculture includes potatoes, hay, fruit, truck crops, livestock, poultry, and dairy products. Its extensive industries produce food products, textiles, chemicals, plastics, concrete, machinery, and wood, paper, and metal products. The county includes parts of Cattaraugus and Tonawanda Indian reservations. **2.** County (1970 pop. 75,909), 264 sq mi (683.8 sq km), N Ohio, bounded on the N by Lake Erie and drained by the Huron and Vermilion rivers; formed 1838; co. seat Sandusky. It has agriculture (wheat, corn, fruit, livestock, and dairy products), limestone quarries, sand pits, and commercial fisheries. Textiles, chemicals, pottery, concrete, and metal and paper products are manufactured. There are a number of summer resorts and recreational areas here. **3.** County (1970 pop. 263,654), 812 sq mi (2,103.1 sq km), extreme NW Pa., bordered on the N by Lake Erie, on the NE by N.Y., on the W by Ohio, and drained by branches of French Creek; formed 1800; co. seat Erie. It is highly industrialized; its manufactures include iron and steel, metal products, furniture, paper, plastics, and concrete. Grapes, cherries, and cabbage are grown.

Erie. 1. City (1970 pop. 1,414), seat of Neosho co., SE Kansas, on the Neosho River NNE of Independence; founded 1867, inc. 1870. It is a trade center and shipping point for an oil and farm area. **2.** City (1970 pop. 129,231), seat of Erie co., NW Pa., on Lake Erie; inc. as a city 1851. Pennsylvania's only port on the Great Lakes, Erie is a busy shipping point for coal, iron, grain, petroleum, heavy machinery, and lumber. Its manufactures include boilers, engines, power shovels, meters, foundry products, plastics, and paper. Fort Presque Isle was built in 1753 by the French, occupied and rebuilt in 1760 by the English, and destroyed during Pontiac's Rebellion in 1763. A peace conference between the British and Indians was held in 1764, but the town was not laid out until 1795. Oliver Hazard Perry's fleet was launched at Crystal Point before his victory over the British during the Battle of Lake Erie in 1813. Presque Isle State Park, located on the long peninsula that helps form Erie's superb harbor, is the area's leading tourist attraction.

Erie, Lake, 9,940 sq mi (25,745 sq km), 241 mi (387.8 km) long and from 30 to 57 mi (48.3–91.7 km) wide, bordered on the N by S Ont., Canada, on the E by W N.Y., on the S by NW Pa. and N Ohio, and on the W by SE Mich.; fourth-largest of the Great Lakes. It is 572 ft (174.5 m) above sea level with a maximum depth of 210 ft (64 m), making it the shallowest of the Great Lakes and the only one with a floor above sea level. It is part of the Great Lakes–St. Lawrence Seaway system and is linked to Lake Huron by the Detroit River, Lake St. Clair, and the St. Clair River, and with Lake Ontario by the Niagara River (Lake Erie's only natural outlet) and the Welland Canal. The New York State Barge Canal links the lake with the Hudson River. Lake Erie is partially icebound in winter and is usually closed to navigation from mid-December to the end of March. Untreated industrial and municipal wastes from lakeshore cities have polluted the waters; a U.S.-Canadian pact (1972) agreed to end the discharge of all contaminating materials into the water. The lake was first seen by Louis Jolliet, the French explorer, in 1669. The British and the French, and later the British and the Americans, fought for its control. The Battle of Lake Erie (Sept. 10, 1813), a naval engagement in the War of 1812, led successfully by U.S. Adm. Oliver H. Perry against the British, was fought at Put-in Bay.

Erie Canal, former artificial waterway, c.360 mi (580 km) long; it extended across central N.Y. from Albany to Buffalo and connected the Hudson River with Lake Erie. Locks overcame the 571-ft (174.2-m) difference between the level of the river and that of the lake. The New York State Barge Canal now follows part of the Erie Canal's course. After the American Revolution, the need for an all-American water route between the Great Lakes and the Atlantic coast became evident, but political and financial considerations delayed the start of work on the canal until 1817. The Erie Canal's middle section (Utica to Salina) was completed in 1820; its eastern section was finished in 1823. The canal was enlarged beginning in 1835. Railroad competition, beginning in the 1850s, eventually destroyed the canal's long-haul advantages. The Erie Canal contributed to New York City's financial development, opened the eastern markets to the farm products of the Midwest and encouraged immigration into that region, and helped create numerous large cities. Its initial success started a wave of canal building in the United States.

Er·in (ĕr′ən, îr′-), ancient and poetic name of Ireland.

Er·i·tre·a (ĕr′ĭ-trē′ə), province (1970 est. pop. 1,836,800), c.48,000 sq mi (124,320 sq km), N Ethiopia, on the Red Sea. Eritrea formed part of the ancient Ethiopian kingdom of Aksum until the 7th cent. Thereafter Ethiopia maintained an intermittent presence in the area until the mid-16th cent., when the Ottoman Empire gained control of much of the coastal region. Beginning in the mid-19th cent. Ethiopia struggled against Egypt and Italy for control of Eritrea. In the 1880s Italy occupied the coastal areas and by 1890 had extended its territory enough to proclaim the colony of Eritrea. The colony was later the main base for Italy's conquest (1935–36) of Ethiopia. In World War II Eritrea was captured (1941) by the British, who occupied it after the war and, beginning in 1949, administered it as a UN trust territory. In 1952 Eritrea became a federated part of Ethiopia; in late 1962 the Eritrean assembly voted to end the federal status and the unify Eritrea with Ethiopia. After 1962 Eritreans who opposed union carried on sporadic guerrilla warfare against Ethiopia.

Er·lang·en (ĕr′läng-ən), city (1974 est. pop. 100,550), Bavaria, S West Germany, at the confluence of the Schwabach and Regnitz rivers. Its manufactures include medical equipment, textiles, leather goods, and beer. Erlangen was sold to Emperor Charles IV in 1361. Chartered in 1398, it passed (1402) to the Franconian branch of the house of Hohenzollern. The city passed to Bavaria in 1810. Rebuilt after a devastating fire in 1706, the present city center has a baroque character; there are also modern industrial and residential sections. Erlangen is well known for its university (founded 1742 at Bayreuth and transferred to Erlangen in 1743).

Er·lang·er (ûr′lăng-gər), city (1970 pop. 12,676), Kenton co., N central Ky.; inc. 1897. Its industries include metal fabrication and the manufacture of clothing, truck bodies, and electrical fixtures.

Er·me·land (ĕr′mə-länt′) or **Erm·land** (ĕrm′länt′), historic region of East Prussia, extending far inland from the Baltic Sea. It was ceded to Poland in 1466 by the Teutonic Knights, passed to Prussia in 1772, and reverted to Poland after World War II.

Er·mine Street (ûr′mĭn), Saxon name for the Roman road in Britain that ran from London to Lincoln and York. It was one of the four main highways of Saxon England.

Er·na·ku·lam (ĕr-nä′ko͞o-ləm), city (1970 est. pop. 213,811), Kerala state, SW India, near Cochin. Manufactures include mats, rope, glycerin, perfume, and soap. Ernakulam was the capital of the former Cochin state and has a Jewish community thought to have been founded in the 2nd or 3rd cent. A.D.

Erne (ûrn), river, 72 mi (115.8 km) long, rising in Lough Gowna, Co. Longford, N Republic of Ireland, and flowing NW through SW Northern Ireland, then back through the Republic before entering the Atlantic Ocean at Donegal Bay. In Northern Ireland the river expands to form two large lakes—Upper Lough Erne (10 mi/16.1 km long) and Lower Lough Erne (18 mi/29 km long).

E·rode (ĭ-rōd′), city (1971 pop. 103,704), Tamil Nadu state, S India, on the Cauvery River. The city is located in a cotton-growing region. Erode was the site of much fighting during the British wars with the Mahrattas.

Er·win (ûr′wĭn), town (1970 pop. 4,715), seat of Unicoi co., NE Tenn., S of Johnson City, in Cherokee National Forest near the Bald Mts. and the N.C. border; settled c.1775, inc. 1903. Erwin has potteries and railroad shops.

Er·y·man·thos, Er·i·man·thos, or **Er·y·man·thus** (all: ĕr′ĭ-măn′thəs), mountain group, S Greece, in NW Peloponnesus, on the border of Achaea, Arcadia, and Elis. The highest peak (c.7,295 ft/2,225 m) is Mt. Erymanthos. In mythology the mountains were the haunt of the Erymanthian boar that Hercules captured.

Er·yx (ĕr′ĭks), ancient city, W Sicily, Italy. Long a source of conflict between Carthage and Syracuse, it was destroyed (c.260 B.C.) by the Carthaginians in the First Punic War. Its temple of Venus Erycina was an important shrine; the temple area was excavated in 1936.

Erz·berg (ĕrts′bĕrkн′), peak, 3,531 ft (1,077 m) high, in Styria, central Austria. Erzberg is almost literally a mountain of iron. Its rich iron ore is mined on the surface in the summer. Some of the mines have been worked for more than 1,000 years.

Erz·ge·bir·ge (ĕrts′gə-bĭr′gə), mountain range, along the Czechoslovakian–East German border, extending c.95 mi (155 km) from the Fichtelgebirge in the SE to the Elbe River in the NE. It reaches its highest point (4,080 ft/1,244.4 m) in Klínovec in Czechoslovakia. From the 14th cent. to the 19th cent. silver and iron were mined extensively in the Erzgebirge. At present the chief ores mined are uranium, lead, zinc, wolframite, tin, copper, bismuth, sulfur, arsenic, and antimony. Coal and lignite mines are also exploited. The Erzgebirge has many famous mineral springs. In 1938 the Czech part of the Erzgebirge was transferred to Germany by the Munich Pact. It was restored to Czechoslovakia in 1945.

Er·zur·um or **Er·zer·um** (both: ĕr′zə-ro͞om′), city (1970 pop. 134,655), capital of Erzurum prov., E Turkey. It is an agricultural trade center. Manufactures include processed foods, cement, and metal goods. Although its origins are obscure, the city was known in the 5th cent. A.D. as an important Byzantine frontier fortress. It was later held by various peoples, before being captured by the Ottoman Turks in the early 16th cent.

Es·bjerg (ĕs′byĕrg′, -byĕrкн′), city (1976 est. pop. 79,160), Ribe co., SW Denmark, a port on the North Sea. It is a commercial and industrial center, with major fisheries.

Es·cam·bi·a (ĕs-kăm′bē-ə). **1.** County (1970 pop. 34,912), 962 sq mi (2,491.6 sq km), S Ala., bordering in the S on Fla. and drained by the Conecuh (Escambia) River; formed 1868; co. seat Brewton. In a coastal-plain timber and agricultural region yielding cotton, fruit, truck crops, poultry, and dairy products, it has oil and natural-gas wells. Textiles, paper products, and chemicals are made. It includes part of Conecuh National Forest. **2.** County (1970 pop. 205,334), 663 sq mi (1,717.2 sq km), extreme NW Fla., bounded on the N by the Ala. border, on the W by the Perdido River (also forming a border with Ala.), on the S by the Gulf of Mexico, on the E by the Escambia River; formed 1821; co. seat Pensacola. Dairying, farming (cotton, corn, peanuts, and vegetables), forestry, and fishing are important. It also has shipyards and some manufacturing.

Escambia, river, c.231 mi (371.7 km) long, rising in SE Ala. (where it is called the Conecuh) and flowing SW through the state, then generally S through the NW corner of Fla., where it empties into Escambia Bay, an arm of Pensacola Bay.

Es·ca·na·ba (ĕs-kə-nä'bə), resort city (1978 est. pop. 13,800), seat of Delta co., W Upper Peninsula, N Mich., on Little Bay de Noc; settled 1852, inc. 1883. It is a railroad and manufacturing center, and from its fine harbor large amounts of ore are shipped. Among its varied manufactures are cranes, paper, wood, and plastics.

Esch-sur-Al·zette (ĕsh'sür-äl-zĕt') or **Esch** (ĕsh), city (1970 est. pop. 27,574), SW Grand Duchy of Luxembourg, on the Alzette River. Its manufactures include iron and steel, cement, and fertilizer.

Esch·wei·ler (ĕsh'vī-lər), city (1974 est. pop. 53,999), North Rhine-Westphalia, W West Germany, near Aachen. The center of a coal basin, it has iron and steel mills. Other manufactures include bricks, chemicals, and textiles. Known in the 9th cent., Eschweiler passed to Prussia in 1815.

Es·con·di·do (ĕs'kən-dē'dō), city (1978 est. pop. 56,200), San Diego co., S Calif.; inc. 1888. Located in a grape-growing valley, Escondido has fruit-packing houses and one of the world's largest avocado-processing plants. Textiles and ball-point ink are among its chief manufactures.

Es·dra·e·lon (ĕz'drə-ē'lən), fertile plain, c.200 sq mi (520 sq km), extending SE c.25 mi (40 km) between the coastal plain, near Mt. Carmel, and the Jordan River valley, N Israel, separating the hills of Galilee on the N from those of Samaria to the S. Once a swampy, malarial lowland, Esdraelon has been drained and turned into one of Israel's most fertile and densely populated regions. Since ancient times the plain has been a battleground.

Es·fa·han (ĕs'fä-hän') or **Is·fa·han** (ĭs'fə-hän'), city (1966 pop. 424,045), capital of Esfahan prov., central Iran, on the Zayandeh River. An ancient and picturesque city, rich in history, Esfahan has long been known for its fine carpets, hand-printed textiles, and metalwork, chiefly silver filigree. It has modern textile and steel mills. A noteworthy city in Sassanid times, Esfahan passed to the Arabs in the mid-7th cent. and served as a provincial capital. In the 11th cent. it was captured by the Seljuk Turks, who made it (1051) the capital of their empire. In the early 13th cent. Esfahan was taken by the Mongols. Tamerlane conquered the city in 1388 and, after its inhabitants rebelled, slaughtered c.70,000 persons in revenge; it is said that he built a large hill with the skulls of the dead. Under Shah Abbas I, who made (1598) Esfahan his capital, the city was embellished with many fine buildings—notably the beautiful imperial mosque, one of the masterpieces of world architecture. At its zenith in the 17th cent. Esfahan had a population of c.600,000, making it one of the world's great cities of the time. However, the city declined rapidly after it was captured (1723) by the Afghans, who massacred most of its inhabitants. Russian troops occupied Esfahan in 1916.

E·sher (ē'shər), urban district (1976 est. pop. 63,970), Surrey, SE England. It is a largely residential suburb of London.

E·skil·stu·na (ā'shĭl-stōō'nä), city (1975 est. pop. 92,663), Södermanland co., SE Sweden, between Hjälmaren and Mälaren lakes. The city was chartered (1659) by Charles X, who founded a gun factory that became the nucleus of an expanding steel and tool industry. Diesel engines and heavy machinery are also manufactured.

Es·ki·şe·hir (ĕs-kē'shĕ-hîr'), city (1975 pop. 258,266), capital of Eskişehir prov., W central Turkey. An industrial center, its manufactures include refined sugar, cement, railroad equipment, textiles, and meerschaum products. It is noted for its hot mineral springs. Godfrey of Bouillon defeated the Seljuk Turks here in 1097.

Es·me·ral·da (ĕz'mə-räl'də), county (1970 pop. 629), 3,570 sq mi (9,246.3 sq km), Nev., in a mountainous region bordering on Calif.; formed 1861; co. seat Goldfield. Mining (gold and silver) and stock grazing are important. It also has some manufacturing.

Es·na (ĕs'nə): *see* Isna, Egypt.

Es·pa·no·la (ĕs'pən-yō'lə), city (1970 pop. 4,528), Rio Arriba and Santa Fe cos., N central N.Mex., on the Rio Grande, in the heart of the Indian pueblo country; founded 1880, inc. 1964. A three-day fiesta every July commemorates the establishment nearby of a settlement by Juan de Oñate on July 11, 1598. There are many scenic, historic, archaeological, and Indian attractions in the area.

Es·pí·ri·to San·to (əs-pē'rē-tōō sän'tōō), state (1975 est. pop. 1,725,-100), 15,200 sq mi (39,368 sq km), E Brazil, on the Atlantic Ocean; capital Vitória.

Es·pi·ri·tu San·to (ĕs-pē'rē-tōō sän'tō) or **Santo**, volcanic island (1969 est. pop. 10,000), 1,485 sq mi (3,846.2 sq km), South Pacific, largest and westernmost island of the New Hebrides.

Es·qui·malt (ĕs-kwī'môlt', ĕ-skwī'-), regional district (1978 est. pop. 15,053), on Vancouver Island, SW British Columbia, Canada, just SW of Victoria. It has the chief naval station and naval dockyard of western Canada. The station was established by the British government in 1855 and taken over by Canada in 1906.

Es·qui·line (ĕs'kwə-līn'), one of the Seven Hills of Rome.

Es·qui·pu·las (ā'skē-pōō'läs), town (1964 pop. 19,164), SE Guatemala, near the Honduras and El Salvador borders. Believed to be a center of Mayan religious ceremonies, Esquipulas was chosen by the Spanish conquerors as a church site. They commissioned (1594) the carving of a figure of Christ from balsam, which became known as the Black Christ of Esquipulas. A series of reputed miracles led to the commissioning of a sanctuary in 1737. Completed in 1758, it is one of the most grandiose colonial churches in Latin America.

Es·sa·oui·ra (ĕs'ə-wîr'ə), city (1971 pop. 30,061), W Morocco, on the Atlantic Ocean. It was founded in 1760. The city declined when Agadir was opened to foreign trade in the 20th cent.

Es·sen (ĕs'ən), city (1975 pop. 684,147), North Rhine-Westphalia, W West Germany, on the Ruhr River. A major industrial center of the Ruhr district, it is the seat of the Krupp steelworks and the chief site for producing electricity in West Germany. Glass, chemicals, textiles, and machinery are also produced, and coal is mined nearby. Essen grew up around a Benedictine convent (founded in the mid-9th cent.). It was a small imperial state from the 13th cent. until 1802, when it passed to Prussia. The city's main industrial growth dates from the second half of the 19th cent. Essen was heavily bombed during World War II, but was rebuilt in modern style after 1945.

Es·se·qui·bo (ĕs'ā-kē'bō), longest river of Guyana, c.600 mi (965 km) long, rising in the Guiana Highlands, S Guyana, and flowing generally N to the Atlantic Ocean. Most of the river's course is broken by rapids and waterfalls.

Es·sex (ĕs'ĭks), one of the early kingdoms of Anglo-Saxon England. It was settled probably in the early 6th cent. by Saxons who traced their royal line back to a continental Saxon god instead of to Woden, as did the rulers of other early kingdoms. The submission of Essex to the overlordship of Wulfhere of Mercia marked the beginning of a long domination by the larger state. In 825 Essex joined other eastern kingdoms in submitting to Egbert of Wessex and became an earldom. Heavily settled by the Danes, it became part of the Danelaw by the treaty of 886, but was retaken by Edward the Elder of Wessex in 917.

Essex, nonmetropolitan county (1976 est. pop. 1,426,200), 1,528 sq mi (3,957.5 sq km), SE England, on the Thames River and the North Sea, one of the Home Counties of London; administrative center Chelmsford. The land rises from the low, irregular coastline to undulating pastoral country. Essex was once part of the kingdom of the East Saxons, and has Roman and Saxon remains.

Essex. 1. County (1970 pop. 637,887), 594 sq mi (1,538.5 sq km), NE Mass., on the Atlantic coast and bounded on the N by N.H.; formed 1643; co. seat Salem. It is interesected by the Merrimack and Ipswich rivers. The county's economy is based on tourism, agriculture, a fishing industry, and diversified manufacturing. Shoes, textiles, and metal products are made here. There are many resorts along the coastline. **2.** County (1970 pop. 932,526), 130 sq mi (336.7 sq km), NE N.J., bounded on the W, N, and E by the Passaic River, on the SE by Newark Bay; formed 1683; co. seat Newark. It is an important part of the N.Y.–N.J. metropolitan area. Primarily industrial, it produces textiles, processed foods, furniture, paper, chemicals, plastics, leather goods, machinery, and metal and wood products. It also has stone quarries, nurseries, and some dairy and truck-crop farms. **3.** County (1970 pop. 34,631), 1,826 sq mi (4,729.3 sq km), NE N.Y., in the Adirondacks bounded on the E by Lake Champlain and drained by the Hudson and Ausable rivers and numerous streams; formed 1799; co. seat Elizabethtown. In a lumbering and agricultural area yielding potatoes, hay, livestock, and dairy products, it also has deposits of metal ores. Lake Placid, Saranac Lake, and other noted resorts are here. **4.** County (1970 pop. 5,416) 664 sq mi (1,719.8 sq km), NE Vt., bounded on the N by the Que., Canada, border, on the E by the Connecticut River (here forming the N.H. border), and drained by the Nulhegan and Moose rivers; formed 1792; co. seat Guildhall. In a lumbering, dairying, and agricultural region, it also has good hunting and fishing and some manufacturing. **5.** County (1970 pop. 7,099), 250 sq mi (647.5 sq km), E Va., in a tidewater region bounded on the NE and E by the Rappahannock River; formed 1692; co. seat Tappahannock. It has agriculture (tobacco, corn, hay, peanuts, truck crops, and livestock) and a lumber industry. Textiles and automobile parts are manufactured.

Essex 1. Industrial town (1970 pop. 4,911), Middlesex co., S Conn., on the Connecticut River W of New London; settled 1690, set off from Saybrook 1852. Essex was once an important shipping and shipbuilding center, and many houses built by sea captains remain. **2.** Uninc. city (1978 est. pop. 42,900), Baltimore co., NE Md., a suburb of Baltimore. **3.** Town (1970 pop. 2,670), Essex co., NE Mass., NE of Salem; settled 1634, set off from Ipswich 1819. Its shipyards were established in 1668. Extensive clam beds are found here. **4.** Town (1970 pop. 10,951), Chittenden co., NW Vt., on the Winooski River; settled 1783. There is some light industry.

Ess·ling·en (ĕs'lĭng-ən), city (1974 est. pop. 97,029), Baden-Württemberg, SW West Germany, on the Neckar River. Manufactures include textiles, metal goods, furniture, and transportation equipment. It is noted for its wines. Founded in the 8th cent., Esslingen was a free imperial city from the 13th cent. to 1802, when it passed to Württemberg. It was (1488) the scene of the founding of the Great Swabian League.

Es·sonne (ĕ-sôn'), department (1975 pop. 923,061), 699 sq mi (1,810.4 sq km), N central France; capital Évry.

Es·ta·dos, Is·la de los (ēz'lə dä dôs ā-stä'dōs, ā-stä'*th*ōs), island, 209 sq mi (541.3 sq km), S Atlantic Ocean, off the E tip of Tierra del Fuego. It belongs to Argentina and has a seismographic station.

Es·tan·cia (ĭ-stän'shə), town (1970 pop. 721), seat of Torrance co., central N.Mex., SE of Albuquerque, in an agricultural and livestock area at an altitude of 6,100 ft (1,860.5 m). Wool and beans are produced. Cibola National Forest and the Manzano Mts. are nearby.

Es·te (ĕs'tā), town (1976 est. pop. 18,277), in Venetia, NE Italy. Manufactures include ceramics, chemicals, and metal goods. The ancient Ateste, it was a center of civilization (10th–2nd cent. B.C.) of which many important remains have been found. It was later a Roman military colony. In 1275 it passed to Padua and in 1405 to Venice. Este has a castle (11th–14th cent.), several fine villas, and an excellent archaeological museum.

Es·ther·ville (ĕs'tər-vĭl'), city (1970 pop. 8,108), seat of Emmet co., N central Iowa, NNW of Emmetsburg on the West Fork of the Des Moines River ESE of Spirit and Okoboji lakes; settled 1857, inc. as a town 1881, as a city 1894. It is a shipping, processing, and trade center for a grain, poultry, and livestock area. The site of Fort Defiance (1862) is preserved in a nearby state park.

Es·till (ĕs'tĭl), county (1970 pop. 12,752), 260 sq mi (673.4 sq km), E Ky., drained by the Kentucky and Red rivers and several creeks; formed 1808; co. seat Irvine. In a hilly agricultural region yielding burley tobacco, corn, truck crops, poultry, livestock, and dairy products, it has coal mines, oil wells, timber, and clothing factories. Part of Daniel Boone National Forest is here.

Es·to·ni·a (ĕ-stō'nē-ə), constituent republic (1970 pop. 1,357,000), 17,413 sq mi (45,100 sq km), W European USSR. It borders on the Baltic Sea in the west; the Gulfs of Riga and Finland (both arms of the Baltic) in the southwest and north, respectively; Latvia in the south; and the Russian SFSR in the east. Tallinn is the capital. Mainly a lowland region, the republic has numerous lakes; Lake Chudskoye (Peipus), the largest, is important for both shipping and fishing. Along Estonia's Baltic coast are more than 800 islands, of which Sarema is the most notable.

Economy. Estonia ranks first among Soviet republics in the extraction of shale and the production of shale oil and gas. Peat, limestone, dolomite, marl, clays (for cement and earthenware), sand (for the glass industry), phosphorite (for fertilizer), and timber are other important natural resources. Estonia is a leading Soviet supplier of various wood products (cellulose, paper, plywood), cotton, linen, and wool textiles, oil industry equipment, chemical fertilizer, electrical and radio apparatus, building materials, window glass, leather footwear, and processed fish. Other industries include shipbuilding and repair, metalworking, and food processing. Fishing, dairy farming, and pig raising are important occupations. The main crops are flax, potatoes, and sugar beets.

History. The Estonians, who are ethnically and linguistically close to the Finns, settled in their present territory before the Christian era. In the 13th cent. the Danes conquered northern Estonia and the German order of the Livonian Brothers of the Sword occupied the southern portion. In 1346 the Danes sold their territory to the order, and Estonia remained under the rule of the knights and the Hanseatic merchants until the order's dissolution in 1561. Northern Estonia then passed to Sweden; the rest was briefly held by Poland but was transferred to the Swedes in 1629. Peter I of Russia conquered Livonia in 1710. Despite some land reforms, the German nobility—the Baltic barons—retained its sway over the Estonian peasantry until the eve of the 1917 Russian Revolution.

Estonian national consciousness began to stir in the mid-19th cent. but was countered by russification, In the aftermath of the 1917 Russian Revolution, Estonia declared itself independent (Feb., 1918), suffered a brief German occupation, and then repulsed an invading Red Army. In 1920 Soviet Russia recognized Estonia's independence. Political stability, however, eluded the republic, which reverted to dictatorship in 1934. In 1939 the USSR secured military bases in Estonia. Complete Soviet military occupation followed, and Estonia was incorporated into the USSR as a constituent republic in 1940. Collectivization of agriculture and nationalization of industry began in the late 1940s, and the Estonian economy was steadily integrated with that of the USSR.

Es·tre·la, Ser·ra da (sĕr'ə dä ĕsh-trā'lə), mountain range, central Portugal. It rises to 6,532 ft (1,992.3 m) in Malhão da Estrela, Portugal's highest peak. The range is an important pastoral region and a year-round resort area.

Es·tre·ma·du·ra (ĕsh'trə-mə-dŏŏr'ə), region, W Portugal, formerly a province, now divided between the prov. of Estremadura (2,064 sq mi/5,345.8 sq km; capital Lisbon) and Ribatejo, with a small part in Beira Litoral.

Estremadura, region (1974 est. pop. 1,162,416), W central Spain, on the border with Portugal. It is a tableland crossed by mountains and by the Tagus and Guadiana rivers. Much of it is poverty-ridden, with poor communications, absentee landlordism, and steady emigration. Reconquered from the Moors in the 12th and 13th cent., the region was frequently a battlefield in the Spanish wars with Portugal and again in the Peninsular War. Most of Estremadura fell to the Nationalists early in the Spanish Civil War.

Esz·ter·gom (ĕs'tĕr-gôm'), city (1977 est. pop. 30,000), N Hungary, on the Danube River and the border of Czechoslovakia. It is a river port with industries producing iron products, machinery, machine tools, textiles, and alcoholic beverages. Situated in an extensive vineyard region, Esztergom carries on trade in wine and grain. Coal and lignite are mined nearby. Its mineral springs make the city popular with tourists. Esztergom was the first royal residence and the capital of Hungary until the 13th cent. King Stephen I, later canonized as Hungary's patron saint, was crowned at Esztergom (his birthplace) in 1001. Mongols sacked Esztergom in 1241, and the Turks occupied it during much of the 16th and 17th cent. Overlooking the Danube is the city's 19th cent. dome-topped cathedral.

E·tah (ē'tə), village, NW Greenland, on Smith Sound, opposite Ellesmere Island. It was often used as a base for arctic expeditions.

E·ta·wah (ĭ-tä'wə), town (1971 pop. 85,900), Uttar Pradesh state, N central India, on the Jumna River. It is a market for grain, oilseed, handloomed fabrics, and leather. The town was held by the Rajputs from the 12th to the 16th cent. It was occupied by rebels during the 1857 Indian Mutiny against the British.

E·thi·o·pi·a (ē-thē-ō'pē-ə), country (1977 est. pop. 29,010,000), 471,776 sq mi (1,221,900 sq km), NE Africa, formerly widely called Abyssinia, bordering on the Red Sea in the N, on the Djibouti Republic in the NE, on Somalia in the E and SE, on Kenya in the S, and on Sudan in the W. Addis Ababa is the capital.

Ethiopia falls into four main geographic regions from west to east—the Ethiopian Plateau, the Great Rift Valley, the Somali Plateau, and the Ogaden Plateau. The Ethiopian Plateau, which is

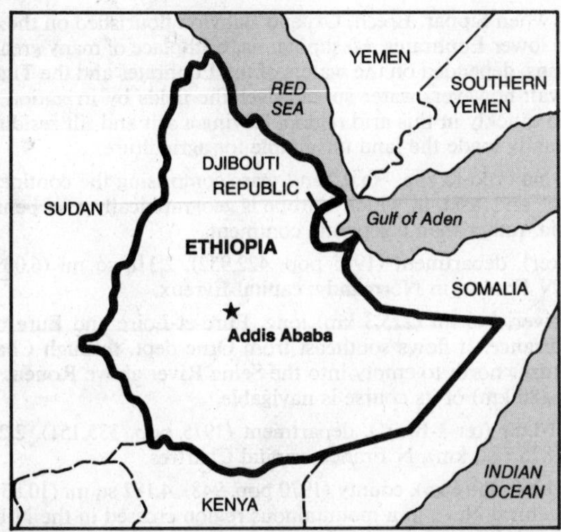

fringed in the west by the Sudan lowlands (made up of savanna and forests), includes more than half the country. It is generally 5,000 to 6,000 ft (1,525–1,830 m) high. The plateau slopes gently from east to west and is cut by numerous deep valleys. The Blue Nile River (in Ethiopia called the Abbai) flows through the center of the plateau from its source, Lake Tana, Ethiopia's largest lake. The Great Rift Valley (which in its entirety runs from southwest Asia to east-central Africa) traverses the country from northeast to southwest and contains the Danakil Desert in the north and several large lakes in the south. The Somali Plateau is generally not as high as the Ethiopian Plateau, but in the Urgoma Mts. it attains heights of more than 14,000 ft (4,270 m). The Awash, Ethiopia's only navigable river, drains the central part of the plateau. The Ogaden Plateau (1,500–3,000 ft/457.5–915 m high) is mostly desert but includes the Wabi-Shabale, Ganale-Dorya (Juba), and Dawa rivers.

The country's inhabitants are equally divided among adherents of Christianity (the great majority of whom belong to the Coptic Ethiopian Orthodox Church), Islam, and traditional religions. There are several distinct ethnic groups. The Amhara and Tigrinya, comprising about 33% of the population, live mostly in the central and northern Ethiopian Plateau. The Galla (40% of the country's population), live in southern Ethiopia. The pastoral Somali live in east and southeastern Ethiopia.

Economy. Ethiopia is an overwhelmingly agricultural country, with the great majority of its economically active population engaged in inefficient subsistence farming. Modern commercial agriculture accounts for less than 10% of annual farm output. The chief farm products are teff and other millets, sorghum, barley, wheat, maize, plantains, peas, potatoes, coffee, groundnuts, cotton, sugar cane, and tobacco. Large numbers of poultry, cattle, sheep, and goats are raised. The leading manufactures include processed food, beverages, textiles and clothing, leather goods, cement, and refined petroleum. No large-scale mineral deposits have been found in Ethiopia; salt, limestone, gold, platinum, iron ore, and sulfur are extracted in small quantities.

History. According to tradition, the Ethiopian kingdom was founded (10th cent. B.C.) by Solomon's first son, Menelik I, allegedly borne by the queen of Sheba. The kingdom of Aksum (Axum) was probably founded in the 1st cent. A.D. by immigrants (mainly traders) from south Arabia who had been settling in northern Ethiopia since about 500 B.C. Aksum controlled much of the Red Sea coast and had links with the Mediterranean world. Under King Ezana, Aksum was converted (4th cent.) to Christianity. In the 6th cent. Jewish influence penetrated Aksum, and some Ethiopians (of whom the modern Falashas are descendants) were converted to Judaism. With the rise of Islam in the 7th cent. Aksum declined, mainly because its land contacts with the Byzantine Empire were severed and its control of the Red Sea trade routes was ended. Aksum soon lost its cohesion, and Ethiopia lapsed into a period of competition among small political units.

A Portuguese embassy reached the Ethiopian court in 1520. In 1530–31 Ahmad Gran, a Moslem Somali leader, conquered much of Ethiopia. The Ethiopian negus (emperor) Lebna Dengel (reigned 1508–40) appealed to Portugal for help against the Somalis, and Portuguese troops were instrumental in defeating them (1543). However, the Somali war exhausted Ethiopia, ending the cultural revival and exposing the empire to incursions by the Galla. For the next two centuries the Ethiopian kingdom, centered at Gondar near Lake Tana, was beset by ruinous civil wars among princes (especially those of Tigre and Amhara), was menaced by the Galla, and was again isolated from the outside world. The reunification of Ethiopia was begun in the 19th cent. by Kasa (c.1818–68), who conquered Amhara, Gojjam, Tigre, and Shoa, and in 1855 had himself crowned negus as Theodore (Tewodros) II. He began to modernize and centralize the legal and administrative systems, despite the opposition of local governors. A brief civil war followed Theodore's death in 1868, and in 1872 a chieftain of Tigre became negus as John (Yohannes) IV. John's attempts to further centralize the government led to revolts by local leaders. The opening (1869) of the Suez Canal increased the strategic importance of Ethiopia, and several European powers sought influence in the area. Menelik II, John's successor, signed a treaty of friendship and cooperation with Italy at Ucciali in 1889. A dispute promptly broke out over the meaning of the treaty, and Italy invaded Ethiopia in 1895. In the war that followed, Menelik decisively defeated the Italians at Aduwa (Adowa). By 1896 Italy recognized the independence of Ethiopia. During his reign, Menelik also greatly expanded the size of Ethiopia, adding the provinces of Harar (in the east), Sidamo (in the south), and Kaffa (in the southwest). Menelik died in 1913 and was succeeded by his grandson Lij Iyasu, who was deposed in 1916. Judith (Zawditu), a daughter of Menelik, was made empress with Ras Tafari Makonnen as regent and heir apparent.

Ras Tafari was given additional powers by the empress in 1928, and on her death in 1930 he was crowned negus as Haile Selassie I. Almost immediately he faced threats from Italy, whose fascist ruler Mussolini was determined to establish an Italian empire and to avenge the defeat at Aduwa. On Oct. 3, 1935, Italy invaded Ethiopia. The League of Nations called (Nov., 1935) for mild economic sanctions against Italy, but they had little effect. In addition, an attempt by the British and French governments to arrange a settlement by giving Italy much of Ethiopia failed. The Italians quickly defeated the ill-equipped Ethiopians. In May, 1936, Addis Ababa was captured, and Haile Selassie fled the country. On June 1, 1936, the king of Italy was also made emperor of Ethiopia, and the country was combined with Eritrea (already held by Italy) and Italian Somaliland to form Italian East Africa. In 1941, during World War II, British and South African forces easily conquered Ethiopia, and Haile Selassie regained his throne. Britain had considerable influence in Ethiopian affairs until the end of the war and administered the small Haud region in the southeast until 1955. Eritrea was federated with Ethiopia in 1952, and in 1962 it was made an integral part of the country.

Ethiopia remained economically underdeveloped, with its wealth concentrated in the hands of a small number of large landlords and the Ethiopian church. On Dec. 14, 1960, a coup d'état was carried out by persons (especially members of the imperial bodyguard) seeking a more equitable distribution of wealth and power in the country. However, the negus was restored to power by the army and air force. Between 1961 and 1967 there were border skirmishes between Ethiopia and Somalia, and in the late 1960s and early 1970s there was considerable fighting between the central government and a guerrilla secessionist movement in Eritrea. Unrest continued among groups (particularly high school and university students) seeking more far-reaching reforms. In a gradual coup d'état that began in Feb., 1974, a group of military officers seized control of the government. After the deposition of the negus, the constitution was suspended and parliament was dissolved, and Lt. Gen. Aman Michael Andom became head of a newly formed cabinet. In Apr., 1974, Prince Zare Yacob was proclaimed crown prince. A provisional military government was established. Following the execution of head of state, Brig. Gen. Teferi Bante, Mengistu Haile Mariam succeeded as chairman of the provisional military administration council. Northern Ethiopia in 1973–74 was ravaged by a long-term drought; Haile Selassie's failure to deal adequately with the widespread starvation there was reportedly a major reason for his downfall. Fighting in the Ogaden region and in Eritrea escalated in 1977–78. Several international powers, including the Soviet Union, Somalia, Cuba, the United States and many Middle Eastern nations, were involved directly or indirectly in the conflicts.

Et·na or **Aet·na** (ĕt′nə), volcano, 11,122 ft (3,392.2 m) high, on the E coast of Sicily, S Italy. It is the highest active volcano in Europe. The shape and height of its central cone have often been changed by eruptions. There are more than 260 lesser craters on the slopes, formed by lateral eruptions. The first known eruption occurred in 475 B.C. Of the numerous later eruptions, often accompanied by earthquakes, those of 1169 and 1669 were the most destructive; the most recent occurred in 1971. The wide base of Mt. Etna, c.93 mi (150 km) in circumference, is encircled by a railroad. There is an observatory at 9,650 ft (2,943.3 m).

E·ton (ēt'n), urban district (1976 est. pop. 4,950), Buckinghamshire, central England, on the Thames River. It is known chiefly for Eton College, largest and most famous of the English public schools, founded by King Henry VI in 1440.

Et·o·wah (ĕt'ə-wä'), county (1970 pop. 94,144), 555 sq mi (1,437.5 sq km), NE Ala., in a hilly region crossed by the Coosa River; formed 1866 as Blaine co., renamed 1868; co. seat Gadsden. Cotton and truck crops are grown. The county has deposits of coal, iron, limestone, fuller's earth, manganese, and barite. Its diverse manufactures include yarn, knit goods, and clothing, lumber and wood products, furniture, tires, pottery, concrete, and machinery.

Etowah, river, 141 mi (226.9 km) long, rising in the Blue Ridge Mts., N Ga., and flowing SW to Rome, Ga., where it joins the Oostanaula River to form the Coosa River. Etowah Mounds, a national historic landmark, is a group of prehistoric Indian earthworks 60 ft (18.3 m) high, located on the river near Cartersville, Ga.

E·tru·ri·a (ĭ-trŏŏr'ē-ə), ancient country, W central Italy, now forming Tuscany and part of Umbria. It was the territory of the Etruscans, who in the 6th cent. B.C. spread Etruscan civilization throughout much of Italy. They were later forced back into Etruria and ultimately dispersed.

Eu·boe·a (yŏŏ-bē'ə): *see* Evvoia, Greece.

Eu·clid (yŏŏ'klĭd), city (1978 est. pop. 60,300), Cuyahoga co., NE Ohio, a suburb adjoining Cleveland, on Lake Erie; settled 1798, inc. 1848. Its many manufactures include airplane and automobile parts, machinery, and machine-shop products. The National Shrine of Our Lady of Lourdes is here.

Eu·fau·la (yŏŏ-fô'lə). **1.** City (1970 pop. 9,102), Barbour co., SE Ala., on the Chattahoochee River ESE of Montgomery; settled in the 1830s, inc. 1857. It ships cotton and produces textiles. Many antebellum plantation houses remain. **2.** City (1970 pop. 2,355), seat of McIntosh co., E Okla., near the confluence of the Canadian and North Canadian rivers NNE of McAlester; founded c.1872 near the site of a Creek settlement.

Eu·gene (yŏŏ-gēn'), city (1978 est. pop. 97,900), seat of Lane co., W Oregon, on the Willamette River; inc. 1862. A processing and shipping center in a farming area, it has huge lumbering and food-processing industries. Eugene is the seat of the Univ. of Oregon.

Eu·nice (yŏŏ'nĭs), city (1970 pop. 11,390), St. Landry and Acadia parishes, S central La.; inc. 1895. It lies in an oil and agricultural region in which rice, cotton, and soybeans are produced. Its manufactures include oil-field equipment and clothing.

Eu·pen (oi'pən), town (1977 pop. 17,346), Liège prov., E Belgium, on the Vesdre River, near the West German border. It is an industrial and commercial center. The districts of Eupen and Malmédy were awarded (1815) to Prussia by the Congress of Vienna. Strategically important for the defense of Belgium, the districts were transferred (1919) to Belgium under the terms of the Treaty of Versailles. During World War II the districts were temporarily annexed (1940–4) by Germany.

Eu·phra·tes (yŏŏ-frā'tēz), river of SW Asia, c.1,700 mi (2,735 km) long, formed by the confluence of the Kara and the Murad rivers, E central Turkey, and flowing generally S through Turkey into Syria, then SE through Iraq, joining with the Tigris River in SE Iraq to form the Shatt al Arab; the united river flows into the Persian Gulf. In its upper course, the Euphrates flows rapidly through deep canyons and narrow gorges. The middle Euphrates traverses a wide flood plain in Syria, where it is used extensively for irrigation. Entering the Syrian Desert and the plains of Iraq, the river loses velocity and becomes a sluggish stream with shifting channels. The river's lower course supplies water to Iraq's great date plantations through a system of dams and canals.

The Euphrates is unnavigable except for shallow-draft vessels; dams have been built on its lower course to control flooding, improve navigation, and supply water for irrigation. The modern waterworks along the Euphrates do not equal in scope those of ancient times when Sippar, Erech, Ur, and Babylon flourished on the banks of the lower Euphrates. Mesopotamia, birthplace of many great civilizations, depended on the waters of the Euphrates and the Tigris for survival; however, water spread over the fields by irrigation evaporated quickly in this arid region, leaving a salt and silt residue that eventually made the land unsuitable for agriculture.

Eu·ra·sia (yŏŏ-rā'zhə, -shə), land mass comprising the continents of Europe and Asia, in which Europe is geographically a W peninsula of Asia, rather than a separate continent.

Eure (œr), department (1975 pop. 422,952), 2,318 sq mi (6,003.6 sq km), N France, in Normandy; capital Évreux.

Eure, river, 140 mi (225.3 km) long, Eure-et-Loire and Eure depts., NW France. It flows southeast from Orne dept. through Chartres, then turns north to empty into the Seine River above Rouen. Some 50 mi (80 km) of its course is navigable.

Eure-et-Loir (œr-ā-lwär'), department (1975 pop. 335,151), 2,269 sq mi (5,876.7 sq km), N France; capital Chartres.

Eu·re·ka (yŏŏ-rē'kə), county (1970 pop. 948), 4,182 sq mi (10,831.4 sq km), central Nev., in a mountainous region crossed in the N by the Humboldt River; formed 1873; co. seat Eureka. The county has livestock ranches and gold mines. Some silver and lead are also mined. Part of Toiyabe National Forest is in the southwest.

Eureka. 1. City (1978 est. pop. 24,100), seat of Humboldt co., NW Calif., on Humboldt Bay; inc. 1856. It is a port of entry. Lumbering and fishing are the chief industries; tourism is also important. A 40-acre (16.2-hectare) redwood park lies within the city limits. **2.** City (1970 pop. 3,028), seat of Woodford co., central Ill., E of Peoria; founded in the 1830s, inc. 1859. It is a farm trade center. **3.** City (1970 pop. 3,576), seat of Greenwood co., SE Kansas, on the Fall River E of Wichita; laid out 1867, inc. 1870. It is a trade center for an oil and cattle region. **4.** Village (1970 pop. 500), seat of Eureka co., central Nev., WNW of Ely at an altitude of 6,837 ft (2,085.3 m). It has gold, silver, and lead mines and stock farms.

Eureka Springs, resort city (1970 pop. 1,670), a seat of Carroll co., NW Ark., NNE of Fort Smith, in the Ozarks near the Mo. line; settled 1879, inc. 1880. It has a large number of mineral springs.

Eur·ope (yŏŏr'əp), sixth-largest continent (1973 est. pop. 640,000,-000), c.4,000,000 sq mi (10,360,000 sq km) including adjacent islands. It is actually a vast peninsula of the great Eurasian land mass. It is separated from Asia by the Urals and the Ural River in the east; by the Caspian Sea and the Caucasus in the southeast; and by the Black Sea, the Bosporus, the Sea of Marmara, and the Dardanelles in the south. The Mediterranean Sea and the Strait of Gibraltar separate it from Africa. Europe is washed in the north by the Arctic Ocean, and in the west by the Atlantic Ocean, with which the North Sea and the Baltic Sea are connected.

The huge Alpine mountain chain, of which the Pyrenees, the Alps, the Carpathians, the Balkans, and the Caucasus are the principal links, traverses the continent from west to east. Between the mountainous Scandinavian peninsula in the north and the Alpine chain in the south extends the great European plain, stretching from the Atlantic coast of France to the Urals. A large part of this plain is fertile agricultural soil; in the east and north there are vast steppe, forest, lake, and tundra regions. South of the Alpine chain extend the Iberian, Italian, and Balkan peninsulas, which are largely mountainous. The Po plain, between the Alps and the Apennines, and the Danubian plain, between the Carpathians and the Alps, are fertile and much-developed regions. Among the chief river systems of Europe are, from east to west, those of the Volga, the Don, the Dnepr, the Danube, the Oder, the Elbe, the Rhine, the Rhône, the Loire, the Garonne, and the Tagus.

The climate of Europe varies from subtropical to polar. The Mediterranean climate of the south is dry and warm. The western and northwestern parts have a mild, generally humid climate, influenced by the North Atlantic Drift. In central and eastern Europe the climate is of the humid continental type with cool summers. In the northeast subarctic and tundra climates are found. All of Europe is subject to the moderating influence of prevailing westerly winds from the Atlantic Ocean.

Indo-European languages predominate in Europe; others spoken include Basque, Maltese, and the languages classified as Finno-Ugric, Samoyedic, Bulgaric, and Turkic. Roman Catholicism is the chief religion of southern and western Europe and the southern part of central Europe; Protestantism is dominant in the United Kingdom, Scandinavia, and the northern part of Europe, and the Orthodox Eastern Church predominates in east and southeast Europe.

Europe is highly industrialized, and manufacturing employs most of the work force. Agriculture, forestry, and fishing are also important. Europe has a large variety of minerals; coal, iron ore, bauxite, and salt are abundant. Oil and gas are found in east and southeast Europe and beneath the North Sea. Coal is used to produce most of Europe's electricity; in Norway and Sweden and in the Alps there are many hydroelectric power plants. The transportation system in Europe is highly developed; interconnecting rivers and canals provide excellent inland water transportation.

The beginnings of civilization in Europe can be traced to ancient times, but they are not as old as the civilizations of Mesopotamia and Egypt. The Roman and Greek cultures flourished in Europe, and European civilization has been spread throughout the world by colonists and immigrants. Once embraced by vast and powerful empires and kingdoms, successful nationalistic uprisings (especially in the 19th cent.) divided the continent into many sovereign states. The political fragmentation led to economic competition and political strife. After World War II Europe became divided into two ideological blocs (Eastern Europe, dominated by the USSR, and Western Europe, dominated by the United States) and became engaged in the Cold War. Tensions eased in the 1960s, and signs of normalization of East-West relations appeared in the 1970s. The expanded European Economic Community, or Common Market, which also has the goal of political unification of member states, is now a strong economic rival to the United States and the USSR.

Eu·ro·pus (yōō-rō'pəs): *see* Dura, Syria.

Eu·taw (yōō'tô), city (1970 pop. 2,805), seat of Greene co., W central Ala., near the Black Warrior River SSW of Tuscaloosa; settled 1818, inc. 1840. A cattle market, it also gins cotton and manufactures textiles and lumber products. Many ante-bellum houses remain.

E·van·ge·line (ĭ-văn'jə-lēn'), parish (1970 pop. 31,932), 672 sq mi (1,740.5 sq km), S central La., drained by Bayou Nexpique; formed 1910; parish seat Ville Platte. In an agricultural area yielding rice, soybeans, and sweet potatoes, it has oil and natural-gas wells, cotton gins, lumber mills, and chemical factories.

Ev·ans (ĕv'ənz), county (1970 pop. 7,290), 186 sq mi (481.7 sq km), SE central Ga., drained by the Canoochee River; formed 1914; co. seat Claxton. It lies in a coastal-plain agricultural (cotton, corn, tobacco, peanuts, and livestock) and timber area. Textiles are manufactured.

Evans, Mount, peak, 14,260 ft. (4,349.3 m) high, N central Colo., in the Front Range of the Rocky Mts. At its summit is the Inter-University High Altitude Laboratory.

Ev·ans·ton (ĕv'ən-stən). **1.** Residential city (1975 pop. 75,300), Cook co., NE Ill., on Lake Michigan; settled 1826, inc. 1892. It has a publishing industry, and photocopying machines, food products, and rust preventatives are among its manufactures. It is the seat of Northwestern Univ. **2.** Town (1970 pop. 4,462), seat of Uinta co., extreme SW Wyo., on the Bear River SW of Green River; settled 1869, inc. 1888. It is the trade center of a farm and dairy area in a region that yields coal and other minerals. The city developed with the arrival of the railroad.

Ev·ans·ville (ĕv'ənz-vĭl'), city (1978 est. pop. 131,100), seat of Vanderburgh co., extreme SW Ind., a port on the Ohio River; inc. 1819. It is the shipping and commercial center for a coal, oil, and farm region. Refrigeration and air-conditioning equipment, aluminum, pharmaceuticals, excavating machinery, and fabricated-metal items are among its manufactures. The Univ. of Evansville and an Indiana State Univ. campus, are here. In an extensive urban renewal program during the 1960s the main street was converted into a seven-block shopping plaza.

E·ven·ki National Okrug (ĕ-vyĕn'kē), administrative division (1970 pop. 13,000), 287,645 sq mi (745,000 sq km), N central Siberian USSR, in the Central Siberian Uplands; capital Tura. The okrug occupies the entire central section of Krasnoyarsk Kray and lies in the forested taiga zone. It was formed in 1930.

Ev·er·est, Mount (ĕv'ə-rĭst, ĕv'rĭst), peak, 29,028 ft (8,853.5 m) high, on the border of Tibet and Nepal, in the central Himalayas. It is the highest elevation in the world. The first eight attempts to scale the peak were unsuccessful, but on May 28, 1953, Sir Edmund Hillary and Tenzing Norkay of Nepal reached the summit.

Ev·er·ett (ĕv'ə-rĭt, ĕv'rĭt). **1.** City (1978 est. pop. 38,100), Middlesex co., E Mass., an industrial suburb of Boston, on the Mystic River; settled c.1643, set off from Malden 1870, inc. as a city 1892. A deep-water port, Everett has coal and petroleum storage facilities, chemical plants, and foundries. **2.** City (1970 pop. 48,600), seat of Snohomish co., NW Wash., on Puget Sound at the mouth of the Snohomish River; inc. 1893. A port of entry with a fine natural harbor, it is an important lumber-shipping center, with pulp and paper mills. Boeing 747 airplanes are manufactured here.

Ev·er·glades (ĕv'ər-glādz'), marshy, low-lying tropical area, c.5,000 sq mi (12,950 sq km), S Fla., extending from Lake Okeechobee S to Florida Bay. Characterized by water, saw grass, hammocks (island-like masses of vegetation), coastal mangrove forests, and solidly packed black muck (resulting from millions of years of vegetable decay in near-stagnant, warm water), the Everglades receives an annual average rainfall of more than 60 in. (152 cm), mainly in the summer. Limestone rims the area, acting as a natural retaining wall against the sea. Colonial expeditions in the 1500s found Indians living in the Everglades. In the late 1830s the Everglades was the site of military operations against the Seminole Indians. Large tracts of land were drained in the late 19th and early 20th cent. After the great fires of 1939 (caused by overdrainage), the first thorough studies of the Everglades concluded that most of the southern part was unfit for cultivation. Construction of retaining walls in the 1960s on the south shore of Lake Okeechobee and land development in Big Cypress Swamp have disrupted the natural flow of water into the Everglades, thus posing the threat of water shortages that can affect its plant and animal life. At the southwestern tip of the state is Everglades National Park, 1,400,533 acres (567,216 hectares), established in 1947. A great variety of flora and fauna is found in the park, which is also a haven for such endangered species as the alligator, egret, and bald eagle.

Ev·er·green (ĕv'ər-grēn'), city (1970 pop. 3,924), seat of Conecuh co., S central Ala., SSW of Montgomery; settled c.1820. A trade and shipping center for a farm area, it makes textile, wood, and metal products.

Evergreen Park, village (1978 est. pop. 24,400), Cook co., NE Ill., a suburb of Chicago; inc. 1893. It is mostly residential.

Eve·sham (ēv'shəm, ē'shəm, ē'səm), municipal borough (1973 est. pop. 14,090), Hereford and Worcester, W central England, on the Avon River. Evesham is the center of the Vale of Evesham, known for its fine market gardens. It is also a popular summer resort. Simon de Montfort, leader of the revolt against Henry III, was killed in 1265 during the Battle of Evesham.

É·vian-les-Bains (ā-vyäN'lā-băN') or **É·vian** (ā-vyäN'), town (1968 pop. 6,052), Haute-Savoie dept., E France, on Lake Geneva. It is a fashionable spa at the foot of the Alps. Évian-les-Bains's mineral water is bottled and exported to all parts of the world. A cease-fire agreement between the French government and the provisional government of Algeria was signed at Évian in 1962.

É·vo·ra (ĕ'vôr-ə), town (1970 pop. 23,665), capital of Évora dist. and of Alto Alentejo, S central Portugal. It is the commercial center of a fertile agricultural area. Évora was an episcopal see early in the Christian era and later a center of trade under the Moors. It was recovered by the Portuguese in 1166. After 1385 it was for many years the favorite seat of the Portuguese court.

É·vreux (ā-vrœ'), town (1975 pop. 47,412), capital of Eure dept., N France, in Normandy. It is an industrial town where metals, textiles, rubber, radio and television parts, and pharmaceuticals are manufactured. Founded in Roman times, the town became (10th cent.) the seat of a county that frequently changed hands throughout the Middle Ages. From 1349 to 1425 the counts of Évreux were also kings of Navarre. The French crown, which had held the county in the 12th cent. and again from 1404 to 1569, acquired it permanently in 1584. Devastated many times during the course of its history, the town was extensively rebuilt following World War II.

É·vry (ā-vrē'), town (1975 pop. 15,354), capital of Essonne dept., N central France. The major industry is the manufacture of aeronautic equipment.

Év·voia (ĕv'yä) or **Eu·boe·a** (yōō-bē'ə), island (1971 pop. 162,986), 1,467 sq mi (3,800 sq km), SE Greece, separated from Boeotia and Attica on the Greek mainland by the Évripos strait. Évvoia is generally mountainous with fertile valleys. Sheep, goats, and cattle are raised, and olives, grapes, and wheat are grown. Magnesite and lignite are mined, and marble is quarried. The island was settled by Ionian and Thracian colonists and was divided among seven independent cities, of which Khalkís and Eretria were the most important. Powerful and prosperous by the 8th cent. B.C., these cities established colonies in Macedonia, southern Italy, and Sicily. The island was under the hegemony of Athens from 506 to 411 B.C. and was taken (c.338 B.C.) by Philip II of Macedon. It was annexed by Rome in 194 B.C. and later passed under Byzantine rule. As a result of the Fourth Crusade, it became (early 13th cent.) a colony of Venice. It was ceded to the Ottoman Turks in 1470. The island rebelled against the Turks in 1821 and in 1830 was incorporated into Greece.

Excelsior Springs, city (1970 pop. 9,411), Clay co., W Mo., NE of Kansas City; founded 1880, inc. 1882. It is a health resort with mineral springs. Frank and Jesse James lived nearby.

Exe (ĕks), river, c.55 mi (90 km) long, rising in the Exmoor, Somerset, SW England, and flowing S across the Cornwall peninsula, past Exeter to the English Channel at Exmouth. Salmon and shellfish are taken from the river; many waterfowl are found along its narrow estuary.

Ex·e·ter (ĕk'sĭ-tər), county borough (1976 est. pop. 93,300), administrative center of Devon, SW England, on the Exe River. It is a market, transportation, and distribution center. There is light manufacturing, with metal and leather goods, paper, and farm implements as the chief products. Because of its strategic location, Exeter was besieged by the Danes in the 9th and 11th cent., by William the Conqueror in 1068, by Yorkists in the 15th cent., and by religious factions in the middle of the 16th cent. There are ruins of the Roman walls and of Rougemont Castle (11th cent.), built by William the Conqueror. Severely damaged by bombing in World War II, the greater part of the city has since been rebuilt.

Exeter, town (1970 pop. 8,892), seat of Rockingham co., SE N.H., SW of Portsmouth. Rev. John Wheelwright settled here in 1638 after his banishment from Mass.; the town was organized in 1639. It served as the capital of N.H. during the Revolutionary War. Phillips Exeter Academy, a 17th cent. garrison house, and several 18th cent. buildings are in Exeter. The sculptor Daniel Chester French and Gen. Lewis Cass were born here.

Ex·moor (ĕks'mŏŏr), high moorland of the Cornwall peninsula, SW England. Underlaid by slate and sandstone, the rugged region is sparsely populated. Sheep and the small Exmoor ponies are grazed on the moorland. Exmoor is a popular vacationland and contains many interesting prehistoric earthworks. It was the home of a legendary nomadic group of brigands called the Doones; Doone Valley became famous after R. D. Blackmore used the legend in his novel *Lorna Doone.*

Ex·mouth (ĕks'mouth), urban district (1973 est. pop. 26,840), Devonshire, SW England, at the mouth of the Exe River. It is a port and a popular summer resort.

Ey·ja·fjör·dur (ā'yä-fyœr'*th*ür), inlet of the Greenland Sea, longest (37 mi/59.5 km) and most scenic fjord in N Iceland.

Eyre, Lake (âr), shallow salt lake, 3,430 sq mi (8,883.7 sq km), central South Australia state, Australia. It is the largest lake in Australia. The lake, 39 ft (12 m) below sea level, is the continent's lowest point. Located in the arid interior of Australia, the lake is frequently dry.

Eyre Peninsula, 200 mi (321.8 km) long, S South Australia state, Australia, between Spencer Gulf and the Great Australian Bight. There are large iron-ore deposits in the Middleback Range at the northeastern base of the peninsula.

E·zi·on-ge·ber (ē'zē-ŏn-gē'bər) ancient port, on the Gulf of Aqaba. The site, near Aqaba, is now some distance from the shoreline. Excavations (1938–40) unearthed the largest copper refineries ever found to have existed in the ancient world.

F

Fa·bri·a·no (fä'brē-ä'nō), town (1971 pop. 27,276), in the Marche, central Italy, in the Apennines. It is an agricultural and industrial center. Paper has been made here since the 13th cent.

Fa·en·za (fä-ĕn'zə, -tsä), city (1976 est. pop. 40,300), in Emilia-Romagna, N central Italy, on the Lamone River. A special kind of richly colored ceramic, called faïence or majolica, has been made here since the 12th cent.; ceramic art flourished from 1450 to 1550 and was revived in the 18th cent.

Faer·oe Islands or **Far·öe Islands** (both: fâr'ō), group of volcanic islands (1970 pop. 36,681), 540 sq mi (1,398.6 sq km), Denmark, in the N Atlantic, between Iceland and the Shetland Islands. There are 18 main islands (one of which is not inhabited) and 4 small, uninhabited islands. The Faeroes are high and rugged and have only sparse vegetation. Their inhabitants depend mainly on fishing and to a lesser extent on farming, whaling, and fowling.

The earliest-known inhabitants were Celtic. In the 8th cent. A.D. the islands were settled by Norsemen. In the early 11th cent. they became part of the kingdom of Norway and were Christianized. The population was nearly wiped out by an outbreak of black plague in the 14th cent. and was soon after replaced by Norwegian settlers. With Norway, the Faeroes passed under Danish rule in 1380, and they remained Danish after the Treaty of Kiel (1814) transferred Norway from the Danish to the Swedish crown. In World War II Great Britain established (1940) a protectorate over the islands after the German occupation of Denmark. In 1948 the Faeroese obtained home rule.

Fair·banks (fâr'băngks'), city (1978 est. pop. 35,100), seat of Fairbanks North Star borough, central Alaska, on the Chena River near its confluence with the Tanana; inc. 1903. The Univ. of Alaska and the U.S. government are the principal employers. Private firms in the retail, wholesale, and construction industries depend heavily on government contracts. Gold was discovered here in 1902, and Fairbanks grew rapidly as a mining camp. The building of the Alaska RR increased the city's importance. The Richardson Highway from Valdez, on the Pacific coast, reaches Fairbanks, which, during World War II, became the terminus of the Alaska Highway.

Fairbanks North Star (nôrth' stär'), borough (1970 pop. 45,864), 7,074 sq mi (18,321.7 sq km), central Alaska; formed after 1961, name changed from North Star in 1967; borough seat Fairbanks. It is a service, supply, and production center for interior and arctic Alaska, and the construction headquarters for the Alaska oil pipeline.

Fair·born (fâr'bôrn'), city (1978 est. pop. 35,300), Greene co., SW Ohio; settled 1799, inc. 1950 with the merging of Osborn and Fairborn. Major employers are Wright State Univ. in nearby Dayton and Wright-Patterson Air Force Base.

Fair·bur·y (fâr'bûr'ē, -bə-rē), city (1970 pop. 5,265), seat of Jefferson co., SE Nebr., near the Kansas border SSW of Lincoln, in a farm area; founded 1869, inc. 1871.

Fair·fax (fâr'făks'), county (1970 pop. 455,032), 399 sq mi (1,033.4 sq km), NE Va., bounded on the E and SE by the Potomac River and the Md. border; formed 1742; co. seat Fairfax (an independent city). It is chiefly residential, with some agriculture (corn, hay, fruit, poultry, and dairy products) and stone, sand, and gravel quarries.

Fairfax, independent city (1978 est. pop. 20,600), NE Va., a residential suburb of Washington, D.C.; inc. 1892.

Fair·field (fâr′fēld′), city (1976 pop. 114,581), New South Wales, SE Australia. It is a suburb of Sydney.

Fair·field 1. County (1970 pop. 792,814), 626 sq mi (1,621.3 sq km), SW Conn., bounded on the S by Long Island Sound, on the W by the N.Y. border, on the NE and E by the Housatonic River; formed 1666; former co. seat Bridgeport. Many residential communities on the southern shore lie within commuting distance from New York City. The county's agriculture includes fruit, truck crops, and dairy products. Amongs its diverse manufactures are electrical equipment, firearms, clothing, tools, machinery, hardware, chemicals, rubber, asbestos, and paper, wood, glass, and metal products. **2.** County (1970 pop. 73,301), 505 sq mi (1,308 sq km), S central Ohio, drained by the Hocking River; formed 1800; co. seat Lancaster. In an agricultural area yielding grain, fruit, livestock, and dairy products, it has timber, oil wells, and sand and gravel pits. Paper, shoes, plastics, and glass and metal products are manufactured. **3.** County (1970 pop. 19,999), 699 sq mi (1,810.4 sq km), N central S.C., bounded on the W and SW by the Broad River, on the NE by Wateree Reservoir; formed 1785; co. seat Winnsboro. It has agriculture (cotton, corn, and cattle), granite quarries, and a lumbering industry.

Fairfield. 1. City (1970 pop. 14,369), Jefferson co., N central Ala., an industrial suburb of Birmingham; inc. 1919. Founded (1910) by the United States Steel Corp., it has steel and chemical industries. **2.** City (1978 est. pop. 54,200), seat of Solano co., W Calif.; founded 1859, inc. 1903. Among its manufactures are aluminum cans and sleeping bags. **3.** Town (1978 est. pop. 59,100), Fairfield co., SW Conn., on Long Island Sound; settled 1639, chartered 1947. It is chiefly residential, but there are diverse light industries. The town was settled on the site of the last battle (1637) in the Pequot War. It was a port of entry and a shipping center until 1890. During the Revolution much of it was burned (1777) by the British. **4.** City (1970 pop. 336), seat of Camas co., S Idaho, SW of Hailey, in a livestock and grain area at an altitude of 5,600 ft (1,708 m). Gold and silver mines are in the area. Sawtooth National Forest is to the north. **5.** City (1970 pop. 5,897), seat of Wayne co., SE Ill., ENE of Mt. Carmel and W of the Little Wabash River, in a farm area; settled c.1819, inc. 1840. Automobile parts are made. **6.** City (1970 pop. 8,715), seat of Jefferson co., SE Iowa, E of Ottumwa; inc. 1847. It is a trade center and manufactures washing machines, farm equipment, and gloves. **7.** Town (1970 pop. 5,684), Somerset co., S Maine, on the Kennebec River above Waterville; settled 1774, inc. 1788. It has woolen and pulp mills, and food is processed. **8.** City (1978 est. pop. 27,400), Butler co., SW Ohio, in the fertile Miami River valley; inc. 1954. Auto body stampings, sulfates, metal products, and construction equipment are manufactured. **9.** Town (1970 pop. 2,074), seat of Freestone co., E central Texas, WSW of Palestine, in a farm, timber, and livestock area; inc. 1933.

Fair Lawn (lôn), borough (1978 est. pop. 35,800), Bergen co., NE N.J., across the Passaic River from Paterson; inc. 1924.

Fair·mont (fâr′mŏnt′). **1.** City (1978 est. pop. 11,700), seat of Martin co., S central Minn.; inc. 1878. It is the trade and manufacturing center for an agricultural area. The courthouse, on the site of a stockade built during the Sioux outbreak of 1862, has pioneer relics. **2.** City (1978 est. pop. 26,200), seat of Marion co., N central W.Va., where the West Fork and the Tygart rivers form the Monongahela, in a rich bituminous-coal area; settled 1793 around Prickett's Fort (1774), inc. 1843. Among the city's manufactures are glassware, light bulbs, and aluminum products. A Union supply depot in the Civil War, Fairmont was raided by Confederate cavalry in 1863.

Fair Oaks (ōks). **1.** Uninc. residential village (1970 pop. 11,256), Sacramento co., N central Calif., on the American River, in a citrus fruit and farm area. **2.** Locality, just E of Richmond, Va. The Battle of Seven Pines was fought here May 31–June 1, 1862. The outcome was the repulsion of Confederate forces and the death of their commander, Joseph Johnston.

Fair·play (fâr′plā′), town (1970 pop. 419), seat of Park co., central Colo., on the South Platte River E of Leadville. It is in a gold-mining area of the Rocky Mts. at an altitude of 9,964 ft (3,039 m).

Fair·view (fâr′vyo͞o′). **1.** Borough (1970 pop. 10,698), Bergen co., NE N.J.; settled 1860, inc. 1894. Embroideries and apparel are the chief manufactures. **2.** Uninc. urban community (1970 pop. 8,517), Dutchess co., SE N.Y., N of Poughkeepsie. **3.** City (1970 pop. 2,894), seat of Major co., NW central Okla., near the Cimarron River NW of Oklahoma City, in a wheat and cattle area; settled 1893, inc. as a town 1901, as a city 1909.

Fairview Park, city (1978 est. pop. 20,100), Cuyahoga co., NE Ohio; inc. 1950. It is a residential suburb of Cleveland.

Fair·weath·er, Mount (fâr′wĕth′ər), peak, 15,300 ft (4,666.5 m) high, on the border between SE Alaska and NW British Columbia, Canada, near Glacier Bay National Monument. It is the highest peak in the Fairweather Range of the St. Elias Mts.

Fai·yum, El (ĕl fī-yo͞om′): *see* Fayyum, Al, Egypt.

Faiz·a·bad (fī′zə-băd, fī′zä-bäd′), city (1969 est. pop. 63,000), capital of Badakhshan prov., NE Afghanistan, on the Kokcha River. The chief commercial center of northeast Afghanistan, Faizabad also has rice and flour mills.

Faizabad, town, Uttar Pradesh state, N central India, on the Gogra River. It is a joint municipality with Ajodhya (total 1971 pop. 102,794). Faizabad trades in local produce and has sugar refineries. It was the capital (1724–75) of the kingdom of Oudh (1724–1856).

Fa·laise (fə-lāz′, fä-), town (1975 pop. 8,368), Calvados dept., N France, in Normandy. Once an important textile center, the town is now an agricultural marketplace and manufactures cheeses and household appliances. William I of England was born here. Falaise was a key point in the Normandy campaign of World War II and was heavily damaged.

Fal·fur·ri·as (făl-fo͝or′ē-əs), city (1970 pop, 6,355), seat of Brooks co., S Texas, SW of Corpus Christi, in a dairying, truck-farming, and ranching area. There are oil wells nearby.

Fal·kirk (fôl′kûrk), burgh (1974 est. pop. 36,589), Central region, central Scotland. The local coal and iron mines have been exhausted, but fire clay is still mined. Concrete, chemicals, metal products, and hosiery are produced. In the first Battle of Falkirk (1298), said to be the first battle in which the longbow was decisive, Edward I and the English defeated the Scots.

Falk·land Islands (fôk′lənd), group of islands (1970 est. pop. 2,045), 4,618 sq mi (11,960.6 sq km), S Atlantic, c.300 mi (480 km) E of the Strait of Magellan. The islands are administered as a British crown colony with the capital at Stanley. There are two large islands (East Falkland and West Falkland) and some 200 small ones. The Falklands are bleak, rocky moorlands, swept by wind and drenched by chill rain. Whales and seals abound in the littoral waters and are hunted for oil and skins; guano is also exported. The British claim is based on probable discovery by the navigator John Davis in 1592; but the islands were claimed and occupied at various times by England, Spain, France, and Argentina. When the seizure of an American sealing vessel in 1832 led to a U.S. punitive expedition, the British occupied the islands.

Fal·lon (făl′ən), county (1970 pop. 4,050), 1,633 sq mi (4,229.5 sq km), E Mont., in an agricultural region bordering in the E on N.Dak. and S.Dak. and drained by O'Fallon Creek; formed 1913; co. seat Baker. Grain and livestock are important to its economy.

Fallon, city (1970 pop. 2,959), seat of Churchill co., W Nev., ENE of Carson City; inc. 1908. It is the trade center of an irrigated farm area served by the Newlands project.

Fall River (fôl), county (1970 pop. 7,505), 1,748 sq mi (4,527.3 sq km), extreme SW S.Dak., bordered on the W by Wyo., on the S by Nebr., and drained by the Cheyenne River; formed 1883; co. seat Hot Springs. In an agricultural area yielding alfalfa, corn, wheat, livestock, and dairy products, it has sandstone quarries, ore deposits, and some manufacturing. Buffalo Gap National Grassland is here.

Fall River, industrial city (1978 est. pop. 101,100), Bristol co., SE Mass., a port of entry on Mt. Hope Bay, at the mouth of the Taunton River; settled 1656, set off from Freetown 1803, inc. as a city 1854. It was once the foremost cotton textile center in the United States, and textiles and clothing are still among its diversified manufactures. In the American Revolution Fall River was the scene (1778) of a skirmish between British and American forces. The first cotton mill was built in 1811. The city was the scene (1892) of the murder trial of Lizzie Borden, who was born and lived in Fall River.

Falls (fôlz), county (1970 pop. 17,300), 761 sq mi (1,971 sq km), E central Texas, in a blackland prairie region drained by the Brazos River; formed 1850; co. seat Marlin. Its agriculture includes cotton, corn, grain sorghums, alfalfa, fruit, truck crops, poultry, livestock, dairy products, and horses. It has some manufacturing.

Falls City, city (1970 pop. 5,444), seat of Richardson co., extreme SE Nebr., near the Kansas line SSE of Lincoln; founded 1857, inc. 1858. It is the trade center of an agricultural region.

Fal·mouth (făl'məth), municipal borough (1973 est. pop. 17,530), Cornwall, SW England, on a small peninsula between Falmouth Bay and Carrick Roads estuary. Falmouth is a port, a resort, and the headquarters of the Royal Cornwall Yacht Club. There are oyster fisheries and engineering and ship repairing industries. The harbor entrance is guarded by Pendennis Castle on the west and St. Mawes Castle on the east (both 16th cent.).

Falmouth. 1. City (1970 pop. 2,593), seat of Pendleton co., N central Ky., on the Licking River SSE of Covington, in a farm area; settled 1776. Pleistocene fossils have been found in the area. **2.** Town (1970 pop. 6,291), Cumberland co., SW Maine, on Casco Bay N of Portland; settled c.1632, inc. 1718. The old town of Falmouth included the site of one of Maine's earliest and largest settlements (c.1632), which suffered greatly during the Indian wars and in 1775, when it was bombarded and nearly destroyed by the British. **3.** Town (1970 pop. 15,942), Barnstable co., SE Mass., on Cape Cod; settled c.1660, inc. 1686. Once a whaling and boatbuilding center, it is now a popular boating and summer resort. Falmouth was attacked by the British in the Revolutionary War and again in the War of 1812. Historic structures include the Ship's Bottom Roof House (1678), the Congregational Church on the town green (1756), and the Julia Wood House (1790), with a historical museum. The town includes the community of Woods Hole, seat of the Oceanographic Institution and Marine Biological Laboratories.

Fal·ster (făl'stər), island (1965 pop. 45,906), 198 sq mi (512.8 sq km), Storstrøm co., SE Denmark, in the Baltic Sea.

Fa·lun (fä'lŭn'), city (1975 est. pop. 38,300), capital of Kopparberg co., S central Sweden; chartered 1614. It is the headquarters of Sweden's oldest company, founded (1347) to operate the copper mines of Falun (now largely exhausted). The company helped finance Gustav II's campaigns during the Thirty Years' War.

Fa·ma·gu·sta (fä'mə-gōō'stə), city (1974 est. pop. 39,400), E Cyprus, on Famagusta Bay. Famagusta occupies the site of ancient Arsinoë, built (3rd cent. B.C.) by Ptolemy II. After the fall (1291) of Acre to the Saracens, Christian refugees greatly increased the city's wealth. The seat (15th–16th cent.) of the Venetian governors of Cyprus, it was strongly fortified by the Venetians. As a British naval base the city was heavily bombed in World War II.

Fan·nin (făn'ən). **1.** County (1970 pop. 13,357), 396 sq mi (1,025.6 sq km), N Ga., in the Blue Ridge bordered on the N by Tenn. and N.C. and drained by the Toccoa River; formed 1854; co. seat Blue Ridge. Farming (corn, hay, potatoes, fruit), cattle ranching, lumbering, and tourism are important. Textiles, clothing, and furniture are manufactured. **2.** County (1970 pop. 22,705), 906 sq mi (2,346.5 sq km), NE Texas, bounded on the N by the Red River, here forming the Okla. border; formed 1837; co. seat Bonham. This rich prairie agricultural area produces cotton, corn, grain, legumes, fruit, truck crops, livestock, and dairy products. It also has stands of timber and deposits of clay and limestone.

Fa·no (fä'nō), city (1976 est. pop. 42,900), in the Marche, central Italy, on the Adriatic Sea. It is a fishing port, a seaside resort, and an agricultural and silk-manufacturing center. An important town in Roman times, it was the scene of a victory by Rome over Carthage (207 B.C.). Fano was destroyed by the Goths in the 6th cent. A.D. but later flourished under the Malatesta family of Rimini. It was under papal control from the mid-15th cent. to 1860. The first printing press in Italy to use Arabic type was set up (1514) in Fano.

Fa·rah (fä-rä'), town (1969 est. pop. 29,000), capital of Farah and Chakhansur prov., W Afghanistan, on the Farah River. Surrounded by a solid earth rampart, it is a market for the products of the surrounding agricultural region. It flourished until Mongols destroyed it in 1221, then revived but suffered renewed devastation by the Persian ruler Nadir Shah in 1737.

Far East (fär' ēst'), in the most restricted sense, region comprising China, Japan, North Korea, South Korea, Mongolia, and the easternmost portion of Soviet Siberia. In a more extended sense, the term includes the Philippines, Vietnam, Cambodia, Laos, Thailand, Malaysia, Singapore, Burma, and Indonesia.

Far·go (fär'gō'), city (1978 est. pop. 58,100), seat of Cass co., E N.Dak., at the head of navigation on the Red River; inc. 1875. A railroad hub and river port, it is the trade and distribution center of a wheat and livestock region. It was founded (1871) with the coming of the Northern Pacific RR and named for William G. Fargo of the Wells-Fargo Express Company.

Far·i·bault (fâr'ə-bō'), county (1970 pop. 20, 896), 713 sq mi (1,846.7 sq km), S Minn., bordering in the S on Iowa and drained by the Blue Earth River; formed 1855; co. seat Blue Earth. It has agriculture (corn, oats, barley, poultry, and livestock) and food-processing and lumbering industries. Concrete is manufactured.

Faribault, city (1978 est. pop. 15,600), seat of Rice co., SE Minn.; founded 1826, inc. 1872. Its manufactures include hand trucks, electrical equipment, metal products, blankets, and food products. Faribault is noted for its peony farms.

Farm·ers Branch (färm'ərz brănch'), city (1978 est. pop. 33,400), Dallas co., N Texas, a suburb adjacent to Dallas; settled 1841, inc. 1946. Insecticides, brooms, and metal products are made.

Farm·er·ville (fär'mər-vĭl'), town (1970 pop. 3,416), seat of Union parish, N central La., NW of Monroe. It is the trade center of a timber and cotton region.

Farm·ing·dale (fär'mĭng-dāl'), village (1970 pop. 9,297), Nassau co., SE N.Y., on central Long Island N of Massapequa; settled 1695, inc. 1904. It has large aircraft factories and plants making tools and electronic equipment.

Far·ming·ton (fär'mĭng-tən). **1.** Town (1970 pop. 14,390), Hartford co., central Conn., on the Farmington River; inc. 1645. It has light tool industries. Points of interest include Miss Porter's School and the Hillstead Museum. **2.** Town (1970 pop. 5,697), seat of Franklin co., W central Maine, on the Sandy River (a tributary of the Kennebec) NW of Augusta; settled c.1776, inc. 1794. The town makes shoes and wood products and processes food. **3.** City (1970 pop. 10,329), Oakland co., SE Mich., a suburb of Detroit; settled 1824 by Quakers, inc. as a city 1925. Automotive parts, tools and dies, building supplies, and computer equipment are among its manufactures. **4.** City (1970 pop. 6,590), seat of St. Francois co., E Mo., SW of St. Louis, in a farm, timber, and lead-mining region; settled c.1799, laid out 1822, inc. 1879. Cobalt and nickel are refined in the area. **5.** City (1978 est. pop. 30,200), San Juan co., NW N.Mex., at the confluence of the San Juan, Animas, and La Plata rivers; inc. 1901. It is the trade center of an oil, natural-gas, and irrigated farm area. Aztec Ruins National Monument and Chaco Canyon National Monument are nearby. **6.** City (1970 pop. 2,526), seat of Davis co., N central Utah, on Great Salt Lake N of Salt Lake City, in an irrigated farm area; settled 1848 by Mormons. Disastrous floods in 1923 and 1930 led to a soil-conservation and reforestation project.

Farm·ville (färm'vĭl'), town (1970 pop. 4,331), seat of Prince Edward co., S central Va., on the Appomatox River ESE of Lynchburg; inc. 1912. It is a processing point in a farm and timber area.

Farn·bor·ough (färn'bûr'ō, -bər-ə), urban district (1971 pop. 41,233), Hampshire, S England. It is the site of the Royal Aircraft Establishment, which does experimental work in aeronautics. Farnborough Hill was the home (1881–1920) of Empress Eugénie.

Farne Islands (färn), group of islets, 2 to 5 mi (3.2–8 km) off the NE coast of Northumberland, N England. Inner Farne, or House Island, sheltered St. Cuthbert in the 7th cent. There is a chapel dedicated to him. The islands are a breeding place of seafowl. Dangerous passages are indicated by lighthouses.

Farn·ham (fär'nəm), urban district (1973 est. pop. 33,140), Surrey, SE England, on the Wey River. It is a market town but is no longer the important grain and wool center it was in the 17th and early 18th cent. A castle, the residence of the bishops of Winchester from 1160 to 1926, now houses the Overseas Service College. The castle was built in the late 17th cent. on the site of a Norman castle that was destroyed in 1648.

Farn·worth (färn'wûrth), municipal borough (1971 pop. 26,841), Greater Manchester, NW England. It has cotton and rayon mills and produces finished textiles, hosiery, knitted goods, machinery, paper, and wood products.

Fa·ro (fä'rōō), town (1970 pop. 20,470), capital of Faro dist. and of Algarve, S Portugal. The southernmost town in Portugal, it was important under the Moors and was retaken in 1249.

Far·öe Islands (fâr'ō): *see* Faeroe Islands.

Far·rukh·a·bad (fə-ro͝okʜ'ä-bäd'), joint municipality with Fategarh (total 1971 pop. 103,282), Uttar Pradesh state, N central India, on the Ganges River. It is a market for grain, fruit, and potatoes. Leather and metal goods are manufactured. A fort was founded (c.1714) on the site of the present-day town. Farrukhabad was captured by the Mahrattas in 1751 and ceded to the British in 1802. The town's British garrison was attacked during the Indian Mutiny.

Fars (färs), province and historic region of Iran, more or less identical with the ancient province of Pars, which was the nucleus of the Persian Empire. The Arabs changed the name Pars to Fars after they conquered the region in the 7th cent.

Far·well (fär'wĕl'), town (1970 pop. 1,185), seat of Parmer co., NW Texas, in the Llano Estacado NW of Lubbock at the N.Mex. border, in a grain-growing region.

Fa·teh·pur Si·kri (fə-tə-po͝or' sĭk'rē), historic city (1971 pop. 13,998), Uttar Pradesh state, N India. It was founded (1569) by the Mogul emperor Akbar and was his capital until 1584. By 1605 it was largely deserted because of the inadequate water supply. A masterpiece of Moslem architecture, the city is unique in India as a nearly intact Mogul city.

Fá·ti·ma (fă'tĭ-mə), hamlet, W central Portugal, in Beira Litoral. At the nearby Cova da Iria is the national shrine of Our Lady of the Rosary of Fátima. This became a center of pilgrimage after the six reported apparitions of the Virgin Mary to three shepherd children, May 13–Oct. 13, 1917.

Faulk (fôk), county (1970 pop. 3,893), 997 sq mi (2,582.2 sq km), N central S.Dak., watered by artificial lakes and intermittent streams; formed 1873; co. seat Faulkton. It is in a farming and ranching area and has some manufacturing.

Faulk·ner (fôk'nər), county (1970 pop. 31,578), 641 sq mi (1,660.2 sq km), central Ark., bounded on the SE by the Arkansas River; formed 1873; co. seat Conway. Its agriculture includes cotton, livestock, poultry, and dairy products. It also has food-processing and lumbering industries. Furniture, paper, shoes, metal products, and machinery are among its manufactures.

Faulk·ton (fôk'tən), city (1970 pop. 955), seat of Faulk co., N central S.Dak., SSW of Aberdeen, in a livestock and grain region.

Fau·quier (fô-kîr'), county (1970 pop. 26,375), 660 sq mi (1,709.4 sq km), N Va., partly in the Blue Ridge and bounded on the SW by the Rappahannock River; formed 1758; co. seat Warrenton. In a rich piedmont agricultural area yielding corn, cattle, hay, tobacco, and grain, it is noted for its thoroughbred horses and has stands of oak, hickory, and pine. Metal products are manufactured.

Fav·er·sham (făv'ər-shəm), municipal borough (1973 est. pop. 15,010), Kent, SE England, a port on a tributary of the Swale River. It is situated in a region where fruit and hops are grown and has shipyards and light industries. There are many Roman and Saxon remains. Faversham was an early member of the federation of the Cinque Ports.

Fa·yal (fä-yäl'), island (1974 est. pop. 17,474), 66 sq mi (170.9 sq km), in the N Atlantic, one of the central Azores, Portugal.

Fay·ette (fā-ĕt', fā'ĕt). **1.** County (1970 pop. 16,252), 627 sq mi (1,624 sq km), NW Ala., drained by the Sipsey and North rivers and crossed by the Fall Line; formed 1824; co. seat Fayette. It has agriculture (cotton, corn, and livestock), a lumbering industry, bituminous-coal mines, and deposits of sandstone and fuller's earth. Clothing, rubber products, sheet metal, and machinery are manufactured. **2.** County (1970 pop. 11,364), 199 sq mi (515.4 sq km), W Ga., bounded on the E by the Flint River; formed 1821; co. seat Fayetteville. In a piedmont agricultural area yielding cotton, corn, truck crops, pecans, peaches, livestock, and dairy products, it has a food-processing industry. Metal products, machinery, and concrete are made. **3.** County (1970 pop. 20,752), 718 sq mi (1,859.6 sq km), S central Ill., drained by the Kaskaskia River; formed 1821; co. seat Vandalia. Its agriculture includes corn, wheat, poultry, livestock, and dairy products. It also has a lumbering industry, petroleum and natural-gas wells, and plants manufacturing wood products and plastics. **4.** County (1970 pop. 26,216), 215 sq mi (556.9 sq km), E Ind., drained by the Whitewater River; formed 1818; co. seat Connersville. Primarily agricultural (livestock and grain), it has some manufacturing (hardware and metal products). **5.** County (1970 pop. 26,898), 728 sq mi (1,885.5 sq km), NE Iowa, in a prairie region drained by the Maquoketa and Turkey rivers and Buffalo Creek; formed 1837; co. seat West Union. In an agricultural area yielding oats, corn, poultry, hogs, cattle, and dairy products, it also has stone quarries and feed mills. Machinery and automobile parts are made. **6.** County (1970 pop. 174,323), 280 sq mi (725.2 sq km), NE central Ky., in a gently rolling upland area of the Bluegrass bounded on the S by the Kentucky River and drained by several creeks; formed 1780; co. seat Lexington. Its agriculture includes burley tobacco, grain, bluegrass seed, cattle, and dairy products. Its horse farms are famous for their thoroughbred race and saddle horses. The county also has bituminous-coal mines, oil and natural-gas wells, and limestone quarries. Clothing, chemicals, glass, concrete, machinery, and wood and metal products are manufactured. **7.** County (1970 pop. 25,461), 406 sq mi (1,051.5 sq km), SW central Ohio, drained by Paint Creek; formed 1810; co. seat Washington Court House. Livestock, dairy products, grain, and fruit are produced. Lumbering and limestone quarrying are also important. Its manufactures include feeds, paper, metal products, and farm machinery. **8.** County (1970 pop. 154,667), 800 sq mi (2,072 sq km), SW Pa., bounded on the S by the W.Va. border, on the W by the Monongahela River, on the SE by Laurel Hill, and drained by the Youghiogheny River; formed 1783; co. seat Uniontown. It is crossed by Chestnut Ridge and the Cumberland Road. Primarily industrial, it has bituminous-coal mines, natural-gas wells, deposits of limestone, and stands of timber. Glass, machinery, clothing, furniture, paper, and metal products are manufactured. The county was the center of the Whiskey Rebellion (1794). **9.** County (1970 pop. 22,692), 704 sq mi (1,823.4 sq km), SW Tenn., bounded on the S by Miss. and drained by the Loosahatchie and Wolf rivers; formed 1824; co. seat Somerville. In an agricultural area yielding soybeans, cotton, corn, and livestock, it has food-processing and lumbering industries. Clothing, wood products, and metalworking machinery are made. **10.** County (1970 pop. 17,650), 936 sq mi (2,424.2 sq km), SE central Texas, drained by the Colorado River and headwaters of the Lavaca and Navidad rivers; formed 1837; co. seat La Grange. This leading poultry-producing county also has oil wells, diversified farming, and a tourism industry. Fertilizer and steel and aluminum products are manufactured. **11.** County (1970 pop. 49,332), 659 sq mi (1,706.8 sq km), S central W.Va., on the Allegheny Plateau, drained by the Kanawha and Meadow rivers; formed 1831; co. seat Fayetteville. Its agriculture includes livestock, dairy products, tobacco, fruit, and truck crops. The county also has bituminous-coal mines and a lumbering industry.

Fayette. **1.** City (1970 pop. 4,568), seat of Fayette co., NW Ala., WNW of Birmingham; inc. 1821. Lumber, textile goods, and livestock feed are produced. There are natural-gas wells nearby. **2.** City (1970 pop. 1,725), seat of Jefferson co., SW Miss., NE of Natchez, in a farm, timber, and oil area. **3.** Commercial city (1970 pop. 3,520), seat of Howard co., N central Mo., near the Missouri River NW of Columbia; founded 1823, inc. 1855.

Fay·ette·ville (fā'ĭt-vĭl'). **1.** City (1978 est. pop. 34,500), seat of Washington co., NW Ark., in the Ozarks; inc. 1836. It is a farm trade center with canneries and woodworking plants. The Univ. of Arkansas is here. **2.** City (1970 pop. 2,160), seat of Fayette co., W central Ga., S of Atlanta, in a farm area. **3.** Village (1970 pop. 4,996), Onondaga co., central N.Y., E of Syracuse; inc. 1844. It has diversified manufacturing. Grover Cleveland lived here as a boy. **4.** City (1978 est. pop. 67,200), seat of Cumberland co., S central N.C., at the head of navigation on the Cape Fear River; inc. 1783. An inland port connected by a channel to the Intracoastal Waterway, Fayetteville is a marketing and shipping center in a farm and timber area. It has large textile and lumber industries and plants making power tools, plastics, tires, and automotive filters. Settled as two towns (1739) by Highland Scots, it became a Tory center during the American Revolution. In 1783 it was renamed for the Marquis de Lafayette. The city was state capital from 1789 to 1793 and the scene (1789) of the state convention that ratified the U.S. Constitution. A great fire in 1831 destroyed c.600 buildings. During the Civil War Gen. W. T. Sherman occupied the town and razed its arsenal (1865). **5.** Town (1970 pop. 7,691), seat of Lincoln co., S central Tenn., SSE of Nashville on the Elk River near the Ala. border, in a farm area; inc. 1809. **6.** Trading town (1970 pop. 1,712), seat of Fayette co., S central W.Va., SE of Charleston, in a farm, coal, and livestock area; settled 1818, inc. 1883.

Fay·yum, Al (äl fī-yo͞om') or **Fai·yum, El** (ĕl fī-yo͞om'), region, N Egypt, W of the Nile River, a depression (entirely below sea level) in the Libyan Desert. It is an irrigated agricultural area made fertile by Nile water and silt. Cereals, fruit, and cotton are produced. Al Fayyum is rich in archaeological finds, including the remains of a Neolithic farm settlement and many papyri written both in ancient Egyptian and in Arabic. The city of Al Fayyum (1970 est. pop. 151,000) is the region's trade, distribution, manufacturing, and transportation center. Industries include cotton ginning, wool and cotton weaving, dyeing, tanning, and cigarette manufacturing.

Faz·zan (fä-zän′) or **Fez·zan** (fĕ-zăn′), historic region, SW Libya. Located on caravan routes connecting the Mediterranean Sea with the Sudan, Fazzan was long important in the trans-Saharan trade. In 19 B.C. Rome conquered the region, and many of its inhabitants were converted to Christianity. After the Vandal invasion of North Africa in the 5th cent. A.D., Fazzan regained its independence. However, in 666 the Arabs conquered the region, and the people were soon converted to Islam. The Arabs held the area until the 10th cent. From the early 16th to the early 19th cent., it was the center of the Bani Mohammed dynasty, which originated in Morocco. Fazzan was annexed by the Ottoman Empire in 1842. Although Italian troops entered the region shortly after their landing in Libya in 1911, the area was not fully pacified by Italy until 1930.

Fear, Cape (fĭr), promontory on Smith Island, off SE N.C., at the mouth of the Cape Fear River. A lighthouse (built 1903) is on the cape, and a lightship is stationed off the dangerous Frying-Pan Shoals, which extend c.20 mi (32 km) to sea.

Fé·camp (fā-käN′), town (1975 pop. 21,910), Seine-Maritime dept., N France. A major port from the 12th to 17th cent., Fécamp is now an important fishing port and a resort on the English Channel. The town also has shipyards, and food, textile, and machine-building industries. Fécamp dates back to Roman times. A monastery founded c.660 became a pilgrimage site. Destroyed by Norsemen, it was rebuilt at the end of the 10th cent. and became the Benedictine Abbey of the Trinity. Fécamp is famous for benedictine liqueur, which was first made by the monks in the 16th cent. and which is now made by a private company on the grounds of the old abbey.

Fed·er·al Hall National Memorial (fĕd′ər-əl): *see* National Parks and Monuments Table.

Fei·ra de San·ta·na (fā′rə dĭ sən-tä′nə), city (1970 pop. 127,105), Bahia state, E Brazil, between the Jacuípe and Pojuca rivers. It is a distribution center for the products of Bahia's interior and is one of the state's leading producers of dried beef.

Feld·kirch (fĕlt′kĭrKH′), town (1971 pop. 21,214), in Voralberg, extreme W Austria, near the Rhine River and the Swiss and Liechtenstein borders; founded c.1190. There are textile mills and breweries in the town.

Fel·ling (fĕl′ĭng), urban district (1971 pop. 38,595), Tyne and Wear, NE England. Felling is part of the Tyneside industrial concentration. There are coal mines nearby.

Fen (fŭn), river, 375 mi (603.4 km) long, N central China, rising in the Wu-t'ai Shan and flowing SW, through a narrow valley, to the Huang Ho. It is navigable for small junks only in its lower course. The wide and fertile lower Fen valley has been irrigated since ancient times; wheat, millet, and cotton are grown.

Fens, the (fĕnz) district, E England, a flat lowland, W and S of The Wash. Extending c.70 mi (115 km) from north to south and c.35 mi (55 km) from east to west, it is traversed by numerous streams. The area was originally the largest swampland in England, formed by the silting up of a bay of the North Sea. The Romans attempted drainage and built a few roads across the Fens; however, the area had become marshy by Anglo-Saxon times. The first effective drainage systems were developed in the 17th cent. by a Dutch engineer. Drainage and construction of dikes and channels continued through the 19th cent., but problems of land sinkage, water accumulation, and frequent flooding existed throughout the period. As a result of flooding in the 20th cent., a drainage-improvement project (completed in the mid-1960s) was undertaken.

Fen·tress (fĕn′trəs), county (1970 pop. 12,593), 499 sq mi (1,292.4 sq km), N Tenn., in the Cumberlands, drained by forks of the Obey and Cumberland rivers; formed 1823; co. seat Jamestown. Its agriculture includes corn, tobacco, fruit, vegetables, livestock, poultry, and dairy products. Lumbering and coal mining are also important. Clothing is manufactured.

Fe·o·do·si·ya (fā′ə-dô′sē-ə), city (1976 est. pop. 75,000), SE European USSR, in the Ukraine, on the Crimean peninsula. It is a major Black Sea port at the western end of the Feodosiya Gulf. Oyster and sturgeon fishing and the production of caviar are important occupations. Industries include printing, steel rolling, and tobacco processing. A popular health resort, Feodosiya occupies the site of ancient Theodosia, which was founded in the 6th cent. B.C. by Greek colonists. Theodosia was destroyed by the Huns in the 4th cent. A.D. It was an insignificant village until the Genoese arrived in the 13th cent., established a flourishing trade colony, and virtually monopolized Black Sea commerce. The khan of Crimea, an ally of the Turks, conquered the city in 1475; it remained under Turko-Tatar control until Russia's annexation of the Crimea in 1783.

Fer·ga·na or **Fer·gha·na** (both: fər-gä′nə), city (1976 est. pop. 132,000), capital of Fergana oblast, Central Asian USSR, in Uzbekistan, in the Fergana Valley. It has silk and cotton industries. Oil, coal, and uranium ore lie in the area.

Fergana Valley or **Ferghana Valley,** region, 8,494 sq mi (22,000 sq km), Central Asian USSR. The Fergana Range (part of the Tien Shan system) rises in the northeast and the Pamir in the south. The narrow Khodzhent or Leninabad pass in the west has historically served as an invasion route into the valley. Chinese Turkistan borders the valley in the southeast. The Fergana Valley consists partly of the very fertile Kara-Kalpak steppe and partly of desert land. A dense irrigation network is linked by the Great Fergana and South Fergana canals. The region is one of Central Asia's most densely populated agricultural and industrial areas and one of the world's oldest cultivated areas. Along the fringes of the valley are deposits of oil, coal, natural gas, and iron ore. According to ancient Chinese sources, the Fergana Valley was a major center of Central Asia as early as the 4th cent. B.C. The introduction of silk raising from China, the development of cotton cultivation, and its favorable location astride the silk route between China and the Mediterranean stimulated the valley's growth. The Arabs, following the path of earlier invaders, occupied the valley in the 8th cent. and introduced Islam. The region was held in the 9th and 10th cent. by the Persian Samanid dynasty, in the 12th cent. by the Seljuk Turks of Khorezm, and in the 14th cent. by the Mongols under Genghis Khan. The valley later belonged to the empire of Tamerlane and his successors. Early in the 16th cent. it was overrun by the Uzbeks. The opening of the sea route to the Orient around that time led to the decline of the prosperous caravan trade through the valley. Russian conquest of the Fergana Valley was completed in 1876.

Fer·gus (fûr′gəs), county (1970 pop. 12,611), 4,250 sq mi (11,007.5 sq km), central Mont., in an irrigated agricultural region drained by the Judith and Box Elder rivers and bounded on the N by the Missouri River; formed 1885; co. seat Lewistown. It is in a grain and livestock area. A section of Lewis and Clark National Forest is in the south.

Fergus Falls, city (1978 est. pop. 12,000), seat of Otter Tail co., W central Minn., on the Otter Tail River; inc. 1872. The chief manufactures are dairy products, packaged meats, clothing, and cabinets.

Fer·man·agh (fər-măn′ə), county (1971 pop. 49,960), 715 sq mi (1,851.9 sq km), SW Northern Ireland; co. town Enniskillen. The Erne River, which widens into the extensive and beautiful Lough Erne, divides Fermanagh into two roughly equal parts. The county's population has declined since the potato famine in the 19th cent.

Fer·mo (fĕr′mō), town (1976 est. pop. 27,000), in the Marche, central Italy, on a hill in the Apennines, near the Adriatic Sea. Leather and cotton goods are manufactured, and silkworms are raised. An ancient town founded by the Sabines, Fermo was held by the papacy from the mid-16th cent. to 1860. Of note are pre-Roman walls, Roman ruins, and a 13th cent. Gothic cathedral.

Fer·nan·di·na Beach (fûr′nən-dē′nə), resort city (1970 pop. 6,955), seat of Nassau co., extreme NE Fla., on Amelia Island at the mouth of the St. Marys River. It is a port of entry and a shrimp-packing, fishing, and pulp-manufacturing center.

Fer·nan·do de No·ro·nha (fĕr-näN′do͝o də nô-rô′nyə), group of 20 islands (1970 pop. 1,239), c.10 sq mi (26 sq km), in the Atlantic Ocean, c.225 mi (360 km) off the NE coast of Brazil. A federal territory of Brazil since 1942, the islands are used as a military base and a penal colony. The main resources of the islands are the guano deposits. The islands were discovered in 1503.

Fern·dale (fûrn′dāl′), city (1978 est. pop. 25,800), Oakland co., SE Mich., a suburb of Detroit; inc. as a city 1927. Its manufactures include automobile and aircraft parts, machinery, and tools.

Fer·ney-Vol·taire (fĕr-nā′vôl-târ′), town (1968 pop. 3,064), Ain dept., E France, on the French-Swiss frontier near Geneva. The town grew after Voltaire bought the seigniory of Ferney in 1758 to escape harassment from both the Genevese and the French. Voltaire lived here until 1778.

Fer·ra·ra (fə-rär'ə, fĕr-rä'rä), city (1976 est. pop. 127,200), capital of Ferrara prov., in Emilia-Romagna, N Italy. It is an industrial and agricultural center, located on a low-lying, marshy plain that has much reclaimed land. Manufactures include chemicals, food products, and refined petroleum. In the early 13th cent. the Este family founded a powerful principality in Ferrara, and during the Renaissance commerce, learning, printing, and the arts flourished about the brilliant court. The religious reformer Savonarola was born here (1452). The city was incorporated into the Papal States in 1558. Among Ferrara's many noteworthy buildings are Este castle (14th cent.), the cathedral (begun 1135), Schifanoia palace (14th–15th cent.), and the Palazzo del Diamanti (15th–16th cent.).

Fer·rol, El (ĕl fĕr-rôl'), officially (since 1939) **El Ferrol del Cau·dil·lo** (dĕl kou-dē'yō), city (1970 pop. 87,736), La Coruña prov., NW Spain, in Galicia. The naval base on the Atlantic was built in the 18th cent. and is one of the most important in Spain. Shipbuilding and ironworks are the main industries.

Fer·ry (fĕr'ē), county (1970 pop. 3,655), 2,202 sq mi (5,703.2 sq km), NE Wash., in a forested mountain area bounded on the N by the Canadian border and watered by the Columbia and Sanpoil rivers; formed 1899; co. seat Republic. It is in a lumbering, mining, and cattle-ranching area and has some agriculture (hay, grain, and dairy products). The county includes parts of Colville Indian Reservation and Colville National Forest.

Fer·tile Cres·cent (fûr'tl krĕs'ənt), historic region of the Middle East. A well-watered and fertile area, it arcs across the northern part of the Syrian desert. It is flanked on the west by the Nile and on the east by the Euphrates and Tigris rivers. From earliest times it was the scene of bloody raids and invasions.

Fes·sen·den (fĕs'ən-dən), city (1970 pop. 815), seat of Wells co., central N.Dak., NE of Bismarck; platted 1893. Wheat, barley, and rye are grown. An annual agricultural exposition is held here.

Fez (fĕz), city (1971 pop. 325,327), N central Morocco. It is located in a rich agricultural region and is noted for its Moslem art and its handicraft industries. Fez was the capital of several dynasties and reached its zenith under the Marinid sultans in the mid-14th cent. Fez consists of the old city (founded A.D. 808) and the new city (founded 1276), connected by walls. The city has more than 100 mosques. The one containing the shrine of Idris II, founder of the old city, is one of the holiest places in Morocco.

Fez·zan (fĕ-zăn'), region, Libya: *see* Fazzan.

Fich·tel·ge·bir·ge (fĭKH'təl-gə-bîr'gə), mountain range, in SE West Germany, between Bayreuth and the Czechoslovakian border. It rises to 3,447 ft (1,051.3 m) in Schneeberg peak. The rugged mountains have dense pine forests and are dotted with resorts. The mountains were once rich in a variety of minerals, but now only lignite and iron are found in large quantities.

Field of the Cloth of Gold, locality between Guines and Ardres, not far from Calais, in France, where in 1520 Henry VIII of England and Francis I of France met for the purpose of arranging an alliance. Both kings brought large retinues, and the name given the meeting place reflects the splendor of the pageantry.

Fie·so·le (fyā'zō-lā), town (1976 est. pop. 14,666), Tuscany, central Italy. The villas and gardens of this tourist center are beautifully situated on a hill overlooking the Arno valley and the city of Florence. An ancient Etruscan town, it was enriched with fine buildings by the Romans. Of note in Fiesole are a well-preserved Roman theater (c.80 B.C.), the ruins of Roman baths, a Romanesque cathedral (11th cent.), and a Franciscan church and convent (on the site of the Roman acropolis). On the lower slopes of the hill is the Church of San Domenico di Fiesole, which has paintings by Fra Angelico.

Fife (fīf), regional authority (1976 est. pop. 338,734), E Scotland, between the Firth of Forth and the Firth of Tay. Fishing villages of great antiquity dot the eastern coast. Fife has pastures and productive farmland in the central valleys and rich coal fields in the west and east. Fife was once a Pictish kingdom.

Fifth Avenue (fĭfth), famous street of the borough of Manhattan, New York City. It begins at Washington Square and ends at the Harlem River. Between 34th and 59th streets, Fifth Ave. is lined with fashionable department stores and specialty shops. From 59th to 110th streets it borders Central Park; on its east side are tall apartment houses, built on the sites of the elegant mansions of 50 years ago.

Fi·gue·ras (fē-gā'räs), town (1970 pop. 22,087), Gerona prov., NE Spain, in Catalonia, near the French border. Traditionally a fortified city because of its strategic location, it is now an important communications center.

Fi·ji (fē'jē), Melanesian island group (1970 est. pop. 524,000), c.7,000 sq mi (18,130 sq km), South Pacific. An independent republic since 1970, Fiji comprises c.320 islands of which some 105 are inhabited. Viti Levu, the largest, constitutes half the land area and is the seat of Suva, the capital of the country. The larger islands are volcanic and mountainous. There are dense tropical forests on the windward sides of the islands and grassy plains and clumps of casuarina and pandanus on the leeward sides; mangrove forests are abundant, and hot springs are common in the mountain regions. The fertile soil yields

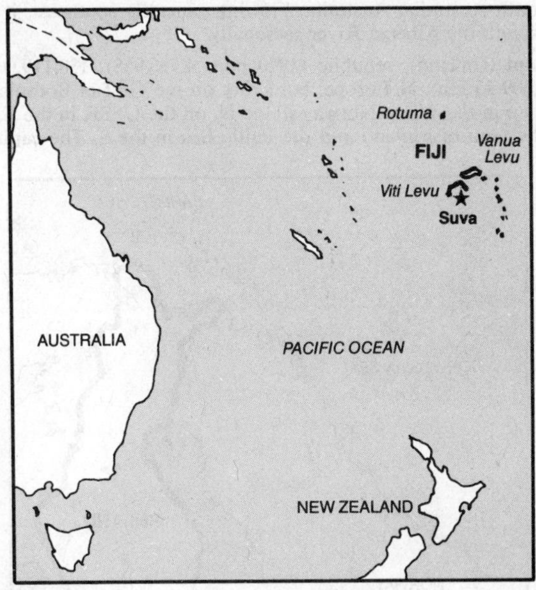

sugar cane, tropical fruits, taro, cotton, pineapples, bananas, and coconuts. Tourism is also important to the economy.

The islands were discovered by the Dutch navigator Abel Tasman in 1643 and were visited by Capt. James Cook in 1774. In 1804 the first European settlement was established. Missionaries arriving in 1835 helped abolish cannibalism. In 1874, after repeated requests by the tribal chiefs of Fiji, Great Britain annexed the islands. During World War II the islands were an important supply point of the South Pacific.

Fill·more (fĭl'môr). **1.** County (1970 pop. 21,916), 859 sq mi (2,224.8 sq km), SE Minn., bordering in the S on Iowa and watered by the Root River; formed 1853; co. seat Preston. In an agricultural area yielding grain, potatoes, poultry, livestock, and dairy products, it also has limestone quarries and a food-processing industry. Textiles, wood products, plastics, and automobile parts are made. **2.** County (1970 pop. 8,137), 577 sq mi (1,494.4 sq km), SE Nebr., drained by branches of the Big Blue River; formed 1856; co. seat Geneva. It is primarily agricultural (corn, wheat, milo, poultry, livestock, and dairy products) and has some manufacturing.

Fillmore. 1. City (1970 pop. 6,621), Ventura co., S Calif., NW of Los Angeles, in a citrus-fruit and oil region; founded 1887 as a railroad town, inc. 1914. **2.** City (1970 pop. 1,411), seat of Millard co., W central Utah, in a dry-farming and ranching area NW of Richfield; settled 1851 by Mormons. Nominally the capital of Utah Territory from 1851 to 1856, it was briefly the seat of the legislature.

Fin·cas·tle (fĭn'kăs'əl), town (1970 pop. 397), seat of Botetourt co., W Va., N of Roanoke, in a fruit-growing, dairying, and timber area.

Find·lay (fĭn'lē, fĭnd'-), city (1970 pop. 35,800), seat of Hancock co., NW Ohio, on the Blanchard River; inc. 1887. Its economy is based upon manufacturing, agriculture, and dairying. Petroleum products, tires, tile brick, washing machines, plastic goods, and food products are among its many manufactures. Gas and oil were discovered here in the 1880s, but by 1900 the supply had greatly diminished.

Fin·gal's Cave (fĭng'gəlz), cavern, 227 ft (69.2 m) long, celebrated for its unusual beauty, on Staffa Island, one of the Inner Hebrides, W Scotland. The entrance is an archway supported by basaltic columns 20 to 40 ft (6.1-12.2 m) high. The cave is inundated by the sea.

Fin·ger Lakes (fĭng'gər), group of 11 long, narrow, glacial lakes in N to S valleys, W central N.Y. Cayuga and Seneca, both more than 35 mi (56 km) long, are the largest and deepest lakes. The region is a major grape and truck-farming and recreation area, with many fine resorts and state parks.

Fin·is·tère (fĭn'ĭs-târ', fē-nē-stâr'), department (1975 pop. 804,088), 2,620 sq mi (6,785.8 sq km), NW France; capital Quimper.

Fin·is·terre, Cape (fĭn'ĭs-târ'), rocky promontory, extreme NW Spain, on the Atlantic coast of Galicia. Off the cape, the English won two naval battles against the French (1747 and 1805).

Finke (fĭngk), river, c.400 mi (645 km) long, S Northern Territory and N South Australia, Australia. Flowing generally southeast, it connects with the Alberga River seasonally.

Fin·land (fĭn'lənd), republic (1970 pop. 4,596,958), 130,119 sq mi (337,009 sq km), N Europe, bordering on the Gulf of Bothnia and Sweden in the W, on Norway in the N, on the USSR in the E, and on the Gulf of Finland and the Baltic Sea in the S. The capital is

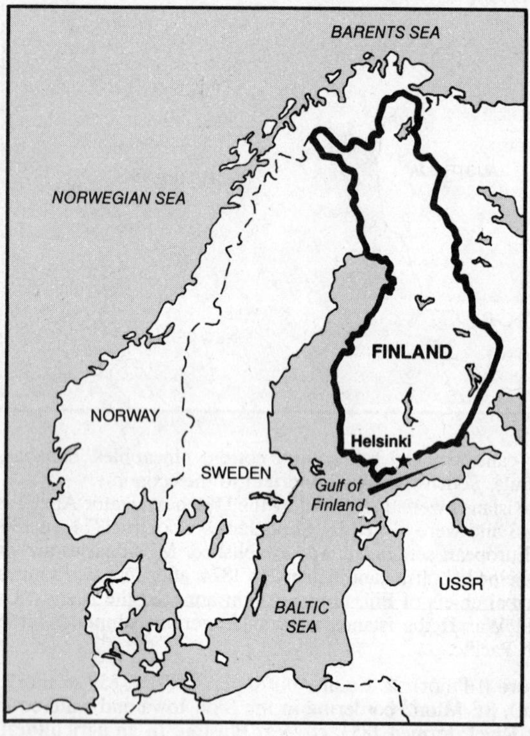

Helsinki. In the south and west is a low-lying coastal strip that includes most of the country's arable land. The coastal strip rises slightly to a vast forested interior plateau that includes about 60,000 lakes, many of which are linked by waterways. The country's third zone lies north of the Arctic Circle and is part of Lapland.

Finland is divided into 12 provinces:

NAME	CAPITAL	NAME	CAPITAL
Ahvenanmaa	Maarianhamina	Mikkeli	Mikkeli
Häme	Hämeenlinna	Oulu	Oulu
Keski-Suomi	Jyväskylä	Pohjois-Karjala	Joensuu
Kuopio	Kuopio	Turku-Pori	Turku
Kymi	Kouvola	Uusimaa	Helsinki
Lappi	Rovaniemi	Vaasa	Vaasa

Economy. Traditionally an agricultural country, Finland accelerated the pace of its industrialization after 1945. By the end of the 1960s manufacturing (plus construction and mining) accounted for about 25% of the annual national product and about 30% of employment; agriculture (plus forestry and fishing) accounted for about 15% of output and about 24% of employment. The leading agricultural commodities produced are hay, oats, barley, potatoes, wheat, rye, sugar beets, and dairy products. Large numbers of poultry, cattle, hogs, reindeer, and sheep are raised. The principal minerals extracted are iron ore, copper, zinc, nickel, titanium, and vanadium. The country's chief manufactures are forest products, processed food, metals, metal products, machinery, chemicals, transportation equipment, textiles, and clothing.

History. Beginning in the 1st cent. A.D., Finnish-speaking people migrated into Finland from the south. By the 8th cent. they had displaced the Lapps who retreated to the far north. The Finns were organized in small-scale political units, with only loose ties beyond the clan level. In the 13th cent. Sweden conquered the country. In the mid-16th cent. Lutheranism was established, and in 1581 the country was raised to the rank of grand duchy. Sweden lost parts of the territory to Russia in 1721 and 1743, and Russia finally conquered and annexed Finland in 1809.

In the 19th cent. the czars allowed the country wide-ranging autonomy, and Finland was able to develop its own democratic system. Finnish nationalism became a powerful movement as the century advanced. Under terms obtained from Russia in 1906, a unicameral parliament (whose members were elected by universal suffrage) was established. Following the Russian Revolution the parliament proclaimed (Dec. 6, 1917) the independence of Finland. In the ensuing civil war (Jan.-May, 1918) the conservative Finnish-nationalist White Guard emerged victorious over the leftist Red Guard. A republic was established and its first president, Kaarlo Juho Stahlberg, elected (1919).

In Nov., 1939, Finland was attacked by Soviet troops, and despite spirited resistance the Finns were defeated and forced to cede about one tenth of their territory (1940). When Germany attacked the USSR in June, 1941, Finland allied itself with Germany. Soviet troops again invaded Finland in 1944 and forced it to sign an armistice that was largely confirmed by a final peace treaty in1947. About 420,000 Finns left the territory ceded to the USSR and were resettled in Finland. Relations with the USSR improved in the postwar period, and treaties of trade and cooperation were signed with both the USSR and Western nations. In the late 1950s and early 1960s the USSR exercised some influence over internal Finnish politics. Communists were included in coalition cabinets from 1966 to 1971 and after 1975.

Government. Under the 1919 constitution as amended, Finland's head of state is the president, who is elected to a six-year term by a 300-member electoral college that is chosen by direct popular vote. Legislation is handled by the unicameral parliament, whose 200 members are elected to four-year terms by a system of proportional representation.

Finland, Gulf of, E arm of the Baltic Sea, c.285 mi (460 km) long and from c.10 to c.75 mi (16-120 km) wide, between Finland and the USSR. The shallow gulf is frozen from Dec. to Mar.

Fin·lay (fĭn'lē), river, c.250 mi (400 km) long, N British Columbia, Canada. It is one of the chief sources of the Peace River, which it joins at Peace River Pass.

Fin·ley (fĭn'lē), city (1970 pop. 809), seat of Steele co., E N.Dak., NW of Fargo, in a dairying, grain, and livestock area.

Fin·ney (fĭn'ē), county (1970 pop. 19,029), 1,302 sq mi (3,372.2 sq km), W Kansas, in an irrigated prairie area drained by the Arkansas and Pawnee rivers; formed 1883; co. seat Garden City. Wheat, livestock, and sugar beets are produced. It has oil and natural-gas wells, food-processing plants, and grain mills. Farm machinery is manufactured.

Fin·ster·aar·horn (fĭn'stər-är'hôrn'), peak, 14,032 ft (4,279.8 m) high, S central Switzerland; highest of the Bernese Alps.

Fire Island (fīr), narrow barrier beach, 32 mi (51.5 km) long, off the S shore of Long Island, SE N.Y., separating Great South Bay from the Atlantic Ocean. Robert Moses State Park and several resort colonies dot the island. Fire Island National Seashore (est. 1964) covers part of the island and includes wooded areas, marshes, and Sunken Forest, an area of unusual plant and animal life.

Fi·ro·za·bad (fə-rō'zə-bäd'), town (1971 pop. 133,945), Uttar Pradesh state, N central India, in a cotton and grain area. Electrical, leather, and cotton goods are manufactured. An ancient city on the site was destroyed by Moslems in the 16th cent. and rebuilt by Mogul emperor Akbar (reigned 1556-1605).

firth (fûrth), Scottish term applied to an arm of the sea, usually an estuary or strait.

Fish·er (fĭsh'ər), county (1970 pop. 6,344), 906 sq mi (2,346.5 sq km), NW central Texas, in a rolling plains area drained by the Double Mountain and Clear forks of the Brazos River; formed 1876; co. seat Roby. In a livestock and agricultural region yielding cotton, grain sorghums, fruit, truck crops, poultry, and dairy products, it also has deposits of oil, natural gas, and gypsum.

Fishers Island, c.8 mi (12.9 km) long and averaging 1 mi (1.6 km) wide, off the NE tip of Long Island, SE N.Y. The island was developed (1925) as a summer resort.

Fitch·burg (fĭch'bûrg'), industrial city (1978 est. pop. 36,600), Worcester co., N Mass., on the N branch of the Nashua River; settled c.1730, inc. as a city 1872. Its important paper industry dates approximately from 1805.

Fitz·ger·ald (fĭts-jĕr'əld), city (1970 pop. 8,187), seat of Ben Hill co., S central Ga., near the Ocmulgee River ENE of Albany; inc. 1896. Founded in 1895 as a colony for Union veterans and their families, it is a processing and shipping center for a farm area producing cotton, peanuts, and tobacco.

Fitzroy. 1. River, 174 mi (280 km) long, formed by the junction of the Dawson and the Mackenzie rivers, E Queensland, Australia, and flowing past Rockhampton to Keppel Bay of the Coral Sea. **2.** River, c.325 mi (525 km) long, rising in the King Leopold Range, N Western Australia state, Australia, and flowing generally W to King Sound of the Indian Ocean.

Fi·u·me (fē-ōō'mē, fyōō'mā): *see* Rijeka, Yugoslavia.

Five Forks (fīv), crossroads near Dinwiddie Courthouse, SW of Petersburg, Va. The last important battle of the Civil War was fought here on Apr. 1, 1865. The Union victory led to the fall of Petersburg, the capture of Richmond, and the surrender of Gen. Robert E. Lee's Confederate army at Appomattox Court House.

fjord or **fiord** (fyŏrd), steep-sided inlet of the sea characteristic of glaciated regions. Fjords probably resulted from the scouring by glaciers of valleys formed by any of several processes, including faulting and erosion by running water. When the regions occupied by these valleys subsided, the valleys were drowned by the sea.

Flag·ler (flăg'lər), county (1970 pop. 4,454), 483 sq mi (1,251 sq km), NE Fla., bounded on the E by the Atlantic Ocean and partially on the W by Crescent Lake; formed 1917; co. seat Bunnell. It is in a lowland agricultural area yielding corn, vegetables, citrus fruit, and livestock.

Flag·staff (flăg'stăf', -stäf'), city (1978 est. pop. 32,000), seat of Coconino co., N Ariz., near the San Francisco Peaks; inc. 1894. Lumbering, ranching, and a lively tourist trade thrive in the region, where many ruined Indian pueblos, numerous state parks, several lakes, and large pine forests are found. Sunset Crater National Monument, Lowell Observatory, and Northern Arizona Univ. are here.

Flam·bor·ough Head (flăm'bûr'ō, -bə-rə), chalk promontory, Humberside, E England, N of Bridlington Bay. There is a lighthouse at the tip. The chalk cliffs are wave-carved into caves and stacks and are inhabited by flocks of seafowl.

Fla·min·i·an Way (flə-mĭn'ē-ən), one of the principal Roman roads, the greatest artery from Rome to Cisalpine Gaul. Construction was begun (220 B.C.) by Caius Flaminius. The original length was 209 mi (336.3 km), but it was increased to 215 mi (346 km) after A.D. 69.

Flan·ders (flăn'dərz), former county in the Low Countries, extending along the North Sea and W of the Scheldt River. It is now divided among Belgium, France, and the Netherlands. In 862 Baldwin Bras-de-Fer, a son-in-law of Emperor Charles II, became the first count of Flanders. Flanders became a fief of the French crown, but its powerful counts enjoyed virtual independence. In the 12th cent. Artois and other areas in western and southern Flanders were lost to the French crown. Prosperity came with the growing cloth industry and the transit trade at such major ports as Bruges (later superseded by Antwerp) and Ghent. The 13th and 14th cent. were turbulent with the struggle between guild workers and patricians and the often violent rivalry among cities. The accession (1322) of the pro-French Louis of Nevers as count of Flanders threw the country into a civil war in which the pro-French party emerged victorious. Duke Philip the Bold of Burgundy succeeded to Flanders in 1384 and subdued a rebellion of weavers at Ghent. The Burgundians and (after 1477) the Hapsburgs kept a firm grip on Flanders, which was a major source of their income.

Flanders, ruled after 1506 by the Spanish line of the house of Hapsburg, joined (1576) in the revolt of the Netherlands against Philip II of Spain, but by 1584 the Spanish had recovered the county. It continued under Spanish rule until 1714, when the Peace of Utrecht awarded it to Austria. Parts of western Flanders were annexed (1668–78) to France. Austria ceded the remainder of Flanders to France in 1797, but the Congress of Vienna awarded (1815) the former Austrian Flanders to the Netherlands.

Flan·dreau (flăn'drōō'), city (1970 pop. 2,027), seat of Moody co., E S.Dak., on the Big Sioux River N of Sioux Falls. The first settlers (1857) were driven away by Indians, but the area was resettled in 1869. In a grain, livestock, and dairy region, the city processes meat, poultry, and dairy products.

Flat·bush (flăt'bŏŏsh'), residential section of central Brooklyn borough of New York City, SE N.Y. In the 17th cent. Flatbush was a Dutch village. Brooklyn College and several 18th cent. buildings are here.

Flat·head (flăt'hĕd'), county (1970 pop. 39,460), 5,137 sq mi (13,304.8 sq km), NW Mont., in an agricultural area bordering in the N on British Columbia and bounded on the E by the Continental Divide; formed 1893; co. seat Kalispell. It is drained by the Flathead River and Flathead Lake. Livestock, grain, fruit, and timber are produced. The county includes sections of Flathead National Forest.

Flathead, river, c.240 mi (385 km) long, rising as the North Fork in SE British Columbia, Canada, and flowing generally SE through NW Mont. to Coram, where it is joined by the Middle Fork (c.85 mi/135 km long) and the South Fork (c.80 mi/130 km long). It continues south through Flathead Lake, then west and south to the Clark Fork River.

Flathead Lake, 197 sq mi (510.2 sq km), NW Mont. Formed by the glacial damming of the Flathead River, Flathead Lake has an irregular shoreline and many small islands. Surrounded by mountains, the lake is a noted recreation area.

Flat Rock, industrial city (1970 pop. 5,643), Wayne co., SE Mich., SW of Detroit; settled 1824, inc. 1923. Metal and paper products are manufactured.

Flat·ter·y, Cape (flăt'ə-rē), NW Wash., at the rocky entrance to Juan de Fuca Strait; discovered in 1778 by Capt. James Cook. A lighthouse and an Indian reservation are on the cape, whose cliffs rise 120 ft (36.6 m) above the Pacific Ocean.

Flèche, La (lä flĕsh'), town (1968 pop. 15,951), Sarthe dept., on the Loir River. Tanning and the manufacture of clothing and paper are the chief industries. The town is famous for its college, the Prytanée, founded by Henry IV in the 16th cent. In 1808 Napoleon I transformed the school into a military academy open only to the sons of officers and members of the Legion of Honor.

Fleet Street (flēt), street in the City of London, England. It is the center of English journalism.

Fleet·wood (flēt'wŏŏd'), municipal borough (1973 est. pop. 30,070), Lancashire, NW England, on Morecambe Bay at the mouth of the Wyre estuary. Fishing is the major industry. Sir Peter Hesketh Fleetwood founded the town in 1836 and developed it into a trading port and seaside resort.

Flem·ing (flĕm' ĭng), county (1970 pop. 11,366), 350 sq mi (906.5 sq km), NE Ky., in a rolling upland area in the outer Bluegrass bounded on the SW by the Licking River and drained by several creeks; formed 1798; co. seat Flemingsburg. Its agriculture includes burley tobacco, corn, alfalfa, and livestock. Shoes and metal products are made.

Flem·ings·burg (flĕm'ĭngs-bûrg'), town (1970 pop. 2,483), seat of Fleming co., NE Ky., NE of Lexington, in a farm and livestock area; settled c.1790. It has flour and feed mills and chicken hatcheries.

Flem·ing·ton (flĕm' ĭng-tən), borough (1970 pop. 3,917), seat of Hunterdon co., W central N.J., W of New Brunswick; settled c.1730, inc. 1910. Its courthouse (1828) was the scene in 1935 of Bruno Hauptmann's trial for the kidnaping and murder of Charles A. Lindbergh, Jr. Chemicals and glass are produced.

Flens·burg (flĕnz'bûrg', flĕns'bŏŏrkh'), city (1974 est. pop. 94,528), Schleswig-Holstein, N West Germany, on the Flensburg Fjord, an arm of the Baltic Sea, at the Danish border. An active Baltic port and commercial center, it has shipyards, rum distilleries, smoked-fish plants, and paper factories. Flensburg was chartered in 1284 and acquired commercial importance in the 16th cent. In 1867 it passed from the Danish crown to Prussia. During World War II it was heavily bombed and lost its large merchant fleet.

Flin·ders (flĭn'dərz), river, 520 mi (836.7 km) long, N Queensland, Australia. It flows northwest from the Eastern Highlands to the Gulf of Carpentaria.

Flinders Ranges, mountain chain, extending 260 mi (418.3 km) between Lake Torrens and Lake Frome, South Australia state, Australia, rising to 3,900 ft (1,189.5 m) at St. Mary's Peak. Uranium and copper are mined here.

Flint (flĭnt), city (1978 est. pop. 164,100), seat of Genesee co., S Mich., on the Flint River; inc. 1855. Since 1902 it has been one of the chief automobile-manufacturing centers of the world. A fur-trading post was established here in 1819. Fur trading was succeeded by lumbering and then by carriage making as Flint's major industry. The city was extensively damaged by a tornado in June, 1953. In the city are a branch of the Univ. of Michigan, the General Motors Institute, an art institute, a planetarium, and a nature preserve.

Flint, municipal borough (1973 est. pop. 15,070), Clwyd, NE Wales, on the Dee estuary. Flint has industries that produce rayon, nylon, paper, and clothing. The castle, built c.1300, was the scene of Richard II's submission to Henry Bolingbroke in 1399.

Flint·shire (flĭnt'shîr', -shər), former county, NE Wales, bounded on the NE by the Dee estuary. The county town was Mold. After the Norman conquest the region was heavily fortified by the border lords against the partially conquered Welsh. In 1974 Flintshire became part of the new nonmetropolitan county of Clwyd.

Flod·den (flŏd'n), field, Northumberland, N England, just across the border from Coldstream, Scotland. It was the scene of the Battle of Flodden Field (1513), in which the English defeated the Scots under James IV, who was killed.

Flor·ence (flôr'əns), city (1971 pop. 461,602), capital of Tuscany and of Firenze prov., central Italy, on the Arno River, at the foot of the Apennines. Florence, the jewel of the Italian Renaissance, is a commercial, industrial, and tourist center. Manufactures include machinery, chemicals, furniture, and pharmaceuticals. The city is noted for its handicraft industry (producing ceramics, mosaics, and metal and leather goods) and is a leading center of women's fashions.

Florence was the site of an Etruscan settlement. It later became a Roman town on the Cassian Way. In the 5th and 6th cent. A.D. the city was controlled, in turn, by the Goths, Byzantines, and Lombards. It became an autonomous commune in the 12th cent. In the 13th cent. the Guelphs defeated the Ghibellines for control of the city. The sale of Florentine silks, tapestries, and jewelry brought great wealth. Florence became a city-state and in the 15th cent. came under the control of Cosimo de' Medici, a wealthy merchant and patron of the arts. Under the Medici and their successors, Florence was for two centuries the golden city, with an unsurpassed flowering of intellectual and artistic life.

The Medici were expelled by a revolution in 1494, the fiery religious reformer Savonarola briefly held power (1494–98), and Machiavelli was a diplomatic representative of the republic. The revolt against the Medici was over by 1512, but another revolution (1527–30) established a new republic, which, however, was forced to surrender to Emperor Charles V after a heroic defense. Under the restored Medici rule, Florence went on expanding and controlled most of Tuscany. In 1569 Cosimo I de' Medici was made grand duke, and Florence became the capital of the grand duchy of Tuscany. The grand duchy, ruled by the house of Hapsburg-Lorraine after the extinction (1737) of the Medici line, was annexed to the kingdom of Sardinia in 1860. Florence was the capital of the newly founded kingdom of Italy from 1865 to 1871. Relatively few of the art treasures of Florence were harmed in World War II; the flooding of the Arno in Nov., 1966, however, caused considerable damage, which art experts sought, with considerable success, to repair.

Most of the city's innumerable monuments date from the 13th to the 15th cent. The Gothic cathedral of Santa Maria del Fiore (begun 1296) has a dome (1420–34) by Brunelleschi; nearby is the slim campanile designed by Giotto. The large Franciscan Church of Santa Croce has frescoes by Giotto, a crucifix by Donatello, and fine works by the Della Robbia family, Rossellino, and others. The Church of Santa Maria Novella (1278–1350) has frescoes by Masaccio, Orcagna, and Ghirlandaio; fine cloisters; and a façade (1470) by Alberti. Some of the best works of Fra Angelico are in the museum of the Monastery of St. Mark. Important frescoes by Masolino, Masaccio, and Filippino Lippi adorn the Church of Santa Maria del Carmine. The Church of San Lorenzo contains Michelangelo's tombs of the Medici, many works by Donatello, and the Laurentian Library. On the Piazza della Signoria is the Palazzo Vecchio, which contains frescoes by Vasari and sculptures by Michelangelo. The Uffizi Museum contains great collections of paintings, especially by Botticelli, Masaccio, and Piero della Francesca. The Pitti Palace (15th–17th cent.) also houses fine paintings, particularly by Raphael, Andrea del Sarto, and Titian. There are many other important museums, churches, and palaces, as well as the Univ. of Florence and the National Library.

Florence. 1. County (1970 pop. 89,636), 805 sq mi (2,085 sq km), E S.C., bounded on the E by the Pee Dee River and drained by the Lynches River; formed 1888; co. seat Florence. It has agriculture (tobacco, cotton, corn, pecans, truck crops, and cattle) and lumbering and food-processing industries. Textiles, concrete, metal products, and machinery are manufactured. **2.** County (1970 pop. 3,298), 489 sq mi (1,266.5 sq km), NE Wis., bounded on the N by the Brule River and on the NE by the Menominee River (both on the Mich. border); formed 1882; co. seat Florence. In a lumbering, dairying, and potato-growing region, it has some manufacturing. The county contains a section of Nicolet National Forest.

Florence. 1. City (1978 est. pop. 34,600), seat of Lauderdale co., NW Ala., on the Tennessee River near Muscle Shoals and adjacent to Wilson Dam; inc. 1818. It is in a cotton and mineral area, and power from Muscle Shoals has stimulated the growth of diversified industries. Aluminum and aluminum products, ceramic tile, textiles, stoves, chemicals, boats, and corrugated boxes are made. The mountain lakes in the area attract many tourists. Florence State Univ. is in the city. Of interest are Pope's Tavern (1811), once a stagecoach stop and later a Civil War hospital, and an Indian mound, with a museum. **2.** Town (1975 est. pop. 2,926), seat of Pinal co., S Ariz., on the Gila River SE of Phoenix; founded 1866, inc. 1906. The surrounding cotton, wheat, and alfalfa area is irrigated by the San Carlos Reservoir. Copper is mined. **3.** Uninc. town (1975 est. pop. 27,200), Los Angeles co., S Calif., a residential and manufacturing suburb SE of Los Angeles. **4.** City (1970 pop. 11,661), Boone co., N central Ky., in a bluegrass farm region; inc. 1830. The city has a race course. **5.** City (1978 est. pop. 33,800), seat of Florence co., NE S.C., in a farm and timber area; inc. 1871. It is an important focal point for railroads (with extensive repair shops and yards) and an industrial and trade center. During the Civil War it was a transportation and supply point and served as the site of a prison camp. **6.** Town (1970 pop. 575), seat of Florence co., extreme NE Wis., NW of Iron Mountain, in a wooded dairying and stock-raising area. Cabinets are made.

Flo·res (flôr'əs), island (1970 pop. 1,108,000), 6,627 sq mi (17,164 sq km), E Indonesia, one of the Lesser Sunda Islands. Flores is mountainous and heavily wooded. Copra, rice, maize, and coffee are grown, and cattle are raised. Flores came under Dutch influence c.1618. Portugal held the eastern end until 1851 and the natives were not completely subjugated until 1907.

Flo·res·ville (flôr'əs-vĭl'), city (1970 pop. 3,707), seat of Wilson co., S Texas, on the San Antonio River SE of San Antonio. It is the market for a farm and cattle area.

Flo·ri·a·nó·po·lis (flôr'ĭ-ə-nŏp'ō-lĭs, flô'ryä-nô'pōō-lēs), city (1970 pop. 115,665), capital of Santa Catarina state, SE Brazil, on Santa Catarina Island. A cultural center and a port city, it is linked with the mainland by a huge suspension bridge (built 1926). The city was founded in 1673 by colonists from São Paulo and became the capital of the captaincy of Santa Catarina in 1739.

Flo·ri·da (flō-rē'dä), city (1970 pop. 32,679), Camagüey prov., E central Cuba. It has good road and rail communications and an economy based on the raising of cattle, citrus fruits, and sugar cane.

Florida (flôr'ə-də, flŏr'-), state (1975 pop., 8,450,000), 58,560 sq mi (151,670 sq km), extreme SE United States, admitted 1845 as the 27th state of the Union. Tallahassee is the capital. A long low peninsula, Florida is bounded on the east by the Atlantic Ocean, on the west by the Gulf of Mexico, and on the north by Ga. and Ala. (where the St. Marys River in the northeast and the Perdido River in the northwest form part of the boundary). Florida is separated from Cuba to the south by the Straits of Florida.

Much of the east coast is shielded from the Atlantic by narrow sandbars and islands that protect the shallow lagoons, rivers, and bays. Immediately inland, pine and palmetto flatlands stretch from the Ga. border almost to the southern tip of the state. The northwest

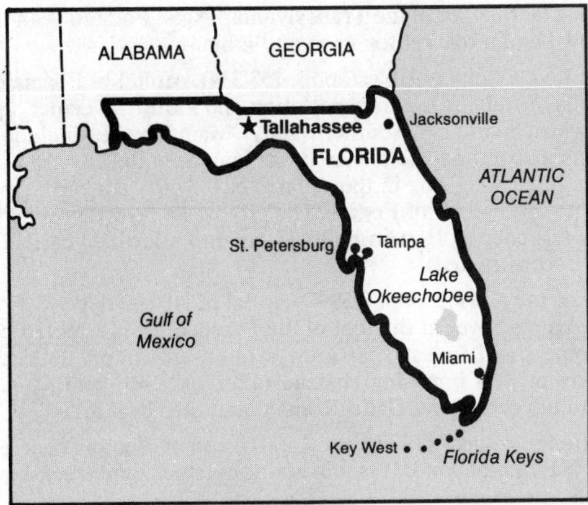

of Florida is a gently rolling panhandle area, cut by deep swamps along the coast. Central Florida abounds in lakes. The Everglades, which includes Big Cypress Swamp, extends over almost the entire southern part of the peninsula. Florida's lower Gulf coast is dotted with tiny islands. The Florida Keys extend south and west from the southern tip of the state.

Warmed by the surrounding subtropical waters and cooled by the trade winds, Florida is famous for its pleasant climate and abundant sunshine. Beautiful beaches attract thousands of vacationers annually. Other attractions include Everglades National Park, with its unusual plant and animal life; Cypress Gardens, near Winter Haven; Palm Beach, with its palatial estates and fine resort facilities; Walt Disney World, an entertainment park near Orlando; and many other picturesque resort areas.

Florida is divided into 67 counties:

NAME	COUNTY SEAT	NAME	COUNTY SEAT
Alachua	Gainesville	Lake	Tavares
Baker	Macclenny	Lee	Fort Myers
Bay	Panama City	Leon	Tallahassee
Bradford	Starke	Levy	Bronson
Brevard	Titusville	Liberty	Bristol
Broward	Fort Lauderdale	Madison	Madison
Calhoun	Blountstown	Manatee	Bradenton
Charlotte	Punta Gorda	Marion	Ocala
Citrus	Inverness	Martin	Stuart
Clay	Green Cove Springs	Monroe	Key West
Collier	East Naples	Nassau	Fernandina Beach
Columbia	Lake City	Okaloosa	Crestview
Dade	Miami	Okeechobee	Okeechobee
De Soto	Arcadia	Orange	Orlando
Dixie	Cross City	Osceola	Kissimmee
Duval	Jacksonville	Palm Beach	West Palm Beach
Escambia	Pensacola	Pasco	Dade City
Flagler	Bunnell	Pinellas	Clearwater
Franklin	Apalachicola	Polk	Bartow
Gadsden	Quincy	Putnam	Palatka
Gilchrist	Trenton	St. Johns	St. Augustine
Glades	Moore Haven	St. Lucie	Fort Pierce
Gulf	Wewahitchka	Santa Rosa	Milton
Hamilton	Jasper	Sarasota	Sarasota
Hardee	Wauchula	Seminole	Sanford
Hendry	La Belle	Sumter	Bushnell
Hernando	Brooksville	Suwannee	Live Oak
Highlands	Sebring	Taylor	Perry
Hillsborough	Tampa	Union	Lake Butler
Holmes	Bonifay	Volusia	De Land
Indian River	Vero Beach	Wakulla	Crawfordville
Jackson	Marianna	Walton	De Funiak Springs
Jefferson	Monticello	Washington	Chipley
Lafayette	Mayo		

Economy. Tourism plays a primary role in Florida's economy. Abundant recreational, boating, and fishing facilities have made tourism a year-round enterprise. Florida leads the nation in the production of oranges, grapefruits, limes, and tangerines. Other important crops raised in the state are tomatoes, sugar cane, and tobacco. Florida also supplies much of the country with many varieties of winter vegetables, and cattle and dairy products are important. The leading manufactured items are food products, chemicals, paper products, electrical equipment, and transportation equipment. Lumber and wood products are also important; most of the state's timber is yellow pine. Florida's mineral resources include phosphate rock, stone, sand, and gravel. Commercial fishing is important.

History. The Spanish explorer Juan Ponce de León is credited with the discovery of the Florida peninsula in 1513. He claimed the area, which he thought was an island, for Spain and named it Florida, probably because it was then the Easter season (*Pascua Florida*). In 1562 the Frenchman Jean Ribaut discovered the St. Johns River, and two years later René de Laudonnière built Fort Caroline at its mouth. Alarmed at this encroachment by the French, Philip II of Spain commissioned Pedro Menéndez de Aviles to drive the French out of the area; this he did ruthlessly. Spanish colonization began when Menéndez founded St. Augustine, the oldest city in the U.S., in 1565. Florida had no precious metals, its soil seemed infertile, and the native Indians were not peaceful. However, the Spanish were compelled to hold Florida because of its strategic location along the Straits of Florida, through which rich treasure ships from the south sailed for Spain.

In 1742 English colonists from Georgia under James E. Oglethorpe, Georgia's founder, defeated the Spanish in the Battle of Bloody Marsh on St. Simons Island, making Florida's northern boundary the St. Marys River. Spain's late entry (1762) into the Seven Years' War cost her Florida, which the British acquired through the Treaty of Paris (1763). Florida, divided into two provinces, prospered under British rule and remained loyal to the mother country during the American Revolution. Under the Treaty of Paris (1783) Florida was returned to Spain. Boundary disputes then developed with the United States. In 1818 Gen. Andrew Jackson again defied Spanish authority and invaded Florida in a punitive attack against the Indians. In 1819, after years of diplomatic wrangling, Spain reluctantly signed the Adams-Onis Treaty ceding Florida to the United States. Official U.S. occupation took place in 1821. Florida, with its present boundaries, was organized as a territory in 1822.

Settlers poured in from neighboring states, especially to the area around the newly founded capital of Tallahassee. Settlement expanded southward and displaced the Seminole Indians, most of whom were transported out of the region by the end of the Second Seminole War (1835–42). However, a small band fled to the wilderness of the Everglades, and today their descendants live on reservations in the Lake Okeechobee area.

Florida was admitted to the Union in 1845 as a slaveholding state. The state seceded from the Union in 1861 and joined the Confederacy. It was readmitted to the Union in 1868. The state sold (1881) 4,000,000 acres (1,620,000 hectares) of land to real-estate promoters, and Florida began to develop.

During the latter half of the 19th cent. Cubans rebelling against Spain received sanctuary and aid in Florida, and the state enthusiastically supported and profited economically from the Spanish-American War (1898), in which Tampa was the chief U.S. base. The drainage of the Everglades, begun in 1906, precipitated one of the state's periodic land booms. The most famous of Florida's land booms started after World War I and reached its peak in 1925. After World War II the state enjoyed phenomenal growth. Manufacturing, particularly industries related to aeronautics, developed at an extraordinary rate.

Government. In 1968 Florida adopted a new constitution. The governor of the state is elected for four years, and the legislature has a house of representatives with 120 members elected for two years and a senate with 40 members elected for four years.

Educational Institutions. These include the Univ. of Florida, at Gainesville; the Univ. of Miami, at Coral Gables; Florida State Univ. and Florida Agricultural and Mechanical Univ., at Tallahassee; Rollins College, at Winter Park; the Univ. of Tampa and the Univ. of South Florida, at Tampa; Florida Southern College, at Lakeland; Stetson Univ., at De Land; Barry College, at Miami; and Bethune-Cookman College, at Daytona Beach.

Florida, Straits of, passage, c.90 mi (144.8 km) wide, between the Florida Keys in the N and Cuba and the Bahamas in the S and SE. It connects the Gulf of Mexico with the Atlantic Ocean.

Florida Keys, chain of small coral and limestone islands and reefs, c.150 mi (240 km) long, extending from Virginia Key, S of Miami Beach, to Key West, and forming the S tip of Fla. state. Many of the islands are habitable and are generally covered by dense growths of low trees and shrubs, with mangrove swamps on the landward side. The Florida Keys are noted for their commercial fisheries, resort areas, and tropical vegetation.

Flor·is·sant (flŏr′ ĭ-sənt), city (1978 est. pop. 74,100), St. Louis co., E Mo., a residential suburb of St. Louis, on the Missouri River; inc.

1829. It was settled by French farmers and fur trappers c.1769, and the first civil government was established in 1786 by the Spanish. The village was predominantly French until the mid-19th cent. Points of interest include Old St. Ferdinand's Shrine and Convent (founded 1789; rebuilt 1820) and historic French and Spanish homes, some dating from 1790.

Florissant Fos·sil Beds National Monument (fŏs′əl bĕdz′): *see* National Parks and Monuments Table.

Floyd (floid). **1.** County (1970 pop. 73,742), 514 sq mi (1,331.3 sq km), NW Ga., in a valley and ridge area bounded on the W by the Ala. border and drained by the Coosa and Etowah rivers; formed 1832; co. seat Rome. Its agriculture includes cotton, corn, hay, sweet potatoes, fruit, poultry, livestock, and dairy products. It also has limestone quarries and a lumbering industry. Wood products, furniture, paper, chemicals, concrete, steel, aluminum, and hardware are manufactured. **2.** County (1970 pop. 55,622), 149 sq mi (385.9 sq km), S Ind., in a hilly region bounded on the SE by the Ohio River and Ky. border; formed 1819; co. seat New Albany. It has agriculture (grain, tobacco, livestock, and dairy products), timber, and sand and gravel pits. Clothing, furniture, paper, chemicals, plastics, leather goods, concrete, and metal products are among its manufactures. **3.** County (1970 pop. 19,860), 503 sq mi (1,302.8 sq km), N Iowa, in a prairie region drained by the Shellrock, Cedar, and Little Cedar rivers; formed 1851; co. seat Charles City. Hogs, cattle, poultry, corn, soybeans, and oats are raised. Food processing is done. Clay products and farm equipment are manufactured. **4.** County (1970 pop. 35,889), 402 sq mi (1,041.2 sq km), E Ky., in the Cumberlands and drained by Levisa Fork; formed 1799; co. seat Prestonsburg. Its agriculture includes white and sweet potatoes, apples, corn, soybeans, cattle, and dairy products. It also has bituminous-coal mines, oil and natural-gas wells, and some manufacturing plants. **5.** County (1970 pop. 11,044), 993 sq mi (2,571.9 sq km), NW Texas, on the Llano Estacado and drained by the White River; formed 1876; co. seat Floydada. In a partially irrigated area, it is primarily agricultural, yielding grain sorghums, wheat, cotton, alfalfa, fruit, truck crops, livestock, poultry, and dairy products. **6.** County (1970 pop. 9,775), 383 sq mi (992 sq km), SW Va., in the Blue Ridge, drained by the Little River and traversed by the Appalachian Trail and Blue Ridge Parkway; formed 1831; co. seat Floyd. It is in a dairying and livestock area and has sawmills and clothing factories. Deposits of arsenopyrite are found here.

Floyd, town (1970 pop. 474), seat of Floyd co., SW Va., SW of Roanoke. It has flour and lumber mills. Clothing is manufactured.

Floyd·a·da (floi-dā′də), town (1970 pop. 4,109), seat of Floyd co., NW Texas, NE of Lubbock in the Llano Estacado; settled c.1889, inc. 1909. The region produces grain sorghums, alfalfa, wheat, cotton, poultry, and dairy products.

Flush·ing (flŭsh′ĭng): *see* Vlissingen, Netherlands.

Flushing. 1. Village (1976 est. pop. 8,313), Genesee co., E central Mich., NW of Flint, in an agricultural area; settled 1833, inc. 1877. There are Indian mounds nearby. **2.** Former village, now in N Queens borough of New York City, SE N.Y.; chartered 1645, inc. into Greater New York City with Queens in 1898. Although chiefly residential, Flushing has gained importance as a trading and manufacturing center. Flushing Meadow (now a park) was the site of two New York World's Fairs (1939–40 and 1964–65) and temporary headquarters of the United Nations (1946–49).

Flu·van·na (flo͞o-văn′ə), county (1970 pop. 7,621), 282 sq mi (730.4 sq km), central Va., bounded on the S by the James River and drained by the Rivanna River; formed 1777; co. seat Palmyra. Its agriculture includes cattle, poultry, hogs, dairy products, and tobacco. Sawmilling, gold mining, and quarrying for quartz are also important. Textiles and metal products are manufactured.

Fly (flī), longest river of the island of New Guinea, c.650 mi (1,045 km) long, rising in the Victor Emmanuel Range and flowing generally SE through Papua New Guinea to the Gulf of Papua.

Foard (fôrd), county (1970 pop. 2,211), 676 sq mi (1,750.8 sq km), N Texas, in an agricultural region drained by the Pease and Wichita rivers; formed 1891; co. seat Crowell. Wheat, cotton, alfalfa, fruit, and truck crops are grown. Cattle, hogs, horses, and poultry are also produced. The county has oil and natural-gas wells and deposits of clay, gypsum, and copper.

Foc·şa·ni (fŏk-shän′, -shä′nē), town (1974 est. pop. 47,189), E central Rumania, at the foot of the Transylvanian Alps. Focşani is an industrial town and a market for wine and grain.

Fog·gia (fôd′jä), city (1976 est. pop. 153,334), capital of Foggia prov., in Apulia, S Italy. It is a transportation and industrial center and the main wheat market of southern Italy. Manufactures include food products, paper, and chemicals. It has long been the custom to store grain in huge holes dug in the squares of the city. An earthquake in 1731 destroyed much of Foggia. The city was a favorite residence of Emperor Frederick II, who built (13th cent.) a fortified castle (since greatly reconstructed).

Foix (fwä), town (1975 pop. 9,599), capital of Ariège dept., S France, on the Ariège River at the foot of the Pyrenees. It is a tourist center with some small industry. It grew around an oratory founded by Charlemagne. Its imposing château (12th and 14th cent.) houses a museum of prehistoric, Gallo-Roman, and medieval art.

Fo·ley (fō′lē), village (1970 pop. 1,271), seat of Benton co., central Minn., NE of St. Cloud. It is in a grain, livestock, and truck-farming area.

Fo·li·gno (fō-lē′nyō), city (1976 est. pop. 45,700), in Umbria, central Italy. Manufactures include chemicals, machinery, paper, and textiles. Foligno was under papal control from the mid-15th cent. until 1860. A local school of art flourished in the 15th cent.

Folke·stone (fōk′stən), municipal borough (1973 est. pop. 45,610), Kent, SE England. Folkestone is a summer resort and an active port. There are vestiges of Roman occupation in the vicinity, and a 13th cent. parish church has been restored. Folkestone was a prosperous port in the Middle Ages and early modern period.

Folk·ston (fōk′stən), city (1970 pop. 2,112), seat of Charlton co., SE Ga., SE of Waycross near the Fla. border and the St. Marys River. It is in a cotton, corn, and timber area.

Fol·som (fōl′səm), city (1971 est. pop. 6,229), Sacramento co., central Calif., on the American River ENE of Sacramento, in a fertile farm area; founded 1855, inc. 1946. The town was on the road to the gold fields. A state prison is here, and Folsom Dam (1956), a unit of the Central Valley project, is nearby.

Fon·da (fŏn′də), village (1970 pop. 1,120), seat of Montgomery co., E central N.Y., on the Mohawk River W of Amsterdam; inc. 1850.

Fond du Lac (fŏn′ də lăk′), county (1970 pop. 84,567), 724 sq mi (1,875.2 sq km), E Wis., drained by the Milwaukee, Rock, Sheboygan, and Fond du Lac rivers; formed 1836; co. seat Fond du Lac. In a dairying and agricultural area yielding corn, alfalfa, and truck crops, it has a thriving tourism industry. Food processing is done.

Fond du Lac, city (1978 est. pop. 36,400), seat of Fond du Lac co., E central Wis., in a resort region at the S end of Lake Winnebago; inc. 1852. Industries include dairy processing and the manufacture of machine tools, leather goods, engines, and snowmobiles. It was a French fur-trading post in the late 18th cent. and grew into a lumbering town in the 19th cent. After the arrival of the railroad, it became an industrial city.

Fon·taine·bleau (fŏn′tən-blō, fôɴ-tĕn-blō′), town (1975 pop. 16,778), Seine-et-Marne dept., N France, SE of Paris. It is a resort and was long a royal residence, chiefly because of the excellent hunting in the vast Forest of Fontainebleau. Louis IV resided in Fontainebleau, and Philip IV and Louis XIII were born here. Francis I built the magnificent French Renaissance palace. In the palace Louis XIV signed (1685) the revocation of the Edict of Nantes, Pope Pius VII was imprisoned (1812–14), and Napoleon signed his first abdication (1814). Fontainebleau also has a military museum. The town was headquarters of the military branch of the North Atlantic Treaty Organization (NATO) from 1945 to 1967.

Fon·tan·a (fŏn-tăn′ə), city (1978 est. pop. 25,000), San Bernardino co., S Calif., at the foot of the San Bernardino Mts.; inc. 1952. Local industries produce steel and gases for industrial and military uses. There is also a large steam plant for the production of electricity. Mormons farmed on the site in the 1850s; in the early 1900s extensive orchards were planted. During World War II the city began its transformation from an agricultural to an industrial community.

Foo·chow (fo͞o′chou′): *see* Fu-chou, China.

Foots·cray (fo͝ots′krā), city (1976 pop. 51,681), Victoria, SE Australia, part of the Melbourne urban agglomeration. It has factories producing munitions, household appliances, and chemicals.

Forbes, Mount (fôrbz), peak, 11,902 ft (3,630.1 m) high, Rocky Mts.,

SW Alta., Canada, in Banff National Park near the British Columbia border.

Ford (fôrd). **1.** County (1970 pop. 16,382), 488 sq mi (1,263.9 sq km), NE central Ill., drained by the Middle Fork of the Vermilion River and a headstream of the Sangamon River; formed 1859; co. seat Paxton. It has agriculture (corn, oats, soybeans, wheat, poultry, livestock, and dairy products), a food-processing industry, and some mining. Clothing is made. **2.** County (1970 pop. 22,587), 1,091 sq mi (2,825.7 sq km), S Kansas, drained by the Arkansas River; formed 1873; co. seat Dodge City. In a gently sloping prairie region yielding livestock and grain, it has a meat-packing industry.

Ford's Theatre National Historic Site (fôrdz): *see* National Parks and Monuments Table.

For·dyce (fôr'dīs'), city (1970 pop. 4,837), seat of Dallas co., S central Ark., SW of Pine Bluff, in a timber and farm area; inc. 1884.

Fore·land, North, and **South Foreland** (both: fôr'lənd), headlands of Kent, SE England, forming parts of the boundary of The Downs (a roadstead). Both are chalk cliff formations, and both have lighthouses. The defeat (1666) of the Dutch off the Forelands was an important battle in the history of British seapower.

For·est (fôr'əst). **1.** County (1970 pop. 4,926), 420 sq mi (1,087.8 sq km), NW Pa., in a forested region drained by the Allegheny and Clarion rivers; formed 1848; co. seat Tionesta. It has agriculture (corn and wheat), a lumbering industry, and natural-gas wells. Glass, hardware, and leather goods are manufactured. A large part of the county is occupied by Allegheny National Forest. **2.** County (1970 pop. 7,691), 1,010 sq mi (2,615.9 sq km), NE Wis., bounded partly on the N by the Brule River and the Mich. border; formed 1885; co. seat Crandon. In a logging area, it also has potato farms. It lies mostly within Nicolet National Forest and has many lakes and streams with good recreational facilities.

Forest, town (1970 pop. 4,085), seat of Scott co., central Miss., ENE of Jackson. It is in Bienville National Forest. Wood products and clothing are made, and poultry is processed.

Forest City. 1. City (1975 est. pop. 4,506), seat of Winnebago co., N central Iowa, NW of Mason City; platted 1856, inc. 1879. Farm produce is processed, and trailers are made. Pilot Knob State Park is nearby. **2.** Textile-mill town (1970 pop. 7,179), Rutherford co., W N.C., SE of Asheville.

Forest Hills. 1. Residential section of central Queens borough of New York City, SE N.Y. Until 1978 the U.S. Open Championship matches were held at the West Side Tennis Club here. **2.** Borough (1970 pop. 9,561), Allegheny co., SW Pa., E of Pittsburgh; inc. 1920.

Forest Park. 1. City (1978 est. pop. 19,600), Clayton co., NW Ga., a suburb of Atlanta; inc. 1908. An army depot and a large state farmers' market are here. **2.** Village (1978 est. pop. 15,400), Cook co., NE Ill., a suburb of Chicago, on the Des Plaines River; inc. 1884 as Harlem, name changed 1907. Its manufactures include packaged meats and office supplies.

For·far (fôr'fər), burgh (1971 pop. 10,500), Tayside region, E Scotland. Jute, linen, processed foods, iron goods, and farm tools are produced. Royalist in the civil wars, Forfar had its charter revoked by Oliver Cromwell in 1651. It was reissued in 1665 by Charles II, who gave the burgh a tower that still stands. Until 1974 Forfar was county town of Angus.

For·lì (fôr-lē'), city (1976 est. pop. 91,800), capital of Forlì prov., Emilia-Romagna, N central Italy. It is an agricultural and industrial center. Manufactures include silk, rayon, clothing, and household appliances. A Roman trade center on the Aemilian Way, Forlì became a free commune in the 11th cent. Of note are the citadel (14th-15th cent.), the clock tower (12th cent.), and the Basilica di San Mercuriale.

For·man (fôr'mən), city (1970 pop. 596), seat of Sargent co., SE N.Dak., W of Lidgerwood, in a grain-farming area.

For·mo·sa Strait (fôr-mō'sə), arm of the Pacific Ocean, between China's Fukien coast and Taiwan, linking the East and South China seas.

For·rest (fôr'əst), county (1970 pop. 57,849), 469 sq mi (1,214.7 sq km), SE Miss., drained by Black Creek; formed 1906; co. seat Hattiesburg. It has agriculture (cotton, corn, and truck crops), lumbering and meat-packing industries, and petroleum and natural-gas wells. Apparel, paper, chemicals, concrete, and metal products are manufactured.

Forrest City, city (1970 pop. 12,521), seat of St. Francis co., E central Ark., at the foot of Crowley's Ridge; inc. 1871. It is a rail and trade center in an area producing timber, peaches, and cotton.

For·syth (fôr'sīth'). **1.** County (1970 pop. 16,928), 218 sq mi (564.6 sq km), N Ga., bounded on the E by the Chattahoochee River; formed 1832; co. seat Cumming. In a piedmont timber and agricultural area yielding cotton, corn, hay, sweet potatoes, and poultry, it has an extensive food-processing industry. Clothing and petroleum products are also made. **2.** County (1970 pop. 215,118), 418 sq mi (1,082.6 sq km), N central N.C., in the Piedmont bounded on the W by the Yadkin River; formed 1849; co. seat Winston-Salem. Its agriculture includes tobacco, corn, hay, and dairy products. Food processing, granite quarrying, and lumbering are also important. Textiles, furniture, electronic equipment, and paper products are made.

Forsyth. 1. City (1970 pop. 3,736), seat of Monroe co., central Ga., NW of Macon, in a farm area; inc. 1822. It processes cotton and lumber. **2.** City (1970 pop. 803), seat of Taney co., S Mo., in the Ozarks NE of Branson. It is in an agricultural and resort area. **3.** City (1976 est. pop. 2,449), seat of Rosebud co., SE Mont., on the Yellowstone River NE of Billings; settled 1882, inc. 1908. It is located in a region of livestock, grain, and sugar-beet farms.

For·ta·le·za (fôr'tə-lā'zə), city (1975 est. pop. 1,109,837), capital of Ceará state, NE Brazil, a port on the Atlantic Ocean. The city is a commercial and industrial city and a processing and shipping center for the products of Brazil's interior (cotton, coffee, hides, carnauba wax, and oiticica oil). Textiles are manufactured, and the city is also known for its traditional handicrafts. Fortaleza was founded in 1609 and became a center of the great sugar plantations of northeast Brazil in colonial times. It was occupied by the Dutch from 1637 to 1654. The city grew after 1808, when the ports of northeast Brazil were opened to international trade.

Fort At·kin·son (ăt'kən-sən), city (1970 pop. 9,164), Jefferson co., S. central Wis., on the Rock River ESE of Madison; settled 1836, inc. as a village 1860, as a city 1878.

Fort Bend (bĕnd), county (1970 pop. 52,314), 867 sq mi (2,245.5 sq km), SE Texas, drained by the Brazos and San Bernard rivers; formed 1837; co. seat Richmond. Its agriculture includes cotton, grain sorghums, rice, corn, soybeans, pecans, fruit, truck crops, livestock, and poultry. Sugar refining is done, and large quantities of oil, natural-gas, and sulfur are produced.

Fort Ben·ning (bĕn' ĭng), U.S. army post, 189,000 acres (76,545 hectares), W Ga., S of Columbus; est. 1918. One of the largest army posts in the United States, it is the nation's largest infantry training center.

Fort Ben·ton (bĕn'tən), city (1970 pop. 1,863), seat of Chouteau co., N central Mont., on the Missouri River NE of Great Falls. Developing around a post founded by the American Fur Company in 1846, it was laid out in 1865 and incorporated in 1883.

Fort Bliss (blĭss), U.S. army post, 1,122,500 acres (454,613 hectares), W Texas, E of El Paso; est. 1849. Originally located near the only ice-free pass through the Rocky Mts., it guarded the U.S.-Mexican border and protected West-bound gold seekers from hostile Indians. The fort's location has changed several times as a result of flooding; its present site, on a mesa, was established in 1890.

Fort Bow·ie National Historic Site (bō'ē, bōō'ē): *see* National Parks and Monuments Table.

Fort Bragg (brăg), U.S. army base, 11,136 acres (4,510 hectares), E N.C., N of Fayetteville; est. 1918. Originally an artillery post, it is now the principal U.S. army airborne training center.

Fort Bridg·er (brĭj'ər), supply post, on Blacks Fork of the Green River, SW Wyo. Founded in 1843, it was an important station on the Oregon Trail. The Mormons held Fort Bridger from 1853 until 1857 but fled into Utah with the approach of Federal troops.

Fort Car·o·line (kăr'ə-līn, -lĭn), settlement near the mouth of the St. Johns River, NE Fla.; est. 1564 by French Huguenots. Angered by the French presence in Florida, a Spanish force attacked the fort in 1565, killed most of the colonists, and renamed the fort San Mateo. In retaliation, the French wiped out the garrison in 1568. Fort Caroline National Memorial includes a replica of the original fort.

Fort Clat·sop National Memorial (klăt'səp): *see* National Parks and Monuments Table.

Fort Col·lins (kŏl'ĭnz), city (1978 est. pop. 58,500), seat of Larimer co., N Colo., on the Cache la Poudre River, at the foot of the Rocky

Mts.; inc. as a city 1883. The area was settled in 1864 around a fortification built to protect a strategic trading post from Indians. The city is a trading, shipping, and processing center of a rich agricultural area raising grain, sugar beets, and livestock.

Fort Da·vis (dā′vəs), uninc. village (1970 pop. 850), seat of Jeff Davis co., W Texas, SE of El Paso in the Davis Mts.

Fort Davis National Historic Site: *see* National Parks and Monuments Table.

Fort Dear·born (dîr′bôrn′), U.S. army post on the Chicago River, NE Ill.; est. 1803. On Aug. 15, 1812, a contingent of troops, militia, women, and children from the fort were attacked by a large Indian force. More than half of the people were killed and most of those remaining were taken prisoner; the fort was destroyed. Fort Dearborn was rebuilt in 1816–17.

Fort-de-France (fôr-də-fräNs′), city (1967 pop. 96,943), capital of Martinique, French West Indies. It is a popular tourist resort and a free port. It was settled in 1762 by the French, who built Fort Royal by the strategically situated harbor. Yellow fever hampered its prosperity, however, and Fort-de-France did not gain importance until after 1902, when the city of St.-Pierre was destroyed by an eruption of Mont Pelée. Drainage of the swamps to control disease further stimulated Fort-de-France's growth.

Fort Dix (dĭks), U.S. army training center, 32,000 acres (12,960 hectares), central N.J., SE of Trenton; est. 1917. During World War II Fort Dix was the largest army training center in the country.

Fort Dodge (dŏj), city (1978 est. pop. 30,200), seat of Webster co., central Iowa, on the Des Moines River; settled c.1846. Fort Clarke, built on the site in 1850, was renamed Fort Dodge the following year, but was abandoned in 1853. The town was laid out in 1854 and incorporated in 1869. Agricultural processing plants and gypsum mills are in the city.

Fort Don·el·son (dŏn′əl-sən), Confederate fortification in the Civil War, on the Cumberland River at Dover, Tenn., commanding the river approach to Nashville. On Feb. 16, 1862, Gen. Ulysses S. Grant captured the fort, opening the way for the advance on Nashville. Fort Donelson National Military Park and National Cemetery are on the site.

Fort Du·quesne (do͞o-kān′, də-), at the junction of the Monongahela and Allegheny rivers, on the site of Pittsburgh, SW Pa. Because of its strategic location, it was a major objective in the last of the French and Indian Wars. The fort was begun by a group of Virginians in 1754. The French drove the Virginians away on Apr. 17, 1754, before George Washington's Va. militia arrived. Fort Duquesne was also the goal of an unsuccessful expedition under English Gen. Edward Braddock in 1755. On Nov. 24, 1758, the French abandoned their position without a fight to advancing British troops and retreated north after burning the fort. The English rebuilt it and renamed it Fort Pitt.

Fort Ed·ward (ĕd′wərd), village (1970 pop. 3,733), Washington co., E central N.Y., on the Hudson River at the confluence of the Champlain division of the Barge Canal and S of Hudson Falls; inc. 1849. The fort was built (1755) to guard the portage between the Hudson River and Lake Champlain; Gen. John Burgoyne occupied it briefly in 1777.

Fort E·rie (îr′ē), town (1976 pop. 24,031), S Ont., Canada, on the Niagara River, opposite Buffalo, N.Y. Fort Erie was built in 1764 and was taken from the British in the War of 1812 by American forces. In Aug., 1814, the Americans withstood successfully a siege by a superior British force but afterward blew up the stronghold and abandoned it. The modern town developed in the 20th cent.

For·tes·cue (fôr′tə-skyo͞o′), river, 340 mi (547.1 km) long, NW Western Australia, flowing intermittently NE and NW from a mountainous region SW of Nullagine to the Indian Ocean near Dampier Archipelago.

Fort Fish·er (fĭsh′ər), Confederate earthwork fortification, built in 1862 to guard the port of Wilmington, N.C. On Jan. 15, 1865, Union forces captured the fort in one of the last large battles of the Civil War. Fort Fisher is now a historic site.

Fort Fred·er·i·ca National Monument (frĕd′ə-rē′kə, frĕd-rē′-): *see* National Parks and Monuments Table.

Fort Gaines (gānz), city (1970 pop. 1,255), seat of Clay co., SW Ga., on a bluff overlooking the Chattahoochee River (here spanned by a cantilever bridge) and S of Columbus; inc. 1830. It is the trade center for a farm area.

Fort Gar·ry (gâr′ē), two trading posts of the Hudson's Bay Company, built on the present-day site of Winnipeg, Man., Canada, at the confluence of the Red and Assiniboine rivers. The first, Upper Fort Garry, was built in 1822 on the site of Fort Gibralter, a post of the North West Company from 1809 to 1816. Damaged by flood, it was replaced by Lower Fort Garry (1831–33) farther down on the Red River. Upper Fort Garry was rebuilt in 1835 and became the center of the Red River fur trade. Fort Garry National Historic Park contains a restoration of Lower Fort Garry.

Fort George G. Meade (mēd), U.S. army post, 13,500 acres (5,467.5 hectares), central Md., between Baltimore and Washington, D.C.; est. 1917 as a World War I induction center. The fort has missile units for the defense of Washington, D.C.

Forth (fôrth), river, c.60 mi (95 km) long, formed by streams that join near Aberfoyle, S central Scotland. It meanders generally eastward past Stirling to the Firth of Forth at Alloa. The Firth of Forth extends c.55 mi (90 km) east from Alloa to the North Sea, reaching widths up to 19 mi (30.6 km).

Fort Hall (hôl), trading post on the Snake River, near Pocatello, SE Idaho; est. 1834. In 1836 it was sold to the Hudson's Bay Company, which occupied the post until 1856. Fort Hall was the main stopping point on the Oregon Trail west of Fort Bridger. The ruins of Fort Hall were destroyed by floods in 1862 and 1864.

Fort Hen·ry (hĕn′rē), Confederate fortification on the Tennessee River, S of the Ky.-Tenn. line. It was the site of the first major Union victory of the Civil War (Feb. 6, 1862).

Fort Hood (ho͝od), U.S. army post, 209,000 acres (84,645 hectares), central Texas, near Killeen; est. 1942 on the site of old Fort Gates. It is the army's main armor training center.

Fort Jef·fer·son National Monument (jĕf′ər-sən): *see* National Parks and Monuments Table.

Fort Knox (nŏks), U.S. military reservation, 110,000 acres (44,550 hectares), Hardin and Meade counties, N Ky.; est. 1917 as a training camp in World War I. In the steel and concrete vaults of the U.S. Depository, built at Fort Knox in 1936 by the Dept. of the Treasury, the bulk of the nation's gold bullion is stored.

Fort-Lamy (fôr′lä-mē′): *see* Ndjamena, Chad.

Fort Lar·a·mie National Historic Site (lăr′ə-mē), 563 acres (228 hectares), SE Wyo.; est. 1938. Founded in 1834 as a fur-trading post, it was bought by the American Fur Company in 1836. In 1849 it became a U.S. army post, which later served as a major stopping place on the Overland Trail. The fort was garrisoned until 1890.

Fort Lar·ned National Historic Site (lär′nĭd): *see* National Parks and Monuments Table.

Fort Lau·der·dale (lô′dər-dāl′), city (1978 est. pop. 153,600), seat of Broward co., SE Fla., on the Atlantic coast; settled around a fort built (c.1837) in the Seminole War, inc. 1911. The city, located on New River and a navigable canal to Lake Okeechobee, is interwoven with more than 270 mi (434.4 km) of natural and artificial waterways. It has one of the largest marinas in the world and one of the most popular beaches in the state. Manufactures include electronic products, boats and yachts, and concrete, fiberglass, metal, and plastic items.

Fort Leav·en·worth (lĕv′ən-wûrth′), U.S. military post, 6,000 acres (2,430 hectares), on the Missouri River, NE Kansas, NW of Leavenworth; est. 1827 to protect travelers on the Santa Fe Trail. The oldest U.S. military prison (est. 1874) and the U.S. Army Command and General Staff College are at the fort.

Fort Lee (lē), borough (1978 est. pop. 32,600), Bergen co., NE N.J., on the Palisades overlooking the Hudson River; settled c.1700, inc. 1904. The fort built here by the Americans to command the Hudson during the Revolution was abandoned on Nov. 20, 1776, after Fort Washington, on the opposite shore, fell to the British. Fort Lee was an early center of the motion-picture industry.

Fort Leon·ard Wood (lĕn′ərd wo͝od′), U.S. army post, 71,000 acres (28,755 hectares), S central Mo.; est. 1940. It is one of the largest basic-training centers in the United States.

Fort Mad·i·son (măd′ĭ-sən), city (1970 pop. 13,996), seat of Lee co., SE Iowa, on the Mississippi River; inc. 1838. Fort Madison, a U.S. trading post, was established here in 1808 as the first fort west of the

Mississippi. The fort was burned by Indians and abandoned in 1813, and the area was resettled in 1833. Fort Madison is a river port and a rail, commercial, and industrial center in a rich agricultural area.

Fort Mc·Hen·ry (mək-hĕn′rē), former U.S. military post in Baltimore harbor; built 1794–1805. In the War of 1812 it was bombarded (Sept. 13–14, 1814) by a British fleet but resisted the attack. Its defense inspired Francis Scott Key to write "The Star-Spangled Banner." During the Civil War the fort was a Union prison camp. It was restored in 1933 and is now a national monument.

Fort Mims (mĭmz), temporary stockade near the confluence of the Tombigbee and Alabama rivers. It was the scene of a massacre (Aug. 30, 1813) in which a large force of Indians led by William Weatherford massacred over 500 whites.

Fort Mon·roe (mən-rō′), SE Va., commanding the entrance to Chesapeake Bay and Hampton Roads. The present fortress (80 acres/32.4 hectares) was built (1819–34) on the site of English fortifications erected in 1609 and 1727. Completely surrounded by a moat, the six-sided fort is the only one of its kind left in the United States. Fort Monroe was held by Union forces throughout the Civil War; Jefferson Davis, president of the Confederacy, was imprisoned here from 1865 until 1867.

Fort Mor·gan (môr′gən), city (1970 pop. 7,594), seat of Morgan co., NE Colo., NE of Denver on the South Platte River; settled on the site of an early trading post (1864); inc. 1887. It is the trading and processing center of a rich farm area yielding sugar beets, grain, dairy products, and livestock.

Fort Moul·trie (mōōl′trē), on Sullivans Island at the entrance to the harbor of Charleston, S.C.; originally called Fort Sullivan. Constructed by Col. William Moultrie, the fort was renamed for him after he repulsed a British naval attack in June, 1776, in one of the most decisive battles of the American Revolution. During the Civil War Confederates held the fort until the evacuation of Charleston in 1865. It was Charleston's chief harbor defense until 1947, when it was abandoned. Fort Moultrie is part of Fort Sumter National Monument.

Fort My·ers (mī′ərz), city (1978 est. pop. 36,600), seat of Lee co., SW Fla., on the Caloosahatchee River, near the Gulf of Mexico; founded 1850, inc. 1905. It has a large tourist industry and is a shipping point for citrus fruits, winter vegetables, flowers (especially gladioli), fish, shrimp, and cattle. Tourist attractions include Thomas A. Edison's estate and boat trips up the Caloosahatchee to the Everglades.

Fort Nas·sau (năs′ô). **1.** Built (1614) on Castle Island, in the Hudson River, S of Albany, N.Y. The fort served as a trading post for the Dutch until 1617, when it was destroyed by flood, and a new fort, Fort Orange, was built on the western bank of the river on the site of Albany. **2.** Built (1623) on the E bank of the Delaware River near Gloucester City, N.J. The Dutch soon abandoned the fort, but after Swedish colonization in the area, the Dutch reoccupied it and for a time united with the Swedes against the English.

Fort Ne·ces·si·ty (nə-sĕs′ĭ-tē), entrenched camp built during the French and Indian Wars by George Washington and his Va. militia at Great Meadows (near the present Uniontown, Pa.). After a large French force attacked Fort Necessity, Washington surrendered on July 4, 1754. Near Fort Necessity National Battlefield is the grave of British Gen. Edward Braddock.

Fort Ni·ag·a·ra (nī-ăg′rə, -ăg′ə-rə), post on the S shore of Lake Ontario, at the mouth of the Niagara River, NW N.Y. It was strategically located on the water route to fur-trapping areas. French explorer Robert LaSalle erected a blockhouse on the river in 1679; in 1726 a stone fort overlooking the river was completed. A British force captured Fort Niagara in 1759 during the French and Indian Wars. The British held the fort until 1796, when it was turned over to the United States by Jay's Treaty. During the War of 1812 the British captured Fort Niagara but returned it to the United States in 1815. The fort remained a U.S. military post until 1946.

Fort Payne (pān), city (1970 pop. 8,435), seat of De Kalb co., NE Ala., NE of Gadsden, in a farm area; settled and inc. 1889. Hosiery is made here, and fuller's earth is found in the region. The site of a Cherokee settlement is nearby.

Fort Pick·ens (pĭk′ənz), fortification on the W end of Santa Rosa Island at the entrance to Pensacola Bay, NW Fla. When Florida joined the Confederacy in Jan., 1861, Fort Barrancas on the mainland was evacuated and its garrison sent to Fort Pickens. Refusing to surrender, the fort was reinforced and repulsed a Confederate attack in Oct., 1861; it remained in Union hands throughout the war.

Fort Pierce (pîrs), city (1978 est. pop. 33,200), seat of St. Lucie co., SE Fla., on Indian River (a lagoon); settled 1860s, inc. 1901. With a good harbor and rail facilities, it is a distributing center for a cattle and farm area yielding citrus fruits and vegetables. Other industries are fishing and tourism. Fort Pierce grew around a fort built in 1838 as protection against the Seminole Indians.

Fort Pierre (pîr), city (1970 pop. 1,448), seat of Stanley co., central S.Dak., on the Missouri River opposite Pierre; founded 1817 as a fur-trading post. It is in a grain and livestock region. There are natural-gas wells nearby.

Fort Pil·low (pĭl′ō), fortification on the Mississippi River, N of Memphis, Tenn.; built by Confederate Gen. Gideon Pillow in 1862. Evacuated by the Confederates after the fall of Island No. 10 to the north, the fort was occupied by Union troops on June 6, 1862. Confederate Gen. Nathan Forrest stormed and captured Fort Pillow on Apr. 12, 1864, killing many Negro defenders. Often called the Fort Pillow Massacre, it became one of the greatest atrocity stories of the Civil War. Charged with ruthless killing, Forrest argued that the soldiers had been killed trying to escape; however, racial animosity on the part of his troops was undoubtedly a factor.

Fort Point National Historic Site (point): *see* National Parks and Monuments Table.

Fort Polk (pōk), U.S. army post, 147,856 acres (59,882 hectares), SW La.; est. 1941. It is a major army warm-weather training center and was used for Southeast Asian infantry assignments in the 1960s because of its similar climate and terrain.

Fort Pu·las·ki (pōō-lăs′kē, pə-), brick fortification on Cockspur Island, SE Ga., at the mouth of the Savannah River; built 1829–47. The fort was seized by Ga. troops during the Civil War in Jan., 1861, but fell to a Union force on Apr. 11, 1862, after a bombardment in which the Federals used rifled cannon for the first time in the war. Fort Pulaski National Monument was established in 1924.

Fort Ri·ley (rī′lē), U.S. military post, 5,760 acres (2,332.8 hectares), NE Kansas, on the Kansas River; est. 1852 to protect travelers on the Santa Fe Trail from Indian attacks. It was a cavalry post and school until 1917, when it became a reserve-officer training center.

Fort Saint John (sānt′ jŏn′), town (1976 pop. 8,947), NE British Columbia, Canada, on the Peace River and the Alaska Highway. A North West Company post established in 1805 is still operated by the Hudson's Bay Company.

Fort Sam Hous·ton (săm′ hyōō′stən), U.S. army base, 3,300 acres (1,336.5 hectares), S Texas, in San Antonio. San Antonio, long a military center, donated land in 1870 for the site of a permanent military post that was constructed from 1876 to 1890. Brooke Army Medical Center is located on the post.

Fort Scott (skŏt), city (1975 est. pop. 8,300), seat of Bourbon co., SE Kansas, near the Mo. border N of Pittsburg; inc. 1860. A military post was established here in 1842. Abandoned in 1855, the fort was rebuilt in the Civil War; some of its buildings are still standing. The city is a farm trade center, and varied products are manufactured.

Fort Scott Historic Area: *see* National Parks and Monuments Table.

Fort Sill (sĭl), U.S. military reservation, 95,000 acres (38,475 hectares), SW Okla.; est. 1869. Fort Sill underwent extensive construction in the early 1870s and became the base of operations for Indian campaigns and for the maintenance of law and order in southwest Okla. The reservation was revitalized by the establishment (1911) of the U.S. army's main field artillery training base.

Fort Smith (smĭth), city (1978 est. pop. 68,600), a seat of Sebastian co., NW Ark., at the Okla. line where the Arkansas and the Poteau rivers join; inc. 1842. It is the rail and trade center of a farm and livestock area and a major industrial hub. The city was founded as a military post in 1817 and was an important supply point during the 1848 gold rush.

Fort Stan·wix (stăn′wĭks′), colonial outpost on the site of Rome, N.Y., controlling a principal route from the Hudson River to Lake Ontario. Originally a French trading center, it was rebuilt by the English in 1758. The fort fell into disrepair until early in the American Revolution, when it was rebuilt by the patriots and called Fort Schuyler. Fort Stanwix is now a national monument.

Fort Stock·ton (stŏk′tən), city (1970 pop. 8,283), seat of Pecos co., W

Texas, on a plateau W of the Pecos River. It is on the site of an army post founded in 1859. Comanche Springs, outside the city, was once a camping ground for Indians and travelers bound (1849–50) for Calif. The area has cotton and truck farms, cattle ranches, and oil and natural-gas wells.

Fort Sum·ner (sŭm'nər), village (1970 pop. 1,615), seat of De Baca co., E N.Mex., NNE of Roswell. It is the trade and shipping center of an irrigated farm and livestock area. The grave of Billy the Kid and the ruins of Fort Sumner (built 1862) are nearby.

Fort Sum·ter (sŭm'tər), fortification, built 1829–60, on a shoal at the entrance to the harbor of Charleston, S.C.; scene of the opening engagement of the Civil War. On Apr. 12, 1861, after a 34-hour Confederate bombardment, the U.S. commander, Maj. Robert Anderson, surrendered the fort. In 1863 Union naval attacks on the fort were thoroughly repulsed. After Gen. W. T. Sherman forced the evacuation of Charleston, the U.S. flag was again raised over the fort by Anderson on Apr. 14, 1865. Fort Sumter became a national monument in 1948.

For·tune Bay (fôr'chən), arm of the Atlantic Ocean, c.80 mi (130 km) long, S N.F., Canada. The French islands of Miquelon and St. Pierre are at its mouth.

Fort Un·ion (yōōn'yən), trading post of the American Fur Company, erected in 1828 near the confluence of the Yellowstone and Missouri rivers, on the Mont.–N.Dak. line. When the U.S. army assumed control in 1867, Fort Union was torn down and Fort Buford, a military post, erected nearby. Fort Union Trading Post is a national historic site.

Fort Union National Monument: *see* National Parks and Monuments Table.

Fort Valley, city (1970 pop. 9,251), seat of Peach co., central Ga., ENE of Columbus and SW of Macon, in a peach-growing area; settled c.1836, inc. 1856.

Fort Van·cou·ver National Historic Site (văn-kōō'vər): *see* National Parks and Monuments Table.

Fort Wal·ton Beach (wôl'tən), city (1970 pop. 19,994), Okaloosa co., NW Fla., on the Gulf of Mexico; inc. 1941. It is a year-round resort, with beaches and freshwater and deep-sea fishing. Its manufactures include sophisticated electronic equipment and small boats. Eglin Air Force Base, on the city's outskirts, contributes significantly to the economy. The city grew around a fort constructed during the Seminole War (1835–42). Its main growth came after 1941, when it developed as a resort center and the air force base was expanded. Indian Temple Mound, a national historic landmark that includes a museum of Indian culture, is here.

Fort Wash·ing·ton (wăsh'ĭng-tən, wôsh'-) military post during the American Revolution, situated on the highest point of Manhattan Island, New York City, overlooking the Hudson River opposite Fort Lee, N.J. It was a hastily built earthwork with no water supply within its walls and no fortifications able to withstand a strong attack. It was, however, strategically located. On Nov. 16, 1776, the fort was attacked and captured by British Gen. William Howe.

Fort Wayne (wān) city (1978 est. pop. 184,800), seat of Allen co., NE Ind., where the St. Joseph and the St. Marys rivers join to form the Maumee River; inc. 1840. It is a major railroad and shipping point, a wholesale and distribution hub, and a manufacturing center, with large electronics and automotive industries. The Miami Indians had their chief town at this strategic water intersection before the French founded (shortly before 1680) a trading post. In 1697 the French built a fort here that remained under their control until 1760, when it was surrendered to the British. The fort was held briefly by Indians in Pontiac's Rebellion. Later, the Miami Indians were subdued by Anthony Wayne, who built (1794) the fort that bore his name. There was further Indian fighting in the War of 1812, but afterward the peaceful fur-trading center began to grow. Industrialization was spurred by the development of the Wabash and Erie Canal and the coming of the railroad (both in the mid-1800s). Of interest are The Landing, the restored main street of the city's original frontier settlement; the sunken gardens at Lakeside Park; and the burial place of John Chapman (Johnny Appleseed).

Fort Wil·liam Hen·ry (wĭl'yəm hĕn'rē) at the S end of Lake George, NE N.Y.; built in 1755 by British colonial leader Sir William Johnson. In 1757, during the French and Indian Wars, it was captured and destroyed by French Gen. Louis Montcalm. Fort William Henry was rebuilt in 1953 and is now a museum.

Fort Worth (wûrth), city (1970 pop. 393,476), seat of Tarrant co., N Texas, 30 mi (48 km) W of Dallas; settled 1843, inc. 1873. An army post was established on the site of Fort Worth in 1847, and after the Civil War the settlement became an Old West cow town. The first railroad (completed 1876) helped establish Fort Worth as a meat-packing and cattle-shipping point. Later in the 19th cent. wheat growing made the city an important center for milling and shipping grain. In 1919 oil was discovered west of Fort Worth, and large refineries and other oil and gas installations were built. Fort Worth is the seat of Texas Christian Univ. and Texas Wesleyan College. Points of interest include Tarrant County Convention Center (1968), Tarrant County Courthouse (1895), Fort Worth Art Center, Amon Carter Museum of Western Art, a museum of science and history, the Health Museum, Greer Island Nature Center, a botanical garden, a zoo, an aquarium, a planetarium, and Heritage Hall. Carswell Air Force Base is nearby.

Fort Yates (yāts), city (1970 pop. 1,153), seat of Sioux co., S N.Dak., on the Missouri River S of Bismarck. It is in a livestock, poultry, and grain area.

Fosse Way (fŏs), Roman road in England. It apparently ran from Exeter northeast past Bath, Cirencester, and Leicester to Lincoln. It intersected Watling Street.

Fos·sil (fŏs'əl), town (1972 est. pop. 533), seat of Wheeler co., N Oregon, S of Condon, in a livestock area. It was named for the prehistoric fossils discovered in the area.

Fossil Butte National Monument: *see* National Parks and Monuments Table.

Fos·ter (fô'stər), county (1970 pop. 4,832), 648 sq mi (1,678.3 sq km), E central N.Dak., in a prairie region drained by the James River; formed 1873; co. seat Carrington. Its agriculture includes dairy products, livestock, wheat, and poultry. There is some manufacturing in the county.

Fos·to·ri·a (fôs-tôr'ē-ə), city (1978 est. pop. 16,800), Hancock, Seneca, and Wood cos., NW Ohio; inc. 1854. A trade and shipping center for a livestock and farm area, the city has grain elevators, livestock markets, and a soybean-processing industry. Its manufactures include carbon products and automotive and electrical equipment.

Foth·er·in·ghay (fŏth'ər-ĭng-gā', fŏth'rĭng-), village, Northamptonshire, central England, on the Nene River. Fotheringhay Castle (12th cent.), now in ruins, was the birthplace of Richard III and the scene of the imprisonment and execution (1587) of Mary Queen of Scots.

Foun·tain (foun'tən), county (1970 pop. 18,257), 397 sq mi (1,028.2 sq km), W Ind., bounded on the W and NW by the Wabash River; formed 1825; co. seat Covington. It has agriculture (grain, soybeans, fruit, truck crops, poultry, livestock, and dairy products), bituminous-coal mines, and sand, gravel, and clay pits. Its manufactures include clothing, rubber, plastics, iron and steel, and metal products.

Fountain Valley, city (1978 est. pop. 54,800), Orange co., S Calif.; inc. 1957. Chiefly residential, Fountain Valley also has industries producing mobile homes, microelectronic components, aerospace fittings, and bathtubs.

Four Lakes, chain of canalized lakes in S Wis. Largest of the four is Mendota, c.6 mi (9.7 km) long, on which the Univ. of Wisconsin campus is located. Indian mounds are nearby.

Fou·ta Djal·lon or **Fu·ta Jal·lon** (both: fōō'tä jä-lôɴ'), highland region, c.30,000 sq mi (77,700 sq km), central Guinea, W Africa. Largely a rolling grassland (average alt. c.3,000 ft/915 m), the region is grazed by cattle of the Fulani. Since the 18th cent. it has been a stronghold of Islam.

Fowl·er (fou'lər), town (1970 pop. 2,643), seat of Benton co., NW Ind., NW of Lafayette, in a farm area; laid out 1871.

Fow·li·ang (fōō'lē-äng'): *see* Ching-te-chen, China.

Fox (fŏks). **1.** River, 176 mi (283.2 km) long, rising in S central Wis. and flowing SW, then NE through Lake Winnebago into Green Bay, an arm of Lake Michigan, at Green Bay, Wis. The river was a well-known route used by early explorers, missionaries, and fur traders to reach the Northwest and the Mississippi River system from the Great Lakes. A barge canal now links the Fox and Wisconsin rivers at Portage, forming a continuous waterway from Lake Michigan to the Mississippi River. **2.** River, c.185 mi (298 km) long, rising in SE

Wis. near Milwaukee and flowing SSW to the Illinois River at Ottawa, Ill.

Fox·bor·ough or **Fox·bor·o** (both: fŏks'bûr'ō), town (1970 pop. 14,218), Norfolk co., SE Mass.; settled 1704, inc. 1778. The chief industrial product is precision instruments. In the town is Schaefer Stadium, home of the New England Patriots football team. During the Revolutionary War cannons and cannonballs were manufactured in Foxborough.

Foxe Basin (fŏks), a widening of the waterway between Baffin Island and the Melville Peninsula, c.340 mi (550 km) long and c.225 mi (360 km) wide, E Franklin dist., Northwest Territories, Canada. The basin is shallow and is ice-clogged most of the year. Foxe Channel (c.200 mi/320 km long and c.90 mi/145 km wide) connects it with Hudson Bay and Hudson Strait.

Foyle (foil), river, c.10 mi (16 km) long, W Northern Ireland, flowing NE through Londonderry to Lough Foyle, a navigable inlet of the Atlantic Ocean c.15 mi (25 km) long. There are valuable freshwater fisheries on the Foyle.

Fra·ming·ham (frā'mĭng-hăm'), town (1978 est. pop. 66,200), Middlesex co., E Mass., on the Sudbury River between Worcester and Boston; settled 1650, inc. 1700. It has varied industries.

France (frăns, fräns), republic (1968 pop., excluding overseas departments and territories, 49,778,540), 211,207 sq mi (547,026 sq km), W Europe. The capital is Paris. France is bordered by the English Channel in the north, the Atlantic Ocean and the Bay of Biscay in the west, and the Mediterranean Sea in the south. The natural land frontiers are the Pyrenees, along the border of Spain, in the southwest; the Jura Mts. and the Alps, along the border of Switzerland and Italy, in the east and southeast; and the Rhine River, which is part of the border of West Germany, in the northeast.

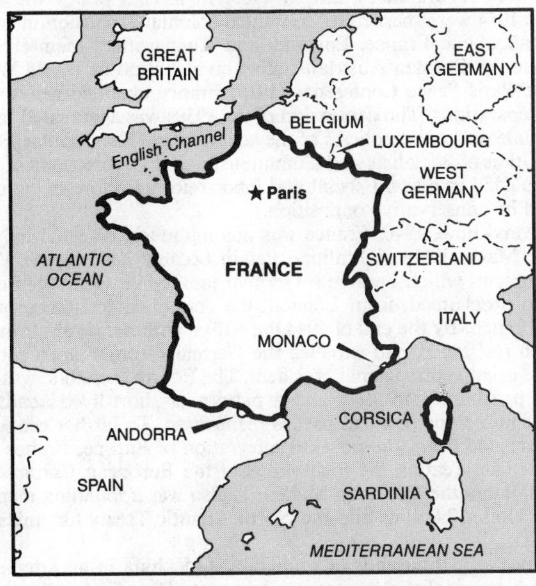

The old historic provinces of France, which were abolished by the Revolution, mirror the natural geographic regions. The heart of France north of the Loire River is the province of Île-de-France, which occupies the greater part of the Paris basin, a fertile depression drained by the Seine and Marne rivers. The basin is surrounded by the provinces of Champagne and Lorraine in the east; Artois, Picardy, French Flanders, and Normandy in the northeast and north; Brittany, Maine, and Anjou in the west; and Touraine, Orléanais, Nivernais, and Burgundy in the south. Further south are Berry and Bourbonnais. Further east, between the Vosges Mts. and the Rhine, is Alsace; south of Alsace, along the Jura, is Franche-Comté. South-central France is occupied by the rugged mountains of the Massif Central. It comprises the provinces of Marche, Limousin, Auvergne, and Lyonnais. East of the Rhône River, which divides the Massif Central from the Alps, are Savoy, Dauphiné, and Provence. The French Alps have some of the highest peaks in Europe, including Mont Blanc. The Rhône valley widens into a plain near its delta on the Mediterranean; part of the coast of Provence forms the celebrated French Riviera. Languedoc extends from the Cevennes Mts. to the Mediterranean coast west of the Rhône. Corsica lies off the Mediterranean coast. The southwestern part of France comprises the small Pyrenean provinces of Roussillon, Foix, Béarn, and French Navarre, and the vast provinces of Gascony and Guienne. The last two constitute the great Aquitainian plain, drained by the Garonne and Dordogne rivers that flow into the Bay of Biscay. The central section of the west coast, between the Gironde estuary and the Loire River, is occupied by the provinces of Saintonge, Angoumois, Aunis, and Poitou.

Metropolitan (European) France is divided into 96 departments:

NAME	CAPITAL	NAME	CAPITAL
Ain	Bourg	Jura	Lons-le-Saunier
Aisne	Laon	Landes	Mont-de-Marsan
Allier	Moulins	Loire	Saint-Étienne
Alpes-de-Haute Provence	Digne	Loire-Atlantique	Nantes
Alpes-Maritimes	Nice	Loiret	Orléans
Ardèche	Privas	Loir-et-Cher	Blois
Ardennes	Mézières	Lot	Cahors
Ariège	Foix	Lot-et-Garonne	Agen
Aube	Troyes	Lozère	Mende
Aude	Carcassonne	Maine-et-Loire	Angers
Aveyron	Rodez	Manche	Saint-Lô
Bas-Rhin	Strasbourg	Marne	Châlons-sur-Marne
Belfort, Territoire de	Belfort	Mayenne	Laval
Bouches-du-Rhône	Marseilles	Meurthe-et-Moselle	Nancy
Calvados	Caen	Meuse	Bar-le-Duc
Cantal	Aurillac	Morbihan	Vannes
Charente	Angoulême	Moselle	Metz
Charente-Maritime	La Rochelle	Nièvre	Nevers
Cher	Bourges	Nord	Lille
Corrèze	Tulle	Oise	Beauvais
Corse-du-Sud	Ajaccio	Orne	Aleçon
Côte-d'Or	Dijon	Paris, Ville-de	Paris
Côtes-du-Nord	Saint-Brieuc	Pas-de-Calais	Arras
Creuse	Guéret	Puy-de-Dôme	Clermont-Ferrand
Deux-Sèvres	Niort	Pyrénées-Atlantiques	Pau
Dordogne	Périgueux	Pyrénées-Orientales	Perpignan
Doubs	Besançon	Rhône	Lyons
Drôme	Valence	Saône-et-Loire	Mâcon
Essonne	Évry	Sarthe	Le Mans
Eure	Évreux	Savoie	Chambréry
Eure-et-Loir	Chartres	Seine-et-Marne	Melun
Finistère	Quimper	Seine-Maritime	Rouen
Gard	Nîmes	Seine-St. Denis	Bobigny
Gers	Auch	Somme	Amiens
Gironde	Bordeaux	Tarn	Albi
Haute-Corse	Bastia	Tarn-et-Garonne	Montauban
Haute-Garonne	Toulouse	Val-de-Marne	Créteil
Haute-Loire	Le Puy	Val-d'Oise	Pontoise
Haute-Marne	Chaumont	Var	Draguignan
Hautes-Alpes	Gap	Vaucluse	Avignon
Haute-Saône	Vesoul	Vendée	La Roche-sur-Yon
Haute-Savoie	Annecy	Viénne	Poitiers
Hautes-Pyrénées	Tabes	Vosges	Épinal
Haute-Vienne	Limoges	Yonne	Auxerre
Haut-Rhin	Colmar	Yvelines	Versailles
Hauts-de-Seine	Nanterre		
Hérault	Montpellier		
Ille-et-Vilaine	Rennes		
Indre	Châteauroux		
Indre-et-Loire	Tours		
Isère	Grenoble		

There are 4 overseas departments, which are legally part of the French Republic:

NAME	CAPITAL	NAME	CAPITAL
French Guiana	Cayenne	Martinique	Fort-de-France
Guadeloupe	Basse-Terre	Réunion	Saint-Denis

Economy. About two thirds of France's land is used for agricultural purposes. Three fifths of the value of total agricultural output derives from livestock. The mountain areas and northwest France are the livestock regions. The leading crops are sugar beets, wheat, corn, potatoes, and barley. Fruit growing is important in the south. The best-known vineyards are in Burgundy, Champagne, the Rhône and Loire valleys, and the Bordeaux region. France is among the world's largest producers of iron ore and bauxite, but has inadequate supplies of coal and petroleum. France's leading industries produce metals, chemicals, foods, and textiles. Tourism is an important industry, and Paris is famous for its luxury goods.

History. Ancient France, known to the Romans as Gaul, was inhabited largely by Celts, or Gauls, and by Basques in present Gascony. The conquest of Gaul by Julius Caesar (58–51 B.C.) became final with the defeat of the rebel Gallic chieftain Vercingetorix. Early in the course of the following five centuries of Roman rule Gaul accepted Latin speech and Roman law. Christianity, introduced in the 1st cent. A.D., spread rapidly. From the 3rd cent., however, the internal decline of the Roman Empire invited barbarian incursions. Among the Germanic tribes that descended upon fertile Gaul, the Visigoths, Franks, and Burgundii were the most important. In 486 the Franks routed Syagrius, last Roman governor of Gaul. Clovis I, who had made himself ruler of all the Franks, then defeated rival tribes, accepted Christianity (496), and founded the dynasty of the Merovingians. Throughout the 6th and 7th cent. Gaul was torn by fractricidal strife between the Merovingian kings of Neustria and of Austrasia, the two realms that ultimately emerged from Clovis' division of Gaul among his sons at his death. Then a more rigorous dynasty, the Carolingians, united (687) Austrasia with Neustria. In 732 the Carolingian Charles Martel decisively defeated the Saracens.

Charles Martel's grandson was Charlemagne. Crowned emperor of the West in 800, Charlemagne gave his subjects an efficient administration, created an admirable legal system, and labored for the rebirth of learning, piety, and the arts. But his heirs could not maintain the empire, which was redivided in 843 into three parts. Charles II was recognized as the ruler of the lands that are now France. Raids by Norsemen, beginning in the late 8th cent., contributed to the decline of royal authority, as did the increasing power of feudal lords.

When the Carolingian dynasty died out in France, the nobles chose (987) Hugh Capet as king. The early Capetians who were dukes of Francia, a small territory around Paris, gradually extended their domain. In the 11th cent. the towns had begun regaining population and wealth. A revival of commerce and theological learning reached its height in the 13th cent. and was aided by the leading role that France played in the Crusades. French courtly poetry and manners became European models.

The fact that the Norman English kings were also French nobles, holding or claiming vast fiefs in France, brought the two nations into centuries of conflict. Philip II of France (1180-1223) soundly defeated the Norman kings, extended the royal domain southward, and rebuilt Paris. Louis IX (1226-70) organized an efficient civil and judicial system. Under Philip IV (1285-1314), the clergy was taxed and the wealthy Knights Templars were destroyed; papal objections to these moves led to the Babylonian Captivity (1309-77) of the popes.

The succession of Philip VI (1328-50) was contested by Edward III of England, who in 1337 proclaimed himself king of France. Thus began the series of wars and truces known as the Hundred Years' War (1337-1453). In 1415 Henry V of England revived the English claim, renewed the war, and crushed the French at Agincourt. In 1428 the English besieged the key city of Orléans. At this hour appeared Joan of Arc, who helped relieve Orléans and in 1429 stood by as Charles VII was crowned at Rheims. In 1453 the English lost their last hold on French soil outside Calais.

It was left for Louis XI (1461-83) to destroy the power of the last great feudal lords and to incorporate into the royal domain almost all of present France. During the 11th cent. the Italian Renaissance spread into France, despite prolonged warfare with the house of Hapsburg, followed by ferocious civil wars between Catholics and Protestants. Henry IV (1589-1610) granted religious toleration to Protestants and made France peaceful and prosperous again. Under his successor, Louis XIII (1610-43), and in the minority of Louis XIV, two great statesmen successively guided the kingdom—Cardinal Richelieu and Cardinal Mazarin. They led France to victory in the Thirty Years' War (1618-48) against the Hapsburg powers, Austria and Spain. Louis XIV (1643-1715) made France the first power in Europe and his court at Versailles the cynosure of Europe. But his many wars undermined French finances. After successful wars against Spain (1667-68) and Holland (1672-78), the War of the Grand Alliance (1688-97) against Louis XIV began to turn the tide; the War of the Spanish Succession (1701-14) marked the end of French expansion in Europe.

Louis XV (1715-74) inherited a France burdened by aristocratic privilege, inequitable taxes, and monopolistic commerce. Rural overpopulation outstripped the stagnant agricultural productivity. The ever expanding bourgeoisie were finding the remnants of feudal dues, services, and other customs increasingly intolerable. Seeking to avenge its defeat by Britain in the Seven Years' War (1756-63), France supported the American Revolution (1775-83). Financially the war was a disaster despite the reforms of minister of finance Jacques Necker. Amid a deepening political crisis, the States-General were convoked in 1789 and transformed into a National Assembly. A constitutional monarchy was created (1791). War with much of Europe began, accompanied by violence and the growth of radical factions in France (1792), and the king and queen were beheaded (1793). Robespierre presided over the Reign of Terror (1793-95) until his own execution, and a reaction ushered in the Directory (1795-99), terminated by Napoleon Bonaparte's coup d'état. Napoleon made himself emperor (1804) and led his armies as far as Moscow. After his defeat at Waterloo (1815) virtually nothing remained for France from the Napoleonic conquests.

The French Revolution and Napoleon established a uniform, modern administrative system, gave land tenure to the peasants, and left to the bourgeoisie a political heritage that they quickly reclaimed. King Louis XVIII (1815-24) granted a moderately liberal charter but his successor, Charles X (1824-30), was the champion of the ultraroyalists. Charles's efforts to restore absolutism led to the July Revolution of 1830, which enthroned Louis Philippe. The increasingly autocratic regime of the "citizen king" fell in the Revolution of 1848. In Dec., 1848, Louis Napoleon Bonaparte, nephew of Napoleon I, was elected president of the Second Republic. In 1852, by a coup d'état, he extended his term and then proclaimed himself emperor as Napoleon III. The Second Empire was a period of colonial expansion (in Senegal and Indochina) and of material prosperity, but it ended disastrously in the Franco-Prussian War (1870-71), in which Alsace and Lorraine were lost to Germany until 1918. After the bloody suppression of the Commune of Paris (1871) by a right-wing provisional government, Marshal MacMahon was elected president of the Third Republic (1873).

The stability of France was once more shaken by the Dreyfus Affair (begun 1894), which discredited monarchists and reactionaries and brought anticlerical, moderate leftists to power. Church and state were separated by law in 1905. In foreign policy the years before 1914 were marked by continued colonial expansion in Africa and Indochina. France, England, and Russia allied themselves to balance the German-Austrian-Italian combination in World War I. At the Paris Peace Conference (1919) France obtained heavy German reparations. The depression of the 1930s was aggravated by the immobile economic policies of the government. The Popular Front, a coalition of Socialists and Communists, won the elections of 1936 and enacted important social and labor reforms before being overturned by conservative opposition.

In May-June, 1940, France was ignominiously defeated by Germany. Marshal Henri Philippe Pétain became head of the Vichy government, which became a German tool, while Gen. Charles de Gaulle proclaimed, from London, the continued resistance of the Free French. By the end of 1944 the Allies, with heroic aid from the French resistance, had expelled the Germans from France and De Gaulle became provisional president. The Fourth Republic was officially proclaimed in 1946 and a pattern of short-lived coalitions among numerous political parties reappeared. To further economic recovery and begin the political integration of Europe, France participated in creating the institutions of the European Community, most notably the Common Market. It also was a founding member of the United Nations and the North Atlantic Treaty Organization (NATO).

France sent thousands of soldiers to Indochina in an attempt to defeat a nationalist-communist movement. The effort collapsed in 1954. A similar war for independence in Algeria destroyed the Fourth Republic. After a right-wing French military coup in Algeria (1958), De Gaulle was invited back to power. He established the Fifth Republic and became its first president. Algerian independence was negotiated. France became a nuclear power (1960). In 1966 De Gaulle withdrew French forces from NATO and forced all U.S. and NATO forces to leave France.

In the spring of 1968 widespread student demonstrations against France's obsolete educational system were joined by striking workers and farmers. De Gaulle won a great electoral victory (June, 1968) but resigned in Apr., 1969. Georges Pompidou was elected president in June and served until his death in 1974. He was succeeded by Valéry Giscard d'Estaing, who restored closer relations with NATO and the United States.

Government. France is governed under the 1958 constitution, amended in 1962, which established the Fifth French Republic. It provides for a strong president, directly elected for a seven-year term. A premier and a cabinet, appointed by the president, are responsible to the national assembly, but are subordinate to the president. Parliament consists of the national assembly and the senate. Deputies to the assembly are elected for five-year terms from single-member districts. The president may dissolve the assembly. Senators are elected for nine-year terms from each department by an electoral college composed of the deputies, the district council members, and the municipal council members from that department.

Franche-Com·té (fräNsh'kôN-tā'), region and former province, E France. The Jura Mts. form the region's eastern border with Switzerland; the Vosges Mts. are in the north. Franche-Comté is largely an agricultural region and has a large dairy industry. Livestock is raised in the Jura district, where there are dense pine forests and extensive grazing lands. Clocks, watches, machines, and plastics are the leading industrial products.

The region was occupied by a Celtic tribe in the 4th cent. B.C. and was conquered by Julius Caesar (52 B.C.). Overrun by the Burgundians (5th cent.), it was included in the First Kingdom of Burgundy and was annexed by the Franks in 534. The territory was united in the 9th cent. as the Free County of Burgundy, or Franche-Comté. It passed to the Holy Roman Empire in 1034; but the allegiance was tenuous, and for six and a half centuries Franche-Comté was perpetually invaded and contested by France, Germany, Burgundy, Switzerland, and Spain. Philip the Bold, duke of Burgundy, acquired Franche-Comté through his marriage to Margaret of Flanders in 1369. After the defeat and death of Charles the Bold (1477), the region passed to Archduke Maximilian of Austria (later Emperor Maximilian I), who in turn gave it to his son Philip I of Spain. At the end of Charles V's reign (1556), Franche-Comté became a possession of the Spanish Hapsburgs. Although some of the region's fortified towns were occupied by France during the Wars of Religion (16th cent.), peace and prosperity continued until the Thirty Years' War (1618–48), when the region was ravaged by both Catholics and Protestants. Louis XIV conquered Franche-Comté in 1668 and again in 1674. After 1676 Franche-Comté became an integral part of France.

Fran·cis·town (frăn'sĭs-toun', frăn'-), town (1971 pop. 18,613), E Botswana; founded 1870. It is the commercial and administrative center for a farming and ranching region. Gold is mined nearby.

Fran·co·ni·a (frăng-kō'nē-ə, -nyə, frăn-), historic region and one of the five basic or stem duchies of medieval Germany, S West Germany. The region was included in the Frankish kingdom of Austrasia, becoming in the 9th cent. a duchy and the center of the East Frankish kingdom. It stretched from the western bank of the Rhine eastward along both banks of the Main River. King Otto I seized the duchy in 939 and partitioned it. Two nominal duchies—Western or Rhenish Franconia and Eastern Franconia—emerged. Rhenish Franconia broke up into the free cities of Frankfurt and Worms, the ecclesiastical states of Mainz and Speyer, the Rhenish Palatinate, the landgraviate of Hesse, and other territories. Eastern Franconia came increasingly under the control of the bishops of Würzburg, who were given legal title by Emperor Frederick I in 1168. The division (16th cent.) of the Holy Roman Empire into circles resulted in the creation of the Franconian circle, which included the bishoprics of Würzburg and Bayreuth, the free imperial city of Nürnberg, and the margraviates of Ansbach and Bayreuth. Most of Eastern Franconia passed to Bavaria between 1803 and 1815.

Franconia Mountains, range in the White Mts., N N.H., rising to 5,249 ft (1,600.9 m) at Mt. Lafayette; part of White Mts. National Forest. Franconia Notch, a scenic, narrow pass (6 mi/9.7 km long), is west of the range. Overlooking the pass is the Old Man of the Mountain, jutting cliffs that form the Great Stone Face.

Frank·fort (frăngk'fərt). **1.** City (1970 pop. 14,956), seat of Clinton co., W central Ind.; laid out in 1830. It is a trade and processing center for a rich farm and livestock region. It has railroad shops, a food-packing plant, and factories making electronic equipment, candy, and automotive parts. **2.** City (1978 est. pop. 23,200), state capital and seat of Franklin co., N central Ky., on both sides of the Kentucky River, in the heart of the Bluegrass; inc. 1796. It is the trade and shipping center for an area yielding tobacco, livestock, and limestone. Among its products are whiskey, metal items, automobile parts, shoes, and wearing apparel. Daniel Boone reached the site in 1770. The city was organized (1786) by the Va. legislature on lands owned by Gen. James Wilkinson and was selected as the capital in 1792. Many old homes and buildings have been preserved. Of interest are the present-day capitol (1909–10), with a giant floral clock in its plaza; the old state house (1827–30), which houses the state historical society; Liberty Hall (1796); the Corner of Celebrities, where the homes of 32 nationally prominent men are in close proximity; and the old cemetery with the graves of Daniel and Rebecca Boone. Kentucky State Univ. is here.

Frank·furt (frängk'fo͝ort') or **Frankfurt am Main** (äm mīn'), city (1974 est. pop. 652,037), Hesse, central West Germany, a port on the Main River. The city is an industrial, commercial, and financial center. Manufactures include chemical and pharmaceutical products, machinery, electrical equipment, leather goods, clothing, printed materials, and motor vehicles. Frankfurt is the site of major international trade fairs, including an annual book fair.

A Roman town founded in the 1st cent. A.D., Frankfurt became (8th cent.) a royal residence under Charlemagne. After the Treaty of Verdun (843) it was briefly the capital of the kingdom of the Eastern Franks. Frankfurt was designated in the Golden Bull (1356) of Emperor Charles IV as the seat of the imperial elections. It was made a free imperial city in 1372. After the emperors ceased to be crowned by the popes, the coronation ceremonies took place (1562–1792) at Frankfurt.

Frankfurt was occupied many times in the wars of the 17th and 18th cent. After the dissolution (1806) of the Holy Roman Empire, Frankfurt was included in the ecclesiastic principality of Regensburg and Aschaffenburg. The principality was converted in 1810 into the grand duchy of Frankfurt. The Congress of Vienna (1814–15) restored Frankfurt to the status of a free city and made it the seat of the diet of the German Confederation. The Frankfurt Parliament, the first German national assembly, met here in 1848–49. After the Austro-Prussian War of 1866, Frankfurt was annexed by Prussia. In 1871 the Treaty of Frankfurt, which ended the Franco-Prussian War, was signed here.

The city was heavily damaged in World War II, but after 1945 many of its historic landmarks were restored and numerous modern structures were built. Points of interest include the city hall, or *Römer* (begun in the 15th cent.), the Gothic Church of St. Bartholomew (13th–15th cent.), the house (now a museum) in which Goethe was born (1749), and the Städel art museum (founded 1816). Frankfurt is the seat of a university (opened 1914) and of the West German central bank and national library.

Frank·furt-an-der-O·der (frăngk'fo͝ort'-än-dər-ō'dər), city (1974 est. pop. 70,817), capital of Frankfurt dist., E East Germany, a port on the Oder River, at the Polish border. It is an industrial center, agricultural market, and rail junction. Manufactures include machinery, wood products, shoes, transistors, chemicals, and food products (notably frankfurter sausages). Lignite is mined nearby. Frankfurt was chartered in 1253. It joined the Hanseatic League in the 14th cent. and became an important commercial center. Frankfurt was frequently besieged, notably in 1631 (during the Thirty Years' War), when it was stormed and sacked by the Swedes under Gustavus II. The city was severely damaged in World War II.

Frank·lin (frăngk'lĭn), district (1971 pop. 7,747), 549,253 sq mi (1,422,565 sq km), Northwest Territories, N Canada. The district was created in 1895 and named for the British explorer Sir John Franklin. It comprises the Boothia and Melville peninsulas and the islands of the Canadian Arctic Archipelago.

Franklin. 1. County (1970 pop. 23,933), 644 sq mi (1,668 sq km), NW Ala., bordering on Miss. in the W and crossed N to S by the Fall Line; formed 1818; co. seat Russellville. It has agriculture (cotton and corn), deposits of iron, coal, limestone, and bauxite, and food-processing and lumbering industries. Clothing, wood, stone, and metal products, and automobile parts are manufactured. **2.** County (1970 pop. 11,301), 615 sq mi (1,592.8 sq km), NW Ark., in the Ozarks, intersected by the Arkansas River and drained by the Mulberry River; formed 1837; co. seats Ozark and Charleston. In an agricultural area yielding corn, cotton, peanuts, livestock, and poultry, it has coal mines, oil and natural-gas fields, and stands of timber. Clothing, furniture, and concrete are produced. Part of Ozark National Forest is here. **3.** County (1970 pop. 7,065), 536 sq mi (1,388.2 sq km), NW Fla., bounded on the S by the Gulf of Mexico, on the W by the Apalachicola River, on the E by the Ochlockonee River; formed 1832; co. seat Apalachicola. In a lowland, partly swampy area, it includes St. Vincent, St. George, and Dog islands, enclosing St. Vincent Sound, Apalachicola Bay, and St. George Sound. Fishing, forestry, and cattle raising are important to its economy. **4.** County (1970 pop. 12,784), 269 sq mi (696.7 sq km), NE Ga., in a piedmont area bordering on S.C. and drained by the Broad River; formed 1784; co. seat Carnesville. Cotton, corn, hay, and sweet potatoes are grown. Its manufactures include textiles, clothing, concrete, and household appliances. **5.** County (1970 pop. 7,373), 664 sq mi (1,719.8 sq km), SE Idaho, bounded on the S by the Utah border and drained by the Bear River; formed 1913; co. seat Preston. Its agriculture includes sugar beets, wheat, alfalfa, and dairy products. Lumbering and some manufacturing are done. Part of Cache National Forest is in the county. **6.** County (1970 pop. 38,329), 434 sq mi (1,124.1 sq km), S Ill., bounded on the NW by the Little Muddy River and drained by the Big Muddy River; formed 1818; co. seat Benton. Corn, wheat, fruit, dairy products, poultry, and livestock are produced. It also has lignite and bituminous-coal mines and oil and natural-gas fields. Flour, machinery, clothing, and leather goods are made. **7.** County (1970 pop. 16,943), 394 sq mi (1,020.5 sq km), E Ind., bounded on the E by the Ohio border and drained by the Whitewater River; formed 1810; co. seat Brookville. It has agriculture (tobacco, grain, and livestock) and factories producing petroleum, coal, and rubber products. **8.** County (1970 pop. 13,255), 586 sq mi (1,517.7 sq km) N central Iowa, in a rolling prairie region; formed 1851; co. seat Hampton. It is in an agricultural area yielding corn, oats, soybeans, cattle, hogs, and poultry and has a food-processing industry. Some quarrying is done. Clay products, farm machinery, and household appliances are among its manufactures. **9.** County (1970 pop. 20,007), 577 sq mi (1,494.4 sq km), E Kansas, in a rolling prairie region drained by the Marais des Cygnes River; formed 1855; co. seat Ottawa. Its diversified agriculture depends primarily on corn and livestock. There are oil and natural-gas fields and plants manufacturing clothing, wood products, and farm machinery. **10.** County (1970 pop. 34,481), 211 sq mi (546.5 sq km), N central Ky., in a gently rolling upland bluegrass area drained by the Kentucky River; formed 1794; co. seat Frankfort. In an agricultural region yielding burley tobacco, grain, dairy products, cattle, hogs, and sheep, it has a food-processing industry, stone quarries, and lead and zinc deposits. Metal products are made. **11.** Parish (1970 pop. 23,946), 648 sq mi (1,678.3 sq km), NE La., bounded on the E by Bayou Macon, on the W by the Boeuf River; formed 1843; parish seat Winnsboro. Its agriculture includes cotton, corn, hay, soybeans, and sweet potatoes. It also has oil and natural-gas fields and lumbering and food-processing industries. Clothing and wood products are manufactured. **12.** County (1970 pop. 22,444), 1,709 sq mi (4,426.3 sq km), W Maine, bordering in the N on Que., Canada, and drained by the Sandy River; formed 1838; co. seat Farmington. Lumbering and agriculture (chiefly apples and dairy products) are important. Shoes, wood products, and paper are manufactured. The county has many recreational areas, including Sugarloaf and Saddleback ski resorts. **13.** County (1970 pop. 58,210), 708 sq mi (1,833.7 sq km), NW Mass., in a rural area bisected by the Connecticut River and drained by the Deerfield and Millers rivers; formed 1811; co. seat Greenfield. It is primarily agricultural, with important ski and tourist areas. **14.** County (1970 pop. 8,011), 568 sq mi (1,471.1 sq km), SW Miss., drained by Homochitto River; formed 1809; co. seat Meadville. It has agriculture (cotton and corn), pine and hardwood timber, and oil fields. Clothing is manufactured in the county. **15.** County (1970 pop. 55,127), 932 sq mi (2,413.9 sq km), E Mo., in the Ozarks, bounded on the N by the Missouri River and drained by the Meramec River; formed 1818; co. seat Union. Its agriculture includes wheat, corn, oats, hay, and livestock. It also has extensive mineral deposits, including copper, lead, zinc, barite, coal, silica, fire clay, and limestone. Shoes, wood and metal products, and textiles are among its manufactures. **16.** County (1970 pop. 4,566), 578 sq mi (1,497 sq km), S Nebr., bounded on the S by Kansas and drained by the Republican River; formed 1867; co. seat Franklin. It is in an agricultural area yielding corn, milo, wheat, cattle, and swine. Some manufacturing is done. **17.** County (1970 pop. 43,931), 1,674 sq mi (4,335.7 sq km), NE N.Y., mostly in the Adirondacks, bounded on the N by the Que., Canada, border and drained by numerous rivers; formed 1808; co. seat Malone. Primarily a dairy region, it has some agriculture (potatoes, hay, hops, fruit, truck crops, and poultry) and industry (clothing, wood products, and leather goods). Tourism is also important; the county has a number of resort lakes and many fine recreational areas. Part of St. Regis Indian Reservation is here. **18.** County (1970 pop. 26,820), 494 sq mi (1,279.5 sq km), N central N.C., in a coastal plain area drained by the Tar River; formed 1779; co. seat Louisburg. It is in an agricultural area yielding tobacco, cotton, corn, soybeans, and livestock. Sawmilling and textile manufacturing are done. **19.** County (1970 pop. 833,249), 538 sq mi (1,393.4 sq km), central Ohio, intersected by the Scioto River and drained by Alum, Darby, and Big Walnut creeks; formed 1803; co. seat Columbus. Its agriculture includes grain, fruit, truck crops, dairy products, and livestock. It also has oil and gas wells, limestone quarries, sand and gravel pits, and lumbering and food-processing industries. Highly industrialized, it produces clothing, shoes, furniture, paper, chemicals, plastics, glass, steel, aluminum, and various metal products. **20.** County (1970 pop. 100,833), 754 sq mi (1,952.9 sq km), S Pa., in a mountainous area bordered on the S by Md. and drained by Conococheague Creek; formed 1784; co. seat Chambersburg. It has agriculture (apples, peaches, and dairy products), stands of timber, and limestone deposits. Machinery, metal products, clothing, furniture, paper, shoes, and leather goods are manufactured. **21.** County (1970 pop. 27,289), 553 sq mi (1,432.3 sq km), S Tenn., partially in the Cumberlands, bounded on the S by the Ala. border and drained by the Elk River; formed 1807; co. seat Winchester. In an agricultural area yielding cotton, corn, hay, potatoes, dairy products, and livestock, it has coal deposits, a lumbering industry, and some manufacturing (primarily clothing). **22.** County (1970 pop. 5,291), 293 sq mi (758.9 sq km), NE Texas, bounded on the N by the Sulphur River and drained by Bayou Cypress; formed 1875; co. seat Mount Vernon. It is mainly agricultural, yielding grain sorghums, cotton, corn, sweet potatoes, fruit, truck crops, dairy products, cattle, hogs, and horses. There are also oil and natural-gas wells. **23.** County (1970 pop. 31,282), 659 sq mi (1,706.8 sq km), NW Vt., bounded on the N by the Que., Canada, border, on the W by Lake Champlain; co. seat St. Albans. In a mountainous resort area, it produces paper, wood, and dairy products, machinery, and maple sugar. It also has deposits of limestone and granite. The Green Mts. are in the east. **24.** County (1970 pop. 28,163), 716 sq mi (1,854.4 sq km), SW central Va., partly in the Piedmont and the Blue Ridge, bounded on the NE by the Roanoke River and drained by the Blackwater River; formed 1785; co. seat Rocky Mount. It has agriculture (tobacco, wheat, and corn), a lumbering industry, and mica mines. Clothing, wood products, furniture, chemicals, and stone products are manufactured. **25.** County (1970 pop. 25,816), 1,253 sq mi (3,245.3 sq km), SE Wash., in the Columbia basin and watered by the Columbia and Snake rivers; formed 1883; co. seat Pasco. In a dryland and irrigated agricultural area yielding wheat, alfalfa, fruit, truck crops, and livestock, it has a food-processing industry and manufactures metal products.

Franklin. **1.** City (1970 pop. 749), seat of Heard co., W Ga., on the Chattahoochee River NNW of La Grange. Tire cords are manufactured. **2.** City (1970 pop. 11,477), seat of Johnson co., S central Ind., inc. 1823. It is a farm trade center. Manufactures include auto parts, electrical components, and copper panels. Franklin College of Indiana is here. **3.** City (1970 pop. 6,553), seat of Simpson co., S central Ky., SSW of Bowling Green near the Tenn. border; founded 1820. It is a trade center and shipping point for an area producing timber, limestone, and farm products (especially strawberries). Clothing is manufactured. **4.** Town (1970 pop. 9,325), seat of St. Mary parish, S La., on Bayou Teche SE of New Iberia, in an oil and sugar-cane region. Carbon black is manufactured. Oaklawn Manor, considered one of the most beautiful ante-bellum houses in the state, is nearby. **5.** Town (1970 pop. 17,830), Norfolk co., SE Mass., near the R.I. line; settled 1660, set off from Wrentham and inc. 1778. Foundry products and rubber goods are manufactured. A memorial marks the birthplace of the educator Horace Mann. **6.** City (1970 pop. 1,193) seat of Franklin co., S central Nebr., on the Republican River SSE of Kearney; inc. 1879. **7.** City (1970 pop. 7,292), Merrimack co., S central N.H., at the junction of the Winnipesaukee and Pemigewasset rivers N of Concord; settled 1764, inc. as a town 1828, as a city 1895. Textiles are among its products. Daniel Webster was born just southwest of the city, and his birthplace is preserved. **8.** Town (1970 pop. 2,336), seat of Macon co., W N.C., on the Little Tennessee River SW of Asheville. It is a mountain resort in a timber, mining (mica and kaolin), and farming area. Nantahala National Forest is nearby. **9.** City (1970 pop. 10,075), Warren co., SW Ohio, on the Great Miami River, in a farm area; inc. 1813. Paper products are manufactured in the city. It was a flourishing river port in the mid-19th cent. **10.** City (1970 pop. 8,629), seat of Venango co., NW Pa., on the Allegheny River SW of Oil City; laid out 1795, inc. 1868. Its manufactures include machinery, steel products, plastics, and oil-well equipment. Dairying is done. Franklin was founded on the site of an Indian town, and several forts were built here during the French and Indian Wars. **11.** Town (1970 pop. 9,497), seat of Williamson co., central Tenn., on the Harpeth River SSW of Nashville; inc. 1815. The surrounding fertile farm area, settled before 1800, has phosphate mines. Franklin was the scene of much fighting in the

Civil War. **12.** Town (1970 pop. 1,063), seat of Robertson co., E central Texas, SE of Waco, in a farm area; settled 1880, inc. 1912. **13.** Independent city (1970 pop. 6,880), SE Va., on the Blackwater River W of Suffolk, in a lumbering area. Paper is manufactured. **14.** Town (1970 pop. 695), seat of Pendleton co., W.Va., in the Eastern Panhandle SE of Elkins near the South Branch of the Potomac River. It is a summer resort in a lumber-milling and dairying region. **15.** City (1970 pop. 12,247), Milwaukee co., W central Wis., a residential suburb of Milwaukee; inc. 1956.

Franklin, State of, government (1784–88) formed by the inhabitants of Washington, Sullivan, and Greene cos. in present-day E Tenn. after N.C. ceded (June, 1784) its W lands to the United States.

Franklin Park, village (1970 pop. 20,348), Cook co., NE Ill., a residential suburb of Chicago; inc. 1892.

Franklin Square, uninc. residential city (1970 pop. 32,600), Nassau co., SE N.Y., on Long Island.

Frank·lin·ton (frăngk'lĭn-tən), town (1970 pop. 3,562), seat of Washington parish, SE La., WNW of Bogalusa; laid out 1821, chartered 1861. It is the trade center of a cotton, sugar, and farm region.

Franz Jo·sef Land (frănts jō'səf, fränts yō'zĕf), archipelago, c.8,000 sq mi (20,720 sq km), in the Arctic Ocean N of Novaya Zemlya, USSR. It consists of 85 islands of volcanic origin. Government observation stations (erected 1929) and settlements are on Hooker and Rudolf islands. Some 90% of Franz Josef Land is covered by ice interspersed with poor lichen vegetation. It was discovered in 1873 by an Austrian expedition. In 1926 the USSR claimed the archipelago as national territory.

Fra·sca·ti (frä-skä'tē), town (1976 est. pop. 19,004), in Latium, central Italy. Beautifully situated in the Alban Hills near the site of ancient Tusculum, it has been a popular summer resort since Roman times. It is famous for its white wine and its villas.

Fra·ser (frā'zər, -zhər), city (1970 pop. 11,868), Macomb co., SE Mich., a suburb of Detroit; inc. as a village 1894, as a city 1957. Automated machine tools and steel products are manufactured.

Fraser, chief river of British Columbia, Canada, c.850 mi (1,370 km) long, rising in the Rocky Mts., at Yellowhead Pass, near the British Columbia–Alta. line and flowing NW through the Rocky Mt. Trench to Prince George, then S and W to the Strait of Georgia at Vancouver. The Fraser River canyon, which begins at Yale, is noted for its scenery; its mountain walls rise more than 3,000 ft (915 m). The river contains the chief spawning grounds in North America for the Pacific salmon. Logging is important along the upper course. The Fraser delta is the most fertile agricultural region of British Columbia; dairying and truck farming are important.

The Fraser River was discovered by Sir Alexander Mackenzie, the Canadian explorer, who followed its upper course on his expedition (1793) to the Pacific Ocean and takes its name from Simon Fraser, the Canadian explorer and fur trader, who followed (1808) the river to its mouth, establishing fur-trading posts along the way. With the discovery of gold (1859) in the Cariboo dist., on the river's upper reaches, the government built a road to serve the valley and settlement of the region followed.

Fra·ser·burgh (frā'zər-bər-ə), burgh (1971 pop. 10,605), Grampian region, NE Scotland, on the North Sea. It is one of Scotland's leading fishing ports.

Frau·en·feld (frou'ən-fĕlt'), city (1977 est. pop. 18,200), capital of Thurgau canton, NE Switzerland, on the Murg River. Although it has aluminum, food, publishing, and textile industries, it is chiefly known for its 11th cent. castle (now a museum), which was the seat (1712–98) of the federal diet.

Fred·er·i·ci·a (frĕd'ə-rĭsh'ē-ə, frĭth-rē'tsē-ä), city (1976 est. pop. 45,320), Vejle co., central Denmark, on the Lille Baelt. It is a port, an industrial center, and an important rail junction. Manufactures include refined petroleum, chemicals, and textiles. Fredericia was built in 1650 by Frederick III as the principal fortress on Jylland and was not permitted to expand beyond its ramparts. In 1849 the Danes defeated the Prussians here. The fortress was closed in 1909, and the city's modern development began.

Fred·er·ick (frĕd'rĭk, frĕd'ə-rĭk). **1.** County (1970 pop. 84,927), 664 sq mi (1,719.8 sq km), N Md., bounded on the N by the Pa. border, on the SW by the Potomac River and the Va. border, on the NE by the Monocacy River; formed 1748; co. seat Frederick. It is in a rolling piedmont area with a section of the Blue Ridge in the extreme west. Its agriculture includes wheat, corn, hay, apples, peaches, truck crops, dairy products, poultry, and livestock. Clothing, wood and metal products, plastics, shoes, machinery, and concrete are manufactured. Tourism is especially important to the county, which has mountain and fishing resorts and a number of state parks. **2.** County (1970 pop. 28,893), 405 sq mi (1,049 sq km), N Va., mainly in the Shenandoah Valley and bordered on the W and NE by W.Va.; formed 1738; co. seat Winchester (an independent city). The leading apple-growing county in the state, it also produces grain, livestock, and dairy products. It has limestone quarries and plants manufacturing clothing, concrete, and machinery.

Frederick. 1. City (1978 est. pop. 25,200), seat of Frederick co., NW Md.; settled 1745, inc. 1817. The processing center of a rich farm area, it has canneries, milk-receiving stations, and plants manufacturing household utensils, aluminum, optical and glass products, leather goods, clothing, and electronic equipment. Frederick was a stop on the road to the Ohio valley and an important grain trading center. During the Battle of Monocacy, Gen. Jubal Early's Confederate forces extracted a $200,000 ransom from Frederick's citizens. Points of interest include the home of Chief Justice Roger Taney, the grave of Francis Scott Key, the house of Barbara Fritchie, a Civil War heroine, and many restored historic buildings. **2.** City (1970 pop. 6,132), seat of Tillman co., SW Okla., SW of Lawton, in a farm area. Cotton processing is the chief industry. Prehistoric implements and fossils have been found nearby.

Frederick Doug·lass Home National Memorial (dŭg'ləs): *see* National Parks and Monuments Table.

Fred·er·icks·burg (frĕd'rĭks-bûrg'). **1.** Town (1970 pop. 5,326), seat of Gillespie co., S central Texas, W of Austin; founded 1846 by Germans, inc. 1928. Many picturesque houses built by the early settlers remain. The town has varied small industries and a prosperous trade in farm produce, wool, mohair, poultry, and cattle. Granite is quarried nearby. **2.** Independent city (1978 est. pop. 17,500), N Va., on the Rappahannock River, midway between Washington, D.C., and Richmond; settled 1671, laid out 1727, inc. as a town 1781, as a city 1879. A city of fine old houses and much historic interest, Fredericksburg attracts many tourists. It is also a farm trade center and an industrial city producing clothing, shoes, and veneers. Its historic buildings include the Rising Sun Tavern (c.1760), a rendezvous for American patriots in the Revolution, and the home of John Paul Jones. Nearby are Wakefield (George Washington's birthplace) and Fredericksburg and Spotsylvania County Battlefields Memorial National Military Park.

Fred·er·ick·town (frĕd'rĭk-toun', frĕd'ə-rĭk-), resort town (1970 pop. 3,799), seat of Madison co., SE Mo., in the Ozarks on a branch of the St. Francis River; founded 1819, inc. 1840. Lead, cobalt, copper, zinc, and iron deposits are found in the area.

Fred·er·ic·ton (frĕd'ə-rĭk-tən), city (1976 pop. 45,248), provincial capital, S central N.B., Canada, on the St. John River. It is a commercial and distribution center where shoes and wood products are manufactured. The city was founded by United Empire Loyalists in 1783. Of interest are the government buildings, the Beaverbrook Art Gallery, and the Playhouse Theatre. The Univ. of New Brunswick and St. Thomas Univ. are in the city.

Fre·de·riks·håb (frĭth'rĭks-hôp'), town (1969 pop. 1,949), in Frederikshåb dist. (1969 pop. 2,496), SW Greenland; founded in 1792. It is an important fishing center.

Fre·de·riks·havn (frĭth'rĭks-houn'), city (1976 est. pop. 34,664), Nordjylland co., N Denmark, a port on the Kattegat; chartered 1818. It is a commercial and industrial center.

Fred·er·ik·sted (frĕd'rĭk-stĕd, frĕd'ə-rĭk-), town (1970 pop. 1,548), chief port and commercial center of St. Croix, U.S. Virgin Islands. Sugar is the principal export.

Fre·do·nia (frĭ-dōn'yə). **1.** City (1970 pop. 3,080), seat of Wilson co., SE Kansas, NNW of Independence, in a farm and petroleum area; founded 1868, inc. 1871. Fall River Reservoir is nearby. **2.** Village (1970 pop. 10,326), Chautauqua co., SW N.Y., near Lake Erie; inc. 1829. Grape juice, wine, canned foods, and seeds are produced here. Fredonia was the site of the first gas well in the United States.

Fred·rik·stad (frĕd'rĭk-stä'), city (1977 est. pop. 28,925), Østfold co., SE Norway, a port on the Oslofjord (an arm of the Skagerrak) at the mouth of the Glåma River. Manufactures include forest products, processed food, and chemicals. Founded by Frederick II in 1567, it was fortified in the 17th cent.

Free·born (frē'bôrn'), county (1970 pop. 38,064), 701 sq mi (1,815.6 sq km), S Minn., bordering on Iowa; formed 1855; co. seat Albert Lea. It lies in an agricultural area that produces livestock, dairy products, grain, and potatoes. It has some mining and manufactures food, processed wood and metal products, and farm machinery.

Free·hold (frē'hōld'), borough (1970 pop. 10,545), seat of Monmouth co., E central N.J.; settled c.1650, called Monmouth Courthouse 1715–1801, inc. as a town 1867, as a borough 1919. Freehold is a farm trade center, with some industry. The Revolutionary War Battle of Monmouth took place nearby in 1778.

Free·port (frē'pôrt'). **1.** City (1978 est. pop. 25,400), seat of Stephenson co., NW Ill., on the Pecatonica River; inc. 1850. It is a trade and manufacturing center in a fertile farm and dairy region. Among its manufactures are farm machinery, plastics, and cosmetics. In 1832 a battle with Black Hawk's Indian forces occurred near here. Freeport was the scene of the second Lincoln-Douglas debate (1858). **2.** Town (1970 pop. 4,781), Cumberland co., SW Maine, on Casco Bay between Portland and Brunswick; settled c.1700, inc. 1789. The papers that established Maine as an independent state were signed here (1820). Freeport grew as a fishing, shipping, and shipbuilding center. Today shoes are manufactured. **3.** Village (1978 est. pop. 39,200), Nassau co., SE N.Y., on the shore of Long Island; settled c.1650, inc. 1892. It is a resort and a deep-sea fishing and oystering center, with access to the Atlantic Ocean through Jones Inlet. Jones Beach State Park is nearby. **4.** City (1978 est. pop. 11,500), Brazoria co., SE Texas, on the Gulf of Mexico at the mouth of the Brazos River on the Intracoastal Waterway; inc. 1913. The center of a thriving industrial area in a ranching, oil, and natural-gas region known as Brazosport, Freeport has large chemical and shrimping industries. Although a port from the 1820s, it became important only with the opening of sulfur mines a century later. Chemical and petrochemical industries developed during World War II. New port facilities were opened in 1955.

Free·stone (frē'stōn'), county (1970 pop. 11,116), 862 sq mi (2,232.6 sq km), E central Texas, bounded on the NE and E by the Trinity River; formed 1850; co. seat Fairfield. Mainly agricultural, it produces cotton, corn, peanuts, fruit, truck crops, poultry, cattle, and hogs. Lumbering is done.

Free·town (frē'toun'), city (1974 est. pop. 274,000), capital of Sierra Leone, W Sierra Leone, a port on the Atlantic Ocean. Freetown is the nation's administrative, communications, and economic center, as well as its main port. Industries include food and beverage processing, petroleum refining, and the manufacture of cigarettes, shoes, and beer. The area was settled in 1787 by freed slaves sent from England by British abolitionists who started the Sierra Leone Company. In 1792 Freetown was founded by former slaves from Nova Scotia sent out by the company. Freetown was used by the British as the base for creating (1808) the Sierra Leone Crown Colony, and from 1808 to 1874 it served as the capital of British West Africa. In 1893 it was made the first British colonial municipality in Africa. During World War II Britain maintained a naval base at Freetown. Freetown is the site of the Univ. of Sierra Leone (1967).

Frei·berg (frī'bo͝orKH'), city (1974 est. pop. 50,815), Karl-Marx-Stadt dist., S East Germany, at the foot of the Erzgebirge. It is an industrial center and a rail junction. Manufactures include precision and optical equipment, leather goods, and textiles. Lead and zinc are mined in the region. Freiberg was for centuries a silver-mining center and was settled by miners in the 12th cent. In the Thirty Years' War it resisted a siege by the Swedes (1642–43), and in the Seven Years' War the Prussians defeated (1762) the Austrians here. Noteworthy buildings include a late Gothic cathedral and numerous Renaissance-style and baroque houses. There is a famous mining academy (founded 1765) in Freiberg.

Frei·burg im Breis·gau (frī'bo͝orKH' ĭm brīs'gou), city (1974 est. pop. 179,196), Baden-Württemberg, SW West Germany, near the Rhine River and at the edge of the Black Forest. Manufactures include textiles, optical and musical instruments, paper, and chemicals; wine and timber are traded. The city is a tourist center. Freiburg was founded in 1120 and passed, with the rest of the Breisgau, to the Hapsburgs in 1368. In the Thirty Years' War (1618–48) the Bavarians and Austrians were defeated here (1644) by the French. The French held Freiburg from 1677 to 1697 and again (1744–48) during the War of the Austrian Succession. In 1805 the city passed to Baden. Freiburg is famous as a cultural center and is the seat of a noted university (founded 1457) and of a number of museums.

Frei·sing (frī'zĭng), city (1974 est. pop. 31,534), Bavaria, S West Germany, on the Isar River. Manufactures include machinery and textiles. Freising was founded in 724. It has a Romanesque cathedral (c.1160), with 18th cent. baroque additions.

Frei·tal (frī'täl'), city (1974 est. pop. 46,061), Dresden dist., SE East Germany; founded 1921. Manufactures of this industrial city include high-quality steel, machinery, optical equipment, and paper.

Fré·jus (frā-zhüs'), town (1975 pop. 28,851), Var dept., SE France, on the Mediterranean. It is a well-known resort of the French Riviera. Corks are made, and masonry is done. Founded by Julius Caesar in 49 B.C., it was an important Roman naval port. Many Roman ruins are preserved, notably the oldest surviving arena of Gaul.

Fre·man·tle (frē'măn'tl), city (1971 pop. 25,990), Western Australia, SW Australia, a suburb of Perth, on the Indian Ocean at the mouth of the Swan River. It is the terminus of the Trans-Australian RR and the chief commercial port of the state.

Fre·mont (frē'mŏnt'). **1.** County (1970 pop. 21,942), 1,562 sq mi (4,045.6 sq km), S central Colo., bounded on the W by the Rocky Mts. and drained by the Arkansas River; formed 1861; co. seat Canon City. In a cattle-grazing area, it has deposits of coal, zinc, and oil. Cement, clay products, concrete, and machinery are manufactured. Sections of San Isabel National Forest are in the county. **2.** County (1970 pop. 8,710), 1,864 sq mi (4,827.8 sq km), E Idaho, in a mountainous area bordered in the N by Mont., in the E by Wyo., and drained by Henrys Fork and the Teton River; formed 1893; co. seat St. Anthony. It has irrigated and dry farming (wheat, oats, sugar beets, potatoes, and dairy products) and cattle ranches. Lumbering is done. Part of Targhee National Forest is here. **3.** County (1970 pop. 9,282), 524 sq mi (1,357.2 sq km), extreme SW Iowa, in a prairie region drained by the Nishnabotna River and bounded on the S by the Mo. border, on the W by the Nebr. border and the Missouri River; formed 1847, annexed part of Otoe co., Nebr., 1943; co. seat Sidney. Cattle, hogs, poultry, grain, fruit, and truck crops are raised. It also has coal deposits and some manufacturing. **4.** County (1970 pop. 28,352), 9,196 sq mi (23,817.6 sq km), central Wyo., bounded on the W and S by the Continental Divide and watered by the Sweetwater, Beaver, and Wind rivers; formed 1884; co. seat Lander. In a cattle, sheep, grain, and sugar-beet area, it has iron deposits, oil and natural-gas wells, and stands of timber. Chemicals are manufactured. The county includes Wind River Indian Reservation, the Wind River Mts., and sections of Shoshone National Forest.

Fremont. **1.** City (1978 est. pop. 117,700), Alameda co., W Calif., on San Francisco Bay; inc. 1956. Long an agricultural center, with champagne vineyards founded (1870) by Leland Stanford, it still has brewing and canning industries and is a shipping point for fruits and vegetables. Its economy was transformed in 1963, however, when General Motors opened an automobile assembly plant. Mission San Jose de Guadalupe (1797) has been restored as a museum. **2.** City (1978 est. pop. 23,000), seat of Dodge co., E central Nebr., on the Platte River; inc. 1858. It is a trade, shipping, and processing center for a grain-growing, dairying, and grazing prairie area. Midland Lutheran College is here. **3.** City (1970 pop. 18,490), seat of Sandusky co., N Ohio, on the Sandusky River; inc. 1849. It is a trade and industrial center in a rich agricultural region. A government trading post was established on the site in 1795. The Battle of Fort Stephenson was fought here (1813) during the War of 1812. The house and tomb of Rutherford B. Hayes are nearby.

Fremont Peak, mountain, 13,730 ft (4,187.5 m) high, W central Wyo., in the Wind River Range.

French Broad (frĕnch' brôd'), river, 210 mi (337.9 km) long, rising in the Blue Ridge, W N.C., and flowing N and then NW to Knoxville, E Tenn., where it joins with the Holston River to form the Tennessee River. The French Broad River was an important route from the southeastern coastal states during the colonial period.

French·burg (frĕnch'bûrg'), city (1970 pop. 467), seat of Menifee co., E central Ky., E of Winchester, in an oil, timber, and farm area. There is good fishing nearby.

French E·qua·tori·al Af·ri·ca (ē'kwə-tôr'ē-əl ăf'rĭ-kə, ĕk'wə-), former French federation in W central Africa. It consisted of Gabon, Middle Congo, Chad, and Ubangi-Shari (now the Central African Empire). The capital was Brazzaville. The federation (originally called French Congo) was officially established in 1910. Until 1920 Chad and Ubangi-Shari were a single territory. About 100,000 sq mi (259,000 sq km) were ceded to Germany in 1911 but were returned to

France by the Treaty of Versailles. During World War II the federation supported the Free French. When the constituent territories voted (1958) to become autonomous republics within the French Community, the federation was dissolved.

French Gui·a·na (gē-ăn′ə, -än′ə), French overseas dept. (1972 est. pop. 50,400), 35,135 sq mi (91,000 sq km), NE South America, on the Atlantic Ocean. Cayenne is the capital. The Oiapoque River on the east and the Tumuc-Humac Mts. on the south separate it from Brazil. The Maroni River on the west forms the border with Surinam. Timber, shrimp, and rum made from local sugar cane are the chief exports. Rice, corn, and bananas are grown for subsistence. There are gold and bauxite deposits.

French settlement dates from 1604. The Portuguese and British occupied it during the Napoleonic Wars, but the Congress of Vienna (1815) restored French authority. French Guiana was used as a penal colony and place of exile during the French Revolution, and under Napoleon III permanent penal camps were established. The colonies were evacuated after World War II. In 1947 French Guiana became an overseas department of the French Republic.

French In·di·a (ĭn′dē-ə), former overseas territory of France in India, composed of the coastal enclaves of Pondicherry, Karikal, Yanaon, and Mahé in the S and the inland trade settlement of Chandernagor near Calcutta.

French Pol·y·ne·sia (pŏl′ə-nē′zhə, -shə), overseas territory of France (1971 pop. 119,918), consisting of 105 islands in the South Pacific. The capital is Papeete, on Tahiti. The territory comprises the Society, Marquesas, Austral, Tuamotu, and Gambier islands. Tropical fruits are grown, and vanilla and copra exported. Phosphate mining, once the major industry, ceased in 1966.

French Union, 1946–58, political entity established by the French constitution of 1946. It comprised metropolitan France (the 90 departments of continental France and Corsica); French overseas departments, territories, settlements, and United Nations trusteeships; French colonies, which became overseas departments of France; and associate states (protectorates), which became autonomous. The union replaced the colonial system. In 1954 the associate states of Vietnam, Laos, and Cambodia withdrew from the union, and in 1956 Morocco and Tunisia, also associate states, became independent. The French Community, established in 1958 by the constitution of the Fifth French Republic, replaced the French Union. Its present members are the French Republic, including metropolitan France (continental France and Corsica), the overseas territories (French Polynesia, New Caledonia, Saint Pierre and Miquelon, the French Southern and Antarctic territories, and the Wallis and Futuna Islands), the overseas departments (French Guiana, Guadeloupe, Martinique, and Réunion), and six independent African republics—the Central African Empire, Chad, the Comoro Islands, the Congo Republic, Gabon, Malagasy Republic, and Senegal.

French West Africa, former federation of eight French overseas territories. The constituent territories were Dahomey (now Benin), French Guinea, French Sudan, Ivory Coast, Mauritania, Niger, Senegal, and Upper Volta. The federation was created in 1895 to consolidate the French holdings in West Africa. During World War II the federation supported the Vichy government until Nov., 1942, when it accepted the authority of the Free French. In 1958 the constituent territories became autonomous republics in the French Community, except for Guinea, which became independent. The federation was dissolved in 1959.

Fres·nil·lo (frāz-nē′yō), city (1970 pop. 101,316), Zacatecas state, N central Mexico. The city is the center of a rich mining area known especially for silver. It has a mining school. Agriculture (cereals, beans) and cattle raising are other important economic activities. Fresnillo was founded in 1554.

Fres·no (frĕz′nō), county (1970 pop. 413,329), 5,966 sq mi (15,452 sq km), S central Calif., stretching across the San Joaquin Valley from the Diablo Range in the W to the crest of the Sierra Nevada in the E; formed 1856; co. seat Fresno. It is drained by the Kings, San Joaquin, and Fresno rivers. Its agriculture includes raisins, grapes, cotton, grain, alfalfa, sugar beets, rice nuts, truck crops, dairy products, livestock, and poultry. It also has oil and natural-gas fields, sand and gravel pits, stone quarries, and extensive lumbering and food-processing industries. Among its manufactures are textiles, clothing, furniture, chemicals, plastics, concrete, metal products, and machinery. There are many recreational areas.

Fresno, city (1978 est. pop. 191,000), seat of Fresno co., S central Calif.; inc. 1885. It is the financial hub of the San Joaquin Valley and an important railroad, processing, and marketing center. Grapes, figs, vegetables, and cotton are grown in the area. Among Fresno's manufactures are boxes, prefabricated buildings, carpets, and wines.

Fri·bourg (frī′bûrg, frē-bo͝or′), canton (1978 est. pop. 181,600), 645 sq mi (1,670.61 sq km), W Switzerland, located on the Swiss Plateau amid the foothills of the Alps. It joined the Swiss Confederation in 1481. The town of Fribourg (1978 est. pop. 39,800), the canton's original settlement and capital, is rich in medieval architecture and picturesquely situated on the Sarine River. It is famous for its chocolate. Founded in 1178, it passed successively to the houses of Kyburg (1218), Hapsburg (1277), and Savoy (1452).

Fri·day Harbor (frī′dē, -dā′), resort town (1977 est. pop. 1,024), seat of San Juan co.; NW Wash.; on San Juan Island SW of Bellingham. It is a port of entry and has salmon canneries.

Frid·ley (frĭd′lē), city (1978 est. pop. 32,300), Anoka co., SE Minn., a suburb of Minneapolis, on the Mississippi River; settled 1847, inc. as a city 1957. A distribution center with railroad yards and warehouses, Fridley produces naval ordnance, pumps, machine tools, transportation equipment, dies and parts, portable generators, electro-medical devices, cosmetics, and linseed oil. In 1965 three tornadoes destroyed substantial parts of the city.

Fried·land (frēt′länt′): *see* Pravdinsk, USSR.

Frie·drichs·ha·fen (frē′drĭĸнs-hä′fən), city (1978 est. pop. 51,930), Baden-Württemberg, S West Germany, a port on the Lake of Constance. Manufactures include textiles, leather goods, machinery, and electrical products. Friedrichshafen was formed in 1811 by the union of the towns of Buchhorn and Hafen. In 1824 it became the summer residence of the kings of Württemberg. As the site of the Zeppelin aircraft works, Friedrichshafen suffered heavy damage in World War II.

Friend·ship (frĕnd′shĭp′), village (1970 pop. 641), seat of Adams co., central Wis., S of Wisconsin Rapids on a tributary of the Wisconsin River. It is in a timber and dairying region.

Fries·land (frēz′lənd, frēs′länt′) or **Fri·si·a** (frĭzh′ē-ə), province (1977 est. pop. 566,042), c.1,325 sq mi (3,430 sq km), N Netherlands; capital Leeuwarden. A great dairying and cattle-raising region, Friesland has fertile land near the coast and sandy heath and fenland in the interior. The Frisians, a Germanic people, were conquered by the Franks in the 8th cent. In 1498 Emperor Maximilian I bestowed all Friesland on Duke Albert of Saxony whose son, for a payment, restored Friesland to Maximilian in 1515. Maximilian's grandson, Emperor Charles V, reduced Friesland by force in 1523. Friesland joined (1579) in the Union of Utrecht against Spanish domination, but it continued to appoint its own stadtholders until 1748, when its stadtholder, Prince William IV of Orange, became sole and hereditary stadtholder of all the United Provinces of the Netherlands.

Frim·ley and Cam·ber·ley (frĭm′lē; kăm′bər-lē), urban district (1973 est. pop. 47,390), Surrey, S England. The Royal Staff College and the Royal Military Academy at Sandhurst are in the district.

Fri·o (frē′ō), county (1970 pop. 11,159), 1,116 sq mi (2,890.4 sq km), S Texas, in an irrigated truck-farming area drained by the Frio River; formed 1858; co. seat Pearsall. Fruit, peanuts, corn, grain sorghums, poultry, livestock, and dairy products are also important. There are oil wells and clay and sand deposits here.

Frio, river, 220 mi (354 km) long, S Texas, flowing S and SE from Edwards Plateau into the Nueces River, below Three Rivers.

Fri·sian Islands (frĭzh′ən, frē′zhən), chain of low-lying islands, off the coasts of the Netherlands, West Germany, and Denmark, in the North Sea. The West Frisian Islands belong to the Netherlands; the East and North Frisians, to West Germany. Fishing and stock raising are pursued on most of the Frisian Islands.

Fri·u·li (frē-o͞o′lē), historic region, now divided between Friuli-Venezia Giulia, NE Italy, and Slovenia, NW Yugoslavia, extending from the E Alps to the Adriatic. Occupied by the Romans (2nd cent. B.C.), it became a Lombard duchy (6th–8th cent.) and a Frankish march (8th cent.). Before A.D. 1000 it was divided into the counties of Gorizia (east) and Friuli (west). The western county passed (11th cent.) to the patriarchs of Aquileia. In 1420 it went to Venice. After the counts of Gorizia became extinct (1500), Emperor Maximilian I incorporated the eastern county into the Hapsburg possessions. By the treaties of Campo Formio (1797) and Paris (1814, 1815) all Friuli became Austrian. After the Austro-Prussian War Austria ceded

(1866) western Friuli to Italy. In 1919 eastern Friuli was also awarded to Italy; with Istria and Trieste it formed the region of Venezia Giulia. The Italian peace treaty of 1947 gave eastern Friuli (but not Gorizia) to Yugoslavia.

Fri·u·li-Ve·ne·zia Giu·lia (frē-ōō'lē-vĕ-nĕt'syä jōō'lyä), region (1971 pop. 1,209,810), 3,031 sq mi (7,850.3 sq km), NE Italy, bordering on Austria in the N and Yugoslavia in the E; capital Trieste. Farming is the chief occupation; industrialization has accelerated since 1945. The region was formed in 1947 by the merger of Udine prov. with that part of the former region of Venezia Giulia not annexed by Yugoslavia. Trieste prov. was added in 1954.

Fro·bish·er Bay (frō'bĭsh-ər, frŏb'ĭsh-), arm of the Atlantic Ocean, 150 mi (241.4 km) long and from 20 to 40 mi (32–64 km) wide, E Franklin dist., Northwest Territories, Canada. Cutting deeply into southeast Baffin Island, it has steep, deeply indented shores and numerous islets. On its southwest side the Grinnell and Southeast ice caps rise to c.3,000 ft (915 m), extending tongues into the bay. The bay was discovered (1576) by Sir Martin Frobisher; until 1860 it was believed to be a strait separating Baffin Island from another island.

Frome (frōōm), urban district (1971 est. pop. 13,384), Somerset, SW England. It has an important woolen industry. Frome is noted for its steep straight streets. The parish church of St. John the Baptist dates from the 14th cent., but is much restored.

Fron·tier (frŭn-tîr'), county (1970 pop. 3,982), 966 sq mi (2,501.9 sq km), S Nebr., watered by Medicine Creek and other branches of the Republican River; formed 1872; co. seat Stockville. It lies in an agricultural region that produces grain and livestock. It has some manufacturing. There are recreational areas at Medicine Creek and Red Willow Reservoir.

Front Range (frŭnt), highest part of the U.S. Rocky Mts., bordering the Great Plains and extending c.300 mi (485 km) S from SE Wyo. to the Arkansas River, central Colo. Part of the Continental Divide, the range has several peaks, including Mt. Elbert and Pikes Peak, that are more than 14,000 ft (4,270 m) high. The Front Range was scouted by explorers Zebulon Pike, in 1806–7, and Stephen Long, in 1819–20. In 1858 gold was discovered at Cripple Creek, Colo., and prospectors rushed into the area. Gold, silver, and beryllium are still mined. Most of the range is in national forests; Rocky Mts. National Park is located in the north.

Front Roy·al (roi'əl), town (1970 pop. 8,211), seat of Warren co., N central Va., on the South Fork of the Shenandoah River S of Winchester, in a resort area; inc. 1788. Silk and rayon are manufactured. Skyline Caverns, with rare calcite formations, are nearby. On May 23, 1862, Stonewall Jackson routed a Union detachment here.

Frun·ze (frōōn'zē), city (1977 est. pop. 511,000), capital of the Kirghiz SSR, Central Asian USSR, on the Chu River and on a branch of the Turkistan-Siberia RR. Industries include metalworking, food processing, and the manufacture of farm machinery, motor vehicles, textiles, building materials, and clothing. The Uzbek khans of Kokand built a fortress on the site in 1846; it was taken by Russian forces in 1862. The city was chartered in 1878.

Fu-chou. 1. (fōō'jō') Or **Foo·chow** (fōō'chou'), city (1970 est. pop. 900,000), capital of Fukien prov., China, a port on the Min River delta. A regional commercial and fishing center, Fuchou has chemical plants, a small integrated iron and steel complex, textile mills, machine shops, food-processing establishments (tea and sugar), and paper mills. The old walled part of the city dates from the T'ang dynasty (A.D. 618–906). After the Opium War (1839–42) Fu-chou was established as a treaty port. By 1850 it was the principal Chinese port and the world's largest tea-exporting center. Its importance declined when the demand for tea decreased and when harbor silting barred large vessels. In the surrounding hills are beautiful pagodas and monasteries, and a summer resort. **2.** Or **Fu·chow** (both: fōō'chou', fōō'jō'), city, N central Kiangsi prov., China. It is an agricultural and commercial center known for its hot springs. Barite deposits are nearby.

Fu·chu (fōō'chōō'). **1.** City (1975 pop. 50,217), Hiroshima prefecture, W Honshu, Japan, on the Ashida River. It is an agricultural and livestock center. **2.** City (1976 est. pop. 184,818), Tokyo metropolis, E central Honshu, Japan. It is a residential suburb of Tokyo and the site of a racetrack. It is noted for its Shinto shrine.

Fue·go (fwā'gō), active volcano, 12,986 ft (3,960.7 m), S central Guatemala, SW of Guatemala City. Among its 20 recorded eruptions are those of 1773 (which destroyed Antigua City), 1880, and 1974.

Fuen·te O·be·ju·na or **Fuen·te·o·ve·ju·na** (both: fwān'tā ō'vā-hōō'nä), town (1970 pop. 9,247), Córdoba prov., S Spain, in Andalusia. An important farm center with livestock-raising and food-processing industries, the town is especially noted for its honey. Lumber, coal, and lead are exploited in the area. In 1430 Fuente Obejuna was given to the Knights of Calatrava. The palace of the Knights, formerly a Moorish castle, is now the parish church.

Fuer·te·ven·tu·ra (fwĕr'tā-vān-tōō'rä), island (1970 pop. 18,192), 668 sq mi (1,730.1 sq km), Canary Islands, Las Palmas prov., Spain, in the Atlantic Ocean ENE of Grand Canary Island. It has many extinct volcanic peaks and often suffers droughts.

Fu·jai·rah (fōō-jī'rä), sheikdom (1968 pop. 9,724), c.450 sq mi (1,165 sq km), part of the federation of United Arab Emirates, E Arabia, on the Gulf of Oman. Although oil has been produced since 1966, agriculture is the most important economic activity. Fujairah was a British protectorate until it joined the United Arab Emirates in 1971.

Fu·ji (fōō'jē), city (1976 est. pop. 200,958), Shizuoka prefecture, S central Honshu, Japan, on Suruga Bay. It is an important communications and industrial center producing paper and paper pulp, machinery, household appliances, and chemical fibers.

Fu·ji·no·mi·ya (fōō'jē-nō'mē-yä), city (1976 est. pop. 79,400), Shizuoka prefecture, central Honshu, Japan, at the foot of Mt. Fuji. It is an important railway junction and point of departure for the Mt. Fuji resort region. The city has a large paper-pulp industry and is noted for its Shinto shrine.

Fu·ji·o·ka (fōō-jē'ō-kä), city (1975 pop. 30,000), Gumma prefecture, central Honshu, Japan, on the Tone River. It is a manufacturing center where silk and soy sauce are produced.

Fu·ji·sa·wa (fōō-jē'sä-wä), city (1976 est. pop. 271,291), Kanagawa prefecture, central Honshu, Japan, on Sagami Bay. It is an industrial and residential suburb of Tokyo and a market for agricultural products from the Sagami plain. Fujisawa is also a resort town.

Fu·ji·ya·ma, Fu·ji-ya·ma (both: fōō'jē-yä'mə, fōō'jē-yä'mä'), **Mount Fu·ji** (fōō'jē), or **Fu·ji-san** (fōō'jē-sän'), volcanic peak, 12,388 ft (3,778.3 m) high, central Honshu, Japan. The highest point on Honshu, it is a sacred mountain and the traditional goal of pilgrimage. According to legend, an earthquake created Fuji in 286 B.C. The beauty of the snow-capped symmetrical cone, ringed by lakes and virgin forests, has inspired Japanese poets and painters throughout the centuries. Its last major eruption was in 1707.

Fu·ji·yo·shi·da (fōō-jē'yō-shē'dä), city (1975 pop. 51,976), Yamanashi prefecture, central Honshu, Japan, on the Katsura River. It is an important communications center for the Mt. Fuji region. The city has a large textile industry.

Fu·ka·ya (fōō-kä'yä), city (1975 pop. 53,100), Saitama prefecture, E central Honshu, Japan. It is an industrial and residential suburb of Tokyo with raw silk, tire, and textile industries.

Fu·kien (fōō'kyĕn'), province (1968 est. pop. 17,000,000), c.46,000 sq mi (119,140 sq km), SE China, on Formosa Strait; capital Fu-chou. About a tenth of the land is arable. Rice, sweet potatoes, wheat, and tea are grown in the uplands, and fruit, silk, and jute are produced in the lowlands. The coastal region is a major sugar-producing area. The chief oil-producing seed is rapeseed, but peanuts and soybeans are also grown. There is some tobacco, and the extensive forests on the mountains provide considerable lumber (fir, pine, bamboo), camphor, and wood oils. Fishing off the island-strewn coast is important. The mineral resources are iron ore, tungsten, and manganese; coal reserves are poor. The industries are light; most important are lumbering and woodworking, tea processing, sugar refining, salt panning, and the manufacturing of textiles, cement, ceramics, and processed foods, including preserved fruits.

Fu·ku·i (fōō-kōō'ē), city (1976 est. pop. 234,001), capital of Fukui prefecture, central Honshu, Japan. A modern textile center, it is especially noted for rayon manufactures. It was an important silk-weaving center in the 10th cent. and became a castle town in the 16th cent. The city suffered a disastrous earthquake in 1948.

Fu·ku·o·ka (fōō-kōō'ô-kä'), city (1976 est. pop. 1,021,623), capital of Fukuoka prefecture, N Kyushu, Japan, on Hakata Bay. A port and textile-producing center, it is the seat of five universities. The ancient port area, Hakata, was in medieval times one of the chief ports of Japan. The Mongols under Kublai Khan were defeated twice (1274 and 1281) at Hakata.

Fu·ku·shi·ma (fōō-kōō'shĭ-mä'), city (1976 est. pop. 250,441), capital of Fukushima prefecture, N Honshu, Japan, on the Kiso plain. A silk-textile center, it is a major commercial city.

Fu·ku·ya·ma (fōō-kōō'yä-mä), city (1976 est. pop. 334,848), Hiroshima prefecture, W Honshu, Japan, on the Ashida River. It is an important commercial, industrial, and communications center with a large electronics industry.

Ful·da (fōōl'dä'), city (1974 est. pop. 59,696), Hesse, E central West Germany, on the Fulda River. It is an agricultural market and an industrial center. Manufactures include textiles, chemicals, rubber products, and carpets. Fulda grew around a Benedictine abbey founded in 744. From the 13th cent. the abbots of Fulda ruled the town and the surrounding area as princes of the Holy Roman Empire, and in 1752 they were raised to the rank of prince-bishops. Fulda was secularized in 1802, and most of it passed to Hesse-Kassel in 1816. A theological seminary is in the city.

Ful·ler·ton (fōōl'ər-tən). **1.** City (1978 est. pop. 95,700), Orange co., S Calif., SE of Los Angeles; founded 1887, inc. 1904. Oil was discovered near Fullerton in 1892, but the city's main growth came with the construction of the Santa Ana Freeway in the 1950s. Fullerton's manufactures include aerospace equipment, food products, electrical and electronic components, ordnance, paper products, musical instruments, and aluminum building products. Muckenthaler Center houses two symphony orchestras, a theater group, and other art associations. **2.** City (1970 pop. 1,444), seat of Nance co., E central Nebr., on the Loup River WSW of Columbus, in a prairie region; platted 1878. Dairy products are made.

Ful·ton (fōōl'tən). **1.** County (1970 pop. 7,699), 611 sq mi (1,582.5 sq km), N Ark., bounded on the N by the Mo. border and drained by the Spring and Strawberry rivers; formed 1842; co. seat Salem. It is mainly agricultural, producing cotton, corn, truck and feed crops, dairy products, livestock, and poultry. Some textile manufacturing is done. **2.** County (1970 pop. 605,210), 530 sq mi (1,372.7 sq km), NW central Ga., bounded on the NW by the Chattahoochee River; formed 1853; co. seat Atlanta. In a piedmont manufacturing, commercial, and agricultural region yielding cotton, corn, fruit, truck crops, dairy products, poultry, and livestock, it has granite quarries, sand and gravel pits, and lumbering and food-processing industries. Among its diversified manufactures are textiles, clothing, furniture, paper, iron and steel, chemicals, plastics, and glass. **3.** County (1970 pop. 41,900), 874 sq mi (2,263.7 sq km), W central Ill., bounded on the SE by the Illinois River and drained by the Spoon River; formed 1823; co. seat Lewiston. Its agriculture includes corn, wheat, soybeans, dairy products, poultry, and livestock. It also has bituminous-coal mines, deposits of lignite and clay, and stands of timber. Food processing is done, and feed, farm equipment, and wood products are manufactured. A group of bayou lakes and resort areas are near the Illinois River. **4.** County (1970 pop. 16,984), 367 sq mi (950.5 sq km), N Ind., drained by the Tippecanoe River; formed 1835; co. seat Rochester. It has agriculture (truck crops, poultry, soybeans, livestock, and dairy products) and timber (sawmills). Its manufactures include dairy and grain products, clothing, and miscellaneous metal products. There are lake resorts in the county. **5.** County (1970 pop. 10,183), 205 sq mi (531 sq km), extreme SW Ky., bounded on the W by the Mississippi River and the Mo. border, on the S by the Tenn. border, and drained by Obion Creek and Bayou de Chien; formed 1845; co. seat Hickman. Cotton, corn, tobacco, and livestock are raised. Food processing and lumbering are done, and some clothing is made. **6.** County (1970 pop. 52,637), 497 sq mi (1,287.2 sq km), E central N.Y., in the Adirondacks, drained by East Canada Creek, the Sacandaga River, and a number of lakes; formed 1838; co. seat Johnstown. Dairying, poultry raising, and general farming are important. It also has lumbering and food-processing industries, textile mills, and plants manufacturing clothing, wood products, and machinery. Gloves have been made here since 1800. **7.** County (1970 pop. 33,071), 407 sq mi (1,054.1 sq km), NW Ohio, bordered on the N by Mich. and drained by the Tiffin River; formed 1850; co. seat Wauseon. Its diversified farm crops include wheat, sugar beets, corn, oats, hay, and tomatoes. It has an extensive food-processing industry. Wood products, furniture, paper, plastics, steel, aluminum, and machinery are manufactured. **8.** County (1970 pop. 10,776), 435 sq mi (1,126.7 sq km), S Pa., in a mountainous region with the Tuscarora Mts. in the E and Sideling Hill in the W; formed 1850; co. seat McConnellsburg. It is in an agricultural area that produces buckwheat, livestock, and gristmill products. It has bituminous-coal and limestone deposits, and there is some clothing and machinery manufacturing. It was settled in the 1740s by the Scotch-Irish.

Fulton. 1. Town (1970 pop. 2,899), seat of Itawamba co., NE Miss., near the East Fork of the Tombigbee River N of Columbus, in a farm, cotton, and timber area; settled 1848, inc. 1850. **2.** City (1970 pop. 12,248), seat of Callaway co., central Mo., in a farm area; inc. 1859. It has printing plants and factories making farm and industrial equipment. During the Civil War the county seceded from the United States and by treaty with the state militia formed the Kingdom of Callaway. On Mar. 5, 1946, Winston Churchill delivered his famous "iron curtain" speech at Westminster College here. The college now houses the Winston Churchill Memorial and Library, including a reconstruction of a Christopher Wren church destroyed in the bombing of London. **3.** City (1970 pop. 14,003), Oswego co., N central N.Y., on the Oswego River; inc. 1835. Machinery, corrugated boxes, thermometers, chocolate, and frozen foods are among its products.

Fu·na·ba·shi (fōō-nä'bä-shē), city (1976 est. pop. 431,940), Chiba prefecture, E central Honshu, Japan, on Tokyo Bay. It is an industrial and residential suburb of Tokyo.

Fu·na·fu·ti (fōō'nä-fōō'tē), chief atoll of the Ellice Islands, S Pacific, a part of the British colony of the Gilbert and Ellice Islands. It comprises 30 islets of a reef 13 mi (21 km) long, with a land area of c.1 sq mi (2.6 sq km). The island was discovered in 1819 and became part of the colony in 1915.

Fun·chal (fōōn-shäl'), city (1975 est. pop. 38,340), capital of Funchal dist., on Madeira island, Portugal. A busy port, it is best known for its beautiful setting and balmy climate, which make it a much-frequented resort. The city was founded in 1421.

Fun·dy, Bay of (fŭn'dē), large inlet of the Atlantic Ocean, c.170 mi (275 km) long and 30 to 50 mi (50–80 km) wide, between N.B. and SW N.S., Canada. In its upper arms, Chignecto Bay and the Mínas Basin, tides reach 40 to 50 ft (12–15 m) in height and create the reversing falls of the St. John River. At low tide, wide flats are laid bare, and the long estuaries of the rivers are drained. Many of the surrounding flats have been reclaimed and transformed into fertile farmland since Acadian settlers built dikes in the early 17th cent.

Fur·nas (fûr'nəs), county (1970 pop. 6,897), 722 sq mi (1,870 sq km), S Nebr., drained by the Republican River and Beaver and Sappa creeks and bordered on the S by Kansas; formed 1873; co. seat Beaver City. In a livestock and grain area, it has some mining and manufacturing.

Fur·neaux Group (fûr'nō grōōp'), about 25 islands, c.900 sq mi (2,330 sq km), Tasmania, SE Australia, in Bass Strait between Tasmania and the Australian mainland. The largest is Flinders Island, and the group forms Flinders municipality (1971 pop. 967). The islands were discovered in 1773 by the British navigator Tobias Furneaux.

Fürth (fürt), city (1974 est. pop. 103,238), Bavaria, S West Germany, at the confluence of the Rednitz and Pegnitz rivers. It is an industrial suburb of Nuremberg; manufactures include toys and glass. Reputedly founded by Charlemagne in the late 8th cent., Fürth rose to importance when the Jews who were denied entrance to Nuremberg settled here (14th cent.). The city passed to Bavaria in 1806 and became part of Middle Franconia.

Fu·ry and Hec·la Strait (fyōōr'ē; hĕk'lə), narrow channel, c.100 mi (160 km) long and from 10 to 15 mi (16–24 km) wide, N Canada, between Baffin Island and Melville Peninsula. It connects Foxe Basin with the Gulf of Boothia.

Fu-shun or **Fu·shun** (both: fōō'shōōn'), city (1970 est. pop. 1,700,000), NE Liaoning prov., China, in a highly industrialized area. It has one of the largest open-cut coal mines in the world. Oil shale deposits, also mined there, are processed in the Fu-shun oil refineries, one of which is the largest in the country. Fu-shun also has a major aluminum reduction plant and factories making automobiles, machinery, chemicals, and synthetic fibers. The city was developed by Russia until 1905 and by Japan until 1945.

Fu·ta Jal·lon (fōō'tä jä-lôN'), region, Guinea: *see* Fouta Djallon.

Fyn (fün), island (1976 est. pop. 446,223), c.1,340 sq mi (3,471 sq km), Fyn co., S central Denmark. It is largely a fertile lowland; dairy goods, sugar beets, and cereals are the chief products. There are many summer residences along the island's coast.

Fyne, Loch (fīn), arm of the Firth of Clyde, Argyllshire, W Scotland, extending 40 mi (64.4 km) N and NE from the Sound of Bute. The loch has long been famous for its herring fisheries.

Ga·bi·i (gā-bē'ī), ancient town of Latium, E of Rome on the road to Praeneste (modern Palestrina). One of the most important of the Latin cities, it was early overshadowed by Rome and had lost all importance even in the days of the republic.

Ga·bon (gä-bôɴ'), republic (1977 pop. 535,000), 103,346 sq mi (267,667 sq km) W central Africa, bordering on the Atlantic Ocean in the W, on Equatorial Guinea and Cameroon in the N, and on the Republic of the Congo in the E and S. Libreville is the capital. Much of Gabon is drained by the Ogooué River, which flows into the Atlantic through a long and broad estuary. The coastline comprises a narrow low-lying strip, which includes a series of lagoons. The interior of the country is made up of mountain ranges and high-lying plateaus.

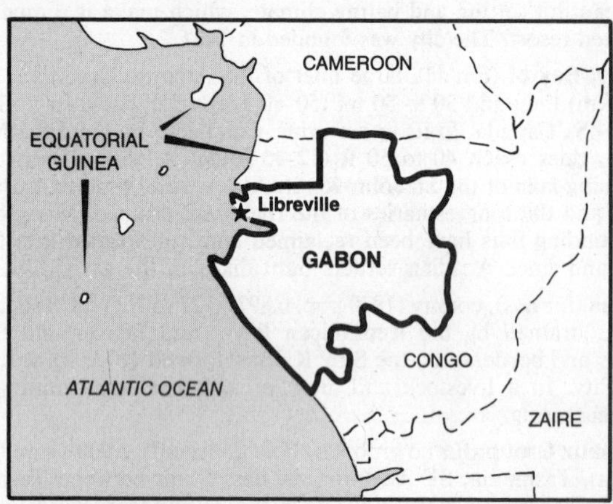

Economy. The majority of the Gabonese workers are engaged in subsistence farming, cassava, plantains, taro, and rice being the chief crops. However, food must be imported to meet the country's needs. Cacao, coffee, and palm products are produced for export. Few animals are raised, partly because of the prevalence of the tsetse fly. Forestry and mining, both largely controlled by European-owned firms, form the backbone of the modern sector of the country's economy. The most important forest products are okoume (a softwood), mahogany, and ebony. The principal minerals extracted are petroleum, manganese and uranium ores, and gold. There are large deposits of iron ore in the east. Gabon's chief manufactures include processed timber, refined petroleum, and agricultural goods.

History. Discoveries by archaeologists indicate that present-day Gabon was inhabited in the late Paleolithic period and that it was later the site of Neolithic cultures. By the 16th cent. A.D. the Omiéné were living along the coast, and in the 18th cent. the Fang entered the region from the north. From the 16th to the 18th cent. Gabon was part of the decentralized Loango empire, between the Ogooué and Congo rivers. In the 1470s Portuguese navigators discovered the Ogooué estuary, and shortly thereafter they began to trade with coastal merchants for black African slaves. The Portuguese were followed by other European traders; by the late 18th cent. the French had gained a dominant position. Slaves continued to be exported from the Gabon coast until the 1880s. Christian missions were established between 1842 and 1844, and additional treaties were signed with nearby African leaders. In 1849 Libreville was founded by the French as a settlement for freed slaves.

In 1885 the Conference of Berlin recognized French rights to the region north of the Congo River that included Gabon. Gabon was initially included in the French Congo and, from 1910 to 1957, formed a part of French Equatorial Africa. The Fang and some other African peoples resisted the imposition of French rule until 1911. In 1946 Gabon was declared an overseas territory of France, and in 1958 the country became internally self-governing. On Aug. 17, 1960, Gabon became an independent republic. Leon Mba, a Fang, was the first president of Gabon. In Feb., 1964, Mba survived a military coup with the help of French troops. Mba died in 1967 and was succeeded by Omar Bongo, who was elected to a new seven-year term as president in 1973.

Ga·bo·ro·ne (gä'bə-rō'nē), city (1971 pop. 18,436), capital of Botswana. The city was founded c.1890. In 1965 it became capital of the Bechuanaland Protectorate; it remained the capital when Bechuanaland became independent as Botswana in 1966.

Ga·bro·vo (gä'brô-vô), town (1969 est. pop. 71,800), N central Bulgaria, in the foothills of the Balkan Mts. It is Bulgaria's chief woolen textile center. The town developed as a strategic point on the northern approaches to Shipka Pass.

Gads·den (gădz'dən), county (1970 pop. 39,184), 508 sq mi (1,315.7 sq km), NW Fla., bounded on the N by the Ga. border, on the E and SE by the Ochlockonee River and Lake Talquin, on the W by the Apalachicola River; formed 1823; co. seat Quincy. Its agriculture includes corn, peanuts, tobacco, vegetables, dairy products, poultry, and hogs. There are large deposits of fuller's earth and clay, and sand and gravel pits. Food and wood products, naval stores, bricks, and cigars are manufactured.

Gadsden, city (1978 est. pop. 49,100), seat of Etowah co., NE Ala., on the Coosa River; inc. 1871. Iron, coal, limestone, sand, clay, and timber are found in the area. Gadsden has metal and textile industries and a large tire and rubber plant.

Gadsden Purchase, strip of land purchased (1853) by the United States from Mexico. The Treaty of Guadalupe Hidalgo (1848) had described the U.S.-Mexico boundary vaguely, and President Franklin Pierce wanted to insure U.S. possession of the Mesilla Valley near the Rio Grande—the most practicable route for a southern railroad to the Pacific. James Gadsden negotiated the purchase, and the U.S. Senate ratified (1854) it by a narrow margin. The area of c.30,000 sq mi (77,700 sq km), purchased for $10 million, now forms extreme southern N.Mex. and Ariz. south of the Gila River.

Ga·e·ta (gä-ā'tä), town (1976 est. pop. 24,151), in Latium, central Italy, a seaport on a high promontory in the Tyrrhenian Sea. It was a favorite resort of the ancient Romans and was a prosperous duchy from the 9th to the 12th cent. Gaeta lost its independence to the Normans (mid-12th cent.) and thereafter shared the fortunes of the kingdom of Naples. The citadel (8th cent.) and the port were strongly fortified (15th–16th cent.). Pope Pius IX took refuge in Gaeta in 1848–49. The fall of the town to Victor Emmanuel II of Sardinia after a siege (1860–61) marked the end of the rule of Francis II of the kingdom of the Two Sicilies.

Gaff·ney (găf'nē), city (1970 pop. 13,253), seat of Cherokee co., NW S.C., near the N.C. line, in a cotton, grain, and peach region; settled in the early 1800s, inc. 1873. Textiles and garments are its major products; the city also has a variety of light manufactures. Cowpens National Battlefield and Kings Mountain National Military Park are nearby.

Gage (gāj), county (1970 pop. 25,731), 858 sq mi (2,222.2 sq km), SE Nebr., in an agricultural region bordered on the S by Kansas and drained by the Big Blue River; formed 1855; co. seat Beatrice. Grain, dairy products, livestock, and poultry are important. It has sand and gravel pits, a food-processing industry, and factories manufacturing clothing, fertilizers, metal products, and farm machinery.

Gail (gāl), town (1970 pop. 150), seat of Borden co., NW Texas, W of

Snyder near the Colorado River and just below the Cap Rock escarpment of the Llano Estacado. It is a trading center for a cattle-ranching area.

Gaines (gānz), county (1970 pop. 11,593), 1,489 sq mi (3,856.5 sq km), NW Texas, on the Llano Estacado in a cattle-ranching area bounded on the W by the N.Mex. border and drained by Mustang Draw; formed 1876; co. seat Seminole. Its agriculture includes grain sorghums, hay, cotton, corn, truck crops, dairy products, poultry, hogs, mules, goats, and sheep. There are natural-gas fields and salt and potash deposits here.

Gaines·bor·o (gānz'bûr'ō), town (1970 pop. 1,101), seat of Jackson co., N central Tenn., on the Cumberland River ENE of Nashville, in a farm, timber, and oil region; inc. 1817.

Gaines·ville (gānz'vĭl'). **1.** City (1978 est. pop. 72,300), seat of Alachua co., N central Fla.; inc. 1869. The Univ. of Florida is a major source of employment in the city. There are also industries manufacturing electronic equipment and wood products. Points of interest are Paynes Prairie State Park, Warrens Cave, and many natural sinkholes. **2.** City (1978 est. pop. 18,900), seat of Hall co., N central Ga., on Lake Lanier, in the foothills of the Blue Ridge Mts.; inc. 1821. It is a trade center and has a textile industry. Riverside Military Academy and Chattahoochee National Forest are nearby. **3.** City (1970 pop. 627), seat of Ozark co., S Mo., in the Ozarks WSW of West Plains, in a cotton-ginning area. Lumber products are made. **4.** Town (1970 pop. 13,830), seat of Cooke co., N Texas, on the Elm Fork of the Trinity River; inc. 1873. It is the commercial and industrial hub of a farm and oil area. Dairy items, oil-field equipment, fishing lures, mobile homes, and furniture are among the manufactures. Gainesville was founded (1850) on the California Trail and later was a riotous cow town, a stopping point on the Chisholm Trail just below a Red River crossing.

Ga·lá·pa·gos Islands (gə-lä'pə-gōs', gä-lä'pä-gôs'), or **Ar·chi·pié·la·go de Co·lón** (är-chē-pyā'lä-gō' dā kō-lōn'), Pacific archipelago belonging to Ecuador (1970 est. pop. 3,000), 3,029 sq mi (7,845.1 sq km), c.650 mi (1,045 km) W of Ecuador, on the equator. They were discovered in 1535 by the Spanish navigator Tomás de Bertanga and named for the gigantic (up to 500 lb/227 kg) land tortoises that are now facing extinction. Ecuador claimed the islands in 1832. The Galápagos are famous for their wildlife. They were visited (1835) by Charles Darwin, who gathered evidence that was used later in support of his theory of natural selection. The Galápagos are now a wildlife sanctuary where naturalists can study living species arrested at various evolutionary stages.

Gal·a·shiels (găl'ə-shēlz'), burgh (1971 pop. 12,605), Borders region, SE Scotland, on the Gala Water. Famous for fine tweeds and woolens, Galashiels is the site of the Scottish College of Textiles. There is also a tanning industry.

Ga·la·ţi (gä-läts', -lä'tsē) or **Ga·latz** (gä-läts'), city (1977 est. pop. 239,306), E Rumania, on the lower Danube. It is an economic center and a major inland port. The city has a large iron and steel plant and shipyards. Of medieval origin, Galaţi became an international trading center in the 18th cent. and was a free port from 1834 to 1883.

Ga·la·tia (gə-lā'shə, -shē-ə), ancient territory of central Asia Minor, in present-day Turkey (around modern Ankara). It was so called from its inhabitants, the Gauls, who invaded from the west and conquered it in the 3rd cent. B.C.

Ga·le·na (gə-lē'nə). **1.** City (1970 pop. 3,930), seat of Jo Daviess co., extreme NW Ill., on the Galena River NW of Stockton; settled c.1820, platted 1826, inc. 1835. The region, rich in lead and zinc deposits, was frequented by French miners as early as the first part of the 18th cent. The city was a prosperous river port until the 1860s, when the mines declined and shipping was diverted to the railroads. Today the city is a trade shipping center for a dairying area. Ulysses S. Grant's home is now a museum. **2.** City (1970 pop. 391), seat of Stone co., SW Mo., in the Ozarks, on the James River SSW of Springfield. Tomatoes, corn, and wheat are grown.

Ga·le·ras (gä-lā'räs), active volcano, 13,997 ft (4,269.1 m) high, SW Colombia, at the S end of the Cordillera Occidental near the Ecuador border.

Gales·burg (gălz'bûrg'), city (1978 est. pop. 34,100), seat of Knox co., W Ill., in a farm, livestock, and coal area; chartered 1841. A trade, rail, and industrial center, it has railroad shops and plants manufacturing power lawn mowers, refrigerators, air conditioners, plastics, containers, and automotive and steel products. Galesburg was founded by Presbyterians from the Mohawk River valley, N.Y. The birthplace of the poet Carl Sandburg has been preserved.

Ga·li·ci·a (gə-lĭsh'ē-ə, -lĭsh'ə), historic region, SE Poland and W Ukraine, covering the slopes of the N Carpathians and plains to the N and bordering on Czechoslovakia in the S. The San River divides Galicia into the western (Polish) and the eastern (Ukrainian) parts. Mainly agricultural, Galicia also has mineral resources.

Originally the duchy of Galich, it was united with the duchy of Vladimir in 1188 and annexed by Casimir III of Poland in the 14th cent. With the First Partition of Poland (1772) most of the region passed to Austria, which made it a crown land. Austria enlarged its holdings with the third Polish partition (1795) and again in 1815. In 1861 Galicia won limited autonomy. In 1918 the Poles wrested western Galicia from Austria and fought the troops of the newly established Ukraine republic in eastern Galicia, forcing them to withdraw. The Paris Peace Conference (1919) assigned eastern Galicia to Poland pending a plebiscite scheduled for 1944. However, in a treaty (1920) with the Ukrainians, upheld by the Polish-Soviet Treaty of Riga (1921), Poland obtained full title to eastern Galicia. In 1939 most of eastern Galicia was incorporated into Ukraine, an act upheld by the Polish-Soviet Treaty of 1945.

Ga·li·ci·a (gə-lĭsh'ē-ə, -lĭsh'ə, gä-lē'thyä), region, NW Spain, on the Atlantic Ocean, S of the Bay of Biscay and N of Portugal. The area is mostly mountainous, with several swift rivers. Fishing and cattle and hog raising are the chief sources of livelihood. The region's mineral resources, chiefly iron and tin, were known to the Romans but are now little exploited. Galicia was (5th–6th cent. A.D.) the center of the kingdom of the German Suevi. Its people's strong spirit of independence was shown in the Middle Ages by the frequent rebellions of the feudal lords against the crown and again in the 19th cent. by the popular resistance to Napoleon I.

Gal·i·lee (găl'ə-lē), region, N Israel, roughly the portion N of the plain of Esdraelon. Galilee was the chief scene of the ministry of Jesus Christ. After the destruction of Jerusalem (A.D. 70), Galilee became the main center of Judaism in Palestine. Zionist colonization of the region began at the end of the 19th cent. Olives and grains are the chief crops of Galilee's fertile farming areas.

Galilee, Sea of, lake, 64 sq mi (165.8 sq km), 14 mi (22.5 km) long, and 3 to 7 mi (4.8–11.3 km) wide, NE Israel; its surface is c.700 ft (215 m) below sea level. The lake is fed and drained by the Jordan River. Mineral springs, some of them hot, discharge into the lake, giving it a saline character. In the time of Christ there was a flourishing fishing industry in the lake; some fishing is still carried on.

Gal·la·tin (găl'ə-tĭn). **1.** County (1970 pop. 7,418), 328 sq mi (849.5 sq km), SE Ill., in a bituminous-coal region bounded on the NE by the Wabash River, on the SE by the Ohio River, and drained by the North Fork of the Saline River; formed 1812; co. seat Shawneetown. It has agriculture (corn, wheat, dairy products, livestock, and poultry) and a lumbering industry. Furniture and wood products are manufactured. Part of Shawnee National Forest is here. **2.** County (1970 pop. 4,134), 100 sq mi (259 sq km), N Ky., in a gently rolling upland area of the outer Bluegrass, bounded on the N by the Ohio River and the Ind. border; formed 1798; co. seat Warsaw. Mainly agricultural (burley tobacco, corn, and livestock), it has some manufacturing. **3.** County (1970 pop. 32,505), 2,517 sq mi (6,519 sq km), SW Mont., in an agricultural area bounded on the S by the Continental Divide and the Idaho border, on the SE by Yellowstone National Park, and drained by the Gallatin River and Hebgen Lake; formed 1856; co. seat Bozeman. The Missouri River is formed in the northwest, near Three Forks, by the junction of the Gallatin River with the Jefferson and Madison rivers. Parts of Gallatin National Forest are in the south and northeast.

Gallatin. 1. City (1970 pop. 1,833), seat of Daviess co., NW Mo., on the Grand River ENE of St. Joseph, in a farm area; laid out 1837, inc. 1856. **2.** City (1970 pop. 13,253), seat of Sumner co., N central Tenn., near Nashville; inc. 1815. It is a livestock and agricultural center that produces tobacco. Its manufactures include tobacco goods, cheese products, furniture, apparel, storm windows and doors, and locks. Nearby is Old Hickory Lake, a popular fishing and recreation area. Andrew Jackson's home, the Hermitage, is nearby.

Gallatin, river, c.120 mi (195 km) long, rising in the Gallatin Range in the NW corner of Yellowstone National Park, NW Wyo., and flowing generally NW to join the Madison and Jefferson rivers at the Three Forks of the Missouri, SW Mont.

Galle (gäl), city (1972 est. pop. 80,000), capital of Southern prov., extreme S Sri Lanka (Ceylon), on the Indian Ocean. An agricultural market center, it exports tea, rubber, coconut oil, cloves, and other products of the surrounding region. Famous as a trade center for Chinese and Arabs by 100 B.C., Galle rose to prominence under Portuguese rule (1057-1640), when it became Sri Lanka's chief port. It was the capital of Sri Lanka under the Dutch (1640-56).

Gal·li·a (găl'ē-ə), county (1970 pop. 25,239), 471 sq mi (1,219.9 sq km), S Ohio, bounded on the E by the Ohio River and the W.Va. line and intersected by Raccoon Creek and small Symmes and Campaign creeks; formed 1803; co. seat Gallipolis. Its agriculture includes livestock, grain, tobacco, and fruit. It has bituminous-coal mines, limestone, sand, and gravel deposits, and plants manufacturing metal products.

Gal·lip·o·li (gə-lĭp'ə-lē), city (1975 pop. 13,426), W Turkey, a port at the E end of the Dardanelles, near the neck of the Gallipoli Peninsula. It has long been a strategic point in the defense of İstanbul (Constantinople) and has numerous historic remains. It was captured by the Ottoman Turks in 1354.

Gallipoli Peninsula, narrow peninsula, c.50 mi (80 km) long, W Turkey, extending SW between the Aegean Sea and the Dardanelles. It was the scene of the Gallipoli campaign of 1915 and was (1920-36) part of the demilitarized Zone of the Straits.

Gal·lo·way (găl'ə-wā'), region, SW Scotland. The Rhinns, or Rinns, of Galloway is a rocky peninsula; its southern extremity is called the Mull of Galloway and is the southernmost point in Scotland. The black, hornless Galloway cattle have long been bred in this region.

Gal·lup (găl'əp), town (1978 est. pop. 18,400), alt. 6,515 ft (1,987 m), seat of McKinley co., NW N.Mex., on the Puerco River near the Ariz. line; inc. 1891. It is a rail and trade center in a mining, timber, and ranching area.

Galt (gôlt), industrial city (1971 pop. 38,897), S Ont., Canada, on the Grand River, NW of Hamilton. Manufactures include textiles, chemicals, and plastics.

Gal·ves·ton (găl'vĭ-stən), county (1970 pop. 169,812), 399 sq mi (1,033.4 sq km), SE coastal Texas, bounded on the E by Galveston Bay, on the S by West Bay and the Gulf of Mexico; formed 1838; co. seat Galveston. The county includes Bolivar Peninsula and Galveston Island. A processing and shipping center on the Intracoastal Waterway, it also has agriculture (cotton, rice, pecans, fruit, truck crops, and livestock), oil and natural-gas fields.

Galveston, city (1978 est. pop. 60,200), seat of Galveston co., on Galveston Island, SE Texas; inc. 1839. The island lies across the entrance to Galveston Bay, an inlet of the Gulf of Mexico. Despite the ship channel to the larger port at Houston, Galveston remains a key port of entry. Oil refining and shipbuilding are major industries, and the city has grain elevators, machine shops, cotton compresses, chemical plants, and large fishing and shrimping fleets. It is also a beach and fishing resort.

The Spanish knew the bay and the island early; it was probably here that Cabeza de Vaca was shipwrecked in 1528. For three hundred years it was visited and occupied by wandering Indian tribes, adventurers, revolutionists, and buccaneers. Settlement began in the 1830s. In 1900 hurricane winds struck Galveston, driving water across the low-lying island. Thousands were killed, and the city left in ruins. When the city was rebuilt, an enormous 10-mi (16-km) protective seawall was built. However, in 1961 another hurricane caused much damage. Fort Crockett, once headquarters for the city's harbor defenses, is now used by the U.S. Army Corps of Engineers. A Coast Guard base and the Texas maritime academy are in Galveston.

Gal·way (gôl'wā'), county (1971 pop. 149,223), 2,293 sq mi (5,939 sq km), W Republic of Ireland; co. town Galway. The county is divided into two sections by Lough Corrib. The mountains of the Connemara region lie to the west; to the east stretches a rolling plain, partially covered with bogs.

Galway, urban district (1971 pop. 27,726), county town of Co. Galway, W Republic of Ireland, on Galway Bay, an inlet of the Atlantic Ocean, near the mouth of the Corrib River. Tourism, food processing, and the production of textiles and furniture are the major industries. Galway was first incorporated by Richard II of England in the late 14th cent. In 1651 the town was taken by parliamentary forces, and in 1691 it was taken by William III after the Battle of Aughrim.

Gam·bi·a (găm'bē-ə), officially **The Gambia,** republic (1977 est. pop. 535,000), 4,361 sq mi (11,295 sq km), W Africa, on the Atlantic Ocean. The capital is Banjul. The smallest country in Africa, Gambia comprises St. Mary's Island and, on the adjacent mainland, a narrow enclave, never more than 30 mi (48 km) wide, in Senegal; this territorial strip borders both banks of the Gambia River for c.200 mi (320 km) above its mouth.

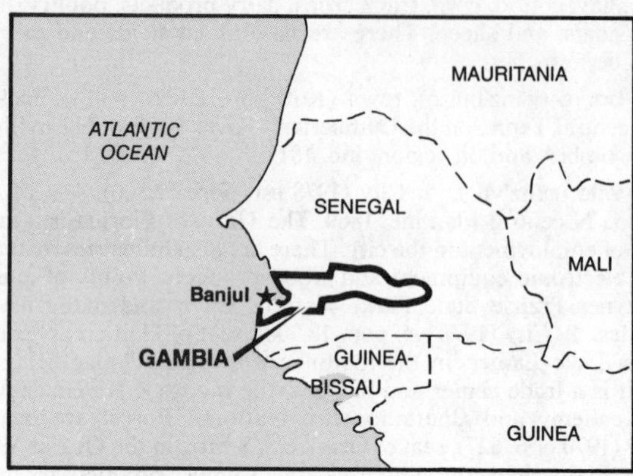

Economy. Despite attempts at diversification, Gambia's economy remains overwhelmingly dependent on the export of peanuts and their by-products. Swamp rice, millet, sorghum, maize, cassava, and beans are grown for subsistence, and cattle are raised.

History. Portuguese explorers reaching the Gambia region in the mid-15th cent. reported a group of small tributary states to the empire of Mali. Englishmen won trading rights from the Portuguese in 1588, but their hold was weak until the early 17th cent. In 1816 the British purchased St. Mary's Island from a local chief and established Banjul (called Bathurst until 1973) as a base against the slave trade. The city remained a colonial backwater under the administration of Sierra Leone until 1843, when it became a separate crown colony. As the French extended their rule over Senegal's interior, trading companies urged Britain to hold on to Gambia. In 1894 the interior was declared a British protectorate. Gambia achieved full self-government in 1963 and independence in 1965 under Dauda Kairaba Jawara and the People's Progressive Party (PPP). The PPP is mainly a party of the predominant Mandingo tribe; the major opposition group is the United Party (UP), dominated by smaller tribes. Following a referendum in 1970, Gambia became a republic.

Government. The president and most members of the House of Representatives are popularly elected; a few representatives are chosen by tribal chiefs from among themselves.

Gambia, river, c.700 mi (1,125 km) long, rising on the Fouta Djallon, N Guinea, W Africa, and flowing generally NW through SE Senegal then W, bisecting Gambia, to the Atlantic Ocean at Banjul. It is navigable for the entire length of Gambia; oceangoing vessels can reach Georgetown, c.175 mi (280 km) upstream.

Gan·der (găn'dər), town (1971 pop. 7,748), NE N.F., Canada. Gander's airport was an important base in World War II.

Gan·dha·ra (gŭn-där'ə), historic region of India, now in NW Pakistan. Situated astride the middle Indus River, the region was originally a province of the Persian Empire and was reached (327 B.C.) by Alexander the Great. It was part of Bactria from the late 3rd cent. to the 1st cent. B.C. In the 1st to 3rd cent. A.D., Gandhara developed a noted school of sculpture, consisting mainly of images of Buddha and reliefs representing scenes from Buddhist texts, but with marked Greco-Roman elements of style. The art form flourished in Gandhara until the 5th cent., when the region was conquered by the Huns.

Gan·dí·a (gän-dē'ä), town (1970 pop. 36,342), Valencia prov., E Spain, in a fertile garden region near the Mediterranean. Large quantities of oranges are grown here.

Gan·ges (găn'jēz) or **Gan·ga** (gŭng'gä), river, c.1,560 mi (2,510 km) long, rising in an ice cave in the Himalayas and flowing generally E through a vast plain to the Bay of Bengal. It is the most sacred river of Hindu India. The fertile Ganges plain is one of the world's most densely populated regions; rice, grains, oilseed, sugar cane, and cotton are the main crops. The upper Ganges supplies water to extensive irrigation works. The river passes the holy bathing sites at Hardwar, Allahabad (where the Jumna River enters the Ganges), and Varanasi. Below Allahabad the Ganges becomes a slow, meandering stream with shifting channels. The lower Ganges is joined by the Brahmaputra River west of Dacca, Bangladesh, to form the Padma, its main channel to the sea. The united rivers branch into many distributaries, forming the vast and fertile Ganges-Brahmaputra delta, which stretches from the Hooghly River on the west to the Meghna on the east. Rice, sugar cane, and jute are the delta's main crops. The delta's southern fringe, a great wilderness of swamp, dense timber forest, small islands, and tidal creeks, has repeatedly suffered great devastation from cyclones and tidal waves.

Gang·tok (gŭng'tŏk'), town (1968 est. pop. 9,000), capital of Sikkim, NE of Darjeeling.

Gann·val·ley (găn-văl'ē), village (1970 pop. 78), seat of Buffalo co., central S.Dak., ESE of Pierre. It is a farm-trading center in a wheat-growing area.

Gap (găp), city (1975 pop. 25,052), capital of Hautes-Alpes dept., SE France, on the Luye River at the foot of the Dauphiné Alps. It has flour mills, creameries, printing plants, and factories making bricks, tile, clothing, and leather products. Founded by Augustus c.14 B.C., it was the capital of medieval Gapençais, which was annexed to the crown of France in 1512. The area was devastated during the Wars of Religion (16th cent.).

Gard (gär), department (1975 pop. 494,575), S France, 2,258 sq mi (5,848.2 sq km), on the Rhône River and the Mediterranean Sea; capital Nîmes.

Gar·da, Lake (gär'dä), largest lake of Italy, 143 sq mi (370.4 sq km), between Lombardy and Venetia, N Italy. It is c.32 mi (52 km) long, with a maximum width of c.11 mi (18 km). The shoreline is dotted with vineyards and resorts.

Gar·den (gär'dn), county (1970 pop. 2,929), 1,678 sq mi (4,346 sq km), W Nebr., watered by the North Platte River; formed 1909; co. seat Oshkosh. It has extensive cattle ranches and is primarily agricultural (wheat, corn, beans, and beets).

Gar·de·na (gär-dē'nə), city (1978 est. pop. 44,900), Los Angeles co., SW Calif., an industrial suburb of Los Angeles; inc. 1930. Among its diverse manufactures are aircraft and missile components, electronic equipment, machinery, tools, chemicals, metal products, clothing, and food products. Gardena is also noted for its plant nurseries.

Gar·de·na, Val (väl gär-dā'nä), Alpine valley, c.15 mi (25 km) long, in the Dolomites, Trentino-Alto Adige, N Italy. Its scenery attracts many tourists. Wood carving (mainly of toys) is a traditional home industry.

Garden City. **1.** Town (1970 pop. 5,790), Chatham co., SE Ga., a suburb NW of Savannah. **2.** City (1978 est. pop. 18,400), seat of Finney co., SW Kansas, on the Arkansas River; inc. 1887. A trade center in an irrigated farm and dairy region producing grain, sugar beets, and alfalfa, it has a gas field and hide-processing and meat-packing plants. Agricultural equipment and concrete and plastic pipes are manufactured. **3.** City (1978 est. pop. 39,000), Wayne co., SE Mich., a suburb of Detroit; inc. as a city 1934. Chiefly residential, the city also manufactures wire cloth, aluminum extrusions, and golf balls. **4.** Village (1978 est. pop. 26,500), Nassau co., SE N.Y., on Long Island; inc. 1919. It is primarily a residential community, with printing and publishing as the major industries. Garden City was founded in 1869. **5.** Village (1970 pop. 300), seat of Glasscock co., W Texas, S of Big Spring. It is a market center for a livestock-ranching area.

Garden Grove (grōv), city (1978 est. pop. 117,300), Orange co., S Calif., a residential suburb of Long Beach and Los Angeles, in a citrus fruit area, on the Santa Ana River; founded 1877, inc. 1956. Many of its residents work in nearby space and defense installations.

Garden of the Gods (gŏdz), park, 770 acres (311.9 hectares), central Colo., near Colorado Springs; noted for its curious, multicolored rock formations. Narrow-crested sandstone hills and ridges have been eroded into grotesque groups with such fanciful names as Kissing Camels.

Gar·di·ner (gärd'nər, gär'dĭ-nər), city (1970 pop. 6,685), Kennebec co., SW Maine, on the Kennebec River S of Augusta; founded 1760, inc. as a town 1803, as a city 1850. Dams, mills, and other works were built here after 1803. In 1823 Benjamin Hale established the first technical school in America here. The poet Edwin Arlington Robinson spent his youth in Gardiner.

Gar·di·ners Island (gärd'nərz, gär'dĭ-nərz), c.3,000 acres (1,215 hectares), in Gardiners Bay between two flukelike peninsulas of E Long Island, SE N.Y. It was settled by Lion Gardiner in 1639 as the first permanent English settlement in what is now N.Y. and was owned for 300 years by his descendants; it is now a game preserve.

Gard·ner (gärd'nər), city (1978 est. pop. 18,900), Worcester co., N central Mass.; settled 1764, inc. as a city 1921. Its furniture industry dates from c.1805.

Gar·field (gär'fēld'). **1.** County (1970 pop. 14,821), 2,994 sq mi (7,754.5 sq km), W Colo., in a plateau area bordering on Utah in the W and drained by the Colorado River and its tributaries; formed 1883; co. seat Glenwood Springs. In a wheat-farming and cattle-grazing area, it has deposits of coal and oil shale. Parts of Grand Mesa, Routt, and White River national forests are here. **2.** County (1970 pop. 1,796), 4,455 sq mi (11,538.5 sq km), E central Mont., in an agricultural area bounded on the W by the Musselshell River, on the N by the Missouri River and Fort Peck Lake; formed 1883; co. seat Jordan. Livestock is raised. Part of Charles M. Russell Wildlife Range is in the north. **3.** County (1970 pop. 2,411), 570 sq mi (1,476.3 sq km), central Nebr., in a partially irrigated agricultural area watered by the Cedar, North Loup, and Calamus rivers; formed 1884; co. seat Burwell. Grain and alfalfa are grown, and cattle raised. It has some mining and a small amount of industry. **4.** County (1970 pop. 56,343), 1,054 sq mi (2,729.9 sq km), N Okla., drained by tributaries of the Cimarron River and by the Salt Fork of the Arkansas River; formed 1907; co. seat Enid. It is in an agricultural area (mainly wheat, also corn, oats, barley, alfalfa, livestock, poultry, and dairy products) with extensive oil and gas wells and refineries. There is waterfowl hunting in the county. **5.** County (1970 pop. 3,157), 5,158 sq mi (13,359.2 sq km), S Utah, in a mountainous area bounded on the E by the Colorado River and drained by the Sevier and Escalante rivers; formed 1882; co. seat Panguitch. Cattle ranching, grain farming, and lumbering are important. The county includes part of Bryce Canyon National Park and sections of Dixie National Forest. **6.** County (1970 pop. 2,911), 714 sq mi (1,849.3 sq km), SE Wash., in a rolling plateau area bounded on the N by the Snake River and rising to the Blue Mts. in the S; formed 1881; co. seat Pomeroy. It is an agricultural county, with wheat, fruit, truck crops, and livestock.

Garfield, industrial city (1978 est. pop. 27,100), Bergen co., NE N.J., on the Passaic River at its confluence with the Saddle River; settled 1679 by the Dutch, inc. 1898.

Ga·ri·glia·no (gä'rē-lyä'nō), lower part of the Liri River, S central Italy, below its junction with the Rapido, or Gari, River near Cassino. It separates Latium from Campania and empties into the Tyrrhenian Sea. A strategic battleground since antiquity, it was the scene (1503) of Gonzalo Fernández de Córdoba's victory over Louis XII of France in the Italian Wars. In World War II heavy fighting occurred (Nov., 1943–May, 1944) near the Garigliano during the Allied drive on Rome.

Gar·land (gär'lənd), county (1970 pop. 54,131), 658 sq mi (1,704.2 sq km), W central Ark., intersected by the Ouachita River; formed 1873; co. seat Hot Springs. It has agriculture (cotton, truck crops, and livestock), ore mines, and lumbering and food-processing industries. Machinery and wood and metal products are made.

Garland, city (1978 est. pop. 127,900), Dallas co., N Texas, a suburb of Dallas; inc. 1891. Since World War II, Garland has grown from an agricultural community into an important center for electronics research and for production of electronic equipment for airplanes and missiles.

Gar·misch-Par·ten·kir·chen (gär'mĭsh-pär'tən-kĭr'KHən), town (1974 est. pop. 26,964), Bavaria, S West Germany, in the Bavarian Alps, at the foot of the Zugspitze. It is an international winter resort and a regional commercial center. The picturesque town has two well-decorated 18th cent. churches.

Gar·ner (gär'nər), town (1970 pop. 2,257), seat of Hancock co., N central Iowa, WSW of Mason City on the Iowa River, in a dairy and farm area; inc. 1881. The Dusenberg brothers, automobile pioneers, began their work here in 1905.

Gar·nett (gär'nĭt, -nĕt), city (1970 pop. 3,169), seat of Anderson co., E Kansas, SSW of Lawrence; founded 1856, inc. as a town 1861, as a city 1870. The poet Edgar Lee Masters was born here.

Ga·ronne (gä-rôn'), river, 402 mi (646.8 km) long, rising in the central

Pyrenees just inside Spain, and flowing generally NE to Toulouse, in SW France, where it swings NW to join the Dordogne River and forms the Gironde estuary. The Garonne is navigable only in its lower course; Bordeaux is the head of oceangoing navigation.

Gar·rard (găr′ərd), county (1970 pop. 9,457), 234 sq mi (606.1 sq km), E central Ky., in a gently rolling upland area of the Bluegrass, bounded on the N by the Kentucky River, on the W by Herrington Lake; formed 1796; co. seat Lancaster. Burley tobacco, corn, wheat, and hay are the principal crops. Some manufacturing is done.

Gar·rett (găr′ĭt), county (1970 pop. 21,476), 659 sq mi (1,706.8 sq km), extreme NW Md., in the Alleghanies bounded on the SE by the North Branch of the Potomac, on the W and SE by the W.Va. border, on the N by the Pa. border; formed 1872; co. seat Oakland. It is drained by the Youghiogheny, Casselman, and Savage rivers. It has agriculture (grain, truck crops, livestock, poultry, and dairy products), bituminous-coal mines, gas wells, and a lumbering industry.

Gar·vin (gär′vĭn), county (1970 pop. 24,874), 814 sq mi (2,108.3 sq km), S central Okla., intersected by the Washita River and Rush and Wildhorse creeks; formed 1907; co. seat Pauls Valley. Its agriculture includes cotton, broomcorn, oats, alfalfa, corn, pecans, fruit, livestock, and poultry.

Gar·wood (gär′wo͝od′), industrial borough (1970 pop. 5,260), Union co., NE N.J., WSW of Elizabeth; inc. 1903. Its manufactures include aluminum and gypsum products.

Gar·y (gâr′ē), city (1978 est. pop. 157,000), Lake co., NW Ind., a port of entry on Lake Michigan; inc. 1906. One of the world's greatest steel centers, Gary was founded by the U.S. Steel Corporation, which bought the land in 1905. Its location midway between the iron-ore beds of the Northwest and the coal areas of the East and Southeast made it ideal for industry. Gary steelworkers were especially active in the nationwide steel strike of 1919, when federal troops occupied the city for several months. The Dunes National Lakeshore Park is adjacent to Gary.

Gar·za (gär′zə), county (1970 pop. 5,289), 914 sq mi (2,367.3 sq km), NW Texas, in rolling plains at the foot of Cap Rock escarpment and drained by the Salt and Double Mountain forks of the Brazos River; formed 1876; co. seat Post. Cattle ranching is done on the lower plains. Its agriculture includes cotton, corn, wheat, grain sorghums, fruit, some truck crops, livestock, and dairy products. There are oil and natural-gas wells and deposits of clay, gypsum, and lignite. Textiles are manufactured.

Gas·con·ade (găs′kə-nād′), county (1970 pop. 11,878), 520 sq mi (1,346.8 sq km), E central Mo., in the Ozarks bounded in the N by the Missouri River and drained by the Gasconade and Bourbeuse rivers; formed 1820; co. seat Hermann. It has agriculture (corn, wheat, and livestock), timber, and deposits of flint, fire clay, diaspore, zinc, and lead. Flour is milled, and wood products, furniture, and shoes manufactured.

Gasconade, river, c.265 mi (426.4 km) long, S central Mo., rising in the Ozarks E of Springfield and flowing NE to the Missouri River E of Jefferson City.

Gas·co·ny (găs′kə-nē), region of SW France. The sandy and swampy Landes along the Atlantic coast, the majestic Pyrenees forming the border with Spain, and the hilly Armagnac region between the Adour and Garonne rivers are the main geographic areas of Gascony. Fishing, stock raising, wine making, brandy distilling, and the tourist trade are the chief industries.

Under the Romans the region was known as Novempopulana and was inhabited by the Vascones, or Basques, who since prehistoric times had lived in the lands north and south of the Pyrenees. Conquered by the Visigoths (5th cent.) and by the Franks (6th cent.), Novempopulana was invaded in turn by the Basque-speaking peoples from south of the Pyrenees, who in 601 set up the duchy of Vasconia or Gascony. The duchy's borders fluctuated as the Basques fought the Visigoths, the Franks, and the Arabs throughout the Merovingian period. Invaded by Norsemen early in the 9th cent., Gascony fell into anarchy. In 1052, with the exception of lower Navarre and Béarn, which continued separate, the remainder of Gascony passed to the duchy of Aquitaine. Gascony fell under English control in 1154 and was a major battleground in the Hundred Years' War (1337–1453); it was completely recovered by France in 1453. Gascony passed, through marriage and inheritance, to Henry of Navarre, who became king of France as Henry IV in 1589. The region was united with the royal domain in 1607.

Gas·coyne (găs′koin′), river, 475 mi (764.3 km) long, W Western Australia, rising in the NE desert region and flowing W to Carnarvon at the Geographe Channel of Shark Bay on the Indian Ocean. Its flow is intermittent except during flood periods.

Gas·pé (găs-pā′, gä-spā′), city (1976 pop. 16,842), E Que., Canada, on Gaspé Bay near the E extremity of the Gaspé Peninsula. It is a resort. Jacques Cartier landed here in 1534.

Gaspé Peninsula, tongue of land, E Que., Canada, between the estuary of the St. Lawrence River on the N and Chaleur Bay on the S, and extending E into the Gulf of St. Lawrence. It is c.150 mi (240 km) long and from 60 to 90 mi (97–145 km) wide. Its backbone is an extension of the Appalachian mountain system. The interior of the peninsula is a mountain wilderness, completely forested, and with numerous mountain streams and lakes, offering excellent hunting and fishing. Settlement is almost wholly confined to the coastal rim, where there is a succession of picturesque villages whose residents live by combining agriculture with fishing (chiefly cod) and lumbering. The coast, with its combination of mountain and sea and its many bold headlands, is famed for its beauty.

Ga·stein (gä-stīn′), valley, Salzburg prov., central Austria, in the N Hohe Tauern range. A popular and beautiful resort area, it has hot radioactive springs. Gold and silver have been mined in the region since Roman times.

Gas·ton (găs′tən), county (1970 pop. 148,415), 356 sq mi (922 sq km), SW N.C., in a piedmont region bounded on the S by the S.C. border, on the E by the Catawba River; formed 1846; co. seat Gastonia. It is mainly agricultural, yielding cotton, wheat, hay, poultry, and dairy products. There is also a lumbering industry.

Gas·to·ni·a (găs-tō′nē-ə), city (1970 pop. 47,142), seat of Gaston co., SW N.C.; inc. 1877. An important textile-mill center, Gastonia is a major producer of fine-combed cotton yarn. Sheeting, tire fabric, and textile machinery are also made.

Gat (gôt): *see* Ghat, Libya.

Ga·ta, Sier·ra de (syĕr′rä dā gä′tä), mountain range, W Spain. Between the valleys of the Douro and Tagus rivers, it separates León from Estremadura. Its highest point is Jálama (5,577 ft/1,701 m).

Gat·chi·na (gä′chē-nə), city (1976 est. pop. 73,000), NW European USSR. Industries include ironworking, sawmilling, and the manufacture of papermaking equipment. The city developed around the imperial palace (built 1766–81; now a museum), which was a favorite residence of the Russian czars during the 19th cent.

Gate City (gāt), trade town (1970 pop. 1,914), seat of Scott co., SW Va., near the Tenn. line WNW of Bristol; settled in the late 18th cent., inc. 1892.

Gates (gāts), county (1970 pop. 8,524), 337 sq mi (872.8 sq km), NE N.C., bounded on the N by the Va. border, on the S by the Chowan River, and on the E by Dismal Swamp; formed 1778; co. seat Gatesville. It is in a coastal-plain agricultural area yielding cotton, corn, peanuts, and soybeans. Sawmilling and some fishing are done.

Gates·head (gāts′hĕd′), county borough (1976 est. pop. 222,000), Tyne and Wear, NE England, on the Tyne River opposite Newcastle upon Tyne. There are locomotive works and railroad shops. Flour milling, shipbuilding, light engineering, packaging, and the manufacture of clothing, glass, iron goods, cables, and chemicals are important. There are coal mines and grindstone quarries in the vicinity. Gateshead is a very old community, probably dating back to Saxon times. The iron and steel industry developed in the 19th cent.

Gates·ville (gāts′vĭl′). **1.** Town (1970 pop. 338), seat of Gates co., NE N.C., WNW of Elizabeth City, in an agricultural and timber area. It has hunting and fishing areas nearby. **2.** City (1970 pop. 4,683), seat of Coryell co., central Texas, on the Leon River WSW of Waco. The area has cotton, grain, and poultry farms. Fort Gates, a long-vanished military post, was established nearby in 1849.

Gate·way National Recreation Area (gāt′wā′): *see* National Parks and Monuments Table.

Gat·i·neau (găt′ĭ-nō), river, c.240 mi (385 km) long, rising in the Laurentians, SW Que., Canada, and flowing S to the Ottawa River at Hull. There are several rapids with hydroelectric power plants.

Ga·tún Lake (gä-to͞on′), artificial lake, 163 sq mi (422.2 sq km), Panama Canal Zone, formed by the impounding of the Chagres River. Gatún Dam (completed 1912), 1.5 mi (2.4 km) long and 115 ft (35 m) high, controls the level of the lake.

Gau·ha·ti (gou-hä'tē), town (1971 pop. 123,783), Assam state, NE India, on the Brahmaputra River. It is a railroad and shipping point for tea, rice, jute, and cotton. The town has an oil refinery and is the site of Gauhati Univ. and Earle Law School.

Gaul (gôl), ancient designation for the land S and W of the Rhine, W of the Alps, and N of the Pyrenees. The name was extended by the Romans to include Italy from Lucca and Rimini northward, excluding Liguria. This extension of the name is derived from its settlers of the 4th and 3rd cent. B.C.—invading Celts, who were called Gauls by the Romans. The Celts in Gaul proper (modern France) probably had been there since 600 B.C. The Italian area of Gaul was called Cisalpine Gaul, as opposed to Transalpine Gaul; Cisalpine Gaul was divided into Cispadane Gaul and Transpadane Gaul. By 121 B.C. Rome had acquired southern Transalpine Gaul.

Julius Caesar conquered Gaul in the Gallic Wars (58 B.C.-51 B.C.). He immortalized its three ethnic divisions, Aquitania (south of the Garonne), Celtic Gaul (modern central France), and Belgica (very roughly Belgium). Aquitania was probably inhabited by the ancestors of the Basques, and the Belgae were probably Celts, like the rest of the Gauls. On the basis of these distinctions, Augustus in 27 B.C. set up great administrative divisions: Narbonensis, under the direct rule of the Roman senate; Aquitania, extending from the Pyrenees to the Loire; Lugdunensis (Celtic Gaul), a central strip mainly between the Loire and the Seine rivers; and Belgica, including most of the rest. The Romanization of Gaul was rapid; the only serious attempt to rebel against Rome was the uprising of Postumus (A.D. 257), but Gallo-Roman civilization was too strong to fall before anything but the Germans of the 5th and 6th cent.

Gaur (gour), ruined city, West Bengal state, India. The city was an ancient Hindu capital of Bengal. It was captured (c.1200) by the Moslems and remained a center of their culture until its abandonment in the late 16th cent.

Ga·var·nie (gä-vär-nē'), village (1968 pop. 167), Hautes-Pyrénées dept., SW France, in the central Pyrenees. Nearby are a celebrated waterfall, 1,385 ft (422.4 m) high, and a gigantic natural amphitheater, the Cirque de Gavarnie. The amphitheater rises in concentric levels from an altitude of c.5,740 ft (1,750 m) and is enclosed by crests reaching altitudes of more than 9,000 ft (2,745 m).

Gäv·le (yäv'lē, yĕv'lə), city (1975 est. pop. 86,911), capital of Gävleborg co., E Sweden, on the Gulf of Bothnia. Although icebound for three months of the year, the port of Gävle has a busy export trade. Manufactures of the city include chemicals, textiles, and beer. Chartered in 1446, Gävle was largely destroyed by a fire in 1869 and was rebuilt in a modern style.

Ga·ya (gä'yə, gī'ə), city (1971 pop. 179,826), Bihar state, E central India. The region is sacred to Buddhist and Hindu pilgrims, who visit the temple of Vishnupad. Bodh Gaya, the site of the Buddha's enlightenment, is to the south. The city processes cotton, jute, sugar, and stone.

Gay·lord (gā'lôrd'). **1.** City (1970 pop. 3,012), seat of Otsego co., N Mich., NE of Traverse City, in a farm and resort area; inc. as a village 1881, as a city 1922. **2.** City (1970 pop. 1,720), seat of Sibley co., S central Minn. SW of Minneapolis; platted 1881. It is a farm trade center.

Ga·za or **Ghaz·zah** (both: gä'zə, găz'ə), town (1968 est. pop. 118,300), SW Asia, on the Philistia plain between the Mediterranean Sea and W Israel. Anciently it was an Egyptian garrison town and later was one of the chief cities of the Philistines. Gaza was besieged for five months by Alexander the Great (332 B.C.) and during the wars of the Maccabees and in the Crusades. The town has long been of commercial importance, the meeting place of caravans between Egypt and Syria. The site of modern Gaza dates from the building programs of Herod the Great. Present-day Gaza is the principal city and administrative center of the Gaza Strip (1971 est. pop. 365,000), a rectangular coastal area (c.140 sq mi/370 sq km), in what was formerly southwest Palestine. It is a densely populated and impoverished region inhabited largely by Arab refugees from Israel. The Gaza Strip has some farming, a modest citrus-fruit industry, and livestock grazing. Between 1917 and 1948 the region was part of Great Britain's Palestine mandate from the League of Nations. After the armistice agreement of 1949 until the 1967 War (with the exception of the Israeli occupation from Nov., 1956, to Mar., 1957), the Gaza Strip was under Egyptian administration; after the 1967 War, Israel occupied the region. Since the late 1940s the Gaza Strip has been the scene of many clashes between Egyptian, Arab guerrilla, and Israeli forces. In the peace treaty signed in 1979 Israel and Egypt agreed to begin negotiations concerning Palestinian autonomy in the area.

gaz·et·teer (găz'ĭ-tîr'), dictionary or encyclopedia listing alphabetically the names of places, political divisions, and physical features of the earth and giving some information about each. The term was first used in its modern sense early in the 18th cent., after the publication (1703) by Lawrence Echard of the *Gazetteer's or Newsman's Interpreter,* a geographic index.

Ga·zi·an·tep (gä'zē-än-tĕp'), formerly **Ain·tab** (īn-täb'), city (1975 pop. 300,801), capital of Gaziantep prov., S Turkey. Gaziantep is an important trading and manufacturing center known for its textiles and pistachio nuts. An ancient Hittite city, it was occupied (8th cent. B.C.) by Sargon of Assyria. It occupied a strategic position in the Crusades and was taken by Saladin in 1183. It was the center of Turkish resistance (1920-21) to the French occupation of the region. After a long siege it was captured by the French, but was returned to Turkey in 1921.

Gdańsk (gə-dänsk', -dănsk') or **Dan·zig** (dăn'sĭg), city (1975 est. pop. 421,000), capital of Gdańsk prov., N Poland, on a branch of the Vistula River and the Gulf of Danzig. One of the chief Polish ports on the Baltic Sea, it is a leading industrial and communications center. Its shipyards are among the world's largest, and it has important mechanical-engineering, machine-building, chemical, and metallurgical industries. Sawmilling, food processing, brewing, distilling, and light manufacturing are also important.

An old Slavic settlement, Gdańsk was first mentioned in 997. After its settlement by German merchants, it joined (13th cent.) the Hanseatic League and developed as an important Baltic trading port. In 1308 it was conquered by the Teutonic Knights and became an object of struggle between their order and Poland. Gdańsk passed to Poland in 1466 and was granted local autonomy under the Polish crown. After the Thirty Years' War the city began to decline. In the War of the Polish Succession, King Stanislaus I took refuge in Gdańsk until it fell (1734) after a heroic defense. The first partition of Poland in 1772 made Gdańsk a free city; the second partition (1793) gave it to Prussia. Napoleon I restored its status as a free city (1807). Reverting to Prussia in 1814, it was fortified and, as Danzig, was the provincial capital of West Prussia until 1919, when by the Treaty of Versailles it once more became a free city.

In order to give Poland a seaport, Gdańsk was included in the Polish customs territory and was placed under a high commissioner appointed by the League of Nations; as the League's authority waned after 1935, Gdańsk came under Nazi control. Hitler's demand (1939) for the city's return to Germany was the principal immediate cause of the German invasion of Poland and thus of World War II. Gdańsk was annexed to Germany from Sept. 1, 1939, until its fall to the Soviet army early in 1945. The Allies returned the city to Poland, which restored the name Gdańk. In 1970 workers' grievances sparked riots in Gdańsk that spread to other cities and led to changes in Poland's national leadership.

Gdy·nia (gə-dĭn'ē-ə, -dĭn'yə), city (1975 est. pop. 221,000), N Poland, a port on the Baltic Sea and the Gulf of Danzig. It is an important rail center with industries producing metals, machinery, and food products. Originally a small German fishing village, it was transferred to Poland after World War I. By 1934 Gdynia was a leading Baltic port. Although the harbor was heavily damaged in World War II, the city suffered relatively little destruction.

Gear·y (gîr'ē), county (1970 pop. 28,111), 399 sq mi (1,033.4 sq km), NE central Kansas, drained in the W by the Smoky Hill and Republican rivers, which join here to form the Kansas River; formed 1855 as Davis co., renamed 1889; co. seat Junction City. It is in a region of low hills that produces wheat, corn, milo, soybeans, and livestock.

Ge·au·ga (jē-ô'gə), county (1970 pop. 62,977), 407 sq mi (1,054.1 sq km), NE Ohio, drained by the Cuyahoga, Chagrin, and Grand rivers and several small lakes; formed 1805; co. seat Chardon. In an agricultural area yielding grain, fruit, truck crops, poultry, and dairy products, it has oil and natural-gas wells, sand and gravel pits, and a food-processing industry.

Gee·long (jĭ-lông'), city (1975 pop. 15,727; urban agglomeration pop. 115,047), Victoria, SE Australia, on an inlet of Port Phillip Bay. Among its industries are wool trading, meat-packing, oil refining, and the manufacture of fertilizers and automobiles.

Ge·la (jā'lä), city (1976 est. pop. 73,268), S Sicily, Italy, on the Mediterranean Sea. It is a port, industrial center, and seaside resort. Petroleum is produced nearby and is refined in the city. Much cotton is grown in the area. The city was founded c.688 B.C. by Greek colonists from Crete and Rhodes and attained its greatest prosperity in the 5th cent. B.C. It was sacked by Carthage in 405 B.C. and never fully recovered. In a necropolis near Gela, Greek vases and other objects have been found; excavations (begun in 1901) have uncovered the ancient Greek wall of Gela (5th–4th cent. B.C.) and two temples. The modern city was founded by Emperor Frederick II in 1230. In World War II Gela was a landing point (July, 1943) for the Allied invasion of Sicily.

Gel·der·land or **Guel·der·land** (both: gĕl'dər-lănd', KHĕl'dər-länt'), province (1977 est. pop. 1,653,516), c.1,940 sq mi (5,025 sq km), E central Netherlands, bordering on West Germany in the E; capital Arnhem. It is an agricultural region. The duchy of Gelderland was conquered (1473) by Charles the Bold of Burgundy, after whose death (1477) it regained its independence. It passed to the House of Hapsburg in 1543 and joined (1579) the Union of Utrecht of the Netherlands against Spain. Part of Gelderland was ceded (1715) by the Netherlands to Prussia.

Gel·li·gaer (gĕl'ĭ-gîr'), urban district (1971 pop. 33,670), Mid Glamorgan, S Wales. Three coal-mining valleys are in the district. Automobiles and rubber products are manufactured.

Gel·sen·kir·chen (gĕl'zən-kĭr'KHən), city (1974 est. pop. 327,591), North Rhine-Westphalia, W West Germany, a port on the Rhine-Herne Canal. It is a major industrial and coal-mining center of the Ruhr district. Manufactures include iron and steel, chemicals, glass, and clothing. Gelsenkirchen was a small village in 1850, but grew rapidly after the opening of the first coal mines in the 1850s.

Gem (jĕm), county (1970 pop. 9,387), 555 sq mi (1,437.5 sq km), SW Idaho, with mountains in the N and an irrigated area along the Payette River in the S; formed 1815; co. seat Emmett. Agriculture (hay, sugar beets, fruit, truck crops, and dairy products) and lumbering are important. Woodworking machinery is made.

Gem·mi (gĕm'ē), pass, 7,620 ft (2,324.1 m) high, S Switzerland, connecting Bern and Valais cantons, in the Bernese Alps.

Gen·er·al Grant National Memorial (jĕn'ər-əl grănt'): *see* National Parks and Monuments Table.

Gen·e·see (jĕn'ĭ-sē'). **1.** County (1970 pop. 445,589), 644 sq mi (1,668 sq km), SE central Mich., drained by the Flint and Shiawassee rivers; organized 1836; co. seat Flint. It has agriculture (grain, hay, sugar beets, truck crops, fruit, livestock, and poultry), marl deposits, and food-processing and tourism industries. Among its extensive manufactures are textiles, wood products, furniture, paper, chemicals, plastics, concrete, metal products, and machinery. **2.** County (1970 pop. 58,722), 501 sq mi (1,297.6 sq km), W N.Y., drained by Tonawanda and Oak Orchard creeks; formed 1802; co. seat Batavia. In an agricultural area yielding wheat, hay, corn, potatoes, truck crops, and dairy products, it has deposits of limestone and gypsum and a food-processing industry. Clothing, paper, concrete, machinery, and metal products are made. Part of Tonawanda Indian Reservation is in the county.

Genesee, river, 158 mi (254.2 km) long, rising in the Allegheny Mts., N Pa., and flowing through W N.Y. to Lake Ontario at Rochester; it is crossed by the New York State Barge Canal. The Genesee valley is noted for its fertility and beauty.

Gen·e·se·o (jĕn'ĭ-sē'ō). **1.** City (1970 pop. 5,840), Henry co., NW Ill., ESE of Moline on the Illinois and Mississippi Canal; settled 1836, inc. 1855. It is a farm trade center. **2.** Farm trade village (1970 pop. 5,714), seat of Livingston co., W central N.Y., on the Genesee River SSW of Rochester; settled c.1790, inc. 1832.

Ge·ne·va (jə-nē'və), canton (1977 est. pop. 400,000), 109 sq mi (282.3 sq km), SW Switzerland, surrounding the SW tip of the Lake of Geneva. Geneva is in the plain between the Jura Mts. and the Alps and is almost entirely surrounded by French territory. Fruit, vegetables, cereals, and wine are produced in the rural areas, while industry—as well as population—is chiefly centered in the city of Geneva (1977 est. pop. 152,600), the capital of the canton. Situated on the Lake of Geneva and divided by the Rhône River, which emerges from the lake, it is a picturesque city joined by numerous bridges. Geneva is a cultural, financial, and administrative center and manufactures watches, jewelry, precision instruments, machinery, automobiles, aluminumware, chocolate, and clothes.

Geneva was an ancient Celtic settlement and was later included in Roman Gaul. It passed successively to the Burgundians (5th cent.), the Franks (6th cent.), Transjurane Burgundy (9th–11th cent.), and the Holy Roman Empire. The bishops of Geneva gradually absorbed the powers of the feudal counts of Geneva and in 1124 became rulers of the city. In 1285 the citizens of Geneva placed themselves under the protection of the counts (later dukes) of Savoy, and by 1387 they had won extensive rights of self-rule. However, by gradually transforming the bishops into their tools, the dukes nearly succeeded in mastering the city by the beginning of the 16th cent. Incensed, the citizens allied themselves with two Swiss cantons—Fribourg and Bern—expelled the bishop (1533), and accepted (1535) the Reformation preached by Guillaume Farel. The arrival (1536) of John Calvin thrust upon Geneva a role of European importance as the focal point of the Reformation. With its population augmented by Protestant refugees, notably Huguenots, Geneva became a cosmopolitan intellectual center. During the 18th cent., when the stern theocracy of Calvin had mellowed into patrician rule, the city's intellectual life reached its zenith.

The city was annexed to France from 1798 to 1813 and joined Switzerland as a canton in 1815. In 1864 Geneva was made the seat of the International Red Cross; it was also the seat of the League of Nations (1920–46). Geneva is headquarters for the International Labor Organization, the World Health Organization, and other international bodies.

Geneva, county (1970 pop. 21,924), 578 sq mi (1,497 sq km), SE Ala., in a coastal-plains area bordering on Fla. and drained by the Pea and Choctawhatchee rivers; formed 1868; co. seat Geneva. Cotton and bees are raised. Lumbering and textile-related industries are also important. Clothing, plastics, and metal products are manufactured.

Geneva. 1. City (1970 pop. 4,398), seat of Geneva co., SE Ala., on the Choctawhatchee River near Fla. border; settled 1836, inc. 1872. It is a trade center, and clothing is made. **2.** City (1970 pop. 10,721), seat of Kane co., NE Ill., on the Fox River W of Chicago; founded c.1833, inc. 1867. Machine tools, electronic equipment, and steel cabinets are manufactured. **3.** City (1970 pop. 2,275), seat of Fillmore co., SE Nebr., WSW of Lincoln; laid out 1858. It is a farm trade center and a livestock-shipping point in a prairie region. **4.** City (1978 est. pop. 16,300), Ontario co., W central N.Y., in the Finger Lakes region; inc. as a city 1897. Located in a farm area, Geneva's manufactures include cans and canning machinery, paper containers, metal products, and water-purification systems. There are also printing plants. It is the seat of Hobart and William Smith colleges. **5.** City (1970 pop. 6,449), Ashtabula co., NE Ohio, WSW of Ashtabula near Lake Erie, in a truck-farming area; settled 1805, inc. 1866. Farm implements and hardware are manufactured.

Geneva, Lake, crescent-shaped lake, 224 sq mi (580.2 sq km), c.45 mi (70 km) long, on the Swiss-French border, between the Alps and the Jura Mts. It has a maximum depth of 1,017 ft (310.2 m). Noted for its deep-blue and remarkably transparent waters, the lake is dotted with numerous resorts and villas. It is subject to seiches, tidal fluctuations that suddenly change the lake's level.

Genk (KHĕngk), city (1977 est. pop. 61,156), Limburg prov., NE Belgium. It is a commercial and industrial center.

Gen·ne·vil·liers (zhĕn-vē-lyā'), town (1975 pop. 50,290), Hauts-de-Seine dept., N central France, on the right bank of the Seine River. It is mainly an industrial community; aircraft equipment, electrical products, radio tubes, and automobiles are manufactured.

Gen·o·a (jĕn'ō-ə), city (1976 est. pop. 800,532), capital of Genoa prov. and of Liguria, NW Italy, on the Ligurian Sea. Genoa's harbor facilities, badly damaged in World War II and by storms in 1954–55, have been rebuilt and greatly modernized. The city is also a commercial and industrial center; manufactures include iron and steel, chemicals, petroleum, airplanes, motor vehicles, and textiles. There are large shipyards in the city.

An ancient town of the Ligures, Genoa flourished under Roman rule. Around the 10th cent. it became a free commune. Helped by Pisa, Genoa drove (11th cent.) the Arabs from Corsica and Sardinia. Rivalry over control of Sardinia resulted in long wars with Pisa; Genoa finally triumphed in the naval battle of Meloria (1284). The Crusades brought Genoa great wealth. The republic acquired possessions and trading privileges in areas from Spain to the Crimea. Genoa's expansion and its military defense were largely financed by a group of merchants, who in 1408 organized a powerful bank, the Banco San Giorgio. Genoese policy in the Orient clashed with the ambitions of Venice, and long wars resulted, ending with the Peace of Turin (1381), which slightly favored Venice. Meanwhile the Genoese republic was weakened by factional strife between the Guelphs and Ghibellines.

From the late 14th to the 16th cent., France and Milan in turn controlled the city, although nominal independence was preserved. The power of Genoa was revived by the seaman and statesman Andrea Doria, who wrote a new constitution in 1528. Later the city came under Spanish, French, and Austrian control. The Austrians were expelled by a popular uprising in 1746, but in 1768 Genoa had to cede Corsica, its last outlying possession, to France. In 1797 French military pressure resulted in the end of aristocratic rule and the formation of the Ligurian Republic, which Napoleon I formally annexed to France in 1805. The Congress of Vienna united (1814) Genoa and Liguria with the kingdom of Sardinia.

Among Genoa's notable buildings are the Cathedral of San Lorenzo (rebuilt in 1100 and frequently restored), the palace of the doges, the richly decorated churches of the Annunciation and of St. Ambrose (both 16th cent.), the medieval Church of San Donato, and many Renaissance palaces. The city is surrounded by old walls and forts, and the streets of the harbor section are very picturesque.

Gen·try (jĕn'trē), county (1970 pop. 8,060), 488 sq mi (1,263.9 sq km), NW Mo., drained by the Grand River; formed 1841; co. seat Albany. It is in an agricultural area that produces corn, wheat, oats, livestock, poultry, and bluegrass seed.

ge·og·ra·phy (jē-ŏg'rə-fē), the science of place, or the study of the surface of the earth, the location and distribution of its physical and cultural features, and the interrelation of these features as they affect man. It integrates data in a study of areal differentiation in the world, making elaborate use of maps as its special tool. Geography was first systematically studied by the ancient Greeks. The Roman contribution to geography was in the exploration and mapping of previously unknown lands. Greek geographic learning was maintained and enhanced by the Arabs during the Middle Ages. The journeys of Marco Polo in the latter part of the Middle Ages began the revival of geographic interest outside the Moslem world. With the Renaissance in Europe came the desire to explore unknown parts of the world that led to the voyages of exploration and to the great discoveries. However, it was mercantile interest rather than a genuine search for knowledge that spurred these endeavors. The 16th and 17th cent. reintroduced sound theoretical geography in the form of textbooks and maps. In the 18th cent. geography began to achieve recognition as a discipline and was taught for the first time at the university level.

Since the end of World War II geography, like other disciplines, has experienced the explosion of knowledge brought on by the new tools of modern technology for the acquisition and manipulation of data; these include aerial photography, remote sensors (including infrared and satellite photography), and the computer (for quantitative analysis and mapping). Today geography is studied by governmental agencies and in many of the world's universities. Research is stimulated by such noted geographic institutions as the Royal Geographical Society (1830, Great Britain), the American Geographical Society (1852, United States), and the Société de Géographie (1821, France).

George (jôrj), county (1970 pop. 12,459), 481 sq mi (1,245.8 sq km), SE Miss., bordered on the E by Ala. and drained by the Pascagoula River; formed 1910; co. seat Lucedale. It has agriculture (cotton, corn, and truck crops) and lumbering.

George, river, c.345 mi (555 km) long, rising in a lake on the Que.-Labrador boundary, E Canada. It flows north through Indian Lake to Ungava Bay (an arm of Hudson Strait).

George, Lake, glacial lake, 33 mi (53 km) long and 1 to 3 mi (1.6–4.8 km) wide, in the foothills of the Adirondack Mts., NE N.Y., draining N into Lake Champlain. The lake was discovered in 1646. During the French and Indian Wars and the American Revolution, the area around Lake George was the scene of many battles. Today the lake is the center of a large resort area.

George Rog·ers Clark National Historical Park (rŏj'ərz klärk'): *see* National Parks and Monuments Table.

George·town (jôrj'toun'), city (1970 pop. 66,070), capital of Guyana, on the Atlantic Ocean at the mouth of the Demerara River. It was known as Stabroek when the Dutch controlled the region and was renamed Georgetown in 1812, after the British had occupied the colony during the Napoleonic Wars.

Georgetown, county (1970 pop. 33,500), 813 sq mi (2,105.7 sq km), E S.C., bounded on the E by the Atlantic Ocean, on the NE by the Pee Dee River, on the S by the Santee River; formed 1798; co. seat Georgetown. It is watered by the Waccamaw and Black rivers. In a lumbering, farming (grain and tobacco), dairying, and fishing area, it has a flourishing tourist industry.

Georgetown. 1. Town (1970 pop. 542), seat of Clear Creek co., N central Colo., in the Front Range on the headstream of Clear Creek W of Denver. It has gold, silver, lead, and copper mines. Prior to 1878 it was one of the most important silver camps in the state. **2.** Town (1970 pop. 1,844), seat of Sussex co., S Del., SSE of Dover, in a farm area; inc. 1869. **3.** Town (1970 pop. 860), seat of Quitman co., SW Ga., on the Chattahoochee River WNW of Cuthbert, in a sawmilling area. **4.** City (1970 pop. 8,629), seat of Scott co., N central Ky., NNW of Lexington; settled 1776 as a pioneer station, inc. 1790. It grew around Royal Spring, which still supplies the city's water. Georgetown is a processing center in a bluegrass farming and horse-breeding region. **5.** Village (1970 pop. 3,087), seat of Brown co., SW Ohio, near the Ohio River SE of Cincinnati; laid out 1819, inc. 1832. It is a trade center for tobacco and other farm products. Shoes are manufactured. The boyhood home of Ulysses S. Grant is here; his birthplace, Point Pleasant, is to the west. **6.** City (1970 pop. 10,449), seat of Georgetown co., E S.C., on the Sampit River at its entrance into Winyah Bay; inc. 1805. It is a port of entry, a resort, and a shipping center, with a steel industry and one of the world's largest paper mills. Rice and indigo plantations were established in the area in the early 1700s, and the city was founded c.1734 as a shipping point for their products. **7.** City (1970 pop. 6,395), seat of Williamson co., central Texas, N of Austin; founded 1848, inc. 1871. It is a market for sheep and goat ranches and processes cotton, dairy products, and poultry.

George Wash·ing·ton Car·ver National Monument (wôsh'ĭng-tən kär'vər): *see* National Parks and Monuments Table.

George Washington Memorial Parkway: *see* National Parks and Monuments Table.

George West (wĕst), city (1970 pop. 2,022), seat of Live Oak co., S Texas, on the Nueces River W of Beeville. It is a market and shipping point for an oil, ranch, and farm area.

Geor·gia (jôr'jə), state (1975 pop. 4,947,000), 58,876 sq mi (152,489 sq km), SE United States, the last of the Thirteen Colonies founded (1733). Atlanta is the capital. Georgia is bounded on the north by Tenn. and N.C., on the east by S.C. and the Atlantic Ocean, on the south by Fla., and on the west by Ala. A number of islands, part of the Sea Islands chain, lie off Georgia's coastline.

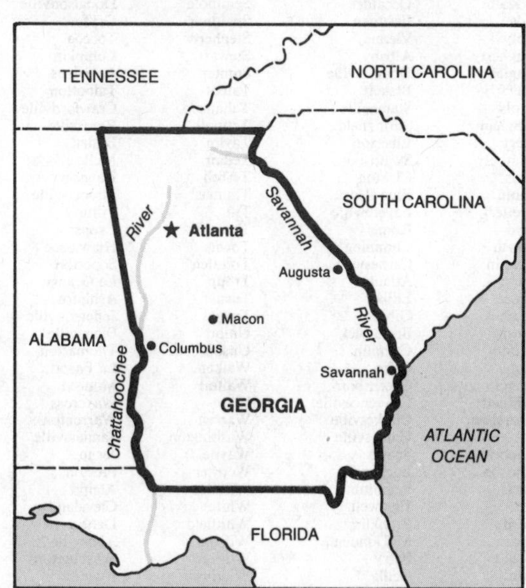

The state has three main topographical areas. Extending inland from the coast is a low coastal plain that covers the southern half of the state. In mountainous north Georgia are the Appalachian Plateau and sections of the Appalachian Valley and the Blue Ridge. Bridging these two sections and embracing about one third of the state is the Piedmont plateau in middle Georgia. Georgia is well drained by many rivers.

With its Southern charm and beauty, the state is a popular vacation area. Georgia's attractions include the Sea Islands; Okefenokee Swamp, a large wilderness area; Chattahoochee and Oconee national forests, with facilities for hunting and fishing; Chickamauga and Chattanooga National Military Park; Kennesaw Mountain National Battlefield Park; and Stone Mountain, near Atlanta, on which is carved a Confederate memorial.

Georgia is divided into 159 counties:

NAME	COUNTY SEAT	NAME	COUNTY SEAT
Appling	Baxley	Jeff Davis	Hazlehurst
Atkinson	Pearson	Jefferson	Louisville
Bacon	Alma	Jenkins	Millen
Baker	Newton	Johnson	Wrightsville
Baldwin	Milledgeville	Jones	Gray
Banks	Homer	Lamar	Barnesville
Barrow	Winder	Lanier	Lakeland
Bartow	Cartersville	Laurens	Dublin
Ben Hill	Fitzgerald	Lee	Leesburg
Berrien	Nashville	Liberty	Hinesville
Bibb	Macon	Lincoln	Lincolnton
Bleckley	Cochran	Long	Ludowici
Brantley	Nahunta	Lowndes	Valdosta
Brooks	Quitman	Lumpkin	Dahlonega
Bryan	Pembroke	McDuffie	Thomson
Bulloch	Statesboro	McIntosh	Darien
Burke	Waynesboro	Macon	Oglethorpe
Butts	Jackson	Madison	Danielsville
Calhoun	Morgan	Marion	Buena Vista
Camden	Woodbine	Meriwether	Greenville
Candler	Metter	Miller	Colquitt
Carroll	Carrollton	Mitchell	Camilla
Catoosa	Ringgold	Monroe	Forsyth
Charlton	Folkston	Montgomery	Mount Vernon
Chatham	Savannah	Morgan	Madison
Chattahoochee	Cusseta	Murray	Chatsworth
Chattooga	Summerville	Newton	Covington
Cherokee	Canton	Oconee	Watkinsville
Clarke	Athens	Oglethorpe	Lexington
Clay	Fort Gaines	Paulding	Dallas
Clayton	Jonesboro	Peach	Fort Valley
Clinch	Homerville	Pickens	Jasper
Cobb	Marietta	Pierce	Blackshear
Coffee	Douglas	Pike	Zebulon
Colquitt	Moultrie	Polk	Cedartown
Columbia	Appling	Pulaski	Hawkinsville
Columbus	Columbus	Putnam	Eatonton
Cook	Adel	Quitman	Georgetown
Coweta	Newnan	Rabun	Clayton
Crawford	Knoxville	Randolph	Cuthbert
Crisp	Cordele	Richmond	Augusta
Dade	Trenton	Rockdale	Conyers
Dawson	Dawsonville	Schley	Ellaville
Decatur	Bainbridge	Screven	Sylvania
De Kalb	Decatur	Seminole	Donalsonville
Dodge	Eastman	Spalding	Griffin
Dooly	Vienna	Stephens	Toccoa
Dougherty	Albany	Stewart	Lumpkin
Douglas	Douglasville	Sumter	Americus
Early	Blakely	Talbot	Talbotton
Echols	Statenville	Taliaferro	Crawfordville
Effingham	Springfield	Tattnall	Reidsville
Elbert	Elberton	Taylor	Butler
Emanuel	Swainsboro	Telfair	McRae
Evans	Claxton	Terrell	Dawson
Fannin	Blue Ridge	Thomas	Thomasville
Fayette	Fayetteville	Tift	Tifton
Floyd	Rome	Toombs	Lyons
Forsyth	Cumming	Towns	Hiawassee
Franklin	Carnesville	Treutlen	Soperton
Fulton	Atlanta	Troup	La Grange
Gilmer	Ellijay	Turner	Ashburn
Glascock	Gibson	Twiggs	Jeffersonville
Glynn	Brunswick	Union	Blairsville
Gordon	Calhoun	Upson	Thomaston
Grady	Cairo	Walker	La Fayette
Greene	Greensboro	Walton	Monroe
Gwinnett	Lawrenceville	Ware	Waycross
Habersham	Clarkesville	Warren	Warrenton
Hall	Gainesville	Washington	Sandersville
Hancock	Sparta	Wayne	Jesup
Haralson	Buchanan	Webster	Preston
Harris	Hamilton	Wheeler	Alamo
Hart	Hartwell	White	Cleveland
Heard	Franklin	Whitfield	Dalton
Henry	McDonough	Wilcox	Abbeville
Houston	Perry	Wilkes	Washington
Irwin	Ocilla	Wilkinson	Irwinton
Jackson	Jefferson	Worth	Sylvester
Jasper	Monticello		

Economy. Cotton, once Georgia's chief crop, has declined in importance; in the early 1970s it ranked in value behind peanuts, tobacco, and corn. Livestock and poultry account for the largest share of farm income; broilers, eggs, and cattle are the most important products. The manufacture of textiles and textile products has long been Georgia's leading industry, centering mainly around Columbus, Augusta, Macon, and Rome. Other major manufactures include transportation equipment, food products, paper products, and chemicals. Much of Georgia is heavily forested with pine, and the state is a leading producer in the South of lumber and pulpwood. Georgia is also a major world supplier of naval stores, especially turpentine and resin. The most valuable minerals produced are clays, stone, kaolin, iron ore, sand, and gravel. Georgia is famous for its fine marble.

History. Creek and Cherokee Indians inhabited the Georgia area when Hernando De Soto and his expedition passed through the region c.1540. The Spanish later established missions and garrisons on the Sea Islands. In 1663 Charles II of England made a grant of land that included Georgia to the eight proprietors of Carolina. An English-Spanish contest for the territory between Charleston (S.C.) and St. Augustine (Fla.) continued intermittently for almost a century. In June, 1732, the English philanthropist James E. Oglethorpe and 19 associates received a charter from George II (for whom the colony was named) to settle and govern the colony of Georgia. The first colonists, led by Oglethorpe, reached the mouth of the Savannah River in Feb., 1733. On a bluff c.18 mi (29 km) upstream, the colonists laid out the first town, Savannah. In 1739 war broke out between Spain and England. Fighting occurred in Georgia, and in 1742, on St. Simons Island, Oglethorpe defeated the Spanish in the Battle of Bloody Marsh, thereby effectively ending Spain's claim to the land north of the St. Marys River.

Georgia's early settlers included Englishmen, Welshmen, Scots Highlanders, Germans, Piedmontese, and Swiss. Jews, Catholics, Negro slaves, and settlers from other American colonies were barred at first. By the time Georgia became a royal colony in 1754, the ban on slaves had been abolished. Georgia fitted well into the British mercantile system, exporting to England rice, indigo, deerskins, lumber, naval stores, beef, and pork and buying there the manufactured articles it needed.

After American independence had been won, Georgia was the first Southern state to ratify (1788) the Constitution. Difficulties with the Federal government stemmed from the related problems of Indian removal and land speculation. Georgia ceded (1802) its western lands to the United States in return for $1,250,000 and a pledge that the Indians would be removed from Georgian lands. By 1826 the Creek Indians had yielded their lands, but in 1827 the Cherokee Indians set themselves up as an independent nation. In 1838 the state forced the Indian chief John Ross to lead his people west.

With the invention of the cotton gin (1793) by Eli Whitney, Georgia began to prosper as a cotton-growing state; by the 1840s a textile industry was established. On Jan. 19, 1861, Georgia seceded from the Union and shortly afterward joined the Confederacy. Georgia became a major Civil War battlefield when, in 1864, Union Gen. William Tecumseh Sherman set fire to Atlanta and marched to the sea, leaving a path of great destruction. After the legislature approved the Fifteenth Amendment (the Thirteenth and Fourteenth having been ratified earlier), Georgia was readmitted (1870) to the Union. The textile industry recovered and was expanding by the 1880s, although agriculture lagged.

After World War I agriculture suffered from boll weevil infestation, soil erosion, and overuse. The state weathered the Depression, but its subsequent history was marked by political and racial conflict. The 1945 constitution contained a provision for Georgia's notorious county-unit system, which led to the political control of urban areas by sparsely populated rural areas. The system was finally abolished by Federal court order in 1962. The integration of public schools, following the 1954 Supreme Court decision, was strenuously opposed by many Georgians. However, in 1961 the legislature abandoned a "massive resistance" policy, and Georgia became the first state in the Deep South to proceed with integration without a major curtailment of its public school system.

Government. Georgia's constitution provides for an elected governor who serves for four years and may not be re-elected to a consecutive term. The legislature, called the general assembly, is made up of a senate with 56 members and a house of representatives with 180 members. Members of both houses are elected for two years.

Educational Institutions. The most prominent are the Univ. of Georgia, at Athens; Georgia Institute of Technology, Emory Univ., Clark College, and Morris Brown College, at Atlanta; Agnes Scott

College, at Decatur; and Mercer Univ. and Wesleyan College, at Macon.

Georgia, Strait of, channel, c.150 mi (240 km) long, between the mainland of British Columbia and Vancouver Island, Canada, linking Puget Sound and Queen Charlotte Sound. It forms part of the inland steamship passage to Alaska.

Geor·gian Bay (jôr'jən), extension of Lake Huron, S Ont., Canada, separated from Lake Huron by Manitoulin Island and by the Bruce Peninsula. Georgian Bay is connected with Lake Ontario by the Severn River and Trent Canal. Many of the well-timbered, rock-bound islands of Georgian Bay are summer resorts.

Georgian Soviet Socialist Republic or **Georgia,** constituent republic (1976 est. pop. 4,954,000), c.26,900 sq mi (69,675 sq km), SE European USSR, in W Transcaucasia; capital Tbilisi. Georgia borders on the Black Sea in the west, on Turkey and the Armenian Republic in the south, on the Azerbaijan Republic in the east, and on the Russian Soviet Federated Socialist Republic in the north. Agriculture is the leading occupation in Georgia. The republic is also rich in minerals, notably manganese.

Georgia developed as a kingdom about the 4th cent. B.C. The Persian Sassanids, who ruled the country from the 3rd cent. A.D., were expelled c.400. In the 12th and 13th cent. Georgia reached its greatest expansion and cultural flowering. Ravaged (13th cent.) by the Mongols, Georgia revived but was again sacked by Tamerlane (c.1386–1403). In the 15th cent. King Alexander I divided Georgia among his three sons, and the period of decline set in. In the 16th cent. Georgia became an object of struggle between Turkey and Persia. In 1555 western Georgia passed under Turkish suzerainty and eastern Georgia under Persian rule. Finally, between 1801 and 1829 Russia acquired all of Georgia. After the Russian Revolution of 1917, the Georgian Menshevik party proclaimed (May, 1918) Georgia's independence. The Soviet government in Moscow recognized (May, 1920) the independence, but in 1921 the Red Army invaded Georgia, and in Feb., 1921, it was proclaimed a soviet republic. It joined the USSR in 1922 and in 1936 became a separate republic.

Geor·gi·na (jôr-jē'nə), intermittent river, c.700 mi (1,125 km) long, E Northern Territory and W Queensland, Australia.

Ge·or·gi·yevsk (gē-ôr'gē-ĭfsk), city (1976 est. pop. 50,000), SE European USSR, in the N foothills of the Caucasus. It is an agricultural center with some industry. It was founded (1777) as a fortress.

Ge·ra (gā'rä), city (1974 est. pop. 113,108), capital of Gera dist., S East Germany, on the White Elster River. It is an industrial center and a rail and road junction. Manufactures include textiles, metal products, and furniture. Gera was chartered in the early 13th cent.

Ge·raards·ber·gen (KHä'rärts-bĕr'KHən), town (1977 est. pop. 30,632), East Flanders prov., W central Belgium, on the Dender River. Manufactures include textiles, watches, and chemicals. Of note is the 15th cent. Gothic town hall.

Ger·a·sa (jĕr'əs-ə), **Ge·rash,** or **Je·rash** (both: jĕr'ăsh), ancient city of the Decapolis, N of Amman, in present-day Jordan. It was a flourishing city in the 2nd and 3rd cent. A.D. The Graeco-Roman city is probably the best-preserved Palestinian city of Roman times. The site is covered with interesting Roman ruins, including a long colonnaded street with more than a hundred columns still standing, a great theater and a smaller one, a triumphal arch, and many temples. It was also important in the development of early Christianity, and several churches of the period have been found here.

Ge·ring (gîr'ĭng), city (1970 pop. 5,639), seat of Scotts Bluff co., W Nebr., on the North Platte River opposite Scottsbluff; founded 1887, inc. 1890. It is the industrial and trade center of an irrigated region producing sugar beets, alfalfa, and livestock and has a beet-sugar factory. Scotts Bluff National Monument is nearby.

Ger·i·zim (gĕr'ĭ-zĭm, gə-rī'-), mountain, 2,890 ft (881.5 m) high, in the Samaritan Hills, W Jordan. Gerizim is sacred to the Samaritans, whose tradition holds that Abraham's offer to sacrifice Isaac occurred here.

Ger·la·chov·ka (gĕr'lä-KHôf'kä), peak, 8,737 ft (2,664.8 m) high, in the Tatra Mts. It is the highest peak of the Carpathian system and of Czechoslovakia.

Ger·man East Africa (jûr'mən), former German colony, E Africa. The German government declared a protectorate over the area in 1885 and by 1898 had conquered all of the territory. Plantations were established and railroad and harbor systems were begun. Discontentment with the administration and with the plantation system led to the Maji Maji rebellion (1905–7). During World War I the Allies captured German East Africa; after the war it was divided into League of Nations mandates. Great Britain was given most of the area, renamed Tanganyika (now Tanzania), while Belgium received Ruanda-Urundi (now Rwanda and Burundi), and Kionga, a village, was ceded to Portugal.

Ger·man·town (jûr'mən-toun'), residential section of Philadelphia, Pa., on Wissahickon Creek. Settled by Dutch and Germans in 1683, Germantown became one of the earliest printing and publishing centers in the country. When the British occupied Philadelphia during the American Revolution, the greater part of their army encamped at Germantown. George Washington's forces unsuccessfully attacked the camp on Oct. 4, 1777, in the last important engagement conducted by Washington before he took the army to Valley Forge for the winter. In 1854 Germantown was annexed to Philadelphia. The Howe House and several other colonial houses, inns, and churches are still standing. Germantown Ave. is a National Historic Landmark.

German Vol·ga Republic (vŏl'gə), former autonomous state, central European USSR, along the lower Volga River. Its largely German population was descended from the German colonists whom Catherine II had invited to settle there in 1762. The autonomous republic was formed in 1924. As a result of the German invasion of the USSR, the republic was dissolved (1941), and the entire German population (c.440,000) was deported to Siberia.

Ger·ma·ny (jûr'mə-nē), country of central Europe. Germany as a whole can be divided into three major geographic regions: the north German plain, the central German uplands, and, in the south, the ranges of the Central Alps and other uplands. West Germany includes parts of all three regions; East Germany is made up largely of the north German plain but includes a small part of the central German uplands.

History. At the end of the 2nd cent. B.C. the German tribes began to expand at the expense of the Celts to the west and south, but they were confined by Roman conquests (1st cent. B.C.–1st cent. A.D.) to the region east of the Rhine and north of the Danube. In a series of great migrations (4th–5th cent.) the German tribes overran most of the Roman Empire, while Slavic tribes occupied Germany east of the Elbe. By the 6th cent. the Anglo-Saxons had established themselves in Britain, and the Franks had taken over nearly all of present-day France, west and south Germany, and Thuringia. Charlemagne conquered the Saxons and extended the Frankish domain in Germany to the Elbe. He was crowned emperor at Rome in 800. In the first division (843) of Charlemagne's empire the kingdom of the Eastern Franks, under Louis the German, emerged as the nucleus of the German state. The Holy Roman Empire came into existence with the imperial coronation (962) of Otto I. Emperor Frederick I (reigned 1152–90; also known as Frederick Barbarossa) of the Hohenstaufen line was one of the most energetic medieval German rulers. He unsuccessfully challenged the power of the pope, being defeated by the Lombard League in 1176. However, Frederick did succeed in partitioning (1180) the domains of Henry the Lion of

Saxony and Bavaria, thus destroying the last great independent German duchy.

Until the dissolution of the Holy Roman Empire in 1806, Germany remained a patchwork of numerous small temporal and ecclesiastical principalities and free cities. The most powerful German state to emerge from the wars of the 17th and 18th cent. was Prussia, which under Frederick II (reigned 1740-86) successfully challenged the military might of Austria and became a European power. The French Revolution and the wars of Napoleon I brought the demise (1806) of the moribund Holy Roman Empire and also forced the German states to accept long-needed social, political, and administrative reforms. By the Congress of Vienna the German map was redrawn in 1814-15, eliminating many petty states and expanding Prussia and Bavaria. German nationalism emerged in the Revolutions of 1848. However, the revolutionists were soon defeated, and William I of Prussia was proclaimed German emperor by the assembled German princes in the Palace of Versailles (1871). The peace treaty with France awarded Alsace and Lorraine to Germany.

The new German empire continued under Otto von Bismarck's autocratic rule and a constitution that favored conservative interests. A master of foreign policy, Bismarck secured Germany against France by maintaining alliances in the east. Reconciliation with Austria led to an alliance (1879), joined in 1882 by Italy. He retained his chancellorship during the brief reign (1888) of Frederick III, but was dismissed in 1890 by William II. By the mid-1880s, Germans had acquired some African territories, but it was only under William II that German colonial expansion began to collide seriously with British and French interests. Equally serious threats to peace were Germany's increasing commercial rivalry with England. Two crises (1905-6 and 1911) over Morocco helped to create and strengthen the Triple Entente of France, Russia, and England, which faced Germany and its allies in World War I (1914-18). Field Marshal Paul von Hindenburg and Chief of Staff Erich Ludendorff controlled Germany until late 1918. Exhausted to the point of collapse but with no enemy troops on its soil, Germany was obliged to accept the Allied armistice terms (Nov., 1918) and, in 1919, the harsh peace terms of the Treaty of Versailles. A democratic and more centralized federal constitution was adopted at Weimar in 1919, and Germany became known as the Weimar Republic.

The economic crisis of the postwar years, marked by mass unemployment and rampant currency inflation, strengthened the extremist parties and wiped out a large portion of the middle class. The assassinations of Matthias Erzberger (1921) and of Walther Rathenau (1922) were symptomatic of the terrorist tactics adopted by the extreme nationalists, many of whom later joined the National Socialist (Nazi) party of Adolf Hitler. As the Nazi and Communist parties gained strength in the Reichstag, Heinrich Brüning and his successors, Franz von Papen and Kurt von Schleicher, failed in their efforts to mold parliamentary majorities without Hitler's support. Hindenburg, old and exhausted, accepted von Papen's assurance that Hitler could be held in check. In Jan., 1933, Hindenburg made Hitler chancellor. In the elections of Mar., 1933, Hitler played upon the electorate's fear of the Communists (especially after the Reichstag building was largely destroyed by fire in Feb., 1933) to win a bare majority of seats in the Reichstag. On Mar. 23 the Enabling Act gave Hitler dictatorial powers.

After the death of Hindenburg (1934), the offices of president and chancellor were combined in the person of the *Führer* (leader) of the Nazi party. Outside Germany, fifth columns were used to undermine the governments of nations that Hitler sought to annex in order to increase the *Lebensraum* (living space) of the Germans. In Oct., 1933, Hitler withdrew from the Geneva Disarmament Conference and from the League of Nations. In Mar., 1936, Germany remilitarized the Rhineland in violation of the Treaty of Versailles and the Locarno Pact. Hitler followed this by concluding an alliance with Fascist Italy by interfering in the Spanish Civil War (1936-39) in support of the Insurgents led by Francisco Franco, and by annexing Austria (Mar., 1938). The Munich Pact (Sept., 1938) marked the culmination of British and French attempts to appease Germany in the hope that Hitler had limited aims. However, in Mar., 1939, Germany marched into Czechoslovakia, and on Aug. 23, 1939, in a surprise move, Germany and the USSR signed a nonaggression pact and other agreements. On Sept. 1, 1939, cutting short negotiations on the status of Danzig (Gdańsk) and the Polish Corridor, Hitler invaded Poland, thus precipitating World War II.

In June, 1941, Hitler launched a vast offensive against the USSR, his former ally. In Dec., 1941, shortly after the Japanese attack on Pearl Harbor, the United States declared war on Germany. In 1942 the tide of the war began to turn against Germany. As its fortunes waned, Germany treated its remaining conquered territories more harshly. Millions of Jews and many other civilians were sent to concentration camps and exterminated, vast slave-labor systems were organized, and many thousands were deported to Germany for forced labor. By early 1945, Germany was being invaded from the west and the east, and most of its cities lay in ruins. On Apr. 30, 1945, with the total collapse of Germany imminent, Hitler committed suicide.

Hitler's successor, Adm. Karl Doenitz, signed (May 7-8, 1945) an unconditional surrender to the Allies, whose military commanders assumed the functions of government in Germany. The agreements of the Yalta Conference (Feb., 1945) were implemented at the Potsdam Conference (July-Aug., 1945). A line formed mostly by the Oder and Neisse rivers was made the eastern boundary of Germany, as East Prussia and Upper and Lower Silesia were placed under Polish administration (except northern East Prussia, which was awarded to the USSR). In the west the Saarland was occupied by French military forces. What remained of Germany was divided into four zones, occupied separately by the armies of Great Britain, France, the United States, and the USSR. Berlin was made the seat of the four-power Allied Control Council. The split between the Western Allies and the USSR, exacerbated by economic and political problems, became complete in 1948. Since 1949 Germany has been divided into the Federal Republic of Germany (West Germany) and the German Democratic Republic (East Germany).

West Germany is a republic (1970 est. pop. 59,214,400), 95,742 sq mi (247,973 sq km); Bonn is the seat of government. Northern West Germany is largely agricultural. Along the northern rim of the Rhenish Slate Mts. lies West Germany's chief mining and industrial region, which includes the Ruhr and Saar basins. The southern part of West Germany consists of plateaus and forested mountains.

West Germany is divided into 10 states:

NAME	CAPITAL	NAME	CAPITAL
Baden-Württemberg	Stuttgart	Lower Saxony	Hanover
Bavaria	Munich	North Rhine-Westphalia	Düsseldorf
Bremen	Bremen	Rhineland-Palatinate	Mainz
Hamburg	Hamburg	Saarland	Saarbrücken
Hesse	Wiesbaden	Schleswig-Holstein	Kiel

Economy. After World War II West Germany recovered quickly, due in large part to the Marshall Plan. The principal West German agricultural products are milk and eggs, potatoes, sugar beets, wheat, barley, rye, and oats. Large numbers of cattle, hogs, and poultry are raised. The chief minerals produced are coal, lignite, potash, petroleum, and iron ore. The leading industrial products include iron and steel, chemicals, motor vehicles, electric and electronic equipment, precision instruments, textiles, refined petroleum, and food products.

History. The states included in the U.S., British, and French occupation zones adopted a constitution in May, 1949, that established the Federal Republic of Germany. Konrad Adenauer became the first chancellor of West Germany; he remained in office until 1963. The first president of West Germany was Theodor Heuss; he was succeeded by Heinrich Lübke (1959), Gustav Heinemann (1969), and Walter Scheel (1974). The occupying powers allowed West Germany considerable autonomy from the start, except in foreign affairs. In 1951 West Germany was given the right to conduct its own foreign relations. In 1952 West Germany, the United States, France, and Great Britain signed the Bonn Convention, in effect a peace treaty, which granted West Germany most of the attributes of national sovereignty. The Paris agreements of 1954 gave West Germany full independence, except that the former occupying powers reserved the right to negotiate with the USSR on matters relating to Berlin and to Germany as a whole. Also, the powers continued to maintain troops in the country. In 1955 West Germany was recognized as an independent country by numerous nations, including the USSR.

National politics in the 1950s and early 1960s were dominated by Adenauer. In 1963 he retired and was replaced as chancellor by Ludwig Erhard. Erhard's government was shaken by a downturn in the economic boom, by controversy over foreign policy, and by a poor showing in the 1965 general election. In 1966 Erhard resigned and was replaced by Kurt Georg Kiesinger, a Christian Democrat.

Willy Brandt assumed the posts of vice chancellor and foreign minister. Under Kiesinger, economic conditions improved, ties with France were strengthened, and talks with the nations of Eastern Europe were initiated. In 1969 Brandt became chancellor. He maintained that neither East nor West Germany were fully sovereign entities and upheld the idea of one German nation. Important milestones in the easing of tensions between the two states were the signing (1970) of treaties of nonaggression and cooperation with the Soviet Union and Poland (ratified in 1972); the signing (1972) of an agreement among the four former occupying powers improving access to West Berlin and permitting West Berliners to visit East Berlin and East Germany more often; a treaty (1973) between East and West Germany that called for increased cooperation between the two states and prepared the groundwork for the establishment of full diplomatic relations; and the initialing (1973) of a treaty between West Germany and Czechoslovakia. Brandt resigned in May, 1974, after it was revealed that an East German spy had been on his personal staff. He was succeeded by Helmut Schmidt, the finance minister.

East Germany is also a republic (1970 est. pop. 17,056,983), 41,610 sq mi (107,771 sq km). East Berlin is the capital. East Germany is largely made up of a low-lying plain, but there are mountains in the west and south.

East Germany is divided into 15 districts:

NAME	CAPITAL	NAME	CAPITAL
Cottbus	Cottbus	Leipzig	Leipzig
Dresden	Dresden	Magdeburg	Magdeburg
East Berlin	East Berlin	Neubrandenburg	Neubrandenburg
Erfurt	Erfurt	Potsdam	Potsdam
Frankfurt	Frankfurt	Rostock	Rostock
Gera	Gera	Schwerin	Schwerin
Halle	Halle	Suhl	Suhl
Karl-Marx-Stadt	Karl-Marx-Stadt		

Economy. Before 1945 the territory that now constitutes East Germany was largely agricultural. Since then industrialization has been greatly accelerated. With very few exceptions, the economy is controlled by the state. The leading agricultural commodities produced in East Germany are wheat, rye, barley, potatoes, and sugar beets. Large numbers of cattle, hogs, sheep, and poultry are also raised. The country's principal manufactures include iron and steel, chemicals, cement, textiles, machinery, precision instruments, footwear, motor vehicles, and electric and electronic equipment. East Germany is a major producer of lignite; coal, potash, uranium, and iron ore are also mined.

History. A congress organized by the Socialist Unity party (SED) in May, 1949, adopted a constitution establishing the German Democratic Republic. The initial constitution, superseded by one adopted in 1968, provided for a president and a bicameral parliament. Wilhelm Pieck became the country's first president and Otto Grotewohl its first prime minister, with Walter Ulbricht as first deputy prime minister. During the 1950s Ulbricht emerged as the leader of East Germany. Under his leadership the country was closely aligned with the USSR, and the liberalizing policies introduced in some of the other East European Communist nations were avoided. In order to reduce the large flow of persons leaving East Germany (about 4 million during 1945–61), many of whom crossed from East to West Berlin, a wall was erected (Aug. 12–13, 1961) between the two parts of the city; it was later reinforced and enlarged. The wall drastically cut the number of emigrants, and gradually this had the effect of solidifying East Germany as an independent country. In 1964 a treaty of friendship and cooperation—in effect a peace treaty—was signed with the USSR; similar treaties with Poland, Czechoslovakia, Hungary, and Bulgaria followed in 1967. In the late 1960s diplomatic contacts with West Germany were initiated; these culminated in 1973 with the signing of a treaty between the two states.

Ger·mis·ton (jûr'mĭ-stən), city (1970 pop. 221,972), Transvaal, NE South Africa, on the Witwatersrand. The chief industries are gold mining and processing and the manufacture of liquid oxygen; other chemicals, machinery, textiles, and clothing are also produced.

Ge·ro·na (hā-rō'nä), city (1975 est. pop. 75,109), capital of Gerona prov., NE Spain, in Catalonia, on the Oñar River. There are food, textile, paper, chemical, and other industries in Gerona. The city dates from pre-Roman times, and the old town has preserved its medieval aspect. The Moors ruled Gerona, with two interruptions, from 714 to 797. In 1808–9, during the Peninsular War, townspeople heroically resisted the French.

Gers (zhĕr), department (1975 pop. 175,366), 2,415 sq mi (6,254.9 sq km), SW France; capital Auch.

Ger·sop·pa, Falls of (jər-sŏp'ə), cataract of the Sharavati River, Karnataka state, SW India. It is one of the most spectacular natural beauties of India. The river cuts through the Western Ghats to fall in four cascades, of which the highest is c.830 ft (255 m).

Ge·ta·fe (hā-tä'fā), town (1970 pop. 69,424), Madrid prov., central Spain, in New Castile, S of Madrid. An industrial and agricultural center, Getafe is located at the exact geographic center of Spain. It has electrical and chemical industries. Cereals, vegetables, grapes, and olives are grown in the surrounding area.

Geth·sem·a·ne (gĕth-sĕm'ə-nē), olive grove or garden, E of Jerusalem, near the foot of the Mount of Olives. It was the scene of the agony and betrayal of Jesus. Ruins of a 4th cent. church and of a church as old as the Crusades were found here.

Get·tys·burg (gĕt'ĭz-bûrg'). **1.** Borough (1970 pop. 7,275), seat of Adams co., S Pa.; inc. 1806. Electrical equipment, food products, and shoes are manufactured here. Gettysburg was settled c.1780 and is named for Gen. James Gettys, to whom its site was granted (17th cent.) by William Penn. The Gettysburg campaign (1863) was a turning point in the Civil War, and President Abraham Lincoln made his famous Gettysburg Address here. Gettysburg National Military Park, Gettysburg National Cemetery, and the farm of President Dwight D. Eisenhower are national historic shrines. **2.** City (1970 pop. 1,915), seat of Potter co., N central S.Dak., SW of Aberdeen; founded 1881 by Civil War veterans. It is a trade and processing center of a dairy, livestock, and grain area.

Gey·sir (gā'sĭr), hot spring, SW Iceland, W of Reykjavík. Although in medieval times it erupted three times daily, weeks now elapse between eruptions. The height and temperature of the jet are variable, reaching up to 200 ft (61 m) and 180°F (82°C).

Ge·zer (gē'zər), ancient city of Canaan, on the coastal plain of Sharon, NW of Jerusalem. Its position guarding the road from Jerusalem to Jaffa has always given it importance. Excavations here (1902–8, 1929, 1934) have made it possible to trace the history of Gezer from Chalcolithic times.

Gha·da·mes (gə-dä'məs): *see* Ghudamis, Libya.

Gha·na (gä'nə), ancient empire, W Africa, in the savanna region of what is now E Senegal, SW Mali, and S Mauritania. The empire was founded c.6th cent. and lay astride the trans-Saharan caravan routes. It prospered from trade—particularly in salt and gold—and tribute. Internal divisions and an Almoravid invasion (1076) contributed to Ghana's decline, and by the 13th cent. it had disintegrated.

Ghana, republic (1977 est. pop. 10,250,000), 92,099 sq mi (238,536 sq km), W Africa, on the Gulf of Guinea, an arm of the Atlantic Ocean. The capital is Accra. Modern Ghana comprises the former British colony of the Gold Coast and the former mandated territory of British Togoland. It is bordered by the Ivory Coast on the west, Upper Volta on the north, and Togo on the east. The coastal region and the far north of Ghana are savanna areas; in between is a forest zone.

Economy. Ghana's economy is predominantly agricultural. The biggest crop and major export is cacao. Coffee, oil palms, and coconuts are also widely grown. Minerals (gold, diamonds, manganese, and bauxite) are found in the north, south, and coastal regions and constitute the second-largest export, followed by timber. The major industries in Ghana are aluminum smelting, the processing of cacao, and the production of lumber and foods and beverages.

History. In precolonial times the area of present-day Ghana comprised a number of independent kingdoms, including Gonja and Dagomba in the north, Ashanti in the interior, and the Fanti states along the coast. Trade was begun, largely in gold and slaves, and intense competition developed among many European nations for trading advantages. In 1874 the British defeated the Ashanti kingdom and organized the coastal region as the colony of the Gold Coast. In 1901 the Northern Territories, a region north of Ashanti, was declared a British protectorate. In the Gold Coast nationalist activity intensified after World War II. Kwame Nkrumah of the Convention People's Party (CPP) emerged as the leading nationalist figure.

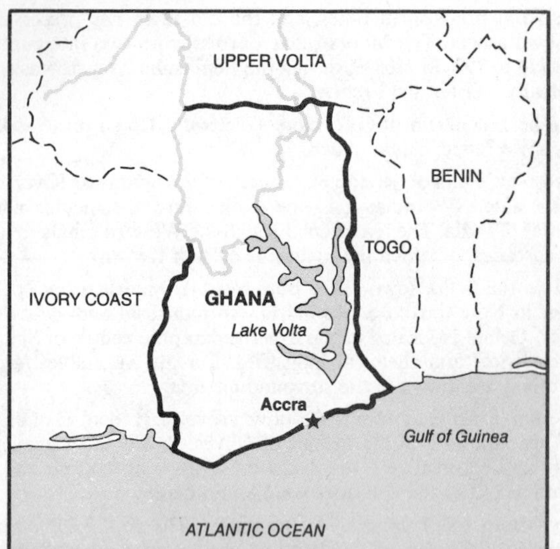

In 1951 the CPP won a general election and Nkrumah became premier. On Mar. 6, 1957, Ghana, named after the medieval West African empire, became an independent country within the Commonwealth of Nations. At the same time the people of British Togoland chose by plebiscite to become part of Ghana. In 1960 Nkrumah transformed Ghana into a republic, with himself as president for life. By 1964 all opposition parties were outlawed, and many critics of the government were subsequently imprisoned. Nkrumah followed an anticolonial, Pan-African policy and grew increasingly less friendly to the West. Falling cacao prices and poorly financed large development projects led to chaotic economic conditions, and in 1966 Nkrumah was overthrown by a military-police coup. The new government, headed by K. A. Busia, was undermined by labor problems, an unpopular currency devaluation, and serious inflation, and in Jan., 1972, it too was overthrown in a bloodless coup led by Col. I. K. Acheampong. The constitution was suspended, and a National Redemption Council (NRC) set up to govern. The NRC pursued a more neutralist course in foreign affairs and concentrated on developing Ghana's economy. The country's large foreign debt was partly repudiated and partly refinanced; imports were curtailed; and the state took controlling interest in foreign mining and timber firms operating in the country. Ghana's social and economic troubles were aggravated by widespread famine (1977–78) in the arid northern regions.

Ghar·da·ïa (gär-dä′yä), town (1974 est. pop. 55,200), N Algeria. It is the chief town of the Mzab, a stony, barren valley of the northern Sahara. Ghardaïa is a center of date production and of the manufacture of rugs and cloth. The city was founded in the 11th cent. First occupied by French troops in 1854, Ghardaïa was not officially annexed to France until 1882.

Ghat or **Gat** (both: gôt), walled town, SW Libya, in an oasis in the Sahara, near the Algerian border. It formerly was an important caravan center. Ghat was captured by the Ottoman Turks in 1875, by the Italians in 1930, and by the French in 1943, during World War II.

Ghats (gôts), two mountain ranges of S India, paralleling the coasts of the Arabian Sea and the Bay of Bengal and forming two sides of the Deccan plateau. Anai Mudi (8,841 ft/2,696.5 m) is the highest peak in the two ranges, which are joined by the Nilgiri Hills in the south. The Western Ghats, c.1,000 mi (1,600 km) long, extend southeast from the Tapti valley, north of Bombay, to Cape Comorin at the southern tip of India. The western side of the range, which receives heavy rainfall from moist monsoon winds, has lush tropical vegetation and dense hardwood forests; the eastern side is relatively dry. The Eastern Ghats, a series of hills c.900 mi (1,450 km) long, extend southwest from the Mahanadi valley to the Nilgiri Hills; the highest elevations are at the northern and southern ends. Numerous rivers cut across the Eastern Ghats and are used for hydroelectric-power generation and irrigation. The range has valuable hardwood trees.

Gha·zi·a·bad (gä′zē-ä-bäd′), city (1971 pop. 119,179), Uttar Pradesh state, N central India. Ghaziabad is an agricultural market. The city was the scene of fighting during the Indian Mutiny (1857).

Gha·zi·pur (gä′zē-po͝or′), town (1971 pop. 45,636), Uttar Pradesh state, N central India, on the Ganges River. Ghazipur has a perfume industry.

Ghaz·ni (gäz′nē), city (1967 pop. 39,900), capital of Ghazni prov., E central Afghanistan, on the Lora River. Ghazni is a market for sheep, wool, camel's hair cloth, corn, and fruit. The famed Afghan sheepskin coats are made in the city. Ghazni was flourishing by the 7th cent. but reached its peak 962 to c.1155, when it was one of Asia's most glorious cities. In 1737 the city became part of the new kingdom of Afghanistan. Ghazni's strong fortress was taken by the British in 1839 and 1842 during the Afghan Wars. The old city of Ghazni, with its numerous bazaars, retains its walls and is topped by a citadel used as a military fort.

Ghaz·zah (gä′zə, găz′ə): *see* Gaza.

Ghent (gĕnt), city (1977 est. pop. 248,671), capital of East Flanders prov., W Belgium, at the confluence of the Scheldt and Leie rivers. Connected with the North Sea by the Ghent-Terneuzen Canal and by a network of other canals, Ghent is a major port and the chief textile, clothing, and steel-manufacturing center of Belgium. Other products of the city include plastics, chemicals, paper, processed food, and motor vehicles. It is also the trade center of a flower- and bulb-producing region.

One of Belgium's oldest cities (first mentioned in the 7th cent.) and the historic capital of Flanders, Ghent developed around a fortress built (early 10th cent.) by the first count of Flanders. By the 13th cent. Ghent had become a major wool-producing center. Its four chief guilds—weavers, fullers, shearers, and dyers—comprised the majority of the working population. There was social conflict between the workers and the rich bourgeoisie; strikes and insurrections were frequent. In 1385 the weavers made a favorable peace with Philip the Bold of Burgundy, who had inherited Flanders in the previous year.

Ghent retained all its liberties and privileges until 1453, when, as a result of an unsuccessful rebellion, they were drastically curtailed by Philip the Good of Burgundy. They were restored by the Great Privilege, promulgated (1477) by Mary of Burgundy. Mary's marriage (1477) to Archduke Maximilian (later Emperor Maximilian I) took place at Ghent, and their children were kept virtual prisoners by the burghers after Mary's death (1482). It was only in 1485 that Maximilian was able to master the rebellious city and obtain the release of his son Philip (later Philip I of Castile). Philip's son, later Emperor Charles V, was born (1500) and reared at Ghent. In 1539 Ghent rose against Charles, who suppressed (1540) the rebellion and established a garrison to prevent further outbreaks. Ghent later joined (1576) William the Silent in the revolt of the Netherlands and Flanders against Spain; the Pacification of Ghent, signed in Nov. of the same year, was an alliance of the provinces of the Netherlands for the purpose of driving the Spanish from the country. For a time Ghent was a city-republic under Calvinist domination, but its capture (1584) by the Spanish under Alessandro Farnese restored it to Hapsburg rule, under which it remained until the French Revolution. The industrialization of Ghent began in the early 19th cent. with the development of its port and textile factories. The city was occupied by the Germans in World Wars I and II.

Ghent is noted for its many beautiful medieval and Renaissance structures. Flemish painting flourished in Ghent under the Burgundian dynasty (15th cent.); Hugo van der Goes worked here most of his life, and the famous masterpiece of the Van Eyck brothers, *The Adoration of the Lamb,* is the altarpiece of the Cathedral of St. Bavon (10th–16th cent.), which also contains a noted Rubens painting.

Ghor (gôr) or **Ghur** (go͝or), mountainous region and province, W central Afghanistan, including a ruined medieval city of the same name. The powerful Moslem Ghorid dynasty was established here in the 12th cent.

Ghor, the, region of the Jordan Valley, c.70 mi (115 km) long, between the Sea of Galilee and the Dead Sea, on the Jordan-Israel border. Entirely below sea level and bordered by steep escarpments, it is part of the Great Rift Valley complex. The Jordan River meanders 160 ft (48.8 m) below the surface through the Ghor. Although the Ghor's flat terraces are fertile, agricultural development is impeded by aridity. In the northern half of the valley, on the Jordan side, is the East Ghor irrigation project (built 1958–66), which makes year-round cultivation possible.

Ghu·da·mis or **Gha·da·mes** (both: gə-dä′məs), town, W Libya, in an oasis in the Sahara, near the borders with Algeria and Tunisia. It was long an important caravan center on the route from Tripoli to the Sudan. The town was held by the Romans and was captured (7th cent.) by the Arabs. The town was occupied by the Italians in 1924 and in 1943, during World War II, by the French.

Gi·ant's Causeway (jī'əntz), headland on the N coast of Co. Antrim, N Northern Ireland, NE of Coleraine. Extending 3 mi (4.8 km) along the coast, it consists of thousands of basaltic columns of volcanic origin, forming three natural platforms. There are several large caves and rock formations. According to legend, the Causeway was built for giants to travel across to Scotland.

Gibbs·town (gĭbz'toun'), uninc. town (1970 pop. 5,676), Gloucester co., SW N.J., near the Delaware River SW of Camden.

Gi·bral·tar (jĭ-brôl'tər), town (1976 est. pop. 30,000), 2.5 sq mi (6.5 sq km), a British crown colony. Gibraltar is located at the northwest end of the Rock of Gibraltar, one of the Pillars of Hercules. The rock itself forms a peninsula of southern Spain at the eastern end of the Strait of Gibraltar, which joins the Mediterranean with the Atlantic. The peninsula is connected with the mainland by a low sandy area of neutral ground. West of the peninsula is the Bay of Gibraltar, an inlet of the strait. There is a safe enclosed harbor of 440 acres (178.2 hectares). The rock, of Jurassic limestone, contains caves in which valuable archaeological finds have been made. It is honeycombed by defense works and arsenals, which are largely concealed. A tunnel bisects the rock from east to west. The town is a free port, with some transit trade. Most of the area is taken up by military installations, and the civilian population is kept small.

Gibraltar was captured (711) by the Moorish leader Tarik. The Spanish took the peninsula in 1309 and held it until 1333, but did not definitively recover it from the Moors until 1462. The English have maintained possession since 1704 despite continual Spanish claims to it. The British post was besieged unsuccessfully by the Spanish and French in 1704, by the Spanish in 1726, and again by the Spanish and French from 1779 to 1783. In World War I Gibraltar served as a naval station. In World War II its fortifications were strengthened, and most of the civilian population was evacuated. It was frequently bombed in 1940-41, but not seriously damaged. After the war Spain again claimed Gibraltar and in the late 1960s closed its border. The residents affirmed (1967) their ties with Britain in a UN-supervised referendum.

Gib·son (gĭb'sən). **1.** County (1970 pop. 30,444), 499 sq mi (1,292.4 sq km), SW Ind., bounded on the W by the Wabash River and drained by the Patoka and Black rivers, and Pigeon Creek; formed 1813; co. seat Princeton. In an agricultural area yielding grain, livestock, poultry, truck crops, and fruit, it has bituminous-coal mines, natural-gas and oil wells, and timber. Its manufactures include food products, concrete blocks, oil-well supplies, and miscellaneous rubber and plastic products. **2.** County (1970 pop. 47,871), 607 sq mi (1,572.1 sq km), NW Tenn., bounded on the NE by the South Fork of the Obion River, on the SW by the Middle Fork of the Forked Deer River; formed 1823; co. seat Trenton. It is in a fertile agricultural area that produces soybeans, cotton, corn, strawberries, vegetables, and livestock. It also has timber and marble and granite deposits. Textile-mill products, clothing, shoes, wood products, furniture, paper, concrete, and metal products are manufactured.

Gibson, city (1970 pop. 701), seat of Glascock co., E Ga., WSW of Augusta, in a cotton and lumber area.

Gibson Desert, central Western Australia, bounded by the Great Sandy Desert on the N and Victoria Desert on the S. Its natural features include salt lakes, scrub, and sand dunes.

Gid·dings (gĭd'ĭngz), city (1970 pop. 2,783), seat of Lee co., S central Texas, E of Austin; founded 1872. It is the processing center of a truck-farm, dairy, poultry, and cotton area and has light industry.

Gies·sen (gē'sən), city (1974 est. pop. 76,217), Hesse, central West Germany, on the Lahn River. Its manufactures include textiles, tobacco products, rubber goods, and machinery. Iron ore is mined nearby. Giessen was chartered by 1248 and became the chief town of Upper Hesse. The city was heavily damaged in World War II.

Gi·fu (gē'fōō'), city (1976 est. pop. 409,404), capital of Gifu prefecture, central Honshu, Japan. A manufacturing and railway center, it has textile and paper industries. It is the seat of three universities and of the Nawa Entomology Institute (founded 1896). The city was reduced to ashes by fires following an earthquake in 1891.

Gi·jón (gē-hôn', hē-), city (1970 pop. 187,612), Oviedo prov., N Spain, in Asturias, on the Bay of Biscay. This major seaport is an industrial and commercial center exporting large quantities of coal and iron. It has steel, iron, chemical, glass, and food and tobacco industries. Of pre-Roman origin, Gijón was one of the first places recaptured from the Moors early in the 8th cent.

Gi·la (hē'lə), county (1970 pop. 29,255), 4,750 sq mi (12,302.5 sq km), E central Ariz., in a mountainous area drained by the San Carlos, Gila, and other rivers; formed 1881; co. seat Globe. Roosevelt Dam and Reservoir on the Salt River and Coolidge Dam and San Carlos Lake on the Gila River provide water for irrigation. Copper, gold, silver, asbestos, and vanadium are mined. The county includes Tonto National Forest.

Gila, river, 630 mi (1,013.7 km) long, rising in the mountains of W N.Mex. and flowing W across Ariz. to the Colorado River at Yuma, Ariz. The Gila valley was occupied by the ancestors of the Pima and Papago Indians, who farmed the region by irrigation. The Gila and its tributaries have many dams to provide flood control, hydroelectricity, and water for irrigation.

Gila Cliff Dwellings National Monument: *see* National Parks and Monuments Table.

Gil·bert (gĭl'bərt), intermittent river, c.320 mi (515 km) long, N Queensland, Australia, rising near the junction of Gregory Range and the Great Dividing Range and flowing NW into the Gulf of Carpentaria.

Gilbert and El·lice Islands (ĕl'əs), former British colony in the central and S Pacific, divided into the Gilbert Islands and Tuvalu.

Gilbert Islands, group of 16 islands (1968 pop. 44,206), a British colony in the central Pacific, formerly (until 1975) a part of the Gilbert and Ellice Islands. The total land area is 102 sq mi (264.2 sq km). The equator runs through the center of the group. There are coconut and breadfruit trees on the islands; fishing and the export of copra are the main economic activities. Nukunau Island was discovered by British Commodore John Byron in 1765; other islands were discovered by captains Thomas Gilbert and John Marshall in 1788, and the remainder were visited between 1799 and 1824. The British made the islands a protectorate in 1892 and a colony in 1915-16. Some of the islands were occupied by the Japanese in 1941 and liberated by U.S. forces in 1943. The Gilberts attained full independence in 1978.

Gil·bo·a (gĭl-bō'ə), range of hills, E spur of the Samarian Hills, located at the SE edge of the Esdraelon plain, NE Israel; rising to 1,630 ft (497.2 m) at Mt. Gilboa.

Gil·christ (gĭl'krĭst'), county (1970 pop. 3,551), 346 sq mi (896.1 sq km), N Fla., bounded on the N by the Santa Fe River, on the W by the Suwannee River; formed 1925; co. seat Trenton. In a flatwoods area with many small lakes, it is dependent on farming (corn, peanuts, and vegetables), cattle raising, and lumbering.

Giles (jīlz). **1.** County (1970 pop. 22,138), 619 sq mi (1,603.2 sq km), S Tenn., bounded on the S by Ala. and drained by the Elk River and Richland Creek; formed 1809; co. seat Pulaski. In a fertile Bluegrass region yielding cotton, tobacco, truck crops, livestock, and dairy products, it has phosphate mines and food-processing and lumbering industries. **2.** County (1970 pop. 16,741), 356 sq mi (922 sq km), SW Va., in the Alleghenies bordered on the N and NW by W.Va. and drained by the New River; formed 1806; co. seat Pearisburg. It lies entirely within Jefferson National Forest. Its agriculture includes corn, wheat, apples, dairy products, livestock, and poultry. Lumbering, leather tanning, and some coal mining are done.

Gil·les·pie (gə-lĕs'pē), county (1970 pop. 10,553), 1,055 sq mi (2,732.5 sq km), S central Texas, in a hilly region of Edwards Plateau and drained by the Pedernales River and several creeks; formed 1848; co. seat Fredericksburg. Ranching, agriculture (corn, wheat, grain sorghums, peanuts, fruit, truck crops, dairy products, and poultry), and granite quarrying are important. Wool and mohair are processed and marketed. The country has good hunting and fishing.

Gil·lette (jĭ-lĕt'), town (1970 pop. 7,194), seat of Campbell co., NE Wyo., SE of Sheridan; inc. 1890. It is a trade center for a grain, livestock, oil, and coal area and a tourist point for a big-game region abounding in deer and antelope.

Gil·li·am (gĭl'ē-əm), county (1970 pop. 2,342), 1,211 sq mi (3,136.5 sq km), N Oregon, in an irrigated agricultural region bounded on the N by the Columbia River and on the W by the John Day River; formed 1885; co. seat Condon. Cattle and wheat are raised.

Gil·ling·ham (gĭl'ĭng-əm), municipal borough (1976 est. pop. 93,900),

Kent, SE England, on the Medway River. Some of the Chatham dockyards (repair and supply facilities) are in Gillingham. There are naval and military establishments (notably the Royal School of Military Engineering) and resort facilities. Manufactures include clothing, chemicals, furniture, and electrical components.

Gil·mer (gĭl'mər). **1.** County (1970 pop. 8,956), 439 sq mi (1,137 sq km), N Ga., in the Blue Ridge and drained by the Coosawattee River; formed 1832; co. seat Ellijay. Its agriculture includes fruit, potatoes, corn, hay, poultry, and livestock. It also has stone quarries, a lumbering industry, and textile mills. In a resort area, the county includes part of Chattahoochee National Forest in the north. **2.** County (1970 pop. 7,782), 339 sq mi (878 sq km), central W.Va., on the Allegheny Plateau drained by the Little Kanawha River; formed 1845; co. seat Glenville. Livestock and poultry are of major importance; fruit, tobacco, potatoes, and corn are also grown. There are oil and natural-gas wells, some coal mines, and stands of timber in the county. Glass and shoes are manufactured.

Gilmer, city (1970 pop. 4,196), seat of Upshur co., NE Texas, WNW of Marshall; settled 1858, inc. 1902. It was transformed from a pinewoods lumbering and farming town into a bustling oil city in the 1930s. Today it is a trade center with some manufacturing.

Gil·pin (gĭl'pən), county (1970 pop. 1,272), 149 sq mi (385.9 sq km), N central Colo.; formed 1861; co. seat Central City. It is in a mining (gold, silver, lead, copper, and zinc) and livestock-grazing region. It includes parts of Arapaho and Roosevelt national forests. The Front Range is in the west.

Gil·roy (gĭl'roi), city (1970 pop. 12,684), Santa Clara co., W Calif.; inc. 1870. Located in the fertile Santa Clara valley, Gilroy supports diversified agriculture, including vineyards, orchards, dairy farms, and plant nurseries. The chief manufactures are farm equipment, processed spices and other foods, wines, paint, fabricated metals, fiber glass, and paper products.

Gipps·land (gĭps'lănd'), geographic area, 13,655 sq mi (35,366.5 sq km), Victoria, SE Australia, E of Melbourne. Dairy and beef cattle are raised, and corn, oats, and sugar beets are grown. The region also supports a timber industry. There are deposits of coal, oil and natural gas (offshore), and limestone. Gold was formerly mined.

Gi·rard (jə-rärd'). **1.** City (1970 pop. 2,591), seat of Crawford co., SE Kansas, NNW of Pittsburg; founded 1868, inc. as a town 1869, as a city 1871. **2.** City (1970 pop. 14,119), Trumbull co., NE Ohio, adjacent to Youngstown, on the Mahoning River; settled c.1800, inc. 1891. Its ironworks date from 1866.

Gi·rar·dot (hē-rär-dôt', -*thôt*'), city (1973 est. pop. 59,165), central Colombia, on the Magdalena River. Girardot is a commercial center and a transportation hub. Coffee, livestock, tobacco, and corn are the principal products. Founded in 1853, Girardot is noted for its vast number of acacia trees.

Gi·re·sun (gē-rĕ-so͞on'), city (1975 pop. 38,205), capital of Giresun prov., NE Turkey, a port on the Black Sea. It is the trade center for a farm region in which maize, filberts, beans, and potatoes are produced. The city was famous in ancient times for its cherry trees.

Gir·ga (gĭr'gə, -gă): *see* Jirja, Egypt.

Gir·nar (gĭr-när'), sacred mountain, 3,666 ft (1,118.1 m) high, Gujarat state, W India, on the Kathiawar Peninsula. A pilgrimage place for adherents of Jainism, it has five peaks, the sides of which are dotted with ancient reservoirs and temple ruins bearing inscriptions that date from 250 B.C.

Gi·ronde (jə-rŏnd', zhē-rôNd'), department (1968 pop. 1,009,390), 3,861 sq mi (10,000 sq km), SW France, on the Bay of Biscay; capital Bordeaux.

Gironde, estuary, c.45 mi (70 km) long and from 2 to 7 mi (3.2–11.3 km) wide, formed by the Garonne and Dordogne rivers N of Bordeaux. Sand banks and a high tidal range hamper navigation. Located between the Médoc and the Côtes vineyards, the Gironde is the great artery of the Bordeaux wine region.

Gis·borne (gĭz'bərn), city (1971 pop. 26,726), E central North Island, New Zealand, on Poverty Bay. It is a resort and a port, exporting wool, frozen meat, and dairy goods. Captain James Cook landed here in 1769.

Giur·giu (jo͝or'jo͞o'), city (1974 est. pop. 47,567), S Rumania, in Walachia, on the Danube River opposite Ruse, Bulgaria, with which it is linked by a bridge. An important inland port, Giurgiu has shipyards and food and other light industries. The city was founded (10th cent.) by Genoese merchants on the site of a Roman settlement. Conquered by the Turks in 1417, it played an important role in the 16th cent. wars between Walachia and Turkey and in the later Russo-Turkish Wars. Remains of the old town walls, the ruins of a medieval fortress, and an old clock tower still stand.

Giv·a·ta·yim (gĭv'ä-tä'yĭm), town (1975 est. pop. 50,000), W central Israel, a residential suburb of Tel Aviv; founded 1942. Industries include printing and food processing.

Gi·za (gē'zə): *see* Jizah, Al, Egypt.

Gji·no·kas·tër (gyē'nô-käs'tər), town (1974 est. pop. 17,400), capital of Gjinokastër prov., S Albania. A commercial center, it produces foodstuffs, leather, and textiles. Dating probably from the 4th cent., Gjinokastër passed to the Turks in the 15th cent. It was captured (1811) by Ali Pasha and was the center (late 1800s) of anti-Turkish resistance. In World War II it was occupied by Greece.

Glace Bay (glās), coal-mining town (1976 pop. 21,836), E Cape Breton Island, N.S., Canada. Exploitation of the mines began toward the end of the 19th cent. The mines extend for several miles under the sea and are among the best equipped in the world. Glace Bay has a good harbor and a large deep-sea fishing fleet. The Marconi wireless tower at Table Head nearby was the transmitter in 1902 of the first transatlantic wireless message.

gla·cier (glā'shər), moving mass of ice, formed in high mountains and in the polar regions by the compacting of snow into névé and then into granular ice and set in motion outward and downward by the pressure of the accumulated mass. Glaciers are of four chief types. Valley, or mountain, glaciers are tongues of moving ice sent out by mountain snowfields into valleys originally formed by streams. They follow the courses of the valleys and are held in by the valley walls. Piedmont glaciers, which occur only in high latitudes, are formed by the spreading of valley glaciers where they emerge from their valleys or by the confluence of several valley glaciers. Small ice sheets known as ice caps are flattened, somewhat dome-shaped glaciers spreading out horizontally in all directions and covering mountains and valleys alike. Continental glaciers are ice sheets of huge extent whose margins may break off to form icebergs.

Glacier, county (1970 pop. 10,783), 2,964 sq mi (7,676.8 sq km), N Mont., in a livestock and grain-growing area bounded on the W by the Continental Divide, on the N by the Alta., Canada, border, and drained by the Milk River and forks of the Marias River; formed 1919; co. seat Cut Bank. It has petroleum and natural-gas wells and refineries. The county includes Blackfeet Indian Reservation and part of Glacier National Park.

Glacier Bay National Monument, 2,803,840 acres (1,135,555 hectares), SE Alaska, near Juneau; est. 1925. Glaciers descending from the towering snow-covered mountains into the bay create one of the world's most spectacular displays of ice. Among the bay's most famous glaciers is Muir Glacier, c.2 mi (3.2 km) wide and rising c.265 ft (80 m) above the water. Many of the glaciers flow into the Pacific Ocean. Wildlife includes bears, deer, mountain goats, porpoises, whales, and waterfowl.

Glacier National Park, 1,013,100 acres (410,306 hectares), NW Mont.; est. 1910. Straddling the Continental Divide, the park contains some of the most beautiful primitive wilderness in the Rocky Mts. There are about 50 glaciers, more than 200 glacier-fed lakes, high peaks, sheer precipices, large forests, waterfalls, much wildlife, and a great variety of wild flowers.

Glacier Peak, mountain, 10,568 ft (3,223.2 m) high, NW central Wash., in the Cascade Range ENE of Everett.

Glad·beck (glät'bĕk), city (1974 est. pop. 81,220), North Rhine-Westphalia, W West Germany, an industrial center of the Ruhr district. Its manufactures include chemicals, clothing, electrical equipment, and metal goods.

Glades (glādz), county (1970 pop. 3,669), 753 sq mi (1,950.3 sq km), S Fla., bounded on the E by Lake Okeechobee and crossed by the Caloosahatchee River; formed 1921; co. seat Moore Haven. In an everglades region, it depends on cattle raising, truck farming, and fishing. It includes a Seminole Indian reservation.

Glade·wa·ter (glād'wô'tər), city (1970 pop. 5,574), Gregg and Upshur cos., NE Texas, on the Sabine River WNW of Longview, in a timber area. It boomed after the East Texas oil field was discovered in 1930.

Glad·stone (glăd'stōn', -stən). **1.** City (1970 pop. 5,237), Delta co., N

Mich., S Upper Peninsula, on an inlet of Green Bay NNE of Escanaba; founded 1887, inc. 1889. It is a resort and a shipping point for an agricultural area. **2.** City (1978 est. pop. 30,000), Clay co., W Mo., a suburb surrounded by Kansas City; founded c.1878, inc. 1952. It has some manufacturing industries.

Glad·win (glăd'wĭn), county (1970 pop. 13,471), 503 sq mi (1,302.8 sq km), central Mich., drained by the Tittabawassee, Tobacco, and Cedar rivers; formed 1831; co. seat Gladwin. Its agriculture includes grain, seed potatoes, dairy products, and livestock. In a resort area, with many small lakes and good hunting and fishing, it also has oil wells and a food-processing industry. Machinery is manufactured.

Gladwin, city (1970 pop. 2,071), seat of Gladwin co., E central Mich., NW of Bay City, in a farm and resort region; settled 1865, inc. as a village 1885, as a city 1893.

Glå·ma or **Glom·ma** (both: glô'mä), longest river of Norway, c.365 mi (590 km) long, rising in the highlands of SE Norway and flowing generally S into the Skagerrak at Fredrikstad. The Glåma's numerous waterfalls are the sites of hydroelectric stations.

Gla·mis (glä'mĭs, glämz), village, Tayside region, E Scotland. King Malcolm II died (1034) nearby, and a sculptured cross in the village is known as King Malcolm's Gravestone. Macbeth was thane of Glamis, and the castle here is erroneously claimed to be the scene of Duncan's murder in Shakespeare's play.

Gla·mor·gan·shire (glə-môr'gən-shîr', -shər), former county, S Wales. The county town was Cardiff. Before the Norman conquest there were several centers of Celtic Christianity in the region. In 1974 Glamorganshire was divided into the nonmetropolitan counties of West Glamorgan, Mid Glamorgan, and South Glamorgan.

Glar·us (glär'əs, -ōōs'), canton (1977 est. pop. 35,600), 264 sq mi (683.8 sq km), E central Switzerland. Located in the basin of the Linth River, it is a mountainous and pastoral region, with forests and meadows in the valleys. Sparsely settled by the Romans after 15 B.C., Glarus was permanently occupied c.500 A.D. by the Alemanni. Glarus joined the Swiss Confederation in 1352. The town of Glarus (1977 est. pop. 6,100), on the Linth, is the capital. Furniture and bleaches are made here.

Glas·cock (glăs'kŏk'), county (1970 pop. 2,280), 142 sq mi (367.8 sq km), E central Ga., bounded on the W by the Ogeechee River; formed 1857; co. seat Gibson. In a coastal-plain agricultural area yielding cotton, corn, peanuts, vegetables, and fruit, it has kaolin deposits and a lumbering industry. Some clothing is manufactured.

Glas·gow (glăs'gō), city (1976 est. pop. 856,012), Strathclyde region, S central Scotland, on the River Clyde. Glasgow is Scotland's leading seaport and center of the great Clydeside industrial belt. Known for its large shipyards, metalworks, and engineering works, Glasgow's manufactured products include electronic equipment, chemicals, carpets, textiles, tobacco, and machine tools. Plagued by widespread slums, the city began a rebuilding program in the late 1950s. Glasgow was founded in the late 6th cent. Its modern commercial growth began with the American tobacco trade in the 18th cent. and the cotton trade in the early 19th cent. Its proximity to the Lanarkshire coal fields and its location on the Clyde (first deepened at Glasgow in 1768) aided its development as a center of heavy industry during the mid-19th cent. Points of interest include St. Mungo's Cathedral (mostly 13th cent.), the Corporation Art Galleries, and the People's Palace museum.

Glasgow 1. City (1970 pop. 11,301), seat of Barren co., S central Ky.; inc. 1799. It is a trade and industrial center for a timber, oil, livestock, and farm area. Spotswood, built in 1795 under the direction of George Washington for his niece, is still occupied. **2.** City (1970 pop. 4,700), seat of Valley co., NE Mont., on the Milk River WNW of Wolf Point; founded 1888, inc. 1911. A lively cow town in the 19th cent., it is today the center of an agricultural and livestock area watered by the Milk River project. There are gas wells nearby.

Glass·bor·o (glăs'bûr'ō), borough (1970 pop. 12,938), Gloucester co., SW N.J.; settled 1775, inc. 1920. It is a trade and processing center for a fruit-growing region and has a large glass industry.

Glass·cock (glăs'kŏk), county (1970 pop. 1,155), 864 sq mi (2,237.8 sq km), W Texas, in a wooded and rolling prairie area drained by tributaries of the Colorado River; formed 1887; co. seat Garden City. Cattle ranching and agriculture (grain sorghums, corn, oats, cotton, and pecans) are important to its economy. It also has some oil and natural-gas wells.

Glass·port (glăs'pôrt'), industrial borough (1970 pop. 7,450), Allegheny co., SW Pa., on the Monongahela River SE of Pittsburgh; inc. 1902 Glass, steel, and tools are produced.

Glas·ton·bur·y (glăs'tən-bĕr'ē), municipal borough (1973 est. pop. 6,580), Somerset, SW England. There is a leather industry. Glastonbury Abbey was a center of learning and an object of pilgrimages in the Middle Ages. Extensive remains of an Iron Age lake village have been found nearby.

Glastonbury, town (1970 pop. 20,651), Hartford co., central Conn., on the Connecticut River; inc. 1690. Its manufactures include aircraft engines, helicopters, typewriters, machine tools, electrical equipment, furniture, leather, and meat products. Poultry, dairy products, fruit, and tobacco are also important.

Glen·coe (glĕn'kō). **1.** Residential village (1970 pop. 10,542), Cook co., NE Ill., on Lake Michigan; inc. 1869. A Nike missile site is here. **2.** City (1970 pop. 4,217), seat of McLeod co., S central Minn., WSW of Minneapolis; platted 1855. It is a farm trade and cattle-shipping center.

Glen Cove, city (1978 est. pop. 25,600), Nassau co., SE N.Y., on the N shore of Long Island, at the entrance to Hempstead harbor; settled 1668, inc. as a city 1918. Although chiefly residential, it has varied light industries. The Webb Institute of Naval Architecture is here.

Glen·dale (glĕn'dāl'). **1.** City (1978 est. pop. 76,700), Maricopa co., S central Ariz., adjacent to Phoenix; inc. 1910. It is located in a rich agricultural region irrigated by the Salt River project. The city has food-processing plants and is a shipping point for fruits and vegetables. Luke Air Force Base, a large jet fighter training center, is in Glendale. **2.** City (1978 est. pop. 135,700), Los Angeles co., S Calif., a suburb of Los Angeles; inc. 1906. It has aerospace and defense-oriented plants, as well as a film industry. The city was founded on part of a ranch that had been the first Spanish land grant in California (1784). Forest Lawn Memorial Park, a large cemetery, is located here. **3.** City (1970 pop. 6,981), St. Louis co., E Mo., a suburb SW of St. Louis; inc. 1950. **4.** City (1970 pop. 13,426), Milwaukee co., SE Wis., a suburb of Milwaukee, on the Milwaukee River; inc. 1950. It has a grain elevator and plants that make electrical switches and appliances, batteries, cans, and plastic products.

Glen·dive (glĕn'dīv'), city (1970 pop. 6,305), seat of Dawson co., E central Mont., on the Yellowstone River NE of Miles City, in a farm, livestock, oil, and natural-gas region; inc. 1902. Laid out in 1880, it was settled in 1881 with the coming of the railroad and grew rapidly as a cattle-shipping center.

Glen·do·ra (glĕn-dôr'ə), city (1978 est. pop. 34,300), Los Angeles co., S Calif., at the base of the San Gabriel Mts.; inc. 1911. Sprinklers and pumps are made in the city. The region was declared open government land in 1869 and a rush of homesteaders began. By the early 1900s it was covered with magnificent groves of orange and lemon trees. The housing boom following World War II converted the city into a residential community.

Glen·elg (glĕn-ĕlg'), city (1971 pop. 15,383), a suburb of Adelaide, S Australia, on an inlet of Gulf St. Vincent. It is a summer resort. In 1836 South Australia's first colonists landed here.

Glenelg, river, 290 mi (466.6 km) long, W Victoria, SE Australia, rising in the Grampians W of Mt. William and flowing generally W and S into Discovery Bay near the border of South Australia. It is frequently dry.

Glen El·lyn (ĕl'ən), village (1978 est. pop. 25,300), Du Page co., NE Ill., a residential suburb of Chicago; inc. 1892. Points of interest include Stacy Tavern, a 19th cent. stagecoach stop on the Chicago-Galena route; a wildlife sanctuary; and an arboretum.

Glen More (mōr, môr): *see* Great Glen, valley, Scotland.

Glenn (glĕn), county (1970 pop. 17,521), 1,317 sq mi (3,411 sq km), N Calif., with the Sacramento Valley to the E and rising in the W to the Coast Ranges, watered by the Sacramento River and its tributaries; formed 1891; co. seat Willows. Cattle raising and irrigated farming (alfalfa, sugar beets, rice, barley, olives, almonds, prunes, pears, apricots, and citrus fruit) are important. Some mining is done, and there are lumbering and food-processing industries. Automobile parts and wood, stone, clay, and glass products are made. The county includes part of Mendocino National Forest.

Glen Rock, borough (1970 pop. 13,011), Bergen co., NE N.J.; settled c.1710, inc. 1896. A residential suburb of New York City, it has a small industrial park. George Washington's army used the area for camping grounds during the Revolutionary War.

Glen Rose (rōz), city (1970 pop. 1,554), seat of Somervell co., N central Texas, SW of Fort Worth; inc. 1926. Cotton and peanuts are grown. Mineral springs and rugged scenery have made Glen Rose a health and vacation resort.

Glen Roy (roi), valley, Inverness-shire, W Scotland, E of Loch Lochy. The Parallel Roads, three terraces on each side of the valley at corresponding heights, are believed to mark receding levels of a lake that once filled the valley.

Glens Falls, city (1978 est. pop. 17,100), Warren co., E central N.Y., in the foothills of the Adirondack Mts. and on the Hudson River; settled 1762, inc. as a city 1908. Paper, clothing, chemicals, and cement are produced.

Glen·view (glĕn'vyōō'), village (1978 est. pop. 31,300), Cook co., NE Ill., a suburb of Chicago; settled 1833, inc. 1899. It is chiefly residential. A dairy research center and a U.S. naval air station are here.

Glen·ville (glĕn'vĭl'), town (1970 pop. 2,183), seat of Gilmer co., central W.Va., on the Kanawha River SW of Clarksburg, in a farm and timber region.

Glen·wood (glĕn'wŏŏd'). **1.** City (1977 est. pop. 5,002), seat of Mills co., SW Iowa, near the Missouri River SSE of Council Bluffs, in a grain-growing area; founded by Mormons, inc. 1857. **2.** City (1970 pop. 2,584), seat of Pope co., W central Minn., S of Alexandria on Lake Minnewaska; platted 1866, inc. as a village 1881, as a city 1912. Two fish hatcheries are nearby. Summer and winter resorts are in the area.

Glenwood Springs, city (1970 pop. 4,106), seat of Garfield co., NW Colo., on the Colorado River NE of Grand Junction; laid out 1883, inc. 1885. At an altitude of c.5,756 ft (1,755.6 m), it has mineral springs and is headquarters for White River National Forest.

Glit·ter·tind·en (glĭt'ər-tĭn'ən), peak, 8,104 ft (2,471.7 m) high, S central Norway, in the Jotunheimen Mts. It is the highest point in Scandinavia.

Gli·wi·ce (glē-vē'tsĕ), city (1975 est. pop. 197,200), SW Poland. A coal-mining and steel-making center of the Katowice region, it also has industries producing machinery and chemicals. Chartered in 1276, Gliwice was ceded by Austria to Prussia in 1742.

globe (glōb), spherical map of the earth (terrestrial globe) or the sky (celestial globe). The terrestrial globe provides the only graphic representation of the areas of the earth without significant distortion or inaccuracy in shape, direction, or relative size. However, the flattening of the earth at the poles and its slight bulge below the equator are normally disregarded in the construction of a globe. Probably the earliest globe was constructed by the Greek geographer Crates of Mallus in the 2nd cent. B.C. Few attempts were made to construct globes in the Middle Ages, although Strabo and Ptolemy, at the beginning of the Christian era, had formulated precise and detailed instructions for doing so. The first globes of modern times were made in the late 15th cent. by Martin Behaim of Nuremberg and Leonardo da Vinci.

Globe, city (1975 est. pop. 6,396), seat of Gila co., E Ariz., in the foothills near the Pinal Mts.; settled 1876, inc. 1907. Though copper had been discovered here earlier (1872), the city grew as a result of a silver boom; only later did it become a big copper center. Copper, silver, gold, and asbestos are still mined.

Glom·ma (glô'mä), river, Norway: *see* Glåma.

Glo·ri·a De·i National Historic Site (glôr'ē-ə dā'ē): *see* National Parks and Monuments Table.

Glos·sop (glŏs'əp), municipal borough (1973 est. pop. 24,820), Derbyshire, central England. A residential suburb of Manchester, Glossop is also the chief cotton-manufacturing city of Derbyshire. Other products are woolens, canned goods, and paper.

Glouces·ter (glŏs'tər, glô'stər), county borough (1976 est. pop. 91,600), administrative center of Gloucestershire, W central England, on the Severn River. It is a market town. Manufactures include aircraft components, agricultural machinery, railroad equipment, and processed foods. The port is still active but has been eclipsed by Bristol since the 15th cent. Gloucester stands upon the site of the Roman city Glevum. In Saxon times it was the capital of Mercia. There is a notable cathedral (begun 1089) in which Edward II is buried.

Gloucester. 1. County (1970 pop. 172,681), 329 sq mi (852.1 sq km), SW N.J., bounded on the W by the Delaware River and drained by the Maurice River and Big Timber Creek; formed 1686; co. seat Woodbury. It is primarily industrial, with some agriculture (truck crops and fruit) and food-processing plants. Among its manufactures are textiles, clothing, chemicals, plastics, concrete, wood and metal products, machinery, and transportation equipment. **2.** County (1970 pop. 14,059), 225 sq mi (582.8 sq km), E Va., bounded on the W and S by the York River, on the E by Mobjack Bay, a section of Chesapeake Bay; formed 1651; co. seat Gloucester. In a tidewater agricultural area yielding truck crops, grain, tobacco, and livestock, it also has fisheries and a lumbering industry.

Gloucester 1. City (1978 est. pop. 26,800), Essex co., NE Mass., on Cape Ann; settled 1623, inc. as a city 1873. It is a port of entry at the head of the excellent Gloucester harbor, which is protected by a breakwater built from Eastern Point. The harbor has been used by fishing ships for over three centuries, and Gloucester is still a great fishing port, with many fish-processing industries and related manufactures. It was once an important shipbuilding center, and the first schooner is said to have been built here in 1713. The picturesque old city is also the center of an extensive summer resort area. Gloucester's development as a resort and artists' colony began late in the 19th cent. Tourist attractions include the famous bronze *Fisherman,* a memorial to the thousands of Gloucestermen lost at sea; Hammond Castle, which houses collections of medieval art; and numerous pre-Revolutionary houses. **2.** Village (1970 pop. 700), seat of Gloucester co., E Va., near Chesapeake Bay E of Richmond, in a fishing area; founded 1769. Winches are manufactured. A courthouse built in 1766 and an old debtors' prison are here.

Gloucester City, city (1970 pop. 14,707), Camden co., SW N.J., on the Delaware River, a suburb adjoining Camden and opposite Philadelphia; site of Fort Nassau (built 1623 by the Dutch); settled c.1682 by Irish Quakers, inc. 1868.

Glouces·ter·shire (glŏs'tər-shîr', -shər), nonmetropolitan county (1971 pop. 1,069,454), 1,258 sq mi (3,258 sq km), W central England; administrative center Gloucester. In the eastern part of the county are the Cotswold Hills, devoted largely to dairy and crop farming; in the center is the fertile valley of the Severn River, devoted to dairy farming and sheep raising; and in the west, on the Welsh border, are the Wye valley and the Forest of Dean, also with sheep raising. The region was part of the Anglo-Saxon kingdom of Mercia.

Glov·ers·ville (glŭv'ərz-vĭl'), city (1978 est. pop. 17,700), Fulton co., E central N.Y.; inc. 1890. Glove making has been important since the late 18th cent. Other industries include tanning and the manufacture of records and toys.

Glynn (glĭn), county (1970 pop. 50,528), 412 sq mi (1,067.1 sq km), SE Ga., bounded on the E by the Atlantic Ocean, on the NE by the Altamaha River, on the SW by the Little Satilla River; formed 1777; co. seat Brunswick. It includes St. Simons and Jekyll islands. In a coastal-plain agricultural area yielding truck crops, dairy products, poultry, and livestock, it also depends on fishing, forestry, and diversified manufacturing. Among its products are clothing, paper, chemicals, and machinery.

Gmünd (gə-münt'): *see* Schwäbisch Gmünd, West Germany.

Gniez·no (gə-nyĕz'nô), city (1975 est. pop. 55,100), central Poland. It is a railway junction with industries producing clothing, leather goods, and metals. Gniezno was the first capital of Poland, whose kings were crowned here until 1320. From 1572 until the early 19th cent. the archbishops of Gniezno acted as protectors of Poland. The city passed to Prussia in 1793 and again in 1815. It was restored to Poland in 1919.

Go·a (gō'ə), **Da·man** (də-män'), and **Di·u** (dē'ōō), union territory (1971 pop. 857,180), c.1,480 sq mi (3,800 sq km), W India, on the Arabian Sea. The union territory is composed of three noncontiguous former Portuguese colonies that were seized by India in 1961. It is administered by the home minister in the central Indian government. There is an elected local assembly. The capital is Panjim.

Goa (1971 pop. 794,530), c.1,430 sq mi (3,700 sq km), on the Malabar Coast, has three principal ports: Agoada, Marmagao, and Panjim. The chief products are rice, cashew nuts, and coconuts. Moslems conquered Goa in 1312. It became part of the Hindu kingdom of Vijayanagar in 1370 but was recaptured by the Moslems 100 years later. The Portuguese annexed it in 1510. Old Goa, the original capital, was a prosperous port city in the late 16th cent. A cathedral, churches, and several palaces survive from this period.

Daman (1971 pop. 38,741), c.30 sq mi (80 sq km), at the mouth of the Daman River on the Gulf of Cambay, was acquired by the Portuguese in 1588. Until the Indians took over, the capital was Daman city, which before the decline of Portuguese power in the 18th cent. enjoyed a large overseas trade, especially with East Africa. Rice, wheat, and tobacco are the chief crops, and fishing is important.

Diu (1971 pop. 23,909), c.20 sq mi (50 sq km), consisting of Diu island and a small area on the coast of the nearby Kathiawar peninsula, was acquired by Portugal in 1535. Diu town has several splendid Catholic churches built before the overseas trade of Diu declined in the late 18th cent. Fishing is the principal occupation, and salt is produced.

Goat Island (gōt), island, W N.Y., in the Niagara River, dividing Niagara Falls into the American and the Canadian falls.

Go·bi (gō'bē), great desert of central Asia, c.500,000 sq mi (1,295,000 sq km), extending c.1,000 mi (1,610 km) from E to W across SE Mongolia and N China from the Great Khingan Mts. to the Tien Shan. The Gobi, located on a plateau from 3,000 to 5,000 ft (915–1,525 m) high, consists of a series of shallow alkaline basins; the western portion of the desert is entirely sandy. Nearly all the region's soil has been removed by the prevailing northwesterly winds and deposited in north-central China as loess; fierce sand and wind storms are common. The Gobi's grassy fringe supports a small population of nomadic Mongolian tribes engaged in sheepherding and goatherding. Many paleontological finds, including dinosaur eggs, have been made in the Gobi. Prehistoric stone implements, some c.100,000 years old, have also been excavated.

Go·da·va·ri (gō-dä'və-rē), river, c.900 mi (1,450 km) long, rising in the Western Ghats in Maharashtra state, W central India, and flowing SE across the Deccan Plateau to the Bay of Bengal. Below Rajahmundry the river divides into two streams that form a huge delta. The delta, site of some of the earliest European settlements in India, has an extensive navigable irrigation-canal system. The Godavari River is sacred to Hindus.

God·havn (gŏd'hä'vən, gôth-houn'), town (1969 pop. 863) in Godhavn dist., W Greenland, on Disko Island; founded in 1773. It is a fishing base and has a geophysical observatory and an arctic research station (established 1906 by the Univ. of Copenhagen).

Godt·håb (gôt'hôp'), town (1975 est. pop. 8,328) in Godthåb dist., W Greenland, on the Godthåb Fjord. The largest town and capital of Greenland, Godthåb is a fishing center. The town was founded in 1721 by Hans Egede and Norwegian missionaries.

God·win-Aus·ten, Mount, (gŏd'wĭn-ô'stĭn), or **K2** (kā'tōō'), peak, 28,250 ft (8,616.3 m) high, in the Karakorum Range, N Pakistan, second-highest peak in the world. It was discovered and measured by the Survey of India in 1856 and named for English topographer Henry Godwin-Austen, who explored and surveyed the region. "K2" is taken from the first letter of Karakorum, and the number indicates that it was the second peak in the range to be measured. An Italian team led by Ardito Desio reached the summit in 1954.

Goffs·town (gôfs'toun'), resort town (1970 pop. 9,284), Hillsborough co., S N.H., WNW of Manchester; inc. 1761. Granted in 1734 by Mass. as Narragansett No. 4, it was regranted in 1748 by the Masonian Proprietor. Wood products are made here.

Go·ge·bic (gō-gē'bĭk), county (1970 pop. 20,676), 1,102 sq mi (2,854.2 sq km), extreme NW Mich., on the W Upper Peninsula bounded on the NW by Lake Superior, on the S and SW by the Wis. border, and drained by the Montreal, Presque Isle, and Ontonagon rivers; formed 1887; co. seat Bessemer. It is crossed by the Gogebic iron range. Agriculture (potatoes, truck crops, grain, hay, fruit, livestock, and dairy products), iron and copper mining, and lumbering are important. Clothing and wood products are manufactured. The county has many small lakes and waterfalls and numerous hunting and fishing resorts.

Gogebic Range, E-W mountain range, 80 mi (128.7 km) long and .5 to 1 mi (.8–1.6 km) wide, extending from the W Upper Peninsula, N Mich., into N Wis. It is known for its iron deposits, discovered in 1848, which lie in some of the world's deepest mines.

Go·gra (gō'grä), river, major tributary of the Ganges, c.640 mi (1,030 km) long, rising in the Himalayas, SW Tibet (China), and flowing generally SE through Nepal (where it is called the Karnali) to join the Ganges in Bihar state, E India.

Goi·â·ni·a (goi-ä'nē-ə, gōō-yä'nyə), city (1970 pop. 388,926), capital of Goiás state, S central Brazil. A modern planned city, it was built to replace the old city of Goiás as state capital and was inaugurated as such in 1937. It is a shipping and processing center for a region producing cattle, minerals, and agricultural commodities.

Goi·ás (goi-äs'), state (1975 est. pop. 3,558,100), 247,912 sq mi (642,092 sq km), central Brazil; capital Goiânia.

Gol·con·da (gŏl-kŏn'də), ruined city, Andhra Pradesh state, SE India. It was the capital (c.1364–1512) of the Bahmani kingdom, but after 1512 it became the capital of the Moslem sultanate of Golconda. The legions of Aurangzeb, the Mogul emperor, captured the city in 1687, after which Golconda gradually fell into ruin. The main feature of the city is its fort on a hill 400 ft (122 m) above the plain. There are also ruins of palaces and mosques. At its peak, the city was famed for the diamonds found to the southeast and cut in Golconda; its name has come to mean "source of great wealth."

Golconda, city (1970 pop. 922), seat of Pope co., extreme SE Ill., on the Ohio River S of Harrisburg, in a rich agricultural area; inc. 1845. Fruit, corn, wheat, livestock, and wood products are produced.

Gold Beach (gōld), city (1971 est. pop. 1,475), seat of Curry co., SW Oregon, on the Pacific coast at the mouth of the Rogue River; inc. 1945. Placer mining for gold was carried on here from c.1853 to 1861. Today it is a lumbering center and fishing resort.

Gold Coast: *see* Ghana, republic.

Gold Coast, city (1971 pop. 66,558), Queensland, E Australia, on the Pacific Ocean. The city, a major resort, stretches for many miles along the coast.

Gold·en (gōl'dən), resort city (1970 pop. 9,817), seat of Jefferson co., N central Colo., just W of Denver, in a coal, gold, and clay area at an altitude of c.5,600 ft (1,708 m). There are farms in the vicinity, and the city manufactures clay products. It was founded in 1859 and served as territorial capital from 1862 to 1867. The Colorado School of Mines is here.

Gold·en·dale (gōl'dən-dāl'), city (1977 est. pop. 3,310), seat of Klickitat co., S central Wash., SSW of Yakima; platted 1871, inc. 1902. It is a trading center for a farm, lumber, and cattle area and has flour and lumber mills. Maryhill Museum of Fine Arts is nearby.

Golden Gate, strait, 4 mi (6.4 km) long and 1 to 2 mi (1.6–3.2 km) wide, linking San Francisco Bay with the Pacific Ocean. It was discovered in 1579 by the English explorer Sir Francis Drake. Known as the Golden Gate before the California gold rush, its name became popular during this period because of its new connotation. The strait is the drowned mouth of the united Sacramento and San Joaquin rivers and forms an excellent channel, c.400 ft (120 m) deep, into San Francisco Bay.

Golden Gate National Recreation Area: *see* National Parks and Monuments Table.

Golden Spike National Historic Site (spīk): *see* National Parks and Monuments Table.

Golden Valley. 1. County (1970 pop. 931), 1,178 sq mi (3,051 sq km), central Mont., in a livestock and grain area drained by the Musselshell River; formed 1920; co. seat Ryegate. Part of Lewis and Clark National Forest is in the northwest. **2.** County (1970 pop. 2,611), 1,014 sq mi (2,626.3 sq km), W N.Dak., on the Mont. line; organized 1912; co. seat Beach. In a fertile agricultural and grazing area yielding livestock, wheat, and grain, it has lignite mines

Gold·field (gōld'fēld'), town (1970 pop. 200), Esmeralda co., SW Nev., a former gold-mining center. Gold was discovered in 1902, and after an early period of disappointment, large yields of high-quality gold were extracted. A rush in 1903 built a lusty city, which was soon remarkable for its exuberant elegance, with a theater, a large hotel (still standing), and various fine residences. A strike by the miners caused Federal troops to be brought to Goldfield in 1907. Production reached its height in 1910, then fell off. The boom ended in 1918, and Goldfield declined almost as fast as it had risen.

Golds·bor·o (gōldz'bûr'ō), city (1978 est. pop. 27,300), seat of Wayne co., E central N.C.; inc. 1847. Goldsboro is a marketplace for bright-leaf tobacco and a shipping center for timber, livestock, and farm products. Furniture, textiles, shoes, leather, metal, and electronic goods are manufactured.

Gold·thwaite (gōld'thwāt'), town (1970 pop. 1,693), seat of Mills co., central Texas, W of Waco, in a hilly region of farms, ranches, and pecan groves.

Gol·go·tha (gŏl′gə-thə): *see* Calvary.

Go·li·ad (gō′lē-ăd′), county (1970 pop. 4,869), 871 sq mi (2,255.9 sq km), S Texas, drained by the San Antonio River; formed 1836; co. seat Goliad. It is primarily agricultural, yielding corn, cotton, grain sorghums, flax, broomcorn, fruit, truck crops, livestock, and dairy products. It also has oil and natural-gas wells.

Goliad, city (1970 pop. 1,709), seat of Goliad co., S Texas, on the San Antonio River, SE of San Antonio. It is a market for the surrounding farm region. A Spanish mission and presidio moved to Goliad in 1749. After the start of the Texas Revolution (1836), Goliad was seized by Texan forces under Col. J. W. Fannin. With the advance of Mexican troops into Texas, Fannin evacuated Goliad with more than 300 men. He was overtaken by the Mexicans and on Mar. 20, 1836, after a hopeless battle, surrendered unconditionally. On Mar. 27 most of the prisoners were shot by the Mexicans. The American settlement grew up across the river, and the restored mission and the ruins of the old presidio are today in a state park.

Go·mel (gô′mĕl), city (1976 est. pop. 349,000), capital of Gomel oblast, W European USSR, on the Sozh River, a tributary of the Dnepr. It is a river port and a large railroad junction in an agricultural area. The city's industries produce machinery, textiles, building materials, food products, electrical equipment, and fertilizers. First mentioned as Gomiy in 1142, it became part of Lithuania in 1537. It was much fought over and passed to Poland by the Treaty of Andrusov (1667) and to Russia in 1772.

Go·na·ïves (gô-nä-ēv′), city (1971 pop. 29,261), W Haiti, a port on the Gulf of Gonaïves. The region's agricultural products (including cotton, sugar, and bananas) are exported from the city's fine harbor. Haiti's independence was proclaimed here in 1804. The Gulf of Gonaïves, situated in the pincers of two mountainous peninsulas, is considered one of the most beautiful in the world.

Gon·bad-e Ka·vus (gōn-bäd′ĕ kä-vo͞os′), city (1966 pop. 40,667), Mazanderan prov., N Iran, on the Gorgan River. It is an agricultural trade center.

Gon·da (gōn′dä), town (1971 pop. 52,647), Uttar Pradesh state, N India, on the Sarayu River. Gonda is a market for maize, sugar cane, pulses, and wheat. It was founded in the early 16th cent.

Gon·dar (gŏn′dər), town (1974 est. pop. 43,040), capital of Begemdir and Simen prov., NW Ethiopia, at an altitude of c.7,300 ft (2,225 m). It is a regional trade center and a tourist spot. Gondar was the capital of Ethiopia from c.1635 to 1867. Gondar is noted for its architectural ruins and for its handicrafts.

Gon·zal·es (gən-zăl′ĭs, -zä′lĭs), county (1970 pop. 16,375), 1,058 sq mi (2,740.2 sq km), S central Texas, drained by the Guadalupe and San Marcos rivers; formed 1836; co. seat Gonzales. It is a leading producer of poultry. Corn, grain sorghums, cotton, hay, pecans, potatoes, fruit, truck crops, livestock, and dairy products are also important. Food-processing and beekeeping are done.

Gonzales, city (1970 pop. 5,854), seat of Gonzales co., S central Texas, near the Guadalupe River, E of San Antonio. Founded in 1825, the city spreads out from squares that were a double plaza of Spanish style. On Oct. 2 1835, the men of Gonzales dispersed Mexican cavalry in the first battle of the Texas Revolution; a state park nearby encloses the battlefield.

Gooch·land (go͞och′lənd), county (1970 pop. 10,069), 289 sq mi (748.5 sq km), central Va., bounded on the S by the James River; formed 1726; co. seat Goochland. In an agricultrual area yielding tobacco, wheat, corn, hay, livestock, and some dairy products, it has deposits of coal and stands of timber.

Goochland, village (1970 pop. 400), seat of Goochland co., central Va., near the James River WNW of Richmond.

Good Hope, Cape of (go͝od′ hōp′), cape, SW Cape Prov., S South Africa, S of Cape Town. It was first circumnavigated in 1488 by Bartolomeu Dias. Vasco da Gama passed it on his voyage to India (1497). The first settlement in the area was made in 1652.

Good·hue (go͝od′hyo͞o), county (1970 pop. 34,804), 758 sq mi (1,963.2 sq km), SE Minn., bounded on the NE by the Mississippi River and the Wis. border and drained by the Cannon River; formed 1853; co. seat Red Wing. Its agriculture includes livestock, dairy products, poultry, corn, oats, and barley. Some mining is done. The county also has lumbering and food-processing industries. Wood and metal products, furniture, leather goods, and shoes are manufactured.

Good·ing (go͝od′ĭng), county (1970 pop. 8,645), 722 sq mi (1,870 sq km), S Idaho, in a tableland area sloping to the Snake River plain in the S; formed 1913; co. seat Gooding. In a cattle-raising and irrigated agricultural area yielding potatoes, beans, sugar beets, alfalfa, and fruit, it has a meat-packing industry. In a winter sports area, the county also has good hunting and fishing.

Gooding, city (1970 pop. 2,599), seat of Gooding co., S central Idaho, on the Little Wood River NW of Twin Falls, in a rich farm and livestock area served by the Minidoka project; founded 1883.

Good·land (go͝od′lənd), city (1970 pop. 5,510), seat of Sherman co., NW Kansas, NW of Hays, near the Colo. border; inc. 1887. Wheat is grown, and livestock raised. Fossils have been found in the area.

Good·win Sands (go͝od′wĭn săndz′), stretch of shoals and sand bars, c.10 mi (15 km) long, lying off the E coast of Kent, SE England. It forms a breakwater east of The Downs, a roadstead. Shipwrecks were formerly frequent on the Sands.

Good·wood (go͝od′wo͝od′), town (1970 pop. 31,592), Cape Prov., SW South Africa, a residential and commercial suburb of Cape Town. The town has some light industry and is on an important railroad line. Goodwood was founded in 1905.

Goole (go͞ol), municipal borough (1973 est. pop. 17,920), Humberside, N England, at the confluence of the Ouse and Don rivers. A significant inland port, it has extensive dockyards. Coal and textiles are exported, and shipbuilding, iron casting, sugar refining, flour milling, and the manufacture of farm machinery, fertilizers, and clothing are leading industries.

Göp·ping·en (gœp′ĭng-ən), city (1975 pop. 55,415), Baden-Württemberg, S West Germany. Its manufactures include machinery, precision instruments, chemicals, pharmaceuticals, wood and leather products, and textiles. Mineral water is bottled and shipped. Göppingen was chartered in the mid-12th cent. The city was twice (1425, 1782) devastated by fire.

Go·rakh·pur (gō′rək-po͝or′), city (1971 pop. 230,701), Uttar Pradesh state, N central India, on the Rapti River. Founded c.1400, it is a conglomeration of farm villages in a densely populated area.

Gor·di·um (gôr′dē-əm), ancient city of Asia Minor, in Phrygia, now in Turkey, SW of Ankara. It was the capital of Phrygia from c.1000 to 800 B.C. Excavations conducted since 1950 have revealed Hittite, Phrygian, Persian, and Graeco-Roman remains. It was here that Alexander the Great is said to have cut the Gordian knot.

Gor·don (gôr′dn), county (1970 pop. 23,570), 358 sq mi (927.2 sq km), NW Ga., in a valley and ridge area drained by the Oostanaula, Conasauga, and Coosawattee rivers; formed 1850; co. seat Calhoun. Its agriculture includes soybeans, cotton, corn, sweet potatoes, hay, fruit, and livestock. Its manufacturing is dominated by a tufted-carpet industry. It also produces other textiles, lumber, and wood products. Part of Chattahoochee National Forest is in the west.

Gor·gan (gôr-gän′), town (1971 est. pop. 55,000), N Iran, E of the Caspian Sea. The surrounding region yields rice, wheat, barley, and nuts and has extensive forests and marshes. The surrounding area, the ancient Hyrcania, was captured by the Arabs (716) and conquered by the Mongols (13th cent.).

Gor·ham (gôr′əm). **1.** Town (1970 pop. 7,839), Cumberland co., SW Maine, just W of Portland; granted 1728, settled 1736, inc. 1764. A garrison house was built on Fort Hill in the French and Indian Wars. **2.** Town (1970 pop. 2,998), Coos co., NE N.H., at the confluence of the Androscoggin and Peabody rivers, on the N edge of White Mountain National Forest; settled c.1805, inc. 1836. In a timber region within sight of the Presidential Range, Gorham is a resort with an annual winter carnival.

Go·ri (gôr′ē), city (1976 est. pop. 54,000), SE European USSR, in Georgia. It has textile plants. Mentioned in the 7th cent. as Tontio, it was later named after a fortress. Gori passed to Russia in 1801. Stalin was born in the city.

Go·rin·chem (gôr′ĭn-KHəm) or **Gor·kum** (gôr′kəm), town (1977 est. pop. 28,539), South Holland prov., W central Netherlands, on the Upper Merwede River. It is a manufacturing center. Gorinchem became a major trade center in the 15th cent. Much of the old town and its walls have been preserved.

Go·ri·zia (gō-rē′tsyä), city (1976 est. pop. 43,375), capital of Gorizia prov., Friuli-Venezia Giulia, NE Italy, on the Isonzo River and the Yugoslav border. It is an industrial, commercial, transport, and tourist center. Manufactures include textiles, leather goods, paper, and machines. Located in the historic region of Friuli, Gorizia was the seat of a duchy from c.1000 to 1500. It passed to the Hapsburgs in 1508 but preserved a remarkable autonomy until the 18th cent. In World War I, Gorizia and the surrounding area were the scene of bloody battles. The Italians took Gorizia in 1916, evacuated it in 1917, and recovered it in 1918.

Gor·ky or **Gor·ki** (both: gôr'kē), formerly **Nizh·ny Nov·go·rod** (nĭsh'nē nŏv'gə-rŏd, nyĕsh'nyĭ nôf'gə-rəd), city (1974 est. pop. 1,319,-000), capital of Gorky oblast, E European USSR, on the Volga and Oka rivers. A major river port and a rail and air center, it is one of the chief industrial cities of the USSR. Heavy machinery, steel, automobiles, chemicals, and textiles are produced. In 1221 a prince of Vladimir founded the city as a frontier post. In 1350 it became the capital of the Suzdal-Nizhny Novgorod principality and was annexed in 1392 by Moscow. From 1608 to 1612 the city was the rallying point for the Russian army that defeated the Polish, Lithuanian, and Cossack armies. Gorky was famous for its annual trade fairs, held from 1817 to 1930, except during the Bolshevik Revolution and the civil war. Its turreted stone kremlin dates from the 13th cent. The city was named for Maxim Gorky, who was born here.

Gör·litz (gœr'lĭts'), city (1974 est. pop. 84,658), Dresden dist., SE East Germany, on the Görlitzer Neisse River, at the Polish border. Manufactures include textiles, metal goods, furniture, and beer. Lignite is mined nearby. Görlitz was founded c.1200 and developed as an important cloth-weaving and trade center. The city passed to the Hapsburgs in 1526 and to Saxony in 1635. In 1815 Görlitz was annexed by Prussia. After World War II a section of the city on the right bank of the Neisse was given to Poland.

Gor·lov·ka (gôr-lôf'kä), city (1976 est. pop. 342,000), S European USSR, in the Ukraine, in the Donets Basin. It is a major coal-mining and industrial center, with a major chemical complex.

Gor·no-Al·tai Autonomous Oblast (gôr'nō-äl-tī'), oblast (1970 pop. 168,000), 35,800 sq mi (92,722 sq km), Altai Kray, SE Siberian USSR. Bordering on Mongolia in the south, it contains most of the Altai Mts. The region is mountainous, and gold, manganese, and mercury are mined. Livestock raising and dairy farming are important, and grain is cultivated. The oblast capital is Gorno-Altaisk (1970 pop. 34,000), a processing center for the agricultural products of the region. A primitive communal society existed in the area from the 3rd millennium B.C., and there is evidence of a Mongolian civilization in the 5th cent. B.C. The Turkish khanate ruled the region from the 6th to the 10th cent. A.D., and the Altaians were under the control of the Mongolian khans from the beginning of the 13th to the 18th cent. In 1756 the Altaians came under Russian hegemony. From 1918 to 1922 there was civil war as the mountain groups fought the Bolshevik forces. Between 1922 and 1948 the state was called the Oirot Autonomous Oblast. It was renamed in 1948.

Gor·no-Ba·dakh·shan Autonomous Oblast (gôr'nō-bä-däKH-shän'), oblast (1970 pop. 98,000), c.24,600 sq mi (63,710 sq km), Central Asian USSR, in Tadzhikistan, in the Pamirs, bordered by China on the E and by Afghanistan on the S and W and separated from Pakistan and Kashmir by a narrow strip of Afghan territory; capital Khorog. The eastern section (East Pamir) is a high plateau, and the western part (West Pamir) is cut by high ranges and deep, narrow valleys. Gold, salt, mica, limestone, and peat are mined. In the east livestock is raised, and in the western valleys grain, vegetables, and beans are grown. Formerly under the control of the Mongols and the Arabs, the region passed to Russian control in 1895. The autonomous oblast was formed in 1925.

Gor·ty·na (gôr-tī'nə), ancient city, S central Crete. Under Rome it was one of the leading cities of the island. Many ancient Greek remains have been discovered on the site. An inscription dating from c.450 B.C. of a code of laws of inheritance, marriage, divorce, and other family matters was found on a wall in 1884.

Go·ryn (gô'rĭn), river, W European USSR, rising E of Lvov and flowing 410 mi (660 km) N into the Pripyat River.

Gor·zów Wiel·ko·pol·ski (gôr'zo͞of vyĕl-kô-pôl'skē), city (1975 est. pop. 87,200), W Poland, on the Warthe River. A transportation and trade center, it also has shoe mills, chemical works, and lignite mines. Chartered in 1257, it was destroyed by the Swedes in the Thirty Years' War.

Gos·forth (gŏz'fərth), urban district (1971 pop. 26,826), Tyne and Wear, NE England. Formerly a coal-mining center, Gosforth is now residential.

Go·shen (gō'shən), county (1970 pop. 10,885), 2,230 sq mi (5,775.7 sq km), SE Wyo., bordering on Nebr. and watered by the North Platte River; formed 1911; co. seat Torrington. Cattle, sheep, and hogs are raised. Sugar beets, corn, alfalfa, beans, and potatoes are grown on irrigated land; dryland wheat and other grains are also important. Beet-sugar refining is done.

Goshen. 1. City (1978 est. pop. 18,900), seat of Elkhart co., N Ind., on the Elkhart River; inc. 1868. Goshen is in a farm and dairy region; poultry is also raised and processed. Its manufactures include rubber goods, electronic controls, wood products, metal products, mobile homes, and boats. There are Amish and Mennonite colonies in the area. **2.** Village (1970 pop. 4,342), seat of Orange co., SE N.Y., SE of Middletown, in a dairy and truck-farm region; settled 1712, inc. 1809. It is a center of trotting races.

Gos·lar (gôs'lär), city (1974 est. pop. 54,041), Lower Saxony, E West Germany, at the foot of the Harz Mts., near the border with East Germany. Since its founding in the 10th cent. Goslar has been a mining center. Today copper, lead, zinc, iron, and sulfur are mined. Manufactures of the city include textiles, clothing, chemicals, and machinery. Goslar was a favorite residence of many early German emperors and was the scene of several imperial diets. It long was a member of the Hanseatic League and was a free imperial city until 1802, when it passed to Prussia. Goslar was awarded to Hanover in 1815 but was regained by Prussia in 1866. The city has preserved much of its medieval character.

Gos·per (gŏs'pər), county (1970 pop. 2,178), 466 sq mi (1,206.9 sq km), S Nebr., watered by Johnson Lake; formed 1873; co. seat Elwood. In an agricultural region yielding grain and livestock, it has a tourist industry and some manufacturing.

Gos·port (gŏs'pôrt'), municipal borough (1976 est. pop. 82,300), Hampshire, S England. There are shipbuilding facilities and various industries. Formerly a victualing station for the Royal Navy, Gosport was an embarkation point for the invasion of France in 1944.

Gö·ta älv (yœ'tä ĕlv'), river, 56 mi (90.1 km) long, SW Sweden, draining Vänern Lake into the Kattegat. It is part of the Göta Canal, a 240-mi (386.2-km) system of rivers, lakes, and canals that crosses southern Sweden from Göteborg to Arkosund on the Baltic Sea and to Stockholm by way of the sea. The canals were opened in 1832.

Gö·te·borg (yœ'tə-bôr'yə) or **Goth·en·burg** (gŏth'ən-bûrg', gŏt'n-), city (1975 pop. 444,651), capital of Göteborg och Bohus co., SW Sweden, on the Kattegat at the mouth of the Göta älv River. It is Sweden's most important seaport and a major commercial and industrial center. Manufactures include iron and steel, ball bearings, motor vehicles, textiles, processed food, and refined petroleum. There are large shipyards and fisheries in the city. Göteborg was founded in 1604 by Charles IX, but was destroyed by the Danes in the Kalmar War. It was rebuilt by Gustavus II in 1619.

Go·tha (gō'tä), city (1974 est. pop. 59,243), Erfurt dist., SW East Germany. It is an industrial, administrative, and cultural center. Manufactures include wood and rubber products, processed food, and printed materials. Gotha was known in the late 12th cent. In 1640 it became the capital of the duchy of Saxe-Gotha (from 1826 to 1918, Saxe-Coburg-Gotha). Gotha has long been a center of geographic research and publishing. The *Almanach de Gotha*, an authoritative reference work on the royal houses and the nobility of numerous countries, was first published here in 1863.

Got·land (gŏt'lənd), region, SE Sweden, in the Baltic Sea, including Gotland Island. Cereals, sugar beets, and vegetables are grown, and sheep are raised. Fishing, cement making, and tourism are the main industries. Archaeological remains indicate that Gotland, inhabited since the Stone Age, had wide commercial contacts from early times, especially under the Vikings (9th-11th cent.). From the 11th to the 14th cent. Gotland prospered as a major trade center of northern Europe, but internal strife between the Hanse merchants and local tradesmen weakened the region. Gotland was conquered by the Swedish King Magnus I in 1280, and later was taken by Waldemar IV of Denmark in 1361 and by the Hanseatic League in 1370. By the Treaty of Stettin in 1570, Gotland passed under Danish rule; by the Peace of Brömsebro in 1645 it was returned to Sweden.

Go·to-ret·to (gō-tō' rĕt'tō), group of more than 125 islands, 249 sq mi

(645 sq km), Nagasaki prefecture, in the East China Sea, off W Kyushu, Japan. Whaling and fishing are the major occupations.

Göt·tin·gen (gœt'ĭng-ən), city (1974 est. pop. 122,428), Lower Saxony, E West Germany, on the Leine River. It is noted for its university, founded in 1734 by Elector George Augustus (George II of England). Manufactures of the city include optical and precision instruments, printed materials, textiles, and aluminum. Known in the 10th cent., Göttingen was granted (1210) a city charter and joined the Hanseatic League. The city was virtually undamaged in World War II and has retained numerous historic buildings, including a 14th cent. town hall, half-timbered houses, and student taverns.

Gott·wald·ov (gôt'väl-dôf'), city (1975 est. pop. 69,788), central Czechoslovakia, in Moravia, on the Dřevnice River. It is one of the world's largest shoe-manufacturing cities. Tires and other rubber goods, machinery, and timber are also important to Gottwaldov's economy.

Gou·da (gou'də, gōō'-, KHou'dä), city (1977 est. pop. 57,773), South Holland prov., W Netherlands, at the confluence of the Gouwe and Hollandsche IJssel rivers. It has an important cheese market. Its manufactures include smoking pipes, textiles, candles, and pottery. Chartered in 1272, Gouda was a center of the medieval cloth trade. Erasmus studied here.

Goul·burn (gōl'bərn), city (1976 pop. 21,735), New South Wales, SE Australia, at the confluence of the Wollondilly and Mulwaree rivers. It is a rail center and wool market.

Goulds (gōōldz), uninc. town (1970 pop. 6,690), Dade co., SE Fla., SW of Miami. Truck crops and citrus fruit are packed and shipped.

Gove (gōv), county (1970 pop. 3,940), 1,070 sq mi (2,771.3 sq km), W Kansas, in a level to sloping area drained in the S by the Smoky Hill and Big Smoky rivers; formed 1868; co. seat Gove. It is primarily agricultural, producing mainly wheat and livestock. Farm machinery is manufactured.

Gove or **Gove City**, city (1970 pop. 172), seat of Gove co., W Kansas, on an affluent of the Smoky Hill River ESE of Oakley. It is in a grain, livestock, and poultry area.

Go·ver·na·dor Va·la·da·res (gô'vĭr-nə-dôr vä'lə-dä'rĭs), city (1970 pop. 125,174), Minas Gerais state, SE Brazil, on the Doce River. Beans, rice, maize, sugar cane, coffee, avocados, and manioc are raised, and cattle are bred. Food processing, lumbering, and the mining of mica and beryl are also carried on.

Gov·er·nors Island (gŭv'ər-nərz), 173 acres (70 hectares), in Upper New York Bay, S of Manhattan Island, SE N.Y. Bought from the Indians by the Dutch in 1637, it was the site of an early New Netherlands settlement. The island received its present name in 1698 (officially 1784), when the British set it aside as the colonial governor's residence. Historic landmarks include Fort Jay (completed c.1800) and Castle William (1811), a military prison. Governors Island is now a U.S. military base.

Gow·er (gou'ər), peninsula, c.15 mi (24 km) long and 5 mi (8 km) wide, S Wales, between Swansea Bay and Carmarthen Bay. Composed of limestone, the peninsula has scenic cliffs and numerous caves, many of which contain Paleolithic and Bronze Age relics.

Graaff Rei·net (gräf rī'nət), town (1970 pop. 22,392), Cape Prov., S South Africa, on the Great Karroo. It is the center of an important farming and stock-raising area. Founded in 1786, it served as the capital of a short-lived Boer republic (1795–96).

Gra·dy (grā'dē). **1.** County (1970 pop. 17,826), 466 sq mi (1,206.9 sq km), SW Ga., bounded on the S by the Fla. border and drained by the Ochlockonee River; formed 1905; co. seat Cairo. In a coastal-plain agricultural area yielding sugar cane, tobacco, corn, truck crops, pecans, tung nuts, and livestock, it has food-processing and lumbering industries. Clothing and farm machinery are made. **2.** County (1970 pop. 29,354), 1,092 sq mi (2,828.3 sq km), central Okla., bounded on the N by the Canadian River and intersected by the Washita River; formed 1907; co. seat Chickasha. Its agriculture includes wheat, oats, hay, alfalfa, livestock, and dairy products. There are oil and natural-gas wells and oil refineries.

Graf·ton (grăf'tən, gräf'-), county (1970 pop. 54,914), 1,732 sq mi (4,485.9 sq km), central and W N.H., in a summer and winter resort region drained by the Connecticut and other rivers; formed 1769; co. seat Woodsville. Lumbering, agriculture (fruit, poultry, and dairy products), mica quarrying, and manufacturing (textiles and wood products) are important here.

Grafton. 1. Town (1970 pop. 11,659), Worcester co., S central Mass.; inc. 1731. Leather products and abrasives are produced here. **2.** City (1970 pop. 5,946), seat of Walsh co., NE N.Dak., on the Park River NNW of Grand Forks, in the Red River valley wheat region; inc. 1883. It is a shipping and processing center for grain, potatoes, sugar beets, and livestock. **3.** City (1970 pop. 6,433), seat of Taylor co., N W.Va., on the Tygart River NE of Clarksburg, in a mining area; settled 1852, inc. 1856. **4.** Village (1970 pop. 5,998), Ozaukee co., E Wis., on the Milwaukee River N of Milwaukee; inc. 1896. Metal products and gasoline engines are made.

Gra·ham (grā'əm). **1.** County (1970 pop. 16,578), 4,618 sq mi (11,960.6 sq km), SE Ariz., in a mountainous region irrigated by the Gila and San Carlos rivers; formed 1909; co. seat Safford. Cotton, alfalfa, citrus fruit, and truck crops are grown. Part of San Carlos Indian Reservation is in the north. **2.** County (1970 pop. 4,751), 891 sq mi (2,307.7 sq km), NW central Kansas, drained by the South Fork of the Solomon River; formed 1867; co. seat Hill City. In a rolling prairie wheat and livestock area, it has oil and natural-gas wells and some manufacturing. **3.** County (1970 pop. 6,562), 292 sq mi (756.3 sq km), extreme W N.C., bounded on the W by the Tenn. border, on the E by the Nantahala River, on the N by the Little Tennessee River; formed 1872; co. seat Robbinsville. It depends on farming (corn, tobacco, and potatoes), cattle raising, and lumbering. Nantahala National Forest is here.

Graham. 1. Town (1970 pop. 8,172), seat of Alamance co., N central N.C., E of Greensboro; est. 1849, inc. 1851. It has textile mills and furniture factories. **2.** City (1970 pop. 7,477), seat of Young co., N Texas S of Wichita Falls, in a ranch, farm, and oil region; founded 1872. It packs meat, makes saddles, mills flour, and refines oil.

Graham Island, 2,485 sq mi (6,436.2 sq km), off the coast of British Columbia, Canada, largest of the Queen Charlotte Islands.

Gra·hams·town (grā'əmz-toun'), city (1970 pop. 41,302), Cape Prov., SE South Africa. It is the commercial center for a rich agricultural region. Founded in 1819 as a military post, Grahamstown was repeatedly attacked in the early 19th cent.

Gra·ian Alps (grā'ən, grī'-), N division of the W Alps, on the border between Savoie dept., France, and NW Piedmont, Italy. Its highest point is Gran Paradiso, 13,324 feet (4,063.8 m).

Grain Coast (grān), W Africa, former name of a part of the Atlantic coast that is roughly identical with the coast of modern Liberia. In the 15th cent. "grains of paradise" (seeds of the melegueta pepper) became a major export item, hence the name Grain Coast.

Grain·ger (grān'jər), county (1970 pop. 13,948), 282 sq mi (730.4 sq km), NE Tenn., bounded on the N by the Clinch River, on the E and S by the Holston River; formed 1769; co. seat Rutledge. In a timber and farm region yielding corn, tobacco, dairy products, and livestock, it has marble quarries and mineral springs. Textiles, clothing, wood products, and furniture are made.

Gram·pi·an (grăm'pē-ən), regional authority (1976 est. pop. 453,829), Scotland. It was established in 1975.

Gram·pi·ans, the (grăm'pē-ənz), highest mountain system of Great Britain, extending NE to SW along the S fringe of the Highlands, central Scotland. Ben Nevis (4,406 ft/1,343.8 m) is the tallest peak. The scenic Grampians, extensively forested, have many lakes and are a popular vacation area.

Gra·na·da (grə-nä'də, grä-nä'*th*ä), city (1971 est. pop. 35,422), W Nicaragua, on Lake Nicaragua. Located in a rich agricultural region, it was founded in 1524 by Francisco Fernández de Córdoba and was the object of repeated Indian raids by French and English pirates. After independence from Spain (1821), Granada became the conservative center, engaging in bloody rivalry with León, the city of the liberals. The struggle led to the founding of Managua (1885).

Granada, city (1975 est. pop. 214,230), capital of Granada prov., S Spain, in Andalusia, at the confluence of the Darro and Genil rivers. Beautifully situated at the foot of the Sierra Nevada, it is a major tourist center. Formerly (17th cent.) a silk center, Granada is now a trade and processing point for an agricultural area that is also rich in minerals. It was originally a Moorish fortress. In 1238 it became the seat of the kingdom of Granada, last refuge of the Moors whom the Christian reconquest had driven south. The concentration of Moorish civilization in Granada gave the city great splendor and made it a center of commerce, industry, art, and science. However, the kingdom was weakened by continuous feuds among noble families. With the surrender (Jan., 1492) of the city of Granada, the Moors lost their last hold in Spain, and the kingdom was united with Castile. Located in Granada is the famous Alhambra, an old Moorish citadel and royal palace, which dominates the city from a hill; on the same hill is the palace of Emperor Charles V. The Palacio del Generalife, summer residence of the Moorish rulers, has celebrated gardens.

Gran·bur·y (grăn'bĕr'ē, -bə-rē), city (1970 pop. 2,473), seat of Hood co., N Texas, on the Brazos River SW of Fort Worth, in a ranch and farm area; settled c.1860, inc. 1873.

Gran·by (grăn'bē, grăm'-), city (1976 pop. 37,132), S Que., Canada, on the North Yamaska River E of Montreal. Located in a farming area, Granby has textile mills and plants that manufacture furniture, tobacco and rubber products, and precision instruments.

Granby. 1. Town (1970 pop. 6,150), Hartford co., N Conn., NW of Hartford, in a dairy region; settled c.1664, inc. 1786. Tobacco and fruit are grown. **2.** Town (1970 pop. 5,473), Hampshire co., W Mass., NE of Holyoke; settled 1727, inc. 1768.

Gran Cha·co (grän' chä'kō), extensive lowland plain, c.250,000 sq mi (647,500 sq km), central South America. It is sparsely populated and is divided among Paraguay, Bolivia, and Argentina. To the north of the Pilcomayo River and to the west of the Paraguay River is the section known as the Chaco Boreal. This is arid land, dotted with swamps in the rainy season and with stretches of dense forest in which the quebracho tree abounds. The Chaco Central, in Argentina south of the Pilcomayo River, has much the same aspect. The eastern part—the Chaco Austral and the region west of the Paraguay River—is the only habitable section of the Gran Chaco. The discovery of oil in a narrow strip of the Chaco Boreal precipitated the Chaco War (1932-35) between Bolivia and Paraguay. After three years of negotiation, Paraguay and Bolivia signed a treaty. Three quarters of the disputed Chaco Boreal went to Paraguay; Bolivia was granted a corridor to the Paraguay River, the privilege of using Puerto Casado, and the right to construct a Bolivian port.

Grand (grănd). **1.** County (1970 pop. 4,107), 1,854 sq mi (4,801.9 sq km), N Colo., in a cattle and truck-farming area bounded in the E by the Front Range and drained by the Colorado River; formed 1874; co. seat Hot Sulphur Springs. The county includes parts of Arapaho and Routt national forests and Rocky Mountain National Park. **2.** County (1970 pop. 6,688), 3,692 sq mi (9,562.3 sq km), E Utah, bounded on the E by the Colo. border, on the W by the Green River, and drained by the Colorado River; formed 1890; co. seat Moab. It has mineral deposits (vanadium, uranium, and copper) and some manufacturing. Arches National Park is here.

Grand, river c.165 mi (265 km) long, rising in the highlands of the Ontario Peninsula, S Ont., Canada, and flowing S past Kitchener and Brantford, then SE to Lake Erie at Port Maitland. The river drains one of the most populated regions of Canada.

Grand. 1. River, c.300 mi (485 km) long, rising SE of Creston, Iowa, and flowing SE across NW Mo. to the Missouri River just below Brunswick. **2.** River, 260 mi (418.3 km) long, rising in S Mich. and flowing N to Lansing, then NW to Lake Michigan at Grand Haven. It is navigable to Grand Rapids. **3.** River, rising in SW N.Dak. and flowing 209 mi (336.3 km) SE through S.Dak. to the Missouri River near Mobridge.

Grand Banks (băngks), submarine plateau rising from the continental shelf, c.36,000 sq mi (93,240 sq km), off SE N.F., Canada. It is c.300 mi (485 km) long and c.400 mi (645 km) wide. The cold Labrador Current flows over most of the banks; the warmer Gulf Stream sweeps along the eastern edge, sometimes crossing the southern part. The mingling of the two currents along with the shallowness of the water forms a favorable environment for plankton upon which cod, haddock, halibut, and other fish feed.

Grand-Bassam (grän-bä-säm'), town (1964 est. pop. 12,330), SE Ivory Coast, a port on the Atlantic Ocean. It is an administrative center in a region where coffee, cacao, pineapples, bananas, palm products, and timber are produced and fishing is important. In 1842-43 French merchants gained treaty rights at Grand-Bassam, which was a center of the gold and palm oil trade.

Grand Blanc (blăngk), city (1970 pop. 5,132), Genesee co., SE central Mich., a suburb SE of Flint; settled 1823, inc. 1930. Automobile bodies and fire-fighting equipment are made.

Grand Canal, longest in the world, extending c.1,000 mi (1,610 m) from Peking to Hangchow, E China, and forming an important N-S waterway on the N China plain. The canal was started in the 6th cent. B.C. and was constructed over a 2,000-year period. Between the 10th and the early 13th cent. the waterway fell into disrepair. Kublai Khan reconstructed the canal in the 13th cent. and extended it to Peking. Improvements were made during the Ming dynasty (1368-1644). The canal is 100 to 200 ft (30.5-61 m) wide and from 2 to 15 ft (.6-4.6 m) deep.

Grand Can·yon (kăn'yən), great gorge of the Colorado River, c.1 mi (1.6 km) deep, from 4 to 18 mi (6.4-29 km) wide, and 217 mi (349.2 km) long, NW Ariz. The canyon shows in its rocks, exposed by more than 8 million years of erosion, the repeated geological sequence of uplift, erosion, submergence, and deposition of materials. Hundreds of ancient Indian pueblos dot the lower canyon walls and the rim. The first white man to see the canyon was the Spanish explorer García López de Cárdenas in 1540. In 1869 U.S. explorer John W. Powell was the first man to lead a party through the canyon in a boat. The Grand Canyon, set aside by the U.S. government in 1908 as a national monument, was expanded in 1919 and designated Grand Canyon National Park (673,575 acres/272,798 hectares). The park contains the most spectacular part of the canyon. Along the forested northern rim and the more accessible southern rim are numerous lookouts; trails wind to the canyon floor. Grand Canyon National Monument (198,280 acres/80,303 hectares), est. 1932, is a primitive area adjoining the park on the west.

Grand Cou·lee (ko͞o'lē), a vertical-walled gorge, c.30 mi (48 km) long, N central Wash., carved by the Columbia River through the Columbia Plateau. It receives water from Grand Coulee Dam, a key unit of the Columbia basin project.

Grande Prai·rie (gräNd prâ-rē'), city (1976 pop. 17,626), W Alta., Canada, NW of Edmonton. It is the chief business center for the Peace River valley farming area.

Grand Forks (fôrks), county (1970 pop. 61,102), 1,438 sq mi (3,724.4 sq km), E N.Dak., bordered on the E by the Red River of the North and watered by the Goose and Turtle rivers; formed 1873; co. seat Grand Forks. It is in a wheat and farming area and has a food-processing industry (dairy products and preserved and frozen fruit and vegetables). Some mining is done. Concrete and machinery are manufactured.

Grand Forks, city (1978 est. pop. 42,900), seat of Grand Forks co., E N.Dak., at the confluence of the Red River of the North and the Red Lake River; inc. 1881. In a spring-wheat, livestock, and farm area, the city has grain elevators, state-operated flour mills, and plants that process and distribute meat, dairy products, sugar beets, and potatoes. The Univ. of North Dakota is here.

Grand Ha·ven (hā'vən), city (1978 est. pop. 12,000), seat of Ottawa co., SW Mich., at the mouth of the Grand River; inc. 1867. It is a port on Lake Michigan that ships sand and gravel. Grand Haven manufactures automobile parts, restaurant equipment, tools, and machinery. The city is also a popular resort.

Grand Island, city (1978 est. pop. 33,400), seat of Hall co., S Nebr., on the Wood River near its junction with the Platte River; settled 1857 on the Platte by Germans, moved 1866 to its present location on the Union Pacific RR, inc. c.1872. The city, which is known as a horse, mule, and cattle market, is also a railroad, manufacturing, and shipping center for a rich, irrigated livestock, grain, and dairy region. The Stuhr Museum of the Prairie Pioneer, on an island in a nearby man-made lake, was designed by Edward Durrell Stone.

Grand Island, 13,000 acres (5,265 hectares), N Mich., in Lake Superior. Heavily wooded, it is a resort area and has a game refuge.

Grand Isle (īl), county (1970 pop. 3,574), 77 sq mi (199.4 sq km), NW Vt., on a peninsula and several islands in Lake Champlain; formed 1802; co. seat North Hero. It includes North Hero Island, Grand Isle, and Isle La Motte, site of the state's first white settlement. In a lumbering, dairying, and fruit-growing area, it has marble quarries and many resorts.

Grand Isle, town (1970 pop. 2,236), Jefferson parish, SE La., on Grand Island at the mouth of Caminada Bay S of New Orleans. It was once the headquarters of Jean Lafitte's corsairs, who engaged in smuggling and piracy in the early 19th cent. Treasure is supposedly buried here. The town is a truck-farming and fishing center.

Grand Junc·tion (jŭngk'shən), city (1978 est. pop. 25,600), seat of Mesa co., W Colo., at the junction of the Gunnison and Colorado rivers; inc. 1882. The shipping and processing center of a large ranch and irrigated farm region, it also serves the area's uranium, oil shale, and coal-mining industries. Electronic equipment is manufactured, and tourism is important to the city. The city is a center of a skiing and hunting region and serves as headquarters for Grand Mesa National Forest. Uranium production was curtailed in 1970 after a radiation hazard arose from the radioactive wastes (tailings) used as construction fill for building projects.

Grand Ledge (lĕj), city (1970 pop. 6,032), Eaton co., S Mich., on the Grand River W of Lansing, in a livestock-raising area; settled c.1848, inc. as a village 1871, as a city 1893. There are soft-coal deposits here, and sandstone found near the river is used for tile and clay products.

Grand Ma·nan (mə-năn'), island c.16 mi (26 km) long and c.7 mi (11.3 km) wide, S N.B. Canada, in the Bay of Fundy. On the north and west sides are bold cliffs, rising from 200 ft to 400 ft (61–122 m). The chief occupation is fishing, and the island is a summer resort. It was settled after the American Revolution by Loyalists; British possession was disputed by the United States until 1817.

Grand Ma·rais (mə-rā'), resort village (1970 pop. 1,301), seat of Cook co., NE Minn., on Lake Superior NE of Duluth. Settled on the site of an Indian village, it was an early fur-trading post and lumber town. Wood products are made.

Grand'Mère (grăN-mâr'), city (1976 pop. 15,999), S Que., Canada, on the St. Maurice River, N of Trois Rivières. The Grand'Mère falls furnish power for paper and pulp mills. The city also has clothing and textile factories.

Grand Port·age National Monument (pôr'tĭj): *see* National Parks and Monuments Table.

Grand Prai·rie (prâr'ē), city (1978 est. pop. 59,000), Dallas and Tarrant cos., N Texas, halfway between Dallas and Fort Worth; inc. 1909. Located in a highly urbanized area, it is a distribution center with a large aerospace industry. Other manufactures include mobile homes, metal goods, plastics, medical supplies, and concrete pipes. An African wildlife preserve and the Six Flags over Texas amusement park are here.

Grand Pré (grăn' prā', grăN'), village, W central N.S., Canada, on an arm of the Bay of Fundy. The area was settled by Acadians, whose expulsion in 1755 is the subject of Longfellow's poem *Evangeline.* Grand Pré National Historic Park contains several remains from the Acadian period.

Grand Rap·ids (răp'ĭdz). **1.** City (1978 est. pop. 182,600), seat of Kent co., W central Mich., on the Grand River; inc. 1850. It is a distribution, wholesale, and industrial center for an area that yields fruit, farm produce, gypsum, and gravel. Furniture manufacturing (begun in 1859) remains an important industry. Among the city's other manufactures are appliances, electronic equipment, automotive parts, and paper products. It has an art gallery, a furniture museum, and a symphony orchestra. **2.** Town (1970 pop. 7,247), seat of Itasca co., N central Minn., near the Mesabi iron range NW of Duluth, in a region of woods, lakes, and streams; settled 1887, inc. 1891. It has paper mills and a fish hatchery.

Grand·son (grăN-sôN'), town (1970 pop. 2,135), Vaud canton, W Switzerland, at the SW end of the Lake of Neuchâtel. Grandson is known chiefly as the scene of the defeat (1476) of Charles the Bold of Burgundy by the Swiss Confederates.

Grand Te·ton National Park (tē'tŏn), 310,443 acres (125,729.4 hectares), NW Wyo.; est. 1929. The park, which includes Jackson Hole and Jackson Lake, embraces the most scenic portion of the glaciated, snow-covered Teton Range; Grand Teton (13,766 ft/4,198.6 m) is the highest peak. The Snake River flows through the park, which is dotted with small lakes and has several glaciers, forests, and a great variety of wildlife.

Grand Trav·erse (trăv'ərs), county (1970 pop. 39,175), 464 sq mi (1,201.8 sq km), NW Mich., bounded on the N by Grand Traverse Bay and drained by the Boardman and Betsie rivers; formed 1840; co. seat Traverse City. In a fruit-growing and agricultural area yielding cherries, corn, potatoes, truck crops, livestock, and dairy products, it has fisheries, mines, and a food-processing industry. Plastics, metal products, and machinery are manufactured. The county has many small lakes and resort areas.

Grand Traverse Bay, arm of Lake Michigan, 32 mi (51.5 km) long and 10 mi (16 km) wide, W central Mich. The bay is known for its fishing and boating.

Grand·view (grănd'vyōō'), city (1978 est. pop. 24,100), Jackson co., W Mo., S of Kansas City; inc. 1912. Grandview is in a farm region. Hardware and electrical equipment are manufactured.

Grandview Heights, city (1970 pop. 8,460), Franklin co., central Ohio, a suburb W of Columbus.

Grand·ville (grănd'vĭl'), city (1970 pop. 10,764), Kent co., W Mich., on the Grand River, in a farm area; settled 1833, inc. as a city 1933. Aircraft parts, boxes, structural steel, die castings, and electrical products are made. Indian mounds, some of which were opened in 1964, are preserved in the northeastern section of the city.

Grange·mouth (grānj'məth, -mouth'), burgh (1974 est. pop. 24,347), Central region, central Scotland, on the Forth River at the E terminus of the Forth and Clyde Canal. Oil refining is important, and there are large chemical and steelworks. The timber trade and sawmilling are also significant. Grangemouth was founded in 1777 to be the terminus of the canal, which opened in 1790.

Grange·ville (grānj'vĭl'), city (1970 pop. 3,636), seat of Idaho co., N central Idaho, SE of Lewiston, in a dry-farming and livestock-raising area; settled 1876, inc. 1897. It was an important outfitting point during the 1898 gold rush. Today it is a tourist center for nearby dude ranches and the headquarters for Nez Percé National Forest.

Gran·ite (grăn'ĭt), county (1970 pop. 2,737), 1,733 sq mi (4,488.5 sq km), W Mont., in a mining and agricultural area drained by the Clark Fork, Rock, and Flint rivers; formed 1893; co. seat Philipsburg. Manganese, silver, and coal are mined. Parts of Deerlodge and Bitterroot national forests are here.

Granite City, city (1978 est. pop. 39,100), Madison co., SW Ill., an industrial suburb of East St. Louis, on the Mississippi; inc. 1896. It is a transportation center, with a port and rail connections.

Granite Falls, city (1970 pop. 3,225), seat of Yellow Medicine co., SW Minn., at the falls of the Minnesota River SE of Montevideo; platted 1872, inc. as a city 1889. It is a farm trade center.

Granite Peak, mountain, 12,799 ft. (3,903.7 m) high, S Mont., in the Absaroka Range NE of Yellowstone National Park. It is the highest point in the state.

Gran·ja, La (lä grăng'hä): *see* San Ildefonso, Spain.

Gran Pa·ra·di·so (grän pä-rä-dē'zō), mountain, 13,323 ft (4,063.5 m) high, in Valle d'Aosta, NW Italy. In the Graian Alps, it is the highest Alpine peak entirely in Italian territory.

Gran Qui·vi·ra National Monument (grăn' kĭ-vîr'ə): *see* National Parks and Monuments Table.

Gran Sas·so d'I·ta·lia (grän säs'sō dē-tä'lyä), mountain group of the central Apennines, in Abruzzi, central Italy. It rises to c.9,560 ft (2,915 m) in Monte Corno, the highest peak in the Apennines.

Grant (grănt). **1.** County (1970 pop. 9,711), 631 sq mi (1,634.3 sq km), S central Ark., drained by the Saline River; formed 1869; co. seat Sheridan. It has agriculture (cotton, corn, peanuts, and livestock), lumbering, some mining, and cotton ginning. Metal products are made. **2.** County (1970 pop. 83,955), 421 sq mi (1,090.4 sq km), N central Ind., drained by the Mississinewa River; formed 1831; co. seat Marion. It is in an agricultural area that produces livestock, grain, poultry, fruit, truck crops, and dairy products. There are natural-gas and oil wells. Its manufactures include processed foods, furniture, paper, plastics, glassware, iron, and metal and wood products. **3.** County (1970 pop. 5,961), 568 sq mi (1,471.1 sq km), SW Kansas, drained by the Cimarron River; formed 1873; co. seat Ulysses. In a gently rolling wheat and cattle area, it has rich natural-gas deposits. Plastics and farm and garden machinery are manufactured. **4.** County (1970 pop. 9,999), 250 sq mi (647.5 sq km), N. Ky., in a gently rolling upland area of the Bluegrass drained by Eagle Creek; formed 1820; co. seat Williamstown. Its agriculture includes burley tobacco, corn, hay, poultry, livestock, and dairy products. There is some manufacturing (mainly rubber products). **5.** Parish (1970 pop. 13,671), 670 sq mi (1,735.3 sq km), central La., bounded on the E by the Little River, on the W and SW by the Red River; formed 1869; parish seat Colfax. Cotton, corn, hay, peanuts, sweet potatoes, truck crops, pecans, and watermelons are grown. It also has oil fields, sand and gravel pits, and a lumbering industry. Clothing is manufactured. Part of Kisatchie National Forest is here. **6.** County (1970 pop. 7,462), 546 sq mi (1,414.1 sq km), W Minn., drained by the Mustinka and Pomme de Terre rivers; formed 1868; co. seat Elbow Lake. In a grain and livestock area, it has some manufacturing. **7.** County

(1970 pop. 1,019), 762 sq mi (1,973.6 sq km), W Nebr., in an agricultural area; formed 1887; co. seat Hyannis. Its economy is based on livestock. **8.** County (1970 pop. 22,030), 3,970 sq mi (10,282.3 sq km), SW N.Mex., in a mountainous region watered by the Gila River and bordering in the W on Ariz.; formed 1868; co. seat Silver City. It is in a copper, lead, and zinc mining area, with livestock grazing and some industry (primarily chemicals). Tourism is also important. Parts of Gila National Forest are here. **9.** County (1970 pop. 5,009), 1,666 sq mi (4,314.9 sq km), S N.Dak., watered by the Heart and Cannonball rivers; formed 1916; co. seat Carson. Its agriculture includes grain, wheat, poultry, livestock, and dairy products. It also has some mining and manufacturing. **10.** County (1970 pop. 7,117), 1,007 sq mi (2,608.1 sq km), N Okla., bounded on the N by the Kansas border and intersected by the Salt Fork of the Arkansas River; formed 1907; co. seat Medford. In an agricultural area yielding wheat, oats, alfalfa, corn, barley, watermelons, cattle, hogs, and dairy products, it has oil and natural-gas wells and petroleum refineries. **11.** County (1970 pop. 6,996), 4,532 sq mi (11,737.9 sq km), E central Oregon, in a mountainous area crossed by the John Day River; formed 1864; co. seat Canyon City. Lumbering and cattle raising are important. Parts of Whitman and Malheur national forests are in the county. **12.** County (1970 pop. 9,005), 684 sq mi (1,771.6 sq km), NE S.Dak., on the Minn. border and watered by intermittent streams; formed 1873; co. seat Milbank. Its agriculture includes wheat, corn, sorghums, alfalfa, soybeans, and dairy products. It has granite quarries and cheese-processing plants. Stone products are made. **13.** County (1970 pop. 41,881), 2,681 sq mi (6,943.8 sq km), E central Wash., in an agricultural region of the Columbia basin; formed 1909; co. seat Ephrata. Fruit, grain, livestock, and alfalfa are raised. There are health resorts in the Soap Lake area. **14.** County (1970 pop. 8,607), 477 sq mi (1,235.4 sq km), NE W.Va., in the Eastern Panhandle bounded on the NW by the North Branch of the Potomac River and the Md. border and drained by the South Branch of the Potomac and Patterson Creek; formed 1866; co. seat Petersburg. It is traversed by the Allegheny Front, Knobly, and Patterson Creek mts. Livestock and poultry are of major importance. Fruit, grain, tobacco, and truck crops are also grown. There are bituminous-coal mines and a lumbering industry. The county includes part of Monongahela National Forest and is a recreational center for hunting, fishing, and white-water boating. **15.** County (1970 pop. 48,398), 1,147 sq mi (2,970.7 sq km), extreme SW Wis., bounded on the S by the Ill. border, on the N by the Wisconsin River, on the W by the Mississippi River (here forming the Iowa border); formed 1836; co. seat Lancaster. It is drained by the Platte, Blue, and Grant rivers. In a farming, dairying, stock-raising, and mining (zinc and lead) region, it has food-processing plants and some manufacturing, especially wood products, furniture, machinery, and electronic equipment.

Grant, village (1970 pop. 1,099), seat of Perkins co., SW Nebr., WSW of North Platte, in a grain and dairy area.

Grant City, city (1970 pop. 1,095), seat of Worth co., NW Mo., near the Iowa border NNE of St. Joseph, in a farm area; settled 1864, inc. 1870.

Gran·tham (grăn'təm, -thəm), municipal borough (1973 est. pop. 27,830), Lincolnshire, E central England, on the Witham River. Grantham is an agricultural center and railroad junction. There are mechanical engineering works. Landmarks include a bronze statue (on St. Peter's Hill) of Sir Isaac Newton, who attended King's School here. At Grantham in 1643 Oliver Cromwell won his first victory over the Royalists.

Grant-Kohrs Ranch National Historic Site (grănt-kôrz'): *see* National Parks and Monuments Table.

Grants (grănts), city (1970 pop. 8,768), Valencia co., W central N.Mex., W of Albuquerque on the San Jose River near the San Mateo Mts.; settled 1882. Once a small farming town, it boomed with the discovery (1950) of uranium reserves.

Grants·burg (grănts'bûrg'), village (1970 pop. 930), seat of Burnett co., NW Wis., N of St. Croix Falls. Dairy products and poultry are processed.

Grants Pass, city (1978 est. pop. 14,700), seat of Josephine co., SW Oregon, on the Rogue River, in a heavily forested area; inc. 1887. It has important lumbering enterprises and is also the commercial center of a large region producing flower bulbs (especially gladioli), fruits, nuts, vegetables, and dairy products.

Grants·ville (grănts'vĭl'), town (1970 pop. 795), seat of Calhoun co., central W.Va., on the Little Kanawha River SE of Parkersburg; laid out 1866. In an agricultural region yielding livestock, fruit, tobacco, and grain, it also has timber and natural-gas and oil wells.

Gran·ville (grăn'vĭl'), county (1970 pop. 32,762), 537 sq mi (1,390.8 sq km), N N.C., in a piedmont area bounded on the N by the Va. border, on the SW by the Neuse River, and drained by the Tar River; formed 1746; co. seat Oxford. It has agriculture (tobacco, corn, soybeans, and wheat) and a lumbering industry. Some manufacturing is done.

Granville Lake, 392 sq mi (1,015.3 sq km), NW Man., Canada. It is one of the network of lakes linked by the Churchill River.

Grape·vine (grāp'vīn'), city (1970 pop. 7,023), Tarrant co., N Texas, NE of Fort Worth, in a blackland farming area; settled 1854, inc. 1907. It is a trade center in a cotton, diversified farming, and oil-refining area.

Gras·mere (grăs'mîr', gräs'-), village, Westmorland, NW England, in the Lake District, near Lake Grasmere. Dove Cottage was the home of William Wordsworth from 1799 to 1808; it now contains a Wordsworth museum. Thomas De Quincey and Samuel Taylor Coleridge also lived in Grasmere.

Grasse (gräs), town (1975 pop. 24,442), Alpes-Maritime dept., SE France. Probably founded in Roman times, Grasse was a commercial center during the Middle Ages. Destroyed many times by the Saracens, it was an independent republic from the 12th cent. until its union with the earldom of Provence in 1226. In 1536 the town was destroyed by Francis I to prevent the advance of Emperor Charles V. Surrounded by fields of flowers and rose gardens, Grasse is a center of the French perfume industry. It has a museum containing paintings by Jean-Honoré Fragonard, who was born in Grasse.

Grass Valley (grăs), city (1970 pop. 5,149), Nevada co., N central Calif., NE of Sacramento; settled 1849, inc. 1861. It has been a gold-mining center since 1850. The house of Lola Montez, who lived here from 1852 to 1854, is preserved.

Gra·tiot (grăsh'ət), county (1970 pop. 39,246), 566 sq mi (1,465.9 sq km), central Mich., drained by the Maple, Pine, and Bad rivers; formed 1831; co. seat Ithaca. Its agriculture includes livestock, poultry, fruit, grain, sugar beets, beans, corn, and dairy products. It has timber, oil wells and refineries, and industry (wood products, plastics, metal products, and machinery). There are mineral springs in the county.

Graves (grāvz), county (1970 pop. 30,939), 560 sq mi (1,450.4 sq km), SW Ky., bounded on the S by Tenn. and drained by the West Fork of Clarks River, Mayfield and Obion creeks, and Bayou de Chien; formed 1821; co. seat Mayfield. In a gently rolling agricultural area that produces dark tobacco and grain, it has clay pits and timber. Its manufactures include tires, clothing, and pottery.

Graves·end (grāvz'ĕnd'), municipal borough (1976 est. pop. 96,000), Kent, SE England, on the Thames River. The town's industries include shipbuilding, metal casting, engineering, papermaking, printing, and the production of tires and rubber products and cement. Known as the "gateway to the Port of London," Gravesend is a station of pilots and customhouse officers. Pocahontas is buried in the parish churchyard.

Gray (grā). **1.** County (1970 pop. 4,516), 869 sq mi (2,250.7 sq km), SW Kansas, drained by the Arkansas River; formed 1887; co. seat Cimarron. It is in a rolling plain agricultural region that produces livestock, wheat, milo, corn, and alfalfa. There is some manufacturing in the county. **2.** County (1970 pop. 26,949), 937 sq mi (2,426.8 sq km), extreme N Texas, in the Panhandle, drained by the North Fork of the Red River and McClellan Creek; formed 1876; co. seat Pampa. One of the state's most productive oil and natural-gas counties, it has huge refineries and manufacturers carbon black, oil-well equipment and supplies, and chemicals. It also has cattle ranches and raises wheat, hay, cotton, fruit, and truck crops. There are clay and caliche deposits.

Gray, city (1970 pop. 2,014), seat of Jones co., central Ga., NNE of Macon, in a farm and cotton area. Peaches are grown here.

Gray·ling (grā'lĭng), city (1970 pop. 2,143), seat of Crawford co., N Mich., on the Au Sable River ESE of Traverse City; settled 1872, inc. as a village 1903, as a city 1935. It is a resort in a timber area. Grayling National Guard Military Reservation is nearby.

Grays Harbor (grāz), county (1970 pop. 59,553), 1,905 sq mi (4,934 sq km), NW Wash., in an area of rolling hills rising to the Olympic Mts. in the E, bounded on the W by the Pacific Ocean, and drained by the Chehalis River; formed 1854 as Chehalis co., renamed 1915; co. seat Montesano. It has some agriculture (dairy products and truck crops).

Gray·son (grā'sən). **1.** County (1970 pop. 16,445), 514 sq mi (1,331.6 sq km), W central Ky., bounded on the N by the Rough River, on the SE by the Nolin River, and drained by Bear Creek; formed 1810; co. seat Leitchfield. In a rolling agricultural area yielding burley tobacco, corn, hay, poultry, and dairy products, it has asphalt and bituminous-coal mines, stone quarries, and a food-processing industry. Clothing, wood products, paper, farm machinery, and hardware are manufactured. **2.** County (1970 pop. 83,225), 940 sq mi (2,434.6 sq km), NE Texas, bounded on the N by the Red River and Lake Texoma (here forming the Okla. border); formed 1846; co. seat Sherman. This rich diversified agricultural, cattle-ranching, and dairying area produces cotton, corn, peanuts, pecans, fruit, truck crops, hogs, sheep, and poultry. It has limited oil and timber resources. Food processing and some manufacturing are done. **3.** County (1970 pop. 15,439), 451 sq mi (1,168.1 sq km), SW Va., bounded on the S by the N.C. border and drained by the New River; co. seat Independence. Its agriculture includes tobacco, corn, hay, beef cattle, and dairy products. Textiles, clothing, furniture, and cheese are made.

Grayson, town (1970 pop. 2,184), seat of Carter co., NE Ky., WSW of Ashland, in an area of farms and clay, iron, and coal deposits.

Graz (gräts), city (1971 pop. 248,500), capital of Styria prov., SE Austria, on the Mur River. It is an industrial, rail, and cultural center. Manufactures include iron and steel, paper, and machinery. Probably founded in the 12th cent., Graz is built around the Schlossberg, a mountain peak, on which are the ruins of a 15th cent. fortress and the famous Uhrturm, or clock tower. The city has several medieval churches (13th–15th cent.) and a twin-naved Gothic parish church that contains Tintoretto's *Assumption of the Virgin.*

Great Ar·te·sian Basin (grāt är-tē'zhən), c.670,000 sq mi (1,735,300 sq km), between the Eastern Highlands and the Western Plateau, E central Australia, extending S from the Gulf of Carpentaria, Queensland, to NE South Australia and N New South Wales. Extremely arid, the basin receives water from the Eastern Highlands as rain is absorbed by porous rock and flows underground toward the center of the saucer-shaped basin. Thousands of wells, some more than 1 mi (1.6 km) deep, tap underground water-bearing rock formations. The rolling surface of the basin supports a pastoral economy based on irrigated grasslands.

Great Au·stra·lian Bight (ô-strāl'yən bīt'), wide bay of the Indian Ocean, indenting the S coast of Australia. An unbroken line of cliffs c.200 ft (60 m) high runs along the coast and extends inland as the arid and desolate Nullarbor Plain.

Great Bar·ri·er Reef (băr'ī-ər), largest coral reef in the world, c.1,250 mi (2,010 km) long, in the Coral Sea, forming a natural breakwater for the coast of Queensland, NE Australia. Composed of several individual reefs, the Great Barrier Reef is separated from the mainland by a shallow lagoon from 10 to 100 mi (16–160.9 km) wide.

Great Bar·ring·ton (băr'ĭng-tən), town (1975 est. pop. 7,068), Berkshire co., SW Mass., on the Housatonic River in the Berkshires near the N.Y. border; settled 1726, set off from Sheffield 1761.

Great Ba·sin (bā'sən), desert region of W United States, bordered by the Sierra Nevada on the W, the Columbia Plateau on the N, the Rocky Mts. on the NE, the Colorado Plateau on the E, and the Mojave Desert on the S. The region is a complex topographic basin, the surface of which is broken by numerous short fault-block mountains, trending mostly north-south and rising sharply in places to more than 10,000 ft (3,050 m) above arid, sediment-floored basins. Death Valley National Monument, 282 ft (86 m) below sea level, is the lowest basin. The region was recognized as an area of interior drainage by J. C. Frémont, who explored (1843–45) and named it. The rivers of the region have no outlet to the sea; they either dry up as they cross the desert, like the Humboldt, or empty into large lakes or into playas that temporarily fill with water after heavy rain. The Great Basin is one of the least populated areas of the United States.

Great Bear Lake, largest lake of Canada and fourth-largest of North America, c.12,275 sq mi (31,795 sq km), c.190 mi (305 km) long and from 25 to 110 mi (40–177 km) wide, N central Mackenzie dist., Northwest Territories, on the edge of the Canadian Shield. It is drained to the west by the Great Bear River (c.100 mi/160 km long), which flows into the Mackenzie River. Even though it is one of North America's deepest (1,356 ft/413.6 m) lakes, its waters are open only about four months each year.

Great Bend, city (1978 est. pop. 16,700), seat of Barton co., central Kansas, on a bend in the Arkansas River; settled and inc. 1871.

Great Brit·ain (brĭt'n), officially the United Kingdom of Great Britain and Northern Ireland, constitutional monarchy (1971 pop. 55,346,-551), 94,226 sq mi (244,044 sq km), on the British Isles, off the Western European continent. It comprises England (1971 pop. 45,870,-062), 50,334 sq mi (130,365 sq km); Wales (1971 pop. 2,723,596), 8,016 sq mi (20,761 sq km); and Scotland (1971 pop. 5,227,706), 30,414 sq mi (78,772 sq km) on the island of Great Britain; and Northern Ireland (1971 pop. 1,525,187), 5,462 sq mi (14,146 sq km) on the island of Ireland. The Isle of Man in the Irish Sea and the Channel Islands in the English Channel are dependencies of the British crown. The capital of Great Britain is London.

Economy. Great Britain is one of the world's leading industrialized nations, despite the lack of most of the raw materials needed for industry. The most important industries are food processing; the manufacture of textiles, vehicles, metals, and chemicals; and mechanical and electrical engineering. Except for foods, the products of these industries are also the leading exports.

Iron ore, coal, tin, zinc, china clay, fluorspar, and oil shale are found in Britain. Of the country's mineral requirements, only coal is present in sufficient quantity. Fishing is important; whitefish, herring, and shellfish make up the bulk of the catch. About half of Great Britain's land area is devoted to agriculture. The widespread dairy industry makes the country self-sufficient in milk, and nearly all the eggs needed are also home-produced. Beef cattle are raised in southeast England, the Midlands, and northeast Scotland. Large numbers of sheep are also raised. Barley, potatoes, and wheat are the main cereal crops. The coal, gas, electricity, railroad, and aviation industries and most of the steel industries are publicly owned.

History. The remains of the dolmens and stone circles built by the earliest inhabitants of Britain are evidence of the developed culture of the prehistoric Britons. They had developed a Bronze Age culture by the time of the first Celtic invaders (early 5th cent. B.C.). In A.D. 43 the emperor Claudius began the Roman conquest of Britain. Under the Roman occupation towns developed, and roads were built to ensure the success of the military occupation. Trade declined with the economic dislocation of the late Roman Empire. As Rome withdrew its legions from Britain, Germanic peoples—the Anglo Saxons and the Jutes—began raids that turned into great waves of invasion and settlement in the later 5th cent. The Celts fell back into remote areas, and the loosely knit tribes of the newcomers gradually coalesced into a heptarchy of kingdoms. Raiding Vikings (known in

English history as Danes) harassed coastal England and finally, in 865, launched a full-scale invasion. They were first confined to the Danelaw, but new Danish invasions late in the 10th cent. overcame ineffective resistance. The Dane Canute ruled all England by 1016. At the expiration of the Scandinavian line in 1042, the Wessex dynasty regained the throne.

A new era began with the conquest of England in 1066 by William, duke of Normandy (William I of England). William I used the feudal system to collect taxes, employed the bureaucracy of the church to strengthen the central government, and made the administration of royal justice more efficient. Henry II's reign (1154-89) was marked by the sharp conflict between king and church that led to the murder of Thomas à Beckett. Henry carried out great judicial reforms that increased the power and scope of the royal courts. As part of his inheritance he brought to the throne Anjou, Normandy, and Aquitaine. The defense and enlargement of these French territories engaged the energies of successive English kings. Conflict between kings and nobles, which had begun under Richard I (1189-99) came to a head under John (1199-1216). A temporary victory of the nobles bore fruit in the most noted of all English constitutional documents, the Magna Carta (1215). The Hundred Years' War with France began (1337) in the reign of Edward III, reached a high tide of English success with the routing of the French at Agincourt (1415) by Henry V, and ended (1453) with French victories that virtually expelled the English from the continent. The plague of the Black Death first arrived in 1348 and had a tremendous effect on economic life, hastening the breakdown of the manorial and feudal systems. At the same time the fast-growing towns and trades gave new prominence to the burgess and artisan classes. Dynastic wars—the Wars of the Roses between the houses of York and Lancaster—ended with the accession of the Tudor family in 1485.

Henry VII (1485-1509) restored political order and the financial solvency of the crown. In 1536 Henry VIII (1509-47) brought about the political union of England and Wales. Henry and his minister Thomas Cromwell greatly expanded the central administration. Many factors, climaxed by the pope's refusal to grant Henry a divorce from Catherine of Aragón, led the king to break with Roman Catholicism and establish the Church of England. As part of the English Reformation (1529-39), Henry suppressed the orders of monks and friars and secularized their property. After a generally hated Roman Catholic revival under Mary I (1553-58), the Roman tie was again cut under Elizabeth I (1558-1603). The Elizabethan age was one of great artistic and intellectual achievement, its most notable figure being William Shakespeare. Overseas trading companies were formed and colonization attempts to the New World were made. A long conflict with Spain culminated in the defeat of the Spanish Armada (1588). The country gentry, enriched by the expansion of the wool trade, became leaders in what, toward the end of Elizabeth's reign, was an increasingly assertive Parliament.

The accession in 1603 of the Stuart James I, who was also James VI of Scotland, united the thrones of England and Scotland. James and his son, Charles I, used extralegal means to obtain money, espoused the divine right of kings, and persecuted opponents and religious minorities, all of which produced a bitter conflict with Parliament that culminated (1642) in the English Civil War. In the war the Parliamentarians, effectively led at the end by Oliver Cromwell, defeated the Royalists. The king was tried for treason and beheaded (1649). The monarchy was abolished, and the country was governed by Parliament until 1653, when Cromwell established the Protectorate. When he died (1658), his son, Richard, succeeded as Lord Protector but governed ineffectively. The threat of anarchy led to an invitation by a newly elected Parliament to Charles, son of Charles I, to become king, ushering in the Restoration (1660).

The Whig and Tory parties developed in the Restoration period. Although Charles II was personally popular, the old issues of religion, money, and the royal prerogative came to the fore again. Charles's brother and successor, James II (1685-88), was an avowed Catholic and aroused fears of a Catholic tyranny. In the Glorious Revolution, Whig and Tory leaders deposed James II and offered the throne to William of Orange (William III). William and Mary were proclaimed king and queen by Parliament in 1689. The Bill of Rights confirmed that sovereignty resided in Parliament. The Act of Toleration (1689) extended religious liberty to all Protestant sects; in subsequent years religious passions slowly subsided. By the Act of Union (1707), the two kingdoms—Scotland and England and Wales—became one.

Britain emerged from the Seven Years' War against France (1756-63) as the possessor of the world's greatest empire, including eastern North America and India. A serious loss was sustained when 13 North American colonies broke away in the American Revolution, but additional colonies were won in the wars against Napoleon I, wars notable for the victories of Horatio Nelson and the duke of Wellington. A vain attempt to solve the centuries-old Irish problem was the union (1801) of Great Britain and Ireland, with Ireland represented in the British Parliament.

The 18th cent. was a time of transition in the growth of the British parliamentary system. The monarch still chose and dismissed ministers as he wished and could dissolve Parliament. Whigs and others called for a reform and reduction of the king's power when George III (1760-1820) arrogantly sought to impose his will on Parliament. The cause of reform, however, was set back by the French Revolution and the ensuring wars with France, which greatly alarmed British society.

The late 18th and early 19th cent. was a time of dynamic economic change—the Industrial Revolution. The growing population provided needed labor for industrial expansion and was accompanied by rapid urbanization. In the 1820s the reform impulse revived: by 1846 political and civil rights had been restored to Catholics, slavery in the Empire was abolished, Parliament was reorganized and the middle class enfranchised, and the cause of free trade was advanced. The Reform Bill of 1867, sponsored by Prime Minister Benjamin Disraeli, enfranchised the urban working classes and was followed in 1884 by a bill extending the vote to agricultural laborers. (Women could not vote until 1918). A coalition of labor and socialist groups, organized in 1900, became the Labour party in 1906.

The reign of Victoria (1837-1901) covered the period of Britain's commercial and industrial leadership of the world and its greatest political influence. Canada was granted self-government (1867), but the Empire was extended in Asia and Africa and wars were fought with China (1839), Russia (the Crimean War of 1853-56), and the South African Boers (1899-1902) to protect imperial interests. Victoria was succeeded by her son Edward VII (1901-10), then by his son, George V (1910-36).

In 1914 Germany's violation of Belgium's neutrality caused Britain to go to war against Germany. World War I drained the nation of wealth and manpower (750,000 were killed) and was followed by a period of moral disillusionment and economic problems. Rebellion in Ireland (1916-21) led to the creation of the Irish Free State and Northern Ireland. In 1926 the country suffered a general strike. Severe economic stress increased during the depression of the early 1930s. George V was succeeded by Edward VIII, after whose abdication (1936) George VI came to the throne. The years prior to the outbreak of World War II were characterized by ineffective attempts to stem the rising tide of German and Italian aggression. Great Britain declared war on Germany on Sept. 3, 1939, after German troops invaded Poland. The nation withstood intensive bombardment during the early years of World War II, but under the leadership of Winston S. Churchill, the British people rose to a supreme war effort. The alliance of Great Britain, the USSR, and the United States led to the formation of the United Nations and brought about the defeat of Germany (May, 1945) and Japan (Sept., 1945).

The war left Britain's economy shattered. Overseas investments had been liquidated to pay the cost of the war. The pound sterling was devalued to make exports more competitive. Most of the colonies, including India (1947), were granted independence. The post-war Labour government pursued a vigorous program of nationalization of industry and extension of social services, including a program of socialized medicine. Elizabeth II succeeded George VI in 1952. Great Britain joined the North Atlantic Treaty Organization (1949) and fought on the United Nations' side in the Korean War (1950-53). Churchill and the Conservatives returned to power in 1951; his successor, Anthony Eden (1955-57), approved an invasion of Egypt with British, French, and Israeli troops after Egypt had nationalized (1956) the Suez Canal.

Great Britain requested membership in the Common Market in 1961 and finally was admitted in 1973. Labour returned to power in 1964 under Harold Wilson. The country faced the compound economic problems of a very unfavorable balance of trade, the instability of the pound sterling, a lagging rate of economic growth, and inflationary wages and prices. Violent confrontations between Catholics and Protestants in Northern Ireland were perennial after 1968. The Conservatives under Edward Heath returned to power in 1970. In 1971 Britain's currency was converted to the decimal system. At the end of 1973 steep increases in the price of oil and a coal shortage caused a severe economic crisis. Elections held in Feb., 1974, resulted in a parliamentary deadlock. Harold Wilson formed a government and called for new elections in Oct., which resulted in a slim Labour majority. Price inflation rose to 26% in 1975, but the first oil from undersea fields off eastern Scotland brightened the nation's economic prospects. James Callaghan became prime minister upon Wilson's retirement in 1976. Growing dissatisfaction with Labour policies, rising inflation, and an increasing number of crippling strikes led to new elections in May, 1979. The Conservatives won a clear majority in Parliament, and Margaret Thatcher took office as prime minister.

Government. Great Britain is a constitutional monarchy. The constitution exists in no one document but is a centuries-old accumulation of statutes, judicial decisions, usage, and tradition. The hereditary monarch is almost entirely limited to exercising ceremonial functions. Sovereignty rests in Parliament, which consists of the House of Commons (with 635 members popularly elected from single-member constituencies), the House of Lords, and the crown. Effective power resides in the Commons; the executive—the cabinet of ministers headed by the prime minister—is usually drawn from the party holding the most seats in the Commons. Elections must be held at least once in five years, but within that period the prime minister may at any time request the crown to dissolve Parliament and call for new elections.

Educational Institutions. There are 44 universities in Great Britain and Northern Ireland, the most famous being those at Oxford, Cambridge, Edinburgh, and London.

Great Di·vid·ing Range (dĭ-vī'dĭng), crest line of the Eastern Highlands of Australia. For the most part it separates rivers draining into the Pacific Ocean from those flowing into the Indian Ocean and the Arafura Sea. Erosion and earth movements have pushed the watershed west of the range in several places.

Greater Man·ches·ter (măn'chĕs-tər, -chə-stər), metropolitan county (1976 est. pop. 2,684,100), W central England, created under the Local Government Act of 1972 (effective 1974). Greater Manchester includes parts of the former counties of Cheshire, Lancashire, and Yorkshire (West Riding).

Great Falls, city (1978 est. pop. 60,300), seat of Cascade co., N central Mont., at the confluence of the Missouri and Sun rivers and near the falls that give the city its name; inc. 1888. As the center of extensive hydroelectric power development, Great Falls is popularly called "Electric City." There are oil and copper refineries, a zinc-reduction plant, and flour mills.

Great Har·wood (här'wŏŏd'), urban district (1971 est. pop. 11,000), Lancashire, N England, NE of Blackburn. It has cotton mills, and there are collieries in the vicinity. It is the birthplace of John Mercer (1791), inventor of the processes of mercerizing textiles and manufacturing parchment paper.

Great In·di·an Desert (ĭn'dē-ən): *see* Thar Desert.

Great Lakes, group of five freshwater lakes, central North America, between the United States and Canada; largest body of fresh water in the world, with a combined surface area of c.95,000 sq mi (246,050 sq km). From west to east they are Lake Superior, Lake Michigan, Lake Huron, Lake Erie, and Lake Ontario, out of which flows the St. Lawrence River. The international boundary passes approximately through the center of all the lakes except Lake Michigan, which lies entirely within the United States. The Great Lakes were formed at the end of the Ice Age when the glacier-carved lake basins were filled with meltwater. The lakes are connected to each other by straits, short rivers, and canals.

French traders were the first Europeans in the region. Étienne Brulé visited Lake Huron c.1612, and in 1614, Brulé and French explorer Samuel de Champlain explored Lake Huron and Lake Ontario. In 1679 French explorer Robert LaSalle sailed from Lake Erie to Lake Michigan. The Great Lakes region, rich in furs, was contested for many years by the French, English, and Americans. The close of the War of 1812 finally ended the struggle for possession of the Great Lakes, and settlement of the region rapidly followed. The opening of the Erie Canal in 1825 accelerated the development of commerce on the Great Lakes. Great quantities of iron ore and grain, much coal, petroleum and steel, and manufactured articles are transported from Apr. until Dec., when ice closes most of the ports and winter storms hinder navigation. The opening of the St. Lawrence Seaway in 1959 has made the Great Lakes a truly international water body. The Illinois Waterway connects the lakes with the Mississippi River and the Gulf of Mexico; the New York State Barge Canal joins the Great Lakes with the Hudson River and the Atlantic Ocean.

Great Mi·am·i (mī-ăm'ē, -ăm'ə), river, Ohio: *see* Miami.

Great Neck, village (1970 pop. 10,798), Nassau co., SE N.Y., on the N shore of Long Island, a primarily residential suburb of New York City; settled c.1634, inc. 1921.

Great Plains, high, extensive grassland region of North America, extending from the Canadian provinces of Alta., Sask., and Man. S through W central United States into Texas. The Great Plains slope gently eastward from the foothills of the Rocky Mts. at an elevation of 6,000 ft (1,830 m) to merge into the prairies at an elevation of 1,500 ft (457.5 m). Much of the Great Plains was once covered by a vast inland sea, and sediments deposited by the sea make up the nearly horizontal rock strata that underlie the area. Intrusive igneous rocks account for sections of higher elevation.

The Great Plains has a semiarid climate; there are wide seasonal temperature ranges and winds of high velocity. In western sections the chinook, a warm winter wind, brings relief from bitterly cold and snowy winters. Grass is the dominant type of vegetation; trees are found in moister areas and along watercourses. The Great Plains are sparsely populated, with huge ranches and farms occupying much of the grassland. Cattle and sheep grazing is an important economic activity in the western plains. The fertile soils of the eastern plains are very productive when water is available. Wheat is the principal farm crop; sorghum and flax are also grown. The westward expansion of the wheat belt during World War I resulted in the creation of the Dust Bowl. A variety of minerals are found on the Great Plains, including oil, natural gas, coal, and gold.

First explored by the Spanish in the 1600s, the Great Plains were called the Great American Desert until well into the 19th cent. They were long inhabited by buffalo and Indians. American pioneers bypassed the region at first in the belief that an area that could not grow trees could not support agriculture. The railroads were largely responsible for populating the region.

Great Rift Valley (rĭft), geologic fault system of SW Asia and E Africa, estending c.3,000 mi (4,825 km) from N Syria to central Mozambique. The valley ranges in elevation from c.1,300 ft (395 m) below sea level (the Dead Sea) to c.6,000 ft (1,830 m) above sea level in southern Kenya. Erosion has concealed some sections, but in places, notably in Kenya, there are sheer cliffs several thousand feet high. The present configuration of the rift, which dates from the mid-Pleistocene epoch, is probably a result of a rifting process associated with thermal currents in the earth's mantle; there is evidence of earlier rift structures.

Great Saint Ber·nard (sānt' bər-närd'), pass: *see* Saint Bernard.

Great Salt Lake, shallow body of salt water, c.1,000 sq mi (2,590 sq km), NW Utah, between the Wasatch Range on the W and the Great Salt Lake Desert on the E; largest salt lake in North America. Fed by the Weber, Jordan, and Bear rivers, the lake varies in size and depth from year to year and with climatic changes; its average depth is 13 ft (4 m). Magnesium chloride, potash, and common table salt have been commercially extracted from the lake. The heavy brine supports no life except brine shrimp and colonial algae. The Great Salt Lake is a remnant of prehistoric Lake Bonneville, which covered an extensive area of the Great Basin and was once c.1,000 ft (305 m) deep. Its various levels are marked by former beachlines on the mountains and by rich soil deposits on the terraces to the east. In 1825 an American fur trader, James Bridger, was probably the first white man to see Great Salt Lake.

Great Sand Dunes National Monument (dōōnz): *see* National Parks and Monuments Table.

Great Slave Lake (slāv), c.10,980 sq mi (28,440 sq km), S Mackenzie dist., Northwest Territories, Canada. It is c.300 mi (485 km) long and from 12 to 68 mi (19.3–109.4 km) wide and is the deepest lake (2,015 ft/614.6 m) of North America. The western shores are wooded, but the long east and north arms reach into tundralike country. Samuel Hearne, a British fur trader, discovered the lake in 1771. Gold was discovered in the 1930s on the northern shore. The area is still important for gold, lead, and zinc.

Great Smok·y Mountains (smō'kē), part of the Appalachian system, on the N.C.–Tenn. border, the highest range E of the Black Hills. The mountains are named for the smokelike haze that envelops them. More than 25 peaks rise over 6,000 ft (1,830 m). The Great

Smokies are noted for their luxuriant vegetation; there are about 100 species of trees and more than 1,300 kinds of flowering plants. Although the region's coves and valleys have been settled since pioneer times, they remained isolated and inaccessible until the 20th cent., when loggers began harvesting the virgin forest. Great Smoky Mountains National Park, 516,626 acres (209,234 hectares), straddles the crest of the Great Smokies for 71 mi (114.2 km); est. 1930. The park includes c.600 mi (965 km) of trails (the Appalachian Trail follows the crest) and many streams and waterfalls.

Great South Bay, arm of the Atlantic Ocean, c.45 mi (72 km) long, between the S shore of Long Island and offshore barrier islands, SE N.Y. With the rapid population growth along its shores, the shallow bay has suffered accelerated deposition and pollution.

Great Vic·to·ri·a Desert (vĭk-tôr'ē-ə), region, c.450 mi (725 km) wide, SE Western Australia and W South Australia, bordered on the N by Gibson Desert, on the S by Nullarbor Plain. Average elevations range from 500 to 1,000 ft (152.5-305 m).

Great Wall of Chi·na (chī'nə): *see* China, Great Wall of.

Greece, republic (1971 pop. 8,745,084), 50,944 sq mi (131,945 sq km), SE Europe, occupying the S part of the Balkan Peninsula and bordering on the Ionian Sea in the W, on the Mediterranean Sea in the S, on the Aegean Sea in the E, on Turkey and Bulgaria in the NE, on Yugoslavia in the N, and on Albania in the NW. Athens is the capital.

About 75% of Greece is mountainous and only about 25% of the land is arable. Northern Greece includes portions of historic Epirus, Macedonia, and Thrace. It takes in part of the Pindus Mts. (which continue into central Greece); low-lying plains along the lower Nestos and Struma rivers; and the Khalkidhikí peninsula. Central Greece, situated north of the Gulf of Corinth, includes the low-lying plains of Thessaly, Attica, and Boeotia and Mt. Olympus. Southern Greece is made up of the Peloponnesus. The fourth region of Greece comprises numerous islands in the Mediterranean, the Ionian Sea, and the Aegean Sea.

Economy. Greece is a relatively poor agricultural country. In the early 1970s farming contributed about 20% of the annual national product and manufacturing about 16%; tourism was an important source of income. The chief agricultural products are wheat, citrus fruits, olives and olive oil, grapes, currants, cotton, tobacco, sugar beets, tomatoes, and potatoes. Large numbers of sheep and goats are raised. The country's principal manufactures are construction materials, textiles, food products, chemicals, refined petroleum, and ships. The chief minerals produced are lignite, bauxite, high-grade iron ore, magnesite, and iron pyrites. Oil in exploitable quantities was discovered off the east coast in the 1970s. Greece has a large merchant fleet. The value of imports (chiefly manufactured goods, food, fuel, and chemicals) is usually considerably higher than the value of exports (chiefly food, wine, and tobacco).

History. Archaeological remains show that Greece had a long prehistory, dating from the Neolithic Age (c.4000 B.C.). By the Bronze Age (c.2800 B.C.) important cultures had developed. The Minoan civilization, centered on Crete, and the Mycenaean civilization, centered on the Peloponnesus, had disappeared by 1100 B.C. The Greek-speaking Achaeans migrated into the area during the 14th cent. B.C., followed successively by the Aeolians, Ionians, and Dorians (11th cent. B.C.). City-states—small settlements that grew into minor kingdoms—developed and established colonies throughout the Mediterranean by the 8th cent. B.C. The city-states varied widely in their political constitutions and cultures. Some were monarchies, some aristocracies or oligarchies, and some, like Athens, were democracies tempered by limited citizenship. Militaristic Sparta had a unique constitutional and social development. A common language and religion helped to give the warring city-states a sense of unity, and they combined effectively to defeat the Persians (449 B.C.) in the Persian Wars.

Athens grew dramatically, and in the age of Pericles (c.495-429 B.C.) developed a culture that left its mark on the course of Western and Eastern civilization. Drama, poetry, sculpture, architecture, and philosophy flourished, and there was a vigorous intellectual life. After the Peloponnesian War (431-404 B.C.), in which Sparta defeated Athens, the hegemony of Greece passed to Corinth and Thebes. Philip II of Macedon conquered Greece in 338 B.C. and paved the way for his son, Alexander the Great, who spread Greek civilization over the known Western world and across Asia to India. After Alexander's death (323 B.C.), his empire was torn apart by his warring generals.

Incessant warfare made Greece increasingly weak, while Rome grew stronger. In 146 B.C., the remnants of the Greek states fell into the hands of Rome. Under Roman rule, the cities had little political or economic importance, but Greek intellectual supremacy continued for many centuries. From the division (A.D. 395) of the Roman Empire into East and West until the conquest (15th cent.) of Greece by the Ottoman Turks, Greece shared the fortunes of the Byzantine Empire. The center of the Greek world was Constantinople, not Greece proper. The Turks completed their conquest (1456) by seizing parts of Greece that were under the rule of French and Italian princes. The Turks practiced religious tolerance, but otherwise their 400-year rule was oppressive.

In 1821 the Greek War of Independence began under the leadership of Alexander and Demetrios Ypsilanti. With Russian, English, and French aid, the Greeks were ultimately successful, and in 1832 Greece became an independent monarchy with Otto I, a Bavarian prince, as king. Forced to abdicate in 1862, Otto was succeeded by a Danish prince, George I (reigned 1863-1913), who introduced (1864) a new constitution establishing a unicameral parliament. England ceded (1864) the Ionian Islands, and in 1881 Greece acquired Thessaly and part of Epirus. Greece was defeated in the Greco-Turkish War over Crete (1897), but because of the pressure of the powers Crete was eventually (1913) incorporated into Greece. After the Balkan Wars (1912-13) Greece obtained southeastern Macedonia, western Thrace, and most of Epirus.

During World War I the neutralist King Constantine I was deposed (1917) by a faction favoring the Allies. For its part in the closing year of the war, Greece received territory from Bulgaria and Turkey. The years between the World Wars were marked by unsettled economic conditions and by violent political strife. A republic, established in 1924, ended with the return of King George II in 1935. In 1936 Premier Ioannis Metaxas, supported by the king, established a dictatorship. The country was occupied by Italian and German troops from May, 1941, to Oct., 1944, when they were expelled by guerrillas and British forces.

After World War II, shaky coalitions held power while Communist-led guerrillas waged civil war from their mountain strongholds. U.S. President Harry S. Truman announced (Mar., 1947) the Truman Doctrine, under which the United States sent military and economic aid to the government. The rebels were defeated by late 1949. Greece was a charter member of the UN, and in 1951 it was admitted to the North Atlantic Treaty Organization (NATO). In 1954 friction with Turkey (and also with Great Britain) arose over the sovereignty of Cyprus, the majority of whose population is ethnically Greek, and it continued after Cyprus became independent in 1960. The moderately liberal Center Union gained a majority of seats in the legislature in 1964 and George Papandreou became Prime Minister. The failure of this government, the weakness of succeeding ones, and the indecisive maneuvering of Constantine II (reigned 1964-73) ended in a rightist army coup d'état (Apr., 1967). A military junta assumed power, imprisoning leftists and placing rigid controls on Greek cultural and political life. These controls were eased somewhat by the early 1970s. In June, 1973, the monarchy was abolished, and Greece became a presidential republic. President George Papadopoulos was ousted in Nov., 1973, by a rival military group that turned over power to a civilian government headed by Constantine Karamanlis in 1974.

Gree·ley (grē'lē). **1.** County (1970 pop. 1,819), 783 sq mi (2,028 sq km), W Kansas, bordered in the W by Colo.; formed 1873; co. seat Tribune. In a farm and cattle-ranching area, it has some manufacturing. **2.** County (1970 pop. 4,000), 570 sq mi (1,476.3 sq km), E central Nebr., drained by the Cedar and North Loup rivers; formed 1871; co. seat Greeley Center. Its economy depends mainly on livestock, grain, and dairy products.

Greeley, city (1978 est. pop. 49,200), seat of Weld co., N Colo., at the base of the Front Range of the Rocky Mts.; inc. 1885. It is a rail, trade, and processing center for a rich irrigated farm area. Greeley was founded (1870) by Horace Greeley through his agent, Nathan C. Meeker, as a cooperative farm and temperance colony. Meeker was killed in the Ute uprising of 1879; his home is now a museum. The city is the seat of the Univ. of Northern Colorado.

Greeley Cen·ter (sĕn'tər), or **Greeley**, village (1970 pop. 580), seat of Greeley co., E central Nebr., N of Grand Island. Dairy products are manufactured.

Green (grēn). **1.** County (1970 pop. 10,350), 282 sq mi (730.4 sq km), central Ky., drained by the Green River and Russell Creek; formed 1792; co. seat Greensburg. In a rolling timber and agricultural area yielding grain, burley tobacco, and livestock, it has oil and natural-gas wells. Clothing and furniture are manufactured. **2.** County (1970 pop. 26,714), 586 sq mi (1,517.7 sq km), S Wis., in a generally hilly area bounded on the S by the Ill. border and drained by the Sugar and Pecatonica rivers; formed 1836; co. seat Monroe. Its economy depends mainly on cheese and other dairy products. Wood products, feed, and hardware are among its manufactures.

Green. **1.** River, 370 mi (595.3 km) long, rising in central Ky. and flowing generally NW, through Mammoth Cave National Park, to the Ohio River near Evansville, Ind. **2.** River, 730 mi (1,174.6 km) long, rising near the Continental Divide, W Wyo., and flowing generally S through W Wyo., NW Colo., and E Utah to the Colorado River in Canyonlands National Park, SE Utah. It is the largest tributary of the Colorado.

Green Bay, city (1978 est. pop. 89,700), seat of Brown co., NE Wis., at the mouth of the Fox River on Green Bay; inc. 1854. One of the best Great Lakes harbors and a railroad center, Green Bay is a port of entry, with heavy shipping and a large wholesale and jobbing trade. Its industries include papermaking, food and dairy processing, and machine building. Jean Nicolet established a trading post on the site of Green Bay in 1634. The permanent settlement, the oldest in the state, dates from 1701. Strategically located, Green Bay became a fur-trading center and was occupied successively by the French (1717), the British (1761), and the Americans (1816). With the settlement of the Old Northwest after the War of 1812 and the decline of the fur trade, Green Bay became the trade center of a lumber and farm area. A branch of the Univ. of Wisconsin is in the city.

Green Bay, arm of Lake Michigan, c.100 mi (160 km) long and from 10 to 20 mi (16–32 km) wide, NE Wis. and NW Mich., separated from the lake by the Door Peninsula. The southern part of the bay is frozen from Dec. to May.

Green·belt (grēn'bĕlt'), city (1978 est. pop. 15,400), Prince Georges co., W central Md., a residential suburb of Washington, D.C.; chartered 1937. Greenbelt was planned and built by the Federal government as an experimental model community.

Green·bri·er (grēn'brī'ər), county (1970 pop. 32,090), 1,026 sq mi (2,657.3 sq km), SE W.Va., bounded on the E by the Va. border and drained by the Greenbrier and Meadow rivers; formed 1778; co. seat Lewisburg. Its agriculture includes livestock, fruit, tobacco, and dairy products. It also has bituminous-coal mines, timber, deposits of limestone, and diversified light industry. Lying partly in the Allegheny Plateau, the county has mineral springs and includes a section of Monongahela National Forest.

Green·cas·tle (grēn'kăs'əl), city (1970 pop. 8,852), seat of Putnam co., W central Ind., NE of Terre Haute; founded c.1823, inc. 1849. It is a farm trade center and seat of DePauw Univ.

Green Cove Springs (kōv), town (1970 pop. 3,857), seat of Clay co., NE Fla., on the W bank of the St. Johns River S of Jacksonville; settled 1830. It is a mineral-springs resort in a farming, lumbering, and sport-fishing area.

Green·dale (grēn'dāl'), village (1970 pop. 15,089), Milwaukee co., SE Wis., a suburb of Milwaukee; inc. 1938. Automotive and machine parts are produced. Greendale is one of three planned communities built by the Federal government in the 1930s.

Greene (grēn). **1.** County (1970 pop. 10,650), 627 sq mi (1,623.9 sq km), W Ala., in a blackbelt region bounded on the N by the Sipsey River, on the W by the Tombigbee River, on the E by the Black Warrior River; formed 1819; co. seat Eutaw. It has agriculture (cotton, corn, pecans, potatoes, and cattle) and a lumbering industry. Clothing, wood products, and paper are manufactured. **2.** County (1970 pop. 24,765), 579 sq mi (1,499.6 sq km), NE Ark., bounded on the E by the St. Francis River, drained by the Cache River, and intersected by Crowley's Ridge; formed 1833; co. seat Paragould. Cotton, corn, hay, and livestock are raised. It also has timber, gravel pits, and a food-processing industry. Textiles, furniture, shoes, sheet metal, and automobile parts are made. **3.** County (1970 pop. 10,212), 404 sq mi (1,046.4 sq km), NE central Ga., in a piedmont area drained by the Oconee, Ogeechee, and Apalachee rivers; formed 1786; co. seat Greensboro. Its agriculture includes cotton, corn, grain, fruit, and livestock. It also has a lumbering industry and textile mills and clothing factories. **4.** County (1970 pop. 17,014), 543 sq mi (1,406.4 sq km), W Ill., bounded on the W by the Illinois River and drained by Macoupin and Apple creeks; formed 1821; co. seat Carrollton. In an agricultural area yielding livestock, corn, wheat, oats, poultry, fruit, and dairy products, it has bituminous-coal mines and deposits of potter's clay. Paper and chemicals are produced. **5.** County (1970 pop. 26,894), 549 sq mi (1,421.9 sq km), SW Ind., drained by the Eel River and the West Fork of the White River; formed 1821; co. seat Bloomfield. It has agriculture (grain, fruit, and livestock), bituminous-coal mines, and lumbering and food-processing industries. Stone, clay and glass products are made. **6.** County (1970 pop. 12,716), 569 sq mi (1,473.7 sq km), central Iowa, in a prairie agricultural area drained by the Raccoon River; formed 1851; co. seat Jefferson. Cattle, hogs, poultry, corn, and soybeans are important. It also has bituminous-coal mines and sand and gravel pits. **7.** County (1970 pop. 8,545), 728 sq mi (1,885.5 sq km), SE Miss., bordered on the E by Ala. and drained by the Leaf and Chickasawhay rivers; formed 1811; co. seat Leakesville. Its agriculture includes cotton, corn, and livestock. It has a lumbering industry and manufactures clothing. The county includes parts of De Soto National Forest. **8.** County (1970 pop. 152,929), 677 sq mi (1,753.4 sq km), SW Mo., in the Ozarks, drained by the James, Sac, Little Sac, and Pomme de Terre rivers; formed 1833; co. seat Springfield. Grain, apples, strawberries, grapes, and tomatoes are grown. It also has food-processing and lumbering industries and limestone, iron, and lead deposits. Among its manufactures are clothing, furniture, chemicals, concrete, steel, and farm machinery. **9.** County (1970 pop. 33,136), 653 sq mi (1,691.3 sq km), SE N.Y., in a mountainous resort area situated mainly in the Catskills, bounded on the E by the Hudson River and drained by Schoharie and Catskill creeks; formed 1800; co. seat Catskill. Dairying, farming (clover, fruit, and truck crops), and poultry raising are important. It also has a lumbering industry and some mining. Clothing, cement, and metal products are made, and foodstuffs are processed. **10.** County (1970 pop. 14,967), 269 sq mi (696.7 sq km), E central N.C., on the coastal plain; formed 1799; co. seat Snow Hill. It has agriculture (tobacco, corn, soybeans, and livestock) and a lumbering industry. Textiles are manufactured.

11. County (1970 pop. 125,057), 416 sq mi (1,077.4 sq km), SW Ohio, intersected by the Little Miami and Miami rivers; formed 1803; co. seat Xenia. Mainly industrial, it also produces corn, wheat, soybeans, and livestock and has stands of timber, sand and gravel pits, and mineral springs. Its diversified manufactures include packaged meats, textiles, wood and metal products, furniture, concrete, iron, and aluminum. **12.** County (1970 pop. 36,090), 577 sq mi (1,494.4 sq km), SW Pa., bounded on the S and W by the W.Va. border, on the E by the Monongahela River; formed 1796; co. seat Waynesburg. Sheep, other livestock, grain, and dairying are important. It also has bituminous-coal mines, oil and natural-gas wells, and timber. There is some clothing manufacturing. **13.** County (1970 pop. 47,630), 617 sq mi (1,598 sq km), NE Tenn., bounded in the SE by the Bald Mts. and the N.C. border and drained by the Nolichucky River; formed 1783; co. seat Greeneville. Its agriculture includes tobacco, corn, hay, fruit, livestock, and dairy products. It also has oak and pine timber and ferroalloy ore deposits. Food processing is done, and textiles, wood and metal products, and furniture are manufactured. Part of Cherokee National Forest is here. **14.** County (1970 pop. 5,248), 153 sq mi (396.3 sq km), N central Va., bounded on the NE by the Rapidan River, on the NW by the Blue Ridge; formed 1838; co. seat Stanardsville. In a diversified agricultural area yielding grain, hay, tobacco, fruit, livestock, and dairy products, it has pine forests and deposits of copper, gneiss, and phyllite. Tourism is also important.

Greene·ville (grēn′vĭl′), town (1978 est. pop. 14,500), seat of Greene co., NE Tenn., in a tobacco, dairy, and cattle area; founded 1783, inc. 1903. It is a leading tobacco market and manufactures televisions and radios, gas pumps, lock sets, and condensed milk. Andrew Johnson's home, tailor shop, and grave are in Andrew Johnson National Monument.

Green·field (grēn′fēld′). **1.** City (1970 pop. 9,986), seat of Hancock co., central Ind., E of Indianapolis; settled 1828, inc. 1850. The poet James Whitcomb Riley was born here, and his house has been restored. **2.** City (1970 pop. 2,212), seat of Adair co., SW Iowa, SW of Des Moines, in a farm and livestock region; settled 1841, inc. 1876. **3.** Town (1978 est. pop. 19,200), seat of Franklin co., NW Mass., at the confluence of the Deerfield and Green rivers, near their junction with the Connecticut River; settled 1686, set off from Deerfield and inc. 1753. It is an industrial center in a prosperous agricultural area. The first cutlery factory in the United States was established here in the early 1800s. Other products include silverware, electronic components, lumber, and paper and wooden boxes. **4.** City (1970 pop. 1,172), seat of Dade co., SW Mo., NNW of Springfield, in a farm area; settled 1808, inc. 1841. **5.** City (1978 est. pop. 34,900), Milwaukee co., SE Wis., a residential suburb of Milwaukee; inc. 1957.

Green·hills (grēn′hĭlz′), city (1970 pop. 6,092), Hamilton co., SW Ohio, a suburb N of Cincinnati; est. 1938 by the Federal Resettlement Administration as a low-cost housing project, inc. 1939.

Green Lake, county (1970 pop. 16,878), 355 sq mi (919.5 sq km), central Wis., in a dairying and farming area intersected by the Fox and Grand rivers; formed 1858; co. seat Green Lake. Peas, corn, soybeans, wheat, barley, and oats are among its crops. It also has tourism and food-processing industries and resorts along Green Lake and Lake Puckaway.

Green Lake, village (1970 pop. 1,109), seat of Green Lake co., central Wis., on Green Lake W of Fond du Lac, in a summer resort and dairying area. Butter, boats, and knit goods are made.

Green·land (grēn′lənd), the largest island in the world (1975 est. pop. 54,000), c.840,000 sq mi (2,175,600 sq km), part of the kingdom of Denmark, lying largely within the Arctic Circle. It is surrounded by the Arctic Ocean in the north; the Greenland Sea in the east; the Denmark Strait in the southeast, which separates it from Iceland; the Atlantic Ocean in the south; and Davis Strait and Baffin Bay in the west, which separate it from Baffin Island, Canada. Greenland is 1,659 mi (2,675.8 km) long from Cape Farewell to Cape Morris Jesup and has a maximum width of about 800 mi (1,290 km). Geologically, the island is part of the Canadian Shield and, therefore, of North America; more than 50% of its ice-free area consists of rocks of the Precambrian era, mostly granites and gneisses. The entire coastline is deeply indented by fjords. There are many offshore islands, of which Disko Island is the largest. Except for about 132,000 sq mi (341,880 sq km) of coastland and coastal islands, an ice sheet covers the island. The extreme northern peninsula (Peary Land) has

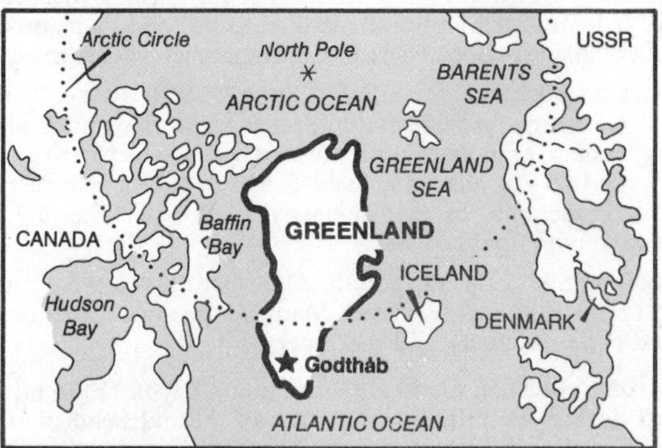

no ice cap. Recent surveys indicate that the thickness of the ice sheet reaches c.14,000 ft (4,300 m) in some places. Cold winds from Greenland's interior make the weather uncertain and foggy. The North Atlantic Drift gives the southwest coast of Greenland a warmer climate and heavy rainfall. There are no forests in Greenland; dwarf trees are found in the southern coastal areas. Natural vegetation also includes mosses, lichens, grasses, and sledges.

Economy. Fishing (cod, shrimp, halibut, and salmon) is the main industry. Sealing is also important along the southeast and northwest coasts. There is extensive sheep breeding in the southern area. The only large-scale mining operation has been for cryolite. Molybdenum, uranium, and coal are also found. The polar bear, musk ox, polar wolf, lemming, Arctic hare, and reindeer are the chief land animals. Sea birds are hunted for their flesh, eggs, and down.

History. Known in ancient times by the Greeks and later by the Irish, Greenland was discovered and colonized (c.982) by Eric the Red, a Norseman, who named it Greenland in order to make it seem attractive to potential settlers. It was in sailing to Greenland (c.1000) that Leif Ericson, the son of Eric the Red, probably reached North America. Greenland became a bishopric c.1110, and ruins of churches of that period remain. By the 12th cent. the population numbered some 10,000. Greenland became self-governing, but failed to achieve political stability. In 1261 the colony came under Norwegian rule, but in the 14th and 15th cent. it was neglected, and the colonists either died out or assimilated with the Eskimos. The British explorers Martin Frobisher and John Davis rediscovered Greenland in the 16th cent. but found no trace of Norsemen.

Modern colonization was begun (1721) by the Norwegian missionary Hans Egede. Danish trading posts were established shortly afterward, and colonization was furthered by deporting undesirable subjects to Greenland. In 1815, at the Congress of Vienna, Denmark retained the colony through an oversight of the delegates. In the 19th and 20th cent. Greenland was explored and mapped by numerous arctic explorers. In World War II, after the German occupation (1940) of Denmark, the United States invoked the Monroe Doctrine for Greenland. A Danish-American agreement for the common defense of Greenland was signed in 1951, and U.S. bases were retained, notably at Thule.

Green·lawn (grēn′lôn′), uninc. urban village (1970 pop. 8,493), Suffolk co., SE N.Y., on the N shore of Long Island E of Huntington.

Green·lee (grēn′lē), county (1970 pop. 10,330), 1,874 sq mi (4,853.7 sq km), SE Ariz., in a mountainous region drained by the Blue, Black, and Gila rivers; formed 1909; co. seat Clifton. Mining (copper and silver) and sheep and cattle raising are of major importance. Cotton, alfalfa, and truck crops are also grown. A large section of Apache National Forest is within the county.

Green Mountains, range of the Appalachian Mts., extending 250 mi (402.3 km) from N to S, mainly in Vt. and extending into W Mass. and S Que., Canada. Mt. Mansfield, 4,393 ft (1,339.9 m) high, in Vermont, is the tallest peak. Composed of some of the oldest rocks in North America, the range has low, rounded peaks, fertile valleys, and streams that have furnished water power for many years.

Green·ock (grē′nək, grĕn′ək), burgh (1971 pop. 69,004), Strathclyde region, W Scotland, on the Firth of Clyde. Greenock is a port, and shipping and shipbuilding are important industries. Others include engineering, textile manufacturing, sugar refining, and the production of office equipment.

Green River, town (1970 pop. 4,196), seat of Sweetwater co., SW Wyo., on the Green River NW of Evanston; founded 1868, inc. 1891. At an altitude of c.6,100 ft (1860.5 m), it is a rail and trade center for a forest, livestock, and farm area.

Greens·bor·o (grēnz′bûr′ō). **1.** Town (1970 pop. 3,371), seat of Hale co., W central Ala., S of Tuscaloosa, in the Black Belt; settled c.1816, inc. 1823. There are many ante-bellum houses here. **2.** City (1970 pop. 2,583), seat of Greene co., NE Ga., SSE of Atlanta; laid out 1786, inc. 1803. It is a processing center (with cotton mills) in a farm and dairy area. **3.** City (1978 est. pop. 161,200), seat of Guilford co., N central N.C.; inc. 1829. It has an important textile industry and manufactures tobacco products and electrical machinery. Greensboro was settled in 1749.

Greens·burg (grēnz′bûrg′). **1.** City (1970 pop. 8,620), seat of Decatur co., SE Ind., SE of Indianapolis; founded 1821, inc. as a town 1837, as a city 1859. It is a shipping point in a farm area. Engine parts and metal products are made. **2.** City (1970 pop. 1,907), seat of Kiowa co., S central Kansas, WSW of Wichita, in a livestock and wheat area; inc. 1886. **3.** Town (1970 pop. 1,990), seat of Green co., S central Ky., SSE of Louisville on the Green River, in a timber, tobacco, and grain area; settled c.1780. **4.** Town (1970 pop. 652), seat of St. Helena parish, SE La., near the Tickfar River NE of Baton Rouge, in an agricultural region. **5.** City (1978 est. pop. 18,900), seat of Westmoreland co., SW Pa.; settled c.1770, inc. as a city 1928. Paper products, building materials, electrical equipment, plate glass, and control systems are made. Col. Henry Bouquet defeated (1763) Indian warriors of Pontiac near here.

Greens·ville (grēnz'vĭl'), county (1970 pop. 9,604), 301 sq mi (779.6 sq km), S Va., bounded on the S by the N.C. border, on the N by the Nottoway River; formed 1781; co. seat Emporia. Peanuts, pork, and pine lumber are its most important products. Food processing and some manufacturing are done.

Green Tree, borough (1970 pop. 6,441), Allegheny co., SW Pa., SW of Pittsburgh; inc. 1885.

Green·up (grēn'əp), county (1970 pop. 33,192), 350 sq mi (906.5 sq km), extreme NE Ky., bounded on the N and E by the Ohio River and the Ohio border and drained by the Little Sandy River; formed 1803; co. seat Greenup. In a hilly agricultural area yielding corn, tobacco, apples, cattle, and truck crops, it has iron deposits. Chemicals and mineral products are manufactured.

Greenup, town (1970 pop. 1,284), seat of Greenup co., NE Ky., on the Ohio River NW of Ashland, in a farm area.

Green·ville (grēn'vĭl'), county (1970 pop. 240,774), 789 sq mi (2,043.5 sq km), NW S.C., bounded on the N by the N.C. line, on the W by the Saluda River; formed 1786; co. seat Greenville. It is drained by the Enoree and Reedy rivers. Agriculture (cotton, corn, and dairy products) and manufacturing, especially of cotton textiles, are important. It also includes a summer-resort area in the Blue Ridge.

Greenville. 1. City (1970 pop. 8,033), seat of Butler co., S central Ala., SSW of Montgomery, in an agricultural area; settled 1819, inc. as a town 1820, as a city 1871. **2.** City (1970 pop. 1,085), seat of Meriwether co., W Ga., E of La Grange, in a truck-farming and cotton area. Lumber products are manufactured. **3.** City (1970 pop. 4,631), seat of Bond co., SW Ill., NE of East St. Louis, in a farm, coal, and natural-gas area; settled 1815, inc. 1855. Evaporated milk and air-conditioning equipment are made. **4.** City (1970 pop. 3,875), seat of Muhlenberg co., W central Ky., S of Owensboro; est. 1799. It is a trade and processing center in an area of farms, coal, oil, clay pits, and timber. **5.** City (1970 pop. 7,493), Montcalm co., S central Mich., NE of Grand Rapids on Flat River, in a farm area; inc. as a village 1867, as a city 1871. It ships potatoes and manufactures refrigerators, auto bearings, and farm implements. **6.** City (1978 est. pop. 42,900), seat of Washington co., W Miss., on Lake Ferguson, a deepwater harbor adjoining the Mississippi River; inc. 1886. It is the trade, processing, and shipping center of the Mississippi-Yazoo delta, a fertile region producing soybeans, oats, corn, timber, and especially cotton. It is also an industrial city, and its many manufactures include saws, metal products, concrete items, and automobile parts. **7.** City (1970 pop. 328), seat of Wayne co., SE Mo., in the Ozarks on the St. Francis River N of Poplar Bluff. It is in an agricultural and timber region. **8.** City (1978 est. pop. 33,400), seat of Pitt co., E N.C., on the Tar River; founded 1786. It grew as a tobacco center, and while still an important tobacco-processing and marketing city, it now has diversified manufacturing industries. **9.** City (1978 est. pop. 12,800), seat of Darke co., W Ohio, in a farm area; settled 1808, inc. as a city 1900. Gen. Anthony Wayne built (1793) a fort on what is now Greenville as a base for his Indian campaign. After his victory at Fallen Timbers, Wayne returned and in 1795 negotiated a treaty with the Indians, who ceded a large part of the Old Northwest to the United States. **10.** Industrial borough (1970 pop. 8,704), Mercer co., NW Pa., near the Ohio border NE of Youngstown, Ohio; settled c.1796, laid out 1798, inc. 1837. A trading center for a truck-farming area, it manufactures aluminum products. **11.** City (1978 est. pop. 56,400), seat of Greenville co., NW S.C., on the Reedy River, in the Piedmont area near the Blue Ridge Mts.; laid out 1797, inc. as a city 1907. It has textile mills, garment factories, farm-produce processing and packing establishments, and plants making pharmaceuticals, fabricated metals, furniture, chemicals, plastics, electronic equipment, and turbine engines. **12.** City (1978 est. pop. 20,900), seat of Hunt co., E Texas, in a prosperous blackland cotton region; inc. 1874. Among its manufactures are electronic systems, clothing, and oil-field equipment.

Green·wich (grĭn'ĭj, -ĭch, grĕn'-), borough (1971 pop. 216,441) of Greater London, SE England, on the Thames River. The system of geographic longitude and time-keeping worked out at the Royal Observatory here have become standard in most countries of the world; the prime meridian, or long. 0°, passes through the observatory.

Green·wich (grĕn'ĭch), residential and resort town (1978 est. pop. 59,700), Fairfield co., SW Conn., on the Mianus and Byram rivers and Long Island Sound; settled 1640, inc. 1955. Settled on land bought from the Indians, Greenwich was long inhabited by farmers and oystermen. In the American Revolution it was plundered (1779) by the British. In the late 19th cent. Greenwich began to attract artists and summer residents.

Greenwich Village, residential district of lower Manhattan, New York City, extending S from 14th St. to Houston St. and W from Washington Square to the Hudson River. A separate village in colonial times, it later became an exclusive residential section. Around 1910, the Village gained renown as the home and workshop of artists, writers, and musicians. Interesting old buildings, many dating from the early and mid-1800s, remain.

Green·wood (grēn'wo͝od'). **1.** County (1970 pop. 9,141), 1,142 sq mi (2,957.8 sq km), SE Kansas, in a rolling flatland to hilly area drained by the Fall River; formed 1855; co. seat Eureka. In a livestock and grain area, it has extensive oil and natural-gas fields. Food processing is done. **2.** County (1970 pop. 49,686), 446 sq mi (1,155.1 sq km), W S.C., in an agricultural area bounded on the NE by the Saluda River; formed 1897; co. seat Greenwood. Textile manufacturing is of primary importance here. The county includes part of Sumter National Forest.

Greenwood. 1. Coal-mining town (1970 pop. 2,156), a seat of Sebastian co., W central Ark., SSE of Fort Smith. It has poultry and dairy farms. **2.** City (1970 pop. 11,408), Johnson co., central Ind.; settled 1822, inc. as a city 1960. Primarily residential, it has some industry. **3.** City (1978 est. pop. 21,000), seat of Leflore co., W central Miss., on the Yazoo River in the Mississippi delta; inc. 1844. It is a retail and trade center for a productive farm region and one of the largest cotton markets in the world. The area's original inhabitants were Choctaw Indians. After the area was ceded to the United States in 1830 by the Treaty of Dancing Rabbit Creek, settlers poured in, carving vast cotton plantations out of the delta swamplands. **4.** City (1978 est. pop. 26,500), seat of Greenwood co., W S.C.; settled 1824, inc. as a city 1927. It is a rail center, with textile and meat-packing industries. A trading post was established here in 1751.

Greer (grîr), county (1970 pop. 7,979), 637 sq mi (1,649.8 sq km), SW Okla., bounded by the North Fork of the Red River (with Altus Dam in the SE) and drained by the Elm and Salt Forks of the Red River; formed 1907; co. seat Mangum. It is in an agricultural area yielding cotton, wheat, corn, livestock, and dairy products. It has granite quarries, and there is some stone manufacturing.

Greer, town (1970 pop. 10,642), Greenville and Spartanburg cos., NW S.C., in a farm region. It has textile mills, food-processing plants, and some light manufacturing.

Gregg (grĕg), county (1970 pop. 75,929), 284 sq mi (735.6 sq km), E Texas, drained by the Sabine River; formed 1873; co. seat Longview. Located in a huge oil field, it also has deposits of natural gas, iron, lignite, and clay. Its agriculture includes cotton, corn, sweet potatoes, fruit, truck crops, livestock, and dairy products.

Greg·o·ry (grĕg'ə-rē), county (1970 pop. 6,710), 997 sq mi (2,582.2 sq km), S S.Dak., on the Nebr. line, watered by intermittent streams and bounded on the E by the Missouri River; formed 1862; co. seat Burke. It is in an agricultural (poultry, grain, and dairy products) and stock-raising region. There is some manufacturing in the county.

Greifs·wald (grīfs'vält), city (1974 est. pop. 53,940), Rostock district, N East Germany, near the Baltic Sea. It is an industrial and commercial center. Manufactures include wood products and machinery. Greifswald was chartered in 1250, and in 1648 it became part of Swedish Pomerania. In 1815 it passed to Prussia. The city has a noted university (founded 1456).

Greiz (grīts), city (1974 est. pop. 37,612), Gera district, S East Germany, on the White Elster River. Manufactures include textiles, paper and paper products, and machinery.

Gre·na·da (grĭ-nā'də), island (1970 est. pop. 87,300), 120 sq mi (310.8 sq km), in the West Indies. It is part of the independent state of Grenada in the Windward Islands. The state includes the southern

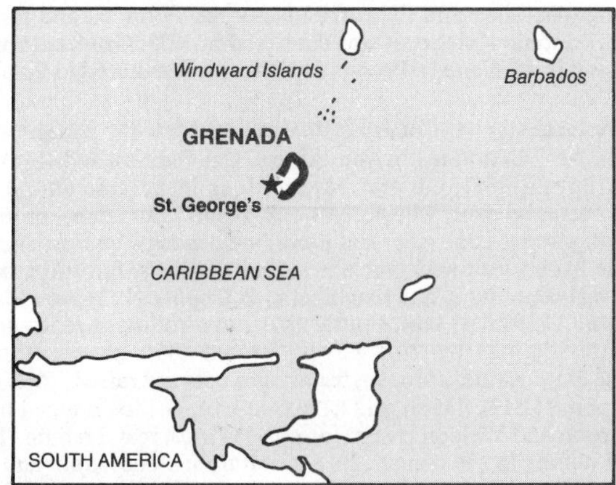

half of the archipelago known as the Grenadines. The capital, main port, and commercial center is Saint George's. Grenada is a volcanic, mountainous island with crater lakes. Its agricultural products include bananas, cacao, nutmeg, mace, sugar, coconuts, cotton, and limes. Tourism is a leading industry. From its discovery by Christopher Columbus in 1498 until French settlement began in 1650, the hostility of the native Carib Indians prevented colonization on Grenada. A point of dispute between England and France, the island became permanently British in 1783. The British colonists imported African slaves and established sugar plantations. In 1967 Grenada became an associated state of Britain with full internal self-government. Complete independence was achieved in Feb., 1974.

Grenada, county (1970 pop. 19,854), 431 sq mi (1,116.3 sq km), N central Miss., drained by the Yalobusha and Skuna rivers; formed 1870; co. seat Grenada. It is a cotton-growing region that also produces corn, livestock, and dairy products. Its timber industry includes logging, sawmills, and wood preserving. Packaged meats, textiles, clothing, and glass products are manufactured.

Grenada, city (1970 pop. 9,944), seat of Grenada co., N central Miss., on the Yalobusha River NE of Greenwood; settled in the early 1830s, inc. 1836. It is a market and processing center for a cotton, livestock, and timber area.

Gren·chen (grĕn'KHən), town (1977 est. pop. 17,800), Solothurn canton, NW Switzerland. It is a watchmaking center.

Gre·no·ble (grə-nō'bəl, grə-nô'blə), city (1975 pop. 166,037), capital of Isère dept., SE France, on the Isère River at the foot of the Alps. It is the hydroelectric center of France and has an important nuclear-research center. Metals, electrical equipment, chemicals, and food products are the chief manufactures. An ancient Roman city, Grenoble came under the Burgundians (5th cent.), the Franks (6th cent.), and the kingdom of Provence (9th–11th cent.). When that kingdom broke up, Grenoble became a possession of the dauphins of Viennois, then passed to the French crown in 1349. In Grenoble are a famous university (founded 1339), the Cathedral of Notre Dame (12–13th cent.), the Church of St. André (13th–14th cent.), the Renaissance palace of the dauphins (now the courthouse), and an art museum. Another museum is devoted to Stendhal, who was born in Grenoble. Grenoble is a noted tourist and skiing center and was the site of the 1968 winter Olympics.

Gresh·am (grĕsh'əm), city (1970 pop. 10,030), Multnomah co., NW Oregon, E of Portland near the Columbia River; settled 1852, inc. as a town 1905, as a city 1921. Aluminum and rubber products are manufactured, and the produce of nearby dairy farms and orchards is processed.

Gret·na (grĕt'nə), city (1978 est. pop. 30,200), seat of Jefferson parish, SE La., on the Mississippi River. Its manufactures include cottonseed oil, alcohol, insecticides, and fertilizer. Founded in the early 19th cent. as Mechanicsham, Gretna merged with McDonoghville in 1913 and is now a suburb of New Orleans.

Gretna Green, village, Dumfriesshire, S Scotland, on the English border. It was famous as a place of runaway marriages from 1754, when English marriage law was tightened, until 1856, when a law was passed requiring that one of the parties to marriage in Scotland must reside in Scotland for at least 21 days before issuance of the license.

Grey·lock, Mount, (grā'lŏk'), peak, 3,491 ft (1,064.8 m), W Mass., highest of the Berkshire Hills and highest point in the state.

Grif·fin (grĭf'ĭn), city (1978 est. pop. 23,900), seat of Spalding co., W central Ga., in a farm and cotton area; inc. 1843. The city has large textile and garment industries. Food processing is also important.

Grif·fith (grĭf'ĭth), residential town (1970 pop. 18,168), Lake co., extreme NW Ind.; inc. 1904.

Griggs (grĭgz), county (1970 pop. 4,184), 714 sq mi (1,849.3 sq km), E N.Dak, drained by the Sheyenne River; formed 1881; co. seat Cooperstown. It is in a wheat-growing area that also produces barley, oats, and flax, and raises beef cattle, hogs, and sheep. Farm machinery is manufactured.

Gri·jal·va (grē-häl'vä), river, c.400 mi (645 km) long, rising in SW Guatemala and flowing NW into S Mexico and N through Chiapas and Tabasco states to the Gulf of Campeche. The Grijalva project was begun in the 1960s along the lower Grijalva and Usumacinta rivers for flood control and sanitation. It is named for the Spanish explorer Juan de Grijalva, who discovered it in 1518.

Grimes (grīmz), county (1970 pop. 11,855), 801 sq mi (2,074.6 sq km), E central Texas, bounded on the W by the Navasota and Brazos rivers; formed 1846; co. seat Anderson. It has diversified agriculture (cotton, corn, grain, watermelons and other fruit, truck crops, livestock, and dairy products). It also has pine forests and good hunting and fishing.

Grims·by (grĭmz'bē), county borough (1976 pop. 93,800), in the Parts of Lindsey, Humberside, E central England, at the mouth of the Humber River. It is one of the largest fishing ports in the world. Grimsby has an extensive trade in fish, coal, grain, and timber and an important frozen food industry. Other industries include the production of rayon, titanium oxide, and chemicals.

Grim·sel (grĭm'zəl), pass, 7,159 ft (2,183.5 m) high, S Switzerland, between the Rhône and Aare valleys. The Grimsel Road (built 1891–94), over the pass, connects Bern and Valais cantons.

Grin·nell (grĭ-nĕl'), city (1970 pop. 8,402), Poweshiek co., central Iowa, NE of Des Moines; founded 1854 by Josiah Bushnell Grinnell, inc. as a town in 1865, as a city in 1882. It manufactures gloves and farm implements, and raises corn, dairy cattle, and poultry.

Gri·sons (grē-sônz', grē-zôN'), canton (1977 est. pop. 162,400), 2,746 sq mi (7,112.1 sq km), E Switzerland, bordering on Italy and Austria; capital Chur. In a region of Alpine peaks and glaciers, forested highlands, and fertile valleys, it has a large tourist industry. A part of Rhaetia under the Roman Empire, the territory preserved Roman laws and customs, although it nominally passed to the Ostrogoths (493) and to the Franks (537). In 1799 Grisons was forced by the French to enter the Helvetic Republic, and in 1803 it became a Swiss canton.

Gris·wold (grĭz'wôld, -wəld), town (1970 pop. 7,763), New London co., SE Conn., bordered on the W by the Quinebaug River in an agriculture and dairy area; settled c.1660, inc. 1815.

Grod·no (grôd'nô), city (1976 est. pop. 176,000), capital of Grodno oblast, W European USSR, in Belorussia, on the Neman River. A river port and an important railway center, it has industries producing machinery, electrical products, textiles, and processed food and tobacco. Dating back to the 10th cent., Grodno was the capital of an independent principality until 1398, when it was included in the grand duchy of Lithuania. It became the second capital of Lithuania and passed to Poland after the union of Lithuania with Poland in 1569. In 1673 it became a seat of Polish diets, the last of which (1793) was forced to consent to the second partition of Poland. Grodno passed to Russia in 1795. It was transferred to Poland in 1920 and was incorporated into the Belorussian Republic in 1939.

Groes·beck (grōs'bĕk'), city (1970 pop. 2,396), seat of Limestone co., E central Texas, E of Waco. It is a market town in a region yielding cotton and other crops, dairy products, clay, timber, and oil. Fort Parker State Park, with a restored stockade, is nearby.

Gro·ning·en (grō'nĭng-ən, KHrō'-), province (1977 est. pop. 544,264), c.900 sq mi (2,330 sq km), NE Netherlands, bordering on West Germany in the E and the North Sea in the N; capital Groningen. The province has a largely agricultural economy. Vast reserves of natural gas were discovered here in 1961.

Groningen, city (1977 est. pop. 161,825), capital of Groningen prov., NE Netherlands. It is an important trade and transportation center. Manufactures include clothing, food products, furniture, and machinery. In the 11th cent. Groningen came under the temporal power of the bishops of Utrecht. It soon rose to prominence and in the 12th cent. supplied ships for the Crusades. In 1284 it joined the Hanseatic League. The city remained loyal to the Hapsburgs at the beginning of the revolt of the Netherlands against Spain, but was captured by the Dutch under Maurice of Nassau in 1594. A picturesque city, Groningen has several fine churches.

Groote Ey·landt (grōōt ī'lənd), 950 sq mi (2,460.5 sq km), Northern Territory, N Australia. It is the largest island in the Gulf of Carpentaria. Manganese ore is mined.

Groo·te Schuur (KHrōō'tə skür), estate, Cape Town, Cape Prov., SW South Africa. The main building of the estate was erected on the site of a large barn dating from 1657. It once was the home of Cecil Rhodes and is now the residence of South Africa's prime minister. The Univ. of Cape Town is on the grounds of the estate; in 1967 Dr. Christiaan Barnard performed the world's first human heart transplant operation here.

Grosse Pointe (grōs' point'), residential city (1970 pop. 6,637), Wayne co., SE Mich., a suburb NE of Detroit, on Lake St. Clair; inc. as a village 1879, as a city 1934. It is a port of entry.

Grosse Pointe Farms (färmz), city (1970 pop. 11,701), Wayne co., SE Mich., a residential suburb of Detroit, on Lake St. Clair; inc. 1893. The mansion of John Dodge and the Alger House, a branch of the Detroit Institute of Arts, are of interest.

Grosse Pointe Park, village (1970 pop. 15,641), Wayne co., SE Mich., a residential suburb of Detroit, on Lake St. Clair; inc. 1907.

Grosse Pointe Woods (wōōdz), village (1978 est. pop. 19,600), Wayne co., SE Mich., a residential suburb of Detroit; inc. 1926.

Gros·se·to (grōs-sā'tō), city (1976 est. pop. 60,300), central Italy, capital of Grosseto prov., Tuscany region, on the Ombrone River near the Tyrrhenian Sea. Situated in the reclaimed Maremma area, it is an agricultural market. Nearby are the ruins of Russellae, an Etruscan town deserted in the 12th cent.

Gross·glock·ner (grōs-glôk'nər), peak, 12,461 ft (3,800.6 m) high, in Tyrol, S Austria, the highest point in the Hohe Tauern range and in Austria. It is traversed by the Grossglocknerstrasse (built 1930–35), a magnificent Alpine highway rising up to 7,770 ft (2,370 m).

Grot·on (grŏt'n). **1.** Town (1970 pop. 38,244), New London co., SE Conn., on the Thames River opposite New London; settled c.1650, inc. 1705. Shipbuilding, the manufacture of chemicals and metal products, and the distribution of fuel oil are among the town's principal industries. A U.S. navy submarine base is here. Groton is the site of Fort Griswold (1775), unsuccessfully defended against the British in 1781. Of interest are guided tours of the submarine base, an annual art festival, and a number of well-maintained colonial homes. **2.** Town (1975 est. pop. 5,497), Middlesex co., N Mass., NW of Boston, in a lumber, dairy, poultry, and apple-orchard area; settled and inc. 1655. It was destroyed in King Philip's War and later rebuilt. Groton School, a preparatory academy, is here.

Grove City (grōv). **1.** Village (1970 pop. 13,911), Franklin co., central Ohio, in a rich agricultural region. It has some manufacturing. **2.** Industrial borough (1970 pop. 8,312), Mercer co., W central Pa., NNW of Pittsburgh, in a coal-mining area; settled 1798, laid out 1844, inc. 1883. It has limestone deposits and manufactures diesel engines.

Grove Hill, town (1970 pop. 1,825), seat of Clarke co., SW Ala., SW of Selma; settled c.1813, inc. 1832. Lumber products are made here.

Grove·land (grōv'lənd), residential town (1975 est. pop. 5,253), Essex co., NE Mass., on the Merrimack River below Haverhill; settled c.1639, set off from Bradford 1850.

Gro·ver City (grō'vər), city (1970 pop. 5,939), San Luis Obispo co., SW Calif., S of San Luis Obispo, in a farm area; inc. 1959.

Groves (grōvz), city (1970 pop. 18,067), Jefferson co., SE Texas, a residential suburb of Port Arthur.

Grove·ton (grōv'tən), city (1970 pop. 1,219), seat of Trinity co., E Texas, SW of Lufkin. It is the market center for a farm, cattle, and timber region.

Groz·ny (grôz'nĭ), city (1976 est. pop. 381,000), capital of the Chechen-Ingush Autonomous SSR, SE European USSR, in the N foothills of the Greater Caucasus. It is the center of one of the USSR's richest oil fields (production began in 1893).

Gru·dziądz (grōō'jôNts), city (1975 est. pop. 82,200), N central Poland, a port on the Vistula River. Industries include flour milling, brewing, distilling, and light manufacturing. Founded and fortified by the Teutonic Knights, it was chartered in 1233. Grudziądz passed to Poland in 1466 and to Prussia in 1772; it was restored to Poland in 1919.

Grun·dy (grŭn'dē). **1.** County (1970 pop. 26,535), 432 sq mi (1,118.9 sq km), NE Ill., drained by the Illinois, Des Plaines, and Kankakee rivers; formed 1841; co. seat Morris. It is in an agricultural area (corn, oats, soybeans, wheat, livestock, poultry, and dairy products) with bituminous-coal, clay, and limestone deposits. Its manufactures include frozen fruit and vegetables, clay products, furniture, paper, chemicals, aluminum, and machinery. **2.** County (1970 pop. 14,119), 501 sq mi (1,297.6 sq km), central Iowa, in a rolling prairie agricultural area drained by Wolf Creek; formed 1851; co. seat Grundy Center. Hogs, cattle, corn, soybeans, and oats are raised. **3.** County (1970 pop. 11,819), 435 sq mi (1,126.6 sq km), N Mo., drained by the Thompson and Weldon rivers; formed 1841; co. seat Trenton. There is coal mining in the county. Its agriculture includes grain and livestock, and it has a meat-processing industry. **4.** County (1970 pop. 10,631), 358 sq mi (927.2 sq km), S central Tenn., in the Cumberlands, drained by the Elk River; formed 1844; co. seat Altamont. In a mining and timber area, it has some agriculture (cattle, hogs, and row crops) and manufacturing (clothing and wood products).

Grundy, town (1970 pop. 2,054), seat of Buchanan co., extreme SW Va., near the Ky. and W.Va. borders NNE of Bristol. It has coal mines and sawmills.

Grundy Cen·ter (sĕn'tər), city (1970 pop. 2,712), seat of Grundy co., central Iowa, SW of Waterloo; inc. 1877. It is a trade center for a rich agricultural area.

Gru·yère (grōō-yâr', grĭ-), district in Fribourg canton, W Switzerland. It is famous for its cattle and for Gruyère cheese.

Gstaad (gə-shtäd'), village, Bern canton, W Switzerland. It is a fashionable winter-sports resort. Cheese and wood products are made in the village.

Gua·da·la·ja·ra (gwä'də-lə-här'ə, gwä'*th*ə-lä-hä'rä), city (1974 est. pop. 1,478,400), capital of Jalisco state, SW Mexico. Guadalajara is a beautiful, spacious city on a plain more than 5,000 ft (1,525 m) high and surrounded by mountains. It is a modern commercial metropolis with many picturesque survivals of the Spanish colonial era. The mild, clear, dry climate has made it a popular health resort. Guadalajara is also an important communications and industrial center. Auto assembly plants and the manufacture of xerographic and photographic equipment are among the leading industries. The region around the city is important for agriculture and livestock raising; some coal is also mined. The most famous products of Guadalajara and its environs are intricately designed and finely worked glassware and pottery.

Founded c.1530, Guadalajara was moved twice, before and during the Mixtón War, because of Indian raids; it was permanently established in 1542. Guadalajara became the seat of the audiencia of Nueva Galicia. Captured in 1810 by Hidalgo y Costilla during the war against Spain, the city was the center of reform activities. Again in 1858, in the War of Reform, it was briefly occupied by the liberals under Benito Juárez.

Guadalajara, town (1975 est. pop. 45,060), capital of Guadalajara prov., central Spain, in New Castile, on the Henares River. It flourished as a Roman colony and belonged to the Moors from the 8th to the 11th cent.

Gua·dal·ca·nal (gwä'dəl-kə-năl'), volcanic island (1970 pop. 23,922), c.2,510 sq mi (6,500 sq km), South Pacific, largest of the Solomon Islands. The island is largely jungle. There are coconut plantations, and some gold has been mined. Discovered by English navigators in 1788, Guadalcanal became a British protectorate in 1893. During World War II the island was occupied by the Japanese. In Aug., 1942, U.S. forces landed, marking the first large-scale invasion of a Japanese-held island; after bitter jungle fighting, the island was conquered (Feb., 1943).

Gua·dal·qui·vir (gwä'dəl-kē-vîr', -*th*äl-), river, c.350 mi (565 km) long, rising in the Sierra de Cazorla, SE Spain, and flowing generally SW past Córdoba and Seville into the Atlantic Ocean near Sanlúcar de Barrameda. There are several hydroelectric plants along its course. In its middle course it flows through a populous fertile region at the foot of the Sierra Morena, where it is used extensively for irrigation. The lower course of the Guadalquivir traverses extensive marshlands that are used for rice cultivation.

Gua·da·lu·pe (gwä'də-lōōp', -lōō'pā, -*thä*-), town (1976 est. pop. 29,100), central Costa Rica. It is an agricultural center and a suburb of San José.

Guadalupe, city (1970 pop. 51,899), Nuevo León state, NE Mexico, on the Santa Catalina River. Its economy is based on agriculture, especially maize, and livestock raising.

Gua·da·lupe (gwä'də-lōōp', -lōō'pā). **1.** County (1970 pop. 4,969), 2,998 sq mi (7,764.8 sq km), E central N.Mex., watered by the Pecos River; formed 1891; co. seat Santa Rosa. It is in a ranching and grain-farming area, with some manufacturing and tourism. **2.** County (1970 pop. 33,554), 715 sq mi (1,851.9 sq km), S central Texas, drained by the San Marcos and Guadalupe rivers; formed 1846; co. seat Seguin. Its agriculture includes cotton, corn, peanuts, grain sorghums, maize, wheat, pecans, vegetables, fruit, livestock, and dairy products. It also has deposits of oil and natural gas, clay pits, some manufacturing, and a food-processing industry. The county has lake resorts and good hunting and fishing.

Guadalupe, river, SE Texas, rising on the Edwards Plateau and flowing c.300 mi (485 km) SE to join the San Antonio River near its mouth in San Antonio Bay.

Guadalupe Mountains National Park, 81,077 acres (32,836 hectares), W Texas; est. 1966. Located in the Guadalupe Mts., the park contains parts of the world's largest and most significant Permian limestone fossil reef.

Gua·dar·ra·ma, Si·er·ra de (sē-ĕr'ə dā gwä'*thä*-rä'mä), mountain range rising from the plateau of central Spain, N of Madrid, and extending c.120 mi (195 km) between the Tagus and Douro rivers. The rugged mountains rise to Peñalara (7,973ft/2,431.8 m), the highest peak. The range is crossed by several passes that link Madrid with Segovia. The passes are usually blocked by heavy winter snowfalls. Largely forested, the mountains yield fine timber. There are skiing resorts on the slopes.

Gua·de·loupe (gwä-də-lōōp'), overseas department of France (1970 est. pop. 327,000), 687 sq mi (1,779.3 sq km), in the Leeward Islands, West Indies. The department comprises the islands of Basse-Terre and Grande-Terre, and the dependencies of Marie-Galante and Les Saintes to the south, Désiderade to the east, and St. Barthélemy and the northern half of Saint Martin to the north. Basse-Terre, on the island of the same name, is the capital. Tourism is a major industry. Basse-Terre, volcanic in origin and extremely rugged, is settled along the coasts and produces bananas, coffee, cacao, and vanilla beans. Grande-Terre has low limestone cliffs and little rainfall; sugar and rum are its chief products. Subsistence farming, livestock raising, and fishing are carried on, and some salt and sulfur are mined.

Discovered by Christopher Columbus in 1493, Guadeloupe was abandoned by Spain in 1604. In 1635 settlement was begun by the French, who imported slaves from Africa for plantation work. By the end of the 17th cent. Guadeloupe was a leading world sugar producer and one of France's most valuable colonies. The islands were hotly contested with the English until they were confirmed as French possessions in 1815. During World War II Guadeloupe at first adhered to the Vichy regime in France, but an accord with the United States in 1942 led to its support of the Free French.

Gua·dia·na (gwä-dyä'nä, -*thy*ä'-), river, 510 mi (820.6 km) long, rising in the La Mancha Plateau, E Spain, and flowing W through central Spain, then S, forming part of the Spanish-Portuguese border, to the Gulf of Cádiz in the Atlantic Ocean. The Guadiana is used to irrigate the fertile Mérida region of Spain and to generate hydroelectric power.

Gua·dix (gwä-dēsh', -*th*ēsh'), town (1970 pop. 15,311), Granada prov., S Spain, in Andalusia. It is the center of a farm area growing olives, flax, wheat, and hemp. Guadix was a Roman colony. Just outside the city are many picturesque caves inhabited by gypsies.

Guam (gwäm), island (1970 pop. 84,996), 209 sq mi (541.3 sq km), W Pacific, an unincorporated territory of the United States; the largest, most populous, and southernmost of the Marianas Islands. Agana is the seat of government, and Apra Harbor, a large U.S. naval base, is nearby. The interior of the island is dense jungle; most of the native villages are on the coast. Discovered in 1521 by Ferdinand Magellan, Guam belonged to Spain until 1898, when it was taken by the United States in the Spanish-American War. From 1917 to 1950, Guam, under the Dept. of the Navy, was governed by a naval officer who was advised by a local congress. The Organic Act of 1950 transferred jurisdiction to the Dept. of the Interior and provided for a governor, appointed every four years by the President of the United States, and a 21-member unicameral legislature elected biennally by the residents. Guamanians are U.S. citizens but cannot vote in U.S. elections. Guam was attacked and captured by Japan in 1941, was retaken by U.S. forces in 1944, and became a major base for assaults on the Japanese mainland. During the Vietnam War in the 1960s Guam was an important base for air assaults on Vietnam and Laos.

Gua·na·ba·co·a (gwä'nä-bä-kō'ä), city (1970 pop. 69,706), La Habana prov., W Cuba, a commercial suburb of Havana. Numerous mineral springs are located near Guanabacoa. The city was founded in 1555 on the site of an Indian settlement.

Gua·na·ba·ra (gwä'nä-bä'rä), state (1970 pop. 4,252,009), 524 sq mi (1,357.2 sq km), SE Brazil, on the Atlantic Ocean; capital Rio de Janeiro. Guanabara Bay, a deep inlet of the Atlantic Ocean, is noted for its beauty.

Gua·na·jua·to (gwä'nä-hwä'tô), city (1970 pop. 36,809), capital of Guanajuato state, W central Mexico. The city, with an altitude of c.6,600 ft (2,015 m), is situated in the Cañada de Marfil, a precipitous ravine encircled by barren hills. Guanajuato has narrow, winding, steep cobblestone streets, sometimes pieced out by stone steps, and the ground underneath is honeycombed with silver-mine shafts. Its geographic position and economic importance as one of Spanish America's chief silver-producing centers gave the city a key role in the wars and revolutions of the 19th and early 20th cent. Guanajuato has recently become a resort city.

Guan·tá·na·mo (gwä-tä'nä-mō), city (1970 pop. 130,100), Oriente prov., SE Cuba, on the Guaso River. It is the processing center for a rich sugar- and coffee-producing region. Founded in the early 19th cent. by Frenchmen fleeing the slave rebellion in Haiti, Guantánamo retains many vestiges of French architecture.

Gua·po·ré (gwä-pōō-rĕ'), river, c.750 mi (1,205 km) long, rising in the mountains of Mato Grosso state, W Brazil. It flows northwest through rain forests, and forms part of the Brazil-Bolivia border before joining the Mamoré River.

Guar·da (gwär'də), city (1970 municipal pop. 40,529), capital of Guarda dist., N central Portugal, in Beira Alta. On the slopes of the Serra da Estrela, it is Portugal's highest city (c.3,400 ft/1,040 m) and is a winter-sports resort. Guarda is also the commercial center of an agricultural region. It has an old fort, a castle, and a Gothic cathedral.

Guá·ri·co (gwä'rē-kō), river, W Venezuela, rising near Valencia and flowing 300 mi (482.7 km) S to join the Apure River. It is navigable for small craft.

Gua·sa·ve (gwä-sä'vā), city (1970 pop. 26,080), Sinaloa state, W Mexico, on the Sinaloa River. The growing of cotton and maize and the raising of livestock are the chief occupations. The city was established in 1595 as a Spanish mission among the Guasave Indians.

Gua·stal·la (gwä-stäl'lä), town (1971 pop. 14,229), Emilia-Romagna, N Italy, on the Po River. It is an agricultural and industrial center. Probably founded in the 7th cent., Guastalla was held by various lords and in 1539 was bought by Ferrante Gonzaga of Mantua.

Gua·te·ma·la (gwä'tə-mä'lə), republic (1970 est. pop. 5,200,000), 42,042 sq mi (108,889 sq km), Central America, bounded on the N and W by Mexico, on the E by Belize and the Caribbean Sea, on the SE by Honduras and El Salvador, and on the S by the Pacific Ocean. The capital is Guatemala City. A highland region, where most of the population lives, cuts across the country from west to east. The rugged main range is flanked on the Pacific side by a string of volcanoes (some active). Volcanic eruptions and floods have plagued Guatemala throughout history. In the center of the range is Lake Atitlán, and south of the highlands is the Pacific coastal lowland. North of them are the Caribbean lowland and the vast tropical forest known as the Petén.

Economy. Coffee, cotton, and bananas are the leading commercial and export crops. There are small manufacturing industries that produce consumer goods. Zinc and lead concentrates are mined and in the north are nickel and petroleum deposits. The leading imports are petroleum, textiles, flour, and machinery.

History. The Maya-Quiché Indians were defeated (1523–24) by the Spaniard Pedro de Alvarado, who became captain general of Guatemala. The conquerors found little of the gold they sought, but cocoa and indigo were raised.

Central America became independent from Spain in 1821. Guatemala was first a part of the Mexican empire of Agustín de Iturbide and then became a nucleus of the Central American Federation. After the federation collapsed, Guatemala became a separate nation (1839). Guatemalan interference in the affairs of other Central American republics during the 19th and early 20th cent. finally led to the Washington Conference of 1907, which established the Central American Court of Justice. The boundary between Guatemala and British Honduras has remained in dispute since 1859; since 1945 Guatemala has actively sought to regain control of British Honduras, over which it claims sovereignty.

After Guatemala declared war on the Axis powers in 1941, the large German-owned coffee holdings were expropriated. In 1944 Juan José Arévalo launched a series of labor and agrarian reforms that were continued by Jacobo Arbenz Guzmán, who succeeded him in 1951. A law expropriating large estates angered foreign plantation owners, particularly the United Fruit Company. As Communist influence in the Arbenz government increased, relations with the United States deteriorated. In 1954 the United States aided the anti-Arbenz military force that placed Col. Carlos Castillo Armas in power. He was assassinated three years later.

In 1963 the prospect of the return to power of Arévalo led to a military coup. However, leftist terrorism mounted and in turn provoked rightist terrorism. In Aug., 1968, the U.S. ambassador was assassinated. In the election of 1970, Col. Carlos Arana Osorio, an extreme conservative, was chosen president. He imposed a one-year state of siege in an attempt to end the violence. In the early 1970s many labor and political leaders were killed and several foreign diplomats were kidnapped.

Government. Guatemala's current constitution was adopted in 1965. It provides for a president, directly elected for a four-year term, who cannot succeed himself. A unicameral legislature is also elected for a four-year term.

Guatemala, city (1971 est. pop. 730,991), S central Guatemala, capital of the republic. Its full name is Santiago de los Caballeros de Guatemala la Nueva. In a broad, fertile, highland valley, c.5,000 ft (1,525 m) high, it enjoys an equable climate the year round. It is the largest city in Central America, with a cosmopolitan atmosphere and many fine public buildings. It is the industrial and commercial center of the republic. To the city's markets come the fruits and vegetables of the tropical coasts and temperate highlands and also native handicrafts, especially textiles. The present city is the third permanent capital of Guatemala and was founded in 1776 after Antigua was abandoned. An earthquake destroyed Guatemala City in 1917–18, but it was rebuilt on the same site. Many interesting remains of Mayan civilization have been unearthed in the vicinity.

Gua·ya·ma (gwä-yä'mä), town (1970 pop. 20,318), SE Puerto Rico. It is the processing and distribution center for a region producing sugar cane, tobacco, coffee, and livestock and is also the headquarters of a major irrigation and electrification project.

Guay·a·quil (gwä'yä-kēl'), city (1974 pop. 814,064), capital of Guayas prov., W Ecuador, on the Guayas River near its mouth on the Gulf of Guayaquil, an inlet of the Pacific Ocean. The chief port of Ecuador, Guayaquil has industries manufacturing textiles, leather goods, cement, alcohol, soap, and iron products. Guayaquil was founded in 1535. It was often subjected to attacks by buccaneers in the 17th cent. and in the 18th and 19th cent. was destroyed repeatedly by fires. The occupation of the city in 1821 by patriot forces under Antonio José de Sucre was the first major step in Ecuador's final liberation from Spain. Guayaquil has several colonial landmarks and a university.

Guay·mas (gwī'mäs), city (1970 pop. 57,492), Sonora state, NW Mexico, on the bay of Guaymas. A port on the Gulf of California, Guaymas stands on a scenic inlet encircled by desert mountains. Its fine beaches, excellent deep-sea fishing, and good transportation facilities have made it a popular tourist resort. Although the surrounding area was explored as early as 1539, the city was not established until the early 18th cent. by Jesuit missionaries. U.S. forces occupied Guaymas in 1846, during the Mexican War, and it was held by the French in 1865–66.

Gu·de·nå (go͞o'*th*ə-nô'), river, 98 mi (157.7 km) long, E Jutland, Denmark. The only Danish river of importance, it flows generally north, traversing several lakes, to Randersfjørd and the Kattegat. It is partly navigable and has salmon fisheries.

Guel·der·land (gĕl'dər-lănd', KHĕl'dər-länt'): *see* Gelderland.

Guelph (gwĕlf), city (1976 pop. 67,538), S Ont., Canada, on the Speed River. It is an industrial city located in a rich farm area. Manufactures include electrical, construction, and farm equipment, textiles, clothing, fiberglass, and tobacco products.

Gué·ret (gā-rĕ'), town (1975 pop. 14,855), capital of Creuse dept., central France. It is a market center and an industrial town. Metals (especially aluminum), shirts, jewelry, and handicrafts are the principal manufactures.

Guer·ni·ca (gwâr'nĭ-kə, gĕr-nē'kä), historic town (1970 pop. 9,977), Vizcaya prov., N Spain, in the Basque Province. It has metallurgical, furniture, and food manufactures. In Apr., 1937, German planes, aiding the insurgents in the Spanish Civil War, bombed and destroyed Guernica. The bombing of Guernica aroused world opinion and became a symbol of fascist brutality. The event inspired one of Picasso's most celebrated paintings.

Guern·sey (gûrn'zē), county (1970 pop. 37,665), 529 sq mi (1,370.1 sq km), E Ohio, drained by Wills Creek; formed 1810; co. seat Cambridge. Its agriculture includes livestock, dairy products, fruit, and grain. It has bituminous-coal mines, clay pits, oil and natural-gas wells, limestone quarries, and stands of timber. Among its manufactures are furniture, plastics, and clay and metal products.

Guernsey, island (1971 pop. 51,458), 25 sq mi (64.8 sq km), in the English Channel, second-largest of the Channel Islands. Guernsey has a low beach in the north and bold rocky cliffs along the south shore. It is a tourist site.

Gui·an·a (gē-ăn'ə, -ä'nə), region, NE South America, facing the Atlantic Ocean on the N and E and enclosed on the W and S within a vast semicircle formed by the linked river systems of the Orinoco, the Rio Negro, and the lower Amazon. The region consists of a cultivated coastal plain, where most of the population lives, and a forested, hilly interior, the Guiana Highlands. The Guiana coast was sighted (1498) by Columbus. The legend of El Dorado drew Sir Walter Raleigh to the region in 1595. The Spanish had also come in search of easy wealth, but, finding none, they left the coast open to exploitation by the Dutch, English, and French. The Dutch were the first to settle, but ownership of territory changed hands many times.

Guiana High·lands (hī'ləndz), mountainous tableland, c.1,200 mi (1,930 km) long and from 200 to 600 mi (321.8–965.4 km) wide, N South America, bounded by the Orinoco and Amazon river basins, and by the coastal lowlands of the Guianas. Geologically, the Guiana Highlands is a shield—a stable mass of Precambrian rock—and is related to the Brazilian Highlands. It consists of vast plateaus of ancient crystalline rocks overlaid by recent sandstone and lava caps. Numerous rivers, fed by heavy rainfall, rise in the highlands and pour over the edges to create deep gorges and magnificent waterfalls. The sparsely populated region is famous for its semideciduous tropical rain forests and rich fauna, including many varieties of brilliantly colored tropical birds. The crystalline rocks of the Guiana Highlands yield gold and diamonds. Large deposits of iron ore, manganese, and bauxite have been made accessible by new roads and railroads, but the enormous potential wealth of the highlands is still largely untapped because of the dense cover of vegetation.

Gui·enne (gē-yĕn'), region of SW France. Guienne as it existed from the time of Henry IV (late 16th–early 17th cent.) to the French

Revolution included part of the Aquitaine basin and part of the Massif Central. It was synonymous with Aquitaine until the Hundred Years' War (1337–1453) and passed to England through the marriage (1152) of Eleanor of Aquitaine to Henry II. In 1453 Guienne was reconquered by France. From the 17th cent. to 1792 it formed part of the vast province of Guienne and Gascony under the jurisdiction of the Parlement of Bordeaux.

Guild·ford (gĭl'fərd), municipal borough (1973 est. pop. 58,470), SE England, on the Wey River. It is a market town and produces knitwear, plastics, and engineered goods.

Guild·hall (gĭld'hôl'), town (1970 pop. 169), seat of Essex co., NE Vt., on the Connecticut River NE of St. Johnsbury; settled 1764. Its town hall dates from 1795.

Guil·ford (gĭl'fərd), county (1970 pop. 288,645), 651 sq mi (1,686.1 sq km), N central N.C., in a piedmont region drained by the Haw and Deep rivers; formed 1770; co. seat Greensboro. Its agriculture includes tobacco, corn, poultry, and dairy products. There is also pine and oak timber. Sawmilling is done.

Guilford (gĭl'fərd), town (1970 pop. 12,033), New Haven co., S Conn., on Long Island Sound; founded 1639. It has fishing and poultry-raising industries. Metal castings and boats are made. Some of the oldest houses in the state are here; the stone Whitfield House (1639–40) was restored in 1936 and is now a state historical museum.

Guilford Court·house (kôrt'hous'), locality, N central N.C., near Greensboro. In the battle of Guilford Courthouse (Mar. 15, 1781), American troops commanded by Nathanael Greene defeated the British led by Gen. Cornwallis. The large British losses compelled Cornwallis to leave the Carolinas.

Gui·ma·rães (gē-mä-ränsh'), city (1975 est. pop. 24,280), Braga dist., NW Portugal, in Minho. It has textile and cutlery manufactures. The town was the seat of Henry of Burgundy and of his son, Alfonso I, first king of Portugal. Alfonso VII of León besieged (1127) Guimarães and forced Alfonso of Portugal to swear fealty, but the Portuguese later established his independence.

Guin·ea (gĭn'ē), an archaic term for the W coast of Africa. In its widest sense it has been applied to the region from Angola to Senegal. Parts of the region bore names originating in early colonial trade, notably Grain Coast, Ivory Coast, Gold Coast, and Slave Coast. Characteristic of the coast are dense tropical forests, heavy rainfall, and a hot, humid climate.

Guinea, republic (1977 est. pop. 4,585,000), 94,925 sq mi (245,856 sq km), W Africa, bounded on the N by Guinea-Bissau, Senegal, and Mali, on the E by the Ivory Coast, on the S by Sierra Leone and Liberia, and on the W by the Atlantic Ocean. Conakry is the capital. A humid and tropical country, Guinea comprises an alluvial coastal plain, the mountainous Fouta Jallon region, a savanna interior, and the forested Guinea Highlands, which rise to c.5,800 ft (1,770 m) in the Nimba Mts.

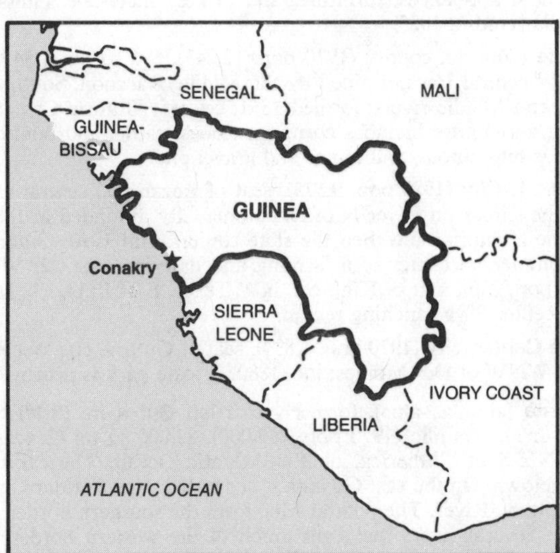

Economy. Predominantly agricultural, Guinea produces rice, millet, manioc, cassava, coffee, bananas, palm kernels, and citrus fruits. Stock raising is important in the highlands. Some of the world's largest bauxite deposits lie in Guinea, and iron ore, gold, and diamonds are also mined. Alumina, made from bauxite, is a leading export; other exports include iron ore and a variety of agricultural products. Guinea has some light industry, but inadequate transportation facilities have hampered industrialization.

History. In the early 18th cent. a Fulani feudal state was established in the Fouta Jallon region. European exploration of the Guinean coast began with the Portuguese in the mid-15th cent.; by the 17th cent. European traders were competing for slaves and by the 19th cent. for palm oil, peanuts, and other products. After a series of wars and agreements with tribal chiefs, France took control of much of Guinea. In 1891 it was constituted as a French colony. Guinean resistance to French rule was not quelled until 1898, however, and sporadic revolts continued into the 20th cent. Little economic development occurred under the colonial regime until just before World War II. The parallel growth of a radical labor movement led to the rise of Sékou Touré, a union leader who also headed the Parti Démocratique de Guinée (PDG). Under his leadership Guinea became the only colony to vote against the constitution of the French Community in 1958 and to opt for complete independence. Touré advocated African unity and steered the country into a union with Ghana and Mali. In the late 1960s Guinea sought improved relations with the West, although its basic international posture has remained one of nonalignment. Touré has continued to foster Pan-Africanism. In 1970 the country was invaded from Guinea-Bissau (then Portuguese Guinea) by a small force that included Guinean exiles opposed to Touré. The invasion was unsuccessful, and several political trials and executions followed. Guinea actively supported the independence movement in Guinea-Bissau, and Conakry was the movement's headquarters. In 1973 Guinea took greater control of the foreign-owned bauxite industry.

Guinea, Gulf of, large open arm of the Atlantic Ocean formed by the great bend of the coast of W Africa. It extends from the western coast of the Ivory Coast to the Gabon estuary and is bounded on the south by the equator.

Guin·ea-Bis·sau (gĭn'ē-bĭ-sou'), independent country (1977 pop. 535,000), 13,948 sq mi (36,125.3 sq km), W Africa, bordering on the Atlantic Ocean in the W, on Senegal in the N, and on the Republic of Guinea in the E and S. The country includes the nearby Bijagós Archipelago and other islands in the Atlantic. The capital and chief city is Bissau. The country is largely a low-lying coastal plain and has many rivers, some with wide swampy estuaries.

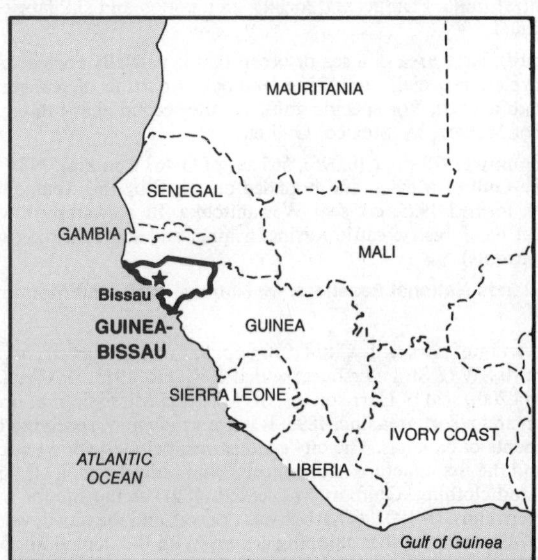

Economy. Farming is by far the leading occupation; rice, palm oil, groundnuts, and coconuts are the main products. Bauxite is mined.

History. The area that became Portuguese Guinea was first visited by the Portuguese in 1446–47, and in the 16th cent. it was an important source of slaves sent to South America. In 1951 it became an overseas province of Portugal. In 1956 the African Party for the Independence of Guinea and Cape Verde (PAIGC) was founded; it was led by Amilcar Cabral until his assassination in 1973. The PAIGC launched a war of independence in Portuguese Guinea in the early 1960s. By 1972 Portugal had about 35,000 troops in the province. In Sept., 1973, the PAIGC declared Guinea-Bissau to be independent of Portugal. Following the coup in Portugal (Apr., 1974), a new Portuguese government initiated negotiations with the PAIGC concerning the political future of the province. The PAIGC contended that the Cape Verde Islands were part of Guinea-Bissau, but Portugal would agree only to hold a future referendum on the status of the islands. Independence was finally achieved on Sept. 10, 1974, and Luis de Almeida Cabral, the PAIGC leader, became president. In the year following independence, some 100,000 refugees of the war with Portugal returned to Guinea-Bissau, thus placing great strains on the country's economy. In 1976, as a step toward eventual union with the Cape Verde Islands, the two countries established a uniform penal and court system.

Güi·nes (gwē'nās), city (1970 pop. 41,407), La Habana prov., W Cuba. It is located in one of the island's most heavily farmed areas. Güines was founded in 1737 as the commercial and financial center of the rich surrounding farm region.

Gu·ja·rat (go͝o'jə-rät'), state (1971 pop. 26,687,186), c.72,000 sq mi (186,480 sq km), W India, on the Arabian Sea and including almost all of the Kathiawar peninsula. Gujarat was constituted in 1960 from the Gujarati-speaking areas in the northern and western portions of the former state of Bombay. The state is the center of the Indian cotton-textile industry. It is a fertile, well-watered region; rice, wheat, and cotton are grown. Salt, limestone, manganese, calcite, and bauxite are mined. Archaeological discoveries have linked Gujarat with the Indus valley civilization (c.3000–1500 B.C.). In 1401 Gujarat became an independent sultanate, but in 1572 was annexed to the Mogul empire. Under the British the region retained its local princely rulers. In 1947 the region was organized into the state of Bombay.

Guj·ran·wal·la (go͝oj'rən-wä'lə), city (1972 est. pop. 360,419), NE Pakistan. It is a commercial center trading in wheat, rice, sugar, oilseed, and oranges. There are varied manufactures.

Guj·rat (go͝oj-rät'), town (1972 pop. 100,581), E Pakistan. It is noted for its furniture, brassware, pottery, and cotton goods. Boots, shawls, and electric fans are also produced. Standing on the site of a fort built by the Mogul emperor Akbar in 1580, Gujrat was the location of the final battle between the British and the Sikhs in 1849.

Gul·bar·ga (go͝ol'bər-gä), town (1971 pop. 145,630), Karnataka state, S central India. Peanuts and locally spun cotton and silk fabrics are marketed.

gulf (gŭlf), large area of a sea or ocean that is partially enclosed. The term refers especially to a long landlocked portion of sea opening through a strait. For specific gulfs, see the second element; e.g., for Gulf of Mexico, *see* Mexico, Gulf of.

Gulf, county (1970 pop. 10,096), 565 sq mi (1,463.4 sq km), NW Fla., on the Gulf of Mexico and bounded on the E by the Apalachicola River; formed 1925; co. seat Wewahitchka. In a swampy lowland area, it has forestry, cattle raising, fishing, and some farming (corn and peanuts).

Gulf Islands National Seashore: *see* National Parks and Monuments Table.

Gulf·port (gŭlf'pôrt'). **1.** City (1970 pop. 9,976), Pinellas co., W Fla., a suburb SW of St. Petersburg; settled 1843, inc. 1913. **2.** City (1970 pop. 44,200), seat of Harrison co., SE Miss., on Mississippi Sound, in a farm and resort area; inc. 1898. It is a port of entry, receiving large shipments of bananas. The city's industries include seafood packaging and the manufacture of chemicals, pharmaceuticals, metal products, and clothing. Gulfport was settled (1891) as the site for a railroad terminus. In 1902 its harbor was opened, and the city developed as an important lumber-shipping center. With the depletion of timber resources, Gulfport extended its shipping facilities and turned to manufacturing and to a growing tourist trade.

Gulf Stream, warm ocean current of the N Atlantic Ocean, off E North America. It was first described (1513) by Ponce de León, the Spanish explorer. The Gulf Stream originates in the Gulf of Mexico and, as the Florida Current, passes through the Straits of Florida and along the coast of southeastern United States with a breadth of c.50 mi (80 km). Off Cape Hatteras it is separated from the coast by a narrow southern extension of the cold Labrador Current and flows northeast into the Atlantic Ocean, where it spreads out and merges with the North Atlantic Drift. It is a common error to confuse the Gulf Stream with the North Atlantic Drift, which is responsible for the warm climate of western Europe.

Gunn·bjørn (go͝on'byôrn'), highest peak, 12,139 ft (3,702.4 m), Greenland, near the SE coast. Glaciers reach from here into the Denmark Strait.

Gun·ni·son (gŭn'ĭ-sən), county (1970 pop. 7,578), 3,220 sq mi (8,339.8 sq km), W central Colo., in a livestock-grazing and coal-mining region drained by the Gunnison River; formed 1877; co. seat Gunnison. Taylor Park reservoir and dam irrigate the county in the east. Ranges of the Rocky Mts. and parts of Grand Mesa, Gunnison, San Isabel, White River, and Uncompahgre national forests are here.

Gunnison, town (1970 pop. 4,613), seat of Gunnison co.; W central Colo., on the Gunnison River in a farm, mine, and resort area; laid out 1879 as a silver-mining town, inc. 1880.

Gunnison, river, 180 mi (289.6 km) long, rising in W central Colo. and flowing SW, W, and NW to the Colorado River at Grand Junction. It flows through magnificent canyons, notably the Black Canyon of the Gunnison, a national monument. Gunnison Tunnel, c.5 mi (8 km) long, was built between 1905 and 1909 to divert the river's water to the Uncompahgre Valley for irrigation.

Gun·ters·ville (gŭn'tərz-vĭl'), town (1970 pop. 6,491), seat of Marshall co., NE Ala., a port on the Tennessee River NNW of Gadsden, in a farm area; founded c.1818 on a Cherokee site, inc. 1847. It has textile mills.

Gun·tur (go͞on-to͝or'), city (1971 pop. 269,941), Andhra Pradesh state, SE India. It is a railroad junction and a cotton and tobacco market. Founded by the French in the 18th cent., Guntur was ceded to Great Britain in 1823.

Gu·ryev (go͝or'yĭf), city (1976 est. pop. 131,000), capital of Guryev oblast, Central Asian USSR, in Kazakhstan, on the Ural River and near the Caspian Sea. A seaport and an industrial center of the Emba oil fields, it has refineries, shipyards, and varied manufactures. Fishing is also important. Founded in 1645 as a military outpost, Guryev was a fishing center until the development of the region's petroleum industry in the 1930s.

Gu·sau (go͞o-zou'), town (1969 est. pop. 80,000), NW Nigeria, on the Sokoto River. It is a regional trade center for peanuts, cotton, and tobacco.

Gus·ta·vo A. Ma·de·ro (go͞os-tä'vô ä' mä-dā'rô), city (1970 pop. 1,182,895), Federal Dist., S central Mexico. Formerly called Guadalupe Hidalgo, it was renamed in 1931. It is a major pilgrimage center. The Treaty of Guadalupe Hidalgo (1848), which ended the Mexican War, was signed here.

Gü·ters·loh (güt'ər-slō), city (1974 est. pop. 78,195), North Rhine-Westphalia, W West Germany. Its manufactures include textiles, household appliances, furniture, and printed materials. Gütersloh was chartered in 1825.

Guth·rie (gŭth'rē), county (1970 pop. 12,243), 596 sq mi (1,543.6 sq km), W central Iowa, drained by the Middle Raccoon, South Raccoon, and Middle rivers; formed 1851; co. seat Guthrie Center. Its prairie agriculture includes corn, oats, hogs, cattle, and poultry. It also has bituminous-coal mines and gravel pits.

Guthrie. **1.** City (1970 pop. 9,575), seat of Logan co., central Okla., near the Cimarron River N of Oklahoma City. Founded in 1889, it was the territorial and then the state capital until 1910. Guthrie is the commercial center of a farming and dairying area. **2.** Village (1970 pop. 250), seat of King co., NW Texas, E of Lubbock. It is a trade center for a ranching region.

Guthrie Center, city (1970 pop. 1,834), seat of Guthrie co., W central Iowa, WNW of Des Moines; inc. 1880. A state park is nearby.

Guy·a·na (gī-ăn'ə, -ä'nə), formerly **Brit·ish Gui·a·na** (brĭt'ĭsh gē-ä'nə, -ăn'ə), republic (1970 pop. 763,000), 83,000 sq mi (214,969 sq km), NE South America, on the Atlantic Ocean. The capital is Georgetown. On the east Guyana is separated from Surinam by the Courantyne River. The Akarai Mts. form the southern border with Brazil. Several rivers make up much of the western border with Brazil and Venezuela, and the Essequibo River flows through the center of the country.

Economy. Agriculture and mining are the principal economic activities. Sugar cane, rice, and coconuts are the leading crops, and

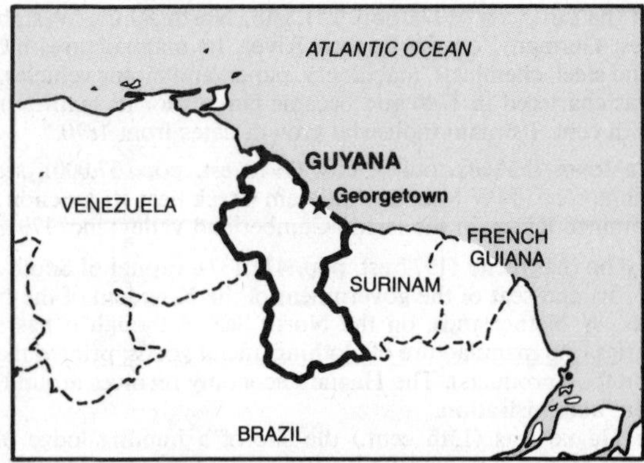

cattle and other livestock are raised. Bauxite, manganese, gold, and diamonds are mined. There are large forest resources (notably greenheart and balata) just beginning to be exploited. The processing of bauxite and sugar cane are the largest industries; the bauxite industry was nationalized in the early 1970s.

History. In the early 17th cent. the Dutch established settlements, and England and France also founded colonies. By the Treaty of Breda (1667) the Dutch gained all the English colonies in Guiana. Possessions continued to change hands until the Congress of Vienna (1815) awarded the settlements of Berbice, Demerara, and Essequibo to Great Britain; they were united as British Guiana in 1831. Slavery was abolished in 1834. In 1879 gold was discovered, thus speeding British expansion toward the Orinoco delta and resulting in the Venezuela Boundary Dispute.

After World War II significant progress toward self-government was begun. Under the 1952 constitution, elections were held and a government formed. However, the British deemed the government pro-Communist and suspended the constitution. Subsequently a new political party emerged. Self-government was granted in 1961. Proportional representation was introduced in 1964 in response to charges that the electoral system was unfair. After the 1964 elections a political coalition was made. Full independence was negotiated in 1966. Antagonism between the East Indians, who control a substantial portion of the nation's commerce, and the blacks led to frequent clashes and bloodshed in the 1960s, but violence subsided by the 1970s. Guyana became a republic in 1970. The boundaries with Venezuela and Surinam became a matter of dispute in the 1960s, with Venezuela laying claim to some 60% of Guyana's territory. Tensions on both fronts eased in 1970 when a 12-year truce was declared with Venezuela and a mutual troop withdrawal agreement was made with Surinam.

Government. Guyana has a parliamentary form of government. The popularly elected national assembly, chosen by proportional representation, elects the president, who is the head of state.

Guy·mon (gī'mən), city (1970 pop. 7,674), seat of Texas co., NW Okla., S of the North Canadian River in the Panhandle E of Boise City; inc. 1905. Wheat is grown, and there are large oil and natural-gas fields in the area.

Guz·mán (gōōs-män'), city (1970 pop. 48,166), Jalisco state, SW Mexico. It is a marketing and processing center, especially for hogs. Guzmán was the site of the pre-Columbian kingdom of Zapotlán, which was conquered by the Spanish in 1526.

Gwa·li·or (gwä'lē-ôr'), city and former princely state, 26,008 sq mi (67,360.7 sq km), central India, part of Madhya Pradesh state since 1956. The state was formed in the mid-18th cent. Forces of Gwalior overran much of central India until they were checked by the British in the early 19th cent., and the state was temporarily annexed to the British domain. When India became independent in 1947, Gwalior and several other princely states were combined into the state of Madhya Bharat. In 1956 Madhya Bharat merged with Madhya Pradesh. The city of Gwalior (1971 pop. 406,755) lies at the foot of Gwalior fort, a stronghold on the Rock of Gwalior. Within the battlemented walls of the fort are elaborately carved palaces and temples. Huge Jain reliefs are carved in the cliffs of the Rock of Gwalior. Among the city's manufactures are textiles, porcelain ware, leather and plastic goods, and processed food.

Gwe·lo (gwĕl'ō), city (1975 est. pop., 20,600), central Rhodesia. It is an industrial and commercial center. Manufactures include footwear, ferroalloys, metal goods, and cement.

Gwent (gwĕnt), nonmetropolitan county (1976 est. pop. 439,600), SE Wales, created under the Local Government Act of 1972 (effective 1974); administrative center Cwmbran. It comprises portions of the former counties of Monmouthshire and Breconshire.

Gwin·nett (gwĭ-nĕt'), county (1970 pop. 72,349), 437 sq mi (1,131.8 sq km), N Ga., in a piedmont area bounded on the NW by the Chattahoochee River and drained by the Apalachee and Yellow rivers; formed 1818; co. seat Lawrenceville. It has agriculture (cotton, corn, hay, fruit, livestock, and poultry), lumber, and granite quarries. Its manufactures include clothing, furniture, paper, glass, steel, and metal products.

Gwyn·edd (gwĭn'ĕ*th*), nonmetropolitan county (1976 est. pop. 225,100), NW Wales, created under the Local Government Act of 1972 (effective 1974); administrative center Caernarvon. It comprises the former counties of Anglesey and Caernarvonshire and portions of the former counties of Merionethshire and Denbighshire.

Gym·pie (gĭm'pē), city (1971 pop. 11,131), Queensland, E Australia. It is an agricultural center. A silver mine is nearby.

Gyön·gyös (dyœn'dyœsh), city (1970 pop. 33,149), N Hungary, at the foot of the Matra Mts. It is the commercial center of a wine-producing and tobacco-growing region. Lead and zinc are mined nearby.

Győr (dyœr), city (1977 est. pop. 120,000), NW Hungary, near the Czechoslovak border at the confluence of the Rába and Danube rivers. Győr is a road and rail hub, a river port, a county administrative center, and a leading industrial city, known especially for its engineering works and textile plants. The site of Győr was a Roman military outpost that was evacuated in the 4th cent. A.D. and later destroyed. The Magyars built fortifications here in the 9th cent. Győr was made a royal free town in 1743. In 1849 Hungarian revolutionary forces were decisively defeated by the Austrians near Győr. The city's industrialization dates from the 19th cent.

Gyu·la (dyōō'lŏ), town (1977 est. pop. 25,000), SE Hungary, on the White Koros River near the Rumanian border. It is an agricultural center.

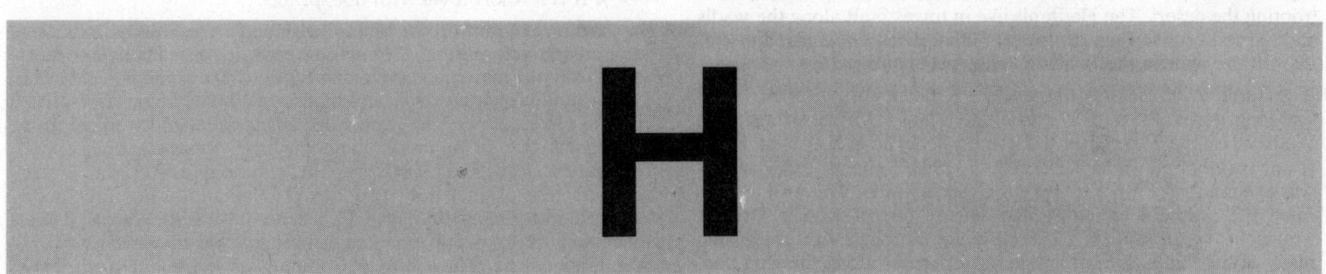

Haa·kon (hăk'ən), county (1970 pop. 2,802), 1,815 sq mi (4,700.9 sq km), W central S.Dak., drained in the S by the Bad River and bounded on the N by the Cheyenne River; formed 1914; co. seat Philip. It has cattle ranches and grain, dairy, and poultry farms.

Haar·lem (här'ləm), city (1977 est. pop. 162,774), capital of North Holland prov., W Netherlands, on the Spaarne River, near the North Sea. Although an industrial center with shipyards, machinery plants, and textile mills, Haarlem is chiefly noted as the center of a famous flower-growing district and the export point for bulbs (especially tulips). Haarlem was chartered in 1245. In 1573 it was sacked by the Spanish during the Revolt of the Netherlands. During the 16th and 17th cent. Haarlem was a center of Dutch painting.

Hab·er·sham (hăb′ər-shăm, -shəm), county (1970 pop. 20,691), 283 sq mi (733 sq km), NE Ga., with the Blue Ridge in the N and a piedmont area in the S; formed 1818; co. seat Clarkesville. Its agriculture includes cotton, hay, sweet potatoes, apples, peaches, and poultry. Lumbering is done, and textiles and clothing are manufactured. Part of Chattahoochee National Forest is in the north.

Ha·chi·no·he (hä-chē′nō-hā), city (1976 est. pop. 227,731), Aomori prefecture, N Honshu, Japan, on the Oirase River and the Pacific Ocean. It is a major fishing and commercial port, agricultural market, and industrial and commercial center. The city is the center of an iron and sand-mining district.

Hack·en·sack (hăk′ən-săk′), city (1978 est. pop. 34,200), seat of Bergen co., NE N.J., on the Hackensack River, a residential and industrial suburb of New York City; settled 1647, inc. as a city 1921. Dutch settlers from Manhattan established a trading post here in 1647. During the Revolution the city served as camping grounds for armies of both sides. It grew as a commercial and shipping center in the early 1800s.

Hack·ney (hăk′nē), borough (1971 pop. 216,659) of Greater London, SE England, on the Lea River. The borough was created in 1965 by the merger of the metropolitan boroughs of Hackney, Shoreditch, and Stoke Newington. Clothing manufacture (in Hackney) and printing and furniture making (in Shoreditch) are the borough's chief industries. London's first theater was built in Shoreditch (c.1575). The parish church of St. Mary in Stoke Newington is one of the few remaining Elizabethan churches.

Ha·da·no (hä-dä′nō), city (1976 est. pop. 106,968), Kanagawa prefecture, E central Honshu, Japan. An important communications center and agricultural market, the city has textile mills and sake and soy-sauce production plants.

Had·ding·ton (hăd′ĭng-tən), burgh (1971 pop. 6,505), Lothian region, SE Scotland. It has a large corn exchange. Haddington was burned by raiding English armies in 1216, 1244, and 1355. Until 1975 Haddington was county town of East Lothian.

Had·don·field (hăd′n-fēld′), borough (1978 est. pop. 11,500), Camden co., SE N.J., a residential suburb of Camden; settled c.1713, inc. 1875. Of interest is Indian King Tavern (1750), where the first state legislature met in 1777.

Ha·de·ra (hə-dā′rä), town (1975 est. pop. 35,700), W Israel, on the Plain of Sharon, near the Mediterranean Sea. Manufactures include tires, paper, and processed foods. Hadera was founded in 1891 by Jewish settlers, who drained vast malarial swamps and planted groves of citrus fruit and fields of grain.

Ha·ders·lev (hă′*th*ərs-lĕv), city (1970 est. pop. 23,200), Sønderjylland co., S Denmark, a seaport on the Haderslev Fjord, an inlet of the Lille Baelt. It is a commercial and industrial center. Haderslev was held by Prussia from 1864 to 1920.

Ha·dhra·maut or **Ha·dra·maut** (both: hä′drä-môt′), region, S Arabia, on the Gulf of Aden and the Arabian Sea, occupying the E part of Southern Yemen. The Hadhramaut consists of a narrow, arid coastal plain, a broad plateau averaging 4,500 ft (1,375 m) high, a region of deeply sunk wadis (watercourses), and an escarpment fronting the desert. The Hadranis live in towns built along the wadis and harvest crops of wheat, maize, millet, dates, coconuts, and coffee. On the plateau the Bedouins raise sheep and goats.

Ha·dri·an's Wall (hā′drē-ənz), ancient Roman wall, 73.5 mi (118.3 km) long, across the narrow part of the island of Great Britain from Wallsend on the Tyne River to Bowness at the head of Solway Firth. It was mainly built from A.D. c.122 to 126 by Emperor Hadrian and was extended by Emperor Severus a century later. The wall demarcated the northern boundary and defense line of Roman Britain. Fragments of the wall, 6 ft (1.8 m) high and 8 ft (2.4 m) thick, and many of the "mile stations" (stone blockhouses along the wall constructed every Roman mile) remain.

Ha-erh-pin (hä′ĕr′bĭn′): *see* Harbin, China.

Ha·fun, Cape (hă-fo͞on′), easternmost point of Africa, coast of NE Somalia, on the Indian Ocean. It is formed by a rocky island and a sandy isthmus connecting it to the mainland.

Ha·gen (hä′gən), city (1975 pop. 231,840), North Rhine-Westphalia, W West Germany, on the Ennepe River. Its manufactures include iron and steel, chemicals, machinery, paper, and motor vehicles. Hagen was chartered in 1746 and became famous for its textiles in the late 18th cent. Its main industrial growth dates from 1870.

Ha·gers·town (hā′gərz-toun′), city (1978 est. pop. 37,000), seat of Washington co., NW Md., on Antietam Creek near its junction with the Potomac River, in the fertile Cumberland valley; inc. 1791.

Hague, The (hāg), city (1977 est. pop. 471,137), capital of South Holland prov. and seat of the government of the Kingdom of the Netherlands, W Netherlands, on the North Sea. Although it has some industries (the manufacture of clothing, metal goods, printed materials, and food products), The Hague's economy revolves around government administration.

The Hague was (13th cent.) the site of a hunting lodge of the counts of Holland. In about 1250 William, count of Holland, began the construction of a palace, around which a town grew in the 14th and 15th cent. In 1586 the States-General of the United Provs. of the Netherlands convened in The Hague, which later (17th cent.) became the residence of the stadtholders and the capital of the Dutch republic. In the early 19th cent., after Amsterdam had become the constitutional capital of the Netherlands, The Hague received its own charter from Louis Bonaparte. It was (1815–30) the alternative meeting place, with Brussels, of the legislature of the United Netherlands. The Dutch royal residence from 1815 to 1948, the city was greatly expanded and beautified in the mid-19th cent. by King William II. Since 1899 the city has been a center for the promotion of international justice and arbitration. Among the numerous landmarks of The Hague are the Binnenhof, which contains the 13th cent. Hall of Knights, where many historic meetings have been held; the Gevangenenpoort, the 14th cent. prison where Jan de Witt and Cornelius de Witt were murdered in 1672; and the Mauritshuis (17th cent.), an art museum with several of the greatest works of Rembrandt and Vermeer. The Peace Palace houses the Permanent Court of Arbitration and, since 1945, the International Court of Justice.

Hahn·ville (hăn′vĭl′), village (1970 pop. 2,522), seat of St. Charles parish, SE La., on the Mississippi River W of New Orleans, in a sugar, truck-farm, and oil region.

Hai·fa (hī′fə), city (1975 est. pop. 227,200), NW Israel, a port on the Mediterranean Sea, at the foot of Mt. Carmel. Haifa is known to have existed by the 3rd cent. A.D. but was of little importance during early Moslem times. Destroyed by Saladin in 1191, it began to revive in the late 18th cent. It was captured by Napoleon in 1799.

Hai-k'ou (hī′kō′) or **Hoi·how** (hoi′hou′), city (1970 est. pop. 500,000), Hainan island, Kwangtung prov., China. A seaport on Hainan Strait, it is an industrial center, with food-processing establishments, cement plants, machine shops, and a large integrated steel complex.

Hail (hīl, hä′ēl), city (1974 pop. 40,502), N central Saudi Arabia, in an oasis. It was the capital of the independent emirate of Jabal Shammar, which Ibn Saud conquered in 1921.

Hai·lar (hī′lär′), city, NW Heilungkiang prov., China, on the Hailar River. It is an agricultural production center. Hailar consists of an old and a new city—the old section, founded in 1734 as a fort, is typically Chinese; the new section is a modern, industrial quarter.

Hai·ley (hā′lē), city (1970 pop. 1,425), seat of Blaine co., S central Idaho, ESE of Boise; laid out 1881, inc. as a village 1903, as a city 1909. It is a resort town with hot springs.

Hai·nan or **Hai-nan** (both: hī′nän′), island in the South China Sea, administratively part of Kwangtung prov., China. Hainan is separated from the mainland by Hainan Strait (c.30 mi/50 km wide). The year-round growing season and monsoon climate favor the cultivation of coffee, rubber, tea, coconuts, sugar cane, and tropical fruit. Hainan is rich in minerals, including iron, tungsten, tin, copper, manganese, lead, silver, coal, graphite, antimony, and crystal. Under Chinese control since the 1st cent. A.D., Hainan was not fully incorporated into China until the 13th cent. In World War II it was occupied (1939) by the Japanese. The island was liberated (1945) by the Nationalists. The Chinese Communists landed in Apr., 1949, and, with the aid of Communist guerrillas from the mountains, gained control in 1950.

Hai·naut (ĕ-nō′), province (1976 est. pop. 1,317,772), 1,437 sq mi (3,721.8 sq km), S Belgium, bordering on France in the S; capital Mons. It is low-lying, except in the southeast, and has considerable productive farmland. The county of Hainaut was created in the late

9th cent., and in the divisions of the Carolingian empire became a fief of Lotharingia. Between 1036 and 1278 Hainaut was frequently joined, through marriage, with Flanders. In 1433 Philip the Good of Burgundy added Hainaut and Holland to his dominions. Hainaut remained under the house of Burgundy until the death (1482) of Mary of Burgundy. By the treaties of the Pyrenees (1659) and of Nijmegen (1678) parts of Hainaut were permanently annexed by France.

Haines (hānz), borough (1970 pop. 1,504), 2,128 sq mi (5,511.5 sq km), SE Alaska; formed after 1961; borough seat Haines. It is in a lumbering, fishing, fish-processing, and fur-farming area.

Haines, city (1970 pop. 1,093), seat of Haines borough, SE Alaska, on Chilkoot Inlet of Lynn Canal SSW of Skagway. A fur-trading post was established here in 1867. Fishing and lumbering are done.

Haines City, city (1970 pop. 8,956), Polk co., central Fla., ENE of Tampa, SW of Orlando; settled in the 1870s. It is a shipping center in a citrus-fruit area.

Hai·phong (hī'fŏng'), city (1960 pop. 182,496), extreme NE Vietnam, on a large branch of the Red River delta near the Gulf of Tonkin. A shipbuilding industry and cement, glass, porcelain, and textile works were established (after 1874) by the French. At the beginning of the French-Indochina War (Nov., 1946), French naval vessels shelled the city, killing c.6,000 Vietnamese. After the French departed (1954), the silted-up harbor was reconstructed with Chinese and Soviet aid and the docks and shipbuilding yards were repaired and modernized. During the Vietnam War, Haiphong was severely bombed from 1965 to 1968 and again from Apr. to Dec., 1972. The harbor was mined by U.S. naval planes in May, 1972, and effectively sealed until the mines were swept by U.S. forces after the cease-fire agreement in early 1973.

Hai·ti (hā'tē), independent republic (1971 pop. 4,314,628), 10,700 sq mi (27,713 sq km), West Indies, on the W third of the island of Hispaniola, bounded on the N by the Atlantic Ocean, on the S by the Caribbean Sea, and on the E by the Dominican Republic. Jamaica lies to the west and Cuba to the northwest. The capital is Port-au-Prince. The offshore islands of Tortuga and Gonâve also belong to Haiti.

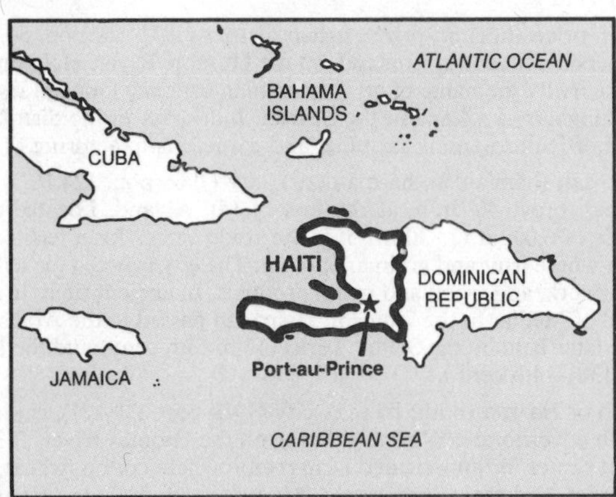

Economy. Agriculture is the principal economic activity in Haiti. Subsistence crops include cassava, rice, sugar cane, sorghum, yams, corn, and plantains. Most Haitians own and farm tiny plots of land, and great population density has caused rural poverty. Haiti's major export is coffee; other exports include cotton, sugar, sisal, bauxite, and essences. Industry in Haiti consists largely of light manufacturing; products include foodstuffs, liquors, essential oils, leather goods, soap, and footwear. Some bauxite and copper are mined but other mineral deposits have barely been tapped.

History. The island of Hispaniola was discovered and named by Columbus in 1492. Disease and ill-treatment by the Spaniards decimated the native Arawak Indians. While establishing plantations in eastern Hispaniola (now the Dominican Republic), however, the Spanish largely ignored the western part of the island, which became a base for French and English buccaneers. Gradually French colonists, importing African slaves, developed sugar plantations on the northern coast. Unable to support its claim to the region, Spain ceded Haiti (then called Saint-Domingue) to France in 1697. It became France's most prosperous colony in the Americas and one of the world's chief coffee and sugar producers.

The pattern of settlement took the French south in the 18th cent. and society became stratified into Frenchmen, Creoles, freed blacks, and black slaves. Between the blacks and whites were the mulattoes, who aspired to the privileges of the whites and who feared the blacks. When French-descended Creole planters sought to prevent mulatto representation in the French National Assembly and in local assemblies in Saint-Domingue, the mulattoes revolted. This rebellion destroyed the entire structure of Haitian society. The blacks formed guerrilla bands led by Toussaint l'Ouverture, a former slave who had been made an officer of the French forces on Hispaniola. In 1795 Spain ceded its part of the island to France, and in 1801 Toussaint conquered it, abolished slavery, and proclaimed himself governor general of Hispaniola. A French punitive force captured Toussaint, who died in a French prison; but the revolt continued and the French troops, already ravaged by yellow fever, were forced to withdraw.

In 1804 Haiti won complete independence. The remaining whites were expelled, and Jean-Jacques Dessalines, an ex-slave, proclaimed himself emperor. His assassination (1806) led to the division of Haiti into a black-controlled north under Emperor Henri Christophe and a mulatto-ruled south under President Alexandre Pétion. After their deaths Haiti was unified by Jean Pierrre Boyer, who also brought (1822–44) Santo Domingo under Haitian control. Seeking to indemnify French planters, Boyer brought financial ruin to Haiti; he was exiled in 1843.

Since the end of his reign, the country has been a republic. Anarchy persisted, intensified by the further decline of the economy. After the dictator Guillaume Sam was killed by a mob in 1915, the United States, troubled over its property and investments in the country and fearing Germany might seize Haiti, landed marines in Port-au-Prince. In 1930 Haiti was allowed to elect a legislature that would, in turn, name a president. The marines were finally withdrawn in 1934, although U.S fiscal control was maintained until 1947.

Political instability persisted in Haiti after World War II. François ("Papa Doc") Duvalier, who was elected president in 1957, used voodoo as an instrument of control over the masses and instilled fear through his paramilitary secret police. In 1964 he proclaimed himself president for life. Upon his death in 1971 his 19-year-old son, Jean-Claude, succeeded him and also became president for life.

Government. Besides the president, Haiti's government consists of a unicameral legislature, which is elected by direct popular vote for a six-year term. The country has one official political party and several unofficial opposition groups, some operating abroad.

Ha·ko·da·te (hä'kô-dä'tē), city (1976 est. pop. 307,453), extreme SW Hokkaido, Japan, on the Tsugaru Strait. Opened (1854) to U.S. ships and a little later (1857) to general foreign trade, it has ironworks, shipyards, and an extensive fishing industry.

Hal·ber·stadt (häl'bər-shtät), city (1974 est. pop. 46,669), Magdeburg district, W East Germany. Manufactures include textiles, paper, processed food, and machinery. Halberstadt was burned (1179) by Henry the Lion; after Henry's fall (1180) the bishopric of Halberstadt was given in temporal fief to the bishops by Emperor Frederick I. The city became (13th–14th cent.) a flourishing trade center. In 1648 it was annexed by Brandenburg.

Hal·den (häl'dən), town (1978 est. pop. 27,054), Østfold co., SE Norway, a port on the Iddefjord (an arm of the Skagerrak), near the Swedish border. Manufactures include forest products, footwear, and textiles.

Hale (hāl). **1.** County (1970 pop. 15,888), 663 sq mi (1,717.2 sq km), W Ala., in the Black Belt bounded on the W by the Black Warrior River; formed 1867; co. seat Greensboro. It has agriculture (cotton, grain, and livestock), timber, and a food-processing industry (especially poultry and fish). Clothing and furniture are manufactured. **2.** County (1970 pop. 34,137), 979 sq mi (2,535.6 sq km), NW Texas, on the Llano Estacado drained by the White River; formed 1876; co. seat Plainview. It is a leading agricultural county, with large irrigated areas producing grain sorghums, alfalfa, potatoes, soybeans, sugar beets, fruit, truck crops, livestock, poultry, and dairy products. There are oil and clay deposits and factories manufacturing fertilizer, seed, and farm machinery.

Ha·le·a·ka·la National Park (hä'lā-ä'kä-lä'), 27,283 acres (11,049.6 hectares), on Maui island, Hawaii; est. 1961. Haleakala Crater,

10,025 ft (3057.6 m) high, has been dormant since the mid-1700s. Its crater, 2,720 ft (829.6 m) deep with an area of 19 sq mi (49.2 sq km), is one of the largest in the world.

Hales·ow·en (hālz'ō-ən), municipal borough (1973 est. pop. 54,120), West Midlands, central England. Listed in the Domesday Book as Hala, Halesowen is a manufacturing town with coal mines and steel plants. It was the birthplace of the printer William Caslon.

Hal·i·car·nas·sus (hăl'ĭ-kär-năs'əs), ancient city of Caria, SW Asia Minor, on the Ceramic Gulf (now the Gulf of Kos) and on the site of the modern city of Bodrum, Turkey. Halicarnassus was Greek in origin. As a Persian vassal it was ruled by tyrants and participated in Xerxes' invasion of Greece (480 B.C.), but after the expulsion of the tyrants (460–455) it joined the Delian League. A dynasty of Carian kings in the 4th cent. B.C. was made famous by Mausolus, whose wife, Artemisia, built him a magnificent tomb, considered one of the Seven Wonders of the World. Alexander the Great conquered the city (c.334 B.C.). It was the birthplace of Herodotus.

Hal·i·fax (hăl'ə-făks'), city (1976 pop. 117,882; metropolitan area pop. 267,991), provincial capital, S central N.S., Canada, on the Atlantic Ocean. It is Canada's principal ice-free Atlantic port and the eastern terminus of its railroad systems and transcontinental highway. Its many industries include commercial fishing, fish processing, shipbuilding, oil refining, and the manufacture of electronics. Halifax was founded in 1749. It served as a naval base for the expedition against Louisburg in 1758, against the American colonies in the American Revolution, and against the United States in the War of 1812. During both World Wars the port was an important naval and air base.

Halifax, county borough (1973 est. pop. 88,580), West Yorkshire, central England, on a small tributary of the Calder River. The borough's industries include the manufacture of carpets, cotton, wool, and worsted goods, and machine tools and boilers.

Halifax. 1. County (1970 pop. 53,884), 734 sq mi (1,901.1 sq km), NE N.C., in a piedmont region bounded on the NW by Lake Gaston, on the N and E by the Roanoke River; formed 1758; co. seat Halifax. It has agriculture (tobacco, peanuts, cotton, and corn), a lumbering industry, and some manufacturing. **2.** County (1970 pop. 30,076), 800 sq mi (2,072 sq km), S Va., bounded on the S by the N.C. border, on the N and E by the Roanoke River, and drained by the Dan, Banister, and Hyco rivers; formed 1752; co. seat Halifax. In a leading tobacco-growing area, it also produces corn, wheat, and livestock and has some manufacturing.

Halifax. 1. Town (1970 pop. 355), seat of Halifax co., NE N.C., on the Roanoke River NE of Raleigh; settled c.1750. The state's first constitutional convention (1776) was held here. **2.** Town (1970 pop. 899), seat of Halifax co., S Va., on the Banister River ENE of Danville; inc. 1875. Tobacco, wheat, and hay are grown.

Hall (hôl). **1.** County (1970 pop. 59,405), 378 sq mi (979 sq km), N Ga., in a piedmont area drained by the Chattahoochee and Oconee rivers; formed 1818; co seat Gainesville. Its agriculture includes cotton, corn, hay, sweet potatoes, and poultry. It has granite quarries. Textiles, clothing, furniture, shoes, chemicals, plastics, concrete, and aluminum are manufactured. **2.** County (1970 pop. 42,851), 536 sq mi (1,388.2 sq km), SE central Nebr., drained by the Platte River and its branches; formed 1858; co. seat Grand Island. In an agricultural area yielding grain, fruit, vegetables, dairy products, livestock, and poultry, it has an extensive food-processing industry. Lumber, wood, and metal products, concrete, and farm machinery are made. **3.** County (1970 pop. 6,015), 885 sq mi (2,292.2 sq km), NW Texas, in a plains region below Cap Rock escarpment of the Llano Estacado and crossed by the Prairie Dog Town Fork of the Red River; formed 1876; co. seat Memphis. In an area devoted to cattle raising and agriculture, its crops include cotton, wheat, grain sorghums, alfalfa, potatoes, truck crops, and some fruit.

Hal·lan·dale (hăl'ən-dāl'), city (1978 est. pop. 34,300), Broward co., SE Fla., on the Atlantic coast and the Intracoastal Waterway; settled 1897, inc. 1927. It has horse and greyhound racetracks.

Hal·le (hä'lə), town (1977 est. pop. 32,138), Brabant prov., central Belgium, on the Charleroi-Brussels Canal. It is a commercial and industrial center. Manufactures include textiles, paper, machines, processed food, and chemicals. Halle's Gothic Church of Our Lady (14th–15th cent.) is a place of pilgrimage.

Halle, city (1974 est. pop. 241,425), capital of Halle district, S central East Germany, on the Saale River. Manufactures include chemicals, refined sugar and other food products, paper, and machinery. Salt and potash are mined in the region. Located on the site of Bronze Age and Iron Age settlements, Halle was first mentioned in the 9th cent. In 968 it was given to the archbishops of Magdeburg. The city was a member (1281–1478) of the Hanseatic League and annexed by Brandenburg in 1648.

Hal·letts·ville (hăl'ĭts-vĭl'), city (1970 pop. 2,712), seat of Lavaca co., S Texas, on the Lavaca River between Houston and San Antonio, in a farm and livestock area.

Hal·lock (hăl'ək), village (1970 pop. 1,477), seat of Kittson co., NW Minn., near the Canadian boundary NNE of Grand Forks, N.Dak.; platted 1879, inc. 1887. It is in a farming and hunting region in the Red River valley.

Hal·ma·he·ra (häl'mä-hĕr'ä), island, c. 7,000 sq mi (18,130 sq km), E Indonesia, between New Guinea and Celebes, on the equator. The largest of the Moluccas and irregular in shape, it consists of two intersecting mountain ranges (rising to c.5,000 ft/1,525 m), which form four rocky peninsulas separated by three deep bays. There are several active volcanos, lush jungles, streams, and a few lakes. Subsistence farming, hunting, and fishing are done. The chief products are spices, resin, sago, rice, tobacco, and coconuts. Known to the Portuguese and the Spaniards as early as 1525, Halmahera came under Dutch influence in 1660. Taken by the Japanese (1942) in World War II, it was frequently bombed by the Allies.

Halm·stad (hälm'städ'), city (1975 est. pop. 64,400), capital of Halland co., SW Sweden, a seaport on the Kattegat at the mouth of the Nissan River. It is an industrial center and summer resort. Manufactures include steel, textiles, and paper, and shipbuilding is carried on. Chartered in 1307, Halmstad was an important fortified city of Denmark before being conquered by Sweden in 1645.

Häl·sing·borg (hĕl'sĭng-bôr'yə), city (1975 est. pop. 101,685), Malmöhus co., S Sweden, a seaport on the Øresund. Manufactures include processed copper, rubber, electrical goods, textiles, and refined sugar. A trade center and stronghold since the 9th cent., Hälsingborg was destroyed during the Danish-Swedish conflicts in the 17th cent. and passed to Denmark. The city was rebuilt after its return to Sweden in 1710.

Hal·tem·price (hôl'təm-prīz'), urban district (1973 est. pop. 54,850), Humberside, E central England, on the Humber River. Haltemprice was originally the name of an Augustinian canonry founded in 1322 at Cottingham, a village in the district. Industries in the district include horticulture, mink farming, and carpet manufacturing.

Ham·a·dan (hăm'ə-dən, hä-mä-dän'), city (1966 pop. 124,167), Kermanshah prov., W Iran, at the foot of Mt. Alvand. Located at an altitude of 6,000 ft (1,830 m), it is the trade center for a fertile farm region where fruit and grain are grown. The city is noted for its rugs, leatherwork, and wood and metal products. In ancient times it was a capital of Media. In the 7th cent. Hamadan passed to the Arabs, and it was later held by the Seljuk Turks (12th–13th cent.) and the Mongols (13th–14th cent.).

Ha·mah or **Ha·ma** (both: hä'mä), city (1970 pop. 137,421), capital of Hamah governorate, W central Syria, on the Orontes River. It is the market center for an irrigated farm region where cotton, wheat, barley, millet, and maize are grown. Manufactures include cotton and woolen textiles, towels, carpets, and dairy products.

The city was settled as far back as the Bronze Age and Iron Age. In the 2nd millennium B.C., it was a center of the Hittites. The Assyrians under Shalmaneser III captured the city in the mid-9th cent. B.C. Later included in the Persian Empire, it was conquered by Alexander the Great and, after his death (323 B.C.), was claimed by the Seleucid kings. The city later came under the control of Rome and of the Byzantine Empire. In A.D. 638 it was captured by the Arabs. Christian Crusaders held Hamah briefly (1108), but in 1188 it was taken by Saladin, in whose family it remained until it passed to Egyptian Mameluke control in 1299. In the early 16th cent. the city came under the Ottoman Empire. After World War I it was made part of the French Levant States League of Nations mandate, and in 1941 it became part of independent Syria.

Ha·ma·mat·su (hä-mä'mät-so͞o), city (1976 est. pop. 472,052), Shizuoka prefecture, S central Honshu, Japan. Its chief products are textiles, musical instruments, motorcycles, and compact cars; the weaving and dyeing industry is also important.

Ham·blen (hăm'blən), county (1970 pop. 38,696), 155 sq mi (401.5 sq km), NE Tenn., in the Great Appalachian Valley, bounded on the N by the Cherokee Reservoir, on the S by the Nolichucky River; formed 1870; co. seat Morristown. It is mainly agricultural (yielding tobacco, corn, potatoes, truck crops, poultry, dairy products, and livestock), with some manufacturing.

Ham·burg (hăm'bûrg', häm'bo͝orkh'), city (1974 est. pop. 1,733,802), coextensive with, and capital of, Hamburg state, 288 sq mi (746 sq km), N West Germany, on the Elbe River near its mouth in the North Sea, and on the Alster River. Hamburg has large shipyards and various industries whose manufactures include machinery, food products, chemicals, metal goods, and printed materials.

Hamburg originated (early 9th cent.) in the Carolingian castle of Hammaburg, probably built by Charlemagne as a defense against the Slavs. It became (834) an archiepiscopal see. The city quickly grew to commercial importance and in 1241 formed an alliance with Lübeck, which later became the basis of the Hanseatic League. In 1558 the first German stock exchange was founded here; with the arrival of Dutch Protestants, Portuguese Jews, and English cloth merchants (expelled from Antwerp), and with the expansion of commercial ties with the United States after 1783, Hamburg continued to prosper. It was occupied by the French in 1806 and in 1815 joined the German Confederation. In 1842 a fire destroyed much of the city. After World War I Hamburg was briefly (1918–19) a socialist republic. During World War II, Hamburg was severely damaged by aerial bombardment. Hamburg today is an elegant, modern city and a cultural center.

Ham·den (hăm'dən), town (1978 est. pop. 50,500), New Haven co., S Conn.; inc. 1786. The town, settled c.1638, is a residential and manufacturing suburb of New Haven, of which it was once a part. Hamden makes firearms, machinery, electrical products, metal goods, rolled steel, and handbags. The town's industrial development dates back to 1798, when Eli Whitney set up an arms factory using techniques of mass production.

Hä·meen·lin·na (hă'mān-lĭn'nä), city (1975 est. pop. 40,761), capital of Häme prov., S Finland. It is a lake port and a manufacturing town with plywood mills and spool mills.

Ha·meln (hä'məln), city (1974 est. pop. 61,722), Lower Saxony, N central West Germany, a port on the Weser River. Its manufactures include carpets, chemicals, and metal products. The city is a tourist center, known as the scene of the legend of the Pied Piper of Hamelin. An ancient Saxon settlement, Hameln became a missionary outpost c.750 and received city rights c.1200. It was a member of the Hanseatic League from 1426 to 1572. The city passed to Hanover in 1814 and to Prussia in 1866.

Ham·hung (häm'ho͝ong'), city (1962 est. pop. 125,000), South Hamgyong prov., E central North Korea. Metalware, cotton textiles, and fertilizers are manufactured. Coal mines are nearby.

Ham·il·ton (hăm'əl-tən), city (1970 pop. 2,060), capital of Bermuda, on Bermuda Island. It is a free port at the head of Great Sound, a lagoon protected by coral reefs. The city is a major tourist resort.

Hamilton, city (1976 pop. 312,003), S Ont., Canada, at the W end of Lake Ontario. Hamilton is an important port, transportation center, and manufacturing city. It is Canada's leading producer of iron and steel; other manufactures include automobiles, heavy machinery, and paper and textile products. The site was settled by United Empire Loyalists in 1778.

Hamilton, city (1976 pop. 87,968), N central North Island, New Zealand, on the Waikato River. Hamilton is the urban center of a densely populated dairy area. It was founded as a military settlement on the site of a deserted Maori village in 1864.

Hamilton, burgh (1976 est. pop. 107,178), Strathclyde region, S central Scotland, at the confluence of the Avon and the Clyde rivers. It is a market town with metal products and other industries. Rudolf Hess landed near Hamilton after his flight from Germany in May, 1941.

Hamilton. 1. County (1970 pop. 7,787), 514 sq mi (1,331.3 sq km), N Fla., bounded on the N by the Ga. border, on the S and E by the Suwannee River, on the W by the Withlacoochee River; formed 1827; co. seat Jasper. In a flatwoods area, it has swamps in the east and is drained by the Alapaha River. Farming (corn, peanuts, cotton, tobacco, and vegetables), stock raising (hogs and cattle), and lumbering are important to its economy. **2.** County (1970 pop. 8,665), 435 sq mi (1,126.7 sq km), SE Ill., drained by the North Fork of the Saline River; formed 1821; co. seat McLeansboro. Its agriculture includes livestock, poultry, fruit, wheat, corn, and redtop seed. It has oil and gas wells and a food-processing industry. Clothing and wood products are manufactured. **3.** County (1970 pop. 54,532), 403 sq mi (1,043.8 sq km), central Ind., drained by the West Fork of the White River and many small creeks; formed 1823; co. seat Noblesville. In a highly industrialized area, it has some mining and agriculture (grain, dairy products, cattle, and draft horses). Its manufactures include packaged foods, clothing, concrete, furniture, and wood and metal products. **4.** County (1970 pop. 18,383), 577 sq mi (1,494.4 sq km), central Iowa, in a prairie agricultural area drained by the Boone and Skunk rivers; formed 1856; co. seat Webster City. Hogs, poultry, corn, soybeans, and oats are raised. It has deposits of bituminous coal. **5.** County (1970 pop. 2,747), 992 sq mi (2,569.3 sq km), W Kansas, in a prairie region bordered on the W by Colo. and drained by the Arkansas River; formed 1873; co. seat Syracuse. Its agriculture includes wheat and livestock, and it has some mining and manufacturing. **6.** County (1970 pop. 8,867), 541 sq mi (1,401.2 sq km), SE central Nebr., bounded on the N and NW by the Platte River; formed 1867; co. seat Aurora. In a grain and livestock area, it has plants manufacturing mobile homes and sporting goods. **7.** County (1970 pop. 4,714), 1,735 sq mi (4,493.7 sq km), NE central N.Y., in the Adirondacks drained by tributaries of the Hudson River and the Raquette and Sacandaga rivers; formed 1816; co. seat Lake Pleasant. In a resort region with many lakes and rivers, it has skiing areas and good hunting and fishing. Some logging is done. Its agriculture includes hay, grain, poultry, livestock, and dairy products. **8.** County (1970 pop. 923,840), 414 sq mi (1,072.3 sq km), extreme SW Ohio, bounded on the W by the Ind. border, on the S by the Ohio River and Ky. border, and drained by the Great Miami, Little Miami, and Whitewater rivers; formed 1790; co. seat Cincinnati. Within the Cincinnati metropolitan area, it has some agriculture (dairy products, poultry, truck crops, and livestock), coal mines, sand and gravel pits, and lumbering and food-processing industries. Textiles, clothing, furniture, paper, chemicals, concrete, and wood and metal products are manufactured. **9.** County (1970 pop. 255,077), 550 sq mi (1,424.5 sq km), SE Tenn., bounded on the S by the Ga. border and crossed by the Tennessee River; formed 1819, absorbed James co. in 1919; co. seat Chattanooga. Its economy is based primarily on manufacturing (textiles, metal products, and chemicals). Farming, mining, and tourism are also important. Part of Chickamauga and Chattanooga National Military Park is in the county. **10.** County (1970 pop. 7,198), 844 sq mi (2,186 sq km), central Texas, in a prairie area drained by the Leon, Bosque, and Lampasas rivers; formed 1842; co. seat Hamilton. Livestock and diversified agriculture (cotton, grain, hay, peanuts, fruit, and potatoes) are important. It markets wool and mohair, has natural-gas fields, and manufactures clothing, wood molding, and steel.

Hamilton. 1. Town (1976 est. pop. 4,196), seat of Marion co., NW Ala., near the Miss. border NW of Birmingham, in a timber and cotton area; settled c.1818. **2.** City (1970 pop. 357), seat of Harris co., W Ga., NNE of Columbus, in an agricultural area. It is a summer resort. **3.** Town (1970 pop. 6,373), Essex co., NE Mass., W of Gloucester; settled 1638, set off from Ipswich 1793. **4.** City (1970 pop. 2,499), seat of Ravalli co., W central Mont., on the Bitterroot River SSW of Missoula; inc. 1894. It is in a farming, dairying, and lumbering area. **5.** City (1978 est. pop. 64,400), seat of Butler co., SW Ohio, on the Great Miami River; inc. 1857. A manufacturing center, Hamilton has paper and pulp mills, blast furnaces, and many factories that make a great variety of products, including safes, machinery, chemicals, and pumps and motors. Hamilton was settled on the site of Fort Hamilton, built in 1791. **6.** City (1970 pop. 2,760), seat of Hamilton co., central Texas, W of Waco, in a cotton, wheat, and livestock area. Deposits of clay, limestone, and natural gas are in the vicinity.

Hamilton, Mount, peak 4,372 ft (1,333.5 m) high, W Calif., in the Coast Ranges, E of San Jose. It is the site of Lick Observatory (built 1876–88), directed by the Univ. of California.

Hamilton Grange National Memorial (grānj): *see* National Parks and Monuments Table.

Ha·mi·na (hä'mĭ-nä), city (1970 pop. 11,028), Kymi prov., SE Finland, on the Gulf of Finland. Hamina is an important port. The Treaty of Fredrikshamn (1809), by which Sweden ceded all of Finland to Russia, was signed in Hamina.

Ham·lin (hăm'lĭn), county (1970 pop. 5,520), 511 sq mi (1,323.5 sq km), E S.Dak., drained by the Big Sioux River and several lakes; formed 1873; co. seat Hayti. It is in a dairying and livestock area, with grain, poultry, and potatoes. Some manufacturing is done.

Hamlin, town (1970 pop. 1,024), seat of Lincoln co., W W.Va., on the Mud River ESE of Huntington, in a coal, natural-gas, and oil region.

Hamm (häm), city (1975 pop. 172,686), North Rhine-Westphalia, W West Germany, on the Lippe River, in the Ruhr district. The city contains iron and steel foundries and manufactures textiles and machinery. Founded in 1226, Hamm was an active member of the Hanseatic League. It passed to Cleves in the 14th cent. and later (1614) to Brandenburg. The city was badly damaged in World War II.

Ham·mer·fest (hä'mər-fĕst'), town (1977 est. pop. 7,451), Finnmark co., N Norway, on Kvaløy island. It is the northernmost town of Europe, but its harbor is always ice-free. The town has sealing and fishing industries and fish-processing plants. Tourists are attracted by its uninterrupted daylight from May 17 to July 29.

Ham·mer·smith (hăm'ər-smĭth'), borough (1971 pop. 184,935), of Greater London, SE England, on the Thames River. The borough was created in 1965 by the merger of the metropolitan boroughs of Hammersmith and Fulham. Hammersmith has pottery works and is the site of the British Broadcasting Corp. Television Centre.

Ham·mond (hăm'ənd). **1.** City (1978 est. pop. 103,000), Lake co., extreme NW Ind.; inc. 1884. Hammond was a meat-packing town until its packinghouse was destroyed by fire in 1901. It now has steel foundries, a publishing industry, and a great variety of manufactures, including petroleum products, soaps and toilet articles, railroad equipment, forgings, valves, and hospital and surgical supplies. **2.** City (1970 pop. 12,478), Tangipahoa parish, SE La.; inc. 1888. It has feed mills and a lumber and wood-products industry. The city is the seat of Southeastern Louisiana Univ.

Ham·monds·port (hăm'əndz-pôrt'), village (1970 pop. 1,066), Steuben co., S N.Y., at the S end of Keuka Lake NE of Bath, in a grape-growing region; inc. 1871. It was the birthplace of Glenn Curtiss and the site of some of his aviation experiments.

Hamp·den (hăm'dən), county (1970 pop. 459,050), 621 sq mi (1,608.4 sq km), SW Mass., bordering on Conn.; formed 1812; co seat Springfield. It is bisected by the Connecticut River, which supplies power to its industrial cities. Tobacco and truck crops are grown; dairying is also important. Its manufactures include food, metal, plastic, wood, and paper products.

Hamp·shire (hămp'shîr', -shər), nonmetropolitan county (1976 est. pop. 1,456,100), 1,503 sq mi (3,892.8 sq km), S central England; adminstrative center Winchester. Hampshire is primarily an agricultural county, devoted to sheep raising and dairy farming. There is much evidence of prehistoric and Roman settlement in the county. Hampshire was part of the Anglo-Saxon kingdom of Wessex and has numerous historical and literary associations. In 1974 a small area in the southwest was transferred to Dorset.

Hampshire. **1.** County (1970 pop. 123,981), 529 sq mi (1,370.1 sq km), W central Mass., in an agricultural and forest area bisected by the Connecticut River and drained by the Westfield River; formed 1662; co. seat Northampton. It has a growing amount of manufacturing and service industries. **2.** County (1970 pop. 11,710), 639 sq mi (1,655 sq km), NE W.Va., in the Eastern Panhandle bounded on the N by the Potomac River, on the E by the Va. border, and traversed by valleys and ridges of the Appalachians; formed 1753; co. seat Romney. It has agriculture (livestock, dairy products, and fruit), timber, limestone quarries, and some manufacturing.

Hamp·ton (hăm'tən, hămp'-), since 1965 part of the Greater London borough of Richmond upon Thames, SE England, on the Thames River. It is the site of Hampton Court Palace, begun by Cardinal Wolsey in 1514 as his private residence. After his downfall it was taken by Henry VIII and remained a royal residence until the time of George II. William III had part of it torn down and rebuilt by Christopher Wren. Much of it is open to the public.

Hampton, county (1970 pop. 15,878), 562 sq mi (1,455.6 sq km), S S.C., bounded on the SW by the Savannah River and drained by the Coosawhatchie River; formed 1878; co. seat Hampton. Its agriculture includes peanuts, strawberries, potatoes, and cotton. The county also has extensive timberland and hunting grounds.

Hampton. **1.** City (1974 est. pop. 4,450), seat of Franklin co., N central Iowa, S of Mason City; founded 1856, inc. 1871. It is a trade and processing center in a farm and livestock area. **2.** Town (1970 pop. 8,011), Rockingham co., SE N.H., on the Atlantic coast SSW of Portsmouth; settled 1638, inc. 1639. Shoes are manufactured here. **3.** Town (1970 pop. 2,966), seat of Hampton co., S S.C., WNW of Charleston, in a cotton and farm area; inc. 1879. **4.** Independent city (1978 est. pop. 130,100), SE Va., a port of Hampton Roads at the mouth of the James River, connected to Norfolk by bridge and by tunnel; settled 1610 by colonists from Jamestown, inc. 1849. It has a large seafood packing and shipping industry (fish, crabs, and oysters) and some light manufacturing. One of the oldest continuous English settlements in the country, Hampton was founded on the site of an Indian village. It was attacked by pirates in the late 17th cent., shelled in the Revolutionary War, sacked by the British in 1813, and burned almost to the ground by Confederates in 1861 to prevent its possession by Union troops. St. John's Episcopal Church (est. 1610), with a Bible dating from 1599 and communion silver that belonged to the Jamestown settlers is here. Langley Air Force Base (est. 1917), the National Aeronautics and Space Administration's Langley Research Center, and historic Fort Monroe are nearby.

Hampton National Historic Site: *see* National Parks and Monuments Table.

Hampton Roads (rōdz), roadstead, 4 mi (6.4 km) long and 40 ft (12.2 m) deep, SE Va., through which the waters of the James, Nansemond, and Elizabeth rivers pass into Chesapeake Bay. One of the finest natural harbors in the world, it has been a major anchorage point since colonial times. The tunnel under the Roads, opened in 1957, is one of the longest vehicular tunnels (7,479 ft/2,281 m) in the United States. Hampton Roads was the site of the Civil War battle (Mar., 1862) between the ironclads *Monitor* and *Merrimack*.

Ham·tramck (hăm-trăm'ĭk), city (1978 est. pop. 21,100), Wayne co., SE Mich., within the confines of Detroit; inc. as a city 1922. Automobiles and automotive parts, paint, plastic products, and sausages are manufactured. The site was settled by Frenchmen in the late 18th cent. The city grew quickly after the coming of the automobile industry c.1910.

Ha·mun-e Hel·mand or **Ha·mun-i-Hel·mand** (both: hä-mo͞on'ē hĕl'-mänd), marshy lake in the Seistan, c.5,000 sq mi (12,950 sq km), on the Iran-Afghanistan border. It is the largest single expanse of fresh water within the central plateau of Iran. The lake varies in size during the year, achieving its maximum extent in the late spring.

Han (hän). **1.** River of S China, 210 mi (338 km) long, rising in W Fukien prov. and flowing S through Kwangtung prov. to the South China Sea at Shan-t'ou. The densely populated delta is a rich agricultural area. **2.** River of central China, c.700 mi (1,125 km) long, rising in SW Shensi prov. and flowing E between the Tsinling and the Ta-pa mts., then SE through Hupeh prov. to join the Yangtze at Wu-han. The river floods its fertile lower valley in summer.

Ha·nau (hä'nou'), city (1974 est. pop. 88,017), Hesse, central West Germany, on the Main and Kinzig rivers. It is a center of the West German jewelry industry. Other manufactures include rubber goods, lamps, and machinery. Hanau was chartered in 1303 and in the 16th cent. accepted refugees from the Low Countries who contributed significantly to the city's economic growth. Hanau passed to Hesse-Kassel in 1736, and with it to Prussia in 1866.

Han-chung (hän'jo͝ong'), city (1970 est. pop. 120,000), SW Shensi prov., China, on the Han River, near the Szechwan border. It is a major agricultural and trade center.

Han·cock (hăn'kŏk'). **1.** County (1970 pop. 9,019), 478 sq mi (1,238 sq km), central Ga., bounded on the E by the Ogeechee River, on the W by the Oconee River, and intersected by the Fall Line; formed 1793; co. seat Sparta. It has agriculture (cotton, corn, pecans, and livestock) and a lumbering industry. Clothing and furniture are manufactured. **2.** County (1970 pop. 23,664), 797 sq mi (2,064.2 sq km), W Ill., bounded on the W by the Mississippi River, here dammed into Lake Keokuk, and drained by the La Moine River and Bear Creek; formed 1825; co. seat Carthage. In an agricultural area yielding corn, wheat, soybeans, fruit, dairy products, livestock, and poultry, it has bituminous-coal mines, limestone quarries, sand pits, lumber, and commercial fisheries. Its manufactures include natural and processed cheese and wood products. **3.** County (1970 pop. 35,096), 305 sq mi (790 sq km), central Ind., drained by Sugar and Brandywine creeks and the Big Blue River; formed 1827; co. seat Greenfield. Livestock, grain, and vegetables are produced. It has lumbering and food-processing industries and plants manufacturing clothing, chemicals, concrete, and metal products. **4.** County (1970 pop. 13,506), 570 sq mi (1,476.3 sq km), N Iowa, drained by branches of the Iowa River and Lime Creek; formed 1851; co. seat Garner. It is in a prairie agricultural area yielding hogs, cattle, poultry, corn, oats, and sugar beets. There are small lakes, peat beds, and sand and gravel pits here. **5.** County (1970 pop. 7,080), 187 sq mi (484.3 sq km), NW central Ky., bounded on the N and NE by the Ohio River and the Ind. border and drained by headstreams of Panther Creek; formed 1829; co. seat Hawesville. It has agriculture (burley tobacco, livestock, corn, grain, and truck crops), coal mines, and oil wells. Paper, clay, and aluminum products are made. **6.** County (1970 pop. 34,590), 1,536 sq mi (3,978.2 sq km), S and SE Maine, on the Atlantic coast and drained by the Penobscot River; formed 1789; co. seat Ellsworth. It has farming and dairying, lead and zinc ores, granite quarrying, lumber (logging, sawmills), hunting, and fishing. Its industry includes food processing and shipbuilding. Its bays, islands, and inland lakes are resort sites. **7.** County (1970 pop. 17,387), 485 sq mi (1,256.2 sq km), S Miss., in a coastal area bounded on the S by Mississippi Sound, on the W by the Pearl River and the La. border; formed 1812; co. seat Bay St. Louis. Its agriculture includes corn, truck crops, pecans, livestock, and poultry. It has lumbering and manufactures metal products and guided missiles. There are shore resorts in the county. **8.** County (1970 pop. 61,217), 532 sq mi (1,377.9 sq km), NW Ohio, intersected by the Blanchard River; formed 1820; co. seat Findlay. It has agriculture (livestock and grain), limestone quarries, and manufacturing (processed foods, tires, concrete, steel, and metal products). **9.** County (1970 pop. 6,719), 231 sq mi (598.3 sq km), NE Tenn., bordering in the N on Va. and drained by the Clinch and Powell rivers; formed 1844; co. seat Sneedville. In an agricultural area yielding livestock, tobacco, and fruit, it is traversed by ridges of the Appalachian Mts. **10.** County (1970 pop. 39,749), 82 sq mi (212.4 sq km), N W.Va. at the tip of the Northern Panhandle, bounded on the N and W by the Ohio River, on the E by Pa.; formed 1848; co. seat New Cumberland. In an industrial region, it has steel milling and manufactures clay products, chemicals, metal alloys, and fabricated products. There is some coal mining. Fruit dominates its agricultural production.

Hand (hănd), county (1970 pop. 5,883), 1,436 sq mi (3,719.2 sq km), E central S.Dak., watered by numerous intermittent streams and artificial lakes; formed 1873; co. seat Miller. It is in a level agricultural region that produces wheat, oats, rye, barley, corn, and dairy products and has cattle ranching. Some manufacturing is done.

Han·da (hän'dä), city (1975 pop. 85,824), Aichi prefecture, S central Honshu, Japan, on the Chita Peninsula and Chita Bay. Handa, a fishing port, is a production center for textiles, sake, and soy sauce.

Han·ford (hăn'fərd), city (1978 est. pop. 18,600), seat of Kings co., central Calif.; inc. 1891. It is a trade and processing center of the San Joaquin Valley.

Hang·chow or **Hang-chou** (both: hăng'chou', häng'jō'), city (1970 est. pop. 1,100,000), capital of Chekiang prov., E China. It is on the Ch'ien-t'ang River at the head of Hangchow Bay, an arm of the East China Sea, and handles river traffic through its port. Long a famous silk-producing center, Hangchow has recently been developed into a major industrial complex. Manufactures include silk and cotton textiles, pig iron and steel products, automobiles, fertilizer, pharmaceuticals, cement, rubber, paper and bamboo products, chemicals, machine tools, electronic equipment, and processed tea.

Hangchow was founded A.D. 606 and was from 907 to 960 the capital of a powerful kingdom. Many of the city's picturesque monasteries and shrines date from this period. It was the capital of the Southern Sung dynasty from 1132 to 1276, when it was sacked by Kublai Khan. In the Southern Sung period Hangchow was a center of art, literature, and scholarship and a cosmopolitan city with a large colony of foreign merchants. Marco Polo described it as the finest and noblest city in the world. It was occupied by the Japanese from 1937 to 1945, and fell to the Chinese Communists in 1949. Hangchow Bay has a spectacular tidal bore, 5 to 15 ft (1.5–4.6 m) high, which menaces shipping.

Han·gö (häng'gö) or **Han·ko** (häng'kō), city (1970 pop. 9,686), Uusimaa prov., SW Finland, on the Baltic Sea. A popular resort and a manufacturing town, it is the most important winter port in Finland.

Han·ni·bal (hăn'ə-bəl), city (1978 est. pop. 17,800), Marion and Ralls cos., NE Mo., on the Mississippi River; inc. 1845. Its industries include meat canning, printing, and the manufacture of boats, candy, cement, and lumber and metal products. The city is famous as the boyhood home of Mark Twain; his house has been preserved, and a museum and statue commemorate him.

Ha·noi (hăn'oi, hə-noi'), city (1970 est. pop. 414,620), N Vietnam, on the right bank of the Red River. It is a transportation hub, a manufacturing center, and an important shipping point for agricultural and industrial products. Hanoi became (7th cent.) the seat of the Chinese rulers of Vietnam. Its Chinese name, Dong Kinh or Tong King, became Tonkin and was applied by Europeans to the entire region. Hanoi was occupied briefly by the French in 1873 and passed to them 10 years later. It became the capital of French Indochina after 1887. Occupied by the Japanese in 1940, Hanoi was liberated in 1945, when it became the seat of the Vietnam government. From 1946 to 1954 it was the scene of heavy fighting between the French and Viet Minh forces. After the French evacuated Hanoi in accordance with the Geneva Conference (July, 1954), the city became the capital of North Vietnam. During the Vietnam War the city remained remarkably intact despite heavy U.S. bombings (1965–68 and 1972).

Han·o·ver (hăn'ō-vər), former independent kingdom and former province of Germany, Lower Saxony, N West Germany. Hanover stretched from the Dutch border and the North Sea in the northwest to the Harz Mts. in the southeast. Most of the territory was included in the duchy of Brunswick, which the house of the Guelphs retained after 1180. In the repeated subdivisions of Brunswick among the various branches of the family, the branch of Brunswick-Lüneburg emerged as the most powerful. In 1692 Duke Ernest Augustus of Calenberg was raised to the rank of elector. His lands became known as the electorate of Hanover. The marriage of Ernest Augustus to Sophia, granddaughter of James I of England, brought (1714) the English throne to his son, Elector George Louis (George I of England). Personal union of Great Britain and Hanover continued under the house of Hanover.

Napoleon I gave the electorate to Prussia in 1805, but in 1807 he assigned part of Hanover to the kingdom of Westphalia under his brother Jérôme Bonaparte, the remainder being divided in 1810 between France and Westphalia. In 1813 Great Britain regained possession, and in 1815 the Congress of Vienna raised Hanover to a kingdom, with membership in the German Confederation. At the accession (1837) of Queen Victoria in England, Hanover was separated from the British crown because of the Salic law of succession. Ernest Augustus, son of George III, became king of Hanover and began his reign by rescinding the constitution of 1833; the Revolution of 1848 forced him to grant a liberal constitution. His son, George V, refused to support Prussia in the Austro-Prussian War (1866) and, as a consequence, lost his kingdom, which was made a Prussian (from 1871 a German) province. After World War II the province was incorporated into Lower Saxony.

Hanover, city (1974 est. pop. 562,951), capital of Lower Saxony, N West Germany, on the Leine River and the Midland Canal. It is a major industrial, commercial, and transshipment center. Manufactures include iron and steel, tires, food products, printed materials, machinery, textiles, and motor vehicles. Hanover was chartered in 1241 and in 1369 passed to Brunswick. In 1386 it joined the Hanseatic League. In 1692 it became the capital of the electorate of Hanover. Hanover was badly damaged in World War II, but after 1945 numerous old buildings were reconstructed and many modern structures were erected. Hanover is the seat of technical, medical, and veterinary universities and several museums. George I of England is buried in Hanover.

Hanover, county (1970 pop. 37,479), 466 sq mi (1,206.9 sq km), E central Va., bounded on the N and NE by the North Anna and Pamunkey rivers, on the S by the Chickahominy; formed 1720; co. seat Hanover. Its eastern section is on the coastal plain; its western section is partially in the Piedmont. Agriculture, especially truck crops, tobacco, livestock, peanuts, and hay, is important.

Hanover. 1. Town (1970 pop. 10,107), Plymouth co., SE Mass., on the North River, in a farm area; settled 1649, set off from Scituate and Abington 1727. It has some light manufacturing. **2.** Town (1970 pop. 8,494), Grafton co., W N.H., near the Connecticut River NW of Lebanon; granted 1761, settled 1765. Dartmouth College is here. **3.** Borough (1978 est. pop. 14,400), York co., SE Pa.; inc. 1815. Industries include food processing and the manufacture of shoes, textiles, twine, and metal products. Standardbred horses are raised. **4.** Village (1970 pop. 300), seat of Hanover co., E central Va., NNE of Richmond. Patrick Henry lived here for many years.

Hans·ford (hănz'fərd), county (1970 pop. 6,351), 906 sq mi (2,349.1 sq km), extreme N Texas, in the high grassy plains of the Panhandle, bordered on the N by Okla.; formed 1876; co. seat Spearman. In a wheat-growing area, it also produces beef cattle, sheep, poultry, hogs, and dairy products. It has oil and natural-gas wells.

Han·son (hăn'sən), county (1970 pop. 3,781), 431 sq mi (1,116.3 sq km), SE S.Dak., drained by the James River; formed 1871; co. seat Alexandria. It is in an agricultural (corn, wheat, and dairy products) and livestock-raising region. It has some manufacturing.

Hanson, town (1970 pop. 7,148), Plymouth co., SE Mass., SSE of Boston; settled 1632, set off from Pembroke 1820. Cranberries are canned here.

Han-tan or **Han·tan** (both: hän'dän'), city (1970 est. pop. 500,000), SW Hopeh prov., China. It is a newly flourishing industrial center, with an iron mine, cotton mills, and chemical and cement plants.

Har·al·son (hăr'əl-sən), county (1970 pop. 15,927), 285 sq mi (738.2 sq km), W Ga., in a piedmont area drained by the Tallapoosa River and bounded on the W by the Ala. border; formed 1856; co. seat Buchanan. Its agriculture includes cotton, corn, grain, fruit, and livestock. Textiles, clothing, and wood products are made.

Ha·ran or **Har·ran** (both: hä-rän'), ancient city of Mesopotamia, now in SE Asiatic Turkey, SE of Urfa. It was an important trade center and the seat of the Assyrian moon god. The Babylonians defeated the Assyrian army at Haran in 609 B.C.

Ha·rar or **Har·rar** (both: hä'rər), city (1974 est. pop. 53,560), capital of Harar prov., E central Ethiopia, at an altitude of c.6,000 ft (1,830 m). It is the trade center for a region where coffee, cereals, and cotton are produced. Harar was probably founded in the 7th cent. The city maintained a precarious independence until its occupation (1875–85) by Egypt. In 1887 it was incorporated into Ethiopia by Menelik II. A walled city, Harar was long a center of Islamic learning.

Har·bin (här'bēn', -bĭn') or **Ha-erh-pin** (hä'ĕr'bĭn'), city (1970 est. pop. 2,750,000), capital of Heilungkiang prov., China, on the Sungari River. Harbin has metallurgical, machinery, chemical, petroleum, and coal industries. There are also railroad shops, food-processing establishments, and plants making aircraft, tractors, ball bearings, precision instruments, cutting tools, electrical and electronic equipment, fertilizer, and lead pencils. Harbin was unimportant until Russia was granted a concession (1896) and built a modern section alongside the old Chinese town. Russia surrendered its concession in 1924.

har·bor (här'bər): *see* Port.

Har·dan·ger·fjord (här-däng'ər-fyôrd'), second-largest fjord of Norway, penetrating 114 mi (183.4 km) from the Atlantic Ocean into Hordaland co., SW Norway. The valleys of the Hardangerfjord are fertile and dotted with picturesque villages. Extending inland from the fjord is the Hardangerfjell, a mountain mass rising to 6,153 ft (1,876.7 m).

Har·dee (här'dē), county (1970 pop. 14,889), 630 sq mi (1,631.7 sq km), central Fla., in a rolling terrain with some swamps and many small lakes; formed 1921; co. seat Wauchula. It is drained by the Peace River. Citrus fruit, strawberries, and truck crops are grown. Cattle and poultry are also important.

Har·de·man (här'də-mən). **1.** County (1970 pop. 22,435), 655 sq mi (1,696.5 sq km), SW Tenn., bounded on the S by the Miss. border and drained by the Hatchie River; formed 1823; co. seat Bolivar. Cotton, corn, fruit, and truck crops are grown, and livestock is raised. **2.** County (1970 pop. 6,795), 685 sq mi (1,774.2 sq km), N Texas, bounded on the S by the Pease River, on the N by the Prairie Dog Fork of the Red River and the Okla. border; formed 1858; co. seat Quanah. Its agriculture includes wheat, grain sorghums, cotton, cattle, poultry, and dairy products. It also has gypsum mines, some oil wells, clay deposits, and hunting and fishing resorts.

Har·din (här'dn, -dĭn). **1.** County (1970 pop. 4,914), 183 sq mi (474 sq km), extreme SE Ill., bounded on the S and E by the Ohio River; formed 1839; co. seat Elizabethtown. It has agriculture (wheat, corn, dairy products, poultry, and livestock), lead, zinc, and fluorspar mines, and limestone quarries. Clothing is manufactured. Part of Shawnee National Forest is here. **2.** County (1970 pop. 22,248), 574 sq mi (1,486.7 sq km), central Iowa, drained by the Iowa River; formed 1851; co. seat Eldora. Its prairie agricultural products include hogs, cattle, poultry, corn, oats, and soybeans. Limestone quarries, sand and gravel pits, and coal deposits are among its natural resources. **3.** County (1970 pop. 78,421), 616 sq mi (1,595.4 sq km), N central Ky., bounded on the N by the Ohio River and the Ind. line, on the E by the Salt River and Rolling Fork, and drained by the Rough and Nolin rivers; formed 1792; co. seat Elizabethtown. It is in a gently rolling agricultural area (livestock, burley tobacco, corn, and wheat) with limestone quarries, sand pits, and asphalt deposits. Its manufactures include plastics and sheet metal. **4.** County (1970 pop. 30,813), 467 sq mi (1,209.5 sq km), NW central Ohio, intersected by the Scioto, Blanchard, and Ottawa rivers; formed 1820; co. seat Keaton. Its agriculture includes livestock, poultry, dairy products, onions, grain, and peppermint. It also has lumbering and food-processing industries, limestone quarries, and gravel pits. Paper, plastics, and metal products are manufactured. **5.** County (1970 pop. 18,212), 587 sq mi (1,520.3 sq km), SW Tenn., bounded on the S by the Miss. and Ala. borders, on the NE by the Tennessee River; formed 1819; co. seat Savannah. In a timber and agricultural area yielding livestock, cotton, corn, and hay, it also has limestone deposits and some manufacturing. Shiloh National Military Park is in the southwest. **6.** County (1970 pop. 29,996), 895 sq mi (2,318.1 sq km), E Texas, bounded on the E by the Neches River and drained by its tributaries; formed 1858; co. seat Kountze. Truck crops, corn, and potatoes are grown. The county also has cattle ranches and oil and natural-gas fields.

Hardin. 1. Village (1970 pop. 1,035), seat of Calhoun co., W Ill., on the Illinois River NW of Alton. It is the trade center of an apple and grain area. **2.** City (1970 pop. 2,733), seat of Big Horn co., S central Mont., on the Bighorn River E of Billings; settled 1906, inc. 1911. A refinery processes sugar beets from the surrounding area.

Har·ding (här'dĭng). **1.** County (1970 pop. 1,348), 2,136 sq mi (5,532.2 sq km), NE N.Mex., watered by branches of the Canadian River; formed 1921; co. seat Mosquero. In a livestock-grazing and grain area, it has oil fields. **2.** County (1970 pop. 1,855), 2,683 sq mi (6,949 sq km), extreme NW S.Dak., bordered on the W by Mont., on the N by N.Dak., and watered by the Little Missouri River; formed 1909; co. seat Buffalo. It is in a cattle-raising area, with large deposits of lignite coal and oil fields.

Har·dins·burg (här'dnz-bûrg'), town (1970 pop. 1,547), seat of Breckenridge co., central Ky., SW of Louisville, in a timber and farm area.

Har·dy (här'dē), county (1970 pop. 8,855), 585 sq mi (1,515.2 sq km), NE W.Va., in the Eastern Panhandle bounded on the E and S by the Va. border and drained by the South Branch of the Potomac River; formed 1785; co. seat Moorefield. Its agriculture includes livestock, dairy products, poultry, fruit, and vegetables. It has an extensive food-processing industry and also manufactures wood products. The county is traversed by ridges and valleys of the Alleghenies. Part of George Washington National Forest is here.

Har·fleur (här-flœr', är-), town (1968 pop. 15,598), Seine-Maritime dept., N France, at the mouth of the Seine River on the English Channel. It was a flourishing port during the later Middle Ages but declined because of silting in the 16th cent. The seige and capture (1415) of Harfleur by the English in the Hundred Years' War is described by Shakespeare in *Henry V.*

Har·ford (här'fərd), county (1970 pop. 115,378), 448 sq mi (1,160.3 sq km), NE Md., bounded on the N by the Pa. border, on the NE by the Susquehanna River, on the SE and S by Chesapeake Bay; formed 1775; co. seat Bel Air. The piedmont agricultural area in the north produces grain, vegetables, fruit, poultry, and dairy products. The coastal-plain area on the southern fringe is occupied by Federal reservations, including Aberdeen Proving Grounds. It has commercial fisheries, shore resorts, lumbering and food-processing industries, and varied manufactures.

Har·ghes·sa or **Har·gei·sa** (both: här-gā'sä), town (1966 est. pop. 42,000), N Somalia. It is a commercial center and watering place for nomadic stock herders. The town was taken in 1870 by Egyptian forces. The British later took control and, in 1941, made Harghessa the capital of British Somaliland.

Har·in·gey (hâr'ĭng-gā'), borough of Greater London (1971 pop. 236,956), SE England. Haringey was created in 1965 by the merger of the municipal boroughs of Hornsey, Tottenham, and Wood Green. It is mainly residential. Tottenham has furniture, light engineering, children's clothing, and printing industries.

Ha·ri Rud (här'ē ro͞od'), river, c.700 mi (1,125 km) long, rising in the Kuh-e Baba range, central Afghanistan, and flowing W and then N into the steppes S of the Kara Kum desert in the Turkmen SSR; its lower course forms part of the Afghanistan–USSR border.

Har·lan (här'lən). **1.** County (1970 pop. 37,370), 469 sq mi (1,214.7 sq km), SE Ky., in the Cumberlands bounded on the S and E by the Va. border and drained by the Cumberland River; formed 1819; co. seat Harlan. In a rich coal-mining area, it has hardwood timber and some agriculture (corn, apples, potatoes, tobacco, poultry, livestock, and dairy products). Labor conflicts between coal operators and miners prompted the nickname "Bloody Harlan" for the county. The mines were unionized in 1941 after 20 years of strife. **2.** County (1970 pop. 4,357), 556 sq mi (1,440 sq km), S Nebr., bounded on the S by Kansas and drained by the Republican River; formed 1871; co. seat Alma. It is primarily agricultural, yielding corn, grain sorghums, alfalfa, wheat, and cattle. Some mining and manufacturing are done.

Harlan. 1. City (1976 est. pop. 5,251), seat of Shelby co., W central Iowa, NE of Council Bluffs, in a farm and livestock area; settled 1858, inc. 1879. **2.** City (1975 est. pop. 2,900), seat of Harlan co., SE Ky., SW of Lynch, in the Cumberlands near the Va. border, in the heart of a coal region; settled 1819. After the railroad's arrival in 1911, Harlan developed as a shipping point for coal and lumber. Indian mounds in the area have yielded many relics.

Har·lech (här'lĕkh, -lĕk, här'lē), village, Merionethshire, W Wales. The ancient village with its 13th cent. castle rests on a cliff 200 ft (61 m) above the modern seaside village. The heroic defense of the castle against the Yorkists in the War of the Roses is the theme of the Welsh battle song, "The March of the Men of Harlech."

Har·lem (här'ləm), residential and business section of upper Manhattan, New York City, bounded roughly by 110th St., the East River and Harlem River (a tidal channel separating Manhattan from the Bronx), 168th St., Amsterdam Ave., and Morningside Park. The Dutch settlement of Nieuw Haarlem was established by Peter Stuyvesant in 1658. To the west of Harlem, near the present site of Columbia Univ., British and Continental forces fought (Sept. 16, 1776) the Battle of Harlem Heights. Harlem remained rural until the 19th cent., when it became a fashionable residential section. With the rapid influx of blacks that began c.1910, Harlem became one of the largest black communities in the United States. After World War II many Puerto Ricans and other Hispanic Americans settled in East Harlem (also called Spanish Harlem).

Har·ling·en (här'lĭng-ən), city (1978 est. pop. 42,900), Cameron co., extreme S Texas; inc. 1910. It is a shipping and processing center in the lower Rio Grande valley, an irrigated farming area yielding citrus and other fruits, truck crops, and cotton. The city, which is linked to the Intracoastal Waterway by a barge channel, has canneries, cotton-processing plants, and factories making wearing apparel, aircraft, and metal and concrete products. Harlingen was founded (c.1904) with the coming of the railroad.

Har·low (här'lō), new town and urban district (1976 est. pop. 80,000), Essex, E England. Harlow was designated one of the new towns in 1946 to alleviate overpopulation in London. Among its industries are furniture making, metallurgy, and printing.

Har·low·ton (här'lō-tən), city (1970 pop. 1,375), seat of Wheatland co., central Mont., on the Musselshell River SE of Great Falls; inc. 1917. It is the trade center of a grain and livestock area.

Har·mon (här'mən), county (1970 pop. 5,136), 545 sq mi (1,411.6 sq km), SW Okla., bounded on the S and W by Texas and drained by the Salt and Prairie Dog Town forks of the Red River; formed 1909; co. seat Hollis. In a hilly agricultural area yielding livestock, cotton, wheat, and sweet potatoes, it has cottonseed-oil mills.

Har·nett (här'nĕt, -nĭt), county (1970 pop. 49,667), 603 sq mi (1,561.8 sq km), central N.C., in a forested area bounded on the SE by the Little River; formed 1855; co. seat Lillington. Tobacco, cotton, and corn are its major crops. Textiles are manufactured.

Har·ney (här'nē), county (1970 pop. 7,215), 10,166 sq mi (26,330 sq km), SE central Oregon, in a livestock-grazing area bounded on the S by the Nev. border; formed 1889; co. seat Burns. Lumbering and agriculture are also important.

Härn·ö·sand (hĕr-nə-sänd'), city (1975 est. pop. 23,900), capital of Västernorrland co., E Sweden, on the Gulf of Bothnia at the mouth of the Ångermanälven River. Its harbor, icebound in winter, is a center of the coastal trade. Manufactures include machinery, processed tobacco, and ships. The city was chartered in 1585. It was burned in 1721 by Russians.

Har·pen·den (här'pən-dən), urban district (1971 pop. 24,161), Hertfordshire, E central England. Mainly residential, it is the site of Rothamsted Experimental Station (for agricultural research).

Har·per (här'pər). **1.** County (1970 pop. 7,871), 801 sq mi (2,074.6 sq km), S Kansas, in a plains region bordered on the S by Okla. and drained by the Chikaskia River; formed 1867; co. seat Anthony. It has agriculture (wheat, livestock, and poultry) and some mining. Farm machinery is made. **2.** County (1970 pop. 5,151), 1,041 sq mi (2,696.2 sq km), NW Okla., bounded in the N by Kansas and intersected by the North Canadian and Cimarron rivers; formed 1907; co. seat Buffalo. In a plains agricultural area yielding livestock, wheat, barley, and broomcorn, it has oil and natural-gas wells.

Har·pers Fer·ry (här'pərz fĕr'ē), town (1970 pop. 423), Jefferson co., E W.Va., at the confluence of the Shenandoah and Potomac rivers; inc. 1763. The town is a tourist attraction, known for its history and its scenic beauty. John Brown seized the U.S. arsenal here on Oct. 16, 1859. Later, in the Civil War, it was held mostly by Union soldiers, but changed hands a number of times. Harpers Ferry never recovered economically. Of particular interest are the fire engine house in which John Brown was captured and the John Brown Museum. Harpers Ferry National Historical Park (est. 1963; 1,530 acres/619.7 hectares) receives many visitors annually.

Har·ran (hä-rän'): *see* Haran, Mesopotamia.

Har·rar (här'ər): *see* Harar, Ethiopia.

Har·ris (hăr'ĭs). **1.** County (1970 pop. 11,520), 465 sq mi (1,204.4 sq km), W Ga., in a piedmont area bounded on the W by the Ala. border and the Chattahoochee River; formed 1827; co. seat Hamilton. Its agriculture includes livestock, poultry, cotton, corn, grain, truck crops, and fruit. It is in a timber area and has furniture and plywood factories and weaving mills. **2.** County (1970 pop. 1,741,-912), 1,723 sq mi (4,462.6 sq km), SE Texas, in the coastal plain bounded on the SE by Galveston Bay and drained by the San Jacinto River and its tributaries; formed 1836; co. seat Houston. It has oil and natural-gas wells and refineries, salt, sulfur, and clay deposits, and agriculture (cattle, poultry, rice, wheat, truck crops, and dairy products). There are many resorts along the bay. Part of Sam Houston National Forest is in the north.

Har·ris·burg (hăr'ĭs-bûrg'). **1.** Town (1970 pop. 1,931), seat of Poinsett co., NE Ark., S of Jonesboro. It is a trade and processing center in a timber, cotton, and rice area. **2.** City (1970 pop. 9,535), seat of Saline co. S central Ill., E of Marion, in a coal, grain, and timber area; platted in the 1850s, inc. 1861. **3.** Village (1970 pop. 100), seat of Banner co., W Nebr., S of Scottsbluff, in a grain area. **4.** City (1978 est. pop. 55,200), state capital and seat of Dauphin co., SE Pa., on the Susquehanna River; settled c.1710, inc. 1791. Rich iron and coal mines are nearby, and the city has a large steel industry. Food processing is also important. Other manufactures include metal products, rails, airplane parts, clothing, textiles, and shoes. Harrisburg became the state capital in 1812 and grew as an inland transportation center with the opening of the Pennsylvania Canal in 1827 and the arrival of the railroad in 1836. Its sprawling Italian Renaissance state capitol (completed 1906) has a 272-ft (83-m) dome modeled after St. Peter's in Rome.

Har·ri·son (hăr'ĭ-sən). **1.** County (1970 pop. 20,423), 479 sq mi (1,240.6 sq km), S Ind., bounded on the E, S, and SW by the Ohio River and the Ky. border, on the W by the Blue River, and drained by Indian and Buck creeks; formed 1808; co. seat Corydon. Its agriculture includes grain and poultry. It also has deposits of natural gas, stone quarries, and lumbering and food-processing industries. Furniture is manufactured. **2.** County (1970 pop. 16,240), 693 sq mi (1,794.9 sq km), W Iowa, bounded on the W by the Missouri River and the Nebr. border and drained by the Boyer and Soldier rivers; formed 1851; co. seat Logan. In a prairie agricultural area yielding wheat, corn, fruit, cattle, and hogs, it also has bituminous-coal mines. **3.** County (1970 pop. 14,158), 308 sq mi (797.7 sq km), N Ky., in a gently rolling upland area of the Bluegrass and bounded on the NE by the Licking River; formed 1793; co. seat Cynthiana. It has agriculture (tobacco, grain, cattle, hogs, poultry, and dairy products), limestone quarries, a lumbering industry, whiskey distilleries, and meat-packing plants. Iron and steel, metal products, and machinery are made. **4.** County (1970 pop. 134,582), 585 sq mi (1,515.2 sq km), SE Miss., bounded on the S by Mississippi Sound and drained by the Biloxi and Wolf rivers; formed 1841; co. seat Gulfport. Its agriculture includes corn, pecans, truck crops, dairy products, and livestock. It also has lumbering, some mining, and an extensive seafood-processing industry. Its manufactures include clothing, chemicals, glass, concrete, wood and metal products, and machinery. Part of De Soto National Forest is here. **5.** County (1970 pop. 10,257), 720 sq mi (1,864.8 sq km), N Mo., bordering on Iowa in the N; formed 1845; co. seat Bethany. Corn, wheat, oats, and livestock are raised. It has some mining and factories producing hats, concrete, and farm machinery. **6.** County (1970 pop. 17,013), 401 sq mi (1,038.6 sq km), E Ohio, drained by the Stillwater and Conotton creeks; formed 1813; co. seat Cadiz. Its economy is based on coal mining, agriculture (cattle, sheep, poultry, grain, and dairy products), limestone quarrying, and some manufacturing. **7.** County (1970 pop. 44,841), 892 sq mi (2,310.3 sq km), NE Texas, bounded on the E by the La. border, on the NE by Caddo Lake, on the SW by the Sabine River, and drained by Little Cypress Bayou; formed 1839; co. seat Marshall. In a hilly wooded region with an extensive lumbering industry, it has agriculture (cotton, corn, forage crops, fruit, truck crops, peanuts, dairy products, and livestock), oil, natural-gas, and lignite deposits, and a clay-products industry. **8.** County (1970 pop. 73,028), 418 sq mi (1,082.6 sq km), N W.Va., on the Allegheny Plateau drained by headstreams of the Monongehela River; formed 1784; co. seat Clarksburg. Its agriculture includes livestock, grain, fruit, and tobacco. Lumber, coal mines, and oil and natural-gas wells contribute to its economy. Glass, pottery, and china wares, concrete, machinery, and food products are made.

Harrison. **1.** City (1970 pop. 7,239), seat of Boone co., NW Ark., in the Ozarks NE of Fayetteville; settled c.1820, inc. as a town 1876, as a city 1909. It is the trade and shipping center of a farming, mining, and lumbering area. **2.** City (1970 pop. 1,460), seat of Clare co., central Mich., N of Mt. Pleasant, in a resort region; inc. as a village 1885, as a city 1891. **3.** Village (1970 pop. 377), seat of Sioux co., NW Nebr., WSW of Chadron near the Wyo. border, in a ranching area. Poultry products, grain, livestock, and potatoes are processed. **4.** Town (1970 pop. 11,811), Hudson co., NE N.J., an industrial suburb on the Passaic River opposite Newark; inc. 1869. **5.** Town (1970 pop. 21,544), Westchester co., SE N.Y., a suburb of New York City between Mamaroneck and Rye.

Har·ri·son·burg (hăr′ĭ-sən-bûrg′). **1.** Village (1970 pop. 626), seat of Catahoula parish, E La., on the Ouachita River NW of Natchez, Miss. Its agricultural products include cotton, corn, sugar cane, grain, livestock, and poultry. **2.** City (1978 est. pop. 20,300), independent city and seat of Rockingham co., NW Va., in the Shenandoah Valley; settled 1739, inc. 1916. It is a processing center in a poultry, dairy, and livestock area. Its manufactures include air conditioners, plastic bottles, clothing, furniture, and fertilizers.

Har·ri·son·ville (hăr′ĭ-sən-vĭl′), city (1970 pop. 5,052), seat of Cass co., W Mo., SSE of Kansas City, in a farm and livestock area; laid out 1837, inc. 1857.

Har·ris·ville (hăr′ĭs-vĭl′). **1.** City (1970 pop. 541), seat of Alcona co., NE Mich., SSE of Alpena on Lake Huron. It is a summer resort with nurseries and commercial fisheries. **2.** Residential town (1970 pop. 1,464), seat of Ritchie co., NW W.Va., ESE of Parkersburg, in a farm area; platted 1822. Textiles are made, and oil and natural-gas wells are nearby.

Har·rods·burg (hăr′ədz-bûrg′), city (1970 pop. 6,741), seat of Mercer co., central Ky., S of Frankfort. It is a trade center in a bluegrass area producing livestock, grain, and tobacco. Clothing, glass, and dairy products are made. It is also a tourist and resort city, with mineral springs. Harrodsburg, the oldest settlement in the state, was founded in 1774.

Har·ro·gate (hăr′ō-gĭt, -gāt), municipal borough (1973 est. pop. 64,620), North Yorkshire, N central England. The borough is a health resort, with more than 80 mineral springs.

Har·row (hăr′ō), borough (1971 pop. 202,718), of Greater London, SE England. Until 1965 Harrow was a municipal borough in the former county of Middlesex. For centuries Harrow grew foodstuffs for London. Now it is mainly residential and contains parts of the Green Belt, areas set aside as parkland. Optical and photographic goods and glass are manufactured. Harrow public school, founded in 1571, is in the borough.

Hart (härt). **1.** County (1970 pop. 15,814), 231 sq mi (598.3 sq km), NE Ga., bounded in the E and N by the S.C. border and the Savannah and Tugaloo rivers; formed 1853; co. seat Hartwell. It is an agricultural county in a piedmont area. Cotton, corn, hay, and sweet potatoes are its major crops. **2.** County (1970 pop. 13,980), 420 sq mi (1,087.8 sq km), central Ky., in a rolling agricultural area bounded on the NE by the Nolin River and drained by the Green River; formed 1819; co. seat Munfordville. Livestock, dairy products, poultry, burley tobacco, corn, and wheat are important. Lumbering and some mining are done. Clothing and wood and metal products are manufactured. The county has resorts and includes part of Mammouth Cave National Park.

Hart, city (1970 pop. 2,139), seat of Oceana co., W central Mich., N of Muskegon, in a fruit and forest area; settled 1856, inc. 1885. Food processing is done.

Hart·ford (härt′fərd). **1.** County (1970 pop. 816,737), 741 sq mi (1,919.2 sq km), N central Conn., bounded on the N by the Mass. border and bisected by the Connecticut River; formed 1666; former co. seat Hartford. It is drained by numerous rivers and is primarily industrial and commercial. Its agriculture includes tobacco, dairy products, poultry, truck crops, fruit, corn, potatoes, nursery products, and feed. Among its extensive manufactures are airplanes and airplane engines and parts, machinery, hardware, tools, electrical equipment and appliances, bricks, sports equipment, textiles, paper, clothing, cutlery, furniture, food products, carpets, silverware, rubber, leather, and wood products, chemicals, automobile parts, paint, and metal products.

Hartford. **1.** City (1978 est. pop. 130,100), state capital, Hartford co., central Conn., on the W bank of the Connecticut River; settled as Newtown 1635–36 on the site of a Dutch trading post (1633; abandoned 1654), inc. 1784. It is a port of entry, an insurance center, and a commercial, industrial, and cultural hub. Its insurance business began in 1794. Manufactures include firearms, typewriters, precision instruments, computers, auto parts, electric equipment, and brushes. Hartford formed (1639) with two other towns the Connecticut Colony, adopting the Fundamental Orders. From 1701 to 1875 it was joint capital with New Haven. It was an important military supply depot during the Revolution, and in 1814–15 it hosted the Hartford Convention. Landmarks include the old statehouse (1796; designed by Charles Bulfinch), where the Hartford Convention met; the capitol (completed 1878; designed by Richard M. Upjohn); and the Travelers Insurance tower. The city has a noted art museum (the Wadsworth Atheneum) and a symphony orchestra. Other attractions are the Harriet Beecher Stowe House (1871). The Hartford *Courant,* founded in 1764, is one of the oldest newspapers in the United States. **2.** City (1970 pop. 1,868), seat of Ohio co., W central Ky., on the Rough River SSE of Owensboro, in a coal, oil, limestone, and farm area; founded c.1790. **3.** Town (1970 pop. 6,477), Windsor co., E Vt., on the Connecticut River at the mouth of the White River E of Rutland; settled 1765. **4.** Industrial city (1970 pop. 6,499), Washington co., E Wis., NW of Milwaukee; platted in the 1830s, settled 1844, inc. 1883. Automobile parts and food products are made.

Hartford City, city (1970 pop. 8,207), seat of Blackford co., E Ind., N of Muncie; settled 1832, laid out 1839, inc. 1857. Office equipment and glass and paper products are made here, and there are oil and natural-gas fields nearby.

Har·ting·ton (här′tĭng-tən), city (1970 pop. 1,581), seat of Cedar co., NE Nebr., near the Missouri River WNW of Sioux City, Iowa, in a corn and grain area; inc. 1883.

Har·tle·pool (härt′lē-po͞ol, här′təl-), county borough (1976 est. pop. 97,100), Cleveland, NE England. The county borough was created in 1966 by the merger of the municipal borough of Hartlepool, the rural district of Stockton, and the county borough of West Hartlepool. There are shipbuilding, iron and steel manufacturing, engineering, and brewing industries. Hartlepool is a seaport.

Hart·ley (härt′lē), county (1970 pop. 2,782), 1,489 sq mi (3,856.5 sq km), extreme N Texas, in the high grassy plains of the Panhandle, bounded on the W by the N.Mex. border and drained by Rita Blanca Creek; formed 1876; co. seat Channing. This large-scale cattle-ranching area produces some grain and has natural-gas wells and glass-sand and clay deposits.

Harts·ville (härts′vĭl′). **1.** Town (1970 pop. 8,017), Darlington co., NE S.C., ENE of Columbia; inc. 1891. Textiles, clothing, and oil, food, chemical, and paper products are made. **2.** City (1970 pop. 2,243), seat of Trousdale co., N central Tenn., near the Cumberland River NE of Nashville; settled in the early 1800s, inc. as a town 1833, as a city 1920. It is in a tobacco-raising area.

Hart·ville (härt′vĭl′), city (1970 pop. 524), seat of Wright co., S central Mo., in the Ozarks, on the Gasconade River E of Springfield. Its economy depends on agriculture.

Hart·well (härt'wĕl'), city (1970 pop. 4,865), seat of Hart co., NE Ga., near the S.C. border NE of Athens; inc. 1856. It is a trade and shipping center for a cotton and dairy area.

Har·vard (här'vərd). **1.** City (1970 pop. 5,177), McHenry co., N central Ill., near the Wis. border NE of Rockford, in a dairy area; inc. 1867. Television sets and farm equipment are made. **2.** Town (1970 pop. 12,494), Worcester co., E central Mass.; inc. 1732. A Shaker house and cemetery, an Indian museum, and a Harvard Univ. observatory are here. Nearby is a museum on the site of Fruitlands, a cooperative vegetarian community founded by Bronson Alcott.

Harvard, Mount, peak, 14,414 ft (4,396.3 m) high, W central Colo., in the Sawatch Range of the Rocky Mts.

Har·vey (här'vē), county (1970 pop. 27,236), 540 sq mi (1,398.6 sq km), SE central Kansas, in a flat to gently rolling prairie region, drained by the Little Arkansas River; formed 1872; co. seat Newton. Its agriculture includes wheat, livestock, and poultry, and it has oil and natural-gas fields. Flour, wood products, furniture, plastics, concrete, metal products, and farm machinery are manufactured.

Har·wich (hăr'ĭj, -ĭch), municipal borough (1976 est. pop. 15,280), Essex, E central England, on the estuary of the Stour and the Orwell rivers. Harwich serves as a port for passenger ships to the Continent. The borough's industries are boatbuilding, fishing, light engineering, and cement manufacture. Harwich, an ancient town, was known in the Middle Ages for its port and had an important shipbuilding industry in the 17th cent.

Ha·ry·a·na (hä'rē-ä'nə), state (1971 pop. 9,971,165), 17,120 sq mi (44,340.8 sq km), N central India; capital Chandigarh. The terrain is generally dry, flat, and barren. Cotton is the main product. There are iron-ore deposits. Haryana was created in 1966 out of the Hindi-speaking portions of Punjab state.

Harz (härts), mountain range, on the East German-West German border, extending c.60 mi (100 km) between the Elbe and Leine rivers. The rugged mountains were once densely forested. They culminate in Brocken peak (3,747 ft/1,142.8 m high). Intensive uranium-ore prospecting began after World War II.

Ha·sa, Al (äl hä'sä, ăl hăs'ə), region, E Saudi Arabia, on the Persian Gulf. Oil, dates, wheat, and rice are produced. It was taken from the Turks in 1914 by Ibn Saud.

Has·kell (hăs'kəl). **1.** County (1970 pop. 3,672), 579 sq mi (1,499.6 sq km), SW Kansas, in a flat to rolling prairie region with sand dunes in the extreme N; formed 1887; co. seat Sublette. Its agriculture includes grain and livestock, and it has small natural-gas fields. Chemical products are made. **2.** County (1970 pop. 9,578), 602 sq mi (1,559.2 sq km), E Okla., bounded on the N by the Canadian and Arkansas rivers; formed 1907; co. seat Stigler. Cattle, cotton, corn, oats, and barley are raised. It has timber, coal mining, and oil and natural-gas wells. There is a canning industry for fruit and vegetables. **3.** County (1970 pop. 8,512), 877 sq mi (2,271.4 sq km), NW central Texas, drained by the Double Mountain Fork of the Brazos River; formed 1858; co. seat Haskell. Primarily agricultural, it produces cotton, grain sorghums, wheat, oats, barley, corn, legumes, dairy products, and livestock. There are oil and natural-gas wells.

Haskell, city (1970 pop. 3,655), seat of Haskell co., W central Texas, N of Abilene; settled 1882, inc. 1907. Long a cow town, it now serves an area largely devoted to farming, dairying, and oil producing.

Has·selt (hä'səlt), city (1977 pop. 62,755), capital of Limburg prov., NE Belgium, in the Campine region, a port on the Albert Canal. Hasselt was chartered in 1232. The Dutch defeated the Belgians here in 1831.

Has·tings (hā'stĭngz), county borough (1976 est. pop. 74,600), East Sussex, SE England. A resort and residential city, it has a marine esplanade, parks, and bathing beaches. The site was occupied in Roman times. The decisive Battle of Hastings took place nearby on Oct. 14, 1066, between the victorious Normans under William, duke of Normandy (later William I), and the Anglo-Saxons under Harold.

Hastings, city (1976 pop. 33,960), SE North Island, New Zealand. It has extensive food-processing industries.

Hastings. 1. City (1970 pop. 6,501), seat of Barry co., S Mich., WSW of Lansing; settled c.1836, inc. as a village 1855, as a city 1871. Aluminum products are made. **2.** City (1970 pop. 12,195), seat of Dakota co., SE Minn., on bluffs above the Mississippi opposite its confluence with the St. Croix River; inc. 1857. It is a farm trade and manufacturing center. **3.** City (1978 est. pop. 22,900), seat of Adams co., S central Nebr.; inc. 1874. Farm equipment is made.

Has·tings-on-Hud·son (hā'stĭngz-ŏn-hŭd'sən), residential and industrial village (1970 pop. 9,479), Westchester co., SE N.Y., on the E bank of the Hudson River NNW of Yonkers; inc. 1879. Copper wire and chemicals are made.

Hatch·ie (hăch'ē), river, c.180 mi (289.6 km) long, rising in N Miss. and flowing NW into the Mississippi River N of Memphis, Tenn.

Hat·field (hăt'fēld'), new town and civil parish (1971 pop. 25,211), Hertfordshire, SE England. Hatfield was designated one of the New Towns in 1948 to alleviate overpopulation in London. Aircraft works are the town's most notable industry.

Hat·ter·as, Cape (hăt'ər-əs), promontory on Hatteras Island, a low, sandy, barrier bar between the Atlantic Ocean and Pamlico Sound, E N.C. Called the "Graveyard of the Atlantic," the cape experiences frequent storms that drive ships landward. Cape Hatteras National Seashore (28,500 acres/11,542.5 hectares; est. 1937), a vast expanse of sand and water, is made up of Hatteras, Bodie, and Ocracoke islands and comprises one of the largest stretches of undeveloped seashore on the U.S. Atlantic coast. Cape Hatteras Lighthouse (built 1870) is the only lighthouse owned by the National Park Service.

Hat·ties·burg (hăt'ēz-bûrg'), city (1978 est. pop. 39,600), seat of Forrest co., SE Miss., on the Leaf River; inc. 1884. It is the rail, trade, and industrial center of a farm and timber area. Once a great lumbering city, it now has many diverse industries.

Hau·ge·sund (hou'gə-sŏŏn'), city (1977 est. pop. 27,211), Rogaland co., S Norway, a port on the North Sea. It has large fisheries and industries producing processed fish and aluminum. Nearby are numerous Viking monuments, including the grave of Harold I (9th cent.).

Hau·ran (hou-rän'): *see* Hawran, Syria.

Haute-Ga·ronne (ōt-gä-rôn'), department (1975 pop. 777,431), 2,433 sq mi (6,301.5 sq km), S France, in Languedoc, bordering Spain; capital Toulouse.

Haute-Loire (ōt-lwär'), department (1975 pop. 205,491), 1,917 sq mi (4,965 sq km), S central France, largely in the Massif Central; capital Le Puy.

Haute-Marne (ōt-märn'), department (1975 pop. 212,304), 2,400 sq mi (6,216 sq km), NE France, largely in Champagne; capital Chaumont.

Hautes-Alpes (ōt-zälp'), department (1975 pop. 97,358), 2,131 sq mi (5,519.3 sq km), SE France, mainly in Dauphiné, bordering on Italy; capital Gap.

Haute-Saône (ōt-sōn'), department (1975 pop. 222,254), 2,063 sq mi (5,343.2 sq km), E France, in Franche-Comté; capital Vesoul.

Haute-Sa·voie (ōt-sä-vwä'), department (1975 pop. 447,795), 1,696 sq mi (4,392.6 sq km), E France, in the N part of the old duchy of Savoy; capital Annecy.

Hautes-Py·ré·nées (ōt-pē-rā-nā'), department (1975 pop. 227,222), 1,740 sq mi (4,506.6 sq km), SW France, in parts of Bigorre, Gascony, and the Basque Provinces; capital Tarbes.

Haute-Vienne (ōt-vyĕn'), department (1975 pop. 352,159), 2,128 sq mi (5,511.5 sq km), central France, in the Massif Central; capital Limoges.

Haut-Rhin (ō-răN'), department (1975 pop. 635,209), 1,360 sq mi (3,522.4 sq km), E France, in lower Alsace; capital Colmar.

Hauts-de-Seine (ōt-də-sĕn), department (1975 pop. 1,438,930), 63 sq mi (163.2 sq km), N central France, W of Paris; capital Nanterre.

Ha·van·a (hə-văn'ə), city (1970 pop. 1,755,400), capital of Cuba and of La Habana prov., W Cuba; largest city and chief port of the West Indies and one of the oldest cities in the Americas. Industries include oil refineries, assembly plants, rum distilleries, sugar refineries, a steel mill, and cigar factories. Founded c.1515, probably by the Spanish explorer Diego de Velázquez, the original settlement was moved from Cuba's southern coast to the site of present-day Havana in 1519. Spanish treasure galleons assembled in Havana's harbor for their return voyage to Spain, and the city tempted many English, French, and Dutch buccaneers. It became the capital of Cuba in the late 16th cent. In 1762, during the French and Indian Wars, Havana fell to Anglo-American forces, but the following year it was returned to Spain in exchange for the Floridas. The blowing up of the U.S. battleship *Maine* in Havana harbor in Feb., 1898, was the immediate cause of the Spanish-American War. In the old city, dominated by Morro Castle and other fortresses, are narrow streets, numerous churches, and fine examples of colonial architecture. The modern section of the city has wide boulevards, impressive public buildings, and magnificent residences.

Havana, city (1970 pop. 4,376), seat of Mason co., W central Ill., SW of Peoria on the Illinois River; founded 1827, inc. 1853. The city is the trade and industrial center for a farm region.

Hav·ant and Wa·ter·loo (hăv'ənt; wô'tər-lo͞o'), urban district (1976 est. pop. 116,400), Hampshire, S England. Manufactures include pharmaceuticals, toys, kitchen equipment, electronic components, and automobiles.

Ha·vel (hä'fəl), river, c.215 mi (345 km) long, rising in the lake region of Mecklenburg, N East Germany; and flowing generally S through West Berlin, then W and NW to enter the Elbe River near Havelberg. During the Soviet blockade of Berlin (1948) the Havel was used as a runway for amphibian aircraft.

Hav·er·ford·west (hăv'ər-fərd-wĕst'), municipal borough (1971 pop. 9,101), Dyfed, SW Wales, on the Western Cleddau River. Until 1974 Haverfordwest was the county town of Pembrokeshire.

Hav·er·hill (hā'vər-ĭl', -vrəl), city (1978 est. pop. 43,100), Essex co., NE Mass., on the Merrimack River; inc. as a town 1641, as a city 1870. Formerly one of the nation's leading shoe producers, Haverhill now processes leather and makes leather products and electronic components. John Greenleaf Whittier's birthplace (the house dates from c.1688) is here.

Hav·er·ing (hăv'ər-ĭng, -nĭng), borough (1971 pop. 246,778), of Greater London, SE England. Havering was created in 1965 by the merger of the municipal borough of Romford and the urban district of Hornchurch. The borough is largely residential but plastics, chemicals, clothing, beer, and other items are manufactured.

Ha·ví·řov (hä'və-rôf'), town (1975 est. pop. 91,653), E Czechoslovakia, in Moravia. An important manufacturing and mining town, Havířov was founded in the 1950s.

Ha·vre (hăv'ər), city (1978 est. pop. 10,500), seat of Hill co., N Mont., on the Milk River; inc. 1892. Founded in 1887 with the coming of the railroad, it is the center of a cattle, sheep, and wheat area served by the Milk River project.

Ha·vre, Le (lə hä'vrə, -vər), city (1975 pop. 217,881), Seine-Maritime dept., N France, in Normandy, at the mouth of the Seine River on the English Channel. It was founded in 1517 by Francis I. Among the city's manufactures are automobiles, cement, synthetic rubber, and fertilizers. During World War II the British bombed the city to prevent its use by the Germans for an invasion of England.

Haw (hô), river, 130 mi (209.2 km) long, N central N.C., rising in Alamance co., and flowing E and SE to join the Deep River and form the Cape Fear River.

Ha·wai·i (hə-wä'ē, -wä'yə, -wī'ē), noncontiguous state of the United States (1975 pop. 853,000), 6,450 sq mi (16,706 sq km), central Pacific, admitted to the Union in 1959 as the 50th state. Hawaii consists of a group of eight major islands and numerous islets in the Pacific Ocean. The capital is Honolulu, on the island of Oahu. Hawaii Island is the largest and geologically the youngest of the group, and Oahu is the most populous and economically important. The other principal islands are Kahoolawe, Kauai, Lanai, Maui, Molokai, and Niihau. The Palmyra atoll and Kingman Reef, which were within the boundaries of Hawaii when it was a U.S. territory, were excluded when statehood was achieved.

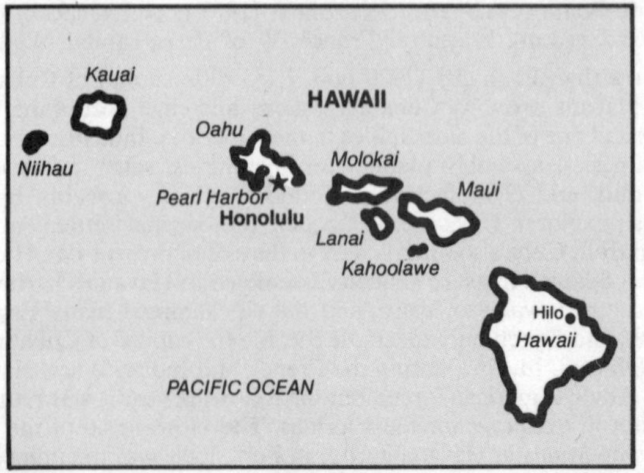

The Hawaiian islands are of volcanic origin and are edged with coral reefs. Generally fertile and with a mild climate, they are sometimes called "the paradise of the Pacific" because of their gaily colored flowers, coral beaches with rolling white surf, and majestic volcanic peaks. Some of the world's largest volcanoes are found on Hawaii and Maui. Vegetation is generally luxuriant throughout the islands, with giant fern forests in Hawaii Volcanoes National Park. Kahoolawe, however, is arid, and Niihau and Molokai have very dry seasons. Although many species of birds and domestic animals have been introduced on the islands, there are few wild animals other than boars and goats, and there are no snakes.

Hawaii is divided into four counties:

NAME	COUNTY SEAT	NAME	COUNTY SEAT
Hawaii	Hilo	Kauai	Lihue
Honolulu	Honolulu	Maui	Wailuku

Economy. Sugar cane and pineapples, grown chiefly on large company-owned plantations, are the major agricultural products and the basis of the islands' principal industry, food processing. Hawaii is one of the world's leading producers of canned pineapple. Other products include coffee, cattle, calves, and dairy products. Commercial fishing is also important, tuna being the principal species. U.S. military defense installations at Pearl Harbor and elsewhere in the state are extremely important to Hawaii's economy. Tourism is another leading source of income.

History. The first known settlers of the Hawaiian Islands were Polynesian voyagers; the date of final migration is believed to be c.750. The islands were opened to Western civilization in 1778 by the English explorer Capt. James Cook, who named them the Sandwich Islands for the English earl of Sandwich. In 1810 Kamehameha I became the sole sovereign of all the islands, and his hospitality enabled American traders to exploit the islands' sandalwood. However, Europeans and Americans brought with them devastating infectious diseases, and over the years the native population was greatly reduced. The adoption of Western ways contributed to the decline of native cultural tradition. Christian missionaries established schools, developed the Hawaiian alphabet, and used it for translating the Bible into Hawaiian. In 1839 Kamehameha III issued a guarantee of religious freedom, and the following year a constitutional monarchy was established.

An American trade treaty was made in 1842, and the supplying of American whaling ships began to replace the waning sandalwood trade. In 1848 the islands' feudal land system was abolished, making private ownership possible and thereby encouraging capital investment in the land. By the time the whaling industry collapsed in the 1860s, the sugar industry, which had been introduced in the 1830s, was well established. The amount of sugar exported to the United States increased, and thousands of Oriental immigrants were brought into Hawaii to work on large sugar plantations.

Toward the end of the 19th cent. agitation for constitutional reform in Hawaii led to the overthrow of the monarchy (1893). A provisional government was established and John L. Stevens, the U.S. minister to Hawaii, proclaimed the country a U.S. protectorate. The provisional government sought U.S. annexation of the islands, but President Grover Cleveland refused after a special investigation revealed that most Hawaiians had not supported the revolution and that Hawaiians and Americans in the sugar industry had played a leading role in overthrowing the monarchy. Cleveland's successor, President William McKinley, favored annexation, which was finally accomplished in 1898. In 1900 the islands were made a territory. In this period Hawaii's pineapple industry expanded.

On Dec. 7, 1941, Japanese aircraft made a surprise attack on Pearl Harbor, plunging the United States into World War II. During the war the Hawaiian islands were the chief Pacific base for U.S. forces and were under martial law (Dec. 7, 1941–Mar., 1943). The postwar years ushered in a dramatic expansion of labor unionism and the tourist trade. After having sought statehood for many decades, Hawaii was finally admitted to the Union on Aug. 21, 1959. In the 1960s and 1970s tourism has continued to grow, creating problems of overcrowding and damage to the islands' natural beauty.

Government. Hawaii's constitution was drafted in 1950 and became effective in 1959 upon attainment of statehood. A governor elected every four years heads the executive. The legislature has a senate with 25 members elected for four-year terms and a house of representatives with 51 members elected for two-year terms.

Educational Institutions. The Univ. of Hawaii is located at Honolulu. In 1961 the Center for Cultural and Technical Interchange between East and West was dedicated at the university.

Hawaii, island (1970 pop. 63,468), 4,037 sq mi (10,455.8 sq km), largest and southernmost island of the state of Hawaii and coextensive with Hawaii co. Geologically the youngest of the Hawaiian group, Hawaii is made up of three volcanic mountain masses rising from the floor of the Pacific Ocean. The only active volcanoes in the United States, outside of Alaska, are found on Hawaii. Lava flows, some of which reach the sea, and volcanic ash cover parts of the island. The north and northeast coasts are rugged with high cliffs; the west and south coasts are generally low, with a few good bathing beaches. Vegetation varies from tropical rain forest to grasslands to barren volcanic areas. Sugar cane and cattle are the island's principal products. It is also known for its health resorts and for its offshore deep-sea fishing.

Hawaii Volcanoes National Park, 229,616 acres (92,994.5 hectares), on Hawaii Island; est. 1916. The park contains two active volcanoes: Kilauea with its fire pit, called Halemaumau, and Mauna Loa with the active Mokuaweoweo crater on its summit.

Haw·ar·den (här'dn), rural district (1971 pop. 42,467), Clwyd, NE Wales. There are ruins of a 13th cent. castle on the grounds of Hawarden Castle (built 1752), which was the home of W. E. Gladstone until his death in 1898.

Hawes·ville (hôz'vĭl'), city (1970 pop. 1,262), seat of Hancock co., NW Ky., on the Ohio River ENE of Owensboro. It is a shipping point for an agricultural area yielding burley tobacco, corn, wheat, and truck crops. There are coal mines and oil wells in the region.

Ha·wick (hô'ĭk), burgh (1974 est. pop. 16,378), Borders region, S Scotland, on the Teviot River. Hawick is famous for its fine woolens and tweeds. Besides the manufacture of knitwear, Hawick's industries include dye works and light engineering plants.

Haw·kins (hô'kĭnz), county (1970 pop. 33,757), 480 sq mi (1,243.2 sq km), NE Tenn., bordered on the N by Va., drained by the Holston River, and traversed by ridges of the Appalachians; formed 1786; co. seat Rogersville. In a hardwood-timber area, its agriculture includes tobacco, corn, hay, livestock, and dairy produce.

Haw·kins·ville (hô'kĭnz-vĭl'), city (1970 pop. 4,077), seat of Pulaski co., central Ga., on the Ocumlgee River SSE of Macon; inc. 1830. It is a trade and processing center in a lumber, cotton, peanut, and dairy area.

Ha·worth (härth, hou'ərth), since 1939 part of the municipal borough of Keighley, West Yorkshire, N England. Patrick Brontë and his family lived in the parsonage, which now houses the Brontë museum and library.

Haw·ran or **Hau·ran** (hou-rän'), district, SW Syria. It is a largely treeless region marked by conical volcanic peaks, barren lava fields, and rich lava soil. Grains and fruits (including grapes) are grown. Most of the inhabitants are Druses, who migrated from Lebanon in the 18th and 19th cent. The district has many ancient towns whose buildings and furniture are made entirely of lava.

Haw·thorne (hô'thôrn'). **1.** City (1978 est. pop. 56,300), Los Angeles co., S Calif., a suburb of Los Angeles; inc. 1922. Toys, cash registers, and defense-related products are made. **2.** Village (1970 pop. 3,539), seat of Mineral co., W Nev., S of Walker Lake. It is a trading center for a mining and ranching region. **3.** Borough (1978 est. pop. 19,000), Passaic co., NE N.J.; settled 1850, inc. 1898. Primarily residential, it produces chemicals and plastics.

Hay (hā), river, c.530 mi (855 km) long, rising in several headstreams in NE British Columbia and NW Alta., Canada, and flowing generally NE through NW Alta. into Great Slave Lake.

Hayes (hāz), county (1970 pop. 1,530), 711 sq mi (1,841.5 sq km), S Nebr., drained by Frenchman Creek and other branches of the Republican River; formed 1877; co. seat Hayes Center. It is in a corn, wheat, and cattle area. There is some mining and manufacturing.

Hayes, river, c.300 mi (485 km) long, rising in a lake NE of Lake Winnipeg, central Man., Canada, and flowing NE to Hudson Bay. It was the chief route used by Hudson's Bay Company traders from Hudson Bay to Lake Winnipeg and the interior.

Hayes, Mount, 13,700 ft (4,178.5 m), E Alaska Range, E Alaska, SSE of Fairbanks.

Hayes Center, village (1970 pop. 237), seat of Hayes co., S Nebr., NW of McCook on a branch of the Republican River. In a livestock and grain area, dairy and poultry products are processed.

Hayes·ville (hāz'vĭl'). Town (1970 pop. 428), seat of Clay co., W N.C., on the Hiwassee River WSW of Asheville, in a resort and sawmilling region.

Hayne·ville (hān'vĭl'), town (1970 pop. 473), seat of Lowndes co., S central Ala., SW of Montgomery.

Hays (hāz), county (1970 pop. 27,642), 650 sq mi (1,683.5 sq km), S central Texas, on the edge of Edwards Plateau and drained by the San Marcos and Blanco rivers; formed 1848; co. seat San Marcos. It has agriculture (cotton, corn, wheat, grain sorghums, hay, flax, livestock, and dairy products) and deposits of limestone and clay.

Hays, city (1978 est. pop. 16,900), seat of Ellis co., W central Kansas; inc. 1885. It is a rail, trade, and medical center in a grain, cattle, and oil area. Medical and hospital supplies and hydraulic cylinders are manufactured. For protection against the Indians, Fort Hays was established (1865) southeast of the present city, on a stagecoach road to Denver. Floods in 1867 forced the fort's relocation to its present site, and the city of Hays grew just east of it. The fort was abandoned in 1889 and the land turned over to the state with the understanding that it be used for a school, an agricultural experiment station, and a state park. Today the school has grown into Fort Hays Kansas State College; the agricultural experiment station (laid out 1901) is one of the world's largest; and the park, a state historic site, contains the fort's surviving buildings.

Hay·ti (hā'tē'), town (1970 pop. 393), seat of Hamlin co., E S.Dak., SSW of Watertown, in a dairying, livestock, and poultry area.

Hay·ward (hā'wərd). **1.** City (1978 est. pop. 98,100), Alameda co., W Calif.; settled 1851, inc. 1876. It is an important commercial and distributing center for an agricultural area. Food processing is its major industry. **2.** City (1975 est. pop. 1,653), seat of Sawyer co., NW Wis., on the Namekagon River SE of Superior, in a lake and woods region; settled 1881, inc. 1915.

Hay·wood (hā'wŏŏd'). **1.** County (1970 pop. 41,710), 561 sq mi (1,453 sq km), W N.C., bounded on the NW by the Great Smoky Mts. and the Tenn. border, on the W by Balsam Mt., and drained by the Pigeon River; formed 1808; co. seat Waynesville. In a largely forested resort area, it has agriculture (tobacco, apples, corn, dairy products, and cattle), a lumbering industry, and some manufacturing. **2.** County (1970 pop. 19,596), 519 sq mi (1,344.2 sq km), W Tenn., drained by the Hatchie River and the South Fork of the Forked Deer River; formed 1823; co. seat Brownsville. It is primarily agricultural (yielding cotton, corn, livestock, and truck crops), with some lumbering and industry.

Haz·ard (hăz'ərd), city (1970 pop. 5,459), seat of Perry co., SE Ky., on the North Fork of the Kentucky River SE of Lexington, in the Cumberland foothills. It is the trade and shipping center for a bituminous-coal area that also yields oil, natural gas, and timber.

Ha·zel Crest (hā'zəl krĕst'), village (1970 pop. 10,329), Cook co., NE Ill., a residential suburb of Chicago; inc. 1911.

Hazel Park, city (1978 est. pop. 20,200), Oakland co., SE Mich., a suburb of Detroit; inc. 1942. Hazel Park has varied light industries and a racetrack. Most of the early settlers were Germans.

Haz·le·hurst (hā'zəl-hûrst'). **1.** City (1970 pop. 4,065), seat of Jeff Davis co., SE Ga., near the Altamaha River SSW of Vidalia; settled in the late 1850s, inc. 1891. It is a tobacco market. **2.** City (1970 pop. 4,577), seat of Copiah co., SW Miss., SSW of Jackson, in a timber and cotton area; founded 1857, inc. 1865.

Healds·burg (hēldz'bûrg'), city (1970 pop. 5,438), Sonoma co., W Calif., NNW of San Francisco; founded 1852, inc. 1867. It is the trading and processing center of a fruit and grape area.

Heard (hûrd), county (1970 pop. 5,354), 297 sq mi (769.2 sq km), W Ga., in a piedmont area bounded on the W by the Ala. border and intersected by the Chattahoochee River; formed 1830; co. seat Franklin. Its economy depends on agriculture (cotton, corn, hay, truck crops, fruit, and livestock) and sawmilling.

Heart (härt), river, 180 mi (290 km) long, rising in the low prairie country near the Little Missouri River, SW N.Dak., and flowing E to the Missouri River at Mandan, N.Dak.

Heaths·ville (hēths′vĭl′), village (1970 pop. 300), seat of Northumberland co., E Va., SE of Fredericksburg, in a grain, truck-farming, and poultry region.

Heb·bron·ville (hĕb′rən-vĭl′), town (1970 pop. 4,079), seat of Jim Hogg co., S Texas E of Laredo. The semiarid plains have oil fields and cattle ranches, and a small cultivated area yields cotton and watermelons.

Heb·burn (hĕb′ərn), urban district (1971 pop. 23,597), Tyne and Wear, NE England, on the Tyne River. It has shipbuilding, electrical industries, and engineering works.

He·ber City (hē′bər), city (1970 pop. 3,245), seat of Wasatch co., N central Utah, SE of Salt Lake City; settled 1859 by Mormons. It is a trade and processing center of a livestock, farming, and mining area.

Heber Springs, resort town (1970 pop. 2,497), seat of Cleburne co., N central Ark., NNE of Little Rock, in a mineral-spring area of the Ozark Mts.; laid out 1881.

Heb·ri·des, the (hĕb′rĭ-dēz′) or **West·ern Islands** (wĕs′tərn), group of more than 500 islands, W and NW Scotland. Less than a fifth of the islands are inhabited. The Outer Hebrides are separated from the mainland and from the Inner Hebrides by the straits of Minch and Little Minch and by the Sea of the Hebrides. Fishing, crop raising, sheep grazing, manufacturing of tweeds and other woolens, quarrying, and tourism are the chief occupations. The original Celtic inhabitants, converted to Christianity by St. Columba (6th cent.), were conquered by the Norwegians (starting in the 8th cent.). They held the islands until 1266. From that time the islands were ruled by various native Scottish chiefs, until the Macdonalds established themselves (1346) as lords of the isles. Since the 16th cent. the Hebrides have belonged to the crown of Scotland. There are interesting prehistoric and ancient historical remains and geologic structures. Under the Local Government Act of 1973, the Outer Hebrides were included in the new Western Isles Island Authority; the Inner Hebrides were included in the new Strathclyde and Highland regions.

He·bron (hē′brən), city (1968 est. pop. 38,000), W Jordan, near Jerusalem. It is in a region where grapes, cereal grains, and vegetables are grown. Tanning, food processing, glassblowing, and the manufacture of sheepskin coats are the major industries. The site of ancient Hebron, which antedates the biblical record, has not been precisely determined. David ruled the Hebrews from Hebron for seven years before moving his capital to Jerusalem. The city has figured in every war in Palestine. It was taken (2nd cent. B.C.) by Judas Maccabeus and destroyed by the Romans. In 636 it was conquered by the Arabs and made an important place of pilgrimage, later to be seized (1099) by the Crusaders and retaken (1187) by Saladin. It later became (16th cent.) part of the Ottoman Empire, was incorporated (1922–48) in the League of Nations Palestine mandate, and in 1948 joined Jordan. Hebron is a sacred place for Moslems and Jews.

Hebron, city (1970 pop. 1,667), seat of Thayer co., SE Nebr., on the Little Blue River SW of Lincoln; founded 1869.

Hec·a·te Strait (hĕk′ə-tē), 160 mi (257.4 km) long, 35 to 80 mi (56.3–128.7 km) wide, separating mainland W British Columbia, Canada, from Queen Charlotte Island.

Heer·len (hār′lən), city (1977 est. pop. 70,928), Limburg prov., SE Netherlands. Its manufactures include textiles and food products. The city was a major coal-mining center from the late 19th cent.

Hef·lin (hĕf′lən), town (1970 pop. 2,872), seat of Cleburne co., NE Ala., near the Tallapoosa River and the Ga. border E of Anniston; settled 1883, inc. 1892. It is in a lumbering area.

Hei·de (hī′də), town (1974 est. pop. 22,154), Schleswig-Holstein, N West Germany, in the center of the Dithmarschen oil fields.

Hei·del·berg (hīd′l-bûrg′, -bĕrKH′), city (1974 est. pop. 129,368), Baden-Württemberg, SW West Germany, picturesquely situated on the Neckar River. Manufactures include printing presses and other machinery, precision instruments, textiles, and leather goods. Heidelberg was first mentioned in the 12th cent. In 1225 it was acquired by the count palatine of the Rhine and until 1720 was the residence of the electors palatine. The Univ. of Heidelberg was founded in 1386. It became a bulwark of the Reformation in the 16th cent., declined after the Thirty Years' War (1618–48), and after the French Revolutionary Wars became the leading university of 19th cent. Germany. Heidelberg is famous for its ruined castle (built mainly in the 16th and early 17th cent.), which was largely destroyed by French troops in the late 17th cent.

Heil·bronn (hīl′brôn′), city (1974 est. pop. 115,924), Baden-Württemberg, S West Germany, a port on the Neckar River. A commercial and industrial center, its manufactures include metal products, textiles, chemicals, and wine. Heilbronn was the site (early 9th cent.) of a Carolingian palace and in the 14th cent. became a free imperial city. In 1802 Heilbronn passed to Württemberg, and later in the 19th cent. it acquired industrial importance. In World War II much of the city was destroyed.

Hei·lung·kiang (hā′lo͝ong′kyäng′, -jē-äng′), province (1967 est. pop. 21,000,000), c.272,000 sq mi (704,480 sq km), NE China; capital Harbin. Heilungkiang is separated from the USSR by the Argun River in the west, the Amur River in the north, and the Ussuri River in the east. Lumbering is a major industry. The northwest was formerly (before 1969) a part of Inner Mongolia; it is predominantly pastureland, with related industries (dairy products and leather). The south is known as the Manchurian plain. It is a great wheat area; millet, kaoliang, soybeans, sugar beets, and flax are also grown. Heilungkiang contains the great Ta-ch'ing oil field, first worked in 1959, and has oil-refining operations. Iron, coal, and magnesite are mined, and aluminum is produced. Manufactures range from processed foods to locomotives. The former provinces of Hingan and Nunkiang were added to Heilungkiang in 1950 and Sungkiang was incorporated in 1954. The northwest section became part of Inner Mongolian Autonomous Region in 1949 and was returned to Heilungkiang in the 1969–70 redistricting.

He·jaz or **Hed·jaz** (both: hē-jăz′), region (1963 est. pop. 2,000,000), c.150,000 sq mi (388,500 sq km), NW Saudi Arabia, on the Gulf of Aqaba and the Red Sea. Extending south to Asir, Hejaz is mainly a dissected highland region lying between the narrow, long coastal strip and the interior desert. There are several oases and some wadis (watercourses) where crops, such as dates, wheat, and millet, and livestock are raised. Hejaz is, however, more important as a place of pilgrimage; it includes the holy cities of Mecca and Medina. Following the fall (1258) of the caliphate of Baghdad, Hejaz came under Egyptian control. In 1517 it came under Turkish suzerainty. The Hejaz was in 1916 proclaimed independent by Husayn ibn Ali, the sherif of Mecca, who with the aid of T. E. Lawrence destroyed Turkish authority. Husayn was defeated in 1924 by Ibn Saud and the formal union of Hejaz into Saudi Arabia was proclaimed in 1932.

Hek·la (hĕk′lä), volcano, 4,747 ft (1,447.8 m) high, SW Iceland; one of the highest in Europe. Since the early 11th cent. more than 20 eruptions have been recorded; the worst occurred in 1766 and the most recent in 1947. Hekla emits steam and has several crater lakes. In medieval Icelandic folklore Hekla was believed to be one of the gates to purgatory; it is also a legendary gathering place for witches.

Hel·e·na (hĕl′ə-nə). **1.** City (1970 pop. 10,415), seat of Phillips co., E central Ark., on the Mississippi River, in the delta cotton country at the S end of Crowley's Ridge; inc. 1833. It is a rail terminus and river port. The city was occupied by Union troops in the Civil War; they were attacked unsuccessfully by Confederates in a battle July 4, 1863. Ante-bellum homes and Indian mounds are in the area. **2.** City (1978 est. pop. 27,700), state capital and seat of Lewis and Clark co., W central Mont., on the E slope of the Continental Divide; inc. 1870. It is a commercial and shipping center in a ranching and mining area. Helena's manufactures include machine parts, concrete, and paints. The city was founded after the discovery of gold (1864) in Last Chance Gulch (now Helena's main street), and grew rapidly. In 1874 a general election ratified the choice of Helena to replace Virginia City as territorial capital. In the 1890s it maintained its position as state capital against the rivalry of Anaconda. Rich silver and lead strikes in the neighborhood kept its mining wealth high.

Hel·go·land (hĕl′gō-länt′) or **Hel·i·go·land** (hĕl′ĭ-gō-lănd′), island (1970 pop. 2,377), c.150 acres (60.8 hectares), Schleswig-Holstein, N West Germany, in the North Sea. Formed of red sandstone, it rises to c.200 ft (60 m) above the sea and is largely covered with grazing land. The island is a popular tourist resort. Strategically located near the mouths of the Weser and the Elbe rivers, Helgoland was captured by the Danes in 1714, was occupied by the English in 1807, and was formally ceded to England by Denmark in 1814. In exchange for rights in Africa, England gave the island to Germany in 1890. The Germans installed fortifications, which were razed after World War I according to the terms of the Treaty of Versailles. However, Germany refortified Helgoland in 1936 and used it as a naval base in World War II. In 1947 British occupation authorities, after evacuating the islanders, blew up the fortifications and part of the island in one of the largest known nonatomic blasts.

Hel·i·con (hĕl'ə-kŏn', -kən), mountain group, c.20 mi (30 km) long, rising to 5,736 ft (1,749.5 m), central Greece, in Boeotia. Helicon formed part of the border between ancient Boeotia and Phocis. In Greek legend it was the abode of the Muses and sacred to Apollo.

He·li·op·o·lis (hē'lē-ŏp'ə-lĭs), ancient city, N Egypt, in the Nile delta below modern Cairo. It was noted as the center of sun worship, and its god Ra or Re was the state deity until Thebes became capital (c.2100 B.C.). Under the New Empire (c.1570 B.C.–c.1085 B.C.), Heliopolis was the seat of the viceroy of northern Egypt. The obelisks called Cleopatra's Needles were erected here. Its schools of philosophy and astronomy declined after the founding of Alexandria in 332 B.C., but the city never wholly lost importance until the Christian era.

Hel·les·pont (hĕl'ĭs-pŏnt'): *see* Dardanelles.

Hell Gate (hĕl), narrow channel of the East River, SE N.Y., between Wards Island and Astoria, Queens, New York City. It was named by the Dutch navigator Adriaen Block, who passed through it into Long Island Sound in 1614.

Hel·mand or **Hel·mund** (hĕl'mənd), river, c.700 mi (1,125 km) long, rising in the Hindu Kush Mts., NE Afghanistan and flowing generally SW to the Seistan basin, SW Afghanistan, where it helps form the Hamun-e Helmand, a marshy lake that extends into Iran. It is the longest river in Afghanistan. The Helmand's ancient irrigation and river-control system was destroyed by Genghis Khan (13th cent.) and Tamerlane (14th cent.).

Hel·mond (hĕl'mônt'), city (1977 est. pop. 58,930), North Brabant prov., SE Netherlands, on the Aa River. Manufactures include textiles and food products.

Helm·stedt (hĕlm'shtĕt'), city (1974 est. pop. 28,786), Lower Saxony, E West Germany, a major frontier station at the East German border. Manufactures include machines and textiles. Helmstedt was founded in the 9th cent. and later (15th–16th cent.) was a member of the Hanseatic League.

Hel·sing·ør (hĕl'sĭng-œr') or **El·si·nore** (ĕl'sə-nôr'), city (1976 est. pop. 55,964), Frederiksborg co., E Denmark, on the Øresund opposite Hälsingborg, Sweden. It is an industrial center, fishing port, and summer resort. Manufactures include ships, machinery, beer, and textiles. Known since the 13th cent., Helsingør is the site of Kronborg castle (1754–85; restored 1925–37), which, although the strongest fortress in Denmark at the time, was taken by the Swedes in 1660. The castle is now a maritime museum and is also used for performances of Shakespeare's *Hamlet*, which is set here.

Hel·sin·ki (hĕl'sĭng'kē), city (1975 est. pop. 502,961), capital of Finland and of Uusimaa prov., S Finland, on the Gulf of Finland. Situated on a peninsula, sheltered by islands, and fortified by Suomenlinna, the city is a natural seaport (blocked by ice from Jan. to May) and the commercial and intellectual center of Finland. It has machine shops, shipyards, food-processing plants, textile mills, clothing and china factories, and chemical plants. The city, founded (1550) by Gustavus I of Sweden, was devastated by a great fire in 1808; it was rebuilt as a well-planned, spacious metropolis. Helsinki grew rapidly after Alexander I of Russia moved (1812) the capital here from Turku. When the Univ. of Helsinki (founded 1640) was moved from Turku in 1828, Helsinki became the center of Finnish nationalism. In the older part of the city is an impressive railway station designed by Eero Saarinen.

Hel·ve·tia (hĕl-vē'shə), region of central Europe, occupying the plateau between the Alps and the Jura mts. The name is derived from the Roman term for its inhabitants, the predominantly Celtic Helvetii, who were defeated (58 B.C.) at Bibracte by Julius Caesar in the Gallic Wars. Helvetia corresponds roughly to the western part of modern Switzerland, and the name is still used in poetic reference to that country and on its postage stamps.

Hem·el Hemp·stead (hĕm'əl hĕm'stəd), municipal borough (1973 est. pop. 71,150), Hertfordshire, SE England. It is a market town and suburb of London. Manufactures include paper, electrical products, office machinery, and photographic apparatus.

Hemp·hill (hĕmp'hĭl'), county (1970 pop. 3,084), 909 sq mi (2,354.3 sq km), extreme N Texas, on the high plains of the Panhandle, bounded on the E by the Okla. border and drained by the Canadian and Washita rivers; formed 1876; co. seat Canadian. Cattle ranching is important, and wheat, grain sorghums, broomcorn, potatoes, milo, alfalfa, and cotton are grown. It has an oil-producing industry, some manufacturing, and good hunting and fishing.

Hemphill, city (1970 pop. 1,005), seat of Sabine co., E Texas, near the La. border E of Lufkin; inc. as a city 1939. It is a trading center in a lumber, cotton, and cattle area.

Hemp·stead (hĕmp'stĕd'), county (1970 pop. 19,308), 735 sq mi (1,903.7 sq km), SW Ark., bounded on the SW by the Red River, on the N by the Little Missouri River; formed 1818; co. seat Hope. In a fertile agricultural area yielding cotton, corn, oats, wheat, soybeans, watermelons, poultry, livestock, and dairy products, it has lumbering and food-processing industries and hunting and fishing areas. Carpets, textiles, bricks, and metal products are made.

Hempstead. **1.** Village (1978 est. pop. 39,300), Nassau co., SE N.Y., on W Long Island; inc. 1853. It is a retail center for the area. Electronic equipment, tools, chemicals, metal products, and furniture are made in the village. Hempstead was settled in 1644 by English colonists. **2.** Town (1970 pop. 1,891), seat of Waller co., S Texas, NW of Houston; founded 1835. It ships watermelons and vegetables.

Hen·der·son (hĕn'dər-sən). **1.** County (1970 pop. 8,451), 381 sq mi (986.8 sq km), W Ill., bounded on the W by the Mississippi River and drained by Henderson Creek; formed 1841; co. seat Oquawka. It is in an agricultural area that produces livestock, corn, wheat, oats, truck crops, poultry, and dairy products. It has limestone deposits, hardwood timber, and some manufacturing. **2.** County (1970 pop. 36,031), 433 sq mi (1,121.5 sq km), NW Ky., bounded on the N by the Ohio River and the Ind. border, on the E and SE by the Green River; formed 1798; co. seat Henderson. In a rolling plateau agricultural area (tobacco, corn, wheat, livestock, and fruit), it has oil wells, coal mines, sand and gravel pits, and stands of timber. Processed foods, knitted goods, textiles, furnitures, plastics, concrete, and wood and aluminum products are made. **3.** County (1970 pop. 42,804), 378 sq mi (979 sq km), W N.C., partly in the Blue Ridge, bounded on the S by the S.C. border and drained by the French Broad and Broad rivers; formed 1838; co. seat Hendersonville. It is chiefly a resort region, with agriculture (apples, vegetables, corn, dairy products, and poultry), textile manufacturing, and sawmilling. **4.** County (1970 pop. 17,360), 515 sq mi (1,333.9 sq km), W Tenn., in an agricultural and lumbering area drained by the Big Sandy River; formed 1821; co. seat Lexington. Cotton, corn, and truck crops are grown. Natchez Trace Forest State Park is in the county. **5.** County (1970 pop. 25,466), 940 sq mi (2,434.6 sq km), E Texas, bounded on the W by the Trinity River, on the E by the Neches River; formed 1846; co. seat Athens. Its agriculture includes truck crops, fruit, cotton, corn, grain, legumes, livestock, and poultry. There are deposits of oil, natural gas, lignite, clay, and glass sand.

Henderson. **1.** City (1978 est. pop. 23,300), seat of Henderson co., NW Ky., on the Ohio River, in an oil, coal, tobacco, corn, and livestock area; founded 1797, inc. as a city 1867. Furniture, clothing, truck axles, and plastics are made. John J. Audubon lived in Henderson from 1810 to 1819; nearby is Audubon Memorial State Park, with a museum and a bird sanctuary. **2.** City (1970 pop. 16,395), Clark co., SE Nev., in a desert area overlooking Las Vegas and surrounded by mountains; inc. 1953. Limestone, titanium, ammonium perchlorate, chlorine, hydrogen, and manganese are produced. The city was founded (1942) to provide houses for employees of a magnesium plant. **3.** City (1978 est. pop. 13,800), seat of Vance co., N N.C.; settled c.1811, inc. 1841. It is an important tobacco market, with food-processing industries. **4.** City (1970 pop. 3,581), seat of Chester co., W central Tenn., SSE of Jackson, in a farm and timber area; inc. 1901. Chickasaw State Park and a number of Indian mounds are nearby. **5.** City (1970 pop. 10,187), seat of Rusk co., NE Texas; inc. 1877. It is a prosperous oil city, with wells and refineries. There are also foundries and lumber and cotton mills.

Hen·der·son·ville (hĕn'dər-sən-vĭl'), city (1975 est. pop. 6,443), seat of Henderson co., W N.C., in the Blue Ridge SSE of Asheville near the S.C. border. It is a mountain resort and a farm trade center.

Hen·dricks (hĕn'drĭks'), county (1970 pop. 53,974), 417 sq mi (1,080 sq km), central Ind., drained by the Eel and Whitelick rivers and Mill Creek; formed 1823; co. seat Danville. Its agriculture includes grain, fruit, and livestock. It also has lumbering and food-processing industries and manufactures metal products.

Hen·dry (hĕn'drē), county (1970 pop. 11,859), 1,187 sq mi (3,074.3 sq km), S Fla., in an everglades region crossed by the Caloosahatchee

River; formed 1923; co. seat La Belle. Sugar cane, truck crops, and livestock are raised.

Heng·e·lo (hĕng'ə-lō'), city (1977 est. pop. 73,084), Overijssel prov., E Netherlands. Manufactures include heavy machinery and metal products.

Heng-yang or **Heng·yang** (both: hŭng'yäng'), city (1970 est. pop. 240,000), central Hunan prov., China, on the Hsiang River, at the mouth of the Lei River. It is the leading transportation center of Hunan. Manufactures include chemicals, machine tools, textiles, paper, processed foods, and fertilizer. Lead and zinc mines are nearby.

Hen·ne·pin (hĕn'ə-pən), county (1970 pop. 960,080), 565 sq mi (1,463.4 sq km), SE central Minn., bounded on the NE by the Mississippi River, on the NW by the Crow River, on the SE by the Minnesota River, with Lake Minnetonka in the SW; formed 1852; co. seat Minneapolis. In a resort and commercial area, it has oil and natural-gas wells, sand and gravel pits, and extensive food-processing and lumbering industries.

Hennepin, village (1976 est. pop. 628), seat of Putnam co., N central Ill., on the Illinois River SW of Spring Valley.

Hen·ri·co (hĕn-rē'kō), county (1970 pop. 154,364), 232 sq mi (600.9 sq km), E central Va., bounded on the S by the James River, on the N and NE by the Chickahominy River; formed 1634; co. seat Richmond (an independent city). In an area divided between the Piedmont and the coastal plain, it has truck and dairy farms. Wood products, chemicals, and electronics are manufactured.

Hen·ri·et·ta (hĕn'rē-ĕt'ə), town (1970 pop. 2,897), seat of Clay co., N Texas, in the valley of the Little Wichita ESE of Wichita Falls. It was founded in ranching country in 1857, abandoned because of Indian attacks, resettled after 1873, and incorporated in 1883.

Hen·ry (hĕn'rē). **1.** County (1970 pop. 13,254), 554 sq mi (1,434.9 sq km), SE Ala., in the coastal plain bounded on the E by the Chattahoochee River and Ga. and drained by the Choctawhatchee River; formed 1819; co. seat Abbeville. Its agriculture includes cotton, peanuts, corn, and poultry. It has lumber (logging and sawmills), bauxite mining, and textile and clothing industries. **2.** County (1970 pop. 23,724), 331 sq mi (857.3 sq km), N central Ga., in a piedmont agricultural area; formed 1821; co. seat McDonough. Cotton, corn, grain, truck crops, fruit, and pecans are grown. Textiles are manufactured. **3.** County (1970 pop. 53,217), 826 sq mi (2,139.3 sq km), NW Ill., bounded on the NW by the Rock River and drained by the Green and Edwards rivers; formed 1825; co. seat Cambridge. The old Illinois and Mississippi Canal crosses the county. Its agriculture includes corn, oats, soybeans, wheat, truck crops, livestock, poultry, and dairy foods. It also has bituminous-coal mines, oil refineries, and some manufacturing. **4.** County (1970 pop. 52,603), 400 sq mi (1,036 sq km), E central Ind., drained by the Big Blue River, Flatrock Creek, and Fall Creek; formed 1821; co. seat New Castle. It is in an agricultural area that produces livestock, grain, truck crops, and poultry. Its diversified manufactures include automobiles and automobile equipment, canned fruit and vegetables, clothing, furniture, and steel products. **5.** County (1970 pop. 18,114), 440 sq mi (1,139.6 sq km), SE Iowa, drained by the Skunk River; formed 1836; co. seat Mount Pleasant. In a prairie agricultural area, it depends on stock raising (hogs, cattle, and poultry) and the growing of corn, oats, and soybeans. There are limestone quarries here. **6.** County (1970 pop. 10,910), 289 sq mi (748.5 sq km), N Ky., in the Bluegrass bounded on the E by the Kentucky River and drained by the Little Kentucky River and Floyd's Fork; formed 1798; co. seat New Castle. Its agriculture includes burley tobacco, corn, hay, dairy products, and livestock. It also has zinc and lead mines and primary metal industries. **7.** County (1970 pop. 18,451), 737 sq mi (1,908.8 sq km), W Mo., drained by the South Grand River; formed 1834 as Rives co., renamed 1841; co. seat Clinton. There is bituminous-coal mining in the county. It has agriculture (corn, wheat, oats, hay, dairy products, and livestock), and its manufactures include clothing, shoes, concrete, and machinery. **8.** County (1970 pop. 27,058), 416 sq mi (1,077.4 sq km), NW Ohio, intersected by the Maumee River; formed 1824; co. seat Napoleon. It has diversified farming (corn, wheat, oats, sugar beets, and truck crops), food-processing plants, and some industry. **9.** County (1970 pop. 23,749), 567 sq mi (1,468.5 sq km), NW Tenn., bounded on the N by the Ky. border and drained by the East Fork of the Clarks River and forks of the Obion River; formed 1821; co. seat Paris. It is in a timber and agricultural area yielding corn, cotton, tobacco, sweet potatoes, and livestock, and has some manufacturing. **10.** County (1970 pop. 50,901), 384 sq mi (994.6 sq km), S Va., in the Piedmont, bounded on the S by the N.C. border and drained by the Smith and Mayo rivers; formed 1777; co. seat Martinsville (an independent city). It has mixed agriculture (especially tobacco and livestock) and manufactures furniture and textiles.

Henry, Cape, SE Va., at the entrance to Chesapeake Bay, E of Norfolk. Cape Henry Memorial marks the approximate spot where the Jamestown settlers landed on Apr. 26, 1607. In 1939 the site was included in Colonial National Historical Park.

Hen·ry·et·ta (hĕn'rē-ĕt'ə), city (1970 pop. 6,430), Okmulgee co., E central Okla., S of Okmulgee; founded c.1900. It has gas and oil wells, coal mines, glassworks, an iron foundry, and a zinc smelter.

Hepp·ner (hĕp'nər), city (1970 pop. 1,429), seat of Morrow co., N Oregon, SW of Pendleton, in a wheat, livestock, and timber area; inc. 1872.

Her·a·cle·a (hĕr'ə-klē'ə), ancient Greek city, in Lucania, S Italy, near the Gulf of Tarentum. Pyrrhus defeated the Romans here in 280 B.C. Bronze tablets giving Roman municipal laws were found here.

Heraclea Pon·ti·ca (pŏn'tĭ-kə), ancient Greek city, a port on the S shore of the Black Sea, on the site of Ereğli, Turkey. Founded in the 6th cent. B.C. by colonists from Megara and Boeotia, it was at its height in the 4th cent. B.C. It was destroyed by the Romans in the wars against Mithridates VI of Pontus.

Her·a·cle·op·o·lis (hĕr'ə-klē-ŏp'ə-lĭs), ancient city, N Egypt, just S of Al Fayyum. One of the oldest Egyptian cities, it was in existence before 3000 B.C.

He·rat (hĕ-rät'), city (1973 est. pop. 108,800), capital of Herat prov., NW Afghanistan, on the Hari Rud. The fertile river valley is renowned for its fruits, especially grapes. Herat has textile weaving and carpet industries and is a market for wool, carpets, dried fruits, and nuts. The city walls are gone, but the great earthwork of the citadel remains. Landmarks include the Great Mosque (first built 12th cent.) and several exquisite minarets. Herat's strategic location on the old trade route from Persia to India and on the caravan road from China and central Asia to Europe has long made it an object of contention. Although taken by various conquerors, it remained under the Persian empire for several centuries. The Mongols under Genghis Khan devastated Herat in 1221. The city was taken by Tamerlane in 1383, and by the Uzbeks in the early 16th cent.; later it was disputed between the Persians and the rulers of an emerging Afghanistan. In the mid-19th cent., British pressure checked Persian claims to Herat, which in 1881 was taken by Abd ar-Rahman and finally confirmed as part of a united Afghanistan.

Hé·rault (ā-rō'), department (1975 pop. 648,202), 2,360 sq mi (6,112.4 sq km), S France, in Languedoc; capital Montpellier.

Her·bert Hoo·ver National Historic Site (hûr'bərt hōō'vər): *see* National Parks and Monuments Table.

Her·cu·la·ne·um (hûr'kyə-lā'nē-əm), ancient city of S Italy, on the Gulf of Naples at the foot of Mt. Vesuvius. Damaged by an earthquake in A.D. 63, it was completely buried, along with Pompeii, by the volcanic eruption of Mt. Vesuvius in A.D. 79. Before the earthquake, it was a popular Roman resort and residential town with fine villas. The first discovery of ruins was made in 1709, and excavations have continued since. Important early finds were the sumptuous so-called Villa of the Papyri (with a large library, and bronze and marble statues), a basilica with fine murals, and a theater.

Her·e·ford (hĕr'ə-fərd), municipal borough (1976 est. pop. 47,800), Hereford and Worcester, W central England. It is a cattle-market town. There is also food processing, brewing, and light manufacturing. Until 1974 Hereford was county town of Herefordshire.

Her·e·ford (hûr'fərd), city (1970 pop. 13,414), seat of Deaf Smith co., N Texas, in the Panhandle; inc. 1906. Fine livestock is raised in Hereford, and cattle feeding is an important industry, along with meat packing and sugar refining. Vegetables and grains are grown on irrigated farms in the semiarid plains.

Her·e·ford and Worces·ter (hĕr'ə-fərd; wŏŏs'tər), nonmetropolitan county, (1976 est. pop. 594,200), W central England, created under the Local Government Act of 1972; administrative center Worcester. It is composed of the county borough of Worcester and most of the former counties of Herefordshire and Worcestershire.

Her·e·ford·shire (hĕr'ə-fərd-shîr', -shər), former county, W central

England, on the Welsh border. The county town was Hereford. In the Middle Ages Herefordshire had a flourishing woolen and cloth trade in the Middle Ages. In 1974 the county was combined with almost all of Worcestershire to form the new nonmetropolitan county of Hereford and Worcester.

Her·ford (hĕr'fôrt), city (1974 est. pop. 64,675), North Rhine-Westphalia, N central West Germany, on the Werre River. Its manufactures include textiles, carpets, furniture, and processed food. Herford passed to Brandenburg in 1647.

He·ri·sau (hā'rĭ-zou'), town (1977 est. pop. 13,900), capital of Appenzell Ausser-Rhoden half canton, NE Switzerland. Embroideries, cotton textiles, machinery, and precision instruments are made.

Her·ki·mer (hûr'kə-mər), county (1970 pop. 67,633), 1,435 sq mi (3,716.7 sq km), NE central N.Y., a long, narrow area with the N part extending into the Adirondacks and the S part in the fertile Mohawk Valley, drained by the Mohawk, Unadilla, Black, and Moose rivers; formed 1791; co. seat Herkimer. It has dairying and general farming, lumbering, some mining, and industry (processed food, textile mill products, clothing, wood products, furniture, paper, leather products, arms, machinery, and equipment). There are many mountain and lake resorts in the north.

Herkimer, industrial village (1970 pop. 8,960), seat of Herkimer co., central N.Y., on the Mohawk River ESE of Utica; settled c.1725 by German immigrants, inc. 1807. Furniture and clothing are made.

Her·mann (hûr'mən), city (1970 pop. 2,658), seat of Gasconade co., E central Mo., on the Missouri River ENE of Jefferson City; settled 1837 by Pennsylvania Germans, inc. 1839. It is in a farm area with diaspore, flint, and clay deposits.

Her·mit·age (hûr'mĭ-tĭj), city (1970 pop. 284), seat of Hickory co., central Mo., in the Ozarks, on the Pomme de Terre River.

Her·mon, Mount (hûr'mən), on the Syria-Lebanon border. The highest of its three peaks (all of which are snow-covered in winter and spring) rises to 9,232 ft (2,815.8 m). Mt. Hermon was sacred to the worshipers of Baal and to the Romans and Druses. It is traditionally designated as the scene of the Transfiguration.

Her·mon·this (hər-mŏn'thĭs), ancient city, N Egypt, S of Thebes near modern Armant. It was founded in prehistoric times and was prominent during the period of Roman supremacy. Hermonthis has a fine temple built c.1500 B.C. and reconstructed by the Ptolemies.

Her·mop·o·lis Mag·na (hûr-mŏp'ə-lĭs măg'nə), ancient city, S Egypt, on the Nile. It was the chief seat for the worship of Thoth. Southwest of the site is a Greco-Egyptian cemetery.

Her·mo·sil·lo (ĕr'mō-sē'yō), city (1974 est. pop. 232,700), capital of Sonora state, NW Mexico, at the entrance to the gorge of the Sonora River. Hermosillo is a transportation and agricultural center in an irrigated area where cereals and cotton are grown and cattle are raised. It was established in 1700 as an Indian town.

Her·mou·po·lis (ĕr-mōō'pô-lĭs) or **Sí·ros** (sē'rôs), city (1971 pop. 13,502), capital of Cyclades prefecture, SE Greece, on the E coast of Síros island. It is the chief city of the Cyclades and a major Aegean port and commercial center.

Her·nan·do (hûr-năn'dō), county (1970 pop. 17,004), 488 sq mi (1,263.9 sq km), W central Fla., between the Withlacoochee River in the E and the Gulf of Mexico in the W; formed 1843; co. seat Brooksville. It is in a lowland area with a marshy coast and small lakes. Agriculture (poultry, cattle, hogs, corn, citrus fruit, and peanuts), sawmilling, and limestone quarrying are important.

Hernando, town (1970 pop. 2,499), seat of De Sota co., extreme NW Miss., S of Memphis, Tenn.; inc. 1837. It is a trade center for a farm region.

Her·ne (hĕr'nə), city (1975 pop. 193,831), North Rhine-Westphalia, W West Germany, a port on the Rhine-Herne Canal. It is an industrial center of the Ruhr district; manufactures include textiles, radio and television sets, and machinery.

Herne Bay (hûrn), urban district (1973 est. pop. 26,510), Kent, SE England. It is a resort with a long coastline and promenades. The town developed after a railroad was built in 1833.

Her·ning (hăr'nĭng), city (1976 est. pop. 46,900), Ringkøbing co., central Denmark. It is an important manufacturing center with textile mills and machine shops.

Her·shey (hûr'shē), uninc. town (1970 pop. 9,000), Dauphin co., SE Pa., ENE of Harrisburg; founded 1903. Owned by the Hershey Company, the village is populated mainly by the company's employees. Chocolate confectionery is made in large quantities.

Her·stal (hĕr'stäl'), city (1977 pop. 41,150), Liège prov., E Belgium, on the Meuse River, an industrial suburb of Liège. It is the center of Belgium's armaments industry; other manufactures include iron and steel, motor vehicles, aircraft engines, and electrical equipment.

Hert·ford (här'fərd, härt'-), municipal borough (1973 est. pop. 20,760), administrative center of Hertfordshire, E central England, on the Lea River. Hertford is an agricultural market. There are several light industries, including brewing, flour milling, and the manufacture of leather goods and stationery. It was important in Saxon times; here, in 672, the archbishop of Canterbury convened the first national church council.

Hert·ford (hûrt'fərd), county (1970 pop. 23,529), 353 sq mi (914.3 sq km), NE N.C., in a coastal-plain and tidewater region bounded on the N by the Va. border, on the E by the Chowan River, and drained by the Meherrin River; formed 1759; co. seat Winton. It has agriculture (peanuts, tobacco, cotton, corn, and livestock), a lumbering industry, and some manufacturing. Recreational opportunities include boating, water-skiing, fishing, and hunting.

Hertford, town (1970 pop. 2,023), seat of Perquimans co., NE N.C., N of Albemarle Sound and SW of Elizabeth City; settled before 1700, inc. 1758. It is a winter resort in a farming and fishing area.

Hert·ford·shire (här'fərd-shĭr', -shər, härt'-), nonmetropolitan county (1976 est. pop. 937,300), 631 sq mi (1,634.3 sq km), E central England; administrative center Hertford. Hertfordshire is primarily an agricultural region. The county figured prominently in the military history of England, particularly during the Wars of the Roses.

Her·zliy·ya (hĕrt-slē'ə), town (1975 est. pop. 47,800), central Israel, near the Mediterranean Sea. It is a resort with fine beaches. It was founded in 1924 and named for Theodor Herzl, the founder of modern Zionism. There are flour mills in the town.

Hesse (hĕs, hĕs'ə), state (1970 pop. 5,382,000), 8,150 sq mi (21,109 sq km), central West Germany, bounded by Baden-Württemberg and Bavaria in the S, Rhineland-Palatinate in the W, North Rhine-Westphalia and Lower Saxony in the N, and East Germany in the E; capital Wiesbaden. It includes the former Prussian province of Hesse-Nassau, which extended from the East German border in the east to the Rhine in the west; it also includes, except for Rhenish Hesse, the former grand duchy (after 1918, state) of Hesse or Hesse-Darmstadt. Nearly all of Hesse is a hilly, agricultural land. Grain, potatoes, and fruit are grown, and cattle are raised here. Along the beautiful Rhine valley some of the finest German wines are produced. The chief manufactures include chemicals, machinery, and metal goods. Lignite, potash, and iron ore are mined.

Het·ting·er (hĕt'ən-jər), county (1970 pop. 5,075), 1,135 sq mi (2,939.7 sq km), SW N.Dak., drained by the Cannonball River; formed 1883; co. seat Mott. It is in an agricultural area that produces wheat, barley, sunflowers, poultry, livestock, and dairy products. There are coal fields in the county, and it has some manufacturing.

Hettinger, city (1970 pop. 1,655), seat of Adams co., SW N.Dak., near the S.Dak. line SW of Mott, in a farm area; inc. 1916. Bricks are made, and flour and dairy products are processed.

Hey·wood (hā'wŏŏd'), municipal borough (1971 pop. 30,418), Greater Manchester, NW England. Products include cotton goods, metal goods, boilers, carpets, paper, rope, and machinery.

Hi·a·le·ah (hī'ə-lē'ə), city (1978 est. pop. 123,700), Dade co., SE Fla., NW of Miami; inc. 1925. Its industries include printing and the manufacture of apparel, furniture and fixtures, metal and plastic goods, transportation equipment, and building supplies. Hialeah Park Race Track, featuring the famed flamingos, is in the city.

Hi·a·was·see (hī'ə-wä'sē), town (1970 pop. 415), seat of Towns co., NE Ga., near the N.C. border NW of Toccoa. It is a resort.

Hi·a·wath·a (hī'ə-wôth'ə), city (1970 pop. 3,365), seat of Brown co., NE Kansas, NW of Atchison; founded 1857, inc. 1859. It is the trade center of an area producing apples, corn, and wheat.

Hib·bing (hĭb'ĭng), village (1978 est. pop. 16,300), St. Louis co., NE Minn., on the Mesabi iron range; inc. 1893. Iron mining, formerly the major industry, has declined, but the manufacture of mining equipment is important. In 1917 Hibbing was moved 2 mi (3.2 km) south to make room for a large open-pit iron mine.

Hi·ber·ni·a (hī-bûr′nē-ə), the Latin name for Ireland.

Hick·man (hĭk′mən). **1.** County (1970 pop. 6,264), 248 sq mi (642.3 sq km), SW Ky., bounded on the W by the Mississippi River and the Mo. border, drained by Obion Creek and Bayou de Chien; formed 1821; co. seat Clinton. It is in a gently rolling agricultural area that produces livestock, grain, dark tobacco, cotton, and fruit. There is some mining in the county. Textiles and shoes are manufactured. **2.** County (1970 pop. 12,096), 613 sq mi (1,587.7 sq km), central Tenn., drained by the Duck River and its tributaries; formed 1807; co. seat Centerville. It has diversified agriculture, phosphate mines, stands of timber, and some manufacturing.

Hickman, town (1970 pop. 3,048), seat of Fulton co., extreme SW Ky., on bluffs above the Mississippi River; settled 1819, inc. 1834. It is a trade, shipping, and processing center for a farm area.

Hick·o·ry (hĭk′ə-rē, hĭk′rē), county (1970 pop. 4,481), 397 sq mi (1,028.2 sq km), SW central Mo., in the Ozarks, drained by the Pomme de Terre and Little Niangua rivers; formed 1845; co. seat Hermitage. Its agriculture includes corn, wheat, oats, hay, and livestock. It has some mining and manufacturing.

Hickory, city (1978 est. pop. 19,100), Burke and Catawba cos., W N.C., at the foot of the Blue Ridge Mts.; inc. 1870. The city's manufactures include hosiery, furniture, porcelain, electrical and electronic products, and textiles.

Hicks·ville (hĭks′vĭl′), uninc. city (1978 est. pop. 50,200), Nassau co., SE N.Y., on Long Island; founded 1648. It is chiefly residential, with some manufacturing.

Hi·dal·go (ē-*th*äl′gō, hĭ-dăl′gō), state (1970 pop. 1,193,845), 8,058 sq mi (20,870 sq km), central Mexico; capital Pachuca de Soto.

Hi·dal·go (hĭ-dăl′gō). **1.** County (1970 pop. 4,734), 3,447 sq mi (8,927.7 sq km), extreme SW N.Mex., bounded on the W by Ariz., on the S and SE by Mexico, and watered by the Gila River; formed 1919; co. seat Lordsburg. It is in a livestock and agricultural area with copper, silver, gold, and lead deposits. Part of Coronado National Forest is here, and ranges of the Continental Divide are in the south. **2.** County (1970 pop. 181,535), 1,541 sq mi (3,991.2 sq km), extreme S Texas, on the Mexican border; formed 1852; co. seat Edinburg. In the south are rich irrigated lands that produce truck crops, cotton, and a large portion of the state's citrus fruits. In the north are cattle, goat, and sheep ranches. The county has an extensive food-processing industry, oil and natural-gas wells and refineries, clay deposits, and winter resorts. Bricks and tile are manufactured.

Hi·dal·go del Par·ral (ē-*th*äl′gō dĕl pä-räl′, hĭ-dăl′gō) or **Par·ral** (pä-räl′), city (1970 pop. 61,729), Chihuahua state, N Mexico, on the Parral River. The city is a mining center, especially for silver, which has been mined in the region since the 16th cent. Pancho Villa was assassinated here (1923).

Hi·er·ap·o·lis (hī-ə-răp′ə-lĭs), ancient city of Phrygia, W Asia Minor, N of Laodicea and on a plateau above the Lycus valley (in present-day Turkey). Devoted to the worship of Leto in ancient times, it became an early seat of Christianity. The Romans greatly enlarged and improved the city, building a large theater and numerous baths about the hot springs for which the site is famous. Today these springs still feed falls that cover the rocks with gleaming white deposits of lime, creating vast falls of crystal incrustations. Extensive ruins survive from the Roman and Christian periods.

Hie·rro (yĕ′rō), volcanic island (1970 pop. 5,503), 107 sq mi (277.1 sq km), westernmost of the Canary Islands, Santa Cruz de Tenerife prov., Spain. It produces wines, figs, potatoes, brandies, tomatoes, and cheese. Ancient geographers marked Hierro as the westernmost point of the world and based longitude here. This reckoning was changed in 1884 to the Greenwich meridian.

Hi·ga·shi-O·sa·ka (hē-gä′shē-ō-sä′kä), city (1976 est. pop. 524,723), Osaka prefecture, W central Honshu, Japan, on the Onii River. It is a residential and industrial suburb of Osaka.

High·land (hī′lənd). **1.** County (1970 pop. 28,996), 549 sq mi (1,421.9 sq km), SW Ohio, drained by the East Fork of the Little Miami River and several creeks; formed 1805; co. seat Hillsboro. It has agriculture (livestock, grain, tobacco, poultry, and dairy products), limestone quarries, and some manufacturing. **2.** County (1970 pop. 2,529), 416 sq mi (1,077.4 sq km), NW Va., in the Alleghenies, bounded on the W and N by the W.Va. border and drained by the Jackson, South Fork of the Potomac, and Cowpasture rivers; formed 1847; co. seat Monterey. It is in a livestock-raising area, with lumbering and some agriculture (grain and potatoes).

Highland. 1. Uninc. community (1970 pop. 12,669), San Bernardino co., SE Calif., in a citrus-grove area at the foot of the San Bernardino Mts. It has citrus-packing plants and some light industry. Norton Air Force Base is adjacent to the community. **2.** City (1970 pop. 5,981), Madison co., SW Ill., ENE of East St. Louis, in a dairy and farm area; founded 1831 by Swiss, inc. 1863. Pipe organs, boxes, and electrical equipment are made. **3.** Residential town (1978 est. pop. 26,400), Lake co., extreme NW Ind., in the Chicago metropolitan area; settled 1850 as Clough Postal Station, inc. 1910.

Highland Park. 1. City (1978 est. pop. 31,300), Lake co., NE Ill., a suburb of Chicago on Lake Michigan; inc. 1869. It is a retail business and medical center for the North Shore area, and the summer home of the Chicago Symphony Orchestra. Fort Sheridan is adjacent to the city. **2.** City (1978 est. pop. 29,400), Wayne co., SE Mich., within the confines of Detroit; laid out 1818, inc. as a city 1917. Tractors, auto parts, packaging equipment, and food products are made, and there are trucking and warehousing industries. Highland Park grew mainly after Henry Ford established a factory here in 1909. **3.** Residential borough (1970 pop. 14,385), Middlesex co., N central N.J., on the Raritan River opposite New Brunswick; inc. 1905. **4.** Town (1970 pop. 10,133), Dallas co., N Texas, a residential suburb within the confines of Dallas; inc. 1913.

High·lands (hī′ləndz), mountain region in the N extremity of Scotland, consisting roughly of the area N of the imaginary line from Dumbarton to Stonehaven excluding the Orkneys, the Shetlands, and the lower coastal area of the N mainland; the Hebrides are usually included. Famous for its rugged beauty, the land is unsuitable for farming. Crofting, fishing, and distilling are the main occupations; in recent years the tourist trade has been an important source of income. The early history of the region is not well-known. By the 11th cent. the Scottish monarchy was definitely centered in the Lowlands, and the Highland lairds were left to run their own affairs. The distinctive marks of the Highlands, the dress and the clan system, now in disuse except as sentimental archaisms, were products of the late Middle Ages. The Highlands were the center of the Jacobite uprisings of 1715 and 1745.

Highlands, county (1970 pop. 29,507), 997 sq mi (2,582.2 sq km), S central Fla., bounded in the E by the Kissimmee River; formed 1921; co. seat Sebring. It has rolling terrain with many lakes, notably Lake Istokpoga, and includes a portion of the Everglades in the southeast. Its economy depends on citrus fruit, cattle, truck crops, and poultry.

High·more (hī′môr′), city (1970 pop. 1,173), seat of Hyde co., central S.Dak., ENE of Pierre. It is in a livestock and grain area.

High Point, city (1978 est. pop. 60,900), Davidson, Guilford, and Randolph cos., N N.C., in a heavily forested piedmont region; settled before 1750, inc. 1859. It is an industrial center noted for the production of furniture and hosiery.

High Wyc·ombe (wĭk′əm), municipal borough (1973 est. pop. 61,190), Buckinghamshire, S England. The town is well-known for its furniture industry and also has papermills, sawmills, and engineering works. Ancient British and Roman remains are found nearby.

Hil·des·heim (hĭl′dĕs-hīm′), city (1974 est. pop. 106,734), Lower Saxony, N central West Germany. Its manufactures include stoves, chemicals, radio and television sets, rubber goods, and motor vehicles. The city received a charter in 1249 and soon afterward joined the Hanseatic League. In 1813 it passed to Hanover, and in 1866 it passed, with Hanover, to Prussia. Among its Romanesque-style buildings are the cathedral (11th cent.), the Church of St. Michael (11th–12th cent.), and the Church of St. Godehard (12th cent.). Many old buildings were badly damaged in World War II.

Hill (hĭl). **1.** County (1970 pop. 17,358), 2,927 sq mi (7,580.9 sq km), N Mont., bordering in the N on Alta. and Sask., Canada, and drained by the Milk and Sage rivers; formed 1912; co. seat Havre. Its economy depends on grain and livestock. **2.** County (1970 pop. 22,596), 1,010 sq mi (2,615.9 sq km), N central Texas, in a rich blackland prairie area bounded on the W by the Brazos River; formed 1853; co. seat Hillsboro. It has agriculture (cotton, corn, grain, clover seed, legumes, fruit, truck crops, livestock, and dairy products), and some manufacturing and processing. Wool is shipped.

Hill City, city (1970 pop. 2,071), seat of Graham co., NW Kansas, NW of Hays on the South Fork of the Solomon River, in a wheat and oil area; laid out 1880, inc. 1888.

Hil·le·rød (hĭl'ə-rœth), city (1976 est. pop. 31,429), capital of Frederiksborg co., E Denmark. It is an industrial and tourist center. The city developed around the famous Renaissance-style Frederiksborg castle (1602-20), which was once a royal residence and today houses the national museum. From 1640 to 1840, Danish kings were crowned in the castle church.

Hil·ling·don (hĭl'ĭng-dən), borough (1971 pop. 234,718) of Greater London, SE England. Hillingdon was created in 1965 by the merger of the municipal borough of Uxbridge and the urban districts of Hayes, Harlington, Ruislip-Northwood, Yiewsley, and West Drayton. Among the borough's industries are printing, motion-picture production, and the manufacture of aircraft, food products, and electrical and musical instruments.

Hills·bor·o (hĭlz'bûr-ō). **1.** City (1970 pop. 4,267), seat of Montgomery co., S central Ill., SSE of Springfield, in a farm and coal area; inc. 1855. **2.** City (1970 pop. 2,730), Marion co., central Kansas, NNE of Wichita, in the Cottonwood Valley; laid out 1879, inc. 1884. It is largely a Mennonite community engaged in winter-wheat and corn production. Tabor College is here. **3.** Town (1970 pop. 831), seat of Jefferson co., E Mo., near the Mississippi River SW of St. Louis, in an agricultural area. **4.** Or **Hillsborough,** town (1970 pop. 2,775), Hillsboro co., S N.H., on the Coontoocook River WSW of Concord; settled 1741, abandoned 1744 in an Indian War, inc. 1772. Electrical equipment is made. Franklin Pierce's house was restored in 1925. **5.** Town (1970 pop. 1,444), seat of Orange co., N central N.C., NW of Durham; settled before 1700, platted 1754. An early capital of the province of North Carolina, it was the scene of disturbances by the Regulators (1768-71). The town produces textiles and furniture and packs meat. **6.** Town (1970 pop. 1,309), seat of Traill co., E N.Dak., on the Goose River and NNW of Fargo, in a grain, livestock, and dairy region; laid out 1880, inc. 1881. **7.** City (1970 pop. 5,584), seat of Highland co., SW Ohio, E of Cincinnati; platted 1807. It is a trade center for a farm, livestock, and dairy area. Several Mound Builders' forts are nearby. **8.** City (1970 pop. 14,675), seat of Washington co., NW Oregon, in the Tualatin valley; inc. 1876. Fruits and vegetables are frozen and packed, and kitchen cabinets and electric organs are manufactured. Settled c.1845, Hillsboro has a pioneer museum and cemetery and a notable old Scottish church. **9.** City (1970 pop. 7,224), seat of Hill co., N central Texas, N of Waco, in a rich black-land region; inc. 1853. It has cotton-processing plants and textile mills. Dairying and livestock breeding is done in the area.

Hills·bor·ough (hĭlz'bûr'ō). **1.** County (1970 pop. 490,265), 1,040 sq mi (2,693.6 sq km), W Fla., on the Gulf coast bounded on the S and W by Tampa Bay; formed 1834; co. seat Tampa. Farming (citrus fruit, truck crops, peanuts, and corn), dairying, poultry raising, and fishing are important. Food, tobacco, and wood products are manufactured. There is some quarrying of phosphate. **2.** County (1970 pop. 223,941), 893 sq mi (2,312.9 sq km), extreme S central N.H., bordered on the S by Mass. and drained by the Contoocook River; co. seat Nashua.

Hillsborough, residential town (1970 pop. 8,753), San Mateo co., W Calif., S of San Francisco; inc. 1910. The founders, hoping to restrict the town to fine residences, banned stores, hotels, and sidewalks.

Hills·dale (hĭlz'dāl'), county (1970 pop. 37,171), 600 sq mi (1,554 sq km), S Mich., bounded on the S by the Ohio border and drained by the St. Joseph River and headstreams of the Kalamazoo River; formed 1829; co. seat Hillsdale. Its agriculture includes grain, fruit, soybeans, livestock, poultry, and dairy products. It has oil and gas wells and lumbering and food-processing industries. Textiles, wood and metal products, furniture, plastics, and machinery are made.

Hillsdale. 1. City (1970 pop. 7,728), seat of Hillsdale co., S central Mich., SW of Jackson; settled 1834, inc. as a village 1847, as a city 1869. Doors and machine parts are made. **2.** Borough (1970 pop. 11,768), Bergen co., NE N.J.; inc. 1923. It is primarily residential.

Hills·ville (hĭls'vĭl'), town (1970 pop. 1,149), seat of Carroll co., SW Va., in the Blue Ridge NE of Galax, in an agricultural and timber area. Grain and cabbage are grown.

Hi·lo (hē'lō), city (1978 est. pop. 29,300), seat of Hawaii co., on Hilo Bay of Hawaii Island; settled by missionaries c.1822, inc. as a city 1911. A port of entry, Hilo is the trade and shipping center for a sugar-cane, orchid, papaya, and macadamia-nut region. The economy is based heavily on sugar and on tourism. Among Hilo's points of interest are the peaks of Mauna Kea and Mauna Loa and the Lyman House (c.1839), now a museum with a collection of Hawaiiana. Hilo was badly damaged by tidal waves in 1946 and 1960.

Hil·ver·sum (hĭl'vər-səm), city (1977 est. pop. 93,951), North Holland prov., central Netherlands. It is the center of Dutch television broadcasting. Its manufactures include chemicals and machinery.

Hi·ma·chal Pra·desh (hĭ-mä'chəl prə-dāsh'), state (1971 pop. 3,424,-332), 21,629 sq mi (56,019 sq km), NW India, in the W Himalayas, bordered by the Tibet region of China on the E; capital Simla. The state is covered with forested mountains, and the valleys are extensively cultivated. The forests supply large quantities of timber, the main source of income. Potatoes and fruits are the chief crops. Salt mining and the making of handicrafts are also practiced here. Himachal Pradesh was formed as a union territory in 1948; it became a state in 1971.

Him·a·la·yas (hĭm'ə-lā'əz, hĭ-mäl'yəz), Asian mountain system, extending c.1,500 mi (2,415 km) E from the Karakorum range through Pakistan, India, China, Nepal, Sikkim, and Bhutan to the S bend of the Brahmaputra River in SE Tibet (China). For most of its length, the Himalayas comprise two nearly parallel ranges separated by a wide valley in which the Indus River flows westward and the Brahmaputra flows eastward. The northern range is called the Trans-Himalayas. The southern range has three parallel zones: the Great Himalayas, the perpetually snow-covered main range in which the highest peaks (average elevation 20,000 ft/6,100 m) are found; the Lesser Himalayas with 7,000 to 15,000 ft (2,135-4,575 m) elevations; and the southernmost Outer Himalayas, 2,000 to 5,000 ft (610-1,525 m) high. A relatively young and still growing system subject to severe earthquakes, the Himalayas' main axis was formed c.25 to 70 million years ago as the earth's crust folded against the northward-moving Indian subcontinent.

Little of the Himalayan region is inhabitable or of great current economic value. The southern approaches are malarial jungle and swamps with many wild animals. Grazing is possible on some of the gentler slopes, and farming is carried on in the valleys; there is some lumbering in the extensive forests found below 12,000 ft (3,660 m). Limited amounts of iron ore, gold, and sapphires are mined.

Hi·me·ji (hĭ-mĕj'ē), city (1976 est. pop. 439,717), Hyogo prefecture, SW Honshu, Japan. A railroad and market center, it manufactures cotton textiles and produces iron and steel.

Him·er·a (hĭm'ər-ə), ancient city on the N coast of Sicily, founded by Greeks in the 7th cent. B.C. Here in 480 B.C. (a traditional date) forces led by Gelon routed the Carthaginians led by Hamilcar. In 409 B.C. the Carthaginians destroyed the city.

Hims (hĭmsh) or **Homs** (hôms), city (1970 pop. 215,423), capital of Hims governorate, W central Syria, on the Orontes River. It is a commercial center located in a fertile plain where wheat, grapes, barley, and onions are grown. Manufactures include refined petroleum, flour, fertilizer, processed foods, and silk, cotton, and woolen textiles. In ancient times Hims was the site of a great temple to Baal. In the early 3rd cent. A.D. a priest of the temple became Roman emperor as Heliogabalus, or Elagabalus. The Arabs took the town in 636. Hims was part of the Ottoman Empire from the 16th cent. until after World War I, when it became part of the French League of Nations mandate.

Hinck·ley (hĭngk'lē), urban district (1973 est. pop. 49,310), Leicestershire, central England, near Watling Street. Hosiery and shoes are the chief manufactures.

Hind·man (hīnd'mən), city (1970 pop. 808), seat of Knott co., E Ky., in the Cumberland foothills ENE of Hazard, in a coal-mining and agricultural area.

Hinds (hīndz), county (1970 pop. 214,973), 877 sq mi (2,271.4 sq km), SW central Miss., bounded on the E by the Pearl River, on the NW and partly on the W by the Big Black River; formed 1821; co. seats Jackson and Raymond. Its agriculture includes cotton, corn, truck crops, cattle, and poultry. It has timber, natural-gas wells, and diversified manufactures (clothing and textiles, wood products, furniture, fertilizer, plastics, glass, brick and clay tile, concrete, iron foundries, metal products, and machinery).

Hin·du Kush (hĭn'dōō kōōsh'), second-highest mountain system in the world, extending c.500 mi (805 km) W from the Pamir Knot, N Pakistan, into NE Afghanistan. It rises to 25,230 ft (7,695.2 m) in Tirich Mir. Glaciated and receiving heavy snowfall, the mountains have permanently snow-covered peaks and little vegetation. Virtually uninhabited, the system is crossed by several high-altitude passes; once followed by Alexander the Great and Tamerlane in their invasions of India, they are now trade routes.

Hin·du·stan (hĭn′dōō-stän′, -stăn′), vague term, usually applied to the Ganges Plain of N India, between the Himalayas in the N and the Deccan plateau in the S. Used variably throughout Indian history, it gradually came to mean the whole of northern India from the Punjab to Assam. The term has also been applied to the whole Indian subcontinent.

Hines·ville (hīnz′vĭl′), city (1970 pop. 4,115), seat of Liberty co., SE Ga., SW of Savannah, in a farm area. Timber and tobacco are produced here.

Hing·ham (hĭng′əm), resort town (1970 pop. 18,845), Plymouth co., E Mass., on the S shore of Hingham Bay; inc. 1635. The Old Ship Church (1681), a fine example of American Gothic architecture, has been in continuous use since it was built.

Hins·dale (hĭnz′dāl′), county (1970 pop. 202), 1,057 sq mi (2,737.6 sq km), SW central Colo., in a tourist and agricultural area, drained by the headwaters of the Rio Grande; formed 1874; co. seat Lake City. Livestock is raised. Parts of the San Juan Mts. and San Juan, Gunnison, Rio Grande, and Uncompahgre national forests are here.

Hin·ton (hĭn′tən), city (1970 pop. 4,530), seat of Summers co., S W.Va., on the New River near its junction with the Greenbrier River, NE of Bluefield; settled 1831, inc. 1880. It is a shipping center for a farm area. There are mineral springs in the area.

Hi·ra·do (hē-rä′dō), town (1970 pop. 32,863), on Hirado island, off NW Kyushu, Japan. It is known for its fine porcelains. The Portuguese traded at its port c.1550, and Dutch and English factories were established in the early 17th cent.

Hi·ra·tsu·ka (hē-rä′tsōō-kä), city (1976 est. pop. 199,021), Kanagawa prefecture, central Honshu, Japan, on Sagami Bay and the Sagami River. It is a commercial and industrial center with industries producing rubber goods, electronic equipment, and fountain pens.

Hi·ro·sa·ki (hē-rō′sä-kē), city (1976 est. pop. 108,600), Aomori prefecture, N Honshu, Japan. A commercial center, it has industries that produce sake and lacquerware. It is a former castle town and has 17th cent. ruins.

Hi·ro·shi·ma (hĭr′ə-shē′mə, hĭ-rō′shĭ-mə), city (1976 est. pop. 863,273), capital of Hiroshima prefecture, SW Honshu, Japan, on Hiroshima Bay. It is an important commercial and industrial center manufacturing textiles, sake, ship components, automobiles, machinery, tools, furniture, and canned foods. The city is also a market for agricultural and marine products. Founded c.1594 as a castle city on the Ota River delta, Hiroshima is divided by the river's seven mouths into six islands connected by 81 bridges. Hiroshima was the target (Aug. 6, 1945) of the first atomic bomb ever dropped on a populated area; almost 130,000 people were killed, injured, or missing, and 90% of the city was leveled.

His·pan·io·la (hĭs′pən-yō′lə), island of the West Indies, 29,530 sq mi (76,482.7 sq km), between Cuba and Puerto Rico. Haiti occupies the western third of the island and the Dominican Republic the remainder. Discovered by Columbus in 1492, the island was first called Española. The later French colony was called Saint-Domingue, after Santo Domingo, the Spanish colony in the eastern part of the island. The terrain, dominated by the Cordillera Central, is rugged. The island's climate is subtropical, and agriculture (coffee, cocoa, and sugar cane) flourishes in the abundant rainfall. Bauxite is mined.

His·sar (hĭ-sär′), town (1971 pop. 89,463), Haryana state, NW India, on the West Jumna Canal. In a well-irrigated area, it is a market for cotton, grain, and oilseed. Cotton and silk fabrics are made. Hissar, founded in 1356, became important under the Mogul empire. It was occupied by the British in 1803.

Hi·ta·chi (hē-tä′chē), city (1976 est. pop. 202,009), Ibaraki prefecture, E central Honshu, Japan, on the Kashima Sea. It is an important industrial city where cement, electrical equipment, and chemicals are produced. It is the site of one of Japan's largest copper mines.

Hitch·cock (hich′kŏk′), county (1970 pop. 4,051), 712 sq mi (1,844.1 sq km), S Nebr., bounded on the S by Kansas and drained by the Republican River; formed 1873; co. seat Trenton. In a corn, wheat, and cattle-raising area, it has oil wells and makes farm machinery.

Hitch·in (hĭch′ən), urban district (1973 est. pop. 29,190), Hertfordshire, SE England. Hitchin appears in the Domesday Book as a royal manor named Hiz. Corn and cattle are traded at a biweekly market. Industries include building contracting, engineering, tanning, parchment making, medicinal distilling, and rose growing.

Hi·va O·a (hē′vä ō′ä), volcanic island (1967 est. pop. 1,000), 154 sq mi (399 sq km), South Pacific, most important of the Marquesas Islands, French Polynesia. The Bay of Traitors, protected by ridges c.3,000 ft (915 m) high, provides a good harbor. Copra is the chief export.

Hi·was·see (hī-wä′sē) or **Hi·a·was·see** (hī′ə-wä′sē), river, c.150 mi (240 km) long, rising in the Blue Ridge of NE Ga. and flowing N through SW N.C. into SE Tenn. to the Tennessee River, NE of Chattanooga.

Hjäl·ma·ren (yĕl′mə-rən), lake, c.190 sq mi (490 sq km), S central Sweden. It is drained into Mälaren Lake by the Eskälven.

Hjør·ring (yœr′ĭng), city (1976 est. pop. 24,400), Nordjylland co., N Denmark. The center of an agricultural region, it has food-processing plants, textile mills, and machine shops. Hjørring dates from the 12th cent. and has retained several medieval churches.

Hka·ka·bo Ra·zi (kä′kä-bō rä′zē), peak, 19,296 ft (5,881 m) high, N Burma, on an outlier of the Himalayan system. It is the highest point in Burma.

Ho·bart (hō′bärt), city (1976 pop. 50,381), capital and principal port of Tasmania, SE Australia, at the foot of Mt. Wellington. Hobart's harbor is one of the finest in the world. The city has diverse industries, including meat packing, food processing, and the making of textiles, chemicals, and glass. It was founded in 1804.

Ho·bart (hō′bərt). **1.** City (1978 est. pop. 22,600), Lake co., extreme NW Ind.; settled c.1849, inc. 1921. Welding machinery and supplies, tools and castings, and aluminum doors and windows are made. **2.** City (1970 pop. 4,638), seat of Kiowa co., SW Okla., SE of Elk City; settled c.1901. It is in a farm area producing livestock, cotton, and some oil.

Hobbs (hŏbz), city (1978 est. pop. 29,900), Lea co., SE N.Mex.; inc. 1929. With the discovery (1927) of oil and natural gas in the area, Hobbs became one of the last great oil-boom towns in the United States. Chemical production is of increasing importance, as are feedlots for livestock and the raising of thoroughbred horses. Cotton and other crops are grown on irrigated farms in the area.

Ho·bo·ken (hō′bō-kən), city (1976 est. pop. 34,097), Antwerp prov., N Belgium, on the Scheldt River, an industrial suburb of Antwerp. It has large shipyards, metal refineries, and wool-processing plants.

Hoboken, city (1978 est. pop. 38,300), Hudson co., NE N.J., on the Hudson River adjoining Jersey City and opposite Manhattan; settled by the Dutch c.1640, inc. as a city 1855. It is a port of entry, a railroad terminal, and a busy seaport. The city has a large food-processing industry. There are also factories making electronic equipment, precision instruments, chemicals, machinery, paper products, and furniture. The site changed title many times before John Stevens gained possession in 1784. He built his home at Castle Point (an unusual rock formation overlooking the river) and laid out the town in 1804. Before the mid-19th cent. Hoboken was a resort and amusement center for New Yorkers.

Hoch·e·lag·a (hŏsh′ə-lăg′ə), former Indian village, Canada, discovered by Jacques Cartier on his second voyage (1535). It was situated at the foot of Mt. Royal in what is now the central part of the city of Montreal. Excavations in Montreal have unearthed hearths, kitchen middens, and burial places of the ancient village.

Hock·ing (hŏk′ĭng), county (1970 pop. 20,322), 431 sq mi (1,090.4 sq km), S central Ohio, intersected by the Hocking River and several creeks; formed 1805; co. seat Logan. Largely industrial, it also has oil wells, coal mines, and some agriculture.

Hock·ley (hŏk′lē), county (1970 pop. 20,396), 906 sq mi (2,346.5 sq km), NW Texas, in a rich agricultural region; formed 1876; co. seat Levelland. In an oil-refining region, it has extensive irrigated fields yielding cotton, corn, wheat, potatoes, truck crops, fruit, livestock, and dairy products. Propane and butane gas are produced.

Hoddes·don (hŏdz′dən), urban district (1971 pop. 26,071), Hertfordshire, E central England. A residential suburb of London, Hoddesdon has light industries and horticultural enterprises.

Ho·dei·da (hōō-dā′dä) or **Al Hu·day·dah** (äl hōō-dā′dä), city (1970 est. pop. 90,000), W Yemen, on the Red Sea. The chief port of the country, it was developed as a seaport in the mid-19th cent. by the Turks. After a disastrous fire in Jan., 1961, destroyed much of the city, it was rebuilt with Soviet aid.

Hodge·man (hŏj′mən), county (1970 pop. 2,662), 860 sq mi (2,227.4 sq km), SW central Kansas, in a rolling prairie region watered by the Pawnee River; formed 1873; co. seat Jetmore. It has agriculture (grain and livestock) and oil and natural-gas fields.

Hodg·en·ville (hŏj'ən-vĭl'), town (1970 pop. 2,562), seat of Larue co., central Ky., S of Louisville near the Nolin River; settled 1789. It is in a farm area that also has limestone quarries and natural-gas wells.

Hód·me·ző·vá·sár·hely (hôd'mĕ-zœ-vä'shär-hä), city (1977 est. pop. 45,200), SE Hungary, near the Tisza River. An agricultural center, it also produces machinery, textiles, and pottery.

Hoek van Hol·land (hŏŏk vän hôl'änt) or **Hook of Holland** (hŏŏk; hŏl'ənd), district of Rotterdam, W Netherlands, on the North Sea. It is an outer port of Rotterdam, with which it is connected by the New Waterway, and a terminus of ships that cross the English Channel from Harwich, England.

Hof (hōf, hôf), city (1974 est. pop. 55,041), Bavaria, E central West Germany, on the Saale River, near the borders with East Germany and Czechoslovakia. The city's industries produce textiles, chemicals and dyes, metal goods, beer, and paper. Hof was first mentioned in the early 13th cent. It went to Bavaria in 1810 and was included in Upper Franconia.

Ho-fei or **Ho·fei** (both: hô'fā'), city (1970 est. pop. 400,000), capital of Anhwei prov., China. A rapidly growing industrial city, it has textile mills, ironworks and steelworks, food and cotton processing plants, and a variety of other manufactures.

Ho·fu (hō'fōō), city (1976 est. pop. 84,000), Yamaguchi prefecture, SW Honshu, Japan, on the Suo Sea. It is a center of chemical industries producing artificial fibers, liquors, medicines, and salt.

Ho·fuf (hōō-fōōf') or **Al Hufuf** (äl), town (1974 pop. 101,271), E Saudi Arabia. Textiles, brass, and copper wire are made in Hofuf. It is also a trade center for dates, wheat, and fruit and has a large mosque.

Hogue, La (lä ōg'), or **La Hougue** (lä ōōg'), cape on the NE coast of the Cotentin peninsula, France, on the English Channel. Off the cape, during the War of the Grand Alliance, a French fleet was defeated (1692) by the English and Dutch.

Ho·hen·frie·de·berg (hō'ən-frē'də-bərKH'), town, SW Poland. In 1745 it was the site of the victory of Frederick II of Prussia over the Austrian and Saxon forces in the War of the Austrian Succession. Hohenfriedeberg was ceded to Poland after World War II.

Ho·hen·wald (hō'ən-wôld'), city (1970 pop. 3,385), seat of Lewis co., central Tenn., SW of Nashville, in a farm and timber area; inc. 1911.

Ho·hen·zol·lern (hō'ən-zŏl'ərn, -tsôl'-), former province of Germany. After 1945 it became part of the temporary state of Württemberg-Hohenzollern, which was included in the state of Baden-Württemberg in 1952.

Ho·he Tau·ern (hō'ə tou'ərn), range of the Eastern Alps, S Austria, extending c.70 mi (115 km) E from the Italian border. It rises to 12,461 ft (3,800.6 m) in the Grossglockner.

Ho·ho·kam Pi·ma National Monument (hō-hō'kəm pē'mə): *see* National Parks and Monuments Table.

Hoi·how (hoi'hou'): *see* Hai-k'ou, China.

Hoke (hōk), county (1970 pop. 16,436), 389 sq mi (1,007.5 sq km), S central N.C.; formed 1911, co. seat Raeford. This forested sand-hills area has farming (cotton, tobacco, grain, corn, and peaches) and sawmills.

Hok·kai·do (hō-kī'dō), island (1976 est. pop. 5,393,587), c.30,130 sq mi (78,035 sq km), N Japan, separated from Honshu by the Tsugaru Strait and from Sakhalin, USSR, by the Soya Strait. It is the northernmost and most sparsely populated of the major islands of Japan. Its rugged interior with many volcanic peaks rises to 7,511 ft (2,290.9 m). Forests, covering most of the island, are a source of lumber, pulp, and paper. Coal, iron, and manganese are mined. Although large areas of the island are unsuited to farming, agriculture is an important occupation. Hokkaido is one of the major fishing centers of the world. The island is also a winter resort and sports area.

Hol·baek (hôl'bĕk'), city (1976 est. pop. 22,100), Vestsjaelland co., E Denmark, a seaport on Holbaek Fjord, an arm of the Isefjord; chartered 1288. It is a commercial and industrial center.

Hol·brook (hŏl'brŏŏk'). **1.** Town (1970 pop. 4,759), seat of Navajo co., E Ariz., on the Little Colorado River SE of Winslow; inc. 1917. It was settled in 1870s as a cow town called Horsehead Crossing and was renamed in 1880. The town's economy is based on livestock, farming, and tourism. The Petrified Forest National Park is nearby. **2.** Town (1970 pop. 11,775), Norfolk co., E Mass.; settled 1710, set off from Randolph and inc. 1872. It has some light manufacturing. **3.** Uninc. town (1970 pop. 13,200), Suffolk co., SE N.Y., on central Long Island, NW of Patchogue; settled 1767.

Hol·den (hōl'dən), town (1970 pop. 12,564), Worcester co., central Mass., a residential suburb of Worcester; settled 1723, set off and inc. 1741. It has some light manufacturing.

Hol·den·ville (hōl'dən-vĭl'), city (1970 pop. 5,181), seat of Hughes co., SE Okla., SE of Shawnee; laid out 1895, inc. 1898. It is the trading point for a farm and oil region. Nearby is the site of historic Fort Holmes (1834).

Hol·drege (hōl'drĭj), city (1970 pop. 5,635), seat of Phelps co., S central Nebr., SW of Kearney, in a grain and dairy area; settled 1883, inc. 1884.

Hol·guín (ōl-gēn'), city (1970 pop. 131,508), Oriente prov., E Cuba. It is a prosperous commercial center and transportation hub in a fertile region where sugar cane, coffee, tobacco, and cattle are raised. The city, founded in 1523, was moved to its present site in the 18th cent.

Hol·land (hŏl'ənd), former county of the Holy Roman Empire and, from 1579 to 1795, chief member of the United Provinces of the Netherlands. Its name is popularly applied to the entire Netherlands. Holland has been divided since 1840 into two provinces, North Holland and South Holland. The county was created in the early 10th cent. and originally controlled not only present North and South Holland, but also Zeeland and part of medieval Friesland. The cloth industry and commerce of Holland, though they developed later than those of Flanders and Brabant, began to rival those of Bruges and Antwerp in the 15th cent. Holland led in the struggle (16th–17th cent.) for Dutch independence.

Holland, city (1978 est. pop. 28,600), Allegan and Ottawa cos., SW Mich., near Lake Michigan, on Lake Macatawa, in a dairy and poultry area; founded 1847 by Dutch settlers, inc. 1867. Furnaces have been made here since 1906. Other products include furniture and boats. Tulip growing is an important industry. The city's many Dutch descendants hold a week-long tulip festival each spring.

Hol·li·days·burg (hŏl'ĭ-dāz-bûrg'), industrial borough (1970 pop. 6,262), seat of Blair co., S central Pa., S of Altoona on the Juniata River; settled 1768, inc. 1836. It has boiler factories and machine works and is a trading center.

Hol·lis (hŏl'ĭs), city (1970 pop. 3,150), seat of Harmon co., extreme SW Okla., near the Texas border W of Altus, in a cotton area; inc. as a town 1905, as a city 1919. It has flour mills.

Hol·li·ster (hŏl'ĭ-stər), city (1970 pop. 7,663), seat of San Benito co., W Calif., E of Monterey Bay, in a fruit and farm area; inc. 1874. Pinnacles National Monument, with numerous rock spires, is nearby.

Hol·lis·ton (hŏl'ə-stən), town (1970 pop. 12,069), Middlesex co., E Mass.; settled c.1659, inc. 1724. Its manufactures include plastics, quartz glass, wood panels, and paper products.

Hol·ly Hill (hŏl'ē), city (1970 pop. 8,191), Volusia co., NE Fla., on the Halifax River (a lagoon) N of Daytona Beach; inc. 1901. It is in a farm area noted for its flowers.

Holly Springs, city (1970 pop. 5,728), seat of Marshall co., N central Miss., near the Tenn. border SE of Memphis, Tenn.; inc. 1837. It is a market town in a cotton area and ships dairy products.

Hol·ly·wood (hŏl'ē-wŏŏd'). **1.** Community (1970 pop. c.194,000), part of the city of Los Angeles, S Calif., on the slopes of the Santa Monica Mts.; inc. 1903, consolidated with Los Angeles 1910. It is the center of the U.S. motion-picture industry and draws many tourists. It is also the home of numerous television, radio, and recording companies. Points of interest include Hollywood Blvd., Sunset Strip, the Hollywood Bowl, Griffith Park (with an observatory and planetarium), and the homes of film celebrities. **2.** City (1970 pop. 106,873), Broward co., SE Fla., on the Atlantic Ocean; inc. 1925. A resort and retirement center, it also produces electronic equipment and building materials.

Holmes (hōmz, hōlmz). **1.** County (1970 pop. 10,720), 483 sq mi (1,251 sq km), NW Fla., bounded on the N by the Ala. border and drained by the Choctawhatchee River; formed 1848; co. seat Bonifay. Agriculture (corn, peanuts, cotton, vegetables, and livestock) and forestry are its major activities. **2.** County (1970 pop. 23,120),

764 sq mi (1,978.8 sq km), W central Miss., bounded on the E by the Big Black River and drained by the Yazoo River; formed 1833; co. seat Lexington. It has cotton and corn farming, timber, sand and gravel pits. clay deposits, and some manufacturing (clothing, wood products, and plastics). **3.** County (1970 pop. 23,024), 424 sq mi (1,098.2 sq km), central Ohio, intersected by Killbuck Creek and the Walhonding River; formed 1825; co. seat Millersburg. Primarily agricultural (yielding corn, wheat, oats, and dairy and poultry products), it also has coal mines and light industry.

Ho·lon (hō-lôn′), city (1975 est. pop. 114,000), W central Israel. Manufactures include textiles, metal and leather goods, processed foods, furniture, glassware, plastics, and construction materials.

Hol·ste·bro (hôl′stə-brō′), city (1976 est. pop. 29,000), Ringkøbing co., W central Denmark, on the Storå River. It is a commercial and industrial center.

Hol·stein (hōl′stīn′, -shtīn′), former duchy, N West Germany, the part of Schleswig-Holstein S of the Eider River. For a time part of the duchy of Saxony, Holstein was created (1111) a county of the Holy Roman Empire. In 1386 the count of Holstein received the duchy of Schleswig as a hereditary fief. Schleswig and Holstein passed (1459) to Christian I of Denmark, who established (1460) the relationship of Denmark, Schleswig, and Holstein as a personal union; the opposition of the German nobles in the two territories to direct Danish rule was a powerful factor in this arrangement, which lasted for four centuries. In 1474 Emperor Frederick III raised Holstein to a duchy under the immediate suzerainty of the Holy Roman Empire (as distinct from Schleswig, which was outside the imperial jurisdiction). Both duchies were divided in the 16th cent. into a royal portion, ruled by the Danish kings; a ducal portion, ruled by the dukes of Holstein-Gottorp, a younger branch of the Danish royal line; and a common portion, ruled jointly by the kings and dukes. Schleswig-Holstein was united under the Danish kings in 1773.

Hol·ston (hōl′stən), river, c.120 mi (195 km) long, formed by the uniting of its N and S forks, NE Tenn., and flowing SW through the Great Appalachian Valley, joining the French Broad River at Knoxville to form the Tennessee River. Settlement along the Holston began before the American Revolution.

Holt (hōlt). **1.** County (1970 pop. 6,654), 456 sq mi (1,181 sq km), NW Mo., between the Missouri and Nodaway rivers and drained by the Tarkio River; formed 1841; co. seat Oregon. It is in an agricultural area that produces corn, soybeans, wheat, oats, apples, cattle, and hogs. There are limestone deposits, and some manufacturing is done. **2.** County (1970 pop. 12,933), 2,408 sq mi (6,236.7 sq km), N Nebr., bounded on the N by the Niobrara River and drained by the Elkhorn River; formed 1860 as West co., renamed 1862; co. seat O'Neill. Corn and cattle are raised. Some mining is done, and farm equipment is manufactured.

Holt, uninc. village (1970 pop. 6,980), Ingham co., S central Mich., SSE of Lansing; est. 1848.

Hol·ton (hōl′tən), city (1970 pop. 3,063), seat of Jackson co., NE Kansas, NNW of Topeka, in a farm region; laid out 1857, inc. 1870.

Ho·ly Island (hō′lē) or **Lin·dis·farne** (lĭn′dĭs-färn′), off the coast of Northumberland, NE England. At low tide the island is connected with the mainland by a stretch of sand. The island is partly cultivated. Tourism and fishing are important. A church and monastery, built in 635 by St. Aidan, represented the first establishment of Celtic Christianity in England. The settlement was burned by the Danes in 793 but rebuilt. When the Danes invaded in 875, the monks fled.

Hol·yoke (hōl′yōk′). **1.** Town (1970 pop. 1,640), seat of Phillips co., NE Colo., E of Sterling near the Nebr. border, in a farm area; inc. 1888. **2.** City (1978 est. pop. 44,400), Hampden co., SW Mass., on the Connecticut River; settled 1745, inc. 1873. The city is an industrial center. Paper and paper products are the leading manufactures; others include printed material, metals, detergents, plastics, machinery, and textiles.

Hom·burg (hôm′bo͝ork): *see* Bad Homburg vor der Höhe, West Germany.

Home of Frank·lin D. Roo·se·velt National Historic Site (frăngk′lĭn; rō′zə-vĕlt′, -vəlt, ro͞o′-): *see* National Parks and Monuments Table.

Ho·mer (hō′mər). **1.** Town (1970 pop. 365), seat of Banks co., NE Ga., E of Gainesville, in an agricultural and lumbering area. **2.** Town (1970 pop. 4,483), seat of Claiborne parish, NW La., ENE of Shreveport; settled 1830, inc. 1850. It is in a cotton region.

Ho·mer·ville (hō′mər-vĭl′), town (1970 pop. 3,025), seat of Clinch co., SE Ga., SW of Waycross near the Okefenokee Swamp; founded 1859, inc. 1869. It is the center of a tobacco and timber region.

Home·stead (hōm′stĕd′). **1.** City (1978 est. pop. 21,300), Dade co., SE Fla.; inc. 1913. It is a trade and shipping center for the redland district, known for its many varieties of citrus and other fruits and vegetables. Nearby Homestead Air Force Base is important to its economy. The city is also a vacation center. **2.** Borough (1978 est. pop. 5,100), Allegheny co., SW Pa., on the Monongahela River just S of Pittsburgh; inc. 1880. It is famous for large ironworks and steelworks, formerly owned by the Carnegie company, where the Homestead strike, one of the most bitterly fought industrial disputes in U.S. labor history, took place in 1892.

Homestead National Monument of America: *see* National Parks and Monuments Table.

Home·town (hōm′toun′), city (1970 pop. 6,729), Cook co., NE Ill., a suburb SW of Chicago; inc. 1953.

Home·wood (hōm′wo͝od′). **1.** City (1978 est. pop. 23,600), Jefferson co., N central Ala., a residential suburb of Birmingham; inc. 1921. **2.** Village (1970 pop. 18,871), Cook co., NE Ill., a residential suburb of Chicago; platted 1852, inc. 1892.

Homs (hôms): *see* Hims, Syria.

Ho·nan or **Ho-nan** (both: hō′năn′), province (1968 est. pop. 50,000,000), c.65,000 sq mi (168,350 sq km), NE China; capital Cheng-chou. It is sparsely settled in the mountainous western region but densely populated and cultivated in the east. Honan is a major wheat- and cotton-producing province; other agricultural products include kaoliang, rice, millet, sweet potatoes, tobacco, fruit, oakleaf silk, and oilseed crops. Coal and iron are mined, and lead and pottery clay are found. Stone Age remains have been discovered in Honan.

Hon·do (hŏn′dō), city (1970 pop. 5,487), seat of Medina co., SW Texas, W of San Antonio, in a farm and livestock area.

Hon·du·ras (hŏn-do͝or′əs), republic (1973 est. pop. 2,800,000), 43,277 sq mi (112,087 sq km), Central America, bounded on the N by the Caribbean Sea, on the E and S by Nicaragua, on the SW by El Salvador, and on the W by Guatemala. Tegucigalpa is the capital. Over 80% of the land is mountainous. In the east are the swamps and forests of the Mosquito Coast.

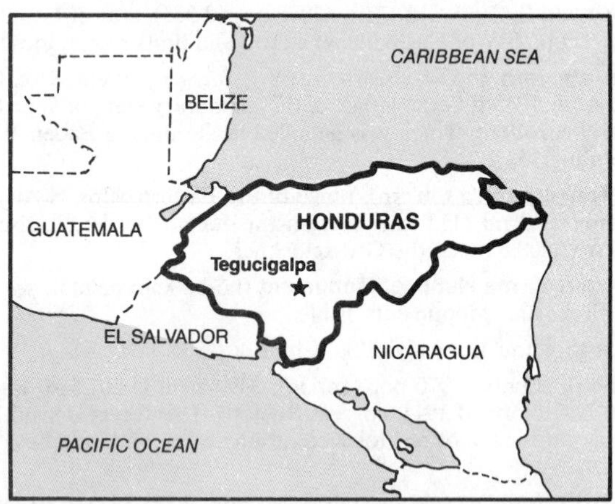

Economy. The economy is based on agriculture; bananas are the most important product, comprising 50% of all exports. Coffee, timber, minerals (silver, lead, zinc), and beef are also exported. Other food crops include corn, beans, rice, and sugar cane. Honduras has rich forest resources and deposits of silver, lead, zinc, gold, cadmium, antimony, and copper, but exploitation is hampered by inadequate road and rail systems. Food processing, especially sugar refining, is done.

History. The restored Mayan ruins of Copán in the west, first discovered by the Spaniards in 1576 and rediscovered in dense jungle in 1839, reflect the great Mayan culture that arose in the 4th cent. It had declined when Columbus sighted the region in 1502. Hernán Cortés arrived in 1524 and ordered Pedro de Alvarado to found settlements along the coast. In 1821 Honduras gained independence from Spain and became part of Iturbide's Mexican Empire; from 1825 to 1838 it was a member of the Central American Federation.

Foreign capital, plantation life, and conservative politics constituted a trio of dominant forces in Honduras from the late 19th cent. to the end of the regime (1933–48) of Tiburcio Carías Andino, when the liberal movement was reawakened. The rights of workers were not effectively defined and protected until a labor code was adopted in 1955 and a new constitution was promulgated in 1957. That year Ramón Villeda Morales became the first liberal president in 25 years. In 1963 Villeda was overthrown and replaced by a military junta under Osvaldo López Arellano. A new constitution was adopted in 1965 (the country's 12th), at which time López was elected president for a six-year term.

The illegal immigration of several hundred thousand Salvadorans across the ill-defined El Salvador-Honduras border and the expulsion of many of them by Honduras led to a war with El Salvador in July, 1969. Although it lasted only five days, its effects were serious. The Organization of American States mediated the dispute and established a demilitarized zone along the border, but border incidents continued for several years. Honduras withdrew from the Central American Common Market as a result of the war. Ramón Ernesto Cruz, who was elected president in 1971 to succeed López, was overthrown (Dec., 1972) by López, who announced that he would govern for not less than five years and promised social and economic reforms. López, in turn, was ousted by the armed forces in 1975, after a scandal involving a banana-exporting company. Col. (now Gen.) Juan Alberto Melgar replaced López as chief of state.

Hong Kong (hŏng' kŏng'), British crown colony (1976 pop. 4,400,000), land area 399 sq mi (1,033.4 sq km), adjacent to Kwangtung prov., SE China, on the estuary of the Canton River E of Macao and SE of Canton. The colony comprises Hong Kong island (29 sq mi/75.1 sq km), ceded by China in 1842 under the Treaty of Nanking; Kowloon peninsula (3.5 sq mi/9 sq km), ceded in 1860 under the Peking Convention; and the New Territories (366 sq mi/948 sq km), a mountainous mainland area adjoining Kowloon, which, with Deep Bay on the west and Mirs Bay on the east and some 235 offshore islands, was leased in 1898 for 99 years. The capital, officially named Victoria but commonly called Hong Kong, is on Hong Kong island, at the foot of Victoria Peak. Hong Kong is a free port, a bustling trade center, and a shipping and banking emporium.

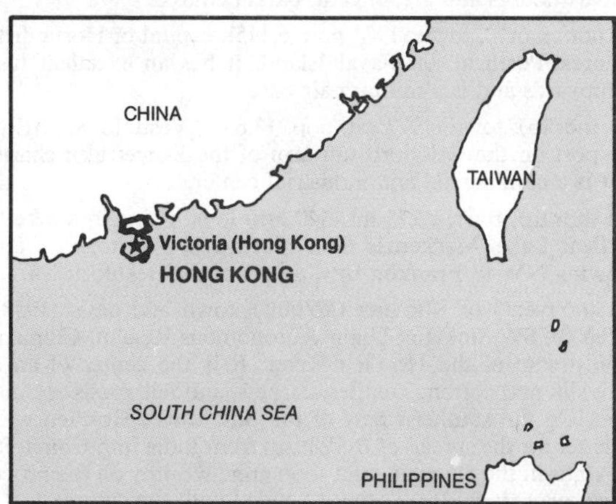

After 1950, when much of its entrepôt trade with Communist China was halted because of UN and U.S. embargoes, Hong Kong began to industrialize. The textile and garment industry is the colony's largest; there are vast complexes of spinning, carding, and weaving mills that run nonstop 24 hours a day. Other industries include shipbuilding, food processing, and the manufacture of plastics, electrical and electronic equipment, rubber products, machinery, chemicals, ceramics, furniture, jewelry, and toys. Tourism is a major source of revenue, and motion-picture production, insurance, and printing and publishing are also important. Because of the mountainous and rocky terrain, only about one seventh of the land is arable. Rice and a variety of vegetables, including cabbage, eggplant, maize, red pepper, leek, and watercress, are grown.

Hones·dale (hōnz'dāl'), industrial borough (1970 pop. 5,224), seat of Wayne co., NE Pa., on the Lackawaxen River NE of Scranton, in an agricultural and resort area; settled 1803, laid out 1826, inc. 1831. Shoes, oil-refining equipment, and textiles are manufactured.

Hon·jo (hōn'jō), city (1975 pop. 51,090), Saitama prefecture, central Honshu, Japan, on the Tone River. It is an agricultural center and a market for raw silk.

Hon·o·lu·lu (hŏn'ə-lōō'lōō), city (1978 est. pop. 328,000), capital of the state of Hawaii and seat of Honolulu co. (596 sq mi/1,543.6 sq km), on the SE coast of the island of Oahu. The city and county are legally coextensive. Honolulu is the economic center and principal port of the Hawaiian Islands. The city lies on a narrow plain between the sea and the Koolau Range. Sugar processing and pineapple canning have long been Honolulu's major industries.

Bypassed by Capt. James Cook when he explored the islands in 1778, Honolulu's harbor was entered in 1794 by William Brown, an English captain. Honolulu became the permanent capital of the kingdom of Hawaii in 1845. In the 19th cent. American and European whalers and sandalwood traders visited its port, and Honolulu was occupied successively by Russian, British, and French forces. The Japanese bombed Pearl Harbor, the U.S. naval base at Honolulu, on Dec. 7, 1941, and during World War II the port became a strategic naval base and a staging area for U.S. forces in the Pacific. Iolani Palace, the former home of Hawaii's kings, is the only royal palace in the United States. The beach at Waikiki is noted for fine bathing and surf riding. The famous Diamond Head crater is nearby.

Hon·shu (hŏn'shōō), island (1975 pop. 88,578,094), c.89,000 sq mi (230,510 sq km), central Japan. It is the largest and most important island of Japan. It is separated from Hokkaido by the Tsugaru Strait, from Kyushu by Shimonoseki Strait, and from Shikoku by the Inland Sea. Honshu is predominantly mountainous and has many volcanoes. It has valuable forests, but a limited amount of arable land. Oil, zinc, and copper are found on the island. Agriculture is varied; rice, other grains, cotton, fruits, and vegetables are grown. The bulk of Japan's tea and silk comes from Honshu. Shipbuilding and metallurgical, chemical, and textile industries are important.

Hood (hŏŏd), county (1970 pop. 6,368), 426 sq mi (1,103.3 sq km), N central Texas, drained by the Brazos River; formed 1866; co. seat Granbury. It is mainly agricultural, yielding peanuts, corn, grain, hay, fruit, truck crops, pecans, livestock, and dairy products. Wool and mohair are marketed.

Hood, Mount, peak, 11,235 ft (3,426.7 m) high, N Oregon, in the Cascade Range, E of Portland. It is the highest point in the state. A symmetrical, extinct volcano with glaciers and forested lower slopes, it is a favorite mountain-climbing and skiing center.

Hood River, county (1970 pop. 13,187), 523 sq mi (1,354.6 sq km), N Oregon, bounded on the N by the Columbia River; formed 1908; co. seat Hood River. Agriculture (particularly fruit) and lumbering are important. The county also has numerous recreational facilities, especially those at Mt. Hood.

Hood River, city (1970 pop. 3,991), seat of Hood River co., N Oregon, on the Columbia River E of Portland; settled 1854, inc. 1895. A processing and shipping point in a fertile fruit-growing valley, it also has a lumbering industry.

Hoogh·ly (hōōg'lē), river, an arm of the Ganges, c.160 mi (260 km) long, formed by the confluence of the Bhagirathi, Jalangi, and Matabhanga rivers, West Bengal state, E India, and flowing S to the Bay of Bengal. It is the major shipping artery through the Hooghlyside industrial area, despite its sandbars and strong tidal currents.

Hoogh·ly-Chin·su·ra (hōōg'lē-chĭn'sōō-rə), town (1971 pop. 105,341), West Bengal state, E India, on the Hooghly River. A center of the Hooghlyside industrial district, the town has many large rice mills. It was founded by the Portuguese in 1537.

Hook·er (hŏŏk'ər), county (1970 pop. 939), 722 sq mi (1,870 sq km), W central Nebr., drained by the Middle Loup and Dismal rivers; formed 1889; co. seat Mullen. It is in a large-scale cattle-raising area.

Hook of Hol·land (hŏŏk; hŏl'ənd): *see* Hoek van Holland.

Hoope·ston (hōōp'stən), city (1970 pop. 6,461), Vermilion co., E central Ill., E of Bloomington near the Ind. border; platted 1871, inc. 1877. It is a canning center and makes metal products.

Hoorn (hôrn), town (1977 est. pop. 27,396), North Holland prov., N central Netherlands, on an inlet of the IJsselmeer. It is a commercial and processing center for a vegetable-growing and dairy-farming region. Hoorn was founded in 1311.

Hoo·sac Range (hōō'sək), S continuation of the Green Mts., NW Mass. and SW Vt., running from N to S. Its maximum height is c.3,000 ft (915 m). A railroad tunnel, c.5 mi (8 km) long, built from 1852 to 1873, crosses the range from east to west.

Ho·pat·cong (hə-păt′kông), resort borough (1970 pop. 9,052), Sussex co., N central N.J., on Lake Hopatcong NNW of Morristown.

Hope (hōp), city (1970 pop. 8,830), seat of Hempstead co., SW Ark., SW of Hot Springs; settled 1874, inc. 1875. The city is the commercial and shipping center for a farm area noted especially for watermelons. Its industries include cotton and poultry processing and metalworking and woodworking.

Ho·peh or **Ho-pei** (both: hō′pā), province (1968 est. pop. 47,000,000), 75,000 sq mi (194,250 sq km), NE China, on the Po Hai, an arm of the Yellow Sea; capital Shih-chia-chuang. Hopeh is mountainous in the north and the west, where rich iron and coal deposits are extensively mined. It is a major cotton-producing province and an important producer of wheat. Other crops include rice, millet, kaoliang, potatoes, sweet potatoes, barley, corn, soybeans, and fruit. Stock raising is important, and fishing and salt production are significant along the coast. Heavy industry (mainly metallurgical, iron and steel, machinery, and textile) is concentrated in and around Peking, Tientsin, and T'ang-shan. One of the earliest regions of Chinese settlement, Hopeh has many prehistoric sites. Parts of the former provinces of Jehol and Chahar were incorporated into Hopeh in 1956.

Hope·town (hōp′toun′) or **Hope Town,** town (1970 pop. 4,210), Cape Prov., central South Africa, on the Orange River; founded 1854. Nearby, diamonds were first discovered (1867) in South Africa. Today the town is mainly an agricultural trade center.

Hope·well (hōp′wĕl′). **1.** Borough (1970 pop. 2,271), Mercer co., W central N.J., N of Trenton; settled before 1700, inc. 1891. Charles A. Lindbergh's home, from which his son was kidnapped, was near Hopewell. The estate was deeded to New Jersey in 1941. **2.** Independent city (1970 pop. 23,471), SE Va., at the confluence of the James and Appomattox rivers; founded 1913, inc. 1916. Hopewell is a port and an industrial center in which chemicals, synthetic textiles, and paper products are manufactured. The city was founded as a munitions center.

Hopewell Village National Historic Site: *see* National Parks and Monuments Table.

Hop·kins (hŏp′kĭnz′). **1.** County (1970 pop. 38,167), 555 sq mi (1,437.5 sq km), W Ky., bounded on the E by the Pond River, on the W by the Tradewater River; formed 1806; co. seat Madisonville. It is in a rolling agricultural area that produces livestock, grain, poultry, burley tobacco, dairy products, and fruit. It has important bituminous-coal mines, oil wells, hardwood timber (sawmills), and manufacturing (food processing, metal products, and machinery). **2.** County (1970 pop. 20,710), 793 sq mi (2,053.9 sq km), NE Texas, bounded on the N by the South Fork of the Sulphur River and drained by White Oak Bayou; formed 1846; co. seat Sulphur Springs. It is a leading dairying county, and also yields cotton, corn, peanuts, grain, sweet potatoes, fruit, truck crops, poultry, and livestock. It has lumbering and food-processing industries, gas and oil wells, and clay mines. Bricks and clothing are made.

Hopkins, city (1978 est. pop. 12,100), Hennepin co., SE Minn., a suburb of Minneapolis; inc. as West Minneapolis 1893, name changed 1928. Foods, especially raspberries, are processed and distributed, and farm machinery, heavy equipment, ordnance, and automotive parts are made in Hopkins.

Hop·kins·ville (hŏp′kĭnz-vĭl′), city (1978 est. pop. 27,200), seat of Christian co., SW Ky.; inc. 1804. Rich agricultural lands surround the city, which is a leading tobacco market. Light fixtures, wire, and clothing are manufactured.

Hop·kin·ton (hŏp′kĭn-tən). **1.** Town (1970 pop. 5,981), Middlesex co., E Mass., WSW of Boston; settled 1715. Daniel Shays was probably born here. **2.** Town (1970 pop. 5,392), Washington co., SW R.I., W of Newport; set off from Westerly and inc. 1757.

Ho·qui·am (hō′kwē-əm), city (1970 pop. 10,466), Grays Harbor co., W Wash., on Grays Harbor; inc. 1890. With its twin city, Aberdeen, it has fishing, lumbering, cranberry, and tourist industries.

Hor·gen (hôr′gən), town (1970 pop. 15,691), Zürich canton, N Switzerland, on the Lake of Zürich. A textile center in the 17th cent., Horgen today has manufactures of machinery, appliances, and electrical goods.

Hor·moz (hôr′mōz′) or **Or·muz** (ôr′mŭz′, ôr-mōōz′), island (1966 est. pop. 600), 5 mi (8 km) long and 3.5 mi (5.6 km) wide, S Iran, in the Strait of Hormoz, between the Persian Gulf and the Gulf of Oman. Salt and red ochre are produced. The town of Hormoz, originally built on the mainland, was moved (c.1300) to the island after repeated attacks by marauding raiders. The new port prospered and served as a center of trade with India and China. It was attacked by the Portuguese under Alfonso de Albuquerque in 1507 and was captured by them in 1514. Its recapture in 1622 by Shah Abbas I with the aid of an English fleet marked the end of the island's prosperity.

Horn, Cape (hôrn), headland, 1,391 ft (424.3 m) high, S Chile, southernmost point of South America, in the archipelago of Tierra del Fuego. It was discovered and first rounded by Willem Schouten, the Dutch navigator, on Jan. 29, 1616. Lashing storms and strong currents made "rounding the Horn" one of the great hazards of sailing-ship days. It is still a formidable challenge to navigation.

Hor·nell (hôr-nĕl′), city (1978 est. pop. 11,100), Steuben co., SW N.Y., on the Canisteo River; settled 1790, inc. 1906. Textiles and steel bearings are manufactured here.

Hor·sens (hôr′səns), city (1976 est. pop. 53,966), Vejle co., central Denmark, a port at the head of the Horsens Fjord, an inlet of the Kattegat. It is a commercial and industrial center. Horsens was a fortified town in the Middle Ages.

Hor·ry (hô′rē), county (1970 pop. 69,992), 1,152 sq mi (2,983.7 sq km), E S.C., bounded on the W by the Little Pee Dee River, on the SE by the Atlantic, and on the NE by the N.C. border; formed 1785 as Kingston, renamed 1801; co. seat Conway. Its economy is based on agriculture (tobacco, corn, soybeans, sweet potatoes, and hogs), industry (including furniture and metal, textile, and wood products), and tourism, with coastal resorts and hunting and fishing areas.

Horse·heads (hôrs′hĕdz′), village (1970 pop. 7,989), Chemung co., S central N.Y., N of Elmira; settled 1789, inc. 1837.

Horse·shoe Bend (hôrs′shōō′), a turn on the Tallapoosa River, near Dadeville, E central Ala., site of a battle on Mar. 27, 1814, in which the Creek Indians, led by William Weatherford, an Indian chief, were defeated by militia under the command of Andrew Jackson. Horseshoe Bend National Military Park is here.

Hor·sham (hôr′shəm), urban district (1973 est. pop. 26,770), West Sussex, SE England. It is a residential area and a market town with tanneries and mills. The Causeway, an old cobbled street, has Tudor and Stuart houses and a 13th cent. parish church.

Hor·ta (hôr′tə, ôr′-), town (1974 pop. 6,145), capital of Horta dist., in the Azores, Portugal, on Fayal Island. It has an excellent harbor with shipyards and is a military air base.

Hor·ten (hôr′tn), town (1977 est. pop. 13,658), Vestfold co., SE Norway, a port on the Oslofjord (an arm of the Skagerrak); chartered 1907. It is a commercial and industrial center.

Hor·ton (hôr′tn), river, c.275 mi (440 km) long, rising in a lake N of Great Bear Lake, Mackenzie dist., Northwest Territories, Canada, and flowing NW to Franklin Bay, a part of the Beaufort Sea.

Ho-t'ien (hô′tyĕn′) or **Kho·tan** (kō′tän′), town and oasis (1958 est. pop. 50,000), SW Sinkiang Uigur Autonomous Region, China, near the headstream of the Ho-t'ien River. It is the center of an area growing silk and cotton. Textiles, carpets, and felt goods are manufactured. On the southern part of the Silk Road, Ho-t'ien was an early center for the spread of Buddhism from India into China. It fell to the Arabs in the 8th cent., and soon grew wealthy on the proceeds of the caravan trade. Its prosperity ended with the conquest of Ho-t'ien by Genghis Khan. After many political changes the region became (1878) permanently part of China.

Hot Spring (hŏt), county (1970 pop. 21,963), 621 sq mi (1,608.4 sq km), SW central Ark., drained by the Ouachita and Caddo rivers; formed 1829; co. seat Malvern. Its agriculture includes cotton, corn, sweet potatoes, and livestock. It has barite mining, lumbering (sawmills and wood products), sand and gravel deposits, and some industry (plastics, bricks, and metal products).

Hot Springs, county (1970 pop. 4,952), 2,022 sq mi (5,237 sq km), NW central Wyo., drained by the Bighorn River, with part of Absaroka Range in the W; formed 1911; co. seat Thermopolis. It has coal mines, oil and gas wells, and agriculture (livestock, grain, and sugar beets). There is some manufacturing in the county.

Hot Springs. 1. City (1978 est. pop. 37,800), seat of Garland co., W central Ark. settled 1807, inc. 1876. The city nearly surrounds Hot Springs National Park. Lumber and metal and electrical products are produced. **2.** City (1976 est. pop. 4,728), seat of Fall River co., SW S.Dak., near the Cheyenne River and the Black Hills SSW of

Rapid City; settled 1879-81, inc. 1882. It is a health resort and tourist center. Angostura Dam, Jewel Cave National Monument, Wind Cave National Park, and Custer State Park are nearby.

Hot Springs National Park, 3,535 acres (1,431.7 hectares), W central Ark.; est. 1921. Long used by the Indians for medicinal purposes, and visited by Spanish explorer Hernando De Soto in 1541, the springs became a Federal Reservation in 1832. More than a million gallons of water a day, with an average temperature of 143°F (62°C), flow from 47 springs. The National Park Service collects, cools, and supplies water to bathhouses in and out of the park.

Hot Sul·phur Springs (sŭl'fər), town (1970 pop. 220), seat of Grand co., N Colo., on the Colorado River in the W foothills of the Front Range WNW of Denver. It has mineral springs.

Hough·ton (hō'tən), county (1970 pop. 34,652), 1,017 sq mi (2,634 sq km), NW Mich., Upper Peninsula, partly bounded on the SE by Keweenaw Bay, drained by the Ontonagon and Sturgeon rivers, intersected on the NE by the Keweenaw Waterway, and traversed by the Copper Range; formed 1845; co. seat Houghton. It is in a copper-mining region. Its agriculture includes potatoes, truck crops, fruit, hay, livestock raising, and dairy products. It has commercial fishing and lumbering, food-processing, and metal-products industries. There are resorts here and several lakes.

Houghton, village (1970 pop. 6,067), seat of Houghton co., extreme N Mich., on the Keweenaw Peninsula and Portage Lake Ship Canal opposite Hancock; settled 1851, inc. 1867. It is a shipping, distributing, and industrial center in the heart of the Lake Superior copper-mining region.

Hough·ton-le-Spring (hō'tn-lə-sprĭng', hou'tən), urban district (1971 pop. 32,666), Tyne and Wear, NE England. It is a market town, with coal mining as the chief industry.

Hougue, La (lä o͞og'), cape: *see* Hogue, La.

Houl·ton (hōl'tən), town (1970 pop. 8,111), seat of Aroostook co., E Maine, on the Meduxnekeag River near the Canadian border; settled 1807, inc. as a plantation 1826, as a town 1831. It is a port of entry and a commercial center.

Hou·ma (ho͞o'mə), city (pop. 30,922), seat of Terrebonne parish, SE La.; inc. 1848. Houma is a port on the Intracoastal Waterway. The processing of seafood, sugar cane, and oil are the leading industries.

Houns·low (hounz'lō), borough (1971 pop. 206,182) of Greater London, SE England, on the Thames River. Hounslow was created in 1965 by the merger of the municipal boroughs of Heston and Isleworth and of Brentford and Chiswick with the urban district of Feltham. Hounslow's manufactures include razor blades, soap, tires, biscuits, precision instruments, pharmaceuticals, and heating equipment. There were prehistoric and Roman settlements here.

Hou·sa·ton·ic (ho͞o'sə-tŏn'ĭk), river rising in the Berkshires, W Mass., and flowing generally S c.130 mi (210 km) through W Conn. to Long Island Sound at Stratford. The river has long been used as a source of power.

Hous·ton (hyo͞o'stən). **1.** County (1970 pop. 56,574), 578 sq mi (1,497 sq km), extreme SE Ala., bounded on the E by the Chattahoochee River and the Ga. border, on the S by the Fla. border; formed 1903; co. seat Dothan. It is in a rich agricultural area that yields peanuts, cotton, corn, potatoes, melons, hogs, and beef cattle. There are oil and natural-gas wells and lumbering and food-processing industries. Its manufactures include cigars, textiles and clothing, furniture, concrete, and farm machinery. **2.** County (1970 pop. 62,924), 379 sq mi (981.6 sq km), central Ga., bounded in the E by the Ocmulgee River; formed 1821; co. seat Perry. Coastal-plain agriculture (cotton, corn, melons, truck crops, peanuts, pecans, peaches, and livestock) and forestry characterize the area. **3.** County (1970 pop. 17,556), 565 sq mi (1,463.4 sq km), extreme SE Minn., drained by the Root River and bounded on the E by the Mississippi River and Wis., on the S by Iowa; formed 1854; co. seat Caledonia. Its agriculture includes livestock, dairy products, grain, and potatoes. It has some mining, lumbering, and dairy processing. Automobile parts are made. **4.** County (1970 pop. 5,853), 201 sq mi (520.6 sq km), NW Tenn., bounded on the W by the Kentucky Reservoir of the Tennessee River; formed 1871; co. seat Erin. It is mainly agricultural, yielding corn, tobacco, sweet potatoes, livestock, and dairy products. **5.** County (1970 pop. 17,855), 1,232 sq mi (3,190.9 sq km), E Texas, bounded on the W by the Trinity River, on the E by the Neches River; formed 1837; co. seat Crockett. In a heavily wooded area, it has agriculture (cotton, corn, pecans, grain, hay, peanuts, fruit, truck crops, livestock, and dairy products), lumbering and food-processing industries, and oil refineries. There are deposits of lignite coal and natural gas. Paper and wood products, furniture, steel, and plastics are manufactured.

Houston. 1. Town (1970 pop. 2,720), a seat of Chickasaw co., NE Miss., NW of Columbus, in a dairy, timber, and farm area; inc. 1837. There are ante-bellum houses here, and nearby are Indian mounds. **2.** City (1970 pop. 2,178), seat of Texas co., S central Mo., SW of Salem, in a dairy region; founded 1845, platted 1846, inc. 1872. Formerly a lumber town, the city now makes shoes and furniture. **3.** City (1978 est. pop. 1,460,000), seat of Harris co., SE Texas, a deepwater port on the Houston Ship Channel; inc. 1837. Houston is a port of entry, a great industrial, commercial, and financial hub, one of the world's major oil centers, and the third-busiest tonnage-handling port in the United States. Houston has numerous space and science research firms, electronics plants, giant oil refineries, one of the world's greatest concentrations of petrochemical works, steel mills, shipyards, grain elevators, breweries, meat-packing houses, paper, rice, and cotton mills, and factories manufacturing and assembling a great variety of products. Harrisburg (now part of Houston) was settled in 1823, and Houston itself, founded in 1836 by J. K. and A. C. Allen and named for Sam Houston, was promoted as a rival to Harrisburg and soon served (1837-39) as capital of the Texas republic. Its phenomenal expansion came after the digging (1912-14) of a ship channel on Buffalo Bayou and Galveston Bay, linking it to the Gulf and making it a deepwater port. The development of the coastal oil fields poured quick wealth into the city; the natural gas, sulfur, salt, and limestone deposits also in the area laid the basis for its great chemical production; shipbuilding during World War II spurred further growth; and the establishment (1961) nearby of the National Aeronautics and Space Administration's Manned Spacecraft Center (renamed the Lyndon B. Johnson Space Center in 1973) brought the aerospace industry. It is the seat of Rice Univ., Texas Southern Univ., the Univ. of Houston, and other schools. Its many parks include the large Hermann Park, which has a zoo, a museum of natural science, and a planetarium. Houston has several notable art museums, an arboretum, and a botanical garden. Other tourist attractions include Old Market Square; Sam Houston Historical Park, which contains restored homes (built 1824-68) and reconstructed buildings; the huge air-conditioned Astrodome (opened 1965) and its adjacent "Astroworld," an amusement center; and the National Space Hall of Fame.

Hove (hōv), municipal borough (1971 pop. 72,659), East Sussex, SE England. It is a modern residential seaside resort.

Ho·ven·weep National Monument (hō'vən-wēp'): *see* National Parks and Monuments Table.

How·ard (hou'ərd). **1.** County (1970 pop. 11,412), 600 sq mi (1,554 sq km), SW Ark., drained by the Saline and Little Missouri rivers; formed 1873; co. seat Nashville. Its agriculture includes fruit, cotton, alfalfa, vegetables, corn, and poultry. It has lumbering, cinnabar mining, cotton ginning, and some manufacturing (concrete, cement, and cutlery). **2.** County (1970 pop. 83,198), 293 sq mi (758.9 sq km), N central Ind., drained by Wildcat Creek; formed 1844; co. seat Kokomo. It is in a rich agricultural area yielding livestock, corn, grain, poultry, soybeans, and truck crops. Its diversified manufactures include food products, plastics, pottery, steel, and miscellaneous metal products. **3.** County (1970 pop. 11,442) 471 sq mi (1,219.9 sq km), NE Iowa, on the Minn. line in the N, in a rolling prairie agricultural area drained by the Upper Iowa, Wapsipinicon, and Turkey rivers; formed 1851; co. seat Cresco. Agricultural products include hogs, cattle, poultry, corn, and oats. Limestone and gravel are its major natural resources. **4.** County (1970 pop. 62,394), 251 sq m (650.1 sq km), central Md., mostly in a piedmont area with part of the coastal plain in the SE, bounded on the NE by the Patapsco River, on the W and SW by the Patuxent River; formed 1851; co. seat Ellicott City. Its agriculture includes dairy products, poultry, truck crops, apples, and some grain. It has lumbering and food-processing industries and factories manufacturing yarn, clothing, furniture, plastics, concrete, metal products, and machinery. **5.** County (1970 pop. 10,561), 469 sq mi (1,214.7 sq km), N central Mo., on the Missouri River; formed 1816; co. seat Fayette. It is in an agricultural region that produces corn, soybeans, milo, tobacco, alfalfa, cattle, and hogs. There is some textile manufacturing in the county. **6.** County (1970 pop. 6,807), 566 sq mi (1,465.9 sq km), E central Nebr., drained by the North Loup and Middle Loup rivers; formed 1871; co. seat St. Paul. It has grain, livestock, dairy, and poultry farming. There is also some mining and manufacturing. **7.** County (1970 pop. 37,796), 912 sq mi (2,362.1 sq km), NW Texas, in rolling plains drained by tributaries of the Colorado River; formed 1876; co. seat Big Spring. Livestock raising, farming (cotton, grain sorghums, fruit, and truck crops), dairying, and oil refining are important to its economy. Clay and farm products are processed.

Howard. 1. City (1970 pop. 918), seat of Elk co., SE Kansas, SE of Wichita; founded 1870, inc. 1877. It is in a farming and oil-producing area. **2.** City (1970 pop. 1,175), seat of Miner co., SE S.Dak., NW of Sioux Falls; platted 1881. It is a trade center of a grain region.

How·ell (hou'əl), county (1970 pop. 23,521), 920 sq mi (2,382.8 sq km), S Mo., in the Ozarks drained by the Eleven Point River; formed 1857; co. seat West Plains. It is in a livestock and farm (corn and hay) region, with pine timber and stone deposits. Its manufactures include wood and metal products, furniture, and shoes.

Howell, city (1970 pop. 5,224), seat of Livingston co., SE Mich., SE of Lansing; settled 1834, inc. as a village 1863, as a city 1915. It is a farm trade center in a dairy region. Metal products are made.

How·land Island (hou'lənd), uninhabited island, .73 sq mi (1.89 sq km), central Pacific Ocean, near the equator, c.1,620 mi (2,605 km) SW of Honolulu. The island was discovered by American traders and was claimed by the United States in 1856, along with Jarvis Island and Baker Island. The three islands were worked for guano deposits by British and American companies during the 19th cent. The guano industry declined, and the islands were forgotten until they became a stop on the air route to Australia. American colonists were brought from Hawaii in 1935 in order to establish U.S. control against British claims, but the colony was disbanded at the outbreak of World War II. While en route to Howland Island in 1937 the aviator Amelia Earhart disappeared.

How·rah (hou'rä), city (1971 pop. 740,622), West Bengal state, E central India, on the Hooghly River opposite Calcutta. A center of the Hooghlyside industrial complex, Howrah produces textiles, jute products, glass, steel, and other items.

Hox·ie (hŏk'sē), city (1970 pop. 1,419), seat of Sheridan co., NW Kansas, NW of Hays, in a farm area; laid out 1886.

Hoy (hoi), island, 13 mi (21 km) long and 6 mi (9.7 km) wide, off N Scotland, in the Orkney Islands. It is located at the southwestern side of the Scapa Flow anchorage. The Old Man of Hoy, a sandstone pinnacle 450 ft (137.3 m) high, is a famous sailors' landmark. The Dwarfie Stone, a huge sandstone block with hollowed rooms inside, is a Viking relic.

Ho·ya (hō'yä), city (1975 pop. 91,546), Tokyo Metropolis, E central Honshu, Japan, on the Shakoji River. It is a residential suburb of Tokyo and an agricultural center where raw silk is produced.

Hoy·ers·wer·da (hoi'ərs-vĕr'dä), city (1974 est. pop. 64,904), Cottbus dist., SE East Germany, on the Black Elster River; chartered 1371. Located in a lignite-mining area, it is an industrial city, manufacturing glass and other products.

Hoy·lake (hoi'lāk'), urban district (1971 pop. 32,196), Merseyside, NW England, on the Wirral Peninsula at the mouth of the Dee River. It is a seaside resort with yachting facilities and a golf course.

Hra·dec Krá·lo·vé (hrä'dĕts krä'lô-və), city (1975 est. pop. 87,217), N Czechoslovakia, in Bohemia, on the Elbe (Labe) River. It is an industrial center, manufacturing machinery, chemicals, photographic equipment, and musical instruments. Founded in the 10th cent., it was a leading town of medieval Bohemia. It suffered heavily in the Hussite and Thirty Years' wars.

Hsi-an (shē'än') or **Si·an** (sē'än', shē'-), city (1970 est. pop. 1,900,000), capital of Shensi prov., China, in the Wei River valley. Situated on the Lung-hai RR, it is an important commercial center in a wheat- and cotton-growing area. It has textile mills, food-processing establishments, and plants making chemicals, cement, motor vehicles, and fertilizer. Hsien-yang, one of several cities which previously occupied this site, was (255–206 B.C.) the capital of the Ch'in dynasty. The present city, then called Chang-an, was the western capital of the T'ang dynasty (618–906). In the Sian Incident (1936) Chiang Kai-shek was kidnaped and kept prisoner until he agreed to form a united front against the Japanese. The city has numerous T'ang dynasty pagodas and is noted for its museum of history, housed in an 11th cent. Confucian temple containing large stone tablets from the T'ang dynasty. The city wall, dating from the Ming dynasty (1368–1644), is still visible in places.

Hsi·ang or **Si·ang** (both: sē'äng', shē'-), river, 715 mi (1,150.4 km) long, rising in NE Kwangsi prov. and flowing N through Hunan prov. to Tung-t'ing lake, SE China. The river is navigable to large vessels for most of its course. The densely populated Hsiang valley is one of China's major industrial regions and yields manganese, lead, and zinc. The delta is an important rice-growing area.

Hsi·ang-t'an (shē-äng'tän') or **Si·ang·tan** (sē-äng'tän', shē-), city (1970 est. pop. 300,000), E central Hunan prov., China, on the Hsiang River. Formerly an agricultural distribution center, it is now industrialized. Products include manganese ore, cement, machine tools, and trucks.

Hsin-chu (shĭn'jo͞o'), city (1973 est. pop. 220,302), NW Taiwan. The city and surrounding area are noted for the production of tea, rice, oranges, and petroleum. Hsin-chu's major industries include petroleum refining and the manufacture of cement, fertilizers, and textiles. Immigrants from the China mainland formed a colony at Hsin-chu in the early 1700s.

Hsin-hui (shĭn'hwē') or **Sun·wui** (so͝on'wē'), town, S Kwangtung prov., SE China, in the Si River delta, near Canton. It has fruit orchards, fruit-processing industries, and machine shops. Tungsten mines are in the area. A treaty port in 1904, it was an important outlet for Chinese emigrants to the United States.

Hsi-ning or **Si·ning** (both: shē'nĭng'), city (1970 est. pop. 250,000), capital of Tsinghai prov., W China, on the Hsi-ning River. For centuries it has been the major commercial hub on the caravan route to Tibet, trading in wool, hides, salt, and timber. More recently it has developed as a processing (flour-milling, wool-spinning, meat-packing) and distribution center for the Tsinghai agricultural basin. Manufactures include chemicals, machinery, motor vehicles, metal products, and textiles. Coal is mined in the area.

Hsin-kao Shan or **Sin·kao Shan** (both: shĭn'gou' shän') or **Mt. Mor·ri·son** (môr'ə-sən), 13,113 ft (3,999.5 m) high, S Taiwan. It is the highest peak on the island and was first ascended in 1896.

Hsin-yang or **Sin·yang** (both: shĭn'yäng'), city (1970 est. pop. 125,000), S Honan prov., China. It is a transportation hub on the Peking-Canton RR. Textiles are produced in the city.

Hsü-chou or **Sü·chow** (both: shü'jō'), city (1970 est. pop. 1,500,000), N Kiangsu prov., E central China. Iron and coal mines are nearby, and the city has a small integrated steel complex. Food is processed, and machine tools are also made.

Huai or **Hwai** (both: hwī), river, c.680 mi (1,095 km) long, rising in the Tung-pai Mts., Honan prov., E China, and flowing E across Anhwei prov., through Hung-tse Lake, to the East China Sea. More than two thirds of the fertile Huai basin is under cultivation; wheat, millet, and kaolin are the main crops.

Huai-nan or **Hwai·nan** (both: hwī'nän'), city (1970 est. pop. 350,000), N central Anhwei prov., China. It was established after 1949 as the center of China's chief coal-mining region.

Hua-lien (hwä'lĕn'), city (1973 est. pop. 96,175), E central Taiwan. It is an important port for the local area and is a market for sugar cane, rice, jute, and camphorwood grown in the surrounding region. The city also produces fertilizers, machinery, and canned foods.

Hual·la·ga (wä-yä'gä), river rising in the Andes of W central Peru and flowing 700 mi (1,126.3 km) generally N to the Marañon River.

Huan·ca·yo (wän-kī'ō), city (1972 pop. 64,777), alt. 10,731 ft (3,721 m), capital of Junín dept., S central Peru. One of Peru's major agricultural centers, it markets and ships the wheat, maize, potatoes, and barley grown in the surrounding area. Silver, copper, and coal are mined in the region. The city is noted for its picturesque colonial architecture.

Huang Ho or **Hwang Ho** (both: hwäng hō) or **Yel·low River** (yĕl'ō), river of N China, c.3,000 mi (4,830 km) long, rising in the Kunlun Mts., NW Tsinghai prov., and flowing generally E into the "great northern bend" (around the Ordos Desert), then E again to the Po Hai, an arm of the Yellow Sea. During the winter dry season the Huang Ho is slow-moving and silt-laden, and occupies only part of its huge bed; with the summer rains, it becomes a raging torrent. Since the 2nd cent. B.C., the lower Huang Ho has inundated the surrounding region some 1,500 times and has made nine major changes in its course. In an attempt to halt the Japanese invasion of China in 1938, the Chinese diverted the Huang Ho south, flooding more than 20,000 sq mi (51,800 sq km) and killing some 900,000 people; it was returned to its present course in 1947. In 1955 a 50-

year construction plan for controlling the river was initiated. Dikes are being repaired and reinforced, and a series of 45 silt-retaining dams, when completed, will control the upper river, produce electricity, and provide water for 20 million acres (8.1 million hectares).

Huang-pu (hwäng'bōō') or **Wham·po·a** (hwäm'pō'ä'), industrial city, S Kwangtung prov., SE China, on an island in the Canton River, SE of Canton, of which it is an outer port.

Huang-p'u (hwäng'pōō'), river, China: *see* Whangpoo.

Huang-shih or **Hwang·shih** (both: hwäng'shē'), city (1970 est. pop. 200,000), E Hupeh prov., China, on the Yangtze River. It is a new industrial center, built after 1950, with a giant iron and steel complex. Other manufactures include cement, building materials, textiles, and processed foods.

Hua·rás or **Hua·raz** (both: wä-räs'), city (1972 pop. 29,719), capital of Ancash dept., W central Peru. It is in a high valley at an altitude of 9,931 ft (3,029 m). Huarás is the center of an agricultural district raising grains and potatoes. Silver, cinnabar, and coal are mined.

Huas·ca·rán (wäs'kä-rän'), extinct volcano, 22,205 ft (6,772.5 m) high, W central Peru, near Huarás. The highest mountain in Peru and one of the highest in the Andes, Huascarán and nearby peaks form an impressive snow-capped rampart. In 1962 an avalanche swept down its slopes and buried the village of Ranrahirca.

Hub·bard (hŭb'ərd), county (1970 pop. 10,583), 932 sq mi (2,413.9 sq km), N central Minn., watered in the S by small lakes; formed 1883; co. seat Park Rapids. In an agricultural area that produces dairy products and potatoes, it has lumbering, some mining, and clothing manufacturing. Its tourist industry centers around state forests in the north and southeast.

Hubbard, industrial village (1978 est. pop. 16,293), Trumbull co., NE Ohio, near the Pa. border NE of Youngstown; settled 1803, inc. as a village 1868. Steel and metal products are made here.

Hub·bell Trad·ing Post National Historic Site (hŭb'əl trā'dĭng): *see* National Parks and Monuments Table.

Hu·bli-Dhar·war (hōōb'lē-där'wär), city (1971 pop. 379,555), Karnataka state, SW India. The cities of Hubli and Dharwar were incorporated as one city in 1961. Dharwar is the district administrative center for a rice- and cotton-growing area. Hubli is a trade and transportation center with cotton and silk factories.

Huck·nall (hŭk'nəl), urban district (1973 est. pop. 27,110), Nottinghamshire, central England. It has coal mines and manufactures hosiery. Lord Byron is buried in the parish church.

Hud·ders·field (hŭd'ərz-fēld'), county borough (1973 est. pop. 130,060), West Yorkshire, N central England, on the Colne River. Its textile industry, including cotton, woolen, and rayon goods, is important. Other products are machinery, iron goods, chemicals, and dyed fabrics. Huddersfield's history dates back to Roman times.

Hud·son (hŭd'sən), county (1970 pop. 607,839), 45 sq mi (116.6 sq km), NE N.J., bounded on the E by the Passaic River and Upper New York Bay and drained by the Hackensack River; formed 1840; co. seat Jersey City. Its diversified manufactures include textiles and clothing, lumber millwork, furniture, paper products, chemicals, plastics, concrete, metal products, machinery and boats.

Hudson. 1. Industrial town (1978 est. pop. 17,300), Middlesex co., E central Mass., on the Assabet River, in an apple-growing region; settled c.1699, inc. 1866. It has various manufacturing industries. **2.** Town (1970 pop. 10,638), Hillsborough co., S N.H., on the Merrimack River opposite Nashua; est. 1673 as part of Dunstable, Mass., included in N.H. in 1746. Synthetic furs, optical equipment, and shoes are among its manufactures. **3.** City (1975 est. pop. 8,700), seat of Columbia co., E N.Y., on the E bank of the Hudson River S of Albany; settled 1662 as Claverack Landing, inc. 1785. Cement and matches are made. **4.** City (1973 est. pop. 5,322), seat of St. Croix co., W central Wis., on the St. Croix River E of St. Paul, Minn.; inc. 1856. A lumber center in the 1860s, it now has railroad shops and plants manufacturing refrigeration equipment.

Hudson, river, c.315 mi (505 km) long, rising in Lake Tear of the Clouds, on Mt. Marcy in the Adirondack Mts., NE N.Y., and flowing generally S to Upper New York Bay at New York City. One of the most important waterways of the world, the Hudson is navigable by ocean vessels to Albany and by smaller vessels to Troy; pleasure boats and self-propelled barges use the canalized section between Troy and Fort Edward, the head of navigation. Divisions of the New York State Barge Canal connect the Hudson with the Great Lakes and with Lake Champlain and the St. Lawrence River. The Hudson is tidal to Troy (c.150 mi/240 km upstream), and this section is considered to be an estuary. The upper course of the river has many waterfalls and rapids. Near Tarrytown the river widens to form the Tappan Zee (which is crossed by a causeway), and from there to its mouth the Hudson is flanked on the west by the sheer cliffs of the Palisades. The Hudson forms part of the N.Y.-N.J. border, and the two states are linked by the George Washington Bridge, vehicular tunnels, and railway tubes.

First sighted by Giovanni da Verrazano in 1524, the river was explored by Henry Hudson in 1609. It was a major route for the Indians and later for the Dutch and English traders and settlers. During the American Revolution both sides fought for control of the Hudson, and there were many battles along its banks. In 1825 the Erie Canal linked the river with the Great Lakes, thus providing the first all-water trans-Appalachian route. A well-known school of painting is associated with the Hudson, and the river is featured in the legend of Rip Van Winkle and other stories by Washington Irving.

Hudson Bay, inland sea of North America, c.475,000 sq mi (1,230,250 sq km), E central Canada. Hudson Bay and James Bay (its southern extension), and all their islands are part of the Northwest Territories. Hudson Strait connects Hudson Bay with the Atlantic Ocean, and Foxe Channel leads to the Arctic Ocean. Hudson Bay occupies the southernmost portion of the Hudson Bay Lowlands, a depression in the Canadian Shield formed during the Pleistocene epoch by the weight of the continental ice sheet. As the ice retreated, the region was flooded by the sea, and sediments were deposited in it. With the burden of ice removed, the floor of the lowlands has been slowly rising and the bay is gradually becoming shallower. Hudson Bay is ice-free and open to navigation from mid-July to Oct.

The bay was explored and named (1610) by Henry Hudson in his search for the Northwest Passage. The surrounding region was a rich source of furs, and France and England struggled for its possession until 1713, when France ceded its claim by the Peace of Utrecht. Hudson's Bay Company set up many trading posts here; some of the posts have operated continuously since 1670.

Hudson Falls, village (1970 pop. 7,917), seat of Washington co., E N.Y., on the Hudson River near Glen Falls; settled 1761, inc. 1810. Papermaking is its chief industry.

Hud·speth (hŭd'spəth), county (1970 pop. 2,392), 4,554 sq mi (11,794.9 sq km), extreme W Texas, in a high plateau region bounded on the N by the N.Mex. border, on the S by the Rio Grande and the Mexican border; formed 1917; co. seat Sierra Blanca. In an irrigated agricultural area yielding cotton, alfalfa, cattle, and sheep, it also has mineral resources, particularly copper.

Hue (hwā), city (1971 est. pop. 199,900), former capital of the historic region of Annam, S Vietnam, in a rich farming area on the Hue River near the South China Sea. It is a market center. A cement plant utilizes the fine limestone deposits in the area. Probably founded in the 3rd cent. A.D., Hue was occupied in turn by the Chams and the Annamese. After the 16th cent. it was the seat of a dynasty that extended its power over southern Annam, modern Cochin China, and parts of Cambodia and Laos. The French occupied the city in 1883. During World War II the Japanese mined iron ore in the area. In the Vietnam War, Hue was the scene of the longest and heaviest fighting of the Tet offensive (Jan.-Mar., 1968); most of the city, including the palaces and tombs of the former Annamese kings, was destroyed.

Huel·va (wĕl'vä), city (1975 est. pop. 112,091), capital of Huelva prov., SW Spain, in Andalusia, on the Odiel River above its junction with the Río Tinto. A busy port with copper, sulfur, and cork exports, it also has large fisheries and summer-resort facilities. A Roman aqueduct supplies the city with water.

Huer·fa·no (wĕr'fōō-nō'), county (1970 pop. 6,590), 1,578 sq mi (4,087 sq km), S Colo., in a coal-mining and livestock-grazing area bounded in the W by the Sangre de Crisco Mts. and drained by the Cucharas and Huerfano rivers; formed 1861; co. seat Walsenburg. Part of San Isabel National Forest in the north.

Hues·ca (wā'skä), town (1975 est. pop. 36,479), capital of Huesca prov., NE Spain, in Aragón, at the foot of the Pyrenees. It is a farm center. After Peter I of Aragón liberated it (1096) from the Moors, Huesca was the residence of the kings of Aragón until 1118. A university, later discontinued, was founded here in 1354.

Hu·ey·town (hyōō'ē-toun'), city (1970 pop. 8,174), Jefferson co., N central Ala., a suburb NW of Bessemer.

Hughes (hyōōz). **1.** County (1970 pop. 13,228), 810 sq mi (2,097.9 sq km), E central Okla., intersected by the Canadian and North Canadian rivers; formed 1907; co. seat Holdenville. It is in an agricultural area yielding corn, peanuts, soybeans, cotton, watermelons, cattle, hogs, and poultry, with oil and natural-gas wells and oil refineries. Clothing, wood products, and plastics are manufactured. **2.** County (1970 pop. 11,632), 748 sq mi (1,937.3 sq km), central S.Dak., bounded on the S by the Missouri River and watered by intermittent streams; formed 1873; co. seat Pierre. Its agriculture includes livestock, wheat, corn, and dairy products. There is some manufacturing in the county. Crow Creek Indian Reservation is in the southeast.

Hu·go (hyōō'gō). **1.** Town (1970 pop. 759), seat of Lincoln co., E Colo., on the Big Sandy Creek SE of Denver, at an altitude of 5,046 ft (1,539 m). It is in a livestock, grain, and dairy-products area. **2.** City (1970 pop. 6,585), seat of Choctaw co., SE Okla, near the Red River E of Durant. It is a railroad and farm center. Creosote is made.

Hu·go·ton (hyōō'gō-tən), city (1970 pop. 2,739), seat of Stevens co., SW Kansas, SW of Dodge City, in the Great Plains near the Okla. border; founded 1885, inc. 1910. The area has large natural-gas fields.

Hu-ho-hao-t'e (hōō'hô-hou'tā) or **Hu·he·hot** (hōō'hā-hōt'), city (1970 est. pop. 700,000), capital of the Inner Mongolian Autonomous Region, N China. The terminus of caravan routes to Sinkiang and to the Mongolian People's Republic, Hu-ho-hao-t'e is also connected by rail with Peking and is a trade center. Manufactures include chemicals, textiles, fertilizers, motor vehicles, and beet sugar and other processed foods. Hu-ho-hao-t'e consists of two sections. The old town is a Mongolian political and religious center dating from the 9th cent. The newer Chinese section, which grew around the railway station after 1921, is the administrative center.

Hu·la, Lake, or **Lake Hu·leh** (both: hōō'lə), near sea-level lake formed by a natural dam of basalt, NE Israel. The Jordan River exits from its southern end. Between 1950 and 1958, c.12,350 acres (5,000 hectares) of the lake's swampy shore were drained; this land, now irrigated by the Jordan, produces rice, cotton, and sugar beets.

Hull (hŭl), city (1976 pop. 61,039), SW Que., Canada, at the confluence of the Ottawa and Gatineau rivers, opposite the city of Ottawa; inc. 1895. Hull has a hydroelectric power station. There are paper, pulp, and lumber mills, iron foundries, and cement and meat-packing plants.

Hull, officially **Kings·ton upon Hull** (kĭngs'tən), county borough (1976 est. pop. 276,600), Humberside, NE England, on the N shore of the Humber estuary at the influx of the small Hull River. Its port is one of the chief outlets for the industrial Yorkshire and Lincolnshire districts, with which there are excellent rail and water connections. Hull is also one of the world's largest fishing ports. Among its many manufactures are processed foods, chemicals, iron and steel products, and machinery. Flour mills and sawmills are nearby. Hull was founded late in the 13th cent. by Edward I, and the construction of docks, which now extend for miles along the Humber, was begun c.1775. Hull's annual fair is one of the largest in England.

Hull, resort town (1970 pop. 9,961), Plymouth co., E Mass., on Nantasket Peninsula extending into Boston Bay SE of Boston; settled 1624, inc. 1647.

Hul·wan (hōōl-wän'), town (1966 pop. 203,500), N Egypt, on the Nile River, opposite the ruins of Memphis. It is a suburb of Cairo. Manufactures include iron and steel, cement, and textiles. The town is a health resort, known for its hot sulfur springs. An ancient burial chamber, one of the largest in the world, was discovered in 1946.

Hu·ma·ca·o (ōō'mä-kä'ō), town (1970 pop. 12,411), E Puerto Rico, on the Humacao River. It is a port of entry in an area growing sugar cane and fruit. The town was founded in 1790.

Hum·ber (hŭm'bər), river, c.75 mi (120 km) long, rising in the Long Mts., W N.F., Canada, and flowing SE then SW, through Deer Lake, to the Bay of Islands at Corner Brook.

Humber, navigable estuary of the Trent and Ouse rivers, c.40 mi (65 km) long and from 1 to 8 mi (1.6-12.9 km) wide, NE England, in Humberside. The shores are generally low, and shoals obstruct shipping in parts.

Hum·ber·side (hŭm'bər-sīd'), nonmetropolitan county (1976 est. pop. 848,600), NE England, created under the Local Government Act of 1972 (effective 1974); administrative center Hull (officially Kingston upon Hull). It is composed of the county boroughs of Grimsby and Kingston upon Hull, and parts of the former counties of Yorkshire and Lincolnshire.

Hum·bolt (hŭm'bōlt'). **1.** County (1970 pop. 99,692), 3,586 sq mi (9,287.7 sq km), NW Calif. on the Pacific Ocean, lying mainly in the Coast Ranges, with part of the Klamath Mts. in the E and NE, and drained by the Klamath, Trinity, Mad, Eel, and Mattole rivers; formed 1853; co. seat Eureka. It is in a logging (redwood, Douglas fir, cedar, and spruce), lumber milling, dairying, and farming (merino sheep, truck crops, apples, berries, nuts, and poultry) region. It has ocean fisheries, and some quarrying and mining. Its manufactures include furniture and concrete. It is a noted recreational area. **2.** County (1970 pop. 12,519), 435 sq mi (1,126.7 sq km), N central Iowa, drained by the Des Moines and East Des Moines rivers; formed 1957; co. seat Dakota City. In a prairie agricultural area yielding hogs, cattle, corn, oats, and soybeans, it has bituminous-coal mines and limestone quarries. **3.** County (1970 pop. 6,375), 9,702 sq mi (25,128.2 sq km), NE Nev., watered by the Quinn, Little Humboldt, and Humboldt rivers, and bordering on Oregon; formed 1861; co. seat Winnemucca. It has agriculture (potatoes, grain, dill, alfalfa, and market seed), cinnabar and opal mines, and cattle ranches.

Humboldt, city (1970 pop. 10,066), Gibson co., W central Tenn.; inc. 1865. It is a trade and processing center in a region yielding fruits (especially strawberries) and vegetables.

Humboldt, river, c.300 mi (485 km) long, rising in several branches in the mountains of NE Nev. and meandering generally through the arid Great Basin to disappear in Humboldt Sink, W Nev. Its length varies with the season, and its volume decreases downstream. Known to early explorers and named by J. C. Frémont, the river was an important route for western settlers.

Humboldt Glacier, NW Greenland. The largest known glacier of the Northern Hemisphere, it debouches into Kane Basin along a front c.60 mi (95 km) wide and 300 ft (91.5 m) high. U.S. explorer E. K. Kane discovered it on his expedition of 1853-55.

Hum·phreys (hŭm'frēz). **1.** County (1970 pop. 14,601), 421 sq mi (1,090.4 sq km), W Miss., bounded partly on the W by the Sunflower River and drained by the Yazoo River and its tributaries; formed 1918; co. seat Belzoni. It has cotton and corn farms, timber, and knitting mills. **2.** County (1970 pop. 13,560), 530 sq mi (1,372.7 sq km), central Tenn., bounded on the W by the Tennessee River and drained by the Duck and Buffalo rivers; formed 1809; co. seat Waverly. Agriculture (corn, peanuts, soybeans, livestock, and dairy products) and forestry are important to its economy.

Humphreys Peak, mountain, 12,633 ft (3,853.1 m) high, N Ariz., near an eroded volcanic crater in the San Francisco Peaks. It is the highest point in the state.

Hu·nan or **Hu-nan** (both: hōō'nän'), province (1968 est. pop. 38,000,000), c.80,000 sq mi (207,200 sq km), S central China, S of Tung-t'ing lake; capital Ch'ang-sha. Largely hilly in the south and west, Hunan becomes an alluvial lowland in the northeast. Rice is the outstanding crop; wheat, corn, sweet potatoes, ramie, tobacco, rapeseed, and tea are also produced. Hunan is famous for its cedar; pine, fir, oak, bamboo, tung and camphorwood are also important. The province abounds in mineral resources, including zinc, antimony, tungsten, manganese, coal, mercury, gold, tin, and sulfur. Under Chinese rule since the 3rd cent. B.C., the region belonged to the kingdom of Wu at the time of the Three Kingdoms (A.D. 220-80) and later became part of the Chu kingdom of the Five Dynasties (907-60). Hunan passed to Communist rule in 1949. Mao Tse-tung was born in Hunan.

Hu·ne·doa·ra (hōō'nā-dwä'rä), city (1974 est. pop. 77,028), W central Rumania, in Transylvania. A major industrial center, it has ironworks and steelworks. Iron ore and coal are mined nearby.

Hun·ga·ry (hŭng'gə-rē), republic (1971 pop. 10,315,597), 35,919 sq mi (93,030 sq km), Central Europe. Budapest is the capital. Hungary borders on Czechoslovakia in the north, on the USSR in the northeast, on Rumania in the east, on Yugoslavia in the south, and on Austria in the west. The Danube River forms the Czechoslovak-Hungarian border from near Bratislava to near Esztergom, then turns sharply south and bisects the country. To the east of the Danube the Great Hungarian Plain extends beyond the Hungarian boundaries to the Carpathians and the Transylvanian Alps.

Economy. Agriculture employs less than one third of the population, although many industrial workers live in rural areas. Nearly all of the arable land is included in collective and state farms. Grains (especially corn and wheat), potatoes, turnips, and grapes and other fruit are grown, and livestock and poultry are raised. Sugar beets, flax, hemp, tobacco, and paprika are the chief industrial crops. Mineral resources are limited; bauxite is the most important. Industry is nationally owned. Products include machinery, textiles, metal products, motor vehicles, and chemicals.

History. The Roman provinces of Pannonia and Dacia embraced part of what was to become Hungary. The Huns and later the Ostrogoths and the Avars settled there for brief periods. In the late 9th cent. the Magyars, a Finno-Ugric people from beyond the Urals, conquered all or most of Hungary and Transylvania. The semilegendary leader, Arpad, founded their first dynasty. The first historic king, St. Stephen (reigned 1001–38), completed the Christianization of the Magyars. Under Bela III (reigned 1172–1196), Hungary came into close contact with Western European culture. In 1222 the lesser nobles forced Andrew II to grant the Golden Bull (the "Magna Carta of Hungary"), which limited the king's power and established the beginnings of a parliament. Mongol invaders occupied the country for a year (1241). When the royal line died out (1301), Charles Robert of Anjou was elected king of Hungary as Charles I (reigned 1308–42), the first of the Angevin line. Under his son, Louis I (reigned 1342–82), Hungary extended its territory into Dalmatia, the Balkans, and Poland.

After the death of Louis I, a series of foreign rulers followed. The advance of the Turks (14th–15th cent.) forced Hungary to retreat from the Balkans. Matthias Corvinus, a Hungarian, reigned (1458–90) over a great period, maintaining a splendid court at Buda and improving the central administration. But under his successors the nobles regained their power. Louis II was defeated and killed by the Turks (1526), an event that launched an era of division, civil war, and Turkish domination of the central plains. Budapest was liberated from the Turks in 1686 by the Hapsburg armies, and the nobles recognized the Hapsburg claim to the Hungarian throne.

Hapsburg rule was unpopular, and in the 1830s a movement that combined Hungarian nationalism with constitutional liberalism gained strength. In 1848 the Hungarian diet established a liberal constitutional monarchy. The Austrians attacked, Hungary was declared an independent republic (Apr., 1849), Russian troops intervened on behalf of the emperor, and the republic collapsed. But after its defeat in the Austro-Prussian War (1866), Austria was obliged to set up the Austro-Hungarian Monarchy, in which Austria and Hungary were nearly equal partners.

After the defeat of the dual monarchy in World War I, Premier Michael Károlyi declared Hungary an independent republic in Jan., 1919. The Communists under Béla Kun seized power in Mar. but were ousted in July by invading Rumanian forces. In Nov. Admiral Horthy de Nagybanya established a government and in 1920 was made regent and head of state. The Treaty of Trianon, signed in 1920, reduced the size and population of Hungary by about two thirds. Frustrated in its attempt to recover the lost territories, the Magyar government turned ultimately to an alliance (1941) with Nazi Germany and participation in World War II. When the Hungarians took steps to withdraw from the war German troops occupied the country (Mar., 1944). The Germans were driven out by Soviet forces (Oct., 1944–Apr., 1945).

National elections were held in 1945 and a new coalition regime instituted long-needed land reforms. Early in 1948 the Communist party seized control of the state. Hungary was proclaimed a People's Republic in 1949. Industry was nationalized, and collectivization of land was ruthlessly pressed. Premier Mátyás Rákosi, the Stalinist in control since 1949, was removed in July, 1953, and Imre Nagy headed a more moderate regime for two years. On Oct. 23, 1956, a popular anti-Soviet revolution broke out in Budapest. A new coalition government under Imre Nagy declared Hungary neutral, withdrew it from the Warsaw Treaty, and appealed to the UN for aid. In brutal fighting Soviet forces suppressed the revolution and set up a new government under Jámos Kádár. Kádár's regime brought increasing liberalization to Hungarian political, cultural, and economic life. Relations with the Catholic Church were improved, and a new primate of Hungary was installed in 1976. Consumer-goods production rose sharply in the 1970s, raising the standard of living.

Government. The country is governed under the 1949 constitution, which was amended in 1972. There is a unicameral parliament, elected every four years, which elects a presidential council. The council chooses the ministers. Actual power resides in the Communist party, officially known as the Hungarian Socialist Workers Party, which heads the People's Patriotic Front.

Hung-tse or **Hung·tse** (both: hōōng'dzŭ'), lake, 65 mi (104.6 km) long, E China, on the border of Anhwei and Kiangsu provs. The San Ho dam, with the largest hydraulic works along the Huai River, is at the outlet of the lake.

Hunt (hŭnt), county (1970 pop. 47,948), 826 sq mi (2,139.3 sq km), NE Texas, in a timber and blackland prairie area drained by the Sabine River and the South Fork of the Sulphur River; formed 1846; co. seat Greenville. Its agriculture includes cotton, corn, clover, grain, pecans, fruit, truck crops, dairy products, poultry, cattle, hogs, and sheep. Some oil refining, manufacturing, and processing are done.

Hun·ter·don (hŭn'tər-dən), county (1970 pop. 69,718), 423 sq mi (1,095.6 sq km), NW central N.J., bounded on the W by the Delaware River and drained by the South Branch of the Raritan River; co. seat Flemington. It has agriculture (poultry, grain, hay, and horses) and manufacturing (clothing, furniture, paper products, chemicals, concrete, and machinery). There are food-processing and tourism industries.

Hun·ting·don (hŭn'tĭng-dən), county (1970 pop. 39,108), 895 sq mi (2,318.1 sq km), S central Pa., in a hilly region drained by the Juniata River; formed 1787; co. seat Huntingdon. It is in a limestone, glass-sand, clay, bituminous-coal, and lumber area. It has some agriculture and manufacturing.

Huntingdon. 1. Industrial borough (1970 pop. 6,987), seat of Huntingdon co., S central Pa., E of Altoona; settled c.1755, laid out 1767, inc. 1796. It manufactures paper and machinery. **2.** Town (1970 pop. 3,661), seat of Carroll co., W central Tenn., NE of Jackson, in a farm area; settled 1921, inc. 1850. Cotton and lumber are shipped.

Huntingdon and God·man·ches·ter (gŏd'mən-chĕs'tər), municipal borough (1973 est. pop. 17,200), Cambridgeshire, E central England, traversed by the Ouse River. The boroughs of Huntingdon and Peterborough were merged in 1961. There are light industries and an agricultural market. Huntingdon has many fine Georgian buildings. Until 1974 the borough was the county town of Huntingdon and Peterborough.

Huntingdon and Pe·ter·bor·ough (pē'tər-bə-rə), former county, E central England. The county town was Huntingdon and Godmanchester. The region is chiefly agricultural. There are remains of two important Roman roads. In Anglo-Saxon times the area came under the control of East Anglia. In 1974 Huntingdon and Peterborough became part of the nonmetropolitan county of Cambridgeshire.

Hun·ting·ton (hŭn'tĭng-tən), county (1970 pop. 34,970), 390 sq mi (1,010.1 sq km), NE Ind., drained by the Wabash, Salamonie, and Little rivers; formed 1832; co. seat Huntington. In an agricultural area yielding grain, truck crops, poultry, livestock, soybeans, and dairy products), it has limestone quarries and timber. Its diversified manufactures include processed food, furniture, chemicals, rubber, concrete, asbestos, aluminum, hardware, and machinery.

Huntington. 1. City (1978 est. pop. 15,000), seat of Huntington co., NE Ind.; inc. 1848. It is a farm trade center and an industrial city. Its manufactures include automotive parts and machinery. **2.** Uninc. town (1970 pop. 12,601), Suffolk co., SE N.Y., on the N shore of Long Island; settled 1653. It is the heart of a township containing 17 contiguous communities, noted for their precision manufactures. The town, which is chiefly residential, has numerous harbors and boatyards. **3.** City (1978 est. pop. 68,200), seat of Cabell co., W W.Va., on the Ohio River; founded 1871 as the W terminus of the Chesapeake & Ohio RR. It is a commercial center and a river port with large shipments of bituminous coal. It has important glass and chemical industries. Other manufactures include electrical goods, wood and metal products, and mattresses.

Huntington Beach, city (1978 est. pop. 163,500), Orange co., S Calif., on the Pacific coast, across from Santa Catalina Island, in a truck-farm, citrus-fruit, and oil area; inc. 1909. It has oil refineries and aerospace, metallurgical, and food-packing industries.

Huntington Park, city (1978 est. pop. 38,400), Los Angeles co., S Calif., a residential and industrial suburb of Los Angeles; founded 1856, inc. 1906.

Huntington Station, uninc. town (1978 est. pop. 29,900), Suffolk co., SE N.Y., on the N shore of Long Island.

Huntington Woods, city (1970 pop. 8,536), Oakland co., SE Mich., a suburb NNW of Detroit; inc. 1932. Metal products are made.

Hunts·ville (hŭnts'vĭl'), town (1976 pop. 11,123), SE Ont., Canada, on the Muskoka River. It is a resort center and has lumber mills and a woodworking plant.

Huntsville. 1. City (1970 pop. 139,282), seat of Madison co., N Ala.; inc. 1811. A major center for U.S. space research, Huntsville has factories manufacturing tires, glass, and electrical equipment. The constitutional convention of the Alabama Territory was held in 1819 in Huntsville, where the first state legislature also met. Numerous ante-bellum buildings remain. **2.** City (1970 pop. 1,287), seat of Madison co., NW Ark., S of Eureka Springs, in a fishing and hunting area. **3.** City (1970 pop. 1,442), seat of Randolph co., N central Mo., W of Moberly, in a farm and coal region; platted 1831. **4.** Town (1970 pop. 337), seat of Scott co., N Tenn., NW of Knoxville, in a lumbering area. **5.** City (1978 est. pop. 23,200), seat of Walker co., E central Texas; inc. 1845. Located in a pine area, it has many sawmills and wood-processing plants. Huntsville, the home of Sam Houston, contains his grave (with an impressive monument), his restored home, and other memorials.

Hun·za (ho͝on'sä), princely state, 3,900 sq mi (10,101 sq km), NW Kashmir, administered by Pakistan. Declared a British protectorate in 1893, Hunza acceded to Pakistan after the partition of British India (1947).

Hu·peh (ho͞o'pā'), province (1974 est. pop. 33,710,000), c.72,000 sq mi (186,480 sq km), central China; capital Wu-han. The central part of Hupeh was once a huge lake and is now a basin, at or below sea level, formed from silt deposited by the Yangtze. Hupeh's lakes and many rivers provide excellent irrigation facilities. Wheat, barley, rapeseed, and beans are raised in the winter, and rice, cotton, tea, soybeans, and corn in the summer.

Hur·ley (hûr'lē), city (1970 pop. 2,418), seat of Iron co., N central Wis., on the Montreal River WSW of Ironwood, Mich.; founded 1885, inc. 1918. The city, in the Gogebic iron range, is a shipping point for iron ore.

Hu·ron (hyo͝or'ən, -ŏn'). **1.** County (1970 pop. 34,083), 822 sq mi (2,129 sq km), E Mich., bounded on the E and N by Lake Huron, on the W by Saginaw Bay, and drained by headwaters of the Cass River; formed 1840; co. seat Bad Axe. Its agriculture includes beans, sugar beets, grain, chicory, livestock, and dairy products. It has commercial fishing, some mining, lumbering, farm-products processing, and wood, metal, and chemical manufacturing. **2.** County (1970 pop. 49,587), 497 sq mi (1,287.2 sq km), N Ohio, drained by the Huron and Vermilion rivers; formed 1815; co. seat Norwalk. In an agricultural area yielding livestock, grain, and garden produce, it also has gravel pits and light industry.

Huron. 1. Village (1970 pop. 6,896), Erie co., N Ohio, SE of Sandusky on Lake Erie at the mouth of the Huron River; settled c.1805. It has a good harbor and is a fishing and tourist center. **2.** City (1978 est. pop. 13,100), seat of Beadle co., E central S.Dak., on the James River; inc. 1883. A shipping and trade center for a large livestock and grain area, it has meat-packing and lumbering industries.

Huron, Lake, 23,010 sq mi (59,596 sq km), between Ont., Canada, and Mich.; second-largest of the Great Lakes. It has a surface elevation of 580 ft (177 m) above sea level and a maximum depth of 750 ft (228.8 m). Lake Huron drains into Lake Erie through the St. Clair River–Lake St. Clair–Detroit River system. It is part of the Great Lakes–St. Lawrence Seaway system and is navigated by oceangoing and lake vessels. Navigation is impeded by ice in the shallower sections from mid-Dec. to early Apr. Samuel de Champlain was the first white man to visit (1615) Lake Huron.

Hurst (hûrst), city (1978 est. pop. 28,500), Tarrant co., N Texas, an industrial and residential suburb of Fort Worth; inc. 1952. Hurst has a large helicopter factory.

Hus·kvar·na (hüs'kvär'nä), city (1970 pop. 14,369), Jönköping co., S Sweden, at the S end of Lake Vättern. Its manufactures include sewing machines, bicycles, textiles, and electrical goods.

Hu·sum (ho͞o'zo͞om'), city (1974 est. pop. 25,030), Schleswig-Holstein, N West Germany, a port on the North Sea. It is a fishing center and major cattle market. First mentioned in the 13th cent., Husum was chartered at the beginning of the 17th cent.

Hutch·in·son (hŭch'ĭn-sən). **1.** County (1970 pop. 10,379), 814 sq mi (2,108.3 sq km), SE S.Dak., in an agricultural and cattle-raising area drained by the James River; formed 1862; co. seat Olivet. Its agriculture includes dairy products, poultry, corn, wheat, barley, and oats. **2.** County (1970 pop. 24,443), 875 sq mi (2,266.3 sq km), extreme N Texas, on the high treeless plains of the Panhandle in an extensive oil and natural-gas area; formed 1876; co. seat Stinnett. Grain sorghums are grown, and beef and dairy cattle, hogs, sheep, horses, and mules are raised. The county has a carbon-black industry and manufactures synthetic rubber, inks, and petroleum products.

Hutchinson. 1. City (1978 est. pop. 41,900), seat of Reno co., S central Kansas, on the Arkansas River; inc. 1872. It is a commercial and industrial center in a grain (especially wheat), livestock, and oil region. Farm equipment and aircraft parts are made, and salt is extracted from great beds beneath the city. **2.** City (1970 pop. 8,031), McLeod co., central Minn., W of Minneapolis, in a dairy and grain region; founded 1855, inc. 1881. Part of the village was burned in the Sioux outbreak in 1862. Cellophane and dairy products are made.

Huy (o͞o-wē'), town (1977 est. pop. 17,792), Liège prov., E Belgium, on the Meuse River; founded in the 9th cent. Its citadel (19th cent.; now a military depot and a prison) dominates the town.

Hvar (KHvär), island (1971 pop. 11,326), 112 sq mi (290.1 sq km), in the Adriatic Sea off the Dalmatian coast, W Yugoslavia. Fruit growing, cattle raising, and fishing are the chief occupations. The island is also a leading tourist center.

Hwai (hwī), river, China: *see* Huai.

Hwai·nan (hwī'nän'): *see* Huai-nan, China.

Hwang Ho (hwäng' hō'), river, China: *see* Huang Ho.

Hwang·poo (hwäng'po͞o'), river, China: *see* Whangpoo.

Hwang·shih (hwäng'shē'): *see* Huang-shih, China.

Hy·an·nis (hī-ăn'ĭs). **1.** Resort village (1970 pop. 7,800), Barnstable co., SE Mass., on Cape Cod; inc. 1639. Major industries are tourism and home construction. President John F. Kennedy and Sen. Robert F. Kennedy had summer homes at nearby Hyannisport; members of the Kennedy family still maintain a compound of houses in the village. **2.** Village (1970 pop. 345), seat of Grant co., W central Nebr., E of Alliance, in a resort, livestock, grain, and poultry area.

Hy·atts·ville (hī'əts-vĭl'), city (1978 est. pop. 13,000), Prince Georges co., W central Md., a suburb of Washington, D.C.; inc. 1886. Hyattsville is a residential community with some light industry.

Hyde (hīd), municipal borough (1971 pop. 37,075), Greater Manchester, NW England. It has iron foundries and factories producing cotton, machinery, rubber, paper, and hats.

Hyde. 1. County (1970 pop. 5,571), 613 sq mi (1,587.7 sq km), E N.C., in a forested and swampy tidewater area bounded on the S by Pamlico Sound; formed 1705; co. seat Swanquarter. Commercial fishing, agriculture (hogs, poultry, soybeans, and cotton), sawmilling, and tourism are important. **2.** County (1970 pop. 2,515), 863 sq mi (2,235.2 sq km), central S.Dak., in an agricultural area drained by the Missouri River and intermittent streams; formed 1873; co. seat Highmore. Corn, wheat, oats, and other grains are grown. Livestock and dairy products are also important.

Hy·den (hī'dən), city (1970 pop. 482), seat of Leslie co., SE Ky., in the Cumberland foothills on the Middle Fork of the Kentucky River WSW of Hazard.

Hyde Park. 1. Village (1970 pop. 2,805), Dutchess co., SE N.Y., on the Hudson River; settled c.1740. It is the site of the Roosevelt estate, where President Franklin D. Roosevelt was born and is buried. The Roosevelt Library (1941) contains historical material dating from 1910 until Roosevelt's death in 1945. The Frederick W. Vanderbilt mansion is also here. Both the Roosevelt and Vanderbilt homes are national historic sites. **2.** Town (1970 pop. 1,347), seat of Lamoille co., N Vt., N of Montpelier; granted 1780, settled 1787. Wood products are made.

Hyde Park, 361 acres (146.2 hectares) in Westminster borough, London, England. Once the manor of Hyde, a part of the old Westminster Abbey property, it became a deer park under Henry VIII. Today's distinctive features of the park are Hyde Park Corner (near the Marble Arch), the meeting place of soapbox orators, and Rotten Row, a famous bridle path.

Hy·der·a·bad (hī'dər-ə-bäd', -băd'), former state, S central India. The region of Hyderabad is now divided among the states of Karnataka, Maharashtra, and Andhra Pradesh. Situated almost entirely within the Deccan plateau, it has abundant crops of cotton and rice in the north, while grains are grown in the heavily irrigated southern area. There are large deposits of coal and iron in the area.

Hyderabad, city (1971 pop. 1,612,276), capital of Andhra Pradesh, central India, was founded in 1589 as the capital of the Golconda kingdom. A commercial center and a transportation hub, the city has fine ancient structures.

Hyderabad, city (1972 pop. 600,000), capital of Sind prov., S Pakistan. It was long noted for its embroideries and cutlery and now has chemical, engineering, food-processing, cotton, cement, cigarette, glass, and match industries. Founded in 1768, Hyderabad was laid out in 1782 and was the capital of the emirs of Sind. The British East India Company occupied Hyderabad in 1839.

Hyères (yĕr), city (1968 pop. 34,875), Var dept., SE France, in Provence, on the Mediterranean Sea. A port in medieval times, Hyères is now a resort. Off the coast is a group of islands of the same name.

Hy·met·tus (hī-mĕt'əs), mountain range, E central Greece, in Attica. Mt. Hymettus (c.3,370 ft/1,030 m) is the highest point. It is noted for its honey. Marble has been quarried here since antiquity.

Hy·sham (hī'shəm), town (1970 pop. 373), seat of Treasure co., S central Mont., near the Yellowstone River NE of Billings. It is a shipping center for a livestock, sugar-beet, and grain area.

Hythe (hīth), municipal borough (1973 est. pop. 12,210), Kent, SE England. A market town, it was one of the Cinque Ports.

Hyu·ga (hyo͞o'gä), city (1975 pop. 40,600), Miyazaki prefecture, E Kyushu, Japan, on the Pacific Ocean. It is a fishing port and agricultural center.

Ia·şi (yä'shē), city (1977 est. pop. 264,947), E Rumania, in Moldavia, near the USSR border. Iaşi is the commercial center of a fertile agricultural region. Textiles, pharmaceuticals, machinery, food products, furniture, plastics, and metal are produced. In 1565 Iaşi became capital of the Rumanian principality of Moldavia, a position it held until Moldavia and Walachia were united in 1859. The city was repeatedly burned and sacked by Tatars, Turks, and Russians. During World War I the city served as Rumania's temporary capital while German forces occupied Walachia. During World War II Iaşi's large Jewish population was massacred by the Nazis in one of the worst pogroms in history. Soviet troops took the city in 1944.

I·ba·dan (ē-bä'dän), city (1971 est. pop. 758,000), SW Nigeria. It is a major commercial and industrial center. Manufactures include canned foods, metal products, furniture, soap, and chemicals. It is also an important market for cacao, which, along with cotton, is produced in the region. Ibadan was founded in the 1830s as a military camp during the Yoruba civil wars and developed into the most powerful Yoruba city-state.

I·ba·gué (ē-bä-gā'), city (1973 pop. 176,223), alt. 4,300 ft (1,311.5 m), capital of Tolima dept., W central Colombia. It is a major commercial center for the Magdalena and Cauca valleys. Coffee, flour, and sugar are produced in the city, and silver and sulfur are shipped. Ibagué grew rapidly as a result of the coffee boom in the 1890s.

I·be·ri·a (ī-bîr'ē-ə), ancient country of Transcaucasia, roughly the E part of present-day Georgian SSR. Iberia was allied to the Romans, ruled by the Sassanids of Persia, and became (6th cent. A.D.) a Byzantine province.

Iberia, parish (1970 pop. 57,397), 588 sq mi (1,522.9 sq km), S La., on the Gulf Coast in a swampland area bounded on the SW by Vermilion Bay, on the E by Grand Lake, and intersected by Bayou Teche; formed 1868; parish seat New Iberia. It is crossed by the Gulf Intracoastal Waterway. In an agricultural area yielding sugar cane, corn, rice, cotton, pepper, truck crops, and dairy products, it has oil and gas wells, salt and limestone deposits, and lumbering and food-processing industries.

I·be·ri·an Mountains (ī-bîr'ē-ən), mountain system, extending c.250 mi (400 km) along the NE edge of the Meseta, NE Spain. Moncayo (7,605 ft/2,320 m) is the highest peak in the system.

Iberian Peninsula, c.230,400 sq mi (596,735 sq km), SW Europe, separated from the rest of Europe by the Pyrenees and washed on the N and W by the Atlantic Ocean and on the S and E by the Mediterranean Sea; the Strait of Gibraltar separates it from Africa. The Iberian Peninsula is dominated by the Meseta, a plateau region ringed and crossed by mountain ranges. Coastal lowlands surround the primarily agrarian-oriented Meseta.

I·ber·ville (ī'bər-vĭl'), parish (1970 pop. 30,746), 627 sq mi (1,623.9 sq km), S La., bounded on the W by the Atchafalaya River, intersected by the Mississippi River, and drained by the Grand River; formed 1805; parish seat Plaquemine. It has agriculture (sugar cane, corn, hay, rice, fruit, and vegetables), oil and gas wells, a lumbering industry, commercial fisheries, cane-sugar refineries, and cotton and moss gins. Chemicals and machinery are manufactured.

I·bi·za (ē-vē'zə), island (1970 pop. 45,075), 221 sq mi (572.4 sq km), Baleares prov., Spain, one of the Balearic Islands, in the W Mediterranean. There are fisheries and saltworks on the island. A picturesque island with Roman, Phoenician, and Carthaginian remains, Ibiza attracts tourists and artists.

I·ca (ē'kə, ē'kä), city (1972 pop. 73,883), capital of Ica dept., SW Peru, on the Pan-American Highway. It is a commercial center for the cotton, wool, and wine produced in the region. There are several summer resorts nearby. Ica is also the archaeological name of the Chincha empire of ancient Peru, which had one of its major centers in the adjacent valley. The empire fell to the Inca in the 15th cent. The Spanish settled the city in 1563.

I·çá (ē-sä'), river, Brazil: *see* Putumayo.

I·car·i·a (ĭ-kâr'ē-ə, ī-kâr'-): *see* Ikaría, Greece.

Ice Age National Scientific Reserve (īs): *see* National Parks and Monuments Table.

I·çel (ē-chĕl'): *see* Mersin, Turkey.

Ice·land (īs'lənd), republic (1976 pop. 220,000), 39,698 sq mi (102,819 sq km), occupying an island in the Atlantic Ocean just S of the Arctic Circle, c.600 mi (970 km) W of Norway and c.180 mi (290 km) SE of Greenland. Reykjavík is the capital. The republic includes several small islands. Deep fjords indent the coasts, particularly in the north and west. The island itself is a geologically young basalt plateau, averaging 2,000 ft (610 m) in height and culminating in vast ice fields. There are about 200 volcanoes, many of them still active. Hot springs abound and are used for inexpensive heating. Only about one fourth of the island is habitable.

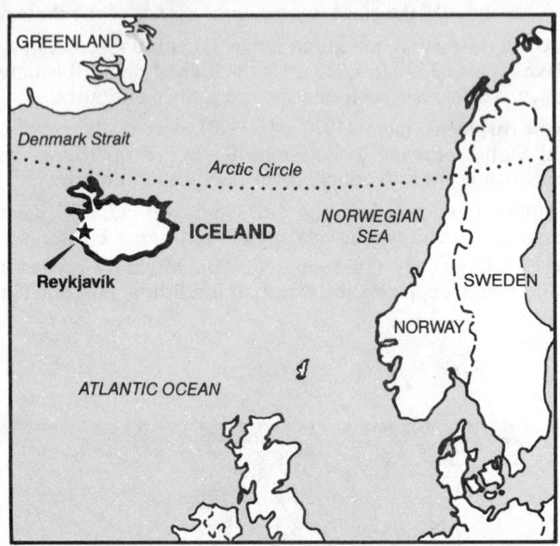

Economy. About 15% of the land is potentially productive, but agriculture, cultivating mainly hay, potatoes, and turnips, is restricted to .5% of the total area. Fruits and vegetables are raised in greenhouses. There are extensive grazing lands, used mainly for sheep raising, but also for horses and cattle. Fishing is the most important industry, with codfish and herring the chief exports. Aside from aluminum smelting, Iceland has little heavy industry.

History. Iceland may be the Ultima Thule of the ancients. Irish monks visited it before the 9th cent. but abandoned it on the arrival (c.850–875) of Norse settlers. In 930 a general assembly, the Althing, was established near Reykjavík at Thingvellir, and Christianity was introduced c.1000 by the Norwegian Olaf I. Politically, Iceland became a feudal state, and the bloody civil wars of rival chieftains facilitated Norwegian intervention. King Haakon IV of Norway obtained suzerainty over Iceland between 1261 and 1264. Norwegian rule brought order, but high taxes and an imposed judicial system caused much discontent. When, with Norway, Iceland passed (1380) under the Danish crown, the Danes showed even less concern for Icelandic welfare.

During the 17th and 18th cent. English, Spanish, and Algerian pirates raided the coasts and ruined trade; epidemics and volcanic eruptions killed a large part of the population; and the creation (1602) of a private trading company at Copenhagen, with exclusive rights to the Iceland trade, caused economic ruin. The trade monopoly was revoked in 1771 and transferred to the Danish crown, and in 1786 trade with Iceland was opened to Danish and Norwegian merchants. The exclusion of foreign traders was lifted in 1854.

The 19th cent. brought a rebirth of national culture and strong agitation for independence, led by Jón Sigurðsson. In 1874 a constitution and limited home rule were granted, and in 1918 Iceland became a sovereign state in personal union with Denmark. After the German occupation (1940) of Denmark in World War II, Great Britain sent a military force to defend the island from possible German attack, and this was replaced after 1941 by U.S. forces. In 1944 Icelanders voted to terminate the union with Denmark; the kingdom of Iceland was proclaimed an independent republic on June 17, 1944. In 1968 Kristjárn Eldjárn was elected president, and he was re-elected in 1972 and 1976.

A major international dispute with Great Britain over fishing rights came to a head in 1975 when Iceland proclaimed a 200-mi (320-km) fishing zone extending into the Atlantic Ocean. Iceland had previously extended the limits of its territorial waters in 1958 and 1972, precipitating several confrontations between the Icelandic coast guard and British warships. When the so-called cod war resumed in 1975, it led to a four-month break in diplomatic relations. An accord reached in 1976 gave Britain limited fishing rights in Icelandic territorial waters.

Government. Iceland is governed by parliamentary democracy, with a cabinet responsible to the Althing and a president elected by popular vote for a four-year term.

I-ch'ang or **I·chang** (both: ē'chäng'), city (1970 est. pop. 160,000), SW Hupeh prov., China, a river port on the Yangtze River. The gorges above the city are noted for their scenic beauty. I-ch'ang has food-processing and cement industries.

I·chi·ha·ra (ē'chē-hä'rə), city (1976 est. pop. 198,896), Chiba prefecture, central Honshu, Japan, on Tokyo Bay. It is an industrial city with important petrochemical, shipbuilding, and steel industries.

I·chi·ka·wa (ē'chē-kä'wə), city (1976 est. pop. 331,019), Chiba prefecture, central Honshu, Japan, on the Edo River. It is an industrial city with metallurgical, chemical, and textile industries.

I·chi·no·mi·ya (ē'chē-nō'mē-yä), city (1978 est. pop. 241,150), Aichi prefecture, central Honshu, Japan. It is an industrial satellite of Nagoya and has spinning, weaving, and metallurgical industries.

Ick·nield Street (ĭk'nēld), name for a prehistoric road in England, extending SW from the Wash, along the line of the Chiltern Hills and Berkshire Downs, to Salisbury Plain.

I·da (ī'də), county (1970 pop. 9,283), 431 sq mi (1,116.3 sq km), W Iowa, drained by the Maple and Sugar rivers; formed 1851; co. seat Ida Grove. In a prairie region, its agriculture includes hogs, cattle, poultry, corn, oats, and barley. It also has bituminous-coal mines and sand and gravel pits.

Ida, Mount, 8,058 ft (2,457.7 m) high, central Crete, Greece. It is the highest mountain on Crete.

I·da·bel (ī'də-bĕl'), city (1970 pop. 5,946), seat of McCurtain co., extreme SE Okla., near the Red River SE of Hugo, in a farm and timber region; inc. 1906. Ouachita National Forest is nearby.

Ida Grove, city (1970 pop. 2,261), seat of Ida co., W central Iowa, ESE of Sioux City on the Maple River, in a farm area; inc. 1878.

I·da·ho (ī'də-hō'), state (1975 pop. 823,000), 83,557 sq mi (216,413 sq km), NW United States, one of the Rocky Mt. states, admitted as the 43rd state of the Union in 1890. The capital is Boise. Idaho is bounded on the north by the Canadian province of British Columbia, on the northeast by Mont., on the east by Wyo., on the south by Utah and Nev., and on the west by Oregon and Wash.

From the northern Panhandle, where Idaho is about 45 mi (72 km) wide, the state broadens south of the Bitterroot Range to 310 mi (499 km) in width. Much of Idaho is a vast expanse of timberland, scenic lakes, wild rivers, cascades, and spectacular gorges. Across southern Idaho the Snake River flows in a great arc; with its tributaries the river has been harnessed to produce hydroelectric power and to reclaim vast areas of dry but fertile land. To the north of the Snake River valley, in central and north-central Idaho, are the massive Sawtooth Mts. and the Salmon River Mts. In the central and north-central regions and in the Panhandle there are tremendous expanses of national forests covering approximately two fifths of the state.

Idaho is divided into 44 counties:

NAME	COUNTY SEAT	NAME	COUNTY SEAT
Ada	Boise	Gem	Emmett
Adams	Council	Gooding	Gooding
Bannock	Pocatello	Idaho	Grangeville
Bear Lake	Paris	Jefferson	Rigby
Benewah	St. Maries	Jerome	Jerome
Bingham	Blackfoot	Kootenai	Coeur d'Alene
Blaine	Hailey	Latah	Moscow
Boise	Idaho City	Lemhi	Salmon
Bonner	Sandpoint	Lewis	Nezperce
Bonneville	Idaho Falls	Lincoln	Shoshone
Boundary	Bonners Ferry	Madison	Rexburg
Butte	Arco	Minidoka	Rupert
Camas	Fairfield	Nez Perce	Lewiston
Canyon	Caldwell	Oneida	Malad City
Caribou	Soda Springs	Owyhee	Murphy
Cassia	Burley	Payette	Payette
Clark	Dubois	Power	American Falls
Clearwater	Orofino	Shoshone	Wallace
Custer	Challis	Teton	Driggs
Elmore	Mountain Home	Twin Falls	Twin Falls
Franklin	Preston	Valley	Cascade
Fremont	St. Anthony	Washington	Weiser

Economy. Agriculture is the most important sector of the state's economy. Cattle and calves are the leading agricultural products; dairy products are also important. Idaho's chief crops are potatoes, hay, wheat, and sugar beets. Food processing is the chief industry; lumber and wood products and chemicals are other major manufactured items. Silver, antimony, phosphate rock, lead, and zinc are the principal minerals produced. Since the 1950s and 1960s tourism has become a major industry in Idaho; about 6 million people visit the state annually.

History. Probably the first white men to enter the area that is now Idaho were members of the Lewis and Clark expedition in 1805. A Canadian, David Thompson of the North West Company, established the first trading post in Idaho in 1809. The next year traders from St. Louis penetrated the mountains, and the first American trading post was established in the area near present-day Rexburg. In this period what is now the state of Idaho was part of Oregon country, held jointly by the United States and Great Britain from 1818 to 1846. By the 1840s rival British and American trappers had severely depleted the region's fur supply. In 1846 the United States gained sole claim to Oregon country south of the 49th parallel by the Oregon Treaty with Great Britain. The area was established as a territory in 1848.

Idaho still had no permanent settlement when Oregon Territory became a state in 1859 and the eastern part of Idaho was added to Washington Territory. It was not until the discovery of gold in the early 1860s that settlers poured in. They soon formed a population large enough to demand new government administration, and Idaho Territory was set up in 1863. Some Indians, upset by the incursion of whites, attacked white settlements. In 1876-77 the Nez Perce Indians, led by Chief Joseph, made their heroic but unsuccessful attempt to flee to Canada while being pursued by U.S. troops.

The late 19th cent. witnessed the growth of cattle and sheep ranching and the strife that developed between the two groups of ranchers over grazing areas. A new mining boom started in 1882 with the discovery of gold, then silver, along the Coeur d'Alene. The coming of the railroads through Idaho in the 1880s and the 1890s brought new settlers and aided in the founding of numerous cities. After statehood was achieved in 1890, farming expanded and private interests developed irrigation projects. Idaho's enormous potential of hydroelectric power is now being tapped by projects along the Snake River.

Government. Idaho's constitution became effective in 1890 upon statehood. The state's chief executive is a governor elected for a term of four years; the governor may succeed himself. The legislature consists of a 70-member house of representatives and a 35-member senate. State representatives and senators are elected every two years.

Educational Institutions. The Univ. of Idaho is located at Moscow.

Idaho, county (1970 pop. 12,891), 8,515 sq mi (22,053.9 sq km), central Idaho, in a mountainous agricultural and mining area bounded in the E by Bitterroot Range and the Mont. border, in the W by the Grand Canyon of the Snake River and the Oregon border, and drained by the Clearwater and Salmon rivers; formed 1861; co. seat Grangeville. Its economy depends largely on wheat and livestock. Copper, gold, silver, and lead are mined. The county contains parts of Nez Perce, Bitterroot, Payette, and Salmon national forests.

Idaho City, village (1970 pop. 164), seat of Boise co., SW Idaho, on a small affluent of the Boise River NE of Boise, in a mountainous area. Lumber milling and placer mining for gold are done. In the 1860s it was an important gold-mining center.

Idaho Falls, city (1970 pop. 35,776), seat of Bonneville co., SE Idaho, traversed by the Snake River; inc. 1900. Idaho Falls is the commercial and processing center of an irrigated livestock, dairy, and farm region producing potatoes, wheat, sugar beets, and seed peas. Concrete, steel, and lumber are manufactured. A nearby nuclear-reactor testing station is also a source of employment. The site of Idaho Falls was originally a miner's fording point over the Snake River, and the first settlers were Mormons.

I·da·li·um (ī-dā'lē-əm), ancient town in Cyprus. An inscription in Phoenician and Cypriote, found on a temple site at Idalium, gave the key to the Cypriote language.

Id·fu (ĭd'fōō) or **Ed·fu** (ĕd'-), town (1966 pop. 27,300), S central Egypt, on the Nile River. It is an agricultural trade center and has paper mills. Idfu was the capital of a predynastic upper Egyptian kingdom that flourished c.3400 B.C. and worshipped Horus. Later, a large sandstone temple of Horus was built here. It is one of the finest extant examples of Egyptian temple architecture. Excavations have yielded mastabas dating from the Old Kingdom, a Roman necropolis, and Coptic and Byzantine remains.

Íd·hra (ē'*th*rä), island (1971 pop. 2,531), 21 sq mi (54.4 sq km), SE Greece, in the Aegean Sea, off the Argolís peninsula of the Peloponnesus. It is mostly barren and rocky. Sponge fishing, shipbuilding, and textile manufacturing are the main occupations. Settled in the 15th cent. by Albanian-speaking Greeks from the Peloponnesus, the island became by the 17th cent. a shipbuilding and commercial center. It declined after the Greek War of Independence (1821-29), in which its seafaring people played an important role.

Id·lib (ĭd'lĭb'), town (1970 pop. 34,515), capital of Idlib governorate, NW Syria. It is the market center for an agricultural region where grains, grapes, olives, sesame, and cotton are grown. Idlib's chief industries are textile manufacture, olive pressing, and fig drying.

I·fe (ē'fā), city (1971 est. pop. 157,000), SW Nigeria. Located in a farm region, the city is an important center for marketing and shipping cacao. According to tradition, Ife is the oldest Yoruba town (founded c.1300) and was the most powerful Yoruba kingdom until the late 17th cent., when Oyo surpassed it. Terracotta and naturalistic bronze sculptures made in the area as early as the 12th cent. are considered among the finest works of West African art.

If·ni (ēf'nē), former Spanish possession (580 sq mi/1,502.2 sq km), SW Morocco, on the Atlantic Ocean. The main industry is fishing. Ifni was ceded by Morocco to Spain in 1860, but Spanish administration was nominal until 1934. Spain returned Ifni to Morocco in 1969.

I·gua·çu Falls or **I·guas·sú Falls** (both: ē'gwä-sōō'), in the Iguaçu River, on the Argentina-Brazilian border near the Paraguay line. Iguaçu Falls has two main sections that are composed of hundreds of waterfalls separated from each other by rocky islands along a 3-mi (4.8-km) escarpment. The highest fall is 210 ft (64 m).

I·gua·la (ē-gwä'lə), city (1970 pop. 60,980), Guerrero state, S Mexico, on the Cocula River. It is the distribution and processing center of the surrounding mining and agricultural region. There are frequent earthquakes. The Plan of Iguala, proclaimed here on Feb. 24, 1821, provided for Roman Catholicism as Mexico's sole religion, absolute independence from Spain, and racial equality.

Ii·zu·ka (ē'zōō-kä), city (1975 pop. 75,417), Fukuoka prefecture, N

Kyushu, Japan, on the Onga River. It is a mining center in a coal-mining area.

IJ or **Y** (both: ī), inlet of the IJsselmeer, North Holland prov., NW Netherlands, on which Amsterdam is located. It is connected by canals with the North Sea and the Lek and Waal rivers.

I·je·bu-O·de (ī-jä'bōō-ō'dä), town (1969 est. pop. 79,000), SW Nigeria. It is a commercial town and a collection point for cacao and palm products. Manufactures include textiles, metal and clay products, processed timber and plywood, canned fruit and juice, and milled rice. Ijebu-Ode was the capital of the Yoruba Ijebu kingdom that was founded by the 15th cent. Long opposed to foreign contacts, the Ijebu kingdom remained closed to Europeans until 1892, when the British seized it in retaliation for the Ijebu's closing of the trade routes to the north during the Yoruba civil wars.

IJ·mui·den (ī-moi'dən), city (1977 est. pop. 63,023), North Holland prov., W Netherlands, on the North Sea. It is a seaport and fishing center at the end of the North Sea Canal.

IJs·sel (ī'səl), river, 72 mi (115.8 km) long, branching from the Neder Rijn (Lower Rhine) River near Arnhem, E Netherlands, and flowing N into the IJsselmeer, near Kampen. The river, canalized in places, passes through an industrial and truck-farming region.

IJs·sel·meer (ī'səl-mâr'), shallow freshwater lake, NW Netherlands. It was formed from the old Zuider Zee by the construction of a dam (completed 1932). The dam, 19 mi (30.6 km) long, has navigation locks and drainage sluices that control the lake's level. Considerable areas have been reclaimed from the former Zuider Zee since 1932. The IJsselmeer is an important freshwater fishing ground. Since 1937 pike, perch, and eels have replaced saltwater fish.

I·ka·ri·a or **I·ca·ri·a** (both: ĭ-kâr'ē-ə, ī-kâr'-), mountainous island (1971 pop. 7,702), c.100 sq mi (260 sq km), SE Greece, one of the Sporades, near Turkey. It has iron-ore deposits and sulfur springs. According to Greek mythology Icarus fell into the sea near here.

I·ke·da (ē-kĕ'dä), city (1976 est. pop. 101,689), Osaka prefecture, S Honshu, Japan, on the Ina River. It is an industrial and residential suburb of Osaka with a major industrial-vehicles industry.

I·lan (ē'län'), city (1973 est. pop. 59,500), NE Taiwan. Located in an agricultural area, it is the largest rice market in Taiwan. Fertilizers and wood and paper products are among the city's manufactures.

Île-aux-Noix (ēl-ō-nwä'), island, 210 acres (85.1 hectares), in the Richelieu River near St. Jean, S Que., Canada. During the French and Indian Wars the French built a fort here to delay the British advance on Montreal but were forced to surrender it in 1760. Named Fort Lennox and occupied by a British garrison, the island fell (1775) to American forces and was used as a base by the American generals Philip Schuyler and Richard Montgomery for attacks on Montreal and Quebec until abandoned in 1776. The British then used the island to supply their operations against the American fleet on Lake Champlain. The present Fort Lennox dates from the 1820s, when the old fortifications were repaired and additions were built. It was a military post until 1870.

Île-de-France (ēl-də-fräNs'), region and former province, N central France, in the center of the Paris basin, a fertile depression where the Marne and Ouse rivers join the Seine. The region has numerous large industrial towns and residential suburbs and supplies the Paris metropolis with fruits, vegetables, and dairy products. The name came into use in the 14th cent., but the region, including the countship of Paris, had been part of the duchy of France since the 10th cent. When Hugh Capet, duke of France and count of Paris, was chosen as the French king in 987, his domains became the nucleus of the ever-growing crown land, which by the time of the death of Louis XI (1483) comprised the major part of present-day France.

I·le·sha (ē-lā'shä), city (1971 est. pop. 200,000), SW Nigeria. Formerly a caravan trade center, Ilesha is today an agricultural and commercial city. Cacao, kola nuts, and yams are shipped. Alluvial gold is found.

I·lhé·us (ē-lyĕ'ōōs), city (1970 pop. 107,738), Bahia state, E Brazil, a port on Ilhéus Bay, an inlet of the Atlantic Ocean. Founded in the mid-16th cent., it is the cacao center of Brazil and ships the agricultural produce of the surrounding hinterland.

I·li (ē'lē'), river of China and the USSR, 590 mi (949.3 km) long, rising in the Tien Shan, NW Sinkiang prov., and flowing W across the China-USSR border, through the sandy Sary-Ishik-Otrau Desert, in Kazakhstan, and into Lake Balkhash.

Il·i·am·na (ĭl'ē-ăm'nə), lake, c.1,000 sq mi (2,590 sq km), SW Alaska, at the base of the Alaska Peninsula. It is the largest lake in Alaska. The lake is noted for sport fishing.

Iliamna Peak, volcano, 10,016 ft (3,054.8 m) high, SW Alaska, W of Cook Inlet and SW of Anchorage.

I·li·gan (ē-lē'gän), city (1970 est. pop. 82,900), capital of Lanao del Norte prov., W central Mindanao, the Philippines, a port on Iligan Bay. The nation's first steel mill was established here in 1964. The city also has chemical and fertilizer plants.

I·lim (ē-lyĕm'), river, c.240 mi (385 km) long, Irkutsk oblast, Russian SFSR, USSR, originating in the Ilim Mts. and flowing N to the Angara River. Half of its course is navigable.

Il·i·pa (ĭl'ĭ-pə), ancient town of Spain, near the modern Seville. Here Scipio Africanus Major defeated (206 B.C.) the Carthaginian forces and paved the way for the defeat of Hannibal at Zama (202 B.C.).

Il·kes·ton (ĭl'kĕs-tən), municipal borough (1973 est. pop. 33,690), Derbyshire, central England. There are iron and coal mines to the south. Rayon, lace, and iron goods are manufactured. Ilkeston is mentioned in the Domesday Book.

Il·lam·pú (ē-yäm-pōō'), peak, 20,873 ft (6,366.3 m) high, in the Cordillera Real of the Bolivian Andes, W Bolivia. Permanently capped with snow, Illampú dominates the scenery visible from La Paz.

Ille-et-Vi·laine (ēl-ā-vē-lĕn'), department (1975 pop. 702,199), 2,609 sq mi (6,757.3 sq km), NW France, in Brittany, on the English Channel; capital Rennes.

Il·li·ma·ni (ē'yē-mä'nē), mountain, 21,151 ft (6,451 m) high, W Bolivia. One of the highest peaks of the Cordillera Real of the Bolivian Andes, it is permanently snow-capped and was first climbed in 1898.

Il·li·nois (ĭl'ə-noi', -noiz'), state (1975 pop. 11,086,000), 56,400 sq mi (146,076 sq km), N central United States, in the Midwest, admitted as the 21st state of the Union in 1818. Springfield is the capital. Illinois is bounded on the north by Wis., on the east by Lake Michigan and Ind., on the southeast and south by Ky. (where the Ohio River forms the boundary), and on the west by Mo. and Iowa (where

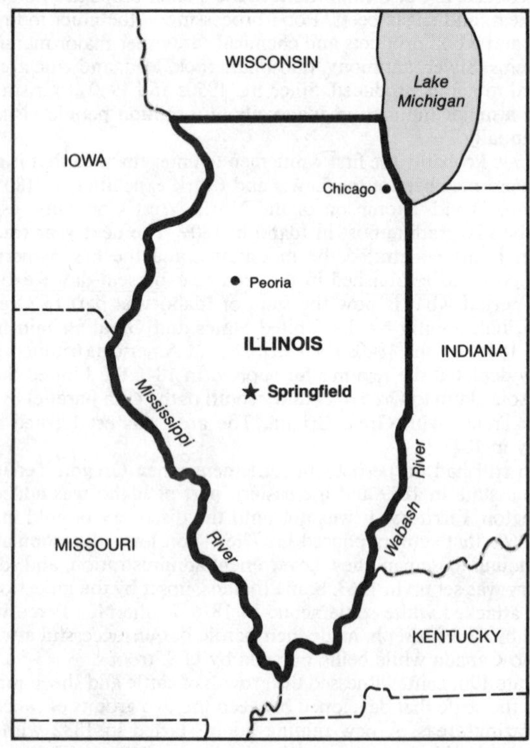

the Mississippi River forms the boundary). The broad level lands that gave Illinois the nickname Prairie State were fashioned by Pleistocene glaciers, which leveled rugged ridges and filled valleys over the northern and central parts of the state. The fertile prairies are drained by more than 275 rivers, most of which flow to the Mississippi-Ohio systems. The St. Lawrence Seaway provides access for oceangoing vessels.

Illinois is divided into 102 counties:

NAME	COUNTY SEAT	NAME	COUNTY SEAT
Adams	Quincy	Lee	Dixon
Alexander	Cairo	Livingston	Pontiac
Bond	Greenville	Logan	Lincoln
Boone	Belvidere	McDonough	Macomb
Brown	Mount Sterling	McHenry	Woodstock
Bureau	Princeton	McLean	Bloomington
Calhoun	Hardin	Macon	Decatur
Carroll	Mount Carroll	Macoupin	Carlinville
Cass	Virginia	Madison	Edwardsville
Champaign	Urbana	Marion	Salem
Christian	Taylorville	Marshall	Lacon
Clark	Marshall	Mason	Havana
Clay	Louisville	Massac	Metropolis
Clinton	Carlyle	Menard	Petersburg
Coles	Charleston	Mercer	Aledo
Cook	Chicago	Monroe	Waterloo
Crawford	Robinson	Montgomery	Hillsboro
Cumberland	Toledo	Morgan	Jacksonville
De Kalb	Sycamore	Moultrie	Sullivan
De Witt	Clinton	Ogle	Oregon
Douglas	Tuscola	Peoria	Peoria
Du Page	Wheaton	Perry	Pinckneyville
Edgar	Paris	Piatt	Monticello
Edwards	Albion	Pike	Pittsfield
Effingham	Effingham	Pope	Golconda
Fayette	Vandalia	Pulaski	Mound City
Ford	Paxton	Putnam	Hennepin
Franklin	Benton	Randolph	Chester
Fulton	Lewistown	Richland	Olney
Gallatin	Shawneetown	Rock Island	Rock Island
Greene	Carrollton	St. Clair	Belleville
Grundy	Morris	Saline	Harrisburg
Hamilton	McLeansboro	Sangamon	Springfield
Hancock	Carthage	Schuyler	Rushville
Hardin	Elizabethtown	Scott	Winchester
Henderson	Oquawka	Shelby	Shelbyville
Henry	Cambridge	Stark	Toulon
Iroquois	Watseka	Stephenson	Freeport
Jackson	Murphysboro	Tazewell	Pekin
Jasper	Newton	Union	Jonesboro
Jefferson	Mount Vernon	Vermilion	Danville
Jersey	Jerseyville	Wabash	Mount Carmel
Jo Daviess	Galena	Warren	Monmouth
Johnson	Vienna	Washington	Nashville
Kane	Geneva	Wayne	Fairfield
Kankakee	Kankakee	White	Carmi
Kendall	Yorkville	Whiteside	Morrison
Knox	Galesburg	Will	Joliet
Lake	Waukegan	Williamson	Marion
La Salle	Ottawa	Winnebago	Rockford
Lawrence	Lawrenceville	Woodford	Eureka

Economy. Corn, hogs, cattle, and soybeans are the principal sources of farm income. Other major crops include hay and wheat. Beneath the fertile topsoil lies mineral wealth, and the state is a leading producer of fluorspar. Bituminous-coal fields and large oil deposits make southern Illinois a major source of fuel. Major industries include the manufacture of electrical and nonelectrical machinery, food products, fabricated and primary metal products, and chemicals; and printing and publishing.

History. At the end of the 18th cent. the Illinois, Sac, Fox, and other Indian tribes were living in the river forests, where many centuries before them the prehistoric Indian Mound Builders had lived. Father Marquette and Louis Jolliet ascended the Illinois River in 1673, and two years later Marquette returned to establish a mission. The French explorer sieur La Salle founded (1680) Fort Creve Coeur and completed (1682–83) Fort St. Louis on Starved Rock cliff. By the Treaty of Paris of 1763, ending the French and Indian Wars, France ceded all of the Illinois country to Great Britain. However, the British did not take possession until Indian resistance, led by the Ottawa Indian chief Pontiac, was quelled (1766). The Illinois region was an integral part of the Old Northwest that came within U.S. boundaries by the 1783 Treaty of Paris ending the Revolution. Under the Ordinance of 1787 the area became the Northwest Territory. Made part of Indiana Territory in 1800, Illinois became a separate territory in 1809 and a state in 1818. In 1820 the capital was moved from Kaskaskia to Vandalia.

The Black Hawk War (1832) practically ended the tenure of the Indians in Illinois and drove them west of the Mississippi. In the 1830s there was heavy and uncontrolled land speculation. Industrial development came with the opening of an agricultural-implements factory by Cyrus H. McCormick at Chicago in 1847 and the building of the railroads in the 1850s. Abraham Lincoln and another Illinois lawyer, Stephen A. Douglas, won national attention with their debates on the slavery issue in the senatorial race of 1858. In 1861 Lincoln became President and fought to preserve the Union in the face of the South's secession. During the Civil War, Illinois supported the Union, but there was much proslavery sentiment in the southern part of the state.

Immediately after the war, industry expanded to tremendous proportions, and the Illinois legislature, by setting aside acreage for stockyards, prepared the way for the development of the meat-packing industry. In the latter part of the 19th cent. farmers in the state revolted against exorbitant freight rates, tariff discrimination, and the high price of manufactured goods. Laborers in factories, railroads, and mines also became restive, and Illinois was the scene of such violent labor incidents as the Haymarket Square riot of 1886 and the Pullman strike of 1894.

In 1937 new oil fields were discovered in southern Illinois, further enhancing the state's industrial development. During World War II the nation's first controlled nuclear reaction was executed at the Univ. of Chicago. World War II spurred the growth of the Chicago metropolitan area, and the opening (1959) of the St. Lawrence Seaway made the city a major port for overseas shipping. Southern Illinois experienced population declines in the 1950s and 1960s as farms in the south became more mechanized.

Government. In 1970 Illinois adopted a new state constitution. The governor is elected for a term of four years. The state legislature, called the general assembly, consists of a house of representatives with 177 members elected to serve for two years and a senate with 59 members elected for two or four years.

Educational Institutions. Among those prominent are the Univ. of Illinois, at Urbana-Champaign; Northwestern Univ., at Evanston; the Univ. of Chicago and the Illinois Institute of Technology, in Chicago; Illinois State Univ., at Normal; and Southern Illinois Univ., at Carbondale.

Illinois, river, 273 mi (439.3 km) long, formed by the confluence of the Des Plaines and Kankakee rivers, NE Ill., and flowing SW to the Mississippi at Grafton, Ill. It is an important commercial and recreational waterway.

Illinois Waterway, 336 mi (540.6 km) long, linking Lake Michigan with the Mississippi River, N Ill.; an important part of the waterway connecting the Great Lakes with the Gulf of Mexico. It extends from the mouth of the Chicago River, on Lake Michigan, following the Chicago Sanitary and Ship Canal, the lower Des Plaines River, and the Illinois River to the Mississippi at Grafton.

Il·lyr·i·a (ĭ-lîr'ē-ə) and **Il·lyr·i·cum** (ĭ-lîr'ĭ-kəm), ancient region of the Balkan Peninsula. In prehistoric times a group of tribes speaking dialects of an Indo-European language swept down to the northern and eastern shores of the Adriatic Sea. Among the Illyrian peoples were the tribes later called the Dalmatians and the Pannonians. Illyria is sometimes taken to include the whole area occupied by the Pannonians, and thus to reach from Epirus north to the Danube River. More usually Illyria is used to mean only the Adriatic coast north of central Albania and west of the Dinaric Alps.

Greek cities were established on the coast in the 6th cent. B.C. but they did not flourish, and generally the Greeks left the Illyrians alone. Philip II of Macedon and later Philip V warred against them, but without permanent results. The Romans established (168–167 B.C.) one of the earliest Roman colonies as Illyricum. The colony was enlarged by the total conquest of Dalmatia in several wars. The southern Illyrians were finally conquered (35–34 B.C.) by Augustus. Illyricum was expanded by conquests (12–11 B.C.) of the Pannonians. In A.D. 6–9 the territory was split into the provinces of Dalmatia and Pannonia, but the term "Illyricum" was later given to a prefecture of the late Roman Empire. Illyricum then included much of the region north of the Adriatic as well as a large part of the Balkan Peninsula. When Napoleon revived (1809) the name for the Illyrian Provs. of his empire, he included much of the region north of the Adriatic and what is today eastern Yugoslavia. Roughly the same region was included in the administrative district of Austria called (1816–49) the Illyrian kingdom.

Il·men (ĭl'mən), shallow lake, varying in size from c.300 to c.800 sq mi (780–2,070 sq km), NW European USSR. It empties through the Volkhov River into Lake Ladoga.

I·lo·i·lo (ē'lō-ē'lō), city (1975 est. pop. 248,000), capital of Iloilo prov. (1970 pop. 1,168,454), SE Panay, the Philippines, on Iloilo Strait. With a fine harbor sheltered by Guimaras island, it is the principal port on Panay. Iloilo is known for its delicate, handwoven fabrics, made from silk and pineapple leaves.

I·lo·rin (ē-lôr′ĕn), city (1971 est. pop. 252,000), SW Nigeria. It is an industrial city and the market center for a wide agricultural region. Manufactures include cigarettes, matches, and sugar. Traditional artisans make woven goods, tin products, wood carvings, and pottery. Ilorin was the capital of a Yoruba kingdom that successfully rebelled against the Oyo empire in 1817 but soon thereafter was incorporated into the Fulani state of Sokoto.

I·ma·ba·ri (ē′mä-bä′rē), city (1976 est. pop. 121,500), Ehime prefecture, N Shikoku, Japan, on the Hiuchi Sea. It is a commercial and fishing port and a manufacturing center with industries producing cotton textiles and food products.

Im·e·ri·tia (ĭm′ə-rĭsh′ə), geographic and historic region, SE European USSR, in Georgia, in the upper Rion River basin. Imeritia is an agricultural region, noted for its mulberry trees and vineyards. Imeritia has been known since 1442. From 1510 it was often invaded by the Turks, to whom it was forced to pay tribute. It was an independent kingdom from the 16th to 18th cent. In 1804 Russia obtained an oath of allegiance from Imeritia, which continued to fight until its annexation to the Russian Empire in 1810.

I·mo·la (ē′mō-lä), city (1976 est. pop. 47,500), Emilia-Romagna, N central Italy, on the Aemilian Way. It is an agricultural and industrial center, known for its ceramics. A Roman town, it later (11th cent.) became a free commune. The city was subsequently ruled by tyrants until it passed to the papacy in the early 16th cent.

Im·pe·ri·a (ēm-pîr′ē-ä, -ē-yä), city (1976 est. pop. 41,837), capital of Imperia prov., Liguria, NW Italy, on the Ligurian Sea. Located on the Italian Riviera, it is a port, industrial center, and winter resort. Manufactures include iron and steel, refined olive oil, and chemicals.

Im·pe·ri·al (ĭm-pîr′ē-əl), county (1970 pop. 74,492), 4,241 sq mi (10,984.2 sq km), SE Calif., bounded on the S by the Mexican border, in the E by the Colorado River and Ariz. border; formed 1907; co. seat El Centro. It is in the Colorado Desert, with mountain ranges enclosing the Imperial Valley, and is irrigated by the All-American Canal and drainage channels carrying water to the Salton Sea. It is the center of a fertile agricultural region yielding lettuce, melons, vegetables, livestock, dairy products, alfalfa, flax, grain, and citrus fruit. It has some quarries and mines and manufactures agricultural chemicals, concrete, machinery, textiles, and clothing.

Imperial, city (1970 pop. 1,589), seat of Chase co., SW Nebr., SW of North Platte in a Great Plains region; settled c.1885.

Imperial Beach, residential and resort city (1978 est. pop. 20,000), San Diego co., S Calif., on the Mexican border; inc. 1956. The southwesternmost city in the continental United States, Imperial Beach has several naval bases and air stations.

Imperial Valley, fertile region in the Colorado Desert, SE Calif., extending S into NW Mexico. Once part of the Gulf of California, most of the region is below sea level. It has one of the longest growing seasons in the United States (more than 300 days) and can, with irrigation, support two crops a year. The valley is an important source of winter fruits and vegetables; cotton, dates, grains, and dairy products are also important. Several disastrous floods on the Colorado River in 1905–6 inundated the area; not until 1935, with the completion of Hoover Dam, was the valley safe from floods.

Imp·hal (ĭmp′hŭl), city (1971 pop. 100,605), capital of Manipur state, NE India, in the Manipur River valley. Industries include weaving and the manufacture of metalware. The inhabitants, of Tibeto-Burman origin, are famous for their music and dance.

Im·roz (ĭm-rôz′), island (1970 pop. 6,786), 108 sq mi (279.7 sq km), NW Turkey, in the Aegean Sea, near the entrance to the Dardanelles. Grain and beans are grown.

I·na·ri, Lake (ĭ′nä-rē), c.500 sq mi (1,295 sq km), N Finland. It empties into the Arctic Ocean. Lake Inari contains more than 3,000 islands and is a tourist attraction.

In·chon (ĭn′chŏn′), city (1975 pop. 799,982), NW South Korea, on the Yellow Sea. Inchon has an ice-free harbor. Iron, steel, coke, light metals, textiles, chemicals, and fertilizers are among its manufactures. Fishing is also an important industry. The city was opened to foreign trade in 1883. It was called Jinsen by the Japanese, who ruled Korea from 1904 to 1945. During the Korean War, U.S. troops landed at Inchon (Sept. 15, 1950) to launch the UN drive northward.

In·dals·älv·en (ĭn′däl-sĕl′vən), river, 261 mi (420 km) long, N central Sweden, rising in the mountains along the Norwegian border and flowing generally SE into the Gulf of Bothnia.

In·de·pend·ence (ĭn′dĭ-pĕn′dəns), county (1970 pop. 22,723), 755 sq mi (1,955.5 sq km), NE central Ark., bounded on the E by the Black River and drained by the White River, with part of the Ozarks in the W; founded 1820; co. seat Batesville. Its agriculture includes cotton, corn, truck crops, fruit, hay, and livestock. It has limestone quarries, manganese, bauxite, and black-marble deposits, and lumbering and food-processing industries. Feed, rubber products, shoes, and leather goods are manufactured.

Independence. 1. Town (1970 pop. 950), seat of Inyo co., E Calif., in the Owens Valley S of Bishop. In a stock-raising and farming area, it has gold, salt, lead, and mercury mines and winter resorts. Mt. Whitney, a large fish hatchery, and old Fort Independence are nearby. **2.** City (1970 pop. 5,910), seat of Buchanan co., NE Iowa, on the Wapsipinicon River E of Waterloo; founded 1847, inc. 1864. Poultry and feed are processed. **3.** City (1970 pop. 10,347), seat of Montgomery co., SE Kansas, on the Verdigris River, near the Okla. line, in an important oil-producing area where corn and wheat are also grown. The town was founded (1869) on a former Osage Indian reservation. It boomed with the discovery of natural gas in 1881 and oil in 1903. **4.** City (1970 pop. 1,784), seat of Kenton co., N Ky., S of Covington. **5.** City (1978 est. pop. 110,700), seat of Jackson co., W Mo., a suburb of Kansas City; inc. 1849. Its manufactures include electrical equipment, dehydrated foods, and oil products. Especially in the 1830s and 1840s it was the starting point for expeditions over the Santa Fe Trail, the Oregon Trail, and the California Trail. A group of Mormons settled here in 1831. It was the home of President Harry S. Truman and is the seat of the Harry S. Truman Library and Museum, on whose grounds the former President is buried. **6.** City (1970 pop. 7,034), Cuyahoga co., NE Ohio, a suburb S of Cleveland; inc. 1950. Chemicals and steel are among its manufactures. **7.** Town (1970 pop. 673), seat of Grayson co., SW Va., in the Blue Ridge WSW of Galax near the N.C. border. Hosiery is manufactured.

Independence Hall, historic building on Independence Square, downtown Philadelphia, Pa., in Independence National Historical Park. Originally constructed as the Pennsylvania colony's statehouse in 1732, the hall was the scene of the proclamation of the U.S. Declaration of Independence (1776) and was the meeting place of the Continental Congress and the Constitutional Convention. The Liberty Bell is on display nearby.

In·de·pend·ent City (ĭn′dĭ-pĕn′dənt), independent city (1975 est. pop. 7,000), SE Va., WSW of Portsmouth, in a farm region; inc. 1876 as Franklin, renamed 1961. Paper and wood products are made.

In·di·a (ĭn′dē-ə), republic (1976 pop. 605,614,000), 1,261,810 sq mi (3,268,090 sq km), S Asia. The second most populous country in the world, it has also been known as the Union of India and as Bharat, its ancient name. New Delhi is the capital. India's land frontier

(c.9,500 mi/15,285 km long) stretches from the Arabian Sea on the west to Burma on the east and touches, from west to east, Pakistan; a small portion of Afghanistan atop the state of Jammu and Kashmir; China; Nepal; Sikkim; Bhutan; and Bangladesh.

The southern half of India is a triangular-shaped peninsula (c.1,300 mi/2,090 km wide at its base) that thrusts into the Indian Ocean between the Bay of Bengal on the east and the Arabian Sea on the west and has a coastline c.3,500 mi (5,630 km) long. In the north, towering above peninsular India, is the Himalayan mountain wall, where rise the three great rivers of the Indian subcontinent—the Indus, the Ganges, and the Brahmaputra. The Indo-Gangetic alluvial plain, which has much of India's arable land, lies between the Himalayas and the dissected uplands of the Deccan, a plateau occupying most of peninsular India. The plain is limited in the west by the Thar Desert of Rajasthan, which merges with the swampy Rann of Kutch to the south. The Narbada River, south of the plain, marks the beginning of the Deccan. The plateau, scarped by the mountains of the Eastern Ghats and Western Ghats, is drained by the Godavari, Kistna, and Cauvery rivers. The much narrower western coast of peninsular India, comprising chiefly the Malabar Coast and the fertile Gujarat plain, curves around the Gulf of Cambay in the north to the Kathiawar peninsula.

India is divided into 21 states and 9 union territories:

NAME	CAPITAL	NAME	CAPITAL
Andaman and Nicobar Islands (territory)	Port Blair	Lakshadweep, or Laccadive, Minicoy, and Amindivi Islands (territory)	Kavaratti
Andhra Pradesh	Hyderabad	Madhya Pradesh	Bhopal
Arunachal Pradesh (territory)	Ziro	Maharashtra	Bombay
Assam	Shillong	Manipur	Imphal
Bihar	Patna	Meghalaya	Shillong
Chandigarh (territory)	Chandigarh	Mizoram (territory)	Aijal
Dadra and Nagar Haveli (territory)	Silvassa	Nagaland	Kohima
Delhi (territory)	Delhi	Orissa	Bhubaneswar
Goa, Daman, and Diu (territory)	Panjim	Pondicherry (territory)	Pondicherry
Gujarat	Ahmedabad	Punjab	Chandigarh
Haryana	Chandigarh	Rajasthan	Jaipur
Himachal Pradesh	Simla	Tamil Nadu	Madras
Jammu and Kashmir	Srinagar	Tripura	Agartala
Karnataka	Bangalore	Uttar Pradesh	Lucknow
Kerala	Trivandrum	West Bengal	Calcutta

Economy. Agriculture supports about 70% of the Indian people. Vast quantities of rice are grown wherever the land is level and water plentiful; other crops are wheat, pulses, jowar and bajra (cereals), and corn. Cotton, tobacco, and jute are the principal nonfood crops. There are large tea plantations. India has perhaps more cattle per capita than any country in the world, but backward stock-raising techniques and the Hindu stricture against the killing of cows detract greatly from their economic value. Sheep are raised by pastoral peoples in grazing areas. Coastal fisheries and, to a lesser extent, inland fisheries as well as pearling grounds are locally important. India has forested mountain slopes, with stands of oak, pine, sal, teak, ebony, palms, and bamboo, and the cutting of timber is a major rural occupation. Aside from mica, manganese, and ilmenite, in which the country ranks high, India's mineral resources, although large, are not as yet fully exploited. Oil fields exist in Assam and Gujarat, but India is deficient in petroleum. Industry in India, traditionally limited to agricultural processing and light manufacture, especially of cotton, woolen, and silk textiles, jute, and leather products, has been greatly expanded in recent years.

History. One of the earliest civilizations of the world, and the most ancient on the Indian subcontinent, was the Indus Valley civilization, which flourished c.2500 B.C. to c.1500 B.C. It was an extensive and highly sophisticated culture, its chief urban centers being Mohenjo-Daro and Harappa. The Aryans invaded through the mountain passes of the northwest, possibly destroyed (c.1500 B.C.) the Indus civilization, and established their homeland in the plains of the Punjab and along the upper Ganges. Over the next 2,000 years the Aryans developed a Brahmanic civilization with a caste system, which evolved into Hinduism. The first important Aryan kingdom was Magadha, with its capital near present-day Patna; it was there, c.540-490 B.C., that the founders of Jainism and Buddhism preached. In 327-325 B.C. Alexander the Great invaded northwest India. The Greek invaders were eventually driven out by Chandragupta of Magadha, founder of the Mauryan empire. The Mauryan emperor Asoka (d. 232 B.C.), perhaps the greatest ruler of the ancient period, unified all of India except the southern tip. Under Asoka, Buddhism was established as the state religion and was also spread to southeast Asia.

During the 200 years of disorder and invasions that followed the collapse of the Mauryan state (c.185 B.C.), Buddhism in India declined. Hindu culture was spread through the Malay Archipelago and Indonesia by Tamil-speaking colonists from the southern Indian kingdoms. Meanwhile, Greeks following Alexander had settled in Bactria (in the area of present-day Afghanistan) and established an Indo-Greek kingdom. After the collapse (1st cent. B.C.) of Bactrian power, fierce tribes from central Asia, the Scythians, Parthians, Afghans, and Kushans, swept into northwest India. There, small states arose and disappeared in quick succession.

In the 4th and 5th cent. A.D., northern India experienced a golden age under the Gupta dynasty, when Hindu art and literature reached a high level. While the Guptas ruled the north, the Pallava kings of Kanchi held sway in the south, and the Chalukyas controlled the Deccan. During the medieval period (8th-13th cent.) several independent kingdoms waxed powerful. In northwest India, beyond the reach of the medieval dynasties, the Rajputs had grown strong and were able to resist the rising forces of Islam.

Islam was first brought to Sind in the 8th cent. by seafaring Arab traders. From 999 to 1026, Mahmud of Ghazni several times breached Rajput defenses and plundered India. In the 11th and 12th cent. Ghaznavid power waned, to be replaced c.1150 by that of the Moslem principality of Ghor. In 1192 the Delhi sultanate was established. The sultanate eventually reduced to vassalage almost every independent kingdom on the subcontinent, except that of Kashmir and the remote kingdoms of the south. Ruling such a vast territory proved impossible; difficulties in the south with the state of Vijayanagar, the last Hindu kingdom in India, and the capture (1398) of the city of Delhi by Tamerlane finally brought the sultanate to an end. The small kingdoms that succeeded it were swept away by a Moslem invader from Afghanistan, Babur, a remote descendant of Tamerlane, who, after the Battle of Panipat in 1526, founded the Mogul empire. The empire was consolidated by Akbar and reached its greatest territorial extent under Aurangzeb (ruled 1659-1707). Islam, however, never supplanted Hinduism as the faith of the majority.

Only a few years before Babur's triumph, Vasco da Gama had landed at Calicut (1498) and the Portuguese had conquered Goa (1510). The splendor and wealth of the Mogul empire attracted British, Dutch, and French competition for the trade that Portugal had at first monopolized. The British East India Company soon became dominant and with its command of the sea drove off the traders of Portugal and Holland. While the Mogul empire remained strong, only peaceful trade relations with it were sought; but in the 18th cent., when an Afghan invasion, dynastic struggles, and incessant revolts of Hindu elements, especially the Mahrattas, were rending the empire, Great Britain and France seized the opportunity to increase trade and capture Indian wealth. From 1746 to 1763 India was a battleground for the forces of the two powers. The victories of Robert Clive, the British commander, ended all French threats to the power of the British company.

Clive's defeat of the Nawab of Bengal at Plassey in 1757 traditionally marks the beginning of the British Empire in India (recognized in the Treaty of Paris of 1763). Warren Hastings, the first governor general of the company's domains to be appointed by Parliament, did much to consolidate Clive's conquests. Only Sind and Punjab remained completely independent. The East India Company administered the rich areas with the populous cities; the rest of British India remained under native princes, with British residents in effective control. British control was extended over Sind in 1843 and Punjab in 1849.

Social unrest, plus the aggrandizing policies of Gov. Gen. Dalhousie, led to the bloody Indian Mutiny of 1857. After the mutiny was suppressed, Great Britain initiated long-needed reforms. The East India Company lost most of its functions, and its governor general became the crown's viceroy. Native rulers were guaranteed the integrity of their domains as long as they recognized the British as paramount. In 1861 the first step was taken toward self-government in British India with the appointment of Indian councilors to advise the viceroy and the establishment of provincial councils with Indian members.

In the early 1900s the British widened Indian participation in legislative councils. Separate Moslem and Hindu constituencies, introduced for the first time, were to be a major factor in the growing split between the two communities. At the outbreak of World War I all elements in India were firmly united behind Britain, but Britain's prestige in India was shattered by the long, costly war. Crop failures and an influenza epidemic that killed millions plagued India in 1918–19. Britain passed the Rowlatt Acts (1919), which enabled authorities to dispense with juries, and even trials, in dealing with agitators. In response Mohandas K. Gandhi organized the first of his many passive resistance campaigns. The massacre of Indians by British troops at Amritsar further inflamed the situation. The Government of India Act (late 1919) set up provincial legislatures in which elected Indian ministers, responsible to the legislatures, had to share power with appointed British governors and ministers. Although the act also provided for periodic revisions, Gandhi felt too little progress had been made, and he organized new protests.

The Government of India Act of 1935 provided for the election of entirely Indian provincial governments and a federal legislature in Delhi. In the first elections (1937) held under the act, the Congress, led by Gandhi and Jawaharlal Nehru, won well over half the seats, mostly in Hindu or general constituencies. The Moslem League won 109 of the 485 Moslem seats. Fearing Hindu domination in a future independent India, Moslem nationalism in India henceforth became a totally separate movement.

In World War II, to secure greater support from India, Britain proposed establishing an Indian interim government, in which Great Britain would maintain control only over defense and foreign policy, to be followed by full self-government after the war. The Congress adamantly demanded that the British leave India and, when the demand was refused, initiated civil disobedience and the Quit India movement. Great Britain's response was to outlaw the Congress and jail Gandhi and other leaders. Mohammed Ali Jinnah, leader of the Moslem League, who offered to support the war, demanded the partition of India into Moslem- and Hindu-majority sections.

The British Labour government in 1946 offered self-government to India, but warned that if no agreement was reached between the Congress and the Moslem League, Great Britain, on withdrawing in June, 1948, would have to determine the apportionment of power between the two groups. Reluctantly the Congress agreed to the creation of Pakistan. The future of Kashmir was never resolved. Nehru became prime minister of India, and Jinnah governor general of Pakistan. Partition left large minorities of Hindus and Sikhs in Pakistan, and Moslems in India. Widespread hostilities erupted in late 1947, in which more than 500,000 people died. Gandhi was killed by a Hindu fanatic in Jan., 1948. The hostility between India and Pakistan was aggravated when warfare broke out (1948) over their conflicting claims to jurisdiction over the rich princely state of Jammu and Kashmir. India became a sovereign republic under a constitution adopted in 1949. The states of the republic were reorganized several times along linguistic lines. India consolidated its territory by acquiring the former French settlements in 1956 and by forcibly annexing the Portuguese enclaves of Goa, Daman, and Diu in Dec., 1961.

A border dispute with China climaxed on Oct. 20, 1962, when the Chinese launched a massive offensive against Ladakh in Kashmir and in areas on the northeastern border. The Chinese announced a cease-fire on Nov. 21 after gaining much territory claimed by India. In Aug., 1965, fighting between India and Pakistan broke out in the Rann of Kutch frontier area and in Kashmir. The United Nations proclaimed a cease-fire in Sept., but clashes continued. India's Prime Minister Lal Bahadur Shastri, who succeeded Nehru after the latter's death in 1964, and Pakistan's President Ayub Khan met (1966) under Soviet auspices in Tashkent, USSR, to negotiate the Kashmir problem. Shastri died in Tashkent and was succeeded, after bitter debate within the Congress party, by Indira Gandhi, Nehru's daughter. The Congress party suffered a setback in the elections of 1967. In 1969 the party split in two: Mrs. Gandhi and her followers formed the New Congress party, and her opponents formed the Old Congress party. In the elections of Mar., 1971, the New Congress party won an overwhelming victory. Rioting and terrorism by Maoists, known as Naxalites, flared in 1970 and 1971.

In Pakistan, attempts by the government (dominated by West Pakistanis) to suppress a Bengali uprising in East Pakistan led in 1971 to the exodus of millions of Bengali refugees (mostly Hindus) from East Pakistan into India. Caring for the refugees imposed a severe drain on India's slender resources. In Dec., 1971, war broke out between India and Pakistan in East Pakistan and in Kashmir. The war ended in two weeks with the creation of independent Bangladesh to replace East Pakistan.

The greatest internal crisis of the postindependence period began in June, 1975, when, after being convicted of illegal campaign practices during the 1971 election, Prime Minister Gandhi imposed a state of emergency, imprisoning 30,000 political opponents and imposing censorship on the press and direct presidential rule on the states. In 1977, in an apparent attempt to seek ratification of the emergency from the electorate, Gandhi called an election. Contrary to expectations, the Congress party was overwhelmingly defeated by a new coalition group, the Janata. The emergency was ended and the Janata leader, Morarji Desai, became prime minister.

Government. India is a federal state with a parliamentary form of government. It is governed under the 1949 constitution (effective since Jan., 1950). The president of India is elected for a five-year term by the elected members of the federal and state parliaments. Executive power is exercised by the prime minister (head of the majority party in the federal parliament) and council of ministers (which includes the cabinet), who are appointed by the president. The federal parliament is bicameral. States have either unicameral or bicameral parliaments and have jurisdiction over police and public order, agriculture, education, public health, and local government. The federal government has jurisdiction over any matter not specifically reserved to the states. In addition the president may intervene in state affairs during emergencies.

In·di·a·na (ĭn'dē-ăn'ə), state (1975 pop. 5,356,000), 36,291 sq mi (93,994 sq km), N central United States, in the Midwest, admitted as the 19th state of the Union in 1816. The capital and largest city is Indianapolis. Indiana is bounded on the north by Mich. and Lake Michigan, on the east by Ohio, on the south by Ky., from which it is separated by the Ohio River, and on the west by Ill. Northern Indiana is a glaciated lake area, separated by the Wabash River from the central agricultural plain, which is rich with deep glacial drift. The southern portion of the state is a succession of bottomlands interspersed with knolls and ridges, gorges and valleys. The unglaciated soil is shallow in southern Indiana, and the cutting of timber has caused erosion, but there is some farming.

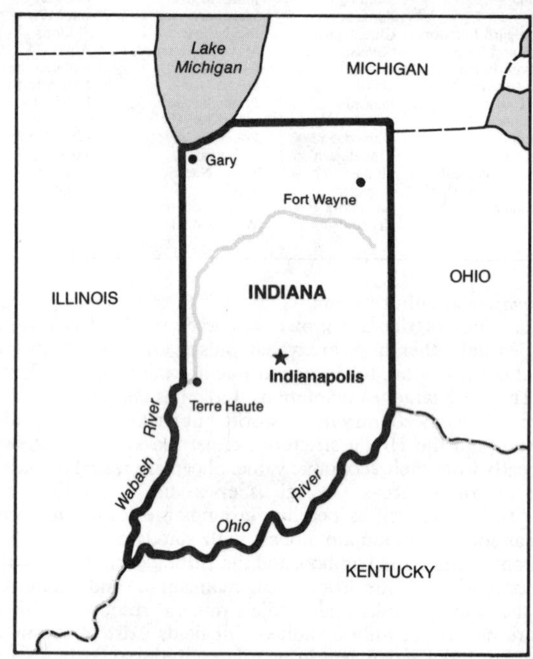

Indiana is divided into 92 counties:

NAME	COUNTY SEAT	NAME	COUNTY SEAT
Adams	Decatur	Daviess	Washington
Allen	Fort Wayne	Dearborn	Lawrenceburg
Bartholomew	Columbus	Decatur	Greensburg
Benton	Fowler	De Kalb	Auburn
Blackford	Hartford City	Delaware	Muncie
Boone	Lebanon	Dubois	Jasper
Brown	Nashville	Elkhart	Goshen
Carroll	Delphi	Fayette	Connersville
Cass	Logansport	Floyd	New Albany
Clark	Jeffersonville	Fountain	Covington
Clay	Brazil	Franklin	Brookville
Clinton	Frankfort	Fulton	Rochester
Crawford	English	Gibson	Princeton

NAME	COUNTY SEAT	NAME	COUNTY SEAT
Grant	Marion	Owen	Spencer
Greene	Bloomfield	Parke	Rockville
Hamilton	Noblesville	Perry	Cannelton
Hancock	Greenfield	Pike	Petersburg
Harrison	Corydon	Porter	Valparaiso
Hendricks	Danville	Posey	Mount Vernon
Henry	New Castle	Pulaski	Winamac
Howard	Kokomo	Putnam	Greencastle
Huntington	Huntington	Randolph	Winchester
Jackson	Brownstown	Ripley	Versailles
Jasper	Rensselaer	Rush	Rushville
Jay	Portland	St. Joseph	South Bend
Jefferson	Madison	Scott	Scottsburg
Jennings	Vernon	Shelby	Shelbyville
Johnson	Franklin	Spencer	Rockport
Knox	Vincennes	Starke	Knox
Kosciusko	Warsaw	Steuben	Angola
Lagrange	Lagrange	Sullivan	Sullivan
Lake	Crown Point	Switzerland	Vevay
La Porte	La Porte	Tippecanoe	Lafayette
Lawrence	Bedford	Tipton	Tipton
Madison	Anderson	Union	Liberty
Marion	Indianapolis	Vanderburgh	Evansville
Marshall	Plymouth	Vermillion	Newport
Martin	Shoals	Vigo	Terre Haute
Miami	Peru	Wabash	Wabash
Monroe	Bloomington	Warren	Williamsport
Montgomery	Crawfordsville	Warrick	Boonville
Morgan	Martinsville	Washington	Salem
Newton	Kentland	Wayne	Richmond
Noble	Albion	Wells	Bluffton
Ohio	Rising Sun	White	Monticello
Orange	Paoli	Whitley	Columbia City

Economy. About three quarters of Indiana is utilized for agriculture. Grain crops, mainly corn and wheat, are important and also support the livestock and dairying industries. Soybeans and hay are also principal crops, and vegetables and fruits are produced in great quantity and variety. Meat-packing is chief among the many industries related to agriculture. Indiana's leading manufactures are iron and steel, electrical equipment, transportation equipment, nonelectrical machinery, chemicals, food products, and fabricated metals. Rich mineral deposits of coal and stone have encouraged construction and industry; petroleum production is also substantial.

History. The region was first explored, notably by the French, in the late 17th cent. At the time of exploration the area was occupied mainly by tribes of Miami, Delaware, and Potawatamie Indians. By the Treaty of Paris of 1763 ending the French and Indian Wars (1689–1763), Indiana passed from French to British control and, along with the rest of the Old Northwest, was united with Canada under the Quebec Act of 1774. By the Treaty of Paris of 1783 ending the Revolutionary War, Great Britain ceded the Old Northwest to the United States. Indiana was still largely unsettled when the Northwest Territory, of which it formed a part, was established in 1787. Indians in the territory resisted white settlement, but they were defeated at Fallen Timbers in 1794 and at Tippecanoe in 1811.

In 1800, Indiana Territory was formed and included the present-day states of Ind., Ill., and Wis., and parts of Mich. and Minn. A constitutional convention met in 1816, and Indiana achieved statehood. Indianapolis was laid out as the state capital. In the 1840s the Wabash and Erie Canal opened between Lafayette and Toledo, Ohio, giving Indiana a water route via Lake Erie to eastern markets. Manufacturing developed rapidly after the Civil War. Factories sprang up, and the old rustic pattern was broken. Industrial development came to the Calumet region in the late 19th cent. with the establishment of an oil refinery at Whiting. The labor movement at Gary figured prominently in the nationwide steel strike just after World War I.

Government. Indiana's constitution dates from 1851 and provides for an elected executive and legislature. A governor serves as the chief executive for a term of four years and may not succeed himself. The legislature, called the general assembly, has a senate with 50 members elected for four years and a house of representatives with 100 members elected for two years.

Educational Institutions. Among the institutions of higher learning in Indiana are Indiana Univ., at Bloomington; Purdue Univ., at Lafayette; the Univ. of Notre Dame, near South Bend; DePauw Univ., at Greencastle; Butler Univ., at Indianapolis; Valparaiso Univ., at Valparaiso; Earlham College, at Richmond; and Goshen College, at Goshen.

Indiana, county (1970 pop. 79,451), 825 sq mi (2,136.8 sq km), W central Pa., bounded on the S by the Conemaugh River and drained by tributaries of the Allegheny River; formed 1803; co. seat Indiana. It has agriculture (grain, potatoes, clover, poultry, livestock, and dairy products), coal mines, oil and natural-gas wells, limestone deposits, and a food-processing industry. Metal, clay, and glass products and tires are manufactured.

Indiana, industrial borough (1978 est. pop. 17,300), seat of Indiana co., W Pa.; inc. 1816. It is the principal supply and trading center for a bituminous-coal area in the Alleghenies and has factories producing rubber goods, scientific instruments, and aluminum products.

Indiana Dunes National Lake·shore (do͞onz; lāk'shôr'): *see* National Parks and Monuments Table.

In·di·an·ap·o·lis (ĭn'dē-ə-năp'ə-lĭs), city (1978 est. pop. 711,000), state capital and seat of Marion co., central Ind., on the White River; selected 1820 as the site of the state capital (which was moved here in 1825), inc. 1847. It is the chief processing point in a rich agricultural region and is a major grain and livestock market. Its many manufactures include chemicals, pharmaceuticals, aircraft and automotive parts, telephone and electronic equipment, and road-building machinery. The city is the seat of many educational institutions, including Butler Univ. The American Legion has its national headquarters here in a building erected as a war memorial. Landmarks are the state capitol (1878–88), the home and burial place of the poet James Whitcomb Riley, the home of President Benjamin Harrison, the Soldiers and Sailors Monument (1902), and the Indianapolis Motor Speedway, site of the famous annual 500-mi (804.5-km) automobile race. Fort Benjamin Harrison is nearby.

In·di·an Ocean (ĭn'dē-ən), third-largest ocean, c.28,350,000 sq mi (73,426,500 sq km), extending from S Asia to Antarctica and from E Africa to SE Australia; it is c.4,000 mi (6,440 km) wide at the equator. The Indian Ocean is connected with the Pacific Ocean by passages through the Malay Archipelago and between Australia and Antarctica; and with the Atlantic Ocean by the expanse between Africa and Antarctica and by the Suez Canal. Its chief arms are the Arabian Sea, the Bay of Bengal, and the Andaman Sea.

The floor of the Indian Ocean has an average depth of c.11,000 ft (3,350 m). The Mid-Oceanic Ridge, a broad submarine mountain range extending from Asia to Antarctica, divides the Indian Ocean into eastern and western sections. The ridge rises to an average height of c.10,000 ft (3,050 m), and a few peaks emerge as islands. A large rift, an extension of the eastern branch of the Great Rift Valley that runs through the Gulf of Aden, extends along most of its length. The Mid-Oceanic Ridge, along with other submarine ridges, encloses a series of deep-sea basins (abyssal plains). The greatest depth (25,344 ft/7,730 m) is in the Java Trench, south of Java, Indonesia.

The surface waters of the ocean are generally warm, although close to Antarctica pack ice and icebergs are found. The Indian Ocean has two water circulation systems—a regular counterclockwise southern system (South Equatorial Current, Mozambique Current, South Pacific Drift, West Australian Current) and a northern system whose currents are directly related to the seasonal shift of the monsoon winds.

In·di·a·no·la (ĭn'dē-ə-nō'lə). **1.** City (1975 est pop. 9,611), seat of Warren co., S central Iowa, S of Des Moines; inc. 1863. Simpson College is here. **2.** City (1970 pop. 8,947), seat of Sunflower co., W Miss., E of Greenville near the Sunflower River, in a cotton area; settled in the mid-19th cent., inc. 1886.

Indian River, county (1970 pop. 35,992), 506 sq mi (1,310.5 sq km), central Fla., on the Atlantic coast bounded on the E by barrier beaches enclosing Indian River lagoon; formed 1925; co. seat Vero Beach. In a coastal-lowland area noted for citrus fruit, especially oranges, it also produces sugar cane and truck crops.

In·di·o (ĭn'dē-ō'), city (1978 est. pop. 18,700), Riverside co., SE Calif., in the Coachella Valley of the Colorado Desert, 22 ft (6.7 m) below sea level; founded 1876, inc. 1930. It is the trade and administrative center for a citrus, grape, and date area.

In·do·chi·na (ĭn'dō-chī'nə), former federation of states, SE Asia. It comprised the French colony of Cochin China and the French protectorates of Tonkin, Annam, Laos, and Cambodia. The federation formed the easternmost region of the Indochinese peninsula and faced east on the South China Sea.

The cultures of Indochina were influenced by China and India. The centuries before European intervention saw the growth and decline of the Khmer Empire in Cambodia, the rise and fall of Champa, and the steady expansion of Annam. European penetration began in the 16th cent.; in the 19th cent. race for a colonial empire, the French took (1862, 1867) Cochin China as a colony and gained protectorates over Cambodia (1863), Annam (1884), and Tonkin (1884). In 1887 they formed those four states into a union of Indochina. Laos was added to the union in 1893.

In World War II, France was forced to accept Japanese intervention in northern Indochina in 1940. Even before the end of the war, the French announced plans for a federation of Indochina within the French Union. The federation was accepted in Cambodia and Laos. Vietnamese nationalists, however, demanded (1945) the complete independence of Annam, Tonkin, and Cochin China as Vietnam, and after Dec., 1946, these regions were plunged into bitter fighting between the French and the extreme nationalists, oftentimes led by Communists. The war in Vietnam dragged on for years, culminating in the French defeat at Dienbienphu. The Geneva Conference in 1954 effectively ended French control of Indochina.

In·do·ne·sia (ĭn'də-nē'zhə, -shə), republic (1975 est. pop. 131,255,000), c.735,000 sq mi (1,903,650 sq km), SE Asia, in the Malay Archipelago. The capital is Djakarta, on Java. Indonesia comprises more than 3,000 islands extending c.3,000 mi (4,830 km) along the equator from the Malaysia mainland toward Australia; the archipelago forms a natural barrier between the Indian and Pacific oceans. The most important islands, culturally and economically, are Java, Bali, and

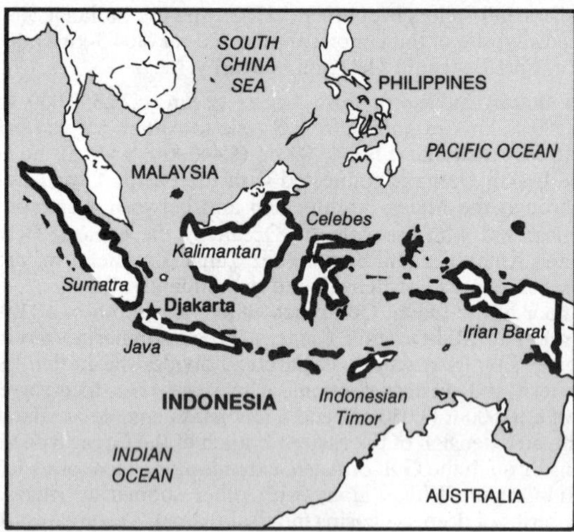

Sumatra. All the larger islands have a central volcanic mountainous area flanked by coastal plains; there are more than 100 active volcanoes. Earthquakes are frequent, although seldom severe. The animal life of Indonesia roughly forms a connecting link between the fauna of Asia and that of Australia. Elephants are found in Sumatra and Borneo, tigers as far south as Java and Bali, and marsupials in Timor and Irian Barat. Crocodiles, snakes, and richly colored birds are everywhere.

Economy. The tropical climate, abundant rainfall, and remarkably fertile volcanic soils permit a rich agricultural yield. Rubber is the most valuable crop; other plantation crops include sugar cane, coffee, tea, tobacco, palm oil, cinchona, cacao, sisal, coconuts, and spices. Rice is the major food crop; cassava, maize, yams, soybeans, peanuts, and fruit are also grown. Fish are abundant, both in the ocean and in inland ponds. Indonesia has great timberlands; vast rain forests of giant trees cover the mountain slopes, and teak, sandalwood, ironwood, camphor, and ebony are cut. Palms and bamboos abound, and a great variety of forest products are manufactured. Petroleum is by far the most important mineral. Indonesia is thought to have about one sixth of all world deposits of tin. Bauxite, nickel, coal, manganese, salt, gold, and silver are also mined. Iron and copper are believed to exist in great quantity, and uranium has been reported. Industry is limited to food, mineral, and wood processing and a variety of light manufactures.

History. Early in the Christian era, Indonesia came under the influence of Indian civilization through the gradual influx of Indian traders and Buddhist and Hindu monks. By the 7th and 8th cent., kingdoms closely connected with India had developed in Sumatra and Java; the spectacular Buddhist temples of Borobudur date from this period. Sumatra was the seat (7th–13th cent.) of the important Buddhist kingdom of Sri Vijaya. In the late 13th cent. the center of power shifted to Java, where the fabulous Hindu kingdom of Majapahit had arisen; for two centuries it held sway over Indonesia and large areas of the Malay Peninsula. Arab traders arrived in the 14th and 15th cent., and by the end of the 16th cent. Islam had replaced Buddhism and Hinduism as the dominant religion.

Early in the 16th cent. the Portuguese, in pursuit of the rich spice trade, began establishing trading posts in Indonesia, after taking (1511) the strategic commercial center of Malacca on the Malay Peninsula. The Dutch followed in 1596 and the English in 1600. By 1610 the Dutch had ousted the Portuguese, who were allowed to retain only the eastern part of Timor, but the English competition remained strong, and it was only after a series of Anglo-Dutch conflicts (1610–23) that the Dutch emerged as the dominant power in Indonesia. Throughout the 17th, 18th, and 19th cent. the Dutch East India Company steadily expanded its control over the entire area. When the company was liquidated in 1799, the Dutch government assumed its holdings, which were thereafter known in English as the Netherlands (or Dutch) East Indies. During the Napoleonic Wars (1811–14) the islands were occupied by the British.

In 1825 Prince Diponegoro of Java launched a long and bloody guerrilla war against the colonists, and in 1906 and again in 1908 the native rulers of Bali led their subjects in suicidal charges against Dutch fortifications. In 1927 the Indonesian Nationalist party arose under the leadership of Sukarno. It received its impetus during World War II, when the Japanese drove out (1942) the Dutch and occupied the islands. In Aug., 1945, immediately after the Japanese surrender, Sukarno and Mohammed Hatta, another nationalist leader, proclaimed Indonesia an independent republic. The Dutch bitterly resisted the nationalists, and four years of intermittent and sometimes heavy fighting followed. Under UN pressure, an agreement was finally reached (Nov., 1949) for the creation of an independent republic of Indonesia. A new constitution provided for a parliamentary form of government. Sukarno was elected president, and Hatta became premier.

The rapid expropriation of Dutch property and the ousting of Dutch citizens (late 1950s) severely dislocated the economy, and soaring inflation and great economic hardship ensued. A widespread native revolt led to increasingly authoritarian rule by Sukarno, who dissolved (1960) the parliament and reinstated the constitution of 1945, which had provided for a strong, independent executive (Hatta had resigned in 1956 following a conflict with Sukarno). The army and the Communist party constituted two important power blocs in Indonesian politics, with Sukarno holding the balance of power between the two. In early 1962 Sukarno dispatched paratroopers to Netherlands New Guinea—territory claimed by Indonesia but firmly held by the Dutch. Netherlands New Guinea was formally annexed by Indonesia in Aug., 1969, and its name was changed to Irian Barat.

Meanwhile, Sukarno made (1963) a major propaganda issue of the newly created Federation of Malaysia and staged guerrilla raids into Malaysian territory on Borneo, beginning a conflict that was waged intermittently for three years. Sukarno began to openly summon Communist leaders for advice, exhibiting hostility toward the United States and cultivating the friendship of Communist China. In 1965 he withdrew Indonesia from the United Nations. The abortive Communist coup against the army that began in Sept., 1965, was swiftly thwarted by army forces under Gen. Suharto, who gradually assumed power (although retaining Sukarno as symbolic leader). The new government steadily increased its power, aided by massive student demonstrations against Sukarno. Gen. Suharto brought an end (1966) to hostilities against Malaysia, re-established close ties with the United States, and re-entered (1966) the United Nations. In 1967 Sukarno was removed from power, and in 1968 Suharto was elected president. In 1975–76 Indonesia annexed Portuguese Timor and incorporated it as a province of the country.

In·dore (ĭn-dôr'), city (1971 pop. 572,622), W central India, on the Saraswati and Khan rivers. It was the capital of the maharajahs of Indore and is the site of their imposing palace. Indore is now a major commercial and industrial center, with chemical, textile, and iron and steel industries.

In·dra·va·ti (ĭn-drä'və-tē), river, 315 mi (506.8 km) long, S India, rising in SW Orissa and flowing W and SW through Madhya Pradesh and into the Godavari River at the Andhra Pradesh boundary.

In·dre (ăɴ'drə), department (1975 pop. 248,523), 2,617 sq mi (6,778 sq km), central France, in parts of Berry, Orléanais, Marche, Touraine, and Poitou; capital Châteauroux.

Indre, river, 165 mi (265.5 km) long, Indre and Indre-et-Loire depts., central France. It flows northwest from the Massif Central foothills to the Loire River below Tours.

In·dre-et-Loire (ăɴ′drā-lwär′), department (1975 pop. 478,601), 2,364 sq mi (6,122.8 sq km), N central France, occupying most of Touraine; capital Tours.

In·dus (ĭn′dəs), chief river of Pakistan, c.1,900 mi (3,060 km) long, rising in the Kailas range in the Tibet region of China and flowing W across Jammu and Kashmir, India, then SW through Pakistan to the Arabian Sea SE of Karachi. The upper Indus, fed by snow and glacial meltwater from the Karakorum, Hindu Kush, and Himalayan mts., flows through deep gorges and scenic valleys; its turbulence makes it unsuitable for navigation. Flowing onto the dry Punjab plains, the Indus becomes a broad, meandering, silt-laden stream and receives the combined waters of the five rivers of the Punjab, its chief affluent. The Indus delta, unlike the deltas of many other rivers, is composed of clay and is infertile. The Indus Valley is Pakistan's most densely populated region and its main agricultural area; wheat, corn, rice, millet, dates, and fruits are the chief crops. In Pakistan the Indus is extensively used for irrigation and hydroelectric power. The use of the Indus and its tributaries has long been a source of conflict between Pakistan and India, although a treaty by which the waters were to be shared was signed in 1960. The river valley was the site of the prehistoric Indus Valley civilization. The river was once considered to be the western boundary of India.

I·nez (ī′nĕz), uninc. city (1970 pop. 470), seat of Martin co., E Ky., in the Cumberland foothills, in a coal, oil, and agricultural area.

In·ger·man·land (ĭng′gər-mən-lănd′), or **In·gri·a** (ĭng′grē-ə), historic region, NW European USSR, along the Neva River on the E bank of the Gulf of Finland. Its name derives from the ancient Finnic inhabitants, the Ingers, some of whose descendants (about 93,000) still live in the area. In medieval times, the region was subject to Great Novgorod, with which it passed in 1478 to the grand duchy of Moscow. Conquered in the early 17th cent. by Sweden, it remained Swedish until Peter I of Russia captured it in 1702. The area was formally ceded to Russia by the Treaty of Nystad (1721), which ended the Northern War between Russia and Sweden.

Ing·ham (ĭng′əm), county (1970 pop. 261,039), 559 sq mi (1,447.8 sq km), S Mich., drained by the Grand and Red Cedar rivers and Sycamore Creek; formed 1829; co. seat Mason. Its agriculture includes livestock, poultry, fruit, grain, sugar beets, corn, hay, beans, truck crops, and dairy products. It has clay and coal deposits and oil and gas wells. Its manufactures include processed foods, lumber and wood products, concrete, and metal products.

In·gle·wood (ĭng′gəl-wo͝od′), city (1978 est. pop. 89,000), Los Angeles co., S Calif., a residential and industrial suburb of Los Angeles, in an oil-producing area; founded 1873, inc. 1908. Its manufactures include machinery, aircraft parts, and electronic equipment.

In·go·da (ĭng′gə-də), river, 360 mi (579.2 km) long, S central Chita oblast, Russian SFSR, USSR, rising near the Mongolian border and flowing NE to join the Onon River, with which it forms the Shilka River.

In·gol·stadt (ĭng′gəl-stăt′, -gôl-shtät′), city (1974 est. pop. 90,357), Bavaria, S West Germany, on the Danube River. It is a commercial and industrial center. Manufactures include engines, machinery, refined oil, textiles, and motor vehicles. Chartered about 1250, Ingolstadt was besieged (1632) by Gustavus II of Sweden during the Thirty Years' War.

In·gul (ĭn-go͞ol′), river, SW European USSR, in the Ukraine, rising N of Kirovograd and flowing S c.210 mi (340 km) to empty into the Bug estuary, an inlet of the Black Sea, at Nikolayev.

In·gu·lets (ĭn-go͞o-lyĕts′), river, c.340 mi (545 km) long, SW European USSR, in the Ukraine. Rising in the Kirovograd region, it flows through the Krivoy Rog iron district and then south to join the Dnepr River above Kherson.

I-ning (ē′nĭng′) or **Kul·dja** (ko͞ol′jä′), city (1970 est. pop. 160,000), W Sinkiang Uigur Autonomous Region, China, on the Ili River in the Dzungarian basin. An old commercial center trading in tea and cattle, it is also an industrial city with manufactures of cotton and wool textiles and carpets. It has fruit orchards, and iron and coal are mined nearby. I-ning was seized by the Russians in 1871 but was restored to China in 1881.

Ink·ster (ĭngk′stər), city (1978 est. pop. 36,300), Wayne co., SE Mich., a suburb of Dearborn, on the Rouge River; settled 1825 as Moulin Rouge, renamed 1863, inc. as a city 1964. The city's residents are primarily employed by nearby automotive plants.

In·land Sea (ĭn′lənd), arm of the Pacific Ocean, c.3,670 sq mi (9,505 sq km), S Japan, between Honshu, Shikoku, and Kyushu islands. It is linked to the Sea of Japan by a narrow channel. The shallow sea is dotted with more than 950 islands. The shores of the Inland Sea are heavily populated and are part of Japan's most important industrial belt. The Inland Sea is famed for its scenic beauty.

Inn (ĭn), river, c.320 mi (515 km) long, rising near the Lake of Sils, SE Switzerland, and flowing NE through the Engadine valley, then through W Austria, past Innsbruck and Solbad Hall, and into SE West Germany. The Inn forms part of the West German-Austrian border before entering the Danube River at Passau.

In·ner Mon·go·li·an Autonomous Region (ĭn′ər mŏng-gō′lē-ən, -gōl′yən), autonomous region (1967 est. pop. 13,000,000), c.164,000 sq mi (424,760 sq km), NE China, bounded on the N by the Mongolian People's Republic. The capital is Hu-ho-hao-t'e. Inner Mongolia is largely steppe country that becomes increasingly arid toward the Gobi Desert in the west. Stock raising, mainly of sheep, goats, horses, and camels, is a major occupation; wool, hides, and skins are important exports. Principal crops are wheat, kaoliang, millet, oats, corn, linseed, soybeans, sugar beets, and rice. There are valuable mineral deposits (coal, lignite, iron ore, lead, zinc, and gold), as yet only partially exploited.

Originally the southern part of Mongolia, Inner Mongolia was settled chiefly by the Tumet and Chahar tribes. After 1635 the Manchus annexed Inner Mongolia. Under Manchu rule southern Mongolia became known as Inner Mongolia; northern Mongolia, conquered by the Manchus at the end of the 17th cent., became known as Outer Mongolia. After the Revolution of 1911 Inner Mongolia became an integral part of the Chinese Republic. In 1928 it was divided among the Chinese provinces of Ninghsia, Suiyuan, and Chahar. After the outbreak (1937) of the Sino-Japanese War, the Mongols of Suiyuan and Chahar established the Japanese-controlled state of Mengkiang or Meng-chiang, with its capital at Kweihwa. The Chinese Communists, after their conquest of Inner Mongolia in 1945, supported the traditional aspirations of the Mongols for autonomy, and in May, 1947, the Inner Mongolian Autonomous Region—with limited powers of self-government within the Communist state—was formally proclaimed. Extensive boundary changes since 1970 have considerably reduced the size of the province.

In·no·shi·ma (ēn-nō′shĭ-mä), city (1975 pop. 41,683), Hiroshima prefecture, on Innoshima Island, Japan, on the Hiuchi Sea. It is a fishing and commercial port and an agricultural center.

Inns·bruck (ĭnz′bro͝ok′), city (1971 pop. 115,200), capital of Tyrol prov., SW Austria, on the Inn River. A famous summer and winter resort, it is also an industrial, commercial, and transport center. Manufactures include textiles, metal products, processed food, and printed materials. Strategically located in the Eastern Alps, Innsbruck grew to early prominence as a transalpine trading post. It was established as a fortified town by 1180 and received city rights in the early 13th cent. The Tyrolese peasants made their heroic stand (1809) against French and Bavarian troops near Innsbruck; a monument in the city commemorates the event. The Hofkirche (built 1553–63), a Franciscan church, is an architectural gem; it contains a large monument to Emperor Maximilian I (d.1519), who often resided in Innsbruck. Equally famous is the Fürstenburg, a 15th cent. castle. The city has several museums and a botanical garden, which has a large collection of Alpine plants. The winter Olympic games were held in Innsbruck in 1964 and 1976.

I·no·wroc·law (ē-nô-vrôts′läf), city (1975 est. pop. 59,700), N central Poland. It is an important railway and industrial center where agricultural machinery, glass, and bricks are produced. It is also a health resort, with saltwater springs. Rock salt is mined nearby. Chartered in 1267, Inowrocław passed to Prussia in 1772 and reverted to Poland in 1919.

In·side Passage (ĭn-sīd′, ĭn′sīd′), natural, protected waterway, c.950 mi (1,530 km) long, threading through the Alexander Archipelago off the coast of British Columbia and SE Alaska. From Seattle, Wash., to Skagway, Alaska, or via Cross Sound to the Gulf of Alaska, the route uses channels and straits between islands and the mainland that afford protection from the storms and open waters of the Pacific Ocean. Snow-capped mountains, waterfalls, glaciers, and narrow channels give the Inside Passage great scenic beauty.

In·ter-A·mer·i·can Highway (ĭn′tər-ə-mĕr′ĭ-kən), section c.3,400 mi

(5,470 km) long, of the Pan-American Highway system, from Nuevo Laredo, Mexico, to Panama City, Panama.

In·ter·la·ken (ĭn'tər-lä'kən), town (1970 pop. 4,735), Bern canton, central Switzerland, between the Lake of Brienz and the Lake of Thun. Interlaken is one of the largest resorts in the Bernese Alps.

In·ter·na·tion·al Date Line (ĭn'tər-năsh'ən-əl), imaginary line on the earth's surface, generally following the 180° meridian of longitude, where, by international agreement, travelers change dates. Traveling eastward across the line, one subtracts one calendar day; traveling westward, one adds a day. The date line bends eastward around the eastern tip of Siberia, westward around the Aleutian Islands, and eastward again around various island groups in the South Pacific in order to avoid a time change in populated areas.

International Falls, city (1970 pop. 6,439), seat of Koochiching co., N Minn., on the Rainy River opposite Fort Frances, Ont., Canada. The city grew after a large paper mill was built in 1904. Today it is a summer resort in a farming and lumbering area.

In·tha·non (ĭn'tə-nôn'), peak, 8,512 ft (2,596.2 m) high, in the Thanon Tong Chai range, NW Thailand. It is the highest point in Thailand.

In·tra·coast·al Wa·ter·way (ĭn'trə-kō'stəl wô'tər-wā'), 2,455 mi (3,950.1 km) long, partly natural, partly man-made, providing sheltered passage for commercial and pleasure boats along the U.S. Atlantic coast from Trenton, N.J., on the Delaware River to Key West, S Fla., and along the Gulf of Mexico coast from the St. Marks River, NW Fla., to Brownsville, Texas, on the Rio Grande. Its total length is 3,100 mi (4,987.9 km), including its mainly open-water extensions north to Boston, Mass., and along the west coast of Florida. The toll-free waterway, authorized by Congress in 1919, is maintained by the Army Corps of Engineers. Many miles of navigable waterways connect with the coastal system, including the Hudson River–New York State Barge Canal, the Savannah River, the Apalachicola River, and the entire Mississippi River system.

In·ver·car·gill (ĭn'vər-kär'gĭl), city (1971 pop. 47,098), extreme S South Island, New Zealand, on the Waiopai River. It is an agricultural center with timber and food-processing industries.

In·ver·ness (ĭn'vər-nĕs'), burgh (1976 pop. 49,738), Highland region, N Scotland, on the Moray Firth at the mouth of the Ness River. It is a seaport and transportation center. There are diverse light industries, including printing, food processing, distilling, wool weaving, and shipbuilding. Electrical and mechanical products and automobile parts are manufactured. There is a herring fishery. Inverness holds an annual cattle and wool market. An ancient town, it is thought to have been a Pict stronghold. Inverness has a museum of Highland relics and hosts an annual Highland Gathering. Until 1975 Inverness was county town of Inverness-shire.

In·ver·ness (ĭn'vər-nĕs'), town (1970 pop. 2,299), seat of Citrus co., central Fla., N of Tampa on Lake Tsala Apopka. It is a trade center in a timber and fruit-growing area.

In·ver·ness-shire (ĭn'vər-nĕs'shîr', -shər), former county, NW Scotland. Inverness was the county town. The county was first settled by Picts and belonged to the independent province of Moray. It came under the Scottish crown in 1078 but was never fully controlled by the central government until the suppression of the Highland clans after 1746. It was a Catholic and Jacobite stronghold in the wars of the 17th and 18th cent. In 1975 Inverness-shire was divided between the Highland and Western Isles regions.

In·ves·ti·ga·tor Strait (ĭn-vĕs'tĭ-gā'tər), strait, 60 mi (96.5 km) long, 35 mi (56.3 km) wide, in the Indian Ocean, separating N Kangaroo Island and the SE coast of S Australia.

In·yo (ĭn'yō'), county (1970 pop. 15,571), 10,130 sq mi (26,236.7 sq km), E Calif., bounded on the W by the crest of the Sierra Nevada, on the E by the Nev. border; formed 1866; co. seat Independence. In the west it rises in many places to more than 14,000 ft (4,270 m); in the east is a large portion of Death Valley National Monument and the lowest point (282 ft/86 m below sea level) in the Western Hemisphere. It has some irrigated land on which stock raising and dairying are done. Its extensive mineral resources include lead, tungsten, talc, molybdenum, zinc, silver, borates, potash, salt, and soda. The county includes sections of Inyo National Forest and China Lake Naval Weapons Center. There are camping, hunting, fishing, and winter sports in the mountains, and winter resorts in Death Valley.

Io·án·ni·na (yô-ä'nē-nä', yä'-), city (1971 pop. 40,130), capital of Ioánnina prefecture, NW Greece, in Epirus, on Lake Ioánnina. The chief city of Epirus, it is the commercial center for an agricultural region that produces cereals, fruits, and wine. Manufactures include textiles and gold and silver products. Founded c.527 by Justinian, Ioánnina was taken (1081) by the Normans and conquered by the Ottoman Turks in 1430. Ioánnina passed to Greece in 1913 as a result of the First Balkan War.

I·o·la (ī-ō'lə), city (1970 pop 6,493), seat of Allen co., SE Kansas, W of Fort Scott, in a dairy region; founded 1859, inc. 1870. It is a trade center, with some industry.

I·o·na (ī-ō'nə), island, 3.5 mi (5.6 km) long and 1.5 mi (2.4 km) wide, Argyllshire, NW Scotland, one of the Inner Hebrides. It is hilly, with shell beaches. Tourism is the main industry. The island is famous as the early center of Celtic Christianity. A group called the Iona Community (est. 1938), dedicated to reviving the spirit of Celtic Christianity, has restored many ancient buildings.

I·o·ni·a (ī-ō'nē-ə), ancient region of Asia Minor, occupying a narrow coastal strip on the E Mediterranean (in present-day W Turkey) as well as the neighboring Aegean Islands. Greek settlers established colonies here before 1000 B.C. A religious league (which reached its full power in the 8th cent. B.C.) was formed, with its center at the temple of Poseidon near Mycale. The fertility of the region and its excellent harbors brought prosperity to the 13 cities of the league. Traders and colonists were sent into the Mediterranean as far west as Spain and up to the shores of the Black Sea. In the 7th cent. B.C. the cities were invaded by the Cimmerians and the Lydians, but it was not until the time of Croesus that their subjugation was completed. When Croesus was conquered (before 546 B.C.) by Cyrus the Great of Persia, the cities came under Persian rule. At the beginning of the 5th cent. B.C. the cities rose in revolt against Darius I. Although the revolt was easily put down, the Persians set out to punish the allies (Athens and Eretria) of the cities. The Persian Wars resulted. Alexander the Great easily captured (c.335) all the Ionian cities. The cities continued to be rich and important through the time of the Roman and Byzantine empires. It was only after the Turkish conquest in the 15th cent. A.D. that their culture was destroyed.

Ionia, county (1970 pop. 45,848), 575 sq mi (1,489.3 sq km), S central Mich., intersected by the Grand River and drained by the Flat, Lookingglass, and Maple rivers; formed 1831; co. seat Ionia. Its agriculture includes livestock, dairy products, poultry, grain, corn, beans, peas, potatoes, and fruit. It has some mining, and its manufactures include processed foods, lumber and wood products, furniture, metal products, and metal-working machinery.

Ionia, city (1970 pop. 6,361), seat of Ionia co., S Mich., E of Grand Rapids; settled 1833, inc. as a village 1865, as a city 1873. Automobile parts are made.

I·o·ni·an Islands (ī-ō'nē-ən), chain of islands (1971 pop. 184,483), c.890 sq mi (2,305 sq km), W Greece, in the Ionian Sea, along the coasts of Epirus and the Peloponnesus. The islands are largely mountainous. Fruits, olives, grains, and vegetables are grown, and sheep, goats, and hogs are raised. Industries include fishing, soapmaking, and boatbuilding. In the 10th cent. A.D. the islands were made a province of the Byzantine Empire. Venice took them in the 14th and 15th cent. and held them until 1797, when the Treaty of Campo Formio gave the islands to France. In 1799 they were seized by a Russo-Turkish fleet and were constituted a republic under Russian protection. In 1807, by the Treaty of Tilsit, Russia returned the islands to France. From 1809 to 1814 the British navy occupied all the islands except Kérkira. In 1815 the islands were placed under British protection. The British ceded the islands to Greece in 1864.

Ionian Sea, part of the Mediterranean Sea, S Europe, between Greece and S Italy. It is connected with the Adriatic Sea by the Strait of Otranto.

I·os (ī'ŏs', ē'ôs'), island, 46 sq mi (119.1 sq km), S central Cyclades, Cyclades dept., Greece, in the Aegean Sea. Its products include cotton, wine, barley, and olive oil. According to legend, Homer is buried here.

I·os·co (ī-ŏs'kō), county (1970 pop. 24,905), 547 sq mi (1,416.7 sq km), NE Mich., bounded on the E by Lake Huron and drained by the Au Sable and Au Gres rivers; organized 1857; co. seat Tawas City. It has some agriculture (livestock, grain, potatoes, corn, and onions), commercial fisheries, cement plants, and food-processing and lumbering industries. The county has hunting and fishing resorts and includes part of Huron National Forest.

Iosh·kar-O·la (yŏsh-kär'ō-lä'): *see* Yoshkar-Ola, USSR.

I·o·wa (ī'ə-wə, ī'ə-wā'), state (1975 pop., 2,847,000), 56,290 sq mi (145,791 sq km), N central United States, in the Midwest, admitted to the Union in 1846 as the 29th state. Des Moines is the capital and largest city. Iowa is bordered on two sides by rivers; the Mississippi separates it on the east from Wis. and Ill., and the Missouri and the Big Sioux separate it on the west from Nebr. and S.Dak. The state is bounded on the north by Minn. and on the south by Mo.

Iowa is an area of rich, rolling plains, interrupted by many rivers. The terrain is low and gently sloping, except for the hills in the unglaciated area of northeast Iowa, the steeply sloping bluffs on the banks of the Mississippi, and the moundlike bluffs on the banks of the Missouri. The rivers of the eastern two thirds of Iowa flow to the Mississippi; those of the west flow to the Missouri. The original woodlands, which included black walnut and hickory, were destroyed by lumbering in the 19th cent., and the present wooded sections are covered only with a second or third growth of timber. Typical of Iowa is the prairie, now covered with fields of corn and other grains.

Iowa is divided into 99 counties:

NAME	COUNTY SEAT	NAME	COUNTY SEAT
Adair	Greenfield	Howard	Cresco
Adams	Corning	Humboldt	Dakota City
Allamakee	Waukon	Ida	Ida Grove
Appanoose	Centerville	Iowa	Marengo
Audubon	Audubon	Jackson	Maquoketa
Benton	Vinton	Jasper	Newton
Black Hawk	Waterloo	Jefferson	Fairfield
Boone	Boone	Johnson	Iowa City
Bremer	Waverly	Jones	Anamosa
Buchanan	Independence	Keokuk	Sigourney
Buena Vista	Storm Lake	Kossuth	Algona
Butler	Allison	Lee	Fort Madison
Calhoun	Rockwell City	Linn	Cedar Rapids
Carroll	Carroll	Louisa	Wapello
Cass	Atlantic	Lucas	Chariton
Cedar	Tipton	Lyon	Rock Rapids
Cerro Gordo	Mason City	Madison	Winterset
Cherokee	Cherokee	Mahaska	Oskaloosa
Chickasaw	New Hampton	Marion	Knoxville
Clarke	Osceola	Marshall	Marshalltown
Clay	Spencer	Mills	Glenwood
Clayton	Elkader	Mitchell	Osage
Clinton	Clinton	Monona	Onawa
Crawford	Denison	Monroe	Albia
Dallas	Adel	Montgomery	Red Oak
Davis	Bloomfield	Muscatine	Muscatine
Decatur	Leon	O'Brien	Primghar
Delaware	Manchester	Osceola	Sibley
Des Moines	Burlington	Page	Clarinda
Dickinson	Spirit Lake	Palo Alto	Emmetsburg
Dubuque	Dubuque	Plymouth	Le Mars
Emmet	Estherville	Pocahontas	Pocahontas
Fayette	West Union	Polk	Des Moines
Floyd	Charles City	Pottawattamie	Council Bluffs
Franklin	Hampton	Poweshiek	Montezuma
Fremont	Sidney	Ringgold	Mount Ayr
Greene	Jefferson	Sac	Sac City
Grundy	Grundy Center	Scott	Davenport
Guthrie	Guthrie Center	Shelby	Harlan
Hamilton	Webster City	Sioux	Orange City
Hancock	Garner	Story	Nevada
Hardin	Eldora	Tama	Toledo
Harrison	Logan	Taylor	Bedford
Henry	Mount Pleasant	Union	Creston
Van Buren	Keosauqua	Winnebago	Forest City
Wapello	Ottumwa	Winneshiek	Decorah
Warren	Indianola	Woodbury	Sioux City
Washington	Washington	Worth	Northwood
Wayne	Corydon	Wright	Clarion
Webster	Fort Dodge		

Economy. Iowa has some of the finest agricultural land in the United States. The deep, porous soil yields corn and other grains in tremendous quantities, and the corn-fed hogs and cattle are nationally known. In 1972 Iowa led the nation in the production of corn, hogs, and pigs and ranked second to Texas in the raising of cattle. In addition to corn, Iowa's other major crops are soybeans, hay, and oats. Agriculture in Iowa also benefits the state's chief industry, food processing. Nonelectrical machinery, electronic equipment, and chemicals are among the other manufactures. Cement is the most important mineral product.

History. In prehistoric times, the Mound Builders, a farming people, lived in the area of present-day Iowa. When white men first came in the 17th cent., various Indian groups, including the Iowa Indians, occupied the land; but it was the belligerent Sioux Indians who dominated the area. In 1673 the French explorers Father Jacques Marquette and Louis Jolliet traveled down the Mississippi River and touched upon the Iowa shores, as did La Salle in 1681–82. A number of Iowa towns developed from trading posts. The United States acquired Iowa as part of the Louisiana Purchase of 1803.

In 1832 the Black Hawk War broke out as Sac and Fox Indians, led by their chief, Black Hawk, fought to regain their former lands in Illinois along the Mississippi River. They were defeated by U.S. troops and were forced to cede to the United States much of their land along the river on the Iowa side. Within two decades after the Black Hawk War, all of the Indians' lands had been ceded to the whites. Meanwhile, a great rush of frontiersmen came to settle the prairies and take over the lead mines. Part of Missouri Territory prior to 1821, Iowa was subsequently part of Michigan Territory and Wisconsin Territory. By 1838 Iowa Territory was organized; the territory achieved statehood in 1846. In 1857 the capital was moved from Iowa City to Des Moines.

Iowa prospered greatly with the beginning of railroad construction in the 1850s. Before and during the Civil War, Iowans, generally owners of small, independent farms, were naturally sympathetic to the antislavery side, and many fought for the Union. During the hard times that afflicted the country in the 1870s, Iowa's farmers found themselves burdened with debts, and many supported the Granger movement, the Greenback party, and the Populist party. Since the end of the 19th cent., except for the depression of the 1930s, Iowa's agricultural community has generally prospered, partly through advances in agricultural science. Farm units grew larger, and mechanization brought great increases in productivity. Organizations such as the Grange, the Farm Bureau, the 4-H clubs, and farmers' cooperatives have promoted improvements in agriculture and, along with county fairs and the state fair, have also contributed to social life in the state.

Government. Iowa's present constitution was adopted in 1857. The governor is elected for a term of two years and may be reelected. The general assembly, or legislature, has a senate with 50 members elected for four-year terms and a house of representatives with 100 members elected for two-year terms.

Educational Institutions. Among those most prominent are Iowa State Univ. of Science and Technology, at Ames; the Univ. of Iowa, at Iowa City; Grinnell College, at Grinnell; Cornell College, at Mount Vernon; Drake Univ., at Des Moines; the Univ. of Northern Iowa, at Cedar Falls; and the Univ. of Dubuque, Loras College, and Clarke College, at Dubuque.

Iowa. 1. County (1970 pop. 15,419), 584 sq mi (1,512.6 sq km), E central Iowa, drained by the Iowa River and forks of the English River; formed 1843; co. seat Marengo. In a rolling agricultural area, it is dependent on stock raising (cattle, hogs, and poultry) and grain growing (corn and oats). **2.** County (1970 pop. 19,306), 761 sq mi (1,971 sq km), S Wis., bordered on the N by the Wisconsin River and drained by the Pecatonica and Blue rivers; formed 1829; co. seat Dodgeville.

Iowa, river, 329 mi (529.4 km) long, rising in the lakes of N Iowa and flowing SE to the Mississippi River, SE Iowa.

Iowa City, city (1978 est. pop. 49,100), seat of Johnson co., E Iowa, on both sides of the Iowa River; founded 1839 as the capital of Iowa Territory, inc. 1853. Its manufactures include foam rubber, animal feed, dentifrices, toilet goods, coated paper, and food products. The city is the seat of the Univ. of Iowa (1855) and is a major center of medical treatment and research. With the arrival of the railroad (1855), Iowa City became an important outfitting center for the westward trails. Nearby are the Herbert Hoover Presidential Library and Herbert Hoover's birthplace.

I-pin or **I·pin** (both: ē'pĭn'), city (1970 est. pop. 275,000), S Szechwan prov., China. It is a commercial and communications center at the junction of the Min and the Yangtze rivers.

I·poh (ē'pō), city (1971 pop. 247,689), capital of Perak state, Malaysia, central Malay Peninsula, in the Kinta River valley. A modern commercial town, it is the greatest tin-mining center of Malaysia. Nearby are rubber plantations and limestone quarries.

Ip·sam·bul (ĭp'säm-bo͞ol'): *see* Abu-Simbel, Egypt.

Ips·wich (ĭps'wĭch), city (1975 est. pop. 67,500), Queensland, E Australia, a suburb of Brisbane. It is the principal coal-mining center of the state and has woolen mills and other industries.

Ipswich, county borough (1976 est. pop. 121,500), administrative center of Suffolk, E England, on the Orwell estuary near its entry into the North Sea. Agricultural machinery and construction vehicles are the chief manufactures of Ipswich, which also has fertilizer, cigarette, malting, milling, brewing, printing, and textile industries. Ipswich was a commercial center and pottery producer from the 7th to 12th cent. It reached the peak of its significance in the woolen trade in the 16th cent. Its port declined with the decline in wool trading but revived with new dock construction in the mid-19th cent. Vestiges of Roman habitation remain in Ipswich. Until 1974 Ipswich was the county town of East Suffolk.

Ipswich. 1. Town (1970 pop. 10,750), Essex co., NE Mass., on the Ipswich River and Ipswich Bay; inc. 1634. Tourism and the production of electrical equipment are important. Points of interest in the town include Choate Bridge, the first stone bridge in the United States (1764), and many historic buildings, notably the John Whipple House (c.1640), with the Ipswich Historical Society collection. **2.** City (1970 pop. 1,187), seat of Edmunds co., N S.Dak., W of Aberdeen, in a farm and livestock area.

I·qui·que (ē-kē'kē), city (1970 pop. 65,040), capital of Tarapacá prov., N Chile, on the Pacific. The city, founded in the 16th cent., was taken (1879) from Peru by Chile during the War of the Pacific. Rock and sand enclose the city on the landward side, and the harbor, although greatly improved, still suffers from Pacific storms. The city has fine beaches and excellent deep-sea fishing.

I·qui·tos (ē-kē'tôs), city (1972 pop. 111,327), capital of Loreto dept., NE Peru, on the Amazon River. With the boom in wild rubber at the beginning of the 20th cent. the city gained prominence, but it declined after the collapse of the market. Today coffee, cotton, timber, balata, and tagua nuts, as well as rubber, are exported. The city was founded in 1863.

I·rá·kli·on (ĭ-rä'klē-ôn') or **Can·di·a** (kăn'dē-ə), city (1971 pop. 77,506), capital of Crete governorate and Iráklion prefecture, N Crete, Greece, a port on the Sea of Crete. It ships wine, olive oil, raisins, and almonds. Iráklion was founded (9th cent.) by the Saracens. In 961 it was conquered by the Byzantines, and in the 13th cent. it became a Venetian colony. In 1669 it was captured by the Ottoman Turks after a two-year siege. It was the capital of Crete until 1841, and in 1913 it passed to Greece. Among Iráklion's historic monuments are a cathedral, several mosques, and remains of Venetian walls and fortifications.

I·ran (ĭ-răn', ē-rän'), kingdom (1971 est. pop. 30,000,000), 636,290 sq mi (1,648,000 sq km), SW Asia. Tehran is the capital. Iran (called Persia before 1935) is bordered on the north by the USSR and the Caspian Sea; on the east by Afghanistan and Pakistan; on the south by the Persian Gulf and the Gulf of Oman; and on the west by Turkey and Iraq.

Iran is composed of a vast central plateau rimmed by mountain ranges and limited lowland regions. The country is subject to numerous and often severe earthquakes. The Iranian Plateau (alt. c.4,000 ft/1,200 m), which extends beyond the low ranges of eastern Iran into Afghanistan, consists of a number of arid basins of salt and sand, such as those of Dasht-e Kavir and Dasht-e Lut, and some marshlands, such as the area around Hamun-e Helmand along the

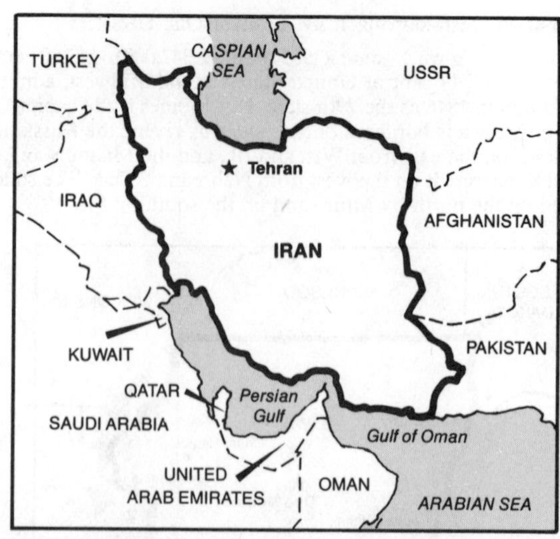

Afghanistan border. The plateau is surrounded by high folded and volcanic mountain chains including the Kopet Mts. in the northwest, the Elburz Mts. in the north, and the complex Zagros Mts. in the west. Narrow coastal plains are found along the shores of the Persian Gulf, Gulf of Oman, and the Caspian Sea; at the head of the Persian Gulf is the Iranian section of the Mesopotamian lowlands.

Economy. Iran's gross national product has been increasing at a rapid rate. Even though revenues from petroleum contribute some 80% of Iran's wealth, agriculture supports 75% of the population. Wheat, the most important crop, is grown mainly in the west and northwest; rice is the major crop in the Caspian littoral. Barley, corn, cotton, tea, hemp, tobacco, sugar beets, fruits (including citrus), nuts, and dates are also grown. Sheep and goats are raised, and silkworms are bred. Petroleum production, Iran's economic mainstay, is concentrated in western Iran. Other important minerals include coal, iron ore, copper, lead, salt, natural gas, manganese, and chromium. Textiles are Iran's second most important industrial product. Other major industries are sugar refining, food processing, and the production of petrochemicals, iron and steel, fertilizers, and machinery. The traditional handicrafts, such as carpet weaving, the manufacture of ceramics, and jewelry, are an important part of Iran's economy.

History. Some of the world's most ancient settlements have been excavated in the Caspian littoral and on the Iranian plateau; village life began there c.4000 B.C. The Aryans came about 2000 B.C. and split into two main groups, the Medes and the Persians. The Persian Empire founded (c.550 B.C.) by Cyrus the Great was succeeded, after Greek and Parthian occupation, by the Sassanid in the early 3rd cent. A.D. Their rule was weakened when Arab invaders took (636) the capital, Ctesiphon, and ended when the Arabs defeated the Sassanid armies at Nahavand in 641. The Arabs brought Islam to Persia, and it was in Persia that the Shiite sect was developed. The Turks began invading Iran in the 10th cent. and soon established several Turkish states. The Turks were followed by the Mongols, led by Genghis Khan in the 13th cent. and Tamerlane in the late 14th cent. The Safavid dynasty (1502–1736), founded by Shah Ismail, restored internal order in Iran and established the Shiite sect as the state religion. The fall of the Safavid dynasty was brought about (1722) by the Afghans, whose brief rule was followed by the prosperous Afshar (1736-50) and Zand (1750-94) dynasties. However, the country was soon again in turmoil, which lasted until the advent of Aga Mohammed Khan, a detested ruler (assassinated 1797), who established the Kajar dynasty (1794–1925). This long period saw Iran steadily lose territory to neighboring countries and fall under the increasing pressure of European nations, particularly czarist Russia and Great Britain.

In 1921 Reza Khan, an army officer, established a military dictatorship. He was subsequently (1925) elected hereditary shah, thus ending the Kajar dynasty and founding the new Pahlevi dynasty. Reza Shah Pahlevi took steps to reduce British commercial influences, introduced many reforms, and encouraged the development of industry and education. In Aug., 1941, British and Soviet forces occupied Iran. On Sept. 16 the shah abdicated in favor of his son Mohammed Reza Shah Pahlevi. At the close of the war, the USSR, dissatisfied with the refusal of the Iranian government to grant it oil concessions, fomented a revolt in the north that led to the establishment (Dec., 1945) of the People's Republic of Azerbaijan and the Kurdish People's Republic. The Soviets finally withdrew (May, 1946) all troops from Iran under international pressure; the Soviet-established governments in the north, lacking popular support, were deposed by Iranian troops late in 1946. In 1951 the National Front movement, headed by Premier Mohammed Mussadegh, a militant nationalist, forced the parliament to nationalize the oil industry and form the National Iranian Oil Company (NIOC). The shah fled Iran but returned when monarchist elements forced Mussadegh from office in Aug., 1953. In 1954 Iran allowed an international consortium of British, American, French, and Dutch oil companies to operate its oil facilities, with profits shared equally between Iran and the consortium.

After 1953 a succession of premiers restored a measure of order to Iran, which established closer relations with the West. Starting in the 1960s and continuing into the 1970s, the Iranian government, at the shah's initiative, undertook a broad program designed to improve economic and social conditions. Land reform was a major priority. A new government-backed political party, the Iran Novin party, was introduced and won an overwhelming majority in the parliament in the 1963 and subsequent elections. Relations with Iraq were strained for much of the late 1960s and early 1970s on the issue of control over the Shatt al Arab and the Persian Gulf. Filling a power vacuum left by the 1971 British withdrawal from the Gulf, Iran increased its defenses with the help of a large U.S. aid program and emerged as the region's strongest military power. The oil industry was fully nationalized in 1973. Meanwhile, protests against the shah were becoming more frequent and more violent. His regime was attacked for being too liberal (for economic and social measures that changed the ancient Moslem order) and too repressive (a major target of protest was the SAVAK, Iran's secret police). In 1978 Iran was plunged into revolution. Finally, the shah went into exile, his hand-picked successor was overthrown, and a government led by the Ayatollah Khomeini, considered a Moslem holy man, took control.

I·ra·pua·to (ē'rä-pwä'tō), city (1974 est. pop. 135,600), Guanajuato state, W central Mexico, on the Irapuato River. It is the commercial and communications center of the surrounding mining and agricultural (cereals and cattle) region.

I·raq or **I·rak** (both: ĭ-răk', ē-räk'), republic (1970 est. pop. 9,440,100), 167,924 sq mi (434,924 sq km), SW Asia. Baghdad is the capital. Iraq is bordered on the south by Kuwait, the Persian Gulf, a neutral zone, and Saudi Arabia, on the west by Jordan and Syria, on the north by Turkey, and on the east by Iran. Basra, on the short Persian Gulf coast, is the only port. Iraq is approximately coextensive with ancient Mesopotamia. The southwest, part of the Syrian Desert, supports a small population of nomadic shepherds. In the rest of the country, life centers on the great southeast-flowing rivers, the Tigris and the Euphrates, which come together in the Shatt al Arab and form a delta with marshes and lakes at the head of the Persian Gulf; Hawr al Hammar is the largest lake. There is very little rainfall except in the northeast.

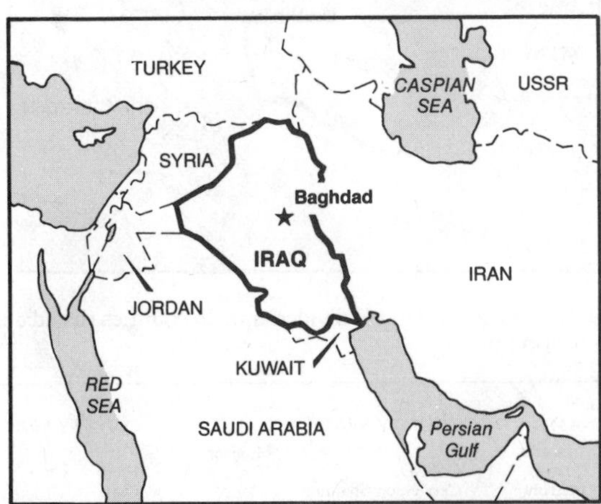

Economy. All agriculture in Iraq depends upon river water. The sandy soil and steady heat of the southeast enable a large date crop and much cotton to be produced. Farther upstream, as the elevation increases, rainfall becomes sufficient to grow diversified crops, including cereals and vegetables. In the mountainous north the economy shifts from agriculture to oil production, notably in the great fields near Mosul and Kirkuk. The oil, which accounts for one quarter of the country's gross national product, is produced mainly by the Iraq Petroleum Company (nationalized in 1972). Aside from petroleum, Iraq has a small, diversified industrial sector, including the production of textiles, cement, food products, leather goods, and a small amount of modern machinery. New industries have been started in electronics products, fertilizers, and refined sugar.

History. Prior to the Arab conquest in the 7th cent. A.D., Iraq had been the site of a number of flourishing civilizations, including the Sumerians, the Akkadians, the Babylonians, and the Assyrians. The capital of the Abbasid caliphate was established at Baghdad in the 8th cent., and the city became a famous center for learning and the arts. Despite fierce resistance, Mesopotamia fell to the Ottoman Turks in the 16th cent. and passed under direct Turkish administration in the 19th cent., when it came to constitute the three Turkish provinces of Basra, Baghdad, and Mosul.

In World War I the British invaded Iraq in their war against the Ottoman Empire; they remained to administer the country under a League of Nations mandate (1920-32). In 1921 the country was made a kingdom headed by Faisal I (reigned 1921-33), and by 1926 an Iraqi parliament and administration was governing the country. Domestic politics were turbulent, with military coups frequent in the 1930s. A pro-Axis group held power briefly in 1941 but was crushed by reinforced British troops.

Iraq, with other members of the Arab League, participated in 1948 in the unsuccessful war against Israel. In the 1950s a national development program, financed mostly by oil royalties, led to flood control measures, improvement of agricultural methods, and the building of a power industry. The USSR's support of Kurdish nationalism caused a break in relations in 1955. Later that year Iraq, Turkey, Pakistan, Iran, and Britain formed the Baghdad Pact for mutual defense. In Feb., 1958, Iraq and Jordan announced the federation of their countries into the Arab Union. In a coup d'état on July 14, 1958, the army led by Gen. Abdul Karim Kassem seized control of Baghdad, killed King Faisal II (reigned 1939-58), and proclaimed a republic. The Arab Union was dissolved, and Iraq's activity in the Baghdad Pact ceased. Diplomatic relations were restored with the USSR, but Iraq pursued a policy of nonalignment in the cold war. In Feb., 1963, Col. Abdul Salam Aref led a coup that overthrew the Kassem regime. Ruling groups since then have been socialist and strongly anti-Israeli. In Apr., 1972, a 15-year friendship treaty with the USSR was signed; the Communist party was legalized in 1972. Iraq took an active part in the 1973 Arab-Israeli War; it also participated in the oil boycott against nations supporting Israel. In 1975, following an agreement under which Iran promised to close its borders to Kurdish rebels, Iraq mounted a massive military offensive against the guerrillas. By Mar., 1975, the rebellion had collapsed, and the Kurdish soldiers had either fled the country or surrendered under an amnesty program offered by the central government.

I·ra·zú (ē'rä-sōō'), active volcano, c.11,260 ft (3,435 m) high, central Costa Rica. It erupted in 1723, destroying Cartago, and in 1963, covering San José with ash.

Ir·bil (ĭr-bĭl'), town (1970 est. pop. 107,400), N Iraq. Irbil is on an artificial mound surmounted by an old Turkish fort.

Ire·dell (īr'dĕl'), county (1970 pop. 72,197), 572 sq mi (1,481.5 sq km), W central N.C., in a piedmont area bounded on the SW by the Catawba River and drained by tributaries of the Yadkin River; formed 1788; co. seat Statesville. Its agriculture includes cotton, corn, wheat, hay, poultry, and dairy products. Lumbering is done, and textiles and furniture are manufactured.

Ire·land (īr'lənd) or *Irish* **Eir·e** (âr'ə), island, 32,598 sq mi (84,429 sq km), second largest of the British Isles. It lies west of the island of Great Britain, from which it is separated by the narrow North Channel, the Irish Sea, and St. George's Channel. A large central plain extending to the Irish Sea between the Mourne Mts. in the north and the mountains of Wicklow in the south is roughly enclosed by a highland rim. The highlands of the north, west, and south, which rise to more than 3,000 ft (915 m), are generally barren, but the central plain is extremely fertile and the climate is temperate and moist, warmed by southwesterly winds. The coastline is irregular, affording many natural harbors. Off the west coast are numerous small islands. The interior is dotted with lakes and wide stretches of river called loughs.

History. The earliest known people in Ireland belonged to the groups that inhabited all of the British Isles in prehistoric times. In the several centuries preceding the birth of Christ, Celtic tribes invaded and established their distinctive culture. The people were organized into clans that in the early period owed allegiance to one of five provincial kings who nominally served the high king of all Ireland at Tara. Parts of Ireland had already been Christianized before St. Patrick completed the process in the 5th cent. The country did not develop a strong central government, however, and it was not united to meet the invasions of the Norsemen who settled on the shores of the island late in the 8th cent. In 1014, at Clontarf, Brian Boru broke the strength of the Norse invaders.

The English conquest of Ireland was begun in the 12th cent. under Henry II. Poynings' Law (1495) provided that future Irish Parliaments and legislation receive prior approval from the English Privy Council. Henry VIII put down a rebellion (1534-37), abolished the monasteries, confiscated lands, and established a Protestant "Church of Ireland" (1537). Under James I, Ulster was settled by Scottish and English Protestants, and many of the Catholic inhabitants were driven off their lands. A rebellion was crushed (1649-50) by Oliver Cromwell with the loss of hundreds of thousands of lives. The English-controlled Irish Parliament passed harsh Penal Laws designed to keep the Catholic Irish powerless. Another unsuccessful rebellion was staged in 1798 by Wolfe Tone, a Protestant who accepted French aid in the uprising. The rebellion convinced British Prime Minister William Pitt that the Irish problem could be solved by abolition of the Irish Parliament, legislative union with Britain, and Catholic emancipation. The first two goals were achieved in 1800, the last in 1829.

After 1829 the Irish representatives in the British Parliament attempted to maintain the Irish question as a major issue in British politics. During the Great Potato Famine (1845-49), caused by a blight that ruined the potato crop, hundreds of thousands perished and many thousands of others emigrated. In the 1870s Irish politicians renewed efforts to achieve Home Rule within the union. Prime Minister William Gladstone twice submitted Home Rule bills (1886 and 1893) that failed. Home Rule was finally enacted in 1914, but the act was suspended for the duration of World War I and never went into effect.

The Irish Republican Brotherhood organized an unsuccessful rebellion on Easter Sunday, 1916. The Catholic nationalist Sinn Fein, linked in the Irish public's mind with the rising, scored a tremendous victory in the parliamentary elections of 1918. Its members refused to take their seats in Westminster, declared themselves the Dáil Éireann (Irish Assembly), and proclaimed an Irish Republic. The British outlawed the Sinn Fein and the Dáil; both went underground and engaged in guerrilla warfare (1919-21) against local Irish authorities representing the union. The British sent troops, the Black and Tans, who inflamed the situation further. A new Home Rule bill was enacted in 1920, establishing separate parliaments for Ulster and Catholic Ireland. This was accepted by Ulster, and Northern Ireland was created. The plan was rejected by the Dáil, but in 1921 a treaty was negotiated granting Dominion status within the British Empire to Catholic Ireland. The Irish Free State was established in Jan., 1922.

Ireland, Northern, division of the United Kingdom of Great Britain and Northern Ireland (1971 pop. 1,527,593), 5,462 sq mi (14,147 sq km), NE Ireland. It is frequently called Ulster. The capital is Belfast. The land is mountainous and has few natural resources.

Northern Ireland is divided into six counties:

NAME	COUNTY SEAT	NAME	COUNTY SEAT
Antrim	Belfast	Fermanagh	Enniskillen
Armagh	Armagh	Londonderry	Londonderry
Down	Downpatrick	Tyrone	Omagh

Economy. Farming (livestock, dairy products, cereals) is the largest single occupation. Northern Ireland's fine linens are famous. Heavy industry is concentrated in and around Belfast, one of the chief ports of the British Isles. Shipbuilding and other engineering, food processing, and the manufacture of textiles are the leading industries; papermaking and furniture manufacturing are also important.

History. In the early 17th cent. much of Northern Ireland was confiscated by the British crown and "planted" with Scottish and English settlers. Ulster took on a Protestant character as compared with the rest of Ireland; but there was no question of political separation until the late 19th cent. The Government of Ireland Act of 1920 established Home Rule separately for the two parts of Ireland, thus creating the province of Northern Ireland. However, the Irish Free State, now the Republic of Ireland, refused to recognize the partition, and violence erupted frequently on both sides of the border.

The late 1960s marked a new stage in the region's troubled history. The Catholic minority, which suffered economic and political discrimination, had grown steadily through immigration from the Republic. In 1968 civil rights protests by Catholics led to widespread violence. In 1969 British troops were called in to help restore order. However, the radical nationalist Irish Republican Army (IRA) and the Ulster Defense Association, a Protestant terrorist group, continued and even intensified their activities. On Mar. 30, 1972, British Prime Minister Edward Heath suspended the government and appointed William Whitelaw secretary of state for Northern Ireland. The parliament was reconstituted, elections were held, and a coalition regime including both Catholics and Protestant Unionists was formed. Extremists of both religions vowed to destroy the coalition Executive; militant Ulster Protestants staged a two-week general strike against the new governmental framework in 1974. The Executive collapsed, and the British government took direct control of the province. Meanwhile, bombings and other terrorist activities had spread to Dublin and London. In 1975 as IRA bombings increased, a constitutional convention charged with devising a new form of government failed. The convention was dissolved in 1976, and London continued to administer the province.

Government. Northern Ireland has 12 representatives in the British Parliament and possesses a large degree of self-government in domestic matters. The Ulster government is known as Stormont, from the hill where the Parliament building stands. About three fifths of the population are Protestant; one third are Catholic. English is the official language.

Ireland, Republic of, country (1971 pop. 2,971,230), 27,136 sq mi (70,282 sq km). From 1922 to 1937 the country was known as the Irish Free State, and from 1937 to 1949 as Eire. Dublin is the capital.

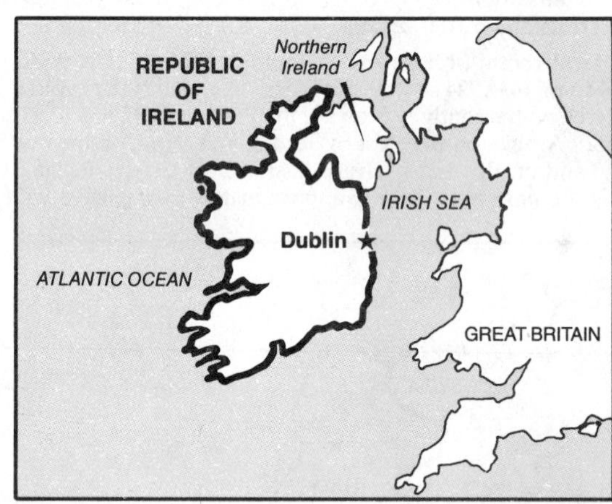

The Republic of Ireland is divided into 26 counties (listed according to historic province):

NAME	COUNTY SEAT	NAME	COUNTY SEAT
Connacht		Munster	
Galway	Galway	Clare	Ennis
Leitrim	Carrick on Shannon	Cork	Cork
Mayo	Castlebar	Kerry	Tralee
Roscommon	Roscommon	Limerick	Limerick
Sligo	Sligo	Tipperary	Clonmel
Leinster		Waterford	Waterford
Carlow	Carlow	Ulster	
Dublin	Dublin	Cavan	Cavan
Kildare	Naas	Donegal	Lifford
Kilkenny	Kilkenny	Monaghan	Monaghan
Laoighis	Maryborough		
Longford	Longford		
Louth	Dundalk		
Meath	Trim		
Offaly	Tullamore		
Westmeath	Mullingar		
Wexford	Wexford		
Wicklow	Wicklow		

Economy. Agriculture is the primary economic activity, engaging about 70% of the land and 30% of the work force. The raising of dairy and beef cattle, sheep, pigs, and poultry is the chief enterprise. Among the leading crops are flax, oats, wheat, turnips, potatoes, sugar beets, and barley. The republic's industries produce such items as linen and laces, food products, textiles, ships, iron products, and handicrafts. Around the free port of Shannon are factories producing electronic equipment, chemicals, plastics, and textiles.

History. After the establishment by treaty with Great Britain of the Irish Free State (Jan., 1922), civil war broke out between supporters of the treaty and opponents, who refused to accept the partition of Ireland and the retention of any ties with Britain. The anti-treaty forces, embodied in the Irish Republican Army (IRA) and led by Eamon de Valera, were defeated, although the IRA continued as a secret terrorist organization. William Cosgrave became the first prime minister. In 1937 a new constitution was promulgated establishing the sovereign nation of Ireland, or Eire, within the British Commonwealth of Nations. During World War II, Eire remained neutral and vigorously protested Allied military activity in Northern Ireland.

The postwar period brought a sharp rise in the cost of living and a decline in population, due in great part to steady emigration to Northern Ireland, Great Britain, and other countries. The Republic of Ireland was proclaimed on Apr. 18, 1949. The country withdrew from the Commonwealth and formally claimed jurisdiction over the Ulster counties. During the 1950s and 60s the republic and Northern Ireland improved their economic relations. But the problem of Northern Ireland flared up again in the late 1960s and 70s with bitter fighting between the Protestant majority and Catholic minority there, aggravated by the actions of the IRA, which was headquartered in the republic. Economic difficulties grew in the 1970s: in 1976 the inflation rate was 21%, unemployment stood at 12%, and a foreign trade deficit was growing. Succeeding years brought only slight improvements.

Government. The republic is governed (under the 1937 constitution) by a two-chamber legislature (the Dáil Éireann and the Śeanad Éireann) and a prime minister and cabinet. The head of state, the president, is popularly elected. The Dáil is chosen by proportional representation.

I·ri (ĭr'ē), city (1975 est. pop. 97,200), SW South Korea. It is an agricultural center. Sake is brewed and rice is processed.

I·ri·an Ba·rat (ĭr'ē-än bär'ät), province (1970 est. pop. 957,000), c.162,000 sq mi (419,580 sq km), Indonesia, comprising the W half of New Guinea and about 12 offshore islands. It is a rugged, densely forested region, with mountains rising to over 16,500 ft (5,032.5 m). The coastal lowlands are swampy and cut by many rivers. Subsistence farming is carried on; taro, bananas, sugar cane, and sweet potatoes are the principal crops. There is fishing along the coast and the rivers. It has deposits of magnetite, oil, nickel, and cobalt.

The Dutch first visited the west coast of the island in 1606. Their claim to the western half of the island was recognized by Great Britain and Germany in treaties of 1885 and 1895. In World War II the northern coastal areas and offshore islands were occupied (1942) by the Japanese but retaken (1944) by the Allies. Following Indonesian independence (1949), the Dutch retained control of the area. Years of dispute over the territory culminated in the landing (early 1962) of Indonesian guerrillas and paratroopers there. The conflict ended in late 1962 when the Netherlands agreed to UN administration of West New Guinea and, after May 1, 1963, transfer of the territory to Indonesian control pending a plebiscite. The plebiscite was held in Aug., 1969; tribal leaders, voting as representatives of their people, chose to remain under Indonesian rule, and Indonesia then formally annexed the territory.

Ir·i·on (ĭr'ē-ən), county (1970 pop. 1,070), 1,073 sq mi (2,779.1 sq km), W Texas, in a broken prairie area drained by the Middle Concho River and spring-fed streams; formed 1889; co. seat Mertzon. In a ranching area (sheep, goats, and cattle), it has some agriculture (grain sorghums, alfalfa, cotton, hogs, and poultry). Wool and mohair are processed and shipped.

I·rish Sea (ī'rĭsh), arm of the Atlantic Ocean, c.40,000 sq mi (103,600 sq km), 130 mi (209.2 km) long and up to c.140 mi (225 km) wide, lying between Ireland and Great Britain. It is connected with the Atlantic by North Channel and St. George's Channel.

Ir·kutsk (ĭr-ko͞otsk'), city (1977 est. pop. 532,000), capital of Irkutsk oblast, S Siberian USSR, at the confluence of the Angara and Irkut rivers. It is an industrial center, a port, and a major stop on the Trans-Siberian RR. Manufactures include aircraft, automobiles, machine tools, textiles, chemicals, food products, and metals. Founded as a Cossack fortress in 1654, Irkutsk became the capital of Eastern Siberia in 1822. It has been a place of exile since the 18th cent.

I·ron (ī'ərn). **1.** County (1970 pop. 13,813), 1,171 sq mi (3,032.9 sq km), N Mich., SW Upper Peninsula, bounded on the S by the Wis. border and drained by the Brule, Michigamme, Paint, and Iron rivers; organized 1885; co. seat Crystal Falls. It has some agriculture (livestock, grain, poultry, potatoes, and fruit) but depends mainly on its lumbering industry. Iron-ore mining is also important. The county includes part of Ottawa National Forest and has many lake and river resorts. **2.** County (1970 pop. 9,529), 554 sq mi (1,434.9 sq km), SE Mo., partly in the St. Francois Mts.; formed 1857; co. seat Ironton. It is in an agricultural and mining (manganese, lead, iron, and granite) area, with oak and pine timber. Its manufactures include mineral products and metal-working machinery. There are resorts in the county, and part of Mark Twain National Forest is here. **3.** County (1970 pop. 12,177), 3,300 sq mi (8,547 sq km), SW Utah, in a mountain and plateau region bordering in the W on Nev., with a desert area in the N; formed 1850; co. seat Parowan. It has large open-pit iron mines, oil and natural-gas wells, and some agriculture (alfalfa and potatoes). Clothing is manufactured. Cedar Breaks National Monument and parts of Dixie National Forest are here. **4.** County (1970 pop. 6,533), 748 sq mi (1,937.3 sq km), N Wis., bounded partly on the N by Lake Superior and the Montreal River (the Mich. border); formed 1893; co. seat Hurley. It is heavily wooded in the south and includes part of the Gogebic iron-ore range. Tourism is of increasing importance to its economy.

I·ron·de·quoit (ĭr-än'də-kwoit'), town (1975 est. pop. 60,800), Monroe co., W N.Y., N of Rochester; settled 1791.

Iron Gate, gorge of the Danube River, c.2 mi (3.2 km) long and c.550 ft (170 m) wide, on the Yugoslav-Rumanian border between Orşova and Turnu-Severin, where the river narrows and swiftly flows through a gap between the Carpathian and Balkan mts. Iron Gate, formerly an obstacle to shipping, was cleared of rock obstructions in the 1860s; the Sip Canal (opened 1896) permits large river craft to get past the gorge. Iron Gate is the site of one of Europe's largest hydroelectric power dams.

Iron Mountain, city (1970 pop. 8,702), seat of Dickinson co., N Mich., S Upper Peninsula, NNE of Kingsford near the Wis. border; settled 1879, inc. 1889. It is a distribution point for ore from the Menominee iron range. Paint, lumber products, and machinery are manufactured. The city is also a summer and winter resort.

I·ron·ton (ī'ərn-tən). **1.** Resort city (1970 pop. 1,452), seat of Iron co., SE Mo., SW of St. Louis, in the Ozarks; founded 1857, inc. 1859. Lead and iron mines are in the area. **2.** Industrial city (1970 pop. 15,030), seat of Lawrence co., S Ohio, on the Ohio River; inc. as a city 1865. Chemicals, dyes, metal pipes, power shovels, industrial tar, and iron products are manufactured, and there is some coal mining. Ironton was a great iron-producing center during the Civil War. From c.1900 to 1910 the city had the largest blast furnace in the world, Big Etna, with a capacity of 100 tons per day. Today the remains of many giant charcoal iron furnaces are local landmarks.

I·ron·wood (ī'ərn-wo͝od'), city (1970 pop. 8,711), Gogebic co., N Mich., extreme W Upper Peninsula, on the Montreal River at the Wis. border; founded 1885, inc. 1889. It is a trading center for the Gogebic iron range.

Ir·o·quois (ĭr'ə-kwoi'), county (1970 pop. 33,532), 1,122 sq mi (2,906 sq km), E Ill., bordering on Ind. in the E and drained by the Iroquois River and Sugar Creek; formed 1833; co. seat Watseka. Its agriculture includes corn, oats, wheat, soybeans, livestock, poultry, and dairy products. Canned foods, clothing, and batteries are made.

Ir·ra·wad·dy (ĭr'ə-wŏd'ē, -wô'dē), chief river of Burma, c.1,000 mi (1,610 km) long, formed by the confluence of the Mali and Nmai rivers in N Burma. The combined stream flows south through gorges strewn with rapids; it receives the Chindwin River, its principal tributary, just below Mandalay. The vast Irrawaddy delta is one of the world's great rice-producing regions.

Ir·tysh (ĭr-tĭsh'), river, c.2,650 mi (4,265 km) long, W Siberian USSR, rising in Sinkiang prov., China, in the Mongolian Altai Mts., and flowing NW through Lake Zaysan (in Kazakhstan) and W Siberia to join the Ob River. The river banks were occupied by Chinese, Kalmyks, and Mongols until the Russians arrived in the late 16th cent.

I·ru·ma (ē-rōō'mä), city (1975 pop. 83,997), Saitama prefecture, central Honshu, Japan, on the Iruma River. It is a residential and industrial suburb of Tokyo.

I·rún (ē-rōōn'), town (1970 pop. 45,060), Guipúzcoa prov., N Spain, in the Basque Provinces near the French border, on the Bidassoa River near the Bay of Biscay. It is a commercial and manufacturing center with a large transit trade. Lead and iron mines are nearby.

Ir·vine (ûr'vən), burgh (1976 est. pop. 50,900), Strathclyde region, SW Scotland, on the Irvine River estuary. There are iron and brass foundries. Other products include chemicals, bottles, and hosiery.

Irvine, city (1970 pop. 2,918), seat of Estill co., central Ky., on the Kentucky River SE of Lexington, in a timber, farm, and oil area.

Ir·ving (ûr'vĭng), city (1978 est. pop. 104,800), Dallas co., N Texas, a suburb of Dallas; inc. as a city 1952. Building supplies, chemicals, electronic equipment, snack foods, and tools are manufactured there. The Univ. of Dallas and the Dallas-Fort Worth Regional Airport (opened 1974) are in Irving.

Ir·ving·ton (ûr'vĭng-tən), town (1978 est. pop. 57,300), Essex co., NE N.J., an industrial suburb of Newark; settled 1692 as Camptown, renamed 1852, inc. 1898. Tools, castings, photographic equipment, paints, building materials, and plastic and paper products are among its manufactures.

Ir·win (ûr'wĭn), county (1970 pop. 8,036), 372 sq mi (963.5 sq km), S central Ga., drained by the Alapaha and Satilla rivers; formed 1818; co. seat Ocilla. It is in a coastal-plain timber and agriculture (cotton, corn, tobacco, peanuts, peaches, and livestock) area.

Ir·win·ton (ûr'wĭn-tən), town (1970 pop. 757), seat of Wilkinson co., central Ga., E of Macon, in an agricultural and kaolin area. Cotton, corn, peanuts, truck crops, and potatoes are grown.

Is·a·bel·a (ĭz'ə-bĕl'ə, ē-sä-bā'lä), ruins of a town on the N shore of Hispaniola, in the Dominican Republic, at the base of Cape Isabela. Believed to have been founded by Columbus (c.1494), it was one of the first Spanish settlements in the New World.

Is·a·bel·la (ĭz'ə-bĕl'ə), county (1970 pop. 44,594), 572 sq mi (1,481.5 sq km), central Mich., drained by the Chippewa and Pine rivers; organized 1859; co. seat Mount Pleasant. Livestock and poultry are raised, as well as sugar beets, beans, grain, corn, and hay. Oil and gas wells and refineries are found in the area.

I·san·ti (ī-săn'tē), county (1970 pop. 16,560), 438 sq mi (1,134.4 sq km), E Minn., drained by the Rum River; formed 1857; co. seat Cambridge. It is in an agricultural area yielding dairy products, grain, livestock, and poultry and has a food-processing industry. Luggage, machinery, and metal products are made.

I·sar (ē'zär'), river, 160 mi (257.4 km) long, rising in the Tyrol, W Austria, and flowing NE through SE West Germany, past Munich, to the Danube River.

I·sau·ri·a (ī-sôr'ē-ə), ancient district of S Asia Minor, on the borders of Pisidia and Cilicia, N of the Taurus range, in present S central Turkey. It was a wild region inhabited by marauding bands. The Isaurians were partially checked (76–75 B.C.) by the Romans but were not completely subdued until the arrival in the 11th cent. A.D. of the Seljuk Turks.

Is·chia (ē'skyä), volcanic island (1971 pop. 14,076), 18 sq mi (46.6 sq km), Campania, S Italy, in the Tyrrhenian Sea between the Gulf of Gaeta and the Bay of Naples. Known as the Emerald Isle, it is a health resort and a tourist center, celebrated for its warm mineral springs and for its scenery. Settled in the 8th cent. B.C., the island was abandoned several times because of volcanic eruptions (the last of which occurred in 1301).

Ischl (ĭsh'əl): *see* Bad Ischl, Austria.

I·se (ē'sā), city (1976 est. pop. 105,252), Mie prefecture, S Honshu, Japan, on Ise Bay. Its three Shinto shrines, set deep in a forest, are said to have been built in 4 B.C. They exhibit an archaic style of architecture, completely without Chinese or Buddhist influence. Ise has several museums.

I·sère (ē-zâr'), department (1975 pop. 860,378), 2,886 sq mi (7,474.7 sq km), SE France, in Dauphiné; capital Grenoble.

Isère, river, 180 mi (289.6 km) long, SE France, rising in the Graian Alps close to the Italian border and flowing W and SW through the Savoy Alps into the Rhone River NW of Valence.

I·ser·lohn (ē'zər-lōn'), city (1975 pop. 97,194), North Rhine-Westphalia, W West Germany. Its manufactures include metal goods, textiles, and electrical products. Iserlohn became an important town in the 13th cent. and was known for the manufacture of armor, chains, and needles.

I·se·yin (ē-sā-äɴ'), town (1971 est. pop. 115,000), SW Nigeria. The city, located in a tobacco-growing region, has an important traditional textile industry. Iseyin was the capital of a small Yoruba kingdom under the Oyo empire. In 1893 it came under British control.

I·se·za·ki (ē'sā-zä'kē), city (1975 pop. 97,841), Gumma prefecture, central Honshu, Japan. It is a center for weaving industries and a market for agricultural products.

Is·fa·han (ĭs'fə-hän): *see* Esfahan, Iran.

Is·fjor·den (ēs-fyôr'dən), inlet of the Greenland Sea and largest fjord of Spitsbergen island, Svalbard, Norway, 65 mi (104.6 km) long and from 8 to 20 mi (12.9–32.2 km) wide.

I·shi·ga·ki (ē'shē-gä'kē), city (1970 pop. 36,554), Okinawa prefecture, Ryukyu Islands, Japan. It is an agricultural center where sake and dried tuna are produced.

I·shi·ka·ri (ē'shē-kä'rē), river, c.225 mi (360 km) long, Japan, rising in the mountainous interior of Hokkaido and flowing generally SW to Ishikari Bay near Otaru. It drains an extensive coal area and waters a fertile agricultural region.

I·shim (ē-shēm'), city (1976 est. pop. 62,000), W Siberian USSR, on the Ishim River and the Trans-Siberian RR. An agricultural center, it produces farm machinery and processes food.

Ishim, river, c.1,130 mi (1,818.2 km) long, W Siberian USSR, rising N of Karaganda in Kazakhstan and flowing W past Atbasar and N past Petropavlovsk to join the Irtysh River at Ust Ishim.

I·shim·bay (ē-shēm-bī'), city (1976 est. pop. 57,000), Bashkir Autonomous Republic, E European USSR, on the Belaya River. Founded in 1932, Ishimbay developed around the first major oil field of the Volga-Ural region, formerly the leading Soviet oil area. Ishimbay's chief industries are oil refining and petrochemical production.

I·shi·no·ma·ki (ē'shē-nō-mä'kē), city (1975 pop. 115,085), Miyagi prefecture, N Honshu, Japan, on Ishinomaki Bay. It is a commercial and fishing port and a center for marine product processing and paper manufacturing.

Ish·pe·ming (ĭsh'pə-mĭng), city (1970 pop. 8,245), Marquette co., N Mich., W Upper Peninsula, W of Marquette; inc. as a village 1871, as a city 1873. It is in the center of the Marquette iron range.

Is·in (ĭs'ĭn), capital of an ancient Semitic kingdom of N Babylonia. The city became important after the third dynasty of Ur fell to the Elamites and the Amorites (c.2025 B.C.). Excavations have brought to light the law code of King Lipit-Ishtar of Isin. This code is one of several codes that predate the stele of Hammurabi.

İs·ken·de·run (ĭs-kĕn'dĕ-rōōn'), formerly **Al·ex·an·dret·ta** (ăl'ĭg-zăn-drĕt'ə), city (1975 pop. 103,164), S Turkey, on the Gulf of Alexandretta, an inlet of the Mediterranean Sea. The principal Turkish port on the Mediterranean, it exports cotton, grain, fruit, wool, and hides. The city was founded by Alexander the Great to commemorate his victory over the Persians at Issus in 333 B.C. In A.D. 1515 the Ottoman Empire captured the city. İskenderun was transferred (1920) to the French Syria League of Nations mandate as part of the sanjak of Alexandretta but was returned to Turkey in 1939.

Is·kŭr (ĭs'kər), river, c.250 mi (400 km) long, rising in the Rhodope Mts., W Bulgaria, and flowing generally NE past Sofia and through the Balkan Mts. to the Danube River. The gorge of the Iskŭr is the chief pass through the Balkans.

Is·lam·a·bad (ĭs-lä'mə-bäd', ĭz-lăm'ə-băd'), city (1972 est. pop. 235,000), capital of Pakistan, NE Pakistan, just NE of Rawalpindi. Construction of Islamabad as the capital, replacing Karachi, began in 1960. There are light manufacturing industries. The nearby Murree Hills serve as a summer resort. Also near the city are the historical ruins of Taxila.

is·land (ī'lənd), relatively small body of land surrounded entirely by water. Depending on their origin, islands are either continental or oceanic. Continental islands are created by the submergence of coastal highlands of which only the summits remain above water or by the sea breaking through an isthmus or peninsula. Oceanic islands can result from the ascending of the ocean floor above water through volcanic action or other earth movements. Oceanic islands that result from coral growth are called atolls. They are always low and occur only in tropical ocean areas.

Island, county (1970 pop. 27,011), 212 sq mi (549.1 sq km), NW Wash., including a number of islands in Puget Sound; formed 1853; co. seat Coupeville. Its agriculture includes grain, poultry, and dairy products. It has fishing and summer resorts. The largest islands in the county are Whidbey and Camano.

Island No. 10, island (no longer extant) in the Mississippi River, between NW Tenn. and SE Mo.; site of an important campaign of the Civil War in which Union troops dug a canal through the swamps to allow their supply barges to bypass the heavily fortified island. After many attempts, Union gunboats finally succeeded in passing the island and reduced its shore batteries. The large Confederate garrison surrendered without a battle on Apr. 7, 1862.

Isle (īl, ēl), river, 145 mi (233.3 km) long, Dordogne and Gironde depts., SW central France, rising in the Haute-Vienne dept. and flowing generally SW, past Périgueux, to the Dordogne River.

Isle La Motte (īl' lə mŏt'), island and town (1970 pop. 262), 6 mi (9.7 km) long and 2 mi (3.2 km) wide, in Lake Champlain, NW Vt. The French chose the island as the site for Fort Ste. Anne (built 1666), the first white settlement in Vt. It remained garrisoned for some time but was abandoned long before permanent settlement began c.1788. Black marble is quarried on the island.

Isle of Wight (wīt), county (1970 pop. 18,285), 321 sq mi (831.4 sq km), SE Va., in a tidewater region bounded on the W by the Blackwater River, on the NE by the James River; formed 1634; co. seat Isle of Wight. Peanuts, corn, soybeans, and some grain are grown. The county is famous for its Smithfield hams. It also has a lumbering industry and a large paper mill.

Isle of Wight, village (1970 pop. 100), seat of Isle of Wight co., SE Va., W of Portsmouth.

Isle Roy·ale National Park (roi'əl), 539,341 acres (218,433 hectares), comprising about 200 islands, in Lake Superior, NW Mich.; est. 1940. Isle Royale, 210 sq mi (543.9 sq km), is the largest island in Lake Superior; Glaciated, the island has many lakes, streams, and inlets and remains a roadless, forested wilderness. Its abundant wildlife includes squirrels, beaver, fox, moose, and many kinds of birds. In prehistoric times, the island's copper was mined by Indians; artifacts from that period have been found. The French, lured by the fur trade, named the island in 1671. Isle Royale originally became U.S. territory in 1783 and was ceded to the United States by the Chippewa Indians in 1843. It was mined for copper from 1843 to 1899 by Americans; large areas of forest were burned to expose the ore and to build settlements. In the early 1900s the island was a popular vacation retreat.

Is·le·ta (ĭz-lā'tə), pueblo (1970 pop. 1,080), Bernalillo co., central N.Mex., on the E bank of the Rio Grande. It is a tourist attraction. According to many experts, the pueblo stands on the site it occupied when discovered in 1540. It was the seat of the Franciscan mission of San Antonio de Isleta from c.1621 until the Pueblo revolt of 1680. The Spanish captured the pueblo in 1681.

Is·ling·ton (ĭz'lĭng-tən), borough (1971 pop. 184,392) of Greater London, SE England. The borough was created in 1965 by the merger of the metropolitan boroughs of Islington and Finsbury. Islington, in the north, is mostly residential, while Finsbury, in the south, is highly industrialized. Industries include special and electrical engineering, printing, clockmaking, and brewing; clothing, furniture, and scientific, surgical, and optical instruments are manufactured. Bunhill Fields in Finsbury contains the graves of several literary figures. John Wesley's chapel and house and the Sadler's Wells Theatre, former home of the Royal Ballet, are also in Finsbury.

I·slip (ī'sləp), town (1970 pop. 278,880), Suffolk co., SE N.Y., on Great South Bay NE of Babylon; settled c. 1665. It is a resort and a residential suburb of New York City.

Is·ma·il·ia (ĭs'mā-ĭl'ē-ə), city (1970 est. pop. 167,500), capital of Ismailia governorate, NE Egypt. Ismailia was founded in 1863 by Ferdinand de Lesseps, who used it as his base of operations during the construction of the Suez Canal.

Is·na (ĭs-nä') or **Es·na** (ĕs'nə), town (1966 pop. 27,400), central Egypt, on the Nile River. It is the center for an agricultural area that is irrigated by the Nile. Isna's manufactures include cotton fabrics and ceramics. Nearby is a Coptic Christian monastery, said to have been founded in the 4th cent. to commemorate those martyred by Diocletian but now believed to date from the 10th or 11th cent.

I·son·zo (ē-zôn'tsô), river, 87 mi (140 km) long, rising in the Julian Alps, NW Yugoslavia, and flowing S through Yugoslavia, then SW through NE Italy before emptying into the Gulf of Trieste. At the entrance to the Venetian plain, the Isonzo valley was the scene of many battles during World Wars I and II.

Is·par·ta (ĭs'pär-tä'), city (1970 pop. 51,107), capital of Isparta prov., W central Turkey. It is a manufacturing center producing cotton, carpets, and attar of roses. A picturesque city, it was severely damaged by an earthquake in 1889.

Is·ra·el (ĭz'rē-əl), republic (1972 pop. 3,164,000), SW Asia, on the Mediterranean Sea. The capital of Israel is Jerusalem. The country is a narrow, irregularly shaped strip of land bounded on the north by Lebanon, on the east by Syria and Jordan, on the west by the Mediterranean Sea, on the southwest by Egypt, and on the south by the Gulf of Aqaba (an arm of the Red Sea).

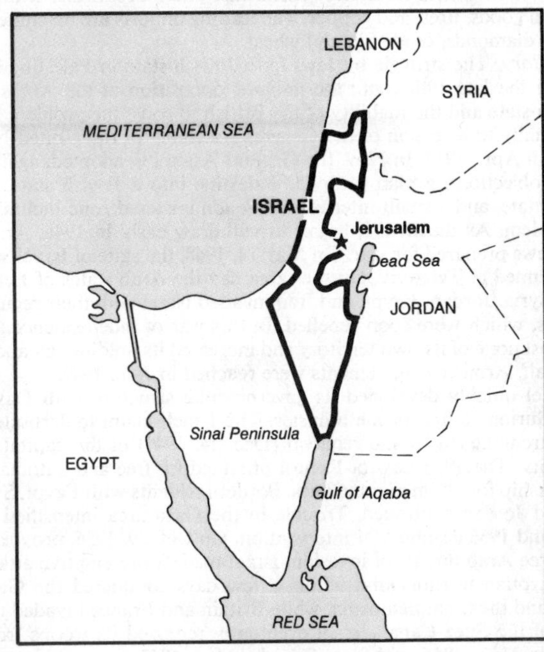

Israel has four principal regions: the plain along the Mediterranean coast; the mountains, which are east of this coastal plain; the Negev, which comprises the southern half of the country; and the portion of Israel that forms part of the Jordan Valley. North of the Negev, Israel enjoys a Mediterranean climate, with long, hot, dry summers and short, cool, rainy winters. This northern half of the country has a limited but adequate supply of water, except in times of drought. The Negev, however, is a semiarid region, having less than 10 in. (25 cm) of rainfall a year. The most important river in Israel is the Jordan. Other, smaller, rivers are the Yarkon, the Kishon, and the Yarmuk. In the southern part of the country are many wadis, or riverbeds, that are dry except in the brief rainy season. As the result of an intensive reforestation program, more than 100 million trees have been planted since 1948, the year the state of Israel came into being.

Israel is divided into six districts:

NAME	CAPITAL	NAME	CAPITAL
Central	Ramla	Northern	Nazareth
Haifa	Haifa	Southern	Beersheba
Jerusalem	Jerusalem	Tel Aviv	Tel Aviv-Jaffa

Economy. The economy of Israel is based on both state and private ownership and operation. Despite adverse conditions, agriculture has been developed successfully. The area of land under cultivation has been greatly increased since the founding of the state in 1948, and extensive irrigation has been provided to develop farm land and compensate for the shortage of water. The chief crop and most important product of Israel is citrus fruit, 95% of which is exported. In addition, flowers, noncitrus fruits, and vegetables are exported to Europe, especially as out-of-season crops in the winter. Other sizable crops are sugar beets, peanuts, cotton, and sisal. Most of the land is held in trust for the people of Israel by the state and the Jewish National Fund. The state and the Fund lease the land to kibbutzim, which are communal agricultural settlements; to moshavim, which are cooperative agricultural communities; and to other agricultural villages.

The Israelis have also made great strides in developing their industries since 1948, despite a comparative scarcity of raw materials. The major industries include the cutting and polishing of diamonds, the manufacture of chemical fertilizers from potash and phosphates, the mining of copper, building construction, and the manufacture of textiles. Besides potash, the Dead Sea yields magnesium and bromine. There are also light industries producing processed foods, precision instruments, shoes, ready-made clothes, and plastic products of various kinds. Diamonds are the largest industrial export, followed by chemical fertilizers, chemicals, pharmaceuticals, textiles, fashion goods, tires, and copper. The leading imports are machinery, rough diamonds, crude oil, and wheat.

History. The struggle by Jews for a Jewish state in Palestine began in the late 19th cent.; the militant opposition of the Arabs to such a state and the inability of the British to solve the problem led eventually to a session of the General Assembly of the United Nations in Apr., 1947. In Nov. the General Assembly adopted, against Arab objections, a plan to divide Palestine into a Jewish state, an Arab state, and a small internationally administered zone including Jerusalem. As the British began to withdraw early in 1948, Arabs and Jews prepared for war. On May 14, 1948, the state of Israel was proclaimed at Tel Aviv. On the same day the Arab states of Lebanon, Syria, Jordan, Egypt, and Iraq invaded Israel with their regular armies, which were soon repelled. In this war of independence, Israel lost none of its own territory and increased its holdings by about one half. Armistice agreements were reached in Jan., 1949.

Israel quickly developed its governmental structure with David Ben-Gurion as first prime minister. The Israeli claim to Jerusalem was strengthened by the removal (Dec. 14, 1949) of the capital to that city. The 1950 Law of Return provided for free and automatic citizenship for all immigrant Jews. Border incidents with Egypt, Syria, and Jordan continued. Trouble in the Gaza area intensified in 1955 and 1956 despite UN intervention. On Oct. 29, 1956, provoked by fierce Arab threats of invasion, Israel made a pre-emptive attack on Egyptian territory and within a few days conquered the Gaza Strip and the Sinai peninsula, while Britain and France invaded the area of the Suez Canal. Israel eventually removed its troops from Sinai in Nov., 1956, and from Gaza by Mar., 1957.

In 1963 Ben-Gurion was succeeded as prime minister by Levi Eshkol, who had to cope with increased guerrilla incursions and bombardments from Syria. The militant Palestine Liberation Organization (PLO) was formed in 1963 by refugees determined to force Israel to give up territory for an independent Arab Palestine. In May, 1967, Egyptian President Gamal Abdel Nasser mobilized the Egyptian army in Sinai and blockaded the Israeli port of Elat (on the Gulf of Aqaba). On June 5, 1967, Israel launched pre-emptive attacks against Egypt, Jordan, and Syria. In six days Israel occupied the Gaza Strip and the Sinai peninsula of Egypt, the Golan Heights of Syria, and the West Bank and Arab sector of Jerusalem (both under Jordanian rule). Cease-fires were obtained on all three fronts by the UN Security Council. On Nov. 22, 1967, the Security Council adopted a resolution calling for the withdrawal of Israeli forces from the Arab territories occupied in the war, the right of all states to live in peace within secure and recognized boundaries, and a just settlement of the Arab refugee problem. Israel continued to occupy the Arab territory taken in the Six-Day War. PLO guerrillas stepped up their incursions, operating largely from Jordan. After Eshkol's death on Feb. 26, 1969, Golda Meir became prime minister.

On Oct. 6, 1973, Egypt and Syria suddenly attacked Israeli positions in the Sinai and the Golan Heights. Egypt succeeded in sending troops in force across the Suez Canal to the east bank before being halted by the Israelis, who toward the end of the fighting managed to send their own troops across the Suez Canal and to drive back the Syrians. A cease-fire was called for by the UN Security Council on Oct. 22 and went into effect shortly thereafter. The Arab nations subsequently halted oil shipments for six months to the United States and other nations that had supported Israel. On Dec. 21, 1973, the first Arab-Israeli peace conference opened in Geneva, Switzerland, under UN auspices. An agreement to disengage Israeli and Egyptian forces was reached in Jan., 1974; a similar agreement was achieved in May, 1974.

In the course of the negotiations, Mrs. Meir resigned and was succeeded (June 3, 1974) by Yitzhak Rabin. The Rabin government was faced with the problem of increased attacks by Palestinian commandos within the borders of Israel. Most of the attacks were launched from refugee camps in south Lebanon. The Israelis retaliated with air and ground attacks on these camps. New Israeli settlements—some authorized and some illegal—were established in the occupied territories. In May, 1977, Rabin resigned and was succeeded by the country's first non-Labor Party prime minister, Menachem Begin, leader of the center-right Likud coalition. Egyptian President Anwar el-Sadat visited Jerusalem in Nov., 1977, urged a peace settlement in a speech to the Knesset, and opened talks with Israeli leaders. Despite increasing diplomatic pressure from the United States, the Begin government refused to agree to relinquish all of the occupied territories, and talks broke off in mid-1978. Israeli forces invaded and briefly occupied south Lebanon in 1978 in retaliation for Palestinian guerrilla attacks. Finally, in 1979, a peace treaty was signed with Egypt.

Is·sa·quah (ĭs'ə-kwô'), city (1977 est. pop. 5,078), King co., W central Wash., ESE of Seattle; settled 1862 as a coal-mining town, inc. 1899. Today it is a trading center for a poultry and dairy region.

Is·sa·que·na (ĭs'ə-kwē'nə), county (1970 pop. 2,737), 415 sq mi (1,074.9 sq km), W Miss., drained by winding bayous, the Mississippi (which forms the W boundary and the La. line), and the Yazoo (which partly forms the S and E boundary); formed 1844; co. seat Mayersville. It has agriculture (cotton, corn, and oats) and timber. It includes part of Delta National Forest and has oxbow lakes along the Mississippi.

Is·sus (ĭs'əs), ancient town of SE Asia Minor, now in Turkey, located near the head of the modern Gulf of Iskenderun. Issus was the scene of three historic battles. In 333 B.C., Alexander defeated the forces of Darius III of Persia. In A.D. 194 Septimius Severus conquered Pescennius Niger, a claimant to the throne of the Roman Empire. In 1622 the Byzantine emperor Heraclius won the first of a series of battles in which the West regained territory formerly lost to the Persians.

Is·syk-Kul (ĭs'ĭk-kool'), lake, E central Asian USSR, in the Kirghiz Republic, in the Ala-Tau Mts. At an altitude of c.5,300 ft (1,615 m) and with an area of c.2,400 sq mi (6,215 sq km), it is one of the largest mountain lakes in the world. It reaches a depth of 2,303 ft (702.4 m), is slightly saline, and is ice-free in winter.

Is·sy-les-Mou·li·neaux (ē-sē'lā-moo'lə-nō'), suburb SW of Paris (1968 pop. 51,666), Hauts-de-Seine dept., N central France. It is an industrial center where metals, aeronautical equipment, chemicals, cartridges, and beer are manufactured.

Is·ta·khr (ĭs-tä'kər), old town, S Iran. Built largely from the ruins of ancient Persepolis, it was a capital of the Sassanid dynasty. Istakhr stubbornly resisted (640–49) the Arabs but soon afterward lost its importance to Shiraz.

İs·tan·bul (ĭs'tăn-bool', -tän-), city (1970 pop. 2,247,630), capital of İstanbul prov., NW Turkey, on both sides of the Bosporus at its entrance into the Sea of Marmara. Its name was officially changed from Constantinople to İstanbul in 1930; before A.D. 330 it was known as Byzantium. One of the great historic cities of the world, İstanbul is the chief city and seaport of Turkey as well as its commercial and financial center. Tobacco is processed, and textiles, glass, shoes, and cement are manufactured. The city is visited by many tourists and is a summer resort. It is the see of the patriarch of the Greek Orthodox Church, of a Latin-rite patriarch of the Roman Catholic Church, and of a patriarch of the Armenian Church. The European part of İstanbul is the terminus of an international rail service (formerly called the Orient Express), and at Haydarpaşa station, on the Asian side, begins the Baghdad Railway. The part of İstanbul corresponding to historic Constantinople is situated entirely on the European side. It rises on both sides of the Golden Horn, an inlet of the Bosporus, on one of the finest sites of the world, and like Rome is built on seven hills. Several miles of its ancient moated and turreted walls are still standing. The chief monument surviving from Byzantine times is the Hagia Sophia, one of the world's noblest works of architecture. Originally a church, it was converted into a mosque after the Ottoman conquest in 1453 and is now a museum. Excavations on the sites of the former Byzantine palaces have brought to light fine works of art, and İstanbul has many monuments of the Byzantine past. The city was destroyed (1509) by an earthquake and was rebuilt by Sultan Beyazid II. Turkish culture reached its height in the 16th cent., and from that period date most of its magnificent mosques, notably those of Beyazid II, Sulayman I, and Ahmed I. They all reflect the influence of the Hagia Sophia—yet are distinctly Turkish—and give the skyline of İstanbul its unique character, a succession of perfectly proportioned domes broken by minarets. In the gardens by the Bosporus stand the buildings of the Seraglio, the former palace of the Ottoman sultans, now a museum. The Seraglio, begun by Mohammed II in 1462, consists of many buildings and kiosks, grouped into three courts, the last of which contained the treasury, the harem, and the private apartments of the ruler. In the 19th cent. the sultans shifted (1853) their residence to the Dolma Bahçe Palace and the Yıldız Kiosk, north of Beyoğlu on the Bosporus. The environs of İstanbul, particularly the villas, gardens, castles, and small communities along the Bosporus, are famed for their beauty. Always a cosmopolitan city, İstanbul has preserved much of its international and polyglot character and contains sizable foreign minorities. In 1973 the European and Asian sections of the city were linked by the opening of the Bosporus Bridge, one of the world's longest (3,524 ft/1,074 m) suspension bridges.

Is·tri·a (ĭs'trē-ə), mountainous peninsula, c.1,500 sq mi (3,885 sq km), NW Yugoslavia, projecting into the N Adriatic between the gulfs of Trieste and Fiume. A section of the northwestern portion belongs to Italy. The area is thickly forested and is predominantly agricultural. Istria was inhabited by Illyrian tribes when it passed (2nd cent. B.C.) to Rome. It remained under nominal Byzantine rule until the 8th cent. A.D. By the 15th cent. Austria and Venice had absorbed parts of the region. The Treaty of Campo Formio (1797) and the Congress of Vienna (1815) added the Venetian part to Austria. In 1919 all Istria passed to Italy, but the Italian peace treaty of 1947 gave most of it to Yugoslavia.

I·ta·bu·na (ē'tə-bōō'nə), city (1970 pop. 113,409), Bahia state, E Brazil, on the Itabuna River. It is a cacao-producing center and has a well-developed cattle industry.

I·tal·ian East Af·ri·ca (ĭ-tăl'yən; ăf'rĭ-kə), former federation of the Italian colonies of Eritrea and Italian Somaliland and the kingdom of Ethiopia. The federation was formed (1936) to consolidate the administration of the three areas. Resistance to Italian rule was particularly strong in Ethiopia, and when British forces invaded the federation in Jan., 1941, they received widespread support. By Dec., 1942, the Italians had been totally defeated. Ethiopia was restored its independence; Eritrea was placed under Ethiopian control in 1952; and Italian Somaliland, after a period as a UN trusteeship, became part of Somalia in 1960.

It·a·ly (ĭt'l-ē) or *Italian* **I·ta·lia** (ē-täl'yä), republic (1973 est. pop. 55,300,000), 116,303 sq mi (301,225 sq km), S Europe, bordering on France in the NW, the Ligurian Sea and the Tyrrhenian Sea in the W, the Ionian Sea in the S, the Adriatic Sea in the E, Yugoslavia in the NE, and Austria and Switzerland in the N. The country includes the large Mediterranean islands of Sicily and Sardinia and several small islands, notably Elba, Capri, Ischia, and the Lipari Islands. Vatican City and San Marino are two independent enclaves on the Italian mainland. Rome is Italy's capital and largest city.

About 75% of Italy is mountainous or hilly, and roughly 20% of the country is forested. There are narrow strips of low-lying land along the Adriatic coast and parts of the Tyrrhenian coast. Northern Italy is made up largely of a vast plain that is contained by the Alps in the north and drained by the Po River and its tributaries. It is the richest part of the country, with the best farm land, the chief port (Genoa), and the largest industrial centers. Northern Italy also has a flourishing tourist trade on the Italian Riviera, in the Alps (including the Dolomites), on the shores of its beautiful lakes, and in Venice. The Italian peninsula, bootlike in shape and traversed in its entire length by the Apennines (which continue on into Sicily), comprises central and southern Italy. Except for the Po and Adige, Italy has only short rivers, among which the Arno and the Tiber are the best known. Most of Italy enjoys a Mediterranean climate; however, that of Sicily is subtropical, and in the Alps there are long and severe winters.

Italy is divided into 20 regions:

NAME	CAPITAL	NAME	CAPITAL
Abruzzi	Aquila	Marche	Ancona
Apulia	Bari	Molise	Campobasso
Basilicata	Potenza	Piedmont	Turin
Calabria	Catanzaro	Sardinia	Cagliari
Campania	Naples	Sicily	Palermo
Emilia-Romagna	Bologna	Trentino-Alto Adige	Trent
Friuli-Venezia Giulia	Trieste	Tuscany	Florence
Latium	Rome	Umbria	Perugia
Liguria	Genoa	Valle d'Aosta	Aosta
Lombardy	Milan	Venetia	Venice

Economy. By the early 1970s industry contributed about 40% of Italy's annual national product and agriculture only about 11%. The principal farm products are wheat, sugar beets, maize, potatoes, tomatoes, citrus fruit, olives, and livestock (especially cattle, pigs, sheep, and goats). In addition, much wine is produced from grapes grown throughout the country. Industry is centered in the north, particularly in the "golden triangle" of Milan-Turin-Genoa. The chief manufactures of the country include iron and steel, refined petroleum, chemicals, textiles, motor vehicles, and machinery. There is a small fishing industry. Italy has only limited mineral resources; the chief minerals produced are petroleum (especially in Sicily), lignite, iron ore, iron pyrites, bauxite, sulfur, and mercury. There are also large deposits of natural gas (methane).

History. The earliest known inhabitants of Italy seem to have been of Ligurian stock. The Etruscans, coming probably from Asia Minor, established themselves in central Italy before 800 B.C. Their expansion was checked in the north by the Celts and in the south by the Samnites in the 4th cent. B.C. The Latins of Latium and their neighbors, the Sabines, were the ancestors of the Romans and defeated the Etruscans and Greek colonists and ruled almost all of Italy by 270 B.C. The history of Italy from the 5th cent. B.C. to the 5th cent. A.D. is largely that of the growth of Rome and of the Roman Empire, of which Italy was the core. By the beginning of the Christian era, all of Italy had been thoroughly latinized, Roman citizenship was extended to all free Italians, an excellent system of roads had been built, and Italy, made tax-exempt, shared fully in the wealth of Rome.

Like the rest of the Roman Empire, Italy in the early 5th cent. A.D. began to be invaded by successive waves of barbarian tribes. The deposition (476) of Romulus Augustulus, the last Roman emperor of the West, is commonly regarded as the end of the Roman Empire. In 488 the Ostrogoth Theodoric the Great invaded Italy, took Ravenna (493), and began a long and beneficent rule. After Theodoric's death (526), Italy was reconquered by Emperor Justinian I of the East. Byzantine rule was soon displaced by that of the Lombards, who under Alboin established (569) a new kingdom. The papacy, which retained control of Rome, emerged as a political power and the chief bulwark of Latin civilization. In the 8th cent., the Franks defeated the Lombards and allied themselves with the pope. Charlemagne was crowned emperor of the West at Rome in 800.

After the breakup of the Carolingian empire (9th cent.), Italy passed to a long succession of weak kings and emperors who struggled with the popes and nobles for power while cities emerged and became increasingly autonomous. To protect their commerce and their industries, the cities grouped together in leagues, which often were at war with each other. The Holy Roman Empire (est. 962) exercised little effective control except in the north. In the 13th cent. the struggle between Emperor Frederick II and the papacy divided the cities and nobles into two strong parties, the Guelphs and Ghibellines. The factional strife led to the rise of despots in some cities, while others retained republican institutions. The most powerful princes and the most powerful republics (e.g., Florence, Venice, and Genoa) tended to increase their territories at the expense of weaker neighbors. By 1500 Italy had fallen into the following chief component parts: in the south, the kingdoms of Sicily and Naples; in central Italy, the Papal States, the republics of Siena, Florence, and Lucca, and the cities of Bologna, Forlì, Rimini, and Faenza; in the north, the duchies of Ferrara and Modena, Mantua, Milan, and Savoy, and the two great merchant republics, Venice and Genoa. Expanding trade with the Middle East brought prosperity, which facilitated the great cultural flowering of the Italian Renaissance.

Italian cultural excellence failed to bring political strength, and the peninsula yielded to foreign domination in the 16th cent. Spain, successful in the Italian Wars (1494–1559) with France, gained the kingdoms of Sicily and Naples and the duchy of Milan. By 1748 the duchies of Milan, Mantua, Tuscany, and Modena had passed to Austria. Remaining independent were the Papal States, the declining republics of Venice, Genoa, and Lucca, and the kingdom of Sardinia (created in 1720). Despite economic decline, Italy continued to have considerable influence on European culture in the 18th cent., especially in architecture and music.

The French Revolution kindled Italian national aspirations, and Napoleon Bonaparte, who defeated Sardinian and Austrian armies in his Italian campaign of 1796–97, was at first acclaimed by most Italians. Napoleon established (1805) a kingdom of Italy (comprising Lombardy, Emilia-Romagna, and Venetia) and directly annexed other regions to France, but failed to unite Italy. The Congress of Vienna (1814–15) generally restored the pre-Napoleonic status quo with increased Austrian influence. The Risorgimento, as the movement for Italian unification was called, achieved its first successes with short-lived revolutionary outbreaks in Naples, Venice, Tuscany, Rome, and the kingdom of Sardinia in 1848–49. Unification was ultimately achieved under the house of Savoy, largely through the efforts of Count Camillo Cavour, Giuseppe Garibaldi, and Victor Emmanuel II, who became king of Italy in 1861. By 1870 the entire peninsula was included in the union, with Rome as the capital.

From 1861 until 1922 Italy was governed under the liberal constitution adopted by Sardinia in 1848. Industrialization proceeded slowly in the north, while in the underdeveloped south, rapid population growth led to mass emigration. Italy acquired part of Somaliland in 1889, Eritrea in 1890, and Libya and the Dodecanese in the Italo-Turkish War (1911–12). At first neutral, Italy entered (1915) World War I on the Allied side and obtained Trieste and other territories at the Paris Peace Conference.

After the war political and social unrest increased, furthering the growth of Fascism. The Fascist leader Benito Mussolini, promising the restoration of social order and of political greatness, became premier in 1922 and was granted dictatorial powers in 1923. He imposed strict controls on political activity, the press, industry, and labor. After 1935 he turned increasingly to militarist and imperialist solutions to Italy's problems. Italy conquered Ethiopia in 1935–36 and Albania in 1939. At the outbreak of World War II, Italy assumed a neutral stance friendly to Germany, but in June, 1940, it declared war on collapsing France and on Great Britain. By July, 1943, its army was shattered, and discontent among Italians culminated in Mussolini's dismissal by King Victor Emmanuel III. A new government surrendered to the Allies while German forces quickly occupied northern and central Italy. Rome fell to the Allies in July, 1944, and in May, 1945, the Germans surrendered. After the war, Italy's borders were established by the peace treaty of 1947, which assigned Trieste, Istria, most of Venezia Giulia, and several Adriatic islands to Yugoslavia and to the Free Territory of Trieste. In 1954 Trieste and its environs were returned to Italy, but the area south of Trieste was reassigned to Yugoslavia in 1975.

In 1946 Italians voted to make the country a republic, and a new constitution went into effect in 1948. The Christian Democrats, Communists, and Socialists emerged from the war as the chief parties, but none could count on majority support and the country was governed by short-lived coalitions headed by Christian Democrats. The Communists, who dominated the local politics of Tuscany, Umbria, and Emilia-Romagna, gradually strengthened their position in parliament through the 1960s and 1970s, and in 1977 were granted a consultative role in a governing coalition. The Italian Communist Party professed independence of the USSR and disowned the radical leftist Red Brigades, whose acts of terrorism culminated in 1978 with the kidnaping and murder of the Christian Democrat leader Aldo Moro. Despite the pervasive instability, Italy's economy has expanded dramatically in the postwar period. The major problems of the 1970s were inflation, trade deficits, and labor unrest.

Government. Under the 1948 constitution, legislative power is vested in a bicameral parliament consisting of the 630-member chamber of deputies, which is popularly elected, and the senate, made up of 315 members elected by region, plus 5 life members nominated by the president of Italy and all living former presidents. The council of ministers, led by the premier, is the country's executive; it must have the confidence of parliament. The head of state is the president, chosen in a joint session by parliament.

I·ta·mi (ē-tä′mē), city (1976 est. pop. 175,051), Hyogo prefecture, S Honshu, Japan, on the Muko River and Osaka Bay. It is a residential suburb of Osaka.

I·ta·pe·cu·ru (ēt′ə-pĕ′ko͞o-ro͞o′), river, c.450 mi (725 km) long, Maranhão, NE Brazil, flowing NE and NW to São José Bay, near Rosario.

I·tas·ca (ī-tăs′kə), county (1970 pop. 35,530), 2,633 sq mi (6,819.5 sq km), N Minn., on the Mississippi River, with numerous lakes; formed 1849; co. seat Grand Rapids. It is in a resort and mining area, with iron-ore deposits in the Mesabi iron range. It has lumber logging and sawmilling and makes wood and paper products. There is some dairy farming in the county.

Itasca, Lake, shallow lake, 2 sq mi (5.2 sq km), in a pine-wooded swampy region, NW Minn. Henry R. Schoolcraft identified it (1832) as the source of the Mississippi. Later geographers consider the source to be above the lake. In 1891 the region was included in a state park, which has a historical and natural history museum.

It·a·wam·ba (ĭt′ə-wäm′bə), county (1970 pop. 16,847), 541 sq mi (1,401.2 sq km), NE Miss., bordering E on Ala. and drained by the East Fork of the Tombigbee River; formed 1836; co. seat Fulton. It has cotton, corn, and sorghum farming, cattle raising, timber sawmilling, and gas wells. Its manufactures include clothing, wood products, furniture, shoes, and copper.

Ith·a·ca (ĭth′ə-kə). **1.** City (1970 pop. 2,749), seat of Gratiot co., central Mich., WSW of Saginaw, in a farm area; inc. 1869. Shoes and gloves are made. **2.** City (1978 est. pop. 30,700), seat of Tompkins co., S central N.Y., at the S end of Cayuga Lake, in the Finger Lakes region; settled 1789, inc. as a city 1888. It is the seat of Cornell Univ. It is also a major producer of salt and an inland shipping point.

I·thá·ki (ē-thä′kē) or **Ithaca**, island (1971 pop. 4,156), c.37 sq mi (96 sq km), W Greece, one of the Ionian Islands. It is mountainous and has little arable land. The chief products are olive oil, currants, and wine. The island is traditionally celebrated as the home of Odysseus. Cyclopean walls and remains of a Corinthian colony (c.8th cent. B.C.) have been found.

I·to (ē′tō), city (1975 pop. 68,072), Shizuoka prefecture, central Honshu, Japan, on the Izu Peninsula and the Sagami Sea. It is a port and hot-springs resort.

I·tsu·ku-shi·ma (ē′tso͞o-ko͞o-shē′mə), sacred island, 12 sq mi (31 sq km), in the Inland Sea, Japan, SW of Hiroshima. It is the site of an ancient Shinto shrine, famous for its magical beauty.

It·u·rae·a (ĭt-yo͞o-rē′ə), ancient country on the N border of Palestine. Ancient geographers are not agreed as to the exact limits of the country. The inhabitants were Arabians with their capital at Chalchis and their religious center at Heliopolis. Ituraea was conquered in 105 B.C. by Aristobulus, king of Judaea, who annexed it to Judaea and converted many of the inhabitants to Judaism. Later, after a brief period of independence, the country was subdued by Pompey. It remained thereafter chiefly in Roman hands, being united (A.D. c.50) to the Roman province of Syria.

I·tu·rup (ē′to͞o-ro͞op), island, 2,587 sq mi (6,700.3 sq km), S Kuril Islands, Russian SFSR, USSR, NE of Hokkaido, Japan. Iturup is noted for fur, whaling, and sulfur mining.

It·ze·hoe (ĭt′sə-hō), city (1974 est. pop. 35,622), Schleswig-Holstein, N West Germany, on the Stör River. It is a commercial center; manufactures include cement and machinery. Itzehoe was founded c.810 by Charlemagne, and it passed to Prussia in 1866.

I·u·ka (ī-yōō'kə), town (1970 pop. 2,389), seat of Tishomingo co., extreme NE Miss., near the Tennessee River SE of Corinth; inc. 1857. It is a resort town.

I·van·hoe (ī'vən-hō'), village (1970 pop. 738), seat of Lincoln co., SW Minn., near the S.Dak. border W of Marshall, in a grain, livestock, and poultry area. Dairy products are processed.

I·va·no-Fran·kovsk (ĭ-vä'nô-frăng'kôfsk'), formerly **Sta·nis·lav** (stə-nyĭ-släf'), city (1976 est. pop. 139,000), capital of Ivano-Frankovsk oblast, extreme SW European USSR, in Ukraine, on the Bystritsa River. It is a rail junction and industrial center situated in a fertile agricultural zone of the Carpathian foothills. The city has oil refineries, railroad repair shops, engineering and food processing plants, and factories that produce farm machinery, metal goods, leather footwear, furniture, cement, clothing, and other items. Oil fields are nearby. An old Ukrainian settlement, Stanislav was chartered in 1662. It passed to Austria in 1772 and to Poland in 1919 and was incorporated into the Ukraine in 1939.

I·va·no·vo (ĭ-vä'nō-vō), city (1976 est. pop. 458,000), capital of Ivanovo oblast, central European USSR, in the Moscow industrial region. A great Soviet textile center, the city was the historic center of Russia's cotton-milling industry. From the 1880s it was a center of labor unrest. During the revolution of 1905, 60,000 workers went on strike and formed one of the first soviets of workers' representatives. After six weeks the strike was crushed.

I·vo·ry Coast (ī'və-rē, ī'vrē), independent republic (1977 est. pop. 7,040,000), 124,503 sq mi (322,463 sq km), W Africa, on the Gulf of Guinea of the Atlantic Ocean. The capital and chief port is Abidjan. The Ivory Coast is bordered by Liberia and Guinea on the west, by Mali and Upper Volta on the north, and by Ghana on the east. The country consists of a coastal lowland in the south, a densely forested plateau in the interior, and a region of high savannas in the north.

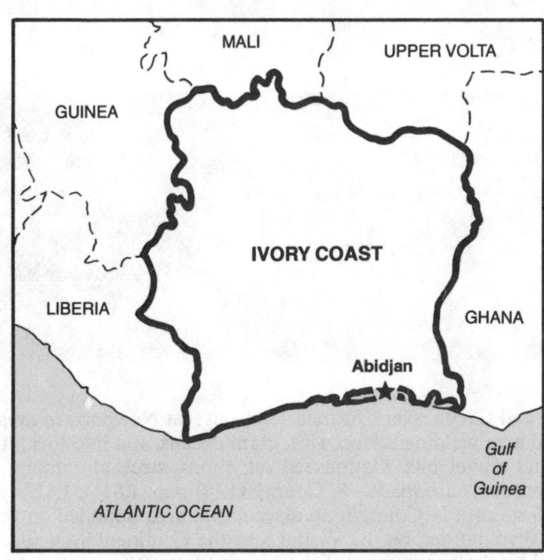

Economy. The wealthiest member of what was formerly French West Africa, the Ivory Coast has enjoyed a high economic growth rate since independence. Despite steady industrialization during the 1960s, the country is still predominantly agricultural. The Ivory Coast is one of the world's largest coffee producers. Cotton, cocoa, bananas, pineapples, and palm kernels are also grown. Mahogany and other hardwood forests provide timber. Livestock is raised in the savannas. Fishing and the canning of tuna are also important occupations. Among the country's industries are the production of flour, palm oil, petroleum, textiles, and cigarettes and the assembly of motor vehicles and bicycles.

History. In precolonial times the Ivory Coast was dominated by native kingdoms. The Portuguese established trading settlements along the coast in the 16th cent., and other Europeans later joined the burgeoning trade in slaves and ivory. After 1870 France undertook a systematic conquest. Due to strong resistance by native tribes, effective French control over the area was not established until after World War I.

Félix Houphouët-Boigny, a planter, formed (1946) the nationalist Parti Démocratique de la Côte d'Ivoire (PDCI). In 1958 the Ivory Coast chose autonomy within the French Community. The following year Houphouët-Boigny played an instrumental role in the formation of the Council of the Entente, a customs union with Benin, Nigeria, and Upper Volta (Togo joined later). In 1960 the Ivory Coast declared itself independent. Houphouët-Boigny has headed the government as well as the PDCI since independence. Despite student and worker unrest, his leadership has not been seriously challenged.

Government. The president is elected for a five-year term by universal suffrage. The national assembly is elected concurrently.

I·vre·a (ē-vrā'ä), city (1976 est. pop. 29,655), Piedmont, NW Italy, on the Dora Baltea River. Manufactures include typewriters and calculating machines, textiles, and silverware. A Roman town, it was later the capital of a Lombard duchy. Berengar II, one of its rulers, was briefly king of Italy (mid-10th cent.). Ivrea passed to the house of Savoy in the 14th cent. The city is dominated by a picturesque castle (14th cent.), which has four red brick towers.

I·vry-sur-Seine (ē-vrē'sür-sĕn'), industrial and commercial suburb SE of Paris (1975 pop. 62,856), Val-de-Marne dept., N central France. Its manufactures include chemicals, metals, pharmaceuticals, and oils. There are churches dating from the 13th, 16th, and 17th cent.

I·wa·ki (ē-wä'kē), city (1976 est. pop. 262,100), Fukushima prefecture, NE Honshu, Japan, on the Iwaki River. It is a major coal-mining center, railway hub, and industrial city where machinery, chemicals, and chemical fertilizers are produced.

I·wa·ku·ni (ē'wə-kōō'nē), city (1976 est. pop. 111,136), Yamaguchi prefecture, SW Honshu, Japan, on the Aki Sea. It is an important industrial center with petroleum refineries and rayon, chemical fiber, paper pulp, and metal machine-tool industries. Iwakuni castle (1603) is an important historical site.

I·wa·ta (ē-wä'tä), city (1975 pop. 67,665), Shizuoka prefecture, central Honshu, Japan, on the estuary of the Tenryu River. It is an agricultural and industrial center.

I·wa·te (ē-wä'tā), dormant volcanic mountain, 6,696 ft (2,042.3 m), Iwate prefecture, N Honshu, Japan. Shrines were built in its ancient crater, which attract thousands of pilgrims every year.

I·wa·tsu·ki (ē-wä'tsōō-kē), city (1975 pop. 60,900), Saitama prefecture, central Honshu, Japan, on the Edo River. It is an industrial center where textiles and dolls are manufactured.

I·wo (ē'wō), city (1971 est. pop. 192,000), SW Nigeria. It is the trade center for a farm region specializing in cacao. Iwo was the capital of a Yoruba kingdom (founded in the 17th cent.) that grew rapidly in the 19th cent. by taking in refugees during the Yoruba civil wars.

I·wo Ji·ma (ē'wə jē'mə, ē'wō), volcanic island, c.8 sq mi (21 sq km), W Pacific, largest and most important of the Volcano Islands. Mt. Suribachi, 546 ft (166.5 m) high, on the south side of the island, is an extinct volcano. The main industries are sulfur mining and sugar refining. During World War II the island, site of a Japanese air base, was taken (Feb.-Mar., 1945), at great cost by U.S. forces.

Ix·elles (ēk-sĕl'), city (1977 est. pop. 80,151), Brabant prov., central Belgium, an industrial suburb of Brussels.

Ix·ta·ci·huatl (ēs'tä-sē'wätl), dormant volcano, 17,342 ft (5,289.3 m) high, central Mexico, on the border between Puebla and Mexico state. Irregular in outline, and snow-capped, it is also popularly known as the Sleeping Woman.

I·za·bal (ē-sä-bäl'), lake, c.30 mi (48 km) long and 15 mi (24 km) wide, E Guatemala, largest lake in the country. It drains into the Caribbean Sea through the Golfete Dulce, a small adjacent lake, and the Rio Dulce, a broad tropical river. In Spanish colonial times Lake Izabal was the scene of lively trading between the seacoast and the highlands, and the small town of Izabal on its south shore was a thriving port, constantly subjected to raids in the 17th cent. by English and Dutch buccaneers. Nearby are many pre-Columbian ruins.

I·zal·co (ē-säl'kō), volcano, 7,828 ft (2,387.5 m) high, W El Salvador. Constantly active and still increasing in height, it is sometimes called the Lighthouse of the Pacific. There have been severe eruptions.

Iz·ard (ĭz'ərd), county (1970 pop. 7,381), 574 sq mi (1,486.7 sq km), N Ark., bounded on the SW by the White River and drained by the Strawberry River; formed 1825; co. seat Melbourne. Its agriculture includes livestock, poultry, cotton, grain, hay, and dairy products. It has lumber (sawmills, hardwood), cotton ginning, and gravel pits.

I·zhevsk (ī-zhĕfsk′), city (1974 est. pop. 534,100), capital of Udmurt Autonomous SSR, E European USSR. A major steel-milling and metallurgical center, Izhevsk has ironworks dating back to 1760.

Iz·ma·il (ēz-mə-ēl′), city (1976 est. pop. 78,000), SW European USSR, in the Ukraine, on an arm of the Danube delta near the Rumanian border. Orchards and vineyards surround the city. Izmail's industries include food and fish processing, winemaking, auto and ship repair, and the manufacture of bricks and tiles. First known in the 16th cent., the city was a Turkish fortress. Russian forces took the city twice (1770, 1790) during the Russo-Turkish Wars. Recaptured by the Russians in 1809, it was ceded to them by the Treaty of Bucharest (1812). At the Congress of Paris in 1856 Izmail was returned to Turkey; but Russia seized the city again in 1878 and held it until 1918, when Rumania took it. Transferred to the USSR in 1940, it was reconquered by the Rumanians the following year but restored to the USSR in 1947.

Iz·mir (ĭz′mēr), formerly **Smyr·na** (smûr′nə), city (1975 pop. 636,078), capital of İzmir prov., W Turkey, on the Gulf of İzmir, an arm of the Aegean Sea. It is an important seaport and commercial and industrial center, whose manufactures include paper, metal goods, dyes, textiles, and processed food and tobacco.

The city was settled during the Bronze Age (c.3000 B.C.). It was colonized (c.1000 B.C.) by Ionians and was destroyed (627 B.C.) by the Lydians. It was rebuilt on a different site in the early 4th cent. B.C. and became one of the largest and most prosperous cities of Asia Minor. Its wealth and splendor increased under Roman rule. The city had a sizable Jewish colony and was an early center of Christianity. It was pillaged by the Arabs in the 7th cent., fell to the Seljuk Turks in the 11th cent., was recaptured for Byzantium during the First Crusade, and formed part of the empire of Nicaea from 1204 to 1261, when the Byzantine Empire was restored. Also in 1261 the Genoese obtained trading privileges, which they retained until the city fell (c.1329) to the Seljuk Turks. The Knights Hospitalers captured the city in 1344, restored Genoese privileges, and held the city until 1402, when it was captured and sacked by Tamerlane. The Mongols were succeeded in 1424 by the Ottoman Turks.

After the collapse of the Ottoman Empire in World War I, the city was occupied (1919) by Greek forces. The Treaty of Sèvres (1920) assigned İzmir and its hinterland to temporary Greek administration, but fighting soon erupted between Greek and Turkish forces. İzmir fell to the Turks in Sept., 1922, and a few days later was destroyed by fire. Thousands of Greek civilian refugees fled from the city. The Treaty of Lausanne (1923) restored İzmir to Turkey. A separate convention between Greece and Turkey provided for the exchange of their minorities, which was carried out under League of Nations supervision. The city suffered greatly from severe earthquakes in 1928 and 1939.

Iz·mit (ĭz′mĭt), city (1975 pop. 164,675), capital of Kocaeli prov., NW Turkey, a port on the Bay of İzmit, at the E end of the Sea of Marmara. It is the center of a rich tobacco- and olive-growing region. Manufactures include beer and paper and paper products.

I·zu·mi (ē-zo͞o′mē), city (1976 est. pop. 120,153), Osaka prefecture, S Honshu, Japan. It is a residential and commercial suburb of Osaka, with numerous textile mills.

I·zu·mi·ot·su (ē-zo͞o′mē-ōt′so͞o), city (1975 pop. 66,250), Osaka prefecture, S Honshu, Japan, on Osaka Bay. It is a commercial port with chemical and textile industries.

I·zu·mi·sa·no (ē-zo͞o′mē-sä′nō), city (1975 pop. 86,139), Osaka prefecture, S Honshu, Japan, on Osaka Bay. It is a fishing port.

I·zu-shi·chi·to (ē′zo͞o-shē-chē′tō), island group, extending c.300 mi (480 km) S of Tokyo Bay, Japan. O-shima is the largest of these volcanic islands, which are now tourist attractions. The islands were formerly used for penal settlements.

Jab·al·pur (jŭb′əl-pôr′), city (1971 pop. 533,751), Madhya Pradesh state, central India, on the Narmada River. It is an important rail junction and military post. Manufactures include weapons, ammunition, and cigarettes.

Ja·blo·nec nad Ni·sou (yä′blô-nĕts nät nĭ′sü), city (1975 est. pop. 37,253), N Czechoslovakia, in Bohemia, on the Lausitzer Neisse River. The glassware center of Czechoslovakia, it also has industries that manufacture automobile equipment, textiles, plastics, and buttons and costume jewelry.

Jack (jăk), county (1970 pop. 6,711), 944 sq mi (2,445 sq km), N Texas, drained by the West Fork of the Trinity River and including part of Lake Bridgeport; formed 1856; co. seat Jacksboro. It is mainly agricultural, yielding livestock, grain, cotton, corn, peanuts, pecans, fruit, and truck crops.

Jacks·bor·o (jăks′bûr-ō). **1.** Village (1974 est. pop. 1,025), seat of Campbell co., NE Tenn., NW of Knoxville. **2.** City (1970 pop. 3,554), seat of Jack co., N Texas, SSE of Wichita Falls; inc. 1899. It grew near a frontier army post, Fort Richardson, founded in 1867.

Jack·son (jăk′sən). **1.** County (1970 pop. 39,202), 1,079 sq mi (2,794.6 sq km), extreme NE Ala., bordering on Ga. and Tenn. and drained by the Tennessee and Paint Rock rivers and Gutersville Lake; formed 1819; co. seat Scottsboro. It is in an agricultural region that produces cotton and corn. Lumber and coal and limestone deposits are also important. Its manufactures include textiles, clothing, plastics, iron and aluminum, and metal products. **2.** County (1970 pop. 20,452), 629 sq mi (1,629.1 sq km), NE Ark., drained by the Black, White, and Cache rivers; formed 1829; co. seat Newport. In an agricultural area yielding cotton, rice, corn, pecans, and livestock, it has sand and gravel pits. Cottonseed oil, shoes, steel, aluminum, and metal products are made. **3.** County (1970 pop. 1,811), 1,623 sq mi (4,203.6 sq km), N Colo., in an agricultural area bounded on the N by the Wyo. border, on the W and S by the Continental Divide, and drained by the headwaters of the North Platte River; formed 1909; co. seat Walden. Lumber and livestock are its major resources. The county includes parts of Routt and Arapaho national forests. **4.** County (1970 pop. 34,434), 935 sq mi (2,421.7 sq km), NW Fla., bordered on the N by Ala., on the E by Ga. and the Chattahoochee River, and drained by the Chipola River; formed 1822; co. seat Marianna. It is in a rolling agricultural area (peanuts, corn, cotton, vegetables, and hogs) with many small lakes. Some manufacturing (food products, lumber, and naval stores) and limestone quarrying are done. **5.** County (1970 pop. 21,039), 346 sq mi (896.1 sq km), NE central Ga., in a piedmont area drained by the Oconee River; formed 1796; co. seat Jefferson. Agriculture (cotton, corn, hay, sweet potatoes, apples, peaches, and poultry) and textile manufacturing are important to its economy. **6.** County (1970 pop. 55,008), 603 sq mi (1,561.8 sq km), SW Ill., bounded in the SW by the Mississippi River and drained by the Big Muddy and Little Muddy rivers; formed 1816; co. seat Murphysboro. The southern section lies within the Ozarks. Its agriculture includes corn, wheat, fruit, truck crops, dairy products, and livestock. It has bituminous-coal mines and plants manufacturing clothing, railroad equipment, beverages, flour, silica, and iron products. Part of Shawnee National Forest is here. **7.** County (1970 pop. 33,187), 520 sq mi (1,346.8 sq km), S Ind., bounded in the S by the Muscatatuck River and drained by the East Fork of the White River; formed 1815; co. seat Brownstown. It has agriculture (grain, truck crops, livestock, and poultry), timber, and diversified manufacturing. **8.** County (1970 pop. 20,839), 644 sq mi (1,668 sq km), E Iowa, bounded on the E by the Mississippi River and the Ill. border and drained by the Maquoketa River; formed 1837; co. seat Maquoketa. Primarily agricultural, it produces hogs, cattle, poultry, corn, oats, and alfalfa. It also has limestone quarries. **9.** County (1970 pop. 10,342), 656 sq mi (1,699 sq km), NE Kansas, in a rolling prairie region watered by headstreams of the Delaware River; formed 1855 as Calhoun co., renamed 1859; co. seat Holton. It has livestock ranches and grain farms. There is some manufacturing. Pottawatomie Indian Reservation is in the west. **10.** County (1970 pop. 10,005), 337 sq mi (872.8 sq km), SE Ky., in the Cumberland foothills, drained by headstreams of the Rockcastle River and several creeks; formed 1858; co. seat McKee. In a mountain agricultural area yielding livestock, fruit, tobacco, hay, and corn, it has coal mines and stands of timber. Miscellaneous manufacturing is done. Part of Daniel Boone National Forest is here. **11.** Parish (1970 pop. 15,963), 583 sq mi (1,510 sq km), N La., drained by the Dugdemona River and Bayou Castor; formed 1845; parish seat Jonesboro. Its agriculture includes cotton, corn, hay, peanuts, white and sweet potatoes, poultry, and dairy products. Some mining, timber logging, and cotton ginning are done. Paper products are manufactured. **12.** County (1970 pop. 143,274), 698 sq mi (1,807.8 sq km), S Mich., drained by the Grand and Raisin rivers and headstreams of the Kalamazoo River; organized 1832; co. seat Jackson. In a recreational area with many small lakes and good fishing, it has diversified agriculture (livestock, poultry, grain, corn, hay, fruit, and dairy products) and some industry. **13.** County (1970 pop. 14,352), 698 sq mi (1,807.8 sq km), S Minn., bordering on Iowa in the S and watered by the West Des Moines River and Heron Lake; formed 1857; co. seat Jackson. Its agriculture includes corn, oats, barley, potatoes, and livestock. Confectionery products and farm machinery are made. **14.** County (1970 pop. 87,975), 736 sq mi (1,906.2 sq km), extreme SE Miss., bounded on the E by the Ala. border, on the S by Mississippi Sound, and drained by the Pascagoula and Escatawpa Rivers; formed 1812; co. seat Pascagoula. It has agriculture (cotton, corn, pecans, livestock, and dairy products), commercial fisheries, and shipbuilding, lumbering, and food-processing industries. Petrochemicals, clothing, wood products, and concrete are made here. The county contains part of De Soto National Forest. **15.** County (1970 pop. 654,178), 603 sq mi (1,561.8 sq km), W Mo., bounded on the N by the Missouri River, on the W by Kansas; formed 1826; co. seat Independence. Its agriculture includes livestock, wheat, corn, oats, and dairy products. It has oil and gas wells, limestone deposits, and timber. Its manufactures include clothing and textiles, wood and metal products, furniture, paper, chemicals, rubber and plastics, cement, concrete, and packaged meats. **16.** County (1970 pop. 21,593), 499 sq mi (1,292.4 sq km), W N.C., partly in the Blue Ridge, bounded on the S by the S.C. border, and drained by the Tuckasegee River; formed 1851; co. seat Sylva. In a resort area, its economy depends on agriculture (livestock, apples, clover, potatoes, and corn), lumbering, and mica and talc mining. Part of Nantahala National Forest is here. **17.** County (1970 pop. 27,124), 420 sq mi (1,087.8 sq km), S Ohio, drained by the Little Scioto River; formed 1816; co. seat Jackson. In an agricultural area yielding livestock, grain, fruit, and dairy products, it also has coal mining and some manufacturing. Buckeye Furnace and Leo Petroglyph state parks are in the county. **18.** County (1970 pop. 30,902), 810 sq mi (2,097.9 sq km), SW Okla., bounded on the S by Texas and drained by the Red River and its North, Salt, and Prairie Dog Town forks; formed 1907; co. seat Altus. It is in an area irrigated by Altus Dam, and has agriculture (cotton, wheat, and alfalfa), oil and natural-gas wells, and copper-ore deposits. Its manufactures include cottonseed oil and clothing. **19.** County (1970 pop. 94,533), 2,817 sq mi (7,296 sq km), SW Oregon, in a mountainous area bordering on Calif. and crossed by the Rogue River; formed 1852; co. seat Medford. Agriculture (poultry, fruit, nuts, and truck crops), lumbering, and tourism are important to its economy. **20.** County (1970 pop. 1,531), 809 sq mi (2,095.3 sq km), SW central S.Dak., in an agricultural area watered by intermittent streams and bounded on the S by the White River; formed 1883 and later absorbed into other counties; reconstituted 1914; co. seat Kadoka. Winter wheat and cattle feed are the most important crops. The county includes part of the Badlands National Monument. **21.** County (1970 pop. 8,141), 327 sq mi (846.9 sq km), N central Tenn., intersected by the Cumberland River; formed 1801; co. seat Gainesboro. Mainly agricultural (tobacco, livestock, and corn), it has oil wells and a lumbering industry. **22.** County (1970 pop. 12,975), 854 sq mi (2,211.9 sq km), SE Texas, on the Gulf coastal plain with Lavaca and Matagorda bays in the S and drained by the Lavaca and Sandy rivers; formed 1836; co. seat Edna. Its agriculture includes cotton, rice, maize, corn, flax, fruit, vegetables, cattle, hogs, poultry, and dairy products. There are oil and natural-gas wells in the county. **23.** County (1970 pop. 20,903), 463 sq mi (1,199.2 sq km), W W.Va., bounded on the NW by the Ohio River and the Ohio border and drained by small creeks; formed 1831; co. seat Ripley. It is in a natural-gas and oil area, and there is some coal. Aluminum is its primary industry. Its agriculture includes livestock, dairy products, fruit, and tobacco. **24.** County (1970 pop. 15,325), 1,000 sq mi (2,590 sq km), W central Wis., intersected by the Black, Buffalo, and Trempealeau rivers; formed 1853; co. seat Black River Falls. Dairy farming is of primary importance.

Jackson. **1.** City (1970 pop. 5,957), Clarke co., SW Ala., on the Tombigbee River N of Mobile, in a lumbering and farming area; inc. 1816. **2.** City (1970 pop. 1,924), seat of Amador co., central Calif., ESE of Sacramento, in a farm, ranch, and mining area; settled c.1848, inc. 1905. **3.** City (1970 pop. 3,778), seat of Butts co., central Ga., SE of Atlanta; inc. 1826. It is a trade and processing center in a farm area. Textiles are produced. **4.** City (1970 pop. 1,887), seat of Breathitt co., E central Ky., on the North Fork of the Kentucky River SE of Lexington, in a mountain region with coal, timber, and farms. **5.** Town (1970 pop. 4,697), East Feliciana parish, SE La., N of Baton Rouge, in a farm region. **6.** City (1978 est. pop. 42,100), seat of Jackson co., S Mich., on the Grand River; inc. 1857. It is an industrial and commercial center in a farm region. Automobile and aircraft parts and accessories, food products, tires, electronic equipment, sheet-metal items, and metal toys are manufactured. **7.** City (1970 pop. 3,550), seat of Jackson co., SW Minn., on the Des Moines River and W of Fairmont, in a flax-growing and poultry region. Settled before 1857, it was destroyed in the Sioux rebellion of 1862, replatted in 1866, and incorporated in 1881. **8.** City (1978 est. pop. 193,000), state capital and seat of Hinds co., W central Miss., on the Pearl River; inc. 1833. It is the state's geographic center, with major rail, warehouse, and distributing operations. Industries include the production of oil and natural gas, food processing, and the manufacture of lumber, metal, glass, and wood products. The site of the city, a trading post near the Natchez Trace, was chosen and laid out as the state capital in 1821. During the Civil War, Jackson was largely destroyed by Gen. W. T. Sherman's forces in 1863. The old capitol (1839) is preserved as a museum; the new capitol was completed in 1903. Among the many points of interest are the governor's mansion (erected 1839); city hall, which was used as a hospital during the Civil War; a 220-acre (89.1-hectare) scale model of the Mississippi River flood-control system; and many ante-bellum homes. The Univ. of Mississippi Medical Center is here. **9.** City (1970 pop. 5,896), seat of Cape Girardeau co., SE Mo., NW of Cape Girardeau; platted 1814, inc. 1819. It is a trade and processing center in a farm and timber area. **10.** Town (1970 pop. 762), seat of Northampton co., NE N.C., SE of Roanoke Rapids. It is a trade center in a farming and sawmilling area. **11.** City (1970 pop. 6,843), seat of Jackson co., S Ohio, SE of Chillicothe; founded 1817. It is in an area of apple orchards, coal mines, and gas wells. Varied products are made. Nearby is Leo Petroglyph State Monument. **12.** City (1978 est. pop. 41,300), seat of Madison co., W Tenn., on the South Fork of the Forked Deer River; inc. 1823. It is a processing and rail shipping point for an extensive farm area. The city has industries producing a great variety of manufactures. Casey Jones is buried in Jackson; his home and railroad museum are of interest. **13.** Town (1970 pop. 2,688), seat of Teton co., NW Wyo., NNE of Afton near the Teton Mts. and the Snake River, at an altitude of c.6,200 ft (1,891 m). It is the tourist and trade center for the Jackson Hole hunting, skiing, dude-ranch, and livestock region.

Jackson, Port, inlet of the Pacific Ocean, 22 sq mi (57 sq km), 12 mi (19.3 km) long and 1.5 mi (2.4 km) wide at its mouth, New South Wales, Australia, forming Australia's finest harbor.

Jackson Hole, fertile Rocky Mt. valley, c.50 mi (80 km) long and 6 to 8 mi (9.6–12.9 km) wide, NW Wyo., in Grand Teton National Park. The valley is popular with hunters. Jackson Hole Wildlife Park, 1,500 acres (607.5 hectares), est. 1948, is the winter home of the largest elk herd in North America.

Jack·son·ville (jăk'sən-vĭl'). **1.** City (1970 pop. 7,715), Calhoun co., NE Ala., N of Anniston, in a cotton area; founded 1833, inc. 1836.

Fort McClellan is nearby. **2.** City (1978 est. pop. 26,300), Pulaski co., central Ark., inc. 1941. The nearby Little Rock Air Force Base, a tactical air command installation, is important to the city's economy. **3.** City (1978 est. pop. 547,000), coextensive (since 1968) with Duval co., NE Fla., on the St. Johns River near its mouth on the Atlantic Ocean; inc. 1832. The second-largest U.S. city in area (c.830 sq mi/2,150 sq km), it is a commercial and financial center, a great rail, air, and highway focal point, and a busy port of entry, with ship-repair yards, extensive freight-handling facilities, and large naval installations. The city is a leading manufacturing center, with lumber, paper, chemicals, food products, and cigars the principal products. It is also a tourist resort, with ocean beaches, fishing and yachting facilities, and inland hunting areas. Jacksonville was settled in 1816 by Lewis Hogan and laid out in 1822. The Seminole War and the Civil War (in which much of the city was destroyed) interrupted its growth, but with the development of a good deepwater harbor and railroads in the late 19th cent., industry and commerce increased. Points of interest include the Confederate monument in Hemming Park and nearby Fort Caroline National Memorial. **4.** City (1978 est. pop. 18,500), seat of Morgan co., W central Ill.; laid out 1825, inc. 1867. Its industries include bookbinding and the manufacture of clothing, plastics, phonograph records, and metal products. Stephen A. Douglas and William Jennings Bryan lived here. **5.** City (1978 est. pop. 19,100), seat of Onslow co., E N.C., on the New River; settled c.1757. It is a trade center in a farm area, with sawmills and plants making clothing, mobile homes, food products, and farm equipment. It is also a summer resort. Camp Lejeune, a U.S. Marine Corps training base, is nearby. **6.** City (1970 pop. 9,734), Cherokee co., E Texas, SE of Dallas; inc. 1916. Founded in 1847, it was moved to its present site in 1872 and was a market for a pine-timber region until the growing of truck crops made it prosperous after the 1920s.

Ja·dot·ville (zhä-dō-vēl'): *see* Likasi, Zaire.

Ja·én (hä-ĕn'), city (1975 est. pop. 82,050), capital of Jaén prov., S Spain, in Andalusia. It is a marketing and distribution center for a fertile area producing olive oil and wine. Nearby lead mines are believed to be among the richest in Europe.

Jaf·fa (jăf'ə), part of Tel Aviv-Jaffa, W central Israel, on the Mediterranean Sea. Originally a Phoenician city, Jaffa has been historically important largely because of its port (which was closed in 1965, when the port of Ashdod was completed). It was captured by Egypt in 1472 B.C. and made a provincial capital. In 701 B.C. the city was besieged by Sennacherib, king of Assyria. After the Captivity in Babylon (6th cent. B.C.) it become Hebrew territory. Alexander the Great took Jaffa in the late 4th cent. B.C. The city changed hands frequently in the fighting between the Maccabees and the Syrians (2nd and 1st cent. B.C.) and was destroyed by Vespasian in A.D. 68. The rebuilt city of Jaffa was conquered by the Arabs in 636. The Crusaders took it in 1126, Saladin recaptured it in 1187, and Richard I retook the city in 1191. In 1196 the Arabs again captured Jaffa, and in the 16th cent. the city, then in decline, was annexed by the Ottoman Empire. In the late 17th cent. Jaffa began to develop again as a seaport. It was captured by Napoleon in 1799. In World War I British troops took Jaffa, which became part of the British-administered Palestine mandate (1922–48). In 1947 and 1948 there was sharp fighting between Jaffa, which was largely inhabited by Arabs, and the adjoining all-Jewish city of Tel Aviv. The Arabs in Jaffa surrendered on the day (May 14, 1948) that the state of Israel was proclaimed. Most of the Arab population soon left. In 1950 the city was incorporated into Tel Aviv.

Jaf·fna (jăf'nə), peninsula, northernmost part of Sri Lanka (Ceylon), separated from India by Palk Strait. There are remains of ancient Tamil culture and of Portuguese and Dutch occupations of the 17th–18th cent. Fishing is an important occupation. The main industries are salt, cement, chemical, and tobacco production. The city of Jaffna (1968 est. pop. 101,000) has a good harbor and an active trade in elephants, peppers, and other commodities.

Jai·pur (jī'po͝or'), city (1971 pop. 613,144), capital of Rajasthan, W India; founded in 1728. Known as the pink city from the color of its houses, it is a transportation junction and a commercial center. It is enclosed by a crenellated wall 20 ft (6.1 m) high. An unusual feature for an Indian city of this size is the system of wide, regular streets.

Ja·kar·ta (jə-kär'tə, jä-kär'tä): *see* Djakarta, Indonesia.

Ja·la·pa En·rí·quez (hä-lä'pä ān-rē'kās), city (1974 est. pop. 161,400), capital of Veracruz state, E central Mexico, on the slopes of the Sierra Madre Oriental. It is located in a rich agricultural region of fertile valleys. The site of a pre-Columbian city, Jalapa was captured by Cortés in 1519. It was an important commercial center during the Spanish colonial era, but declined in the late 18th cent.

Ja·lis·co (hä-lē'skō), state (1970 pop. 3,296,587), 31,152 sq mi (80,684 sq km), W Mexico, bounded on the west by the Pacific; capital Guadalajara.

Jal·u·it (jăl'o͞o-ĭt), atoll (1970 pop. 492), c.40 mi (65 km) long and c.20 mi (30 km) wide, central Pacific, one of the Ralik Chain in the U.S. Marshall Islands. It comprises some 85 islets. In World War II it was the headquarters of the Japanese admiralty for the Marshall Islands. U.S. forces captured the atoll in 1944.

Ja·mai·ca (jə-mā'kə), republic (1973 est. pop. 2,000,000), 4,232 sq mi (10,961 sq km), coextensive with the island of Jamaica, West Indies, S of Cuba and W of Haiti, in the Caribbean Sea. The capital is Kingston. Although largely a limestone plateau more than 3,000 ft (915 m) above sea level, Jamaica has a mountainous backbone that extends across the island from the west and rises to the Blue Mts. in the east. A narrow plain along the northern coast and several larger plains near the south shore are Jamaica's major agricultural zones. The north coast also has fine beaches.

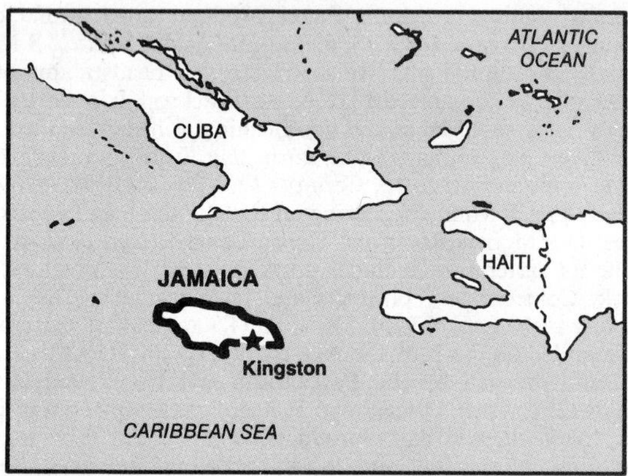

Economy. Jamaica's major crops are coffee, sugar cane, from which rum and molasses are also made, bananas, ginger, citrus fruits, cocoa, pimento, and tobacco. Most of these crops are grown on large plantations. Small peasant farms produce some ginger, bananas, and sugar cane for export but mainly raise such subsistence crops as yams, breadfruit, and cassava. Since large, easily accessible deposits of bauxite were discovered in 1942, Jamaica has become one of the world's leading suppliers of this ore. Along with the alumina made from it, bauxite accounts for about half of Jamaica's foreign exchange. Tourism is the second-biggest earner of exchange.

History. Discovered by Christopher Columbus in 1494, Jamaica was conquered and settled in 1509 by Spaniards under a license from Columbus's son. Spanish exploitation decimated the native Arawaks. The island remained Spanish until 1655, when Adm. William Penn and Robert Venables captured it; it was formally ceded to England in 1670, but the local white population obtained a degree of autonomy. Jamaica prospered from the wealth brought by buccaneers, notably Sir Henry Morgan, to Port Royal, the capital; in 1692, however, the city sank into the sea during an earthquake, and Spanish Town became the new capital.

A huge, mostly black, slave population grew up around the sugar-cane plantations in the 18th cent., when Jamaica was a leading world sugar producer. Freed slaves and runaways, sometimes aided by the Maroons (slaves who had escaped to remote areas after Spain lost control of Jamaica), collaborated in fomenting frequent rebellions against the white colonials. The sugar industry declined in the 19th cent., partly because of the abolition of slavery in 1833 (effective 1838) and partly because of the elimination in 1846 of the imperial preference tariff for colonial products entering the British market. Economic hardship was the prime motive behind the Morant Bay rebellion by freedmen in 1865. The British ruthlessly quelled the uprising and also forced the frightened white legislature to surrender its powers; Jamaica then became a crown colony.

A new constitution in 1884 marked the initial revival of local autonomy for Jamaica. Despite labor and other reforms, riots recurred, notably those of 1938, which were caused mainly by unemployment and resentment against British racial policies. A royal commission investigating the 1938 riots recommended an increase of economic development funds and a faster restoration of representative government for Jamaica. In 1944 universal suffrage was introduced, and a new constitution provided for a popularly elected house of representatives. In 1962 Jamaica won complete independence.

Jamaica Bay, c.20 sq mi (50 sq km), SW Long Island, SE N.Y., separated from the Atlantic Ocean by Rockaway Peninsula; the Rockaway Inlet links it to the sea. Nearly all of the bay is in New York City; since 1950 much of the adjacent area has been reclaimed for housing. John F. Kennedy International Airport extends into the bay. Part of Gateway National Recreation Area, the shallow bay is used for boating and fishing and is a wildlife refuge.

Jam•bi (jäm′bē): *see* Djambi, Indonesia.

Jam•bol (yäm′bôl): *see* Yambol, Bulgaria.

James (jāmz). **1.** Unnavigable river, 710 mi (1,142.4 km) long, rising in central N.Dak. and flowing across S.Dak. to the Missouri River at Yankton, S.Dak. Jamestown Dam on the river is an irrigation and flood-control unit of the Missouri River basin project. **2.** River, 340 mi (547.1 km) long, formed in W central Va. by the union of the Jackson and Cowpasture rivers and winding E across Va. to enter Chesapeake Bay through Hampton Roads. One of Virginia's chief rivers, it is navigable for large ships to Richmond, c.100 mi (160 km) upstream. The James's upper course flows through scenic gorges in the Blue Ridge Mts. and the Piedmont.

James Bay, shallow southern arm of Hudson Bay, c.300 mi (480 km) long and 140 mi (225 km) wide, E central Canada, in the Northwest Territories between Ont. and Que. The bay was discovered (1610) by Henry Hudson but was named for Capt. Thomas James, an Englishman who explored much of it in 1631.

James City, county (1970 pop. 17,853), 150 sq mi (388.5 sq km), SE Va., in a tidewater region bounded on the N by the York River, on the W by the Chickahominy River; formed 1634; co. seat Williamsburg (an independent city). Its agriculture includes truck crops, tobacco, corn, potatoes, fruit, dairy products, and livestock. Extensive fishing and oystering and some lumbering are done. Colonial National Historical Park is in the county.

James•town (jāmz′toun′). **1.** City (1970 pop. 1,027), seat of Russell co., S Ky., in the Cumberland foothills WSW of Somerset. Wood products, woolens, and flour are manufactured. **2.** City (1978 est. pop. 36,600), Chautauqua co., W N.Y., on Chautauqua Lake; founded c.1806, inc. as a city 1886. It is the business and financial center of a dairy, livestock, and vineyard area, and its chief industries are food processing and furniture making. **3.** City (1978 est. pop. 15,200), seat of Stutsman co., SE N.Dak., on the James River, in a farm area; settled 1872 when Fort Seward was established to protect railroad workers, inc. 1896. It is the trade center for an agricultural area. Fort Seward State Monument and a restored frontier village are on the outskirts. **4.** Resort town (1970 pop. 2,911), Newport co., comprising Conanicut Island and Dutch and Gould islands, S R.I., in Narragansett Bay; inc. 1678. Primarily rural, Jamestown has palatial estates and some historic buildings. **5.** Town (1970 pop. 1,899), seat of Fentress co., N Tenn., NW of Knoxville, in a farm area; settled 1827, inc. 1835. **6.** Former village, SE Va., first permanent English settlement in America; est. May 14, 1607, by the London Company on a marshy peninsula (now an island) in the James River and named for the reigning English monarch, James I. Disease, starvation, and Indian attacks wiped out most of the colony, but the London Company continually sent more men and supplies, and John Smith briefly provided efficient leadership. After the severe winter of 1609–10 (the "starving time"), the survivors prepared to return to England but were stopped by the timely arrival of Lord De la Warr with supplies. John Rolfe cultivated the first tobacco here in 1612, and in 1614 he assured peace with the Indians by marrying Pocahontas, daughter of the Indian chief Powhatan. In 1619 the first representative government in the New World met at Jamestown, and Jamestown remained the capital of Virginia throughout the 17th cent. The village was almost entirely destroyed during Bacon's Rebellion; it was partially rebuilt but fell into decay with the removal of the capital to Williamsburg (1698–1700). Of the 17th cent. settlement, only the old church tower (built c.1639) and a few gravestones were visible when National Park Service excavations began in 1934. Except for the land owned by the Association for the Preservation of Virginia Antiquities, Jamestown is today the property of the U.S. government.

Jam•mu (jŭm′ōō′), town (1971 pop. 155,249), Jammu and Kashmir state, N India, on the Tawi River in the Himalayan foothills. It is strategically important as the southern terminus of a highway linking the Vale of Kashmir with the North Indian plain. On one bank of the river is Jammu's old Fort of Bahu; on the other bank is the maharajah's palace, which dominates the city.

Jammu and Kash•mir (kăsh′mēr, kăsh-mēr′): *see* Kashmir.

Jam•na•gar (jäm′nə-gər), city (1971 pop. 214,853), Gujarat state, W central India. A port on the Gulf of Kutch, an arm of the Arabian Sea, Jamnagar has naval and aeronautical schools and is known for its silk, embroidery, and marble. There are cotton-textile mills.

Jam•shed•pur (jăm′shĕd-pōōr′), city (1971 pop. 355,783), Bihar state, E central India, at the confluence of the Subarnarekha and Kharkai rivers. A great steel-producing center, it was built in the early 20th cent. Nearby are extensive coal and iron deposits.

Janes•ville (jānz′vĭl′), city (1978 est. pop. 49,200), seat of Rock co., S central Wis., on the Rock River; inc. 1853. It is an industrial and commercial center in a grain, dairy-farm, and tobacco area.

Jan May•en (yän mī′ən), island, c.145 sq mi (375 sq km), in the Arctic Ocean, c.300 mi (485 km) E of Scoresby Sound, E Greenland. It was annexed by Norway in 1929. The island is barren tundra land and, except for a meteorological and wireless station, is uninhabited. It was discovered (1607) by Henry Hudson.

Ja•pan (jə-păn′), country (1976 pop. 112,200,000), 142,811 sq mi (369,881 sq km), occupying an archipelago off the coast of E Asia; capital Tokyo. Japan proper has four main islands: Hokkaido, Honshu, Shikoku, and Kyushu. There are also many smaller islands stretched in an arc between the Sea of Japan and the East China Sea and the Pacific Ocean. Honshu, Shikoku, and Kyushu enclose the Inland Sea. The general features of the four main islands are mountains, sometimes snowcapped, the highest and most famous being the sacred Fujiyama; rushing short rivers; forested slopes; irregular and lovely lakes; and small, rich plains.

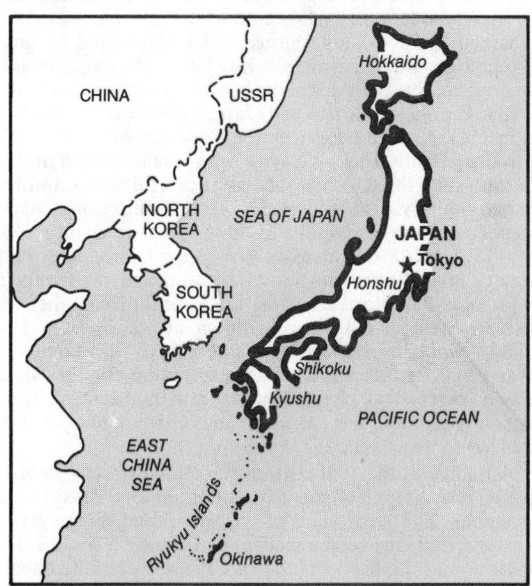

Economy. Arable land in Japan is intensively cultivated; farmers use irrigation, terracing, and multiple cropping to coax rich crops from the overworked soil. Rice and other cereals are the main crops; some vegetables and industrial crops, such as mulberry trees (for feeding silkworms), are also grown, and livestock is raised. Fishing is highly developed, and the annual catch is one of the biggest in the world. In the late 19th cent. Japan was rapidly and thoroughly industrialized. Textiles were a leading item, and vast quantities of light manufactures were also produced. In the 1950s and 1960s textiles became less important in Japanese industry while the production of heavy machinery expanded. Japan is the world's leading producer of ships and also ranks high in the production of cars and trucks, steel, and textiles. The manufacture of electronic equipment is also important. Japanese industry depends heavily on imported raw materials.

History. Japan's early history is lost in legend. The divine design of the empire—supposedly founded in 660 B.C. by the emperor Jimmu, a lineal descendant of the sun goddess and alleged ancestor of the present emperor—was held as official dogma until 1945. Actually, reliable records date back only to about A.D. 400. In the first centuries of the Christian era the country was inhabited by numerous clans or tribal kingdoms ruled by priest-chiefs. Contacts with Korea were close, and bronze and iron implements were probably introduced by invaders from Korea around the 1st cent. By the 5th cent. the Yamato clan had settled near modern Kyoto and had established a loose control over the other clans of central and western Japan, laying the foundation of the Japanese state. From the 6th to the 8th cent. the rapidly developing society gained much in the arts of civilization under the strong cultural influence of China, then flourishing in the splendor of the T'ang dynasty. Buddhism was introduced, and the Japanese upper classes assiduously studied Chinese language, literature, philosophy, art, science, and government, creating their own forms adapted from Chinese models. The Yamato priest-chief assumed the dignity of an emperor, and an imposing capital city, modeled on the T'ang capital, was erected at Nara, to be succeeded by an equally imposing capital at Kyoto. By the 9th cent., however, the powerful Fujiwara family had established a firm control over the imperial court. The Fujiwara influence and the power of the Buddhist priesthood undermined the authority of the imperial government. Provincial gentry—particularly the great clans who opposed the Fujiwara—evaded imperial taxes and grew strong. A feudal system developed. Civil warfare was almost continuous in the 12th cent. The Minamoto family defeated their rivals, the Taira, and became masters of Japan. Their great leader, Yoritomo, took the title of shogun, established his capital at Kamakura, and set up a military dictatorship.

For the next 700 years Japan was ruled by warriors. The old civil administration gradually decayed, and the imperial court at Kyoto fell into obscurity. In 1274 and again in 1281 the Mongols under Kublai Khan tried unsuccessfully to invade the country. In 1331 the emperor Daigo II attempted to restore imperial rule. He failed, but the revolt brought about the downfall of the Kamakura regime. The Ashikaga family took over the shogunate in 1338 and settled at Kyoto, but were unable to consolidate their power. The next 250 years were marked by civil wars, during which the feudal barons (the daimyo) and the Buddhist monasteries built up local domains and private armies. This period saw the birth of a middle class.

The first European contact with Japan was made by Portuguese sailors in 1542. A small trade with the West developed. Christianity was introduced by St. Francis Xavier, who reached Japan in 1549. In the late 16th cent. three warriors, Nobunaga, Hideyoshi, and Ieyasu, established military control over the whole country and succeeded one another in the dictatorship. Hideyoshi unsuccessfully invaded Korea in 1592 and 1596 in an effort to conquer China. After Hideyoshi's death, Ieyasu took the title of shogun, and his family ruled Japan for over 250 years. They set up a centralized, efficient, but repressive system of feudal government at Yedo (later Tokyo). Christianity was suppressed, and all intercourse with foreign countries was prohibited except for a Dutch trading post at Nagasaki. Tokugawa society was rigidly divided into the daimyo, samurai, peasants, artisans, and merchants, in that order. Oppression of the peasants led to many sporadic uprisings.

By the middle of the 19th cent. most daimyo were in debt to the merchants, and discontent was rife among impoverished but ambitious samurai. The great clans of western Japan, notably Choshu and Satsuma, had long been impatient of Tokugawa control. In 1854 an American naval officer, Matthew C. Perry, forced the opening of trade with the West. Japan was compelled to admit foreign merchants and to sign unequal treaties. Attacks on foreigners were answered by the bombardment of Kagoshima and Shimonoseki. Threatened from within and without, the shogunate collapsed. After brief fighting, in 1868 the boy emperor Meiji was "restored" to power, and the imperial capital was transferred to Tokyo.

Although the Meiji restoration was originally inspired by antiforeign sentiment, Japan's new rulers quickly realized the impossibility of expelling the foreigners. Instead they strove to strengthen Japan by adopting the techniques of Western civilization. Under the leadership of an exceptionally able group of statesmen, Japan was rapidly transformed into a modern industrial state and a great military power. Feudalism was abolished in 1871. In 1889 the emperor granted a constitution, modeled in part on that of Prussia. Supreme authority was vested in the emperor, who in practice was largely a figurehead controlled by the clan oligarchy. Universal manhood suffrage was not granted until 1925.

After the Meiji restoration nationalistic feeling ran high. At first concerned with defending Japanese independence against the Western powers, Japan soon joined them in the competition for an empire in the Orient. By 1899 Japan cast off the shackles of extraterritoriality, but not until 1911 was full tariff autonomy gained. The first Sino-Japanese War (1894–95) marked the real emergence of imperial Japan. An alliance with Great Britain in 1902 increased Japanese prestige, which reached a peak as a result of the Russo-Japanese War in 1904–5. Unexpectedly the Japanese smashed the might of Russia with speed and efficiency. The Treaty of Portsmouth recognized Japan as a world power. A territorial foothold had been gained in Manchuria, and in 1910 Japan was able to annex Korea. During World War I the Japanese secured the German interests in Shantung (later restored to China) and received the German-owned islands in the Pacific as mandates. In 1915 Japan presented the Twenty-One Demands designed to reduce China to a protectorate. The other world powers opposed the demands giving Japan policy control in Chinese affairs and prevented their execution, but China accepted the rest of the demands.

During the next decade the expansionist drive abated in Japan, and liberal and democratic forces gained ground. Trade and industry, stimulated by World War I, continued to expand, though interrupted by the earthquake of 1923, which destroyed much of Tokyo and Yokohama. Japan pursued a moderate policy toward China, relying chiefly on economic penetration and diplomacy to advance Japanese interests. This and other foreign policies pursued by the government displeased more extreme militarist and nationalist elements developing in Japan, some of whom disliked capitalism and advocated state socialism. Militarist propaganda was aided by the depression of 1929, which ruined Japan's silk trade. In 1931 the Kwantung army precipitated an incident at Mukden and promptly overran all of Manchuria, which was set up as the puppet state of Manchukuo. When the League of Nations criticized Japan's action, Japan withdrew from the organization.

During the 1930s the military party gradually extended its control over the government. Military extremists instigated the assassination of Prime Minister Ki Inukai in 1932 and an attempted coup d'état in 1936. In July, 1937, after an incident at Peking, Japanese troops invaded the northern provinces. Chinese resistance led to full-scale though undeclared war. A puppet Chinese government was installed at Nanking in 1940. Meanwhile relations with the Soviet Union were tense and worsened after Japan and Germany joined together against the Soviet Union in the Anti-Comintern Pact of 1936. In 1938 and 1939 armed clashes took place on the Manchurian border. Japan then stepped up an armament program, extended state control over industry (1938), and intensified police repression of dissident elements.

After World War II erupted (1939) in Europe, Japan signed a military alliance with Germany and Italy, sent troops to Indochina (1940), and announced the intention of creating a Greater East Asia Co-Prosperity Sphere under Japan's leadership. In Apr., 1941, a neutrality treaty with Russia was concluded. In Oct., 1941, the militarists achieved complete control in Japan, when Gen. Hideki Tojo succeeded a civilian, Prince Fumimaro Konoye, as prime minister. Unable to neutralize U.S. opposition to its actions in Southeast Asia, Japan opened hostilities against the United States and Great Britain on Dec. 7, 1941, by striking at Pearl Harbor, Singapore, and other Pacific possessions. By the end of 1942 the spread of Japanese military might over the Pacific to the doors of India and of Alaska was prodigious. Then the tide turned; territory was lost to the Allies island by island; Japan itself was intensively bombed; and finally in 1945, following the explosion of atomic bombs by the United States over Hiroshima and Nagasaki, Japan surrendered on Aug. 14.

The surrender was unconditional; the empire was dissolved, and Japan was deprived of all territories it had seized by force. The country was demilitarized, and war-potential industries were forbidden. Until these conditions were fulfilled Japan was to be under Allied military occupation. The occupation force, under the command of Gen. Douglas MacArthur, controlled Japan through the existing machinery of the Japanese government. A new constitution was adopted in 1946 and went into effect in 1947; the emperor publicly disclaimed his divinity. The general conservative trend in politics was tempered by the elections of 1947, which made the Social Democratic Party the dominant force in a two-party coalition government, but in 1949 it lost power completely when the conservatives took full charge under Shigeru Yoshida. Economic revival proceeded slowly with much unemployment and a low level of production. On Apr. 28, 1952, Japan again assumed full sovereignty. The elections in 1952 kept the conservative Liberal Party in power. In Nov., 1954, the Japan Democratic Party was founded. This new group attacked governmental corruption and advocated stable relations with the USSR and Communist China. In Dec., 1954, Yoshida resigned, and Ichiro Hatoyama, leader of the opposition, succeeded him. The Liberal and Japan Democratic parties merged in Nov., 1955, to become the Liberal Democratic Party (LDP).

In 1951 Japan signed a security treaty with the United States, providing for U.S. defense of Japan against external attack and allowing the United States to station troops in the country. New security treaties with the United States were negotiated in 1960 and 1970. Many Japanese felt that military ties with the United States would draw them into another war. Student groups and labor unions, often led by Communists, demonstrated during the 1950s and 1960s against military alliances and nuclear testing. Prime Minister Nobusuke Kishi was forced to resign in 1960 following the diet's acceptance, under pressure, of the U.S.-Japanese security treaty. He was succeeded by Hayato Ikeda, also of the Liberal Democratic Party. Ikeda led his party to two resounding victories in 1960 and 1963. He resigned in 1964 and was replaced by Eisaku Sato, under whose administration an agreement was reached whereby the United States returned (1972) the Ryukyu Islands. Sato was succeeded (1972) by Kakuei Tanaka, also a Liberal Democrat. Tanaka resigned in 1974 because of alleged irregularities in his personal finances and was replaced by Takeo Miki. It was during Miki's term that Japan's greatest postwar political scandal erupted when it was revealed that various high-ranking Japanese officials had received more than $12 million in bribes from the Lockheed Aircraft Corp., an American concern. Former Premier Tanaka was arrested, jailed, and indicted (1976) for having accepted a $1.6-million payment. In the midst of the scandal an election was held for the house of representatives; the LDP, although retaining a bare majority, suffered a major setback; Miki resigned and was succeeded by Takeo Fukuda.

Government. The emperor is the symbolic head of state but sovereignty rests with the people. The bicameral national diet has sole legislative power. Executive power is vested in an 18-member cabinet appointed and headed by the prime minister, who is elected by the diet.

Japan, Sea of, enclosed arm of the Pacific Ocean, c.405,000 sq mi (1,048,950 sq km), located between Japan and the Asian mainland, connecting with the East China Sea, the Pacific Ocean, and the Sea of Okhotsk through several straits.

Japan Cur·rent (kûr'ənt), warm ocean current of the Pacific Ocean, off E Asia. A northward flowing branch of the North Equatorial Current, it runs east of Taiwan and Japan; the Tsushima Current separates from the main current and flows into the Sea of Japan. At about lat. 35° N it divides to form an eastern branch flowing nearly to the Hawaiian Islands and a northern branch that skirts the coast of Asia and merges with the waters of the cold Oyashio Current to form the North Pacific Current.

Ja·pu·rá (zhä'po͞o-rä'), river, c.1,300 mi (2,090 km) long, rising as the Caquetá in the Andes, SW Colombia and flowing SE into Brazil, where it enters the Amazon through a network of channels.

Ja·ro·sław (yä-rô'släf), town (1975 est. pop. 31,300), SE Poland, on the San River. Primarily an agricultural and trading center, it has food-processing plants and flour mills. The town was founded in the 11th cent. It passed to Poland in 1382. Despite continuous Tatar raids, it developed as an important trade center in the 15th and 16th cent. It passed to Austria in 1772 and was restored to Poland in 1919.

Jar·row (jăr'ō), municipal borough (1971 pop. 28,779), Tyne and Wear, NE England, on the Tyne estuary. Industries include the manufacture of iron and steel products and shipbuilding and repairing. St. Paul's Church and an adjacent Benedictine monastery (now in ruins) were both founded in the 7th cent. The Venerable Bede lived, worked, and died in the monastery.

Jas·per (jăs'pər). **1.** County (1970 pop. 5,760), 373 sq mi (966.1 sq km), central Ga., bounded in the W by the Ocmulgee River and drained by the Little River; formed 1807; co. seat Monticello. It is in a piedmont agricultural area yielding cotton, corn, truck crops, and livestock. The county includes part of Oconee National Forest. **2.** County (1970 pop. 10,741), 495 sq mi (1,282.1 sq km), SE central Ill., drained by the Embarrass River; formed 1831; co. seat Newton. It is primarily agricultural (livestock, poultry, redtop seed, corn, wheat, and dairy products), with some manufacturing (wood products, brooms, and beverages). **3.** County (1970 pop. 20,429), 562 sq mi (1,455.6 sq km), NW Ind., bounded in the N by the Kankakee River and drained by the Iroquois River; formed 1835; co. seat Rensselaer. Farming, dairying, stock raising, and lumber and flour milling are the major occupations. **4.** County (1970 pop. 35,425), 736 sq mi (1,906.2 sq km), central Iowa, drained by the Skunk and North Skunk rivers; formed 1846; co. seat Newton. It is in a prairie agricultural area yielding hogs, cattle, poultry, corn, oats, and wheat, with bituminous-coal deposits in the southwest. **5.** County (1970 pop. 15,994), 683 sq mi (1,769 sq km), SE central Miss., drained by Tallahala and Tallahoma creeks, and by affluents of the Leaf River; formed 1833; co. seats Bay Springs and Paulding. It has corn and cotton farming, lumbering (logging and sawmills), and oil wells. The county includes part of Bienville National Forest. **6.** County (1970 pop. 79,852), 642 sq mi (1,662.8 sq km), SW Mo., in an Ozark region bounded on the W by the Kansas border and drained by the Spring River; formed 1841; co. seat Carthage. Its agriculture includes grain, poultry, strawberries, and dairy products. It also has lead, zinc, marble, and limestone deposits and stands of timber. Processed foods, clothing, shoes, and stone, metal, and wood products are made. **7.** County (1970 pop. 11,885), 652 sq mi (1,688.7 sq km), extreme S S.C., bounded on the W by the Savannah River, on the NE by the Coosawhatchie River, and on the SE by the Broad River; formed 1912; co. seat Ridgeland. It is in an agricultural area yielding corn, potatoes, tomatoes, and livestock and has extensive stands of timber. **8.** County (1970 pop. 24,692), 907 sq mi (2,349.1 sq km), E Texas, bounded on the W by the Neches River; formed 1836; co. seat Jasper. Lumbering is its chief industry. It also has diversified agriculture (including livestock and dairy products).

Jasper. 1. City (1978 est. pop. 11,700), seat of Walker co., NW central Ala., in a coal-mining area; inc. 1889. The city's industries produce coal, lumber, and textiles. **2.** City (1970 pop. 394), seat of Newton co., NW Ark., in the Ozarks SSW of Harrison, in a lumber-milling area. Lead and zinc are mined. **3.** City (1970 pop. 2,221), seat of Hamilton co., N Fla., NW of Lake City near the Ga. border; settled c.1825. It is a trade and processing center in a tobacco, timber, and truck-farming area. **4.** Town (1970 pop. 1,202), seat of Pickens co., N Ga., N of Atlanta, in a farm and lumber area. **5.** City (1970 pop. 8,641), seat of Dubois co., SW Ind., NE of Evansville; platted 1830, inc. as a town 1866, as a city 1915. Wood products and furniture are its chief manufactures. **6.** Town (1970 pop. 2,009), seat of Marion co., SE Tenn., near the Ala. and Ga. borders, W of Chattanooga, in the fertile valley of the Sequatchie River. **7.** City (1970 pop. 6,251), seat of Jasper co., E Texas, N of Beaumont; settled 1824, inc. 1926. It is a lumber town with related industries.

Ja·va (jä'və), island (1974 est. pop. 75,000,000), c.51,000 sq mi (132,090 sq km), Indonesia, S of Borneo, from which it is separated by the Java Sea, and SE of Sumatra across Sunda Strait. For centuries Java has been the cultural, political, and economic center of the area. A chain of volcanic mountains, most of them densely forested with teak, palms, and other woods, traverses the length of the island from east to west. There are almost 2 million acres of planted teak forests. The climate is warm and humid, and the volcanic soil is exceptionally fertile. Most of Indonesia's sugar cane and kapok are grown in Java. Rubber, tea, coffee, tobacco, cacao, and cinchona are produced in highland plantations. Rice is the chief small-farm crop. Cattle are raised in the east. In the northeast are important oil fields, and tin, gold, silver, coal, manganese, phosphate, and sulfur are also mined.

Early in the Christian era Indians began colonizing Java, and by the 7th cent. "Indianized" kingdoms were dominant in both Java and Sumatra. Islam was introduced in the 13th cent., and the Moslem state of Mataram emerged in the 16th cent. Following the Portuguese, the Dutch arrived in 1596, and in 1619 the Dutch East India Company established its chief post in Batavia, thence gradually absorbing the native states. Between 1811 and 1815 Java was briefly under British rule. When the Dutch returned to power, they resorted to a system of enforced labor and harsh methods of exploitation, which led to a native uprising (1825-30). In World War II Java was left unprotected by the disastrous Allied defeat in the battle of the Java Sea in Feb., 1942; the island was occupied by the Japanese until the war ended.

Ja·va·ri (zhä'vä-rē'), river, c.500 mi (805 km) long, rising in the Cerro de Canchyuaya, E Peru, and flowing NE, forming part of the boundary between Brazil and Peru, before entering the Amazon River near Tabatinga. It is navigable for most of its length.

Jay (jā), county (1970 pop. 23,575), 386 sq mi (999.7 sq km), E Ind., bordered in the E by Ohio and drained by the Salamonie River; formed 1835; co. seat Portland. It has agriculture (livestock, poultry, grain, soybeans, truck crops, and dairy products) and lumbering and food-processing industries. Its manufactures include automobile parts and stone, clay, and glass products.

Jay, town (1970 pop. 1,594), seat of Delaware co., NE Okla., near the Ark. border SE of Vinita, in a timber and farm region.

Jed·da (jĕd'ə): *see* Jidda, Saudi Arabia.

Jeff Da·vis (jĕf dā'vəs). **1.** County (1970 pop. 9,425), 330 sq mi (854.7 sq km), SE central Ga., bounded in the NW by the Ocmulgee River, in the N by the Altamaha River, and drained by the Satilla River; formed 1905; co. seat Hazlehurst. In a coastal-plain agricultural area yielding tobacco, corn, truck crops, sugar cane, peanuts, pecans, and livestock, it also has a lumbering industry and textile mills. **2.** County (1970 pop. 1,527), 2,258 sq mi (5,848.2 sq km), extreme W Texas, on a high plateau rising to the scenic Davis Mts. and touching on the Mexico border in the W; formed 1887; co. seat Fort Davis. It is drained by Salt Draw and Toyah creeks. Cattle ranching is the major activity. It also produces sheep, goats, horses, and fruit.

Jef·fer·son (jĕf'ər-sən). **1.** County (1970 pop. 644,991), 1,117 sq mi (2,893 sq km), central Ala., crossed by Locust Fork; formed 1819; co. seat Birmingham. It has bituminous-coal and iron mines, limestone quarries, clay pits, and timber. Largely industrial, its manufactures include food products, clothing and textiles, furniture, paper, chemicals, clay products, cement, concrete, iron and steel, metal products, and machinery. **2.** County (1970 pop. 85,329), 873 sq mi (2,261.1 sq km), SE central Ark., intersected by the Arkansas River and drained by Plum Bayou; formed 1829; co. seat Pine Bluff. It has agriculture (cotton, hay, grain, rice, poultry, and livestock), lumbering and shipbuilding industries, sand and gravel pits, cottonseed-oil mills, and iron and steel foundries. Other manufactures include textiles, paper, and farm machinery. **3.** County (1970 pop. 235,300), 783 sq mi (2,028 sq km), central Colo., in a coal-mining and irrigated agricultural area bounded in the E by the South Platte River; formed 1861; co. seat Golden. Sugar beets, beans, livestock, and fur-bearing animals are raised. Part of Pike National Forest is here. **4.** County (1970 pop. 8,778), 605 sq mi (1,567 sq mi), NW Fla., in a lowland, partly swampy area of the Panhandle bounded on the N by the Ga. border, on the S by the Gulf of Mexico, on the E by the Aucilla River; formed 1827; co. seat Monticello. Its economy depends on forestry and agriculture (corn, peanuts, cotton, vegetables, tung nuts, hogs, and cattle). **5.** County (1970 pop. 17,174), 532 sq mi (1,377.9 sq km), E Ga., drained by the Ogeechee River; formed 1796; co. seat Louisville. Coastal-plain agriculture (cotton, corn, and peanuts) and sawmilling are the principal occupations. **6.** County (1970 pop. 11,740), 1,096 sq mi (2,838.6 sq km), E Idaho, in an irrigated area of the Snake River valley; formed 1913; co. seat Rigby. Livestock is raised, and clover, legumes, sugar beets, potatoes, and fruit are grown. **7.** County (1970 pop. 31,848), 574 sq mi (1,486.7 sq km), S Ill., watered by the Big Muddy River and Rend Lake; formed 1819; co. seat Mount Vernon. It has agriculture (livestock, fruit, poultry, clover seed, corn, and wheat) and bituminous-coal mines. Among its manufactures are railroad cars, clothing, shoes, feed, and food products. **8.** County (1970 pop. 27,006), 366 sq mi (947.9 sq km), SE Ind., bounded in the S by a bend in the Ohio River (here forming the Ky. border) and drained by Big Creek; formed 1810; co. seat Madison. Agriculture (tobacco, grain, and livestock), diversified manufacturing, lumbering, and oil refining are important to its economy. **9.** County (1970 pop. 15,774), 436 sq mi (1,129.2 sq km), SE Iowa, drained by the Skunk River; formed 1839; co. seat Fairfield. In a prairie agricultural area yielding hogs, cattle, poultry, corn, soybeans, and hay, it also has coal mines and limestone quarries. **10.** County (1970 pop. 11,945), 549 sq mi (1,421.9 sq km), NE Kansas, in a hilly area crossed by the Delaware River and bounded on the S by the Kansas River; formed 1855; co. seat Oskaloosa. It is a rural area that produces corn and dairy products and raises livestock and poultry. There is some manufacturing. **11.** County (1970 pop. 695,055), 375 sq mi (971.3 sq km), N central Ky., bounded on the W by the Ohio River and the Ind. border; formed 1780 from Kentucky co., a Va. district; co. seat Louisville. It is heavily industrial and has coal, phosphate, and fluorspar mines, oil and natural-gas wells, timber, and deposits of limestone, sand, and gravel. Among its manufactures are whiskey, tobacco products, clothing and textiles, furniture, paper, chemicals, plastics, concrete, iron and steel, and metal products. Agriculture (potatoes, onions, burley tobacco, livestock, grain, and fruit) is also important. **12.** Parish (1970 pop. 338,229), 369 sq mi (955.7 sq km), extreme SE La., in the Mississippi delta bounded on the S by the Gulf of Mexico, on the N by Lake Pontchartrain, and traversed by the Gulf Intracoastal Waterway; formed 1825; parish seat Gretna. It has a wide variety of industries, including food processing, shipbuilding, petroleum refining, fishing, mining, and lumbering. Textiles, paper products, chemicals, glassware, concrete, asbestos, machinery, and metal products are manufactured. It also has oil and natural-gas wells. **13.** County (1970 pop. 9,295), 520 sq mi (1,346.8 sq km), SW Miss., bounded on the W by the Mississippi River and the La. border; formed 1799 as Pickering co., renamed 1802; co. seat Fayette. Cotton and corn are grown. Lumbering and the production of oil and natural gas are also important. Part of Homochitto National Forest is here. **14.** County (1970 pop. 105,248), 667 sq mi (1,727.5 sq km), E Mo., bounded on the E by the Mississippi River and the Ill. border; formed 1818; co. seat Hillsboro. Its agriculture includes corn, wheat, barley, and livestock. There are silica and barite mines. Among its manufactures are textiles and clothing, stone, clay, glass, and metal products, and leather goods. **15.** County (1970 pop. 5,238), 1,651 sq mi (4,276.1 sq km), SW central Mont., in a stock-raising area bounded on the S by the Big Hole Jefferson River, on the W by the Continental Divide, and drained by the Boulder River; formed 1865; co. seat Boulder. Gold, silver, lead, and zinc are mined. The county contains portions of Deerlodge and Helena national forests. **16.** County (1970 pop. 10,436), 577 sq mi (1,494 sq km), SE Nebr., bounded on the S by Kansas and drained by the Little Blue River; formed 1856; co. seat Fairbury. It lies in a grain, livestock, dairy, and poultry-farming area. It has some mining, and its industry includes food processing (meat packing) and clothing, brick, and clay-tile manufacturing. **17.** County (1970 pop. 88,508), 1,293 sq mi (3,348.9 sq km), N N.Y., bounded on the W by Lake Ontario, on the NW by the St. Lawrence River, and drained by the Black and Indian rivers; formed 1805; co. seat Watertown. It is in a dairying and iron-producing area, with limestone and talc deposits and fisheries. Its industry includes lumber sawmilling, food processing, and clothing and machinery manufacturing. **18.** County (1970 pop. 96,193), 411 sq mi (1,064.5 sq km), E Ohio, bounded on the E by the Ohio River, here forming the W.Va. line, and drained by Yellow and Cross creeks; formed 1797; co. seat Steubenville. Coal mining and the production of steel are important. Clay products, concrete blocks, and building materials are manufactured. **19.** County (1970 pop. 7,125), 780 sq mi (2,020 sq km), S Okla., bounded on the S by the Red River, here forming the Texas border, and drained by Beaver and Mud creeks; formed 1907; co. seat Waurika. Cattle raising and agriculture (cotton, grain, corn, poultry, and dairy products) are important. It also has some oil and natural-gas wells. **20.** County (1970 pop. 8,548), 1,794 sq mi (4,646.5 sq km), central Oregon, drained by the Deschutes River; formed 1914; co. seat Madras. Its economy depends on lumbering, agriculture (wheat, hay, potatoes, and livestock), and tourism. The county, which includes Mt. Jefferson, has many recreational facilities. **21.** County (1970 pop. 43,695), 652 sq mi (1,688.7 sq km), W central Pa., bounded on the N by the Clarion River and drained by tributaries of the Allegheny River; formed 1804; co. seat Brookville. In a stock-raising and wheat-growing region, it has food-processing and lumbering industries, bituminous-coal, clay, and glass-sand mines, and oil and natural-gas wells. Glass, clay, and metal products are made. **22.** County (1970 pop. 24,940), 274 sq mi (709.7 sq km), E Tenn., in the Great Appalachian Valley bounded on the NW by the Holston River and drained by the French Broad River; formed 1792; co. seat Dandridge. Corn, tobacco, and soybeans are grown. The county has zinc mines and factories producing canned vegetables, furniture, and leather goods. **23.** County (1970 pop. 246,402), 951 sq mi (2,463.1 sq km), SE Texas, bounded on the E by Sabine Lake and the La. border, on the S by the Gulf of Mexico, crossed by the Gulf Intracoastal Waterway and drained by the Neches river; formed 1836; co. seat Beaumont. Oil refining and shipping are of major importance here. It also has natural-gas fields, cattle ranches, and general agriculture (rice, wheat, fruit, truck crops, and dairy products). **24.** County (1970 pop. 10,661), 1,805 sq mi (4,675 sq km), W Wash., bounded on the W by the Pacific, on the E and N by Hood Canal and Puget Sound; formed 1852; co. seat Port Townsend. Its economy depends on dairy farms, logging, and paper manufacturing. The Olympic Mts. occupy the interior of the county, which includes parts of Olympic National Forest and Olympic National Park. **25.** County (1970 pop. 21,280), 211 sq mi (546.5 sq km), extreme NE W.Va., at the tip of the Eastern Panhandle in the S part of the Great

Appalachian Valley, with the Blue Ridge along its SE border, bounded on the NE by the Potomac River and the Md. border, on the SE and SW by the Va. border; formed 1801; co. seat Charles Town. It has agriculture (livestock, especially horses, dairy products, and fruit) and nonmetallic mineral mining. There is some lumbering and machinery manufacturing. It is in a scenic resort region. **26.** County (1970 pop. 60,060), 564 sq mi (1,460.8 sq km), S Wis., drained by the Rock, Bark, and Crawfish rivers; formed 1836; co. seat Jefferson. In a dairy-farming area, it also has general agriculture and diversified manufacturing.

Jefferson. 1. City (1970 pop. 1,647), seat of Jackson co., NE Ga., NE of Atlanta. It is a farm trade center. **2.** City (1970 pop. 4,735), seat of Greene co., central Iowa, NW of Des Moines on the Raccoon River; settled c.1854, inc. 1872. It is a trade center of a farm area. Dairy products are made. **3.** Town (1970 pop. 943), seat of Ashe co., NW N.C., in the Blue Ridge NW of North Wilkesboro. It is in Yadkin National Forest. **4.** Village (1970 pop. 2,472), seat of Ashtabula co., NE Ohio, S of Ashtabula, in a farm, dairy, and livestock area; founded 1804. Chemicals are produced. **5.** Borough (1970 pop. 8,512), Allegheny co., SW Pa., SE of Pittsburgh; inc. 1952. Bituminous coal is mined. **6.** City (1970 pop. 2,866), seat of Marion co., E Texas, on Big Cypress Bayou N of Marshall. The region was settled in the 1830s. Jefferson grew as a river port. It was also a large lumbering and industrial center. Today the region produces timber, cotton, truck crops, and cattle, as well as oil and natural gas. **7.** City (1970 pop. 5,429), seat of Jefferson co., S Wis., ESE of Madison; settled 1836, inc. 1878. Meat products and furniture are made.

Jefferson, Mount, 10,499 ft (3,202.2 m) high, in the Cascade Range, NW central Oregon, ESE of Salem.

Jefferson City, city (1978 est. pop. 35,900), state capital and seat of Cole co., central Mo., on the S bank of the Missouri River, near the mouth of the Osage River; inc. 1825. The state government is the major employer. The city has printing and publishing houses; other industries produce shoes, clothing, electrical appliances, and steel products. It was a small river village when it was chosen (1821) for the state capital. Because of divided loyalties, Jefferson City was occupied by Union troops during the Civil War.

Jefferson Da·vis (dā'vĭs). **1.** Parish (1970 pop. 29,554), 658 sq mi (1,704.2 sq km), SW La., bounded on the E and SE by Bayou Nezpique; formed 1910; parish seat Jennings. Agricultural crops include rice, corn, cotton, and sweet potatoes. The area is noted for lumber, oil, and natural gas. Industries include rice milling, cotton ginning, metalworking, and oil refining. The county includes Lake Arthur, which has fishing and camping. **2.** County (1970 pop. 12,936), 414 sq mi (1,072.3 sq km), S Miss., drained by tributaries of the Leaf and Pearl Rivers; formed 1906; co. seat Prentiss. It has cotton and corn farming, lumbering, oil and gas wells, and clothing manufacturing.

Jef·fer·son·ville (jĕf'ər-sən-vĭl'). **1.** Town (1970 pop. 1,302), seat of Twiggs co., central Ga., SE of Macon, in a farm and timber area. There are kaolin deposits here. **2.** City (1978 est. pop. 22,300), seat of Clark co., S Ind., at the falls of the Ohio River opposite Louisville, Ky.; inc. 1817. Its shipbuilding industry dates from the 19th cent. Jeffersonville was founded (1802) on the site of Fort Steuben (originally Fort Finney) by veterans of George Rogers Clark's expedition, who were given the land in gratitude for their services.

Jeh·lam (jā'ləm), river, Kashmir: *see* Jhelum.

Je·hol (jə-hôl', -hōl'), former province, NE China, divided (1955) between the Inner Mongolian Autonomous Region and the provinces of Hopeh and Liaoning. Jehol was the traditional gateway to Mongolia and from time to time was overrun by Tatars, Huns, and Khitan Mongols. Taken by the Japanese early in 1933 and included in Manchukuo, it was not restored to China until the end of World War II. From 1945 to 1955 it was administered as part of Manchuria.

Je·le·nia Gó·ra (yĕ-lĕ'nyä go͞o'rä), city (1975 est. pop. 58,800), SW Poland. It is an industrial and commercial center known especially for its woolen textiles. Chartered in 1312, the city passed to Bohemia in 1368 and to Prussia in 1741.

Je·mappes (zhə-mäp'), town (1970 pop. 12,455), Hainaut prov., S Belgium. It is a coal-mining center; manufactures include iron and steel. At Jemappes in 1792 the French defeated the Austrians in one of the first important battles of the French Revolutionary Wars. The victory opened the way to Brussels for the French.

Je·na (yā'nä), city (1974 est. pop. 99,431), Gera dist., S East Germany, on the Saale River. Manufactures include pharmaceuticals, glass, and optical and precision instruments. Jena was known in the 9th cent. and was chartered in the 13th cent. In 1806 Napoleon I decisively defeated the Prussians at Jena. The Univ. of Jena was founded in 1557–58 and reached its height in the late 18th and early 19th cent. At that time the dramatist Friedrich von Schiller, the philosophers Georg Hegel, Johann Fichte, and Friedrich Schelling, and the poet August Wilhelm von Schlegel taught here.

Je·na (jē'nə), town (1970 pop. 2,431), seat of La Salle parish, central La., NE of Alexandria, in a timber, farm, and oil region.

Jen·kins (jĕng'kĭns), county (1970 pop. 8,332), 351 sq mi (909.1 sq km), E Ga., intersected by the Ogeechee River formed 1905; co. seat Millen. Coastal-plain agriculture (cotton, corn, truck crops, and livestock) and timber abound in this area.

Jen·né (jĕn'ā): *see* Djenné, Mali.

Jen·nings (jĕn'ĭngz), county (1970 pop. 19,454), 377 sq mi (976.4 sq km), SE Ind., drained by Vernon, Graham, and Sand creeks; formed 1816; co. seat Vernon. Bituminous-coal mining and agriculture (livestock, grain, and tobacco) characterize the region, which also has timber and limestone quarries. Some manufacturing is done.

Jennings, city (1970 pop. 11,783), seat of Jefferson Davis parish, SW La., on the Mermentau River; inc. 1888.

Jer·auld (jə-rôld'), county (1970 pop. 3,310), 528 sq mi (1,367.5 sq km), SE central S.Dak., in an agricultural area watered by intermittent streams; formed 1883; co. seat Wessington Springs. Its agriculture includes corn, hay, poultry, livestock, and dairy produce.

Jer·ba (jĕr'bə), island (1966 pop. 62,445), 197 sq mi (510.2 sq km), SE Tunisia, in the Mediterranean Sea. Fruits are grown on the island, once identified as the land of the lotus eaters. It has extensive Roman remains.

Je·rez de la Fron·te·ra (hā-rāz' dā lä frôn-tā'rä), city (1970 pop. 149,867), Cádiz prov., SW Spain, in Andalusia. Jerez de la Frontera is an important commercial center noted for its sherry and cognac. Its horses of mixed Spanish, Arab, and English blood are world-famous. Captured by the Moors in 711, the city was recovered by Alfonso X of Castile in 1264.

Jer·i·cho (jĕr'ə-kō), ancient city, Palestine, in the Jordan valley N of the Dead Sea, near modern Ariha, Jordan. Jericho was captured from the Canaanities by Joshua, according to the biblical account, and was destroyed, an event several times repeated in its history. One of its conquerors was Herod the Great, who sacked and rebuilt it. Later it fell to the Moslems. Excavations of the mound of Tell es Sultan, the original site, were begun early in the 20th cent. and have revealed the oldest known settlement in the world, dating perhaps from c.8000 B.C.

Jer·ome (jə-rōm'), county (1970 pop. 10,253), 593 sq mi (1,535.9 sq km), S Idaho, bounded in the S by the Snake River; formed 1919; co. seat Jerome. It is a livestock-raising and irrigated agricultural area, producing potatoes, apples, sugar beets, alfalfa, and dairy products.

Jerome, city (1970 pop. 4,183), seat of Jerome co., S Idaho, near the Snake River NNW of Twin Falls; laid out 1907, inc. 1909. It is a processing center in an area irrigated by the Minidoka project.

Jer·sey (jûr'zē), county (1970 pop. 18,492), 374 sq mi (968.7 sq km), W Ill., bounded in the S by the Mississippi River, in the W by the Illinois River, and drained by Macoupin Creek; formed 1839; co. seat Jerseyville. Agriculture (apples, corn, wheat, and livestock), manufacturing (shoes, leather, concrete products, flour, cigars, beverages, and lumber), and bituminous-coal mining are the major activities here. There are resorts on the Illinois River.

Jersey, island (1971 pop. 72,532), 45 sq mi (116.6 sq km), in the English Channel, largest of the Channel Islands, which are dependencies of the British Crown. It is 15 mi (24.1 km) from the Normandy coast of France and southeast of Guernsey. Jersey, like the other Channel Islands, is a vacation resort. Large quantities of vegetables (especially potatoes, tomatoes, and broccoli) and fruits are raised; cattle raising and dairying are important.

Jersey City, city (1978 est. pop. 236,900), seat of Hudson co., NE N.J., a port on a peninsula formed by the Hudson and Hackensack rivers and Upper New York Bay, opposite lower Manhattan; settled before 1650, inc. as Jersey City 1836. It is a port of entry, a great shipping and manufacturing center, and a major transportation terminal point and distribution center. It has railroad shops, oil refineries, warehouses, and more than 600 plants manufacturing a great variety of products. The area was acquired by Michiel Pauw c.1629 as the patroonship of Pavonia. The Dutch soon set up the trading posts of Paulus Hook, Communipaw, and Horsimus. In 1674 the site came permanently under British rule. The fort at Paulus Hook was captured by Light-Horse Harry Lee on Aug. 19, 1779. The city's industrial growth began in the 1840s with the arrival of the railroad and the improvement of its water transport system.

Jer·sey·ville (jĕr'zē-vĭl'), city (1970 pop. 7,446), seat of Jersey co., W Ill., NE of the junction of the Illinois and the Mississippi rivers; platted 1834, inc. 1855. A shipping and trade center in a fruit and farm area, it also produces conveyors.

Je·ru·sa·lem (jə-ro͞o'sə-ləm), city (1972 pop. 304,500), capital of Israel, situated on a ridge 2,500 ft (760 m) high that lies W of the Dead Sea and the Jordan River. Jerusalem is an administrative, religious, and cultural center. Construction and tourism are the city's major industries. Manufactures include cut diamonds, plastics, and shoes. Jerusalem is a holy city for Jews, Christians, and Moslems.

The eastern part of Jerusalem is the Old City, a quadrangular area built on two hills and surrounded by a wall completed in 1542 by Sulayman I. The Moslem quarter, in the east, contains a sacred enclosure, the Haram esh-Sherif, within which are the Dome of the Rock (completed 691), or Mosque of Omar, and the Mosque of al-Aksa. The wall of the Haram incorporates the only extant piece of the Temple of Solomon; this, the Wailing Wall, is a holy place for Jews. Nearby is the Jewish quarter, with several famous old synagogues. Largely destroyed in previous Arab-Israeli fighting, it was recaptured in 1967 by the Israelis, who began to rebuild and renovate it. To the west of the Jewish quarter is the Armenian quarter, site of the Gulbenkian Library. The Christian quarter occupies the northern and northwestern parts of the Old City. Its greatest monument is the Church of the Holy Sepulcher. Through the area runs the Via Dolorosa, where Jesus is said to have carried his cross. The New City, extending west and southwest of the Old City, has largely developed since the 19th cent. It is the site of several educational institutions, as well as the Knesset (the Israeli parliament) and other government buildings. To the east of the Old City is the Valley of the Kidron, across which lie the Garden of Gethsemane and the Mount of Olives.

Despite incomplete archaeological work, it is evident that Jerusalem was occupied as far back as the 4th millenium B.C. In the late Bronze Age (2000–1550 B.C.) it was a Jebusite (Canaanite) stronghold. David captured it (c.1000 B.C.) from the Jebusites and walled the city. After Solomon built a temple on Mt. Moriah in the 10th cent. B.C., Jerusalem became the spiritual and political capital of the Hebrews. In 586 B.C. it fell to the Babylonians, and the temple was destroyed. Under Persian rule the temple was rebuilt (538–515 B.C.). The city was the capital of the Maccabees in the 2nd and 1st cent. B.C., and then of the Roman puppet kings, the Herods. The Roman emperor Titus ruined the city and destroyed the temple (A.D. 70), and Hadrian forbade the Jews to live on the site (132).

With the imperial toleration of Christianity (from 313), Jerusalem underwent a revival. The Moslems, who believe that the city was visited by Mohammad, captured it in 637, making it their chief shrine after Mecca. Jerusalem was conquered by the Crusaders in 1099 and was the capital of the Latin Kingdom of Jerusalem until 1187, when Moslems under Saladin recaptured the city. Thereafter, under Mameluke and then Ottoman rule, it was rebuilt and restored. In the early 19th cent. the flow of Christian pilgrims and Jewish immigrants increased. After World War I Jerusalem was made the capital of the British-held League of Nations Palestine mandate (1922–48). As the end of the mandate approached, Arabs and Jews both sought to hold sole possession of the city, and fighting broke out. The Old City and all areas held by the Arab Legion were annexed by Jordan in Apr., 1949. The New City, held by the Israelis, was made the capital of Israel on Dec. 14, 1949. In the Arab-Israeli War of 1967, Israeli forces took the Old City. Late in June of that year the Israeli government formally annexed the Old City and placed all of Jerusalem under a unified administration.

Jer·vis Bay (jär'vĭs, jûr'-), sheltered inlet of the Pacific Ocean, 10 mi (16.1 km) long and 6 mi (9.7 km) wide, SE Australia. In 1915 the harbor and part of the coast were transferred to the federal government by New South Wales. Jervis Bay then became the port of the landlocked Australian Capital Territory. The area around the bay is a popular summer resort.

Jes·sa·mine (jĕs'ə-mĭn), county (1970 pop. 17,430), 177 sq mi (458.4 sq km), E central Ky., in a gently rolling upland area of the Bluegrass region; bounded on the SW and SE by the Kentucky River; formed 1798; co. seat Nicholasville. Its agriculture includes dairy products, livestock, poultry, burley tobacco, corn, wheat, and truck crops. It has limestone quarries, and clothing and paper factories.

Jes·sore (jĕ-sōr'), city (1974 pop. 76,168), SW Bangladesh, on the Bhairab River. Modern Jessore, a market town for rice and sugar, also has rice and oilseed mills and celluloid and plastics industries.

Jes·up (jĕs'əp), city (1970 pop. 9,091), seat of Wayne co., SE Ga., near the Altamaha River SW of Savannah, in a hunting and fishing region; inc. 1870. It is a trade center for an area yielding tobacco, and timber, and agricultural products.

Jet·more (jĕt'môr'), city (1970 pop. 936), seat of Hodgeman co., SW Kansas, NNE of Dodge City, in a farm area; settled 1879 as Buckner, name changed 1880, inc. 1887.

Jew·el Cave National Monument (jo͞o'əl): *see* National Parks and Monuments Table.

Jew·ell (jo͞o'əl), county (1970 pop. 6,099), 915 sq mi (2,369.9 sq km), N Kansas, in a rolling plains area bordering in the N on Nebr.; formed 1867; co. seat Mankato. It is an agricultural region that produces grain and livestock. There is a meat-packing industry.

Jew·ish Autonomous Oblast (jo͞o'ĭsh) or **Bi·ro·bi·dzhan** (bē'rō-bē-jän'), autonomous region (1976 pop. 190,000), c.13,800 sq mi (35,740 sq km), Khabarovsk Kray, Far Eastern USSR, in the basins of the Biro and Bidzhan rivers, tributaries of the Amur; capital Birobidzhan. The region is bounded on the south by Mongolia and on the north by the Bureya and Khingan mts., which yield gold, tin, iron ore, and graphite. Mining, agriculture, lumbering, and light manufacturing are the major economic activities. Formed in 1928 to give Soviet Jews a home territory and to increase settlement along the vulnerable borders of the Soviet Far East, the region was raised to the status of an autonomous oblast in 1934.

Jhang-Ma·ghi·a·na (jəng-mə-gyä'nə), twin cities (1972 est. pop. 135,722), central Pakistan, on the Chenab River.

Jhan·si (jän'sē), city (1971 pop. 173,255), Uttar Pradesh state, N central India. An agricultural market and small industrial center, it has iron and steel mills and manufactures brassware. The city grew around a fort built in 1613. It reverted to Great Britain in 1853, when the ruling prince died without heirs. British residents in Jhansi were massacred during the Indian Mutiny (1857).

Jhe·lum or **Jeh·lam** (both: jā'ləm), westernmost of the five rivers of the Punjab, 480 mi (772.3 km) long, rising in W Kashmir and flowing W through the Vale of Kashmir, S across the Punjab, where it forms part of the India-Pakistan border, then SW across NE Pakistan to the Chenab River. The Jhelum was crossed in 326 B.C. by Alexander the Great, who defeated the Indian king Porus.

Jid·da (jĭd'ə) or **Jed·da** (jĕd'ə), city (1974 pop. 561,104), Hejaz, W Saudi Arabia, on the Red Sea. Jidda is the port of Mecca and annually receives a hugh influx of pilgrims. Jidda was ruled by the Turks until 1916, when it became part of the independent Hejaz. In 1925 it was conquered by Ibn Saud.

Ji·hla·va (yĭ'hlä-vä), city (1975 est. pop. 45,362), W central Czechoslovakia, in Moravia, on the Jihlava River. Jihlava has industries manufacturing linen and woolen cloth, machinery, footwear, and tobacco. Chartered in 1227, it was the site of the signing (1436) of the Compactata—the Magna Carta of the Hussites.

Jim Hogg (jĭm hôg), county (1970 pop. 4,654), 1,139 sq mi (2,950 sq km), extreme S Texas; formed 1913; co. seat Hebbronville. In a petroleum-producing and cattle-ranching area, it also has some general agriculture (peanuts, corn, cotton, and dairy products).

Jim Thorpe (thôrp): *see* Mauch Chunk, Pa.

Jim Wells (wĕlz), county (1970 pop. 33,032), 846 sq mi (2,191.1 sq km), S Texas, bounded on the NE by the Nueces River; formed 1911; co. seat Alice. A major producer of oil and natural gas, the county also has diversified agriculture (including grain, cotton, corn, fruit, peanuts, livestock, and dairy products) and sulfur mines.

Jin·ja (jĭn'jə), city (1969 pop. 52,509), SE Uganda, on the Victoria Nile River, near Victoria Nyanza. It is an industrial city and the commercial and processing center for a region where cotton, sugar cane, maize, and groundnuts are grown. Manufactures include refined copper, metal goods, forest products, textiles, soap, and processed food. Jinja was founded in 1901 as a trading post.

Ji·pi·ja·pa (hē'pē-hä'pä), city (1974 pop. 19,719), W Ecuador, on the equatorial lowlands. A few miles inland from the Pacific, Jipijapa is famous for the manufacture of high-grade Panama hats. It is the trade center for an agricultural region.

Jir·ja (jĭr′jă) or **Gir·ga** (gĭr′gă), town (1966 pop. 44,300), central Egypt, on the Nile. It is noted for its pottery. The town is the seat of a Coptic bishop. A Roman Catholic monastery, said to be the oldest in Egypt, is in Jirja. Nearby is the ancient city of Abydos.

Ji·zah, Al (äl jē′zə) or **Gi·za** (gē′zə), city (1976 pop. 1,246,054), capital of Al Jizah governorate, N Egypt, surburb of Cairo. It is a manufacturing and agricultural trade center. Products include cotton textiles, cigarettes, and footwear. Nearby are the Great Sphinx and the pyramid of Khufu (Cheops).

João Pes·so·a (zhwouɴ pə-sō′ə), city (1970 pop. 197,398), capital of Paraíba state, NE Brazil, at the confluence of the Sanhauá and Paraíba do Norte rivers. Industries in the city produce tobacco, shoes, and cement. The city was established in the late 16th cent. Its Franciscan convent and church are excellent examples of colonial architecture. Nearby are several resort areas.

Jo Da·viess (jō dā′vəs), county (1970 pop. 21,766), 606 sq mi (1,569.5 sq km), extreme NW Ill., bounded in the N by the Wis. border, in the W by the Mississippi River and the Iowa border; formed 1827; co. seat Galena. It is drained by the Apple, Plum, and Galena rivers. Livestock, poultry, corn, oats, barley, and alfalfa are among its agricultural products. Its resources include timber and deposits of lead and zinc. A wide variety of goods, such as beverages, gloves, and thermometers, are manufactured.

Jodh·pur (jŏd′po͝or) or **Mar·war** (mär′wär), city and former principality, Rajasthan state, NW India. The state was founded in the 13th cent. and was later a vassal of the Mogul empire. The British brought it under their control in 1818, and in 1949 it was merged with the state of Rajasthan. The city (1971 pop. 318,894) was founded in 1459. It is surrounded by a wall nearly 6 mi (9.7 km) long. Jodhpur is an important marketplace for wool. Its manufactures include textiles and electrical and leather goods. Towering above the city on a rock 400 ft (122 m) high is an old fortress housing several palaces and the treasury of the maharajah, which contains a famous gem collection.

Jog·ja·kar·ta (jŏg′yä-kär′tä) or **Djok·ja·kar·ta** (jŏk′-), city (1961 pop. 312,698), S Java, Indonesia, at the foot of volcanic Mt. Merapi, capital of the autonomous district of Jogjakarta, a former sultanate. It is the cultural center of Java, known for its artistic life, particularly its drama and dance festivals and handicraft industries. It is also the trade hub of a major rice-producing region, and there is some manufacturing. The town was founded in 1749. It was the focus of the revolt against the Dutch (1825–30) and was the stronghold of the Indonesian independence movement from 1946 to 1950.

Jo·han·nes·burg (jō-hăn′ĭs-bûrg′), city (1970 metropolitan pop. 1,407,963), Transvaal, NE South Africa, on the S slopes of the Witwatersrand at an altitude of 5,750 ft (1,753.8 m). Johannesburg is the center of an important gold-mining industry, and a manufacturing and commercial center. Manufactures include cut diamonds, industrial chemicals, plastics, cement, electrical and mining equipment, paper and paper products, glass, food products, and beer. Johannesburg was founded as a mining settlement in 1886, when gold was found on the Witwatersrand.

John D. Rock·e·fel·ler, Jr., Memorial Parkway (rŏk′ə-fĕl′ər): *see* National Parks and Monuments Table.

John Fitz·ger·ald Ken·ne·dy National Historic Site (fĭts-jĕr′əld kĕn′ĭ-dē): *see* National Parks and Monuments Table.

John Muir National Historic Site (myo͝or): *see* National Parks and Monuments Table.

John·son (jŏn′sən). **1.** County (1970 pop. 13,630), 676 sq mi (1,750.8 sq km), NW Ark., bounded on the S by the Arkansas River and drained by small Mulberry River and Piney Creek; formed 1833; co. seat Clarksville. It is in an agricultural area that produces cotton, fruit, livestock, and poultry. It has natural-gas wells, coal mines, timber (sawmills and millwork), and some manufactures (including plastic products). Part of Ozark National Forest is in the north. **2.** County (1970 pop. 7,727), 313 sq mi (810.7 sq km), E central Ga., bounded in the W by the Oconee River and drained by the Little Ohoopee River; formed 1858; co. seat Wrightsville. Its coastal-plain agriculture includes cotton, corn, potatoes, truck crops, and fruit. **3.** County (1970 pop. 7,550), 345 sq mi (893.6 sq km), S Ill., in the Ozarks and drained by the Cache River; formed 1812; co. seat Vienna. Agriculture (fruit, corn, wheat, dairy products, and livestock) and lumbering (wood products) predominate in the area. Part of Shawnee National Forest is here. **4.** County (1970 pop. 61,138), 315 sq mi (815.9 sq km), central Ind., drained by the West Fork of the White River; formed 1815; co. seat Franklin. It has agriculture (wheat, corn, livestock, vegetables, and dairy products) and some manufacturing. **5.** County (1970 pop. 72,127), 620 sq mi (1,605.8 sq km), E Iowa, drained by the Iowa River; formed 1839; co. seat Iowa City. It is characterized by prairie agriculture (corn, hogs, cattle, and poultry). It also has limestone quarries. **6.** County (1970 pop. 220,073), 476 sq mi (1,232.8 sq km), E Kansas, in a rolling plains area with low hills, bounded on the N by the Kansas River, on the E by the Mo. border; formed 1855; co. seat Olathe. It has agriculture (stock raising, corn growing, and dairying), and scattered oil and gas fields. Its diversified manufactures include dairy products, clothing, lumber millwork, drugs, metal products, and farm and miscellaneous machinery. **7.** County (1970 pop. 17,539), 264 sq mi (683.8 sq km), E Ky., in a mountainous agricultural area in the foothills of the Cumberlands and drained by the Levisa Fork; formed 1843; co. seat Paintsville. Livestock and poultry are raised, and corn, soybeans, potatoes, apples and tobacco grown. It also has coal mines and oil wells. **8.** County (1970 pop. 34,172), 826 sq mi (2,139.3 sq km), W Mo., drained by the Blackwater River; formed 1834; co. seat Warrensburg. Its agriculture includes corn, wheat, oats, hay and livestock. It has coal mines, stone quarries, clay pits, and clothing and metal manufacturing. **9.** County (1970 pop. 5,743), 377 sq mi (976.4 sq km), SE Nebr., drained by branches of the Nemaha River; formed 1855; co. seat Tecumseh. It has feed, livestock, fruit, grain, dairy, and poultry farming. Its industry includes poultry and egg processing. **10.** County (1970 pop. 11,569), 293 sq mi (758.9 sq km), extreme NE Tenn., bounded on the N by the Va. border, on the E and SE by the N.C. border, drained by the Watauga River and traversed by the Iron Mts. It is in a lumbering and agricultural (tobacco, truck crops, fruit, and livestock) area within Cherokee National Forest. **11.** County (1970 pop. 45,769), 740 sq mi (1,916.6 sq km), N central Texas, bounded on the SW by the Brazos River and drained by tributaries of the Trinity River; formed 1854; co. seat Cleburne. This rich agricultural area produces cotton, corn, oats, peanuts, pecans, fruit, truck crops, poultry, livestock, and dairy products. It also has deposits of clay, limestone, and gravel and some manufacturing and processing. **12.** County (1970 pop. 5,587), 4,175 sq mi (10,813.3 sq km), N Wyo., watered by the Powder River; formed 1875 as Pease co., renamed 1879; co. seat Buffalo. It is in an agricultural (cattle, sheep, grain, and sugar beets) and mining (coal, uranium, oil, and gas) region, with some manufacturing.

Johnson, city (1970 pop. 1,038), seat of Stanton co., SW Kansas, SW of Garden City, in a wheat-growing area.

Johnson City. 1. Village (1970 pop. 18,025), Broome co., S N.Y., in a tri-city area including Endicott and Binghamton; inc. 1892. **2.** City (1978 est. pop. 39,500), Washington co., NE Tenn., in a mountainous region; settled before 1800, inc. 1869. It is an important burley tobacco and dairy market and a railroad center. **3.** City (1970 pop. 767), seat of Blanco co., central Texas. It is the site of the LBJ Ranch, known as the Texas White House when Lyndon B. Johnson was President. The Lyndon B. Johnson National Historic Site includes the former President's boyhood home in the town and his birthplace 13 mi (20.9 km) to the west.

John·ston (jŏn′stən.) **1.** County (1970 pop. 61,737), 797 sq mi (2,064.2 sq km), central N.C., on the coastal plain drained by the Neuse River; formed 1746; co. seat Smithfield. It has agriculture (tobacco, wheat, oats, potatoes, cotton, and corn) and timber (pine and gum). Cotton and lumber milling and tobacco processing are done. **2.** County (1970 pop. 7,870), 638 sq mi (1,652.4 sq km), S Okla., bounded on Lake Texoma and the Washita River and drained by the Blue River; formed 1907; co. seat Tishomingo. In a cattle-raising area, it also produces peanuts and vegetables and has some industry.

Johnston, town (1978 est. pop. 24,200), Providence co., N central R.I., a suburb of Providence; inc. 1759. Among its manufactures are jewelry, textiles, and fabricated metals.

Johnston City, village (1970 pop. 767), seat of Blanco co., S central Texas, W of Austin on the Pedernales River in a agricultural area.

John·stone (jŏn′stən), burgh (1974 est. pop. 23,603), Strathclyde region, W Scotland. There are cotton mills and engineering works. Chemicals, machine tools, and shoelaces are manufactured.

Johnston Island, central Pacific, c.3,000 ft (915 m) long and c.600 ft

(185 m) wide, c.700 mi (1,126 km) SW of Honolulu. It was discovered by the British in 1807 and claimed by the United States in 1858. The island became a U.S. naval base in 1941.

Johns·town (jŏnz'toun'). **1.** City (1970 pop. 10,045), seat of Fulton co., E central N.Y.; founded 1772, inc. 1895. Its leather glove industry began in 1800. Knitted goods, boats, gelatin, and chemicals are also made. The last Revolutionary battle in New York was fought in Johnstown on Oct. 25, 1781. **2.** Industrial city (1978 est. pop. 38,300), Cambria co., SW Pa., on the Conemaugh River at the mouth of Stony Creek; settled 1770, inc. as a city 1936. Situated in a beautiful mountain region, it is a center of heavy industry. Manufactures include iron, steel, coal products, refractories, chemicals, wearing apparel, and mining, telegraph, railroad, and industrial equipment. The city expanded with the rapid growth of iron and steel industries after the Civil War. On May 31, 1889, the dam across the river c.12 mi (19 km) above the city broke as a result of heavy rains, and the city was flooded, with the loss of about 2,200 lives. Flooding occurred again in 1936. Johnstown Flood National Memorial and Allegheny Portage Railroad National Historic Site are nearby.

Jo·hor or **Jo·hore** (jō-hôr', jə-), state (1972 est. pop. 1,365,018), 7,360 sq mi (19,062 sq km), at the S extremity of the Malay Peninsula, Malaysia, opposite Singapore. It is largely covered with rain forests and swamps. Johor has extensive rubber plantations; other agricultural products are rice, copra, pineapples, gambier, and palm products. Tin and bauxite are mined. In 1914 Johor became a British protectorate. Until 1948, when it entered the Federation of Malaya, Johor was classified as one of the Unfederated Malay States.

Johor Ba·ha·ru (bə-här'ōō), **Johore Bha·ru**, or **Johore Bah·ru** (both: bär'ōō), city (1971 pop. 135,936), capital of Johor, Malaysia, S Malay Peninsula, opposite Singapore. The city is connected with Singapore by a causeway across Johore Strait, an arm of the Singapore Strait. It is a trade center for rubber and tropical produce.

Jö·kuls·á (yœ'küls-ou'), name of several Icelandic rivers formed by glaciers. The best known is the Jökulsá á Fjöllum, which rises on the north slope of the Vatnajökull in southeast Iceland and flows c.130 mi (210 km) north into the Axarfjörður, forming the Dettifoss c.30 mi (50 km) from its mouth.

Jo·li·et (jō'lē-ĕt'), city (1978 est. pop. 70,900), seat of Will co., NE Ill., on the Des Plaines River; inc. 1857. It is an important river port, with limestone quarries and coal mines in the area. Earth-moving equipment, wire, radio and television parts, wallpaper, chemicals, and paper and metal products are made in the city.

Jo·li·ette (zhō'lē-ĕt'), city (1976 pop. 18,118), S Que., Canada, on L'Assomption River, NE of Montreal. Its industries include steel, paper, and textile manufacturing, tobacco processing, and limestone quarrying.

Jo·lo (hō'lō, hō-lō'), island (345 sq mi/893.6 sq km), Sulu Archipelago, the Philippines. The seaport city, Jolo (1970 pop. 46,586), on the northwest coast of the island, is the the trading and shipping hub of the archipelago. An ancient walled city, it was once a pirate base and served as the residence of a sultan until the sultanate was abolished in 1940.

Jones (jōnz). **1.** County (1970 pop. 12,270), 402 sq mi (1,041.2 sq km), central Ga., bounded in the SW by the Ocmulgee River; formed 1807; co. seat Gray. It is a farming (peaches, cotton, corn, pimientos, truck crops, and livestock) and sawmilling area. Part of Oconee National Forest is here. **2.** County (1970 pop. 19,868), 585 sq mi (1,515.2 sq km), E Iowa, drained by the Wapsipinicon and Maquoketa rivers; formed 1837; co. seat Anamosa. Cattle, hogs, poultry, corn, oats, and wheat characterize its prairie agriculture. There are limestone quarries and sand and gravel pits. **3.** County (1970 pop. 56,357), 706 sq mi (1,828.5 sq km), SE Miss., drained by Leaf River and Tallahala Creek; formed 1826; co. seats Laurel and Ellisville. Its agriculture includes cotton, corn, sweet potatoes, and poultry. It has lumber and food-processing industries, oil and gas wells, and diversified manufactures (clothing, brick and clay tile, concrete, and machinery). **4.** County (1970 pop. 9,779), 467 sq mi (1,209.5 sq km), E N.C., in a forested and swampy tidewater area drained by the Trent River; formed 1778; co. seat Trenton. Tobacco, corn, cotton, and soybeans are grown, and livestock (especially hogs) is raised. Sawmilling is done. **5.** County (1970 pop. 1,882), 973 sq mi (2,520.1 sq km), S central S.Dak., in an agricultural area drained by the Bad River and numerous creeks; formed 1916; co. seat Murdo. Livestock, poultry, and grain are important. **6.** County (1970 pop. 16,106), 959 sq mi (2,483.8 sq km), W central Texas, in the rolling plains drained by the Clear Fork of the Brazos River; formed 1858; co. seat Anson. This rich agricultural area produces cotton, corn, grain sorghums, some fruit, truck crops, and livestock.

Jones·bor·o (jōnz'bûr-ō, bŭr-ō). **1.** City (1978 est. pop. 30,700), a seat of Craighead co., NE Ark., on Crowley's Ridge; founded 1859, inc. 1883. It is the trade, distributing, and industrial center for a large farm area. Arkansas State Univ. is here. Parts of the city were devastated by tornadoes in May, 1973. **2.** City (1970 pop. 4,105), seat of Clayton co., W central Ga., S of Atlanta, in a farm area; settled 1823, inc. 1859. Lumber and textile products are made. A nearby historical museum is on the site of the Battle of Jonesboro (1864). **3.** City (1970 pop. 1,676), seat of Union co., S Ill., N of Cairo, in a farm area; laid out 1816, inc. 1857. **4.** Town (1970 pop. 5,072), seat of Jackson parish, N central La., SSW of Ruston. It is in a timber and farm region. **5.** Town (1970 pop. 1,510), seat of Washington co., NE Tenn., WSW of Johnson City; laid out 1779, inc. 1815.

Jones·ville (jōnz'vĭl'), town (1970 pop. 700), seat of Lee co., extreme SW Va., near the Powell River SE of Harlan, Ky.

Jön·kö·ping (yœn'chœ-pĭng), city (1975 est. pop. 89,300), capital of Jönköping co., S Sweden, at the S end of Lake Vättern. It is a commercial and industrial center. The safety match was developed here, and the city has large match factories (founded 1844). Other manufactures include machinery, paper, textiles, and airplanes. Jönköping was chartered in 1284 by Magnus I. Gustavus Adolphus gave (1620) it special privileges after its citizens had burned the city to prevent the Danes from sacking it.

Jon·quière (zhôɴ-kyĕr'), city (1976 pop. 60,691), S Que., Canada, on the Saguenay River, adjacent to its twin city, Kénogami, W of Chicoutimi. Its chief industries produce paper and pulp.

Jop·lin (jŏp'lĭn), city (1978 est. pop. 40,200), Jasper and Newton cos., SW Mo., at the edge of the Ozarks; settled c.1839, inc. 1873. It is the shipping and processing point of a grain and livestock region.

Jor·dan (jôr'dn), kingdom (1970 est. pop. 2,348,000), 37,737 sq mi (97,740 sq km), SW Asia, bordering on Israel in the W, on Syria in the N, on Iraq in the NE, and on Saudi Arabia in the E and S. The capital is Amman. East Jordan, which encompasses about 92% of the country's land area, is made up of a section of the Arabian Plateau that in the northeast includes part of the Syrian Desert. In the western part of the plateau are the Jordanian Highlands. Central Jordan includes the Jordan River, the Dead Sea, and the Arabah (a dry riverbed). West Jordan, which is part of historic Palestine, is composed of the hilly regions of Samaria (in the north) and Judaea (in the south). Samaria has abundant fertile soil, and Judaea is largely stony and barren.

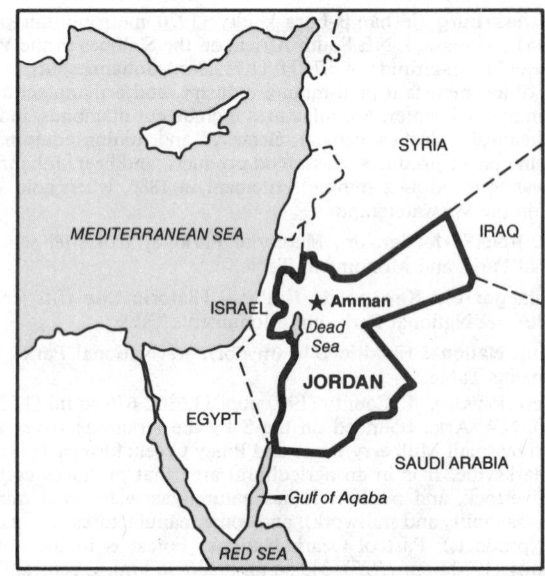

Economy. Jordan's economy is largely agricultural. Only about 10% of the country's land is arable, and farm output is further limited by the small size of most farms, inefficient methods of tilling the soil, inadequate irrigation, and the dislocations caused by the Arab-Israeli Wars. The principal crops are wheat, barley, lentils, tomatoes, eggplants, citrus fruits, and grapes. Many Jordanians support themselves by raising sheep and goats. Manufactures are largely limited to foodstuffs, beverages, clothing, construction materials (especially cement), soap, dairy products, and cigarettes. Numerous artisans make items of leather, wood, and metal. Phosphate rock and potash are the only minerals produced in significant quantities.

History. The region of present-day Jordan roughly corresponds to the biblical lands of Ammon, Bashan, Edom, and Moab. The area was conquered by the Seleucids in the 4th cent. B.C. and was part of the Nabatean empire from the 1st cent. B.C. to the mid-1st cent. A.D., when it was captured by the Romans. In the period between the 6th and 7th cent. it was the scene of fighting between the Byzantine Empire and Persia. In the early 7th cent. the region was invaded by the Moslem Arabs, and after the Crusaders captured Jerusalem in 1099, it became part of the Latin Kingdom of Jerusalem. In 1516 the Ottoman Turks gained control of what is now Jordan, and it remained part of the Ottoman Empire until the 20th cent.

After the fall of the Ottoman Empire in World War I the region came under (1919) the government of Faisal I centered at Damascus. When Faisal was defeated by the French, Transjordan (as Jordan was then known) was made (1920) part of the British League of Nations mandate of Palestine. In 1921 Faisal's brother, Abdullah ibn Husain, was made head of Transjordan, which was administered separately from Palestine and was specifically exempted from being part of a Jewish national home. In 1928 Transjordan became a constitutional monarchy, and in 1946 it became independent as the Hashemite Kingdom of Transjordan.

When Palestine was partitioned and the state of Israel was established in 1948, Transjordan sent forces to fight Israel. The troops of the Arab Legion (the British-trained Jordanian army) were unsuccessful against the Israeli forces, but they did gain control of most of that part of west-central Palestine that the United Nations had designated as Arab territory. In Dec., 1949, Jordan (so renamed in Apr., 1949) concluded an armistice with Israel, and early in 1950, it formally annexed the West Bank. The annexation increased Jordan's population by about 450,000 persons, many of them homeless refugees from Israel. In 1951 Abdullah was assassinated and was succeeded by his son Talal. Talal, however, was mentally ill, and in 1952 parliament replaced him with his son Hussein I. The new king, under pressure from nationalists, ended Jordan's treaty relationship with Great Britain in 1956. Relations with Egypt and Iraq deteriorated in the late 1950s, but improved by the late 1960s.

In the brief Arab-Israeli War of 1967, Jordanian forces were routed by Israel, and Jordan lost the West Bank. In 1968–69 there were clashes along the frontier with Israel, but of greater significance was the growing hostility between the Jordanian government and the Palestinian guerrilla organizations operating in Jordan. The guerrillas sought to establish an independent Palestinian state, a goal that conflicted with Hussein's intention of reestablishing Jordan's control over the West Bank. In Sept., 1970, the country was engulfed in a bloody 10-day civil war, which ended when other Arab countries arranged a cease fire. In July, 1971, the army carried out a successful offensive that destroyed the guerrillas' bases. Jordan played a minor role in the Arab-Israeli War of Oct., 1973, sending a small number of troops to fight on the Syrian front. Bowing to pressure from other Arab states, Hussein formally renounced all Jordanian claims to the West Bank in 1974. He also agreed to recognize the Palestine Liberation Organization as the sole negotiating representative of the Palestinian people. Although the policy of "Jordanization" (the removal of Palestinian influence in Jordanian affairs) remained in effect, Hussein moved (1976) to reassert some influence over the West Bank. In 1975–76, Jordan and Syria agreed to coordinate their military forces in order to achieve a common "eastern front" policy against Israel.

Government. Under the 1952 constitution as amended, the most powerful political and military figure in the country is the king. He appoints a cabinet (headed by a prime minister), which is responsible to the bicameral parliament that consists of a 30-member senate (appointed by the king) and a 60-member house of representatives (popularly elected to 4-year terms).

Jordan, town (1970 pop. 529), seat of Garfield co., E central Mont., NW of Miles City. It is a trading center in an irrigated grain and livestock area.

Jordan, river, c.200 mi (320 km) long, formed in the Hula basin, N Israel, by the confluence of three headwater streams and meandering S through the Sea of Galilee to the Dead Sea. Deep and turbulent during the rainy season, the Jordan is reduced to a sluggish, shallow stream during the summer. As it nears the Dead Sea, its salinity increases. Although the river is not navigable, its waters are potentially valuable for irrigation.

Jos (jôs), city (1969 est. pop. 105,000), central Nigeria, on the Jos Plateau. It is a mining center for tin ore, which is processed in the city, and a collection point for hides and skins and for market-garden produce. It is also a resort.

Jo·se·phine (jō'zə-fēn', -sə-), county (1970 pop. 35,746), 1,625 sq mi (4,208.8 sq km), SW Oregon, in a mountainous area bordering on Calif. and crossed in the N by the Rogue River; formed 1856; co. seat Grants Pass. Lumbering, fruit-growing, and poultry raising are done here. The county includes Oregon Caves National Monument and part of Siskiyou National Forest.

Josh·u·a Tree National Monument (jŏsh'o͞o-ə trē'): *see* National Parks and Monuments Table.

Jos·te·dals·breen (yô'stə-däls-brā'ən), largest glacier of the European mainland, 315 sq mi (815.9 sq km), SW Norway. The glacier is 60 mi (96.5 km) long and 15 mi (24.1 km) wide, with its head c.6,700 ft (2,045 m) above sea level.

Jo·tun·hei·men (yō'to͝on-hā'mən), mountain group, S central Norway; highest of Scandinavia. It culminates in Galdhøpiggen (8,097 ft/2,469.6 m) and Glittertinden (8,104 ft/2,471.7 m). Sparsely inhabited, the region is used for summer pasture.

Jour·dan·ton (jərd'ən-tən), city (1970 pop. 1,841), seat of Atascosa co., S Texas, S of San Antonio, in a ranch and farm area. It is also the headquarters for nearby oil fields.

Ju·ab (jo͞o'ăb), county (1970 pop. 4,574), 3,412 sq mi (8,837.1 sq km), W Utah, in a mining and agricultural area bounded on the W by the Nev. border and watered by the Sevier River; formed 1852; co. seat Nephi. Gypsum, quartz, and silver are among its mineral resources. Cattle and alfalfa are its principal agricultural products.

Juan de Fu·ca Strait (wän də fyo͞o'kə), inlet of the Pacific Ocean, 100 mi (161 km) long and 11 to 17 mi (17.7–27.4 km) wide, between Vancouver Island, British Columbia, and Wash. state, linking the Strait of Georgia and Puget Sound with the Pacific. It forms part of the U.S.-Canada border. Discovered by the English captain Charles W. Barkley in 1787, the strait was named for a sailor, Juan de Fuca, who reputedly had discovered it for Spain in 1592.

Juan Fer·nán·dez (hwän fər-năn'dĕz), group of small islands, S Pacific, c.400 mi (645 km) W of Valparaiso, Chile. They belong to Chile. Volcanic in origin, they have a pleasant climate and are rugged and heavily wooded. The chief occupation is lobster fishing. Discovered in 1563, the islands achieved fame with the publication of Daniel Defoe's *Robinson Crusoe* (1719), generally acknowledged to have been inspired by the confinement (1704–9) of Alexander Selkirk, a Scottish sailor. Occupied by the Spanish in 1750, the islands passed to Chile when it won independence.

Juá·rez (wär'ĕz), city (1970 pop. 436,054) Chihuahua state, N Mexico, on the Rio Grande opposite El Paso, Texas. It is a shipping point and the commercial and processing center for the surrounding cotton-growing area. Developing (1659) as the focal point for Spanish colonial expansion to the north, it was originally called El Paso del Norte and included settlements on both sides of the river, until they were split by the Treaty of Guadalupe Hidalgo (1848), which ended the Mexican War.

Ju·ba (jo͞o'bä), city (1973 pop. 56,737), S Sudan, a port on the White Nile. At a conference in Juba in 1947, representatives of the northern and southern parts of Sudan agreed to unify the country, thus dashing Britain's hopes of adding the south to Uganda. Beginning with a mutiny of southern troops at Juba in 1955, southern unrest led to a Sudanese civil war that was not settled until 1969.

Juba, river, c.1,000 mi (1,610 km) long, formed at the Ethiopia-Somali Republic border, E Africa, and meandering S through SW Somali Republic to the Indian Ocean near Kismayu. The river is extensively used for irrigation.

Ju·dae·a or **Ju·de·a** (both: jo͞o-dē'ə), Greco-Roman name for S Palestine. In the time of Christ it was both part of the province of Syria and a kingdom ruled by the Herods. It was the southernmost of the Roman divisions of Palestine.

Ju·dah (jo͞o'də), the southern of the two kingdoms remaining after the division of the kingdom of the Jews that occurred under Rehoboam. The northern kingdom, Israel, was continually at war with Judah. Judah lasted from 931 B.C. to 586 B.C.

Ju·dith (jo͞o'dĭth), river, 124 mi (199.5 km) long, central Mont., rising in the Little Belt Mts. and flowing NE to the Missouri River NW of Winifred.

Judith Basin, county (1970 pop. 2,667), 1,880 sq mi (4,869.2 sq km), central Mont., in an agricultural region drained by branches of the Judith River; formed 1920; co. seat Stanford. Grain and livestock are raised. Part of Lewis and Clark National Forest is in the county.

Juiz de Fora (zhwēzh dĭ fô'rə), city (1970 pop. 238,052), Minas Gerais state, SE Brazil. It is an industrial and commercial city with more than half of the labor force engaged in textile production. Foodstuffs are also produced. The city was founded at the end of the 18th cent. In the 19th cent., coffee cultivation was the main economic activity.

Ju·juy (ho͞o-hwē'), city (1960 pop. 44,188), capital of Jujuy prov., NW Argentina, on the Bermejo River. In the scenic foothill region of the eastern Andes, it is the center of an agricultural, mining, and cattle-raising area. There are interesting Indian ruins nearby.

Jules·burg (jo͞olz'bûrg'), town (1970 pop. 1,578), seat of Sedgwick co., extreme NE Colo., on the South Platte River near the Nebr. border, in a farm area; founded 1881, inc. 1886. Concrete and farm machinery are produced.

Jul·ian Alps (jo͞ol'yən), mountain range, NE Italy and NW Yugoslavia, between the Carnic Alps and the Dinaric Alps, rising to 9,396 ft (2,865.8 m) in Triglav, the highest peak in Yugoslavia. The forested, glacier-scoured region is a popular resort area.

Ju·li·a·ne·håb (yo͞ol-yä'nə-hôp'), town (1975 est. pop. 2,923), in Julianehåb dist. (1969 pop. 3,213), SW Greenland. It is a fishing port with canneries. Sheep are raised in the surrounding region.

Jü·lich (yü'lĭKH'), former duchy, West Germany, between Cologne and Aachen. At first a county, Jülich was raised to a duchy in 1356, and in 1423 it was united with the county of Berg. Both Jülich and Berg passed in 1521 to duke John III of Cleves and in 1666 to the Palatinate-Neuburg branch of the Bavarian house of Wittelsbach. Occupied by the French from 1794 to 1814, the territory was assigned (1815) to Prussia at the Congress of Vienna.

Jülich, town (1974 est. pop. 32,002), North Rhine–Westphalia, W West Germany. It has some light industry. Originally a Roman settlement, Jülich was chartered in the mid-13th cent. The town was almost totally destroyed in World War II.

Ju·lier (zhü-lyā'), pass, 7,504 ft (2,288.7 m) high, Grisons canton, SE Switzerland, connecting the Upper Engadine Valley to the Oberhalbstein Valley. It has been used since ancient times.

Jul·lun·dur (jŭl'ən-dər), city (1971 pop. 296,103), Punjab state, NW India. It has flour and silk mills. Jullundur was the capital of Punjab from the time of India's independence (1947) until Chandigarh was built in 1953.

Ju·met (zhü-mā'), city (1970 pop. 28,029), Hainaut prov., S Belgium. Manufactures include metal products, glass, and beer.

Jum·na (jŭm'nə) or **Ya·mu·na** (yä'mə-nə), river, c.850 mi (1,370 km) long, rising in the Himalayas, N India, and flowing generally SE, past Delhi, to the Ganges River at Allahabad. The Jumna's confluence with the Ganges is sacred to Hindus. Along the Jumna's banks are many historic monuments including the Taj Mahal at Agra.

Ju·na·gadh (jo͞o'nə-gäd') or **Ju·na·garh** (-gär'), town (1971 pop. 95,945) Gujarat, W India. It is a market for gold and silver embroidery, perfume, and copper and brass vessels. The town has ancient Buddhist caves and Rajput forts.

Junc·tion (jŭngk'shən). **1.** City (1970 pop. 2,654), seat of Kimble co., W central Texas, NW of San Antonio at the junction of two forks that form the Llano River; settled 1876, inc. 1928. Once a cow town, Junction now ships livestock and pecans as well as immense quantities of wool and mohair from surrounding ranches. **2.** Town (1970 pop. 135), seat of Piute co., S Utah, E of Beaver on the Piute Reservoir, at an altitude of 6,250 ft (1,906.3 m). It is in an irrigated agricultural and cattle area.

Junction City, city (1978 est. pop. 20,400), seat of Geary co., NE Kansas, at the confluence of the Republican and Smoky Hill rivers; inc. 1859. The rail and trade center of an agricultural and dairy area, it grew as the supply point for nearby Fort Riley.

Jun·diaí (zho͞oN-dyī'), city (1970 pop. 169,096), São Paulo state, S Brazil, on the Jundiaí River. It is an agricultural and industrial center. Among its products are textiles, ceramics, furniture, soap, wines, foodstuffs, brushes, shoes, paper, matches, chemicals, and agricultural tools. The city was established in the 17th cent.

Ju·neau (jo͞o'nō), county (1970 pop. 18,455), 774 sq mi (2,004.7 sq km), central Wis., in a dairying region bounded on the E by the Wisconsin River and drained by the Yellow and Baraboo rivers; formed 1856; co. seat Mauston. Livestock and general crops are also important. The county has a thriving tourism industry.

Juneau. 1. City (1978 est. pop. 18,400), state capital, coextensive with Juneau borough, SE Alaska, in the Alaska Panhandle; settled by gold miners 1880, inc. 1900. A port on Gastineau Channel, Juneau is a trade center for the Panhandle area, with an ice-free harbor. Salmon and halibut fishing, lumbering, and tourism are important economic activities. Joseph Juneau and a partner discovered gold nearby in 1880, and the city developed as a gold rush town. It was officially designated as capital of the Territory of Alaska in 1900 but did not function as such until the government offices were moved from Sitka in 1906. In 1959 it became state capital. Glacier Bay National Monument is to the northwest. **2.** City (1970 pop. 2,043), seat of Dodge co., E central Wis., NW of Milwaukee; inc. 1887. The area produces cheese, fertilizer, and peas.

Jung·frau (yo͝ong'frou'), peak, 13,668 ft (4,168.7 m) high, S central Switzerland, in the Bernese Alps. It was first ascended in 1811. The Jungfraujoch is a mountain saddle 11,333 ft (3,456.6 m) high, the highest point in Europe reached by rail.

Ju·ni·a·ta (jo͞o'nē-ăt'ə), county (1970 pop. 16,712), 387 sq mi (1,002.3 sq km), central Pa., in a mountainous agricultural area drained by the Juniata River; formed 1831; co. seat Mifflintown. Dairying, fruit-growing, manufacturing (shirts and textiles), and sawmilling are major occupations.

Ju·nín (ho͞o-nēn'), city (1970 pop. 69,731), Buenos Aires prov., E Argentina, on the Salado River. It is a busy commercial center for an agricultural and livestock area. Junín began as a frontier fort (est. 1827) during the struggle against the Indians of Pampa.

Ju·ra (zhü-rä'), department (1975 pop. 238,856), 1,934 sq mi (5,009.1 sq km), E France, in Franche-Comté, bordering on Switzerland; capital Lons-le-Saunier.

Ju·ra (jo͝or'ə, zhü-rä', yo͞o'rä), mountain range, part of the Alpine system, E France and NW Switzerland. It extends in narrow, parallel ridges c.160 mi (260 km) from the Rhine River at Basel to the Rhône River southwest of Geneva; Crêt de la Neige (5,652 ft/1,723 m), in France, is the highest peak. The Jura's rounded crests and summits are covered with dense pine forests and good pasture lands. Hydroelectric plants in the Jura supply power to pulp and paper, textile, watchmaking, and woodworking industries. The Jura Mts. are a popular year-round resort region.

Ju·ruá (zho͞o-rwä'), river, c.1,500 mi (2,415 km) long, rising in the Cerros de Canchyuaya, E Peru, and flowing in a winding course generally NE through Acre and Amazonas states, W Brazil, to the Amazon River E of Fonte Boa. One of the Amazon's longer tributaries, it is navigable along one third of its course and was important for transport during the wild-rubber boom.

Jus·tice (jŭs'tĭs), village (1970 pop. 9,473), Cook co., NE Ill., a suburb W of Chicago; inc. 1911.

Ju·taí (zho͞o'tā), river, c.800 mi (1,287.2 km) long, Brazil, rising in W Amazonas and flowing NE to the Amazon River. Its lower course is navigable.

Jut·land (jŭt'lənd), peninsula, c.250 mi (400 km) long and up to 110 mi (177 km) wide, N Europe, comprising continental Denmark and N Schleswig-Holstein state, West Germany. It is bounded by the Skagerrak in the north, the North Sea in the west, the Kattegat and Lille Baelt in the east, and the Eider River in the south. The term usually is applied only to the Danish territory. Western Jutland is windswept and sandy and has poor soil. Its coast is marshy, with many lagoons. The east coast of Jutland is fertile and densely populated. Dairying and livestock raising are the main occupations. In 1916, off the coast of western Jutland, British and German fleets engaged in the largest naval battle of World War I.

Jy·väs·ky·lä (yo͞o'väs-ky'lă), city (1975 est. pop. 61,209), capital of Keski-Suomi prov., S central Finland. Situated on Lake Päijänne, it is an important port. Paper and wood products are made. The city was chartered in 1837.

K2 (kā'to͞o'), peak, Kashmir: *see* Godwin-Austen, Mount.

Kab·ar·di·no-Bal·kar Autonomous Soviet Socialist Republic (kăb'ər-dē'nō-bôl-kär'), autonomous republic (1970 pop. 589,000), c.4,800 sq mi (12,430 sq km), SE European USSR, in the N part of the Caucasus Mts.; capital Nalchik. The area is a largely unsettled, roadless mountain wilderness. Livestock and poultry are raised, and wheat, corn, hemp, and fruit are grown. Much of the republic's industry is related to agricultural processing. Lumbering, metallurgy, and mining are also important. The Kabardins occupied the land in the foothills of the central Caucasus between the 13th and 15th cent. It is not known when the Balkars settled. The Kabardin area became a Muscovite protectorate in 1557. Its annexation by Russia began with the treaty of Kuchuk Kainarji (1774) and was completed in 1827. The area was organized as an oblast in 1922 and became an autonomous republic in 1936.

Ka·bul (kä'bo͝ol), city (1971 pop. 318,094), capital of Afghanistan in E Afghanistan on the Kabul River. It is strategically located in a high narrow valley, wedged between two mountain ranges that command the main approaches to the Khyber Pass. The city's chief products are woolen and cotton cloth, beet sugar, plastics, leather goods, furniture, glass, matches, soap, and machinery. Kabul's history dates back more than 3,000 years, although the city has been destroyed and rebuilt on several different sites. It was conquered by Arabs in the 7th cent. Babur made it the capital (1504–26) of the Mogul empire, and it remained under Mogul rule until its capture (1738) by Nadir Shah of Persia. It became Afghanistan's capital in 1773. During the Afghan Wars a British army took (1839) Kabul. In 1842 the withdrawing British troops were ambushed and almost annihilated after the Afghans had promised them safe conduct; in retaliation another British force partly burned Kabul. The British again occupied the city in 1879. Kabul's old section, with its narrow, crooked streets, contains extensive bazaars; the modern section has administrative and commercial buildings.

Ka·bwe (kä'bwā), city (1976 est. pop., 115,000), central Zambia. It is a lead and zinc mining center.

Ka·desh (kā'dĕsh), ancient city of Syria, on the Orontes River. Here Ramses II fought (c.1300 B.C.) the Hittites in a great battle that ended in a truce.

Ka·do·ka (kə-dō'kə), city (1970 pop. 815), seat of Jackson co., SW central S.Dak., SW of Pierre. It is a tourist center in an agricultural and livestock area. Badlands National Monument is nearby.

Ka·do·ma (kä-dō'mä), city (1976 est. pop. 142, 993), Osaka prefecture, Honshu, Japan, on the Furu River. It is an industrial and residential suburb of Osaka, with mechanical and textile industries.

Ka·du·na (kə-do͞o'nə), town (1969 est. pop. 174,000), N Nigeria. A leading commercial and industrial center, Kaduna has cotton textile, beverage, and furniture factories. It is also a trade center for the surrounding agricultural area. The city was founded by the British in 1913 and became the capital of Nigeria's Northern Region in 1917.

Kae·song (kā'sông'), city (1967 est. pop. 140,000), S North Korea. A longtime commercial center, it is important chiefly for its exports of ginseng, a valuable medicinal root. There is also active trade in rice, barley, and wheat. Fine porcelain is made in the city, and there is some heavy industry. Intersected by the 38th parallel, Kaesong served as the main contact point between North and South Korea from 1945 to 1951 and passed from United Nations to North Korean forces several times during the Korean War.

Ka·fu·e (kä-fo͞o'ā), river, c.600 mi (965 km) long, rising along the Zambia-Zaire border, S central Africa, near Lubumbashi, and meandering through central Zambia to the Zambezi River. The river has a good hydroelectricity-generating potential.

Ka·ga (kä'gä), city (1975 pop. 47,400), Ishikawa prefecture, W Honshu, Japan. It is an agricultural market, hot spring resort, and industrial center with mechanical and textile industries.

Ka·ge·ra (kä-gâr'ə), river, c.250 mi (400 km) long, formed on the Rwanda-Tanzania border, E central Africa, and flowing N and E forming part of Tanzania's borders with Rwanda and Uganda, before emptying into Victoria Nyanza. The Kagera's headwaters are the remotest sources of the Nile.

Ka·go·shi·ma (kä'gō-shē'mä), city (1976 est. pop. 468,649), Kagoshima prefecture, extreme S Kyushu, Japan, on Satsuma Peninsula and Kagoshima Bay. An important port, it has a navy yard. The city's industries produce Satsuma porcelain ware, silk and cotton clothing, tinware, and wood products. Kagoshima is the site (since 1961) of a major Japanese rocket base.

Ka·ho·ka (kə-hō'kə), city (1970 pop. 2,207), seat of Clark co., NE Mo., WNW of Keokuk, Iowa; laid out 1851, inc. 1886. It is a trade center of a farm and timber region.

Ka·ho·o·la·we (kä-hō'ō-lä'vā, -wā, kä'ho͞o-), uninhabited island, 45 sq mi (116.6 sq km), central Hawaii; separated from Maui island to the NE by Alalakeiki Channel. The island, low and unfertile, has served as a prison and as a military target range.

Ka·hu·lu·i (kä'ho͞o-lo͞o'ē), city (1970 pop. 8,280), Maui co., Hawaii, on N Maui SE of Wailuku. It is the principal port of the island.

Kai-e-teur Falls (kī'ĕ-to͞or), waterfall, 741 ft (226 m) high, in the Potaro River, W Guyana.

K'ai-feng or **Kai·feng** (both: kī'fŭng'), city (1970 est. pop. 330,000), NE Honan prov., China, on the Lunghai RR. It is a commercial, agricultural, and industrial center. Manufactures include agricultural machinery, motor vehicles, electrical and electronic equipment, fertilizer, chemicals, and processed foods. The Huang Ho, just to the north, has frequently flooded the city. Founded in the 3rd cent. B.C., it was capital of the Five Dynasties (906–59) and then capital of the northern Sung dynasty (960–1127). The city fell to the Mongols in the 13th cent.

Kai Islands or **Kei Islands** (both: kī), island group, c.550 sq mi (1,425 sq km), E Indonesia, SE of Ceram, in the Banda Sea, in the Moluccas. It is densely forested with valuable timber.

Kai·las (kī-läs'), peak, c.22,280 ft (6,795 m) high, SW Tibet (China), highest point of the Kailas Range, in the Himalayas. The dwelling place of the Hindu god Shiva, Kailas is the goal of pilgrimages.

Kai·lu·a (kī-lo͞o'ä), uninc. city (1978 est. pop. 39,000), Honolulu co., Hawaii, on the SE coast of Oahu, on Kailua Bay. An agricultural experiment station is here.

Kai·nan (kī-nän'), city (1975 pop. 53,250), Wakayama prefecture, S Honshu, Japan, on the Kii Sound. It is a port and industrial center with spinning, textile, and print-dyeing industries.

Kair·ouan (kĕr-wän'): *see* Al Qayrawan, Tunisia.

Kai·sers·lau·tern (kī'zərs-lou'tərn), city (1974 est. pop. 102,119), Rhineland-Palatinate, W West Germany, on the Lauter River. It is a commercial, industrial, and cultural center. There are ironworks, textile mills, and sewing-machine, furniture, and automobile factories. Charlemagne built a castle in Kaiserslautern that was later enlarged (1153–58) by Emperor Frederick I; some ruins of the castle remain today. The city was repeatedly devastated by warring armies.

Kai·zu·ka (kī-zo͞o'kä), city (1975 pop. 79,506), Osaka prefecture, S Honshu, Japan, on Osaka Bay. It is a commercial port and industrial center where textiles and flour are produced.

Ka·jaa·ni (kä'yä-nē), city (1975 est. pop. 20,583), Oulu prov., central Finland, on the Kajaaninjoki River. Forest products and sports equipment are manufactured. Kajaani was chartered in 1651. The Kajaneborg fortress, around which the city grew, was taken by the Russians in 1716. Restored in 1937, the fortress is a tourist attraction.

Ka·ka·mi·ga·ha·ra (kä-kä'mē-gä-hä'rä), city (1975 pop. 94,192), Gifu prefecture, central Honshu, Japan. It is an agricultural and commercial center.

Ka·ki·na·da (kä-kĭ-nä'də), town (1971 pop. 164,172), Andhra Pradesh state, SE India, on the Godavari River delta. Formerly an important port on the Bay of Bengal, it is now a market for sugar cane, oilseed, cotton, rice, jute, and iron ore.

Ka·ko·ga·wa (kä-kō'gä-wä), city (1976 est. pop. 176,558), Hyogo prefecture, S Honshu, Japan, on the Kako River. It is an industrial center producing woolen and rubber goods and fertilizers.

Ka·la·ha·ri (kä'lä-hä'rē), arid plateau region, c.100,000 sq mi (259,000 sq km), in Botswana, Namibia, and the Republic of South Africa. The Kalahari, covered largely by reddish sand, lies between the Orange and Zambezi rivers and is studded with dry lake beds. Grass grows throughout the Kalahari in the rainy season, and some parts also support low thorn scrub and forest. Grazing and a little agriculture are possible in certain areas.

Ka·lakh (kä'läKH), ancient city: *see* Calah.

Ka·lá·mai (kä-lä'mā) or **Ka·la·ma·ta** (kä'lə-mä'tə, kăl'ə-), city (1971 pop. 39,133), capital of Messinia prefecture, S Greece, in the Peloponnesus; a port on the Gulf of Messinia. It is an agricultural trade center. Silk and flour are manufactured. The city developed after c.1205, when it became a fief of the Villehardouin family. It later came under the rule of Venice and the Ottoman Turks.

Kal·a·ma·zoo (kăl'ə-mə-zōō'), county (1970 pop. 201,550), 567 sq mi (1,468.5 sq km), SW Mich., drained by the Kalamazoo and Portage rivers; organized 1830; co. seat Kalamazoo. Farm crops include corns, wheat, soybeans, hay, and fruit. Livestock is raised, and dairy products are made. Among its manufactures are lumber, wood products, furniture, pharmaceuticals, and automotive equipment.

Kalamazoo, city (1978 est. pop. 77,200), seat of Kalamazoo co., SW Mich., on the Kalamazoo River at its confluence with Portage Creek; inc. 1838. It is an industrial and commercial center in a fertile farm area that produces celery, peppermint, and fruit. Kalamazoo has a large paper industry.

Ka·le·mi (kä-lā'mē), formerly **Al·bert·ville** (ăl'bərt-vĭl'), city (1967 est. pop. 87,000), Shaba region, SE Zaire, on Lake Tanganyika at the mouth of the Lukuga River. It is a commercial center. Manufactures include textiles and cement. The city was founded in 1892 by Belgians as a military post.

Kal·goor·lie (kăl-gōōr'lē), town (1976 pop. 9,064), Western Australia, SW Australia. It is the chief gold-mining town of the state. Gold was found at nearby Coolgardie in 1892; nickel is also mined.

Ka·li·nin (kä-lē'nĭn), formerly **Tver** (tə-vĕr'), city (1976 est. pop. 395,000), capital of Kalinin oblast, central European USSR, at the confluence of the Volga and Tver rivers. A major port on the upper Volga as well as an industrial center, it has industries producing linen textiles, heavy machinery, and rolling stock. The city grew around a fort established in the late 12th cent. From the mid-13th cent. until the late 14th cent. it was the seat of a powerful principality that rivaled Moscow.

Ka·li·nin·grad (kä-lē'nĭn-gräd', kə-lyē'nyĭn-grät'), formerly **Kö·nigs·berg** (kōōn'ĭgz-bûrg', koe'nĭKHs-bĕrKH), city (1976 est. pop. 345,000), capital of Kaliningrad oblast, W European USSR, on the Pregolya River near its mouth on the Vislinski Zalev, which empties into the Gulf of Kaliningrad on the Baltic Sea. A major ice-free Baltic seaport and naval base and an important commercial center, Kaliningrad has industries that produce ships, machinery, food products, metals, automobile parts, and textiles. The city was founded (1255) as a fortress of the Teutonic Knights by King Ottocar II of Bohemia. It joined (1340) the Hanseatic League and became (1457) the seat of the grand master of the Teutonic Order. It was the residence of the dukes of Prussia from 1525 until the union (1618) of Prussia and Brandenburg and became (1701) the coronation city of the kings of Prussia. The Univ. of Königsberg (founded 1544) reached its greatest fame when Immanuel Kant taught here. The city was transferred to the USSR in 1945. The new Soviet city was laid out after 1945.

Kal·i·spell (kăl'ĭ-spĕl'), city (1978 est. pop. 11,700), seat of Flathead co., NW Mont., at the head of Flathead Lake near Glacier National Park; inc. 1892. It is the tourist and trade center of a rich agricultural, fruit, and timber region.

Ka·lisz (kä'lēsh, -lĭsh), city (1975 est. pop. 87,300), central Poland. An industrial center, it has industries producing textiles, machinery, metals, and chemicals. One of the oldest Polish towns, it has been identified as the Slavic settlement of Calissia mentioned in the 2nd cent. A.D. by Ptolemy. It flourished as a trade center from the 13th cent. The city passed to Prussia in 1793, was transferred to Russia in 1815, and was restored to Poland in 1919.

Kal·ka·ska (kăl-kăs'kə), county (1970 pop. 5,372), 564 sq mi (1,460.8 sq km), N Mich., drained by the Manistee and Boardman rivers; organized 1871; co. seat Kalkaska. Logging is a major industry. There are oil and gas wells here, and automotive parts and equipment are manufactured. There is some stock raising, dairying, and farming of potatoes, grain, and fruit.

Kalkaska, village (1970 pop. 1,475), seat of Kalkaska co., N Mich., E of Traverse City, in a farm and lake region; inc. 1887.

Kal·mar (käl'mär), city (1975 est. pop. 44,500), capital of Kalmar co., SE Sweden, on the Kalmarsund (an arm of the Baltic Sea) opposite Öland Island. It is a commercial, industrial, and tourist center. Manufactures include matches, glass, processed food, and ships. It has been an important trade center since the 8th cent.

Kal·myk Autonomous Soviet Socialist Republic (kăl'mĭk), autonomous republic (1970 pop. 268,000), c.29,400 sq mi (76,145 sq km), SE European USSR, on the Caspian Sea; capital Elista. Lying mostly in the vast depression of the northern Caspian lowland, the republic is largely a steppe and desert area. Stock raising is by far the leading economic activity, and fishing is important. Irrigation has made limited agriculture possible. A seminomadic branch of the Oirat Mongols, the Kalmyks migrated from Chinese Turkistan to the steppe west of the Volga's mouth in the mid-17th cent. They became allies of the Russians and were charged by Peter I with guarding the eastern frontier of the Russian Empire. Under Catherine II, however, the Kalmyks became vassals. The Kalmyk Autonomous Oblast was established in 1920; it became an autonomous republic in 1936. During World War II, Kalmyk units fought the Russians in collaboration with the Germans. As a result, the Kalmyks were deported to Siberia in 1943, and their republic was dissolved. The Kalmyk Autonomous SSR was officially re-established in 1958.

Ka·lo·csa (kŏ'lô-chŏ), town (1970 pop. 16,004), S Hungary, near the Danube River. It is famed for its embroidery.

Ka·lu·ga (kä-lōō'gä), city (1976 est. pop. 255,000), capital of Kaluga oblast, central European USSR, on the Oka River. Known since 1389 as a Muscovite outpost, it is a river port and an industrial center producing machinery, electrical equipment, and textiles.

Ka·lund·borg (kä'lōōn-bôr'), city (1970 pop. 19,216), Vestsjaelland co., central Denmark, a port on the Kalundborg Fjord, an arm of the Store Baelt. It is a commercial, industrial, and communications center. Manufactures include chemicals and machinery.

Ka·ma (kä'mä), river, c.1,260 mi (2,030 km) long, E European USSR, the chief left tributary of the Volga, rising in the foothills of the central Urals and flowing N, E, and SW to join the Volga.

Ka·ma·ku·ra (kä-mä'kōō-rä), city (1976 est. pop. 168,183), Kanagawa prefecture, central Honshu, Japan, on Sagami Bay and at the base of the Miura Peninsula. It is a resort and residential area but is chiefly noted as a religious center, the site of more than 80 shrines and temples. An earthquake in 1923 severely damaged the city.

Ka·mar·ha·ti (kä'mär-hä'tē), city (1971 pop. 169,222), West Bengal state, NE India. It is a suburb of Calcutta.

Kam·chat·ka (kăm-chăt'kə), peninsula, 104,200 sq mi (269,878 sq km), Far Eastern USSR, separating the Sea of Okhotsk in the W from the Bering Sea and the Pacific Ocean in the E. The peninsula's central valley, drained by the Kamchatka River, is enclosed by two parallel volcanic ranges. Kamchatka is covered with mountain vegetation, except in the central valley and on the west coast, which has peat marshes and tundralike moss. There are numerous forests, mineral springs, and geysers. Kamchatka's mineral resources include oil, coal, gold, mica, pyrites, sulfur, and tufa. Fishing, sealing, hunting, and lumbering are the main occupations. Cattle breeding is carried on in the south, and farming (rye, oats, potatoes, vegetables) in the Kamchatka valley. Reindeer are also raised on the peninsula. Industries include fish processing, shipbuilding, and woodworking. Kamchatka was first sighted in 1697. Its exploration and development continued in the early 18th cent. under Czar Peter I. Russian conquest was complete by 1732.

Ka·me·nets-Po·dol·ski (kä'mĭn-yĭts-pə-dôl'skē), city (1976 est. pop. 77,000), SW European USSR, in the Ukraine. It is a rail terminus and has industries that produce foodstuffs, tobacco, machinery, machine tools, and automobile parts. Kamenets-Podolski passed to Poland in the 14th cent. It came under Russian control in 1793.

Ka·mensk-Shakh·tin·ski (kä'myĭnsk-shäkh'tyĭn-skē), city (1976 est. pop. 75,000), SE European USSR, on the Donets River. A mining center of the Donets coal basin, the city is also an important producer of artificial fibers.

Ka·met (kŭm'ät'), peak, 25,447 ft (7,761.3 m) high, on the border of India and Tibet, in the Himalayas.

Ka·mi·na (kä-mē'nä), city (1967 est. pop. 115,000), Shaba region, S Zaire. Kamina was used by the Belgians as a center for interventionist actions in the early months of Zaire's independence (1960).

Kam·loops (kăm'lo͞ops), city (1976 pop. 58,311), S British Columbia, Canada, at the junction of the North Thompson and South Thompson rivers. A trading post was first established on the site in 1812. A village grew up at the time of the Cariboo gold rush (1860). It is now a tourist and supply center for an extensive lumbering, mining, and farming district.

Kam·pa·la (käm-pä'lä), city (1969 pop. 331,889), capital of Uganda, on Victoria Nyanza. It is Uganda's communications, economic, and transportation center. Manufactures include processed foods, beverages, shoes, enamelware, furniture, and machine parts. Kampala grew up around a fort constructed (1890) for the British East Africa Company. Despite its proximity (20 mi/32.2 km) to the equator, the city has a moderate climate, largely because of its altitude (c.4,000 ft/1,220 m). The city is built on and around six hills and has wide avenues that fan out toward the surrounding suburbs.

Kam·pen (käm'pən), town (1977 est. pop. 29,584), Overijssel prov., central Netherlands, on the IJssel River, near the IJsselmeer. It is a trade and industrial center. Kampen was first mentioned in the 13th cent., and in the 15th cent. it was a member of the Hanseatic League. Notable structures in the town include several churches and buildings dating from the 14th and 15th cent.

Kam·pot (käm'pôt'), town (1962 est. pop. 12,558), capital of Kampot prov., S Cambodia, on the Gulf of Siam. It is a seaport and the center of the Cambodian pepper culture.

Kan (gän), river, c.550 mi (885 km) long, flowing N through the plain of central Kiangsi prov., SE China, past Nan-ch'ang to P'o-yang lake. The lower Kan valley is fertile; rice and tea are the main crops.

Ka·nab (kə-năb'), city (1970 pop. 1,381), seat of Kane co., extreme S Utah, ESE of St. George near the Ariz. border on Kanab Creek; settled 1864. In a region of plateaus and cliffs north of the Grand Canyon and south of Zion National Park, it is a tourist resort and a ranching center.

Ka·na·bec (kə-nā'bĕk), county (1970 pop. 9,775), 525 sq mi (1,359.8 sq km), E Minn., drained by the Snake River; formed 1858; co. seat Mora. It is in an agricultural area that produces dairy products, livestock, poultry, and grain. It has some mining and food processing, and manufactures plastic products and machinery.

Ka·nan·ga (kə-näng'gə), formerly **Lu·lua·bourg** (lo͞o-lwä-bo͝or'), city (1971 est. pop. 483,400), capital of Kasai-Occidental prov., S central Zaire, on the Lulua River. It is the commercial and transportation center of an agricultural region where cotton is grown. The city was founded in 1884. Kananga grew rapidly in the early 20th cent. with the coming of the railroad.

Ka·na·wha (kə-nô'wə), county (1970 pop. 229,515), 908 sq mi (2,351.7 sq km), W central W.Va., on the Allegheny Plateau bounded on the W by the Coal River and drained by the Kanawha, Elk, and Pocatalico rivers; formed 1788; co. seat Charleston. Its agriculture includes corn, dairy products, livestock, poultry, and tobacco. It also has bituminous-coal mines. oil and natural-gas wells, and food-processing and lumbering industries. Chemicals, concrete, and glass and metal products are manufactured.

Ka·na·za·wa (kä-nä'zä-wä), city (1976 est. pop. 400,411), capital of Ishikawa prefecture, central Honshu, Japan, on the Sea of Japan. It produces cotton and silk textiles, machinery, rolling stock, iron, and fine porcelain and lacquer ware. The city was the seat of the Maeda clan (16th–19th cent.) and gradually became an industrial center.

Kan·chen·jun·ga (kän'chən-jo͝ong'gə), mountain, on the Sikkim-Nepal border, E Himalayas; geologically regarded as the main axis of the Himalayan range. The third-highest mountain in the world, it has five peaks, of which the tallest is 28,208 ft (8,603.4 m). In 1955 a British expedition under Charles Evans climbed the mountain, but in deference to the wishes of Sikkimese authorities the party stopped a few yards short of the summit.

Kan·chi·pu·ram (kŭn'chē-po͝or-əm), formerly **Con·jee·ve·ram** (kən-jē'və-rəm), city (1971 pop. 110,505), Tamil Nadu state, S India. The patterns and texture of its saris are famous. Sacred to Hindus, it is known as the Golden City. Several temples in the Dravidian style survive from the period (3rd–8th cent.) when it was a center of Brahmanical and Buddhist culture.

Kan-chou or **Kan·chow** (both: gän'jō'), city, SW Kiangsi prov., China, on the Kan River. It is a transportation, distribution, and commercial center. Fertilizer is manufactured, and tungsten mines are nearby.

Kan·da·har (kŭn'də-här'), city (1973 est. pop. 140,000), capital of Kandahar prov., S Afghanistan. Kandahar is a market for sheep, wool, cotton, food grains, fresh and dried fruit, and tobacco. Woolen cloth, felt, and silk are manufactured. The surrounding irrigated region produces fine fruits, especially grapes, and the city has plants for canning, drying, and packing fruit. Kandahar may have been founded by Alexander the Great (4th cent. B.C.). India and Persia long fought over the city, which was strategically located on the trade routes of central Asia. It was conquered by Arabs in the 7th cent. Genghis Khan sacked it in the 12th cent., after which it became a major city of the Karts (Mongol clients) until their defeat by Tamerlane in 1383. Babur, founder of the Mogul empire of India, took Kandahar in the 16th cent. It was later contested by the Persians and by the rulers of emerging Afghanistan, who made it the capital (1748–73) of their newly independent kingdom. British forces occupied Kandahar during the First Afghan War (1839–42). The British again held the city from 1879 to 1881.

Kan·da·lak·sha (kən-də-läk'shə), city (1974 est. pop. 43,000), NW European USSR, on the Kandalaksha Bay of the White Sea. It is a seaport and has aluminum plants and hydroelectric stations. A settlement at the present site was known to the Vikings.

Kan·di·yo·hi (kăn'də-yō-hī'), county (1970 pop. 30,548), 783 sq mi (2,028 sq km), SW central Minn., watered by several lakes; formed 1858; co. seat Willmar. It is in an agricultural (corn, oats, barley, livestock, dairy products, and poultry) and food-processing area. It manufactures metal products and has a tourist industry.

Kan·dy (kăn'dē, kän'-), city (1972 est. pop. 85,000), capital of Central prov., Sri Lanka (Ceylon), on the Kandy Plateau. It is a mountain resort and the market center for an area producing tea, rubber, rice, and cacao. The main part of the city overlooks a scenic artificial lake built by the last king of Kandy in 1806. Near the lake is the Temple of the Tooth, said to house one of Buddha's teeth.

Kane (kān). **1.** County (1970 pop. 251,005), 516 sq mi (1,336.4 sq km), NW Ill., drained by the Fox River and small Mill Creek; formed 1836; co. seat Geneva. In an agricultural area yielding dairy products, livestock, and grain, it has limestone quarries and industrial centers producing varied manufactures. **2.** County (1970 pop. 2,421), 4,016 sq mi (10,401.4 sq km), S Utah, in a grazing and agricultural area bordering on Ariz. and bounded in the E and SE by the Colorado River; formed 1864; co. seat Kanab. It is drained by the Paria and Virgin rivers and includes part of Bryce Canyon National Park.

Ka·ne·o·he (kä'nā-ō'hā), uninc. city (1978 est. pop. 35,000), Honolulu co., Hawaii, on the E coast of Oahu, on Kaneohe Bay. Once the site of a pineapple plantation and cannery, it is now a residential seaside community. A missile-tracking station is here.

Kan·ga·roo Island (kăng'gə-ro͞o'), small island, South Australia, S Australia, at the entrance to Gulf St. Vincent. The chief products are barley, sheep, salt, gypsum, and eucalyptus oil. At its west end is Flinders Chase, a large reservation for native flora and fauna.

Kang·hwa or **Kang·hoa** (both: käng'hwä'), island, 163 sq mi (422.2 sq km), off SW South Korea, in the Yellow Sea. Farming and fishing are important occupations. Kanghwa was briefly the site of the Korean capital in the 13th cent. It was stormed by the French in 1866 and by the Americans in 1871.

Kang·nung (käng'no͝ong'), city (1975 pop. 55,900), NE South Korea, a port on the Sea of Japan. It is also an agricultural center and is famed for its beautiful scenery.

K'ang-ting or **Kang·ting** (both: käng'dĭng'), city, W Szechwan prov., China, in the Kan-tzu Tibetan Autonomous Region. It is a transportation center on the main road from Ch'eng-tu to Lhasa, Tibet.

Kan·ia·pis·kau or **Can·la·pis·cau** (both: kăn'yə-pĭs'kou), river, c.575 mi (925 km) long, issuing from Kaniapiskau Lake, NE Que., Canada, and flowing generally NW past Fort Mackenzie to the Koksoak River, which then flows NE to Ungava Bay at Fort Chimo.

Ka·nin (kä'nyĭn), peninsula, N European USSR, projecting into the Barents Sea between the White Sea in the W and Chesha Bay in the E. Its northernmost cape is called Kanin Nos.

Kan·ka·kee (kăng'kə-kē'), county (1970 pop. 97,250), 680 sq mi (1,761.2 sq km), NE Ill., bordered on the E by Ind., formed 1853; co. seat Kankakee. The Kankakee and Iroquois rivers drain this county. Corn, oats, soybeans, wheat, livestock, and poultry are its major products. Its manufactures include farm machinery, wood products, brick, tile, clothing, stoves, paint, fiber drums, hosiery, and dairy and food products. There are limestone deposits.

Kankakee, city (1978 est. pop. 27,700), seat of Kankakee co., E Ill., on the Kankakee River; inc. 1855. It is an industrial and shipping center for a farm area. Kankakee's varied manufactures include ranges, water heaters, furniture, tractors, farm and garden equipment, biochemicals, and pharmaceuticals. Limestone quarries are nearby.

Kan·kan (käN-käN'), city (1967 est. pop. 50,000), E Guinea, a port on the Milo River, a tributary of the Niger. It is the commercial center for a farm area where rice, sesame, maize, tomatoes, oranges, mangoes, and pineapples are grown. Diamonds are mined. Bricks and fruit juices are made. Kankan was probably founded in the 18th cent. as a trade center that linked the Sudan region with the forest belt and the Atlantic coast. The French occupied the city in 1891.

Kan·nap·o·lis (kə-năp'ə-lĭs), uninc. city (1978 est. pop. 37,000), Cabarrus and Rowan cos., W central N.C.; founded c.1905. It is a planned company town owned by Cannon Mills, known for its production of household linens.

Kan·nauj (kə-nouj'), town (1971 pop. 28,189), Uttar Pradesh state, N central India, on the Ganges River. It is a market center for food grains, oilseed, fruit, perfume, and rose water. Kannauj declined after being conquered by Turkish tribes in 1018.

Ka·no (kä'nō), city (1971 est. pop. 357,000), N Nigeria. It is the trade and shipping center for an agricultural region where cotton, cattle, and groundnuts are raised. Peanut flour and oil, cotton textiles, steel furniture, processed meat, concrete blocks, shoes, and soap are the chief manufactures. The city has long been known for its leatherwork. Kano's written history dates back to A.D. 999, when the city was already several hundred years old. In the early 16th cent. Kano accepted Islam, and c.1600 it was temporarily held by the Moslem state of Bornu. Kano reached the height of its power in the 17th and 18th cent. In 1903 a British force captured the city.

Ka·non·ji (kä-nōn'jē), city (1975 pop. 31,700), Kagawa prefecture, E Shikoku, Japan, on the Hiuchi Sea. It is a religious center and agricultural market noted for its Buddhist temple.

Ka·no·ya (kä-nō'yä), city (1975 pop. 38,500), Kagoshima prefecture, S Kyushu, Japan, on the Osumi Peninsula. It is an agricultural market with a silk-rayon weaving industry.

Kan·pur (kän'po͝or), city (1971 pop. 1,151,975), Uttar Pradesh state, N central India, on the Ganges River. A major industrial center, it produces chemicals, textiles, leather goods, and food products. During the Indian Mutiny (1857), Nana Sahib, whose claim to a pension had been rejected, slaughtered the entire British garrison, including women and children.

Kan·sas (kăn'zəs), state (1975 pop. 2,276,000), 82,264 sq mi (213,064 sq km), central United States, admitted to the Union in 1861 as the 34th state. Topeka is the capital. Almost rectangular in shape, Kansas is bounded on the north by Nebr., on the east by Mo., on the south by Okla., and on the west by Colo.

Part of the Great Plains, Kansas is famous for its seemingly endless fields of ripe golden wheat. The land rises more than 3,000 ft (915 m) from the eastern alluvial prairies to western semiarid high plains, which stretch toward the foothills of the Rocky Mts. The rise is so gradual, however, that it is imperceptible, although the terrains of the east and the west are markedly different. The state is drained by the Kansas and Arkansas rivers, both of which generally run

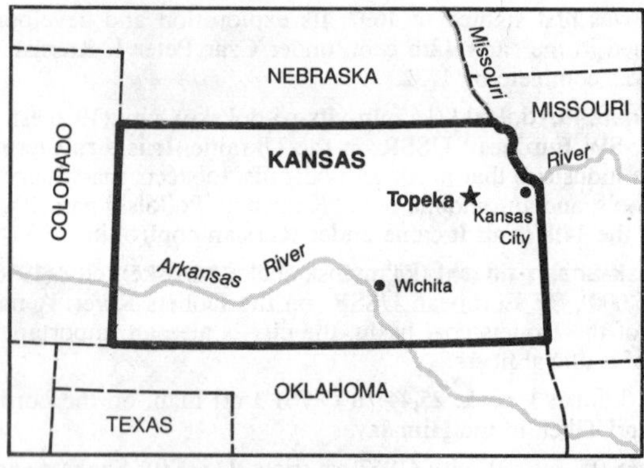

from west to east. Occasional dust storms plague farmers and ranchers in the west. The climate is continental, with wide extremes—cold winters with blizzards and hot summers with tornadoes. Floods also wreak havoc in the state.

Kansas is divided into 105 counties:

NAME	COUNTY SEAT	NAME	COUNTY SEAT
Allen	Iola	Linn	Mound City
Anderson	Garnett	Logan	Oakley
Atchison	Atchison	Lyon	Emporia
Barber	Medicine Lodge	McPherson	McPherson
Barton	Great Bend	Marion	Marion
Bourbon	Fort Scott	Marshall	Marysville
Brown	Hiawatha	Meade	Meade
Butler	El Dorado	Miami	Paola
Chase	Cottonwood Falls	Mitchell	Beloit
Chautauqua	Sedan	Montgomery	Independence
Cherokee	Columbus	Morris	Council Grove
Cheyenne	St. Francis	Morton	Elkhart
Clark	Ashland	Nemaha	Seneca
Clay	Clay Center	Neosho	Erie
Cloud	Concordia	Ness	Ness City
Coffey	Burlington	Norton	Norton
Comanche	Coldwater	Osage	Lyndon
Cowley	Winfield	Osborne	Osborne
Crawford	Girard	Ottawa	Minneapolis
Decatur	Oberlin	Pawnee	Larned
Dickinson	Abilene	Phillips	Phillipsburg
Doniphan	Troy	Pottawatomie	Westmoreland
Douglas	Lawrence	Pratt	Pratt
Edwards	Kinsley	Rawlins	Atwood
Elk	Howard	Reno	Hutchinson
Ellis	Hays	Republic	Belleville
Ellsworth	Ellsworth	Rice	Lyons
Finney	Garden City	Riley	Manhattan
Ford	Dodge City	Rooks	Stockton
Franklin	Ottawa	Rush	La Crosse
Geary	Junction City	Russell	Russell
Gove	Gove	Saline	Salina
Graham	Hill City	Scott	Scott City
Grant	Ulysses	Sedgwick	Wichita
Gray	Cimarron	Seward	Liberal
Greeley	Tribune	Shawnee	Topeka
Greenwood	Eureka	Sheridan	Hoxie
Hamilton	Syracuse	Sherman	Goodland
Harper	Anthony	Smith	Smith Center
Harvey	Newton	Stafford	St. John
Haskell	Sublette	Stanton	Johnson
Hodgeman	Jetmore	Stevens	Hugoton
Jackson	Holton	Sumner	Wellington
Jefferson	Oskaloosa	Thomas	Colby
Jewell	Mankato	Trego	Wakeeney
Johnson	Olathe	Wabaunsee	Alma
Kearny	Lakin	Wallace	Sharon Springs
Kingman	Kingman	Washington	Washington
Kiowa	Greensburg	Wichita	Leoti
Labette	Oswego	Wilson	Fredonia
Lane	Dighton	Woodson	Yates Center
Leavenworth	Leavenworth	Wyandotte	Kansas City
Lincoln	Lincoln		

Economy. Farming, though now surpassed by manufacturing as a source of income, is still important to the state's economy, and Kansas is the nation's leading producer of wheat and second-largest producer of sorghum for grain. Corn and hay are also major crops. Cattle and calves are raised and constitute the single most-valuable agricultural item. Meat-packing and dairy industries are major economic activities, and the Kansas City stockyards are among the nation's largest. Food processing ranked as the state's third-largest industry in the early 1970s. The two leading industries are the manufacture of transportation equipment and of chemicals. Other important manufactured items are aircraft, petroleum and coal products, and nonelectrical machinery. The state is a major producer of crude petroleum and has large reserves of natural gas and helium.

History. When the Spanish explorer Francisco Vásquez de Coronado visited (1541) the Kansas area, it was occupied by various Plains Indian tribes, notably the Kansas, the Wichita, and the Pawnee. French traders were active among the Indians during most of the 18th cent. By the Treaty of Paris of 1763, ending the French and Indian Wars, France ceded the territory of western Louisiana (including Kansas) to Spain. In 1800 Spain secretly retroceded the territory to France, from whom the United States acquired it in the Louisiana Purchase in 1803. Most of the territory that eventually became Kansas was considered unsuitable for settlement because of its apparent barrenness. In the 1830s the region was designated permanent Indian country, and northern and eastern tribes were relocated here. Forts were constructed for frontier defense and for the protection of the growing trade along the Santa Fe Trail. Kansas was organized as a territory in 1854.

The Kansas-Nebraska Act (1854), an attempted compromise on the extension of slavery, repealed the Missouri Compromise and reopened the issue of extending slavery by allowing settlers of territories to decide the matter themselves. Towns were established by hostile free staters and proslavery settlers. Violence soon came to "bleeding Kansas." On May 21, 1856, proslavery groups raided Lawrence. A few days later a band led by the abolitionist crusader John Brown murdered five proslavery men in the Pottawatomie massacre. The Kansas conflict and the issue of statehood for the territory became a national issue. Kansas became a state in 1861, with the capital at Topeka. In the Civil War Kansas fought with the North and suffered the highest rate of fatal casualties of any state in the Union.

With peace came the development of the prairie lands. The construction of railroads made cow towns such as Abilene and Dodge City the shipping point for large herds of cattle driven overland from Texas. Winter wheat replaced spring wheat on an ever-increasing scale. Corn, too, soon became a major money crop. Over the years improved agricultural methods and machines increased crop yield. Irrigation proved practicable in some areas, and winter wheat and alfalfa could be cultivated in dry regions. As part of the Dust Bowl, Kansas sustained serious land erosion during the long drought of the 1930s. During World War II agriculture thrived and industry expanded rapidly. Kansas has become increasingly industrialized and urbanized, and industrial production has surpassed farm production in economic importance.

Government. Government in Kansas is based on the constitution of 1859, adopted just before Kansas attained statehood. An elected governor heads the executive and serves a term of two years. The legislature has a house of representatives and a senate, with the 125 members of the house elected for two-year terms and the 40 members of the senate elected for four-year terms.

Educational Institutions. Prominent are the Univ. of Kansas, at Lawrence; Kansas State Univ., at Manhattan; Wichita State Univ., at Wichita; and Washburn Univ. of Topeka, at Topeka.

Kansas or **Kaw** (kô), river, 170 mi (273.5 km) long, formed by the junction of the Smoky Hill and Republican rivers in NE Kansas and flowing E to the Missouri River at Kansas City. Heavy floods (especially in 1951) on the Kansas and its tributaries caused great damage in this primarily agricultural region.

Kansas City, two adjacent cities of the same name, one (1978 est. pop. 166,800), seat of Wyandotte co., NE Kansas (inc. 1859), the other (1978 est. pop. 454,000), Clay, Jackson, and Platte cos., NW Mo. (inc. 1850). They are at the junction of the Missouri and Kansas rivers and together form a large commercial, industrial, and cultural center. They are a port of entry, the focus of many transportation lines, and a huge market for wheat, hay, poultry, and seed. Both cities have large stockyards, grain elevators, food-processing establishments, oil refineries, steel mills, soap and farm-machinery factories, automobile-assembly plants, and railroad shops. The area was the starting place of many Western expeditions. Several historic settlements of the early 19th cent. were predecessors to the present-day cities. Kansas City, Kansas, is the seat of the Univ. of Kansas Medical Center, and a state school for the blind (est. 1868). A 19th cent. Indian cemetery here is part of a unique center city mall. Kansas City, Mo., with its fine parks and residential districts, is the site of the noted Nelson Art Gallery and the Atkins Museum of Fine Arts. Among its educational institutions are the Univ. of Missouri-Kansas City and Kansas City Art Institute. Extensive flood damage in 1951 led to several river-control projects in the region.

Kan·su or **Kan-su** (both: kăn'sōō', gän'-), province (1968 est. pop. 13,000,000), NW China, bordered by the Mongolian People's Republic on the N; capital Lan-Chou. Winter wheat, kaoliang, millet, corn, rice, cotton, and tobacco are grown. Livestock are raised in the mountainous areas. Kansu's mineral resources include coal, copper, gold, and large deposits of iron ore and oil. Oil refining is done. Long isolated from the center of Chinese power, the Kansu area has traditionally been independent of all but the strongest central governments. After the 13th cent., Moslem strength grew, and fierce Moslem rebellions often plagued the central government. Kansu's boundaries have been changed several times in recent years.

Ka·nu·ma (kä-nōō'mä), city (1975 pop. 55,800), Tochigi prefecture, central Honshu, Japan. It is an industrial center where brooms, hemp yarn and rope, and wood fittings are produced.

Kan·ye (kän'yə), town (1971 pop. 10,664), SE Botswana. It is a commercial and administrative center. Asbestos is mined nearby.

Kao-hsiung or **Kao·hiung** (both: gou'shyōōng'), city (1975 est. pop. 998,900), S Taiwan. It is a leading port and major industrial center. Its industries produce sugar, petroleum products, cement, aluminum, wood and paper products, fertilizers, metals, and machinery; shipbuilding is also carried on. The city was developed as a manufacturing center and port by the Japanese.

Ka·o·lack (kä'ō-lăk, kou'-), city (1973 est. pop. 114,000), W Senegal, a port on the Saloum River. Lying in a farm area, Kaolack is a major peanut marketing and exporting center and has a large peanut oil factory. Brewing, leather tanning, cotton ginning, and fish processing are also important industries.

Ka·pa·a (kä-pä'ä), city (1970 pop. 3,794), Kauai co., Hawaii, on E Kauai, NNE of Lihue, in a pineapple region.

Ka·pi·la·vas·tu (kä'pĭ-lə-vä'stōō), ancient town, S Nepal. According to legend, the Buddha, whose father ruled the state of Kapilavastu, was born nearby.

Kap·lan (kăp'lən), rice-milling town (1970 pop. 5,540), Vermilion parish, S central La., SW of Lafayette, in an oil and natural-gas region; inc. 1902. Rice, cotton, poultry, and eggs are shipped.

Ka·pos·vár (kŏ'pôsh-vär'), city (1977 est. pop. 70,000), SW Hungary, on the Kapos River. It is a market for agricultural goods and livestock, and an industrial center.

Ka·pu·as (kä'pōō-äs), river, c.710 mi (1,140 km) long, rising in the mountains of central Borneo and flowing SW through W Kalimantan, Indonesia, to the South China Sea near Pontianak. Its valley is intensively cultivated; rice is the chief crop.

Kap·us·ka·sing (kăp'əs-kā'sĭng), town (1976 pop. 12,676), central Ont., Canada, on the Kapuskasing River, N of Timmins. It has lumbering and pulp and paper mills.

Ka·ra (kä'rə), river, c.140 mi (225 km) long, NE European and NW Siberian USSR, flowing N from the N Urals into the Kara Sea. It forms part of the traditional border between European and Asian Russia.

Ka·ra·bakh (kə-rə-bäKH'): *see* Nagorno-Karabakh, USSR.

Ka·ra-Bo·gaz-Gol (kä'rə-bə-gäz'gôl), shallow bay, c.7,000 sq mi (18,130 sq km), Central Asian USSR, in Turkmenistan. An arm of the Caspian Sea, it acts as a natural evaporating basin, drawing off the water of the Caspian and depositing salts along its shores.

Ka·ra·bük (kä-rä'bük), city (1975 est. pop. 69,070), N Turkey. It was built in the 1930s as the seat of the iron and steel industry of Turkey. Nearby are the Zonguldak coal fields.

Ka·ra·chay-Cher·kess Autonomous Oblast (kä-rä-chī'chĕr-kĕs'), administrative division (1970 pop. 345,000), c.5,500 sq mi (14,245 sq km), Stavropol Kray, SE European USSR, in the Greater Caucasus, along the upper Kuban River; capital Cherkessk. The oblast consists of lowland steppe in the north and the Caucasian foothills in the south. Grains, fruits, and vegetables are grown and livestock is raised. The oblast has coal, lead, zinc, copper, and gold mines. Industrial products include building materials, foodstuffs, and machinery. The Karachay, Turkic-speaking Moslems who arrived in the region in the 14th cent., were conquered by the Russians in 1828. The region was included (1921) in the Mountain People's Republic, but in 1922 it became the Karachay-Cherkess Autonomous Oblast. In 1924 it was divided into the Karachay Autonomous Oblast and the Cherkess National Okrug; the latter became an autonomous oblast in 1928. In 1943 the Karachay, accused of collaborating with the Germans in World War II, were deported to Siberia and their autonomous oblast was abolished. The Karachay-Cherkess Autonomous Oblast was re-established in 1957, when the "rehabilitation" of deported peoples was decreed.

Ka·ra·chi (kə-rä'chē), city (1972 pop. 2,850,000), former capital of Pakistan, SE Pakistan, on the Arabian Sea near the Indus River delta. It is Pakistan's chief seaport and industrial center, as well as a transportation, commercial, and financial hub and a military headquarters. It has a large automobile assembly plant, an oil refinery, a steel mill, shipbuilding and repair and railroad yards, jute and textile factories, printing and publishing plants, food processing plants, and chemical and engineering works. Filmmaking and fishing are also important. An old settlement, Karachi was developed as a port and trading center by Hindu merchants in the early 18th cent. In 1843 it passed to the British, who made it the seat of the Sind government and a military center. Karachi served as Pakistan's capital from 1947, when the country gained independence, until 1959.

Ka·ra·gan·da (kä'rä-gän'dä), city (1977 est. pop. 576,000), capital of Karaganda oblast, Central Asian USSR, in Kazakhstan, on the Trans-Kazakhstan RR. It consists of about 50 coal-mining settlements scattered around the central part of the city. Its industries include iron and steel foundries, flour mills, food and beverage plants, ship repair yards, and factories that produce mining equipment, building materials, machinery, and footwear. Karaganda was founded in 1857 as a copper-mining settlement. The Karaganda coal basin, developed in the late 1920s, is one of the USSR's largest producers of bituminous coal.

Ka·raj (kä-räj'), city (1966 pop. 44,243), Tehran prov., N Iran, on the Karaj River. It is an agricultural market and a transportation center. Chemicals are manufactured.

Ka·ra-Kal·pak Autonomous Soviet Socialist Republic (kär'ə-kŭl-päk', kə-rä'-), autonomous republic (1970 pop. 702,000), c.61,000 sq mi (158,000 sq km), Central Asian USSR, on the Amu Darya River; capital Nukus. The republic comprises parts of the Ustyurt plateau, the Kyzyl-Kum desert, and the Amu Darya delta on the Aral Sea. The republic is the USSR's chief producer of alfalfa; other crops are cotton, rice, corn, and jute. Livestock raising and silkworm breeding are widespread. There are many light industries. In the 18th cent. the Kara-Kalpak migrated to their present homeland and in the 19th cent. came under the rule of the khanate of Khiva. The khanate passed under Russian control at the end of the 19th cent. and under Bolshevik control by 1920. The Kara-Kalpak Autonomous Oblast was formed in 1925 and became an autonomous republic in 1932.

Ka·ra·ko·rum (kär'ə-kôr'əm), ruined city, central Mongolian People's Republic, near the Orkhon River, SW of Ulan Bator. The area around Karakorum had been inhabited by nomadic Turkic tribes from the 1st cent. A.D., but the city itself was not laid out until c.1220, when Genghis Khan established his residence there. The city was abandoned (and later destroyed) after Kublai Khan transferred (1267) the Mongol capital to Khanbaliq (modern Peking). The ruins of the ancient Mongol city were discovered in 1889.

Karakorum, mountain system, extending c.300 mi (485 km), between the Indus and Yarkand rivers, N Kashmir, S central Asia; SE extension of the Hindu Kush. Karakorum's main range has some of the world's highest peaks, including Mt. Godwin-Austen. Karakorum also has several of the world's largest glaciers. Karakorum Pass (alt. 18,290 ft/5,578.5 m), the chief pass of the system, is on the main Kashmir-China route.

Ka·ra-Kul (kär'ə-kōōl'), mountain lake, c.140 sq mi (365 sq km), Gorno-Badakhshan Autonomous Oblast, Central Asian USSR, in the Pamir, near the Chinese border. It is c.12,840 ft (3,915 m) above sea level, and its greatest depth is 780 ft (237.9 m).

Ka·ra-Kum (kär'ə-kōōm'), two deserts, S USSR. The Caspian Kara-Kum, the larger desert (c.115,000 sq mi/297,850 sq km), is west of the Amu Darya River and includes most of the Turkmen Republic. The Kara-Kum Canal (c.500 mi/805 km long), permits irrigated agriculture (mainly cotton) and industry to flourish along the southern margin of the desert. Natural gas deposits have been discovered. The Aral Kara-Kum desert (c.15,440 sq mi/40,000 sq km) lies northeast of the Aral Sea in the Kazakh Republic.

Ka·ra·mai (kä-rä-mī'), city, N Sinkiang Uigur Autonomous Region, China, in the Dzungarian basin. Since the discovery (1955) of one of the largest oil fields in China, it has grown into an oil-producing and refining center.

Ka·ra·man (kä-rä-män'), town (1975 pop. 43,735), S central Turkey, at the N foot of the Taurus Mts. The ancient Laranda, Karaman was renamed after the chieftain of a Turkic tribe who conquered the city c.1250 and set up the independent Moslem state of Karamania, which at one time comprised most of Asia Minor. Karamania existed until its subjugation by the Ottoman Turks in the late 15th cent.

Kara Sea, shallow section of the Arctic Ocean, off N USSR, between Severnaya Zemlya and Novaya Zemlya. No deeper than 650 ft (198.3 m), it is important as a fishing ground. The ice-locked sea is navigable only during August and September.

Ka·ra·tsu (kä-rä'tsōō), city (1975 pop. 75,224), Saga prefecture, NW Kyushu, Japan, on Karatsu Bay. It is a summer resort and fishing port important historically as Japan's ancient communications point with Korea.

Kar·ba·la (kär'bə-lə), city (1970 est. pop. 107,500), central Iraq, at the edge of the Syrian Desert. The city's trade is in religious objects, hides, wool, and dates. Karbala is second only to Mecca in being a holy place visited by Shiite pilgrims.

Kar·cag (kŏr'tsŏg), city (1977 est. pop. 21,700), E Hungary. A road and rail junction, Karcag is an important communications point.

Ka·re·li·an Autonomous Soviet Socialist Republic (kə-rē'lē-ən, -rēl'yən) or **Ka·re·li·a** (kə-rē'lē-ə, -rēl'yə), autonomous region (1970 pop. 714,000), c.66,540 sq mi (172,340 sq km), NW European USSR, extending from the Finnish border in the W to the White Sea in the E and from the Kola Peninsula in the N to Lakes Ladoga and Onega in the S; capital Petrozavodsk. A glaciated plateau, Karelia is covered by about 50,000 lakes and by coniferous forests; fishing and lumbering are major industries. Agriculture is possible only in the south; dairy farming and livestock raising are also carried on. Karelia has valuable deposits of iron ore, magnetite, lead, zinc, copper, titanium, marble, and pyrite.

The Karelians, a major division of the Finns, were first mentioned in the 9th cent. and formed a strong medieval state. Karelia, the region north and east of Lake Onega, was conquered in the 12th-13th cent. by the Swedes, who took the west, and by Novgorod, which took the east. The eastern part was taken from Russia by Sweden in 1617 but restored in 1721 by the Treaty of Nystad. The western part shared the history of Finland until 1940. In 1920 an autonomous oblast, known as the Karelian Workers' Commune, was set up in eastern Karelia; in 1923 it was made into the Karelian Autonomous SSR, which, after the Soviet-Finnish War of 1939–40, incorporated most of the territory ceded by Finland to the USSR. In Mar., 1940, the region's status was raised to that of a constituent republic, called the Karelo-Finnish SSR. During World War II, the Finns (allies of the Axis powers) occupied most of Karelia; but it was returned to the USSR in 1944. Karelia reverted to the status of an autonomous republic in 1956.

Karelian Isthmus, land bridge, NW European USSR, connecting Russia and Finland. Situated between the Gulf of Finland in the west and Lake Ladoga in the east, it is 25 to 70 mi (40.2–112.6 km) wide and 90 mi (144.8 km) long. Originally part of the Grand Duchy of Sweden, the isthmus passed to Russia in 1721, and—except for its southernmost section—became part of Finland in 1917. The isthmus was formally ceded to the USSR in 1944.

Ka·ri·ya (kä-rē'yä), city (1975 pop. 96,152), Aichi prefecture, central Honshu, Japan. It is an industrial center with textile, mechanical, and food-processing industries.

Kar·kheh (kär'kĕ), river, c.350 mi (565 km) long, rising in the Zagros Mts., W Iran, and flowing S into the Khuzistan lowland, where it forms a swamp bordering the Tigris River. An ancient storage dam on the river at Shush made Khuzistan one of the most prosperous agricultural regions of Asia until the system fell into disrepair and the irrigated area reverted to desert.

Kar·li (kär'lē), village, Maharashtra state, W India. Nearby are Buddhist caves that may have been excavated as early as the 2nd cent. B.C. The most famous of them measures 124 ft by 45 ft (37.8 m by 13.7 m) and is India's largest cave temple.

Karl-Marx-Stadt (kärl'märks'shtät'), formerly **Chem·nitz** (kĕm'nĭts), city (1974 est. pop. 303,811), capital of Karl-Marx-Stadt dist., S East Germany, on the Chemnitz River. It is a major industrial center and a road and rail junction. Manufactures include machine tools, machinery, chemicals, optical instruments, furniture, and textiles. Nearby is a large open-pit lignite mine. Of Wendish origin, the city was chartered in 1143. It was devastated in the Thirty Years' War (1618–48) and recovered its prosperity after the introduction (late 17th cent.) of cotton milling.

Kar·lo·vy Va·ry (kär'lô-vĭ vä'rĭ) or *German* **Karls·bad** (kärls'bät', kärlz'băd), city (1975 est. pop. 50,386), NW Czechoslovakia, in Bohemia, at the confluence of the Teplá and Ohře rivers. Karlovy Vary is one of the best-known spas of Europe. The medicinal springs, known for centuries, attracted European aristocrats until World War I. Karlovy Vary is also noted for its china, glass, and porcelain industries. The city was chartered in the 14th cent. by Emperor Charles IV.

Karls·hamm (kärls'hä'mən), city (1975 est. pop. 14,900), Blekinge co., SE Sweden, a busy port on the Baltic Sea; chartered 1664. It is the seat of a large fishing fleet.

Karl·sko·ga (kärl-sko͞o'gä), city (1975 est. pop. 38,103), Örebro co., S Sweden; chartered 1940. An industrial center, it is the seat of the Bofors iron and armaments works and has other industries that manufacture steel, machines, explosives, chemicals, and clothing.

Karls·kro·na (kärls-kro͞o'nä), city (1975 est. pop. 50,100), capital of Blekinge co., SE Sweden, on the Baltic Sea. It is a seaport and fishing center with a large modern port. The headquarters of the Swedish navy since 1679, the city has many service-connected industries. Manufactures include metal goods, canned food, and porcelain.

Karls·ru·he (kärls'ro͞o-ə, kärlz'-), city (1975 est. pop. 280,448), Baden-Württemberg, SW West Germany, on the N fringes of the Black Forest, connected by canal with a port on the nearby Rhine River. It is a transportation, industrial, and cultural center and is the seat of the federal constitutional court and the federal court of justice. Manufactures include textiles, jewelry, pharmaceuticals, machinery, and refined oil. Karlsruhe was founded in 1715. After 1771 it was the capital of the duchy (later grand duchy and, after 1919, state) of Baden. The old part of Karlsruhe, badly damaged in World War II, was laid out as a vast semicircle with the streets converging radially upon the ducal palace (1752–85; restored after 1945).

Karl·stad (kärl'städ), city (1975 est. pop. 72,369), capital of Värmland co., S Sweden, on Lake Vänern. It has ironworks and machine shops and other industries that manufacture forest products and heavy machinery. It was chartered by Charles IX in 1584. A fire in 1865 destroyed much of the city. The treaty that severed the union of Norway and Sweden was negotiated and signed here in 1905.

Kar·nak (kär'năk), village, central Egypt, on the Nile, E of Luxor. It occupies part of the site of Thebes. Remains of the pharaohs abound at Karnak. Most notable is the Great Temple of Amon. Although there was an older foundation, the temple was largely conceived and accomplished in the XVIII dynasty, and it is often considered the finest example of New Empire religious architecture.

Kar·nal (kər-näl'), town (1971 pop. 92,835), Haryana state, N central India. It is a market for rice, wheat, and maize, a cattle-breeding center. The British occupied the town in 1805.

Kar·na·ta·ka (kär-nä'tə-kə), formerly **My·sore** (mī-sôr'), state (1971 pop. 29,263,334), 74,122 sq mi (191,976 sq km), SW India, bordering on the Arabian Sea; capital Bangalore. Coffee is the major crop, but cotton, millet, sugar cane, rice, and fodder are also grown. The state has the most valuable sandalwood forests in India. Karnataka produces nearly all of India's gold and chromite and has considerable deposits of iron ore and manganese. The region was part of the empire of the Mauryas (c.325–185 B.C.). From the 3rd to the 11th cent. it was ruled by the Gangas and Chalukyas. In 1313 it was conquered by the Delhi Sultanate. In the late 18th cent. the Moslem leaders Haider Ali and his son, Tippoo Sahib, conquered the Hindu rulers of Karnataka. They fought the British but were finally defeated in 1799. The state acceded to the Indian Union in 1947.

Karnes (kärnz), county (1970 pop. 13,462), 758 sq mi (1,963.2 sq km), S Texas, drained by the San Antonio River; formed 1858; co. seat Karnes City. It is a leading producer of flax and also yields corn, grain sorghums, cotton, peanuts, fruit, truck crops, livestock, and poultry. There are oil and natural-gas wells and deposits of clay and fuller's earth here.

Karnes City, town (1970 pop. 2,926), seat of Karnes co., S Texas, SE of San Antonio; settled 1885, inc. 1914. In a farm region, it also has oil refineries, a uranium-processing plant, and gas wells.

Kár·pa·thos (kär'pä-thôs), island (1971 pop. 5,420), c.110 sq mi (285 sq km), SE Greece, in the Aegean Sea, one of the Dodecanese.

Kar·roo (kə-ro͞o'), semiarid plateaus of W Cape Prov., Republic of South Africa. The Little Karroo is located north of the Cape Ranges and extends c.200 mi (320 km) from east to west at an altitude of from 1,000 to 2,000 ft (305–610 m). It is separated from the Great Karroo (c.300 mi/480 km long; alt. 2,000–3,000 ft/610–915 m) by the Zwartberg Mts. The Karroo, where irrigated, is very fertile. Livestock grazing is important, and citrus fruits and grains are raised. The name is also applied to the low scrub vegetation found in semiarid regions and also to a system of rocks laid down over central and southern Africa during the late Paleozoic and early Mesozoic eras.

Kars (kärs), city (1975 pop. 54,787), capital of Kars prov., E Turkey, in Armenia, on the Kars River, near the Soviet border. Its manufactures include textiles, carpets, and food products. An old fortified city, well situated in the mountains, Kars was the capital of an Armenian state in the 9th and 10th cent. It was destroyed by Tamerlane in 1386 and was captured and rebuilt by the Ottoman Turks in the 16th cent. In 1828, 1855, and 1877 the city was occupied by Russia and together with the surrounding region was ceded to Russia by the Congress of Berlin in 1878. It was returned to Turkey in 1921.

Kar·shi (kär'shē), city (1976 est. pop. 91,000), S Central Asian USSR, in Uzbekistan, on the Kashka-Darya River. It is the center of a fertile oasis that produces wheat, cotton, and silk. Karshi was founded in the 14th cent.

Karst (kärst), limestone plateau, in the Dinaric Alps, NW Yugoslavia, N of Istria. It is characterized by deep gullies, caves, sinkholes, and underground drainage—all the result of carbonation-solution. Rough pasture or forest covers much of the surface, and there is little arable land.

Ka·run (kä-ro͞on'), river, c.450 mi (725 km) long, rising in the Zagros Mts., W Iran, and flowing S to the Shatt al Arab on the Iraq border. The river was opened to foreign trade in 1888.

Ka·rur (kə-ro͞or'), city (1971 pop. 65,246), Tamil Nadu state, S central India. Milled rice, cotton fabrics, and brassware are the city's chief products.

Kar·vi·ná (kär'vĭ-nä), city (1975 est. pop. 81,394), N central Czechoslovakia, in Moravia, near the Polish border. It is an industrial center for a coal-mining region. Formerly in Austria, the city became (after 1918) an object of dispute between Poland and Czechoslovakia; after World War I a conference of Allied ambassadors awarded (1920) it to Czechoslovakia despite Polish claims. The city was seized by Poland in Oct., 1938, but was restored to Czechoslovakia in 1945.

Ka·sai (kä-sī'), river, c.1,100 mi (1,770 km) long, rising in central Angola, S central Africa, flowing E, N, and NW through W Zaire to the Congo River. The Kasai, navigable for c.475 mi (765 km) above its mouth, is an important trade artery.

Ka·shan (kä-shän'), city (1966 pop. 58,468), Tehran prov., central Iran. The city has long been noted for its silk textiles, carpets, ceramics, copperware, and rose water. Kashan's skyline is dominated by a 13th cent. minaret that is 150 ft (45.8 m) high.

Kash·gar (kăsh'gär), city (1970 est. pop. 175,000), SW Sinkiang Uigur Autonomous Region, China, on the Kashgar River (a tributary of the Tarim). It is the hub of an important commercial district and a center for caravan trade with India, Afghanistan, and the USSR. Cotton and wool cloth, rugs, and gold and silver jewelry are manufactured. From Kashgar a mountain pass provides a route to Samarkand and thence to the Middle East. The city first came under Chinese rule in the period of the Han dynasty (206 B.C.–A.D. 221). Romans traded here in the 6th cent. Visited by Marco Polo in 1275, Kashgar was soon after conquered by Genghis Khan. The city passed definitively to China in 1760, but since then there have been uprisings and periods of contested control.

Ka·shing (kä'shĭng'): *see* Chia-hsing, China.

Ka·shi·wa·za·ki (kä-shē-wä'zä-kē), city (1975 pop. 53,500), Niigata prefecture, central Honshu, Japan, on the Japan Sea. It is an agricultural center and a resort with hot springs.

Kash·ka-Dar·ya (kăsh'kə-dər-yä'), river, c.200 mi (320 km) long, Central Asian USSR. It is the basis of a wide network of irrigation canals.

Kash·mir (kăsh'mîr, kăsh-mîr'), officially **Jam·mu and Kashmir** (jŭm'o͞o), former princely state, c.86,000 sq mi (222,800 sq km), NW India and NE Pakistan, bordered on the W by Pakistan, on the S by India, and on the N and E by China. The region is administered in two sections: the Indian state of Jammu and Kashmir (1971 pop. 4,615,176), c.54,000 sq mi (139,900 sq km), with its capital at Srinagar, the historic capital of the state; and the Pakistani-controlled Azad Kashmir, c.32,000 sq mi (82,900 sq km), with Muzaffarabad as its capital. Kashmir is covered with lofty, rugged mountains, including sections of the Himalayan and Karakorum ranges. The valley of the Jhelum River, the celebrated Vale of Kashmir, produces abundant crops of wheat and rice. The handicraft industry, particularly the making of woolen cloth and shawls (cashmeres), for which the state was renowned, has declined.

In the late 14th cent., after years of Buddhist and Hindu rule, Kashmir was conquered by Moslems who converted most of the population to Islam. It became part of the Mogul empire in 1586, but by 1751 the local ruler was independent. After a century of disorder the British pacified Kashmir in 1846 and installed a Hindu prince as ruler. When India was partitioned in 1947, a Moslem revolt, supported by tribesmen from Pakistan, flared up. The Hindu ruler fled to Delhi and there signed an agreement that placed Kashmir under the dominion of India. Pakistan, backing the rebels, later dispatched troops to oppose the Indian forces.

The fighting was ended by a UN cease-fire in 1949, but the region was divided between India and Pakistan along the cease-fire line. A constituent assembly in Indian Kashmir voted in 1953 for incorporation into India, but this move was delayed by continued Pakistani-Indian disagreement and disapproval by the United Nations of annexation without a plebiscite. In 1955 an outbreak of fighting ended in an agreement between India and Pakistan to keep their respective forces in Kashmir 6 mi (10 km) apart. A new vote by the assembly in Indian Kashmir in 1956 led to the integration of Kashmir as an Indian state; Azad Kashmir remained, however, under the control of Pakistan. India ignored subsequent Pakistani protests and UN resolutions calling for a plebiscite. The situation was further complicated in 1959, when Chinese troops occupied the district of Ladakh and neighboring areas. Indian-Pakistani relations became more inflamed in 1963 when a Sino-Pakistani agreement defined the Chinese border with Pakistani Kashmir. Serious fighting between India and Pakistan broke out again in Aug., 1965. A UN cease-fire took effect in September. In Jan., 1966, an agreement was reached providing for the mutual troop withdrawal to the positions held before the latest outbreak of fighting. In the Dec., 1971, war between India and Pakistan, however, there was further fighting in Kashmir in which India made some gains. In Dec., 1972, a new cease-fire line along the current positions was agreed to by India and Pakistan.

Kas·kas·ki·a (kăs-kăs'kē-ə), village (1970 pop. 79), Randolph co., SE Ill., on Kaskaskia Island in the Mississippi River where it is joined by the Kaskaskia River. The settlement was established by Jesuit missionaries in 1703. The French built a fort in 1721 and occupied it until 1755. In 1778 George Rogers Clark took possession of the village for the United States. Kaskaskia thrived as the capital of Illinois Territory (1809–18) and state capital (1818–20). The community declined after the capital was shifted (1820) to Vandalia, and periodic floods discouraged further growth.

Kas·sa·la (kä'sä-lä'), city (1973 pop. 98,751), NE Sudan. It is a cotton market and rail transport center and has extensive fruit gardens. Founded in 1840, Kassala was captured by the Mahdists in 1885 and by the Italians in 1894. Restored to Egyptian sovereignty in 1897, it became part of the Anglo-Egyptian Sudan.

Kas·sel (kä'səl), city (1974 est. pop. 210,042), Hesse, E West Germany, on the Fulda River. Manufactures include machinery, chemicals, textiles, optical and precision instruments, locomotives, and motor vehicles. Kassel was mentioned in 913 and was chartered in 1198. As a center of German airplane and tank production in World War II, Kassel was severely damaged by Allied air raids. Many historic buildings were destroyed.

Kas·se·rine Pass (kăs'ə-rēn), gap, 2 mi (3.2 km) wide, central Tunisia, in the Grand Dorsal chain (an extension of the Atlas Mts.). A key point in World War II, the pass was the scene of an Axis breakthrough (Feb. 20, 1943), but was retaken with heavy losses by U.S. forces on Feb. 25.

Kas·ta·mo·nu (kä'stä-mō-nōō'), city (1975 pop. 29,839), capital of Kastamonu prov., N Turkey. It is a manufacturing center, noted for its textiles and copper utensils, and is the chief city of a region rich in minerals. Kastamonu was captured by the Ottoman Turks in 1393, was taken by Tamerlane in 1403, and was regained by the Ottomans in 1460.

Kas·to·rí·a (käs-tô-rē'ə), city (1971 pop. 15,407), capital of Kastoría prefecture, N Greece, in Macedonia, on a peninsula extending into Lake Kastoría. It is a market for farm produce and has fisheries. In the 17th and 18th cent. it was a major fur-trading center.

Ka·su·gai (kä-sōō'gī), city (1976 est. pop. 222,335), Aichi prefecture, central Honshu, Japan. It is a suburb of Nagoya and the site of silk and textile industries.

Ka·tah·din, Mount (kə-tä'dĭn), mountain, 5,267 ft (1,606.4 m) high, between branches of the Penobscot River in N central Maine. It is the highest point in Maine and the northern terminus of the Appalachian Trail.

Ka·tan·ga (kə-täng'gə, -tăng'-): *see* Shaba, province, Zaire.

Ka·te·ri·ni (kä-tə-rē'nē), city (1971 pop. 28,808), capital of Piería prefecture, N Greece, in Macedonia. It is the commercial center for a productive tobacco-growing region.

Ka·thi·a·war (kä'tē-ə-wär'), peninsula, c.25,000 sq mi (64,750 sq km), W India, between the Gulf of Kutch and the Gulf of Cambay. The region, mostly level, produces much cotton and has stone quarries and cement and chemical industries.

Kat·mai National Monument (kăt'mī), 2,792,137 acres (1,130,815 hectares), at the N end of the Alaska Peninsula, S Alaska; est. 1918. Mt. Katmai and Novarupta volcanoes and the Valley of the Ten Thousand Smokes are located in this dying volcanic region, which is the site of one of the greatest eruptions in history, that of Novarupta in 1912. As lava beneath Mt. Katmai drained west to Novarupta, its top collapsed, forming a crater, 8 mi (12.9 km) in circumference and 3,700 ft (1,128.5 m) deep, in which a lake has formed. The Valley of the Ten Thousand Smokes (72 sq mi/186.5 sq km) has countless holes and cracks through which hot gases passed to the surface; all but a few are now extinct. The region is inaccessible except to specially equipped expeditions. The national monument also includes glacier-covered peaks and crater lakes.

Kat·man·du (kät'män-dōō'), city (1971 pop. 105,402), capital of Nepal, central Nepal, c.4,500 ft (1,375 m) above sea level, in a fertile valley of the E Himalayas. It lies astride an ancient trade and pilgrim route from India to Tibet, China, and Mongolia. Katmandu became independent in the 15th cent. and was captured in 1768 by the Gurkhas, who made it their capital. In the late 18th cent. the city became the seat of a British resident. Landmarks include the elaborate royal palace, several pagoda-shaped temples, and many Sanskrit libraries.

Ka·to·wi·ce (kä'tô-vē'tsĕ), city (1975 est. pop. 343,700), S Poland. One of the chief mining and industrial centers of Poland, it has industries producing heavy machinery and chemicals; mines in the region yield coal, iron, zinc, and lead. The city was chartered in 1865 and passed from Germany to Poland in 1921.

Ka·tri·ne·holm (kä'trē-nə-hôlm'), city (1975 est. pop. 24,900), Södermanland co., S Sweden; chartered 1917. It is a commercial, industrial, and transportation center. The city has one of Europe's largest dairies. Ball bearings, automobile bodies, furniture, and machinery are manufactured.

Ka·tsi·na (kä-tsē'nə, kä'tsĭ-nə), city (1971 est. pop. 109,000), N Nigeria, near the Niger frontier. The city is the trade center for an agricultural region where guinea corn and millet are grown and groundnuts, cotton, and hides are produced. Leather handicrafts are made. In the 17th and 18th cent. it was the largest of the seven Hausa city-states and the cultural and commercial center of Hausaland. In 1807 Katsina was conquered by the Fulani.

Ka·tsu·ta (kä-tsōō'tä), city (1975 pop. 79,996), Ibaraki prefecture, central Honshu, Japan, on the Naka River. It is a commercial center with mechanical, automotive, and electronics industries.

Kat·te·gat (kăt'ĭ-găt'), strait, c.140 mi (225 km) long and from 40 to 100 mi (64.4–160.9 km) wide, between Sweden and Denmark. It is connected with the North Sea through the Skagerrak and with the Baltic Sea by way of the Øresund, Store Baelt, and Lille Baelt.

Ka·tun (kə-tōōn'), river, c.415 mi (670 km) long, Altai Kray, S Siberian USSR. It rises in the Katun Alps and flows generally north to join the Biya, with which it forms the Ob River.

Ka·tyn (kə-tĭn'), village, W central European USSR. It was occupied by the Germans in Aug., 1941, during World War II. In 1943 the German government announced that the mass grave of some 4,250 Polish officers had been found in a forest near Katyn and accused the Soviets of having massacred them. The Soviet government asserted that the Poles, war prisoners, had been captured and executed by invading German units in 1941. The Soviets refused to permit an investigation by the International Red Cross.

Kau·ai (kou'ī'), county (1970 pop. 29,761), 619 sq mi (1,603.2 sq km), Hawaii, including Kauai and Niihau islands; co. seat Lihue. The islands are mountainous and heavily forested. Sugar cane is the principal crop.

Kauai, circular island (1970 pop. 29,524), 549 sq mi (1,421.9 sq km), N Hawaii, separated from Oahu to the SE by Kauai Channel. Geologically, Kauai is the oldest of the Hawaiian Islands. It was formed by now extinct volcanoes. High annual rainfall has eroded deep valleys in Kauai's central mountain mass. An independent kingdom when visited by English Capt. James Cook in 1778, Kauai became part of the Kingdom of Hawaii in 1810. The first sugar plantation in the islands was established here in 1835. Agriculture is the main industry, with sugar cane, rice, and pineapples the chief crops; ranching and tourism are also important.

Kauf·man (kôf'mən), county (1970 pop. 32,392), 816 sq mi (2,113.4 sq km), NE Texas mainly in an area bounded on the W by the Trinity River; formed 1848; co. seat Kaufman. It has agriculture (cotton, corn, grain sorghums, hay, peanuts, fruit, truck crops, livestock, and dairy products), lumbering and food-processing industries, and some manufacturing.

Kaufman, city (1970 pop. 4,012), seat of Kaufman co., N Texas, ESE of Dallas; founded 1848, inc. 1873. This market center processes cotton and other products of the rich blackland area.

Kau·kau·na (kô-kô'nə), industrial city (1970 pop. 11,308), Outagamie co., E Wis., on the Fox River; settled 1793, inc. 1885. The city has a large paper plant; dairy items, foundry products, machine tools, and farm equipment are also manufactured. A fur-trading post was established on the site in 1760.

Kau·nas (kou'näs), city (1976 est. pop. 352,000), W European USSR, in Lithuania, on the Neman River. It is a river port and an industrial center with industries producing machinery, iron and steel, chemicals, plastics, and textiles. Probably founded as a fortress at the end of the 10th cent., Kaunas was a medieval trading center and a Lithuanian stronghold against the Teutonic Knights. It passed to a united Lithuanian-Polish state in 1569 and to Russia in the third partition of Poland (1795). Although strongly fortified by the Russians, it was captured (1915) by the Germans in World War I. From 1918 to 1940 Kaunas was the provisional capital of Lithuania. Kaunas was occupied by German forces from 1941 to 1944.

Ka·vál·la (kä-vä'lä, kə-văl'ə), city (1971 pop. 46,234), capital of Kaválla prefecture, NE Greece, in Macedonia; a port on the Gulf of Kaválla, an inlet of the Aegean Sea. Surrounded by a rich tobacco-growing hinterland, it is a leading Greek city for processing and exporting tobacco. Fish and manganese are also shipped, and flour is manufactured. Kaválla was held by the Ottoman Turks from 1387 to 1913, when it passed to Greece.

Kaw (kô), river: *see* Kansas, river.

Ka·wa·go·e (kä-wä'gō-ā), city (1976 est. pop. 233,974), Saitama prefecture, central Honshu, Japan. Silk textiles are manufactured in the city. Kawagoe is the site of Kitain Temple (built 830), famed for its images of the 500 disciples of Buddha.

Ka·wa·gu·chi (kä-wä'gōō-chē), city (1976 est. pop. 351,802), Saitama prefecture, central Honshu, Japan, on the Ajikawa and Kizagawa rivers. A Tokyo suburb, it has ironworks and textile mills.

Ka·wa·ni·shi (kä-wä'nē-shē), city (1976 est. pop. 119,562), Hyogo prefecture, central Honshu, Japan, on the Ina River. It is an agricultural and commercial center with a hat-manufacturing industry.

Ka·war·tha Lakes (kə-wôr'thə), group of 14 lakes, in a region c.50 mi (80 km) long and c.25 mi (40 km) wide, S Ont., Canada. They are popular as summer resorts.

Ka·wa·sa·ki (kä-wä'sä-kē), city (1976 est. pop. 1,025,455), Kanagawa prefecture, central Honshu, Japan, on Tokyo Bay. It has steel mills, shipyards, oil refineries, engineering works, and factories that produce motors, electrical machinery, petrochemicals, and cement.

Kay (kā), county (1970 pop. 48,791), 950 sq mi (2,460.5 sq km), N Okla., bounded on the N by the Kansas line and intersected by the Arkansas and Chikaskia rivers; formed 1893; co. seat Newkirk. In an agricultural area yielding wheat, barley, alfalfa, and livestock, it also has oil and natural-gas wells, oil refineries, and some manufacturing.

Kayes (kāz), town (1972 est. pop. 37,000), W Mali, a port on the Senegal River. It is the commercial center for a region where peanuts and gum arabic are produced. The town has tanneries.

Kay·se·ri (kī'sĕ-rē'), city (1975 pop. 207,039), capital of Kayseri prov., central Turkey, at the foot of Mt. Erciyas. It is an important commercial center and has textile mills, sugar refineries, and cement and carpet factories. It was taken by the Seljuk Turks in the mid-11th cent., briefly held (1097) by the Crusaders, and captured (1243) by the Mongols. The city was occupied by the Mamelukes of Egypt in 1419. Sultan Selim I incorporated Kayseri into the Ottoman Empire in 1515. The city has numerous historical remains.

Kays·ville (kāz'vĭl'), city (1970 pop. 6,192), Davis co., N Utah, between Salt Lake City and Ogden, in an irrigated farm area; settled 1849 by Mormons, inc. 1868.

Ka·zakh Soviet Socialist Republic (kä-zäk', kə-) or **Ka·zakh·stan** (kä'zäk-stän', kə-zək-hstän'), constituent republic (1970 pop. 12,850,000), c.1,050,000 sq mi (2,719,500 sq km), S USSR. It borders on Siberia in the north, China in the east, the Kirghiz, Uzbek, and Turkmen republics in the south, and the Caspian Sea in the west. Alma-Ata is the capital. Kazakhstan is largely lowland in the north and west, hilly in the center, and mountainous in the south and east. Most of the region is desert or has limited and irregular rainfall; however, dry farming along the northern borders of Kazakhstan has been expanding considerably since cultivation began in 1954. Kazakhstan produces much of the USSR's wool and cattle and a very great part of its wheat. The Kazakh Plateau covers the core of the region, and has important mineral resources, including coal, oil, iron ore, copper, lead, zinc, nickel, chromium, and silver.

The original Turkic tribes were conquered by the Mongols in the 13th cent. and ruled by various khanates until the Russian conquest (1730–1840). In 1916 the Kazakhs rebelled against Russian domination and were establishing a Western-style state at the time of the 1917 Bolshevik Revolution. Organized as the Kirghiz Autonomous SSR in 1920, it was renamed the Kazakh Autonomous SSR in 1925 and became a constituent republic in 1936.

Ka·zan (kə-zăn', -zän', kə-zä'nyə), city (1977 est. pop. 970,000), capital of the Tatar Autonomous Soviet Socialist Republic, E European USSR, on the Volga. It is a major historic, cultural, shipping, industrial, and commercial center. Manufactures include aircraft, machines and machine tools, chemicals, explosives, electrical equipment, building materials, food products, and furs. Founded in 1401, Kazan became the capital of a powerful, independent Tatar khanate (1445), which emerged from the empire of the Golden Horde. The khanate was conquered and the city sacked in 1552 by Ivan IV. It became the capital of the Volga region in 1708 and was an outpost (18th cent.) of Russian colonization in the east. It was burned in 1774 and rebuilt by Catherine II.

Ka·zan·lik (kä'zän-lĭk'), town (1969 est. pop. 50,300), central Bulgaria, in the Kazanlik valley, a region famous for its rose fields. Kazanlik developed in the 17th cent. as a manufacturing center for attar of roses.

Kaz·bek, Mount (käz-bĕk'), peak, 16,558 ft (5,050.2 m) high, SE European USSR, in Georgia, in the Greater Caucasus. An extinct volcano, it rises above the Daryal gorge and the Georgian Military Road. Mt. Kazbek was first scaled in 1868.

Kaz Da·ğı (käz dä-ï'), range, NW Turkey, SE of the location of ancient Troy. Mt. Gargarus (5,797 ft/1,768 m) is the highest point. The mountain was dedicated in ancient times to the worship of Cybele.

Ka·ze·run (kä'zĕ-rōōn'), city (1971 est. pop. 42,000), Fars prov., SW Iran. It is an agricultural trade center.

Ké·a (kē'ä) or **Ke·os** (kē'ŏs), island (1971 pop. 1,666), c.61 sq mi (160 sq km), SE Greece, in the Aegean Sea; one of the Cyclades. Fruits, barley, and silk are produced.

Ke·a·la·ke·ku·a Bay (kā'ə-lä-kä-kōō'ə), on the Kona coast of the island of Hawaii. Capt. James Cook, who discovered the islands in 1776, stopped here on his second voyage to Hawaii and was killed during a beach fight with the natives on Feb. 14, 1779. A monument to him stands on the shore.

Keans·burg (kēnz'bûrg'), resort borough (1970 pop. 9,720), Monmouth co., E N.J., on Raritan Bay ESE of Perth Amboy, inc. 1917.

Kear·ney (kär'nē), county (1970 pop. 6,707), 512 sq mi (1,326.1 sq km), S Nebr., bounded on the N by the Platte River; formed 1860; co. seat Minden. Its agriculture includes wheat, corn, and cattle.

Kearney, city (1978 est. pop. 20,600), seat of Buffalo co., S central Nebr., on the Platte River; inc. 1873. It is a commercial, industrial, and transportation center in an agricultural area. Farm and irrigation equipment are among its many products. Fort Kearny, established nearby in 1848 to protect the Oregon Trail, was abandoned in 1871.

Kearns (kûrnz), uninc. town (1970 pop. 17,071), Salt Lake co., N Utah, a suburb of Salt Lake City. There are dairy farms in the area, and sugar beets are grown.

Kear·ny (kär'nē), county (1970 pop. 3,047), 853 sq mi (2,209.3 sq km), W Kansas, in a gently rolling plain drained by the Arkansas River; formed 1873; co. seat Lakin. Grain and livestock farming and some mining and manufacturing are done.

Kearny, town (1978 est. pop. 40,000), Hudson co., NE N.J.; inc. 1899. The town is the site of shipyards (greatly enlarged in 1941) and dry docks. Its chief product is communications equipment. Kearny contains much of the tidal wastelands between the Passaic and the Hackensack rivers that is being reclaimed for industrial and recreational purposes.

Keb·ne·kai·se (kĕb'nə-kī'sə), mountain peak, 6,962 ft (2,123.4 m) high, Norrbotten prov., N Sweden. It is the highest point in Sweden. There are 16 small glaciers on the slopes.

Kecs·ke·mét (kĕch'kĕ-māt'), city (1977 est. pop. 73,400), central Hungary, in a fruit-growing region. It is a a road and rail hub and a manufacturing city whose industries produce food products, alcoholic beverages, textiles, and furniture. Known since the 4th cent., the city has a museum, and a law school with a large library.

Kee·ling Islands (kē'lĭng): *see* Cocos Islands.

Kee·lung (kē'lo͝ong'): *see* Chi-lung, Taiwan.

Keene (kēn), city (1978 est. pop. 21,300), seat of Cheshire co., SW N.H., on the Ashuelot River; settled 1736, inc. as a city 1873. It is a trade and manufacturing center in a farming and resort area. Mt. Monadnock, a popular ski site, is to the east.

Keet·mans·hoop (kēt'mäns-ho͝op'), town (1970 pop. 10,297), S Namibia. The trade center for a region where karakul sheep are raised, was founded in 1866 as a German missionary station.

Kee·wa·tin (kē-wä'tĭn, -wā'-), administrative district (228,160 sq mi/ 590,934 sq km), Northwest Territories, Canada, N of Manitoba and W of Hudson Bay. Its boundaries include all of Hudson and James bays and all of the mainland of the Northwest Territories east of long. 102° W, except for the Boothia and Melville peninsulas.

Ke·fal·li·ní·a (kĕf-ä-lē-nē'ä) or **Ceph·a·lo·nia** (sĕf-ə-lō'nyə), island (1971 pop. 31,787), c.300 sq mi (780 sq km), W Greece, the largest of the Ionian Islands. It has an irregular coastline and is largely mountainous, rising to c.5,340 ft (1,630 m) at Mt. Ainos, which in ancient times was crowned by a temple to Zeus. Sheep raising and fishing are important occupations on the island. Kefallinía was an ally of Athens in the Peloponnesian War and later was a member of the Aetolian League. The island was taken by Rome in 189 B.C. After the division of the Roman Empire (A.D. 395), it was held by the Byzantine Empire until its occupation (1126) by Venice. It subsequently was ruled by several Italian families, was seized by the Ottoman Turks (1479), and was ceded (1499) to Venice, which held it until the Treaty of Campo Formio (1797). In 1953 the island was devastated by earthquakes.

Kef·la·vík (kĕp'lä-vēk'), town (1970 pop. 5,663), SW Iceland, on the Faxaflói, W of Reykjavík. It is a major fishing port, best known for its airport, built by the United States during World War II.

Keigh·ley (kēth'lē), municipal borough (1973 est. pop. 56,040), West Yorkshire, N central England, at the junction of the Aire and Worth rivers. Keighley's products include woolen, silk, and rayon goods; spinning machinery and looms; and sewing and washing machines.

Kei Islands (kī), Indonesia: *see* Kai Islands.

Keith (kēth), county (1970 pop. 8,487), 1,032 sq mi (2,672.9 sq km), W Nebr., drained by the North and South Platte rivers; formed 1873; co. seat Ogallala. It is in an agricultural region that produces livestock and grain. There is some mining. Electronic equipment is manufactured.

Kei·zer (kī'zər), uninc. town (1970 pop. 11,405), Marion co., NW Oregon, a suburb of Salem.

Kells (kĕlz): *see* Ceanannus Mor, Republic of Ireland.

Kel·logg (kĕl'ôg), city (1976 est. pop. 3,613), Shoshone co., NE Idaho, in the Coeur d'Alene Mts. mining district ESE of Coeur d'Alene; inc. 1913. It grew around the Bunker Hill and Sullivan mines (discovered in 1885), now combined as one of the world's largest lead mines. The Sunshine mine, a leading producer of silver, is also here, and gold and zinc are mined in the area.

Ke·low·na (kĭ-lō'nə), city (1976 pop. 51,955), S British Columbia, Canada, on Okanagan Lake. It is a tourist resort and serves as a trade center for a fruit-growing and lumbering area.

Kel·so (kĕl'sō), city (1970 pop. 10,296), seat of Cowlitz co., SW Wash., on the Cowlitz River near the Columbia, in a rich farm area; inc. 1889. Boatbuilding, meat-packing, and the manufacture of cement are the major industries. Settled in 1847, Kelso was a stopping place for early steamboat travel along the Cowlitz River.

Kem (kĕm), river, c.240 mi (385 km) long, Karelian Autonomous Republic, NW European USSR rising SE of Kuusamo, Finland, and flowing E into the White Sea. The first hydroelectric station along the Kem went into operation in 1967.

Ke·me·ro·vo (kĕ'mĕ-rō'vō), city (1976 est. pop. 446,000), capital of Kemerovo oblast, central Siberian USSR, on the Tom River. It is a coal-mining center, with chemical and synthetic fiber industries.

Ke·mi (kĕ'mē), city (1975 est. pop. 27,893), Lappi prov., W central Finland, on the Gulf of Bothnia at the mouth of the Kemijoki River. An old trading post, it was chartered in 1869. Kemi is a port and has large sawmills and pulp mills and a power station.

Ke·mi·jo·ki (kĕ'mē-yô-kē), longest river of Finland, c.345 mi (555 km) long, rising near Sokosti peak, NE Finland, and flowing generally SW, then W into the Gulf of Bothnia at Kemi. It is an important logging route.

Kem·mer·er (kĕm'ər), town (1970 pop. 2,292), seat of Lincoln co., SW Wyo., NNE of Evanston, on Hams Fork near the Utah border, in a coal area; settled 1897, inc. 1899. It is a trade center for a phosphate-mining and livestock-ranching region. Fossil fish beds are nearby.

Kem·pen·land (kĕm'pən-länd), region, Limburg and Antwerp provs., NE Belgium, and North Brabant prov., S Netherlands. It is a coal-mining and manufacturing area. Once covered by moors and marshes, it has been partially reclaimed.

Kem·per (kĕm'pər), county (1970 pop. 10,233), 757 sq mi (1,960.6 sq km), E Miss., bounded on the E by the Ala. border; formed 1833; co. seat De Kalb. It has cotton, corn, and cattle farming, lumber sawmilling, and some manufacturing (farm machinery and equipment).

Kemp·ten (kĕmp'tən), city (1974 est. pop. 57,022), Bavaria, S West Germany, on the Iller River, in the Allgäu. It is the center of a dairying region. Among the city's manufactures are textiles, paper, and machinery. Of Celtic origin, Kempten became a flourishing Roman colony. A free imperial city from the late 13th cent., it was sacked (1632) by the Swedes in the Thirty Years' War.

Ke·nai Peninsula (kē'nī'), borough (1970 pop. 16,586), 12,474 sq mi (32,307.7 sq km), S central Alaska; formed 1964; borough seat Soldotna. Fishing, lumbering, and food processing are done. There are oil wells in the borough. It includes Chugach National Forest and Kenai National Moose Range.

Kenai Peninsula, S Alaska, jutting c.150 mi (240 km) into the Gulf of Alaska, between Prince William Sound and Cook Inlet. The Kenai Mts., c.7,000 ft (2,135 m) high, occupy most of the peninsula. There are forest, mineral, and fishing resources in the east and, in the western section, good farmland.

Ke·nans·ville (kē'nənz-vĭl'), town (1970 pop. 762), seat of Duplin co., SE N.C., SW of Kinston. Sawmilling is done.

Ken·dall (kĕn'dəl). **1.** County (1970 pop. 26,374) 320 sq mi (828.8 sq km), NE Ill., drained by the Fox River; formed 1841; co. seat Yorkville. It is in a rich dairying and farming area yielding corn, oats, soybeans, wheat, livestock, and poultry. Metal and Bakelite products, batteries, plumbing supplies, and tin cans are manufactured. **2.** County (1970 pop. 6,964), 670 sq mi (1,735.3 sq km), S central Texas, on the S edge of Edwards Plateau and drained by the Guadalupe and Blanco rivers; formed 1862; co. seat Boerne. It is particularly noted for its sheep and goat ranches. Cattle, poultry, corn, fruit, and grain sorghums are also important. The county has timber, clay and limestone deposits and good hunting and fishing.

Ken·dall·ville (kĕn'dəl-vĭl'), city (1970 pop. 6,838), Noble co., NE

Ind., NNW of Fort Wayne, in a lake region; settled 1833. A shipping center for grain and onions, it also manufactures refrigerators, farm equipment, boats and furniture.

Ken·e·dy (kĕn'ə-dē), county (1970 pop. 678), 1,394 sq mi (3,610.5 sq km), extreme S Texas, on a flat coastal plain; formed 1921 from sections of Willacy, Hidalgo, and Cameron cos.; co. seat Sarita. In a large-scale ranching area (most of the huge King Ranch is here), it has some general agriculture and dairying.

Ken·il·worth (kĕn'əl-wûrth'), urban district (1973 est. pop. 19,730), Warwickshire, central England. A market town, it is famous for the ruins of Kenilworth Castle, celebrated in Sir Walter Scott's novel *Kenilworth* and founded c.1120. In the castle's Great Hall, Edward II was forced to relinquish his crown in 1327.

Kenilworth, borough (1970 pop. 9,165), Union co., NE N.J., NW of Elizabeth; inc. 1907. Its diversified manufactures include plastics and pharmaceuticals.

Ke·ni·tra (kə-nē'trə), city (1971 pop. 139,206), NW Morocco, on the Sebou River. It is a busy port exporting agricultural products. The city was built by the French and called by them Port Lyautey. American troops landed here in Nov., 1942, during World War II.

Ken·more (kĕn'môr'), village (1978 est. pop. 21,000), Erie co., NW N.Y., a residential suburb adjacent to Buffalo; inc. 1899.

Ken·ne·bec (kĕn'ə-bĕk'), county (1970 pop. 95,306), 872 sq mi (2,258.5 sq km), SW Maine, watered by the Kennebec River; formed 1799; co. seat Augusta. In a lumbering, resort, and shipping area, it has diversified manufactures, including textiles, paper and metal products, shoes, and machinery.

Kennebec, town (1970 pop. 372), seat of Lyman co., S central S.Dak., SE of Pierre on Medicine Creek. It is a trade center for a farming region yielding dairy produce, livestock, poultry, and grain.

Kennebec, river, 164 mi (263.9 km) long, rising in Moosehead Lake, NW Maine, and flowing S to the Atlantic. French explorer Samuel de Champlain explored it in 1604 and 1605; in 1607 English colonist George Popham established a short-lived colony, Fort St. George, at its mouth. Trading posts were established shortly after 1625.

Ken·ne·bunk (kĕn'ə-bŭngk'), town (1970 pop. 4,546), York co., S Maine; inc. 1820. The first settlement (c.1650) grew as a trading and, later, a shipbuilding and shipping center. The Wedding Cake House at Kennebunk is known for its scroll-saw architecture.

Ken·ne·dy, Cape (kĕn'ĭ-dē): *see* Canaveral, Cape.

Kennedy, Mount, 13,905 ft (4,241 m) high, SW Yukon Territory, Canada, in the St. Elias Mts. near the Alaskan border. It was named in honor of U.S. President John F. Kennedy in 1965. Although discovered in 1935, the mountain was climbed for the first time in 1965 by a team that included Robert F. Kennedy, the President's brother.

Ken·ner (kĕn'ər), city (1978 est. pop. 48,800), Jefferson parish, SE La., a suburb of New Orleans; inc. 1952.

Ken·ne·saw (kĕn'ə-sô'), city (1970 pop. 3,548), Cobb co., NW Ga., NW of Atlanta; inc. 1908. Kennesaw Mountain National Battlefield Park is nearby.

Ken·nett (kĕn'ĭt), city (1970 pop. 10,090), seat of Dunklin co., SE Mo., near the St. Francis River WNW of Caruthersville, in a cotton and timber region; platted 1846, inc. 1896.

Ken·ne·wick (kĕn'ĭ-wĭk'), city (1970 pop. 15,212), Benton co., SE Wash., on the Columbia River near the influx of the Snake River, in an irrigated farm and vineyard region; inc. 1904. Food processing is the chief industry. The Atomic Energy Commission's nearby Hanford Works (established during World War II) is a major employer.

Ké·nog·a·mi (kə-nŏg'ə-mē), city (1971 pop. 10,970), SE Que., Canada, on the Saguenay River, adjacent to its twin city, Jonquière. It has pulp and paper mills and a hydroelectric station.

Ke·no·ra (kə-nôr'ə), town (1976 pop. 10,565), W Ont., Canada, at the N end of the Lake of the Woods. There are fish-processing plants and lumber, flour, pulp, and paper mills in the town. Kenora serves as a base for fishing, hunting, and canoe trips.

Ke·no·sha (kə-nō'shə), county (1970 pop. 117,917), 273 sq mi (707.1 sq km), extreme SE Wis., bounded on the E by Lake Michigan, on the S by the Ill. border, and drained by the Des Plaines and Fox rivers; formed 1850; co. seat Kenosha. In a dairying and agricultural (livestock, oats, corn) region, it also has varied manufacturing and numerous lake resorts.

Kenosha, industrial city (1978 est. pop. 81,900), seat of Kenosha co., SE Wis., a port of entry on Lake Michigan; inc. 1850. Clothing, automobiles, electronic equipment, and metal products are among its many manufactures. The first public school in the state was opened here in 1849.

Ken·sing·ton and Chel·sea (kĕn'zĭng-tən; chĕl'sē), borough (1971 pop. 184,392) of Greater London, SE England. It was created in 1965 by the merger of the metropolitan London boroughs of Kensington and Chelsea. Kensington is a largely residential district with fashionable shopping streets and is the site of the Victoria and Albert Museum, the Science Museum, the Royal College of Art, and the Royal College of Science. Chelsea is a literary and artistic quarter. Sir Thomas More, D. G. Rossetti, James Whistler, Charles Dickens, and many others were associated with it. Thomas Carlyle's house has been restored.

Kent (kĕnt), nonmetropolitan county (1976 est. pop. 1,448,100), 1,525 sq mi (3,949.8 sq km), SE England, between the Thames estuary and the Strait of Dover; administrative center Maidstone. The Isle of Sheppey is separated from the north coast by the narrow Swale channel. The region, largely agricultural, is a market-gardening center. Crops include fruit, grain, and hops. Sheep and cattle grazing, fishing, and dairying are also done. Julius Caesar landed at Kent in 55 B.C. Kent later became one of the seven Anglo-Saxon kingdoms.

Kent. 1. County (1970 pop. 81,892), 595 sq mi (1,541.1 sq km), central Del. bounded on the N by the Smryna River, on the W by the Md. border, on the E by Delaware Bay and River; formed 1683; co. seat Dover. It has agriculture (corn, wheat, vegetables, fruit, dairy products, and poultry), commercial fisheries, a food-processing industry, and some manufacturing. **2.** County (1970 pop. 16,146), 284 sq mi (735.6 sq km), NE Md., a peninsula on the Eastern Shore, bounded on the E by the Del. border, on the W by Chesapeake Bay; formed 1642; co. seat Chestertown. In a coastal-plain agricultural area yielding vegetables, fruit, corn, wheat, livestock, poultry, and dairy products, it has a large seafood industry. Clothing and chemical products are made. It has lumbering and hunting, and there are summer resorts here. Many beautiful old churches and houses are in the county. **3.** County (1970 pop. 411,044), 857 sq mi (2,219.6 sq km), SW Mich., intersected by the Grand River and drained by the Flat, Rogue, and Thornapple rivers; organized 1836; co. seat Grand Rapids. Its agriculture includes livestock, poultry, grain, corn, potatoes, beans, and fruit. Dairy products are made. Gypsum quarries and gravel pits are found in the area, which has many lake resorts. **4.** County (1970 pop. 142,382), 172 sq mi (445.5 sq km), W central R.I., bounded on the E by Narragansett Bay, on the W by the Conn. border, and drained by the Moosup, Flat, and Wood rivers; formed 1750; co. seat East Greenwich. In an agricultural area yielding dairy products, poultry, corn, potatoes, fruit, mushrooms, and truck crops, it also has textile factories, fishing and lumbering industries, and many coastal resorts. **5.** County (1970 pop. 1,434), 880 sq mi (2,279.2 sq km), NW Texas, in a rolling-plains area drained by the Salt and Double Mountain forks of the Brazos River; formed 1876; co. seat Jayton. Cattle ranching is the main occupation. Sheep, hogs, horses, and poultry are also raised, and corn grain sorghums, oat, wheat, and peanuts grown. It has deposits of oil, clay, sand, and gravel.

Kent. 1. Industrial city (1978 est. pop. 25,500), Portage co., NE Ohio; settled in 1805 as Franklin Mills, combined with Carthage and renamed 1863, inc. as a city 1920. Electric motors, compressors, drilling rigs, fasteners, and locks are made here. **2.** City (1978 est. pop. 18,300), King co., W central Wash., near Puget Sound; inc. 1890. Formerly a farm area, Kent is now rapidly urbanizing. It has a large aerospace industry. Food is processed, and electrical supplies and chemical products are made.

Kent·land (kĕnt'lənd), town (1970 pop. 1,864), seat of Newton co., NW Ind., near the Ill. border NW of Lafayette; settled 1860. There are grain and dairy farms. Feed and fertilizer are produced.

Ken·ton (kĕn'tən), county (1970 pop. 129,440), 165 sq mi (427.4 sq km), extreme N Ky., in a gently rolling upland area in the outer Bluegrass bounded on the N by the Ohio River and the Ohio border, on the E by the Licking River; formed 1840; co. seat Independence. The manufacturing area in the north produces canned fruit and vegetables, textiles, paper, plastics, metal products, and machinery. Its agriculture consists of dairy products, poultry, livestock, corn, burley tobacco, and alfalfa.

Kenton, city (1970 pop. 8,315), seat of Hardin co., W central Ohio,

WNW of Marion on the Scioto River, in a farm area; platted 1833. Toys and machine tools are manufactured here.

Kent's Cavern or **Kent's Hole,** limestone cave, Devonshire, SW England, near Torquay. The floor is composed of several strata, with remains indicating the prehistoric coexistence of man and now extinct animals. The cave was extensively explored from 1865 to 1880.

Ken·tuck·y (kən-tŭk'ē), state (1975 pop. 3,386,000), 40,395 sq mi (104,623 sq km), S central United States, admitted as the 15th state of the Union in 1792. Frankfort is the capital. The northern boundary is formed by the Ohio River, separating Kentucky from Ohio, Ind., and Ill. The river runs generally southwest below Covington until it joins the Mississippi River, which forms the western border with Mo. At the southwest tip of the state about 5 sq mi (13 sq km) of Kentucky territory, created by a double hairpin turn in the Mississippi River, protrudes north from Tenn. into Mo. Tenn. borders Kentucky in a straight line on the south. In the east, the boundary with W.Va. is formed by the Big Sandy River and its tributary, the Tug Fork, while the Va. border runs through the Cumberland Mts.

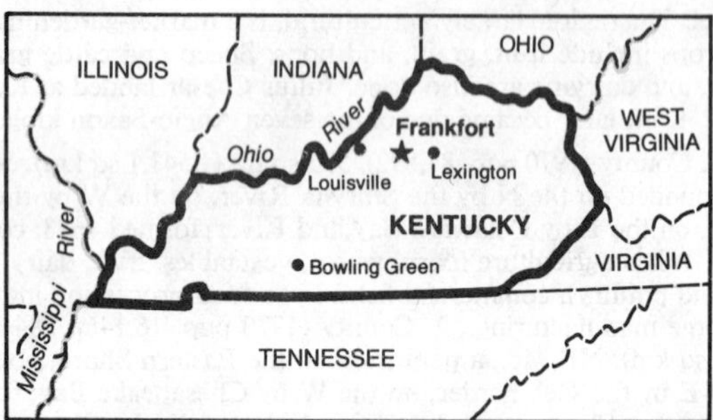

From elevations of about 2,000 ft (610 m) on the Cumberland Plateau in the southeast, where Black Mt. (4,145 ft/1,263 m) marks the state's highest point, Kentucky slopes to elevations of less than 800 ft (244 m) along the western rim. The narrow valleys and sharp ridges of the mountain region are noted for forests of giant hardwoods and scented pine. To the west, the plateau breaks in a series of escarpments, bordering a narrow plains region interrupted by many single conical peaks called knobs. Surrounded by the knobs region on the south, west, and east and extending as far west as Louisville is the Bluegrass country, the heart and trademark of the state. To the south and west lie the rolling plains and rocky hillsides of the Pennyroyal, a section that takes its name from a species of mint that grows abundantly in the area. Here underground streams have washed through limestone to form miles of subterranean passages, some of the notable ones being in Mammoth Cave National Park. Northwest Kentucky is generally rough, rolling terrain, with scattered but important coal deposits. The isolated far-western region, bounded by the Mississippi, Ohio, and Tennessee rivers, is referred to as the Purchase, or Jackson Purchase (for Andrew Jackson, who was a prominent member of the commission that bought it from the Chickasaw Indians in 1818). Consisting of flood plains and rolling uplands, it is the largest migratory bird route in the United States. Kentucky's climate is generally mild, with few extremes of heat and cold.

Kentucky is divided into 120 counties:

NAME	COUNTY SEAT	NAME	COUNTY SEAT
Adair	Columbia	Carlisle	Bardwell
Allen	Scottsville	Carroll	Carrollton
Anderson	Lawrenceburg	Carter	Grayson
Ballard	Wickliffe	Casey	Liberty
Barren	Glasgow	Christian	Hopkinsville
Bath	Owingsville	Clark	Winchester
Bell	Pineville	Clay	Manchester
Boone	Burlington	Clinton	Albany
Bourbon	Paris	Crittenden	Marion
Boyd	Catlettsburg	Cumberland	Burkesville
Boyle	Danville	Daviess	Owensboro
Bracken	Brooksville	Edmonson	Brownsville
Breathitt	Jackson	Elliott	Sandy Hook
Breckinridge	Hardinsburg	Estill	Irvine
Bullitt	Shepherdsville	Fayette	Lexington
Butler	Morgantown	Fleming	Flemingsburg
Caldwell	Princeton	Floyd	Prestonsburg
Calloway	Murray	Franklin	Frankfort
Campbell	Alexandria	Fulton	Hickman

NAME	COUNTY SEAT	NAME	COUNTY SEAT
Gallatin	Warsaw	Martin	Inez
Garrard	Lancaster	Mason	Maysville
Grant	Williamstown	Meade	Brandenburg
Graves	Mayfield	Menifee	Frenchburg
Grayson	Leitchfield	Mercer	Harrodsburg
Green	Greensburg	Metcalfe	Edmonton
Greenup	Greenup	Monroe	Tompkinsville
Hancock	Hawesville	Montgomery	Mount Sterling
Hardin	Elizabethtown	Morgan	West Liberty
Harlan	Harlan	Muhlenberg	Greenville
Harrison	Cynthiana	Nelson	Bardstown
Hart	Munfordville	Nicholas	Carlisle
Henderson	Henderson	Ohio	Hartford
Henry	New Castle	Oldham	La Grange
Hickman	Clinton	Owen	Owenton
Hopkins	Madisonville	Owsley	Booneville
Jackson	McKee	Pendleton	Falmouth
Jefferson	Louisville	Perry	Hazard
Jessamine	Nicholasville	Pike	Pikeville
Johnson	Paintsville	Powell	Stanton
Kenton	Independence	Pulaski	Somerset
Knott	Hindman	Robertson	Mount Olivet
Knox	Barbourville	Rockcastle	Mount Vernon
Larue	Hodgenville	Rowan	Morehead
Laurel	London	Russell	Jamestown
Lawrence	Louisa	Scott	Georgetown
Lee	Beattyville	Shelby	Shelbyville
Leslie	Hyden	Simpson	Franklin
Letcher	Whitesburg	Spencer	Taylorsville
Lewis	Vanceburg	Taylor	Campbellsville
Lincoln	Stanford	Todd	Elkton
Livingston	Smithland	Trigg	Cadiz
Logan	Russellville	Trimble	Bedford
Lyon	Eddyville	Union	Morganfield
McCracken	Paducah	Warren	Bowling Green
McCreary	Whitely City	Washington	Springfield
McLean	Calhoun	Wayne	Monticello
Madison	Richmond	Webster	Dixon
Magoffin	Salyersville	Whitley	Williamsburg
Marion	Lebanon	Wolfe	Campton
Marshall	Benton	Woodford	Versailles

Economy. The state is noted for the distilling of Bourbon whiskey and for the breeding of thoroughbred racehorses. Tobacco has long been the state's chief crop, and it is also the chief farm product, followed by cattle, dairy products, and hogs. Hay, corn, and soybeans are other major crops. The state's chief industries manufacture electrical equipment, food products, nonelectrical machinery, chemicals, and fabricated and primary metals. Kentucky is one of the country's major producers of coal, the state's most valuable mineral. Other mineral products include stone, petroleum, and natural gas.

History. The first major expedition to the region was led by Dr. Thomas Walker in 1750. Walker was soon followed by hunters and scouts including Christopher Gist. Further exploration was interrupted by the French and Indian War (1754–63) and Pontiac's Rebellion, an Indian uprising (1763–66), but, with the British victorious in both, settlers soon began to enter Kentucky. A surveying party under James Harrod established the first permanent settlement at Harrodsburg in 1774, and the next year Daniel Boone blazed the Wilderness Road from Tennessee into the Kentucky region and founded Boonesboro. Kentucky was made (1776) a county of Virginia, and new settlers came through the Cumberland Gap and over the Wilderness Road or down the Ohio River. In 1792 a constitution was framed and accepted, and in the same year the Commonwealth of Kentucky was admitted to the Union. The state grew fast as trade and shipping centers developed and river traffic down the Ohio and Mississippi increased.

In the first half of the 19th cent., Kentucky was primarily a state of small farms rather than large plantations and was not adaptable to extensive use of slave labor. After 1850, however, the state was converted into a huge slave market for the lower south. Soon Kentucky, like other border states, was torn by conflict over the slavery issue. At the outbreak of the Civil War, Kentucky attempted to remain neutral. Confederate forces invaded and occupied part of southern Kentucky, but were forced to retreat when Ulysses S. Grant crossed the Ohio and took Paducah. The state was thus secured for the Union. After the war many in the state opposed federal Reconstruction policies, and Kentucky refused to ratify the Thirteenth and Fourteenth amendments to the U.S. Constitution.

After the turn of the century, the depressed price of tobacco gave rise to a feud between buyers and growers, resulting in the Black Patch War. For more than a year general lawlessness prevailed until the state militia forced an agreement in 1908. Coal mining, which began on a large scale in the 1870s, was well established in mountainous eastern Kentucky by the early 20th cent. The mines boomed during World War I, but after the war intense labor troubles developed. The attempt of the United Mine Workers of America (U.M.W.) to organize the coal industry in Harlan co. in the 1930s resulted in outbreaks of violence. It was not until 1939 that the U.M.W. was finally recognized as a bargaining agent for most of the state's miners. Since World War II construction of turnpikes, extensive development of state parks, and a marked rise in tourism have all contributed to the development of the state.

Government. Kentucky's present constitution was adopted in 1891. The governor of the state is elected for a term of four years. The general assembly, or legislature, is bicameral with a senate of 38 members and a house of representatives of 100 members. State senators are elected to serve for terms of four years and representatives for two years.

Educational Institutions. Those prominent in the state include the Univ. of Kentucky and Transylvania Univ., at Lexington; the Univ. of Louisville, at Louisville; Kentucky Wesleyan College, at Owensboro; Union College, at Barbourville, and Kentucky State Univ., at Frankfort.

Kentucky, river, 259 mi (416.7 km) long, formed by the junction of the North Fork and the Middle Fork rivers, central Ky., and flowing NW to the Ohio River at Carrollton. The river is navigable for its entire length by means of locks. The Kentucky's upper course flows through a coal-mining district and the middle course through a deep gorge before entering the fertile bluegrass region.

Ken·ya (kēn'yə, kĕn'-), republic (1977 est. pop. 14,115,000), 224,960 sq mi (582,646 sq km), E Africa; capital Nairobi. Kenya is bordered by the Somali Democratic Republic on the east, the Indian Ocean on the southeast, Tanzania on the south, Victoria Nyanza on the southwest, Uganda on the west, the Sudan on the northwest, and Ethiopia on the north. The country, which lies astride the equator, is made up of several geographical regions. The first is a narrow, dry coastal

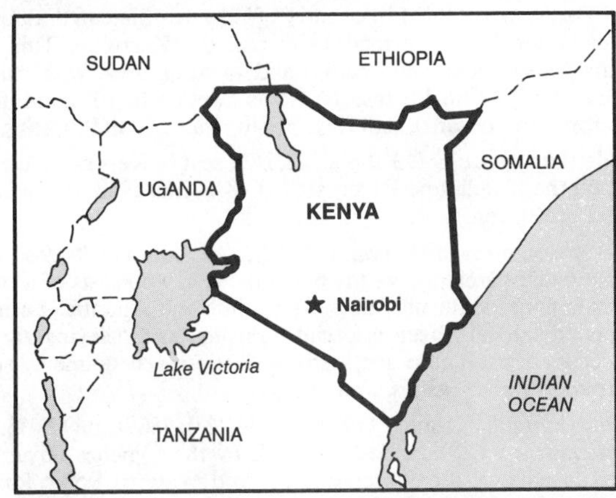

strip that is low lying except for the Taita Hills in the south. The second, an inland region of bush-covered plains, constitutes most of the country's land area. In the northwest, straddling Lake Rudolf and the Kulal Mts., are high-lying scrublands. In the southwest are the fertile grasslands, and forests of the Kenya highlands. In the west is the Great Rift Valley, an irregular depression that cuts through western Kenya from north to south in two branches. It is also the location of some of the country's highest mountains.

Economy. The great majority of Kenyans are engaged in subsistence farming, but industry is growing. Coffee, tea, sisal, pyrethrum, maize, and wheat are grown in the highlands. Coconuts, cashew nuts, cotton, sugar cane, sisal, and maize are grown in the lower-lying areas. Much of the country remains grassland, where large numbers of cattle are pastured. Kenya's leading manufactures include refined petroleum, processed food, cement, textiles, leather goods, and metal products. The chief minerals produced are limestone, soda ash, gold, and salt. Kenya attracts many tourists.

History. During the 1950s and 1960s, the anthropologist L. S. B. Leakey discovered in northern Tanzania the remains of men who lived c.2 million years ago. These persons, perhaps the earliest men on earth, most likely also inhabited southern Kenya. In the Kenya highlands, the existence of farming and domestic herds can be dated to c.1000 B.C. Trade between the Kenya coast and Arabia was brisk by A.D. 100. Arabs settled on the coast by the 8th cent., and they soon established several autonomous city-states (including Mombasa, Malindi, and Pate). Around the year 1000, ironworking reached the interior of Kenya, at about the same time that the first Bantu-speaking people arrived there. The Portuguese first visited the Kenyan shores in 1498, and by the end of the 16th cent. they controlled much of the coast. The Portuguese were replaced as the leading power on the coast by two Arab dynasties: the Busaidi dynasty, based first at Masqat (in Oman) and from 1832 on Zanzibar, and the Mazrui dynasty, based at Mombasa. From the early 19th cent. there was long-distance caravan trading between Mombasa and Victoria Nyanza. The British and German governments agreed upon spheres of influence in East Africa in 1886, with most of present-day Kenya passing to the British. In 1888 a British concessionary association was given a royal charter as the Imperial British East Africa Company, but severe financial difficulties soon led to its takeover by the British government. A railroad was built (1895-1901) from Mombasa to Kisumu on Victoria Nyanza in order to facilitate trade with the interior and with Uganda. In 1903 the first European settlers established themselves as large-scale farmers, taking land from the Kikuyu, Masai, and others. At the same time, Indian merchants moved inland from the coast.

From the 1920s to the 1940s European settlers controlled the government and owned extensive farmlands; Indians were small traders and lower-level government employees; and black Africans grew cash crops (coffee and cotton) on a small scale, were subsistence farmers, or were laborers in the towns. In the 1920s black Africans began to protest their inferior status. Protest reached a peak between 1952 and 1956 with the so-called Mau-Mau Emergency, an armed revolt led by the Kikuyu against British rule. The British declared a state of emergency and imprisoned many of the colony's nationalist leaders, including Jomo Kenyatta (1893-1978). After the revolt, Britain increased black African representation in the colony's legislative council. On Dec. 12, 1963, Kenya became independent. In 1964 the country became a republic, with Kenyatta as president. The first decade of independence was characterized by disputes among ethnic groups (especially between the Kikuyu and the Luo), by economic growth and diversification, and by the end of European predominance. Boundary disputes with the Somali Democratic Republic resulted in sporadic fighting (1963-68). More than 70% of the country was affected by the sub-Saharan drought of the early 1970s.

Government. Kenya is run by a president, who is assisted by a vice president and a cabinet. There is a unicameral legislature. The only legal political party is the Kenya African National Union.

Kenya, Mount, extinct volcano, central Kenya, just S of the equator. Its highest peak, Batian, reaches 17,058 ft (5,202.7 m) making Mt. Kenya the second-highest mountain in Africa. The Kikuyu people cultivate Mt. Kenya's fertile lower slopes. From 5,000 to 15,000 ft (1,525-4,575 m) are dense woodlands inhabited by elephants, buffalo, and leopards.

Ke·o·kuk (kē'ə-kŭk'), county (1970 pop. 13,943), 579 sq mi (1,499.6 sq km), SE Iowa, drained by the Skunk, North Skunk, and South Fork English rivers; formed 1837; co. seat Sigourney. Its prairie agriculture includes hogs, cattle, poultry, corn, oats, and wheat. There are limestone quarries and clay pits in the area.

Keokuk, city (1978 est. pop. 13,500), extreme SE Iowa, on the Mississippi River at the foot of the Des Moines River rapids and in a farm area; inc. 1847. A gravity dam (built 1910-13) and a power plant on the Mississippi furnish hydroelectric power for Keokuk's industries, which include food processing and packaging and the manufacture of sponge rubber goods, trucks, corrugated cartons, and a great variety of metals and metal products. The first cabin was erected here in 1820, and a trading post founded in 1829 was named for Keokuk, a Sac Indian chief (who is buried beneath an impressive statue in the city's Rand Park). Tourist attractions include Keokuk Dam, the Unknown Soldier Monument in the national cemetery, and many old homes.

Ke·os (kē'ŏs): *see* Kéa, Greece.

Ke·o·sau·qua (kē'ə-sô'kwə), town (1970 pop. 1,018), seat of Van Buren co., SE Iowa, in a bend of the Des Moines River SE of Ottumwa; settled 1836, inc. 1851. Dairy equipment, dairy products, and feed are manufactured.

Ke·ra·la (kā'rä-lä), state (1971 pop. 21,280,397), 15,003 sq mi (38,858 sq km), SW India, on the Arabian Sea; capital Trivandrum. The most densely populated Indian state, Kerala was created in 1956 from the Malayalam-speaking former princely states of Cochin and Travancore and Malayalam-speaking areas formerly in Madras state. In 1957 India's first Communist state administration was elected in Kerala. Kerala takes its name from the ancient Tamil kingdom of Kerala (Chera), which traded with the Phoenicians, Greeks, and Romans.

Ke·ra·ma Islands (kä-rä'mä, kə-rä'mə), volcanic and coral island group, 18 sq mi (46.6 sq km), off the SW coast of Okinawa Islands, Ryukyu Islands, Japan, in the East China Sea. Heavily wooded and mountainous, the islands are important for fishing, sweet potatoes, and sugar cane.

Kerch (kĕrch), city (1976 est. pop. 152,000), SE European USSR, in the Crimea. It lies on the Kerch Strait of the Black Sea and at the eastern end of the Kerch Peninsula, a land strip between the Sea of Azov and the Black Sea. A seaport and major industrial center, it has iron and steel mills, machine, chemical, and coking plants, shipyards, fisheries, and canneries. Iron ore, vanadium, and natural gas are extracted nearby. The city was founded (6th cent. B.C.) by Greek colonists from Miletus. It became (5th cent. B.C. to 4th cent. A.D.) the capital of the European part of the Kingdom of Bosporus. Conquered (c.110 B.C.) by Mithridates VI of Pontus, it then passed under Roman and Byzantine rule, and was taken by Novogorod in the 9th cent. Later (13th cent.) it became a Genoese trade center and was conquered (1475) by the Crimean Tatars. It was captured (1771) by the Russians in the first Russo-Turkish War (1768–74). There are ruins of the ancient acropolis on top of the steep hill of Mithridates. Archaeological remains, discovered in catacombs and burial mounds near the city, are in the archaeological museum (founded 1826), which is famous for its Greco-Scythian antiquities.

Ker·gue·len (kûr'gə-lĕn'), subantarctic island of volcanic origin, 1,318 sq mi (3,413.6 sq km), in the S Indian Ocean, c.3,300 mi (5,310 km) SE of the S tip of Africa; largest of the 300 Kerguelen Islands (total area c.2,700 sq mi/6,995 sq km), part of the French Southern and Antarctic Territories. Glacial lakes, peat marshes, lignite, and guano deposits are found on the island. The island is used mainly as a research station and a seal-hunting and whaling base. Kerguelen was discovered in 1772. It has belonged to France since 1893.

Ke·rin·tji (kə-rĭn'chē), peak, 12,467 ft (3,802.4 m) high, in the Pegunungan Barisan, W central Sumatra, Indonesia.

Kér·ki·ra (kĕr'kē-rä) or **Cor·fu** (kôr'fōō), city (1971 pop. 28,630), capital of Kérkira prefecture, NW Greece, on Kérkira Island. Olive oil, wine, and citrus fruits are shipped, and textiles are manufactured. It is also a commercial and tourist center and has been the summer home of the Greek royal family.

Kérkira or **Corfu,** island (1971 pop. 89,578), 229 sq mi (593.1 sq km), NW Greece, in the Ionian Sea, the second largest of the Ionian Islands, separated by a narrow channel from the Albanian and Greek coasts. Kérkira is largely a fertile lowland producing olive oil, figs, wine, and citrus fruit. Livestock raising and fishing are important sources of livelihood. The island was settled c.730 B.C. by Corinthian colonists and became the competitor of Corinth in the Adriatic Sea. The two rivals fought the first recorded (by Thucydides) naval battle in 665 B.C. In 433 B.C. Kérkira concluded an alliance with Athens; this alliance helped precipitate (431) the Peloponnesian War. The island passed under Roman rule in 229 B.C. and in A.D. 336 became part of the Byzantine Empire. It was seized from the Byzantines by the Normans of Sicily in the 1080s and 1150s, by Venice (1206), and later by Epirus (1214–59) and the Angevins of Naples. In 1386 the Venetians obtained a hold that ended only with the fall of the Venetian republic in 1797. Under Venetian rule, the island successfully resisted two Turkish sieges (1537, 1716). The island was under the protection of Great Britain from 1815 to 1864, when it was ceded to Greece.

Kerk·ra·de (kĕrk'rä'də), city (1977 est. pop. 46,732), Limburg prov., SE Netherlands, on the West German border. Coal mining began here in the 12th cent.

Ker·man (kĕr-män', kər-), city (1971 est. pop. 88,000), capital of Kerman prov., E central Iran. It is noted for making and exporting carpets. Cotton textiles and goats-wool shawls are also manufactured. Kerman was under the Seljuk Turks in the 11th and 12th cent., but remained virtually independent. Marco Polo visited (late 13th cent.) and described the city. Kerman changed hands many times in ensuing years, prospering under the Safavid dynasty (16th cent.) and suffering under the Afghans (17th cent.). In 1794, Aga Mohammed Khan, shah of Persia, ravaged the city by selling 20,000 of its inhabitants into slavery and by blinding another 20,000.

Ker·man·shah (kĕr-män'shä, kər-), city (1971 est. pop. 190,000), capital of Kermanshah prov., W Iran. It is the trade center for a rich agricultural region that produces grain, fruits, and sugar beets. Manufactures include carpets, canvas shoes, textiles, and refined petroleum. Kermanshah was founded in the 4th cent. A.D. It was captured by the Arabs in the 7th cent. and was later a frontier fortress against the Ottoman Turks, who occupied it a number of times including the period from 1915 to 1917.

Ker·mit (kûr'mĭt), city (1970 pop. 7,884), seat of Winkler co., W Texas, near the N.Mex. border S of Odessa; inc. 1938. The center of a thriving oil region, it has refineries, carbon-black plants, and similar enterprises.

Kern (kûrn), county (1970 pop. 330,234), 8,152 sq mi (21,113.7 sq km), S Calif., in the S end of the San Joaquin Valley, walled in by the Tehachapi Mts., the Sierra Nevada, and the Coast Ranges, with part of the Mojave Desert in the E and SE; formed 1866; co. seat Bakersfield. It is the state's leading petroleum-producing county, with oil and natural-gas fields along the western side of San Joaquin Valley. There is extensive mining in the Mojave desert (borax, tungsten, silver, and gold), as well as sand, clay, and gravel pits and gypsum and stone deposits. It is an important irrigated agricultural county, yielding potatoes, cotton, grain, alfalfa, grapes, fruit, truck crops, cattle, sheep, hogs, poultry, and dairy products. It has a food-processing industry and varied manufactures. Part of Sequoia National Forest is in the north.

Kern, river, 155 mi (249.4 km) long, rising in the S Sierra Nevada Mts., E Calif., and flowing S, then SW to a reservoir in the extreme S part of the San Joaquin Valley. U.S. explorer John Frémont named the river in honor of Edward M. Kern, the topographer of his third expedition. Gold was discovered along the river in 1853.

Kerr (kûr), county (1970 pop. 19,454), 1,101 sq mi (2,851.6 sq km), SW central Texas, in a scenic hill country of Edwards Plateau drained by the Guadalupe River; formed 1856; co. seat Kerrville. This is a ranching (sheep, goats, and cattle) and vacation area, with camps, guest ranches, and health resorts. It has farming (corn, grain, fruit, truck crops, and pecans), dairying, poultry raising, and hunting.

Kerr·ville (kûr'vĭl'), city (1970 pop. 12,672), seat of Kerr co., S central Texas, on the Guadalupe River; settled 1846, inc. 1942. It is a vacation and health resort.

Ker·ry (kĕr'ē), county (1971 pop. 112,772), 1,815 sq mi (4,700.9 sq km), SW Republic of Ireland; co. town Tralee. Kerry consists of a series of mountainous peninsulas that extend into the Atlantic. Farming (oats and potatoes), sheep and cattle raising, and dairying are the chief occupations. There are many well-preserved dolmens, stone forts, round towers, castles, and abbeys.

Ker·shaw (kûr-shô'), county (1970 pop. 34,727), 786 sq mi (2,035.7 sq km), N central S.C., bounded on the E by the Lynches River and watered in part by the Wateree River and Wateree Pond; formed 1791; co. seat Camden. It has timber, some agriculture (peaches, pecans, and sweet potatoes), and a tourist area in the Sand Hills.

Ker·u·len (kĕr'ōō-lĕn), river, 785 mi (1,263 km) long, E Mongolian People's Republic, rising in the Kentei Mts., NE of Ulan Bator, and flowing S, then E to Hu-lun Lake, NE China.

Ke·sen·nu·ma (kā-sā-nōō'mä), city (1975 pop. 66,616), Miyagi prefecture, NE Honshu, Japan, on Kesennuma Bay. It is a fishing port.

Ke·she·na (kə-shē'nə), uninc. village (1971 est. pop. 950), seat of Menominee co., E central Wis.

Ket (kĕt), river, c.845 mi (1,359.6 km) long, W central Siberian USSR, rising just N of Krasnoyarsk and flowing NW and W into the Ob. The Ket is navigable c.410 mi (660 km).

Ketch·i·kan (kĕch'ĭ-kăn'), city (1978 est. pop. 7,200), seat of Ketchikan Gateway borough, SE Alaska, a port of entry on Revillagigedo Island in the Alexander Archipelago. A supply point for miners in the gold rush of the 1890s, it has become a center of Alaska's fishing and pulp industries and a tourist hub. Its excellent ice-free harbor on Tongass Narrows makes it an important port on the Inside Passage and a distribution point for a large area.

Ketchikan Gate·way (gāt'wā'), borough (1970 pop. 10,041), 1,345 sq mi (3,483.6 sq km), extreme SE Alaska, in the Alexander Archipelago; formed after 1961, name changed from Gateway in 1967;

borough seat Ketchikan. It is in a lumbering area, with some mining and food-processing industries.

Ket·ter·ing (kĕt'ər-ĭng), municipal borough (1973 est. pop. 44,480), Northamptonshire, central England. It is a center for the manufacture of shoes and textiles. There are also iron mining and smelting, engineering, and cardboard and brush-making industries.

Kettering, city (1978 est. pop. 68,700), Montgomery co., SW Ohio, a suburb of Dayton; settled c.1812, inc. 1952. Its manufactures include shock absorbers, electric motors, and tool and die products.

Ket·tle (kĕt'l), river, c.160 mi (257.4 km) long, British Columbia, Canada, and NE Wash., rising in the Monashee Mts. W of Upper Arrow Lake and flowing S to the Wash. border, finally joining the Columbia River NW of Colville.

Keu·ka Lake (kyōō'kə), 18 mi (28.9 km) long and .5 to 2 mi (.8-3.2 km) wide, W central N.Y., one of the Finger Lakes, draining NE into Seneca Lake. It is surrounded by a resort, grape-growing, and wine-making region.

Ke·wa·nee (kē-wô'nē), industrial city (1978 est. pop. 15,300), Henry co., NW Ill., in a farm and livestock area; inc. 1855. Its manufactures include gloves, trailers and trucks, boilers, steel doors and windows, farm machinery, foundry equipment, valves, and pipe fittings.

Ke·wau·nee (kē-wô'nē), county (1970 pop. 18,961), 331 sq mi (857.3 sq km), E Wis., in partly wooded, hilly terrain bounded on the E by Lake Michigan, on the NW by Green Bay, and drained by several small streams; formed 1852; co. seat Kewaunee. Farming, dairying, lumbering, and woodworking are done.

Kewaunee, city (1970 pop. 2,901), seat of Kewaunee co., E Wis., on Lake Michigan E of the city of Green Bay; inc. 1883. It is in a dairying area and produces cheese, butter, and furniture.

Ke·wee·naw (kē'wĭ-nô'), county (1970 pop. 2,264), 538 sq mi (1,393.4 sq km), NW Upper Peninsula, Mich., on the NE Keweenaw Peninsula in Lake Superior; formed 1861; co. seat Eagle River. It is in a copper-mining region traversed by the Copper Range. There is lumbering and some farming (potatoes and dairy products). Hunting and fishing resorts, several lakes, and two state parks are in the county.

Ke·ya Pa·ha (kē'ə pä'hä), county (1970 pop. 1,340), 769 sq mi (1,991.7 sq km), N Nebr., bounded on the N by the S.Dak. border, on the S by the Niobrara River, and drained by the Keya Paha River; formed 1884; co. seat Springview. It is in a livestock-raising and grain-farming area, with some manufacturing.

Key Lar·go (kē' lär'gō), narrow island, c.30 mi (48 km) long, off S Fla., largest of the Florida Keys.

Key·port (kē'pôrt'), borough (1970 pop. 7,205), Monmouth co., E central N.J., on Raritan Bay SSE of Perth Amboy; settled before 1700, inc. 1908. It is a resort, farming, and fishing center. Ceramic tiles are made here.

Key·ser (kī'zər), city (1970 pop. 6,586), seat of Mineral co., NE W.Va., in the Eastern Panhandle on the North Branch of the Potomac River W of Martinsburg; settled 1802, inc. 1923. It has pulp and paper mills and manufactures textiles and electrical equipment.

Keytes·ville (kēts'vĭl'), city (1970 pop. 730), seat of Chariton co., N central Mo., near the Missouri River W of Moberly; platted 1830. Grain and lumber products are processed.

Key West, city (1978 est. pop. 24,500), seat of Monroe co., S Fla., on an island at the SW extremity of the Florida Keys; inc. 1828. It is the southernmost city of the continental United States, a port of entry, a winter and fishing resort, a shrimping and fishing center, and an artists' colony. Its military installations include a major U.S. naval air station, a naval base, and a U.S. Coast Guard base (at Fort Taylor, built 1844–46). Early Spanish sailors called the site Cayo Hueso (Bone Island), because of the human bones they found. A railroad (completed 1912) linked the Keys with the mainland. It was abandoned after being damaged by a hurricane in 1935 and was replaced by a highway (completed in 1938). After a severe decline in industry, the Federal government took over (1934) the bankrupt city.

Kha·ba·rovsk (KHä-bä'rŏfsk), city (1977 est. pop. 524,000), capital of Khabarovsk Kray, Far Eastern USSR, on the Amur River near its junction with the Ussuri. The city has oil refineries, shipyards, and factories that produce farm machinery, trucks, aircraft, diesel engines, and machine tools. Khabarovsk, formerly a fortified trading post, prospered greatly after the coming of the railroad in 1905.

Khabarovsk Kray (krī) or **Khabarovsk Territory,** administrative division (1970 pop. 1,346,000), 305,000 sq mi (789,950 sq km), Far Eastern USSR, in Siberia, bounded by the Sea of Okhotsk in the E, Primorsky Kray and Manchuria in the S, and the Kolyma Range in the N; capital Khabarovsk. The mountainous territory yields gold, oil, tin, and coal. Grain and potatoes are grown in the Amur valley, and in the north there are reindeer herds and fur trappers. Herring, flounder, and salmon are caught along the coast. The territory was founded in 1938 and reorganized in 1953 and 1957.

Kha·bur (KHä-bŏŏr'), river, c.200 mi (320 km) long, rising in SE Turkey, and flowing generally S through NE Syria to enter the Euphrates River.

Khair·pur (KHīr'pŏŏr), city (1961 pop. 34,144), SE Pakistan, in Sind prov. It trades in wheat, cotton, tobacco, and dates. Manufactures include textiles, armaments, and pharmaceuticals. The city was the capital of the former princely state of Khairpur, which was founded in 1783 and merged into Pakistan in 1955.

Kha·kass Autonomous Oblast (KHä-käs'), administrative division (1970 pop. 446,000) 23,900 sq mi (61,901 sq km), S central Siberian USSR, in Krasnoyarsk Kray; capital Abakan. The oblast, largely consisting of black-earth steppe, is bounded by the upper Yenisei River on the east and by the wooded Kuznetsk Ala-Tau and Sayan ranges on the west and south, respectively. The oblast's swift-flowing rivers provide hydroelectric power, and many of the numerous lakes are sources of therapeutic mineral waters. Mining, forestry, and food processing are the main industries. The region, known for mining and trade from the 8th to the 11th cent., came under Russian control in the 17th cent. The autonomous oblast was formed in 1930.

Khal·ki·dhi·kí (KHäl'kə-*the*'kē) or **Chal·ci·di·ce** (kăl-sĭd'ĭ-sē), peninsula, NE Greece, projecting into the Aegean Sea from SE Macedonia. The region is largely mountainous, dry, and agricultural. Olive oil, wine, wheat, and tobacco are produced; magnesite is mined. In antiquity the peninsula was famous for its timber. The peninsula was named for Khalkís, which established colonies in the 8th and 7th cent. B.C. In the 4th cent. B.C. the peninsula was conquered by Philip II of Macedon, and in the 2nd cent. B.C. by Rome.

Khal·kís (KHäl-kēs') or **Chal·cis** (kăl'sĭs), city (1971 pop. 36,300), capital of Évvoia (Euboea) prefecture, E Greece, on the island of Évvoia. Connected to the mainland by a bridge, the city is a trade center for wine, cotton, and citrus fruits. Soap and cement are manufactured. The chief city of ancient Euboea, Khalkís was settled by the Ionians and early became a commercial and colonizing center. The city was subdued by Athens (c.506 B.C.) and led the revolt of Euboea against Athens in 446 B.C. Again defeated, it came under Athenian rule until 411 B.C. In 338 B.C. it passed to Macedonia. Aristotle died here (322 B.C.). In succeeding centuries the city was used as a base for invading Greece. It was occupied by the Venetians in 1209, passed to the Ottoman Turks in 1470, and in 1830 became part of Greece.

Khan·a·bad (KHän'ä-bäd'), city (1967 est. pop. 30,000), NE Afghanistan, near the USSR border. It is a market town for wool and silk.

Kha·na·qin (KHä'nä-kēn'), town (1965 pop. 23,527), E Iraq, on a tributary of the Diyala. It is located in an oil-producing region.

Khand·wa (kŭnd'wä), town (1971 pop. 85,513), Madhya Pradesh state, central India. Khandwa is a market for cotton, timber, and grain. There are cotton gins and oilseed mills.

Kha·niá (KHä-nyä') or **Ca·ne·a** (kə-nē'ə), city (1971 pop. 40,564), capital of Khaniá prefecture, NW Crete, Greece, a port on the Gulf of Khaniá, an arm of the Sea of Crete. Olives, citrus fruits, and wine are shipped. One of the oldest Cretan cities, it was conquered in 69 B.C. by the Romans and in A.D. 826 fell under Arab rule. Reconquered (961) by the Byzantine Empire, it became (13th cent.) a Venetian colony. The Ottoman Empire took the city in 1645. It was the capital of Crete from 1841 to the mid-20th cent.

Khan·ty-Man·si National Okrug (KHän'tē-män'sē), administrative division (1970 pop. 272,000), 201,969 sq mi (523,100 sq km), W Siberian USSR; capital Khanty-Mansisk. The region is mostly forest and swamp with numerous lakes and peat bogs and is sparsely populated. Lumbering, fishing, fur farming and trading, and reindeer breeding are the okrug's chief occupations. The okrug, formed in 1930, was known until 1940 as the Ostyak-Vogul National Okrug.

Kha·rag·pur (kä-räg'pŏŏr), city (1971 pop. 161,911), West Bengal state, E central India. It is an industrial city and has a scientific-research center.

Kha·ri·jah, Al (äl khär'ēn-jä) or **Khar·ga** (khär'gə), large oasis, S central Egypt, in the Libyan Desert. Populated chiefly by Bedouins and Berbers, the irrigated oasis produces cereals, vegetables, dates, citrus fruits, and alfalfa. Cattle and poultry are also raised. The oasis was prosperous in ancient times, and there are ruins of temples built by the Achaemenids of ancient Persia and by the Romans.

Khar·kov (kär'kŏf, -kŏv), city (1977 est. pop. 1,405,000), capital of Kharkov oblast, S European USSR, in the Ukraine, at the confluence of the Kharkov, Lopan, and Udy rivers in the upper Donets valley. Kharkov is one of the country's main rail junctions and economic and cultural centers. Proximity to the iron mines of Krivoy Rog and the coal of the Donets Basin has provided the basis for engineering industries that produce a wide variety of other heavy metal items. Kharkov's industries also include food and tobacco processing, printing, and the manufacture of chemicals. Founded in 1656, it became an important frontier headquarters of the Ukrainian Cossacks, who kept the city loyal to the czar during the Cossack uprisings of the late 17th cent. Russia's annexation of the Crimea in 1783 and colonization of the steppes stimulated Kharkov's economic growth. The coal and metallurgical industries developed after the 1860s, and railroads were built in the late 19th cent. The city became the capital of the Ukraine in 1919 but was superseded by Kiev in 1934. Kharkov's landmarks include the cathedral of the Protectoress (1686), the cathedral of the Assumption (1771), and a bell tower that was built to celebrate Napoleon's defeat in 1812.

Khar·toum (kär-to͞om'), city (1973 pop. 333,921), capital of the Sudan, a port at the confluence of the Blue Nile and White Nile rivers. Food, beverages, cotton, gum, and oil seeds are processed in the city. Manufactures include cotton textiles, knitwear, glass, and tiles. Founded in 1821 as an Egyptian army camp, Khartoum developed as a trade center and slave market. In the war between Great Britain and the forces of the Mahdi, Gen. Charles Gordon was killed here (1885) after resisting a long siege, during which the city was severely damaged. Khartoum was retaken by H. H. Kitchener in 1898.

Khas·ko·vo (KHäs'kŏ-vŏ), city (1969 est. pop. 66,900), S Bulgaria, in an agricultural region noted for its tobacco. The city has one of Bulgaria's largest cigarette factories.

Kha·tan·ga (kə-täng'gə, -tăng'-), river, Krasnoyarsk Kray, N central Siberian USSR, formed by the union of the Kotui and the Kheta rivers. From the Kotui it is c.715 mi (1,150 km) long and flows north through the central Siberian Plateau past Khatanga village and northeast into the Khatanga Gulf of the Laptev Sea.

Kher·son (KHĕr-sôn'), city (1976 est. pop. 315,000), capital of Kherson oblast, S European USSR, in the Ukraine, on the Dnepr River near its mouth on the Black Sea. It is a rail junction and a sea and river port, exporting grain, timber, and manganese ore and importing oil from the Caucasus. Kherson has one of the Ukraine's largest cotton textile mills; the city's other industries include shipbuilding and food processing. Kherson was founded in 1778 by Potemkin as a naval station, fortress, and shipbuilding center. Kherson's landmarks include the fortress with earthen ramparts and stone gates and the 18th cent. cathedral that contains Potemkin's tomb.

Khing·an, Great (khĭng'än', shĭng'än'), mountain range, Heilungkiang, Kirin, and Liaoning prov., NE China, extending c.750 mi (1,205 km) from the Amur River S to the Liao River; the highest point is 5,657 ft (1,724 m). The range forms the eastern edge of the Mongolian Plateau.

Khí·os (KHē'ôs, kī'ŏs) or **Chi·os** (kī'ŏs), island (1971 pop. 52,487), c.350 sq mi (905 sq km), E Greece, in the Aegean Sea, just W of Asia Minor. It is mountainous and is famous for its scenic beauty and good climate. The island produces olives, figs, and mastic and has marble quarries, lignite deposits, and sulfur springs. Sheep and goats are raised. Khíos was colonized by Ionians and later held (494–479 B.C.) by the Persians. In 479 B.C. it recovered its independence and joined the Delian League. The island maintained its independence from Rome until the reign of Vespasian (1st cent. A.D.). It became part of the Byzantine Empire and later passed (1204) to the Latin emperors of Constantinople and then (1261) to the Genoese. The Ottoman Turks conquered the island in 1566 and held it until the First Balkan War (1912), when it was taken by Greece. A rebellion against Turkish rule resulted (1822) in a ruthless massacre of the population. Khíos claims to be the birthplace of Homer.

Khi·va (KHē'və, KHē-vä'), city (1974 est. pop. 26,000), Central Asian USSR, in Uzbekistan, in the Khiva oasis and on the Amu Darya River. Industries include metalworking, cotton and silk spinning, wood carving, and carpetmaking. The city, in existence by the 6th cent., passed to Russia in 1873.

Khmel·nit·sky (KHmĕl-nēt'skē), city (1976 est. pop. 161,000), capital of Khmelnitsky oblast, SW European USSR, in the Ukraine, on the Southern Bug River. It has metal forges, food-processing plants, and factories that produce machine tools, equipment for power stations, reinforced concrete items, clothing, and footwear. Known since the 15th cent., the city was a fortress by the late 16th cent. and became part of Russia in 1795.

Khmer Republic (kmĕr): *see* Cambodia.

Kho·kand (kŏ-känt'): *see* Kokand, USSR.

Kho·per (KHə-pyôr'), river, c.625 mi (1,005 km) long, S European USSR, rising SW of Penza and flowing SW, then S into the Don. It is partly navigable.

Kho·rezm (kə-rĕz'əm) or **Khwa·razm** (kwä-răz'əm), ancient and medieval state of central Asia, situated in and around the basin of the lower Amu Darya River; now an oblast, NW Uzbek Republic, USSR. It was a part of the Achaemenid empire of Cyrus the Great in the 6th cent. B.C. and became independent in the 4th cent. B.C. It was later inhabited by Persians who adhered to Zoroastrianism. Khorezm was conquered by the Arabs in the 7th cent. and was converted to Islam. In the late 12th cent., Khorezm gained independence from the Seljuk Turks. At the height of its power in the early 13th cent. it ruled from the Caspian Sea to Bukhara and Samarkand. It was conquered in 1221 by Genghis Khan and was included in the Golden Horde. In the late 14th cent., Khorezm was destroyed by Tamerlane. A century of struggle over Khorezm between the Timurids, the descendants of Tamerlane, and the Golden Horde was followed by the Uzbek conquest in the early 16th cent. Khorezm became an independent Uzbek state and was known as the khanate of Khiva after Khiva became the capital.

Kho·rog (KHô-rōg'), city (1975 est. pop. 15,000), capital of Gorno-Badakhshan Autonomous Oblast, Central Asian USSR, in Tadzhikistan, in the Pamir. Khorog has shoe factories and metal-working plants, and building materials are produced.

Khor·ra·ma·bad (KHō-räm'ä-bäd'), city (1971 est. pop. 62,000), capital of Luristan governorate, Khuzistan prov., W Iran. It is the trade center of a mountainous region yielding fruit, grain, and wool.

Khor·ram·shahr (KHo͝or'äm-shär'), city (1971 est. pop. 90,000), Khuzistan prov., SW Iran, at the confluence of the Karun River and the Shatt al Arab, near the Persian Gulf. Its development dates to the late 19th cent., when steam navigation on the Karun was started.

Khors·a·bad (KHôr'sä-bäd'), village, NE Iraq, near the Tigris River NE of Mosul. It is built on the site of Dur Sharrukin, an Assyrian city founded 8th cent. B.C. by Sargon. Its mounds were excavated in 1842 and in 1851, and statues of Sargon and of huge, winged bulls that guarded the gates of the royal palace were taken to the Louvre. In 1932 hundreds of cuneiform tablets in the Elamite language and a list of kings ruling from c.2200 B.C. to 730 B.C. were discovered.

Kho·tan (kō'tän'): *see* Ho-t'ien, China.

Kho·tin (KHə-tyēn'), city, SW European USSR, in the Ukraine, on the Dnestr River. It is in an agricultural district and has agricultural and food-processing industries. Its strategic location at an important Dnestr River crossing caused the city to change hands frequently from the 16th to 18th cent. Seized by Russia in 1739, Khotin was incorporated into the Russian Empire in 1812 as part of Bessarabia. The city was under Rumanian rule from 1918 to 1940 and under German occupation from 1941 to 1944. Khotin has remains of an imposing fortified castle that was built (13th cent.) by the Genoese, enlarged (14th–15th cent.) by the Moldavians, and restored (18th cent.) by the Turks.

Khul·na (ko͝ol'nə), town (1974 pop. 437,304), SW Bangladesh, near the Ganges delta. It is the trade and processing center for the products of the Sundarbans, a swampy, forested region. Agricultural products, especially rice and jute, are processed, and there is some textile manufacturing and shipbuilding.

Khvoy (kə-voi'), city (1971 est. pop. 51,000), West Azerbaijan prov., NW Iran. It is the trade center for a fertile farm region. Khvoy was attacked by Russia in 1827, occupied by Turkey in 1911, and held by the Soviet Union during World War II.

Khwa·razm (kwä-răz'əm), ancient state: *see* Khorezm.

Khy·ber Pass (kīʹbər), narrow, steep-sided pass, 28 mi (45 km) long, winding through the Safed Koh Mts., on the Pakistan-Afghanistan border; highest point 3,500 ft (1,067.5 m). It links the cities of Peshawar, Pakistan, and Kabul, Afghanistan. For centuries a trade and invasion route from central Asia, the Khyber Pass was one of the principal approaches of the armies of Alexander the Great, Tamerlane, Babur, Mahmud of Ghazni, and Nadir Shah in their invasions of India. The pass was also important in the Afghan Wars fought by the British in the 19th cent. The Khyber Pass is now traversed by a modern road, an old caravan route, and a railroad (built 1920–25), which passes through 34 tunnels and over 92 bridges and culverts.

Kiakh·ta (kyäKHʹtə): *see* Kyakhta, USSR.

Kia·ling (kyäʹlĭngʹ): river, China: *see* Chia-ling.

Ki·a·mich·i (kīʹə-mĭshʹē), river, c.100 mi (160 km) long, SE Okla., rising near the Ark. border and flowing SW and S, then SE to the Red River S of Fort Towson.

Kia·mus·ze (kyäʹmo͞oʹso͞oʹ, jē-äʹ-): *see* Chia-mu-ssu, China.

Ki·an (jēʹänʹ): *see* Chi'an, China.

Kiang·si (kyăngʹsēʹ, jē-ängʹsēʹ), province (1968 est. pop. 22,000,000), c.66,000 sq mi (170,940 sq km), SE China; capital Nan-ch'ang. Kiangsi is one of China's leading rice producers; other food crops include wheat, sweet potatoes, barley, and corn. Commerical crops are cotton, oil-bearing plants, ramie, tea, sugar cane, tobacco, and oranges. Mineral resources include tungsten, high-grade coking coal, kaolin, uranium, manganese, tin, and antimony. Traditionally known as Kan, Kiangsi was ruled by the Chou dynasty (722–481 B.C.); it received its present name under the Southern Sung dynasty (A.D. 1127–1280). The province, whose present boundaries date from the Ming dynasty, passed under Manchu rule in 1650. The Chinese Communist movement began (1927) in Kiangsi; the famous long march began here. Following World War II, during which Kiangsi was largely free of Japanese forces, the province passed (1949) to the Communists.

Kiang·su (kyăngʹso͞oʹ, jē-ängʹso͞oʹ), province (1968 est. pop. 47,000,000), c.41,000 sq mi (106,190 sq km), E China, on the Yellow Sea; capital Nanking. Kiangsu consists largely of the alluvial plain of the Yangtze River. The province straddles two agricultural zones, with wheat, millet, koaliang, corn, soybeans, and peanuts cultivated in the north and rice, tea, sugar cane, and barley raised in the south. Cotton is grown along the coast (north and south) in the saline soil, which is not suited for other crops. Tea is planted in the western hills. Kiangsu, which is known to the Chinese as "the land of rice and fish," is rich in marine products. It is also a major salt-producing area. Textile, food-processing, cement, and fertilizer industries are found throughout the province. Kiangsu was originally part of the Wu kingdom, and the name Wu is still its traditional name. The capture of Kiangsu in 1937 was an important phase of Japan's effort to conquer China. Liberated by the Chinese Nationalists in 1945, Kiangsu fell to the Communists in 1949. Many archaeological sites have been excavated since 1956.

Kiao·chow (kyouʹchouʹ, jē-ouʹjōʹ), former German territory, area c.200 sq mi (520 sq km), along the S coast of Shantung prov., China. Germany leased Kiaochow in 1898 for 99 years, but in 1914 Japan seized it. Through agreements reached at the Washington Conference in 1922, Kiaochow was returned to China.

Kick·a·poo (kĭkʹə-po͞oʹ), river, c.100 mi (160 km) long, SW Wis., rising S of Tomah and flowing SSW into the Wisconsin River near its junction with the Mississippi E of Prairie du Chien.

Kick·ing Horse (kĭkʹĭng hôrsʹ), river of SE British Columbia, Canada, rising in the Rocky Mts., and flowing SW and NW to Golden, where it enters the Columbia River. Its course is rapid, with several high falls. Kicking Horse Pass, 5,339 ft (1,628.4 m) high, northwest of Lake Louise, in Banff National Park, connects the Bow River with the Kicking Horse and is one of the principal rail and highway passes over the Continental Divide.

Kid·der (kĭdʹər), county (1970 pop. 4,362), 1,358 sq mi (3,517.2 sq km), S central N.Dak., watered by several small lakes including Horsehead Lake; formed 1873; co. seat Steele. It is in an agricultural area that produces duram wheat (for macaroni), oats, corn, rye, sunflowers, livestock, and poultry. There is a grain-milling industry.

Kid·der·min·ster (kĭdʹər-mĭnʹstər), municipal borough (1973 est. pop. 49,960), Hereford and Worcester, W central England. It is a market town. Kidderminster carpets have been produced since 1735; other industries include spinning, dyeing, metal forging, and the production of beet sugar.

Kiel (kēl), city (1974 est. pop. 264,290), capital of Schleswig-Holstein, N West Germany, on Kiel Bay, an arm of the Baltic Sea. Situated at the head of the Kiel Canal, the city was Germany's chief naval base from 1871 to 1945, when the naval installations were dismantled. Kiel is now a shipping and industrial center, with large shipyards and factories that manufacture textiles, processed foods, and printed materials. Chartered in 1242, Kiel joined the Hanseatic League in 1284. Kiel passed to Denmark in 1773; it was annexed by Prussia in 1866. The sailors' mutiny that began at Kiel at the end of World War I touched off a socialist revolution in Germany.

Kiel Canal, artificial waterway, 61 mi (98.1 km) long, in Schleswig-Holstein, N West Germany, connecting the North Sea with the Baltic Sea. Built (1887–95) to facilitate movement of the German fleet, the Kiel Canal was widened and deepened from 1905 to 1914.

Kiel·ce (kyĕlʹtsĕ), city (1975 est. pop. 151,200), S central Poland. It is a railway junction and manufacturing center where metals, agricultural machinery, and chemicals are produced. It also has marble quarries. Founded in 1173, Kielce obtained municipal rights in the 14th cent. The city passed to Austria in 1795 and to Russia in 1815 and reverted to Poland in 1919.

Ki·ev (kēʹĕf, kēʹyəf), city (1977 est. pop. 2,079,000), capital of the Ukrainian Soviet Socialist Republic and of Kiev oblast, SW European USSR, a port on the Dnepr River. Kiev is a leading industrial, commercial, and cultural center. Food processing, metallurgy, and the manufacture of machinery, machine tools, rolling stock, chemicals, building materials, and textiles are the major industries.

Kiev probably existed as a commercial center as early as the 5th cent. A Slavic settlement on the great trade route between Scandinavia and Constantinople, Kiev was tributary to the Khazars when the Varangians under Oleg established themselves in 882. Under Oleg's successors it became the capital of medieval Kievan Russia (the first Russian state) and was a leading European cultural and commercial center. It was also an early seat of Russian Christianity. The city reached its apogee in the 11th cent., but by the late 12th cent. it had begun to decline. From 1240, when it was devastated by the Mongols, until the 14th cent., the city paid tribute to the Golden Horde. Kiev then passed under the control of Lithuania, which in 1569 was united with Poland. In 1648, when the Ukrainian Cossacks rose against Poland, Kiev became for a brief period the center of a Ukrainian state. After the Ukraine's union with Russia in 1654, however, the city was acquired (1686) by Moscow. In Jan., 1918, Kiev became the capital of the newly proclaimed Ukrainian republic; but in the ensuing civil war (1918–20), it was occupied in succession by German, White Russian, Polish, and Soviet troops. In 1934 the capital of the Ukrainian SSR was transferred from Kharkov to Kiev. German forces held the city during World War II and massacred thousands of its inhabitants, including 50,000 Jews. Postwar reconstruction of the heavily damaged city was not completed until c.1960.

Ki·ga·li (kĭ-gäʹlē), town (1970 est. pop. 60,000), capital of Kigali prefecture and of Rwanda, central Rwanda. It is the country's main administrative and economic center.

Ki·go·ma-U·ji·ji (kē-gōʹmä-o͞o-jēʹjē), municipality (1967 pop. 21,369), capital of Kigoma prov., W Tanzania, a port on Lake Tanganyika. There are fisheries. Ujiji and Kigoma were important settlements of Arab and Swahili ivory and slave traders between c.1850 and c.1890. The explorer Henry M. Stanley successfully ended his search for David Livingstone at Ujiji on Nov. 10, 1871. The region was occupied by the Germans in the 1890s. Kigoma and Ujiji were combined into a single municipality in the 1960s.

Ki·lau·e·a (kēʹlou-āʹä), crater, 3,646 ft (1,112 m) deep, central Hawaii Island, Hawaii, on the SE slope of Mauna Loa in Hawaii Volcanoes National Park. One of the largest active craters in the world, Kilauea has a circumference of c.8 mi (13 km) and is surrounded by a wall of volcanic rock 200 to 500 ft (61–152.5 m) high. The usual level of the lake of molten lava is c.740 ft (225 m) below the pit's rim.

Kil·dare (kĭl-dârʹ), county (1971 pop. 71,977), 654 sq mi (1,693.9 sq km), E central Republic of Ireland; co. town Naas, the ancient seat of the kings of Leinster. The region is a flat plain, containing the greater portion of the Bog of Allen and the Curragh. Agriculture is the chief occupation; the breeding of racehorses is significant. There are many pre-Christian and early Christian remains.

Kil·gore (kĭlʹgôrʹ), city (1970 pop. 9,495), Gregg and Rusk cos., NE

Texas, W of Shreveport, La. near the Sabine River; settled 1872, inc. 1931. Transformed from a sleepy pinewoods town to a roaring boom town after the East Texas Oil Field was discovered in 1930, Kilgore has many oil wells crowding the city's stores and houses.

Kil·i·man·ja·ro (kĭl'ə-mən-jär'ō), highest mountain of Africa, NE Tanzania. An extinct volcano, it rises in two snow-capped peaks, Kibo (19,340 ft/5,898.7 m) and Mawenzi (17,564 ft/5,357 m), which are joined by a broad saddle (alt. c.15,000 ft/4,575 m). Coffee and plantains are the chief crops raised on Kilimanjaro's intensively cultivated lower southern slopes.

Kil·ken·ny (kĭl-kĕn'ē), county (1971 pop. 61,473), 796 sq mi (2,061.6 sq km), S Republic of Ireland; co. town Kilkenny. It is mainly a rolling plain, with low hills to the south. Grains and vegetables are grown, and livestock is raised. There are food-processing and brewing industries. In the northeast is a large anthracite coal field. Kilkenny is roughly coextensive with the ancient kingdom of Ossory. It is rich in antiquities.

Kilkenny, urban district (1971 pop. 9,838), co. town of Co. Kilkenny, S Republic of Ireland, on the Nore River. Kilkenny was the seat of the kings of Ossory. The Statute of Kilkenny (1366) forbade the English settlers from marrying the Irish inhabitants, speaking Irish, or wearing Irish dress.

Kil·lar·ney (kĭ-lär'nē),urban district (1971 pop. 7,184), Co. Kerry, SW Republic of Ireland. The town, which has footwear and other industries, is a tourist center for the three Lakes of Killarney. They occupy a wooded valley stretching south between the mountains. Lough Leane or Lower Lake is the largest; it has about 30 islands. On the island are the ruins of an abbey founded c.600 by St. Finian.

Kill Devil Hills, town (1970 pop. 357), Dare co., NE N.C., on a sandy peninsula separating Albemarle Sound from the Atlantic SSE of Kitty Hawk. The Wright Brothers National Memorial is here.

Kil·leen (kĭ-lēn'), city (1978 est. pop. 54,700), Bell co., central Texas, in a ranching and cotton region; founded 1882, inc. 1893. The city has some varied light manufacturing, but adjacent Fort Hood is the major source of employment.

Kil·lie·cran·kie, Pass of (kĭl'ē-krăng'ke), wooded pass, Tayside region, central Scotland, through which the river Garry flows, near Pitlochry. In 1689 Jacobite Highlanders defeated a large government force under Hugh MacKay and the Jacobite leader, Viscount Dundee, was killed.

Kil·ling·ly (kĭl'ĭng-lē), town (1970 pop. 13,573), Windham co., NE Conn., on the Quinebaug River and the R.I. border, in a farm area; settled 1693, inc. 1708. It was once a great textile town and still has some textile manufactures.

Kill Van Kull (kĭl' văn kŭl'), channel, 4 mi (6.4 km) long and .5 mi (.8 km) wide, connecting Upper New York Bay with Newark Bay, between Bayonne, N.J., and Staten Island, N.Y. It is the main route for ships docking at Port Elizabeth and Port Newark, N.J.

Kil·mar·nock (kĭl-mär'nək), burgh (1974 est. pop. 50,318), Strathclyde region, SW Scotland. An industrial town in a mining region, it has industries that manufacture carpets, hosiery, farm and hydraulic machinery, whiskey, and shoes. Its textile industry dates from 1603. Robert Burns's first poems were published here in 1786; the Burns Monument has a museum.

Ki·lung (kē'lo͝ong'): *see* Chi-lung, Taiwan.

Kim·ball (kĭm'bəl), county (1970 pop. 6,009), 953 sq mi (2,468.3 sq km), W Nebr., bordering on Colo. and Wyo. and drained by Lodgepole Creek; formed 1888; co. seat Kimball. It is in a livestock, wheat, small grains, and potato farming area, with oil and gas wells and some manufacturing.

Kimball, city (1970 pop. 3,680), seat of Kimball co., W central Nebr., S of Scottsbluff near the Wyo. border; inc. 1885. It is a trade center for a potato, wheat, and bean-growing area. There are dairy farms and oil wells in the vicinity.

Kim·ber·ley (kĭm'bər-lē), town (1976 pop. 7,111), SE British Columbia, Canada. At an elevation of more than 3,000 ft (915 m), it is the site of the Sullivan mine, where large quantities of silver, lead, and zinc are mined.

Kimberley, city (1970 pop. 105,258), Cape Prov., central South Africa. The city is primarily a diamond-mining center, although textiles, construction materials, and machinery are manufactured. Kimberley was founded in 1871 when diamonds were discovered on a nearby farm. The De Beers Consolidated Mines, organized by Cecil Rhodes, assumed control of the diamond fields in 1888.

Kim·ber·ly (kĭm'bər-lē), village (1970 pop. 6,131), Outagamie co., E central Wis., E of Appleton; inc. 1910. Paper products are made.

Kim·ble (kĭm'bəl), county (1970 pop. 3,904), 1,274 sq mi (3,299.7 sq km), W central Texas, in the scenic Edwards Plateau drained by the North and South Llano rivers; formed 1858; co. seat Junction. It has ranching (especially goats, also sheep and cattle) and some agriculture (corn, oats, grain sorghums, wheat, and pecans). It is a leading producer of wool and mohair. Hunting and fishing attract tourists.

Kin·a·ba·lu or **Kin·i·ba·lu, Mount** (both: kĭn'ə-bə-lo͞o'), peak, 13,455 ft (4,103.8 m) high, N Sabah, Malaysia, NE of Kota Kinabalu. It is the highest peak on Borneo.

Kin·car·dine·shire (kĭn-kär'dn-shîr', -shər), or earlier **the Mearns** (mûrnz), former county, E Scotland. Stonehaven was the county town. Kincardineshire, long inhabited by Picts, was occupied briefly by the Romans. Remains of their forts and of Pictish castles are found in the area. In 1975 Kincardineshire became part of the Grampian region.

Kin·der·hook (kĭn'dər-ho͝ok'), village (1970 pop. 1,233), Columbia co., SE N.Y.; settled before the Revolution, inc. 1838. Richard Upjohn designed St. Paul's Church (1851) here. Martin Van Buren was born and is buried in Kinderhook; the Van Buren homestead, Lindenwald, is south of the village. The House of History, maintained by the county historical society, occupies an early 19th cent. mansion.

Kin·di·a (kĭn'dē-ə, -dyə), town (1967 est. pop. 45,000), W Guinea. It is the trade center for an area where bananas, manioc, rice, fruits, and vegetables are grown and bauxite is mined. Tonic water and soap are also produced.

Ki·nesh·ma (kē'nyĭsh-mə), city (1976 est. pop. 100,000), N central European USSR, on the Volga River. A river port and a rail terminus, it is an old textile center with sawmills and chemical plants.

King (kĭng). **1.** County (1970 pop. 464), 944 sq mi (2,445 sq km), NW Texas, in a rolling plains area drained by tributaries of the Wichita and Brazos rivers; formed 1876; co. seat Guthrie. The county has huge cattle ranches, some agriculture (grain sorghums, wheat, cotton, poultry, and other livestock), oil wells, and deposits of lime and copper. **2.** County (1970 pop. 1,159,369), 2,136 sq mi (5,532.2 sq km), W central Wash., bounded on the W by Puget Sound and rising to the Cascade Range in the E; formed 1852; co. seat Seattle. It is heavily industrial, with many manufactures (especially aircraft and airplane and aerospace components). Lumber is processed, and plywood and paper manufactured. The area is mainly residential around Seattle, but truck farming and dairying are still important.

King and Queen (kwēn), county (1970 pop. 5,491), 318 sq mi (823.6 sq km), E Va., in a tidewater region bounded on the SW by the Mattaponi and York rivers; formed 1691; co. seat King and Queen Court House. It is primarily agricultural, yielding tomatoes, corn, wheat, tobacco, legumes, and livestock. Oystering is done.

King and Queen Court House (kôrt'hous', kōrt'-), village (1970 pop. 50), seat of King and Queen co., E Va., near the Mattaponi River ENE of Richmond.

King·fish·er (kĭng'fĭsh'ər), county (1970 pop. 12,857), 904 sq mi (2,341.4 sq km), central Okla., in a diversified agricultural area intersected by the Cimarron River; formed 1890; co. seat Kingfisher. Wheat, oats, barley, and milo are the most important crops. It also has oil wells and refineries.

Kingfisher, city (1970 pop. 4,042), seat of Kingfisher co., central Okla., NW of Oklahoma City; founded 1889. It is a trade center for a grain, livestock, and oil-refining area.

King George (jôrj), county (1970 pop. 8,039), 178 sq mi (461 sq km), E Va., in a rolling dairying and agricultural area bounded on the N by the Potomac River, on the S by the Rappahannock River; formed 1720; co. seat King George. Truck crops, tobacco, and hay are grown, and cattle and sheep raised. It also has commercial fisheries.

King George, village (1970 pop. 200), seat of King George co., E Va., E of Fredericksburg, in an agricultural area.

King George Mount, peak, 11,226 ft (3,423.9 m) high, SE British Columbia, Canada, in the Rockies near the Alta. border S of Banff.

Kin·gi·sepp (kĭn'gĭ-syĕp'), city, NW European USSR, SW of Leningrad, near the Estonian border, on the Luga River. A river port, it has leather and shoe industries. The site was settled in the 9th cent., and the fortress of Yam was founded there in 1384 as a frontier post of Novgorod. The fortress was taken by Sweden in 1585 and passed to Russia in 1703.

King Island, fertile island, 424 sq mi (1,098.2 sq km), at the W end of Bass Strait, NW of Tasmania, Australia. Its mines yield tungsten, tin, and zircon.

King Island, small rocky island, off the W coast of Seward Peninsula, Alaska, at the S end of Bering Strait. Its highest peak is 700 ft (213.5 m). Captain Cook discovered the island in 1778.

King·man (kĭng'mən), county (1970 pop. 8,886), 865 sq mi (2,240.4 sq km), S central Kansas, in a plains region watered by the Chikaskia River and the South Fork of the Ninnescah River; formed 1886; co. seat Kingman. In a wheat and cattle area, it has oil wells and some manufacturing (cutlery, hardware, and farm machinery).

Kingman. 1. City (1975 est. pop. 7,397), seat of Mohave co., NW Ariz., founded 1882, inc. 1952. Gold, silver, lead, and other minerals are mined near here, and livestock is raised. **2.** City (1970 pop. 3,622), seat of Kingman co., S central Kansas, WSW of Wichita; founded c.1872, inc. 1883. It is a trade center in a grain and livestock region.

Kings (kĭngz). **1.** County (1970 pop. 66,717), 1,395 sq mi (3,613.1 sq km), SW central Calif., in the level irrigated farmland of the San Joaquin Valley, drained by the Kings River, with Tulare Lake in the central part and Kettleman Hills in the SW; formed 1893; co. seat Hanford. Its agriculture includes cotton, grain, dairy products, beef cattle, raisin and table grapes, other fruit, nuts, olives, hay, flax, and poultry. It has rich oil and natural-gas fields, gypsum deposits, petroleum refineries, and a food-processing industry. Textiles, chemical products, and concrete are manufactured. **2.** County (1970 pop. 2,602,012), 71 sq mi (183.9 sq km), SE N.Y., coextensive with Brooklyn borough of New York City; formed 1683; co. seat Brooklyn.

Kings, river, 125 mi (201.1 km) long, rising in three forks in the Sierra Nevada, E Calif., and flowing SW to Tulare Lake in the San Joaquin Valley. Its middle and southern forks flow through the great gorges of Kings Canyon National Park.

Kings·bur·y (kĭngz'bĕr'ē), county (1970 pop. 7,657), 819 sq mi (2,121.2 sq km), E central S.Dak., in an agricultural area watered by natural and artificial lakes; formed 1873; co. seat De Smet. Its agriculture includes corn, wheat, alfalfa, barley, oats, and livestock. The county also offers good hunting and fishing opportunities and has some light industry.

King's Lynn (lĭn), municipal borough (1973 est. pop. 29,990), Norfolk, E England, on the Great Ouse River near its influx into The Wash. The town's harbor is the base for a fishing fleet. A farm market, it is also a center of fertilizer production, canning, flour milling, beet-sugar refining, shipbuilding, and metalworking. King's Lynn dates from Saxon times.

Kings Mountain National Military Park: *see* National Parks and Monuments Table.

Kings Peak, mountain, 13,528 ft. (4,126 m) high, NE Utah, in the Uinta Mts., a range of the Rockies, near the Wyo. border and E of Salt Lake City. It is the highest point in the state.

Kings Point, residential village (1970 pop. 5,614), Nassau co., SE N.Y., on NW Long Island, N of Great Neck; inc. 1924. The U.S. Merchant Marine Academy was founded here in 1938.

Kings·port (kĭngz'pôrt'), city (1978 est. pop. 33,000), Hawkins and Sullivan cos., NE Tenn., on the Holston River near the Va. border; inc. 1917. Industries include printing and bookbinding and the manufacture of film, textiles, and plastics. The city, which is encircled by mountains, stands on the site of forts Robinson (1761) and Patrick Henry (1775) on the old Wilderness Road.

Kings·ton (kĭngz'tən), city (1976 pop. 56,032), S Ont., Canada, on Lake Ontario, near the head of the St. Lawrence River at the end of Rideau Canal. Industries include the manufacture of locomotives, textiles, aluminum products, synthetic yarn, and ceramics. On the site stood Fort Frontenac, of great importance in the French and Indian War. The present city was founded by United Empire Loyalists in 1783 and prospered during the War of 1812. From 1841 to 1844 it served as the capital of Canada. Fort Henry, built during the War of 1812 and rebuilt from 1832 to 1836, is now a museum.

Kingston, city (1973 est. pop. 603,717), capital of Jamaica, SE Jamaica. The country's chief port, it exports sugar, rum, molasses, and bananas. The city's industries include tourism, food processing, and oil refining. Kingston was founded in 1693 on a deep, landlocked harbor. The former capital, Port Royal, at the tip of the long, narrow peninsula forming the harbor, was inundated after an earthquake in 1692; the capital was then moved to Spanish Town and, in 1872, to Kingston. After fire destroyed the new Port Royal in 1703, Kingston became Jamaica's leading commercial city. It has suffered from severe hurricanes and was leveled by an earthquake in 1907.

Kingston. 1. Town (1970 pop. 5,999), Plymouth co., SE Mass., on the Atlantic coast between Plymouth and Duxbury; set off from Plymouth 1726. Textiles and concrete products are made. **2.** City (1970 pop. 291), seat of Caldwell co., NW Mo., NE of Kansas City. Corn and oats are grown. **3.** City (1978 est. pop. 23,900), seat of Ulster co., SE N.Y., on the Hudson River; inc. as a village 1805, and as a city through the union (1872) of Kingston and Rondout. The eastern gateway to the Catskill-Shawangunk vacationland and the center of an expanding industrial region, it has plants making electronic computers, farm machinery, and apparel. Fur-trading posts were built here between 1611 and 1615. The first permanent settlement was established in 1652. Kingston served as the first capital of New York state until it was burned by the British in Oct., 1777. Its growth in the early 19th cent. was stimulated by the Delaware and Hudson Canal. Among notable landmarks are many old Dutch stone houses; the senate house (1676), meeting place of the first New York state legislature and now a museum; and the old Dutch church (1659) and cemetery (1661). **4.** Borough (1978 est. pop. 15,800), Luzerne co., NE Pa., on the Susquehanna River opposite Wilkes-Barre; settled 1769, inc. 1857. Although chiefly residential, it has railroad shops and varied manufactures. It was devastated by a flood in June, 1972. **5.** Town (1970 pop. 4,142), seat of Roane co., E central Tenn., WSW of Knoxville between the Tennessee and the Clinch rivers, in a farming area; settled c.1800, inc. 1820.

Kingston upon Hull (hŭl): *see* Hull, England.

Kingston upon Thames (tĕmz), borough (1971 pop. 140,210) of Greater London, SE England. The borough was created in 1965 by the merger of the municipal boroughs of Kingston upon Thames, Malden and Coombe, and Surbiton. Mainly residential, it has light-engineering works and manufactures electronic equipment. In the 10th cent. several Anglo-Saxon kings were crowned here.

Kings·town (kĭngz'toun'), town (1970 pop. 17,258), capital of St. Vincent, British West Indies. The chief port of entry, it is also a popular winter resort.

Kings·tree (kĭngz'trē'), resort town (1970 pop. 3,381), seat of Williamsburg co., E central S.C., on the Black River NNE of Charleston, in a lumber and tobacco area; settled c.1736, inc. 1866.

Kings·ville (kĭngz'vĭl'), city (1978 est. pop. 29,200), seat of Kleberg co., S Texas; inc. 1911. It is headquarters of the gigantic King Ranch, part of which is nearby. The city is a processing center for a farm, oil, and gas area.

Kings·wood (kĭngz'wŏŏd'), urban district (1976 est. pop. 78,800), Avon, SW England. A residential suburb, Kingswood has a footwear industry. It is noted for its open-air chapel.

King Wil·liam (wĭl'yəm), county (1970 pop. 7,497), 278 sq mi (720 sq km), E Va., in a tidewater region bounded on the SW by the Pamunkey River, on the NE by the Mattaponi and York rivers; formed 1702; co. seat King William. It has an extensive lumbering industry, agriculture (soybeans, corn, wheat, and livestock), and deposits of sand, gravel, clay, and marl. Fish and shellfish are processed.

King William, village (1970 pop. 200), seat of King William co., E Va., ENE of Richmond.

King William Island, part of the Arctic Archipelago, in the Arctic Ocean, central Northwest Territories, Canada, between Boothia Peninsula and Victoria Island. The island was discovered (1831) by Sir James C. Ross, who also explored the northern coast. In 1837 Thomas Simpson of the Hudson's Bay Company traced the southern coast. The ships of the expedition of Sir John Franklin were wrecked off the west coast, and the island was further explored by searchers for Franklin, notably John Rae and Sir Francis L. McClintock. Roald Amundsen wintered here in 1903–4 while on his way through the Northwest Passage.

King·wood (kĭng'wŏŏd'), town (1970 pop. 2,550), seat of Preston co., N W.Va., SE of Morgantown, in a coal-mining, timber, and farm area; settled 1811, inc. 1853. Limestone is quarried here.

Kin·hwa (jĭn'hwä'): *see* Chin-hua, China.

Kin·i·ba·lu, Mount (kĭn'ə-bə-lo͞o'), Malaysia: *see* Kinabalu, Mount.

Kin·ney (kĭn'ē), county (1970 pop. 2,006), 1,391 sq mi (3,602.7 sq km), SW Texas, on the S edge of Edwards Plateau bounded on the SW by the Rio Grande and the Mexico border and drained by tributaries of the Nueces River and the Rio Grande; formed 1850; co. seat Brackettville. Cattle, sheep, and goats are raised in this ranching area. Wool and mohair are marketed.

Kin·ross (kĭn-rôs'), burgh (1971 pop. 2,418), Tayside region, E Scotland, on Loch Leven. Kinross House, in the style of an Italian Renaissance mansion, was built for James II of England in 1685. Until 1975 Kinross was the county town of Kinross-shire.

Kin·ross-shire (kĭn-rôs'shĭr', -shər), former county, E Scotland. Kinross was the county town. In 1975 Kinross-shire became part of the Tayside region.

Kin·sha·sa (kĭn-shä'sə), formerly **Le·o·pold·ville** (lē'ə-pōld'vĭl'), city (1974 est. pop. 2,008,000), capital of Zaire, W Zaire, a port on Stanley Pool of the Congo River. Major industries are food and beverage processing, tanning, construction, ship repairing, and the manufacture of chemicals, mineral oils, textiles, and cement. Kinshasa was founded in 1881 by Henry M. Stanley, the Anglo-American explorer, who named it after his patron, Leopold II, king of the Belgians. In 1898 the rail link with Matadi was completed, and in 1926 the city succeeded Boma as the capital of the Belgian Congo. A major anti-Belgian rebellion that took place in Jan., 1959, started the country on the road to independence (June, 1960). In 1966 the city's name was changed to Kinshasa, the name of one of the African villages that occupied the site in 1881.

Kins·ley (kĭnz'lē), city (1970 pop. 2,212), seat of Edwards co., SW central Kansas, on the Arkansas River NE of Dodge City; founded c.1875, inc. 1878. The city is the center of a wheat-growing area. The Santa Fe Trail passed through here, and a battle with the Indians was fought nearby in 1858.

Kin·ston (kĭn'stən), city (1978 est. pop. 25,000), seat of Lenoir co., E N.C., on the Neuse River; settled c.1740, inc. 1849. It is a market for bright leaf tobacco. Lumber, textiles, and fertilizers are produced.

Kin·tyre (kĭn-tīr'), peninsula, 42 mi (67.6 km) long and 10 mi (16 km) wide, Argyllshire, W Scotland, joined to the mainland at the isthmus of Tarbert between East Loch Tarbert and West Loch Tarbert. The terrain is hilly and uncultivated.

Kio·ga (kyō'gə), lake, Uganda: *see* Kyoga.

Ki·o·wa (kī'ə-wə). **1.** County (1970 pop. 2,209), 1,767 sq mi (4,576.5 sq km), E Colo., in a wheat and livestock area bordering on Kansas; formed 1889; co. seat Eads. It is watered by Big Sandy Creek and reservoirs in the south. **2.** County (1970 pop. 4,088), 720 sq mi (1,864.8 sq km), S Kansas, in a rolling plain watered by Rattlesnake Creek and by headstreams of the Salt Fork of the Arkansas River and the Medicine Lodge River; formed 1886; co. seat Greensburg. It is in a grain and livestock area, with some mining and manufacturing. **3.** County (1970 pop. 12,532), 1,032 sq mi (2,672.9 sq km), SW Okla., bounded on the W by the North Fork of the Red River and drained by Elk Creek; formed 1901; co. seat Hobart. It has agriculture (wheat, cotton, oats, barley, livestock, and poultry), granite and marble quarries, oil wells, and some manufacturing.

Kiowa, town (1970 pop. 235), seat of Elbert co., central Colo., SE of Denver, at an altitude of 6,400 ft (1,952 m). Dairy products, livestock, grain, and beans are processed.

Kir·ghiz Soviet Socialist Republic (kĭr-gēz'), **Kir·ghi·zia** (kĭr-gē'-zhə), or **Kir·ghiz·stan** (kĭr-gē-stän'), constituent republic (1970 pop. 3,368,000), c.76,600 sq mi (198,395 sq km), Central Asian USSR. It borders on China in the southeast and on the Kazakh SSR, the Uzbek SSR, and the Tadzhik SSR in the north, west, and southwest. Frunze is the capital. Kirghizia has rich pasturage for goats, sheep, cattle, and horses. There are coal, antimony, lead, tungsten, mercury, uranium, petroleum, and natural gas deposits. The area was formed into the Kara-Kirghiz Autonomous Oblast in 1924, an autonomous republic in 1926, and a constituent republic in 1936.

Ki·rin (kē'rĭn'), province (1968 est. pop. 17,000,000), NE China; capital Ch'ang-ch'un. It is bordered by the USSR on the northeast, by North Korea on the southeast, and by the Inner Mongolian Autonomous Region on the west. Soybeans, wheat, upland rice, sweet potatoes, and beans are grown. Vast timberlands, among the best in China, are exploited, and iron, coal, gold, and lead are extracted.

Kirin, city: *see* Chi-lin, China.

Kirk·cal·dy (kûr-kôl'dē, -kô'dē, -kä'dē), burgh (1974 est. pop. 50,063), Fife, E Scotland, on the Firth of Forth. It is one of the largest producers of linoleum and oilcloth in Great Britain. Other industries include textile printing and the manufacture of farm machinery. The city is the birthplace of the economist Adam Smith.

Kirk·cud·bright (kûr-ko͞o'brē), burgh (1971 pop. 2,506), Strathclyde region, SW Scotland, at the head of the Dee estuary. It has granaries and creameries and is a market town and artists' colony. There are traces of an ancient wall and moat. Until 1975 Kirkcudbright was county town of Kirkcudbrightshire.

Kirk·cud·bright·shire (kûr-ko͞o'brē-shĭr', -shər), former county, SW Scotland, in the Galloway district. Kirkcudbright was the county town. In 1975 Kirkcudbrightshire became part of the Strathclyde region.

Kir·kin·til·loch (kûr'kĭn-tĭl'ŏKH), burgh (1974 est. pop. 26,845), Strathclyde region, W Scotland, on the Forth and Clyde Canal. An engineering center, the burgh has factories that produce mining machinery and valves. Chartered in the 13th cent., the burgh is located on the line of the Roman Antonine wall.

Kirk·land (kûrk'lənd), city (1970 pop. 14,970), King co., W Wash., a suburb of Seattle on Lake Washington; inc. 1905. Furniture is the principal manufacture.

Kirkland Lake, mining town, E Ont., Canada. It is one of Canada's largest gold-mining centers. Gold was discovered here in 1911.

Kırk·lar·e·li (kərk-lär'ĕ-lĕ'), city (1975 pop. 33,260), capital of Kırklareli prov., NW Turkey. It is a transportation hub and a trade center for butter and cheese. During the First Balkan War the Bulgarians defeated (1912) the Turks here.

Kirks·ville (kûrks'vĭl'), city (1978 est. pop. 14,800), seat of Adair co., N Mo.; inc. 1857. Among its manufactures are shoes, gloves, machinery, and hospital equipment.

Kir·kuk (kĭr-ko͝ok'), city (1970 est. pop. 207,900), NE Iraq. It is the center of Iraq's oil industry and a market for the region's produce. Kirkuk is built on a mound containing the remains of a settlement dating back to 3000 B.C.

Kirk·wall (kûrk'wôl', -wəl), burgh (1974 est. pop. 4,814), Orkney Islands, N Scotland, on the E coast of Mainland Island. It is a trading and boatbuilding center. Kirkwall was founded prior to 1046 and became important as a port on the northern trade route to Scandinavia and the Baltic states. St. Magnus Cathedral dates from 1137.

Kirk·wood (kûrk'wo͝od'), city (1978 est. pop. 29,800), St. Louis co., E Mo., a suburb of St. Louis; inc. 1865. Lime, cement, and lumber products are made.

Ki·rov (kē'rôf', -rôv', -rəf), formerly **Vyat·ka** (vyät'kə), city (1976 est. pop. 376,000), capital of Kirov oblast, central European USSR, on the Vyatka River. It is a river port and an industrial center with sawmills and machine and metalworking plants. Founded in 1174, it soon became the capital of an independent republic that was annexed to Moscow by Ivan III in 1489. In the 19th cent. it was used as a place of political exile.

Ki·ro·va·bad (kĭ-rō'və-băd'), city (1976 est. pop. 211,000), SE European USSR, in Azerbaijan, on the Gandzha River. Kirovabad produces cotton and silk textiles, building materials, carpets, cottonseed oil, agricultural implements, copper sulfate, and wine. It was founded in the 6th cent., c.4 mi (6 km) east of the modern city, but was demolished by earthquake in 1139, after which the survivors settled on the present site. The medieval city was an important textile and wine center. It was destroyed by the Mongols in 1231 and recovered slowly. It was the seat of a khanate under Persian suzerainty from the 17th cent. until its conquest (1804) by the Russians.

Ki·rov·o·grad (kĭ-rō'və-grăd'), city (1976 est. pop. 224,000), capital of Kirovograd oblast, S central European USSR, in the Ukraine, on the Ingul River. It is an agricultural trade center, with one of the USSR's largest farm machinery plants. Other industries include metallurgy, food processing, and the manufacture of building materials. It was founded as a fortress in 1754.

Ki·rovsk (kē'rəfsk), city (1974 est. pop. 40,000), N European USSR, on the Kola Peninsula. The city is the center of a mining complex that produces apatite and nepheline, raw materials for the superphosphate and aluminum industries.

Kır·şe·hir (kĭr'shĕ-hēr'), city (1975 pop. 41,325), capital of Kırşehir prov., central Turkey. It is noted for its carpets.

Kirt·land Hills (kûrt'lənd), village (1970 pop. 452), Lake co., NE Ohio, NE of Cleveland; settled 1808–9, inc. 1926. The first Mormon temple was built here (1833–36) by Joseph Smith and his followers.

Ki·ru·na (kē'ro͞o-nä'), city (1975 est. pop. 26,542), Norrbotten co., N Sweden. The northernmost city in Sweden, it is the center of the Lapland iron-mining region. Kiruna became the most extensive city (c.5,500 sq mi/14,245 sq km) in the world in 1948, when several distant mining villages were incorporated into it.

Kir·yu (kĭr-yo͞o'), city (1976 est. pop. 134,137), Gumma prefecture, central Honshu, Japan. A major center of silk production since the 8th cent., it now manufactures rayon as well.

Ki·san·ga·ni (kē'sən-gä'nē), formerly **Stan·ley·ville** (stăn'lē-vĭl'), city (1974 est. pop. 311,000), N central Zaire, a port on the Congo River. Manufactures include metal goods and beer, and cotton and rice are shipped from the city. Founded in 1883 by the explorer Henry M. Stanley and originally located on a nearby island in the river, the city became the stronghold of Patrice Lumumba in the late 1950s. After the assassination of Lumumba in 1961, Antoine Gizenga set up a government that rivaled the central government in Leopoldville (now Kinshasa). Gizenga's regime was quashed in 1962, but in 1964, 1966, and 1967 the city was the site of temporarily successful revolts against the central government.

Ki·sa·ra·zu (kē-sä-rä'zo͞o), city (1976 est. pop. 100,131), Chiba prefecture, E central Honshu, Japan, on Tokyo Bay. It is a residential and industrial suburb of Tokyo.

Ki·se·levsk (kĭ-sĕ-lyôfsk'), city (1974 est. pop. 125,000), S Siberian USSR. It is a major coal-mining center in the Kuznetsk Basin and also manufactures mining machinery.

Kish (kĭsh), ancient city of Mesopotamia, in the Euphrates valley, E of the modern city of Hillah, Iraq. It was occupied from very ancient times, and its remains go back as far as the protoliterate period in Mesopotamia. In the early 3rd millennium B.C., Kish was a Semitic city. There is an excavated palace of Sargon I of Agade and a great temple built by Nebuchadnezzar and Nabonidus in the later Babylonian period. The site has also yielded a complete sequence of pottery from the Sumerian period to that of Nebuchadnezzar.

Ki·shi·nev (kĭ-shĭ-nyôf', -nĕf'), city (1977 est. pop. 489,000), capital of the Moldavian Soviet Socialist Republic, SW European USSR, on the Byk River, a tributary of the Dnestr. Major industries include food and tobacco processing, metalworking, and the manufacture of building materials, machinery, plastics, rubber, and textiles. Founded in the early 15th cent. as a monastery town, Kishinev was taken in the 16th cent. by the Turks and in 1812 by the Russians, who made it the center of Bessarabia. Rumania held the city from 1918 to 1940, when it was seized by the USSR.

Ki·shi·wa·da (kē-shē-wä'dä), city (1976 est. pop. 175,899), Osaka prefecture, SW Honshu, Japan, on Osaka Bay. It is an industrial and residential suburb of Osaka.

Kis·kun·fé·legy·há·za (kĭsh'ko͝on-fā'lĕd-yə-hä'zŏ), city (1977 est. pop. 27,300), S central Hungary. It is a road and rail junction; trade and industry are based on agricultural products.

Kis·lo·vodsk (kĭs-lə-vôtsk'), city (1976 est. pop. 97,000), S European USSR, in the N Caucasus mts. It is a famous health resort.

Kis·ma·yu (kĭs-mī'yo͞o), town (1968 est. pop. 18,000), SW Somalia, on the Indian Ocean. Kismayu was founded in 1872 by the sultan of Zanzibar and passed to Great Britain in 1887. In 1924 it was transferred to Italian control. The town has several mosques and a palace that was constructed by the sultan.

Kis·si·dou·gou (kē-sē-do͞o'go͞o), town (1961 est. pop. 12,000), S Guinea. It is a market town for an agricultural area that produces coffee, rice, palm products, kola nuts, and other crops. There are sawmills in the town, and diamonds are mined nearby.

Kis·sim·mee (kĭs-ĭm'ē), city (1975 est. pop. 12,500), seat of Osceola co., central Fla., on Lake Tohopekaliga S of Orlando. Among its industries are boatbuilding, fruit packing, sawmilling, and the manufacture of boxes and plastics.

Kissimmee, Lake, 55 sq mi (142.5 sq km), central Fla. The Kissimmee River, 140 mi (225.3 km) long, rises in small lakes and flows south through Lake Kissimmee to Lake Okeechobee. The lake and river region is a major U.S. cattle-raising area.

Kist·na (kĭst'nə) or **Krish·na** (krĭsh'nə), river, c.800 mi (1,290 km) long, rising in Maharashtra state, central India, in the Western Ghats, and flowing SE through Andhra Pradesh state to the Bay of Bengal. The river's source is sacred to Hindus.

Ki·su·mu (kē-so͞o'mo͞o), city (1971 est. pop. 35,000), capital of Nyanza prov., SW Kenya, on Kavirondo Gulf (an arm of Victoria Nyanza). It is the principal lake port of Kenya and the commercial center of a prosperous farm region. Manufactures include refined sugar, frozen fish, textiles, and processed sisal.

Ki·ta·kyu·shu (kē-tä'kyo͞o-sho͞o), city (1976 est. pop. 1,063,990), Fukuoka prefecture, N Kyushu, Japan, on the Shimonoseki Strait between the Inland Sea and the Korea Strait. Kitakyushu is one of Japan's most important manufacturing regions and one of its chief ports and railroad centers. It has a great variety of industries, the chief of which produce iron and steel, textiles, chemicals, machinery, ships, porcelain, and glass.

Ki·ta·mi (kē-tä'mē), city (1975 pop. 73,000), Hokkaido prefecture, NE Hokkaido, Japan, on the Tokoro River. It is an agricultural market and a major center for the production of peppermint.

Kit Car·son (kĭt kär'sən), county (1970 pop. 7,530), 2,171 sq mi (5,622.9 sq km), E Colo., in a grain and livestock area bordering on Kansas; formed 1889; co. seat Burlington. It is drained by branches of the Republican and Smoky Hill rivers.

Kitch·e·ner (kĭch'ə-nər), city (1971 pop. 111,804), S Ont., Canada, in the Grand River valley; settled by Mennonites in 1806. Its products include packaged meats, metal goods, and rubber products.

Kí·thi·ra (kē'thîr-ä) or **Cy·the·ra** (sə-thîr'ə), island (1971 pop. 3,961), c.109 sq mi (282 sq km), S Greece, in the Mediterranean Sea, southernmost of the Ionian Islands, off the S Peloponnesus. Mostly rocky with many streams, it produces wine, goat cheese, olives, corn, and flax. Ancient Kíthira was a center of the cult of Aphrodite. The island passed to Greece in 1864.

Kit·i·mat (kĭt'ĭ-măt), town (1971 pop. 11,803), W British Columbia, Canada, at the head of Douglas Channel. It is the site of a huge aluminum smelter. There are also pulp and paper mills.

Kit·sap (kĭt'săp'), county (1970 pop. 101,732), 393 sq mi (1,017.9 sq km), W Wash., on a peninsula bounded on the W by Hood Canal, on the E by Puget Sound; formed 1857; co. seat Port Orchard. It has commercial fisheries and diversified agriculture (fruit, nuts, dairy products, poultry, and truck crops).

Kit·tan·ning (kĭ-tăn'ĭng), industrial borough (1975 est. pop. 5,500), seat of Armstrong co., W Pa., on the Allegheny River NE of Pittsburgh; settled 1796, laid out 1804, inc. 1821. It has gas and oil wells, brick and clay works, and diversified agriculture.

Kit·ta·tin·ny Mountain (kĭt'ə-tĭn'ē), ridge of the Appalachian system, extending across NW N.J. from Shawangunk Mt., SE N.Y., to Blue Mt., E Pa. Kittatinny Mt. is a major resort and recreation area; the Appalachian Trail follows the ridge. The Delaware River cuts through the western part of the ridge forming Delaware Water Gap.

Kit·ter·y (kĭt'ə-rē), town (1970 pop. 11,028), York co., extreme SW Maine, at the mouth of the Piscataqua River opposite Portsmouth, N.H.; inc. 1647. Its economy centers around tourism and the Portsmouth Naval Shipyard, which is located on two islands and connected with Kittery by two bridges. The oldest town in Maine (settled c.1623), it grew as a trading, fishing, lumber-shipping, and shipbuilding center. John Paul Jones's ship *Ranger* (built in 1777), and the *Kearsarge* of Civil War fame were both built in Kittery. There are several 18th cent. houses in the town.

Kit·ti·tas (kĭt'ĭ-təs), county (1970 pop. 25,039), 2,315 sq mi (5,995.9 sq km), central Wash., in a mountainous area drained by the Yakima River; formed 1883; co. seat Ellensburg. It is primarily agricultural, yielding beef cattle, hay, and corn. Some dairying is done. There are deposits of coal, gold, and silica.

Kitt·son (kĭt'sən), county (1970 pop. 6,853), 1,124 sq mi (2,911.2 sq km), extreme NW Minn., bounded on the W by the Red River and the N.Dak. border, on the N by the Man., Canada, border; formed 1862; co. seat Hallock. Its agriculture includes wheat, grain, potatoes, livestock, and dairy products. Clothing is manufactured, and some mining is done.

Kit·ty Hawk (kĭt′ē hôk′), sandy peninsula, NE N.C., E of Albemarle Sound. Nearby is Kill Devil Hill, where the Wright brothers experimented successfully (1900–1903) with gliders and airplanes. Wright Brothers National Memorial, commemorating their first successful flight, is here.

Ki·twe (kē′twā), city (1976 est. pop. 281,000), N central Zambia; founded 1936. Copper is mined, and food products, clothing, and plastics are manufactured.

Kiu·kiang (jē-ō′jē-äng′): *see* Chiu-chiang, China.

Ki·vu (kē′vo͞o), lake, 1,042 sq mi (2,698.8 sq km), on the Zaire-Rwanda border, E central Africa; highest lake in Africa (4,788 ft/1,460.3 m). It is drained by the Ruzizi River, which flows south into Lake Tanganyika.

Ki·zel (kĭ-zyĕl′), city (1974 est. pop. 42,000), E European USSR, on the Kizel River and the W slopes of the Urals. It is a coal-mining and industrial center with coal-concentrating factories and plants that produce mining equipment, clothing, and food products. It was founded in the late 18th cent. when the coal mines of the Kizel basin were being developed.

Ki·zil (kĭ-zĭl′): *see* Kyzyl, USSR.

Kı·zıl A·da·lar (kĭ-zĭl′ ä-dä-lär′) or **Princ·es Islands** (prĭn′sĭz), group of nine small islands (1970 pop. 15,244), c.4 sq mi (10.4 sq km), NW Turkey, in the Sea of Marmara, near İstanbul. The islands were used as places of exile in Byzantine times.

Kızıl Ir·mak (ĭr-mäk′), longest river of Turkey, c.715 mi (1,150 km) long, rising in the Kızıl Dağ, N central Turkey, and flowing in a wide arc SW, then N, and then NE into the Black Sea. The river is an important source of hydroelectric power.

Kizil Kum (ko͝om), desert, USSR: *see* Kyzyl-Kum.

Klad·no (kläd′nô), city (1975 est. pop. 62,019), NW Czechoslovakia, in Bohemia. An industrial center of the Kladno coal-mining region, it has large iron and steel plants, and manufactures chemicals and machinery. Known in 973, Kladno grew rapidly with the opening of its first coal mine in 1846.

Kla·gen·furt (klä′gən-fo͝ort′), city (1973 pop. 82,512), capital of Carinthia prov., S Austria, on the Glan River. Situated in a mountain lake region, it is a noted winter sports center. Manufactures include machinery, textiles, and leather goods. Klagenfurt was chartered about the mid-13th cent. The city has a cathedral (16th cent.), a theological seminary, and several museums.

Klam·ath (klăm′əth), county (1970 pop. 50,021), 5,973 sq mi (15,470.1 sq km), S Oregon, in the Cascade foothills bordering in the S on Calif. and drained by the Klamath River; formed 1882; co. seat Klamath Falls. It is in a lumbering and irrigated agricultural area yielding potatoes, hay, alfalfa, and livestock. Crater Lake National Monument is here.

Klamath, river, c.265 mi (425 km) long, rising in Upper Klamath Lake in the Klamath Mts., SW Oregon and flowing generally SW across NW Calif. to the Pacific Ocean. Most of its course passes through national forests and wildlife refuges. The river is used for irrigation and power production.

Klamath Falls, city (1978 est. pop. 16,400), seat of Klamath co., SW Oregon, at the S tip of Upper Klamath Lake; settled 1867, inc. 1905. It is the processing and distributing center of a lumber, livestock, and farm area, and is a resort center. There is some manufacturing. The Klamath irrigation project (1900) and the coming of the railroad (1909) stimulated its growth from a hamlet to a thriving city.

Klay·pe·da (klī′pē-dä′), formerly **Me·mel** (mā′məl), city (1976 est. pop. 169,000), NW European USSR, in Lithuania, on the Baltic Sea, at the entrance to the Kursky Zaliv. An ice-free seaport and an industrial center, it has shipyards and industries producing textiles, fertilizers, and wood products. One of the oldest cities of Lithuania, Klaypeda was the site of a settlement as early as the 7th cent. It was conquered and burned in 1252 by the Teutonic Knights. The city was ceded (1629) by Prussia to Sweden but reverted to Prussia in 1635. In the Napoleonic Wars the city was (1807) the refuge and residence of Frederick William III of Prussia, who signed here the edict emancipating the serfs in his kingdom.

Kle·berg (klā′bûrg′), county (1970 pop. 33,166), 851 sq mi (2,204.1 sq km), S Texas, on the Gulf coast protected by Padre Island and indented by Baffin Bay; formed 1913; co. seat Kingsville. This cattle-ranching area includes part of the King Ranch and has deposits of oil and natural gas. Peanuts, cotton, truck crops, grain sorghums, and vegetables are grown here. There are fishing and bathing resorts along the coastline.

Klerks·dorp (klĕrks′dôrp), town (1970 pop. 63,558), Transvaal, NE Republic of South Africa, on the Schoonspruit River. The town, which has grain elevators, lumberyards, and food-processing and beverage-making industries, is the mining and processing center for major gold and uranium deposits and is also the distribution center for neighboring farms. Klerksdorp was founded in 1837 by Boer farmers. Gold mining began in 1886 but declined in the late 1890s. Heavy fighting occurred in the area during the South African War (1899–1902). Gold mining revived in 1932, and the town underwent an economic revival, which accelerated after World War II.

Klick·i·tat (klĭk′ə-tăt′), county (1970 pop. 12,138), 1,912 sq mi (4,952.1 sq km), S Wash., bounded on the S by the Columbia River and rising to the Cascades in the N; formed 1860; co. seat Goldendale. It is drained by the Klickitat River. Primarily agricultural (wheat, alfalfa, hay, and cattle), it also has a lumbering industry and an aluminum-producing plant.

Kłodz·ko (klôts′kô), town (1975 est. pop. 28,000), SW Poland. It is a commercial center with textile mills, metalworks, slate quarries, and sugar refineries. Founded in the 10th cent., it was capital of a county created in 1462. It was seized by Frederick II of Prussia in the War of the Austrian Succession and was ceded to Prussia in 1745.

Klon·dike (klŏn′dīk), region of Yukon Territory, NW Canada, just E of the Alaska border. It surrounds the Klondike River, a small stream that enters the Yukon River from the east at Dawson. The discovery in 1896 of rich placer gold deposits in Bonanza Creek, a tributary of the Klondike, caused the Klondike stampede of 1897–98. News of the discovery reached the United States in July, 1897, and within a month thousands of people were rushing north. Most landed at Skagway at the head of Lynn Canal and crossed by Chilkoot or White Pass to the upper Yukon, which they descended to Dawson. Others went in by the Copper River Trail or over the Teslin Trail by Stikine River and Teslin Lake, and some by the all-Canadian Ashcroft and Edmonton trails. The other main access route was up the Yukon River by steamer. With unexpected thousands in the region a food famine threatened, and supplies were commandeered and rationed. The population in the Klondike in 1898 was c.25,000. Thousands of others who did not find claims drifted down the Yukon and found placer gold in Alaskan streams, notably at Nome, to which there was a new rush. Others went back to the United States. Gold is still mined in the area.

Klo·ster·neu·burg (klô′stər-noi′bo͝orkh), city (1971 pop. 21,912), Lower Austria prov., NE Austria, on the Danube River and the N slope of the Wienerwald, near Vienna. Klosterneuburg was formed in 1938 through the merger of seven towns. It is the site of the oldest Augustinian monastery (consecrated 1136) in Austria.

Knife (nīf), river, c. 165 mi (265 km long), W central N.Dak., rising N of Fairfield and flowing E to the Missouri River just N of Stanton.

Knos·sos (nŏs′əs), ancient city: *see* Cnossus.

Knott (nŏt), county (1970 pop. 14,698), 356 sq mi (922 sq km), SE Ky., in the Cumberland foothills drained by Troublesome Creek; formed 1884; co. seat Hindman. It is in a mountainous coal-mining and agricultural area, with some manufacturing.

Knox (nŏks). **1.** County (1970 pop. 60,939), 728 sq mi (1,885.5 sq km), NW central Ill., drained by the Spoon River and Pope and Henderson creeks; formed 1825; co. seat Galesburg. It has agriculture (corn, wheat, oats, livestock, poultry, and dairy products), deposits of bituminous coal, clay, and gravel, and some manufacturing. **2.** County (1970 pop. 41,546), 516 sq mi (1,336.4 sq km), SW Ind., bounded on the W by the Wabash River and the Ill. border, on the E by the West Fork of the White River, on the S by the White River; formed 1790; co. seat Vincennes. In an agricultural area yielding fruit and grain, it has bituminous-coal mines, creameries, fruit-packing plants, and nurseries. **3.** County (1970 pop. 23,689), 373 sq mi (966.1 sq km), SE Ky., in the Cumberlands drained by the Cumberland River and several creeks; formed 1799; co. seat Barbourville. It has bituminous-coal mines, agriculture (livestock, fruit, vegetables, grain sorghums, corn, potatoes, hay, and tobacco), sawmills, and oil wells. **4.** County (1970 pop. 29,013), 369 sq mi (955.7 sq km), S Me., on Penobscot Bay in a coastal and island area drained by the St. George River; formed 1860; co. seat Rockland. The inland agricultural and lake region produces poultry, eggs, and apples. Limestone quarrying, fishing, packaging and shipping of fish, and manufacturing (textile-mill products, lumber products, machinery, and boats) are done. **5.** County (1970 pop. 5,692), 512 sq mi (1,326.1 sq km), NE Mo., drained by the Salt and Middle and South Fabius rivers; formed 1845; co. seat Edina. Its agriculture includes corn, oats, hay, and livestock. It has lumbering and some mining. Leather goods are manufactured. **6.** County (1970 pop. 11,723), 1,107 sq mi (2,867.1 sq km), NE Nebr., bounded on the N by the Missouri River and drained by the Niobrara River; formed 1857; co. seat Center. Grain, livestock, dairy products, and poultry are important to its economy. It also has some mining and manufacturing. **7.** County (1970 pop. 41,795), 532 sq mi (1,377.9 sq km), central Ohio, drained by the Kokosing and Mohican rivers and the North Fork of the Licking River; formed 1808; co. seat Mount Vernon. Basically rural, its economy is based on agriculture (livestock, dairy products, grain, and fruit) and light industry. **8.** County (1970 pop. 276,293), 508 sq mi (1,315.7 sq km), E Tenn., in the Great Appalachian Valley drained by the Tennessee and Clinch rivers; formed 1792; co. seat Knoxville. It has agriculture (corn, hay, tobacco, fruit, livestock, and dairy products), deposits of bituminous coal and zinc, marble quarries, a food-processing industry, and plants manufacturing clothing, chemicals, and metal products. **9.** County (1970 pop. 5,972), 854 sq mi (2,211.9 sq km), N Texas, in a plains area drained by the Brazos River and the North and South forks of the Wichita River; formed 1858; co. seat Benjamin. Its economy is based on agriculture, including cotton, truck crops, cattle, hogs, poultry, and dairy products.

Knox, city (1970 pop. 3,519), seat of Starke co., NW Ind., on the Yellow River SW of South Bend, in an agricultural area.

Knox·ville (nŏks'vĭl'). **1.** Town (1970 pop. 75), seat of Crawford co., central Ga., WSW of Macon, in an agricultural area. **2.** City (1970 pop. 7,755), seat of Marion co., S central Iowa, SE of Des Moines, in a farm and coal area; platted 1845, inc. 1853. **3.** City (1978 est. pop. 186,200), seat of Knox co., E Tenn., on the Tennessee River; inc. 1876. A port of entry, it is a major trade and shipping center for a farm, bituminous-coal, and marble area. Its industries include meat packing, tobacco marketing, and the manufacture of seat belts, plastics, textiles, and marble, wood, and metal products. The city is surrounded by mountains and lakes. A house was built on this site c.1785, followed by a fort and then a town. Knoxville twice (1796–1812 and 1817–18) served as the state capital. During the Civil War the area was torn by divided loyalties. Knoxville is headquarters of the Tennessee Valley Authority, and Norris Dam, from which the city procures its power, is nearby.

Ko·be (kō'bē), city (1976 est. pop. 1,363,992), capital of Hyogo prefecture, S Honshu, Japan, on Osaka Bay. A port city, it has shipbuilding yards, vehicle factories, iron and steel mills, sugar refineries, and chemical, rubber, and food-processing plants. Kobe has seven colleges and universities and many temples and shrines. Kobe was heavily bombed during World War II but has since been rebuilt.

Ko·blenz (kō'blĕnts), city (1974 est. pop. 119,295), Rhineland-Palatinate, W West Germany, at the confluence of the Rhine and the Moselle rivers. Its manufactures include machines, furniture, pianos, textiles, and printed materials. It is an important trade center for Rhine and Moselle wines. The city was founded (9 B.C.) by Drusus. It was prominent in Carolingian times as a residence of the Frankish kings. Koblenz was held by the archbishops of Trier from 1018 to the late 18th cent. In 1794 it was occupied by French troops and in 1798 was annexed by France. The city passed to Prussia in 1815. After World War I it was occupied by Allied troops from 1919 to 1929.

Ko·chi (kō'chē), city (1976 est. pop. 286,629), capital of Kochi prefecture, S Shikoku, Japan. The city exports dried bonito, ornamental coral, cement, and paper.

Ko·dai·ra (kō-dī'rä), city (1976 est. pop. 157,540), Tokyo Metropolis, central Honshu, Japan. It is a suburb of Tokyo.

Ko·di·ak (kō'dē-ăk'), city (1970 pop. 3,798), seat of Kodiak Island borough, S central Alaska, on the NE coast of Kodiak Island SSW of Anchorage. A salmon-fishing center, the area also produces cattle and timber.

Kodiak Island, borough (1970 pop. 9,409), 5,375 sq mi (13,921.3 sq km), S Alaska, in a generally hilly, grassy area; formed after 1961; borough seat Kodiak. It is in a fishing and seafood-processing region, with some agriculture, mining, and manufacturing.

Kodiak Island, 5,363 sq mi (13,890.2 sq km), off S Alaska, separated from the Alaska Peninsula by Shelikof Strait. Alaska's largest island, Kodiak is mountainous and heavily forested in the north and east; the native grasses in the south offer good pasturage for cattle and sheep. The island has many ice-free, deeply penetrating bays that provide sheltered anchorages and transportation routes. The Kodiak bear and the Kodiak king crab are native to the island. Most of the island is a national wildlife refuge. In 1912 the eruption of Mt. Katmai on the mainland blanketed the island with volcanic ash, causing widespread destruction and loss of life. Discovered in 1763 by a Russian fur trader, the island was the scene of the first permanent Russian settlement in Alaska (1784).

Ko·dok (kō'dŏk), town, SE Sudan, on the White Nile. In 1898 it was the scene of the Fashoda Incident, which brought Britain and France to the brink of war and resulted, in 1899, in an agreement establishing the frontier between the Sudan and the French Congo along the watershed between the Congo and Nile basins.

Ko·fo·ri·du·a (kō-fō-rē-do͞o'ä), town (1970 pop. 46,235), capital of the Eastern region, S Ghana. It is the commercial center for a region producing palm oil, cassava, and corn. Koforidua was founded (c.1875) by refugees from Ashanti.

Ko·fu (kō'fo͞o), city (1976 est. pop. 195,009), capital of Yamanashi prefecture, central Honshu, Japan. It is an industrial center, with manufactures of silk textiles and crystal ware, as well as a collection point for silk cocoons and raw silk.

Ko·ga·nei (kō-gä'nā), city (1976 est. pop. 102,489), Tokyo Metropolis, central Honshu, Japan. It is a suburb of Tokyo.

Ko·hat (kō-hät'), town (1972 pop. 48,091), N Pakistan, on the Kohat Toi River. The town, enclosed by a wall with 14 gates, is noted for its cotton fabrics and sarongs. Kohat contains a 19th cent. British fort built on the site of an old Sikh fortress.

Ko·kand or **Kho·kand** (kō-känt'), city (1976 est. pop. 152,000), Central Asian USSR, in Uzbekistan, in the Fergana Valley. It is a center for the manufacture of fertilizers, chemicals, machinery, and cotton and food products. Important since the 10th cent., Kokand became the capital of an Uzbek khanate that became independent of the emirate of Bukhara in the middle of the 18th cent. Kokand was taken by the Russians in 1876 and became part of Russian Turkistan. It was the capital (1917–18) of the anti-Bolshevik autonomous government of Turkistan.

Kok·ko·la (kôk'kô-lä), city (1975 est. pop. 22,096), Vaasa prov., W Finland, on the Gulf of Bothnia. It is a port with steel, engineering, and lumber industries. It was chartered in 1620.

Ko·ko·mo (kō'kə-mō'), city (1978 est. pop. 52,300), seat of Howard co., N central Ind., on Wildcat Creek; inc. 1865. Radios, automobile parts, and metal products are manufactured here.

Ko·ko Nor (kō'kō' nôr'), salt lake, c.1,625 sq mi (4,210 sq km), in the Tibetan highlands, NE Tsinghai prov., China; one of the largest lakes in China. At an altitude of 10,515 ft (3,207 m), it is shallow and brackish and of little economic value.

Ko·ku·bun·ji (kō-ko͞o'bo͞on-jē), city (1975 pop. 88,159), Tokyo Metropolis, central Honshu, Japan. It is a suburb of Tokyo and is noted for its Buddhist temple founded in 1588.

Ko·la Peninsula (kō'lə, kô'-), peninsula, c.50,000 sq mi (129,500 sq km), NW European USSR, in Murmansk oblast. Forming an eastern extension of the Scandinavian peninsula, it lies between the Barents Sea to the north and the White Sea to the south. In the northeastern part are tundras; the southwestern area is forested. The peninsula has rich mineral deposits.

Ko·lar (kō-lär'), city (1971 pop. 43,345), Karnataka state, SW India. Founded in the late 19th cent., it is the center of the Indian gold-mining industry. The first hydroelectric project in southern India was built in 1902 to provide electricity for the gold fields.

Kol·ding (kôl'dĭng), city (1976 est. pop. 54,434), Vejle co., S central Denmark, a port on Kolding Fjord, an arm of the Lille Baelt. It is a commercial, industrial, and fishing center. Of note in the city are Koldinghus, a royal castle built in 1248, and the oldest stone church (built in the 13th cent.) in Denmark.

Kol·guy·ev (kəl-go͞o'yĭf), island, 1,350 sq mi (3,496.5 sq km), off NE European USSR, in the Barents Sea, E of the Kanin peninsula. It is a tundra region, and the inhabitants engage in fishing, seal hunting, reindeer raising, and trapping.

Kol·ha·pur (kōl'hä-po͝or'), former princely state, 3,219 sq mi (8,337.2 sq km), Maharashtra state, SW India. Largely agricultural, the region produces cotton and textiles. It also has large bauxite deposits. A center of the Mahrattas, Kolhapur was an important state of the Deccan. It was transferred to Maharashtra state in 1960. The city of Kolhapur (1971 pop. 259,068) was the capital of the former state. It occupies the site of an ancient Buddhist center.

Ko·lín (kô'lēn), city (1975 est. pop. 29,808), N central Czechoslovakia, in Bohemia, on the Elbe River. It is a river port and manufactures railroad cars, chemicals, light machinery, and metal products. The city also has a petroleum refinery and a hydroelectric station. Founded in the 13th cent., Kolín grew rapidly after the construction (19th cent.) of the Vienna-Prague railway.

Kol·mar (kōl'mär, kôl-mär'): *see* Colmar, France.

Köln (kœln): *see* Cologne, West Germany.

Ko·łob·rzeg (kô-lôb'zhĕk) or **Kol·berg** (kôl'bĕrKH), town (1975 est. pop. 31,800), NW Poland, on the Baltic Sea at the mouth of the Prośnica River. It is a seaport, seaside resort, and rail junction. A salt-trading center in the Middle Ages, it was chartered in 1255. It was besieged three times by the Russians in the Seven Years' War before it fell in 1761.

Ko·lom·na (kŏ-lôm'nä), city (1976 est. pop. 144,000), central European USSR, at the confluence of the Moskva and Oka rivers. Locomotives and machine tools are produced. Known in 1177, the city became a Muscovite outpost in 1301 and has been an industrial center since 1863.

Ko·lo·my·ya (kə-lə-mĭ'yə), city (1976 est. pop. 50,000), SW European USSR, in the Ukraine, on the Prut River in the Carpathian foothills. Industries include food processing, woodworking, oil refining, and the manufacture of building materials, farm machinery, and textiles. First mentioned in 1240, Kolomyya passed in the 14th cent. to the Poles. It was taken by Austria during the Polish partition of 1772 and became part of the newly independent republic of Ukraine in 1918 but reverted to Poland in 1920. It was incorporated into the Ukrainian Soviet Socialist Republic in 1939.

Kol·we·zi (kōl-wĕz'ē), city (1968 est. pop. 71,000), Shaba region, SE Zaire. It is a center for copper and cobalt mining. There are copper-ore concentration plants, a zinc refinery, and a brewery.

Ko·ly·ma (kŏ-lĭ-mä'), river, c.1,500 mi (2,415 km) long, rising in several headstreams in the Kolyma and Cherskogo ranges, Far Eastern USSR. It flows generally north to the Arctic Ocean at Nizhniye Kresty. Its upper course crosses the rich Kolyma Gold Fields. Gold mining was begun in the 1930s, and both the fields and the surrounding area were developed with the use of forced labor. The Kolyma Range, east of the Kolyma River, extends northeast from Magadan and rises to c.6,000 ft (1,830 m).

Ko·ma·ki (kō-mä'kē), city (1975 pop. 97,445), Aichi prefecture, Honshu, Japan, on the Nobi Plain. It is a suburb of Nagoya and an agricultural market.

Ko·man·dor·ski Islands (kŏm'ən-dôr'skē) or **Com·mand·er Islands** (kə-măn'dər), group of treeless islands, off E Kamchatka Peninsula, E Far Eastern USSR, in SW Bering Sea. These hilly, foggy islands often have earthquakes. Their inhabitants, Russians and Aleuts, are engaged in fishing, hunting, and whaling.

Ko·már·no (kô'mär-nô) or **Ko·már·om** (kô'mä-rôm), city of Czechoslovakia and Hungary, on both sides of the Danube, at its confluence with the Nitra and Váh rivers. Komárno (1970 pop. 27,031) is located on the left bank and belongs to Czechoslovakia. It is a shipbuilding center and has flour mills and machinery and textile plants. Hungarian Komárom (1968 est. pop. 26,800), on the right bank, has lumber yards, sawmills, and textile plants. The site of Komárno was fortified by the Romans. It became a free city in 1331. Later a part of the Austro-Hungarian Monarchy, it was partitioned in 1920 between Hungary and Czechoslovakia.

Ko·ma·tsu (kō-mä'tso͞o), city (1976 est. pop. 101,232), Ishikawa prefecture, central Honshu, Japan. It is a flourishing market town noted for its production of silk, rayon, and pottery.

Ko·mi Autonomous Soviet Socialist Republic (kō'mē), autonomous region (1970 pop. 965,000), c.160,000 sq mi (414,400 sq km), NE European USSR; capital Syktyvkar. The region is a wooded lowland, stretching across the Pechora and the Vychegda river basins and the upper reaches of the Mezen River. The northern part is permanently frozen, wooded tundra. Mining is the most important economic activity. There is also extensive lumbering, stock raising, fishing, and hunting. The area belonged to the Novgorod Republic from the 13th cent. The Zyrian Autonomous Oblast was constituted in 1921; it became an autonomous republic in 1936.

Ko·mi-Perm·yak National Okrug (kō'mē-pĭrm'yäk), administrative division (1970 pop. 212,000), 12,664 sq mi (32,800 sq km), E central European USSR, in the basin of the upper Kama River. The terrain is slightly hilly and heavily forested. Lumbering is the major industry of the okrug. Among the crops grown are rye, oats, spring wheat, and flax. The territory is the oldest and most populous of the USSR's national okrugs, which was established in 1925.

Ko·mo·ti·ní (kô-mə-tĭ-nē'), city (1971 pop. 28,896), capital of Rodhópi prefecture, NE Greece, in Thrace. It is the commercial center for a region that produces grains, silk, and tobacco.

Kom·pong Cham (käm'pông' chäm), city (1967 est. pop. 31,000), capital of Kompong Cham prov., SE Cambodia, a port on the Mekong River. It has a large textile factory, built with aid from China. In Sept., 1973, it was the scene of heavy fighting as government forces, reinforced and supplied via the Mekong River, withstood a massive attack by the Khmer Rouge.

Kompong Som (sôm), formerly **Si·ha·nouk·ville** (sē-hä'nək-vĭl'), city and seaport (1962 pop. 6,578), located in, but politically independent of Kampot prov., S Cambodia, on the Gulf of Siam. Although a new city (completed 1960), it is the principal deepwater port and commercial outlet of Cambodia.

Kom·so·molsk (kŏm-sŏ-mŏlsk'), city (1976 est. pop. 246,000), Khabarovsk Kray, S Far Eastern USSR, on the Amur River. It is a manufacturing center producing steel, refined oil, and wood products. Tin mines are nearby. The city was founded (1932) by the Komsomol, the Communist youth organization.

Ko·na (kō'nə), district, along the W coast of the island of Hawaii. It is Hawaii's coffee belt and the only coffee-producing area in the United States. The Kona coast, with fine deep-sea fishing offshore, is a favorite tourist spot.

Kong Karls Land (kông kärlz), island group, 128 sq mi (331.5 sq km), in the Barents Sea, part of the Norwegian possession of Svalbard, W of Spitsbergen.

Kon·go, kingdom of the (kŏng'gō), former state of W central Africa, founded in the 14th cent. In the 15th cent. the kingdom stretched from the Congo River in the north to the Loje River in the south and from the Atlantic Ocean in the west to beyond the Kwango River in the east. In 1491 Portuguese missionaries, soldiers, and artisans were welcomed at Mbanza, the capital of the kingdom. The missionaries soon gained converts, including Nzinga Nkuwu (who took the name João I), and the soldiers helped him defeat an internal rebellion. The next ruler, Afonso I (reigned 1505–43), was raised as a Christian and attempted to convert the kingdom to Christianity and European ways. However, the Portuguese residents in the Kongo were primarily interested in increasing their private fortunes (especially through capturing black Africans and selling them into slavery). After the death of Afonso, the Kongo declined rapidly and suffered major civil wars. In 1641 Manikongo Garcia II allied himself with the Dutch in an attempt to control Portuguese slave traders, but in 1665 a Portuguese force decisively defeated the army of the Kongo. The kingdom disintegrated into a number of small states, all controlled to varying degrees by the Portuguese. The area of the Kongo was incorporated mostly into Angola and partly into the Independent State of the Congo in the late 19th cent.

Kongs·berg (kôngs'bĕr), city (1977 est. pop. 20,192), Buskerud co., SE Norway, on the Lågen River. It is a commercial, industrial, and winter sports center and has a hydroelectric power plant. Kongsberg has old silver mines that are tourist attractions.

Kö·nigs·berg (ko͞o'nĭgz-bûrg', kœ'nĭKHs-bĕrKH): *see* Kaliningrad, USSR.

Kö·niz (kœ'nĭts), town (1977 est. pop. 34,200), Bern canton, W central Switzerland. It is a suburb of Bern. The Romanesque-Gothic church, founded in the 10th cent. by Rudolph II of Burgundy, has noteworthy frescoes.

Ko·no·top (kŏn'ə-tŏp'), city (1976 est. pop. 76,000), central European USSR, in Ukraine, on the Ezuch River. It is a rail junction and agricultural center, with food-processing plants, an electromechanics industry, and factories that produce clothing and mining equipment. Konotop was founded in 1634 by Poles.

Kon·stan·ti·nov·ka (kŏn'stən-tyē'nəf-kə), city (1976 est. pop. 111,000), S central European USSR, in the Donets Basin of the Ukraine. It is a zinc-refining and superphosphate-producing center.

Kon·stanz (kôn'stänts): *see* Constance, West Germany.

Kon·ya (kôn'yä), city (1975 pop. 246,381), capital of Konya prov., S central Turkey. It is the trade center of a rich agricultural and livestock-raising region. Manufactures include cement, carpets, and leather, cotton, and silk goods. The city was important in Roman times, but it reached its peak after the victory (1071) of Alp Arslan over the Byzantines at Manzikert, which resulted in the establishment (1099) of the sultanate of Iconium or Rum (so called after Rome), a powerful state of the Seljuk Turks. In the late 13th cent. the Seljuks were defeated by the Mongols, and their territories subsequently passed to Karamania. In the 15th cent. the whole region was annexed to the Ottoman Empire. Konya lost its political importance but remained a religious center as the chief seat of the whirling dervishes, whose order was founded here in the 13th cent. The tomb of the founder, several medieval mosques, and the old city walls have been preserved.

Koo·chi·ching (ko͞o'chə-chĭng), county (1970 pop. 17,131), 3,129 sq mi (8,104.1 sq km), N Minn., in a marshland area bounded on the N by Ont., Canada, and the Rainy River and drained by the Big Fork and Little Fork rivers; formed 1906; co. seat International Falls. It is in a resort, lumbering, and agricultural (cattle, dairy products, and truck crops) area. Its main industry is paper milling.

Ko·o·lau Range (kō'ə-lou'), mountain chain, extending NW to SE, E Oahu Island, Hawaii, rising to 3,105 ft (947 m) in Konahuanui. It is cut by two scenic passes, Nuuanu Pali and Waimanalo Pali.

Koo·te·nai (ko͞ot'n-ā'), county (1970 pop. 35,332), 1,249 sq mi (3,234.9 sq km), N Idaho, in a rolling, wooded area bordering on Wash. and watered by Coeur d'Alene Lake and the Spokane and Coeur d'Alene rivers; formed 1864 co. seat Coeur d'Alene. Lumbering, dairying, and agriculture (pears, cherries, apples, and wheat) are the primary industries.

Kootenai or **Koo·te·nay** (ko͞ot'n-ā'), river, 407 mi (654.9 km) long, rising in the Rocky Mts., SE British Columbia, Canada, and flowing S into NW Mont., NW through N Idaho, then N into Canada. There it flows through Kootenay Lake (64 mi/103 km long; 191 sq mi/494.7 sq km), an expansion of the river, before joining the Columbia River at Castlegar.

Ko·per (kô'pĕr), town (1971 pop. 35,407), NW Yugoslavia, in Slovenia, on the Istrian peninsula in the Gulf of Trieste. It is a fishing port and has small shipyards. From 1278 until 1797 the town was the capital of Istria under Venetian rule. The Treaty of Campo Formio transferred Koper to Austria. The town passed to Italy after World War I and became part of the Free Territory of Trieste in 1947. In 1954 Koper was annexed to Yugoslavia.

Kö·ping (chœ'pĭng'), city (1975 est. pop. 21,700), Västmanland co., S central Sweden, at the W end of Lake Mälaren. It is an important lake port and a commercial and industrial center. Manufactures include machinery, textiles, and cement.

Ko·rat (kō-rät'): *see* Nakhon Ratchasima, Thailand.

Kor·çë (kôr'chə), city (1971 est. pop. 47,900), capital of Korçë prov., SE Albania, near the Greek border. Located in an agricultural region, it is a commercial and industrial center producing leather, tobacco and glass products, and knitwear. There are lignite, copper, and iron ore deposits nearby. Known in 1280, it was destroyed (1440) by the Turks but developed again after the 16th cent. Ever since Albania gained independence in the Balkan Wars, Korçë has been claimed by Greece. Greek troops occupied it in 1912-13 during the Balkan Wars and again early in World War I. From 1916 to 1920 it was occupied and administered by the French, and in World War II it was held (Nov., 1940-Apr., 1941) by the Greeks.

Kor·ču·la (kôr'cho͞o-lä), island (1971 pop. 20,176), 105 sq mi (272 sq km), in the Adriatic Sea, off Dalmatia, W Yugoslavia. It is covered with pine forests, pastures, and vineyards. Most of the inhabitants are sailors, farmers, or fishermen. The island was colonized by the Greeks in the 4th cent. B.C.

Ko·re·a (kō-rē'ə, kô-), historic region (85,049 sq mi/220,277 sq km), E Asia. Seoul was the traditional capital. A peninsula, 600 mi (965.4 km) long, Korea separates the Yellow Sea (and Korea Bay, a northern arm of the Yellow Sea) on the west from the Sea of Japan on the east. On the south it is bounded by Korea Strait (connecting the

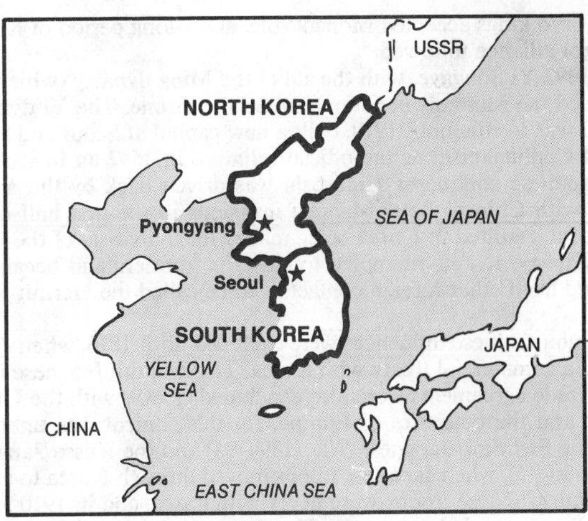

Yellow Sea and the Sea of Japan), and on the north its land boundaries with China and with the USSR are marked chiefly by the Yalu and Tumen rivers. The Korean peninsula is largely mountainous. Off the heavily indented coast lie some 3,420 islands, most of them rocky and uninhabited.

Economy. The country once had large timber resources. Most of the remaining stands are in the north. The south, on the other hand, is largely deforested—the result of illegal cutting after 1945 and damage during the Korean War (1950-53). Korea has great mineral wealth, most of it concentrated in the north. Of the peninsula's five major minerals—gold, iron ore, coal, tungsten, and graphite—only the tungsten and amorphous graphite are found principally in the south. North Korea is especially rich in iron and coal and also has deposits of graphite, gold, tungsten, magnesite, zinc, barites, magnatite, molybdenum, limestone, mica, and fluorite. Because of the mountainous and rocky terrain, only about 20% of Korean land is arable. Rice is the chief crop. Barley, wheat, corn, soybeans, and grain sorghums are also extensively cultivated, especially in the uplands; other crops include cotton, tobacco, fruits, potatoes, beans, and sweet potatoes. Livestock plays a minor role in Korean agriculture, especially in the north. In the south, cattle are used largely as beasts of burden. The fishing waters off Korea are among the best in the world. Deep-sea fishing is expanding, and Korean ships now range into the Atlantic and Arctic Oceans. Almost all of the deep-sea catch (consisting largely of tuna) is canned and exported.

The Korean economy was shattered by the war of 1950 to 1953. Postwar reconstruction was abetted by enormous amounts of foreign aid and intensive government economic development programs. Major North Korean products include iron, steel, and other metals, machinery, textiles (synthetics, wool, cotton, silk), and chemicals. In the south the traditional consumer goods industries (textiles, garments, food-processing) are still dominant, but heavy industry has been established and a great variety of products are now manufactured; these include electrical and electronic equipment, chemicals, ceramic goods, and plywood.

History. Chinese and Japanese influences have been strong throughout Korean history, but the Koreans, descended from Tungusic tribal peoples, are a distinct racial and cultural group. The documented history of Korea begins in the 12th cent. B.C., when a Chinese scholar, Ki-tze (Kija), founded a colony at Pyongyang. After 100 B.C. the Chinese colony of Lolang, established near Pyongyang, exerted a strong cultural influence on the Korean tribes settled in the peninsula. The kingdom of Koguryo, the first native Korean state, arose in the north near the Yalu River in the 1st cent. A.D., and by the 4th cent. it had conquered Lolang. In the south, two kingdoms emerged, that of Paekche (A.D. c.250) and the powerful kingdom of Silla (A.D. c.350). With Chinese support, the kingdom of Silla conquered Koguryo and Paekche in the 7th cent. and unified the peninsula. Under Silla rule, Korea prospered and the arts flourished; Buddhism, which had entered Korea in the 4th cent., became dominant in this period. In 935 the Silla dynasty was peacefully overthrown by Wang Kon, who established the Koryo dynasty. During the Koryo period, literature was cultivated, and although Buddhism remained the state religion, Confucianism—introduced from China during the Silla years—controlled the pattern of government. In 1231 Mongol forces invaded from China, initiating a war that was waged intermittently for some 30 years. Peace came when the Koryo kings accepted Mongol rule, and a long period of Koryo-Mongol alliance followed.

In 1392 Yi Songgye, with the aid of the Ming dynasty (which had replaced the Mongols in China), seized the throne. The Yi dynasty, which was to rule until 1910, built a new capital at Seoul and established Confucianism as the official religion. In 1592 an invasion of the Japanese conqueror Hideyoshi was driven back by the Yi dynasty with Chinese help. Manchu invasions in the first half of the 17th cent. resulted in Korea being made (1637) a vassal of the Manchu dynasty. Korea attempted to close its frontiers and became so isolated from other foreign contact as to be called the Hermit Kingdom.

All non-Chinese influences were excluded until 1876, when Japan forced a commercial treaty with Korea. To offset the Japanese influence, trade agreements were also concluded (1880s) with the United States and the countries of Europe. Japan's control was tightened after the first Sino-Japanese War (1894–95) and the Russo-Japanese War (1904–5), when Japanese troops moved through Korea to attack Manchuria. These troops were never withdrawn, and in 1910 Japan formally annexed the country. The Japanese instituted vast social and economic changes, building modern industries and railroads, but their rule (1910–45) was harsh and exploitative.

In World War II, at the Cairo Conference (1943), the United States, Great Britain, and China promised Korea independence. At the end of the war Korea was arbitrarily divided into two zones as a temporary expedient; Soviet troops were north and Americans south of the line of lat. 38° N. The Soviet Union thwarted all UN efforts to hold elections and reunite the country under one government. When relations between the Soviet Union and the United States worsened, trade between the two zones ceased. In 1948 two separate regimes were formally established—the Republic of Korea (1975 pop. 34,688,079; capital Seoul) in the south, and the Democratic People's Republic (1976 est. pop. 17,000,000; capital Pyongyang) in the north. By mid-1949 all Soviet and American troops were withdrawn. In June, 1950, the North Korean army launched a surprise attack against South Korea, initiating the Korean War, and with it, severe hardship, loss of life, and enormous devastation. After the war the boundary was stabilized along a line running from the Han estuary northeast across the 38th parallel, with a "no-man's land," 1.24 mi (2 km) wide and occupying a total of 487 sq mi (1,261.3 sq km), on either side of the boundary. In 1971 negotiations between North and South Korea provided the first hope for peaceful reunification; in Nov., 1972, an agreement was reached for the establishment of joint machinery to work toward unification.

Kor·ho·go (kôr-hō'gō), town (1967 est. pop. 30,000), N Ivory Coast. It is a processing center for a mountainous region where cotton, kapok, rice, millet, groundnuts, maize, and yams are grown and sheep and goats are raised. Diamonds are mined in the area.

Kó·rin·thos (kôr'ĭn-thôs): *see* Corinth, Greece.

Ko·ri·ya·ma (kō-rē'yä-mä), city (1976 est. pop. 183,700), Fukushima prefecture, N Honshu, Japan, on the Abukuma River. It is a major commercial and communications center with industries producing textiles, electrical appliances, and food products.

Kö·rös (kœ'rœsh), river, c.345 mi (555 km) long, formed in E Hungary by the junction of three headstreams that rise in Transylvania, NW Rumania. It meanders west through farmland to the Tisza River at Csongrád. The Körös is used for irrigation.

Kor·sør (kôr-sœr'), city (1970 com. pop. 19,864), Vestsjaelland co., S central Denmark, a seaport on the Store Baelt. In the city are fisheries and factories producing glass and processed food.

Kort·rijk (kôrt'rīk), city (1977 pop. 75,535), West Flanders prov., SW Belgium, on the Leie River. It is an important linen and textile-manufacturing center. Kortrijk was one of the earliest (14th cent.) and most important cloth-manufacturing towns of medieval Flanders. The Church of Notre Dame (13th cent.) in the city contains Anthony Van Dyck's *Elevation of the Cross* (1631).

Kós (kŏs, kôs), island (1971 pop. 16,650), 111 sq mi (287.5 sq km), SE Greece, in the Aegean Sea, one of the Dodecanese Islands. Fishing and sponge diving are important occupations. Grain, tobacco, olive oil, and wine are produced, and cattle, horses, and goats are raised. Kós has mineral deposits and several sulfur springs. In ancient times the island was controlled in turn by Athens, Macedon, Syria, and Egypt. A cultural center, it was the site of a school of medicine founded in the 5th cent. B.C. by Hippocrates. Kós later enjoyed great prosperity as a result of its alliance with the Ptolemaic dynasty of Egypt, which valued the island as a naval base. The island became part of modern Greece in 1947.

Kos·ci·us·ko (kŏs'ē-ŭs'kō), county (1970 pop. 48,127), 538 sq mi (1,393.4 sq km), N Ind., drained by the Tippecanoe and Eel rivers; formed 1835; co. seat Warsaw. It is the center of the state's lake region. Agriculture (livestock, poultry, dairy foods, grain, truck crops, and soybeans) and lumbering are its major occupations.

Kosciusko, city (1970 pop. 7,266), seat of Attala co., central Miss., NE of Jackson, in a cotton, dairy, and timber area; settled in the early 1830s on the old Natchez Trace, inc. 1836. It is a processing center and manufactures textiles, metal products, and flour.

Kosciusko, Mount, 7,316 ft (2,231.4 m) high, SE New South Wales, Australia, in the Australian Alps. It is the highest peak of Australia.

Ko·shi·ga·ya (kō-shē'gä-yä), city (1976 est. pop. 202,000), Saitama prefecture, central Honshu, Japan, on the Motoara River. It is a suburb of Tokyo and is noted for its peach orchards.

Kosh·tan-Tau (kôsh'tän-tou'), peak, c.16,880 ft (5,150 m) high, Kabardino-Balkar Autonomous SSR, S European USSR, in the central Greater Caucasus.

Ko·ši·ce (kô'shĭ-tsĕ), city (1975 est. pop. 174,388), E Czechoslovakia, in Slovakia. It is a major industrial center and transportation hub and a market for the surrounding agricultural area. The city's industries include food processing, brewing and distilling, printing, and the manufacture of machinery, cement, and ceramics. Originally a fortress town, Košice was chartered in 1241 and became an important trade center during the Middle Ages. It was frequently occupied by Austrian, Hungarian, and Turkish forces. By the Treaty of Trianon (1920) the city passed from Hungary to Czechoslovakia.

Kos·suth (kə-sōōth'), county (1970 pop. 22,937), 979 sq mi (2,535.6 sq km), N Iowa, on the Minn. border in a rolling prairie agricultural area drained by the East Des Moines River; formed 1851; co. seat Algona. Cattle, hogs, poultry, corn, oats, and soybeans are major products. It has sand and gravel pits.

Ko·stro·ma (kŏ-strŏ-mä'), city (1976 est. pop. 247,000), capital of Kostroma oblast, E European USSR, on the Volga at the mouth of the Kostroma River. It is a major linen-milling center. Metallurgy, ship repair, and the production of machinery, plywood, and footwear are also important. Founded in 1152, it became an important commercial center after it was annexed by Moscow in 1364.

Ko·sza·lin (kô-shä'lēn), city (1975 est. pop. 77,600), NW Poland, near the Baltic Sea. Its industries produce canned fish, metal products, and chemicals. Founded in 1188, it prospered from the 14th to the 16th cent., but suffered greatly in the Thirty Years' War. The city was transferred from Germany to Poland in 1945.

Ko·ta (kō'tə), city (1971 pop. 213,005), Rajasthan state, NW India, on the Chambal River. Kota, enclosed by a massive wall, is a market for sugar cane, oilseed, and building stone. The Mathureshi temple is the most famous of Kota's many temples.

Kota Ba·ha·ru (bə-här'ōō) or **Kota Bah·ru** (bär'ōō), city (1971 pop. 55,052), capital of Kelantan state, Malaysia, central Malay Peninsula, on the South China Sea at the mouth of the Kelantan River. It is a modern city with an important power station. It was seized by Japan (Dec. 10, 1941) early in the campaign against Singapore during World War II.

Kota Ki·na·ba·lu (kĭn'ə-bə-lōō'), town (1971 pop. 41,830), capital of Sabah, Malaysia, in N Borneo on a small inlet of the South China Sea. It was founded in 1899 and in 1947 replaced Sandakan as the capital of what was then British North Borneo.

Ko·tel·ny Island (kō-tĕl'nē), largest island of the Anjou group of the New Siberian Islands, c.100 mi (160 km) long and c.60 mi (100 km) wide, off N Siberian USSR. The island was discovered in 1773.

Kö·then (kœ'tən), city (1974 est. pop. 35,451), Halle district, central East Germany. There are lignite mines, sugar refineries, textile mills, and chemical factories in the city. Köthen was from 1603 to 1847 the residence of the dukes of Anhalt-Köthen, at whose court Johann Sebastian Bach was musical director from 1717 to 1723.

Kot·ka (kŏt'kä), city (1975 est. pop. 34,026), Kymi prov., SE Finland, on the Gulf of Finland. It is a major export center for paper, pulp, and timber and has chemical industries. It was chartered in 1878.

Ko·tor (kô'tôr), town (1971 pop. 32,439), SW Yugoslavia, in Montenegro, on the Bay of Kotor, an inlet of the Adriatic. It is a seaport and a tourist center. The town was colonized by Greeks (3rd cent. B.C.) and later belonged to the Roman and Byzantine empires. In 1797 it passed to Austria and became an important naval base; in 1918 it was transferred to Yugoslavia.

Kott·bus (kŏt'bəs, kôt'bo͝os): *see* Cottbus, East Germany.

Kot·ze·bue (kŏt'sə-bo͞o'), city (1970 pop. 1,696), NW Alaska, on Kotzebue Sound at the tip of Baldwin Peninsula; inc. 1958. A regional trade center, Kotzebue has a tourist industry. The city, set on a tundra, began in the 18th cent. as an Eskimo trading post for arctic Alaska and part of Siberia.

Kountze (ko͞onts), city (1970 pop. 2,173), seat of Hardin co., E Texas, NNW of Beaumont, in the pine-forested coastal plain. It is primarily a lumber-milling center, though rice farming and cattle raising are also important.

Ko·vel (kō'vəl, kô'vĭl), city (1974 est. pop. 40,000), W European USSR, in the Ukraine, on the Tura River. A rail junction and agriculture center, it has food and peat processing plants, railroad shops, and sewing, flax, and woodworking industries. First mentioned in the 14th cent., Kovel belonged to Lithuania and passed to Poland when the two states were united in 1569. The city was taken by Russia in 1795. It was again under Polish rule from 1921 to 1945, when it reverted to the USSR.

Kov·rov (kŏv-rôf'), city (1976 est. pop. 138,000), central European USSR, on the Klyazma River. Kovrov is an industrial center that produces excavating machines, linen textiles, and machine tools.

Ko·ya (kō'yä), peak, 2,858 ft (871.7 m) high, S Honshu, Japan. On its summit is a Buddhist monastery, founded in 816. The monastery has 120 temples and is visited by more than a million pilgrims annually.

Koy·u·kuk (kī'ə-kŭk'), river, rising on the S slopes of the Brooks Range, N Alaska, and flowing c.500 mi (805 km) SW to the Yukon River near the village of Koyukuk.

Ko·zhi·kode (kō'zhə-kōd'): *see* Calicut, India.

Kra, Isthmus of (krä), narrow neck of the Malay Peninsula, c.40 mi (60 km) wide, SW Thailand, between the Bay of Bengal and the Gulf of Siam. It has long been the proposed site of a ship canal that would bypass the congested Straits of Malacca.

Kra·ka·to·a (krä'kə-tō'ə), island, c.5 sq mi (13 sq km), W Indonesia, in Sunda Strait between Java and Sumatra. A volcanic explosion in 1883 blew up most of the island and altered the configuration of the strait; the accompanying tsunami caused great destruction and loss of life along the nearby coasts of Java and Sumatra. So great was the outpouring of ashes and lava that new islands were formed and debris was scattered across the Indian Ocean as far as Madagascar. Since then there have been lesser eruptions.

Kra·ków (krä'kou'), city (1975 est. pop. 684,600), S Poland, on the Vistula. A river port and industrial center, it has varied manufactures including metals, machinery, electrical equipment, and chemicals. One of eastern Europe's largest iron and steel plants is in the city. Founded c.700, Kraków became (1320) the residence of the kings of Poland. The Kraków fire (1595) caused the transfer (1596) of the royal residence to Warsaw, but the kings were crowned and buried in Kraków until the 18th cent. The city passed to Austria in 1795 and was included (1809) in the grand duchy of Warsaw. In 1815 the Congress of Vienna made the city and its vicinity into the republic of Kraków, a protectorate of Russia, Prussia, and Austria, and in 1846 it was included in Austria. The city reverted to Poland in 1919. Its university, founded in 1364, has long been a leading European center of learning; Copernicus was one of its students. The city has some 50 old churches, many of which contain works of art.

Kra·ma·torsk (krä-mä-tôrsk'), city (1976 est. pop. 167,000), S central European USSR, in the Donets Basin of the Ukraine. It is an iron and steel center with factories that produce equipment for coal-mining and chemical industries.

Kras·no·dar (krăs'nə-där'), city (1977 est. pop. 552,000), capital of Krasnodar Kray, SE European USSR, on the Kuban River. A river port and railroad junction, it has petroleum refineries and machinery, metalworking, textile, chemical, and food-processing plants. Founded in 1794 by Cossacks upon orders from Catherine II, it served as a military center protecting Russia's Caucasian frontier.

Krasnodar Kray, administrative division (1970 pop. 4,511,000), 32,317 sq mi (83,701 sq km), SE European USSR, extending E from the Sea of Azov and the Black Sea into the Kuban steppe and straddling the northwestern end of the Greater Caucasus; capital Krasnodar. It is one of the USSR's principal tobacco-growing regions. The subtropical Black Sea littoral produces fruit, tea, and wine and is dotted with health resorts. There are petroleum, gas, machinery, cement, and lumber industries. The area north of the Kuban belonged to the Crimean Khanate and was annexed by Russia in 1783. The Black Sea littoral was ceded to Russia by Turkey in the Treaty of Adrianople (1829). The remainder, known as Circassia, was annexed in 1864. Krasnodar Kray was formed in 1937.

Kras·no·vodsk (krăs'nə-vôtsk'), city (1976 est. pop. 54,000), S Central Asian USSR, in the Turkmen Republic, on the Krasnovodsk Gulf of the Caspian Sea. It is the western terminus of oil and natural gas pipelines. The city was founded in 1869.

Kras·no·yarsk (kräs'nə-yärsk'), city (1977 est. pop. 769,000), capital of Krasnoyarsk Kray, W Siberian USSR, on the Yenisei River. A major river port and rail center, it has industries producing heavy equipment for the Trans-Siberian RR, building and mining equipment, and farm and shipbuilding machinery. There are also plants producing cement, aluminum, and textiles. One of the world's largest hydroelectric plants is on the Yenisei at Krasnoyarsk. Founded in 1628, it grew rapidly after the discovery of gold and the construction of the Trans-Siberian RR.

Krasnoyarsk Kray, administrative division (1970 pop. 2,962,000), c.928,000 sq mi (2,403,520 sq km), central Siberian USSR, extending from the Sayan Mts. and the Minusinsk basin in the S across the Siberian wooded steppe, taiga, and tundra to the Arctic Ocean; capital Krasnoyarsk. The territory stretches along the entire course of the Yenisei, comprising parts of the West Siberian lowland on the left bank and the central Siberian Plateau on the right bank. There are deposits of brown coal, graphite, iron ore, manganese, gold, copper, nickel, aluminum, uranium, and mica. In the north is an extensive lumber industry. Grain is grown, cattle and reindeer are raised, and fur trapping is carried on. The territory was organized in 1934.

Kras·no·ye Se·lo (kräs'nə-yə syĭ-lô'), city (1969 est. pop. 22,000), NW European USSR. It is a rail terminus and has industries producing paper and plastics. Krasnoye Selo was a favorite summer resort of St. Petersburg before the Russian Revolution.

kray (krī), administrative and territorial unit of the USSR. There are six krays, or territories, within the Russian Soviet Federated Socialist Republic (RSFSR). Historically, these areas were frontier zones at the edges of the Russian Empire.

Kre·feld (krā'fĕlt'), city (1975 pop. 231,642), North Rhine-Westphalia, W West Germany, a port on the Rhine River. It is the center of the West German silk and velvet industry. Other manufactures include quality steels, machinery, and dyes. Krefeld was chartered in 1373 and was an important linen-weaving center until it passed (1702) to Prussia. The silk industry, encouraged by a monopoly given to the city by Frederick II of Prussia, soon replaced linen weaving; and in the 20th cent. the manufacture of artificial silk became important. The city was heavily damaged in World War II.

Krem·en·chug (krĕm'ĕn-cho͝ok'), city (1976 est. pop. 202,000), S central European USSR, in the Ukraine, on the Dnepr River. It is the center of an industrial complex based on a hydroelectric plant. Kremenchug was founded in 1571 as a fortress.

Kre·me·nets (krĕm'ə-nĕts'), city (1967 est. pop. 20,000), W European USSR, in the Ukraine. It is a rail terminus, highway hub, and agriculture trade center. Food and tobacco processing and the manufacture of milling machinery, tiles, cement, and hats are important industries. Founded in the 11th cent., Kremenets was part of the Kievan duchy. After the Polish-Lithuanian union in 1569, it served as a royal residence. The city passed to Russia in 1795. It was under Polish rule from 1919 to 1945, when it reverted to the USSR.

Kreuz·ling·en (kroits'lĭng-ən), town (1977 est. pop. 16,100), Thurgau canton, NE Switzerland, on the Lake of Constance. It is an industrial center with the oldest aluminum-rolling mill in Switzerland. Foodstuffs, chemicals, and motor vehicles are also manufactured.

Kri·ens (krē-ĕns'), town (1970 pop. 20,409), Lucerne canton, central Switzerland, at the foot of Mt. Pilatus. It is a suburb of Lucerne.

Krish·na (krĭsh'nə), river: *see* Kistna.

Krish·na·gar (krĭsh'nə-gər), town (1971 pop. 86,354), West Bengal state, E central India, on the Jalangi River. The main products of the area are rice, jute, sugar, ceramics, and plywood.

Kris·tian·sand (krĭs'chən-sănd'), city (1970 pop. 56,914), capital of Vest-Agder co., S Norway, a commercial and passenger port on the Skagerrak. Manufactures include ships, textiles, canned fish, and beer. The city was founded (1641) by Christian IV.

Kris·tian·stad (krĭs'chən-stăd'), city (1975 est. pop. 44,300), capital of Kristianstad co., SE Sweden, on the Helgaän River. It is a commercial and industrial center, located in a fertile agricultural region. Manufactures include textiles, machinery, and processed food. Founded (1614) by Christian IV of Denmark, Kristianstad changed hands frequently, but passed definitively to Sweden in 1678.

Kris·tian·sund (krĭs'chən-so͝ond'), city (1977 est. pop. 18,701), Møre og Romsdal co., W Norway, a port on the Atlantic Ocean. It is the site of a large trawler fleet and has industries that produce ships and fish and forest products. Chartered in 1742, Kristiansund was destroyed (1940) by bombardment in World War II and has since been rebuilt on three islands enclosing the harbor.

Kris·ti·ne·hamn (krĭs'tē-nə-hăm'ən), city (1975 est. pop. 22,124), Värmland co., S central Sweden, a port on Lake Vänern. The city was first chartered in 1582.

Kri·voy Rog (krĭv'oi rōg, rôk), city (1977 est. pop. 641,000), SW European USSR, in the Ukraine, at the confluence of the Ingulets and Saksagan rivers. It is a rail junction, an industrial center, and a metallurgical and coking center of one of the world's richest iron-mining regions. Burial mounds in the area indicate that Scythians inhabited it and used the iron deposits. Founded in the 17th cent., Krivoy Rog's industrial growth dates from 1881, when French, Belgian, and other foreign interests founded a mining syndicate.

Krk (kûrk), island (1971 pop. 13,078), 157 sq mi (406.6 sq km), in the Adriatic, off the Dalmatian coast, NW Yugoslavia. It has several small seaside resorts.

Kr·nov (kər-nôf'), city (1975 est. pop. 25,160), N Czechoslovakia, in Moravia, on the Opava River, near the Polish border. It manufactures machinery, textiles (especially woolens), and musical instruments (notably organs). The city was founded in 1221 and served as the capital of an independent duchy from 1377 to 1523.

Kro·mě·říž (krô'myĕr-zhĕsh), city (1975 est. pop. 23,687), central Czechoslovakia, in Moravia, on the Morava River. An agricultural center, it manufactures farm machinery and machine tools and has sugar refineries. Kroměříž was chartered in 1290.

Kron·shtadt (krŏn-shtät'), city, NW European USSR, on the small island of Kotlin in the Gulf of Finland. It is the chief naval base for the Soviet Baltic fleet. The harbor is icebound for several months each year. It was founded (1703) by Peter I as a port and a fortress to protect the site of St. Petersburg. The visit (1891) of a French naval squadron to Kronshtadt was followed by a Franco-Russian military agreement heralding the formation of the Triple Entente of France, England, and Russia. Mutinies of the naval garrison took place in 1825 and 1882 and played a part in the revolutions of 1905 and 1917.

Kroon·stad (kro͞on'stät), town (1970 pop. 51,988), Orange Free State, E central South Africa, on the Vals River. It is an agricultural and industrial center. The town's chief industries are clothing manufacture and mineral processing and the production of machine parts. Kroonstad was founded in 1855. Its growth was stimulated by the discovery of gold in the region in the late 19th cent.

Kru·ger National Park (kro͞o'gər), game reserve, c.8,000 sq mi (20,720 sq km), Transvaal, NE Republic of South Africa. One of the world's largest wildlife sanctuaries, it has almost every species of game in southern Africa.

Kru·gers·dorp (kro͞o'gərz-dôrp'), city (1970 pop. 92,725), Transvaal, NE South Africa. Krugersdorp is the center for a region where gold, manganese, asbestos, lime, and uranium are mined. The city has uranium extraction plants. It also serves as the trade center for the surrounding farming area.

Kru·še·vac (kro͞o'shĕ-väts), town (1971 pop. 29,469), E Yugoslavia, in Serbia. A commercial center, it has chemical and munitions industries. The seat of the kings of Serbia until 1389, it has retained the ruins of a medieval castle.

Kua·la Lum·pur (kwä'lə lo͝om'po͝or), city (1971 pop. 451,278), capital of the Federation of Malaysia, S Malay Peninsula, at the confluence of the Klang and Gombak rivers. Kuala Lumpur is the commercial center of a tin-mining and rubber-growing district and is a transportation hub. It was founded in 1857 by Chinese tin miners. In 1896 it became the capital of the Federated Malay States.

Kuala Te·reng·ga·nu (tə-rĕng-gä'no͞o) or **Kuala Treng·ga·nu** (trĕng-gä'no͞o), city (1971 pop. 53,353), capital of Terengganu state, Malaysia, central Malay Peninsula, on the South China Sea at the mouth of the Terengganu River. It is a port and has a weaving industry.

Ku·ban (ko͞o-băn'), river, c.570 mi (920 km) long, rising in the Greater Caucasus on the W slopes of Mt. Elbrus, S European USSR, and flowing N in a wide arc past Karachayevsk, Cherkessk, and Armavir, then W past Krasnodar, entering the Sea of Azov through two arms. Its upper course is precipitous and leads through several gorges.

Ku·ching (ko͞o'chĭng), city (1971 pop. 63,491), capital of Sarawak, Malaysia, in W Borneo and on the Sarawak River; founded 1839.

Kuei-lin or **Kwei·lin** (both: gwā'lĭn'), city (1970 est. pop. 235,000), N Kwangsi Chuang Autonomous Region, S China, on the Kuei River. It is a transportation center. Paper products are manufactured in the city. A large tin mine is nearby, and tungsten, manganese, and antimony are also found in the area. Kuei-lin is known for its beautiful scenery, often pictured by Chinese landscape painters.

Kuei-yang or **Kwei·yang** (both: gwā'yäng'), city (1970 est. pop. 1,500,000), capital of Kweichow prov., SW China. Textiles, chemical fertilizers, machine tools, petroleum products, cement, and paper are among its manufactures. Important coal fields are nearby.

Kuf·stein (ko͝of'shtīn), city (1971 pop. 12,766), in Tyrol prov., W Austria, on the Inn River, near the West German border. It is a summer and winter resort. Manufactures include skis and chemicals.

Kul·dja (ko͞ol'jä'): *see* I-ning, China.

Kulm·bach (ko͝olm'bäkh), town (1975 pop. 25,939), Bavaria, E central West Germany, on the White Main River. It has breweries, textile and paper mills, and canneries. Known in 1035, Kulmbach passed to Prussia in 1791. In 1807 it was taken by France, and in 1810 it was annexed by Bavaria and made part of Upper Franconia. On a nearby hill is the fortress (now a museum) of Plassenburg (12th cent.; rebuilt in Renaissance style 1560–70), which served as a prison from 1808 to the early 20th cent.

Ku·ma·mo·to (ko͞o-mä'mō-tō), city (1976 est. pop. 495,811), capital of Kumamoto prefecture, W Kyushu, Japan. An agricultural market town, it has manufactures of bamboo ware and pottery. It was an important castle town in the 17th cent.

Ku·ma·no·vo (ko͞o'mä-nô-vô), town (1971 pop. 46,406), S Yugoslavia, in Macedonia. It is the center of a tobacco-growing region. The Serbs won a decisive victory over the Turks at Kumanovo in 1912.

Ku·ma·si (ko͞o-mä'sə), city (1970 pop. 345,117), capital of the Ashanti Region, central Ghana. It is a commercial and transportation center in a cocoa-producing region, and has a large central market. Kumasi was founded c.1700.

Kum·ba·ko·nam (ko͝om-bə-kō'nəm), town (1971 pop. 112,971), Tamil Nadu state, SE India, on the Cauvery River. The area is known for its betel vines. Manufactures include brassware, textiles, and jewelry. The town is a Brahmanic cultural center. The many Hindu temples along the river are visited by pilgrims every 12 years.

Kum·gang San (ko͞om'gäng' sän), mountain range, SE North Korea, rising to 5,374 ft (1,639.1 m). There are scenic ravines and caverns and many ancient Buddhist temples.

Kun·gälv (kŭng'ĕlv'), town (1970 pop. 11,500), Göteborg och Bohus co., SW Sweden, on the Götaälv River. Founded in the 10th cent., Kungälv was one of the chief cities of medieval Norway. The town was plundered (1135) by Wends, seized (1368) by Hansa merchants, and ceded (1658) to Sweden.

Kun·lun (ko͞on'lo͞on'), mountain system of central Asia, between the Himalayas and the Tien Shan, extending c.1,000 mi (1,610 km) E from the Pamir mts., along the Tibet-Sinkiang prov. border of W China and into Tsinghai prov., where it branches into the mountain ranges of central China. It rises to 25,340 ft (7,728.7 m). The Kunlun system acts as a natural barrier between northern Tibet and the Tarim basin of Sinkiang. Great sections of the system are inaccessible and uninhabited; there is a very small nomad population, and yaks are the beasts of burden in the high mountain passes.

K'un-ming or **Kun·ming** (both: ko͞on'mĭng'), city (1970 est. pop. 1,700,000), capital of Yünnan prov., S China, on the N shore of Tien Ch'ih. It is a major commercial and cultural center and leading transportation hub. Coal is mined, and the city has an iron and steel complex. Other manufactures include phosphorus, chemicals, machinery, textiles, paper, and cement. K'un-ming has long been noted for its scenic beauty and equable climate. It consists of an old walled city, a modern commercial suburb, and a residential and university section. Although it was often the seat of kings in ancient times, K'un-ming's modern prosperity dates only from 1910, when the railroad from Hanoi was built. In World War II K'un-ming was important as the Chinese terminus of the Burma Road.

Kun·san (ko͝on'sän'), city (1975 pop. 154,485), SW South Korea, on the Yellow Sea at the Kum River estuary. It is a major port, especially for rice shipments, and is a commercial center for the rice grown in the Kum basin. Rice processing, fishing and fish processing, shipbuilding, and the production of salt, rubber, and alcohol are the chief industries. Kunsan gained importance with the development of its port, which was opened to foreign trade in 1899.

Kuo·pio (kwô'pyô), city (1975 est. pop. 71,684), capital of Kuopio prov., central Finland, on Lake Kallavesi. Situated in a large forest region, its industries are based on timber.

Ku·ra (ko͞o-rä'), river, c.950 mi (1,530 km) long, the chief river of Georgian SSR and Azerbaijan SSR, S European USSR, rising in NE Turkey, NW of Kars, and flowing NE into the USSR, then SE, parallel to the Caucasus Mts., to the Caspian Sea. There are hydroelectric plants on the river.

Ku·ra·yo·shi (ko͞o-rä'yō-shē), city (1975 pop. 34,800), Tottori prefecture, W Honshu, Japan, on the Tenjin River. It is an agricultural and communications center with a textile industry.

Kur·di·stan (kûr'dĭs-tăn'), an extensive plateau and mountain region in SW Asia (c.74,000 sq mi/191,660 sq km), including parts of E Turkey, NE Iraq, NW Iran and smaller sections of NE Syria and Soviet Armenia. The region lies astride the Zagros Mts. and the eastern extension of the Taurus Mts. and extends in the south across the Mesopotamian plain.

Kurdistan was conquered by the Arabs and converted to Islam in the 7th cent. The region was held by the Seljuk Turks in the 11th cent., by the Mongols from the 13th to 15th cent., and then by the Ottoman Empire. The Treaty of Sèvres (1920) provided for the creation of an autonomous Kurdish state. Because of Turkey's military revival under Kemal Atatürk, however, the Treaty of Lausanne (1923), which superseded Sèvres, failed to mention Kurdistan. Revolts by the Kurds of Turkey in 1925 and 1930 were forcibly quelled. Agitation among Iraq's Kurds for a unified and autonomous Kurdistan led in the 1960s to prolonged warfare between Iraqi troops and the Kurds. In 1970 Iraq finally promised local self-rule to the Kurds. The Kurds refused to accept the terms of the agreement, contending that the president of Iraq would retain real authority.

Ku·re (ko͞o'rĕ), city (1976 est. pop. 241,961), Hiroshima prefecture, SW Honshu, Japan, on Hiroshima Bay. It is a major naval base and port, with shipbuilding yards. Steel, pulp, files, machinery, and tools are manufactured.

Kure Island, circular atoll, c.15 mi (24 km) in circumference, in the NW part of the Hawaiian group, NW of Midway Island. Kure is uninhabited but has a large variety of sea birds. The island was annexed in 1886 by the Kingdom of Hawaii.

Kur·gan (ko͝or-gän'), city (1976 est. pop. 297,000), capital of Kurgan oblast, W Siberian USSR, on the Tobol River. Its factories produce agricultural and road-building equipment, machine tools, and food products. Kurgan was founded in the 17th cent.

Ku·ril Islands (ko͝or'ĭl, ko͞o'rēl), island chain, c.6,020 sq mi (15,590 sq km), Sakhalin oblast, E USSR, stretching c.775 mi (1,245 km) between S Kamchatka Peninsula and NE Hokkaido, Japan, and separating the Sea of Okhotsk from the Pacific Ocean. Active volcanoes are present and earthquakes are frequent. The low temperature, high humidity, and persistent fog make the islands unpleasant for human habitation. There are, however, communities engaged in sulfur mining, hunting, and fishing. In 1875 Japan gave up Sakhalin in return for Russian withdrawal from the Kuriles, and the Japanese held the islands until the end of World War II. The Yalta Conference ceded the islands to the USSR, and Soviet forces occupied the chain in Sept., 1945. Japan has challenged the Soviet right to the Kuriles, and failure to resolve the impasse has been a major obstacle to the signing of a peace treaty between Japan and the USSR.

Kur·land (ko͝or'lənd, ko͝or'länt): *see* Courland, USSR.

Kur·nool (kər-no͞ol'), town (1971 pop. 136,682), Andhra Pradesh state, S central India, at the confluence of the Tungabhadra and Hindri rivers. Kurnool is a market for grain, hides, and cotton. There are ruins of a fort built in the 16th cent. The town was overrun by Moslems in 1565 and was ceded to the British in 1800.

Kursk (ko͝orsk), city (1976 est. pop. 363,000), capital of Kursk oblast, central European USSR, at the confluence of the Tuskor and Seim rivers. An important rail junction, it has machine, chemical, and synthetic fiber plants. A large iron deposit is south of the city. Known since 1095, Kursk was destroyed by the Mongols in 1240 and was rebuilt as a Muscovite fortress in 1586.

Kur·sky Za·liv (ko͝or'skē zä'lēv) or **Cour·land Lagoon** (ko͝or'lănd), lagoon, 56 mi (90.1 km) long and 28 mi (45 km) wide, W USSR, in the Lithuanian and Russian republics. It is separated from the Baltic Sea by Courland Spit, a sandspit that leaves only a narrow opening at the Klaypeda Channel in the north.

Ku·ru·me (ko͞o-ro͞o'mā), city (1976 est. pop. 207,586), Fukuoka prefecture, W Kyushu, Japan, on the Chikugo Plain. It is a commercial and agricultural center and manufactures rubber and cotton goods.

Ku·ru·ne·ga·la (ko͝or'o͞o-nā'gə-lə), town (1972 est. pop. 28,000), W central Sri Lanka. It is an important road junction and the commercial center of a coconut, rice, and rubber plantation district. Overlooking the town is Elephant Hill, a stronghold in the 14th cent.

Ku·sa·tsu (ko͞o-sä'tso͞o), town (1975 est. pop. 64,873), Gumma prefecture, central Honshu, Japan. As early as the 12th cent. its hot sulfur springs were known for their medicinal properties.

Ku·shi·ro (ko͞o-shē'rō), city (1975 pop. 206,840), SE Hokkaido, Japan, on the Pacific Ocean. The island's only ice-free trading port, it exports timber, fish, and coal. Kushiro is also a major base for fishermen. The city is the center of the huge Kushiro coal field, which extends far out to sea; mining is carried on in the sea. Industrialization has made Kushiro important in the production of marine products, dairy products, lumber, paper, pulp, and fertilizer.

Kus·ko·kwim (kŭs'kō-kwĭm), river, c.800 mi (1,290 km) long, rising on the NW slopes of the Alaska Range, central Alaska, and flowing SW to the Bering Sea.

Ku·sta·nay (ko͞o-stə-nī'), city (1976 est. pop. 151,000), capital of Kustanay oblast, NW Central Asian USSR, in Kazakhstan, on the Tobol River. It is an agricultural center and producer of chemical fibers. Rich iron deposits are nearby.

Kü·tah·ya (kü-tä'yä), city (1975 pop. 82,400), capital of Kütahya prov., W central Turkey. An agricultural market center, the city has been famous since the 16th cent. for the manufacture of ceramics. It was occupied by the Seljuk Turks soon after the battle of Manzikert (1071). In the 15th cent. it passed to the Ottomans.

Ku·ta·i·si (ko͞o-tä-ē'sĭ), city (1976 est. pop. 177,000), SE European USSR, in Georgia, on the Rioni River. It has industries producing trucks, mining and transport equipment, textiles, chemicals, and food products. Kutaisi was the capital of ancient Colchis (8th cent. B.C.), and the capital of Imeritia in the 13th, 15th, and 16th cent. A.D. It was taken by the Russians in 1810.

Kutch or **Cutch** (both: kŭch), district, 17,000 sq mi (44,030 sq km), Gujarat state, W India, bounded on the N by Pakistan. It is largely barren except for a fertile band along the Arabian Sea. There is some horse and camel breeding. Formerly a princely state, Kutch was established in the 14th cent. by Rajputs, was often invaded from Sind, and passed under British rule in 1815. Kutch was incorporated into Gujarat in 1960. The Rann of Kutch, a salt waste (9,000 sq mi/23,310 sq km) mainly in the north of the district, was the scene of Indo-Pakistani fighting in 1965.

Kut·ná Ho·ra (ko͝ot'nä hô'rä), city (1975 est. pop. 19,529), NW Czechoslovakia, in Bohemia. Now an agricultural center, it was an important silver-mining center in the Middle Ages. The city lost its importance after the silver mines closed in the 17th cent. Kutná Hora is rich in medieval architecture.

Ku·wait (ko͞o-wāt', -wĭt'), independent sheikdom (1970 pop. 738,663), 6,177 sq mi (15,998 sq km), NE Arabian peninsula, at the head of the Persian Gulf. A low, sandy region, generally barren and sparsely settled, Kuwait is bounded by Saudi Arabia on the south and by Iraq on the north and west. The capital is Al-Kuwait, or Kuwait (1970 pop. 80,008).

Economy. Kuwait's traditional exports were pearls and hides, but since 1946 it has become a major petroleum producer. It is estimated that Kuwait possesses about one fifth of the world's oil reserves. The

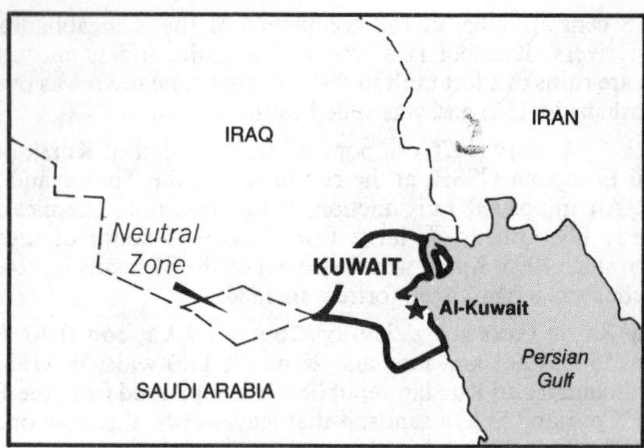

main concession for oil exploitation was held by a joint British-American firm until 1974 when Kuwait took control of most of the operations. Recently expanded industries produce chemicals, fertilizer, building materials, and processed food. Wheat, dates, and vegetables are the chief farm products, but most food is imported.

History. Kuwait was settled by Arab tribes in the early 18th cent. The present ruling dynasty was founded by Sabah abu Abdullah (ruled 1756–72). In 1897 the sheik, fearing that the Turks intended to make their nominal authority effective, made Kuwait a British protectorate. In June, 1961, the British ended their protectorate but supplied troops in July at the request of the sheik when Iraq claimed sovereignty over Kuwait. The country is a large donor of financial aid to the other Arab countries.

Government. Kuwait is governed under a constitution promulgated in 1963. The sheik and the council of ministers appointed by him constitute the executive. A 50-member national assembly is elected every four years by the male citizens.

Ku·wa·na (kōō-wä'nä), city (1975 pop. 83,440), Mie prefecture, S Honshu, Japan, on Ise Bay. It is an important port and industrial center with large metal and textile industries.

Kuy·by·shev (kwē'bĭ-shĕf, kōō'ē-bĭ-shĭf), formerly **Sa·ma·ra** (sə-mâr'ə), city (1977 est. pop. 1,204,000), capital of Kuybyshev oblast, E central European USSR, on the left bank of the Volga and at the mouth of the Samara River. It is a major river port and rail center and has industries producing automobiles, aircraft, locomotives, machinery, ball bearings, synthetic rubber, chemicals, textiles, and petroleum. Founded in 1586 as a Muscovite stronghold for the defense of the Volga trade route and of Russia's eastern frontier, Samara was attacked by the Nogai Tatars (1615) and the Kalmyks (1644) and opened its gates to the Cossack rebels in 1670. Its industrial expansion dates from the early 20th cent., when railroads to Siberia and central Asia were built. Samara was (1918) the seat of the anti-Bolshevik provisional government and constituent assembly of Russia. During World War II the central government of the USSR was transferred to Kuybyshev (1941–43) from Moscow.

Kuz·netsk Ala-Tau (kōōz-nĕtsk' ə-lä'tou), mountain range, S Siberian USSR, E of Novokuznetsk, rising to c.6,900 ft (2,105 m). Part of the great mountain system of central Asia, the range is composed mainly of metamorphic rocks and yields such minerals as iron, manganese, and gold.

Kuznetsk Basin, coal basin, c.10,000 sq mi (25,900 sq km), W Siberian USSR, between the Kuznetsk Ala-Tau and the Salair Ridge. It has extensive coal deposits, particularly of high-grade coking coal. The first iron-smelting works were founded in 1697. Coal deposits were discovered in 1721 and first mined in 1851. The Kuznetsk industrial region produces iron and steel, zinc, aluminum, heavy machinery, and chemicals.

Kwa·ja·lein (kwŏj'ə-lən, -lān'), coral atoll, 6.5 sq mi (16.8 sq km), central Pacific, in the Ralik Chain of the Marshall Islands. Kwajalein, a group of 97 islets surrounding a lagoon, is a district headquarters of the U.S. Trust Territory of the Pacific Islands. A large Japanese naval and air base was located here during World War II, and after the U.S. conquest of the Marshalls (1944) U.S. military bases were established.

Kwang·ju (kwäng'jōō'), city (1970 est. pop. 503,000), capital of South Cholla prov., SW South Korea, in the Yongsan River lowland. A regional agricultural and commercial center built on the site of an ancient market, Kwangju has rice mills and industries that produce rayon and cotton textiles and beer. In the hills around Kwangju are ancient tombs and temples.

Kwang·si Chuang Autonomous Region (kwäng'sē' chwäng', kwăng'-), province (1968 est. pop. 24,000,000), c.85,000 sq mi (220,150 sq km), S China, bordering on Vietnam; capital Nan-ning. Rice is an important crop.Wheat, corn, vegetables, peanuts, tropical and subtropical fruit, sesame, rapeseed, jute, sugar cane, and tobacco are also grown. Kwangsi is a major producer of manganese ore and a significant source of fluorspar, tin, tungsten, and antimony. The Kwangsi Chuang Autonomous Region was created in 1958 from Kwangsi prov.

Kwang·tung (kwäng'tŏŏng', kwăng'-), province (1968 est. pop. 40,000,000), c.89,400 sq mi (231,545 sq km), S China; capital Canton. The province includes some 730 islands. In the canals and off the coast there is considerable fishing. Its crops include sugar cane, rice, hemp, tobacco, tea, tropical and subtropical fruits, and peanuts. Kwangtung has tungsten, iron, manganese, titanium, tin, gold, and silver deposits. There is offshore drilling for oil. Kwangtung came under Chinese suzerainty during the unification under the Ch'in dynasty (c.211 B.C.), and was more firmly absorbed during the Han dynasty. Kwangtung was the main scene of China's early foreign contact, chiefly through Canton; there was trade with the west during the Roman Empire, trade with the Arabs during the T'ang dynasty, and European trade that originated during the 16th cent. with the Portuguese. Kwangtung has been a center of revolutionary activity; the Kuomintang was formed here (1912) under the leadership of Sun Yat-sen.

Kwan·za (kwän'zə), river: *see* Cuanza.

Kwei·chow (kwā'chou', gwā'jō'), province (1968 est. pop. 17,000,000), c.66,000 sq mi (170,940 sq km), SW China; capital Kuei-yang. Kweichow is almost entirely a high plateau, and its sheer limestone hills form some of the most spectacular karst scenery in the world. Rice is the major crop. Soybeans, corn, wheat, millet, barely, kaoliang, and beans are also raised. Commerical crops include rapeseed, tobacco, tea, oakleaf silk, sugar cane, and indigo. Cotton is being developed. Kweichow has rich forests, and lumber, tung oil, lacquer, and paint are produced. Mineral resources include mercury, coal, iron, phosphorous, copper, manganese, and silver. Chinese settlement of the region began around 2,000 years ago, but it was only in the 10th cent. that it passed under the suzerainty of China. Kweichow became a province in the 17th cent. under the Ming dynasty.

Kwei·lin (gwā'lĭn'): *see* Kuei-lin, China.

Kwei·yang (gwā'yäng'): *see* Kuei-yang, China.

Kwi·na·na (kwī-nä'nə), city (1971 pop. 12,208), Western Australia, SW Australia, a suburb of Perth. A new industrial city, Kwinana has oil refineries and steelworks.

Kyakh·ta or **Kiakh·ta** (both: kyäKH'tə), city, Buryat Autonomous SSR, S Siberian USSR, near the Soviet-Mongolian border. It has textile, lumber, and food-processing plants. Founded in 1728, it was a trading point between Russia and Western Europe and China; it was then a trading center between Russia and Outer Mongolia.

Kyo·ga or **Kio·ga** (kyō'gə), lake, c.100 mi (160 km) long, formed by the Victoria Nile, S central Uganda, E Africa. It occupies part of the same depression as Victoria Nyanza, to which it was once joined. The shallow lake has large areas of papyrus swamp.

Kyo·to (kyō'tō, kē-ō'-), city (1976 est. pop. 1,461,567), capital of Kyoto prefecture, S Honshu, Japan, on the Kamo River. Yodo is its port. Industries include copper rolling, food processing, and the manufacture of electrical equipment, spinning and dyeing machinery, precision tools, chemicals, and cameras. The city is famous for its cloisonné, bronzes, damascene work, porcelain, and lacquer ware, and its renowned silk industry dates from 794. Founded in the 8th cent., it lost its political power to Tokyo after 1192. For centuries it has been the cultural heart of Japan. Rich in historic interest, Kyoto is the site of the tombs of many famous Japanese; the old imperial palace as well as Nijd Castle (former palace of the shoguns), with their fine parks and gardens, are also in the city. In addition, Kyoto is a religious center, noted especially for its ancient Buddhist temples, its Heian shrine (a Shinto holy place), and its 59-ft (18-m) statue of Buddha.

Kyu·shu (kyōō'shōō, kē-lōō'-), island (1976 est. pop. 13,600,207), c.13,760 sq mi (35,640 sq km), S Japan. It is the third-largest, southernmost, and most densely populated of the major islands of Japan.

Mainly of volcanic origin, the island has a mountainous interior and there are many hot springs. Rice, tea, tobacco, sweet potatoes, fruits, wheat, and soybeans are major crops. Coal, zinc, and copper are mined in Kyushu, and raw silk is extensively produced. The island is noted for its porcelain (Satsuma and Hizen ware).

Kyu·sten·dil (kyo͝o-stĕn-dēl′), city (1969 est. pop. 43,700), SW Bulgaria, near the Yugoslav border. Famous for its mineral springs used to heat hothouses, Kyustendil is a market city for fruit and other agricultural produce. There are varied light industries. The city's history dates to Roman times. It was the capital of an independent Bulgarian principality when the Turks took it in the 14th cent. The city remained under Turkish rule until 1878.

Ky·zyl or **Ki·zil** (both: kĭ-zĭl′), city (1976 est. pop. 57,000), capital of Tuva Autonomous Republic, S Siberian USSR, on the Yenisei River. It services motor transport and has brickyards, sawmills, furniture factories, and food-processing plants.

Ky·zyl-Kum or **Ki·zil Kum** (both: kĭ-zĭl′ko͞om′), desert, c.115,000 sq mi (297,850 sq km), Central Asian USSR, in Kazakhstan and Uzbekistan. This vast region southeast of the Aral Sea between the Amu Darya and Syr Darya rivers consists mainly of rocky areas covered by sparse vegetation and shifting sand dunes. Cotton, rice, and wheat are grown in river valleys and irrigated oases. Important gold deposits have been discovered in the desert.

Laa·land (lô län): *see* Lolland, Denmark.

La·bé (lă-bā′), town (1967 est. pop. 26,000), W central Guinea, in the Fouta Jallon. It is the market center for a farm region where citrus fruit, bananas, vegetables, and rice are grown and cattle are raised. Labé was incorporated in the Mali empire in the early 13th cent. The Fulani settled here in the second half of the 18th cent., displacing the original inhabitants.

La Belle (lə bĕl′), city (1970 pop. 1,823), seat of Hendry co., S central Fla., ENE of Fort Myers. It is a shipping point for sugar cane, cotton, cattle, and watermelon. There are oil wells in the vicinity.

La·bette (lə-bĕt′), county (1970 pop. 25,775), 654 sq mi (1,693.9 sq km), SE Kansas, in a level area bordering in the S on Okla. and drained in the E by the Neosho River; formed 1867; co. seat Oswego. Its agriculture includes grain, livestock, and dairy products. It also has oil and gas fields and coal deposits. Clothing, wood products, furniture, plastics, and metal products are made.

Lab·ra·dor-Un·ga·va (lăb′rə-dôr′ŭng-gā′və, -gä′-), peninsular region of E Canada, c.550,000 sq mi (1,424,500 sq km), bounded on the W by Hudson Bay, on the N by Hudson Strait and Ungava Bay, on the E by the Atlantic Ocean, and on the S by the St. Lawrence River. It is almost completely unpopulated. The western four fifths of the peninsula belongs to Quebec prov. The eastern fifth, called simply Labrador, is part of Newfoundland prov. The region south of Ungava Bay, originally a possession of the Hudson's Bay Company, was made a part of the Northwest Territories in 1869 and later (1895) became a separate district. In 1912 it was added to Quebec prov., but in 1927 the eastern coast was awarded to Newfoundland by the British Privy Council. The northern part of the region is a cold, barren tundra; the southern part is covered by coniferous forests. Geologically part of the Canadian Shield, the glaciated peninsula has many lakes and streams.

La Bre·a (lə brā′ə), area, S Calif., formerly in Rancho La Brea. The La Brea asphalt pits, which yielded prehistoric animal and plant remains, are in Hancock Park, Los Angeles. The first fossils were found in 1875; since 1906 the pits have been extensively explored.

La·bu·an (lä′bo͞o-än′), island, 38 sq mi (98.4 sq km), part of Sabah, Malaysia, off N Borneo, in the South China Sea. Coconuts, rubber, and rice are the main products. Labuan was ceded to Great Britain by the sultan of Brunei in 1846 and became a crown colony in 1848. It was included in the Straits Settlements in 1906 and in 1946 was joined to British North Borneo, which became Sabah in 1963.

La Ca·na·da (lä kən-yä′də), uninc. residential town (1970 pop. 20,652), Los Angeles co., S Calif., NW of Pasadena.

Lac·ca·dive, Min·i·coy, and A·min·di·vi Islands (lăk′ə-dīv′, mĭn′ĭ-koi′, ä′mĭn-dē′vē), island group (1971 pop. 31,798), 11 sq mi (28.5 sq km), SW India, in the Arabian Sea off the coast of Kerala state. The capital of this group of 26 islands (10 are inhabited) is Kavarrati Island. The population engages in fishing and copra production. The islands compose the Union Territory of Lakshadweep.

La·cey (lā′sē), city (1970 pop. 9,696), Thurston co., W central Wash., E of Olympia, in a farming area.

La·chine (lă-shēn′), city (1976 pop. 41,503), S Que., Canada, on Montreal island, at the E end of Lake St. Louis just SW of Montreal. Its industries include iron and steel foundries and the manufacture of tires, wire, and tiles. Lachine was first settled in 1675 and in 1689 was the scene of a massacre by the Iroquois Indians. The city is the southwest terminal of the Lachine Canal, connecting Lake St. Louis with the St. Lawrence River at Montreal.

La·chish (lā′kĭsh), ancient city, S Palestine, SW of Jerusalem. Excavations (begun in 1935) show that Lachish had been populated since c.3200 B.C. and was a thriving community as early as the 17th cent. B.C. The finds include 21 ostraca, or potsherds, written in ink. They were written (c.589 B.C.) in Hebrew by local commanders to their officers when Lachish was being threatened by the Babylonians under Nebuchadnezzar.

Lach·lan (läk′lən), river, SW New South Wales, Australia, rising in the Eastern Highlands near Forbes and flowing 922 mi (1,483.5 km) NW and then SW to join the Murrumbidgee River.

La·chute (lă-sho͞ot′), town (1976 pop. 11,928), S Que., Canada, on the North River, W of Montreal. It is at the foot of the Laurentian Mts.

Lack·a·wan·na (lăk′ə-wŏn′ə), county (1970 pop. 234,131), 454 sq mi (1,175.9 sq km), NE Pa., in an anthracite-mining and industrial region drained by the Lackawanna and Susquehanna rivers; formed 1878; co. seat Scranton. The Pocono Plateau, south of the Lackawanna River valley, is the anthracite center of the world. Manufacturing (textiles, clothing, food products, and paints), printing, and agriculture (vegetables, poultry, and dairy products) are major industries.

Lackawanna, city (1978 est. pop. 23,800), Erie co., W N.Y., on Lake Erie; inc. 1909. Lackawanna is one of the major steelmaking centers in the United States. The spectacular Basilica of Our Lady of Victory is a Roman Catholic shrine.

La·clede (lə-klēd′), county (1970 pop. 19,944), 770 sq mi (1,994.3 sq km), S central Mo., in the Ozarks, drained by the Gasconade River; formed 1849; co. seat Lebanon. It is in an agricultural region yielding livestock, dairy products, wheat, corn, and hay. It has oak timber and a food-processing industry. Clothing, metal and wood products, furniture, and machinery are manufactured. Part of Mark Twain National Forest is in the county.

La·con (lā′kən), city (1970 pop. 2,147), seat of Marshall co., N central

Ill., on the Illinois River SW of La Salle, in a grain and livestock region; laid out 1826, inc. 1839. Bricks are made.

La·co·ni·a (lə-kō′nē-ə) or **Lac·e·dae·mon** (lăs′ĭ-dē′mən), ancient region, S Peloponnesus, Greece, bounded on the W by Messenia and on the N by Arcadia and Argolis. On the Eurotas (now Evrotás), the principal river, stood Sparta, which dominated the region until the rise of the second Achaean League in the 3rd and 2nd cent. B.C.

Laconia, city (1978 est. pop. 15,000), seat of Belknap co., central N.H., near Lakes Winnisquam and Winnipesaukee on the Winnipesaukee River; settled c.1761, inc. as a city 1893. It is a popular summer and winter resort and the industrial and trade center of a farm region.

Lac qui Parle (lăk-ē pärl′), county (1970 pop. 11,164), 768 sq mi (1,989.1 sq km), W Minn., bordering on S.Dak., bounded on the N and E by the Minnesota River (flowing through Lac qui Parle), and drained in the SE by the Lac qui Parle River; formed 1871; co. seat Madison. It is in an agricultural area that produces corn, oats, barley, and livestock and has soybean-oil mills.

La Crosse (lə krôs′), county (1970 pop. 80,468), 451 sq mi (1,168.1 sq km), W Wis., in a fertile farming, dairying, and stock-raising area drained by the La Crosse and Black rivers; formed 1851; co. seat La Crosse. It has lumbering and food-processing industries.

La Crosse. 1. City (1970 pop. 1,583), seat of Rush co., W central Kansas, NW of Great Bend, in a farm and oil region; founded 1876, inc. 1886. **2.** City (1978 est. pop. 48,900), seat of La Crosse co., W Wis., at the foot of high bluffs on the Mississippi, where the La Crosse and Black rivers meet; inc. 1856. Air-conditioning systems, rubber footwear, clothing, metal products, and vending machines are made. A French fur-trading post was here in the late 18th cent., and later the city had a thriving lumber industry.

La·dakh (lə-däk′), region (1971 pop. 105,001), 45,762 sq mi (118,524 sq km), E Kashmir, on the border of China. Although allied ethnologically and geographically with Tibet, the region has a predominantly Moslem population. It was nominally a dependency of Tibet, but after 1531 it was invaded periodically by Moslems from Kashmir. It was annexed to Kashmir in the mid-19th cent. The region is now claimed by China, which in 1962 occupied part of the area despite Indian opposition.

La·do·ga, Lake (lä′də-gə, lăd′ə-), c.7,000 sq mi (18,130 sq km), NW European USSR, in Karelia, NE of Leningrad. The largest lake in Europe, it is c.130 mi (210 km) long and c.80 mi (130 km) wide and has a maximum depth of 738 ft (225 m). Located on the heavily glaciated Baltic Shield, the lake has shores that are low and marshy in the south, rocky and indented in the north. It is subject to autumn storms and freezes every year for two months in the north and four months in the south. Until the Finnish-Russian War of 1939–40, the northern part of the lake belonged to Finland; cession of the Finnish shore to the USSR was confirmed by the peace treaty of 1947. During the defense of Leningrad against the Germans in World War II, the frozen Lake Ladoga was the lifeline by which Leningrad was supplied in the winters from 1941 to 1943.

La·dy·smith (lā′dē-smĭth′), town (1970 pop. 28,920), Natal, E South Africa. The town has railroad yards and food-processing, textile, and tire factories. It is the distribution center for the surrounding agricultural and coal-mining region. Ladysmith was founded in 1851 and grew after a railroad to Durban was opened in 1886. During the South African War, Sir George White's British forces at Ladysmith were under siege by Boers from Nov., 1899, to Feb., 1900, when British reinforcements arrived.

Ladysmith, city (1970 pop. 3,674), seat of Rusk co., N central Wis., NNE of Eau Claire on the Flambeau River, in a lake and woods region; inc. 1905. Paper and dairy products are made.

La·e (lä′ē, lī), town (1971 pop. 38,707), Papua New Guinea, on NE New Guinea island, at the head of the Huon Gulf. It was founded in 1927 to serve air transport into the Morobe gold fields in the mountainous interior.

La·fa·yette (lăf′ē-ĕt′, lä′fē-). **1.** County (1970 pop. 10,018), 523 sq mi (1,354.6 sq km), SW Ark., bounded on the S by the La. border, on the W by the Red River; formed 1827; co. seat Lewisville. It has agriculture (cotton, corn, potatoes, and hay), oil and natural-gas wells, and a lumbering industry. Cotton ginning and some manufacturing (carpets and furniture) are done. **2.** County (1970 pop. 2,892), 549 sq mi (1,421.9 sq km), N Fla., bounded on the E and NE by the Suwannee River; formed 1856; co. seat Mayo. It is a swampy flatwoods area with many small lakes. Farming (corn, peanuts, and tobacco), cattle raising, and lumbering are the major industries. There are limestone and phosphate deposits. **3.** Parish (1970 pop. 11,643), 283 sq mi (733 sq km), S La., bounded on the N by Bayou Carencro and drained by the Vermilion River; formed 1823; parish seat Lafayette. Its agriculture includes cotton, sugar cane, rice, sweet potatoes, corn, hay, and livestock. It has oil and natural-gas wells and refineries, fishing, and lumbering and food-processing industries. Machinery and hardware are made. **4.** County (1970 pop. 24,181), 668 sq mi (1,730.1 sq km), N Miss., drained by the Yocono and Tallahatchie rivers and Sardis Reservoir; formed 1836; co. seat Oxford. It is in an agricultural area that produces cotton, corn, dairy products, poultry, and truck crops. It has pine and hardwood timber, and manufactures clothing and wood products. **5.** County (1970 pop. 26,626), 634 sq mi (1,642.1 sq km), W Mo., bounded on the N by the Missouri River; formed 1820, name changed from Lillard in 1825; co. seat Lexington. In a livestock (cattle, dairying, poultry, and hogs) and farming (wheat, corn, oats, and apples) area, it has coal mines and rock quarries. Its manufactures include clothing, wood products, and machinery. **6.** County (1970 pop. 17,456), 643 sq mi (1,665.4 sq km), S Wis., bordered on the S by Ill. and drained by the Pecatonica and Galena rivers; formed 1846; co. seat Darlington. It is primarily agricultural, especially dairying and livestock raising. There are lead and zinc mines here.

Lafayette. 1. City (1970 pop. 3,530), seat of Chambers co., E Ala., near the Ga. border E of Alexander City, in a cotton, corn, and potato area; settled 1883; inc. 1907. Clothing and cement products are made. There are pulpwood mills in the area. **2.** City (1970 pop. 20,484), Contra Costa co., NW Calif., a residential suburb in the San Francisco-Oakland area; settled 1848, inc. 1968. **3.** City (1978 est. pop. 48,800), seat of Tippecanoe co., W central Ind., on the Wabash River; inc. 1853. A manufacturing city in a grain, livestock, and dairy area, it has railroad shops, meat-packing houses, and plants making aluminum and rubber goods. It is the seat of Purdue Univ. The nearby site of the battle of Tippecanoe (Nov., 1811) is a state memorial. **4.** City (1978 est. pop. 77,700), seat of Lafayette parish, S central La., on the Vermilion River (which is linked to the Intracoastal Waterway); settled 1770s by Acadians, inc. 1836. It is a commercial, shipping, and medical center for an area producing sugar cane, rice, cotton, dairy cattle, livestock, and petroleum. Manufactures include building materials, electrical appliances, auto parts, furniture, and metal products. The city retains a colorful Cajun atmosphere. **5.** City (1970 pop. 2,583), seat of Macon co., N Tenn., NE of Nashville; settled c.1842, inc. 1843.

La Fay·ette (lə fĕt′), city (1970 pop. 6,044), seat of Walker co., extreme NW Ga., S of Chattanooga, Tenn.; founded 1835. It is a farm trade center and produces textiles.

La Fol·lette (lə fŏl′ĭt), city (1970 pop. 6,902), Campbell co., NE Tenn., on Norris Lake NNW of Knoxville, at the E base of the Cumberland Plateau; inc. 1897. The city is a coal-mining center.

La·fourche (lə-fo͞osh), parish (1970 pop. 68,941), 1,141 sq mi (2,955.2 sq km), extreme SE La., bounded on the S by the Gulf of Mexico, on the SW by Bayou Pointe au Chien, on the E partly by Barataria Bay, intersected by the navigable Bayou Lafourche, and crossed by the Gulf Intracoastal Waterway and Southwestern Louisiana Canal; formed 1805; parish seat Thibodaux. Sugar cane, corn, potatoes, and hay are grown. Natural-gas and oil wells are found here. There is some manufacturing, including sugar refining.

Lag·an (lăg′ən), river, c.40 mi (65 km) long, rising in Slieve Croob, Co. Down, SE Northern Ireland, flowing NW, then NE past Lisburn to Belfast Lough at Belfast.

La·gash (lā′găsh), ancient city of Sumer, S Mesopotamia. Lagash was flourishing by c.2400 B.C., but traces of habitation go back at least to the 4th millennium B.C. After the fall of Akkad (2180 B.C.), when the rest of Mesopotamia was in a state of chaos, Lagash was able to maintain peace and prosperity under its ruler Gudea. Thousands of inscribed tablets have been found at the site.

La·ghou·at (lä-gwät′), town (1974 est. pop. 41,900), N Algeria, an oasis on the N edge of the Sahara Desert. It is an important marketplace and is known for rug and tapestry weaving. There are natural-gas deposits in the region. Laghouat traces its history at least to the 11th cent. The Turks captured it in 1786, and the French conquered the city in 1852.

La·gos (lā′gŏs, lä′gôs), city (1971 est. pop. 1,112,000), capital of Nigeria, SW Nigeria, on the Gulf of Guinea. It comprises four islands and four mainland sections that are interconnected by causeways and bridges. Lagos is Nigeria's economic center and chief port. Industries include railroad repair, motor vehicle assembly, food processing, and the manufacture of metal products, textiles, beverages, chemicals, pharmaceuticals, soap, and furniture.

An old Yoruba town, Lagos, beginning in the 15th cent., grew as a trade center and seaport. From the 1820s until it became a British colony, Lagos was a notorious center of the slave trade. Britain annexed the city in 1861, both to tap the trade in palm products and other goods with the interior and to suppress the slave trade. In 1906, Lagos was joined with the British protectorate of Southern Nigeria, and, in 1914, when Southern and Northern Nigeria were amalgamated, it became part of the small coastal Colony of Nigeria. In 1954 most of the colony was merged with the rest of Nigeria, but Lagos was made a separate federal territory. From the late 19th cent. to independence in 1960, Lagos was the center of the Nigerian nationalist movement.

La·gos (lä'gōōsh'), city (1970 municipal pop.16,610), Faro dist., S Portugal, in Algarve, on the Atlantic Ocean. The excellent harbor shelters much coastwise trade and an important sardine and tuna fishing fleet. Sancho I with the help of bands of Crusaders captured (1189) the city from the Moors; in 1191 it was recaptured by the Moors but was soon (c.1250) restored to the Portuguese. Lagos was a starting port for Portuguese navigators in the time of Prince Henry the Navigator.

La Grande (lə grănd'), city (1970 pop. 9,645), seat of Union co., NE Oregon, SE of Pendleton, in the Grande Ronde valley; founded on the Oregon Trail 1861, inc. 1885. The city is the processing center of a timber, livestock, and farm area.

La·grange (lə-grānj'), county (1970 pop. 20,890), 379 sq mi (981.6 sq km), NE Ind., bounded in the N by the Mich. line and drained by the Pigeon and Elkhart rivers; formed 1832; co. seat Lagrange. Dairying, farming (soybeans and grain), and stock raising are done.

Lagrange, town (1970 pop. 2,053), seat of Lagrange co., NE Ind., E of Elkhart, in a livestock and dairying area; platted 1836, inc. 1855. Condensed milk is produced.

La Grange (lə grānj'). **1.** City (1978 est. pop. 23,100), seat of Troup co., W central Ga., inc. 1828. It is an industrial center that has retained its charm as a residential city, containing a number of classic revival houses. **2.** Village (1978 est. pop. 17,000), Cook co., NE Ill., a suburb of Chicago; settled 1830s, inc. 1879. It is primarily residential. **3.** City (1970 pop. 1,713), seat of Oldham co., N central Ky., NW of Frankfort, in a farm area. **4.** City (1970 pop. 3,092), seat of Fayette co., S central Texas, on the Colorado River SE of Austin; settled 1828. Early a wealthy market for blackland cotton plantations, it now serves a farm region and processes dairy products.

La Grange Park, village (1970 pop. 15,459), Cook co., NE Ill., a suburb of Chicago; inc. 1892.

La Gran·ja (lä gräng'hä): *see* San Ildefonso, Spain.

La Guai·ra (lä gwī'rä), city, Federal Dist., N Venezuela, on the Caribbean Sea NW of Caracas. It is the principal international port of Venezuela and a seaside resort. Founded in the 16th cent. as an outlet for Caracas, the city was sacked by English pirates in 1743.

La·gu·na, La (lä lä-gōō'nä), city (1970 pop. 79,963), on Teneriffe island, Canary Islands. The center of a fertile farm area producing cereals, grapes, fruits, and vegetables, it is also a tourist resort.

La·gu·na Beach (lə-gōō'nə), city (1970 pop. 14,550), Orange co., S Calif., on the Pacific coast; founded 1887, inc. 1927. It is a residential and resort city with a noted art colony.

La·gu·na District (lä-gōō'nä), irrigated area in E Durango and W Coahuila states, N central Mexico. Originally a 900,000-acre (364,500-hectare) tract, consisting of large estates, the land was reapportioned (1936) and distributed to Mexican farmers. It was a successful experiment in agrarian reform until 1952, when a severe drought scorched more than half the district, turning 200,000 acres (81,000 hectares) of wheat and cotton fields into a dust bowl. Settlement has continued here, but on a greatly reduced scale.

La Ha·bra (lə hä'brə), city (1978 est. pop. 43,900), Orange co., S Calif.; inc. 1925. A suburb of Los Angeles, it is a center for an area where citrus fruits, avocados, and vegetables are grown and oil is produced. La Habra was settled in the 1860s by Basque sheepherders.

La·hai·na (lə-hī'nə, lä-hī'nä), city (1970 pop. 3,718), Maui co., on the W coast of Maui Island, Hawaii, in a sugar-cane and pineapple region. It was the scene of the first white settlement in the islands and served as capital from 1810 until the seat of government was moved (1845) to Honolulu. Hawaii's first school was established here in 1831.

La·hon·tan, Lake (lə-hŏn'tən), extinct lake of W Nev. and NE Calif. It was formed by heavy precipitation caused by the Pleistocene glaciers and with Lake Bonneville occupied a part of the Great Basin region. Lake Lahontan vanished shortly after the Pleistocene epoch, but Pyramid, Winnemucca, and Walker lakes and Carson Sink are its remnants. The area is rich in Pleistocene fossils.

La·hore (lə-hôr'), city (1972 pop. 2,050,000), capital of Punjab prov., E central Pakistan, on the Ravi River. A railway center near the Indo-Pakistani border, Lahore is a banking and commercial city that markets the products of the surrounding fertile agricultural area. Its diverse industries include motion-picture production, food processing, engineering, metalworking, sawmilling, and the manufacture of textiles, chemicals, pharmaceuticals, iron and steel, electrical goods, rubber and leather items, farm implements, sewing machines, surgical instruments, bicycles, carpets, glass, and matches.

According to Hindu legend, Lahore was founded by Loh, or Lava, son of Rama, the hero of the Sanskrit epic *Ramayana*. In 1036 it was conquered from a Brahman dynasty by the Moslem Turkish Ghaznivids, who made it the capital of their empire in 1106. It passed in 1186 to the Ghori sultans, also from Afghanistan. The city, which suffered Mongol raids in the 13th and 14th cent., entered the period of its greatest glory in the 16th cent., when it became one of the capitals of the Mogul empire. Lahore was annexed in 1767 by the Sikhs. It passed to the British in 1849. When Pakistan won independence in 1947, Lahore became the capital of its West Punjab; from 1955 to 1970 it was the capital of the province of West Pakistan, and upon the province's dissolution it became the capital of Punjab prov.

The architectural remains of the Mogul period, although imperfectly preserved, are among the most splendid of Mogul art. Especially notable are the palace and mausoleum of emperor Jahangir and the Shalimar gardens, just outside the city; only three sections of the gardens remain of the original seven that had symbolized the divisions of the Islamic paradise. Lahore's museum of Indian antiquities is among the most noted in the East.

Lah·ti (lä'tē), city (1975 est. pop. 94,864), Häme prov., S central Finland. It is an important lake port as well as a transportation center. It has many large factories and is a center of the Finnish wood-products industry. The city was chartered in 1905.

Lai·e (lī'ā), town (1970 pop. 3,009), Honolulu co., Hawaii, on the NE coast of Oahu, NNW of Honolulu, in a sugar-cane region. Mormons came here in 1864 and started a large sugar plantation. In 1919 the largest Mormon temple outside of Salt Lake City was built here.

La Jol·la (lə hoi'ə), resort on the Pacific Ocean, S Calif., an uninc. district within the confines of San Diego; founded 1869. The beautiful ocean beaches and sea-washed caves attract visitors and year-round residents. The Scripps Institution of Oceanography and the Univ. of California at San Diego are located in La Jolla.

La Jun·ta (lə hŭn'tə), city (1970 pop. 7,938), seat of Otero co., SE Colo., on the Arkansas River below Pueblo, in a farm and livestock area; founded 1875, inc. 1881. It has railroad shops, a creamery, a grain elevator, and a meat-packing plant.

lake (lāk), body of standing water occupying a hollow in the earth's surface. Ponds are generally small, shallow lakes, although no clear size criterion exists to differentiate them. Although lakes are usually thought to be freshwater bodies, many lakes, especially in arid regions, become quite salty because a high rate of evaporation concentrates inflowing salts. Most modern lakes were formed as a result of glacial action of the Pleistocene ice sheets.

Lake. 1. County (1970 pop. 19,548), 1,256 sq mi (3,253 sq km), W Calif., in a mountain and valley region in the Coast Ranges, drained by Cache Creek and headstreams of Eel River; formed 1861; co. seat Lakeport. It is in a fruit-growing (especially Bartlett pears) region, with dairying, stock raising, and alfalfa, grain, vegetable, and nut farming. It is a leading quick-silver producing county, with sand and gravel quarrying and mineral-water bottling. It is a scenic recreational region, with fishing, hunting, camping, and mineral and hot spring resorts. Part of Mendocino National Forest is in the north. **2.** County (1970 pop. 8,282), 380 sq mi (984.2 sq km), central Colo., in a mining, dairying, and livestock-grazing area drained by the headwaters of the Arkansas River; formed 1861; co. seat Leadville. Gold, silver, lead, copper, molybdenum, and zinc are mined. It contains a large part of San Isabel National Forest. **3.** County (1970 pop. 69,305), 960 sq mi (2,486.4 sq km), central Fla., bounded on the NE by the St. Johns River; formed 1887; co. seat Tavares. It has rolling terrain, with hundreds of lakes. Citrus-fruit growing predominates in the area, which also has canneries and packing houses. Truck crops, watermelon, corn, peanuts, cotton, and poultry are produced. The county also has concrete plants, sawmills, and deposits of diatomite, sand, and peat. **4.** County (1970 pop. 382,638), 457 sq mi (1,183.6 sq km), extreme NE Ill., bounded in the E by Mich., in the N by Wis., and drained by the Fox and Des Plaines rivers; formed 1839; co. seat Waukegan. Dairying and agriculture (livestock, poultry, corn, oats, and wheat) are major occupations. Sand, gravel, and stone deposits are found. The county includes many suburbs of Chicago and numerous lakes where sport fishing and duck hunting are done. **5.** County (1970 pop. 546,253), 514 sq mi (1,331.3 sq km), extreme NW Ind., bounded in the N by Lake Michigan, in the W by the Ill. border, in the S by the Kankakee River, and traversed by the Grand Calumet and Little Calumet rivers; formed 1836; co. seat Crown Point. The heavily industrialized Calumet region is part of the Chicago metropolitan area and is one of the world's most important steel-manufacturing centers. Agricultural areas of the county produce dairy foods, truck crops, poultry, soybeans, and corn. **6.** County (1970 pop. 5,661), 572 sq mi (1,481.5 sq km), W central Mich., drained by the Pere Marquette and Little Manistee rivers, and the small South Branch of the Manistee River; formed 1871; co. seat Baldwin. There is some agriculture, particularly livestock, poultry, dairy products, potatoes, and grain. It is a resort area with hunting and fishing. Part of Manistee National Forest is in the north. **7.** County (1970 pop. 13,351), 2,062 sq mi (5,340.6 sq km), NE Minn., in an extensively watered area bounded on the S by Lake Superior, on the N by a chain of lakes along the Ont. border; formed 1856; co. seat Two Harbors. It has iron-ore mining, lumbering, agriculture (dairy products, poultry, grain, and potatoes), and farm machinery manufacturing. The county lies within Superior National Forest and is part of the recreational region known as Arrowhead Country. **8.** County (1970 pop. 14,445), 1,494 sq mi (3,869.5 sq km), NW Mont., in a mountainous region drained by the Flathead and Swan rivers; formed 1893; co. seat Polson. Livestock, sugar beets, and grain are its main products. Flathead Indian Reservation runs throughout the county. **9.** County (1970 pop. 197,200), 232 sq mi (600.9 sq km), NE Ohio, bounded on the N by Lake Erie and drained by the Grand and Chagrin rivers; formed 1840; co. seat Painesville. It has commercial fisheries, agriculture (fruit, truck crops, and poultry), manufacturing, and a number of lakeshore resorts. **10.** County (1970 pop. 6,343), 8,231 sq mi (21,318.3 sq km), S Oregon, in a mountainous area bordering on Calif. and Nev.; formed 1874; co. seat Lakeview. Lumbering, livestock raising, and dairying are important. The county contains many lakes, Fremont National Forest, and part of the Great Sandy Desert. **11.** County (1970 pop. 11,456), 571 sq mi (1,478.9 sq km), E S.Dak., in an agricultural area watered by Lakes Madison and Herman; formed 1873; co. seat Madison. Cattle, hogs, sheep, corn, and grain crops are raised. The county also has some manufacturing. **12.** County (1970 pop. 8,074), 164 sq mi (424.8 sq km), extreme NW Tenn., bounded on the N by the Ky. border, on the W by the Mississippi River and the Mo. border; formed 1870; co. seat Tiptonville. Cotton, corn, and soybeans are its principal crops.

Lake An·des (ăn'dēz), city (1970 pop. 948), seat of Charles Mix co., SE S.Dak., on Lake Andes SW of Mitchell. In a livestock and wheat region, it is a fishing resort and trade center.

Lake But·ler (bŭt'lər), city (1970 pop. 1,598), seat of Union co., N Fla., N of Gainesville. Lumber and truck crops are produced.

Lake Charles (chärlz), city (1978 est. pop. 75,700), seat of Calcasieu parish, SW La.; inc. 1867. It is located on Lake Charles at the mouth of the Calcasieu River in a rice, timber, oil, and natural-gas region. The city is a leading producer of petrochemicals. Other industries include the manufacture of rubber and tires, plastics, and aluminum, the refining of petroleum, and aircraft maintenance. Lake Charles is an important deepwater port and port of entry.

Lake Che·lan National Recreation Area (shə-lăn'): *see* National Parks and Monuments Table.

Lake City. 1. Town (1974 est. pop. 1,606), a seat of Craighead co., NE Ark., E of Jonesboro near the St. Francis River, in an area of sunken lands caused by earthquakes (1811). **2.** Town (1970 pop. 91), seat of Hinsdale co., SW central Colo., on a branch of the Gunnison River in the San Juan Mts. SE of Montrose. At an altitude of 8,500 ft (2,592.5 m), it is a resort. **3.** City (1970 pop. 10,575), seat of Columbia co., N Fla.; inc. 1921. It was founded in the 1830s as a military post. Lake City is located in a farm and cattle area and produces tobacco, lumber, and naval stores. **4.** City (1970 pop. 704), seat of Missaukee co., N central Mich., NE of Cadillac on Lake Missaukee, in a resort and farm area; inc. 1932. Indian earthworks are nearby. **5.** Town (1970 pop. 6,247), Florence co., E central S.C., S of Florence; settled in the mid-18th cent., inc. 1874. It has livestock and truck farms. Clothing is made, and tobacco is processed.

Lake District, region of mountains and lakes, c.30 mi (50 km) in diameter, NW England. It includes the Cumbrian Mts. and part of the Furness peninsula. There are 15 lakes, several beautiful falls, and some of England's highest peaks. Numerous ancient relics remain, such as the stone circle near Keswick, and there are ruins of old castles and churches and remains of Roman occupation. This scenic district is a favorite resort of artists and writers. William Wordsworth, Samuel Taylor Coleridge, and Robert Southey were known as the Lake Poets. Tourism is a major source of income.

Lake For·est (fôr'ĕst, -ĭst), city (1978 est. pop. 15,200), Lake co., NE Ill., a residential suburb of Chicago, on Lake Michigan; inc. 1861.

Lake Ge·ne·va (jə-nē'və), resort city (1970 pop. 4,890), Walworth co., SE Wis., on Lake Geneva SW of Milwaukee; settled c.1800, inc. as a village 1844, as a city 1883. Many Chicago residents have estates here, and there is a hotel designed by Frank Lloyd Wright.

Lake George (jôrj), village (1970 pop. 1,046), seat of Warren co., E N.Y.; inc. 1903. Situated on the southern tip of Lake George in the foothills of the Adirondack Mts., it is a year-round tourist and sports center. Vestiges of Fort William Henry, built by Sir William Johnson, and of Fort George, begun by Jeffrey Amherst, are here.

Lake·hurst (lāk'hûrst), borough (1970 pop. 2,641), Ocean co., E central N.J.; inc. 1921. It is important as the site of the Lakehurst Naval Air Station (est. 1919), which until 1962 accommodated dirigibles. The burning of the *Hindenburg,* which took 36 lives, occurred here (May 6, 1937) as the hydrogen-filled zeppelin was being moored.

Lake Jack·son (jăk'sən), city (1978 est. pop. 15,900), Brazoria co., SE Texas, on a branch of the Brazos River, near the Gulf of Mexico; founded 1941. It is mainly residential.

Lake·land (lāk'lənd). **1.** Resort city (1978 est. pop. 48,800), Polk co., central Fla., in the highland region; inc. 1885. It is an important processing and shipping center for a citrus-fruit region and is a retail trade center serving a large area. It has many diverse manufactures. **2.** City (1970 pop. 2,569), seat of Lanier co., S central Ga., NE of Valdosta, in a farm area.

Lake Mead National Recreation Area (mēd): *see* National Parks and Monuments Table.

Lake Mer·e·dith National Recreation Area (mĕr'ĭ-dĭth): *see* National Parks and Monuments Table.

Lake of the Woods (wŏŏdz), county (1970 pop. 3,987), 1,311 sq mi (3,395.5 sq km), N Minn., bounded on the NE by the Rainy River, and extending N between the Canadian provs. of Man. and Ont; formed 1922; co. seat Baudette. It includes the Northwest Angle, the northernmost land in the continental United States. It has agriculture (seeds, small grains, potatoes, hay, and livestock), forestry, pulp and paper factories, a pharmaceutical industry, and resorts.

Lake of the Woods, 1,485 sq mi (3,846.2 sq km), c.70 mi (115 km) long, on the U.S.-Canada border in the pine forest region of N Minn., SE Man., and SW Ont. More than two thirds of the lake is in Canada. A remnant of former glacial Lake Agassiz, it has a very irregular shoreline and approximately 14,000 islands. Abundant in fish and game, the region is a resort area.

Lake Park, town (1970 pop. 6,993), Palm Beach co., SE Fla., on Lake Worth (a lagoon) N of West Palm Beach.

Lake Pla·cid (plăs'ĭd), village (1970 pop. 2,731), Essex co., NE N.Y.; settled 1850, inc. 1900. In the Adirondack Mts. at an altitude of 1,800 ft (549 m), the village surrounds Mirror Lake. It is a famous resort and sports center. The 1932 Winter Olympics were held in the village; the 1980 games will also take place here. Lake Placid has a summer theater and music festival and a figure-skating school. The burial place of the abolitionist John Brown is nearby.

Lake Pleas·ant (plĕz′ənt), village (1970 pop. 100), seat of Hamilton co., E central N.Y., on Lake Pleasant NE of Utica. In the Adirondacks, it has summer and winter resorts and a lumbering industry.

Lake·port (lāk′pôrt′), city (1970 pop. 3,005), seat of Lake co., W Calif., on Clear Lake NNW of San Francisco; settled 1859, inc. 1888. It is a summer resort and farm trade center.

Lake Prov·i·dence (prŏv′ĭ-dəns, -dĕns′), town (1970 pop. 6,183), seat of East Carroll parish, NE La., at the foot of Lake Providence near the Mississippi and ENE of Monroe, in a cotton and farm area; settled c.1812, inc. 1876. It was a refuge for pirates and outlaws.

Lake Ron·kon·ko·ma (rŏng-kŏng′kə-mə, rŏn-), resort village (1970 pop. 8,250), Suffolk co., SE N.Y., on central Long Island, NW of Patchogue on Lake Ronkonkoma.

Lake Suc·cess (sək-sĕs′), village (1970 pop. 3,254), Nassau co., SE N.Y., on NW Long Island; settled c.1730, inc. 1926. A residential suburb of New York City, Lake Success was the temporary home of the United Nations from 1946 to 1950.

Lake·view (lāk′vyōō′). **1.** Uninc. town (1970 pop. 16,222), Calhoun co., S central Mich., a suburb of Battle Creek. **2.** Town (1970 pop. 2,705), seat of Lake co., S Oregon, near the Calif. border E of Klamath Falls; founded 1876, inc. 1889. It is the trade center for a lumbering and cattle- and sheep-raising area. Uranium from nearby mines is processed here.

Lake Village, resort city (1970 pop. 3,310), seat of Chicot co., SE Ark., on Lake Chicot, near the Mississippi River SW of Greenville, Miss.; founded in the 1850s, inc. 1901. It has cotton gins, woodworking plants, and commercial fisheries.

Lake Wales (wālz), city (1970 pop. 8,240), Polk co., central Fla., in the highlands W of Lake Kissimmee and SE of Lakeland; platted 1911. A noted lake resort, it is also a lumber-milling center.

Lake·wood (lāk′wŏŏd′). **1.** City (1978 est. pop. 80,900), Los Angeles co., S Calif., a residential and industrial suburb of Long Beach; inc. 1954. **2.** City (1978 est. pop. 125,800), Jefferson co., N central Colo., a suburb of Denver; inc. 1969. **3.** Township (1978 est. pop. 31,700), Ocean co., E central N.J., on the Metedeconk River, a health resort in a pine forest and lake region, near the Atlantic coast; settled 1800, inc. 1892. It has poultry farms and plants making a variety of products. It was the site of early ironworks. Georgian Court College is in the town, and the former Rockefeller estate (now a state reserve), a naval air station, and a parachuting center are nearby. **4.** City (1978 est. pop. 63,800), Cuyahoga co., NE Ohio, a suburb of Cleveland, on Lake Erie; inc. 1911. It has many varied industries.

Lake Worth (wûrth), city (1978 est. pop. 27,600), Palm Beach co., SE Fla., on Lake Worth (a lagoon); inc. 1913. It is a resort center popular for its bathing and fishing facilities. Sports equipment, clothing, and food products are among its manufactures.

La·kin (lā′kən), city (1970 pop. 1,570), seat of Kearny co., SW Kansas, W of Garden City on the Arkansas River, in a farm area; founded 1872, inc. 1887. There are gas and oil wells in the vicinity.

La·ko·ta (lə-kō′tə), city (1970 pop. 964), seat of Nelson co., NE N.Dak., ESE of Devils Lake, in a wheat, livestock, and dairy region.

Lak·shad·weep (läk′shäd-wēp′), union territory, India: *see* Laccadive, Minicoy, and Amindivi Islands.

La·mar (lə-mär′). **1.** County (1970 pop. 14,335), 605 sq mi (1,567 sq km), NW Ala., in a level region bordering on Miss. and drained by the Buttahatchee River; formed 1867 as Jones co., re-established as Sanford co. 1868, name changed to Lamar 1877; co. seat Vernon. It has cotton and bee farming, mule raising, lumbering, and clothing manufacturing. **2.** County (1970 pop. 10,688), 181 sq mi (468.8 sq km), W central Ga.; formed 1920; co. seat Barnesville. In a piedmont agricultural area yielding peaches, pecans, cotton, corn, truck crops, and livestock, it also has a lumbering industry. Textiles are manufactured. **3.** County (1970 pop. 15,209), 500 sq mi (1,295 sq km), S Miss., drained by Black and Red creeks; formed 1904; co. seat Purvis. It has extensive lumbering, corn, cotton, and pecan farms, and oil and gas wells. Its manufactures include clothing, wood products, furniture, and concrete. Petroleum refining is done. **4.** County (1970 pop. 36,062), 984 sq mi (2,548.6 sq km), NE Texas, bounded on the N by the Red River and the Okla. border, on the S by the North Fork of the Sulphur River; formed 1840; co. seat Paris. It has diversified agriculture (cotton, corn, grain, pecans and peanuts, fruit, truck crops, dairy products, livestock, and poultry). Food processing, lumbering, and some manufacturing are done.

Lamar. 1. City (1970 pop. 7,797), seat of Prowers co., SE Colo., on the Arkansas River E of Pueblo; inc. 1886. It is a processing center in a farm area. **2.** City (1970 pop. 3,760), seat of Barton co., SW Mo., NNE of Joplin, in a coal and farm area; founded c.1856. President Harry S. Truman was born here in 1884.

La Marque (lə märk′), city (1970 pop. 16,131), Galveston co., SE Texas, in an agricultural and oil area; settled c.1860, inc. 1953. Originally a farm settlement, it later became a railroad shipping point. Today it is primarily a residential suburb for workers in Texas City and other nearby industrial centers.

Lamb (lăm), county (1970 pop. 17,770), 1,022 sq mi (2,647 sq km), NW Texas, on the Llano Estacado drained by the Double Mountain Fork of the Brazos River; formed 1876; co. seat Littlefield. This rich agricultural and livestock region, with much irrigated land, leads the state in production of grain sorghums. It also raises cotton, corn, some fruit, truck crops, dairy products, poultry, beef cattle, and hogs.

Lam·ba·ré·né (läm-bə-rā′nā), town (1970 est. pop. 18,000), W Gabon, on the Ogooué River. It is a river port and trade center. The mission hospital founded by Albert Schweitzer is here.

Lam·bèse (läm-bĕz′): *see* Tazoult, Algeria.

Lam·beth (lăm′bəth), borough (1971 pop. 302,616) of Greater London, SE England, on the Thames River. The borough was created in 1965 by the merger of the metropolitan borough of Lambeth with part of the metropolitan borough of Wandsworth. It is largely residential but is important as an area of governmental and commercial offices. The National Theatre (the Old Vic), the National Film Theatre, and the Royal Festival Hall are in Lambeth, as are the Imperial War Museum, Morley College, and eight hospitals, two of which (St. Thomas's and King's College hospitals) have medical schools.

La·me·sa (lə-mē′sə), city (1970 pop. 11,559), seat of Dawson co., NW Texas, in the Llano Estacado; inc. 1917. Lamesa is a processing and shipping center for an irrigated area where cattle and poultry are raised and cotton, grains, black-eyed peas, and soybeans are grown.

La Me·sa (lä mā′sə), city (1978 est. pop. 45,400), San Diego co., S Calif., a suburb of San Diego; inc. 1912. It is a retail-trade center for an area of truck and poultry farms.

La·mí·a (lä-mē′ä, lā′mē-ə), city (1971 pop. 37,872), capital of Fthiótis prefecture, E central Greece. It is an agricultural center. Founded about the 5th cent. B.C., it was the chief city of the small region of Malis and developed as an ally of Athens. It gave its name to the Lamian War (323–322 B.C.), waged by the confederate Greeks against Antipater, the Macedonian general, who took refuge in the city and was besieged here for several months. Antipater conquered (322 B.C.) the confederates at Crannon, near Lárisa.

La Mi·ra·da (lä mə-rä′də), city (1978 est. pop. 41,000), Los Angeles co., S Calif.; inc. 1960.

Lam·mer·muir Hills (lăm′ər-myŏŏr′), range of hills, Lothian and Borders regions, SE Scotland. Meikle Says Law (1,749 ft/533.4 m) is the highest point. Sheep are grazed in the hills.

La·moille (lə-moil′), county (1970 pop. 13,309), 475 sq mi (1,230.3 sq km), N central Vt., in a lumbering and dairy area drained by the Lamoille River; formed 1835; co. seat Hyde Park. It has deposits of asbestos and granite quarries and manufactures machinery, wood products, and textiles. The Green Mts. are in the north.

La·mon Bay (lä-môn′, lə-), landlocked bay, c.60 mi (96 km) wide, Philippines, bounded on the W by Luzon, on the N by the Polillo Islands, on the S by Alabat Island, on the SE by Camarines Norte. It merges with the Philippine Sea.

La·mont (lə-mŏnt′), village (1970 pop. 7,007), Kern co., S Calif., SSE of Bakersfield, in a farm and fruit region.

La Moure (lə mŏŏr′), county (1970 pop. 7,117), 1,137 sq mi (2,944.8 sq km), SE N.Dak., drained by the James River; formed 1873; co. seat La Moure. It is in a rich agricultural area that produces grain, livestock, and dairy products. There is some manufacturing.

La Moure, city (1970 pop. 951), seat of La Moure co., SE N.Dak., on the James River SSE of Jamestown, in a grain, livestock, and dairy area; settled 1883, inc. 1906.

Lam·pas·as (lăm-păs′əs), county (1970 pop. 9,323), 726 sq mi (1,880.3 sq km), central Texas, bounded on the W by the Colorado River; formed 1856; co. seat Lampasas. In a livestock and poultry area, it has general agriculture (cotton, grain sorghums, corn, oats, barley, wheat, pecans, and fruit) and deposits of clay and glass sand.

Lampasas, city (1970 pop. 5,922), seat of Lampasas co., central Texas, NNW of Austin; settled 1854, inc. 1874. An early cow town, it now manufactures leather goods, plastic products, and machinery. It is also a shipping point for livestock, pecans, wool, and mohair.

Lam·pe·du·sa (lăm'pĭ-dōō'zä), island, 8 sq mi (20.7 sq km), S Sicily, Italy, in the Mediterranean Sea between Malta and Tunisia, the largest of the Pelagie Islands. Sponge and sardine fishing are the main occupations. Lampedusa was settled in the 18th cent.

Lamp·sa·cus (lămp'sə-kəs), ancient Greek city of NW Asia Minor, on the Hellespont (now Dardanelles) opposite Callipolis (now Gallipoli). It was colonized in the 7th cent. B.C. by Greeks. After the Battle of Mycale (479 B.C.) the citizens joined with the Athenians, and the city continued to flourish under the Greeks and the Romans.

La·na·i (lä-nä'ē, lə-nī'), island (1970 pop. 2,204), 141 sq mi (365.2 sq km), central Hawaii, W of Maui Island across the Auau Channel. For many years the island was used for sugar-cane raising and cattle grazing. The entire island was purchased in 1922 by a pineapple company and developed as a pineapple-growing center.

Lan·ark (lăn'ərk), burgh (1971 pop. 8,701), Strathclyde region, S central Scotland, on the Clyde River. It has cattle markets and textile mills. Robert Owen conducted industrial and social experiments at the nearby New Lanark mills, founded by his father-in-law, David Dale, in 1785. Until 1975 Lanark was county town of Lanarkshire.

Lan·ark·shire (lăn'ərk-shîr', -shər), former county, central Scotland. The county town was Lanark. The region, which includes the valley of the Clyde and is sometimes referred to as Clydesdale, has a varied terrain, rising from the level valley in the north to more than 2,000 ft (610 m) in the mountainous southern portion. In 1975 Lanarkshire became part of the Strathclyde region.

Lan·ca·shire (lăng'kə-shîr', -shər), nonmetropolitan county (1976 est. pop. 1,375,500), 1,878 sq mi (4,864 sq km), N England, on the Irish Sea; administrative center Lancaster. The northwestern portion of the county is part of the picturesque Lake District; in the west and south are lowlands (the Lancashire plain) and occasional moors, with rich deposits of coal, slate, and sandstone. In Anglo-Saxon times Lancashire was part of the kingdom of Northumbria. In 1351 it was made a county palatine, and in 1399 the palatine rights were vested in the king. Lancashire's economic growth began in medieval times with the introduction of the woolen industry. The process was accelerated by the Industrial Revolution, and the population increased rapidly in the 19th and early 20th cent.

Lan·cas·ter (lăng'kə-stər), municipal borough (1973 est. pop. 50,570), administrative center of Lancashire, NW England, on the Lune River. The city's products include furniture, textiles, farm machinery, linoleum, soap, and flour. Lancaster Castle occupies the site of a Roman station. It has a Norman keep and tower (built 1170) with a turret called John o' Gaunt's Chair.

Lan·cas·ter (lăng'kăs-tər, lăng'kə-stər). **1.** County (1970 pop. 167,972), 845 sq mi (2,188.6 sq km), SE Nebr., drained by small tributaries of the Missouri River; formed 1855; co. seat Lincoln. In a commercial and agricultural area yielding corn, milo, wheat, and livestock, it has some mining, and an extensive food-processing industry. Furniture, wood, paper, and metal products, machinery, chemicals, plastics, and concrete are manufactured. **2.** County (1970 pop. 320,079), 945 sq mi (2,447.6 sq km), SE Pa., in one of the richest agricultural areas of the United States, bounded in the W by the Susquehanna River; formed 1729; co. seat Lancaster. Tobacco, cattle, and small grains are of major importance. Manufacturing (linoleum, textiles, lumber, and leather), limestone quarrying, and slate mining are also done. **3.** County (1970 pop. 43,328), 504 sq mi (1,305.4 sq km), N S.C., bounded on the W by the Catawba River and on the N by the N.C. border; formed 1785; co. seat Lancaster. It is in a timber-processing and agricultural area yielding cotton, corn, wheat, oats, hay, and tobacco. **4.** County (1970 pop. 9,126), 142 sq mi (367.8 sq km), E Va., on the S shore of the Northern Neck peninsula, bounded on the W and S by the Rappahannock River; formed 1652; co. seat Lancaster. It has agriculture (truck crops, tobacco, poultry, and livestock), commercial fisheries specializing in oysters and crab, and lumbering and food-processing industries. There are many summer resorts along the heavily indented shoreline.

Lancaster. 1. Uninc. city (1978 est. pop. 37,300), Los Angeles co., S Calif., in Antelope Valley and the Mojave Desert; laid out 1894. It is a trade center for an irrigated farming area. An Indian museum has prehistoric artifacts. Edwards Air Force Base is nearby. **2.** City (1970 pop. 3,230), seat of Garrard co., central Ky., SSW of Lexington, in a bluegrass region; settled 1798. Burley tobacco, corn, wheat, and hay are grown. **3.** Town (1970 pop. 6,095), Worcester co., central Mass., NE of Wachusett Reservoir; settled 1643, inc. 1653. Destroyed in King Philip's War, it was later attacked again. The town has one of Charles Bulfinch's finest churches (1816-17). **4.** City (1970 pop. 821), seat of Schuyler co., N Mo., near the Chariton River N of Kirsville, in a grain and poultry area; inc. 1856. **5.** Town (1970 pop. 3,166), seat of Coos co., NW N.H., at the influx of the Israel River into the Connecticut River, ENE of St. Johnsbury, Vt.; inc. 1764. It is a trade center for White Mt. resorts. **6.** Village (1970 pop. 13,365), Erie co., W N.Y.; inc. 1849. It has lumber mills, dairy farms, and stone quarries. **7.** City (1978 est. pop. 39,700), seat of Fairfield co., S central Ohio, on the Hocking River, in a livestock and dairy area; founded 1800, inc. as a village 1831. Its manufactures include glassware, shoes, and automotive parts. The birthplace of the brothers Gen. William T. Sherman and Senator John Sherman has been preserved. In the area are many covered bridges and an Indian mound in the form of a cross. **8.** City (1978 est. pop. 53,300), seat of Lancaster co., SE Pa., on the Conestoga River, in the heart of the Pennsylvania Dutch country; inc. as a city 1818. It is the commercial center for one of the most productive agricultural counties in the United States. Chief products are tobacco, small grains, and livestock. Lancaster has a huge farmers' market and one of the largest stockyards east of Chicago. Manufactures include linoleum, watches, radio tubes, cigars, razors, tools, and metal products. The area was settled by German Mennonites c.1709, and the famous Conestoga wagon was developed here shortly thereafter. A munitions center during the Revolution, it was briefly (1777) a meeting place of the Continental Congress and served as capital of the state for more than 10 years before 1812. Points of interest include Wheatland, the home of James Buchanan (built in 1828; a national shrine since 1962); homes of several Revolutionary War patriots; and the Fulton Opera House (1854), one of the oldest continuously operating theaters in the country and now a historic monument. **9.** Town (1970 pop. 9,186), seat of Lancaster co., N S.C., near the N.C. border and the Catawba River NNE of Columbia; settled c.1795, inc. 1830. Textiles and concrete, wood, metal, and grain products are made. **10.** City (1970 pop. 10,522), Dallas co., NE Texas, in a blackland farming area; settled 1846, inc. 1886. Clothing is made in the city. **11.** Village (1970 pop. 125), seat of Lancaster co., E Va., N of Newport News. Seafood and vegetables are canned here. **12.** City (1970 pop. 3,756), seat of Grant co., SW Wis., N of Dubuque, Iowa, in a farm and dairy area; settled before 1840, inc. 1878.

Lan·cas·ter Sound (lăng'kə-stər), arm of Baffin Bay, c.200 mi (320 km) long and 40 mi (64.4 km) wide, E Franklin dist., Northwest Territories, Canada. It extends west between Devon and Baffin islands and is part of the shortest water route across north Canada to the Beaufort Sea. It was discovered in 1616.

Lan-chou or **Lan·chow** (both: län'jō'), city (1970 est. pop. 1,500,000), capital of Kansu prov., W China, on the Huang Ho at its confluence with the Wei. It has one of the largest oil refineries in the country, a gas-diffusion plant for processing plutonium, textile mills, and petrochemical, plastic, chemical fertilizer, and machine manufactures.

Lan·der (lăn'dər), county (1970 pop. 2,666), 5,621 sq mi (14,558.4 sq km), central Nev., with the Reese River flowing N between Toiyabe Range and the Shoshone Mts., and the Humboldt River flowing across the N part of the county; formed 1862; co. seat Austin. It is in a copper and barite mining area (with some gold, silver, and lead deposits), and has cattle ranches.

Lander, town (1970 pop. 7,125), seat of Fremont co., W central Wyo., SW of Riverton on the Popo Agie River; settled c.1870, inc. 1890. At an altitude of c.5,563 ft (1,696.7 m), it is a tourist center in a dude-ranch, livestock, farm, and mineral (oil, iron ore, gold, uranium, and coal) area. Wind River Indian Reservation is nearby.

Landes (läɴd), region, SW France. It is a vast, flat, nearly triangular tract of sand and marshland, stretching along the Atlantic coast for more than 100 mi (160.9 km) between the Adour River and the Médoc region. Formerly, sheep grazing was the only occupation in this insalubrious region, but much of the land has been reclaimed through drainage and the planting of pine forests.

Landes, department (1975 pop. 288,323), 3,566 sq mi (9,235.9 sq km), SW France, in Gascony, on the Atlantic coast; capital Mont-de-Marsan.

Lands·berg am Lech (länts'bĕrKH äm lĕKH') or **Landsberg,** town

(1970 pop. 14,205), Bavaria, S West Germany, on the Lech River. Textiles, metal goods, and paper are manufactured. Its fortress served as a political prison; Adolf Hitler wrote *Mein Kampf* while imprisoned here in 1923–24, and numerous convicted Nazi war criminals were held here after 1945.

Lands End (lăndz' ĕnd'), promontory, Cornwall, SW England, forming the westernmost extremity of the English mainland. Of wave-carved granite, it has cliffs c.60 ft (20 m) high. Offshore are reefs and rocky islets, on one of which is Longships Lighthouse.

Lands·hut (länts'ho͝ot'), city (1974 est. pop. 56,405), Bavaria, SE West Germany, on the Isar River. Once the capital of Lower Bavaria, it is now a transportation and industrial center. Manufactures include glass, ceramics, chemicals, and machinery. Founded in 1204, the city suffered heavily in the Thirty Years' War (1618–48).

Lands·kro·na (länts-kro͞o'nä), city (1975 est. pop. 38,409), Malmöhus co., SW Sweden, a seaport on the Øresund. It is a commercial and industrial center. Manufactures include refined sugar, metal goods, rubber, and textiles. Chartered in 1413, Landskrona was devastated in the Danish-Swedish wars of the 16th–17th cent. The city was largely rebuilt in the mid-18th cent.

Lane (lān). **1.** County (1970 pop. 2,707), 720 sq mi (1,864.8 sq km), W Kansas, in a rolling plains region with low hills in the E and NE; formed 1886; co. seat Dighton. It is in an agricultural area that produces grain and livestock. There is some manufacturing in the county. **2.** County (1970 pop. 215, 401), 4,562 sq mi (11,815.6 sq km), W Oregon, on the Pacific coast in a mountainous area crossed by the Willamette and Siuslaw rivers; formed 1851; co. seat Eugene. Its economy is based on timber and forest products, agriculture (fruit, grain, poultry, truck crops, livestock, and dairy products), and recreational and tourist facilities.

La·nett (lə-nĕt'), cotton-milling city (1970 pop. 6,908), Chambers co., E Ala., on the Chattahoochee River ENE of Montgomery; inc. 1893.

Lang·don (lăng'dən), city (1970 pop. 2,182), seat of Cavalier co., NE N.Dak., NW of Grand Forks; inc. 1888. It is a shipping point for durum wheat and cattle.

Lang·e·land (läng'ə-län'), narrow island (1965 pop. 17,745), 110 sq mi (284.9 sq km), S Denmark, between Fyn and Lolland. The island is largely agricultural, and grain is the chief product.

Lang·en·sal·za, Bad (bät läng'ən-zäl'tsä), town (1970 pop. 16,951), Erfurt district, SW East Germany, on the Unstrut River. It is an industrial and horticultural center. Manufactures include textiles, paper, processed food, and beer. Bad Langensalza was an early seat (13th cent.) of the Teutonic Knights. The town was annexed by Prussia in 1815. The town has retained parts of its medieval walls.

Lang·lade (lăng'lăd'), county (1970 pop. 19,220), 858 sq mi (2,222.2 sq km), NE Wis., in a wooded lake region drained by the Wolf and Eau Claire rivers; formed 1879; co. seat Antigo. Lumbering and dairying are its chief industries. Part of Nicolet National Forest is here.

Lang·ley Park (lăng'lē), uninc. town (1970 pop. 11,564), Prince Georges co., W central Md., a suburb of Washington, D.C.

Lan·gres (läɴ'grě), town (1968 pop. 11,835), Haute-Marne dept., NE France. It has an old and famous cutlery industry. Langres has preserved a large part of its ancient fortifications.

Lang·side (lăng'sīd'), district of Glasgow, S central Scotland. At the Battle of Langside (1568) the 1st earl of Murray defeated the forces of Mary Queen of Scots led by Archibald Campbell, 5th earl of Argyll. As a result Mary fled to England.

Lan·gue·doc (läɴg-dôk'), region and former province, S France, bounded by the foot of the Pyrenees, the upper Garonne River, the Auvergne Mts., the Rhône, and the Mediterranean. The Garonne plains are fertile farming and wine-producing districts. The name was derived from the language of its inhabitants. It now generally refers to Lower Languedoc, an alluvial plain along the Mediterranean, with a warm climate; wine is the chief product. Historically, Languedoc roughly corresponds to Narbonensis prov. of Roman Gaul; Lower Languedoc was the later Septimania. It was conquered by the Franks in the 8th cent. and incorporated into the French royal domain in 1271.

La·nier (lə-nîr'), county (1970 pop. 5,031), 177 sq mi (458.4 sq km), S Ga., in a coastal-plain area intersected by the Alapaha River; formed 1919; co. seat Lakeland. Agriculture (corn, sweet potatoes, tobacco, fruit, and livestock) and forestry are the major industries.

Lans·dale (lănz'dāl'), borough (1978 est. pop. 21,200), Montgomery co., SE Pa.; inc. 1872. The Jenkins House here dates from 1702.

Lans·downe (lănz'doun'). **1.** Uninc. town (1970 pop. 16,976), Baltimore co., NE Md., a suburb of Baltimore. **2.** Borough (1970 pop. 14,090), Delaware co., SE Pa., a suburb of Philadelphia; inc. 1893.

L'Anse (lăns), village (1970 pop. 2,538), seat of Baraga co., N Mich., W Upper Peninsula, at the head of Keweenaw Bay; inc. 1873. A fur-trading post and a mission were on the site. Today it is a resort where wood products are made.

Lans·ford (lăns'fərd, lănz'-), industrial borough (1970 pop. 5,168), Carbon co., E Pa., S of Hazleton; founded 1846, inc. 1877. It has coal mines and machine works.

Lan·sing (lăn'sĭng). **1.** Village (1978 est. pop. 30,200), Cook co., NE Ill., a suburb of Chicago, near the Ind. border; inc. 1893. **2.** City (1978 est. pop. 124,100), state capital, Clinton, Eaton, and Ingham cos., S Mich., on the Grand River at its confluence with the Red Cedar River; inc. 1859. Automobiles and automobile parts are the major manufactures. The city grew after it was made the state capital (1847), and industrial development came with the railroads (1870s) and the automobile industry (1897).

Lan·tan·a (lăn-tăn'ə), town (1970 pop. 7,126), Palm Beach co., SE Fla., on Lake Worth (a lagoon) S of West Palm Beach; inc. 1921.

La·nús (lä-no͞os'), city (1970 pop. 449,824), Buenos Aires prov., E Argentina. It is an administrative center in the Greater Buenos Aires area.

La·nu·vi·um (lə-no͞o'vē-əm, -nyo͞o'-), ancient city of Latium, Italy, S of Rome, in the Alban Hills near the Appian Way. It was celebrated for its temple of Juno. There are ruins of a temple and Roman walls.

La·od·i·ce·a (lā-ŏd'ĭ-sē'ə, lā'ə-dĭ-), name of several Greek cities of Asia and Asia Minor built by the Seleucids in the 3rd cent. B.C. The most important, Laodicea ad Lycum, was north of Colossae near the present Denizli. On the trade route from the East, the city prospered, particularly under Rome. Extensive Roman ruins include theaters, an aqueduct, a gymnasium, and sarcophagi. Laodicea ad Mare, a seaport of Syria south of Antioch, flourished under the Romans.

Laoigh·is or **Lao·is** (both: lā'ĭsh), or **Leix** (lāks), county (1971 pop. 45,349), 664 sq mi (1,719.8 sq km), central Republic of Ireland; co. town Port Laoise. A part of the central plain of Ireland, Laoighis is generally level, except for the Slieve Bloom Mts. in the northwest. Agriculture and dairy farming are the main occupations.

Laon (läɴ), commercial town (1975 pop. 27,914), capital of Aisne dept., N France. It has forges, a printing plant, and factories that make heating equipment and metal goods. Situated on a rocky height c.300 ft (90 km) above the plain, it was fortified as early as Roman times. During the Middle Ages it was torn by bitter struggles against the bishops by the burghers, who ultimately succeeded (12th cent.) in obtaining recognition of their charter.

La·os (lä'ōs, lā'ŏs), country (1976 est. pop. 3,500,000), 91,428 sq mi (236,800 sq km), SE Asia. The capital is Vientiane; until 1975 the royal capital was Luang Prabang. A landlocked region, Laos is bordered by China on the north, by Vietnam on the east, by Cambodia on the south, and by Thailand and Burma on the west. In general, the Mekong River, most of which flows in a broad valley, forms the

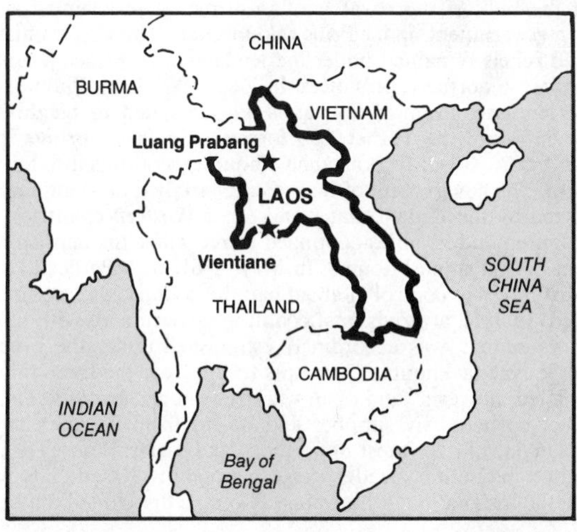

boundaries with Burma and Thailand. Except for the Mekong lowlands and three major plateaus, the terrain of Laos is rugged, mountainous, and heavily forested; crests in the north tower over 9,000 ft (2,745 m).

Economy. Laos is one of the regions of Southeast Asia least touched by modern civilization. There are no railroads; roads and trails are limited; and use of the country's main communications artery, the Mekong River, is impeded by many falls and rapids. More than half of the population lives along the Mekong and its tributaries. Most are subsistence farmers, who even weave their own cloth. Rice is by far the chief crop; corn and vegetables are also grown. Commercial crops include coffee, tobacco, sugar cane, and cotton. Fish from the rivers supplement the diet. Forests cover about two thirds of the country; teak is cut and lac is extracted, but poor transportation and the lack of industry limit production. Although tin is mined, mineral resources are practically undeveloped.

History. The Laotians are descendants of Thai tribes that were pushed southward from Yünnan, China, in the 13th cent. and gradually infiltrated the territory of the Khmer Empire. In the mid-14th cent. a powerful kingdom called Lan Xang was founded in Laos by Fa Ngoun (1353–73), who is also credited with the introduction of Hinayana Buddhism and much of Khmer civilization into Laos. Lan Xang waged intermittent wars with the Khmer, Burmese, Vietnamese, and Thai, and by the 17th cent. it held sway over sections of Yünnan, southern Burma, the Vietnamese and Cambodian plateaus, and large stretches of northern Thailand. In 1707, however, internal dissensions brought about a split of Lan Xang into two kingdoms: Luang Prabang in upper (northern) Laos and Vientiane in lower (southern) Laos. In the early 19th cent. Siam was dominant over the two Laotian kingdoms, although Siamese claims were disputed by Annam. After French explorations in the late 19th cent. Siam was forced (1893) to recognize a French protectorate over Laos, which was incorporated into the union of Indochina.

During World War II, Laos was gradually occupied by the Japanese, who in 1945 persuaded the king of Luang Prabang to declare the country's independence. The French nevertheless re-established (1946) dominion over Laos, recognizing the king as constitutional monarch of the entire country. In 1949 Laos became a semiautonomous state within the French Union. In 1951 a Communist Laotian nationalist movement, the Pathet Lao, was formed by Prince Souphanouvong in North Vietnam. In 1953 Pathet Lao guerrillas accompanied a Viet Minh invasion of Laos from Vietnam and established a government at Samneua in northern Laos. That year Laos attained full sovereignty.

The new country faced immediate civil war as Pathet Lao forces, supported by the Viet Minh, soon occupied sizable portions of the country. Agreements reached at the Geneva Conference of 1954 provided for the withdrawal of foreign troops and the establishment of the Pathet Lao in two northern provinces. An agreement between the royal forces and the Pathet Lao signed in 1957 provided for the re-establishment of government authority in the north, partial integration of Pathet Lao troops into the Laotian royal army, and Pathet Lao participation in the government. In 1959, however, hostilities were renewed. A succession of coups resulted (1960) in a three-way struggle for power. The neutralists were headed by Premier Souvanna Phouma, who remained in the administrative capital of Vientiane. The rightists were led by Gen. Phoumi Nosavan, who controlled the bulk of the royal Laotian army; he proclaimed a pro-Western government under Prince Boun Oum. The Communist Pathet Lao rebels remained under the leadership of Prince Souphanouvong in the northern provinces. In Dec., 1960, Gen. Phoumi captured Vientiane; Premier Souvanna Phouma fled to neighboring Cambodia while the Pathet Lao forces, allied with proneutralist Laotian troops (loyal to Souvanna Phouma), continued fighting in the north. The government of Boun Oum, installed in Vientiane, was recognized by the United States and other Western countries. The Soviet Union and its allies continued to recognize the deposed government of Souvanna Phouma. In May, 1961, with Pathet Lao and neutralist forces in control of about half the country, a cease-fire was arranged. In 1962 a provisional coalition government, with all factions represented, was accordingly established under the premiership of Souvanna Phouma. Attempts to integrate the three military forces failed, however, and open warfare resumed in 1963. The Pathet Lao, bolstered by supplies and troops from North Vietnam, solidified control over most of northern and eastern Laos. Disgruntled right-wing military leaders staged a coup in 1964 and attempted to force the resignation of Souvanna Phouma; the United States and the Soviet Union emphasized their support of the premier, however, and he remained in office. Pathet Lao guerrilla activity decreased after the start (1965) of U.S. bombings of North Vietnamese military bases and communications routes.

Communist pressure increased during 1969, and early in 1970 the Pathet Lao launched several major offensives. North Vietnamese use of the Ho Chi Minh Trail increased greatly after a U.S.-South Vietnamese invasion of Cambodia in the spring of 1970 closed alternate routes. The attack drove the North Vietnamese deeper into Laos, and Laos became another battleground of the Vietnam War, with heavy U.S. aerial bombardments. The United States extended enormous military and economic aid to the Laotian government, but the Pathet Lao, supported by North Vietnamese troops, scored major gains, consolidating their control over more than two thirds of Laotian territory (but over only one third of the population). A cease-fire was finally declared in Feb., 1973. A coalition government (inaugurated in early 1974) was established with Souvanna Phouma as premier. In effect, the country was partitioned between the Pathet Lao and the royalists, with each side administering its own zone. This arrangement lasted until April, 1975, at which time Cambodia and South Vietnam fell to the Communists. With the collapse of pro-Western regimes in neighboring states, the Pathet Lao moved gradually and, in general, peacefully, to consolidate its control over all Laos. By Aug., 1975, the transformation was completed and most of the pro-Western middle class had fled to Thailand. On Dec. 3, 1975, the monarchy was abolished and the abdication of King Savang Vatthana announced. (The king and his family were arrested in 1977 for antigovernment activities.) Prince Souphanouvong became chief of state and Kaysone Phomvihan premier.

La Paz (lä päs′, lə päz′), city (1976 pop. 654,713), W Bolivia, administrative capital (since 1898) and largest city of Bolivia. La Paz, the highest capital in the world, lies at an altitude of c.12,000 ft (3,660 m) and is crowded into a long, narrow valley cut by the La Paz River. The site, where there was an Inca village, was chosen in 1548 because it offered protection in winter from the wind and cold of the barren high plateau c.1,400 ft (425 m) above. The Plaza Murillo, named after the independence leader Pedro Domingo Murillo, with the national palace, cathedral, and other buildings, is small; there are only a few broad, long avenues, and the streets ascend steeply on either side. La Paz's location on colonial trade routes made it the commercial and political focus of colonial life. It is an agricultural market and has light manufacturing industries. There are extraordinary tourist attractions in the region, notably the Andean peaks Illimani and Illampú and Lake Titicaca.

La Paz, town (1970 pop. 46,011), capital of Baja California Sur state, NW Mexico. It is a fishing and pearling center located on a bay near the entrance to the Gulf of California. Maize, cotton, and cattle are raised in the surrounding area. Its warm, dry climate has made La Paz a winter resort.

La·peer (lə-pēr′), county (1970 pop. 52,361), 659 sq mi (1,706.8 sq km), E Mich., drained by the Flint and Belle rivers; organized 1833; co. seat Lapeer. There is much farming of grain, sugar beets, beans, potatoes, celery, onions, and fruit, as well as stock raising and dairying. Industries include the manufacture of foundry products, aircraft parts, and furniture. There are numerous small lakes in the area.

Lapeer, city (1970 pop. 6,314), seat of Lapeer co., SE Mich., E of Flint, in a farm and lake region; settled 1831, inc. 1869.

La Pé·rouse Strait (lä pā-ro͞oz′) or **So·ya Strait** (sō′yä), channel, 25 mi (40.2 km) wide, separating N Hokkaido Island, Japan, from S Sakhalin Island, USSR, and connecting the Sea of Japan on the W with the Sea of Okhotsk on the E.

La·place (lə-pläs′), village (1970 pop. 5,953), St. John the Baptist parish, SE La., on the Mississippi River WNW of New Orleans, in a truck-farm and sugar-cane region.

Lap·land (lăp′lănd′, -lənd), vast region of N Europe, largely within the Arctic Circle. Lapland is mountainous in north Norway and Sweden and consists largely of tundra in the northeast. There are also extensive forests and many lakes and rivers. The climate is arctic and the vegetation is generally sparse, except in the forested southern zone. Lapland is very rich in mineral resources, particularly in high-grade iron ore, copper, nickel, and apatite. The region abounds in sea and river fisheries and in aquatic and land fowl. Reindeer are essential to the economy; there is a growing tourist industry in the region.

The Lapps or Laplanders, who constitute the indigenous population, number about 31,500 and are concentrated mainly in Norway. Largely nomadic, the majority of the Lapps follow their reindeer herds, wintering in the lowlands and summering in the western mountains. Little is known of their early history; it is believed that they came from central Asia and were pushed to the northern extremity of Europe by the migrations of the Finns, Goths, and Slavs. Though mainly conquered by Sweden and Norway in the Middle Ages, the Lapps long resisted Christianization, which was completed only in the 18th cent. by Russian and Scandinavian missionaries.

La Pla·ta (lä plä'tä), city (1970 pop. 408,300), capital of Buenos Aires prov., E central Argentina, inland from Ensenada, its port on the Río de la Plata. Industrial growth has been steady, and large quantities of oil, grain, and refrigerated meat products are exported. La Plata is Argentina's main oil-refining center. It is also a major cultural center, with fine museums and colleges and a national university. The city was founded in 1882. During the dictatorship of Juan Perón (1946-55) both city and province were renamed Eva Perón, in honor of his wife. The name La Plata was restored when Perón's regime was overthrown (1955).

La Pla·ta (lə plăt'ə, plä'tə), county (1970 pop. 19,199), 1,683 sq mi (4,359 sq km), SW Colo., in a livestock-grazing area bordering in the S on N.Mex., bounded on the W by the La Plata Mts., and drained by the Animas River and branches of the San Juan River; formed 1874; co. seat Durango. Gold, silver, lead, and coal are mined. Part of San Juan National Forest is here.

La Pla·ta (lə plăt'ə), town (1970 pop. 1,561), seat of Charles co., SW Md., S of Washington, D.C.; inc. 1888. It is a tobacco market.

La·porte (lə-pôrt'), borough and mountain resort (1970 pop. 207), seat of Sullivan co., NE Pa., ENE of Williamsport.

La Porte (lə pôrt'), county (1970 pop. 105,342), 608 sq mi (1,574.7 sq km), NW Ind., bounded in the NW by Lake Michigan, on the N by the Mich. border, and partly in the S by the Kankakee River; formed 1832; co. seat La Porte. It has agriculture (grain and livestock), lake shipping, fisheries, a lumbering industry, and some manufacturing.

La Porte. **1.** City (1978 est. pop. 20,600), seat of La Porte co., NW Ind.; inc. 1835. It is a manufacturing center in a fertile farmland on the edge of the Calumet industrial region. **2.** City (1970 pop. 7,149), Harris co., S Texas, on Galveston Bay ESE of Houston; settled 1889. It is a summer resort with spacious beaches.

Lap·peen·ran·ta (läp'pĕn-rän'tä), city (1975 est. pop. 52,682), Kymi prov., SE Finland, on Lake Saimaa. It is an important trade and industrial center. The city was chartered in 1649 and became an important border fortress after the Treaty of Nystad (1721).

Lap·tev Sea (läp'tĕv, -tyĭf), section of the Arctic Ocean, c.250,900 sq mi (649,830 sq km), N Siberian USSR, between the Taymyr Peninsula and the New Siberian Islands. The Laptev Sea, part of the Northern Sea Route, is navigable only during Aug. and Sept.

La Puen·te (lä pwĕn'tē), city (1978 est. pop. 31,000), Los Angeles co., S Calif., a residential suburb of Los Angeles; laid out 1841, inc. 1956.

L'Aq·ui·la (lä'kwē-lä), or **L'Aquila de·gli A·bruz·zi** (dā'lyē ä-brōōt'-tsē), city (1976 est. pop. 65,451), capital of L'Aquila prov. and of Abruzzi, central Italy, on the Pescara River. It is an agricultural and industrial center. Manufactures include building materials, textiles, and electronic equipment. L'Aquila is situated at the foot of the Gran Sasso d'Italia mountain group and is a popular base for mountain climbing.

La·rache (lä-räsh'), city (1971 pop. 45,710), N Morocco, on the Atlantic Ocean. Vegetables, cork, and timber are exported. The Phoenicians founded a trading post on the site, which was later captured by the Romans. Spain held the city twice (1610-91 and 1911-56).

Lar·a·mie (lăr'ə-mē), county (1970 pop. 56,360), 2,703 sq mi (7,000.8 sq km), SE Wyo., bordering on Colo. and Nebr. and watered by Chugwater, Horse, and Lodgepole creeks; formed 1867; co. seat Cheyenne. In an agricultural area yielding livestock, grain, and sugar beets, it has granite quarrying and light manufacturing (processed foods, textiles, chemical products, concrete, and machinery).

Laramie, city (1978 est. pop. 24,900), seat of Albany co., SE Wyo., on the Laramie River; inc. 1874. It is a commercial and industrial center for a livestock, mining, and timber region. Tourism is an important economic activity; the city is surrounded by mountain ranges and many nearby ski, hunting, and fishing areas. It is the seat of the Univ. of Wyoming. Laramie was settled in 1868 with the arrival of the railroad. Nearby is the site of Fort Sanders, established in 1866 to protect the Overland Trail and workers on the Union Pacific RR.

Larch·mont (lärch'mŏnt), suburban village (1975 est. pop. 7,000), Westchester co., SE N.Y., on Long Island Sound between New Rochelle and Mamaroneck; developed c.1845, inc. 1891.

La·re·do (lə-rā'dō), city (1978 est. pop. 80,000), seat of Webb co., S Texas, on the Rio Grande; founded 1755, inc. 1852. It is the major port of entry on the U.S.-Mexican border, with a thriving export-import trade and a busy tourist industry. It is also a wholesale and retail center for a large area on both sides of the Rio Grande. Important to its economy are cattle ranching, irrigated farming, oil production, and mining and smelting. A wide variety of products are manufactured, including clothing, electronic equipment, ceramics, medical supplies, and leather goods. Laredo was founded by the Spanish and retains a semi-Mexican flavor. It grew as a post on the road to San Antonio and other Texas cities. After the Texas Revolution its ownership remained in doubt until the southern boundary of Texas was definitively established by the Mexican War. Laredo's growth was aided by the arrival of the railroads (1880s), the development of irrigated farming, the discovery of oil and natural gas, and the opening (1936) of an excellent highway to Mexico City. Laredo Air Force Base, a large jet-pilot training center, is on the outskirts.

Lar·go (lär'gō), town (1978 est. pop. 47,000), Pinellas co., W Fla., on the Pinellas peninsula and the Gulf Coast, across the bay from Tampa; settled 1853, inc. 1905. It is a packing, canning, and shipping center in a citrus fruit area. Its beautiful beaches and many recreational facilities make it a popular resort spot.

Lar·i·mer (lăr'ə-mər), county (1970 pop. 89,900), 2,610 sq mi (6,759.9 sq km), N Colo., in an irrigated agricultural area bordering on Wyo. and drained by the Cache la Poudre and Laramie rivers; formed 1861; co. seat Fort Collins. Sugar beets, beans, and livestock are its major products. The county includes parts of the Front Range and Rocky Mountain National Park.

Lá·ri·sa (lä'rē sä) or **La·ris·sa** (lə-rĭs'ə), city (1971 pop. 72,336), capital of Lárisa prefecture, E Greece, in Thessaly on the Piniós River. It is an agricultural trade center and a transportation hub. The chief city of ancient Thessaly, it was annexed (4th cent. B.C.) by Philip II of Macedon and in 196 B.C. became an ally of Rome. It was taken from the Byzantine Empire by Bulgaria and later was held by Serbia, with which it passed (15th cent.) under the rule of the Ottoman Turks. In the Greek War of Independence the city was (1821) the headquarters of Ali Pasha. Turkey ceded the city to Greece in 1881.

Lar·ka·na (lär-kä'nə), town (1972 pop. 71,943), S central Pakistan, on the Ghar canal. Famous for the quality of its rice, it is an important grain market and a trading center for silk and cotton goods. Brass and other metalware are manufactured. Remains of the ancient city of Mohenjo-Daro have been uncovered nearby.

Lark·spur (lärk'spûr'), city (1970 pop. 10,487), Marin co., W Calif., a residential suburb of San Francisco near Mt. Tamalpais; inc. 1908. Nearby Larkspur Canyon has a redwood grove.

Lar·na·ca (lär'nə-kə), town (1974 est. pop. 19,800), SE Cyprus, on Larnaca Bay. It is a port and district administrative center. Salt and umber are mined in the district. There is a tradition that Lazarus settled in Larnaca after his resurrection and became its first bishop. In the town is a fort built by the Turks in 1625.

Larne (lärn), municipal borough (1971 pop. 18,242), Co. Antrim, NE Northern Ireland, on an inlet of the North Channel. It is a seaport and a tourist center.

Lar·ned (lär'nĭd), city (1970 pop. 4,567), seat of Pawnee co., central Kansas, SW of Great Bend, on the Arkansas River and the old Santa Fe Trail; laid out 1873, inc. 1876. It is a trade center, producing feed and dairy and metal goods. Pawnee Rock, nearby, was the scene of bitter Indian battles; it is included in a state park.

La Ro·ma·na (lä rô-mä'nä), city (1970 pop. 36,722), SE Dominican Republic, on the Caribbean Sea. It is a provincial capital and port.

Lar·sa (lär'sə), ancient city of S Babylonia, in modern Iraq. In 1763 B.C. Hammurabi defeated Larsa and succeeded in uniting Babylonia under his power. Temple libraries have been found in the ruins.

La·rue (lə-rōō'), county (1970 pop. 10,672), 260 sq mi (673.4 sq km), central Ky., bounded on the E and NE by the Rolling Fork and drained by the Nolin River; formed 1843; co. seat Hodgenville. It is in a rolling agricultural area (corn, burley tobacco, livestock, and dairy products), with timber, gas wells, and limestone deposits. There is some clothing manufacturing.

La-sa (lä'sə, -sä, lăs'ə): *see* Lhasa, China.

La·salle (lə-săl'), city (1976 pop. 76,713), S Que., Canada, SW of Montreal on the St. Lawrence River at the head of the Lachine Rapids. It is a suburb of Montreal.

La Salle (lə săl'). **1.** County (1970 pop. 111,409), 1,153 sq mi (2,986.3 sq km), N Ill., drained by the Illinois, Fox, Vermilion, and Little Vermilion rivers; formed 1831; co. seat Ottawa. Agricultural products include corn, oats, soybeans, wheat, livestock, poultry, and dairy foods. The area has deposits of bituminous coal, limestone, silica, clay, shale, and cement rock. **2.** Parish (1970 pop. 13,295), 643 sq mi (1,665.4 sq km), central La., bounded on the W by the Little River; formed 1908; parish seat Jena. Sugar cane, rice, cotton, corn, and hay are grown. There are oil fields here, much timber, and some industry, including cotton ginning and lumber milling. **3.** County (1970 pop. 5,014), 1,501 sq mi (3,887.6 sq km), S Texas, in a cattle-ranching area drained by the Nueces and Frio rivers; formed 1858; co. seat Cotulla. The county contains part of the irrigated Winter Garden truck-farming and citrus-growing region and has some dairying. There are oil and natural-gas wells.

La Salle, city (1978 est. pop. 9,900), La Salle co., N Ill., on the Illinois River, in an area where coal and other minerals are found; settled 1830, inc. 1852. It forms a tri-city unit with Peru and Oglesby. Manufactures include zinc products, cement, electrical and electronic equipment, steel items, clocks, and motors. Nearby is Starved Rock State Park on the site of a fort begun in 1680 by Henri de Tonti and completed by sieur La Salle and Tonti in 1682–83.

Las An·i·mas (läs ăn'ə-məs), county (1970 pop. 15,744), 4,794 sq mi (12,416.5 sq km), SE Colo., in a coal-mining and livestock-grazing area bordering on N.Mex. and drained by the Purgatoire and Apishapa rivers; formed 1866; co. seat Trinidad. Part of the Sangre de Cristo Mts. and San Isabel National Forest are in the west.

Las Animas, city (1970 pop. 3,148), seat of Bent co., SE Colo., on the Arkansas River E of Pueblo; founded 1869, inc. 1886. It is a processing center in a farm region.

Las Cru·ces (läs kro͞o'sĭs), city (1978 est. pop. 41,500), seat of Dona Ana co., SW N.Mex., on the Rio Grande, in a farm area irrigated by the Elephant Butte system; founded 1848, inc. 1907. Its economy is based chiefly upon agriculture and the nearby White Sands Missile Range, testing grounds for the first atomic bomb and now a major military and NASA testing site. The city also has textile and canning industries. The name, Spanish for "the crosses," refers to a massacre (1830) of some 40 travelers by Apache Indians on this site. Nearby are the historic village of Mesilla, the picturesque Organ Mts., Fort Fillmore (1851), Fort Seldon (1865), and the Indian village Tortugas.

La Se·re·na (lä sā-rā'nä), city (1970 pop. 71,898), capital of Coquimbo prov., N central Chile, on the Elqui River. A commercial and agricultural center in a region of orchards and vineyards, it is a popular resort. La Serena was founded in 1544, destroyed by Indians in 1549, and sacked by the English in 1680. It was the site of Chile's declaration of independence in 1818.

Lash·kar (lŭsh'kər), city, Madhya Pradesh state, central India. Formerly the capital of Madhya Bharat state, it adjoins Gwalior town and is a modern commercial center and transportation hub.

Las·sen (lăs'ən), county (1970 pop. 16,796), 4,561 sq mi (11,813 sq km), NE Calif., on a high volcanic plateau extending E from the Cascade Range to the Sierra Nevada along the S and SW boundaries; formed 1864; co. seat Susanville. In a densely forested area, it has extensive logging and lumber-milling operations. Stock grazing and irrigated farming (hay, potatoes, fruit, and dairy products) are also done, chiefly in the Honey Lake valley. Fishing, hunting, camping, and winter sports attract vacationers.

Lassen Volcanic National Park, 106,934 acres (43,308 hectares), N Calif., at the S tip of the Cascade Range; est. 1916. The park contains volcanic peaks, lava flows, vents, and hot springs. Lassen Peak, 10,457 ft (3,189.4 m) high, is the only active volcano in the United States excluding Alaska and Hawaii. It erupted in 1914 and was intermittently active until 1921.

Las Ve·gas (läs vā'gəs). **1.** City (1978 est. pop. 155,800), seat of Clark co., S Nev.; inc. 1911. Revenue from hotels, gambling, entertainment, and other tourist-oriented industries forms the backbone of Las Vegas's economy. The city is also the commercial hub of a ranching and mining area. In the 19th cent. Las Vegas was a watering place for travelers to southern Calif. In 1855–57 the Mormons maintained a fort here, and in 1864 Fort Baker was built by the U.S. army. **2.** City (1970 pop. 13,835), seat of San Miguel co., N N.Mex., ESE of Santa Fe; settled c.1835, inc. as a town 1888, as a city 1897. It is a mountain and health resort in a dude-ranch area, and a shipping center for a sheep, cattle, and farm region.

La·ta·cun·ga (lä'tä-ko͝ong'gä), city (1974 pop. 22,106), capital of Cotopaxi prov., N central Ecuador. A town of the ancient Incas, it is in a high mountain basin between the east and west Andean cordilleras. The city has suffered severe earthquakes.

La·tah (lā'tä', lə-tä'), county (1970 pop. 24,898), 1,090 sq mi (2,823.1 sq km), N Idaho, with long, rolling hills, drained by the Palouse and Potlatch rivers and bordering on Wash.; formed 1888; co. seat Moscow. Lumber, livestock, wheat, cherries, and pears are its major products. It includes part of St. Joe National Forest in the north.

Lat·a·ki·a (lăt-ə-kē'ə, lä'tä-kē'ä), city (1970 pop. 125,716), capital of Latakia governorate, W Syria, a port city on the Mediterranean Sea. Industries include sponge fishing, vegetable-oil milling, and cotton ginning. Formerly an ancient Phoenician city, it was rebuilt (c.290 B.C.) by Seleucus I and later prospered as the Roman Laodicea ad Mare. Byzantines and Arabs fought over it from the 7th to 11th cent. A.D. The city was captured in 1098 by the Crusaders and flourished in the 12th cent. until after its capture in 1188 by Saladin. From the 16th cent. to World War I it was part of the Ottoman Empire.

Lat·ga·le (lät'gä-lĕ) or **Lat·gal·li·a** (lăt-găl'ē-ə, -yə), region and former province, NW European USSR, in Latvia, N of the Western Dvina River. The region was settled in the early Middle Ages. Latgale formed the southern part of Livonia until 1561, when it passed to Poland. The area was ceded to Russia during the Polish partition of 1772. In 1918 it became part of Latvia.

Lat·i·mer (lăt'ə-mər), county (1970 pop. 8,601), 737 sq mi (1,908.8 sq km), SE Okla., in a cattle-raising area drained by Gaines Creek; formed 1907; co. seat Wilburton. It also has extensive gas deposits.

La·ti·na (lä-tē'nä), city (1976 est. pop. 79,400), capital of Latina prov., in Latium, central Italy, near the Tyrrhenian Sea. It is an industrial, commercial, and agricultural center.

Lat·in A·mer·i·ca (lăt'n ə-mĕr'ĭ-kə), the Spanish-speaking, Portuguese-speaking, and French-speaking countries (except Canada) of North America, South America, Central America, and the West Indies. The term is also used to include Puerto Rico, the French West Indies, and other islands of the West Indies where a Romance tongue is spoken. Occasionally the term is used to include British Honduras, Guyana, French Guiana, and Surinam.

lat·i·tude (lăt'ĭ-to͞od', -tyo͞od'), angular distance of any point on the surface of the earth N or S of the equator. The equator is latitude 0°, and the North Pole and South Pole are latitudes 90° N and 90° S, respectively. The length of one degree of latitude averages about 69 mi (111 km); it increases slightly from the equator to the poles as a result of the earth's polar flattening. An imaginary line on the earth's surface connecting all points equidistant from the equator is called a parallel of latitude. Because of their special meanings, four fractional parallels are also shown. These are the Tropic of Cancer (23½° N) and the Tropic of Capricorn (23½° S), marking the farthest points north and south of the equator where the sun's rays fall vertically, and the Arctic Circle (66½° N) and the Antarctic Circle (66½° S), marking the farthest points north and south of the equator where the sun appears above the horizon each day of the year. Parallels of latitude and meridians of longitude together form a grid by which any point on the earth's surface can be specified.

La·ti·um (lā'shē-əm), region (1971 pop. 4,754,484), 6,642 sq mi (17,203 sq km), central Italy, extending from the Apennines W to the Tyrrhenian Sea; capital Rome. The region is mostly hilly and mountainous, with a narrow coastal plain. Agriculture forms the backbone of the economy. Industry in the region has been spurred (mid-20th cent.) by the construction of hydroelectric facilities on the Aniene and Liri rivers and a nuclear power plant at Latina. There is a large tourist industry, and fishing is pursued along the coast. In ancient times Latium comprised a limited area east and south of the Tiber River that extended to the Alban Hills; only after it became part of Italy in 1870 did it approximately reach its present limits. In early Roman times Latium was inhabited by the Latins, the Etruscans, and several Italic tribes. In the 3rd cent. B.C., Rome subdued all of Latium. After the fall of Rome, Latium was invaded in turn by the Visigoths, the Vandals, and the Lombards. From the 8th cent. the duchy of Rome, including most of modern Latium, belonged to the popes. Except for the area south of Terracina, which belonged to the kingdom of Naples, Latium remained a part of the Papal States until 1870. In World War II southern Latium was the scene of bloody battles during the Allied drive on Rome.

La·trobe (lə-trōb′), industrial borough (1970 pop. 11,749), Westmoreland co., SW Pa., in the foothills of the Alleghenies; inc. 1854. Among its varied manufactures are steel, castings, beer, ceramics, forgings, carbides, and aluminum siding.

La Tuque (lä tük′), town (1976 pop. 12,067), S Que., Canada, on the St. Maurice River, NW of Quebec. La Tuque, in a lumbering and farming region, was established as a trading post in the French period; it grew after the coming of the railroad in 1908.

Lat·vi·a (lăt′vē-ə), constituent republic (1976 est. pop. 2,497,000), 24,590 sq mi (63,688 sq km), NW European USSR; capital Riga. It borders on Estonia in the north, Lithuania in the south, the Baltic Sea with the Gulf of Riga in the west, the Russian Republic in the east, and Belorussia in the southeast. Latvia falls into four historic regions: north of the Western Dvina River are Vidzeme and Latgale, which were parts of Livonia; south of the Dvina are Kurzeme and Zemgale, which belonged to the former duchy of Courland. Latvia is largely a fertile lowland, with numerous lakes and hills rising to the east. Dairying and stock raising are extensively carried on. Latvia also has valuable timber resources. Machinery, metals, electrical equipment, and textiles are among the chief industrial products.

The Letts were conquered and Christianized by the Livonian Brothers of the Sword in the 13th cent. Their country formed the southern part of Livonia until 1561, when Courland became a vassal duchy under Polish suzerainty and Livonia passed to Poland. In 1629 Sweden conquered Livonia (save Latgale), which it lost in turn to Russia in 1721. With the first (1772) and third (1795) partitions of Poland, Latgale and Courland also passed to Russia. Between 1817 and 1819 the serfs were emancipated, and in the middle of the 19th cent. a national revival began. In the revolution of 1905 the Letts played a prominent role, and bloody reprisals were meted out. Latvia was devastated in World War I. However, the collapse of Russia and Germany made Latvian independence possible in 1918. Peace with Russia followed in 1920. The Latvian constitution of 1920 provided for a democratic republic. However, there was no political stability, and in 1934 the constituent assembly was dissolved. Soviet troops occupied Latvia in 1940, and subsequent elections held under Soviet auspices resulted in the absorption of Latvia into the USSR as a constituent republic. Occupied (1941–44) by German troops in World War II, it was reconquered by the Soviet Union.

Lau·der·dale (lô′dər-dāl′). **1.** County (1970 pop. 68,111), 662 sq mi (1,714.6 sq km), extreme NW Ala., bordering on Miss. and Tenn. and drained in the S by Pickwick Landing Reservoir, Lake Wilson, and Wheeler Reservoir; formed 1818; co. seat Florence. It is in a cotton-growing area, with diversified manufacturing (clothing, furniture, and ceramic and metal products) and a food-processing industry. **2.** County (1970 pop. 67,087), 708 sq mi (1,833.7 sq km), E Miss., drained by affluents of the Chickasawhay River; formed 1833; co. seat Meridian. Its agriculture includes cotton, corn, sweet potatoes, dairy products, beef cattle, and hogs. It has lumbering and food-processing industries, textile mills, and factories manufacturing clothing, wood and metal products, furniture, asphalt, ceramic tiles, and machinery. **3.** County (1970 pop. 20,271), 477 sq mi (1,235.4 sq km), W Tenn., bounded on the W by the Mississippi River, on the N by the Forked Deer River, on the S by the Hatchie River; formed 1835; co. seat Ripley. Its economy is based on lumbering and agriculture (cotton, corn, livestock, and truck crops).

Lau·en·burg (lou′ən-bûrg′, -bŏŏrκн′), former duchy, NE West Germany, on the right bank of the lower Elbe. The duchy belonged to a branch of the house of Saxony from the 12th to the late 17th cent., when it passed to the house of Hanover. Lauenburg was occupied by France from 1803 to 1813. The Congress of Vienna awarded (1815) it to Prussia and made it a member state of the German Confederation, but Prussia ceded it to the Danish crown in exchange for western Pomerania. In the Danish War of 1864 the duchy was seized by Prussia and Austria, and Austria soon afterward ceded its rights to Prussia. Lauenburg was incorporated into the province of Schleswig-Holstein in 1876 and ceased to be a duchy in 1918.

Laun·ces·ton (lôn′sĕs-tən, län′-), city (1976 pop. 32,947; urban agglomeration pop. 62,181), on Tasmania, SE Australia, where the North Esk and South Esk rivers join to form the Tamar estuary; founded 1806. Launceston is the main port for trade with the Australian mainland. There are woolen mills and brass works.

Lau·rel (lôr′əl), county (1970 pop. 27,386), 448 sq mi (1,160.3 sq km), SE Ky., in a mountainous agricultural area of the Cumberland foothills bounded on the W by the Rockcastle River, on the S by the Laurel River, and drained by several creeks; formed 1825; co. seat London. It produces livestock, poultry, dairy foods, tobacco, corn, wheat, and fruit and has bituminous-coal mines, lumbering and food-processing industries, and some manufacturing.

Laurel. 1. Town (1970 pop. 10,525), Prince Georges co., W central Md., a residential suburb between Washington, D.C., and Baltimore; patented in the late 1600s, inc. 1870. In the area are three racetracks, the Patuxent Wildlife Research Refuge, and Fort George G. Meade (est. 1917). **2.** City (1978 est. pop. 23,200), a seat of Jones co., SE Miss., on Tallahala Creek; founded 1882, inc. 1892. Industries include petroleum exploration and production and meat and poultry processing. Oil was discovered in the vicinity in 1944.

Lau·rens (lôr′ənz). **1.** County (1970 pop. 32,738), 811 sq mi (2,100.5 sq km), central Ga., in a coastal-plain area intersected by the Oconee River; formed 1807; co. seat Dublin. Agriculture (cotton, corn, peanuts, and livestock) and lumbering are the major occupations. **2.** County (1970 pop. 49,713), 713 sq mi (1,846.7 sq km), NW central S.C., bounded on the SW by the Saluda River and on the NE by the Enoree River; formed 1785; co. seat Laurens. Agriculture (cotton, grain, vegetables, peanuts, and dairy products) and textile manufacturing are important. Part of Sumter National Forest is here.

Laurens, city (1970 pop. 10,298), seat of Laurens co., NW S.C.; inc. 1875. Textiles and glass products are made, and grain is processed.

Lau·ren·tian Mountains (lô-rĕn′shən), S Que., Canada, N of the St. Lawrence and Ottawa rivers, rising to 3,150 ft (960.8 m) in Mt. Tremblant. The region is a popular year-round recreational area.

Laurentian Plateau: *see* Canadian Shield.

Lau·rin·burg (lô′rĭn-bûrg′, lŏr′-), city (1970 pop. 8,859), seat of Scotland co., S N.C., near the S.C. border SSW of Fayetteville, in a cotton and fruit area; inc. 1877.

Lau·sanne (lō-zăn′, -zän′), city (1977 est. pop. 132,800), capital of Vaud canton, W Switzerland, on the Lake of Geneva. An important rail junction and lake port, it is the trade and commercial center of a rich agricultural region. Food and tobacco products are produced. Lausanne is also a well-known resort city. Originally a Celtic settlement and a Roman military camp, it was ruled by prince-bishops until 1536, when it was conquered by Bern. Bernese rule ended in 1798, and Lausanne became (1803) the capital of the newly formed canton of Vaud. In the 18th cent., Lausanne was the residence of Gibbon, Rousseau, and Voltaire. The Univ. of Lausanne was founded as a Protestant school of theology in 1537 and became famous as a center of Calvinism.

Lau·zon (lō-zôɴ′), city (1976 pop. 12,663), S Que., Canada, on the St. Lawrence River adjoining Lévis and opposite Île d'Orléans. Settled in 1647, it is a shipbuilding center.

La·va Beds National Monument (lä′və, lăv′ə): *see* National Parks and Monuments Table.

La·va·ca (lə-văk′ə), county (1970 pop. 17,903), 975 sq mi (2,525.3 sq km), S Texas, in a coastal-plain region drained by the Lavaca and Navidad rivers; formed 1846; co. seat Hallettsville. It has agriculture (cotton, corn, grain sorghums, peanuts, hay, fruit, truck crops, poultry, cattle, and hogs), oil and natural-gas wells, and deposits of sand, gravel, and clay.

La·val (lə-văl′), city (1976 pop. 246,243), coextensive with Île-Jésus (94 sq mi/243.5 sq km), S Que., Canada, between the Rivière des Mille Îles and the Rivière des Prairies, NW of Montreal. It is a residential suburb of Montreal, with summer tourist facilities.

Laval, town (1975 pop. 51,544), capital of Mayenne dept., NW France, in Maine. It has been noted for its linen products since the 14th cent. It was founded in the 9th cent.

La Ve·ga (lä vā′gä), city (1970 pop. 31,085), central Dominican Republic, on the Camú River. La Vega is a communications center near a religious sanctuary erected on the site of an important battle in the colonial period. The city was founded in 1495.

La Verne (lə vûrn′), city (1970 pop. 12,965), Los Angeles co., S Calif., in a citrus-fruit area; inc. 1906. It is chiefly residential.

La·von·gai (lə-vŏng'gī'), volcanic island, c.460 sq mi (1,190 sq km), in the Bismarck Archipelago, part of Papua New Guinea. Lavongai is mountainous and densely forested. There are several coconut plantations. Germany held the island from 1884 until World War I.

Lá·vri·on (lä'vrē-ôn) or **Lau·ri·um** (lôr'ē-əm), town, E central Greece, in Attica, on the Aegean Sea. It is a mining, smelting, and shipping center for lead, manganese, cadmium, and silver ores. Silver was mined here from the 6th to the 2nd cent. B.C. and was one of the chief sources of Athenian revenue in the 5th cent. B.C.

Lawn·dale (lôn'dāl'), city (1978 est. pop. 23,900), Los Angeles co., S Calif., in the Centinela valley; inc. 1959.

Law·rence (lôr'əns, lŏr'-). **1.** County (1970 pop. 27,281), 686 sq mi (1,776.7 sq km), N Ala., drained in the N by Wheeler Reservoir (in the Tennessee River); formed 1818; co. seat Moulton. It is in a cotton, corn, and poultry-farming area, with deposits of coal, limestone, and asphalt. It has lumber logging and clothing manufacturing. Part of William B. Bankhead National Forest is in the south. **2.** County (1970 pop. 16,320), 592 sq mi (1,533.3 sq km), NE Ark., bounded on the E by the Cache River, intersected by the Black River, and drained by the Spring and Strawberry rivers and Village Creek; formed 1815; co. seat Walnut Ridge. Its agriculture includes cotton, corn, rice, hay, and livestock. It has lumbering, some mining, and manufacturing (textiles, wood products, shoes and leather goods, and sporting goods). **3.** County (1970 pop. 17,522), 374 sq mi (968.7 sq km), SE Ill., bounded in the E by the Wabash River and drained by the Embarrass River; formed 1821; co. seat Lawrenceville. In an agricultural area yielding poultry, livestock, soybeans, corn, and wheat, it has oil and natural-gas wells and refineries. Some manufacturing is done. **4.** County (1970 pop. 38,038), 459 sq mi (1,188.8 sq km), S Ind., drained by the Salt Creek and the East Fork of the White River; formed 1818; co. seat Bedford. Large limestone quarries and agriculture (fruit and grain) characterize the area. **5.** County (1970 pop. 10,726), 425 sq mi (1,100.8 sq km), E Ky., bounded on the E by the Big Sandy River and its Tug Fork (both forming the W.Va. line here), drained by Levisa Fork and small Blaine Creek; formed 1821; co. seat Louisa. In a mountain agricultural area (dairy products, livestock, poultry, corn, sorghum, tobacco, and fruit), it has coal mines, oil and gas wells, fire-clay and sand pits, and timber. Its manufactures include some textiles and wood products. **6.** County (1970 pop. 11,137), 433 sq mi (1,121.5 sq km), S Miss., drained by the Pearl River; formed 1814; co. seat Monticello. It has cotton and corn farming, lumber logging, and clothing manufacturing. **7.** County (1970 pop. 24,585), 619 sq mi (1,603.2 sq km), SW Mo., in the Ozarks, drained by the Spring River; formed 1845; co. seat Mount Vernon. It is in an agricultural area that produces wheat, corn, oats, strawberries, poultry, and dairy products. It also has lead and zinc mines, limestone deposits, and factories manufacturing clothing, wood and metal products, leather, and machinery. **8.** County (1970 pop. 56,868), 456 sq mi (1,181 sq km), S Ohio, bounded on the S by the Ohio River, here forming a boundary with Ky. and W.Va., and drained by Symmes Creek; formed 1816; co. seat Ironton. Coal mining, agriculture (dairy products, grain, livestock, fruit, tobacco, hay, and truck crops), and diversified industry are important. **9.** County (1970 pop. 107,374), 367 sq mi (950.5 sq km), W Pa., in a manufacturing and mining area drained by the Shenango, Hahoning, and Beaver rivers; formed 1849; co. seat New Castle. An iron center in the mid-19th cent., it now produces metal, glass, clay, stone, and cement. It is also a dairying region. **10.** County (1970 pop. 17,453), 800 sq mi (2,072 sq km), W S.Dak., in a farming and mining region of the Black Hills; and bordering on Wyo.; formed 1875; co. seat Deadwood. Agriculture (livestock and grain), lumbering, mining, and tourism are important. The Homestake Gold Mine and part of Black Hills National Forest are in the county. **11.** County (1970 pop. 29,097), 634 sq mi (1,642.1 sq km), S Tenn., in an upland agricultural area bordered on the S by Ala. and drained by the Buffalo River; formed 1817; co. seat Lawrenceburg. Its farms yield cotton, corn, livestock, and dairy products.

Lawrence. 1. Town (1978 est. pop. 25,700), Marion co., central Ind., a residential suburb of Indianapolis, on the West Fork of the White River. **2.** City (1978 est. pop. 51,900), seat of Douglas co., NE Kansas, on the Kansas River; inc. 1858. Manufactures include farm chemicals and corrugated boxes. Major employers are the Univ. of Kansas and Haskell Institute. Lawrence was founded in 1854 by the New England Emigrant Aid Company. In 1856 there was a proslavery raid on the town that instigated the retaliatory Pottawatamie killings by John Brown. **3.** City (1978 est. pop. 67,000), Essex co., NE Mass., on the Merrimack River; settled 1655, set off from Andover and Methuen 1847, inc. as a city 1853. It is a port of entry. Textiles, textile machinery, leather goods, wearing apparel, electrical equipment, and rubber and paper products are manufactured. Boston capitalists laid out an industrial town in 1845 and built a granite dam on the Merrimack River. They also built mills and workers' dwellings, which were soon crowded with laborers, mainly from Europe, and Lawrence became one of the world's greatest centers for woolen textiles. Several disasters have occurred here—the collapse and burning of the Pemberton Mill in 1860, when over 500 trapped workers were killed or injured; the tornado of 1890; and the labor strife of 1912, when the strikers (members of the International Workers of the World) finally won some of their demands. **4.** Residential village (1975 est. pop. 6,700), Nassau co., SE N.Y., on the S shore of Long Island, near Lynbrook; inc. 1897. It is in a resort area.

Law·rence·burg (lôr'əns-bûrg', lŏr'-). **1.** City (1970 pop. 4,636), seat of Dearborn co., SE Ind., on the Ohio River near the Ohio border S of Greendale; laid out 1802. It is a port of entry, and whiskey is distilled here. Prehistoric fortifications have been found near here. The city was inundated by a flood in 1937. **2.** City (1970 pop. 3,579), seat of Anderson co., N central Ky., just S of Frankfort, in a bluegrass region; settled 1776 as Coffman's Station, inc. 1820. It is a farm trade center, with a distillery and several small factories. **3.** City (1970 pop. 8,889), seat of Lawrence co., S Tenn., SSW of Columbia, in a dairy, livestock, cotton, fruit, and grain area; founded c.1815, inc. 1825. Cheese, clothing, bicycles, and lumber products are made.

Law·rence·ville (lôr'əns-vĭl', lŏr'-). **1.** City (1970 pop. 5,115), seat of Gwinnett co., N central Ga., NE of Atlanta; inc. 1821. Apparel, leather goods, aluminum products, and lumber are produced. **2.** Industrial city (1970 pop. 5,863), seat of Lawrence co., SE Ill., on the Embarrass River near the Wabash River, in an oil, natural-gas, and farm region; founded 1821, inc. 1835. The city has oil refineries and manufactures oil-well and telephone equipment, asphalt products, chemicals, and gasoline. **3.** Town (1970 pop. 1,636), seat of Brunswick co., SE Va., SSW of Richmond near the N.C. border; founded 1814, inc. 1874. Wood products are made.

Law·ton (lôt'n), city (1978 est. pop. 83,200), seat of Comanche co., SW Okla.; inc. 1901. It is a commercial and trade center for the surrounding cotton, wheat, and cattle area and for Fort Sill, a U.S. field artillery center. Nearby is a large limestone quarry and the Wichita Mts. Wildlife Refuge.

Lay·ton (lāt'n), city (1970 pop. 13,603), Davis co., N Utah, between the Wasatch Range and Great Salt Lake. In an irrigated farm area served by the Weber basin project, it has a beet-sugar refinery, canning and packing plants, and a flour mill.

Lea (lē), county (1970 pop. 49,554), 4,393 sq mi (11,377.9 sq km), extreme SE N.Mex., in the Llano Estacado bounded on the S and E by the Texas border; formed 1917; co. seat Lovington. This livestock-grazing region also has oil and natural-gas fields.

Lead (lēd), city (1970 pop. 5,420), Lawrence co., W S.Dak., in the Black Hills; laid out 1876 after the discovery of gold, inc. 1890. It is the site of the Homestake Mine, in operation since 1877.

Lead·ville (lĕd'vĭl'), mining city (1970 pop. 4,314), alt. c.10,200 ft (3,110 m), seat of Lake co., central Colo., near the headwaters of the Arkansas River, in the Rocky Mts.; inc. 1878. Farming, ranching, and the tourist trade have kept this famous mining city from becoming another ghost town. Rich placer gold deposits were discovered c.1860 in California Gulch. Oro City, the principal camp, flourished for about two years until the diggings were exhausted. The camps were virtually deserted until 1877, when the discovery of carbonates of lead with a high silver content again transformed Oro City into a boom town. Leadville became one of the greatest silver camps in the world, with a population estimated at 40,000. In 1893, with the repeal of the Sherman Silver Act, silver mining collapsed; in the 1890s, with the discovery of gold nearby, Leadville revived.

Leaf (lēf), river, c.180 mi (290 km) long, S central and SE Miss., rising just S of Forest and flowing S to Hattiesburg, then SE to join the Chickasawhay River and form the Pascagoula.

League City (lēg), city (1970 pop. 10,818), Galveston co., SE Texas; inc. 1961. The aeronautics industry is important; the Lyndon B. Johnson Space Center is nearby. Other industries in League City produce petroleum and petroleum products and paints.

Leake (lēk), county (1970 pop. 17,085), 586 sq mi (1,517.7 sq km), central Miss., drained by the Pearl and Yockanookany rivers;

formed 1833; co. seat Carthage. It is in a cattle, hogs, poultry, cotton, corn, and soybean farming area, with timber and pulpwood. Clothing, feed, gloves, and wood products are made.

Leakes·ville (lēks′vĭl′), city (1970 pop. 1,090), seat of Greene co., SE Miss., ESE of Hattiesburg on the Chickasawhay River, in a lumbering area.

Lea·key (lē′kē), city (1970 pop. 393), seat of Real co., SW Texas, N of Uvalde on the Frio River. It is a trade center for a ranching area. Frio Canyon, a scenic tourist attraction, is here.

Lea·ming·ton (lē′mĭng-tən), town (1971 pop. 10,435), S Ont., Canada, on Lake Erie. In a farm area, it has large canneries.

Leav·en·worth (lĕv′ən-wûrth′), county (1970 pop. 53,340), 465 sq mi (1,204.4 sq km), NE Kansas, in a gently rolling to hilly area bounded on the E by the Missouri River and the Mo. border, on the S by the Kansas River; formed 1855; co. seat Leavenworth. In a grain-growing, stock-raising, and dairying region, it has some mining. Flour, paper, steel, and metal products are made.

Leavenworth, city (1978 est. pop. 24,500), seat of Leavenworth co., NE Kansas, on the Missouri River; inc. 1855. It is the commercial center of a farm and livestock region, with a flour mill, a shipyard, and plants that make automobile batteries and machinery. Nearby Fort Leavenworth, with its various institutions (including the Federal penitentiary, which is located on the grounds although operated by the Justice Dept.), is an important factor in the city's economy. Leavenworth was settled (1854) near the fort by proslavery Missourians and flourished as a supply point on the westward travel routes.

Lea·wood (lē′wŏŏd′), city (1970 pop. 10,645), Johnson co., NE Kansas, a residential suburb of Kansas City; inc. 1948.

Leb·a·non (lĕb′ə-nən), republic (1973 est. pop. 3,000,000), 4,015 sq mi (10,400 sq km), SW Asia, on the Mediterranean Sea; capital Beirut. The country faces the eastern shore of the Mediterranean and is bordered on the north and east by Syria and on the south by Israel. Much of the terrain is mountainous. The Lebanon Mts. run parallel to the coast; on the eastern border is the Anti-Lebanon range. Between the two mountain ranges lies the fertile valley of Al Biqa.

Economy. Prior to the onset of the 1975 civil war between Moslems and Christians, Lebanon had been the traditional commercial and distribution center of the Middle East. After commerce, the nation's chief industries were engaged in food processing and the production of textiles and tobacco products. Lebanon's agricultural commodities included citrus fruits, sugar beets, potatoes, and grapes; wheat was the most important grain.

History. In ancient times the area of Lebanon and Syria was occupied by the Canaanites, who founded the great Phoenician cities. The Phoenician cities occupied a favored position in the Persian Empire and were conquered by Alexander the Great. The region came under Roman dominion starting in 64 B.C. and was Christianized before the Arab conquest in the 7th cent. By then the Maronites had established themselves. Later (11th cent.) the Druses settled in southern Lebanon and in adjacent regions of Syria, and trouble between them and the Christians was to become a constant theme in regional history. After the Crusaders (12th cent.), Lebanon was loosely ruled by the Mamelukes (c.1300). Invasions by Mongols and others contributed to the decline of trade until the reunification of the Middle East under the Ottoman Turks (early 16th cent.).

Conflict among the religious communities, culminating in massacres of the Maronites by the Druses in 1860, led to intervention by France (1861), and the Ottoman sultan was forced to appoint a Christian governor for Lebanon. After World War I, Lebanon and Moslem areas not previously part of it became a French mandate known as Greater Lebanon. In 1926 the mandate was given a republican constitution. In World War II the Free French conquered the coast (1941) and proclaimed Lebanon an independent republic. Elections were held in 1943, and, after considerable controversy, Lebanon became independent on Jan. 1, 1945. In that year it became a member of the United Nations, and soon afterward all British and French troops were evacuated.

As a member of the Arab League, Lebanon declared war on Israel in 1948 but took little part in the conflict. After 1952 Lebanon formed closer ties with the West. In 1958 U.S. forces were called in by the pro-Western government of President Camille Chamoun to suppress a rebellion. During the Arab-Israeli wars of 1967 and 1973 Lebanon gave verbal support to the Arab effort against Israel but did not become involved in any military action. However, Lebanon's position became increasingly difficult because of the activities against Israel of Palestinian terrorists based in Lebanon. These commandos, drawn from among the 200,000 Palestinian refugees living in Lebanon, engaged in frequent raids into Israeli territory and gradually became a disruptive force in internal Lebanese affairs. In 1969, 1972, and again in 1973, the Lebanese army clashed with the Palestinians but failed to rout them. Israel, in turn, repeatedly accused Lebanon of not doing enough to prevent Palestinian terrorist activity and frequently engaged in retaliatory raids into Lebanon. During the 1973 Arab-Israeli war, the Lebanese army did not interfere with Palestinian guerrilla bases in southern Lebanon.

The civil war that erupted in 1975 was in many ways a result of these external events and a culmination of the complex internal tensions that existed in Lebanese society: that of rich against poor, Moslems against Christians, left against right, and Palestinians against the Lebanese elite. The war was carried on by paramilitary forces that far outnumbered the small Lebanese army. Christian troops consisted of the right-wing Phalangists led by Pierre Gemayel and the National Union party of former president Camille Chamoun. Moslem forces, generally left wing in political orientation, were often aided by the Palestine Liberation Army (PLA). The fighting intensified throughout 1975 into 1976 despite several attempts to arrange a cease-fire. By early 1976 the Moslem side, and particularly the PLA, appeared to have a military advantage, with Christians isolated in several coastal enclaves. Fearing that a Palestinian victory would lead to war with Israel, Syria dispatched troops (April, 1976) into Lebanon to bolster the Christian position and isolate the PLA from other Moslem units. By the fall of 1976 a Syrian army of some 20,000 controlled roughly half of the country, including most of Beirut. A conference of Arab states (1976) sponsored by Saudi Arabia called for an Arab League peace-keeping force to be sent to Lebanon and for a withdrawal of Palestinian forces to predetermined bases. By mid-1977 this force, largely Syrian in composition, had imposed a tenuous cease-fire but had failed to remove the Palestinian presence from the country. In the meantime, the political situation in Lebanon remained fluid. Elias Sarkis, a moderate Christian, was elected president (1976) by parliament; but political activity had, in effect, ceased, and the country remained in a state of de facto partition under the rule of opposing armies.

Government. Political life in Lebanon has been profoundly affected by the country's religious diversity. Traditionally, the president was a Christian and the prime minister a Moslem. The sects were represented proportionally in the legislature, cabinet, and civil service. The legislature consisted of a 99-member chamber of deputies elected every four years. One of the underlying causes of the 1975 civil war was a belief on the part of many Moslems that they were discriminated against by the Christians.

Lebanon, county (1970 pop. 99,665), 363 sq mi (940.2 sq km), SE central Pa., in an industrial, mining, and agricultural area with Blue Mt. in the N and Lebanon Valley in the S; formed 1813; co. seat Lebanon. It has poultry, corn, and dairy farms, a food-processing industry, deposits of magnetite, and limestone quarries.

Lebanon. 1. Town (1970 pop. 3,804), New London co., E Conn., NW of Norwich, in a dairy region; inc. 1700. It has many old buildings, including the Revolutionary War Office (1727) and the Governor Trumbull House (1740). **2.** City (1970 pop. 9,766), seat of Boone co., central Ind., NW of Indianapolis; founded 1831, inc. as a town 1853, as a city 1875. Meat processing and some manufacturing are done. **3.** City (1970 pop. 5,528), seat of Marion co., central Ky., SE of Louisville, in a bluegrass region. It has distilleries, garment factories, and a meat-packing plant. A national cemetery is nearby. **4.** City (1970 pop. 8,616), seat of Laclede co., S central Mo., NE of Springfield, in a farm and recreational area of the Ozarks; founded c.1849. Clothing is made. **5.** City (1975 est. pop. 9,900), Grafton co., W N.H., on the Mascoma River near its confluence with the Connecticut River; granted and chartered 1761. Clothing, metal and wood products, and electrical equipment are manufactured. **6.** City (1970 pop. 7,934), seat of Warren co., SW Ohio, NE of Cincinnati, in a rich farm area; founded 1803. Several old buildings, including The Golden Lamb Hotel (1816), remain. Nearby is Fort Ancient State Memorial, with a prehistoric earthwork fortification, burial mounds, and remains of village sites. **7.** City (1970 pop. 6,636), Linn co., NW Oregon, in the Willamette Valley SSE of Salem; platted 1851, inc. 1878. In a lumbering and farming area, the city makes wood and paper products. **8.** City (1978 est. pop. 27,900), seat of Lebanon co., SE Pa., in the Pennsylvania Dutch farm country; founded 1753, inc. as a city 1868. It has steel and steel-fabricating industries. Lebanon was a flourishing town before 1790. **9.** City (1970 pop. 12,492), seat of Wilson co., N central Tenn., in a timber and livestock area of the Cumberland River basin; inc. 1819. Steering gears, clocks, boots, luggage, and plastic products are manufactured. Points of interest include an early log meeting house and ante-bellum houses. **10.** Town (1970 pop. 2,272), seat of Russell co., SW Va., NNE of Bristol, in a farm and cattle area; settled c.1816.

Lebanon, mountain range, c.100 mi (160 km) long, paralleling the Mediterranean Sea from S Lebanon N into Syria and rising steeply from the coast; Qurnet as Sawda (10,131 ft/3,090 m) is the highest peak. A great fault line, site of the fertile Al Biqa valley, separates the Lebanon from the Anti-Lebanon Mts. to the east. The mountains were famed in ancient times for the huge, old cedars that extended in a narrow strip for 85 mi (136.8 km) along the upper western slope of the range; however, these trees were depleted by long use as a building material and a fuel, and only 10 small, isolated groves remain.

Lec·ce (lĕt'chĕ), city (1976 est. pop. 88,499), capital of Lecce prov., Apulia region, S Italy. It is an industrial and agricultural center. Manufactures include ceramics, toys, food products, and wine. A Greek and later a Roman town, Lecce was from 1053 to 1463 a semi-independent county under various lords. There are many fine churches and palaces in a characteristic baroque style.

Le Cen·ter (lə sĕn'tər), village (1970 pop. 1,890), seat of Le Sueur co., S Minn., SSW of Minneapolis; settled 1864, inc. 1876. It is a farm trade center.

Lech (lĕKH), river, c.175 mi (280 km) long, rising in Vorarlberg, W Austria, and flowing NE into S West Germany past Augsburg to the Danube River. There are some 20 hydroelectric stations on the river.

Lech·feld (lĕKH'fĕlt'), plain near Augsburg, S West Germany, drained by the Lech River. In 955 Otto I defeated the Magyars here and stopped their expansion into central Europe.

Le·comp·ton (lə-kŏmp'tən), city (1970 pop. 434), Douglas co., NE Kansas, on the Kansas River; a residential suburb between Lawrence and Topeka. The pro-slavery Lecompton Constitution was formulated (Sept., 1857) in the city, and was ratified (Dec., 1857) after an election in which voters were given a choice only between limited or unlimited slavery. At a subsequent election (Aug., 1858), Kansans decisively rejected the Lecompton Constitution. Kansas was later (1861) admitted as a free state.

Le·duc (lə-do͝ok'), town (1976 pop. 8,576), central Alta., Canada, S of Edmonton. It is the center of the Leduc oil field (discovered 1947).

Led·yard (lĕd'yərd, lĕj'ərd), town (1970 pop. 14,837), New London co., SE Conn., on the Thames River; settled c.1653, inc. 1836.

Lee (lē). **1.** County (1970 pop. 61,268), 612 sq mi (1,585 sq km), E Ala., in a piedmont area leveling off to flatlands below the Fall Line, bounded on the E by the Chattahoochee River and the Ga. border; formed 1866; co. seat Opelika. It is in a cotton, corn, and truck-farming area, with granite, dolomite, and manganese deposits. Lumbering, food processing, and textile weaving are done. Its factories produce clothing, paper products, plastics, concrete, iron, steel, aluminum, and machinery. **2.** County (1970 pop. 18,884), 608 sq mi (1,574.7 sq km), E Ark., bounded on the E by the Mississippi and drained by the St. Francis and L'Anguille rivers; formed 1873; co. seat Marianna. In an agricultural area yielding cotton, corn, small grains, grasses, soybeans, and some rice, it has oak, gum, poplar, walnut, pecan, and ash timber. Some mining and manufacturing are done. **3.** County (1970 pop. 105,216), 786 sq mi (2,035.7 sq km), SW Fla., in a lowland area bounded on the W by the Gulf of Mexico and drained by the Caloosahatchee River; formed 1887; co. seat Fort Myers. A chain of barrier islands border the county, sheltering several lagoons and Pine Island. Farming (citrus fruit and vegetables), flower growing, cattle raising, and fishing are the main occupations. **4.** County (1970 pop. 7,044), 354 sq mi (916.9 sq km), SW central Ga., bounded in the E by the Flint River and drained by the Kinchafoonee River; formed 1826; co. seat Leesburg. It is in a coastal-plain agriculture (peanuts, pecans, corn, and livestock) and timber area. **5.** County (1970 pop. 37,947), 729 sq mi (1,888.1 sq km), N Ill., drained by the Rock and Green rivers; formed 1839; co. seat Dixon. It has agriculture (corn, wheat, oats, soybeans, poultry, and livestock) and sand and gravel pits. Dairy products, farm machinery, cement and clay products, shoes, metal and wire goods, caskets, and gasoline are produced. **6.** County (1970 pop. 42,996), 527 sq mi (1,364.9 sq km), extreme SE Iowa, bounded in the NE by the Skunk River, in the E by the Mississippi River and the Ill. border, in the S by the Des Moines River and the Mo. border; formed 1836; co. seat Fort Madison. In a prairie agricultural area yielding hogs, cattle, poultry, corn, oats, and soybeans, it has limestone quarries, deposits of coal, and some industry. **7.** County (1970 pop. 6,587), 210 sq mi (543.9 sq km), E Ky., in the Cumberlands, drained by the Kentucky River and its North, Middle, and South forks; formed 1870; co. seat Beattyville. It is in a mountain agricultural area (livestock, fruit, tobacco, corn, and potatoes), with coal mines and oil wells. It has lumbering and some manufacturing (shoes). It includes part of Daniel Boone National Forest. **8.** County (1970 pop. 46,148), 455 sq mi (1,178.5 sq km), NE Miss., drained by tributaries of the Tombigbee River; formed 1866; co. seat Tupelo. It has cotton, corn, dairy, and cattle farming and lumbering. Its manufactures include food products, clothing and textiles, wood products, furniture, plastics, and machinery. The county includes Ackia Battleground National Monument, Brices Cross Roads National Battlefield Site, and Tupelo National Battlefield Site. **9.** County (1970 pop. 30,467), 256 sq mi (663 sq km), central N.C., in a forested piedmont and sand-hill area bounded on the NW by the Deep River, on the N by the Cape Fear River; formed 1907; co. seat Sanford. Farming (tobacco, cotton, and corn), sawmilling, and manufacturing are done. **10.** County (1970 pop. 18,323), 409 sq mi (1,059.3 sq km), NE central S.C., drained by the Lynches and Black rivers; formed 1902; co. seat Bishopville. It is in a timber and agricultural area yielding cotton, sweet potatoes, peanuts, and cucumbers. The county includes Lee State Park. **11.** County (1970 pop. 8,048), 637 sq mi (1,649.8 sq km), S central Texas, drained by Yegua Creek; formed 1874; co. seat Giddings. This diversified agricultural and livestock area yields cotton, peanuts, corn, grain sorghums, hay, pecans, fruit, truck crops, cattle, goats, sheep, hogs, poultry, and dairy products. **12.** County (1970 pop. 20,321), 434 sq mi (1,124.1 sq km), extreme SW Va., in an angle formed by Ky. on the N and Tenn. on the S; formed 1792; co. seat Jonesville. It is in a mountain and valley region drained by the Powell River. Agriculture (barley, tobacco, fruit, truck crops, and livestock) and coal mining are important. It also has limestone quarries, some oil wells, and a lumbering industry. Recreational facilities include limestone caves, the Cumberland Mts., and a section of Jefferson National Forest.

Lee, resort town (1970 pop. 6,426), Berkshire co., W Mass., in the Berkshires, on the Housatonic River S of Pittsfield; settled 1760, set off from Great Barrington and Washington 1777. It has paper mills and marble quarries.

Leeds (lēdz), county borough (1976 est. pop. 744,500), West Yorkshire, N central England, on the Aire River. Leeds is a communications center and an important transportation junction. Manufactures include woolen goods and clothing, metal goods (locomotives, machinery, farm implements, and airplane parts), leather goods, chemicals, and glass. Leeds has a classical town hall (1858) in which triennial musical festivals are held; of interest also are St. Peter's Church, the Cathedral of St. Anne, St. John's Church, and the City Art Gallery. Kirkstall Abbey, founded in the 12th cent., is near the city.

Leeds, city (1970 pop. 6,991), Jefferson, Shelby, and St. Clair cos., N central Ala., E of Birmingham, in a lumber, coal, iron, and limestone area; founded 1881. Textile and steel products are made.

Lee·la·nau (lē'lə-nô'), county (1970 pop. 10,872), 349 sq mi (903.9 sq km), NW Mich., a peninsula bounded on the W by Lake Michigan and on the E by Grand Traverse Bay; formed 1863; co. seat Leland. Fruit is grown, especially cherries and strawberries. There is also poultry, truck, and potato farming, and dairy products are made. Sawmills and fisheries are found here. There are resorts on Leelanau and Glen lakes, as well as hunting, skiing, and recreational areas.

Leer·dam (lâr-däm'), town (1970 pop. 13,282), South Holland prov., S central Netherlands. It is famous for its glassware and ceramics.

Lees·burg (lēz'bûrg'). **1.** City (1970 pop. 11,869), Lake co., N central Fla., in a hill and lake region; inc. 1875. It is a processing and shipping center in a citrus-fruit and truck-farm area. Cattle raising and the manufacture of such items as crates, boxes, concrete products, fertilizer, athletic equipment, and mobile homes are also important. **2.** City (1970 pop. 996), seat of Lee co., SW central Ga., N of Albany near the Kinchafoonee River, in an agricultural area. **3.** Town (1970 pop. 4,821), seat of Loudoun co., NE Va., NW of Washington, D.C.; settled 1749, inc. 1758. The town has many colonial houses and is in a rich farm area noted for its cattle and thoroughbred horses.

Lee's Sum·mit (lēz sŭm'ĭt), city (1970 pop. 16,230), Jackson co., W Mo., in the Kansas City metropolitan area; inc. 1868. Its manufactures include tools, machinery, and plastics.

Lees·ville (lēz'vĭl'), town (1970 pop. 8,928), seat of Vernon parish, W central La., N of De Ridder, in a timber and livestock area; inc. 1899. Kisatchie National Forest and Fort Polk are nearby.

Leeu·war·den (lā'wärd'n, -värd'n), city (1977 est. pop. 85,435), capital of Friesland prov., N Netherlands. It is the center of an agricultural and dairying region and has a noted cattle market. Manufactures include food products, clothing, and artificial silk. Chartered in 1435, Leeuwarden was (16th–18th cent.) the center of a goldworking and silverworking industry.

Lee·ward Islands (lē'wərd), N group of the Lesser Antilles in the West Indies, extending SE from Puerto Rico to the Windward Islands. Largely volcanic in origin, the Leeward Islands have lush, subtropical vegetation, rich soil, and abundant rainfall. The warm, delightful climate is tempered by the surrounding water so that there is little variation in temperature. Most of the islands have become popular winter resorts. Products for the most part are agricultural—fruits, vegetables, sugar, cotton, coffee, and tobacco. Columbus discovered the Leeward Islands in 1493, but settlement began only after the British arrived in the 17th cent. In 1623 both the English and the French established colonies. By 1632 the sharp, three-way colonial conflict of England, France, and Spain had begun. The Spanish were forced from the struggle, but for nearly two centuries the islands were pawns in the Anglo-French worldwide wars. Their final disposition—a division of the islands between the competing powers—did not come until the end of the Napoleonic Wars in 1815.

Le·flore (lə-flōr'), county (1970 pop. 42,111), 588 sq mi (1,522.9 sq km), W Miss., drained by the Tallahatchie and Yalobusha rivers, which unite here to form the Yazoo River; formed 1871; co. seat Greenwood. It is in a cotton-growing area, with a lumbering industry and cottonseed-oil mills. Farm equipment is manufactured.

Le Flore (lə flōr'), county (1970 pop. 32,137), 1,560 sq mi (4,040.4 sq km), SE Okla., bounded on the N by the Arkansas River and on the E by the Ark. border; formed 1907; co. seat Poteau. It is drained by the Poteau and Kiamichi rivers and includes part of the Ouachita Mts. Agriculture (corn, cotton, livestock, and vegetables), lumbering, and coal mining are important.

Le·gas·pi (lə-găs'pē, -gäs'pē), city (1970 pop. 35,911), capital of Albay prov., SE Luzon, the Philippines, on Albay Gulf. Founded c.1639, it is a seaport shipping copra and hemp. In World War II it was the scene (Dec. 12, 1941) of a large Japanese landing. Towering directly behind the city is the volcano Mt. Mayon. Its eruption in 1814 severely damaged the town and killed more than 1,000 people.

Leg·horn (lĕg'hôrn') or *Italian* **Li·vor·no** (lē-vôr'nô), city (1971 pop. 173,774), capital of Livorno prov., Tuscany, central Italy, on the Ligurian Sea and on the Aurelian Way. It is a busy port and commercial, industrial, and tourist center. Manufactures include refined petroleum, iron and steel, chemicals, and electrical equipment. The city has major shipyards and a fishing industry. A fortified castle in the Middle Ages, Leghorn was developed (16th cent.) into a flourishing city by the Medici. In 1590 Ferdinand I, grand duke of Tuscany, made it a free port and opened it to all religious and political refugees. The city was badly damaged in World War II.

Le·gna·no (lā-nyä'nō), city (1975 est. pop. 48,347), Lombardy, NW Italy, near Milan. Manufactures of this important industrial center include chemicals, plastics, steel, and textiles. Near Legnano the Lombard League defeated (1176) Emperor Frederick I.

Leg·ni·ca (lĕg-nē'tsä), city (1975 est. pop. 82,100), SW Poland, on the Kaczawa River. A center of a vegetable-growing region, it also has manufactures of textiles, machinery, and chemicals. Chartered in 1252, it was acquired (1742) by Prussia. The city was heavily damaged in World War II, but it has retained its 11th cent. castle (rebuilt 1835) and parts of its medieval walls and towers.

Le·high (lē'hī'), county (1970 pop. 255,304), 347 sq mi (898.7 sq km), E Pa., in an industrial and farm area drained by the Lehigh River; formed 1812; co. seat Allentown. Potatoes are grown here. Limestone and slate are among its natural resources. Manufactures include cement, motor vehicles, and textiles.

Lehigh, river, 103 mi (165.7 km) long, rising in NE Pa. and flowing generally SE to the Delaware River at Easton. The river flows through a major industrial and anthracite-coal area.

Le·high·ton (lē-hīt'n), industrial borough (1970 pop. 6,095), Carbon co., E Pa., on the Lehigh River NNW of Allentown, in a coal-mining area; settled c.1746, inc. 1866. It was destroyed in the French and Indian Wars and then rebuilt.

Leh·man Caves National Monument: (lē'mən): *see* National Parks and Monuments Table.

Leices·ter (lĕs'tər), county borough (1976 est. pop. 289,400), administrative center of Leicestershire, central England. Leicester was of industrial importance as early as the 14th cent.; the making of hosiery and shoes are long-established industries. Other manufactures are chemicals, aniline dyes, textiles, textile and woodworking machinery, and light-metal products. Leicester was the Ratae Coritanorum, or Ratae, of the Romans, whose Fosse Way passes nearby. It was also a town of the ancient Britons and was one of the Five Boroughs of the Danes. Its antiquities include the Jewry Wall, a Roman structure 18 ft (5.5 m) high and 70 ft (21.4 m) long (near which extensive Roman remains have been found); remains of the ancient Norman castle in whose banquet hall county assizes are now held; and ruins of an abbey founded in 1143.

Leicester, town (1970 pop. 9,140), Worcester co., central Mass., W of Worcester; settled 1713, inc. 1714. Textiles are produced.

Leices·ter·shire (lĕs'tər-shîr', -shər), nonmetropolitan county (1976 est. pop. 837,900), 832 sq mi (2,154.9 sq km), central England; administrative center Leicester. There is good farming land in the uplands of the east, while the west is devoted primarily to mining and industry. Its agriculture includes sheep, dairy cattle, wheat, and barley. Stilton cheese is a well-known dairy product of the region. Leicestershire was part of the Anglo-Saxon kingdom of Mercia. At Bosworth Field, in 1485, Richard III was slain by the forces of Henry Tudor, who ascended the throne as Henry VII.

Lei·den or **Ley·den** (both: līd'n), city (1971 pop. 100,135), South Holland prov., W Netherlands, on the Old Rhine (Oude Rijn) River. Its manufactures include textiles, medical equipment, machinery, and food products. The city is famous for its university (founded 1575), which was a center for the study of Protestant theology, classical and oriental languages, science, and medicine in the 17th and 18th cent.

Dating from Roman times, Leiden has had an important textile industry since the 16th cent., when there was an influx of weavers from Flanders. The city took a prominent part in the revolt (late 16th cent.) of the Netherlands against Spanish rule. Besieged and reduced to starvation in 1574, it was saved from surrender when William the Silent ordered the flooding of the surrounding land by cutting the dikes, thus enabling the fleet of the Beggars of the Sea to sail to its relief across the countryside. Leiden became famous as a center of printing after 1580 and was the birthplace of the painters Jan van Goyen, Jan Steen, Lucas van Leyden, and Rembrandt.

Leigh (lē), municipal borough (1971 pop. 46,117), Greater Manchester, NW England. Coal mines and cotton mills are here.

Lein·ster (lĕn'stər, līn'-), province (1971 pop. 1,498,400), 7,580 sq mi (19,632 sq km), E Republic of Ireland. Its wealth and accessibility made the province subject to Danish and Anglo-Norman invasions.

Leip·zig (līp'sĭg, -sĭk, -tsĭKH), city (1974 est. pop. 570,972), capital of Leipzig district, S central East Germany, at the confluence of the Pleisse, White Elster, and Parthe rivers. It is an industrial, commercial, and transportation center. Manufactures include textiles, steel, machinery, chemicals, paper, toys, and motor vehicles. Important international trade and industrial fairs have been held in the city since the Middle Ages.

Originally a Slavic settlement called Lipsk, Leipzig was chartered at the end of the 12th cent. The city was the scene of the famous religious debate between Martin Luther, Carlstadt, and Johann Eck in 1519. In 1539 it accepted the Reformation. The city was one of the leading cultural centers of Europe in the age of the philosopher and mathematician Leibnitz, who was born here in 1646, and of the composer Johann Sebastian Bach, who was cantor at the Church of St. Thomas from 1723 until his death. The Univ. of Leipzig (founded 1409; renamed Karl Marx Univ. in 1953) became one of the most important in Germany. In the 18th cent. Gottsched, Gellert, Schiller, and many others made Leipzig a literary center; the young Goethe studied here in 1765. The city's musical reputation reached its peak in the 19th and early 20th cent. Felix Mendelssohn, who died here in 1847, made the Gewandhaus concerts internationally famous. Robert Schumann worked in Leipzig, Richard Wagner was born here in 1813, and the Leipzig Conservatory (founded by Mendelssohn in 1842–43) became one of the world's best-known musical academies.

The Battle of Leipzig, Oct. 16–19, 1813, also called the Battle of the Nations, was a decisive victory of the Austrian, Russian, and Prussian forces over Napoleon I. It is estimated that 120,000 men were killed or wounded in the battle. Until World War II Leipzig was the center of the German book and music publishing industry, and the center of the European trade in furs and smoked foods. The city was badly damaged in World War II. Noteworthy buildings include the Church of St. Thomas (late 15th cent.), which has housed the tomb of Bach since 1950; the Gewandhaus, built in 1884 to replace an earlier structure; the old city hall (1558); the old stock exchange (1682); the Church of St. John (17th cent.); and the former German supreme court building (now an art museum).

Lei·ri·a (lā-rē'ə), town (1970 pop. 10,286), capital of Leiria dist., W central Portugal, in Beira Litoral. It is an agricultural trade center. Here Alfonso I erected (beginning 1135) a castle on a cliff above the present city; it was taken and retaken in the wars with the Moors.

Lei·sure City (lē'zhər), uninc. town (1970 pop. 5,600), Dade co., SE Fla., SW of Miami.

Leitch·field (lĭch'fēld'), town (1970 pop. 2,983), seat of Grayson co., W central Ky., SW of Louisville, in a farm area.

Leith (lēth), former burgh, Lothian region, SE Scotland, on the S shore of the Firth of Forth. It was incorporated into Edinburgh in 1920. As a strategically located port, Leith was sacked by the English in 1544 and 1547 and burned in the Jacobite uprising of 1715.

Lei·tha (lī'tä), river, 112 mi (180.2 km) long, formed in E Austria by the confluence of the Schwarza and Pitten rivers and flowing generally E to an arm of the Danube River. It was long a historic boundary between Austria and Hungary.

Leith Hill, 965 ft (294.3 m) high, Surrey, SE England; highest point of the North Downs. On the summit is a tower, with a view on clear days of London and the English Channel.

Lei·trim (lē'trĭm), county (1971 pop. 28,360), 589 sq mi (1,525.5 sq km), N Republic of Ireland; Co. town Carrick-on-Shannon. The county is divided into two parts by Lough Allen; the northern part is mountainous, the southern part level. Cattle and sheep raising and tillage are the chief occupations.

Leix (lāks), county, Ireland: *see* Laoighis.

Le·land (lē'lənd). **1.** Village (1970 pop. 400), seat of Leelanau co., NW Mich., NNW of Traverse City between Lake Leelanau and Lake Michigan, in a dairy and fruit area. It is a resort and has commercial fisheries. **2.** City (1970 pop. 6,000), Washington co., W central Miss., just E of Greenville, in a rich cotton, rice, and truck-farm area; settled 1847, laid out 1884. Cottonseed oil, farm implements, and furniture are made here.

Le Mars (lə märz'), city (1970 pop. 8,159), seat of Plymouth co., NW Iowa, on the Floyd River NNE of Sioux City; founded in the 1870s, inc. 1881. It is a trade center for a farm and dairy region.

Lem·hi (lĕm'hī'), county (1970 pop. 5,566), 4,585 sq mi (11,875.2 sq km), E Idaho, in a mountain and valley area drained by the Salmon River and its tributaries and bounded on the E by the Bitterroot Range and Mont.; formed 1869; co. seat Salmon. Livestock raising and mining (gold, tungsten, lignite, and coal) are major occupations. Manganese deposits are also found.

Lem·nos (lĕm'nŏs), island: *see* Límnos.

Lem·on Grove (lĕm'ən), uninc. town (1970 pop. 20,300), San Diego co., S Calif. It is residential and agricultural, with some industry.

Le·moyne (lə-moin'), residential borough (1975 est. pop. 4,400), Cumberland co., S Pa., on the Susquehanna River, in a grain-farming area; inc. 1905. The site of Fort Washington marks the northernmost point of Confederate invasion (1863) during the Civil War.

Lem·pa (lĕm'pə, läm'pä), river, c.200 mi (320 km) long, rising in Guatemala and flowing S through Honduras into El Salvador, then generally S to the Pacific Ocean. The fertile Lempa valley, supports a dense agricultural population.

Le·na (lē'nə, lyĕ'-), river, easternmost of the great rivers of Siberia, c.2,670 mi (4,295 km) long, rising near Lake Baykal, SE Siberian USSR. It flows northeast, then north along the east side of the central Siberian uplands and empties through a c.250-mi (400-km) wide delta into the Arctic Ocean. It is navigable for 2,135 mi (3,435.2 km). Coal, oil, and gold are found along the Lena and its tributaries.

Le·na·wee (lĕn'ə-wē'), county (1970 pop. 81,951), 754 sq mi (1,952.9 sq km), SE Mich., bounded on the S by the Ohio border, drained by the Raisin and Tiffin rivers; formed 1826; co. seat Adrian. It is chiefly an agricultural area dealing in grain, corn, beans, truck crops, and livestock. Dairy products are made. There is some manufacturing. Hatcheries and sand and gravel pits are found here.

Len·in·a·bad (lĕn'ĭn-ə-băd', lyĕ'nyĭ-nə-bät'), city (1976 est. pop. 121,000), capital of Leninabad oblast, Central-Asian USSR, in Tadzhikistan, on the Syr Darya River at its exit from the Fergana Valley. It is the major Soviet center for silk production; other industries produce clothing, footwear, and food products. Leninabad, located on an ancient caravan route from China to the Mediterranean, was a famous town marking the farthest expansion of Alexander the Great. The fortress he founded, called Alexandria Eskhat, was plundered (711) by the Arabs and later (1220) was razed by Genghis Khan. As part of the Kokand khanate (early 19th cent.), Leninbad was annexed (1866) by Russia. The city and surrounding area belonged to Uzbekistan from 1924 to 1929.

Len·in·a·kan (lĕn'ĭn-ə-kän', lyĕ'nyĭ-nə-kän'), city (1976 est. pop. 188,000), SE European USSR, in Armenia, near the Turkish border. It has textile and metalworking plants. Leninakan was founded (1837) on the site of a Turkish fortress.

Len·in·grad (lĕn'ĭn-grăd'), city (1970 pop. 3,950,000), capital of Leningrad oblast, NW European USSR, at the head of the Gulf of Finland on both banks of the Neva River and on the islands of its delta. Leningrad's port is linked by deepwater canal with Kotlin Island, where the outer port and the Kronshtadt naval base are located. Leningrad is a major seaport, rail junction, and industrial, cultural, and scientific center. Although the harbor is frozen for three or four months annually, icebreakers have prolonged the navigation season. The seaport is one of the world's largest, but it handles relatively little traffic because the volume of foreign trade for the USSR is small. The river port, one of the most important in the country, stands at the end of two artificial waterways, the Volga-Baltic and the White Sea-Baltic. A series of canals within the city carries considerable cargo. The city's diverse industries include shipbuilding, metallurgy, oil refining, printing, woodworking, food and tobacco processing, and the manufacture of machinery, electrical equipment, chemicals, pharmaceuticals, and textiles.

Originally called St. Petersburg, the city was built in 1703 by Peter I (Peter the Great), who sought a Russian outlet to the sea. The fortress of Peter and Paul was erected to defend the new city, which became the Russian capital in 1712. St. Petersburg soon replaced Arkhangelsk as Russia's leading seaport and became an important commercial center. From the second half of the 18th cent. it was also the country's principal industrial center. Its apex as an international center of literature, music, theater, and ballet was reached in the late 19th and early 20th cent.

The city, renamed Petrograd in 1914, was the scene of the revolutions of 1905 and of Feb. and Oct., 1917. Although it lost much of its former glamour, the city remained the economic and cultural rival of Moscow, which replaced it as capital in 1918. Petrograd was renamed Leningrad in 1924. During World War II the city was besieged by German armies for more than two years, during which many thousands died of famine and disease.

The city's main thoroughfare is the celebrated Nevsky Prospekt. On it are the high-spired admiralty building, the Winter Palace, the Hermitage museum, the huge-domed Cathedral of St. Isaac (1858), and the equestrian statue of Peter the Great. The city's oldest building is the fortress of Peter and Paul (1703). Leningrad has a university and numerous theaters, museums, scientific and medical institutes, and libraries.

Len·in·o·gorsk (lĕn′ĭn-ə-gôrsk′, lyĕ′nyĭ-nə-gôrsk′), city (1976 est. pop. 69,000), E Central Asian USSR, in Kazakhstan, in the Altai Mts. It is a mining center in an area that is the chief source of Soviet zinc and lead. Silver, copper, and gold are also mined. The first mines were opened in the late 18th cent. Leninogorsk's metallurgical factories produce lead, copper, and zinc concentrates and cadmium.

Len·in Peak (lĕn′ĭn), 23,405 ft (7,138.5 m) high, on the border of Tadzhikistan and Kirghizia, Central Asian USSR. It is the highest point in the Trans-Alai range.

Len·insk-Kuz·nets·ki (lĕn′ĭnsk-ko͞oz-nĕts′kĭ, lyĕ′nĭnsk-ko͞oz-nyĕt′-skē), city (1976 est. pop. 131,000), S central Siberian USSR, on the Inya River. It is a coal center in the Kuznetsk Basin. Founded in 1864 as a mining settlement, it developed rapidly in the 1930s.

Len·ko·ran (lĕng′kə-rän′, lyĕn-kə-rän′yə), city (1974 est. pop. 38,000), SE European USSR, in the Azerbaijan Republic, near the Iranian border, on the Caspian Sea. It is a port and an important food-processing center for fish and tea. Lenkoran, known since the 17th cent., was ceded to Russia by Persia in 1813. The Lenkoran Lowland, a coastal strip c.40 mi (65 km) long, has a humid, subtropical climate. Citrus fruit, tea, and rice are grown here.

Len·nox (lĕn′əks), uninc. city (1970 pop. 16,121), Los Angeles co., S Calif., an industrial suburb of Los Angeles.

Le·noir (lə-nôr′), county (1970 pop. 55,204), 400 sq mi (1,036 sq km), E central N.C., on the coastal plain and drained by the Neuse River; formed 1791; co. seat Kinston. Farming (tobacco, corn, and cotton), sawmilling, and manufacturing are the major occupations.

Lenoir, city (1978 est. pop. 15,500), seat of Caldwell co., W N.C.; inc. 1851. It is a resort in the eastern foothills of the Blue Ridge Mts.

Len·ox (lĕn′əks), resort town (1970 pop. 5,804), Berkshire co., W Mass., in the Berkshires S of Pittsfield; settled c.1750, set off from Richmond 1767. The annual Berkshire Festival is held at Tanglewood, a former estate mainly in the adjoining town of Stockbridge. Fanny Kemble, Henry Ward Beecher, and Edith Wharton lived in Lenox. It is noted for its many beautiful estates.

Lens (läns), city (1975 pop. 40,199), Pas-de-Calais dept., N France. Since the 19th cent. it has been an important coal center. It is also a manufacturing center with metallurgical and textile industries. The victory (1648) of the French under Louis II de Condé was the last important battle of the Thirty Years' War. Lens was occupied and devastated by the Germans in both world wars.

Le·o·ben (lā-ō′bən), city (1971 pop. 35,153), Styria prov., S central Austria, on the Mur River. An industrial center in a coal-mining region, it has large ironworks, textile mills, and breweries. An armistice between France and Austria was signed (1797) at Leoben to conclude Napoleon I's victorious Italian campaign.

Le·o·la (lē-ō′lə), city (1970 pop. 787), seat of McPherson co., N S.Dak., NW of Aberdeen. It is a shipping center for a grain and hog-raising area.

Leom·in·ster (lĕm′ĭn-stər), city (1978 est. pop. 36,600), Worcester co., N central Mass.; set off from Lancaster 1740, inc. as a city 1915. Plastics are made here.

Le·on (lē′ŏn). **1.** County (1970 pop. 103,047), 670 sq mi (1,735.3 sq km), NW Fla., on the Ga. border in the N and bounded in the W by the Ochlockonee River and Lake Talquin; formed 1824; co. seat Tallahassee. Rolling terrain in the north and coastal plains in the south characterize this region, which also has many lakes and part of Apalachicola National Forest. Agriculture (corn, peanuts, cotton, vegetables, cattle, hogs, poultry, and dairy products) and forestry are the main occupations. **2.** County (1970 pop. 8,738), 1,102 sq mi (2,854.2 sq km), E central Texas, bounded on the W by the Navasota River, on the E by the Trinity River; formed 1846; co. seat Centerville. In an agricultural area yielding cotton, corn, grain sorghums, peanuts, potatoes, truck crops, fruit, livestock, and dairy products, it has food-processing and lumbering industries.

Leon, city (1970 pop. 2,142), seat of Decatur co., S Iowa, SSW of Des Moines, near the Mo. border, in a farm area; settled 1840, inc. 1858. It is a shipping center for livestock and dairy products.

Le·ón (lā-ōn′), region and former kingdom (1974 est. pop. 1,199,214), NW Spain, E of Portugal and Galicia. Northern León, which is crossed by the Cantabrian Mts., has coal mines, forests, and mountain pastures; the rest of the region is a dry plateau. Livestock are raised, and cereals, hops, and flax are grown.

Early in the Christian reconquest, the kings of Asturias gained control over León (8th–9th cent.); their territory became the kingdom of Asturias and León. The power of the kings also extended over Galicia and part of Castile, Navarre, and the Basque Province, but it was too weak to prevent the rise of the independent kingdoms of Navarre and Castile. León was conquered (1037) by Ferdinand I of Castile, on whose death (1065) the kingdoms again became separate. Reunited in 1072 under Alfonso VI, León and Castile were again separated in the 12th cent. and remained so until Ferdinand III accomplished the final reunion in 1230.

León, city (1974 est. pop. 468,900), Guanajuato state, central Mexico. It is located in a fertile river valley c.5,600 ft (1,710 m) high and is a commercial, agricultural, and mining center. It is also one of Mexico's leading shoe manufacturers and has gained fame for the knives and iron goods produced by local artisans. The city's mines yield gold, copper, silver, lead, and tin. León was founded in 1577.

León, city (1971 pop. 54,841), capital of Léon dept., W Nicaragua. It is a rail and commercial center. León was founded in 1524 on Lake Managua and moved westward to its present site in 1610 after a severe earthquake. It became the stronghold of the liberal forces after independence from Spain (1821) and engaged in bitter rivalry with conservative Granada.

León, city (1975 est. pop. 113,339), capital of León prov., NW Spain, at the foot of the Cantabrian Mts. at the confluence of the Bernesga and Torio rivers. It is an agricultural and commercial center. Dating from Roman times, it was reconquered from the Moors in 882 by Alfonso III of Asturias. The city flourished in the 12th and 13th cent. but declined after the kings of León and Castile made Valladolid their favored residence. It still retains a medieval atmosphere, and its many historic monuments attract tourists.

Leon·ard·town (lĕn′ərd-toun′), town (1970 pop. 1,406), seat of St. Marys co., SW Md., SSE of Washington, D.C.; laid out 1708, inc. 1858. It is a trade center for tobacco, farm produce, and lumber.

Le·o·ni·a (lē-ō′nē-ə, -ōn′yə), residential borough (1970 pop. 8,847), Bergen co., NE N.J., SSW of Englewood; inc. 1894.

Le·on·ti·ni (lē′ən-tī′nī′), ancient city, E Sicily, S of Catania. It was (729 B.C.) a colony of Chalcidians from the island of Naxos and passed (5th cent. B.C.) under the rule of Syracuse. It was sacked (A.D. 848) by Saracens and destroyed (1693) by an earthquake.

Le·o·pold·ville (lē′ə-pōld-vĭl′, lā′ô-pôld-vēl′): *see* Kinshasa, Zaire.

Le·o·ti (lē-ō′tī′), city (1970 pop. 1,916), seat of Wichita co., W central Kansas, NW of Dodge City, in a grain and dairy area; settled 1885, inc. 1887.

Le·pon·tine Alps (lə-pŏn′tən), section of the Central Alps, S Switzerland, along the Swiss-Italian border, extending from the Pennine Alps in the WSW to the Rhaetian Alps in the ENE. Its highest peak is Monte Leone (11,683 ft/3,563.3 m).

Lep·tis (lĕp′tĭs), ancient city of Libya, E of Tripoli. It was founded (c.600 B.C.) by Phoenicians. Annexed (46 B.C.) to the Roman province of Africa, it flourished as an important port under the Romans. Some of the most impressive ruins of Roman Africa are here, including walls, baths, arches, temples, and forums. The city is sometimes called Leptis Magna.

Lé·ri·da (lā′rĭ-də, lā′rē-*th*ä), city (1975 est. pop. 81,200), capital of Lérida prov., NE Spain, in Catalonia, on the Segre River. Lérida is the center of a fertile farm area and manufactures arms and chemicals. It was taken (49 B.C.) by Julius Caesar, who defeated Pompey's generals here. Lérida fell to the Moors in A.D. 714 and was liberated (1149) by Raymond Berengar IV of Barcelona. Traditionally a strategic, fortified city, Lérida was a key defense point for Barcelona in the Spanish Civil War; it fell (April, 1938) after a nine-month battle.

Ler·ma (lĕr′mə, -mä), river, c.350 mi (565 km) long, rising in Mexico state, central Mexico, and flowing NW and W through Guanajuato state to Lake Chapala, crossing the part of the central plateau known as the Anáhuac. The river draining the lake and flowing NW through Jalisco state to the Pacific Ocean is generally called the Río Grande de Santiago (c.200 mi/320 km long) but it is considered a continuation of the Lerma. The river system is extensively used for irrigation and hydroelectric power.

Ler·wick (lûr'wĭk, lĕr'ĭk), burgh (1971 pop. 6,107), Shetland Islands Authority, extreme N Scotland. On the southeastern coast of Mainland island, Lerwick is the central market town of the Shetlands.

Les·bos (lĕz'bŏs, -bōs) or **Lés·vos** (lāz'vôs), island (1971 pop. 114,797), c.630 sq mi (1,630 sq km), E Greece, in the Aegean Sea near Turkey. A fertile island, it has vast olive groves and also produces wheat, wine, and citrus fruit. Sardines are caught. Lesbos was a center of Bronze Age civilization and later (c.1000 B.C.) was settled by Aeolians. The island was a brilliant cultural center from the 7th to the 6th cent. B.C. Lesbos joined the Delian League and revolted unsuccessfully against Athens in 428–427 B.C. Later, Lesbos passed to Macedonia, Rome, and the Byzantine Empire. It became part of Greece in 1913.

Les Cayes (lā kā'): *see* Aux Cayes, Haiti.

Les·lie (lĕz'lē), county (1970 pop. 11,623), 412 sq mi (1,067.1 sq km), SE Ky., in the Cumberlands drained by the Middle Fork of the Kentucky River; formed 1878; co. seat Hyden. It is in a mountain agricultural area that produces corn, hay, truck crops, livestock, fruit, and tobacco. It has bituminous-coal mines.

Le·so·tho (lə-sō'tō), formerly **Ba·su·to·land** (bə-sōō'tō-lănd'), kingdom (1977 est. pop. 1,075,000), 11,720 sq mi (30,355 sq km), S Africa, enclave within the Republic of South Africa. Maseru is the capital. The Draksenberg range occupies the eastern part of the country. The rest of the country is a rocky tableland with a dry climate.

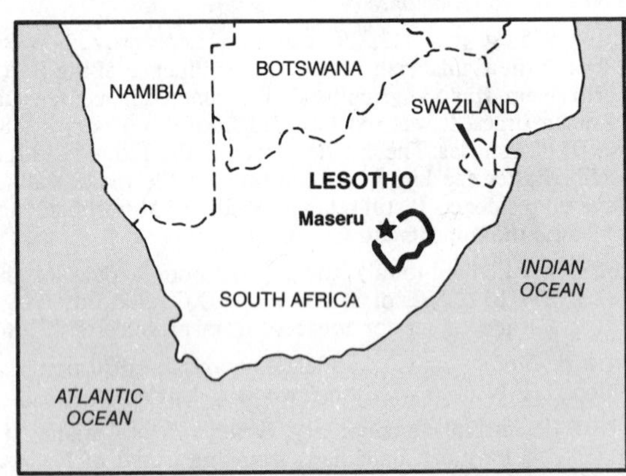

Economy. Only a small percentage of Lesotho's land is arable. Maize, sorghum, and wheat are extensively cultivated. Sheep are bred for wool, and cattle and Angora goats are raised. All land in Lesotho is held by the king in trust for the Sotho nation and is apportioned on his behalf by local chiefs. Lesotho's mineral resources are limited to diamonds, and it has varied light industries.

History. The Sotho are made up of remnants of ethnic groups that were scattered during the disturbances accompanying the rise of the Zulu (1816–30). Moshesh, a paramount chief, not only defended his people from Zulu raids but preserved their independence against Boer and British interlopers. Following wars with the Boers, Moshesh put the Sotho under British protection (1868). The protectorate was annexed to Cape Colony in 1871 without Sotho consent; but in 1884 it was placed under the direct control of Britain. A resident commissioner was established at Maseru. When the Union of South Africa was forged in 1910, Basutoland came under the jurisdiction of the British High Commissioner in South Africa. Sotho opposition prevented annexation by the Republic of South Africa. In 1960 the British granted Basutoland a new constitution, and on Oct. 4, 1966, Basutoland became independent as Lesotho. Following general elections in early 1970 Prime Minister Leabua Jonathan declared a state of emergency and suspended the constitution; King Moshoeshoe II went into exile but returned at the end of the year. In 1973 an interim assembly began work on a new constitution, but the Congress party, led by Ntsu Mokhehle, refused to participate. In Jan., 1974, Chief Leabua Jonathan accused the Congress party of attempting to stage a coup d'état, and in the months that followed hundreds of its members reportedly were killed.

Les·ser Slave Lake (lĕs'ər slāv'), 60 mi (96.5 km) long and from 3 to 10 mi (4.8–16.1 km) wide, central Alta., Canada, NW of Edmonton. It drains into the Athabasca River by the Lesser Slave River. Commercial fishing, lumbering, and farming are done.

Le Sueur (lə sōōr'), county (1970 pop. 21,332), 441 sq mi (1,142.2 sq km), S Minn., bordered on the W by the Minnesota River; formed 1853; co. seat Le Center. It is in an agricultural area that produces livestock, corn, oats, barley, and dairy products. It has some mining and a food-processing industry. Clothing, pharmaceuticals, concrete, tools, and auto parts are manufactured.

Lesz·no (lĕsh'nô), town (1975 est. pop. 37,500), SW Poland. A railway junction, it has industries producing synthetic fibers, chemicals, and carpets. Chartered in 1547, it passed to Prussia in 1793 and again in 1815. It reverted to Poland in 1919. Leszno was a center of the Protestant Reformation of the 16th cent.

Letch·er (lĕch'ər), county (1970 pop. 23,165), 339 sq mi (878 sq km), SE Ky., in the Cumberlands bounded on the E and SE by Va. and drained by the North Fork of the Kentucky River and Poor Fork of the Cumberland River; formed 1842; co. seat Whitesburg. It is in an important bituminous-coal area, with clay, sand, and gravel pits and limestone quarries. It has lumber sawmilling, and its agriculture includes dairy products, poultry, livestock, corn, soybeans, apples, sweet and Irish potatoes, and tobacco.

Letch·worth (lĕch'wûrth'), urban district (1973 est. pop. 31,520), Hertfordshire, E central England; founded 1903. The main industries are printing and the manufacture of printing machinery.

Leth·bridge (lĕth'brĭj'), city (1976 pop. 46,752), S Alta., Canada, on the Oldman River. Formerly a coal-mining center, Lethbridge is now a commercial center for an irrigated farming and ranching district. Industries include sugar refining, food processing, brewing, steel fabricating, and the manufacture of electronic equipment.

Le·ti·ci·a (lə-tē'sē-ə, lā-tē'sē-ä), town (1968 est. pop. 4,600), capital of Amazonas commissary, SE Colombia, on the upper Amazon River. The Leticia region, a narrow strip of land extending south of the Putumayo River to the Amazon, was disputed, at times violently, between Colombia and Peru (1932–34). The region was awarded to Colombia by the League of Nations in 1934.

Leu·ca·di·a (lōō-kā'dē-ə), uninc. town (1970 pop. 6,500), San Diego co., S Calif., N of San Diego; founded 1888.

Leu·cas (lōō'kəs), island: *see* Levkás.

Leuc·tra (lōōk'trə), village of ancient Greece, in Boeotia, SW of Thebes. The Spartans were defeated (371 B.C.) by the Thebans here. The battle dealt a severe blow to Spartan hegemony.

Leu·na (loi'nä), city (1974 est. pop. 10,550), Halle dist., S central East Germany. The city grew as the center of the German synthetic chemical industry, and it is today the seat of the largest chemical works in East Germany.

Le·val·lois-Per·ret (lə-vä'lwä'pĕ-rā'), residential and industrial suburb of Paris (1968 pop. 59,212), Hauts-de-Seine dept., N central France, on the Seine River. Automobiles, electrical and radio equipment, and perfume are manufactured. The town also has foundries, distilleries, and food-processing plants.

Le·vant (lə-vănt'), collective name for the countries of the E shore of the Mediterranean from Egypt to, and including, Turkey. The divisions of the French mandate over Syria and Lebanon were called the Levant States.

Lev·el·land (lĕv'əl-lănd'), city (1970 pop. 11,445), seat of Hockley co., NW Texas, on the Llano Estacado; inc. 1926. The economy is based chiefly on oil, agriculture, and the manufacture of mobile homes. Reese Air Force Base is nearby.

Le·ven (lē'vən), burgh (1971 pop. 9,454), Fife region, E Scotland, at the mouth of the Leven River on the Firth of Forth. It is a summer resort, famous for its golf links and beaches.

Leven, Loch, lake, 3.5 mi (5.6 km) long, Tayside region, E Scotland. Its several islands include Castle Island, with the ruins of the castle in which Mary Queen of Scots was imprisoned in 1567–68.

Le·ver·ku·sen (lā'vər-kōō'zən), city (1974 est. pop. 167,671), North Rhine-Westphalia, W West Germany, on the Rhine River. It is an industrial center and a road and rail junction.

Lé·vis (lē'vĭs, lā-vē'), city (1976 pop. 17,819), S Que., Canada, on the St. Lawrence River opposite Quebec. Settled in 1647, it was a base

(1759) for Wolfe's siege of Quebec. Lévis is a port with shipbuilding and other industries.

Lev·it·town (lĕv'ət-toun'), uninc. residential city (1978 est. pop. 65,500), Nassau co., SE N.Y., on Long Island; founded 1947 as a private, low-cost housing development for veterans.

Lev·kás (lĕf-käs') or **Leu·cas** (lo͞o'kəs), mountainous island (1971 pop. 22,917), c.115 sq mi (300 sq km), W Greece, in the Ionian Sea, one of the Ionian Islands. Olive oil, currants, wine, and tobacco are produced. The island was colonized (7th cent. B.C.) by Corinthians, and Corinth and Levkás were allies in the Peloponnesian War. Levkás later was the capital of the Acarnanian League (3rd cent. B.C.). The island was captured (1697) from the Ottoman Turks by Venice, which held it until 1797.

Le·vy (lē'vē), county (1970 pop. 12,756), 1,082 sq mi (2,802.4 sq km), N Fla., in a flatwoods, partly swampy area bounded on the SW by the Gulf of Mexico, on the W by the Suwannee River, on the S by the Withlacoochee River and Lake Rousseau; formed 1845; co. seat Bronson. Stock raising (hogs and cattle), farming (corn, vegetables, and peanuts), lumbering, fishing, and quarrying (limestone and dolomite) are the major occupations.

Lew·es (lo͞o'ĭs), municipal borough (1973 est. pop. 14,170), administrative center of East Sussex, SE England. Lewes is a farm market with light manufactures. In 1264, Lewes was the scene of a victory by Simon de Montfort, earl of Leicester, over Henry III.

Lewes, resort and deep-sea fishing town (1970 pop. 2,563), Sussex co., SE Del., at the mouth of Delaware Bay NW of Rehoboth Beach; inc. 1857. The first white settlement along the Delaware was made here by the Dutch in 1631 and eventually passed to the English.

Lewes, river, 338 mi (543.8 km) long, rising in N British Columbia, Canada, and flowing NW to the Pelly River at Fort Selkirk. It is the main tributary of the Yukon River. Discovered in 1843, it was an important route during the Klondike gold rush (1897–98).

Lew·is (lo͞o'ĭs). **1.** County (1970 pop. 3,867), 478 sq mi (1,238 sq km), W Idaho, in an agricultural area bounded on the E by the Clearwater River; formed 1911; co. seat Nezperce. Wheat, potatoes, grain, and beans are grown. It also has a lumbering industry. **2.** County (1970 pop. 12,355), 485 sq mi (1,256.2 sq km), NE Ky., bounded on the N by the Ohio River and the Ohio border and drained by the North Fork of the Licking River; formed 1806; co. seat Vanceburg. It is in a rolling agricultural area yielding dairy products, livestock, burley tobacco, grain, and truck crops. Sawmilling and manufacturing of wood and leather products are done. **3.** County (1970 pop. 10,993), 505 sq mi (1,308 sq km), NE Mo., bounded on the E by the Mississippi River and drained by the Wyaconda River and North and Middle Fabius rivers; formed 1833; co. seat Monticello. Its agriculture includes corn, wheat, oats, soybeans, and livestock. It has a lumbering industry and iron foundries. **4.** County (1970 pop. 23,644), 1,293 sq mi (3,348.9 sq km), N central N.Y., rising to the foothills of the Adirondacks in the E and drained by the Black River; formed 1805; co. seat Lowville. It is in a dairying and farming area. Sawmilling, cheese processing, and manufacturing (clothing and wood and paper products) are also important. **5.** County (1970 pop. 6,761), 285 sq mi (738.2 sq km), central Tenn., drained by the Buffalo River; formed 1843; co. seat Hohenwald. It has some farming (corn, hay, and cotton), but depends primarily on lumbering, dairying, and livestock raising. **6.** County (1970 pop. 45,467), 2,447 sq mi (6,337.7 sq km), SW Wash., drained by the Cowlitz and Chehalis rivers; formed 1845; co. seat Chehalis. Dairy products, poultry, grain, and truck crops are produced. It also has lumbering and food-processing industries. Parts of Snoqualmie and Gifford Pinchot national forests are here. **7.** County (1970 pop. 17,847), 392 sq mi (1,015.3 sq km), N central W.Va., on the Allegheny Plateau drained by a headstream of the Monongahela River; formed 1816; co. seat Weston. Its agriculture includes poultry, livestock, fruit, and tobacco. It also has timber and has oil and natural-gas wells.

Lewis and Clark (klärk), county (1970 pop. 33,281), 3,478 sq mi (9,008 sq km), W central Mont., in a mountainous region crossed by the Continental Divide and drained by the Missouri and Blackfoot rivers; formed 1876; co. seat Helena. In a livestock and grain region, it also has gold mines. Sections of Helena National Forest are in the southeast and northwest.

Lew·is·burg (lo͞o'ĭs-bûrg'). **1.** Borough (1970 pop. 5,718), seat of Union co., central Pa., on the West Branch of the Susquehanna River NW of Sunbury; laid out 1785, inc. 1795. Its manufactures include furniture and textiles. **2.** Town (1970 pop. 7,207), seat of Marshall co., S central Tenn., S of Nashville, in a timber and limestone area; inc. 1837. It has poultry, livestock, dairy, and grain farms and makes metal and lumber products. **3.** Trading town (1970 pop. 2,407), seat of Greenbrier co., SE W.Va., ENE of Hinton, in a farm and livestock area; settled 1769, inc. 1782.

Lew·i·sham (lo͞o'ĭ-shəm), borough (1971 pop. 264,800) of Greater London, SE England, on the Thames. The borough was created in 1965 by the merger of the metropolitan boroughs of Lewisham and Deptford. It is mainly residential, but there is some light engineering. Deptford was noted in Elizabethan times for its cattle market and royal dockyard.

Lew·is·ton (lo͞o'ĭ-stən). **1.** City (1978 est. pop. 26,700), seat of Nez Perce co., NW Idaho, at the Wash. line and at the junction of the Snake and Clearwater rivers; founded 1861. It is the commercial and industrial center of a timber, grain, and livestock region that also has lime, clay, and silica deposits. The city has food-processing plants, a large pulp and paper mill, and factories making ammunition primers and concrete products. Lewiston grew as a supply and shipping center after gold was discovered on the Clearwater. It was the first capital (1863–64) of Idaho Territory. **2.** Industrial city (1978 est. pop. 40,300), Androscoggin co., SW Maine, on the Androscoggin River opposite Auburn; inc. 1795. A 50-ft (15.3-m) waterfall has supplied power for textile mills since the early 19th cent. Lewiston also has printing and poultry-hatching industries.

Lew·is·town (lo͞o'ĭs-toun'). **1.** City (1970 pop. 2,706), seat of Fulton co., central Ill., SW of Peoria, in a farm and coal area; settled 1821, inc. 1857. It was the home of Edgar Lee Masters, and the territory, its people, and the local legends are sources of his *Spoon River Anthology*. **2.** City (1970 pop. 6,437), seat of Fergus co., central Mont., SE of Great Falls; laid out 1882, inc. 1899. It is the trading center of the Judith Basin, which had a gold rush after 1880, later became a prosperous cattle-ranching region, and after 1910 was peopled by wheat-farming pioneers. Lewistown also has a petroleum refinery. **3.** Borough (1978 est. pop. 10,000), seat of Mifflin co., central Pa., on the Juniata River, in a beautiful farm and dairy area; inc. 1795. Many Amish live and farm in the surrounding area.

Lew·is·ville (lo͞o'ĭs-vĭl'). **1.** Town (1970 pop. 1,653), seat of Lafayette co., SW Ark., near the Red River NW of El Dorado, in a timber area; settled c.1825, inc. 1889. **2.** City (1970 pop. 9,264), Denton co., N Texas, NW of Dallas; settled 1844. Long a farm center in a cotton and grain area, it is today primarily a residential city for workers in the Dallas–Fort Worth industrial area. It also has industries manufacturing aluminum products and children's clothing.

Lewis with Har·ris (hăr'ĭs), island, 825 sq mi (2,136.8 sq km), largest and northernmost of the Outer Hebrides, Western Isles Authority, NW Scotland, 24 mi (38.6 km) from the mainland across the Minch. Harris has hilly terrain. Central Lewis is a vast, wet moor, uninhabited and unproductive. Crofting, fishing, and stock raising are the main occupations. The thriving Harris tweed industry utilizes home looms throughout the island.

Lex·ing·ton (lĕk'sĭng-tən), county (1970 pop. 89,012), 716 sq mi (1,854.4 sq km), central S.C., bounded on the NE by the Congaree River and on the SW by the North Fork of the Edisto River; formed 1785; co. seat Lexington. In the Sand Hill belt, it has some agriculture (cotton, asparagus, and peaches) and manufacturing.

Lexington. 1. Town (1970 pop. 322), seat of Oglethorpe co., NE Ga., ESE of Athens, in a farm and timber area. **2.** City (1978 est. pop. 190,900), seat of Fayette co., N central Ky., in the heart of a bluegrass region; inc. 1832. The outstanding center in the United States for the raising of thoroughbred horses, it is also an important market for tobacco and bluegrass seed and a railroad shipping point for oil, coal, farm produce, and quarry products. Lexington has railroad shops, meat-packing plants, distilleries, and plants making electronic equipment, electric typewriters, and paper products. The city was named in 1775 by a group of hunters who were encamped on the site when they heard the news of the Battle of Lexington. The city is the seat of the Univ. of Kentucky. Places of interest include Ashland, the home of Henry Clay (designed by Latrobe in 1806 and rebuilt with the original materials in the 1850s); the Thomas Hart house (1794); the home of Mary Todd Lincoln; and the library, which has a file of the *Kentucky Gazette,* founded by John Bradford in 1787. **3.** Town (1978 est. pop. 32,500), Middlesex co., E Mass., a residential suburb of Boston; settled c.1640, inc. 1713. On April 19, 1775, the first battle of the Revolution was fought here. The site is marked by a monument on the triangular green, around which are several 17th cent. buildings, including Buckman Tavern (1710), where the minutemen assembled; and an old burying ground. Other attractions include Monroe Tavern (1695), British headquarters during the battle; and the Hancock-Clarke House (1698), where John Hancock and Samuel Adams were awakened by Paul Revere's alarm. The first state normal school in the country was established here in 1839. **4.** City (1970 pop. 2,756), seat of Holmes co., central Miss., NNE of Jackson; inc. 1836. It has sawmills and sand quarries and ships cotton and dairy products. **5.** City (1970 pop. 5,388), seat of Lafayette co., W central Mo., on the Missouri River E of Kansas City, in a farm and coal area; laid out 1822, inc. 1845. In the Civil War the city, with its Union garrison, was besieged and captured (Sept., 1861) by Mo. militia. The courthouse, dating from 1847, and other ante-bellum buildings are among the points of interest. **6.** City (1970 pop. 5,654), seat of Dawson co., S central Nebr., on the Platte River SE of North Platte, in a farm region. Laid out in 1872 and named Plum Creek to commemorate a camping spot on the Oregon Trail, it was renamed in 1889. **7.** City (1978 est. pop. 16,700), seat of Davidson co., central N.C., in the Yadkin valley; inc. 1827. Major industries are food processing and the manufacture of furniture and textiles. **8.** Town (1970 pop. 969), seat of Lexington co., W central S.C., W of Columbia; settled c.1818, inc. 1861. It makes textiles and lumber products and processes poultry. **9.** Town (1970 pop. 5,024), seat of Henderson co., W Tenn., E of Jackson, in a corn, cotton, and truck-farm area; inc. 1824. Textiles are made. **10.** Town (1970 pop. 7,597), independent city and seat of Rockbridge co., W central Va., in the Shenandoah valley, in a farm area not far from Natural Bridge; laid out 1777, inc. 1841. It is the seat of Virginia Military Institute and Washington and Lee Univ. and the burial place of Robert E. Lee and Stonewall Jackson. The town was bombarded and partially burned in the Civil War.

Lexington Park, uninc. town (1970 pop. 9,136), St. Marys co., SW Md., SSE of Washington, D.C., in a farm region.

Ley·den (līd'n): *see* Leiden, Netherlands.

Ley·te (lā'tē, -tā), island (1970 est. pop. 1,340,000), 2,785 sq mi (7,213.2 sq km), one of the Visayan Islands, the Philippines, between Luzon and Mindanao. A fertile agricultural land, it is the nation's leading producer of sweet potatoes and bananas and a major producer of corn and peanuts. It has commercial coconut plantations and extensive forest reserves; lumbering is an important industry. In World War II Leyte was occupied by the Japanese in early 1942. It was the scene of the first American landing (Oct. 20, 1944) in the campaign to recover the Philippines. The Battle of Leyte Gulf followed, in which American naval forces destroyed the Japanese fleet.

Lha·sa or **La-sa** (lä'sə, -sä, lăs'ə), city (1970 est. pop. 175,000), capital of Tibet Autonomous Region, SW China. It is on a tributary of the Tsangpo (Brahmaputra) River at an altitude of c.11,800 ft (3,600 m). Chemicals are manufactured, and copper and gold, which are mined nearby, are processed. Because of the remoteness of the city and the traditional hostility of the Tibetan clergy toward foreigners, Lhasa has long been called the Forbidden City. Prior to the Chinese occupation (1951) of Tibet, Lhasa was the center of Lamaism. On a nearby hill, backed by lofty mountains in the distance, stands the magnificent Potala, the former palace of the Dalai Lama, a gigantic block of buildings nine stories high, whitewashed save for the central portion, which is red, and surmounted by gilded roofs and towers. A smaller palace of the Dalai Lama is set in the beautifully wooded grounds of Jewel Park. The holiest temple in Lhasa is the Jokang. Several of the religious edifices were damaged during the Tibetan revolt (1959-60) against the Chinese.

Liao (lyou), principal river of NE China, c.900 mi (1,450 km) long, rising in Inner Mongolia and flowing E then S through the fertile Liao alluvial plain to the Gulf of Liaotung. The shallow, silt-laden Liao is navigable for light junks c.400 mi (645 km) upstream.

Liao·ning or **Liao-ning** (both: lyou'nĭng'), province (1968 est. pop. 28,000,000), c.89,000 sq mi (230,510 sq km), NE China, on the Po Hai and West Korea Bay; capital Shen-yang. A part of Manchuria, it encompasses the Liaotung peninsula and the plain of the Liao River. Soybeans are the major crop, and millet, kaoliang, wheat, rice, sweet potatoes, beans, cotton, fruit, and oakleaf silk (pongee silk) are also produced. Liaoning is a major coal-producing area and contains more than half of China's iron ore reserves; there are large deposits of magnesite and smaller ones of copper, lead, and molybdenum. Important manufactures include aircraft, locomotives, tractors, a wide range of heavy equipment, paper, and brick and tile. Along the coast, salt production and fishing are important.

Japan acquired (1895) the Liaotung peninsula after the first Sino-Japanese War, but was forced by Russia, Germany, and France to return it to China that same year. In 1898 Russia received the southern portion of the Liaotung peninsula as a 25-year leasehold. After the Russo-Japanese War (1904-5), Japan took this territory (which it called Kwantung) and held it until the end of World War II, when approximately the same area was made the Port Arthur Naval Base District, under joint Soviet and Chinese operation. The district has been under sole Chinese administration since 1955. The eastern part of Jehol prov. became part of Liaoning in 1956, and in 1970 more than 30,000 sq mi (77,700 sq km) of territory from the Inner Mongolian Autonomous Region was added to Liaoning in the west.

Liao·pei or **Liao-pei** (lyou'bā'), former province, NE China. It was one of nine provinces created in Manchuria in 1945 by the Chinese Nationalist government. However, since the Nationalists never gained effective control of Manchuria after World War II, the province existed only on paper. It was later divided between the Inner Mongolian Autonomous Region and the provinces of Heilungkiang, Kirin, and Liaoning.

Liao-yang or **Liao·yang** (both: lyou'yäng'), city (1970 est. pop. 250,000), E Liaoning prov., China, on a tributary of the Hun River. Iron and coal are mined, and there are textile and other light industries. One of the oldest cities of Manchuria, Liao-yang contains several Buddhist temples built in the 11th cent. In the Russo-Japanese War it was the site of a battle (Aug. 23-Sept. 3, 1904) in which the Russians were forced to retreat.

Liao-yüan or **Liao·yüan** (lyou'ywän'), city, SW Kirin prov., China. It is a coal-mining center with iron and steel works.

Li·ard (lē'ärd, lē-ärd'), river, 755 mi (1,214.8 km) long, rising in the Pelly Mts., SE Yukon Territory, Canada, and flowing SE into N British Columbia, passing through the main range of the Rocky Mts., thence NE through densely wooded country to the Mackenzie River at Fort Simpson, SW Mackenzie dist., Northwest Territories. It is navigable to Fort Liard, an old Hudson's Bay Company post, c.165 mi (265 km) from its mouth.

Lib·by (lĭb'ē), city (1970 pop. 3,286), seat of Lincoln co., extreme NW Mont., on the Kootenai River near the Cabinet Mts. and W of Whitefish; inc. 1909. Founded as a gold-mining town in the 1860s, it is primarily a dairy-farming and lumbering center.

Lib·er·al (lĭb'rəl, lĭb'ər-əl), city (1970 pop. 13,789), seat of Seward co., SW Kansas; founded 1888, inc. 1945. It is the trade center for a grazing and farm area. Beef processing and the cattle and feed-grain industries are important to the economy. Oil and natural gas are extracted, and helium is processed in the city. Aircraft, fabricated metals, and handling equipment are also manufactured in Liberal. The International Pancake Race between the housewives of Liberal and Olney, England, is held annually on Shrove Tuesday.

Li·be·rec (lĭ'bĕ-rĕts), city (1975 est. pop. 76,441), N Czechoslovakia, in Bohemia, on the Lausitzer Neisse River near the East German and Polish borders. The city is a textile center known especially for its woolens; textile machinery, electrical equipment, and automobiles are also produced there. Founded c.1350, Liberec has enjoyed prosperity since the 16th cent., when cloth making was introduced; the first textile factories were built in the 18th cent.

Li·be·ri·a (lī-bîr'ē-ə), republic (1970 est. pop. 1,225,000), 43,000 sq mi (111,370 sq km), W Africa, bordered on the NW by Sierra Leone, on the N by Guinea, and on the E by the Ivory Coast. Monrovia is the capital. Liberia can be divided into three distinct topographical areas: a flat coastal plain of some 10 to 50 mi (16-80 km), with creeks, lagoons, and mangrove swamps; an area of broken, forested hills and mountain ranges, with altitudes from 600 to 1,200 ft (185-365 m), which covers most of the country; and an area of mountains in the northern highlands, with elevations reaching 4,540 ft (1,384.7 m) in the Nimba Mts. and 4,528 ft (1,381 m) in the Wutivi Mts. Vegetation in much of the country is dense forest growth.

Economy. Until the 1950s Liberia's economy was almost totally dependent upon subsistence farming and the production of rubber. The American-owned Firestone plantation was the country's largest employer. With the discovery of high-grade iron ore, the production and export of minerals became the country's major cash-earning economic activity. Other important minerals include gold, diamonds, barite, and kyanite. Some three quarters of the population

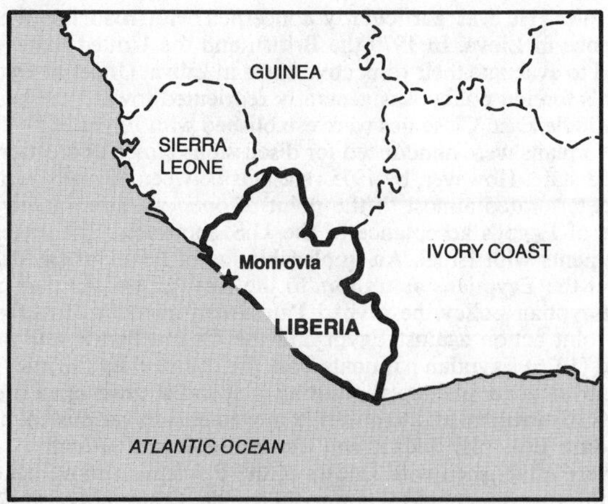

engage in the subsistence farming, producing rice, cassava, yams, and okra. Iron ore, rubber, diamonds, and timber provide the bulk of the export earnings. Much of the country's industry is directed toward the production of consumer goods. The government derives a sizable income from registering ships; low fees and lack of control over shipping operations have made the Liberian merchant marine one of the world's largest. Internal communications are poor, with few paved roads and only a few short, freight-carrying rail lines.

History. Liberia was founded in 1821, when officials of the American Colonization Society were granted possession of Cape Mesurado by local De chiefs. American Negro settlers, most of them freed slaves, were landed in 1822, the first of some 15,000 to settle in Liberia. In 1847, primarily due to British pressures, the colony was declared independent. The immigration of American Negroes virtually came to an end with the American Civil War.

Efforts to modernize the economy led to the raising of a sizable foreign debt, resulting in the overthrow of the government in 1871. Conflicts over territorial claims led to the loss of large areas of claimed, but uncontrolled, lands to Britain and France. By 1909 the Liberian government was bankrupt, and a series of international loans were floated.

In 1930 scandals broke out over the exportation of forced labor from Liberia. The president and his associates were pressured to resign, and international control of the republic was proposed. Under the leadership of presidents Edwin Barclay (1930–44) and William V. S. Tubman (1944–71), however, Liberia avoided such control. Under Tubman, new policies to open the country to international investment and to allow the tribal peoples a greater say in Liberian affairs were undertaken. These new approaches gradually improved Liberia's social, political, and economic affairs. Upon Tubman's death in 1971, Vice President W. R. Tolbert took charge, and in 1972 he was elected to the presidency.

Government. The Liberian constitution is modeled on that of the United States. The president may be elected once to an eight-year term. Legislative authority is divided between a Senate and a House of Representatives.

Lib·er·ty (lĭb'ər-tē). **1.** County (1970 pop. 3,379), 838 sq mi (2,170.4 sq km), NW Fla., bounded in the E by the Ochlockonee River, in the W by the Apalachicola River; formed 1855; co. seat Bristol. Forestry and agriculture (livestock, corn, and peanuts) are the major occupations. Apalachicola National Forest is in the south. **2.** County (1970 pop. 17,569), 510 sq mi (1,320.9 sq km), SE Ga., bounded on the SE by the Atlantic Ocean and St. Catherines Island, on the NE by the Canoochee River; formed 1777; co. seat Hinesville. In a coastal-plain agricultural area yielding corn, sugar cane, rice, truck crops, and livestock, it has fishing and lumbering industries. Fort Stewart is in the northwest. **3.** County (1970 pop. 2,359), 1,439 sq mi (3,727 sq km), N Mont., bordering on Alta., Canada, in the N and drained by the Marias River; formed 1920; co. seat Chester. Its economy is based on grain and livestock. **4.** County (1970 pop. 33,014), 1,180 sq mi (3,056.2 sq km), E Texas, with the coastal plain in the S and rolling, wooded hills in the N, drained by the Trinity River; formed 1836; co. seat Liberty. It has diversified agriculture (rice, wheat, cotton, corn, truck crops, pecans, livestock, and dairy products), and a lumbering industry.

Liberty. 1. Town (1970 pop. 1,831), seat of Union co., SE Ind., E of Connersville, in a farm area; settled 1822. **2.** Town (1970 pop. 1,765), seat of Casey co., central Ky., SW of Lexington, in a farm area; settled 1791. **3.** Town (1970 pop. 612), seat of Amite co., SW Miss., SE of Natchez, in a lumber-milling area. **4.** City (1970 pop. 13,704), seat of Clay co., W central Mo., in a grain, tobacco, and livestock area; laid out 1822. It has railroad yards and grain elevators. **5.** City (1970 pop. 5,591), seat of Liberty co., E Texas, on the Trinity River NE of Houston; founded c.1830. It is the center of an area producing oil, cattle, rice, and timber.

Liberty Island, c.10 acres (4 hectares), in Upper New York Bay, SW of Manhattan island, SE N.Y.; part of Statue of Liberty National Monument. In the mid-1700s, John Bard, a physician, established New York City's first quarantine station here. The Statue of Liberty was placed on the island in 1885, using star-shaped Fort Wood (built in 1841 for harbor defense) as a base. Formerly called Bedloe's Island, Liberty Island was renamed by Congress in 1956.

Lib·er·ty·ville (lĭb'ər-tē-vĭl'), village (1970 pop. 11,684), Lake co., NE Ill., in a lake area; inc. 1882. Earth-moving equipment, outdoor rugs, and pressure hoses are made.

Li·bre·ville (lē'brĕ-vēl'), city (1969 pop. 73,000), capital of Gabon, a port on the Gabon River estuary, near the Gulf of Guinea. It is also a trade center for a lumbering region. The city was founded in 1843 as a French trading station. It was the chief port of French Equatorial Africa before the development (1934–46) of Pointe Noire.

Lib·y·a (lĭb'ē-ə), country (1977 est. pop. 2,580,000), 679,358 sq mi (1,759,537 sq km), N Africa, bordered on the W by Algeria, on the NW by Tunisia, on the N by the Mediterranean Sea, and on the S by Chad and Niger. Tripoli is the capital.

Libya falls into three main geographic regions—Tripolitania in the west, Fazzan in the southwest, and Cyrenaica in the east. Tripolitania in turn can be divided into three zones. In the north is a low-lying coastal plain called the Gefara, which, although mainly arid, has several irrigated areas. South of the Gefara is a mountainous zone known as the Jabal; it is mostly arid and barren, but has scattered areas of cultivation. South of the Jabal is an upland plateau, largely desert, but crossed by a string of oases in the south. South of Tripolitania is the Fazzan region, which is largely made up of sandy desert but has a number of scattered oases. In the north along the Mediterranean is a narrow upland plateau called the Jabal al Akhdar. In the west the Jabal al Akhdar drops abruptly to the shore of the Gulf of Sidra, which deeply indents Libya's Mediterranean coastline, and in the east it falls gradually toward the Egyptian border, where there is another upland region. South of the Jabal al Akhdar is a vast region of sandy desert, which in the east includes part of the Libyan Desert. Cyrenaica is fringed in the southwest by the Tibesti Mts. (located mostly in Chad).

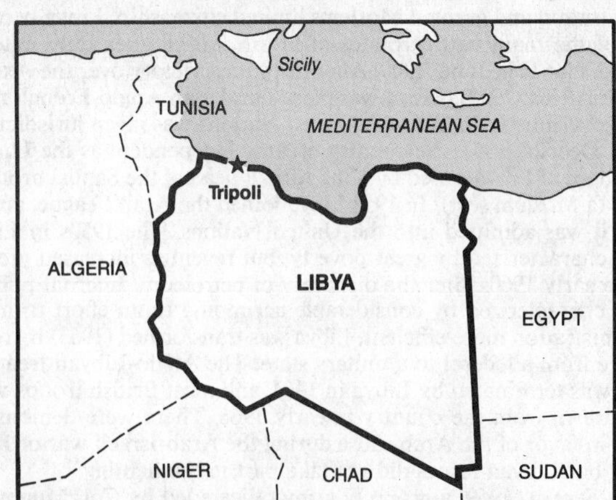

Economy. Until the late 1950s Libya was a poor agricultural country. However, in 1958 petroleum was discovered in the region south and southeast of the Gulf of Sidra, and since 1961 crude petroleum has been exported on an increasingly significant scale. Much of the oil has a low sulfur content, which is desirable because it causes less pollution when burned. The petroleum was located and extracted by a large number of foreign firms. As production increased, so did Libya's receipts of royalties and taxes. The Libyan government nationalized (1972–73) 51% ownership in most of the subsidiaries of foreign petroleum firms operating in the country. The remaining subsidiaries were completely nationalized. All phases of the petroleum industry are regulated by the Libyan National Oil Corporation (LINOCO), founded in 1970. In 1962 Libya joined the Organization of Petroleum Exporting Countries (OPEC). Libya is in addition a major producer of natural gas and has several large gas liquefication plants. Gypsum, salt, and limestone are also produced.

Despite the petroleum boom, the majority of Libya's workers are still employed in agriculture. Farming is severely limited by the small amount of fertile soil and by inadequate rainfall. It is estimated that only 8% of the country's land area is cultivable; most of this area is used for pasturing livestock, leaving very little land for raising crops. Overall, Libya's chief agricultural products are wheat, barley, olives, dates, tobacco, citrus fruit, tomatoes, millet, maize, groundnuts, and almonds. Large numbers of sheep and goats are raised. Libya has little industry. The principal manufactures are refined petroleum, liquefied natural gas, construction materials (especially cement), and basic consumer items such as processed food, beverages, clothing, footwear, soap, and cigarettes. Handicraftsmen produce woven goods and items of metal, leather, and wood.

History. Throughout most of its history the territory that constitutes modern Libya has been held by foreign powers. The Ottomans gained control of most of North Africa in the 16th cent., dividing it into three regencies—Algeria, Tunisia, and Tripoli (which also included Cyrenaica). Ottoman rule in Tripoli was limited largely to the coastal region, where taxes were regularly collected. The Janissaries, professional soldiers of slave origins, became a military caste, wielding considerable influence over the Ottoman governor. From the early 1600s the Janissaries chose a leader, called the dey. Numerous pirates who preyed on the shipping of Christian nations in the Mediterranean were based at Tripoli's ports. In 1711 Ahmad Karamanli, a Janissary, became dey, killed the Ottoman governor, and was made governor. The post of governor remained hereditary in the Karamanli family until 1835. In the 18th cent. and during the Napoleonic Wars, the dey took in great revenues from the pirates and also extended the central government's control to much of the interior. After 1815 England, France, and the kingdom of the Two Sicilies undertook a successful campaign against the pirates, which undermined the finances of the dey and thus facilitated the re-establishment of direct Ottoman rule in Tripoli in 1835.

During the Turko-Italian War of 1911–12 Italy conquered northern Tripoli, but Turkey granted Tripoli autonomy. The Libyans continued to fight the Italians, and Italy was forced to undertake a long series of wars of pacification. Until the 1930s little was done to develop the country. In 1934 Tripolitania and Cyrenaica were formally united to form the colony of Libya; Fazzan was administered as part of Tripolitania. In 1939 Libya was made an integral part of Italy, and governor-general Italo Balbo reorganized the colonial administration and granted Moslems limited citizenship. Libya became one of the main battlegrounds of North Africa after Italy entered World War II in June, 1940. After the Allied victory over the Axis in North Africa (1943), Libya was placed under an Anglo-French military government. In 1949 the United Nations was given jurisdiction.

On Dec. 24, 1951, the country became independent as the United Kingdom of Libya, ruled by King Idris I, head of the Sanusi brotherhood (a Moslem sect). In 1953 Libya joined the Arab League, and in 1955 it was admitted into the United Nations. The 1950s in Libya were characterized by great poverty, but revenues increased greatly in the early 1960s after the discovery of petroleum. Internal politics were characterized by considerable acrimony. In an effort to make administration more efficient, Libya was transformed (1963) by royal decree from a federal to a unitary state. The Anglo-Libyan treaty of 1953 was terminated by Libya in 1964, and most British troops were withdrawn from the country in early 1966. There were demonstrations in favor of the Arab cause during the Arab-Israeli war of June, 1967, but Libyan forces did not take part in the fighting.

On Sept. 1, 1969, a group of army officers led by Col. Muammar al-Qaddafi staged a successful coup d'état, ousting Idris. Government was placed in the hands of a 12-member Revolutionary Command Council (RCC) headed by al-Qaddafi. The RCC appointed a cabinet headed by a prime minister, and al-Qaddafi served in that office from early 1970 to mid-1972, when he turned the post over to Abdul Salam Jallud. Al-Qaddafi retained the office of president of the RCC, the most important political and military office in the country. (In 1977 the RCC was renamed the General Secretariat of the People's Congress.) Al-Qaddafi pursued a policy of Arab nationalism and strict adherence to Islamic law; he also espoused socialist principles. He was particularly concerned with reducing Western influence in Libya. In 1970 the British and the United States were forced to evacuate their respective bases in Libya. Under al-Qaddafi, Libya's foreign policy was generally reoriented toward the heart of the Middle East. Close ties were established with Egypt in 1971, and in 1973 plans were announced for discussions aimed at creation of a unified state. However, by 1975 relations between the two countries had deteriorated almost to the point of open warfare, largely as a result of Egypt's acceptance of the U.S.-sponsored disengagement agreements with Israel. An implacable foe of Israel, al-Qaddafi regarded the Egyptians as traitors to the Arab cause. As part of his anti-Egyptian policy, he invited Palestinian guerrillas to Libya to plan joint action against Egypt and the United States, and he expelled (1976) Egyptian nationals from the country. By the mid-1970s the Libyan government was granting tacit and at times open support to acts of terrorism and to guerrilla movements in various countries, including Ethiopia, Sudan, and the Philippines. Particularly close ties were established with leaders of the Palestinian movement and with the government of Idi Amin of Uganda. Coups aimed at deposing al-Qaddafi were thwarted in 1969, 1970, and Aug., 1975. The official name of the country became the People's Socialist Libyan Arab Republic in 1977.

Lib·y·an Desert (lĭb'ē-ən), NE part of the Sahara Desert, NE Africa, in SW Egypt, E Libya, and NW Sudan. It is a region of sand dunes, stony plains, and rocky plateaus.

Li·can·cá·bur (lē'käng-kä'bər), volcano, 19,455 ft (5,933.8 m) high, N Chile, near the Bolivian border NE of Antofagasta.

Li·ca·ta (lē-kä'tä), city (1976 est. pop. 42,292), S Sicily, Italy, on the Mediterranean Sea at the mouth of the Salso River. Licata is a seaport, seaside resort, and commercial and industrial center. It was founded in the early 3rd cent. B.C.

Lich·field (lĭch'fēld'), municipal borough (1973 est. pop. 23,690), Staffordshire, W central England. It is a market town with light industries, famous for its three-spired cathedral and its close associations with Dr. Samuel Johnson. The cathedral, dating from the 13th and 14th cent., replaced the original church built by St. Chad, who first founded the see in the 7th cent. It suffered considerable damage at the hands of the parliamentary forces during the English Civil War and was not completely restored until the 19th cent. The house where Dr. Johnson was born and lived is now a museum containing many relics of his life and works.

Lick·ing (lĭk'ĭng), county (1970 pop. 107,799), 686 sq mi (1,776.7 sq km), central Ohio, drained by the Licking River and Raccoon Creek; formed 1808; co. seat Newark. It is in an agricultural area yielding grain, livestock, and dairy products and has sand and gravel pits and some manufacturing.

Licking, river, c.320 mi (515 km) long, rising in E Ky. and flowing NW to the Ohio River opposite Cincinnati. It was an important means of travel for Indians and pioneers and a busy trade route.

Li·di·ce (lē'də-chā', lĭ'dyĭ-tsĕ), village, NW Czechoslovakia, in Bohemia. In reprisal for the assassination of Reinhard Heydrich, the Germans "liquidated" (1942) Lidice by killing all the men, deporting all women and children, and razing the village to the ground. After World War II a new village was built near the site of old Lidice, which is now a national park and memorial.

Li·ding·ö (lē'dĭng-œ'), city (1975 est. pop. 36,727), Stockholm co., SE Sweden, on Lidingö Island in the Baltic Sea; chartered 1926. It is a residential suburb of Stockholm and a resort.

Lid·kö·ping (lēd'chœ'pĭng), city (1975 est. pop. 26,400), Skaraborg co., S Sweden, a port on Lake Vänern; chartered 1446. It has machine shops, match factories, and porcelain works.

Li·do (lē'dō), long, narrow, sandy island in Venetia, Italy, separating the lagoon of Venice from the Adriatic. It is a fashionable resort.

Liech·ten·stein (lĭkн'tən-shtīn'), principality (1970 pop. 21,350), 61 sq mi (157 sq km), W central Europe. It is situated in the Alps between Austria and Switzerland and is bounded in the west by the Rhine River. Vaduz is the capital.

Economy. Traditionally agricultural, Liechtenstein has been increasingly industrialized in recent years. Only a fraction of the population still engages in agriculture (dairying, wine production, and the raising of livestock and cereals). The leading manufactured products are machinery and other metal goods, ceramics, textiles, and foodstuffs. Tourism is an increasingly important industry. Much revenue

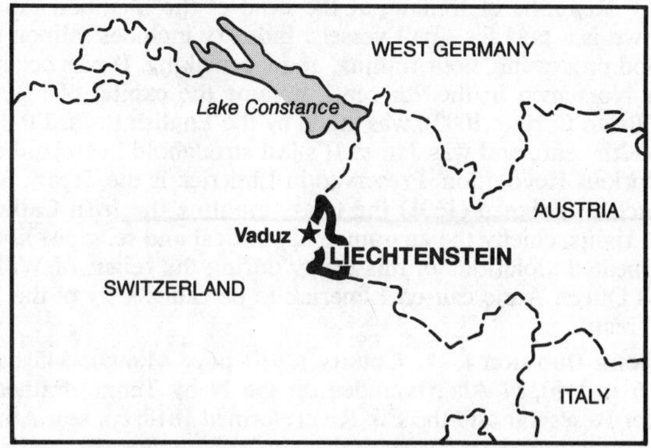

is derived from the sale of postage stamps and from the minimal taxes imposed on international corporations.

History. The principality was created in 1719 by uniting the county of Vaduz with the barony of Schellenburg. The princes, vassals of the Holy Roman emperors, also owned huge estates (many times larger than their principality) in Austria and adjacent territories; they rarely visited their country but were active in the service of the Hapsburg monarchy. Liechtenstein became independent in 1866, after having been a member of the German Confederation from 1815 to 1866. The principality escaped the major upheavals of the 19th and 20th cent. A parliament-approved proposal granting women the right to vote was decisively defeated in two referendums (1971, 1973), but since then, communes have been given the authority to allow women to vote in local elections. Liechtenstein remains the only Western European country to deny women suffrage.

Government. The 1921 constitution, amended in 1972, establishes a parliament of 21 members, elected by male suffrage. Since 1919, Liechtenstein has been represented abroad through Switzerland.

Li·ège (lē-āzh'), province (1976 est. pop. 1,010,237), 1,526 sq mi (3,952.3 sq km), E Belgium, bordering on West Germany in the E; capital Liège.

Liège, city (1977 pop. 221,404), capital of Liège prov., E Belgium, at the confluence of the Meuse and Ourthe rivers, near the Dutch and West German borders. The commercial center of the industrial Meuse valley and itself a major industrial center, Liège is also an important transportation hub. Manufactures include metal goods, armaments, motor vehicles, electrical and electronics equipment, chemicals, textiles, clothing, and furniture.

A growing trade center by the 10th cent., Liège became the capital of the prince-bishopric of Liège, which was part of the Holy Roman Empire and lasted until 1792. Liège city, the strongly fortified key to the Meuse valley, suffered numerous sieges in its history. In the late Middle Ages Liège was torn by bitter social strife. In the 14th cent. the workers (organized in guilds) won far-reaching concessions from the nobles and the wealthy bourgeoisie and began to take part in the government of the city. The episcopal functionaries were placed (1373) under the supervision of a tribunal of 22 persons, 14 of whom were burgesses. This Peace of the Twenty-Two remained the basic guarantee of the constitutional liberty of the inhabitants of Liège until 1792. In 1465 the city became a protectorate of Burgundy, and in 1467 Charles the Bold, duke of Burgundy, abolished the citizens' communal liberties. The citizens of Liège, encouraged by Louis XI of France, rose in rebellion, but Charles forced Louis to assist him in suppressing the revolt and then sacked the city (1468).

Liège remained technically a sovereign member of the Holy Roman Empire after the Netherlands passed (1477) under Hapsburg rule; however, the prince-bishops were dependent on the Spanish kings and, after 1714, the emperors. In 1792 the French entered Liège. In the 19th cent. the city was a center of Walloon particularism, of rapid industrial growth, and of social unrest. In World War I its fortifications, reputed to be among the strongest in Europe, fell (1914) to the Germans after a 12-day siege. In World War II Liège was again taken (May, 1940) by the Germans. It was liberated (May, 1944) by U.S. forces, but during the Battle of the Bulge (Dec., 1944–Jan., 1945) it suffered considerable destruction from German rockets. Liège retains some historic buildings, including a cathedral (founded 971), the Church of the Holy Cross (10th cent.), and the Church of St. Denis (10th–11th cent.).

Lie·pa·ya (lyĕ'pä-yä), city (1976 est. pop. 103,000), W European USSR, in Latvia. An ice-free port on the Baltic Sea, it is located at the end of an isthmus separating the Baltic from Lake Liepaya. Metallurgy is the leading industry; others include shipbuilding, food and fish processing, and sugar refining. Founded by the Teutonic Knights in 1263, the city passed to Russia in 1795. In the late 19th and early 20th cent. Liepaya acquired great commercial importance and became one of the main Russian emigration ports with a direct shipping line to the United States. The city was under German occupation during most of World War I. It was briefly the site of the provisional Latvian government when Bolshevik forces attacked Riga in 1918. Held by the Germans from 1941–45, Liepaya suffered heavy damage. After World War II it was annexed by the USSR along with the rest of Latvia.

Liè·vre (lyä'vrə), river, c.200 mi (320 km) long, rising in Kempt Lake, S Que., Canada, and flowing generally SW into the Ottawa River near Buckingham. Parts of it are navigable.

Lif·fey (lĭf'ē), river, c.50 mi (80 km) long, rising in the Wicklow Mts., E Republic of Ireland, and flowing W, NE, and then E through Dublin to Dublin Bay.

Light·house Point (līt'hous'), city (1975 est. pop. 11,900), Broward co., SE Fla., just NE of Pompano Beach.

Li·gny (lē-nyē'), village (1969 est. pop. 2,000), Namur prov., central Belgium, near Namur. At Ligny, on June 16, 1815, Napoleon defeated the Prussians early in the Waterloo campaign.

Li·gu·ri·a (lĭ-gyŏŏr'ē-ə, lē-gōō'ryä), region (1975 est. pop. 1,867,383), 2,098 sq mi (5,433.8 sq km), NW Italy, extending along the Ligurian Sea and bordering France on the W; capital Genoa. The generally mountainous region has a steep, narrow coastal strip that includes the Italian Riviera. In the interior, the Ligurian Alps rise in the west and the Ligurian Apennines in the east. Flowers (mostly for use in making perfume), olives, wine grapes, citrus fruit, mushrooms, and cereals are grown. Chestnuts are gathered in the mountains, where there are extensive pastures, timberland, and marble, slate, quartz, and limestone quarries. Fishing is pursued along the coast. Manufactures of the region include iron and steel, ships, machinery, textiles, chemicals, processed food, and forest products.

Liguria derives its name from the ancient Ligurii, who occupied the Mediterranean coast from the Rhône River to the Arno River. In the 4th cent. the Ligurii were driven from the Alpine regions by Celtic immigrants, while Phoenicians, Greeks, and Carthaginians colonized the coast. In the 2nd cent. B.C. the entire region was subdued by the Romans. Throughout the Middle Ages, Genoa struggled with local feudal lords (and at times with Venice) for control of the area. By the 16th cent. it controlled virtually all of present-day Liguria, and from that time until its annexation (1815) by the kingdom of Sardinia, Liguria shared the history of Genoa.

Li·gu·ri·an Sea (lĭ-gyŏŏr'ē-ən, lē-gōō'ryän), arm of the Mediterranean Sea, between the Ligurian coast and the islands of Corsica and Elba. The Gulf of Genoa is its northernmost part. The sea's northwest coast is noted for its favorable climate and scenic beauty.

Li·hu·e (lē-hōō'ā), city (1970 pop. 3,124), seat of Kauai co., Hawaii, on the SE coast of Kauai, SSW of Kapaa, in a sugar-cane region.

Li·ka·si (lĭ-kä'sē), formerly **Ja·dot·ville** (zhä-dō-vēl'), city (1970 pop. 146,394), Shaba region, SE Zaire. It is a major industrial, mining, and transportation center. Copper and cobalt are mined and refined, and cement, chemicals, and beverages are manufactured.

Lille (lēl), city (1975 pop. 172,280), capital of Nord dept., N France, near the Belgian border. It is a commercial, cultural, and manufacturing center, long known for its textile products. Lille was the chief city of the county of Flanders, a brilliant residence of the 16th cent. dukes of Burgundy and (after 1668) the capital of French Flanders. Taken (1708) after a costly siege by Eugene of Savoy and the duke of Marlborough, it was restored to France in the Peace of Utrecht (1713). Among its principal buildings are the citadel, the old stock exchange (17th cent.), several fine churches, and the unfinished cathedral (begun 1854). Lille has one of the most important art museums in Europe; its paintings include many of the best works of the Flemish, Dutch, French, and Spanish masters.

Lil·le·ham·mer (lĭl'ə-hä'mər), town (1977 est. pop. 21,497), capital of Oppland co., S Norway, at the N end of Lake Mjøsa. It is a commercial center and a popular summer and winter resort. Its open-air museum (founded 1887) features complete farms, peasant cottages, workshops, and handicrafts of the region.

Lil·ling·ton (lĭl'ĭng-tən), town (1970 pop. 1,155), seat of Harnett co., central N.C., SSW of Raleigh, on the Cape Fear River.

Li·long·we (lē-lông′gwā), city (1975 est. pop. 102,000), S central Malawi, in a fertile agricultural area. The capital of Malawi since 1966, it was founded in 1947 as an agricultural marketing center.

Li·ly·bae·um (lĭ-lĭ-bē′əm), ancient city of Sicily, on the extreme W coast. It is the modern Marsala. It was founded (396 B.C.) by Carthage and became a stronghold. In the First Punic War it resisted a long Roman siege (250–242). Later Rome acquired Sicily by treaty, and Lilybaeum became a subject of the empire.

Li·ma (lē′mə, -mä), city (1972 metropolitan area pop. 3,350,000), W Peru, capital and largest city of Peru. Its port is Callao. The Lima urban area is Peru's economic center and the site of oil-refining and diversified manufacturing industries.

The city was founded on Jan. 18, 1535, by Francisco Pizarro and was the capital of Spain's New World empire until the 19th cent. Rebuilt several times, Lima reflects the architectural styles prevalent in various periods; much of the city is characterized by modern steel and concrete buildings. Although many streets are narrow and preserve a colonial atmosphere, spacious boulevards traverse the entire metropolitan area. Small squares, statues of national heroes, parks, and gardens are common. The focal point of the city's life is the central square, the Plaza de las Armas. It is dominated by the huge national palace and cathedral. The cathedral, begun by Pizarro and containing what are claimed to be his remains, was almost totally destroyed by earthquakes in 1687 and 1746, along with much of the city. Notable public buildings include the National Library, founded in 1821 by José de San Martín, and the Univ. of San Marcos, founded in 1551. The library, which contained priceless documents of the Spanish Conquest and rare European books, was looted by Chilean soldiers during Chile's occupation of Lima (1881–83) in the War of the Pacific.

Li·ma (lī′mə), city (1978 est. pop. 50,000), seat of Allen co., NW Ohio; settled 1831, inc. 1842. Located in a rich farm area, it is a processing and marketing center for grain, dairy, and meat products. Auto engines, school buses, electric signs and motors, cranes and power shovels, petroleum products, cigars, steel castings, machine tools, plastics, chemicals, and fertilizers are produced in the city.

Li·ma·sa·wa (lē′mə-sä′wə), island, 3 sq mi (7.8 sq km), Philippines, off the S coast of Leyte Island. Limasawa was Magellan's second landing point in the Philippines (Mar. 25, 1521).

Li·mas·sol (lĭm′ə-sôl′), city (1974 est. pop. 55,000), S Cyprus, on Akrotiri Bay. It is a port and a resort. Wine and agricultural goods are exported. Umber is mined in the district.

Lim·burg (lĭm′bûrg′, -bûrkh′), province (1976 est. pop. 692,127), 930 sq mi (2,408.7 sq km), NE Belgium, bordering on the Netherlands in the N; capital Hasselt. It is largely agricultural. Most of Limburg was included in the prince-bishopric of Liège until 1792. It became (1815) part of the Dutch province of Limburg, which was divided between Belgium and the Netherlands in 1839.

Limburg, province (1977 est. pop. 1,055,619), c.850 sq mi (2,200 sq km), SE Netherlands, bordering on Belgium in the W and S and West Germany in the E; capital Maastricht. Rich in historic antiquities, the province takes its name from the former duchy of Limburg, which comprised the southern part of the modern province and an eastern portion of modern Liège prov. in Belgium. Founded in the 11th cent., the duchy was divided in the Peace of Westphalia (1648) between the United Netherlands and the Spanish Netherlands. The duchy was united (1815) under the kingdom of the Netherlands. Limburg prov., established in 1815, was contested after the establishment (1831) of an independent Belgium. The Dutch-Belgian treaty of 1839 divided the territory, which was incorporated, respectively, with the Dutch and Belgian provinces of Limburg. There was some Belgian separatist feeling in the Netherlands' Limburg prov. in the 19th cent., and the province was not fully integrated into the Dutch national structure until the early 20th cent.

Lim·er·ick (lĭm′ər-ĭk), county (1971 pop. 140,370), 1,037 sq mi (2,685.8 sq km), SW Republic of Ireland; co. town Limerick. The region is an agricultural plain lying south of the Shannon estuary. Dairy farming and salmon fishing are the chief occupations. There are food-processing, wool, and paper industries. After the Anglo-Norman invasion and the organization of Limerick as a shire (c.1200), it was controlled by the earls of Desmond.

Limerick, county borough (1971 pop. 57,161), co. town of Co. Limerick, SW Republic of Ireland, at the head of the Shannon estuary. The town is a port for small vessels. Industry includes salmon fishing, food processing, flour milling, and lacemaking. It was occupied by the Norsemen in the 9th cent., became the capital of Munster under Brian Boru (c.1000), was taken by the English toward the end of the 12th cent., and was James II's last stronghold in Ireland after the Glorious Revolution. Preserved in Limerick is the Treaty Stone on which was signed (1691) the treaty granting the Irish Catholics certain rights, chiefly the guarantee of political and religious liberty. The repeated violations of this treaty during the reigns of William III and Queen Anne caused Limerick to be called City of the Violated Treaty.

Lime·stone (līm′stōn′). **1.** County (1970 pop. 41,699), 545 sq mi (1,411.6 sq km), N Ala., bounded on the N by Tenn., drained by Wheeler Reservoir and the Elk River; formed 1818; co. seat Athens. It has cotton and corn farms, timber, phosphate deposits, and some industry (food processing and clothing manufacturing). Electrical and transportation equipment are made. **2.** County (1970 pop. 18,100), 932 sq mi (2,413.9 sq km), E central Texas, drained by the Navasota River; formed 1846; co. seat Groesbeck. It has agriculture (cotton, corn, grain sorghums, legumes, hay, fruit, truck crops, pecans, dairy products, and livestock). There are oil and natural-gas deposits and clay pits.

Limestone, town (1970 pop. 10,360), Aroostook co., NE Maine, ENE of Caribou near the N.B., Canada, boundary, in a potato-growing region; settled 1849, inc. 1869. It is a port of entry.

Lim·fjord (lĭm′fyœrd′), waterway, c.110 mi (180 km) long, cutting across N Jutland, Denmark, and connecting the North Sea with the Kattegat. It is irregular in shape, forming Løgstør, a lagoon 15 mi (24.1 km) wide in its middle section. Before 1825, when the fjord cut through to the North Sea, its western part consisted of several freshwater lakes that drained eastward into the Kattegat.

Lím·nos (lēm′nôs) or **Lem·nos** (lĕm′nŏs), island (1971 pop. 17,367), 186 sq mi (481.7 sq km), NE Greece, in the Aegean Sea near Turkey. It is largely mountainous, with areas of fertile lava soil. Fruits, wine, silk, and wheat are produced, sheep and goats are raised, and fish are caught. A medicinal earth, used in treating open wounds and snake bites, has been produced here since ancient times. The island became a colony of Athens c.500 B.C. After the fall (1204) of the Byzantine Empire, Límnos was captured by the Genoese, who held the island until 1464, when it passed to Venice. It was seized by the Ottoman Turks in 1479 and became part of Greece in 1913.

Li·moges (lĭ-môzh′), city (1975 pop. 143,689), capital of Haute-Vienne dept., W central France, on the Vienne River. It is famous for its ceramics industry, which uses the abundant kaolin in the area. The shoe industry is also large. An ancient town, Limoges became (12th cent.) the seat of the viscounty of Limoges and (1589) the capital of Limousin prov. Richard I of England was killed in battle near Limoges (1199). In 1370 Edward the Black Prince burned the city and massacred its inhabitants. The famous Limoges enamel industry was fully developed by the 13th cent., but it declined when Limoges was once more devastated in the Wars of Religion. Prosperity returned with the introduction (1771) of china manufactures. Limoges has an art gallery with many works by Renoir, who was born here.

Li·món (lē-môn′), city (1976 est. pop. 31,900), capital of Limón prov., Costa Rica, on the Caribbean Sea. The leading port of Costa Rica, it is also a tourist resort. Limón was founded (1874) during the construction of the railroad to San José.

Li·mou·sin (lē-mōō-zăN′), region and former province, S central France, in the arid, hilly country W of the Auvergne Mts. In 918 Limousin was enfeoffed to the duchy of Aquitaine. Ravaged by Edward the Black Prince in the Hundred Years' War, Limousin was reconquered for France (1370–74) by Bertrand du Guesclin.

Lim·po·po (lĭm-pō′pō), river, c.1,100 mi (1,770 km) long, rising in Transvaal prov., Republic of South Africa. It flows in a great arc, first north (forming part of the South Africa–Botswana border), then east (forming the South Africa–Rhodesia border), and finally southeast through Mozambique to the Indian Ocean. The lower Limpopo waters a fertile and heavily populated region.

Li·na·res (lē-nä′räs), city (1974 est. pop. 46,330), Jaén prov., S Spain, in Andalusia. The rich silver and lead mines nearby have brought prosperity to the city, which now has many metallurgical industries. Powder and dynamite are the chief products.

Lin·coln (lĭng′kən), county borough (1976 est. pop. 73,700), administrative center of Lincolnshire, E England, on the Witham River. Located at the junction of the Roman Fosse Way and Ermine Street, it is a center of road and rail transportation. Manufactures include heavy machinery, light-metal products, automobile parts, radios, and food products. Lincoln was an ancient British settlement and one of the Five Boroughs of the Danes. Parliamentarians captured it in 1644. Lincoln Cathedral, first built from 1075 to 1501, has a central tower containing the famous bell Great Tom of Lincoln. One of the few extant copies of the Magna Carta is in the cathedral.

Lincoln. **1.** County (1970 pop. 12,913), 565 sq mi (1,463.4 sq km), SE Ark., bounded on the NE by the Arkansas River and drained by Bayou Bartholomew; formed 1871; co. seat Star City. Its agriculture includes cotton, truck crops, corn, and fruit. Lumbering and cotton ginning are done, and textiles and metal products manufactured. **2.** County (1970 pop. 4,836), 2,593 sq mi (6,715.9 sq km), E Colo., drained by Big Sandy Creek; formed 1889; co. seat Hugo. It is in a wheat and livestock area and produces glass. **3.** County (1970 pop. 5,895), 192 sq mi (497.3 sq km), NE Ga., bounded in the E by the S.C. line, formed here by the Savannah River, and in the S by the Little River; formed 1796; co. seat Lincolnton. In a piedmont agricultural area yielding cotton, corn, hay, truck crops, and fruit, it also has sawmills. **4.** County (1970 pop. 3,057), 1,203 sq mi (3,115.8 sq km), S Idaho, in a livestock-raising and farming area in the Snake River Plain, watered by the Big Wood and Little Wood rivers; formed 1895; co. seat Shoshone. An irrigated region in the southwest produces sugar beets, potatoes, and dry beans. **5.** County (1970 pop. 4,582), 726 sq mi (1,880.3 sq km), central Kansas, in a rolling prairie region drained by the Saline River; formed 1867; co. seat Lincoln. It is in a wheat and cattle region and has quartz quarries, lumber, and some manufacturing. **6.** County (1970 pop. 16,663), 340 sq mi (880.6 sq km), E central Ky., in a rolling upland area of the outer Bluegrass drained by the Dix and Green rivers and small Fishing Creek; formed 1780 from Kentucky co., Va.; co. seat Stanford. In an agricultural (dairy products, livestock, burley tobacco, corn, wheat, and fruit) and timber region, it has plants manufacturing clothing and metal products. **7.** Parish (1970 pop. 33,800), 469 sq mi (1,214.7 sq km), N La., drained by the Middle Fork of Bayou D'Arbonne; formed 1873; parish seat Ruston. Agricultural products include cotton, corn, hay, fruit, peanuts, and sweet potatoes. Dairying is done, and there is some manufacturing. Natural-gas wells are found here. **8.** County (1970 pop. 20,537), 457 sq mi (1,183.6 sq km), S Maine, on the Atlantic coast, watered by the Sheepscot and Eastern rivers; formed 1760; co. seat Wicasset. It is in a fishing and resort area, with some agriculture in the north, and has lumbering, food-processing, and boatbuilding industries. Wood products and electrical equipment are made. Resorts dot its rugged coastline and offshore islands. **9.** County (1970 pop. 8,143), 531 sq mi (1,375.3 sq km), SW Minn., bordering in the W on S.Dak., and drained by tributaries of the Yellow Medicine River; formed 1866; co. seat Ivanhoe. In an agricultural area yielding corn, oats, barley, and livestock, it has some manufacturing. **10.** County (1970 pop. 26,198), 586 sq mi (1,517.7 sq km), SW Miss., watered by Bogue Chitto; formed 1870; co. seat Brookhaven. Cotton and corn farming, dairying, and cattle raising are major occupations. There are also oil and natural-gas wells, stands of timber, and factories producing brick and clay tile, wood products, and machinery. **11.** County (1970 pop. 18,041), 629 sq mi (1,629.1 sq km), E Mo., bounded on the E by the Mississippi River and drained by the Cuivre River; formed 1818; co. seat Troy. It is in an agricultural area that produces wheat, corn, apples, and livestock. It has coal mines and limestone deposits, and manufactures clothing. **12.** County (1970 pop. 18,063), 3,715 sq mi (9,621.9 sq km), extreme NW Mont., in a mountainous region bordering in the N on British Columbia, in the W by Idaho, and watered by the Kootenai River and Lake Koocanusa; formed 1909; co. seat Libby. Livestock, dairy products, and lumber are important to its economy. It also has lead, silver, and gold mines and a tourism industry. Most of the county is within Kootenai National Forest. **13.** County (1970 pop. 29,538), 2,525 sq mi (6,539.8 sq km), SW central Nebr., where the South Platte and North Platte rivers join to form the Platte River; formed 1860; co. seat North Platte. Livestock and grain are its major products. **14.** County (1970 pop. 2,557), 10,649 sq mi (27,580.9 sq km), E Nev., in a mountainous region bordering on Ariz. and Utah, watered by Meadow Valley Wash, with part of Egan Range in the N; formed 1866; co. seat Pioche. It is in a mining (lead and zinc) and ranching area. **15.** County (1970 pop. 7,560), 4,859 sq mi (12,584.8 sq km), S central N.Mex., drained by Rio Hondo and Salt Creek; formed 1869; co. seat Carrizozo. Livestock grazing, farming, and coal mining are done. The county is a center for quarter-horse racing. It includes parts of Lincoln and Cibola national forests and White Sands Missile Range. **16.** County (1970 pop. 32,682), 297 sq mi (769.2 sq km), W central N.C., in a piedmont area bounded on the E by the Catawba River; formed 1779; co. seat Lincolnton. Its agriculture includes cotton, corn, wheat, hay, soybeans, oats, barley, vegetables, apples, livestock, poultry, and dairy products. Among its manufactures are textiles, furniture, electronic components, and wood and metal products. **17.** County (1970 pop. 19,482), 973 sq mi (2,520.1 sq km), central Okla., in a diversified agricultural area intersected by the Deep Fork River; formed 1890; co. seat Chandler. It also has oil and natural-gas wells, oil refineries, and a growing amount of light industry. **18.** County (1970 pop. 25,755), 986 sq mi (2,553.7 sq km), W Oregon, bounded on the W by the Pacific Ocean and drained by the Alsea and Yaquina rivers; formed 1893; co. seat Newport. Fishing, lumbering, and tourism are important to its economy. **19.** County (1970 pop. 11,761), 576 sq mi (1,491.8 sq km), SE S.Dak., in a rolling prairie area bounded on the W by the Idaho border and on the E by the Big Sioux River; formed 1862; co. seat Canton. It is mainly agricultural, yielding corn, oats, soybeans, alfalfa, cattle, hogs, and sheep. **20.** County (1970 pop. 24,318), 581 sq mi (1,504.8 sq km), S Tenn., bounded on the S by the Ala. border and crossed by the Elk River; formed 1809; co. seat Fayetteville. In a lumbering area, it has agriculture (livestock, corn, grain, tobacco, and dairy products) and some industry. **21.** County (1970 pop. 9,572), 2,306 sq mi (5,972.5 sq km), E Wash., bounded on the N by the Spokane and Columbia rivers; formed 1883; co. seat Davenport. Wheat, livestock, and fruit are important to its economy. Some lumber milling is done. Grand Coulee Dam is at its northwest corner. **22.** County (1970 pop. 18,912), 438 sq mi (1,134.4 sq km), W W.Va., on the Allegheny Plateau drained by the Guyandot, Mud, and Coal rivers and their tributaries; formed 1867; co. seat Hamlin. Tobacco is the main farm product, though the county also raises livestock and produces fruit. Its major resources are bituminous coal, oil, and natural gas. **23.** County (1970 pop. 23,499), 892 sq mi (2,310.3 sq km), N central Wis., drained by the Wisconsin River; formed 1874; co. seat Merrill. The wooded lake region in the north is a resort area; dairying and farming are done in the south. It also has paper mills and some manufacturing plants. **24.** County (1970 pop. 8,640), 4,101 sq mi (10,621.6 sq km), W Wyo., in a grain and livestock area bordering on Utah and Idaho and watered by the Salt, Greys, and Green rivers; formed 1911; co. seat Kemmerer. It also has deposits of coal and oil.

Lincoln. **1.** City (1978 est. pop. 15,200), seat of Logan co., central Ill., in a farm area; inc. 1865. It is a shipping and industrial center. The city was platted and promoted (1853) with the aid of Abraham Lincoln. Lincoln practiced law here from 1847 to 1859, and buildings and places associated with him have been preserved or reconstructed. **2.** City (1970 pop. 1,582), seat of Lincoln co., N central Kansas, on the Saline River NW of Salina; laid out 1871, inc. 1879. It is a shipping point for wheat and livestock. **3.** Residential town (1970 pop. 7,567), Middlesex co., E Mass., SE of Concord; settled c.1650, set off from Concord, Lexington, and Weston 1754. **4.** City (1978 est. pop. 165,100), state capital, and seat of Lancaster co., SE Nebr.; inc. 1869. It is the railroad, trade, and industrial center for a large grain and livestock area. Rubber products, candy, sports and industrial vehicles, and circuit breakers are among its manufactures. Founded in 1864 as Lancaster, the city was chosen as the site of the capital in 1867 and renamed. It is the seat of the Univ. of Nebraska, Union College, and Nebraska Wesleyan Univ. It has a planetarium, an art gallery and sculpture garden, and several parks. William Jennings Bryan lived in Lincoln from 1887 to 1916, and his home is preserved. **5.** Town (1970 pop. 16,182), Providence co., NE R.I.; set off from Smithfield and inc. 1871. Once a textile town, its manufactures now include wire, tubing, metal parts, and thread. Limestone has been quarried here since colonial times. Many pre-Revolutionary houses are in the town.

Lincoln, Mount, peak, 14,284 ft (4,356.6 m) high, central Colo., highest elevation of the Park Range of the Rocky Mts.

Lincoln Boyhood National Memorial: *see* National Parks and Monuments Table.

Lincoln Heights (hīts), city (1970 pop. 6,099), Hamilton co., SW Ohio, a suburb N of Cincinnati; inc. 1946.

Lincoln Home National Historic Site: *see* National Parks and Monuments Table.

Lincoln Memorial, monument, 164 acres (66.4 hectares), in Potomac

Park, Washington, D.C.; built 1914–17. The building, designed by Henry Bacon and styled after a Greek temple, has 36 Doric columns representing the states of the Union at the time of President Abraham Lincoln's death. Inside the building is a heroic statue of Lincoln by Daniel Chester French and two murals by Jules Guerin.

Lincoln Park. 1. City (1978 est. pop. 47,700), Wayne co., SE Mich., a suburb adjacent to Detroit, on the Detroit River; inc. 1921. It is a residential community in a highly industrialized area. **2.** Borough (1970 pop. 9,034), Morris co., N central N.J., W of Paterson; inc. 1922. It is a summer resort in a truck-farming area.

Lin·coln·shire (lĭng'kən-shîr', -shər), nonmetropolitan county (1976 est. pop. 524,500), 2,662 sq mi (6,894.6 sq km), E England, on the Humber estuary, the North Sea, and The Wash; administrative center Lincoln. It was formerly divided into the Parts of Holland in the southeast, the Parts of Kesteven in the southwest, and the Parts of Lindsey in the north. Lincolnshire is an important agricultural, fishing, and industrial area. In Anglo-Saxon times Lincolnshire was variously under the control of Mercia and Northumberland. In 1974 the parts of Holland, Kesteven, and Lindsey were abolished and a considerable area in the north was attached to the new nonmetropolitan county of Humberside.

Lin·coln·ton (lĭng'kən-tən). **1.** City (1970 pop. 1,442), seat of Lincoln co., NE Ga., NW of Augusta, in a farm, livestock, and timber region. **2.** Town (1970 pop. 5,293), seat of Lincoln co., W central N.C., NW of Charlotte; inc. 1785. Textiles and furniture are made.

Lin·coln·wood (lĭng'kən-wŏŏd'), village (1970 pop. 12,929), Cook co., NE Ill., a suburb of Chicago; settled in the 1840s, inc. 1911.

Lin·dau (lĭn'dou), town (1974 est. pop. 24,375), Bavaria, S West Germany, on an island in the Lake of Constance. Connected by bridges with the mainland, it is a summer resort. Lindau was an imperial city from 1275 to 1803 and passed to Bavaria in 1805.

Lin·den (lĭn'dən). **1.** Town (1970 pop. 2,697), seat of Marengo co., W central Ala., W of Montgomery, in a corn, cotton, and timber area; founded 1823 by Bonapartist exiles. **2.** City (1978 est. pop. 39,500), Union co., NE N.J., in the New York metropolitan area; inc. 1925. During the first half of the 20th cent. Linden changed from an agricultural district to a city of diverse manufactures, chief among which are chemicals, petroleum products, and automobiles. **3.** Town (1970 pop. 1,062), seat of Perry co., W central Tenn., on the Buffalo River SW of Nashville, in a corn, hay, and peanut area; inc. 1849. **4.** Town (1970 pop. 2,264), seat of Cass co., NE Texas, NE of Longview, near the point where Texas, Ark., and La. meet; founded c.1850. The red sandy land supports lumbering and truck farming.

Lin·den·hurst (lĭn'dən-hûrst'), village (1978 est. pop. 30,200), Suffolk co., SE N.Y., on S Long Island; inc. 1923. It is a residential area.

Lin·den·wold (lĭn'dən-wōld'), borough (1970 pop. 12,199), Camden co., SW N.J.; settled 1742, inc. 1929.

Lin·des·nes (lĭn'dĭs-nĕs') or **the Naze** (nāz), cape, in Vest-Agder co., southernmost point of the Norwegian mainland. An old lighthouse (1655) is here.

Lin·di (lĭn'dē), river, NE Zaire, rising W of Lake Edward and the Zaire-Uganda border and flowing c.375 mi (605 km) NW, W, and SW to enter the upper Congo River WNW of Kisangani.

Lin·dis·farne (lĭn'dĭs-färn): *see* Holy Island.

Lind·say (lĭn'zē), town (1976 pop. 13,062), SE Ont., Canada, on the Scugog River, NE of Toronto. It is an industrial town, with woolen, flour, and lumber mills, in a scenic lake district.

Lí·ne·a, La (lä lē'nā-ä), city (1970 pop. 52,127), Cádiz prov., S Spain, on the Strait of Gibraltar. Situated on the Spanish border north of the neutral zone that separates the city from the British colony, La Línea supplies the British colony with fresh fruits and vegetables.

Line Islands (līn) or **E·qua·to·ri·al Islands** (ē'kwə-tôr'ē-əl, ĕk'wə-), coral group, 43 sq mi (111.4 sq km), central and S Pacific. Once valuable for guano deposits, the islands now have coconut groves, airfields, and meteorological stations. The islands were uninhabited when discovered by American sailors in 1798, although a few show evidence of ancient Polynesian culture.

Ling·ga Archipelago (lĭng'gə): *see* Riau Archipelago.

Lin·kö·ping (lĭn'chœ'pĭng), city (1975 est. pop. 109,236), capital of Östergötland co., S Sweden, near Lake Roxen. Manufactures include motor vehicles, railroad cars, airplanes, electrical appliances, and processed food. Linköping flourished in the Middle Ages as an intellectual and religious center. In 1598 Sigismund III, king of Sweden, was defeated by the future Charles IX at nearby Stangebrö and deposed. Linköping has a Romanesque cathedral (12th cent.).

Lin·lith·gow (lĭn-lĭth'gō), burgh (1971 pop. 2,777), Lothian region, central Scotland. Manufactures include chemicals, paper, whiskey, and soap. Linlithgow Palace, now a ruin, was the birthplace of James V and Mary Queen of Scots.

Linn (lĭn). **1.** County (1970 pop. 163,213), 713 sq mi (1,846.7 sq km), E Iowa, drained by the Cedar and Wapsipinicon rivers; formed 1837; co. seat Cedar Rapids. Its prairie agriculture includes hogs, cattle, poultry, corn, and oats. There are limestone quarries and sand and gravel pits. **2.** County (1970 pop. 7,770), 607 sq mi (1,572.1 sq km), E Kansas, in a gently sloping to rolling hills area bordering in the E on Mo. and drained by the Marais des Cygnes River; formed 1855; co. seat Mound City. In a livestock and general farming area, it has oil and gas fields and coal, lead, and zinc deposits. Clothing and farm machinery are manufactured. **3.** County (1970 pop. 15,125), 624 sq mi (1,616.2 sq km), N Mo.; formed 1837; co. seat Linneus. It has agriculture (corn, soybeans, wheat, milo, hay, cattle, and hogs), coal mines, and factories manufacturing clothing, shoes, and automobile parts. **4.** County (1970 pop. 71,914), 2,294 sq mi (5,941.5 sq km), W Oregon, in a level agricultural area bounded on the W by the Willamette River and rising to the Cascade Range in the E; formed 1847; co. seat Albany. Its agriculture includes fruit, truck crops, grain, seeds, hay, and dairy products. Logging is also important. Part of Willamette National Forest is in the county.

Linn, city (1970 pop. 1,289), seat of Osage co., central Mo., near the Missouri River ESE of Jefferson City, in a farm and clay region; founded 1842, inc. 1911.

Lin·ne·us (lĭn'ē-əs), city (1970 pop. 400), seat of Linn co., N central Mo., NW of Brookfield, in a grain and livestock area.

Lin·ton (lĭn'tən), city (1970 pop. 1,695), seat of Emmons co., S N.Dak., on Beaver Creek SSE of Bismarck, in a grain, livestock, and dairy region; inc. 1916.

Lin·wood (lĭn'wŏŏd'), city (1970 pop. 6,159), Atlantic co., SE N.J., WSW of Atlantic City; inc. 1931.

Linz (lĭnts), city (1971 pop. 202,874), capital of Upper Austria, NW Austria, a major port on the Danube River. It is a commercial and industrial center and a rail junction. Manufactures include iron and steel, machinery, and textiles. Originally a Roman settlement, Linz was made a provincial capital of the Holy Roman Empire in the late 15th cent. The city has numerous historic structures, including the Romanesque Church of St. Martin (8th cent.) and the baroque cathedral (17th cent.), where Anton Bruckner was organist (1856–68).

Lip·a·ri Islands (lĭp'ə-rē), formerly **Ae·o·li·an Islands** (ē-ō'lē-ən), volcanic island group (1971 pop. 10,043), 44 sq mi (114 sq km), Messina prov., NE Sicily, Italy, in the Tyrrhenian Sea. The group includes Vulcano, the site in former times of the worship of the mythical fire god, with a high volcano that emits hot sulfurous vapors; and Stromboli, with an active volcano (3,040 ft/927.2 m) that has several craters.The islands were colonized by the Greeks in the 6th cent. B.C. Under the Roman Empire and the Fascist regime in Italy (20th cent.) the island group served as a place of exile.

Li·petsk (lē'pĕtsk, lyĕ'pyĭtsk), city (1976 est. pop. 363,000), capital of Lipetsk oblast, E central European USSR, on the Voronezh River. It is the center of an iron-ore-mining area. Industrial products include steel, silicates, tractors, food products, cement, and metal goods. The city has mineral springs and since the 18th cent. has been a health resort center. It was founded in the 13th cent., completely destroyed by the Tatars at the end of the 13th cent., and rebuilt (1707) by Peter the Great as a metallurgical center.

Lip·pe (lĭp'ə), former state, N central West Germany, between the Teutoburg Forest and the Weser River. It was incorporated in 1947 into the state of North Rhine-Westphalia. Originally included in the duchy of Saxony, Lippe became a lordship in the 12th cent., a county in 1529, and a principality in 1720. It sided with Prussia in the Austro-Prussian War (1866) and joined the German Empire in 1871. A local electoral victory (Jan., 1933) of the National Socialists in Lippe helped Adolf Hitler gain power.

Lippe, river, c.150 mi (241.4 km) long, rising in the Teutoburg Forest, W West Germany and flowing W into the Rhine River. It is canalized to permit barge navigation.

Lips·comb (lĭp'skəm), county (1970 pop. 3,486), 934 sq mi (2,419.1 sq km), extreme N Texas, in the NE corner of the Panhandle, with high plains in the W and high rolling hills in the E; formed 1876; co. seat Lipscomb. It is mainly agricultural, yielding wheat, grain sorghums, barley, oats, corn, alfalfa, some fruit, truck crops, livestock, and poultry. Quail, wild turkey, and deer are hunted.

Lipscomb, village (1970 pop. 125), seat of Lipscomb co., extreme N Texas, on the edge of the high plains of the Panhandle, WSW of Woodward. It is a trade center for a farm and livestock area.

Li·ri (lē′rē), river, 98 mi (157.7 km) long, rising in the Apennines, in Latium, central Italy, and flowing generally SE to the Tyrrhenian Sea. In World War II the area around the river was the scene of heavy fighting between Allied and German troops.

Lis·bon (lĭz′bən), city (1975 est. pop. 829,900), W Portugal, capital of Portugal, of Estremadura prov., and of Lisboa dist., on the Tagus River where it broadens to enter the Atlantic Ocean. Lisbon has one of the best harbors in Europe. The city's industries include the production of textiles, chemicals, and steel; oil and sugar refining; and shipbuilding. A large tourist trade is drawn to Lisbon, which is set on seven terraced hills. The Castelo de São Jorge, a fort that dominates the city, may have been built by the Romans on the site of the citadel of the early inhabitants, who traded with Phoenician and Carthaginian navigators. The Romans occupied the town in 205 B.C. It was conquered by the Moors in 714. The city's true importance dates from 1147, when King Alfonso I, with the help of Crusaders, drove out the Moors. Alfonso III transferred (c.1260) his court here from Coimbra, and the city rose to great prosperity in the 16th cent. with the establishment of Portugal's empire in Africa and India. Although many of the old buildings were destroyed by earthquakes, particularly the disastrous earthquake of 1755, some of the medieval buildings remain. The old quarter, the picturesque and crowded Alfama, surrounds the 12th cent. cathedral (rebuilt later). The new quarter, built by the marqués de Pombal after the great earthquake, centers about a large square, the Terreiro do Paço. Some of the well-known buildings in and near Lisbon are the Renaissance Monastery of São Vicente de Fora, with the tombs of the Braganza kings; the Church of St. Roque, with the Chapel of St. John (built by John V in the 18th cent.); and the monastery at Belém, on the Tagus facing the sea, built by Manuel I to commemorate the discovery of the route to India by Vasco da Gama.

Lisbon. **1.** Town (1970 pop. 6,544), Androscoggin co., SW Maine, on the Androscoggin River SE of Lewiston; inc. 1799. Textiles and gypsum and plastic products are made. **2.** City (1970 pop. 2,090), seat of Ransom co., SE N.Dak., on the Sheyenne River SW of Fargo, in a livestock, dairy, and farm region; settled 1878, platted 1880, inc. 1883. **3.** Industrial village (1970 pop. 3,521), seat of Columbiana co., NE Ohio, SSW of Youngstown; founded 1802 as New Lisbon, inc. 1814. Ceramics and leather goods are made.

Lis·burn (lĭz′bərn), municipal borough (1971 pop. 31,836), Co. Antrim, E Northern Ireland, on the Lagan River. Its chief industry, linen manufacture, was introduced by the Huguenots after the revocation of the Edict of Nantes (1685).

Li·sieux (lē-zyœ′), town (1975 pop. 25,521), Calvados dept., N France. It is one of the oldest towns in Normandy. Its modern importance dates from the canonization (1925) of St. Thérèse, whose shrine attracts many pilgrims.

Lis·more (lĭz′môr), city (1976 pop. 22,080), New South Wales, E Australia, on the North Arm of the Richmond River. An important industrial city, Lismore is a leading producer of butter.

Lis·more (lĭz′môr, lĭz-môr′), island, 9.5 mi (15.3 km) long and 1.5 mi (2.4 km) wide, Argyll dist., W Scotland, in Loch Linnhe. There are ruins of several old castles, one of which was a 9th cent. Viking fortress, another the residence of the bishops of Argyll.

Litch·field (lĭch′fēld′), county (1970 pop. 144,091), 925 sq mi (2,395.8 sq km), NW Conn., bordered on the N by Mass., on the W by N.Y., and drained by the Housatonic River and numerous other rivers and lakes; constituted 1751; former co. seats Winsted and Litchfield. Its agriculture includes dairy products, truck crops, fruit, tobacco, and poultry. Among its diversified manufactures are hardware, electrical equipment, typewriters, silverware, machinery, glass and plastic products, sports equipment, textiles and clothing, tools, and furniture. The county includes many lake and mountain resorts.

Litchfield. **1.** Town (1970 pop. 7,399), Litchfield co., NW Conn., SW of Torrington; inc. 1719. It is a summer resort in an agricultural area that also has light industry. The town preserves its early 19th cent. appearance. Ethan Allen, Henry Ward Beecher, and Harriet Beecher Stowe were born here. **2.** City (1970 pop. 7,190), Montgomery co., S central Ill., S of Springfield, in a coal-mining and dairy-farming area; inc. 1859. Shoes, clothing, and metal products are made. **3.** City (1970 pop. 5,262), seat of Meeker co., central Minn., W of Minneapolis; settled 1856, platted 1869, inc. 1872. It is a trade center of a farm and dairying area.

Lith·gow (lĭth′gō), town (1971 pop. 12,814), New South Wales, SE Australia, in the Blue Mts. It is a coal-mining center.

Lith·u·a·ni·a (lĭth′o͞o ā′nē ə), constituent republic (1970 pop. 3,129,000), 25,174 sq mi (65,201 sq km), W European USSR. Vilnius is the capital. Lithuania borders on the Baltic Sea in the west, Latvia in the north, Belorussia in the east, Poland in the south, and Kaliningrad oblast (formerly East Prussia) in the southwest.

Economy. Dairy farming and stock raising are carried on extensively, and grains, flax, sugar beets, potatoes, and vegetables are grown. Primarily agricultural before 1940, Lithuania has since developed considerable industry, including food processing, shipbuilding, and the manufacture of textiles, machinery, metal products, chemicals, and electrical equipment.

History. The pagan Liths, or Lithuanians, may have settled along the Nemen as early as 1500 B.C. In the 13th cent. the Livonian Brothers of the Sword and the Teutonic Knights conquered the region now comprising Estonia, Latvia, and parts of Lithuania. To protect themselves against the Knights, the Lithuanians formed (13th cent.) a strong unified state. The grand dukes Gedimin (1316–41) and Olgerd (1345–77) expanded their territories at the expense of the neighboring Russian principalities. Lithuania became one of the largest states of medieval Europe, including all Belorussia, a large part of the Ukraine, and sections of Great Russia. Olgerd's son, Jagiello, became king of Poland in 1386 as Ladislaus II. In 1569 Lithuania fully merged with Poland by the Union of Lublin. By the three successive partitions of Poland (1772, 1793, 1795) Lithuania disappeared as a national unit and passed to Russia.

A Lithuanian cultural revival began in the 19th cent. Proclaimed (Feb., 1918) an independent kingdom under German protection, Lithuania became (Nov., 1918) an independent republic. In 1940 Soviet forces occupied the country and made it a constituent republic of the USSR. During the German occupation (1941–44) in World War II the considerable Jewish minority was largely exterminated. In 1944 the Communist government returned.

Lit·itz (lĭt′ĭts), borough (1970 pop. 7,072), Lancaster co., SE Pa., N of Lancaster; settled c.1740 by Moravians, laid out 1757, inc. 1759. Shoes, textiles, and chocolate are manufactured.

Lit·tle A·mer·i·ca (lĭt′l ə-mĕr′ĭ-kə), base for Antarctic exploring expeditions, Antarctica, on the Ross Ice Shelf, S of the Bay of Whales. Richard E. Byrd, a U.S. explorer, established and named Little America in 1929 and built bases on the same site in 1933–35 and 1939–41 for subsequent expeditions.

Little Big·horn (bĭg′hôrn′), river, c.90 mi (145 km) long, rising in the Bighorn Mts., N Wyo., and flowing N to join the Bighorn River in S Mont. On June 25–26, 1876, Sioux and Cheyenne warriors defeated the forces of Col. George Custer in the Little Bighorn valley. Custer Battlefield National Monument occupies the site of the battle. The graves of those killed in the battle are located around a granite monument marking the spot of Custer's "last stand."

Little Col·o·ra·do (kŏl′ə-rä′dō, -răd′ō), river, rising in the mountains near the Ariz.-N.Mex. border and flowing generally NW 315 mi (506.8 km) to the Colorado River just above the Grand Canyon.

Little Falls. **1.** City (1970 pop. 7,467), seat of Morrison co., central Minn., NNW of St. Cloud; settled 1855, inc. as a village 1879, as a city 1889. It is a resort and trade center for an agricultural area. Charles A. Lindbergh Memorial Park is nearby. **2.** City (1970 pop. 7,629), Herkimer co., E central N.Y., SE of Utica, at the falls of the Mohawk River on the Barge Canal, in a dairy region; settled c.1722, inc. as a village 1811, as a city 1895.

Little Fer·ry (fĕr′ē), borough (1970 pop. 9,064), Bergen co., NE N.J., on the Hackensack River ESE of Passaic; settled 1636, inc. 1894.

Lit·tle·field (lĭt′l-fēld′), city (1970 pop. 6,738), seat of Lamb co., NW Texas, in the Llano Estacado NW of Lubbock, in an area yielding cotton, sorghum, and other grains; settled 1911, inc. 1925.

Little Fork (fôrk), river, c.132 mi (210 km) long, N Minn., formed by the confluence of three forks NW of the Mesabi iron range and flowing NW to enter the Rainy River on the U.S.-Canada boundary.

Little Ka·na·wha (kə-nô'wə), river, c.160 mi (255 km) long, rising in central W.Va. and flowing NW to the Ohio River at Parkersburg.

Little Mis·sou·ri (mĭ-zŏŏr'ē). **1.** River, c.145 mi (235 km) long, rising in the Ouachita Mts., SW Ark., and flowing generally SE to join the Ouachita River N of Camden. **2.** River, c.560 mi (900 km) long, rising in NE Wyo. and flowing NE into Garrison Reservoir on the Missouri River, W N.Dak.

Little Pee Dee (pē' dē'), river, 145 mi (233.3 km) long, rising in S N.C. and flowing S across the border into S.C., then SE and S to the Pee Dee River W of Myrtle Beach.

Little Red River (rĕd), 105 mi (169 km) long, rising in the Boston Mts., NW Ark., and flowing SE to the White River. Greers Dam and reservoir (completed 1964) provide flood control and power.

Little River, county (1970 pop. 11,194), 544 sq mi (1,409 sq km), extreme SW Ark., bounded on the W by the Texas border, on the S by the Red River, on the N and E by the Little River; formed 1867; co. seat Ashdown. Its agriculture includes fruit, cotton, truck crops, corn, hay, and livestock. It also has sawmills, sand and gravel pits, and plants manufacturing paper and transportation equipment.

Little Rock (rŏk), city (1978 est. pop. 142,700), state capital and seat of Pulaski co., central Ark., on the Arkansas River; inc. 1831. It is a river port and the administrative, commercial, transportation, and cultural center of the state. The city's industries process agricultural products, fish, beef, poultry, and bauxite and timber. The settlement was a well-known river crossing when Arkansas Territory was established in 1819. It became territorial capital in 1821 and state capital in 1836. The city became a center of world attention in 1957, when Federal troops were sent to enforce a 1954 U.S. Supreme Court ruling against segregation in the public schools. Of interest are the old statehouse (which served as capitol from 1836 to 1910) and several museums. The present capitol building was built in 1911.

Little Sil·ver (sĭl'vər), borough (1970 pop. 6,010), Monmouth co., E central N.J., SE of Red Bank; inc. 1923.

Little Sioux (sōō), river, 221 mi (355.6 km) long, rising in SW Minn. and flowing generally SW across NW Iowa to the Missouri River S of Sioux City. Flowing through a rich agricultural area in the Corn Belt, the river is used extensively for irrigation.

Little Ten·nes·see (tĕn-ə-sē'), river, c.135 mi (220 km) long, rising in the Blue Ridge, NE Ga., and flowing generally NW across SW N.C. and through E Tenn. to the Tennessee River opposite Lenoir City.

Lit·tle·ton (lĭt'l-tən). **1.** City (1978 est. pop. 27,300), seat of Arapahoe co., N central Colo.; platted 1812, inc. 1890. It is a suburb south of Denver in an irrigated farm area. Its industries include petroleum research and aerospace firms as well as the manufacture of tires, precision castings, photographic equipment, and tow trucks. **2.** Town (1970 pop. 6,380), Middlesex co., NE Mass., NW of Boston; settled c.1686 on the site of an Indian village, inc. 1715. Electronic equipment is made. **3.** Town (1970 pop. 5,290), Grafton co., NW N.H., near the Connecticut River SE of St. Johnsbury, Vt., in a timber and resort region; settled 1769, inc. 1784.

Little Val·ley (văl'ē), village (1970 pop. 1,340), seat of Cattaraugus co., W N.Y., NW of Olean, in a dairying region; inc. 1876.

Little Wa·bash (wô'băsh), river, c.200 mi (320 km) long, E Ill., rising near Mattoon and flowing SE into the Wabash River.

Liu-chou or **Liu·chow** (both: lyōō'jō'), city, N central Kwangsi Chuang Autonomous Region, S China, on the Liu River. At the intersection of highways and three railroads, it is a manufacturing town with important paper and wood-product industries, a large integrated iron and steel complex, machine shops, chemical plants, textile mills, and food-processing establishments.

Li·vad·i·ya (lĭ-văd'ē-ə, lyĭ-vä'dyē-ə), town, SE European USSR, in the Ukraine, in the Crimea, on the Black Sea. It produces wine and is a noted health resort. Dating from medieval times, Livadiya became a summer residence of the Russian czars in 1861. The Livadiya palace, built in 1910–11, is now a sanatorium. It was the site of the Yalta Conference in 1945.

Live Oak (lĭv' ōk'), county (1970 pop. 6,697), 1,056 sq mi (2,735 sq km), S Texas, drained by the Frio, Atascosa, and Nueces rivers; formed 1856; co. seat George West. Cattle are raised, and grain sorghums, broomcorn, cotton, corn, flax, truck crops, and citrus fruits are grown. There are oil and natural-gas wells here.

Live Oak. 1. Uninc. town (1970 pop. 5,400), Santa Cruz co., W Calif., NW of Santa Cruz, in an area of seaside resorts. **2.** City (1970 pop. 6,830), seat of Suwannee co., N Fla., near the Suwannee River W of Jacksonville. Its principal crops are tobacco and watermelons.

Liv·er·more (lĭv'ər-mōr'), city (1978 est. pop. 48,600), Alameda co., W central Calif.; inc. 1876. The major sources of employment are wineries and the Livermore Radiation Laboratory of the Univ. of California, which conducts nuclear research.

Liv·er·pool (lĭv'ər-pōōl'), county borough (1976 est. pop. 539,700), Merseyside, NW England, on the Mersey River near its mouth. It is one of Britain's greatest ports and largest cities and the country's major outlet for industrial exports. A large center for food processing (especially flour and sugar), Liverpool has a variety of industries, including the manufacture of electrical equipment, chemicals, and rubber. Liverpool was once famous for its pottery, and its textile industry was also prosperous; however, since World War II its cotton market has declined considerably.

In 1207 King John granted Liverpool its first charter. In 1644, during the English Civil War, Liverpool surrendered to the royalists under Prince Rupert after several sieges. Air raids during World War II caused heavy damage and casualties. Liverpool Cathedral, designed by Sir George Gilbert Scott, was begun in 1904 and completed in 1978. St. George's Hall is an imposing building in a group that includes libraries and art galleries. The Walker Gallery has a fine collection of Italian and Flemish paintings, as well as more modern works. The statesman William Gladstone and the artist George Stubbs were born in Liverpool.

Liv·ing·ston (lĭv'ĭng-stən). **1.** County (1970 pop. 40,690), 1,043 sq mi (2,701.4 sq km), E central Ill., drained by the Vermilion River; formed 1837; co. seat Pontiac. Its agriculture includes corn, oats, wheat, soybeans, livestock, poultry, and dairy products. There are bituminous-coal mines, clay pits, and stone quarries in the county. **2.** County (1970 pop. 7,596), 312 sq mi (808.1 sq km), W Ky., bounded on the W and N by the Ohio River and the Ill. border, on the S by the Tennessee River, and crossed by the Cumberland River; formed 1798; co. seat Smithland. In a gently rolling agricultural area yielding burley tobacco, livestock, and corn, it has fluorite mines and limestone quarries. Some manufacturing is done. **3.** Parish (1970 pop. 36,511), 654 sq mi (1,693.9 sq km), SE La., bounded on the W and S by the Amite River, on the SE by Lake Maurepas; formed 1832; parish seat Livingston. Its agriculture includes strawberries, vegetables, corn, cotton, sweet potatoes, hay, and livestock. There is much pine timber and pulpwood, as well as sand and gravel pits. Hunting and fishing areas are found in the parish. **4.** County (1970 pop. 58,967), 571 sq mi (1,478.9 sq km), SE Mich., drained by the Huron and Shiawassee rivers; organized 1836; co. seat Howell. It is primarily agricultural, yielding grain, beans, sugar beets, potatoes, livestock, poultry, and dairy products. The county has some manufacturing and numerous small lakes and summer resorts. **5.** County (1970 pop. 15,368), 533 sq mi (1,380.5 sq km), N Mo., drained by the Grand River; formed 1837; co. seat Chillicothe. Corn, wheat, and oats are grown, and livestock is raised. It has limestone quarries and food-processing and lumbering industries. Textiles, wood and clay products, machinery, and leather goods are manufactured. **6.** County (1970 pop. 54,041), 638 sq mi (1,652.4 sq km), W N.Y., in the Finger Lakes region drained by the Genesee River; formed 1821; co. seat Geneseo. In a dairying and farming region yielding fruit, truck crops, grain, potatoes, hay, and poultry, it has deposits of gypsum and limestone and a food-processing industry. Clothing, chemicals, concrete, metal products, and machinery are made.

Livingston. 1. Town (1970 pop. 2,358), seat of Sumter co., W Ala., between the Tombigbee River and the Miss. border, in the Black Belt; founded c.1833. **2.** Village (1970 pop. 1,398), seat of Livingston parish, SE La., ENE of Baton Rouge, in a timber region. **3.** City (1970 pop. 6,883), seat of Park co., S Mont., on the Yellowstone River N of Yellowstone National Park; founded 1882 by the Northern Pacific RR, inc. 1889. Tourism, mining, and livestock raising are important to its economy. **4.** Town (1970 pop. 3,050), seat of Overton co., N central Tenn., NW of Knoxville, in a farm and timber area; inc. 1835. **5.** Town (1970 pop. 3,965), seat of Polk co., E Texas, near the Trinity River NE of Houston, in an oil-producing region. There are truck and poultry farms in the area.

Liv·ing·stone (lĭv'ĭng-stən), city (1976 est. pop., with suburbs, 65,000), S Zambia, on the Zambezi River, which forms the border with Rhodesia. It is an industrial, commercial, and transportation center. Manufactures include clothing, textiles, and food products.

Founded in 1905, the city was named for David Livingstone, the Scots explorer. From 1911 to 1935 it served as capital of the British protectorate of Northern Rhodesia. The Livingstone Museum contains archaeological, ethnological, and historical materials, including letters and relics of Livingstone. Victoria Falls is nearby.

Li·vo·ni·a (lĭ-vō'nē-ə, -nyə), region and former Russian province, comprising present Estonia and parts of Latvia. It borders on the Baltic Sea and its arms, the Gulf of Riga and the Gulf of Finland, and extends east to Lake Chudskoye and the Narva. Livonia, also known as Livland, was named after the Livs, a Finnic tribe that inhabited the coast when, in the 13th cent., the Livonian Brothers of the Sword conquered the entire region. The knights formed a strong state and threatened Lithuania and Novgorod in the 13th and 14th cent. After the dissolution (1561) of the Livonian Order, Livonia was contested by Poland, Russia, and Sweden, finally passing to Russia in 1772. In 1783 Livonia was constituted a Russian province, and in 1918 it was divided between Estonia and Latvia.

Livonia, city (1978 est. pop. 115,600), Wayne co., SE Mich., a suburb of Detroit; founded 1835, inc. 1950. Among its manufactures are auto bodies and parts, tools and dies, and paints.

Li·vor·no (lē-vôr'nō): *see* Leghorn, Italy.

Liz·ard, The (lĭz'ərd), peninsula, Cornwall, SW England. Its southern extremity (the southernmost point of Great Britain) is called Lizard Point or Lizard Head. The coast has colored serpentine rocks, small coves and bays, wave-hollowed caves, islets, and dangerous reefs.

Lju·blja·na (lyōō'blyä-nä), city (1971 pop. 173,662), capital of Slovenia, NW Yugoslavia, on the Sava River. An industrial and transportation center, it has industries that manufacture machinery, optical instruments, textiles, and chemicals. Known as Emona in Roman times, Ljubljana passed in 1277 to the Hapsburgs. The city passed to Yugoslavia in 1919 and was made the capital of Slovenia in 1946.

Llan·dud·no (lăn-dŭd'nō, -dĭd'-), urban district (1973 est. pop. 17,700), Gwynedd, NW Wales, on a point of land jutting into the Irish Sea. Llandudno is a popular seaside resort with a mild climate.

Lla·nel·li (lă-nĕl'ē'), municipal borough (1973 est. pop. 25,870), Dyfed, S Wales, on the estuary of the Burry River. There are tinplate works and steelworks. Pottery and chemicals are made.

Llan·gol·len (lăn-gŏl'ĭn), urban district (1971 pop. 3,108), Clwyd, NE Wales, at the head of the Vale of Llangollen on the Dee River. Antiquities in the vicinity include the castle Dinas Bran (13th cent.), Eliseg's Pillar (a shaft of a cross dating probably from the 9th cent.), and Valle Crucis Abbey (1200). The International Musical Eisteddfod has been held here since 1949.

Lla·no (lă'nō), county (1970 pop. 6,979), 941 sq mi (2,437.2 sq km), central Texas, in a hilly area on the E side of Edwards Plateau, bounded on the E by the Colorado River and drained by the Llano River; formed 1856; co. seat Llano. Ranching, farming (peanuts, corn, pecans, grain sorghums, oats, and fruit), and granite quarrying are done. The scenic area has good hunting and fishing.

Llano, town (1970 pop. 2,608), seat of Llano co., central Texas, on the Llano River NW of Austin; inc. 1901. The rough country here has cattle, goat, and sheep ranches and some farms. Llano has mineral-processing plants and a granite-quarrying industry.

Llano Es·ta·ca·do (ĕs'tə-kä'dō) or **Staked Plain** (stākt), level, semi-arid, plateaulike region of the S Great Plains, c.40,000 sq mi (103,600 sq km), E N.Mex. and W Texas, between the Pecos River and the Cap Rock escarpment. The region's wind-swept grasslands, formerly used for cattle ranching, are now dotted with dry-land and irrigated farms and oil and natural-gas fields.

lla·nos (lăn'ōs, lä'nōs), Spanish-American term for prairies, specifically those of the Orinoco River basin of northern South America, in Venezuela and eastern Colombia. Shunned by man before the Spanish came, the llanos of the Orinoco are a vast, hot region of rolling savanna broken by low-lying mesas, scrub forest, and scattered palms. Elevation above sea level never reaches more than a few hundred feet. During the dry season (Nov. to Apr.) the land is dry, the grass brown, brittle, and inedible; during the rainy season much of the area is inundated. The sparsely populated llanos support a pastoral economy. Cattle raising is dominant. With flood-control and water-storage projects in the region, sections of the llanos have been turned into fertile agricultural land.

Lloyd·min·ster (loid'mĭn-stər), city (1976 pop. 10,311), on the Alta.-Sask. boundary, Canada. The city is chartered by both provinces. Farming and ranching are the chief activities of the region, which has oil and natural-gas deposits.

Llul·lail·la·co (yōō'yī-yä'kō), extinct volcano, 22,057 ft (6,727.4 m) high, on the border of Chile and Argentina. One of the highest peaks in the Andes and perpetually snowcapped, it overlooks a pass used for rail and highway traffic.

Lo·a (lō'ə), town (1970 pop. 324), seat of Wayne co., S central Utah, SE of Richfield on the Fremont River, at an altitude of 7,000 ft (2,135 m).

Lo·a (lō'ä), longest river of Chile, 275 mi (442.5 km) long, flowing S from the Andes, N Chile, then W and N through the Atacama Desert, before turning W to the Pacific Ocean. It is not navigable but affords some water supply and hydroelectric power for nitrate-mining communities in its vicinity.

Lo·bi·to (lō-bē'tō), city (1970 pop. 59,528), W central Angola, on the Atlantic Ocean. Angola's chief port, it is also a road and railroad hub. Exports include minerals, grain, coffee, sisal, sugar, fish, salt, and beans. Among the city's industries are shipbuilding, food processing, and the manufacture of cement and building materials. Lobito was founded in 1843 on orders from Queen Maria II in response to requests by the Portuguese inhabitants of Benguela for a healthier and strategically more favorable living area. It is built mainly on reclaimed land.

Lo·car·no (lō-kär'nō), town (1977 est. pop. 15,300), Ticino canton, S Switzerland, at the N end of Lago Maggiore. In a beautiful resort region with a mild climate, Locarno attracts a great number of tourists. It has an annual film festival. Jewelry, motors, soap, and food products are made. In 1512 it was taken from Milan by the Swiss cantons, and in 1803 it was included in Ticino canton.

Loch (lŏKH, läk), for names of Scottish lakes and inlets beginning thus, see second element; e.g., for Loch Awe, *see* Awe, Loch.

Loches (lōsh), town (1975 pop. 6,738), Indre-et-Loire dept., W central France, in Touraine, on the Indre River. It is famous for its medieval buildings, especially the ancient château that dominates the town. Originally established by the counts of Anjou, it later became (mid-13th cent.) a royal residence and a state prison.

Lock·hart (lŏk'härt'), city (1970 pop. 6,489), seat of Caldwell co., S central Texas, S of Austin; founded 1848, inc. 1870. In a region of cotton and dairy farms, the city has a clothing factory and other small industries. There are oil fields nearby.

Lock Ha·ven (lŏk hā'vən), industrial city (1978 est. pop. 9,900), seat of Clinton co., N central Pa., on the West Branch of the Susquehanna River at the junction of Bald Eagle Creek, in a rich agricultural area; settled 1769, inc. as a city 1870. The city was a lumber center in the 19th cent., and it now has varied industries.

Lock·port (lŏk'pōrt'). **1.** City (1970 pop. 9,985), Will co., NE Ill., SW of Chicago; settled 1830, laid out 1837, inc. 1853. It is near locks of the old Illinois and Michigan Canal and of the modern Chicago Sanitary and Ship Canal. There is a large oil refinery here. **2.** Industrial city (1978 est. pop. 26,300), seat of Niagara co., W N.Y., on the New York State Barge Canal, in a rich fruit and dairy region; settled 1821, inc. 1865. Automotive radiators, wood, metal, and paper products are among the many manufactures. The city was built around a series of locks on the old Erie Canal.

Lo·cle, Le (lə lôk'l), town (1977 est. pop. 13,100), Neuchâtel canton, NW Switzerland, in the Jura Mts. near the French border. It has been a watchmaking center since the 17th cent.

Lo·cris (lō'krĭs), region of central Greece. The state was probably in existence before the arrival of the Phocians. Largely hemmed in by stronger states, the Locrians played a minor role in Greek history. However, they founded (c.700 B.C.) one of the earliest Greek colonies in southern Italy.

Lo·cust Grove (lō'kəst grōv'), uninc. town (1970 pop. 11,626), Nassau co., SE N.Y., on Long Island.

Lod (lōd), city (1975 est. pop. 35,500), central Israel. Its manufactures include chemicals, oil products, electronic equipment, and cigarettes. Lod was probably of Hebrew foundation and is frequently mentioned in the Bible. It was destroyed (A.D. 66-70) by the Romans in the Jewish-Roman war and, after the destruction of the Temple in Jerusalem (A.D. 70), became the temporary seat of many famous Jewish teachers. Hadrian rebuilt the city. It is the traditional home and place of burial of St. George (4th cent.?), England's patron saint, and has a church in his honor. Lod was occupied by the Crusaders in 1099, destroyed by Saladin in 1191, and rebuilt by King Richard I of England. After the Arab-Israeli War of 1948 most Arabs left the city, which was then settled by Jewish immigrants.

Lodge·pole Creek (lŏj'pōl'), river, 212 mi (341.1 km) long, rising SE of Laramie, Wyo., and flowing generally SE into the SW corner of Nebr., where it turns S and joins the South Platte River at Julesburg in the NE corner of Colo.

Lo·di (lô'dē), city (1976 est. pop. 44,332), Lombardy, N Italy, on the Adda River, near Milan. It is an important dairy center. Machines, electrical goods, and ceramics are also produced. At Lodi on May 10, 1796, Napoleon Bonaparte defeated the Austrians.

Lo·di (lō'dī). **1.** City (1978 est. pop. 32,400), San Joaquin co., central Calif., on the Mokelumne River, in a rich farm area; inc. 1906. The city has foundries, a cannery, and a meat plant. Wine is made, as well as other diverse manufactures. Lodi was founded in 1869 and settled by wheat farmers from the Dakotas, mostly of German descent. **2.** Industrial borough (1978 est. pop. 23,300), Bergen co., NE N.J.; inc. 1894. It has chemical industries.

Łódź (lo͞oj), city (1975 est. pop. 798,300), central Poland. An important industrial city, Łódź is the center of the Polish textile industry. Other manufactures include machinery, electrical equipment, chemicals, and metals. Chartered in 1423, the city passed to Prussia in 1793 and to Russia in 1815. It reverted to Poland in 1919. In World War II the city was incorporated into Germany.

Lo·fo·ten (lō'fōt'n) and **Ves·ter·å·len** (vĕs'tə-rô'lən), two contiguous island groups (1976 est. pop. 88,151), NW Norway, in the Norwegian Sea. Situated within the Arctic Circle, the islands extend c.150 mi (240 km) from northeast to southwest and are from 1 to 50 mi (1.6–80.5 km) off the mainland. The North Atlantic Drift gives these northern islands a temperate climate. The chief economic importance of these island groups lies in their cod and herring fisheries, which are among the richest in the world.

Lo·gan (lō'gən). **1.** County (1970 pop. 16,789), 718 sq mi (1,859.6 sq km), W Ark., bounded on the N by the Arkansas River and drained by the Petit Jean River; formed 1871 as Sarber co., renamed 1875; co. seats Paris and Booneville. Its agriculture includes fruit, truck crops, cotton, corn, potatoes, livestock, and dairy products. Coal mining, lumbering, and cotton ginning are done. Among its manufactures are textiles, rubber, shoes, wood products, gears, and machinery. Part of Ozark National Forest is here. **2.** County (1970 pop. 18,852), 1,827 sq mi (4,731.9 sq km), NE Colo., bordering on Nebr. and drained by the South Platte River; formed 1887; co. seat Sterling. In an irrigated agricultural area yielding wheat, sugar beets, and cattle, it has some oil and natural-gas wells. Processed foods and machinery are manufactured. **3.** County (1970 pop. 33,538), 622 sq mi (1,611 sq km), central Ill., drained by the Salt and Kickapoo creeks and small Sugar Creek; formed 1839; co. seat Lincoln. Agriculture (corn, wheat, soybeans, hay, oats, livestock, and poultry) and some manufacturing (china, cigars, dairy products, and caskets) characterize the area. Bituminous-coal mining and lumbering are also major occupations. **4.** County (1970 pop. 3,814), 1,073 sq mi (2,779.1 sq km), W Kansas, in a farming and grazing area drained by the North Fork of the Smoky Hill River; formed 1881 as St. John co., renamed 1887; co. seat Oakley. Some mining and manufacturing are done. **5.** County (1970 pop. 21,793), 563 sq mi (1,458.2 sq km), S Ky., bounded on the S by Tenn. and drained by the Mud and Red rivers and several creeks; formed 1792; co. seat Russellville. It is in a rolling agricultural area (dark tobacco, corn, wheat, livestock, poultry, and dairy products), with bituminous-coal and asphalt mines, lumber, and stone quarries. Its manufactures include flour, packaged meats, hosiery, clothing, chemicals, and wood and metal products. **6.** County (1970 pop. 991), 570 sq mi (1,476.3 sq km), central Nebr., in a farming area; formed 1885; co. seat Stapleton. Livestock and grain are its major products. **7.** County (1970 pop. 4,245), 1,003 sq mi (2,599.8 sq km), S N.Dak., watered by Beaver Creek; formed 1873; co. seat Napoleon. It is in an agricultural area that produces wheat, oats, flax, cattle, and dairy products. There is some manufacturing in the county. **8.** County (1970 pop. 35,072), 461 sq mi (1,194 sq km), W central Ohio, drained by the Great Miami and Mad rivers; formed 1817; co. seat Bellefontaine. It has agriculture (livestock, dairy products, and grain), limestone quarries, sand and gravel pits, and diversified manufacturing. **9.** County (1970 pop. 19,645), 747 sq mi (1,934.7 sq km), central Okla., intersected by the Cimarron River; formed 1890; co. seat Guthrie. It has diversified agriculture (grain, cotton, fruit, livestock, and dairy products), oil and gas wells, and some manufacturing. **10.** County (1970 pop. 46,269), 456 sq mi (1,181 sq km), SW W.Va., on the Allegheny Plateau, drained by the Guyandot River; formed 1824; co. seat Logan. It is in a bituminous-coal region and also has timber, natural-gas fields and some livestock ranches. Fruit and tobacco are grown.

Logan. 1. Town (1970 pop. 1,526), seat of Harrison co., SW Iowa, on the Boyer River NNE of Council Bluffs, in a farm area; settled as Boyer Falls, renamed 1864, inc. 1876. **2.** City (1970 pop. 6,269), seat of Hocking co., S central Ohio, on the Hocking River SE of Columbus; founded 1816, inc. 1839. Pottery is made from local clay. **3.** City (1978 est. pop. 24,100), seat of Cache co., N Utah, on the Logan River; inc. 1859. It is the center of an irrigated dairy and farm area, with huge cheese plants, other food-processing facilities, and factories making farm machinery, pianos, plastics, and knitted goods. Logan was founded (1859) by Mormons. Utah State Univ. is here. **4.** Trading city (1970 pop. 3,311), seat of Logan co., S W.Va., SW of Charleston, in a mining and timber area; settled c.1765.

Logan, Mount, 19,850 ft (6,054.3 m) high, extreme SW Yukon Territory, Canada, just E of Alaska. It is the highest mountain in Canada. It caps an immense tableland and is the center of the greatest glacial expanse in North America. The first ascent was made in 1925.

Lo·gans·port (lō'gənz-pôrt'), city (1978 est. pop. 17,700), seat of Cass co., N central Ind., at the confluence of the Wabash and the Eel rivers; inc. 1838. In a fertile farm area, it has diversified industries.

Lo·gro·ño (lō-grō'nyō), city (1975 est. pop. 96,546), capital of Logroño prov., N Spain, in Old Castile, on the Ebro River. It is a farm-processing center noted for its Rioja wine; wood products and textiles are also made. Navarre and Castile fought over Logroño from the 10th cent. until its final annexation (1173) to Castile.

Loir (lwär), river, 193 mi (310.5 km) long, rising S of Chartres, N central France, and flowing generally SW through a fertile agricultural region to join the Sarthe River N of Angers.

Loire (lwär), department (1975 pop. 742,396), 1,843 sq mi (4,773.4 sq km), E central France, in part of Beaujolais and Lyonnais; capital Saint-Étienne.

Loire, longest river of France, c.630 mi (1,015 km) long, rising in the Cévennes Mts., SE France, and flowing in an arc through central and W France to the Atlantic Ocean at Saint-Nazaire. The upper Loire swiftly flows northwestward through numerous gorges in the Massif Central. At Orléans it swings southwest and enters a wide fertile valley. Silting, shallowness, and seasonal volume fluctuations limit the use of the Loire for navigation.

Loire-At·lan·tique (lwär' ät-läɴ-tēk'), department (1975 pop. 934,499), 2,661 sq mi (6,892 sq km), NW France, in S Brittany, on the Atlantic coast; capital Nantes.

Loi·ret (lwä-rā'), department (1975 pop. 490,189), 2,603 sq mi (6,741.8 sq km), N central France, partly in Orléanais; capital Orléans.

Loir-et-Cher (lwär' ā-shĕr'), department (1975 pop. 283,686), N central France, in Orléanais; capital Blois.

Lo·ja (lô'hä), city (1974 pop. 343,153), capital of Loja prov., S Ecuador, on the Zamora River. It is a center for the agricultural and mineral resources (gold, silver, and copper) of the area. Founded in 1546, the city was the site of the Ecuadorian declaration of independence in 1820.

Lo·ko·ja (lō-kō'jə), town (1963 pop. 25,000), central Nigeria, at the junction of the Niger and Benue rivers. Lokoja is the trade and distribution center for an agricultural (chiefly cotton) region and has food-processing industries. In 1859 a British trading and missionary settlement was founded in Lokoja. In 1900 Lokoja served as the staging point for the British conquest of northern Nigeria and became the temporary capital of the protectorate of Northern Nigeria.

Lol·land (lŏl'ənd, lô'län) or **Laa·land** (lô'län), island (1970 pop. 74,819), 479 sq mi (1,240.6 sq km), SE Denmark, in the Baltic Sea, E of Langeland, S of Sjaelland, and W of Falster. The island is low-lying and agricultural; sugar beets are the main crop. There are numerous summer resorts on the island's southwest coast.

Lom·bard (lŏm'bärd), village (1978 est. pop. 35,600), Du Page co., NE Ill., a residential suburb of Chicago; inc. 1869.

Lom·bar·dy (lŏm'bər-dē, lŭm'-), region (1975 est. pop. 8,837,656), c.9,200 sq mi (23,830 sq km), N Italy, bordering on Switzerland in the N; capital Milan. Lombardy has Alpine peaks and glaciers in the north, several picturesque lakes, and upland pastures that slope to the rich, irrigated Po valley in the south. Rice, cereals, forage, flax, and sugar beets are the main crops of Lombardy, and the mulberry is extensively cultivated for use in sericulture. Manufactures include textiles, clothing, iron and steel, machinery, motor vehicles, chemicals, and wine.

The Lombard plain, located in the central part of Lombardy at the confluence of several Alpine passes, has for centuries been a much coveted and frequently invaded area, and it has been a battlefield in many wars. First inhabited by a Gallic people, the region became (3rd cent. B.C.) part of the Roman province of Cisalpine Gaul. It suffered heavily during the barbarian invasions. In A.D. 569 the region was made the center of the kingdom of the Lombards. Lombardy was united in 774 with the empire of Charlemagne. After a period of confusion (10th cent.), power gradually passed (11th cent.) from feudal lords to autonomous communes, and a general economic revival occurred. In the 12th cent. several cities united in the Lombard League in order to defy Emperor Frederick I and defeated him at Legnano (1176). The 13th cent. was marked by struggles between Guelphs (pro-papal) and Ghibellines (pro-imperial), which resulted in wars among cities and rivalries between families within cities. After the end (mid-16th cent.) of the Italian Wars, the rest of Lombardy followed the fortunes of Milan. Spanish rule (1535–1713) was followed by that of Austria (1713–96) and of France (1796–1814). The Lombardo-Venetian kingdom was established under Austrian rule in 1815. Lombardy ousted the Austrians in 1859, and the kingdom was dissolved.

Lom·bok (lŏm-bŏk′), island (1970 est. pop. 1,602,000), c.1,825 sq mi (4,725 sq km), E Indonesia, one of the Lesser Sundas, separated from Bali by the Strait of Lombok. Its southern area is a fertile plain producing maize, rice, coffee, cotton, and tobacco. The inhabitants are skilled weavers and metalworkers. First visited by the Dutch in 1674, Lombok became part of the Netherlands East Indies in 1894.

Lo·mé (lô-mā′), city (1971 est. pop. 144,300), capital of Togo, on the Gulf of Guinea. It is the country's administrative, communications, and industrial center. Lomé was a small village until 1897, when it became the capital of the German colony of Togo.

Lo·mi·ta (lō-mē′tə), city (1970 pop. 19,784), Los Angeles co., S Calif., a residential suburb of Los Angeles; inc. 1964.

Lo·mond, Loch (lō′mənd, -mən), largest lake of Scotland, 23 mi (37 km) long and from 1 to 5 mi (1.6–8 km) wide, between Strathclyde and Central regions, W Scotland. At the southern end of the lake there are numerous wooded islands.

Lo·mo·no·sov (lə-mə-nô′səf), city (1974 est. pop. 43,000), NW European USSR, on the Gulf of Finland. It is a rail terminus and summer resort and has foundries and brick factories. In Lomonosov are a palace built (1710–25) by Peter I and the Chinese Palace, built (1762–68) by Catherine the Great. The city was part of the Soviet bridgehead on the Gulf of Finland during the German siege of Leningrad in World War II.

Lom·poc (lŏm′pōk), city (1978 est. pop. 25,300), Santa Barbara co., S Calif., in an oil area; inc. 1888. It has a huge flower-seed industry and two large diatomaceous earth mines. Petroleum and food are processed, and phonograph records are manufactured.

Łom·ża (lôm′zhä), town (1975 est. pop. 29,000), NE Poland, on the Narew River. Manufactures include cotton cloth, agricultural machinery, and bricks. There are clay deposits in the vicinity. Łomża dates from c.1000; it passed to Prussia in 1795 and to Russia in 1815 and reverted to Poland in 1921.

Lon·don (lŭn′dən), city (1976 pop. 240,392), SE Ont., Canada, on the Thames River. London was settled in 1826. Its streets and bridges are named for those of old London in England. In a rich agricultural district, it is a notable industrial, commercial, and financial center.

London, capital of Great Britain, chief city of the Commonwealth of Nations, SE England, on both sides of the Thames River, since the 1965 London Government Act is officially called Greater London (1971 pop. 7,379,014), c.620 sq mi (1,605 sq km). It consists of the Corporation of the City of London and 32 boroughs.

London is a financial, commercial, industrial, and cultural center and one of the world's greatest ports. It exports manufactured goods and imports petroleum, tea, wool, raw sugar, timber, butter, metals, and meat. London is also a great manufacturing city. Clothing, furniture, precision instruments, jewelry, cement, chemicals, and stationery are produced. Engineering and scientific research are also important. London is rich in artistic and cultural activity, with numerous theaters, cinemas, museums, galleries, and opera and concert halls. Municipal parks include Hyde Park, Kensington Gardens, and Regent's Park. Besides the British Museum, the art galleries and museums include the Victoria and Albert Museum, the National Gallery, and the Tate Gallery. The Univ. of London is the largest in Great Britain.

London was an early Roman outpost known as Londinium. After the withdrawal of the Roman legions in the 5th cent., London was lost in obscurity until 886, when it again emerged as an important town under the control of King Alfred. Under the Normans and Plantagenets, the city grew commercially and politically and during the reign of Richard I (1189–99) obtained a form of municipal government from which the modern City Corporation developed. Medieval London saw the foundation of the Inns of Court and the construction of Westminster Abbey. By the 14th cent. London had become the political capital of England.

The reign of Elizabeth I (1558–1603) brought London to a level of great wealth, power, and influence—the undisputed center of England's Renaissance culture. In 1665 the great plague took some 75,000 lives. A great fire in Sept., 1666, lasted five days and virtually destroyed the city. Sir Christopher Wren played a large role in rebuilding the city. He designed more than 51 churches, notably the rebuilt St. Paul's Cathedral. Until 1750, when Westminster Bridge was opened, London Bridge, first built in the 10th cent., was the only bridge to span the Thames. The area of present-day Greater London had about 1.1 million people in 1801; by 1851 the population had increased to 2.7 million, and by 1901 to 6.6 million. During the Victorian era London acquired tremendous prestige as the capital of the British Empire and as a cultural and intellectual center.

Many buildings of central London were destroyed or damaged in air raids during World War II. Today there are numerous blocks of new office buildings and districts of apartment dwellings constructed by the government authorities.

London. 1. City (1970 pop. 4,337), seat of Laurel co., SE Ky., NNW of Middlesboro. It is the trade center for a coal, corn, and tobacco area. **2.** Rural village (1970 pop. 6,481), seat of Madison co., SW central Ohio, WSW of Columbus; founded 1811, inc. 1831. Metal products are made here.

Lon·don·der·ry (lŭn′dən-dĕr′ē, lŭn′dən-dĕr′ē) or **Der·ry** (dĕr′ē), county (1971 pop. 130,296), 804 sq mi (2,082.4 sq km), NW Northern Ireland; co. town Londonderry. Much of the county is mountainous. Agriculture is the main occupation. The district was dominated for many centuries by the O'Neill family, whose confiscated estates were granted, in 1609, to the city companies of London.

Londonderry or **Derry**, county borough (1971 pop. 66,645), co. town of Co. Londonderry, NW Northern Ireland, on the Foyle River near the head of Lough Foyle. Londonderry is a naval base and seaport with extensive exporting of livestock. The staple industry is the manufacture of linen shirts and collars, but a program of industrial development and diversification was begun in 1969. Londonderry grew up around an abbey founded in 546 by St. Columba. The town was burned by the Danes in 812. In 1311 it was granted to Richard de Burgh, earl of Ulster. It was turned over (1613) to the corporations of the City of London. In the siege of Londonderry by the forces of James II (beginning in Apr., 1689), the town was held for 105 days under the leadership of George Walker; a triumphal arch, a column, and one of the town gates commemorate the siege.

Long (lông), county (1970 pop. 3,746), 403 sq mi (1,043.8 sq km), SE Ga., bounded in the SW by the Altamaha River; formed 1920; co. seat Ludowici. It is in a coastal-plain agricultural (corn, truck crops, and livestock) and lumbering area.

Long Beach. 1. City (1978 est. pop. 341,200), Los Angeles co., S Calif., on San Pedro Bay; inc. 1888. It is a port and a year-round resort noted for its long, wide beaches and active marina. The city has a large oil industry; oil (discovered in 1921) is found both underground and offshore. Manufactures include aircraft, automobile and missile parts, electronic equipment, building materials, canned seafoods, liquid gas, and metal, rubber, and chemical products. Points of interest in the city include an adobe ranch house (1844) that is now a museum and the ocean liner ***Queen Mary***, which was purchased in 1967 and converted into a museum, hotel, and tourist center. **2.** Resort city (1970 pop. 6,170), Harrison co., SE Miss., on Mississippi Sound just WSW of Gulfport, in a truck-farm area; inc. 1905. Furniture is made. **3.** City (1978 est. pop. 33,100), Nassau co., SE N.Y., on Long Island; inc. 1922. It is a beach resort on the Atlantic Ocean. Labels, clothing, and umbrellas are manufactured.

Long Branch (brănch), city (1978 est. pop. 29,800), Monmouth co., E central N.J., on the Atlantic coast; settled 1740, inc. 1904. It has garment and electronics industries. Long Branch has been a popular ocean resort since the early 19th cent.

Long Ea·ton (ēt'n), urban district (1973 est. pop. 33,560), Derbyshire, central England. A large number of products are manufactured, including synthetic fabrics and apparel, and railroad carriages.

Long·fel·low National Historic Site (lông'fĕl'ō): *see* National Parks and Monuments Table.

Long·ford (lông'fərd), county (1971 pop. 28,227), 403 sq mi (1,043.8 sq km), N central Republic of Ireland; co. town Longford. A part of the central plain of Ireland, it has level land with numerous small lakes, bogs, and marshes. Raising beef cattle is the principal occupation; oats and potatoes are the chief crops.

Long·horn Cavern (lông'hôrn'), limestone cave, central Texas, at Burnet. On the northern edge of the Edwards Plateau, the cave (explored length c.8 mi/13 km) lies beneath a triangular ridge rising above the valley of the Colorado River.

Long Island, 1,723 sq mi (4,462.6 sq km), 118 mi (190 km) long, and from 12 to 20 mi (19.3–32.2 km) wide, SE N.Y. It is the fourth-largest island of the United States and the largest outside of Alaska and Hawaii. It is separated from Staten Island by the Narrows, from Manhattan and the Bronx by the East River, and from Conn. by Long Island Sound, an arm of the Atlantic Ocean; on the south is the Atlantic Ocean. Eastern Long Island has two flukelike peninsulas that are separated by Peconic Bay. The northern fluke, terminating in Orient Point, follows part of the Harbor Hill moraine, a hilly ridge that extends west along northern Long Island to the Narrows and was deposited by melting ice during the last stage of the Ice Age. The southern fluke, terminating in Montauk Point, follows the Ronkonkoma moraine, a somewhat older morainal ridge that extends west to join the Harbor Hill moraine at Lake Success. Low, wooded hills capped by glacial deposits lie north of the moraines and contrast with a broad, low-lying outwash plain to the south. Long beaches, backed by dunes and shallow lagoons, fringe the south shore; the north shore has low cliffs and is deeply indented by bays.

Both the Dutch and the English established farming, whaling, and fishing settlements on Long Island, but it remained sparsely settled until bridges, highways, and railroads provided easy access to New York City. Industrial and residential growth has been especially rapid since 1945. Farming has declined in importance but continues to be profitable in eastern Long Island. Sand and gravel are quarried from the island's glacial deposits. Sport and commercial fishing is important on the south and east coasts.

lon·gi·tude (lŏn'jĭ-to͞od', -tyo͞od'), angular distance on the earth's surface measured along the equator E or W of the prime meridian. A meridian of longitude is an imaginary line on the earth's surface from pole to pole; two opposite meridians form a great circle dividing the earth into two hemispheres. By international agreement, the meridian passing through the original site of the Royal Greenwich Observatory at Greenwich, England, is designated the prime meridian, and all points along it are at 0° longitude. All other points on the earth have longitudes ranging from 0° to 180° E or from 0° to 180° W. Meridians of longitude and parallels of latitude form a grid by which any position on the earth's surface can be specified.

Long·mead·ow (lông'mĕd-ō), town (1970 pop. 15,630), Hampden co., SW Mass., a residential suburb adjoining Springfield, on the Connecticut River; settled 1644, set off and inc. 1783.

Long·mont (lông'mŏnt'), city (1978 est. pop. 34,800), Boulder co., N Colo.; inc. 1885. It is the trading and processing center for a rich farm area irrigated by the Colorado–Big Thompson project. Sugar is produced, and campers, trailers, and batteries are made.

Long Prai·rie (prâr'ē), village (1970 pop. 2,416), seat of Todd co., central Minn., between Little Falls and Alexandria; platted 1867, inc. 1883. It is a farm trade center.

Long Range, mountain range, extending c.300 mi (485 km) along the W coast of Newfoundland Island, Canada, and rising to 2,672 ft (815 m) in the Lewis Hills. Part of the Appalachian system, the range consists of parallel ridges that rise steeply from the coast and slope gently eastward. The range is economically important for timber.

Longs Peak (lôngz), 14,255 ft (4,347.8 m) high, N Colo., in the Front Range of the Rocky Mts. From the east side of its snow-capped peak there is a 2,000 ft (610 m) drop to Chasm Lake.

Lon·gueuil (lôɴ-gâl'), city (1976 est. pop. 122,429), S Que., Canada, on the St. Lawrence River opposite Montreal.

Long·view (lông'vyo͞o'). **1.** City (1978 est. pop. 53,000), seat of Gregg co., E Texas; inc. 1872. The city has oil and natural-gas wells, oil refineries, and plants making a wide variety of products. It is also a livestock center. **2.** City (1978 est. pop. 29,500), Cowlitz co., SW Wash., a port of entry at the junction of the Columbia and the Cowlitz rivers; inc. 1924. Its manufactures include aluminum, metal, paper, pulp, and wood products. The city was founded in 1922 as a lumber town on the site of the historic settlement Monticello, which had been swept away by a flood in 1867.

Long·wy (lôɴ-wē'), town (1975 pop. 20,131), Meurthe-et-Moselle dept., NE France, near the Belgian and Luxembourg borders. It is a center of the Lorraine iron and steel industry.

Long·year·by·en (lông'yîr-bü'ən), town and administrative center of Svalbard, on Isfjorden, Spitsbergen Island. It is a coal-mining settlement, founded (1905) by an American company. It was destroyed (Sept., 1943) by German battleships but was quickly rebuilt.

Lo·noke (lō'nōk'), county (1970 pop. 26,249), 800 sq mi (2,072 sq km), central Ark., bounded on the N by Cypress Bayou; formed 1873; co. seat Lonoke. Its agriculture includes rice, cotton, truck crops, fruit, and livestock. Wood products, cheese and other dairy products, feed, textiles, furniture, and metal products are made. Cotton ginning, sawmilling, and rice milling are done.

Lonoke, trading city (1970 pop. 3,140), seat of Lonoke co., central Ark., E of Little Rock, in an area producing rice and cotton.

Lons-le-Sau·nier (lôɴ'lə-sō-nyā'), town (1975 pop. 20,942), capital of Jura dept., E France, at the foot of the Jura Mts. A saltwater spa since Roman times, the town has food and textile industries.

Look·out, Cape (lo͝ok'out'), point of a sandy reef off E N.C., SW of Cape Hatteras. A lighthouse on the point was built in 1859 and is included in Cape Lookout National Seashore.

Lop Bu·ri (lŭp bo͞o-rē'), town (1972 est. pop. 33,302), capital of Lop Buri prov., S Thailand. It was ruled by the Mons in the 7th and 8th cent. and by the Khmers from the 10th to the 13th cent.

Lop Nor (lôp' nôr'), salt basin, SE Sinkiang Uigur Autonomous Region, China, in the Tarim River basin. Since 1964 it has been used by the Chinese government for nuclear test explosions. Once a large salt lake (as mapped by ancient Chinese geographers), it is now largely dried up, with marshes and small, shifting lakes.

Lo·rain (lō-rān', lô-), county (1970 pop. 256,843), 495 sq mi (1,282.1 sq km), N Ohio, bounded on the N by Lake Erie and drained by the Black and Vermilion rivers; formed 1824; co. seat Elyria. It is in a highly industrial area specializing in steel and automobile parts. Truck farming is also important.

Lorain, city (1978 est. pop. 81,000), Lorain co., N Ohio, on Lake Erie at the mouth of the Black River; inc. 1834. It is an important ore-shipping port, with shipyards, steel works, automobile-assembly plants, and commercial fisheries. Power equipment, building materials, navigation equipment, and toys are among the manufactures.

Lor·ca (lôr'kä), city (1970 pop. 25,208), Murcia prov., SE Spain, in Murcia, on the Guadalentín River. It is a market center for a fertile, irrigated region producing cereals and livestock. Hemp sandals and woolen products are made in Lorca. Nearby are gypsum quarries and sulfur and iron mines. Taken by the Moors in the 8th cent., the city was liberated in 1234.

Lord Howe Island (hou), volcanic island (1969 pop. 280), 5 sq mi (13 sq km), S Pacific, a dependency of New South Wales, Australia. It is a resort c.300 mi (485 km) east of the Australian coast. The island was discovered in 1788 by the British and was settled in 1834.

Lords·burg (lôrdz'bûrg'), village (1970 pop. 3,429), seat of Hidalgo co., SW N.Mex., SW of Silver City near the Ariz. border. It is the trade center of a region yielding copper, gold, and silver, cattle and sheep, and truck crops.

Lor·e·lei (lôr'ə-lī), cliff, 433 ft (132 m) high, on the right bank of the Rhine River, near St. Goarshausen, W West Germany, about midway between Koblenz and Bingen. Here the Rhine forms a dangerous narrows, and in German legend a fairy lived on the rock and by her singing lured the sailors to their deaths.

Lo·re·to (lə-rā'tō), town (1976 est. pop. 5,900), in the Marche, central Italy, on a hill overlooking the Adriatic Sea. It has silk industries and is a famous place of pilgrimage. According to legend, the Holy

House of the Virgin in Nazareth was brought to Loreto through the air by angels in 1294.

Lo·rient (lôr′ē-äN, lô-ryäN′), town (1975 pop. 69,769), Morbihan dept., NW France, a port and naval station on the Atlantic Ocean. It is a great shipbuilding center. Established (17th cent.) as a port to serve the French East India Company, it was developed as a naval base by Napoleon I and became the country's chief naval yard. In World War II it was the Germans' major submarine base on the Atlantic and was destroyed by Allied bombs in 1942–43.

Lor·raine (lō-rān′, lô-rân′), region and former province, NE France, bordering in the N on Belgium, Luxembourg, and West Germany, in the E on Alsace, in the S on Franche-Comté, and in the W on Champagne. Except for the Vosges Mts. in the southeast and the ridges paralleling the Moselle and Meuse rivers, Lorraine is a slightly rolling plateau with pastures and some agricultural districts. Hops are grown, and there are numerous vineyards. In the east salt is mined; some coal is found in the north. The principal wealth of Lorraine lies in its iron deposits; these deposits constitute the richest fields of iron in Europe outside the USSR and Sweden.

Lorraine was in the 9th cent. part of the kingdom of Lotharingia; it became a duchy under the Holy Roman Empire. It passed in 1048 to the house of Alsace, which then became the house of Lorraine and controlled the duchy until 1738. Under Duke Charles II (1559–1608) Lorraine enjoyed a period of relative order and prosperity amidst a Europe torn by religious and imperialistic strife. Lorraine was occupied by France in the Thirty Years' War (1618–48). In the Treaty of Ryswick (1697), Leopold I was given possession of the duchy. His heir, Francis I, by an arrangement (1735) with Louis XV, exchanged the duchies of Lorraine and Bar for Tuscany; Lorraine and Bar were then given to Louis XV's father-in-law, Stanislaus I, ex-king of Poland, upon whose death (1766) they passed to France. As a French province, Lorraine continued to enjoy certain exemptions and privileges. In 1871, as a result of the Franco-Prussian War, the eastern part of Lorraine was ceded to Germany and united with Alsace as the imperial land (Reichsland) of Alsace-Lorraine. After World War I Alsace-Lorraine was returned to France, but it was again annexed (1940–44) by Germany during World War II.

Los Al·a·mi·tos (lôs ăl′ə-mē′tōs), city (1970 pop. 11,346), Orange co., S Calif., in a farm area; inc. 1960. It has a racecourse.

Los Al·a·mos (lôs ăl′ə-mōs′), county (1970 pop. 15,198), 110 sq mi (284.9 sq km), N N.Mex., in a high plateau area largely within the Valle Grande Mts.; formed 1949 from parts of Sandoval and Santa Fe cos.; co. seat Los Alamos. The region is a center for nuclear fission research and atomic weapons development. It includes Bandolier National Monument.

Los Alamos, uninc. town (1978 est. pop. 16,800), seat of Los Alamos co., N central N.Mex. It is on a long mesa extending from the Jemez Mts. The U.S. government chose the site in 1942 for atomic research, and the first atomic bombs were produced here. In 1947 the Atomic Energy Commission took over the town. In 1962 government control ended and Los Alamos became a self-governing community. The Los Alamos Scientific Laboratory, operated by the Univ. of California, is a national historic landmark.

Los Al·tos (lôs ăl′təs), residential city (1978 est. pop. 25,400), Santa Clara co., W Calif.; inc. 1952.

Los Altos Hills, residential town (1970 pop. 6,871), Santa Clara co., W Calif., WNW of San Jose; inc. 1956.

Los An·ge·les (lôs ăn′jə-ləs, -lēz′), county (1970 pop. 7,041,980), 4,071 sq mi (10,543.9 sq km), SW Calif., in the fertile Los Angeles basin bounded on the W by the Pacific Ocean and nearly surrounded by mountains; formed 1850; co. seat Los Angeles. The county has valuable irrigated farmland, producing fruit, nuts, vegetables, poultry, dairy products, field crops, and livestock. Motion pictures, television programs, and records are among its best-known products. It also has a wide variety of manufactures, oil and gas fields, granite and stone quarries, lumber, and ocean fisheries.

Los Angeles, city (1978 est. pop. 2,765,000), seat of Los Angeles co., S Calif.; inc. 1850. It is a port of entry on the Pacific coast, with a fine harbor at San Pedro Bay. Two mountain ranges, the Santa Monica and Verdugo, cut across the center of the city. Los Angeles is the shipping, industrial, communication, financial, and distribution hub of an agricultural area that produces citrus fruit, vegetables, grains, nuts, and dairy products. It is also a leading producer of aircraft, military ordnance, glass, furniture, lumber and wood products, electrical machinery, transportation equipment, and fabricated metal.

The Los Angeles metropolitan area covers five counties and encompasses 34,000 sq mi (88,060 sq km) and nearly 10 million people. The region is a great industrial and urban complex; residues of industry, combined with a high motor-vehicle density, have created a serious problem of smog pollution. Los Angeles is the only major U.S. city without a public transportation system. Water for the city is obtained from the Colorado and Owens rivers and the Mono Basin, at distances of c.240 to c.400 mi (385–645 km).

The site of the city was visited by the Spanish explorer Gaspar de Portolá in 1769, and in 1781 El Pueblo de Nuestra Señora de los Angeles de Porciuncula was founded. The city served several times as the capital of the Spanish colonial province of Alta California and was a cattle-ranching center under Spanish and Mexican rule. In 1846 Los Angeles was captured from the Mexicans by U.S. forces. The railroads (Southern Pacific in 1876; Santa Fe in 1885) and the discovery of oil in the 1890s stimulated expansion, as did the development of the motion-picture industry in the early 20th cent.

In Los Angeles are botanical gardens; art, history, movie, industrial, and science museums; and many parks, including Griffith Park, the largest urban park in the world, with a zoo and a planetarium. The La Brea Tar Pits are one of the world's biggest sources of Ice Age fossils. There are large ethnic communities from Mexico and Latin America, India, Japan, and China, as well as a substantial American Indian population. Los Angeles has a symphony orchestra and professional baseball, football, basketball, and hockey teams. The motion-picture and television industries, the proximity of many resorts, the fine beaches, and the climate attract thousands of tourists annually. Other attractions in the region include the Santa Anita and Hollywood racetracks and Disneyland (at Anaheim). Among the city's many educational institutions are the Univ. of Southern California; the Univ. of California, Los Angeles; Occidental College; Loyola Marymount Univ.; and Pepperdine Univ.

Los Ba·nos (lôs băn′əs), city (1970 pop. 9,188), Merced co., central Calif., in the San Joaquin Valley NW of Fresno; laid out 1889, inc. 1907. It is a diversified agricultural area.

Los Ga·tos (lôs găt′əs, gä′tōs), city (1978 est. pop. 25,300), Santa Clara co., W Calif.; inc. 1887. It is a residential community in a fruit and poultry area. Wine and electronic equipment are produced.

Lo-shan or **Lo·shan** (both: lō′shän′), city, central Szechwan prov., China, just S of Ch'eng-tu, on the Min River. Machine tools and textiles are manufactured. Nearby are decorated grottoes, a colossal stone Buddha, and the sacred peak Omei.

Los Lu·nas (lôs lo͞o′nəs), village (1970 pop. 973), seat of Valencia co., W central N.Mex., SW of Albuquerque near the Rio Grande. It is a trade center in an area yielding alfalfa, grains, and sheep.

Lot (lôt), department (1975 pop. 150,725), 2,019 sq mi (5,229.2 sq km), S central France, in Quercy; capital Cahors.

Lot, river, c.300 mi (485 km) long, rising in the Cévennes Mts., SE France, and flowing W past Mende and Cahors to join the Garonne River. The limestone plateaus through which the Lot winds are intersected by fertile valleys and vineyards.

Lo·ta (lō′tä), city (1970 pop. 48,166), S central Chile, a port on the Gulf of Arauco, an inlet of the Pacific Ocean. Founded in the 17th cent., the city grew rapidly after coal was discovered in the region (1837). There are also industries producing copper and ceramics.

Lot-et-Ga·ronne (lô-tā-gä-rôn′), department (1975 pop. 292,616), 2,069 sq mi (5,358.7 sq km), SW France, in Agenais and parts of Bazadais and Bezaume; capital Agen.

Lo·thi·an (lō′*th*ē-ən), region (1976 est. pop. 755,293), S Scotland. It was organized in 1975.

Lou·don (loud′n), county (1970 pop. 24,266), 240 sq mi (621.6 sq km), E Tenn., in the Great Appalachian Valley bounded on the NW by the Clinch River and crossed by the Tennessee and Little Tennessee rivers; formed 1870; co. seat Loudon. It has agriculture (corn, tobacco, hay, fruit, livestock, and dairy products), large tracts of timber, and some manufacturing.

Loudon, farm trade town (1970 pop. 3,728), seat of Loudon co., E Tenn., SW of Knoxville on the Tennessee River; settled 1828, inc. 1860. Hosiery, furniture, and dairy and plastic products are made. Nearby is the site of Fort Loudoun (1756), which fell to the Cherokee in 1760 after a long siege.

Lou·doun (loud′n), county (1970 pop. 37,150), 517 sq mi (1,339 sq km), N Va., in a rolling piedmont area rising to the Blue Ridge in the NW and bounded on the NE by the Potomac River and the Md. border; formed 1757; co. seat Leesburg. Its agriculture includes horses, cattle, wheat, corn, tobacco, apples, dairy products, and poultry. Tourism and light industry are also important.

Lough·bor·ough (lŭf'bər-ə), municipal borough (1973 est. pop. 49,010), Leicestershire, central England, on the Soar River. It is a market town with engineering works; its products include hosiery, shoes, pharmaceuticals, boilers, and pottery. Bell foundries were built in 1840; the great bell of St. Paul's Cathedral in London was cast here in 1881.

Lou·i·sa (lōō-ē'zə). **1.** County (1970 pop. 10,682), 403 sq mi (1,043.8 sq km), SE Iowa, drained by the Iowa River; formed 1836; co. seat Wapello. Its prairie agriculture includes cattle, hogs, poultry, corn, oats, and wheat. A fertile section in the east is artificially drained by pumping ditches. It has limestone quarries. **2.** County (1970 pop. 14,004), 514 sq mi (1,331.3 sq km), central Va., bounded on the N and NE by the North Anna River; formed 1742; co. seat Louisa. It is primarily agricultural, yielding tobacco, grain, hay, dairy products, livestock, and poultry. It also has some timber.

Louisa. 1. City (1970 pop. 1,781), seat of Lawrence co., NE Ky., at the W.Va., border where the junction of the Levisa and Tug forks forms the Big Sandy River; settled c.1790. It is the trade and shipping center for a mountain farm area producing oil, natural gas, coal, fire-clay, and sand. **2.** Town (1970 pop. 633), seat of Louisa co., central Va., E of Charlottesville. It is a trade center in an agricultural and timber area. Lumber is processed and clothing manufactured.

Lou·is·burg (lōō'ĭs-bûrg'), town (1976 pop. 1,519), E Cape Breton Island, N.S., Canada. The town, an ice-free port, is near the site of the great fortress of Louisbourg, built (1720–40) by France as its Gibraltar in America. French privateers, using the harbor as a base, preyed on New England fishermen working the Grand Banks, until 1745, when a small force of New Englanders under William Pepperrell attacked Louisbourg and forced its surrender. Three years later it was returned to France by the Treaty of Aix-la-Chapelle in exchange for Madras, India, but it fell (1758) to a British land and sea attack, which reduced it to ruins. The site is a national historic park, and the buildings have been reconstructed.

Louisburg, town (1970 pop. 2,941), seat of Franklin co., NE N.C., NE of Raleigh; settled 1758, inc. 1764. It is a tobacco market.

Lou·ise, Lake (lōō-ēz'), 1.5 mi (2.4 km) long, alt. 5,680 ft (1,732.4 m), SW Alta., Canada, in the Rocky Mts., in Banff National Park. Noted for its scenic beauty, it is surrounded by high peaks, glaciers, and snow fields. The lake was discovered in 1882. It has become a popular year-round tourist and mountain-climbing center.

Lou·i·si·ade Archipelago (lōō-ē'zē-äd', -ăd'), SW Pacific, part of Papua New Guinea. The archipelago comprises c.10 volcanic islands and numerous coral reefs. Most of the islands had gold reserves, but mining largely ceased after World War II.

Lou·i·si·an·a (lōō-ē'zē-ăn'ə, lōō'ə-, lōō'ē-), state (1975 pop. 3,782,000), 48,523 sq mi (125,675 sq km), S central United States, admitted

to the Union in 1812 as the 18th state. Baton Rouge is the capital. Louisiana is bounded on the north by Ark., on the east by Miss. (the line is formed by the Mississippi River in the north, by the Pearl in the extreme south), on the south by the Gulf of Mexico, and on the west by Texas (the Sabine River marks most of the boundary).

A low country on the Gulf coastal plain and the Mississippi alluvial plain, Louisiana rises in only to some 535 ft (163.2 m). The rainy coast country contains marshes and fertile delta lands; inland are rolling pine hills and prairies. The Mississippi dominates the many waterways, including the Red, Ouachita, Atchafalaya, and Calcasieu rivers, and the coast is threaded by many slow bayous. There are lagoons such as Lake Ponchartrain, oxbow lakes made by Mississippi River cutoffs, and other lakes where the slow streams are clogged. A variety of recreational facilities makes the state an excellent vacationland.

Louisiana is divided into 64 parishes:

NAME	PARISH SEAT	NAME	PARISH SEAT
Acadia	Crowley	Madison	Tallulah
Allen	Oberlin	Morehouse	Bastrop
Ascension	Donaldsonville	Natchitoches	Natchitoches
Assumption	Napoleonville	Orleans	New Orleans
Avoyelles	Marksville	Ouachita	Monroe
Beauregard	De Ridder	Plaquemines	Pointe a la Hache
Bienville	Arcadia	Pointe Coupee	New Roads
Bossier	Benton	Rapides	Alexandria
Caddo	Shreveport	Red River	Coushatta
Calcasieu	Lake Charles	Richland	Rayville
Caldwell	Columbia	Sabine	Many
Cameron	Cameron	St. Bernard	Chalmette
Catahoula	Harrisonburg	St. Charles	Hahnville
Claiborne	Homer	St. Helena	Greensburg
Concordia	Vidalia	St. James	Convent
De Soto	Mansfield	St. John the Baptist	Edgard
East Baton Rouge	Baton Rouge	St. Landry	Opelousas
East Carroll	Lake Providence	St. Martin	St. Martinville
East Feliciana	Clinton	St. Mary	Franklin
Evangeline	Ville Platte	St. Tammany	Covington
Franklin	Winnsboro	Tangipahoa	Amite
Grant	Colfax	Tensas	St. Joseph
Iberia	New Iberia	Terrebonne	Houma
Iberville	Plaquemine	Union	Farmerville
Jackson	Jonesboro	Vermilion	Abbeville
Jefferson	Gretna	Vernon	Leesville
Jefferson Davis	Jennings	Washington	Franklinton
Lafayette	Lafayette	Webster	Minden
Lafourche	Thibodaux	West Baton Rouge	Port Allen
La Salle	Jena	West Carroll	Oak Grove
Lincoln	Ruston	West Feliciana	St. Francisville
Livingston	Livingston	Winn	Winnfield

Economy. The climate (subtropical in the south and temperate in the north), together with the rich alluvial soil, makes the state one of the nation's leading producers of sweet potatoes, rice, and sugar cane. Other major commodities are soybeans, cotton, cattle, and dairy products. Strawberries, corn, hay, pecans, and truck vegetables are also produced in quantity. Fishing is a major industry; shrimp, menhaden, and oysters are principal catches. Louisiana is a leading fur-trapping state; its marshes supply most of the country's muskrat furs. Pelts are also obtained from mink, coypus, opossums, otter, and raccoon. The state leads the nation in the production of salt and sulfur; it ranks second in the production of crude petroleum (of which many deposits are off shore), natural gas, and natural-gas liquids. Timber is plentiful; forests cover more than 50% of the land area. The state is rapidly industrializing. It has giant oil refineries, petrochemical plants, metal foundries, and sawmills and paper mills. Other industries produce foods, clay, glass, and transportation equipment. Tourism brings in more than $600 million a year.

History. Louisiana is rich in tradition and legend. Three different groups have contributed to its unique heritage: the Creoles, descendants of the original Spanish and French colonists; the Cajuns, whose French ancestors were expelled from Nova Scotia by the British in 1755; and the American cotton planters. The region was possibly visited by Cabeza de Vaca and his fellow survivors of a Spanish expedition of 1528, and it was certainly seen by some of De Soto's men (1541–42). In 1682 La Salle reached the mouth of the Missisippi and claimed for France all of the land drained by that river and its tributaries, naming it Louisiana after Louis XIV. Europeans did not permanently settle here until 1699, when Pierre le Moyne, sieur d'Iberville, founded a settlement near Biloxi. This settlement became the seat of government for Louisiana, an enormous territory embracing the entire Mississippi drainage basin. Natchitoches (the oldest settlement within the present boundaries of the state) grew from a French military and trading post established (c.1714) to protect the Red River area from the Spanish. New Orleans was founded in 1718, and in 1723 the capital was transferred here. Large numbers of Negroes were brought in as slaves. Indigo plantations and fur trading brought some wealth, but the colony did not prosper.

In order to keep the entire Louisiana territory from falling into the hands of the British at the end of the French and Indian Wars, the French secretly ceded the area west of the Mississippi River and the Isle of Orleans to Spain. By the Treaty of Paris (1763), Great Britain gained control of all Louisiana east of the Mississippi except the Isle of Orleans. Spanish control was not confirmed until 1769. During the Spanish years agriculture flourished with the cultivation of rice and sugar cane. The Spanish government welcomed thousands of Acadians, who settled what came to be known as the Cajun country.

During the American Revolution New Orleans was a center for Spanish aid to the colonies. After the war Louisiana's control of the great inland trade route, the Mississippi, led to heated controversy with the Americans. Conflicts over shipping rights were partly resolved by the Pinckney Treaty (1795), in which Spain granted Americans free navigation of the river. In 1802 it became known that Napoleon I had forced the retrocession of the territory to France, as confirmed in the secret Treaty of San Ildefonso (1800). To the surprise of the American representatives in France, Napoleon decided to sell all of Louisiana to the United States. Americans took possession in 1803, and in 1804 the territory was divided into two parts. That north of lat. 33° N (the present northern boundary of the state of Louisiana) was first called the District of Louisiana (1804–5), then the Territory of Louisiana (1805–12), and finally Missouri Territory. The southern part, which was called the Territory of Orleans, was admitted to the Union in 1812 as the state of Louisiana. Settlement (1819) of the West Florida Controversy gave Louisiana the area between the Mississippi and Pearl rivers.

After statehood French and Spanish influence remained, not only in the Creole and Cajun societies but also in the civil law (based on French and Spanish codes) and in the division of the state into parishes rather than counties. However, the whole population united behind Andrew Jackson to defeat (1815) the British at the Battle of New Orleans during the War of 1812. With settlers pouring in from other Southern states, great sugar and cotton plantations developed rapidly in the fertile lowlands, and the less productive uplands were also settled. The advent of steam propulsion on the Mississippi (1812) was a boon to the state's economy. Plantation owners, with their large landholdings and many slaves dominated politics and largely controlled the state.

On Jan. 26, 1861, Louisiana seceded from the Union and six weeks later joined the Confederacy. The fall of New Orleans to David G. Farragut in 1862 prefaced the detested military occupation under Gen. B. F. Butler. Western Louisiana remained in the hands of the Confederates. The state was readmitted to the Union in 1868. The next few years were characterized by corruption and great economic distress. Reconstruction finally ended with the disputed presidential election of 1876, when Louisiana's electoral votes were "traded" to the Republicans (whose candidate was Rutherford B. Hayes) in exchange for the withdrawal of Federal troops from the state.

Economic recovery was slow. The disrupted plantation system was largely replaced by farm tenancy and sharecropping. The pattern of Louisiana's economy was changed by the discovery of oil and natural gas in the early 1900s, and industries began to grow on the basis of cheap fuel and cheap labor. In 1928 a virtual revolution occurred when Huey P. Long was elected governor. His almost dictatorial rule brought material progress at the cost of official corruption. After his assassination in 1935 his political heirs made their peace with the New Deal, and Federal funds poured into the state. The effects of growing industrialization and natural disasters (including hurricanes and floods on the Mississippi River), are the major problems confronting Louisiana today.

Government. The state's present constitution (effective Jan., 1975) replaced the constitution of 1921, which had been amended more than 500 times. The executive branch is headed by a governor elected for a four-year term and permitted one re-election. The bicameral legislature has a senate with 39 members and a house of representatives with 105 members, all elected for four-year terms.

Educational Institutions. Among the state's more prominent are Tulane Univ., Dillard Univ., and Loyola Univ., all at New Orleans; Louisiana State Univ. and Agricultural and Mechanical College, mainly at Baton Rouge; and Louisiana Tech Univ., at Ruston.

Louisiana Purchase, extending from the Mississippi River to the Rocky Mts. and from the Gulf of Mexico to British North America. Acquired by an 1803 treaty with France, it included the former Spanish region of Louisiana (ceded secretly to France in 1800). The acquisition doubled the national domain, increasing it c.828,000 sq mi (c.2,144,520 sq km).

Lou·is·ville (lo͞o'ē-vĭl'). **1.** City (1970 pop. 2,691), seat of Jefferson co., E central Ga., SW of Augusta on the Ogeechee River, in a farm area; laid out 1786 as the prospective capital of the state. The statehouse was completed in 1796, and the legislature held its last session in Louisville in 1805. **2.** Village (1970 pop. 1,020), seat of Clay co., S central Ill., W of Olney, in an agricultural, oil, and natural-gas area. Corn, wheat, poultry, fruit, and truck crops are grown. **3.** City (1978 est. pop. 319,800), seat of Jefferson co., NW Ky., at the Falls of the Ohio; inc. 1780. It is a port of entry and an important industrial, financial, marketing, and shipping center. The city has some of the nation's largest whiskey distilleries and cigarette factories. There are also railroad shops, sawmills, meat-packing houses, glassworks, chemical plants, and a wide variety of other industries. A settlement grew after George Rogers Clark built (1778) a fort as a base of operations against the British and the Indians. The city was chartered by the Va. legislature in 1780. During the Civil War it was a center of pro-Union activity and a supply base for Federal forces. The Univ. of Louisville is here. Churchill Downs, a noted racetrack and scene of the famous annual Kentucky Derby (first held in 1875), is in Louisville. **4.** City (1970 pop. 6,626), seat of Winston co., E central Miss., SW of Columbus; inc. 1836. It is a processing center for a farm, dairy, and timber area. **5.** Village (1970 pop. 6,298), Stark co., NE Ohio, ENE of Canton, in a dairying area; inc. 1860. Foundry products, clothing, and structural steel are among its manufactures.

Loup (lo͞op), county (1970 pop. 854), 574 sq mi (1,486.7 sq km), central Nebr., drained by the Calamus and North Loup rivers; formed 1883; co. seat Taylor. Livestock and grain are its major products.

Loup, river, formed in E central Nebr. by the junction of the North Loup, 212 mi (341.1 km) long, and the Middle Loup, c.221 mi (355 km) long. It flows c.68 mi (110 km) east to the Platte River at Columbus and is used for power, recreation, and irrigation.

Loup City, city (1970 pop. 1,456), seat of Sherman co., central Nebr., NW of Grand Island on the Middle Loup River; settled 1873. Dairy products are made here.

Lourdes (lo͝ord, lo͝ordz), town (1975 pop. 17,870), Hautes-Pyrénées dept., SW France, at the foot of the Pyrenees. It is famous for its Roman Catholic shrine where Our Lady of Lourdes is believed to have appeared (1858) to St. Bernadette. Millions of people make the pilgrimage to Lourdes each year, drawn by their faith in the miraculous cures attributed to the waters of the shrine.

Lou·ren·ço Mar·ques (lō-rĕn'sō mär'kĕs, lô-), city (1972 est. pop. 230,000), capital of Mozambique, a port on the Indian Ocean. The economy is dominated by the modern port on Delagoa Bay; coal, cotton, sugar, chrome, ore, sisal, copra, and hardwood are the chief exports. The city's main manufactures are food products, beverages, cement, pottery, furniture, shoes, and rubber. Founded in the late 18th cent., the city is named for a Portuguese trader, who first explored the area in 1544. Its main growth dates from 1895, when a railroad to Pretoria, South Africa, was completed.

Louth (louth, lou*th*), county (1971 pop. 74,941), 317 sq mi (821 sq km), NE Republic of Ireland; co. town Dundalk. It borders on the Irish Sea from the mouth of the Boyne River to Carlingford Lough. The terrain is an undulating plain except for a hilly district in the north. Among the industries are cotton and linen manufacturing, brewing, and food processing.

Louth (louth), municipal borough (1971 pop. 11,746), Lincolnshire, E England, on the Lud River. The town's industries include trading and processing of farm produce and the manufacture of agricultural implements, malt, and lime. There are ruins of a Cistercian abbey founded in 1139.

Lou·vain (lo͞o-văɴ'), city (1970 pop. 30,623), Brabant prov., central Belgium, on the Dijle River. It is a commercial, industrial, and cultural center and a rail junction. Mentioned in the 9th cent., Louvain was a center of the wool trade and of the cloth industry in the Middle Ages. In the 14th cent. there was also much strife between the nobles and the weavers, and after the nobles gained sway (1383) most of the weavers emigrated to Holland and England, and the city declined. In 1426 Duke John IV of Brabant founded a famous Roman Catholic university here. In 1968, as a result of a long-standing dispute between Flemish and French-speaking sectors, the university was divided into two autonomous units.

Lou·vière, La (lä lo͞o-vyâr′), town (1977 pop. 76,167), Hainaut prov., S Belgium. It is an industrial center of the Bassin du Centre coal-mining region. Manufactures include steel and machinery.

Love (lŭv), county (1970 pop. 5,637), 513 sq mi (1,328.7 sq km), S Okla., bounded on the S by the Red River, here forming the Texas border; formed 1907; co. seat Marietta. In an irrigated agricultural area, its economy is based mainly on cattle ranching. It also has a number of oil and gas wells.

Love·land (lŏv′lənd). **1.** City (1978 est. pop. 28,400), Larimer co., N Colo.; inc. 1881. It is a food-processing center in a fertile farm area irrigated by the Colorado-Big Thompson project, as well as a growing industrial hub. **2.** City (1970 pop. 7,126), Clermont, Hamilton, and Warren cos., SW Ohio, on the Little Miami River NE of Cincinnati; settled c.1825 as Paxton, inc. 1876 as Loveland.

Love·lock (lŭv′lŏk′), city (1970 pop. 1,571), seat of Pershing co., W Nev., on the Humboldt River NE of Reno; settled c.1860, inc. 1919. A station on the old trail along the Humboldt, it is now the center of a ranch and mine area.

Loves Park (lŭvz), city (1970 pop. 12,390), Winnebago co., N Ill., on the Rock River; inc. 1947. It is chiefly residential.

Lov·ing (lŭv′ĭng), county (1970 pop. 164), 647 sq mi (1,675.7 sq km), W Texas, in the high rolling prairies bordered on the N by N.Mex., on the W by the Pecos River; formed 1887; co. seat Mentone. Cattle, horses, mules, and poultry are raised.

Lov·ing·ston (lŭv′ĭng-stən), village (1970 pop. 400), seat of Nelson co., central Va., NE of Lynchburg, in an apple-growing area.

Lov·ing·ton (lŭv′ĭng-tən), town (1970 pop. 8,915), seat of Lea co., SE N.Mex., near the Texas border NE of Carlsbad; founded 1908. It is a rail and trade center for an oil, cattle, and farm area.

Low Archipelago (lō): *see* Tuamotu islands.

Low Countries, region of NW Europe comprising the Netherlands, Belgium, and the grand duchy of Luxembourg. The northern parts of the Netherlands and Belgium form a low plain bordering on the North Sea, but southern Belgium and Luxembourg are part of the Ardennes plateau. One of the wealthiest areas of medieval and modern Europe, it also has been chronically a theater of war.

Low·ell (lō′əl), city (1978 est. pop. 90,100), Middlesex co., NE Mass., at the confluence of the Merrimack and Concord rivers; settled 1653, set off from Chelmsford 1826, inc. as a city 1836. Its manufactures include electronic and electrical equipment, textiles, rubber products, chemicals, machine parts, foodstuffs, shoes, and plastics. The city developed after textile mills were built at Pawtucket Falls, and it became one of the great textile centers of the country.

Low·er Aus·tri·a (lō′ər ô′strē-ə), province (1971 pop. 1,414,161), c.7,400 sq mi (19,165 sq km), NE Austria, bordering on Czechoslovakia in the N and NE; prov. seat Vienna. It is a picturesque, hilly region, drained by the Danube River and containing peaks of the Eastern Alps and the Wienerwald. The province is noted for its grain production and its wines.

Lower Bur·rell (bûr′əl, bə-rĕl′), city (1970 pop. 13,654), Westmoreland co., SW Pa., NE of Pittsburgh; inc. 1959. Steel is produced.

Lower Cal·i·for·ni·a (kăl′ə-fôr′nyə, -fôr′nē-ə): *see* Baja California.

Lower Hutt (hŭt), city (1976 pop. 64,553), S North Island, New Zealand, at the mouth of the Hutt River, near Wellington. It is a manufacturing city with automobile assembly plants.

Lower Saint Croix National Scenic River (sānt kroi′): *see* National Parks and Monuments Table.

Lower Sax·o·ny (săk′sə-nē), state (1974 est. pop. 7,264,800), 18,295 sq mi (47,384 sq km), N West Germany; capital Hanover. Situated on the North German plain, it is mountainous in the south; heaths and moors form the central belt. Farming and cattle raising are important occupations. Industry (including the manufacture of iron and steel, textiles, machinery, food products, and chemicals) is well developed. Lower Saxony has had no historic unity since 1180, when Emperor Frederick I broke up the duchy of Henry the Lion of Saxony, of which it was a part. The current state was formed in 1946 by the merger of the Prussian province of Hanover and the states of Brunswick, Oldenburg, and Schaumburg-Lippe.

Lowe·stoft (lōs′tôft′, -təf), municipal borough (1973 est. pop. 53,260), Suffolk, the easternmost town in England. It is a popular seaside resort and has fishing, shipbuilding, food-processing, and other light industries. Oliver Cromwell took the town in 1643, and in 1665 the coastal waters were the scene of a naval victory of the English under the duke of York (later James II) over the Dutch. Most of the old houses were destroyed by a fire in the 17th cent., but St. Margaret's Church, from the 15th cent., has an older tower. Fine bone china was produced in Lowestoft from about 1750 to 1800.

Lowndes (loundz). **1.** County (1970 pop. 12,897), 716 sq mi (1,854.4 sq km), S central Ala., in the Black Belt, bounded on the N by the Alabama River; formed 1830; co. seat Hayneville. It is in a cotton, corn, and dairy-farming region, with lumber logging, some mining, and textile manufacturing. **2.** County (1970 pop. 55,112), 506 sq mi (1,310.5 km), S Ga., bounded in the S by the Fla. border and drained by the Withlacoochee River; formed 1825; co. seat Valdosta. Coastal-plain agriculture (tobacco, cotton, peanuts, corn, watermelons, and livestock) and forestry are important to its economy. **3.** County (1970 pop. 49,700), 508 sq mi (1,315.7 sq km), E Miss., bordered on the E by Ala. and drained by the Tombigbee River; formed 1830; co. seat Columbus. It has cotton, corn, and hay farming, dairying, and stock raising. Food processing and lumbering are done. Textiles, clothing, wood products, furniture, chemicals, cement, concrete, and farm machinery are manufactured.

Low·ville (lou′vĭl′), village (1970 pop. 3,671), seat of Lewis co., N N.Y., SE of Watertown, in a farm area; inc. 1871. Cheese and wood products are made.

Loy·al·ty Islands (loi′əl-tē), coral group, S Pacific, a part of the French overseas territory of New Caledonia. The group comprises three islands and many islets, and has a total land area of c.800 sq mi (2,070 sq km). The chief exports are copra, rubber, and sugar cane.

Lo-yang or **Lo·yang** (both: lō′yäng′), city (1970 est. pop. 750,000), NW Honan prov., China, on the Lo River. Manufactures include ball and roller bearings, tractors, heavy machinery, glass, cement, and textiles. Lo-yang, a major Chinese cultural center, was the capital of several ancient dynasties, particularly that of the Eastern Chou kingdom (770–256 B.C.) and the T'ang dynasty (A.D. 618–906). The nearby Lung-men grottoes, embellished in the 6th cent. A.D., contain colossal carvings of Buddha.

Lo·zère (lô-zĕr′), department (1975 pop. 74,825), 1,995 sq mi (5,167 sq km), S central France; capital Mende.

Lu·an·da (lo͞o-ăn′də), city (1970 pop. 475,328), capital of Angola, a port on the Atlantic Ocean. Manufactures include processed foods, beverages, textiles, cement and other construction materials, plastic products, metalware, cigarettes, and shoes. Petroleum, found nearby, is refined in the city. Luanda has a natural harbor, with a modern port. The chief exports are coffee, cotton, sugar, palm products, and manganese ore. Founded in 1575 by the Portuguese, the city has been the administrative center of Angola since 1627 (except for 1640–48). From c.1550 to c.1850 it was the center of a large slave trade to Brazil.

Luang Pra·bang (lwäng′ prä-bäng′), city (1973 est. pop. 43,000), capital of Luang Prabang prov. and the royal capital of Laos, NW Laos, on the Mekong River. It is a river port and a market for rubber, rice, teak, and fish. Zinc is mined nearby. According to tradition, Luang Prabang was founded by Indian Buddhist missionaries. For several centuries it was the center of a Laotian-Thai kingdom that controlled most of Laos and parts of Siam.

Lu·ang·wa (lo͞o-äng′wä), river, E Zambia, S central Africa, rising in several headstreams near the Mozambique border and flowing c.500 mi (805 km) SSW to the Zambezi River. It forms part of the border between Zambia and Mozambique.

Lu·an·shya (lo͞o-än′shä), city (1976 est. pop. 135,000), N central Zambia, near Zaire. It is a copper-mining center.

Lu·bang (lo͞o-bäng′), island group, 95 sq mi (246 sq km), Mindoro prov., S China Sea, Philippines, separated from Luzon by the Verde Island Passage. Rice growing and stock raising are the major economic endeavors by the native Tagalogs.

Lub·bock (lŭb′ək), county (1970 pop. 179,295), 892 sq mi (2,310.3 sq km), NW Texas, drained by the intermittent Double Mountain Fork of the Brazos River; formed 1876; co. seat Lubbock. One of the state's leading agricultural counties, it produces cotton, sweet and grain sorghums, wheat, forage, fruit, truck crops, sugar beets, dairy products, poultry, beef cattle, sheep, and hogs.

Lubbock, city (1978 est. pop. 169,000), seat of Lubbock co., NW Texas; inc. 1909. In the Llano Estacado on a branch of the Brazos River, Lubbock was settled in 1879 by Quakers. It is the trade center for a cotton- and grain-growing region. Its manufactures include sprinkler heads, earth-moving equipment, mobile and modular homes, sausage casings, and sheets. Lubbock Lake Site is an important geological formation. Reese Air Force Base is nearby.

Lü·beck (lü'bĕk'), city (1974 est. pop. 234,510), Schleswig-Holstein, NE West Germany, on the Trave River near its mouth on the Baltic Sea. It is a major port and a commercial and industrial center. The city contains foundries, machinery plants, textile mills, and large shipyards; it is also noted for the manufacture of marzipan.

Known in the 11th cent., Lübeck was destroyed by fire in 1138 but was refounded in 1143. It was acquired and chartered by Henry the Lion c.1158; the charter, which granted far-reaching communal rights, was copied by more than 100 other German cities. In 1226 Frederick II made Lübeck a free imperial city. Ruled by a merchant aristocracy, it soon rose to great commercial prosperity, acquired hegemony over the Baltic trade, and headed the Hanseatic League. However, the rise of the maritime powers of Denmark and Sweden and the revolution in commerce caused by the development of the Americas resulted in the decline of the League and, with it, of Lübeck. In the French Revolutionary Wars Lübeck was sacked by French troops in 1803, and occupied by them in 1806. Lübeck joined the North German Confederation and later the German Empire as a free Hanseatic city; it retained that status until 1937, when it was incorporated into Schleswig-Holstein. Despite heavy damage by bombing in World War II, the inner city of Lübeck remains one of the finest examples of medieval Gothic architecture in northern Europe. Among the buildings that have been restored are the city hall (13th–15th cent.), the churches of St. Catherine and St. Jacob (both 14th cent.), the Hospital and Church of the Holy Ghost (13th cent.), the Holstentor (completed 1477), an imposing city gate flanked by two round towers, the cathedral (founded in 1173), and the large brick Church of St. Mary (13th–14th cent.).

Lub·lin (lo͞o'blēn), city (1975 est. pop. 272,000), SE Poland. Manufactures include textiles, automobiles, trucks, agricultural machinery, and electrical products. One of the oldest Polish towns, Lublin was the meeting place of several diets (16th–18th cent.), one of which united (1569) Poland with Lithuania. Lublin passed to Austria in 1795 and to Russia in 1815. In 1944 it was the seat of a provisional government rivaling the Polish government-in-exile in London.

Lu·bum·ba·shi (lo͞o'bo͝om-bä'shē), formerly **E·lis·a·beth·ville** (ĭ-lĭz'ə-bəth-vĭl'), city (1974 est. pop. 404,000), capital of Shaba region, SE Zaire, near the border with Zambia. Copper is smelted, and textiles, food products and beverages, printed materials, and bricks are manufactured. Founded in 1910, Lubumbashi was the capital of the secessionist state of Katanga (1960–63) and was the scene of bloody strife between UN troops and Katangan forces.

Lu·ca·ni·a (lo͞o-kā'nē-ə), ancient region of S Italy. It was bounded on the east by the Gulf of Tarentum (now Taranto) and by Apulia, on the north by Samnium and Campania, on the west by the Tyrrhenian Sea, and on the south by Bruttium. Italic tribes and Greek colonists lived in the region before the Roman conquest in the 3rd cent. B.C.

Lucania, Mount, 17,147 ft (5,229.8 m) high, in the St. Elias Mts., SW Yukon Territory, Canada, near the Alaska line.

Lu·cas (lo͞o'kəs). **1.** County (1970 pop. 10,163), 434 sq mi (1,124.1 sq km), S Iowa, drained by the Chariton River; formed 1846; co. seat Chariton. It is in a prairie agricultural area yielding hogs, cattle, poultry, corn, and hay. Bituminous-coal deposits are in the eastern section. **2.** County (1970 pop. 483,594), 343 sq mi (888.4 sq km), NW Ohio, bounded on the W by the Mich. border, on the SE by the Maumee River, and on the NE by Lake Erie; formed 1835; co. seat Toledo. It has agriculture (wheat, corn, oats, and soybeans), extensive transportation facilities, and diversified manufacturing (glass, spark plugs, scales, and elevators). Fallen Timbers State Park and the site of Fort Meigs are within the county.

Luc·ca (lo͞o'kä), city (1976 est. pop. 91,658), capital of Lucca prov., Tuscany, N central Italy, near the Ligurian Sea. It is a commercial and industrial center and an agricultural market (olive oil, wine, and tobacco). Manufactures include textiles, paper, and food products. A Ligurian settlement, later a Roman town, Lucca became (6th cent.) the capital of a Lombard duchy and (12th cent.) a free commune, which soon developed into a republic. In spite of ruthless strife between Guelphs and Ghibellines and frequent wars (especially with Pisa and Florence) the city prospered. Numerous churches, showing Pisan influence, were built from the 12th to the 14th cent. Lucca remained an independent republic until Napoleon I made it a principality (1805). In 1817 Lucca became part of the duchy of Parma and in 1847 of the grand duchy of Tuscany; in 1860 it was annexed to the kingdom of Sardinia.

Luce (lo͞os), county (1970 pop. 6,789), 906 sq mi (2,346.5 sq km), NE Upper Peninsula, Mich., bounded on the N by Lake Superior and drained by the Tahquamenon River; formed 1887; co. seat Newberry. It is in a forest and farm area, growing potatoes, celery, beans, and strawberries. There are sawmills and hunting and fishing areas here. Some manufacturing is done.

Luce·dale (lo͞os'dāl'), town (1970 pop. 2,083), seat of George co., SE Miss., near the Ala. border SE of Hattiesburg, in a farming area.

Lu·ce·ra (lo͞o-chĕr'ə), town (1976 est. pop. 27,700), Apulia, S Italy. It is an agricultural and industrial center. Already important in the 4th cent. B.C., the town was destroyed by the Byzantines in the 7th cent. A.D. It was revived (13th cent.) by Emperor Frederick II, who built a great castle (now in ruins).

Lu·cerne (lo͞o-sûrn'), canton (1977 est. pop. 291,100), 576 sq mi (1,491.8 sq km), central Switzerland. Lucerne is mainly an agricultural and pastoral region, with orchards and large forested areas. Its capital is Lucerne (1970 pop. 69,879), which is on both banks of the Reuss River where it flows out of the Lake of Lucerne. One of the largest resorts in Switzerland, Lucerne has manufactures of textiles, metal goods, and chemicals, and has a printing industry. The city grew around the monastery of St. Leodegar, founded in the 8th cent. An important trade center on the St. Gotthard route, it became a Hapsburg possession in 1291. Lucerne joined the Swiss Confederation in 1332 and gained full freedom after the battle of Sempach (1386). It became capital of the Helvetic Republic in 1798.

Lucerne, Lake of, irregular-shaped lake, 44 sq mi (114 sq km), central Switzerland. Surrounded by mountains, the Lake of Lucerne is noted for its scenic beauty; many resort towns are along its shores.

Luck·now (lŭk'nou'), city (1971 pop. 750,512), capital of Uttar Pradesh state, N central India. An educational and cultural center, it also has varied manufactures. It was the capital of the kingdom of Oudh (1775–1856) and then of Oudh prov. It became the capital of the United Provinces when Agra and Oudh merged in 1877. In the Indian Mutiny the British garrison in Lucknow suffered heavy casualties during a five-month siege (June–Nov., 1857). Lucknow was a focus of the movement (1942–47) for an independent Pakistan.

Lü·den·scheid (lü'dən-shīt), city (1974 est. pop. 78,002), North Rhine-Westphalia, W West Germany. It is an industrial center; manufactures include metal products and household goods. Lüdenscheid was chartered in 1287.

Lü·de·ritz (lü'dər-ĭts), town (1970 pop. 6,642), SW Namibia, on Lüderitz Bay, an arm of the Atlantic Ocean. Fish and lobsters are processed in the city, and diamonds are mined nearby.

Lu·dhi·a·na (lo͞o'dē-ä'nä), city (1971 pop. 401,124), Punjab state, NW India. It was founded in the late 15th cent.; today it is an industrial center. Hosiery, cotton textiles, bicycle parts, and sewing machines are the important manufactures.

Lud·ing·ton (lŭd'ĭng-tən), city (1970 pop. 9,021), seat of Mason co., N Mich., on Lake Michigan N of Muskegon; settled 1847, inc. 1873. It is a car-ferry port in a lake resort region and has light industry.

Lud·low (lŭd'lō). **1.** City (1970 pop. 5,815), Kenton co., N central Ky., a suburb W of Covington on the Ohio River; settled c.1790, inc. as a village 1864, as a city c.1920. Among its products are furniture and machinery. **2.** Town (1970 pop. 17,580), Hampden co., SW Mass., on the Chicopee River; settled c.1750, set off from Springfield 1774, inc. 1775. Its manufactures include brassieres, shoelaces, and electronics and television components.

Lu·do·wic·i (lo͞o'də-wĭs'ē), city (1970 pop. 1,419), seat of Long co., SE Ga., SW of Savannah, in an agricultural area. Naval stores and lumber are produced here.

Lud·wigs·burg (lo͞od'vĭgz-bûrg', lo͞ot'vĭkhs-bo͝orkh'), city (1970 pop. 78,019), Baden-Württemberg, SW West Germany, near the Neckar River. It is a transportation and industrial center. Manufactures include machine tools, metal goods, chemicals, porcelain, and organs. Ludwigsburg grew around the large baroque castle built 1704–33 in imitation of Versailles.

Lud·wigs·ha·fen am Rhein (lo͝od'vĭgz-hä'fən äm rīn', lo͞ot' vĭKHs-), city (1970 pop. 176,031), Rhineland Palatinate, W West Germany, a port on the left bank of the Rhine River. The city is a major transshipment point and is a leading center of the West German chemical industry. Machinery and motor vehicles are also produced. Founded as a small fortress in the 17th cent., Ludwigshafen was badly damaged in World War II and was the scene (1948) of a disastrous explosion of several chemical plants.

Luf·kin (lŭf'kən), city (1978 est. pop. 28,100), seat of Angelina co., E Texas; inc. 1890. Situated in the deep pine woods, it is the core of a region of forest industries with many sawmills and the first plant to make newsprint from native pine. There are numerous other manufactures, including oil-field equipment, engines, gears, and iron and steel castings. Fuller's earth is found in the region.

Lu·ga·no (lo͞o-gä'nō), city (1977 est. pop. 28,400), Ticino canton, S Switzerland, near the Italian border, on the Lake of Lugano. A commercial center in the Middle Ages, Lugano today is a popular resort noted for its scenery and climate.

Lu·go (lo͞o'gō), city (1975 est. pop. 56,900), capital of Lugo prov., NW Spain, in Galicia, on the Miño River. The city is the processing and trade center for a fertile farm area. It has well-preserved Roman walls (3rd cent. B.C.) and a 12th cent. cathedral (restored).

Lu·le·å (lo͞o'lĕ-ô'), city (1975 est. pop. 66,290), capital of Norrbotten co., NE Sweden, a port on the Gulf of Bothnia at the mouth of the Luleälv River. Although its harbor is icebound most of the winter, large quantities of iron ore and timber are exported. The city has a large smelting plant and railroad shops. Luleå was chartered by Gustavus II in 1621. It was destroyed by fire in 1887, and was rebuilt in a modern style.

Lu·le·älv (lo͞o'lĕ-ôlv'), river, c.275 mi (445 km) long, rising near the Norwegian border, Norrbotten prov., N Sweden, and flowing SE to the Gulf of Bothnia at Luleå. It has spectacular falls.

Lu·lua·bourg (lo͞o-lwä-bo͝or'): *see* Kananga, Zaire.

Lum·ber (lŭm'bər), river, 125 mi (201.1 km) long, rising in central N.C., and flowing SE past Lumberton, then SSW into S.C. to join the Little Pee Dee River.

Lum·ber·ton (lŭm'bər-tən), city (1970 pop. 16,961), seat of Robeson co., S N.C., on the Lumber River; founded 1787, inc. 1852. It is a tobacco market and has textile and lumber mills.

Lump·kin (lŭmp'kən), county (1970 pop. 8,728), 292 sq mi (756.3 sq km), N Ga., in the Blue Ridge, drained by the Chestatee and Etowah rivers; formed 1832; co. seat Dahlonega. Farming (cotton, corn, hay, and potatoes), lumbering, and gold mining are important.

Lumpkin, city (1970 pop. 1,431), seat of Stewart co., SW Ga., near the Chattahoochee River SSE of Columbus; inc. 1831. Nearby are the Providence Caverns, wide, eroded gullies cut through layers of sedimentary soil.

Lu·na (lo͞o'nə), county (1970 pop. 11,706), 2,957 sq mi (7,658.6 sq km), SW N.Mex., in a mountainous area bounded on the S by the Mexican border; formed 1901; co. seat Deming. Its economy depends on livestock and grain.

Lund (lŭnd), city (1975 est. pop. 76,284), Malmöhus co., S Sweden. It is a commercial and industrial center and a rail junction. Manufactures include paper, printed materials, and clothing. Mentioned (c.920) in the sagas as Lunda, it became the Roman Catholic archiepiscopal see for Scandinavia in 1103–4. The city declined after it became (1536) a Lutheran bishopric, and it was devastated during the Danish-Swedish wars of the 17th cent. It passed definitively to Sweden in 1658. Lund has a fine 11th cent. Romanesque cathedral and a museum of folk customs.

Lun·dy Isle (lŭn'dē), 3 mi (4.8 km) long, off Devonshire, SW England, at the mouth of the Bristol Channel. Inhabited in prehistoric times, the island was a stronghold of pirates and smugglers from the Middle Ages until the 18th cent.

Lü·ne·burg (lo͞o'nə-bûrg', lü'nə-bo͝orKH'), city (1974 est. pop. 65,301), Lower Saxony, N West Germany, on the Ilmenau River. It is a rail junction and river port. There are large saltworks and chemical and textile industries in the city. Its hot salt springs and mud baths have long been frequented. Dating from the 10th cent., Lüneburg was an important member of the Hanseatic League. Predominately built in the late-Gothic and Renaissance styles, the city has several fine churches, a large city hall (begun 13th cent., additions as late as the 18th cent.), and many gabled houses.

Lü·nen (lü'nən), city (1975 pop. 85,876), North Rhine-Westphalia, N central West Germany, on the Lippe River. Its manufactures include machinery, metal fittings, motor vehicles, textiles, and aluminum. Coal is mined. Lünen was chartered in 1265.

Lun·en·burg (lŭn'ən-bûrg'), county (1970 pop. 11,687), 443 sq mi (1,147.4 sq km), S Va., in a rolling agricultural region bounded on the S by the Meherrin River, on the N by the Nottoway River; formed 1746; co. seat Lunenburg. Tobacco, grain, cattle, and poultry are important to its economy. It also has some industry.

Lunenburg. 1. Residential town (1970 pop. 7,419), Worcester co., N Mass., E of Fitchburg; settled 1721, inc. 1728. **2.** Village, seat of Lunenburg co., S Va., SW of Victoria and WSW of Petersburg.

Lu·né·ville (lü-nā-vēl'), town (1975 pop. 22,709), Meurthe-et-Moselle dept., NE France, on the Meurthe River in Lorraine. It is known for its crockery. Railroad equipment, textiles, and wooden toys are also made. The treaty signed here in 1801 between France and Austria supplemented the terms of the Treaty of Campo Formio.

Lu·ray (lo͞o-rā'), tourist town (1970 pop. 3,612), seat of Page co., N Va., in the Shenandoah valley, in a farm area; inc. 1812. The Luray Caverns, discovered in 1878, are remarkable for the beauty and color of their large stalagmite and stalactite formations and pools.

Lur·gan (lûr'gən), municipal borough (1971 pop. 24,055), Co. Armagh, central Northern Ireland, near Lough Neagh. A textile center since the 17th cent., Lurgan also has factories for the production of machinery. The town was founded by settlers from England.

Lu·sa·ka (lo͞o-sä'kə), city (1976 est. pop. 483,000), alt. 4,200 ft (1,281 m), capital of Zambia, S central Zambia. A sprawling city located in a productive farm area, Lusaka is an administrative, financial, and commercial center. Manufactures include foodstuffs, beverages, clothing, and cement (made from limestone quarried nearby). Lusaka was founded by Europeans in 1905 and was named after the headman of a nearby African village. Its main growth occurred after 1935, when it replaced Livingstone as the capital of the British colony of Northern Rhodesia. The Univ. of Zambia (1965) is here.

Lu·sa·ti·a (lo͞o-sā'shē-ə, -shə), region of E East Germany and SW Poland. It extends north from the Lusatian Mts., at the Czechoslovak border, and west from the Oder River. The hilly and fertile southern section is known as Upper Lusatia, the sandy and forested northern part as Lower Lusatia. Forestry, farming, and stock raising are the chief occupations. There are lignite mines, textile mills, and glass-making factories. The Lusatians are descended from the Slavic Wends, and part of the population still speaks Wendish and has preserved traditional dress and customs. The region was colonized by the Germans beginning in the 10th cent. and was constituted into the margraviates of Upper and Lower Lusatia. Both margraviates changed hands frequently among Saxony, Bohemia, and Brandenburg. In 1346 several towns of the region formed the Lusatian League and preserved considerable independence. Under the Treaty of Prague (1635) all of Lusatia passed to Saxony. The Congress of Vienna awarded (1815) Lower Lusatia and a large part of Upper Lusatia to Prussia.

Lü-shun (lo͞o'sho͞on'), formerly **Port Ar·thur** (pôrt är'thər), city, SW Liaoning prov., China, at the tip of the Liao-tung peninsula. It has been combined with Ta-lien into the joint municipality of Lü-ta. Lüshun is an important naval base dominating the entrance to the Po Hai. The city was the administrative center of the Liao-tung leasehold from 1898 to 1945. As a Russian base (1898–1905), it was the site on Feb. 8, 1904, of the surprise Japanese naval attack that precipitated the Russo-Japanese War. The city passed to Japan by the Treaty of Portsmouth (1905). In 1945 it became the headquarters of the Port Arthur Naval Base District under joint Sino-Soviet administration. China regained exclusive control in 1955.

Lu·si·ta·ni·a (lo͞o'sĭ-tā'nē-ə), Roman province in the Iberian Peninsula. As constituted (A.D. c.5) by Augustus it included all of modern central Portugal as well as much of western Spain. The province took its name from the Lusitani, a group of warlike tribes who resisted Roman domination until their leader, Viriatus, was killed (139 B.C.) by treachery.

Lusk (lŭsk), town (1970 pop. 1,495), seat of Niobrara co., E central Wyo., on the Niobrara River near the Nebr. border E of Casper; settled in the late 1880s, inc. 1898. It is the trade and processing center of a ranching, dry-farming, and oil region.

Lü-ta (lo͞o'dä'), city (1970 est. pop. 4,000,000), S Liaoning prov., China, at the tip of the Liaotung peninsula. It comprises the municipalities of Lü-shun and Ta-lien. Lü-ta has notable vineyards and fields that produce winter wheat and cotton.

Lu·te·tia (lo͞o-tē'shə), ancient name for Paris, France.

Lu·ton (lo͞ot'n), county borough (1976 est. pop. 164,500), Bedfordshire, S central England on the Lea River. Automobiles, ball bearings, and aircraft parts are among the products manufactured in Luton. It is also the center of the English millinery industry, established in Luton in the time of James I.

Lutsk (lo͞otsk), city (1976 est. pop. 128,000), capital of Volyn oblast, SW European USSR, in the Ukraine, on the Styr River. A river port, it has industries producing machinery, food products, textiles, and shoes. First mentioned in 1085, Lutsk was part of Kievan Russia until 1154, when it became the capital of the Lutsk independent principality. It was taken by Lithuania in the 14th cent., and was an important trade city from the 14th to the 16th cent. Lutsk was part of Poland from the second half of the 16th cent., was taken by Russia in 1791, was Polish again from 1919 to 1939, and was ceded to the Ukrainian SSR in 1939.

Lüt·zen (lüt'sən), town (1965 est. pop. 4,800), Leipzig district, S central East Germany. In the Thirty Years' War Gustavus II of Sweden defeated (1632) Gen. Albrecht Wallenstein, but died in battle.

Lu·verne (lo͞o'vərn). **1.** Town (1970 pop. 2,440), seat of Crenshaw co., S central Ala., NNE of Andalusia, in a cotton and lumber area; inc. 1891. **2.** City (1970 pop. 4,703), seat of Rock co., extreme SW Minn., SSW of Pipestone; settled 1867, platted 1870, inc. as a village 1877, as a city 1904. It is in a farm area, and there are granite quarries nearby.

Lux·em·bourg (lŭk'səm-bûrg', lük-säɴ-bo͝or') or **Lux·em·burg** (lŭk'səm-bûrg', lo͝ok'səm-bo͝ork'), grand duchy (1972 est. pop. 343,000), 998 sq mi (2,584.8 sq km), W Europe. Roughly triangular, it borders on Belgium in the west and north, West Germany in the east, and France in the south. The city of Luxembourg is the capital. The grand duchy is a constitutional monarchy with a bicameral legislature, the upper chamber being appointed by the sovereign, the lower chamber elected by direct universal suffrage. Luxembourg is drained by the Sauer and Alzette rivers, both tributaries of the Moselle, which forms part of its eastern border. The Ardennes Mts. extend into northern Luxembourg.

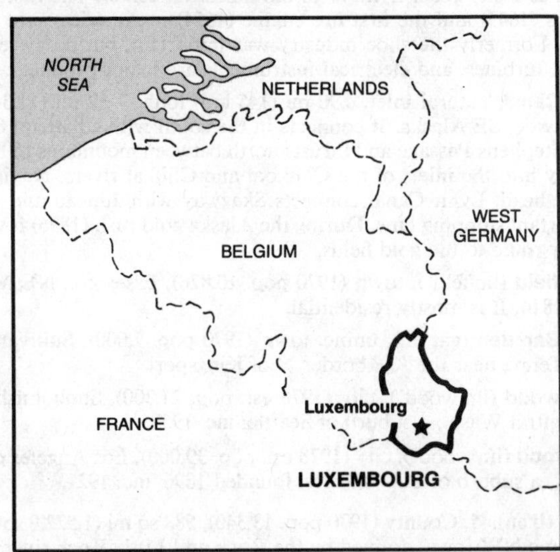

Economy. Luxembourg is a major iron and steel producer. Other industries are food processing, tanning, and the production of textiles, chemicals, and cement. Grains and potatoes are grown and livestock are raised. Iron and steel products are the main exports; imports include fuel, food, cloth, and manufactured goods.

History. The county of Luxembourg was one of the largest fiefs in the Holy Roman Empire and rose to prominence when its ruler was elected emperor as Henry VII in 1308. Made a duchy in 1354, it passed to the House of Burgundy in 1443 and to the House of Hapsburg in 1482. For the following three centuries it shared the history of the southern Netherlands, passing from Spanish to Austrian rule in 1714. The duchy was formally ceded to France in 1797 and made a grand duchy in union with the Netherlands in 1815. The major part of the grand duchy was obtained by newly independent Belgium in 1839; the remainder became an autonomous member of the German Confederation while remaining formally a possession of the Dutch royal family. In 1867 the European powers declared Luxembourg a neutral territory, and Prussian forces withdrew.

Germany occupied Luxembourg during World Wars I and II. In 1922 the grand duchy formed an economic union with Belgium, which was widened to include the Netherlands in the Benelux Economic Union of 1958. In 1961 Prince Jean (b.1921) was made head of state; he became Grand Duke in 1964.

Luxembourg, province (1976 est. pop. 220,259), 1,706 sq mi (4,418.5 sq km), SE Belgium, in the Ardennes, bordering on the grand duchy of Luxembourg in the E and on France in the S; capital Arlon. It is mainly agricultural, producing grain, livestock, and dairy goods. The province was detached from the grand duchy of Luxembourg in 1839. In World War II it was a major battleground in the Battle of the Bulge (Dec., 1944–Jan., 1945).

Luxembourg or **Luxemburg**, city (1973 est. pop. 78,300), capital of the grand duchy of Luxembourg, S Luxembourg, at the confluence of the Alzette and Pétrusse rivers. It is a commercial, industrial, administrative, and cultural center and a rail junction. Manufactures include iron and steel, furniture, leather goods, machinery, textiles, beer, and processed food. A picturesque city, Luxembourg developed around a 10th cent. castle that was one of Europe's strongest fortresses until its fortifications were dismantled according to the terms of the Treaty of London (1867). Of note are the Cathedral of Notre Dame and the city hall (both 16th cent.).

Lux·or (lŭk'sôr, lo͝ok'-), city (1970 est. pop. 84,600), central Egypt, on the Nile. It is west of Karnak and occupies part of the site of Thebes. The temple of Luxor, the greatest monument of antiquity in the city, was built in the reign of Amenhotep III as a temple to Amon. The temple, 623 ft (190 m) long, was much altered by succeeding pharaohs, especially by Ramses II, who had many colossal statues of himself erected on the grounds. In early Christian times the temple was made into a church. Numerous temples and burial grounds, including the Valley of the Tombs of the Kings, are nearby.

Lu·zerne (lo͞o'zərn), county (1970 pop. 342,329), 886 sq mi (2,294.7 sq km), E central Pa., in an anthracite-mining and industrial region drained by the Susquehanna and Lehigh rivers; formed 1786; co. seat Wilkes-Barre. It is a hilly area, except for the Wyoming Valley. Anthracite coal and sandstone are the major resources. Industries include manufacturing (textiles, clothing, food products, beverages, wire, and tobacco), printing, and agriculture (vegetables, fruit, grain, dairy products, and poultry).

Lu·zon (lo͞o-zŏn'), island (1970 pop. 16,669,724), 40,420 sq mi (104,688 sq km), largest, most populous, and most important of the Philippine Islands. The irregular coastline provides several fine bays, most notably Manila Bay. Northern Luzon, which is drained by the Cagayan River, is mountainous. In the east the great Sierra Madre range so closely parallels the shore that almost no coastal plain exists. In the west the Zambales range runs from Lingayen Gulf south to Bataan peninsula. Between the rugged coastal mountains, in central Luzon, lies a great fertile plain. It supplies food for almost the entire Manila area and is a major rice and sugar-cane region. Other major crops are hemp, tobacco, corn, fruits, vegetables, and cacao. Luzon has important lumbering and mining industries; there are gold, manganese, chromite, nickel, copper, and iron deposits, and the mountain pines and the bamboo on Bataan peninsula have many commercial uses. Manufacturing is centered in the Manila area, where the major industries produce textiles, chemicals, metal products, and automobiles.

Luzon has played a leading role in the nation's history. Manila harbor has been important since the arrival of the Spanish in the late 16th cent. It was on Luzon that the Filipino revolt against Spanish rule began (1896), that U.S. forces wrested control of the islands from Spain (1898), and that the Philippine insurrection against U.S. rule broke out (1899). In World War II the island was invaded by Japanese forces in several places on Dec. 10, 1941, and in early 1942 the Allied forces made their last stand on Bataan peninsula and Corregidor. Luzon was recovered (1945) after a major landing from Lingayen Gulf (Jan.) and a bloody fight for Manila (Feb.). After the war, Luzon's central plain was the focus of a Communist-dominated land reform movement, the Hukbalahap, which resorted to terrorism and caused much civil strife until it was finally brought under control (c.1954). In 1968 the movement became active again; mounting violence spurred President Ferdinand Marcos to initiate (Aug., 1969) an intensive military campaign to restore order. In mid-1972 Luzon experienced the worst flood in the nation's history; more than 400 died and there was extensive property damage.

Lvov (lə-vôf'), city (1977 est. pop. 642,000), capital of Lvov oblast, SW European USSR, in the Ukraine, at the watershed of the Western Bug and Dnestr rivers in the N foothills of the Carpathian Mts. Lvov is a major rail and highway junction and an industrial and commercial center. Machine building, food processing, oil refining, and the manufacture of chemicals and pharmaceuticals, motor vehicles, radio and electrical apparatus, and textiles are the leading industries. Founded c.1256, the city developed as a commercial center on the trade route from Vienna to Kiev and served as an outpost against Tatar invasions. Lvov was captured by the Poles in the 1340s, the Turks in 1672, and the Swedes in 1704. During the first partition of Poland (1772) it passed to Austria, and became the capital of Galicia. Lvov was the chief center of the Ukrainian national movement in Galicia after 1848. The city was taken by Poland in 1919 and confirmed as Polish by the Soviet-Polish Treaty of Riga (1921). Lvov was annexed to the Ukraine by the USSR in 1939. German forces held the city during much of World War II and exterminated most of the Jewish population. In 1945 Poland formally ceded Lvov to the USSR.

Lya·khov Islands (lyä'KHəf), c.2,700 sq mi (6,995 sq km), S group of the New Siberian Islands, N Siberian USSR, between the Laptev Sea and the East Siberian Sea, Yakut Autonomous SSR. They were discovered (1770) by Ivan Lyakhov, a Russian merchant. In 1928 the Soviet government established a geophysical station here.

Ly·all·pur (lī'əl-po͝or'), city (1972 pop. 822,263), NE Pakistan, in a cotton- and wheat-growing area. It is an important commercial center, especially for grains, cloth, and ghee (clarified butter). Manufactures include textiles, pharmaceuticals, bicycles, textile machinery, hosiery, flour, sugar, vegetable oil, and soap.

Lyc·a·o·ni·a (lĭk'ā-ō'nē-ə, lĭ'kā-), ancient country of S Asia Minor (now in Turkey), between Galatia and Cilicia on the N and S and Phrygia and Cappadocia on the W and E. Passing successively to the Persians, Syrians, and Romans, it was divided by the Romans between Galatia and Cappadocia.

Ly·ci·a (lĭsh'ē-ə), ancient country, SW Asia Minor. A mountainous promontory, it was held by the Persians, the Seleucids, and the Romans (from 189 B.C.). Ruins include rock-cut tombs and Grecian sculptures dating from the 5th cent. B.C.

Ly·com·ing (lī-kō'mĭng), county (1970 pop. 113,296), 1,215 sq mi (3,146.9 sq km), N central Pa., in a hilly agricultural and forest region drained by the West Branch of the Susquehanna River; formed 1795; co. seat Williamsport. Grain, clover, and poultry are raised. The county also has dairying, limestone quarrying, and anthracite-coal mining. Textiles and leather goods are manufactured.

Lydd (lĭd), municipal borough (1971 pop. 4,301), Kent, SE England. A military training center, the borough gave its name to lyddite (picric acid), an explosive that was tested at the military camp here in 1888. Lydd was a member of the Cinque Ports but is no longer a seaport because of changes in the shoreline.

Lyd·i·a (lĭd'ē-ə), ancient country, W Asia Minor, N of Caria and S of Mysia (now NW Turkey). The tyrant Gyges was the founder of the Mermnadae dynasty, which lasted from c.700 B.C. to 550 B.C. The little kingdom grew to an empire in the chaos that had been left after the fall of the Neo-Hittite kingdom. Lydia was proverbially golden with wealth, and the capital, Sardis, was magnificent. To Lydian rulers is ascribed the first use of coined money in the 7th cent. B.C. Lydia had close ties with the Greek cities of Asia, which were for a time within the Lydian empire. The last ruler was Croesus, who was defeated (c.546 B.C.) by Cyrus the Great of Persia. Lydia was then absorbed into the Persian Empire.

Ly·ell, Mount (lī'əl), peak, 13,095 ft (3,934 m) high, E central Calif., in the Sierra Nevada near the boundary of Yosemite National Park SSW of Mono Lake. It has a large glacier.

Ly·man (lī'mən), county (1970 pop. 4,060), 1,685 sq mi (4,364.2 sq km), S central S.Dak., in an agricultural and cattle-raising area bounded on the S by the White River, on the E by the Missouri River; formed 1890; co. seat Kennebec. Spring and winter wheat are among its major crops.

Lyme Re·gis (lĭm' rē'jĭs), municipal borough (1971 pop. 3,394), Dorset, SW England. It is a tourist resort. Paleontological discoveries have been made in the blue Lias rocks quarried near Lyme Regis.

Lym·ing·ton (lĭm'ĭng-tən), municipal borough (1976 est. pop. 164,500), Hampshire, S England, on the Solent channel at the mouth of the Lymington River. It is a market town, resort, and port; coast trading and yacht building are pursued and piston rings are produced. A Roman camp was in the vicinity. Henry II landed at Lymington in 1154 on the way to his coronation.

Lyn·brook (lĭn'bro͝ok'), village (1978 est. pop. 22,700), Nassau co., SE N.Y., a suburb of New York City on the S shore of Long Island; settled 1785, inc. 1911. There is some light manufacturing.

Lynch·burg (lĭnch'bûrg'). **1.** Town (1970 pop. 538), seat of Moore co., S Tenn., SSE of Shelbyville, in a timber and agricultural area. Whiskey is distilled. **2.** Independent city (1978 est. pop. 65,400), central Va., on the James River; settled 1757, inc. as a city 1852. It is a trade center and tobacco market in the foothills of the Blue Ridge Mts. Its manufactures include shoes, foundry and fabricated metal products, clothing, electronic equipment, machinery, tools, printed materials, furniture, wood products, and medical supplies. There is a nuclear research facility. Lynchburg was a Confederate supply base in the Civil War. Randolph-Macon Woman's College, Lynchburg College, and other schools are here.

Lynd·hurst (lĭnd'hûrst'), city (1970 pop. 19,749), Cuyahoga co., NE Ohio; inc. 1917. It is a residential suburb of Cleveland.

Lyn·don (lĭn'dən), city (1970 pop. 958), seat of Osage co., E Kansas, on a small affluent of the Marais des Cygnes River S of Topeka. It is a trade center in a livestock and grain region.

Lyndon B. John·son National Historic Site (jŏn'sən): *see* National Parks and Monuments Table.

Lynn (lĭn), county (1970 pop. 9,107), 915 sq mi (2,369.9 sq km), NW Texas, on the Llano Estacado; formed 1876; co. seat Tahoka. It is in an agricultural area yielding large crops of grain sorghums, also legumes, cotton, wheat, watermelons, livestock, dairy products, and poultry. There are deposits of sodium sulphate, magnesium sulphate, silica, and potash. The county includes Tahoka Lake and other intermittently dry lakes.

Lynn, city (1978 est. pop. 77,400), Essex co., E Mass.; inc. as a town 1631, as a city 1850. Lynn is an old industrial center. The first ironworks (1643) and the first fire engine (1654) in America were built here. Formerly the shoe industry was important, but today jet engines, turbines, and electrical instruments are major products.

Lynn Canal, natural inlet, c.90 mi (145 km) long, 7–12 mi (11.3–19.3 km) wide, SE Alaska. It connects in the south with Chatham Strait and Stephens Passage and thrusts north between mountains to break finally into the inlets of the Chilkoot and Chilkat rivers. Navigable to its head, Lynn Canal connects Skagway with Juneau and is an important shipping lane. During the Alaska gold rush (1896) it was a major route to the gold fields.

Lynn·field (lĭn'fēld'), town (1970 pop. 10,826), Essex co., NE Mass.; inc. 1814. It is mostly residential.

Lynn Gar·den (gär'dn), uninc. town (1970 pop. 7,000), Sullivan co., NE Tenn., near the Ky. border N of Kingsport.

Lynn·wood (lĭn'wo͝od'), city (1978 est. pop. 21,300), Snohomish co., W central Wash., a suburb of Seattle; inc. 1959.

Lyn·wood (lĭn'wo͝od'), city (1978 est. pop. 39,000), Los Angeles co., S Calif., a suburb of Los Angeles; founded 1896, inc. 1921.

Ly·on (lī'ən). **1.** County (1970 pop. 13,340), 588 sq mi (1,522.9 sq km), extreme NW Iowa, drained by the Rock and Little Rock rivers and bounded in the N by the Minn. border, in the W by the Big Sioux River (here forming the S.Dak. border); formed 1851; co. seat Rock Rapids. A prairie agricultural area, it is known for stock raising (hogs, cattle, and poultry) and for the growing of corn and oats. **2.** County (1970 pop. 32,071), 841 sq mi (2,178.2 sq km), E Kansas, in a level to hilly area drained by the Neosho and Cottonwood rivers; formed 1857 as Breckinridge co., renamed 1862; co. seat Emporia. In a grain and livestock area, it has lumbering, some mining, and a food-processing industry. Wood and metal products and machinery are among its manufactures. **3.** County (1970 pop. 5,562), 216 sq mi (559.4 sq km), W Ky., bounded on the SW by the Kentucky Reservoir (Tennessee River) and crossed by the Cumberland River; formed 1854; co. seat Eddyville. In a gently rolling agricultural area yielding livestock, burley tobacco, and grain, it has limestone quarries and hardwood timber. Its manufactures include plastic materials and resins. Part of Kentucky Woodlands Wildlife Refuge is here. **4.** County (1970 pop. 24,273), 713 sq mi (1,854.4 sq km), SW Minn., drained by the Yellow Medicine, Cottonwood, and Redwood rivers; formed 1868; co. seat Marshall. Its agriculture includes corn, soybeans, small grains, cattle, and hogs. Food processing and lumber milling are done, and plastics and concrete are manufactured. **5.** County (1970 pop. 8,221), 2,030 sq mi (5,257.7 sq km), W Nev., with the East and West Walker rivers forming the Walker River below Yerington, and Lahontan Reservoir, on the Carson River, supplying water for irrigation; formed 1861; co. seat Yerington. It is in a livestock and dairying region, with copper mining and some light industry (meat packing and cement). Part of Mono National Forest is in the south in the Sierra Nevada.

Ly·on·nais (lē-ô-nĕ'), region and former province, E central France. It included Lyonnais proper (the region around Lyons, its capital), which Philip IV acquired c.1307; the former counties of Forez and Beaujolais, annexed in 1531; and the tiny dependency of Franc-Lyonnais. It is primarily a grazing region, with great industrial centers noted especially for their textile production.

Ly·ons or *French* **Ly·on** (both: lē-ôɴ'), city (1975 pop. 456,716), capital of Rhône dept., E central France, at the confluence of the Rhône and Saône rivers. It leads Europe in silk and rayon production. The city has important metal, machine, clothing, and chemical industries; a river port; a stock exchange (founded 1506, the oldest in France); a university (founded 1808); and several fine museums. It is also a gastronomic capital. Founded in 43 B.C. as a Roman colony, ancient Lugdunum soon became the principal city of Gaul. Christianity was first introduced into Gaul here, and the importance of Lyons until c.1300 was chiefly religious. Lyons (which after the breakup of the Carolingian empire passed to the kingdom of Arles) was ruled by its archbishops until c.1307, when Philip IV incorporated the city and Lyonnais proper into the French crownlands. In 1793 Lyons was devastated by French Revolutionary troops after an insurrection, but it recovered quickly because of the invention of the Jacquard loom. During the German occupation in World War II (1940–44), Lyons was the capital of the French resistance movement.

Ly·ons (lī'ənz). **1.** City (1970 pop. 3,739), seat of Toombs co., SE Ga., WNW of Savannah, in a farm and timber area; inc. 1897. **2.** Village (1970 pop. 11,124), Cook co., NE Ill., a residential suburb of Chicago, on the Des Plaines River; inc. 1888. It was settled at the edge of an early travel route, the portage between the Chicago and the Des Plaines rivers. **3.** City (1970 pop. 4,355), seat of Rice co., central Kansas, NW of Hutchinson, on the old Santa Fe Trail, in an oil and wheat region; laid out 1876, inc. 1880. **4.** Village (1970 pop. 4,496), seat of Wayne co., W central N.Y., on the Barge Canal SE of Rochester, in a fruit-growing region; settled 1800, inc. 1831.

Lys (lēs), river, c.135 mi (220 km) long, rising in the hills of Artois, N France, and flowing NE, forming the Franco-Belgian border between Armentières and Menen. It continues into Belgium past Kortrijk to empty into the Scheldt River at Ghent. The Lys valley is known for flax spinning and weaving.

Lyth·am Saint Anne's (lĭth'əm sānt' ănz'), municipal borough (1973 est. pop. 42,120), Lancashire, NW England, on the N shore of the Ribble estuary. It is a seaside resort. Lytham Saint Anne's was founded in the 12th cent. by Benedictine monks.

Ma·an (mə-än'), town (1973 est. pop. 9,500), S Jordan. It is the terminus of the country's main rail line. Important since Biblical times.

Maas (mäs), river: *see* Meuse.

Maas·tricht (mäs'trĭKHt'), city (1977 est. pop. 110,191), capital of Limburg prov., SE Netherlands, on the Maas River and the Albert Canal system. It is an important transportation point and an industrial center. Its manufactures include textiles, ceramics, glass, paper, printed materials, and chemicals. An episcopal see from 382 to 721, Maastricht has the oldest church in the Netherlands, the Cathedral of St. Servatius, founded in the 6th cent. In 1284 the city came under the dual domination of the dukes of Brabant and the prince-bishop of Liège. It was for many years a strategic fortress and suffered many sieges.

Ma·cao (mə-kou'), Portuguese overseas province (1970 est. pop. 248,316), 6 sq mi (15.5 sq km), adjoining Kwangtung prov., SE China, on the estuary of the Canton River, W of Hong Kong and S of Canton. It consists of a rocky, hilly peninsula (c.2 sq mi/5 sq km), connected by a sandy isthmus to China's Chung-Shan Island, and two small islands. The city of Macao is approximately coextensive with the peninsula and contains almost the entire population of the province, which is overwhelmingly Chinese. Macao, a free port, is a leading trade, tourist, and fishing center, with gambling casinos and a recently established textile industry. Other leading products are fresh and salted fish and handicraft items, especially firecrackers and matches. Macao is separated from China by a barrier gate (built 1849, replacing one erected by the Chinese in 1573).

Macao is the oldest permanent European settlement in the Far East. First visited by Vasco da Gama in 1497, it was a desolate spot when the Portuguese established a trading post here in 1557. For nearly 300 years the Portuguese paid China an annual tribute for using the peninsula, but in 1849 Portugal proclaimed it a free port. With the gradual silting up of its harbor and the rise (19th cent.) of Hong Kong, Macao lost its pre-eminent position and became identified to a large extent with smuggling and gambling interests. Since 1949 the population has been greatly swelled by an influx of Chinese refugees from the Communist mainland. Macao's historic structures include the remaining façade of St. Paul's Basilica (built 1635 by Roman Catholic Japanese artisans; burned 1835), a fascinating example of late Italian Renaissance architecture, with mixed Western and Oriental motifs; St. Domingo's church and convent (founded c.1670); the fort and chapel of Guia (1626); the fort of São Paulo de Monte (16th cent.); and statues to da Gama and Luís de Camões, who wrote (1558–59) part of *The Lusiads* here.

Ma·ca·pá (mə-kə-pä'), city (1970 pop. 51,563), capital of Amapá federal territory, extreme N Brazil, on the Amazon River. Its economy is based on mining (manganese and iron) and tropical forest products. Macapá's rubber resources have not been fully exploited.

Mac·clen·ny (mə-klĕn'ē), town (1970 pop. 2,733), seat of Baker co., NE Fla., W of Jacksonville.

Mac·cles·field (măk'əlz-fēld'), municipal borough (1973 est. pop. 45,420), Cheshire, W England. Silk manufacture, of which Macclesfield is the principal center in England, was introduced in the town in 1756; other industries manufacture clothing, shoes, electrical appliances, and paper. The Church of St. Michael dates from 1278.

Mac·e·don (măs'ĭ-dŏn), ancient country, N Greece, the modern Macedonia. Macedon proper constituted the coast plain northwest, north, and northeast of the Chalcidice (now Khalkidhikí) peninsula; Upper Macedon was the highland to the west and the north of the plain. The plain was fertile and productive, and there were important silver mines in the eastern part.

The first influence of Greek culture in Macedon came from the colonies along the shore founded in the 8th cent. B.C. and after. Alexander I (d. 450 B.C.) was the first Macedonian king to enter into Greek politics; for the next century the Hellenic influences grew and the state became stronger. With Philip II (reigned 359–336 B.C.) these processes reached their culmination, for by annexing Upper Macedon, Chalcidice, and Thrace he made himself the strongest power in Greece. He created an excellent army with which his son, Alexander the Great, forged his empire. When the Macedonian generals carved the empire up after Alexander's death (323 B.C.), Macedon, with Greece as a dependency, was one of the states formed. Antigonus II (reigned 277–239 B.C.) restored Macedon economically. Antigonus III (reigned 229–221 B.C.) re-established Macedonian hegemony over the Greek city-states. The Romans took control of the area and eventually (146 B.C.) annexed Macedonia, making it the first Roman province. In the history of Greek culture Macedon had its single significance in producing the conquerors and armies who created the Hellenistic empires and civilizations.

Mac·e·do·ni·a (măs'ĭ-dō'nē-ə, -dōn'yə), region, SE Europe, on the Balkan Peninsula, divided among Greece, Yugoslavia, and Bulgaria. Corresponding roughly with ancient Macedon, it extends from the Aegean Sea northward between Epirus in the west and Thrace in the east. The region is predominately mountainous, encompassing parts of the Pindus and Rhodope mts. Tobacco is the main crop; grains and cotton are also grown, and sheep and goats are raised. The mining of iron, copper, lead, and chromite is important.

With the division (395) of the Roman Empire, Macedonia came under Byzantine rule. It was settled (6th cent.) by the Slavs and wrested from the Byzantine Empire by Bulgaria in the 9th cent. It again became part of the Byzantine Empire in 1261, but in the 14th cent. Stephen Dushan of Serbia conquered all Macedonia except for present-day Thessaloníki. The fall of the Serbian empire in the late 14th cent. brought Macedonia under the rule of the Ottoman Turks, which lasted for five centuries. In the 19th cent. the Ottoman Empire lost control over the major sections of Greece, Serbia, and Bulgaria, each of which claimed Macedonia on historical or ethnic grounds. A secret, Bulgarian-supported terrorist organization working for Macedonian independence sprang up. The Balkan Wars (1912–13) left Bulgaria with only a small share of Macedonia, the rest of which was divided roughly along the present lines. Thousands of Macedonians fled to Bulgaria. Relations were strained among the Balkan States over the Macedonian question, and border skirmishes and terrorist attacks were frequent. In World War II all Macedonia was occupied (1941–44) by Bulgaria, but the prewar boundaries were confirmed in the peace treaty of 1947.

Ma·cei·ó (mä'sā-ô'), city (1970 pop. 242,867), capital of Alagoas state, E Brazil, on a narrow strip of land between a lagoon and the Atlantic Ocean. A commercial and distribution center, Maceió exports sugar and textiles. On the outskirts are coconut plantations. The city grew around a sugar mill following the Dutch occupation during the early 17th cent.

Ma·ce·ra·ta (mä'chā-rä'tä), town (1976 est. pop. 37,500), capital of Macerata prov., in the Marche, central Italy. It is an agricultural and industrial center. Manufactures include musical instruments, furniture, and construction materials. Macerata was ruled by the papacy from the mid-15th cent. to 1797. It retains its medieval walls and has a university that was founded in 1290.

Mac·gil·li·cud·dy's Reeks (mə-gĭl'ə-kŭd'ēz rēks'), highest mountain range of Ireland, Co. Kerry, SW Republic of Ireland. It includes Carrantuohill and other peaks more than 3,000 ft (915 m) high.

Ma·chi·as (mə-chī'əs), town (1970 pop. 2,441), seat of Washington co., SE Maine, on the Machias River, in a granite and truck-farm area. The English established a trading post near the mouth of the Machias River in 1633, but the French, who claimed the area, later destroyed the post. Permanent settlement began in 1763, and the town was incorporated in 1784.

Ma·chi·da (mä-chē'dä), city (1976 est. pop. 263,758), Tokyo Metropolis, E central Honshu, Japan, on the Tsurumi River. It is an industrial and residential suburb of Tokyo.

Ma·chu Pic·chu (mä'cho͞o pēk'cho͞o), fortress city of the ancient Incas, Peru, NW of Cuzco. It is perched high upon a rock in a narrow saddle between two sharp mountain peaks and overlooks the Urubamba River 2,000 ft (610 m) below; it was unknown to Spanish explorers. Discovered in 1911 by the American explorer Hiram Bingham, the imposing city is one of the few urban centers of pre-Columbian America found virtually intact. Perhaps the most extraordinary ruin in the Americas, Machu Picchu contains 5 sq mi (13 sq km) of terrace and construction, with more than 3,000 steps linking it to many levels. It shows admirable architectural design and execution, although the stonework is not always as refined as in other Inca sites. Legend indicates that the city may have been the home of the Incas prior to their migration to Cuzco as well as their last stronghold after the Spanish Conquest.

Mac·kay (mə-kī'), city (1975 est. pop. 20,550), Queensland, NE Australia on the Pioneer River. A port city, Mackay exports sugar.

Mac·ken·zie (mə-kĕn'zē), district, 527,490 sq mi (1,366,199 sq km), one of the three districts of the Northwest Territories, Canada. Established in 1920, it lies between Yukon Territory (on the west) and Keewatin dist. (on the east) and includes the lower two thirds of the Mackenzie valley, Great Bear Lake, Great Slave Lake, and many smaller lakes.

Mackenzie, river, c.1,120 mi (1,800 km) long, issuing from Great Slave Lake, S Mackenzie dist., Northwest Territories, Canada, and flowing generally NW to the Arctic Ocean through a great delta. Between Great Slave Lake and Lake Athabasca it is known as the Slave River. At Lake Athabasca, the Finlay-Peace river system and the Athabasca River join the Mackenzie. The Finlay-Peace-Mackenzie system (c.2,600 mi/4,185 km long) is the second-longest continuous stream in North America. The river is navigable from the Arctic Ocean to Great Slave Lake between June and Oct. Between Great Slave Lake and Lake Athabasca there are rapids (14 mi/22.5 km) that must be portaged; above the rapids are more than 400 mi (643.6 km) of navigable waters. The Mackenzie basin, flanked by the Rocky Mts. and the Canadian Shield, is the northern portion of the Great Plains of North America; arctic air masses follow the valley south into the interior of the continent. Numerous trading posts were established along the Mackenzie in the early part of the 19th cent. and fur trapping is still an important activity. In the early 1970s large natural-gas fields were discovered in the region. Sir Alexander Mackenzie, the Canadian explorer, was the first to descend (1789) the river to the Arctic Ocean.

Mackenzie Mountains, range, N Rocky Mts., E Yukon Territory and W Mackenzie dist., Canada. The range stretches c.500 mi. (805 km), reaching its highest elevation at Keele Peak (9,750 ft/2,973.8 m). The Mackenzie Mts. Preserve was established in 1938 in the southern part of the range.

Mac·ki·nac (măk'ə-nô'), historic region of the Old Northwest. The name, in the past, was variously applied to different areas: to Mackinac Island; to the whole fur-trading region supplied from the island; to the northern mainland shore; and to the southern mainland shore, where Mackinaw City, Mich., is today and where a fort called Old Mackinac once stood.

The Straits of Mackinac, a passage between the Upper and Lower peninsulas of northern Mich., served for many years as an important Indian gathering place. In 1634 the French explorer Jean Nicolet was the first white man to pass through the straits. Father Jacques Marquette established a mission at St. Ignace in 1671. A fort was later built there, and it became the headquarters of French trade operations in New France and an important military post in the Old Northwest; its importance declined in 1701 when Detroit was founded. The region passed into British hands in 1761 during the French and Indian Wars. In 1763 the British garrison at Old Mackinac was massacred by Ottawa Indians during Pontiac's Rebellion.

The island and the straits were awarded to the United States in 1783 by the Treaty of Paris but remained in British hands until 1794. One of the first events of the War of 1812 was the British capture of Mackinac; it was returned to U.S. control by the Treaty of Ghent in 1814. After the war, Mackinac Island became the center of operations of John Jacob Astor's American Fur Company and thrived until the 1830s, when fur trading declined. After the 1840s the straits area changed from an important crossroads to an out-of-the-way shipping point. Iron-ore mining revitalized the area in the early 20th cent., but the ore was depleted after World War II. The Mackinac Straits Bridge, the third longest suspension bridge in the world (3,800 ft/1,159 m long; opened 1957) spans the straits and links St. Ignace with Mackinaw City. The straits are an important link in the Great Lakes–St. Lawrence waterway.

Mackinac, county (1970 pop. 9,660), 1,014 sq mi (2,626.3 sq km), SE Upper Peninsula, Mich., bounded on the S by Lakes Michigan and Huron, and by their connection, the Straits of Mackinac, drained by the Carp and Pine rivers, and including Bois Blanc and Mackinac islands; formed 1818; co. seat St. Ignace. It is in a forest, resort, and agricultural area where fruit and potatoes are grown and dairy products are made. There is some mining, as well as commercial fishing. Part of Marquette National Forest and several lakes are in the county.

Mac·mil·lan (mək-mĭl'ən), river, c.200 mi (320 km) long, rising in two main forks in the Selwyn Mts., E Yukon Territory, Canada, and flowing generally W to the Pelly River. It was an important route to the gold fields from c.1890 to 1900.

Ma·comb (mə-kōm'), county (1970 pop. 625,309), 481 sq mi (1,245.8 sq km), SE Mich., bounded on the SE by Lake St. Clair and Anchor Bay, drained by the Clinton River and its affluents; formed 1818; co. seat Mount Clemens. It is a truck-farm, grain, and poultry area, as well as a fast-growing industrial region. There are several major automobile installations.

Ma·comb (mə-kōm'), city (1978 est. pop. 24,500), seat of McDonough co., W Ill.; inc. as a city 1856. A trade and manufacturing center in a rich farm, clay, and coal region, the city is known for its artistic clay products.

Ma·con (mā'kən). **1.** County (1970 pop. 24,841), 616 sq mi (1,595.4 sq km), E Ala., in the Black Belt, drained by the Tallapoosa River and its branches; formed 1832; co. seat Tuskegee. It is in an agricultural area that produces cotton, corn, peanuts, sweet potatoes, cattle, and dairy products. Clothing and concrete are manufactured. **2.** County (1970 pop. 12,933), 399 sq mi (1,033.4 sq km), central Ga., in a coastal-plain agricultural area drained by the Flint River; formed 1837; co. seat Oglethorpe. Peaches, cotton, corn, truck crops, peanuts, and pecans are grown. **3.** County (1970 pop. 125,010), 577 sq mi (1,494.4 sq km), central Ill., drained by the Sangamon River and Lake Decatur; formed 1829; co. seat Decatur. Agriculture (corn, wheat, soybeans, oats, livestock, poultry, and dairy products), diversified manufacturing, and bituminous-coal mining are the major occupations. The county has a number of recreational areas. **4.** County (1970 pop. 15,432), 798 sq mi (2,066.8 sq km), N Mo., drained by the Chariton River; formed 1837; co. seat Macon. It is in an agricultural (corn, wheat, oats, hay, and livestock) and bituminous-coal area. Fruit and vegetables are preserved. **5.** County (1970 pop. 15,788), 513 sq mi (1,328.7 sq km), W N.C., in the Blue Ridge, crossed by the Nantahala Mts. and drained by the Nantahala and Little Tennessee rivers; formed 1828; co. seat Franklin. A resort area included in Nantahala National Forest, it has lumbering, farming (vegetables, apples, and corn), dairying, poultry raising, and mica mining. **6.** County (1970 pop. 12,315), 304 sq mi (787.4 sq km), N Tenn., bordered in the N by Ky. and drained by affluents of the Barren and Cumberland rivers; formed 1842; co. seat Lafayette. Its agriculture includes corn, tobacco, and livestock. It also has some oil wells. There are garment, lumbering, and machine and transportation-equipment industries here.

Macon. 1. City (1978 est. pop. 120,700), seat of Bibb co., central Ga., at the head of navigation on the Ocmulgee River; inc. 1823. It is the industrial, processing, and shipping center for an extensive farm area. Textiles, clay products, insulation board, tile brick, rockets, explosives, and fabricated steel are among its manufactures. Fort Hawkins was established on the east side of the river in 1806 and renamed Newtown in 1821. Macon was laid out on the west side in 1823, and Newtown was annexed in 1829. Wesleyan College and Mercer Univ. are here. Also in Macon are the birthplace of Sidney Lanier, a restored grand-opera house (1884), Fort Hawkins (1806; partially restored), a museum of arts and sciences, and a planetarium. **2.** City (1970 pop. 2,612), seat of Noxubee co., E Miss., SSW of Columbus on the Noxubee River; inc. 1836. It is a processing center for a cotton, timber, and dairy area. **3.** City (1970 pop. 5,301), seat of Macon co., N Mo., S of Kirksville; founded c.1856, inc. 1856. It is a shipping point for farm produce and livestock.

Macon, Bayou, c.145 mi (235 km) long, rising in SE Ark. and flowing S into NE La. to the Tensas River. It was used as a rendezvous by the bandits Frank and Jesse James.

Mâ·con (mä-kôN'), town (1975 pop. 39,344), capital of Saône-et-Loire dept., E central France, in Burgundy, on the Saône River. It is famous for its quality wines. The town also has foundries and plants that manufacture motorcycles, electrical equipment, and clothing. Mâcon was acquired by the French crown in 1238, passed to Burgundy by the Treaty of Arras (1435), and was recovered by France in 1477. In the 16th cent. it was a Huguenot stronghold.

Ma·cou·pin (mə-ko͞o'pən), county (1970 pop. 44,557), 872 sq mi (2,258.5 sq km), SW central Ill., drained by Macoupin, Cahokia, and Otter creeks; formed 1829; co. seat Carlinville. It has agriculture (livestock, corn, wheat, oats, soybeans, poultry, and dairy products), bituminous-coal mines, and some manufacturing (brick, tile, and gloves). There are clay pits and tracts of timber.

Mac·quar·ie (mə-kwŏr'ē), river, 590 mi (949.3 km) long, rising in the Blue Mts., E New South Wales, Australia, and flowing NW to the Darling River. It flows through an important sheep- and wheat-raising area.

Mac·quar·ie Island (mə-kwŏr'ē), uninhabited volcanic island, S Pacific, SE of Tasmania, Australia, to which it belongs. Mountainous, with small glacial lakes, it is 21 mi (33.8 km) long and 3 mi (4.8 km) wide. Sea elephants, seals, and penguins are found here. It has a meteorological station.

Mac·tan (mäk-tän'), coral island (1970 pop. 70,729), 24 sq mi (62.2 sq km), Cebu prov., the Philippines, just off the coast of Cebu Island. The spot where Magellan was killed by natives in 1521 is marked by a monument.

Mad·a·gas·car (măd'ə-găs'kər): *see* Malagasy Republic.

Ma·dei·ra (mə-dîr'ə, -dĕr'ə), river, c.900 mi (1,450 km) long, formed by the junction of the Beni and Mamoré rivers on the Bolivia-Brazil border. It flows north along the border for c.60 mi (95 km), then northeast in a winding course through the Rondônia and Amazonas sections of northwest Brazil into the Amazon River. At its mouth is an extensive marshy region formed by the Madeira's distributaries. The river is navigable by ocean vessels to the falls and rapids near Pôrto Velho, Brazil.

Ma·dei·ra Islands (mə-dîr'ə, -dĕr'ə), archipelago (1975 est. pop. 265,600), 308 sq mi (797.7 sq km), coextensive with Funchal dist., Portugal, in the Atlantic Ocean c.350 mi (565 km) off Morocco. Madeira, the largest island, and Porto Santo are inhabited. Sugar cane and Madeira wine are produced on the islands, and there are light industries such as embroidering and the manufacture of reed furniture and baskets. Madeira is a year-round resort. Mountain peaks, which descend steeply into deep, green valleys and advance to the sea as precipitous basalt cliffs, give the island unusual scenic beauty. The islands were known to the Romans as the Purple Islands and were rediscovered by the Portuguese (1418–20). Settlement took place rapidly under the orders of Prince Henry the Navigator. Madeira was occupied by the British in the early 19th cent.

Ma·de·ra (mə-dĕr'ə), county (1970 pop. 41,519), 2,148 sq mi (5,563.3 sq km), central Calif., stretching from the level San Joaquin Valley in the W to the crest of the Sierra Nevada in the NE; formed 1893; co. seat Madera. It is watered by the San Joaquin, Chowchilla, and Fresno rivers and the Madera Canal. The rich irrigated valley area yields cotton, alfalfa, barley, potatoes, truck crops, dairy products, livestock, poultry, almonds, pistachios, and grapes. The county also has lumbering and food-processing industries, mining (pumice, gold, and copper), and deposits of natural gas. Its manufactures include glassware, concrete, and machinery. Devils Postpile National Monument and parts of Yosemite National Park and Sierra National Forest are here.

Madera, city (1978 est. pop. 18,600), seat of Madera co., central Calif., in the San Joaquin valley; inc. 1907. It is known for its wines; other products include olives, packed meats, lumber, paper, glass, and air-cooling and farm equipment.

Madh·ya Bha·rat (mŭd'yə bä'rŭt, -rət), former state, W central India. It comprised 25 former princely states. In 1956 it was incorporated into Madhya Pradesh state.

Madhya Pra·desh (prə-dāsh', prä'dĕsh), state (1971 pop. 41,650,684), 171,210 sq mi (443,434 sq km), central India, between the Deccan and the Ganges plains; capital Bhopal. Madhya Pradesh consists of upland zones separated by plains. Adequate rainfall and plentiful good soil permit a prosperous, predominantly agricultural economy. Grains, especially wheat, are the main crops of the north. In the southeast, rice is the largest crop; in the southwest, cotton. Spinning and weaving are the chief industries. The state is rich in such minerals as manganese, bauxite, iron ore, and coal. Nominally within the Mogul empire, the area was ruled during the 16th and 17th cent. by the Gonds and in the 18th cent. by the Mahrattas. The British occupied it in 1820. From 1903 until 1950 the state was called Central Provinces and Berar.

Ma·dill (mə-dĭl'), city (1970 pop. 2,875), seat of Marshall co., S Okla.,

ESE of Ardmore, in a farm, ranch, and oil area; settled c.1900. Lake Texoma, a resort area, is nearby.

Ma·di·nat ash Shab (mä'dē-nät' äsh shäb'), town (1967 est. pop. 20,000), SW Southern Yemen, just N of Aden. It was built in the 1960s as the federal capital of the Federation of South Arabia. From 1967 to 1970 it was the capital of Southern Yemen along with Aden.

Mad·i·son (măd'ĭ-sən). **1.** County (1970 pop. 186,540), 803 sq mi (2,079.8 sq km), N Ala., bounded on the N by the Tenn. border and drained by Wheeler Reservoir and the Flint River; formed 1808; co. seat Huntsville. Its agriculture includes cotton, corn, tobacco, mules, cattle, and hay. Mining, lumbering, and food processing are done. Among its manufactures are clothing, plastics, concrete, wood and metal products, and machinery. **2.** County (1970 pop. 9,453), 832 sq mi (2,154.9 sq km), NW Ark., in the Ozarks, drained by the White and King rivers and War Eagle Creek; formed 1836; co. seat Huntsville. It is in an agricultural area that produces truck crops, hay, grain, livestock, and dairy products. It has timber and some textile manufacturing. Part of Ozark National Forest is in the south. **3.** County (1970 pop. 13,481), 702 sq mi (1,818.2 sq km), N Fla., in a flatwoods region, with swamps and many small lakes, bordered in the N by Ga. and bounded in the W by the Aucilla River, in the NE by the Withlacoochee River; in the SE by the Suwannee River; formed 1827; co. seat Madison. Agriculture (corn, peanuts, cotton, tobacco, hogs, and poultry) and forestry are the major occupations. **4.** County (1970 pop. 13,517), 281 sq mi (727.8 sq km), NE Ga., in a piedmont region drained by the Broad River; formed 1811; co. seat Danielsville. Cotton, corn, sweet potatoes, and hay are grown. Furniture and yarn, thread, and clothing are manufactured. **5.** County (1970 pop. 13,452), 473 sq mi (1,225.1 sq km), E Idaho, in an irrigated agricultural area in the Snake River Plain, watered by the Teton and Henrys Fork rivers; formed 1913; co. seat Rexburg. The area yields sugar beets, seed peas, dry beans, potatoes, wheat, and livestock. **6.** County (1970 pop. 250,911), 732 sq mi (1,895.9 sq km), SW Ill., bounded on the W by the Mississippi River; formed 1812; co. seat Edwardsville. Largely industrial, the county also has agriculture (corn, wheat, dairy products, poultry, and livestock), bituminous-coal mines, and oil and natural-gas wells in its eastern section. **7.** County (1970 pop. 138,522), 453 sq mi (1,173.3 sq km), central Ind., drained by the West Fork of the White River and by numerous small creeks; formed 1823; co. seat Anderson. It is in a rich agricultural area that produces corn, hogs, cattle, tomatoes, soybeans, and poultry. It has lumbering, limestone quarrying, and a food-processing industry. Wood products, furniture, glassware, concrete, metal products, and machinery are manufactured. **8.** County (1970 pop. 11,558), 565 sq mi (1,463.4 sq km), S central Iowa, drained by the North and Middle rivers; formed 1846; co. seat Winterset. Its prairie agriculture includes hogs, cattle, corn, oats, soybeans, and apples. There are bituminous-coal and limestone deposits here. **9.** County (1970 pop. 42,730), 446 sq mi (1,155.1 sq km), E central Ky., partly in the outer Bluegrass, bounded on the N, NE, and NW by the Kentucky River and drained by several creeks; formed 1785; co. seat Richmond. In a rolling agricultural area yielding burley tobacco, livestock, corn, hay, poultry, and dairy products, it has bituminous-coal mines, clay pits, and a food-processing industry. Its manufactures includes asbestos, aluminum, metal products, and machinery. **10.** Parish (1970 pop. 15,065), 662 sq mi (1,714.6 sq km), NE La., bounded on the W by Bayou Macon, on the E by the Mississippi River, and intersected by the Tensas River; formed 1838; parish seat Tallulah. It is in a fertile lowland agricultural area, growing cotton, oats, hay corn, and soybeans. Industries include cotton ginning, lumber milling, and the manufacture of cottonseed and soybean products. There are fishing and hunting areas on the oxbow lakes formed by the Mississippi. **11.** County (1970 pop. 29,737) 727 sq mi (1,882.9 sq km), central Miss., bounded on the SE by the Pearl River, on the NW by the Big Black River; formed 1828; co. seat Canton. Cotton, corn, hay, truck crops, and fruit are grown. Timber sawmilling and food processing are done. Among its manufactures are clothing, furniture, wood and metal products, machinery, and chemicals. **12.** County (1970 pop. 8,641), 496 sq mi (1,284.6 sq km), SE Mo., partly in the St. Francois Mts. and drained by the St. Francis and Castor rivers; formed 1818; co. seat Fredericktown. It has ranching and farming (wheat, corn, and hay), lumbering, and extensive mining (tungsten, manganese, lead, zinc, iron, cobalt, copper, antimony, nickel, and granite). Clothing and shoes are made. Part of Mark Twain National Forest is here. **13.** County (1970 pop. 5,014), 3,530 sq mi (9,142.7 sq km), SW Mont., in an agricultural and mining region drained by the Madison, Ruby, Beaverhead, and Jefferson rivers; formed 1865; co. seat Virginia City. Livestock is raised, and gold, silver, lead, and copper mined. Parts of Deerlodge, Gallatin, and Beaverhead national forests are in the county, which is heavily mountainous in the east. **14.** County (1970 pop. 27,402), 572 sq mi (1,481.5 sq km), NE central Nebr., drained by the Elkhorn River; formed 1865; co. seat Madison. It is primarily agricultural, yielding livestock, grain, dairy products and poultry. **15.** County (1970 pop. 62,864), 661 sq mi (1,712 sq km), central N.Y., drained by the Chenango and Unadilla rivers and several creeks, and including part of Oneida Lake and many small lakes and reservoirs; formed 1806; co. seat Wampsville. It is in a dairying, truck-farming (particularly onions and cabbage), grain-growing, and poultry-raising area. Lumbering and food processing are done, and wood products, furniture, plastics, and machinery manufactured. The county has many lake resorts. **16.** County (1970 pop. 16,003), 450 sq mi (1,165.5 sq km), W N.C., in a mountainous region bounded on the N by the Bald Mts. and the Tenn. border and drained by the French Broad River; formed 1851; co. seat Marshall. A resort area, partly in Pisgah National Forest, it has sawmilling, farming (tobacco and corn), and cattle raising. **17.** County (1970 pop. 28,318), 464 sq mi 1,201.8 sq km), central Ohio, in an agricultural area drained by Deer, Paint, and Darby creeks; formed 1810; co. seat London. Corn, soybeans, and wheat are its major crops. Some manufacturing is done. **18.** County (1970 pop. 65,774), 561 sq mi (1,453 sq km), W Tenn., drained by the Middle and South forks of the Forked Deer River; formed 1821; co. seat Jackson. Its agriculture includes truck crops, cotton, and livestock. Processed foods, textiles, paper, plastics, concrete, machine, and transportation equipment are manufactured. **19.** County (1970 pop. 7,693), 478 sq mi (1,238 sq km), E central Texas, bounded on the W by the Navasota River, on the E by the Trinity River; formed 1853; co. seat Madisonville. It is primarily agricultural, yielding cotton, corn, legumes, potatoes, fruit, truck crops, livestock, horses, poultry, and some dairy products. Lumbering is also done. **20.** County (1970 pop. 8,638), 327 sq mi (846.9 sq km), N Va., with the Blue Ridge in the NW, bounded in the SW and S by the Rapidan River and drained by the Robertson River; formed 1792; co. seat Madison. Its agriculture includes grain, fruit, tobacco, livestock, and dairy products. It also has stands of timber, good trout fishing, and plants manufacturing clothing, wood products, and furniture. Part of Shenandoah National Park is here.

Madison. 1. Resort town (1970 pop. 9,768), New Haven co., S Conn., on Long Island Sound; set off from Guilford 1826. The Graves House dates from 1675. **2.** City (1970 pop. 3,737), seat of Madison co., N Fla., near the Ga. border E of Tallahassee; settled in the 1830s. It is a trade center for a farm area and makes wood products. **3.** City (1970 pop. 2,890), seat of Morgan co., N central Ga., ESE of Atlanta; inc. 1809. It is the trade center for a farm and lumber area. **4.** Industrial city (1970 pop. 7,042), Madison co., SW Ill., on the Mississippi River N of East St. Louis; inc. 1891. It is a steel and railroad center. Paper, magnesium, and asphalt products are made. **5.** City (1970 pop. 13,081), seat of Jefferson co., SE Ind., on the Ohio River; settled c.1806, inc. 1838. It is a port of entry and a major tobacco marketing center. Among its manufactures are machinery, electric motors, organs, and metal products. The city has fine examples of Federal, Regency, Gothic, Georgian, Classic Revival, and Italianate architecture. **6.** City (1970 pop. 2,242), seat of Lac qui Parle co., W Minn., SSE of Ortonville; settled c.1875, platted 1884, inc. as a city 1902. It is a trade center for a livestock, dairy, and grain region. **7.** City (1970 pop. 1,595), seat of Madison co., NE Nebr., S of Norfolk and NW of Omaha, in a farm area; settled 1868, inc. 1873. **8.** Borough (1970 pop. 16,710), Morris co., NE N.J., a residential suburb of the N.Y.-N.J. area; settled 1685, inc. 1889. Drew Univ. and part of Fairleigh Dickinson Univ. are here. Originally called Bottle Hill, it was renamed in 1834. Sayre House (1745) was Anthony Wayne's headquarters during the Revolution. The borough is noted for its roses. **9.** City (1970 pop. 6,315), seat of Lake co., SE S.Dak., NW of Sioux Falls, in a resort and farm area; platted 1880, inc. 1885. It is a trade and processing center. **10.** Town (1970 pop. 299), seat of Madison co., N Va., near the E foot of the Blue Ridge NNE of Charlottesville. Clothing is manufactured. **11.** Town (1970 pop. 2,342), seat of Boone co., SW W.Va., SW of Charleston on the Little Coal River; inc. 1906. There are natural-gas wells and coal mines here. **12.** City (1978 est. pop. 168,700), state capital, and seat of Dane co., S central Wis., on an isthmus between Lakes Monona and Mendota; inc. 1856. It is a trading and manufacturing center in a fertile agricultural region. Meat products, dairy machinery, batteries, and medical equipment are made. Madison was founded in 1836, and was chosen territorial capital before it was settled. It is the seat of the Univ. of Wisconsin. Many parks dotting the wooded lake shores make it an attractive residential city. Among its points of interest are the elaborate capitol, which houses the legislative library; a Unitarian church designed by Frank Lloyd Wright; a large arboretum; and Vilas Park, which contains a zoo.

Madison, river, 183 mi (294.4 km) long, rising in Yellowstone National Park, NW Wyo., and flowing W then N through SW Mont. to join the Jefferson and Gallatin rivers at the Three Forks of the Missouri River. The river is used for irrigation.

Mad·i·son·ville (măd'ĭ-sən-vĭl'). **1.** City (1978 est. pop. 18,400), seat of Hopkins co., W Ky., in a coal and farm area; inc. 1807. Food is processed and canned, and there is some light manufacturing. **2.** Town (1970 pop. 2,614), seat of Monroe co., SE Tenn., SW of Knoxville, in a timber, cotton, and tobacco area; inc.1865. **3.** City (1970 pop. 2,881), seat of Madison co., E central Texas, NNW of Houston, in a timber and oil region; settled 1851.

Ma·dras (mə-drăs', -dräs'), city (1971 pop. 2,470,288), capital of Tamil Nadu state, SE India, on the Bay of Bengal. A commercial and manufacturing center, it has large textile mills, chemical plants, and tanneries. Leather, peanuts, and cotton are exported. Largely built around Fort St. George, a British outpost completed in 1640, the city soon became an important British trading center. The French captured it in 1746, but the British recovered it two years later. A cultural center, the city is the seat of the Univ. of Madras (1857).There are many large public buildings and a famous shore drive, the Marina. Near the city is Mt. St. Thomas, the traditional site of the martyrdom (A.D. 68) of the apostle Thomas. He is supposedly buried in Madras at the Cathedral of St. Thomé.

Mad·ras (măd'rəs), city (1970 pop. 1,689), seat of Jefferson co., N central Oregon, N of Bend, in a farm and ranch area; inc. 1910. It is the trade center of the Deschutes River valley.

Ma·dre de Di·os (mä'drē dā dē-ōs', mä'thrā *thā* thyōs'), river, c. 700 mi (1,125 km) long, rising in the Andes of SE Peru and flowing NE through NW Bolivia to the Beni River. It is a major artery of northwest Bolivia, but frequent rapids make its lower course only partly navigable. It drains a rubber-producing area.

Ma·drid (mə-drĭd'), city (1974 est. pop. 3,274,043), capital of Spain and of Madrid prov., central Spain, in New Castile, on the Manzanares River. The newest of the great Spanish cities, it lacks the traditions of the ancient Castilian and Andalusian towns. Lying on a vast open plateau, it is subject to extremes of temperature. Madrid is almost in the exact geographic center of Spain and is Spain's chief transportation and administrative center. Its commercial and industrial life developed very rapidly after the 1890s. Besides its many manufacturing industries, Madrid is foremost as a banking, education, printing, publishing, and motion-picture center.

Madrid was first mentioned in the 10th cent. as a Moorish fortress. Alfonso VI of Castile drove out the Moors in 1083. The Cortes of Castile met in Madrid several times, but Madrid became the capital of Spain only in 1561, in the reign of Philip II. The city expanded rapidly in the 18th cent. under the Bourbon kings (especially Charles III). From that period date the royal palace and the Prado, which houses one of the finest art collections in the world. At the beginning of the Peninsular War a popular uprising against the French took place at Madrid on May 2, 1808, and a fierce battle was fought in the Puerta del Sol, the city's central square. Madrid again played a heroic role in the Spanish Civil War (1936–39), when it resisted 29 months of siege by the Insurgents, surrendering, thus ending the war, only late in Mar., 1939.

Among the many landmarks are the huge royal palace; the Buen Retiro park, opened in 1631; and the imposing 19th cent. building containing the national library (founded 1712), the national archives, a museum of Spanish modern art, and an archaeological museum.

Ma·du·ra or **Ma·doe·ra** (both: mä-do͝or'ä), island (1970 est. pop. 2,650,000), c.1,760 sq mi (4,560 sq km), Indonesia, near the NE coast of Java, from which it is separated by Madura Strait. Principal products are salt, obtained from pans along the coast, and fish. From the 11th to the 18th cent. Madura was dominated by the rulers of Java.

Ma·du·rai (măd'yo͞o-rī'), city (1971 pop. 548,298), Tamil Nadu state, S India, on the Vaigai River. It is known as the "city of festivals and temples." The Meenakshi temple, which has 1,000 carved pillars, is especially famous. Madurai is also an educational and cultural center and a market for tea, coffee, and cardamom. Important industries are the weaving and dyeing of silk and muslin cloth, the making of brassware, and wood carving. As Mathurai, the city was the capital of the Pandya kingdom from the 5th cent. B.C. until the 11th cent. A.D. In the 14th cent. it was captured by Moslem invaders, who held it until 1378, when it became part of the Hindu Vijayanagar kingdom. From c.1550 until 1736 the city was the capital of the Nayak kingdom. The Carnatic Nawabs then gained control and in 1801 ceded it to the British.

Mae·an·der (mē-ăn'dər), ancient name of the Büyük Menderes River, c.250 mi (400 km) long, W Turkey. It rises in three branches west of Afyonkarahisar and flows generally west into the Aegean Sea. Its valley is extremely fertile. Its winding and wandering course gave rise to the word "meander."

Ma·e·ba·shi (mä-ā'bä-shē), city (1976 est. pop. 253,552), capital of Gumma prefecture, central Honshu, Japan, on the Tone River. It is a silk textile center.

Mae Nam Chao Phra·ya (mă näm' chou prä-yä'), river, Thailand: *see* Chao Phraya.

Maes·teg (mī-stäg'), urban district (1971 pop. 20,970), Mid Glamorgan, S Wales. It is a coal-mining town.

Maf·e·king (măf'ə-kĭng', mä'fə-), town (1970 pop. 6,515), Cape Prov., N central South Africa. It is the market for the surrounding cattle-raising and dairy-farming area and is an important railroad depot. Mafeking was founded in 1885 on the site of a black African settlement. In the South African War (1899–1902) the British garrison withstood a Boer siege for 217 days; the fort is now a national monument. Mafeking was the extraterritorial capital of the Bechuanaland protectorate until it became independent as Botswana in 1965.

Ma·ga·dan (mä'gə-dăn', mə-gə-dän'), city (1976 est. pop. 112,000), capital of Magadan oblast, Far Eastern USSR, a port on the Sea of Okhotsk. It has shipyards and canning factories.

Ma·ga·dha (mä'gə-də), ancient Indian kingdom, situated within the area of modern Bihar. The kingdom rose to prominence in the mid-7th cent. B.C. It fell (c.325) to Chandragupta, who made the kingdom the nucleus of the Mauryan empire. After a period of obscurity, it became (4th cent. A.D.) the power-base of the Gupta dynasty. Buddhism and Jainism first developed in Magadha.

Ma·ga·di, Lake (mə-gä'dē), c.20 mi (30 km) long and 2 mi (3.2 km) wide, S Kenya, in the Great Rift Valley. Formed and constantly resupplied by volcanic springs, the lake has a thick crust of carbonate of soda. The crust is removed by a floating dredge and then pumped to refineries, where it is processed into soda ash.

Mag·da·le·na (măg'də-lā'nə, mäg'dä-lā'nä), river, c.1,000 mi (1,610 km) long, rising in the Cordillera Central, SW Colombia and flowing N to the Caribbean Sea near Barranquilla. It flows through the Andes to a broad, swampy, alluvial plain where the Cauca River joins its lower course. The Magdalena links the interior highlands with the coastal lowlands. Its navigability is hampered by sandbars, rapids, and fluctuating water levels. Its tropical valley is thinly populated. Economic development has been retarded except for the oil industry. Coffee is the chief crop along the river's upper course.

Mag·de·burg (măg'də-bûrg', mäk'də-bo͝orкн'), city (1974 est. pop. 276,089), capital of Magdeburg district, W East Germany, on the Elbe River. It is a large inland port, an industrial center, and a rail and road junction. Manufactures include steel, paper, textiles, machines, and chemicals. There are lignite and potash mines nearby.

Known in 805, Magdeburg became, under Emperor Otto I, an outpost for the colonization of the Wendish territories. In 968 it was made an archiepiscopal see. The city of Magdeburg obtained (13th cent.) a charter that was the model of hundreds of medieval town charters in Germany, Austria, Bohemia, and Poland. Under this Magdeburg Law a town governed itself through an elected council, had its own courts of justice, and was exempt from all duties except the payment of rent to the prince of the land. Magdeburg prospered and became one of the chief members of the Hanseatic League. It accepted (1524) the Reformation, joined (1531) the Schmalkaldic League, and continued its resistance against Emperor Charles V until its fall (1551) to Maurice of Saxony.

During the Thirty Years' War the imperial forces laid siege to Magdeburg in 1630. On May 20, 1631, troops stormed the city and put the garrison to the sword. Fires mysteriously broke out in various quarters, and by the following day virtually the entire city had burned down. Roughly 25,000 persons perished. The city was rebuilt and its trade revived after the Peace of Westphalia (1648), which transferred the city to the electorate of Brandenburg. From the late 17th cent. Magdeburg was an important Prussian fortress. The city was severely damaged in World War II.

Ma·gel·lan, Strait of (mə-jĕl'ən), c.330 mi (530 km) long and 2.5 to 15 mi (4–24.1 km) wide, separating South America from Tierra del Fuego and other islands S of the continent. The strait, discovered by Ferdinand Magellan in 1520, was important in the days of sailing ships, especially before the building of the Panama Canal, and is still used by ships rounding South America. One of the most scenic waterways in the world, it affords an inland passage protected from almost continuous ocean storms.

Ma·gen·ta (mə-jĕn'tə, mä-jān'tä), town (1976 est. pop. 23,780), Lombardy, N Italy, near Milan. Manufactures include matches, textiles, and plastics. At the Ticino River nearby, the French and the Sardinians won a decisive victory (1859) over the Austrians, which opened the way to Milan.

Mag·gio·re, La·go (lä'gō mə-jôr'ē, mäd-jôr'ā), second-largest lake in Italy, 82 sq mi (212.4 sq km), in the Alpine foothills of Piedmont and Lombardy. It is formed by the Ticino River and lies partly in Switzerland. Along part of its western shore run the Simplon Road (built by Napoleon) and a railroad. The lake has many villas and resorts.

Ma·ghreb or **Ma·grib** (both: măg'rəb, mä'grĭb), Arabic term for NW Africa. It is generally applied to all of Morocco, Algeria, and Tunisia but actually pertains only to the area of the three countries between the high ranges of the Atlas Mts. and the Mediterranean Sea.

Mag·na Grae·ci·a (măg'nə grē'shē-ə, -shə), Greek colonies of S Italy. The Greek overseas expansion of the 8th cent. B.C. founded a number of towns, on both coasts from the Bay of Naples and the Gulf of Taranto southward, that became the centers of a new, thriving Greek territory. Unlike Greek Sicily, Magna Graecia began to decline by 500 B.C., probably because of malaria and endless warfare among the colonies. Only Tarentum and Cumae remained individually very significant. Through Cumae especially, the Etruscans of Capua and the Romans came into early contact with Greek civilization.

Mag·ne·si·a (măg-nē'shē-ə, -zhē-ə), two ancient cities of Lydia, W Asia Minor. One city (Magnesia ad Maeandrum), southeast of Smyrna, was later colonized by Ionians and given by Artaxerxes I to Themistocles, who died there. There are important ruins on the site, including the celebrated temple of Artemis Leucophryene, built in the 2nd cent. B.C. Magnesia ad Sipylum, on the Hermus River at the foot of Mt. Sipylus, northeast of Smyrna, was (190 B.C.) the scene of the defeat of Antiochus III by the Romans.

Mag·ni·to·gorsk (măg-nē'tō-gôrsk'), city (1976 est. pop. 393,000), SW Siberian USSR, on the slopes of Mt. Magnitnaya in the S Urals, on the Ural River. Built (1929-31) under the first five-year plan on the site of iron and magnetite deposits, the city became an important symbol of Soviet industrial growth. Coking coal for steel production comes from the Kuznetsk and Karaganda basins; there are also coke and chemical plants.

Mag·no·lia (măg-nōl'yə, -nō'lē-ə). **1.** City (1970 pop. 11,303), seat of Columbia co., SW Ark.; inc. 1855. Its oil industry has been important since 1938. Textiles, chemicals, lumber, and metal and plastic products are also produced. **2.** City (1970 pop. 1,970), seat of Pike co., SW Miss., near the La. line S of McComb, in a farm, dairy, and timber area. **3.** Borough (1970 pop. 5,893), Camden co., SW N.J., SE of Camden; inc. 1915.

Ma·gof·fin (mə-gä'fən), county (1970 pop. 10,443), 303 sq mi (784.8 sq km), E Ky., in the Cumberlands, drained by the Licking River and by several creeks; formed 1860; co. seat Salyersville. It is in an agricultural area that produces livestock, fruit, tobacco, potatoes, and corn. It has bituminous-coal mines, oil wells, and timber.

Ma·gog (mā'gŏg), city (1976 pop. 13,290), S Que., Canada, on Lake Memphremagog, SW of Sherbrooke. Founded by Loyalist emigrants from the United States after 1776, Magog is a resort and trade center, with textile mills and dairying.

Ma·ha·ba·li·pu·ram (mə-hä'bə-lĭ-pŏŏr-əm), village, Tamil Nadu state, SE India, on the Coromandel Coast. Archaeological remains represent some of the earliest-known examples of Dravidian architecture (c.7th cent. A.D.) in India. The site is often called the Seven Pagodas because of the high pinnacles of seven of its temples.

Ma·ha·na·di (mə-hä'nə-dē), river, c.550 mi (885 km) long, rising in SE Madhya Pradesh state, central India, and flowing N then E through a gorge in the Eastern Ghats, across Orissa state, and entering the Bay of Bengal near Cuttack.

Ma·ha·rash·tra (mä'hə-räsh'trə), state (1971 pop. 50,335,492), 118,530 sq mi (306,993 sq km), W India, on the Arabian Sea; capital Bombay. The mountains of the Western Ghats run parallel to the coast of the state, leaving a narrow strip known as the Konkan between the Arabian Sea and the lofty mountain barrier. Beyond the Western Ghats is a vast plateau including the Tapti trough, a fertile belt where cotton is cultivated. Rice, grown in the coastal area, is the principal food crop. The state is rich in minerals; manganese, iron ore, bauxite, coal, and salt are mined. Industry includes the manufacture of textiles, electrical products, and chemicals.

The Moslem rulers of India controlled the area of Maharashtra from the early 14th cent. to the mid-17th cent. In the 16th cent. Portugal was the leading foreign power in the region, but Great Britain gradually gained influence and by the early 19th cent. had incorporated the Maharashtran area into the Bombay presidency, which later became a province of British India. The state was formed in 1960, when the old state of Bombay was split along linguistic lines lines into two new states, Maharashtra and Gujarat.

Ma·has·ka (mə-hăs'kə), county (1970 pop. 22,177), 572 sq mi (1,481.5 sq km), S central Iowa, drained by the Des Moines, Skunk, and North Skunk rivers; formed 1843; co. seat Oskaloosa. Its rolling prairie agriculture includes hogs, cattle, sheep, poultry, corn, oats, and hay. Bituminous coal is mined in the southwest.

Mah·di·a (mə-dē'ə): *see* Al Mahdiyah, Tunisia.

Ma·hi (mä'hē), river, 350 mi (563.2 km) long, W India, rising in the Vindhya Mts. and flowing NW and SW past Bombay into the Gulf of Cambay.

Mah·no·men (mô-nō'mən), county (1970 pop. 5,638), 563 sq mi (1,458.2 sq km), NW Minn., on the Wild Rice River, in White Earth Indian Reservation; formed 1906; co. seat Mahnomen. It is a trading point for livestock, poultry, dairy products, flour, and lumber.

Mahnomen, village (1970 pop. 1,313), seat of Mahnomen co., NW Minn., NE of Moorhead.

Ma·hón (mä-ōn'), town (1970 pop. 19,279), capital and chief town of Minorca island, Baleares prov., Spain, in the W Mediterranean Sea.

Ma·ho·ning (mə-hō'nĭng), county (1970 pop. 304,545), 419 sq mi (1,085.2 sq km), E Ohio, bounded on the E by the Pa. line and intersected by the Mahoning and Little Beaver rivers; formed 1846; co. seat Youngstown. The eastern part has an extensive steel industry, and the western portion is mainly agricultural, yielding corn, oats, and livestock.

Mahoning, river, c.90 mi (145 km) long, rising in NE Ohio, E of Canton, and flowing NW to Alliance, then NE past Warren, where it turns SE to flow past Youngstown into NW Pa. and joins the Shenango River to form the Beaver River.

Mai·den Cas·tle (mā'dn kăs'əl), prehistoric fortress, Dorset, S England, near Dorchester. The finest earthwork in the British Isles, c.115 acres (45 hectares) in area, is here. Excavations in 1934-37 revealed evidences of a Neolithic village, with a two-ditch irrigation system, indicating occupation c.2000 B.C. On the same site are remains of an Iron Age fortified village (300 B.C.), which was eventually taken by the Romans. The inhabitants ceased to occupy Maiden Castle about A.D. 70 when they moved to a town in the nearby valley.

Maid·en·head (măd'n-hĕd'), municipal borough (1973 est. pop. 48,210), Berkshire, S central England, on the Thames River. It is a residential town and a resort. There are also brewing and milling industries. The 13th cent. stone bridge was rebuilt in the 1770's.

Maid·stone (măd'stōn', -stən), municipal borough (1973 est. pop. 72,110), administrative center of Kent, SE England, on the Medway River. It is a market town. There are paper, printing, quarrying, brewing, engineering, and agricultural industries. There is evidence of a Roman station. Chillington Manor (Elizabethan) contains the Maidstone Museum, the public library, and the headquarters of the Kent Archaeological Society. The grammar school dates from 1549.

Mai·du·gu·ri (mī-dŏŏ'gŏŏr-ē), town (1971 est. pop. 169,000), NE Nigeria. Leather goods made from the hides of crocodiles caught in Lake Chad are a leading product of the town.

Mai·kop (mī-kôp'), city (1976 est. pop. 127,000), capital of Adyge

Autonomous Oblast, Krasnodar Kray, S European USSR, at the foot of the Greater Caucasus and on the Belaya River. It has machinery, lumber, and food-processing industries. Nearby are the important Maikop oil fields, discovered in 1900-1901. Maikop was founded in 1857 as a Russian fortress. It was captured by German troops in 1942 and retaken by Soviet forces in 1943.

Main (mīn, mān), river, c.310 mi (500 km) long, formed near Kulmbach, E West Germany, by the confluence of the Roter Main and the Weisser Main rivers, both of which rise in the Fichtelgebirge. It then winds generally west through central West Germany and past a heavily industrial area to join the Rhine River at Mainz.

Maine (mān, mĕn), region and former province, NW France, S of Normandy and E of Brittany. Maine is primarily agricultural, with stock raising in the hilly Perche. Important during Roman times, Maine was Christianized between the 4th and 6th cent. Made a county in the 10th cent., it passed (1126) to Anjou and was held for long periods by England. It frequently reverted to the French crown, or to members of the royal family, until it was finally united with the crown in 1584 upon the death of the duke of Alençon.

Maine (mān), state (1975 pop. 1,055,000), 33,215 sq mi (86,027 sq km), in the extreme NE corner of the United States, largest of the New England states; admitted as the 23rd state of the Union in 1820. Augusta is the capital. The Canadian provinces of Que. and N.B. border Maine from the northwest around to the southeast coast, with the St. John and St. Croix rivers forming part of the international boundary with N.B. To the south is the Atlantic Ocean. New Hampshire (to the west) is the only state bordering Maine.

Geologic action laid down a bedrock of sandstone, shale, and limestone. Much of the soft rock eroded into tableland valleys, while the more resistant rock remained, forming the generally mountainous west, the mountains of Mt. Desert Island in the east, and isolated peaks including Katahdin (5,268 ft/1,606 m), the highest point in the state. Receding glaciers left more than 2,200 lakes (Moosehead Lake is the largest) and watercourses for more than 5,000 streams. The major rivers are the St. John, the Penobscot, the Kennebec, the Androscoggin, and the Saco. The jigsawed coastline of 2,500 mi (3,219 km) is rugged and wild east of the Kennebec, but west of the river the shoreline has sandy beaches and marshy lowlands. Four fifths of the state is still forested with hemlock, spruce, fir, and hardwoods. In the shelter of lakes and woods, particularly in the north counties, wildlife has found refuge. Moose, deer, black bears, and smaller game are still found; fish and fowl are plentiful.

Maine is divided into 16 counties:

NAME	COUNTY SEAT	NAME	COUNTY SEAT
Androscoggin	Auburn	Oxford	South Paris
Aroostook	Houlton	Penobscot	Bangor
Cumberland	Portland	Piscataquis	Dover-Foxcroft
Franklin	Farmington	Sagadahoc	Bath
Hancock	Ellsworth	Somerset	Skowhegan
Kennebec	Augusta	Waldo	Belfast
Knox	Rockland	Washington	Machias
Lincoln	Wiscasset	York	Alfred

Economy. Much of Maine's abundant natural and industrial resources remain undeveloped. The mineral wealth of the state includes superior varieties of granite, as well as sand and gravel, zinc, and peat. Gold was discovered at Pembroke in 1965. Maine is the third-largest producer of beryllium concentrate in the United States. The population of Maine is centered on the cleared land along the coast and major rivers. With many factors operating against prosperity—a generally poor soil and a short growing season, geographic remoteness, an inadequate water distribution system, and a lack of coal and steel—Maine has had a very low population increase in the last century. However, the picturesque coastal and island resorts of Maine hold a strong appeal for visitors and, combined with abundant wildlife to attract sportsmen, make the tourist trade a most important feature of Maine's economy. Maine's other economic assets are its protected harbors, which serve as fishing ports; rapid rivers, which provide power for mills and factories; extensive lumber and fishing resources; and farming regions. Fishing, one of the state's earliest industries, remains important and Maine lobsters are nationally famous.

Manufacturing is the largest economic sector, accounting for one third of all production. Maine is a leading producer of paper and paper products, which account for about one third of the value of all manufactures in the state. Leather goods, food products, textiles, and transportation equipment are also produced. Agriculture, which occupies approximately one third of the population, has been developed despite adverse soil and climatic conditions. Dairying, poultry raising, and market gardening predominate. Potatoes, hay, apples, and oats are the chief crops.

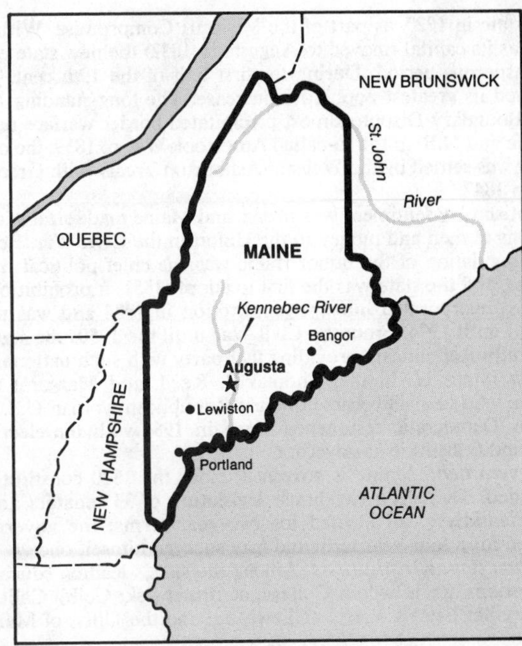

History. At the time of settlement by white men the friendly Abnaki Indians were scattered along the coast of Maine and in some inland areas. The coast may have been visited by the Norsemen and was known to British, French, and Spanish mariners before the sieur de Monts and Samuel de Champlain established a short-lived French colony in 1604 at the mouth of the St. Croix River. English colonists of the Plymouth company founded Fort St. George, on the present site of Phippsburg, at the mouth of the Kennebec in 1607, but the settlement failed, and the colonists returned to England in 1608. In 1620 the Council for New England (successor to the Plymouth Company) granted Ferdinando Gorges and Captain John Mason the territory between the Kennebec and Merrimack rivers extending 60 mi (97 km) inland. At this time the region became known as Maine, either to honor Henrietta Maria, queen of Charles I, who was feudal proprietor of the province in France called Maine, or to distinguish the mainland from the offshore islands. Gorges and Mason divided (1629) their grant, with Gorges taking the area east of the Piscataqua. Permanent settlements were established at Monhegan, Saco, and York. Neglected after Gorges's death in 1647, Maine settlers came under the jurisdiction of the Massachusetts Bay Colony in 1652.

King Philip's War (1675-76) was the first of many struggles between the British on one side and the French and Indians on the other that slowed down the further settlement of Maine. French influence declined rapidly after 1688, and local government and institutions in the Massachusetts tradition took root. Because it was on the frontier, the province was repeatedly ravaged by Indians, but their strength was broken during Queen Anne's War (1702-3). Maine nevertheless recovered quickly and soon had prosperous fishing, lumbering, and shipbuilding industries.

Dissatisfaction with British rule was first expressed openly after Parliament passed the Stamp Act in 1765; in protest, a mob at Falmouth (Portland) seized a quantity of the hated stamps. During the Revolution, Falmouth paid dearly for its defiance; it was devastated by a British fleet in 1775. During the war supplies were cut off, and Indian attacks were frequent, but with American independence won, economic development was rapid in what was then called the Dist. of Maine, one of the three admiralty districts of Massachusetts set up by the Continental Congress in 1775.

Agitation for statehood, which had been growing since the Revolution, became widespread after the War of 1812. Equality of power between North and South was preserved by admitting Maine as a free state in 1820, as part of the Missouri Compromise. With Portland as its capital (moved to Augusta in 1832) the new state entered a prosperous period. During the first half of the 19th cent. Maine enjoyed its greatest population increase. The long-standing Northeast Boundary Dispute almost precipitated border warfare between Maine and N.B. in the so-called Aroostook War of 1839; the controversy was settled by the Webster-Ashburton Treaty with Great Britain in 1842.

Antislavery sentiment was strong, and Maine made sizable contributions of men and money to the Union in the Civil War. For decades regulation of the liquor traffic was the chief political issue in Maine, and the state was the first to adopt (1851) a prohibition law. It was incorporated into the constitution in 1884 and was not repealed until 1934. From the Civil War until the 1950s the state was generally Republican, providing that party with such national leaders as James G. Blaine, Thomas B. Reed, and Margaret Chase Smith, who in 1948 became the first Republican woman U.S. Senator. A Democratic resurgence began in 1954 with the election of Edmund S. Muskie as governor.

Government. Maine is governed under the 1820 constitution as amended. There is a two-house legislature of 33 senators and 151 representatives, all elected for two-year terms; the governor is elected for a four-year term and may succeed himself once.

Educational Institutions. Among the state's leading educational institutions are Bowdoin College, at Brunswick; Colby College, at Waterville; Bates College, at Lewiston; and the Univ. of Maine, at Orono.

Maine, Gulf of, part of the Atlantic Ocean, between SE Maine and SW N.S., at the entrance of the Bay of Fundy.

Maine-et-Loire (mā'nā-lwär', mĕ'-), department (1975 pop. 629,849), 2,753 sq mi (7,130.3 sq km), NW France, roughly coextensive with Anjou; capital Angers.

Main·land (măn'lănd', -lənd). **1.** Island (1971 pop. 6,502), 178 sq mi (461 sq km), N Scotland. It is the largest of the Orkney Islands. Kirkwall Bay and Scapa Flow deeply indent its shores. The interior has hills, moors, several lakes, and fertile valleys. Cattle and sheep are raised; eggs are a leading product. There is also a distilling industry. Local customs in some districts reveal the Norse ancestry of many of the inhabitants. There are numerous Pictish remains—mounds, underground dwellings, circles, and standing stones. **2.** Island, 375 sq mi (971.3 sq km), extreme N Scotland. It is the largest of the Shetland Islands. There are remains of a prehistoric village.

Mainz (mīnts), city (1974 est. pop. 184,030), capital of Rhineland-Palatinate, W central West Germany, a port on the left bank of the Rhine River opposite the mouth of the Main River. The city is an industrial, commercial, and transportation center. Chemicals, motor vehicles, machinery, cement, champagne, and printed materials are produced; the city is also a trade center for Rhine wines.

Mainz grew on the site of the Roman camp (founded 1st cent. B.C.). The city was made (746–47) the seat of the first German archbishop, St. Boniface (c.675–754). The later archbishops acquired considerable territory around Mainz and in Franconia, on both sides of the Main, which they ruled as princes of the Holy Roman Empire. Under their rule Mainz flourished as a commercial and cultural center. Jews, who had one of their oldest settlements in Germany at Mainz, played an important part in the prosperity of the city. Johann Gutenberg (c.1397–1468) lived in Mainz, which he made the first printing center of Europe. Occupied in 1792 by the French, the city was ceded to France by the treaties of Campo Formio (1797) and Lunéville (1801), and the archbishopric was secularized and reduced to a diocese in 1803.

The Congress of Vienna made (1815) Mainz a federal fortress of the German Confederation and awarded it, with Rhenish Hesse, to the grand duchy of Hesse-Darmstadt. Mainz was severely damaged during World War II, but was largely restored and rebuilt after 1945. Noteworthy structures in the old inner city include the six-towered Romanesque cathedral (consecrated 1009; restored 19th cent.); the Renaissance-style electoral (archiepiscopal) palace (17th–18th cent.), which houses an art gallery and a museum of Roman and Germanic antiquities; and the Church of St. Peter (18th cent.). The Univ. of Mainz was founded in 1477, was discontinued in 1816, and was re-established in 1946 as the Johannes Gutenberg Univ.

Mai·po (mī-pō͞o'), volcano, 17,464 ft (5,326.5 m) high, in the Andes on the Chile-Argentina border SE of Santiago, Chile.

Mai·pú (mī-pō͞o'), battlefield, central Chile, a few miles S of Santiago. On Apr. 5, 1818, San Martín routed the Spanish royalist army at Maipú and assured Chilean independence.

Mai·sons-Al·fort (mā-zôNz-äl-fôr'), suburb SE of Paris (1975 pop. 54,146), Val-de-Marne dept., N central France. With some agriculture, it is mainly an industrial town producing chemicals and metals.

Mait·land (māt'lənd), city (1976 pop. 35,996), New South Wales, SE Australia, on the Hunter River. It is a railroad junction and agricultural center with light manufacturing. Maitland began as a convict settlement in 1824. The river has flooded in 1893, 1949, and 1955.

Mai·zu·ru (mī'zō͞o-rō͞o), city (1975 pop. 82,600), Kyoto prefecture, SW Honshu, Japan, on Maizuru Bay. An important port and naval base with the best natural harbor on the Japan Sea coast.

Ma·jor (mā'jər), county (1970 pop. 7,529), 963 sq mi (2,494.2 sq km), NW Okla., drained by the Cimarron River; formed 1907; co. seat Fairview. It is in an agricultural area yielding wheat, barley, broomcorn, oats, peanuts, dairy products, and livestock.

Ma·jor·ca (mə-jôr'kə, -yôr'-), island (1970 pop. 460,030), 1,405 sq mi (3,639 sq km), Spain, largest of the Balearic Islands, in the W Mediterranean. Majorca is mountainous in the northwest, rising to 4,739 ft (1,445.4 m); the south and east form a gently rolling, fertile region. Its mild climate and beautiful scenery have long made Majorca a popular resort; tourism is its major industry. Cereals, flax, grapes, and olives are grown, a light wine is produced, hogs and sheep are raised, and lead, marble, and copper are mined. In 1276 the kingdom of Majorca was formed from the inheritance of James I of Majorca. It comprised the Balearic Islands, Roussillon and Cerdagne (between France and Spain), and several fiefs in southern France. In 1343 Peter IV of Aragón took the kingdom from James II and reunited it with the crown of Aragón. The island's commerce declined, partly because of the warfare between the native peasantry and the Aragonese nobles and Catalan traders, but mainly because of the change in trade routes after the discovery of America. Majorca is known for its stalagmite caves and for its architectural treasures and prehistoric monuments. The abandoned monastery where Frédéric Chopin and George Sand lived is a landmark.

Ma·jun·ga (mə-jŭng'gə), city (1972 est. pop. 67,456), NW Malagasy Republic, on Madagascar, on the Mozambique Channel. Despite its shallow harbor, Majunga is one of the nation's chief ports.

Ma·kas·ar (mə-kăs'ər), city (1971 pop. 434,766), SW Celebes, capital of South Sulawesi prov., Indonesia. It is an important seaport. Exports include coffee, teak, spices, copra, rubber, rattan, and gums and resins. The city is a commercial center, with a large central market. Industries include the manufacture of cement and paper, and automobile assembly. Once a center of spice smuggling, Makasar was a thriving port when the Portuguese arrived (16th cent.). The Dutch supplanted the Portuguese, triumphing over the indigenous sultan in 1667. The city became a free port in 1848.

Ma·ke·yev·ka (mä-kĕ'yəf-kä'), city (1976 est. pop. 437,000), S European USSR, in the Ukraine, in the Donets Basin. It is a metallurgical and coal-mining center and has machinery and coking plants. Makeyevka was founded (1899) as a metallurgical settlement.

Ma·khach·ka·la (mə-KHăch'kə-lä'), city (1976 est. pop. 231,000), capital of Dagestan Autonomous Republic, SE European USSR, a port on the Caspian Sea. It is an important commercial and industrial center with oil refineries that are linked by pipeline with the Grozny fields. Aircraft and textiles are manufactured.

Ma·kin (mä'kĭn, mä'-), atoll: *see* Butaritari.

Mal·a·bar Coast (măl'ə-bär'), SW coast of India stretching c.525 mi (845 km) from Goa to the S tip of the peninsula at Cape Comorin. It is a narrow coastal plain bounded by the Western Ghats. Monsoon rains make the coast a fertile rice-growing region. It was the scene of trade struggles in the 16th and early 17th cent. between the Portuguese and their European and Indian rivals.

Mal·a·bo (măl'ə-bō'), formerly **San·ta I·sa·bel** (săn'tə ĭz'ə-bĕl', sän'tä ē'sä-vĕl'), city (1970 est. pop. 20,000), capital of Equatorial Guinea, on Fernando Po island, in the Gulf of Guinea. Fish processing is the city's main industry, and cacao and coffee are the leading exports. The city was founded in 1827 by the British as a base for the suppression of the slave trade.

Ma·lac·ca (mə-lăk'ə): *see* Melaka, Malaysia.

Malacca, Strait of, c.500 mi (805 km) long and from c.30 to 200 mi (50–320 km) wide, between Sumatra and the Malay Peninsula. Linking the Indian Ocean with the South China Sea, it is one of the world's most important sea passages.

Ma·lad City (mə-lăd'), city (1970 pop. 1,848), seat of Oneida co., SE Idaho, SE of Pocatello. Founded by Mormons in 1864, it was a pony-express station. Today it is a shipping center for a livestock, sugar-beet, wheat, and potato area.

Ma·la·det·ta Mountains (mä'lə-dĕt'ə), massif of the central Pyrenees, NE Spain, near the French border. Its highest point, Pico de Aneto (11,168 ft/3,406.2 m), is also the highest in the Pyrenees.

Má·la·ga (măl'ə-gə, mä'lä-gä), city (1975 est. pop. 408,458), capital of Málaga prov., S Spain, in Andalusia, on the Guadalmedina River and the Costa del Sol. Picturesquely situated on the Bay of Málaga, it is one of the best Spanish Mediterranean ports. Olives, almonds, and dried fruits are exported. Málaga's mild climate and luxurious flora, as well as the beautiful beaches nearby, make it also a popular resort. Founded (12th cent. B.C.) by the Phoenicians, the city passed to the Carthaginians, the Romans, the Visigoths, and finally (711) the Moors. It flourished from the 13th cent. as a seaport of the Moorish kingdom of Granada, until it fell to Ferdinand and Isabella in 1487. Picasso was born in Málaga.

Mal·a·gas·y Republic (măl'ə-găs'ē), republic (1976 pop. 7,700,000), 226,658 sq mi (587,044 sq km), in the Indian Ocean, separated from E Africa by the Mozambique Channel; capital Tananarive. The nation is made up of Madagascar, the world's fourth-largest island, and several small islands including Sainte-Marie, Nossi-Bé (Nosy-Bé), Juan de Nova, Europa, and Bassas da India. Madagascar is made up of a highland plateau fringed by a lowland coastal strip, narrow (c.30 mi/50 km) in the east and considerably wider (c.60–125 mi/95–200 km) in the west. The plateau rises in the north, where Mount Maromokotro (9,450 ft/2,882.2 m), the loftiest point in the country, is located, and in the center, where the Ankaratra Mts. reach c.8,670 ft (2,645 m). Once heavily wooded, the plateau is now largely deforested. Lagoons along much of the east coast are connected in part by the Pangalanes Canal, which can accommodate small boats.

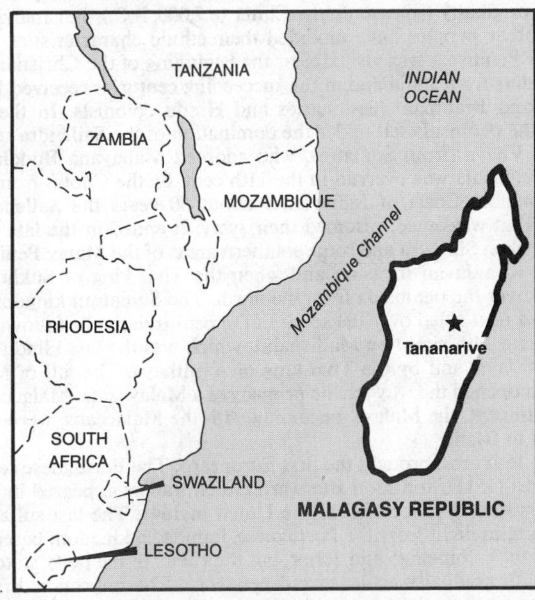

Economy. The economy of the Malagasy Republic is overwhelmingly agricultural, largely of a subsistence type; the best farmland is in the east and northwest. The principal crops are rice, manioc, millet, pulses, sugar cane, groundnuts, coffee, raffia, tobacco, cloves, and vanilla. In addition, large numbers of poultry, cattle, goats, sheep, and hogs are raised. Manufactures are mostly confined to agricultural products, beverages, and basic consumer goods like clothing. Refined petroleum, cement, paper, and radio and television receivers are also produced, and motor vehicles are assembled. The country has a small but growing mining industry; the chief minerals extracted are chromite, graphite, phosphates, ilmenite, mica, zircon, and industrial beryl and garnets.

History. The earliest history of Madagascar is unclear. Black Africans and Indonesians reached the island about 2,000 years ago, the Indonesian immigration continuing until the 15th cent. From the 9th cent. Moslem traders (including some Arabs) from East Africa and the Comoro Islands settled in northwest and southeast Madagascar. Probably the first European to see Madagascar was Diogo Dias, a Portuguese navigator, in 1500. Between 1600 and 1619, Portuguese Roman Catholic missionaries tried unsuccessfully to convert the Malagasy. From 1642 until the late 18th cent. the French maintained footholds, first at Fort-Dauphin in the southeast and finally on Sainte Marie Island off the east coast. By the beginning of the 17th cent. there were a number of small Malagasy kingdoms. Later in the century the Sakalawa under Andriandahifotsi conquered west and north Madagascar, but the kingdom disintegrated in the 18th cent. The Merina people of the interior were united under King Andrianampoinimerina (reigned 1787–1810), who also subjected the Bétsiléo. Radama I (reigned 1810–28), in return for agreeing to end the slave trade, received British aid in modernizing and equipping his army, which helped him to conquer the Betsimisáraka kingdom. The Protestant London Missionary Society was welcomed, and it gained many converts, opened schools, and helped to transcribe the Merina language. Merina culture began to spread over Madagascar. Radama was succeeded by his wife Ranavalona I (reigned 1828–1861), who, suspicious of foreigners, declared (1835) Christianity illegal and halted most foreign trade. During her rule the Merina kingdom was wracked by intermittent civil war. Under her successors the anti-European policy was reversed and missionaries (including Roman Catholics) and traders were welcomed again. By the end of the century the Merina kingdom included all Madagascar except the south and part of the west. In 1883 the French bombarded and occupied Tamatave, and in 1885 they established a protectorate over Madagascar, which was recognized by Great Britain in 1890. Resistance to the French, caused heavy fighting from 1894 to 1896. In 1896 French troops under J. S. Gallieni defeated the Merina and abolished the monarchy. By 1904 the French fully controlled the island.

Under the French, who governed the Malagasy through a divide-and-rule policy, the Merina benefited most from colonial rule. Merina nationalism developed early in the 20th cent., and in 1916 a Merina secret society was suppressed after a plot against the colonialists was discovered. During World War II Madagascar was aligned with Vichy France until 1942, when it was conquered by the British; in 1943 the Free French regime assumed control. From 1947 to 1948 there was a major uprising against the French, who crushed the rebellion, killing between 11,000 and 80,000 (estimates vary) Malagasy in the process. As in other French colonies, indigenous political activity increased in 1956, and the Social Democratic Party (PSD), led by Philibert Tsiranana (a Tsimihety), gained predominance in Madagascar. On Oct. 14, 1958, the country—renamed the Malagasy Republic—became autonomous within the French Community and Tsiranana was elected president. On June 26, 1960, it became fully independent. Under Tsiranana (re-elected in 1965 and 1972), an autocratic ruler whose PSD controlled parliament, government was centralized, the coastal peoples were favored over those of the interior (especially the Merina), and French economic and cultural influence remained strong. In a controversial move beginning in 1967, Tsiranana cultivated economic relations with white-ruled South Africa. After his re-election in Jan., 1972, students and workers staged a series of protest demonstrations. At the height of the crisis, Tsiranana handed power over to Gabriel Ramanantsoa; following a national referendum approving a plan to allow Ramanantsoa to rule for 5 years without parliament, Tsiranana resigned (Oct., 1972). Ramanantsoa resigned in Feb., 1975; his successor, Richard Ratsimandrava, was assassinated by paramilitary police after 6 days in power. In June, 1975, Didier Ratsiraka emerged as head of state. Under Ratsiraka parliament remained suspended, and a supreme executive council was created to direct policy. In 1977 the country became known officially as Madagascar.

Mä·lar·en (mĕ'lä-rən), lake, c.440 sq mi (1,140 sq km), E central Sweden. The lake's scenic shores and more than 1,000 islands have many villas and historic castles and ruins.

Mal·a·spi·na (măl'ə-spē'nə), glacier, c.1,500 sq mi (3,885 sq km), SE Alaska, between Yakutat Bay and Icy Bay and flowing into the Gulf of Alaska. The glacier was named for an Italian navigator who explored this region for Spain in 1791.

Ma·la·tya (mä'lə-tyä'), city (1975 pop. 154,056), capital of Malatya prov., E central Turkey, in the E Taurus mts. It is the commercial center for a rich farm region that produces apricots, grapes, and grains. Manufactures of the city include cement, cotton textiles, and sugar. Situated at a strategic crossroads in ancient times, the city was the capital of a small Hittite kingdom c.1100 B.C. An important city of Cappadocia, it became a metropolitan see in early Christian times. The city frequently suffered from attack and changed hands many times. In 1516 it was annexed by the Ottoman Empire.

Ma·la·wi (mä-lä'wē), formerly **Ny·as·a·land** (nī-ăs'ə-lănd'), republic (1977 est. pop. 5,225,000), 45,200 sq mi (117,068 sq km), E central Africa, bordering on Zambia in the W, on Tanzania in the N, and on Mozambique in the E, S, and SW; capital Lilongwe. Malawi is long and narrow, and about 20% of its total area is made up of Lake Nyasa, within the Great Rift Valley. Much of the rest of the country is made up of a plateau that averages 2,500 to 4,500 ft (762.5–1,372.5 m) in height, but reaches elevations of c.8,000 ft (2,440 m) in the north and almost 10,000 ft (3,050 m) in the south.

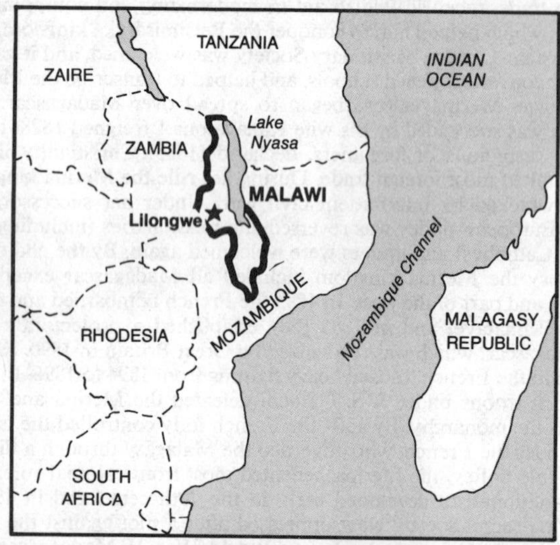

Economy. Malawi is an overwhelmingly agricultural country, with a very low per capita income. About 85% of the cultivated land is made up of small farms held under traditional terms of tenure; the principal crops raised are maize, pulses, millet, sorghum, groundnuts, cassava, and potatoes. The rest of the farmland is included in large estates, where tea, tobacco, sugar cane, and tung oil are produced. Few minerals are extracted, but there are substantial deposits of bauxite.

History. The first inhabitants of present-day Malawi were probably Pygmylike hunter-gatherers. In the 15th cent. Bantu-speaking persons migrated from the west and north, eventually forming the Malawi kingdom (late 15th–late 18th cent.), centered in the Shire River valley. In the 18th cent. the kingdom conquered much of modern Rhodesia and Mozambique. However, shortly thereafter it declined as a result of internal rivalries and incursions by the Yao, who sold their Malawi captives as slaves. In the 1840s the region was thrown into further turmoil by the arrival from southern Africa of the predatory Ngoni. Great Britain sent a consul to the area in 1883 and proclaimed the Shire Highlands Protectorate in 1889. In 1891 the British Central African Protectorate (known from 1907 until 1964 as Nyasaland), which included most of present-day Malawi, was established. During the 1890s British forces ended the slave trade in the protectorate. Europeans established coffee-growing estates in the Shire region; they were worked by Africans who thereby earned the cash necessary to pay taxes.

In 1915 John Chilembwe, a Yao Christian missionary aggrieved by British rule led a small-scale revolt. The revolt was easily suppressed, but it was long remembered by other Africans intent upon ending foreign control. In 1944 the protectorate's first political movement, the moderate Nyasaland African Congress, was formed. In 1953 the Federation of Rhodesia and Nyasaland (linking Nyasaland, Northern Rhodesia, and Southern Rhodesia) was formed, over the strong opposition of Nyasaland's black Africans, who feared the imposition of more aggressively European-oriented policies of Southern Rhodesia. In 1958 Dr. Hastings Kamuzu Banda became the leader of the nationalist movement, which was renamed the Malawi Congress Party in 1959. Banda organized protests that led to the declaration of a state of emergency in 1959–60. The Federation of Rhodesia and Nyasaland was ended in 1963, and on July 6, 1964, Nyasaland became independent as Malawi. Banda led the country in the era of independence, first as prime minister and, in 1966, as president; he was made president for life in 1971. He quickly alienated other leaders and crushed a revolt led by H. B. M. Chipembere in 1965 and one led by Yatuta Chisiza in 1967. Arguing that the country's economic well-being depended on friendly relations with the white-run governments in southern Africa, Banda established diplomatic ties between Malawi and South Africa in 1967.

Malawi, Lake: *see* Nyasa, Lake.

Ma·lay Archipelago (mə-lā', mā'lā), great island group of SE Asia. It lies between the Asian mainland and Australia, separating the Pacific Ocean from the Indian Ocean.

Malay Peninsula, southern extremity (c.70,000 sq mi/181,300 sq km) of the continent of Asia, lying between the Andaman Sea of the Indian Ocean and the Strait of Malacca on the W and the Gulf of Siam and the South China Sea on the E. It stretches south for c.700 mi (1,125 km) from the Isthmus of Kra, where it is narrowest, to Singapore. The northern part of the peninsula forms a part of Thailand; the southern part constitutes West Malaysia, the Malayan part of the Federation of Malaysia. A mountain range forms the backbone of the peninsula; from it numerous short, swift rivers flow east and west. More than half of the land surface is covered with tropical rain forest. The region is one of the richest of the world in the production of tin and rubber; other products include timber, copra and coconut oil, palm oil, tapioca, peanuts, pineapples, and bananas. Rice is the chief foodstuff. The peninsula forms a physical and cultural link between the mainland of Asia and the islands of Indonesia (often included in the Malay Archipelago).

The Malays, historically the dominant cultural group, probably came originally from southern China (c.2,000 B.C.), but marriages with other peoples have modified their ethnic characteristics. The Malay Peninsula was visited near the beginning of the Christian era by traders from India and in the succeeding centuries received Buddhist and Brahman missionaries and Hindu colonists. In the 8th cent. the peninsula fell under the domination of the Sailendra rulers of Sri Vijaya (from Sumatra), who adopted Mahayana Buddhism. The peninsula was overrun in the 11th cent. by the Cholas from the Coromandel Coast of India; after about 50 years the Sailendras, somewhat weakened, resumed their sway. It ended in the late 13th cent., when Sumatra and some southern areas of the Malay Peninsula fell to a Javan invasion and when the Thai king of Sukhothai swept over the peninsula from the north. The Sumatran kingdom of Melayu next ruled over the south of the peninsula, to be followed in turn (late 14th cent.) by Madjapahit, which was the last Hindu empire of Java, and by the Thai king of Ayutthaya. The fall of Madjapahit opened the way for the primacy of a Malay state, Malacca. In the 15th cent. the Malays, beginning with the Malaccans, were converted to Islam.

The 16th cent. brought the first Europeans. The Portuguese seized Malacca (1511), and soon afterward Dutch traders appeared in Malayan waters. Malacca fell to the Dutch in 1641. The last sultan of Malacca, in flight from the Portuguese, founded a kingdom based on the Riau Archipelago and Johor, but the rulers of the petty states in the south gradually achieved independence. The important British role began with the founding of settlements at Pinang (1786) and Singapore (1819). The rising power of Siam and an increasingly imperial Britain became rivals. The British established protectorates over several Malay states, and in 1909 the boundary between Siam and Malaya was fixed by Siam's transfer to Great Britain of suzerainty over Kedah, Perlis, Kelantan, and Terengganu.

Ma·lay·sia, Federation of (mə-lā'zhə, -shə), country (1975 pop. 12,368,000), 128,430 sq mi (332,634 sq km), Southeast Asia. Malaysia consists of two parts: West Malaysia (1971 pop. 8,791,690), 50,700 sq mi (131,313 sq km), on the Malay Peninsula and coextensive with the former Federation of Malaya; and East Malaysia (1971 pop. 1,632,635), 77,730 sq mi (201,321 sq km), on the island of Borneo. The two parts are separated by the South China Sea. West Malaysia is bordered on the north by Thailand, on the east by the South China Sea, on the south by Singapore (separated by the narrow Johore Strait), and on the west by the Strait of Malacca and the Andaman Sea. East Malaysia is bordered on the north by the South China Sea and the Sulu Sea, on the east by the Celebes Sea, and on the south and west by Kalimantan (Indonesian Borneo). Along the coast within Sarawak are the two small portions of the British protectorate of Brunei. The capital of Malaysia is Kuala Lumpur. Both East and West Malaysia have mountainous interiors and coastal plains. Lying close to the equator, Malaysia has a tropical rainy climate.

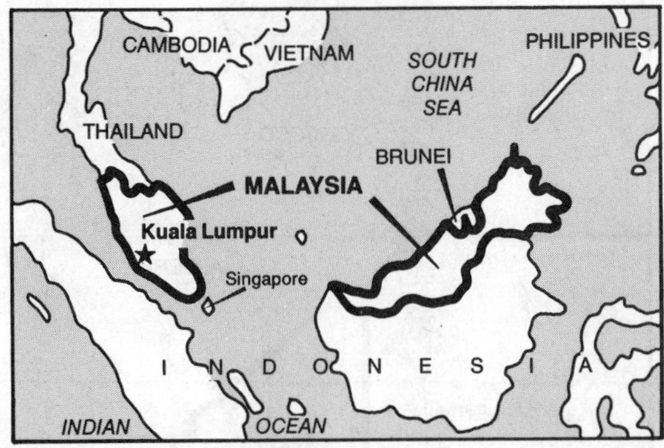

Nearly three fourths of the land area is forested, and many parts of the country have not yet been explored.

Malaysia is a country of vast ethnic diversity. Of the total population a little more than two fifths are Malays, a little less than two fifths are Chinese, about one tenth are Indians and Pakistanis, and the remainder belong to sixteen indigenous ethnic groups. Conflict between the ethnic groups, particularly between Malays and Chinese, has played a large role in Malaysian history.

Economy. The federation is one of the world's leading suppliers of tin and rubber. Other large exports are timber and forest products, palm oil, and iron ore. Other minerals found in Malaysia are petroleum, bauxite, ilmenite, copper, and gold. Subsistence agriculture, however, remains the basis of livelihood for most Malaysians; rice is the staple food, while fish supply most of the protein. Industry, mainly processing and light manufacturing, is largely concentrated in West Malaysia.

History. When the Portuguese captured Malacca (1511), its sultan fled first to Pahang and then to Johor and the Riau Archipelago. From both Johor and Acheh in Sumatra unsuccessful attacks were made on Malacca. Acheh and Johor also fought each other. The main issue in these struggles was control of trade through the Strait of Malacca. Kedah, Kelantan, and Terengganu, north of Malacca, became nominal subjects of Siam. By 1619 the Dutch had established themselves in Batavia (Djakarta), and in 1641, allied with Johor, they captured Malacca after a six-month siege. In the late 17th cent. when the Bugis from Celebes, a Malay people economically pressured by the Dutch, began settling near Selangor on the west coast of the peninsula, where they traded in tin. The Bugis captured Johor and Riau in 1721 and, with a few interruptions, maintained control there for about a century. The Bugis were also active in Perak and Kedah. Earlier, in the 15th and 16th cent., another Malay people, the Minangkabaus from Sumatra, had peacefully settled inland from Malacca. Their settlements eventually became the state of Negeri Sembilan.

The British role on the peninsula began in 1786, when Francis Light of the British East India Company, searching for a site for trade and a naval base, obtained the cession of the island of Pinang from the sultan of Kedah. In 1819 the British founded Singapore, and in 1824 they formally (actual control had been exercised since 1795) acquired Malacca from the Dutch. A joint administration was formed for Pinang, Malacca, and Singapore, which became known as the Straits Settlements. During this period Siam was asserting its influence southward on the peninsula. The Anglo-Siamese treaty of 1821 recognized Siamese control of Kedah but left the status of Perak, Kelantan, and Terengganu ambiguous. In 1841 the sultan of Kedah was restored, but Perlis was carved out of the territory of Kedah and put under Siamese protection.

Later in the century there was conflict between Chinese settlers, who worked in the tin mines, and Malays; there were civil wars among the Malays; and there was an increase in piracy in the western part of the peninsula. Merchants asked the British to restore order. The British were also concerned that Dutch, French, and German interest in the area was increasing. In each state a British "resident" was installed to advise the sultan (who received a stipend) and to supervise administration. The Pangkor Treaty of 1874 with Perak served as a model for subsequent treaties. In 1896 the four states were grouped together as the Federated Malay States with a British resident general. Johor, which had signed a treaty of alliance with Britain in 1885, accepted a British adviser in 1914. British control of the four remaining Malayan states was acquired in 1909, when, by treaty, Siam relinquished its claims to sovereignty over Kedah, Kelantan, Perlis, and Terengganu. These four, along with Johor, became known as the Unfederated Malay States.

In the latter half of the 19th cent. Malaya's economy assumed essentially its present character. The output of tin, mined for centuries, increased greatly with the utilization of modern methods. Rubber trees were introduced (Indian laborers were imported to work the rubber plantations), and Malaya became a leading rubber producer. Malaya's economic character and its geographic position gave it great strategic importance, and the peninsula was quickly overrun by the Japanese at the start of World War II and held by them for the duration of the war. Malaya's Chinese population received particularly harsh treatment during the Japanese occupation.

When the British returned they arranged (1946) a centralized colony, called the Malayan Union, comprising all their peninsula possessions. Influential Malays vehemently opposed the new organization; they feared that the admission of the large Chinese and Indian populations of Pinang and Malacca to Malayan citizenship would end the special position Malays had always enjoyed. The British backed down and established the Federation of Malaya (1948) headed by a British high commissioner. The Federation was an expansion of the former Federated Malay States. Pinang and Malacca became members in addition to the nine Malay states, but there was no common citizenship. In that same year a Communist insurrection began that was to last more than a decade. In combatting the uprising the British resettled nearly 500,000 Chinese. "The Emergency," as it was called, was declared ended in 1960, although outbreaks of terrorism have continued sporadically.

The insurrection spurred the movement for Malayan independence. In 1957 the Federation became an independent state within the Commonwealth of Nations and was admitted to the United Nations. The constitution guaranteed special privileges for Malays. In 1963 Singapore, Sabah, and Sarawak were added to the federation, creating the Federation of Malaysia. Since Singapore has a large Chinese population, the latter two states were included to maintain a non-Chinese majority. Brunei was also included in the plan but declined to join. Malaysia retained Malaya's place in the United Nations and the Commonwealth. Indonesia, which described the federation as a British imperialist subterfuge, waged an undeclared war against it. In the struggle Malaysia received military aid from Great Britain and other Commonwealth nations. Hostilities continued until President Sukarno's fall from power in Indonesia (1965). Friction soon developed between Malay leaders and Singapore's prime minister, Lee Kuan Yew, who worked to improve the position of Chinese in the Federation. In 1965 Singapore peacefully seceded from Malaysia. Inter-communal tension continued, however, and serious violence broke out in 1969 following general elections in West Malaysia in which the Chinese made gains at the expense of the Malays. Parliament was suspended for 22 months. General elections in Aug., 1974, resulted in an overwhelming victory for the National Front Coalition government.

Government. Malaysia is a constitutional monarchy with parliamentary democracy. The sovereign appoints the cabinet, headed by the prime minister, who must be a member and have the confidence of the house of representatives. The parliament has two chambers. The house consists of 154 members, all elected by popular vote in single-member districts. The senate consists of 58 members chosen for six-year terms.

Mal·bork (mäl'bôrk'), town (1975 est. pop. 32,500), N Poland, on the Nogat River. It is a rail junction with sugar refineries and manufactures of rubber and pharmaceuticals. Originally a castle founded (1274) by the Teutonic Knights, Malbork became the seat of their grand master in 1309. It successfully withstood sieges by the Poles in 1410 and 1454, but in 1457 Malbork was sold to Poland by mercenaries whose pay was in arrears. The town passed to Prussia in 1772. The castle (rebuilt in the 14th and 19th cent.) is one of the finest examples of German secular medieval architecture.

Mal·den (môl'dən), city (1978 est. pop. 55,700), Middlesex co., E Mass., a suburb of Boston, in the Mystic valley; settled 1640, inc. 1882. Among its varied manufactures are processed foods, cans, aluminum products, and tools. A number of historic churches are here.

Mal·dives (măl'dīvz), formerly **Mal·dive Islands** (măl'dīv'), republic (1975 est. pop. 136,000), 115 sq mi (297.9 sq km), S Asia, stretching c.500 mi (805 km) from N to S in the N Indian Ocean, SW of Sri Lanka. The Maldives consist of 19 atolls made up of nearly 2,000 coral islands that are the exposed tops of a submarine ridge. They have a tropical monsoon climate modified by their marine location.

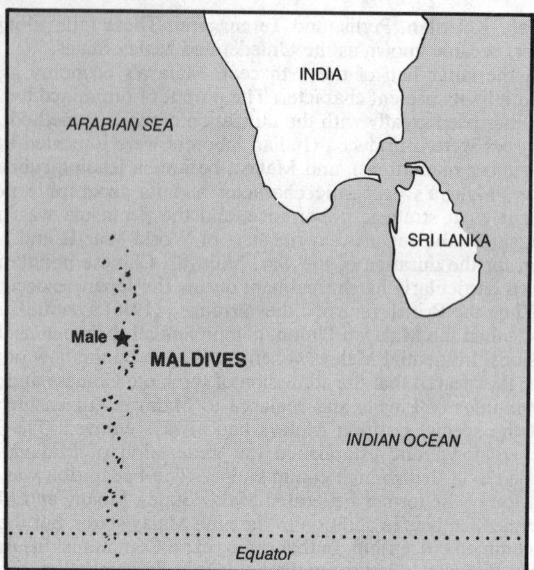

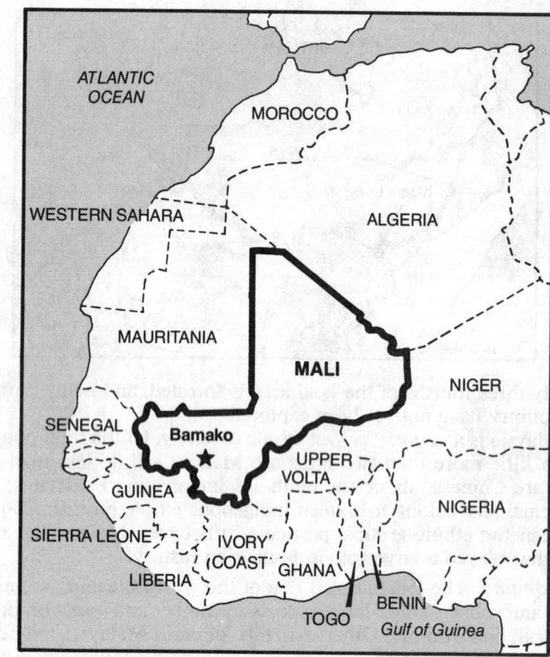

The islands are covered with tropical vegetation, particularly coconut palms. About 200 of the islands are inhabited, and some have freshwater lagoons. Tropical fruit and corn are raised; fish, coconuts, and coconut products (especially copra) are the nation's chief sources of income.

The Maldives were orginally settled by peoples who came from southern Asia. In the 12th cent. Islam was brought to the islands. With the coming of the Portuguese in the 16th cent., the Maldives were intermittently under European influence. In 1887 they became a British protectorate and military base but retained internal self-government. The Maldives obtained independence as a sultanate in 1965, but in 1968 the ad-Din dynasty, which had ruled since the 14th cent., was ended and a republic was declared.

Mal·don (môl'dən), municipal borough (1971 pop. 13,840), Essex, E England, on the Blackwater estuary. It is a market town with iron foundries and other small industries. Prehistoric traces have been found in the vicinity and there may have been a Saxon settlement. A battle against Danish raiders was fought near Maldon in 991.

Ma·le·gaon (mä'lə-goun'), town (1971 pop. 197,784), Maharashtra state, W central India, at the confluence of the Girna and Masam rivers. It is a weaving center for saris. Malegaon was captured by the British in 1818 during the war with the Pindaris, marauding tribes who were often mercenaries of Mahratta leaders.

Mal·heur (măl'hŏŏr), county (1970 pop. 23,169), 9,859 sq mi (25,534.8 sq km), SE corner of Oregon, drained by the Malheur and Owyhee rivers, which flow to the Snake River, and bordering on Nev. and Idaho; formed 1887; co. seat Vale. It is in a dairying and agricultural area yielding sugar beets, fruit, truck crops, and hay. There is a food-processing industry, and concrete is manufactured.

Ma·li (mä'lē), independent republic (1977 est. pop. 5,925,000), 478,764 sq mi (1,240,000 sq km), W Africa, bordered on the N by Algeria, on the E and SE by Niger, on the S by Upper Volta and Ivory Coast, and on the W by Guinea, Senegal, and Mauritania; capital Bamako. In the south, traversed by the Niger and Senegal rivers, are fertile areas. Elsewhere the country is arid desert or semidesert and barely supports grazing (mainly cattle, sheep, and goats).

Economy. The Niger River serves as an important transportation artery and a source of fish. Peanuts and cotton are Mali's only significant cash crops, but subsistence agriculture is also practiced. Live animals and preserved fish are the most important exports. Mali has varied light industries, including canning and preserving, cotton ginning, peanut-oil extraction, brickmaking, and the production of textiles, cigarettes, matches, and hardware. Some salt and gold are mined for local trade, but the country's extensive mineral resources (bauxite, manganese, iron ore, phosphates, lithium, diamonds) remain largely unexploited. There is hope of finding oil in the desert regions.

History. The Mali region has been the seat of extensive empires and kingdoms, notably those of Ghana (4th–11th cent.), Mali, and Gao. The medieval empire of Mali was a powerful state and one of the world's chief gold suppliers; it attained its peak in the early 14th cent. under Emperor Mansu Musa (reigned c.1312–1337). The Mali empire was followed by the Songhai empire of Gao, which rose to great power in the late 15th cent. In 1590 the weakening empire was shattered by a Moroccan army attack. The Moroccans, however, could not effectively dominate the vast region. By the late 18th cent. the area was in a semianarchic condition and was subject to incursions by the Tuareg and Fulani. The 19th cent. witnessed a great resurgence of Islam; Moslem states opposed French invasion of the region. By 1898, however, the French conquest was virtually complete; Mali, called French Sudan, became part of the Federation of French West Africa. A nationalist movement blossomed during the period between the two World Wars. The Sudanese Union, a militantly anticolonial party, became the leading political force. Its leader, Modibo Keita, was a descendant of the Mali emperors. In the French constitutional referendum of 1958, French Sudan voted to join the French Community as the autonomous Sudanese Republic. In 1960 the Sudanese Republic, renamed the Republic of Mali, obtained full independence. Under Keita's presidency Mali became a one-party state committed to socialist policies. In 1962 the country withdrew from the franc zone and adopted a nonconvertible national currency. The resulting economic and financial difficulties forced an accommodation with France in 1967. Militant elements in the Sudanese Union opposed this rapprochement, however, and Keita formed a people's militia to destroy opposition. In 1968 a bloodless military coup overthrew the Keita regime. The new rulers governed through the Military Committee of National Liberation, with Lt. Moussa Traoré as president.

In the early 1970s Mali suffered from the effects of a prolonged drought that desiccated the Sahel region of Africa. The drought shattered Mali's agriculture economy by killing thousands of head of livestock and hindering crop production. The resulting famine, disease, and poverty contributed to the deaths of untold thousands and forced the southward migration of many tribes. Mali received emergency aid from UN-supervised international relief programs. The drought may have permanently extended desert conditions into central and southern Mali.

Ma·li·bu Beach (măl'ə-bōō), residential area (1970 pop. 7,000), Los Angeles co., S Calif., W of Los Angeles and near Santa Monica.

Ma·lin·di (mə-lĭn'dē), town (1962 pop. 5,818), SE Kenya, on the Indian Ocean; a resort and commercial center. Probably founded in the 10th cent. by Arab traders, Malindi became an important city-state and a major port. The Portuguese navigator Vasco da Gama landed here in 1498 and erected a monument that still stands.

Mal·lor·ca (mä-lyôr'kä): *see* Majorca.

Mal·mé·dy (mäl-mā-dē'), town (1970 pop. 6,464), Liège prov., E Belgium, near the West German border. It is a manufacturing and tourist center. The town passed (1815) to Prussia and to Belgium by the

Treaty of Versailles after World War I. In World War II there was heavy fighting here during the Battle of the Bulge (Dec., 1944).

Malmes·bur·y (mämz'bĕr-ē, -bə-rē), municipal borough (1971 pop. 2,526), Wiltshire, S England. It is famous for its magnificent Benedictine abbey, founded in the 12th cent., of which only the nave remains.

Mal·mö (măl'mō, mäl'mœ), city (1975 est. pop. 243,591), capital of Malmöhus co., S Sweden, on the Øresund opposite Copenhagen. It is a major naval and commercial port and an industrial center. Manufactures include textiles, clothing, metal goods, processed food, and cement. There are also shipyards and machine shops. Founded in the 12th cent., Malmö was an important trade and shipping center during the Hanseatic period. It was usually a Danish possession until it passed to Sweden in 1658.

Ma·lo·los (mə-lō'ləs, mä-lō'lōs), town (1970 pop. 73,996), capital of Bulacan prov., SW Luzon, the Philippines, N of Manila. It is a marketing center for surrounding farms. The Spanish settled here in 1580. Malolos was the capital of the Philippine republic proclaimed (June, 1898) by the insurrectionary leader Emilio Aguinaldo; U.S. forces captured the town in Mar., 1899.

Ma·lone (mə-lōn'), village (1975 est. pop. 7,800), seat of Franklin co., N N.Y., near the Canadian boundary ESE of Massena. It is a port of entry and manufactures clothing and footwear. Members of the Fenian movement gathered here (1866) to attack Canada.

Mal·ta (môl'tə), independent state (1967 pop. 315,765), 122 sq mi (316 sq km), in the Mediterranean Sea S of Sicily. It comprises the islands of Malta (95 sq mi/246 sq km), Gozo (26 sq mi/67.3 sq km), and Comino (1 sq mi/2.6 sq km), as well as two uninhabited rocks. The group is sometimes called the Maltese Islands. Valletta is the capital. Malta has no rivers or lakes, no natural resources, and very few trees. Nevertheless it is of great strategic value and has long been an important British military base.

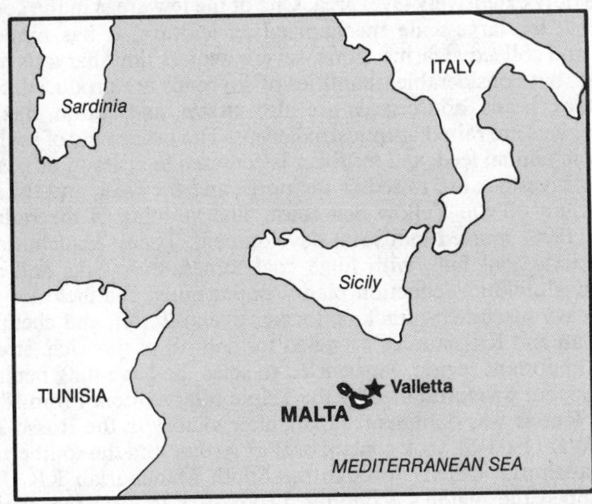

Economy. The decline in activity at the base, beginning in the late 1960s, created serious economic problems for the country. Although the soil is poor, agriculture is the principal occupation in Malta. The chief crops are wheat, barley, potatoes, and other vegetables and fruits. Tourism is of increasing importance, and efforts have been made to stimulate food processing and other light industries; manufactured goods such as textile threads and rubber products are the leading exports.

History. The island of Malta (ancient Melita) belonged successively to the Phoenicians, Greeks, Carthaginians, Romans, and Saracens. The Normans of Sicily occupied it c.1090. In 1530 the Hapsburg Charles V granted Malta to the Knights Hospitalers, who held it until 1798, when it was surrendered to Napoleon. The British ousted the French in 1800, and for most of the 19th cent. Malta was ruled by a military governor. During World War II Malta was subjected to extremely heavy bombing by Italian and German planes, and in 1942 King George VI awarded the entire population the George Cross for bravery.

Malta became fully independent in 1964 and chose to remain in the Commonwealth of Nations. In 1965 it joined the United Nations. In 1973 Malta initiated a seven-year economic development plan intended to free the country, after 1979, from its dependence on British rental payments for the military base.

Malta, city (1970 pop. 2,195), seat of Phillips co., NE Mont., E of Havre on Milk River; inc. 1909. A cattle town in the late 19th cent., it is the chief town of the Milk River reclamation project.

Mal·va·si·a (măl-və-sē'ə) or **Mon·em·va·sí·a** (mô'nĕm-və-sē'ə), village, S Greece, in the Peloponnesus, on a rocky island joined to the mainland by a mole. In the Middle Ages it was a fortress and an important commercial port, exporting Malvasian or malmsey wine. It was (1821) the seat of the first Greek national assembly.

Mal·vern (môl'vərn, mô'-), urban district (1973 est. pop. 30,420), Hereford and Worchester, W central England, on the E slopes of the scenic Malvern Hills. Occupying the site of the medieval Chase of Malvern, a royal forest, Malvern today is primarily a health and holiday resort. The priory church of Great Malvern dates from 1085; the Norman arches of the interior remain intact.

Mal·vern (măl'vərn), city (1974 est. pop. 9,848), seat of Hot Spring co., S central Ark., near the Ouachita River SW of Little Rock; laid out 1873, inc. 1876. Lumber and metal products, cotton goods, and bricks are made.

Mal·vern Hills (môl'vərn, mô'-), range of hills, c.9 mi (14.5 km) long, W central England, in Hereford and Worcester. The highest points are the Worcester Beacon (1,395 ft/425.5 m) and the Hereford Beacon (1,114 ft/339.8 m); on the latter was an ancient British camp.

Ma·mar·o·neck (mə-măr'ə-nĕk'), village (1978 est. pop. 18,100), Westchester co., SE N.Y., a suburb of New York City, on Long Island Sound; settled 1661, inc. 1895. It is a boating center, with a fine marina. Although it is primarily residential, there is considerable industry.

Mam·moth Cave National Park (măm'əth), 51,354 acres (20,798.4 hectares), S Ky.; est. 1936. Located in a hilly and forested region, the park offers numerous outdoor activities. It is the site of Mammoth Cave, one of the largest known caves in the world. Composed of a series of subterranean chambers and narrow passages formed by the dissolution of limestone, the cave has five separate levels. Its full extent is still unexplored, but the known passages extend c.150 mi (240 km), disclosing limestone formations (stalactites, stalagmites, and columns), lakes, and rivers. Echo River, c.360 ft (110 m) below the surface, flows through the cave's lowest level. Hanson's Lost River, an underground stream, joins Mammoth Cave with the extensive Flint Ridge cave system; this long-sought link was discovered in 1972. The cave contains the mummified body of a man believed to date from the pre-Columbian period. Eyeless fish, bats, and insects are also found. Mammoth Cave was an Indian habitation long before it was visited by Ky. pioneers in 1799.

Ma·mo·ré (mä'mô-rā'), river, c.600 mi (965 km) long, formed by tributaries rising in the Andes and plains of central Bolivia. It flows north, past Trinidad, to the Brazilian border. After forming part of the Bolivia-Brazil border, the Mamoré joins with the Beni River to form the Madeira River.

Man, Isle of (măn), island (1971 pop. 49,743), 227 sq mi (588 sq km), off Great Britain, in the Irish Sea. The coast is rocky with precipitous cliffs; the Calf of Man is a detached rocky islet off the southwest coast. The scenery is varied and beautiful, the climate very mild (subtropical plants are grown without protection), and the island is a popular resort. Oats, barley, turnips, and potatoes are grown, and sheep are raised. Dairying and fishing are carried on, and Manx tweeds are made from locally produced wool. There is some light industry. Traces of occupants of the isle from Neolithic times exist; there are ancient crosses and other stone monuments, a round tower, an old fort, and castles.

Occupied in the 9th cent. by Vikings, the island was a dependency of Norway until 1266, when it passed to Scotland, but from the 14th to the 18th cent. (except for brief periods when it reverted to the English crown) it belonged to the earls of Salisbury and of Derby. Since 1765, when Parliament purchased it from the Duke of Atholl, the Isle has been a dependency of the crown, but it is not subject to acts of the British Parliament.

Ma·na·do (mä-nä'dō), town (1971 pop. 169,684), capital of North Sulawesi prov., on the NE coast of Celebes, Indonesia.

Ma·na·gua (mä-nä'gwä), city (1971 pop. 375,278), W Nicaragua, capital of Nicaragua, on the S shore of Lake Managua. It is the commercial and industrial center of the country. Managua was made permanent capital in 1855 to end the bitter feud between Granada and León. During periods of disorder (1912–25 and 1926–33) it was occupied by U.S. troops. Managua is generally hot and sultry. A fairly constant wind blows from nearby Lake Managua. Managua was damaged by earthquake and fire in 1931 and by fire in 1936. On Dec. 23, 1972, it was almost completely destroyed in an earthquake that took more than 10,000 lives.

Ma·nas·sas (mə-năs'əs), town (1970 pop. 9,164), seat of Prince William co., N Va., in a farm area; inc. 1873, rechartered 1938. It was a key railroad junction during the Civil War, and the Battles of Bull Run were fought nearby. Today its population is growing rapidly as a suburb of Washington, D.C.

Manassas National Battlefield Park: *see* Bull Run.

Man·a·tee (măn'ə-tē', măn'ə-tē'), county (1970 pop. 97,115), 739 sq mi (1,914 sq km), SW Fla., on the Gulf of Mexico and Tampa Bay and drained by the Manatee and Myakka rivers; formed 1855; co. seat Bradenton. It includes scattered lakes, part of Sarasota Bay, and small offshore islands.

Ma·naus (mə-nous'), city (1970 pop. 284,118), capital of Amazonas state, NW Brazil, on the Rio Negro. It is the chief commercial and cultural center of the upper Amazon region and an important river port, with floating docks that can accommodate oceangoing vessels. Brazil nuts, rubber, hardwoods, and animal skins are exported. Founded in 1669, Manaus grew slowly until the late 19th cent., when the wild-rubber boom brought prosperity and short-lived splendor. In recent years, renewed interest in the Amazon basin and the discovery of oil nearby has brought new importance to Manaus.

Man·cha, La (lä män'chä), region of central Spain, in New Castile. This high, barren plateau, dotted with windmills, was made famous as the scene of most of the adventures of Don Quixote in the novel by Cervantes.

Manche (mäɴsh), department (1975 pop. 451,662), 2,296 sq mi (5,946.6 sq km), NW France, in Normandy, on the English Channel, coextensive with the Cotentin peninsula; capital Saint-Lô.

Man·ches·ter (măn'chĕs'tər, -chĭ-stər), county borough (1976 est. pop. 490,000), Greater Manchester, NW England, on the Irwell, Medlock, Irk, and Tib rivers. Manchester is the center of the most densely populated area of England. It has long been the leading textile city (its textile industry dates back to the 14th cent.) of England and among the world's foremost cotton cities, serving as a distribution point for the mills of surrounding towns. It is also a center of printing and publishing. The Manchester Ship Canal (35.5 mi/57.1 km; opened in 1894) gave the city access to the sea. After World War I the artificial-silk industry tended to balance losses in the cotton market.

A Celtic settlement is believed to have existed on the site of Manchester. The Romans called the town Mancunium, and there are remains of their occupation. Manchester's first charter was granted in 1301. Representation in Parliament was achieved in 1832, and in 1838, thanks to the efforts of Richard Cobden, Manchester was incorporated as a borough. It was the center of the Manchester school of economics and the Anti-Corn Law League, led by Cobden and John Bright. The influential liberal Manchester *Guardian* was founded in 1821. During World War II Manchester suffered extensively from air raids.

The borough has several libraries, including the John Rylands Library (founded 1899) and the Chetham Library (founded 1653), one of Europe's first free public libraries. The Victoria Univ. of Manchester, formerly Owens College, opened in 1851. John Dalton, Lord Rutherford, and Niels Bohr, among others, did significant work in nuclear physics in Manchester. At Jodrell Bank, nearby, is the world's largest radio telescope.

Man·ches·ter (măn'chĕs'tər). **1.** Town (1978 est. pop. 51,300), Hartford co., central Conn.; settled c.1672, inc. 1823. Among its many manufactures are fiberboard, electrical goods, and textiles. **2.** City (1976 est. pop. 4,914), seat of Delaware co., NE Iowa, on the Maquoketa River W of Dubuque; settled 1850, inc. 1886. It is a trade center for a livestock, farm, and dairy area. **3.** City (1970 pop. 1,664), seat of Clay co., SE Ky., NNW of Middlesboro, in an agricultural area of the Cumberland foothills. There are coal mines in the area. **4.** Resort town (1975 est. pop. 5,534), Essex co., NE Mass., on the Atlantic coast between Salem and Gloucester; settled 1626, set off from Salem 1645. It has a boatbuilding industry. **5.** Village (1975 est. pop. 8,632), St. Louis co., E Mo., W of St. Louis; settled in the 1790s, inc. 1950. **6.** City (1978 est. pop. 80,600), Hillsboro co., S N.H., on both sides of the Merrimack River; settled 1722, inc. as a city 1846. Among its various manufactures are textiles, shoes, machinery, and electrical and electronic products. In 1838 textile interests founded the city and established a giant textile-manufacturing company. Until the 1930s and the moving of much of the textile industry to the South, Manchester was heavily dependent on this industry. **7.** City (1970 pop. 6,208), seat of Coffee co., central Tenn., near the Duck River SSE of Nashville, in a farm and timber area; settled 1836, inc. 1838. It has a cannery and makes clothing. **8.** Resort town (1970 pop. 2,919), a seat of Bennington co., SW Vt., N of Bennington and E of Mt. Equinox; chartered 1761, settled c.1764, laid out 1784. Wood and plastic products are made.

Man·chu·kuo (măn'chōō'kwō'), former country, comprising Manchuria and Jehol prov., China. The Japanese invaded Manchuria in 1931 and founded Manchukuo in 1932. Manchukuo, ostensibly an independent Manchu territory, was actually a puppet-state. The Japanese military kept strict control of the administration and fought a continuing guerrilla war with native resistance groups. To develop Manchukuo as a war base, the Japanese greatly expanded industry and railroads. After World War II Chinese sovereignty was reasserted over the area.

Man·chu·ri·a (măn-chŏŏr'ē-ə), region, c.600,000 sq mi (1,554,000 sq km), NE China. It is separated from the USSR largely by the Amur, Argun, and Ussuri rivers, from North Korea by the Yalu and Tumen rivers, and from Mongolia by the Khingan Mts. It includes the Liao-tung peninsula. Until 1860 it included territory now in Siberia and until 1955 territory now in the Inner Mongolian Autonomous Region. Much of the region is hilly to mountainous. The Great and Lesser Khingan in the north and the Ch'ang-pai in the east are the greatest ranges.

Manchuria has vast timber reserves; the country's finest timber is found in the eastern Manchurian highlands. Mineral resources, chiefly coal and iron, are concentrated in the southwest. Magnesite, oil, uranium, and gold are also important.

The great Manchurian plain (average elevation c.1,000 ft/305 m) is the only extensively level area. One of the few areas in the country suitable for large-scale mechanized agriculture, it has numerous state and collective farms. Long, severe winters limit harvests to one a year, but considerable quantities of soybeans are produced. Sweet potatoes, beans, and cereals are also grown, and cotton, flax, and sugar beets are raised as industrial crops. The processing of soybeans into oil, animal feed, and fertilizer is centered in cities in or near the plain. Livestock are raised in the north and the west, and fishing is important off the Yellow Sea coast. The building of the railroads (after 1896) spurred industrial development. Today Manchuria is a great industrial hub, with huge coal mines, ironworks and steelworks, aluminum reduction plants, paper mills, and factories making heavy machinery, tractors, locomotives, aircraft, and chemicals.

Japan and Russia long struggled for control of this rich, strategically important region. Japan tried to seize the Liao-tung peninsula in 1895, but was forestalled by the Triple Intervention. From 1898 to 1904, Russia was dominant. Japan, after victory in the Russo-Japanese War (1904–5), took control of Port Arthur and the southern half of Manchuria. Chiefly through the South Manchurian RR, Japan developed the region's economy. From 1918 to 1931 the warlords Chang Tso-lin and Chang Hsueh-liang controlled Chinese military power. Japan occupied Manchuria in 1931–32, when Chinese military resistance, sapped by civil war, was weak. It was, in effect, an unofficial declaration of war on China. Manchuria was a base for Japanese aggression in northern China and a buffer region for Japanese-controlled Korea. In 1932, under the aegis of Japan, Manchuria with Jehol prov. was constituted Manchukuo, a nominally independent state. During World War II the Japanese developed a huge industrial complex of metallurgical, coal, petroleum, and chemical industries. Soviet forces, which occupied Manchuria from July, 1945, to May, 1946, dismantled and removed over half of the Manchurian industrial plant. The Chinese Communists were strongly established in Manchuria and by 1948 had captured the major cities and inflicted devastating losses on the Nationalist army. From 1949 to 1954 Manchuria, ruled by Kao Kang, was the most staunch of the Communist areas in China. With the help of Soviet technicians the Communists rapidly restored Manchuria's large industrial capacity. Since the Sino-Soviet rift in the 1960s there has been a massive Soviet military buildup along the border and several border incidents have occurred.

Man·da·lay (măn'də-lā'), city (1970 est. pop. 402,000), Mandalay division, central Burma, a railroad terminus on the Irrawaddy River. As a city it dates from c.1850. It was the capital of the Burman kingdom from 1860 to 1885, when it was annexed to British Burma. A center of Burmese Buddhism, the city is noted for its Arakan pagoda, built around an ancient shrine. The group of sacred buildings known as the Seven Hundred and Thirty Pagodas was erected in the reign (1853–78) of King Mindon. Mandalay was heavily damaged in World War II.

Man·dan (măn′dăn), city (1970 pop. 11,093), seat of Morton co., S N.Dak., on the Missouri River opposite Bismarck; inc. 1881. It is the distributing center for a grain, livestock, and dairy region.

Man·ga·lore (măng′gə-lôr′), city (1971 metropolitan area pop. 214,093), Karnataka state, SW India, on the Arabian Sea. A port, it trades in spices, rice, coffee, nuts, and timber.

Man·gum (măng′gəm), city (1970 pop. 4,066), seat of Greer co., SW Okla., NNW of Altus, in an irrigated farm region; laid out 1883, inc. 1900. Quartz Mountain State Park is nearby.

Man·gysh·lak Peninsula (mäng′gĭsh-läk′), E Central Asian USSR, extending into the NE Caspian Sea. Except for the Kara-Tau range, the peninsula is below sea level; Batyr Sink is c.430 ft (130 m) below sea level. Oil, manganese, and coal are found on the peninsula.

Man·has·set (măn-hăs′ət), uninc. town (1970 pop. 8,541), Nassau co., SE N.Y., a suburb on the N shore of Long Island.

Man·hat·tan (măn-hăt′n). **1.** City (1978 est. pop. 33,600), seat of Riley co., NE Kansas, at the confluence of the Big Blue and Kansas rivers; inc. 1857. It is the trade and processing center of a farm area. Dress patterns are manufactured. Much of the economy is dependent upon Kansas State Univ. and nearby Fort Riley. **2.** Borough (1970 pop. 1,539,233), 22 sq mi (57 sq km), New York City, SE N.Y., coextensive with New York co. It is composed chiefly of Manhattan Island, and is bounded by the Hudson River on the west, New York Bay on the south, the East River on the east, and the Harlem River and Spuyten Duyvil Creek on the northeast and north. Many bridges, tunnels, and ferries link it to the other boroughs and to New Jersey. Manhattan is the cultural and commercial heart of the city, and its dramatic skyline symbolizes New York City around the world. The Manhattan Indians sold (1626) the island to Peter Minuit of the Dutch West India Company, supposedly for some $24 worth of merchandise. A town built at the tip of the island was called New Amsterdam and served as the capital of the colony of New Netherland during the Dutch domination. In 1664 the English captured New Netherland and renamed it New York. The boundary of New York City first extended beyond Manhattan Island when some Westchester co. towns were annexed in 1874. In 1898 Manhattan became one of the five boroughs established by the Greater New York Charter.

Manhattan Beach, city (1978 est. pop. 33,300), Los Angeles co., S Calif., on Santa Monica Bay; inc. 1912. It is a residential and beach community with an oil refinery and factories that produce aircraft and missile parts, electrical equipment, and pottery.

Man·i·coua·gan (măn′ĭ-kwŏg′ən), river, 310 mi (498.8 km) long, rising in E central Que., Canada, and flowing S to the St. Lawrence River near Baie Comeau. The river is a source of hydroelectricity.

Ma·nil·a (mə-nĭl′ə), city (1975 est. pop. 1,438,300), former capital of the Philippines, SW Luzon, on Manila Bay. Manila is still the metropolitan center of the country, its chief port, and focus of all governmental, commercial, industrial, and cultural activities. It is the manufacturing center of the Philippines, with large metal fabrication, automobile assembly, and textile and garment industries. It also has food- and hemp-processing plants, cigarette factories, and establishments making toilet articles, pharmaceuticals, and other chemical products.

A fortified walled colony was established here in 1571 and developed mainly by Spanish missionaries. Except for two years (1762–64) when the city was in British hands, it remained under Spanish control until the Spanish-American War (1898), when it was seized by U.S. forces after the Battle of Manila Bay. Filipino uprisings occurred for several years, and not until 1901 was a civil government definitely established. In World War II the city was occupied by the Japanese (Jan. 2, 1942). Its recovery (Feb., 1945) involved fierce house-to-house fighting, which reduced the old walled city to rubble, destroying many fine examples of 17th cent. Spanish architecture. Only the Church of San Agustin (1606) survived. Reconstruction of the Manila Cathedral began in 1958. In 1968 Manila was shaken by a severe earthquake, which killed more than 300 people and caused extensive property damage. In 1972 the city was damaged by floodwaters resulting from more than three weeks of torrential rains.

Manila, town (1970 pop. 226), seat of Daggett co., NE Utah, NNW of Vernal near the Wyo. border and the Uinta Mts.

Manila Bay, nearly landlocked inlet of the South China Sea, SW Luzon, the Philippines. About 35 mi (56 km) wide at its broadest point and 30 mi (48 km) long, it is the best natural harbor in the Orient and one of the finest in the world. The entrance to Manila Bay (c.11 mi/18 km wide) is divided by the island of Corregidor into two channels; the northern channel, between Corregidor and Bataan peninsula, is only c.2 mi (3.2 km) wide. During the Spanish-American War, in the Battle of Manila Bay (May 1, 1898), an American squadron destroyed the Spanish fleet off Cavite within a few hours. The Manila Bay area was the focus, during the early phase of World War II, of a desperate attempt to save the Philippines from Japanese conquest. In the Allied recovery of the Philippines (1944–45), many Japanese ships were sunk in the bay.

Ma·ni·pur (mŭn′ĭ-po͝or′), state (1971 pop. 1,069,555), 8,628 sq mi (22,346.5 sq km), NE India, bordered by Burma on the S and E; capital Imphal. The terrain, mostly jungle, is on a high plateau. The area was administered from Assam state until 1947, when it became a union territory. Manipur became a state in 1972.

Man·is·tee (măn′ə-stē′, măn′əs-tē′), county (1970 pop. 20,393), 553 sq mi (1,432.3 sq km), NW Mich., bounded on the W by Lake Michigan and drained by the Manistee and Little Manistee rivers; formed 1855; co. seat Manistee. Its agriculture includes fruit, potatoes, cucumbers, livestock, and dairy products. It also has salt mines, fisheries, some manufacturing, and resorts. Part of Manistee National Forest is in the county.

Manistee, city (1970 pop. 7,723), seat of Manistee co., NW Mich., on Lake Michigan W of Cadillac, in a fruit and dairy region; inc. 1869. It is a resort and shipping center and is known for its salt plants.

Man·is·tique (măn′ə-stēk′, măn′ə-stēk′), city (1970 pop. 4,324), seat of Schoolcraft co., N Mich., E Upper Peninsula, on Lake Michigan NE of Escanaba; settled 1833, inc. as a village 1885, as a city 1901.

Man·i·to·ba (măn′ĭ-tō′bə), province (1975 pop. 1,019,000), 246,512 sq mi (638,466 sq km), including 27,239 sq mi (70,549 sq km) of water surface, W central Canada. Winnipeg is the capital. Easternmost of the Prairie Provinces, Manitoba is bounded on the north by Keewatin dist. of the Northwest Territories (with a northeast shore line on Hudson Bay), on the east by Ont., on the south by the U.S. borders of Minn. and N.Dak., and on the west by Sask.

Numerous lakes (Winnipeg, Manitoba, and Winnipegosis) and rivers (Nelson, Churchill, and Hayes) flow northeast across Manitoba into Hudson Bay. Miles of almost uninhabited treeless tundra surround the port of Churchill. Extending south from Churchill and east from Lake Winnipeg, the topography is that of the Canadian Shield. The southern part of Manitoba is dominated by lakes, with Lake Winnipeg paralleled in the west by Lakes Winnipegosis and Manitoba. To the west and north of the Red River valley, the land rises in an escarpment extending into plateaus. Much of this heavily forested area has been set off as reserves, and the Riding Mt. area is a national park. To the south, where most of the population is concentrated, are fields of wheat, barley, oats, and flax.

Economy. Canada's wheat industry originated in Manitoba, and Manitoba's bread wheat has set the standards for the world. Grain is shipped in quantity from Churchill (the only port in the Prairie Provinces) during the three ice-free months of the year. Although agriculture has been continually extended, manufacturing has nevertheless displaced it as the leading industry in the province. Foods, minerals, clothing, electrical items, chemicals, furniture, leather, fabricated metals, and transportation equipment are major products. In the southwest, near Brandon, are large oil reserves, and the municipal districts of Flin Flon and The Pas, on the Saskatchewan River, are gateways to the rich mineral deposits (chiefly nickel, copper, and zinc) and timberlands of the central west. Beluga whales are still caught by native fishermen at Churchill, and there are about 500 licensed fur farms.

History. The history of Manitoba began along Hudson Bay. To exploit the fur trade, King Charles II granted (1670) the Hudson's Bay Company propriety over a vast area including the present-day province of Manitoba, then occupied by the Assiniboin, Ojibwa, and Cree Indians. Manitoba was explored and posts were established by the French as well as by the British; their rival claims were resolved when England's conquest of Canada in the French and Indian Wars was confirmed by the Treaty of Paris in 1763. Scotsmen took over much of the French fur trade, organized the North West Company, and challenged the monopoly of the Hudson's Bay Company. The resulting violent rivalry deterred colonization until the merger of the two companies in 1821. From then until 1870, when the Hudson's Bay Company sold its vast domain to the newly created confederation of Canada, that company was in sole control, and settlement of the area increased. Despite a rebellion (1869) of Indians who feared the loss of their land, Manitoba was organized as a province in 1870. Its area was enlarged in 1881, and in 1912 it was given its present extension to Hudson Bay.

During the last part of the 19th cent. and the first part of the 20th, the Canadian government advertised for immigrants to settle the prairies, and huge numbers of Russians, Poles, Estonians, Scandinavians, and Hungarians came from Europe. The largest single group was from the Ukraine. Today the Ukrainians are an important part of Manitoban culture.

Manitoba, Lake, 1,817 sq mi (4,706 sq km), SW Man., Canada; one of the largest lakes of North America. A remnant of glacial Lake Agassiz, it has commercial fisheries.

Man·i·tou·lin Islands (măn′ĭ-to͞o′lĭn), archipelago consisting of three large islands and several smaller ones, in N Lake Huron, NW of Georgian Bay. The islands, in a noted fishing region, are popular resorts. Dairying, lumbering, mixed farming, and tourism are the major activities. Manitoulin, c.80 mi (130 km) long and from 2 to 30 mi (3.2–48.3 km) wide, is the world's largest lake island. It encloses more than 100 lakes and has a much-indented, rugged coast.

Ma·ni·to·woc (măn′ĭ-tə-wŏk′), county (1970 pop. 82,294), 589 sq mi (1,525.5 sq km), E Wis., bounded in the E by Lake Michigan and drained by the Manitowoc River; formed 1836; co. seat Manitowoc. It has oats and alfalfa farming, dairying, and stock raising. Its heavy industries include aluminum, construction machinery, and electrical equipment.

Manitowoc, industrial city (1975 est. pop. 32,500), seat of Manitowoc co., E Wis., a port of entry on Lake Michigan at the mouth of the Manitowoc River; inc. 1870. Its shipbuilding industry dates from 1847; submarines were made here in World War II. Among the city's many other products are aluminum ware and soap. The North West Company established a trading post here in 1795. Manitowoc and its twin city, Two Rivers, were founded in 1836.

Ma·ni·za·les (mä′nē-sä′lĕs), city (1973 pop. 231,066), alt. 7,063 ft (2,153 m), capital of Caldas dept., W central Colombia, on the slopes of the Cordillera Central. It is a commercial and agricultural center in a region that produces a large share of Colombia's coffee.

Man·ka·to (măn-kā′tō). **1.** City (1970 pop. 1,287), seat of Jewell co., N central Kansas, NNW of Salina, in a grain and livestock area; laid out 1872, inc. 1880. **2.** City (1978 est. pop. 28,400), seat of Blue Earth co., S Minn., at the confluence of the Blue Earth and Minnesota rivers; inc. 1865. It is a trade and processing center for a farm and dairy region. Mankato stone has been quarried here for over 100 years. Sibley Park in Mankato was the site of Camp Lincoln, where 38 Sioux were hanged, after their revolt in 1862.

Man·ly (măn′lē), municipality (1976 pop. 36,705), New South Wales, SE Australia, a suburb of Sydney, on Port Jackson, an inlet of the Pacific Ocean. It is a resort.

Mann·heim (măn′hīm′, män′-), city (1974 est. pop. 320,508), Baden-Württemberg, central West Germany, on the right bank of the Rhine River at the mouth of the Neckar River. It is a major inland port and an industrial center. Manufactures include precision instruments, chemicals, building materials, textiles, agricultural machinery, and motor vehicles. Mannheim was mentioned in the 8th cent. as a small fishing village. It was fortified and chartered in 1606–7. In 1720 the city became the residence of the electors palatine, who built (1720–60) a large palace and held a brilliant court. Elector Charles Theodore made (late 18th cent.) Mannheim one of the great musical and theatrical centers of Europe. W. A. Mozart lived (1777–78) here and Johann Schiller began (1782–83) his career at the Mannheim theater. Mannheim was awarded to Baden in 1802. Although many of the historic buildings were heavily damaged in World War II, the city has restored the chateau and the regularly laid-out 18th cent. baroque buildings of the inner city, including the Jesuit church (1733–60) and the city hall (1700–23).

Man·ning (măn′ĭng). **1.** Village (1970 pop. 45), seat of Dunn co., W central N.Dak, N of Dickinson on the Knife River. **2.** Town (1970 pop. 4,025), seat of Clarendon co., E central S.C., ESE of Columbia; settled c.1855, inc. 1861. Tobacco, lumber, and food are processed.

Man·re·sa (män-rā′sä), city (1970 pop. 57,846), Barcelona prov., NE Spain, in Catalonia, on the Cardoner River. It is an industrial center with textile, metallurgical, and glass industries.

Mans, Le (lə mäɴ), city (1975 pop. 152,285), capital of Sarthe dept., NW France, on the Sarthe River. The historical capital of Maine, it is also an important manufacturing, commercial, educational, and communications center. Le Mans, which dates from pre-Roman times and before Charlemagne was a Merovingian capital, has witnessed frequent sieges and battles throughout its history. The Cathedral of St. Julien du Mans (11th–13th cent.) is partly Romanesque; its Gothic part has perhaps the most daring system of flying buttresses of any Gothic cathedral. Le Mans was the birthplace of Henry II of England and John II of France. Today, Le Mans is famous for its annual international auto race.

Mans·field (mănz′fēld′), municipal borough (1973 est. pop. 58,450), Nottinghamshire, central England, on the W border of Sherwood Forest. It is in a coal district, with manufactures of hosiery, shoes, and metal products. Limestone and sandstone are quarried nearby.

Mansfield. 1. Town (1970 pop. 19,994), Tolland co., NE Conn.; settled c.1692, inc. 1702. The Univ. of Connecticut is in Storrs, which is included within Mansfield. The town also includes Mansfield Hollow, the site of a large flood-control project. **2.** City (1970 pop. 6,432), seat of De Soto parish, NW La., S of Shreveport; inc. 1847. It is in an area yielding oil and natural gas, farm products, cotton, and timber. Mansfield Battle Park, site of the Civil War Battle of Sabine Crossroads, is nearby. **3.** Town (1975 est. pop. 12,586), Bristol co., SE Mass., SSW of Boston; settled 1659, set off from Norton 1770. Machine parts are produced. **4.** Industrial city (1978 est. pop. 55,700), seat of Richland co., N central Ohio, in a hilly region surrounded by fertile farmlands; inc. 1828. It is a manufacturing, commercial, and insurance center. Among its diverse products are tires, automobile bodies, electrical appliances, sports vehicles, and brass goods. South Park, with a reconstructed blockhouse of the War of 1812, and Kingwood Center and Gardens, with landscaped displays and a pre-Civil War French-provincial mansion are here.

Mansfield, Mount, peak, 4,393 ft (1,339.9 m) high, N central Vt. It is the highest peak in the Green Mts. and the highest point in the state. At the foot of the mountain is a deep gorge called Smugglers Notch. The area is a winter-sports center.

Man·su·rah, Al (ăl măn-so͝or′ə), city (1970 est. pop. 212,000), N Egypt, a port in the Nile River delta. It is an agricultural market and industrial center. Manufactures include ginned cotton, cottonseed oil, and textiles. Al Mansurah was founded in 1221. In 1250 Crusaders under Louis IX of France suffered a crushing defeat here at the hands of the Mamelukes.

Man·te·o (măn′tē-ō′), resort and fishing town (1970 pop. 547), seat of Dare co., NE N.C., on Roanoke Island SE of Elizabeth City.

Man·ti (măn′tī′), city (1970 pop. 1,803), seat of Sanpete co., central Utah, in the Sanpete Valley W of the Wasatch Mts. and S of Provo, at an altitude of 5,550 ft (1,692.8 m); founded 1849 by Mormons.

Man·ti·ne·a (măn'tə-nē'ə), city of ancient Greece, in E central Arcadia. In the Peloponnesian War a coalition of Mantinea and Argos, urged on by Athens, was defeated (418 B.C.) by Sparta here. It was also the scene of the victory of Thebes over Sparta (362 B.C.).

Man·tor·ville (măn'tər-vĭl'), village (1970 pop. 479), seat of Dodge co., SE Minn., on a branch of the Zumbro River WNW of Rochester, in a grain, potato, and livestock area.

Man·tu·a (măn'choo-ə, -too-ə), city (1976 est. pop. 65,574), capital of Mantova prov., Lombardy, N Italy, bordered on three sides by lakes formed by the Mincio River. It is an agricultural, industrial, and tourist center. Manufactures include machinery, furniture, and petroleum. Originally an Etruscan settlement, Mantua was later a Roman town and afterward a free commune (12th–13th cent.). It flourished under the Gonzaga family (1328–1708), passed to Austria in 1708, was taken by Napoleon I in 1797, was retaken by Austria in 1815, and was returned to Italy in 1866. The Gonzaga palace (13th–18th cent.), among the finest in Europe, has frescoes by Mantegna and Giulio Romano and numerous other works of art.

Man·y (măn'ē), town (1970 pop. 3,112), seat of Sabine parish, W La., NW of Alexandria, in a timber, farm, and oil district; settled in the early 19th cent. Fort Jesup, a frontier post, is nearby.

Ma·nych (mə-nĭch'), two rivers, SE European USSR. The Western Manych, c.200 mi (320 km) long, rises near Stavropol in the northern Caucasus and flows northwest through Lake Manych-Gudilo into the lower Don River. The Eastern Manych rises in a marshy area and flows c.100 mi (160 km) east to a system of salt lakes and marshes west of the Caspian Sea.

Man·za·la, Lake (măn-zä'lə), salt water lagoon, c.660 sq mi (1,710 sq km), NE Egypt, near Port Said, partly separated from the Mediterranean Sea by a narrow peninsula. The Suez Canal cuts through the eastern part of the lake's basin.

Man·za·na·res (män'*th*ä-nä'räs,-sä-), river, c.55 mi (90 km) long; rising in the Sierra de Guadarrama, central Spain, and flowing S past Madrid (where it is canalized) into the Jarama River. The Manzanares is used for irrigation and hydroelectric-power generation.

Man·za·nil·lo (män'sä-nē'yô), city (1970 pop. 77,880), Oriente prov., SE Cuba, a port on the Guacanayabo Gulf of the Caribbean Sea.

Manzanillo, city (1970 pop. 36,982), Colima state, SW Mexico. The nation's chief Pacific port.

Man·zi·kert (măn'zə-kərt), village, E Turkey, SE of Erzurum. It was an important town of ancient Armenia. A council held in A.D. 726 reasserted the independence of the Armenian Church from the Orthodox Eastern Church. In 1071, the Seljuk Turks routed the troops of Byzantine Emperor Romanus IV in a decisive battle that resulted in the fall of Asia Minor to the Seljuks.

Ma·ple·wood (mā'pəl-wood'). **1.** Village (1978 est. pop. 27,800), Ramsey co., SE Minn., a residential suburb of St. Paul; inc. 1957. **2.** City (1978 est. pop. 10,800), St. Louis co., E Mo., a suburb of St. Louis; settled 1825, inc. 1908. Manufactures include structural steel products and honing tools. **3.** Residential township (1970 pop. 24,932), Essex co., NE N.J., SW of Orange; inc. 1922.

Ma·quo·ke·ta (mə-kō'kə-tə), city (1970 pop. 5,677), seat of Jackson co., E Iowa, S of Dubuque on the Maquoketo River; inc. 1853. There are limestone quarries and dairy farms in the area.

Maquoketa, river, rising in E Iowa and flowing c.130 mi (209.2 km) SE to the Mississippi River near Bellevue.

Mar·a·cai·bo (măr'ə-kī'bō, mä'rä-), city (1971 pop. 651,574), capital of Zulia state, NW Venezuela, at the outlet of Lake Maracaibo. It is a commercial and industrial center and the oil capital of South America. Other exports include coffee, cacao, sugar, hardwoods, and some minerals found in the surrounding region. Maracaibo was founded in 1571. In the 17th cent. it was sacked five times, notably by Sir Henry Morgan in 1669. After 1918 exploitation of the vast petroleum resources of the Maracaibo basin resulted in a rapid expansion and modernization of the city.

Maracaibo, Lake, largest lake of South America, c.5,100 sq mi (13,210 sq km), NW Venezuela, extending c.110 mi (175 km) inland. Lake Maracaibo is a major artery of communication for products of the adjacent region and those of the Colombian-Venezuelan highlands. A dredged channel gives oceangoing vessels access to the lake. Discovered in 1499 by Alonso de Ojeda, the Spanish explorer, the lake lies in the extremely hot, humid, and disease-ridden lowlands of the Maracaibo basin, a region which, almost enclosed by mountains, is semiarid in the north but has an average annual rainfall of 50 in. (127 cm) in the south. By far the most vital activity is production of petroleum. Developed since 1918 by foreign concerns, the region is one of the greatest oil-producing areas in the world.

Ma·ra·cay (mä-rä-kī'), city (1971 pop. 255,134), capital of Aragua state, N Venezuela, at the E end of Lake Valencia. It is a commercial, agricultural, and industrial city.

Ma·ra·gheh (mä-rä-gä'), city (1966 pop. 54,106), East Azerbaijan prov., NW Iran, on the S slopes of Mt. Sahand. It is the trade and transportation center of a fertile fruit-growing region. After the Arab conquest in the 7th cent. Maragheh developed rapidly as a provincial capital. In 1029 it was seized by the Oghuz Turks, but they were driven out by a Kurdish chief who established a local dynasty. The city was destroyed by the Mongols in 1221 and was temporarily occupied by Russia in 1828. Maragheh's celebrated observatory (13th cent.) is now in ruins.

Ma·rais (mä-rā'), old quarter of Paris, on the right bank of the Seine. Until the 18th cent. it was the most aristocratic section of Paris. The Marais park, surrounded by uniform houses in pink brick and gray slate, remains a perfect ensemble of 17th cent. architecture. Nearby is the Carnavalet, once the home of Mme de Sévigné, which now houses the municipal museum of Paris. Although the quarter is now very populous, it still has fine mansions dating from its most prosperous period. Among its many renovated and restored buildings is the sumptuous Hotel Sully (17th cent.).

Mar·ais des Cygnes (mĕr'ē də sēn'), river, c.140 mi (225 km) long, rising in E central Kansas, SW of Topeka, and flowing SE into W Mo. to join the Little Osage River and form the Osage River.

Ma·ra·jó (mä'rä-zhô'), island, c.150 mi (240 km) long and c.100 mi (160 km) wide, N Brazil, at the mouth of the Amazon River. It divides the river into the Amazon proper and the Pará. Cattle are raised on the extensive eastern grasslands, and water buffaloes are bred in the low, swampy west. The island is famous for its prehistoric mounds.

Ma·ra·nhão (mə-rə-nyouN'), state (1975 est. pop. 3,330,000), 126,897 sq mi (328,663 sq km), NE Brazil, on the Atlantic Ocean; capital São Luís.

Ma·ra·ñón (mä'rä-nyôn'), river, c.1,000 mi (1,610 km) long, rising in Lake Lauricaucha in the Cordillera Occidental, W central Peru, flowing generally NW, then E across the Andes to join the Ucayali River in NE Peru where it forms the Amazon River. Some consider the Marañón to be the authentic headwater of the Amazon. Pedro de Ursúa, the Spanish explorer, descended the Marañón in 1560.

Ma·raş (mä-räsh'), city (1975 pop. 128,891), capital of Maraş prov., S central Turkey, in the Taurus Mts. It is an agricultural trade center and a transportation hub. Ancient inscriptions indicate that Maraş was a Hittite city-state c.1000 B.C. The city was captured by the Arabs in A.D. 638 and was annexed by the Ottoman Empire in the early 16th cent.

Mar·a·thon (măr'ə-thŏn'), village and plain, ancient Greece, NE of Athens. Here the Athenians and Plataeans under Miltiades defeated a Persian army in 490 B.C.

Marathon, county (1970 pop. 97,457), 1,586 sq mi (4,107.7 sq km), central Wis., drained by the Wisconsin River and its tributaries (Eau Claire and Eau Pleine); formed 1850; co. seat Wausau. In a dairying and lumbering region, it is a major producer of both natural and processed cheese, and wood and paper products are among its leading industries.

Mar·bel·la (mär-bā'lyä), city (1970 pop. 19,648), Málaga prov., S Spain, in Andalusia, on the Mediterranean Sea. The city is a noted tourist resort.

Mar·ble Canyon National Monument (mär'bəl): *see* National Parks and Monuments Table.

Mar·ble·head (mär'bəl-hĕd'), town (1978 est. pop. 21,700), Essex co., NE Mass., on the Atlantic coast; inc. 1649. A fishing village for many years, Marblehead became a resort in the 19th cent.; it is famous especially for yachting. In Burial Hill cemetery are the graves of hundreds of Revolutionary soldiers.

Marble Hill, city (1970 pop. 589), seat of Billinger co., SE Mo., W of Cape Giradeau. Corn and wheat are grown.

Mar·burg an der Lahn (mär'bûrg' än dĕr län', -boorKH') or **Marburg,**

city (1974 est. pop. 71,604), Hesse, central West Germany, on the Lahn River. It is chiefly known for its Protestant university, founded in 1527 by Philip of Hesse. Manufactures include chemicals, pharmaceuticals, machinery, and surgical and optical instruments. Marburg grew in the 12th cent. around a castle; it was chartered in 1227 and, at intervals during the 13th to 17th cent., served as the residence of the landgraves of Hesse. Marburg became part of the Prussian province of Hesse-Nassau in 1866. The castle, which still dominates the picturesque city, was the scene of the famous Marburg Colloquy, held (1529) under the auspices of Philip of Hesse; it failed to bring about agreement between Luther and Melanchthon on the one side and Zwingli on the other. St. Elizabeth of Hungary is buried in the fine Gothic church (13th–14th cent.) dedicated to her; the remains of Field Marshal Hindenburg and of Frederick William I and Frederick II of Prussia were transferred to the church in 1946.

Marche (märsh), region and former province, central France, on the NW margin of the Massif Central. Marche is primarily an agricultural region that also specializes in sheep raising. The name of the region derived from its location as a northern border fief (march) of the duchy of Aquitaine. Marche passed (13th cent.) to the house of Lusignan but was seized (early 14th cent.) by Philip IV of France. Briefly united with the crown lands, it ultimately became an appanage of the house of Bourbon. It came definitively to France in 1531.

Mar·che (mär'kā) or **the March·es** (mär'chĭz), region (1975 est. pop. 1,390,388), 3,742 sq mi (9,691.8 sq km), E central Italy, extending from the E slopes of the Apennines to the Adriatic Sea; capital Ancona. The Marche is mostly hilly or mountainous, except for a narrow coastal strip. Farming is the chief occupation; cereals, grapes, vegetables, and tobacco are the main products, and livestock is raised. Industry has expanded in the 20th cent. with the construction of hydroelectric facilities. Manufactures include textiles, chemicals, fertilizer, and refined petroleum.

The Umbri and the Picentes (Greek colonists for whom part of the region was called Picenum) lived in the region when it was colonized (3rd cent. B.C.) by Rome. In the 6th cent. the northern section, including four of the cities of the Pentapolis and adjoining territories, came under Byzantine rule; the southern section became a part of the Lombard duchy of Spoleto. In the 8th cent. the region passed, as part of the donations of Pepin the Short (754) and Charlemagne (774), under the nominal rule of the papacy, but later emperors granted fiefs in the area until the 13th cent. From the 13th to the 16th cent. the popes gradually established their rule in the Marche and ended local autonomy. The region was occupied by the French from 1797 to 1815, when it was restored to the papacy. The Marche was united with the kingdom of Sardinia in 1860.

March·feld (märkн'fĕlt'), plain, NE Austria, NE of Vienna, between the Danube and the Morava rivers, on the border of Czechoslovakia. A strategic approach to Vienna, it was the site of several important battles. In 1260, Ottocar II of Bohemia defeated Bela IV of Hungary on the Marchfeld, and in 1278, Ottocar was defeated and slain by the forces of Rudolf I of the house of Hapsburg. In 1809, Napoleon I was defeated on the Marchfeld at Aspern by Archduke Charles, but was victorious at Wagram.

Mar·cy, Mount (mär'sē), 5,344 ft (1,629.9 m) high, NE N.Y., in the Adirondack Mts.; highest peak in the state. Lake Tear of the Clouds, on its southern slope, is the source of the Hudson River.

Mar del Pla·ta (mär dĕl plä'tä), city (1970 pop. 302,282), E central Argentina, on the Atlantic Ocean. It is one of the most popular seaside resorts in South America. Fishing and fish processing are also important industries. The city was founded in the 1850s.

Ma·ree, Loch (mə-rē'), lake, 13 mi (21 km) long and 1 to 3 mi (1.6–4.8 km) wide, Highland region, NW Scotland. It drains into the Minch through the Ewe River and Loch Ewe. Set in the Highlands, Loch Maree is known for its scenery.

Ma·ren·go (mä-rĕng'gô), village, Piedmont, NW Italy, near Alessandria. It was the site of a famous battle (June 14, 1800) between the French under Napoleon Bonaparte and the Austrians in which the Austrians were completely defeated.

Ma·ren·go (mə-rĕng'gō), county (1970 pop. 23,819), 978 sq mi (2,533 sq km), W Ala., in the Black Belt, bounded on the W by the Tombigbee River; formed 1818; co. seat Linden. It is in an agricultural area that produces cotton, livestock and dairy products, velvet beans, peanuts, small grains, hay and clovers, and truck crops. Lumbering and food processing are done, and clothing, cement, and plastics manufactured.

Marengo, city (1970 pop. 2,235), seat of Iowa co., E central Iowa, on the Iowa River SW of Cedar Rapids, in a farm area; laid out 1847, inc. 1859. Amana Society settlements are nearby.

Mar·e·o·tis (măr'ē-ō'tĭs) or **Mar·yut** (mər-yo͞ot'), salt lake, c.95 sq mi (245 sq km) excluding marshes, N Egypt, in the Nile delta. It is separated from the Mediterranean Sea by the narrow isthmus on which Alexandria is situated.

Mar·fa (mär'fə), city (1970 pop. 2,647), seat of Presidio co., W Texas, N of the Rio Grande and S of the Davis Mts.; founded 1884, inc. 1887. The city is a summer resort, and there are truck farms in the region.

Mar·ga·ri·ta (mär-gä-rē'tä, -gə-rē'tə), island (1961 est. pop. 75,000), 444 sq mi (1,150 sq km), in the Caribbean Sea off the coast of Venezuela. Island industries produce canned fish, salt, fishing boats, ceramics, tiles, shoes, and sisal hats. Margarita is also a popular tourist resort. The island was discovered by Columbus in 1498. During colonial times it was an important pearl-fishing center.

Mar·gate (mär'gĭt), municipal borough (1973 est. pop. 50,290), in the Isle of Thanet, Kent, SE England. It is a seaport with light industries and, since the late 18th cent., a popular resort. Of interest is the Church of St. John the Baptist (partly Norman).

Ma·ri (mä'rē), ancient city of Mesopotamia (modern Syria), on the middle Euphrates. The site was discovered by chance in the early 1930s by Arabs digging graves and has subsequently been excavated by the French. The earliest evidence of habitation goes back to the Jemdet Nasr period in the 3rd millenium B.C., and Mari remained prosperous throughout the early dynastic period. The temple of Ishtar and other works of art show that Mari was at this time an artistic center with a highly developed style of its own. As the commercial and political focus of western Asia c.1800 B.C., its power extended over 300 mi (485 km) from the frontier of Babylon proper, up the Euphrates, to the border of Syria. The archives of the great King Zimri-lim, a contemporary of Hammurabi in the 18th cent. B.C., were discovered in 1937. They contain more than 20,000 clay documents, which have helped fix the dates of events in Mesopotamia in the 2nd millennium B.C. Hammurabi conquered Mari c.1700 B.C., and it never regained its former status.

Ma·ri·a·na·o (mä'rē-ä-nä'ô), city (1970 pop. 368,747), La Habana prov., W Cuba, a suburb of Havana. Marianao encloses the military base of Ciudad Columbia. Chemicals, beer, and textiles are produced in the city, which also has a fine beach.

Mar·i·a·nas Islands (mâr'ē-ăn'əz, măr'-), island group (1970 pop. 9,640), W Pacific, comprising one of the six districts of the U.S. Trust Territory of the Pacific Islands and the island of Guam (1970 pop. 84,996). The Marianas lie east of the Philippines and south of Japan and extend 350 mi (563.2 km) from north to south. The northern islands are composed of volcanic rock; the southern islands of madrepore limestone covering a volcanic base. Sugar cane, coffee, and coconuts are the chief products. There are deposits of phosphate, sulfur, and manganese ore. The islands were discovered in 1521 by Ferdinand Magellan, who named them the Ladrones Islands (Thieves Islands). They were renamed the Marianas by Spanish Jesuits who arrived in 1668. Nominally a possession of Spain until 1898, the islands were sold to Germany in 1899, except for Guam, which was ceded to the United States. The German islands were seized by Japan in 1914 and were mandated to Japan by the League of Nations in 1920. U.S. forces occupied the Marianas (1944) during World War II, and in 1947 the group (exclusive of Guam) was included in the U.S. Trust Territory of the Pacific Islands.

Marianas trench, Marianas trough, or **Marianas deep,** elongated depression on the floor of the Pacific Ocean, 210 mi (337.9 km) SW of Guam. It is the deepest (36,198 ft/11,040.4 m) known depression on the earth's surface. The trench was first sounded (1959) by Soviet scientists; its bottom was reached (1960) by two men in a U.S. Navy bathyscaph.

Mar·i·an·na (mâr'ē-ăn'ə, măr'-). **1.** City (1970 pop. 6,196), seat of Lee co., E Ark., near the Mississippi River S of Forrest City; inc. 1877. Lumber products and clothing are made. **2.** City (1970 pop. 6,741), seat of Jackson co., NW Fla., near the Ala. border NW of Tallahassee. In a dairy and farm area yielding fruit, nuts, cotton, and corn, it also has clothing factories and limestone quarries.

Ma·ri·as (mə-rī'əs), river, c.210 mi (340 km) long, rising in several branches in NW Mont. near the Continental Divide and flowing SE to the Missouri River near Fort Benton.

Ma·ri Autonomous Soviet Socialist Republic (mä'rē), autonomous republic (1970 pop. 685,000), c.8,900 sq mi (23,050 sq km), E central European USSR, in the middle Volga valley; capital Yoshkar-Ola. The region is a rolling plain, heavily forested with fir and pine. There is an extensive lumbering industry. In the nonforested agricultural areas, grain and flax are grown, and there is dairy farming and livestock raising. Ruled by the eastern Bulgars from the 9th to the 12th cent., the Mari were then conquered (1236) by the Golden Horde. The Russians under Ivan IV assumed control in 1552. The autonomous republic was organized in 1936.

Ma·rib (mä'rĭb), ancient city, Yemen, SW Arabia, 140 mi (225.3 km) inland at an altitude of 3,900 ft (1,189.5 m). It was one of the chief cities, perhaps the capital, of Sheba. The site has yielded numerous inscriptions and ruins, including the famous dam that was built in the 6th cent. B.C. and was one of the great engineering feats of antiquity. The dam collapsed in the 6th cent. A.D., flooding the countryside.

Ma·ri·bor (mä'rĭ-bôr), city (1971 pop. 97,167), NW Yugoslavia, in Slovenia, on the Drava River. It has industries that produce machinery, armaments, automobiles, airplanes, chemicals, and textiles. Known as early as the 12th cent., it was an important city of Styria until its transfer (1919) to Yugoslavia.

Mar·i·co·pa (măr'ə-kō'pə), county (1970 pop. 969,425), 9,155 sq mi (23,711.5 sq km), SW central Ariz., with irrigated regions along the Salt, Gila, Santa Cruz, Verde, and Agua Fria rivers; formed 1871; co. seat Phoenix. Long-staple cotton, citrus fruit, truck crops, figs, alfalfa, and lettuce are grown. Cattle ranches and health resorts are found. The county includes Gila Bend and Salt River Indian reservations and ranges of mountains in the northeast.

Ma·rie Byrd Land (mə-rē' bûrd'), area of W Antarctica, E of the Ross Shelf Ice and the Ross Sea and S of the Amundsen Sea. The Ford Ranges lie in the northwest part. The region was discovered and claimed for the United States by Richard E. Byrd in 1929. Much of this region was explored during the second Byrd expedition (1933–35) and the U.S. Antarctic Service Expedition (1939–41).

Mar·ies (măr'ēz), county (1970 pop. 6,851), 526 sq mi (1,362.3 sq km), S central Mo., in the Ozarks, drained by the Gasconade River; formed 1855; co. seat Vienna. It is in an agricultural area that produces wheat, corn, oats, and livestock.

Mar·i·et·ta (mâr'ē-ĕt'ə). **1.** City (1978 est. pop. 31,700), seat of Cobb co., NW Ga.; inc. 1834. Aircraft are manufactured. A summer resort area, Marietta is at the foot of Kennesaw Mt., the scene of a Union defeat in the Civil War. Kennesaw Mountain National Battlefield Park marks the site. Many Civil War dead are buried in the city. **2.** City (1978 est. pop. 15,600), seat of Washington co., SE Ohio, at the confluence of the Muskingum and Ohio rivers; inc. 1801. It is a trading center for an agricultural and dairying area. Among the city's varied manufactures are office equipment, alloys, plastics, ventilators, and paints. It was the first planned, permanent settlement in Ohio and the Northwest Territory. Founded in 1788 by the Ohio Company of Associates, it grew as a shipbuilding and shipping center for a farm area. The first houses were in a stockaded enclosure called Campus Martius. Points of interest include the Ohio River Museum (est. 1972), and Mound Cemetery, named for a large Indian mound within its enclosure, where numerous Revolutionary officers are buried. **3.** City (1970 pop. 2,013), seat of Love co., S Okla., S of Ardmore near the Red River, in a farm and resort area; settled c.1887. There are oil and gas wells in the area.

Ma·rin (mə-rĭn'), county (1970 pop. 206,758), 521 sq mi (1,349.4 sq km), W Calif., in a wooded hilly peninsula containing many residential suburbs of San Francisco, reaching S to the Golden Gate, and lying between San Pablo and San Francisco bays on the E and the Pacific Ocean on the W; formed 1893; co. seat San Rafael. Its agriculture includes dairying, poultry and stock raising (cattle, sheep, and hogs), and hay, truck, and fruit farming. It has stone, sand, gravel, and clay quarries, timber, and a food-processing industry. Mt. Tamalpais and Muir Woods National Monument are here.

Mar·i·nette (măr'ə-nĕt', mĕr'-), county (1970 pop. 35,810), 1,378 sq mi (3,569 sq km), NE Wis., bounded in the E by the Menominee River, which forms the Mich. border, in the SE by Green Bay, and drained by the Peshtigo River; formed 1879; co. seat Marinette. Its agriculture includes dairying and potato growing. It has stone quarries, lumbering, and diversified manufactures (food, furniture, paper, chemical, wood, and metal products, machinery, and ships).

Marinette, city (1970 pop. 12,696), seat of Marinette co., NE Wis., on Green Bay at the mouth of the Menominee River; inc. 1887. A port of entry, it is the center of a tri-city area embracing Peshtigo, Wis., and Menominee, Mich. Among the city's manufactures are pulp and paper, machinery, and aluminum castings. Fur trading began here c.1795 and gave way to lumbering, which flourished until the 1930s.

Ma·rin·gá (măr'ən-gä'), city (1970 pop. 51,620), Paraná state, SE Brazil. Coffee processing is the main industry.

Mar·i·on (măr'ē-ən, mâr'-). **1.** County (1970 pop. 23,788), 743 sq mi (1,924.4 sq km), NW Ala., bordering on Miss., drained by the Buttahatchee River, and crossed N to S by the Fall Line; formed 1818; co. seat Hamilton. It has cotton, poultry, and bee farming, coal mines, lumber sawmilling, and some industry (weaving and knitting mills and clothing and machinery factories). **2.** County (1970 pop. 7,000), 584 sq mi (1,512.6 sq km), N Ark., in the Ozarks, bounded on the N by the Mo. border, intersected by the White River, and drained by the Buffalo River and Crooked Creek; formed as Searcy in 1835, name changed in 1836; co. seat Yellville. Its agriculture includes cotton, corn, truck crops, grain, and livestock. It has timber, lead and zinc mines, and a boatbuilding industry. Buffalo River State Park, a fishing area, is in the county. **3.** County (1970 pop. 69,030), 1,599 sq mi (4,141.4 sq km), N central Fla., drained by the Ochlawaha River; formed 1844; co. seat Ocala. It is in a flatwoods area with scattered lakes. Farming (citrus fruit, vegetables, corn, and peanuts), stock raising (cattle and hogs), forestry (lumber and naval stores), and quarrying (limestone and phosphate) are its major activities. **4.** County (1970 pop. 5,099), 365 sq mi (945.4 sq km), W Ga., in a coastal-plain area drained by the Kinchafoone River; formed 1827; co. seat Buena Vista. Cotton, corn, peanuts, pecans, truck crops, and fruit are grown. Its industry includes lumbering, poultry dressing, and furniture manufacturing. **5.** County (1970 pop. 38,986), 580 sq mi (1,502.2 sq km), S central Ill., drained by the Skillet Fork, Crooked Creek, and a small headstream of the Kaskaskia River; formed 1823; co. seat Salem. It has agriculture (corn, wheat, fruit, livestock, poultry, and dairy products) and some manufacturing of metal products, shoes, clothing, and lumber. **6.** County (1970 pop. 793,769), 392 sq mi (1,015.3 sq km), central Ind., drained by the West Fork of the White River; formed 1821; co. seat Indianapolis. Its agriculture includes wheat, corn, truck crops, soybeans, cattle, hogs, and dairy products. It also has coal mines, lumber, sand and gravel pits, and iron, steel, and aluminum foundries. Among its manufactures are textiles, furniture, paper, wood, chemical, and metal products, glassware, concrete, and machinery. **7.** County (1970 pop. 26,352), 498 sq mi (1,289.8 sq km), S central Iowa, drained by the Skunk and Des Moines rivers and by Whitebreast Creek; formed 1845; co. seat Knoxville. It is in a rolling prairie agricultural area (hogs, cattle, poultry, and corn), with bituminous-coal mines and some limestone quarries. **8.** County (1970 pop. 13,935), 945 sq mi (2,447.6 sq km), E central Kansas, in a gently rolling to hilly area drained by the Cottonwood River; formed 1855; co. seat Marion. It is in a wheat and livestock region, with oil fields in the north and south. It produces natural and processed cheese and manufactures machinery. **9.** County (1970 pop. 16,714), 343 sq mi (888.4 sq km), central Ky., in a rolling upland area partly in the outer Bluegrass; formed 1834; co. seat Lebanon. An agricultural county yielding burley tobacco, corn, and hay, it has sawmills, stone quarries, whiskey distilleries, and plants manufacturing clothing and hardwood flooring. **10.** County (1970 pop. 22,871), 550 sq mi (1,424.5 sq km), S Miss., partly bordered on the S by La. and drained by the Pearl River; formed 1811; co. seat Columbia. It has cotton, corn, and truck farms, timber, oil and gas wells, and industry (clothing, furniture, plastics, and concrete manufacturing). **11.** County (1970 pop. 28,121), 440 sq mi (1,139.6 sq km), NE Mo., bounded on the E by the Mississippi River and drained by the South Fabius River; formed 1826; co. seat Palmyra. It is in an agricultural area that produces corn, wheat, oats, and dairy products. Its industry includes food processing and the manufacture of shoes, fertilizer, and machinery. **12.** County (1970 pop. 64,724), 405 sq mi (1,049 sq km), central Ohio, in an agricultural area intersected by the Scioto River; formed 1823; co. seat Marion. It has limestone quarries, sand and gravel pits, and some manufacturing. **13.** County (1970 pop. 151,309), 1,166 sq mi (3,019.9 sq km), NW Oregon, bounded in the W by the Willamette River and in the S by the North Santiam River, with the Cascade Range in the E; formed 1843 as Champoick co., renamed 1849; co. seat Salem. In an agricultural area yielding dairy products, poultry, fruit, truck crops, grain, seeds, and hay, it has mineral mining and lumbering. Textiles, millwork, paper products, chemicals, plastics, concrete, metal products, and machinery are manufactured. Part of Willamette National Forest is here. **14.** County (1970 pop. 30,270), 487 sq mi (1,261.3 sq km), E S.C., bounded in the W by the Pee Dee River and in the E by the Little Pee Dee River; formed 1785; co. seat Marion. Tobacco, cotton, and truck farming are done. It has lumbering, food-processing, and textile industries and manufactures furniture and machinery. **15.** County (1970 pop. 20,577), 507 sq mi (1,313.1 sq km), S Tenn., bordered in the S by Ala. and Ga. and drained by the Tennessee and Sequatchie rivers; formed 1817; co. seat Jasper. It is in a bituminous-coal and iron-ore mining area, with dairying, livestock raising, and some agriculture (corn, hay, soybeans, cotton, and tobacco). Its manufactures include textiles, wood products, and cement. **16.** County (1970 pop. 8,517), 380 sq mi (984.2 sq km), E Texas, bounded on the E by the La. border and drained by Cypress Bayou; formed 1860; co. seat Jefferson. Cattle, poultry, and hogs are raised, and cotton, corn, sweet potatoes, peanuts, fruit, and truck crops grown. The county also has stands of timber, oil and natural-gas wells, and deposits of clay and iron. Part of Caddo Lake, a hunting and fishing area, is here. **17.** County (1970 pop. 61,356), 309 sq mi (800.3 sq km), N W. Va., on the Allegheny Plateau, drained by the Monongahela and Tygart rivers; formed 1842; co. seat Fairmont. It is in a bituminous-coal region, with lumber and gas and oil fields. Its manufactures include glass, paper and metal products, and machinery.

Marion. 1. City (1970 pop. 4,289), seat of Perry co., W central Ala., NW of Selma, in a farm area; settled 1817, inc. 1835. Judson College and a military academy are here. Many ante-bellum houses remain. **2.** City (1975 est. pop. 2,280), seat of Crittenden co., E Ark., WNW of Memphis, Tenn., in a cotton-growing area. **3.** City (1970 pop. 11,724), seat of Williamson co., S Ill.; inc. 1841. It is the commercial and retail center of a farm and coal area and has a large soft drink bottling plant. **4.** City (1978 est. pop. 40,200), seat of Grant co., E central Ind., on the Mississinewa River; settled 1826, inc. 1889. It is a trade, processing, and industrial center in a farm area. Its diversified manufactures include auto parts and television and radio tubes. The city developed with the discovery of gas and oil in the late 1880s. **5.** City (1970 pop. 18,028), Linn co., E central Iowa, adjoining Cedar Rapids; inc. 1865. It is chiefly residential. Home construction and mobile home manufacturing are its main industries. **6.** City (1970 pop. 2,052), seat of Marion co., E central Kansas, NNE of Wichita, in a farm region; settled 1860, laid out 1866, inc. 1875. **7.** City (1970 pop. 3,008), seat of Crittenden co., W Ky., NE of Paducah; inc. 1844. It is a shipping point for fluorspar. **8.** Town (1970 pop. 3,335), seat of McDowell co., W N.C., in the Blue Ridge ENE of Asheville. Textile products and furniture are made. It is in a resort area near Mt. Mitchell. **9.** City (1978 est. pop. 39,200), seat of Marion co., central Ohio; inc. 1830. A rail, industrial, and agricultural center, it is noted for its production of power shovels, cranes, and road-building equipment. Limestone quarries are in the area. Marion was the home of President Warren G. Harding; his house is preserved as a museum, and his burial place is marked by a circular marble monument. **10.** Town (1970 pop. 7,435), seat of Marion co., E S.C., E of Florence, in a farming area; founded c.1800, inc. 1847. Textile and wood products are made. **11.** Town (1970 pop. 8,158), seat of Smyth co., SW Va., in the Holston River valley NE of Bristol; inc. 1832. Textiles and wood products are made.

Mar·i·po·sa (măr'ə-pō'sə, -zə), county (1970 pop. 6,015), 1,455 sq mi (3,768.5 sq km), central Calif., on the W slope of the Sierra Nevada and drained by the Merced, Tuolumne, and Chowchilla rivers; formed 1850; co. seat Mariposa. It has gold and quartz mining, lumbering (pine, fir, and spruce), stock raising (cattle, hogs, sheep, and poultry), and sand and gravel quarrying. The eastern portion of the county is in Yosemite National Park, and it also includes parts of Sierra and Stanislaus national forests.

Mariposa, former town, (1970 pop. 950), central Calif., E of Turlock. A boom town of the Mother Lode in the gold rush, it is today the seat of Mariposa co. and a tourist center for Yosemite National Park. Its courthouse (1854) is said to be the oldest in the state.

Mar·i·time Alps (măr'ə-tīm'), mountain range, S part of the Western Alps, stretching 120 mi (193.1 km) along the French-Italian border near the Mediterranean. Its highest point is 10,817 ft (3,299.2 m).

Maritime Provinces, Canada, term applied to Nova Scotia, New Brunswick, and Prince Edward Island, which before the formation of the Canadian confederation (1867) were politically distinct from Canada proper.

Ma·ri·tsa (mä-rē'tsä), river, c.300 mi (485 km) long, rising in the Rila Mts., W Bulgaria, and flowing SE between the Balkans and Rhodope Mts., to Edirne, Turkey, where it turns S to enter the Aegean Sea near Enez.

Markham, Mount, peak, 14,272 ft (4,353 m), Victoria Land, Antarctica, between Mt. Albert Markham and Mt. Kirkpatrick, S of Shackleton Inlet. It was discovered in 1902.

Mark·lee·ville (märk'lē-vĭl'), town (1970 pop. c.125), seat of Alpine co., E Calif., in the Sierras S of Carson City, Nev.

Marks (märks), town (1970 pop. 2,609), seat of Quitman co., NW Miss., on the Coldwater River S of Memphis, Tenn., in a cotton area; inc. 1906. Chemicals are manufactured.

Marks·ville (märks'vĭl'), town (1970 pop. 4,519), seat of Avoyelles parish, E central La., SE of Alexandria, in a farm area; settled in the late 18th cent. by Acadians. A state park containing Indian mounds is nearby.

Marl (märl), city (1975 pop. 91,779), North Rhine-Westphalia, W West Germany. It is an industrial and mining (coal, lead, and zinc) center. Now a modern city, Marl was first mentioned in the 9th cent. and was chartered in 1936.

Marl·bo·ro (märl'bûr'ō, -bŭr'ō), county (1970 pop. 27,151), 482 sq mi (1,248.4 sq km), NE S.C., bounded in the SW by the Pee Dee River and in the N and NE by the N.C. border; formed 1785; co. seat Bennettsville. Its agriculture includes cotton, soybeans, corn, peaches, and tobacco. It has sand and gravel pits and textile, food-processing, lumbering, and machinery-manufacturing industries.

Marl·bor·ough (märl'bûr'ō, -bŭr'ō) or **Marlboro,** city (1978 est. pop. 31,500), Middlesex co., E Mass.; settled on the site of an Indian village 1657, inc. as a city 1890. It has been a shoe-manufacturing center for many years. The community was almost destroyed (1676) in King Philip's War.

Mar·lin (mär'lĭn), city (1970 pop. 6,351), seat of Falls co., E central Texas, SE of Waco near the Brazos River; founded 1850. It is a health resort with mineral springs.

Mar·lin·ton (mär'lĭn-tən), town (1970 pop. 1,286), seat of Pocahontas co., E W.Va., on the Greenbrier River E of Richwood, in a resort area; settled 1749.

Mar·ly-le-Roi (mär-lē'lə-rwä'), town (1968 pop. 12,016), Yvelines dept., N France, on the Seine River near Versailles. Nearby is the hamlet of Marly-la-Machine, where in 1682 a huge hydraulic engine was built to supply the fountains of Versailles. Considered one of the wonders of the world, the engine was in use until 1804.

Mar·ma·ra, Sea of (mär'mər-ə), c.4,430 sq mi (11,473.7 sq km), NW Turkey, between Europe in the N and Asia in the S. It is connected on the east with the Black Sea through the Bosporus and on the west with the Aegean Sea through the Dardanelles. The sea has no strong currents and the tidal range is minimal.

Marne (märn), department (1975 pop. 530,399), NE France, in Champagne; capital Châlons-sur-Marne.

Marne, river, c.325 mi (525 km) long, rising in the Langres plateau, NE France, and flowing in an arc generally NW to the Seine River near Paris. During World War I and World War II, the Marne region was the scene of much fighting.

Ma·ro·ni (mə-rō'nē), river of South America, rising in the Tumuc-Humac Mts. of French Guiana near the Brazilian border and flowing c.450 mi (725 km) N along the border between Surinam and French Guiana to the Atlantic Ocean.

Mar·que·sas Islands (mär-kā'zəz, -səz), volcanic group (1971 pop. 5,593), South Pacific, a part of French Polynesia. There are 12 islands in the group, which lies c.740 mi (1,190 km) northeast of Tahiti. The Marquesas, famous for their rugged beauty, are fertile and mountainous, rising to c.4,130 ft (1,260 m) on Hiva Oa. The islands are divided into two groups. The southern cluster (sometimes called the Mendaña Islands) was discovered in 1595 by the Spanish navigator Alvaro de Mendaña de Neira; the northern group (sometimes called the Washington Islands) was discovered in 1791 by the American navigator Captain Joseph Ingraham. In 1813, Commodore David Porter claimed Nuku Hiva for the United States, naming it Madison Island, but the U.S. Congress never ratified the claim. France took possession of the islands in 1842 and established a settlement on Nuku Hiva, which was abandoned in 1859. In 1870 the French administration over the Marquesas was reinstated. The islands are the setting for Herman Melville's novel *Typee.*

Mar·quette (mär-kĕt'). **1.** County (1970 pop. 64,686), 1,828 sq mi (4,734.5 sq km), NW Upper Peninsula, Mich., bounded on the N by Lake Superior, drained by the Dead and Michigamme rivers and by several branches of the Escanaba River; formed 1851; co. seat Marquette. It includes the Huron Mts. and the Marquette iron range, where extensive mining is done. Marble quarrying, lumbering, and woodworking are major occupations. There is some livestock, poultry, and truck farming. Hunting, fishing, and camping resorts are found in the county, as well as a national experimental forest, a fish hatchery, a state park, and several lakes. **2.** County (1970 pop. 8,865), 457 sq mi (1,183.6 sq km), central Wis., drained by the Fox River and its tributaries and containing Buffalo Lake; formed 1836; co. seat Montello. It is in a stock-raising, dairying, and farming (potatoes and vegetables) area. It has granite quarries and some manufacturing (processed food, metal products, and electrical equipment).

Marquette, city (1978 est. pop. 23,900), seat of Marquette co., N Mich., Upper Peninsula, on Lake Superior; settled 1849, inc. as a city 1871. It is a center of iron-ore shipping and of industry and trade for a mining, lumbering, farming, and resort region. Chemicals, wood products, and machinery are manufactured. Marquette is the seat of Northern Michigan Univ.

Mar·ra·kesh (mə-rä'kĕsh, mär'ə-kĕsh'), city (1971 pop. 332,741), W central Morocco. The city, renowned for leather goods, is one of the principal commercial centers of Morocco. It was founded in 1062 and was the capital of Morocco from then until 1147 and again from 1550 to 1660. It was captured by the French in 1912. Beautifully situated near the Atlas Mts., Marrakesh has extensive gardens, a 14th cent. palace, and a former palace of the sultan that is now a museum of Moroccan art. The 220-ft (67.1-m) minaret (completed 1195) of the Koutoubya mosque dominates the city.

Mar·sa·la (mär-sä'lä), city (1976 est. pop. 49,200), W Sicily, Italy, a port on the Mediterranean Sea, located on Cape Boeo. It is noted for its sweet wine. The ancient Lilybaeum, it was later renamed Marsah al Allah, or port of God, by the Arabs. In 1860, Garibaldi landed here at the start of his successful campaign to conquer the kingdom of the Two Sicilies.

Mar·seilles (mär-sā'), city (1975 pop. 908,600), capital of Bouches-du-Rhône dept., SE France, on the Gulf of Lions, an arm of the Mediterranean Sea. It is the second largest city of France and an important seaport; an underground canal links it with the Rhône River. Marseilles is a major industrial city where flour, vegetable oil, soap, cement, sugar, sulfur, chemicals, and processed foods are produced. The oldest town of France, it was settled by Phocaean Greeks from Asia Minor c.600 B.C. Known as Massilia, it became an ally of Rome, which annexed it (49 B.C.) after it supported Pompey against Caesar in the Roman civil war. Although the city retained its internal autonomy, it was of secondary importance during the Middle Ages. The upper city was ruled by its bishops from A.D. 539 until 1288, when it was reunited with the lower city, which had been governed independently by a city council since 1214. During the Crusades (11th–14th cent.) Marseilles was a commercial center and a transit port for the Holy Land. The city declined commercially in the first half of the 14th cent.

Marseilles was taken by Charles I of Anjou (13th cent.) and then absorbed by Provence and bequeathed (with Provence) to the French crown in 1481. In the 1700s commerce revived, mainly with the Levant and the Barbary States; although the plague wiped out almost half its population in 1720, Marseilles continued to enjoy prosperity until the French Revolution, when it was torn by civil strife. In the 19th cent. the French conquest of Algeria and the opening of the Suez Canal led to a tremendous expansion of the port of Marseilles and to the city's industrialization. Despite its long history, Marseilles has few buildings dating back further than the 18th cent. A landmark of Marseilles harbor is the Chateau d'If, a castle built on a small, rocky isle. Excavations in 1966–67 uncovered what are believed to be vestiges of the ramparts of old Massilia.

Mar·shall (mär'shəl). **1.** County (1970 pop. 54,211), 571 sq mi (1,478.9 sq km), NE Ala., bounded on the N by the Paint Rock River, with Wheeler and Guntersville Reservoirs on the Tennessee River; formed 1836; co. seat Guntersville. It is in an agricultural area that produces cotton and bees. It has some mining, timber, and a food-processing industry. Textiles and clothing, wood products, concrete, metal products, and machinery are made here. **2.** County (1970 pop. 13,302), 395 sq mi (1,023.1 sq km), N central Ill., drained by the Illinois River and Sandy Creek; formed 1839; co. seat Lacon. It has agriculture (corn, oats, wheat, soybeans, fruit, livestock, poultry, and dairy products), bituminous-coal mines, timber, and manufacturing plants, woolen goods, clothing, cigars, and feed. **3.** County (1970 pop. 34,986), 444 sq mi (1,150 sq km), N Ind., drained by the Yellow and Tippecanoe rivers; formed 1835; co. seat Plymouth. In an agricultural area yielding livestock, grain, soybeans, poultry, fruit, truck crops, and dairy products, it is especially noted for the production of spearmint and peppermint oil. Its manufactures include textiles, lumber and wood products, plastics, concrete, metal products, and machinery. There are resort lakes and recreation areas in the county. **4.** County (1970 pop. 41,076), 574 sq mi (1,486.7 sq km), central Iowa, drained by the Iowa and North Skunk rivers; formed 1846; co. seat Marshalltown. Its prairie agriculture includes cattle, hogs, poultry, corn, oats, and wheat. Bituminous-coal mines are found in the west, and limestone quarries are in the east. **5.** County (1970 pop. 13,139), 883 sq mi (2,287 sq km), NE Kansas, in a gently rolling to hilly area bordered on the N by Nebr. and drained by the Big Blue and Little Blue rivers; formed 1855, name changed from Lykins in 1861; co. seat Marysville. It raises wheat, milo, corn, and livestock. There are limestone deposits, and some manufacturing is done. **6.** County (1970 pop. 20,381), 303 sq mi (784.8 sq km), W Ky., bounded on the N by the Tennessee River, on the E by its Kentucky Reservoir, and drained by the East and West forks of the Clarks River; formed 1842; co. seat Benton. Its agriculture includes dairy products, poultry, livestock, corn, dark tobacco, strawberries, and lespedeza. It also has hardwood timber, limestone deposits, and clay pits. Furniture and chemicals are made. **7.** County (1970 pop. 13,060), 1,789 sq mi (4,635.5 sq km), NW Minn., bounded on the W by N.Dak. and the Red River, and drained by the Snake, Thief, and Middle rivers; formed 1879; co. seat Warren. Small grains, potatoes, sugar beets, and sunflowers are grown. It has some manufacturing and tourism. Agassiz National Wildlife Refuge is in a marshy area in the east. **8.** County (1970 pop. 24,027), 710 sq mi (1,838.9 sq km), N Miss., on the Tenn. border, partly bounded on the S by the Tallahatchie River and drained by the Coldwater River; formed 1836; co. seat Holly Springs. In a hilly agricultural area that produces cotton, corn, hay, and dairy products, it has lumber sawmilling and plants manufacturing textiles, plastics, bricks, and clay tiles. The county includes part of Holly Springs National Forest. **9.** County (1970 pop. 7,682), 366 sq mi (947.9 sq km), S Okla., bounded in the E and S by Lake Texoma, which is formed by Denison Dam on the Red River; formed 1907; co. seat Madill. It is in an agricultural and cattle-raising area yielding alfalfa, corn, barley, cotton, pecans, and dairy products. It has oil and gas fields and some lumbering. Textiles and trailers are manufactured. **10.** County (1970 pop. 5,965), 848 sq mi (2,196.3 sq km), NE S.Dak., on the N.Dak. border and watered by numerous lakes; formed 1885; co. seat Britton. It is in a rich farming and livestock-raising region and manufactures miscellaneous textile products. Fort Sisseton, a historic military outpost, is here. **11.** County (1970 pop. 17,319), 377 sq mi (976.4 sq km), central Tenn., drained by the Duck River and its tributaries; formed 1836; co. seat Lewisburg. Its agriculture includes livestock, dairy products, grain, and tobacco. Among its diversified manufactures are wood and metal products, furniture, paper, leather, iron and aluminum, and machinery. **12.** County (1970 pop. 37,598), 306 sq mi (792.5 sq km), N W.Va., the southernmost county of the Northern Panhandle, bounded on the W by the Ohio River and the Ohio border, on the E by Pa., and drained by Wheeling, Fish, and Grave creeks; formed 1835; co. seat Moundsville. It is in an industrial area; the diversified manufacturing is based on the region's coal, natural-gas, oil, glass sand, clay, and timber. There is some livestock, dairy, feed crop, tobacco, and truck farming.

Marshall. 1. City (1970 pop. 1,397), seat of Searcy co., N central Ark., SE of Harrison, in a timber, fruit, and farm area. **2.** City (1970 pop. 3,468), seat of Clark co., E Ill., WSW of Terre Haute, Ind., in a farm and oil area; inc. 1853. Chemicals and electronic equipment are made. **3.** City (1970 pop. 7,253), seat of Calhoun co., S Mich., ESE of Battle Creek, in a farm region; settled 1831, inc. as a village 1836, as a city 1859. Pumps, paper products, and refrigerators are made. **4.** City (1970 pop. 9,886), seat of Lyon co., SW Minn., SSW of Granite Falls; settled 1871, platted 1872, inc. 1901. It is a farm trade center and processes poultry. **5.** City (1970 pop. 12,051), seat of Saline co., N central Mo.; inc. 1839. It is the processing center of a farm area, with meat-packing houses, grain and seed mills, and cold storage facilities. Shoes are also manufactured. **6.** Town (1970 pop. 982), seat of Madison co., W N.C., NNW of Asheville on the French

Broad River, in a burley-tobacco region; settled 1816, inc. 1852. It has cotton and lumber mills. **7.** City (1978 est. pop. 21,300), seat of Harrison co., E Texas, in a pine-covered hill and lake area; inc. 1844. Live-oak-shaded streets and mansions recall the plantation past of the city, which now has railroad shops and chemical and aluminum industries.

Marshall Islands, archipelago (1970 pop. 22,888), central Pacific, an administrative district of the U.S. Trust Territory of the Pacific Islands. The Marshalls extend over a 700-mi (1,125-km) area and comprise two major groups: the Ratak Chain in the east, and the Ralik Chain in the west, with a total of 34 atolls, c.900 reefs, and a land area of 70 sq mi (181.3 sq km). The chief industry is coconut planting; copra, sugar, and coffee are the major exports. Some of the islands were discovered by Spanish explorers in the early 16th cent. and were named after a British captain who visited them in 1788. Much mapping was done on Russian expeditions under Adam Johann von Krusenstern (1803) and Otto von Kotzebue (1815 and 1823). Germany annexed the group in 1885 and tried with little success to establish a colony. The administrative affairs of the islands continued to be managed largely by private German and Australian interests. In 1914, Japan seized the Marshalls and in 1920 received a League of Nations mandate over them. In World War II the islands were taken by U.S. forces (1943–44); they were included in the Trust Territory of the Pacific Islands in 1947.

Mar·shall·town (mär'shəl-toun'), city (1978 est. pop. 27,100), seat of Marshall co., central Iowa, on the Iowa River; inc. 1863. It is the rail and trade center of a rich grain and livestock area and a busy manufacturing city with the nickname of "Little Pittsburgh."

Marsh·field (märsh'fēld'). **1.** Resort town (1970 pop. 15,223), Plymouth co., SE Mass., on the Atlantic coast; settled 1632, inc. 1640. Sand and gravel are the major industrial products. Points of interest include Winslow House, home of Edward Winslow, and several other colonial buildings. Daniel Webster lived in Marshfield and is buried here. **2.** City (1970 pop. 2,961), seat of Webster co., S Mo., in the Ozarks ENE of Springfield; settled c.1830, platted 1856, inc. 1877. The area has diversified agriculture.

Marston Moor, battlefield, Yorkshire, N England, near York. The battle fought here on July 2, 1644, between the Royalists, under Prince Rupert and the duke of Newcastle, and the Parliamentarians, under Lord Fairfax of Cameron, Oliver Cromwell, and the earl of Leven, resulted in the first major victory for the Parliamentarians in the English Civil War.

Mar·tha's Vine·yard (mär'thəz vĭn'yərd), island (1970 est. pop. 6,000), c.100 sq mi (260 sq km), SE Mass., separated from the Elizabeth Islands and Cape Cod by Vineyard and Nantucket sounds. As a result of glaciation, the island has morainal hills composed of boulders and clay deposits in the north, and low, sandy plains in the south. The English were the first to settle the island (1642); they engaged in farming, brickmaking, salt production, and fishing. Martha's Vineyard became an important commercial center, with whaling and fishing as the main occupations, in the 18th and early 19th cent. In the late 1800s the island, with its harbors, beaches, and scenic attractions, developed into a summer resort. Much of the island's interior is set aside as a state forest.

Mar·tin (mär'tn). **1.** County (1970 pop. 28,035), 559 sq mi (1,447.8 sq km), SE Fla., between Lake Okeechobee in the W and the Atlantic in the E, and partly sheltered in the E by Jupiter Island's barrier beach; formed 1925; co. seat Stuart. It is in a lowland area with swamps and many small lakes in the west. Truck crops and citrus fruit are produced. Some cattle raising and fishing are done. **2.** County (1970 pop. 10,969), 345 sq mi (893.6 sq km), SW Ind., drained by the Lost River and the East Fork of the White River; formed 1820; co. seat Shoals. In an agricultural area yielding grain, corn, hay, and livestock, it has lumber, coal deposits, and oil and gas fields. Its manufactures include textiles, wood products, and concrete. **3.** County (1970 pop. 9,377), 231 sq mi (598.3 sq km), E Ky., in the Cumberlands, bounded on the E by Tug Fork and the W.Va. border and drained by several creeks; formed 1870; co. seat Inez. It is in a mountain agricultural area that produces livestock, fruit, and tobacco. It also has bituminous-coal mines, oil wells, and some manufacturing. **4.** County (1970 pop. 24,316), 707 sq mi (1,831.1 sq km), S Minn., bordering on Iowa and watered by Tuttle Lake and by Middle Chain of Lakes; formed 1857; co. seat Fairmont. It lies in an agricultural area yielding corn, oats, barley, potatoes, and livestock. Textile mill products, stone, clay, and glass products, machinery, and travel trailers and campers are manufactured. **5.** County (1970 pop. 24,730), 455 sq mi (1,178.5 sq km), E N.C., in the coastal plain, bounded on the N by the Roanoke River; formed 1774; co. seat Williamston. Peanuts and tobacco are grown and processed here. Lumbering and fishing are also important. **6.** County (1970 pop. 4,774), 911 sq mi (2,359.5 sq km), W Texas, on the S Llano Estacado with Cap Rock escarpment in the NW; formed 1876; co. seat Stanton. Cattle ranching is the major occupation. Other agricultural products include cotton, grain sorghums, grain, fruit, truck crops, poultry, hogs, sheep, and dairy productions. There are clay deposits and some oil wells.

Martin, city (1970 pop. 1,248), seat of Bennett co., SW S.Dak., near the Nebr. border SE of Rapid City, in a cattle and wheat area; founded 1912, inc. 1926. Pine Ridge Indian Reservation and a migratory bird refuge are nearby.

Mar·ti·nez (mär-tē'nəs), city (1970 pop. 16,506), seat of Contra Costa co., W Calif., on Carquinez Strait between San Pablo and Suisun bays, in a farm area; inc. 1884.

Mar·ti·nique (mär'tə-nēk'), overseas department of France (1973 est. pop. 343,100), 425 sq mi (1,100.8 sq km), in the Windward Islands, West Indies; capital Fort-de-France. The department and the island of Martinique are coterminous. Of volcanic origin, the island is rugged and mountainous and reaches its greatest height in Pelée volcano. Most agriculture is carried on in the hot valleys and along the coastal strips; about 80% of this area is devoted to sugar cane, introduced from Brazil in 1654. The island's industries consist mainly of sugar and rum production and pineapple canning. Tourism is a growing source of revenue. Discovered by Columbus, probably in 1502, the island was ignored by the Spanish; colonization began in 1635, when the French, who had promised the native Carib Indians the western half of the island, established a settlement. The French proceeded to eliminate the Caribs and later imported African slaves as sugar plantation workers. In the 18th cent. Martinique's sugar exports made it one of France's most valuable colonies; although slavery was abolished in 1848, sugar continued to hold a dominant position in the economy. A target of dispute during the Anglo-French worldwide colonial struggles, Martinique was finally confirmed as a French possession after the Napoleonic wars. Martinique supported the Vichy regime after France's collapse in World War II, but in 1943 a U.S. naval blockade forced the island to transfer its allegiance to the Free French. It became a department of France in 1946.

Mar·tins·burg (mär'tənz-bûrg'). **1.** Industrial borough (1970 pop. 2,088), Blair co., S Pa., SSE of Altoona; settled c.1793 by Quakers, laid out 1815, inc. 1832. Shoes and canned goods are made. **2.** Industrial city (1978 est. pop. 13,300), seat of Berkeley co., NE W.Va., in the Eastern Panhandle; settled 1732, inc. as a city 1859. It is a railroad center in a region that grows apples and peaches. Manufactures include textiles, hosiery, glassware, and cement. Limestone is quarried nearby. In the Civil War the city's strategic location on the railroad made it a frequent military objective. Nearby Bunker Hill, settled c.1729, is the oldest recorded settlement in the state.

Mar·tins Ferry (mär'tənz), industrial city (1970 pop. 10,757), Belmont co., E Ohio, on the Ohio River opposite Wheeling, W.Va.; settled 1780, inc. as a city 1885. It is a coal-mining and steel-manufacturing city. The novelist William Dean Howells was born here.

Mar·tins·ville (mär'tənz-vĭl'). **1.** City (1970 pop. 9,723), seat of Morgan co., central Ind., on the White River SW of Indianapolis; settled 1822. A health resort with mineral springs, the city also makes wood products and has diversified agriculture. **2.** City (1978 est. pop. 18,600), independent city and seat of Henry co., S Va., in the Blue Ridge foothills near the N.C. border; founded 1793, inc. as a city 1928. Tobacco is processed, and furniture, chemicals, textiles, and textile products are manufactured.

Mar·war (mär'wär): *see* Jodhpur, India.

Ma·ry (mä-rē'), city (1976 est. pop. 70,000), capital of Mary oblast, Central Asian USSR, in Turkmenistan. Lying in a large oasis of the Kara-Kum desert, on the Murgab River delta, Mary is the center of a rich cotton-growing area. It is a rail junction and carries on extensive trade in cotton, wool, grain, and hides. Mary is also a major textile center. Mary arose in 1884 as a Russian military center c.20 mi (30 km) from the site of ancient Merv and was itself called Merv until 1937.

Mar·y·bor·ough (mâr'ə-bûr'ō -bŭr'ō), city (1971 pop. 19,304),

Queensland, E Australia, on the Mary River. Sugar, fruit, coal, and timber are exported, and there are shipyards and locomotive works in the city.

Mar·y·land (mĕr′ə-lənd), state (1975 pop. 4,114,000), 10,577 sq mi (27,394 sq km), E United States, in the Middle Atlantic region, one of the original Thirteen Colonies. Annapolis is the capital. A seaboard state, Maryland is divided by Chesapeake Bay, which runs almost to the northern border; thus the section of Maryland called the Eastern Shore is separated from the main part of the state. Maryland is bounded on the north by Pennsylvania and on the east by Delaware and the Atlantic Ocean. For the most part, the Potomac River separates the main part of Maryland from Virginia (to the south) and the long, narrow western handle from West Virginia (to the south and west). The District of Columbia cuts a rectangular indentation into the state at the estuary of the Potomac.

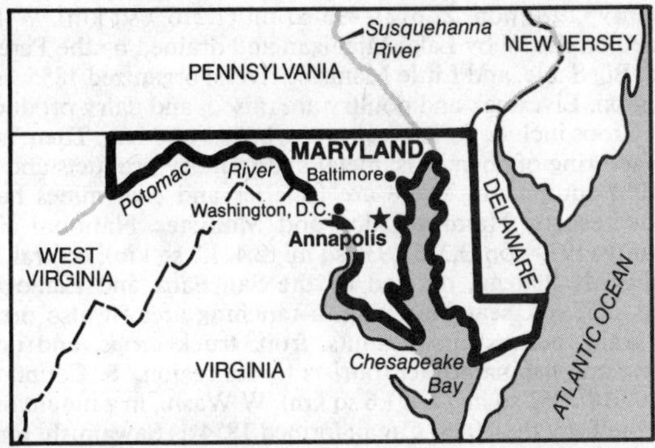

The main part of the state is divided by the Fall Line, which runs between Baltimore and Washington, D.C.; to the north and west is the rolling Piedmont, rising to the Blue Ridge and to the Pennsylvania hills. The heavily indented shores of Chesapeake Bay fringe the land with bays and estuaries, which helped in the development of a farm economy relying on water transport. Maryland has nearly 3 million acres (1.2 million hectares) of forest land.

The state has become increasingly popular as a vacation area—Ocean City is a popular seashore resort, and both sides of Chesapeake Bay are lined with beaches and small fishing towns. The Eastern Shore is still noted for its unique rural beauty and architecture. Annapolis, with its well-preserved Colonial architecture and 18th cent. waterfront, is the site of the U.S. Naval Academy.

Maryland is divided into 23 counties and 1 independent city:

NAME	COUNTY SEAT	NAME	COUNTY SEAT
Allegany	Cumberland	Garrett	Oakland
Anne Arundel	Annapolis	Hartford	Bel Air
Baltimore	Towson	Howard	Ellicott City
Baltimore (Independent City)		Kent	Chestertown
		Montgomery	Rockville
Calvert	Prince Frederick	Prince Georges	Upper Marlboro
		Queen Annes	Centreville
Caroline	Denton	St. Marys	Leonardtown
Carroll	Westminster	Somerset	Princess Anne
Cecil	Elkton	Talbot	Easton
Charles	La Plata	Washington	Hagerstown
Dorchester	Cambridge	Wicomico	Salisbury
Frederick	Frederick	Worcester	Snow Hill

Economy. While the iron mines of western Maryland have declined along with other mining activity, iron and steel production is still a major industry, and Maryland is an important mineral processor. Shipbuilding flourished early and in later years continued, but on more massive lines. Other important industries include the manufacture of primary metals, food products, transportation equipment, electrical machinery, chemicals, and apparel. Shipping (Baltimore is a major port), tourism (especially in the Chesapeake Bay cities), and printing and publishing are also big industries. Manufacturing is the largest sector of the economy except for government work. Although manufacturing well exceeds agriculture as a source of income, Maryland's farms yield corn, hay, tobacco, soybeans, and other crops. Income from livestock and livestock products is almost twice that from crops; dairy and poultry farms thrive, and Maryland is famous for breeding horses.

History. John Cabot was probably the first white man to see what is today Maryland when he sailed along the eastern coast in 1498. The Chesapeake region was later explored (1574) by Pedro Menéndez Marqués, governor of Spanish Florida. In 1608 the region was charted by Capt. John Smith. William Claiborne of Virginia, under license from Charles I of England, set up a fur-trading post on Kent Island in 1631. The next year Charles granted a charter to George Calvert, 1st Baron Baltimore, yielding him feudal rights to the region between lat. 40° N and the Potomac River. The territory was named Maryland in honor of Henrietta Maria, queen consort of Charles I. George Calvert's son, Cecilius Calvert, 2nd Baron Baltimore, undertook development of the colony as a haven for his persecuted fellow Catholics and also as a source of income. In 1634 a settlement called St. Mary's was set up with Leonard Calvert, brother of the proprietor, as governor. The Algonquian-speaking Indian tribes withdrew gradually and for the most part peacefully from the area during the colonial period. By 1650 a bicameral legislature existed.

Religious conflict was strong in ensuing years as the Puritans set out to destroy the religious freedom guaranteed with the founding of the colony. A toleration act (1649) was passed in an attempt to save the Catholic settlers from persecution, but it was repealed (1654) after the Puritans seized control. A brief civil war ensued (1655), from which the Puritans emerged triumphant. After England's Glorious Revolution of 1688 the government of the colony passed to the Crown; the Church of England was made the established Church, and Maryland became (1691) a royal province. In 1694, when the capital was moved from St. Mary's to Annapolis, those were the only towns in the province, but the next century saw the emergence of commercially oriented Baltimore. Public education began in 1723 when the general assembly authorized a free school in each county. Tobacco became the basis of the economy by 1730.

Economic and religious grievances led Maryland to support the growing colonial agitation against England. In 1776 Maryland adopted a declaration of rights and a state constitution and sent soldiers and supplies to aid the war for independence; supposedly the high quality of its regular "troops of the line" earned Maryland its nickname, the Old Line State. At Annapolis Congress ratified the Treaty of Paris, ending the Revolutionary War in 1783. In 1791 Maryland and Virginia contributed land and money for the new national capital in the District of Columbia.

After the War of 1812 the state entered a period of great commercial and industrial expansion. This was accelerated by the building of the National Road; the opening of the Chesapeake and Delaware Canal (1829); and the opening (1830) of the Baltimore & Ohio RR, the first railroad in the United States open for public traffic. Southern ways and sympathies persisted, however, among the plantation owners, and as the rift between North and South widened, Maryland was torn by conflicting interests. In 1860 there were 87,000 slaves in Maryland, but industrialists and businessmen successfully exerted their influence to keep the state in the Union. At the beginning of the Civil War a pro-Southern mob in Baltimore attacked units of the Massachusetts militia. President Lincoln suspended habeas corpus and sent troops to Maryland who imprisoned large numbers of secessionists.

With the end of the Civil War, industry was quickly revived and became a dominant force, economically and politically. New railroad lines traversed the state, making it more than ever a crossing point between North and South. Labor troubles hit Maryland with the Panic of 1873, and four years later railroad wage disputes resulted in large-scale rioting in Cumberland and Baltimore.

The great influx of population into the state during World War I was repeated and accelerated in World War II—war workers poured into Baltimore, where vital shipbuilding and aircraft plants were in operation, and military and other government employees moved into the area around Washington, D.C. The opening of the Chesapeake Bay Bridge in 1952 spurred significant industrial expansion on the Eastern Shore; a parallel bridge was opened in 1973.

Government. Maryland is governed under a constitution adopted in 1867. The general assembly consists of 43 senators and 123 delegates, all elected for four-year terms. The governor, also elected for a four-year term, may succeed himself once.

Educational Institutions. Among the more prominent in the state are Johns Hopkins Univ., at Baltimore; St. John's College, at Annapolis; Towson State College, at Towson; and the Univ. of Maryland, at College Park.

Mar·ys·ville (mâr′ēz-vĭl′). **1.** City (1970 pop. 9,353), seat of Yuba co.,

N central Calif., at the confluence of the Yuba and Feather rivers N of Sacramento; settled 1842, inc. 1851. It is located in a rich farm area (rice, barley, and fruit) and since 1849 has been a supply point for a gold-mining region. **2.** City (1970 pop. 3,588), seat of Marshall co., NE Kansas, on the Big Blue River NW of Topeka; settled 1851, inc. 1861. Once a ferry crossing (1849) on the Oregon Trail, it is the shipping center of a grain, livestock, and dairy area. **3.** Village (1970 pop. 5,744), seat of Union co., W central Ohio, NW of Columbus; settled 1816, laid out 1820, inc. 1838. Primarily a farm community, it has some manufacturing.

Mar·yut (mər-yo͞ot'), salt lake, Egypt: *see* Mareotis.

Mar·y·ville (mâr'ē-vĭl'). **1.** City (1970 pop. 9,970), seat of Nodaway co., NW Mo., N of St. Joseph, in a farming region; settled c.1845, inc. 1869. Northwest Missouri State Univ. is here. **2.** City (1978 est. pop. 17,800), seat of Blount co., E Tenn.; settled around Fort Craig (built 1785), inc. as a town 1830, as a city 1927. With its twin city, Alcoa, it is an important center for the production of aluminum and aluminum products. Textile, rubber, and plastic products are also manufactured. Great Smoky Mountains National Park and the Tuckaleechee Caverns are in the area.

Ma·sa·da (mə-sā'də), ancient mountaintop fortress in Israel, the final outpost of the Zealot Jews in their rebellion against Roman authority (A.D. 66–73). Located in the Judaean Desert, the fortress sits atop a mesa-shaped rock that towers some 1,300 ft (395 m) above the western shore of the Dead Sea. According to the ancient historian Josephus, Masada was first fortified sometime during the 1st or 2nd cent. B.C. Between 37 and 31 B.C. Herod the Great, king of Judaea, further strengthened Masada, building two ornate palaces, a bathhouse, aqueducts, and surrounding siege walls. In A.D. 66, with the outbreak of the Jewish war against Rome, the zealots, an extremist Jewish sect, seized the fortress in a surprise attack and massacred its Roman garrison. Masada remained under Zealot control until A.D. 73, when, after a siege of almost two years, the 15,000 soldiers of Rome's tenth legion finally subdued the 1,000 men, women, and children holding the fortress. In a final act of defiance, however, almost all of the Jewish defenders had killed themselves. Only two women and five children survived. Excavated (1963–65) by an international team of volunteer archaeologists, Masada is now a major tourist site and an Israeli historical shrine.

Mas·ba·te (mäs-bä'tə), island (1970 pop. 492,868), 1,262 sq mi (3,268.6 sq km), Philippines, one of the Visayan Islands, S of SE Luzon. Gold, which has been mined here for centuries, is still produced in quantity; copper is also mined.

Mas·ca·ra (măs'kər-ə), town (1974 est. pop. 70,600), NW Algeria. It is an administrative center, a garrison town, and a marketplace noted for its white wine and for its trade in cereals and tobacco. Mascara occupies the site of a Roman settlement.

Mas·ca·rene Islands (măs-kə-rēn'), in the Indian Ocean, E of Madagascar. They include Mauritius, Réunion, and Rodriguez. Apparently known to the Arabs, they were rediscovered by the Portuguese at the beginning of the 16th cent.

Mas·e·ru (măz'ə-ro͞o'), city (1972 est. pop. 17,000), capital of the Kingdom of Lesotho, on the Caledon River near the border with the Republic of South Africa. A trade and transportation hub, it lies on Lesotho's main road and is linked with South Africa's rail network. The city's few manufactures include candles, carpets, and retreaded tires. Maseru was a small trading town when it was made the capital of the Basuto people by Moshesh I, their paramount chief, in 1869. It was the capital of the British Basutoland protectorate from 1869 to 1871 and from 1884 to 1966, when Lesotho achieved independence.

Mash·had (mäsh-häd'), city (1966 pop. 409,616), capital of Khorasan prov., NE Iran. It is an industrial and trade center and a transportation hub. Manufactures include carpets, textiles, pharmaceuticals, and processed foods. It is the site of the beautiful shrine of the Imam Ali Riza, a Shiite holy person. Imam Riza died (819) in the city after visiting the grave of Caliph Harun ar-Rashid, who had died there 10 years before; he was buried next to Harun, and the shrine was built over both graves. The city was attacked by the Oghuz Turks (12th cent.) and by the Mongols (13th cent.), but recovered by the 14th cent. It prospered under the Safavids, who were devout Shiite Moslems, and reached its greatest glory in the 18th cent., when Nadir Shah made Mashhad the capital of Persia. The city took on strategic importance in the late 19th cent. because of its proximity to the Russian and Afghan borders. The bombing of the sanctuary of the Imam Riza by the Russians in 1912 caused widespread resentment.

Mas·jed So·ley·man (mäs-jĕd' so͞o-lā-män'), city (1966 pop. 64,488), Khuzistan prov., SW Iran, on the Karun River. The site of the first discovery of petroleum in Iran (1908), it is an oil-refining center.

Ma·son (mā'sən). **1.** County (1970 pop. 16,180), 541 sq mi (1,401.2 sq km), central Ill., bounded on the W by the Illinois River and its bayou lakes, on the S by Sangamon River and Salt Creek; formed 1841; co. seat Havana. It has agriculture (corn, wheat, soybeans, livestock, poultry, truck crops, dairy products, and watermelons), diversified manufacturing, commercial fisheries, and river and rail shipping. **2.** County (1970 pop. 17,273), 239 sq mi (619 sq km), NE Ky., in the outer Bluegrass bounded on the N by the Ohio River and the Ohio border, and drained by the North Fork of the Licking River; formed 1788; co. seat Maysville. It is in a gently rolling upland agricultural area that produces burley tobacco, corn, wheat, apples, livestock, and milk. Some mining, food and tobacco processing, and manufacturing (yarn, clothing, and machinery) are done. **3.** County (1970 pop. 22,612), 493 sq mi (1,276.9 sq km), W Mich., bounded on the W by Lake Michigan and drained by the Pere Marquette, Big Sable, and Little Manistee rivers; organized 1855; co. seat Ludington. Livestock and poultry are raised, and dairy products are made. Crops include fruit, corn, grain, beans, and hay. There is some manufacturing of chemicals, metal and foundry products and transportation equipment. There are fisheries and coal mines here, as well as resorts, Hamlin Lake, and Manistee National Forest. **4.** County (1970 pop. 3,356), 935 sq mi (2,421.7 sq km), central Texas, on Edwards Plateau, drained by the San Saba and Llano rivers; formed 1858; co. seat Mason. In a ranching area, it also produces corn, grain, pecans and peanuts, fruit, truck crops, and poultry. Hunting and fishing attract tourists to the region. **5.** County (1970 pop. 20,918), 962 sq mi (2,491.6 sq km), W Wash., in a mountain area cut in the E by the Hood Canal; formed 1854 as Sawamish, renamed 1864; co. seat Shelton. It is in a lumbering area, with fishing, clamming, and oystering. It manufactures wood products and has other heavy industries. Part of Olympic National Forest and Skokomish Indian Reservation are here. **6.** County (1970 pop. 24,306), 432 sq mi (1,118.9 sq km), W W.Va., bounded on the N and W by the Ohio River and the Ohio border and drained by the Kanawha River; formed 1804; co. seat Point Pleasant. It is in a bituminous-coal region, and there are some natural-gas fields. Among its diversified industries are shipbuilding, chemical manufacturing, plastic and synthetic production, and metal industries.

Mason. 1. City (1970 pop. 5,468), seat of Ingham co., SE of Lansing; settled 1836, inc. as a village 1865, as a city 1875. Milk products are made. **2.** Village (1970 pop. 5,677), Warren co., SW Ohio, NNE of Cincinnati, in a farm area; inc. 1837. **3.** City (1970 pop. 1,806), seat of Mason co., W central Texas, on Comanche Creek in the Llano River valley NW of San Antonio. Settled by Germans before the Civil War, the stone-built city and nearby farms have a continental air. Mason handles the wool, mohair, and cattle from ranches in the rugged, sparsely settled hills of the Edwards Plateau.

Mason City, city (1978 est. pop. 30,200), seat of Cerro Gordo co., N central Iowa; inc. 1874. It is the rail, trade, and industrial center of a large agricultural area. The major industries are food processing and the manufacture of cement, brick, tile, doors and windows, fertilizers, feeds, apparel, and mobile homes.

Ma·son-Dix·on Line, (mā'sən-dĭk'sən), boundary between Pa. and Md. (running between lat. 39°43'26.3" N and lat. 39°43'17.6" N), surveyed by the English astronomers Charles Mason and Jeremiah Dixon between 1763 and 1767. The ambiguous description of the boundaries in the Md. and Pa. charters led to a protracted disagreement between the proprietors of the two colonies; the dispute was submitted to the English court of chancery in 1735. A compromise between the Penn and Calvert families in 1760 resulted in the appointment of Mason and Dixon. By 1767 the surveyors had run their line 244 mi (392.6 km) west from the Del. border, every fifth milestone bearing the Penn and Calvert arms. The survey was completed to the western limit of Md. in 1773; in 1779 the line was extended to mark the southern boundary of Pa. with Va. Before the Civil War the term "Mason-Dixon Line" popularly designated the boundary dividing the slave states from the free states.

Ma·so·vi·a (mə-sō'vē-ə), historic region, almost coextensive with Warsaw prov., central Poland. At the death (1138) of Boleslaus III, Masovia became an independent duchy under the Piast dynasty. It became a suzerainty of Great Poland in 1351 and was finally united with it in 1526. Masovia passed to Prussia during the 18th cent.

partitions of Poland and was later a part of the Russian Empire. It reverted to Poland in 1918.

Mas·sa (mäs'-ä), city (1976 est. pop. 65,332), capital of Massa-Carrara prov., Tuscany, N central Italy, near the Ligurian Sea. Marble is quarried, and chemicals are produced here. From the 15th to the 19th cent. Massa was the capital of the independent principality, later duchy, of Massa and Carrara.

Mas·sac (măs'ăk', -ək), county (1970 pop. 13,889), 246 sq mi (637.1 sq km), extreme S Ill., bounded on the S by the Ohio River, on the NW by the Cache River; formed in 1843; co. seat Metropolis. It has agriculture (corn, wheat, livestock, cotton, and truck crops), lumbering, and manufacturing of clothing and wood products.

Mas·sa·chu·setts (măs'ə-cho͞o'sĭts), state (1975 pop. 5,800,000), 8,257 sq mi (21,386 sq km), NE United States, in New England, one of the Thirteen Colonies. Boston is the capital and largest city. Massachusetts is bounded on the north by Vt. and N.H., on the east and south by the Atlantic Ocean, further on the south by R.I. and Conn., and on the west by N.Y.

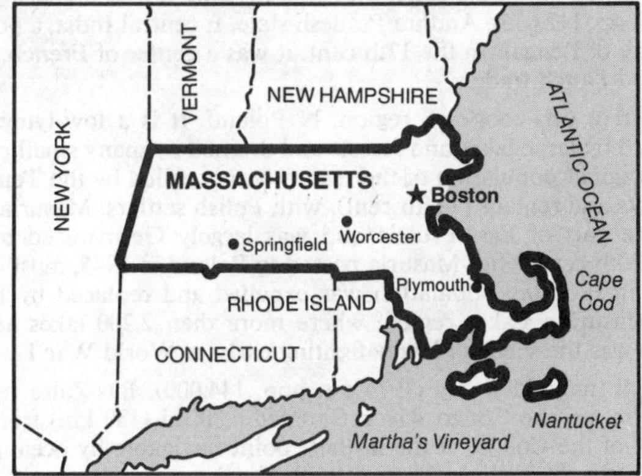

The eastern part of the commonwealth (its official designation), including the Cape Cod peninsula and the islands lying off it to the south—the Elizabeth Islands, Martha's Vineyard, and Nantucket—is a low coastal plain. In this area short, swift rivers such as the Merrimack have long supplied industry with power, and an indented coastline provides many good natural harbors, with Boston a major port. In the interior rise uplands separated by the rich Connecticut River valley, and farther west lies the Berkshire valley, surrounded by the Berkshire Hills, part of the Taconic Mts. The western streams feed both the Hudson and the Housatonic rivers. The state has a mean altitude of c.500 ft (155 m). As a recreation and vacation land, Massachusetts has great stretches of seashore in the east and many lakes and streams in the wooded Berkshire Hills in the west. There are numerous state parks, forests, and beaches, and Cape Cod is the site of a national seashore.

Massachusetts is divided into 14 counties:

NAME	COUNTY SEAT	NAME	COUNTY SEAT
Barnstable	Barnstable	Hampshire	Northampton
Berkshire	Pittsfield	Middlesex	Cambridge
Bristol	Taunton	Nantucket	Nantucket
Dukes	Edgartown	Norfolk	Dedham
Essex	Salem	Plymouth	Plymouth
Franklin	Greenfield	Suffolk	Boston
Hampden	Springfield	Worcester	Worcester

Economy. Massachusetts is an overwhelmingly industrial state, and, with its predominantly urban population, one of the most densely settled in the nation. It has many diverse manufactures; chief among them are electrical and electronic equipment, plastic products, shoes and leather goods, clothing and textiles, firearms, paper and paper products, machinery, tools, and metal and rubber products. Shipping, printing, and publishing are important, and the jewelry industry dates from before the American Revolution. Leading agricultural products include cranberries, tobacco, hay, apples, vegetables, greenhouse and nursery items, and milk and other dairy goods; poultry is also raised. The fishing fleets of Gloucester and New Bedford still bring in a large and varied catch, and the coastal waters abound in shellfish.

History. The coast of what is now Massachusetts was probably skirted by Norsemen in the 11th cent., and Europeans (mostly English) sailed offshore in the late 16th and early 17th cent. Settlement began when the Pilgrims arrived on the Mayflower and landed (1620) at a point which they named Plymouth. Weathering early difficulties, the colony eventually prospered. Other Englishmen soon established fishing and trading posts nearby. In 1626 Roger Conant founded Naumkeag (Salem), which in 1628 became the nucleus of a Puritan colony led by John Endecott of the New England Company; in 1629 the New England Company was reorganized as the Massachusetts Bay Company after receiving a more secure patent from the crown. In 1630, John Winthrop led the first large Puritan migration from England (900 settlers on 11 ships). Boston supplanted Salem as capital of the colony, and Winthrop replaced Endecott as governor.

The "Bay Colony" continued to be governed as a private company for the next 50 years. It was a thoroughgoing Puritan theocracy. The status of freeman was restricted (until 1664) to church members, and the state was regarded as an agency of God's will on earth. When Roger Williams spoke out in favor of religious toleration he was expelled (1635), as were other dissenters. A steady stream of newcomers from England more than offset these departures.

The early Puritans were primarily agricultural people, although a merchant class soon formed. Most of the inhabitants lived in villages, beyond which lay their privately owned fields. The typical village was composed of houses grouped around the common—a plot of land held in common by the community. In the meetinghouse of the chief village of a town was also held the town meeting, traditionally regarded as a foundation of American democracy. In practice the town meeting served to enforce unanimity and conformity, and participation was as a rule restricted to male property holders and church members. The Puritans, however, were not invariably grim and dictatorial; like 17th cent. Englishmen in general, they had a gregarious side and they also zealously promoted the development of educational facilities.

Massachusetts Bay Colony naturally rejoiced at the triumph of the Puritan Revolution in England, but with the restoration in 1660 the colony's happy prospects faded. Its charter was revoked in 1684 for violation of royal orders. The Massachusetts Bay and Plymouth colonies became part of the Dominion of New England under the governorship of Sir Edmund Andros in 1686. In 1689 aroused Bostonians ended Andros's arbitrary rule. Meanwhile, Increase Mather had gone (1688) to England to complain of Andros's administration. Mather managed to obtain a new charter uniting Massachusetts Bay, Plymouth, and Maine into the one royal colony of Massachusetts. This charter abolished church membership as a test for voting, although Congregationalism remained the established religion.

By the mid-18th cent. the Massachusetts colony had come a long way from its humble agricultural beginnings. Fish and lumber were exported along with farm products in a lively trade carried by ships built in Massachusetts and manned by local seamen. But the increasing British tendency to regulate colonial affairs, especially trade without colonial advice, was most unwelcome. In the 1760s a bitter struggle developed between the governor, Thomas Hutchinson, and the anti-British party in the legislature led by Samuel Adams, John Adams, James Otis, and John Hancock. The Stamp Act (1765) and the Townshend Acts (1767) preceded the Boston Massacre (1770), and the Tea Act (1773) brought on the Boston Tea Party. Through Committees of Correspondence, Massachusetts and the other colonies had been sharing their grievances, and in 1774 they called the First Continental Congress at Philadelphia for united action. The mounting tension in Massachusetts exploded in April, 1775, when the Massachusetts militia engaged a British force at Lexington and Concord. Patriot militia from other colonies hurried to Massachusetts, where, after the battle of Bunker Hill (June 17, 1775), George Washington took command of the patriot forces.

Victorious in the Revolution, the colonies faced depressing economic conditions. Nowhere were those conditions worse than in W Massachusetts, where discontented Berkshire farmers erupted in Shays's Rebellion in 1786. The uprising was promptly quelled, but it frightened conservatives into support of a new, stronger national constitution; this constitution was ratified by Massachusetts in 1788. Independence had closed the old trade routes within the British Empire, but newer ones were soon opened up, and trade with China became especially lucrative. European wars at the beginning of the 19th cent. at first further stimulated the carrying trade but then led to interference with American shipping. War with Great Britain in 1812 was violently unpopular in New England. There was talk of secession at the abortive Hartford Convention of New England Federalists, over which George Cabot presided.

After 1816, new industries, financed by money made in shipping and shielded from foreign competition by protective tariffs grew rapidly, transforming the character of the commonwealth and its people. Labor was plentiful and often ruthlessly exploited. The power loom made Massachusetts an early center of the American textile industry. Agriculture, on the other hand, went into a sharp decline because Massachusetts could not compete with the new agricultural states of the West.

In 1820 Maine was separated from Massachusetts and admitted to the Union as a separate state under the terms of the Missouri Compromise. In the same year the Massachusetts constitution was considerably liberalized by the adoption of amendments which abolished all property qualifications for voting and removed religious tests for officeholders. In the 1830s and 40s the state became the center of religious and social reform movements. The intellectual and cultural center of the nation, Massachusetts gave to the nation such writers and poets as Ralph Waldo Emerson, Henry Thoreau, Nathaniel Hawthorne, Richard Henry Dana, Emily Dickinson, Oliver Wendell Holmes, Henry Wadsworth Longfellow, James Russell Lowell, and John Greenleaf Whittier. In the 1830s reformers began to devote energy to the antislavery crusade. With the Whig party broken on the slavery issue, Massachusetts turned to the new Republican party and voted for John C. Frémont in 1856 and Abraham Lincoln in 1860.

After the war Massachusetts, with other northern states, experienced rapid industrial expansion. Labor remained cheap and plentiful as European immigrants streamed into the state. The Irish began arriving in droves in the 1840s, and they continued to land in Boston for years to come. After them came French Canadians, Portuguese, Italians, Poles and other Slavs, and Scandinavians. Of all the immigrant groups, the Irish came to be the most influential, especially in politics. Their religion (Roman Catholic) and their political faith (Democratic) definitely set them apart from the old native Yankee stock. Practically all of the immigrants went to work in the factories. The rise of industrialism was accompanied by a growth of cities, although the small mill town remained important.

World War I caused a vast increase in industrial production. Industry spurted forward again during World War II, and in the postwar era the state has continued to develop. The decline of the textile industry has been offset by the growth of the electronics industry, attracted by the skilled labor in the Boston area. In 1973 the state's economy was dealt a severe blow when several U.S. military installations, including the 172-year-old Boston Naval Shipyard and Westover Air Force Base, were closed by the Federal government in an economy move.

Government. The governor of Massachusetts is elected for a four-year term. The legislature (the General Court) has a senate of 40 members and a house of representatives with 240 members, all of whom serve two-year terms.

Educational Institutions. Besides Harvard Univ. and the Massachusetts Institute of Technology, at Cambridge, educational institutions include Radcliffe College, also at Cambridge; Amherst College and the Univ. of Massachusetts, at Amherst; Boston College, at Chestnut Hill; Boston Univ., Simmons College, and Northeastern Univ., at Boston; Brandeis Univ., at Waltham; Clark Univ., College of the Holy Cross, and Worcester Polytechnic Institute, at Worcester; Lowell Technological Institute, at Lowell; Mount Holyoke College, at South Hadley; Smith College, at Northampton; Tufts Univ., at Medford; Wellesley College, at Wellesley; Wheaton College, at Norton; and Williams College, at Williamstown.

Massachusetts Bay, inlet of the Atlantic Ocean. The bay, with its arms (Boston, Cape Cod, and Plymouth bays), extends 65 mi (104.6 km) from Cape Ann on the north to Cape Cod on the south. Its coastline varies from the irregular, rocky shore of the north to the sandy beaches of the south.

Mas·sa·pe·qua (măs-ə-pē'kwə), uninc. city (1978 est. pop. 27,400), Nassau co., SE N.Y., on the S shore of Long Island. It is chiefly residential.

Mas·sa·wa or **Me·se·wa** (both: mə-sä'wə), city (1974 est. pop. 23,880), Eritrea prov., N Ethiopia, a port on the Red Sea. It is the main port for northern Ethiopia and is linked by rail with Asmara.

Mas·se·na (mə-sē'nə), village (1978 est. pop. 13,100), St. Lawrence co., extreme N N.Y., on the St. Lawrence River; settled 1792, inc. 1886.

Mas·sif Cen·tral (mă-sēf' sĕn-träl'), great mountainous plateau, c.33,000 sq mi (85,470 sq km), S central France, covering almost a sixth of the surface of the country. The chief water divide of France, it borders on the Paris basin in the north, the Rhône valley and basin in the east and south, and the Aquitanian basin in the west. The core of the Massif is the volcanic mass of the Auvergne Mts. that rises to the Massif's highest point, Puy de Sancy (6,188 ft/1,887.3 m). The Cévennes limit the Massif Central on the southeast and the Causses form its southwest border. The Massif Central is the most rugged and geologically diverse region within France.

Mas·sil·lon (măs'ə-lŏn'), city (1978 est. pop. 34,700), Stark co., NE Ohio, on the Tuscarawas River; inc. 1853. It is an industrial city; among its manufactures are surgical gloves, food products, aluminum cans, and plastics. Jacob S. Coxey, the social reformer, was mayor of the city in the early 1930s.

Mas·sive, Mount, (măs'ĭv), peak, 14,418 ft (4,397.5 m) high, W central Colo., in the Sawatch Mts. It is the second-highest peak in the U.S. Rocky Mts.

Ma·su·li·pa·tam (mə-sōō'lə-pŭt'əm) or **Ban·dar** (bŭn'dər), town (1971 pop. 112,636), Andhra Pradesh state, E central India, a port on the Bay of Bengal. In the 17th cent. it was a center of French, British, and Dutch trade.

Ma·su·ri·a (mə-zōōr'ē-ə), region, N Poland. It is a low-lying area covered by large lakes and forests and drained by many small rivers. The original population of the region was expelled by the Teutonic Knights and replaced (14th cent). with Polish settlers. Masuria later became part of East Prussia and was largely Germanized by the early 20th cent. After Masuria passed to Poland in 1945, most of the German-speaking population was expelled and replaced by Poles. The Masurian Lakes region, where more than 2,700 lakes are located, was the scene of heavy fighting early in World War I.

Ma·ta·di (mə-tä'dē), city (1974 est. pop. 144,000), Bas-Zaire region, W Zaïre, on the Congo River. Situated c.80 mi (130 km) from the mouth of the Congo, at the farthest point navigable by oceangoing vessels, the city is linked by rail with Kinshasa.

Mat·a·dor (măt'ə-dôr'), town (1970 pop. 1,091), seat of Motley co., NW Texas, NE of Lubbock.

Mat·a·gor·da (măt'ə-gôr'də), county (1970 pop. 27,913), 1,157 sq mi (2,996.6 sq km), SE coastal Texas, on Matagorda Bay, traversed by the Gulf Intracoastal Waterway and drained by the Colorado River; formed 1836; co. seat Bay City. A leading cattle-raising county, it also has diversified agriculture (rice, cotton, grain, corn, truck crops, fruit, dairy products, livestock, and poultry) and fisheries.

Matagorda Bay, inlet of the Gulf of Mexico, c.50 mi (80 km) long and from 3 to 12 mi (4.8–19.3 km) wide, SE Texas, protected by a long sandspit, Matagorda Peninsula. The bay, with its arm, Lavaca Bay, was probably visited (1685) by the French explorer Robert La Salle on his last expedition. Matagorda Island is a sandbar at the entrance of San Antonio Bay. The area has often hurricanes.

Mat·a·mo·ros (măt'ə-môr'əs), city (1974 est. pop. 165,100), Tamaulipas state, NE Mexico, near the mouth of the Rio Grande, opposite Brownsville, Texas. Matamoros, linked by rail and highway with the United States, is an international trading center and a point of entry. Fishing is an important industry, and maize, cotton, and cattle are raised in the surrounding area. Noted for its heroic defense against numerous U.S. adventurers during the 19th cent., the city fell to the forces of Zachary Taylor in the Mexican War in 1846.

Ma·tane (mə-tăn'), town (1976 pop. 12,726), SE Que., Canada, on the St. Lawrence River at the mouth of the Matane River at the beginning of the Gaspé Peninsula.

Mat·a·nus·ka-Su·sit·na (măt'ə-nōōs'kə-sōō-sĭt'nə), borough (1970 pop. 6,509), 25,730 sq mi (66,640.7 sq km), S central Alaska; formed after 1961; borough seat Palmer. It is in a farming and coal-mining region, with some industry.

Ma·tan·zas (mə-tăn'zəs, mä-tän'säs), city (1970 pop. 85,376), capital of Matanzas prov., W central Cuba. A port with a large, deep harbor, it exports sugar, fruits, and sisal. Industries in the city include sugar refineries and textile mills. Matanzas is a popular stopover for vacationers, who explore the picturesque Yumurí River valley and the caves of Bellamar, famous for their calcite crystal formations. Founded in 1693, it was once a pirate haven but by the early 19th cent. had become Cuba's second city, mainly because of the growth of the sugar industry.

Mat·a·pan, Cape (măt'ə-păn'), or **Cape Taí·na·ron** (tâ'nä-rôn), S Greece, southern extremity of the Greek mainland, of the Peloponnesus, and of the Taygetus Mts., projecting into the Ionian Sea. In World War II the British won an important naval battle (1941) over the Italians off Cape Matapan.

Ma·ta·pé·di·a, Lake (măt'ə-pē'dē-ə), 14 mi (22.5 km) long and 2 mi (3.2 km) wide, E Que., Canada, at the base of the Gaspé Peninsula.

Ma·ta·ró (mä'tə-rō'), city (1970 pop. 73,129), Barcelona prov., NE Spain, in Catalonia. It is a Mediterranean port and a manufacturing center, especially of knitted goods. The first railroad in Spain was built (1848) from Barcelona to Mataró.

Ma·te·ra (mə-tĕr'ə), city (1976 est. pop. 48,785), capital of Matera prov., in Basilicata, S Italy, in the Apennines. It is an agricultural and industrial center with woolen textile mills and food-processing factories.

Math·ews (măth'yo͞oz), county (1970 pop. 7,168), 89 sq mi (230.5 sq km), E Va., at the tip of the tidewater peninsula, bounded in the E by Chesapeake Bay and in the N and S by inlets of bay; formed 1790; co. seat Mathews. It produces truck crops and poultry, bulbs, and some tobacco, peanuts, and grain.

Mathews, village (1970 pop. 500), seat of Mathews co., E Va., near Chesapeake Bay ESE of West Point in a farming area.

Ma·thu·ra (mŭt'o͝o-rə) or **Mut·tra** (mŭ'trə), city (1971 pop. 131,813), Uttar Pradesh state, N central India, on the Jumna River. An agricultural market town, it is best known as a Hindu pilgrimage site, the reputed birthplace of the god Krishna. The region is rich in archaeological remains.

Ma·to Gros·so (mä'to͞o grô'so͞o) state (1975 est. pop. 2,006,900), 475,501 sq mi (1,231,548 sq km), central and W Brazil; capital Cuiabá.

Ma-tsu (mä'tso͞o'), island, in the East China Sea, off Fukien prov., China, E of Fu-chou, and c.100 mi (160 km) from Taiwan. It remained a Chinese Nationalist-held outpost after the Communist takeover of the mainland in 1949.

Ma·tsu·do (mä-tso͞o'dō), city (1976 est. pop. 358,139), Chiba prefecture, E central Honshu, Japan. It is a suburb of Tokyo.

Ma·tsu·e (mä-tso͞o'ā), city (1976 est. pop. 129,277), capital of Shimane prefecture, SW Honshu, Japan, a port on the Sea of Japan. It is a distribution center and popular tourist spot.

Ma·tsu·mo·to (mä-tso͞o-mō'tō), city (1976 est. pop. 187,225), Nagano prefecture, central Honshu, Japan. It is a market for silkworms and raw silk. Industries include food processing, machine-building, and textile manufacturing.

Ma·tsu·ya·ma (mä-tso͞o-yä'mä), city (1976 est. pop. 374,492), capital of Ehime prefecture, NW Shikoku, Japan, a port on the Inland Sea. It is an important agricultural distribution point and fishing port. Cotton textiles and paper products are manufactured. Matsuyama has two universities. Its feudal castle (built 1603), one of the best preserved in Japan, stands in a magnificent park.

Mat·ta·ga·mi (mə-tăg'ə-mē), river, 275 mi (442.5 km) long, rising in the lake district, E Ont., Canada, SW of Timmins, flowing N to join the Missinaibi River, with which it forms the Moose River.

Mat·ter·horn (măt'ər-hôrn, mä'tər-), peak, c.14,685 ft (4,480 m) high, in the Pennine Alps, on the Swiss-Italian border, near Zermatt. Its distinctive pyramidal peak was formed by the enlargement of several cirques. It was first scaled in 1865 by Edward Whymper, the English mountaineer.

Mat·toon (mə-to͞on'), city (1978 est. pop. 18,800), Coles co., E central Ill.; inc. 1859. It is a processing, rail, and industrial center for a rich farm and dairy region. Among its manufactures are road-building equipment, paper and brass products, and springs. Nearby are many oil wells, a fish hatchery, and Paradise Lake. Gen. Ulysses S. Grant took command of his first Civil War troops in Mattoon.

Mau·beuge (mō-bœzh'), city (1975 pop. 35,399), Nord dept., N France, on the Sambre River near the Belgian border. Iron and steel products, glass, and china are major manufactures. An abbey was founded on the site of Maubeuge by St. Aldegone in the 7th cent. Still standing are remains of fortifications built in 1685 by the famous military engineer Sébastien Vauban.

Mauch Chunk (mouch' chŭngk'), borough (1970 pop. 5,456), seat of Carbon co., E Pa., on the Lehigh River SE of Hazleton, in a mountain resort area; settled 1815, inc. 1956. Created in 1954 by a merger of the boroughs of Mauch Chunk and East Mauch Chunk, it was named first for the athlete Jim Thorpe, who is buried here.

Mau·i (mou'ē), county (1970 pop. 46,156), 1,173 sq mi (3,038.1 sq km), Hawaii, including Kahoolawe, Lanai, Maui, and Molokai islands; co. seat Wailuku.

Maui, island (1970 pop. 38,691), 728 sq mi (1,885.5 sq km), second-largest island in the state of Hawaii, separated from the island of Hawaii by the Alenuihaha Channel and from Molokai by the Pailolo Channel. Maui is made up of two mountain masses, which constitute the east and west peninsulas, connected by an isthmus. The highest point on the island is the Haleakala volcano (10,025 ft/3,057.6 m). The island's chief industries are the cultivation of sugar cane and pineapples.

Mau·mee (mô-mē', mô'mē), village (1978 est. pop. 17,100), Lucas co., NW Ohio, on the Maumee River; inc. 1838. It is largely residential. Maumee was the site of Fort Miami, a British post surrendered to the Americans during the War of 1812.

Mau·na Ke·a (mou'nə kā'ə), dormant volcano, 13,796 ft (4,207.8 m) high, in the S central part of the island of Hawaii. It is the loftiest peak in the Hawaiian Islands and the highest island mountain in the world, rising c.32,000 ft (9,760 m) from the Pacific Ocean floor. It has many cinder cones on its flanks and a great crater at the summit. Its fertile lower slopes are used for agriculture, especially the growing of coffee beans. The upper slopes are snow-covered in winter.

Mauna Lo·a (lō'ə), mountain, 13,680 ft (4,172.4 m) high, in the S central part of the island of Hawaii, in Hawaii Volcanoes National Park. Its many craters include Kilauea and Mokuaweoweo, two of the world's largest active craters. Mauna Loa has erupted twice (1942, 1949) since its period of greatest activity in 1881.

Mau·re·ta·ni·a (môr'ĭ-tā'nē-ə), ancient district of Africa in Roman times. In a vague sense it meant only "the land of the Moors" and lay west of Numidia, but more specifically it usually included most of present-day northern Morocco and western Algeria. It was a complex of native tribal units, but after the 2nd cent. B.C. Roman influence became paramount. Revolts later occurred, and Mauretania was subdued A.D. 41–42. Roman influence was never complete, however, and native chieftains remained powerful. With the onset of the barbarian invasions, Roman control weakened, and by the end of the 5th cent. A.D. the district had disappeared.

Mau·ri·ta·ni·a, Islamic Republic of (môr'ĭ-tā'nē-ə), republic (1975 est. pop. 1,318,000), 397,953 sq mi (1,030,700 sq km), NW Africa, bordered on the W by the Atlantic Ocean, on the N by Morocco, on the NE by Algeria, on the E and SE by Mali, and on the SW by Senegal. Nouakchott is the capital. The area figure used above does not include that portion of the former colony of Spanish (Western) Sahara annexed by Mauritania in 1976. Most of Mauritania is made up of low-lying desert, which forms part of the Sahara. Along the Senegal River (which forms the border with Senegal) in the southwest is the semiarid Sahel with some fertile alluvial soil. A wide

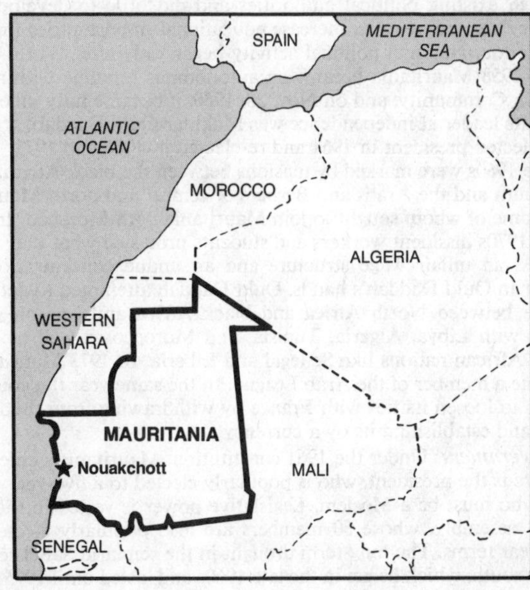

sandstone plateau runs through the center of the country from north to south. In the southeast is the Hodh, a large basin in the desert.

History. The economy is sharply divided between a traditional agricultural sector and a modern mining industry. The great majority of the country's workers are engaged either in raising crops or pasturing livestock and are largely unaffected by the mining industry. The principal agricultural products, produced chiefly near the Senegal and in scattered oases, are millet, pulses, dates, maize, groundnuts, gum arabic, rice, and wheat. Large numbers of sheep, goats, cattle, and camels are raised. There is a small but growing fishing industry based in the Atlantic and on the Senegal River. A large deposit of high-grade iron ore was discovered in northern Mauritania in the late 1950s and production for export began in 1963. Mining was controlled by the Iron Mining Co. of Mauritania, a concern jointly owned by several West European companies until it was nationalized in 1974. Foreign sales of iron ore account for a majority of the country's export earnings. In the 1970s an oil refinery and a sugar plant were built and financing for other projects was obtained from Kuwait. Copper mining began in the early 1970s; salt is also produced and there are untapped deposits of gypsum and titanium. Mauritania has a small network of all-weather roads that serve mainly the south. The country is a charter member of the West African Economic Community, founded in 1974.

History. The remains of paleolithic and neolithic cultures have been discovered in northern Mauritania. By the 1st millennium A.D. Sanhaja Berbers had migrated into Mauritania, pushing the black African inhabitants (especially the Soninké) southward toward the Senegal River. The Hodh region, which became desert only in the 11th cent., was the center of the ancient empire of Ghana (700–1200), whose capital, Kumbi-Saleh, has been located near the present-day border with Mali. Until the 13th cent., Oualata, Awdaghost, and Kumbi-Saleh, all in southeast Mauritania, were major centers along the trans-Saharan caravan routes linking Morocco with the region along the upper Niger River. In the 11th cent. the Almoravid movement was founded among the Moslem Berbers of Mauritania. In the 14th and 15th cent., southeast Mauritania was part of the empire of Mali, centered along the upper Niger. By this time the Sahara had encroached on much of Mauritania, consequently limiting agriculture and reducing the population.

In the 1440s Portuguese navigators explored the Mauritanian coast and established a fishing base on Arguin Island, located near the present-day boundary with Spanish Sahara. From the 17th cent., Dutch, British, and French traders were active along the southern Mauritanian coast; they were primarily interested in the gum arabic gathered near the Senegal. Under Louis Faidherbe, governor of Senegal (1854–61; 1863–65), France gained control of southern Mauritania. The region was declared a protectorate in 1903, but parts of the north were not pacified until the 1930s.

Until 1920, when it became a separate colony in French West Africa, Mauritania was administered as part of Senegal. Saint-Louis, in Senegal, continued to be Mauritania's administrative center until 1957, when it was replaced by Nouakchott. The French ruled through existing political authorities and did little to develop the country's economy or to increase educational opportunities for the population. National political activity began only after World War II. In 1958 Mauritania became an autonomous republic within the French Community, and on Nov. 28, 1960, it became fully independent. Its leader at independence was Makhtar Ould Daddah, and he was elected president in 1961 and re-elected in 1966 and 1971.

The 1960s were marked by tensions between the black Africans of the south and the Arabs and Berbers of central and north Mauritania, some of whom sought to join Mauritania with Morocco. In the early 1970s dissident workers and students protested what they considered an unfair wage structure and an undue concentration of power in Ould Daddah's hands. Ould Daddah attempted to act as a bridge between North Africa and black Africa and was on good terms with Libya, Algeria, Tunisia, and Morocco as well as with black African nations like Senegal and Liberia. In 1973 Mauritania became a member of the Arab League. In the same year the country began to loosen its ties with France by withdrawing from the franc zone and establishing its own currency.

Government. Under the 1961 constitution, Mauritania's chief executive is the president, who is popularly elected to a five-year term and who must be a Moslem. Legislative power is vested in the national assembly, whose 50 members are also popularly elected to five-year terms. The long-term drought in the semiarid Sahel region in the south, which began in the late 1960s and lasted until 1975, was estimated to have caused the death of about 80% of the country's livestock. In 1975, Mauritania became involved in a dispute with Morocco and Algeria over possession of the adjacent colony of Spanish (Western) Sahara. An agreement between Spain, Morocco, and Mauritania was reached (1975) providing for a partition of the former colony, with Mauritania receiving slightly less than one third of the 102,703-square-mile (266,000-square-kilometer) territory. The Polisario Front, a pro-independence guerrilla group based in the colony, refused to recognize the division of Spanish Sahara, and in June, 1976, its forces shelled the capital city of Nouakchott. Algeria, which also refused to recognize the partition, broke diplomatic relations with Morocco and Mauritania and was believed to be providing arms to the Polisario Front. By 1977 the guerrillas were in possession of large areas of NE Mauritania.

Mau·ri·tius (mô-rĭsh'əs), island country (1975 est. pop. 899,000), 790 sq mi (2,046.1 sq km), in the SW Indian Ocean, part of the Mascarene Island group, c.500 mi (805 km) E of the Malagasy Republic. The island of Rodriguez and two groups of small islands, Agalega and Cargados Carajos, are dependencies of Mauritius. The capital is Port Louis. Mauritius is surrounded by coral reefs. A central plateau is ringed by mountains of volcanic origin which rise to c.2,700 ft (825 m) in the southwest. The island has a tropical rainy climate.

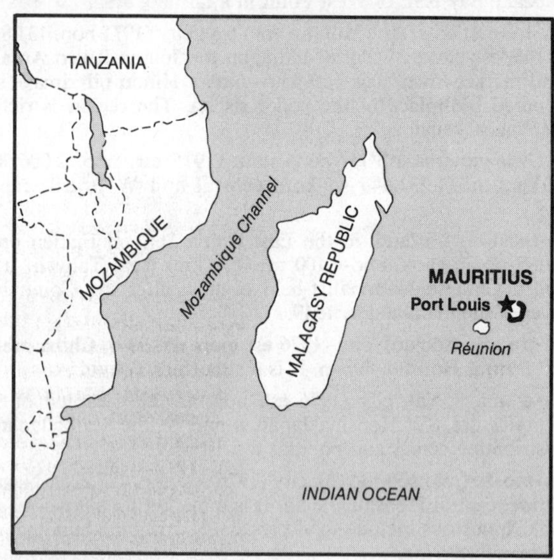

Economy. Sugar cane is the chief crop raised and sugar and molasses are the major exports; tea and food crop production have been encouraged to diversify the economy and reduce the need to import food. The fishing industry is of some importance.

History. Mauritius was probably visited by Arabs and Malays in the Middle Ages. Portuguese sailors visited it in the 16th cent. The island was occupied by the Dutch from 1598 to 1710 and named after Prince Maurice of Nassau. The French settled the island in 1722 and called it Île de France. It became an important way station on the route to India. The French introduced the cultivation of sugar cane and imported large numbers of African slaves to work the plantations. The British captured the island in 1810 and restored the Dutch name. After the abolition of slavery in the British Empire in 1833, indentured laborers were brought from India; their descendents constitute a majority of the population today.

Politics on Mauritius was long the preserve of the French and the Creoles, but the extension of the franchise under the 1947 constitution gave the Indians political power. Indian leaders in the 1950s and 1960s favored independence, while the French and Creoles wanted continuing association with Britain, fearing domination by the Hindu Indian majority. Independence, however, was granted in 1968. Clashes among the island's ethnic groups have frequently resulted in bloodshed. High unemployment and overpopulation—a result of the eradication of malaria in the 1960s—are serious problems.

Mau·ry (môr'ē), county (1970 pop. 44,028), 614 sq mi (1,590.3 sq km), central Tenn., drained by the Duck River; formed 1807; co. seat Columbia. It has livestock raising (cattle and mules), dairying, and agriculture (barley, tobacco, corn, soybeans, and wheat). There are phosphate rock deposits here.

Maus·ton (mô'stən), city (1970 pop. 3,466), seat of Juneau co., central Wis., on the Lemonweir River E of La Crosse; settled c.1840, inc. 1883.

Mav·er·ick (măv'ə-rĭk, măv'rĭk), county (1970 pop. 18,093), 1,289 sq mi (3,338.5 sq km), SW Texas, bounded on the SW by the Rio Grande and the Mexican border; formed 1856; co. seat Eagle Pass. This rich agricultural area (part of the Winter Garden region irrigated by the Rio Grande), produces a large part of the state's spinach crop, other truck-farming crops, and cotton and grain sorghums.

May, Cape (mā), peninsula, extreme S N.J. It has a lighthouse on Cape May Point at the entrance to Delaware Bay. The cape is bisected by a canal, c.3 mi (4.8 km) above the point, which was constructed by the U.S. government in 1942–43 as a war emergency measure to provide an alternative to the longer, more hazardous route around the cape.

Ma·ya·güez (mä'yä-gwĕs'), city (1970 pop. 68,872), W Puerto Rico, on Mona Passage. It is a port of entry as well as a shipping and manufacturing center in an area where sugar cane, coffee, tobacco, and livestock are raised.

Ma·yenne (mä-yĕn'), department (1968 pop. 252,762), 1,997 sq mi (5,172.2 sq km), NW France, in parts of Maine and Anjou; capital Laval.

May·er·ling (mī'ər-lĭng), village, Lower Austria prov., E Austria, on the Schwechat River, in the Wienerwald. It is the site of the hunting lodge (now a convent) where Crown Prince Rudolf and Baroness Maria Vetsera died mysteriously in Jan., 1889.

May·ers·ville (mā'ərz-vĭl'), village (1970 pop. 500), seat of Issaquena co., W Miss., on the Mississippi River S of Greenville.

Mayes (māz), county (1970 pop. 23,302), 648 sq mi (1,678.3 sq km), NE Okla., intersected by the Neosho River, which is impounded here by Grand River Dam; formed 1907; co. seat Pryor. Its agriculture includes corn, cotton, oats, and livestock. There are some oil wells and timber.

May·field (mā'fēld'), city (1978 est. pop. 9,700), seat of Graves co., SW Ky., in an area of farms and clay deposits; founded 1823. It is an agricultural trade center with a large tobacco market. In a plot at the local cemetery are the curious Wooldridge monuments—stone figures of an eccentric aristocrat (buried there in 1899), his family, friends, and animal pets.

May·nard (mā'nərd), industrial town (1970 pop. 9,710), Middlesex co., NE Mass., near Concord; settled 1638, set off from Sudbury and Stow 1871. It has an electronics laboratory.

May·nard·ville (mā'nərd-vĭl'), city (1970 pop. 702), seat of Union co., NE Tenn., NNE of Knoxville, in a fertile farm area.

May·nooth (mā-nōōth'), town, Co. Kildare, E Republic of Ireland. It is the seat of St. Patrick's College (1795), the principal institution in Ireland for training Roman Catholic clergy, now a constituent college of the National Univ. of Ireland. Near the college are the ruins of Maynooth Castle, founded c.1176. It was besieged in the reign of Henry VIII and dismantled in the 17th cent.

May·o (mā'ō), county (1971 pop. 109,497), 2,084 sq mi (5,397.6 sq km), W Republic of Ireland; co. town Castlebar. The western portion, including large Achill island, is mountainous; the eastern part is more level. There are numerous lakes and the irregular coast line is deeply indented by bays. Oats and potatoes are grown and cattle, sheep, pigs, and poultry are raised. The county was not brought fully under English control until the late 16th cent.

Mayo, town (1970 pop. 793), seat of Lafayette co., N Fla., WNW of Gainesville, in a lumbering area. There are limestone quarries here.

Ma·yon, Mount (mä-yōn'), active volcano, 8,070 ft (2,461.4 m) high, SE Luzon, Philippines. It is considered one of the world's most perfect cones. The last major eruption took place in 1947.

Ma·yotte (mä-yôt'), island (1970 pop. 48,000), 144 sq mi (373 sq km), Comoro Islands, in Mozambique Channel, Indian Ocean. Rising to 2,165 ft (660.3 m), it is a producer of rum, vanilla, oils, and sugar. The Comoros declared independence from France in 1975, but Mayotte remained under French jurisdiction.

Mays Landing (māz), town (1970 pop. 1,272), seat of Atlantic co., SE N.J., on Great Egg Harbor River NW of Atlantic City; settled c. 1710. It is an inland resort.

Mays·ville (māz'vĭl'). **1.** City (1970 pop. 7,411), seat of Mason co., N Ky., on the Ohio River SE of Cincinnati, Ohio; est. 1787, inc. 1833. It is a trade, transportation, and industrial center making textiles, shoes, and metal products and processing tobacco. **2.** City (1970 pop. 1,045), seat of De Kalb co., NW Mo., ENE of St. Joseph; settled 1845. Corn, wheat, and poultry are raised.

Ma·zar-i-Sha·rif (mä-zär'ē-shä-rēf'), city (1969 est. pop. 43,197), capital of Mazar-i-Sharif prov., N Afghanistan, near the USSR border. It is held sacred as the alleged burial place of Ali, son-in-law and cousin of Mohammed. Mazar-i-Sharif, the center of Afghanistan's rug and carpet industry, also has cotton and silk industries.

Ma·za·ru·ni (mä-zə-rōō'nē), river, c.350 mi (565 km) long, rising in the Guiana Highlands, NW Guyana, and flowing generally E to the Essequibo River at Bartica. The river is the center of Guyana's diamond industry.

Ma·zat·lán (mä'sät-län'), city (1974 est. pop. 147,000), Sinaloa state, W Mexico, on the Pacific coast. Mazatlán is a major seaport and a popular resort with a beautiful setting. Spanish colonial trade with the Philippines stimulated the development of the port.

Mba·ba·ne (əm-bä-bä'nā), town, (1976 pop. 22,000), capital of Swaziland, NW Swaziland, in the Mdimba Mts. It serves as a commercial hub for the surrounding agricultural region.

Mban·da·ka (əm-bän-dä'kä), formerly **Co·quil·hat·ville** (kô-kē-yä-vēl'), city (1974 est. pop. 134,000), capital of Equateur region, W Zaire, a port on the Congo River. It is a commercial and transportation center and has tanning and fishing industries. The city was founded in 1883 by the explorer Henry M. Stanley.

Mbu·ji-Ma·yi (əm-bōō'jē-mī'yē), formerly **Bak·wan·ga** (bäk-wäng'-gä), city (1970 pop. 256,154), S central Zaire, on the Sankuru River. A commercial center in Luba country, it handles most of the industrial diamonds produced in Zaire. After Zaire attained independence (1960) the city's population grew rapidly with the immigration of Luba people from other parts of the country. From 1960 to 1962 it was the capital of the secessionist Mining State of South Kasai.

Mc·Al·es·ter (mə-kăl'ĭ-stər), city (1978 est. pop. 16,200), seat of Pittsburg co., SE Okla.; inc. 1899. Once a coal-mining and farming community, it is now a regional distribution center with a busy stockyard. Aircraft and truck parts, boats, and clothing are manufactured. Nearby is a state penitentiary, the site of an annual rodeo.

Mc·Al·len (mə-kăl'ən), city (1978 est. pop. 52,000), Hidalgo co., extreme S Texas, on the Rio Grande; inc. 1911. It is a port of entry and a packing and processing center for the citrus fruit, truck crops, and other produce of the lower Rio Grande valley. The city has oil refineries, chemical plants, and other manufacturing industries.

Mc·Ar·thur (mə-kär'thər), village (1970 pop. 1,543), seat of Vinton co., S central Ohio, ESE of Chillicothe; platted 1815, inc. 1851. In an area of coal mines, oil and natural-gas wells, and limestone quarries, the village makes wood products and brick.

Mc·Clain (mə-klān'), county (1970 pop. 14,157), 573 sq mi (1,484.1 sq km), central Okla., bounded in the NE by the Canadian River and drained by small creeks; formed 1907; co. seat Purcell. It is in an agricultural area that produces cotton, alfalfa, oats, peaches, corn, broomcorn, dairy products, and livestock.

Mc·Clure Strait (mə-klŏŏr'), arm of the Beaufort Sea, c.170 mi (275 km) long and 60 mi (95 km) wide, W Franklin dist., Northwest Territories, Canada. It extends west from Viscount Melville Sound, between Melville and Eglinton islands on the north and Banks Island on the south. In 1954, U.S. icebreakers cut through the strait for the first time, opening the last obstacle to the shortest water route across the Canadian arctic region.

Mc·Clus·ky (mə-klŭs'kē), city (1970 pop. 664), seat of Sheridan co., central N.Dak., NNE of Bismarck. Livestock, dairy products, and poultry are produced.

Mc·Cone (mə-kōn'), county (1970 pop. 2,875), 2,607 sq mi (6,752.1 sq km), NE Mont., in an agricultural region bounded on the N by the Missouri River and drained by Redwater Creek; formed 1919; co. seat Circle. Its economy depends on grain and livestock.

Mc·Con·nells·burg (mə-kŏn'lz-bûrg'), borough (1970 pop. 1,228), seat of Fulton co., S Pa., W of Chambersburg, in a farm and resort area; settled c.1730, laid out 1786, inc. 1814.

Mc·Con·nels·ville (mə-kŏn'lz-vĭl'), village (1970 pop. 2,107), seat of Morgan co., SE Ohio, on the Muskingum River SSE of Zanesville, in a farm area; platted 1817, inc. 1836.

Mc·Cook (mə-ko͝ok'), county (1970 pop. 7,246), 577 sq mi (1,494.4 sq km), SE S.Dak., in an agricultural area drained by branches of the Vermillion River; formed 1873; co. seat Salem.

McCook, city (1970 pop. 8,285), seat of Red Willow co., S Nebr., on the Republican River near the Kansas border; founded 1881, inc. 1883. In a rich farming area, it has processing plants.

Mc·Cor·mick (mə-kôr'mĭk), county (1970 pop. 7,955), 306 sq mi (792.5 sq km), W S.C., in an agricultural area bounded on the W by the Savannah River; formed 1916; co. seat McCormick. Forestry, tourism, and textile milling are also important.

McCormick, town (1970 pop. 1,864), seat of McCormick co., W S.C., near the Savannah River W of Columbia; inc. 1882. Textile products are made, and poultry is processed.

Mc·Crack·en (mə-krăk'ən), county (1970 pop. 58,281), 251 sq mi (650.1 sq km), W Ky., bounded on the N by the Ohio River and the Ill. border, on the NE by the Tennessee River, and drained by the Clarks River; formed 1824; co. seat Paducah. In a gently rolling agricultural area yielding dark tobacco, corn, dairy products, poultry, livestock, and fruit, it has clay, fluorspar, and coal mines and lumbering and food-processing industries. Textiles and clothing, concrete, and metal products are manufactured.

Mc·Crea·ry (mə-krîr'ē), county (1970 pop. 12,548), 418 sq mi (1,082.6 sq km), SE Ky., in the Cumberlands, bounded on the S by Tenn., on the N and E by the Cumberland River, and drained by the South Fork of the Cumberland River; formed 1912; co. seat Whitley City. It lies in a bituminous-coal and timber region. Its agriculture includes dairy products, poultry, livestock, apples, Irish and sweet potatoes, corn, lespedeza, and tobacco. Textiles and wood products are made. Part of Daniel Boone National Forest is in the county.

Mc·Cul·loch (mə-kŭl'ə), county (1970 pop. 8,571), 1,066 sq mi (2,760.9 sq km), central Texas, on Edwards Plateau, bounded on the N by the Colorado River and drained by the San Saba River and Brady Creek; formed 1856; co. seat Brady. It has diversified agriculture (oats, corn, grain sorghums, cotton, barley, wheat, milo, fruit, pecans and peanuts, livestock, and poultry).

Mc·Cur·tain (mə-kûr'tn), county (1970 pop. 28,642), 1,800 sq mi (4,662 sq km), extreme SE Okla., bounded in the E by the Ark. border, in the S by the Red River, and drained by the Little River and its Mountain Fork; formed 1907; co. seat Idabel. It is in a lumbering area, with sawmills and paper mills. There is agriculture (truck crops, soybeans, grain, fruit, cotton, corn, alfalfa, hay, and cattle and some mining and manufacturing.

Mc·Don·ald (mək-dŏn'ld), county (1970 pop. 12,357), 540 sq mi (1,398.6 sq km), extreme SW Mo., in the Ozarks drained by the Elk River; formed 1849; co. seat Pineville. In an agricultural area yielding fruit, grain, dairy products, and poultry.

Mc·Don·ough (mək-dä'nə), county (1970 pop. 36,653), 582 sq mi (1,507.4 sq km), W Ill., drained by the La Moine River and its branches; formed 1826; co. seat Macomb. Agriculture (livestock, corn, wheat, oats, soybeans, hay, poultry, and dairy products) and bituminous-coal mining are the major occupations.

McDonough, city (1970 pop. 2,675), seat of Henry co., NW central Ga., SE of Atlanta; inc. 1823. Clothing and tools are manufactured here.

Mc·Dow·ell (mək-dou'əl). **1.** County (1970 pop. 30,648), 436 sq mi (1,129.2 sq km), W central N.C., in the Blue Ridge and drained by the Catawba River; formed 1842; co. seat Martion. In a resort area, it has farming (corn, apples, soybeans, dairy products, and poultry), livestock raising, lumbering, and textile manufacturing. Pisgah National Forest is in the north. **2.** County (1970 pop. 50,666), 533 sq mi (1,380.5 sq km), S W.Va., on the Allegheny Plateau, bordered on the W, S, and SE by Va. and drained by Tug and Dry forks of the Big Sandy River; formed 1858; co. seat Welch. There is extensive bituminous-coal mining and some lumbering.

Mc·Duf·fie (mək-dŭf'ē), county (1970 pop. 15,276), 253 sq mi (655.3 sq km), E Ga., bounded in the N by the Little River and intersected by the Fall Line; formed 1870; co. seat Thomson. It is in a farming (cotton, corn, grain, truck crops, and fruit) and sawmilling area.

Mc·Hen·ry (mək-hĕn'rē). **1.** County (1970 pop. 111,555), 611 sq mi (1,582.5 sq km), NE Ill., on the Wis. border and drained by the Fox and Kishwaukee rivers; formed 1836; co. seat Woodstock. It is in a dairying region, with livestock and poultry raising and corn and wheat farming. The county has many lake resorts, with good fishing and duck hunting. **2.** County (1970 pop. 8,977), 1,871 sq mi (4,845.9 sq km), N central N.Dak., in an agricultural area drained by the Souris River; formed 1873; co. seat Towner. Livestock, dairy produce, poultry, and wheat are among its products.

Mc·In·tosh (măk'ĭn-tŏsh'). **1.** County (1970 pop. 7,371), 431 sq mi (1,116.3 sq km), SE Ga., in a coastal plain area bounded on the SE by the Atlantic, on the SW by the Altamaha River; formed 1793; co. seat Darien. Dairying, truck farming, fishing and seafood processing, and sawmilling are done here. **2.** County (1970 pop. 5,545), 993 sq mi (2,571.9 sq km), S N.Dak., on the S.Dak. border in a rich prairie area watered by Beaver Creek; formed 1883; co. seat Ashley. Its agriculture includes dairy products, flour, grain, livestock, and poultry. **3.** County (1970 pop. 12,472), 608 sq mi (1,574.7 sq km), E Okla., bounded in the S by the Canadian River and intersected by the North Canadian River; formed 1907; co. seat Eufaula. It has agriculture (cotton, potatoes, peanuts, corn, grain, dairy products, and livestock), oil and gas wells, and textile and metal industries.

McIntosh, city (1970 pop. 563), seat of Corson co., N S.Dak., NNW of Pierre, in a livestock and grain area. Lignite is mined.

Mc·Kean (mə-kēn'), county (1970 pop. 51,915), 992 sq mi (2,569.3 sq km), N Pa., in a plateau area bordered in the N by N.Y. and drained by the Allegheny River; formed 1804; co. seat Smethport. It is a leading producer of lubricating oils. Other industries include dairying and the manufacture of oil-well supplies, explosives, chemicals, and glass, clay, and wood products.

Mc·Kee (mə-kē'), city (1970 pop. 255), seat of Jackson co., central Ky., in the Cumberlands SSE of Winchester, in a coal-mining and timber region. Corn, hay, and tobacco are grown.

Mc·Kees·port (mə-kēz'pôrt'), city (1978 est. pop. 31,800), Allegheny co., SW Pa., in hilly terrain at the confluence of the Monongahela and Youghiogheny rivers; settled 1755, inc. as a city 1890. It is primarily an industrial city, with factories that manufacture steel, machine parts, and automobile bodies.

Mc·Ken·zie (mə-kĕn'zē), county (1970 pop. 6,127), 2,735 sq mi (7,083.7 sq km), W N.Dak., on the Mont. border in an agricultural area watered by the Missouri and Yellowstone rivers; formed 1883; co. seat Watford City. It is rich in lignite, coal, and oil deposits, and has wheat and livestock ranches and stands of timber. There are irrigation projects in the northwest along the Missouri River.

Mc·Kin·ley (mə-kĭn'lē), county (1970 pop. 43,208), 5,456 sq mi (14,131 sq km), NW N.Mex., bounded on the W by the Ariz. border and drained by Rio Puerco; formed 1899; co. seat Gallup. In an area specializing in cattle and sheep grazing, it also has a lumbering industry and deposits of coal, uranium, bentonite, oil, helium, lignite, and natural gas. Cibola National Forest is within the county.

McKinley, Mount, peak, 20,320 ft (6,197.6 m) high, S central Alaska, in the Alaska Range. It is the highest point in North America. Permanent snowfields cover more than half the mountain and feed numerous glaciers. Mt. McKinley was first successfully scaled by the American explorer Hudson Stuck in 1913.

Mc·Kin·ney (mə-kĭn'ē), city (1978 est. pop. 13,000), seat of Collin co., N Texas; inc. 1849. It is a shipping point for cotton and grains and has small industries.

Mc·Lean (mə-klān'). **1.** County (1970 pop. 104,389), 1,173 sq mi (3,038.1 sq km), central Ill., drained by the Sangamon and Mackinaw rivers and by Kickapoo, Salt, Money, and Sugar creeks; formed 1930; co. seat Bloomington. Agriculture (corn, oats, wheat, soybeans, livestock, poultry, and dairy products), bituminous-coal mining, limestone quarrying, manufacturing, and shipping are the major occupations. **2.** County (1970 pop. 9,062), 257 sq mi (665.6 sq km), W Ky., crossed by the Green River, bounded on the W by the Green and Pond rivers, and drained by the Rough River; formed 1854; co. seat Calhoun. In an agricultural area yielding soybeans, corn, wheat, livestock, and tobacco, it has bituminous-coal mines, oil and gas wells, timber, and some industry (food processing and wood products and furniture manufacturing). **3.** County (1970 pop. 11,251), 2,065 sq mi (5,348.4 sq km), central N.Dak., in an agricultural area bounded on the W by Fort Berthold Indian Reservation and on the W and S by the Missouri River; formed 1883; co. seat Washburn. It has coal and lignite deposits. Livestock, poultry, dairy cattle, and various grains are raised.

McLean, city (1970 pop. 17,698), Fairfax co., N Va., a suburb of Washington, D.C.

Mc·Leans·bor·o (mə-klănz'bûr-ō, -bŭr-ō), city (1970 pop. 2,630), seat of Hamilton co., S Ill., SE of Mt. Vernon, in a farm area; inc. 1840. Clothing is made here.

Mc·Len·nan (mə-klĕn'ən), county (1970 pop. 147,553), 1,000 sq mi (2,590 sq km), E central Texas, drained by the Brazos River and branches of the Bosque River; formed 1850; co. seat Waco. This rich agricultural area produces oats, cotton, corn, grain sorghums, peanuts and pecans, fruit, truck crops, dairy products, poultry, beef cattle, hogs, sheep, horses, and mules. There are deposits of limestone, clay, and cement rock.

Mc·Leod (mə-kloud'), county (1970 pop. 27,662), 488 sq mi (1,264 sq km), S central Minn., watered by forks of the Crow River; formed 1856; co. seat Glencoe. Its agriculture includes corn, oats, barley, potatoes, livestock, and dairy products. Among its manufactures are furniture, paper and metal products, and farm machinery.

Mc·Lough·lin House National Historic Site (mə-klŏk'lĭn): *see* National Parks and Monuments Table.

Mc·Lough·lin, Mount (mə-glä'klən), peak, 9,510 ft (2,900.6 m), Cascade Range, E Jackson co., SW Oregon, NW of Klamath Falls.

Mc·Minn (mək-mĭn'), county (1970 pop. 35,462), 435 sq mi (1,126.7 sq km), SE Tenn., in the Great Appalachian Valley bounded on the SW by the Hiwassee River; formed 1819; co. seat Athens. A leading dairying county, it produces corn, cotton, tobacco, hay, and livestock. There is lumbering and diversified manufacturing.

Mc·Minn·ville (mək-mĭn'vĭl'). **1.** City (1970 pop. 10,125), seat of Yamhill co., NW Oregon; inc. 1876. It is a trade and processing center in the fertile Willamette River valley and also has a large lumber industry. **2.** Town (1970 pop. 10,662), seat of Warren co., central Tenn.; inc. 1809. It has various manufacturing industries.

Mc·Mul·len (mək-mŭl'ən), county (1970 pop. 1,095), 1,159 sq mi (3,001.8 sq km), S Texas, drained by the Frio and Nueces rivers; formed 1858; co. seat Tilden. It has cattle ranching, agriculture (grain sorghums, corn, peanuts, some cotton, hogs, poultry, and dairy products), and oil and natural-gas wells.

Mc·Mur·ray (mək-mûr'ē, -mŭr'e), town, (1976 pop. 15,424), NE Alta., Canada, on the Athabasca River. It is an important river port and transshipment point for the Northwest Territories.

Mc·Nair·y (mək-nâr'ē), county (1970 pop. 18,369), 569 sq mi (1,473.7 sq km), SW Tenn., bounded in the S by the Miss. border and drained by tributaries of the South Fork of the Forked Deer River and the Hatchie and Tennessee rivers; formed 1823; co. seat Selmer. Its agriculture includes cotton, corn, and hay. It has lumbering and textile industries.

Mc·Pher·son (mək-fûr'sən). **1.** County (1970 pop. 24,778), 895 sq mi (2,321 sq km), central Kansas, in a rolling plain area drained in the NW by the Smoky Hill River; formed 1870; co. seat McPherson. It is in a wheat and livestock region, with oil and gas fields and lumber. Its manufactures include flour, petroleum, plastics, metal products, and machinery. **2.** County (1970 pop. 623) 855 sq mi (2,214.5 sq mi), W central Nebr., in an agricultural area; formed 1887; co. seat Tryon. Livestock and grain are important to its economy. **3.** County (1970 pop. 5,022), 1,151 sq mi (2,981.1 sq km), N S.Dak., on the N.Dak. border, watered by intermittent streams in the E and by small lakes; formed 1873; co. seat Leola. It is in an agricultural and cattle-raising area that produces wheat, flax, corn, and oats. It has some manufacturing.

McPherson, city (1970 pop. 10,851), seat of McPherson co., central Kansas, in a farm area on the old Santa Fe Trail; inc. 1874. The city has an oil refinery, a flour mill, and factories that make a variety of products.

Mc·Rae (mə-krā'), city (1970 pop. 3,151), seat of Telfair co., S central Ga., SE of Macon on the Little Ocmulgee River, in a farm and fruit area; settled in the mid-19th cent., inc. 1874. Lumber and clothing are manufactured here.

Mead, Lake (mēd), 247 sq mi (639.7 sq km), on the Nev.-Ariz. border, formed by Hoover Dam across the Colorado River. The lake is 115 mi (185 km) long, from 1 to 8 mi (1.6-12.9 km) wide, and 589 ft (179.6 m) at its maximum depth; it has the greatest capacity of any reservoir in the United States and is one of the largest in the world. Lake Mead, with its 550 mi (885 km) shoreline, is the focal point of Lake Mead National Recreation Area.

Meade (mēd). **1.** County (1970 pop. 4,912), 976 sq mi (2,527.8 sq km), SW Kansas, in a rolling prairie region bordered on the S by Okla. and drained by Crooked Creek; formed 1885; co. seat Meade. It has grain and livestock farming, oil and gas wells, volcanic ash deposits, and some manufacturing. **2.** County (1970 pop. 18,796), 308 sq mi (797.7 sq km), NW Ky., bounded on the N and NW by the Ohio River and the Ind. border; formed 1823; co. seat Brandenburg. In an agricultural area yielding tobacco, corn, soybeans, and livestock, it has lumbering, limestone quarrying, and a chemical industry. Otter Creek Recreational Area is in the east. **3.** County (1970 pop. 17,020), 3,465 sq mi (8,974.4 sq km), W S.Dak., drained by the Belle Fourche River and bounded in the E by the Cheyenne River; formed 1889; co. seat Sturgis. It is in a ranching (cattle and sheep) and agricultural (wheat and oats) area that is rich in mineral resources (gold, manganese, lignite, bentonite, and fuller's earth). It has some manufacturing.

Meade, city (1970 pop. 1,899), seat of Meade co., SW Kansas, SSW of Dodge City, in a cattle area; inc. 1885.

Mead·ville (mēd'vĭl'). **1.** Town (1970 pop. 594), seat of Franklin co., SW Miss., ESE of Natchez, in a lumber-milling area. **2.** City (1978 est. pop. 16,400), seat of Crawford co., NW Pa.; settled 1788, inc. 1866. It is an industrial city in a rich agricultural region. There is a railroad shop and factories that manufacture zippers, acetate yarn, flat glass, furnaces, and machines and machine parts.

Mea·gher (mē'gər), county (1970 pop. 2,122), 2,354 sq mi (6,096.9 sq km), central Mont., in a mountainous region drained by the Smith River and branches of the Musselshell River; formed 1866; co. seat White Sulphur Springs. Livestock is raised.

Meath (mēth, mē*th*), county (1971 pop. 71,729, 903 sq mi (2,338.8 sq km), E Republic of Ireland; co. town Navan. The land is mostly level, with extensive fertile areas near the Boyne and Blackwater rivers. Grain and potato cultivation and cattle raising support the bulk of the population, and there is some manufacturing.

Meaux (mō), city (1975 pop. 42,243), Seine-et-Marne dept., N France, in Brie, on the Marne River. It is an industrial center where metals, flour, chemicals, carbon paper, candy, and cheeses are manufactured. In the massacre of Meaux (1358), thousands of peasants who had participated in the Jacquerie were slain.

Mec·ca (mĕk'ə), city (1974 pop. 366,801), capital of the Hejaz, W Saudi Arabia. The birthplace A.D. c.570 of Mohammed the Prophet, it is the holiest city of Islam. It is in a narrow valley overlooked by hills crowned with castles. The city was an ancient center of commerce and a place of great sanctity for idolatrous Arab sects before the rise of Mohammed. His flight (the hegira) from Mecca in 622 is the beginning of the Moslem era. He captured the city shortly after. Although Mecca never lost its sanctity, it declined rapidly in commercial importance after its capture by the Umayyads in 692. It was sacked in 930 by the Karmathians and taken by the Ottoman Turks in 1517. In Mecca, in 1916, Husayn ibn Ali proclaimed his independence from Turkey and maintained himself as king of the Hejaz until Mecca fell to Ibn Saud in 1924. At the center of Mecca is the Great Mosque, the Haram, which encloses the Kaaba, the chief goal of Moslem pilgrimage. The bazaar outside the mosque is noted for its silks, beadwork, and perfumes. The commerce of the city depends almost wholly on the pilgrims. Despite the ban against unbelievers, the holy city was visited and described in the 19th cent. by Richard Burton and others.

Me·che·len (mĕkʜ'ə-lən) or **Mech·lin** (mĕk'lĭn), city (1977 pop. 77,868), Antwerp prov., N central Belgium, on the Dijle River. It is a commercial, industrial, and transportation center and was formerly a famous lace-making center. Manufactures include textiles, steel, motor vehicles, and processed food. Founded in the early Middle Ages, Mechelen was until 1356 a fief of the prince-bishops of Liège. It then passed to Louis de Mâle and the dukes of Burgundy. The city was damaged often in the many wars that were fought in the Low Countries. However, Mechelen retains many noteworthy buildings, including the Gothic Cathedral of St. Rombaut (13th cent.), which has a 319-ft (97.3-m) tower and a famous carillon and which contains Anthony Van Dyck's great painting, the *Crucifixion,* and the churches of Notre Dame and of St. John, both of which have paintings by Rubens.

Meck·len·burg (mĕk'lən-bûrg', -bŏŏrkʜ'), former state, N East Germany, bordering on the Baltic Sea. As constituted in 1947 under Soviet military occupation, Mecklenburg consisted of the former states of Mecklenburg-Schwerin and Mecklenburg-Strelitz and part of the former Prussian province of Pomerania. In 1952 it was abolished as an administrative unit, and its territory was included in the districts of Schwerin, Rostock, and Neubrandenburg.

The region of Mecklenburg was occupied (6th cent. A.D.) by the Wends. Later awarded as a march to the dukes of Saxony, it was subdued (12th cent.) by Henry the Lion, and the Wendish prince Pribislaw became a vassal of the Holy Roman Empire. In 1348 the princes were raised to ducal rank. In 1621 the duchy was divided into Mecklenburg-Schwerin and Mecklenburg-Güstrow, but during the Thirty Years War both dukes were deposed (1628). However, Gustavus II of Sweden restored the region (1631) to its former rulers. The line of Mecklenburg-Güstrow died out in 1701, and the line of Mecklenburg-Strelitz took its place. At the Congress of Vienna both divisions of Mecklenburg were raised (1815) to grand duchies. They both joined the German Confederation, sided with Prussia in the Austro-Prussian War of 1866, and joined the German Empire at its founding in 1871. The grand dukes were deposed in 1918. In 1934 the separate states of Mecklenburg-Schwerin and Mecklenburg-Strelitz were united.

Meck·len·burg (měk'lən-bûrg'). **1.** County (1970 pop. 354,656), 530 sq mi (1,372.7 sq km), S N.C., in a piedmont region bounded on the SW by the S.C. border, on the W by the Catawba River; formed 1762; co. seat Charlotte. Its economy is based on agriculture (cotton, corn, hay, dairy products, and poultry), lumbering, and manufacturing. **2.** County (1970 pop. 29,426), 612 sq mi (1,585.1 sq km), S Va., bounded in the S by N.C., in the N by the Meherrin River, and drained by the Roanoke River, which is joined here by the Dan River; formed 1764; co. seat Boydton. It is an important tobacco growing county that also produces corn, cotton, hay, peanuts, livestock, and poultry. Its industries include food processing, lumbering, and manufacturing of clothing and textiles, and wood products.

Me·co·sta (mĭ-kôs'tə), county (1970 pop. 27,992), 563 sq mi (1,458.2 sq km), central Mich., drained by the Muskegon, Little Muskegon, Chippewa, and Pine rivers; organized 1859; co. seat Big Rapids. It is a partially agricultural area where livestock and poultry are raised, and grain, potatoes, fruit, and corn are grown. There is some manufacturing.

Me·dan (mə-dän'), city (1971 pop. 635,562), capital of North Sumatra prov., NE Sumatra, Indonesia, on the Deli River.

Me·del·lín (mĕd'l-ēn', mä'dā-yēn'), city (1973 pop. 1,100,082), capital of Antioquia dept., W central Colombia. Textiles, steel, sugar, and coffee are the principal products. Coal, gold, and silver are mined in the surrounding region. The city, which was founded in 1675, is located in a small intermontane valley at an altitude of c.5,000 ft (1,525 m). Until the development of transportation in the 19th cent., it was practically isolated.

Med·ford (mĕd'fərd). **1.** City (1978 est. pop. 58,900), Middlesex co., E Mass., a residential and industrial suburb of Boston, on the Mystic River; settled 1630, inc. as a city 1892. Truck bodies, valves, wax, paper, and furniture are among its products. A shipping and shipbuilding center from the 17th to the 19th cent., Medford was also known for its rum. It is the seat of Tufts Univ. **2.** Town (1970 pop. 1,304), seat of Grant co., N Okla., NNE of Enid, in a wheat, livestock, and oil area. **3.** City (1978 est. pop. 33,100), seat of Jackson co., SW Oregon, on Bear Creek; inc. 1884. It is a trade, shipping, and medical center in an agricultural area. There are fruit-packing plants and lumber mills in the city. Between 1836 and 1856 the area was the scene of a number of bloody conflicts between white settlers and the Rogue River Indians. Gold was discovered nearby in 1851. Medford is the headquarters for Crater Lake National Park and Rogue River National Forest. **4.** City (1970 pop. 3,454), seat of Taylor co., N central Wis., on the Black River NW of Wausau; inc. 1889. Wood products are made.

Me·di·a (mē'dē-ə), ancient country of W Asia, extending from the Caspian Sea to the Zagros Mts. The Medes were an Indo-European people who spoke an Iranian language closely akin to old Persian. They extended their rule over Persia during the reign of Sargon (d. 705 B.C.). The dynasty continued until the rule of Astyages, when it was overthrown (c.550 B.C.) by Cyrus the Great and united with the Persian Empire. In the 2nd cent. B.C. Media became part of the Parthian kingdom. It was divided by the Romans into Media Atropatene in the north and Media Magna in the south.

Media, borough (1970 pop. 6,444), seat of Delaware co., SE Pa., W of Philadelphia; settled c.1682, laid out c.1848, inc. 1850. It is in a farming area, with diversified light industries.

Med·i·cine Bow (mĕd'ĭ-sən bō), river, c.120 mi (193.1 km) long, S Wyo., rising in the N Medicine Bow Mts. and flowing N and W into the Seminoe Reservoir, where it joins the North Platte River.

Medicine Bow Mountains, outlying E range of the Rocky Mts., SE Wyo. and N Colo., extending S from Medicine Bow, Wyo., c.100 mi (160 km) to Cameron Pass, Colo. Peaks include Medicine Bow Peak (12,013 ft/3,664 m) and Elk Mt. (11,156 ft/3,402.6 m). Much of the area is in Medicine Bow National Forest.

Medicine Hat (hăt), city (1976 pop. 32,811), SE Alta., Canada, on the South Saskatchewan River. It is the center of a farming and ranching area. Natural-gas deposits are exploited.

Medicine Lodge (lŏj), city (1970 pop. 2,545), seat of Barber co., S Kansas, SW of Wichita, in a wheat, natural-gas, and oil area; founded 1873, inc. 1879. In 1867 a peace treaty was concluded near here with the five tribes of the Plains Indians—the Apache, Arapaho, Cheyenne, Comanche, and Kiowa; every five years since 1927 a pageant has commemorated the event.

Me·di·na (mə-dē'nə), city (1974 pop. 198,186), Hejaz, W Saudi Arabia. It is situated c.110 mi (180 km) inland from the Red Sea in a well-watered oasis where much fruit is raised. Before the flight (hegira) of Mohammed from Mecca to the city in 622, Medina was called Yathrib. Mohammed quickly gained control of Medina and used it as the base for converting and conquering Arabia. Medina grew rapidly until 661, when the Umayyad dynasty transferred the capital of the caliphate to Damascus. Thereafter Medina was reduced to a provincial town. Local warfare drained the city's prosperity. It came under the sway of the Ottoman Turks in 1517. The Wahabis captured it in 1804, but it was retaken for the Turks by Muhammad Ali in 1812. In World War I the forces of Husayn ibn Ali captured Medina. In 1924 it fell to Ibn Saud, after a 15-month siege. The city is surrounded by double walls flanked by bastions and pierced by nine gates. The chief building is the large mosque, which contains the tombs of Mohammed, his daughter Fatima, and the caliph Omar. A pilgrimage to Mecca usually includes a visit to Medina.

Me·di·na. 1. (mə-dĭ'nə) County (1970 pop. 82,717), 424 sq mi (1,098.2 sq km), N Ohio, drained by the Rocky and Black rivers; formed 1818; co. seat Medina. It has some agriculture (dairy products, grain, fruit, and truck crops), but light manufacturing is becoming increasingly important. **2.** (mə-dē'nə) County (1970 pop. 20,249), 1,353 sq mi (3,504.3 sq km), SW Texas, crossed E to W by the Balcones Escarpment, separating Edwards Plateau in the N from the plains of the S, and drained by the Medina River; formed 1848; co. seat Hondo. Its agriculture includes cattle, sheep, goats, poultry, corn, grain sorghums, hay, vegetables, pecans, and peanuts. There are deposits of oil, natural gas, and clay.

Me·di·na (mə-dĭ'nə), city (1970 pop. 10,913), seat of Medina co., N Ohio; laid out 1818, inc. as a city 1950. It is a processing point in a farm area, with a bee industry.

Med·i·ter·ra·ne·an Sea (mĕd'ĭ-tə-rā'nē-ən), the world's largest inland sea, c.965,000 sq mi (2,499,350 sq km), surrounded by Europe, Asia, and Africa. It is c.2,400 mi (3,860 km) long with a maximum width of c.1,000 mi (1,610 km); its greatest depth is c.14,450 ft (4,405 m), off Cape Matapan, Greece. It connects with the Atlantic Ocean through the Strait of Gibraltar; with the Black Sea through the Dardanelles, the Sea of Marmara, and the Bosporus; and with the Red Sea through the Suez Canal. Its chief divisions are the Tyrrhenian, Adriatic, Ionian, and Aegean seas. Shallows (Adventure Bank) between Sicily and Cape Bon, Tunisia, divide the Mediterranean into two main basins. It is of higher salinity than the Atlantic and has little variation in tides. The shores are chiefly mountainous. Earthquakes and volcanic disturbances are frequent. Strong local winds, such as the hot, dry sirocco from the south and the cold, dry mistral and bora from the north, blow across the sea. Fish (about 400 species), sponges, and corals are plentiful.

Some of the most ancient civilizations flourished around the Mediterranean. It was opened as a highway for commerce by merchants trading from Phoenicia. Carthage, Greece, Sicily, and Rome were rivals for dominance of its shores and trade; under the Roman Empire it became virtually a Roman lake. Later, the Byzantine Empire and the Arabs dominated the Mediterranean. Products of the Orient passed to Europe over Mediterranean trade routes until the establishment of a route around Africa (late 15th cent.). With the opening of the Suez Canal (1869) the Mediterranean resumed its importance as a link on the route to the East. Since World War II the Mediterranean region has been of strategic importance to both the United States and the USSR.

Mé·doc (mā-dôk′), region, SW France, a peninsula extending NW of Bordeaux between the Bay of Biscay and the Gironde River estuary. The region is covered with some of France's most famous vineyards.

Me·do·ra (mĭ-dôr′ə), village (1970 pop. 129), seat of Billings co., W N.Dak., W of Dickinson, in a cattle-raising and grain area. Nearby is the site of Chimney Butte Ranch, where Theodore Roosevelt ranched for a time; it is now a national park.

Meek·er (mē′kər), county (1970 pop. 18,387), 620 sq mi (1,605.8 sq km), S central Minn., drained by the Crow River; formed 1856; co. seat Litchfield. It is in an agricultural area that produces corn, oats, barley, potatoes, livestock, and dairy products. Its industry includes food processing (dairy and grain mills), textile weaving and finishing, and the manufacture of metal products and tools.

Meeker, town (1970 pop. 1,597), seat of Rio Blanco co., NW Colo., on the White River, in a resort, farm and mine area; inc. 1885.

Mee·rut (mē′rət), city (1971 pop. 271,325), Uttar Pradesh state, N central India. An agricultural market, it processes flour and vegetable oil. Meerut was conquered by Moslems in 1192, ravaged by Tamerlane in 1399, and became part of the Mogul empire. The first outbreak of the Indian Mutiny occurred in Meerut in May, 1857, but the British held the city.

Mé·ga·ra (mĕg′ər-ə, mâ′gä-rä), town (1971 pop. 17,294), E central Greece, on the Saronic Gulf. Wine, olive oil, and flour are produced. It is the site of the ancient town of Mégara, the capital of Mégaris, a small district between the Gulf of Corinth and the Saronic Gulf. The Dorians made Mégara a wealthy city by means of maritime trade. After the Persian Wars the citizens of Mégara summoned the aid of Athens against Corinth (459 B.C.), but soon thereafter expelled the Athenians. The mathematician Euclid was probably born here.

Me·gha·la·ya (mā′gə-lā′yə), state (1971 pop. 983,336), c.8,700 sq mi (22,535 sq km), NE India, bordered on the S by Bangladesh; capital Shillong. Meghalaya is in the Garo, Khasi, and Jaintia hills, at an elevation of 4,000–6,000 ft (1,220–1,830 m). Meghalaya was formerly part of Assam state; it became a separate state in 1972.

Megh·na (māg′nə), river, c.130 mi (210 km) long, formed at the W end of the Surma valley, NE Bangladesh, by the branches of the Surma River. The Meghna is an important inland waterway, navigable throughout its length by river steamers. In the springtime, at high tide, tidal bores, c.20 ft (6 m) high, rush upstream with great destructive force.

Me·gid·do (mə-gĭd′ō), ancient city, Palestine, by the Kishon River on the S edge of the plain of Esdraelon, N of Samaria. It was inhabited from the 4th millennium B.C. to c.450 B.C. Situated in a strategic position, controlling the route that connected Egypt with Mesopotamia, it has been the scene of many battles throughout history. Excavations have unearthed 20 strata of settlements.

Meigs (mĕgz). **1.** County (1970 pop. 19,799), 434 sq mi (1,124.1 sq km), SE Ohio, bounded on the SE by the Ohio River, here forming the W.Va. line; formed 1819; co. seat Pomeroy. It is in an industrial area, with coal mining and some agriculture (livestock, dairy products, fruit, grain, and truck crops). **2.** County (1970 pop. 5,219), 191 sq mi (494.7 sq km), SE Tenn., in the Great Appalachian Valley, bounded in the NW by the Tennessee River and drained by the Hiwassee River; formed 1836; co. seat Decatur. Its agriculture includes tobacco, corn, soybeans, wheat, fruit, and cattle.

Mei·ners Oaks (mī′nərz ōks), uninc. town (1970 pop. 5,600), Ventura co., S Calif., NW of Los Angeles, in a farm area; founded 1925.

Mei·ning·en (mī′nĭng-ən), city (1974 est. pop. 26,134), Suhl district, SW East Germany, on the Werra River. Manufactures include machinery, textiles, lumber, and metal products. Meiningen was first mentioned in the 10th cent. and was the capital of the duchy of Saxe-Meiningen from 1680 to 1918.

Meis·sen (mī′sən), city (1974 est. pop. 43,561), Dresden district, SE East Germany, on the Elbe River. A porcelain manufacturing center since 1710, Meissen is famous for its figurines (often called Dresden china). Other manufactures include metal products and leather goods. Meissen was founded (929) by Henry of Saxony and became (965) the seat of the margraviate of Meissen, where the Wettin dynasty of Saxony originated. The diocese of Meissen was founded in 968, was suppressed in 1581, and was restored in 1921. The Albrechtsburg (15th cent.), a large castle, dominates the city; it housed (1710–1864) the royal porcelain manufacture, begun under the patronage of Elector Frederick Augustus I.

Mek·nès (mĕk-nĕs′), city (1971 pop. 248,369), N central Morocco. It has a noted carpet-weaving industry. There are also woolen mills, cement and metal works, oil distilleries, and food-processing plants. Meknès was founded (c.1672) by Sultan Ismail, who undertook such palatial building operations that the city was called the Versailles of Morocco. Little of his construction has survived.

Me·kong (mā′kŏng′, mē′-), one of the great rivers of SE Asia, c.2,600 mi (4,185 km) long. It rises in the Tibetan highlands of China as the Dza Chu and flows generally south through Yünnan prov. in deep gorges and over rapids. Leaving Yünnan, the Mekong forms the Burma-Laos border, then curves east and south through northwest Laos before marking part of the Laos-Thailand border. From southwestern Laos the river descends onto the Cambodian plain, receiving water from Tônlé Sap during the dry season by way of the Tônlé Sap River; during the rainy season, however, the floodwaters of the Mekong reverse the direction of the Tônlé Sap River and flow into Tônlé Sap, a lake that is a natural reservoir. The Mekong River finally flows into the South China Sea through many distributaries in the vast Mekong delta (c.75,000 sq mi/194,250 sq km). The delta, crisscrossed by many channels and canals, is one of the greatest rice-growing areas of Asia. It was the scene of heavy fighting in the Vietnam War.

Me·la·ka or **Ma·lac·ca** (both: mə-lăk′ə), (1971 pop. 86,357), capital of Melaka state, Malaysia, S Malay Peninsula. It was founded c.1400 by a Malay prince who had been driven from Singapore after a brief reign there. The city quickly gained wealth as a center of trade with China, Indonesia, India, and the Middle East. Gujarati traders introduced Islam to the Malay world through Malacca. In 1511 Malacca was captured by the Portuguese In the early 17th cent. the Dutch entered the region and captured Malacca in 1641 after a long siege. They utilized the city more as a fortress guarding the strait than as a trading port. The Dutch retained nominal control until 1824, although during the wars of the French Revolution and the Napoleonic period (1795–1818) the British occupied Malacca at the request of the Dutch government-in-exile. In 1824 the Dutch formally transferred Malacca to Great Britain.

Mel·a·ne·sia (mĕl′ə-nē′zhə, -shə), one of the three main divisions of Oceania, in the SW Pacific Ocean, NE of Australia and S of the equator. Melanesia includes the Solomon Islands, New Hebrides, New Caledonia, the Bismarck Archipelago, and the Admiralty and Fiji islands. New Guinea is sometimes included in Melanesia.

Mel·bourne (mĕl′bərn), city (1976 pop. 65, 065, capital of Victoria, SE Australia, on Port Phillip Bay at the mouth of the Yarra River. Melbourne is a rail hub and financial and commercial center. Wool and raw and processed agricultural goods are exported. The city is heavily industrialized; industries include shipbuilding and the manufacture of automobiles, farm machinery, textiles, and electrical goods. Settled in 1835, it was named (1837) for Lord Melbourne, the British prime minister. From 1901 to 1927 the city was the seat of the Australian federal government. The Univ. of Melbourne, Melbourne Technical College, the Australian Ballet School, and the National Art Gallery are in the city.

Melbourne, city (1976 est. pop. 1,286), seat of Izard co., N Ark., NW of Batesville, in a stock-raising and dairying area. Cotton, corn, and hay are grown.

Mel·fi (mĕl′fē), town (1971 pop. 15,194), in Basilicata, S Italy. It is an agricultural and manufacturing center noted for its wine. In 1041 it was made the first capital of the Norman county of Apulia. At Melfi Emperor Frederick II promulgated (c.1231) his important code, the Constitutions of Melfi, or *Liber Augustalis.* In 1528 the town was sacked by the French, and it never recovered its position as a flourishing commercial center.

Me·lil·la (mā-lē′lyä), city (1970 pop. 60,843), Spanish possession, on the Mediterranean coast of Morocco, NW Africa. It is a fishing port and an export point for iron ore. Spain has held the city since 1496 despite many attacks by Moroccans. The revolt that began (1936) the Spanish Civil War broke out in Melilla.

Me·li·to·pol (mĕl′ə-tô′pəl), city (1976 est. pop. 155,000), S European USSR, in the Ukraine, on the Molochnoy River.

Melk (mĕlk), town (1971 pop. 5,100), Lower Austria province, N central Austria, on the Danube River. A noted tourist spot, it was one of the earliest residences of the Austrian rulers. The large Benedictine abbey, founded in 1089, has a library whose holdings include about 2,000 old manuscripts and 80,000 volumes.

Mel·lette (mə-lĕt′), county (1970 pop. 2,420), 1,306 sq mi (3,382.5 sq km), S S. Dak., bounded in the N by the White River; formed 1909; co. seat White River. It is in a cattle-ranching and agricultural area that produces wheat, oats, barley, milo, and alfalfa.

Mel·rose (mĕl′rōz′), burgh (1971 pop. 2,188), Borders region, S Scotland, on the Tweed River. It is the site of one of the finest ruins in Scotland—Melrose Abbey, founded for Cistercians by David I in 1136 and now owned by the nation.

Melrose, city (1978 est. pop. 31,500), Middlesex co., E Mass., a suburb of Boston; settled c.1629, set off from Malden and inc. 1850.

Me·lun (mə-lœɴ′), town (1975 pop. 37,705), capital of Seine-et-Marne dept., N central France, SE of Paris. It is an important industrial center where automobile bodies, airplane engines, leather products, pharmaceuticals, and elastics are produced.

Mel·ville, Lake (mĕl′vĭl′), saltwater lake, 1,133 sq mi (2,934.5 sq km), SE Labrador, Canada, extending c.120 mi (195 km) inland from Hamilton Inlet, an arm of the Atlantic Ocean. It receives the Churchill River in Goose Bay, its southwest arm.

Melville Bay, broad indentation of the W coast of Greenland, opening to the SW into Baffin Bay. The inland ice cap comes down to the coast, and glaciers discharge much ice into its waters.

Melville Island, 2,240 sq mi (5,801.6 sq km), Northern Territory, N Australia, in the Timor Sea 16 mi (26 km) off the coast. It is 65 mi (104.6 km) long and 45 mi (72.4 km) wide and is separated from Bathurst Island by Apsley Strait. It consists largely of mangrove jungle with sandy soil.

Melville Island, c.16,400 sq mi (42,475 sq km), W Franklin dist., Northwest Territories, Canada, N of Victoria Island; largest of the Parry Islands. Generally hilly (rising to c.1,500 ft/460 m), it has several ice-covered areas in the interior. There are musk oxen on the island. Sir William Parry, the British explorer, discovered Melville Island in 1819.

Melville Peninsula, 24,156 sq mi (62,564 sq km), c.250 mi (400 km) long and from 70 to 135 mi (112.6–217.2 km) wide, S Franklin dist., Northwest Territories, Canada, between the Gulf of Bothnia and Foxe Basin, and separated from Baffin Island to the N by the Fury and Hecla Strait. It is joined to the mainland by the Rae Isthmus. Numerous streams radiate from the peninsula's central hilly section, which rises to 1,850 ft (564.3 m). The tundra-covered region is virtually uninhabited and unimportant economically.

Me·mel (mā′məl): *see* Klaypeda, USSR.

Memel Territory, name applied to the district (1,092 sq mi/2,828.3 sq km) of former East Prussia situated on the E coast of the Baltic Sea and the right bank of the Neman River. In 1919 the Treaty of Versailles placed the district under League of Nations-sponsored French administration. Lithuanian troops occupied the area in 1923, forcing the French garrison to withdraw. The Allied council of ambassadors then drew up a new status for the territory, which in 1924 became an autonomous region within Lithuania with its own legislature. The 1938 electoral victory of the National Socialists here was followed in March, 1939, by a German ultimatum demanding the district's return. In 1945 Soviet forces took the area and restored it to Lithuania, by then a part of the USSR.

Mem·ming·en (mĕm′ĭng-ən), city (1974 est. pop. 34,799), Bavaria, S West Germany. Manufactures include metal products, textiles, machinery, beer, and chemicals. Historically a Swabian town, Memmingen was first mentioned in the early 12th cent. and became a free imperial city in 1286.

Mem·phis (mĕm′fĭs), ancient city of Egypt, capital of the Old Kingdom (c.3100–c.2258 B.C.), at the apex of the Nile delta near Cairo. It was reputedly founded by Menes, the first king of united Egypt. Its god was Ptah. The temple of Ptah, the palace of Apries, and two huge statues of Ramses II are among the most important monuments found at the site. The necropolis of Sakkara, near Memphis, was a favorite burial place for pharaohs of the Old Kingdom. Across the Nile are the great pyramids. Memphis remained important during the long dominance by Thebes and became the seat of the Persian satraps (525 B.C.). Second only to Alexandria under the Ptolemies and under Rome, it finally declined with the founding of nearby Fustat by the Arabs, and its ruins were largely removed for building in the new city and, later, in Cairo.

Memphis. 1. City (1970 pop. 2,081), seat of Scotland co., NE Mo., near the North Fabius River WNW of Keokuk, Iowa; settled 1838, inc. 1880. It ships livestock and grain. **2.** City (1978 est. pop. 655,000), seat of Shelby co., SW Tenn., on the Fourth, or Lower, Chickasaw Bluff above the Mississippi, at the mouth of the Wolf River; inc. 1826. An important river port, Memphis is a port of entry, a rail center, and a leading hardwood lumber, cotton, poultry, and livestock market. Its wide variety of manufactures includes textiles, heating equipment, pianos, and automobile and truck parts. The area was strategically important during the time of the British, French, and Spanish rivalries in the 18th cent. A U.S. fort was erected here in 1797. The city was established (1819) by Andrew Jackson (who named it), Marcus Winchester, and John Overton. In the Civil War it fell, on June 6, 1862, to a Union force and was an important Federal base for the rest of the war. Severe yellow-fever epidemics occurred in the 1870s; thousands died, and so many people fled the city that its charter had to be surrendered (1879); the charter was not restored until 1891. E. H. "Boss" Crump ruled Memphis from 1909 until his political hold was broken after 1948. The city is the seat of Memphis State Univ., the Univ. of Tennessee Medical Units, and other schools. It has a museum of natural history, a planetarium, an art gallery, a notable park system, botanical gardens, a zoo, an aquarium, a coliseum, a speedway, and a greyhound park. It is the seat of a large medical center and St. Jude Children's Research Hospital. A number of antebellum homes in the city have been restored. A leading tourist attraction is Beale St., made famous by W. C. Handy, the black composer and compiler of the blues. **3.** City (1970 pop. 3,227), seat of Hall co., NW Texas, ESE of Amarillo; founded 1889, inc. 1906. A center of farms producing grains, cotton, forage, and other crops, Memphis has processing plants and small industries.

Me·na (mē′nə), city (1970 pop. 4,530), seat of Polk co., W Ark., SSE of Fort Smith, in a rich farm section of the Ouachita Mts.; founded 1896. Clothing, leather goods, and wood products are made.

Me·nam Chao Phra·ya (mä-näm′ chou′prä-yä′), river, Thailand: *see* Chao Phraya.

Me·nard (mə-närd′). **1.** County (1970 pop. 9,685), 312 sq mi (808.1 sq km), central Ill., bounded on the N by the Sangamon River and Salt Creek; formed 1839; co. seat Petersburg. It has agriculture (corn, wheat, soybeans, livestock, poultry, and dairy products) and the manufacturing of radiator guards, canned foods, beverages, brick, and tile. **2.** County (1970 pop. 2,646), 914 sq mi (2,367.3 sq km), W central Texas, on Edwards Plateau, drained by the San Saba River; formed 1858; co. seat Menard. In a ranching area, it has some irrigated farming (truck crops, grain, and hay), poultry raising, and dairying. Wool and mohair are marketed.

Menard, town (1970 pop. 1,740), seat of Menard co., W central Texas, on the San Saba River SE of San Angelo. It handles wool, mohair, and cattle from surrounding ranches.

Me·nash·a (mə-năsh′ə), city (1978 est. pop. 15,000), Winnebago co., E Wis., on Lake Winnebago and the Fox River, adjacent to its twin city of Neenah; settled 1840s, inc. 1874. Menasha is a great papermaking center.

Mende (mäɴd), city (1975 pop. 10,451), capital of Lozère dept., S France, on the Lot River. Mende is a tourist resort. It was originally a small Gallo-Roman city that became an episcopal see in the 5th cent. Bishops ruled the town until 1306, when they were forced to cede a portion of it to Philip the Handsome.

Men·den·hall (mĕn′dən-hôl′), town (1970 pop. 2,402), seat of Simpson co., S central Miss., SE of Jackson, in a timber area.

Men·de·res (mĕn′də-rĕs′), name of several rivers in Turkey. The Büyük Menderes is the ancient Maeander; the Küçük Menderes is the ancient Scamander.

Men·dip Hills (mĕn′dĭp′), range of hills, c.25 mi (40 km) long, across N Somerset, SW England, extending SE from the vicinity of Hutton to the Frome valley. Composed primarily of limestone, the hills have numerous caves, some of which show signs of prehistoric occupation. In the hills are Roman ruins.

Men·do·ci·no (mĕn′də-sē′nō), county (1970 pop. 51,101), 3,510 sq mi (9,090.9 sq km), W Calif., on the coast, in a mountain and valley region traversed by several of the Coast Ranges and drained by the Eel, Russian, Big, Noyo, and Navarro rivers; formed 1850; co. seat Ukiah. It has extensive lumbering and sawmilling, stock raising (cattle and sheep), and farming in the valleys (apples, pears, prunes, nuts, grapes, hops, berries, hay, poultry, and dairy products). Its manufactures include machinery, furniture, and wood products. There are hot springs resorts. Trout and steelhead fishing and deer hunting are done. The county lies partly in Medocino National Forest, and Round Valley Indian Reservation is in the northeast.

Mendocino, Cape, westernmost point of Calif., N of San Francisco. It was discovered in 1542 by Juan Rodríguez Cabrillo.

Men·do·za (mĕn-dō'zə), city (1970 pop. 118,568), capital of Mendoza prov., W Argentina. With a backdrop of snow-capped mountains, Mendoza is surrounded by a fertile oasis, known as the Garden of the Andes, irrigated by the Mendoza River. It is the center of a rich wine-producing region, largely settled by Italian immigrants. Mendoza was founded in 1561 and belonged to Chile until the creation of the viceroyalty of Río de la Plata (1776). Destroyed by earthquake in 1861, the town was rebuilt and expanded rapidly after the completion of the railroad to Buenos Aires late in the 19th cent.

Me·nen (mā'nən), city (1977 pop. 34,302), West Flanders prov., SW Belgium, on the Leie River, near the French border. Manufactures include machinery, textiles, and tobacco products. Founded in 1578, Menen was strongly fortified in the 17th cent.

Men·i·fee (mĕn'ə-fē), county (1970 pop. 4,050), 210 sq mi (543.9 sq km), E Ky., bounded on the NE by the Licking River, on the S by the Red River, and drained by several creeks; formed 1869; co. seat Frenchburg. In a rolling agricultural area yielding livestock, grain, and burley tobacco, it has oil and gas wells and some timber sawmilling. Part of Daniel Boone National Forest is in the county.

Men·lo Park (mĕn'lō), residential city (1978 est. pop. 26,500), San Mateo co., W Calif.; inc. 1874. Space products are manufactured in the city. Menlo College and a Stanford Univ. research institute are here.

Me·nom·i·nee (mə-nŏm'ə-nē). **1.** County (1970 pop. 24,587), 1,038 sq mi (2,688.4 sq km), SW Upper Peninsula, Mich., bounded on the SE by Green Bay, on the SW by Wis., and drained by the Menominee, Cedar, and Little Cedar rivers; organized 1863; co. seat Menominee. Livestock, poultry, potatoes, and sugar beets are raised. Industries include lumbering, commercial fishing, and manufacturing. There are resorts here, and a state park on Green Bay. **2.** County (1970 pop. 2,607), 360 sq mi (932.4 sq km); NE central Wis., drained by the Wolf River; formed 1959 from parts of Oconto and Shawano cos.; co. seat Keshena. Its major industry is lumber milling.

Menominee, city (1970 pop. 10,748), seat of Menominee co., N Mich., W Upper Peninsula, on Green Bay at the mouth of the Menominee River; inc. 1883. It is a car-ferry port and the marketplace for a cheese-producing area. A fur-trading post was established near here in 1796, and several sawmills were operating along the river by 1832. Menominee became a great shipping point for lumber; at one time it was known as the white-pine capital of the world. Today its diversified manufacturing is augmented by a growing tourist industry. Of interest is the "mystery ship," raised (1969) from the bottom of Green Bay, where it sank in 1864.

Menominee, river, 118 mi (190 km) long, formed by the union of the Brule and the Michigamme rivers above Iron Mountain, W Upper Peninsula, N Mich., and flowing SE into Green Bay at Menominee. It passes through an iron-ore region. Once used for lumbering, it now furnishes water power.

Me·nom·o·nie (mə-nŏm'ə-nē), city (1970 pop. 11,112), seat of Dunn co., W Wis., on the Red Cedar River; platted 1859, inc. 1882. Once a lumber town, it is now a trade center in an area of poultry and dairy farms.

Men·teith (mĕn-tēth'), lake, up to 1.5 mi (2.4 km) across, SW Perthshire, central Scotland, near Stirling. Mary Queen of Scots, as a child of five, was hidden at Inchmahome priory on the largest of the lake's three islands.

Men·ton (mäN-tôN'), town (1975 pop. 25,129), Alpes-Maritime dept., SE France, near the Italian border on the Mediterranean Sea. A popular resort of the Riviera, it was a part of Monaco until 1848 when it declared itself a free city under the protection of Sardinia. It passed to France after a plebiscite in 1860.

Men·tone (mĕn'tōn'), village (1970 pop. 50), seat of Loving co., W Texas, NNW of Pecos on the Pecos River.

Men·tor (mĕn'tər), residential village (1978 est. pop. 41,300), Lake co., NE Ohio, on Lake Erie; founded 1799, inc. 1855. James Garfield was living here when he was elected President.

Mer·a·mec (mĕr'ə-măk'), river, rising in S central Mo., E of Salem, and flowing 207 mi (333.1 km) N, NE, and SE to the Mississippi below St. Louis.

Mer·ced (mər-sĕd'), county (1970 pop. 104,629), 1,958 sq mi (5,071.2 sq km), central Calif., extending across the San Joaquin Valley from Diablo Range to the foothills of the Sierra Nevada, irrigated by the Merced, San Joaquin, and Chowchilla rivers; formed 1855; co. seat Merced. It is in a fertile agricultural area yielding dairy products, poultry, cotton, grain, nuts, cattle, grapes, figs, peaches, apricots, sweet potatoes, tomatoes, melons, and alfalfa. It has mining (gold, platinum, sand and gravel, and silver), lumber milling, a food-processing industry, and plants manufacturing textiles, concrete, plastics, metal and wood products, and machinery.

Merced, city (1978 est. pop. 33,100), seat of Merced co., central Calif.; inc. 1889. It is a center for tourism and farm trade in a cotton, fruit, and dairy region.

Mer·ce·des (mĕr-sā'dəs), city (1975 pop. 34,667), capital of Soriano dept., SW Uruguay, a port on the Río Negro. An agricultural and livestock center, the city has a shipyard and several fine beaches and resorts. Mercedes was founded in 1781.

Mer·ce·des (mĕr-sā'dēz), city (1970 pop. 9,355), Hidalgo co., extreme S Texas, WNW of Brownsville; founded 1907, inc. 1909. A processing center for the citrus fruit and vegetables of the irrigated lower Rio Grande valley, it packs meat and has small factories.

Mer·cer (mûr'sər). **1.** County (1970 pop. 17,294), 556 sq mi (1,440 sq km), NW Ill., bounded on the W by the Mississippi River and drained by the Edwards River and Pope Creek; formed 1825; county seat Aledo. It has agriculture (livestock, corn, poultry, wheat, soybeans, hay, alfalfa, and dairy products), bituminous-coal mining, and manufacturing of brick, feed, and pearl buttons. **2.** County (1970 pop. 15,960), 256 sq mi (663 sq km), central Ky., bounded on the E by the Kentucky River and drained by Salt River and Beech Fork; formed 1785; co. seat Harrodsville. In a rolling bluegrass agricultural area yielding livestock, poultry, grain, and burley tobacco, it has calcite mines and limestone quarries. Distilled liquors, clothing, glassware, metal products, and machinery are made. **3.** County (1970 pop. 4,910), 456 sq mi (1,181 sq km), N Mo., drained by the Weldon River; formed 1845; co. seat Princeton. It is in an agricultural area that produces livestock and corn. There is some mining and manufacturing in the county. **4.** County (1970 pop. 304,116), 228 sq mi (590.5 sq km), W central N.J., bounded on the W by the Delaware River, crossed by Delaware and Raritan Canal, and drained by Millstone River and Crosswicks Creek; formed 1838; co. seat Trenton. It has mineral mining and some agriculture (truck crops, poultry, dairy products, and fruit). Its industry includes food processing and the manufacture of clothing, furniture, paper products, chemicals, plastics, concrete, metal products, and machinery. **5.** County (1970 pop. 6,175), 1,042 sq mi (15,993.3 sq km), central N.Dak., in an agricultural area drained by the Knife River and bounded on the N and E by the Missouri River; formed 1873; co. seat Stanton. Wheat and corn are grown, and lignite is mined. **6.** County (1970 pop. 35,558), 454 sq mi (1,175.9 sq km), W Ohio, bounded on the W by the Ind. border and drained by the Wabash and St. Marys rivers; formed 1824; co. seat Celina. It has agriculture (livestock, grain, and poultry), limestone quarries, and manufacturing. Fort Recovery State Park is here. **7.** County (1970 pop. 127,225), 670 sq mi (1,735.3 sq km), NW Pa., in a manufacturing area drained by the Shenango River and tributaries of the Allegheny River and bounded in the W by the Ohio border; formed 1800; co. seat Mercer. Iron, steel, tin, and electrical products come from this county, along with engines and railroad cars. Potatoes and dairy foods are among its agricultural products. There are bituminous-coal, sandstone, and limestone deposits. **8.** County (1970 pop. 63,206), 417 sq mi (1,080 sq km), S W. Va., on the Allegheny Plateau, bordered on the S by Va. and drained by the Bluestone River; formed 1837; co. seat Princeton. Its agriculture includes livestock, fruit, and tobacco. It also has semibituminous-coal mines, limestone deposits and various industries.

Mercer, borough (1970 pop. 2,773), seat of Mercer co., NW Pa., NNE of New Castle, in a farm and bituminous-coal region; settled 1795, laid out 1803, inc. 1814.

Mer·ci·a (mûr'shē-ə), one of the kingdoms of Anglo-Saxon England, consisting generally of the region of the Midlands. It was settled by Angles c.500, probably first along the Trent valley. Its history emerges from obscurity with the reign of Penda, who extended his power over Wessex (645) and East Anglia (650). After his death Mercia suffered a three-year loss of ascendancy and was converted to Christianity. Penda's son, Wulfhere, then re-established a Greater Mercia that finally under Æthelbald in the 8th cent. extended over all of southern England. This hegemony was strengthened by Offa (reigned 757–96), who controlled East Anglia, Kent, and Sussex and maintained superiority of a sort over Wessex and Northumbria. After his death, Mercian power gradually gave way before that of Wessex. In 874 Mercia succumbed to the invading Danish army, and ultimately the eastern part became (886) a portion of the Danelaw, while the western part was controlled by Alfred of Wessex.

Mer de Glace (mĕr də gläs'), glacier (3.5 mi/5.6 km long; 16 sq mi/41.4 sq km), Haute-Savoie dept., E France, on the N slope of Mont Blanc. It is formed by the junction of three smaller glaciers. Famous for its majestic beauty, it is a tourist attraction.

Mé·ri·da (mä'rē-dä), city (1974 est. pop. 233,900), capital of Yucatán state, SE Mexico. It is the chief commercial, communications, and cultural center of the Yucatán peninsula. Founded (1542) on the site of a ruined Mayan city, Mérida has many fine examples of Spanish colonial architecture, notably the 16th cent. cathedral.

Mérida, city (1970 pop. 40,059), Badajoz prov., SW Spain, in Estremadura, on the Guadiana River. It is a rail hub and agricultural center. The colony Emerita Augusta, founded by the Romans in the 1st cent. B.C., it became the capital of Lusitania. Its Roman remains, among the most important in Spain, include a magnificent bridge, a triumphal arch, a theater with marble columns, an aqueduct, a temple, an imposing circus, and an amphitheater. Mérida was later the chief city of Visigothic Lusitania. It fell (713) to the Moors, under whom it prospered. Conquered (1228) by Alfonso IX of León, it was given to the Knights of Santiago but quickly declined.

Mer·i·den (mĕr'ĭ-dən), city (1978 est. pop. 58,100), New Haven co., S central Conn.; settled 1661, inc. as a town 1806, as a city 1867, town and city consolidated 1922. Meriden is known for its large silver industry. Silverware and pewter were made here in the 18th cent. by Samuel Yale and later by the Rogers Brothers and a forerunner of the International Silver Company. Jewelry and bathroom and lighting fixtures are also manufactured.

Me·rid·i·an (mə-rĭd'ē-ən). **1.** City (1978 est. pop. 44,500), seat of Lauderdale co., E Miss., near the Ala. line; settled 1831, inc. 1860. It is the trade, shipping, and industrial center for a farm, livestock, and timber area. In the Civil War the city was the temporary state capital (1863); it was destroyed by Gen. W.T. Sherman in Feb., 1864. **2.** City (1970 pop. 1,162), seat of Bosque co., central Texas, on the Bosque River NW of Waco; settled 1854, inc. 1886. It is a trade and shipping center for a diversified agricultural area.

Mer·i·on·eth·shire (mĕr'ē-ŏn'ĭth-shîr', -shər), former county, NW Wales, on Cardigan Bay. The county town was Dolgellau. Merionethshire was organized as a shire by the English in the 13th cent., but the remoteness of the region made it one of the last districts in Wales to submit to English influence. In 1974 Merionethshire became part of the new nonmetropolitan county of Gwynedd.

Mer·i·weth·er (mĕr'ē-weth'ər), county (1970 pop. 19,461), 499 sq mi (1,292.4 sq km), W Ga., in a piedmont area bounded on the E by the Flint River; formed 1827; co. seat Greenville. It is in a peach-growing region that also produces pecans, melons, peppers, cotton, and livestock. It has lumbering and textile industries and processes fruit, vegetables, and soups.

Mer·o·ë (mĕr'ō-ē'), ancient city N Sudan, on the E bank of the Nile N of Khartoum. In the mid-6th cent. B.C., Meroë replaced Napata as the central city of the Cushite dynasty and from 530 B.C. until A.D. 350 served as the capital of the dynasty. By the 1st cent. B.C., Meroë was a major center for iron smelting. It is believed that knowledge of iron casting was carried (7th-10th cent.) from the middle Nile to the middle Niger by a great African overland route. Among Meroë's extensive ruins are royal palaces (6th cent. B.C.) and a temple of Amon. Nearby are three groups of pyramids.

Mer·rick (mĕr'ĭk), county (1970 pop. 8,571), 480 sq mi (1,243.2 sq km), E central Nebr., in an agricultural area bounded in the S by the Platte River and drained by its branches; formed 1858; co. seat Central City. Flour, livestock, and grain are its major products.

Mer·rill (mĕr'əl), city (1970 pop. 9,502), seat of Lincoln co., N central Wis., at the confluence of the Wisconsin and Prairie rivers, N of Wausau; settled c.1847, inc. 1883. Once a lumber town, it now makes varied products.

Mer·ri·mack (mĕr'ə-măk'), county (1970 pop. 80,925), 931 sq mi (2,411.3 sq km), S central N.H., in a hilly region drained by the Merrimack, Contoocook, Suncook, and other rivers; formed 1823; co. seat Concord. Its rivers yield water power for industries producing textiles, shoes, wood, leather, and metal products, paper, machinery, and electrical instruments. Granite quarrying, mica mining and processing, and flour milling are also important. Its agriculture includes dairy products, poultry, truck crops, and fruit.

Merrimack, river, c.110 mi (175 km) long, formed at Franklin, S central N.H., by the junction of the Pemigewasset and Winnipesaukee rivers and flowing S past Concord and Manchester into NE Mass., where it flows NE past Lowell and Lawrence to the Atlantic Ocean at Newburyport. It has long been used as a source of power for textile and other mills.

Mer·se·burg (mĕr'zə-bûrg', -bŏŏrkh'), city (1974 est. pop. 54,269), Halle district, S central East Germany, on the Saale River. It is an industrial city and a lignite-mining center. Manufactures include chemicals, paper, steel, bricks, and beer. A fortress in the 9th cent., Merseburg was a favorite residence of Henry I (Henry the Fowler) and of Emperor Otto I.

Mers-el-Ke·bir (mĕrs'-ĕl-kə-bîr'), town, NW Algeria, on the Gulf of Oran. Originally a Roman port, it has a long history of maritime importance. During the 15th cent. it was a center of activity for corsairs and was twice occupied by the Portuguese. The Spanish held the town from 1505 to 1792; the French arrived in the 19th cent. After France's defeat by Germany in June, 1940, the French fleet sought refuge at Mers-el-Kebir, but the British navy sank or damaged most of the ships. The French naval base at Mers-el-Kebir came to include subterranean installations where atomic tests were held. In 1962 the Evian Agreement, by which Algerian independence was acknowledged, allowed France to maintain the Mers-el-Kebir base for 15 years; however, the French evacuated the base in 1967.

Mer·sey (mûr'zē), river, c.70 mi (115 km) long, formed at Stockport, W England, by the confluence of the Etherow and Goyt rivers. It flows east to the Irish Sea near Liverpool. The estuary of the Mersey is navigable for oceangoing vessels. Mersey Tunnel or Queensway, a vehicular tunnel (opened 1934) with a length of 2.3 mi (3.7 km), is the longest subaqueous tunnel in the world; it connects Liverpool and Birkenhead. The Mersey River is of great commercial importance to the cities served by it, especially Liverpool and Manchester. Shipbuilding, milling, and oil refining are important industries along the river.

Mer·sey·side (mûr'zē-sīd'), metropolitan county (1976 est. pop. 1,578,000), NW England, created under the Local Government Act of 1972 (effective 1974). It is subdivided into five metropolitan districts. Merseyside is composed of the county boroughs of Birkenhead, Bootle, Liverpool, Saint Helens, Southport, and Wallasey, and parts of the former counties of Cheshire and Lancashire. It is mainly industrial, with shipbuilding, flour milling, steel manufacturing, oil and sugar refining, and diversified manufacturing.

Mer·sin (mĕr-sēn'), formerly **İ·çel** (ē-chĕl'), city (1975 pop. 152,186), capital of İçel prov., S Turkey, on the Mediterranean Sea. A rail terminus and modern seaport, it exports cotton, chrome, copper, and agricultural produce.

Mer·thyr Tyd·fil (mûr'thər tĭd'vĭl), county borough (1973 est. pop. 53,680), Mid Glamorgan, S Wales, on the Taff River. In the center of the great coal field of Glamorganshire, it has ironworks and steelworks. After World War II, light industries were stressed to revive the economy. Textiles, clothing, and leather goods are also made.

Mer·ton (mûr'tn), borough (1971 pop. 176,524) of Greater London, SE England. Merton was created in 1965 by the merger of the metropolitan London boroughs of Mitcham and Wimbledon and the urban districts of Merton and Morden. The area is largely residential with some industry, including tanning and the manufacture of silk and calico prints, varnish and paint, and toys. An annual fair dating from Elizabethan times is held in Mitcham. Wimbledon is England's tennis headquarters; the first Wimbledon Championship match took place in 1877. Merton has remains of a priory that was founded in 1115 and destroyed by Oliver Cromwell.

Mert·zon (mûrt'sən), town (1970 pop. 513), seat of Irion co., W Texas, SW of San Angelo on a tributary of the Concho River.

Merv (myĕrf), ancient city, Central Asian USSR, in Turkmenistan, in a large oasis of the Kara-Kum desert, on the Murgab River. The city, known in antiquity as Margiana, or Antiochia Margiana, was founded in the 3rd cent. B.C. on the site of an earlier settlement. Its periods of greatness were from A.D. 651 to 821, when it was the seat of the Arab rulers of Khorasan and Transoxania and one of the main centers of Islamic learning, and from 1118 to 1157, when it was the capital of the Seljuk Empire under the last sultan, Sandzhar. The Mongols destroyed the city early in the 13th cent., but it was slowly rebuilt, to be destroyed again by the Bukharans in 1790. The Russians conquered the area in 1884. Several mausoleums, mosques, and castles of the 11th and 12th cent. are preserved.

Me·sa (mā'sə), county (1970 pop. 54,374), 3,301 sq mi (8,549.6 sq km), W Colo., bordering on Utah and drained by the Colorado and Gunnison rivers; formed 1883; co. seat Grand Junction. It is in an extensively irrigated farming area that produces fruit, beans, grain, sugar beets, and potatoes. It has a large cattle industry, and there are rich coal and uranium deposits, as well as oil and gas wells. Its industry includes food processing, petroleum refining, and the manufacture of concrete, iron, metal products, and electronic equipment. Colorado National Monument and parts of Grand Mesa, White River, and Uncompahgre national forests are here.

Mesa, city (1978 est. pop. 112,400), Maricopa co., S central Ariz., in the lush Salt River valley; inc. 1883. Electronic components, fabricated metals, aircraft, and machine tools are among the manufactures of this city, which almost doubled in population between 1960 and 1970.

Me·sa·bi (mə-sä'bē), range of low hills, NE Minn., famous for its extensive iron ore deposits. The ores are found in a belt c.110 mi (175 km) long and from 1 to 3 mi (1.6–4.8 km) wide between Babbitt and Grand Rapids; they occur in horizontal layers (up to 500 ft/152.5 m thick) near the surface and are mined by the open-pit method. Reserves of high-grade hematite iron are nearly exhausted, and lower-grade taconite deposits are now being worked. The taconite contains mostly chert and magnetite (an iron-bearing mineral) and must undergo a costly and complex beneficiation process before being shipped in the form of pellets containing c.60% iron. The Mesabi iron-ore deposits were first discovered in 1887 by Leonidas Merritt and his brothers, who organized the Mountain Iron Company in 1890 to mine the ore; John D. Rockefeller gained control of the company in the Panic of 1893.

Mesa Verde National Park (vûrd'), 52,074 acres (21,090 hectares), SW Colo.; est. 1906. It includes the most notable and best-preserved cliff dwellings and relics in the United States, covering four archaeological periods.

Me·se·wa (mə-sä'wə), *see* Massawa, Ethiopia.

Me·sil·la (mā-sē'yə), town (1970 pop. 1,713), Dona Ana co., SW N.Mex., on the Rio Grande and near Las Cruces; settled c.1850. The whole Mesilla Valley became part of the United States under the Gadsden Purchase (1853). During the Civil War it changed hands several times. From July, 1861, to Aug., 1862, it was headquarters for Col. John R. Baylor of the Confederate army, who proclaimed Mesilla the capital of the new Confederate territory. A museum commemorates Billy the Kid, who once stood trial here.

Me·so·lón·gi·on (mĕs'ə-lông'gē-ôn') or **Mis·so·lon·ghi** (mĭs'ə-lông'gē), town (1971 pop. 11,614), capital of Aetolia and Acarnania prefecture, W central Greece, a port on the Gulf of Pátrai. It trades in fish, wine, and tobacco. Mesolóngion was a major stronghold of the Greek insurgents in the Greek War of Independence. Its inhabitants successfully resisted a siege by forces of the Ottoman Empire in 1822–23 and held out heroically against a second siege from 1825 to 1826, when the Ottoman forces captured the town. Lord Byron, the English poet who supported the Greek insurgents, died here in 1824.

Mes·o·po·ta·mi·a (mĕs'ə-pə-tā'mē-ə), ancient country of Asia, the region about the Tigris and Euphrates rivers, included in modern Iraq. The region extends from the Persian Gulf north to the mountains of Armenia and from the Zagros and Kurdish mts. on the east to the Syrian Desert. From the mountainous north, Mesopotamia slopes down through grassy steppes to a central alluvial plain, which was once rendered exceedingly fertile by a network of canals. The south was long thought to be the cradle of civilization until earlier settlements (which probably date from about 5000 B.C.) were found in northern Mesopotamia. Tell Halaf, the most advanced of these early cultures, is famous for Halaf ware, the finest prehistoric pottery in Mesopotamia. The Al Ubaid culture that followed flourished in both northern and southern Mesopotamia.

During the next period the transformation of the village culture into an urban civilization took place. Erech (modern Warka), the foremost site at the beginning of this period, has yielded such monumental architecture as the temple of Inanna and the ziggurat of Anu. Also found at Erech were tablets including the earliest pictographic writing. The early dynastic phase that followed saw the development of city-states all over the Middle East. The Sumerians, the inhabitants of the city states of southern Mesopotamia, were unified at Nippur, where they gathered together to worship Enlil, the wind god. The famous first dynasty of Ur came at the end of the early dynastic period. Sargon founded (c.2340) the Akkadian dynasty, the first empire in Mesopotamia, whose example of empire building was later followed by the old Babylonian dynasty and late Assyrian Empire.

There was also a great cultural exchange between the Mesopotamians and the Elamites (and other Iranians), who for centuries had threatened each other. Mesopotamia still had prestige at the time of Alexander the Great. The Arabs took it from the Byzantine Empire, and it rose to great prominence after Baghdad was made (A.D. 762) the capital of the Abbasid caliphate. This glory was destroyed when the Mongols under Hulagu Khan devastated the area in 1258, destroying the ancient irrigation system. In the centuries following, Mesopotamia never regained its former prominence. In World War I, however, it was an important battlefield.

Mes·se·ne (mə-sē'nē), ancient city, central Messenia, Greece. It was founded (c.369 B.C.) under Theban auspices to be a capital and fort for the Messenians, whom the Battle of Leuctra had just freed from the Spartans.

Mes·se·ni·a (mə-sē'nē-ə, -sēn'yə), ancient region of SW Greece, in the Peloponnesus. Excavation has revealed an important center of Mycenaean culture at Pylos dating from the 13th cent. B.C. From the 8th cent. B.C. the Messenians were engaged in a series of revolts against expanding Sparta. After the First Messenian War the Spartans annexed (c.700 B.C.) the eastern part of Messenia. With the Second Messenian War the remaining inhabitants were reduced (7th cent. B.C.) to helots. The Third Messenian War (464–459 B.C.) was a failure for Messenia, but very costly to Sparta. The Battle of leuctra (371 B.C.) freed Messenia. The region gave its name to Messina, Sicily, because of an influx of Messenian colonists (c.490 B.C.).

Mes·si·na (mə-sē'nə, mās-sē'nä), city (1976 est. pop. 265,318), capital of Messina prov., NE Sicily, Italy, on the Strait of Messina, opposite the Italian mainland. It is a busy seaport and a commercial and industrial center. Manufactures include processed food, chemicals, and pharmaceuticals. Founded (late 8th cent. B.C.) by Greek colonists, the city was captured (5th cent. B.C.) by Anaxilas of Rhegium. It became involved in several wars, particularly against Syracuse and Carthage, and was taken in 282 B.C. by mercenaries called Mamertines. The Romans answered an appeal for help from the Mamertines and intervened in Sicily, thus precipitating the first of the Punic Wars. Messina was subsequently allied with Rome.

The city was conquered by the Arabs in the late 9th cent. A.D. but was liberated by the Normans in 1061. It developed a thriving silk industry (which declined in the 18th cent.). Messina later came under the rule of the Angevins, the Aragonese, and the Spanish Bourbons. A heroic insurrection against the Bourbons took place from 1774 to 1778. Garibaldi took Messina in July, 1860, but the Bourbon garrison resisted in the citadel until March, 1861. The city suffered a severe plague in 1743 and major earthquakes in 1783 and 1908. The earthquake of Dec. 28, 1908, destroyed 90% of Messina's buildings, including fine churches and palaces, and cost about 80,000 lives; afterward the city was completely rebuilt in conformity with standards for quake-resistant construction. In World War II, the Sicilian campaign ended with the fall of Messina to the Allies on Aug. 17, 1943.

Messina, Strait of, channel, c.20 mi (32 km) long and from 2 to 10 mi (3.2-16.1 km) wide, separating the Italian peninsula from Sicily and connecting the Ionian and Tyrrhenian seas. The currents, whirlpools, and winds of the strait, which still hamper navigation, gave rise in ancient times to many legends about its dangers to navigators.

Me·ta (mā'tə), river, c.650 mi (1,045 km) long, Colombia, rising in headstreams in the Cordillera Oriental and flowing generally NE, forming part of the Colombia-Venezuela border before joining the Orinoco at Puerto Carreño.

Met·a·pon·tum (mĕt'ə-pŏn'təm), ancient city of Magna Graecia, on the Gulf of Taranto, SE Italy. Settled by Greeks, c.7th cent. B.C., it flourished and gave refuge to Pythagoreans expelled from Crotona. Pythagoras taught and died here.

Met·calfe (mĕt′kăf′), county (1970 pop. 8,177), 296 sq mi (766.6 sq km), S Ky., drained by several creeks; formed 1860; co. seat Edmonton. In a rolling agricultural area that produces livestock, corn, wheat, and burley tobacco, it has lumbering, some mining, clothing manufacturing, and other industry.

Me·thu·en (mə-thōō′ən), town (1978 est. pop. 35,400), Essex co., NE Mass., a suburb of Boston; settled c.1642, set off from Haverhill 1725.

Me·trop·o·lis (mə-trŏp′ə-lĭs), city (1970 pop. 6,940), seat of Massac co., S Ill., on the Ohio River NW of Paducah, Ky., in a farm and timber region; laid out 1839, inc. 1845. It processes uranium and fluorine. In nearby Fort Massac State Park is the site of Fort Ascension (later Fort Massiac). The fort, established by the French in 1757, was later held by the English and then by the Americans.

Met·ter (mĕt′ər), city (1970 pop. 2,912), seat of Candler co., E Ga., near the Canoochee River WNW of Savannah. It is a tobacco market and a processing center for a timber and farm area.

Me·tuch·en (mĭ-tŭch′ən), borough (1978 est. pop. 13,900), Middlesex co., NE N.J.; settled before 1700, inc. 1900. Although chiefly residential, it has manufactures of electronic and electrical equipment and cleaning compounds. In June, 1777, a brief but bloody skirmish occurred here between British troops and a small American force.

Metz (mĕts, mĕs), city (1975 pop. 111,869), capital of Moselle dept., NE France, on the Moselle River. It is a cultural and commercial center of Lorraine and an industrial city producing metals, machinery, tobacco, clothing, and food products. Of pre-Roman origin, the city was the capital of the Mediomatrici, a Gallic people. One of the most important cities of Roman Gaul, it was invaded and destroyed by the Vandals (406) and the Huns (451). Metz was an early episcopal see and became the capital of Austrasia (the eastern portion of the Merovingian Frankish empire) in the 6th cent. After the division of the Frankish empire (8th cent.) the bishops of Metz greatly increased their power, ruling a relatively vast area as a fief of the Holy Roman Empire. Metz was a major cultural center of the Carolingian Renaissance (8th cent.) and was later (10th cent.) a prosperous commercial city with an important Jewish community. Metz became a free imperial city in the 12th cent. and was then one of the richest and most populous cities of the empire.

During the Reformation the bourgeoisie of Metz welcomed Protestantism, but the city never became a bastion of Calvinism, and the uneasy bourgeoisie accepted the protection of the French crown. In 1552 Henry II annexed the three bishoprics of Lorraine (Metz, Toul, and Verdun), and soon after, Metz, under the command of François de Guise, resisted a long siege (1552–53) by Emperor Charles V. The Peace of Westphalia (1648), ending the Thirty Years War, confirmed the three bishoprics in French possession. An important fortress and garrison town, Metz was besieged (1870) by the Germans in the Franco-Prussian War, and after a two-month siege, 179,000 French soldiers capitulated. During the German annexation of eastern Lorraine (1871–1918), Metz, largely French-speaking, was a center of pro-French sentiment. During World War II the city suffered greatly under German occupation. There are many Gallo-Roman ruins in Metz, including an aqueduct, thermal baths, and part of an amphitheater. Much has also been preserved from the medieval period. The celebrated Cathedral of St. Étienne was built from c.1221 to 1516. The Place Sainte-Croix is a square surrounded by medieval houses (13th–15th cent.).

Meu·don (mœ-dôɴ′), town (1975 pop. 52,806), Hauts-de-Seine dept., N central France, a suburb SW of Paris. Metal products, automobile bodies, and explosives are the chief manufactures. The astrophysics department of the Paris Observatory is located in the pavilion of an 18th cent. château, which commands a magnificent view of Paris. Richard Wagner and Auguste Rodin lived in Meudon.

Meurthe (mœrt), river, c.105 mi (170 km) long, rising in the Vosges, NE France, and flowing NW past Lunéville to join the Moselle River just N of Nancy. Its very irregular level has necessitated an intricate system of controls.

Meurthe-et-Mo·selle (mœrt′ā-mō-zĕl′), department (1975 pop. 722,587), 2,021 sq mi (5,234.4 sq km), NE France, in Lorraine, bordering on Belgium and Luxembourg; capital Nancy.

Meuse (myōōz, mœz), department (1975 pop. 203,904), 2,042 sq mi (5,288.8 sq km), NE France, in Lorraine, bordering on Belgium; capital Bar-le-Duc.

Meuse, river, c.560 mi (900 km) long, rising in the Langres Plateau, NE France and flowing N past Sedan (the head of navigation) and Charleville-Mézières into S Belgium. From Namur the Meuse winds eastward skirting the Ardennes, passes Liège, and turns north, where it forms part of the Belgian-Dutch border before swinging westward through southeastern Netherlands (where it is called the Maas). Near 's Hertogenbosch it branches out to form a common delta with the Rhine River. The Meuse is linked with the Belgian port of Antwerp by the Albert Canal and with Rotterdam and other Dutch ports by the intricate system of Dutch waterways; it is thus one of the chief thoroughfares of Europe. The Belgian section of the Meuse valley, especially around Namur and Liège, is an important industrial and mining region. A strategic line of defense, particularly in Belgium and France, the valley has been a battleground in many wars, and most of the cities along its course have been strongly fortified since the Middle Ages.

Me·war (mĕ-wär′): *see* Udaipur, India.

Mex·i·cal·i (mĕk′sə-kăl′ē, mā-hē-kä′lē), city (1974 est. pop. 317,200), capital of Baja California state, NW Mexico, across the border from Calexico, Calif.

Mex·i·co (mĕk′sə-kō′), officially United States of Mexico, republic (1970 pop. 48,377,363), 761,600 sq mi (1,972,544 sq km), S North America, bordering on the United States in the N, on the Gulf of Mexico (including its arm, the Bay of Campeche) and the Caribbean Sea in the E, on Belize and Guatemala in the SE, and on the Pacific Ocean in the S and W; capital Mexico City.

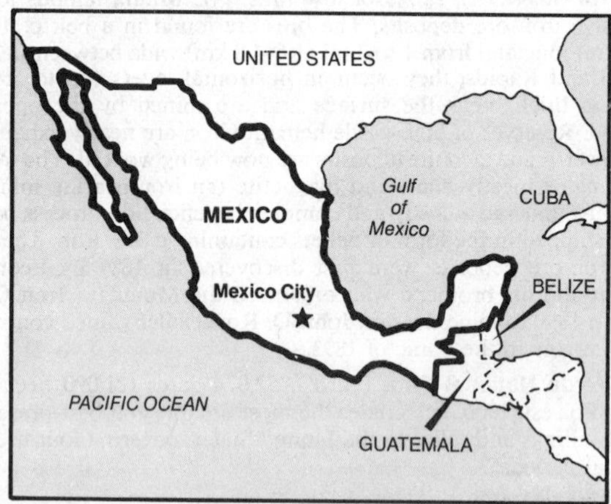

Most of Mexico is highland or mountainous and only about 15% of the land is arable; about 20% of the country is forested. Most of the Yucatán peninsula and the Isthmus of Tehuantepec in the southeast is lowland, and there are lowlying strips of land along the Gulf of Mexico, the Pacific Ocean, and the Gulf of California (which separates the Baja, or Lower, California peninsula from the rest of the country). The heart of Mexico is made up of the Mexican Plateau (c.700 mi/1,125 km long and c.4,000–8,000 ft/1,220–2,440 m high), which is broken by mountain ranges and segmented by deep rifts. The plateau is fringed by two mountain ranges, the Sierra Madre Oriental (in the east) and the Sierra Madre Occidental (in the west), which converge just south of the plateau. Within the plateau are drainage basins, which have no outlet to the sea and which contain some of the country's major cities. The Laguna District, one of the drainage basins, was (1936) the scene of a major experiment in land reapportionment. In the north the plateau is arid except for irrigated areas and is used principally for raising livestock. In the south the deserts yield to the broad, shallow lakes of a region, comprising the Valley of Mexico, known as the Anáhuac and famous for its rich cultural heritage. South of the Anáhuac is a chain of extinct volcanoes. To the south are jumbled masses of mountains and the Sierra Madre del Sur. The climate of the country varies with the altitude, so that there are hot, temperate, and cool regions.

Mexico is divided into 31 states and the Federal District:

NAME	CAPITAL	NAME	CAPITAL
Aguascalientes	Aguascalientes	Baja California	Mexicali

NAME	CAPITAL	NAME	CAPITAL
Baja California Sur	La Paz	Nayarit	Tepic
Campeche	Campeche	Nuevo León	Monterrey
Chiapas	San Angrés Tuxtla	Oaxaca	Oaxaca
		Puebla	Puebla
Chihuahua	Chihuahua	Querétaro	Querétaro
Coahuila	Saltillo	Quintana Roo	Chetunal
Colima	Colima	San Luis Potosí	San Luis Potosí
Durango	Durango	Sinaloa	Culiacán
Federal District	Mexico City	Sonora	Hermosillo
Guanajuato	Guanajuato	Tabasco	Villahermosa
Guerrero	Chilpancingo	Tamaulipas	Ciudad Victoria
Hidalgo	Pachuca	Tlaxcala	Tlaxcala
Jalisco	Guadalajara	Veracruz	Jalapa
México	Toluca	Yucatán	Mérida
Michoacán	Morelia	Zacatecas	Zacatecas
Morelos	Cuernavaca		

Economy. Since 1945, Mexico has enjoyed considerable economic growth, especially of its industrial plant. By 1970 manufacturing and commerce each contributed about 26% of the annual national product and agriculture (including forestry and fishing) about 16%. The Mexican government plays a major role in the economy and owns and operates basic industries and means of transport. About half of the country's workers (including those largely outside the money economy) are engaged in farming, which is still largely run by inefficient, outmoded methods. Because rainfall is not adequate outside the coastal regions, agriculture depends largely on irrigation, which began on a large scale only in the 1940s. The leading crops raised are maize, wheat, sugar cane, beans, citrus fruits, tomatoes, plantains, cotton, rice, coffee, sisal, and chili peppers. Maguey is widely grown and is processed into the alcoholic beverages pulque and mescal. Mexico has considerable mineral resources. The great majority of operating mines are foreign-owned, but laws passed in the early 1960s provide for a gradual transition to Mexican control. The chief minerals produced are antimony, barite, copper, graphite, gypsum, and lead. Large deposits of petroleum were discovered in the 1970s; Mexico's reserves are estimated to be as vast as Iran's and Kuwait's. The leading manufactures include iron and steel, motor vehicles, processed food, cement, refined petroleum and petrochemicals, chemical fertilizers, rubber products, forest products, textiles, and aluminum. Mexico is also known for its handicrafts, especially pottery, woven goods, and silverwork. There is a large tourist industry.

History. Before the arrival of the Spanish conquistadores in the early 16th cent., great Indian civilizations developed and flourished in Mexico. The Maya, the Aztec, the Toltec, the Mixtec, the Zapotec, and the Olmec all left behind impressive remains of great interest. The first European to visit Mexico was Francisco Fernández de Córdoba in 1517. The conquest was begun from Cuba in 1519 by Hernán Cortés, who managed to conquer the Aztec capital, Tenochtitlán; to capture Montezuma, the Aztec ruler, and to bring down his empire. In 1528 the first audiencia (royal court) was set up The territory was constituted the viceroyalty of New Spain in 1535, and the process of Christianizing and Europeanizing the Indians went forward. Nevertheless, the Spanish remained a small minority and establishing control over the vast Indian population remained an arduous task, interrupted by serious revolts like the Mixtón War (1541). The population developed slowly into three groups—whites, Indians, and mestizos (mixed white and Indian). The groups did not coalesce easily, despite the efforts of able viceroys like Luis de Velasco (both father and son) and the younger conde de Revilla Gigedo. The efforts of the church resulted in at least nominal conversion of the Indians to Roman Catholicism. However, concentration of land and political power in the church's hands did nothing to close the gap in status between the wealthy, almost exclusively Spanish landowning class and the depressed laboring class on the land, in the mines, and in the small factories. The growth of an underprivileged mestizo class and the antagonism between those whites born in Spain (*gachupines*) and those born in America (*criollos,* or creoles) added to the stress. The mercantilist system, under which manufacturing was largely forbidden in New Spain, drained the wealth of the country to Spain. Lesser officials often were corrupt. At the same time, new territory was conquered. In the 16th cent. California was explored, but it was not until the middle and late 18th cent. that northeast Mexico and Texas were occupied in any large degree. Many of the administrative evils were ended by the reforms (especially that of 1786) of José de Gálvez, but discontent with Spanish rule continued to grow among the creoles.

The establishment of the United States and the ideas of the French Revolution had considerable influence on Mexicans. The occupation (1808) of Spain by Napoleon I, who placed his brother Joseph Bonaparte on the Spanish throne, opened the way for a revolt in Mexico. The priest Hidalgo y Costilla began the rebellion by issuing (Sept. 16, 1810) a revolutionary tract calling for racial equality and the redistribution of land. Armies, made up mostly of lower-class mestizos and Indians and shunned by the creoles, sprang up. Hidalgo was at first successful, but lost (1811) the decisive battle of Calderón Bridge. By 1815, the other commanders had been defeated or dispersed. When the liberals came to power in Spain in 1820, the more conservative elements in Mexico (primarily the higher clergy and the creoles) sought independence as a means of maintaining the status quo. Spain accepted Mexican independence in Sept., 1821, and a short-lived empire with the royalist general Augustin de Iturbide at its head was established (1822). In 1823 the republican leaders Santa Anna and Guadalupe Victoria drove out Iturbide and a republic was set up with Guadalupe Victoria as its first president. There was a frequent turnover of governments, and the national budget usually ran a deficit. In 1832 the ambitious Santa Anna, who had a great influence over Mexican politics until 1855, became president. Waste, corruption, and inefficiency were rampant at the time. The war with Texas led to an all-out war with the United States, the Mexican War (1846–48), which was ended by the Treaty of Guadalupe Hidalgo, by which Mexico lost a large block of territory. After the war, Santa Anna returned to power as "perpetual dictator," but he was overthrown in 1855. A group of reform-minded leaders, especially Benito Juárez—drafted the liberal constitution of 1857, which secularized church property and reduced the privileges of the army. Conservative opposition was bitter, and civil war ensued; the liberals were victorious in the War of Reform (1858–61). The conservatives then sought foreign aid and received it from Napoleon III of France. French intervention followed and led to a brief and ill-starred interlude of empire (1864–67) under Maximilian, a Hapsburg prince. With the end of French aid the empire collapsed and Juárez again ruled Mexico, but political disturbances prevented the accomplishment of his reform program. Porfirio Díaz led a successful armed revolt in 1876 and, except for the period from 1880 to 1884, firmly held the reins of power as president until 1911. It was a period of considerable economic growth, but social inequality was increased by the favoritism shown the great landowners and foreign investors; the Indians sank deeper into peonage.

In Nov., 1910, an idealistic liberal leader, Francisco I. Madero, began an armed revolt against Díaz, who had gone back on his word not to seek reelection. Madero was quickly successful, and in May, 1911, Díaz resigned and went into exile. Madero was elected president in Nov., 1911. Well-meaning but ineffectual, he was attacked by conservatives and revolutionaries alike. In Feb., 1913, Madero was overthrown by his general, Victoriano Huerta, and was murdered. President Huerta's regime was dictatorial and repressive, and revolts soon broke out under the leadership of Venustiano Carranza, Francisco "Pancho" Villa, and Emiliano Zapata. In 1914, Huerta resigned, partly because of U.S. military intervention, and Carranza became president. Civil war broke out again, but by the end of 1915 Carranza had established control over the country, although Villa and Zapata maintained opposition bands for a number of years. In 1916, Villa led a raid into the United States, which resulted in an unsuccessful U.S. expedition into Mexico. Carranza sponsored the constitution of 1917, which was similar to the 1857 constitution, but which in addition provided for the nationalization of mineral resources, for the restoration of communal lands to the Indians, for the separation of church and state, and for educational, agrarian, and labor reforms. However, most provisions of the constitution were not implemented, and in 1920 Carranza was deposed by General Álvaro Obregón, his former military chief, who was subsequently elected president. Obregón was succeeded by Plutarco Elías Calles in 1924, who continued the agrarian and educational programs, but who became embroiled in serious controversies with the United States over rights to petroleum and with the church over the separation of church and state. In some regions Catholic militants were in open revolt, and in the country as a whole from 1926 to 1929 church schools were closed and no church services were held. Both controversies subsided, partly because of the intervention of the U.S. ambassador, Dwight Morrow. Re-elected in 1928, Obregón was assassinated before taking office. Calles remained the most powerful person in Mexico during the next six years. In 1929 he organized the National Revolutionary party (in 1946 renamed the Institutional Revolutionary party), the chief political party in Mexico. Calles's hegemony ended, however, with the inauguration (1934) of Lázaro Cárdenas. Vigorous and idealistic, Cárdenas instituted reforms to improve the lot of the underprivileged and to make the Indian an organic part of the state. He redistributed much land and supported the Mexican labor movement. Railroads were nationalized, and foreign holdings, particularly in petroleum fields, were expropriated with compensation. Educational opportunities were increased and illiteracy reduced, medical facilities were extended, transport and communications were improved, and plans were drawn up for land reclamation and for hydroelectric and industrial projects. A settlement with the church was reached. The pace of reform slowed under Manuel Avila Camacho, who became president in 1940. Relations with the United States improved. In World War II, Mexico declared war (1942) on the Axis powers; it made substantial contributions to the Allied cause and also received considerable U.S. economic aid.

Since World War II, Mexico has enjoyed considerable economic development, but most of the benefits have accrued to the middle and upper classes. Under President Miguel Alemán (1946-52) vast irrigation projects and hydroelectric plants were constructed, and industrialization advanced rapidly. Under the moderate presidents, the government continued to play a dominant role in national affairs, and attempts were made to improve the conditions of the lower classes. The tax structure was reformed somewhat, some large estates were confiscated and the land redistributed, and educational opportunities in rural areas were increased. In foreign affairs, Mexico maintained friendly relations with the United States, ratifying treaties settling long-standing border disputes. Unlike most other American nations, Mexico maintained diplomatic relations with revolutionary Cuba, but it supported the United States during the Cuban missile crisis (1962). In 1970 Luis Echeverría Álvarez became president. He took significant steps toward reforming the government. Mexico's support of a UN resolution condemning Zionism as racism created some strain in relations with the United States. In the first years of the Echeverría administration, guerrilla activity increased. In 1976, Echeverría was succeeded by José López Portillo, who was expected to pursue a more moderate version of his predecessor's policies.

Government. Under the constitution of 1917 as amended, Mexico is a federal republic whose chief executive and head of state is the president, directly elected to a nonrenewable six-year term and assisted by a cabinet. The bicameral legislature is made up of the Senate, comprising 64 members directly elected to six-year terms, and the Chamber of Deputies, consisting of 219 members serving three-year terms (194 of the deputies are directly elected, and 25 are chosen by a system of proportional representation).

Mexico, city (1974 est. pop. 8,299,200), central Mexico, capital and largest city of Mexico, near the S end of the plateau of Anáhuac, at an altitude of c.7,800 ft (2,380 m). The horizons of the city are almost obscured by mountain barriers, and the peaks of Popocatépetl and Ixtacihuatl are nearby. The climate is cool, dry, and healthful. Much of the surrounding valley is a lake basin with no outlet, and in the past during the rainy seasons mountain floods swelled the lakes.

Drainage and artesian wells have lowered the water table so that the surface crust, formerly supported by subsoil water, can no longer sustain the heavier buildings of the city, which are sinking some 4 to 12 in. (10.2-30.5 cm) a year. Some of Mexico's finest buildings have been damaged, among them the old cathedral (begun in 1553 on the site of an Aztec temple) and the Palace of Fine Arts. Nevertheless, many monuments of Spanish colonial architecture remain. The cathedral and the National Palace are on the great central square, or Plaza de la Constitución, where the streets of the old town crisscross in a rough gridiron. From the Plaza the great avenues span out to the far sections of the capital. Many colonial churches are to be found, notably on the Paseo de la Reforma, which cuts across the city to Chapultepec. Public buildings of the 19th cent. have a ponderous grandeur that shows French influence, but the recent edifices are starkly modern. The National Univ. of Mexico, founded in the 16th cent., is now housed in University City (opened 1952), built on a lava outcrop in the outskirts.

The city has been the metropolis of Mexico since New Spain was created. It was taken in 1847 by Winfield Scott's American army in the Mexican War. The French army captured Mexico City in 1863, and Emperor Maximilian, crowned in 1864, did much to beautify it before it was recaptured by Mexicans under Benito Juárez. In the years of revolution after 1910 it was a magnet for divergent insurrectionary forces. Perhaps the most spectacular incidents were the occupations (1914-15) by Francisco Villa and Emiliano Zapata. Today Mexico City forms the core of the Federal Dist. and is the commercial, industrial, financial, political, and cultural center of the nation. Among its important manufactures are iron and steel, petroleum, food products, textiles, glassware, machinery, chemicals, and consumer items. The city, with its rich local color and extraordinary cultural attractions, has become a focal point for tourists. The Olympics were held in Mexico City in 1968.

Mexico, city (1978 est. pop. 12,100), seat of Audrain co., central Mo., in a farm area; inc. 1857. Firebrick and shoes are manufactured, and there are livestock markets and horse stables in the city.

Mexico, Gulf of, arm of the Atlantic Ocean, c.700,000 sq mi (1,813,000 sq km), SE North America. The Gulf stretches more than 1,100 mi (1,770 km) from west to east and c.800 mi (1,290 km) from north to south. It is bordered by the southeast coast of the United States from Fla. to Texas, and the east coast of Mexico from Tamaulipas to Yucatán. At the entrance of the Gulf is the island of Cuba. On the northern side of Cuba the Gulf is connected with the Atlantic Ocean by the Straits of Florida (through which the Gulf Stream passes); on the southern side of Cuba it is connected with the Caribbean Sea by the Yucatán Channel. The Bay of Campeche and Apalachee Bay are the Gulf's largest arms. Sigsbee Deep (12,714 ft/3,877.8 m), the deepest part of the Gulf, lies off the Mexican coast. The shoreline is low, sandy, and marshy, with many lagoons and deltas. The U.S. Intracoastal Waterway follows the Gulf's northern coast. Oil deposits from the continental shelf are tapped by offshore wells.

Mi·am·i (mī-ăm'ē, -ăm'ə). **1.** County (1970 pop. 39,246), 377 sq mi (976.4 sq km), N central Ind., intersected by the Wabash, Mississinewa, and Eel rivers; formed 1832; co. seat Peru. It is in an agricultural area that produces grain, fruit, livestock, poultry, and dairy products. Its manufactures include natural and processed cheese, lumber and wood products, furniture, chemicals, plastics, and farm machinery. **2.** County (1970 pop. 19,254), 592 sq mi (1,533.3 sq km), E Kansas, in a rolling plain region bordering in the E on Mo. and drained by the Marais des Cygnes River; formed 1855; co. seat Paola. It has livestock and grain farming, oil and gas fields, plants and manufacturing clothing and metal products. **3.** County (1970 pop. 84,342), 407 sq mi (1,054.1 sq km), W Ohio, intersected by the Great Miami and Stillwater rivers; formed 1807; co. seat Troy. It has agriculture (mainly corn, soybeans, and livestock) and highly diversified manufacturing, including clothing, transportation equipment, and metal, wood, paper, and meat products.

Miami. **1.** City (1978 est. pop. 348,000), seat of Dade co., SE Fla., on Biscayne Bay at the mouth of the Miami River; inc. 1896. A port of entry, it is also one of the most popular and famous resorts of the United States. Tourism is its major industry, and there are extensive recreational facilities for fishing, swimming, golf, yachting, and horse and dog racing, with many related businesses and enterprises. The city is also the processing and shipping hub of a large agricultural region and a center for rebuilding and repairing aircraft. Its manufactures include aluminum products, clothing, furniture, transportation equipment, machinery, stone, clay, glass, lumber, and wood products. Other industries are printing and publishing, fishing, and shellfishing. The first settlement was made here in the 1870s near the site of Fort Dallas, built in 1836 during the Seminole War. In 1895 Henry M. Flagler became interested in the area. He made Miami a railroad terminus in 1896, dredged the harbor, and began the development of a recreation center. The city received its greatest impetus during the Florida land boom of the mid-1920s. Biscayne Boulevard, with its park and causeways spanning Biscayne Bay and leading to Miami Beach, is well known, as is the Dade County Art Museum. The Orange Bowl is the home of the city's major league football team, the Miami Dolphins, and site of a traditional New Year's festival and college bowl game. **2.** City (1978 est. pop. 13,700), seat of Ottawa co., extreme NE Okla., in the foothills of the Ozarks on the headwaters of Grand Lake, which provides both electric power and recreation. It is a trade, shipping, and marketing center for a tri-state region where lead and zinc are mined. **3.** City (1970 pop. 611), seat of Roberts co., extreme N Texas, in the high plains of the Panhandle, NE of Pampa. It is a market and shipping point for a livestock, grain, and cotton region.

Miami or **Great Miami,** river, c.160 mi (260 km) long, formed in W Ohio near Indian Lake and flowing generally SW past Dayton to the Ohio River at the Ind. line. The Miami River system has large-scale flood-control projects. The Miami and Erie Canal (c.240 mi/385 km long; opened in the 1830s) linked the upper Miami River with Lake Erie and was the principal transportation route of western Ohio until the 1850s. The Little Miami River (95 mi/152.9 km long) to the east and generally parallel, rises southeast of Springfield and enters the Ohio River at Cincinnati.

Miami Beach, city (1978 est. pop. 90,800), Dade co., SE Fla., on an island between Biscayne Bay and the Atlantic Ocean; inc. 1915. It is connected to Miami by four causeways. Miami Beach is a popular year-round resort, world famous for its "gold coast" hotel strip (with over 350 hotels and motels), palatial estates, and recreational facilities. The city's chief source of income is from tourism.

Mi·an·wa·li (mē-än'wä-lē), town (1972 pop. 48,370), N Pakistan, on the Indus River. It is the market for a district that produces food grains, oilseed, hides, and wool.

Mich·i·gan (mĭsh'ĭ-gən), state (1975 est. pop. 9,135,000), 58,216 sq mi (150,779 sq km), not including c.38,575 sq mi (99,909 sq km) of the Great Lakes, N United States, admitted to the Union in 1837 as the 26th state. Lansing is the capital. The Lower Peninsula, shaped like a mitten, thrusts northward from Ind. and Ohio. On the east it is separated from Ont., Canada, by Lake Erie and Lake Huron, and by the Detroit and St. Clair rivers, which together link the two Great Lakes; on the west it is separated from Wis. by Lake Michigan. Across Lake Michigan, northeast of Wis., the Upper Peninsula stretches eastward, separating Lake Michigan from Lake Superior, and itself separated from Ontario only by the narrow St. Marys River. The Upper Peninsula is separated from the Lower Peninsula by the Straits of Mackinac; a bridge connecting the two peninsulas was opened in 1957 and has spurred the development of the Upper Peninsula.

The eastern portion of the Upper Peninsula has swampy flats and limestone hills on the Lake Michigan shore, while sandstone ridges rise abruptly from the rough waters of Lake Superior; in the west the land rises to forested mountains, still rich in copper and iron. The whole of the Upper Peninsula is northern woods country. Deer, bears, and other big and small game in the forests, as well as abundant fish in streams and the lakes, make the area a sportsman's delight. The scarred timberlands are being reforested with second growth. The Lower Peninsula is a different sort of country, less wild but in parts no less beautiful. Its forests were also cut over in the lumber boom of the late 19th cent. Today its great wealth lies in the many farms and factories. The surrounding waters temper the climate, allowing a long growing season.

The northern Michigan wilds, numerous inland lakes, and some 3,000 mi (4,800 km) of shoreline, combined with a pleasantly cool summer climate, have long attracted thousands of vacationers. In the winter Michigan's snow-covered hills bring skiers from all over the Midwest. Places of interest in the state include Greenfield Village, a re-creation of a 19th cent. American village, and the Henry Ford Museum, both at Dearborn; Pictured Rocks and Sleeping Bear Dunes national lake shores; and Isle Royal National Park.

Michigan is divided into 83 countries:

NAME	COUNTY SEAT	NAME	COUNTY SEAT
Alcona	Harrisville	Lake	Baldwin
Alger	Munising	Lapeer	Lapeer
Allegan	Allegan	Leelanau	Leland
Alpena	Alpena	Lenawee	Adrian
Antrim	Bellaire	Livingston	Howell
Arenac	Standish	Luce	Newberry
Baraga	L'Anse	Mackinac	St. Ignace
Barry	Hastings	Macomb	Mount Clemens
Bay	Bay City	Manistee	Manistee
Benzie	Beulah	Marquette	Marquette
Berrien	St. Joseph	Mason	Ludington
Branch	Coldwater	Mecosta	Big Rapids
Calhoun	Marshall	Menominee	Menominee
Cass	Cassopolis	Midland	Midland
Charlevoix	Charlevoix	Missaukee	Lake City
Cheboygan	Cheboygan	Monroe	Monroe
Chippewa	Sault Ste. Marie	Montcalm	Stanton
Clare	Harrison	Montmorency	Atlanta
Clinton	St. Johns	Muskegon	Muskegon
Crawford	Grayling	Newaygo	White Cloud
Delta	Escanaba	Oakland	Pontiac
Dickinson	Iron Mountain	Oceana	Hart
Eaton	Charlotte	Ogemaw	West Branch
Emmet	Petoskey	Ontonagon	Ontonagon
Genesee	Flint	Osceola	Reed City
Gladwin	Gladwin	Oscoda	Mio
Gogebic	Bessemer	Otsego	Gaylord
Grand Traverse	Traverse City	Ottawa	Grand Haven
Gratiot	Ithaca	Presque Isle	Rogers City
Hillsdale	Hillsdale	Roscommon	Roscommon
Houghton	Houghton	Saginaw	Saginaw
Huron	Bad Axe	St. Clair	Port Huron
Ingham	Mason	St. Joseph	Centreville
Ionia	Ionia	Sanilac	Sandusky
Iosco	Tawas City	Schoolcraft	Manistique
Iron	Crystal Falls	Shiawassee	Corunna
Isabella	Mount Pleasant	Tuscola	Caro
Jackson	Jackson	Van Buren	Paw Paw
Kalamazoo	Kalamazoo	Washtenaw	Ann Arbor
Kalkaska	Kalkaska	Wayne	Detroit
Kent	Grand Rapids	Wexford	Cadillac
Keweenaw	Eagle River		

Economy. Livestock raising and dairying are of great importance, but crops account for almost half of farm income. Corn is the chief crop, followed by hay, dry beans, wheat, oats, and soybeans. The manufacture of transportation equipment is by far the state's chief industry, and Detroit, Dearborn, Flint, Pontiac, and Lansing are centers for automobile manufacturing. Michigan's other leading industries produce nonelectrical machinery, fabricated metal products, primary metals, chemicals, and food products. Industrial centers include Saginaw, Bay City, Muskegon, and Jackson. Although mining contributes less to income in the state than either agriculture or manufacturing, Michigan in 1972 was the nation's 13th leading state in mineral production. The chief minerals produced are iron ore, cement, copper, and sand and gravel. In 1972 Michigan ranked first among the states in production of peat, bromine, calcium-magnesium chloride, gypsum, and magnesium compounds, and ranked second to Minnesota in iron-ore production and second to California in sand and gravel. Fishing is also important, both commercially and as a resource attracting visitors to the state.

History. The Ojibwa, the Ottawa, the Potawatomi, and other Algonquian-speaking Indian tribes were living in what is now Michigan when the French explorer Étienne Brulé landed at the narrows of Sault Sainte Marie in 1618. Later French explorers, traders, and missionaries came, including Jean Nicolet, Jacques Marquette, and Robert Cavelier, sieur de La Salle. Mackinac Island (in the Straits of Mackinac) became a center of the fur trade. Fort Pontchartrain, later Detroit, was founded in 1701 by Antoine de la Mothe Cadillac. All this vast region was weakly held by France until lost to Great Britain in the French and Indian Wars (1754–63). The Indians of Michigan, who had lived in peace with the French, resented the coming of the British, who were the allies of the hated Iroquois. Under Pontiac the Indians rose in bloody revolt (1763–66) and terrorized the British in their newly acquired territory. Despite provisions of the Treaty of Paris (1783), the British held stubbornly to Detroit and Mackinac until 1796, when they evacuated the area under the terms of Jay's Treaty. After passage of the Northwest Ordinance in 1787, Michigan became part of the Northwest Territory. After the Northwest Territory was broken up, Detroit was made (1805) capital of Michigan Territory.

Michigan remained in British hands through most of the War of

1812 until William Henry Harrison in the Battle of Thames and Oliver Hazard Perry in the Battle of Lake Erie restored U.S. control. After peace came, pioneers moved into Michigan. Treaties were negotiated with the Indians, in which they ceded their lands to the whites. Steamboat navigation on the Great Lakes and sale of public lands in Detroit both began in 1818, and the Erie Canal was opened in 1825. Farmers came to the Michigan fields, and the first sawmills were built along the rivers. The Michigan electorate, impatient for statehood, organized a government without U.S. sanction and in 1836 operated as a state, although legally outside the Union until 1837. Detroit served as the capital until 1847, when it was replaced by Lansing.

After statehood Michigan promptly adopted a program of internal improvement through the building of railroads, roads, and canals, including the Soo Ship Canal at Sault Sainte Marie. A survey of mineral resources made in 1837 prefaced the development of the copper and iron mines. At the same time lumbering was expanding, and the population grew as German, Irish, and Dutch immigrants arrived. In the 1880s farmers joined with workers in the mines and lumber camps to support measures for agrarian improvement and public welfare. Reforms influenced by the labor movement were the creation of a state board of labor (1883), a law enforcing a 10-hr day (1885), and a moderate child-labor law (1887).

With the invention of the automobile, industry in Michigan was altered radically. Henry Ford established the Ford Motor Company in 1903 and introduced conveyor-belt assembly lines in 1918. Along with the development of mass-production methods came the growth of the labor movement. In the 1930s, when the automobile industry was well established in the state, labor unions struggled for recognition. The conflict between labor and the automotive industry, which continued into the 1940s, included sit-down strikes and was sometimes violent. In World War II Michigan produced large numbers of tanks, airplanes, and other war materiel. In the early 1960s and again in the mid-1970s economic growth lagged and unemployment became a problem in the state. Detroit was shaken by severe race riots in 1967 that left 43 persons dead and many injured. Resistance to busing was a major political issue in the state in the early 1970s.

Government. Michigan's constitution, adopted in 1963, provides for an elected governor as the state's chief executive. The governor serves for a term of four years and may succeed himself in the office. The state legislature is made up of a senate and house of representatives. The senate has 38 members elected for terms of four years and the house of representatives has 110 members elected for two-year terms.

Educational Institutions. Higher education is well represented by such institutions as the Univ. of Michigan, at Ann Arbor; Michigan State Univ., at East Lansing; the Univ. of Detroit and Wayne State Univ., at Detroit; and many other private and state colleges.

Michigan, Lake, 22,178 sq mi (57,441 sq km), 307 mi (494 km) long and 30 to 120 mi (48.3–193.1 km) wide, bordered by Mich., Ind., Ill., and Wis.; third largest of the Great Lakes and the largest freshwater lake entirely within the United States. Its surface is 581 ft (177.2 m) above sea level, and the lake is 923 ft (281.5 m) deep. The Straits of Mackinac, its only natural outlet, connect the lake with Lake Huron to the northeast; the Illinois Waterway links Lake Michigan with the Mississippi River and Gulf of Mexico. Many islands are found in the northern part of the lake; the northern shoreline is indented, with Green Bay and Grand Traverse Bay the largest bays. The southern part of Lake Michigan has a regular shoreline necessitating the building of man-made harbors. Sand dunes border the eastern and southern shores of the lake. The forested northern region of Lake Michigan is generally sparsely populated. The southern portion, located near the heart of the Midwest, is one of the most important urban industrial areas in the United States.

Lake Michigan was discovered in 1634 by the French explorer Jean Nicolet and was later explored by the French traders Marquette and Jolliet. French missionary and trade centers thrived here by the late 1600s. As part of the bitterly contested Northwest Territory, the area passed to England in 1763 and then, in 1796, to the United States. The area was isolated until the 1830s, when improvements in transportation brought settlers here.

Michigan City, city (1978 est. pop. 40,800), La Porte co., NW Ind., on Lake Michigan; inc. 1836. A resort area with sand dune beaches and a state park, it also has many industries.

Mi·cho·a·cán (mē′chō-ä-kän′), state (1970 pop. 2,341,556), 23,202 sq mi (60,093 sq km), S Mexico; capital Morelia.

Mi·cro·ne·sia (mī′krō-nē′zhə, -shə), one of the three main divisions of Oceania, in W Pacific Ocean, N of the equator. Micronesia includes the Caroline Islands, Marshall Islands, Marianas Islands, Gilbert Islands, and Nauru.

Mid·del·burg (mĭd′l-bûrg′), city (1977 est. pop. 37,327), capital of Zeeland prov., SW Netherlands, on the former island of Walcheren. It is a trade, manufacturing, and tourist center. Chartered in 1217, Middelburg developed into an important medieval trade center. The last Spanish fortress in Zeeland, it was captured (1574) by the Beggars of the Sea. Although heavily damaged in World War II, Middelburg retains many beautiful old buildings.

Mid·dle·bor·o or **Mid·dle·bor·ough** (both: mĭd′l-bûr′ō), town (1970 pop. 13,607), Plymouth co., SE Mass.; inc. 1669. Cranberry-processing is a major industry in the town. Fire apparatus, brass products, and shoes are also made. The town was destroyed by Indians in King Philip's War and later rebuilt. Points of interest include an Indian site believed to date from 2500 B.C.; restored Revolutionary industries, such as a slitting mill and an iron foundry; and the Tom Thumb historical museum.

Mid·dle·bourne (mĭd′l-bôrn′), town (1970 pop. 814), seat of Tyler co., NW W.Va., NE of Parkersburg, in a farming, gas, and oil region.

Mid·dle·burg (mĭd′l-bûrg′), borough (1970 pop. 1,369); seat of Snyder co., central Pa., SW of Sunbury; settled c.1760, laid out 1800, inc. 1862. It makes textile products.

Mid·dle·bur·y (mĭd′l-bĕr′ē, -bə-rə), town (1970 pop. 6,532), seat of Addison co., W Vt., on both sides of Otter Creek; chartered 1761, first settled 1773, permanently settled 1783.

Mid·dle East (mĭd′l), term applied to the countries of SW Asia and NE Africa lying W of Afghanistan, Pakistan, and India. Thus defined it includes the Asian part of Turkey, Syria, Israel, Jordan, Iraq, Iran, Lebanon, the countries of the Arabian peninsula (Saudi Arabia, Yemen, Southern Yemen, Oman, United Arab Emirates, Qatar, Bahrain, Kuwait), and Egypt and Libya. The term is sometimes used in a cultural sense to mean the group of lands in that part of the world predominantly Islamic in culture, thus including the remaining states of North Africa as well as Afghanistan and Pakistan.

Mid·dles·brough (mĭd′lz-brə), county borough (1976 est. pop. 153,900), administrative center of Cleveland, NE England, on the S bank of the Tees estuary. It grew from a village in the mid-19th cent. after the opening of a railroad and the discovery of iron in the Cleveland Hills nearby. It has huge iron, steel, and chemical plants.

Mid·dle·sex (mĭd′l-sĕks′), former county adjoining London, SE England. In 1965 most of the county was reorganized into the Greater London boroughs of Barnet, Brent, Ealing, Enfield, Haringey, Harrow, Hillingdon, Hounslow, and Richmond-upon-Thames. The remainder became part of the counties of Hertfordshire and Surrey. Middlesex has been important from Roman times, when the Roman road known as Watling Street traversed the district.

Middlesex. 1. County (1970 pop. 115,018), 371 sq mi (960.9 sq km), S Conn., on Long Island Sound, drained by the Connecticut River; constituted 1785; former co. seat Middletown. Its agriculture includes tobacco, potatoes, truck crops, dairy products, fruit, and poultry. Among its many diversified manufactures are tools, hardware, electrical equipment, boats, textiles, clothing, metal and paper products, cosmetics, rubber goods, piano and automobile parts, farm implements, chemicals, asbestos, and cigars. There are also fisheries, sandstone and feldspar quarries, and a growing tourism industry. **2.** County (1970 pop. 1,398,397), 829 sq mi (2,147.1 sq km), NE Mass., bordered on the N by N.H. and drained by the Charles, Concord, Sudbury, and Assabet rivers; formed 1643; co. seat Cambridge. Its rivers furnish water power for industrial towns that produce shoes, textiles, machinery and other metal products, watches, food products, wood products, rubber goods, and agricultural produce. **3.** County (1970 pop. 583,813), 312 sq mi (808.1 sq km), central N.J., bounded on the E by Raritan Bay and Authur Kill and drained by the Raritan, Millstone, and South rivers; formed 1683; co. seat New Brunswick. It is in an industrial, agricultural, and residential area, with lumber and sand and gravel pits. Its agriculture includes truck crops, poultry, dairy products, and fruit. Textiles, clothing, wood, paper, and metal products, furniture, chemicals, concrete, and machinery are manufactured. It has shipyards and dry docks on Raritan Bay. **4.** County (1970 pop. 6,295), 130 sq mi (336.7 sq km), E Va., in the tidewater region, bounded in the N by the Rappahannock River and in the S by Dragon Run and the Piankatank River;

formed 1674; co. seat Saluda. It is in a farming (truck crops, tobacco, corn, and melons) and livestock-raising (cattle, hogs, poultry, and sheep) area. It has fishing and oystering, and its industries include lumbering, seafood processing, and boatbuilding.

Mid·dle·ton (mĭd'l-tən), municipal borough (1973 est. pop. 53,340), Greater Manchester, NW England, on the Irk River. Manufactures include cotton and silk textiles, soap, plastics, and chemicals.

Middletown. 1. Industrial city (1978 est. pop. 38,700), Middlesex co., central Conn., on the W bank of the Connecticut River; settled 1650, inc. 1784, town and city consolidated 1923. Its manufactures include brake linings, marine hardware, rubber footwear, clothing, and textiles. Shipping brought early prosperity to Middletown, and during colonial days it was the state's leading shipping, commercial, and cultural center. It is the seat of Wesleyan Univ. **2.** Township (1975 est. pop. 57,000), Monmouth co., E N.J., NW of Red Bank; settled 1665. The first Baptist church in the state was built here in 1668. Marlpit Hall (c.1684) is kept as a museum. **3.** Industrial city (1978 est. pop. 25,700), Orange co., SE N.Y., on the Walkill River; settled 1756, inc. as a city 1888. It is a farm trade center with railroad shops and foundries. Among its products are clothing and leather goods. **4.** Industrial city (1978 est. pop. 46,600), Butler co., SW Ohio, on the Great Miami River, in a farm area; inc. 1866. Its major industry is steel production. Miami Univ. has a branch here. **5.** Borough (1970 pop. 9,080), Dauphin co., SE Pa., on the Susquehanna River below Harrisburg; laid out 1755, inc. 1829. A farm trade center, it also makes metal products and clothing. **6.** Rural and resort town (1978 est. pop. 17,000), Newport co., SE R.I., on Rhode Island and Narragansett Bay; set off from Newport and inc. 1743. Its name is derived from its location between Newport and Portsmouth. During the Revolution the town was pillaged (1776) by the British.

Middle West or **Mid·west** (mĭd'wĕst'), section of the United States about the Great Lakes and the upper Mississippi valley. As commonly used, the term refers to the states of Ohio, Indiana, Illinois, Michigan, Wisconsin, Minnesota, Iowa, Missouri, Kansas, and Nebraska. The area has some of the richest farming land in the world and is known for its corn and hogs. Parts of the Middle West also have an enormous amount of industry.

Mid Gla·mor·gan (mĭd glə-môr'gən), nonmetropolitan county (1976 est. pop. 540,000), S Wales, created under the Local Government Act of 1972 (effective 1974); administrative center Cardiff. It comprises the county borough of Merthyr Tydfil, and parts of the former counties of Glamorganshire, Breconshire, and Monmouthshire.

Mid·land (mĭd'lənd), town (1976 pop. 11,568), S Ont., Canada, on Georgian Bay, NW of Toronto. Midland is a port, and has grain elevators and plants that manufacture textiles, cameras, optical goods, and other products. The Martyrs' Shrine, commemorating the deaths of five Jesuit priests who were among the eight North American martyrs canonized in 1930, and other remembrances of the early colonial period are nearby.

Midland. 1. County (1970 pop. 63,769), 520 sq mi (1,346.8 sq km), central Mich., drained by the Tittabawassee, Pine, and Chippewa rivers; organized 1855; co. seat Midland. There is some agriculture (livestock, poultry, grain, beans, and dairy products). Oil and gas wells, salt deposits, and coal mines are found here. **2.** County (1970 pop. 65,433), 938 sq mi (2,429.4 sq km), W Texas, on the S Llano Estacado, drained by tributaries of the Colorado River; formed 1885; co. seat Midland. In an oil-producing region, it also has cattle ranches, grain, cotton, corn, and dairy farms, and extensive deposits of potash.

Midland. 1. City (1978 est. pop. 37,700), seat of Midland co., central Mich., in the Saginaw valley at the confluence of the Tittabawassee and Chippewa rivers; inc. 1887. Midland owes its development after 1890 to the Dow Chemical Company, whose corporate headquarters are here. Silicone products, chemicals, cans, magnesium, and plastics are among its manufactures. Oil, coal, and salt are found in the area. **2.** City (1978 est. pop. 66,500), seat of Midland co., W Texas, on the S border of the Llano Estacado; inc. 1906. Midland has prospered partly because of its cattle ranches, but the city's reputation for spectacular wealth and its great spurt in population after 1940 resulted from the location of many oil-company offices here. Clothing, mobile homes, aircraft, fabricated steel, plastic products, and tools for mining and drilling are manufactured.

Midland Canal, artificial waterway system of West Germany and East Germany, extending c.200 mi (320 km) along the North German plain from the Dortmund-Ems Canal, West Germany, to Magdeburg, East Germany, on the Elbe River. An eastward extension of the Midland Canal passes through Berlin and connects with the Oder River. The canal facilitates east-west transportation of raw materials and manufactured goods.

Mid·lands (mĭd'ləndz), region of central England. It is usually considered to include the present counties of Derbyshire, eastern Hereford and Worcester, Leicestershire, Northamptonshire, Nottinghamshire, Staffordshire and Warwickshire. The region is highly industrialized.

Mid·lo·thi·an (mĭd-lō'*th*ē-ən), formerly **Ed·in·burgh·shire** (ĕd'n-bûr-shîr', -shər), former county, SE Scotland, on the S shore of the Firth of Forth. The county town was Edinburgh. The Heart of Midlothian was a popular name for the former Tolbooth prison at Edinburgh and became the title of a novel by Sir Walter Scott. Under the Local Government Act of 1973, Midlothian was divided between the Lothian and Borders regions.

Mid·way (mĭd'wā'), island group (2 sq mi/5.2 sq km), central Pacific, c.1,150 mi (1,850 km) NW of Honolulu, comprising Sand and Eastern islands with the surrounding atoll. It is a U.S. military base, with no indigenous population. Discovered by Americans in 1859, Midway was annexed in 1867. A cable station was opened in 1903. In 1935 Midway became a commercial air station of Pan American Airways, and in 1941 a U.S. naval base was opened. The Battle of Midway (June 3–6, 1942), one of the decisive Allied victories of World War II, occurred nearby. The battle, fought mostly with aircraft, resulted in the destruction of three Japanese aircraft carriers, crippling the Japanese navy. The islands are now administered by the U.S. Dept. of the Interior.

Mid·west City (mĭd'wĕst'), city (1978 est. pop. 51,500), Oklahoma co., central Okla., a residential suburb of Oklahoma City; founded 1942 with the activation of adjoining Tinker Air Force Base, a logistics center.

Mi·e·res (mē-ĕr'əs), city (1970 pop. 22,790), Oviedo prov., N Spain, in Asturias, on the Lena River. It is an important mining center for coal, sulfur, and cinnabar and has iron and steel plants.

Mif·flin (mĭf'lĭn), county (1970 pop. 45,268), 431 sq mi (1,116.3 sq km), central Pa., in an agricultural region bisected by the Jacks Mts. and drained by the Juanita River; formed 1789; co. seat Lewistown. Grain and fruit are grown, and textiles, metal products, and bricks manufactured. There are limestone and sand deposits in the county.

Mif·flin·town (mĭf'lĭn-toun'), borough (1970 pop. 828), seat of Juniata co., S central Pa., NW of Harrisburg on the Juniata River; laid out 1791, inc. 1833. In an agricultural area, it has limestone and shale quarries. Shirts are manufactured.

Mik·ke·li (mĭk'ə-lē), city (1975 est. pop. 27,112), capital of Mikkeli prov., S central Finland. It is in the Saimaa lake region and is an important lake port, commercial center, and transportation hub. It was chartered in 1838.

Mí·ko·nos (mē'kô-nôs) or **Myk·o·nos** (mĭk'ə-nŏs'), mountainous island (1971 pop. 3,823), c.35 sq mi (90 sq km), SE Greece, in the Aegean Sea; one of the Cyclades. It is a tourist resort.

Mi·ku·lov (mĭk'o͝o-lôf'), town (1970 pop. 6,267), S central Czechoslovakia, in Moravia, near the Austrian border. It is an agricultural market and has textile and food-processing industries. Armistice agreements ending the Franco-Austrian War (1805) and the Austro-Prussian War (1866) were signed at Mikulov.

Mi·lac·a (mə-lăk'ə), village (1970 pop. 1,940), seat of Mille Lacs co., E Minn., NNW of Minneapolis; settled in the 1880s, inc. 1897. Originally a lumber town, it is now a dairy and farm trade center.

Mi·lam (mī'ləm), county (1970 pop. 20,028), 1,027 sq mi (2,659.9 sq km), central Texas, bounded on the E by the Brazos River and drained by the Little River; formed 1836; co. seat Cameron. It has diversified agriculture (cotton, corn, grain sorghums, truck crops, livestock, poultry, and dairy products). There are oil wells and deposits of lignite, peat, clay, and mineral salts.

Mi·lan (mĭ-lăn', -län'), city (1971 pop. 1,724,173), capital of Lombardy and of Milano prov., N Italy, at the heart of the Po basin. Because of its strategic position at the intersection of several major transportation routes, it has been since the Middle Ages an international commercial, financial, and industrial center. Manufactures include textiles, clothing, machinery, chemicals, printed materials, motor vehicles, airplanes, and rubber goods. The city has a large construction industry, and it is one of the most important silk markets in

Europe. Probably of Celtic origin, Milan was conquered by Rome in 222 B.C. In later Roman times it was the capital (A.D. 305–402) of the Western Empire and the religious center of N Italy. In 313 Constantine I issued the Edict of Milan, which granted religious toleration. Milan was conquered by the Lombards in 569. In the 12th cent. it became a free commune and gradually gained supremacy over the cities of Lombardy. From the 11th to the 13th cent. it suffered from internal warfare and from the enmity of rival cities. As a member of the Lombard League, Milan contributed to the defeat of Frederick I at Legnano (1176). The city's independence was recognized in the Peace of Constance (1183). Galeazzo Visconti received (1395) the title of duke of Milan from the emperor, and under him the duchy became one of the most important states in Italy. After the death of the last Visconti (1447) the Sforza became dukes of Milan. The city passed under Spanish domination (1535), then Austrian rule made it the capital of the Cisalpine Republic (1797) and of the kingdom of Italy (1805–14). In 1815, Milan again came under Austria. In 1859 the city was united with the kingdom of Sardinia. Its industrial importance grew after it was incorporated (1861) into Italy.

The most striking feature of the city is the large, white-marble cathedral (1386–1813). Other points of interest include Brera Palace and Picture Gallery (17th cent.); the Castello Sforzesco (15th cent., with 19th cent. additions), which houses a museum of art; the Church of Santa Maria delle Grazie (1465–90), containing the famous fresco, the *Last Supper,* by Leonardo da Vinci; and the Basilica of Sant' Ambrogio (founded in the 4th cent., rebuilt in the 11th–12th cent.). Milan has three universities, a music conservatory, and an opera house, Teatro alla Scala (opened in 1778).

Mi·lan (mī'lən), city (1970 pop. 1,794), seat of Sullivan co., N Mo., W of Kirksville, in a livestock and truck-farm region; laid out 1845, inc. 1869.

Mil·bank (mĭl'băngk'), city (1970 pop. 3,727), seat of Grant co., NE S.Dak., NE of Watertown near the Minn. border; settled 1880. A shipping center for a livestock and grain region, the city processes meat and dairy products, and quarries granite.

Miles City (mīlz), city (1970 pop. 9,023), seat of Custer co., SE Mont., ENE of Billings on the Yellowstone River at the mouth of the Tongue River; inc. 1887. It grew up around Fort Keogh (built 1877), was reached by the Northern Pacific RR in 1881, and became the leading town of a cattle and sheep region.

Mi·le·tus (mī-lē'təs), ancient seaport of W Asia Minor, in Caria, on the mainland not far from Sámos. It was occupied by Greeks c. 1000 B.C. and became one of the principal cities of Ionia. From the 8th cent. B.C. it led in colonization, especially on the Black Sea. The Milesians were strong enough to resist the Lydian kings and were not molested by the Persians. In 499 B.C., however, they stirred up the revolt of Ionian Greeks against Persia; the Persians sacked the city (494 B.C.). It remained an important seaport until the harbor silted up early in the Christian era. Miletus produced some of the earliest Greek philosophers, including Thales and Anaximander.

Mil·ford (mĭl'fərd). **1.** City (1978 est. pop. 52,200), New Haven co., SW Conn., on Long Island Sound; settled 1639, inc. as a city 1959. Writing pens, electrical products, thermostats, and rivets are produced. **2.** City (1970 pop. 5,314), Kent and Sussex cos., S central Del., SSE of Dover; laid out 1787, inc. 1867. A trade, processing, and shipping center for a farm area, it also makes dental supplies. **3.** Industrial town (1978 est. pop. 25,400), Worcester co., S Mass., on the Charles River, in a farm area; settled 1662, set off from Mendon and inc. 1780. Pink granite has been quarried here since the mid-1800s. **4.** Town (1970 pop. 6,622), Hillsborough co., S N.H., on the Souhegan River NW of Nashua; formed 1794 from parts of Hollis and Amherst. It quarries granite and makes furniture and machinery. **5.** Borough (1970 pop. 1,190), seat of Pike co., NE Pa., on the Delaware River SW of Port Jervis, N.Y., in a farm area; settled 1733, inc. 1874. It is a summer resort.

Milford Ha·ven (hā'vən), urban district (1971 pop. 13,745), Dyfed, SW Wales, a seaport on the N side of the estuary called Milford Haven. The bay forms a splendid natural harbor that can handle large oil tankers, making the town a key oil port and refining center. From early times the town was a port for trade with Ireland. Henry II invaded Ireland from Milford Haven in 1172; Henry Tudor (who became Henry VII) landed here from France in 1485.

Milk (mĭlk), river, 729 mi (1,173 km) long, rising in the Rocky Mts., NW Mont., and flowing N into Alta., Canada, then in long curves E, S into Mont. again, and generally SE to the Missouri River, entering just below Fort Peck Dam. The Milk River reclamation project (est. 1911) irrigates c.134,000 acres (54,270 hectares).

Mil·lard (mĭl'ərd), county (1970 pop. 6,988), 6,793 sq mi (17,598.9 sq km), W Utah, bordering on Nev. and watered by the Sevier River and Sevier Lake; formed 1851; co. seat Fillmore. The irrigated lands produce alfalfa seed and wheat.

Mil·lau (mē-yō'), town (1975 pop. 21,907), Aveyron dept., S France, on the Tarn River. The center of the French glove industry, the town also has tanning and dyeing industries. At the nearby village of Roquefort, the famous cheese is manufactured from sheep's milk. Millau was a Huguenot stronghold in the 16th cent.

Mil·ledge·ville (mĭl'ĭj-vĭl'), city (1970 pop. 11,601), seat of Baldwin co., central Ga., on the Oconee River, in a fertile farm area; inc. 1836. Among the industries are the manufacture of textiles, pharmaceuticals, prefabricated homes, and food products. Laid out in 1803 as the site of the state capital, Milledgeville was the seat of government from 1807 to 1868. Many ante-bellum homes survive; among them are the old executive mansion (1838) and the old state capitol (1807), now part of Georgia Military College.

Mille Lacs (mĭl lăks), county (1970 pop. 15,703), 568 sq mi (1,471.1 sq km), E central Minn., drained by the Rum River; formed 1857; co. seat Milaca. It is in a dairying, livestock, poultry, and potato-farming region. There is some mining, lumber sawmilling, and its industry includes food processing and the manufacture of clothing, wood products, furniture, machinery, and auto parts. Mille Lacs Lake Indian Reservation is in the north.

Mille Lacs Lake, 207 sq mi (5361 sq km), E central Minn., N of Minneapolis. Sieur Duluth, a French explorer, visited Ojibwa Indians living on its shore in 1679.

Mil·len (mĭl'ən), city (1970 pop. 3,713), seat of Jenkins co., E Ga., S of Augusta on the Ogeechee River; settled in the 1830s, inc. 1881. It is a railroad junction in a farm area and processes lumber and cotton.

Mil·ler (mĭl'ər). **1.** County (1970 pop. 33,385), 627 sq mi (1,623.9 sq km), extreme SW Ark., bounded on the W by the Texas border, on the S by La., on the E and N by the Red River, and drained by the Sulphur River; formed 1820; co. seat Texarkana. Its agriculture includes cotton, oats, truck crops, livestock, and dairy products. There are also oil and natural-gas wells, sand and gravel pits, stands of timber, and plants manufacturing clay and metal products. **2.** County (1970 pop. 6,424), 287 sq mi (743.3 sq km), SW Ga., in a coastal plain area drained by Spring Creek; formed 1856; co. seat Colquitt. It is in a farming (corn, peanuts, sugar cane, truck crops, and livestock) and timber region. There is clothing manufacturing in the county. **3.** County (1970 pop. 15,026), 603 sq mi (1,561.8 sq km), central Mo., in the Ozarks, drained by the Osage River; formed 1837; co. seat Tuscumbia. In a tourist and agricultural (corn, wheat, livestock, and dairy products) area, it has barite mining, lumbering, and clothing and shoe manufacturing.

Miller, city (1970 pop. 2,148), seat of Hand co., E central S.Dak., ENE of Pierre; settled 1882, inc. 1910. It is a trade center in a grain and livestock region.

Mill·ers·burg (mĭl'ərz-bûrg'), village (1970 pop. 2,979), seat of Holmes co., NE central Ohio, SW of Canton; settled 1816, laid out 1824, inc. 1835. There are gas and oil wells, coal mines, and dairy farms in the area.

Mills (mĭlz). **1.** County (1970 pop. 11,832), 446 sq mi (1,155.1 sq km), SW Iowa, on the Nebr. line (here formed by the Missouri River) and drained by the West Nishnabotna River and by Keg and Silver creeks; formed 1851; co. seat Glenwood. The region is characterized by prairie agriculture (hogs, cattle, poultry, corn, wheat, and oats) and has bituminous-coal deposits. **2.** County (1970 pop. 4,212), 734 sq mi (1,901.1 sq km), central Texas, bounded on the SW by the Colorado River and drained by Pecan Bayou; formed 1887; co seat Goldthwaite. This diversified agricultural and ranching area yields sheep, goats, beef and dairy cattle, poultry, oats, grain sorghums, barley, peanuts, cotton, and fruit.

Mill Springs (mĭl), village, Wayne co., on the Cumberland River, S of Frankfort, SE Ky. It was the site of the opening battle (Jan. 19, 1862) of the Ky.-Tenn. campaign of the Civil War and the first important Union victory in the West.

Mill Valley, city (1970 pop. 12,942), Marin co., W Calif., a suburb on Richardson Bay, an inlet of San Francisco Bay; inc. 1900. It is a residential community set in heavily timbered hills and valleys; redwood trees predominate. Mt. Tamalpais State Park is adjacent to the city, and Muir Woods National Monument is nearby.

Mill·ville (mĭl′vĭl′), city (1978 est. pop. 26,600), Cumberland co., S N.J., on the Maurice River, in a poultry, fruit, and truck-farm area; settled 1756, inc. 1866.

Mí·los (mē′lôs) or **Mi·lo** (mē′lō, mī′-), mountainous island (1971 pop. 4,499), 58 sq mi (150.2 sq km), SE Greece, in the Aegean Sea, one of the Cyclades. The island's products include grain, cotton, fruits, and olive oil. Mílos flourished as a center of early Aegean civilization because of its deposits of obsidian and its strategic location between the Greek mainland and Crete. It lost importance when bronze replaced obsidian as a material for tools and weapons. Despite its neutrality in the Peloponnesian War, Mílos fell victim to Athens, which conquered the island in 416 B.C. and then massacred the men, enslaved the remaining persons, and founded an Athenian colony. Much excavation has been done on Mílos. The most famous find is the Venus of Milo (now in the Louvre, Paris), discovered in 1820.

Mil·pi·tas (mĭl′pē′təs), city (1978 est. pop. 33,000), Santa Clara co., W Calif., a suburb of San Jose, in a truck-farm and citrus-fruit area; inc. 1954. Industries are automobile assembling, food distributing, and the installation and maintenance of irrigation systems.

Mil·ton (mĭl′tən). **1.** Town (1970 pop. 5,360), on the Blackwater River; seat of Santa Rosa co., extreme NW Fla., NE of Pensacola; founded c.1825. Textiles, chemicals, and metal products are made. **2.** Town (1978 est. pop. 27,300), Norfolk co., E Mass., a residential suburb of Boston, on the Neponset River; settled 1636, set off from Dorchester and inc. 1662. Granite quarries are nearby. Milton is the seat of Curry College and several preparatory schools, including Milton Academy (1798). Harvard's meteorological observatory is on Blue Hill.

Milton Keyes (kēz), new town (1971 pop. 46,473), Buckinghamshire, S central England. Milton Keyes was designated one of the New Towns in 1967 to alleviate overpopulation in London.

Mil·wau·kee (mĭl-wô′kē), county (1970 pop. 1,054,249), 237 sq mi (613.8 sq km), SE Wis., bounded in the E by Lake Michigan and drained by the Wilwaukee, Menomonee, and Root rivers; formed 1834; co. seat Milwaukee. Its agriculture includes dairying, stock raising, truck farming, and horticulture. It has stone quarries. In a highly industrialized region that manufactures electrical and nonelectrical machinery, iron and steel, furniture, paper products, chemicals, plastics, concrete, and metal products, it also has food-processing, textile, and lumbering industries.

Milwaukee, city (1978 est. pop. 641,000), seat of Milwaukee co., SE Wis., at the point where the Milwaukee, Menomonee, and Kinnickinnic rivers enter Lake Michigan; inc. 1846. It is a port of entry, shipping heavy cargo via the St. Lawrence Seaway, a major producer of heavy machinery and electrical equipment, and one of the world's leading manufacturers of diesel and gasoline engines, tractors, and beer. Motorcycles, refrigeration equipment, chocolate, and electronic products are also produced.

In 1673 Father Jacques Marquette visited the site, which was then an Indian gathering and trading center. In 1795 the North West Company established a fur-trading post. Solomon Juneau, the fur trader, arrived in 1818, and in 1838 several settlements merged to form Milwaukee village. Milwaukee grew as a shipping center and became famous for its numerous industries, notably brewing and meat packing. German refugees arrived in large numbers after 1848, stimulating the city's political, economic, and social growth. Among the city's educational institutions are Marquette Univ. and the Univ. of Wisconsin at Milwaukee. Local attractions include the breweries, with their guided tours, a public library and museum, an art center, and a church built by Frank Lloyd Wright.

Mil·wau·kie (mĭl-wô′kē), city (1970 pop. 16,444), Clackamas and Multnomah cos., NW Oregon, on the Willamette River; inc. 1903. The city is a distribution center for farms and orchards of the Willamette valley and has numerous warehouse facilities. Chief among its varied manufactures are tools and textiles.

Min (mĭn). **1.** Chief river of Fukien prov., SE China, c.350 mi (565 km) long, rising in Wu-i Shan and flowing SE to the South China Sea near Fu-chou. **2.** River, W Szechwan prov., central China, c.500 mi (805 km) long, rising in the Min Shan and flowing S through the Ch'eng-tu Plain to the Yangtze River at I-pin. In the 2nd cent. B.C. the Min's water was diverted into numerous irrigation channels. The system is still used to water the fertile Ch'eng-tu Plain.

Mi·nas Ge·rais (mē′nəs zhĭ-rīs′), state (1975 est. pop. 12,550,600), 226,707 sq mi (587,171 sq km), E Brazil; capital Belo Horizonte.

Minch (mĭnch), strait, 20 to 45 mi (32.2-72.4 km) wide, separating the N Outer Hebrides from the mainland of Scotland. Little Minch, to the southwest, 14 to 20 mi (22.5-32.2 km) wide, separates Skye from the middle Outer Hebrides islands.

Min·cio (mēn′chō), river, c.47 mi (75 km) long, in Lombardy, N Italy. It flows generally south from the southern end of Lake Garda through Mantua (where it forms three lakes) to the Po River.

Min·da·na·o (mĭn′də-nä′ō, -nou′), island (1970 pop. 7,292,691), 36,537 sq mi (94,631 sq km), Philippines, NE of Borneo. The island is generally mountainous, heavily forested, and indented by several deep bays. Off the northeast coast in the Philippine Sea is the Mindanao Trench (c.35,000 ft/10,675 m deep), one of the greatest known ocean depths. Mindanao is the country's major pineapple- and hemp-producing island. Coffee and rice are also grown in abundance. There are commercial coconut and rubber plantations, fish culture areas, and crocodile grounds. Lumbering and mining are major industries. Mindanao is a prime source of Philippine mahogany, and iron, gold, and coal are produced in quantity.

In the middle of the 14th cent. Islam spread from Malaya and Borneo to the Sulu Archipelago, and from there to Mindanao. The arrival of the Spanish in the late 16th cent. united the various Moslem groups in a holy war against the conquerors that lasted some 300 years. Although many of the Philippine Islands suffered extensive damage in World War II, Mindanao emerged relatively unscathed. During the 1960s it experienced a phenomenal population increase and rapid development. The native Moros, finding themselves outnumbered and in many cases pushed off their lands, retaliated with terrorist activities. When the Philippine army attempted to restore order, fierce fighting often resulted. In 1969 and the early 1970s several thousand people were killed and hundreds of villages were burned. The discovery in 1971 of a Stone Age people, the Tasaday, living in an area now under rapid exploitation, strengthened the government's desire to protect such peoples against encroaching lumbering, mining, and ranching interests.

Min·den (mĭn′dən), city (1974 est. pop. 79,737), North Rhine-Westphalia, N West Germany, a port on the Weser River and the Midland Canal. It is an industrial center and rail junction. Manufactures include ships, machinery, textiles, clothing, coffee filters, chemicals, beer, and foundry products. Minden was founded c.800 by Charlemagne. In the 13th cent. it joined the Hanseatic League, and in 1530 it accepted the Reformation. Minden passed to Brandenburg in the Peace of Westphalia (1648). In the Seven Years War the English and the Hanoverians defeated (1759) the French at Minden. The city passed to Prussia in 1814.

Minden. 1. City (1970 pop. 13,996), seat of Webster parish, NW La.; inc. 1850. It is the shipping center of an area rich in timber, oil, and natural gas. Industries include lumbering, gas and oil production, steel fabrication, and the manufacture of plywood and of sand and gravel. **2.** City (1970 pop. 2,669), seat of Kearney co., S Nebr. SSE of Kearney, in a farm area; founded 1876. A replica of a pioneer village is here. **3.** Village (1970 pop. 1,200), seat of Douglas co., W Nev., on the East Carson River S of Carson City, at an altitude of 4,700 ft (1433.5 m). Alfalfa, cattle, grain, poultry, and potatoes are raised. Mono National Forest is nearby.

Min·do·ro (mĭn-dôr′ō), island (1970 pop. 473,940), 3,759 sq mi (9,735.8 sq km), Philippines, SW of Luzon. Although there is relatively little arable land, subsistence farming is carried on. Coal is mined, lumbering is an important industry, and there are major crocodile grounds.

Min·e·o·la (mĭn′ē-ō′lə), village (1978 est. pop. 20,700), seat of Nassau co., SE N.Y., on Long Island, a suburb of New York City; inc. 1906.

Min·er (mī′nər), county (1970 pop. 4,454), 571 sq mi (1,478.9 sq km), E S.Dak., watered by several streams and artificial lakes; formed 1873; co. seat Howard. Its agriculture includes dairy products, corn, grain, poultry, cattle, and hogs. There is some manufacturing.

Min·er·al (mĭn′ər-əl). **1.** County (1970 pop. 786), 921 sq mi (2,385.4 sq km), S Colo., drained by headwaters of the Rio Grande; formed 1893; co. seat Creede. It is a silver-mining and livestock-grazing region. It includes part of Rio Grande and San Juan National forests, with Wheeler National Monument in the northeast. **2.** County (1970 pop. 2,958), 1,223 sq mi (3,167.6 sq km), W Mont., in an agricultural region bordering on Idaho and drained by the Clark Fork:

formed 1914; co. seat Superior. Livestock and dairy products are of major importance. Gold and silver mining is also done. Parts of Lolo and Cabinet national forests are here. **3.** County (1970 pop. 7,051), 3,765 sq mi (9,751.4 sq km), SW Nev., in a mountainous region bordering on Calif., with Walker Lake and Wassuck Range in the W.; formed 1911; co. seat Hawthorne. It has some mining (gold, silver, and tungsten), cattle ranching, and a tourist trade. Part of Walker River Indian Reservation is in the northwest. **4.** County (1970 pop. 23,109), 330 sq mi (854.7 sq km), NE W.Va., in the Eastern Panhandle, bounded on the N and NW by the North Branch of the Potomac, drained by Patterson Creek, and traversed by the Allegheny Mts.; formed 1866; co. seat Keyser. its agriculture includes livestock, dairy products, fruit, and grain. Bituminous coal, limestone, and timber are its major natural resources.

Mineral Wells (wĕlz), city (1970 pop. 18,411), Palo Pinto and Parker cos., N Texas; inc. 1882. Among its industrial products are clay pipe, electronic equipment, and craft instruments. The mineral water made this hill city a popular health resort in the late 19th and early 20th cent., and oil activity in the area also spurred the city's growth. Tourists and sportsmen are attracted by Possum Kingdom Lake and State Park to the northwest.

Min·gan Islands (mĭng'gən), group of 15 small islands and many islets, E Que., Canada, in the St. Lawrence River, N of Anticosti island. They were discovered (1535) by Jacques Cartier. In 1836 the islands were acquired by the Hudson's Bay Company.

Min·go (mĭng'gō), county (1970 pop. 32,780), 423 sq mi (1,095.6 sq km), SW W. Va., bounded on the W by Tug Fork and the Ky. border; formed 1895; co. seat Williamson. It has extensive bituminous-coal fields, oil and natural-gas wells, and timber. Its agriculture includes livestock, fruit, and tobacco. Some food processing is done.

Min·gre·li·a (mĭn-grē'lē-ə, mĭng-grē'-), lowland region, SE European USSR, in Georgia, bordering the Black Sea. Tea and grapes are the chief products. Mingrelia was a vassal principality under the Ottoman Empire. It was annexed to Russia in 1803.

Mi·nho (mē'nyōō), river, c.210 mi (340 km) long, rising in Galicia, NW Spain, and flowing generally SW to the Atlantic Ocean. The lower part of the Minho forms a section of the border between Spain and Portugal.

Min·i·do·ka (mĭn'ə-dō'kə), county (1970 pop. 15,731), 750 sq mi (1,942.5 sq km), S Idaho, in an irrigated farm region; formed 1913; co. seat Rupert. Potatoes, sugar beets, dry beans, and alfalfa are grown, and livestock is raised.

Minidoka project, S Idaho, in the Snake River valley. Developed by the U.S. Bureau of Reclamation, it irrigates more than one million acres (405,000 hectares) of land. The Snake is impounded by Minidoka Dam (completed 1906), forming Lake Walcott, and by American Falls Dam (1927), which forms American Falls Reservoir.

Min·ne·ap·o·lis (mĭn'ē-ăp'ə-lĭs). **1.** City (1970 pop. 1,971), seat of Ottawa co., N central Kansas, N. of Salina; laid out 1866, inc. 1871. It is a trade center for an agricultural and livestock region. **2.** City (1978 est. pop. 359,000), seat of Hennepin co., E Minn., at the head of navigation on the Mississippi River, at St. Anthony Falls; inc. 1856. A port of entry, it is also a major industrial and rail hub. With adjacent St. Paul (the two are known as the Twin Cities), it is the processing, distributing, and trade center for a vast grain and cattle area. Chief among its many manufactures are computers and electronic equipment, instruments, graphic art products, machinery, fabricated metals, and textiles and garments. The falls were visited by Louis Hennepin in 1683; Fort Snelling was established in 1819; and a sawmill was built at the falls in 1821. The village of St. Anthony was settled c.1839 on the east side of the river near the falls. Minneapolis originated on the west side of the river c.1847 and included much of the reservation of Fort Snelling. It annexed St. Anthony in 1872. The city became the country's foremost lumber center, and after the plains were planted with wheat and the railroads were built, flour milling developed. The city was laid out with wide streets and numerous lakes and parks. In Minnehaha Park is the Stevens House (1849), the first frame house in the state. Also of interest are Fort Snelling State Park, several art galleries and museums (including the American Swedish Institute), The Guthrie Theater, and the Minneapolis Grain Exchange.

Min·ne·ha·ha (mĭn'ē-hä'hä), county (1970 pop, 95,209), 815 sq mi (2,110.9 sq km), SE S.Dak., on the Minn. border, drained by the Big Sioux River; formed 1862; co. seat Sioux Falls. Its agriculture includes dairy produce, grain, livestock, and poultry. It has a food-processing, industry and manufactures textiles, plastics, concrete, wood and metal products and farm machinery.

Minnehaha Falls, 53 ft (16.2 m) high, SE Minn., in Minnehaha Creek, which flows from Lake Minnetonka (23 sq mi/59.6 sq km) SE to the Mississippi River. The surrounding area, including the gorge cut by the receding falls, is a state park. Most of the year only a thin trickle of water passes over the falls.

Min·ne·so·ta (mĭn'ĭ-sō'tə), state (1975 pop. 3,944,000), 84,068 sq mi (217,736 sq km), N central United States, in the Great Lakes region, admitted as the 32nd state of the Union in 1858. Saint Paul, the capital, and its twin city Minneapolis, together have almost a third of the state's population. Minnesota is bounded on the north by the Canadian provs. of Man. and Ont., on the east by Lake Superior and Wisconsin (the St. Croix and Mississippi rivers forming much of the border with that state), on the south by Iowa, and on the west by South Dakota and North Dakota. The climate is humid continental. Winter locks the land in snow, and spring is brief; summers are hot. Prehistoric glaciers left marshes, boulder-strewn hills, and rich, gray drift soil stretching from the northern pine wilderness to the broad southern prairies. In the eastern part of the state are mountains from which iron ore is extracted. South of the iron country lie rolling hills. In the south and west are prairies. The state has more than 19 million acres (7.7 million hectares) of woodland and two national forests. Another great resource of Minnesota is its water. The state has more than 11,000 lakes and numerous streams and rivers. The rivers feed three great river systems: the Red River of the north and its tributaries in the west run north to Hudson Bay; the streams that run east into Lake Superior eventually help to supply the St. Lawrence; and the Mississippi flows south from Lake Itasca, gathering volume from the waters of the St. Croix and Minnesota rivers before leaving the state. Locks and other improvements enable barge traffic to pass over the Falls of St. Anthony into Minneapolis. Duluth, at the western tip of Lake Superior, has the largest inland harbor in the United States. With the completion of the Saint Lawrence Seaway (1959) and a marine terminal, the city became a key port for overseas trade.

The beauty of Minnesota's lakes and dense green forests, as seen in Voyageurs National Park, has long attracted vacationers, and the abundant fish in the state's many rivers, lakes, and streams provide excellent fishing. Also of interest to tourists are the Grand Portage and Pipestone national monuments, Itasca State Park (site of the headwaters of the Mississippi River), the Minnesota Museum of Mining (near Chisholm), and the world's largest open-pit iron mine at Hibbing. The University of Minnesota, with its many associated campuses, exerts a great influence on the state.

Minnesota is divided into 87 counties:

NAME	COUNTY SEAT	NAME	COUNTY SEAT
Aitkin	Aitkin	Kandiyohi	Willmar
Anoka	Anoka	Kittson	Hallock
Becker	Detroit Lakes	Koochiching	International Falls
Beltrami	Bemidji	Lac qui Parle	Madison
Benton	Foley	Lake	Two Harbors
Big Stone	Ortonville	Lake of the Woods	Baudette
Blue Earth	Mankato	Le Sueur	Le Center
Brown	New Ulm	Lincoln	Ivanhoe
Carlton	Carlton	Lyon	Marshall
Carver	Chaska	McLeod	Glencoe
Cass	Walker	Mahnomen	Mahnomen
Chippewa	Montevideo	Marshall	Warren
Chisago	Center City	Martin	Fairmont
Clay	Moorhead	Meeker	Litchfield
Clearwater	Bagley	Mille Lacs	Milaca
Cook	Grand Marais	Morrison	Little Falls
Cottonwood	Windom	Mower	Austin
Crow Wing	Brainerd	Murray	Slayton
Dakota	Hastings	Nicollet	St. Peter
Dodge	Mantorville	Nobles	Worthington
Douglas	Alexandria	Norman	Ada
Faribault	Blue Earth	Olmsted	Rochester
Fillmore	Preston	Otter Tail	Fergus Falls
Freeborn	Albert Lea	Pennington	Thief River Falls
Goodhue	Red Wing	Pine	Pine City
Grant	Elbow Lake	Pipestone	Pipestone
Hennepin	Minneapolis	Polk	Crookston
Houston	Caledonia	Pope	Glenwood
Hubbard	Park Rapids	Ramsey	St. Paul
Isanti	Cambridge	Red Lake	Red Lake Falls
Itasca	Grand Rapids	Redwood	Redwood Falls
Jackson	Jackson	Renville	Olivia
Kanabec	Mora	Rice	Faribault

NAME	COUNTY SEAT	NAME	COUNTY SEAT
Rock	Luverne	Traverse	Wheaton
Roseau	Roseau	Wabasha	Wabasha
St. Louis	Duluth	Wadena	Wadena
Scott	Shakopee	Waseca	Waseca
Sherburne	Elk River	Washington	Stillwater
Sibley	Gaylord	Watonwan	St. James
Stearns	St. Cloud	Wilkin	Breckenridge
Steele	Owatonna	Winona	Winona
Stevens	Morris	Wright	Buffalo
Swift	Benson	Yellow Medicine	Granite Falls
Todd	Long Prairie		

Economy. Minnesota ranks first in the nation in the production of iron and manganiferous ore. In the Vermilion and Cuyuna ranges the iron (discovered in 1884 and 1911) is mined underground, but in the rich Mesabi iron range (1890) open-pit methods are used. Granite (from St. Cloud) and sand and gravel production are also among the largest in the country. Wheat, once paramount in the fields, has yielded its preeminence to corn and livestock. The state is a leader in the production of creamery butter, dry milk, cheese, sweet corn, and soybeans. In the early 1950s manufacturing displaced agriculture as the major source of income in Minnesota. Major industries in the state include the manufacture of processed foods, electronic equipment, machinery, paper products, chemicals, and stone, clay, and glass products. St. Paul is one of the nation's largest meat-packing centers. Printing and publishing are also important.

History. Archaeological evidence indicates that Minnesota was inhabited as early as c.20,000 years ago. Some experts argue on the basis of the Kensington Rune Stone and other evidence that the first white men to reach Minnesota were the Norsemen, but others refute this. That French fur traders came in the mid-17th cent. is undeniable. At the time the French arrived, the dominant Indians were the Ojibwa in the east and the Sioux in the west. Minnesota remained excellent country for fur trade throughout the British regime that followed the French and Indian Wars (1763). After the American Revolution the eastern part of Minnesota was included in the Northwest Territory and was governed under the Ordinance of 1787; the western part was joined to the United States by the Louisiana Purchase. Only after the War of 1812 did settlement begin in earnest. In 1820 Fort St. Anthony (later Fort Snelling) was founded as a guardian of the frontier. Treaties (1837, 1845, 1851, and 1855) with the Ojibwa and the Sioux, by which the U.S. government acquired Indian lands, and the opening of a land office at St. Croix Falls in 1848 initiated a period of real expansion. Territorial status was granted in 1849. The Missouri and White Earth rivers were the western boundary.

The building (1851–53) of the Soo Ship Canal at Sault Sainte Marie, Mich., opened a water route for lake shipping eastward. The Panic of 1857 hit Minnesota particularly hard because of land speculation, but difficult times did not prevent the achievement of statehood in 1858, with St. Paul as the capital. The population swelled from 6,000 in 1850 to nearly 440,000 in 1870. During the Civil War years and afterward the Sioux reacted to broken promises, fraudulent dealings, and the encroachment of the white men on their lands with massacres and raids. Later in the century came immigrants from Scandinavia. Lumbering, which had begun in 1839, became paramount, and logging camps were established. A boom in wheat made the Minnesota flour mills famous across the world. The opening of the iron mines gave new impetus to Minnesota's economy but also created discontent among the laborers. They joined forces with the farmers in the 1890s in the Populist party.

In the 20th cent. the state has been notable for experimentation in novel features of local government and has also been a leader in the use of cooperatives. Credit unions, cooperative creameries, grain elevators, and purchasing associations were supported by legislation in 1919 that protected and encouraged the institutions. Today there are several thousand cooperative associations in Minnesota serving diversified needs. The state has become progressively more urban; in 1970 the urban population was two thirds of the total.

Government. The state is governed under the 1858 constitution. The legislature has 67 senators elected for four-year terms and 134 representatives elected for two-year terms. The governor is elected for a four-year term and may succeed himself.

Minnesota, river, 332 mi (534.2 km) long, rising in Big Stone Lake at the W boundary of Minnesota and flowing SE to Mankato, then NE to the Mississippi S of Minneapolis. It was an important route of explorers and fur traders. The river follows the valley of the prehistoric Warren River, the outlet of Lake Agassiz.

Min·ne·ton·ka (mĭn'ĭ'tŏng'kə), village (1978 est. pop. 42,700), Hennepin co., SE Minn., a residential suburb of Minneapolis, near Lake Minnetonka; inc. 1956.

Min·ne·wau·kan (mĭn'ə-wô'kən), city (1970 pop. 496), seat of Benson co., N central N.Dak., W of Devils Lake, in a grain area.

Mi·nor·ca (mĭ-nôr'kə), Spanish island (1970 pop. 50,217), 271 sq mi (701.9 km), in the W Mediterranean Sea, the second largest of the Balearic Islands. The terrain is mostly low but has a hilly center. Cereals, wine, olive oil, and flax are the chief products. Much of the agriculture is irrigated. Lobster fishing, the export of livestock, and shoe making add to the economy. Tourism is also important. A great number of megalithic monuments have been found. Minorca shared the history of the other Balearic Islands until 1708, when it was occupied by the English during the War of the Spanish Succession. England retained it until the Seven Years' War, when it was seized by the French. The Treaty of Paris (1763) restored Minorca to Britain, but the French and Spanish again seized it (1782) in the American Revolution. In 1798, in the French Revolutionary Wars, England regained control; the Peace of Amiens (1802) awarded Minorca to Spain. In the Spanish Civil War of 1936–39 Minorca remained in Loyalist hands until Feb., 1939.

Mi·not (mī'nət), city (1978 est. pop. 33,300), seat of Ward co., NW N.Dak., on the Souris River; inc. 1887. It is a commercial and transportation center for an extensive agricultural area.

Minsk (mĭnsk), city (1977 est. pop. 1,215,000), capital of the Belorussian Soviet Socialist Republic and of Minsk oblast, W European USSR, on a tributary of the Berezina. It is a cultural and industrial center and a railroad junction with machine, machine-tool, tractor, automobile, textile, and food-processing factories. First mentioned in 1067, it was an outpost on the road from Kiev to Polotsk. It became the capital of the Minsk principality in 1101 and part of Lithuania in 1326. At the end of the 15th cent. it became a great craft and trade center. Magdeburg Law was introduced into the local government of the city in 1499. Joined to Poland in 1569, it passed to Russia in the Second Partition of Poland (1793). The city's industrial development began in the 1870s. From 1941 to 1943 Minsk was a concentration center for Jews prior to their extermination by the Nazis. Although the city was heavily damaged in the war, several monuments remain, including a former 17th cent. Bernardine convent and the 17th cent. Ekaterin Cathedral.

Min·ute Man National Historical Park (mĭn'ət): *see* National Parks and Monuments Table.

Min·ya, Al (ăl mĭn'yə), city (1970 est. pop. 122,000), capital of Al Minya governorate, N central Egypt, on the Nile River. It is an agricultural trade center. Products include ginned cotton, flour, and rugs.

Min·ya Kon·ka (mĭn'yə kŏng'kə), peak, 24,900 ft (7,594.5 m) high, SW

Szechwan prov., central China, in the Himalayas. It was climbed (1932) by an American expedition.

Mi·o (mī'ō'), village (1970 pop. 500), seat of Oscoda co., NE central Mich., SW of Alpena on the Au Sable River. It is in Huron National Forest.

Miq·ue·lon (mĭk'ə-lŏn'), island: *see* Saint Pierre and Miquelon.

Mir·a·mi·chi (mĭr'ə-mĭ-shē', -mē'shē), river, c.135 mi (220 km) long, rising in several forks and tributaries in N.B., Canada, and flowing E past Newcastle into the Gulf of St. Lawrence at Miramichi Bay. The bay was visited (1534) by Jacques Cartier. Several Acadian fishing villages are on its shores.

Mir·ny (mĭr'nē), town (1974 est. pop. 27,000), NE Siberian USSR, in Yakut Autonomous Republic. Founded in 1956, when diamonds were discovered, Mirny is now the center of the Soviet diamond-mining industry.

Mir·za·pur (mĭr'zä-pŏŏr'), town (1971 pop. 105,920), Uttar Pradesh state, N central India. Shellac and cement are manufactured. Many Hindu pilgrims visit the shrine of the goddess Vindhyeshwari.

Mish·a·wa·ka (mĭsh'ə-wô'kə), city (1978 est. pop. 39,300), St. Joseph co., N Ind., on both banks of the St. Joseph River and adjacent to South Bend; settled c.1830, inc. 1899.

Mis·kolc (mĭsh'kôlts'), city (1977 est. pop. 203,000), NE Hungary, on the Sajó River. Miskolc has large iron and steel mills, lime and cement works, and machinery and motor vehicle factories. Iron ore and lignite are mined nearby, and the region's numerous limestone caves are used as cellars by local winemakers. Miskolc also has an important trade in metal products and agricultural goods. Miskolc was granted the status of a free city in the 15th cent. Frequent invasions (by Mongols in the 13th cent., Turks in the 16th and 17th cent., and German imperial forces in the 17th and 18th cent.) marked the city's history. Industrialization began in the second half of the 19th cent. Present-day landmarks include the Avas Reformed Church (15th cent.) and a museum containing Scythian art.

Mis·sau·kee (mĭ-sô'kē), county (1970 pop. 7,126), 565 sq mi (1,463.4 sq km), N central Mich., drained by the Muskegon River and its affluents; organized 1871; co. seat Lake City. It is partially agricultural, yielding livestock, poultry, potatoes, hay, and grain.There are many resorts and small lakes suitable for fishing. Part of Manistee National Forest is within the county.

Mis·si·nai·bi (mĭs'ə-nī'bē), river, c.265 mi (425 km) long, rising in Missinaibi Lake, central Ont., Canada, and flowing N and NE to the Mattagami River, SW of Moosonee, to form the Moose River.

Mis·sis·sip·pi (mĭs'ĭ-sĭp'ē), state (1975 pop. 2,331,000), 47,716 sq mi (123,584 sq km), S United States, admitted as the 20th state of the Union in 1817. Jackson is the capital and largest city. The Mississippi River, which gives the state its name, forms most of the state's western boundary. Across the Mississippi River lie Ark. and La. The latter, also bordering a stretch of Mississippi on the south, cuts eastward like the toe of a boot until it reaches the Pearl River, which forms the lower part of Mississippi's southern line; Ala. is on the east, Tenn. on the north.

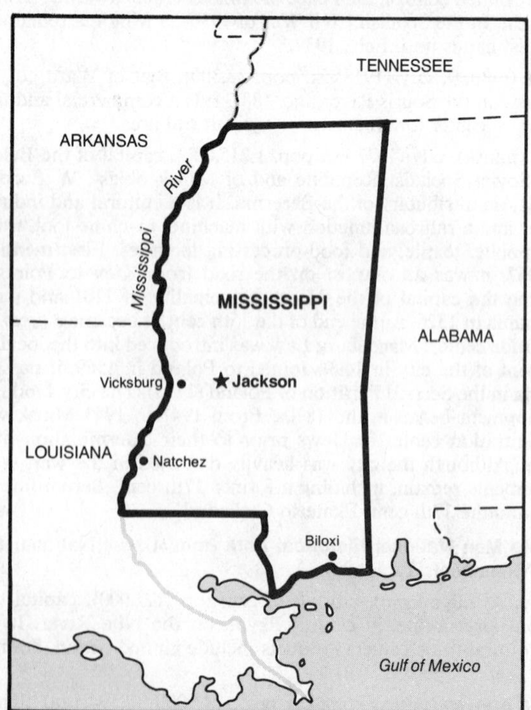

The generally hilly land reaches its highest point (806 ft/246 m) in the northeastern corner along the Tennessee River. The most distinctive region in the state's varied topography is the Delta, a flat alluvial plain between the Mississippi and the Yazoo rivers. The Delta is a highly productive cotton area. A wide belt of longleaf yellow pine (the piny woods) covers most of southern Mississippi to within a few miles of the coastal-plain meadows. Important there are lumbering and allied industries. Most of the state's rivers belong to either the Mississippi or the Alabama river systems, with the Pontotoc Ridge the divide. The climate of Mississippi is subtropical in the southern part and temperate in the northern part; the average annual rainfall is more than 50 in. (127 cm). Historical sites in the state include Old Spanish Fort, the oldest house on the Mississippi River, near Pascagoula, and Vicksburg National Military Park, Brices Cross Roads National Battlefield Site, and Tupelo National Battlefield Site. Mississippi, in the path of waterfowl migrations down the Mississippi valley, is noted for its duck and quail hunting. Along the Gulf Coast, a favorite fishing area, are several resort cities and part of Gulf Islands National Seashore. In Natchez and Biloxi are many fine ante-bellum mansions.

Mississippi is divided into 82 counties:

NAME	COUNTY SEAT	NAME	COUNTY SEAT
Adams	Natchez	Leflore	Greenwood
Alcorn	Corinth	Lincoln	Brookhaven
Amite	Liberty	Lowndes	Columbus
Attala	Kosciusko	Madison	Canton
Benton	Ashland	Marion	Columbia
Bolivar	Cleveland	Marshall	Holly Springs
Calhoun	Pittsboro	Monroe	Aberdeen
Carroll	Vaiden and Carrollton	Montgomery	Winona
Chickasaw	Houston and Okolona	Neshoba	Philadelphia
Choctaw	Ackerman	Newton	Decatur
Claiborne	Port Gibson	Noxubee	Macon
Clarke	Quitman	Oktibbeha	Starkville
Clay	West Point	Panola	Sardis and Batesville
Coahoma	Clarksdale	Pearl River	Poplarville
Copiah	Hazlehurst	Perry	New Augusta
Covington	Collins	Pike	Magnolia
De Soto	Hernando	Pontotoc	Pontotoc
Forrest	Hattiesburg	Prentiss	Booneville
Franklin	Meadville	Quitman	Marks
George	Lucedale	Rankin	Brandon
Greene	Leakesville	Scott	Forest
Grenada	Grenada	Sharkey	Rolling Fork
Hancock	Bay St. Louis	Simpson	Mendenhall
Harrison	Gulfport and Biloxi	Smith	Raliegh
Hinds	Jackson and Raymond	Stone	Wiggins
Holmes	Lexington	Sunflower	Indianola
Humphreys	Belzoni	Tallahatchie	Charleston and Sumner
Issaquena	Mayersville	Tate	Senatobia
Itawamba	Fulton	Tippah	Ripley
Jackson	Pascagoula	Tishomingo	Iuka
Jasper	Bay Springs and Paulding	Tunica	Tunica
		Union	New Albany
Jefferson	Fayette	Walthall	Tylertown
Jefferson Davis	Prentiss	Warren	Vicksburg
Jones	Ellisville and Laurel	Washington	Greenville
Kemper	De Kalb	Wayne	Waynesboro
Lafayette	Oxford	Webster	Walthall
Lamar	Purvis	Wilkinson	Woodville
Lauderdale	Meridian	Winston	Louisville
Lawrence	Monticello	Yalobusha	Coffeeville and Water Valley
Leake	Carthage		
Lee	Tupelo	Yazoo	Yazoo City

Economy. One of the more rural states in the Union, Mississippi is a leader in the production of cotton. The most important crops are cotton lint, soybeans, cottonseed, and hay. There has been a great rise in livestock raising and, especially, dairying. The state's most important and valuable mineral resources, petroleum and natural gas, have been developed only since the 1930s. More than one third of the state's land is subject to oil and gas development. Sand and gravel and clays are also produced. Industry has grown rapidly since oil development began and has been helped by the Tennessee Valley Authority and the state's program to balance agriculture with industry. Industrial products, including clothing, wood products, foods, and chemicals, have exceeded in value those of agriculture in recent years. On the Gulf coast there is a profitable fishing and seafood processing industry. Despite modernization efforts, however, the state's per capita income is one of the lowest in the nation.

History. Hernando De Soto's expedition undoubtedly passed (1540-42) through the region, then inhabited by the Choctaw, Chickasaw, and Natchez Indians, but the first permanent white settlement was not made until 1699, when Pierre le Moyne, sieur d'Iberville, established a French colony on Biloxi Bay. The region was part of Louisiana until 1763, when, by the Treaty of Paris, England received practically all the French territory east of the Mississippi River. English colonists had made the Natchez district a thriving agricultural community by the time Bernardo de Gálvez captured it for Spain in 1779. By the Treaty of Paris of 1783 the United States, with English approval, claimed most of the area of the present-day state of Mississippi. Spain denied this claim, and the long, involved West Florida Controversy ensued. In the Pinckney Treaty (1795) Spain accepted lat. 31°N as the northern boundary of its territory but did not evacuate Natchez until the arrival of American troops in 1798. Congress immediately created the Mississippi Territory, with Natchez as the capital.

In 1817 Mississippi became a state, with substantially its present-day boundaries; the eastern section of the Mississippi Territory was organized as Alabama Territory. Land hunger increased as new settlers arrived, lured by the continuing cotton boom. By a series of Indian treaties (1820, 1830, 1832) all the Indians in the state were pushed westward across the Mississippi. After 1840 slaves in the state outnumbered whites. On Jan. 9, 1861, Mississippi became the second state to secede from the Union. Civil War fighting did not reach Mississippi until Apr., 1862, when Union forces were victorious at Corinth and Iuka. Grant's brilliant Vicksburg campaign ended large-scale fighting in the state. Mississippi was readmitted to the Union early in 1870 after ratifying the Fourteenth and Fifteenth Amendments and meeting other Congressional requirements.

After Reconstruction the blacks were virtually disenfranchised. White supremacy was bolstered by the Constitution of 1890, later used as a model by other Southern states; under its terms a prospective voter could be required to read and interpret any of the constitution's provisions. On the ruins of the shattered plantation economy rose the sharecropping system. The landowners maintained their hold on politics until 1904, when the small farmers, still the dominant voting group, elected James K. Vardaman governor. Nevertheless this agrarian revolt did not alter a deep-seated obscurantism that was reflected in the Jim Crow laws (1904) and the ban on the teaching of evolution in the public schools (1926). Prohibition was put into effect in 1908 and not repealed until 1959.

Since the disastrous flood of 1927 the Federal government has taken over flood-control work. Navigation, too, has not been neglected; the Intracoastal Waterway provides a protected channel along the entire Mississippi coastline and links the state's ports with all others along the Gulf coast. The state has made attempts to wipe out illiteracy but its per capita expenditure on education was among the lowest of any state in 1972. Mississippi is still plagued by racial problems. The 1954 Supreme Court ruling against racial segregation in public schools brought massive resistance. There was not even token integration of public schools in Mississippi until 1962. Racial antagonisms during the 1960s resulted in many acts of violence, despite which some progress has been made. After the Federal Voting Rights Act of 1965, many blacks succeeded in registering and voting. In 1967, for the first time since 1890, a black was elected to the legislature.

Government. Mississippi is governed under the 1890 constitution. The bicameral legislature consists of 52 senators and 122 representatives, all elected for four-year terms. The governor is also elected for a four-year term.

Educational Institutions. Besides the Univ. of Mississippi the institutions of higher education in the state include Mississippi State Univ., at State College, and Mississippi State College for Women, at Columbus.

Mississippi. **1.** County (1970 pop. 62,060), 904 sq mi (2,341.4 sq km), NE Ark., bounded on the N by the Mo. border, on the E by the Mississippi River; formed 1833; co. seats Blytheville and Osceola. Its agriculture includes cotton, soybeans, corn, alfalfa, and vegetables. It has a diversified food-processing industry and manufactures wood and metal products, furniture, shoes, and machinery. **2.** County (1970 pop. 16,647), 411 sq mi (1,064.5 sq km), SE Mo., on the Mississippi River, with drainage canals; formed 1845; co. seat Charleston. It is in an agricultural region yielding cotton, corn, wheat, potatoes, and livestock. Lumbering is done, and clothing, shoes, and farm machinery are manufactured.

Mississippi, river, principal river of the United States, c.2,350 mi (3,780 km) long, exceeded in length only by the Missouri River, the chief of its numerous tributaries. The combined Missouri-Mississippi system (from the Missouri's headwaters in the Rocky Mts. to the mouth of the Mississippi River) is c.3,740 mi (6,020 km) long and ranks as the world's third-longest river system after the Nile and the Amazon. With its tributaries, the Mississippi drains c.1,231,000 sq mi (3,188,290 sq km) of the central United States, including all or part of 31 states and c.13,000 sq mi (33,670 sq km) of Alta. and Sask., Canada.

The Mississippi River rises in small streams that feed Lake Itasca in north Minn. and flows generally south to enter the Gulf of Mexico through a huge delta in southeast La. A major economic waterway, the river is navigable from the sediment-free channel maintained through South Pass in the delta to the Falls of St. Anthony in Minneapolis, with canals circumventing the rapids near Rock Island, Ill., and Keokuk, Iowa. The Mississippi connects with the Intracoastal Waterway in the south and with the Great Lakes-St. Lawrence Seaway system in the north by way of the Illinois Waterway. The Mississippi River receives the Missouri River north of St. Louis and expands to a width of c.3,500 ft (1,070 m); it swells to c.4,500 ft (1,375 m) at Cairo, where it receives the Ohio River. The lower Mississippi meanders in great loops across a broad alluvial plain (25–125 mi/40–201 km wide) that stretches from Cape Girardeau, Mo., to the delta section south of Natchez, Miss. The Mississippi enters a birdsfoot-type delta, where it discharges into the Gulf of Mexico through a number of distributaries, the most important being the Atchafalaya River and Bayou Lafourche. The main stream continues southeast through the delta to enter the gulf through several mouths, including Southeast Pass, South Pass, and Pass à Loutre. The flow of the river is greatest in the spring, when heavy rainfall and melting snow on the tributaries (especially the Missouri and the Ohio) cause the main stream to rise and frequently overflow its banks and levees, inundating vast areas of the plain.

The Spanish explorer Hernando De Soto is credited with the European discovery of the Mississippi River in 1541. The French explorers Jacques Marquette and Louis Jolliet reached it through the Wisconsin River in 1673, and in 1682 La Salle traveled down the river to the Gulf of Mexico and claimed the entire territory for France. France ceded the river to Spain in 1763 but regained it in 1800; the United States acquired the Mississippi River as part of the Louisiana Purchase in 1803. A major artery for the Indians and the fur-trading French, the river became in the 19th cent. the principal outlet for the newly settled areas of mid-America; exports were floated downstream with the current, and imports were poled or dragged upstream on rafts and keelboats. The first steamboat plied the river in 1811, and successors became increasingly luxurious as river trade increased in profitability and importance. Traffic from the north ceased after the outbreak of the Civil War. During the Civil War the Mississippi was an invasion route for Union armies and the scene of many important battles. Especially decisive were the capture of New Orleans (1862) by Adm. David Farragut, the Union naval commander, and the victory of Union forces under Gen. Grant at Vicksburg in 1863.

River traffic resumed after the end of the war; it is colorfully described in Mark Twain's *Life on the Mississippi* (1883). However, much of the trade was lost to the railroads in the mid-1800s, and the river ports declined in importance. With modern improvements in the channels of the river there has been a great increase in traffic, especially since the mid-1950s, with bulky items such as petroleum products, chemicals, sand, gravel, and limestone being the principal items of freight. Cotton and rice are important crops in the lower Mississippi valley; sugar cane is raised in the delta. The Mississippi is rich in freshwater fish; shrimp are taken from the briny delta waters. The delta also yields sulfur, oil, and gas.

Mississippi Sound, arm of the Gulf of Mexico, c.100 mi (160 km) long and from 7 to 15 mi (11–24 km) wide, extending from Lake Borgne in Louisiana on the W to Mobile Bay in Alabama on the E. It is part of the Intracoastal Waterway and is separated from the Gulf by a series of narrow islands and sand bars.

Mis·so·lon·ghi (mĭs'ə-lông'gē): *see* Mesolóngion, Greece.

Mis·sou·la (mĭ-zōō'lə), county (1970 pop. 58,263), 2,612 sq mi (6,765.1 sq km), W Mont., in an irrigated agricultural area drained by the Clark Fork and Bitterroot and Blackfoot rivers; formed 1865;

co. seat Missoula. Its major products are livestock, sugar beets, grain, and lumber. Part of Lolo National Forest is in the northwest.

Missoula, city (1978 est. pop. 29,800), seat of Missoula co., W Mont., on the Clark Fork of the Columbia River; inc. 1889. In the midst of five watered valleys, large forests, and an extensive dairy and cattle area, Missoula is a commercial and medical center with a busy lumber industry. The Treaty of Hell Gate in 1855 between the warring Salish (Flathead) and Blackfoot Indian nations brought peace to the area and opened it to settlement. Hell Gate town was founded nearby in 1860 and moved to the present Missoula site six years later. The coming (1883, 1908) of the railroads stimulated Missoula's growth. The Univ. of Montana and a regional headquarters of the U.S. Forest Service are here.

Mis·sou·ri (mĭ-zo͝or'ē, -zo͝or'ə), state (1975 est. pop. 4,786,000), 69,686 sq mi (180,487 sq km), central United States, admitted as the 24th state of the Union in 1821. The capital is Jefferson City. Missouri is bounded on the north by Iowa; on the west by Nebr., Kansas, and Okla.; on the south by Ark.; and on the east, where the Mississippi River forms the border, by Ill., Ky., and Tenn.

Two great rivers, the Mississippi and the Missouri, have had a great influence on the development of Missouri. The Mississippi tied the region to the South, particularly to New Orleans, and the Missouri, which crosses the state from west to east and enters the Mississippi near St. Louis, was the greatest avenue of pioneer advance westward across the continent. The region north of the Missouri River is largely prairie land, where corn and livestock are raised. Most of the region south of the Missouri is covered by foothills and by the plateau of the Ozark Mts., a unique region of hill scenery populated by a relatively isolated, self-reliant people. The rough, heavily forested eastern section of the Ozarks extends into the less hilly farming plateau in the west and encompasses the irregular, twisting Lake of the Ozarks to the northwest. In southwestern Missouri there is a long, narrow area of flat land, part of the Great Plains, where livestock and forage crops are raised. In the southeast below Cape Girardeau are the cotton fields of the Mississippi flood plain.

Places of cultural and historic interest in Missouri include the Jefferson National Expansion Memorial, a national historic site, in St. Louis; George Washington Carver National Monument, in Diamond; Wilson's Creek National Battlefield, near Springfield; the William Rockhill Nelson Gallery of Art, in Kansas City; the Harry S. Truman Memorial Library, in Independence; and the Museum of the American Indian, in St. Joseph.

Missouri is divided into 114 counties:

NAME	COUNTY SEAT	NAME	COUNTY SEAT
Adair	Kirksville	Benton	Warsaw
Andrew	Savannah	Bollinger	Marble Hill
Atchison	Rockport	Boone	Columbia
Audrain	Mexico	Buchanan	St. Joseph
Barry	Cassville	Butler	Poplar Bluff
Barton	Lamar	Caldwell	Kingston
Bates	Butler	Callaway	Fulton
Camden	Camdenton	Mississippi	Charleston
Cape Girardeau	Jackson	Moniteau	California
Carroll	Carrollton	Monroe	Paris
Carter	Van Buren	Montgomery	Montgomery City
Cass	Harrisonville	Morgan	Versailles
Cedar	Stockton	New Madrid	New Madrid
Chariton	Keytesville	Newton	Neosho
Christian	Ozark	Nodaway	Maryville
Clark	Kahoka	Oregon	Alton
Clay	Liberty	Osage	Linn
Clinton	Plattsburg	Ozark	Gainesville
Cole	Jefferson City	Pemiscot	Caruthersville
Cooper	Boonville	Perry	Perryville
Crawford	Steelville	Pettis	Sedalia
Dade	Greenfield	Phelps	Rolla
Dallas	Buffalo	Pike	Bowling Green
Daviess	Gallatin	Platte	Platte City
De Kalb	Maysville	Polk	Bolivar
Dent	Salem	Pulaski	Waynesville
Douglas	Ava	Putnam	Unionville
Dunklin	Kennett	Ralls	New London
Franklin	Union	Randolph	Huntsville
Gasconade	Hermann	Ray	Richmond
Gentry	Albany	Reynolds	Centerville
Greene	Springfield	Ripley	Doniphan
Grundy	Trenton	St Charles	St. Charles
Harrison	Bethany	St. Clair	Osceola
Henry	Clinton	St. Francois	Farmington
Hickory	Hermitage	St. Louis	Clayton
Holt	Oregon	St Louis (Independent City)	
Howard	Fayette	Ste. Genevieve	Ste. Genevieve
Howell	West Plains	Saline	Marshall
Iron	Ironton	Schuyler	Lancaster
Jackson	Independence	Scotland	Memphis
Jasper	Carthage	Scott	Benton
Jefferson	Hillsboro	Shannon	Eminence
Johnson	Warrensburg	Shelby	Shelbyville
Knox	Edina	Stoddard	Bloomfield
Laclede	Lebanon	Stone	Galena
Lafayette	Lexington	Sullivan	Milan
Lawrence	Mount Vernon	Taney	Forsyth
Lewis	Monticello	Texas	Houston
Lincoln	Troy	Vernon	Nevada
Linn	Linneus	Warren	Warrenton
Livingston	Chillicothe	Washington	Potosi
McDonald	Pineville	Wayne	Greenville
Macon	Macon	Webster	Marshfield
Madison	Fredericktown	Worth	Grant City
Maries	Vienna	Wright	Hartville
Marion	Palmyra		
Mercer	Princeton		
Miller	Tuscumbia		

Economy. Missouri's economy rests chiefly on industry. The manufacture of transportation equipment is the major industry; food products and chemicals are next in commercial importance, followed by printing and publishing. Nonelectrical machinery, fabricated metals, and electrical equipment are also produced. St. Louis is an important center for the manufacture of metals and chemicals. In Kansas City, long a leading market for livestock and wheat, the manufacture of vending machines and of cars and trucks are leading industries. Lead, cement, stone, iron ore, coal, and zinc are the chief minerals produced, and in 1972 Missouri was the leading state in the production of lead. Outside of its major cities, Missouri remains predominantly agricultural. The most valuable farm products are cattle, hogs, soybeans, and dairy items. Missouri ranked fourth in the production of hogs in 1972 and eighth in the production of cattle. After soybeans, the chief crops are corn, hay, and wheat. Springfield is a major agricultural center; Sedalia is the site of the annual state fair; and every fall Kansas City is host to the Midwest's largest horse and cattle show, the American Royal.

History. Missouri's recorded history begins in the latter half of the 17th cent. when the French explorers Jacques Marquette and Louis Jolliet descended the Mississippi River, followed by sieur de La Salle, who claimed the whole area drained by the Mississippi River for France and called the territory Louisiana. When the French explorers came the area was inhabited by the Osage and the Missouri Indians. Trade down the Mississippi prompted the settlement of Ste. Genevieve about 1735 and the founding of St. Louis in 1764. After a period of Spanish possession (1762–1800) the Louisiana Territory (including the Missouri area) was retroceded to France, but in 1803 it passed to the United States as part of the Louisiana Purchase.

By the time of the Lewis and Clark expedition (1803–6), St. Louis was already known as the gateway to the Far West. The U.S. Territory of Missouri was set up in 1812. The coming of the steamboat increased trade on the Mississippi, and settlement progressed. Planters from the South had introduced slavery; the question of admitting the Missouri Territory as a state became a burning national issue because it involved the question of extending slavery into the territories. The dispute was resolved by the Missouri Compromise, which admitted (1821) Missouri to the Union as a slave state, but excluded slavery from lands of the Louisiana Purchase north of lat. 36°30′. Slaveholding interests became politically powerful, but the state remained principally a fur-trading center. From Missouri traders established a thriving commerce over the Santa Fe Trail with the inhabitants of N.Mex., and pioneers followed the Oregon Trail to settle the Northwest. Franklin, Westport, Independence, and St. Joseph became famous as the points of origin of these expeditions. The present boundaries of the state were formed after the Indians gave up their claim to Platte co. in 1836, and this strip of land in the northwest corner of present-day Missouri was added to the state.

During the Civil War most Missourians remained loyal to the Federal government. A state convention, which met in Mar., 1861, voted against secession. Confederate forces challenged Federal authority in the State in 1861, 1862, and 1864, and guerrilla activities persisted. A new Missouri rose out of the war—the river life and steamboating began to decline. The coming of the railroads brought the decay of many of Missouri's river towns and tied the state more closely to the East and North. Urbanization and industrialization progressed, and the Louisiana Purchase Exposition, held at St. Louis in 1904, dramatically revealed Missouri's economic growth.

The depression years of the 1930s sent farm values crashing down, and many banks failed. Prosperity returned, however, during World War II, when both St. Louis and Kansas City served as vital mid-continental transportation centers. After the war Missouri's industrialization increased enormously.

Government. In 1945 Missouri adopted a new state constitution that is still in effect today. The governor of the state is elected for a term of four years. The general assembly, or legislature, has a senate with 34 members elected for four years and a house of representatives with 163 members elected for two years.

Educational Institutions. Institutes of higher learning include the Univ. of Missouri, at Columbia; St. Louis Univ., Washington Univ., and Webster College, at St. Louis; Rockhurst College, at Kansas City; and Westminster College, at Fulton.

Missouri, river, c.2,565 mi (4,130 km) long (including its Jefferson-Beaverhead-Red Rock headstream), the longest river of the United States and the principal tributary of the Mississippi River. The length of the combined Missouri-Mississippi system from the headwaters of the Missouri to the mouth of the Mississippi is c.3,740 mi (6,020 km), making it the world's third-longest river. The Missouri drains an area of c.580,000 sq mi (1,502,200 sq km), including 2,550 sq mi (6,605 sq km) in Canada.

The principal headwaters of the Missouri are the Jefferson, Madison, and Gallatin rivers, which rise high in the Rocky Mts. and join to form the Missouri near Three Forks, Mont. The Missouri's upper course flows north through scenic mountain terrain including Gate of the Mountains, a deep gorge. At Great Falls, Mont., the river enters a 10-mi (16-km) stretch of cataracts that prevented navigation to the upper river and effectively established Fort Benton, Mont., as the head of navigation for 19th cent. riverboats. Below Fort Benton, the Missouri follows a meandering course east then southeast across the Great Plains, crossing Mont., N.Dak., and S.Dak., and forming part of the boundaries of Nebr., Kansas, and Iowa before crossing Mo. and entering the Mississippi River 17 mi (27.4 km) north of St. Louis. Nicknamed "Big Muddy" for its heavy load of silt, the brown waters of the Missouri do not readily mix with the gray waters of the Mississippi until c.100 mi (160 km) downstream.

Above Sioux City, Iowa, the Missouri's fluctuating flow is regulated by seven major dams and more than 80 other dams on tributary streams. These dams, with their reservoirs, are part of the coordinated, basin-wide Missouri River Basin project (authorized by the U.S. Congress in 1944), which provides for flood control, hydroelectric power, irrigation water, and recreational facilities. The dams serve to impound for later use the spring rains and snow melt that swell the volume of the river in Mar. and Apr. and also the second flood stage that frequently occurs in June as the snow melts in the remoter mountain regions. Silt, fertilizers, and pesticides, which are contained in the runoff from agricultural lands, pollute the water above Sioux City, but wastes from industrial plants and from inadequately treated municipal sewage create a more serious level of pollution downstream.

The Missouri River was an important artery of commerce for the village Indians of the Plains culture long before the French explorers Jacques Marquette and Louis Jolliet passed the mouth of the river in 1683 and the Canadian explorer Vérendrye visited the upper reaches of the river in 1738. David Thompson, a Canadian fur trader, explored part of the river in 1797. Meriwether Lewis and William Clark followed the Missouri on their journey (1803–6) to the Pacific Ocean and described it at length. The first steamboat ascended the river in 1819 and hundreds more later navigated the uncertain waters to Fort Benton. Mormons bound for Utah and pioneers bound for Oregon and California followed the Missouri valley and that of the Platte overland to the West. River traffic declined with the loss of freight to the railroads after the Civil War, but it has been revitalized in the 20th cent. in the section below Sioux City.

Mis·tas·si·ni, Lake (mĭs'tə-sē'nē), c.840 sq mi (2,175 sq km), S Que., Canada, NW of Lake St. John, in sparsely settled country. It drains west to James Bay by way of the Rupert River.

Mis·ti, El (ĕl mē'stē), dormant volcano, 19,098 ft (5,824.9 m) high, in the Cordillera Occidental, S Peru, rising over the city of Arequipa. El Misti, with its perfect snow-capped cone, apparently achieved significance in the Inca religion, and has often figured in Peruvian legends and poetry.

Mi·tan·ni (mĭ-tăn'ē), ancient kingdom established in the 2nd millennium B.C. in NW Mesopotamia. It was founded by Aryans but was later made up predominantly of Hurrians. Mitanni controlled Assyria for a period and was engaged in military efforts to hold back Egyptian forces intent on conquering Syria. In c.1450 B.C. the army of Thutmose III of Egypt successfully advanced as far as Euphrates; the king of Mitanni surrendered, sending tribute to Egypt, which halted its invasion. Friendly relations later developed between the two powers. In the 14th cent. B.C., Mitanni became involved in struggles with the Hittites and c.1335 fell to the Hittites as well as to resurgent Assyrian forces.

Mitch·ell (mĭch'əl). **1.** County (1970 pop. 18,956), 510 sq mi (1,320.9 sq km), SW Ga., in a coastal-plain area bounded on the N by Flint River; formed 1857; co. seat Camilla. It is in an agricultural (cotton, corn, pecans, peanuts, and livestock) and timber (logging, sawmills, and wood products) region. Its industry includes poultry dressing, yarn milling, and clothing manufacturing. **2.** County (1970 pop. 13,108), 467 sq mi (1,209.5 sq km), N Iowa, on the Minn. border and drained by the Wapsipinicon, Cedar, and Little Cedar rivers; formed 1851; co. seat Osage. The principle industry is prairie agriculture (cattle, hogs, corn, and hay). The area also has many limestone quarries and sand and gravel pits. **3.** County (1970 pop. 8,010), 714 sq mi (1,849.3 sq km), N central Kansas, in a plains region drained by the Solomon River; formed 1867; co. seat Beloit. In a wheat and livestock area, it has farm-machinery manufacturing. **4.** County (1970 pop. 13, 447), 215 sq mi (556.9 sq km), W N.C., bounded on the N by the Tenn. border, on the W by the Nolichucky River, with the Unaka Mts. in the N and the Blue Ridge in the S; formed 1861; co. seat Bakersville. Its economy depends on tourism, farming (tobacco, corn, potatoes, apples, and clover), dairying, livestock raising, mining (mica, feldspar, and kaolin), and sawmilling. Pisgah National Forest occupies a large portion of the county. **5.** County (1970 pop. 9,073), 922 sq mi (2,388 sq km), W Texas, in a rolling prairie area watered by the Colorado River; formed 1876; co. seat Colorado City. Ranching, agriculture (cotton, grain sorghums, oats, wheat, alfalfa, and peanuts), and dairying are important. It also has oil wells.

Mitchell, city (1978 est. pop. 14,100), seat of Davison co., SE S.Dak.; inc. 1881. Mitchell is a trade, distribution, and shipping center for a grain, dairy, and livestock area. It has meat-packing and food-processing plants. Its huge Corn Palace, which has murals of colored corn along its entire exterior, is redecorated annually. Harvest festivals are held every Sept.

Mitchell, river, 300 mi (482.7 km) long, Cape York Peninsula, Queensland, Australia, rising in the Great Dividing Range and flowing WNW into the Gulf of Carpentaria.

Mitchell, Mount, peak, 6,684 ft (2,038.6 m) high, W N.C., in the Black Mts. of the Appalachian system. It is the highest peak east of the Mississippi River.

Mit·i·lí·ni or **Myt·i·le·ne** (both: mĭt'l-ē'nē), city (1971 pop. 23,426), capital of Lesbos prefecture, E Greece, a port on the island of Lesbos in the Aegean Sea. There are Roman remains.

Mi·to (mē'tō), city (1976 pop. 201,787), capital of Ibaraki prefecture, central Honshu, Japan, on the Naka River. It is chiefly a communications center. From 1606 Mito was the seat of a branch of the Tokugawa family. The city's Tokiwa Park is one of the greatest landscape gardens of Japan.

Mi·ya·za·ki (mē-yä'zä'kē), city (1975 pop. 240,001), capital of Miyazaki prefecture, SE Kyushu, Japan, on the Hyuga Sea. It is a popular tourist and resort center and the seat of a Shinto shrine dedicated to Jimmu, first emperor of Japan.

Mi·ya·zu (mē-yä'zo͞o), town (1970 pop. 31,602), Kyoto prefecture, S Honshu, Japan, on Miyazu Bay. It is a fishing port and processes marine products. Nearby is Ama-no-hashidate, or "heaven's bridge," a long promontory covered with pine trees whose fantastic shapes are reflected in the waters of the bay.

Mi·zo·ram (mĭ-zô'rəm), union territory (1971 pop. 321,686), c.8,000 sq mi (20,720 sq km), NE India, in the Mizo Hills, bordered on the E by Burma; capital Aijal. Mizoram became a union territory in 1972. Formerly, its area was part of Assam state.

Mjø·sa (myœ'sä), largest lake of Norway, 141 sq mi (365.2 sq km), and 1,453 ft (443.2 m) deep, on the Oppland-Hedmark border, SE Norway.

Mo·ab (mō'ăb), ancient nation located in the uplands E of the Dead Sea, now part of Jordan. Archaeological exploration in Moab has shown that settlements first occurred in the 13th cent. B.C.

Moab, city (1970 pop. 4,793), seat of Grand co., E Utah, on the Colorado River NNW of Monticello near the La Salt Mts.; inc. 1902. Originally settled by Mormons, it was abandoned in 1855 and resettled as a cow town and farm center after 1876. Today it is a tourist center in an area yielding large amounts of uranium, oil, and potash.

Mo·bile (mō-bēl'), county (1970 pop. 317,308), 1,240 sq mi (3,211.6 sq km), extreme SW Ala., in the coastal plain, bounded on the S by Mississippi Sound, on the E by Mobile Bay and Mobile River, on the W by the Miss. border; formed 1812; co. seat Mobile. It has agriculture (truck crops, fruit, and dairy products), commercial fisheries, a seafood-processing industry, and oil and gas refineries. Lumbering is done, and paper, chemicals, clothing, concrete, metal products, and machinery are manufactured.

Mobile, city (1978 est. pop. 208,100), seat of Mobile co., SW Ala., at the head of Mobile Bay; inc. 1814. One of the country's major ports, Mobile is an important shipping and shipbuilding center. There are oil refineries and industries that produce paper, textiles, aluminum, and chemicals. Mobile was founded at its present site in 1710 by the sieur de Bienville. It was the capital of French Louisiana from 1710 to 1719. The British held it from 1763 to 1780, when Bernardo de Gálvez took it for Spain. Mobile was seized for the Americans by Gen. James Wilkinson in 1813. During the Civil War ships from Mobile evaded the Federal blockade until Adm. David Farragut's victory at Mobile Bay (Aug. 5, 1864). Gen. E. R. S. Canby captured the city in Apr., 1865. Mobile has many beautiful ante-bellum homes and magnificent gardens. Of historical interest are the homes of Adm. Raphael Semmes and Gen. Braxton Bragg, the headquarters of Gen. E. R. S. Canby, and Forts Morgan and Gaines at the entrance to Mobile Bay. Mobile is the seat of Spring Hill College (the oldest in the state), Mobile College, and the Univ. of South Alabama. The Azalea Trail Festival dates from 1929.

Mobile Bay, arm of the Gulf of Mexico, SW Ala., from 8 to 18 mi (12.9–29 km) wide, extending c.35 mi (56 km) from the Gulf to the mouth of the Mobile River. A ship channel connects Mobile Bay with the Gulf. The Intracoastal Waterway passes through the southern part of the bay.

Mo·cha (mō'kə), town, S Yemen, a port on the Red Sea. It was noted for the export of the coffee to which it gave its name but declined as a trading port in the late 19th cent.

Mocks·ville (mŏks'vĭl'), town (1970 pop. 2,529), seat of Davie co., W central N.C., SW of Winston-Salem; settled before 1750. Clothing and furniture are made.

Mo·de·na (mô'dā-nä), city (1976 est. pop. 178,530), capital of Modena prov., Emilia-Romagna, N central Italy, on the Panaro River. It is an agricultural, commercial, and industrial center. Manufactures include motor vehicles, shoes, and machine tools. An Etruscan settlement, the city was the site of a Roman colony founded in the early 2nd cent. B.C. and located on the Aemilian Way. Modena became a free commune in the 12th cent. and in 1288 permanently passed to the Este family of Ferrara. The duchy of Modena, established in 1452, became the seat of the Este family after it lost (1598) Ferrara.

Mo·des·to (mə-dĕs'tō), city (1978 est. pop. 94,800), seat of Stanislaus co., central Calif., on the Tuolumne River, near the N end of the San Joaquin Valley; inc. 1884. The center of a farming and fruit-growing area, it has food-processing plants and companies that manufacture paper cartons and cans.

Mo·di·ca (mô'dĭ-kä), city (1976 est. pop. 30,800), SE Sicily, Italy. It is the center of an agricultural region where livestock is raised. Known in ancient times as Motyca, it was a feudal county in the 12th cent. and enjoyed a high degree of independence from the 14th to 18th cent. Nearby are the Cava d'Ispica (a series of limestone grottoes containing cave dwellings) and prehistoric and early Christian tombs.

Mo·doc (mō'dŏk), county (1970 pop. 7,469), 4,094 sq mi (10,603.5 sq km), NE Calif., on a high semiarid volcanic plateau bounded on the N by the Oregon border, on the E by the Nev. border, and watered by the Pit River and Clear Lake Reservoir; formed 1874; co. seat Alturas. In a stock-raising and farming (barley, potatoes, and hay) area, it has a lumbering industry, sand and gravel pits, and gold deposits. Fishing and hunting attract tourists. The county includes part of Modoc National Forest.

Moe·ris (mĭr'əs), ancient name of Lake Karun, c.90 sq mi (235 sq km), NE Egypt, in Al Fayyum. The size of the lake is much reduced from that described by ancient travelers, such as Herodotus. Ancient irrigation works were excavated in the late 1920s.

Moe·sia (mē'shə), ancient region of SE Europe, S of the lower Danube River. Inhabited by Thracians, it was captured by the Romans in 29 B.C. and organized as a Roman province, comprising roughly what is now Serbia (Upper Moesia) and Bulgaria (Lower Moesia).

Mof·fat (mŏf'ət), county (1970 pop. 6,525), 4,743 sq mi (12,284.4 sq km), extreme NW Colo., bordering on Utah and Wyo. and drained by the Yampa, Little Snake, and Green rivers; formed 1911; co. seat Craig. It is in a livestock-grazing area, with oil and gas wells and some manufacturing. Part of Dinosaur National Monument is here.

Mog·a·dish·o (mŏg'ə-dĭsh'o͞o), city (1972 est. pop. 230,000), capital of the Somali Republic, on the Indian Ocean. It is a port and a commercial and financial center. Mogadisho has little industry except for food and beverage processing. Uranium ore has been discovered nearby. Mogadisho was settled by Arab colonists c.900, and by the early 12th cent. it had become an important trade center for East Africa. During the 16th cent. it was controlled by Portugal. In 1871 the city was occupied by the sultan of Zanzibar, who leased it to the Italians in 1892. In 1905 Italy purchased the city and made it the capital of its colony of Italian Somaliland. Mogadisho was captured and occupied during World War II by British forces.

Mo·gi·lev (mŏg'ə-lĕf'), city (1976 est. pop. 264,000), capital of Mogilev oblast, W European USSR, on the Dnepr River. It is an important rail and highway junction, a river port, and an industrial center where metal products, machinery, chemicals, and artificial fibers are produced. The city grew around a castle dating from 1267. It was part of the grand duchy of Lithuania (united with Poland in 1569), was later held by Sweden, and passed to Russia during the First Partition of Poland (1772). Mogilev was occupied and heavily damaged by the Germans during World War II.

Mo·gi·lev-Po·dol·ski (mŏg'ə-lĕf' pə-dôl'skē, mə-gə-lyôf'-), city (1976 est. pop. 264,000), SW European USSR, in the Ukraine, at the confluence of the Dnestr and Derlo rivers. A river port, it has machine and food-processing plants. There are limestone quarries in the area.

Mo·gol·lon Plateau (mō'gə-yōn'), tableland, part of the Colorado Plateau, from 7,000 to 8,000 ft (2,135–2,440 m) high, E central Ariz. It is covered by pine forests, parts of which are included in Coconino, Tonto, and Sitgreaves national forests. Its southern edge is a rugged escarpment called the Mogollon Rim.

Mo·hács (mō'hăch', mô'häch'), town (1970 pop. 19,583), S Hungary, on the Danube River near the Yugoslavia border. It is an important river port and railroad terminus and has a modern metallurgical industry. Leather and silk goods, foodstuffs, textiles, and hemp are also produced in the city. Mohács is best known for the crushing defeat (Aug. 29, 1526) here of Louis II of Hungary and Bohemia by Sulayman I of Turkey. The defeat brought with it more than 150 years of Ottoman domination in Hungary. At Mohács are monuments to the slain and to Hungarian independence. Mohács was also the scene (1687) of a Turkish defeat by Charles V of Lorraine, which hastened the end of Turkish rule in Hungary.

Mo·hall (mō'hôl'), city (1970 pop. 950), seat of Renville co., N N.Dak., N of Minot, in a dairying, livestock, and poultry area.

Mo·ha·ve (mō-hä'vē), county (1970 pop. 25,857), 13,217 sq mi (34,232 sq km), NW Ariz., bounded on the N by Lake Mead and the Colorado River, on the S by the Bill Williams River, on the W by the Nev. and Calif. borders; formed 1864; co. seat Kingman. Mining (lead, silver, zinc, gold, and copper) and cattle grazing are the major occupations. The county includes sections of Grand Canyon National Park, Lake Mead Recreational Area, and Hualapai and Fort Mohave Indian reservations.

Mohawk, river, c.140 mi (225 km) long, rising in central N.Y. and flowing S then SE past Utica and Schenectady to enter the Hudson River at Cohoes. The Mohawk is canalized from Rome to its mouth (completed 1918) as part of the New York State Barge Canal, which links the Hudson River with the Great Lakes; it is now mainly used by pleasure craft. The beautiful and fertile Mohawk valley was the scene of many battles and raids in the French and Indian Wars and in the American Revolution. The valley was long an important route to the West.

Mohawk Trail, old road (c.100 mi/160 km long) in central N.Y. following the Mohawk River. It was an easy route through the Appalachians by which thousands of settlers emigrated from the Eastern seaboard to the Midwest. The Erie Canal rendered the road less important, and when the railroads were built its value was further diminished.

Moi·sie (mwä-zē'), river, 210 mi (337.9 km) long, rising in E Que., Canada, near the Labrador border, and flowing S to the St. Lawrence. The Hudson's Bay Company has an important trading post at the village of Moisie near the river's mouth.

Mojave, river, c.100 mi (160 km) long, rising in the San Bernardino Mts., S Calif., and flowing generally N to disappear in the Mojave Desert. Due to the porous soil and rapid evaporation, much of its course is underground except during the short wet season.

Mojave or **Mohave Desert,** c.15,000 sq mi (38,850 sq km), region of low, barren mountains and flat valleys, 2,000 to 5,000 ft (610–1,525 m) high, S Calif., part of the Great Basin, bordered on the N and W by the Sierra Nevada and the Tehachapi, San Gabriel, and San Bernardino mts. and merging with the Colorado Desert in the SE. Once a part of an ancient interior sea, the desert was formed by volcanic action and by material deposited by the Colorado River. Located in the rain shadow of the Coast Ranges, the Mojave receives an average annual rainfall of 5 in. (12.7 cm), mostly in winter. Juniper and Joshua trees are found on the higher, outer mountain slopes; desert-type vegetation and numerous intermittent lakes and streams are present in the valleys. Minerals found in the desert include borax and other salines, gold, silver, and iron. Military installations were established in the Mojave during World War II. Death Valley National Monument and Joshua Tree National Monument are located in the region.

Mo·kel·um·ne (mō-kĕl'ə-mē), river, c.140 mi (225 km) long, central Calif. flowing from the Sierra Nevada to the San Joaquin River NW of Stockton.

Mo·ku·a·we·o·we·o (mō-kōō'ə-wā'ō-wā'ō), volcanic crater at the summit of Mauna Loa, S central part of the island of Hawaii. The second-largest active crater in the world, it has a depth of c.800 ft (245 m) and is 3.7 mi (6 km) in circumference and 1.7 mi (2.7 km) wide. During active periods, lava streams flow from Mokuaweoweo down the slopes of Mauna Loa.

Mol (môl), city (1977 pop. 29,373), Antwerp prov., N Belgium, near the Dutch border; founded in the 9th cent. It is a manufacturing city and the center of nuclear research in Belgium.

Mold (mōld), urban district (1971 pop. 8,239), administrative center of Clwyd, NE Wales. It is in a farming and coal-mining area. Until 1974 Mold was the county town of Flintshire.

Mol·da·vi·a (mŏl-dā'vē-ə), historic province (c.14,700 sq mi/38,075 sq km), E Rumania, separated in the E from the Moldavian SSR by the Prut River and the W from Transylvania by the Carpathians. Moldavia borders on the Ukraine in the north and on Walachia in the south. It comprises roughly the modern administrative divisions of Bacău, Galaţi, and Iaşi. Suceava and Iaşi, its historic capitals, and Galaţi, its port on the Danube, are the chief cities. Moldavia, a fertile plain drained by the Siretul, is the granary of Rumania. Besides farming there is livestock raising, and orchards and vineyards dot the countryside. Lumbering and petroleum extraction are the main industries. The region was part of the Roman province of Dacia and has retained its Latin speech despite the centuries of invasion and foreign rule. Greek, Slavic, Turkish, Jewish, and other elements have influenced its culture. Moldavia was part of the Kievan state from the 9th to the 11th cent. In the 13th cent. the Cumans, who then held Moldavia, were expelled by the Mongols. When the Mongols withdrew, Moldavia became (early 14th cent.) a principality under native rulers. It then included Bukovina and Bessarabia. Like its sister principality, Walachia, it was torn by strife among the boyars—the great landowners and officeholders—and among rival claimants to the throne. The rural population was reduced to misery and virtual slavery (which lasted well into the 19th cent.) by the princes, who ruled with Oriental absolutism and cruelty. Moldavia reached its height under Stephen the Great (1457–1504), who in 1475 routed the Turks, but in 1504 it became tributary to the sultans. Although it was frequently occupied by foreign powers in the continuous wars among Turkey, Austria, Transylvania, Poland, and Russia, Moldavia remained under the Ottoman Empire. Southern Bessarabia early passed under the rule of the khans of Crimea. Early in the 18th cent. the Turks ended the rule by native princes—who had sided with the enemy as often as with Turkey—and appointed governors (hospodars), mostly Greek Phanariots. The Greeks surpassed their predecessors in avarice, while the nobility fell into total decay and corruption. Their rule was ended (1822) after the Greek insurrection instigated by Alexander Ypsilanti, and native hospodars were appointed. Meanwhile, Bukovina was taken (1775) by Austria and Bessarabia by Russia (1812). After the Russo-Turkish War of 1828–29, Moldavia and Walachia were made virtual protectorates of Russia, although they continued to pay tribute to the sultan. A Rumanian national uprising (1848–49) was suppressed by Russian intervention. In the Crimean War, Moldavia was again occupied by Russia, but in 1856 the two Danubian principalities, Walachia and Moldavia, were guaranteed independence under the nominal suzerainty of Turkey. With the accession (1859) of Alexander John Cuza as prince of both Moldavia and Walachia the history of modern Rumania began.

Mol·da·vi·an Soviet Socialist Republic (mŏl-dā'vē-ən), constituent republic (1976 est. pop. 3,850,000), c.13,000 sq mi (33,670 sq km), SW European USSR; capital Kishinev. The Prut River separates it from Rumania in the west. In the north and east, the Dnestr River forms its approximate boundary with the Ukraine, on which it also borders in the south. The fertile soil supports wheat, corn, barley, tobacco, sugar beets, soybeans, and sunflowers, as well as extensive fruit orchards, vineyards, and walnut groves. Beef and dairy cattle are raised, and beekeeping and silk breeding are widespread. Food processing is the main industry; others include metalworking, engineering, and the manufacture of electrical equipment.

A historic passageway between Asia and southern Europe, Moldavia was often subject to invasion and warfare. The main part of Moldavia was an independent principality in the 14th cent. and came under Ottoman Turkish rule in the 16th cent. It became a highly fortified Turkish border region and was a frequent target in Russo-Turkish wars. Eastern Moldavia passed to Russia in 1791. Russia acquired further Moldavian territory in 1793 and especially in 1812, when the Russians received all of Bessarabia (the name for the area of Moldavia between the Prut and Dnestr rivers). The rest of Moldavia remained with the Turks and later passed to Rumania, which seized Bessarabia in 1918. In 1924 the USSR, refusing to sanction the seizure, established the Moldavian ASSR in the Ukraine. Rumania was forced to cede Bessarabia to the USSR in 1940. The predominantly Ukrainian districts in the south and around Khotin in the north were incorporated into the Ukraine, as were parts of the Moldavian ASSR; the rest was merged with what remained of the Moldavian ASSR and made a constituent republic. Taken by Rumania in 1941, the republic was reconquered by the USSR in 1944.

Mo·line (mō-lēn'), city (1978 est. pop. 43,800), Rock Island co., NW Ill., on the Mississippi River, in a coal area; inc. 1848. It is a transportation and industrial center with railroad repair shops, and has been a major producer of farm machinery since John Deere moved here in 1847. Moline, with East Moline, Rock Island, and Davenport, Iowa, is part of an economic unit called the Quad Cities.

Mo·li·se (mō-lē'zā), region (1975 est. pop. 329,705), 1,714 sq mi (4,439.3 sq km), S central Italy, bordering on the Adriatic Sea in the E; capital Campobasso. Mostly mountainous, Molise is crossed by the Apennines. The main occupation in the generally poor region is farming. Molise's few industries include the processing of food and the manufacture of clothing. Molise was conquered by the Romans in the 4th cent. B.C. After the fall of Rome it came under the Lombard duchy of Benevento (6th–11th cent.). From 1948 to 1965 Molise was included in the region of Abruzzi e Molise.

Mo·lo·kai (mŏl'ə-kī', mō'lō-), island (1970 pop. 5,261), 261 sq mi (676 sq km), Maui co., Hawaii, between Oahu and Maui islands. Molokai is generally mountainous. On the north coast, separated by a rocky mountain wall from the rest of the island and accessible only over a 2,000-ft (610-m) pass, is the Kalaupapa peninsula, the site of a government leper colony established (1860) by Father Damien, the Belgian missionary, who worked here until his death. Molokai has many cattle ranches and pineapple plantations.

Mo·luc·cas (mə-lŭk'əz, mō-) or **Spice Islands** (spīs), island group and province (1971 pop. 2,012,385), c.32,300 sq mi (83,660 sq km), E Indonesia, between Celebes and New Guinea. Of volcanic origin, the Moluccas are mountainous, fertile, and humid. They are the original source of nutmeg and cloves. Other spices, copra, and forest products are also produced. Sago is the staple food. The islands were explored by Magellan in 1511–12 and thereafter settled by the Portuguese. In the 17th cent. they were taken by the Dutch, who secured a monopoly in the clove trade. Twice the British gained a foothold in the islands, which passed definitively to the Dutch in the first quarter of the 19th cent.

Mom·ba·sa (mŏm-bä'sä, -băs'ə), city (1973 est. pop. 301,000), capital of Coast prov., SE Kenya, mostly on Mombasa Island in the Indian Ocean and partly on the mainland (with which it is connected by a causeway). It is an important commercial and industrial center. Manufactures include processed food, cement, and glass. From the 8th to the 16th cent. Mombasa was a center of the Arab trade in ivory and slaves. The city was visited (1498) by Vasco da Gama on his first voyage to India. The Portuguese controlled the city until 1698, when it was regained by the Arabs; the Portuguese briefly held the city again in 1729. It came under Zanzibar in the mid-19th cent. and passed to Great Britain in 1887. Mombasa was the capital of the British East Africa Protectorate from 1887 to 1907.

Mon·a·co (mŏn'ə-kō, mə-nä'kō), independent principality (1968 pop. 23,035), c.370 acres (150 hectares), on the Mediterranean Sea, an enclave within Alpes-Maritimes dept., SE France, near the Italian border. It consists of three adjoining sections—La Condamine, the business district; Monte Carlo, the site of the famous casino; and Monaco-Ville, the capital, atop a rocky promontory. Its beautiful location, natural harbor, exceptionally mild climate, and the gambling tables of Monte Carlo make Monaco one of the best known resorts of the Riviera. Other attractions include a 16th cent. palace, a 19th cent. cathedral, and a noted oceanographic museum.

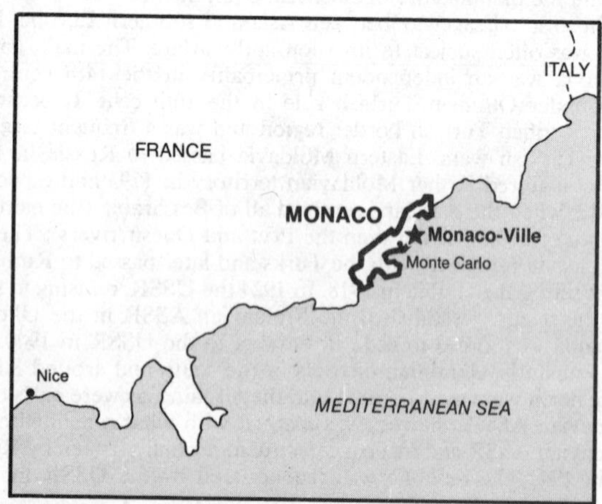

In addition to tourism and the foreign businesses attracted to Monaco by freedom from taxation, shipping and the manufacture of perfumes are important. Only about 2,500 of the total population are citizens of Monaco (Monegasques), and they are not admitted to the gambling tables. Monaco has a customs union with France, and its currency is interchangeable with the French. There are neither income nor corporation taxes; the chief sources of state revenue are excise, stamp, transfer, and estate taxes.

History. Probably settled by Phoenicians in ancient times, Monaco was annexed by Marseilles and Christianized in the 1st cent. A.D. In the 7th cent. it was part of the kingdom of the Lombards, and in the 8th cent. of the kingdom of Arles. It was under Moslem domination (8th cent.) after the Saracens invaded France. Monaco was ruled by the Genovese Grimaldi family from the 13th cent. In 1731 the male line died out, but the French Goyon-Matignon family, which succeeded by marriage, assumed the name Grimaldi. Monaco was under Spanish proection from 1542 to 1641, under French protection from 1641 to 1793, annexed to France in 1793, under Sardinian protection from 1815 to 1861, and again under French protection thereafter. Until 1911, when the first constitution was promulgated, the prince was an absolute ruler. Rainier III, who succeeded his grandfather Louis II in 1949, married Grace Kelly in 1956, and a male heir was born in 1958. In 1963 new fiscal agreements severely curtailed the right of French citizens to use Monaco as a tax haven.

Government. In accordance with the 1962 constitution, Monaco is governed by the prince, who is assisted by a minister of state a cabinet and the National Council, which is elected by universal suffrage every five years. The prince may initiate legislation, but all laws must be approved by the National Council. Monaco has a police force and a Royal Guard. By a treaty of 1918, the succession to the throne must be approved by the French government. Should the throne become vacant for any reason, Monaco would become an autonomous state under French protection.

Mo·nadh·li·ath Mountains (mō'nə-lē'ə), Inverness-shire, N central Scotland, between the Spey River and Loch Ness. Carn Ban (3,087 ft/941.5 m) is the highest point.

Mo·nad·nock (mə-năd'nŏk), isolated peak, 3,165 ft (965.3 m) high, SW N.H. It is much visited for its fine view. The peak lends its name to the geomorphic term "monadnock," an isolated mountain remnant standing above the general level of the land because of its greater resistance to erosion.

Mon·a·ghan (mŏn'ə-gən), county (1971 pop. 46,242), 498 sq mi (1,289.8 sq km), N Republic of Ireland, bordered on the N by Northern Ireland; co. town Monaghan. The northwest portion of the county is a part of the fertile central plain of Ireland; to the south and east are hilly sections. It is primarily an agricultural county.

Monaghan, urban district (1971 pop. 5,255), county town of Co. Monaghan, N Republic of Ireland. It is a farm market with some manufacturing.

Mon·a·hans (mŏn'ə-hănz'), city (1970 pop. 8,333), seat of Ward co., W Texas, N of the Pecos River; inc. 1928. It developed from a ranching center into an oil town in the 1930s.

Mo·na Passage (mō'nə), strait, c.80 mi (130 km) wide, between Puerto Rico and the Dominican Republic, connecting the N Atlantic Ocean with the Caribbean Sea. In it is Mona Island c.20 sq mi (50 sq km), part of Puerto Rico. The island was discovered by Columbus in 1493, and in 1508 Ponce de León stopped here.

Mön·chen·glad·bach (mŭn'kən-gläd'bäk, mœn'ᴋʜən-glät'bäᴋʜ'), city (1974 est. pop. 263,356), North Rhine-Westphalia, W West Germany. Cotton textiles and machinery are produced. Mönchengladbach developed around a Benedictine abbey (founded c.972), which was rebuilt several times between the 14th and the 18th cent. and which now serves as the city hall.

Moncks Cor·ner (mŭngkz kôr'nər), town (1977 est. pop. 3,213), seat of Berkeley co., SE S.C., N of Charleston, in a diversified argricultural area; settled c.1800, inc. 1885. Lumber products are made.

Monc·ton (mŭngk'tən), city (1976 pop. 55,934), SE N.B., Canada, on the Petitcodiac River. Textiles as well as wood, metal, meat, and petroleum products are manufactured, and wood and meat are processed. Magnetic Hill, an optical illusion, and the Tidal Bore, a high tide occurring twice daily, are features of the city.

Mon·em·va·si·a (mô'nĕm-və-sē'ə): *see* Malvasia, Greece.

Mon·ghyr (mŭng-gîr'), city (1971 pop. 102,462), Bihar state, NE India, on the Ganges River. The city has one of India's largest railroad workshops and a firearms industry that dates back to the 18th cent.

Mon·go·li·a (mŏng-gō'lē-ə, mŏn-), Asian region (c.906,000 sq mi/2,346,540 sq km), bordered roughly by Sinkiang Uigur Autonomous Region, China, on the W, the Manchurian provinces of China on the E, Siberia on the N, and the Great Wall of China on the S. It now comprises the Mongolian People's Republic (Outer Mongolia) and the Inner Mongolian Autonomous Region of China. Mongolia is chiefly a region of desert and of steppe plateau from c.3,000 to

5,000 ft (910–1,520 m) high. The climate is cold and dry and the population sparse. The Gobi desert, which is entirely wasteland, is in the central section. To the west are the Altai Mts., which rise to 15,266 ft (4,653 m). Rivers include a section of the Huang Ho (Yellow River) in the south and the Selenga, Orkhon, and Kerulen in the north. Rainfall averages less than 15 in. (38.1 cm) a year, but irrigation has made some cultivation possible; wheat and oats are the chief crops. Mongolia has traditionally been a land of pastoral nomadism; livestock raising and the processing of animal products are the main industries. Wool, hides, meat, cloth, and leather goods are exported. Coal, iron ore, gold, and oil are important mineral resources. Mongolia is crossed north to south by a railroad linking Peking with the USSR. Camels and horses are the chief means of transportation, although the use of jeeps and trucks is increasing.

In the 1st cent. A.D. Mongolia was inhabited by various Turkic tribes who dwelt mainly along the upper course of the Orkhon River. It was also the home of the the Huns, who ravaged (1st–5th cent.) north China. The Uigur Turks founded their first empire (744–856) with its capital near Karakorum in western Mongolia. Many smaller territorial states followed until (c.1205) Genghis Khan conquered all Mongolia, united its tribes, and from his capital at Karakorum led the Mongols in creating one of the greatest empires of all time. His successors established the Golden Horde in southeastern Russia and founded the Hulagid dynasty of Persia and the Yüan dynasty (1260–1368) of China. After the decline of the Mongol empire, Mongolia intruded less in world affairs. China, which earlier had gained control of Inner Mongolia, subjugated Outer Mongolia in the late 17th cent., but in the succeeding years struggled with Russia for control. Outer Mongolia finally broke away in 1921 to form the Mongolian People's Republic. Inner Mongolia remained under Chinese control, although the Japanese conquered Jehol (1933), which they included in Manchukuo, and Chahar and Suiyuan (1937), which they formed into Menchiang. These areas were returned to China after World War II. In 1949 the Chinese Communists joined most of Inner Mongolia to north Jehol prov. and west Heilungkiang prov. to form the Inner Mongolian Autonomous Region. In 1945 Tannu Tuva, long recognized as part of Mongolia, was incorporated within the USSR.

Mon·go·li·an People's Republic (mŏng-gō'lē-ən, mŏn-), country (1976 est. pop. 1,500,000), 604,247 sq mi (1,565,000 sq km), N central Asia. It is unofficially called Outer Mongolia. The capital is Ulan Bator. Bordered on the west, south, and east by China, and on the north by the USSR, it comprises more than half the region historically known as Mongolia. A high country, its average elevation exceeds 5,100 ft (1,555.6 m); the central, northern, western, and southwestern areas are covered with hills, high plateaus, and mountain ranges. Much of the Gobi desert lies to the south and east. Numerous beautiful lakes fill the depressions between the mountains; the largest, Uvs Nuur (c.1,300 sq mi/3,370 sq km) is saltwater. The main rivers are in the north and include the Selenga, with its long tributary the Orkhon, which flows into Lake Baykal in the USSR, and the Kerulen. Navigability is limited. The country's climate is dry continental, with little rain or snow and great extremes in temperature. Winters are severe, causing the ground to freeze to unusual depths; summers can be very hot.

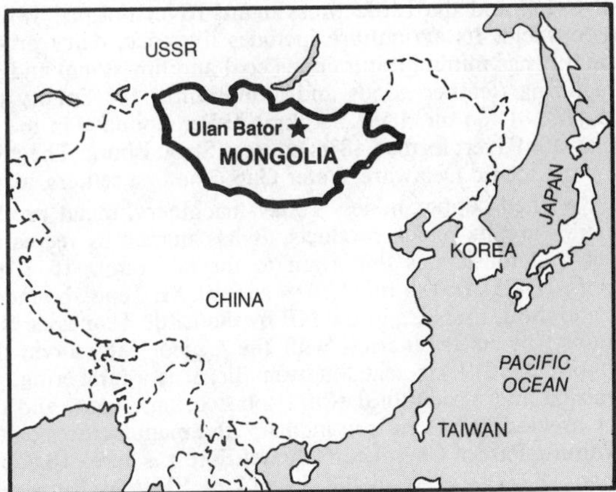

The paucity of snow permits year-round grazing, and nomadic herding has been the major occupation for centuries. Animal husbandry, now generally organized under state cooperatives, is still the mainstay of the Mongolian economy. Sheep and goats constitute most of the livestock, followed by cattle and horses; yaks are raised in the higher altitudes, and camels are extremely important in the desert and semidesert areas. Agriculture is limited. Wheat is the chief crop, followed by oats. Barley, corn, millet, rye, legumes, and potatoes are also grown. Hunting is a source of revenue; the country abounds in wildlife, and sable, ermine, fox, lynx, woodchuck, marmot, snow leopard, squirrel, and wolf are all trapped for their furs. Mongolia has valuable timberlands, especially in the northern mountainous area; logs are shipped down the Selenga, Orkhon, and Kerulen rivers. Mineral resources are abundant. The extensive coal deposits have been exploited since 1913. Gold, fluorspar, wolframite, iron ore, tungsten, molybdenum, lead, silver, and salt are also mined, and oil was pumped from fields in the eastern Gobi region from 1951 to 1971. Industry, developed with Soviet aid, is centered chiefly in Ulan Bator. It is based largely on the country's livestock resources, with dairy products, packed meats, leather and leather goods, and woolen textiles and related items (clothing, blankets, carpets) the chief manufactures. The building-material and lumber industries are also important, and cement, glassware, woodwork, furniture, paper, and matches are manufactured. Mongolia's main exports are livestock, wool, hides, meat, butter, and furs; more than 90% of the trade is with the USSR and countries of eastern Europe. The country has c.1,000 mi (1,610 km) of railroads. Beasts of burden are still used, notably in the south, where camel caravans are common.

The population is predominantly Khalkha Mongol. Minorities include Oirat Mongols, Kazakhs, Tuvinians, Chinese, and Russians. Khalkha Mongolian, the official language, was until 1946 written in the old Uigur Turkic script; it now uses the Cyrillic alphabet. The dominant religion was long Lamaist Buddhism, but the government undermined its influence, and religion is now openly practiced by very few, mostly the old.

The area was under Chinese control from 1691 until the collapse of the Manchu dynasty in China in 1911, when a group of Mongol princes ousted the Manchu governor and proclaimed an autonomous Mongolia. The new state was reoccupied by the Chinese in 1919. The Chinese were driven out by White Russian forces in early 1921, and the Whites in turn were ousted by Red Army troops and Mongolian units under the Mongolian Communist leaders. Mongolia was proclaimed an independent state in July, 1921, and remained a monarchy until 1924. The establishment (Nov., 1924) of the Communist-led Mongolian People's Republic was followed by a struggle to divest the privileged classes of their capital (largely in the form of land and livestock) and persecution of the Lama priests; this in turn led to the Lama Rebellion of 1932, when priests led thousands of people, with some 7 million head of livestock, across the border to Inner Mongolia. In 1936 the USSR signed a mutual aid pact with the republic, thus formalizing the existing close relations between the two countries. A new constitution, adopted in 1940, consolidated the power of the Communist regime. During World War II the Mongolian army joined the USSR in Manchuria in the last, brief stage of the war against Japan. In 1945 a plebiscite was held under a Sino-Soviet agreement, and the republic overwhelmingly voted for continued independence. A Soviet-sponsored application for Outer Mongolia's admission to the United Nations was rejected in 1947, but approved in 1961. The border between Outer Mongolia and Communist China was fixed by treaty in Dec., 1962. A 20-year treaty of friendship and cooperation was signed with the USSR in 1966.

Mon·i·teau (măn'ə-tō'), county (1970 pop. 10,742), 418 sq mi (1,082.6 sq km), central Mo., bounded on the NE by the Missouri River; formed 1845; co. seat California. It is in an agricultural region that produces wheat, corn, and oats and raises cattle, poultry, mules, sheep, and saddle horses.

Mon·mouth (mŏn'məth), municipal borough (1973 est. pop. 7,000), Gwent, SE Wales, at the junction of the Monnow and Wye rivers. It is a popular tourist and agricultural center with flourishing cattle and produce markets. There are food-processing and paper industries. Remains of a 12th cent. castle in which Henry V was born and a Norman church are here. Until 1974 Monmouth was the county town of Monmouthshire.

Monmouth, county (1970 pop. 461,849), 477 sq mi (1,235.4 sq km), E central N.J., bounded on the E by the Atlantic, on the N by Raritan and Sandy Hook bays, and drained by Metedeconk, Manasquan, and Shark rivers, with Navesink Highlands and Sandy Hook in the NE; formed 1683; co. seat Freehold. The inland agricultural area produces potatoes, soybeans, truck crops, grain, wheat, and corn. The commercial and industrial areas process and manufacture food products, clothing, lumber millwork and wood products, furniture, paper products, chemicals, plastics, concrete, metal industries, metal products, and machinery. There are many coastal resorts.

Monmouth. **1.** City (1970 pop. 11,022), seat of Warren co., W Ill.; inc. 1852. It is a trade center in a farm area. The city has a packing plant and companies that manufacture pottery, hobby kits, and dog food. **2.** Town (1970 pop. 5,237), Polk co., NW Oregon, SW of Salem in the Willamette River valley, in a lumber, dairy, and fruit region; laid out 1855, inc. 1880. Oregon College of Education is here.

Mon·mouth·shire (mŏn′məth-shîr′, -shər), former county SE Wales. The county town was Newport. In 1974 most of Monmouthshire was reorganized as the nonmetropolitan county of Gwent; small areas in western Monmouthshire became part of the new nonmetropolitan counties of Mid Glamorgan and South Glamorgan.

Mo·no (mō′nō), county (1970 pop. 4,016), 3,027 sq km), E Calif., in a rugged area of the Sierra Nevadas, bounded on the E by the Nev. border and drained by the Owens, East Walker, and West Walker rivers; formed 1861; co. seat Bridgeport. Stock raising, mining (pumice, gold, lead, silver, and andalusite), and lumbering are important. The county includes sections of Mono and Inyo national forests, part of the John Muir Trail, many small lakes, and other scenic and recreastional areas.

Mo·noc·a·cy (mə-nŏk′ə-sē), river, c.60 mi (95 km) long, rising in S Pa., and flowing S across Md. to join the Potomac River near Frederick, Md. On its banks, just east of Frederick, was fought the Battle of Monocacy (July 9, 1864). Although the Union forces under Gen. Lew Wallace were defeated, they delayed the Confederate forces under Gen. J. A. Early long enough to give Gen. U. S. Grant time to dispatch troops to defend Washington and drive Early back into Va.

Mo·no·na (mə-nō′nə), county (1970 pop. 12, 069), 699 sq mi (1,810.4 sq km), W Iowa, bordered on the W by Nebr. and the Missouri River and drained by the Little Sioux, Maple, and Soldier rivers; formed 1851; co. seat Onawa. Its prairie agriculture yields corn, hogs, cattle, and poultry. There are sand and gravel pits and bituminous-coal deposits.

Monona, city (1970 pop. 10,420), Dane co., S Wis.; inc. 1938.

Mo·non·ga·he·la (mə-nŏn′gə-hē′lə), city (1970 pop. 7,113), Washington co., SW Pa., on the Monongahela River S of Pittsburgh, in an industrial and agricultural region; settled 1770, inc. as a borough 1833, as a city 1873. There are coal mines in the area.

Monongahela, river, 128 mi (206 km) long, formed at Fairmont, N W.Va., by the junction of the West Fork and Tygart rivers. It flows north through a highly industrialized valley into southwestern Pa., where it joins the Allegheny River and forms the Ohio River at Pittsburgh. The canalized river is navigable for most of its length.

Mon·on·ga·li·a (mŏn′ən-gā′lē-ə), county (1970 pop. 63,714), 365 sq mi (945.4 sq km), N W.Va., on the Allegheny Plateau, bordered on the N by Pa. and drained by the Monongahela and Cheat rivers; formed 1776; co. seat Morgantown. Its natural resources include bituminous coal, oil and natural gas, limestone, and sand. Dairy processing is done, and glass, concrete, bricks, and metal products are manufactured.

Mon·re·a·le (mōn′rā-ä′lā), town (1976 est. pop. 24,988), NW Sicily, Italy, near Palermo. An agricultural market and tourist center, it commands a magnificent view of the fertile Conca d'Oro plain. Its cathedral, one of the masterpieces of Norman-Sicilian architecture, was begun (1174) by William II of Sicily. The cathedral has fine copper doors by Bonanno Pisano; its interior is decorated with exceptional Byzantine mosaics.

Mon·roe (mən-rō′). **1.** County (1970 pop. 20,883), 1,035 sq mi (2,680.7 sq km), SW Ala., in a coastal-plain area bounded on the SW by the Alabama River, on the S by the Little River; formed 1815; co. seat Monroeville. Cotton and peanut farming, beekeeping, logging, and clothing manufacturing are done. **2.** County (1970 pop. 15,657), 607 sq mi (1,572.1 sq km), E Ark., drained by the White and Cache rivers; formed 1829; co. seat Clarendon. In an agricultural area yielding cotton, rice, soybeans, wheat, and livestock, it has textile and metal-products manufacturing. There is commercial fishing in the county, which includes a state game refuge. **3.** County (1970 pop. 52,586), 1,034 sq mi (2,678.1 sq km), extreme S Fla., bordered on the W by the Gulf of Mexico, on the S by the Florida Keys, enclosing Florida Bay; formed 1824; co. seat Key West. Fishing, dairying, and poultry raising are the major occupations. Citrus fruits (especially limes) are grown on the Florida Keys. Much of the county is included in Everglades National Park. **4.** County (1970 pop. 10,991), 399 sq mi (1,033.4 sq km), central Ga., bounded on the E by the Ocmulgee River; formed 1821; co. seat Forsyth. In a piedmont agricultural area yielding corn, truck crops, pecans, fruit, and livestock, it has lumbering and textile industries. **5.** County (1970 pop. 18,831), 380 sq mi (984.2 sq km), SW Ill., bounded on the W by the Mississippi River and drained by the Kaskaskia River; formed 1816; co. seat Waterloo. It has agriculture (corn, wheat, hay, fruit, livestock, poultry, and dairy products) and limestone quarries. **6.** County (1970 pop. 85,221), 386 sq mi (999.7 sq km), S central Ind., drained by the West Fork of the White River and Salt, Bean Blossom, and Clear creeks; formed 1818; co. seat Bloomington. Its agriculture includes grain, corn, tobacco, livestock, and dairy products. It has timber, limestone quarries, and plants manufacturing electronic equipment, concrete, and stone products. Lake Monroe and several state and national forests make this a recreational area. **7.** County (1970 pop. 9,357), 435 sq mi (1,126.7 sq km), S Iowa, in a prairie agricultural area; formed 1843; co. seat Albia. Corn, oats, and hay are grown. There are also coal mines in the county. **8.** County (1970 pop. 11,642), 334 sq mi (865.1 sq km), S Ky., bordered on the S by Tenn. and drained by the Cumberland River and several creeks; formed 1820; co. seat Tompkinsville. In a hilly agricultural area yielding corn, wheat, hay, burley tobacco, and livestock, it has oil wells, timber, and limestone quarries. Its industry includes knitting mills and the manufacture of clothing and wood products. **9.** County (1970 pop. 119,172), 562 sq mi (1,455.6 sq km), extreme SE Mich., bounded on the S by the Ohio border, on the E by Lake Erie, on the NE by the Huron River, and drained by the Raisin River; formed 1817; co. seat Monroe. It is in a stock-raising, dairying, and agricultural area growing corn, wheat, soybeans, and hay. There are nurseries, limestone quarries, and salt beds here. Among its manufactures are paper products, furniture, iron and steel castings, and pollution-control equipment. **10.** County (1970 pop. 34,043), 769 sq mi (1,991.7 sq km), NE Miss., bordered on the E by Ala. and drained by Buttahatchee River and the East Fork of the Tombigbee River; formed 1821; co. seat Aberdeen. It lies in a cotton, corn, and dairy region, with timber, clay pits, and a food-processing industry. Clothing, wood products, chemicals, and motor vehicle parts are made. **11.** County (1970 pop. 9,542), 669 sq mi (1,732.7 sq km), NE central Mo., in a timber and coal area drained by the Salt River; formed 1831; co. seat Paris. Farming (corn, oats, and wheat) and livestock raising (poultry and saddle horses) are the major occupations. **12.** County (1970 pop. 711,917), 673 sq mi (1,743.1 sq km), W N.Y., bounded on the N by lake Ontario, crossed by the Barge Canal, and drained by the Benesee River and several creeks; formed 1821; co. seat Rochester. In an apple-growing area, it also yields truck crops, dairy products, grain, and potatoes and has stands of timber, limestone and gypsum quarries, and sand and gravel pits. It has food-processing and metalworking industries and diverse manufactures (textiles and clothing, wood and paper products, furniture, chemicals, plastics, concrete, and machinery). **13.** County (1970 pop. 15,739), 455 sq mi (1,178.5 sq km), E Ohio, bounded on the SE by the Ohio River, which forms the W.Va. border here, and drained by Sunfish Creek and the Little Muskingum River; formed 1813; co. seat Woodsfield. Its agriculture includes livestock, dairy products and grain. It has mining (bituminous coal and limestone) and some manufacturing (leather goods and aluminum). **14.** County (1970 pop. 45,422), 611 sq mi (1,582.5 sq km), E Pa., bounded in the E by the Delaware River; formed 1836; co. seat Stroudsburg. The county is cut by the scenic Delaware Water Gap. The first settlers, arriving in 1725, worked copper mines. Today machinery, metal products, and textiles are its major products. It has numerous recreational facilities, and tourism is important to the economy. **15.** County (1970 pop. 23,475), 660 sq mi (1,709.4 sq km), SE Tenn., bounded in the SE and the E by N.C., in the NE by the Little Tennessee River, and drained by its tributaries, with the Unicoi Mts. along the S border; formed 1819; co. seat Madisonville. It is a lumbering., livestock-raising and agricultural (fruit, tobacco, hay, corn, and dairy products) region. It has an iron industry and manufactures clothing and furniture. Part of Cherokee National Forest is here. **16.** County (1970 pop. 11,272), 473 sq mi (1,225.1 sq km), SE W.Va., in a mountainous region bordered on the S and E by Va. and drained by tributaries of the New and Greenbrier rivers; formed 1799; co. seat Union. In a livestock-raising, dairying, and fruit-growing region, it has deposits of limestone, iron ore, and natural gas. Some manufacturing of rubber and plastic products is done. **17.** County (1970 pop. 31,610), 915 sq mi (2,369.9 sq km), W central Wis., drained by the Black, La Crosse, and Kickapoo rivers; formed 1854; co. seat Sparta. It is in a dairying and farming (tobacco, grain, livestock, and poultry) region. Its industries include lumbering, food processing, and the manufacture of wood products, furniture, paper, iron foundries, metal products, and farm machinery.

Monroe. 1. Residential town (1970 pop. 12,047), Fairfield co., SW Conn.; settled c.1755, inc. 1823. Electronic components are among the town's manufactures. **2.** City (1970 pop. 8,071), seat of Walton co., N central Ga., E of Atlanta; inc. 1821. Monroe is a trade and processing center in a cotton and farm area. **3.** Industrial city (1978 est. pop. 63,200), seat of Ouachita parish, SE La., on the Ouachita River; founded c.1785, inc. as a city 1900. The center of a large natural-gas field (discovered 1916), it has important chemical and carbon-black plants, as well as pulp, paper, and lumber mills. The first settlers founded (c.1785) Fort Miró. The community was renamed in 1819 after the *James Monroe*, the first steamship to come up the Ouachita. **4.** City (1978 est. pop. 24,600), seat of Monroe co., SE Mich., on Lake Erie; settled 1778, inc. 1837. Paper products, fuel-burning equipment, chairs, and auto parts are made. The city has large nurseries and is the shipping point for a farm region. Limestone quarries and an atomic breeder reactor plant are nearby. Monroe was the scene of the River Raisin massacre during the War of 1812. George A. Custer lived here, and the local museum has a large collection of Custer memorabilia. **5.** City (1970 pop. 11,282), seat of Union co., S N.C., in the Piedmont; settled 1751, inc. 1844. Poultry is processed, and textiles, metal alloys, surgical equipment, and aircraft parts are made. **6.** City (1975 est. pop. 9,200), seat of Green co., S Wis., N of Freeport, Ill., in a dairy region; inc. as a village 1859, as a city 1882. Cheese products are made.

Mon·roe·ville (mən-rō'vĭl'), city (1970 pop. 4,846), seat of Monroe co., SW Ala., near the Alabama River NNE of Montgomery; settled c.1815. It is a trade center for a farm and timber area.

Mon·ro·vi·a (mən-rō'vē-ə), city (1970 est. pop. 100,000), capital of the Republic of Liberia, NW Liberia, a port on the Atlantic Ocean at the mouth of the St. Paul River. The city's economy revolves around its harbor, which was substantially improved by U.S. forces under lend-lease during World War II. In 1948 the first port capable of handling oceangoing vessels was opened; there are now several ports, including a free port. The main exports are latex and iron ore. The city also has extensive storage and ship-repair facilities. Manufactures include cement, refined petroleum, food products, bricks and tiles, furniture, and pharmaceuticals. Roads and railroads connect Monrovia with Liberia's interior. Monrovia was founded in 1822 by the American Colonization Society as a haven for freed slaves from the United States and the British West Indies and was named for James Monroe, then U.S. President. Descendants of those early settlers still control the city. The Univ. of Liberia (1862) and Cuttington College and Divinity School (1889; Episcopal) are here.

Monrovia, city (1978 est. pop. 30,000), Los Angeles co., S Calif., in the foothills of the San Gabriel Mts.; inc. 1886. The city has industries that manufacture electronic equipment, plastics, chemicals, machinery, and printed materials.

Mons (mônꜱ), city (1977 est. pop. 93,332), capital of Hainaut prov., SW Belgium, near the French border. It is the processing and shipping center of the Borinage coal-mining district and is also a manufacturing center. Known since the 7th cent., Mons became (1295) the seat of the counts of Hainaut. In the wars of the 16th to 18th cent. it was often attacked and occupied by Dutch, Spanish, and French forces. In World Wars I and II the city was the site of several battles. Of note in Mons are the Gothic Church of St. Waltrude (15th–16th cent.), the city hall (15th cent.), and many beautiful houses of the 16th to 18th cent.

Mon·tague (mŏn-tāg'), county (1970 pop. 15,326), 932 sq mi (2,413.9 sq km), N Texas, bounded on the N by the Red River and the Okla. border and drained by tributaries of the Red and Trinity rivers; formed 1857; co. seat Montague. It has diversified agriculture (peanuts, cotton, corn, grain, fruit, truck crops, poultry, dairy products, and cattle), oil and natural-gas fields, and some manufacturing and processing plants.

Mon·ta·gue (mŏn'tə-gyōō'), uninc. village (1970 pop. 350), seat of Montague co., N Texas, ESE of Wichita Falls. It is a market center in an agricultural area.

Mon·tan·a (mŏn-tăn'ə), state (1975 pop. 740,000), 147,138 sq mi (381,087 sq km), NW United States, in the Rocky Mt. region, admitted as the 41st state of the Union in 1889. Helena is the capital. The state lies on the northwest border of the United States, south of the Canadian provinces of British Columbia, Alta., and Sask. It is bounded by N.Dak. and S.Dak. on the east, by Wyo. and Idaho on the south, and by Idaho on the west.

Montana is thinly populated and has many remote areas. In the eastern half of the state are broad plains, drained by the Missouri River, which originates in southwest Montana, and by its tributaries, the Milk, the Marias, the Sun, and especially the Yellowstone. Much of Montana's western boundary is marked by the crest of the Bitterroot Range, part of the Rocky Mts., which dominate the western section of the state and through which runs the Continental Divide.

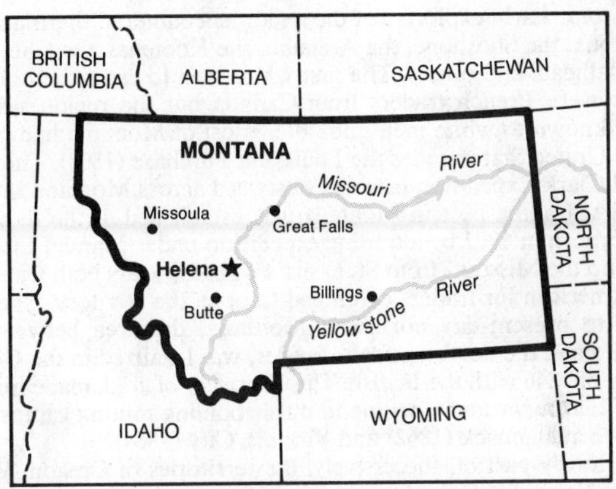

Montana's very name is derived from the Spanish word *montaña*, meaning mountain country. High granite peaks, green forests, blue lakes, and such natural wonders as those of Glacier National Park have helped make tourism a growing industry and a major source of income. Other places of interest include Custer Battlefield National Monument, Big Hole National Battlefield, Grant-Kohrs Ranch National Historic Site, and the National Bison Range, near Ravalli, where herds of buffalo may still be seen. Strips of Yellowstone National Park, including the north and west entrances, are also in Montana, as are such Indian reservations as the Blackfoot, the Fort Belknap, the Fort Peck, and the Crow. The many kinds of fish found in the rushing mountain streams and innumerable lakes bring fishermen to the state, and the abundant wildlife—elk, deer, bear, moose, and waterfowl—attract hunters. The state's outstanding recreational areas also include facilities for skiing, hiking, boating, and swimming.

Montana is divided into 56 counties:

NAME	COUNTY SEAT	NAME	COUNTY SEAT
Beaverhead	Dillon	Meagher	White Sulphur Springs
Big Horn	Hardin		
Blaine	Chinook	Mineral	Superior
Broadwater	Townsend	Missoula	Missoula
Carbon	Red Lodge	Musselshell	Roundup
Carter	Ekalaka	Park	Livingston
Cascade	Great Falls	Petroleum	Winnett
Chouteau	Fort Benton	Phillips	Malta
Custer	Miles City	Pondera	Conrad
Daniels	Scobey	Powder River	Broadus
Dawson	Glendive	Powell	Deer Lodge
Deer Lodge	Anaconda	Prairie	Terry
Fallon	Baker	Ravalli	Hamilton
Fergus	Lewistown	Richland	Sidney
Flathead	Kalispell	Roosevelt	Wolf Point
Gallatin	Bozeman	Rosebud	Forsyth
Garfield	Jordan	Sanders	Thompson Falls
Glacier	Cut Bank	Sheridan	Plentywood
Golden Valley	Ryegate	Silver Bow	Butte
Granite	Philipsburg	Stillwater	Columbus
Hill	Havre	Sweet Grass	Big Timber
Jefferson	Boulder	Teton	Choteau
Judith Basin	Stanford	Toole	Shelby
Lake	Polson	Treasure	Hysham
Lewis and Clark	Helena	Valley	Glasgow
Liberty	Chester	Wheatland	Harlowton
Lincoln	Libby	Wibaux	Wibaux
McCone	Circle	Yellowstone	Billings
Madison	Virginia City		

Economy. In and around the mountainous western region are the large mineral deposits for which Montana is famous—copper, silver, gold, zinc, lead, and manganese. The eastern part of the state is noted for its petroleum and natural gas, and there are also vast coal deposits. Montana also mines vermiculite, chromite, fluorspar, tungsten, uranium, and phosphate rock. In 1972 the most valuable minerals produced were petroleum, copper, sand and gravel, and gold and silver. Leading industries manufacture forest products, processed food, refined petroleum, and coal products. In eastern Montana much of the high grass of the Great Plains is gone, but the cattle and sheep remain, grazing mainly on short grass. Agriculture, with the aid of irrigation, provides a major share of Montana's income. Cattle are the most valuable farm item, and dairy products are also important. The principal crops raised are wheat, hay, barley, and sugar beets.

History. Early explorers of the region encountered the Blackfoot, the Sioux, the Shoshone, the Arapaho, the Kootenai, the Cheyenne, the Flathead, and others. The first white men to cross the northern plains were French traders from Canada but the region was not really known to white men until after most of Montana had passed to the United States under the Louisiana Purchase (1803). The Lewis and Clark Expedition traveled westward across Montana in 1805. The first trading post in Montana was established at the mouth of the Bighorn in 1807 by a trading expedition under Manuel Lisa that came up the Missouri from St. Louis. For some years both Canadian and American fur traders continued to open the territory. The U.S. claim to present-day northwest Montana, the area between the Rockies and the northern Idaho border, was legalized in the Oregon Treaty of 1846 with the British. The discovery of gold, made initially in 1852, brought many people to mushrooming mining camps such as those at Bannack (1862) and Virginia City (1864).

Previously part of, successively, the territories of Oregon, Washington, Nebraska, Dakota, and Idaho, Montana itself became a territory in 1864. The territory was still a rough frontier, however, and the first governor, Sidney Edgerton, was driven out of the region. After the Civil War the grasslands attracted ranchers, and the first cattle were brought in from Texas in 1866 over the Bozeman Trail, to the area east of the Bighorn Mts. The Sioux did not tamely submit to having their lands taken from them, and in 1876 at the Battle of the Little Bighorn they won against Gen. George A. Custer and his force one of the greatest of all Indian victories. The Sioux were, however, subdued, and the gallant attempt of Chief Joseph of the Nez Perce Indians to lead his people into Canada to escape pursuing U.S. troops had its pitiful end in Montana. Great ranches spread out across the plains, and cow towns that were to grow into cities such as Billings and Missoula sprang up as the railroads were built in the West (c.1880–c.1910). Achievement of statehood in 1889 and the building of the railroads put an end to the era of the open range.

The discovery of silver at Butte (1875) had been followed (c.1880) by discovery of copper at that same "richest hill on earth." Montana's fate was subsequently linked to copper, and the Amalgamated Copper Company (later renamed Anaconda Copper Mining Company) came to to exercise control over state politics. There were struggles between the company and the workingmen that led to strikes, disorder, and bloodshed, but they led also to enactment of some early measures for social security. Farmers, brought out by the trainload to develop the lands of eastern Montana, planted their fields in the second decade of the 20th cent. The initial yield of wheat was great, but the calamitous drought of 1919 and the consequent dust storms seared the fields. Montana was already accustomed to depression by 1929. By the late 1940s Federal dam and irrigation projects had opened many acres to cultivation. The eastern Montana oil boom of the early 1950s stimulated Montana's economy, and diversified manufacturing has recently been growing.

Government. In 1973 Montana implemented a new constitution, which replaced the one adopted in 1889. The governor of the state is elected for a term of four years and may be re-elected. The Legislative Assembly is made up of a senate with 50 members and a house of representatives with 100 members. State senators are elected for terms of four years and representatives for terms of two years.

Educational Institutions. The state's major institutions of higher learning are included in the Univ. of Montana and Montana State Univ. systems.

Mon·tau·ban (môN-tō-bäN'), city (1975 pop. 35,940), capital of Tarn-et-Garonne dept., S France, on the Tarn River. It is a commercial and industrial center where aeronautic and electrical equipment, food products, textiles, shoes, and tiles are produced. Founded in 1144, Montauban was a stronghold of the Albigenses in the 13th cent. and of the Huguenots in the 16th cent. It enjoyed prosperity until the time of Louis XIV's religious persecutions (17th cent.).

Mon·tauk Point (mŏn'tôk'), E extremity of the S peninsula of Long Island, SE N.Y. Approximately 115 mi (185 km) E of Manhattan, it is the easternmost point of the state. It has been the site of a lighthouse since 1795.

Mont·bé·liard (môN-bā-lyär'), industrial town (1975 pop. 30,425), Doubs dept., E France, on the Rhône-Rhine Canal. Among its manufactures are clocks, textiles, and wood and metal products. With its surrounding countryside it constituted a county (after the 12th cent.) of the Holy Roman Empire. The county passed (1397) to the counts (later dukes) of Württemberg, who held it, with interruptions, until its capture by French Revolutionary troops in 1793. The town was a Huguenot refuge during the Reformation. It was formally ceded to France by the Treaty of Lunéville (1801).

Mont Blanc (môN bläN'), Alpine massif, on the French-Italian border, SE of Geneva. One of its several peaks, also called Mont Blanc (15,771 ft/4,810.2 m), is the highest peak in France and the second highest in Europe. The southeastern (Italian) face is a massive wall; on the northwestern slopes are numerous glaciers. There are many hotels and hostels along the base of Mont Blanc. The first successful ascent of Mont Blanc was made in 1786. In 1965 a highway tunnel (7 mi/11.3 km long) under Mont Blanc, linking Chamonix with Courmayeur, Italy, was opened to traffic. It provides a short, year-round route between Paris and Rome.

Mont·calm (mŏnt-käm'), county (1970 pop. 39,660), 712 sq mi (1,844.1 sq km), central Mich., drained by the Flat and Pine rivers and Fish Creek; organized 1850; co. seat Stanton. There is much general farming (potatoes, grain, beans, corn, and livestock) and some manufacturing. Oil and gas wells and refineries are found here. The county has several lake resorts.

Mont·clair (mŏnt-klâr'). **1.** Residential city (1978 est. pop. 21,300), San Bernardino co., SE Calif., in a citrus fruit area; inc. 1956. It has some light manufacturing. **2.** Town (1978 est. pop. 42,300), Essex co., NE N.J., a suburb of Newark and New York City, on a slope of the Watchung Mts.; settled c.1666 as part of Newark, set off from Newark 1812, set off from Bloomfield and inc. 1868.

Mont-de-Mar·san (môN-də-mär-säN'), town (1975 pop. 26,166), capital of Landes dept., SW France. It is a commercial center where important fairs are held.

Mon·te Al·bán (mōn'tā äl-bän'), ancient city, c.7 mi (11.3 km) from Oaxaca, SW Mexico, capital of the Zapotec. Monte Albán was built on an artificially leveled, rocky promontory above the Valley of Oaxaca. Located around an enormous plaza are long, low buildings set off by sunken courts and stairways. The tombs have yielded great archaeological treasure—jewelry of gold, copper, jade, rock crystal, obsidian, and turquoise mosaic and bone and wood carving showing elaborate religious symbolism. Excavation was begun (1931) by the Mexican archaeologist Alfonso Caso. The Zapotec apparently had an advanced culture here c.200 B.C. and already were using the bar and dot system of numerals used by the Maya. The final epoch (c.1300–1521), terminated by the Spanish Conquest, covers the ascendancy of the Mixtec, when the Zapotec were driven from Monte Albán and Mitla.

Mon·te·bel·lo (mŏn'tə-bĕl'ō), city (1970 pop. 42,807), Los Angeles co., S Calif., a residential and industrial suburb of Los Angeles; inc. 1920.

Mon·te Car·lo (mŏn'tē kär'lō, môN-tā' kär-lō'), town (1968 pop. 9,948), principality of Monaco, on the Mediterranean Sea and the French Riviera. It is a tourist center noted for its gambling casino (built 1858) and for its scenery, fine villas, and luxurious hotels.

Mon·te Cris·to (mŏn'tē krĭs'tō, môN'tā krēs'tō), unpopulated, rocky island, 6 sq mi (15.5 sq km), belonging to Italy, in the Tyrrhenian Sea between Corsica and the Italian coast. It owes its fame to the novel by Alexandre Dumas père, *The Count of Monte Cristo.*

Mon·te·go Bay (mŏn-tē'gō), city (1970 pop. 43,754), NW Jamaica. A port, railroad terminus, and commercial center, it is also one of the most popular resorts in the Caribbean.

Mon·tel·lo (mŏn-tĕl'ō), city (1970 pop. 1,082), seat of Marquette co., central Wis., W of Fond du Lac on the Fox River; settled 1849, inc. 1938. John Muir lived here as a boy.

Mon·te·ne·gro (mŏn'tə-nē'grō), constituent republic of Yugoslavia

(1971 pop. 530,361), 5,332 sq mi (13,810 sq km), SW Yugoslavia. Its name means "black mountain." Titograd is the capital. Situated at the southern end of the Dinaric Alps, Montenegro is almost entirely mountainous and its access is difficult. The barren karst of Montenegro proper, on the west, is separated by the Zeta River and its plain from the higher Brda region, on the east, which has forests and pastures. Sheep and goat raising are important occupations. Only about 6% of the area is cultivated.

The Montenegrin people are Serbs, but they are recognized as a separate ethnic nationality. They belong mostly to the Orthodox faith. From the 14th to the 19th cent. their principal activity was fighting the Turks, who never entirely conquered their mountain stronghold. The region constituting present Montenegro was in the 14th cent. the virtually independent principality of Zeta in the Serbian empire. After Serbia was defeated by the Turks in the Battle of Kossovo (1389), Montenegro continued to resist and became a refuge for Serbian nobles who fled Turkish rule. However, the princes of Montenegro ruled only a small part of the present republic, the rest being governed by Turkey after 1499 and by Venice, which held Kotor, the main port. From 1515 until 1851 the rule of Montenegro was vested in the prince-bishops of Cetinje; these were assisted by civil governors.

With Danilo I, who ruled from 1696 to 1735, the episcopal succession was made hereditary in the Niegosh family, the office passing ordinarily from uncle to nephew. Danilo I also inaugurated (1715) the traditional alliance of Montenegro with Russia. Peter I, who reigned from 1782 to 1830, instituted internal reforms and sought to end the blood feuds and endemic lawlessness. Peter II (reigned 1830–51), continued his predecessor's work of reform and fostered a revival of learning and culture; aside from occasional border warfare, he lived in relative peace with his neighbors, Turkey and Austria. Danilo II, who succeeded him, secularized his principality in 1852 and transferred his ecclesiastic functions to an archbishop. Under Nicholas I (reigned 1860–1918) Montenegro was formally recognized as an independent state at the Congress of Berlin (1878), which increased its territory and gave it a narrow outlet on the Adriatic. In 1910 Nicholas proclaimed himself king. He fought Turkey in the Balkan Wars and took Shkodër in 1913, but was forced by the pressure of the European powers to evacuate the city. Montenegro did, however, receive part of the territory claimed by newly independent Albania, and when World War I broke out (1914), the Montenegrins invaded Albania. Montenegro declared war on Austria in Aug., 1914, but late in 1915 it was overrun by Austro-German forces. In Nov., 1918, a national assembly declared Nicholas deposed and effected the union of Montenegro with Serbia. In 1946 Montenegro became one of the six republics of Yugoslavia, and its territory was enlarged with the addition of part of the Dalmatian coast.

Mon·te·rey (mŏn'tə-rā'), county (1970 pop. 247,450), 3,324 sq mi (8,609.2 sq km), W Calif., on the Pacific and Monterey Bay, bounded on the N by the Pajaro River valley, with the Salinas River valley in its center and flanked on the E by the Gabilan and Diablo ranges, on the W by the Santa Lucia Range; formed 1850; co. seat Salinas. The Salinas valley is a rich agricultural area that produces lettuce, other vegetables, sugar beets, beans, grain, fruit, dairy products, livestock, and poultry. It has oil and gas wells, sand and gravel pits, and magnesite mines. Much of its industry handles food products and it also manufactures clothing, wood products, chemicals, concrete, minerals, and machinery. Monterey Peninsula is a famed scenic resort region. The county includes Big Sur (redwoods), old Soledad, Carmel, and San Antonio de Padua missions, U.S. Fort Ord and Hunter Liggett Military Reservation, and part of Los Padres National Forest in the east.

Monterey. 1. City (1978 est. pop. 28,000), Monterey co., W Calif., a port on Monterey Bay; founded 1770, inc. 1850. It is a popular resort and the home of many artists and writers. One of the oldest cities in California, it is rich in historic tradition. The bay was discovered by Juan Cabrillo in 1542 and entered and named by Sebastián Vizcaíno in 1602. In 1770 an expedition under Gaspar de Portolá arrived and established a presidio. Junípero Serra remained to found a Franciscan mission (which a year later was moved near what is now Carmel). Monterey was the capital of Alta California during many of the years between 1775 and 1846. In 1846 the city was taken by a U.S. naval force under Com. John D. Sloat, and in 1849 the state constitutional convention met here. Monterey became a whaling and fishing center. California's first theater (1844) and first brick building (1847) are still standing, and it was in Monterey that California's first newspaper was established in 1846. The many old structures include the customs house (1827) and the jail (1854). There are numerous museums. The Presidio of Monterey (1770) is the home of a branch of the U.S. army language school. **2.** Town (1970 pop. 223), seat of Highland co., NW Va., in the Alleghenies NW of Stauton, in an agricultural and lumber-milling area. Good fishing attracts visitors to this scenic area.

Monterey Park, city (1978 est. pop. 52,400), Los Angeles co., S Calif., a residential suburb of Los Angeles; inc. 1916.

Mon·ter·rey (mŏn-tə-rā', mōn-tā-rā'), city (1974 est. pop. 1,006,200), capital of Nuevo León state, NE Mexico. Located in a valley surrounded by mountains, Monterrey is a rail and highway hub, an important industrial center, and the site of the nation's largest iron and steel foundries. Monterrey's numerous industries also include breweries, glass factories, metalworks, paper plants, cotton and flour mills, and factories manufacturing construction materials, electrical equipment, furniture, and textiles. Mexico's chief lead smelting center, Monterrey also produces silver, gold, copper, antimony, and bismuth. Its moderate, dry climate, cool mountains, and hot springs make Monterrey a popular resort. The city was founded in 1579. During the Mexican War, it was captured by Zachary Taylor after a courageous defense (Sept. 19–24, 1846) by the besieged Mexicans. The Obispado chapel and its cathedral (18th cent.) are good examples of colonial architecture.

Mon·te·sa·no (mŏn'tə-sä'nō), city (1977 est. pop. 2,790), seat of Grays Harbor co., W Wash., W of Olympia; settled 1852, platted 1869, inc. 1883. It is a processing center in a farm and timber area.

Mon·te·vi·de·o (mŏn'tə-vĭ-dā'ō, môn'tā-vē-*th*ā'ō), city (1975 pop. 1,229,748), S Uruguay, capital of Uruguay, on the Río de La Plata. It is one of the major ports of South America and the governmental, financial, and commercial center of Uruguay. Much of the South Atlantic fishing fleet is based in Montevideo. The city has industries producing textiles, dairy items, wines, and packaged meats. Tourism is also important. Montevideo's origins lay in the colonial rivalry of the Spanish and Portuguese. The Portuguese constructed (1717) a fort on top of the hill that overlooks the harbor. Captured by the Spanish in 1724, the fort became the nucleus of the settlement founded in 1724 by the governor of Buenos Aires. Montevideo became the capital of Uruguay in 1828. It suffered during Uruguay's 19th cent. civil wars and was besieged from 1843 to 1851. Today Montevideo is spacious, modern, and attractive, with broad, tree-lined boulevards, numerous beautiful parks, and fine buildings and residences.

Mon·te·vid·e·o (mŏn'tə-vĭd'ē-ō), city (1970 pop. 5,661), seat of Chippewa co., SW Minn., on the Minnesota River NW of Granite Falls; platted 1870, inc. as a village 1879, as a city 1908. It is a farm trade center and makes dairy products.

Mon·te·zu·ma (mŏn'tĭ-zo͞o'mə), county (1970 pop. 12,952), 2,095 sq mi (5,426.1 sq km), extreme SW Colo., bordering on N.Mex., and Utah, bounded on the E by La Plata Mts., and drained by the Dolores River; formed 1889; co. seat Cortez. It is in a livestock-raising region, with oil and gas wells, sand and gravel pits, and a lumbering industry. It includes Mesa Verde National Park, Yucca House National Monument, Hovenweep National Monument in the west, and part of San Juan National Forest in northeast.

Montezuma, city (1970 pop. 1,353), seat of Poweshiek co., SE central Iowa, E of Des Moines; inc. 1868.

Montezuma Castle National Monument, 842 acres (341 hectares), central Ariz.; est. 1906. Montezuma Castle, built c.1250, is a 5-story, 20-room apartment house perched high in the cavity of a cliff. It was named by early settlers who believed it had been built by the Aztecs.

Mont·fer·rat (mŏnt-fə-răt', -rät'), historic region of Piedmont, NW Italy, south of the Po River. In the late 10th cent. Montferrat was a marquisate held by the Aleramo family, and its rulers played an important role in the Crusades. In 1310 it passed to the Paleologo family. With the extinction of the Paleologo line, Emperor Charles V gave (1536) Montferrat to the Gonzaga family of Mantua, despite the claims of the house of Savoy. After Francesco Gonzaga's death in 1612, Savoy renewed its claims on Montferrat and invaded (1613) the region. Spain and France intervened. The Treaty of Cherasco (1631) assigned parts of Montferrat to the house of Savoy, and the rest passed to the Nevers (French) branch of the Gonzaga family. All of Montferrat was recognized by the Peace of Utrecht (1713) as belonging to the house of Savoy.

Mont·gom·er·y (mŏnt-gŭm′ə-rē, -gŭm′rē). **1.** County (1970 pop. 167,790), 790 sq mi (2,046.1 sq km), SE central Ala., in the Black Belt bounded on the NW by the Alabama River, on the N by the Tallapoosa River; formed 1816; co. seat Montgomery. It has agriculture (cotton, grain, livestock, and bees), lumber, sand and gravel pits, a food-processing industry, and plants manufacturing clothing, furniture, concrete, metal products, and machinery. **2.** County (1970 pop. 5,821), 775 sq mi (2,007.3 sq km), W Ark., drained by the Ouachita and Caddo rivers; formed 1842; co. seat Mount Ida. Its agriculture includes cotton, grain, livestock, poultry, and dairy products. It has stone quarrying, lumber sawmilling, cotton ginning, and food processing, and manufactures shoes and leather gloves. Part of Ouachita National Forest is in the county. **3.** County (1970 pop. 6,099), 235 sq mi (608.7 sq km), SE central Ga., bounded on the W by the Oconee River, on the S by the Altamaha River; formed 1793; co. seat Mount Vernon. It is in a coastal-plain agricultural area that produces cotton, corn, tobacco, peanuts, and livestock. Logging is done, and clothing is manufactured. **4.** County (1970 pop. 30,260), 706 sq mi (1,828.5 sq km), S central Ill., drained by Shoal and Macoupin creeks; formed 1821; co. seat Hillsboro. Its agriculture includes corn, wheat, oats, soybeans, livestock, poultry, and dairy products. There are zinc smelters and bituminous-coal mines here. Among its manufactures are paper boxes, glass jars, concrete blocks, shoes, radiators, metal products, and beverages. **5.** County (1970 pop. 33,930), 507 sq mi (1,313.1 sq km), W central Ind., drained by Sugar and Raccoon creeks; formed 1822; co. seat Crawfordsville. It has agriculture (grain, truck crops, livestock, poultry, and dairy products), clay pits, and timber. Wood products, steel and copper, sheet metal, miscellaneous metals products, and machinery are produced. **6.** County (1970 pop. 12,781), 422 sq mi (1,093 sq km), SW Iowa, drained by the East Nishnabotna, West Nodaway, and Tarkio rivers; formed 1851; co. seat Red Oak. Its prairie agriculture includes hogs, cattle, corn, wheat, and oats. There are bituminous-coal deposits in the area. **7.** County (1970 pop. 39,949), 628 sq mi (1,626.5 sq km), SE Kansas, in a level to hilly area bordering in the S on Okla. and drained by the Verdigris and Elk rivers; formed 1867; co. seat Independence. It lies in a livestock, grain, poultry, and dairy region, with numerous oil and gas wells. It has food-processing, lumbering, and iron and aluminum industries and manufactures clothing, furniture, plastics, cement, metal products, and machinery. **8.** County (1970 pop. 15,364), 204 sq mi (528.4 sq km), E Ky., drained by several creeks; formed 1796; co. seat Mount Sterling. It is in a rolling upland agricultural area yielding livestock, poultry, burley tobacco, corn, wheat, and dairy products. Clothing and machinery are manufactured. **9.** County (1970 pop. 522,809), 494 sq mi (1,279.5 sq km), central Md., bounded on the NE by the Patuxent River, on the S by the District of Columbia, on the W and SW by the Potomac River and the Va. border; formed 1776; co. seat Rockville. This rolling piedmont area contains many residential suburbs. The agricultural hinterland produces truck crops, apples, corn, wheat, hay, cattle, poultry, and dairy products. Textiles, processed foods, lumber millwork, furniture, chemicals, plastics, and machinery are manufactured. **10.** County (1970 pop. 12,918), 403 sq mi (1,043.8 sq km), N central Miss., drained by the Big Black River; formed 1871; co. seat Winona. It is in a cotton, corn, livestock, and dairy-farming area, with timber sawmilling and manufacturing (wood products, concrete, and machinery). **11.** County (1970 pop. 11,000), 533 sq mi (1,380.5 sq km), E central Mo., on the Missouri River and drained by the Loutre River; formed 1818; co. seat Montgomery City. Farming (corn, wheat, and oats) and livestock raising are the major occupations. It also has limestone quarries and fire-clay pits. **12.** County (1970 pop. 55,883), 409 sq mi (1,059.3 sq km), E central N.Y., in the fertile Mohawk River Valley, traversed by the Barge Canal and drained by Scholarie Creek; formed 1772, name changed from Tryon in 1784; co. seat Fonda. It is in a dairying and farming region, with diversified manufactures (textiles, carpets, clothing, paper products, chemicals, and machinery). **13.** County (1970 pop. 19,267), 488 sq mi (1,263.9 sq km), central N.C., bounded on the W by the Yadkin River and drained by the Uharie River; formed 1778; co. seat Troy. This forested piedmont area specializes in farming (peaches, cotton, tobacco, and corn), dairying, poultry raising, lumbering, and textile manufacturing. **14.** County (1970 pop. 608,413), 459 sq mi (1,188.8 sq km), W Ohio, intersected by the Great Miami, Stillwater, and Mad rivers and by the small Bear, Wolf, and Twin creeks; formed 1803; co. seat Dayton. Its agriculture includes livestock, grain, tobacco, dairy products, and truck crops, and it has sand and gravel pits. There is extensive manufacturing in the county, with food-processing, textile, lumbering, and metal industries. Paper, concrete, chemical and plastic products are made. **15.** County (1970 pop. 624,080), 492 sq mi (1,274.3 sq km), SE Pa., in an agricultural and manufacturing area bounded in the SE by the city of Philadelphia and drained by the Schuylkill River; formed 1783; co. seat Norristown. Its manufactures include textiles, metal products, and rubber goods. There are limestone deposits in the area. **16.** County (1970 pop. 62,721), 539 sq mi (1,396 sq km), N Tenn., bordered in the N by Ky. and drained by the Cumberland and Red rivers; formed 1796; co. seat Clarksville. It has agriculture (especially dark tobacco), livestock ranches, and iron-ore and limestone deposits. Its industries include food and tobacco processing, lumbering, and the manufacture of clothing, paper products, leather, metal products, and machinery. **17.** County (1970 pop. 49,479), 1,090 sq mi (2,823.1 sq km), E Texas, drained by tributaries of the San Jacinto River; formed 1837; co. seat Conroe. Livestock raising, dairying, and farming (peanuts, sweet potatoes, truck crops, fruit, some corn, and cotton) are done. Lumbering and the production and processing of oil and natural gas are also important. **18.** County (1970 pop. 47,157), 394 sq mi (1,020.5 sq km), SW Va., mainly in the Great Appalachian Valley, with the Alleghenies in the N and NW and the Blue Ridge in the S and SE, bounded in the W by the New River and drained by the Little and Roanoke rivers; formed 1776; co. seat Christiansburg. It has farming (grain, hay, and fruit), livestock raising, and dairying. There is some coal mining, and its industries include food processing, and the manufacture of furniture, machinery, and electrical equipment. Part of Jefferson National Forest is here.

Montgomery, city (1978 est. pop. 161,000), state capital and seat of Montgomery co., E central Ala., near the head of navigation on the Alabama River just below the confluence of the Coosa and Tallapoosa rivers, and in the rich Black Belt; inc. 1819. It is an industrial city and an important market center for agricultural goods, especially cotton, livestock, and dairy products. Manufactures include machinery, glass products, textiles, refrigeration equipment, axles, furniture, food items, and paper. Montgomery became the state capital in 1847 and boomed as a river port and cotton market. In the capitol building (erected 1857) the convention met (Feb., 1861) that formed the Confederate States of America. Jefferson Davis was inaugurated president on the capitol steps, and the city served as the Confederate capital until the seat was moved to Richmond in May, 1861. The city was occupied by Federal troops in the spring of 1865. It is the seat of Alabama State Univ. Maxwell Air Force Base and Gunter Air Force Base are nearby. Points of interest include the "first White House of the Confederacy" (built c.1825), preserved as a Confederate museum; a planetarium; a museum of fine arts; the state archives and history museum; and many ante-bellum homes and buildings.

Montgomery City, city (1970 pop. 2,187), seat of Montgomery co., E central Mo., WNW of St. Louis; laid out 1853, inc. 1859.

Mont·gom·er·y·shire (mŏnt-gŭm′ə-rē-shîr′, -shər, -gŭm′rē-), former county, central Wales. The county town was Montgomery. In medieval times Montgomeryshire was important as the heavily fortified border district of the Norman lords of the Welsh Marches. In 1974 Montgomeryshire became part of the new nonmetropolitan county of Powys.

Mon·ti·cel·lo (mŏn′tĭ-sĕl′ō), 640 acres (259.2 hectares), central Va., near Charlottesville; home of Thomas Jefferson for 56 years. The mansion, which he designed, was begun in 1770 on property inherited from his father. The house is one of the earliest examples of the American classic revival. Not long after Jefferson's death, his daughter, unable to maintain the property, sold it, retaining only the family burial plot in which Jefferson is interred. In 1923 Jefferson P. Levy sold Monticello to the Thomas Jefferson Memorial Foundation. Dedicated as a national shrine in 1926, and extensively renovated during the next 30 years, the estate was opened to the public in 1954.

Monticello. 1. City (1972 est. pop. 7,034), seat of Drew co., SE Ark., SE of Pine Bluff. A processing center for a farm and timber area, it makes boats, hardware, furniture, and rugs. **2.** Town (1970 pop. 2,473), seat of Jefferson co., NW Fla., near Lake Miccosukee ENE of Tallahassee; settled in the early 19th cent. A trade center in a farm and dairy area, it also makes tung oil and plywood. **3.** City (1970 pop. 2,132), seat of Jasper co., central Ga., N of Macon; inc. 1810. Peanuts and peaches are grown. **4.** City (1973 est. pop. 4,360, seat of Piatt co., central Ill.; on the Sagamon River NE of Decatur, in an agricultural region; inc. 1841. **5.** City (1970 pop. 4,869), seat of

White co., NW Ind., on the Tippecanoe River W of Logansport; settled 1831, laid out 1834, inc. 1853. It is a resort center in an agricultural area. **6.** City (1970 pop. 3,618), seat of Wayne co., S Ky., in rolling country between Lake Cumberland and Cumberland National Forest; settled before 1800. It is in a farm and timber area, with oil fields in the vicinity. **7.** Town (1970 pop. 1,790), seat of Lawrence co., S central Miss., S of Jackson on the Pearl River, in a pine-timber and poultry area; founded 1798. Lumber products are made. **8.** Town (1970 pop. 157), seat of Lewis co., NE Mo., on North Fabius River, near the Mississippi NW of Quincy, Ill. **9.** Resort village (1970 pop. 5,991), seat of Sullivan co., SE N.Y., NW of Middletown, in a lake region; inc. 1830. **10.** City (1970 pop. 1,431), seat of San Juan co., SE Utah, in the Abajo Mts. SSE of Moab, at an altitude of 7,066 ft (2,155.1 m); founded 1887 by Mormons. Ores from nearby uranium and vanadium mines are processed here.

Mon·til·la (môn-tē'yä), town (1970 pop. 22,059), Córdoba prov., S Spain, in Andalusia. It is the center of an agricultural district famous for wines. Deriving from Montilla wines, the term "amontillado" now designates a type of sherry wine.

Mont·lu·çon (môN-lü-sôN'), town (1975 pop. 56,468), Allier dept., central France, on the Cher River. Industry developed in the 19th cent. because of nearby coal fields and iron-ore deposits. Today there are metallurgy, rubber, chemical, clothing, synthetic-textile, and wax industries.

Mont·mar·tre (môN-mär'trə), hill in Paris, on the right bank of the Seine River. The highest point of Paris, it is topped by the Church of Sacré-Coeur. Until the 20th cent. Montmartre retained a rural look and provided material for Van Gogh, Pissarro, Utrillo, and other artists. Montmartre is also famed for its night life. The cemetery of Montmartre contains the tombs of Stendhal, Renan, Heine, Berlioz, and Alfred de Vigny. The town of Montmartre was annexed to Paris in 1860. The hill, a natural fortress, played a military role during the Paris Commune (1871) and other periods.

Mont·mo·ren·cy (môN-mô-räN-sē'), town (1975 pop. 20,860), Val d'Oise dept., N France, a suburb N of Paris. Jean Jacques Rousseau lived here (1756–62), first at Hermitage, a cottage on the estate of his friend Mme d'Épinay, and after his quarrel with her, in Montmorency itself.

Mont·mo·ren·cy (mŏnt'mə-rĕn'sē), county (1970 pop. 5,247), 555 sq mi (1,437.5 sq km), NE Mich., drained by the Thunder Bay, Rainy, and Black rivers; organized 1881; co. seat Atlanta. It is a dairying and potato-raising area, as well as a hunting and fishing region, with a state forest, a game refuge, and several small lakes.

Mon·tour (mŏn-tŏŏr'), county (1970 pop. 16,508), 130 sq mi (336.7 sq km), central Pa., in an agricultural and industrial area drained by the Susquehanna River; formed 1850; co. seat Danville. It is known for the manufacture of metal products and clothing, and for its limestone deposits.

Mont·par·nasse (môN-pär-näs'), quarter of Paris, on the left bank of the Seine River. Its famous cafés were long centers of the Parisian artistic and intellectual world. The quarter contains the Pasteur Institute, the ancient catacombs, and the Montparnasse cemetery, with the tombs of Saint-Saëns, Houdon, Baudelaire, Poincaré, César Franck, and Maupassant.

Mont·pel·ier (mŏnt-pēl'yər), city (1978 est. pop. 8,200), state capital (since 1805) and seat of Washington co., central Vt., at the junction of the Winooski and North Branch rivers; inc. 1855. Its economy is dominated by state government and insurance industries. It is also a trading center in a lumber, granite, and winter resort area. There are granite plants in the city. Other manufactures include bakery products, clothespins, plastic products, and sawmill machinery. Of interest are the state capitol and an art gallery for wood sculpture. The city, which is surrounded by mountains, has an excellent view of Mt. Mansfield, the highest point in the state.

Mont·pel·lier (môN-pĕ-lyā'), city (1975 pop. 191,354), capital of Hérault dept., S France, near the Mediterranean coast. Its industries include food processing, salt working, textile milling, printing, and the manufacture of metal items and chemicals. Dating from the 8th cent., Montpellier was the center of a fief under the counts of Toulouse; it passed (13th cent.) to the kings of Majorca, from whom it was purchased (1349) by Philip VI of France. A Huguenot center, it was besieged and taken by Louis XIII in 1622. Montpellier's fame rests principally on its university, founded in 1289. Its noted medical faculty is traced to the 10th cent. Rabelais was a student.

Mon·tre·al (mŏn'trē-ôl', môN'rä-äl'), city (1976 pop. 1,080,546; metropolitan pop. 2,802,485), S Que., Canada, on Montreal Island in the St. Lawrence River. Montreal is a cultural, commercial, financial, and industrial center. It lies at the foot of Mt. Royal—the source of its name—and has an excellent harbor on the St. Lawrence Seaway, which connects the city to the industrial centers of the Great Lakes. Montreal is a transshipment point for oil, grain, sugar, machinery, and manufactured goods. Its manufactures include steel, electronic equipment, aircraft, ships, raw textiles, clothing, and tobacco.

A stockaded Indian village, Hochelaga, was found on the site (1535) by Jacques Cartier, and the island was visited in 1603 by Champlain, but it was not settled by the French until 1642, when a band of priests, nuns, and settlers under Paul de Chomedey, sieur de Maisonneuve, founded the Ville Marie de Montréal. The settlement grew to become an important center of the fur trade and the starting point for the western expeditions of Louis Jolliet, Jacques Marquette, La Salle, La Vérendrye, and Daniel Duluth. It was fortified in 1725 and remained in French possession until 1760, when Vaudreuil de Cavagnal surrendered it to British forces under Lord Jeffrey Amherst. Americans under Richard Montgomery occupied it briefly (1775-76) during the American Revolution. The city's growth was aided by the opening in 1825 of the Lachine Canal, making possible water communications with the Great Lakes. From 1844 to 1849, Montreal was the capital of United Canada. The area of Old Montreal has undergone extensive restoration. Among the city's notable buildings are the Gothic Church of Notre Dame (c.1820), St. Sulpice Seminary (1685), and the Château de Ramezay (1705). McGill Univ., the Univ. of Montreal, and Loyola College are here.

Mon·treuil (môN-trœ'yə), town (1975 pop. 96,587), Seine-Saint-Denis dept., N central France, a suburb of Paris. Long famous for its peaches and pears, Montreuil is now an important industrial center. Among its products are metals, furniture, porcelain and glassware, electrical equipment, dresses, and toys. Montreuil was founded before A.D. 1000.

Mon·treux (môN-trœ'), resort area (1977 est. pop. 20,400), on the NE shore of the Lake of Geneva, Vaud canton, W Switzerland.

Mon·trose (mŏn-trōz'), burgh (1974 est. pop. 10,112), Angus, NE Scotland, on the North Sea at the mouth of the South Esk River. Open to water on three sides, it is a spacious resort town, with flax and jute mills, boat yards, fruit canneries, and a fishing industry. Montrose was the scene of John de Baliol's surrender of the Scottish throne to Edward I of England in 1296.

Montrose, county (1970 pop. 18,366), 2,239 sq mi (5,799 sq km), W Colo., bordering on Utah and drained by the Dolores, San Miguel, and Uncompahgre Rivers; formed 1883; co. seat Montrose. It is in an agricultural region that produces fruit, beans, hay, and livestock and has uranium, radium, coal, silver, and copper mines. Lumber sawmilling and candy processing are done. The county includes parts of Uncompahgre, Gunnison, and Manti-La Sal national forests. Black Canyon of the Gunnison National Monument is in the northeast.

Montrose. 1. City (1970 pop. 6,496), seat of Montrose co., SW Colo., on the Uncompahgre River SE of Grand Junction, in a farm area at an altitude of 5,820 ft (1,775.1 m); founded and inc. 1882. Meat is packed, and sawmilling done. Carnotite deposits nearby are a source of radium. **2.** Borough (1970 pop. 2,058), seat of Susquehanna co., NE Pa., NNW of Scranton; settled 1799, laid out 1811, inc. 1824.

Mon·tross (mŏn'trôs'), town (1970 pop. 419), seat of Westmoreland co., E Va., SE of Fredericksburg. It has canneries. George Washington Birthplace National Monument and Stratford Hall, home of the Lee family, are nearby.

Mont-Saint-Mi·chel (môN-săN-mē-shĕl'), rocky isle (1968 pop. 105) in the Gulf of Saint-Malo, an arm of the English Channel, Manche dept., NW France, 1 mi (1.6 km) off the coast, near Avranches. The isle, accessible by land at low tide, is also linked with the mainland by a causeway (built 1875). The Benedictine abbey of Mont-Saint-Michel was founded in 708. A gigantic group of buildings, rising three stories high, serves, with the summit of the cone-shaped rock, as a base for the great abbey church. Six of these structures on the side facing the sea form the unit called La Merveille, constructed from 1203 to 1228. Strongly fortified, the abbey was frequently assaulted by the English in the Hundred Years' War but was never captured. It remains one of the major tourist attractions of Europe.

Mont·ser·rat (mŏnt'sə-răt'), island (1970 pop. 12,302), 38 sq mi (98.4

sq km), British West Indies, one of the Leeward Islands. It is a rugged, scenic island of volcanic origin. Plymouth is the capital and only outlet for the cotton and other agricultural products of the island. Montserrat was discovered in 1493 by Columbus and colonized by the English in 1632. After changing hands several times between France and Britain, it was definitively awarded to Great Britain in 1783.

Montserrat, mountain, 4,054 ft (1,236.5 m) high, NE Spain, rising abruptly from a plain in Catalonia, NW of Barcelona. On a narrow terrace, more than halfway up its precipitous cliffs, is a celebrated Benedictine monastery. Only ruins are left of the old monastery (11th cent.). The present monastery was built in the 18th cent. and restored after being destroyed by French troops in 1812. The Renaissance church (16th cent.; largely restored in the 19th and 20th cent.) contains the black wooden image of the Virgin which, according to tradition, was carved by St. Luke, brought to Spain by St. Peter, and hidden in a cave near Montserrat during the Moorish occupation.

Mon·za (mōn'tsä), city (1976 est. pop. 120,574), Lombardy, N Italy. Manufactures of this industrial center include felt hats, carpets, textiles, and machinery. The cathedral, founded (6th cent.) by the Lombard queen Theodolinda, contains the iron crown of Lombardy, which was made, according to tradition, from a nail of Christ's cross, and which was used to crown Charlemagne, Charles V, Napoleon I, and other emperors as kings of Lombardy or of Italy.

Moo·dy (mo͝o'dē), county (1970 pop. 7,622), 523 sq mi (1,354.6 sq km), E S.Dak., on the Minn. border and drained by the Big Sioux River; formed 1873; co. seat Flandreau. It is in a rich farming and livestock-raising region that produces corn, soybeans, flax, alfalfa, barley, oats, and wheat. Its manufactures include textiles and paper products.

Moore (mo͝or, môr). **1.** County (1970 pop. 39,048), 704 sq mi (1,823.4 sq km), central N.C., in the Piedmont, with forested sand hills and drained by the Deep River; formed 1784; co. seat Carthage. Its economy is based on farming (tobacco, peaches, corn), poultry raising, sawmilling, and textile manufacturing. **2.** County (1970 pop. 3,568), 122 sq mi (316 sq km), S Tenn., bounded in the SE by the Elk River; formed 1871; co. seat Lynchburg. Livestock raising and grain and tobacco farming are done. Its industries include lumbering and liquor distilling. **3.** County (1970 pop. 14,060), 912 sq mi (2,362.1 sq km), extreme NW Texas, in the high plains of the Panhandle, drained by the Canadian River; formed 1876; co. seat Dumas. One of the richest areas of the Panhandle natural-gas and oil field, it also has processing plants for carbon black, gasoline, nitrate, and helium. Wheat, corn, and milo are grown, and cattle are raised.

Moore·field (mo͝or'fēld', môr'-), town (1970 pop. 2,124), seat of Hardy co., NE W.Va., in the Eastern Panhandle, SE of Cumberland on the South Branch of the Potomac River, in a hunting and fishing area; chartered 1777.

Moore Ha·ven (hā'vən), city (1970 pop. 974), seat of Glades co., S Fla., ENE of Fort Myers, on the W shore of Lake Okeechobee near the entrance of the Caloosahatchee Canal, in a truck-farming and fishing area.

Moores Creek National Military Park (mo͝orz, môrz), 50 acres (20.3 hectares), SE N.C.; est. 1926. The patriot victory over the Loyalists at Moores Creek Bridge on Feb. 27, 1776, prevented the intended British invasion of N.C.

Moor·head (mo͝or'hĕd', môr'-), city (1978 est. pop. 28,700), seat of Clay co., NW Minn., on the Red River; inc. 1881. It is a shipping and processing center for a dairy and farm area. Farm and oil equipment and furniture are manufactured. Moorhead's population is predominantly of Scandinavian origin.

Moose Fac·to·ry (mo͞os făk'tə-rē), trading post, NE Ont., Canada, near the mouth of the Moose River on James Bay. A fort was built here by the Hudson's Bay Company in the early 1670s. In the struggle between the English and French in Canada, the fort changed hands several times and shortly after 1696 was destroyed. In 1730 the company built a post close to the ruins of the original fort. This post has been in continuous operation to the present day.

Moose·head Lake (mo͞os'hĕd'), 35 mi (56.3 km) long, from 2 to 10 mi (3.2–16.1 km) wide, with an area of 120 sq mi (310.8 sq km), W Maine, N of Augusta. It has an irregular shoreline and numerous islands. The region around the lake is a picturesque resort area.

Moose Jaw (jô), city (1976 pop. 32,581), S central Sask., Canada. It has oil refineries, meat-packing and dairy-processing plants, flour, lumber, and woolen mills, grain elevators, and stockyards.

Mop·ti (môp'tē), city (1972 est. pop. 43,000), central Mali, a port at the confluence of the Niger and Bani rivers. Mopti is the market center for a region where rice, cotton, peanuts, and manioc are produced and cattle are raised.

Mo·ra (môr'ə), county (1970 pop. 4,673), 1,942 sq mi (5,029.8 sq km), NE N.Mex., bounded on the E by the Canadian River and drained by the Mora River; formed 1860; co. seat Mora. Its economy depends on livestock raising, farming (hay, alfalfa, grain, and fruit), and lumbering. Santa Fe National Forest is in the southwest.

Mora. 1. Village (1970 pop. 2,582), seat of Kanabec co., E Minn., N of Minneapolis; settled 1881. It is a farm trade center and makes dairy equipment. **2.** Village (1970 pop. 900), seat of Mora co., N N.Mex., on the Mora River in the Sangre de Cristo Mts. N of Las Vegas, at an altitude of c.7,200 ft (2,195 m).

Mo·ra·da·bad (môr'ə-də-băd', mōr'ä-dä-bäd'), city (1971 pop. 258,590), Uttar Pradesh state, N central India. It is an important rail junction and an agricultural market center.

Mo·ra·va (mô'rä-vä), river, c.240 mi (385 km) long, rising in the Sudetes, N Czechoslovakia, and flowing generally S into the Danube River W of Bratislava. Its lower course forms part of the Austria-Czechoslovakia border. Sugar beets, grains, and tobacco are raised in the fertile valley.

Mo·ra·vi·a (mə-rā'vē-ə, mō-), region, central Czechoslovakia. The region is bordered on the west by Bohemia, on the east by the Little and White Carpathian Mts., which divide it from Slovakia, and on the north by the Sudetes Mts., which separate it from Silesia and include the Moravian Gate, a historically strategic north-south route. Central Moravia is a valley, opening in the S on Austria and drained by the Morava River and its tributaries.

A fertile agricultural area that encompasses the Haná region, Moravia is also highly industrialized. Diverse mineral resources, such as lignite, coal, oil, iron, copper, silver, and lead, spurred industrialization in the 20th cent. Among the region's other products are machinery, machine tools, armaments, automobiles, beer, liquor, clothing, furniture, and lumber. Brno, the former Moravian capital, is one of Europe's leading textile centers, and Gottwaldov is famous for its shoe industry; and Ostrava is a coal-mining center with a large iron and steel industry.

History. With Bohemia and Czech Silesia, Moravia makes up the portion of Czechoslovakia traditionally occupied by the Czechs, a branch of the Western Slavs, who displaced Germanic tribes that occupied the region (1st to the 5th cent. A.D.). By the 9th cent. the Moravians formed a great empire that included Bohemia, Silesia, Slovakia, southern Poland, and northern Hungary. The empire reached its height under Svatopluk, but after his death (A.D. 894) it broke apart and (early 10th cent.) fell to the Magyars. When Emperor Otto I defeated (955) the Magyars, Moravia became a march of the Holy Roman Empire. From the early 11th cent. it was a crown land of the kingdom of Bohemia, with which it passed (1526) under Austrian rule. Moravia, generally more tolerant of Hapsburg authority than Bohemia, suffered less in the religious and civil strife of the 16th cent. Hapsburg rule was finally overthrown in 1918, and Moravia was incorporated into Czechoslovakia.

Mor·ay·shire (mûr'ē-shîr', -shər) or **El·gin·shire** (ĕl'gĭn-shîr', -shər), former county NE Scotland, on the Moray Firth. Elgin was the county town. Morayshire was colonized by Northern Picts and came under crown dominion in the time of Malcolm III of Scotland (late 11th cent.). In 1975 Morayshire was divided between the Highland and Grampian regions.

Mor·bi·han (môr-bē-äɴ'), department (1975 pop. 563,588), 2,611 sq mi (6,N762.5 sq km), W France, in Brittany, on the Atlantic coast; capital Vannes.

Mord·vin·i·an Autonomous Soviet Socialist Republic (môrd-vĭn'ē-ən), autonomous republic (1976 est. pop. 985,000), c.10,000 sq mi (25,900 sq km), E European USSR; capital Saransk. A densely forested steppe, it consists of the Volga upland in the east and the Oka-Don lowland in the west. Lumbering, agricultural processing, and the manufacture of automobiles, machinery, furniture, paper, and wood chemicals are the major industries. Mordvinians, an ancient Finno-Ugric ethnic group, were first mentioned in the 6th cent. A.D. In the mid-13th cent. they fell under the Golden Horde and, when it disintegrated, passed to the Kazan khanate. Russia annexed the territory of the Mordvinians in 1552. The Mordvinian Autonomous SSR was formed in 1934.

More·cambe and Hey·sham (môr′kəm; hā′shəm), municipal borough (1973 est. pop. 42,010), Lancashire, NW England, on Morecambe Bay. Morecambe, a seaside resort, and Heysham, a port with service to Belfast, were joined in 1928.

More·head (mōr′hĕd′), city (1970 pop. 7,161), seat of Rowan co., N Ky., ENE of Lexington. The town processes local clay, has sawmills, and makes clothing.

More·house (môr′hous′), parish (1970 pop. 32,463), 804 sq mi (2,082.4 sq km), N La., bounded on the E by the Boeuf River, on the W by the Ouachita River, on the N by the Ark. border; formed 1844; parish seat Bastrop. It has agriculture (cotton, corn, hay, fruit, and livestock), a large natural-gas field, lumbering and food-processing industries, paper mills, and clothing and bottling factories.

Mo·re·lia (mô-rā′lyä), city (1974 est. pop. 199,100), capital of Michoacán state, W Mexico. It is the commercial and processing center of an irrigated agricultural and cattle-raising area. Founded in 1541, Morelia is built on a rocky hill and is surrounded by a fertile valley at the western edge of the central plateau. High peaks border the valley on three sides.

Mo·re·los (mô-rā′lôs), state (1970 pop. 620,392), 1,917 sq mi (4,965 sq km), S Mexico; capital Cuernavaca.

Mo·re·no (mə-rĕn′ō), city (1970 pop. 114,041), Buenos Aires prov., E Argentina.

Mores·by (mōrz′bē, môrz′-), island, 991 sq mi (2,567 sq km), Queen Charlotte Islands, off the coast of W British Columbia, Canada, in the Pacific Ocean.

Mo·res·net (mô′rĕz-nā′), district, 1.5 sq mi (3.9 sq km), Liège prov., E Belgium, near the West German border. Under joint Prussian and Dutch (after 1830, Belgian) suzerainty from 1816, it was awarded (1919) to Belgium under the Treaty of Versailles.

More·ton Bay (môr′tn), inlet of the Pacific Ocean, 65 mi (104.6 km) long and 20 mi (32.2 km) wide, Queensland, E Australia, nearly enclosed by Moreton and Stradbroke islands. The bay is the entrance to the port of Brisbane.

Mor·gan (môr′gən). **1.** County (1970 pop. 77,306), 574 sq mi (1,486.7 sq km), N Ala., drained in the N by Wheeler Reservoir; formed 1818; co. seat Decatur. It is in a timber and agricultural area that produces cotton and hogs. There are deposits of limestone, coal, fuller's earth, and asphalt. Textiles, chemicals, concrete, wood and metal products, and machinery are manufactured. **2.** County (1970 pop. 20,105), 1,282 sq mi (3,320.4 sq km), NE Colo., drained by the South Platte River; formed 1889; co. seat Fort Morgan. In an irrigated agricultural region yielding sugar beets, beans, and livestock, it has oil and natural-gas wells and a food-processing industry. Metal products are made. **3.** County (1970 pop. 9,904), 356 sq mi (922 sq km), N central Ga., in a piedmont region drained by the Little River; formed 1807; co. seat Madison. Its agriculture includes cotton, corn, grain sorghums, peaches, and livestock. It has a lumbering industry and manufactures twine, clothing, and furniture. **4.** County (1970 pop. 36,174), 565 sq mi (1,463.4 sq km), W central Ill., bounded on the NW by the Illinois River and drained by Apple, Sandy, and Indian creeks; formed 1823; co. seat Jacksonville. It has agriculture (corn, wheat, oats, livestock, poultry, and dairy products), meatpacking plants, and factories manufacturing clothing, steel products, shoes, wire novelties, cigars, and books. **5.** County (1970 pop. 44,176), 406 sq mi (1,051.5 sq km), central Ind., drained by the West Fork of the White River; formed 1821; co. seat Martinsville. In an agricultural area yielding hogs, grain, fruit, poultry, and dairy products, it has timber, clay deposits, and artesian springs. Its manufactures include wood products, furniture, brick and tile, and machinery. **6.** County (1970 pop. 10,019), 369 sq mi (955.7 sq km), E Ky., drained by the Licking River and several creeks; formed 1822; co. seat West Liberty. It lies in a hilly agricultural area in the Cumberland foothills. Corn, tobacco, livestock, poultry, and fruit are important products. It has bituminous-coal mines, timber, and oil and gas wells. Shoes are manufactured here. Part of Cumberland National Forest is in the county. **7.** County (1970 pop. 10,083), 596 sq mi (1,543.6 sq km), central Mo., in the Ozarks, with part of the Lake of the Ozarks in the S; formed 1833; co. seat Versailles. Its economy is based on tourism, farming (wheat, corn, and oats), dairying, and poultry raising. There are stands of timber and deposits of coal and barite in the area. **8.** County (1970 pop. 12,375), 418 sq mi (1,082.6 sq km), SE Ohio, intersected by the Muskingum River and Meigs and Wolf creeks; formed 1817; co. seat McConnelsville. It is in an agricultural area that produces livestock, dairy products, corn, wheat, and cabbages, with bituminous-coal mines and limestone quarries. Some manufacturing (millwork, furniture, metal products, and machinery) is done. **9.** County (1970 pop. 13,619), 539 sq mi (1,396 sq km), NE central Tenn., on the Cumberland Plateau; formed 1817; co. seat Wartburg. Livestock raising, dairying, and farming (corn, hay, tobacco, fruit, and vegetables) are important. It has bituminous-coal mining and lumbering and textile industries. **10.** County (1970 pop. 3,983), 603 sq mi (1,561.8 sq km), N Utah, watered by the Weber River, with the Wasatch Range throughout the county; formed 1862; co. seat Morgan. It is in an irrigated agricultural and dairying region that produces hay, sugar beets, fruit, truck crops, livestock, and poultry. Sporting goods and cement are manufactured. It is a recreation area, with hunting, water and snow skiing, boating, and fishing. **11.** County (1970 pop. 8,547), 233 sq mi (603.5 sq km), NE W.Va. in the Eastern Panhandle, bounded on the N by the Potomac River and the Md. border and partly on the S by Va.; formed 1820; co. seat Berkeley Springs. It has agriculture (livestock, dairy products, and fruit), glass-sand pits, and timber. Berkeley Springs is a health resort.

Morgan 1. City (1970 pop. 280), seat of Calhoun co., SW Ga., W of Albany on Ichawaynochaway Creek, in a farm area. **2.** Or **Morgan City**, city (1970 pop. 1,586), seat of Morgan co., N Utah, on the Weber River SE of Ogden; settled 1860 by Mormons. It is a trade and processing center in an irrigated region.

Morgan City, city (1978 est. pop. 16,500), St. Mary parish, S La., a fishing port on the Atchafalaya River; inc. 1860. The city is headquarters for offshore petroleum drilling and a trade center for a bayou area where rice and sugar cane are grown. It has shipyards, a large shrimp fleet, and an oyster industry.

Mor·gan·field (môr′gən-fēld′), city (1970 pop. 3,563), seat of Union co., W Ky., near the Ohio River SW of Henderson, in a coal, oil, and farm area.

Mor·gan·ton (môr′gən-tən), town (1970 pop. 13,625), seat of Burke co., W N.C., on the Catawba River in the foothills of the Blue Ridge Mts.; founded 1784, inc. 1885. A lake resort town, it has industries that manufacture furniture, textiles, apparel, and electrical parts.

Mor·gan·town (môr′gən-toun′). **1.** City (1970 pop. 1,394), seat of Butler co., W central Ky., NW of Bowling Green, in an agricultural, timber, and coal area. **2.** City (1978 est. pop. 30,100), seat of Monongalia co., N W.Va., near the Pa. line, on the Monongahela River; inc. 1785. A shipping point for a coal-mining region, it also has glass, textile, and chemical industries. Fort Morgan was built here in 1772, and the first settlers arrived the same year. Iron, discovered in 1789, was the principal industry until the Civil War.

Mor·gar·ten (môr′gär′tn), mountain, 4,084 ft (1,245.6 m) high, N central Switzerland. Here, on Nov. 15, 1315, a small Swiss force decisively defeated the Austrians, thus paving the way for Swiss independence.

Mo·ri·o·ka (môr′ē-ō′kə), city (1976 est. pop. 220,051), capital of Iwate prefecture, N Honshu, Japan, on the Kitakami River. An industrial and commercial center, it is noted for the production of ironware.

Mor·ley (môr′lē), municipal borough (1973 est. pop. 44,790), West Yorkshire, N England. Woolen textiles and other products are made. Coal is mined in the borough. It was besieged by Royalists in the English civil war.

Mo·ro (mōr′ō, môr′ō), city (1970 pop. 290), seat of Sherman co., N Oregon, ESE of The Dalles, in a wheat-growing region.

Mo·roc·co (mə-rŏk′ō), kingdom (1973 est. pop. 15,600,000), 171,834 sq mi (445,050 sq km), NW Africa, on the Mediterranean Sea and the Atlantic Ocean. Rabat is the capital. Morocco is bordered by Western Sahara on the south and by Algeria on the south and east. Ifni, formerly a Spanish-held enclave on the Atlantic coast, was ceded to Morocco in 1969. Two cities, Ceuta and Melilla, and several small islands off the Mediterranean coast remain part of metropolitan Spain. The population of Morocco is concentrated in the coastal regions, where rainfall is most plentiful. Central Morocco is largely occupied by the Atlas Mts. Southern Morocco lies in the Sahara Desert.

Economy. In parts of the Rif Mts. in the northeast wheat and other cereals can be raised without irrigation. On the Atlantic coast, where there are extensive plains, olives, citrus fruits, and grapes are grown, largely with water supplied by artesian wells. Fishing is also

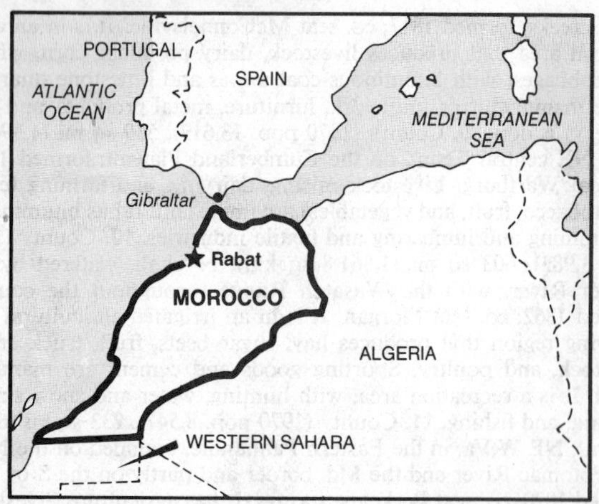

important. In the northern foothills of the Atlas Mts. there are large mineral deposits; phosphates are the most important, but there are also iron, zinc, copper, lead, molybdenum, cobalt, and the only sizable coal deposits in all North Africa. Petroleum is also found.

History. Berbers inhabited Morocco at the beginning of the historic era. In Roman times Morocco was roughly coextensive with the province of Mauretania Tingitania. The Vandals were the earliest (5th cent.) of barbarian peoples to take the area as the Roman Empire declined. The Arabs first swept into Morocco c.685, bringing with them Islam. Morocco became an independent state in 788 under the royal line founded by Idris I. After 900 the country again broke into small tribal states. The Almoravids overran (c.1062) Morocco and established a kingdom stretching from Spain to Senegal. They were succeeded by the Almohads (c.1174) and the Merinids (1259–1550), but the country was rarely completely unified, and conflict between Arabs and Berbers was incessant. Beginning with the capture of Ceuta in 1415, Portugal took all the chief Moroccan ports except Melilla and Larache, both of which fell to Spain. The present ruling dynasty, the Alawite, or second Sherifian, dynasty, came to power in 1660 and recaptured many European-held strongholds.

In the 19th cent. the strategic importance and economic potential of Morocco excited the interest of the European powers. In 1880 the major European nations and the United States decided to preserve the territorial integrity of Morocco and to maintain equal trade opportunities for all. Political and commercial rivalries soon disrupted this cordial arrangement and brought on several international crises. France, Spain, Great Britain, and Germany all maneuvered to gain the upper hand. The French steadily annexed territory, and in 1912 the sultan agreed to a French protectorate over nine-tenths of the country, with Spain administering two zones in the south.

A strong threat to European rule was posed (1921–26) by the revolt (the Rif War) of Adb-el-Krim. In 1937 the French crushed another nationalist revolt. During World War II, French Morocco was seized from the Vichy French by the Allies in 1942. After the war the nationalist movement gained strength and received the active support of the sultan, Sidi Mohammed, whom the French deposed in 1953. In 1956 France relinquished its rights in Morocco, the Spanish surrendered their protectorate, and Tangier was given to the newly independent Morocco by international agreement.

The sultan became (1957) King Mohammed V (Sidi Mohammed). After his death (1961) his son Hassan II ascended the throne. Border hostilities with Algeria erupted in the 1960s but were settled in a 1970 agreement. King Hassan assumed absolute powers in 1965. The country returned to a modified form of parliamentary democracy in 1970. Political unrest, repression of dissenters, and attempts on the king's life marked the 1970s. In 1975 Spanish Sahara, relinquished by Spain, was divided between Morocco and Mauritania and renamed Western Sahara. An Algerian-supported Saharan independence movement opposed Moroccan administration, and prolonged guerrilla warfare ensued.

Government. Morocco is governed under a 1972 constitution. The king holds effective power and appoints the ministers, and there is a unicameral parliament. Islam is the state religion and Arabic is the official language, but French and Spanish are also spoken. There are universities at Rabat, Fez, and Marrakesh.

Mo·rón (mō-rôn'), city (1970 pop. 485,983), Buenos Aires prov., E Argentina, in the Greater Buenos Aires area. Settled in the early 16th cent., Morón became an outpost on the route between Buenos Aires and Chile and Peru.

Mo·ro·ni (mō-rō'nē), town (1966 pop. 11,515), capital of the French overseas territory of the Comoro Islands, on Grand Comoro island, at the N end of the Mozambique Channel, an arm of the Indian Ocean.

Mo·ro·tai (mō-rō-tī'), island (c.695 sq mi/1,800 sq km), E Indonesia, one of the Moluccas. Heavily wooded, it produces timber and resin.

Mor·rill (môr'ĭl, mŏr'-), county (1970 pop. 5,813), 1,403 sq mi (3,633.8 sq km), W Nebr., in an irrigated farm area drained by the North Platte River; formed 1909; co. seat Bridgeport. Sugar beets, beans, and grain are grown.

Mor·ril·ton (môr'ĭl-tən, mŏr'-), city (1970 pop. 6,814), seat of Conway co., W central Ark., on the Arkansas River NW of Little Rock; founded c.1876, inc. 1879. A trade center, it also processes cotton, lumber, and meat and makes clothing. Cattle are bred here.

Mor·ris (môr'ĭs, mŏr'-). **1.** County (1970 pop. 6,432), 697 sq mi (1,805.2 sq km), E central Kansas, in a rolling plains region watered by the Neosho River; formed 1855, co seat Council Grove. In a livestock and grain area, it has industries producing plastics, metal products, and farm machinery and equipment. **2.** County (1970 pop. 383,454), 468 sq mi (1,212.1 sq km), N N.J., in a hilly resort area with many lakes and mountain ridges, bounded on the SE and E by the Passaic River and drained by numerous rivers and streams; formed 1739; co. seat Morristown. It is primarily residential, with some dairy and truck farming and light industries. Among its diverse manufactures are pharmaceuticals, textiles and clothing, lumber and wood products, chemicals, plastics, glass, concrete, and paper. **3.** County (1970 pop. 12,310), 263 sq mi (681.2 sq km), NE Texas, bounded on the N by the Sulphur River, on the S by Cypress Bayou; formed 1875; co. seat Daingerfield. Farming (cotton, corn, peanuts, fruit, and vegetables), livestock raising, and dairying are important to its economy. It also has iron-ore deposits and pig-iron and lumbering industries.

Morris. 1. City (1977 est. pop. 8,563), seat of Grundy co., NE Ill., on the Illinois River SW of Joliet, in a coal area; platted 1842, inc. 1853. It processes local clay and makes paper products. **2.** City (1970 pop. 5,366), seat of Stevens co., W Minn., W of Glenwood, in a farm region; platted 1869, inc. as a village 1878, as a city 1903. The Univ. of Minnesota has an agricultural experiment station here.

Morris Jes·up, Cape (jĕs'əp), northernmost land point in the world, N Greenland. At lat. 83°39′ N, it is 440 mi (708 km) from the North Pole. U.S. explorer Robert Peary reached the cape in 1892.

Mor·ri·son (môr'ĭ-sən, mŏr'-), county (1970 pop. 26,949), 1,127 sq mi (2,918.9 sq km), central Minn., drained by the Mississippi River, with Lake Alexander in NW; formed 1856; co. seat Little Falls. It is in an agricultural area yielding dairy products, livestock, and grain. Its industry includes food processing, some clothing manufacturing, lumbering, papermaking, and boatbuilding.

Morrison, city (1970 pop. 4,387), seat of Whiteside co., NW Ill., NE of Rock Island, in a farm and dairy area; founded 1855, inc. 1867. Electrical equipment is made.

Morrison, Mount, Taiwan: *see* Hsin-Kao Shan.

Morristown (môr'ĭs-toun', mŏr'-). **1.** Town (1978 est. pop. 16,800), seat of Morris co., N N.J., on the Whippany River; settled c.1710, inc. 1865. Although chiefly residential, it has stone quarries and plants that make a wide variety of products. This quiet village was a center of Revolutionary activity, particularly in the winters of 1777 and 1779–80, when the Continental Army encamped here. Benedict Arnold was court-martialed in the town. Alfred Vail and S. F. B. Morse perfected (c.1837) the telegraph here. Morristown grew with the area's iron industry. The Seeing Eye School (est. 1929) for training dogs to aid the blind is nearby. Of interest are the Schuyler-Hamilton House (1760), the courthouse (1826), and the municipal building, which was built in 1918. Morristown National Historical Park includes the Ford Mansion, which was Washington's headquarters in 1779–80, a historical museum, and the reconstructed sites of encampment of the Continental Army at Fort Nonsense and at Jockey Hollow. **2.** City (1978 est. pop. 19,700), seat of Hamblen co., NE Tenn., in a fertile valley of a mountainous region; settled 1783, inc. 1867. Furniture is made. **3.** Town (1970 pop. 4,052), Lamoille co., N central Vt., N of Waterbury, in a farming and lumbering region; chartered 1781.

Mor·ris·ville (môr'ĭs-vĭl', mŏr'-), borough (1970 pop. 11,309), Bucks co., SE Pa., on the Delaware River opposite Trenton, N.J.; settled c.1624 by the Dutch West India Company, inc. 1804. Tiles and rubber and plastic products are the principal manufactures. Nearby is William Penn's manor, Pennsbury, which has been reconstructed.

Mor·ro Castle (môr'ō), fort at the entrance to the harbor of Havana, Cuba. It was erected by the Spanish in 1589 to protect the city from buccaneers and captured by the British in 1762. The fort at the entrance to the harbor of Santiago de Cuba is also called Morro Castle. It was taken by the American forces in the Spanish-American War (1898).

Mor·row (môr'ō, mŏr'ō). **1.** County (1970 pop. 21,348), 404 sq mi (1,046.4 sq km), central Ohio, drained by the Kokosing River and Whetstone and Big Walnut creeks; formed 1848; co. seat Mount Gilead. Its agriculture includes livestock, dairy products, grain, fruit, and soybeans. It has oil and gas fields, and its industries produce fertilizers, plastic products, machinery, and electrical equipment. **2.** County (1970 pop. 4,465), 2,059 sq mi (5,332.8 sq km), N Oregon, bounded in the N by the Columbia River and by Wash., with the Blue Mts. in the S; formed 1885; co. seat Heppner. It has grain and livestock farming, and its industry includes food processing and lumbering. Part of Umatilla National Forest is in the southeast.

Mor·ton (môr'tn). **1.** County (1970 pop. 3,576), 725 sq mi (1,877.8 sq km), extreme SW Kansas, in a rolling plains region bordered on the W by Colo., on the S by Okla., and drained by the Cimarron River; formed 1886; co. seat Elkhart. Wheat and grain sorghums are grown. It also has gas fields and some manufacturing. **2.** County (1970 pop. 20,310), 1,920 sq mi (4,972.8 sq km), central N.Dak., in an agricultural area bounded on the E by the Missouri River and drained by Muddy Creek and the Heart River; formed 1873; co. seat Mandan. Manufacturing and oil refining are also important here.

Morton, town (1970 pop. 2,738), seat of Cochran co., NW Texas, W of Lubbock on the Llano Estacado; inc. 1934. It is in an area of cattle ranches and cotton farms and has an oil refinery.

Mor·van (môr-väɴ'), mountainous region, E central France, in Nivernais and Burgundy. The northernmost part of the Massif Central, this heavily forested region rises to 2,959 ft (902.5 m). Cattle raising and the production of charcoal are the chief industries.

Mos·cow (mŏs'kou, -kō), city (1970 pop. 7,061,000), capital of the USSR, of the Russian Soviet Federated Socialist Republic, and of Moscow oblast, W central European USSR, on the Moskva River near its junction with the Moscow Canal. Moscow is the USSR's leading economic and cultural center. The hub of the Soviet railroad network, it is also an inland port. Moscow's major industries include machine building, metalworking, oil refining, publishing, brewing, filmmaking, and the manufacture of machine tools, precision instruments, building materials, automobiles, trucks, aircraft, chemicals, wood and paper products, textiles, clothing, and footwear.

The major sections of Moscow form concentric circles, of which the innermost is the Kremlin whose walls represent the city limits as of the late 15th cent. Adjoining the Kremlin in the east is the huge Red Square, on which are the Lenin Mausoleum, the historical museum, and the imposing cathedral of Basil the Beatified (16th cent.), now an antireligious museum. To the east of Red Square extends the old district of Kitaigorod, once the merchant's quarter and now an administrative hub with various government offices and ministries. Encircling the Kremlin and Kitaigorod are the Bely Gorod, traditionally the most elegant part of Moscow and now a commercial and cultural area; the Zemlyanoy Gorod, named for the earthen and wooden ramparts that once surrounded it; and the inner suburbs.

Among Moscow's many cultural and scientific institutions are the Univ. of Moscow (founded 1755), the Academy of Sciences of the USSR, a conservatory (1866), the Tretyakov art gallery (opened in the 1880s), the Museum of Oriental Cultures, the State Historical Museum, the Lenin Museum and Lenin Library, the Agricultural Exhibition, and the People's Friendship Univ. (1960) for foreign students. Theaters include the Moscow Art Theater, the Bolshoi (opera and ballet), and the Maly Theater (drama). A new Palace of Congresses was built (1961) inside the Kremlin walls for meetings of the Supreme Soviet. Moscow's numerous large parks and recreation areas include Gorky Central Park, the forested Izmailovo and Sokolniki parks, and Ostankino Park. The Moscow subway system opened in 1935.

Although archaeological evidence indicates that the site has been occupied since Neolithic times, the village of Moscow was first mentioned in the Russian chronicles in 1147. Moscow became (c.1271) the seat of the grand dukes of Suzdal-Vladimir, who later assumed the title of grand dukes of Moscow. The first stone walls of the Kremlin were built in 1367. Moscow, or Muscovy, achieved dominance over the Russian lands by virtue of its strategic location, its leadership in the struggle against the Tatars, and its gathering of neighboring principalities under Muscovite suzerainty. By the 15th cent. Moscow had become the capital of the Russian national state and also the seat of the Metropolitan (later Patriarch) of the Russian Orthodox Church.

The Russian capital was transferred to St. Petersburg (now Leningrad) in 1712; but Moscow's cultural and social life continued uninterrupted, and the city never ceased to be the religious center of Russia. Built largely of wood until the 19th cent., Moscow suffered from numerous fires, the most notable of which occurred in the wake of Napoleon I's occupation in 1812. Rebuilt, the city developed from the 1830s as a major textile and metallurgical center. In 1918 the Soviet government transferred the capital back to Moscow and fostered spectacular economic growth in the city, whose population doubled between 1926 and 1939. During World War II the German columns were stopped only 20 to 25 mi (32-40 km) from the city's center, but Moscow suffered virtually no war damage.

Mos·cow (mŏs'kō), city (1978 est. pop. 15,200), seat of Latah co., NW Idaho, at the Wash. border; inc. 1887. It is a trade center for a lumber and farm area that grows wheat and peas. There are sawmills, food-processing plants, and factories that manufacture a variety of products. The Univ. of Idaho is here.

Moscow Basin, lignite basin, c.200 mi (320 km) long and 50 mi (80 km) wide, central European USSR, S of Moscow. Lowgrade bituminous coal is mined here.

Mo·selle (mō-zĕl'), department (1975 pop. 1,006,373), 2,399 sq mi (6,213.4 sq km), NE France, bordering on Luxembourg and Germany; capital Metz.

Moselle, river, 320 mi (514.9 km) long, rising in the Vosges Mts., NE France, and winding generally N past Épinal and Metz. Leaving France, it forms part of the border between Luxembourg and West Germany, then enters West Germany, passes Trier, and cuts between the Eifel and the Hunsrück ranges to reach the Rhine River at Koblenz. The West German section of the Moselle valley has numerous old castles and celebrated vineyards.

Mo·shi (mō'shē), city (1967 pop. 26,864), NE Tanzania, on the S slope of Mt. Kilimanjaro, near Kenya. It is the center of a rich coffee-growing region and is an industrial, tourist, and transportation center. Manufactures include ginned cotton, cured coffee, beverages, and clothing. The original town was the capital of a 19th cent. kingdom of the Chagga people and became (late 19th cent.) an administrative center under the Germans. In the 20th cent. the British moved Moshi to its present site.

Mos·kva (mŏs-kvä'), river, c.310 mi (500 km) long, rising in the hills W of Moscow, in central European USSR, and meandering generally E past Moscow to join the Oka River near Kolomna. It is connected with the upper Volga River by the Moscow-Volga Canal (80 mi/130 km long), built between 1932 and 1937.

Mos·que·ro (mŏs-kĕr'ō), village (1970 pop. 244), seat of Harding co., NE N.Mex., NNW of Tucumcari.

Mos·qui·to Coast (mə-skē'tō), region, E coast of Nicaragua and Honduras, extending from the San Juan River N into NE Honduras. It is sultry and swampy, rising to low hills in the west. Banana cultivation is the main economic activity. In the early colonial period English and Dutch buccaneers preyed on Spanish shipping from coastal bases, and English loggers exploited the forest products. England established a protective kingdom at Bluefields in 1678. In 1848 the British claimed and took San Juan del Norte to offset U.S. interest in a transisthmian route. The Clayton-Bulwer Treaty (1850) between the United States and Great Britain checked British expansion, but relinquishment of the coast was delayed until a separate treaty was concluded with Nicaragua (1860), which established the autonomy of the so-called Mosquito Kingdom. In 1894 José Santos Zelaya ended the anomalous position of the territory by forcibly incorporating it into Nicaragua. The northern part was awarded to Honduras in 1960 by the International Court of Justice.

Moss (môs), city (1977 est. pop. 25,715), capital of Østfold co., SE Norway, a port on the Oslofjord. It is a commercial, industrial, and tourist center, with shipyards and sawmills. On Aug. 14, 1814, the convention establishing the personal union of Sweden and Norway was signed here.

Most (môst), city (1975 est. pop. 59,909), NW Czechoslovakia, in Bohemia, near the East German border. It is a railway junction and industrial city in a lignite-mining area and has pipelines that carry gas to Prague. Chemicals, steel, and ceramics are the major products. The city, whose history dates at least to the 11th cent., has several medieval churches and an old town hall.

Mos·ta·ga·nem (môs'tä-gä-hĕm'), city (1974 est. pop. 101,780), capital of Mostaganem dept., NW Algeria, a port on the Mediterranean Sea. It was founded in the 11th cent. and reached its commercial height in the 16th cent. under the Turks. Its population had declined to about 3,000 when the French arrived in 1833. Wine, grain, meat, and wool are exported.

Mos·tar (mō'stär), city (1971 pop. 89,405), SW Yugoslavia, in Bosnia and Hercegovina, on the Neretva River. It has industries that produce tobacco, wine, and aluminum products. Bauxite and lignite are mined nearby. Known in 1442, it passed to Austria in 1878 and to Yugoslavia in 1918. The city has numerous Turkish mosques.

Mo·sul (mō-so͞ol', mō'səl), city (1970 est. pop. 293,100), N Iraq, on the Tigris River, opposite the ruins of Nineveh. Trade in agricultural goods and exploitation of oil are the main occupations. Mosul was the chief city of northern Mesopotamia from the 8th to 13th cent., when it was devastated by the Mongols. The city was occupied by the Persians (1508) and the Turks (1534–1918). Under the British occupation and mandate (1918–32) it regained its stature as the chief city of the region. Its possession by Iraq was disputed by Turkey (1923–25) but was confirmed by the League of Nations (1926).

Mo·ta·gua (mō-tä'gwä), river, c.250 mi (400 km) long, rising in S central Guatemala and flowing NE to the Gulf of Honduras. The longest river within Guatemala, it waters a valley where hemp and bananas are raised.

Mo·ta·la (mo͞o'tə-lä'), city (1975 est. pop. 32,400), Östergötland co., S Sweden, on Lake Vättern. It is an important lake port. An industrial center, it manufactures locomotives and radio and television sets.

Moth·er Lode (mŭ*th*'ər lōd), belt of gold-bearing quartz veins, central Calif., along the W foothills of the Sierra Nevada. The discovery of alluvial gold on the South Fork of the American River led to the 1848 gold rush. Mark Twain and Bret Harte helped make the Mother Lode famous.

Moth·er·well and Wish·aw (mŭ*th*'ər-wĕl'; wĭsh'ô), burgh (1971 pop. 74,184), Strathclyde region, S central Scotland. The two parts of the burgh were united in 1920. In a coal and iron region, it is a center of steel and other heavy industry.

Mot·ley (mŏt'lē), county (1970 pop. 2,178), 980 sq mi (2,538.2 sq km), NW Texas, in broken plains just below the Cap Rock escarpment of the Llano Estacado and drained by the North, South, and Middle Pease rivers; formed 1876; co. seat Matador. In a chiefly cattle-ranching area, it also yields cotton, grain sorghums, wheat, some fruit, truck crops, hogs, sheep, horses, and poultry. There are deposits of clay, lignite, sand, gravel, and caliche.

Mott (mŏt), city (1970 pop. 1,368), seat of Hettinger co. SW N.Dak., on the Cannonball River SW of Bismarck; inc. 1928. It is a trade and processing center of a wheat and dairy region. Lignite and coal are mined.

Mou·lins (mo͞o-lăɴ'), city (1975 pop. 26,067), capital of Allier dept., central France, on the Allier River. Clothing, shoes, dyes, automobile parts, and household products are manufactured. Moulins has remarkable artistic and historic treasures. The cathedral contains a 15th cent. triptych, considered one of the finest French paintings of the period. In 1566 Charles IX held a great assembly at Moulins at which important administrative and legal reforms were adopted.

Moul·mein (mo͞ol-mān', mōl-), city (1970 est. pop. 173,000), SE Burma, near the mouth of the Salween River. A river port and commercial center, it has teak mills and shipyards. From 1826 to 1852, Moulmein was the chief town of British Burma.

Moul·ton (mōlt'n), town (1970 pop. 2,470), seat of Lawrence co., NW Ala., SW of Decatur; inc. 1818. Cotton is ginned and lumber is milled.

Moul·trie (mōl'trē), county (1970 pop. 13,263), 326 sq mi (844.3 sq km), central Ill., bounded on the SW by Lake Shelbyville; formed in 1843; co. seat Sullivan. It has agriculture (corn, wheat, soybeans, broomcorn, livestock, and poultry) and manufacturing of shoes, concrete products, and cheese and other dairy products.

Moultrie, city (1978 est. pop. 13,900), seat of Colquitt co., SW Ga., on the Ochlockonee River; inc. 1890. The town grew as a lumbering and naval stores center; when the timber was depleted the area turned to livestock raising and diversified farming.

Mound City. 1. City (1970 pop. 1,177), seat of Pulaski co., S Ill., on the Ohio River NNE of Cairo, in a rich agricultural region; inc. 1857. During the Civil War it was a important Union naval base. **2.** City (1970 pop. 714), seat of Linn co., E Kansas, NNW of Fort Scott, in a livestock and fruit region. Oil and natural-gas wells and coal mines are in the vicinity. **3.** Town (1970 pop. 164), seat of Campbell co., N S.Dak., WNW of Aberdeen. It produces wheat and dairy products.

Mound City Group National Monument: *see* National Parks and Monuments Table.

Mounds·ville (moundz'vĭl'), city (1970 pop. 13,560), seat of Marshall co., W.Va., in the Northern Panhandle on the Ohio River; settled 1771, inc. 1865. In a coal-mining region, it has sawmills and factories producing various manufactures. Grave Creek Indian Burial Mound is nearby.

Moun·tain Ash (moun'tən), urban district (1971 pop. 27,806), Mid Glamorgan, S Wales. It is dependent upon the coal mines nearby, which were developed in the 19th cent.

Mountain City, resort town (1970 pop. 1,883), seat of Johnson co., NE Tenn., SE of Bristol, in a mountainous region; inc. 1905. It is a farm trade center and processes lumber.

Mountain Home. 1. Resort town (1970 pop. 3,936) seat of Baxter co., N Ark., ENE of Harrison. **2.** City (1970 pop. 6,451), seat of Elmore co., SW Idaho, on a plain near the Snake River SE of Boise; inc. 1896. The city grew as a wool-shipping center on the railroad. Today there is irrigated farming (potatoes and sugar beets), stock raising, lumbering, and mining in the area.

Mountain View (vyo͞o). **1.** City (1970 pop. 1,866), seat of Stone co., N Ark., WNW of Batesville, in the Ozarks. It is a summer resort in an agricultural, woodworking, and cotton-ginning area. **2.** City (1978 est. pop. 55,400), Santa Clara co., W Calif., on San Francisco Bay; inc. 1902. It has publishing and printing firms, research organizations, and diverse manufacturing industries. Moffet Naval Air Station adjoins the city.

Mount Ayr (âr), town (1970 pop. 1,762), seat of Ringgold co., SW Iowa, near the Mo. border SE of Creston; founded c.1855, inc. 1875. It is in a livestock-raising area.

Mount Car·mel (kär'məl) **1.** City (1970 pop. 8,096), seat of Wabash co., SE Ill., on the Wabash River SW of Vincennes, Ind.; laid out 1818, inc. 1825. **2.** Borough (1970 pop. 9,317), Northumberland co., E central Pa., SW of Hazelton; settled c.1775, inc. 1862. Anthracite coal is mined, and apparel, chemicals, and cigars are made.

Mount Car·roll (kăr'əl), city (1970 pop. 2,143), seat of Carroll co., NW Ill., SW of Freeport near the Mississippi, in a farm and dairy region; founded 1843, inc. 1867.

Mount Clem·ens (klĕm'ənz), city (1978 est. pop. 21,000), seat of Macomb co., SE Mich., on the Clinton River; settled c.1798, inc. as a city 1879. It is a health resort, known for its mineral waters. It also has a large floral industry, an automobile paint and vinyl plant, and factories making pottery, chemicals, and seat belts. Nearby is Selfridge Air Force Base.

Mount Des·ert Island (dĕz'ərt), c.100 sq mi (260 sq km), largest island off the coast of Maine; separated from the mainland by Frenchman Bay, Mt. Desert Narrows, and Western Bay. The island's rugged topography is a result of glacial action. Numerous lakes and streams are found on the island. It is almost equally divided into east and west halves by Somes Sound. A chain of rounded granite peaks dominates the island. They were named *Monts Deserts,* meaning "wilderness mountains," by the French explorer Samuel de Champlain, who landed on the island in 1604. The first French Jesuit mission and colony in America was established here in 1613. The French relinquished their claims in 1713, and the first permanent English settlement began in 1762. The island developed as a fishing and lumbering center, and by the end of the 19th cent. it had become a famous resort area. A forest fire in 1947 damaged much of the eastern half of the island. The major part of the island is in Acadia National Park.

Mount Gil·e·ad (gĭl'ē-əd), village (1970 pop. 2,971) seat of Morrow co., N central Ohio, E of Marion, in a farm area; settled 1811, inc. 1864. Hydraulic presses are made here.

Mount Hol·ly (hŏl'ē). **1.** Township (1970 pop. 12,713), seat of Burlington co., W N.J., S of Trenton and E of Camden; settled c.1680 by Friends. It is a trade center. Of interest are a Friend's meetinghouse (1775) that was used by the British as a commissary when they occupied the town in 1776; the courthouse (1796); the old firehouse (1752); and Stephen Girard's house (1777). **2.** Industrial city (1970 pop. 5,107), Gaston co., S N.C., NW of Charlotte on the Catawba River, in a corn and wheat region; inc. 1889. It is a textile center with cotton, woolen, and hosiery mills.

Mount I·da (ī'də), city (1970 pop. 819), seat of Montgomery co., W Ark., W of Hot Springs, in a dairying and sawmilling area.

Mount Joy (joi), borough (1970 pop. 5,041), Lancaster co., SE Pa., NW of Lancaster; settled 1768, laid out 1812, inc. 1851. Its chief manufacture is shoes.

Mount Kis·co (kĭs'kō'), residential village (1970 pop. 8,172), Westchester co., SE N.Y., N of White Plains; inc. 1874. New Croton Reservoir is nearby.

Mount Mc·Kin·ley National Park (mə-kĭn'lē), 1,939,493 acres (785,495 hectares), in the Alaska Range, S central Alaska; est. 1917. Located in a region of spectacular mountain scenery, the park contains Mt. McKinley, the highest point in North America. The park includes glaciers, tundra, and abundant wildlife (caribou, mountain sheep, bears, and wolves).

Mount Ol·i·vet (ŏl'ə-vĕt'), city (1970 pop. 442), seat of Robertson co., N Ky., WSW of Maysville, in a bluegrass agricultural region.

Mount Pleas·ant (plĕz'ənt). **1.** City (1970 pop. 7,007), seat of Henry co., SE Iowa, NW of Burlington; platted 1837, inc. 1842. Iowa Wesleyan is here. **2.** City (1978 est pop. 24,800), seat of Isabella co., central Mich., on the Chippewa River; settled before 1860, inc. as a city 1889. The city grew after oil was found nearby in 1928. There are oil wells and refineries here. Mount Pleasant is the seat of Central Michigan Univ. **3.** Borough (1970 pop. 5,895), Westmoreland co., SW Pa., SE of Pittsburgh, in a coal area; laid out c.1797, inc. 1828. Glass products, cement, and mining machinery are made. **4.** Resort town (1970 pop. 6,879), Charleston co., SE S.C., on Charleston Harbor E of Charleston, with which it is connected by a bridge; inc. 1837. There is shrimp fishing. **5.** City (1970 pop. 9,459), seat of Titus co., E Texas, SW of Texarkana; inc. 1900. Settled before the middle of the 19th cent., it became a pine-woods lumbering center. Today the region also produces oil, sweet potatoes, truck crops, peanuts, cotton, and livestock.

Mount·rail (mount'rāl'), county (1970 pop. 8,437), 1,819 sq mi (4,711.2 sq km), NW central N.Dak., in a rich agricultural area on the Missouri River and drained by the White Earth River; formed 1908; co. seat Stanley. Its agricultural products include livestock, dairy produce, grain, flax, and vegetables. Verendrye National Monument is in the southern part of the county.

Mount Rain·ier (rā-nîr', rə-, rā'nîr), city (1970 pop. 8,180), Prince Georges co., central Md., NE of Washington, D.C.; inc. 1910.

Mount Rainier National Park, 241,992 acres (98,007 hectares), SW Wash., in the Cascade Range; est. 1899. The area is dominated by Mt. Rainier, a volcanic peak, 14;410 ft (4,395.1 m) high. The mountain is snow-crowned and has 26 glaciers; its heavily forested lower slopes are popular with mountain climbers.

Mount Rush·more National Memorial (rŭsh'môr'), 1,278 acres (517.6 hectares), SW S.Dak., in the Black Hills; est. 1925, dedicated 1927. There, carved on the face of the mountain and visible for 60 mi (96.5 km), are the enormous busts of four U.S. Presidents—Washington, Jefferson, Lincoln, and Theodore Roosevelt. The sculpture, nearly completed when the sculptor, Gutzon Borglum, died (1941), was finished later that year by his son Lincoln. It took 14 years to complete the figures.

Mount Ster·ling (stûr'lĭng). **1.** City (1970 pop. 2,182), seat of Brown co., W central Ill., E of Quincy; settled 1830, inc. 1837. It is a processing center in a farm area. **2.** City (1970 pop. 5,083), seat of Montgomery co., E central Ky., E of Lexington, in a bluegrass region; platted 1793. A farm trade center, it also processes food and makes clothing. In the Civil War Mt. Sterling was captured and sacked (1863) by Gen. John H. Morgan. Indian mounds are nearby.

Mount Ver·non (vûr'nən), NE Va., overlooking the Potomac River near Alexandria, S of Washington, D.C.; home of George Washington from 1747 until his death in 1799. The land was patented in 1674, and the house was built in 1743 by Lawrence Washington, George Washington's half brother. George Washington inherited it in 1754 and made additions that were not completed until after the Revolution. The mansion is a wooded structure of Georgian design, two and on-half stories high, with a broad, columned portico; wide lawns, fine gardens, and subsidiary buildings surround it. The mansion has been restored, after Washington's detailed notes, with much of the original furniture, family relics, and duplicate pieces of the period. The estate was purchased in 1860 by the Mount Vernon Ladies' Association (organized in 1856), its permanent custodian. In the tomb (built 1831-37) are the sarcophagi of George and Martha Washington and the bodies of other members of the family.

Mount Vernon. 1. Town (1970 pop. 1,579), seat of Montgomery co., SE central Ga., near the Oconee River W of Savannah, in a farm and timber area. **2.** City (1978 est. pop. 17,000), seat of Jefferson co., SE Ill.; settled 1819, inc. 1872. It is a trade, rail, and industrial center in a farm and coal region. **3.** City (1970 pop. 6,770), seat of Posey co., extreme SW Ind., on the Ohio River W of Evansville; settled 1816, inc. 1865. It is a processing and shipping point for a farming and oil-producing area. **4.** Town (1970 pop. 1,639), seat of Rockcastle co., central Ky., S of Richmond and on the old Wilderness Road, in a coal-mining area; settled 1810, inc. 1818. **5.** City (1970 pop. 2,600), seat of Lawrence co., SW Mo., E of Joplin, in a farm region; laid out 1845. **6.** City (1978 est. pop. 66,600), Westchester co., SE N.Y., between the Bronx and Hutchinson rivers and adjacent to the Bronx; settled 1664, inc. 1892. Although primarily a residential suburb of New York City, it has many industries. The area was settled as part of Eastchester township. John Peter Zenger was arrested for libel here in 1733. The city itself was not founded until 1851 when a cooperative group bought the land and built a planned community. St. Paul's Church (c.1761), a national historic site, is here. **7.** City (1978 est. pop. 14.400), seat of Knox co., central Ohio, on the Kokosing River; laid out 1805, inc. as a city 1880. It is a trade and manufacturing center for a fertile farm and livestock area. **8.** Town (1970 pop. 1,806), seat of Franklin co., E Texas, WSW of Texarkana. The county has pine woods, corn fields, truck farms, and oil wells. **9.** City (1970 pop. 8,804), seat of Skagit co., NW Wash., on the Skagit River SSE of Bellingham; laid out 1877, inc. 1890. It is a trade center in a farm and dairy area.

Mourne Mountains (môrn), in S Co. Down, SE Northern Ireland; Slieve Donard (2,796 ft/852.8 m) is the highest peak in Northern Ireland. The district is barren and sparsely populated. Granite and sand and gravel are quarried.

Mow·er (mou'ər), county (1970 pop. 44,919), 703 sq mi (1,820.8 sq km), S Minn., bordering on Iowa and drained by headwaters of the Cedar River; formed 1855; co. seat Austin. It is a corn, soybean, cattle, and hog farming area, with a food-processing industry (meat packing and dairy products). There is some mining. Paper products and concrete are made.

Mo·zam·bique (mō'zəm-bēk'), country (1977 est. pop. 9,560,000), 302,328 sq mi (783,030 sq km), SE Africa, bordering on the Indian Ocean in the E, on South Africa and Swaziland in the S, on Rhodesia, Zambia, and Malawi in the W, and on Tanzania in the N. Lourenço Marques is the capital.

The Mozambique Channel (an arm of the Indian Ocean) separates the country from the island of Madagascar. South of the Zambezi estuary the coastal belt is very narrow, and in the far north the coastline is made up of rocky cliffs. Along the northern coast are numerous islets and lagoons; in the far south is Delagoa Bay. The northern and central interior is mountainous. About one third of Lake Nyasa falls within Mozambique's boundaries. Much of the country is covered with savanna; there are also extensive hardwood forests, and palms grow widely along the coast and near rivers.

More than 95% of the population is made up of black Africans, virtually all of whom speak a Bantu language. The principal ethnic groups are the Yao, Makonde, Makua, Thonga, Chewa, Nyanja, Sena, Shona, and Tonga. Small numbers of Swahili live along the coast.

Economy. Mozambique is an overwhelming agricultural country, with the majority of its workers engaged in subsistence cultivation. Many black Africans grow cash crops; about 150,000 work on plantations; and approximately 80,000 are employed as migratory laborers in South African mines. The principal food crops are maize, cassava, pulses, rice, potatoes, plantains, groundnuts, and sesame.

Cotton (grown mainly in the north) and cashew nuts are the chief cash crops produced on private black African plots. The leading plantation crops are sugar cane, tea, copra, and sisal. Large numbers of cattle and goats are raised. There are small forestry and fishing industries. The territory's mineral wealth has not been determined fully, and mining is a minor factor in Mozambique's economy. Mozambique's rudimentary industrial plant is devoted largely to the processing of raw materials. In addition, refined petroleum, construction materials (particularly cement), steel, chemical fertilizer, clothing, and footwear are produced.

History. Bantu-speaking black Africans began to migrate into the region of Mozambique in the middle of the 1st millennium A.D. From 1000, Arab and Swahili traders settled along parts of the coast. The traders had contact with the interior, and Sofala, an early settlement, was particularly noted as a gold- and ivory-exporting center closely linked with Kilwa (on the coast of modern Tanzania). In 1498 Vasco da Gama, a Portuguese navigator, visited Quelimane and Moçambique. In 1505 the Portuguese under Francisco de Almeida occupied Moçambique and established a settlement at Sofala. The Mwanamutapa kingdom controlled the region between the Zambezi and Save rivers and was the source of much of the gold exported at Sofala. Swahili traders resident in Mwanamutapa began to redirect the kingdom's gold trade away from Portuguese-controlled Sofala and toward more northern ports. Thus, Portugal became interested in directly controlling the interior. However, the Swahili traders who lived there, fearing for their commercial position, resisted the Portugese. Between 1569 and 1574 two Portugese armies marched into the interior, but most of the men were killed in fighting with black Africans. In the late 16th and early 17th cent. the official Portuguese presence in the interior was limited to small trading colonies along the Zambezi. At the same time Portuguese adventurers began to establish control over large estates (*prazos*), which resembled feudal kingdoms. Black Africans were forced to work on plantations, and considerable slave-raiding was undertaken (especially after 1650). Some of the plantation owners (*prazeros*) maintained private armies, and they were generally independent of the Portuguese crown. From about 1628 the Portuguese gained increasing influence in Mwanamutapa. Mozambique was ruled as part of Goa in India until 1752, when it was given its own administration headed by a captain-general. Although the Portuguese helped introduce several American crops (notably maize and cashew nuts), their impact was mainly destructive. From the mid-18th to the mid-19th cent. large numbers of black Africans were exported as slaves. In the 1820s and 1830s groups of Nguni-speaking people from South Africa invaded Mozambique; in southern Mozambique they held effective control until the late 19th cent.

When the scramble for African territory among the European powers began in the 1880s, the Portuguese government had only an insecure hold on Mozambique. Nevertheless, Portugal tried to increase its nominal holdings, partly in an attempt to connect by land its territory in Mozambique and in Angola. Beginning in the 1890s and ending only around 1920, the Portuguese established their authority in Mozambique by force of arms against determined black African resistance. In the 1890s several private companies were founded to develop and administer most of Mozambique. In 1910 the status of the territory was changed from province to colony. After the 1926 revolution in Portugal, the Portuguese government took a more direct interest in Mozambique. The government furthered economic development by building railroads and by systematically forcing black Africans to work on white-owned land. Portuguese colonial policy was based on the egalitarian theory of "assimilation": a culturally assimilated African was to be given the same legal status as a white Portuguese. In practice, however, very few black Africans qualified for citizenship and they were directed to work for whites or to grow export crops.

In 1951 the status of Mozambique was changed to "overseas province," and in 1972 Mozambique was declared to be a "self-governing state." In both instances, however, Portugal maintained firm control over the territory. Between 1961 and 1963 several laws (one of which abolished forced labor) were passed to improve the living conditions of the black Africans. At the same time, nationalist sentiment was growing in Mozambique. In 1962 several nationalist groups were united to form the Mozambique Liberation Front (Frelimo), headed by Eduardo Mondlane. The Portuguese adamantly refused to give the territory independence, and in 1964 Frelimo initiated guerrilla warfare. In 1969 Mondlane was killed in Dar es Salaam; he was succeeded by Uria Simango (1969) and by Samora Moisès Machel (1970). By the early 1970s, Frelimo controlled much of central and northern Mozambique. Frelimo's efficacy was hurt somewhat by internal dissension and by the defection of some of its leaders to the Portuguese side. Frelimo received aid from several foreign sources, including the government of Sweden (beginning in 1969). Roman Catholic missionaries accused the Portuguese of massacring about 400 inhabitants of the village of Wiriyamu (near Tete) in Dec., 1972; Portugal denied the charge. It was reported that beginning in mid-1973 Portugal had resettled about 1 million black Africans in fortified villages to insulate them from Frelimo activities.

On Apr. 25 the government of Portugal was overthrown by the military. The new regime made an effort to resolve the conflict in Mozambique by implementing a number of reforms, and by entering (June) into negotiations with Frelimo. The talks resulted in a mutual cease-fire (July 29) and an agreement (Sept. 7) for Mozambique to become independent in June, 1975. In reaction to the agreement, a group of white rebels attempted to seize control of the Mozambique government but were quickly subdued. On Sept. 20 an interim black government took office with Joaquim Chissano of Frelimo as premier. As black rule of Mozambique became a reality and as increased racial violence erupted, there was an exodus of Europeans from Mozambique.

Mo·zhaisk (mə-zhīsk′), city (1967 est. pop. 21,000), W central European USSR, on the Moskva River. It is a rail terminus and has clothing and brick factories. First mentioned in 1231, Mozhaisk was joined with Moscow principality in 1303 and became an important fortress and commercial center. The city was taken by the Germans on Oct. 15, 1941, in their furthest advance directly west of Moscow. It was recaptured during the Soviet winter offensive of 1941–42.

Msta (əm-stä′), river, c.280 mi (450 km) long, rising N of Vyshne Volochek, NW European USSR, and flowing generally NW into Lake Ilmen near Novgorod.

Mtwa·ra-Mi·kin·da·ni (əm-twä′rä-mē-kēn-dä′nē), municipality (1967 pop. 20,413), capital of Mtwara prov., SE Tanzania, a port on the Indian Ocean. Cashew nuts and lime are processed.

Mu·fu·li·ra (mo͞o-fo͞o-lē′rä), city (1976 est. pop., with suburbs, 154,100), N central Zambia, on the border with Zaire. It is a copper-mining center.

Muh·len·berg (myo͞o′lən-bûrg′), county (1970 pop. 27,537), 482 sq mi (1,248.4 sq km), W Ky., bounded on the NE by the Green River, on the E by the Mud River, on the W by the Pond River; formed 1798; co. seat Greenville. It is in an important bituminous-coal area, with timber, oil wells, and clay pits. Its agriculture includes dairy products, poultry, livestock, soybeans, burley tobacco, grain, hay, and truck crops. Clothing and wood products are manufactured.

Mühl·hau·sen (mül-hou′zən), city (1974 est. pop. 44,106), Erfurt dist., SW East Germany, on the Unstrut River. It is a major center for the manufacture of textiles and clothing. Other products include paper, machinery, and furniture. Barite is mined nearby. Fortified (10th cent.) by Henry I, Mühlhausen was a favorite residence of the German rulers. It was made a free imperial city in 1180 and later (13th cent.) joined the Hanseatic League. It became (16th cent.) an Anabaptist center. Mühlhausen changed hands several times before passing in 1815 to Prussia. Noteworthy structures of the city include several Gothic churches and many houses dating from the 16th, 17th, and 18th cent.

Muir Woods National Monument: (myo͝or): *see* National Parks and Monuments Table.

Mu·ka·che·vo (mo͞o'kä-chĕv-ô), city (1976 est. pop. 69,000), extreme SW European USSR, in the Ukraine. It is a rail terminus and highway junction and has food, tobacco, beer, wine, furniture, textile, and timber industries. From the 9th to the 11th cent., Mukachevo was part of the Kievan state. Taken by the Hungarians in 1018, it later (15th cent.) developed as a prominent trade and craft center. Part of the Transylvanian duchy from the 16th cent., Mukachevo then came under Austrian control and was made a key fortress of the Austro-Hungarian empire. Mukachevo passed to Czechoslovakia in 1919, was under German-Hungarian occupation from 1938 to 1944, and was ceded to the Ukrainian SSR in 1945.

Mu·kal·la (mo͞o-kăl'ə), town (1970 est. pop. 65,000), Southern Yemen, a port on the Gulf of Aden. Fish products, tobacco, and coffee are exported.

Muk·den (mo͝ok'dĕn', mo͝ok'-): *see* Shen-yang, China.

Mu·la·tas (mo͞o-lä'täs), archipelago off the NE coast of Panama. It consists of 332 coral islands. The inhabitants are almost pure-blooded aborigines of Carib origin; fishing and coconut gathering are the chief occupations. Protected by a treaty with the government of Panama, the Indians did not consent to scientific observation of their culture and visits of tourists until the late 1940s.

Mule·shoe (myo͞ol'sho͞o'), town (1970 pop. 4,525), seat of Bailey co., NW Texas, NW of Lubdock near the N.Mex. border; settled 1913, inc. 1926. The area has ranches, but it derives its income primarily from the cotton, grain, and vegetable farms of the region.

Mul·ha·cén (mo͞ol'lä-*thān'*) or **Mu·ley-Ha·cén** (mo͞o-lā'ä-*thān'*), peak, 11,424 ft (3,484.3 m) high, S Spain. It is the highest point of the Sierra Nevada and of Spain.

Mül·heim an der Ruhr (mül'hīm än dĕr ro͝or'), city (1975 pop. 190,689), North Rhine-Westphalia, W West Germany, on the Ruhr River. It is an industrial center of the Ruhr district. The city formerly produced mainly coal and steel, but in the mid-20th cent. its products include machinery, precision and optical instruments, clothing, and chemicals. Mülheim was chartered in 1808.

Mul·house (mü-lo͞oz'), city (1975 pop. 117,013), Haut-Rhin dept., E France, in Alsace, on the Ill River and the Rhône-Rhine canal. Cotton, wool, and clothing are the chief manufactures; machinery, chemicals, automobile parts, and steel pipes are also produced. Nearby are the only important potash mines in Western Europe. Mulhouse became a free imperial city in the 13th cent. In 1515 it became an allied member of the Swiss Confederation, and in 1586 it became a neutral republic. In 1798 Mulhouse voted to unite with France. After the Franco-Prussian War (1871), the city was made a part of Germany until 1918.

Mull (mŭl), island (1961 pop. 3,185), 351 sq mi (909.1 sq km), Strathclyde region, NW Scotland, one of the Inner Hebrides, separated from the mainland by the Sound of Mull and the Firth of Lorn. The land is mountainous, rising from the deeply indented coast line to 3,169 ft (966.5 m) at Ben More. Mull has gardens and farms. There are several medieval castles.

Mul·len (mŭl'ən), village (1970 pop. 667), seat of Hooker co., central Nebr.,NNW of North Platte near the Middle Loup River, in a livestock and grain area. Dairy and poultry products are made.

Mull of Gal·lo·way (găl'ə-wā), headland, 239 ft (73 m) high, Wigtownshire, SW Scotland, the southernmost extremity of Scotland.

Mul·tan (mo͞ol-tän'), city (1972 metropolitan area pop. 542,195), E central Pakistan, in the Punjab. It is an important road and rail junction, an agricultural center, and a market for textiles, leather goods, and other products. The city's industries include metalworking, flour and oil milling, and the manufacture of cotton textiles, shoes, carpets, and glass. Multan is also known for its handicrafts, especially pottery and enamel work. The city was conquered (c.326 B.C.) by Alexander the Great, visited (A.D. 641) by the Chinese Buddhist scholar Hsüan-tsang, taken (8th cent.) by the Arabs, and captured by Tamerlane in 1398. In the 16th and 17th cent., Multan enjoyed peace under the early Mogul emperors. In 1818 the city was seized by Ranjit Singh, leader of the Sikhs. The British held it from 1848 until Pakistan achieved independence in 1947.

Mult·no·mah (mŭlt-nō'mə), county (1970 pop. 554,668), 424 sq mi (1,098.2 sq km), NW Oregon, drained by the Columbia and Willamette rivers, and bordering on Wash.; formed 1854; co. seat Portland. It has coal mining and agriculture (dairying, poultry, fruit, truck crops, and grain). Its industry includes food processing and lumbering and the manufacture of paper products, furniture, textiles, chemicals, concrete, metal products, and transportation equipment.

Mun·cie (mŭn'sē), city (1978 est. pop. 76,700), seat of Delaware co., E Ind., on the White River; inc. 1854. It is a trade, processing, and manufacturing center. The city is in a rich agricultural area that has dairying, livestock raising, and grain, soybean, fruit, and truck crops. Machine tools, metal goods, glass, electrical components, wire, and automotive and marine parts are among the many manufactures. The town was first established by the Delaware Indians. White settlers were in the area before the land passed (1818) by treaty to the U.S. government. Industrialization came after the discovery (1886) of natural gas in the county.

Mün·den (mün'dən), town (1974 est. pop. 27,326), Lower Saxony, E West Germany, where the Fulda and Werra rivers flow together to form the Weser River. Its manufactures include machine tools, chemicals, and lead, aluminum, and rubber goods. Münden was founded in the late 12th cent. on the site of a Carolingian palace. Noteworthy structures of the picturesque town include the palace (renovated in the 16th cent. and now a museum), a stone bridge (built c.1400), a Renaissance-style town hall, and numerous half-timber houses.

Mun·ford·ville (mŭn'fərd-vĭl'), town (1970 pop. 1,233) seat of Hart co., W central Ky., NE of Bowling Green.

Mu·nich (myo͞o'nĭk), or *German* **Mün·chen** (mün'KHən), city (1970 pop. 1,293,590), capital of Bavaria, S West Germany, on the Isar River near the Bavarian Alps. It is a commercial, industrial, transportation, communications, and cultural center. Its industries produce machinery, chemicals, pharmaceuticals, processed foods, precision and optical instruments, textiles, electrical appliances, and beer; vehicles and airplanes are assembled. The city is a major tourist and convention center and has radio, television, and film studios.

Situated near a settlement that was established in Carolingian times, Munich was founded (1158) by Henry the Lion, duke of Saxony and of Bavaria. In 1255 it was chosen as the residence of the Wittelsbach dukes of Bavaria; it later became (1506) the capital of the dukedom. In 1806 the city was made capital of the kingdom of Bavaria. Under the kings Louis I (1825–48), Maximilian II (1848–64), and Louis II (1864–86), Munich became a cultural and artistic center. After World War I National Socialism (Nazism) was founded here, and on Nov. 8, 1923, Adolf Hitler failed in his attempted Munich "beer-hall putsch"—a coup aimed at the Bavarian government. Munich was badly damaged during World War II, but after 1945 it was extensively rebuilt and many modern buildings were constructed.

Among the city's chief attractions are the Frauenkirche (Church of Our Lady), a twin-towered cathedral built from 1468 to 1488; the Renaissance-style St. Michael's Church (1583–97); the Theatinerkirche (17th–18th cent.), a baroque church; Nymphenburg castle (1664–1728); and the large English Garden (laid out 1789–1832). The city also has several leading museums, a famous university, a conservatory of music, an opera, numerous theaters, and many publishing houses. Munich is noted for its lively Fasching (Shrove Tuesday) and Oktoberfest (October festival) celebrations.

Mu·ni·sing (myo͞o'nĭ-sĭng), city (1970 pop. 3,677), seat of Alger co., N Mich., N Upper Peninsula, SE of Marquette; inc. as a village 1897, as a city 1916. Paper is made. The region, heavily wooded, and among cliffs and falls, attracts tourists and fishermen. Grand Island is offshore, and a ski area is nearby.

Mun·ster (mŭn'stər), province (1971 pop. 882,002), 9,315 sq mi (24,126 sq km), SW Republic of Ireland. The largest of the Irish provinces, it comprises the counties of Clare, Cork, Kerry, Limerick, Tipperary, and Waterford. It was of the ancient kingdoms of Ireland.

Mün·ster (mün'stər), city (1970 pop. 262,567), North Rhine-Westphalia, W West Germany, a port and industrial center on the Dortmund-Ems Canal. Its manufactures include heavy machinery, textiles, metal products, and beer. The city is also a trade center for grain and lumber. Münster was founded (c.800) as a Carolingian episcopal see. Its bishops ruled a large part of Westphalia as princes of the Holy Roman Empire from the 12th cent. until 1803, when the bishopric was secularized. From the 14th cent. the city was a prominent member of the Hanseatic League. In 1534-35 it was the scene of the Anabaptist experimental government under John of Leiden. Münster passed to Prussia in 1816. It was severely damaged in World War II but was rebuilt after 1945. Münster retains some of its medieval character. Its historic buildings include the cathedral (13th cent.), the Lambertikirche (14th-15th cent.), the Liebfrauenkirche (14th cent.), a Gothic city hall (14th cent.), and gabled houses.

Mur (mŏŏr), river, c.300 mi (480 km) long, rising in the Hohe Tauern, S central Austria, and flowing NE to Bruck, where it receives the Mürz River, its chief tributary. Turning southeast, it flows past Graz (the head of navigation) and through northwest Yugoslavia to the Drava River.

Mu·ra·no (mōō-rä'nō), suburb of Venice, NE Italy, on five small islands in the Lagoon of Venice. From the late 13th cent. it was the center of the Venetian glass industry, which reached a peak in the 16th cent. and was revived in the 19th cent. With its old houses, canals, and bridges, Murano has the same quaint charm as Venice. Of note are a Venetian-Byzantine basilica (7th-12th cent.) and a museum of old and new Venetian glass.

Mur·chi·son (mûr'chĭ-sən), intermittent river, 440 mi (708 km) long, W Western Australia. It flows southwest to Gantheaume Bay on the Indian Ocean.

Mur·cia (mŏŏr'shə, mŏŏr'*th*yä), region and former Moorish kingdom, SE Spain, on the Mediterranean Sea. The area has a generally rugged terrain, except along its coastal plain, and is one of the hottest and driest regions of Europe. However, an irrigation system (dating from Moorish times) and several fertile valleys permit the growing of large crops of citrus and other fruits, vegetables, almonds, olives, and grapes. The region was settled by the Carthaginians. It was taken (8th cent. A.D.) by the Moors and emerged as an independent kingdom after the fall (11th cent.) of the caliphate of Córdoba. In 1243 it became a vassal state of Castile, which in 1266 annexed it outright.

Murcia, city (1975 est. pop. 172,000), capital of Murcia prov., SE Spain, on the Segura River. The city lies in one of the finest garden regions in Spain. The silk industry, a traditional occupation for many years, has declined. There are food-processing and other light industries. Lead, silver, sulfur, and iron are mined nearby. Murcia rose to prominence under the Moors. The Gothic cathedral (14th-15th cent.) and the episcopal palace are landmarks.

Mur·do (mûr'dō), city (1970 pop. 865), seat of Jones co., S central S.Dak., SSW of Pierre, in a livestock, poultry, and grain area.

Mu·re·şul (mōō'rĕ-shōōl), river, c.470 mi (755 km) long, rising in the Carpathian Mts., N central Rumania. It flows generally west into southern Hungary, where it joins the Tisza River at Szeged.

Mu·ret (mü-rā'), town (1968 pop. 13,598), Haute-Garonne dept., S France. It is an agricultural market and produces foundry products, surgical instruments, and bricks. In 1213 Simon de Montfort, leader of the Albigensian Crusade, defeated the nobles of southern France at Muret, thus ending their independence.

Mur·frees·bor·o (mûr'frēz-bûr'ō, -bŭr'ō). **1.** Town (1970 pop. 1,350), seat of Pike co., SW Ark., near the Little Missouri River SW of Hot Springs, in a farm area. There are mercury and cinnabar mines here. **2.** City (1978 est. pop. 31,900), seat of Rutherford co., central Tenn., on Stones River; inc. 1817. It is the processing center of a dairy, livestock, and farm area, and its manufactures include electrical equipment, furniture, and rubber tires. Murfreesboro was the capital of the state from 1819 to 1826. Andrew Jackson and Thomas Hart Benton practiced law here. The Civil War Battle of Murfreesboro (or Stones River) was fought here (Dec. 31, 1862-Jan. 2, 1863). Stones River National Battlefield commemorates the battle, and Civil War dead are buried in Stones River National Cemetery. Another historic attraction is Oakland Mansion, scene of the surrender (July, 1862) of a Federal garrison to Gen. N. B. Forrest.

Mur·gab (mŏŏr-gäb'), river, 530 mi (852.8 km) long, rising in the Paropamisus range, NE Afghanistan, flowing NW into the USSR, through the Merv oasis, and disappearing into the Kara Kum desert, SE Turkmen SSR. It forms part of the Afghanistan-USSR border.

Mur·mansk (mŏŏr-mänsk'), city (1976 est. pop. 369,000), capital of Murmansk oblast, NW European USSR, on the Kola Gulf of the Barents Sea. The terminus of the Northeast Passage, it is a leading Soviet freight port, a naval base, and a base for fishing fleets and is the world's largest city north of the Arctic Circle. The port at Murmansk is ice free. The city is also a railroad terminus. Murmansk has fish canneries, shipyards, textile factories, breweries, and sawmills. Lumber, fish, and apatite are exported, and machinery and coal are imported. Murmansk was only a small village before World War I. The port and its rail line inland from Leningrad were built in 1915-16, when the Central Powers cut off the Russian Baltic and Black Sea supply routes. Allied forces occupied the Murmansk area from 1918 to 1920, during the Russian civil war. A major World War II supply base and port for Anglo-American convoys, Murmansk was bombarded by the Germans.

Mu·rom (mōō'rəm), city (1976 est. pop. 111,000), W central European USSR, on the Oka River. It is a port and a rail junction, with railroad repair shops and machinery, woodworking, and textile industries. First mentioned in the chronicles in 862, the city was ravaged by the Mongols in the 13th cent., and in 1393 it passed to the grand duchy of Moscow.

Mu·ro·ran (mŏŏr'ə-rän'), city (1975 pop. 158,715), SW Hokkaido, Japan, on Uchiura Bay. It is a major industrial center and port, with iron, steel, and cement works and an oil refinery. Hot spring resorts are nearby.

Mur·phy (mûr'fē), town (1970 pop. 614), seat of Owyhee co., SW Idaho, S of Caldwell.

Mur·phys·bor·o (mûr'fēz-bûr'ō, -bŭr'ō), city (1970 pop. 10,013), seat of Jackson co., S Ill., on the Big Muddy River; inc. 1867. It is a trade and distributing center for a rich farm area. Shoes, clothing, and aluminum products are made. An apple festival is held here.

Mur·ray (mûr'ē, mŭr'ē). **1.** County (1970 pop. 12,986), 342 sq mi (885.8 sq km), N Ga., bounded on the N by the Tenn. border, on the W by the Conasauga River; formed 1832; co. seat Chatsworth. In an agricultural area yielding corn, cotton, hay, fruit, and livestock, it has talc mines, timber and a textile industry. Part of Chattahoochee National Forest is in the east. **2.** County (1970 pop. 12,508), 708 sq mi (1,833.7 sq km), SW Minn., drained by headwaters of the Des Moines River, with Lake Shetek in the N; formed 1857; co. seat Slayton. It is in an agricultural area that produces corn, oats, barley, and livestock. It has some mining and makes mobile homes and wood products. **3.** County (1970 pop. 10,669), 423 sq mi (1,095.6 sq km), S Okla., intersected by the Washita River and including part of the Arbuckle Mts.; formed 1907; co. seat Sulphur. It has farming (corn, cotton, wheat, and fruit), livestock and poultry raising, and dairying. There are sand and gravel deposits, and oil wells.

Murray, city (1970 pop. 13,537), seat of Calloway co., SW Ky., near the Tenn. border; inc. 1844. There are some manufacturing industries in the city, and agriculture and tourism are also important. Murray is located near a recreational area operated by the Tennessee Valley Authority.

Murray, principal river of Australia, 1,609 mi (2,589 km) long, rising in the Australian Alps, SE New South Wales, and flowing W to form the New South Wales-Victoria boundary. It then flows southwest across South Australia state through Lake Alexandrina, a lagoon, into the Indian Ocean. It receives its main tributary, the Darling River, at Wentworth. The Murray-Darling watercourse is 2,911 mi (4,683.8 km) long but is of little use for navigation except in the lower reaches. The Murray valley contains most of Australia's irrigated land; vines, fruits, and vegetables are grown.

Mur·rum·bidg·ee (mûr'əm-bĭj'ē), river, c.1,050 mi (1,690 km) long, rising in the Australian Alps, SE New South Wales, Australia, and flowing generally W to the Murray River on the Victoria border.

Mur·ten (mŏŏr'tn), town (1970 pop. 4,256), Fribourg canton, W Switzerland, on the Lake of Murten. It is known chiefly as the scene of the defeat (1476) of Charles the Bold of Burgundy by the Swiss.

Muş (mōōsh), city (1975 pop. 27,730), capital of Muş prov., E Turkey. Grapes are grown nearby. Founded c.400 B.C., it was an important town of Armenia. Muş was captured by the Seljuk Turks, the Mongols, and Tamerlane before being annexed by the Ottoman Empire in 1515.

Mu·sa·la (mōō'sä-lä'), mountain, 9,592 ft (2,925.6 m) high, SW Bulgaria. It is the highest peak of Bulgaria and of the Rhodope range.

Mu·sa·shi·no (mōō-sä'shē-nō'), city (1976 est. pop. 139,204), Tokyo Metropolis, E central Honshu, Japan, on the Sumida River. It is a suburb of Tokyo.

Mus·cat (mŭs-kăt'), city (1969 est. pop. 9,980), capital of Oman, SE Arabia, on the Gulf of Oman. It is flanked by rugged mountains. Muscat, which has a fine harbor, was seized by the Portuguese in 1508 and kept by Portugal until 1648. Persian princes held it until 1741, when it became the capital of Oman. Dates, dried fish, and mother-of-pearl are exported.

Mus·ca·tine (mŭs'kə-tēn', mŭs'kə-tēn'), county (1970 pop. 37,181), 442 sq mi (1,144.8 sq km), SE Iowa, bounded in the SE by the Mississippi River (here forming the Ill. border); formed 1836; co. seat Muscatine. A prairie agricultural area yielding hogs, cattle, poultry, grain, and truck crops, it has limestone quarries, sand and gravel pits, and bituminous-coal deposits.

Muscatine, city (1978 est. pop. 22,700), seat of Muscatine co., SE Iowa, on the Mississippi River; inc. 1851. An early center of river traffic and lumbering, Muscatine today is the shipping and processing center of a rich agricultural area. Food products, grains, industrial alcohol, and vitamins are among the city's products.

Mus·cle Shoals (mŭs'əl shōlz'), town (1970 pop. 6,907), Colbert co., NW Ala., on the Tennessee River opposite Florence; inc. 1923. It is the center of experimental development of phosphate and nitrate fertilizers and animal foods. The river formerly descended at that point in a series of rapids called Muscle Shoals (more than 35 mi/56 km long with a drop of more than 130 ft/40 m). They were unnavigable, and early in the 19th cent. plans were made to construct a canal around them. However, a canal built by the state in the 1830s silted up and was abandoned. In 1890 a canal was successfully built by U.S. army engineers. In World War I a project for extracting nitrate was undertaken, but the nitrate works built in 1916 remained unused until the area was bought (1933) by the Tennessee Valley Authority. The building of Wilson Dam (1925) and of Wheeler Dam (1936) resulted in the submersion of the shoals.

Mu·shin (mōō'shĭn'), city (1971 est. pop. 176,000), SW Nigeria, an industrial and residential suburb of Lagos. Manufactures include textiles, furniture, printed materials, metal products, plastics, milk products, and shoes. Mushin grew mainly after World War II.

Mu·si (mōō'sē), river, c.325 mi (525 km) long, rising in the Pegunungan Barisan, S Sumatra, Indonesia. It flows southeast to Palembang (head of oceangoing navigation), then northeast through the swampy coastal plain to the Bangka Strait. Rubber and a variety of tropical crops are raised in the valley.

Mus·ke·gon (mŭs-kē'gən), county (1970 pop. 157,426), 504 sq mi (1,305.4 sq km), SW Mich., bounded on the W by Lake Michigan and drained by the Muskegon and White rivers; formed 1859; co. seat Muskegon. There is some agriculture (livestock, grain, and dairy products). Industries include paper mills, as well as the manufacture of chemicals and office furniture and equipment. There are oil wells and refineries, commercial fisheries, and resorts. It includes part of Manistee National Forest in the north.

Muskegon, city (1978 est. pop. 44,200), seat of Muskegon co., W Mich., on Lake Michigan; inc. as a city 1869. A port of entry, with a large landlocked harbor, the city is a car-ferry terminus and a shipping point for a farm, fruit, and industrial region. Among its many manufactures are automobile parts and engines, foundry products, pistons, bowling and billiard supplies, copper wire, gasoline pumps, and heavy machinery. Agriculture and tourism are also important. A fur-trading post was established here c.1810. The first sawmill was built in 1837, and the lumber industry thrived until 1890, when the city was swept by fire.

Muskegon, river, 227 mi (365.2 km) long, rising in Houghton Lake, N central Mich., and flowing SW to Lake Michigan at Muskegon. At its mouth the river widens into Muskegon Lake, forming a harbor c.2.5 mi (4 km) wide and c.5.5 mi (8.9 km) long.

Mus·king·um (mə-skĭng'əm), county (1970 pop. 77,826), 651 sq mi (1,686.1 sq km), central Ohio, intersected by the Muskingum and Licking rivers and by small Salt and Jonathan creeks; formed 1804; co. seat Zanesville. It has agriculture (livestock, dairy products, fruit, and grain), bituminous-coal mines, oil and gas fields, limestone quarries, and sand, gravel, and clay pits. Its industries include food processing and metalworking. Wood, paper, chemical, plastic, and pottery products and machinery are made.

Muskingum, river, 111 mi (178.6 km) long, formed in NE Ohio, at Coshocton, by the union of the Walhonding and Tuscarawas rivers and flowing S through Zanesville, then SE to the Ohio River at Marietta. The Muskingum River system has extensive flood control projects.

Mus·ko·gee (mŭs-kō'gē), county (1970 pop. 59,542), 822 sq mi (2,129 sq km), E Okla., bounded in the N by the Arkansas River and in the S by the Canadian River; formed 1907; co. seat Muskogee. It has agriculture (livestock, dairy products, soybeans, wheat, grain, sorghums, cotton, and potatoes), oil and natural-gas fields, and mineral mines. Its industry includes food processing, paper milling, iron and steel working, and the manufacture of furniture, glass and metal products, and machinery.

Muskogee, city (1978 est. pop. 38,900), seat of Muskogee co., E Okla., near the junction of the Arkansas, Verdigris, and Grand rivers; inc. 1898. It is an important transportation, trade, and industrial center in the agricultural Arkansas valley, with a new port (opened 1971). Muskogee has food-processing plants, meat-packing houses, seed mills, and numerous manufacturing industries. Of interest are the Five Civilized Tribes Museum and the beautiful flower gardens in Honor Heights Park; an antique car museum; and nearby Fort Gibson (1824; restored), with its national cemetery. A state fair and an azalea festival are held annually.

Mus·sel·shell (mŭs'əl-shĕl'), county (1970 pop. 3,734), 1,886 sq mi (4,884.7 sq km), central Mont., in an agricultural region drained by the Musselshell River; formed 1911; co. seat Roundup. Livestock, grain, and coal are important to its economy.

Musselshell, river c.292 mi (470 km) long, rising in central Mont. in the Crazy Mts. and flowing generally E, then N to join the Missouri River.

Mu-tan-chiang or **Mu·tan·kiang** (both: mōō'dän'jē-äng'), city (1970 est. pop. 400,000), SE Heilungkiang prov., China. It is a railroad junction and a lumbering center in a rich timber region. Manufactures include wood pulp, paper, compound medicines, tires, aluminum, and cement.

Mut·tra (mŭ'trə): *see* Mathura, India.

Mu·zaf·far·na·gar (mōō-zäf'ər-nə-gär'), town (1971 pop. 114,859), Uttar Pradesh state, N central India. It is the center of an area growing wheat and sugar cane.

Mu·zaf·far·pur (mōō-zäf'ə-pōōr'), city (1971 pop. 127,045), Bihar state, NE India, on the Gandak River. It is the site of Bihar Univ.

Muz·tagh (mōōs-täg'), mountain, 24,757 ft (7,550.9 m), SW Singkiang Uighur Autonomous Region, W China, near the Tadzhik SSR border. It is the second-highest peak of the Kunlun Mts.

Mwan·za (mwän'zä), city (1967 pop. 34,861), capital of Mwanza prov., NW Tanzania, a port on Victoria Nyanza. Industries include meat-packing, fishing, and the manufacture of textiles and soap. There is an institute for research in tropical diseases.

Mwe·ru, Lake (mwä'rōō), c.70 mi (115 km) long and 30 mi (50 km) wide, alt. c.3,000 ft (910 m), central Africa, on the Zaire-Zambia border. The lake has large fisheries.

Myc·a·le (mĭk'ə-lē), promontory, W Asia Minor, opposite Samós Island. The center of the Ionian League was in the temple of Poseidon. In 479 B.C. the Greeks destroyed the Persian fleet here.

My·ce·nae (mī-sē'nē), ancient city of Greece, in Argolis. In historical times it had little importance and was usually dependent on Argos. Its significance is in its remote past as a center of Mycenaean civilization. The famous Lion Gate, which led into the city, and the Treasury of Atreus, the largest of the beehive tombs outside the walls of the city, are the most notable of its ancient remains.

Myit·kyi·na (myĭ'chē-nä), town (1964 est. pop. 14,000), N Burma, on the Irrawaddy River. It is a trade center (including teak and jade) and a railroad terminus. In World War II its capture (Aug., 1944) by Allied troops after a siege of 78 days marked an important stage in the liberation of Burma from the Japanese.

My·ko·nos (mĭk'ə-nōs'): *see* Mikonos, Greece.

My·lae (mī'lē), ancient port, NE Sicily, now Milazzo. It was settled by colonists from Messina. Here in 260 B.C. the Romans in a newly built fleet defeated the Carthaginians in the First Punic War.

My·men·singh (mī'mĕn-sĭng'), town (1974 pop. 76,036), N central Bangladesh, on a channel of the Brahmaputra River. It is a trading center for rice, jute, sugar cane, oilseeds, tobacco, mustard, and pulses. Once noted for the manufacture of glass bangles, Mymensingh now has jute-pressing and electrical-supply industries.

My·ra (mī'rə), ancient city and seaport of Lycia, S Asia Minor. Ruins of a theater are on the acropolis.

Myr·tle Beach (mûr'tl), city (1970 pop. 9,035), Horry co., E S.C., on the Atlantic coast SE of Conway; inc. 1938. It is a seaside resort. There is some farming and fishing. Metal products and electronic equipment are made.

My·si·a (mĭsh'ē-ə), ancient region, NW Asia Minor, N of Lydia. Mysia passed successively to Lydia, Persia, Macedon, Syria, Pergamum, and Rome.

My·sore (mī-sôr'): *see* Karnataka, state, India.

Mys·tic (mĭs'tĭk), river, c.10 mi (16 km) long, rising in SE Conn. and flowing S past Old Mystic and Mystic villages to Long Island Sound. Mystic Seaport, a maritime museum, is at its mouth.

Myt·i·le·ne (mĭt'l-ē'nē): *see* Mitilíni, Greece.

Mzab (əm-zäb'), stony, barren valley, Algeria, in the N Sahara. It was settled c.1000. The Mzabites dug wells, created date-palm oases, and built seven towns, united in a confederation. Their aptitude for trade made the area a caravan junction. France occupied the Mzab in 1853 and annexed it formally in 1882.

Nab·a·tae·a (năb'ə-tē'ə), ancient kingdom of Arabia, S of Edom, in present-day Jordan. It flourished from the 4th cent. B.C. to A.D. 106, when it was conquered by Rome. The history of Nabataea consists mainly of the struggle to control the trade routes between the Orient and the Mediterranean Sea. Petra, the capital city, is noted for its unique rock-cut monuments, tombs, and temples.

Na·blus (nä-bloōs'), city (1967 pop. 44,223), W Jordan. It is the market center for a region where wheat and olives are grown and sheep and goats are grazed. Manufactures include soap made from olive oil and colorful shepherds' coats. Nablus, an ancient Canaanite town, has remains dating from c.2000 B.C., about the time when the city was held by Egypt. The Samaritans made it their capital and built a temple to rival that of Jerusalem. The city was destroyed (129 B.C.) by John Hyrcanus I. Under Hadrian it was rebuilt and named Neapolis. Nearby are the reputed sites of the tomb of Joseph and the well of Jacob. The city came under Israeli occupation following the Arab-Israeli War of 1967.

Nack·a (nä'kä), city (1975 est. pop. 55,321), Stockholm co., E Sweden, on the Baltic Sea, a suburb of Stockholm. It has radio and television stations and shipyards. Manufactures include steam turbines and processed food.

Nac·og·do·es (năk'ə-dō'chəz), county (1970 pop. 36,362), 902 sq mi (2,336.2 sq km), E Texas, bounded on the W and S by the Angelina River and partly on the E by Attoyac Bayou; formed 1836; co. seat Nacogdoches. Its agriculture includes dairy products, cotton, corn, hay, legumes, sugar cane, fruit, truck crops, and livestock. There are also deposits of oil and natural gas, clay and lignite mines, lumber mills, and some manufacturing plants.

Nacogdoches, city (1978 est. pop. 27,800), seat of Nacogdoches co., E Texas, in a pine and hardwood forest area containing clays and sandy loams; settled 1779. Industries in this rapidly growing city include lumbering, clay refining, meat-packing, poultry raising and processing, and the manufacture of feed and fertilizer, brass valves, and wood products. Tourism is also important; the city is in the center of a large recreational area that contains Sam Rayburn Reservoir and many lakes. A Spanish mission was founded here in 1716, but permanent settlers did not arrive until 1779. The settlement was an eastern bastion of the Spanish colony against the French in Louisiana. After the Louisiana Purchase it was twice (1812, 1819) seized by raiding expeditions from the United States. In 1820 about 100 American families were issued land grants here. Their settlement led to the Fredonian Rebellion in 1826. The city was active in the Texas Revolution (1835–36) and later developed into a market for cotton plantations. The state's first oil wells were drilled near the city in 1859. On the campus of Stephen F. Austin State Univ. is the Old Stone Fort, a Spanish presidio built in 1779.

Na·di·ad (nə-dē-äd'), city (1971 pop. 108,268), Gujurat state, W central India. It is a market for agricultural products. Metal utensils are made. The city formerly had walls and a moat.

Naest·ved (nĕst'vĕth'), city (1976 est. pop. 38,700), Storstrøm co., SE Denmark. It is a seaport, linked (since 1938) with the Karrebaek Fjord (an arm of the Store Baelt) by a 5-mi (8.1-km) canal. It is also an industrial center and a rail junction.

Na·fud (nä-foōd'), desert area in the N part of the Arabian Peninsula, occupying a great oval depression. It is 180 mi (290 km) long and 140 mi (225.3 km) wide. This area of red sand is surrounded by sandstone outcrops that have eroded into grotesque shapes. The Nafud is noted for its sudden violent winds, which have formed great crescent-shaped dunes, rising up to 600 ft (183 m). Rainfall occurs once or twice a year. In some lowland basins, especially those near the Hejaz Mts., there are oases where dates, vegetables, barley, and fruits are raised. The Nafud is connected to the Rub al Khali, the great desert of southern Arabia, by the Dahna, a corridor of gravel plains and sand dunes.

Na·ga·land (nä'gə-lănd'), state (1971 pop. 515,561), 6,365 sq mi (16,485 sq km), NE India. Formerly the Naga Hills–Tuensang area in Assam state, it became a separate state in 1961. It is a forested and undeveloped region bounded on the east by Burma. There is a strong movement for independence, and the region has been the scene of clashes between Indian troops and the Naga tribes.

Na·ga·no (nä-gä'nō), city (1976 est. pop. 234,000), capital of Nagano prefecture, central Honshu, Japan, on the Tenryu River. It has printing, food-processing, machine-building, and textile-manufacturing industries. Nagano is a religious center, the site of Zenkoji, a 7th cent. Buddhist temple with statues sent from the king of Korea in 552.

Na·ga·o·ka (nä'gä-ō'kä), city (1976 est. pop. 173,375), Niigata prefecture, central Honshu, Japan. An industrial center, it has oil refineries, chemical plants, engineering works, and machine-building factories. Nagaoka is also a distribution point for agricultural products.

Na·ga·sa·ki (nä'gə-sä'kə, năg'ə-săk'ē), city (1976 est. pop. 449,371), capital of Nagasaki prefecture, W Kyushu, Japan, on Nagasaki Bay. It is one of Japan's leading ports. Shipbuilding is the chief industry; steelworks, collieries, fisheries, and electrical machinery plants are also important. Nagasaki's port, the first to receive Western trade, was known to Portuguese and Spanish traders before it was opened to the Dutch in 1567. After the Portuguese and Spanish merchants were forced to leave Japan in 1637, the Dutch traders were restricted (1641–1858) to De-Jima, an island in the harbor. Nagasaki was gradually reopened to general foreign trade during the 1850s. Long a center of Christianity, the city had until 1945 Japan's largest Roman Catholic cathedral. During World War II, on Aug. 9, 1945, Nagasaki became the target of the second atomic bomb ever detonated on a populated area; about 75,000 people were killed or wounded, and more than one third of the city was devastated.

Na·ger·coil (nä'gər-koil), city (1971 pop. 141,207), Tamil Nadu state, S India. Nagercoil is the southernmost city in India. It is in a region where rice is grown and ilmenite and monazite are mined.

Na·gor·no-Ka·ra·bakh Autonomous Oblast (nä-gôr'nō-kä-rä-bäk'), autonomous region (1970 pop. 149,000), 1,699 sq mi (4,400.4 sq km), SE European USSR, in Azerbaijan, between the Caucasus and the Karabakh range; capital Stepanakert. The region has numerous mineral springs as well as deposits of polymetallic ore, lithographic stone, marble, and limestone. Farming and grazing are important and there are various light industries. The area was taken by Armenia in the 1st cent. A.D. and by the Arabs in the 7th cent. In the early 17th cent. it passed to the Persians, who permitted local autonomy, and in the mid-18th cent. the Karabakh khanate was formed. Karabakh alone was ceded to Russia in 1805; the khanate passed to the Russians by the Treaty of Gulistan in 1813. In 1822 the area became a Russian province. The Nagorno-Karabakh Autonomous Oblast was established in 1923.

Na·go·ya (nä'gô-yä'), city (1976 est. pop. 2,080,050), capital of Aichi prefecture, central Honshu, Japan, on Ise Bay. A major port, transportation hub, and industrial center, it has iron and steel works, textile mills, aircraft factories, automotive works, and chemical, plastics, and fertilizer plants. Fine porcelain, pottery, and cloisonné are also produced. The city has nine universities. Nagoya has two famous shrines: the Atsuta (founded in the 2nd cent.), where the sacred imperial sword is housed; and the Higashi Honganji, which was built in 1692. A fortress town in the 16th cent., Nagoya retains a castle built by Ieyasu in 1612 and reconstructed in 1959.

Nag·pur (năg-pŏŏr'), city (1971 pop. 866,144), Maharashtra state, central India, on the Nag River. Formerly the capital of Berar and Madyha Pradesh states, it is a transportation center and a leading industrial and commercial city. Founded in the 18th cent., it passed in 1853 to the British.

Nagy·ka·ni·zsa (nŏd'yə-kŏ-nĭ'zhŏ), city (1977 est. pop. 46,000), SW Hungary. It is an industrial center producing oil derricks and other heavy equipment, glass, and beer. There are oil fields nearby. Founded c.1300 as a fortress, Nagykanizsa was captured by the Turks in 1600. It has a museum, an 18th cent. Franciscan church, and the ruins of the old fortress.

Nagy·kő·rös (nŏd'yə-kœ'rœsh), town (1977 est. pop. 22,300), central Hungary. It is the center of a fruit-growing region and has an old Reformed church, a college, and a town hall.

Na·ha (nä'hä), city (1976 est. pop. 295,547), on Okinawa island, in the Ryukyu Islands, Japan. A port on the southwest coast, it is also the chief manufacturing center of the island. In 1853 Com. Matthew C. Perry chose Naha as his first base for the penetration of Japan. The city was virtually destroyed during World War II. In 1945 it became the headquarters of the U.S. military governor of the Ryukyus, and when the island was returned to Japan in 1972, it became the capital of Okinawa prefecture.

Na·ha·vand (nä'hä-vänd'), city (1966 pop. 24,000), Hamadan governorate, Kermanshah prov., W Iran. It is an agricultural trade center. Nahavand was the scene of a decisive victory of the Arabs over the Persians in 641 or 642.

Na·huel Hua·pí (nä-wĕl' wä-pē'), lake, c.210 sq mi (545 sq km), in Río Negro and Neuquén provs., W central Argentina. The 45-mi (72.4-km) lake is a popular resort area.

Na·hun·ta (nā-hŭn'tə), city (1970 pop. 974), seat of Brantley co., SE Ga., E of Waycross. Sawmilling is done.

Nairn (nârn), burgh (1971 pop. 8,038), Highland region, N Scotland, at the mouth of the Nairn River on Moray Firth. It is a resort and fishing center with a good harbor. Until 1975 Nairn was county town of Nairnshire.

Nairn·shire (nârn'shĭr', -shər), former county, NE Scotland. Nairn was the county town. The county was settled by Northern Picts and was part of the province of Moray before it came under the Scottish crown. In 1975 Nairnshire became part of the Highland region.

Nai·ro·bi (nī-rō'bē), city (1973 est. pop. 630,000), capital of Kenya, S Kenya, in the E African highlands. A modern city with broad boulevards, Nairobi is Kenya's administrative, communications, and economic center. It is the trade and distribution center for a productive agricultural area specializing in coffee and cattle. The chief manufactures are food products, beverages, construction materials, cigarettes, chemicals, textiles, clothing, glass, and furniture. Although Nairobi is only 90 mi (144.8 km) south of the equator, it has a moderate climate, largely because of its high altitude (c.5,500 ft/1,680 m). Many tourists are attracted to Nairobi National Park, a large wildlife sanctuary and to nearby scenic and hunting areas.

Nairobi was founded in 1899 on the site of a waterhole of the pastoral Masai as a railhead camp on the Mombasa-Uganda line. In 1905 it replaced Mombasa as the capital of the British East Africa Protectorate and became the center of the prosperous European-dominated highlands farming area. In the 1950s the Mau Mau rebellion flared among Kikuyu people near Nairobi; there were related disturbances in the city. The National Univ., several medical and technical schools, Coryndon Memorial Museum, which has collections on Kenya's prehistory and natural history, and the Sorsbie Art Gallery are in Nairobi.

Nai·va·sha (nī-vä'shä), lake, 12 mi (19.3 km) long and 9 mi (14.5 km) wide, W central Kenya, E Africa, in the Great Rift Valley. Located near Nairobi, the lake is a popular resort. Fish, waterfowl, and hippopotamuses abound here.

Na·ja·fa·bad (nə-jä'fə-bäd'), city (1971 est. pop. 46,000), Esfahan prov., W central Iran, near Esfahan. It is the trade center for an agricultural region noted for its pomegranates.

Najd (näjd), region, Arabia: *see* Nejd.

Na·ka·tsu (nä-kä'tsōō), city (1975 est. pop. 44,200), Oita prefecture, NE Kyushu, Japan, on the Suo Sea at the mouth of the Yamakuni River. It is a commercial center and port.

Na·khi·che·van (nə-khē-chĭ-vän'), city (1975 est. pop. 37,000), capital of Nakhichevan Autonomous SSR, SE European USSR, in Azerbaijan. Its industries include food processing, wine making, cotton ginning, and the production of furniture, leather, and building materials. The city was ruled by Armenians, Persians, Arabs, Mongols, and Turks, became a flourishing Armenian trade center in the 15th cent., and was ceded by Persia to Russia by the Treaty of Turkmanchai (1828). In the 19th cent. it was an important trading post. It has Greek and Roman remains and two 12th cent. mausoleums.

Nakhichevan Autonomous Soviet Socialist Republic, autonomous republic (1976 est. pop. 227,000), 2,124 sq mi (5,501.2 sq km), SE European USSR, in Azerbaijan, bordering on Iran and Turkey in the south; capital Nakhichevan. The lowlands are irrigated and produce cotton, tobacco, rice, winter wheat, and fruits. In the foothills grapes are grown for the wine industry, and silkworms are raised. There are salt, molybdenum, lead, and zinc deposits. The republic's industries include food processing, cotton cleaning, and the bottling of mineral water. The republic was founded in 1924.

Na·khod·ka (nə-kôt'kə), city (1976 est. pop. 127,000), Far Eastern USSR, E of Vladivostok, on the Sea of Japan. A port city with fewer winter ice problems than Vladivostok, Nakhodka has assumed an increasingly large share of shipping from the Soviet Far East.

Na·khon Pa·thom (nä'kôn pä'təm), town (1972 est. pop. 37,807), capital of Nakhon Pathom prov., SW Thailand, on the Mekong River. It is a transportation and commercial center. Phra Pathom, the largest Buddhist stupa in Thailand, is here.

Nakhon Rat·cha·si·ma (rä-chä'sĭ-mä') or **Korat** (kō-rät'), city (1972 est. pop. 77,397), capital of Nakhon Ratchasima prov., S central Thailand, on the Mun River. Nakhon Ratchasima is the administrative, economic, and transportation center of the Korat plateau. Copper deposits are nearby. Founded in the 17th cent., the city grew rapidly after the construction (1890) of the RR from Bangkok.

Nak·nek (năk'nĕk'), village (1970 pop. 318), seat of Bristol Bay borough, S Alaska, near the head of the Alaska Peninsula, on Kvichak Bay of Bristol Bay at the mouth of the Naknek River. Fishing and fish processing are done.

Nak·skov (näk'skou), city (1976 est. pop. 17,291), Storstrøm co., SE Denmark, a seaport at the head of Nakskov Fjord (an arm of the Langelands Baelt). It has large sugar refineries.

Na·ku·ru (nä-kŏŏr'ōō), city (1971 est. pop. 45,000), capital of Rift Valley prov., W central Kenya. Founded in the early 20th cent. as a center of European settlement, Nakuru is a growing commercial and industrial city. Manufactures include textiles, processed food, and pyrethrum extract. Nearby is Lake Nakuru (c.35 sq mi/90 sq km), noted for its flamingo haunts.

Nal·chik (näl'chĭk), city (1974 est. pop. 195,000), capital of the Kabardino-Balkar Autonomous Republic, S European USSR, on the N slope of the Greater Caucasus. A health and tourist resort, it also has considerable industry, notably a molybdenum-tungsten mill. Nalchik was founded in 1817 as a Russian stronghold.

Na·man·gan (nä'män-gän'), city (1976 est. pop. 217,000), capital of Namangan oblast, Central Asian USSR, in Uzbekistan, in the Fergana Basin. A center for the production of cotton and silk, it also has food-processing plants. Russian forces captured Namangan in 1875.

Na·ma·qua·land (nə-mä'kwə-lănd') or **Na·ma·land** (nä'mə-lănd'), region, c.150,000 sq mi (388,500 sq km), SW Africa. It extends from Windhoek, Namibia, in the north to northwest Cape Prov., Republic of South Africa, in the south and from the Namib Desert in the west to the Kalahari Desert in the east. The Orange River divides the region into Great Namaqualand (in Namibia) and Little Namaqualand (in Cape Prov.). An arid region, Namaqualand is populated chiefly by the pastoral-agricultural Nama (Hottentots), who speak a Khoikhoi language. Near the Atlantic Ocean are extensive alluvial diamond beds; copper is mined in Little Namaqualand. Karakul pelts are a major export of the region.

Nam·hoi (näm'hoi'): *see* Fo-shan, China.

Na·mib (nä'mĭb), desert, c.800 mi (1,290 km) long and from 30 to 100 mi (50–160 km) wide, SW Africa, along the coast of Namibia. It occupies a rocky platform between the Atlantic Ocean and the escarpment of the interior plateau. Isolated mountains rise from the desert and sand dunes cover its southern portion. Tungsten, salt, and alluvial diamonds are mined.

Na·mib·i·a (nə-mĭb'ē-ə), formerly **South West Af·ri·ca** (ăf'rĭ-kə), country (1977 pop. 900,000), c.318,000 sq mi (823,620 sq km), SW Africa. It is bordered by Angola in the north, by Zambia in the northeast, by Botswana in the east, by South Africa in the southeast and south, and by the Atlantic Ocean in the west. The Orange River forms the southern boundary, and the Kunene, Okavango, and Zambezi rivers form parts of the northern and northeastern borders. The country includes the Caprivi Strip in the northeast, and the South African enclave of Walvis Bay in the west. The capital is Windhoek.

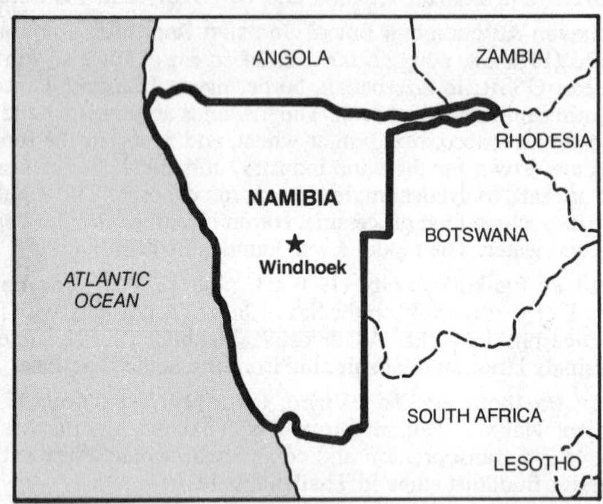

The country has four main geographic regions: the arid and barren Namib Desert, which runs along the entire Atlantic coast; an extensive central plateau that averages c.3,600 ft (1,100m) in elevation; the western fringes of the Kalahari Desert in the east; and an alluvial plain in the north that includes the Etosha Pan, a large salt marsh. The country has an ethnically diverse population.

Economy. Because of inadequate rainfall, crops are not widely raised and pastoralism forms the backbone of the economy. Goats and sheep are raised mainly in the south, and cattle are herded chiefly in the north. Agricultural income is derived mainly from Karakul pelts, livestock, and dairy goods. The country's few manufactures are made up mostly of processed food. There is an extensive mining industry, run principally by foreign-owned companies. The chief minerals are diamonds, copper, lead, manganese ore, zinc concentrate, salt, amethyst, germanium, and vanadium. Fishing fleets operate in the Atlantic.

History. The earliest inhabitants were San hunters and gatherers, who lived here as early as 2,000 years ago. By A.D. c.500, Nama herders had entered the region. The Herero and Ovambo migrated into the area after about 1800. Portuguese navigators landed on the coast in the early 15th cent. In the late 18th cent. Dutch and British captains laid claim to parts of the coast. These claims, however, were disallowed by their governments. In the 18th cent. English and German missionaries arrived. In 1884 the German government proclaimed a protectorate over Angra Pequeña (now Lüderitz), to which the rest of the area (called South West Africa) was soon added. Conflicts between the indigenous population and the Europeans led to outbreaks of violence. In 1903 the Nama began a revolt, joined by the Herero in 1904. The Germans pursued an uncompromising military campaign that by 1908 had resulted in the death of about 54,000 Herero (out of a total Herero population of about 70,000), many of whom were driven into the Kalahari Desert, where they perished; 30,000 other black Africans also died in the revolt. In 1908 diamonds were discovered near Lüderitz, and a large influx of Europeans began. After World War I South Africa began (1920) to administer South West Africa as a mandate. In 1921–22 the Bondelzwarts, a small Nama group, revolted against South African rule, but they were crushed by South African forces.

After 1945 South Africa refused to surrender its mandate and place South West Africa under the UN trusteeship system. However, the UN General Assembly in Oct., 1966, passed a resolution terminating the mandate, and in 1968 it resolved that the country be known as Namibia. However, the South African government maintained that the United Nations had no authority over South West Africa and proceeded with plans for establishing 10 African homelands (Bantustans) in the country and for tying it more closely to South Africa. In 1968 the legislative council of the Ovambo was opened, followed by that of the Okavango in 1970.

Nam·oi (năm'oi'), river, 526 mi (846.3 km) long, N central New South Wales, Australia. Rising as the Peel River in the Liverpool Range, it flows northwest to the Barwon River, a tributary of the Darling River.

Nam·pa (năm'pə), city (1978 est. pop. 24,900), Canyon co., SW Idaho, in the fertile Treasure Valley; inc. 1890. It is the commercial, processing, and shipping center for an irrigated agricultural, orchard, and dairy region and is included in the Boise project. It has food-processing plants and a huge sugar factory. Camp trailers, mobile homes, tin cans, and wood products are also manufactured.

Nam·po (näm'pō'), formerly **Chin·nam·po** (chē'näm'pō'), city, W North Korea, on Korea Bay. It is the port city for Pyongyang and is also a leading metallurgical center. Other industries include rice and flour milling.

Nam·pu·la (năm-po͞o'lə), city (1970 pop. 120,188), NE Mozambique. It is an agricultural trade center and railroad junction.

Nams·os (näm'sōs'), town (1977 est. pop. 11,481), Nord-Trøndelag co., W Norway, a port at the mouth of the Namsen River on the Namsenfjord. In World War II Namsos was the scene (1940) of heavy fighting between the British and the Germans.

Na-mu Hu (nä'mo͞o' ho͞o') or **Nam Tso** (näm' tsō'), salt lake, 950 sq mi (2,460.5 sq km), central Tibet, SW China. The largest lake in Tibet, it is at an altitude of 15,180 ft (4,629.9 m).

Na·mur (nä-mo͝or'), province (1977 est. pop. 398,916), S Belgium, bordering on France in the S; capital Namur. The province is generally hilly and largely agricultural. There are also extensive marble, chalk, and stone quarries, coal and iron mines, and glass and cutlery factories. The province, which is mainly French-speaking, includes the former county of Namur, part of the former prince-bishopric of Liège, and part of Hainaut.

Namur, city (1977 pop. 100,039), capital of Namur prov., S central Belgium, at the confluence of the Meuse and Sambre rivers. It is a commercial and industrial center and a rail junction. Manufactures include machinery, leather goods, and porcelain. It is an episcopal see and a tourist spot. Namur was a Merovingian fortress (7th cent.) and later (10th cent.) became the seat of a county. The county fell to the counts of Flanders in 1262 and in 1421 was bought by Philip the Good of Burgundy. It later shared the history of the Austrian and Spanish Netherlands. Because of its strategic location, Namur was frequently besieged. In the War of the Grand Alliance it fell (1692) to the French but was retaken by the Dutch in 1695. The first Barrier Treaty (1709) gave the Netherlands the right to garrison Namur, a right confirmed by two further treaties (1713, 1715) supplementing the Peace of Utrecht. Refortified in 1887, it was part of the Belgian defenses on the Meuse at the start of World War I.

Na·nai·mo (nə-nī'mō), city (1976 pop. 40,336), SW British Columbia, Canada, on Vancouver Island. It is a port, the base of a herring-fishing fleet, and the trade center for a farm and lumbering region. There are several sawmills and a pulp-producing plant in the city.

Na·na·ku·li (nä'nä-kōō'lē), city (1970 pop. 6,506), Honolulu co., Hawaii, on the W coast of Oahu, in a sugar-cane region.

Nance (năns), county (1970 pop. 5,142), 438 sq mi (1,134.4 sq km), E central Nebr., in an agricultural area drained by the Loup River; formed 1879; co. seat Fullerton. Grain, livestock, and dairy products are important to its economy.

Nan-ch'ang or **Nan·chang** (both: nän'chäng'), city (1970 est. pop. 900,000), capital of Kiangsi prov., China, on the Kan River, near the S end of P'o-yang Lake. A major transportation center, it is a large economic and industrial center with machine shops, food-processing establishments, the country's largest integrated silk complex, and plants making fertilizer, tractors, cement, tires, and pharmaceuticals. An old walled city, Nan-ch'ang dates from the Sung dynasty (12th cent.). Nan-ch'ang is considered the birthplace of the People's Liberation Army. Here, in 1927, a force of 30,000 Communist troops, led by Chu Teh, rose against the Kuomintang government and briefly established the first Soviet republic in China. Occupied by the Japanese (1939–45) in World War II, Nan-ch'ang was reoccupied by the Nationalists in 1945 but fell to the Communists in 1949.

Nan·cy (năn'sē, nän-sē'), city (1975 pop. 107,902), capital of Meurthe-et-Moselle dept., NE France, on the Meurthe River and the Marne-Rhine Canal. It is the administrative, economic, and educational center of Lorraine. Situated at the edge of the huge Lorraine iron fields, Nancy is an industrial city manufacturing foundry products, boilers, electrical equipment, machine tools, and textiles. In the city are a noted fine arts museum, an academy of fine arts, and a large university (founded 1854). Nancy grew around a castle of the dukes of Lorraine and became the duchy capital in the 12th cent. In 1477 Charles the Bold of Burgundy was defeated and killed at the gates of Nancy by Swiss troops and the forces of René II of Lorraine. The major part of the center of Nancy, a model of urban planning and a gem of 18th cent. architecture, was built during the liberal reign of Stanislaus I, duke of Lorraine (reigned 1738–66) and ex-king of Poland. Nancy passed to the French crown in 1766. In 1848 it was one of the first cities to proclaim the republic. From 1870 to 1873 it was occupied by the Germans following the Franco-Prussian War, and it was partially destroyed in World War I. Points of interest include the Place Stanislas, the Place de la Carrière, and the Church of Cordeliers (15th cent.), which houses the magnificent tombs of the princes of Lorraine.

Nan·da De·vi (nŭn'dä dā'vē), peak, 25,645 ft (7,821.8 m) high, Uttar Pradesh state, N India, in the Himalayas. Hindus believe that the goddess Nanda, wife of Siva, lives here. The peak was scaled in 1936 by an Anglo-American expedition.

Nan·der (nän'dər), town (1971 pop. 126,400), Maharashtra state, S central India, on the Godavari River. Nander is known for its fine muslin. It is also a market for cattle, grain, and cotton.

Nan·ga Par·bat (nŭng'gə pŭr'bət), peak, 26,660 ft (8,131.3 m) high, in the Punjab Himalayas, Pakistan, the seventh-highest peak in the world. A German-Austrian team reached the peak in 1953.

Nan·king (năn'kĭng') or **Nan-ching** (nän'jĭng'), city (1970 est. pop. 2,000,000), capital of Kiangsu prov., E central China, in a bend of the Yangtze River. Nanking is at the intersection of three major railroad lines. Industry, which once centered around "nankeen" cloth (unbleached cotton goods), has been vigorously developed under the Communist government. The city now has an integrated iron-steel complex, an oil refinery, food-processing establishments, and some 600 plants making chemicals, textiles, cement, fertilizers, machinery, electronic equipment, optical instruments, photographic equipment, and trucks. Nanking has long been a literary and political center.

Nanking was the capital of China from the 3rd to the 6th cent. and again from 1368 to 1421. The Treaty of Nanking, signed in 1842 at the end of the Opium War, opened China to foreign trade. During the Taiping Rebellion insurgents held the city from 1853 to 1864. It was captured by the revolutionists in 1911, and in 1912 it became the capital of China's first president, Sun Yat-sen. In 1927 the city fell to the Communists. The Kuomintang under Chiang Kai-shek retook the city, and it became (1928) the Nationalist capital. In 1932 when the Japanese were threatening to attack the city, the government was temporarily removed to Lo-yang, and on Nov. 21, 1937, just before Nanking fell to the Japanese, it was moved to Chungking. The Japanese entry into the city, which was accompanied by widespread killing and brutality, became known as the "rape of Nanking." The Japanese established (1938) a puppet regime in Nanking. Chinese forces reoccupied the city Sept. 5, 1945, and the capitulation of the Japanese armies in China was signed here on Sept. 9. Nanking again fell to the Communists in Apr., 1949, and from 1950 until 1952, when it became the provincial capital, Nanking was administered as part of an autonomous region.

The city is the seat of numerous institutions of higher learning, and is also noted for its large library. Both its astronomical observatory and its botanical gardens are among the largest in the country. The original city wall (70 ft/21.4 m high), most of which still stands, dates from the Ming dynasty and encircles most of the modern city. The tomb of the first Ming emperor is approached by an avenue lined with colossal images of men and animals. Also of interest are the tomb of Sun Yat-sen, a memorial to China's war dead (a steel pagoda), and the Taiping museum.

Nan Ling (nän' lĭng'), mountain range of Kwangtung, Hunan, and Kwangsi provs., S China, rising to c.6,900 ft (2,105 m). The Nan Ling forms the geographic boundary between central and southern China. The mountains separate the Yangtze and Si basins, protect southern China from cold northern air masses, and divide the Cantonese civilization and linguistic area from that of northern China.

Nan-ning (nän'nĭng'), city (1970 est. pop. 375,000), capital of the Kwangsi Chuang Autonomous Region, S China, on the Si River in a fertile farming area. The city has an iron and steel complex, a sugar refinery, other food-processing plants, and factories making fertilizer, machine tools, cement, and farm machinery.

Nan·terre (nän-târ'), city (1975 pop. 95,032), capital of Hauts-de-Seine dept., N central France, on the right bank of the Seine River. It is an industrial center where metals, automobiles, electrical equipment, machine tools, and rolling stock are made. In May, 1968, the Nanterre branch of the Univ. of Paris was the scene of student protests that spread and led to a national political crisis.

Nantes (nănts, nänt), city (1968 pop. 265,009), capital of Loire-Atlantique dept., W France, on the Loire River. It is an important industrial and shipping center with its ocean port at Saint-Nazaire. Food products (especially biscuits), naval equipment, metals, dyes, clothing, bicycles, and agricultural equipment are the leading manufactures. Nantes became an important trade and administrative center under the Romans. Nantes was ravaged and held (843–936) by Norsemen and later (10th cent.) fell to the dukes of Brittany, who resided here until Brittany became part of France in 1524. During the French Revolution, Nantes was the scene of massacres by the revolutionaries in 1793. Nantes was a center of resistance to the German occupation in World War II. Points of interest include a 10th cent. castle and a 15th cent. cathedral with tombs of dukes of Brittany.

Nan·ti·coke (năn'tĭ-kōk'), city (1970 pop. 14,632), Luzerne co., NE Pa., on the Susquehanna River; founded 1793, inc. as a city 1926. It is largely residential. Its manufactures include chemicals, clothing, and shoes. Nanticoke was formerly the heart of an anthracite-coal mining region, but mechanization and reduced mining operations have diminished the importance of coal to its economy.

Nan·tuck·et (năn-tŭk'ĭt), island, c.14 mi (23 km) long, from 3 to 6 mi (4.8–9.7 km) wide, SE Mass., S of Cape Cod, from which it is separated by Nantucket Sound. Muskeget Channel is located between Nantucket and Martha's Vineyard to the west. Exhibiting evidence of glaciation (terminal moraine, outwash plain), Nantucket has sandy beaches and low, rolling hills composed of sand and gravel. It is sparsely vegetated; wild cranberries, heather, and wild roses predominate. Nantucket and the small adjacent islands constitute Nantucket co. (1970 pop. 3,774), 46 sq mi (119.1 sq km). Settled in 1659, the island was part of N.Y. from 1660 to 1692, when it was ceded to Mass. Nantucket was a major whaling port until the decline of the industry (c.1850), and it later developed into a resort and artists' colony. The village of Nantucket is the trade center of the island and is known for its many old houses. The island has a whaling museum. The first U.S. lightship station (est. 1856) is located near Nantucket.

Nan-t'ung or **Nan·tung** (both: nän'tōōng'), town (1970 est. pop. 300,000), N Kiangsu prov., E central China, on the Yangtze River near the East China Sea. The center of an important cotton-growing area, it is an old town that has become industrialized. Manufactures include textiles, cottonseed oil, processed foods, and chemicals.

Na·pa (năp'ə), county (1970 pop. 79,140), 790 sq mi (2,046.1 sq km), W Calif., in a mountainous region bounded on the S by San Pablo Bay, with the Napa River valley extending SE from the base of Mt. St. Helena; formed 1850; co. seat Napa. Its agriculture includes prunes, wine grapes, poultry, dairy products, livestock, nuts, fruit, grain, and hay. Quicksilver, pumice, and sand and gravel are quarried. Wine making has been the principal industry since the 1850s. Clothing, furniture, leather goods, concrete, and metal products are manufactured.

Napa, city (1970 pop. 36,103), seat of Napa co., W Calif., on the Napa River; inc. 1872. Grapes and other fruits are grown in the adjacent Napa valley, which is well known for its wines. Napa's manufactures include concrete and steel products and leather goods.

Na·pa·ta (năp'ə-tə), ancient city of Nubia, just below the Fourth Cataract of the Nile. From about the 8th cent. B.C., Napata was the capital of the kingdom of Cush. Many great temples like those of Thebes were built here.

Na·per·ville (nā'pər-vĭl'), city (1970 pop. 22,617), Du Page co., NE Ill., on the Du Page River; settled 1831–32, inc. as a city 1890.

Na·pi·er (nā'pē-ər), city (1971 pop. 40,186), E central North Island, New Zealand, on Hawke Bay. It is a major wool-exporting port as well as a fishing port and wool market. Napier suffered a ruinous earthquake in 1931.

Na·ples (nā'pəlz) or *Italian* **Na·po·li** (nä'pō-lē), city (1976 est. pop. 1,223,927), capital of Campania and of Naples prov., S central Italy, on the Bay of Naples, an arm of the Tyrrhenian Sea. It is a major seaport, with shipyards, and a commercial, industrial, and tourist center. Manufactures include iron and steel, petroleum, textiles, food products, chemicals, and machinery.

An ancient Greek colony, Naples was known variously as Parthenope, Palaepolis, and Neapolis. It was conquered (4th cent. B.C.) by the Romans, who favored it because of its Greek culture, its scenic beauty, and its baths. The Roman poet Virgil, who often stayed there, is buried nearby. In the 6th cent. A.D. Naples passed under Byzantine rule; in the 8th cent. it became an independent duchy. In 1139 the Norman Roger II added the duchy to the kingdom of Sicily. Emperor Frederick II embellished the city and founded its university (1224). The execution (1268) of Conradin left Charles of Anjou undisputed master of the kingdom. He transferred the capital from Palermo to Naples. After the Sicilian Vespers insurrection (1282), Sicily proper passed to the house of Aragón, and the Italian peninsula south of the Papal States became known as the kingdom of Naples. Naples was its capital until it fell to Garibaldi and was annexed to the kingdom of Sardinia (1860). The city suffered severe damage in World War II.

Naples is beautifully situated at the base and on the slopes of the hills enclosing the Bay of Naples. The bay, dominated by Mt. Vesuvius, extends from Cape Misena in the north to the Sorrento peninsula in the south and is dotted with towns and villas. Especially interesting parts of the city are the Old Spacca Quarter (the heart of Old Naples) and the seaside Santa Lucia sector. Noteworthy structures in Naples include the Castel Nuovo (1282), the Castel dell'Ovo (rebuilt by the Angevins in 1274), the Renaissance-style Palazzo Cuomo (late 15th cent.), the large Carthusian Monastery of St. Martin (remodeled in the 16th and 17th cent.), the neoclassic Villa Floridiana, which houses a museum of porcelain, china, and Neopolitan paintings, the Church of Santa Chiara, which contains the tombs of Robert the Wise and other Angevin kings, the Cathedral of St. Januarius (14th cent.), with numerous later additions, including a 17th cent. baroque chapel, the Royal Palace (early 17th cent.), and the Church of Santa Maria Donna Regina. Naples has several museums including the National Museum, which holds the Farnese collection and most of the objects excavated at nearby Pompeii and Herculaneum. The Teatro San Carlo, a famous opera house, was opened in 1737. The city has a conservatory and several art academies.

Naples, resort city (1970 pop. 12,042), Collier co., SW Fla., on the Gulf of Mexico; inc. 1927. It is noted for its beach.

Na·po·le·on (nə-pō'lē-ən, -pōl'yən). **1.** City (1970 pop. 1,036), seat of Logan co., S N.Dak., SE of Bismarck. In a livestock and dairy region, it has wheat farms. **2.** City (1970 pop. 7,791), seat of Henry co., NW Ohio, on the Maumee River SW of Toledo, in a rich farm area; platted 1832. Metal products are manufactured.

Na·po·le·on·ville (nə-pōl'yən-vĭl', -pōl'ē-ən-), town (1970 pop. 1,008), seat of Assumption parish, SE La., S of Baton Rouge, in a sugar-cane and petroleum region; founded c.1818, inc. 1879. Antebellum houses are in the area.

Na·ra (nä'rä), city (1976 est. pop. 264,739), capital of Nara prefecture, S Honshu, Japan. An ancient cultural and religious center, it was founded in 706 by imperial decree and was modeled after Ch'ang-an, the capital of T'ang China. Nara was (710–84) the first permanent capital of Japan. The noted temple, Todai-ji, has a 53.5-ft (16.3 m) high image of Buddha, said to be one of the largest bronze figures in the world. Nara Park includes the celebrated Imperial Museum, which houses ancient art treasures and relics. Near the city is wooded Mt. Kasuga, the traditional home of the gods; its trees are never cut. Also nearby is Horyu-ji, founded in 607, the oldest Buddhist temple in Japan, with the grave of Jimmu, the first emperor.

Na·ra·shi·no (nä'rə-shē'nō), city (1976 est. pop. 119,476), Chiba prefecture, E central Honshu, Japan, on Tokyo Bay. It is a newly developed suburb of Tokyo and has a large metals industry.

Na·ra·yan·ganj (nä-rä'yən-gənj), city (1974 pop. 176,459), E central Bangladesh, at the confluence of the Lakhya and Dhaleshwari rivers. It is the river port for Dacca and one of Bangladesh's busiest trade centers. The city is a collection center for jute and hides and skins. Narayanganj and Dacca together make up the principal industrial region of Bangladesh. There are jute presses, cotton textile mills, and leather, glass, footwear manufactures.

Nar·ba·da (nər-bŭd'ə), river, c.775 mi (1,245 km) long, rising in Madhya Pradesh state, central India, and flowing W between the Satpura and Vindhya ranges to the Gulf of Cambay. Because the river is turbulent and confined between steep banks, it is unsuitable for navigation or irrigation. The Narbada, sacred to Hindus, is said to have sprung from the body of the god Shiva.

Nar·bonne (när-bôn'), city (1975 pop. 39,342), Aude dept., S France, near the Mediterranean coast. It is the commercial center of a wine-growing region and an industrial city producing sulfur, copper, and clothing. It was the first Roman colony established in Transalpine Gaul (118 B.C.). It later became the capital of the Roman province of Gallia Narbonensis. Narbonne was occupied by the Visigoths in A.D. 413 and taken by the Saracens in 719 and the Franks in 759. It later became the seat of the viscounts of Narbonne, vassals of the counts of Toulouse, and was united to the French crown in 1507. Its port, silted up in 1320, brought great wealth to the city, especially during the Middle Ages. Narbonne was an important center of the Jews in the Middle Ages. Their expulsion (late 13th cent.) and the Black Death (1310), which is said to have taken 30,000 lives, were severe blows to the city's prosperity. In Narbonne are the remains of a Roman amphitheater and bridge, the splendid St. Just Cathedral (13th–14th cent.), and an archiepiscopal palace (13th cent.), now the town hall and museum.

Na·rew (nä'rĕf), river, c.275 mi (445 km) long, rising in the Białowieza Forest, W European USSR, near the border with Poland, and flowing generally NW through NE Poland past Łomża, the head of navigation, then SW to the Vistula River near Warsaw. During World Wars I and II, several major battles took place along the Narew River.

Na·rod·na·ya (nä-rô'dnə-yə), peak, 6,214 ft (1,895.3 m) high, NE European USSR, in the N Urals. It is the highest peak of the Urals.

Nar·ra·gan·sett (năr'ə-găn'sĭt), town (1970 pop. 7,138), Washington co., S R.I., on the W shore of Narragansett Bay N of Point Judith; settled in the mid-17th cent., set off from South Kingston 1888, inc. 1901. It is a summer resort. The coastal area was severely damaged by hurricanes in 1938 and 1954.

Narragansett Bay, arm of the Atlantic Ocean, 30 mi (48 km) long and from 3 to 12 mi (4.8–19.3 km) wide, deeply indenting the state of Rhode Island. Its many inlets provided harbors that were advantageous to colonial trade and later to resort development.

Nar·va (när'vä), city (1976 est. pop. 71,000), NW European USSR, in Estonia, on the left bank of the Narva River. A leading textile center, it has machinery plants, sawmills, flax and jute factories, fisheries, and food-processing industries. The city is also an important producer of electric power. Founded by the Danes in 1223, Narva passed to the Livonian Knights in 1346 and was a member of the Hanseatic League. In 1492 Ivan III of Russia built the fortress Ivangorod on the right bank of the Narva, facing the Hermann fortress of the knights. After the dissolution (1561) of the Livonian Order, the city was first seized by the Russians, then taken (1581) by the Swedes; it continued to be contested by the two nations. In 1700

Charles XII of Sweden, with inferior forces, resoundingly defeated Peter I of Russia at Narva in the first battle of the Northern War (1700–1721). Peter captured the city in 1704, and it remained part of Russia until 1919, when it was incorporated into newly independent Estonia. German forces occupied the city in World War II. In 1945 all Estonian territory east of the Narva River was ceded to the USSR. The city is dominated by two old fortresses, and it has a 14th cent. Eastern Orthodox cathedral (originally Roman Catholic), and a 17th cent. town hall and exchange buildings.

Nar·vik (när'vĭk), city (1977 est. pop. 19,468), Nordland co., N Norway, an ice-free port on the Ofotfjord opposite the Lofoten Islands. It was founded (1887) as the Atlantic port for the Kiruna and Gällivare iron mines in Sweden. The city is now a tourist center. In World War II Narvik fell to the Germans when they invaded Norway on Apr. 9, 1940. To prevent the Germans from using Narvik as a shipping base for Swedish iron ore, a British expeditionary force briefly occupied (May 28–June 9, 1940) the port.

Na·ryn (nə-rĭn'), river, c.450 mi (725 km) long, rising in several branches in the Tien Shan mountain system, SW Kirghizia and SE Uzbekistan, Central Asian USSR, and flowing generally W through the Fergana Valley where it joins with the Kara Darya to form the Syr Darya. Its lower course is used for irrigating a cotton-growing area. The city of Naryn (1975 est. pop. 26,000) on the upper course of the river, at an altitude of 6,610 ft (2,016 m), is the center of a wheat-growing and sheep-grazing district.

Nase·by (nāz'bē), village, Northamptonshire, central England, near Northampton. Nearby, on June 14, 1645, the Parliamentarians under Oliver Cromwell defeated the Royalists under Charles I in a decisive battle of the English Civil War.

Nash (năsh), county (1970 pop. 59,122), 544 sq mi (1,410 sq km), E central N.C., in a coastal-plain area bounded on the NE by Fishing Creek and crossed by the Tar River; formed 1777; co. seat Nashville. It has agriculture (tobacco, cotton, and corn), stands of pine timber, and a tourist industry.

Na·sho·ba (nə-shō'bə), former community, SW Tenn., on the Wolf River near Memphis. It was founded by Frances Wright and others as a place in which black slaves, who were purchased especially for the purpose, might be educated for freedom. Influenced by the example of Robert Owen's cooperative colony at New Harmony, Ind., Frances Wright bought the land in 1825. She and the other trustees were impractical administrators, and the venture was unsuccessful. The difficulties of pioneering were increased by swamp fever and by poor management during her frequent absences. The community was denounced as a center of free love and miscegenation and by 1829 was practically deserted by its white members. The following year the slaves were transported to Haiti.

Nash·u·a (năsh'ōō-ə), city (1978 est. pop. 63,300), seat of Hillsborough co., S N.H., on the Merrimack and Nashua rivers near the Mass. line; settled c.1655, inc. as a city 1853. Because of the availability of water power, Nashua developed (early 19th cent.) as a textile mill town. Its chief manufactures include electronic equipment, paper, shoes and leather, and machinery.

Nash·ville (năsh'vĭl'). **1.** City (1970 pop. 4,016), seat of Howard co., SW Ark., NW of Hope, in a timber, fruit, and farm area. It is noted as a peach-shipping center. **2.** City (1970 pop. 4,323), seat of Berrien co., S Ga., N of Valdosta, in a tobacco and farm area; inc. 1892. Lumber is produced. **3.** City (1970 pop. 3,027), seat of Washington co., S Ill., W of Mt. Vernon, in a farm and coal region; founded 1824, inc. 1853. **4.** Town (1970 pop. 527), seat of Brown co., S central Ind., S of Indianapolis, in an agricultural and timber area. **5.** Town (1970 pop. 1,670), seat of Nash co., NE N.C., W of Rocky Mount, in a tobacco and farm area. **6.** City (1978 est. pop. 459,000), state capital, coextensive with Davidson co., central Tenn., on the Cumberland River, in a fertile farm area; inc. as a city 1806, merged with Davidson co. 1963. It is a port of entry and an important commercial and industrial center. The city has railroad shops and factories making a great variety of goods including automobile glass, wearing apparel, footwear, food products, tires, and commercial, industrial, and agricultural chemicals. Nashville is noted for its music industry. It also has many publishing houses producing religious materials, school annuals, magazines, and telephone directories. Two large insurance companies have their headquarters in Nashville. The city was founded (1779) by a group of pioneers under James Robertson (who is buried here). Fort Nashborough was built on the banks of the river, and the next year 60 families arrived to settle the area. The settlement developed early as a cotton center and river port and later as a railroad hub. It became permanent capital of the state in 1843. After the fall of Fort Donelson in Feb., 1862, Nashville was abandoned to Union troops and became an important Union base for the remainder of the Civil War. Sometimes called the Athens of the South, Nashville has many buildings of classical design (including a replica of the Parthenon, built in 1897) and is the seat of Vanderbilt Univ., Fisk Univ., Tennessee State Univ., the Univ. of Tennessee at Nashville, and many other educational institutions. Among the points of interest are the capitol, with the tomb of James K. Polk, the war memorial building, the country music hall of fame and museum, "Opryland, U.S.A." (a family entertainment complex), and several old churches and ante-bellum homes.

Na·sik (nä'sĭk), town (1971 pop. 176,187), Maharashtra state, W central India. It is a center of brassware manufacture and cattle and poultry breeding. It is holy as the site of the Ramayana exile of the Hindu god Rama and Sita, his wife.

Nass (năs), river, 236 mi (380 km) long, rising in the Coast Mts., W British Columbia, Canada, and flowing SW to Portland Inlet of the Pacific Ocean. It has valuable salmon fisheries.

Nas·sau (nä'sou), former duchy, central West Germany, situated N and E of the Main and Rhine rivers. The region takes its name from the small town of Nassau, on the Lahn River east of Ems, where the original castle of the house of Nassau was built in the early 12th cent. by a count of Laurenburg. His descendants took the title of count of Nassau. In 1255 the dynasty split into two main lines. The ruling line was called the Walramian line for Count Walram II; his younger brother, Otto, founded the Ottonian line.

In 1806 Nassau, which had received some territorial additions, joined the Confederation of the Rhine and was raised to a duchy. In 1816 the territories belonging to the various branches of the Walramian line were united by Duke William (1816–39). His successor, Adolf, sided against Prussia in the Austro-Prussian War (1866) and as a result lost his duchy to Prussia. Nassau was then united with the former Electoral Hesse to form the Prussian province of Hesse-Nassau. Duke Adolf of Nassau, however, succeeded in 1890 to the grand duchy of Luxembourg, where his descendants continue to rule.

The Ottonian line of Nassau acquired (15th cent.) the lordship of Breda and settled in the Netherlands. It came into European prominence in the 16th cent. with William the Silent, who inherited the principality of Orange in southern France and became stadtholder of the Netherlands. His sons, Maurice of Nassau and Frederick Henry, succeeded him as princes of Orange and as stadtholders; these titles then passed to Frederick Henry's son, William II of Orange, and to William's son William III, who also became king of England. William III died (1702) without direct heirs, and the principality of Orange (which had become purely titular) passed to John William Friso, of the collateral branch of Nassau-Dietz. His son, Prince William IV, became (1748) hereditary stadtholder of the Netherlands, and from him all subsequent rulers of the Netherlands (except Louis Bonaparte) are descended in direct line. The Dutch line of the Nassau family is known as the house of Orange.

Nas·sau (năs'ô), city (1970 pop. 101,503), capital of the Bahama Islands. A port on New Providence island, it has a large and beautiful harbor and is the commercial and social center of the islands. Its warm, healthful climate and colorful atmosphere have made it a favorite winter resort. In the 18th cent. it was a rendezvous for pirates. Three forts, Nassau (1697), Charlotte (1787–94), and Fincastle (1793), were built to ward off the numerous Spanish invasions. American revolutionists in 1776 captured and held it a short time.

Nas·sau (năs'ô). **1.** County (1970 pop. 20,626), 650 sq mi (1,683.5 sq km), extreme NE Fla., bounded on the E by the Atlantic Ocean, on the W and N by the St. Marys River, here forming the Ga. border; formed 1824; co. seat Fernandina Beach. It is in a lowland area protected by Amelia Island, a barrier island. Agriculture (corn, poultry, and dairy products), forestry, paper making, and fishing are the major occupations. **2.** County (1970 pop. 1,428,838), 289 sq mi (748.5 sq km), SE N.Y., on W Long Island, bounded on the N by Long Island Sound, on the S by the Atlantic Ocean; formed 1898; co. seat Mineola. It is part of the New York City metropolitan area and is chiefly residential. It has some agriculture (mainly truck crops and flowers), commercial fisheries, sand and gravel pits, and a food-processing industry. Among its diverse manufactures are clothing, furniture, paper products, chemicals, plastics, machinery, and aircraft equipment. The deeply indented northern shore has many old estates and summer communities; yachting and fishing are popular off the shore. The southern shore has numerous resorts, with its bays sheltered from the Atlantic by fine barrier beaches that are linked to Long Island by causeways.

Nas·ser, Lake (nä'sər), c.1,550 sq mi (4,015 sq km), on the Nile River, SE Egypt and N Sudan. It extends c.350 mi (565 km) behind Aswan High Dam to the Second Cataract at Wadi Halfa (now submerged). The lake's rising waters forced more than 80,000 people to relocate and submerged many historic sites.

Na·tal (nə-tăl', -täl'), province (1970 pop. 4,245,675), 33,578 sq mi (86,967 sq km), E South Africa, on the Indian Ocean, bounded in the N by Mozambique and Swaziland and in the W by the Orange Free State and Lesotho; capital Pietermaritzburg. The province rises from a narrow (except in the north) coastal belt to an inland region fringed in the west by the Drakensberg Range. Sugar refining is the main industry. Sheep, cattle, citrus fruits, maize, sorghum, cotton, bananas, and pineapples are also raised. Industries include textile, clothing, rubber, fertilizer, paper, and food-processing plants, tanneries, and oil refineries. Natal produces considerable coal (especially coking coal) and timber.

In the early 19th cent. Natal was inhabited primarily by Bantu-speaking Zulu people. In the 1820s and 1830s the British acquired much of Natal from the Zulu chiefs Chaka and Dingane. Boer farmers arrived in 1837 and, after battles with the Zulu, established (1838–39) a republic. In 1843 Britain annexed Natal to Cape Colony, and a Boer exodus followed. In 1856 Natal became a separate colony. Sugar-cane cultivation began c.1860, and many Indians (mostly indentured laborers) came to work in the sugar industry. Many Indians remained in Natal as free men after their term of indenture expired; and by 1900 they outnumbered whites. In 1893 Natal was given internal self-government, and in 1910 it became a founding province of the Union of South Africa.

Na·tal (nə-täl', -tôl'), city (1970 pop. 264,567), capital of Rio Grande do Norte state, NE Brazil, just above the mouth of the Potengi River. Its port is important in the handling of coastal shipping and in the export of the state's tungsten. There is also some light industry in the city. Beautifully situated among white, palm-studded beaches, Natal is a modern city that has retained its colonial flavor. Natal was founded on Christmas Day, 1599. It was occupied by the Dutch from 1633 to 1654 and in 1817 was briefly the seat of a republican government until it was suppressed by imperial authorities. It grew rapidly during World War II, when an airport was built for flights to Africa.

Na·tash·kwan (nə-tăsh'kwən), river, 241 mi (387.8 km) long, rising in S Labrador, Canada, and flowing S across E Que. to the Gulf of St. Lawrence. It is noted for trout and salmon fishing. Iron-bearing sands found along its banks are mined.

Natch·ez (năch'ĭz), city (1978 est. pop. 23,400), seat of Adams co., SW Miss., on bluffs above the Mississippi River; settled 1716, inc. 1803. It is the trade, shipping, and processing center for a cotton, livestock, and timber area where oil and natural gas are found. One of the oldest towns on the Mississippi River, Natchez was founded in 1716 when Fort Rosalie was established here, but Natchez Indians annihilated the garrison in 1729. The area passed to England (1763), to Spain (1779), and to the United States (1798). Natchez was capital of the Mississippi Territory from 1798 to 1802. Its strategic location at the junction of the Mississippi and the southern terminus of the Natchez Trace brought prosperity to the area. Natchez served as state capital from 1817 to 1821. In the Civil War it was taken by Federal forces in 1863. The city has preserved its ante-bellum charm.

Natchez Trace (trās), road, from Natchez, Miss., to Nashville, Tenn., of great commercial and military importance from the 1780s to the 1830s. It grew from a series of Indian trails used in the 18th cent. by the French, English, and Spanish successively. After U.S. expansion into the Old Southwest, it was improved by the army. Andrew Jackson marched over the Trace to New Orleans in the War of 1812 and later used it in his Indian campaigns. With the coming of steamboat transportation, it passed into decline. The Natchez Trace Parkway generally follows the old Natchez Trace.

Natch·i·toches (năk'ĭ-tŏsh'), parish (1970 pop. 35,219), 1,297 sq mi (3.359.2 sq km), NW central La., intersected by the Red and Cane rivers; formed 1805; parish seat Natchitoches. Agriculture consists mainly of cotton, as well as corn, hay, and peanuts. There is some manufacturing, including the processing of timber and farm products. The county contains part of Kisatchie National Forest.

Natchitoches, city (1970 pop. 15,974), seat of Natchitoches parish, NW La.; inc. 1819. Its industry is centered on farm products, chiefly cotton. Natchitoches was founded c.1714 as a French military and trading post. It was the dividing line between French and Spanish territory and an important port on the Red River until the river changed its course in the early 1800s (the riverbed has since filled and is now known as Cane River Lake). The city was occupied by the Union army during the Civil War. A tour of old homes and plantations every Oct. attracts tourists.

Na·tick (nā'tĭk), town (1978 est. pop. 30,900), Middlesex co., E Mass., a residential and industrial suburb of Boston, on Lake Cochituate; founded 1651 as a "praying Indian" village by John Eliot, settled by whites 1718, inc. 1781. Manufactures include shoes, electronic components, and clothing.

Na·tion·al Cap·i·tal Parks (năsh'ən-əl kăp'ĭ-təl, năsh'nəl): *see* National Parks and Monuments Table.

National City, city (1978 est. pop. 52,100), San Diego co., S Calif., on San Diego Bay; inc. 1887. Citrus fruits and vegetables are packed, and there are defense-related industries.

National Forest System, federally owned reserves, c.187,000,000 acres (75,680,000 hectares), administered by the Forest Service of the U.S. Dept. of Agriculture. The system is made up of 155 national forests and 19 national grasslands in 44 states, Puerto Rico, and the Virgin Islands. Most of the acreage is found in Western states, with Alaska, Idaho, and Calif. having the largest holdings. In the East large national forests are in the Green, White, Allegheny, and Blue Ridge mts. The national grasslands are on the Great Plains. By law the reserves must be used for timber production, watershed land, wildlife preservation, livestock grazing, mining, and recreation. In 1891 Congress authorized the President to set aside forest reserves; Yellowstone Park Timber Reserve (now Shoshone National Forest) in Wyo. was the first (1891) to be established. The forest reserves were administered by the General Land Office of the Dept. of the Interior until 1905, when they were transferred to the Forest Service and designated national forests in 1907.

national parks and monuments. The National Park Service, a bureau of the U.S. Dept. of the Interior, was established in 1916 to correlate the administration of 37 national parks and monuments under the charge of the department. By 1975 it was administering the more than 300 areas of scenic, historic, or scientific interest that make up the National Park System. The areas are classified into natural, historic, recreational, and cultural groupings to facilitate park management and to identify areas by their prominent characteristics. Instructed by an act of Congress to "conserve the natural and historic objects in such manner as will leave them unimpaired for the enjoyment of future generations," the National Park Service directs a wide program of construction and of educational and protective work. Congress laid the foundation of the National Park System in 1872 when it established Yellowstone National Park "as a pleasuring ground for the benefit and enjoyment of the people." Congress accelerated expansion of the National Park System in 1906 with the passage of the Antiquities Act, which permitted the President to proclaim national historic landmarks, structures, and "other objects of historic and scientific interest" on Federal lands. Until 1925, when an act was passed authorizing acceptance of donated land, nearly all of the National Park System was carved out of public lands. In 1933 the National Park Service was given trusteeship over areas hitherto under the jurisdiction of the Agriculture and War Depts. Since then Congress has also authorized the preservation of significant historic sites and the establishment of national parkways, national seashores, national recreational areas, national lakeshores, national wild and scenic rivers, national scenic trails, and national preserves.

National Road, U.S. highway built in the early 19th cent. It finally extended from Cumberland, Md., to St. Louis and was the great highway of Western migration. Agitation for a road to the West began c.1800. Congress approved the route, but construction did not begin until 1815. The first section (called the Cumberland Road) was built of crushed stone. Opened in 1818, it ran from Cumberland to Wheeling, W.Va., following in part the Indian trail known as Nemacolin's Path. Largely through the efforts of Henry Clay it was continued (1825–33) westward through Ohio, using part of the road built by Ebenezer Zane. By this time the older part of the road needed repairs. Control of the road was therefore turned over to the states through which it passed, where tolls for maintenance were collected. It was carried on to Vandalia, Ill., and finally to St. Louis.

NATIONAL PARKS

Name	*Location*	*Date Authorized*	*Special Characteristics*
*Acadia	S Maine	1919	Mountain and coast scenery.
*Arches	E Utah	1929	Giant arches formed by erosion.
*Big Bend	W Texas	1935	Canyons and desert plain on the Rio Grande.
*Bryce Canyon	SW Utah	1923	Canyon with colored walls and rock formations.
*Canyonlands	SE Utah	1964	Rocks, spires, and mesas; Indian petroglyphs.
Capitol Reef	S Utah	1937	Highly colored sandstone cliffs dissected by gorges; named for a white, dome-shaped rock.
*Carlsbad Caverns	SE N.Mex.	1923	Great limestone caverns.
*Crater Lake	SW Oregon	1902	Blue lake in a crater.
*Everglades	S Fla.	1934	Subtropical wilderness.
*Glacier	NW Mont.	1910	Region of glaciers, forests, and lakes.
*Grand Canyon	NW Ariz.	1908	Great gorge of the Colorado River.
*Grand Teton	NW Wyo.	1929	Scenic portion of the Teton Range.
*Great Smoky Mountains	N.C., Tenn.	1926	Wild, beautiful area in the Great Smoky Mts.
*Guadalupe Mountains	W Texas	1966	Mountain region; contains a limestone fossil reef.
*Haleakala	Hawaii, on Maui	1960	Largest inactive crater in the world.
*Hawaii Volcanoes	Hawaii, on Hawaii	1916	Volcanic region.
*Hot Springs	Central Ark.	1921	Mineral springs.
*Isle Royale	NW Mich.	1931	Forested islands in Lake Superior.
Kings Canyon	E Calif.	1890	Canyons, peaks, sequoias. See Sequoia National Park.
*Lassen Volcanic	N Calif.	1907	Volcanic peaks and lava formations.
*Mammoth Cave	Central Ky.	1926	Extensive underground passages.
*Mesa Verde	SW Colo.	1906	Prehistoric cliff dwellings.
*Mount McKinley	Central Alaska	1917	Highest peak in North America.
*Mount Rainier	SW Wash.	1899	Volcanic peak and glaciers.
North Cascades	N Wash.	1968	Area of great alpine scenery in the Cascade Range; bisected by Ross Lake National Recreation Area.
*Olympic	NW Wash.	1938	Rain forests and glaciers in the Olympic Mts.
*Petrified Forest	E Ariz.	1906	Petrified logs; part of the Painted Desert.
Platt	S Okla.	1906	Cold mineral springs in the Arbuckle Mts.
*Redwood	NW Calif.	1968	Coast redwood forests.
Rocky Mountain	Central Colo.	1915	Scenic Rocky Mts. region on the continental divide; many high snow-capped peaks.
*Sequoia	E Calif.	1890	Groves of giant sequoias.
*Shenandoah	N Va.	1926	Forested region of the Blue Ridge Mts.
*Virgin Islands	Virgin Islands, on St. John	1956	Unusual scenery, marine life, coral gardens; ruins of the Danish colony.
Voyageurs	N Minn.	1971	Scenic northern lakes region; interesting glacial features and history.
*Wind Cave	SW S.Dak.	1903	Limestone caverns in the Black Hills.
*Yellowstone	Wyo., Mont., Idaho	1872	Geysers, Yellowstone canyon, falls; first and largest U.S. national park.
*Yosemite	Central Calif.	1890	Mountain region with Yosemite Valley.
*Zion	SW Utah	1909	Multicolored canyon in a desert region.

NATIONAL MONUMENTS

Name	*Location*	*Date Authorized*	*Special Characteristics*
Agate Fossil Beds	NW Nebr.	1965	World-famous quarries containing numerous well-preserved Miocene mammal fossils.
Alibates Flint Quarries and Texas Panhandle Pueblo Culture	NW Texas	1965	Flint quarries, first worked by Indians c.10,000 years ago; rich archaeological and historic area.
*Aztec Ruins	NW N.Mex.	1923	Ruins of a Pueblo Indian town.
Badlands	SW S.Dak.	1929	See Badlands.
Bandelier	N N.Mex.	1916	Ruins of prehistoric Pueblo Indian homes.
Biscayne	S Fla.	1968	Example of a living coral reef; includes part of Biscayne Bay.
Black Canyon of the Gunnison	W Colo.	1933	Deep, narrow canyon of the Gunnison River, named for its dark-colored walls, which are always in shadow.
Booker T. Washington	Central Va.	1956	Birthplace and childhood home of Booker T. Washington.
Buck Island Reef	Virgin Islands, on Buck Island	1961	One of the finest marine gardens in the Caribbean; bird rookeries and grottoes.
Cabrillo	S Calif.	1913	Memorial to Juan Rodriguez Cabrillo.
*Canyon de Chelly	NE Ariz.	1931	Ruins of prehistoric Indian villages.
Capulin Mountain	NE N.Mex.	1916	Huge cinder cone of extinct volcano.
Casa Grande Ruins	S Ariz.	1892	Huge building built c.600 years ago in the ruins of an Indian pueblo.
Castillo de San Marcos	NE Fla.	1924	Old Spanish masonry fort in Saint Augustine, Fla.
Castle Clinton	SE N.Y.	1946	See Battery, The.
Cedar Breaks	SW Utah	1933	Amphitheater (2,000 ft/610 m deep) formed by erosion.
Chaco Canyon	NW N.Mex.	1907	Ruins representing the highest point of pueblo Indian prehistoric civilization (A.D. 900–1000).
Channel Islands	SW Calif.	1938	Part of the Santa Barbara Islands; sea lions, fossils.

*See separate article for additional information. For example, for Acadia, see Acadia National Park.

NATIONAL MONUMENTS (Continued)

Name	Location	Date Autho-rized	Special Characteristics
Chiricahua	SE Ariz.	1924	Unusually shaped rock formations.
Colorado	W Colo.	1911	Huge monoliths and other unusual erosional features.
*Craters of the Moon	S Idaho	1924	Volcanic cones, craters, fissures, lava flows.
Custer Battlefield	SE Mont.	1879	Site of the Custer massacre. See Little Bighorn, river.
*Death Valley	Calif., Nev.	1933	Lowest point in North America; desert environment.
Devils Postpile	E Calif.	1911	Basaltic columns, some 60 ft (18 m) high.
*Devils Tower	NE Wyo.	1906	Volcanic rock tower; first national monument.
Dinosaur	Utah, Colo.	1915	Rich quarries of well-preserved fossils.
Effigy Mounds	NE Iowa	1949	Outstanding examples of Indian mounds.
El Morro	W N.Mex.	1906	Sandstone monolith with inscriptions of Spanish explorers and American pioneers.
Florissant Fossil Beds	Central Colo.	1969	Well-preserved insect, seed, and leaf fossils of the Oligocene period; petrified sequoia tree stumps.
Fort Frederica	SE Ga.	1936	Ruins of a fort built by James Oglethorpe on one of the Sea Islands.
Fort McHenry	N Md.	1925	Historic shrine; place where the *Star-spangled Banner* was written. See Fort McHenry.
Fort Jefferson	S Fla.	1935	In the Dry Tortugas Islands; the largest all-masonry fort in the Western Hemisphere; built 1846.
Fort Matanzas	NE Fla.	1924	Spanish fort in Saint Augustine, Fla.
Fort Pulaski	SE Ga.	1924	Fort on Cockspur Island. See Fort Pulaski.
Fort Stanwix	Central N.Y.	1935	See Fort Stanwix.
Fort Sumter	SE S.C.	1948	Scene of the engagement that opened the Civil War. See Fort Sumter.
Fort Union	NW N.Mex.	1954	Ruins of a U.S. army fort on the Santa Fe Trail.
Fossil Butte	W Wyo.	1972	Area containing Paleocene-Eocene fossil fish.
George Washington Birthplace	E Va.	1930	Estate and reconstructed mansion.
George Washington Carver	SW Mo.	1943	Birthplace and boyhood home of George Washington Carver.
Gila Cliff Dwellings	SW N.Mex.	1907	Well-preserved dwellings built by the Pueblo Indians into a 150-ft (46-m) cliff.
*Glacier Bay	SE Alaska	1925	Glaciers, ice displays; largest unit of the National Park System.
Grand Canyon	NW Ariz.	1932	Part of the Grand Canyon.
Grand Portage	NE Minn.	1951	9-mi (14-km) portage on the route to the Northwest used by explorers, missionaries, and fur traders.
Gran Quivira	Central N.Mex.	1909	Ruins of a Spanish mission and Indian pueblos.
Great Sand Dunes	S Colo.	1932	Large, high sand dunes in the Sangre de Cristo Mts.
Hohokam Pima	Central Ariz.	1972	Archaeological remains of the Hohokam culture.
Homestead	SE Nebr.	1936	Site of the first farm claimed under the Homestead Act.
Hovenweep	Utah, Colo.	1923	Prehistoric Indian pueblos and cliff dwellings.
Jewel Cave	SW S.Dak.	1908	Limestone caves with chambers connected by narrow passages; in the Black Hills.
Joshua Tree	S Calif.	1936	Rare Joshua trees, or "praying plant"; named by Mormons because of upstretched arms.
*Katmai	SW Alaska	1918	Volcanic area; second-largest unit of the National Park System.
Lava Beds	N Calif.	1925	Examples of volcanism; scene of Modoc Indian uprising.
Lehman Caves	E Nev.	1922	Honeycombed limestone caves; rock formations.
Marble Canyon	N Ariz.	1969	Canyon of the Colorado River with high vertical walls of red sandstone and white limestone.
*Montezuma Castle	Central Ariz.	1906	Well-preserved prehistoric cliff dwellings.
Mound City Group	S Ohio	1923	Prehistoric Indian mounds.
Muir Woods	W Calif.	1908	Grove of virgin redwood trees.
*Natural Bridges	SE Utah	1908	Three huge natural bridges.
Navajo	NE Ariz.	1909	Ruins of large cliff dwellings.
Ocmulgee	Central Ga.	1934	Remains of mounds and prehistoric towns.
Oregon Caves	SW Oregon	1909	Limestone caverns with four levels; rock formations.
Organ Pipe Cactus	S Ariz.	1937	Organ pipe cactus and other unique desert growth.
Pecos	N N.Mex.	1965	15th cent. ruins of Pecos Pueblo, once the largest Indian settlement in the Southwest.
Pinnacles	W Calif.	1908	Rock spires from 500 to 1,200 ft (152–366 m) high; caves.
Pipe Spring	NW Ariz.	1923	Spring first visited by the Mormons; old fort.
Pipestone	SW Minn.	1937	Quarry that was a source for Indian peace pipes.
*Rainbow Bridge	S Utah	1910	Pink sandstone arch.
Russell Cave	NE Ala.	1961	Cave containing a nearly continuous archaeological record of human habitation from about 6000 B.C. to A.D. 1650.
Saguaro	SE Ariz.	1933	Saguaro, other cacti, varied desert growth.
Saint Croix Island	E Maine	1949	Commemorates the French settlement on the island in the Saint Croix River.
Scotts Bluff	W Nebr.	1919	Landmark on the Oregon Trail.
Statue of Liberty	SE N.Y.	1924	See Ellis Island.
Sunset Crater	N Ariz.	1930	Volcanic cinder cone with multicolored crater; lava flows; ice cave.
Timpanogos Cave	N Utah	1922	Limestone cavern on Mt. Timpanogos.

*See separate article for additional information. For example, for Acadia, see Acadia National Park.

NATIONAL MONUMENTS (Continued)

Name	Location	Date Autho-rized	Special Characteristics
Tonto	Central Ariz.	1907	Well-preserved 14th cent. cliff dwellings built by the Salado Indians in the Salt River valley.
Tumacacori	S Ariz.	1908	Mission founded by Father Eusebio F. Kino; rebuilt by the Franciscans.
Tuzigoot	Central Ariz.	1939	Excavated ruins of a large Indian pueblo.
Walnut Canyon	N Ariz.	1915	12th cent. Sinagua Indian cliff dwellings.
White Sands	S N.Mex.	1933	Wind-drifted gypsum sands.
Wupatki	N Ariz.	1924	Several prehistoric pueblos.
Yucca House	SW Colo.	1919	Remains of a prehistoric Indian village.

NATIONAL HISTORIC SITES

Name	Location	Date Autho-rized	Special Characteristics
*Abraham Lincoln Birthplace	Central Ky.	1916	Traditional birthplace cabin in memorial building on site of Lincoln's birthplace.
Adams	E Mass.	1946	Home of Presidents John Adams and John Quincy Adams and other members of the family.
Allegheny Portage Railroad	SW Pa.	1964	Inclined-plane railroad that lifted passengers and cargoes of boats on the Pennsylvania Canal over the Allegheny Mts.
Andersonville	SW Ga.	1970	Civil War prison camp. See under Andersonville.
Andrew Johnson	NE Tenn.	1935	Home, shop, and grave of President Andrew Johnson; site includes Andrew Johnson National Cemetery.
Bent's Old Fort	SE Colo.	1960	Fur-trading post, Indian rendezvous, and rest-station on the Santa Fe Trail; built c.1830 by Charles and William Bent. See Bent's Fort.
Carl Sandburg Home	SW N.C.	1968	Farm home of author Carl Sandburg.
†Chicago Portage Railroad	NE Ill.	1952	Portion of a portage discovered (1673) by Jacques Marquette and Louis Jolliet; later used as a link between the Great Lakes and the Mississippi River.
†Chimney Rock	W Nebr.	1956	500-ft (150-m) landmark on the Oregon Trail.
Christiansted	Virgin Islands, on St. Croix	1952	Commemorates the Virgin Islands' colonial development, especially under Danish rule in the 18th and 19th cent.
†Dorchester Heights	E Mass.	1951	Site of American batteries that helped force the evacuation of British forces from Boston (1776) during the Revolution.
Edison	NE N.J.	1962	Buildings and equipment used by Thomas A. Edison.
Eisenhower	S Pa.	1969	Home and farm of President Dwight D. Eisenhower.
Ford's Theatre	Washington, D.C.	1970	Sites of President Abraham Lincoln's assassination and death; includes the Lincoln Museum.
Fort Bowie	SE Ariz.	1964	Ruins of a fort (est. 1862) that was the base of military operations against Geronimo and his followers.
Fort Davis	W Texas	1961	Key post in the defensive system of W Texas, guarding (1854-91) the San Antonio-El Paso road through the Davis Mts; troops attached here fought the Comanche and Apache Indians.
*Fort Laramie	SE Wyo.	1938	Buildings of an old fort on the Oregon Trail.
Fort Larned	Central Kansas	1964	Protected the Santa Fe Trail; served as a military base during the Plains War (1860s) and later as an Indian Bureau administrative center.
Fort Point	W Calif.	1970	Largest brick and granite mid-19th cent. coastal fortification on the west coast of North America.
Fort Raleigh	NE N.C.	1941	Site of the first attempted settlement by the English in North America. See Roanoke Island.
†Fort Scott	SE Kansas	1965	A historic area; commemorates historic events in Kansas prior to and during the Civil War.
Fort Smith	NW Ark.	1961	One of the first U.S. military posts in the Louisiana Purchase; maintained law and order in the Oklahoma Territory. See Fort Smith, Ark.
Fort Union Trading Post	N.Dak., Mont.	1966	American Fur Company trading post. See Fort Union.
Fort Vancouver	SW Wash.	1948	Site of a Hudson's Bay Company post (1825–49) and later of a U.S. army fort.
†Gloria Dei	SE Pa.	1942	Second-oldest Swedish church in the United States; founded 1677, present building erected c.1700.
Golden Spike	N Utah	1957	Site where the Union Pacific RR and the Central Pacific RR joined to form the first transcontinental railroad.
Grant-Kohrs Ranch	W Mont.	1972	Headquarters of one of the largest 19th cent. range ranches.
Hampton	NE Md.	1948	Late 18th cent. Georgian mansion.
Herbert Hoover	E Iowa	1965	Birthplace, childhood home, and burial place of President Herbert Hoover.
Home of Franklin D. Roosevelt	SE N.Y.	1944	Home, "Summer White House," and burial place of Franklin D. and Eleanor Roosevelt. See Hyde Park.
Hopewell Village	SE Pa.	1938	Restored 19th cent. iron-making village.
Hubbell Trading Post	NE Ariz.	1965	Example of a late 19th cent. trading post in the Southwest.
†Jamestown	SE Va.	1940	Site of the first permanent English settlement in America. See Jamestown, Va.
Jefferson National Expansion Memorial	E Mo.	1935	Area commemorating westward exploration and settlement; includes Gateway Arch. See St. Louis, Mo.
John Fitzgerald Kennedy	E Mass.	1967	Birthplace and early boyhood home of President John F. Kennedy.

*See separate article for additional information. For example, for Acadia, see Acadia National Park.
†Not owned by the Federal government.

NATIONAL HISTORIC SITES (Continued)

Name	*Location*	*Date Authorized*	*Special Characteristics*
John Muir	W Calif.	1964	John Muir House and Martínez Adobe, commemorating contributions of John Muir to conservation and literature.
Lincoln Home	Central Ill.	1971	Only private home owned by Abraham Lincoln; he was living here when he was elected President.
Longfellow	E Mass.	1972	Home of Henry Wadsworth Longfellow (1837–82) in Cambridge; also George Washington's headquarters during the siege of Boston (1775–76).
Lyndon B. Johnson	SE Texas	1969	Sites of the birthplace and boyhood home of President Lyndon B. Johnson.
†McLoughlin House	NW Oregon	1941	Home of the fur trader Dr. John McLoughlin.
Pennsylvania Avenue	Washington, D.C.	1965	Portion of Pennsylvania Ave. and adjacent area between the Capitol and the White House.
Puukohola Heiau	Hawaii, on Hawaii	1972	Hill of Whale Temple built (1791) by King Kamehameha the Great; ruins of John Young's house.
Sagamore Hill	SE N.Y.	1962	Estate and Victorian-style home of President Theodore Roosevelt (1858–1919).
Saint-Gaudens	W N.H.	1964	Memorial to the American sculptor Augustus Saint-Gaudens; contains his home, studios, gardens.
†Saint Paul's Church	SE N.Y.	1943	18th cent. church associated with the events leading to the arrest of John Peter Zenger; link in American architectural history.
St. Thomas	Virgin Islands, on St. Thomas	1960	Fort Christian (1680), the oldest standing structure in the Virgin Islands and the center of early Danish settlement.
Salem Maritime	NE Mass.	1938	Wharf and buildings important during Salem's seafaring days.
†San Jose Mission	S Texas	1941	Restored Spanish frontier mission (est. 1720).
San Juan	NE Puerto Rico	1949	Oldest fortification within the limits of U.S. territory, built (16th cent.) by the Spanish to protect the harbor guarding the sea lanes to the New World.
Saugus Iron Works	E Mass.	1968	Reconstruction of the 17th cent. Colonial ironworks.
The Mar-A-Lago	E Fla.	1969	Private mansion illustrating the affluent society's way of life in the 1920s.
Theodore Roosevelt Birthplace	SE N.Y.	1962	Birthplace and boyhood home of President Theodore Roosevelt.
Theodore Roosevelt Inaugural	W N.Y.	1966	Ansley Wilcox House, where Theodore Roosevelt took the oath of office (1901) as President.
†Touro Synagogue	SE R.I.	1946	Fine example of Colonial architecture; one of the oldest synagogues in the country.
Vanderbilt Mansion	E N.Y.	1940	19th cent. palatial Victorian residence of a grandson of Cornelius Vanderbilt.
Whitman Mission	SW Wash.	1936	Site of the mission of Dr. Marcus Whitman.
William Howard Taft	SW Ohio	1969	Birthplace and early home of President William Howard Taft.

NATIONAL HISTORICAL PARKS

Name	*Location*	*Date Authorized*	*Special Characteristics*
Appomattox Court House	S central Va.	1930	Site of Lee's surrender to Grant. See Appomattox, Va.
Chalmette	SE La.	1939	Scene of part of the battle of New Orleans in the War of 1812.
Chesapeake and Ohio Canal	D.C., Md., W.Va.	1938	See Chesapeake and Ohio Canal.
City of Refuge	Hawaii, on Hawaii	1955	Ancient burial ground and place of refuge.
*Colonial	SE Va.	1930	Historic Yorktown, Jamestown, and Cape Henry. Colonial Parkway connects some sites with Williamsburg.
Cumberland Gap	Ky., Tenn., Va.	1940	Mountain pass of the Wilderness Road. See Cumberland Gap.
George Rogers Clark	SW Ind.	1966	Memorial near the site of old Fort Sackville, seized from British by Gen. G. R. Clark in 1779.
Harpers Ferry	Md., W.Va.	1944	See Harpers Ferry.
Independence	SE Pa.	1948	Historic points of interest; site of the signing of the Declaration of Independence. See Independence Hall.
Minute Man	E Mass.	1959	Scene of fighting on the opening day of the Revolutionary War; includes North Bridge, Minute Man statue, Battle Road, and the home of Nathaniel Hawthorne.
*Morristown	N N.J.	1933	Site of military encampments during the Revolution; Washington's headquarters, 1779–80.
Nez Perce	NW Idaho	1965	22 sites that preserve and commemorate the history and culture of the Nez Perce Indians.
San Juan Island	NW Wash.	1966	Dedicated to the peaceful relationship between the United States, Britain, and Canada since the San Juan Boundary Dispute.
Saratoga	E N.Y.	1938	Scene of a famous battle in the Revolution.
Sitka	SE Alaska	1910	Site of the Tlingit Indians' defeat by Russian settlers in 1804.

NATIONAL MEMORIALS

Name	*Location*	*Date Authorized*	*Special Characteristics*
Arkansas Post	SE Ark.	1960	Site of the first permanent French settlement in the lower Mississippi valley. See Arkansas Post.
*Arlington House	NE Va.	1925	Former home of the Custis and Lee families; memorial to Robert E. Lee.
Benjamin Franklin	SE Pa.	1972	Statue of Benjamin Franklin in the rotunda of the Franklin Institute.
Chamizal	W Texas	1966	Memorializes the peaceful settlement of the 99-year border dispute between the United States and Mexico.

*See separate article for additional information. For example, for Acadia, see Acadia National Park.
†Not owned by the Federal government.

NATIONAL MEMORIALS (Continued)

Name	Location	Date Authorized	Special Characteristics
Coronado	SE Ariz.	1952	Area near Francisco Vásquez de Coronado's point of entry (1540) into the United States.
De Soto	W Fla.	1948	Commemorates the landing (1539) of Hernando De Soto in Florida and his exploration of S United States.
Federal Hall	SE N.Y.	1939	Site of the first seat of the Federal government and George Washington's inauguration (1789).
Fort Caroline	NE Fla.	1950	Area overlooking the site of Fort Caroline.
Fort Clatsop	NW Oregon	1958	Site of the winter encampment of the Lewis and Clark Expedition.
Frederick Douglass Home	Washington, D.C.	1962	Home of Frederick Douglass from 1877 to 1895.
General Grant	SE N.Y.	1958	Tomb of President Ulysses S. Grant and his wife, Julia.
Hamilton Grange	SE N.Y.	1962	Home of Alexander Hamilton.
Johnstown Flood	SE Pa.	1964	Memorializes the Johnstown flood of 1889. See Johnstown, Pa.
Lincoln Boyhood	SW Ind.	1962	Site of the farm where Abraham Lincoln was raised and the burial place of his mother, Mary Hanks Lincoln.
*Lincoln Memorial	Washington, D.C.	1911	Classical structure with a heroic statue of Lincoln.
*Mount Rushmore	SW S.Dak.	1925	Carvings of Washington, Jefferson, Lincoln, and Theodore Roosevelt on the granite face of Mt. Rushmore.
Perry's Victory and International Peace Memorial	N Ohio	1936	Scene of the victory near Put in Bay of Oliver H. Perry in the War of 1812.
Roger Williams	E R.I.	1965	Memorial to Roger Williams, the founder of the Rhode Island colony and a religious freedom pioneer.
Thaddeus Kosciuszko	SE Pa.	1972	Commemorates the life and work of Thaddeus Kosciuszko.
*Thomas Jefferson	Washington, D.C.	1934	Classical structure with a statue of Jefferson.
*Washington Monument	Washington, D.C.	1848	555-ft (169-m) high obelisk honoring George Washington.
Wright Brothers	NE N.C.	1927	Scene of the first (1903) successful flight of Wilbur and Orville Wright.

NATIONAL MEMORIAL PARK

Name	Location	Date Authorized	Special Characteristics
*Theodore Roosevelt	W N.Dak.	1947	Part of Roosevelt's Elkhorn Ranch; badlands along the Little Missouri River.

NATIONAL MILITARY PARKS

Name	Location	Date Authorized	Special Characteristics
Chickamauga and Chattanooga	Ga., Tenn.	1890	See Chattanooga, Tenn.
Fort Donelson	NW Tenn.	1928	See Fort Donelson.
Fredericksburg and Spotsylvania County Battlefields Memorial	NE Va.	1927	See Fredericksburg, Va.
Gettysburg	S Pa.	1895	See Gettysburg, Pa.
Guilford Courthouse	N N.C.	1917	See Guilford Courthouse, N.C.
Horseshoe Bend	E Ala.	1956	See Horseshoe Bend.
Kings Mountain	N S.C.	1931	Site of an American victory over the British at a critical point during the revolution (Oct. 7, 1780).
*Moores Creek	SE N.C.	1926	Site of a battle between Patriots and Loyalists.
Pea Ridge	NW Ark.	1956	Site of the Civil War battle of Pea Ridge, which saved Mo. for the Union.
Shiloh	SW Tenn.	1894	Site of the Civil War battle of Shiloh.
Vicksburg	W Miss.	1899	Site of the Vicksburg Campaign of the Civil War.

NATIONAL BATTLEFIELD PARKS

Name	Location	Date Authorized	Special Characteristics
Kennesaw Mountain	NW Ga.	1917	Site of Gen. W.T. Sherman's attack on the Confederate forces in the Atlanta Campaign.
Manassas	NE Va.	1940	See Bull Run, Va.
Richmond	E Va.	1936	See Richmond, Va.

NATIONAL BATTLEFIELDS

Name	Location	Date Authorized	Special Characteristics
Big Hole	SW Mont.	1910	Scene of an attack by U.S. soldiers on Chief Joseph.
Cowpens	NW S.C.	1929	Site of an American militia victory over British infantry and cavalry forces in the Revolutionary War Battle of Cowpens (Jan. 17, 1781).
Fort Necessity	SW Pa.	1931	See Fort Necessity.
Petersburg	SE Va.	1926	Scene of the Battle of the Crater and a 10-month Union campaign (1864–65) to seize Petersburg, Va., a railroad center supplying Richmond and Gen. R.E. Lee.
Stones River	Central Tenn.	1927	See Murfreesboro, Tenn.
Tupelo	NE Miss.	1929	See Tupelo, Miss.
Wilson's Creek	SW Mo.	1960	Site of Civil War battle for control of Missouri (Aug. 10, 1861).

NATIONAL BATTLEFIELD SITES

Name	Location	Date Authorized	Special Characteristics
Antietam	Central Md.	1890	See Antietam, Md.
Brices Cross Roads	NE Miss.	1929	Site of a rout of Union troops by Confederate cavalry under Gen. N.B. Forrest (June 10, 1864).

*See separate article for additional information. For example, for Acadia, see Acadia National Park.

NATIONAL CEMETERIES

Name	Location	Date Authorized	Special Characteristics
Antietam (Sharpsburg)	Central Md.	c.1862	Civil War cemetery.
Battleground	Washington, D.C.	1864	Civil War cemetery.
Fort Donelson (Dover)	NW Tenn.	c.1867	Civil War cemetery.
Fredericksburg	NE Va.	c.1865	Civil War cemetery.
Gettysburg	S Pa.	1863	Civil War cemetery; site of President Lincoln's Gettysburg Address.
Poplar Grove (Petersburg)	SE Va.	c.1866	Civil War cemetery.
Shiloh (Pittsburg Landing)	SW Tenn.	c.1866	Civil War cemetery.
Stones River (Murfreesboro)	Central Tenn.	c.1865	Civil War cemetery.
Vicksburg	W Miss.	c.1865	Civil War cemetery.
Yorktown	SE Va.	c.1866	Civil War cemetery.

NATIONAL RECREATION AREAS

Name	Location	Date Authorized	Special Characteristics
Amistad	S Texas	1965	U.S. part of Amistad Reservoir, on the Rio Grande.
Arbuckle	S Okla.	1965	See Arbuckle Mts.
Bighorn Canyon	Mont., Wyo.	1964	Yellowstone Reservoir and spectacular Bighorn Canyon, on the Bighorn River.
Coulee Dam	NE Wash.	1946	Franklin D. Roosevelt Lake, formed by the Grand Coulee Dam in the Columbia River; interesting geology.
Curecanti	E Colo.	1965	Blue Mesa, Morrow Point, and Crystal reservoirs in the upper Black Canyon of the Gunnison.
Delaware Water Gap	N.J., Pa.	1965	Scenic Delaware Water Gap.
Gateway	N.Y., N.J.	1972	Beaches, marshes, islands, and waters in and around New York City. One of the first two national urban recreation areas.
Glen Canyon	Ariz., Utah	1958	Lake Powell, formed by the Glen Canyon Dam.
Golden Gate	W Calif.	1972	Offers a variety of recreation in and around San Francisco. One of the first two national urban recreation areas.
Lake Chelan	N Wash.	1968	Located in the Stehekin Valley and in the northern part of fjordlike Lake Chelan.
Lake Mead	Ariz., Nev.	1936	Lake Mead, formed by Hoover Dam, and Lake Mohave, formed by Davis Dam; the first national recreation area established by Congress.
Lake Meredith	NW Texas	1965	Includes Lake Meredith, on the Canadian River, a popular water-sports area in the Southwest.
Ross Lake	N Wash.	1968	Extends along the Skagit River canyon; bisects North Cascades National Park.
Shadow Mountain	N Colo.	1952	Shadow Mountain Lake and Lake Granby; part of the Colorado–Big Thompson project.
Whiskeytown-Shasta-Trinity	N Calif.	1965	Reservoirs and forestland; the National Park Service runs the Whiskeytown unit, and the Forest Service administers the Shasta and Trinity units.

NATIONAL LAKESHORES

Name	Location	Date Authorized	Special Characteristics
Apostle Islands	NW Wis.	1970	Apostle Islands and a strip of the Bayfield Peninsula, on the south shore of Lake Superior.
Indiana Dunes	NW Ind.	1966	200-ft (60-m) sand dunes, beaches, and marshes along the south shore of Lake Michigan.
Pictured Rocks	N Mich.	1966	Sandstone cliffs, sand dunes, beaches, marshes, waterfalls, and inland lakes along Lake Superior; the first national lakeshore.
Sleeping Bear Dunes	W central Mich.	1970	Section of the Lake Michigan shoreline and the North and South Manitoulin islands; beaches, sand dunes, forests, and lakes.

NATIONAL SEASHORES

Name	Location	Date Authorized	Special Characteristics
Assateague Island	Md., Va.	1965	35-mi (56-km) barrier island; beaches; wildlife refuge including the wild Chincoteague ponies.
Cape Cod	SE Mass.	1961	See Cape Cod.
Cape Hatteras	E N.C.	1937	The first national seashore. See under Cape Hatteras.
Cape Lookout	E N.C.	1966	Three barrier islands with beaches, sand dunes, and salt marshes; Cape Lookout Lighthouse.
Cumberland Island	SE Ga.	1972	Largest island off Ga.; beaches, sand dunes, marshes, and lakes.
*Fire Island	SE N.Y.	1964	Barrier beach.
Gulf Islands	Fla., Miss.	1971	Historic forts and white sand beaches near Pensacola, Fla.; Fort Massachusetts and primitive offshore islands in S Miss.
Padre Island	S Texas	1962	See Padre Island, Texas.
*Point Reyes	W Calif.	1962	Coastal area with beaches and steep bluffs.

NATIONAL PARKWAYS

Name	Location	Date Authorized	Special Characteristics
Baltimore-Washington	Central Md.	1950	Approach to the nation's capital from the northeast; includes Greenbelt Park, a natural woodland.

*See separate article for additional information. For example, for Acadia, see Acadia National Park.

NATIONAL PARKWAYS (Continued)

Name	Location	Date Authorized	Special Characteristics
Blue Ridge	Va., N.C.	1936	Scenic route in the Blue Ridge Mts. between Shenandoah and Great Smoky Mts. national parks; many roadside parks, lookouts, and trails; the first national parkway.
George Washington Memorial	Va., Md.	1930	Parkway connecting landmarks associated with the life of George Washington along both sides of the Potomac River from Mt. Vernon to Great Falls, then S to Chain Bridge.
John D. Rockefeller, Jr., Memorial	NW Wyo.	1972	Scenic corridor between Yellowstone and Grand Teton national parks commemorating Rockefeller's role in the creation of many national parks.
Natchez Trace	Miss., Ala., Tenn.	1938	Parkway following the general location of the old trail known as Natchez Trace.
Suitland	Md., D.C.	1949	Landscaped parkway between Washington, D.C., and Suitland, Md. (Andrews Air Force Base).
NATIONAL SCENIC TRAIL			
Appalachian	Maine, N.H., Vt., Mass., Conn., N.Y., N.J., Pa., Md., W.Va., Va., N.C., Tenn., Ga.	1968	See Appalachian Trail.
NATIONAL SCENIC RIVER			
Lower Saint Croix	Minn., Wis.	1972	Scenic lower course of the St. Croix River; part of the Wild and Scenic Rivers System.
NATIONAL SCENIC RIVERWAYS			
*Ozark	Mo.	1964	Scenic parts of the Current and Jacks Fork rivers; the first national scenic riverway.
Saint Croix	Minn., Wis.	1968	200 mi (320 km) of the St. Croix River and its Namekagon tributary; trails, camping, boating.
Wolf	Wis.	1968	Scenic 24-mi (38-km) stretch of fast water.
NATIONAL RIVER			
Buffalo	NW Ark.	1972	130-mi (209-km) stretch of the Buffalo River and its valley; the first national river.
INTERNATIONAL PARK			
†Roosevelt-Campobello	SW N.B., Canada	1964	Summer home of President Franklin D. Roosevelt on Campobello island; first international park to be administered by a joint U.S.– Canadian commission.
NATIONAL SCIENTIFIC RESERVE			
†Ice Age	Wis.	1964	Contains features of continental glaciation; first national scientific reserve.
NATIONAL CAPITAL PARKS			
National Capital Parks	D.C., Va., Md.	1790	More than 700 parks in and around Washington, D.C.
WHITE HOUSE			
*White House	Washington, D.C.	1943	Official residence of the President.
PARKS (other)			
Catoctin Mountain Park	NW Md.	1936	Campgrounds, trails, and scenic drive located in the Catoctin Mts.; Camp David, the presidential retreat, is here.
John F. Kennedy Center for the Performing Arts	Washington, D.C.	1958	Site of cultural performances in its theater, concert hall, and opera house.
Piscataway Park	S Md.	1961	Preserves the view from Mt. Vernon of the opposite shore of the Potomac River.
Prince William Forest Park	NE Va.	1936	Woodland with 89 species of trees in pure stands.
Theodore Roosevelt Island	Washington, D.C.	1932	Wilderness preserve in the Potomac River; a tribute to the "conservationist President."
Wolf Trap Farm Park for the Performing Arts	N Va.	1966	Set in a rolling, wooded landscaped area to provide artistic enjoyment and recreation; the first national park for the performing arts.

*See separate article for additional information. For example, for Acadia, see Acadia National Park.
†Not owned by the Federal government.

Na·tro·na (nə-trō'nə), county (1970 pop. 51,264), 5,342 sq mi (13,835.8 sq km), central Wyo., watered by the North Platte River, the Sweetwater River, headstreams of the Powder River, and part of Pathfinder Reservoir; formed 1888; co. seat Casper. It is in a livestock and mining region that produces oil, coal, and uranium, radium, and vanadium ores. Its industries include food processing, petroleum refining, and the manufacture of concrete.

Nat·u·ral Bridge (năch'ər-əl, năch'rəl), village (1970 est. pop. 200), Rockbridge co., W Va., in the Shenandoah valley; founded 1774. Nearby is the famous Natural Bridge over the gorge of Cedar Creek. It is a limestone arch 215 ft (65.6 m) high with a span of 90 ft (27.5 m) and was once owned by Thomas Jefferson, who built a cabin for visitors. Today a public highway crosses the bridge.

Natural Bridges National Monument, 7,600 acres (3,078 hectares), SE Utah; est. 1908. Located in an area of colored cliffs and box canyons, the monument contains three natural sandstone bridges: Owachomo (also called Rock Mound), 106 ft (32.3 m) high with a span of 180 ft (54.9 m); Kachina, 210 ft (64 m) high with a span of 206 ft (62.8 m); and Sipapu, 220 ft (67.1 m) high with a span of 268 ft (81.7 m).

Nau·cal·pan (nou-käl'pən), city (1970 pop. 373,605), Mexico state, S central Mexico, on the Hondo River. It is an industrial extension of Mexico City.

Nau·cra·tis (nô'krə-tĭs), ancient city of Egypt, on the Canopic branch of the Nile, SE of Alexandria. It was probably given (7th cent. B.C.) to Greek colonists from Miletus and was the first Greek settlement in Egypt. The rise of Alexandria and the shifting of the Nile caused its decline. The site has been excavated, revealing pottery of a Greek type and ruins of Greek temples.

Nau·ga·tuck (nô'gə-tŭk'), borough (1978 est. pop. 26,200), New Haven co., SW Conn., on the Naugatuck River; settled 1704, inc. 1844. Rubber products have been made since Henry Goodyear established a rubber plant here in 1843. Other manufactures are candy, machinery, iron castings, and electrical and metal products.

Naugatuck, river, 65 mi (104.6 km) long, rising in NW Conn. and flowing S, past Waterbury, to the Housatonic River at Derby. Many manufacturing centers are along its course.

Naum·burg (noum'bûrg', -bŏŏrKH') or **Naumburg an der Saa·le** (än dər zä'lə), city (1974 est. pop. 36,358), Halle dist., S East Germany, on the Saale River. Manufactures of this industrial city include machine tools, processed food, textiles, and toys. Founded in the 11th cent., Naumburg developed as a trade center and joined the Hanseatic League. It passed to Saxony in 1564 and to Prussia in 1815. The city has retained parts of its medieval walls and a beautiful cathedral (13th-14th cent.), with some of the finest sculptures of the German Gothic period.

Na·u·ru (nä-ōō'rōō), atoll and parliamentary republic (1975 est. pop. 8,000), c.8 sq mi (20.7 sq km), central Pacific, just S of the equator and W of the Gilbert Islands. Until 1968, when it became one of the world's smallest independent states, Nauru was administered by Australia under United Nations Trusteeship. Australia and New Zealand are heavily dependent on its high-grade phosphate deposits.

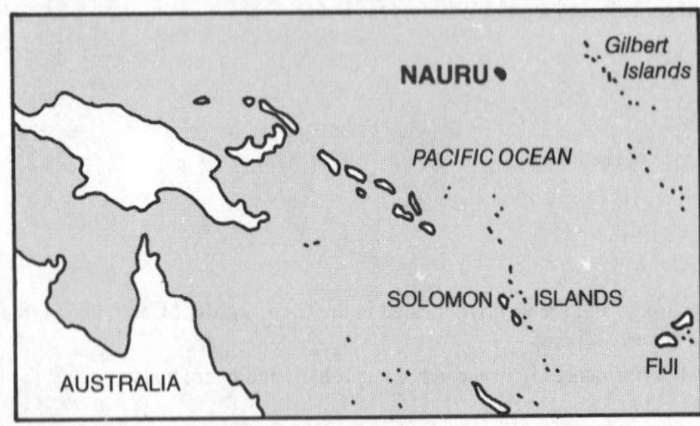

Almost 2,000,000 tons of phosphates are produced annually. After independence Nauru took control of the deposits from the British Phosphate Commission. Nauru was discovered in 1798 by the British and was annexed in 1888 by Germany. Occupied during World War I by Australian forces, it was placed (1920) under a League of Nations mandate to Australia. Throughout World War II the island was occupied by the Japanese.

Nau·voo (nô-vōō', nô'vōō), city (1970 pop. 1,047), Hancock co., W Ill., on heights overlooking a bend in the Mississippi River NE of Keokuk, Iowa. Settled shortly after 1830 as Commerce, it was occupied and renamed in 1839 by the Mormons under Joseph Smith and incorporated in 1841. The city reached its most prosperous state under the Mormons, with about 20,000 inhabitants. After Smith and his brother were killed by a mob in Carthage jail (1844), the group left (1846) Ill. for Utah. Smith's house, part of an old hotel, and other old buildings are still standing. The city is a farm trade center and makes wine and cheese.

Nav·a·jo (năv'ə-hō', nä'və-), county (1970 pop. 47,559), 9,911 sq mi (25,669.5 sq km), E Ariz., in a mountainous area bordering on Utah and crossed by the Little Colorado River in the S; formed 1895; co. seat Holbrook. It is a sheep and alfalfa district, known for Indian handicrafts and its tourist trade. Black Mesa and parts of Navajo and Hopi Indian reservations are in the north. Navajo National Monument is near the Utah line.

Navajo National Monument: *see* National Parks and Monuments Table.

Nav·an (năv'ən), urban district (1971 pop. 4,607), county town of Co. Meath, E Republic of Ireland, at the confluence of the Boyne and Blackwater rivers. It produces woolens and has sawmills. Clothing, furniture, and carpets are also made. Nearby are some medieval ruins and Navan Mote, an imposing earthwork.

Na·varre (nə-vär'), province (1974 est. pop. 486,607), N Spain, bordering on France, between the W Pyrenees and the Ebro River. Pamplona is the capital. The mountain slopes have extensive cattle pastures and vast forests that yield hardwoods, which are economically important. The fertile valleys produce sugar beets, cereals, and vegetables; vineyards are important. The establishment of hydroelectric plants has permitted some industrial development.

The population of Navarre is largely of Basque stock. The pass of Roncesvalles, which leads from France to Navarre, made the region strategically important early in its history. The Basques defended themselves successfully against the Moorish invaders and against the Franks; the domination of Charlemagne, who conquered Navarre in 778, was short-lived. In 824 the Basque Iñigo Aritza was chosen king of Pamplona, which expanded under his successors and became known as the kingdom of Navarre. It reached its zenith under Sancho III (reigned 1000–1035), who married the heiress of Castile and ruled nearly all of Christian Spain. On his death the Spanish kingdoms were divided (into Navarre, Aragón, and Castile). In 1234 Navarre passed through inheritance to the house of Champagne and in 1305 to King Philip IV of France. Navarre stayed with the French crown until the death (1328) of Charles IV. In 1479 Navarre passed, through marriage, to the counts of Foix and then to the house of Albret. Ferdinand V annexed most of Navarre in 1515. The area north of the Pyrenees (Lower Navarre) remained an independent kingdom until it was incorporated (1589) into the French crown when Henry III of Navarre became King Henry IV of France. It was united with Béarn into a French province. Until the French Revolution the kings of France carried the additional title king of Navarre. Since the rest of Navarre was in Spanish hands, the kings of Spain also carried (until 1833) the title king of Navarre. In 1833 Navarre sided with the Carlists but recognized Isabella II as queen in 1839.

Na·var·ro (nə-văr'ō, -vĕr'ō), county (1970 pop. 31,150), 1,070 sq mi (2,771.3 sq km), E central Texas, in a rich blackland prairie area bounded on the NE by the Trinity River; formed 1846; co. seat Corsicana. It has diversified agriculture (cotton, corn, grain, fruit, truck crops, pecans, livestock), oil and natural-gas wells, and clay mines. Food processing, oil refining, and manufacturing are done.

Nav·a·so·ta (năv'ə-sō'tə), city (1970 pop. 5,111), E central Texas, near the confluence of the Navasota and Brazos Rivers NW of Houston. The place was known to the Spanish. Later a town grew about a stagecoach stop here and was laid out in 1858. The city is a shipping center for a farming and lumbering region and has small industries. A statue of sieur La Salle commemorates the tradition that he was killed near Navasota.

Navasota, river, rising in E central Texas and flowing c.130 mi (209.2 km) SE and S to the Brazos River SW of Navasota.

Náv·pak·tos (näf'päk-tôs), town (1971 pop. 8,170), central Greece, a port on the Gulf of Corinth near the Gulf of Pátrai. The town was captured by Athens in 456 B.C. and was an important Athenian naval base in the Peloponnesian War. It later declined but rose to commercial importance as part of the Byzantine Empire. Known also as Lepanto in medieval and modern times, it came under the rule of Venice in 1407 and under the Ottoman Turks in 1499. In 1571 Christian forces of Europe won a major naval battle over the Ottoman Empire near the town. In 1828 Greek insurgents took the town from Ottoman control.

Náv·pli·on (näf'plē-ôn), town (1971 pop. 9,281), capital of Argolís prefecture, S Greece, in the Peloponnesus, a port on the Gulf of Argolís. It is a commercial center that ships tobacco, cotton, and fruits. In 1715 Návplion was captured by the Ottoman Turks from Venice. The town was taken in 1822 by Greek insurgents and was (1830–34) the first capital of independent Greece.

Na·vy Island (nā'vē), in the Niagara River, just above Niagara Falls, S Ont., Canada. It is famous as the scene of the last stand made by William Lyon Mackenzie and some of his fellow rebels in the Upper Canadian Rebellion of 1837.

Náx·os (näk'sôs, năk'səs), island (1971 pop. 14,201), c.160 sq mi (415 sq km), SE Greece, in the Aegean Sea; largest of the Cyclades. The fertile island produces fruits, olive oil, and a noted white wine. It has been a source of white marble, emery, and granite since ancient times. Náxos is famous in mythology as the place where Theseus abandoned Ariadne. It was a center of the worship of Dionysius. The island was colonized by the Ionians and in 490 B.C. was captured and sacked by the Persians. It was a member of the Delian League, but after an unsuccessful attempt to secede was captured (c.470 B.C.) and became a tributary to Athens. Náxos passed to Venice in 1207 and was the seat of a Venetian duchy until 1566, when it fell to the Ottoman Turks. It became part of independent Greece in 1829.

Na·ya·rit (nä'yä-rēt'), state (1970 pop. 547,992), 10,547 sq mi (27,316.7 sq km), W Mexico, on the Pacific Ocean; capital Tepic. The Nayarit region was known to the Spanish early in the 16th cent. Spain did not finally conquer the area until the early 17th cent., however. Shortly afterward Nayarit became a dependency of Guadalajara and, upon Mexican independence, part of Jalisco. Continued turbulence led to Nayarit's separation as a territory in 1884. It became a state in 1917.

Naz·a·reth (năz'ər-əth, -ə-rĭth), town, N Israel, in Galilee. As the home of Jesus Christ, it is a great pilgrimage and tourist center. The town is also the trade center for an agricultural region. Nazareth is first mentioned in the New Testament, although its settlement antedates historic times. It was captured by Crusaders in 1099, taken by Saladin in 1187, and retaken in 1229 by Frederick II. In 1263 Moslems conquered Nazareth and massacred its Christian population. In 1517 Nazareth was annexed by the Ottoman Empire. The town was part of the British-administered Palestine mandate (1922–48) and was captured by Israeli forces in the 1948 War.

Nazareth, borough (1970 pop. 5,815), Northamptom co., E Pa., NNE of Bethlehem; settled c.1740 by Moravians, inc. 1856. Textile and paper products, musical instruments, and cement are made. Whitefield House, begun in 1740, is an excellent example of Moravian architecture.

Na·zas (nä'säs'), river, c.180 mi (290 km) long, rising in the Sierra Madre Occidental, Durango state, N Mexico, and flowing generally E to disappear into the ground near Torreón. During the wet season it usually inundates a vast desert basin.

Naze, the (nāz), cape, Norway: *see* Lindesnes.

Ndja·me·na (ən-jä'mä-nä), formerly **Fort-La·my** (fôr'lä-mē'), city (1972 est. pop. 179,000), capital of Chad and of Chari-Baguirmi prefecture, SW Chad. It is a port on the Chari River and a transportation hub leading to Nigeria, Sudan, and the Central African Empire. Ndjamena is also a major regional market for livestock, salt, dates, and grains. Meat processing is the chief industry. The city was founded by the French in 1900. Its name was changed in 1973.

Ndo·la (ən-dō'lä), city (1976 est. pop., with suburbs, 260,000), N central Zambia, near Zaire. It is a commercial, mining, and manufacturing center, located on the Copperbelt. Copper mining in Ndola long antedates the coming of the Europeans (c.1900). Manufactures include cement, footwear, and soap; motor vehicles are assembled.

Neagh, Lough (nā), lake, 153 sq mi (396.3 sq km), 18 mi (29 km) long and 11 mi (18 km) wide, central Northern Ireland. It is the largest freshwater body in the British Isles but is not a deep lake. Fed by the Upper Bann, Blackwater, and other streams, it is noted for pollan, trout, and eel fisheries. Mesolithic man is believed to have first appeared in Ireland (c.6000 B.C.) near the lake.

Ne·an·der·thal (nā-än'dər-täl', nē-ăn'dər-thôl'), small valley, W West Germany, E of Düsseldorf. In 1856 the remains of Neanderthal man were discovered here.

Ne·ap·o·lis (nē-ăp'ə-ləs), name of many cities in ancient Greek and Roman times. The most important is the modern Naples, Italy.

Neath (nēth), municipal borough (1973 est. pop. 27,280), West Glamorgan, S Wales, on the Neath River. It is a market and industrial town. Metallurgy is the main industry. The ruins of Neath Abbey, founded c.1130 are here.

Ne·bit-Dag (nyĕ-bēt'däk'), city (1976 est. pop. 65,000), Central Asian USSR, in Turkmenistan, at the S foot of the Greater Balkhan Range. It is in an industrial region yielding oil and natural gas.

Ne·bras·ka (nə-brăs'kə), state (1975 pop. 1,553,000), 77,227 sq mi (200,018 sq km), central United States, in the Great Plains region, admitted as the 37th state of the union in 1867. Lincoln is the capital. The state is roughly rectangular, except in the east where the border is formed by the irregular course of the Missouri River and in the southwest where the state of Colorado cuts out a squared corner. Elsewhere Nebraska is bounded by Wyo. on the west, by S.Dak. on the north, by Iowa and Mo. on the east, and by Kansas on the south.

The land rises gradually from 840 ft (256 m) in the east to 5,300 ft (1,615 m) in the west. The Platte River, formed in west Nebraska by the junction of the North Platte and the South Platte, flows across the state to join the Missouri south of Omaha. The Platte and the Missouri, with their tributaries, give Nebraska all-important water sources that are still essential to farming in this agrarian state. From the Missouri westward over about half the state stretch undulating farm lands, where the fertile silt is underlaid by deep loess soil. Nebraska's population is concentrated here, many being farmers who produce grains for the market or for feeding hogs and dairy cattle. To the west and northwest the sand hills of Nebraska fan out, their wind-eroded contours now more or less stabilized by grass coverage. Cattle graze on the slopes and tablelands, protected in the severe winters by the sand bluffs and the valleys. The climate is severely continental throughout Nebraska. A low of −40°F (−40°C) in the winter is not unusual, and during the short intense summers temperatures may easily reach 110°F (43°C). In the far west the land rises to the foothills of the Rocky Mts.

Points of interest to the traveler include Father Flanagan's Boys Town, near Omaha; the Oglala National Grassland; the Fort Niobrara National Wildlife Refuge, near Valentine; and the Homestead National Monument, near Beatrice. The pioneers' migration west over the Oregon Trail is commemorated by the Scotts Bluff National Monument and the Chimney Rock National Historic Site. Hundreds of fresh and alkali lakes in the state attract sportsmen and campers.

Nebraska is divided into 93 counties:

NAME	COUNTY SEAT	NAME	COUNTY SEAT
Adams	Hastings	Boyd	Butte
Antelope	Neligh	Brown	Ainsworth
Arthur	Arthur	Buffalo	Kearney
Banner	Harrisburg	Burt	Tekamah
Blaine	Brewster	Butler	David City
Boone	Albion	Cass	Plattsmouth
Box Butte	Alliance	Cedar	Hartington

NAME	COUNTY SEAT	NAME	COUNTY SEAT
Chase	Imperial	Lancaster	Lincoln
Cherry	Valentine	Lincoln	North Platte
Cheyenne	Sidney	Logan	Stapleton
Clay	Clay Center	Loup	Taylor
Colfax	Schuyler	McPherson	Tryon
Cuming	West Point	Madison	Madison
Custer	Broken Bow	Merrick	Central City
Dakota	Dakota City	Morrill	Bridgeport
Dawes	Chadron	Nance	Fullerton
Dawson	Lexington	Nemaha	Auburn
Deuel	Chappell	Nuckolls	Nelson
Dixon	Ponca	Otoe	Nebraska City
Dodge	Fremont	Pawnee	Pawnee City
Douglas	Omaha	Perkins	Grant
Dundy	Benkelman	Phelps	Holdrege
Fillmore	Geneva	Pierce	Pierce
Franklin	Franklin	Platte	Columbus
Frontier	Stockville	Polk	Osceola
Furnas	Beaver City	Red Willow	McCook
Gage	Beatrice	Richardson	Falls City
Garden	Oshkosh	Rock	Bassett
Garfield	Burwell	Saline	Wilber
Gosper	Elwood	Sarpy	Papillion
Grant	Hyannis	Saunders	Wahoo
Greeley	Greeley Center	Scotts Bluff	Gering
Hall	Grand Island	Seward	Seward
Hamilton	Aurora	Sheridan	Rushville
Harlan	Alma	Sherman	Loup City
Hayes	Hayes Center	Sioux	Harrison
Hitchcock	Trenton	Stanton	Stanton
Holt	O'Neill	Thayer	Hebron
Hooker	Mullen	Thomas	Thedford
Howard	St. Paul	Thurston	Pender
Jefferson	Fairbury	Valley	Ord
Johnson	Tecumseh	Washington	Blair
Kearney	Minden	Wayne	Wayne
Keith	Ogallala	Webster	Red Cloud
Keya Paha	Springview	Wheeler	Bartlett
Kimball	Kimball	York	York
Knox	Center		

Economy. Mineral deposits of oil, sand and gravel, and stone contribute to the state's economy, but agriculture remains the dominant occupational pursuit. Chief products are cattle, corn, hogs, wheat, hay, and grain sorghum. Nebraska ranked fourth among the states in cattle production in 1972. Nebraska's largest industry is food processing. The state has diversified its industries since World War II, and the manufacture of electrical machinery and chemical products is also important.

History. The Spanish explorer Francisco Vásquez de Coronado and his men probably came through Nebraska in 1541. Development began only after the area passed from France to the United States in the Louisiana Purchase of 1803. Manuel Lisa, a fur trader, probably established the first trading post in the Nebraska area in 1813. Bellevue, the first permanent settlement, first developed as a trading post. Steamboating on the Missouri River, initiated in 1819, brought much business to the river ports of Omaha and Brownville. Military posts, notably Fort Atkinson (1819-27), were founded to protect developing commerce from the Indians. The natural highway formed by the Platte valley became deeply rutted in the 1840s and 1850s by the wagons of the pioneers going west over the Oregon Trail and also the California and Mormon trails. Nebraska settlers made money supplying the wagon trains with fresh mounts and pack animals as well as food.

When Nebraska became a territory (1854), it extended from lat. 40° N to the Canadian border. In 1863 the territory was reduced to its present-day size by the creation of the territories of Dakota and Colorado, and statehood was granted in 1867. In that year the Union Pacific RR was built across the state, and the land boom, already vigorous, became a rush. Farmers settled on free land obtained under the Homestead Act of 1862, and eastern Nebraska took on a settled look. The population rose from 28,841 in 1860 to 122,993 in 1870. The Pawnee Indians were subdued in 1859, and by 1880 war with the Sioux and other Indian disturbances were over.

Many farmers joined the Granger movement in the lean 1870s and the Farmers' Alliances of the 1880s. The first national convention of the Populist party was held at Omaha in 1892, and Nebraska's most famous son, William Jennings Bryan, headed the Populist and Democratic tickets in the presidential election of 1896. The return of prosperous days in the early 1900s was marked by progressive legislation, the building of highways, and conservation measures. Overexpansion of credits and overconfidence made the depression of the 1920s and 1930s all the more disastrous. Better weather and the huge food demands of World War II renewed prosperity in Nebraska. Since the war, efforts have continued to make the best use of the water supply, notably in such Federal plans as the Missouri river basin project, a vast dam and water-diversion scheme.

Government. Nebraska's present constitution was adopted in 1875. The executive branch is headed by a governor elected for a four-year term. The unicameral legislature has 49 members elected on a nonpartisan basis for terms of four years.

Educational Institutions. The state's leading institution of higher education is the Univ. of Nebraska, mainly at Lincoln.

Nebraska City, city (1970 pop. 7,441), seat of Otoe co., SE Nebr., on the Missouri River ESE of Lincoln; founded 1854, inc. 1855, consolidated 1857 with two contiguous settlements. It is a trade and processing point in a fertile farm area yielding grain, livestock, and apples.

Ne·chak·o (nĭ-chăk'ō), river, 287 mi (461.8 km) long, rising in Tetachuck and Ootsa lakes, central British Columbia, Canada, and flowing NE, then E to the Fraser River at Prince George. Kenney Dam (325 ft/99.1 m high; completed 1952) and Kemano Dam (320 ft/97.6 m high; completed 1954) are among the highest dams in Canada.

Nech·es (nĕch'ĭz), river, 416 mi (669.3 km) long, E Texas. Rising in Van Zandt co., it flows southeast, past Beaumont, to the head of Sabine Lake.

Neck·ar (nĕk'ər), river, 228 mi (366.9 km) long, rising in the Black Forest, SW West Germany. It flows generally north past Tübingen, Stuttgart, and Heilbronn, then west past Heidelberg before joining the Rhine River at Mannheim. The Neckar is celebrated for its scenic charm; its hilly banks are covered with fine vineyards, orchards, and woods.

Ne·der·land (nē'dər-lănd'), city (1978 est. pop. 18,500), Jefferson co., SE Texas; founded by Dutch settlers in 1897, inc. 1940. Primarily a residential suburb, it has oil companies and industries that produce butadiene, plastics, and synthetic rubber.

Need·ham (nē'dəm), town (1978 est. pop. 29,700), Norfolk co., E Mass., a suburb of Boston; founded 1680, set off from Dedham and inc. 1711. Although it is largely residential, textiles, paper products, electronic equipment, and other items are manufactured.

Nee·nah (nē'nə), city (1978 est. pop. 25,200), Winnebago co., E Wis., on Lake Winnebago at the mouth of the Fox River; settled c.1835 on the site of a Winnebago Indian village, inc. as a city 1873. Located in a dairy-farming region, Neenah is known, with its twin city Menasha, as a center for the manufacture of paper and paper products. Neenah's industrial development began c.1850, when flour mills serving the surrounding farming area were opened.

Neer·win·den (nîr-vĭn'dən), village, Liège prov., E Belgium. In the War of the Grand Alliance the French under Marshal Luxembourg defeated (1693) William III of England here.

Ne·gau·nee (nə-gô'nē), city (1970 pop. 5,248), Marquette co., N Mich., W Upper Peninsula, on the Marquette iron range W of Ishpeming, in a resort region; settled 1846, inc. as a village 1862, as a city 1873. Iron ore was discovered in the region in 1844.

Ne·gev (nĕg'ĕv) or **Neg·eb** (nĕg'ĕb), hilly desert region of S Israel, c.5,140 sq mi (13,315 sq km), bordered by the Judaean Hills, the Wadi Arabah, the Sinai peninsula, and the narrow Mediterranean coastal plain. The Negev receives c.2 to 4 in. (5-10 cm) of rain annually. In the Beersheba basin there are fertile loess deposits, but the region's aridity prevented cultivation until irrigation was provided by the National Water Carrier Project, which taps the Sea of Galilee. The Negev region also has a good mineral potential; copper, phosphates, and natural gas are commercially extracted. The Negev was the scene of much fighting between Egyptian and Israeli forces after the partition of Palestine in 1948.

Ne·goi·u (nə-goi'ōō), peak, 8,344 ft (2,544.9 m) high, central Rumania, NW of Cîmpulung. It is the second-highest peak of the Transylvanian Alps.

Ne·gom·bo (nĭ-gŏm'bō), town (1971 pop. 57,115), W Sri Lanka, at the mouth of the Negombo Lagoon. Chiefly noted for its ceramics and brassware, it is also a fishing center and a market for coconut products and cinnamon. Many 17th cent. Dutch buildings remain.

Ne·gro, Río (rē'ō nā'grō), river, c.400 mi (645 km) long, formed in central Argentina by the confluence of the Neuquén and the Limay rivers, and flowing E across Río Negro prov. to the Atlantic Ocean. The river is used for irrigation.

Negro, Río, river, c.1,400 mi (2,255 km) long, rising as the Guainía River in E Colombia, where it flows NE before turning S to form part of the Colombia-Venezuela border. It then flows southeast through Amazonas state, Brazil, to the Amazon near Manaus. The river is filled with islands and has many secondary channels. An important commercial channel (rubber and nuts are shipped on it), the Río Negro was discovered in 1638 and was named for its black color, which results from vegetal debris, not sediment.

Negro, Río, principal river of Uruguay, c.500 mi (805 km) long, rising in S Brazil and flowing SW across central Uruguay to the Uruguay River. It traverses a sheep-raising region; there is agriculture along its lower course. On the river is Embalse del Río Negro (c.4,000 sq mi/10,360 sq km), the largest artificial lake in South America.

Ne·gros (nā'grōs'), island (1970 est. pop. 2,300,000), 4,905 sq mi (12,704 sq km), one of the Visayan Islands, Philippines, between Panay and Cebu. Although mountainous, Negros has extensive arable lowlands; they are intensively cultivated and densely populated. Two thirds of the nation's sugar cane is grown here, and sugar processing is a major industry. The island is also a leading banana- and corn-producing region. It has coconut plantations, copper and coal deposits, and a lumber industry.

Nei-chi·ang (nā'jē-äng'), city (1970 est. pop. 240,000), central Szechwan prov., China, on the To River. It is a port and railroad center with sugar-refining and food-processing industries.

Neills·ville (nēlz'vĭl'), city (1970 pop. 2,750), seat of Clark co., central Wis., SE of Eau Claire on the Black River; settled 1844, inc. 1882. It is a commercial center for dairying, stock-raising, and farming.

Neis·se (nī'sə), two rivers of SW Poland. The Glatzer Neisse, c.120 mi (195 km) long, rises in the Sudetes, southwest Poland, and winds generally northeast past Kłodzko to the Oder River near Brzeg. A large dam at Otmuchow serves hydroelectric and irrigation projects. The Lausitzer or Lusatian Neisse, c.140 mi (225 km) long, rises in the Sudetes, northwest Czechoslovakia, and flows generally north to the Oder River near Guben, East Germany. Since 1945 it has formed part of the border between East Germany and Poland.

Nejd (nĕjd) or **Najd** (năjd), region, central Saudi Arabia. It is a vast plateau from 2,500 to 5,000 ft (762.5–1,525 m) high. There is a chain of oasis settlements in the eastern section; elsewhere the area is roamed by nomadic Bedouins. The Nejd, the stronghold of the Wahabi movement, was gradually conquered (1899–1912) from Turkey by the Wahabi leader, Ibn Saud. From here he completed his conquest of the Hejaz and Al Hasa. In 1932 the Nejd became part of his newly constituted domain, Saudi Arabia.

Ne·ligh (nē'lē), city (1970 pop. 1,764), seat of Antelope co., NE Nebr., on the Elkhorn River WNW of Norfolk, in a farm area; inc. 1873.

Nel·lore (nĕ-lôr', lōr'), city (1971 pop. 133,607), Andhra Pradesh state, SE India, on the Pennar River. Nellore is a market for cotton and oilseed. It also has milling and processing industries.

Nel·son (nĕl'sən), city (1976 pop. 9,235), SE British Columbia, on the Kootenay River. It is a center for a lumbering and farming region.

Nelson, municipal borough (1973 est. pop. 31,220), Lancashire, N England. It has cotton and rayon factories.

Nelson, city (1976 pop. 32,793), N South Island, New Zealand, at the head of Tasman Bay. It is a port with light industries. The Cawthron Institute for scientific research is in the city.

Nelson. 1. County (1970 pop. 23,477), 437 sq mi (1,131.8 sq km), central Ky., in the outer Bluegrass region, bounded on the SW by Rolling Fork, on the E by Beech Fork; formed 1784; co. seat Bardstown. It is in a rolling agricultural area that produces livestock, dairy products, grain, and burley tobacco. It has hardwood timber and some mining and manufactures distilled liquor, clothing, and stone, clay, and glass products. **2.** County (1970 pop. 5,807), 997 sq mi (2,582.2 sq km), E central N.Dak., in an agricultural area watered by the Sheyenne and Goose rivers and Stump Lake; formed 1883; co. seat Lakota. Dairy products, livestock, and grain are its principal sources of income. **3.** County (1970 pop. 11,702), 471 sq mi (1,219.9 sq km), central Va., in a piedmont region with the Blue Ridge in the W and NW, bounded on the SE by the James River, and drained by the Rockfish River; formed 1807; co. seat Lovingston. Its agriculture includes fruit, tobacco, corn, and livestock. It has some mining (titanium ores, apatite, and soapstone) and clothing and lumbering industries. Parts of the Blue Ridge Parkway, the Appalachian Trail, and George Washington National Forest are in the county.

Nelson, city (1970 pop. 746), seat of Nuckolls co., S Nebr., SE of Hastings on a branch of the Little Blue River, in a grain area.

Nelson, river, c.400 mi (645 km) long, issuing from the NE end of Lake Winnipeg, central Man., Canada, and flowing NE to Hudson Bay at Port Nelson. With the Bow-South Saskatchewan-Saskatchewan river system, the Nelson is part of a 1,600-mi (2,575-km) continuous stream from western Alberta to Hudson Bay. The Nelson is being developed to make use of its great hydroelectric power potential. Kettle Rapids Dam, the second-largest hydroelectric facility in Canada, is on the river. The Nelson's mouth was discovered (1612) by Sir Thomas Button. The river was long followed by fur traders.

Nem·a·ha (nĕm'ə-hô'). **1.** County (1970 pop. 11,825), 709 sq mi (1,836.3 sq km), NE Kansas, in a gently sloping terrain bordered on the N by Nebr. and watered by the South Fork of the Nemaha River; formed 1855; co. seat Seneca. It is in an agricultural area that produces cattle, hogs, sheep, dairy products, milo, wheat, oats, corn, alfalfa, and red clover hay. Machinery is manufactured. **2.** County (1970 pop. 8,976), 399 sq mi (1,033.4 sq km), SE Nebr., in an agricultural area bounded in the E by the Missouri River and the Mo. border; formed 1854; co. seat Auburn. It is drained by the Little Nemaha River. Its economy depends on fruit, grain, livestock, dairy products, and poultry.

Nem·an (nĕm'ən), river, c.580 mi (935 km) long, rising in central Belorussia, W European USSR, SW of Minsk, and flowing generally W to Grodno, then N and W through S Lithuania to form part of the Lithuania-Kaliningrad Oblast border before entering the Kursky Zalev of the Baltic Sea through a small delta. The meeting of Napoleon I and Czar Alexander I, which resulted in the Treaty of Tilsit (1807), took place on a raft in the middle of the river.

Ne·me·a (nē'mē-ə), city of ancient Greece, in N Argolis. At the temple of Zeus were held the Nemean games, which from 573 B.C. were one of the four Panhellenic festivals; the games were held in the second and fourth years of each Olympiad. The temple and palaestra have been excavated.

Ne·mi, Lake (nā'mē, nĕm'ē), small picturesque crater lake, c.1 mi (1.6 km) long, in the Alban Hills, central Italy, SE of Rome. The sacred wood and the ruins of the celebrated temple of Diana are here. Two pleasure ships of the Roman emperor Caligula that were lying at the bottom of Lake Nemi for almost 2,000 years were raised (1930–31) from the lake after its level had been lowered c.70 ft (21 m). No valuables were found, but objects interesting from the artistic and technical point of view were recovered. During World War II the ships were destroyed (1944) by the retreating German forces.

Nen Chi·ang (nŭn' jē-äng') or **Non·ni** (nŏn'ē), river, 740 mi (1,190.7 km) long, rising in the I-lo-hu-li Shan, N Heilingkiang prov., NE China, and flowing S along the E side of the Great Khingan Range to the Sungari River. It forms part of the Heilungkiang-Kirin prov. border. The river and its valley form an important trade artery.

Nene (nĕn, nĕn) or **Nen** (nĕn), river, c.90 mi (145 km) long, rising in the Northampton Uplands, central England, and flowing NE past Northampton and Peterborough to the Wash. It is navigable to Peterborough and drains part of the Fens.

Ne·nets National Okrug (nyĕ-nyĭts'), administrative division (1970 pop. 39,000), 68,224 sq mi (176,700 sq km), extreme NE European USSR; capital Naryan-Mar. Formed in 1929, the okrug forms the northern part of Arkhangelsk oblast and extends along the tundra coast of the Barents, White, and Kara seas. Reindeer raising, fishing, fur trapping, and seal hunting are the chief occupations. Fish canning, sawmilling, and hide processing are important. The Nentsy, previously known as Samoyeds, speak a Finno-Ugric language. They were first mentioned in the 11th cent., became tributaries of the grand duchy of Moscow c.1500, and gained their independence in the 17th cent. after prolonged warfare.

Ne·o·sho (nē-ō'shō, -shə), county (1970 pop. 18,812), 587 sq mi (1,520.3 sq km), SE Kansas, in a sloping to gently rolling area drained by the Neosho River; formed 1855, name changed from Dorn in 1861; co. seat Erie. It is in a grain, poultry, and dairy region, with oil, gas, and coal mining. Its industry includes lumber sawmilling, petroleum refining, and the manufacture of clothing, cement, metal products, and farm machinery.

Neosho, city (1970 pop. 7,517), seat of Newton co., SW Mo., in the Ozarks SE of Joplin; inc. 1855. It is in an agricultural area yielding strawberries, raspberries, and dairy products.

Neosho, river, c.460 mi (740 km) long, rising in E central Kansas and flowing SE into NE Okla. (where it is generally known as the Grand River) then S to join the Arkansas River near Muskogee, Okla. Pensacola Dam and Fort Gibson dam and reservoir are on the river.

Ne·pal (nə-pôl', -päl', -păl'), independent kingdom (1976 est. pop. 12,900,000), c.54,000 sq mi (139,860 sq km), central Asia, bordered on the N by the People's Republic of China (Tibet region) and on the W, S, and E by India.

The south, known as the Terai, is a comparatively low region of cultivable land, swamps, and forests that provide valuable timber. In the north is the main section of the Himalayas, including Mt. Everest. Central Nepal, an area of moderately high mountains, contains the Katmandu valley, or Valley of Nepal, the country's most densely populated region and its administrative, economic, and cultural center. Nepal's railroads, connecting with lines in India, do not reach the valley, which is served by a bridgelike cable line. The population of Nepal represents a long intermingling of Mongolians, who migrated from the north (especially Tibet), and Indo-Aryans, who came from the Ganges plain in the south. The chief ethnic group, the Newars, were probably the original inhabitants of the Katmandu valley. Several ethnic groups are classified together as Bhotias; among them are the Sherpas, famous for guiding mountain-climbing expeditions, and the Gurkhas, a term sometimes loosely applied to the fighting castes, who achieved fame in the British Indian army.

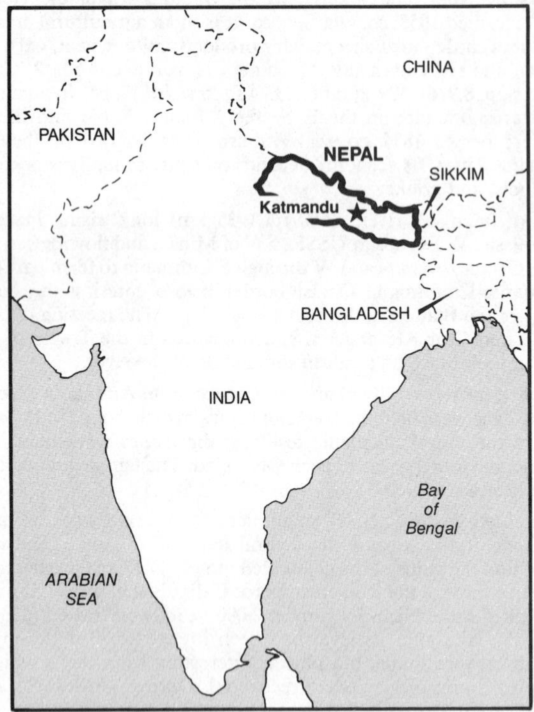

Economy. The overwhelming majority of Nepal's people engage in agriculture, which contributes about two thirds of the national income. In the Terai, the main agricultural region, rice is the chief crop; other food crops include pulses, wheat, barley, and oilseeds. Jute, tobacco, cotton, indigo, and opium are also grown in the Terai, whose forests provide sal wood and commercially valuable bamboo and rattan. In the lower mountain valleys rice is produced during the summer, and wheat, barley, oilseeds, potatoes, and vegetables are grown in the winter. Corn, wheat, and potatoes are raised at higher altitudes, and terraced hillsides are also used for agriculture. Large quantities of medicinal herbs, grown on the Himalayan slopes, are sold worldwide. Livestock raising is second to farming in Nepal's economy; oxen predominate in the lower valleys, yaks in the higher, and sheep, goats, and poultry are plentiful everywhere. Transportation and communication difficulties have hindered the growth of industry and trade. Significant quantities of mica are found in the hills of Nepal. Tourism is the chief source of foreign income (along with subsidies from the Indian government and Gurkha pensions).

History. By the 4th cent. A.D. the Newars of the central Katmandu valley had apparently developed a flourishing Hindu-Buddhist culture. From the 8th to 11th cent. many Buddhists fled to Nepal from India, which was being invaded by Moslems, and a group of Hindu Rajput warriors set up the principality of Gurkha just west of the Katmandu valley. Although a Newar dynasty, the Mallas, ruled the valley from the 14th to 18th cent., there were internecine quarrels among local rulers. These were exploited by the Gurkha king Prithur Narayan Shah, who conquered the Katmandu valley in 1768. Gurkha armies seized territories far beyond present-day Nepal; but their invasion of Tibet, over which China claimed sovereignty, was defeated in 1792 by Chinese forces. An ensuing peace treaty forced Nepal to pay China an annual tribute, which continued until 1910. Also in 1792 Nepal first entered into treaty relations with Great Britain. Gurkha expansion into northern India, however, led to a border war (1814–16) and to British victory over the Gurkhas, who were forced by treaty to retreat into roughly the present borders of Nepal and to receive a British envoy at Katmandu.

The struggle for power among the Nepalese nobility culminated in 1846 with the rise to political dominance of the Rana family. Jung Bahadur Rana established a line of hereditary prime ministers, who controlled the government until 1950. In 1854 Nepal again invaded Tibet, which was forced to pay tribute from then until 1953. Under the Ranas, Nepal was deliberately isolated from foreign influences; this policy helped to maintain independence during the colonial period but prevented economic and social modernization. Relations with Britain were cordial, however, and in 1923 a British-Nepalese treaty expressly affirmed Nepal's full sovereignty.

The successful Indian movement for independence (1947) stimulated democratic sentiment in Nepal. The newly formed Congress party of Nepal precipitated a revolt in 1950 that forced the autocratic Ranas to share power in a new cabinet. In 1959 a democratic constitution was promulgated, and parliamentary elections gave the Congress party a clear majority; the following year, however, King Mahendra (reigned 1952–72) cited alleged inefficiency and corruption in government as evidence that Nepal was not ready for Western-style democracy. He dissolved parliament, detained many political leaders, and in 1962 inaugurated a system of "basic democracy," based on the elected village council (panchayat) and working up to district and zonal panchayats and national panchayat.

King Mahendra modernized Nepal through such programs as a land reform that distributed large holdings to landless families and a law removing the legal sanctions for caste discrimination. Charges that India had sheltered antiroyalist Nepalese politicians caused friction with India in the late 1960s; in 1969 Nepal canceled an arms agreement with India and ordered the Indians to withdraw their military mission from Katmandu and their wireless operators from the Tibet-Nepal frontier. Meanwhile Nepal has received U.S., Soviet, and Chinese economic assistance; a Chinese-financed highway linking Katmandu with Tibet is a particularly important aid project. Crown Prince Birenda succeeded to the throne (1972) upon his father's death; like previous Nepalese monarchs, he married a member of the Rana family in order to assure political peace. Nepal has maintained a position of nonalignment in foreign affairs.

Ne·phi (nē'fī'), city (1970 pop. 2,699), seat of Juab co., central Utah, SSW of Provo; settled 1851 by Mormons. It is a processing and shipping center for a wheat, poultry, and livestock area. Rubber products are made, and gypsum is mined and made into plaster.

Nep·tune City (nĕp'to͞on', -tyo͞on'), resort borough (1970 pop. 5,502), Monmouth co., E N.J., SSW of Asbury Park; inc. 1881.

Ner·chinsk (nyĕr'chĭnsk'), city, SE Siberian USSR. Founded in 1654, the city was a Russian outpost in the Far East from the 17th to the 19th cent. A Russo-Chinese border treaty signed at Nerchinsk in 1689 granted the Transbaikalia area to Russia and left the Amur valley to China. The treaty also permitted Russian trading caravans to go to Peking; Nerchinsk became an important customs and trade center on the caravan route.

Nes·con·set (nĕs-kŏn'sət), uninc. village (1970 pop. 7,500), Suffolk co., SE N.Y., on Long Island.

Ne·sho·ba (nə-shō'bə), county (1970 pop. 20,802), 568 sq mi (1,471.1 sq km), E central Miss., drained by the Pearl River and its tributaries; formed 1833; co. seat Philadelphia. It is in a cotton, corn, and fruit area, with lumbering and industry (wood products, clothing, and paper products). Choctaw Indian Reservation is within the county.

Ness (nĕs), county (1970 pop. 4,791), 1,081 sq mi (2,799.8 sq km), W central Kansas, in a rolling prairie region watered by Walnut Creek; formed 1880; co. seat Ness City. It is in a wheat and cattle region, with oil and gas wells. There is some manufacturing in the county.

Ness, Loch, lake, 22 mi (35.4 km) long, Inverness-shire, N central Scotland, in the Great Glen. More than 700 ft (213.5 m) deep and ice-free, it is fed by the Oich and other streams and drained by the Ness to the Moray Firth. It forms part of the Caledonian Canal. Since Dec., 1933, when newspapers published accounts of a "monster" 40 to 50 ft (12.2–15.3 m) long, said to have been seen in the loch, there have been several alleged sightings.

Ness City, city (1970 pop. 1,756), seat of Ness co., W central Kansas, N of Dodge City, in an oil and wheat area; founded 1878, inc. 1886.

Ne·tan·ya (nə-tän'yə), city (1975 est. pop. 82,400), W central Israel, on the Mediterranean Sea. It is a beach resort and the trade center for agricultural settlements in the region. Diamond cutting and polishing and citrus packing are the chief industries. Netanya, founded in 1929, was named for the U.S. philanthropist Nathan Straus, who contributed funds to educational and social agencies in Palestine. Wingate Institute for Physical Education and Zichron Ya'akov, one of the first modern Jewish settlements (1882) in Palestine, are nearby.

Neth·er·lands (nĕ*th*'ər-ləndz), kingdom (1971 pop. 13,182,800), 15,963 sq mi (41,344 sq km), NW Europe, bounded by the North Sea on the N and W, by Belgium on the S, and by West Germany on the E. It is popularly known also as Holland. Amsterdam is the constitutional capital; The Hague is the seat of government.

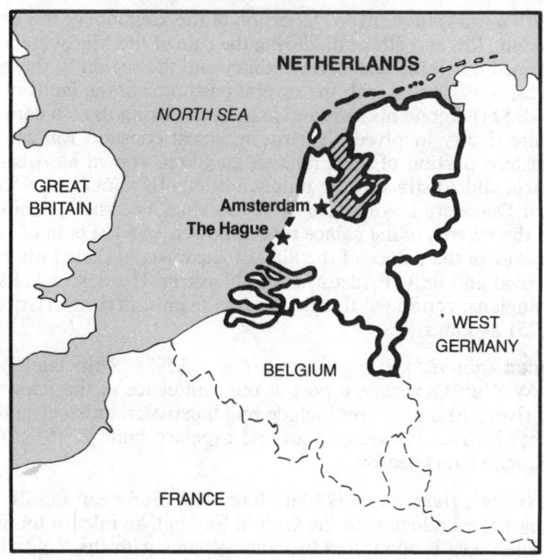

About 40% of the land is situated below sea level and is made up of territory reclaimed from the sea since the 13th cent. The kingdom includes one overseas territory, the Netherlands Antilles in the Caribbean. Surinam, in NE South America, received its independence in 1975. The West Frisian Islands are located off the northern coast of the Netherlands. The country is crossed by drainage canals, and the main rivers are canalized and interconnected by artificial waterways that are linked with the river and canal systems of Belgium and West Germany. The population density of the Netherlands is one of the highest in the world, and great skill is necessary to maintain the high standard of living the country enjoys.

The Netherlands is divided into 11 provinces:

NAME	CAPITAL	NAME	CAPITAL
Drenthe	Assen	North Holland	Haarlem
Friesland	Leeuwarden	Overijssel	Zwolle
Gelderland	Arnhem	South Holland	Hague, The
Groningen	Groningen	Utrecht	Utrecht
Limburg	Maastricht	Zeeland	Middelburg
North Brabant	's Hertogenbosch		

Economy. The Netherlands is highly industrialized, with industry contributing some 40% to national income in the early 1970s. The chief manufactures are textiles, machinery, electrical equipment, iron and steel, refined petroleum, ships, processed foods, plastics, and chemicals. Agriculture is specialized, mechanized, and efficient. Dairy farming is especially important. Cattle are widely raised, and there is a large poultry industry. The major crops are truck-farm commodities, beets, and potatoes; relatively little grain is raised. Fishing contributes significantly, although less than in the past, to the economy. The country's few natural resources include coal, natural gas, and petroleum.

The Netherlands carries on a large foreign trade; the main exports are machinery, textiles, petroleum products, fruits and vegetables, and meat. A considerable amount of the country's wealth is contributed annually by financial and transportation services.

History. One of the Low Countries, the Netherlands did not have a unified history until the late 16th cent. With the decline of Roman rule nearly the entire area was taken (4th–8th cent.) by the Franks, and with the breakup of the Carolingian empire, most of it passed (9th cent.) to the east Frankish (i.e., German) kingdom and thus to the Holy Roman Empire. The counts of Holland emerged as the most powerful medieval lords of the region. In the 14th and 15th cent. Flanders, Holland, Zeeland, Gelderland, and Brabant passed to the dukes of Burgundy. The Dutch towns nearly all belonged to the Hanseatic League, and they enjoyed vast autonomous privileges. In 1477 Mary of Burgundy's marriage to the Archduke Maximilian (later Emperor Maximilian I) brought the Low Countries into the house of Hapsburg. Emperor Charles V gave them (1555) to his son Philip II of Spain. The northern provinces (i.e., the present Netherlands) succeeded (1572–74) in expelling the Spanish garrisons, formed (1579) the Union of Utrecht, and declared (1581) their independence.

Fighting with Spain continued intermittently until 1609 and was resumed in the Thirty Years' War (1618–48), at the end of which the independence of the United Provinces—as the Netherlands was called—was recognized in the Peace of Westphalia (1648). While involved in the religious struggle between Calvinists and Remonstrants, the Dutch laid the foundation of their commercial and colonial empire. The Dutch East India Company was founded in 1602, the Dutch West India Company in 1621. The United Provinces opened their doors to religious refugees, notably to Portuguese and Spanish Jews and to French Huguenots; they contributed vastly to the immense prosperity and cultural excellence of 17th cent. Holland.

Opponents of the ruling house of Orange reasserted the rights of the provinces and the States-General in 1650. Jan de Witt was chosen (1652) grand pensionary of Holland and directed the fate of the Dutch republic for the following 20 years. De Witt's administration was largely taken up by the Dutch wars with England (1652–54, 1664–67) and with France (1672). Power was restored in 1672 to William III, who conducted two more wars with France before his death in 1702. The States-General then resumed control until 1747, when William IV became hereditary stadtholder (chief of state). In the 18th cent. the relative commercial, military, and cultural positions of the United Provinces declined as those of England and France ascended. In the French Revolutionary Wars the French overran (1794–95) the Netherlands and set up first a republic, then a kingdom, under French protection.

At the Congress of Vienna (1814-15) the former United Provinces and the former Austrian Netherlands were united. In 1830, however, the former Austrian provinces (Belgium) rebelled against Dutch rule and declared their independence. From 1849 to 1890 the Netherlands enjoyed a period of commercial expansion and internal development. Industrialization and trade unionism grew in the late 19th cent., and considerable national social-welfare legislation was passed. In 1890 Queen Wilhelmina began her reign of almost 60 years. The Netherlands was neutral in World War I. In World War II Germany invaded (May, 1940) the Netherlands, crushed Dutch resistance, and established a harsh occupation, exterminating most Dutch Jews. The German collapse in May, 1945, was followed by the immediate return of the queen and the cabinet and by a relatively speedy recovery.

The Netherlands gave Indonesia independence in 1949, and in 1962 relinquished Netherlands New Guinea. The Netherlands, which enjoyed a traditional friendship with Israel, suffered considerably from the ban on the sale of crude petroleum imposed (1973–74) by Arab nations in the wake of the 1973 Arab-Israeli War. There were racial tensions in the country in the late 1970s. Extremists from among the 40,000 South Moluccan immigrants living in the Netherlands committed several acts of terrorism (1976–77), demanding Dutch support for South Moluccan independence from Indonesia.

In 1976 the monarchy was severely shaken by charges that Prince Bernhard, the consort to Queen Juliana, had taken bribes from the Lockheed Aircraft Corp. to promote the sale of Lockheed planes to the Dutch government. No evidence was found that the prince had accepted bribes, but he was criticized for the impropriety of many of his business ventures. Following the disclosures Bernhard resigned all his official posts but retained the title of prince. A left-center coalition led by Jorp den Uyl, head of the Labor party, came to power in 1973.

Government. The Netherlands is a constitutional monarchy governed under a constitution promulgated in 1814 and since frequently revised. Executive power rests formally with the crown and in practice with the premier and his cabinet. Legislative power is vested in the bicameral States-General. The deliberative upper, or first, chamber is elected by the 11 provincial estates, and the more powerful lower, or second, chamber is chosen by direct universal suffrage. The royal succession is settled on the house of Orange.

Net·til·ling Lake (nĕch'ə-lĭng'), freshwater lake, 1,956 sq mi (5,066 sq km), S Baffin Island, Franklin dist., Northwest Territories, Canada. It is located in an arctic lowland region and is fed by Amadjuak Lake and by numerous streams that drain the tundra. The lake is frozen most of the year.

Net·tu·no (nā-tōō'nō), town (1976 est. pop. 24,110), in Latium, central Italy, on the Tyrrhenian Sea. It is an agricultural center and a seaside resort. With nearby Anzio it was the site of an Allied landing (Jan. 22, 1944) in World War II.

Ne·tza·hual·có·yotl (nā-tsä'wäl-kō'yōt'l), city (1970 pop. 580,438), Mexico state, S central Mexico. It is a communications center whose importance lies chiefly in its proximity to Mexico City.

Neu·bran·den·burg (noi-brän'dən-bûrg'), city (1974 est. pop. 59,971), capital of Neubrandenburg dist., N East Germany, on the Tollensesee. Manufactures include machinery, chemicals, and food products. It was founded in 1248 by the margraves of Brandenburg.

Neu·châ·tel (nœ'shə-tĕl', nōō'shə-tĕl'), canton (1977 est. pop. 162,900), 309 sq mi (800.3 sq km), NW Switzerland, in the Jura Mts. A part of Burgundy by the 10th cent., Neuchâtel was later governed by counts under the Holy Roman Empire. The county passed (1504) to the French house of Orléans-Longueville and in 1648 became independent. In 1707 it chose Frederick I of Prussia as its prince. It remained an autonomous principality, although in 1815 it became a canton of the Swiss Confederation, with which it had been allied since the 15th cent. In 1848 a revolution abolished the monarchy within Neuchâtel, and in 1857, after some complications, the king of Prussia renounced his claim to the canton. Its capital, Neuchâtel (1970 pop. 38,784), has industries that produce watches, jewelry, machinery, and chocolate. The town still retains a medieval aspect with its numerous statues, fountains, and old structures. It has an old church (12th–13th cent.), a castle (12th–17th cent.), and a noted university (founded 1838). The town is on the northern shore of the Lake of Neuchâtel, 24 mi (38.6 km) long and 4 to 5 mi (6.4–8 km) wide. The lake is surrounded by valuable vineyards and picturesque settlements. There are many remains of lake dwellings.

Neu·hau·sen am Rhein·fall (noi'hou'zən äm rīn'fäl'), town (1970 pop. 12,103), Schaffhausen canton, N Switzerland, on the right bank of the Rhine River. It is a manufacturing center.

Neuil·ly-sur-Seine (nœ-yē'sür-sĕn', -sōōr-sān'), city (1975 pop. 65,983), Hauts-de-Seine dept., N central France. A wealthy suburb of Paris, Neuilly-sur-Seine also manufactures machines, boilers, and precision instruments. The American Hospital of Paris is here.

Neu·mün·ster (noi'mün'stər), city (1974 est. pop. 85,645), Schleswig-Holstein, N West Germany. It is a transportation and industrial center; manufactures include machinery, textiles, and clothing. Known in the 12th cent., Neumünster was chartered in 1870.

Neun·kir·chen (noin'kĭr'KHən), city (1971 pop. 10,922), Lower Austria province, E Austria, on the Schwarza River. Metal goods, furniture, and building materials are manufactured.

Neunkirchen, city (1974 est. pop. 55,884), Saarland, W West Germany. Manufactures include iron and steel, machinery, chemicals, and textiles. Neunkirchen was first mentioned in the 13th cent.

Neuse (nōōs, nyōōs), river, formed in N N.C. NE of Durham by the junction of three rivers and flowing c.275 mi (442.5 km) generally SE past Smithfield and Kingston to New Bern at the head of its estuary and then into Pamlico Sound.

Neu·sie·dler Lake (noi'zēd'lər), c.130 sq mi (335 sq km), on the Austria-Hungary border SE of Vienna. The lake's area and depth (average 5 ft/1.5 m) vary considerably with the seasons. The heavy growth of lake reeds supplies the Austrian cellulose industry. Carp fisheries are on the lake. Its lonely and desolate salt marshes attract a variety of wildlife and have been protected since 1935. The Neusiedler region has noted resorts. There are remains of prehistoric lake dwellers in the vicinity.

Neuss (nois), city (1975 pop. 147,833), North Rhine-Westphalia, W West Germany. It is a rail junction and canal port, near the left bank of the Rhine opposite Düsseldorf. Its industries produce metal goods, heavy and light machinery, paper, and food products. Built on the site of a Roman camp, Neuss was chartered in the 12th cent. It belonged to the archbishopric of Cologne until the French Revolutionary Wars. In 1474–75 Charles the Bold of Burgundy, supporting the archbishop in a quarrel with the chapter of Neuss, unsuccessfully besieged the city for 11 months. It passed to Prussia in 1815.

Neu·stadt an der Wein·stras·se (noi'shtät' än dər vīn'shtrā'sə), city (1974 est. pop. 51,124), Rhineland-Palatinate, W West Germany; chartered 1275. It is the center of the Rhenish Palatinate wine trade; manufactures include metal products and textiles.

Neu·stre·litz (noi-shtrā'lĭts), city (1974 est. pop. 27,074), Neubrandenburg district, N East Germany. It is a transportation center and has metalworks, publishing houses, and wood mills. Neustrelitz was founded (1733) as the capital of Mecklenburg-Strelitz.

Neus·tri·a (nōō'strē-ə, nyōō'-), portion of the kingdom of the Franks in the 6th, 7th, and 8th cent., during the rule of the Merovingians. It comprised the Seine and Loire country and the region to the north. The realm originated with the several partitions of the lands of Clovis I (d. 511) among his sons and grandsons during the 6th cent. The dynastic rivalry involved Neustria in almost constant warfare with the eastern portion of the Frankish kingdom, known as Austrasia. Neustria and Austrasia were reunited briefly by Clotaire I, Clotaire II, and Dagobert I. After Dagobert the kings became insignificant, while the mayors of the palace rose in power. In 687 Pepin of Heristal, mayor of the palace of the king of Austrasia, defeated his Neustrian rival and united Austrasia and Neustria. His descendants, the Carolingians, continued to rule the two realms, first as mayors and after 751 as kings.

Neu·wied (noi-vēt'), city (1974 est. pop. 62,598), Rhineland-Palatinate, W West Germany, a port at the confluence of the Rhine and Wied rivers. Manufactures include building materials, steel, and machinery. Neuwied developed around a palace built in 1648. There are Roman ruins nearby.

Ne·va (nē'və), river, 46 mi (74 km) long, NW European USSR, connecting Lake Ladoga with the Gulf of Finland, an inlet of the Baltic Sea. The Neva is connected by canal systems with the Volga River and with the White Sea. It freezes in winter.

Ne·vad·a (nə-văd'ə, –vä'də), state (1975 pop. 596,000), 110,540 sq mi (286,299 sq km), W United States, admitted as the 36th state of the Union in 1864. Carson City is the capital. Nevada is bounded on the west by Calif., on the north by Oregon and Idaho, on the east by Utah, and on the southeast by Ariz. (with the Colorado River marking most of the border).

Most of the state lies within the Great Basin. The rivers in the southeast belong to the Colorado River system, while those of the extreme north drain into the Snake. Like the Humboldt, most Nevada rivers go nowhere, ending instead in desolate alkali sinks—except where they have been diverted for irrigation and reclamation. About a half million acres (202,345 hectares) of land are being reclaimed by the Humboldt project, the Newlands project, and the Truckee River storage project. The alkali sinks and great arid stretches clothed with sagebrush and creosote bush typify Nevada's landscape. Its mountain chains generally run north and south, segmenting the state. On the Calif. border stand the lofty Sierra Nevada. The days and nights in this dry country are generally clear, and the temperature varies with the season as well as the altitude. The mean elevation is c.5,500 ft (1,676 m). In the north and west the winters reach extreme cold, while in parts of the south the summers approach ovenlike heat.

Besides Reno and Las Vegas, there are many points of interest. Hoover Dam impounds Lake Mead, one of the largest artificial lakes in the world. Lake Mead Recreational Area has facilities for fishing, swimming, and boating. Other attractions include Lake Tahoe, on

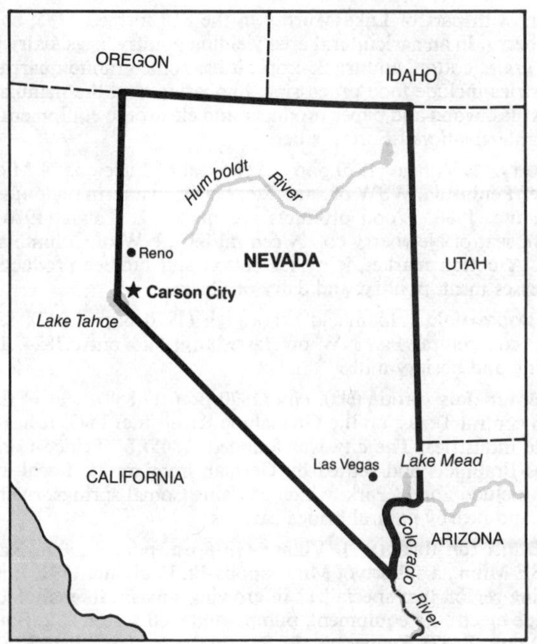

the Nevada-Calif. line, Lehman Caves National Monument, Death Valley National Monument, and restored mining ghost towns like Virginia City.

Nevada is divided into 16 counties and independent Carson City:

NAME	COUNTY SEAT	NAME	COUNTY SEAT
Churchill	Fallon	Lyon	Yerington
Clark	Las Vegas	Mineral	Hawthorne
Douglas	Minden	Nye	Tonopah
Elko	Elko	Pershing	Lovelock
Esmeralda	Goldfield	Storey	Virginia City
Eureka	Eureka	Washoe	Reno
Humboldt	Winnemucca	White Pine	Ely
Lander	Austin	Carson City	
Lincoln	Pioche		

Economy. Cattle and sheep raising, chiefly on the high plateaus, is one of the important industries of the state. Because of the prevailing dryness and the steep slopes, agriculture is not highly developed but is devoted mainly to growing hay and alfalfa as winter food for the cattle; however, wheat and barley are also grown. The fortune of Nevada is in the wealth below the surface of the land—lead, silver, gold, zinc, antimony, arsenic, and tungsten. In recent years the state has been a leading producer of copper, gold, sand and gravel, iron ore, and mercury. Petroleum was discovered in 1954. There is also some manufacturing; principal industries include the manufacture of stone, clay, and glass products, chemicals, food products, lumber, electrical machinery, and fabricated metals. In the 20th cent. resorts have been developed — notably at Reno, "the divorce capital of the world," and at Las Vegas, renowned for its gambling (legalized in 1931) and night life. Gambling taxes are a primary source of state revenue.

History. In the 1770s several Spanish explorers came near the area of present-day Nevada. A half century later fur traders thrusting into the Rocky Mts. for beaver pelts publicized the region. Jedediah S. Smith came across southern Nevada on his way to California in 1827. Later many wagon trains crossed Nevada on the way to California, especially during and after the gold rush of 1849. Travelers going to Calif. over the Old Spanish Trail also crossed southern Nevada, and Las Vegas became a station on the route. The United States acquired the area from Mexico in the Mexican War and included almost all of it in the Mormon-ruled Utah Territory (1850). Trouble between the U.S. government and the Mormons led to an army campaign against the Mormons in 1857, and Brigham Young recalled back to Salt Lake City most of the few Mormons who had settled in the region that is now Nevada.

After gold was found in 1859, a rush from Calif. began and multiplied manyfold as news of the Comstock Lode silver strike spread. Partly to impose order on the lawless, wide-open mining towns, Congress made Nevada into a territory in 1861. It was rushed into statehood in 1864, with Carson City as its capital, so that President Lincoln could get more votes to pass the Thirteenth Amendment. In 1866 Nevada acquired its present-day boundaries when the southern tip was gained and more eastern land was added to a strip acquired in 1862. Communications with the East, which had been spectacularly but briefly maintained by the pony express, were firmly established by the completion of the transcontinental RR in 1869. The state continued to be dependent on its precious ores, and its fate was affected by new strikes such as the "big bonanza" (1873) and the discovery (1900) of silver deposits at Tonopah, of copper at Ely, and of gold at Goldfield (1902). The political leaders of Nevada were vociferous in favor of the free coinage of silver.

In the 20th cent. the Federal government has played an active role in Nevada. The Newlands Irrigation Project (1907) was the first built by the Federal government. The Hoover Dam was completed in 1936. The U.S. Atomic Energy Commission began conducting nuclear tests in Nevada at Frenchman Flat and Yucca Flat in the 1950s. Nevada's constitution was adopted in 1864.

Government. The legislature is composed of 20 senators elected for four-year terms and 40 assemblymen elected for two-year terms. The governor is elected for a four-year term.

Educational Institutions. The most prominent in the state is the Univ. of Nevada, at Reno.

Ne·vad·a. 1. (nə-vā'də) County (1970 pop. 10,111), 616 sq mi (1,595.4 sq km), SW Ark., bounded on the N by the Little Missouri River and drained by Bayou Dorcheat; formed 1871; co. seat Prescott. Its agriculture includes poultry, livestock, dairy products, soybeans, cucumbers, watermelons, cotton, and corn. It has oil wells, iron-ore deposits, and timber. Textiles, clothing, and asphalt products are made. **2.** (nə-văd'ə, -vä'də) County (1970 pop. 26,346), 973 sq mi (2,520.1 sq km), E Calif., in a wooded mountainous area rising to the crest of the Sierra Nevada and drained by the Bear River and forks of the Yuba River; formed 1851; co. seat Nevada City. Lumbering is its primary industry. Stock raising, dairying, and some fruit growing are done in the lower foothills. There are also gold and silver mines and sand and gravel pits. The county is a popular recreational area, with many beautiful lakes, winter resorts in the Donner Pass region, and opportunities for camping, hiking, hunting, and fishing.

Ne·vad·a (nə-vā'də). **1.** City (1970 pop. 4,952), seat of Story co., central Iowa, E of Ames, in a farm area; founded c.1853, inc. 1869. **2.** City (1970 pop. 9,736), seat of Vernon co., SW Mo., E of Fort Scott, Kansas; founded 1855, inc. 1869. It is a shipping center for a farm area that also yields oil, asphalt, and coal.

Ne·vad·a City (nə-văd'ə, -vä'də), city (1970 pop. 2,314), seat of Nevada co., N central Calif., NNE of Sacramento, in a gold-mine, orchard, and resort area of the Sacramento valley; laid out 1849, inc. 1851. Numerous landmarks from the gold-rush days are here.

Ne·vers (nə-vâr'), city (1975 pop. 45,480), capital of Nièvre dept., central France, on the Loire and Nièvre rivers. It is the center of an iron and steel district and has important pottery and china industries. Other manufactures include metal and foundry products, mechanical and electrical equipment, refrigerators, chemicals, textiles, and printed matter. Nevers became the seat of a bishopric in the 6th cent. and was long the capital of the duchy and province of Nivernais. Among the points of interest are the ducal palace (15th-16th cent.), now a courthouse; the Church of St. Étienne (11th cent.), a gem of Romanesque architecture; the cathedral (13th-16th cent.); and the Church of St. Bernadette-du-Banlay (1966).

Ne·vis (nē'vĭs, nĕv'ĭs), island, British West Indies: *see* Saint Kitts-Nevis.

New Al·ba·ny (nōō ôl'bə-nē, nyōō). **1.** City (1978 est. pop. 36,900), seat of Floyd co., S Ind., near the falls of the Ohio River opposite Louisville, Ky.; inc. 1819. It was a shipbuilding center in the 19th cent. Today the city's industries produce plywood, men's suits, machine parts, chemicals, and many other products. **2.** City (1970 pop. 6,426), seat of Union co., N Miss., on the Tallahatchie River NW of Tupelo; settled c.1840. It is a processing center for a cotton, dairy, and timber area and manufactures shirts and furniture.

New Am·ster·dam (ăm'stər-dăm'), Dutch settlement at the mouth of the Hudson River on the S end of Manhattan Island; est. 1624. It was the capital of the colony of New Netherland from 1626 to 1664, when it was captured by the British and renamed New York.

New·ark (nōō'ərk, nyōō'-). **1.** City (1978 est. pop. 29,900), Alameda co., W Calif., on the E side of San Francisco Bay; inc. 1955. **2.** City (1978 est. pop. 28,100), New Castle co., NW Del.; settled before 1700, inc. 1852. It has a huge automobile assembly plant, several research laboratories, and a variety of light manufactures. The only Revolutionary battle on Delaware soil was fought (Sept., 1777) at nearby Cooch's Bridge. The Univ. of Delaware is here. **3.** City (1978 est. pop. 324,000), seat of Essex co., NE N.J., on the Passaic River and Newark Bay; settled 1666, inc. as a city 1836. Only 8 mi (13 km) west of New York City, Newark is a port of entry and a major transportation, industrial, commercial, and manufacturing center. Its leather industry dates from the 17th cent., and its jewelry manufactures and insurance businesses were started in the early 19th cent. Among the city's many other products are beer, cutlery, electronic equipment, pharmaceuticals, fabricated metal items, and paints and varnishes. Newark was settled (1666) by Puritans from Connecticut under the leadership of Robert Treat. The city's industrial growth began after the Revolution, aided by new inventions and the development of transportation facilities. The Morris Canal was opened in 1832, and the railroads arrived in 1834 and 1835. Newark Port opened in 1915, and the city's shipbuilding played an important role in World War I. Newark's landmarks include Trinity Cathedral (1810, with the spire of a church built in 1743), the Sacred Heart Cathedral (begun 1898, completed 1953), the First Presbyterian Church (1791), the Newark Public Library (founded 1888), the Newark Museum (1909), and the county courthouse (1906), designed by Cass Gilbert. Aaron Burr and Stephen Crane were born in Newark. **4.** Village (1978 est. pop. 10,200), Wayne co., W central N.Y., on the Barge Canal, in a farm area. Food is processed, and jewelry, furniture, and cartons are manufactured. **5.** City (1970 pop. 41,836), seat of Licking co., central Ohio, on the Licking River, in a livestock area; inc. 1826. It is a farm trade and processing center, a transportation hub, and an industrial city. An outstanding group of Indian mounds and the Museum of Ohio Indian Art are here.

New Au·gus·ta (ô-gŭs'tə), town (1970 pop. 511), seat of Perry co., SE Miss., ESE of Hattiesburg, in a lumber-milling region.

Ne·way·go (nĭ-wā'gō), county (1970 pop. 27,992), 849 sq mi (2,198.9 sq km), W central Mich., drained by the Muskegon, Pere Marquette, and White rivers; organized 1851; co. seat White Cloud. There is some stock and poultry raising, dairying, and farming (potatoes, corn, grain, and hay). Light manufacturing and sawmilling are done. The county contains hunting and fishing resorts and Manistee National Forest.

New Bed·ford (bĕd'fərd), city (1978 est. pop. 98,200), Bristol co., SE Mass., at the mouth of the Acushnet River on Buzzard's Bay; settled 1640, set off from Dartmouth 1787, inc. as a city 1847. Formerly one of the world's greatest whaling ports, it has become a leading port for the fishing and scalloping industries. During the Revolution the harbor was a haven for American privateers, prompting the British to invade and burn the town in 1778. The whaling industry boomed after the Revolution, reaching a peak in the 1850s. The first cotton-textile mill in the city dates from 1846, but the textile industry declined in the 1920s. Today New Bedford's manufactures include clothing, textiles, electrical machinery, electronic components, rubber products, and tools and dies. The Bourne Whaling Museum, the Old Dartmouth Historical Society, Friends' Academy (1810), and the Swain School of Design are here. The Free Public Library holds a large collection of material on whaling. There is a sizable Portuguese-speaking colony in the city.

New·berg (nōō'bûrg', nyōō'-), city (1970 pop. 6,507), Yamhill co., NW Oregon, in the Willamette Valley SW of Portland; founded by Quakers, inc. 1893. The trade and packing center of a fruit- and nut-growing area, it also has flour, pulp, and lumber mills.

New Ber·lin (bûr'lĭn), city (1978 est. pop. 31,200), Waukesha co., SE Wis., a suburb of Milwaukee; founded 1840, inc. 1959.

New Bern (bûrn), city (1978 est. pop. 18,300), seat of Craven co., E N.C., a port and trading center at the junction of the Neuse and the Trent rivers; inc. 1723. Settled in 1710 by Swiss and German colonists, New Bern was an early colonial capital; in 1774 it was the seat of the first provincial convention. Notable among the old buildings is the beautiful Tryon Palace (1767–70), which was the colonial capitol and governor's mansion; it was badly burned in 1798 and was not reconstructed until the 1950s.

New·ber·ry (nōō'bĕr'ē, -bə-rē, nyōō'-), county (1970 pop. 29,273), 635 sq mi (1,644.7 sq km), N central S.C., bounded on the E by the Broad River, on the S by the Saluda River, and on the N by the Enoree River, with part of Lake Murray in the SE; formed 1785; co. seat Newberry. In an agricultural area yielding poultry, eggs, dairy products, grain, cotton, and truck crops, it has some granite quarries. Its industries include food processing, lumbering, and the manufacture of textiles, wood and paper products, and electronic equipment. Part of Sumter National Forest is here.

Newberry. 1. Village (1970 pop. 2,334), seat of Luce co., N Mich., E Upper Peninsula, WSW of Sault Ste. Marie, in a farm region; settled 1882, inc. 1886. Wood products are made. **2.** Town (1970 pop. 9,218), seat of Newberry co., N central S.C., NW of Columbia; inc. 1832. A cotton market, it makes cotton and lumber products and processes meat, poultry, and dairy products.

New Bloom·field (blōōm'fēld'), borough (1970 pop. 1,032), seat of Perry co., central Pa., NW of Harrisburg; laid out c.1824. It has planing and hosiery mills.

New Braun·fels (broun'fəlz), city (1970 pop. 17,859), seat of Comal co., S central Texas, on the Guadalupe River; inc. 1847. It has large textile industries. The city was founded (1845) by Prince Carl von Solms-Braunfels and settled by German immigrants. Local attractions include Landa Park, which contains Comal springs, river, and lake, and nearby natural bridge caverns.

New Brigh·ton (brīt'n). **1.** Village (1978 est. pop. 22,500), Ramsey co., SE Minn., a suburb of Minneapolis–St. Paul; inc. 1891. It is in a farming region that specializes in growing squash. Its manufactures include electronic equipment, pumps, and well screens. **2.** Borough (1970 pop. 7,637), Beaver co., W Pa., on the Beaver River S of New Castle, in an industrial area; settled c.1801, inc. 1838. Metal and clay products are made.

New Brit·ain (brĭt'n), industrial city (1978 est. pop. 75,900), Hartford co., central Conn.; settled c.1686, inc. 1871. The tin shops and brassworks in the city were established in the 18th cent. Of interest are the city hall (1884) and a park designed by Frederick Law Olmsted in the center of the city.

New Britain, volcanic island (1970 est. pop. 154,000), c.14,600 sq mi (37,815 sq km), SW Pacific, largest island of the Bismarck Archipelago and part of Papua New Guinea. The island is mountainous, with active volcanoes, hot springs, and peaks more than 7,000 ft (2,135 m) high. There are many European plantations on New Britain; the major export is copra, and some copper, gold, iron, and coal are mined. Discovered by the English explorer William Dampier in 1700, New Britain became part of German New Guinea in 1884. In 1920 it was mandated to Australia by the League of Nations and in 1947 was made a UN trust territory under Australian control.

New Bruns·wick (brŭnz'wĭk'), province (1975 pop. 675,000), 27,985 sq mi (72,481 sq km), including 512 sq mi (1,326 sq km) of water surface, E Canada. Fredericton is the capital. One of the Maritime Provs., New Brunswick is bounded on the north by Chaleur Bay and Que. prov., on the east by the Gulf of St. Lawrence, Northumberland Strait, and N.S., on the south by the Bay of Fundy, and on the west by Maine. Its irregular coastline provides excellent facilities for fishing and shipping enterprises. The largest river, the St. John, crosses the province from northwest to southeast, and the Miramichi River flows northeasterly and drains the central lowlands of the province.

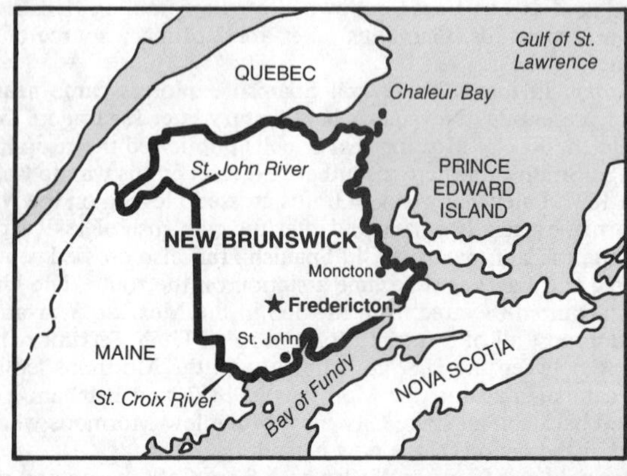

Economy. Dairying thrives on fine pasturage, and the major crops are hay, clover, oats, potatoes, berries, and fruit. A careful conservation program maintains a supply of second-growth hardwoods and softwoods; forests cover about three fourths of the total area, and lumbering is New Brunswick's most important industry. Great quantities of pulpwood and paper are produced, and the mills at Dalhousie on the north coast are among the largest in the world. Manufacturing has greatly expanded since World War II; in addition to wood items and pulp and paper, products include stoves and heating equipment, shoes, and confectionery. Industry is generally run by hydroelectric power, and fuel resources include coal and much untapped water power, which is being developed. In the northeast there are valuable deposits of lead, copper, silver, and pyrite. The major mineral product is zinc; mined in quantity since the late 1950s, it now accounts for half of all of New Brunswick's mineral income. New Brunswick's fisheries are among the most valuable in Canada, with a variety of freshwater and saltwater fish (cod, salmon, herring, and sardines) as well as shellfish (lobsters, oysters, and clams). Trade flows in and out of the ports of St. John and Moncton, facilitated by railroad connections throughout the province, eastward to N.S. and westward to Que. New Brunswick derives considerable revenue from sportsmen and tourists. Its forests are still filled with bear, deer, and moose, and the rivers abound in trout and silver salmon. Easy accessibility from the United States has made Woodstock the gateway to the province. Permanent summer residences are concentrated around Passamaquoddy Bay. Natural attractions include the Grand Falls on the upper reaches of the St. John River as well as the spectacular Bay of Fundy tides—the highest in the world—sometimes surging to over 50 ft (15 m). The tides in turn cause the Reversing Falls of St. John and the "Bore," a twice-daily tidal wave coming up the Petitcodiac River. They have also sculpted the famous Hopewell Rocks, another tourist attraction.

History. The first white man said to have sailed along the New Brunswick coast was a Portuguese navigator, Estevão Gomes (1525), although there is evidence of Basque fishermen at an earlier date. Jacques Cartier landed at Point Escuminac in 1534 and skirted the shores of Miramichi Bay. The first white settlement in New Brunswick was made in 1604 at the mouth of the St. Croix River by Champlain and the sieur de Monts. During this period the present province of Nova Scotia and the coast of New Brunswick were considered one region, called Acadia by the French and Nova Scotia by the British. British control of this region was confirmed by the Peace of Utrecht (1713–14). Doubting the loyalty of the Acadians, the British expelled them in 1755, although many fled into the interior, which was still effectively controlled by the French. (Today about 40% of the people of New Brunswick are Acadians.) Great Britain gained possession of the rest of New Brunswick when it gained all of Canada after the French and Indian Wars. When the population of New Brunswick was increased by many thousands of Loyalists who fled New England after the American Revolution, that area was organized (1784) into a separate colony.

By the middle of the 19th cent. lumbering and farming were extending into the interior, and St. John was a busy port and shipbuilding town. In 1867 under the British North America Act, federation with the other provinces into the dominion of Canada was somewhat reluctantly accepted. At present the population is about equally distributed between farm, nonfarm rural, and urban areas, although urbanization is noticeably increasing.

New Brunswick, city (1978 est. pop. 45,700), seat of Middlesex co., central N.J., on the Raritan River; settled 1681, inc. as a city 1784. Originally developed as a commercial center (especially for collecting and shipping grain), New Brunswick now manufactures pharmaceutical, medical, and surgical supplies. The city is the seat of Rutgers Univ. Joyce Kilmer was born here; his birthplace is now an American Legion post. Nearby Camp Kilmer was an important U.S. army reserve base during World War II and the Korean War.

New·burgh (no͞o'bûrg', nyo͞o'-), city (1978 est. pop. 25,200), Orange co., SE N.Y., on the W bank of the Hudson River, opposite Beacon; settled 1709 by Palatines, inc. 1800. Once an important river port and whaling town, the city now has textile and garment industries. Other important manufactures are handbags, floor tile, and varied metal products. A major thermoelectric plant is in Newburgh. The city has many old houses, and the streets run sharply to the river. At Hasbrouck House (1750; now a museum), Washington made his headquarters from Apr., 1782, to Aug., 1783. It was in Newburgh that the Continental Army was disbanded.

New·burn (no͞o'bərn, nyo͞o'-), urban district (1971 pop. 39,379), Northumberland, NE England, on the Tyne River.

New·bur·y (no͞o'bĕr'ē, -bə-rē, nyo͞o'-), municipal borough (1973 est. pop. 24,850), Berkshire, S central England. In a farming region, it trades in wool, malt, and farm products. Paper, furniture, and metal products are made. In the Middle Ages the town became an important textile-manufacturing center. The 16th cent. cloth hall now contains a museum.

New·bur·y·port (no͞o'bĕr'ē-pôrt', -bə-rē-, nyo͞o'-), city (1978 est. pop. 16,600), Essex co., NE Mass., at the mouth of the Merrimack River; settled 1635, set off from Newbury and inc. 1764. Its silverware and rum industries date from colonial times. An early shipbuilding, whaling, and shipping center, it declined after the Embargo of 1808 and the War of 1812, although ships continued to be built here throughout the clipper-ship era. Its many notable old houses include the Coffin House (c.1651), the Swett-Isley House (c.1671), and the Short House (c.1732). A fire in 1811 destroyed much of the village. William Lloyd Garrison was born in Newburyport.

New Cal·e·do·ni·a (kăl'ə-dō'nē-ə, -dōn'yə), overseas territory of France (1975 est. pop. 136,000), South Pacific, c.700 mi (1,125 km) E of Australia. It comprises the island of New Caledonia, the Isle of Pines, the Loyalty Islands, Walpole Island, and the Huon, Chesterfield, and Belep groups; the total land area is 7,082 sq mi (18,342.4 sq km). The capital is Nouméa on New Caledonia island. The principal industries are the mining and refining of nickel, iron mining, and the production of coffee and copra. Cattle and poultry are raised, but many foodstuffs must still be imported from Australia. The government consists of a governor appointed by France, an elected territorial assembly, and a council. L'Union calédonienne, a largely and indigenous party, controls the assembly. The inhabitants are largely Melanesians, with Polynesians in the outlying islands. Capt. James Cook sighted and named the main island in 1774; the French annexed it in 1853. New Caledonia Island, the largest island of the territory (6,223 sq mi/16,118 sq km) is mountainous and temperate in climate. It is rich in mineral resources, especially nickel, iron, manganese, cobalt, gold, and silver. The island is densely forested in some places, but almost all the kauri pine that was once an important export has been cut down.

New Ca·naan (kā'nən), town (1970 pop. 17,451), Fairfield co., SW Conn.; settled c.1700, inc. 1801. It is mainly a residential town and summer resort. Dairy products are made.

New Car·lisle (kär'līl', kär-līl'), village (1970 pop. 6,112), Clark co., W central Ohio, W of Springfield, in a rich farm area; founded 1810, inc. 1831.

New·cas·tle (no͞o'kăs'əl, -kä'səl, nyo͞o'-), city (1971 pop. 138,696), New South Wales, SE Australia, on the Pacific Ocean. It is a port and the center of the largest coal-mining area in the country. The city has steel mills and shipyards; chemicals, glass, fertilizer, and textiles are also produced. The first permanent settlement on the site was made in 1804.

Newcastle, town (1976 pop. 6,423), E central N.B., Canada, on the Miramichi River. Located in a lumbering region, it has sawmills and a large pulp mill.

Newcastle, city (1970 pop. 3,432), seat of Weston co., NE Wyo., near the S.Dak. border SSW of Rapid City, S.Dak.; founded and inc. 1889 with the coming of the railroad. A ranching, oil, and timber area, it has an oil refinery, a meat-packing plant, and sawmills.

New Cas·tle (kăs'əl, kä'səl), county (1970 pop. 385,856), 438 sq mi (1,134.4 sq km), N Del., bounded on the N by the Pa. border, on the W by the Md. border, on the S by the Smyrna River, and on the E by the Delaware River; formed 1683; co. seat Wilmington. It is crossed by the Chesapeake and Delaware Canal and drained by a number of creeks. Agriculture (dairy products, poultry, corn, wheat, and livestock) and shipping are important to its economy.

New Castle. **1.** City (1970 pop. 4,814), New Castle co., NE Del., on the Delaware River S of Wilmington; inc. 1878. Peter Stuyvesant built (1651) Fort Casimir near here. It was held successively by the Swedes, the Dutch, and the English, who changed the name of the fort and settlement from New Amstel to New Castle. Nylon products, steel castings, aircraft parts, and chemicals are made here. **2.** City (1978 est. pop. 20,000), seat of Henry co., E Ind.; inc. 1839. It is a farm trade center, and its manufactures include auto and truck parts, doors, and metal products. The city has a number of prehistoric Indian mounds. **3.** City (1970 pop. 755), seat of Henry co., N Ky., NW of Frankfort, in a bluegrass agricultural area. **4.** City (1978 est. pop. 34,400), seat of Lawrence co., W Pa., at the junction of the

Shenango and Neshannock rivers, in a fertile farm area; inc. 1825. Coal, limestone, iron ore, and clay deposits found in the region contribute to the city's economy. Manufactures include bronze tools and parts, pottery and china, and rolling mill and steel plant equipment. The Hoyt Institute of Fine Arts is here **5.** Town (1970 pop. 225), seat of Craig co., W Va., in the Alleghenies on Craig Creek NNW of Roanoke.

New·cas·tle-un·der-Lyme (nōō'kăs'əl-ŭn'dər-līm', -kä'səl-, nyōō'-), municipal borough (1973 est. pop. 75,940), Staffordshire, W central England, on the Lyme River. It is partly in the Potteries district. Among the industries are coal mining and brick, tile, and clothing manufacturing. There are ruins of a castle built (12th cent.) by Ranulf, earl of Chester. Chesterton, a section of the borough, has extensive Roman remains.

Newcastle upon Tyne (tīn'), county borough (1976 est. pop. 295,800), administrative center of Northumberland (but located within the metropolitan county of Tyne and Wear since 1974), NE England, on the Tyne River. It is an important shipping and trade center. Its coal-shipping industry began in the 13th cent., but coal exports were exceeded by wool exports until the 16th cent. There are many heavy industries in the area, and the city is one of the chief shipbuilding centers of England. Newcastle stands on the site of the Roman military station Pons Aelii, at Hadrian's Wall. Later the site was occupied by the Angles until the Norman conquest. In 1080 Robert II, duke of Normandy and eldest son of William the Conqueror, built a fortified castle from which the town took its name. The castle was besieged and repaired several times; the oldest parts now standing date from 1177. The town walls, of which traces and towers remain, are attributed to Edward I. There are several notable old buildings including Trinity Almshouse (1492). The Royal Grammar School was founded in the 16th cent.

New City, uninc. village (1970 pop. 30,400), seat of Rockland co., SE N.Y., a suburb of New York City. Situated in a farming region, New City is mostly residential.

New Cum·ber·land (kŭm'bər-lənd). **1.** Borough (1970 pop. 9,803), Cumberland co., S Pa., on the Susquehanna River just S of Harrisburg; laid out c.1810, inc. 1831. **2.** City (1970 pop. 1,865), seat of Hancock co., W.Va., in the Northern Panhandle, on the Ohio River N of Weirton; platted 1839. There are coal mines in the area.

New Del·hi (dĕl'ē), city (1971 pop. 292,857), capital of India, Delhi union territory, N central India, on the right bank of the Jumna River. It was constructed between 1912 and 1929 to replace Calcutta as capital of British India; New Delhi was officially inaugurated in 1931. The city is a transportation hub and trade center with textile mills, printing plants, and light industrial facilities. New Delhi has broad, symmetrically aligned streets that provide vistas of historic monuments. Between the main government buildings a broad boulevard, the Raj Path, leads from a massive war memorial arch (built 1921) through a great court to the resplendent sandstone and marble government house (formerly the viceroy's palace). In the southern section of the city is the prayer ground where Mahatma Gandhi was assassinated (1948).

New Eng·land (ĭng'glənd), name applied to the region comprising six states of the NE United States—Maine, New Hampshire, Vermont, Massachusetts, Rhode Island, and Connecticut. Topographically it is partly cut off from the rest of the nation by the Appalachian Mts. on the west, from which the land slopes gradually toward the Atlantic Ocean. Because of the generally poor soil, agriculture was never a major part of the region's economy. However, excellent harbors and nearby shallow banks teeming with fish made New England a fishing and commercial center. Shipbuilding was important until the end (mid-1800s) of the era of wooden ships. The War of 1812 had an adverse effect on the region's trade, and opposition to the war was so great that New England threatened secession. After the war the growth of manufacturing (especially of cotton textiles) was rapid, and the region became highly industrialized. Agriculture dwindled with migration to the West. After World War II the character of New England industry changed. Traditional industries such as textiles have been superseded by such modern industries as electronics. Tourism, long a source of income for the region, has greatly increased, and people come to New England from all parts of the country for both winter and summer vacations. Stone quarrying, dairying, and potato farming are important.

New·fane (nōō'fān', nyōō'-), village (1970 pop. 183), seat of Windham co., SE Vt., on the West River NW of Brattleboro. It is a resort.

New·found·land (nōō'fən-lənd, -lănd', nyōō'-, nōō-found'lənd, nyōō-), province (1975 pop. 549,000), 156,185 sq mi (404,519 sq km), E Canada. It is sometimes called Newfoundland and Labrador. The province consists of the island of Newfoundland and adjacent islands (1971 pop. 493,938), 43,359 sq mi (112,300 sq km), and the mainland area of Labrador and adjacent islands (1971 pop. 28,166), 112,826 sq mi (292,219 sq km). The capital is St. John's. Newfoundland island lies at the mouth of the Gulf of St. Lawrence and is bounded on the north, east, and south by the Atlantic Ocean and separated on the northwest from Labrador by the Strait of Belle Isle. Labrador, part of the Labrador-Ungava peninsula, forms the northeastern tip of the Canadian mainland. It is bounded on the east by the Atlantic Ocean down to the Strait of Belle Isle and on the south and west by Quebec. Cape Chidley, Labrador's northernmost point, is on the Hudson Strait.

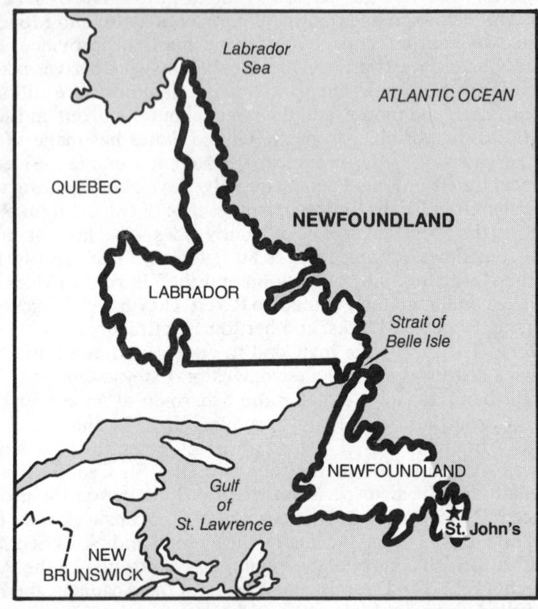

Newfoundland has a rocky, irregular coast, indented with numerous inlets. The major portion of the island is a plateau, with many lakes and marshes and with forests covering less than half the area. Throughout the province the inland wilderness has an abundance of fur-bearing animals, waterfowl, and fish, while caribou graze on the tundra of the northern wasteland. The cod-fishing area of the Grand Banks is probably the best in the world. Cod, lobster, herring, and salmon are caught throughout the coastal waters. The province has a generally cool and moist climate. Nearly half the population lives in St. John's or the surrounding Avalon Peninsula. Corner Brook is the second-largest city. Most of the inhabitants are of English or Irish descent, but in Labrador there are small numbers of Indians and Eskimo.

Economy. Labrador is rich in mineral resources (iron, copper, graphite, nickel, zinc), timber, and waterpower. Exploitation of the tremendous iron reserves in the southwest lake district, begun in the 1950s, and the growth of the logging industry have brought new towns and roads. A giant hydroelectric project has been built at Churchill Falls. Mining is the main industry, and Newfoundland is the leading Canadian province in iron production. The processing of fish and the manufacture of wood products are also important. There are large pulp and paper mills at Grand Falls and Corner Brook, both on Newfoundland. Agriculture in the province is limited by the unfavorable soil and climate, and much of the food supply must be imported.

History. Vikings visited the area c.1000 and briefly established a settlement on Newfoundland. After the two voyages of John Cabot at the end of the 15th cent., fishermen and explorers from several European countries came to the area. Sir Humphrey Gilbert claimed Newfoundland for England in 1583. The first settlers arrived in 1610. France contested England's claims, and Newfoundland changed hands several times. The Treaty of Paris of 1763 definitively awarded Newfoundland and Labrador (where the French had established trading posts) to Great Britain, although France retained fishing rights on the northwest coast of Newfoundland until 1904. Representative government was introduced in 1832 and parliamentary government in 1855. In 1869 the voters of Newfoundland rejected union with Canada. In 1895 iron ore was discovered in the Grand Falls (now Churchill Falls) region of Labrador. Possession of Labrador was disputed between Quebec and Newfoundland until 1927, when the British Privy Council confirmed Newfoundland's title to it. After World War II Newfoundland voted to join Canada, and in 1949 it became Canada's tenth province.

New Glas·gow (glăs'gō, glăz'-), town (1976 pop. 10,672), N N.S., Canada, on East River. It is an industrial town in a coal region. Steel products are manufactured.

New Gra·na·da (grə-nä'də), former Spanish colony, N South America. It included at its greatest extent present Colombia, Ecuador, Panama, and Venezuela.

New Guin·ea (gĭn'ē), island (1970 est. pop. 3,200,000), c.342,000 sq mi (885,780 sq km), SW Pacific, N of Australia. It is the world's second-largest island after Greenland. Politically it is divided into two sections: the Indonesian province of Irian Barat (West Irian or West New Guinea; formerly Netherlands New Guinea) in the west, and the self-governing country of Papua New Guinea in the east. The island is c.1,500 mi (2,415 km) long and c.400 mi (645 km) wide at the center. Largely tropical, it has vast mountain ranges such as the Owen Stanley and the Bismarck mts. The lower courses of the large rivers are generally swampy, with a few grassy plains. The inhabitants of New Guinea are Melanesians, Negritos, and Papuans, some of whom, in the more inaccessible regions, still practice headhunting and cannibalism. The fauna, generally similar to that of Australia, consists largely of marsupials and monotremes, with venomous snakes among the reptiles. The island is known for its many unique species of butterflies and birds of paradise. There are mangrove and sandalwood forests. Agriculture, largely for subsistence, forms the basis of New Guinea's economy. Agricultural products include sweet potatoes, copra, cocoa, coffee, pyrethrum, sisal hemp, rubber, kapok, sago, sugar cane, coconuts, nutmeg, and tobacco. Pearl-shell culture and tortoise fishing are carried on along the coasts. Although some gold, silver, and manganese are mined and oil is extracted in Irian Barat, much of the area remains unexploited.

New Guinea was probably first sighted by the Portuguese explorer Antonio d'Abreu in 1511 and was named for its resemblance to the Guinea coast of West Africa. In 1828 the Dutch formally annexed the western half of the island, and in 1885 the British proclaimed a protectorate over the southeastern coast and the adjacent islands under the name of British New Guinea; in the same year the Germans took possession of the northeast. Australia obtained control of British New Guinea in 1905 and renamed it the Territory of Papua. During World War I Australian forces occupied the German-controlled region in the northeast, which was mandated to Australia by the League of Nations in 1920. Renamed the Territory of New Guinea, this area became a UN trust territory under Australian control after World War II. In 1949 the territories of Papua and New Guinea were merged administratively, and in 1973 they were united into a self-governing country that later gained complete independence from Australia. Netherlands New Guinea was transferred to Indonesian administration in 1963 and became a province in 1969.

New·hall (nōō'hôl', nyōō'-), uninc. town (1970 pop. 9,651), Los Angeles co., S Calif., N of Los Angeles; founded 1876. There are oil and natural-gas wells in the area.

New·ham (nōō'əm, nyōō'-), borough (1971 pop. 235,700) of Greater London, SE England, on the Thames River. Newham was created in 1965 by the merger of the county boroughs of East Ham and West Ham, part of the metropolitan London borough of Woolwich, and part of the municipal borough of Barking. Newham is residential in the northeast. The Royal Docks and associated industries are in the south; chemical factories and railroad yards predominate in the northwest. Few buildings in the borough are more than a century old, because the area's growth stemmed largely from London's 19th cent. industrial expansion. The southwest especially suffered from slum conditions; much of it was destroyed during World War II and was rebuilt in the 1960s.

New Hamp·shire (hămp'shər, hăm'shər, -shîr'), state (1975 pop. 822,000), 9,304 sq mi (24,097 sq km), NE United States, in New England, one of the Thirteen Colonies. It is bounded on the north by the Canadian province of Quebec, on the east by Maine and the Atlantic Ocean, on the south by Mass., and on the west by the Connecticut River and the Vt. border. Concord is the capital.

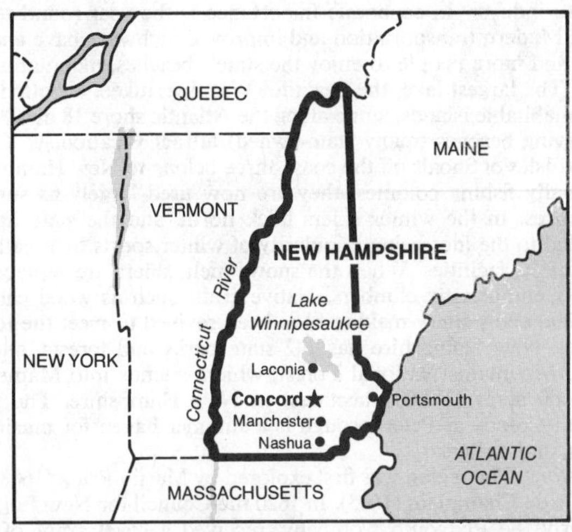

The continental ice sheet once covered the entire state; in receding it scraped the mountains, peneplained the intervening upland areas, and rerouted the water courses into precipitous streams and beautiful lakes. Across the north-central part of the state the residual White Mountains of the Appalachian chain form ranges abruptly broken by notches cut into their rocky walls. Between the Carter-Moriah Range and the Presidential Range in the east the Ellis River drops 80 ft (24.4 m) through Pinkham Notch. West of the Presidential Range the cascading courses of the Ammonoosuc and Saco rivers divide it from the Franconia Mts. at Crawford Notch. To the southwest Franconia Notch overlooks the famous Old Man of the Mountain, beneath which the Pemigewasset tumbles on its way to join the Merrimack. The northernmost gap, Dixville Notch, is surrounded by rocky pinnacles that look down upon a wild, fir-covered country abounding in lakes and streams. South of the mountains the lake and upland area is frequently interrupted by isolated peaks called monadnocks. Practically every part of the state is within sight of, and identifies itself with, some peak. Along the coast the ocean tempers the climate, but inland there are great temperature extremes. Occasional high winds and violent storms roar through the narrow valleys and rebound off the rocky walls.

New Hampshire is divided into 10 counties:

NAME	COUNTY SEAT	NAME	COUNTY SEAT
Belknap	Laconia	Hillsborough	Nashua
Carroll	Ossipee	Merrimack	Concord
Cheshire	Keene	Rockingham	Exeter
Coos	Lancaster	Strafford	Dover
Grafton	Woodsville	Sullivan	Newport

Economy. Intensive agriculture is hampered by the mountainous topography and by extensive areas of unfertile and stony soil, but the upper Connecticut valley (known as Coos country) is pleasantly pastoral. Farmers' main sources of income are dairy products, eggs, cattle, and greenhouse products. Hay, apples, sweet corn, and potatoes are the chief crops. However, since the late 1800s manufacturing has been predominant. Industry is concentrated in the intervales of the rivers, where the abundant water power has been harnessed. Leading products are leather and leather goods (such as shoes and boots), electrical and other machinery, textiles and related products, and paper and paper products. The production of lumber, maple sugar, and pulp and paper are major industries, especially in the north. Printing and publishing are also important enterprises. The state's only port, Portsmouth, is situated on the estuary of the Piscataqua River. Its busy naval base, which has been building ships since 1800, is a commercial center in the state.

Although New Hampshire has long been known as the Granite State, its large deposits of granite—used for building as early as 1623—are no longer extensively quarried. Today sand and gravel, stone, clays, and feldspar are the state's leading minerals. In 1962 the White Mts. were discovered to be an important potential source of thorium, an expensive nuclear fuel. Nevertheless, mineral production remains a minor factor in New Hampshire's economy. Second only to industry in economic importance is the year-round tourist trade. Modern transportation and improved highways have enabled more and more people to enjoy the state's beaches, mountains, and lakes. The largest lake, the beautiful Winnipesaukee, is dotted with 274 inhabitable islands, while along the Atlantic shore 18 mi (29 km) of curving beaches (many state-owned) attract vacationists. Of the rugged Isles of Shoals off the coast, three belong to New Hampshire. Originally fishing colonies, they are now used largely as summer residences. In the winter skiers flock north, and the state has responded to the increasing popularity of winter sports by greatly expanding its facilities. When the snows melt, skiers are replaced by equally enthusiastic climbers. Native crafts such as wood carving, weaving, and pottery making have been revived to meet the tourist market. New Hampshire has 142 state parks and forests, and the White Mountains National Forest, which extends into Maine, has c.724,000 acres (293,000 hectares) in New Hampshire. The MacDowell Colony at Peterborough is a summer haven for musicians, artists, and writers.

History. The region was first explored by Martin Pring (1603) and Samuel de Champlain (1605). In 1620 the Council for New England, formerly the Plymouth Company, received a royal grant of land between lat. 40° N and 48° N. In 1629 Capt. John Mason obtained rights to the area between the Piscataqua and the Merrimack, naming it New Hampshire. The first permanent colony had been established at Dover some time before 1628. Portsmouth was founded by Anglican farmers and fishermen in 1630. Although New Hampshire was proclaimed a royal colony in 1679, Massachusetts continued to press land claims until the two colonies finally agreed on the eastern and southern boundaries (1739–41). Although they were technically independent of each other, the crown habitually appointed a single man to govern both colonies until 1741, when Benning Wentworth was made the first governor of New Hampshire alone. Wentworth and his friends purchased the land rights in 1746, laying claim to lands east of the Hudson River and thereby provoking a protracted controversy with New York. Although a royal order in 1764 established the Connecticut River as the western boundary of New Hampshire, the dispute flared up again in the American Revolution and was only settled when Vermont became a state. The French and Indian Wars had prevented colonization of the inland areas, but once the hostile Indians were thoroughly defeated a land rush began. Lumber camps were set up and sawmills were built along the streams. A textile industry was started by Scottish-Irish settlers. By the time of the Revolution many of the inhabitants had tired of British rule and were eager for independence. In Dec., 1774, a band of patriots overpowered Fort William and Mary (later Fort Constitution) and secured the arms and ammunition for their cause.

New Hampshire was the first colony to declare its independence from Great Britain and to establish its own government (Jan., 1776). After the war an economic depression, marked by severe deflation, a shortage of currency, and unequal distribution of taxation, moved the people to open protest against the state legislature. Forces under Gen. John Sullivan put down the protest without bloodshed (1786). New Hampshire became the ninth and last necessary state to ratify (1788) the new Constitution of the United States. In 1819 New Hampshire abolished state support of the Congregational Church through the Toleration Act. New Hampshire's northern boundary was fixed in 1842, when the Webster-Ashburton Treaty set the international line between Canada and the United States.

During the Civil War, New Hampshire was a strong supporter of the Northern cause and contributed many troops to the Union forces. After the war its economy began to emerge as primarily industrial, and population growth was steady. The production of woolen and cotton goods and the manufacturing of shoes led all other enterprises. The forests were rapidly and ruthlessly exploited, but in 1911 a bill was passed to protect big rivers by creating forest reserves at their headwaters. The Great Depression of the 1930s severely dislocated the state's economy, especially in the one-industry towns. The recent establishment of important new industries such as electronics has successfully counterbalanced the departure to other states of older industries such as textiles.

Government. New Hampshire's present constitution was adopted in 1784; it is the second-oldest in the country. The state's executive branch is headed by a governor and five powerful administrative officers called councillors. The governor is elected for a two-year term and is traditionally limited to two successive terms. Perhaps the most unusual feature of New Hampshire politics is the size of its bicameral legislature (General Court), with 24 senators and from 375 to 400 representatives, all elected for two years.

Educational Institutions. Among the state's more prominent institutions of higher learning are the Univ. of New Hampshire, at Durham, and Dartmouth College, at Hanover.

New Hamp·ton (hămp'tən), city (1970 pop. 3,621), seat of Chickasaw co., NE Iowa, E of Charles City, in a farm area; founded 1854, platted 1867, inc. 1873.

New Han·o·ver (hăn'ō'vər, hăn'ə-), county (1970 pop. 82,996), 185 sq mi (479.2 sq km), SE N.C., in a forested tidewater area on the Atlantic Ocean, bounded on the E by Onslow Bay, on the W by the Cape Fear River, on the N by the Northeast Cape Fear River; formed 1729; co. seat Wilmington. It has dairying, poultry and stock raising, manufacturing, and shipping. There are resorts along the coast.

New Har·mo·ny (här'mə-nē), town (1970 pop. 971), Posey co., SW Ind., on the Wabash River; founded 1814 by the Harmony Society under George Rapp. In 1825 the Harmonists sold their holdings to Robert Owen and moved to Economy, Pa., where their sect survived for another 78 years. Owen established a communistic colony in New Harmony that gained prominence as a cultural and scientific center and attracted many noted scientists, educators, and writers. Dissension arose, and in 1828 the community ceased to exist as a distinct enterprise, although the town remained an intellectual center. The nation's first kindergarten, first free public school, first free library, and first school with equal education for boys and girls were all established here. Many of the old buildings have been restored and are now open to visitors.

New Ha·ven (hā'vən), county (1970 pop. 744,948), 604 sq mi (1,564.4 sq km), S Conn., on Long Island Sound; constituted 1666; former co. seats New Haven and Waterbury. Industry produces a wide variety of goods, especially brass and other metal products.

New Haven. 1. City (1978 est. pop. 121,200), New Haven co., S Conn., a port of entry where the Quinnipiac and other small rivers enter Long Island Sound; inc. 1784. Firearms and ammunition, prestressed concrete, shirts, tools, pyrotechnic devices, rubber products, and door locks are among the many manufactures. The city is the seat of Yale Univ. New Haven was founded in 1637–38 by Puritans led by Theophilus Eaton and John Davenport. It was one of the first planned communities in America and was the chief town of a colony that later included Milford, Guilford, Stamford, Branford, and Southold (on Long Island). Its government was theocratic; religion was a test for citizenship, and life was regulated by strict rules. In 1665 the colony was reluctantly united with Conn.; it was joint capital with Hartford from 1701 to 1875. In the late 18th and early 19th cent. New Haven was a thriving port. Manufacturing grew, and New Haven firearms, hardware, and coaches and carriages became famous products. The world's first commercial telephone exchange was established here in 1879. The city today centers upon a large public green, dating from 1680, on which stand three churches built between 1812 and 1816. Many old buildings have been preserved, and there is a historic district. Landmarks in the city are two traprock cliffs—West Rock, with the Judges' Cave, and East Rock. Since the 1950s New Haven has received national attention for its pioneering urban-renewal projects. The nation's first antipoverty program began here in 1962. Despite these improvements the city suffered a serious racial riot in 1967. Noah Webster and Eli Whitney lived in New Haven and are buried in the city. **2.** City (1970 pop. 5,728), Allen co., NE Ind., on the Maumee River E of Fort Wayne; platted 1839. A farm trade center, it also has some manufacturing.

New Heb·ri·des (hĕb'rə-dēz'), island group (1974 est. pop. 93,000), c.5,700 sq mi (14,765 sq km), South Pacific, E of Australia. The islands are jointly governed by France and Great Britain. New Hebrides is a 450-mi (724-km) chain of 80 islands. The administrative center, Vila, is on Efate. The chief industries are copra production, tuna fishing, manganese mining, and cattle raising. The natives are predominantly Melanesians, with some Polynesians. Because of the many native tongues Pidgin English has become the lingua franca.

The New Hebrides were discovered by the Portuguese navigator Pedro Fernandez de Queiros in 1606. Capt. James Cook made the first systematic exploration of the islands in 1774. English missionaries began arriving in the early 19th cent. With them came the "sandalwooders," who, once the local sources of sandalwood ran out, began kidnapping natives for the sugar and cotton plantations in Queensland, Australia. British attempts to halt the decimation of the native population met success in 1887, when the islands were placed under an Anglo-French naval commission. The commission was replaced by a condominium in 1906. Each power has sovereignty over its own nationals but no territorial sovereignty. Natives owe allegiance to neither Great Britain nor France. There is joint administration on all levels, with resident commissioners (representatives of the British and French high commissioners) stationed at various places in the island group.

New Hyde Park (hīd), village (1978 est. pop. 9,700), Nassau co., SE N.Y., on Long Island; inc. 1927. It is a residential community with some manufacturing.

New I·be·ri·a (ī-bîr'ē-ə), city (1978 est. pop. 34,500), seat of Iberia parish, S La., on Bayou Teche, which is connected to the Intracoastal Waterway by a canal; inc. 1836. It is a processing center for a sugar cane, oil, dairy, vegetable farm, rock salt, and fishing area. The city has carbon plants, salt mines, canneries, shipyards, and factories making microwave ovens and custom-built homes. Acadian refugees settled here beginning c.1765, and French is still spoken by many of the inhabitants. Numerous old houses are in the area; among them are Justine (1822) and Shadows on the Teche (1834), a classic example of Greek revival architecture.

New·ing·ton (no͞o'ĭng-tən, nyo͞o'-), town (1978 est. pop. 30,700), Hartford co., central Conn., a suburb of Hartford; settled 1670, inc. 1871. Although chiefly residential, it has some manufacturing.

New Ire·land (īr'lənd), volcanic island (1970 pop. 50,600), c.3,340 sq mi (8,650 sq km), SW Pacific, in the Bismarck Archipelago, part of Papua New Guinea. New Ireland is largely mountainous, rising to c.4,000 ft (1,220 m). Much of the island is under cultivation, especially the east coast. The island was first sighted in 1616 but until 1797 was thought to be part of New Britain, from which it is separated by a 20-mi (32-km) channel. The island was a German protectorate from 1884 to 1914.

New Jer·sey (jûr'zē), state (1975 pop. 7,335,000), 7,836 sq mi (20,295 sq km), E United States, one of the Middle Atlantic states and one of the Thirteen Colonies. The capital is Trenton. Surrounded by water except along the northern border with N.Y., New Jersey is bounded on the east by the Hudson River, New York Bay, and the Atlantic Ocean and on the south and west by Delaware Bay and the Delaware River.

The northern third of New Jersey lies within the Appalachian Highland region, where ridges running northeast and southwest shelter valleys containing pleasant streams and glacial lakes. Beyond the crest of wooded slopes are long-established farms given over to dairying and field crops. Southeast of the Highlands lie the Triassic lowlands or piedmont plains, extending from the northeastern border to Trenton and encompassing every major city of the state except Camden and Atlantic City. East of Newark and Hackensack acres of tidal marshes have been converted to industrial use. Drainage is provided by the state's major rivers, the Passaic, the Raritan, and the Hackensack. The busy lowlands give way in the south to the coastal plains, which cover more than half the state. The coast itself is highly developed as a resort area. Sandbars make large harbors impractical but provide sheltered waterways.

New Jersey is divided into 21 counties:

NAME	COUNTY SEAT	NAME	COUNTY SEAT
Atlantic	Mays Landing	Middlesex	New Brunswick
Bergen	Hackensack	Monmouth	Freehold
Burlington	Mount Holly	Morris	Morristown
Camden	Camden	Ocean	Toms River
Cape May	Cape May Court House	Passaic	Paterson
Cumberland	Bridgeton	Salem	Salem
Essex	Newark	Somerset	Somerville
Gloucester	Woodbury	Sussex	Newton
Hudson	Jersey City	Union	Elizabeth
Hunterdon	Flemington	Warren	Belvidere
Mercer	Trenton		

Economy. New Jersey is an industrial giant, a major transportation terminus, a long-established playground for summer vacationers, and a year-round commuter area pouring thousands daily into New York City and Philadelphia. The state is noted for its output of chemicals and pharmaceuticals, machinery, processed foods, and a host of other products, including electronic equipment and missile components. Rubber and textiles are also produced, as well as silk goods and synthetics. Although stone, zinc, and sand and gravel are the state's only native mineral resources of consequence, it is a center for copper smelting and oil refining as well as the nation's third-largest producer of titanium concentrate. New Jersey is also a leading state in agricultural income per acre. The scrub pine area of the southern inland region is used for cranberry and blueberry culture. North of the pine belt the soil is extremely fertile and supports a variety of crops, most notably potatoes, corn, hay, peaches, and vegetables. Dairy products, cattle, eggs, and poultry are also important.

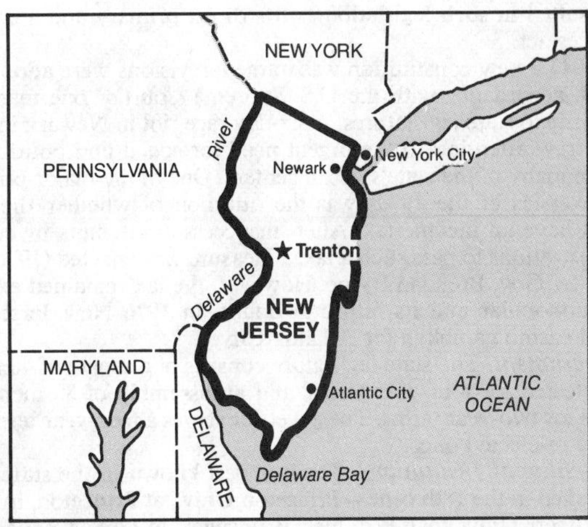

A tremendous transportation system, concentrated in the industrial lowlands, funnels state products and a huge volume of interstate traffic to the seaports of Newark, Hoboken, Jersey City, and Perth Amboy and to the New York area. New Jersey has a massive concentration of railroad trackage and a large network of toll roads and freeways. Traffic to and from New York is served by railway tunnels and by the George Washington Bridge, the Lincoln and Holland vehicular tunnels, and three bridges to Staten Island. Because of this extensive transportation network New Jersey's ocean beaches, inland lakes, forests, and mountain areas have become the basis for a thriving vacation industry.

History. The history of New Jersey goes back to Dutch and Swedish communities established in the 1620s and 1630s. Small Dutch colonies were located on the present sites of Hoboken, Jersey City, and Gloucester City. New Sweden, a group of Scandinavian communities in the Delaware valley, was annexed by the New Netherland colony in 1655. In 1664 Richard Nicolls, acting for the future King James II, seized New Netherland for the English, and New Jersey remained under proprietary rule until it was returned to the crown in 1702. During that period New Jersey's history was marked by land-title disputes, dubious business transactions, and frequent changes of authority. The governor of New York administered the province from 1702 until 1738, when Lewis Morris was appointed governor of New Jersey alone. Until Trenton became the capital in 1790, the legislature met in alternate years at Perth Amboy and Burlington. Under the royal governors the same problems persisted—land titles were in dispute and opposition to the proprietors culminated in riots in the 1740s.

Because of its strategic position New Jersey was of major concern in the American Revolution. The patriot cause was generally accepted, and in June, 1776, the provincial congress declared New Jersey a state. Altogether about 90 engagements were fought in the state, including Washington's victories at Trenton and Princeton in 1776; Washington moved his army across the state four times, wintering twice at Morristown. New Jersey was the third state to ratify (Dec., 1787) the Constitution of the United States.

During the next 50 years, a period of enormous economic expansion, the dominance of the landed aristocracy gave way to industrial growth and to a more democratic state government. Textile mills, potteries, shoe factories, and brickworks were built. Prior to the Civil War an era of reform resulted in the framing of a new state constitution (1844) in which property qualifications for suffrage were abolished. During and after the war, population and industry showed rapid and steady growth. Large economic interests grasped control of political power, giving rise to sporadic but unsustained popular movements for reform. After the 1870s easy incorporation laws and low corporation tax rates attracted new trusts to incorporate through "dummy" offices in the state. A general reform movement sponsored by Woodrow Wilson when he was governor (1910-

12) resulted in such legislation as the direct primary and a corrupt practices act.

In 1947 a new constitution was framed; revisions were adopted in 1966 in accordance with the U.S. Supreme Court's "one man, one vote" rule to state legislatures. A six-day race riot in Newark in July, 1967, drew attention to the urgent need for social and political reform in many of the state's urban centers. One of the major political controversies of the 1970s was the question of whether the state should have an income tax. After unsuccessful attempts by several administrations to pass such a tax, a measure was enacted (1976) and signed by Gov. Brendan Byrne. However, the tax remained exceedingly unpopular and its future in doubt. In 1976 New Jersey approved casino gambling for Atlantic City.

Government. The state legislature consists of a senate of 40 members, elected for four-year terms, and an assembly of 80 members, elected for two-year terms. The governor serves a four-year term and may be re-elected once.

Educational Institutions. The two best known in the state were established in the 18th cent.—Princeton Univ., at Princeton, in 1746; and Rutgers Univ., mainly at New Brunswick, in 1766. Among other educational institutions are Fairleigh Dickinson Univ., with three campuses; Seton Hall Univ., mainly at South Orange; and Stevens Institute of Technology, at Hoboken. The Institute for Advanced Study, at Princeton, is one of the leading research centers of the country.

New Ken·sing·ton (kĕn'zĭng-tən), city (1978 est. pop. 17,300), Westmoreland co., SW Pa., on the Allegheny River, in a coal-mining area; laid out 1891 on the site of Fort Crawford (1778), inc. as a city 1933. Aluminum products have been made here since 1892.

New Kent (kĕnt), county (1970 pop. 5,300), 210 sq mi (543.9 sq km), E Va., in a tidewater region on the peninsula between the Chickahominy and Pamunkey rivers in the N and the York River in the NE; formed 1654; co. seat New Kent. It is in an agricultural area yielding truck crops, sweet and white potatoes, corn, soybeans, livestock, and poultry. There are timberlands and lumber milling here.

New Kent, village (1970 pop. c.50), seat of New Kent co., E Va., E of Richmond.

New·kirk (nōō'kûrk', nyōō'-), city (1970 pop. 2,173), seat of Kay co., N Okla., NNE of Ponca City, in a wheat and oil area; settled c.1893.

New·land (nōō'lənd, nyōō'-), resort town (1970 pop. 524), seat of Avery co., NW N.C., NNW of Morganton, in the Blue Ridge Mts.

New·lands project (nōō'ləndz, nyōō'-), on the Carson and Truckee rivers, W Nev., one of the first projects built by the U.S. Bureau of Reclamation (1903-8). The project irrigates c.71,500 acres (28,960 hectares); grains and truck crops are grown and livestock is raised. Lahontan Dam (completed 1915) produces electricity for the project.

New Lex·ing·ton (lĕk'sĭng-tən), village (1970 pop. 4,921), seat of Perry co., E central Ohio, SSW of Zanesville, in a coal, oil, and clay area; laid out 1817, inc. 1841.

New Lon·don (lŭn'dən), county (1970 pop. 230,654), 667 sq mi (1,727.5 sq km), SE Conn., on Long Island Sound and the R.I. border, bounded on the W by the Connecticut River; constituted 1666; former co. seats New London and Norwich. The county has diversified manufactures (textiles, metal products, hospital supplies, machinery, printing presses, paper products, chemicals, clothing, thermos bottles, bedding, thread, silverware, boats, and leather goods). Its agriculture includes dairy products, poultry, fruit, and truck crops. Resorts line the coast.

New London. 1. City (1978 est. pop. 30,100), New London co., SE Conn., on the Thames River near its mouth on Long Island Sound; laid out 1646 by John Winthrop, inc. 1784. It is a deepwater port of entry, with shipbuilding, textile, and food-processing industries. A privateers' rendezvous during the Revolution, New London survived a partial burning by Benedict Arnold in 1781 and a British blockade during the War of 1812. The city reached the height of its maritime prosperity in the 19th cent., when it flourished as a shipping, shipbuilding, and whaling and sealing port. The last whaler ceased operations in 1909, and the excellent harbor is now used mainly by the U.S. navy as a submarine base and by yachtsmen and students of the U.S. Coast Guard Academy (located in the city). The city also has a whaling museum, an art museum, and many old buildings, including the Hempsted House (1678) and the old town mill (1650; built by John Winthrop). Old Fort Trumbull, built in 1849 on the site of a Revolutionary fort, now houses a U.S. navy underwater sound laboratory. **2.** City (1970 pop. 967), seat of Ralls co., NE Mo., on the Salt River near the Mississippi and S of Hannibal. It is the trade center of an agricultural region. There is a fine courthouse (built 1857-58) here. **3.** City (1970 pop. 5,801), Outagamie and Waupaca cos., central Wis., NW of Appleton; founded 1853, inc. 1877. Wood products are made.

New Ma·drid (mə-drĭd'), county (1970 pop. 23,420), 679 sq mi (1,758.6 sq km), extreme SE Mo., on the Mississippi River and crossed by the Little River and drainage canals; formed c.1788; co. seat New Madrid. Cotton, corn, wheat, and lumber are its major products.

New Madrid, city (1970 pop. 2,719), seat of New Madrid co., SE Mo., on the Mississippi River SW of Cairo, Ill.; laid out 1789 on land granted by the Spanish, inc. 1868. It is a cotton center and river port in a farm region. In the Civil War Federal troops captured New Madrid from the Confederates before taking (1862) Island No. 10 in the Mississippi.

New·mar·ket (nōō'mär'kĭt, nyōō'-), urban district (1973 est. pop. 13,370), Suffolk, E England. It has been a racing center since early in the 17th cent. One of the courses on Newmarket Heath is crossed by an ancient earthwork known as the Devil's Dyke.

New Mar·tins·ville (mär'tnz-vĭl'), city (1970 pop. 6,528), seat of Wetzel co., NW W.Va., on the Ohio River SSW of Moundsville; platted 1838. Chemicals and glassware are produced.

New Mex·i·co (mĕk'sĭ-kō'), state (1975 pop. 1,145,000), 121,666 sq mi (315,115 sq km), SW United States, admitted to the Union in 1912 as the 47th state. The capital is Santa Fe. The state is bounded on the north by Colo., on the east by Okla. and Texas, on the south by Texas and Mexico, and on the west by Ariz.

New Mexico is roughly bisected by the Rio Grande and has an approximate mean altitude of 5,700 ft (1,738.5 m). The topography of the state is marked by broken mesas, wide deserts, heavily forested mountain wildernesses, and high, bare peaks. The mountain ranges, part of the Rocky Mts., are in broken groups, running north to south and flanking the Rio Grande. In the southwest is the Gila Wilderness. Broad, semiarid plains, particularly prominent in southern New Mexico, are covered with cactus, yucca, creosote bush, sagebrush, and desert grasses.

Millions of acres of the wild and beautiful country of New Mexico are under Federal control as national forests and monuments and, together with the attractive climate of the state, make tourism a chief source of income. Best known of the state's attractions are the Carlsbad Caverns National Park and the Aztec Ruins, White Sands, Bandelier, Capulin Mountain, Chaco Canyon, El Morro, Fort Union, Gila Cliff Dwellings, and Gran Quivira national monuments.

New Mexico is divided into 32 counties:

NAME	COUNTY SEAT	NAME	COUNTY SEAT
Bernalillo	Albuquerque	McKinley	Gallup
Catron	Reserve	Mora	Mora
Chaves	Roswell	Otero	Alamogordo
Colfax	Raton	Quay	Tucumcari
Curry	Clovis	Rio Arriba	Tierra Amarilla
De Baca	Fort Sumner	Roosevelt	Portales
Dona Ana	Las Cruces	Sandoval	Bernalillo
Eddy	Carlsbad	San Juan	Aztec
Grant	Silver City	San Miguel	Las Vegas
Guadalupe	Santa Rosa	Santa Fe	Santa Fe
Harding	Mosquero	Sierra	Truth or Consequences
Hidalgo	Lordsburg	Socorro	Socorro
Lea	Lovington	Taos	Taos
Lincoln	Carrizozo	Torrance	Estancia
Los Alamos	Los Alamos	Union	Clayton
Luna	Deming	Valencia	Los Lunas

Economy. Because rainfall is scanty and irrigation opportunities are few, most of the state's farmland is given over to grazing. There are many large ranches, and cattle and sheep graze year-round on the open range. The two notable rivers besides the Rio Grande—the Pecos and the San Juan—are used for some irrigation; however, the Rio Grande, harnessed by the Elephant Butte Dam, remains the major irrigation source for the area of most extensive farming. Cotton lint, hay, wheat, and sorghum grains are the major crops. Dairy products are also very important.

The growth of military establishments and atomic-energy centers has greatly contributed to the economic advance of New Mexico in recent years. In the early 1970s about one quarter of the personal

income in the state came from government payrolls. Manufacturing, centered especially around Albuquerque, includes food and mineral processing and the production of chemicals, electrical equipment, and ordnance.

Much of the state's income is derived from its considerable mineral wealth. In 1972 the state ranked first nationally in the production of uranium ore, manganese ore, potash, salt, and perlite; third in copper ore; and fourth in natural gas, beryllium, and tin concentrates. Petroleum and coal are also found in large quantities. Silver and turquoise have been used in making Indian jewelry since long before the coming of the white man. About one fourth of the land is forested. Pinewood is the chief commercial wood.

History. Use of the land and minerals goes back to the prehistoric Indian cultures in the Southwest that long preceded the flourishing sedentary civilization of the Pueblo Indians that the Spanish found along the Rio Grande and its tributaries. Word of the pueblos reached the Spanish through Cabeza de Vaca, who may have wandered across southern New Mexico between 1528 and 1536; they were enthusiastically identified by Fray Marcos de Niza as the fabulously rich Seven Cities of Cibola. A full-scale expedition (1540–42) to find the cities was dispatched from New Spain, under the leadership of Francisco Vásquez de Coronado. The treatment of the Indians by Coronado and his men led to the long-standing hostility between the Indians and the Spanish. Reports of precious minerals and grazing lands led to the founding of the first regular colony at San Juan by Juan de Oñate in 1598. In 1609 Pedro de Peralta was made governor of the "Kingdom and Provinces of New Mexico," and a year later he founded his capital at Santa Fe. The little colony did not prosper greatly, although some of the missions flourished and haciendas were founded. Subjecting the Indians to forced labor caused further trouble. The fierce Apache Indians rose in 1676, and in 1680 came the great Pueblo revolt led by Popé. The Indians fell upon the Spanish and wiped out their settlements, and the survivors were driven entirely out of New Mexico. The Spanish did not return until the stern campaign of Diego de Vargas Zapata re-established their control in 1692.

In the 18th cent., despite sporadic Indian warfare, the development of ranching and of some farming and mining was more thorough, laying the foundations for the Spanish culture in New Mexico that still persists. When Mexico achieved its independence from Spain in 1821, New Mexico became a province of Mexico, and trade was opened with the United States. By the following year the Santa Fe Trail was being traveled by the wagon trains of American traders. The Mexican War marked the coming of the Anglo-American culture to New Mexico. The Treaty of Guadalupe Hidalgo (1848) ceded New Mexico to the United States. The territory, which included Arizona and other territories, was enlarged by the Gadsden Purchase (1853). The Compromise of 1850 organized New Mexico as a territory without restriction on slavery. Conflict with the Apache and Navaho Indians continued sporadically until the Apache chief Geronimo surrendered in 1886.

The coming of the Santa Fe RR in 1879 encouraged the great cattle boom of the 1880s. There were typical cow towns, feuds among cattlemen as well as between cattlemen and the authorities, and the activities of such outlaws as Billy the Kid. The cattlemen were unable to keep out the sheepherders and were overwhelmed by the homesteaders and squatters. Land claims gave rise to bitter quarrels among the homesteaders, the ranchers, and the old Spanish families, who made claims under the original grants. Statehood was granted in 1912. Pancho Villa raided Columbus, N.Mex., in March, 1916. In 1943 the U.S. government built Los Alamos as a center for atomic research. The first atomic bomb was exploded at the White Sands Proving Grounds in July, 1945.

About one third of the population today is of mixed Spanish descent (including many recent immigrants from Mexico). In many isolated communities Spanish is still the dominant tongue. The Apache, Navaho, and Ute Indians live on Federal reservations within the state—the Navaho reservation, with over 16 million acres (6.5 million hectares), is the largest in the country—and the Pueblo Indians, settled, agricultural people, live in pueblos scattered throughout the state.

Government. New Mexico is governed under the constitution of 1912. The legislature has a senate of 42 members elected for four-year terms and a house of representatives with 70 members elected for two-year terms. The governor is elected for four years.

Educational Institutions. The most prominent in the state is the Univ. of New Mexico, at Albuquerque.

New Mil·ford (mĭl'fərd). **1.** Town (1970 pop. 14,601), Litchfield co., W Conn., on the Housatonic River; inc. 1712. It is situated in a dairy-cattle and poultry region. Its manufactures include concentrated foods, paper products, brass and copper, electronic equipment, and precision instruments. The present town hall is on the homesite of Roger Sherman, a signer of the Declaration of Independence. **2.** Borough (1970 pop. 19,149), Bergen co., NE N.J., on the Hackensack River; inc. 1922. It is primarily residential. New Milford was settled in 1695 by French Huguenots. One of the original houses still stands. A French Huguenot cemetery is also in the borough. In 1776 George Washington's forces crossed the Hackensack River here during their retreat from Fort Lee to Trenton, N.J.

New·nan (no͞o'nən, nyo͞o'-), city (1970 pop. 11,205), seat of Coweta co., W Ga., in a rich farm and livestock area; inc. 1828.

New Neth·er·land (nĕ*th*'ər-lənd), territory included in a commercial grant by the government of Holland to the Dutch West India Company in 1621. Colonists were settled along the Hudson River region; in 1624 the first permanent settlement was established at Fort Orange (now Albany, N.Y.). The principal settlement in the tract after 1625 was New Amsterdam (later New York City), at the southern end of Manhattan Island, which was purchased from the Indians in 1626. Colonization proceeded slowly, hampered by trouble with the Indians, poor administration, and rivalry with New England settlers. In 1664 the territory was taken by the English, who divided it into the colonies of New York and New Jersey.

New Or·le·ans (ôr'lē-ənz, ôr'lənz, ôr-lēnz'), city (1970 pop. 593,471), seat of Orleans parish, SE La., between the Mississippi River and Lake Pontchartrain, 107 mi (172.2 km) by water from the river mouth; founded 1718 by the sieur de Bienville, inc. 1805. It was built within a great bend of the Mississippi (and is therefore called the Crescent City) on subtropical lowlands, now protected from flooding by levees.

New Orleans has long been one of the busiest international ports in the country. Coastwise traffic is heavy (the city is at the junction of the Intracoastal Waterway with the Mississippi River), and New Orleans is a major rail, highway, air, and river focus. The bountiful natural resources in the area have made the city one of the leading industrial centers in the South. Food processing is a major enterprise. The city has huge oil and chemical industries, great shipbuilding and repair yards, and plants manufacturing a wide variety of products.

Most of the larger industries have been developed recently, but soon after the sieur de Bienville had the city platted in 1718, it took prominence as a port, and in 1722 it became the capital of the French colony. The transfer of Louisiana to Spain by the secret Treaty of Fontainebleau (1762) was confirmed by the Treaty of Paris (1763). New Orleans—deeply involved in the struggle for control of the Mississippi—was returned to French hands only briefly before passing to the United States with the Louisiana Purchase (1803). Nevertheless, the tone of the city's life was dominated by Creole culture until late in the 19th cent., and the French influence is still seen even today. The westward movement in the United States propelled the queen city of the Mississippi to almost fabulous heights as a port and market for cotton and slaves. New Orleans then was stamped with its lasting reputation for glamour, elegance, and wickedness. Jazz had its origin in the late 19th cent. among the black musicians of New Orleans. The golden era ended in the Civil War when the city fell (1862) to Adm. David G. Farragut and suffered under the occupation by Union troops. New Orleans recovered from Reconstruction and passed through the end of the river-steamboat era to emerge as a modern city.

Its past, however, continues to attract visitors. The picturesque French quarter (Vieux Carré) of the old city, north of broad Canal St., is a major tourist attraction. In the heart of the quarter is Jackson Square (the former Place d'Armes); fronting upon the square are the Cabildo (1795; formerly the government building and now housing part of the state museum); St. Louis Cathedral (1794); and other 18th and 19th cent. structures. The annual Mardi Gras is perhaps the best-known festival in the United States. Adding to the color of the city are the many parks, museums (including a jazz museum and the Isaac Delgado Museum of Art), and gardens. New Orleans is also an educational center, the seat of Dillard Univ., Loyola Univ., Tulane Univ., and other schools.

New Paltz (pôlts), resort village (1970 pop. 6,058), Ulster co., SE N.Y., SSW of Kingston, in a farm area; settled 1677 by Huguenots, inc. 1887. Several Huguenot houses remain.

New Phil·a·del·phi·a (fĭl'ə-dĕl'fē-ə), city (1978 est. pop. 15,500), seat of Tuscarawas co., E Ohio, on the Tuscarawas River, in a coal and clay area; founded 1804, inc. 1833. Foundry products, machinery, and pottery are made. Nearby is the Schoenbrunn Village State Memorial, a reconstruction of the first settlement in Ohio.

New Ply·mouth (plĭm'əth), city (1976 pop. 37,711), W central North Island, New Zealand, on the Tasman Sea. It is a port and has iron and copper industries. Nearby is an oil field.

New·port (no͞o'pôrt', nyo͞o'-), municipal borough (1973 est. pop. 22,430), administrative center of the Isle of Wight, S England. It is also a port and the commercial center of the island, with agricultural markets and light industry (plastics, soft drinks, and woodwork). In the 17th cent. King Charles I was imprisoned in nearby Carisbrooke Castle. The town grammar school dates from the early 17th cent. There are remains of a Roman villa in Newport.

Newport, county (1970 pop. 94,228), 115 sq mi (297.9 sq km), SE R.I., on the Mass. border and the Atlantic Ocean, and including Block, Conanicut, Prudence, and Rhode (or Aquidneck) islands in Narragansett Bay; formed 1703 as Rhode Island co., renamed 1729; co. seat Newport. It is in a resort and agricultural area that also has fisheries. Its industries include food processing (especially seafood), boatbuilding, and the manufacture of metal products and electrical equipment.

Newport. 1. City (1970 pop. 7,725), seat of Jackson co., NE Ark., on the White River; settled c.1873, inc. 1875. It is a rail, trade, and processing center for a livestock area yielding cotton and rice. **2.** Town (1970 pop. 708), seat of Vermillion co., W Ind., on the Little Vermilion River near its mouth on the Wabash River and N of Terre Haute. Clay products are made. **3.** City (1978 est. pop. 21,100), a seat of Campbell co., N Ky., on the Ohio River opposite Cincinnati, Ohio, and on the E bank of the Licking River opposite Covington; laid out 1791, inc. as a city 1835. It has a large steel-rolling mill, several clothing factories, a brewery, and a lumber mill. **4.** Town (1970 pop. 5,899), seat of Sullivan co., SW N.H., on the Sugar River E of Claremont; inc. 1761. The Congregational church (1822) is a fine example of colonial architecture. Shoes and woolen goods are made. **5.** City (1970 pop. 5,188), seat of Lincoln co., W Oregon, a port of entry on Yaquina Bay SW of Salem; settled c.1855, inc. 1882. A lumber-shipping center, it is also a resort, with commercial and sports fishing. **6.** City (1978 est. pop. 28,100), seat of Newport co., SE R.I., on Aquidneck (also called Rhode) Island; settled 1639, inc. 1784. The economy of this historic city, a port of entry, revolves chiefly around its many naval installations. Also important are the summer tourist industry, educational facilities, fishing, and agriculture. William Coddington and John Clarke founded Newport in 1639. Newport and Portsmouth united in 1640 and entered a permanent federation with Providence and Warwick in 1654. Shipbuilding, dating from 1646, and foreign commerce, especially trade in black slaves, pineapples, rum, and molasses, brought prosperity. The town early harbored refugees of various groups—Friends and Jews first arrived in the 1650s, and the Seventh-Day Baptists organized a church here in 1671. Jewish merchants contributed greatly to Newport's pre-Revolutionary prosperity, and it became the leading town of the colony. In the American Revolution the British occupied the town (1776–79); many buildings were destroyed, most of the citizens moved away, and Newport never fully regained its former economic prestige. It was replaced in importance by Providence, with which it was joint state capital until 1900. In the 19th cent. Newport developed as a fashionable resort of the very rich, and many palatial mansions were built. Outstanding tourist attractions are The Breakers, the former summer house of Cornelius Vanderbilt, Belcourt Castle, The Elms, Marble House, and Château-sur-Mer. Cliff Walk and Ocean Drive are known for their spectacular views of the ocean and the coastline. Of historic interest are the Wanton-Lyman-Hazard House (c.1675; scene of a Stamp Act riot in 1765); Trinity Church (1726); the beautiful old colony house or statehouse (1739); Touro Synagogue (1763), oldest in the country and since 1946 a national historic site; the Redwood Library and Athenaeum (1747); and the brick market house or city hall (1762). Newport is host to yacht races (including the America's Cup) and tennis tournaments (tennis was popularized here, and the National Tennis Hall of Fame is in the Newport casino). Notable jazz and folk music festivals were held here in the 1950s and 1960s. The U.S. navy base in Newport was closed in early 1974. Newport was hard hit by hurricanes in 1938 and 1954; its historic waterfront is undergoing major redevelopment. **7.** Town (1970 pop. 7,328), seat of Cocke co., E Tenn., on the Pigeon River E of Knoxville, in a farm and timber area; inc. 1832. It has a cannery and manufactures automotive parts and furniture. **8.** City (1970 pop. 4,664), seat of Orleans co., N Vt., on Lake Memphremagog; settled 1793, chartered 1803, inc. 1917. A port of entry, it is a resort and trade center. The city has long been important as a gateway between Canada and the United States. **9.** Town (1970 pop. 1,418), seat of Pend Oreille co., NE Wash., on the Pend Oreille River at the Idaho border; founded 1890, inc. 1903. Lumber, potatoes, and hay are processed.

Newport, county borough (1976 est. pop. 134,100), Gwent, SE Wales, on the Usk River. Newport has large steel works. Coal is exported, and iron ore is imported. Aircraft are made, and tin plate, iron, steel, aluminum, and other metal goods are manufactured. In 1839 Newport was the scene of Chartist riots.

Newport Beach, residential and resort city (1978 est. pop. 65,000), Orange co., S Calif., on Newport Bay and the Pacific Ocean; inc. 1906. It is a popular seaside resort and yachting center. Manufactures include electronic equipment, plastics, and fiberglass boats.

Newport News (no͞oz, nyo͞oz), independent city (1978 est. pop. 141,400), SE Va., at the mouth of the James River, off Hampton Roads, near Norfolk; inc. 1896. It is a major port for transatlantic and intracoastal shipping. Newport News is also one of the world's largest shipbuilding and repair centers. Its manufactures include metal products, building materials, and processed seafood. Newport News was settled by Irish colonists c.1620 but did not grow appreciably until 1880, when it became the terminus of the Chesapeake and Ohio RR; the shipbuilding industry began in 1886. During the Civil War the U.S. army captured Newport News and established a fortified base and prison camp. In 1862 the famous battle between the ironclad ships *Monitor* and *Merrimack* took place off Newport News. The city's points of interest include the Mariners Museum, the War Memorial Museum of Virginia, the Peninsula Junior Nature Museum and Planetarium, and the Victory Arch.

New Port Rich·ey (rĭch'ē), resort city (1973 est. pop. 7,137), Pasco co., W Fla., near the Gulf coast NW of Tampa, in a cattle and citrus-fruit area. Anclote National Wildlife Refuge is nearby.

New Prov·i·dence (prŏv'ĭ-dəns), borough (1970 pop. 13,796), Union co., NE N.J.; settled c.1720, set off and inc. 1899. It is largely residential but has some light industry, research laboratories, and insurance-company offices. Roses are grown commercially.

New River, c.320 mi (515 km) long, rising in the Blue Ridge, NW N.C., and flowing NE through SW Va., then NW into W.Va., where it joins with the Gauley River to form the Kanawha River. Bluestone Dam (completed 1952), near Hinton, W.Va., provides flood control and power.

New Roads (rōdz), town (1970 pop. 3,945), seat of Pointe Coupee parish, S central La., near the Mississippi River NW of Baton Rouge, on the shore of a lake. It is in a cotton and sugar-cane area.

New Ro·chelle (rō-shĕl', rə-), city (1978 est. pop. 70,500), Westchester co., SE N.Y., on Long Island Sound; settled by Huguenots 1688, inc. as a village 1858, as a city 1899. Although mainly a residential suburb of New York City, it has some light industry. The house where Thomas Paine lived has been preserved.

New Rock·ford (rŏk′fərd), city (1970 pop. 1,969), seat of Eddy co., E central N.Dak., on the James River NNW of Jamestown; inc. 1912. It is in a dairy, livestock, and grain region.

New Rom·ney (rŏm′nē), municipal borough (1971 pop. 3,414), Kent, SE England, in Romney Marsh. Until the sea receded, it lay on the coast and was one of the Cinque Ports; many documents concerning the Cinque Ports are kept in the town guildhall. A famous sheep fair is held in New Romney in Aug. Of several ancient churches, only the partly Norman Church of St. Nicholas remains.

New·ry (nŏŏr′ē, nyŏŏr′ē), urban district (1971 pop. 20,279), Co. Down, SE Northern Ireland, on the Clanrye River and the Newry Canal. Newry is a seaport with linen mills, tobacco and food processing, and varied manufactures. In the 12th cent. Maurice McLoughlin, king of Ireland, founded an abbey on the site; the abbey became in 1543 a collegiate church of secular priests and was later dissolved. The town's castle was taken by Edward Bruce in 1315. Newry contains St. Patrick's parish church (1578), the first Protestant church built in Ireland.

New Sa·lem (sā′ləm), former town, Sangamon co., central Ill., NW of Springfield on the Sangamon River. Lincoln's home was here from 1831 to 1837. The site is now a state park; the buildings that were here in Lincoln's day have been restored so that the town appears as it did then. Settled in 1828, the town declined after 1839 as nearby Petersburg grew. Among the old places restored are the Rutledge Tavern, Denton Offut's store, and the Lincoln-Berry Store.

New Sar·um (săr′əm): *see* Salisbury, England.

New Si·be·ri·an Islands (sī-bîr′ē-ən), archipelago, c.10,900 sq mi (28,230 sq km), N Siberian USSR, in the Arctic Ocean between the Laptev and East Siberian seas, part of the Yakut Autonomous Republic. The archipelago is separated into two groups by the Sannikov Strait. The northern group is called the New Siberian or Anjou islands (c.8,200 sq mi/21,240 sq km); the southern group consists of the Lyakhov Islands (c.2,700 sq mi/6,990 sq km). The islands are almost always covered by snow and ice and have a very scant tundra. They were discovered (1773) by Ivan Lyakhov, a Russian merchant. Mammoth fossils have been found in the islands. The islands were neglected until 1927, when meteorological stations were set up.

New Smyr·na Beach (smûr′nə), city (1970 pop. 10,580), Volusia co., NE Fla., on Indian River (a lagoon) and Ponce de Leon Inlet of the Atlantic Ocean; inc. 1903. It is a center for citrus-fruit packing and has commercial fishing and seafood-processing industries and varied light manufacturing. It is also a tourist city, with 8 mi (12.8 km) of white sand beaches. A Spanish Franciscan mission was established here in 1696 on the site of an Indian village. Colonists arrived in 1767, but the settlement did not prosper until the advent of the railroad in the mid-19th cent. In the area are a huge Indian mound made of shells and artifacts, and the ruins of a Spanish fort (c.1565).

New South Wales (wālz), state (1976 pop. 4,776,258), 309,443 sq mi (801,457 sq km), SE Australia. It is bounded on the east by the Pacific Ocean. Sydney is the capital. There are four main geographic regions: the coastal lowlands; the eastern highlands, culminating in Mt. Kosciusko; the western slopes; and the western plains, which cover about two thirds of the state. The Sydney-Newcastle-Wollongong area is the greatest industrial region in the commonwealth, with steel the principal product. Agriculture is also important: wheat, wool, and meat are produced, and there is considerable dairy farming. Tropical fruits and sugar cane are grown in the northeast. The state's rich mineral resources include coal, gold, iron, copper, silver, lead, and zinc.

The area was first visited in 1770 by Capt. James Cook, who proclaimed British sovereignty over the east coast of Australia. Sydney, the first Australian settlement, was founded in 1788 as a prison farm. During the 1820s and 1830s the character of New South Wales changed as the wool industry grew and the importation of convicts ceased. In the early 19th cent. the colony included Tasmania, South Australia, Victoria, Queensland, the Northern Territory, and New Zealand. These territories were separated and made colonies in their own right between 1825 and 1863. In 1901 New South Wales was federated as a state of the Commonwealth of Australia. The Australian Capital Territory (site of Canberra, the federal capital), an enclave in New South Wales, was ceded to the commonwealth in 1911.

New Swe·den (swēd′n), Swedish colony (1638–55), on the Delaware River, including parts of what are now Pa., N.J., and Del. The New Sweden Company was organized in Sweden in 1633. Two ships, commanded by Peter Minuit, reached the Delaware River in Mar., 1638. Minuit immediately bought land from the Indians and founded Fort Christina, where Wilmington, Del., now stands. Peter Stuyvesant, with a Dutch force larger than the population of New Sweden, took the colony in 1655.

New·ton (nŏŏt′n, nyŏŏt′n). **1.** County (1970 pop. 5,844), 822 sq mi (2,129 sq km), NW Ark., in the Ozarks, drained by the Buffalo River and its tributaries; formed 1842; co. seat Jasper. Its agriculture includes livestock, poultry, fruit, truck crops, and dairy products. It has lead and zinc mining and some lumber milling. Part of Ozark National Forest is in the south. **2.** County (1970 pop. 26,282), 273 sq mi (707.1 sq km), N central Ga., drained by the Alcovy and Yellow rivers; formed 1821; co. seat Covington. It is in a piedmont agricultural area (cotton, corn, truck crops, fruit, and livestock) and has a lumbering industry. Its manufactures include textiles and clothing, chemicals, plastics, and wire products. **3.** County (1970 pop. 11,606), 413 sq mi (1,069.7 sq km), NW Ind., bounded on the W by the Ill. border, on the N by the Kankakee River, and drained by Iroquois River; formed 1835; co. seat Kentland. In a rich agricultural area yielding poultry, grain, livestock, and dairy products, it has limestone quarrying and ships grain and seed. There is some manufacturing (feed, cheese, metal products, and electrical equipment). **4.** County (1970 pop. 18,983), 580 sq mi (1,502.2 sq km), E central Miss., drained by Chunky and Tuscolameta creeks; formed 1836; co. seat Decatur. It is in a corn, cotton, and dairy farming area, with timber sawmilling, cheese processing, and clothing and machinery manufacturing. **5.** County (1970 pop. 32,981), 629 sq mi (1,629.1 sq km), SW Mo., in the Ozarks; formed 1854; co. seat Neosho. Agricultural products include berries, hay, fruit, and truck crops. Zinc, lead, and tripoli are mined. There are stands of oak timber here. **6.** County (1970 pop. 11,657), 949 sq mi (2,437.2 sq km), E Texas, bounded on the E by the Sabine River and drained by its tributaries; formed 1846; co. seat Newton. This pine-forest region, in which lumbering is the main industry, also has agriculture (cotton, corn, hay, fruit, truck crops, and livestock) and some oil and gas deposits.

Newton. 1. City (1970 pop. 624), seat of Baker co., SW Ga., SSW of Albany on the Flint River. **2.** City (1970 pop. 3,024), seat of Jasper co., SE Ill., on the Embarrass River SSE of Mattoon, in a farm area; settled 1828, inc. 1831. **3.** City (1978 est. pop. 15,500), seat of Jasper co., central Iowa; inc. 1857. It is an industrial city where washing machines are produced. **4.** City (1978 est. pop. 16,300), seat of Harvey co., S central Kansas, in a wheat area; inc. 1872. It is an important railroad division point and has a large mobile-home industry. Farm equipment is also made. The Chisholm Trail passed through the site. In the early 1870s Mennonites from Russia brought seed for what became the first hard winter wheat in Kansas. The city still has a large Mennonite population, and there is a monument to their ancestors. **5.** City (1978 est. pop. 87,200), Middlesex co., E Mass., a suburb of Boston on the Charles River; settled before 1640, inc. as a city 1873. It comprises a large number of beautiful residential villages and several industrial villages. The city is the seat of Newton College, Mount Alvernia College, and other schools. Horace Mann, Nathaniel Hawthorne, Mary Baker Eddy, and Samuel Francis Smith lived in Newton. **6.** Town (1970 pop. 7,297), seat of Sussex co., NW N.J., WNW of Morristown; settled c.1760, inc. 1864. A dairy center, it also makes photographic materials and textile products. Nearby lakes offer recreational facilities. **7.** Town (1970 pop. 7,857), seat of Catawba co., W central N.C., NW of Charlotte; settled c.1748. Hosiery, furniture, and gloves are manufactured. **8.** Town (1970 pop. 1,529), seat of Newton co., E Texas, near the Sabine River NNE of Beaumont, in a farm, timber, and livestock area.

Newton Falls, city (1970 pop. 5,378), Trumbull co., NE Ohio, on the Mahoning River NW of Youngstown. Automobile parts are made.

New·town (nŏŏ′toun′, nyŏŏ′-), town (1970 pop. 16,942), Fairfield co., SW Conn., on the Housatonic River; inc. 1711. Fabric rubber hose and plastic products are made. There are dairy and fruit farms in the area.

New·town·ards (nŏŏt′n-ärdz′, nyŏŏt′-), municipal borough (1971 pop. 15,387), Co. Down, E Northern Ireland, near the head of Strangford Lough. There are textile and other industries. A Dominican monastery (now in ruins) was founded here in 1244.

new towns, planned urban communities in Great Britain, developed by long-term loans from the central government and first authorized by the New Towns Act of 1946. The chief purpose of the act was to reduce congestion in the great cities (or at least prevent its increase) through the creation of attractive, healthful urban units that would provide local employment for their residents. The idea goes back to the book by Ebenezer Howard on "garden cities" (1898). It was given impetus by the example of the "new towns" of Letchworth (1903) and Welwyn Garden City (1919-20), both established with private capital. The act of 1946 empowered the government to designate areas (which might or might not already contain an existing municipality) as new towns, to appoint development corporations, and to approve their plans. New towns in Northern Ireland have development commissions, established and governed under a separate act (1965). Most of the 28 new towns were intended to alleviate the growth problems of Greater London, Manchester, Merseyside, Tyneside-Wearside, Birmingham and the Black Country, and Clydeside. New towns have also been designated to stimulate economic growth (Craigavon) and to provide needed housing and community services for industrial areas (Corby, Glenrothes, Cwmbran). More recently the instrument of the new town development corporation and the machinery of the new towns legislation have been used to decentralize population through the expansion of already large towns (Peterborough, Northampton, and Ipswich). Central Lancashire New Town, designated in 1970, represents yet another variation, the "clustertown." The area is comparatively large, containing a county borough (Preston) and two other towns (Leyland and Chorley). Its total population of some 250,000 is projected to increase to 435,000 by 1993. Its purpose is to relieve congestion in Liverpool and Manchester and create a more balanced distribution of economic growth for Lancashire.

New Ulm (ŭlm), city (1978 est. pop. 14,600), seat of Brown co., S Minn., at the confluence of the Minnesota and Cottonwood rivers; inc. as a city 1876. New Ulm, a processing and trade center for a grain and cattle region, has a huge rye mill. The city's manufactures include dairy products, flour, beer, tools, textiles, electronic equipment, and mobile homes. New Ulm was settled in 1854 by Germans, who named it after Ulm, Germany.

New Wa·ter·ford (wô'tər-fərd, wŏt'ər-), town (1976 pop. 9,223), on NE Cape Breton Island, N.S., Canada, NE of Sydney. It is chiefly a coal-mining center.

New West·min·ster (wĕst'mĭn'stər), city (1976 pop. 38,393), SW British Columbia, Canada, on the Fraser River, part of metropolitan Vancouver. Founded in 1859, it was the capital of British Columbia until Victoria was made capital after the union of British Columbia and Vancouver Island in 1866. New Westminster is a year-round port, with an excellent harbor that is the base of the Fraser River fishing fleet. Among the city's industries are salmon, fruit, and vegetable canneries, foundries, oil refineries, paper, lumber, and flour mills, and meat-packing plants.

New Wind·sor (wĭn'zər): *see* Windsor, England.

New Windsor, uninc. village (1970 pop. 8,803), Orange co., SE N.Y., on the Hudson River S of Newburgh. Washington had his headquarters here in 1779 and 1780-81.

New York (yôrk), state (1975 pop. 18,009,000), 49,576 sq mi (128,402 sq km), E United States, one of the Middle Atlantic states and one of the Thirteen Colonies. Albany is the capital. The state is bounded on the north by the Canadian provinces of Que. and Ont., with the St. Lawrence River and Lake Ontario marking the Ont. border. In the northwest the Niagara River, with magnificently scenic Niagara Falls, forms the border with Ont. between Lake Ontario and Lake Erie. Lake Erie itself and a minute part of Pa. constitute the rest of the western border. Pa. and N.J. are to the south, except where the state extends into the Atlantic Ocean at New York City and Long Island. To the east New York borders on Conn., Mass., and Vt.

Lake Champlain, stretching past the Canadian border, is the chief northern feature of the great valley (including the Hudson River and its west tributary, the Mohawk River) that dominates all eastern New York. The Hudson is noted for its beauty, as are Lake Champlain and neighboring Lake George, which have many resorts. West of the lakes are the wild and rugged Adirondack Mts., another major vacationland, with woods in the north and sports and health centers. The rest of northeastern New York is hilly, sloping gradually to the valleys of the St. Lawrence and Lake Ontario. The Mohawk River, which flows from Rome to the Hudson, is part of the New York State Barge Canal, a major route to the Great Lakes and the mid-

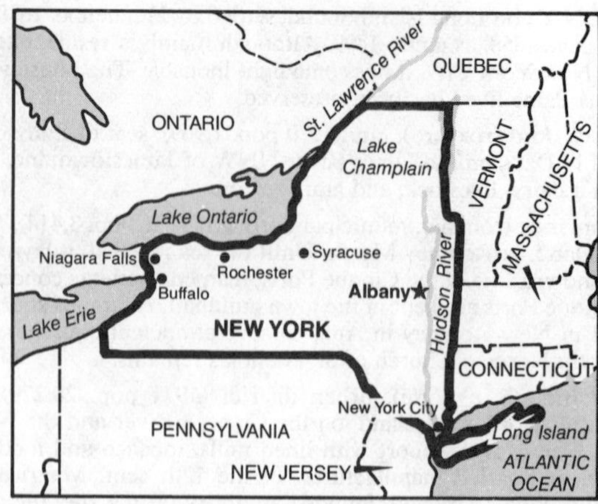

western United States. Most of the southern part of the state is on the Allegheny plateau, which rises in the southeast to the Catskill Mts. The western extension of the state to Lakes Ontario and Erie is hilly land with many bodies of water, notably Oneida Lake and the celebrated Finger Lakes.

In addition to the great forest preserves of the Adirondack and Catskill mts., New York has many state parks, among which Jones Beach State Park and Allegany State Park are well known. Part of Fire Island is a national seashore. The racetrack and mineral waters of Saratoga Springs make it both a pleasure and health resort, and the Thousand Islands are popular with summer vacationers. Among the several places of historic interest in the state under Federal administration are those at Hyde Park, with the burial place of Eleanor and Franklin D. Roosevelt, and the Vanderbilt Mansion.

New York is divided into 62 counties:

NAME	COUNTY SEAT	NAME	COUNTY SEAT
Albany	Albany	Niagara	Lockport
Allegany	Belmont	Oneida	Utica
Bronx	Bronx	Onondaga	Syracuse
Broome	Binghamton	Ontario	Canandaigua
Cattaraugus	Little Valley	Orange	Goshen
Cayuga	Auburn	Orleans	Albion
Chautauqua	Mayville	Oswego	Oswego
Chemung	Elmira	Otsego	Cooperstown
Chenango	Norwich	Putnam	Carmel
Clinton	Plattsburg	Queens	Jamaica
Columbia	Hudson	Rensselaer	Troy
Cortland	Cortland	Richmond	Staten Island
Delaware	Delhi	Rockland	New City
Dutchess	Poughkeepsie	St. Lawrence	Canton
Erie	Buffalo	Saratoga	Ballston Spa
Essex	Elizabethtown	Schenectady	Schenectady
Franklin	Malone	Schoharie	Schoharie
Fulton	Johnstown	Schuyler	Watkins Glen
Genesee	Batavia	Seneca	Waterloo
Greene	Catskill	Steuben	Bath
Hamilton	Lake Pleasant	Suffolk	Riverhead
Herkimer	Herkimer	Sullivan	Monticello
Jefferson	Watertown	Tioga	Owego
Kings	Brooklyn	Tompkins	Ithaca
Lewis	Lowville	Ulster	Kingston
Livingston	Geneseo	Warren	Lake George
Madison	Wampsville	Washington	Hudson Falls
Monroe	Rochester	Wayne	Lyons
Montgomery	Fonda	Westchester	White Plains
Nassau	Mineola	Wyoming	Warsaw
New York	New York	Yates	Penn Yan

Economy. Except in the mountain regions, the state is rich agriculturally. The Finger Lakes region has apple orchards; in 1973 New York along with Wash. state led the nation in apple production. New York has vineyards and is famous for its champagnes. Other areas of the state produce diverse crops, especially grains, truck crops, hay, and potatoes. The state is a leading producer of milk and dairy products.

The state has mineral resources—emery, garnet, talc, titanium concentrate, zinc, lead, salt, gypsum, stone, sand and gravel, iron ore, petroleum, and natural gas—but most of its great industries depend on imported raw materials. The state has a complex system of railroads, air routes, and modern highways that serve industry. The rivers and the New York State Barge Canal, an improvement of the old Erie Canal, also carry much freight, although they are not as important as they once were. Ocean shipping is handled by New York City and by Buffalo, made accessible to oceangoing shipping by the St. Lawrence Seaway, which opened in 1959.

New York is the nation's chief manufacturing state and, by virtue of New York City, its commercial and financial leader as well. Its manufactures include wearing apparel, food products, machinery, chemicals, paper, electrical equipment, optical instruments and cameras, and transportation equipment. Printing and publishing, mass communications, advertising, and entertainment are among New York City's notable industries.

History. Before Europeans began to arrive in the 16th cent., New York was inhabited mainly by Algonquian- and Iroquoian-speaking Indians. The Algonquians lived chiefly in the Hudson valley and on Long Island. The Iroquoians, living in the central and western parts of the state, included the Cayuga, Mohawk, Oneida, Onondaga, and Seneca tribes, who joined c.1570 to form the Iroquois Confederacy.

Giovanni da Verrazano, a Florentine in the service of France, visited (1524) New York Bay but did little exploring. In 1609 Samuel de Champlain, a Frenchman, went down Lake Champlain from Canada, and Henry Hudson, an Englishman in the service of the Dutch, went up the Hudson nearly to Albany. The French from Canada continued to penetrate northern and western New York, but they came up against the uncompromising hostility of the Iroquois Confederacy. The Dutch early claimed the Hudson region, and the Dutch West India Company planted (1624) their colony of New Netherland, with its chief settlement at New Amsterdam on the lower tip of present-day Manhattan island (purchased in 1626 from the Canarsie Indians for trinkets worth about $24).

The last and most able of the Dutch administrators, Peter Stuyvesant (in office 1647-64), was unable to meet the threat of the English, who had been penetrating Long Island and southeastern New York ever since Lion Gardiner came in 1639. The English, claiming the whole region on the basis of the explorations of John Cabot, made good their claim in the Second Dutch War (1664-67). New Netherland then became the colonies of New York and New Jersey, granted by King Charles II to his brother, the duke of York (later James II). Except for brief recapture (1673-74) by the Dutch, New York remained English until the American Revolution.

After the early days of the colony the popular governor Thomas Dongan (1683-88) put New York on a firm basis and began to establish the alliance of the English with the Iroquois, which later played an important part in New York history. The rebellious New York City merchant Jacob Leisler gained sufficient popular support to grasp brief (1689-91) control of the provincial government and run it in the interest of the popular antiaristocratic party. Leisler's government was overthrown by royal authority, and he was executed for treason. After this the rift continued to widen between the great landholders and the small farmers and artisans, who were allied with the increasingly powerful merchant group.

The threat of the French was continuous, and New York was involved in a number of the French and Indian Wars (1689-1763). Raids and counterraids of the wars and sporadic Indian massacres hindered growth in central New York, and much of western New York remained unsettled by colonists throughout the 18th cent. Colonial self-assertiveness grew after the warfare with the French ended; there was considerable objection to the restrictive commercial laws, and the Navigation Acts were flouted by smuggling. When the Stamp Act was passed, New York was a leader of the opposition.

About one third of all the military engagements of the American Revolution took place in New York state. The first major military action in the state was the capture (May, 1775) of Ticonderoga by Ethan Allen and his Green Mountain Boys and Benedict Arnold. Crown Point was also taken. In Aug., 1776, however, George Washington was unable to hold lower New York against the British, who besieged New York City and held it to the war's end. The state had, however, declared independence and functioned with Kingston as its capital. The British plan of taking the entire state and thus separating New England from the South failed finally (Oct., 1777) in the battles near the present-day resort of Saratoga Springs.

After the war speculation in western New York land rose to dizzying heights. The eastern boundary of the state was established after long wrangles and violence when Vt. was admitted as a state in 1791. The influence of Alexander Hamilton was paramount in bringing New York to accept (1788) the Constitution of the United States at a convention in Poughkeepsie. New York City was briefly (1789-90) the capital of the new nation and was also the state capital until 1797, when Albany succeeded it.

From the 1780s increased commerce (somewhat slowed by the Embargo Act of 1807) and industry, especially textile milling, marked the turn away from the old, primarily agricultural, order. The state's development was quickened and broadened by the building of the Erie Canal. The canal, completed in 1825, and railroad lines constructed (from 1831) parallel to it made New York the major east-west commercial route in the 19th cent. and helped to account for the growth and prosperity of the port of New York. Albany grew, and New York City became the financial capital of the nation. New constitutions broadened the suffrage in 1821 and again in 1846; slavery was abolished in 1827. New York was a leader in numerous 19th cent. reform groups. Antislavery groups made their headquarters in New York, as did the first woman's rights leaders. Migrants from New England had been settling on the western frontier, and in the 1840s famine and revolution in Europe resulted in a great wave of Irish and German immigrants, whose first stop in America was usually New York City.

Despite the Draft Riots (1863) in New York City and the activities of the Peace Democrats, New York state strongly favored the Union and contributed much to its cause in the Civil War. Industrial development was stimulated by the needs of the military, and railroads increased their capacity. New York City's newspapers, notably the *Tribune* under the guidance of Horace Greeley, had considerable national influence, and after the war book publishing centered more and more in the city, whose libraries expanded. From 1867 to 1869 Cornelius Vanderbilt consolidated the New York Central RR system. As economic growth accelerated, political corruption became rampant. Samuel J. Tilden won a national reputation in 1871 for prosecuting the Tweed Ring of New York City, headed by William Marcy Tweed, but Tammany soon recovered much of its prestige and influence as the Democratic city organization.

The inpouring after 1880 of immigrants from Ireland, Italy, and Eastern Europe brought workers for the old industries, which were expanding, and for the new ones, including the electrical and chemical industries. Labor conditions worsened but were challenged by the growing labor movement, whose targets included sweatshops (particularly notorious in New York City). A fire in 1911 at the Triangle Shirt Waist Company in New York City that killed 140 workers resulted in the passage of some early labor laws like the Widowed Mothers Pension Act. The reform programs enacted in the 1920s and 1930s emphasized public works, conservation, reorganization of state finances, social welfare, and extensive labor laws. Fiorello LaGuardia, Republican mayor of New York City (1934-45), enthusiastically supported President Franklin D. Roosevelt's social and economic reforms. Gov. Thomas E. Dewey (1942-54) had the immense task of coordinating state activities with national efforts in World War II, which strained New York's resources to the utmost. Gov. Nelson Rockefeller (1958-73) increased the state's social welfare programs, greatly expanded the State Univ. (est. 1948), and began construction of a large state-office and cultural complex in Albany.

Government. Under its present constitution (adopted 1894), New York is run by a governor, who is elected to a four-year term and may succeed himself, and by a bicameral legislature made up of a 60-member senate and a 150-member assembly. Members of both branches are elected to two-year terms.

Educational Institutions. Apart from New York City, the institutions of higher education in the state include Alfred Univ., Bard College, Colgate Univ., Cornell Univ., Hobart College, Long Island Univ., Rensselaer Polytechnic Institute, Sarah Lawrence College, Skidmore College, State Univ. of New York, Syracuse Univ., U.S. Military Academy, Univ. of Rochester, Vassar College, and Wells College.

New York, county (1970 pop. 1,539,233), 22 sq mi (57 sq km), SE N.Y., coextensive with Manhattan borough of New York City and Manhattan Island; formed 1683; co. seat New York.

New York, city (1970 pop. 7,895,563), area with water surface c.365 sq mi (945 sq km), SE N.Y., largest city in the United States and one of the three largest in the world, on New York Bay at the mouth of the Hudson River. It comprises five boroughs, each coextensive with a county: Manhattan (New York co.), the heart of the city, an island; the Bronx (Bronx co.), on the mainland, northeast of Manhattan and separated from it by the Harlem River; Queens (Queens co.), on Long Island, east of Manhattan across the East River; Brooklyn (Kings co.), also on Long Island, on the East River adjoining Queens and on New York Bay; and Richmond (Richmond co.), on

Staten Island, southwest of Manhattan and separated from it by Upper Bay. The metropolitan area (1970 est. pop. 11,600,000) encompasses parts of southeastern N.Y. state, northeastern N.J., and southwestern Conn. and includes both industrial and residential areas.

New York is the largest port in the United States and one of the leading ports in the world. It is the trade center of the nation and the financial center of the entire world. Manufacturing—mostly of small but diverse types—accounts for a large but declining amount of employment. Clothing, chemicals, metal products, and processed foods are some of the main manufactures. New York is also the principal center of U.S. television and radio broadcasting, book publishing, and other mass communications. The most celebrated newspaper is the New York *Times.* New York was the site of two World's Fairs (1939–40; 1964–65). The city is served by three major airports; railroads converge upon it from all points. Since the mid-19th cent. various waves of immigration have widened the already broad spectrum of New York City life.

Although Giovanni da Verrazano may have been the first European to explore the region and Henry Hudson certainly visited the area, it was with Dutch settlement that the city truly began to emerge. In 1624 the colony of New Netherland was established, with the town of New Amsterdam on the lower tip of Manhattan as its capital. Under the Dutch, schools were opened and the Dutch Reformed Church was established. In 1664 the English, at war with the Netherlands, seized the colony for the duke of York, for whom it was renamed, and New York became the capital of the British province of New York. The Dutch returned to power only briefly (1673–74). English rule was not without dissension; the autocratic rule of British governors was one of the causes of an insurrection that broke out in 1689 under the leadership of Jacob Leisler; the insurrection was put down in 1691 by British authorities. Meanwhile, despite such outbreaks, New York was growing, and its expanding commerce was attracting settlers from many nations. The first newspaper, the New York *Gazette,* appeared in 1725, and 10 years later the trial of John Peter Zenger helped establish the principle of a free press. Kings College (now Columbia Univ.) was founded in 1754.

New York was active in the colonial opposition to British measures after trouble in 1765 over the Stamp Act. As revolutionary sentiments increased, the New York Sons of Liberty forced (1775) Gov. William Tryon and the British colonial government from the city. Although many New Yorkers were Loyalists, Continental forces commanded by George Washington defended (1776) the city. After the patriot defeat in the Battle of Long Island and the succeeding actions at Harlem Heights and White Plains, Washington gave up New York, and the British occupied the city until the end of the war for independence. After the Revolution, New York was briefly (1789–90) the first capital of the United States and was the state capital until 1797. President Washington was inaugurated (Apr. 30, 1789) at Federal Hall, where the Subtreasury Building now stands.

By 1790 New York was the largest city in the United States, with over 33,000 inhabitants; by 1800 the number had risen to 60,515. The opening of the Erie Canal (1825), which made New York the seaboard gateway for the Great Lakes region, ushered in another era of commercial expansion. During the Civil War the majority of New Yorkers supported the Union. However, in 1863 the Draft Riots broke out in protest against the Federal Conscription Act. Extensive immigration had begun before the Civil War, and after 1865, with the acceleration of industrial development, another wave of immigration began that reached its height in the late 19th and early 20th cent.

Until 1874, when portions of Westchester were annexed, the city's boundaries were those of present-day Manhattan. With the adoption of a new charter in 1898 New York became Greater New York, a metropolis of five boroughs—New York City was split into the present Manhattan and Bronx boroughs, and the independent city of Brooklyn was annexed, as were Queens co. and Staten Island.

In New York Bay are Liberty Island (with the Statue of Liberty), Governors Island, and Ellis Island, which along with Castle Garden served as the point of entry for many immigrants. New York City is the seat of the United Nations. It is also a cultural capital. Lincoln Center for the Performing Arts houses the Metropolitan Opera Company, the New York Philharmonic-Symphony Orchestra, the New York City Ballet, the New York City Opera, the Juilliard School, and the Vivian Beaumont Theatre and Library Building. Also in the city are Carnegie Hall and New York City Center. Among the best known of the city's many museums and scientific collections are the Metropolitan Museum of Art, the Museum of Modern Art, the Solomon R. Guggenheim Museum (designed by Frank Lloyd Wright), the Frick Collection (housed in the Frick mansion), the Whitney Museum of American Art, the Museum of Primitive Art, the Museum of the City of New York, the American Museum of Natural History (with the Hayden Planetarium), the museum and library of the New-York Historical Society, and the Brooklyn Museum. The New York Public Library is the administrative hub for many smaller public libraries.

Major educational institutions include City Univ. of New York, Columbia Univ., Cooper Union, Fordham Univ., General Theological Seminary, Jewish Theological Seminary, New School for Social Research, New York Univ., and Union Theological Seminary. Among New York's noted houses of worship are Trinity Church, St. Paul's Chapel (dedicated 1776), Saint Patrick's Cathedral, the Cathedral of St. John the Divine, Riverside Church, and Temple Emanu-El. New York's parks and recreation centers include parts of Gateway National Recreation Area; Central Park, the Battery, Washington Square Park, Riverside Park, and Fort Tryon Park in Manhattan; the New York Zoological Park (Bronx Zoo), the New York Botanical Garden, and Van Cortlandt Park in the Bronx; and Coney Island and Prospect Park in Brooklyn. Among the many other places of interest are Rockefeller Center, Greenwich Village, and Times Square. Of historic interest are Fraunces Tavern (built 1719), where Washington said farewell to his officers after the Revolution; Gracie Mansion (built late 18th cent.), now the official residence of the mayor; the Edgar Allan Poe Cottage; and Grant's Tomb.

New York Bay, arm of the Atlantic Ocean at the mouth of the Hudson River, SE N.Y. and NE N.J., enclosed by the shores of NE N.J., E Staten Island, S Manhattan, and W Long Island (Brooklyn) and opening on the SE to the Atlantic Ocean between Sandy Hook, N.J., and Rockaway Point, N.Y. It is a sheltered deep harbor able to accommodate the largest ships. The tidal range of the bay is very small and it is ice-free. New York Bay is divided into Upper and Lower Bay, which are connected by the Narrows, a strait (c.3 mi/4.8 km long; 1 mi/1.6 km wide) separating Staten Island from Brooklyn. The Verrazano-Narrows Bridge spans the strait between Fort Wadsworth and Fort Hamilton. Upper Bay, c.5.5 mi (8.8 km) in diameter, is one of the world's busiest harbors, with extensive port facilities on all shores. The larger Lower Bay includes Raritan Bay on the west and Gravesend Bay on the northeast. Ambrose Channel, federally maintained, crosses Sandy Hook bar at the bay's entrance and extends north to the piers of Upper Bay, where it is 2,000 ft (610 m) wide. Giovanni da Verrazano, the Italian explorer, was the first European to enter (1524) the bay. Henry Hudson later explored it (1609) and claimed the region for the Dutch East India Company.

New York State Barge Canal, waterway system, 525 mi (845 km) long, traversing New York state and connecting the Great Lakes with the Hudson River and Lake Champlain. The canal, a modification and improvement of the old Erie Canal, was authorized (1903) by public vote, was begun in 1905, and was completed in 1918. Its main sections are the Erie Canal, extending from Troy to Tonawanda; the Champlain Canal, joining the Erie Canal at Waterford and extending north (via the Hudson as far as Fort Edward) to Whitehall on Lake Champlain; the Oswego Canal, connecting the Erie Canal with Oswego on Lake Ontario; and the Cayuga and Seneca Canal, joining the Erie Canal with Cayuga and Seneca lakes. The Barge Canal (12 ft/3.7 m deep), with 57 electrically operated locks, accommodates 2,000-ton vessels and is toll-free. Pleasure craft are the most numerous vessels on the canal.

New Zea·land (zē'lənd), country (1976 pop. 3,121,904), 103,736 sq mi (268,676 sq km), in the S Pacific Ocean. The capital is Wellington. New Zealand comprises North Island and South Island, Stewart Island, the Chatham Islands, and a number of small outlying islands. The Cook Islands, internally self-governing, are "free associates" of New Zealand.

North Island is known for its active volcanic mountains, its hot springs, and its mineral deposits. On South Island are the massive Southern Alps, many beautiful fjords, and the largest areas of virgin forest. Among the unusual native animals are the kiwi, the albatross, and the tuatara (survivor of a prehistoric order of reptiles, Rhynchocephalia); there are no land snakes.

Economy. Agriculture is the mainstay of the economy, although industry employs a larger number of people. The principal exports are dairy products, meat, and wool. Other important products include wheat, fruits, and kauri gum. Small amounts of coal, gold,

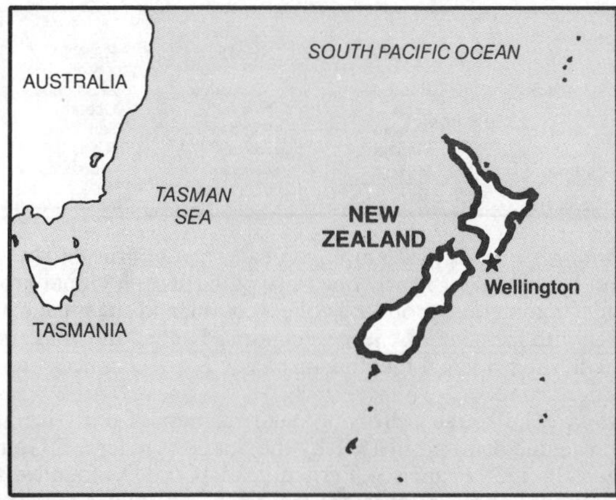

iron, and oil are also produced. Large oyster beds are found in the Foveaux Strait between Stewart Island and South Island. Food processing is the largest manufacturing industry.

History. The Dutch navigator A. J. Tasman was the first European to reach New Zealand, which was already inhabited by Polynesian Maoris. New Zealand was visited by Capt. James Cook four times between 1769 and 1777. The first missionary, Samuel Marsden, arrived in 1814. In 1840 the first permanent settlement was made at Wellington by a group sent by the New Zealand Company. In that year the Treaty of Waitangi guaranteed the Maoris the full possession of their land in exchange for their recognition of the sovereignty of the British crown. Nevertheless, the white settlers continued until 1870 to wage bloody territorial battles against the Maoris. Of the present Maori population of about 225,000, most live on North Island.

Originally part of the New South Wales colony of Australia, New Zealand was made a separate colony in 1841. In 1852 the British Parliament granted it self-government, and by the Statute of Westminster of 1931 the colony became completely independent (although the New Zealand Parliament did not confirm the statute until 1947, preferring instead to have Great Britain conduct its foreign affairs). New Zealand was the first country to grant (1893) women over 21 the right to vote. A comprehensive social-security system was begun in 1898 with the enactment of an old-age pension law.

Government. The New Zealand government consists of the governor general (representing the British crown), a prime minister and cabinet (the effective executive), and a unicameral parliament.

Ne·ya·ga·wa (nā'yä-gä'wä), city (1976 est. pop. 259,456), Osaka prefecture, SW Honshu, Japan, on the Shinyodo River. It is a suburb of Osaka.

Ney·sha·bur (nā'shä-bo͝or'), city (1966 pop. 33,482), Khorasan prov., NE Iran. It is the trade center for a farm region where cotton, fruit, and grain are grown. Manufactures of the city include food products and leather goods. Neyshabur was founded by the Sassanid ruler Shapur I in the 3rd cent. A.D. Under the Seljuk Turks (11th–12th cent.) the city became an important cultural center. Al-Ghazali, the noted philosopher of the 11th–12th cent., studied in Neyshabur, and his famous contemporary Omar Khayyam, the poet and mathematician, was born in the city and is buried here. The tomb of Omar was rebuilt in 1934. Near Neyshabur archaeologists have made important finds of glazed pottery and stucco work from the 9th and 10th cent.

Ne·zhin (nyā'zhĭn), city (1976 est. pop. 68,000), S central European USSR, in the Ukraine, on the Oster River. It is a rail terminus and an agricultural trade center. Industries include engineering, food processing, and the manufacture of machinery, railroad cars, lacquers, paints, clothing, building materials, and whaling equipment. Known in the 11th cent., the city became an important trading center in the 17th and 18th cent. after Greek merchants received permission to settle here in 1657.

Nez·perce (nĕz'pûrs'), village (1970 pop. 555), seat of Lewis co., W Idaho, ESE of Lewiston, in a wheat and lumber-milling area.

Nez Perce (nĕz' pûrs'), county (1970 pop. 30,376), 847 sq mi (2,193.7 sq km), W Idaho, in a stock-raising and agricultural area bounded in the W by the Snake River and the Wash. border and drained by the Clearwater River; formed 1861; co. seat Lewiston. Wheat, fruit, and vegetables are grown. Lumber milling is also important. Part of Nez Perce Indian Reservation is in the north.

Nez Perce National Historical Park: *see* National Parks and Monuments Table.

Nga·mi, Lake (əng-gä'mē), reedy marsh, c.40 mi (65 km) long and from 4 to 8 mi (6.4–12.9 km) wide, NW Botswana. During the Pleistocene epoch the lake covered an extensive area. Since the late 1880s, when papyrus growth blocked the mouth of its main tributary, the lake has greatly shrunk in size. David Livingstone, the Scottish explorer, was the first European to visit (1849) the lake.

Nha Trang (nyä' träng'), city (1968 est. pop. 102,000), S central Vietnam, a port on the South China Sea. During the Vietnam War, Nha Trang was the site of a major U.S. military installation that extended 15 mi (24 km) south to the harbor of Cam Ranh Bay. The city is situated on the main highway that runs north from Saigon to Hue.

Ni·ag·a·ra (nī-ăg'rə, -ăg'ər-ə), county (1970 pop. 235,720), 533 sq mi (1,380.5 sq km), W N.Y., bounded on the W by the Niagara River and Lake Erie, on the N by Lake Ontario, and drained by Tonawanda Creek; formed 1808; co. seat Lockport. Its agriculture includes apples, peaches, pears, prunes, cherries, dairy products, and truck crops. It also has limestone quarries, commercial fisheries, and industry (food processing, lumber millwork, and the manufacture of paper products, chemicals, plastics, and machinery). The Niagara Falls resort area is here.

Niagara, river, 34 mi (54.7 km) long, issuing from Lake Erie between Buffalo, N.Y., and Fort Erie, Ont., Canada. It flows north around Grand Island (American) and over Niagara Falls to Lake Ontario; the river forms part of the U.S.-Canadian border. The upper section of the river is navigable for c.20 mi (30 km) to a series of rapids above the falls; in its last 7 mi (11.3 km) it is again navigable, from Lewiston, N.Y., to Lake Ontario. The New York State Barge Canal enters the river at Tonawanda, N.Y.; the Welland Canal, on the Ontario side, is a lake-freighter route around the falls. Hydroelectric power is generated by diverting water from the river above Niagara Falls to generating plants.

Niagara Falls, city (1976 pop. 69,423), S Ont., Canada, on the Niagara River opposite Niagara Falls, N.Y. It is a port of entry and an important industrial city, the home of Canadian factories for many well-known U.S. firms. Electric power is supplied by the falls.

Niagara Falls, city (1978 est. pop. 78,400), Niagara co., W N.Y., at the falls of the Niagara River; inc. 1892. Tourism is one of its oldest industries, and there are many state parks in the area, including New York State Niagara Reservation. The city is also a port of entry; its manufactures include chemical and mechanical products, rocket parts, and food products. One of the world's first hydroelectric power plants was built here. Settled by Indians, the site was occupied by the French in the 1680s, captured by the British in 1759, and settled by Americans in 1805. Lost to the British during the War of 1812, it was regained by the United States after the Treaty of Ghent in Dec., 1814.

Niagara Falls, in the Niagara River, W N.Y. and S Ont., Canada. The falls are on the international line between the cities of Niagara Falls, N.Y., and Niagara Falls, Ont. Goat Island splits the cataract into the American Falls (167 ft/51 m high and 1,060 ft/323.3 m wide) and the Horseshoe, or Canadian, Falls (158 ft/48.2 m high and 2,600 ft/793 m wide). The falls were formed c.10,000 years ago as the retreating glaciers exposed the Niagara escarpment, thus permitting the waters of Lake Erie to flow north, over the scarp, to Lake Ontario. The escarpment has been gradually eroded back toward Lake Erie, a process that has formed the Niagara Gorge (c.7 mi/11 km long). Horseshoe Falls is eroding upstream at a faster rate than the American Falls because of the greater volume of water passing over it. A great rock slide occurred (1954) at the American Falls and formed a huge talus slope at its base. Water was diverted from the American Falls for several months in 1969 by the U.S. Corps of Engineers to study the bedrock and to remove some of the talus. International agreements control the diversion of water for hydroelectric power; weirs divert part of the flow above the deeper Canadian Falls to supplement the flow in the shallower American Falls. Hydroelectric-power developments were authorized under the Niagara Diversion Treaty (1950), which stipulated a minimum flow to be reserved for the falls and the equal division of the remaining flow between the United States and Canada. The governments of the two countries also control the appearance of the surrounding area, much of which has been included in parks since 1885.

Ni·ag·a·ra-on-the-Lake (nī-ăg′rə-ŏn-thə-lāk′, -ăg′ər-ə-), or **Niagara,** town (1976 pop. 12,485), S Ont., Canada, on Lake Ontario at the mouth of the Niagara River. It was settled (1784) by American Loyalists, and in 1792 Lt. Gov. John Simcoe made the town the capital of Upper Canada. The legislature met here until 1796. Fort George, built (1796–99) to defend the settlement, was taken in 1813 by the United States but retaken in the same year.

Nia·mey (nyä-mā′), city (1975 est. pop. 130,000), capital of Niger and its Niamey dept., SW Niger, a port on the Niger River. Much of its importance stems from its location on the Niger River at the crossroads of the country's two main highways. The city is the trade center for an agricultural region that specializes in growing groundnuts. Manufactures include bricks, food products, beverages, ceramic goods, cement, and shoes. Niamey was a small town when the French colonized the area in the late 19th cent., but it grew after it became the capital of Niger in 1926. It is the site of the National Museum, which has ethnological and zoological collections.

Ni·as (nē′əs), volcanic island (1970 pop. 388,000), 1,569 sq mi (4,063.7 sq km), Indonesia, in the Indian Ocean off Sumatra. Rice and other food crops are grown, livestock are raised, and there is hunting and fishing. The Dutch began trading here in 1669. The island is subject to severe earthquakes.

Ni·cae·a (nī-sē′ə), city of Bithnyia, N Asia Minor. Built in the 4th cent. B.C., it flourished under the Romans and was the scene of the ecumenical council called in A.D. 325 by Constantine I. Another council held in 787 sanctioned the devotional use of images. The city, captured by the Turks in 1078 and by the Crusaders in 1097, passed finally to the Turks in 1330.

Nic·a·ra·gua (nĭk′ə-rä′gwə), republic (1970 pop. 1,974,924), 49,579 sq mi (128,410 sq km), Central America, bordered on the N and NW by Honduras, on the E by the Caribbean Sea, on the S by Costa Rica, and on the SW by the Pacific Ocean. The capital is Managua. The northwestern highlands have peaks as high as 8,000 ft (2,440 m). On

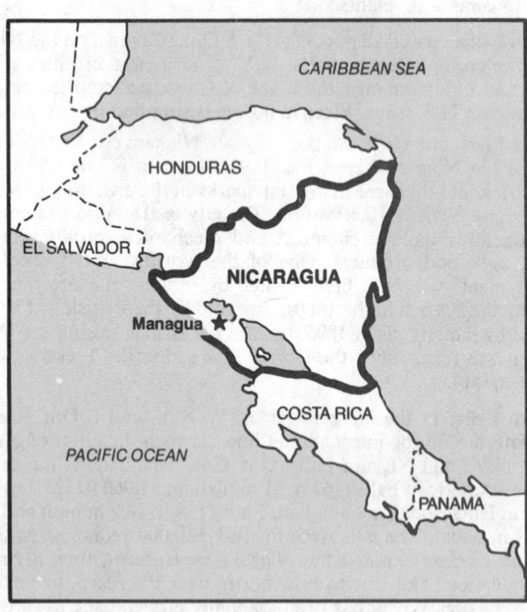

the Caribbean is the torrid Mosquito Coast, with the historic port of Bluefields. A lowland belt running northwest to southeast contains lakes Managua and Nicaragua. A narrow volcanic belt is squeezed between the lakes and the Pacific; in this region the productive wealth and the population (almost entirely mestizo) are concentrated.

Nicaragua is divided into 16 departments:

NAME	CAPITAL	NAME	CAPITAL
Boaca	Boaca	Chontales	Juigalpa
Carazo	Jinotepe	Esteli	Esteli
Chinandega	Chinandega	Granada	Granada
Jinotega	Jinotega	Matagalpa	Matagalpa
León	León	Nueva Segovia	Ocotal
Madriz	Somoto	Río San Juan	San Carlos
Managua	Managua	Rivas	Rivas
Masaya	Masaya	Zelaya	Bluefields

Economy. Agriculture employs nearly two thirds of the work force and accounts for nearly one third of the gross national product. The chief commercial crops are coffee, cotton, and sugar cane; these, together with meat, are the largest exports. Timber and gold are also exported. The principal manufactured goods are chemicals, textiles, and processed foods.

History. The country probably takes its name from Nicarao, an Indian cacique defeated in 1522 by the Spanish under Gil González de Ávila. In 1524 Francisco Fernández de Córdoba founded León and Granada. León became the political and intellectual capital, Granada the stronghold of the aristocracy. Under Spanish rule Nicaragua was part of the captaincy general of Guatemala. After declaring independence from Spain (1821) Nicaragua was briefly part of the Mexican Empire of Agustín de Iturbide and then (1825–38) a member of the Central American Federation.

Nicaraguan politics were wracked by conflict between Liberals and Conservatives, centered respectively in León and Granada; Managua was founded as the capital in 1855 as a compromise. British influence had been established along the east coast in the 17th cent., and in 1848 the British seizure of San Juan del Norte opened a period of conflict over control of the Mosquito Coast. The United States was interested in a transisthmian canal, and its interest was heightened by the discovery of gold in California. In 1851 Cornelius Vanderbilt opened a transisthmian route through Nicaragua for the gold seekers. The Clayton-Bulwer Treaty (1850) settled some of the issues between Great Britain and the United States concerning the proposed canal, but Nicaragua remained in a state of disorder until 1857.

There was a long period of quiet under Conservative control until the Liberal leader, José Santos Zelaya, became president in 1894. He instituted a vigorous dictatorship, extended Nicaraguan authority over the Mosquito Coast, promoted economic development, and interfered in the affairs of neighboring countries. His financial dealings with Britain aroused U.S. apprehensions and helped bring about his downfall (1909). In 1912 U.S. marines were landed to support the provisional president, Adolfo Díaz, in a civil war. The Bryan-Chamorro Treaty, giving the United States exclusive rights for a Nicaraguan canal and other privileges, was ratified in 1916. (It was terminated in 1970.) The Liberals opposed the U.S. intervention, and there was guerrilla warfare against the U.S.-supported regime for years. The U.S. diplomat Henry L. Stimson succeeded in getting most factions to agree (1927) to binding elections. The marines were withdrawn in 1933.

Three years later Anastasio Somoza emerged as the strong man in Nicaragua. He officially became president in 1937 and ruled for 20 years. Nicaragua virtually became Somoza's private estate; relations with other Central American republics were poor. Somoza was assassinated in 1956, and his son Luis Somoza Debayle became president. Another son, Anastasio Somoza Debayle, headed the armed forces. René Schick Gutiérrez was chosen by the Somoza family to be elected president in 1963. Anastasio Somoza Debayle was elected president in 1967. In Aug., 1971, following an agreement between the ruling National Liberal party (NLP) and the opposition Conservative party, the national assembly voted to abrogate the 1950 constitution. Although Somoza resigned from office in May, 1972, handing power to the new governing council, he retained effective control of the country as head of the armed forces, leader of the NLP, and, after the earthquake (Dec., 1972) that devastated Managua, as director of the emergency relief operations. Somoza resigned from the armed forces in Feb., 1974, to become a presidential candidate and easily won the September elections. In the early 1970s sporadic terrorism and guerrilla incidents were directed against the Somoza regime.

Nicaragua, Lake, 3,089 sq mi (8,000.5 sq km), c.100 mi (160 km) long and up to 45 mi (72.4 km) wide, SW Nicaragua; the largest lake of Central America. It is drained into the Caribbean Sea by the San Juan River. Lake Nicaragua occupies part of the Nicaragua Depression, an extensive lowland region stretching across the isthmus.

Once part of the sea, the lake was formed when the land rose. There are several islands in the lake, and small volcanoes rise above its surface. The fresh water of Lake Nicaragua contains fish usually associated with salt water, including tuna and sharks, which have adapted to the environmental change. Located only 110 ft (33.6 m) above sea level, the lake reaches a depth of 84 ft (25.6 m).

Nice (nēs), city (1975 pop. 344,481), capital of Alpes-Maritimes dept., SE France, on the Mediterranean Sea. Nice is the most famous resort on the French Riviera. Although the economy depends mainly on the tourist trade, the electronics industry as well as other manufactures are important. The port of Nice handles both commercial and tourist traffic. Probably a Greek colony established in the 5th cent. B.C., Nice was pillaged and burned by the Saracens in 859 and 880. In the 13th and 14th cent. the city belonged to the counts of Provence and Savoy. In 1543 the united forces of Francis I and Barbarossa attacked and burned Nice. It was annexed to France in 1793, restored to Sardinia in 1814, and again ceded to France in 1860 after a plebiscite. Nice has several churches dating from the 12th through the 17th cent.

Ni·chi·nan (nē'chē-nän'), city (1975 pop. 38,200), Miyazaki prefecture, SE Kyushu, Japan, on the Pacific Ocean. It is an important fishing port and manufacturing center for pulp and paper products.

Nich·o·las (nĭk'ə-ləs). **1.** County (1970 pop. 6,508), 204 sq mi (528.4 sq km), NE Ky., in the Bluegrass, bounded on the NE by the Licking River; formed 1799; co. seat Carlisle. It is in a gently rolling upland agricultural area that produces burley tobacco and grain. There is some manufacturing. **2.** County (1970 pop. 22,552), 642 sq mi (1,662.8 sq km), central W.Va., on the Allegheny Plateau, bounded on the SW by the Gauley and Meadow rivers and drained by the Cherry River; formed 1818; co. seat Summersville. Its agriculture includes livestock, fruit, and tobacco. Bituminous-coal mining, limestone quarrying, lumbering, food processing, and paper and rubber manufacturing are done. Carnifex Ferry Battlefield State Park and part of Monongahela National Forest are here.

Nich·o·las·ville (nĭk'ə-ləs-vĭl'), city (1970 pop. 5,829), seat of Jessamine co., central Ky., just SSW of Lexington, in a fertile bluegrass region.

Nic·o·bar Islands (nĭk'ə-bär'): *see* Andaman and Nicobar Islands.

Nic·ol·let (nĭk'ə-lĕt'), county (1970 pop. 24,518), 432 sq mi (1,118.9 sq km), S Minn., bounded on the S and E by the Minnesota River, with Swan Lake (once a large body of water but now mainly dry) near the center of the county; formed 1853; co. seat St. Peter. It is in an agricultural area that produces livestock, dairy products, corn, oats, barley, and potatoes. There is a food-processing industry. Boatbuilding and some mining are done.

Nic·o·me·di·a (nĭk'ə-mē'dē-ə), ancient city, NW Asia Minor, near the Bosporus, in present-day Turkey. Refounded (264 B.C.) by Nicomedes I of Bithynia to replace Astacus as his capital, it flourished for centuries. The Goths sacked the city in A.D. 258. Diocletian chose it for the eastern imperial capital, but it was soon superseded by Byzantium (Constantinople).

Ni·cop·o·lis (nĭ-kŏp'ə-lĭs), ancient city, NW Greece, in Epirus. It was founded by Octavian (later Augustus) to celebrate the victory (31 B.C.) at Actium, which is nearby.

Nic·o·si·a (nĭk'ə-sē'ə), city (1974 est. pop. 51,000), capital of Cyprus, on the Pedias River in the central plain of the island. It is an agricultural trade center and has textile, brandy, cigarette, leather, pottery, and other manufactures. Known as Ledra or Ledrae in antiquity, it was the residence of the Lusignan kings of Cyprus from 1192, became a Venetian possession in 1489, and fell to the Turks in 1571. There are remnants of the Venetian fortifications and museums with notable collections of antiquities. Nicosia was the scene of bitter strife just prior to Cypriot independence (1960) and after the Turkish invasion of Cyprus (1974).

Ni·co·ya, Gulf of (nĭ-kō'yə), inlet of the Pacific Ocean, Central America, between the Nicoya Peninsula and the NW mainland of Costa Rica. The village of Nicoya on the peninsula was probably the first Spanish settlement (c.1530) in Costa Rica.

Ni·da·ros (nē'də-rōs'): *see* Trondheim, Norway.

Nieuw·poort or **Nieu·port** (both: nōō'pôrt', nyōō'-), town (1970 pop. 8,273), West Flanders prov., W Belgium, on the North Sea at the mouth of the Yser River. It is a fishing port, an industrial center, and a beach resort.

Niè·vre (nyĕ'vrə), department (1975 pop. 245,212), 2,640 sq mi (6,837.6 sq km), central France; capital Nevers.

Ni·gel (nī'jəl), town (1970 pop. 41,179), Transvaal, NE South Africa. Gold mining and processing are the main industries. Founded in 1909, the town is the site of Sub-Nigel, formerly one of the world's richest gold mines.

Ni·ger (nī'jər), republic (1977 est. pop. 4,805,000), 489,189 sq mi (1,267,000 sq km), W Africa, bordering on Upper Volta and Mali in the W, on Algeria and Libya in the N, on Chad in the E, and on Nigeria and Benin in the S; capital Niamey.

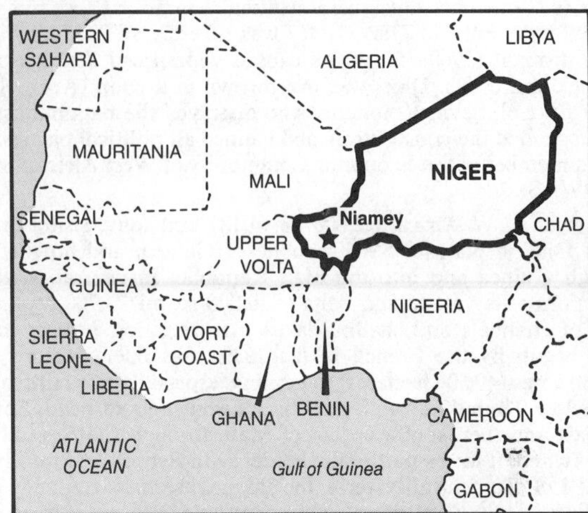

Niger is extremely arid except along the Niger River in the southwest and near the border with Nigeria in the south, where there are strips of savanna. Most of the rest of the country is either semidesert or part of the Sahara. The main ethnic groups are the Hausa (who make up about 55% of the population), the closely related Djerma and Songhai (together 24%), the Fulani (11%), and the Tuareg (3%).

Economy. The economy of Niger is overwhelmingly agricultural, with about 90% of its work force engaged in farming (largely of a subsistence type). The Hausa, Djerma, and Songhai are mainly sedentary farmers, and the Fulani and Tuareg are principally nomadic and seminomadic pastoralists. The leading farm crops are millet, sorghum, cassava, groundnuts, rice, cotton, sugar cane, and dates. Poultry, goats, cattle, sheep, and camels are also raised. Most of the country's few manufactures are basic consumer goods. In addition, groundnut oil, ginned cotton, and construction materials (mainly bricks and cement) are produced. In the early 1970s large high-grade uranium-ore deposits in the Aïr Mts. began to be worked by a Franco-Niger company, and a uranium-ore concentrating plant was opened. Small quantities of cassiterite (tin ore), low-grade iron ore, gypsum, phosphates, natron, and salt also are extracted in the country. There is a small but growing fishing industry, operating mainly in the Niger River and in Lake Chad (in the southeast). Niger has a very limited transportation network.

History. Numerous Neolithic remains have been found in the desert areas of Niger. The region was probably known to the Egyptians before the beginning of recorded history. In the 11th cent. A.D. Tuareg migrated from the desert to the Aïr region, where they later (c.1300) established a state centered at Agadès (Agadez). Agadès was situated on a major trans-Saharan caravan route. In eastern Niger, Bilma, a salt-mining center, was on another important trans-Saharan route that linked North Africa with the state of Bornu. In the 14th cent. the Hausa founded several city-states in southern Niger. In the early 16th cent. much of Niger came under the Songhai empire and passed to Bornu. In the 17th cent. the Djerma people settled in southwest Niger near the Niger River. In the early 19th cent. Fulani gained control of southern Niger as a result of the holy war waged against the Hausa by the Moslem reformer Usuman dan Fodio. At the Conference of Berlin (1884-85) the territory of Niger was placed within the French sphere of influence.

In 1900 Niger was made a military territory within Upper Senegal-Niger, and in 1922 it was constituted a separate colony within French West Africa. The French undertook little economic development and provided few new educational opportunities. National political activity began in 1946. The first important political organization was the Niger Progressive party (PPN). In the mid-1950s a leftist party (later called Sawaba) headed by Bakary Djibo became predominant in the colony. However, in 1958 the PPN (which favored autonomy for Niger within the French community) regained power.

Niger achieved full independence on Aug. 3, 1960, and PPN leader Hamani Diori became its first president; he was re-elected president in 1965 and 1970. In the early years of independence sporadic terrorist campaigns were waged by the outlawed Sawaba party. Otherwise Niger enjoyed political stability, despite its weak economy and occasional ethnic conflicts (especially between the Tuareg and the central government); the PPN maintained firm control of the country's 50-member unicameral national assembly. Close ties were retained with France. The country was severely affected by the Sahelian drought (1968–75), which caused widespread destruction of livestock and crops. Diori was overthrown in a coup (April, 1974) led by Lt.-Col. Seyni Kountche, who dissolved the national assembly, suspended the constitution, and banned all political parties. Niger is a member of the Economic Community of West African States (ECOWAS).

Niger, river of W Africa, c.2,600 mi (4,185 km) long, rising on the Fouta Djallon plateau, SW Republic of Guinea, and flowing NE through Guinea and into the Mali Republic. In central Mali the Niger forms its vast inland delta (c.30,000 sq mi/77,700 sq km), a maze of channels and shallow lakes. An irrigation project in the delta, begun by the French in the 1930s, has opened more than 100,000 acres (40,500 hectares) to farming, especially rice cultivation. Just below Timbuktu, Mali, the Niger begins a great bend, flowing first east and then southeast out of Mali, through the Republic of Niger (where it forms part of the border with Benin), and into Nigeria. At Lokoja, central Nigeria, the Benue, its chief tributary, joins the Niger, which then flows south, emptying through a great delta into the Gulf of Guinea. The delta (c.14,000 sq mi/36,260 sq km)—the largest in Africa—is characterized by swamps, lagoons, and navigable channels. Much of the Niger is seasonally navigable, and below Lokoja it is open to ships virtually all year. The Niger is a major source of fish, especially perch and tiger fish.

The course of the Niger long puzzled Western geographers; only from 1795 to 1797 did Mungo Park, the Scottish explorer, correctly establish the eastern flow of the upper Niger, and it was not until 1830 that Richard and John Lander, English explorers, proved that the river emptied into the Gulf of Guinea. The water level of the Niger was substantially lowered as a result of the long-term West African drought that began in the late 1960s.

Ni·ge·ri·a, Federation of (nī-jîr′ē-ə), republic (1977 est. pop. 83,800,000), 356,667 sq mi (923,768 sq km), W Africa, bordering on the Gulf of Guinea (an arm of the Atlantic Ocean) in the S, on Benin in the W, on Niger in the NW and N, on Chad in the NE, and on Cameroon in the E; capital Lagos.

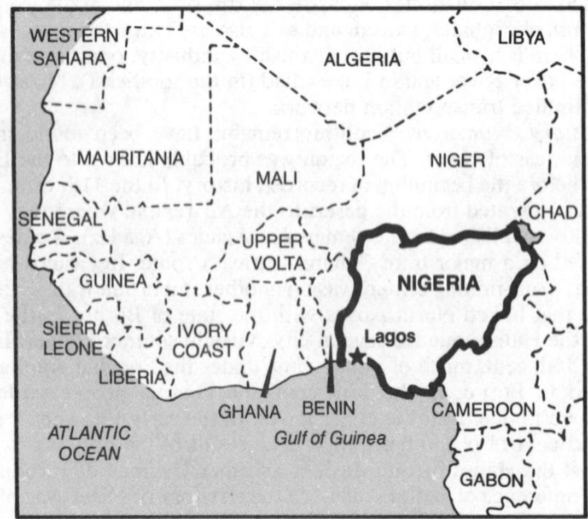

The Niger River and its tributaries drain most of the country. Nigeria has a 500-mi (805-km) coastline, for the most part made up of sandy beaches, behind which lies a belt of mangrove swamps and lagoons that averages 10 mi (16 km) in width but increases to c.60 mi (95 km) wide in the great Niger delta in the east. North of the coastal lowlands is a broad hilly region, with rainforest in the south and covered with savanna in the north. Behind the hills is the great plateau of Nigeria (average elevation 2,000 ft/610 m), a region of plains covered largely with savanna but merging into scrubland in the north. Greater altitudes are attained on the Bauchi and Jos plateaus in the center and in the Adamawa Massif (which continues into Cameroon) in the east.

Economy. The economy of Nigeria is mainly agricultural, with about 70% of the work force engaged in farming. The chief crops grown in the north are sorghum, millet, soybeans, groundnuts, and cotton; the principal agricultural commodities produced in the south are maize, yams, rice, palm products, cacao, and rubber. In addition, large numbers of poultry, goats, sheep, and cattle are raised in the country. Petroleum is the leading mineral produced in Nigeria. The low sulfur content of much of Nigeria's petroleum makes it especially desirable in a pollution-conscious world. Other minerals extracted include tin, coal, limestone, salt, columbite, tantalite, low-grade iron ore, and gold. Industry in Nigeria is largely confined to the processing of agricultural goods and to the manufacturing of basic consumer goods. Other manufactures include refined petroleum, cement, and metal goods. Traditional craftsmen produce woven goods, pottery, metal objects, and carved wood and ivory.

History. By c.2000 B.C. most of Nigeria was sparsely inhabited by persons who had a rudimentary knowledge of raising domesticated food plants and of herding animals. From c.800 B.C. to A.D. c.200 the Neolithic Nok culture flourished on the Jos Plateau; the Nok people made fine terra-cotta sculptures and probably knew how to work tin and iron. The first important centralized state to influence Nigeria was Kanem-Bornu (approximately 8th cent.), north of Lake Chad (outside modern Nigeria). In the 11th cent. Kanem-Bornu expanded south of Lake Chad into present-day Nigeria. Kano and Katsina, Hausa city-states, competed for the lucrative trans-Saharan trade with Kanem-Bornu. In the late 16th cent. Kanem-Bornu replaced Songhai as the leading power in northern Nigeria.

In southwestern Nigeria two Yoruba states—Oyo and Benin—had developed by the 14th cent. Benin was the leading state in the 15th cent. but began to decline in the 17th cent., and by the 18th cent. Oyo controlled Yorubaland. The Ibo people in the southeast lived in small communities. In the late 15th cent. the Portuguese began to purchase slaves and agricultural produce from coastal middlemen. The Portuguese were followed by British, French, and Dutch traders. Among the Ibo and Ibibio a number of city-states were established by individuals who had become wealthy by engaging in the slave trade. In 1804 Usuman dan Fodio (1754–1817), a Fulani and a pious Moslem, began a holy war to reform the practice of Islam in the north. He soon conquered the Hausa city-states, but Bornu maintained its independence. In 1817 Usuman dan Fodio's son, Mohammed Bello (d.1837), established a state centered at Sokoto, which controlled most of northern Nigeria until the coming of the British (1900–1906). Under both Usuman dan Fodio and Mohammed Bello, Moslem culture and trade flourished in the Fulani empire. The Bornu empire disintegrated, partly because of the ruler's disinterest in affairs of state and partly because of devastating civil wars. In 1807 Great Britain abolished the slave trade; however, the trade continued until about 1875. Many African middlemen gave up the slave trade and turned to selling palm products. In 1817 a long series of civil wars began in the Oyo Empire, lasting until 1893.

In the 19th cent. a number of British and Scottish explorers visited Nigeria and determined the course of the Niger River. Britain annexed Lagos in 1861. By 1885 Sir George Goldie had gained control of all the British and French companies trading on the Niger. Largely because of Goldie's efforts Great Britain was able to claim southern Nigeria at the Conference of Berlin held in 1884–85. In the following years the British established their rule in southwestern Nigeria, partly by signing treaties and partly by using force. Jaja, a leading African trader based at Opobo in the Niger delta and strongly opposed to European competition, was captured in 1897. Goldie's firm, given (1886) a British royal charter as the Royal Niger Company, antagonized Europeans and Africans alike by its monopoly of trade on the Niger. In 1900 its charter was revoked, and British forces under Frederick Lugard began to conquer the north.

By 1906 Britain controlled Nigeria, which was divided into Southern and Northern Nigeria. In 1914 the two regions were amalgamated, and the Colony and Protectorate of Nigeria was established. The administration of Nigeria was based on "indirect rule"; under this system Britain ruled through existing political institutions. All important decisions were made by the British governor, and the

African rulers, partly by being associated with the colonialists, soon lost most of their traditional authority.

Under the British, Nigeria's economy grew considerably. From 1922 African representatives were elected to the legislative council of Southern Nigeria; otherwise Africans continued to have no role in the higher levels of government. Occasionally discontent with colonial rule flared into open protest. A small Western-educated elite developed in Lagos and a few other southern cities. In 1947 Great Britain promulgated a constitution that gave the traditional authorities a greater voice in national affairs. The Western-educated elite was excluded. Three major political parties emerged—the National Council of Nigeria and the Cameroons (NCNC; from 1960 known as the National Convention of Nigerian Citizens), led by Nnamdi Azikiwe and largely based among the Ibo; the Action Group, led by Obafemi Awolowo and with a mostly Yoruba membership; and the Northern People's Congress (NPC), led by Ahmadu Bello and based in the north. With Nigerian independence scheduled for 1960, elections were held in 1959; no party won a majority, and the NPC combined with the NCNC to form a government. Nigeria attained independence on Oct. 1, 1960, with Abubakar Tafawa Balewa of the NPC as prime minister and Azikiwe of the NCNC as governor-general; when Nigeria became a republic in 1963, Azikiwe was made president. The first years of independence were characterized by severe conflicts within and between regions.

In1964 there was great controversy over the 1963 population census and regional imbalances in representation in the federal parliament. National elections late in 1964 were hotly contested, and in Oct., 1965, elections in the western region were marred by widespread irregularities. In Jan., 1966, Ibo army officers staged a successful coup d'état, which resulted in the deaths of Federal Prime Minister Balewa, Northern Prime Minister Ahmadu Bello, and Western Prime Minister S. I. Akintola. Maj. Gen. Johnson T. U. Aguiyi-Ironsi, an Ibo, became head of a military government and suspended the national and regional constitutions. In July, 1966, a coup led by Hausa army officers ousted Ironsi (who was killed) and placed Lt. Col. Yakubu Gowon at the head of a new military regime. In Sept., 1966, many Ibos living in the north were massacred. In May, 1967, the eastern parliament gave Lt. Col. Chukwuemeka O. Ojukwu, the region's leader, authority to declare the region an independent republic.

On May 30 Ojukwu proclaimed the independent Republic of Biafra, and in July fighting broke out between Biafra and Nigeria. Biafra made some advances early in the war, but soon federal forces gained the initiative. Biafra capitulated on Jan. 15, 1970, and the secession ended. The early 1970s were marked by reconstruction in areas that were formerly part of Biafra, by the gradual reintegration of Ibos into national life, and by a slow return to civilian rule. In July, 1975, Gen. Gowon was overthrown and Brigadier (later Gen.) Murtala Ramat Mohammed became Nigeria's leader. On Feb. 13, 1976, Mohammed was assassinated; Lt. Gen. Olusegun Obasanjo was named head of state.

Spurred by the booming petroleum industry, the Nigerian economy quickly recovered from the effects of the civil war, although inflation, high unemployment, and a decline in the world price of peanuts and cocoa caused problems. The Sahelian drought (1968–75) affected northern Nigeria, destroying livestock and the fishing industry on Lake Chad. In 1975 Gowon was overthrown in a military coup led by Gen. Murtala Rufai Mohammed. He, in turn, was assassinated in an unsuccessful coup attempted by junior officers (1976), 30 of whom were later executed. As Mohammed's successor the military junta chose Gen. Olusegun Obasanjo. In 1976 the government increased the number of states from 12 to 19.

Ni·i·ga·ta (nē-gä'tə), city (1970 pop. 383,869), capital of Niigata prefecture, N Honshu, Japan, on the Sea of Japan at the mouth of the Shinano River. It is a port city and has an important chemical industry based on the area's coal and natural-gas deposits. The city is traversed by many canals and is the site of one of the largest flower farms in the Orient. Niigata was opened to foreign trade in 1869.

Ni·i·ha·ma (nē-hä'mə), city (1976 est. pop. 132,693), Ehime prefecture, N Shikoku, Japan. It is a commercial and fishing port and a manufacturing and mining center.

Ni·i·ha·u (nē'hou'), island (1970 pop. 237), 70 sq mi (181.3 sq km), in Kauai co., Hawaii, W of Kauai island. Niihau is suitable only for cattle grazing, and the island, privately owned since 1864 when it was purchased from the Hawaiian kingdom by an American family, is operated as a ranch.

Ni·i·tsu (nē'tsōō'), city (1975 pop. 42,900), Niigata prefecture, W central Honshu, Japan. It is a center for mechanical and chemical industries in a region that produces oil and natural gas.

Ni·i·za (nē'zə), city (1976 est. pop. 111,922), Saitama prefecture, E central Honshu, Japan, on the Yonase River; a suburb of Tokyo.

Nij·me·gen (nī'mā'gən), city (1977 est. pop. 148,073), Gelderland prov., E Netherlands, on the Waal River, near the West German border. It is a rail and water transportation point and an industrial center. Its manufactures include metal products, paper, clothing, and soap. One of the oldest cities in the Netherlands, Nijmegen was founded in Roman times and flourished under Charlemagne (8th-early 9th cent.). It was chartered in 1184, became a free imperial city, and later joined the Hanseatic League. It subscribed (1579) to the Union of Utrecht, formed as a defensive measure against Philip II of Spain. The treaties of Nijmegen (1678–79), which ended the Dutch War (1672–78) of Louis XIV of France, were signed here. Nijmegen has a 13th cent. church (the Groote Kerk), a 16th cent. town hall, and the remains of a palace built (c.777) by Charlemagne and rebuilt by Frederick Barbarossa in 1165.

Nik·ko (nēk'kō), town (1975 pop. 26,279), Tochigi prefecture, central Honshu, Japan. It is a tourist resort and religious center, famous for its ornate temples and shrines dating from the Yedo period (1600–1868). Within the shrine of Ieyasu is the Yomeimon (Gate of Sunlight), perhaps the most beautiful gate in Japan.

Ni·ko·la·yev (nĭk'ə-lä'yəf), city (1976 est. pop. 436,000), capital of Nikolayev oblast, S European USSR, in the Ukraine, at the confluence of the Bug and Ingul rivers. Nikolayev has shipyards, machinery plants, and cast-iron works. Founded in 1784 as a fortress near the site of the ancient Greek colony of Olbia, the city was named Nikolayev in 1788 when it became a shipbuilding center.

Ni·ko·pol (nĭ-kô'pəl), town (1963 est. pop. 5,763), N Bulgaria, a port on the Danube River bordering Rumania. Farming, viticulture, and fishing are the chief occupations. Nikopol is one of the world's major sources of manganese. The town's industries also include steel rolling, flour milling, tanning, shipbuilding, and the manufacture of construction materials. Founded in 629 by Byzantine emperor Heraclius, Nikopol (then Nicopolis) became a flourishing trade and cultural center of the second Bulgarian kingdom. In 1396 at Nikopol the Ottoman Turks under Bayazid I defeated an army of Crusaders. The Turkish victory removed the last serious obstacle to a Turkish advance on Christian Europe. However, when Tamerlane defeated Bayazid (1402), Europe gained a respite. The Turks strongly fortified Nikopol, which was strategically important during the Russo-Turkish wars (18th–19th cent.), but the city later declined.

Ni·ko·pol (nĭ-kô'pəl), city (1976 est. pop. 143,000), S European USSR, in the Ukraine, on the Dnepr River. It is the industrial center of one of the world's richest manganese-mining areas. The city has steel and machinery plants, shipyards, flour mills, and food-processing and brewing industries. Nikopol stands on the site of one of the earliest trade routes and strategic crossing points over the Dnepr. The city was founded in the 18th cent. on the site of a fortified camp of the Zaporozhe Cossacks.

Nik·šić (nĭk'shĭch'), town (1971 pop. 28,547), SW Yugoslavia, in Montenegro. It is the commercial center of an agricultural region. Founded in the early Middle Ages, Nikšić was under Turkish rule until 1878, when it passed to Montenegro.

Nile (nīl), longest river in the world, c.4,160 mi (6,695 km) long from its remotest headstream, the Luvironza River in Burundi, central Africa, to its delta on the Mediterranean Sea, NE Egypt. The Nile flows northward and drains c.1,100,000 sq mi (2,850,000 sq km), about one tenth of Africa, including parts of Egypt, Sudan, Ethiopia, Kenya, Uganda, Rwanda, Burundi, and Zaire. Its waters support practically all agriculture in the most densely populated parts of Egypt, furnish water for more than 20% of Sudan's total crop area, and are widely used throughout the basin for navigation and hydroelectric power.

The trunk stream of the Nile is formed at Khartoum, Sudan, 1,857 mi (2,988 km) from the sea, by the junction of the Blue Nile (c.1,000 mi/1,610 km long) and the White Nile (c.2,300 mi/3,700 km long). The Blue Nile rises in the headwaters of Lake Tana, northwest Ethiopia, a region of heavy summer rains, and is the source of floodwaters that reach Egypt in Sept.; the Blue Nile contributes more than half of all Nile waters throughout the year. During floodtime it also carries great quantities of silt from the highlands of Ethiopia. The

White Nile rises in the headwaters of Victoria Nyanza in a region of heavy, year-round rainfall; unlike the Blue Nile, it has a constant flow, owing in part to its source area and in part to the regulating effects of its passage through lakes Victoria and Albert and the Sudd swamps.

The Gezira, or "island," formed between the Blue Nile and the White Nile as they come together at Khartoum is Sudan's principal agricultural area and the only large tract of land outside Egypt irrigated with Nile waters. From Khartoum to the Egyptian border at Wadi Halfa (now submerged) and on to Aswan in Egypt, the Nile occupies a narrow entrenched valley with little floodplain for cultivation; in this stretch it is interrupted by six cataracts (rapids). From Aswan the river flows north 550 mi (885 km) to Cairo, bordered by a floodplain that gradually widens to c.12 mi (20 km); irrigated by the river, this intensively cultivated valley contrasts with the barren desert on either side. North of Cairo is the great Nile delta (c.100 mi/160 km long and up to 115 mi/185 km wide), which contains 60% of Egypt's cultivated land and extensive areas of swamps and shallow lakes. Two distributaries, the Damietta (Dumyat) on the east and the Rosetta (Rashid) on the west, each c.150 mi (240 km) long, carry the river's remaining water to the Mediterranean Sea. Regular steamship service is maintained on the Nile between Alexandria (reached by canal) and Aswan; the Blue Nile is navigable from June through Dec. from Suki (above Sennar Dam) to Roseires Dam; the White Nile is navigable all year between Khartoum and Juba in Sudan and between Nimule and Murchison Falls on the Victoria Nile.

The use of the Nile for irrigation, now regulated by the Nile Waters Treaty of 1959, dates back to at least 4000 B.C. in Egypt. The traditional system of basin irrigation—in which Nile floods were trapped in basins and crops were grown in soaked and silt-replenished soil—has been replaced in the last 150 years by a system of perennial irrigation and the production of two or three crops a year. The delta barrages, just below Cairo, channel water into a system of feeder canals for the delta, and other barrages keep the level of the Nile high enough all year for perennial irrigation in the valley of Upper Egypt. Nile water is also used for irrigation in the Fayyum Basin. The Aswan Dam (completed 1902 and raised twice since then) was the first dam built on the Nile to store part of the autumn flood for later use; it is now supplemented by the Aswan High Dam (completed 1971), 5 mi (8 km) upstream, with a storage capacity of 48 billion cu m, sufficient (with existing dams) to hold back the entire flood for later use.

Construction of the Aswan High Dam has added c.1,800,000 acres (729,000 hectares) of irrigated land to Egypt's cultivable area and converted c.730,000 acres (295,650 hectares) from basin to perennial irrigation. Other important storage dams, all outside Egypt but built with Egypt's help or cooperation, are the Owen Falls Dam (1954) and Jebel Aulia Dam (1937) on the White Nile; the Sennar (1927) and Roseires (1966) on the Blue Nile; and the Kashm-el-Girba Dam (1964) on the Atbara River. A number of other schemes to increase the available waters of the Nile have been proposed from time to time; they include the Equatorial Nile Project, or Jonglei Diversion Canal, to carry the White Nile around the Sudd and reduce evaporation losses; the construction of a dam at Lake Tana on the Blue Nile; and the construction of dams at Nimule on the Bahr-el-Jebel, on the Albert Nile, and below Lake Kyoga on the Victoria Nile.

The source of the Nile and its life-giving floods was a mystery for centuries. Ptolemy held that the source was the "Mountains of the Moon." James Bruce, the Scottish explorer, identified (1770) Lake Tana as the source of the Blue Nile, and John Speke, the British explorer, is credited with the identification (1861–62) of Lake Victoria and Ripon Falls as the source of the White Nile.

Niles (nīlz). **1.** Village (1978 est. pop. 29,700), Cook co., NE Ill., a suburb adjacent to Chicago, on the Chicago River; settled 1832, inc. 1899. It has large plants making duplicating machines, electronic equipment, and tools and dies. The village has a replica (half-size) of the leaning tower of Pisa. **2.** City (1978 est. pop. 14,200), Berrien co., SW Mich., on the St. Joseph River, in a farm and fruit area; inc. 1829. It was the site of a Jesuit mission (1690) and of Fort St. Joseph, built by the French (1697). The fort fell to the British (1761), to the Indians (Pontiac's Rebellion, 1763), and to the Spanish and Indians (1780, 1781); it was later abandoned. Permanent settlement began in 1827, and as a station on the stagecoach route between Detroit and Chicago, Niles grew as a commercial and industrial center. **3.** City (1978 est. pop. 24,300), Trumbull co., NE Ohio, on the Mahoning River; settled 1806, inc. as a city 1895. It is an iron and steel center. The city has a memorial to President William McKinley.

Nîmes (nēm), city (1975 pop. 127,933), capital of Gard dept., S France, in Cévennes. Its products include machinery, textiles and clothing, and tinware. An old Gallic town, it became Roman c.120 B.C. United to the French crown in 1258, it later became a stronghold of the Huguenots but suffered greatly from the revocation of the Edict of Nantes (1685). Nîmes is famous for its remarkable collection of Roman relics. The magnificent Roman arena (1st cent. A.D.), seating up to 24,000, is still in use. The well-preserved Maison Carrée, a Roman temple (1st or 2nd cent. A.D.), one of the finest examples of Roman architecture, houses a museum of Roman antiquities. Other Roman relics are the temple of Diana (2nd cent. A.D.), a watchtower, and the nearby Pont du Gard.

Nin·e·veh (nĭn'ə-və), ancient city, capital of the Assyrian Empire, on the Tigris River opposite the site of modern Mosul, Iraq. The old capital, Assur, was replaced by Calah, which seems to have been replaced by Nineveh. Nineveh was thereafter generally the capital and reached its full glory under Sennacherib and Assurbanipal. It continued to be the leader of the ancient world until it fell to a coalition of Babylonians, Medes, and Scythians in 612 B.C. and the Assyrian Empire came to an end. Excavations, begun in the middle of the 19th cent., have revealed an Assyrian city wall with a perimeter of c.7.5 mi (12 km). The palaces of Sennacherib and Assurbanipal, containing magnificent sculptures, have been discovered, as well as Assurbanipal's library, including over 20,000 cuneiform tablets.

Ning·hsia Hui Autonomous Region (nĭng'shyä' hwē'), autonomous region (1968 est. pop. 2,000,000), c.105,800 sq mi (247,025 sq km), N China; capital Ying-ch'uan. Ninghsia is part of the Inner Mongolian plateau, and desert and grazing land make up most of the area. Extensive land reclamation and irrigation projects have increased cultivation. Wheat, kaoliang, rice, beans, fruit, and vegetables are grown. Wools, furs, hides, and rugs are exported, and there is some gold and silver mining. Desert lakes yield salt and soda. Formerly a province, Ninghsia was incorporated into Kansu in 1954 but was detached and reconstituted as an autonomous region for the Hui people in 1958. In 1969 Ninghsia Hui received a part of the Inner Mongolian Autonomous Region.

Ning-po or **Ning·po** (nĭng'pô'), city (1970 est. pop. 350,000), NE Chekiang prov., SE China, at the confluence of the Yung (or Ning-po) and Yao rivers. It is one of China's leading fishing ports. Industries include salt panning, food canning, and the production of textiles. Long a center of culture and religion, Ning-po has many temples and Buddhist monasteries. The present site of Ning-po has been occupied since the 8th cent. A.D. From 1433 to 1549 it served as the port of entry for Japanese missions to the Chinese court. The Portuguese established a trading settlement here in the 16th cent. In the Opium War (1841), British forces occupied the city. The Treaty of Nanking (1842), ending hostilities, made Ning-po a treaty port.

Ni·o·brar·a (nī'ə-brâr'ə), county (1970 pop. 2,924), 2,613 sq mi (6,767.7 sq km), E Wyo., bordering on S.Dak. and Nebr. and watered by branches of the Cheyenne River; formed 1911; co. seat Lusk. In a grain and livestock region, it has petroleum and natural-gas fields. There is some manufacturing.

Niobrara, river, c.430 mi (690 km) long, rising in the High Plains, E Wyo., and flowing E across N Nebr. to the Missouri River. The Mirage Flats irrigation project uses water impounded by Box Butte Dam (completed 1946) to irrigate c.35,000 acres (14,160 hectares).

Ni·o·ro (nē-ō'rō), town (1967 est. pop. 11,000), W Mali. The market center for an agricultural region where millet, cotton, and gum arabic are produced, Nioro has a cotton ginning and treatment plant. Cattle raising is also important, and the town has an experimental station for raising Karakul sheep. In the 16th cent. Nioro, located at the convergence of several trade routes, was part of the Songhai empire. By the early 17th cent. it was able to form the independent state of Kaarta. In 1854 the state fell to Al-hajj Umar, the militant Moslem reformer, and in 1891 Nioro was taken by the French.

Niort (nyôr), city (1968 pop. 50,079), capital of Deux-Sèvres dept., W France, in Poitou. An old agricultural marketplace, it now has plywood, chemical, metallurgy, clothing, tobacco, and printing industries. Niort was originally a Gallo-Roman town. During the 16th and 17th cent. it was a stronghold of the Huguenots. Of the old fortress (12th–13th cent.), two huge towers remain; there are also several fine Renaissance buildings, including a town hall (16th cent.) and a church (15th–17th cent.).

Nip·i·gon, Lake (nĭp'ĭ-gŏn'), c.1,870 sq mi (4,840 sq km), central

Ont., Canada. It has many islands. Its outlet, the Nipigon River (40 mi/64 km long), flows south into Lake Superior.

Nip·is·sing, Lake (nĭp'ĭ-sĭng'), c.350 sq mi (910 sq km), S Ont., Canada, between the Ottawa River and Georgian Bay.

Nip·pon (nĭ-pŏn', nĭp'ŏn), name for Japan, derived from Dai Nippon, meaning Great Japan. It comes from the Chinese ideograph for the place where the sun comes from, or Land of the Rising Sun.

Nip·pur (nĭ-po͝or'), ancient city of Babylonia, a N Sumerian settlement on the Euphrates. It was the seat of the important cult of the god Enlil, or Bel. Excavations at Nippur have yielded the remains of several temples that date from the middle of the 3rd millennium B.C. and were later rebuilt and restored many times. Clay tablets found here serve as a primary source of information on Sumerian civilization. Relics of the Persian and Parthian periods have also been unearthed at the site.

Niš or **Nish** (both: nēsh), city (1971 pop. 193,320), E Yugoslavia, in Serbia, on the Nišava River. An important railway and industrial center, it has industries that manufacture machinery, leather and tobacco products, and armaments. It was the site of a victory (A.D. 269) of Claudius II over the Ostrogoths and was the birthplace of Constantine I. In 441 it was destroyed by the Huns but was rebuilt (6th cent.) by Emperor Justinian I. In the Middle Ages the city passed back and forth between the Bulgarian and Serbian empires. The Turks captured it c.1386, were defeated in 1443 by John Hunyadi, and recaptured it again in 1456. It became (until 1878) their most important military stronghold in the Balkans. It passed to Serbia in 1878. The Tower of Skulls was built to commemorate the Serbs massacred by the Turks in the uprising of 1809.

Ni·shi·no·mi·ya (nē'shē-nō'mē-yä), city (1976 est. pop. 403,756), Hyogo prefecture, S Honshu, Japan, on Osaka Bay. Nishinomiya is a resort, as well as the site of several temples that were founded in the 7th and 8th cent.

Ni·shi·o (nē-shē'ō), city (1975 pop. 62,600), Aichi prefecture, central Honshu, Japan. It is an agricultural market and manufacturing center for textiles and sake.

Ni·te·rói (nē'tə-roi'), city (1975 est. pop. 376,033), Rio de Janeiro state, SE Brazil, on Guanabara Bay opposite the city of Rio de Janeiro. It is a residential suburb of Rio and an important industrial center. Foodstuffs, transportation equipment, textiles, pharmaceuticals, and metals are the principal products. The area was settled by Indians in 1573 on land granted by the king of Portugal. By 1819 the Indian community was extinct.

Ni·tra (nyĭ'trä), city (1975 est. pop. 57,105), S Czechoslovakia, in Slovakia, on the Nitra River, a tributary of the Danube. It is an agricultural market center and has sugar refineries, breweries, and food-processing industries. Dating from Roman times, Nitra was important from the 9th cent. onward as a religious center and fortress. It became a free city by royal decree in 1248. Nitra's Roman Catholic bishopric church and a castle (founded c.830) are the oldest structures in Slovakia.

Ni·tro (nī'trō), city (1970 pop. 8,019), Kanawha and Putnam cos., W W.Va., on the Kanawha River NW of Charleston; inc. 1932. It developed around a government explosives plant, which was built in World War I and later abandoned. Chemicals and rayon are made.

Ni·u·e (nē-o͞o'ā), coral island (1971 est. pop. 5,100), c.100 sq mi (260 sq km), South Pacific, a territory of New Zealand. The island has fertile soil and exports copra and bananas.

Ni·velles (nē-vĕl'), city (1977 pop. 20,606), Brabant prov., central Belgium. It is an industrial center and a rail junction. Manufactures include machinery, paper, and railroad equipment. Of note are a 7th cent. convent and a Romanesque church (11th cent; rebuilt in the 18th cent.).

Ni·ver·nais (nē-vĕr-nā'), region and former province, central France. Drained by the Loire and Yonne rivers, it is a hilly plateau, rising to the Morvan Mts. in the east. It has metallurgical, chemical, and livestock industries. A county after the 10th cent., it passed (1384) through inheritance to Philip the Bold of Burgundy, and later, as a duchy, passed (1565) through a complicated succession to the house of Gonzaga. Cardinal Mazarin bought (1659) the title, which remained with his family even after Louis XIV incorporated (1669) Nivernais into the royal domain.

Ni·zam·a·bad (nĭ-zäm'ə-bäd', -zăm'-), town (1971 pop. 114,868), Adhra Pradesh state, S central India. It is a market for grain, sugar, and vegetable oil. The district is irrigated by the Nizamsagar hydroelectric project.

Nizh·ne·var·tovsk (nyēzh'nyĭ-vär'təfsk), city (1976 est. pop. 63,000), N Siberian USSR. The discovery of a huge oil field at nearby Lake Samotlor in 1965 quickly transformed the small village of Nizhnevartovsk into an oil center. The field is one of the world's largest.

Nizh·ni Ta·gil (nĭzh'nē tə-gĭl', nyēzh'nyē tə-gēl'), city (1976 est. pop. 396,000), E European USSR, in the central Urals, on the Tagil River. It is a leading metallurgical and heavy industry center. Railroad cars, machinery, and chemicals are manufactured, and copper, iron, and gold are mined in the district.

No (nō), lake, S central Sudan, in the swampy Sudd region. It is formed by the flood waters of the White Nile and varies in size seasonally. Its maximum area is c.40 sq mi (105 sq km). Much papyrus grows in the lake.

No·a·tak (nō'o͞o-tăk), river, c.400 mi (640 km) long, NW Alaska, rising in Brooks Range and flowing W to Kotzebue Sound opposite Kotzebue. There are mineral deposits in its basin. It was first explored in 1885–86.

No·be·o·ka (nō'bē-ō'kə), city (1976 est. pop. 135,882), Miyazaki prefecture, E Kyushu, Japan, on the mouth of the Gokase River. It is a commercial and fishing port and a production center for chemicals, foodstuffs, and textiles.

No·ble (nō'bəl). **1.** County (1970 pop. 31,382), 410 sq mi (1,061.9 sq km), NE Ind., drained by the Elkhart River; formed 1835; co. seat Albion. In an agricultural area yielding livestock, poultry, fruit, grain, soybeans, truck crops, and dairy products, it has gravel pits and plants processing iron and aluminum. Among its diverse manufactures are confectionery, wood and paper products, plastics, and machinery. There are many small lakes in the county. **2.** County (1970 pop. 10,428), 398 sq mi (1,030.8 sq km), E Ohio, drained by Wills, Duck, and Seneca creeks; formed 1851; co. seat Caldwell. Its agriculture includes livestock, grain, tobacco, poultry, and dairy products. It has coal mines, oil wells, clay pits, limestone quarries, and some manufacturing (wood products and auto parts). **3.** County (1970 pop. 10,043), 744 sq mi (1,925 sq km), N Okla., bounded in the NE by the Arkansas River and drained by Black Bear Creek; formed 1907; co. seat Perry. It has agriculture (wheat, oats, cotton, livestock, and dairy products), and oil and natural-gas wells.

No·bles (nō'bəlz), county (1970 pop. 23,208), 712 sq mi (1,844.1 sq km), SW Minn., bordering on Iowa and watered by headwaters of the Little Rock River; formed 1857; co. seat Worthington. It is in an agricultural area that produces corn, oats, barley, potatoes, and livestock. It has some mining and a food-processing industry. Wood and plastic products and mobile homes are manufactured.

No·bles·ville (nō'bəlz-vĭl'), city (1970 pop. 7,548), seat of Hamilton co., central Ind., NNE of Indianapolis, in a grain and livestock area; settled 1823, inc. as a town 1851, as a city 1887. Rubber goods are made, and racing horses are bred in the area.

No·da (nō'də), city (1975 pop. 78,193), Chiba prefecture, E central Honshu, Japan, on the Edo River. It is an industrial center.

Nod·a·way (nŏd'ə-wā'), county (1970 pop. 22,476), 877 sq mi (2,271.4 sq km), NW Mo., drained by the Nodaway and Little Platte rivers; formed 1845; co. seat Maryville. It is primarily agricultural (corn and wheat), with some stock raising (especially hogs).

No·gal·es (nō-gä'lās), city (1970 pop. 56,865), Sonora state, NW Mexico, on the Ariz. border, contiguous to Nogales, Ariz. Nogales derives its importance chiefly from international trade. A brief but bloody border dispute led to the city's occupation by Americans from Nogales, Ariz., in 1918.

No·gal·es (nə-găl'ĭs), city (1970 pop. 8,946), seat of Santa Cruz co., S Ariz., on the Mexican border; founded 1880, inc. 1893. In a rich mining and ranching area, it is important chiefly as a port of entry.

No·ginsk (nō-gĭnsk', nə-gēnsk'), city (1976 est. pop. 111,000), central European USSR, on the Klyazma River. It is a major textile center, processing cotton, silk, and wool.

No·ki·a (nō'kē-ä), town (1975 est. pop. 22,308), SW Finland, on Lake Näsijärvi. It is an industrial community where wood and rubber products are manufactured.

No·la (nô'lä), town (1976 est. pop. 21,700), in Campania, S Italy. It is an agricultural center with food-processing industries. An Etruscan stronghold as early as 500 B.C., Nola flourished after passing (c.316

B.C.) to Rome and was an important center of early Christianity. Nearby are Roman ruins (an amphitheater and tombs) and an old cemetery where Christian martyrs were buried.

No·lan (nō'lən), county (1970 pop. 16,220), 921 sq mi (2,385.4 sq km), W Texas, drained by the Colorado and Brazos rivers; formed 1876; co. seat Sweetwater. This ranching, agricultural, and dairying area yields cattle, sheep, goats, horses, hogs, poultry, cotton, corn, grain sorghums, oats, wheat, hay, some fruit, and vegetables. It has oil wells and gypsum quarries. There are recreation areas at Sweetwater and Trammel lakes.

Nol·i·chuck·y (nŏl'ə-chŭk'ē), river, c.150 mi (240 km) long, rising in the Blue Ridge, W N.C., and flowing NW and W to the French Broad River W of Greeneville, Tenn. The first settlement on the river was made in 1772.

Nome (nōm), city (1970 pop. 2,488), W Alaska, on the S side of Seward Peninsula, on Norton Sound; founded c.1898, when gold was discovered on the beach here. It is a commercial and supply center, with an airport and steamer connections to Seattle. Major economic mainstays are tourism, fishing, and fur trapping. The city is a center of Eskimo handicrafts. Oil deposits have been found in the area. Nome was a gold rush town from 1899 to 1903; it attracted some 20,000 prospectors, but many died or left because of the hardships. Dredging, which replaced older methods of mining, ceased in 1962. A U.S. air force base is here.

Non·ni (nŏn'ē), river, China: *see* Nen Chiang.

Noot·ka Sound (no͞ot'kə, no͝ot'-), inlet of the Pacific Ocean and natural harbor on the W coast of Vancouver Island, SW British Columbia, Canada, lying between the mainland and Nootka Island (206 sq mi/533.5 sq km). The mouth of the sound was sighted (1774) by Juan Pérez, the Spanish explorer. The sound itself was discovered by Capt. James Cook (1778), who was the first European to land in the region. John Meares, the British explorer, established a trading post on Nootka Sound in 1788. Its seizure by Spaniards in 1789 became the subject of a controversy between Spain and England over claims in the region. The Nootka Convention (1790) resolved the dispute and opened the Northern Pacific coast to British settlement.

Nor·co (nôr'kō), city (1970 pop. 14,511), Riverside co., S Calif.; inc. 1964. Norco is in a farm region of the Santa Ana mountain plains. Poultry and eggs are the city's major products. Its industries are associated with those of adjacent Corona.

Nord (nôr), department (1975 pop. 2,510,738), 2,216 sq mi (5,739.4 sq km), N France, bordering on the North Sea and Belgium; capital Lille.

Nord·fjord (nôr'fyôr'), inlet, c.50 mi (80 km) long, Sogn og Fjordane co., SW Norway. To the south, between Nordfjord and Sognafjord, is Jostedalsbreen glacier. The Nordfjord's several branches, cutting deeply into the mountains and celebrated for their scenery, are favored tourist spots.

Nord·hau·sen (nôrt'hou'zən), city (1974 est. pop. 44,442), Erfurt dist., W East Germany, at the S foot of the Harz Mts. It is an industrial center and rail junction. Manufactures include cotton and linen textiles, clothing, heavy machinery, and construction materials. Nearby are potash mines. Known in the early 10th cent., Nordhausen was chartered in the 12th cent. and was a free imperial city from 1253 to 1803. In 1815 it passed to Prussia. The city was severely damaged in World War II.

Nord·kapp (nōr'käp'), promontory, Norway: *see* North Cape.

Nord·kyn, Cape (nōr'kən), northernmost point of the European mainland, Finnmark co., N Norway, E of North Cape.

Nörd·ling·en (nœrt'lĭng-ən), town (1974 est. pop. 16,028), Bavaria, S West Germany. It is a manufacturing center and a rail junction. Historically a Swabian town, Nördlingen was founded in the 9th cent. and became a free imperial city c.1217. In the Thirty Years' War an imperial army defeated (1634) troops at Nördlingen led by Duke Bernhard of Saxe-Weimar; the victory was a major reason for France's entry into the war in 1635. In 1645 the town was the scene of a German defeat by the French. It passed to Bavaria in 1803. The picturesque town retains its walls (14th–16th cent.), a town hall (14th cent.), the late-Gothic Church of St. George (1427–1505), and numerous 16th and 17th cent. houses.

Nore (nôr), river, c.70 mi (115 km) long, rising in NE Co. Tipperary, Republic of Ireland. It flows northeast, then southeast through a rich agricultural region to the Barrow River near New Ross.

Nore, the, sandbank in the Thames estuary, SE England, 3 mi (4.8 km) E of Sheerness. At the east end is Nore Lightship. The name is also applied to part of the Thames estuary, a famous anchorage.

Nor·folk (nôr'fək), nonmetropolitan county (1976 est. pop. 662,500), E England; administrative center Norwich. The region is one of flat, fertile farmlands, with a long, low coast bordering on the North Sea and the Wash. Norfolk produces excellent cereal and root crops and supports extensive breeding of cattle and poultry. Fishing, the manufacture of agricultural machinery, and light industries are also important. There are numerous vestiges of habitation dating from prehistoric times. After the Anglo-Saxon invasion of England, Norfolk became a part of the kingdom of East Anglia, the home of the "north folk" of that region, whence its name.

Norfolk, county (1970 pop. 604,854), 398 sq mi (1,030.8 sq km), E Mass., S of Boston and bounded on the SW by R.I.; formed 1793; co. seat Dedham. It is drained by the Charles and Neponset rivers and has agriculture in the southwest and heavily populated suburban areas in the northeast. Shoes, textiles, metal, wood, and paper products, building materials, and machinery are manufactured. It also has granite quarries and a printing and publishing industry.

Norfolk. 1. City (1978 est. pop. 17,900), Madison co., NE Nebr., on the Elkhorn River; inc. 1881. A trade and railroad center in a fertile farming region, it has a livestock market, feed mills, creameries, bottling companies, and meat-packing and food-processing plants. **2.** City (1978 est. pop. 266,400), independent and in no county, SE Va., on the Elizabeth River and the S side of Hampton Roads; founded 1682, inc. as a city 1845. It is a port of entry and a major commercial, industrial, shipping, and distribution center. With Portsmouth and Newport News, it forms the Port of Hampton Roads, one of the best natural harbors in the world. Industries include shipbuilding, meat and seafood processing, automobile assembling, and the manufacture of fertilizers, farm implements, chemicals, textiles, and peanut oil. Norfolk is also a major military center; with Portsmouth the city forms an extensive naval complex. The huge operating base (largest in the United States) includes a major naval air station, a supply center, and numerous other facilities. A rallying point for Tory forces at the start of the American Revolution, Norfolk was attacked (1776) by Americans and in the ensuing battle caught fire and was nearly destroyed. In the Civil War it was first a Confederate naval base; the battle between the *Monitor* and the *Merrimack* was fought in Hampton Roads. Norfolk fell to Union forces in May, 1862. Of interest in the city are St. Paul's Church (1738; only building to survive the burning of 1776); Fort Norfolk (1794); the Gen. Douglas MacArthur Memorial, where the general is buried; and many old homes. The Chesapeake Bay Bridge–Tunnel links Norfolk with the Delmarva Peninsula, and the Hampton Roads Bridge–Tunnel links it with Hampton, Va.

Norfolk Island, island (1970 est. pop. 1,380), 13 sq mi (33.7 sq km), South Pacific, a territory of Australia, c.1,035 mi (1,665 km) NE of Sydney. A resort, Norfolk has luxuriant vegetation and is known for its pine trees. Bean and palm seeds are exported, and livestock is raised. Discovered in 1774 by Capt. James Cook, the island was claimed by Great Britain in the hope that the pines would provide masts for the navy. When the wood proved unsatisfactory, Norfolk was made into a prison island (1788–1855). In 1856 the prisoners were removed and some descendants of the *Bounty* mutineers were moved here from Pitcairn Island. Norfolk Island was annexed to Tasmania in 1844, became a dependency of New South Wales in 1896, and was transferred to Australia in 1913.

Nor·i·cum (nôr'ĭ-kəm), province of the Roman Empire, corresponding roughly to modern Austria S of the Danube and W of Vienna. Noricum was incorporated into the Roman Empire in 16 B.C. It prospered as a frontier colony for centuries, then declined and was overrun by German tribes in the 5th cent.

No·rilsk (nŏ-rēlsk', nə-rēlsk'), city (1976 est. pop. 168,000), Krasnoyarsk Kray, N Siberian USSR. The northernmost major city in the Soviet Union, Norilsk is the center of a region where nickel, copper, cobalt, platinum, and coal are mined. Founded in 1935, Norilsk was reportedly the site of forced labor camps during the Stalin era.

Nor·mal (nôr'məl), town (1978 est. pop. 36,600), McLean co., central Ill.; inc. 1865. It lies in a productive farming area noted for fruits and nursery stock. Paper products are manufactured. The town grew around Illinois State Univ. (formerly Illinois State Normal Univ.), which remains a major contributor to its economy.

Nor·man (nôr'mən), county (1970 pop. 10,008), 885 sq mi (2,292.2 sq km), NW Minn., bounded on the W by the Red River of the North and N.Dak. and drained by the Wild Rice River; formed 1881; co. seat Ada. It is in an agricultural area that produces sugar beets, barley, wheat, potatoes, and livestock. Some mining is done.

Norman, city (1978 est. pop. 62,500), seat of Cleveland co., central Okla.; inc. 1891. Air conditioners, packaged foods, and airplanes are among the city's manufactures. The Univ. of Oklahoma is here.

Norman, intermittent river, 260 mi (418.3 km) long, N Queensland, Australia, rising in SE Gregory Range and flowing WNW and NNW, past Normanton, to the Gulf of Carpentaria.

Nor·man·dy (nôr'mən-dē), region and former province, NW France, bordering on the English Channel. It now includes five departments—Manche, Calvados, Eure, Seine-Maritime, and Orne. Rouen was the historic capital. Normandy is a region of flat farmland, forests, and gentle hills. The economy is based on cattle raising, fishing, and tourism. There are also shipbuilding, metalworking, oil-refining, and textile industries. Normandy has outstanding beach resorts, notably Deauville, Granville, and Étretat. Mont-Saint-Michel lies off the coast where Normandy and Brittany meet.

History. Part of ancient Gaul, the region was conquered by Julius Caesar and became part of the province of Lugdunensis. It was conquered by the Franks in the 5th cent. Repeatedly devastated (9th cent.) by the Norsemen, it finally was ceded (911) to their chief, Rollo, 1st duke of Normandy, by Charles III of France. The Norsemen (or Normans), for whom the region was named, acquired neighboring territories in a series of wars. In 1066 Duke William invaded England, where he became king as William I (William the Conqueror). In 1144 Geoffrey IV of Anjou conquered Normandy; branches of the Angevin dynasty came to rule England, as well as vast territories in France, Sicily, and southern Italy, where the Normans had begun to establish colonies in the 11th cent. After falling to the English in the Hundred Years' War (1337–1453), Normandy was permanently restored to France in 1450. In 1790 the province, with others in France, was abolished and replaced by the present-day departments.

Nor·ris·town (nôr'ĭs-toun'), borough (1978 est. pop. 33,800), seat of Montgomery co., SE Pa., on the Schuylkill River; settled c.1712, laid out 1784, inc. 1812. It is a regional trade center. Its manufactures include clothing, woolen textiles, metal products, electrical machinery, and asbestos products.

Norr·kö·ping (nôr'chœ'pĭng), city (1975 est. pop. 119,169), Östergötland co., SE Sweden, a seaport at the head of the Bråviken, a narrow inlet of the Baltic Sea. A major textile center, it also has industries producing paper, rubber, furniture, electrical goods, and processed food. Norrköping was founded in the 14th cent. and was burned (1719) by the Russians in the Northern War.

North Ad·ams (nôrth ăd'əmz), city (1978 est. pop. 19,400), Berkshire co., NW Mass., in the Berkshire Hills, on the Hoosic River near the Vt. border; settled c.1737, set off from Adams and inc. 1878. It is a commercial and industrial center in a summer resort and winter ski area. Manufactures include electronic and electrical components, paper products, and chemicals.

North·al·ler·ton (nôr-thăl'ər-tən), urban district (1971 pop. 10,508), administrative center of North Yorkshire, N England. Until 1974 it was included in the North Riding of Yorkshire.

North America, third-largest continent (1971 est. pop. 327,000,000), c.9,400,000 sq mi (24,346,000 sq km), the northern of the two continents of the Western Hemisphere. North America is usually considered to include all of the mainland and related offshore islands lying north of the Isthmus of Panama, which connects it with South America. However, other definitions exclude Central America (1971 est. pop. 19,000,000), 202,200 sq mi (523,698 sq km) and the Caribbean islands (1971 est. pop. 26,000,000), c.91,000 sq mi (235,690 sq km). Greenland, the French islands of St. Pierre and Miquelon (off Canada), and Hawaii (formerly considered part of Oceania) are categorized as parts of North America. The continent is bounded on the north by the Arctic Ocean, on the west by the Pacific Ocean and the Bering Sea, and on the east by the Atlantic Ocean, the Gulf of Mexico, and the Caribbean Sea. Its coastline is long and irregular. Hudson Bay is by far the largest body of water indenting the continent; others include the Gulf of St. Lawrence and the Gulf of California. The Arctic Archipelago, the West Indies, the Alexander Archipelago, and the Aleutian Islands are the principal islands off the continent's coasts. Mt. McKinley (20,320 ft/6,194 m), Alaska, is the highest point on the continent; the lowest point (282 ft/86 m below sea level) is in Death Valley, Calif. The Red Rock-Missouri-Mississippi river system (c.3,740 mi/6,020 km long) is the longest of North America. Together with the Ohio River and numerous other tributaries, it forms the world's greatest inland waterway system. Other major rivers include the Colorado, Columbia, Delaware, Mackenzie, Nelson, Rio Grande, St. Lawrence, Susquehanna, and Yukon. Lake Superior (31,820 sq mi/82,414 sq km), the westernmost of the Great Lakes, is the continent's largest lake.

Physiographically, the Anglo-American section of the continent may be divided into five major regions: the Canadian Shield, an area of ancient rock that occupies most of the northeastern quadrant, including Greenland; the Appalachian Mts., a geologically old and eroded system that extends from Newfoundland, Canada, to Alabama; the Coastal Plain, a belt of lowlands that extends from southern New England to Mexico; the Central Plains, which extend down the middle of the continent from the Mackenzie valley to the Gulf Coastal Plain; and the North American Cordillera, a complex belt of geologically young mountains and associated plateaus and basins, which extend from Alaska into Mexico. The Coastal Plain and the main belts of the North American Cordillera continue south into Mexico (where the Mexican Plateau, bordered by the Sierra Madre Oriental and the Sierra Madre Occidental, is a continuation of the intermontane system) to join the Transverse Volcanic Range, a zone of high and active volcanic peaks south of Mexico City.

North America, extending to within 10° of latitude of both the equator and the North Pole, embraces every climatic zone, from tropical rain forest and savanna on the lowlands of Central America to areas of permanent ice cap in central Greenland. During the Ice Age of the Pleistocene epoch, a continental ice sheet, centered on Hudson Bay, covered most of northern North America; glaciers descended the slopes of the Rocky Mts. and those of the Pacific Margin. Extensive glacial lakes, such as Bonneville, Lahontan, Agassiz, and Algonquin, were formed by glacial meltwater; their remnants are still visible in the Great Basin and along the edge of the Canadian Shield in the form of the Great Salt Lake, the Great Lakes, and the large lakes of west-central Canada.

The first human inhabitants of North America crossed to Alaska from northeastern Asia more than 48,000 years ago, moved southward along the Pacific coast, and then eastward. European discovery and settlement of North America dates from the 10th cent., when Norsemen settled (986) in Greenland. Although evidence is fragmentary, they probably reached eastern Canada c.1000 at the latest. Of greater impact were Christopher Columbus's discovery of the Bahamas in 1492, and John Cabot's explorations of eastern Canada (1497), which established English claims to the continent. Spanish and French expeditions also explored much of North America. Today the population of Canada and the United States is largely of European and African origin and is highly urbanized (about 74% live in urban areas); Mexico's population, about 60% mestizo (of mixed European and Indian origin), is moderately urbanized (about 59%) and clusters around Mexico City. People of European descent are a minority in most Central American and Caribbean countries.

North America's extensive agricultural lands (especially in Canada and the United States) are a result of the interrelationship of favorable climatic conditions, fertile soils, and technology. North America produces most of the world's corn, meat, cotton, soybeans, tobacco, and wheat, along with a variety of other food and industrial raw material crops. Mineral resources are also abundant; the large variety includes coal, iron ore, bauxite, copper, natural gas, petroleum, mercury, nickel, potash, and silver. Much of North America's great hydroelectric potential is being developed. The factories of North America provide an abundance of basic and manufactured products that provide a high standard of living for the people of Canada and the United States.

North·amp·ton (nôr-thămp'tən, nôrth-hămp'-), county borough (1976 est. pop. 142,000), administrative center of Northamptonshire, central England, on the Nene River. Shoemaking has long been the chief industry; engineering is second (roller bearings, earth-moving equipment, and motor vehicle components). Northampton was an important settlement of the Angles and of the Danes, and its Norman castle was the scene of parliaments from the 12th to the 14th cent. and of many sieges. In 1460 Henry VI was defeated by the Yorkists in Northampton. In 1675 fire destroyed much of the town. There are Roman and ancient British remains in the vicinity. The Church of St. Giles has a Norman doorway; All Saints' has a 14th cent. tower; and St. Peter's (12th cent.) has a Norman interior. The

12th cent. St. Sepulchre's is one of the four round churches in England. St. John's Hospital was founded in 1138.

Northampton. **1.** County (1970 pop. 24,009), 536 sq mi (1,388.2 sq km), NE N.C., in a piedmont region bounded on the N by the Va. border, on the SW by the Roanoke River, and drained by the Meherrin River; formed 1741; co. seat Jackson. Farming (peanuts, cotton, and corn) and lumbering are the mainstays of its economy. **2.** County (1970 pop. 214,545), 374 sq mi (968.7 sq km), E Pa., in an industrial region bounded in the E by the Delaware River and in the NW by the Lehigh River; formed 1752; co. seat Easton. It was first settled by English and Scots-Irish, then by the Germans. Manufacturing (steel, cement, metal products, textiles, and clothing) and slate and limestone quarrying characterize the region. **3.** County (1970 pop. 14,442), 220 sq mi (569.8 sq km), E Va., at the S end of the Eastern Shore, with Cape Charles at the S tip, and at the N side of the entrance from the Atlantic Ocean to Chesapeake Bay; formed 1634; co. seat Eastville. It is in a fertile coastal-plain truck-farming area, known for white and sweet potatoes, tomatoes, cabbage, beans, and other vegetables. It also grows strawberries and has poultry raising and dairying. There are ocean fisheries, and several canneries process seafoods and vegetables.

Northampton. **1.** City (1978 est. pop. 30,300), seat of Hampshire co., W Mass., on the Connecticut River; inc. as a town 1656, as a city 1883. Cutlery, brushes, dinnerware, wire, optical devices, plastic products, and caskets are made in Northampton. It is the seat of Smith College and Clarke School for the Deaf. Calvin Coolidge lived in Northampton and was mayor of the city. His papers and mementos are preserved in the Forbes Library. **2.** Industrial borough (1970 pop. 8,389), Northampton co., E Pa., on the Lehigh River N of Allentown; settled c.1763, inc. 1901. Textiles, clothing, and cement are made.

North·amp·ton·shire (nôr-thămp'tən-shîr', -shər), nonmetropolitan county (1976 est. pop. 505,900), 914 sq mi (2,367.3 sq km), central England; administrative center Northampton. The county is undulating agricultural country, devoted to pasture and forests. In Anglo-Saxon times the area was part of the kingdom of Mercia and was probably organized as a shire in Danish times.

North An·do·ver (ăn'dō-vər), town (1970 pop. 16,284), Essex co., NE Mass., on the Merrimack River, in a dairy and farm area; settled c.1644, set off from Andover and inc. 1855. A former textile town, its manufactures include telephone equipment, chemicals, plastics, and textile machinery.

North Ar·ling·ton (är'lĭng-tən), borough (1970 pop. 18,096), Bergen co., NE N.J., a residential and industrial suburb of Newark, on the Passaic River; settled 1700s, inc. 1896.

North At·lan·tic Drift (ăt-lăn'tĭk drĭft'), warm ocean current in the N part of the Atlantic Ocean. It is a continuation of the Gulf Stream, the merging point being at lat. 40° N and long. 60° W. Off the British Isles it splits into two branches, one going south as the Canary Current and the other going north along the coast of western and northern Europe, where it exerts considerable influence upon the climate as far as the north coast of western USSR.

North At·tle·bor·o (ăt'l-bûr'ō, -bŭr'ō), industrial town (1978 est. pop. 19,100), Bristol co., SE Mass., near the R.I. border; settled 1669, set off from Attleboro and inc. 1887. Jewelry has been made here since 1807. The Woodcock Tavern dates from 1670.

North Au·gus·ta (ô-gŭs'tə), city (1970 pop. 12,883), Aiken co., SW S.C., on the Savannah River opposite Augusta, Ga.; settled c.1860, inc. 1906. Located in a dairy-farming and poultry-raising region, it is mostly residential. Veneer, bricks and tiles, paper products, and textiles are manufactured.

North Bat·tle·ford (băt'l-fərd), city (1976 pop. 13,158), W Sask., Canada, at the confluence of the North Saskatchewan and Battle rivers, opposite Battleford. It has grain elevators and tanneries.

North Bay, city (1976 pop. 51,639), SE Ont., Canada, on Lake Nipissing. It is the transportation and commercial center of lumbering and mining districts and an outfitting point for hunting and fishing parties. Mining equipment is manufactured.

North Bell·more (bĕl'môr'), uninc. town (1978 est. pop. 23,600), Nassau co., SE N.Y., on Long Island.

North Bend (bĕnd), city (1970 pop. 8,553), Coos co., SW Oregon, on Coos Bay; settled 1853, inc. 1903. In a dairying and cranberry-growing area, it also makes wood products and has an important fishing industry. Its resort activities include sports fishing.

North·bor·ough (nôrth'bûr'ō, bŭr'ō), rural town (1970 pop. 9,218), Worcester co., E central Mass., ENE of Worcester; settled c.1672, set off from Westborough 1766. Eli Whitney was born here.

North Bra·bant (brə-bănt', brä'bənt), province (1977 est. pop. 1,991,-176), c.1,920 sq mi (4,975 sq km), S Netherlands, bordering on Belgium in the S and on West Germany in the E; capital 's Hertogenbosch. Wheat and sugar beets are grown, cattle are raised, and dairying is pursued. The history of the province was that of Brabant until the late 16th cent., when the Dutch revolted against the harsh Spanish rule. As a result of the Spanish reconquest of the larger part of the duchy, Brabant was divided by the Peace of Westphalia (1648) between the Spanish (later Austrian) Netherlands and the United Provinces of the Netherlands. North Brabant, the smaller part occupied by the United Provinces, was administered by the United Provinces as a territory and was not granted a seat in the States-General. In 1795 North Brabant became a province of the Netherlands.

North Brad·dock (brăd'ək), borough (1970 pop. 10,838), Allegheny co., W Pa., a suburb of Pittsburgh, on the Monongahela River; inc. 1897. Andrew Carnegie's first steel plant was built here in 1875. The borough was the site of Gen. Edward Braddock's defeat in the French and Indian Wars and of a mass meeting of farmers instituting the Whiskey Rebellion.

North Bran·ford (brăn'fərd), town (1970 pop. 10,778), New Haven co., S Conn., on the Branford River; settled c.1680, inc. 1831. A large rock quarry is here, and there is some light industry.

North·bridge (nôrth'brĭj'), town (1970 pop. 11,795), Worcester co., S Mass., on the Blackstone River; settled 1704, set off from Uxbridge and inc. 1772.

North·brook (nôrth'bro͝ok'), village (1978 est. pop. 30,800), Cook co., NE Ill., a suburb of Chicago; settled 1836. It is largely residential, but has some industry and research laboratories and is an insurance center. Originally a farming community, Northbrook developed industry after the coming of the railroad in 1871. Botanical gardens and a forest preserve are nearby.

North Cald·well (käl'dwĕl'), borough (1970 pop. 6,733), Essex co., NE N.J., SW of Paterson; inc. 1898.

North Ca·na·di·an (kə-nā'dē-ən), river, 760 mi (1,223 km) long, rising in NE N.Mex., and flowing SE through Okla. to join the Canadian River in the Eufaula reservoir, E Okla. Federal dams and reservoirs on the river are part of the Arkansas River basin project.

North Can·ton (kăn'tən), city (1970 pop. 15,228), Stark co., NE Ohio, a suburb of Canton; settled c.1815, inc. as a city 1961. Vacuum cleaners and die castings are among the city's manufactures.

North Cape or **Nord·kapp** (nōr'käp), promontory, rising steeply c.1,000 ft (305 m) from the Arctic Ocean, near the N end of Magerøya Island, N Norway. North Cape, at lat. 71° 10′ N, is considered to be the northernmost important point of the European continent. The point actually situated farthest north on the mainland is Cape Nordkyn.

North Carolina, state (1975 pop. 5,425,000), 52,586 sq mi (136,198 sq km), SE United States, one of the Thirteen Colonies; capital Raleigh. It is bounded on the north by Va., on the east by the Atlantic Ocean, on the south by S.C. and Ga., and on the west by Tenn.

Serving as a buffer against the Atlantic is a long chain of islands, with constantly shifting sand dunes, from which project three famous capes—Hatteras, Lookout, and Fear. Between the islands and the shoreline stretch the lagoons—Albemarle Sound and Pamlico

Sound are the largest—that receive the Chowan, Roanoke, Tar, and Neuse rivers as well as Cape Fear River (Wilmington, the chief port, is at the head of its broad estuary). The mainland bordering the sounds is low, flat tidewater country, often swampy, even beyond the Dismal Swamp. In the upper coastal plain the land rises gradually from the tidewater level, reaching 500 ft (152 m) at the fall line. There begins the Piedmont, a rolling hill country with many swift streams such as the Broad River; the Catawba, or Wateree; and the Yadkin, or Pee Dee, with its three large dams. The hydroelectric power these rivers generate has made this a great manufacturing area, and the Piedmont supports most of the state's population and has its largest cities. At the western edge of the Piedmont the land rises abruptly in the Blue Ridge, then dips down to several basins, and rises again in the Great Smoky Mts., with Mt. Mitchell (6,684 ft/2,037 m) the highest peak east of the Mississippi River. The French Broad River, the Watauga, and other rivers rising west of the Blue Ridge flow into the Mississippi system, almost all via the Tennessee River. North Carolina, in the warm temperate zone, has a mild, generally uniform climate, and the rainfall is abundant and well distributed.

The climate, the many miles of beaches, and the spectacularly beautiful mountains attract large numbers of visitors each year. Chief among the tourist attractions are the Cape Hatteras National Seashore, the Cape Lookout National Seashore, the Blue Ridge Parkway, and the Great Smoky Mts. National Park. Wildlife abounds in the national forests (the state has four) and in the Dismal Swamp. Places of historic interest include Fort Raleigh National Historic Site, on Roanoke Island; the Wright Brothers National Memorial, at Kitty Hawk; Carl Sandburg Home National Historic Site, at Flatrock; and Guilford Courthouse and Moores Creek national military parks. One of the largest military reservations in the nation is at Fort Bragg, near Fayetteville, and the huge U.S. Marine Corps amphibious training base is at Camp Lejeune, near the mouth of the New River.

North Carolina is divided into 100 counties:

NAME	COUNTY SEAT	NAME	COUNTY SEAT
Alamance	Graham	Johnston	Smithfield
Alexander	Taylorsville	Jones	Trenton
Alleghany	Sparta	Lee	Sanford
Anson	Wadesboro	Lenoir	Kinston
Ashe	Jefferson	Lincoln	Lincolnton
Avery	Newland	McDowell	Marion
Beaufort	Washington	Macon	Franklin
Bertie	Windsor	Madison	Marshall
Bladen	Elizabethtown	Martin	Williamston
Brunswick	Southport	Mecklenburg	Charlotte
Buncombe	Asheville	Mitchell	Bakersville
Burke	Morganton	Montgomery	Troy
Cabarrus	Concord	Moore	Carthage
Caldwell	Lenoir	Nash	Nashville
Camden	Camden	New Hanover	Wilmington
Carteret	Beaufort	Northampton	Jackson
Caswell	Yanceyville	Onslow	Jacksonville
Catawba	Newton	Orange	Hillsboro
Chatham	Pittsboro	Pamlico	Bayboro
Cherokee	Murphy	Pasquotank	Elizabeth City
Chowan	Edenton	Pender	Burgaw
Clay	Hayesville	Perquimans	Hertford
Cleveland	Shelby	Person	Roxboro
Columbus	Whiteville	Pitt	Greenville
Craven	New Bern	Polk	Columbus
Cumberland	Fayetteville	Randolph	Asheboro
Currituck	Currituck	Richmond	Rockingham
Dare	Manteo	Robeson	Lumberton
Davidson	Lexington	Rockingham	Wentworth
Davie	Mocksville	Rowan	Salisbury
Durham	Durham	Rutherford	Rutherfordton
Duplin	Tarboro	Sampson	Clinton
Edgecombe	Tarboro	Scotland	Laurinburg
Forsyth	Winston-Salem	Stanly	Albermarle
Franklin	Louisburg	Stokes	Danbury
Gaston	Bastonia	Surry	Dobson
Gates	Gatesville	Swain	Bryson City
Graham	Robbinville	Transylvania	Brevard
Granville	Oxford	Tyrrell	Columbia
Greene	Snow Hill	Union	Monroe
Guilford	Greensboro	Vance	Henderson
Halifax	Halifax	Wake	Raleigh
Harnett	Lillington	Warren	Warrenton
Haywood	Waynesville	Washington	Plymouth
Henderson	Hendersonville	Watauga	Boone
Hertford	Winton	Wayne	Goldsboro
Hoke	Raeford	Wilkes	Wilkesboro
Hyde	Swanquarter	Wilson	Wilson
Iredell	Statesville	Yadkin	Yadkinville
Jackson	Sylva	Yancey	Burnsville

Economy. The state leads the nation in the production of tobacco, textiles, and furniture. It grows 40% of all U.S. tobacco. Broilers (North Carolina ranks fourth in their production), dairy items, corn, soybeans, peanuts, hogs, and eggs are also important. Plentiful forests supply the thriving lumber and furniture industries. North Carolina has long been a major textile manufacturer, producing cotton, knit, synthetic, and silk goods. Other leading manufactures are electrical machinery and chemicals. The state also has mineral resources: it leads the nation in the production of feldspar, mica, and lithium materials; is second in olivine and crushed granite; third in talc; and fourth in asbestos, clays, and phosphate rock. There is valuable coastal fishing, with shrimp, menhaden, and crab the principal catches.

History. North Carolina's treacherous coast was explored by Giovanni da Verrazano in 1524 and possibly by some Spanish navigators. The first permanent settlements were made (c.1653) around Albemarle Sound by colonials from Va. Charles II of England assigned the territory south of Va. betweem the 36th and 31st parallels to eight court favorites, who became the "true and absolute Lords Proprietors" of Carolina. After 1691 the province was known as North Carolina. Deputy governors, appointed from Charleston, S.C., ruled North Carolina from 1691 to 1711. Their failure to provide a stable and efficient government severely retarded the growth of the colony. By 1700 there were only some 4,000 people, predominantly of English stock, along Albemarle Sound. But by 1711, when North Carolina was made a separate colony with its own governor, there were three towns—Bath, Edenton, and New Bern. The destructive war with the Tuscarora Indians broke out that year. The Tuscarora were defeated, and in 1714 the remnants of the tribe moved north to join the Iroquois Confederacy. A long, bitter boundary dispute with Va. was partially settled in 1728 when a joint commission ran the boundary line 240 mi (386 km) inland.

When the British government became dissatisfied with the work of the proprietors, North Carolina was made (1729) a royal colony. Thereafter the region developed more rapidly. The Indians were gradually pushed back over the Appalachians as the Piedmont was increasingly occupied by Germans, Scots-Irish, Swiss, French, and Welsh immigrants. In 1768 the back-country farmers, enraged by the excessive taxes imposed by a legislature dominated by the eastern aristocracy, organized the Regulator movement in an attempt to effect reforms. The insurgents were suppressed at Alamance in 1771. Settlements had been established beyond the mountains before the Revolution and were increased after the war. In 1784 North Carolina ceded its western lands to the United States, spurring the transmontane people to organize a new, short-lived government, the state of Franklin. The territory became (1796) the state of Tennessee.

In 1835 the western part of the state, long suppressed by the tidewater aristocracy, finally succeeded in enacting a constitution that abolished the property and religious qualifications for voting and holding office (except for Jews) and provided for the popular election of governors. In the same year began the final forced removal of most of the Cherokee Indians.

Few North Carolinians held black slaves, and not until after the firing on Fort Sumter did the state secede and join (May, 1861) the Confederacy. The coast was ideal for blockade-running, and the last important Confederate port to fall (Jan., 1865) was Wilmington. The period of Reconstruction saw the beginning of the modern state, with a tremendous rise in industry in the Piedmont. Increased use of tobacco in the Civil War stimulated the growth of tobacco manufacturing, first centered at Durham. Agriculture, however, was in a critical condition. The old plantation system was replaced by farm tenancy, which long remained the dominant system of holding land (in the early 1970s at least one quarter of the farms were still being operated by tenants). The nationwide agrarian revolt reached North Carolina in the Granger movement (1875), the Farmers' Alliance (1887), and the Populist party, which united with the Republicans to carry the state elections in 1894 and 1896.

The turn of the century marked the beginning of a new progressive era, typified by the successful airplane experiments of the Wright Brothers near Kitty Hawk. But one old pattern was strengthened when a suffrage amendment, the "grandfather clause" assuring white supremacy, was added (1900) to the state constitution. A huge highway development program, begun by the counties in 1921, was assumed by the state (1931). Industrialization grew rapidly after World War II, and in the 1950s the value of manufactured goods surpassed that of agriculture for the first time. This industrialization continued during the 1960s and early 1970s, increasing at a rate unmatched by any other Southern state. North Carolina, more than many other Southern states, was able to make a peaceful adjustment to integration in the public schools following the Supreme Court's desegregation ruling in 1954.

Government. North Carolina's present constitution dates from 1868 but was thoroughly revised in 1875-76. The state's executive branch is headed by a governor elected for a four-year term and not permitted immediate re-election. North Carolina's bicameral general assembly has a senate with 50 members and a house with 120 members, all elected for two-year terms.

Educational Institutions. The main institutions of higher learning include the Univ. of North Carolina; Duke Univ., at Durham; East Carolina Univ., at Greenville; Appalachian State Univ., at Boone; and Wake Forest Univ., at Winston-Salem.

North Cas·cades National Park (kăs-kādz'): *see* National Parks and Monuments Table.

North Channel, strait, c.75 mi (120 km) long, between Northern Ireland and Scotland, connecting the Irish Sea with the Atlantic Ocean. It is 13 mi (21 km) across at its narrowest point.

North Chi·ca·go (shĭ-kä'gō, -kô'-), industrial city (1978 est. pop. 39,800), Lake co., NE Ill.; inc. 1909. Its economy is closely intertwined with the neighboring city of Waukegan, which has a good harbor on Lake Michigan. Pharmaceuticals, chemicals, and iron, steel, and wood products are among the many manufactures.

North Col·lege Hill (kŏl'ĭj), city (1970 pop. 12,363), Hamilton co., SW Ohio, a suburb of Cincinnati; inc. as a city 1940.

North Da·ko·ta (də-kō'tə), state (1975 pop. 640,000), 70,665 sq mi (183,022 sq km), N central United States, admitted to the Union in 1889 simultaneously with S.Dak. (they are the 39th and 40th states). Bismarck, on the eastern bank of the Missouri River, is the capital of the state; North Dakota is bounded on the north by the Canadian provinces of Sask. and Man., on the east by the Red River, which separates it from Minn., on the south by S.Dak., and on the west by Mont.

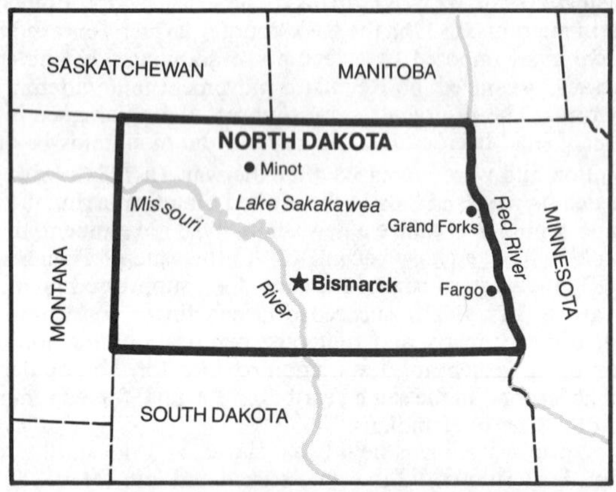

Situated in the geographical center of North America, North Dakota is subject to the extremes of a continental climate. Semiarid conditions prevail in the western half of the state, but in the east an average annual rainfall of 22 in. (55 cm), much of it falling in the crop-growing months, enables the rich soil to yield abundantly. North Dakota is one of the most rural states in the nation (the 1970 census classified 44.3% of its population as urban).

The eastern half of the state is in the central lowlands, a belt of fertile black earth. Along the banks of the Red River there is a wedge of land, c.40 mi (60 km) wide at the Canadian border and tapering to 10 mi (16 km) in the south, that is the floor of the former glacier Lake Agassiz. Treeless, except along the riversides, and without rocks, this flat land was transformed into the bonanza wheat fields of the 1870s and 1880s. To the west of the valley a series of escarpments rises some 300 ft (91 m) to meet the drift prairies, where rolling hills, scattered lakes, and occasional moraines form a pleasant and fertile countryside.

In the western part of the state a combination of unfavorable topography and scant rainfall precludes intensive cultivation except in the river valleys. An area some 50 mi (80 km) east of the Missouri River is a farm and grazing belt, divided from the drift prairies by the Missouri escarpment. Westward from the Missouri rolls an irregular plateau, covered with short prairie grasses and cut by deep coulees. Where wind and rain have eroded the hillsides there are unusual formations of sand and clay; along the Little Missouri this section is called the Badlands. Situated there are the three units of the Theodore Roosevelt National Memorial Park. Other tourist attractions in the state include the International Peace Garden on the Canadian border and recreational facilities provided by reservoirs.

North Dakota is divided into 53 counties:

NAME	COUNTY SEAT	NAME	COUNTY SEAT
Adams	Hettinger	McLean	Washburn
Barnes	Valley City	Mercer	Stanton
Benson	Minnewaukan	Morton	Mandan
Billings	Medora	Mountrail	Stanley
Bottineau	Bottineau	Nelson	Lakota
Bowman	Bowman	Oliver	Center
Burke	Bowbells	Pembina	Cavalier
Burleigh	Bismarck	Pierce	Rugby
Cass	Fargo	Ramsey	Devils Lake
Cavalier	Langdon	Ransom	Lisbon
Dickey	Ellendale	Renville	Mohall
Divide	Crosby	Richland	Wahpeton
Dunn	Manning	Rolette	Rolla
Eddy	New Rockford	Sargent	Forman
Emmons	Linton	Sheridan	McClusky
Foster	Carrington	Sioux	Fort Yates
Golden Valley	Beach	Slope	Amidon
Grand Forks	Grand Forks	Stark	Dickinson
Grant	Carson	Steele	Finley
Griggs	Cooperstown	Stutsman	Jamestown
Hettinger	Mott	Towner	Cando
Kidder	Steele	Traill	Hillsboro
La Moure	La Moure	Walsh	Grafton
Logan	Bapoleon	Ward	Minot
McHenry	Towner	Wells	Fessenden
McIntosh	Ashley	Williams	Williston
McKenzie	Watford City		

Economy. Agriculture continues to be North Dakota's principal pursuit, and the processing of grain, meat, and dairy products is vital to the state. The productivity of the soil makes North Dakota a leader in wheat, flaxseed, barley, rye, and oats. However, cattle and cattle products exceed all the crops except wheat in income earned. In the northwestern area of the state oil was discovered in 1951, and petroleum is now North Dakota's leading mineral product, ahead of lignite, sand, and gravel. There are about 2,000 producing oil wells. Underlying the western cos. are lignite reserves estimated at 350 billion tons, a large part of the nation's coal reserves. The Missouri and the Red River, once the major transportation routes, are more important now for their irrigation potential. Several dams have been built, notably Garrison Dam, and a number of Federal reclamation projects have been completed as part of the Missouri River basin project.

History. The Mandan Indians were found on the banks of the Missouri River in 1738 by the French explorer Pierre de la Vérendrye. Other tribes were the Arikara, Hidatsa, Cheyenne, Cree, Sioux, Assiniboin, Crow, and Ojibwa (Chippewa). Subsequent explorations were concerned with the fur trade. With the Louisiana Purchase of 1803 the northwestern half of North Dakota became part of the United States. The southeastern half was acquired from Great Britain in 1818 when the international line with Canada was fixed at the 49th parallel. From its post at Fort Union, established in 1828, John Jacob Astor's American Fur Company gradually gained monopolistic control for a time over the trade of the region. Supply and transport were greatly facilitated when a paddlewheel steamer, the *Yellowstone,* inaugurated steamboat travel on the turbulent upper Missouri River in 1832.

The first permanent farming community was established (1851) at Pembina. Dakota Territory was organized in 1861 to include what eventually became present-day N.Dak., S.Dak., Mont., and Wyo. Several military posts had been established, starting in 1857, to protect travelers and railroad workers from Indians. Retributive campaigns against the Sioux were waged in 1863-66. A treaty was signed in 1868. In 1876, after gold was discovered on Indian land in the Black Hills, the unwillingness of the whites to respect treaty agreements led to further war with the Indians. Ultimately, the Sioux were returned to reservations. The first cattle ranch in North Dakota was established in 1878. With the construction of railroads in the 1870s and 1880s, thousands of European immigrants arrived. Local politics was rapidly reduced to a struggle between the agrarian groups and the corporate interests. Agrarian discontent was focused on marketing practices of the large grain interests. In 1919 a reform legislature established an industrial commission to manage state-owned enterprises and created the Bank of North Dakota to handle public funds and provide low-cost rural credit.

Government. The state is governed under the 1889 constitution. The legislature consists of 51 senators elected to four-year terms and 102 representatives elected to two-year terms. The governor is elected for a four-year term.

Educational Institutions. Those prominent in the state include the Univ. of North Dakota, at Grand Forks; North Dakota State Univ. of Agriculture and Applied Science, at Fargo; and Jamestown College, at Jamestown.

North·east Passage (nôrth'ēst'), water route along the N coast of Europe and Asia, between the Atlantic and Pacific oceans. Beginning in the 15th cent., efforts were made to find a new all-water route to India and China. Most of these attempts were directed at seeking a northwest passage. However, English, Dutch, and Russian navigators did try to seek a northeast route by sailing along the northern coast of Russia and far into the arctic seas. In the 1550s English ships made the first attempt to find the passage. Willem Barentz, the Dutch navigator, made several futile voyages in the 1590s, as did Henry Hudson in the early 17th cent. The decline of Dutch shipping in the 1700s left the exploration mainly to the Russians. The Northeast Passage was not, however, traversed until Nils A. E. Nordenskjöld of Sweden accomplished the feat in 1878–79. In the early 1900s icebreakers sailed through the passage, and in the 1930s the Northern Sea Route, a shipping lane, was established by the USSR. Since World War II the USSR has maintained a regular highway for shipping along this passage.

North·ern Dvi·na (nôr'*th*ərn dvē-nä'), river: *see* Dvina.

Northern Ire·land (īr'lənd): *see* Ireland, Northern.

Northern Territory, territory (1970 est. pop. 71,400), 520,280 sq mi (1,347,525 sq km), N central Australia, bounded on the N by the Timor Sea, the Arafura Sea, and the Gulf of Carpentaria. Darwin is the territorial capital. In the north are lowlands, in the southeast are low plains sloping toward the Lake Eyre depression, and in the southwest are the MacDonald Ranges. Some 20,000 aborigines live in Northern Territory on 15 reservations with a total area of 94,000 sq mi (243,460 sq km). Gold is worked to a very small extent; uranium, bauxite, manganese, iron, lead, and zinc deposits are increasingly exploited. Stockbreeding is the major rural activity. There is very little farming. Northern Territory's first settlement was established at Port Essington in 1824, in an attempt to forestall French colonization. The settlement failed, and permanent settlement did not resume until 1869. Northern Territory was part of New South Wales from 1825 to 1863 and of South Australia from 1863 to 1911. Transferred to direct rule by the commonwealth in 1911, it was divided in 1926 but was reunited in 1931.

North·field (nôrth'fēld'). **1.** City (1970 pop. 10,235), Rice co., SE Minn., near Minneapolis–St. Paul, on the Cannon River; inc. 1875. It is the trade center for a dairy and farming region. Northfield's manufactures include plastic, wood, food, and electrical products, and woodworking and conveyor machinery. Color film is processed, packaged, and printed. On Sept. 7, 1876, Jesse and Frank James and their bandit gang attempted a bank robbery here. In the ensuing gun battle several gang members and two Northfield citizens were killed. The gang broke up after the capture of several of its members. Each Sept., Northfield celebrates the event by holding a festival during which the robbery attempt is re-enacted. **2.** City (1970 pop. 8,646), Atlantic co., SE N.J., W of Atlantic City; inc. 1905.

North·fleet (nôrth'flēt'), urban district (1971 pop. 26,679), Kent, SE England. Shipbuilding and the production of cement and paper are the main industries.

North Fork (fôrk), river, c.100 mi (160 km) long, rising in the Ozarks, S Mo., and flowing S to the White River in N Ark. Near its mouth is Norfolk Dam (completed 1944), which impounds Norfolk Lake.

North Hale·don (hāl'dən), borough (1970 pop. 7,614), Passaic co., NE N.J., N of Paterson; inc. 1901.

North Ha·ven (hā'vən), town (1978 est. pop. 23,800), New Haven co., S Conn., on the Quinnipiac River; settled c.1650, set off from New Haven 1786. Although chiefly residential, it has some manufactures.

North He·ro (hîr'ō), town (1970 pop. 364), seat of Grand Isle co., NW Vt., on North Hero Island in Lake Champlain W of St. Albans.

North High·lands (hī'ləndz), uninc. town (1978 est. pop. 36,100), Sacramento co., N central Calif., a residential suburb of Sacramento, in the Sacramento valley.

North Hol·land (hŏl'ənd), province (1977 est. pop. 2,300,595), c.1,080 sq mi (2,800 sq km), NW Netherlands, a peninsula between the North Sea in the W and the IJsselmeer in the E; capital Haarlem. North Holland is largely made up of low-lying fenland. Since the 1940s manufacturing has formed the backbone of its economy.

North Island (1976 pop. 2,268,393), 44,281 sq mi (114,688 sq km), New Zealand. It is the smaller but more populous of the two principal islands of the country. Separated from South Island by Cook Strait, North Island is irregularly shaped with a long peninsula projecting northwest. The island contains most of New Zealand's dairy and wine industries. Oil, iron, and coal are found here. Near the center of the island is a hot springs resort area.

North Kings·town (kĭng'stən, kĭngz'toun'), town (1970 pop. 29,793), Washington co., S central R.I., on Narragansett Bay; inc. as Kings Towne 1674, divided into North Kingstown and South Kingstown 1723. The site of North Kingstown was settled in 1641 by Roger Williams. North Kingstown is a regional trade center and fishing port and attracts many tourists. Its manufactures include machine tools, primary metals, printed materials, chemicals, plastics, and textiles. Points of interest in North Kingstown include Smith's Castle (1678), now a museum containing 18th and 19th cent. American furnishings; Casey House (1725), which retains the bullet holes made during skirmishes between British troops and minutemen during the Revolutionary War; Old Narragansett Church (1707); the birthplace (now a museum) of Gilbert Stuart (1755–1828), the portrait painter; and South County Museum, containing antiques and collections of early New England firearms, vehicles, and furniture.

North·lake (nôrth'lāk'), city (1970 pop. 14,212), Cook and Du Page cos., NE Ill., near Chicago; inc. 1949. Its varied manufactures include food, electrical, and paper products. St. John Vianney Roman Catholic Church, which is shaped like a fish, has the largest mosaic-tile mural in the Western Hemisphere.

North Las Ve·gas (läs vā'gəs), city (1978 est. pop. 39,500), Clark co., SE Nev., a suburb of Las Vegas; inc. 1946. The Garden of Cities features trees and plants from cities of the United States.

North Lit·tle Rock (lĭt'l rŏk'), city (1978 est. pop. 61,700), Pulaski co., central Ark., on the Arkansas River opposite Little Rock; settled c.1856, inc. as a city 1903. North Little Rock lies in a cotton, rice, soybean, dairy-cattle, and truck-farm area. Its manufactures include food products, lumber and preserved wood, insecticides and fertilizers, nonwoven fabric, metal products, and mattresses. In the early 19th cent. the discovery of a small silver vein drew settlers to the area, which was called Silver City in the futile hope that more silver would be found. Most of the area later became part of Little Rock, but in 1903 local citizens pushed a bill through the Ark. legislature permitting a part of Little Rock to secede and join the small village of North Little Rock.

North Loup (lo͞op), river, 212 mi (341.1 km) long, Nebr., rising in the center of Cherry co. and flowing SE to unite with the Middle Loup and South Loup rivers and form the Loup River.

North Man·ches·ter (măn'chĕs'tər, măn'chəs-tər), town (1970 pop. 5,791), Wabash co., NE Ind., WSW of Fort Wayne; settled 1834, inc. 1872. Furniture is manufactured.

North Man·ka·to (măn-kā'tō), city (1970 pop. 7,347), Nicollet co., S Minn., on the Minnesota River NW of Mankato; inc. 1922.

North Mer·rick (mĕr'ĭk), uninc. residential village (1970 pop. 13,650), Nassau co., SE N.Y., on Long Island.

North Mi·am·i (mī-ăm'ē, -ăm'ə), city (1978 est. pop. 45,400), Dade co., SE Fla., a suburb of Miami, on Biscayne Bay; inc. 1926. It is mainly a residential and resort city. Its manufactures include boats, wooden furniture, and aluminum products.

North Miami Beach, resort city (1978 est. pop. 37,900), Dade co., SE Fla., on the Atlantic coast; inc. 1931.

North New Hyde Park (hīd), uninc. town (1970 pop. 15,000), Nassau co., SE N.Y., on Long Island.

North Olm·sted (ŏm'stĕd), city (1978 est. pop. 38,700), Cuyahoga co., NE Ohio, a suburb of Cleveland; inc. as a city 1951. It is mainly residential; the chief industry produces printed materials. The first U.S. municipal bus line began operations in the city in 1931.

North Os·se·ti·a (ŏ-sē'shē-ə): *see* Ossetia, USSR.

North Palm Beach (päm), town (1970 pop. 9,035), Palm Beach co., SE Fla., on Lake Worth (a lagoon) N of West Palm Beach.

North Plain·field (plăn'fēld'), residential borough (1978 est. pop. 19,100), Somerset co., NE N.J.; settled 1736, inc. 1885.

North Platte (plăt), city (1978 est. pop. 22,500), seat of Lincoln co., W central Nebr., at the confluence of the North Platte and South Platte rivers; inc. 1873. A processing and shipping point for grain and livestock, it has meat-packing plants and a railroad repair shop. Scouts Rest Ranch (once a home of Buffalo Bill, who lived here for 30 years) is nearby, and there is an annual rodeo.

North Platte, river, c.680 mi (1,095 km) long, rising in the Park Range, N Colo., and flowing in a great bend N through SE Wyo., then E across W central Nebr. to join the South Platte River at North Platte city and form the Platte River. The North Platte project and the Kendrick project utilize the North Platte for power and irrigation. The valley of the North Platte, followed by the Overland Trail, was a chief route used by pioneers.

North Platte project, unit of the U.S. Bureau of Reclamation in the North Platte River valley, W Nebr. and E Wyo. It supplies power to many towns and industries and provides irrigation for c.335,000 acres (135,675 hectares) of land extending 111 mi (178.6 km) along the valley from Guernsey, Wyo., to below Bridgeport, Nebr.

North Pole, N end of the earth's axis, lat. 90° and long. 0°. It is distinguished from the north magnetic pole. The North Pole was first reached (1909) by U.S. explorer Robert E. Peary.

North·port (nôrth'pôrt'). **1.** City (1970 pop. 9,435), Tuscaloosa co., W central Ala., on the Black Warrior River opposite Tuscaloosa; inc. 1871. **2.** Resort village (1970 pop. 7,494), Suffolk co., SE N.Y., on the N shore of Long Island; settled c.1683, inc. 1894.

North Prov·i·dence (prŏv'ĭ-dəns), town (1978 est. pop. 26,900), Providence co., NE R.I.; set off from Providence and inc. 1765. Once a large textile town, it is now mainly suburban. A major portion of Rhode Island College is within the town's limits.

North Read·ing (rĕd'ĭng), residential town (1970 pop. 11,264), Middlesex co., NE Mass., on the Ipswich River; settled 1651, set off from Reading and inc. 1853.

North Rhine-West·pha·li·a (rīn' wĕst-fā'lē-ə, -fāl'yə), state (1974 est. pop. 17,217,700), 13,111 sq mi (33,957.5 sq km), W West Germany, bounded by Belgium and the Netherlands in the W, Lower Saxony in the N and E, Hesse in the SE, and Rhineland-Palatinate in the S; capital Düsseldorf. A highly industrialized state, its manufactures include chemicals, machines, processed foods, textiles, clothing, and iron and steel. The state was formed in 1946 through the union of the former Prussian province of Westphalia, the northern part of the former Prussian Rhine Prov., and the former state of Lippe.

North Rich·land Hills (rĭch'lənd), town (1970 pop. 16,514), Tarrant co., N Texas, a residential suburb of Fort Worth; inc. 1953.

North Riv·er·side (rĭv'ər-sīd'), village (1970 pop. 8,097), Cook co., NE Ill., a suburb W of Chicago; inc. 1923.

North Roy·al·ton (roi'əl-tən), city (1970 pop. 12,807), Cuyahoga co., NE Ohio, a suburb of Cleveland; settled 1811, inc. as a village 1927, as a city 1960.

North Saint Paul (sānt' pôl'), village (1970 pop. 11,950), Ramsey co., SE Minn., a suburb of St. Paul, in a lake resort region; inc. 1888. Conveyor systems and masonry products are manufactured.

North Sea, arm of the Atlantic Ocean, c.222,000 sq mi (574,980 sq km), c.600 mi (965 km) long and c.400 mi (645 km) wide, NW of Central Europe. It washes the shores of Great Britain, Norway, Denmark, West Germany, the Netherlands, Belgium, and the northern tip of France. In the south the Strait of Dover connects it with the English Channel. The North Sea is deepest (c.2,165 ft/660 m) along the coast of Norway and contains several shallows, the largest of which is the Dogger Bank, midway between England and Denmark. The cod and herring fisheries of the North Sea are of great value. In 1970 oil was discovered under the sea floor.

North Syd·ney (sĭd'nē), town (1976 pop. 8,319), NE Cape Breton Island, N.S., Canada, on Sydney Harbour. It is the coal-shipping port for the nearby Sydney Mines and a winter base for the Cape Breton fisheries.

North Syr·a·cuse (sĭr'ə-kyo͞os', -kyo͞oz', sĕr'-), village (1970 pop. 8,687), Onondaga co., central N.Y.; inc. 1925.

North Tar·ry·town (tăr'ē-toun'), village (1970 pop. 8,334), Westchester co., SE N.Y., on the E bank of the Hudson River N of Tarrytown; inc. 1875. Philipsburg Manor and the Old Dutch Church (c.1697) are here. In the Sleepy Hollow Cemetery are the graves of Washington Irving, Andrew Carnegie, and Whitelaw Reid.

North Ton·a·wan·da (tŏn'ə-wŏn'də), industrial and commercial city (1978 est. pop. 40,600), Niagara co., W N.Y., on the Niagara River at the terminus of the Barge Canal; settled c.1802, inc. as a city 1897.

North·um·ber·land (nôr-thŭm'bər-lənd), nonmetropolitan county (1976 est. pop. 287,300), NE England; administrative center Newcastle upon Tyne. Northernmost of the English counties, it is separated from Scotland by the Cheviot Hills and the Tweed River, and borders on the North Sea. The terrain is level along the rugged coast line and hilly in the interior, where high moorlands alternate with fertile valleys. Chief industries are coal mining, shipping, and ship building and repairing. Sheep and cattle are raised. In the 6th cent. the Angles established themselves in the region, which later became the kingdom of Northumbria. The area suffered severely during the border wars between England and Scotland. In 1974 a small but populous area in the southeast became part of the new metropolitan county of Tyne and Wear.

Northumberland. 1. County (1970 pop. 99,190), 454 sq mi (1,175.9 sq km), E central Pa., in an anthracite-coal region drained by the Susquehanna River and its West Branch; formed 1772; co. seat Sunbury. Its agriculture includes truck crops, grain, poultry, livestock, and dairy products. It has a food-processing industry and manufactures textiles and metal products. **2.** County (1970 pop. 9,239), 190 sq mi (492.1 sq km), E Va., on the NE shore of the Northern Neck peninsula, bounded in the E by Chesapeake Bay, with many bays and inlets on its coast; formed 1648; co. seat Heathsville. Its agriculture includes grain, potatoes, truck crops (especially tomatoes), tobacco, livestock, and poultry. It has fishing and oystering, and a seafood-processing industry. It also manufactures wood products. There are shore resorts in the county.

Northumberland Strait, arm of the Gulf of St. Lawrence, Canada, c.200 mi (320 km) long and from 9 to 30 mi (14.5–48.3 km) wide, separating P.E.I. from N.B. and N.S.

North·um·bri·a, kingdom of (nôr-thŭm'brē-ə), one of the Anglo-Saxon kingdoms in England. It was originally composed of two independent kingdoms divided by the Tees River: Bernicia and Deira, both settled by invading Angles c.500. Ethelfrith of Bernicia (593–616) united the kingdoms to form Northumbria and added Scottish and Welsh territory. He was defeated by Edwin of Deira (616–32), who accepted (627) Roman Christianity and established Northumbrian supremacy in England. Edwin was killed by Cadwallon of the Welsh kingdom of Gwynned and, after a year of anarchy, was succeeded by Oswald of Bernicia (633–41), who brought in St. Aidan to introduce Celtic Christianity. Under Oswald's successors Northumbria's power declined as that of Mercia increased. The late 7th and 8th cent. saw almost constant political discord but also the golden age of arts, scholarship, and literature in Northumbria. The invading Danes with their smashing victory at York in 867 soon wrecked all culture. They occupied southern Northumbria, and the Angles were able to keep only a small kingdom stretching from the Tees north to the Firth of Forth; however, all Northumbria acknowledged Edward the Elder of Wessex as overlord in 920. The conquering Canute (1015) and his successors installed Danish earls, the last of whom was expelled in 1065. Harold II of England came north and defeated later invaders before the coming of William the Conqueror.

North Val·ley Stream (văl'ē strēm'), uninc. residential town (1970 pop. 14,881), Nassau co., SE N.Y., on Long Island.

North Van·cou·ver (văn-ko͞o'vər), city (1976 pop. 31,934), SW British Columbia, Canada, on Burrard Inlet of the Strait of Georgia, opposite Vancouver, of which it is a suburb. Shipbuilding, woodworking, and chemical manufacturing are the chief industries.

North Vi·et·nam (vē'ĕt-näm', -năm'): *see* Vietnam.

North-West Fron·tier Province (nôrth'wĕst' frŭn-tîr'), province and historic region (1972 pop. 10,909,000), c.41,000 sq mi (106,190 sq km), NW Pakistan, bounded on the N and W by Afghanistan; capital Peshawar. An area of high, barren mountains dissected by fertile valleys, it is predominantly agricultural. The region has been historically and strategically influenced by its proximity to the Khyber Pass, through which invaders came from central Asia. Alexander the Great conquered the region c.326 B.C., but his garrisons were routed by Chandragupta, founder of the Maurya empire of India. In the early centuries A.D., Kanishka and his Kushan dynasty ruled the area. The Pathans arrived in the 7th cent., and by the 10th cent. conquerors from Afghanistan had made Islam the dominant religion. Under local Pathan rulers from the late 12th cent. until Babur annexed it to his Mogul empire, the region paid nominal allegiance to the Moguls in the 16th and 17th cent. After Nadir Shah's invasion (1738), it became a feudatory of the Afghan Durrani kingdom. The Sikhs later held the area, which passed to Great Britain in 1849. The British separated the region from the North-West Provinces of India in 1901 and constituted the North-West Frontier Province, whose people voted to join newly independent Pakistan in 1947. From 1955 to 1970 the North-West Frontier Province was a section of the consolidated province of West Pakistan.

North·west Pas·sage (nôrth'wĕst' păs'ĭj), water routes through the Arctic Archipelago, N Canada, and along the N coast of Alaska between the Pacific and Atlantic oceans. Even though the explorers of the 16th cent. demonstrated that the American continents were a true barrier to a short route to the Orient, there still remained hope that a natural passage would be found leading directly through the barrier. During the same period, the idea of reaching China and India by sailing over the North Pole or by sailing through a passage north of Europe and Asia—the Northeast Passage—also became popular. However, the Northwest Passage remained the most important goal, and the search for the passage continued even though at that time such a route had no commercial value.

Sir Martin Frobisher, the English explorer, was the first European to explore (1576-78) the eastern approaches of the passage. John Davis also explored (1585-87) this area, and in 1610 Henry Hudson sailed north and discovered Hudson Bay while seeking a short route to the Orient. Soon afterward, William Baffin, an English explorer, discovered (1616) Baffin Bay, through which the passage was finally discovered. Although one of the avowed goals of Hudson's Bay Company was to find the Northwest Passage, little was accomplished until a century after its charter. British, Spanish, and Americans pushed explorations on the Pacific coast; and the explorations of the Russians about Kamchatka and Alaska, together with the voyages of Alexander Mackenzie, the Canadian explorer, showed the contours of the continental barrier. Wars between Britain and France interrupted the search for the Northwest Passage, and later explorations were made in the interests of science, not commerce.

The desire to extend man's knowledge was the chief motive in arctic exploration after the expeditions of British explorers John Ross and David Buchan were sent out in 1818. The last tragic expedition of Sir John Franklin indirectly had more effect than any other voyage, because of the many expeditions sent out to discover his fate. In his expedition (1850-54) Robert J. Le M. McClure penetrated the passage from the west along the northern coast of the continent and by a land expedition reached Viscount Melville Sound, which had been reached (1819-20) by Sir William Edward Parry from the east. The actual existence of the Northwest Passage had thus been proved. It was many years, however, before a transit of the passage was made. This feat, which had been attempted by so many men, was first accomplished (1903-6) by the Norwegian explorer Roald Amundsen. Interest in the Northwest Passage slackened until the 1960s, when oil was discovered in northern Alaska, provoking a desire for a short water route to the east coast of the United States. In 1969 the SS *Manhattan*, an ice-breaking tanker, became the first commercial ship to transit the Northwest Passage.

Northwest Territories, region (1975 pop. 38,000), 1,304,903 sq mi (3,379,699 sq km), NW Canada. The Northwest Territories lie west of Hudson Bay, north of lat. 60° N, and east of Yukon, and occupy more than one third of Canada's area. Yellowknife has been the territorial capital since 1967; before 1967 the government was conducted from Ottawa. The region is divided into three administrative districts—Keewatin, west of Hudson Bay and including the islands in Hudson Strait, Hudson Bay, and James Bay; Mackenzie, east of Yukon; and Franklin in the northern section, consisting of the Arctic Archipelago and the Boothia and Melville peninsulas.

Geographically the region is separated by the tree line, which runs roughly northwest to southeast. The tundras extend over much of the north and east. Most of the development of the Northwest Territories has taken place in Mackenzie dist., an area well-covered with soft woods and rich in minerals. In this district are two of the largest lakes in the world, Great Slave and Great Bear, which are linked to the Arctic Ocean by one of the world's longest rivers, the Mackenzie, flowing 2,635 mi (4,241 km) from its source in British Columbia.

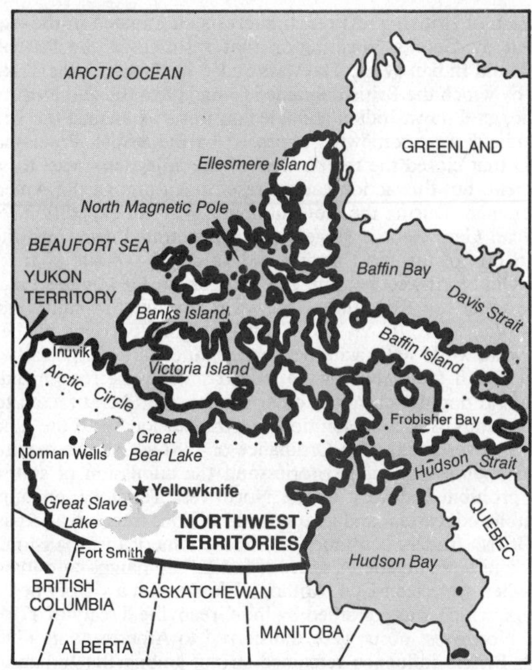

Economy. Agriculture in the Northwest Territories is virtually impossible except for limited cultivation south of the Mackenzie River region. Trapping, the region's oldest industry, ranks third after mining and fishing. A thriving commercial fishing industry, based on whitefish and lake trout, is centered on the village of Hay River, on Great Slave Lake. Leading minerals are lead and zinc, both mined at Pine Point on Great Slave Lake. Oil is pumped and refined at Fort Norman and Norman Wells on the Mackenzie River; gold and copper are extracted, and the region also has iron-ore, tungsten, silver, cadmium, and nickel deposits. Important hydroelectric developments are on the Yellowknife and Snare rivers. Transportation and communication in the Northwest Territories are difficult. Long winters close the rivers to navigation for all but two months of the year. Most commerce, supply, and travel continue to be airborne while an extensive northern roads program is in progress. Nahanni and Baffin Islands national parks are here.

History. Sir Martin Frobisher was the first European explorer to touch some part of the area, but it was Henry Hudson who discovered Hudson Bay in 1610. Samuel Hearne (1771), Alexander Mackenzie (1789), and Sir John Franklin (1818-25) mapped the region's coasts and rivers. The Northwest Territories were sold by the Hudson's Bay Company to the Canadian confederation in 1870. Some lands were added to the provinces of Que. and Ont. The province of Man. was carved from them in 1870, and Alta. and Sask. in 1905. The present boundaries of the Northwest Territories were set in 1912.

Government. Today the territory is administered by a commissioner and a 14-member council sitting in Yellowknife, backed up by the Royal Canadian Mounted Police. The government operates a school system, and, with the aid of scattered missions throughout the vast region, provides extensive health and welfare services. The Territories are represented in the Canadian Parliament by one elected representative in the House of Commons.

Northwest Territory, first possession of the United States, comprising the geographic region known as the Old Northwest, S and W of the Great Lakes, NW of the Ohio River, and E of the Mississippi River. Men from New France began to penetrate this rich fur country in the 17th cent.; in 1634 the French explorer Jean Nicolet became the first to enter the region. He was followed by explorers, traders, and missionaries. The two chief posts of the Old Northwest were Detroit and Mackinac (Michilimackinac), but French influence spread among the Indian tribes east to the Iroquois country.

In the 18th cent. the Northwest was coveted not only by the British colonists in Canada, but also by those in the American seaboard colonies, who organized the Ohio Company in 1747 for the purpose of extending settlement westward. At the same time, the French sought to strengthen their hold on the Northwest by building forts.

The clash of British and French interests culminated in the expedition led by George Washington that resulted in the last of the French and Indian Wars. The wars ended in 1763 with the Treaty of Paris, by which the British obtained Canada and the Old Northwest. Pontiac, an Ottawa Indian chief, led an uprising against the British. The Indians were somewhat appeased by the British Proclamation of 1763 that closed the region west of the Allegheny Mts. to white settlement, but this action caused resentment among the American frontiersmen. During the Revolutionary War an expedition led by American Gen. George Rogers Clark penetrated deep into the region in one of the most daring and valuable exploits of the war.

The Old Northwest became U.S. territory in 1783 by the Treaty of Paris ending the Revolution. However, the four so-called landed states—Va., Mass., N.Y., and Conn.—claimed portions of the Old Northwest, while states with no western land claims, especially Md., argued that if the claims of the landed states were recognized, the wealth and population of the other states would be attracted to the western lands. The final solution was the cession of all the lands to the U.S. government. The Ordinance of 1787 set up the machinery for the organization of territories and the admission of states. Its terms prohibited slavery in the Northwest Territory, encouraged free public education, and guaranteed religious freedom and trial by jury. British traders continued to oppose American expansion, and the Indians were hostile. A series of Indian campaigns culminated in 1794, when American Gen. Anthony Wayne won a victory at Fallen Timbers, which was solidified by the Greenville Treaty of 1795.

The Northwest posts were transferred to Americans in 1796, although British influence remained strong among the Indians. Settlers poured into the southern part of the Territory, and in 1799 a legislature was organized. In 1800 the western part was split off as Indiana Territory, and by 1803 the eastern portion was populated enough to be admitted to the Union as Ohio. Other territories were then formed—Michigan in 1805, Illinois in 1809, and Wisconsin in 1836. The British traders, however, wanted the Northwest set aside as Indian land. Indian unrest led Tecumseh and Shawnee Prophet to seek a permanent foothold for the natives. Some Western Americans, meanwhile, sought to extend the Northwest to Canada. The quarrel over the Northwest was a major cause of the War of 1812. The Treaty of Ghent, which ended the war, irrevocably gave the Northwest to the United States.

North·wich (nôrth′wĭch′), urban district (1973 est. pop. 17,710), Cheshire, W central England, at the confluence of the Weaver and Dane rivers. Northwich has long been the center of England's salt production, but the manufacture of chemicals is now its leading occupation. Its rock-salt fields were in use before the Christian era.

North·wood (nôrth′wo͝od′), town (1970 pop. 1,950), seat of Worth co., N Iowa, on the Shellrock River N of Mason City; settled 1853, inc. 1875.

North York·shire (yôrk′shîr′, -shər), nonmetropolitan county (1976 est. pop. 653,000), N England, created under the Local Government Act of 1972 (effective 1974); administrative center Northallerton. It is composed of the county borough of York, and parts of the former county of Yorkshire.

Nor·ton (nôr′tən), county (1970 pop. 7,279), 872 sq mi (2,258.5 sq km), N Kansas, in a rolling plain region bordering in the N on Nebr. and watered by Prairie Dog Creek and the North Fork of the Solomon River; formed 1867; co. seat Norton. It is in a grain and livestock area, with some mining and metal-products manufacturing.

Norton. 1. City (1970 pop. 3,627), seat of Norton co., NW Kansas, NW of Salina near the Nebr. border, in a farm area; founded 1872, inc. 1885. The area has silica mines. **2.** Town (1970 pop. 9,487), Bristol co., SE Mass., NW of Taunton; settled 1669, set off from Taunton 1711. Textiles are produced. Wheaton College is here. **3.** City (1970 pop. 12,308), Summit co., NE Ohio, a residential suburb of Akron.

Norton Sound, inlet of the Bering Sea, c.150 mi (240 km) long and 125 mi (200 km) across at its widest point, W Alaska, S of the Seward Peninsula. Norton Bay is its northeast arm.

Nor·um·be·ga (nôr′əm-bē′gə), name vaguely used, especially on European maps of the 16th and 17th cent., to indicate a region, a river, or a city on the E coast of North America. Fabulous tales were told of the city, but its location and its identity are uncertain. Probably the word is an Indian version of the old form of Norway. In the late 19th cent. Professor E.N. Horsford identified Norumbega as the site of a Norse settlement in America, claiming to have discovered its position on the Charles River at Watertown, Mass. No conclusive results have been reached on the matter, and it is generally considered that Norumbega is purely mythical.

Nor·walk (nôr′wôk′). **1.** City (1978 est. pop. 84,900), Los Angeles co., S Calif.; settled in the 1850s, inc. 1957. Cerritos College is here. **2.** City (1978 est. pop. 76,100), Fairfield co., SW Conn., at the mouth of the Norwalk River, on Long Island Sound; settled 1640, inc. 1913. Electronic and electrical equipment, pumps, and food products are manufactured, and aircraft research is carried on. Norwalk was burned by the British in the Revolution. The city includes numerous small islands in the harbor and the village of Silvermine, an artists' colony. **3.** City (1970 pop. 13,386), seat of Huron co., N Ohio; inc. 1881. It is a trade and processing center for a farm area, with factories that make rubber products, packaging materials, and machine parts. The city was settled (c.1817) by "Fire Sufferers" from Norwalk, Conn., whose homes had been burned by the British in the Revolution.

Nor·way (nôr′wā′), kingdom (1973 est. pop. 3,960,000), 125,181 sq mi (324,219 sq km), N Europe, occupying the W part of the Scandinavian peninsula. Oslo is the capital. Extending from the Skagerrak (an arm of the North Sea) c.1,100 mi (1,770 km) northeast to North Cape on the Arctic Ocean, the country forms a narrow mountainous strip along the North Sea and the Atlantic Ocean. It has a long land frontier with Sweden and in the north borders on Finland and the USSR. The coastline, c.1,700 mi (2,740 km) long, is fringed with islands and is deeply indented by numerous fjords. From the coast the land rises sharply to high plateaus. Galdhøpiggen, in the Jotunheimen range, is the highest point (8,098 ft/2,468 m); west of it lies Jostedalsbreen, the largest glacier field in Europe. The mountains

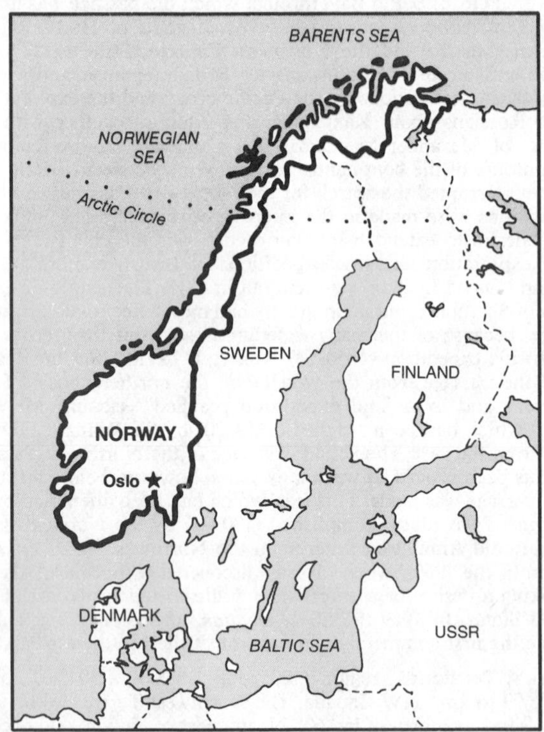

and plateaus are intersected by fertile valleys and by rapid rivers, which furnish hydroelectric power and are used for logging. The Glåma, in the south, is the most important river. Most of the population is concentrated along the southern coast and valleys, where the chief cities are located. The beautiful Norwegian fjords and the midnight sun of the far north attract many tourists. Because of the North Atlantic Drift, Norway has a mild and humid climate for a northern country. The majority of Norwegians are of Scandinavian stock, but in the northern county of Finnmark, Lapps and Finns predominate. The literary language of Norway for many years was Danish, from which Rigsmål (officially Bokmål), one of the two official idioms of Norway, is derived. Landsmål (officially Nynorsk), the other official idiom, is similar.

Norway is divided into 20 counties:

NAME	COUNTY SEAT	NAME	COUNTY SEAT
Akershus	Oslo	Oppland	Lillehammer
Aust-Agder	Arendal	Oslo	Oslo
Bergen	Bergen	Østfold	Moss
Buskerud	Drammen	Rogaland	Stavanger
Finnmark	Vadsø	Sogn og Fjordane	Hermansverk
Hedmark	Hamar	Sør-Trøndelag	Trondheim
Hordaland	Bergen	Telemark	Skien
Møre og Romsdal	Molde	Troms	Tromsø
Nordland	Bodø	Vest-Agder	Kristiansand
Nord-Trøndelag	Steinkjer	Verstold	Tønsberg

Economy. Almost three quarters of the land is unproductive; less than 4% is under cultivation. The vast mountain pastures are used for the grazing of cattle and sheep, and, in the north, for reindeer raising. About one quarter of Norway is forested. The chief industries are shipping, the production of oil from the North Sea, and trading. (Norway's oil reserves may be equivalent to those of the Persian Gulf.) Norway has one of the great merchant fleets of the world. Fishing is important, and fresh, canned, and salted fish from Norway are exported throughout the world. The pulp, paper, and electrochemical industries are important to the economy. Mineral resources include large offshore reserves of oil and natural gas; pyrites and iron ore, which are heavily mined; and some coal, copper, zinc, and lead.

History. Harold I, of the Yngling or Scilfing dynasty, defeated (c.900) the numerous petty kings who had divided Norway and conquered the Shetland and the Orkney islands, but failed to establish permanent unity. In the next two centuries Norsemen raided widely in western Europe and established the Norse duchy of Normandy. Christianity gained a foothold under Olaf I (c.1000) and was established by Olaf II (reigned 1015–28). Olaf II was driven out of Norway by King Canute of England and Denmark, but his son, Magnus I, was restored (1035) to the Norwegian throne. After Harold III died while invading England (1066), Norway entered a period of decline and civil war. Sverre (reigned 1184–1202), who created a new nobility grounded in commerce, consolidated the royal power. Under Magnus VI (reigned 1263–80) medieval Norway reached its greatest flowering and enjoyed peace and prosperity. During this time Iceland and Greenland recognized Norwegian rule.

The separate development of Norway was halted by the accession (1319) of Magnus VII, who was also king of Sweden. Margaret of Denmark united Norway, Sweden, and Denmark and in 1397 had the Kalmar Union drawn up. Norway was ruled by Danish governors for the following four centuries, losing territory to Sweden but developing economically. In 1814 Denmark, which had sided with France in the Napoleonic Wars, was obliged to cede Norway to the Swedish crown. The Norwegians failed in an attempt to set up a separate kingdom and were forced to accept Swedish King Charles XIII as their sovereign. Despite some Swedish concessions to growing Norwegian nationalism, Swedish-Norwegian relations were strained throughout the 19th cent. Finally, in 1905, the Norwegian parliament declared the dissolution of the union, Sweden acquiesced, and Norway chose in a plebiscite to become an independent monarchy.

Norway remained neutral in World War I and attempted to remain neutral in World War II, but in Apr., 1940, German troops invaded and occupied the country until 1945. Although half of the Norwegian fleet was sunk assisting the allies during the war, Norway quickly recovered its commercial position. Postwar economic policy included a degree of socialization under the ruling Labor party. Norway was one of the original members of the United Nations, and it became a member of the North Atlantic Treaty Organization (NATO) in 1949. King Olaf V succeeded Haakon VII (reigned 1905–57) in 1957 as the second king of independent Norway. Norwegian voters rejected membership in the Common Market in 1972, but trade agreements with the market were made the next year. New wealth derived from its North Sea oil fields promised to make Norway affluent, but delays in oil development and a large foreign-trade deficit caused an economic crisis in 1978.

Government. Norway is a constitutional monarchy; the legislative power is vested in the parliament or Storting. The two main parties are the moderately socialist Labor party and the pro-free enterprise Conservative party; there are also five other parties.

Nor·we·gian Sea (nôr-wē'jən), part of the Atlantic Ocean, NW of Norway, between the Greenland Sea and the North Sea. It is separated from the Atlantic by a submarine ridge. The warm Norwegian Current gives the sea generally ice-free conditions.

Nor·well (nôr'wĕl'), town (1970 pop. 7,796), Plymouth co., SE Mass., on the North River ENE of Rockland; settled 1634, set off from Scituate 1888.

Nor·wich (nôr'ĭch, -ĭj,), county borough (1976 est. pop. 119,200), administrative center of Norfolk, E England, on the Wensum River just above its confluence with the Yare. It is a market center for cattle and grain. Farm machinery, textiles, chocolate, shoes, and mustard are among its manufactures. Norwich was sacked by the Danes in the 11th cent. and devastated by the Black Death in 1348. There are many medieval churches and a great cathedral founded in 1096 by the first bishop of Norwich. Norwich Castle, part of which dates from Norman times, was made into a museum for collections of natural history and local antiquities in 1894. There are many other old buildings, including St. Giles's Hospital (13th cent.), Suckling House (14th cent.), Strangers Hall (15th cent.; now a museum), the guildhall (15th cent.), and St. Andrew's Hall (15th cent.; formerly a Dominican church). The Maddermarket Theatre is a reconstruction of a Shakespearean theater.

Nor·wich (nôr'wĭch). **1.** Industrial city (1978 est. pop. 40,300), SE Conn., New London co., on hilly ground, where the Yantic and Shetucket form the Thames; settled 1659, inc. 1784, town and city consolidated 1952. It has various manufacturing industries. The last battle between the Mohegans and Narragansetts took place on the site in 1643, and the Indian chiefs are buried here. Norwich was a leading colonial industrial city; Thomas Danforth began making pewterware here in 1733. The many historic structures include the Leffingwell Inn (1675) and the home of Benedict Arnold, who was born here. **2.** City (1970 pop. 8,843), seat of Chenango co., S central N.Y., NNE of Binghamton; settled 1788, inc. 1915. Clothing, pharmaceuticals, and electronic equipment are among its products.

Nor·wood (nôr'wo͝od'). **1.** Town (1978 est. pop. 31,200), Norfolk co., E Mass.; settled 1678, set off from Dedham and Walpole and inc. 1872. It is chiefly residential. **2.** City (1978 est. pop. 24,000), Hamilton co., SW Ohio, a suburb of Cincinnati; settled early 1800s, inc. 1888. It has varied light industries. **3.** Borough (1970 pop. 7,229), Delaware co., SE Pa., NE of Chester; inc. 1893.

No·teć (nô'tĕch'), river, c.270 mi (435 km) long, NW Poland. It rises south of Inowrocław and flows generally west into the Warta River near Gorzów Wielkopolski. The Noteć is connected by canal with the Vistula River and is navigable for almost its entire length.

No·to (nō'tō), peninsula, c.45 mi (70 km) long and from 6 to 17 mi (9.7–27.4 km) wide, Ishikawa prefecture, W central Honshu, Japan, between the Sea of Japan and Toyama Bay. The rugged peninsula has a deeply indented east coast. Farming, lumbering and fishing are major economic activities.

Not·od·den (nō'tôd'n), town (1977 est. pop. 13,012), Telemark co., SE Norway. The world's first nitrate factory was built here in 1905. Today the town also has paper mills, iron foundries, and several large hydroelectric stations.

No·tre Dame Bay (nō'tər dām'), arm of the Atlantic Ocean, c.40 mi (65 km) long and 50 mi (80 km) wide, E N.F., Canada. The bay has an irregular shoreline and contains many islands. There are numerous fishing settlements along the coast, many of which have fish-processing plants.

Notre Dame Mountains, section of the Appalachian system, extending c.500 mi (800 km) from the Green Mts. of Vt. into the Gaspé Peninsula, Canada. Worn low by erosion, the ancient mountains have an average elevation of c.2,000 ft (610 m).

Not·ta·way (nŏt'ə-wā), river, c.140 mi (225 km) long, issuing from Mattagami Lake, W Que., Canada, and flowing NW into S James Bay. It is noted for sturgeon.

Not·ting·ham (nŏt'ĭng-əm), county borough (1976 est. pop. 280,300), administrative center of Nottinghamshire, central England, on the Trent River. It is a center of rail and road transportation. The most important industries are the manufacture of lace, hosiery, cotton, and silk. Coal, cigarettes, pharmaceuticals, bicycles, and electronic equipment are among Nottingham's many other products. In the 9th cent. it was one of the Danish Five Boroughs. In the 12th cent. much of the city was destroyed by fire. In 1642 the town was the scene of the raising of the standard of Charles I, marking the beginning of the civil war. Early in the 19th cent. Luddites were active in the city. The present castle (17th cent.) overlooking the Trent River was burned in 1831 during Reform Bill riots, was restored in 1878, and now houses an art museum. The earlier Norman castle on the same site was once the prison of David II of Scotland and the headquarters of Richard III before the Battle of Bosworth Field. According to tradition Robin Hood was born in Nottingham. William Booth, founder of the Salvation Army, was born here in 1829.

Not·ting·ham·shire (nŏt'ĭng-əm-shîr', -shər), nonmetropolitan county (1976 est. pop. 977,500), central England; administrative center Nottingham. The land, partially reclaimed fenland, is low-lying and fertile. An area of moors devoted to pasturage, in the south, is known as the Wolds. Coal fields extend along the western border and there are small oil fields. The county was a part of the Anglo-Saxon kingdom of Mercia. In 1974 a small area in the northwest was assigned to the new metropolitan county of South Yorkshire.

Not·to·way (nŏt'ə-wā'), county (1970 pop. 14,260), 308 sq mi (797.7 sq km), central Va., bounded on the S by the Nottoway River; formed 1788; co. seat Nottoway. Its agriculture includes tobacco, fruit, hay, grains, dairy products, livestock, and poultry. Among its manufactures are clothing, textiles, millwork, and metal products.

Nottoway, village (1970 pop. 100), seat of Nottoway co., central Va., W of Petersburg.

Nottoway, river, 175 mi (281.6 km) long, S Va., rising in N Lunenburg co. and flowing SE to the N.C. border, where it joins the Blackwater River and forms the Chowan River.

Nouak·chott (nwäk'shŏt'), city (1972 est. pop. 37,000), capital of the Islamic Republic of Mauritania, Nouakchott dist., W Mauritania, a port on the Atlantic Ocean. Nouakchott was a small village until 1957, when it was chosen as capital. A large-scale construction program began in 1958. Its ocean port has modern storage facilities, especially for petroleum. Handicrafts are made, and light industry is carried on in the city. Some historians believe that nearby stood the ribat (monastery) from which the Moslem Almoravides set out on their conquests of Africa and Spain in the 11th cent.

Nou·mé·a (no͞o-mā'ə), town (1975 est. pop. 60,200), capital of the French overseas territory of New Caledonia, on New Caledonia Island, South Pacific. The site of a U.S. airfield in World War II, it is an important air base on transpacific flight routes. Nouméa is the seat of the South Pacific Commission, an international body formed in 1947 to promote the economic and social welfare of island people. The town was a French penal colony from 1864 to 1897.

No·va Lis·bo·a (nô'və lēzh-vō'ə), city (1970 pop. 61,885), W central Angola. Nova Lisboa stands on a high plateau and serves as a road, rail, and air transport hub and as a commercial and shipping center for a rich agricultural region. Milling and the production of lime are carried on in the city. Nova Lisboa was founded in 1912.

No·va·ra (nō-vä'rä), city (1976 est. pop. 102,132), capital of Novara prov., Piedmont, N Italy. It is an agricultural and industrial center and a rail junction. Manufactures include textiles, chemicals, machinery, and printed materials. Several battles were fought (1500, 1513) near Novara during the Italian Wars. At Novara, in Mar., 1849, the Austrians defeated the Piedmontese under Charles Albert. The Church of San Gaudenzio (16th–17th cent.) has an impressive campanile (19th cent.).

No·va Sco·tia (nō'və skō'shə), province (1975 pop. 822,000), 21,425 sq mi (55,491 sq km), E Canada. One of the Maritime Provinces, it comprises a mainland peninsula and the adjacent Cape Breton Island. The capital is the port of Halifax. It is bounded on the north by the Gulf of St. Lawrence, on the east and south by the Atlantic Ocean, and on the west by N.B., from which it is largely separated by the Bay of Fundy. The climate is moderate and the rainfall abundant. The east coast is rocky, with numerous bays and coves, and is dotted with many fishing villages. Off the beautiful south shore is Sable Island, called the graveyard of the Atlantic; on the west coast huge Fundy tides wash the shores.

A system of railroads and highways interlaces the province, providing an additional lure to tourists. Frequently visited historical spots include the Alexander Graham Bell Museum at Banneck, the Shrine of Evangeline at Grand Pré, and the town of Annapolis Royal, site of the first permanent Canadian settlement (1610). Cape Breton Island (est. 1936) and Kejimkujik (est. 1968) national parks are in Nova Scotia. Sportsmen are attracted by abundant game and all types of fishing and some of the best sailing on the continent.

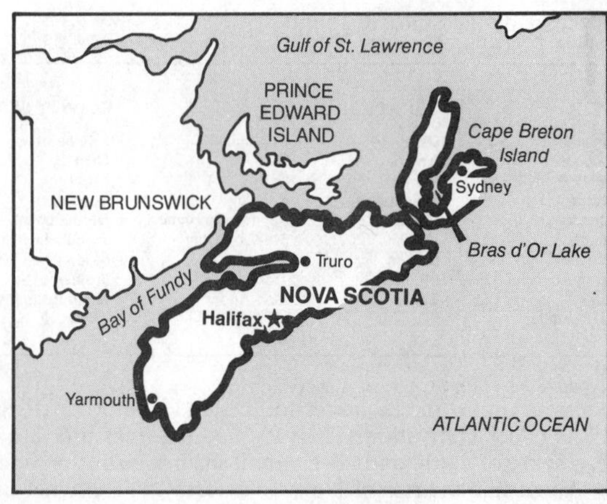

Economy. Coal is mined principally in the Sydney-Glace Bay area of Cape Breton Island. Gypsum, barite, and salt are also mined. Fishing is next in importance; cod, lobster, and haddock are the biggest catches. Inland, the forests yield spruce lumber, and the province's industries produce much pulp and paper. In the northwest there is dairying, and the region of Annapolis and Cornwallis supports valuable apple orchards. There are also important hay, grain, fruit, and vegetable crops. The iron and steel industry is centered at Sydney, and the province has a variety of manufacturing industries, including those processing food and fish. In addition to its all-year port facilities, Halifax is a railroad terminus and a center of shipbuilding and sugar refining. The rivers of Nova Scotia have a number of small hydroelectric stations that help support the economy.

History. Two Algonquian tribes, the Abnaki and the Micmac, inhabited the area before white men arrived. John Cabot may have landed (1497) on the tip of Cape Breton Island. A successful French settlement was made in 1610 at Port Royal (now Annapolis Royal). For the next century and a half France and England contested bitterly for colonial rights to Acadia, which included present-day Nova Scotia, New Brunswick, and Prince Edward Island. The Treaty of Paris (1763) gave all of French North America to England. Prince Edward Island, joined to Nova Scotia in 1763, became separate in 1769. In 1784 New Brunswick and Cape Breton also became separate colonies; Cape Breton rejoined Nova Scotia in 1820. During the early 19th cent. thousands of Scots and Irish emigrated to Nova Scotia. Under the leadership of Joseph Howe, Nova Scotia became the first colony to achieve (1848) responsible (or cabinet) government. It acceded to the Canadian confederation as one of the four original members in 1867. Nova Scotia has pioneered in Canadian history with the first newspaper (Halifax *Gazette*, 1752), the first printing press (1751), and the first university (King's College, Windsor, 1788–89).

No·va·to (nō-vä'tō), residential city (1970 pop. 31,006), Marin co., W Calif., on San Pablo Bay; inc. 1960.

No·va·ya Zem·lya (nô'vä-yä zĕm'lē-ä'), archipelago, c.35,000 sq mi (90,650 sq km), in the Arctic Ocean between the Barents and Kara seas, NW USSR. It consists of two main islands and many smaller ones. In the north the archipelago is glaciated and is covered by arctic desert; the southern part is tundra. Copper, lead, zinc, and asphaltite are found here. Fishing, sealing, and trapping are the chief occupations of the region's small population. The islands have been used by the Russians for thermonuclear testing. Discovered by Novgorodians in the 11th or 12th cent., the islands were sighted by explorers searching for the Northeast Passage in the 1500s.

Nov·go·rod (nŏv'gə-rŏd'), city (1970 pop. 128,000), capital of Novgorod oblast, NW European USSR, on the Volkhov River near the point where it leaves Lake Ilmen. Novgorod's industries produce china, furniture, bricks, and wood and food products. Its magnificent architectural monuments earned it the name the "museum city" until World War II, when it was held by the Germans (1941–44) and suffered great damage. Chief among the losses was the 12th cent. kremlin, containing the Cathedral of St. Sophia (founded 1045).

History. Novgorod is one of the oldest Russian cities, believed to have been founded in antiquity by the Slovenes. A major commercial and cultural center of medieval Europe, it lay on the chief trade routes of Eastern Europe. Rurik, legendary founder of the Russian state, made it his capital in 862. The capital was transferred to Kiev in 886, but Novgorod broke away from Kiev in 1136, when it became the capital of an independent republic—Sovereign Great Novgorod, an area that embraced the whole of northern Russia to the Urals. Situated on the great trade route to the Volga valley, it became, with London, Bruges, and Bergen, one of the four chief trade centers of the Hanseatic League. At its height, in the 14th cent., its population rose to c.400,000. Its colorful splendor during that period, its hundreds of churches, its great shops, and its huge fairs have furnished rich themes for later Russian art and folklore. The city fell to Muscovite invasions in 1456 and 1470 and came under Moscow's complete control in 1478. It lost its commercial dominance to the newly built St. Petersburg in 1703. Since the devastations of World War II, many of Novgorod's historic buildings have been restored.

No·vi (nō'vī'), village (1975 est. pop. 17,100), Oakland co., SE Mich., NW of Detroit; inc. 1958. Metal products are made.

No·vi Li·gu·re (nô'vē lē'gōōr-ā), town (1976 est. pop. 32,268), Piedmont, NW Italy. It is an industrial and transportation center. In 1799 Austrian and Russian forces defeated the French here.

No·vi Pa·zar (nō'vē pə-zär'), town (1971 pop. 29,072), E Yugoslavia, in Serbia, on the Raška River. It is a trade and industrial center where textiles, carpets, and copper ware are produced. Known in the 9th cent., it was the capital of Serbia from the 12th to the 14th cent. The town was captured by the Turks in 1456. It was occupied by Austria from 1879 to 1908 but remained under Turkish civil administration until 1913, when it passed to Serbia. It became part of Yugoslavia after World War I. The town retains much of its Turkish architecture.

No·vi Sad (nō'vē säd'), city (1971 pop. 141,712), NE Yugoslavia, in Serbia, on the Danube River. It has industries that produce agricultural machinery, electrical equipment, and munitions. Known in the 16th cent., it rapidly developed as a commercial center and was made (1748) a royal free city of Austria-Hungary. In the 18th and early 19th cent. Novi Sad was the center of the Serbian literary revival. It was incorporated into Yugoslavia in 1918.

No·vo·cher·kassk (nō'vō-chĕr-käsk'), city (1976 est. pop. 183,000), SE European USSR, on the Aksai River, a tributary of the Don. It manufactures locomotives, machine tools, mining and building equipment, electrical apparatus, oil field machinery, and chemicals. Founded in 1805, it remained the administrative center of the Don Cossacks until 1920.

No·vo·kuz·netsk (nô'vŏ-kōōz-nĕtsk'), city (1977 est. pop. 537,000), S central Siberian USSR, on the Tom River. Iron, steel, mining equipment, chemicals, aluminum, and iron alloys are produced. The old town of Kuznetsk was founded by Cossacks in 1617 and was a trading center until the 20th cent. It was developed in the 1930s as an iron and steel center of the Kuznetsk Basin and was merged with its newer industrial section in 1932.

No·vo·mos·kovsk (nô'vŏ-mŏs-kôfsk'), city (1976 est. pop. 68,000), W central European USSR; founded 1930. An industrial center in the Moscow coal basin, it has coal mines and chemical plants.

No·vo·ros·siysk (nô'vŏ-rŏ-sēsk'), city (1976 est. pop. 150,000), Krasnodar Kray, SE European USSR, on the Black Sea. A major port and a naval base, it exports grain, has shipyards, and is a major center of the Soviet cement industry. The city stands on the site of a Genoese colony (13th–14th cent.) and of a Turkish fortress, captured by the Russians in 1808. The present city was founded in 1838, and the first cement factory was opened in 1882.

No·vo·si·birsk (nô'vŏ-sĭ-bîrsk'), city (1976 est. pop. 1,304,000), capital of Novosibirsk oblast, S Siberian USSR, on the Ob River and the Trans-Siberian RR. Novosibirsk has machine, textile, chemical, and metallurgical industries. Founded in 1893, during the construction of the Trans-Siberian RR, it grew as a trade center. Its growth is largely due to the proximity of the Kuznetsk Basin. The region forming Novosibirsk oblast is predominantly agricultural, although there is coal mining to the east.

No·wa·ta (nə-wä'tə), county (1970 pop. 9,773), 537 sq mi (1,390.8 sq km), NE Okla., bounded in the N by the Kansas border and drained by the Verdigris River; formed 1907; co. seat Nowata. In a stock-raising and agricultural area that produces cotton, corn, wheat, and oats, it has petroleum and natural-gas fields and refineries. Textiles are manufactured.

Nowata, city (1970 pop. 3,679), seat of Nowata co., NE Okla., NNE of Tulsa, in a livestock and oil area; settled 1888, inc. 1895.

No·wy Sącz (nô'vĭ sôNch'), city (1975 est. pop. 48,600), SE Poland, on the Dunajec. It is a railway junction and industrial center producing machinery, agricultural tools, chemicals, and footwear. There are deposits of lignite and petroleum in the vicinity. Chartered in 1298, it passed to Austria in 1772 and was included in Poland in 1919.

Nox·u·bee (nŏk-shə-bē), county (1970 pop. 14,288), 695 sq mi (1,800 sq km), E Miss., bordering in the E on Ala. and drained by Noxubee River; formed 1833; co. seat Macon. It is in a soybean, cotton, corn, dairy, and cattle-raising region, with timber logging and clothing, brick, and clay-tile manufacturing.

No·yon (nwä-yôN'), town (1975 pop. 13,889), Oise dept., N France. It has foundries; metalworks; and machine, asbestos, clothing, furniture, and food-processing industries. In 768 at Noyon, Charlemagne was crowned king of the Franks. The town was devastated in both World Wars, but the Cathedral of Notre Dame (12th–13th cent.) has survived. The house where John Calvin was born is now a museum.

Nu·bi·a (nōō'bē-ə, nyōō'-), ancient state of NE Africa. At the height of its political power Nubia extended from the First Cataract of the Nile to Khartoum, in the Sudan. It early came under the influence of the pharaohs, and in the 20th cent. B.C. Seti I completed the occupation of the area. Many centuries later Egypt itself was ruled (8th and 7th cent. B.C.) by conquering Nubians of the Cush (Kush) kingdom. Nubia later was attacked by Assyrians and Ethiopians. After c.350 A.D. the region came under the sway of the Nobatae, a Negro tribe that formed a powerful kingdom with its capital at Dongola. The kingdom was converted to Christianity in the 6th cent. A.D. Joined with the Christian kingdom of Ethiopia, it long resisted Moslem encroachment, but in the 14th cent. it finally collapsed. Nubia was then broken up into many petty states. Mohammed Ali of Egypt conquered (1820–22) Nubia, and in the late 19th cent. much of the area was held by supporters of the Mahdi.

Nu·bi·an Desert (nōō'bē-ən, nyōō'-), region of the Sahara Desert, c.157,000 sq mi (406,630 sq km), NE Republic of Sudan, NE Africa, between the Nile and the Red Sea. The arid region, largely a sandstone plateau, has numerous wadis flowing toward the Nile, whose great bends are entrenched in the western part of the region.

Nuck·olls (nŭk'əlz), county (1970 pop. 7,404), 579 sq mi (1,499.6 sq km), S Nebr., in a farming region bordered in the S by Kansas and drained by the Republican and Little Blue rivers; formed 1871; co. seat Nelson. It is in a livestock and grain area.

Nu·e·ces (nōō-ā'səs, nyōō'-), county (1970 pop. 237,544), 841 sq mi (2,178.2 sq km), S Texas, on the Gulf coastal plain bounded on the N by the Nueces River and Nueces Bay, on the NE and E by Corpus Christi Bay and Laguna Madre, and sheltered from the Gulf of Mexico by Mustang and Padre islands; formed 1846; co. seat Corpus Christi. It has agriculture (cotton, grain sorghums, truck crops, citrus fruit, livestock, poultry, and dairy products), oil and natural-gas wells and refineries, extensive shipping facilities, and plants manufacturing metal, stone, clay, and glass products.

Nueces, river, 315 mi (506.8 km) long, Texas, rising on Edwards Plateau and flowing SE to Nueces Bay. It receives the combined Frio and Atascosca Rivers at Three Rivers.

Nue·vi·tas (nwā-vē'täs), city (1970 pop. 20,734), Camagüey prov., E Cuba, on the Guincho peninsula. Nuevitas is sheltered by a huge harbor, has two auxiliary ports, and is a major shipping point for Cuban sugar. The large bay was discovered by Columbus in 1492. Founded in 1775, the city was moved to its present site in 1828.

Nue·vo La·re·do (nwā'vō lä-rā'dō, *-thō*), city (1974 est. pop. 184,600), Tamaulipas state, NE Mexico, across the Rio Grande River from Laredo, Texas. Nuevo Laredo is the northern terminus of the Inter-American Highway, as well as the chief point of entry for U.S. tourists driving to Mexico. It is also a center of international trade and the distribution point for an agricultural (mainly cotton) and livestock-raising area. Founded in 1755, the city was part of Laredo until the end of the Mexican War in 1848.

Nuevo Le·ón (lā-ōn'), state (1970 pop. 1,694,689), 25,136 sq mi (65,102 sq km), N Mexico; capital Monterrey.

Nu·ku·a·lo·fa (nōō'kwä-lô'fä), town (1966 pop. 15,685), capital of the Kingdom of Tonga, on the N coast of Tongatabu Island.

Nu·ku Hi·va (nōō'kə hē'və), volcanic island, 127 sq mi (329 sq km), South Pacific, largest of the Marquesas Islands, French Polynesia. The island has fertile, well-watered valleys. Copra is exported.

Nu·kus (no͞o-ko͞os'), city (1976 est. pop. 96,000), capital of Karakalpak Autonomous Republic, Central Asian USSR, on the Amu Darya River. It has alfalfa and food-processing industries and is a center for repairing machines and motors.

Null·ar·bor Plain (nŭl'ə-bôr', nŭl'är-bər), c.300 mi (480 km) wide, SW South Australia state, Australia. Bordered on the north by the Great Victoria Desert and on the south by the Great Australian Bight, its maximum height is 1,000 ft (305 m). It is the site of a major research and rocket center.

Nu·man·ti·a (no͞o-măn'shē-ə, -shə, nyo͞o'-), ancient settlement, Spain, near the Durius (now Douro) River N of modern Soria. Its inhabitants withstood repeated Roman attacks from the time of Cato the Elder's campaign (195 B.C.) until Scipio Aemilianus finally took the city in 133 B.C., after an eight-month blockade, thus completing the conquest of Spain. Archaeologists have uncovered the remains of Roman camps and evidence of settlement dating back to the Bronze Age.

Nu·mid·i·a (no͞o-mĭd'ē-ə, nyo͞o'-), ancient country of NW Africa, very roughly the modern Algeria. It was part of the Carthaginian empire until Masinissa, ruler of eastern Numidia, allied himself (c.206 B.C.) with Rome in the Punic Wars. After the Roman victory over Carthage led to peace in 201 B.C., Masinissa was awarded rule of all Numidia. Numidia's encroachments on reviving Carthage furnished Rome with a pretext for the Third Punic War (149–146 B.C.). Later, in the Roman civil war, King Juba I sided with Pompey, and Numidia lost (46 B.C.) all independence with Julius Caesar's victory. Numidia was invaded by the Vandals in the 5th cent. A.D. and by the Arabs in the 8th cent.

Nun·ea·ton (nŭn-ē'tən), municipal borough (1976 est. pop. 111,100), Warwickshire, central England. There are coal mines in the vicinity, and iron goods, hats, and cotton, silk, rayon, and woolen textiles are manufactured. There are remains of a 12th cent. nunnery. George Eliot was born in the borough.

Nu·ni·vak (no͞o'nə-văk'), island, c.1,700 sq mi (4,405 sq km), off W Alaska, in the Bering Sea. Fogbound most of the year, Nunivak is covered with low vegetation and has a small Eskimo population engaged in hunting and fishing. Reindeer and musk oxen have been introduced. The island was discovered in 1821 by Russian explorers.

Nu·rem·berg (no͝or'əm-bûrg', nyo͝or'-), or *German* **Nürn·berg** (nœrn'bĕrкн), city (1974 est. pop. 509,813), Bavaria, S West Germany, on the Pegnitz River. One of the great historic cities of Germany, Nuremberg is an important commercial, industrial, and transportation center. Its manufactures include electrical equipment, heavy machinery, precision instruments, chemicals, textiles, printed materials, beer, and liquor.

First mentioned in 1050, Nuremberg received a charter in 1219 and was made a free imperial city by the end of the 13th cent. The city was independent of the burgraviate of Nuremberg, which came (1192) under the control of the Hohenzollern family. The cultural flowering of Nuremberg in the 15th and 16th cent. made it the center of the German Renaissance. In 1525 Nuremberg accepted the Reformation, and the religious Peace of Nuremberg, by which the Lutherans gained important concessions, was signed here (1532). In the Thirty Years' War, Gustavus II was besieged (1632) in Nuremberg by Wallenstein. The city then declined, recovering its importance only in the 19th cent. when it grew as an industrial center. In 1806 Nuremberg passed to Bavaria.

After Adolf Hitler came to power, Nuremberg was made a national shrine by the Nazis, who held their annual party congresses nearby from 1933 through 1938. At the party congress of 1935 the so-called Nuremberg Laws were promulgated; they deprived German Jews of civic rights, forbade intermarriage between Jews and non-Jews, and deprived persons of partly Jewish descent of certain rights. Until 1945 Nuremberg was the site of roughly half the total German production of airplane, submarine, and tank engines; as a consequence, the city was heavily bombed by the Allies during World War II and was largely destroyed. After the war, Nuremberg was the seat of the international tribunal for war crimes. Since 1945 much of the city's architectural beauty has been restored. Among the historic buildings are the churches of St. Sebald (1225–73), St. Lorenz (13th–14th cent.), St. Jacob (14th cent.), and Our Lady (1352–61); the Hohenzollern castle (11th–16th cent.); the old city hall (1616–22); and the house (now a museum) where Albrecht Dürer lived from 1509 to 1528. A large portion of the city walls (14th–17th cent.) still stands.

Nu·ri·stan (no͝or'ĭ-stăn'), region on the S slopes of the Hindu Kush, NE Afghanistan, bordered on the E by Pakistan. It is inhabited by an ethnically distinctive people (numbering c.60,000), who practiced animism until their forcible conversion to Islam in 1895–96. Inhabiting relatively isolated villages in deep, narrow mountain valleys, they grow wheat, barley, millet, peas, wine grapes, and other fruit and raise livestock (chiefly goats). A special artisan caste specializes in woodcarving, pottery making, weaving, and metalwork.

Nu·say·bin (no͞os'ĭ-bēn'), town (1970 pop. 13,941), SE Turkey, near the Syrian border. It is a commercial and transportation center. It has ruins of the ancient Nisibis, the residence of early (2nd cent. B.C.–1st cent. A.D.) Armenian kings. In early Christian times it was a center of Nestorianism.

Nut·ley (nŭt'lē), town (1978 est. pop. 31,500), Essex co., NE N.J., a residential suburb of Newark, on the Passaic River; settled 1680, inc. 1902. Pharmaceuticals, textiles, and paper products are made. After the Civil War the town was a center for writers and artists.

Nu·u·a·nu Pa·li (no͞o'o͞o-ä'no͞o pä'lē), sheer cliff and mountain pass, alt. 1,200 ft (366 m), Koolau Range, SE Oahu island, Hawaii. The pass is the principal route between Honolulu and eastern Oahu.

Nu·wa·ra E·li·ya (no͞o'wər-ə ā'lē-yə), town (1971 pop. 16,347), S Sri Lanka. A hill resort and health center in a tea-growing area, it was first settled in 1827 by the British.

Nu·zi (no͞o'zē), site near Kirkuk, N Iraq. Thousands of clay tablets unearthed here bear Akkadian inscriptions that reveal much about ancient laws and customs.

Ny·ack (nī'ăk'), residential village (1970 pop. 6,659), Rockland co., SE N.Y., on the W bank of the Hudson River opposite Tarrytown; settled 1684 by Dutch, inc. 1833.

Ny·as·a, Lake (nī-ăs'ə), or **Lake Ma·la·wi** (mä-lä'wē), c.11,600 sq mi (30,040 sq km), c.360 mi (580 km) long and from 15 to 50 mi (24–80 km) wide, E central Africa, in the Great Rift Valley. Lake Nyasa is bordered by Tanzania in the north and northeast, by Mozambique in the east, and by Malawi in the south and west. The lake is bounded by steep mountains, except in the south. First discovered by the Portuguese explorer Caspar Boccaro c.1616, Lake Nyasa was visited and named by the Scottish missionary David Livingstone in 1859.

Ny·as·a·land (nī-ăs'ə-lănd'): *see* Malawi.

Ny·borg (nü'bôr'), port city (1970 com. pop. 17,775), Fyn co., S central Denmark, at the head of Nybord Fjord (an arm of the Store Baelt). It is an industrial center, with shipyards and plants manufacturing textiles and tobacco products.

Nye (nī), county (1970 pop. 5,599), 18,064 sq mi (46,785.8 sq km), central and S Nev., in a mountain region bordering on Calif; formed 1864; co. seat Tonopah. It is in a farming and cattle-ranching area, with oil and gas wells, mineral mines, and mineral-products industry. Sections of Toiyabe National Forest are in the north. Pahute Mesa, Amargosa Desert, and part of Death Valley National Monument are in the south.

Nyí·regy·há·za (nyē'rĕj-hä'zô), city (1977 est. pop. 75,000), NE Hungary. It is a road and rail junction and the market for an extensive agricultural region. Known in the 13th cent., the city was destroyed during the Turkish occupation (16th cent.) of Hungary but was rebuilt in the 18th cent. Its museum contains valuable gold relics dating from Avar times.

Ny·kø·bing (nü'kœ'bĭng), city (1976 est. pop. 21,800), capital of Storstrøm co., SE Denmark, on Falster Island and on the Guldborg Sund, connected by bridge with Lolland. It is a seaport and has sugar refineries, breweries, and textile mills. Of note are a Gothic church (until 1532 a Franciscan monastery) and the ruins of a 12th cent. castle.

Ny·kö·ping (nü'chœ'pĭng), city (1975 est. pop. 38,900), capital of Södermanland co., SE Sweden, a port on the Baltic Sea. It is a commercial and industrial center. An atomic research center is nearby. Nyköping was founded in the 13th cent. on the site of a former trading town. It was destroyed by fire in 1665, was rebuilt, and was sacked by the Russians in 1719. Historic structures in the city include St. Nicholas Church (13th–18th cent.), the city hall (17th cent.), and ruins of Nyköpingshus castle (13th cent.).

Nys·sa (nĭs'ə), name of several ancient cities devoted to the worship of Dionysus. The best known of them is a town of Cappadocia, Asia Minor, near the Halys (now the Kizil Irmak) River. It was the residence of St. Gregory of Nyssa.

O·a·hu (ō-ä'hōō), island (1970 pop. 629,145), 593 sq mi (1,535.9 sq km), chief island of Hawaii, part of Honolulu co., between Molokai and Kauai. Oahu is composed of two parallel mountain ranges that are separated by a rolling plain dissected by deep gorges. Oahu has no active volcanoes, but there are many extinct craters. Pearl Harbor indents the island's southern coast. The island is an important defense area that includes the headquarters of the U.S. Pacific Command and the Pearl Harbor naval base. There are many bathing beaches (including Waikiki), some of which have coral gardens. Large pineapple and sugar-cane plantations cover the rural areas of the island, and their products form Oahu's chief agricultural exports. Dairy farming and fishing are also important economic activities.

Oak Creek (ōk), city (1970 pop. 13,928), Milwaukee co., SE Wis., a suburb of Milwaukee, on Lake Michigan; inc. 1955. Electronic equipment, heavy machinery, and aluminum products are made.

Oak·dale (ōk'dāl'). **1.** City (1970 pop. 6,594), Stanislaus co., central Calif., near the Stanislaus River SE of Stockton; founded 1871, inc. 1906. Farm products are canned. **2.** City (1970 pop. 7,301), Allen parish, SW La., near the Calcasieu River SSW of Alexandria, in a timber area; laid out 1886.

Oak Forest, village (1978 est. pop. 23,900), Cook co., NE Ill., a residential suburb of Chicago, in a diversified farming area; inc. 1947.

Oak Grove (grōv), town (1970 pop. 1,980), seat of West Carroll parish, extreme NE La., NE of Monroe, in a cotton area; inc. as a village 1908, as a town 1928.

Oak Har·bor (här'bər), town (1977 est. pop. 11,618), Island co., NW Wash., on Whidbey Island NW of Everett; settled 1851, inc. 1915. It is a recreation and trade center. A U.S. naval air station is nearby.

Oak·land (ōk'lənd), county (1970 pop. 907,871), 867 sq mi (2,245.5 sq km), SE Mich., drained by the Shiawassee, Huron, and Clinton rivers and by the River Rouge; organized 1820; co. seat Pontiac. There is some farming of grain, corn, hay, and livestock. Nurseries and poultry hatcheries are found here. Automotive parts and equipment are manufactured. There are many small lakes and resorts.

Oakland. 1. City (1978 est. pop. 326,000), seat of Alameda co., W Calif., on the E side of San Francisco Bay; inc. 1852. A containerized shipping port and a major rail terminus, it has shipyards, chemical plants, glassworks, and food-processing establishments. The San Francisco-Oakland Bay Bridge was opened in 1936, and several tunnels connect Oakland with other nearby cities. Of interest are the new Oakland Museum, Chabot Observatory, the Morcom Rose Garden, and Jack London Square (a restaurant and entertainment area). The city is the seat of Mills College. A large U.S. naval supply center and a U.S. army terminal and depot are in the city. **2.** Town (1970 pop. 1,786), seat of Garrett co., NW Md., SW of Cumberland, in a farm region; settled 1790, inc. 1861. It has canneries and grain mills. **3.** Residential borough (1970 pop. 14,420), Bergen co., NE N.J.; settled early 1700s, inc. 1902.

Oakland Park, city (1978 est. pop. 20,300), Broward co., SE Fla., on the Atlantic coast.

Oak Lawn (lôn), village (1978 est. pop. 63,100), Cook co., NE Ill., a suburb of Chicago; inc. 1909. It is residential with some light manufacturing industries. Products include metalwork, machine tools and parts, kitchen cabinets, and musical instruments.

Oak·mont (ōk'mŏnt'), borough (1970 pop. 7,550), Allegheny co., SW Pa., on the Allegheny River NE of Pittsburgh; inc. 1889. It is in a farming and industrial area.

Oak Park. 1. Village (1978 est. pop. 58,400), Cook co., NE Ill., a residential suburb adjacent to Chicago; settled 1833, inc. 1901. Some 25 houses in the village were designed by Frank Lloyd Wright. Ernest Hemingway was born here. **2.** City (1978 est. pop. 32,300), Oakland co., SE Mich., a suburb of Detroit; inc. 1927. It is chiefly residential.

Oak Ridge (rĭj), city (1978 est. pop. 26,700), Anderson and Roane cos., E Tenn., on Black Oak Ridge and the Clinch River; founded by the U.S. government 1942, inc. as an independent city 1959. Many activities in the fields of atomic energy and nuclear physics are pursued; manufactures include complex nuclear instruments, electronic instrumentation, irradiated products, and nuclear fuel. The site was chosen (1942) for what was then called the Clinton Engineer Works, and the city was built by the Federal government to house the workers who developed the uranium-235 and plutonium-239 for the atomic bomb. The existence and purpose of the community were kept secret from most of the country until the summer of 1945. The project was under the control of the Atomic Energy Commission, but the city has since (1955–59) been turned over to its residents. The former Clinton National Laboratory for nuclear research became (1948) the Oak Ridge National Laboratory. The Oak Ridge Institute of Nuclear Studies (1948), composed of many sponsoring educational institutions, and the Univ. of Tennessee Biomedical Science graduate school are also here. Tourist attractions include the American Museum of Atomic Energy; a nearby nuclear graphite reactor; the K-25 overlook, from which can be seen the Oak Ridge gaseous diffusion plant; and an arboretum.

Oak·wood (ōk'wŏŏd'), city (1970 pop. 10,095), Montgomery co., SW Ohio, a residential suburb adjacent to Dayton; inc. 1907.

Oa·xa·ca (wä-hä'kä), city (1974 est. pop. 114,900), capital of Oaxaca state (36,375 sq mi/94,211.3 sq km), S Mexico. Situated in a valley encircled by low mountains, Oaxaca is a commercial and tourist center with gardens and many fine examples of colonial church architecture. The church and monastery of Santo Domingo is a national monument. Oaxaca is noted for hand-wrought gold and silver filigree, pottery, and sarapes. The city is subject to severe earthquakes. According to Aztec tradition, Oaxaca was founded in 1486, during the brief ascendancy of the Aztec over the Mixtec and Zapotec Indians. Prominent in the Mexican revolution against Spain, the city also joined in the War of the Reform and in resistance to the French intervention.

Ob (ŏb, ôb), river, c.2,300 mi (3,700 km) long, W Siberian USSR. With the Irtysh River, its chief tributary, it is c.3,460 mi (5,570 km) long and is the world's fourth-longest river. Formed by the junction of the Biya and Katun rivers southwest of Biysk, the upper Ob flows northwest, then northeast through the west Siberian lowlands to be joined by the Tom River. The middle Ob flows northwest through the swampy forests in the Tomsk and Narym regions and then is joined by the Chulym, Ket, and Irtysh rivers. The lower Ob consists of the Great Ob and the Small Ob and flows north, then east into Ob Bay, an estuary and shallow arm of the Arctic Ocean between the Yamal and Gyda peninsulas. The valley of the middle Ob is subject to flooding each spring as the thaw occurs in the upper Ob basin before the ice in the lower course of the Ob has melted. Although frozen from five to six months of the year, the Ob is an important trade and transport route. The Ob is rich in sturgeon, salmon, and carp. There are major deposits of oil and natural gas in the basin of the middle and lower Ob.

O·beid, El (ĕl ō-bād'): *see* Al Ubayyid, Sudan.

O·ber·alp (ō'bər-älp'), Alpine pass, 6,733 ft (2,053.6 m) high, between Grisons and Uri cantons, S central Switzerland. The peak Oberalpstock, 10,926 ft (3,332.4 m) high, is northeast of the pass.

O·ber·am·mer·gau (ō'bər-ä'mər-gou'), town (1970 est. pop. 4,700), Bavaria, S West Germany, in the Bavarian Alps. It has been a noted center of woodcarving since the 12th cent. Oberammergau is famous for the Passion play performed here every 10 years (last in 1970), originally (1634) in fulfillment of a vow made during a plague in 1633. Tourism is the town's major source of income.

O·ber·hau·sen (ō′bər-hou′zən), city (1974 est. pop. 239,309), North Rhine-Westphalia, W West Germany, an industrial center of the Ruhr district. It is a port and a rail junction. Manufactures include iron and steel, machinery, and chemicals. Oberhausen was chartered in 1874.

O·ber·lin (ō′bər-lĭn). **1.** City (1970 pop. 2,291), seat of Decatur co., NW Kansas, NNE of Garden City near the Nebr. border, in a farm and oil area; laid out 1878, inc. 1885. In the cemetery is a monument to the settlers killed (1878) in a Cheyenne raid. **2.** City (1970 pop. 1,857), seat of Allen parish, SW La., NE of Lake Charles, in a timber region. **3.** City (1970 pop. 8,761), Lorain co., N Ohio, SSW of Lorain. Oberlin College (1833) is here.

O·bi·on (ō-bī′ən), county (1970 pop. 30,247), 556 sq mi (1,440 sq km), NW Tenn., bounded in the N by Ky., in the NW by Reelfoot Lake, and drained by the Obion River and its tributaries; formed 1823; co. seat Union City. It is in a fertile farm region that produces cotton and livestock. Its industries include food processing, lumbering, and the manufacture of plastics, metal products, and transportation equipment.

o·blast (ô′bləst, ŏ′bläst′), administrative and territorial division in the USSR. The boundaries of oblasts are usually based on economic and administrative considerations; as a result, oblasts vary considerably in size and population.

O'·Bri·en (ō-brī′ən), county (1970 pop. 17,522), 575 sq mi (1,489.3 sq km), NW Iowa, in a prairie agricultural area drained by the Little Sioux and Floyd rivers; formed 1851; co. seat Primghar. Stock raising (hogs, cattle, and poultry) and grain growing (corn, oats, and barley) are major occupations here.

O·bua·si (ō-bwä′sē), town (1970 pop. 31,005), S central Ghana. Highly concentrated gold ore is mined, and there are gold-extraction plants. Gold was mined in Obuasi as early as the 17th cent. From the late 1890s it was developed by Europeans into a modern mining town.

Ob·wal·den (ôp′väl′dən), half canton, Switzerland: *see* Unterwalden.

O·cal·a (ō-kăl′ə), city (1978 est. pop. 33,500), seat of Marion co., N central Fla.; inc. 1868. It is a trade, processing, and transportation hub for a major citrus region also known for its thoroughbred horses, cattle, lumber, phosphatic limestone, and fuller's earth. Manufactures include wood products, clothing, concrete goods, and metalware. Ocala is also a resort city; fish and game abound in the many nearby lakes and streams and in Ocala National Forest.

o·cean (ō′shən), interconnected mass of water covering 70.78% of the surface of the earth, often called the world ocean. It is subdivided into four major units that are separated from each other by the continental masses. The Antarctic Ocean is sometimes considered a separate ocean, extending from the shores of Antarctica northward to c.40° S lat. The Atlantic, Indian, and Pacific oceans extend northward from Antarctica as huge "gulfs" separating the continents. The Arctic Ocean, nearly landlocked by Eurasia and North America and nearly circular in outline, caps the north polar region. The major oceans are further subdivided into smaller regions loosely called seas, gulfs, or bays. Some of these seas, such as the Sargasso Sea of the North Atlantic Ocean, are only vaguely defined, while others, such as the Mediterranean Sea and the Black Sea, are almost totally surrounded by land areas. Large totally landlocked saltwater bodies, such as the Caspian Sea and Salton Sea, are actually salt lakes.

Ocean, county (1970 pop. 208,470), 639 sq mi (1,655 sq km), E N.J., bounded on the E by Barnegat Bay, with Long Beach Island and Island Beach peninsula lying between the bay and the Atlantic, and drained by the Toms and Metedeconk rivers; formed 1850; co. seat Toms River. The inland agricultural area produces truck crops, poultry, dairy products, and fruit. Part of the county is in a pine barrens region (with timber, cranberries, and huckleberries), and it has metal-ore mining and sand and gravel pits. Its industry includes food processing and the manufacture of clothing, lumber millwork, furniture, concrete, metal products, and machinery. There are many popular summer resorts.

O·ce·an·a (ō′shē-ăn′ə), county (1970 pop. 17,984), 536 sq mi (1,388.2 sq km), W Mich., bounded on the W by Lake Michigan and drained by the White River; formed 1855; co. seat Hart. Livestock and poultry raising, dairying, and farming (truck crops, potatoes, fruit, and beans) are important. Some manufacturing is done. There are fisheries, resorts, and part of Manistee National Forest in the area.

Ocean City, city (1970 pop. 10,575), Cape May co., SE N.J., a resort on the Atlantic coast; inc. 1897. Ocean City is on an island between the Atlantic Ocean and Great Egg Harbor Bay; it is linked to the mainland by a 2-mi (3.2-km) causeway.

O·ce·an·i·a (ō′shē-ăn′ē-ə, -ä′nē-ə), collective name for the approximately 25,000 islands of the Pacific, usually excluding such nontropical areas as the Ryukyu and Aleutian islands and Japan, as well as Formosa, Indonesia, and the Philippines, whose populations are more closely related to mainland Asia. Oceania is generally considered synonymous with the South Sea Islands and is divided ethnologically into Melanesia, Micronesia, and Polynesia.

Ocean Island, also known as **Ba·na·ba** (bə-nä′bə), island (1968 pop. 2,192), 2.2 sq mi (5.7 sq km), central Pacific, a part of the British colony of the Gilbert and Ellice Islands. The island was discovered by the British in 1804, annexed in 1900, and was the administrative capital of the colony until World War II. Ocean Island has important phosphate deposits mined by the British Phosphate Commission, but the deposits are expected to be depleted by about 1980.

O·cean·port (ō′shən-pôrt′, -pōrt′), borough (1970 pop. 7,503), Monmouth co., E N.J., near the Shrewsbury River SE of Red Bank.

O·cean·side (ō′shən-sīd′). **1.** City (1978 est. pop. 65,000), San Diego co., S Calif., on the Gulf of Santa Catalina; inc. 1888. It is a commercial and trading center for a rich inland farm area and for nearby Camp Pendleton, a U.S. Marine Corps base. The city has a large flower and bulb industry, and its principal manufactures are electric and electronic components. Deep-sea fishing and tourism are also important. **2.** Uninc. resort city (1978 est. pop. 36,400), Nassau co., SE N.Y., on the S shore of Long Island.

Ocean Springs (sprĭngz), resort town (1975 est. pop. 16,800), Jackson co., SE Miss., on Biloxi Bay E of Biloxi. The town is on the site of Old Biloxi, the first white settlement in the lower Mississippi valley, made by the French under sieur d'Iberville in 1699.

O·cha·kov (ə-chä′kəf), city, SW European USSR, on the Dnepr-Bug estuary and on the Black Sea. It is the center of an agricultural district and a seaport with fishing industries. In the 7th and 6th cent. B.C., there were several Greek colonies in the area. The city fell to the Russians (1788) during the Russo-Turkish War from 1787 to 1792. In the Crimean War it was occupied (1855) by the allies.

Och·il·tree (ŏk′əl-trē′), county (1970 pop. 9,704), 905 sq mi (2,344 sq km), extreme N Texas, on the high plains of the Panhandle, bounded on the N by the Okla. border and drained by tributaries of the Canadian and North Canadian rivers; formed 1876; co. seat Perryton. A leading wheat-producing county, it also has cattle ranching.

Och·lock·o·nee (ŏk-lŏk′ə-nē), river, c.135 mi (217.2 km) long, rising in SW Worth co., Ga., and flowing S to Moultrie, where it turns SW into NW Fla., through Lake Talquin, and SE into Apalachee Bay, an arm of the Gulf of Mexico.

Oc·mul·gee (ŏk-mŭl′gē), river, c.255 mi (410 km) long, formed SE of Atlanta, NW Ga., by the confluence of the Yellow, South, and Alcovy rivers and flowing SE past Macon to join the Oconee River and form the Altamaha River near Lumber City. The river passes the remains of prehistoric Indian villages preserved in Ocmulgee National Monument.

O·co·nee (ō-kō′nē). **1.** County (1970 pop. 7,915), 186 sq mi (481.7 sq km), N central Ga., in a piedmont area drained by the Apalachee and Oconee rivers; formed 1875; co. seat Watkinsville. In an agricultural region yielding cotton, corn, grain, sorghum, and fruit, it has some manufacturing. **2.** County (1970 pop. 40,728), 654 sq mi (1,693.9 sq km), extreme NW S.C., in the Blue Ridge, bounded in the NW by the Chattooga River, in the SW by the Tugaloo River, and in the E by the Keowee and Seneca rivers; formed 1768; co. seat Walhalla. Its agriculture includes cattle, apples, cotton, corn, and wheat. It has lumbering and textile industries. This summer-resort area includes part of Sumter National Forest.

Oconee, river, 282 mi (453.7 km) long, rising in the Appalachian Mts., N Ga., and flowing SE to the Ocmulgee River to form the Altamaha River.

O·con·o·mo·woc (ə-kŏn′ə-mə-wôk′), city (1970 pop. 8,741), Waukesha co., SE Wis., W of Milwaukee; inc. 1875. It is a resort in a lake and mineral-springs area.

O·con·to (ō-kŏn′tō), county (1970 pop. 25,553), 1,001 sq mi (2,592.6 sq km), NE Wis., bounded in the E by Green Bay and drained by the Oconto River; formed 1851; co. seat Oconto. It is in a dairying and lumbering region. Its industries include cheese processing, millwork, boatbuilding, and the manufacture of paper, steel, and machinery.

Oconto, city (1970 pop. 4,667), seat of Oconto co., NW Wis., on the W shore of Green Bay at the mouth of the Oconto River; inc. 1869. An early lumber center, it is now a summer resort and makes motorboats, food products, and apparel.

O·den·se (ōd′n-sə), city (1976 est. pop. 168,206), capital of Fyn co., S central Denmark, a seaport linked by canal with the Odense Fjord (an arm of the Kattegat). It is an important commercial, industrial, and cultural center and a rail junction. There are large shipyards and plants manufacturing metal goods, motor vehicles, machinery, dairy products, and processed food. Odense was founded in the 10th cent. Of note are a 12th cent. church and the 13th cent. Cathedral of St. Knud, one of the finest examples of Danish Gothic architecture. The house of the writer Hans Christian Andersen, who was born in Odense in 1805, is now a museum.

O·den·ton (ōd′n-tən), uninc. village (1970 pop. 5,989), Anne Arundel co., central Md., S of Baltimore.

O·den·wald (ōd′n-vält′), hilly, forested region, S central West Germany, bordering on the Neckar and Main rivers and the Rhine plain. Fruit and grapes are grown, and there are porphyry quarries.

O·der (ō′dər), river, 562 mi (904.3 km) long, rising in the E Sudetes, N central Czechoslovakia, and flowing generally NW through SW Poland, then N along the Poland–East Germany border to the Baltic Sea. The Oder is an important waterway connecting the industrial region of Silesia with the sea.

O·des·sa (ō-dĕs′ə), city (1970 pop. 892,000), capital of Odessa oblast, SW European USSR, in the Ukraine, a port on Odessa Bay of the Black Sea. Odessa is one of the USSR's major industrial, cultural, scientific, and resort centers. Grain, sugar, machinery, coal, petroleum products, cement, metals, jute, and timber are the chief items of trade at the port, which is the leading Soviet Black Sea port and is kept open all year with the aid of icebreakers. Odessa is also a naval base and the home port of a fishing and an Antarctic whaling fleet. The city's industries include shipbuilding, oil refining, machine building, metalworking, food processing, and the manufacture of chemicals, machine tools, movie equipment, clothing, and products made of wood, jute, and silk. Large health resorts are located nearby. Ukrainians, Russians, Jews, and Greeks predominate in Odessa's cosmopolitan population. It has a university (est. 1865), an opera and ballet theater (1809), a historical museum (1825), a municipal library (1830), an astronomical observatory (1871), an opera house (1883–87), and a picture gallery (1898).

The city is said to occupy the site of an ancient Miletian Greek colony that disappeared between the 3rd and 4th cent. In the 14th cent. the site, then under Lithuanian control, became a Crimean Tatar fortress and trade center called Khadzhi-Bei. In 1764 it passed to the Turks, from whom it was captured by the Russians in 1789. They rebuilt Odessa as a fort, commercial port, and naval base. It was a free port from 1819 to 1849, and in 1866 it was linked by rail with Kiev, Kharkov, and the Rumanian city of Jassy.

Industrialization began in the latter part of the 19th cent. Odessa was the scene in 1905 of the workers' outbreak led by sailors from the battleship *Potemkin.* Following the 1917 Bolshevik Revolution, the city was successively occupied by the Central Powers, the French, the Reds, and the Whites until the Red Army definitively took it in 1920 and united it with the Ukrainian SSR. The city fell to German and Rumanian forces in Oct., 1941. It was under Rumanian administration as the capital of Transnistra until its liberation (Apr., 1944) by the Soviet Army.

Odessa, city (1978 est. pop. 90,400), seat of Ector co., W Texas; founded 1881, inc. 1927. Great oil deposits to the south changed Odessa from a small ranch town into a large oil center with refineries and plants producing fuels, chemicals, plastics, synthetic rubber, industrial gas, and oil-field equipment. The region is underlaid with potash deposits.

Oel·wein (ōl′wīn′), city (1975 est. pop. 7,300), Fayette co., NE Iowa, NE of Waterloo; inc. 1888. It is a trade and rail center for a farm and livestock area.

Oe·ta (ē′tə), mountain range, central Greece, stretching c.15 mi (25 km) W from Thermopylae on the Gulf of Lamía. Mt. Oeta (c.7,060 ft/2,155 m) is the highest peak. In legend Hercules died here on a pyre after being poisoned by Nessus' robe.

O'·Fal·lon (ō-făl′ən). **1.** City (1973 est. pop. 10,045), St. Clair co., SW Ill., E of East St. Louis; inc. 1865. Coal is mined, and apparel is made. **2.** City (1970 pop. 7,018), St. Charles co., E Mo., WNW of St. Louis; founded 1857, inc. 1912.

O·fen (ō′fən), Alpine pass, 7,070 ft (2,156.5 m) high, Grisons canton, E Switzerland. The Ofen Pass Road links the Engadine Valley with the Italian Tyrol.

Of·fa·ly (ô′fə-lē, ŏf′ə-), county (1971 pop. 51,834), 771 sq mi (1,996.9 sq km), central Republic of Ireland. The county town is Tullamore. A part of the central plain of Ireland, the county is for the most part flat, and sections are covered by the Bog of Allen. Agriculture is the chief occupation. Among the light industries is distilling. The region formed part of the kingdom of Offaly in ancient Ireland. It was known as King's County until the establishment of the Irish Free State.

Of·fen·bach am Main (ô′fən-bäкн′ äm mīn′), city (1970 pop. 117,306), Hesse, central West Germany, on the Main River. It is an industrial center long famous for the manufacture of leather goods; chemicals, metal products, and machinery are also produced. Offenbach was first mentioned in the late 10th cent.; it was annexed by Hesse-Darmstadt in 1816. A Renaissance-style palace (1564–78) and museums of leathercraft and typography are located in the city.

O·ga·den (ō-gä′dän), region, Harar prov., SE Ethiopia, bordering on the Somali Democratic Republic. It is an arid region, inhabited mainly by Somali pastoral nomads. The region was conquered by Menelik II of Ethiopia in 1891. A clash (Dec. 5, 1934) between Italian and Ethiopian troops at the watering hole of Walwal in Ogaden was a pretext by Italy to begin a war (1935–36) against Ethiopia.

O·gal·lal·a (ō′gə-lä′lə), city (1970 pop. 4,976), seat of Keith co., W central Nebr., on the South Platte River W of North Platte city, in a wheat and cattle region; laid out 1875, inc. 1930.

Og·bo·mo·sho (ŏg′bə-mō′shō), city (1971 est. pop. 387,000), SW Nigeria. It is the trade center for a farming region. Cotton cloth is woven. Ogbomosho was founded in the 17th cent.

Og·den (ôg′dən, ŏg′-), city (1978 est. pop. 66,900), seat of Weber co., N Utah, at the confluence of the Ogden and Weber rivers; inc. 1851. Aerospace industries, Hill Air Force Base, and the Ogden Defense Depot are the major employers. The site of a trading post in the 1820s, the area was settled by Mormons in 1847. Two Mormon tabernacles are in Ogden. The city is surrounded by mountains, with major ski resorts.

Ogden, river, 35 mi (56.3 km) long, rising in the Wasatch Range in N Utah and flowing SW to join the Weber River at Ogden. The river has been used for irrigation for nearly a century, and it now irrigates c.23,000 acres (9,315 hectares).

Og·dens·burg (ôg′dənz-bûrg′, ŏg′-), city (1978 est. pop. 12,800), St. Lawrence co., N N.Y., on the St. Lawrence River at the mouth of the Oswegatchie River, in a resort area opposite Prescott, Ont.; settled by French missionaries and trappers 1749, inc. as a city 1868. An art gallery contains works of Frederic Remington, who lived here.

O·gee·chee (ō-gē′chē), river, c.250 mi (402.3 km) long, NE Ga., rising E of Greensboro and flowing SE past Louisville to the Atlantic Ocean S of Savannah.

O·ge·maw (ō′gə-mô′), county (1970 pop. 11,903), 574 sq mi (1,486.7 sq km), NE central Mich., drained by the Au Gres and Rifle rivers; formed 1875; co. seat West Branch. Livestock, poultry, potatoes, grain, and fruit are raised. Oil wells are found here. The county includes part of Huron National Forest and many small lakes with hunting and fishing areas and resorts.

O·gle (ō′gəl), county (1970 pop. 42,867), 757 sq mi (1,960.6 sq km), N Ill., drained by the Rock, Leaf, and Kyte rivers; formed in 1836; co. seat Oregon. It has agriculture (livestock, poultry, dairy products, corn, oats, barley, hay, and truck crops), food-processing plants, and some manufacturing.

O·gle·thorpe (ō′gəl-thôrp′), county (1970 pop. 7,598), 432 sq mi (1,118.9 sq km), NE Ga., in a piedmont region; formed 1793; co. seat Lexington. It is in an agricultural (cotton, corn, grain, and fruit) and timber area. It has some mineral mining and a textile industry (weaving mills and clothing manufacturing).

Oglethorpe, city (1970 pop. 1,286), seat of Macon co., W central Ga., SSW of Macon, in a farm area; inc. 1849.

O·go·ki (ō-gō′kē), river, c.300 mi (480 km) long, rising in lakes W of Lake Nipigon, W central Ont., Canada, and flowing NE to the Albany River.

O·go·oué or **O·go·we** (both: ō′gə-wā′), river, c.560 mi (900 km) long, rising on the Batéké Plateau, SW Congo Republic, and flowing NW and W across Gabon to the Gulf of Guinea, near Port-Gentil, where it forms a large delta.

O·hi·o (ō-hī′ō), state (1975 pop. 10,731,000), 41,222 sq mi (106,765 sq km), N United States, in the Great Lakes region of the Middle West, admitted as the 17th state of the Union in 1803. Columbus is the capital. The Ohio River separates it in the southeast from W.Va. and in the south from Ky.; Ohio is bounded on the west by Ind., on the

north by Mich. and Lake Erie, and on the east by Pa. From the dunes on Lake Erie to the gorge-cut plateau along the Ohio River, the land is fairly flat, with some pleasant rolling country and in the southeast rugged little hills leading to the mountains of West Virginia.

Ohio is divided into 88 counties:

NAME	COUNTY SEAT	NAME	COUNTY SEAT
Adams	West Union	Holmes	Millersburg
Allen	Lima	Huron	Norwalk
Ashland	Ashland	Jackson	Jackson
Ashtabula	Jefferson	Jefferson	Steubenville
Athens	Athens	Knox	Mount Vernon
Auglaize	Wapakoneta	Lake	Painesville
Belmont	St. Clairsville	Lawrence	Ironton
Brown	Georgetown	Licking	Newark
Butler	Hamilton	Logan	Bellefontaine
Carroll	Carrollton	Lorain	Elyria
Champaign	Urbana	Lucas	Toledo
Clark	Springfield	Madison	London
Clermont	Batavia	Mahoning	Youngstown
Clinton	Wilmington	Marion	Marion
Columbiana	Lisbon	Medina	Medina
Coshocton	Coshocton	Meigs	Pomeroy
Crawford	Bucyrus	Mercer	Celina
Cuyahoga	Cleveland	Miami	Troy
Darke	Greenville	Monroe	Woodsfield
Defiance	Defiance	Montgomery	Dayton
Delaware	Delaware	Morgan	McConnelsville
Erie	Sandusky	Morrow	Mount Gilead
Fairfield	Lancaster	Muskingum	Zanesville
Fayette	Washington Court House	Noble	Caldwell
		Ottawa	Port Clinton
Franklin	Columbus	Paulding	Paulding
Fulton	Wauseon	Perry	New Lexington
Gallia	Gallipolis	Pickaway	Circleville
Geauga	Chardon	Pike	Waverly
Greene	Xenia	Portage	Ravenna
Guernsey	Cambridge	Preble	Easton
Hamilton	Cincinnati	Putnam	Ottawa
Hancock	Findlay	Richland	Mansfield
Hardin	Kenton	Ross	Chillicothe
Harrison	Cadiz	Sandusky	Fremont
Henry	Napoleon	Scioto	Portsmouth
Highland	Hillsboro	Seneca	Tiffin
Hocking	Logan	Shelby	Sidney

NAME	COUNTY SEAT	NAME	COUNTY SEAT
Stark	Canton	Warren	Lebanon
Summit	Akron	Washington	Marietta
Trumbull	Warren	Wayne	Wooster
Tuscarawas	New Philadelphia	Williams	Bryan
Union	Marysville	Wood	Bowling Green
Van Wert	Van Wert	Wyandot	Upper Sandusky
Vinton	McArthur		

Economy. The state has many manufacturing centers, with an emphasis on heavy industry. Its leading products are transportation equipment, primary and fabricated metals, machinery, and plastic and rubber goods. Ohio leads the nation in the production of lime, is second in the production of clays, and ranks third in the production of salt, sand and gravel, and stone. It also ranks high in the exploitation of coal. Although most of the working population is engaged in industry and most of the state's income is derived from commerce and manufacturing, Ohio has extensive farms, and large amounts of corn, soybeans, hay, wheat, cattle, hogs, and dairy items are produced. Railroads and highways crisscross the state, bearing a tremendous traffic of raw materials and manufactures. The Lake Erie ports—Toledo, Cleveland, and Sandusky—handle much iron and copper ore, coal, oil, and finished materials (including steel and automobile parts).

History. In prehistoric times Ohio was inhabited by the mound builders, many of whose mounds are preserved in state parks and in the Mound City Group National Monument. Before the arrival of Europeans, eastern Ohio was the scene of bloody Indian warfare when the Iroquois exterminated (1655) the Erie Indians. La Salle began his explorations of the Ohio valley in 1669 and claimed the entire area for France. The Ohio River became a magnet for fur traders and landseekers, and rivalry for control of the forks of the Ohio River led to the outbreak (1754) of the last of the French and Indian Wars. The defeat of the French gave the land to the British, but British possession was disturbed by Pontiac's Rebellion and the Indian flare-ups that followed. In 1774 the British placed the region between the Ohio River and the Great Lakes within the boundaries of Canada. During the American Revolution George Rogers Clark conducted military operations in the Ohio country. Ohio was part of the vast area ceded to the United States by the Treaty of Paris (1783). It was the first region developed under the provisions of the Ordinance of 1787, which established the Old Northwest. Marietta, founded in 1788, was the first permanent American settlement in the Old Northwest. In the years that followed, various land companies were formed, and settlers poured in. The Indians, supported by the British, resisted American settlement. They were decisively defeated by Anthony Wayne in the battle of Fallen Timbers (1794). The British thereafter (1796) withdrew their outposts from the Northwest under the terms of Jay's Treaty, and the area was pacified.

Ohio became a territory in 1799 and a state in 1803, with Chillicothe as its capital. Columbus became the permanent capital in 1816. In the War of 1812 the Americans lost many of the first battles of the war in the Old Northwest. The area was secured with O. H. Perry's naval victory on Lake Erie near Put-in-Bay, Ohio, and W. H. Harrison's victory in the battle of the Thames on Canadian soil. After the war Ohio's growth was spurred by the building of the Erie Canal, other canals, and toll roads. Railroads gradually succeeded canals.

After the Civil War industrial development increased rapidly when the shipment of ore from the upper Great Lakes region was intensified and the development of the petroleum industry in northeast Ohio shifted the center of economic activity from the banks of the Ohio River to the shores of Lake Erie, particularly around Cleveland. Immigrants began to swell the population, and huge fortunes were made. Ohio became very important politically, contributing seven of the eleven presidents from Grant to Harding.

Both farms and industries in Ohio were hard hit by the Great Depression. In the 1930s the state was wracked by major strikes such as the sit-down strikes in Akron (1935–36) and the so-called Little Steel strike (1937). World War II brought great prosperity to Ohio, but labor strife was later resumed, as in the steel strikes of 1949 and 1959. Industrialization has continued, and Ohio has become an important center for industrial research.

Government. Ohio's present constitution was adopted in 1851. The state's executive branch is headed by a governor elected for a four-year term and permitted two successive terms. Ohio's bicameral general assembly has a senate with 33 members, elected for four-year terms (half each two years) and a house with 99 members elected for two-year terms.

Educational Institutions. An unusually large number of colleges and universities are located in the state. Among them are Antioch College, at Yellow Springs; Bowling Green State Univ., at Bowling Green; Case Western Reserve Univ. (formerly Western Reserve Univ. and Case Institute of Technology), at Cleveland; Kent State Univ., at Kent; Kenyon College, at Gambier; Miami Univ., at Oxford; Oberlin College, at Oberlin; the Ohio State Univ., at Columbus; Ohio Univ., at Athens; Ohio Wesleyan Univ., at Delaware; Univ. of Cincinnati; and Univ. of Toledo.

Ohio. **1.** County (1970 pop. 4,289), 87 sq mi (225.3 sq km), SE Ind., bounded on the E by the Ky. border and the Ohio River and drained by Laughery Creek; formed 1844; co. seat Rising Sun. Its agriculture includes livestock, truck crops, tobacco, grain, and dairy products. Furniture and construction machinery and materials are manufactured. **2.** County (1970 pop. 18,790), 596 sq mi (1,543.6 sq km), W central Ky., bounded on the W and S by the Green River and crossed by the Rough River and South Fork of Panther Creek; formed 1798; co. seat Hartford. In a rolling agricultural area yielding livestock, grain, burley tobacco, hay, and strawberries, it has bituminous-coal mines, oil wells, limestone quarries, and timber. Its manufactures include textiles, furniture, and metal products. **3.** County (1970 pop. 64,197), 107 sq mi (277.1 sq km), W W.Va., in the Northern Panhandle, bounded on the W by the Ohio River and the Ohio border, on the E by the Pa. border, and drained by Wheeling Creek; formed 1776; co. seat Wheeling. Its agriculture includes livestock, dairy products, tobacco, and truck crops. It has meat-packing, dairy-processing, and iron, steel, and metalworking industries.

Ohio, river, 981 mi (1,578.4 km) long, formed by the confluence of the Allegheny and Monongahela rivers in W Pa., at Pittsburgh, and flowing generally NW, then SW and W to enter the Mississippi River at Cairo, Ill. The Ohio's course follows a portion of the southern edge of the region covered by continental ice during the Pleistocene epoch; glacial meltwater probably cut its original channel. The Ohio River basin covers c.204,000 sq mi (528,360 sq km). The Ohio is prone to spring flooding, and extensive flood control and protection devices have been constructed along the river and its tributaries. These devices also improve the river's navigability; a 9-ft (2.7-m) channel is maintained along its entire length. A system of modern locks and dams, constructed since 1955 to replace older structures, speeds the transit of barges and pleasure craft.

The French explorer La Salle reportedly reached the Ohio River in 1669, but there was no significant interest in the valley until the French and the British began to struggle for control of the river in the 1750s. An early settlement was established at the forks of the Ohio (modern Pittsburgh) by the Ohio Company in 1749, but it was captured by the French in 1754. It was recaptured by the British and named Fort Pitt in 1758. Britain gained control of the river by the treaty of 1763, but settlement of the area was prohibited. Britain ceded the region to the United States at the end of the Revolutionary War (1783), and it was opened to settlement by the Ordinance of 1787, which established the Northwest Territory. Thereafter, until the opening of the Erie Canal in 1825, the Ohio River was the principal route to the newly opened West. Traffic declined on the river after the railroads were built in the mid-1800s, but it began to revive after World War II.

Ohio and E·rie Canal (îr'ē), former waterway of Ohio, 307 mi (494 km) long, between Lake Erie at Cleveland and the Ohio River at Portsmouth; built 1825–32. It flourished until the advent of the railroad era in the 1850s.

O·hrid (ō'krēd'), town (1971 pop. 26,370), extreme S Yugoslavia, in Macedonia, on a rock above Lake Ohrid, on the Yugoslav-Albanian frontier. It is a tourist and commercial center, as well as a railroad terminus. Fishing, tanning, sericulture, and the manufacture of silk textiles are also economically important. Ohrid is near the site of the Greek colony of Lychnidos, founded in the 3rd cent. B.C. It was captured by the Romans in A.D. 168. In the 9th cent. Ohrid was incorporated into the first Bulgarian empire, and in the 10th cent. it flourished as the political and cultural center of Bulgaria. Ohrid was captured by the Serbs in 1334 and fell to the Turks in 1394. After World War I Ohrid was joined to Yugoslavia. The town is also noted for its museums, galleries, and educational institutions.

Ohrid, Lake, deepest lake of the Balkans, c.130 sq mi (335 sq km), on the Yugoslav-Albanian border. It is connected with Lake Prepa by underground channels. On its shores stand several monasteries.

Oil City (oil), city (1978 est. pop. 14,900), Venango co., NW Pa., on the Allegheny River; inc. 1871. The city was founded after Edwin L. Drake struck oil nearby in 1859. Today it is a major refining and shipping center and a producer of oil-field equipment.

Oise (wäz), department (1975 pop. 606,320), 2,261 sq mi (5,856 sq km), N France, in Picardy; capital Beauvais.

Oise, river, 186 mi (299.3 km) long, rising in the Ardennes Mts., S Belgium, and flowing through N France generally SW to join the Seine River near Pontoise. Navigable for most of its length, the Oise is an important transportation route.

O·i·ta (ō-ē'tə), city (1976 est. pop. 329,456), capital of Oita prefecture, NE Kyushu, Japan, a port on Beppu Bay. It is a rail hub, a manufacturing center, and a distribution point for agricultural products. Oita was a castle town in the 16th cent. and traded with the Portuguese.

O·jai (ō'hī'), mountain resort city (1970 pop. 5,591), Ventura co., S Calif., NW of Los Angeles; laid out 1874, inc. 1921.

O·jos del Sa·la·do (ō'hōs *th*ĕl sä-lä'*th*ō), peak, 22,539 ft (6,874.4 m) high, on the border between Argentina and Chile, in the Andes. In 1956 a Chilean expedition reported its height to be 23,239 ft (7,087.9 m), thus making it the tallest peak in the Western Hemisphere, but this report has been unconfirmed.

O·ka (ō-kä'). **1.** River, c.925 mi (1,490 km) long, rising S of Orel, central European USSR, and flowing N past Orel and Kaluga, E past Serpukhov and Ryazan, and then NE past Murom to join the Volga River at Gorky. It is navigable by large vessels below Kolomna, c.550 mi (885 km) upstream, and traverses densely populated agricultural and industrial areas. **2.** River, c.600 mi (970 km) long, rising in the Sayan Mts., Buryat Autonomous Republic, S central Siberian USSR. It flows north through Irkutsk oblast to join the Angara River below Bratsk.

O·ka·loo·sa (ō'kə-lo͞o'sə), county (1970 pop. 88,187), 944 sq mi (2,445 sq km), NW Fla., bounded by the Ala. line in the N and the Gulf of Mexico in the S; formed 1915; co. seat Crestview. It is in a rolling agricultural area (corn, peanuts, cotton, hogs, cattle, and poultry) drained by the Blackwater, Yellow, and Shoal rivers. Forestry and some fishing are done. Part of Eglin Air Force Base is in the south.

O·ka·na·gan Lake (ō'kə-nä'gən), 69 mi (111 km) long and from 2 to 4 mi (3.2–6.4 km) wide, S British Columbia, Canada. The lake is in a prosperous fruit-growing region.

O·ka·nog·an (ō'kə-nŏg'ən), county (1970 pop. 25,867), 5,301 sq mi (13,729.6 sq km), N Wash., on the British Columbia line, bounded in the S by the Columbia River and drained by the Okanogan and Methow rivers, with the Cascade Mts. in the W; formed 1888; co. seat Okanogan. It is in a lumbering and mining (gold, silver, and copper) region. Its agriculture includes apples, wheat, hay, and livestock (especially sheep), and it has some food-processing plants. It contains parts of Chelan and Colville national forests and the Colville Indian Reservation.

Okanogan, town (1977 est. pop. 2,250), seat of Okanogan co., N central Wash., on the Okanogan River; settled 1886, inc. 1907. It is a trade and marketing center for a ranching and fruit-growing area.

O·ka·ra (ō-kä'rə), city (1972 pop. 101,791), N central Pakistan. It is a market for food grains, oilseed, and cotton. There is also a factory for spinning and weaving cotton.

O·ka·van·go (ō'kə-văng'gō), river, c.1,000 mi (1,610 km) long, rising in the highlands of central Angola, W central Africa, and flowing SE, across the Caprivi Strip, to the Okavango Swamp, N Botswana. The Okavango Swamp (c.4,000 sq mi/10,360 sq km) occupies a depression that contained a large prehistoric lake. The northern part of the swamp has papyrus growth and is wet throughout the year; the rest of the swamp fills with water as the seasonal cycle progresses.

O·ka·ya·ma (ō'kə-yä'mə), city (1976 est. pop. 520,469), capital of Okayama prefecture, SW Honshu, Japan, on an inlet of the Inland Sea. It is a railroad hub, an industrial center, and an important market for peaches and fancy matting. Stoneware, cotton textiles, machinery, chemicals, and rubber goods are produced in Okayama.

O·kee·cho·bee (ō'kə-chō'bē), county (1970 pop. 11,233), 780 sq mi (2,020.2 sq km), central Fla., bounded on the W by the Kissimmee River and on the S by Lake Okeechobee; formed 1917; co. seat Okeechobee. Cattle raising predominates in this grassy plains area, which has many small lakes and some swamps. Poultry raising and truck farming are also done.

Okeechobee, city (1970 pop. 3,715), seat of Okeechobee co., S central Fla., on the N shore of Lake Okeechobee. A Seminole War battle took place in the vicinity on Dec. 25, 1837.

Okeechobee, Lake, c.700 sq mi (1,815 sq km), SE Fla., N of the Everglades; third-largest freshwater lake and fourth-largest lake wholly within the United States. It is c.35 mi (55 km) long and up to 25 mi (40 km) wide, with a maximum depth of 15 ft (4.6 m). In reclaiming the Everglades and adjacent lands, many canals (also used for transportation) were built extending from the southern part of the lake, itself a link in the Okeechobee Waterway. A levee, built after the disastrous hurricane of 1926, rims the lake's southern shore and protects the region from flood waters. The levees and canals have impeded the flow of water from the lake into the Everglades, which now suffers from saltwater intrusion. The drained lands bordering the lake produce vegetables and sugar cane.

Okeechobee Waterway or **Cross-Flor·i·da Waterway** (krôs′flôr′ĭ-də), 155 mi (249.4 km) long, across S Fla., from Stuart on the Atlantic Ocean to Fort Myers on the Gulf of Mexico. Its main segments are the St. Lucia Canal, Lake Okeechobee, Lake Hicpochee, and Caloosahatchee River. The shallow (6 ft/1.8 m) waterway has four locks and is used by small commercial and pleasure craft.

O·ke·fe·no·kee Swamp (ō′kə-fə-nō′kē, -nŏk′ē), c.600 sq mi (1,555 sq km), c.40 mi (65 km) long and averaging 20 mi (32 km) in width, SE Ga., extending into N Fla. It is a saucer-shaped depression with low ridges and small islands rising above the water and vegetation cover. It abounds in varied wildlife.

O·ke·mah (ō-kē′mə), city (1970 pop. 2,913), seat of Okfuskee co., E central Okla., E of Oklahoma City, in an agricultural and oil region; settled 1902.

Ok·fu·skee (ōk-fŭs′kē), county (1970 pop. 10,683), 638 sq mi (1,652.4 sq km), central Okla., intersected by the North Canadian River; formed 1907; co. seat Okemah. Its agriculture includes peaches, apples, cotton, grain, pecans, peanuts, livestock, and dairy products. It has oil and natural-gas fields and textile and machine industries.

O·khotsk, Sea of (ō-kŏtsk′), 590,000 sq mi (1,528,100 sq km), NW arm of the Pacific Ocean, W of the Kamchatka peninsula and the Kuril Islands. It is connected with the Sea of Japan by Tatar and La Pérouse straits and with the Pacific Ocean by passages through the Kuril Islands. The sea is generally less than 5,000 ft (1,525 m) deep; its deepest point, near the Kuriles, is 11,033 ft (3,365 m). The sea is icebound from Nov. to June.

O·ki·na·wa (ō′kə-nä′wə), island (1975 pop. 924,540), 454 sq mi (1,175.9 sq km), W Pacific Ocean, SW of Kyushu, and part of Okinawa prefecture, Japan. It is the largest of the Okinawa Islands in the Ryukyu Islands archipelago. Okinawa is a long, narrow, irregularly shaped island of volcanic origin with coral formations in the southern part. The northern part is mountainous, rising to 1,657 ft (505 m), and has a dense vegetation cover. Sugar cane, sweet potatoes, and rice are grown and fishing is important. In World War II. U.S. army and marine forces landed here on Apr. 1, 1945, and fought one of the bloodiest campaigns of the war, while the navy offshore suffered heavy damage in resisting attacks by suicide planes. The Japanese garrison ended organized resistance on June 21, 1945. Okinawa was placed in Aug., 1945, under a U.S. military governor and remained under U.S. control until May, 1972, when it was returned to Japan. U.S. military bases were allowed to remain on the island.

Okla. Abbreviation for Oklahoma.

O·kla·ho·ma (ō′klə-hō′mə), state (1975 pop. 2,750,000), 69,919 sq mi (181,090 sq km), SW United States, admitted as the 46th state of the Union in 1907. Oklahoma City is the capital. The state is bounded on the north by Kansas and on the east by Mo. and Ark.; the Red River marks the southern border with Texas except in the west, where the Texas Panhandle thrusts north and reduces Oklahoma to a thin panhandle bounded in the west by N.Mex. and touching Colo. in the north.

The high, short-grass plains of western Oklahoma are part of the Great Plains and, like the rest of that area, are frozen by north winds in the winter and baked by soaring heat in the summer. There are wide grazing lands and broad wheat fields. The plains are broken here and there, notably by Black Mesa in the Panhandle and by the Wichita Mts. in the southwest, but the general slope is downward to the east, and central and eastern Oklahoma is mostly prairie, rising

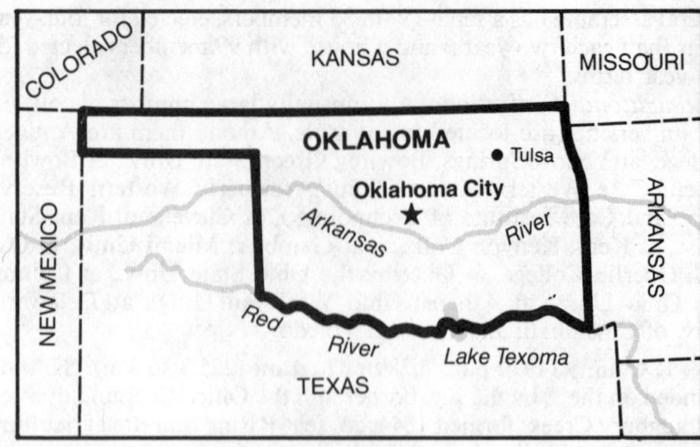

in the northeast to the Ozark Mts. and in the southeast to the Ouachita Mts. The rivers that flow from west to east across the state—the Arkansas and its tributaries, the Cimarron and the Canadian (with the North Canadian) in the north, the Red River with the Washita and other tributaries in the south—are much more prominent in the east. Platt National Park, the site of extensive mineral springs, and Arbuckle National Recreation Area are in southern Oklahoma.

Oklahoma is divided into 77 counties:

NAME	COUNTY SEAT	NAME	COUNTY SEAT
Adair	Stilwell	Le Flore	Poteau
Alfalfa	Cherokee	Lincoln	Chandler
Atoka	Atoka	Logan	Guthrie
Beaver	Beaver	Love	Marietta
Beckham	Sayre	McClain	Purcell
Blaine	Watonga	McCurtain	Idabel
Bryan	Durant	McIntosh	Eufaula
Caddo	Anadarko	Major	Fairview
Canadian	El Reno	Marshall	Madill
Carter	Ardmore	Mayes	Pryor Creek
Cherokee	Tahlequah	Murray	Sulphur
Choctaw	Hugo	Muskogee	Muskogee
Cimarron	Boise City	Noble	Perry
Cleveland	Norman	Nowata	Nowata
Coal	Coalgate	Okfuskee	Okemah
Comanche	Lawton	Oklahoma	Oklahoma City
Cotton	Walters	Okmulgee	Okmulgee
Craig	Vinita	Osage	Pawhuska
Creek	Sapulpa	Ottawa	Miami
Custer	Arapaho	Pawnee	Pawnee
Delaware	Jay	Payne	Stillwater
Dewey	Taloga	Pittsburg	McAlester
Ellis	Arnett	Pontotoc	Ada
Garfield	Enid	Pottawatomie	Shawnee
Garvin	Pauls Valley	Pushmataha	Antlers
Grady	Chickasha	Roger Mills	Cheyenne
Grant	Medford	Rogers	Claremore
Greer	Mangum	Seminole	Wewoka
Harmon	Hollis	Sequoyah	Sallisaw
Harper	Buffalo	Stephens	Duncan
Haskell	Stigler	Texas	Guymon
Hughes	Holdenville	Tillman	Frederick
Jackson	Altus	Tulsa	Tulsa
Jefferson	Waurika	Wagoner	Wagoner
Johnston	Tishomingo	Washington	Bartlesville
Kay	Newkirk	Washita	Cordell
Kingfisher	Kingfisher	Woods	Alva
Kiowa	Hobart	Woodward	Woodward
Latimer	Wilburton		

Economy. Formerly the major crop of Oklahoma was cotton, but now wheat is the leading cash crop; however, income from livestock exceeds that from crops. Many minerals are found in the state, including coal, but the mineral that has given the state its wealth is oil. After the first well was drilled in 1888, the petroleum industry grew to enormous proportions. Many of Oklahoma's factories process raw materials found in the state. There is also an aviation industry.

History. Several Indian cultures, some highly advanced materially, existed in the area before the first European, Francisco Coronado, visited here in 1541. The Indians roamed over the land, tribes of the Plains cultures—Osage, Kiowa, Comanche, and Apache—in the west, and the Wichita and other relatively sedentary tribes farther east. The first European trading post was established at Salina by the Chouteau family of St. Louis before the territory was transferred to the United States by the Louisiana Purchase in 1803.

In the 1830s intense white pressure for their lands forced the Five Civilized Tribes (the Cherokee, the Choctaw, the Chickasaw, the Creek, and the Seminole) to abandon their old homes east of the

Mississippi and to take up residence in what was to become the Indian Territory. Their tragic removal is known as the Trail of Tears. They settled on the hills and little prairies of the eastern section and built separate organized states and communities. The Five Civilized Tribes clashed briefly with the Plains Indians, particularly the Osage, but they were for a time free from white interference, and they were able to establish a civilization that strongly affected the whole history of the region. As a punishment for taking the Confederate side during the Civil War, the Five Civilized Tribes lost the western part of the Indian Territory, and the Federal government began assigning lands here to such landless eastern tribes as the Delaware and the Shawnee, as well as to nomadic Plains tribes. The territory was victimized by lawlessness and served as a hideout for white outlaws.

Immediately after the Civil War the long drives of cattle from Texas to the Kansas railroad began to cross Oklahoma, traveling over the cattle trails that became part of Western folklore. The first railroad to cross Oklahoma was built between 1870 and 1872, and thereafter it was not possible to keep white settlers out, despite laws and Indian treaties. In the 1880s, land-hungry frontier farmers, the boomers, agitated to obtain the "unassigned" lands in the western section—the lands not given to any Indian tribe. The agitation succeeded, and a large strip was opened for settlement on Apr. 22, 1889. Those who illegally entered ahead of the set time were the sooners. Later other strips of territory were opened, and settlers poured in from the Middle West and the South.

The western section of what is now the state of Oklahoma became the Oklahoma Territory in 1890; it included the Panhandle, that tiny strip of territory that, taken from Texas by the Compromise of 1850, had become a no-man's-land where settlers came in undisturbed. In 1893 the Dawes Commission was appointed to implement a policy of dividing the tribal lands into individual holdings; the Indians resisted, but the policy was finally enforced in 1906. The wide lands of the Indian Territory were thus made available to white men. The Indian Territory and Oklahoma Territory were united in 1907 to form the state of Oklahoma.

Already the oil boom had reached major proportions, and the young state was on the verge of great economic development. At the same time, cotton, wheat, and corn were major money crops, and cattle-land holdings, although shrinking, were still enormous. Recurrent drought in the 1920s burned the wheat in the fields, and overplanting, overgrazing, and unscientific cropping aided the weather in making northwestern Oklahoma part of the Dust Bowl of the 1930s. A great number of tenant farmers were compelled to leave their dust-stricken farms and went west as migrant laborers; the tragic plight of these Okies is the theme of John Steinbeck's *Grapes of Wrath*. The demands for food in World War II and Federal price supports for agricultural products after the war aided farm prosperity. Large state and Federal programs for conserving the water of rivers and for supplying irrigation have resulted in the construction of many large dams and reservoirs. In 1971 the opening of the Oklahoma portion of the Arkansas River Navigation System gave the cities of Muskogee and Tulsa direct access to the sea. The Indian population of the state, exceeded only by that of Ariz., was almost 65,000 in 1970.

Government. The original 1907 constitution is still in effect. Oklahoma has a legislature of 48 senators and 99 representatives elected for two-year terms. The governor is elected for a four-year term.

Educational Institutions. The most important in the state are the Univ. of Oklahoma and Oklahoma State Univ.

Oklahoma, county (1970 pop. 527,717), 700 sq mi (1,813 sq km), central Okla., intersected by the North Canadian River and the Deep Fork; formed 1907; co. seat Oklahoma City. It has dairying, stock raising, and agriculture (wheat, poultry, and oats). There are petroleum and natural-gas fields. Its diversified industries produce textiles, millwork, iron, steel and aluminum, furniture, tires, and construction machinery. Paper, chemicals, plastic, brick and concrete, and metal products are also manufactured.

Oklahoma City (1978 est. pop. 371,000), state capital, and seat of Oklahoma co., central Okla., on the North Canadian River; inc. 1890. It is an important livestock market, the state's wholesale and distributing center, and a farm trade and processing point. Oil is a major product; the city is situated in the middle of an oil field (opened 1928), and there are wells even on the capitol grounds. The city has large stockyards and meat-packing houses, grain mills, and cotton-processing plants. Nearby Tinker Air Force Base, a logistics center with one of the world's largest air depots, is an important source of civilian employment. Oklahoma City was settled in a land rush after the area was opened to homesteaders on Apr. 22, 1889. It became the state capital in 1910. Of interest are the capitol, the state historical museum, the National Cowboy Hall of Fame and Western Heritage Center, the civic center buildings and monuments, a theater complex, the Oklahoma Health Sciences Center, and a zoo.

Ok·la·wa·ha (ŏk′lə-wô′hô), river, c.120 mi (195 km) long, N central Fla. Rising in the large lake system of central Fla., it flows north, receiving the waters of Silver Springs and Orange Lake, then turns east, converging with the St. Johns River near Palatka.

Ok·mul·gee (ōk-mŭl′gē), county (1970 pop. 35,358), 700 sq mi (1,813 sq km), E central Okla., intersected by the Deep Fork and including Okmulgee and Henryetta lakes; formed 1907; co. seat Okmulgee. Its agriculture includes grain, cotton, livestock, pecans, peanuts, corn, and dairy products. It has petroleum and natural-gas fields.

Okmulgee, city (1970 pop. 15,180), seat of Okmulgee co., E central Okla., in an oil and farm area; inc. 1900. It is an agricultural processing center and has glass plants and an oil refinery. It was founded on the site of the Creek capital (1868–1907) and boomed with the discovery of oil in 1907.

O·ko·lo·na (ō′kə-lō′nə). **1.** Town (1978 est. pop. 23,800), Jefferson co., NW Ky., a suburb of Louisville. **2.** City (1970 pop. 3,002), a seat of Chickasaw co., NE Miss., NNW of Columbus, in a dairy, livestock, and farm area; founded 1848. Brick, tile, and textiles are made here. The town was burned in the Civil War.

o·krug (ō′kro͞og′), administrative division in the USSR. Okrugs are sparsely settled ethnic minority areas that are smaller than a kray (territory), and that usually have populations of less than 100,000. There are currently ten national okrugs; all were formed between 1929 and 1937. Originally conceived by sympathetic ethnographers as a kind of reservation, the national okrugs have become an instrument for integrating the lives of the natives into the Soviet system.

Ok·tib·be·ha (ŏk-tĭb′ə-hô′), county (1970 pop. 28,752), 454 sq mi (1,175.9 sq km), NE central Miss., drained by the Noxubee and Oktibbeha rivers; formed 1833; co. seat Starkville. It is in a dairy, cattle, soybean, cotton, corn, and hay region, with lumber sawmilling and industry (food processing, textile finishing, and the manufacturing of clothing, wood products, and furniture).

Ö·land (œ′länd′, ûr′lănd′), narrow island (1970 pop. 22,561), 520 sq mi (1,346.8 sq km), Kalmar co., SE Sweden, in the Baltic Sea, separated from mainland Sweden by the Kalmarsund. There are many summer resorts on the island. Sugar beets, cereals, and vegetables are grown, and cattle are raised. The island also has a fishing fleet. Öland has numerous monuments dating from the Stone Age. It has often been a battleground in the wars among the Scandinavian countries.

O·la·the (ō-lā′thə), city (1978 est. pop. 25,100), seat of Johnson co., NE Kansas, near Kansas City; inc. 1858. Its manufactures include aircraft communication and guidance systems, batteries, machinery, and plastic components.

Ol·bi·a (ōl′bē-ə), Ionic Greek colony of Miletus, founded at the beginning of the 6th cent. B.C. It is on the right bank of the Bug River in south-central Ukrainian SSR. The leading Milesian colony and later a republic, its economy centered around handicrafts and trade. Its prosperity resulted especially from the exportation of wheat. It flowered for 300 years, ending in the 3rd cent. B.C. In the 2nd cent. B.C., Olbia was incorporated into the Scythian state of the Crimea. Excavations have unearthed towers and city gates from the Hellenic period and part of a temple of Apollo from the Roman period.

Ol·den·burg (ōl′dən-bûrg′, ôl′dən-bo͝ork′), former state, NW West Germany. It is now included in the state of Lower Saxony. The former state was divided into Oldenburg proper, stretching south from the North Sea, west of the Weser River; Birkenfeld; and the district (but not the city) of Lübeck. Originally a part of Saxony, the county of Oldenburg came into prominence in the 12th cent., when the counts became princes of the empire. In 1448 Count Christian became king of Denmark as Christian I, while his younger brother, Gerard, and his successors continued to rule Oldenburg. Oldenburg passed (1676) to Christian V of Denmark. In 1773 Christian VII exchanged Oldenburg for ducal Holstein. Peter I of Oldenburg lost his duchy to Napoleon I but recovered Oldenburg and the bishopric of Lübeck in 1813. A member of the German Confederation from 1815, Oldenburg sided (1866) with Prussia in the Austro-Prussian

War and joined (1871) the German Empire. The last grand duke abdicated in 1918, and Oldenburg joined the Weimar Republic.

Oldenburg, city (1974 est. pop. 134,280), Lower Saxony, NW West Germany, on the Hunte River and the Küstenkanal (Coast Canal). It is a rail junction, transshipment point, agricultural market, and industrial center. Manufactures include machinery, glass, chemicals, and textiles. Oldenburg was first mentioned in 1108 and was chartered in 1345. It was the seat of the counts of Oldenburg until 1667, when it passed to Denmark. Noteworthy buildings include the former ducal palace (17th–18th cent.) and the Gothic Lambertikirche, a church built in the 13th cent. (rebuilt 18th–19th cent.).

Old Faith·ful (fāth'fəl), geyser: *see* Yellowstone National Park.

Old Forge (fôrj, fōrj), borough (1970 pop. 9,522), Lackawanna co., NE Pa., SW of Scranton on the Lackawanna River; inc. 1899. Apparel is made, and there are anthracite-coal mines.

Old·ham (ōl'dəm), county borough (1976 pop. 227,500), Greater Manchester, NW England. It is a center of cotton spinning and has numerous mills, foundries, and engineering works. The town hall, the art gallery, the museum, and Alexandra Park are noteworthy.

Oldham. 1. County (1970 pop. 14,687), 184 sq mi (476.6 sq km), N Ky., in the outer Bluegrass, bounded on the W by the Ohio River and the Ind. border and drained by Floyds Fork; formed 1823; co. seat La Grange. It is in a rolling upland agricultural area that produces burley tobacco, dairy products, oats, and corn. It has stone quarries, and its industry includes food processing. **2.** County (1970 pop. 2,258), 1,478 sq mi (3,828 sq km), extreme NW Texas, in the high plains of the Panhandle, bounded on the W by the N.Mex. border and drained by the Canadian River and its tributaries; formed 1876; co. seat Vega. This wheat and cattle-ranching area also produces truck crops, hogs, sheep, poultry, and dairy products.

Old·man (ōld'mən), river, c.250 mi (400 km) long, rising in the Rocky Mts., SW Alta., Canada, and flowing generally E past Lethbridge to join the Bow River W of Medicine Hat and form the South Saskatchewan River. Bituminous coal is mined along the river.

Old Or·chard Beach (ôr'chərd), town (1970 pop. 5,404), York co., SW Maine, on the Atlantic coast; settled c.1631, inc. 1883. For many years a popular resort, it has a beach and amusement facilities. There was a trading post nearby before 1630.

Old Sar·um (sâr'əm), site of a former city, Wiltshire, S England, just N of Salisbury (New Sarum). Excavations in the mound on which the settlement stood have revealed remains of an ancient British camp, the Roman station Sorbiodunum, a still later Saxon town, and a Norman town. Old Sarum was important until internal strife caused the decay of the old city; shortage of water and exposure to winds may also have been causes of its decline.

Old Say·brook (sā'brŏŏk'), resort town (1970 pop. 8,468), Middlesex co., S Conn., on the Connecticut River and Long Island Sound; settled 1635, set off from Saybrook 1852, inc. 1854.

Old Town, city (1970 pop. 8,741), Penobscot co., S central Maine, on the Penobscot River above Bangor; settled 1774, inc. as a town 1840, as a city 1891. Maine's first railroad (1836) connected Old Town with Bangor. Canoes, shoes, woolen goods, and wood products are made.

Old West·bur·y (wĕst'bûr'ē, -bə-rē), residential village (1970 pop. 2,667), Nassau co., SE N.Y., on W Long Island NE of Mineola; inc. 1924. The Old Westbury Gardens, a park with formal gardens after the style of 18th cent. England, is here.

O·le·an (ō'lē-ăn', ō'lē-ăn'), city (1978 est. pop. 18,400), Cattaraugus co., W N.Y., on the Allegheny River near the Pa. line; settled 1804, inc. 1893. Manufactures include diesel engines, compressors, tile, plastics, and electrical and electronic parts.

O·lek·ma (ō-lĕk'mə), river, c.820 mi (1,320 km) long, rising in the Yablonovy range, SE Siberian USSR. It flows north through Amursk oblast and the Yakut Autonomous Republic to the Lena River below Olekminsk.

O·len·ek (ŏl'ən-yôk'), river, c.1,350 mi (2,175 km) long, rising in the central Siberian plateau, Krasnoyarsk Kray, E Siberian USSR. It winds east then north through northwestern Yakut Autonomous Republic to the Laptev Sea. It is navigable for c.600 mi (965 km) upstream and abounds in fish.

O·lé·ron (ô-lā-rôɴ'), island (1968 est. pop. 16,355), 68 sq mi (176.1 sq km), Charente-Maritime dept., W France, in the Bay of Biscay. It is an oystering, farming, and ranching area and a summer vacation spot. Oléron was a stronghold of Protestantism in the 16th cent. A bridge (1966) links it with the mainland.

O·lin·da (ō-lĭn'də, ŏŏ-lēn'də), city (1970 pop. 187,553), Pernambuco state, E Brazil, on the Atlantic Ocean. Founded in 1537, it was captured by the Dutch in the 1630s and burned to the ground. Olinda's reputation as a center of learning dates from 1796, when a Jesuit seminary was founded here.

Ol·ive·hurst (ŏl'ĭv-hûrst'), uninc. town (1970 pop. 8,100), Yuba co., N central Calif., S of Marysville.

Ol·i·ver (ŏl'ə-vər), county (1970 pop. 2,322), 720 sq mi (1,864.8 sq km), central N.Dak., in an agricultural area watered by the Square Butte Creek and bounded on the E by the Missouri River; formed 1885; co. seat Center. Its agriculture includes livestock, poultry, grain, and potatoes. Deposits of lignite coal are plentiful here.

Ol·i·vet (ŏ'lə-vĕt'), town (1970 pop. 103), seat of Hutchinson co., SE S.Dak., SSE of Mitchell.

Ol·i·vette (ŏl'ə-vĕt'), city (1970 pop. 9,156), St. Louis co., E Mo., a suburb W of St. Louis; settled c.1828, inc. 1930. Carburetors, conveyors, and whiskey are made.

O·liv·i·a (ō-lĭv'ē-ə), village (1970 pop. 2,553), seat of Renville co., SW Minn., E of Granite Falls, in a farm area; platted 1878, inc. 1881.

Olm·sted (ŏm'stĕd', ōm'-), county (1970 pop. 84,104), 655 sq mi (1,696.5 sq km), SE Minn., drained by the Root River and branches of the Zumbro River; formed 1855; co. seat Rochester. In an agricultural area yielding livestock, dairy products, corn, oats, barley, and potatoes, it has limestone quarries and food-processing and metalworking industries.

Ol·ney (ŏl'nē), city (1975 est. pop. 8,500), seat of Richland co., SE Ill., W of Lawrenceville; inc. 1841. A trade and shipping center of a farm area, it also makes lubricants and tricycles.

O·lo·mouc (ô'lô-mōts'), city (1975 est. pop. 96,207), N central Czechoslovakia, in Moravia, on the Morava River. Olomouc is an industrial city, with factories producing steel, machinery, electrical equipment, and food products, especially candy and chocolate. An ancient town, it was once the leading city of Moravia and was strongly fortified. In 1242 Wenceslaus II of Bohemia defeated the Mongol invaders here. An agreement between Austria and Prussia was signed here (1850), dissolving the German Union under Prussia's presidency and restoring the German Confederation, headed by Austria. Prussia smarted under the "humiliation of Olmütz" until 1866, when it defeated Austria in war. Present-day landmarks include the Cathedral of St. Wenceslaus (begun 12th cent.), the city hall (rebuilt 13th cent.), and two Gothic churches.

Olsz·tyn (ôl'shtĭn), city (1975 est. pop. 112,700), N Poland. It is a trade, manufacturing, and railroad center. Founded (1348) by the Teutonic Knights, who built its impressive castle, it was ceded to Poland in 1466 and to Prussia in 1772. The city was retained by Germany after a plebiscite in 1920. It suffered heavy damage in World War II and reverted to Poland in 1945.

Ol·ten (ôlt'n), town (1977 est. pop. 19,900), Solothurn canton, N Switzerland, on the Aare River. It is an important rail center and has manufactures of aluminumware, shoes, and chemicals.

O·lym·pi·a (ō-lĭm'pē-ə), important center of the worship of Zeus in ancient Greece, in Elis near the Alpheus (now Alfiós) River. It was the scene of the Olympic Games. The great temple of Zeus, situated between the Cladeus and Alpheus rivers, was especially celebrated for the ivory, gold-adorned statue of Zeus by Phidias—one of the Seven Wonders of the World. Excavation, which revealed the great temple, also uncovered the Hermes of Praxiteles, several other temples within the sacred enclosure, and the stadium.

Olympia, city (1978 est. pop. 25,800), state capital, and seat of Thurston co., W Wash., at the S tip of Puget Sound, on Budd Inlet; inc. 1859. It is a port of entry. Lumber products and beer are manufactured and there are oyster fisheries. Settled in 1846, it was made capital of the newly created Washington Territory in 1853. Of interest are a state historical museum, the state library, the fine old capitol building (1893), and the newer, imposing group of white sandstone capitol buildings. The Olympic Mts. may be seen to the north, and Mt. Rainier to the east.

O·lym·pic Mountains (ō-lĭm'pĭk), highest part of the Coast Ranges, on the Olympic Peninsula, NW Wash. Mt. Olympus (7,954 ft/2,426 m) is the highest point in the mountains, which are composed mainly of sedimentary rock. The west side of the mountains is in one of the areas of greatest precipitation in the United States, with an annual rainfall of 130 in. (330.2 cm); the northeast side is in one of the driest areas on the West Coast. On the upper slopes are 60 small glaciers fed by heavy winter snows. The greater part of the Olympic Mts. is included in Olympic National Park (est. 1938). Rugged mountains, alpine meadows, coniferous rain forests, glaciers, lakes, and streams characterize this area. The national park includes a 50-mi (80-km) stretch of shoreline along the Pacific Ocean that contains scenic seascapes and wildlife sanctuaries.

O·lym·pus (ō-lĭm′pəs), mountain range, N Greece, on the border of Thessaly and Macedonia, near the Aegean coast. It rises to c.9,570 ft (2,920 m) at Mt. Olympus, the highest point in Greece. In Greek mythology the summit, shut from the sight of men on earth by clouds, was the home of the Olympian gods. Later the name Olympus was given to the remote heavenly palace of the gods.

O·lyn·thus (ō-lĭn′thəs), ancient city of Greece, on the peninsula of Chalcidice (now Khalkidhiki), NE of Potidaea. A league of Chalcidic cities grew up in the late 5th cent. B.C., and Olynthus, as the head of this Chalcidian League, vigorously opposed the threats of Athens and Sparta. Athens captured the city and held it for a brief time. In 379 B.C. Sparta defeated Olynthus and dissolved the league, which was, however, re-formed after the fall of Sparta. Olynthus had been allied with Philip II of Macedon against Athens, but, fearing Philip's power, sought Athenian aid. Philip attacked, and Demosthenes in his Olynthiac orations eloquently urged his fellow Athenians to aid the threatened city. Philip destroyed (348 B.C.) the city despite Athenian aid. Excavations at Olynthus have revealed the layout of the city.

Ol·y·phant (ŏl′ə-fənt), industrial borough (1970 pop. 5,422), Lackawanna co., NE Pa., on the Lackawanna River NE of Scranton; settled 1858, inc. 1877. Anthracite coal is mined.

O·magh (ō′mä′), urban district (1971 pop. 27,998), county town of Co. Tyrone, W Northern Ireland, on the Strule River. It is a farm market. Dairy products are processed and shirts are manufactured.

O·ma·ha (ō′mə-hô′, -hä′), city (1978 est. pop. 382,900), seat of Douglas co., E Nebr., on the W bank of the Missouri River; inc. 1857. It is a busy port of entry and a major transportation center. Located in the heart of the country's great farming region, it is one of the largest livestock markets and meat-processing centers in the world and a market for agricultural products. Much of the city's industry is devoted to food processing. Among its many manufactures are farm machinery, fertilizers, computer components, telephone equipment, furniture, clothing, insecticides, soap, cans, chemicals, paints, and airplane and automobile parts. Omaha is the home of many insurance companies and a center for medical treatment and research. Founded when the Nebraska Territory was opened to settlement in 1854, it grew as a supply point for westward migration and became a thriving transportation and industrial center after the arrival of the railroad in 1869. It was the territorial capital from 1855 to 1867. The city has fine park and school systems and is the seat of Creighton Univ. and the Univ. of Nebraska at Omaha. Of interest are the Joslyn Art Museum, an aerospace museum, a Mormon cemetery, and Fontenelle Forest. Fort Omaha (built 1868) now serves as headquarters of the naval reserve training command.

Omaha, beach, W central section, Normandy beaches, NW France, on both sides of the Vire River at Saint-Laurent-sur-Mer. On June 6, 1944, troops led by Gen. Omar Bradley established a beachhead here despite strong resistance.

O·man (ō-män′), sultanate (1970 est. pop. 750,000), c.82,000 sq mi (212,380 sq km), SE Arabian peninsula, on the Gulf of Oman and the Arabian Sea, bordered on the W by Southern Yemen and Saudi Arabia and on the N by the United Arab Emirates, which separate the major portion of the sultanate from a small area on the Strait of Hormuz. The capital is Muscat. Oman comprises a narrow coastal plain backed by hill ranges and an interior desert plateau. In the extreme north dates are cultivated, and in the southwest there is an abundance of sugar cane and cattle. Fishing is an important industry. The major product, however, is oil, which began to be exported in 1967.

History. Much of the coast of Oman was controlled by Portugal from 1508 to 1659, when Turkey took possession. The Turks were driven out in 1741 by Ahmad ibn Said of Yemen, who founded the present royal line. In the late 18th cent. Oman began its close ties with Great Britain, which have continued to the present. In the early

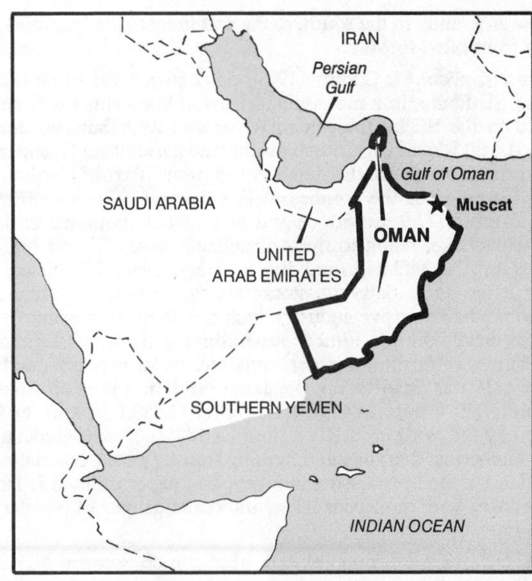

19th cent. Oman was the most powerful state in Arabia, controlling Zanzibar and much of the coast of Iran and Baluchistan. The sultan of Oman has had frequent clashes with the imam (leader) of the interior tribes. In 1957 the tribes revolted but were suppressed with British aid. In 1970 sultan Said bin Timur was deposed by his son, Qabus bin Said, who promised to use oil revenues for modernization. Rebel activity was a problem in Dhofar prov. until 1976, when it was quelled. A general amnesty was granted and more aid to the province was promised. The sultan exercises absolute power.

Om·bos (ŏm′bōs′), ancient city, S Egypt, on the Nile, S of Idfu. It was strategically located on top of a hill. The city attained great importance under the Ptolemies, who built a mighty temple complex dedicated to the crocodile-headed god Suchos and the falcon-headed Haroeris.

Om·dur·man (ŏm′dŏŏr-män′), city (1973 pop. 299,401), central Sudan, on the White Nile opposite Khartoum. Industries include leather tanning and furniture and pottery making. In 1884 the Mahdi made his military headquarters at the village of Omdurman. The Battle of Karari, which took place (1898) near Omdurman, marked the defeat of the Mahdist state in the Sudan by the Anglo-Egyptian army of Lord Kitchener.

O-mei or **O·mei** (both: ō′mā′), peak, c.10,000 ft (3,050 m) high, SW Szechwan prov., central China. With many Buddhist images and temples and monasteries, it is one of China's sacred peaks.

Omsk (ômsk, ŏmsk), city (1977 est. pop. 1,026,000), capital of Omsk oblast, W Siberian USSR, at the confluence of the Irtysh and Om rivers and on the Trans-Siberian RR. It is a major river port and produces agricultural machinery and railway equipment. There are also oil refineries, grain mills, and textile plants. During the civil war that followed the Revolution of 1917, Omsk served as headquarters of anti-Bolshevik forces.

O·mu·ta (ō′mōō-tä′, ō-mōō′tə), city (1976 est. pop. 165,435), Fukuoka prefecture, W Kyushu, Japan, a port on the Amakusa Sea. Coal is exported, and there is a large chemical industry.

On·a·wa (ŏn′ə-wə, -wä′), city (1970 pop. 3,154), seat of Monona co., W Iowa, near the Missouri SSE of Sioux City; platted 1857, inc. 1858. It is a trade center of a grain and livestock area.

On·do (ŏn′dō), city (1969 est. pop. 86,000), SW Nigeria. It is the market center for a cacao and timber region and has rice mills and sawmills. Formerly the capital of a Yoruba kingdom, Ondo came under British protection in 1893.

O·ne·ga (ō-nĕg′ə), river, c.260 mi (420 km) long, rising in Lake Lacha, NW European USSR, and flowing N into the Onega Gulf of the White Sea, SW of Arkhangelsk.

Onega, Lake, c.3,800 sq mi (9,840 sq km), NW European USSR, in Karelia, between Lake Ladoga and the White Sea. The second-largest lake in Europe, it is c.150 mi (240 km) long with a maximum width of c.60 mi (95 km) and a maximum depth of c.360 ft (110 m). The lake is located on the heavily glaciated Baltic Shield. Its shores are low and sandy in the south, rocky and indented in the north. It is frozen from Nov. to May.

O·nei·da (ō-nī′də). **1.** County (1970 pop. 2,864), 1,191 sq mi (3,084.7 sq km), SE Idaho, in a mountainous region bordering on Utah and crossed in the SE by the Malad River and its tributaries; formed 1864; co. seat Malad City. Stock raising and agriculture (wheat, alfalfa, and sugar beets) are the major occupations. Part of Caribou National Forest is in the southeast. **2.** County (1970 pop. 273,037), 1,227 sq mi (3,177.9 sq km), central N.Y., partly bounded in the W by Oneida Lake, rising to the Adirondacks in the E and NE, and drained by the Mohawk and Black rivers; formed 1798; co. seat Utica. It has some dairying, stock raising, farming, and limestone quarrying. Its extensive industry includes food processing, textile milling, metalworking, lumber sawmilling and millwork, and the manufacture of furniture, paper, concrete, metal products, and machinery. It has resorts on the lakes and in the Adirondacks. **3.** County (1970 pop. 24,427), 1,112 sq mi (2,880.1 sq km), N Wis., drained by the Wisconsin River; formed 1885; co. seat Rhinelander. It has lumbering, dairying, and farming (mostly potatoes) on cutover forest land, a food-processing industry, and paper mills. It is largely a resort area with numerous lakes, and contains part of Nicolet National Forest.

Oneida, city (1978 est. pop. 10,700), Madison co., central N.Y.; inc. 1901. It has milk-processing plants and factories manufacturing paper and plastic products. Nearby was the Oneida Community, a religious society of Perfectionists that was established (1848) by John Humphrey Noyes. Members of the sect held all property in common and practiced complex marriage and common care of the children. The community prospered by making steel traps and silverware. In 1881 it was reorganized as a joint stock company, and the social experiments were abandoned.

Oneida Lake, c.80 sq mi (210 sq km), central N.Y., NE of Syracuse. The New York State Barge Canal links the eastern end of the lake with the Mohawk River and also follows part of the Oneida River, which flows from the western end of the lake c.20 mi (30 km) into the Oswego River.

O'·Neill (ō-nēl′), city (1970 pop. 3,753), seat of Holt co., NE Nebr., on the Elkhorn River ESE of Atkinson; settled 1874 by Irish. It is a shipping point for a farming area.

O·ne·on·ta (ō′nē-ŏn′tə). **1.** City (1970 pop. 4,390), seat of Blount co., N central Ala., NE of Birmingham, in a mountainous region; inc. 1891. Clothing and lumber and cement products are made. **2.** City (1978 est. pop. 16,300), Otsego co., E central N.Y., on the Susquehanna River, in a farm area of the Catskills; settled c.1780, inc. as a city 1909. Oneonta grew after the coming of the railroad in 1865. Feed and grain, apparel, truck bodies, and electronic items are among the local manufactures. Many Indian artifacts have been found in the area.

O·ni·da (ō-nī′də), city (1970 pop. 785), seat of Sully co., central S.Dak., NNE of Pierre, in a dairying, livestock, and poultry area.

O·nit·sha (ō-nĭch′ə), city (1971 est. pop. 197,000), SE Nigeria, a port on the Niger River. The city's manufactures include textiles, beverages, shoes, lumber, phonograph records, and printed materials. Fishing and canoe building are traditional local industries. Onitsha was probably founded in the 16th cent. by immigrants from Benin. In the 17th cent. Ibo tribesmen arrived. In 1857 a British trading station and a Christian mission were established in the city, and in 1884 Onitsha came under British protection.

O·no·mi·chi (ō′nə-mē′chē), city (1976 est. pop. 103,105), Hiroshima prefecture, SW Honshu, Japan, on the Inland Sea. It is a shipping center and the site of several Buddhist temples.

On·on·da·ga (ŏn′ən-dô′gə, -dä′-), county (1970 pop. 472,835), 794 sq mi (2,056.5 sq km), central N.Y., in the Finger Lakes region, drained by the Seneca and Oswego rivers and crossed by the Barge Canal; formed 1794; co. seat Syracuse. It has dairying, farming (poultry, corn, truck crops, potatoes, and hay), extensive manufacturing, and saltworks. There are resorts on Oneida and Skaneateles lakes.

Ons·low (ŏnz′lō), county (1970 pop. 103,126), 756 sq mi (1,958 sq km), SE N.C., bounded on the S by Onslow Bay and drained by the New River; formed 1734; co. seat Jacksonville. This heavily forested and partly swampy tidewater area has farming (tobacco and corn), sawmilling, and fishing.

On·ta·ke (ōn-tä′kē), mountain, 10,049 ft (3,064.9 m), central Honshu, Japan. Its summit has a 14th cent. Shinto shrine.

On·tar·i·o (ŏn-târ′ē-ō), province (1975 pop. 8,226,000), 412,582 sq mi (1,068,587 sq km), E central Canada. Toronto is the capital. Ontario is bounded on the north by Hudson Bay and James Bay; on the east by Que.; on the south by the St. Lawrence River, lakes Ontario, Erie, Huron, and Superior, and by the United States; and on the west by Man.

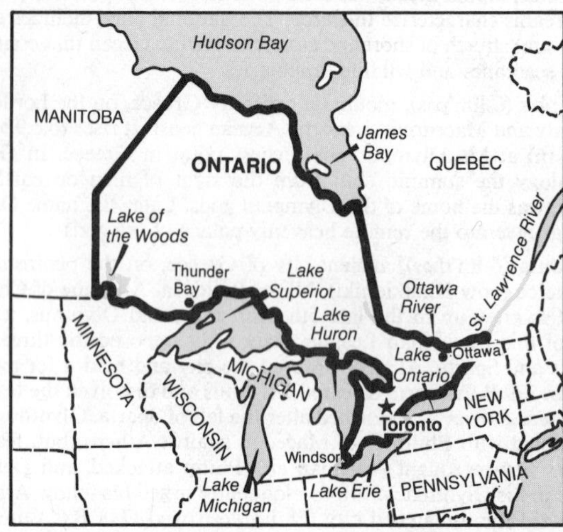

In the western and central portion of Ontario is the Canadian Shield, a region of mineral-rich rock covered with forests and broken by a labyrinth of rivers and lakes. In the north is the Hudson Bay Lowlands, an area of marshes, swampland, and forest. In the south and east are the Great Lakes-St. Lawrence lowlands, where nine tenths of the population lives and where industry and agriculture are concentrated. Climate varies among the regions. The far north has subarctic conditions, while the west has a temperate climate. Around the Great Lakes the weather is moderate and summers are longer than in other parts of the province. The St. Lawrence River gives Ontario access to the Atlantic. Other important rivers are the Ottawa, the St. Clair, the Detroit, and the St. Marys. Several of the province's rivers are used to generate hydroelectric power, among them the Niagara, with its famous falls.

Economy. Ontario has four national parks and numerous tourist attractions, notably Niagara Falls and the annual Shakespeare Festival at Stratford. The most important economic activity in Ontario is manufacturing. Major industrial products include transportation equipment, foods and beverages, metals and metal products, electrical goods, machinery, chemicals, and paper products. Agriculture is also important, with cattle, dairy products, and hogs producing the most income. Other major crops are corn, wheat, potatoes, and soybeans. Mining is important in the Canadian Shield region, where iron ore, copper, nickel, zinc, gold, silver, and uranium are found. Ontario is also a major producer of lumber, pulp, and paper.

History. Before the arrival of the white man the area of Ontario was inhabited by several Indian tribes, the largest of which was the Huron. Étienne Brulé explored southern Ontario in 1610–12. Henry Hudson sailed into Hudson Bay in 1611 and claimed the region for England. Within a few years Samuel de Champlain reached (1615) the eastern shores of Lake Huron, and French explorers, missionaries, and trappers had established posts at several points. In the late 17th cent. the Anglo-French struggle for control of Ontario began. The conflict was resolved by the Treaty of Paris of 1763, which gave Great Britain all of France's mainland North American territory. In 1774 the British attached Ontario to Quebec, which had a predominantly French culture. The Constitutional Act of 1791 split Quebec into Lower Canada (present-day Quebec) and Upper Canada (present-day Ontario), with the Ottawa River as the dividing line. The two colonies were rejoined in 1841 and granted parliamentary self-government in 1849. However, conflict between French and English made the united province unworkable, and in 1867 when the confederation of Canada was formed, Ontario and Quebec became separate provinces. With the construction of the transcontinental railroad in the 1880s, settlement increased in western Canada and Ontario's commerce and industry flourished. The exploitation of the minerals in the Canadian Shield region began in the early 20th cent.

Educational Institutions. Among the province's universities are the Univ. of Toronto, the Univ. of Ottawa, McMaster Univ.,

Queen's Univ., and the Univ. of Western Ontario.

Ontario, county (1970 pop. 78,849), 649 sq mi (1,680.9 sq km), W central N.Y., in the Finger Lakes region, partly bounded on the E by Seneca Lake; formed 1789; co. seat Canandaigua. This fruit-growing and truck-farming area has diversified manufacturing, nurseries, agriculture (grain, hay, potatoes, dairy products, and poultry), and wineries using locally grown grapes. It includes Canandaigua, Honeoye, and Canadice lakes.

Ontario. 1. City (1978 est. pop. 69,200), San Bernardino co., SE Calif., near Los Angeles, in a region of vineyards; inc. 1891. Citrus fruits are grown. Also important to the economy are the manufacture of iron products, clothing, and mobile homes and the overhauling of jet engines. **2.** City (1970 pop. 6,523), Malheur co., E central Oregon, on the Idaho border; inc. 1899. A railroad and highway center in the Owyhee reclamation project, it processes and ships farm products.

Ontario, Lake, 7,540 sq mi (19,529 sq km), 193 mi (310.5 km) long and 53 mi (85.3 km) at its greatest width, between SE Ont., Canada, and NW N.Y.; smallest and lowest of the Great Lakes. It has a surface elevation of 246 ft (75 m) above sea level and a maximum depth of 778 ft (237.3 m). Lake Ontario is fed chiefly by the waters of Lake Erie by way of the Niagara River. The lake is drained to the northeast by the St. Lawrence River. Oceangoing vessels reach the lake through the St. Lawrence Seaway and use the Welland Canal to bypass Niagara Falls and reach Lake Erie. Navigation on the lake is not usually impeded by ice in winter. Commercial fishing is important, but pollution and infestation of lampreys have reduced the value of the catch. A U.S.-Canadian pact (1972) established that water quality would be improved and further pollution ended. The first European to see (1615) Lake Ontario was Étienne Brulé, the French explorer; later that year Samuel de Champlain visited it.

On·to·na·gon (ŏn'tə-nô'gən, -nŏg'ən), county (1970 pop. 10,548), 1,321 sq mi (3,421.4 sq km), NW Upper Peninsula, Mich., bounded on the N by Lake Superior, drained by the Ontonagon River and by the Iron and Firesteel rivers; formed 1848; co. seat Ontonagon. The Porcupine Mts. are in the northwest, where much copper is mined. It is a lumber area with some agriculture (livestock, hay, grain, and corn). Commercial fishing is also done. The county has many resorts and includes part of Ottawa National Forest.

Ontonagon, village (1970 pop. 2,432), seat of Ontonagon co., N Mich., W Upper Peninsula, on Lake Superior at the mouth of the Ontonagon River; inc. 1885. Paper and wood products are made.

Oost·en·de (ō-stĕn'də), city (1976 est. pop. 71,446), West Flanders prov., W Belgium, on the North Sea. It is a major commercial and fishing port, connected by canals with Bruges and Ghent. It is also an industrial center and a seaside resort. Manufactures include processed food, ships, and chemicals. A port by the time of the First Crusade (11th cent.), Oostende was fortified (1583) by William the Silent and played a leading role in the Dutch struggle for independence. The city was taken (1604) by the Spaniards after a three-year siege, was sacked again in 1745 by the French, and suffered heavy Allied bombardment in World War II.

Oo·ta·ca·mund (ōō'tə-kə-mŭnd'), town (1971 pop. 63,003), Tamil Nadu state, SW India, in the Nilgiri Hills. At an altitude of c.7,000 ft (2,135 m), the town is a hot-weather resort. It is also a market for tea, coffee, teak, and sandalwood. There are tea plantations, a munitions factory, and breweries.

O·pal Cliffs (ō'pəl), uninc. town (1970 pop. 5,425), Santa Cruz co., W Calif., E of Santa Cruz.

O·pa-Lock·a (ō'pə-lŏk'ə), city (1970 pop. 11,902), Dade co., SE Fla.; inc. 1926. Diverse industrial plants are in the city. Its city hall is patterned after a Moorish castle, and other buildings are also of unusual Arabian architecture.

O·pa·va (ô'pə-və), city (1975 est. pop. 54,873), N central Czechoslovakia, in Moravia, on the Opava River near the Polish border. A prosperous market center in a fertile agricultural region, it has food-processing plants and industries producing machinery, mining equipment, metal goods, textiles, pharmaceuticals, and timber. Opava was founded in the 12th cent. and later became the capital of Austrian Silesia. In 1820 representatives of the European great powers met here, at the Congress of Troppau, to discuss problems arising after the settlement of the Napoleonic Wars.

O·pe·li·ka (ō'pə-lī'kə), city (1978 est. pop. 22,600), seat of Lee co., E Ala., near the Chattahoochee River, in a farm area; inc. 1854. It is a trade center, with textile, lumber, and metallurgical industries.

Op·e·lou·sas (ŏp'ə-lōō'səs), city (1978 est. pop. 20,000), seat of St. Landry parish, S central La.; inc. 1821. Its industries are based chiefly on the agricultural products and livestock of the surrounding region. Opelousas still retains some of its early French and Spanish flavor. It was founded c.1765 by French traders and served (1863) as state capital during part of the Civil War.

O·po·le (ô-pô'lə), city (1975 est. pop. 106,000), S Poland, on the Oder River. A river port and rail junction, it is also an important trade center, with manufactures of cement, metals, and furniture. Originally a Slavic settlement, it passed (1532) to the house of Hapsburg and (1742) to Prussia. It was the capital (1919–45) of the Prussian province of Upper Silesia.

O·por·to (ō-pōr'tō, ō-pôr'-) or **Por·to** (pōr'tōō), city (1975 est. pop. 335,700), capital of Porto dist. and Douro Litoral, NW Portugal, near the mouth of the Douro River. It is an important Atlantic port. Oporto's most famous export is port wine, to which the city gives its name, but cork, fruits, olive oil, and building materials are also exported. Cotton, silk, and wool textiles are milled; clothing and leather goods are made. An ancient settlement, probably of pre-Roman origin, Oporto was captured by the Moors in 716 and retaken in 1092. Oporto was for some time the chief city, although not the capital, of little Portugal. Wine exports increased after the Methuen Treaty (1703) with England. After the French conquest of Portugal in the Peninsular War, Oporto was the first city to revolt (1808). It was retaken by the French but liberated (1809) by Wellington. In 1832, in the Miguelist Wars, Dom Pedro I of Brazil long withstood a siege of the city by his brother, Dom Miguel. The city's most conspicuous landmark is the Torre dos Clérigos, a baroque tower.

Opp (ŏp), city (1970 pop. 6,493), Covington co., S central Ala., near the Fla. border; inc. 1902. It is a processing and shipping center for a pine, cotton, and peanut area.

Op·por·tu·ni·ty (ŏp'ər-tōō'nĭ-tē, -tyōō'nĭ-tē), uninc. town (1970 pop. 16,604), Spokane co., E Wash., a residential suburb of Spokane.

O·quaw·ka (ō-kwô'kə), resort village (1970 pop. 1,352), seat of Henderson co., W Ill., on the Mississippi River NNE of Burlington, Iowa; inc. 1857.

O·ra·dea (ô-rä'dyä), city (1977 est. pop. 171,258), W Rumania, in Crişana-Maramureş, near the Hungarian border. It is the marketing and shipping center for a livestock and agricultural region. Oradea is also an important industrial city. Destroyed (1241) by the Tatars, it was rebuilt in the 15th cent. Oradea was held by the Turks from 1660 to 1692. Hungary ceded it (1919) to Rumania after World War I, but it was occupied by Hungarian forces during World War II.

Or·a·dell (ŏr'ə-dĕl', ôr'-), borough (1970 pop. 8,903), Bergen co., NE N.J., N of Hackensack; settled by the Dutch in the 1700s, inc. 1894.

Ör·ae·fa·jö·kull (œr'ī-və-yœ'kyōōt'l), mountain, SE Iceland, rising from the Vatnajökull glacier. Öraefajökull is an ice-covered, three-peaked volcano. The largest of its four recorded eruptions occurred in 1362. Its highest peak, Hvannadalshnúkur (6,950 ft/2,119.8 m), is also the highest point in Iceland.

O·ran (ō-răn', ō-rän'), city (1974 est. pop. 485,139), capital of Oran dept., NW Algeria, a port on the Gulf of Oran of the Mediterranean Sea. One of the country's leading ports, it ships wheat, wine, alcohol, vegetables, meat, wool, cigarettes, and iron ore. The city, surrounded by vineyards and market gardens, is a commercial, industrial, and financial center. The site of modern Oran has been inhabited since prehistoric times. Spanish forces captured and fortified the city in 1509 and held it until the Turks arrived in 1708. Spain recovered Oran in 1732. The city was successfully besieged (1791) and was made a provincial capital of the Ottoman Empire. French troops captured Oran in 1831 and began to develop it as a naval base. Held by Vichy France during World War II, Oran fell to the Allied forces in Nov., 1942. Civil strife ravaged the city in the late 1950s. Oran consists of a modern, French-style section and an old Spanish-type quarter with a casbah (fortress) and an 18th cent. mosque.

O·range (ô-räNzh'), town (1968 pop. 20,779), Vaucluse dept., SE France. An agricultural market center, the town also produces refined sugar, pâtés, preserves, wool, and shoes. Tourism is also important. Orange was an earldom probably founded by Charlemagne. It became the capital of a principality (12th cent.) and was passed eventually (1554), through inheritance, to William the Silent, of the house of Nassau. Orange was conquered (1672) by Louis XIV and confirmed in French possession by the Treaty of Ryswick (1697) and the Peace of Utrecht (1713), although the title remained with the Dutch princes of Orange. The town has important Roman ruins, notably a triumphal arch (1st cent. A.D.) and an amphitheater (c.120 A.D.) that is still in use.

Or·ange (ôr'ĭnj, ŏr'-). **1.** County (1970 pop. 1,421,233), 782 sq mi (2,025.4 sq km), SW Calif., in a coastal plain and foothill region, drained by the Santa Ana River and rising to the Santa Ana Mts. along the E border; formed 1889; co. seat Santa Ana. It is a leading citrus-fruit county, which also raises avocados, truck crops, lima beans, alfalfa, sugar beets, dairy products, and poultry. There are extensive petroleum and natural-gas fields and sand, gravel, and clay pits. It also has food processing, textile, and metalworking industries. Furniture, paper, chemicals, plastics, concrete, and metal products are made. The county includes part of Cleveland National Forest and has coastal resorts. **2.** County (1970 pop. 344,311), 910 sq mi (2,356.9 sq km), central Fla., bounded in the E by the St. Johns River; formed 1824; co. seat Orlando. The hilly lake region in the west contains part of Lake Apopka. A major citrus-growing area, it also produces truck crops, poultry, and dairy products. There is some manufacturing. **3.** County (1970 pop. 16,968), 405 sq mi (1,049 sq km), S Ind., drained by the Lick, Lost, and Patoka rivers; formed 1815; co. seat Paoli. In an agricultural area yielding grain, fruit, livestock, poultry, and dairy products, it has limestone quarries, timber sawmills, and bituminous-coal mines. It manufactures furniture. The mineral springs in the county (notably French Lick) are resort areas. **4.** County (1970 pop. 221,657), 829 sq mi (2,147.1 sq km), SE N.Y., bounded on the E by the Hudson River, on the SW by the Delaware River and the N.J. and Pa. borders, and including parts of the Hudson Highlands; formed 1683; co. seat Goshen. It is an important dairying region and has general farming (truck crops, fruit, hay, and poultry) and manufacturing. There are many mountain and lake resorts here. The county includes West Point and part of Palisades Interstate Park. **5.** County (1970 pop. 57,707), 400 sq mi (1,036 sq km), N central N.C., in a piedmont area bounded on the SW by the Haw River; formed 1753; co. seat Hillsboro. It has agriculture (tobacco, corn, dairy products, and poultry), timber, and textile and furniture factories. **6.** County (1970 pop. 71,170), 356 sq mi (922 sq km), SE Texas, bounded on the E by the Sabine River and the La. border, on the W and SW by the Neches River, on the S by Sabine Lake; formed 1852; co. seat Orange. It has agriculture (rice, wheat, potatoes, fruit, truck crops, pecans, livestock, and dairy products), lumbering in the north, oil and natural-gas wells, and a chemicals industry. Hunting and fishing are popular recreational activities. **7.** County (1970 pop. 17,676), 690 sq mi (1,787.1 sq km), E central Vt., bounded in the E by the Connecticut River and drained by the White, Waits, and Ompompanoosuc rivers; formed 1781; co. seat Chelsea. It is in a dairying and lumbering region that also produces maple sugar. Its manufacturing includes textiles, furniture, and metal and plastic products. **8.** County (1970 pop. 13,792), 354 sq mi (916.9 sq km), N central Va., in the Piedmont, bounded in the N by the Rapidan River and in the S by the North Anna River; formed 1734; co. seat Orange. Its agriculture includes grain, fruit, legumes, sweet potatoes, tobacco, truck crops, livestock, poultry, and dairy products. It has some mining and lumbering, textile, furniture, and brick-manufacturing industries.

Orange. 1. City (1978 est. pop. 84,800), Orange co., S Calif.; inc. 1888. Citrus fruits are packed and processed, and rubber products, doors, and industrial furnaces are manufactured. **2.** Town (1970 pop. 13,524), New Haven co., SW Conn., a residential suburb of New Haven, on the Housatonic; settled 1720, set off from Milford 1822, inc. 1921. Tools, furniture, and clothing are made. The first house (1720) still stands. **3.** Industrial town (1970 pop. 6,104), Franklin co., N Mass., on Millers River E of Greenfield; settled c.1746, inc. 1810. Machinery and shoes are produced. **4.** City (1978 est. pop. 29,400), Essex co., NE N.J.; settled c.1675, set off from Newark 1806, inc. as a city 1872. Orange and the surrounding municipalities of East Orange, West Orange, South Orange, and Maplewood are known as The Oranges, a single suburb of Newark and New York City. Although chiefly residential, Orange has plants making office machines, clothing, aircraft parts, and pharmaceuticals. **5.** City (1978 est. pop. 25,100), seat of Orange co., SE Texas, a deep-water port on the Sabine River at its junction with the Intracoastal Waterway; settled c.1800, inc. 1858. It is a port of entry, with shipyards, oil and gas wells, and major petrochemical plants. **6.** Town (1970 pop. 2,768), seat of Orange co., central Va., NE of Charlottesville, in the Piedmont; settled c.1810, inc. 1856. It is in a farm area and has textile and lumber mills. Montpelier, home of James Madison, is nearby.

Orange, chief river of S Africa, c.1,300 mi (2,090 km) long, rising in the Maluti Mts., N Lesotho, flowing SW through Lesotho, then meandering NW and W through central Republic of South Africa and entering the Atlantic Ocean at Oranjemund. The lower Orange River flows through the southern part of the Kalahari and Namib deserts; in very dry years it does not reach the sea. At the mouth of the river are rich alluvial diamond beds. The river is used extensively for irrigation.

Or·ange·burg (ôr'ĭnj-bûrg', ŏr'-), county (1970 pop. 69,789), 1,106 sq mi (2,864.5 sq km), S central S.C., bounded in the SW by the South Fork of the Edisto River, in the E by the Marion River, and drained by the North Fork of the Edisto; formed 1785; co. seat Orangeburg. In a major agriculture and dairy area that produces corn, soybeans, cotton, sweet potatoes, and truck crops, it has lumbering, food-processing, and textile industries.

Orangeburg, city (1970 pop. 13,252), seat of Orangeburg co., central S.C., on the North Fork of the Edisto River; settled 1732, inc. as a city 1883. It is the trade and processing center of a cotton area and has large textile and garment industries. Tools, office machines, wood, and chemicals are also made here. Orangeburg was a planned settlement est. by German-Swiss immigrants who had free grants of land. South Carolina State College and Claflin College are here. Points of interest include Edisto gardens and the Donald Bruce House (c.1735). A U.S. fish hatchery, an aircraft and automobile museum, and a state park are nearby.

Orange City, city (1975 est. pop. 4,017), seat of Sioux co., NW Iowa, NNE of Sioux City; founded 1869 by Dutch, inc. 1884.

Orange Free State, province (1970 pop. 1,715,589), 49,866 sq mi (129,153 sq km), E central South Africa. Bloemfontein is the capital and largest city. The province is chiefly a plateau, rising gradually from c.4,000 ft (1,220 m) in the west to c.6,000 ft (1,830 m) in the east; there are higher elevations in the Drakensberg Range in the southeast. The economy is mainly agricultural; maize, sorghum, potatoes, wheat, sheep, and cattle are raised. Gold mining is important, and uranium oxide, diamonds, and coal are mined. Synthetic rubber, fertilizers, plastics, textiles, and processed foods are manufactured, and oil is refined from coal.

In the early 19th cent. the sparsely populated Orange Free State was inhabited mainly by the Bantu-speaking Tswana people. Boer farmers entered the territory from the 1820s; after 1835 their immigration accelerated. In 1848 the British, who then held Cape Colony and Natal, annexed the region as the Orange River Sovereignty. After conflicts with the Boers and failure to establish an orderly administration, Britain, by the Bloemfontein Convention (1854), granted the territory independence as the Orange Free State. With the increased tension following the raid into the Transvaal (1895–96), led by L. S. Jameson, the Free State was drawn into the conflict between Britons and Boers that resulted in the South African War (1899–1902). The British again annexed the Free State, as the Orange River Colony, in 1900. In 1907 the colony was granted self-government, and in 1910 it became a founding province of the Union of South Africa. The Univ. of the Orange Free State in Bloemfontein is the chief institution of higher education.

Orange Park, town (1970 pop. 7,619), Clay co., NE Fla., on the St. Johns River SSW of Jacksonville; founded 1790 on a Spanish land grant, inc. 1879.

O·ra·ni·en·burg (ō-rä'nē-ən-bûrg', -bo͝ork'), city (1974 est. pop. 24,452), Potsdam district, central East Germany, on the Havel River. It is a center of a fruit-growing region. Manufactures include chemicals, machinery, and processed food.

Or·chard Me·sa (ôr'chərd mā'sə), uninc. town (1970 pop. 5,824), Mesa co., W central Colo., near Grand Junction, in an irrigated area producing fruit and vegetables.

Or·chom·e·nus (ôr-kŏm'ə-nəs), ancient city of Boeotia, central Greece, NW of Lake Copaïs. After 1600 B.C. it was a center of the Mycenaean civilization. The city was later eclipsed by Thebes.

Ord (ôrd), city (1970 pop. 2,439), seat of Valley co., central Nebr., on the North Loup River NNW of Grand Island, in a farm area; surveyed 1874.

Or·dos (ôr'dəs), sandy desert plateau region, c.35,000 sq mi (90,650 sq km), Inner Mongolian Autonomous Region, N China. The Great Wall of China separates the Ordos from the fertile loess land to the south and east. The region has many salt lakes and intermittent streams. Large soda deposits are mined. There are some grasslands.

Or·du (ôr-dōō'), city (1975 pop. 47,302), capital of Ordu prov., N Turkey, a port on the Black Sea. Hazelnuts are grown and exported. It is the site of Cotyora, founded by Greek colonists c.500 B.C.

Ord·way (ôrd'wā'), town (1970 pop. 1,017), seat of Crowley co., SE Colo., near the Arkansas River E of Pueblo, in a farm area; inc. 1900.

Or·dzho·ni·kid·ze (ôr-jä'nə-kĭd'zə), city (1976 est. pop. 276,000), capital of the North Ossetian Autonomous Soviet Socialist Republic, SE European USSR, on the Terek River at the foot of the Caucasus. It has an electric zinc smelter, lead and silver refineries, chemical plants, food-processing factories, and industries producing chemicals, motors, tractor equipment, clothing, and textiles. Founded in 1784 as a fortress, it was long the military and political center of Russia in the Caucasus. It was the capital of Gorskaya ASSR in 1921, which in 1936 became North Ossetian ASSR.

Ör·e·bro (œr'ə-brōō'), city (1975 est. pop. 117,837), capital of Örebro co., S central Sweden, W of Lake Hjälmaren. It is a commercial, industrial, and transportation center. Manufactures include shoes, paper, and processed food. Known since the 11th cent., it was the site of 15 national diets, notably the one in 1529, which brought the Reformation to Sweden.

Or·e·gon (ôr'ə-gən, -gŏn', ŏr'-), state (1975 pop. 2,314,000), 96,981 sq mi (251,181 sq km), NW United States, in the Pacific Northwest, admitted 1859 as the 33rd state. Salem is the capital. Oregon is bounded on the north by Wash., from which it is largely separated by the Columbia River; on the east by Idaho, with the Snake River forming the boundary in the northern half; on the south by Nev. and Calif.; and on the west by the Pacific Ocean.

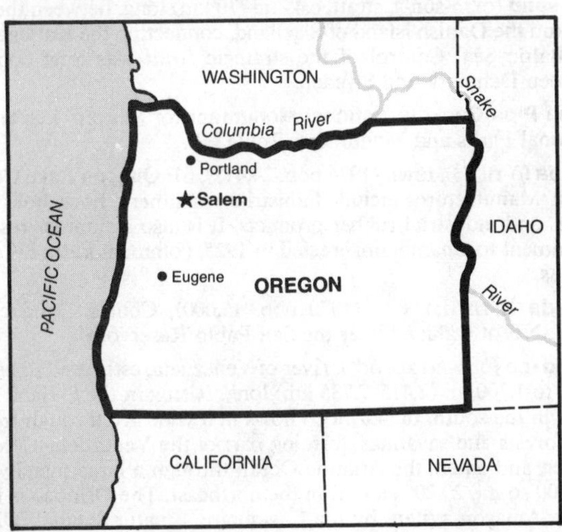

The state's contrasting physical features are characterized by great forested mountain slopes and treeless basins, rushing rivers and barren playas, lush valleys and extensive wastelands. The major determinant for these unusual climatic differences is the Cascade Range, a rugged mountain chain running north to south c.100 mi (160 km) inland. As the eastward-moving air masses rise and meet the cooler mountain temperatures, rain is precipitated over the western third of Oregon. Dry air and continental climate prevail over the eastern two thirds of the state. The western shoreline is bordered by narrow coastal plains of sandy beaches, luxuriant pastures, and occasional jutting promontories. About 25 mi (40 km) inland, the rugged Coast Range rises to heights of 4,000 ft (1,220 m) to serve as the western wall of the Willamette valley. In the valley lies the agricultural, commercial, and industrial center of the state. The Cascades are on the east, with beautiful Mt. Hood rising to the state's highest elevation (11,235 ft/3,426.7 m). Mighty stands of timber cover the slopes. Eastward the Cascades level out into plateaus drained in the north by the Deschutes and the John Day rivers. To the south a variegated pattern of marshland and mountain merges in the east into the semiarid Great Basin. North of this area rise the pine-covered Blue Mts. and Wallowa Mts., which in some places extend to the Snake River to form precipitous gorges. Other parts of the region where the Snake cuts through the plateau are more level and have been made productive through irrigation. Oregon's irrigation projects include the Deschutes, the Umatilla, and the Vale; the Klamath, shared with California; and the Boise and the Owyhee, shared with Idaho.

Oregon's beautiful ocean beaches, lakes, and mountains draw thousands of visitors annually. Major attractions are the Oregon Caves National Monument, Fort Clatsop National Memorial, and McLoughlin House National Historic Site; Crater Lake National Park is a vacationer's paradise. There are 13 national forests, 1 national grassland, and more than 220 state parks.

Oregon is divided into 36 counties:

NAME	COUNTY SEAT	NAME	COUNTY SEAT
Baker	Baker	Lake	Lakeview
Benton	Corvallis	Lane	Eugene
Clackamas	Oregon City	Lincoln	Newport
Clatsop	Astoria	Linn	Albany
Columbia	St. Helens	Malheur	Vale
Coos	Coquille	Marion	Salem
Crook	Prineville	Morrow	Heppner
Curry	Gold Beach	Multnomah	Portland
Deschutes	Bend	Polk	Dallas
Douglas	Roseburg	Sherman	Moro
Gilliam	Condon	Tillamook	Tillamook
Grant	Canyon City	Umatilla	Pendleton
Harney	Burns	Union	La Grande
Hood River	Hood River	Wallowa	Enterprise
Jackson	Medford	Wasco	The Dalles
Jefferson	Madras	Washington	Hillsboro
Josephine	Grants Pass	Wheeler	Fossil
Klamath	Klamath Falls	Yamhill	McMinnville

Economy. The state's major sources of farm income are cattle, dairy items, wheat, and greenhouse products. Chief crops in terms of quantity produced are hay, wheat, potatoes, and barley. In 1970 Oregon was the nation's leading producer of snap beans, peppermint, and sweet cherries and the second-largest producer of broccoli and strawberries. The state's 30,739,000 acres (12,450,000 hectares) of rich forest lands (about one half the area of the state) comprise the country's greatest reserves of standing timber; huge areas have been set aside for conservation. Oregon has been the nation's foremost lumber state since 1950; it produces about 25% of the nation's lumber and about 70% of its plywood; wood processing is Oregon's major industry. Douglas fir predominates in the Cascades and western pine in the eastern regions. Other major products are food, paper and paper items, machinery, and fabricated metals. Printing and publishing are important businesses. Abundant, cheap electric power is supplied by numerous dams, most notably those on the Columbia River—Bonneville Dam, The Dalles Dam, McNary Dam, and the John Day Dam. Oregon's river resources are one of its greatest assets. Its salmon-fishing industry is one of the world's largest; other catches are tuna and crabs. Although mining is still underdeveloped, Oregon leads the nation in the production of nickel. Sand and gravel, stone, and cement are also major sources of mineral income. Tourism is the state's third-largest industry (after wood processing and agriculture).

History. Spanish seamen skirted the Oregon coast in the 16th cent., and the Englishman Sir Francis Drake may have sailed as far as Oregon in 1579. Capt. James Cook charted some of the coastline (1778). By this time the Russians were pushing southward from posts in Alaska and the British fur companies were exploring the West. The Oregon coast was soon active with the vessels of several nations engaged in fur trade with the Indians. An American, Robert Gray, first sailed up the Columbia River (1792), thus establishing U.S. claim to the areas that it drained. Canadian traders of the North West Company were approaching the Columbia River country when the overland Lewis and Clark expedition arrived in 1805. John Jacob Astor's agents (in the Pacific Fur Company) founded Astoria, the first permanent settlement in the Oregon country. In the War of 1812 the post was sold (1813) to the North West Company, but in 1818 a treaty provided for joint rights for the United States and Great Britain in Oregon (i.e., the whole Columbia River area).

The North West Company merged with the Hudson's Bay Company in 1821, and soon the region was dominated by the Hudson's Bay men. American trappers rivaled the Hudson's Bay men on the southeastern edge. They were the harbingers of the "great migration" westward over the Oregon Trail beginning with enormous wagon trains in 1842 and 1843. Trouble between the settlers and the British followed. Americans demanded the ousting of the British from the whole of the Columbia River country up to lat. 54°40′ N;

one of the slogans of the 1844 election was "Fifty-four forty or fight." War with Britain was a threat momentarily, but diplomacy prevailed. In 1846 the boundary was set at the line of lat. 49° N.

The Oregon Territory was created in 1848, embracing the area west of the Rockies from the 42nd to the 49th parallel. The area was reduced with the creation of the Washington Territory in 1853, and Oregon became a state in 1859. Wheat farming prospered and in 1867–68 a surplus crop was shipped to England—the beginning of Oregon's great wheat export trade. Cattle and sheep were driven up from California to graze on the tall grass of the semiarid plateaus. The 1850s to 1870s were plagued by Indian uprisings, but by 1880 Indian troubles were over, and the next few decades brought increasing settlement and internal improvements. Lumbering, which had long been important, became a leading industry. The huge stands of Douglas fir and cedar brought fortunes to the lumbering kings, and the threat to natural resources led ultimately to the creation of national forests. In the 1880s the influx of Chinese threatened the labor market and brought violent anti-Chinese sentiment.

In the 1930s one of the most disputed issues was the question of public or private development of power. Federal power and irrigation projects have had a profound effect on the economy of the entire Pacific Northwest. Many acres have been opened to irrigated farming, and the tremendous industrial expansion of World War II was to a large extent dependent on Bonneville power.

Government. Oregon has been a leader in social and political reforms: it was the first state to ban nonrecyclable containers and to ease the laws against marijuana use. Oregon still operates under its original constitution, drawn and ratified in 1857. Its executive branch is headed by a governor elected for a four-year term. Its bicameral legislature has a senate with 30 members elected for four-year terms and an assembly with 60 members elected for two years.

Educational Institutions. Among the more prominent are the Univ. of Oregon, at Eugene; Oregon State Univ. at Corvallis; Reed College, and the Univ. of Portland, at Portland; and Willamette Univ. at Salem.

Oregon, county (1970 pop. 9,180), 784 sq mi (2,030.6 sq km), S Mo., in the Ozarks, drained by the Eleven Point and Spring rivers; formed 1845; co. seat Alton. Livestock raising and dairying are the major agricultural activities. Its timber resources include oak, hickory, and walnut. Part of Clark National Forest is here.

Oregon. 1. City (1970 pop. 3,539), seat of Ogle co., N Ill., on the Rock River SW of Rockford; settled 1829, inc. 1843. Its products include road equipment and evaporated milk. **2.** City (1970 pop. 789), seat of Holt co., NW Mo., near the Missouri River NW of St. Joseph, in an agricultural and livestock area. Corn, apples, and wheat are grown. Big Lake State Park is nearby. **3.** City (1970 pop. 16,563), Lucas co., NW Ohio, a suburb adjacent to Toledo, on Lake Erie; inc. 1958. It is a port with railroad-owned docks. The city has industries producing oil, chemicals, and wire products. About two thirds of the city's area is open farmland. The chief crops are tomatoes, soybeans, greenhouse vegetables, fruits, and grains.

Oregon Caves National Monument: *see* National Parks and Monuments Table.

Oregon City, city (1975 est. pop. 11,000), seat of Clackamas co., NW Oregon, at the falls of the Willamette River S of Portland. The Hudson's Bay Company founded a sawmill here and had the city platted in 1842, but the company's rights were lost to American immigrants. Incorporated in 1849, the city was the territorial capital until 1851. Today it is the processing center of a lumbering, dairying, and fruit-growing area.

Oregon Trail, overland emigrant route in the United States from the Missouri River to the Columbia River country (all of which was then called Oregon). The trail originated at various places on the Missouri, although Independence, St. Joseph, and Westport (now part of Kansas City, Mo.) were favorite starting points. In open country the different wagon trains might spread out over a large area, only to converge again for river crossings, mountain passes, and other natural constrictions. The end of the trail, at first in the Willamette River valley, shifted as settlement spread. The first train of emigrants to reach Oregon (1842) was led by Elijah White. In 1843 occurred the "great emigration" of more than 900 persons and more than 1,000 head of stock. Four trains made the journey in 1844, and by 1845 the emigrants reached a total of more than 3,000. It took the average emigrant train six months to traverse the c.2,000-mi (3,200-km) trail. Travel upon the trail gradually declined with the coming of the railroads, and it was abandoned in the 1870s.

O·re·kho·vo-Zu·ye·vo (ôr′ə-kô′və-zo͞o-yĕ′vō), city (1976 est. pop. 128,000), W central European USSR, on the Klyazma River. There is a large textile industry dating from the 18th cent.

O·rel (ô-rĕl′, ô-ryôl′), city (1976 est. pop. 282,000), capital of Orel oblast, central European USSR, on the Oka River. It is a large railroad junction, an agricultural trade center, and an industrial city producing machinery, textiles, construction equipment, automobile parts, and clothing. It was founded in 1564 by Ivan IV as a fortified settlement to protect the southern border of Muscovy from Crimean Tatar attacks. The city was almost totally destroyed in World War II. The author Ivan Turgenev was born in Orel.

O·rem (ôr′əm, ōr′-), city (1978 est. pop. 40,400), Utah co., N central Utah, near Provo; settled 1861, inc. 1919. Orem is located in an irrigated truck-farming and fruit-growing area. It has a large steel mill; other manufactures include electronic components and canned foods.

O·ren·burg (ōr′ən-bûrg′, ôr′-), formerly **Chka·lov** (chəkä′ləf), city (1976 est. pop. 435,000), capital of Orenburg oblast, Central Asian USSR, on the Ural River. A rail junction, it is a major food-processing and agricultural machine center. Other industries produce elevator equipment, leather goods, clothing, and silk. Founded in 1735 as a fortress, Orenburg became a center for Russian trade with Kazakhstan and central Asia.

O·ren·se (ō-rĕn′sā), city (1975 est. pop. 69,500), capital of Orense prov., NW Spain, in Galicia, on the Miño River. It is the center of an agricultural region with extensive vineyards. There are some light industries. A Roman settlement, it reached its greatest importance as the capital of the Suebi (5th–6th cent.). There are hot sulfur springs, known since Roman times.

Ø·re·sund (œ′rə-sŭn′), strait, c.45 mi (70 km) long, between the Sweden and the Danish island of Sjaelland, connecting the Kattegat with the Baltic Sea. Control of the strategic strait was long contested between Denmark and Sweden.

Or·gan Pipe Cac·tus National Monument (ôr′gən pīp′ kăk′təs): *see* National Parks and Monuments Table.

O·ril·lia (ō-rĭl′yə), town (1976 pop. 24,412), SE Ont., on Lake Couchiching. Manufactures include industrial machinery, household appliances, and industrial rubber products. It is also a summer resort. A monument to Champlain, erected in 1925, commemorates his explorations.

O·rin·da (ō-rĭn′də), city (1970 pop. 15,000), Contra Costa co., W Calif., NE of Oakland near the San Pablo Reservoir.

O·ri·no·co (ôr′ə-nō′kō, ōr′-), river of Venezuela, estimated to be from 1,500 to 1,700 mi (2,415–2,735 km) long. Rising in the Guiana Highlands in the south, the Orinoco flows in a wide arc through tropical rain forests and savannas, forming part of the Venezuela-Colombia border, and enters the Atlantic Ocean through a large marshy delta (c.7,800 sq mi/20,200 sq km) in the northeast. The Orinoco is joined to the Amazon system by the Casiquiare, a natural canal. Divided into upper and lower courses by the Ature and Maipures cataracts, the river is navigable for most of its length. Dredging permits ocean-going vessels to reach Ciudad Bolívar, c.270 mi (435 km) upstream. Christopher Columbus probably discovered the mouth of the Orinoco in 1498, and Lope de Aguirre, the Spanish adventurer, seems to have traveled most of its length in 1560. In 1799 Alexander von Humboldt, the German naturalist, explored the upper reaches, but it was not until the 1950s that its headwaters were located.

O·ris·sa (ô-rĭs′ə), state (1971 pop. 21,934,827), 60,162 sq mi (155,820 sq km), E India, on the Bay of Bengal. Bhubaneswar is the capital. The state depends mainly upon agriculture and fishing for its livelihood. In ancient times the region of Orissa was the center of the strong Kalinga kingdom. With its gradual decline, several Hindu dynasties arose. After long resistance to the Moslems, the region was finally overcome (1568) by Afghan invaders and soon after passed to the Mogul empire. In 1803 it was conquered by the British. The coastal section became in 1936 the province of Orissa. In 1948 and 1949 the area of Orissa was almost doubled with the addition of 24 former princely states. In 1950 Orissa became a constituent state.

O·ri·za·ba (ōr′ə-zä′bə, ō′rē-sä′vä), city (1974 est. pop. 105,200), Veracruz state, E central Mexico. It is the commercial center of a prosperous bean, citrus, and tropical fruit-growing region. The development of waterpower has stimulated manufacturing industries, especially cotton and wool textile factories. A favorite resort of Emperor Maximilian, Orizaba has remained a popular vacation spot. Mineral springs are nearby, and the majestic cone of Mt. Orizaba (or Citlaltépetl; 18,701 ft/5,703.8 m) rises in the distance. French forces used the city as a base for their invasion of Mexico.

Or·khon (ôr′kŏn′), river, c.300 mi (480 km) long, rising in the Khangai Mts., N central Mongolian People's Republic, and flowing E, then N, past the site of ancient Karakorum, and then NE to join the Selenga River just S of the USSR border. It is navigable for shallow-draft vessels only during July and Aug. The Orkhon Inscriptions, discovered in 1889, date from the 8th cent. They comprise minor Chinese texts and the oldest known material in a Turkic language.

Ork·ney (ôrk′nē), island authority (1976 est. pop. 17,748), 376 sq mi (973.8 sq km), N Scotland, consisting of the Orkney Islands, an archipelago of about 70 islands in the Atlantic Ocean and the North Sea, N of Caithness across the Pentland Firth. Less than half the islands are inhabited. Orkney is one of Scotland's richest farming regions. Beef cattle and eggs are the most important products. Sheep and pigs are also raised. Some fishing, mainly for lobster, is carried on in Scapa Flow and in the north. Orkney was settled by Picts. From 875 to 1231 it was a Viking earldom under the Norwegian crown. Orkney passed to the Scottish earls of Angus on the death of the last Viking earl. It became a possession of the Scottish crown in 1472, but the long Norse occupation left marked Scandinavian traces in the people and their culture.

Or·lan·do (ôr-lăn′dō), city (1978 est. pop. 114,000), seat of Orange co., central Fla., in a lake region; inc. 1875. It has resort activities and is the trade and shipping center for a citrus-fruit and vegetable area. It also has aerospace and electronic industries. Orlando was settled near Fort Gatlin, a post established (c.1837) during the Seminole War. Of interest are Eola Park, on the shores of Lake Eola in downtown Orlando, botanical gardens, a museum, and a planetarium. Walt Disney World is nearby.

Orland Park (ôr′lənd), village (1970 pop. 6,391), Cook co., NE Ill., a suburb SW of Chicago; inc. 1892. Radio and television equipment is made.

Or·lé·a·nais (ôr′lā-ə-nā′), region and former province, N central France, on both sides of the Loire River. The ancient forest of Orléans occupies the center of the region. The fertile Loire valley yields fruits, vegetables, and grapes and is dotted by many fine châteaux. South of the Loire bend is the swampy Sologne Plain, which has been considerably improved by drainage. The nucleus of the Orléanais has been part of the royal domain since the time of Hugh Capet (10th cent.). There are abundant ruins of fortresses and churches from the Carolingian period (c.7th cent.).

Or·le·ans. 1. (ôr′lē-ənz, ôr′lənz) Parish (1970 pop. 593,471), 197 sq mi (510.2 sq km), SE La., coextensive with New Orleans, on the E bank of the Mississippi and extending N and NE to the shores of Lake Pontchartrain; formed 1805; parish seat New Orleans. A commercial, metropolitan area, it is the south terminus of river navigation on the Mississippi, and a center for traffic on the Gulf Intracoastal Waterway. It is a leading U.S. cotton market and exports petroleum, cotton, lumber, machinery, food products, iron, and steel. Industries include sugar milling, petroleum refining, cottonseed processing, shipbuilding, and textile manufacturing. The area is of great historical importance. **2.** (ôr′lēnz′) County (1970 pop. 37,305), 396 sq mi (1,025.6 sq km), W N.Y., bounded on the N by Lake Ontario, crossed by the Barge Canal, and drained by Oak Orchard Creek; formed 1824; co. seat Albion. This fruit-growing area also has truck farming, dairying, and diversified manufacturing. **3.** (ôr-lēnz′) County (1970 pop. 20,153), 715 sq mi (1,851.9 sq km), N Vt., on the Que. border, drained by the Barton, Missisquoi, Black, and Clyde rivers; formed 1792; co. seat Newport. It is in a dairying region that produces natural and processed cheese. Its industries include clothing, manufacturing, lumbering, and metalworking. There are many lake resorts.

Or·lé·ans (ôr′lā-äN′), city (1975 pop. 106,249), capital of Loiret dept., N central France, on the Loire River. A commercial and transportation center, it has food-processing, tobacco, machine-building, electrical, pharmaceutical, chemical, and textile industries. The old city is surrounded by sprawling modern suburbs. Orléans was first known as Genabum, a commercial city of the Carnutes, a Celtic tribe. The city revolted against Julius Caesar (52 B.C.), was burned, and was rebuilt and called Aurelianum. Unsuccessfully attacked by Attila the Hun (451), it was taken by Clovis I (498), after which it became (511) the capital of the Frankish kingdom of Orléans. The kingdom was united with Neustria in the 7th cent. Under the Capetians, the first kings of France, the city became (10th cent.) a principal residence of the French kings. The siege of Orléans (1428–29) by the English threatened to bring all of France under England's rule, and its lifting by Joan of Arc turned the tide of the Hundred Years' War (1337–1453). In the Wars of Religion (16th cent.) the city was briefly the headquarters of the Huguenots and was besieged in 1563 by Catholic forces. Orléans remained in Catholic hands until the Edict of Nantes (1598). During the 17th and 18th cent. the city was a prosperous industrial and commercial center, and its university (founded 14th cent.) was famous throughout Europe. The advent of railroads in the 19th cent. somewhat reduced the city's importance as a trade center dependent on the Loire River port. Orléans was severely damaged during the German invasion of France in 1940, and many irreplaceable historic buildings were destroyed. Several fine structures remain, including the Cathedral of Sainte-Croix, rebuilt (17th–19th cent.) after its destruction by the Huguenots in 1568, and the Renaissance town hall, where Francis II died in 1560. The feast of Joan of Arc is celebrated with particular splendor each May.

Or·lé·ans, Île d' (ēl′ dôr-lā-äN′), island, 20 mi (32 km) long and 5 mi (8 km) wide, S Que., Canada, in the St. Lawrence NE of Quebec. It is a popular tourist attraction. Potatoes, strawberries, cheese, and poultry are the chief products of the island. Settled (1651) by the French, it was the site of one of James Wolfe's camps in his attack on Quebec in 1759.

Or·ly (ôr-lē′), city (1975 pop. 26,109), Val-de-Marne dept., N central France, a suburb SE of Paris. It is the site of Orly Field, one of the chief airports of Paris.

Or·mond Beach (ôr′mənd), resort and residential city (1970 pop. 14,063), Volusia co., NE Fla., on Halifax River (a lagoon) and the Atlantic Ocean; inc. 1880. It was founded (1873) as a health resort and was the winter home of several famous people, including John D. Rockefeller, who died here in 1937.

Orms·kirk (ôrmz′kûrk′), urban district (1973 est. pop. 28, 860), Lancashire, N England. Silk and cotton textiles and metal goods are made. Nearby are ruins of Burscough Abbey (12th cent.).

Or·muz (ôr′mŭz′, ôr-mōōz′): *see* Hormoz, Iran.

Orne (ôrn), department (1975 pop. 293,523), 2,355 sq mi (6,099.5 sq km), NW France, in Normandy and part of Perche; capital Alençon.

O·ro·fi·no (ôr′ə-fē′nō), city (1970 pop. 3,883), seat of Clearwater co., N Idaho, on the Clearwater River E of Lewiston; founded 1898, inc. as a village 1906, as a city 1927. The name was taken from a mining camp established nearby in the gold rush of 1860. It is a lumber town in an area of white-pine forests.

O·ro·moc·to (ôr′ə-mŏk′tō), town (1976 pop. 10,276), S central N.B., Canada, on the St. John River. The town developed because of its proximity to Camp Gagetown, the largest (436 sq mi/1,129.2 sq km) military camp in Canada.

Or·o·no (ôr′ə-nō′, ŏr′-). **1.** Town (1970 pop. 9,989), Penobscot co., S Maine, on the Penobscot River NNE of Bangor; settled c.1775, inc. 1806. Wood products, pulp, and wool are produced. The Univ. of Maine (est. 1865) is here. **2.** Village (1970 pop. 6,787), Hennepin co., SE Minn., on Lake Minnetonka W of Minneapolis; inc. 1954.

O·ron·say (ôr′ən-zā′, -sā′), island, 3 sq mi (7.8 sq km), NW Scotland, one of the Inner Hebrides. The island contains ruins of a 14th cent. priory, a sculptured cross from 1510, and many carved stones unearthed (1882) from Viking graves.

O·ron·tes (ō-rŏn′tēz), river, c.250 mi (400 km) long, rising in the N part of the Al Biqa valley, Lebanon, and flowing generally N through Syria, then W into S Turkey and into the Mediterranean Sea. The river is unnavigable but is important for irrigation. Marshes on its middle course have been drained and the land reclaimed for farming.

O·ro·ville (ôr′ə-vĭl′, ŏr′-), city (1975 est. pop. 8,100), seat of Butte co., N Calif., N of Sacramento on the Feather River; inc. 1906. Settled in 1849, it boomed as a gold camp and river town. Today it is the center of an olive-growing and orchard region in the Sacramento Valley.

Or·re·fors (ôr′ə-fôrs′, -fôsh′), town, Kronoberg co., SE Sweden. It is noted for the manufacture of fine crystal and glassware.

Orr·ville (ôr′vĭl′, ŏr′-), city (1970 pop. 7,408), Wayne co., NE Ohio,

SW of Akron, in a farm area; settled c.1850, inc. 1864. It has light industry.

Or·sha (ôr'shə), city (1976 est. pop. 114,000), W central European USSR, a port at the confluence of the Dnepr and Orshitsa rivers. The city's industries include machine building, metalworking, food processing, and the production of machine tools, reinforced concrete, silicone, and linen. First mentioned in 1067, the city passed to Lithuania in the 13th cent. It was an important Polish fortress from the 16th cent. until its annexation by Russia in 1772.

Orsk (ôrsk), city (1976 est. pop. 243,000), Orenburg oblast, in the foothills of the S Ural Mts., E European USSR, on the Ural River. It has rich iron, copper, nickel, and coal deposits. There are metallurgical plants, machine works, and oil refineries. Orsk was founded in 1735.

Or·te·gal, Cape (ôr'tā-gäl'), NW Spain, in Galicia, extending into the Atlantic. It is usually considered the southwestern limit of the Bay of Biscay.

Ort·les (ôrt'lās'), range of the Ötztal Alps, in Trentino-Alto Adige, N Italy. It has many glaciers. Ortles peak, 12,792 ft (3,901.6 m) high, the highest peak, was first ascended in 1804.

Or·to·na (ôr-tō'nə), town (1971 pop. 20,894), Abruzzi, central Italy, on the Adriatic Sea. Now a small fishing port and a seaside resort, it was a major port from the 11th cent. to 1447, when its fleet and arsenal were destroyed by the Venetians. The 12th cent. cathedral (now restored) and the Aragonese castle (15th cent.) were heavily damaged in World War II.

Or·ton·ville (ôr'tn-vĭl'), city (1970 pop. 2,665), seat of Big Stone co., W Minn., at the S.Dak. border on Big Stone Lake and NW of Montevideo; settled 1872. It is a resort and a farm trade center, with granite quarries nearby.

O·ru·ro (ō-ro͝or'ō), city (1976 pop. 124,121), capital of Oruro dept., W Bolivia. Oruro's economy is based on the silver, wolfram, copper, and tin found in the area. Because of the altitude (12,159 ft/3,708.5 m), agriculture is almost nonexistent. Oruro was founded in 1595.

Or·vi·e·to (ôr'vē-ā'tō), city (1976 est. pop. 17,600), in Umbria, central Italy, on the Poglia River. Situated at the top of a rocky hill, it is a tourist and pilgrimage center. Orvieto is probably located on the site of an Etruscan town (sacked by the Romans in 280 B.C. and later rebuilt). It became a free commune by the 12th cent. but was at the mercy of indigenous and foreign tyrants until it passed to the popes in 1448. There are notable Romanesque, Gothic, and Renaissance buildings, but the fame of the city is due mainly to its beautiful cathedral (begun in 1290). The cathedral's white and black marble façade is decorated with delicate sculptures and colorful mosaics.

O·sage (ō'sāj', ō-sāj'). **1.** County (1970 pop. 13,352), 707 sq mi (1,831.1 sq km), E Kansas, in a gently rolling plains area drained by the Marais des Cygnes River; formed 1855, name changed from Weller in 1859; co. seat Lyndon. It is in a livestock-raising and grain-farming area, with bituminous-coal mining, clothing manufacturing, and a paper industry. **2.** County (1970 pop. 10,994), 608 sq mi (1,574.7 sq km), central Mo., in the Ozarks, with the Missouri River in the N and the Osage River in the W and drained by the Gasconade River; formed 1841; co. seat Linn. Corn, wheat, hay, potatoes, and livestock are its major agricultural products. Fire-clay and lead mines are found throughout the area. **3.** County (1970 pop. 29,750), 2,272 sq mi (5,884.5 sq km), N Okla., bounded in the N by the Kansas border, in the SW by the Arkansas River, and drained by the Caney River; formed 1907; co. seat Pawhuska. It is in a cattle-ranching and petroleum-producing area with some agriculture (wheat, cotton, and dairy products) and limestone quarries. Its manufacturing includes metal goods and machinery, textiles, and wood and concrete products.

Osage, city (1970 pop. 3,815), seat of Mitchell co., NE Iowa, NE of Mason City, in a farm area; settled 1853, inc. 1871.

Osage, river, c.360 mi (580 km) long, formed by the confluence of the Marais des Cygnes and Little Osage rivers, W Mo., and flowing NE to join the Missouri River near Jefferson City. The Osage River basin project provides for flood control and power production.

O·sa·ka (ō-sä'kə, ō'sä-kä), city (1976 est. pop. 2,750,418), capital of Osaka prefecture, S Honshu, Japan, on Osaka Bay, at the mouth of the Yodo River. Osaka is the focal point of a chain of industrial cities stretching to Kobe. Food processing, printing, and the manufacture of steel, chemicals, and textiles are among the chief industries. The city is also a major port and transportation hub. Osaka is known for its puppet and other theaters, and its parks and gardens are noted for their beauty. Landmarks include the Buddhist temple of Shitennoji, founded in 593, and Temmangu, a Shinto shrine founded in 949. The city was the site of imperial palaces as early as the 4th cent. Its importance as a commercial center dates from the 16th cent. Hideyoshi's huge castle, reconstructed in 1931, still dominates the city.

O·sa·wat·o·mie (ō'sə-wŏt'ə-mē, ŏs'ə-), city (1970 pop. 4,294), Miami co., E Kansas, on the Marais des Cygnes River; founded 1855 by the New England Emigrant Aid Company, inc. 1883. It is a farm trade center in an area of oil and natural-gas fields. The town has a memorial park that contains the cabin where John Brown lived in 1856.

Os·borne (ŏz'bərn), county (1970 pop. 6,416), 886 sq mi (2,294.7 sq km), N central Kansas, in a rolling plains region drained by the South Fork of the Solomon River; formed 1827; co. seat Osborne. In a livestock and grain area, it has meat-packing plants, some mining, and a metalworking industry.

Osborne, city (1970 pop. 1,980), seat of Osborne co., N central Kansas, on the South Fork of the Solomon River NW of Salina; founded 1871, inc. 1879. It is a shipping point for a livestock and grain area.

Os·ce·o·la (ŏs'ē-ō'lə). **1.** County (1970 pop. 25,267), 1,313 sq mi (3,400.7 sq km), central Fla., in a lowland area with many lakes and tributaries of the Kissimmee River; formed 1887; co. seat Kissimmee. It is in a cattle and citrus-fruit region, with some truck farming. Some mining and lumbering are done. Furniture, wood and plastic products, concrete, and engines are manufactured. **2.** County (1970 pop. 8,555), 398 sq mi (1,030.8 sq km), NW Iowa, on the Minn. border and drained by the Ocheyedan River; formed 1851; co. seat Sibley. It is in a prairie agricultural area yielding hogs, cattle, poultry, corn, oats, and wheat. **3.** County (1970 pop. 14,838), 581 sq mi (1,504.8 sq km), central Mich., intersected by the Muskegon River and drained by the South Branch of the Manistee River; formed 1869; co. seat Reed City. Livestock, grain, potatoes, corn, and hay are raised. Some manufacturing is done. The county contains resorts, hunting and fishing areas, and part of Manistee National Forest in the west.

Osceola. 1. City (1975 est. pop. 8,371), a seat of Mississippi co., NE Ark., near the Mississippi River N of Memphis, Tenn. In a rich cotton area, it makes cottonseed oil and allied products. **2.** City (1975 est. pop. 3,375), seat of Clarke co., SSW of Des Moines; settled 1850, inc. 1859. **3.** City (1970 pop. 874), seat of St. Clair co., W Mo., on the Osage River SSE of Kansas City, in a farm and coal region; settled in the 1830s, inc. 1883. It is a resort center. **4.** City (1970 pop. 923), seat of Polk co., E central Nebr., NW of Lincoln on a branch of the Big Blue River, in a livestock, dairy, poultry, and grain area.

Os·co·da (ŏs-kō'də), county (1970 pop. 4,726), 565 sq mi (1,463.4 sq km), NE Mich., intersected by the Au Sable River and drained by the Upper South Branch of the Thunder Bay River; organized 1881; co. seat Mio. There is some stock raising and farming. Dairy products and hardwood timber are made. Hunting and fishing areas, part of Huron National Forest, and several small lakes are found here.

Osh (ôsh), city (1976 est. pop. 155,000), capital of Osh oblast, Central Asian USSR, in Kirghizia, in the Fergana Valley. One of the oldest settlements of Central Asia, Osh was for centuries a major silk-production center, strategically situated on a trade route to India.

Osh·a·wa (ŏsh'ə-wə, -wä', -wô'), city (1976 pop. 107,023), SE Ont., Canada, on Lake Ontario. The production of automobiles, begun in 1907, is the leading industry. Many other products are made, notably leather goods. The town is on the site of a French trading post.

O-shi·ma (ō-shē'mə), island, c.35 sq mi (90 sq km), near the entrance to Tokyo Bay, E Japan. Agriculture and fishing are chief activities here. The island is a recreation area for Tokyo. It was visited (17th cent.) by Maarten Vries, the Dutch navigator.

Osh·kosh (ŏsh'kŏsh'). **1.** City (1970 pop. 1,067), seat of Garden co., W Nebr., SE of Alliance, on the North Platte River; settled 1855. **2.** City (1978 est. pop. 49,600), seat of Winnebago co., E Wis., on Lake Winnebago; inc. 1846. It is a resort center and has varied industries. Father Allouez visited the site in 1670; French explorers traveled here in the 18th cent.; and a French fur-trading post was set up in the early 19th cent. Oshkosh grew as a lumber town. The downtown area was destroyed by fire in 1875.

O·shog·bo (ō-shŏg'bō), city (1971 est. pop. 253,000), SW Nigeria, on the Oshun River. Primarily a farming and commercial city, it has cotton gins, a textiles industry, and cigarette factories. Oshogbo was probably founded in the 17th cent. In 1839 it was the site of a decisive battle in which Ibadan, a Yoruba city-state, defeated Ilorin, an expansionist Fulani state, thus halting Ilorin's southward advance.

O·si·jek (ō'sē-yěk'), city (1971 pop. 93,912), N Yugoslavia, in Croatia, on the Drava River. It is a river port and industrial center. Metal and wood items, textiles, and furniture are among its industrial products. Osijek grew around a castle built in 1091 on the site of the Roman colony and fortress of Mursa. It was under Turkish rule from 1526 to 1687 and passed to Yugoslavia in 1918.

O·si·pen·ko (ŏs'ə-pěng'kō): *see* Berdyansk, USSR.

Os·ka·loo·sa (ŏs'kə-lōō' sə). **1.** City (1978 est. pop. 10,200), seat of Mahaska co., SE Iowa, on the North and South Skunk rivers; settled 1844 by Quakers, inc. 1852. It is the trade and processing center of a rich farm and livestock area. Coal has been mined here for more than 100 years; a huge strip mine is still in operation. Manufactures include farm equipment, trousers, fire hydrants, and feeds and seeds. A small fort was established here in 1835 and became a post on a much traveled westward trail. **2.** City (1970 pop. 955), seat of Jefferson co., NE Kansas, NE of Topeka. It is a trading center in a grain, livestock, and dairy region.

Os·kars·hamn (ôs'kərs-hä'mən), city (1975 est. pop. 22,700), Kalmar co., SE Sweden, a seaport on the Kalmarsund (an arm of the Baltic Sea); chartered 1856. Manufactures of this industrial center include processed copper, paper, and machinery.

Os·lo (ŏz'lō, ŏs'-), city (1977 est. pop. 461,881), capital of Norway, of Akershus co., and of Oslo co., SE Norway, at the head of the Oslofjord (a deep inlet of the Skagerrak). Oslo is Norway's largest city, its main port, and its chief commercial, industrial, and transportation center. Manufactures include ships, processed food, textiles, forest products, machines, and printed materials. Founded c.1050 by Harold III, Oslo became (1299) the national capital. In the 14th cent. it came under the dominance of the Hanseatic League. After a great fire (1624) the city was rebuilt by Christian IV and was renamed Christiania (or Kristiania); in 1925 the name Oslo again became official. The city's modern growth dates from the late 19th cent. In World War II Oslo fell (Apr. 9, 1940) to the Germans and was occupied until the surrender (May, 1945) of the German forces in Norway. The city's chief public buildings include the royal palace (1848), the Storting (parliament), and the city hall (1950), decorated by many Norwegian artists. Surviving medieval structures include the Akerskirke (12th cent.) and the Akershus fortress (13th cent.). The Univ. of Oslo (founded 1811), the national theater (1899), the national gallery, a Nobel Institute, and a college of architecture are among the city's cultural institutions. The Folk Museum has reconstructions of old Norwegian timber houses and of a 12th cent. stave church, and the Kon-Tiki Museum has mementos of Thor Heyerdahl's trip (1947) across the Pacific Ocean.

Os·na·brück (ŏz'nə-brŏŏk', ôs'nä-brük'), city (1974 est. pop. 163,674), Lower Saxony, N West Germany, on the Hase River, linked by canal with the Midland Canal. It is an industrial center, with iron and steel mills, machinery plants, and factories that manufacture textiles, paper, and motor vehicles. Located on the site of an ancient Saxon settlement, Osnabrück was made (783) an episcopal see by Charlemagne. The city became a member of the Hanseatic League and a center of the linen trade. It accepted the Reformation in 1543; however, under the Peace of Westphalia, the see was occupied alternately by Catholic and Lutheran bishops. The bishopric was secularized in 1803, and the city passed (1815) to Hanover at the Congress of Vienna. The Catholic diocese was restored in 1857. Osnabrück was badly damaged in World War II.

O·sor·no (ō-sôr'nō), city (1970 pop. 68,815), capital of Osorno prov., S central Chile, in the heart of the lake district. Osorno is chiefly an agricultural processing and distributing center. Founded in 1553, it was later destroyed by the Araucanian Indians and was re-established in 1796. An influx of immigrants in the latter half of the 19th cent. has given Osorno the atmosphere of a German town.

Os·ro·e·ne (ŏz'rō-ē'nē), ancient kingdom of NW Mesopotamia, in present-day SE Turkey and NE Syria. Edessa was its capital. It broke away (2nd cent. B.C.) from the Seleucid empire to form a separate kingdom. It came under Roman rule (late 2nd cent. A.D.).

Oss (ôs), city (1977 est. pop. 45,582), North Brabant prov., S Netherlands; chartered 1399. It is an industrial center; manufactures include meat products, chemicals, and pharmaceuticals.

Os·se·tia (ŏ-sē'shə), region of the central Caucasus, S European USSR. On the northern slope is the North Ossetian Autonomous Soviet Socialist Republic (1970 pop. 552,000), 3,100 sq mi (8,029 sq km), in the Russian Republic; capital Ordzhonikidze. On the southern slope is the South Ossetian Autonomous Oblast (1970 pop. 99,000), 1,500 sq mi (3,885 sq km), in Georgia; capital Tskhinvali. Both sections have valleys that produce fruit, wine, grain, and cotton. Lumbering and livestock raising are important in the mountains. North Ossetia has lead, silver, zinc, and oil deposits. The Ossetians, an Iranian-speaking people, are descended from medieval Alans. During the 17th cent. the northern Ossetians were subject to Karbada princelings. Russian influence became strong in the 18th cent., and between 1801 and 1806 Ossetia was annexed to Russia.

Os·si·ning (ŏs'ə-nĭng), village (1978 est. pop. 19,300), Westchester co., SE N.Y., on the Hudson River; settled c.1750, inc. 1813 as Sing Sing, renamed 1901. Fine wire, surgical instruments, heart pumps, and maps are among the village's manufactures. Ossining is the site of Sing Sing state prison (built 1825–28).

Os·si·pee (ŏs'ə-pē), rural and resort town (1970 pop. 1,647), seat of Carroll co., E N.H., E of Lake Winnipesaukee; inc. 1785. The town makes wood products.

Os·so·ry (ôs'ə-rē), ancient kingdom of Ireland. An independent state on the borders of Leinster and Munster, its overlordship was long disputed. It became part of Leinster under the Normans in the 12th cent., and by the middle of the 14th cent. was part of the earldom of Ormonde.

Ö·ster·sund (œ'stər-sŭnd'), city (1975 est. pop. 47,100), capital of Jämtland co., central Sweden, on Lake Storsjön; founded 1786. It is a commercial, industrial, and transportation center.

Os·ti·a (ŏs'tē-ə), ancient city of Italy, at the mouth of the Tiber. It was founded (4th cent. B.C.) as a protection for Rome, then developed (from the 1st cent. B.C.) as a port. Augustus, Claudius I, Trajan, and Hadrian expanded the city and harbor. From the 3rd cent. A.D. the city began to decline. The ruins rival those of Pompeii in showing the layout of an ancient Italian city.

O·stra·va (ô'strə-və), city (1975 est. pop. 300,945), N central Czechoslovakia, in Moravia, near the junction of the Oder and Ostravice rivers. Anthracite and bituminous coal, iron and steel, rolling stock, machinery, and ship and bridge parts are the major products of the city, which also has a large chemical industry. Ostrava is also a road and rail hub and the site of several hydroelectric stations. Well-known in the Middle Ages, it was later important because of its strategic location guarding the Moravian Gate, the entrance to the Moravian lowlands. The city's industrial prominence dates from the opening of its first coal mine and the coming of the railroad in the late 19th cent. German forces occupied Ostrava from 1939 to 1945.

O·stro·łe·ka (ô'strə-lěɴ'kə), town (1975 est. pop. 28,000), NE Poland, on the Narew River. Pulp and paper, lumber, and bricks are produced. Chartered in 1427, the town passed to Prussia in 1795 and to Russia in 1815. It reverted to Poland in 1920.

Os·we·go (ŏs-wē'gō), county (1970 pop. 100,897), 968 sq mi (2,507.1 sq km), N central N.Y., bounded on the NW by Lake Ontario, on the S by Oneida Lake and the Oneida River, crossed by the Barge Canal and drained by the Oswego and Salmon rivers; formed 1816; co. seat Oswego. It is in a dairying area, with diversified manufacturing, gas wells, farming (strawberries and other truck crops), poultry and stock raising, and shipping facilities. There are lake resorts.

Oswego. 1. City (1970 pop. 2,200), seat of Labette co., SE Kansas, near the Neosho River SE of Parsons, in a farm and livestock area; founded 1866, inc. 1870. Clothing is made. **2.** City (1978 est. pop. 22,200), seat of Oswego co., N central N.Y., on Lake Ontario and the Oswego River; founded 1722, inc. as a city 1848. It is a port of entry. The city's manufactures include aluminum, textiles, and paper products. A trading post established here after the English founded Oswego (1722) became a vital outlet for the Albany fur trade. The strategic location prompted the building of Fort Oswego (1727), Fort George (1755), and present Fort Ontario (1755; an active U.S. army post until 1946, and a state historic site since 1951). These fortifications were much contested in the colonial wars. The city's importance as a lake port came with the completion of the Barge Canal (1917) and the St. Lawrence Seaway (1959).

Os·wes·try (ŏz'wə-strē), rural district (1971 pop. 30,320), Salop, W central England. It is a market town with plastics, clothing, and printing industries. The district is named for St. Oswald, a Northumbrian king who was killed here in a battle (7th cent.) against King Penda of Mercia.

Oś·wię·cim (ôsh-vyĕɴ'tsēm), formerly **Ausch·witz** (oush'vĭtz'), town (1975 est. pop. 42,700), SE Poland. It is a railway junction and industrial center producing chemicals, leather, and agricultural implements. In World War II the Germans organized a concentration camp system here. Some 4,000,000 prisoners, mostly Jews, were annihilated here.

O·ta·ru (ō-tä'rōō), city (1975 pop. 184,406), SW Hokkaido, Japan, on Ishikari Bay. It is the main coal-exporting port of the island and a center of herring fisheries.

O·ter·o (ō-tĕr'ō). **1.** County (1970 pop. 23,523), 1,254 sq mi (3,247.9 sq km), SE Colo., drained by the Purgatoire, Apishapa, and Arkansas rivers; formed 1889; co. seat La Junta. It is in an irrigated agricultural area that produces livestock, sugar beets, grain, truck crops, and feed. It has a food-processing industry and manufactures metal products. **2.** County (1970 pop. 41,097), 6,638 sq mi (17,192.4 sq km), S N.Mex., bounded on the S by the Texas border and drained by Rio Penasco; formed 1899; co. seat Alamogordo. Livestock grazing and lumbering are done. Ranges of the Sacramento and Guadalupe mts. are in the east and southeast. The county includes Mescalero Indian Reservation and portions of White Sands National Monument, White Sands Missile Range, and Fort Bliss Military Reservation.

O·toe (ō'tō), county (1970 pop. 15,576), 619 sq mi (1,603.2 sq km), SE Nebr., bounded in the E by the Missouri River at the Iowa-Mo. border; formed 1855; co. seat Nebraska City. It is drained by the Little Nemaha River and its branches. Livestock, feed, grain, fruit, dairy products, and poultry characterize its agriculture. Some manufacturing is done.

Ot·se·go (ŏt-sē'gō). **1.** County (1970 pop. 10,422), 530 sq mi (1,372.7 sq km), N Mich., drained by the Sturgeon and Black rivers and by the North Branch of the Au Sable River; formed 1875; co. seat Gaylord. Its agriculture includes livestock, grain, and hay. Industries include sawmilling and the manufacture of hardwood timber. It is a year-round resort area for hunting, fishing, and boating. The county includes many small lakes. **2.** County (1970 pop. 56,181), 1,013 sq mi (2,623.7 sq km), central N.Y., bounded on the W by the Unadilla River and drained by the Susquehanna River, issuing here from Otsego Lake; formed 1791; co. seat Cooperstown. This dairying area also yields poultry, livestock, grain, and hay. There are resorts by several small lakes and mineral springs.

O·tsu (ō'tsōō), city (1976 est. pop. 195,414), capital of Shiga prefecture, S Honshu, Japan. It is a tourist center and a port for excursion steamers on Lake Biwa. Nylon, cotton yarn, textiles, polypropylene, electrical appliances, precision instruments, and machinery are among the city's industrial products. A former imperial seat (2nd and 7th cent.), Otsu is the site of a 7th cent. Buddhist temple.

Ot·ta·wa (ŏt'ə-wə, -wä', -wô'), city (1976 pop. 304,462). capital of Canada, SE Ont., at the confluence of the Ottawa and Rideau rivers. The Rideau Canal separates the city into upper and lower towns; along its banks and those of the rivers are many landscaped drives. Ottawa has industries that produce paper and paper products, printed materials, telecommunications equipment, and electrical products. Ottawa proper was founded in 1827 by Col. John By, an engineer in charge of construction of the Rideau Canal. Formerly Bytown, the present name was adopted in 1854. In 1858 Ottawa was chosen by Queen Victoria to be the capital of the United Provinces of Canada, and in 1867 it became capital of the Dominion of Canada. The government buildings, built between 1859 and 1865, were burned in 1916 but were immediately rebuilt on an enlarged scale. Other notable buildings are Rideau Hall, the residence of the governor general, the Anglican and Roman Catholic cathedrals, the National Museum, the National Art Gallery, the National Arts Centre, the Dominion Observatory, and the Royal Mint.

Ottawa. 1. County (1970 pop. 6,183), 723 sq mi (1,872.6 sq km), NE central Kansas, in a rolling prairie region intersected by the Solomon River and drained in the SW by the Saline River; formed 1860; co. seat Minneapolis. It is in a wheat-farming and cattle-raising area that also produces small grains and hay. There is some mining and a cheese-processing industry. Motor homes and campers are manufactured here. **2.** County (1970 pop. 128,181), 564 sq mi (1,460.8 sq km), SW Mich., bounded on the W by Lake Michigan and drained by the Grand and Black rivers; formed 1837; co. seat Grand Haven. Its agriculture includes livestock, poultry, fruit, truck crops, grain, celery, and potatoes. Tulips are grown. The county has oil refineries and commercial fisheries, as well as resorts. **3.** County (1970 pop. 37,099), 261 sq mi (676 sq km), N Ohio, bounded in the NE by Lake Erie and drained by the Portage River and Toussaint and Packer creeks; formed 1840; co. seat Port Clinton. It has agriculture (grain, fruit, sugar beets, truck crops, and dairy products) and limestone quarries. Its industries include food processing, textile milling, metal-working, and the manufacture of machinery and wood, paper, chemical, and plastic products. Perry's Victory and International Peace Memorial is here. The Bass Islands are resorts. **4.** County (1970 pop. 29,800), 464 sq mi (1,201.8 sq km), extreme NE Okla., bordered in the N by Kansas, in the E by Mo., with part of the Ozarks in the E; formed 1907; co. seat Miami. It is drained by the Neosho and Spring rivers and includes a section of the Lake of the Cherokees. In a rich lead and zinc region, it has stock raising, dairying, and agriculture (corn, grain, and poultry). Its manufactures include textiles, tires, leather, concrete, metals, wood and metal products, and machinery.

Ottawa. 1. City (1978 est. pop. 17,600), seat of La Salle co., N central Ill., at the confluence of the Fox and Illinois rivers, in a fertile farm area; inc. as a city 1853. Rich deposits of silica in the area are used in the production of glass and sand products. The city has a marble manufacturing industry; other products are machines, tools, and plastic products. Points of interest include the site of the first Lincoln-Douglas debate (1858) and Fort Johnson (1832). **2.** City (1970 pop. 11,036), seat of Franklin co., E Kansas, on the Marais des Cygnes River; inc. 1867. It is the trade center of a farm area, and its industries produce mobile homes, plastic products, clothing, cabinets, and building blocks. The city is named for the Ottawa Indians, who moved here (1832) after ceding their Ohio lands to the United States; they were subsequently removed (1867) to Oklahoma. **3.** Village (1970 pop. 3,622), seat of Putnam co., NW Ohio, on the Blanchard River N of Lima, in a farm area; founded 1833, inc. 1861. Television tubes are made.

Ottawa, river, c.700 mi (1,125 km) long, largest tributary of the St. Lawrence River, Canada. It rises in the Laurentian Highlands, SW Que., and flows generally west through La Vérendrye Provincial Park to Lake Timiskaming, then southeast, forming part of the Que.-Ont. border, past Ottawa, and into the St. Lawrence River near Montreal. The river is navigable for large vessels as far as Ottawa; it is connected with Lake Ontario by the Rideau Canal system. Lumbering is the chief industry along the lower river. Samuel de Champlain was the first European to visit (1613–15) the valley. The river became an important highway for fur traders and missionaries.

Ot·ter Creek (ŏt'ər), river, c.100 mi (160 km) long, W Vt., rising in the Green Mts. near Dorset and flowing N past Rutland, Middlebury, and Vergennes to Lake Champlain.

Otter Tail (tāl), county (1970 pop. 46,097), 1,962 sq mi (5,081.6 sq km), W central Minn., in an extensively watered area drained by the Pelican, Pomme de Terre, and Otter Tail rivers and numerous lakes; formed 1858; co. seat Fergus Falls. In an agricultural area yielding livestock, dairy products, poultry, grain, and potatoes, it has deposits of marl and peat and food-processing and boatbuilding industries. Clothing, wood products, and mobile homes are manufactured.

Ot·to·man Empire (ŏt'ə-mən), vast state founded in the late 13th cent. by the Ottoman or Osmanli Turks (the last of the Turkish peoples to invade the Near East) and ruled by the descendants of Osman I until its dissolution in 1918. Modern Turkey formed only part of the empire, but the terms "Turkey" and "Ottoman Empire" were often used interchangeably.

The Ottoman state began as one of many small Turkish states that emerged in Asia Minor during the breakdown of the empire of the Seljuk Turks. The Ottoman Turks began to absorb the other states, and during the reign (1451–81) of Mohammed II they ended all other local Turkish dynasties. The early phase of Ottoman expansion took place at the expense of the Byzantine Empire, Bulgaria, and Serbia. The great Ottoman victories of Kossovo (1389) and Nikopol (1396) placed large parts of the Balkan Peninsula under Ottoman rule. After a setback at the hands of Tamerlane, who captured Beyazid in 1402, the Ottomans rallied and defeated the Byzantine Greeks at Constantinople (1453). Their success was due partly to the weakness and disunity of their adversaries and partly to their superior military organization. Turkish expansion reached its peak in the 16th cent.

under Selim I and Sulayman I (Sulayman the Magnificent). The Hungarian defeat (1526) at Mohacs prepared the way for the capture (1541) of Buda and the absorption of the major part of Hungary; Transylvania became a tributary principality, as did Walachia and Moldavia. The Asiatic borders of the empire were pushed deep into Persia and Arabia. Selim I defeated the Mamelukes of Egypt and Syria, took Cairo in 1517, and assumed the succession to the Caliphate. Algiers was taken in 1518, and Mediterranean commerce was threatened by corsairs who sailed under Turkish auspices. Sulayman reorganized the Turkish judicial system, and his reign saw the flowering of Turkish literature, art, and architecture.

After Sulayman's death the court at Constantinople became corrupt. Ottoman administration was inept and exploitive, though tolerant of non-Moslems. In 1683 a huge Turkish army surrounded Vienna. The relief of Vienna by John III of Poland and subsequent Turkish losses ended in negotiations in 1699 that cost Turkey Hungary and other territories and from which the disintegration of the Ottoman Empire may be dated. The breakup of the state gained impetus with the Russo-Turkish Wars in the 18th cent., in which Turkey lost the northern and northeastern coasts of the Black Sea. Greece and Egypt followed early in the 19th cent. Drastic reforms were introduced, but they came too late. By the 19th cent. Turkey was known as the Sick Man of Europe, increasingly dependent on the European powers that loaned it money in return for commercial privileges and markets.

The rebellion (1875) of Bosnia and Hercegovina precipitated the Russo-Turkish War of 1877–78, after which Rumania, Serbia, and Montenegro were declared fully independent, Bulgaria became virtually independent, and Bosnia and Hercegovina passed under Austrian administration. The Armenian massacres of the late 19th cent. turned world public opinion against Turkey. In 1909 the Young Turks, a reformist and nationalist group, forced the deposition of the despotic sultan Abd Al-Hamid II. A war with Italy (1911–12) resulted in the loss of Libya; and in the two successive Balkan Wars (1912–13) Turkey lost nearly its entire territory in Europe to Bulgaria, Serbia, Greece, and newly independent Albania. The outbreak of World War I found Turkey solidly lined up with the Central Powers. In 1917 British forces occupied Baghdad and Jerusalem. Arabia rose against Turkish rule, and in 1918 Turkish resistance collapsed in both Asia and Europe. An armistice was concluded in Oct., and the Ottoman Empire came to an end. With the victory of the Turkish nationalists, who overthrew the sultan in 1922, the history of modern Turkey began.

Ot·tum·wa (ə-tŭm′wə), city (1978 est. pop. 26,100), seat of Wapello co., SE Iowa, on the Des Moines River, in a farm and coal area; inc. 1851. A commercial and industrial center, Ottumwa has a large meat-packing plant and a farm-machinery industry.

Ötz·tal Alps (œts′täl′), mountain group, in the Tyrol, W Austria, S of the Inn River. It rises to 12,382 ft (3,776.5 m) in the Wildspitze, the highest peak in the Tyrol.

Ouach·i·ta (wŏsh′ĭ-tô′). **1.** County (1970 pop. 30,896), 738 sq mi (1,911.4 sq km), S Ark., drained by the Ouachita and Little Missouri rivers; formed 1842; co. seat Camden. Its agriculture includes cotton, corn, and poultry. It has petroleum and natural-gas wells, lumber, and a food-processing industry. Clothing, furniture, paper, and machinery are manufactured. **2.** Parish (1970 pop. 115,387), 638 sq mi (1,652.4 sq km), NE central La., bounded on the E by Bayou Lafourche and intersected by the Ouachita River and Bayou D'Arbonne; formed 1805; parish seat Monroe. In the west there is stock raising, fruit growing, and lumbering. In the east cotton and corn are grown and dairying is done. Large natural-gas fields are found here. There is some industry, including the processing of farm products.

Ouachita, river, c.600 mi (965 km) long, rising in the Ouachita Mts., W Ark., and flowing E, SE, and S through a rich cotton-producing region of S Ark. and NE La. and into the Red River system. It is joined by the Tensas River at Jonesville, La., below which it is called the Black River. The river is navigable for shallow-draft vessels below Arkadelphia.

Ouachita Mountains, range of ridges between the Arkansas and Red rivers, extending c.200 mi (320 km) from central Ark. into SE Okla. Magazine Mt. (c.2,800 ft/855 m high) is the tallest peak. The Ouachita Mts. are geologically considered offshoots of the Appalachian Mts. They are composed of strongly folded and faulted sedimentary rocks. Mineral springs, lakes, and wooded areas attract tourists.

Oua·ga·dou·gou (wä′gə-dōō′gōō), city (1975 pop. 172,661), capital of Upper Volta. Ouagadougou is the trade and distribution center for an agricultural region whose main crop is peanuts. The city's industry is limited to handicrafts and the processing of food and beverages. Ouagadougou was founded in the late 11th cent. as the capital of a Mossi empire. It was captured by French forces in 1896.

Ouarg·la (wär′glə), town and oasis (1974 est. pop. 26,200), E Algeria. Ouargla lies in the heart of a huge palm grove. There are oil fields nearby. The oasis was settled A.D. c.1000 by Moslems who were fleeing religious persecution. It became a small city-state that from the 16th cent. paid tribute to the Turks. The town was conquered by French forces in 1853.

Ou·de·naar·de (ou′də-när′də), town (1977 pop. 27,177), East Flanders prov., W Belgium, on the Scheldt River. It is a textile center and a rail junction. At Oudenaarde, in 1708, the allies under the duke of Marlborough and Eugene of Savoy defeated the French under the dukes of Burgundy and of Vendôme in the War of the Spanish Succession.

Oudh (oud), historic region of N central India. The region passed under Gupta rule in the 4th cent. A.D. and later became (11th–12th cent.) the center of the Rajput state of Kanauj. In the 13th cent. it was conquered by the legions of the Delhi Sultanate. It became (16th cent.) a province of the Mogul empire and was subsequently governed by the nawabs of Oudh. The annexation (1856) of Oudh as a British province was a major cause of the Indian Mutiny (1857–58). In 1877 Oudh was joined with the presidency of Agra to form the United Provinces, now the constituent state of Uttar Pradesh.

Ou·grée (ōō-grā′), city (1970 pop. 20,574), Liège prov., E Belgium, on the Meuse River, a suburb of Liège. It is a center of heavy industry.

Oui·dah (wē′də), town (1967 est. pop. 19,887), S Benin, a port on the Gulf of Guinea. Palm products, copra, coffee, and citrus fruit are processed nearby. Fishing as well as curing and drying is carried on in the town. Ouidah was the capital of a small state founded about the 16th cent. From the early 17th cent., Portuguese, French, and Dutch traders were intermittently active here. In the 18th and early 19th cent. Ouidah was an important export point for black African slaves. The town was annexed by France in 1886. Ouidah has a Portuguese fort (1788) that contains a museum.

Ouj·da or **Oudj·da** (both: ōōj-dä′), city (1971 pop. 175,532), NE Morocco, near the Algerian border. It is a railroad junction, agricultural market, and commercial center. It was occupied by the French in 1844, 1859, and 1907.

Ou·lu (ou′lōō), city (1975 est. pop. 93,707), capital of Oulu prov., W central Finland, at the mouth of the Oulu River on the Gulf of Bothnia. It is a seaport and has metal shops, leather plants, and wood-processing and other industries. The city grew around a castle founded in 1590 and was chartered in 1610.

Ou·ray (ōō-rā′), county (1970 pop. 1,546), 540 sq mi (1,398.6 sq km), SW Colo., drained by the Uncompahgre River; formed 1877; co. seat Ouray. In a livestock-ranching area, it has lead and zinc mines and some manufacturing. Part of Uncompahgre National Forest and ranges of the Rocky Mts. are here.

Ouray, city (1970 pop. 741), seat of Ouray co., SW Colo., SW of Grand Junction, at an altitude of c.7,800 ft (2,379 m); settled 1875, inc. 1884. In a mining and farming area, it has hot springs and is a health resort.

Ou·ri·que (ō-rē′kə), town (1970 pop. 3,482), Beja dist., S Portugal, in Baixo Alentejo. Although tradition says Alfonso I defeated the Moors here in 1139, the Battle of Ourique was actually fought at some undetermined place nearby.

Ou·ro Prê·to (ō′rōō prā′tōō), city (1970 pop. 46,166), Minas Gerais state, E Brazil. Founded near the end of the 17th cent., it was a prosperous gold-mining town and cultural center in the 18th cent. Since 1933 the city has been a national historic site, preserving the 18th cent. atmosphere of narrow, twisting, cobbled streets, the colonial mint and treasury, the old houses, the theater (oldest in South America), the governor's mansion, and the old churches, most notably the Church of São Francisco, decorated with magnificent carvings. A mining school (est. 1875) is in Ouro Prêto.

Ouse (ōōz). **1.** Or **Great Ouse,** river, c.155 mi (250 km) long, rising in the Northampton Highlands, Northamptonshire, S central England. The Great Ouse flows generally northeast past Bedford and Ely to the Wash near King's Lynn, Norfolk, and drains the eastern Midlands and the west Fens. **2.** River, c.60 mi (100 km) long, formed by the confluence of the Ure and Swale rivers near Boroughbridge, North Yorkshire, NE England. It flows generally southeast past York to join with the Trent River and form the Humber River. Navigable to York, the Ouse is an important commercial waterway.

Ou·ta·gam·ie (ou'tə-găm'ē), county (1970 pop. 119,398), 634 sq mi (1,642.1 sq km), E Wis., drained by the Wolf, Fox, and Embarrass rivers; formed 1851; co. seat Appleton. It is in a dairying and paper-milling region. Its agriculture includes corn, alfalfa, cabbage, oats, and livestock. Textiles, wood products, plastics, concrete, iron and steel, and machinery and electrical equipment are produced.

Out·er Bar·ri·er (ou'tər băr'ē-ər), series of sandy barrier islands or offshore bars, extending c.75 mi (120 km) along the S shore of Long Island, SE N.Y., from Rockaway Beach to the E end of Shinnecock Bay and separating a series of lagoons (Great South Bay, Moriches Bay, and Shinnecock Bay) from the Atlantic Ocean. East Rockaway, Jones, Fire Island, Moriches, and Shinnecock inlets pierce the barrier, forming narrow, sandy islands. The sparsely settled and largely undeveloped low-lying islands suffer from wave erosion.

O·ver·ijs·sel (ō'vər-ī'səl), province (1977 est. pop. 992,953), c,1,500 sq mi (3,885 sq km), E central Netherlands, between the IJsselmeer and West Germany; capital Zwolle. The province is generally sandy but supports extensive stock raising and dairying. The lordship of Overijssel belonged in the Middle Ages to the bishop of Utrecht, but was sold (1527) to Emperor Charles V. It joined (1579) the Union of Utrecht and became one of the United Provs. of the Netherlands.

O·ver·land (ō'vər-lənd, -lănd'), city (1978 est. pop. 20,100), St. Louis co., E Mo., a suburb of St. Louis; inc. 1939.

Overland Park, city (1978 est. pop. 83,100), Johnson co., NE Kansas, a residential suburb of Kansas City; inc. 1960.

Overland Trail, any of several trails of westward migration in the United States. The term is sometimes used to mean all the trails westward from the Missouri to the Pacific.

O·ver·lea (ō'vər-lē'), uninc. town (1970 pop. 13,086), Baltimore co., N Md., a residential suburb of Baltimore.

O·ver·ton (ō'vər-tən), county (1970 pop. 14,866), 442 sq mi (1,144.8 sq km), N Tenn., in the Cumberlands, drained by affluents of the Obey and Cumberland rivers; formed 1806; co. seat Livingston. It has agriculture (corn, hay, tobacco, livestock, poultry, and dairy products) and some bituminous-coal mining. Its manufactures include metal-working machinery, clothing, and furniture.

O·vie·do (ō-vyā'*th*ō), city (1975 est. pop. 159,652), capital of Oviedo prov., NW Spain, near the great mining district of the Cantabrian Mts. Ordnance, firearms, gunpowder, textiles, and many other products are manufactured. Founded c.760, Oviedo flourished in the 9th cent. as the capital of the Asturian kings but declined after the capital was transferred to León early in the 10th cent. The cathedral, begun in 1388, contains the tombs of the Asturian kings and has a high square tower. The city suffered severely during the revolt of the Asturian miners in 1934 and in a siege during the civil war.

O·wa·ton·na (ō'wə-tŏn'ə), city (1978 est. pop. 17,000), seat of Steele co., SE Minn.; inc. 1854. It has many diversified industries.

O·we·go (ō-wē'gō), resort village (1974 est. pop. 4,686), seat of Tioga co., S N.Y., on the Susquehanna River E of Elmira; settled 1787, inc. 1827. Furniture is made.

Ow·en (ō'ĭn). **1.** County (1970 pop. 12,163), 391 sq mi (1,012.7 sq km), SW central Ind., drained by the West Fork of the White River and Mill Creek; formed 1818; co. seat Spencer. Its agriculture includes grain, fruit, and livestock. It has timber and limestone quarries. Cement products, drugs, and typewriter ribbons are manufactured. **2.** County (1970 pop. 7,470), 351 sq mi (909.1 sq km), N Ky., in the Bluegrass, bounded on the W by the Kentucky River and drained by Eagle Creek; formed 1819; co. seat Owenton. In a rolling upland agricultural area yielding burley tobacco, corn, and wheat, it has lead and zinc mines, limestone quarries, and industry (cheese processing and shoe manufacturing).

Ow·ens (ō'ĭnz), river, c.120 mi (195 km) long, rising in the Sierra Nevada, E Calif., SE of Yosemite National Park, and flowing SE, nominally to enter Owens Lake (now dry), near Mt. Whitney. An aqueduct diverts most of the river's water to Los Angeles.

O·wens·bor·o (ō'ĭnz-bûr'ō, -bŭr'ō), city (1978 est. pop. 51,100), seat of Daviess co., W Ky., on the Ohio River; settled c.1800, inc. as a city 1866. It is an important tobacco market and a shipping point for a farm and oil region. Its varied manufactures include radio tubes, whiskey, chemicals, electrical equipment, steel, cigars, and furniture.

Owen Sound, city (1976 pop. 19,525), SE Ont., Canada, on Owen Sound. It is a port and railroad terminal in a farming region, and it has large grain elevators. There are printing and other industries.

Owen Stan·ley Range (stăn'lē), mountain chain, c.300 mi (480 km) long, SE Papua New Guinea, on New Guinea island. It rises to Mt. Victoria (13,363 ft/4,075.7 m). The region, drained by several small rivers, is largely jungle.

Ow·en·ton (ō'ĭn-tən), town (1970 pop. 1,280), seat of Owen co., N Ky., N of Frankfort, in a farm area.

Ow·ings Mill (ō'ĭngz mĭl'), uninc. town (1970 pop. 7,360), Baltimore co., N Md., NW of Baltimore.

Ow·ings·ville (ō'ĭngz-vĭl'), town (1970 pop. 1,381), seat of Bath co., NE Ky., ENE of Lexington, in a farm area.

O·wo (ō'wō), city (1969 est. pop. 93,000), S Nigeria. It is primarily a farming and commercial city, located in an area producing cacao and timber. Owo was the capital of a Yoruba state founded in the 14th cent. and came under British protection in 1893.

O·wos·so (ō-wŏs'ō), city (1978 est. pop. 17,500), Shiawassee co., S Mich., on the Shiawassee River, in a farming region; inc. 1859.

Ows·ley (ouz'lē), county (1970 pop. 5,023), 197 sq mi (510.2 sq km), E Ky., in the Cumberlands, drained by the South Fork of the Kentucky River and several creeks; formed 1843; co. seat Booneville. It is in a mountain agricultural region yielding livestock, fruit, and tobacco and has bituminous-coal mines and timber. It includes part of Cumberland National Forest.

O·wy·hee (ō-wī'ē), county (1970 pop. 6,422), 7,648 sq mi (19,808.3 sq km), SW Idaho, in a hilly region bordering on Oregon and Nev. and bounded on the N by the Snake River; formed 1863; co. seat Murphy. Irrigated areas are in the east along the Bruneau River and in the southwest along forks of the Owyhee River. Stock raising, dairying, agriculture (hay, sugar beets, fruit, and truck crops), and mining (lignite and quartz) are major occupations.

Owyhee, river, c.300 mi (480 km) long, rising in several branches in SW Idaho, N Nev., and SE Oregon and flowing N across NE Oregon to the Snake River.The Owyhee reclamation project of the U.S. Bureau of Reclamation irrigates c.118,000 acres (47,790 hectares) west of the Snake River.

Ox·ford (ŏks'fərd), county borough (1976 est. pop. 117,400), administrative center of Oxfordshire, S central England. In addition to its importance as the seat of Oxford Univ., Oxford has significant industries, including the manufacture of automobiles and steel products. A trading town and frontier fort, it was raided by Danes in the 10th and 11th cent. By the 12th cent. it was the site of a castle, an abbey, and the university. During the 13th cent. there were frequent conflicts between the town and the university in which the latter, supported by the church and the king, was usually victorious. During the civil wars Oxford was the Royalist headquarters; it was besieged but not damaged by the Parliamentarians. Among its famous historic buildings (apart from the colleges) are the Radcliffe Camera (1737), the Observatory (1772), and the Sheldonian Theatre (designed by Christopher Wren); the churches of St. Mary the Virgin (13th cent.) and St. Michael (11th cent.); and several old inns. The Ashmolean Museum and the Bodleian Library are notable.

Oxford, county (1970 pop. 43,457), 2,080 sq mi (5,387.2 sq km), W Me., bordering on N.H. and Que. and drained by the Androscoggin and Little Androscoggin rivers; formed 1805; co. seat South Paris. It is in a lumbering, mining (mica, feldspar, quartz, and semiprecious stones), farming, and dairying area, with some manufacturing (paper, wood products, and shoes and moccasins). There are winter sports areas, and summer resorts in the Rangeley Lakes region.

Oxford. 1. Town (1970 pop. 10,345), Worcester co., S Mass.; settled 1687 by French Protestants, inc. 1693. It is chiefly residential, with some light manufacturing. Clara Barton was born here. **2.** City (1970 pop. 13,846), seat of Lafayette co., N central Miss.; inc. 1837. It is principally a university town, the seat of the Univ. of Mississippi ("Ole Miss"). Household appliances, motors, clothing, and lumber products are manufactured in the city. Oxford was the home of the novelist William Faulkner and the setting for some of his works. An annual pilgrimage takes tourists to the many ante-bellum buildings in the area. In 1962 Oxford was the scene of rioting and conflict when the first black student was enrolled in the university. **3.** Town

(1970 pop. 7,178), seat of Granville co., N N.C., N of Raleigh; settled 1760, laid out 1811, inc. 1816. It is a processing center in a tobacco and farm area. **4.** Village (1970 pop. 15,868), Butler co., SW Ohio, near the Ind. border, in a farm area; laid out 1810, inc. 1830. It is the seat of Miami Univ. There are many old houses.

Ox·ford·shire (ŏks'fərd-shîr', -shər) or **Ox·on** (ŏk'sŏn'), nonmetropolitan county (1976 est. pop. 541,800), 749 sq mi (1,940 sq km), S central England; administrative center Oxford. The chief occupation is farming (wheat, barley, and oats), with some dairying and sheep raising. Ironstone and limestone are found. In the Middle Ages Oxfordshire was a part of the Anglo-Saxon kingdom of Mercia. During the English civil war it was a stronghold of Royalist resistance.

Ox·nard (ŏks'närd'), city (1978 est. pop. 95,900), Ventura co., S Calif., on the Pacific coast; inc. 1903. Its economy, formerly based on agriculture, mining, and nearby military bases, has been broadened to include large industrial and commercial operations. Oxnard is the gateway for visitors to the Santa Barbara Islands and to Los Padres National Forest.

O·yo (ō'yō), city (1971 est. pop. 136,000), SW Nigeria. It is primarily a farming town, producing tobacco, yams, and cassava. Traditional artisans make textiles and leather goods and carve utensils from shells of the calabash gourd. Oyo was founded c.1835 as the successor of Old Oyo (Katunga), the capital of the Yoruba empire of Oyo, which was destroyed in the Yoruba civil wars of the early 19th cent. The city came under British protection in 1893.

Oys·ter Bay (oi'stər), uninc. area (1970 pop. 6,800) of the town of Oyster Bay, Nassau co., SE N.Y., on N Long Island, on Long Island Sound; settled 1653. It is chiefly residential. Nearby is Theodore Roosevelt's estate, Sagamore Hill, which was made a national shrine in 1953 and a national historic site in 1963. Also of interest in Oyster Bay are several 18th cent. homes and the Theodore Roosevelt memorial bird sanctuary (owned by the National Audubon Society), which adjoins Roosevelt's grave.

O·zark (ō'zärk'), county (1970 pop. 6,226), 731 sq mi (1,893.3 sq km), S Mo., in the Ozarks, drained by the North Fork of the White River; formed 1841; co. seat Gainesville. Agriculture is the major activity, producing corn, wheat, oats, cotton, and livestock. Parts of Mark Twain National Forest are in the county.

Ozark. 1. City (1970 pop. 13,555), seat of Dale co., SE Ala., in a timber and farm area; settled 1820, inc. 1870. Textiles, wearing apparel, and farm equipment are made. **2.** City (1976 est. pop. 2,984), a seat of Franklin co., NW Ark., in the Ozarks on the Arkansas River; settled 1836. It has coal mines and processes timber and food. **3.** City (1970 pop. 2,384), seat of Christian co., S Mo., S of Springfield, in a farm region; laid out 1840, inc. 1855. Clothing is made.

Ozark National Scenic Riverways, 72,101 acres (29,201 hectares), along the Current and Jacks Fork rivers, SE Mo.; est. 1964 as the first national riverway. Many large caves with interesting dripstone formations are found along the rivers. Forests cover about 75% of the riverways. Wildlife and fish are abundant in the area.

O·zarks, Lake of the (ō'zärks'), man-made lake, 93 sq mi (240.9 sq km), c.130 mi (210 km) long, central Mo., largest reservoir in the state, formed by the impounding of the Osage River by Bagnell Dam.

Ozarks, the, or **Ozark Plateau,** upland region, actually a dissected plateau, c.50,000 sq mi (129,500 sq km), chiefly in S Mo. and N Ark., but partly in Okla. and Kansas, between the Arkansas and Missouri rivers. The Ozarks, which rise from the surrounding plains, are locally referred to as mountains. Composed of igneous rock overlain by limestone and dolomite, the ancient land form has been worn down by erosion. Summits (knobs) are found wherever there is a resistant rock outcrop; the Boston Mts. are the highest and most rugged section, with several peaks more than 2,000 ft (610 m) high. The Ozarks are rich in lead and zinc, and there are good fruit-growing areas. Subsistence farming and household crafts are found in the more isolated regions. The scenic Ozarks, with forests, streams, and mineral springs, are a popular tourist region.

O·zau·kee (ō-zô'kē), county (1970 pop. 54,461), 235 sq mi (608.7 sq km), E Wis., bounded in the E by Lake Michigan and drained by the Milwaukee River; formed 1853; co. seat Port Washington. It is in a dairying, stock-raising, and truck-farming area. It has fisheries, and its industries include food-processing plants and iron and aluminum foundries. Textiles, paper products, chemicals, plastics, leather goods, and machinery and electrical equipment are manufactured.

Ózd (ōzd), city (1976 est. pop. 40,000), NE Hungary, near the Czechoslovak border. It is an industrial center with ironworks, steelworks, and lignite mines.

O·zo·na (ō-zō'nə), town (1970 pop. 2,864), seat of Crockett co., W Texas, SW of San Angelo, in a ranch and oil area.

Paarl (pärl), town (1970 pop. 48,585), Cape Prov., S South Africa, on the Berg River. It is the center of South Africa's wine industry and of a tobacco-growing region. Canned foods, textiles, and cigarettes are important products. Paarl was founded in 1687.

Pa·bia·ni·ce (päb'yä-nē'tsĕ), city (1975 est. pop. 66,800), central Poland. A textile center, it has industries producing chemicals, machine tools, and electric bulbs. Founded in the 13th cent., the city passed to Russia in 1815. It reverted to Poland in 1919.

Pa·cha·ca·mac (pə-chä'kə-mäk'), ruins of a walled Indian city, Peru, SE of Lima. Built before the time of the Incas in one of the irrigable valleys of the coastal desert, Pachacamac is noted for its great pyramidal temple and for the remains of polychrome frescoes adorning its adobe walls.

Pa·chu·ca de So·to (pä-chōō'kə də-sō'tō), city (1970 pop. 84,543), capital of Hidalgo state, central Mexico, at the head of a ravine surrounded by foothills of the Sierra Madre Oriental. Pachuca was founded in 1534 on the site of an ancient Toltec city. The region is extremely rich in ore deposits, especially silver, which has been mined since Aztec times.

Pa·cif·ic (pə-sif'ĭk), county (1970 pop. 15,796), 908 sq mi (2,351.7 sq km), extreme SW Wash., bounded in the W by the Pacific Ocean and Willapa Bay; formed 1851; co. seat South Bend. Its agriculture includes fruit and dairy products.

Pacific Grove, residential and resort city (1970 pop. 13,505), Monterey co., W central Calif., on a point where Monterey Bay meets the Pacific Ocean; inc. 1889. Among the natural attractions of the area are the millions of Monarch butterflies that arrive each fall to spend the winter.

Pacific Islands, Trust Territory of the (1973 pop. 114,773), consisting of the Caroline Islands, the Marshall Islands, and the Marianas Islands, held by the United States under United Nations trusteeship. The territory, covering a vast area of the Pacific, includes more than 2,000 islands and islets. The combined land area is 717 sq mi (1,857 sq km). The islands were acquired by Germany but were seized by

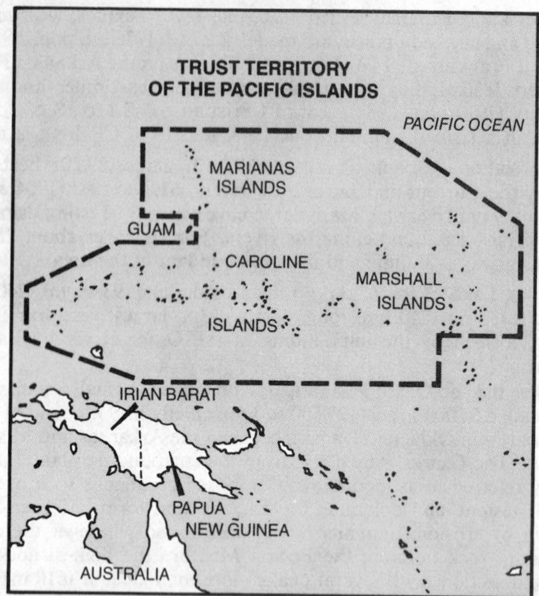

Japan in 1914. In 1922 they were mandated to Japan by the League of Nations. During World War II the islands were occupied (1944) by U.S. forces and administered by the naval government on Guam. In 1947 the United Nations approved U.S. trusteeship of the islands under the administration of the U.S. Dept. of the Interior.

Pacific Ocean, largest and deepest ocean, c.70,000,000 sq mi (181,300,000 sq km), occupying about one third of the earth's surface; named by the Spanish explorer, Ferdinand Magellan; the southern part is also known as the South Sea. It extends from the arctic to antarctic regions between North and South America on the east and Asia and Australia on the west. It is connected with the Arctic Ocean by the Bering Strait; with the Atlantic Ocean by the Drake Passage, Straits of Magellan, and the Panama Canal; and with the Indian Ocean by passages in the Malay Archipelago and between Australia and Antarctica. Its maximum length is c.9,000 mi (14,500 km), and its greatest width c.11,000 mi (17,700 km), between the Isthmus of Panama and the Malay Peninsula. The principal arms of the Pacific Ocean are (in the north) the Bering Sea; (in the east) the Gulf of California; (in the south) Ross Sea; and (in the west) the Sea of Okhotsk, the Sea of Japan, and the Yellow, East China, South China, Philippine, Coral, and Tasman seas.

Along the eastern Pacific shore the coast generally rises abruptly from a deep sea floor to mountain heights on land, and there is a narrow continental shelf. The Asian coast is generally low and indented and is fringed with islands rising from a wide continental shelf. A series of volcanoes, the Circum-Pacific Ring of Fire, rims the Pacific basin. The approximately 20,000 islands in the Pacific Ocean are concentrated in the south and west. Most of the larger islands are structurally part of the continent and rise from the continental shelf. Scattered around the Pacific and rising from the ocean floor are high volcanic islands (such as the Hawaiian Islands) and low coral islands (such as those of Oceania). The floor of the Pacific Ocean, which has an average depth of c.14,000 ft (4,300 m), is largely a deep sea plain. The greatest known depth (36,198 ft/11,040 m) is in the Challenger Deep in the Marianas Trench. Huge whirls, formed by the major ocean currents, are found roughly north and south of the equator; the Equatorial Counter Current separates them.

The Pacific islands of the south and west were populated by Asian migrants who crossed long distances of open sea in primitive boats. European travelers including Marco Polo had reported an ocean off Asia, and in the late 15th cent. trading ships had sailed around Africa to the western rim of the Pacific, but recognition of the Pacific as distinct from the Atlantic Ocean dates from Balboa's sighting of its eastern shore (1513). Magellan's crossing of the Philippines (1520–21) initiated a series of explorations, including those of Drake, Tasman, Dampier, Cook, Bering, and Vancouver, which by the end of the 18th cent. had disclosed the coastline and the major islands. In the 16th cent. supremacy in the Pacific area was shared by Spain and Portugal. The English and the Dutch established footholds in the 17th cent., France and Russia in the 18th, and Germany, Japan, and the United States in the 19th.

Pa·dang (pä-däng'), town (1971 pop. 196,339), capital of West Sumatra prov., on W Sumatra, Indonesia, on the Indian Ocean at the mouth of the small Padang River. An important port, it has a large trade in coffee, copra, rubber, spices, tobacco, and cement.

Pa·der·born (pä'dər-bôrn'), city (1975 pop. 103,230), North Rhine-Westphalia, N central West Germany. It is an agricultural market and industrial center; manufactures include chemicals, machinery, and textiles. Paderborn was made (805) an episcopal see by Charlemagne. The city grew rapidly in the 11th cent. and in the 13th cent. joined the Hanseatic League.

Pa·dre Island (pä'drā, -drē), low, sandy island, c.115 mi (185 km) long, less than 3 mi (4.8 km) wide, S Texas. It is the longest barrier beach in the United States and is characterized by large, irregular sand dunes, sparse vegetation, and a strong prevailing wind off the gulf. Padre Island was discovered and charted in 1519. It became infamous as a ship's graveyard; during a hurricane in 1553, most of a Spanish treasure fleet of 20 ships broke up on the island. Padre Island National Seashore is located in the undeveloped central part of the island, where more than 350 kinds of birds, and many small animals, reptiles, and varied marine life are found.

Pad·u·a (păj'ōō-ə, păd'yōō-ə), city (1975 est. pop. 242,186), capital of Padova prov., in Venetia, NE Italy. It is an agricultural, commercial, and industrial center and a rail junction. Manufactures include machinery, motor vehicles, leather goods, and processed food. The city was destroyed by the Lombards in A.D. 601 but recovered quickly. Padua was from the 12th to the 14th cent. a free commune of great political and economic importance. It subdued neighboring cities and became an artistic center, where Giotto painted his masterpiece, a series of frescoes (1304–6) in the Capella degli Scrovegni. Under the rule of the munificent Carrara family (1318–1405) and under the domination of Venice (1405–1797), Padua continued to flourish. Mantegna (1431–1506), a native of Padua, produced much work here; parts of frescoes executed by him are preserved in the 13th cent. Eremitani church. The Univ. of Padua, the oldest in Italy after that of Bologna, was founded in 1222 by teachers and students who had fled from Bologna. Galileo taught (1592–1610) at the university, and Dante, Petrarch, and Tasso were students here.

Pa·du·cah (pə-dōō'kə, -dyōō'-). **1.** City (1978 est. pop. 30,100), seat of McCracken co., SW Ky., on the Ohio River at the mouth of the Tennessee River; inc. as a city 1856. It is an important tobacco market, a farm trade and shipping point, and a river port. It also has railroad shops, boat yards, and a shoe factory. The city suffered serious floods in 1884, 1913, and 1937. The city hall was designed by Edward Durrell Stone. **2.** City (1970 pop. 2,052), seat of Cottle co., NW Texas, S of Childress, in a livestock and cotton area; settled 1885, inc. 1910.

Paes·tum (pĕs'təm), ancient city of Lucania, S Italy. It was a colony of the Greek city of Sybaris (c.600 B.C.). It flourished with the rest of Magna Graecia through the 6th cent. B.C. The Romans took the city in 273 B.C. The ruins include some of the finest and best-preserved Doric temples in existence.

Pa·gan (pə-gän'), ruined city, central Burma, on the Irrawaddy River. It is one of the great archaeological treasures of Southeast Asia and a holy place of pilgrimage. Founded c.849, it has thousands of Buddhist shrines and temples, principally in stone and brick. Occupied by the Mongols in 1287, Pagan was sacked and burned by the Shans in 1299.

Page (pāj). **1.** County (1970 pop. 18,537), 535 sq mi (1,385.7 sq km), SW Iowa, drained by the Nodaway, East Nodaway, Tarkio, and East Nishnabotna rivers; formed 1847; co. seat Clarinda. Its prairie agriculture includes corn, hogs, cattle, and poultry. Bituminous-coal deposits are found in the east. **2.** County (1970 pop. 16,581), 316 sq mi (818.5 sq km), N Va., in the Shenandoah Valley, with Massanutten Mt. in the W and the Blue Ridge in the E, drained by the South Fork of the Shenandoah River; formed 1831; co. seat Luray. It is in a fertile agricultural area yielding grain, hay, fruit, poultry, livestock, and dairy products. Its industries includes food processing and lumbering.

Pa·go Pa·go (päng'ō päng'ō), village (1970 pop. 2,451) and capital of American Samoa, on Tutuila island. Pago Pago has an excellent landlocked harbor. From 1878 to 1951 it was a coaling and repair station for the U.S. navy.

Pa·go·sa Springs (pə-gō'sə), town (1970 pop. 1,360), seat of Archuleta co., SW Colo., E of Durango; platted 1880, inc. 1891.

Päi·jän·ne (pä'yăn'ä), lake c.560 sq mi (1,450 sq km), S central Finland. One of the largest lakes of the Finnish plateau, it consists of several long rocky basins dotted with numerous islands.

Paines·ville (pānz'vĭl'), city (1970 pop. 16,536), seat of Lake co., NE Ohio, on the Grand River, in a farm area; laid out c.1805, inc. as a city 1902. It has railroad shops and plants that manufacture chemicals and machinery.

Paint·ed Desert (pān'tĭd), badlands on the NE bank of the Little Colorado River, NE Ariz., stretching c.200 mi (320 km) SE from the Grand Canyon. Striking bands of color result from irregularly eroded layers of red and yellow sediment and bentonite clay.

Paint Rock, town (1970 pop. 193), seat of Concho co., W central Texas, on the Concho River E of San Angelo. It is a wool and mohair market in a sheep, goat, and cattle ranching region. Prehistoric rock paintings and good hunting and fishing areas are nearby.

Paints·ville (pānts'vĭl'), city (1970 pop. 3,868), seat of Johnson co., E Ky., on Paint Creek and SSW of Ashland, in a coal, timber, and livestock area.

Pais·ley (pāz'lē), burgh (1974 est. pop. 94,025), Strathclyde region, W Scotland. It has a thriving textile industry and manufactures boilers, chemicals, and soap. Patterned Paisley shawls were famous in the 19th cent.

Pak·i·stan (păk'ĭ-stăn', pä'kĭ-stän'), republic (1976 est. pop. 72,370,000), 310,403 sq mi (803,944 sq km), S Asia. Islamabad is the capital. Pakistan is bordered by India on the east, the Arabian Sea on the south, Iran on the southwest, and Afghanistan on the west and north; in the northeast is the disputed territory of Jammu and Kashmir, of which the part occupied by Pakistan borders on China. Pakistan is composed of four provinces—Baluchistan, North-West Frontier Prov., Punjab, and Sind, all of which closely coincide with the historic regions—and a federal district that is the site of the capital. The country has a generally hot and dry climate, with desert conditions prevailing throughout most of the area. The Indus River is the nation's lifeline. It flows the length of the country and is fed by the combined waters of the five rivers of Punjab.

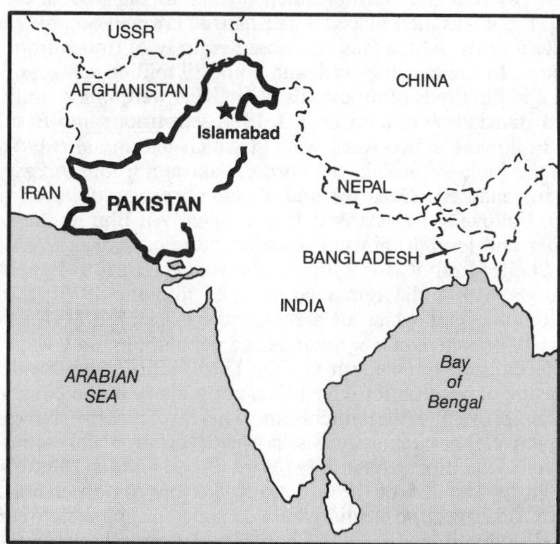

The plateau region of western Pakistan, roughly coextensive with Baluchistan prov., is an arid region with relatively wetter conditions in its northern sections. Numerous low mountain ranges rise from the plateau. Large portions of the region are unfit for agriculture, and although some cotton is raised, nomadic sheep grazing is the principal activity. Coal, chromite, and natural gas are found in this area, and fishing and salt trading are carried on along the rugged Makran coast. Quetta, the chief city, is an important railroad center on the line between Afghanistan and the Indus valley. East of the plateau region are the extensive alluvial plains of eastern Pakistan, through which flow the Indus and its tributaries. The region, closely coinciding with Sind and Punjab provs., is hot and dry and is occupied in its eastern part by the Thar desert. A variety of crops (especially wheat, rice, and cotton) are raised here. The Indus delta, however, is not fertile and supports little agriculture. The irrigated portions of the plain are densely populated. The higher parts of the plain, in the north, have a favorable subtropical climate and have become Pakistan's resort center. In northwestern Pakistan is a region of low hills and plateaus interspersed with fertile valleys. The elevation of the region tempers the arid climate. It is a predominantly agricultural area, with wheat the chief crop; fruit trees and livestock are also raised. Oil and coal are the chief minerals. Peshawar and Rawalpindi, the largest cities of this area, are major manufacturing centers. In the northern section of the North-West Frontier Prov. and in the Pakistani-occupied sector of Kashmir are the rugged ranges and the high, snow-capped mountains of the Hindu Kush, Himalaya, and Karakorum mts.

The people of Pakistan are a mixture of many ethnic groups, a result of the occupation of the region by groups passing through on their way to India. The Pathan tribesmen of the North-West Frontier Prov. are a large, indigenous group that has long resisted advances by invaders and would like to establish an autonomous state within Pakistan. Baluchi tribesmen have also pressed for the creation of a state that would incorporate parts of Afghanistan and Iran. Despite improved health standards, sickness, disease, and malnutrition are still prevalent. A widespread effort to educate the population has raised the literacy rate to more than 15%.

Economy. Agriculture is the mainstay of Pakistan's economy, employing some 70% of the population. Most of Pakistan's agricultural output comes from the Indus basin. Wheat, rice, cotton, sugar cane, and tobacco are the chief crops. The country is not self-sufficient in food, even though vast irrigation schemes have extended farming into arid areas, and fertilizers and new varieties of crops have increased yields. The rapidly increasing industrial base is able to supply most of the country's needs in consumer goods. Pakistan's natural resources provide materials for such industries as textile production (the biggest earner of foreign exchange), oil refining, metal processing, and cement and fertilizer production. Since the mid-1950s electric power output has greatly increased through the development of hydroelectric power potential and the use of nuclear power plants. Pakistan's chief imports are machinery, food, manufactured goods, and mineral fuels.

History. The northwest of the Indian subcontinent, which now constitutes Pakistan, lies athwart the historic invasion routes through the Khyber, Gumal, and Bolan passes from central Asia to India, and for thousands of years invaders and adventurers swept down upon the settlements there. The earliest-known culture was the prehistoric Indus Valley civilization, which flourished until overrun by Aryan invaders c.1500 B.C. The Aryans were followed by the Persians of the Achaemenid empire, who by c.500 B.C. reached the Indus River. Alexander the Great, conqueror of the Persian empire, invaded the Punjab in 326 B.C. The Seleucid empire, heir to Alexander's Indian conquest, was checked by the Mauryas, who by 305 B.C. occupied the Indus plain and much of Afghanistan. After the fall of the Mauryas (2nd cent. B.C.) the Indo-Greek Bactrian kingdom rose to power, but was in turn overrun (c.97 B.C.) by Scythian nomads called Saka and then by the Parthians (A.D. c.7). The Parthians, of Persian stock, were replaced by the Kushans; the Kushan Kanishka ruled (2nd cent. A.D.) all of what is now Pakistan from his capital at Peshawar. In 712 the Moslem Arabs appeared in force and conquered Sind, and by 900 they controlled most of northwest India. They were followed by the Ghaznavid and Ghorid Turks. The first Turki invaders reached Bengal c.1200 and an important Moslem center was established there, principally through conversion of the Hindus.

Although the northeast of the Indian subcontinent (now Bangladesh) remained, with interruptions, part of a united Mogul empire in India from the early 16th cent. to 1857, the northwest changed hands many times before it became (1857) part of imperial British India. It was overrun by Persians in the late 1730s; by the Afghans, who held Sind and the Punjab during the latter half of the 18th cent.; and by the Sikhs, who rose to power in the Punjab under Ranjit Singh (1780–1839). The British attempted to subdue the anarchic northwest during the First Afghan War (1839–42) and succeeded in conquering Sind in 1843 and the Punjab in 1849. The turbulence of the region was intensified by the fierce forays of Baluchi and Pathan tribesmen from the mountainous hinterlands. The British occupied Quetta in 1876 and again attempted to conquer the tribesmen in the Second Afghan War (1878–80) but were still unsuccessful. With the creation of the North-West Frontier Prov. in 1901 the British shifted from a policy of conquest to one of containment.

Unlike previous settlers in India, the Moslem immigrants were not absorbed into Hindu society. Their ranks were augmented by the millions of Hindus who had been converted to Islam during the declining years of the Mogul empire; there was cultural interchange between Hindu and Moslem, but no homogeneity emerged. After the Indian Mutiny (1857), a rising Hindu middle class began to assume dominant positions in industry, education, the professions, and the civil service. Although, in these early decades of the Indian National Congress, vigorous efforts were made to include Moslems in the nationalist movement, concern for Moslem political rights led to the formation of the Moslem League in 1906; in the ensuing years Hindu-Moslem conflict became increasingly acute. The idea of a Moslem nation, distinct from Hindu India, was introduced in 1930 by the poet Mohammed Iqbal and was ardently supported by a group of Indian Moslem students in England, who were the first to use the name Pakistan. It gained wide support in 1940 when the Moslem League, led by Mohammed Ali Jinnah, demanded the establishment of a Moslem state in the areas of India where Moslems were in the majority. The league received nearly all of the Moslem vote in the 1946 elections, and Britain, faced with a united Moslem voice, reluctantly agreed to the formation of Pakistan as a separate dominion under the provisions of the Indian Independence Act, which went into effect on Aug. 15, 1947.

Jinnah became the governor-general of the new nation and Liaquat Ali Khan the first prime minister. While India inherited most of the British administrative machinery, Pakistan had to start with practically nothing; records and Moslem administrators were transferred from New Delhi to a chaotic, makeshift capital at Karachi. Disturbances in Delhi between the Hindus and the Moslems were only a prelude to the slaughter in the Punjab, where the Gurdaspur district had been partitioned to give India access to Kashmir. Although there was some violence in Calcutta, the efforts of Mohandas K. Gandhi prevented widespread killing in partitioned Bengal. The communal strife took more than a million lives; 7.5 million Moslem refugees fled to both parts of Pakistan from India, and 10 million Hindus left Pakistan for India. In the princely states of Junagadh, and Hyderabad, Moslem rulers held sway over a Hindu majority but India forcibly joined both states to the Union, dismissing the wishes of the rulers and basing its claims instead on the wishes of the people and the facts of geography. In Kashmir the situation was precisely the opposite; a Hindu ruler held sway over a Moslem majority in a country that was geographically and economically tied to West Pakistan. The ruler signed over Kashmir to India in Oct., 1947, but Pakistan refused to accept the move. Fighting broke out and continued until Jan., 1948, when India and Pakistan both appealed to the United Nations, each accusing the other of aggression. A cease-fire was agreed upon and a temporary demarcation line partitioned (1949) the disputed state.

In the meantime Pakistan faced serious internal problems. A liberal statement of constitutional principles was promulgated in 1949, but parts of the proposed constitution ran into orthodox Moslem opposition. On Oct. 16, 1951, Prime Minister Liaquat Ali Khan was assassinated by an Afghan fanatic. His death left a leadership void that his successors failed to fill. In East Bengal, which had over half of the nation's population, there was increasing dissatisfaction with the federal government in West Pakistan. In 1954, faced with growing crises, the government dissolved the constituent assembly and declared a state of emergency. In 1955 the existing provinces and princely states of West Pakistan were merged into a single province made up of 12 divisions, and the name of East Bengal was changed to East Pakistan, thus giving it at least the appearance of parity with West Pakistan. In Feb., 1956, a new constitution was finally adopted, and Pakistan formally became a republic within the British Commonwealth; Gen. Iskander Mirza became the first president. Economic conditions remained precarious, even though large shipments of grain from the United States after 1953 had helped to relieve famine. In foreign relations, Pakistan's conflict with India over Kashmir remained unresolved, and Afghanistan continued its agitation for the formation of an autonomous Pushtunistan nation made up of the Pathan tribesmen along the northwest frontier.

After 1956 the threat to the stability of the Pakistan government gradually increased, stemming from continuing economic difficulties, frequent cabinet crises, and widespread political corruption. In Oct., 1958, President Mirza abrogated the constitution and granted power to the army under Gen. Mohammed Ayub Khan. Under his dictatorship a vigorous land reform and economic development program was begun, and a new constitution, which provided for a federal Islamic republic with two provinces (East and West Pakistan) and two official languages (Bengali and Urdu), went into effect in 1962. The new city of Islamabad, north of Rawalpindi, which had been interim capital since 1959, became the national capital, and Dacca, in East Pakistan, became the legislative capital. Communal strife was constantly present in the subcontinent. Diplomatic relations between Pakistan and Afghanistan were severed (1961-63) after some border clashes and continued Afghan agitation, supported by the USSR, for an independent Pushtunistan. A series of conferences on Kashmir was held (Dec., 1962-Feb., 1963) between India and Pakistan following the Chinese assault (Oct., 1962) on India; both nations offered important concessions and solution of the long-standing dispute seemed imminent. Pakistan signed a bilateral border agreement with China that involved the boundaries of the disputed state, and relations with India again became strained. Pakistan's continuing conflict with India over Kashmir erupted in fighting (Apr.-June, 1965) in northwest India and southeastern West Pakistan and in an outbreak of warfare (Aug.-Sept.) in Kashmir. In 1966 President Ayub Khan and Prime Minister Lal Bahadur Shastri of India reached an accord in the Declaration of Tashkent at a meeting sponsored by the USSR. However, the basic dispute over Kashmir remained unsettled. In an effort to gain support in the conflict with India, Pakistan somewhat modified its pro-Western policy after 1963.

East Pakistan's long-standing discontent with the federal government was expressed in 1966 by a movement for increased autonomy, supported by a general strike. Following disastrous riots in late 1968 and early 1969, Ayub resigned and handed the government over to Gen. Agha Mohammed Yahya Khan, the head of the army. The first direct universal voting since independence was held in Dec., 1970, to elect a National Assembly that would draft a new constitution and restore federal parliamentary government. The Awami League, under sheik Mujibur Rahman, in a campaign for full autonomy in East Pakistan, won an overwhelming majority in the national assembly by taking 153 of the 163 seats allotted to East Pakistan. The opening session of the National Assembly, scheduled to meet in Dacca in Mar., 1971, was twice postponed by Yahya, who then canceled the election results, banned the Awami League, and imprisoned Mujib in West Pakistan on charges of treason.

East Pakistan declared its independence as Bangladesh on Mar. 26, 1971, but was then placed under martial law and occupied by the Pakistani army, which was composed entirely of troops from West Pakistan. In the ensuing civil war, some 10 million refugees fled to India and hundreds of thousands of civilians were killed. India supported Bangladesh and on Dec. 3, 1971, sent troops into East Pakistan. Following a two-week war, in which fighting also broke out along the India-West Pakistan border, Pakistani troops in East Pakistan surrendered (Dec. 16) and a cease-fire was declared on all fronts. Following Pakistan's defeat, Zulfikar Ali Bhutto, the deputy premier and foreign minister, took absolute control in West Pakistan. Sheik Mujib was eventually allowed to return to Bangladesh. Relations with India remained strained into the 1970s; they improved somewhat when an agreement was reached (1972) on exchange of prisoners of war and hostage populations, and when Pakistan recognized Bangladesh (1974). The Bhutto government faced increasing opposition into the 1970s, particularly in the North-West Frontier prov. and in Baluchistan, where secessionist movements were active. The greatest crisis, however, occurred following elections for the National Assembly (Mar., 1977) in which Bhutto's People's Party won 155 of the 210 seats. The opposition claimed that Bhutto had rigged the election and demanded his immediate resignation. He refused, and over the next weeks rioting broke out across Pakistan. By June, more than 300 were dead and many opposition leaders were under arrest. In an attempt to quell the disturbances, Bhutto offered to hold a referendum on his leadership.

Government. Pakistan is governed by the constitution of Apr., 1973, which provides for a federal parliamentary form of government. There is a bicameral Parliament whose 210 members in the national assembly and 63 members in the senate (14 from each province, 5 from federally administered tribal areas, and 2 from the federal capital) are elected by popular vote. The president is the head of state, and the prime minister, in whom most power is vested, is the chief executive. There is an independent judiciary branch.

The government and the opposition leaders reached an agreement on July 2 that provided for the dissolution of the assemblies and the scheduling of new elections. This agreement was repudiated, and the army led a coup on July 5 that resulted in the arrest of many politicians, including prime minister Bhutto. Martial law was imposed, and power was assumed by Gen. Mohammad Zia ul-Hag. New elections were announced but were postponed until the outcome of several political trials. Criminal charges were lodged against former president Bhutto, and amid worldwide protest and consternation he was hanged on Apr. 4, 1979.

Pa•kok•ku (pə-kō'ko͞o), town (1962 est. pop. 157,000), central Burma, a port on the Irrawaddy River. It is a trading and shipping center.

Pa•lat•i•nate (pə-lăt'ə-nāt'), two regions of West Germany. They are related historically, but not geographically.

The Rhenish or Lower Palatinate is a district (c.2,100 sq mi/5,440 sq km) of the state of Rhineland-Palatinate. The Rhenish Palatinate extends from the left bank of the Rhine River and borders in the south on France and in the west on Saarland and Luxembourg. Neustadt an der Weinstrasse is the capital. It is a rich agricultural region, famed for its wines.

The Upper Palatinate is a district (c.3,725 sq mi/9,650 sq km) of northeastern Bavaria, separated in the east from Czechoslovakia by the Bohemian Forest. Regensburg is the capital. Agriculture and cattle raising are the chief occupations.

The name of the two regions came from the office known as count palatine, a title used by high-ranking nobles with judicial functions in the Roman, Byzantine, and Holy Roman empires. Emperor Frederick I bestowed (1156) the title count palatine on his half-brother Conrad, who was in possession of territories on both sides of the Rhine. When Conrad's line died out, the Palatinate passed (1214) to the Bavarian Wittelsbach dynasty. The Wittelsbachs enlarged their holdings along the Bohemian border, which were constituted as the Upper Palatinate. After 1356 the Wittelsbach counts in possession of the two palatinates were designated electors palatine, and their territories were called the Electoral Palatinate. The direct line was succeeded to (1559) by the counts palatine of Simmern, who in turn were succeeded by the dukes palatine of Neuburg (1685) and the counts palatine of Sulzbach (1742). The Rhenish Palatinate flourished in the 15th and 16th cent., and its capital, Heidelberg, was a center of the German Renaissance and Reformation.

After the territorial reshufflings of the French Revolutionary and Napoleonic wars (1789–1814), part of the Rhenish Palatinate had been dividing among neighboring states and the rest, along with the Upper Palatinate, was incorporated into an enlarged Bavaria. After World War II the Rhenish Palatinate became (1946) a district of the newly created state of Rhineland-Palatinate.

Pal•a•tine (păl'ə-tīn), one of the Seven Hills of ancient Rome.

Pa•lat•ka (pə-lăt'kə), city (1975 est. pop. 8,800), seat of Putnam co., NE Fla., on the St. Johns River S of Jacksonville: inc. 1872. Founded as a trading post in 1821, it was destroyed in the Seminole War and was later the scene of several Civil War skirmishes. Palatka is a commercial center and river port manufacturing paper products and furniture. The city is noted for its azalea gardens.

Pa•lau (pä-lou') or **Pe•lew** (pə-lo͞o'), island group (1970 pop. 11,210), c.192 sq mi (497 sq km), W Pacific, in the W Caroline Islands. Palau is one of six administrative districts of the U.S. Trust Territory of the Pacific Islands. Palau consists of about 200 islands and islets, of which Babelthuap, Koror (the administrative center of the group), Arakabesan, and Malakal are the most important. Trochus shells and scrap metal are the only notable exports. Spain held the islands for about 300 years before selling them to Germany in 1899. Japan seized them in 1914 and was given a mandate over them by the League of Nations in 1920. A major Japanese naval base in World War II, Palau was seized by U.S. forces in 1944.

Pa•la•wan (pä-lä'wän), island (1970 pop. 232,322), 4,550 sq mi (11,785 sq km), fifth-largest of the Philippines, N of Borneo and between the Sulu Archipelago and the South China Sea. Lumbering is an important industry on Palawan; it is a leading wood-products manufacturing area.

Pa•lem•bang (pä'lĕm-bäng'), city (1971 pop. 582,961), capital of South Sumatra prov., on SE Sumatra, Indonesia. It is a deepwater port on both banks of the Musi River and the trade and shipping center for nearby oil fields. There are large oil refineries, rubber plants, textile mills, fertilizer factories, and food-processing plants. Palembang in the 8th cent. was the capital of the powerful Hindu-Sumatran kingdom of Sri Vijaya. The Dutch began trading here in 1617, and later it was intermittently under British rule.

Pa•len•cia (pə-lĕn'chə, -chē-ə, pä-lān'*th*yä), city (1975 est. pop. 62,186), capital of Palencia prov., N central Spain, in León. An industrial center with iron foundries, textile mills, and chemical plants, it was formerly noted for its woolen industry. Palencia was occupied by the Romans and sacked (6th cent.) by the Visigoths. It was recovered from the Moors in the 10th cent. and was in the 12th and 13th cent. a favorite residence of the kings of León. The first university in Spain was founded here (1212 or 1214) but was removed to Salamanca in 1238. There is a notable Gothic cathedral (14th–16th cent.) containing a fine collection of old Flemish tapestries and paintings by El Greco.

Pa•len•que (pə-lĕng'kā, pä-lāng'kĕ), ancient city of the Maya in Chiapas, S Mexico, in the Usumacinta valley. Its architectural elegance was a high point in the art of the Classic period. The Temple of Inscriptions, noted for its hieroglyphic tablets, is one of the best-preserved Mayan temples.

Pa•ler•mo (pə-lûr'mō), city (1975 est. pop. 673,163), capital of Palermo prov. and of Sicily, NW Sicily, Italy, on the Tyrrhenian Sea. Situated on the edge of the Conca d'Oro a beautiful and fertile plain, it is Sicily's chief seaport. Manufactures include textiles, food products, chemicals, and cement. There are also shipyards in the city. An ancient Phoenician community founded between the 8th and 6th cent. B.C., it later became a Carthaginian military base and was conquered by the Romans in 254 B.C.–253 B.C. Palermo was under Byzantine rule from A.D. 535 to A.D. 831, when it fell to the Arabs, who held it until 1072. The city's prosperity dates from the Arab domination and continued when, under the Normans, it served (1072–1194) as the capital of the kingdom of Sicily. Under king Roger II (1130–54) and later under emperor Frederick II (1220–50), Palermo attained its main artistic, cultural, and commercial flowering. The French Angevin dynasty transferred the capital to Naples; its misrule led to the Sicilian Vespers insurrection (1282), which began in Palermo. The city is rich in works of art; Byzantine, Arab, and Norman influence are blended in many buildings. Points of interest include the Arab-Norman Palatine Chapel (1130–40) in the large palace of the Normans (today also the seat of the Sicilian parliament); the cathedral (founded in the late 12th cent.), which contains the tombs of Frederick II and other rulers; the Church of St. John of the Hermits (1132); the Palazzo Abbatellis (15th cent.), which houses the National Gallery of Sicily; the Gothic Palazzo Chiaramonte (1307); and the Capuchin catacombs. The city has a university.

Pal•es•tine (păl'ə-stīn'), historic region on the E shore of the Mediterranean Sea, comprising parts of modern Israel, Jordan, and Egypt; also known as the Holy Land.

Palestine's boundaries, never constant, always included at least the land between the Mediterranean and the Jordan River. So defined, the region is c.140 mi (225 km) long and c.30 to c.70 mi (50–115 km) wide. Outside these bounds were such biblical lands as Edom, Gilead, Moab, and Hauran. The British mandate of Palestine (1920–48) included also the Negev, a c.100-mile-long (160-km) desert stretching south to the Gulf of Aqaba. From east to west, Palestine proper comprises three geographic zones: the depression—northernmost extension of the Great Rift Valley—in which lie the Jordan River, Lake Hula, the Sea of Galilee (Lake Tiberias), the Dead Sea, and the Arabah, a dry valley south of the Dead Sea; a ridge rising steeply to the west of this cleft; and a coastal plain c.12 mi (20 km) wide. In northern Palestine the ridge is interrupted by the Plain of Esdraelon (Jezreel) and the connecting valley of Bet Shean (Beisan), the most fertile part of the region. The highland area to the north is called Galilee, its chief centers being Zefat and Nazareth, near which rises Mt. Tabor. To the south of the Plain of Esdraelon the broad ridge stretches unbroken to the Negev. First there are the hills of Samaria, at the center of which is Nablus; it lies between Mt. Ebal and Mt. Gerizim. The mountains of Judaea are west of the Dead Sea. In Judaea are Jerusalem, Bethlehem, and Hebron. Well to the south, in the Negev, lies Beersheba. The towns of the coastal plain are Akko (Acre), Haifa, Netanya, and the twin cities of Tel Aviv-Jaffa. To the south is Gaza. The various sections of the plain are named the Valley of Zebulun or Plain of Acre, south of Akko; Sharon, south of Mt. Carmel; and the Shephelah, or Philistia, in the extreme south. Agriculture in the Jordan valley centers around Lake Hula and the Sea of Galilee. The chief town is Tiberias. Farther south the valley is too narrow to be of much use, except for providing water power, and there is only one city, Jericho, east of Jerusalem. The surface—c.1,300 ft (400 m) below sea level—of the Dead Sea, into which the Jordan River empties, is the lowest spot on the earth's surface.

At the start of the Zionist colonization of Palestine in the late 19th cent., the rural people were Arab peasants (fellahin). Most of the population were Moslems, but in the urban areas there were important groups of both Christians and Jews. The Holy Land derives its special character from being a place of pilgrimage. Shrines, shared in common by several religions, cluster most numerously about Jerusalem, Bethlehem, Nazareth, and Hebron.

History. The earliest known inhabitants of Palestine were of the same group as the Neanderthal inhabitants of Europe. By the 4th millennium B.C. they were herders and farmers. During the 2nd millennium, Palestine was ruled by the Hyksos and by the Egyptians. Toward the end of this period Moses led the Hebrew people out of Egypt, across the Sinai, and into Palestine. The Philistines invaded (c.1200 B.C.) the southern coastland and established a powerful kingdom. The Hebrews were subject to the Philistines until c.1000 B.C., when an independent Hebrew kingdom was established under Saul, who was succeeded by David and then by Solomon. After the expansionist reign of Solomon (c.950 B.C.), the kingdom broke up into two states, Israel, with its capital at Samaria, and Judah, under the house of David, with its capital at Jerusalem. The two kingdoms were later conquered by expanding Mesopotamian states, Israel by Assyria (c.720 B.C.) and Judah by Babylonia (586 B.C.). In 539 B.C. the Persians conquered the Babylonians. Under Persian rule Palestine enjoyed considerable autonomy. The Macedonian Greeks (333 B.C.) and the Seleucids then held the region. Jewish revolt under the Maccabees established a new Jewish state in 142 B.C. The state lasted until 63 B.C., when Pompey conquered Palestine for Rome.

When the Jews revolted in A.D. 66, the Romans destroyed the Temple (A.D. 70). When Emperor Constantine converted to Christianity (A.D. 312), Palestine became a center of Christian pilgrimage. It was conquered in 614 by the Persians, was recovered briefly by the Byzantine Romans, but fell to the Moslem Arabs by the year 640. Under the Fatimid caliph al Hakim (996–1021), the Christians and Jews were harshly suppressed, and many churches were destroyed. In 1099 Palestine was captured by the Crusaders who established the Latin Kingdom of Jerusalem. The Crusaders were driven out of Palestine by the Mamelukes in 1291. Under Mameluke rule Palestine declined. In 1516 the Mamelukes were ousted by the Ottoman Turks, whose rule lasted four centuries.

In the late 19th cent. the Zionist movement was founded with the goal of establishing a Jewish homeland in Palestine. Dozens of Zionist colonies were founded there. At the same time Arab nationalism was developing in the Middle East in opposition to Turkish rule. In World War I the British, with Arab aid, gained control of Palestine, which was assigned to the British as a League of Nations mandate in 1922. In the Balfour Declaration (1917) they promised Zionist leaders to aid the establishment of a Jewish "national home" in Palestine, with due regard for the rights of non-Jewish Palestinians. In 1919 there were about 568,000 Moslem Arabs, 74,000 Christians, and 58,000 Jews in Palestine.

The rise of Nazism in Europe during the 1930s led to a great increase in immigration. Arabs, fearful of the consequences of continued Jewish immigration with its attendant land purchases, conducted strikes and boycotts. During World War II the British outraged the Zionists by limiting immigration and land purchases. Secret Jewish terrorist groups became active. Illegal immigration, involving many survivors of Hitler's death camps, took place on a large scale. The British, despairing of finding a solution, turned the Palestine problem over to the United Nations (Feb., 1947). At this time there were about 1,091,000 Moslem Arabs, 614,000 Jews, and 146,000 Christians in Palestine.

Pal·es·tine (păl'ə-stēn'), city (1970 pop. 14,525), seat of Anderson co., E Texas; inc. 1871. It is a market, processing, and rail center for a rich oil area and for the truck crops, livestock, and other produce of the rolling red hills.

Pal·es·tri·na (păl'ə-strē'nə), town (1971 pop. 11,432), in Latium, central Italy. It is an agricultural market located on the site of Praeneste, a town founded by c.800 B.C. and later destroyed and rebuilt by the Romans in the 1st cent. B.C. Of note are the ruins of a temple of Fortuna (8th cent. B.C.), celebrated for its oracles.

Pal·i·sades (păl'ĭ-sādz'), cliffs along the W bank of the Hudson River, NE N.J. and SE N.Y., with a general altitude of from 350 ft to 550 ft (107–168 m). A large part of the most scenic section is embraced in Palisades Interstate Park (c.47,000 acres/19,035 hectares). The park, on the west bank of the river, includes a chain of wooded recreational areas between Fort Lee, N.J., and Newburgh, N.Y.

Palk Strait (pôk, pôlk), 40 to 85 mi (64.4–136.8 km) wide, between India and Sri Lanka. It is studded with shoal reefs called Adam's Bridge and by small islands. The strait's treacherous waters are avoided by most ships.

Pall Mall (pĕl mĕl, păl măl, pôl môl), street in the City of Westminster borough, London, England. St. James's Palace, Marlborough House, and a number of private clubs are on Pall Mall. The name derives from the game pall mall, or paille maille, which was played in front of the palace in the 17th cent.

Pal·ma (päl'mä) or **Palma de Mal·lor·ca** (*th*ā mä-lyôr'kä), city (1975 est. pop. 262,948), capital of Majorca island and of Baleares prov., Spain, on the Bay of Palma. It is the chief port and commercial center of the Balearic Islands. Picturesquely situated along the bay and into the surrounding hills, it is one of Europe's most renowned resorts. Craft industries supplement tourism. Stone Age remains have been found. The imposing Gothic cathedral, founded after James I of Aragón wrested (1229) Palma from the Moors, was finished in the 17th cent. There are several ancient churches and fine private homes.

Pal·mas, Las (läs päl'mäs), city (1975 est. pop. 327,489), capital of Las Palmas prov., Spain, on Grand Canary. Industries include fishing, fish processing, and tourism. The city was founded in 1478; still standing is a house where Columbus stayed in 1492.

Palm Beach (päm), county (1970 pop. 348,993), 2,023 sq mi (5,239.6 sq km), SE coast of Fla., on the Atlantic in the Everglades, including Lake Okeechobee in the NW and crossed by several drainage canals; formed 1909; co. seat West Palm Beach. In a truck-farm, dairy, and poultry area, with citrus groves along the coast, it has fisheries and mines and manufactures food, metal, and wood products, clothing, and concrete. It has a noted resort section.

Palm Beach, town (1978 est. pop. 9,800), Palm Beach co., SE Fla., on a barrier beach between the Atlantic Ocean and Lake Worth (a lagoon); settled c.1870, inc. 1911. It is a well-known resort, with many fine estates and luxurious hotels. After the arrival of Henry M. Flagler in 1893, Palm Beach was rapidly developed.

Pal·mer (pä'mər). **1.** City (1970 pop. 1,140), seat of Matanuska-Susitna borough, S Alaska, NE of Anchorage; inc. 1951. It is the agricultural market of Matanuska Valley. **2.** Town (1970 pop. 11,680), Hampden co., S Mass.; settled 1716, inc. 1775. It is an industrial and trade center.

Pal·mer·ston North (pä'mər-stən), city (1976 pop. 57,931), S North Island, New Zealand. It is a transportation and farm-marketing center with diverse industries.

Pal·met·to (păl-mĕt'ō), city (1970 pop. 7,422), Manatee co., SW Fla., S of Tampa near the entrance to Tampa Bay at the mouth of the Manatee River; settled 1866, inc. 1894.

Pal·mi·ra (päl-mē'rä), city (1973 pop. 140,481), W Colombia, on the Pan American Highway. An agricultural center in the Cauca valley, Palmira gave its name to the tobacco of the region. Sugar cane, coffee, rice, and corn are also grown. The city was founded in 1705.

Palm Springs, city (1978 est. pop. 32,200), Riverside co., S Calif.; founded 1876, inc. 1938. It is a verdant desert oasis and a fashionable resort. It was known to the Spanish as early as 1774 as Agua Caliente because of its hot springs. By 1872 it was a regular stop on the stagecoach run between Prescott, Ariz., and Los Angeles. Nearby are Mt. San Jacinto, with a cable run almost to the top, and Palm Canyon, containing forests of Washingtonia palms estimated to be over 1,000 years old.

Pal·my·ra (păl-mī'rə), ancient city of central Syria, an oasis N of the Syrian Desert and NE of Damascus. Palmyra was important in Syrian-Babylonian trade by the 1st cent. B.C. Tradition says it was founded by Solomon. Palmyra became of true importance only after Roman control was established (A.D. c.30). Local tribes vied for control, which fell to the Septimii by the 3rd cent. A.D. Septimius Odenathus built Palmyra into a strong autonomous state that practically embraced the Eastern Empire, including Syria, northwest Mesopotamia, and western Armenia. After his death the territory was conquered by Aurelian, who partly destroyed (273) the city. Palmyra was taken by the Arabs and sacked by Tamerlane. It fell into ruins, and even the ruins were forgotten until the 17th cent. The temple dedicated to Baal and other remains show the Oriental splendor of Palmyra at its prime.

Palmyra. 1. City (1970 pop. 3,188), seat of Marion co., NE Mo., on the North River near the Mississippi and NW of Hannibal; laid out

1819, inc. 1855. Shoes are made. **2.** Borough (1970 pop. 6,969), Burlington co., SW N.J., on the Delaware River NE of Camden; inc. 1923. **3.** Village (1975 est. pop. 3,672), Wayne co., W central N.Y., on the Barge Canal SE of Rochester; inc. 1819. Joseph Smith lived in Palmyra and published the *Book of Mormon* here. **4.** Borough (1970 pop. 7,615), Lebanon co., SE Pa., ENE of Harrisburg; settled 1749, inc. 1913. There are limestone quarries and shoe factories here. **5.** Village (1970 pop. 200), seat of Fluvanna co., central Va., on the Rivanna River SE of Charlottesville.

Pal·o Al·to (păl'ō ăl'tō), county (1970 pop. 13,289), 560 sq mi (1,450.4 sq km), NW Iowa, drained by the West Des Moines River; formed 1851; co. seat Emmetsburg. It is in a prairie agricultural area (cattle, hogs, poultry, corn, oats, and soybeans), with a resort lake region.

Palo Alto, city (1978 est. pop. 54,600), Santa Clara co., W Calif.; inc. 1894. Although primarily residential, Palo Alto has an electronics industry. A local attraction is "El Palo Alto," a tree that is more than 1,000 years old. Stanford Univ. is nearby.

Pal·o·mar, Mount (păl'ō-mär'), peak, 6,126 ft (1,868.4 m) high, S Calif., NE of San Diego, in Cleveland National Forest. It is the site of Mount Palomar Observatory, operated jointly by the California Institute of Technology and Carnegie-Mellon Univ. It has the world's largest reflecting telescope.

Palo Pin·to (pĭn'tō), county (1970 pop. 28,962), 948 sq mi (2,455.3 sq km), N central Texas, drained by the Brazos River; formed 1856; co. seat Palo Pinto. Its agriculture includes peanuts, corn, grain sorghums, wheat, fruit, truck crops, livestock, poultry, and dairy products. There are also oil and natural-gas wells.

Palo Pinto, village (1970 pop. 525), seat of Palo Pinto co., N central Texas, W of Fort Worth, in a farm and ranch area.

Pa·los de la Fron·te·ra (pä'lôs dā lä frôn-tā'rä), town (1970 pop. 4,390), Huelva prov., SW Spain, in Andalusia, on the Tinto River near its mouth. From its port (now silted up), Columbus sailed on his first voyage of discovery (1492).

Pa·louse (pə-lo͞os'), river, c.140 mi (225 km) long, rising in NW Idaho and flowing W into Wash., where it turns S to enter the Snake River NW of Dayton.

Pa·mir (pä-mîr') or **Pa·mirs** (pä-mîrz'), mountainous region of central Asia, located mainly in the Tadzhik SSR and extending into NE Afghanistan and SW Sinkiang, China. Many peaks rise to more than 20,000 ft (6,100 m); Mount Communism (24,590 ft/7,480 m) and Lenin Peak (23,508 ft/7,170 m) are the highest peaks in the Pamir. The region forms a geologic structural knot from which the great Tien Shan, Karakorum, Kunlun, and Hindu Kush mountain systems radiate. Snow-capped throughout the year, the Pamir experiences long cold winters and cool summers. Coal is mined in the west, but nomadic sheep herding in the upland meadows is the main economic activity. Terak Pass, used by Marco Polo on his way to China in 1271, is one of several high passes through the Pamir.

Pam·li·co (păm'lə-kō'), county (1970 pop. 9,467), 338 sq mi (875.4 sq km), E N.C., in a forested and swampy tidewater area bounded in the E by Pamlico Sound, in the S by the Neuse River; formed 1872; co. seat Bayboro. Farming (potatoes, corn, soybeans, cabbage, and tobacco), fishing, and lumbering are important. It has many resort and retirement communities.

Pamlico Sound, lagoon, 80 mi (128.7 km) long and 15 to 30 mi (24–48 km) wide, E N.C., separated from the Atlantic Ocean by a row of low, sandy barrier islands. It is linked on the north with Albemarle Sound. Cape Hatteras National Seashore is located on the barrier islands. Fish, oysters, and waterfowl abound.

Pam·pa (păm'pə), city (1978 est. pop. 20,000), seat of Gray co., extreme N Texas. This cow town on the Panhandle ships cattle and wheat and packs meat, but discovery of oil and gas has also made it an industrial center with refineries, carbon-black plants, and other oil-based industries.

pam·pas (păm'pəz, päm'päs), wide, flat, grassy plains of temperate S South America, c.300,000 sq mi (777,000 sq km), particularly in Argentina and extending into Uruguay. The Pampa of central and northern Argentina occupies c.250,000 sq mi (647,500 sq km). Cattle was first introduced to the region by the Portuguese in the 1550s. Throughout the colonial period under Spain, only a small part of the Pampa was used. Herds of cattle roamed freely over the Pampa and the gaucho, the Argentine cowboy, was the region's dominant figure in the 18th and early 19th cent. A new economic era was initiated in the second half of the 19th cent., when a growing European market for agricultural products brought immigrant farmers (mostly Italian, Spanish, French and German) to the Pampa. Settlement spread into the interior and land was brought under the plow as unfriendly Indians were driven out of the region and the gaucho yielded to the farmer. In the 20th cent. agriculture remains the chief economic activity of the Pampa.

Pam·phyl·i·a (păm-fĭl'ē-ə), ancient region of S Asia Minor, on the coast between Lycia and Cilicia, in present S Turkey. It passed to Rome after the surrender (188 B.C.) of Antiochus III.

Pam·plo·na (păm-plō'nə, päm-plō'nä), city (1975 est. pop. 163,197), capital of Spanish Navarre, N Spain, on the Arga River. It is an important communications, agricultural, and industrial center, manufacturing kitchenware and chemicals. An ancient city of the Basques, it was repeatedly captured (5th–9th cent.) by the Visigoths, the Franks, and the Moors. In 824 the Basque kingdom of Pamplona, later called the kingdom of Navarre, was founded. Pamplona remained the capital of Navarre until 1512, when Ferdinand V united the major part of Navarre with Castile. In the Peninsular War, Pamplona was taken (1808) by the French and (1813) by the English. The celebration of the feast of San Fermin, described in Hemingway's *The Sun Also Rises,* is marked by running bulls through the streets of the city to the bullring. The city is still surrounded by old walls and fortifications and has retained its Gothic cathedral (14th–15th cent.).

Pan·a·ma (păn'ə-mä'), republic (1970 pop. 1,428,082), 29,209 sq mi (75,650 sq km), occupying the Isthmus of Panama, which connects Central and South America. To the west and east of Panama respectively are Costa Rica and Colombia; the Panama Canal Zone bisects the country. The capital and largest city is Panama City. In the west are rugged mountains (Chiriquí is 11,410 ft/3,478 m high) of volcanic origin, which yield in the middle of the country to low hills; there is a low mountain range in the east. Lowlands line both the Caribbean and Pacific coasts and there are numerous offshore islands. The climate is generally tropical with abundant rainfall.

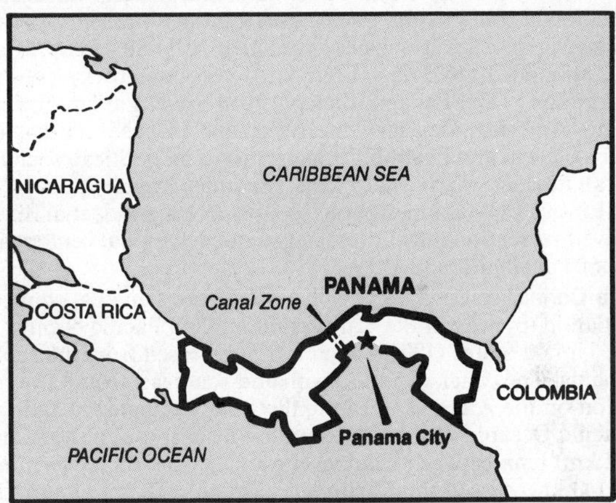

Panama is divided into 9 provinces:

NAME	CAPITAL	NAME	CAPITAL
Bocas del Toro	Bocas del Toro	Herrera	Chitré
Chiriquí	David	Los Santos	Las Tablas
Coclé	Penonomé	Panama	Panama
Colón	Colón	Veraguas	Santiago
Darién	La Palma		

Economy. Only about a quarter of the land is used for agriculture. On the upland savannas cattle are grazed and subsistence crops (notably rice), sugar cane, cocoa, and coffee are grown. Bananas are grown on the Pacific coast. The country has various light industries. Bananas are the leading export, followed by shrimp and fish products, sugar, and coffee. Manufactured goods, raw materials, and foodstuffs are imported. Much of the trade is with the United States. In 1977 the per capita gross national product was $1,060.

History. The coast of Panama was first sighted in 1501 by the Spaniard Rodrigo de Bastidas, and Columbus dropped anchor off the present-day Portobelo in 1502. Vasco Núñez de Balboa became governor of the region, and in 1513 he made his momentous voyage across the isthmus to the Pacific Ocean. Under the governorship of Pedro Arias de Ávila, Panama City was founded (1519). Panama was subordinated to the viceroyalty of Peru and remained in this status until 1717, when it was transferred to New Granada. It became a part of independent Colombia in 1821. Its significance as a crossroads was enhanced when U.S. settlers bound for the west passed through Panama. W. H. Aspinall built (1848–55) the Panama RR, and the question of a canal across the isthmus became paramount. The project ultimately led to a revolution against Colombian sovereignty and the establishment of Panama as a separate republic. The new state, proclaimed in Nov., 1903, was under the aegis of the United States, and the canal and U.S. interests in it became the determinants of Panama's history.

The internal politics of the republic have been stormy, with frequent changes of administration. U.S. forces were landed in 1908, 1912, and 1918. In 1958 and again in 1960 steps were taken by the United States to assuage Panamanian discontent by establishing uniform wages and employment opportunities in the Canal Zone and by reaffirming Panama's titular sovereignty over the zone. Gen. Omar Torrijos Herrera emerged as the dominant political figure in 1968. A series of negotiations aimed at resolving the conflict over the canal was begun in 1973. In 1977 the United States and Panama signed a treaty under which the canal would revert to Panama by the year 2000. The agreement appeased nationalist sentiment and increased Torrijos's prestige.

Panama, city (1970 est. pop. 420,000), central Panama, capital and largest city of Panama, on the Gulf of Panama, bounded on the W by the Panama Canal Zone. Panama city is no longer a port; commerce is handled through neighboring Balboa.

Founded in 1519 by Pedro Arias de Ávila, the city flourished in early colonial times as the Pacific port of transshipment of Andean riches to Spain. After it was destroyed in 1671 by Sir Henry Morgan, it was refounded (1673) 5 mi (8.1 km) west on a rocky peninsula. Construction of the Panama Canal (1914) brought assured prosperity, and American sanitary measures and disease control made Panama a clean and healthful tropical city. The political, social, and cultural nucleus of the nation, it expanded rapidly after World War II into a polyglot metropolis, boasting new residential districts, improved recreational facilities, and such educational centers as the Univ. of Panama (founded 1935).

Panama Canal, waterway across the Isthmus of Panama, connecting the Atlantic (by way of the Caribbean Sea) and Pacific oceans, built by the United States (1904–14) on territory leased from the republic of Panama. The canal, running south and southeast from Limón Bay at Colón on the Atlantic Ocean to the Bay of Panama at Balboa on the Pacific Ocean, is 40 mi (64 km) long from shore to shore and 51 mi (82 km) long between channel entrances. The Pacific terminus is 27 mi (43 km) east of the Caribbean terminus. The minimum depth is 41 ft (12.5 m). From Limón Bay a ship is raised by Gatun Locks (a set of three) to an elevation 85 ft (25.9 m) above sea level, traverses Gatun Lake, then crosses the Continental Divide through Gaillard (formerly Culebra) Cut and is lowered by Pedro Miguel Lock to Miraflores Lake and then by the Miraflores Locks (a set of two) to sea level. Passage requires 7 to 8 hours. The locks (1000 ft/305 m long and 110 ft/33.5 m wide) cannot accommodate the largest modern merchant ships and warships.

History. The United States became concerned with plans for a canal after settlers had begun to pour into Oregon and Calif. Active negotiations led in 1846 to a treaty, by which the republic of New Granada (consisting of present-day Panama and Colombia) granted the United States transit rights across the Isthmus of Panama. A French company under Ferdinand de Lesseps began to build a sea-level canal in Panama in 1881, but poor planning, disease among the workers, construction troubles, and inadequate financing drove the company into bankruptcy in 1889. In 1901 the U.S. congress authorized purchase of the French rights to construct a Panama canal. A treaty signed (Jan., 1903) with Colombia would have given the United States a strip of land across the Isthmus of Panama but the Colombian senate refused to ratify. An insurrection was encouraged by the United States, and Panama rose in revolt on Nov. 3, 1903, declaring itself independent of Colombia. On Nov. 17 the Hay-Bunau-Varilla Treaty was signed, granting to the United States, in return for a cash payment of $10 million and an annuity, exclusive control of a canal zone in perpetuity. Construction of a lock canal was decided on in 1906. Most of the actual construction work was supervised by Col. G. W. Goethals. The canal was informally opened Aug. 15, 1914; formal dedication took place on July 12, 1920. The total cost was $336,650,000, and c.240 million cu yd (184 million cu m) of earth were evacuated. A reservoir was added in 1935, a bridge in 1962, and a widening of the Gaillard Cut in 1969. In the 1960s there was increasing agitation in Panama to achieve greater Panamanian control over the canal. In 1977 two new treaties were signed replacing the 1903 document; under their terms the canal will revert to full Panamanian control by the year 2000, with the United States retaining rights of priority use for its warships and of defense against any threat to the canal's neutrality.

Panama Canal Zone, area within Panama (1970 pop. 44,198), 553 sq mi (1,432 sq km), administered by the United States under a 1977 treaty with Panama. The zone extends 5 mi (8.1 km) on either side of the Panama Canal. About three quarters of the civilian population are U.S. citizens. The zone is administered by a governor appointed by the President of the United States. The governor is also the president of the Panama Canal Company.

Panama City, city (1978 est. pop. 39,700), seat of Bay co., NW Fla., on St. Andrews Bay; inc. 1909. A Gulf Coast resort with amusement parks and excellent fishing, it is also a port of entry. The city's industries produce paper, clothing, chemicals, and plastics. Tyndall Air Force Base and the U.S. Navy Mine Defense Laboratory are nearby.

Pan-A·mer·i·can Highway (păn′ə-mĕr′ĭ-kən), system of roads, c.16,000 mi (25,745 km) long, linking the nations of the Western Hemisphere and extending from Alaska to Chile. The highway has many branches that connect with large cities off the main north-south route. Climatic zones along the highway vary from lush jungle to cold mountain passes nearly 15,000 ft (4,575 m) high. The scenery is often spectacular.

Pan·a·mint Range (păn′ə-mĭnt′), rugged fault-block mountains, SE Calif., near the Nev. border. Telescope Peak (11,045 ft/3,368.7 m high) is the tallest peak. The range forms the western boundary of Death Valley; to the west of the range is Panamint Valley.

Pa·nay (pä-nī′), island (1970 pop. 2,116,545), 4,446 sq mi (11,515.1 sq km), one of the Visayan Islands, Philippines, NW of Negros. Primarily agricultural with extensive lowlands, it is a major rice- and corn-producing area. There are also sugar-processing, lumbering, and fishing industries.

Pan·čev·o (pän′chə-vō′), city (1971 pop. 54,269), NE Yugoslavia, in the Vojvodina region of Serbia, on the Tamiš River near its confluence with the Danube. It has industries that produce aircraft, electrical equipment, and glass.

Pan·dhar·pur (pŭn′dər-pŏŏr′), town (1971 pop. 53,634), Maharashtra state, W India, on the Bhima River. Many Hindu pilgrims attend the festivals held three times a year at the temple of Vishnu.

Pan·guitch (păn′gwĭch), city (1970 pop. 1,318), seat of Garfield co., SW Utah, NE of Cedar City, at an altitude of 6,624 ft (2,020.3 m). Settled by Mormons c.1865, it was briefly abandoned (1867–71) and later became a center of grain farming and ranching.

pan·han·dle (păn′hăn′dl), a strip of land projecting from the main body of an area and shaped like the handle of a pan, such as the panhandles of W.Va., Okla., Texas, Idaho, and Alaska.

Panhandle, city (1970 pop. 2,141), seat of Carson co., extreme N Texas, on the high plains of the Panhandle ENE of Amarillo; settled 1896, inc. 1909. It is a marketing and shipping center for an oil, wheat, and cattle region.

Pa·ni·pat (pä′nə-pət), town (1971 pop. 88,017), Haryana state, NW India, on the Western Jumna Canal. It has sugar-processing plants and blanket and brassware industries. On a plain astride the easiest route from Afghanistan to central India, Panipat has seen several great battles. In 1526 Babur defeated the Delhi Sultanate here, thus paving the way for the formation of the Mogul empire. In 1556 Akbar defeated the Afghans at Panipat and thus secured Mogul rule. In 1761 Panipat was the site of an Afghan victory over the Mahrattas.

Pan·mun·jom (păn′mŏŏn-jŏm′), village, N South Korea. It lies south of the 38th parallel, the military demarcation line that partitions Korea. In the Korean War the truce negotiations, begun at nearby Kaesong, were moved in Oct., 1951, to Panmunjom, where the truce was signed on July 27, 1953.

Pan·no·ni·a (pə-nō′nē-ə), ancient Roman province, central Europe, SW of the Danube, including parts of modern Austria, Hungary, and Yugoslavia. Its natives, the warlike Pannonians, were Illyrians.

Their final subjugation by Rome took place in A.D. 9. Pannonia was divided A.D. c.103 into the provinces of Upper Pannonia and Lower Pannonia and abandoned by the Romans after 395.

Pa·no·la (pə-nō'lə). **1.** County (1970 pop. 26,829), 693 sq mi (1,794.9 sq km), NW Miss., drained by the Tallahatchie and Yocona Rivers; formed 1836; co. seat Batesville and Sardis. Its agriculture includes cotton, corn, hay, livestock, poultry, and dairy products. It has sawmills, clay and gravel deposits, and a textile industry. Clothing and wood and metal products are made. **2.** County (1970 pop. 15,894), 869 sq mi (2,250.7 sq km), E Texas, bounded on the E by the La. border and drained by the Sabine River; formed 1846; co. seat Carthage. This rolling wooded area has an extensive lumbering industry, agriculture (cotton, truck crops, corn, peanuts and pecans, fruit, livestock, and poultry), and deposits of clay and iron lignites. There is a huge natural-gas field tapped by interstate pipelines.

Pan·tel·le·ri·a (păn-tĕl'lĕ-rē'ä), volcanic island, 32 sq mi (82.9 sq km), S Italy, in the Mediterranean Sea between Sicily and Tunisia. Sweet wine, capers, raisins, and dried figs are exported. A colony of the Phoenicians and then of the Carthaginians, it passed to the Romans in 217 B.C. The island was later taken by the Arabs (8th cent. A.D.) and by the Normans (12th cent.). During World War II, Pantelleria was bombed into surrender by the Allies in 1943. On the island are extinct cones (the highest rising to 2,743 ft/836.6 m), numerous fumaroles, and hot mineral springs.

Pá·nu·co (pä'nōō-kō'), river, c.315 mi (505 km) long, rising as the Santa María River in San Luis Potosí state, N central Mexico, and flowing generally E to empty into the Gulf of Mexico near Tampico. It is navigable for c.200 mi (320 km).

Pao-chi (bou'jē', pou'chē'), city (1970 est. pop. 275,000), SW Shensi prov., China, on the Wei River. On the Lunghai RR, it is a newly flourishing industrial center with manufactures of textiles and paper.

Pa·o·la (pā-ō'lə, pī-), city (1970 pop. 4,622), seat of Miami co., E Kansas, SSW of Kansas City; laid out 1885, inc. as a village 1855, as a city 1860. It is a rail center for an oil and farm area.

Pa·o·li (pā-ō'lē, pī-), town (1970 pop. 3,281), seat of Orange co., S Ind., NW of Louisville, Ky., in a farm area; settled 1807, inc. 1869. Wood products (especially furniture) are made.

Pao-ting or **Pao·ting** (both: bou'dĭng'), city (1970 est. pop. 350,000), central Hopeh prov., China. It is a port on the Fu River and an agricultural distribution center, with food-processing industries. Chemicals, fertilizer, cotton cloth, synthetic fibers, and medicines are also manufactured.

Pao-t'ou or **Pao·tow** (both: bou'tō'), city (1970 est. pop. 800,000), Inner Mongolian Autonomous Region, China, on the Huang Ho. Vigorous industrialization in recent years has made it a major manufacturing center. Iron and coal are mined in the vicinity, and the city has a large integrated iron and steel complex as well as sugar refineries, textile mills, and plants making motor vehicles, chemicals, fertilizers, and aluminum. Two nuclear reactors are nearby.

Pá·pa (pä'pŏ), town (1976 est. pop. 32,000), W Hungary, in a grain- and beet-growing area. It is an industrial town; cotton spinning, tobacco processing, and the manufacture of textiles and electric appliances are the major industries. Pápa has an 18th cent. chateau built by Count Maurice Esterházy.

Pa·pal States (pā'pəl), independent territory under the temporal rule of the popes from 754 to 1870. The territory varied in size at different times; in 1859 it included c.16,000 sq mi (41,440 sq km) extending north-south on the Italian peninsula, from the Adriatic Sea and lower course of the Po River to the Tyrrhenian Sea, thus including the present regions of Latium, Umbria, Marche, and eastern Emilia-Romagna.

The nucleus of the states consisted of endowments given to the popes from the 4th cent. in and around Rome, in other areas of the Italian mainland, and in Sicily, Sardinia, and other lands. The popes gradually lost their more distant lands, but in the duchy of Rome papal power became stronger and increasingly independent of the Eastern emperors and of the other states in Italy. In 754 Pepin the Short gave the exarchate of Ravenna and the Pentapolis to Pope Stephen II. Over these vast territories the popes were unable to exercise effective temporal sovereignty. In Rome itself, the popes' temporal power, remained greatly limited by the interference of the emperors, by the power of the nobles, and by the ambitions of the commune of Rome.

Actual control by the papacy of its territories began in the 16th cent., when Cesare Borgia, son of Pope Alexander VI, conquered the petty states of the Romagna and Marche; after his fall (1503) most of them passed directly under papal rule. The last principalities to lose their autonomy to the popes were Ferrara (1598) and Urbino (1631). Papal troops, mostly Swiss and other mercenaries, offered almost no resistance to the French invaders under Napoleon Bonaparte in 1796. The Congress of Vienna fully restored (1815) the states of the papacy and placed them under Austrian protection. During the Risorgimento, only French intervention at Rome prevented the total absorption of the Papal States. After the Austrians left (1859) Bologna and the Romagna, both united (1860) with the kingdom of Sardinia, as did Marche and Umbria. Victor Emmanuel II of Italy seized what remained of the Papal States—Rome—in 1870. The popes, however, did not acquiesce to their loss of temporal power until 1929, when the Lateran Treaty established the Vatican City.

Pa·pan·tla (pä-pän'tlä), town (1970 pop. 26,773), Veracruz state, E central Mexico. It is known for the nearby ruins of Tajín, thought to be the remains of a pre-Columbian civilization (fl. 400?-900?) known as Classic Veracruz. The most impressive relic is a spectacular pyramid in seven rectangular tiers, lined with niches that contained small idols.

Pa·pe·e·te (pä'pā-ā'tā), town (1971 pop. 25,342), capital of Tahiti and of French Polynesia, South Pacific. Papeete ships copra, vanilla, and mother-of-pearl. It has an important nuclear laboratory.

Pa·phos (pā'fŏs), two ancient cities, SW Cyprus, on the coast. Old Paphos was probably founded in the Mycenaean period by colonists from Greece or Phoenicia. Modern excavations have revealed ruins dating from 3000 B.C. New Paphos, to the northwest, was an important seaport and the capital of Cyprus from the middle of the Hellenistic period until the time of Constantine.

Pa·pil·lion (pə-pĭl'yən), village (1974 est. pop. 6,493), seat of Sarpy co., E Nebr., just SW of Omaha.

Pap·u·a New Guin·ea (păp'yōō-ə nōō gĭn'ē, nyōō, pä'pōō-ä'), country (1971 pop. 2,490,037), 183,540 sq mi (475,369 sq km), SW Pacific, encompassing the E half of the island of New Guinea, as well as the Bismarck Archipelago, the Trobriand Islands, Samarai Island, Woodlark Island, D'Entrecasteaux Islands, the Louisiade Archipelago, and the northernmost Solomon Islands of Buka and Bougainville. The capital is Port Moresby.

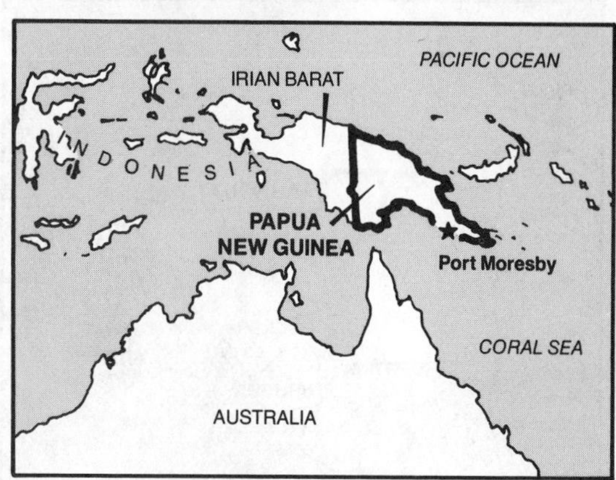

Economy. Agriculture is the mainstay of the economy. Sweet potatoes constitute the main food crop. Agricultural exports (notably coconut products, rubber, coffee, cocoa, and tea) generally represent the surplus from subsistence farming and the crops from European-owned plantations. Timber is also exported. Silver, manganese, copper, and gold are mined, but the gold reserves have steadily declined. Pearl-shell and tortoise fisheries dot the coast. Exploration for oil is underway.

History. Papua, the southern section of the country, was annexed by Queensland in 1883 and the following year became a British protectorate called British New Guinea. It passed to Australia in 1905 as the Territory of Papua. The northern section of the country formed part of German New Guinea from 1884 to 1914 and was called Kaiser-Wilhelmsland. Occupied by Australian forces during World War I, it was mandated to Australia by the League of Nations in 1920 and became known as the Territory of New Guinea. Australian rule was reconfirmed by the United Nations in 1947. In 1949 the territories of Papua and New Guinea were merged administratively, but they remained constitutionally distinct. They were combined on Dec. 1, 1973, as the self-governing country of Papua New Guinea, with Australia retaining control of defense and foreign relations. Full independence was attained on Sept. 16, 1975; Michael Somare became the country's first prime minister. The threat of a separatist movement in Bougainville was removed in 1976 when the central government granted the island a greater degree of autonomy.

Pa·rá (pä-rä'), state (1975 est. pop. 2,544,300), 474,896 sq mi (1,229,-981 sq km), N Brazil, in the lower Amazon River basin bordering on the Guianas and the Atlantic Ocean; capital Belém.

Pará, city: *see* Belém, Brazil.

Pará, river, c.200 mi (320 km) long, N Brazil. It is actually the southeastern arm or estuary of the Amazon, divided from the rest of the river by Marajó Island.

Pa·ra·cel Islands (pä-rä-sĕl'), group of low coral islands and reefs in the South China Sea, c.175 mi (280 km) SE of Hainan island. They are rich in guano and are thought to be underlain by oil deposits.

Par·a·dise (păr'ə-dīs'), uninc. town (1978 est. pop. 24,600), Butte co., N central Calif., in the foothills of the Sierra Nevada.

Par·a·gould (păr'ə-go͞old'), city (1970 pop. 10,639), seat of Greene co., NE Ark.; inc. 1882. The processing and trade center of an agricultural region, the city also has railroad shops and a wide variety of manufactures, including electric motors and automobile parts.

Pa·ra·gua·çu or **Pa·ra·guas·su** (both: păr'ə-gwə-so͞o'), river, c.300 mi (480 km) long, Bahia state, E Brazil. Rising in the Serra do Sincorá, it flows east into the Atlantic Ocean at All Saints Bay.

Par·a·guay (păr'ə-gwā', -gwī'), republic (1970 pop. 2,395,614), 157,047 sq mi (406,752 sq km), S central South America. The capital is Asunción. Paraguay and Bolivia are the two landlocked nations of the continent. Paraguay is enclosed by Bolivia, Brazil, and Argentina. The eastern part of the country, between the Paraguay and Paraná rivers, where most of the population live, is a lowland, rising in the east and north to a plateau region. The Paraná River, south of the Iguaçu River, separates Paraguay from Argentina. The Paraguay River also forms a border with Argentina from its confluence with the Paraná River north to the Pilcomayo River. The section west of the Paraguay River is a dry plain, part of the Chaco.

The population is largely mestizo, a mixture of Spanish and Guaraní Indian strains. There has been considerable emigration to neighboring countries, especially Argentina. Spanish is the official language, but Guaraní is widely spoken. The Jesuit missions made it possible for the Guaraní culture to blend with the Spanish. In later days European immigrants added new elements to the distinctive civilization of Paraguay. The established religion is Roman Catholicism. Most of the small number of Protestants are Mennonites.

Economy. More than half of Paraguay's labor force are engaged in agriculture and forestry; less than 15% work in industry and mining. The principal crops are rice, corn, soybeans, wheat, cotton, and tobacco. Orange groves furnish a large part of the world's supply of petitgrain, used in perfumes and flavorings. Maté (Paraguay tea) is used domestically and is also exported. In addition to quebracho, hardwoods and cedars are commercially exploited. Meat packing, vegetable-oil processing, and textile manufacturing are the main industries. Sandstone, clay, and limestone are found. The leading exports are meats, timber, and oils. The leading imports are foodstuffs, vehicles and machinery, chemicals, and fuels. Customs duties furnish an important part of Paraguay's revenues. River transportation is a vital supplement to Paraguay's inadequate roads and railroads.

History. European influence in Paraguay began with the early explorations of the Río de la Plata. Juan Díaz de Solís was the first to come (1516), and Sebastian Cabot followed him (1527) to the Paraguay River, which was thought to offer access to Peru. A colony grew up, as Asunción became the nucleus of the La Plata region. At the end of the 16th cent. Hernando Arias de Saavedra, called Hernandarias, became governor of Río de la Plata prov., of which Paraguay was a part; it was through his efforts that the administrations of present Argentina and Paraguay were separated (1617). The Jesuit missions were founded in the days of Hernandarias (most of them in the trans-Paraná area, now in Argentina). Real independence from Spain was asserted when in 1721 José de Antequera led a successful revolt and governed independently for some 10 years. In 1776 the region was made part of the viceroyalty of the Río de la Plata. Manuel Belgrano was unsuccessful in carrying the Argentinian revolution against Spain into Paraguay in 1810, but the next year the colonial officials there were quietly overthrown.

In 1814 the first of the three great dictators who were to mold Paraguay came to power. He was José Gaspar Rodríguez Francia, the incorruptible, harsh, and autocratic dictator known as "El Supremo," who ruled Paraguay until his death in 1840. He was succeeded by Carlos Antonio López, who held absolute power from 1844 to 1862. His son, Francisco Solano López, succeeded him and brought on disaster by involving Paraguay in war with Brazil, Argentina, and Uruguay (1865-70). The Paraguayans fought heroically and sustained the loss of more than half the population. Recovery from the catastrophe was slow, and the desperate state of the economy was matched by political confusion, as warring caudillos established short-lived dictatorships. In the late 19th and early 20th cent. conditions improved. Trade increased as Paraguayan products found markets, immigration was encouraged, and farming and small industries prospered fitfully. In 1932 the unsettled boundary with Bolivia plunged Paraguay into another major war—the Chaco War, which lasted until 1935. From it the little country emerged victorious but exhausted. A rapid succession of governments was broken by the years when Higinio Morínigo was in power (1940-48). Signs of recovery from the Chaco War appeared in improvements in education, public health, and roads, but the oppressive dictatorship of Morínigo was challenged by numerous uprisings. He was overthrown in 1948. After another series of short-lived governments. Gen. Alfredo Stroessner engineered a successful coup in 1954. Stroessner repeatedly suppressed opposition. He was re-elected in 1958 and 1963. The 1967 constitution permitted him to be re-elected again in 1968 and 1973.

Government. Paraguay is governed under a 1967 constitution. The president, popularly elected, serves a five-year term. The legislature has two houses, a 30-member senate and a 60-member chamber of deputies.

Paraguay, river, c.1,300 mi (2,090 km) long, rising in the highlands of central Mato Grosso state, Brazil. Flowing generally southward, it forms the border between Brazil and Paraguay, then crosses the center of Paraguay, dividing the Gran Chaco from eastern Paraguay, and continues its flow southwest to the Paraná River, forming part of the Paraguay-Argentina border. Navigable for most of its course, the Paraguay is a major artery of the Río de la Plata system.

Pa·ra·í·ba (pä'rə-ē'bə), state (1975 est. pop. 2,675,100), 21,765 sq mi (56,371 sq km), NE Brazil, on the Atlantic; capital João Pessoa.

Paraíba or **Paraíba do Sul** (do͞o so͞ol'), river, c.650 mi (1,045 km) long, rising as the Paraitinga in the Serra do Mar, São Paulo state, SE Brazil, and flowing SW to a point NE of São Paulo city, where it makes a hairpin turn and continues NE through the state of Rio de Janeiro to the Atlantic Ocean near Campos. Its beautiful valley forms a rich agricultural region (site of Brazil's first commercial coffee plantation) as well as a major industrial center.

Par·a·mar·i·bo (păr'ə-măr'ə-bō'), city (1971 pop. 102,300), capital of Surinam (Suriname), on the Surinam River near the Atlantic Ocean. It exports bauxite, sugar cane, rice, cacao, coffee, rum, and tropical woods. Paramaribo has canals that are reminiscent of the Netherlands. The area was settled by the British from Barbados in 1630, and in 1650 the city became the capital of the new English colony. Paramaribo changed hands often between the British and Dutch but finally came under Dutch rule in 1815.

Par·a·mount (păr'ə-mount'), city (1978 est. pop. 31,200), Los Angeles co., S Calif.; inc. 1957. Originally a dairy region, it has become highly industrialized since the 1950s. It has an oil refinery and industries that produce metal and plastic products, automotive parts, sports equipment, and furniture.

Pa·ram·us (pə-răm'əs), borough (1978 est. pop. 28,300), Bergen co., NE N.J.; settled 1668, inc. 1922. It is a large retail-trade center.

Pa·ra·ná (păr'ə-nä'), state (1975 est. pop. 8,449,200), 77,048 sq mi (199,554 sq km), S Brazil, on the borders of Paraguay and Argentina and on the Atlantic Ocean; capital Curitiba.

Pa·ra·ná (pä-rä-nä'), city (1970 pop. 189,537), capital of Entre Ríos prov., NE Argentina, a port on the Paraná River. It is the center of a grain and cattle district; there is an agricultural school nearby. Founded in 1730, Paraná was the capital of the Argentine confederation from 1853 to 1862.

Paraná, river, c.2,000 mi (3,200 km) long, formed by the junction of the Paranaíba and the Rio Grande, SE Brazil. It has the second-largest drainage system in South America. It flows generally southwest to its confluence with the Paraguay River, forming the southern border of Paraguay, then south and east through northeast Argentina to join the Uruguay River in a huge delta at the head of the Río de la Plata. The lower Paraná is hampered by shifting channels, sandbars, and fluctuating river flow, and is subject to flooding. The Brazilian stretch flows in a deep bed and is broken by many waterfalls. The Paraná was first ascended (1526) by Sebastian Cabot, the English explorer in the service of Spain.

Pa·ra·na·guá (pä'rə-nə-gwä'), city (1975 est. pop. 51,510), Paraná state, SE Brazil, on the Atlantic Ocean. Founded c.1600, the city has fine port facilities and has become increasingly important with the development of coffee cultivation in the area.

Pa·ra·na·í·ba (pä'rə-nə-ē'bə), river, c.500 mi (805 km) long, rising in W Minas Gerais state, Brazil. It flows generally westward through an agricultural region before joining the Rio Grande to form the Paraná. Diamonds are found along its course.

Pa·ra·na·pa·ne·ma (păr'ə-năp'ə-nā'mə), river, c.500 mi (805 km) long, SE Brazil, rising in SE São Paulo state and flowing W along the Parana–São Paulo border to the Paraná River.

Pa·ray-le-Mo·nial (pä-rā'lə-mô-nyäl'), town (1975 pop. 11,545), Saône-et-Loire dept., E central France. Ceramics and hosiery are produced. In the 17th cent. St. Margaret Mary founded the cult of the Sacred Heart of Jesus here. The town's Romanesque Church of Notre Dame dates from the 12th cent.

Par·do (pär'dōō'). **1.** River, c.400 mi (645 km) long, E Brazil, rising in the Serra do Espinhaco, NE Minas Gerais state, and flowing NE and E to the Atlantic Ocean. **2.** River, 230 mi (370.1 km) long, S Mato Grasso, SW Brazil. It flows from north of Campo Grande southeast to the Paraná River. **3.** River, c.300 mi (480 km) long, S Brazil. Rising in southwestern Minas Gerais state, it flows northwest into the Rio Grande in São Paulo state, north of Barretos.

Par·du·bi·ce (pär'dōō-bĭ'tsĕ), city (1975 est. pop. 79,869), N central Czechoslovakia, in Bohemia, on the Elbe River.

Pa·rí·cu·tin (pä-rē'kōō-tēn), active volcano, c.8,200 ft (2,500 m) high, Michoacán state, W central Mexico. In one of the most spectacular eruptions of modern times, Parícutin burst forth from a cornfield on Feb. 20, 1943, and grew discontinuously until 1952, spewing forth over a billion tons of lava. The cone (c.2,000 ft/610 m high) is a remarkable example of volcanic growth, and its development was closely studied by international scientific teams.

Par·is (păr'ĭs), city (1968 pop. 2,590,771; metropolitan area pop. 8,196,746), N central France, capital of the country, on the Seine River. It is the commercial and industrial focus of France and a cultural and intellectual center of international renown. Situated in the center of the Paris basin, and only 90 mi (145 km) from the English Channel, the city handles a great volume of shipping. Three airports and seven major railroad stations make Paris one of the great transportation centers of western Europe. The Paris metro (subway) was built in 1900. Elegant stores and hotels, lavish night clubs, theaters, and gourmet restaurants help make tourism the biggest industry in Paris. Other leading industries manufacture luxury articles, high-fashion clothing, perfume, and jewelry. Heavy industry, notably automobile manufacture, is located in the suburbs. About one quarter of the French labor force is concentrated in the Paris area.

Paris is divided into 20 arrondissements (boroughs). In 1976 the national government re-established the office of mayor of Paris, which had been abolished following the uprising of the Commune of Paris in 1871. The office of city mayor, appointive after 1871, was again made elective in 1977. Geographically, Paris is divided into roughly equal sections by the Seine River. On the right (northern) bank are the Bois de Boulogne, Arc de Triomphe, Bibliothèque nationale, Élysée Palace, Place de la Concorde, Opéra, Comédie française, Louvre, Palais de Chaillot, and the Champs Élysées and the other great streets and boulevards. To the north of the right bank is Montmartre, the highest area in Paris, topped by the Church of Sacré-Coeur. Much of the right bank, which has many of the most fashionable streets and shops, has a stately, formal air. The left bank, with the Sorbonne, the French Academy, the Panthéon, the Luxembourg Palace and Gardens, the Chamber of Deputies, the Quai d'Orsay, and the Hotel des Invalides, is the governmental and to a large extent the intellectual section. The old Latin Quarter, for nearly a thousand years the preserve of university students and faculty; the Faubourg Saint-Germain section; and Montparnasse are the most celebrated left-bank districts. The Eiffel Tower, chief landmark of Paris, stands by the Seine on the Champ-de-Mars. The historical nucleus of Paris is the Île de la Cité, a small boat-shaped island largely occupied by the huge Palais de Justice and the Cathedral of Notre Dame de Paris.

History. Julius Caesar conquered Paris in 52 B.C. It was then a fishing village, called Lutetia Parisiorum (the Parisii were a Gallic tribe), on the Île de la Cité. Under the Romans the town spread to the left bank and acquired considerable importance. Clovis I and several other Merovingian kings made Paris their capital; under Charlemagne it became a center of learning. In 987 Paris was firmly established as the French capital.

In the 11th cent. the city spread to the right bank. During the next two centuries streets were paved and the city walls enlarged; the first Louvre (a fortress) and several churches, including Notre Dame, were constructed. The Sorbonne became a fountainhead of theological learning. During the period of the Hundred Years' War the city suffered civil strife, occupation by the English (1419–36), famine, and the Black Death. In the 16th cent. the Louvre was transformed from a fortress to a Renaissance palace. In the Wars of Religion (1562–98), Parisian Catholics took part in the massacre of St. Bartholomew's Day (1572). Cardinal Richelieu established the French Academy (1635) and built the Palais Royal (1629).

During the late 17th and the 18th cent. Paris acquired further glory as the scene of many of France's greatest cultural achievements. At the same time, growing industries resulted in the creation of such suburbs (faubourgs) as Saint-Antoine and Saint-Denis. Throughout the turbulent period of the Revolution the city played a central role. Napoleon (emperor, 1804–15) began a large construction program (including the building of the Arc de Triomphe). In the first half of the 19th cent. Paris grew rapidly. In 1801 it had 547,000 people and 1,696,000 in 1861. The great boulevards and parks of modern Paris were the work of Baron Georges Haussmann, who was appointed prefect by Napoleon III. During the Franco-Prussian War (1870–71), Paris was besieged and occupied by the Germans. After the Germans withdrew, Parisian workers rebelled against the French government and established the Commune of Paris, which was bloodily suppressed.

In World War I the Germans failed to reach Paris. During the 1920s Paris was home to many artists and writers from the United States and elsewhere. German troops occupied the city during World War II from June 14, 1940, to Aug. 25, 1944. In 1969 Les Halles, Paris's famous central market, was dismantled.

Paris (păr'ĭs). **1.** City (1975 est. pop. 3,947), a seat of Logan co., Ark., ESE of Fort Smith, in a coal-mining and farming region of the Ouachita foothills. **2.** City (1970 pop. 615), seat of Bear Lake co., SE Idaho, ENE of Preston, at an altitude of 5,967 ft (1,819.9 m). Its agriculture includes sugar beets, grain, livestock, and dairy products. **3.** City (1975 est. pop. 9,600), seat of Edgar co., E Ill., NE of Terre Haute, Ind.; inc. 1853. An industrial and rail center in a farm and coal area, it mills corn and makes electrical equipment and metal products. **4.** City (1970 pop. 7,823), seat of Bourbon co., N Ky., NE of Lexington on the South Fork of the Licking River, in a fertile bluegrass region; founded 1774. A bluegrass-seed market, it also has tobacco storehouses. An early distillery (1790) here was one of the first in the state. Thouroughbred horses are raised nearby. **5.** City (1970 pop. 1,442), seat of Monroe co., NE Mo., WSW of Hannibal, in a farm area; laid out 1831, inc. 1880. **6.** City (1970 pop. 9,892), seat of Henry co., NW Tenn., WNW of Nashville, in an area of clay pits and timber; settled 1821, inc. 1823. It manufactures carburetors, shirts, and clay products. **7.** City (1978 est. pop. 23,400), seat of Lamar co., E Texas, in the Red River valley; settled 1824. It is a processing center for the rich farms of the blackland region. There are various manufactures.

Park (pärk). **1.** County (1970 pop. 2,185), 2,162 sq mi (5,599.6 sq km), central Colo., drained by headwaters of the South Platte River; formed 1861; co. seat Fairplay. It lies in farming, livestock-grazing, and mining (gold, silver, and lead) region, with some manufacturing. It includes part of Park Range and Pike national forests. **2.** County (1970 pop. 11,197), 2,627 sq mi (6,803.9 sq km), S Mont., in an agricultural region drained by the Yellowstone River and bordering on Wyo. and Yellowstone National Park; formed 1887; co. seat Livingston. Livestock are raised. Absaroka National Forest and Absaroka Range are in the south. **3.** County (1970 pop. 17,752), 6,959 sq mi (18,023.8 sq km), NW Wyo., bordering on Mont. and Yellowstone National Park, and watered by the Shoshone and Greybull rivers and by Shoshone Reservoir; formed 1909; co. seat Cody. Its irrigated agriculture includes sugar beets, grain, beans, and livestock. It has oil and gas wells, petroleum refineries, and factories manufacturing plastics and concrete. Shoshone National Forest, part of Absaroka Range, and Shoshone Cavern National Monument are here.

Park City, city (1970 pop. 1,193), Summit co., N central Utah, SE of Salt Lake City. It developed as a mining camp after gold, silver, and lead were discovered in 1869 and flourished in the late 19th cent. Mining is still carried on in the area, which is a resort center for nearby ski resorts.

Parke (pärk), county (1970 pop. 14,628), 445 sq mi (1,152.6 sq km), W Ind., bounded on the W by the Wabash River and drained by Sugar and Raccoon creeks; formed 1821; co. seat Rockville. In an agricultural and bituminous-coal region, it has some manufacturing (plastics and electronic equipment), timber, clay and gravel pits, mineral springs, and fisheries.

Par·ker (pär'kər), county (1970 pop. 33,888), 904 sq mi (2,341.4 sq km), N central Texas, drained by the Brazos River and the Clear Fork of the Trinity River; formed 1855; co. seat Weatherford. This rich diversified agricultural, ranching, and dairying area yields melons, peaches, other fruits, truck crops, peanuts, corn, grains, pecans, beef cattle, poultry, and horses.

Parker, city (1970 pop. 1,005), seat of Turner co., SE S.Dak., SW of Sioux Falls; settled 1879. It is in a livestock and dairy region.

Par·kers·burg (pär'kərz-bûrg'), city (1978 est. pop. 36,300), seat of Wood co., NW W.Va., at the confluence of the Little Kanawha and Ohio rivers; settled 1785, inc. 1820. An industrial and shipping center in a coal region, it has industries producing synthetic fibers, plastics, and glass.

Parkes (pärks), town (1971 pop. 8,849), New South Wales, SE Australia. It is the site of a radiotelescope (opened 1961) capable of receiving radio waves from a distance of 1 billion light-years.

Park Range, part of the Rocky Mts., central Colo. and S Wyo., extending N from the Colorado River. Mt. Lincoln (14,284 ft/4,356.6 m) is the highest peak.

Park Rap·ids (răp'ĭdz), village (1970 pop. 2,772), seat of Hubbard co., N central Minn., ENE of Detroit Lakes; founded 1880, inc. 1891. It is a resort in a dairy and timber region.

Park Ridge (rĭj), city (1978 est. pop. 43,000), Cook co., NE Ill., a suburb adjacent to Chicago, on the Des Plaines River; inc. 1873. It is chiefly residential. O'Hare International Airport is nearby.

Par·ma (pär'mə), city (1975 est. pop. 177,894), capital of Parma prov., in Emilia-Romagna, N Italy, on the Parma River and the Aemilian Way. It is an agricultural market and an industrial center. Manufactures include textiles, watches, footwear, and fertilizer. Parmesan cheese is also produced. Parma was the site of a Roman colony (founded 183 B.C.) and became a free commune by the 12th cent. It later was ruled by outside powers (particularly Milan and France) and in 1513 was added to the Papal States by Pope Julius II. In 1545 Pope Paul III created the duchy of Parma and Piacenza and bestowed it on his son, Pier Luigi Farnese, whose descendants ruled it (with interruptions) until 1731. The duchy then passed, through the female line, to the Spanish Bourbons; the cadet line of Bourbon-Parma began in 1748. It was displaced in 1802, when Napoleon I annexed the duchy to France. The Congress of Vienna (1814–15) awarded it to Marie Louise, who ruled it from 1816 to 1847; it was then restored to the Bourbons. In 1860 the duchy was incorporated into the kingdom of Sardinia. The Parma school of painting flourished here in the 16th cent.; its leading artists were Correggio (who executed frescoes for the Convent of St. Paul and for the Romanesque cathedral) and Parmigiano. Points of interest in the city include an octagonal Romanesque baptistry (13th cent.) and the Palazzo della Pilotta (1583–1622; damaged in World War II), which contains the National Museum of Antiquities, the National Gallery, and the Farnese Theatre.

Par·mer (pär'mər), county (1970 pop. 10,509), 859 sq mi (2,224.8 sq km), NW Texas, on the Llano Estacado bounded on the W by the N.Mex. border and drained by the White River; formed 1876; co. seat Farwell. It has agriculture (especially wheat, also grain sorghums, hay, barley, peanuts, and cotton), livestock raising (cattle, hogs, and sheep), and some dairying.

Par·na·í·ba (pär'nə-ē'bə), river c.800 mi (1,290 km) long, rising in the highlands of NE Brazil. It flows generally north, forming the boundary between Maranhão and Piauí states, and enters the Atlantic Ocean through a delta near the town of Parnaíba, which is the shipping center for the valley. The river is filled with rapids.

Par·nas·sós (pär-năs'əs, -nə-sôs'), mountain, c.8,060 ft (2,460 m) high, Phocis, central Greece. In ancient Greece it was sacred to Apollo, Dionysus, and the Muses. The fountain of Castalia was on its slopes; at the foot of the mountain lay Delphi. Bauxite is mined on the slopes.

Par·nu (pär'no͞o), city (1974 est. pop. 49,000), W European USSR, in Estonia, on the Gulf of Riga. A seaport, it exports timber and flax and is also a beach and health resort. It was founded c.1250 by the Livonian Knights and became a city of the Hanseatic League. After the dissolution (1561) of the Livonian Order it was contested by Sweden, Russia, and Poland. Parnu was incorporated into newly independent Estonia in 1918.

Par·o·pa·mi·sus (păr'ə-pə-mī'səs), mountain range, N Afghanistan, stretching c.300 mi (480 km) W from the Hindu Kush toward the Elburz Mts. in Iran. It rises to c.11,000 ft (3,355 m). Silver and lead crystal deposits are found here.

Pá·ros (pä'rôs, pâr'ŏs), island (1971 pop. 6,776), c.81 sq mi (210 sq km), SE Greece, in the Aegean Sea; one of the Cyclades. Wine, tobacco, figs, and grains are produced on the island. The beautiful white, semitransparent Parian marble, used by sculptors and architects as early as the 6th cent. B.C., is quarried on Mt. Hagios. Páros was settled by Ionians and became a maritime power and a center of Aegean trade. In the 7th cent. B.C. it established colonies in Thásos and on the Sea of Marmara. During the Persian Wars, Athens accused Páros of aiding the Persians and captured the island in 479 B.C. Páros was held by the Ottoman Turks from 1537 to 1832, when it joined Greece. Two marble fragments of a historical inscription, called the Parian Chronicle, have been found on the island. The larger fragment (covering 1581–354 B.C.) is one of the Arundel Marbles, housed at Oxford, England; the smaller (covering 356–299 B.C.) is in a museum on Páros.

Pa·row (pă-rou'), town (1970 pop. 60,146), Cape Prov., SW South Africa, near Cape Town. It is an industrial center whose manufactures include printed materials, processed timber, and metal goods.

Par·o·wan (păr'ə-wăn,' păr'ə-wăn'), city (1970 pop. 1,423), seat of Iron co., SW Utah, near the Parowan Mts. NE of Cedar City, at an altitude of 5,990 ft (1,827 m); settled 1851.

Par·ral (pä-räl'): *see* Hidalgo del Parral, Mexico.

Par·ra·mat·ta (păr'ə-măt'ə), city (1976 pop. 131,665), New South Wales, SE Australia, a suburb of Sydney, on the Parramatta River. It has an automobile parts industry. Founded in 1788, it is the second-oldest settlement in Australia.

Par·ras de la Fuen·te (pä'räs dā lä fwān'tā), city (1970 pop. 32,664), Coahuila state, N Mexico. It is a road and rail junction located in a well-watered valley of a semiarid region. Parras, an agricultural center, has orchards and vineyards that make the city famous for wines and brandies. Cattle raising and cotton and flour milling are also important. In 1846, during the Mexican War, Parras was held by U.S. troops.

Par·snip (pär'snĭp), river, c.150 mi (240 km) long, rising in central British Columbia, Canada, and flowing NW to join the Finlay River at Williston Lake and form the Peace River. Discovered by Sir

Alexander Mackenzie in 1793, it was an important fur-trade route.

Par·sons (pär'sənz). **1.** City (1978 est. pop. 11,900), Labette co., SE Kansas; inc. 1871. It is a shipping point for dairy products, grain, and livestock. There are various manufacturing industries. **2.** City (1970 pop. 1,784), seat of Tucker co., N W.Va., NE of Elkins; inc. 1893. It has tanneries and woolen mills. The Battle of Corrick's Ford was fought here in 1861.

Par·thi·a (pär'thē-ə), ancient country of Asia, SE of the Caspian Sea. In its narrowest limits it consisted of a mountainous region intersected with fertile valleys. It was included in the Assyrian and Persian empires, the Macedonian empire of Alexander the Great, and the Syrian empire. The Parthians were famous horsemen and archers and may have been of Scythian stock. In 250 B.C. they freed themselves from the rule of the Seleucidae and founded the Parthian empire. At its height, in the 1st cent. B.C., this empire extended from the Euphrates across Afghanistan to the Indus and from the Oxus to the Indian Ocean. Defeating Marcus Licinius Crassus in 53 B.C., the Parthians threatened Syria and Asia Minor, but they were turned back by Ventidius in 39–38 B.C. Then began the decline of the empire, which in A.D. 226 was conquered by Ardashir I (Artaxerxes), the founder of the Persian dynasty of the Sassanidae.

Pas·a·de·na (păs'ə-dē'nə). **1.** City (1978 est. pop. 106,200), Los Angeles co., S Calif., at the base of the San Gabriel Mts.; inc. 1866. Among the city's many manufactures are cosmetics, electronic equipment and systems, ceramics, plastics, and aircraft and missile components. Pasadena is the scene of the annual Tournament of Roses and of the postseason college football game held (Jan. 1) in the famous Rose Bowl (seating 100,000 spectators). The city is also the seat of the California Institute of Technology (with its noted NASA jet propulsion laboratory). The Huntington Library and Art Gallery, four museums, and several gardens noted for their rare flora are here. **2.** City (1978 est. pop. 98,100), Harris co., S Texas, on the Houston ship channel; inc. 1928. It is an industrial suburb of Houston located in a highly productive oil area. The city has oil refineries, chemical plants, iron and steel works, paper and grain mills, food-processing establishments, and factories making oil field equipment, machinery and tools, building materials, paint, and clothing. The Lyndon B. Johnson Manned Space Center is just south, on Clear Lake. The city is near the old San Jacinto battlefield (Apr., 1836); the San Jacinto Monument (built 1936–39; 570 ft/173.9 m high) commemorates the event.

Pa·sar·ga·dae (pə-sär'gə-dē'), capital of ancient Persia under Cyrus the Great, NE of Persepolis, in present Iran. The buildings of Cyrus include a temple in the form of a tower, the remains of his palace, and his tomb. According to Greek historians, Alexander the Great found Cyrus' tomb already rifled, and he sealed its entrance. The tomb has long since been reopened.

Pas·ca·gou·la (păs'kə-gōō'lə), city (1978 est. pop. 31,300), seat of Jackson co., extreme SE Miss. A port of entry on Mississippi Sound at the mouth of the Pascagoula River, it is a resort and a fishing and shipbuilding center, with paper mills, an oil refinery, and factories producing chemicals and pet foods. It grew around a Spanish fort, built in 1718 and still extant.

Pas·co (păs'kō), county (1970 pop. 75,955), 742 sq mi (1,921.8 sq km), W central Fla., on the Gulf of Mexico and watered by many small lakes; formed 1887; co. seat Dade City. It is in a citrus-fruit, poultry, and cattle area, with lumber sawmilling. It processes fruit and manufactures concrete and metal cans and lids.

Pasco, city (1978 est. pop. 15,600), seat of Franklin co., SE Wash., on the Columbia River near its confluence with the Snake and Yakima rivers. It is a trade and shipping center for the Columbia basin project; its industries manufacture paper, container board, and machine parts. Pasco was an early railroad division point. With Kennewick and Richland it forms a tri-city area that grew during World War II, when the Nuclear Regulatory Commission's Hanford Works were constructed nearby.

Pas-de-Ca·lais (pä-də-kä-lā'), department (1975 pop. 1,403,035), 2,563 sq mi (6,638.2 sq km), N France, on the Strait of Dover; capital Arras.

Pas·o Ro·bles (pä'sō rō'blās), resort city (1970 pop. 7,168), San Luis Obispo co., S Calif., on the Salinas River N of San Luis Obispo, in a farm and grain area; inc. 1889. It has hot mineral springs. San Miguel Arcangel Mission (founded 1797) is nearby.

Pas·quo·tank (păs'kwō-tăngk'), county (1970 pop. 26,824), 228 sq mi (590.5 sq km), NE N.C., bounded on the S by Albemarle Sound, on the E by the Pasquotank River, and partly in the Dismal Swamp; formed 1672; co. seat Elizabeth City. This forested tidewater area has farming (truck crops, soybeans, corn, potatoes, cabbage, tomatoes, and snap beans), livestock raising (cattle and hogs), sawmilling, fishing, and textile and furniture manufacturing.

pass (păs), opening or way by which a natural or artificial barrier can be crossed. The term "pass" is usually applied to a relatively narrow passage through a mountainous region. A pass, like an isthmus, may have great strategic and economic importance; the history of a nation has often been determined by its success or failure in defending a pass, and land trade routes must necessarily cross passes.

Pas·sa·ic (pə-sā'ĭk), county (1970 pop. 460,782), 194 sq mi (502.5 sq km), N N.J., on the N.Y. line, watered by many lakes and the Passaic, Ramapo, and Pequannock rivers; formed 1837; co. seat Paterson. The highly industrialized region in the southeast processes and manufactures food products, textile-mill products, clothing and textiles, lumber millwork, furniture, paper products, chemicals, plastics, concrete, metal products, and machinery. It has metal-ore mining, some agriculture (dairy products, poultry, truck crops, and fruit), and resort areas.

Passaic, city (1978 est. pop. 47,100), Passaic co., NE N.J., a port on the Passaic River; settled 1678 by Dutch traders as Acquackanonk, named Passaic 1854, inc. as a city 1873. Formerly a great textile center, it now has highly diversified industries.

Pas·sa·ma·quod·dy Bay (păs'ə-mə-kwŏd'ē), inlet of the Bay of Fundy, between Maine and N.B., at the mouth of the St. Croix River. Most of it (including Campobello island) is within Canada's border.

Pas·sau (päs'ou), city (1974 est. pop. 50,669), Bavaria, SE West Germany, at the confluence of the Danube, Inn, and Ilz rivers, near the border with Austria. It is a river port, rail junction, and industrial center; manufactures include machinery, textiles, optical equipment, and printed materials. A Roman frontier outpost known as Castra Batava, Passau was made (738–39) an episcopal see by St. Boniface. The bishops of Passau were temporal lords of a substantial territory until 1803, when the bishopric was secularized and awarded to Bavaria; the diocese was restored in 1817. The Treaty of Passau (1552) helped pave the way for the religious peace of 1555. Noteworthy buildings in Passau include the cathedral (15th–17th cent.), the Gothic city hall (begun 1398), the baroque episcopal palace, the Oberhaus fortress (13th–16th cent.), and a former Benedictine monastery (founded in the 8th cent.).

Pas·to (päs'tô), city (1973 pop. 119,339), alt. 8,510 ft (2,595.6 m), capital of Nariña dept., SW Colombia. It is a distributing and processing center for the agricultural products of the surrounding region. Varnish and woolen goods are produced in the city. Occupied for a short period in 1831 by Ecuadorian forces, Pasto was the scene of the treaty (1832) by which Colombia (then called New Granada) and Ecuador became separate states. The city retains a colonial appearance, with narrow streets and many old churches.

Pa·su·ru·an (pä'sōō-rōō-än'), city (1961 pop. 63,408), E Java, Indonesia, on Madura Strait. A port, it exports sugar, rubber, and coffee.

Pat·a·go·ni·a (păt'ə-gō'nē-ə, -gōn'yə), region, c.300,000 sq mi (777,000 sq km), primarily in S Argentina, S of the Río Colorado and E of the Andes, but including extreme SE Chile and N Tierra del Fuego. Patagonia, except for the far southern plains, the sub-Andean region, and the Andes, is a vast, wind-swept semiarid plateau, sloping gently toward the east and terminating in cliffs along the Atlantic Ocean. Until recently sheep raising (mainly for wool) was the major industry of Patagonia, but oil production has become increasingly important. There are also deposits of coal and iron ore. Studies have revealed the presence of vast untrapped mineral wealth. Tourist resorts in the lake region are very popular. Cattle are raised, and agriculture is practiced in irrigated oases. A rich field for paleontologists, Patagonia has been visited by many scientific expeditions since the days of Charles Darwin. Of the original inhabitants, the Tehuelches (the "Patagonian giants") are the most important. Among the native animals are the guanaco, the rhea, the puma, and the deer.

Probably first visited (1501) by Amerigo Vespucci, the Patagonian coast was explored (1520) by Ferdinand Magellan. Settlements were attempted in the 16th and 17th cent., but the inhospitable country and natives discouraged colonization. It was not until after Julio A. Roca, an Argentine general, campaigned against the Indians that

Argentine ranchers began entering the territory in the late 19th cent. Chileans had been coming in for some time, and despite efforts to exclude them during and after the Argentine-Chilean boundary dispute in the early 20th cent., many continued to immigrate.

Pa·tan (pä'tən), city (1971 pop. 59,049), central Nepal, in the Katmandu valley, c.4,000 ft (1,220 m) above sea level. Agriculture and grazing are important in the surrounding area, and wool and leather are exported. The city is the center of the Banra sect of goldsmiths and silversmiths. Founded in the 7th cent., Patan is the oldest of Nepal's chief cities. It was the capital of a Nepali kingdom from the 17th cent. until captured by the Gurkhas in 1768. According to legend, the Indian emperor Asoka visited the area c.250 B.C. and built the four stupas that still stand on the four sides of Patan.

Pa·tay (pä-tā'), village (1968 pop. 1,948), Loiret dept., N central France. At Patay, in 1429, Joan of Arc defeated the English—one of the most serious English defeats in the Hundred Years' War.

Pat·chogue (păt'chôg'), village (1978 est. pop. 11,000), Suffolk co., SE N.Y., on Long Island, on Great South Bay; inc. 1893.

Pa·ter·nò (pä'tār-nô'), city (1975 est. pop. 47,954), E Sicily, Italy, at the foot of Mt. Etna, probably the ancient Hybla. It is an agricultural market and a food-processing center.

Pat·er·son (păt'ər-sən), city (1978 est. pop. 131,100), seat of Passaic co., NE N.J., at the falls of the Passaic River; inc. 1851. Founded in 1791 by Alexander Hamilton and others of the Society for Establishing Useful Manufactures, Paterson was a planned attempt to promote industrial independence in the newly-formed United States. In 1792 and 1794 cotton-spinning mills, forerunners of the city's textile industry, were established. In 1835 Samuel Colt began his manufacture of the Colt revolver. Shortly thereafter the silk industry was established. The iron industry, which initially supplied Paterson with textile machinery, was producing locomotives in great numbers by 1880. After World War I the aeronautics industry moved to Paterson. Industrial diversification has steadily increased. Today the silk industry is gone, but textiles and transportation equipment are still manufactured, and there is a large garment industry. Among the many other manufactures are electronic equipment, paper and food products, fabricated metals, rubber, and plastics.

Paterson has suffered many of the urban problems that plague large cities. During the first half of the 20th cent., notably in 1912–13, 1933, and 1936, many bitter strikes arose from bad labor conditions in the silk industry. More recently Paterson has attracted large numbers of immigrants, and the shortage of housing has become a major problem. Today about one third of the population is Spanish-speaking. Of special interest is the historic district that centers around the roaring falls of the river. Designated a national historic site in 1970, it is a unique display of industrial history, with old cobblestone streets and stone bridges; the abandoned houses of workmen and mill owners; and a great variety of industrial works, including several locomotive factories (one dating back to 1830), the Colt gun factory (1835), and historic spinning mills and waterworks.

Pa·ti·a·la and East Pun·jab States Union (păt'ē-äl'ə; pŭn-jäb', -jăb'), former union of princely states, NW India. It was the only area in India in which the Sikhs had a majority. It was merged with Punjab state in 1956.

Pát·mos (păt'məs, pät'môs), island (1971 pop. 2,432), c.13 sq mi (34 sq km), SE Greece, in the Aegean Sea; one of the Dodecanese, near Turkey. The Monastery of St. John, founded here in the 11th cent., holds a valuable manuscript collection.

Pat·na (păt'nə, pŭt'-), city (1971 pop. 474,349), capital of Bihar state, NE India, on the Ganges River. It is the hub of a rice-growing region. The ancient Pataliputra, it was an imperial city during the Mauryan (c.325–185 B.C.) and Gupta (c.320–545 A.D.) eras. Asoka (270–230 B.C.) built a large palace here, and numerous ruins of the period remain.

Pa·to·ka (pə-tō'kə), river, 138 mi (222 km) long, SW Ind., rising in SE Orange co. and flowing generally W into the Wabash River opposite Mt. Carmel, Ill.

Pa·tos, La·go·a dos (lə-gô'ə dōōs pä'tōōs), shallow tidal lagoon, c.150 mi (240 km) long and up to 30 mi (48 km) wide, Rio Grande do Sul state, SE Brazil. A wide sandbar separates it from the Atlantic Ocean. The lagoon is an important fishing ground.

Pá·trai (pä'trā) or **Pa·tras** (pə-träs', păt'rəs), city (1971 pop. 111,607), capital of Akhaía prefecture, central Greece, in the Peloponnesus. It is a port on the Gulf of Pátrai, which connects the Gulf of Corinth with the Ionian Sea. Pátrai is a commercial, industrial, and transportation center that ships currants, tobacco, wine, olive oil, and sheepskins. It was allied with Athens in the Peloponnesian War and became (3rd cent. B.C.) a leading member of the Second Achaean League. It led a revolt against the Macedonians in 218 B.C. but sank into insignificance before the Roman conquest (146 B.C.) of Greece; it was revived (late 1st cent. B.C.) as a Roman military colony by Augustus and soon flourished as a port. The city was conquered by the French nobleman Geoffroi I de Villehardouin in 1205, captured by the Ottoman Turks in 1458, passed to Venice in 1687, and was retaken by the Turks in 1715. The city was destroyed (1821) in the Greek War of Independence and was rebuilt on a rectangular pattern by Count Capo d'Istria in 1829.

Pat·rick (păt'rĭk), county (1970 pop. 15,282), 464 sq mi (1,201.8 sq km), S Va., partly in the Piedmont and rising to the Blue Ridge in the W and NW, drained by the Mayo, Smith, and Dan rivers; formed 1790; co. seat Stuart. Its agriculture includes apples, peaches, corn, tobacco, and livestock. It has lumbering, and manufactures textiles, clothing, and furniture. Fairy Stone Mt. and the Pinnacles of Dan are here. The Blue Ridge Parkway and the Appalachian Trail traverse the county.

Pat·ta·ni (pät'tä-nē'), city (1972 est. pop. 26,243), capital of Pattani prov., S Thailand, on the E coast of the Malay Peninsula, near the mouth of the Gulf of Siam. It is a port and the center of a region producing most of the spices grown in Thailand, as well as rubber and coconuts. Pattani was one of the first places in Siam opened to the Portuguese in the 16th cent.

Pátz·cua·ro (päts'kwä-rō), lake, c.100 sq mi (260 sq km) Michoacán state, W Mexico. Its indented shores, dotted with Tarascan Indian villages, green islands, and the curious native sailboats help make Lake Pátzcuaro popular as a resort.

Pau (pō), city (1975 pop. 83,498), capital of Pyrénées-Atlantiques dept., SW France, at the foot of the Pyrenees. It is a major year-round tourist center, renowned for its scenery. It has metallurgical and wool industries, and shoes and clothing are manufactured. Founded in the 11th cent., it became the capital of Béarn in the 15th cent. and the residence of the kings of Navarre in 1512.

Paul·ding (pôl'dĭng). **1.** County (1970 pop. 17,520), 318 sq mi (823.6 sq km), NW Ga., in a piedmont region; formed 1832; co. seat Dallas. It is in an agricultural (cotton, corn, grain, and fruit) and timber area, with yarn mills and clothing factories. **2.** County (1970 pop. 19,329), 416 sq mi (1,077.4 sq km), NW Ohio, bounded in the W by the Ind. border and drained by the Auglaize and Maumee rivers; formed 1820; co. seat Paulding. In an agricultural region yielding corn, soybeans, wheat, sugar beets, oats, and livestock, it has limestone quarries and clay pits. Cement products, construction machinery, and auto parts are manufactured.

Paulding. 1. Town (1970 pop. 200), a seat of Jasper co., E central Miss., SW of Meridian. **2.** Village (1970 pop. 2,983), seat of Paulding co., NW Ohio, SSW of Bryan, in a farm region; inc. 1872.

Pauls Valley (pôlz), city (1970 pop. 5,769), seat of Garvin co., S central Okla., on the Washita River SSE of Oklahoma City; settled c.1847. It is the center of a farm area producing alfalfa, cotton, broomcorn, and pecans.

Pa·vi·a (pä-vē'ä), city (1975 est. pop. 87,909), capital of Pavia prov., Lombardy, N Italy, on the Ticino River near its confluence with the Po. Pavia has long been an agricultural center and is now also an industrial and transportation center. Manufactures include textiles, machinery, and food products. Known as Ticinum in Roman times, it was an important stronghold of the empire and later served as the capital of the Lombard kings. From the 9th to the 12th cent. the Italian kings, and several German kings, received the Iron Crown of Lombardy at Pavia. In the 12th cent. the city became a free commune, loyal, however, to the emperor. It was the last Lombard city to fall to the Visconti (1359). Pavia suffered heavily during the Italian wars, and near here, in 1525, Emperor Charles V defeated and captured Francis I of France. The city came successively under Spanish, French, and Austrian domination, and was liberated in 1859. Among Pavia's notable structures are the Romanesque St. Michael's Church (12th cent.) and the Lombard-Romanesque St. Peter's Church (12th cent.), where St. Augustine is buried. There is a university, which was established (1361) around a celebrated law school (founded in the 9th cent.).

Pav·lovsk (păv'lŏfsk'), town, NW European USSR, a summer resort near Leningrad. Founded by Catherine the Great in 1777, it became the royal summer residence in 1796, and in the 19th cent. it also served as a summer residence for the nobility of St. Petersburg. Pavlovsk contains English gardens, villas, mansions, a palace (1782–86) in the Russian classical style, several park pavilions (1780–83), and the mausoleum of Paul I (early 19th cent.).

Paw·hus·ka (pô-hŭs'kə), city (1970 pop. 4,238), seat of Osage co., NE Okla., NW of Tulsa; settled 1872, inc. 1906. It is in a grazing area that also yields oil and natural gas.

Paw·nee (pô-nē'). **1.** County (1970 pop. 8,484), 755 sq mi (1,955.5 sq km), central Kansas, in a rolling plains area drained by the Arkansas and Pawnee rivers; formed 1867; co. seat Larned. It is in a wheat and livestock area, with some mining and metal-products manufacturing. **2.** County (1970 pop. 4,473), 433 sq mi (1,121.5 sq km), SE Nebr., in a farm area bounded in the S by Kansas and drained by branches of the Nemaha River; formed 1857; co. seat Pawnee City. Farming yields livestock, grain, poultry, and dairy products. **3.** County (1970 pop. 11,338), 561 sq mi (1,453 sq km), N Okla., bounded in the NE by the Arkansas River and drained by the Cimarron River; formed 1907; co. seat Pawnee. Its agriculture includes cotton, barley, broomcorn, corn, fruit, livestock, and dairy products. It has oil and natural-gas wells and manufactures clothing.

Pawnee, city (1970 pop. 2,443), seat of Pawnee co., N central Okla., near the Arkansas River NW of Tulsa; founded c.1893 on the site of a trading post and Indian agency.

Pawnee City, city (1970 pop. 1,267), seat of Pawnee co., SE Nebr., WNW of Falls City; inc. 1858.

Paw Paw (pô' pô'), village (1970 pop. 3,160), seat of Van Buren co., SW Mich., WSW of Kalamazoo, in a resort and fruit-growing region. Wine and grape juice are produced.

Paw·tuck·et (pô-tŭk'ĭt), city (1978 est. pop. 68,100), Providence co., NE R.I., on the Blackstone River at Pawtucket Falls; settled 1671, inc. 1885 after the E section (which was part of Mass. until 1862) was merged with the W section into a R.I. town. Pawtucket has been a textile center since Samuel Slater built the nation's first successful water-powered cotton mill here in 1793. Among the city's many other manufactures are industrial fasteners, jewelry, Christmas ornaments, paper, and tires. The area, deeded to Roger Williams in 1638, was a haven for religious freedom in New England. Pawtucket's first settler was an ironworker who established (1671) a forge at the falls. After World War II, when much textile manufacturing moved south, Pawtucket shared the decline of many New England towns, but in the late 1950s its industries were revitalized with state aid. Two major urban renewal programs have greatly modernized the area. The city's chief point of interest is the 1793 Slater mill, now a museum.

Pa·xoí (pä-ksē') or **Pax·os** (păk'səs), island (1971 pop. 2,227), c.7 sq mi (18 sq km), NW Greece, in the Ionian Sea; one of the Ionian Islands. Olive oil, citrus fruits, and almonds are produced.

Pax·ton (păks'tən), city (1970 pop. 4,373), seat of Ford co., E central Ill., NNE of Champaign, in a farm area; settled in the 1850s by Swedes, inc. 1865. Hosiery and electrical equipment are made.

Pay·ette (pā-ĕt'), county (1970 pop. 12,401), 403 sq mi (1,043.8 sq km), W Idaho, in a farming area bounded in the W by the Snake River and Oregon and irrigated by the Payette River; formed 1917; co. seat Payette. Dairying and agriculture (hay, sugar beets, fruit, and truck crops) are the major occupations.

Payette, city (1975 est. pop. 5,235), seat of Payette co., W Idaho, near the junction of the Payette and Snake rivers; settled 1884, inc. 1891. It is a trade and processing center for an irrigated farm, dairy, and fruit region.

Payne (pān), county (1970 pop. 50,654), 697 sq mi (1,805.2 sq km), N central Okla., intersected by the Cimarron River; formed 1907; co. seat Stillwater. It has stock raising, dairying, and diversified agriculture (grain, cotton, and corn). There are oil and gas fields, and its industry includes food-processing plants, textile mills, petroleum refineries, and iron foundries.

Pay·san·dú (pī'sän-dōō'), city (1975 pop. 98,733), capital of Paysandú dept., W Uruguay, a port on the Uruguay River. It is the commercial center for a rich stock-raising and farming district. In the city are meat-packing plants. Paysandú was settled in 1772 by a missionary and his Christianized Indian followers.

Pa·zar·dzhik (pä-zär-jĭk'), city (1972 est. pop. 61,400), S central Bulgaria, on the Maritsa River. Pazardzhik, a commercial center, was under Turkish rule from the 15th to the 19th cent.

Pea·bod·y (pē'bŏd'ē, -bə-dē), city (1978 est. pop. 43,500), Essex co., NE Mass., a suburb of Boston, on the Danvers River; settled c.1633, inc. as South Danvers 1855, name changed 1868. Its tanning industry dates from early in the 18th cent. Leather goods, chemicals, electronic equipment, and machine tools are also produced. The Peabody Institute library contains much of the memorabilia of George Peabody, the philanthropist, for whom the city was named. There are many old houses here.

Peace (pēs), river, 945 mi (1,520.5 km) long, formed by the junction of the Finlay and Parsnip rivers at Williston Lake, N central British Columbia, Canada. It flows east through the Rocky Mts., then generally northeast across northern Alta. and onto the Northern Plains where it meanders to the Slave River at Lake Athabasca. From the head of the Finlay River the Peace River is 1,195 mi (1,922.8 km) long; it is one of the chief headstreams of the Mackenzie River. The valley of the middle Peace is fertile, with wheat the chief crop. Large natural-gas reserves are tapped along the river; oil, coal, salt, and gypsum deposits are also worked. The Peace River was probably discovered (1775–78) by Peter Pond, the American fur trader, and was first explored (1792–93) by Sir Alexander Mackenzie, the Canadian explorer. It was long an important route of fur traders. Settlement in the valley began in the early 1900s.

Peach (pēch), county (1970 pop. 15,990), 151 sq mi (391.1 sq km), central Ga., in a coastal plain region bounded on the W by the Flint River; formed 1924; co. seat Fort Valley. It is in a peach-growing and timber area that also produces cotton, corn, truck crops, peanuts, and pecans. Its manufactures include prepared feeds, wood products, and chemicals.

Peak District or **The Peak** (pēk), dissected plateau, c.30 mi (50 km) long and 22 mi (35 km) wide, Derbyshire, central England, forming the S extremity of the Pennines. Kinderscout (2,088 ft/636.8 m) is the highest peak. Peak District has many caves.

Pea Ridge National Military Park (pē): *see* National Parks and Monuments Table.

Pear·is·burg (pâr'ĭs-bûrg'), town (1970 pop. 2,169), seat of Giles co., SW Va., W of Roanoke in the Alleghenies near the W.Va. line; settled 1782, inc. 1914. It is a farm trade center.

Pearl (pûrl), river, 485 mi (780.4 km) long, rising in E Miss. and flowing S to Lake Borgne, an inlet of the Gulf of Mexico; its lower section forms the Miss.-La. boundary. The lower Pearl valley accounts for about one half of U.S. tung-oil production.

Pearl Harbor, land-locked harbor, on the S coast of Oahu island, Hawaii, W of Honolulu; one of the largest and best natural harbors in the E Pacific Ocean. In the Pearl Harbor vicinity are many U.S. military installations. The United States first gained rights here in 1887, when the Hawaiian monarchy granted permission for the maintenance of a coaling and repair station. After the United States annexed Hawaii in 1900, Pearl Harbor was made a naval base. On Dec. 7, 1941, while negotiations were going on with Japanese representatives in Washington, Japanese carrier-based planes swept in without warning over Oahu and attacked the bulk of the U.S. Pacific fleet, moored in Pearl Harbor. Nineteen naval vessels, including eight battleships, were sunk or severely damaged; 188 U.S. aircraft were destroyed. On Dec. 8, the United States declared war on Japan. Pearl Harbor is now a national historic landmark; a memorial has been built over the sunken hulk of the USS *Arizona.*

Pearl River, county (1970 pop. 27,802), 828 sq mi (2,144.5 sq km), S Miss., bounded on the W by the Pearl River and the La. border and drained by the Wolf River; formed 1890; co. seat Poplarville. It has cotton and corn farming, tung and pecan groves, timber, and industry (wood products, clothing, chemicals, and farm machinery).

Pearl River, uninc. village (1970 pop. 17,146), Rockland co., SE N.Y., near the N.J. line, a residential suburb of New York City.

Pear·sall (pîr'sôl'), city (1976 est. pop. 6,495), seat of Frio co., SW Texas, SW of San Antonio; settled 1881, inc. 1909. It is a processing center in a winter-garden area producing peanuts, watermelons, and other truck crops. Oil fields and cattle ranches are also in the region.

Pear·son (pîr'sən), city (1970 pop. 1,700), seat of Atkinson co., S Ga., WNW of Waycross. Naval stores and lumber are produced.

Pea·ry Land (pîr'ē), peninsula, N Greenland, extending into the Arctic Ocean. It terminates in Cape Bridgman in the northeast and Cape

Morris Jesup in the north, the most northerly point of land yet discovered. The area is mountainous (rising to c.6,400 ft/1,950 m) and is free of the inland icecap. Sparse vegetation supports musk oxen and caribou. The peninsula is named for Robert E. Peary who first explored it in his expedition of 1891-92.

Peć (pěch), town (1971 pop. 42,113), S Yugoslavia, in Serbia, in the Kossovo-Metohija region. A trade center, it has industries that produce jewelry, carpets, and small arms. The town is noted for its 13th cent. patriarchal cathedral and several Turkish mosques and houses.

Pe•cho•ra (pə-chō'rə, pyĭ-chô'rə), river, c.1,120 mi (1,800 km) long, rising in the N Urals, N European USSR. It flows generally north through the forest and tundra regions of the Komi Autonomous Republic and the Nenets National Okrug into Pechora Bay (an inlet of the Barents Sea), forming a vast delta at Naryan-Mar. The Pechora coal basin extends eastward to Vorkuta from the middle course of the river.

Pe•cos (pā'kəs, -kōs), county (1970 pop. 13,748), 4,740 sq mi (12,276.6 sq km), extreme W Texas, extending from the Glass Mts. in the SW to the Pecos River in the NE; formed 1871; co. seat Fort Stockton. It has large-scale ranching (cattle, sheep, and goats) and some irrigated agriculture (alfalfa seed, melons, and truck crops). Oil production and tourism are important to its economy.

Pecos, city (1970 pop. 12,682), seat of Reeves co., W Texas, on the Pecos River; inc. 1903. It is the market for an extensive ranch and farm area. It is also a sulfur, gas, and oil center. There are cattle feed lots, vegetable-packing houses, large automotive proving grounds, and a garment industry. Pecos was founded in the 1880s as a cow town at a crossing of the river. The annual rodeo, held here since 1883, was the world's first.

Pecos, river, 926 mi (1,490 km) long, rising in N N.Mex. near the Truchas peaks and flowing SE across E N.Mex. and W Texas to the Rio Grande. It drains c.38,300 sq mi (99,200 sq km).

Pecos National Monument: *see* National Parks and Monuments Table.

Pécs (pāch), city (1976 est. pop. 164,000), SW Hungary, near the Yugoslav border. Pécs is the industrial center of Hungary's chief coal-mining region. Coke, metals, agricultural machinery, tobacco, and leather goods are produced in the city, which is also famous for its pottery. There are extensive vineyards in the surrounding area. One of Hungary's oldest cities, Pécs was the site of a Celtic settlement and became the capital of the Roman province of Lower Pannonia under Emperor Hadrian. Pécs was under Turkish rule from 1543 to 1686. Many German miners and colonists settled here during the 18th cent., and in 1780 it became a free city. The 11th cent. cathedral (rebuilt in the late 19th cent.) is the most notable historic building in Pécs; the city also has an episcopal palace, a Turkish minaret, and several churches that were formerly mosques.

Pee•bles•shire (pē'bəlz-shîr', -shər), former county SE Scotland. Peebles was the county town. In 1975 Peeblesshire became part of the Borders region.

Pee Dee (pē' dē') or **Great Pee Dee,** river, c.435 mi (700 km) long, rising in the Blue Ridge, W N.C., and flowing NE then SE to Winyah Bay, S.C. It is called the Yadkin until it is joined by the Uharie River west of Troy, N.C.

Peeks•kill (pēks'kĭl), city (1978 est. pop. 21,300), Westchester co., SE N.Y., on the Hudson River; settled 1665, inc. as a village 1816, as a city 1940. Clothing, optical instruments, lighting fixtures, and office equipment are manufactured there. Peekskill was a prominent trade center after the Revolutionary War. St. Peter's Church, dedicated in 1767, has been restored.

Pe•gu (pĕ-gōō'), city (1970 est. pop. 125,000), capital of Pegu div., S Burma, on the Pegu River. It is a port and railway junction. Founded c.825 by the Mons, it became their capital when their king, Binnya U, established his palace here. Pegu was the center of one of the three chief states of Burma from the 14th to the late 15th cent.; in the 16th cent. it was the capital of a united Burmese kingdom. In the 18th cent. the Talaings rebelled against the Burmese and set up their capital at Pegu; it was destroyed by the Burmese in 1757, but was later rebuilt as the center of a Burmese province.

Pei (bā), river, c.200 mi (320 km) long, formed by the union of two headstreams in the Nan Ling Mts., N Kwangtung prov., S China, and flowing S into the Si River, E of Canton, to form the Canton River delta. Its entire course is navigable.

Pei-hai (bā'hī'), town (1970 est. pop. 175,000), Kwangsi Chuang Autonomous Region, SE China, a port on the Gulf of Tonkin. The chief pearl grounds of China are in the waters near Pei-hai. The town became a treaty port in 1877.

Pei•pus, Lake (pī'pəs): *see* Chudskoye, Lake, USSR.

Pe•ka•long•an (pə-kä-lông'än'), city (1971 pop. 111,537), N central Java, Indonesia, on the Java Sea. It is a textile and batik center and the principal port for central Java; sugar, rubber, and tea are exported. A Dutch fort was built here in 1753.

Pe•kin (pē'kĭn), city (1978 est. pop. 33,200), seat of Tazewell co., central Ill., a port on the Illinois River; inc. 1839. A processing, rail, and shipping point in a grain, livestock, and coal area, Pekin has a large food industry. Cereals, liquor, yeast, and malt are produced.

Pe•king (pē'kĭng', pā'-), city (1970 est. pop. 8,000,000), capital of the People's Republic of China. It is in central Hopeh prov., but constitutes an independent unit (6,564 sq mi/17,000 sq km) administered directly by the national government. The second-largest city in China (after Shanghai), Peking is the political, cultural, financial, educational, and transportation center of the country. It has also become a great industrial area, the heart of a vast complex of textile mills, ironworks, steelworks, railroad repair shops, machine shops, chemical plants, and factories manufacturing heavy machinery, electronic equipment, aircraft, plastics, synthetic fibers, and rolling stock. It is a rail hub, receiving lines from all sections of the country. It has air connections with all major Chinese cities and with numerous foreign countries.

Since 723 B.C. several cities, bearing various names, have existed at this site. The nucleus of the present city was Kublai Khan's capital, Cambuluc (constructed 1260-90). Under the name Peking the city was the capital of China from 1421 until 1911. The gateway to Mongolia and Manchuria, it was often the prize of contending armies. In 1860 Great Britain and France captured it after the battle of Pa-li-ch'iao and forced the Chinese government to concede the Legation Quarter for foreign settlements. This cession was among the factors responsible for the Boxer Rebellion (1900), in which the foreign colony was besieged until relieved by a combined expeditionary force of American, Japanese, and European troops. The city repeatedly changed hands during the civil wars that followed the establishment of the Chinese Republic in 1911-12. From 1912 to 1927 Peking, Canton, and Han-k'ou alternated as centers of government. In 1928 when the seat of government was transferred to Nanking, the name Peiping was adopted. Japan occupied the city after the famous Marco Polo Bridge incident in 1937 during the Second Sino-Japanese War. The Japanese made the city the capital of a puppet state. With the end of World War II and the abolition of the last foreign concessions (1946), the city was entirely restored to Chinese sovereignty. In Jan., 1949, it fell to the Communists, who later that year designated it the capital of the newly founded People's Republic of China and restored the name Peking.

Peking in the main consists of two formerly walled districts, the Outer or Chinese City and the Inner or Tatar City. The 25 mi (40 km) of ramparts and monumental gates that once surrounded the cities have been razed by the Communist government and replaced by wide avenues. Within the Tatar City is the Forbidden City (formerly the emperor's residence), the Imperial City (where his retinue was housed), and the Legation Quarter. The Imperial City is now the seat of the Communist government. On the southern edge of the Tatar City is the T'ien An Men Square, which contains the monument to the heroes of the revolution, the Great Hall of the People, and the museum of history and revolution. Celebrations held in the square include May Day and the founding date (Oct. 1) of the People's Republic. Peking is known for its artificial lakes and for its parks and temples. It contains many of the greatest examples of architecture of the Ming and Ch'ing dynasties. The Temple of Heaven (15th cent.) is set in a large park and has a massive altar of white marble before which the emperors prayed at the summer solstice. In the temple of Confucius, built by Kublai Khan, are guarded incised boulders that date from the Chou dynasty. The Forbidden City, now a vast museum, contains the imperial palaces (two groups of three each) and smaller palaces, all filled with art treasures. Just outside Peking, rivaling the beauties within, is the imperial summer palace with its lovely parks. After 1949 Peking began to spread well beyond its two core cities, and hundreds of new buildings, hotels, and cultural centers now dot the suburbs. A subway was completed in 1969. Peking has an opera, a ballet, and the impressive national library. It is the seat of many learned societies, research organizations, academies of fine arts, drama, dance, and music, and more than 25 institutions of higher learning. The Peking zoo is famous for its collection of pandas. In addition to the many tourist attractions in the city, the Great Wall and the gigantic Ming tombs are easily accessible. At nearby Chou-k'ou-tien were discovered several fossil bones of *Sinanthropus pekingensis* (Peking man), an early example of prehistoric man.

Pe·lée (pə-lā'), volcano, 4,800 ft (464 m) high, on N Martinique Island, French West Indies. On May 8, 1902, Pelée erupted, engulfing Saint-Pierre at its base and killing c.40,000 people in the city and adjacent area. Because the chemical composition of the volcanic ash prevents plant life, the thick layer deposited over a wide expanse is almost complete wasteland. Pelée also erupted in 1792 and 1851.

Pe·lee Island (pē'lē), 18 sq mi (46.6 sq km), S Ont., Canada, in W Lake Erie. Ferry service connects the island with the Canadian and U.S. mainland.

Pel·ham Man·or (pĕl'əm măn'ər), village (1970 pop. 6,673), Westchester co., SE N.Y., a suburb NE of New York City on the N shore of Long Island Sound; settled in mid-17th cent., inc. 1891. Anne Hutchinson was murdered by Indians near here.

Pe·li·on (pē'lē-ən), mountain, 5,252 ft (1,601.9 m) high, N Greece, E Thessaly, on the Aegean coast. In ancient legend, the centaur Chiron lived on the mountain.

Pel·la (pĕl'ə), ancient city of Greek Macedonia, NW of Thessalonica (now Thessaloniki). It became the capital of the Macedonian kingdom in the 4th cent. B.C. It declined after the Roman conquest of Macedonia (168 B.C.). Alexander the Great was born at Pella. Modern excavations have revealed many ancient buildings.

Pella, city (1970 pop. 6,688), Marion co., S central Iowa, ESE of Des Moines; settled 1847 by Dutch, inc. 1855. A trade center in a farm and livestock area, it also has some manufacturing. An annual tulip festival is held.

Pel·ly (pĕl'ē), river, c.330 mi (530 km) long, rising W of the Mackenzie Mts., S central Yukon Territory, Canada, and flowing generally NW to join the Yukon River at Fort Selkirk. The Pelly was discovered (1840) by Robert Campbell of the Hudson's Bay Company.

Pelly Lake, 331 sq mi (857.3 sq km), NW Keewatin dist. and NE Mackenzie dist., Northwest Territories, Canada, W of and connected to Lake Garry. Back River is its outlet on the east.

Pel·o·pon·ne·sus (pĕl'ə-pə-nē'səs), peninsula (1971 pop. 986,912), c.8,300 sq mi (21,500 sq km), S Greece. It is linked with central Greece by the Isthmus of Corinth, and is washed by the Aegean Sea on the east and southeast, by the Ionian Sea on the southwest and west, and by the gulfs of Pátrai and Corinth on the north. Its deeply indented south coast terminates in Cape Matapan. Predominately agricultural and pastoral, the Peloponnesus produces currants, grapes, figs, citrus fruit, olives, tobacco, and wheat. The most fertile parts of the peninsula are the coastal strips in the north and west. Sheep and goat raising, textile manufacturing, fishing, and sericulture are major sources of income.

The chief ancient divisions of the Peloponnesus were Elis, Achaea, Argolis, and the city-state of Corinth in the north; Arcadia in the center; and Lacedaemonia in the south. Sparta, Corinth, Argos, and Megalopolis were among its chief cities in ancient times. Originally populated by Leleges and Pelasgians, the peninsula was later occupied by the Achaeans and then by the Dorians, who dominated the Peloponnesus in historic times. At the time of the Peloponnesian War (5th cent. B.C.) almost the entire peninsula was dominated by Sparta. Spartan hegemony was broken in the 4th cent. B.C. by Epaminondas of Thebes, who thus prepared the way for the establishment of Macedonian supremacy over the Peloponnesus.

Under Roman (from 146 B.C.) and Byzantine rule the Peloponnesus was reduced to provincial status and in the centuries that followed was repeatedly raided and invaded by Slavs, Bulgars, and Petchenegs. In 1204 the French Villehardouin family received the principality of Achaia or Achaea (i.e., the Peloponnesus) as fief, except for several ports, which passed to Venice. A French feudal state was created and enjoyed a period of great prosperity and chivalrous culture. The principality passed first to the Angevin dynasty of Naples (1278) later to various nobles, then to the Byzantine Greeks (1432), and finally to the Ottoman Turks (1460). As a result of the Greek War of Independence (1821–29) the peninsula passed to independent Greece.

Pe·lo·tas (pə-lô'təsh), city (1970 pop. 150,278), Rio Grande do Sul state, S Brazil, an inland port on the São Gonçalo canal. It is a major export point for a stock-raising region and a transfer point for ocean-to-lake shipping. The city is Brazil's main producer of dried beef. Lard, shoes, furniture, candles, and soap are also manufactured.

Pe·lu·si·um (pə-lo͞o'shē-əm), ancient city of Egypt, on the easternmost branch of the Nile (long since silted up) and E of Modern Port Said. It was an important fortress against attacks from the east. The Assyrians under Sennacherib were supposedly struck by pestilence at Pelusium, and in 525 B.C. the Persians under Cambyses overthrew Psamtik III here. There are Roman remains on the site.

Pem·ba (pĕm'bə), island (1967 pop. 164,321), c.380 sq mi (985 sq km), NE Tanzania, in the Indian Ocean just off the E African mainland. The island is the world's leading producer of cloves. Coconuts are also exported, and fishing is an important industry. Traders from the Persian Gulf region settled on the island beginning in the 10th cent. The Portuguese occupied the island in the 16th cent. but were displaced by Omani Arabs in 1698. In 1822 the island was conquered by Sayyid Said (later the sultan of Zanzibar). Along with Zanzibar, Pemba passed under British rule in 1890, became independent in 1963, and merged with Tanganyika to form Tanzania in 1964.

Pem·bi·na (pĕm'bə-nə, -nô'), county (1970 pop. 10,728), 1,124 sq mi (2,911.2 sq km), extreme NE N.Dak., in an agricultural area drained by the Tongue and Pembina rivers and bounded on the E by the Red River of the North and on the N by the Canadian border; formed 1867; co. seat Cavalier. Its agriculture includes livestock, grains, sugar beets, and potatoes. Retail trade, food processing, and manufacturing are also important.

Pem·bi·na (pĕm'bĭ-no͞o'), river, c.210 mi (337.9 km) long, central Alta., Canada, rising in the Rocky Mts. near the E boundary of Jasper National Park and flowing generally NE and N into Athabaska River W of Athabaska.

Pem·broke (pĕm'brōk', -bro͝ok'), town (1976 pop. 14,927), SE Ont., Canada, NW of Ottawa, on the Ottawa River. It is a lumbering center and has steel and electric-products factories.

Pembroke, city (1970 pop. 1,361), seat of Bryan co., SE Ga., W of Savannah, in a farm and timber area.

Pembroke, municipal borough (1976 est. pop. 14,570), Dyfed, SW Wales, on an inlet of Milford Haven bay. The town is an agricultural market with tourism, ship repairing, and light industries; it was formerly the site of several military bases. Pembroke contains a 10th cent. priory and 12th cent. castle. Henry VII was born here.

Pem·broke·shire (pĕm'brŏk-shĭr', -shər, -bro͝ok-), former county, SW Wales. The county town was Haverfordwest. Pembrokeshire is rich in megalithic remains. The region was harried by the Norsemen in the early Middle Ages and conquered by the Normans in the 11th cent. In the 12th cent. Flemish settlers were brought into the district. In 1974 Pembrokeshire became part of the new nonmetropolitan county of Dyfed.

Pem·i·scot (pĕm'ĭ-skät', -skō'), county (1970 pop. 26,373), 493 sq mi (1,276.9 sq km), extreme SE Mo., on the Mississippi River; formed 1861; co. seat Caruthersville. Agricultural products include corn, wheat, oats, hay, and dairy goods. Cotton and gum and oak timber are grown and processed here.

Pe·nang (pə-năng'): *see* Pinang, Malaysia.

Pen·arth (pĕ-närth'), urban district (1971 pop. 23,965), South Glamorgan, S Wales. A suburb of Cardiff and a seaside resort, it also has cement works and is a coal port.

Pen-ch'i (bĕn'chē'), city (1970 est. pop. 750,000), S Liaoning prov., China. It is an important heavy industrial center with rich iron and coal mines. It was founded as a metallurgical center in 1915.

Pen·del·i·kón (pĕn'dĕl-ĭ-kôn'), mountain, c.3,670 ft (1,120 m) high, central Greece, NE of Athens. The white marble quarried here was used for many buildings of ancient Athens.

Pen·der (pĕn'dər), county (1970 pop. 18,149), 871 sq mi (2,255.9 sq km), SE N.C., on the Atlantic, bounded on the SE by Onslow Bay and drained by a branch of the Cape Fear River; formed 1875; county seat Burgaw. This forested tidewater area has farming (tobacco, corn, and vegetables), sawmilling, and fishing.

Pender, village (1970 pop. 1,229), seat of Thurston co., NE Nebr., SE of Wayne; inc. 1886.

Pen·dle·ton (pĕn'dl-tən). **1.** County (1970 pop. 9,949), 279 sq mi

(722.6 sq km), N Ky., mostly in the outer Bluegrass, bounded in the NE by the Ohio River and the Ohio line, and drained by the Licking River and its South Fork; formed 1787; co. seat Falmouth. It is in a gently rolling upland agricultural area that produces burley tobacco, dairy products, poultry, hay, corn, and honey. Stone quarrying and food processing are done, and shoes, metal products, and auto parts are made. **2.** County (1970 pop. 7,031), 695 sq mi (1,800 sq km), E W.Va. in the Eastern Panhandle, bounded on the E, S and SW by Va., and traversed by the Allegheny Front in the W; formed 1788; co. seat Franklin. Its agriculture includes livestock, dairy products, and fruit, and it has crushed stone and limestone industries and timber (sawmills). The county includes Seneca Caverns, Smoke Hole Caverns, Seneca Rocks and parts of George Washington and Monongahela National Forests.

Pendleton, city (1978 est. pop. 14,000), seat of Umatilla co., NE Oregon, on the Umatilla River, in the foothills of the Blue Mts.; founded 1869 on the old Oregon Trail, inc. 1889. A distribution and trade center for an extensive wheat, livestock, and timber region, Pendleton also has woolen and pine mills.

Pend O·reille (pän' də-rā'), county (1970 pop. 6,025), 1,406 sq mi (3,641.5 sq km), extreme NE Wash., on the Idaho and British Columbia borders, drained by the Clark Fork River; formed 1911; co. seat Newport. It has lumbering, mining (lead and zinc ores, gold, silver, and copper), and agriculture. Part of Kaniksu National Forest is here.

Pend Oreille Lake, 148 sq mi (383.3 sq km), 65 mi (104.6 km) long, and 1,200 ft (366 m) deep, N Idaho. It is the largest lake in Idaho and one of the largest and deepest lakes in the United States. The lake, with the surrounding national forests, is a scenic landmark in a farming, lumbering, and mining region.

Pe·ne·us (pə-nē'əs), river, Greece: *see* Piniós.

Pen·ner (pĕn'ər), river, 350 mi (563.2 km) long, rising in the Eastern Ghats, Karnataka state, S India, and flowing N into Andhra Pradesh state, then E to the Bay of Bengal, near Nellore. The river is used for irrigation.

Pen·nine Alps (pĕn'īn'), mountain range, SW of Central Alps, stretching SW-NE along the Swiss-Italian border from Great St. Bernard Pass to Simplon Pass. It rises to 15,203 ft (4,637 m).

Pen·nines (pĕn'īnz') or **Pennine Chain,** mountain range, sometimes called the "backbone of England," extending c.160 mi (260 km) from the Cheviot Hills on the Scottish border to the Peak District in Derbyshire. The range consists of a series of upland blocks, separated by transverse valleys. There are caverns, and several chasms are more than 300 ft (91.5 m) in depth. Cross Fell (2,930 ft/893.7 m) is the highest peak. The range is sparsely populated; sheep raising, quarrying, and tourism are important economic activities.

Pen·ning·ton. 1. County (1970 pop. 13,266), 622 sq mi (1,611 sq km), NW Minn., drained by the Red Lake and Thief rivers; formed 1910; co. seat Thief River Falls. In an agricultural area yielding dairy products, small grains, and livestock, it has a food-processing industry and manufactures transportation equipment. **2.** County (1970 pop. 59,349), 2,776 sq mi (7,189.8 sq km), SW S.Dak., on the Wyo. border, crossed in the W by the Black Hills and traversed in the E by the Cheyenne River; formed 1875; co. seat Rapid City. It is in a grain-farming and ranching region that has gold mines and granite and limestone quarries. Its industries include food processing, lumbering, and the manufacture of wood products, concrete, and machinery. Mt. Rushmore National Memorial, Custer State Park, Black Hills National Forest, and Badlands National Monument are in the county.

Penn·syl·va·nia (pĕn'səl-vān'yə, -vā'nē-ə), state (1975 pop. 11,808,-000), 45,333 sq mi (117,412 sq km), E United States, one of the Middle Atlantic states and one of the Thirteen Colonies. It is bounded on the north by Lake Erie and New York; on the east by the winding Delaware River, which separates it from N.Y. and N.J.; on the south by Del., Md., and W.Va.; and on the west by W.Va. and Ohio. Harrisburg, midway between the metropolitan areas of Philadelphia and Pittsburgh, is the capital.

Except for coastal plains in northwestern and southeastern Pennsylvania, the state is a succession of mountains and rolling hills with narrow valleys. Mountain ridges rise from the Delaware River into the Piedmont, with the parallel ranges of the Blue Mts. and Allegheny Mts. occupying the central portion of the state. Uplands of the plateau extend to the western border. In the east and center Pennsylvania is drained by the Delaware and the Susquehanna river systems, and in the west by the Allegheny and the Monongahela rivers, which join at Pittsburgh to form the Ohio River. The great forests and lush vegetation that once covered the entire state were transformed during the Carboniferous period into tremendous deposits of anthracite coal in the east and extensive bituminous beds in the west. Large areas of woodland remain and, in some isolated sections, have retained an almost primitive wildness.

Pennsylvania is divided into 67 counties:

NAME	COUNTY SEAT	NAME	COUNTY SEAT
Adams	Gettysburg	Lackawanna	Scranton
Allegheny	Pittsburgh	Lancaster	Lancaster
Armstrong	Kittanning	Lawrence	New Castle
Beaver	Beaver	Lebanon	Lebanon
Bedford	Bedford	Lehigh	Allentown
Berks	Reading	Luzerne	Wilkes-Barre
Blair	Hollidaysburg	Lycoming	Williamsport
Bradford	Towanda	McKean	Smethport
Bucks	Doylestown	Mercer	Mercer
Butler	Butler	Mifflin	Lewistown
Cambria	Ebensburg	Monroe	Stroudsburg
Cameron	Emporium	Montgomery	Norristown
Carbon	Mauch Chunk	Montour	Danville
Centre	Bellefonte	Northampton	Easton
Chester	West Chester	Northumberland	Sunbury
Clarion	Clarion	Perry	New Bloomfield
Clearfield	Clearfield	Philadelphia	Philadelphia
Clinton	Lock Haven	Pike	Milford
Columbia	Bloomsburg	Potter	Coudersport
Crawford	Meadville	Schuylkill	Pottsville
Cumberland	Carlisle	Snyder	Middleburg
Dauphin	Harrisburg	Somerset	Somerset
Delaware	Media	Sullivan	Laporte
Elk	Ridgway	Susquehanna	Montrose
Erie	Erie	Tioga	Wellsboro
Fayette	Uniontown	Union	Lewisburg
Forest	Tionesta	Venango	Franklin
Franklin	Chambersburg	Warren	Warren
Fulton	McConnellsburg	Washington	Washington
Greene	Waynesburg	Wayne	Honesdale
Huntingdon	Huntingdon	Westmoreland	Greensburg
Indiana	Indiana	Wyoming	Tunkhannock
Jefferson	Brookville	York	York
Juniata	Mifflintown		

Economy. Iron smelting became important in the 18th cent., made possible by abundant supplies of ore and hardwoods for the furnaces. In the 19th cent. Pennsylvania emerged as the nation's leading steel producer, and its giant mills now account for about one fourth of the total U.S. output. The state is also one of the leaders in nickel production, and oil is refined. Heavily industrialized as well, the state has factories that manufacture metal products, foodstuffs, machinery, chemicals, and wearing apparel. Agriculture is concentrated in the fertile counties of the southeast, the principal crops being hay, corn, wheat, oats, tobacco, barley, rye, potatoes, and fruit. Transportation facilities have kept pace with the state's rapid development. The accessible and landlocked harbor of Philadelphia has been supplemented by the Great Lakes port of Erie and access to the South and Midwest by Pittsburgh and the Ohio River.

History. In the early 1600s the English, Dutch, and Swedes disputed right to the region. The original permanent settlement was made at Tinicum Island in the Schuylkill River (1643) by Johan Printz, governor of New Sweden. Swedish jurisdiction was short-lived as the Dutch (1655) and then the British (1664) took over the Delaware area. The duke of York remained in control until 1681, when, in payment of a royal debt, William Penn was granted proprietary rights to almost the whole of what is now Pennsylvania. A devout Quaker, who had suffered for his beliefs, Penn viewed his colony as a holy experiment, designed as an asylum for the persecuted. In 1681 he sent William Markham as deputy to establish a government at Uppland and sent commissioners to plot the City of Brotherly Love (Philadelphia). Penn carefully constructed a constitution that gave Pennsylvania the most liberal government in the colonies. In 1682 Penn arrived at Uppland (renamed Chester). Shortly thereafter he met with the chiefs of the Delaware tribes and a treaty was signed promoting long-lasting good will between the Indians and the white settlers. After William and Mary ascended (1689) the throne of England, the province was taken away from Penn, but in 1694 full proprietary rights were restored.

By the time of Penn's death in 1718, Pennsylvania had developed into a dynamic and growing colony, enriched by the continuous immigration of numerous different peoples. The Quakers, English, and Welsh were concentrated in Philadelphia and the eastern counties. The Germans (Pennsylvania Dutch) settled in the farming areas of southeastern Pennsylvania, where they retained their cohesion and to a considerable extent their culture. After 1718 the Scotch-Irish began colonizing in the Cumberland valley and gradually pushed the frontiers toward western Pennsylvania. Resentful of encroachment on their lands, the Indians allied themselves with the French, who were then fortifying positions in the Ohio valley. The frontier settlements were severely ravaged until the French abandoned (1758) Fort Duquesne and Pontiac's Indian rebellion (1763) was suppressed.

A growing struggle between those favoring and those opposing rule by proprietors was overshadowed by the opposition to British imperial policy that culminated in the American Revolution. In 1776 a provincial convention created the Commonwealth of Pennsylvania under one of the most democratic of the new state constitutions (replaced in 1790). The state was invaded, Philadelphia was occupied by the British, and Valley Forge witnessed the heroic endurance of Washington's troops in the winter of 1777–78. Philadelphia, host to the First and Second Continental Congresses and scene of the signing of the Declaration of Independence, was the site of the Constitutional Convention of 1787 and served as the seat of the new Federal government from 1790 to 1800. But opposition to Federal taxation in rural Pennsylvania led to violence in the Whiskey Rebellion of 1794 and the Fries Rebellion of 1798, while anti-Eastern sentiment forced removal of the state capital to Lancaster in 1799, then to Harrisburg in 1812.

The economic and social development of western Pennsylvania encouraged programs of internal improvements. Turnpikes, canals, and eventually railroads hastened transportation and communication across the wide state. Free public education emerged in the Free School Act of 1834.

Pennsylvania was the scene of several battles in the Civil War, notably the Gettysburg campaign of 1863. With the close of the war came the rapid emergence of the state as an industrial commonwealth. Supported by high protective tariffs, the industries found favorable markets and a constant supply of immigrant labor. The first oil well was dug at Titusville in 1859. But it was steel that became the basic industry, using iron ore brought across the Great Lakes and the native Pennsylvania coal.

In the face of increasing concentration of power, labor struggled to achieve safer working conditions, higher wages, and shorter hours. This brought bloodshed during the fight between mine owners and the radical Molly Maguires and reached a climax in the strike at Homestead in 1892. During the 1930s the Congress of Industrial Organizations (CIO) successfully promoted unionization in the steel industry, while the United Mine Workers acquired increasing strength in the coal fields.

Government. Pennsylvania is governed under the constitution adopted in 1873 and amended extensively since then. The governor serves a four-year term and may succeed himself for one additional term. The state legislature, called the general assembly, consists of a senate of 50 members and a house of representatives of 203 members.

Educational Institutions. Among the state's many institutions of higher education are Bryn Mawr College, at Bryn Mawr; Bucknell Univ., at Lewisburg; Carnegie-Mellon Univ., the Univ. of Pittsburgh, and Duquesne Univ., at Pittsburgh; Dickinson College, at Carlisle; Drexel Univ., Temple Univ., the Univ. of Pennsylvania, and La Salle College, at Philadelphia; Haverford College, at Haverford; Lehigh Univ., at Bethlehem; Pennsylvania State Univ., at University Park; Swarthmore College, at Swarthmore; Villanova Univ., at Villanova; and Lafayette College, at Easton.

Pennsylvania Avenue National Historic Site: *see* National Parks and Monuments Table.

Penn Yan (pĕn' yăn'), village (1970 pop. 5,293), seat of Yates co., W central N.Y., S of Geneva on an arm of Keuka Lake, in the Finger Lakes grape region; inc. 1883. It was settled by Pennsylvanians and New Englanders ("Yankees").

Pe·nob·scot (pə-nŏb'skət, -skŏt'), county (1970 pop. 125,393), 3,390 sq mi (8,780.1 sq km), E central Me., drained by the Penobscot River, with numerous lakes; formed 1816; co. seat Bangor. It is in a diversified agricultural, lumber, and industrial area yielding paper and allied products, textiles, wood products, shoes, concrete, electrical machinery and equipment, and boats. There is hunting and fishing in the north.

Penobscot, river, 350 mi (563.2 km) long, rising in numerous lakes in central Maine and flowing generally E in four branches, uniting, then flowing S into Penobscot Bay. The river, navigable to Bangor, is an important source of power for pulpwood and paper mills. The Penobscot's upper course is in a wooded region famous for hunting, fishing, and canoeing. Pulpwood and petroleum products are the principal freight on the river.

Penobscot Bay, inlet of the Atlantic Ocean, 35 mi (56.3 km) long and 27 mi (43.4 km) wide, S Maine. The bay was entered by the English explorer Martin Pring in 1603; the French explorer Samuel de Champlain claimed the area for France in 1604. Trading posts and missions were established, and for many years the possession of the area was disputed among the French, English, and Americans.

Pen·sa·co·la (pĕn'sə-kō'lə), city (1978 est. pop. 65,900), seat of Escambia co., extreme NW Fla., on Pensacola Bay, an inlet of the Gulf of Mexico; inc. 1822. It is a port of entry with a good natural harbor and shipping and fishing industries. The city has industries that produce synthetic fibers, paper products, chemicals, naval stores, and nuclear-reactor parts. The Spanish established a short-lived settlement (1559–61) here. In 1698 a new Spanish colony was founded. Between 1719 and 1723 possession of Pensacola shifted between the Spanish and the French. The Spanish then held it until 1763, when it passed to the British. It again became Spanish in 1783 and was the capital of West Florida until 1822. Although still Spanish, Pensacola was a British base in the War of 1812 until it was captured (1814) by Andrew Jackson. The United States took formal possession in 1821 after the purchase of Florida. During the Civil War the city was abandoned (1862) to Union forces, but Fort Pickens had remained in Federal hands from the beginning. Much of the life of the city is related to the U.S. naval air station, established here in 1914. Of interest are several historical museums and the naval-aviation museum at the air station. The ruins of old Fort Barrancas and of Forts San Carlos (built in the 1780s), Pickens, and McRae (built in the 1830s) are on the shores of Pensacola Bay. The eastern section of Gulf Islands National Seashore is here.

Pen·tic·ton (pĕn-tĭk'tən), city (1976 est. pop. 21,344), S British Columbia, Canada, located where the Okanagan River flows into Okanagan Lake. It is a service and trade center for a resort and fruit-growing area.

Pent·land Firth (pĕnt'lənd), channel, 6 to 8 mi (9.7–12.9 km) wide and c.14 mi (23 km) long, N Scotland. Connecting the North Sea with the Atlantic Ocean, it separates Caithness on the Scottish mainland from the Orkney Islands.

Pen·za (pĕn'zä, pyĕn'zə), city (1976 est. pop. 436,000), capital of Penza oblast, S central European USSR, on the Sura River. It is a large railroad junction and the center of an extensive and fertile black-earth district. There are machine, paper, and food-processing industries. It was founded in 1666 as a fortress.

Pen·zance (pĕn-zăns'), municipal borough (1973 est. pop. 19,360), Cornwall, SW England, at the head of Mounts Bay. Penzance is the westernmost borough in England. It is a resort and a port for the Scilly Islands and has flour mills. The borough was sacked by the Spanish in 1595 and until the 18th cent. was subject to raids by Mediterranean pirates.

Pe·o·ria (pē-ōr'ē-ə, -ôr'-), county (1970 pop. 195,318), 624 sq mi (1,616.2 sq km), central Ill., bounded on the E and S by the Illinois River and Lake Peoria and drained by the Spoon River and Kickapoo Creek; formed 1825; co. seat Peoria. It has agriculture (corn, wheat, oats, soybeans, livestock, poultry, truck crops, fruit, and dairy products), bituminous-coal mines, sand and gravel deposits, timber, commercial fisheries, diversified manufacturing, and a food-processing industry.

Peoria, city (1978 est. pop. 126,000), seat of Peoria co., central Ill., on Lake Peoria and the Illinois River; inc. as a city 1845. A busy port of entry, it is a regional trade and transportation point; grain, livestock, and coal from the area are marketed, processed, and shipped in Peoria. It has large distilleries, a brewery, and factories producing numerous heavy and light industrial goods. La Salle established Fort Creve Coeur in the region in 1680, and the spot later became a French trading post. It was twice visited by military expeditions in the Revolutionary War. During the War of 1812 it was the scene of Indian depredations, and Fort Clark was built here in 1813. The first permanent American settlement was established in 1819.

Pep·in (pĭp'ən, pĕp'-), county (1970 pop. 7,319), 237 sq mi (613.8 sq km), W Wis., in a hilly region bounded in the SW by Lake Pepin and drained by the Chippewa River; formed 1858; co. seat Durand. Its agriculture includes corn, beans, hay, dairy products, and livestock. It has a food-processing industry and manufactures tires and farm machinery and equipment.

Pepin, Lake, a widening of the Mississippi River, 21 mi (34 km) long and c.3 mi (5 km) wide, SE Minn., between Wabasha and Red Wing; formed by a natural dam of silt dropped by the Chippewa River. It is a popular recreation area.

Pep·per·ell (pĕp'rəl, pĕp'ə-rĕl), town (1970 pop. 5,887), Middlesex co., N Mass., near the N.H. border W of Lowell; settled 1720, set off from Groton 1753. Paper products, shoe laces, and yarn are made.

Per·cé Rock (pĕr-sā'), E Que., Canada, just off the tip of the Gaspé Peninsula. It is a massive rock (1,420 ft/433 m long; c.300 ft/90 m wide; and 290 ft/88 m at the highest point), rising sheer from the Atlantic. It takes its name from an arch 50 ft (15 m) high near its seaward end. With nearby Bonaventure Island, it is a well-known tourist attraction.

Perche (pĕrsh), region and former county, NW France. Horse breeding is a significant industry, and Perche has given its name to the Percheron horse. Much of the region is forested, and there are many apple orchards and grasslands. Perche was attached to the French crown in 1525.

Per·di·do, Mon·te (môn'tĕ pĕr-*thē'thô*), peak, 11,007 ft (3,357.2 m) high, NE Spain, near the French border. It is one of the highest peaks of the Pyrenees.

Pe·rei·ra (pĕ-rā'rä), city (1973 pop. 174,128), capital of Risaralda dept., W central Colombia, in the upper Cauca valley. It is a major distribution center for coffee and cattle and for the mineral resources (gold and silver) of the region.

Pe·re·kop, Isthmus of (pĕr'ĭ-kôp'), c.19 mi (30 km) long and from 5 to 14 mi (8–23 km) wide, SW European USSR, in the Ukraine, connecting the Crimea with the Ukrainian mainland. It separates the Gulf of Perekop (an arm of the Black Sea) in the west from the Sivash Sea (an inlet of the Sea of Azov) in the east. Because of its strategic position and economic importance (salt extraction from the lakes in the southern part), the Greeks and Tatars fortified the isthmus with moats and ramparts. Before the 15th cent. there was a Genoese colony here. The isthmus passed to Russia in 1783.

Pe·re·slavl-Za·les·ski (pĕr'ə-slăv'əl-zə-lĕs'kē), city (1974 est. pop. 33,000), central European USSR. It has industries that produce textiles, film, and food products. The city was founded in 1152, was included in the Suzdal principality, and from 1175 to 1302 was the capital of an independent principality; it passed to Moscow in 1302. The city is on Lake Pleshcheyevo (19 sq mi/49.2 sq km), where Peter I built the first ships of the Russian navy. Remains of the flotilla are in a nearby museum.

Per·ga·mum (pûr'gə-məm), ancient city of NW Asia Minor, in Mysia (modern Turkey), in the fertile valley of the Caicus. It became important c.300 B.C., after the breakup of the Macedonian empire. The independence of Pergamum ended when Attalus III (d.133) bequeathed the kingdom to the Roman people. The chief glory of Pergamum was its sculpture, including the Dying Gaul and a frieze for a great altar of Zeus, glorifying the defeat (190) of Antiochus III of Syria at Magnesia. The cultured Pergamene rulers also built up a library second only to the one at Alexandria. One of the library's specialties was the use of parchment, which takes its name from the city. Eventually the library was given by Antony to Cleopatra. Under Rome, Pergamum was reconstituted as the province of Asia, and Ephesus rapidly eclipsed Pergamum as the chief city of Asia Minor. Pergamum accepted Christianity early; it was one of the Seven Churches of Asia.

Per·i·bon·ca (pĕr'ə-bŏng'kə), river, c.280 mi (450 km) long, rising in the Otish Mts., central Que., Canada, and flowing S through Peribonca Lake to Lake St. John. It is an important source of hydroelectric power.

Pé·ri·gord (pā-rē-gôr'), region of SW France. The region consists of low, arid limestone plateaus, the deep and fertile valleys of the Lot and Dordogne rivers, and extensive oak forests. Périgord is noted for its truffles and goose livers. There are numerous cave dwellings from the Paleolithic period. Occupied during Gallic and Roman eras by the Petrocorii, Périgord became a county under the Merovingians (9th cent). First enfeoffed to the dukes of Aquitaine, it later passed to England, was returned to France c.1370 as a fief of the French crown, and passed eventually to the house of Bourbon (1574). It was inherited by Henry of Navarre, and after he became king of France as Henry IV (1589), Périgord was incorporated (1607) into the royal domain as part of the province of Guienne.

Pé·ri·gueux (pā-rē-gœ'), city (1975 pop. 35,120), capital of Dordogne dept., SW France, on the Isle River. A commercial center, it is famous for its pâtés The city's major manufactures are tobacco products, chemicals, and leather goods. Périgueux was the ancient Vesumna or Vésona of Gallic Petrocorii and became the capital of Périgord in the 9th cent. It passed to Philip II of France in 1204, was taken by the English in 1356, was regained by France in 1454, and became (16th cent.) a Protestant stronghold. Remains from Roman times include large arenas and an amphitheater.

Pe·rim (pə-rĭm'), island, c.5 sq mi (13 sq km), off SW Arabian Peninsula in the Bab el Mandeb strait. A rocky and barren island, it is strategically located at the southern entrance to the Red Sea. Perim was occupied by France in the 18th cent., and then briefly by Britain in 1799. It was reoccupied by the British in 1857 and later connected administratively with Aden. Perim flourished between 1883 and 1936 as a coaling station. In 1967 Perim's small population voted to become part of Southern Yemen.

Per·kins (pûr'kĭnz). **1.** County (1970 pop. 3,423), 885 sq mi (2,292.2 sq km), SW central Nebr., in an agricultural area bordered in the W by Colo.; formed 1887; co. seat Grant. Wheat, grain, and livestock are raised. **2.** County (1970 pop. 4,769), 2,860 sq mi (7,407.4 sq km), NW S.Dak., on the N.Dak. border, drained by the North Fork of the Grand River, Moreau River, and several artificial lakes; formed 1909; co. seat Bison. It is in a grain-farming and ranching (cattle and sheep) area, with lignite deposits. Silver jewelry is manufactured.

Perm (pĕrm), city (1974 est. pop. 972,000), capital of Perm oblast, NE European USSR, on the Kama River. It is a transfer center for rail and river traffic and a major producer of machinery. Perm also has chemical plants and oil refineries. It was founded in 1780 and underwent rapid industrial growth in the 19th cent. Perm oblast is a major mining region.

Per·nam·bu·co (pər-nəm-bōō'kōō), state (1975 est. pop. 5,843,400), 37,946 sq mi (98,280 sq km), NE Brazil, on the Atlantic Ocean; capital Recife.

Per·nik (pĕr'nĭk), city (1972 est. pop. 79,900), W Bulgaria, on the Struma River. The industrial center of a coal-mining region, it has iron smelters, glassworks, and power plants.

Pé·ronne (pā-rôn'), town (1968 pop. 7,701), Somme dept., N France, in Picardy, on the Somme River. It is a farm trade center, and its manufactures include wool, bricks, furniture, and plastics. It was a residence (10th cent.) of the Frankish kings and was united to the French crown in 1477. It was here that Charles the Bold of Burgundy forced on Louis XI of France a humiliating treaty in 1468. In World War I the town was almost entirely destroyed during the five months of bloody fighting (1916) along the Somme.

Per·pi·gnan (pĕr-pē-nyäN'), city (1975 pop. 106,426), capital of Pyrénées-Orientales dept., S France, near the Spanish border and the Mediterranean. It is a farm trade center, handling wine, fruits, and vegetables. It has distilleries, canneries, and factories making chocolate, clothing, paper, and toys. Tourism is also important. Founded c.10th cent., Perpignan was the fortified capital of the Spanish kingdom of Roussillon. Among its notable buildings are the Loge (14th cent.), built to house the merchants' exchange; the Gothic Cathedral of St. Jean (14th–15th cent.); and the castle of the kings of Majorca (13th–15th cent.), which forms part of the old citadel.

Per·quim·ans (pûr-kwĭm'ənz), county (1970 pop. 8,351), 246 sq mi (637.1 sq km), NE N.C., bounded on the S by Albemarle Sound and drained by the Perquimans River; formed 1672; co. seat Hertford. This partly forested tidewater area, with Dismal Swamp in the northeast, has farming (peanuts, cotton, and corn), fishing, sawmilling.

Per·ry (pĕr'ē). **1.** County (1970 pop. 15,388), 734 sq mi (1,901.1 sq km), W central Ala., in the Black Belt, drained by the Cahaba River; formed 1819; co. seat Marion. It has cotton and livestock farming, food processing, clothing manufacturing, and other industry. Part of Talladega National Forest is in the northeast. **2.** County (1970 pop. 5,634), 556 sq mi (1,440 sq km), central Ark., bounded on the NE by the Arkansas River; formed 1840; co. seat Perryville. In an agricultural area yielding cotton, corn, hay, truck crops, and livestock, it has oil and gas wells and lumber mills. **3.** County (1970 pop. 19,757), 439 sq mi (1,137 sq km), SW Ill., bounded partly on the E by the Little Muddy River and drained by Beaucoup and Galum creeks; formed 1827; co. seat Pinckneyville. It has agriculture (corn, wheat, dairy products, livestock, and poultry), bituminous-coal mines, oil wells, and plants manufacturing flour, explosives, cigars, and machinery. **4.** County (1970 pop. 19,075), 384 sq mi (994.6 sq km), S Ind., bounded on the S and E by the Ohio River, here forming the Ky. line, and drained by a tributary of the Ohio River; formed 1814; co. seat Cannelton. In a bituminous-coal-mining and agricultural (dairy products, poultry, and livestock) area, it has a lumbering industry and sandstone quarries. Furniture, clay and metal products, and machinery are manufactured. **5.** County (1970 pop. 26,259), 341 sq mi (883.2 sq km), SE Ky., in the Cumberlands, drained by the North Fork of the Kentucky River and several creeks; formed 1820; co. seat Hazard. It is in an important coal-mining area, with oil and gas wells and some timber. Its agriculture includes dairy products, livestock, poultry, apples, corn, sweet and Irish potatoes, soybeans, and tobacco. **6.** County (1970 pop. 9,065), 653 sq mi (1,691.3 sq km), SE Miss., drained by the Leaf River and several creeks; formed 1820; co. seat New Augusta. Agriculture (cotton and corn) and lumbering are important. Part of De Soto National Forest is here. **7.** County (1970 pop. 14,393), 476 sq mi (1,232.8 sq km), E Mo., bounded in the E by the Mississippi River; formed 1820; co. seat Perryville. Grain, livestock, lumber, and iron deposits characterize the area. **8.** County (1970 pop. 27,434), 409 sq mi (1,059.3 sq km), central Ohio, drained by Rush, Sunday, Jonathan, and Moxahala creeks; formed 1817; co. seat New Lexington. Its agriculture includes livestock, grain, truck crops, and fruit. It has coal mines, oil and gas wells, and sand and gravel pits. Tile, brick, pottery, and metal products, and machinery are made. It includes part of Buckeye Lake, which is a resort area. **9.** County (1970 pop. 28,615), 550 sq mi (1,424.5 sq km), S central Pa., in a mountainous area bounded in the E by the Susquehanna River and drained by the Juniata River; formed 1820; co. seat New Bloomfield. Grist-mill products, clothing, hardware, flour, dairy goods, and textiles are produced. **10.** County (1970 pop. 5,238), 411 sq mi (1,064.5 sq km), W central Tenn., bounded in the W by the Tenn. River and drained by the Buffalo River; formed 1821; co. seat Linden. Its agriculture includes dairy products, livestock, corn, hay, and peanuts. It has lumbering and other industries.

Perry. 1. Town (1970 pop. 7,701), seat of Taylor co., NW Fla., SE of Tallahassee, in a fishing and hunting region. It is a center for the processing of timber, especially cypress. **2.** City (1970 pop. 7,771), seat of Houston co., central Ga., S of Macon, in a farm and quarry area; settled c.1820, inc. 1824. **3.** City (1970 pop. 6,906), Dallas co., central Iowa, NW of Des Moines; settled c.1848, inc. 1875. Poultry is processed and farm machinery is made. **4.** City (1970 pop. 5,341), seat of Noble co., N central Okla., NNE of Oklahoma City, in a farm and oil area; settled 1893, inc. 1894.

Per·rys·burg (pĕr'ēz-bûrg'), town (1970 pop. 7,693), Wood co., NW Ohio, on the Maumee River SW of Toledo; laid out 1816, inc. 1833. A busy river port from 1828 to 1840, it was noted for its shipbuilding.

Perry's Victory and International Peace Memorial National Memorial: *see* National Parks and Monuments Table.

Per·ry·ton (pĕr'ē-tən, -tn), city (1970 pop. 7,810), seat of Ochiltree co., extreme N Texas, NE of Amarillo near the Okla. border, founded 1919, inc. 1920. It is the shipping and manufacturing center of a wheat and cattle area.

Per·ry·ville (pĕr'ē-vĭl'). **1.** City (1974 est. pop. 948), seat of Perry co., central Ark., WNW of Little Rock on Fourche La Fave River, in a stock-raising and agricultural area. Sawmilling is done. **2.** City (1970 pop. 5,149), seat of Perry co., SE Mo., NNW of Cape Girardeau, near the Mississippi River, in a farm region; founded c.1821, inc. 1831.

Per·sep·o·lis (pər-sĕp'ə-lĭs), ancient city of Persia, ceremonial capital of the Achaemenid empire under Darius I and his successors. The ruins of Persepolis lie northeast of Shiraz in a fertile plain of the Pulvar River, with strong natural mountain defenses. There are ruins of the palaces of Darius I, Xerxes, and later kings as well as the citadel that contained the treasury looted by Alexander; the ruins lie on a huge platform constructed of limestone from the adjacent mountain. A few miles distant are the rock-hewn tombs of Achaemenid kings and monuments of the Sassanids. Nearby excavations have disclosed a village of the Neolithic period, with mural decorations in red ocher that date back to about 4000 B.C.

Per·shing (pûr'shĭng), county (1970 pop. 2,670), 6,001 sq mi (15,542.6 sq km), NW central Nev., in a mountainous region crossed by the Humboldt River; formed 1919; co. seat Lovelock. Tungsten and quicksilver are mined, and alfalfa, other grains, sugar beets, and livestock are raised.

Per·sia (pûr'zhə, -shə): *see* Iran.

Per·sian Gulf (pûr'zhən, -shən), arm of the Arabian Sea, 90,000 sq mi (233,100 sq km), between Arabia and Iran, extending c.600 mi (965 km) from the Shatt al Arab delta to the Strait of Hormuz, which links it with the Gulf of Oman. The Persian Gulf, called the Arabian Gulf by the Arabs, is very shallow and has many islands. It was generally thought that the gulf extended farther north and that sediment dropped by the Tigris, Euphrates, Karun, and Karkheh rivers filled the northern part of the gulf to create a great delta; but geologic investigations now indicate that the marshlands of the delta represent a sinking of the earth's crust.

The Persian Gulf was an important transportation route in antiquity but declined with the fall of Mesopotamia. In succeeding centuries control of the region was contested by Arabs, Persians, Turks, Russians, and Western Europeans. In 1853 Britain and the Arab sheikdoms of the Persian Gulf signed the Perpetual Maritime Truce, formalizing the temporary truces of 1820 and 1835; the sheiks agreed to stop harassing British shipping in the Arabian Sea and to recognize Britain as the dominant power in the gulf. These sheikdoms thus became known as the Trucial States. An international agreement among the major powers in 1907 placed the gulf in the British sphere of influence. Although oil was discovered in the gulf in 1908, it was not until the 1930s, when major finds were made, that keen international interest in the region revived.

Since World War II the Persian Gulf oil fields, among the most productive in the world, have been extensively developed, and modern port facilities have been constructed. In the late 1960s, following British military withdrawal from the area, the United States and the USSR sought to fill the vacuum. In 1971 the first U.S. military installation in the gulf was established at Bahrain. The long-standing Arab-Persian conflict in the gulf, combined with the desire of all the states for control of large oil reserves, has led to numerous boundary disputes.

Per·son (pûr'sən), county (1970 pop. 25,914), 401 sq mi (1,038.6 sq km), N N.C., on the Va. border; formed 1791; co. seat Roxboro. This piedmont tobacco and timber area also has textile manufacturing and sawmilling.

Perth (pûrth), city (1976 pop. 87,576; urban agglomeration pop. 805,489), capital of Western Australia, SW Australia, on the estuary of the Swan River. Perth was founded in 1829 but did not gain importance until the Coolgardie gold rush (1890s), the development of the port at Fremantle, and the construction of rail lines to the east (early 20th cent.). The Univ. of Western Australia and Murdoch Univ. are in Perth.

Perth, burgh (1974 est. pop. 44,066), Tayside region, central Scotland, on the Tay River. Perth is famous for dye works and cattle markets and has linen and wool factories. Strategically located between the Highlands and the Lowlands, Perth was the capital of Scotland from the 11th to the mid-15th cent. James I of Scotland was murdered here in 1437. John Knox preached his famous sermon against idolatry in the Church of St. John in 1559. Gowrie House (no longer standing) was the scene (1600) of a plot to seize James VI (James I of England). James I in 1618 issued the Five Articles of Perth, which opened the battle between crown and church. The earl of Montrose took the city after the Battle of Tippermuir in 1644; Oliver Cromwell seized it again in 1651. It was held by Jacobites in 1689, 1715, and 1745. Until 1975 Perth was the county town of Perthshire.

Perth Am·boy (ăm'boi), city (1978 est. pop. 34,300), Middlesex co., NE N.J., with a harbor on Arthur Kill at the mouth of the Raritan River; settled 1683, inc. as a city 1718. A port of entry, Perth Amboy is a shipping center with industries that include the smelting and refining of lead, copper, and silver; oil refining; printing; and the manufacture of chemicals. The city was the capital of East Jersey from 1684 until the union of East and West Jersey in 1702 and was alternate capital with Burlington until 1790. Perth Amboy grew after connections with the interior were established, particularly after it became the tidewater terminal of the Lehigh Valley RR in 1876.

Perth·shire (pûrth'shîr', -shər), former county, central Scotland. The county town was Perth. It is largely a mountainous region. Wild forests and rolling moor cover much of the area, which is noted for its fine scenery. In 1975 Perthshire was divided between the Tayside and Central regions.

Pe·ru (pə-rōō'), republic (1973 est. pop. 14,640,000), 496,220 sq mi (1,285,210 km), W South America, bordering on the Pacific Ocean in the W, on Ecuador and Colombia in the N, on Brazil and Bolivia in the E, and on Chile in the S. Lima is the capital.

Peru varies greatly in climate and topography and falls into three main geographic regions—a narrow strip of desert along the coast, a region of high mountains in the center, and a large area of forested mountains and lowlands in the east. The desert region stretches the length of Peru's Pacific coastline and is extremely arid because of the effects of the cold Humboldt, or Peru, Current. The coast and also the mountains are frequently shaken by severe earthquakes. Off the coast are small islands, notably the Lobos and Chincha islands, where guano (used as fertilizer) is harvested. The central region is made up mostly of three ranges of the Andes Mts., the Cordillera

Occidental in the west and the Cordillera Central and its continuation, the Cordillera Real, in the east. The rugged eastern ranges receive considerable rainfall and are drained by numerous rivers, which have cut deep canyons. Between the eastern and western ranges of the Andes in the south, and extending into Bolivia, is the Altiplano Plateau, which includes small, scattered basins of cultivable soil and pastureland and also part of Lake Titicaca. The eastern region, called La Montaña, includes more than half of the country's land area. It is made up of the highly forested Cordillera Oriental of the Andes and low-lying tropical plains, covered by rain forests and drained by the Amazon River and its tributaries. The region is generally inaccessible and sparsely inhabited, but contains much fertile soil that has not yet been cleared and cultivated.

Peru is divided into a constitutional province (Callao) and 23 departments:

NAME	CAPITAL	NAME	CAPITAL
Amazonas	Chachapoyas	Ayacucho	Ayacucho
Ancash	Huarás	Cajamarca	Cajamarca
Apurímac	Abancay	Callao	Callao
Arequipa	Arequipa	Cuzco	Cuzco
Huancavelica	Huancavelica	Madre de Dios	Puerto Maldonado
Huánuco	Huánuco	Moquegua	Moquegua
Ica	Ica	Pasco	Cerrc de Pasco
Junin	Huancayo	Piura	Piura
La Libertad	Trujillo	Puno	Puno
Lambayeque	Chiclayo	San Martin	Moyobamba
Lima	Lima	Tacna	Tacna
Loreto	Iquitos	Tumbes	Tumbes

Economy. Farming provides the livelihood for the majority of Peruvians. However, industry was being developed at a high rate in the 1960s and early 1970s. The chief farm commodities produced are cotton, sugar cane, coffee, cacao, wheat, rice, maize, and barley. Large numbers of poultry, sheep, cattle, llamas, alpacas, and hogs are raised. Peru is one of the world's foremost fishing countries in terms of the value and weight of the annual catch. Peru has a large mining industry, and the most valuable minerals produced are copper and silver; other minerals extracted include gold, iron ore, mercury, phosphate rock, salt, tin ore, and zinc. Petroleum is produced along the northern coast and in the Amazon basin. Peru's principal manufactures include iron and steel, processed food, cement, refined minerals (especially copper, zinc, and lead), textiles, consumer goods (including clothing, footwear, and household appliances), and processed fish (mainly fish meal and fertilizer). There is a substantial tourist industry.

History. Peru has been inhabited since at least the 9th millennium B.C. It was later the center of several developed Indian cultures, including the Chavín, the Chimu, the Nazca, and the Aymara. In the 12th cent. A.D. the Quechua-speaking Inca Indians settled around Cuzco, and in the mid-15th cent. they established by conquest a large, well-organized empire that included most of present-day Peru and Ecuador and parts of Bolivia, Chile, Argentina, and Colombia. Around 1530 the empire was weakened by civil war initiated by Atahualpa and Huascar, who had been designated as dual heirs by their father, Huayna Capac. Atahualpa had emerged victorious by 1532, when Francisco Pizarro, a Spaniard, arrived on the coast of Peru with a small band of adventurers. Sensing no danger to his empire, Atahualpa agreed to meet Pizarro at Cajamarca. However, Pizarro, whose horses and firearms (both unknown to the Incas) gave him an overriding advantage, imprisoned Atahualpa after he had refused to accept Spanish suzerainty and Christianity. The emperor's followers collected a huge ransom in gold and silver for his release; nevertheless, the Spaniards executed him in mid-1533. By late 1533 Pizarro had captured Cuzco, the Inca capital, and the empire had disintegrated. In 1535 Pizarro founded Lima, which in 1542 became the center of Spanish rule in South America. From 1536 to 1544 Manco Capac, who had succeeded Atahualpa as emperor, led several unsuccessful uprisings against the Spaniards. At the same time, Pizarro and his brothers and companions were unsuccessfully challenged by other Spaniards for control of the area. Pizarro forced the Indians to work in the mines, on the lands held in encomienda from Spanish landlords, and in the small textile mills. The New Laws of Bartolomé de Las Casas, which would have ended these abuses, caused Gonzalo Pizarro to revolt (1544). He defeated the viceroy, Blasco Núñez Vela, but was in turn defeated (and executed) by Pedro de la Gasca in 1548. However, the New Laws were never administered for the benefit of the Indians.

In the late 16th cent., the viceroyalty of Peru was expanded to include all of Spanish-ruled South America except Venezuela, and the mining of silver and gold increased. Lima was the administrative, religious, economic, and cultural center of the viceroyalty. In the 18th cent. Peru was drastically reduced in size by the creation of the viceroyalty of New Granada and a viceroyalty centered at Buenos Aires; as a result, Lima lost control over considerable trade and mineral wealth. At the same time, government in Peru was reformed. However, Spaniards retained control in the viceroyalty, and the Indians and creoles (persons of Spanish descent born in Peru) remained powerless and poor, working mostly as laborers or subsistence farmers. There were a few uprisings by Indians and creoles in the late 18th and early 19th cents.

The ideas of the French Revolution and Napoleon I's conquest (1808) of Spain led to strong independence movements in each of Spain's Latin-American holdings except Peru. Peru's loyalty to Spain was due to the relatively large number of Spaniards who resided there, to the concentration of Spanish power at Lima, and to the efficiency of the government in the viceroyalty. As a result, Peru achieved independence (1821) largely because of the efforts of outsiders, notably José de San Martín and Simón Bolívar. However, Spanish forces remained in the interior. Bolívar took over the leadership of the liberation movement from the self-effacing San Martín after a mysterious conference (July, 1822) between the two at Guayaquil. In 1824 Bolívar and his lieutenant Antonio José de Sucre assured Peru's independence by defeating Spain at the battles of Junín and Ayacucho. Government in Peru became confused as several military leaders vied for power. Peru continued to be torn by civil strife until the emergence of Gen. Ramón Castilla, who was president from 1844 to 1850 and from 1855 to 1862. Under Castilla, Peru enjoyed stability and economic development. Peruvian society, however, remained sharply divided between the wealthy oligarchy (made up mostly of creoles) and the great majority of inhabitants (mostly Indians). A republican constitution was promulgated in 1860 and remained in effect until 1920.

After Castilla, Peruvian politics again were in turmoil because of corruption, growing foreign indebtedness, and an attempt by Spain to regain Peru. Claiming that Peru had not met its financial obligations, Spain seized the guano-rich Chincha Islands in 1863. Aided by Chile, Bolivia, and Ecuador, Peru defeated the Spaniards at Callao in 1866; a truce was signed in 1871 and in 1879 Spain formally recognized Peru's independence. The first civilian president of Peru, Manuel Pardo (1872-76), tried to better the country's financial position, but was seriously hampered by the declining international price of guano, one of Peru's major resources. In 1873 Peru signed a secret defensive alliance with Bolivia, whose valuable coastal nitrate deposits (especially in Atacama) were worked by Chileans. When disagreements over the mining led to war between Bolivia and Chile, Peru tried to mediate but refused to declare its neutrality. Therefore, Chile declared war (1879) on Peru. Chile badly defeated the allies and by the Treaty of Ancón (1883) Peru had to yield the nitrate-rich province of Tarapacá and also to surrender the other southern coastal provinces of Tacna and Arica to Chilean administration until a plebiscite would be held 10 years later. The plebiscite was never carried out, and there ensued the Tacna-Arica Controversy, which was not resolved until 1929. Peru emerged nearly bankrupt from the War of the Pacific. President A. A. Cáceres (1886-90) created a syndicate of foreign capitalists to manage the guano deposits and the railroads, and thus foreign influence and holdings increased.

The first third of the century was dominated by President Augusto B. Leguía (1908-12, 1919-30), who mostly governed as a virtual dictator and successfully promoted economic development in the interest of the country's dominant oligarchy. Peru benefited, in turn, from a rubber boom in the Putumayo River region, from the opening (1914) of the Panama Canal, and from a large increase in its exports that began during World War I. In 1924 a new political party, the Alianza Popular Revolucionaria Americana (APRA), was founded; it called for radical reform, especially of the condition of the Indians. The party was banned into the 1930s. Those years were marked by a bitter rivalry between leftists and rightists, with the latter dominating politics. Peru was involved in a serious boundary dispute with Ecuador in 1941 and sided with the Allies in World War II. APRA was allowed to take part in the 1945 elections and backed the victorious moderate, José Luís Bustamante y Rivero. However, APRA split with Bustamante in 1947, and the resulting disputes led to a military coup by Manuel Odría in 1948. Odría, a conservative, was president until 1956, when Prado was again elected president, this time with APRA support. Peru's economic situation improved because of increased sales and U.S. loans. However, the rate of inflation increased, causing much labor unrest. In the 1962 presidential elections Haya de la Torre won by a small plurality, but did not receive the required one third of the total vote. The military seized power and conducted elections in 1963. The new regime was plagued by budgetary deficits, reduced sales abroad, spiraling inflation, and poor relations with foreign companies active in Peru. In 1968 a military junta installed Gen. Juan Velasco Alvarado as president at the head of a revolutionary government. Velasco suspended the constitution and assumed dictatorial powers. His declared intention was to democratize Peruvian society and to give the masses a greater voice in government, but few steps were taken toward fulfilling these goals. He also sought (and with more concrete results) to diversify the country's economy by exploiting systematically its natural resources, especially petroleum. Velasco was ousted in a military coup in Aug., 1975. His successor, Gen. Francisco Morales Bermúdez, set up a government that shifted to the right.

Government. Under the 1933 constitution as amended, Peru's chief executive and head of state is the president, who is directly elected to a six-year term. The president is assisted by a cabinet led by a prime minister. Legislative power is vested in a bicameral parliament, consisting of the 45-member senate and the 140-member chamber of deputies. Members of both are directly elected to six-year terms. The constitution was suspended in 1968 and the parliament dissolved.

Peru. 1. City (1978 est. pop. 11,000), La Salle co., N Ill., on the Illinois River; inc. 1835. Clocks, watches, and metal products are made here. **2.** City (1978 est. pop. 13,400), seat of Miami co., N Ind., on the Wabash River; inc. 1847. It is a processing and rail center for a fertile farm area. Among its products are furniture, plastic items, and electrical equipment.

Pe·ru·gia (pĕ-rōō'jä, -jə), city (1975 est. pop. 136,799), capital of Umbria and of Perugia prov., central Italy, situated on a hill overlooking the valley of the Tiber River. It is a commercial and industrial center. Manufactures include chocolate and textiles. Perugia was inhabited by the Umbrians and the Etruscans before it came under the control of Rome (c.310 B.C.). It became a Lombard duchy in the late 6th cent. A.D. In the 12th cent. it attained the status of a free commune and gradually gained hegemony over other Umbrian cities. Although nominally under papal control, it was in fact ruled by strong tyrants until 1540, when it was conquered by Pope Paul III. To help control the city Pope Paul built an imposing citadel (designed by Antonio da San Gallo and dismantled in 1860). Perugia was the artistic center of Umbria. Points of interest in the city include the imposing Palazzo dei Priori (13th-15th cent.), which houses the National Gallery of Umbria, the marble Great Fountain (13th cent.), the Collegio del Cambio, the Gothic cathedral (14th-15th cent., with later baroque additions), the Gothic Church of San Domenico, which houses an archaeological museum, and well-preserved medieval quarters.

Pe·sa·ro (pā'zä-rō), city (1975 est. pop. 89,890), capital of Pesaro e Urbino prov., in the Marche region, central Italy, on the Adriatic Sea at the mouth of the Foglia River. It is an agricultural and industrial center and a seaside resort. Manufactures include musical instruments, motor vehicles, and ceramics. A Roman colony, Pesaro was later one of the cities of the Pentapolis (5th-11th cent.). The house of Malatesta gained power here in the 13th cent.; it was succeeded by the Sforza (15th-16th cent.) and by the dukes of Urbino (16th-17th cent.) In 1631 the city passed directly under the Holy See.

Pes·ca·do·res (pĕs'kə-dôr'ĭs, -ēz), group of 64 small islands, c.50 sq mi (130 sq km), in Formosa Strait, off the W coast of Taiwan. Fishing is the main occupation, and peanuts and sweet potatoes are grown; coral is an important product. The group was named the Pescadores, or fishermen's islands, by the Portuguese in the 16th cent. China ceded them (1895) to Japan after the First Sino-Japanese War, and they were returned to China after World War II.

Pe·sca·ra (pĕs-kä'rä), city (1976 est. pop. 135,167), capital of Pescara prov., in Abruzzi, central Italy, on the Adriatic Sea at the mouth of the Pescara River. It is a fishing port and a seaside resort. Cement and textiles are produced.

Pe·sha·war (pĕ-shä'wər), city (1972 est. pop. 210,000), capital of the North-West Frontier Prov., NW Pakistan. A road and rail center near the famed Khyber Pass, Peshawar is an important military and communications center. Local handicrafts and farm produce from the surrounding fertile agricultural valley are sold at colorful bazaars in the city. Industries include food processing, and the manufacture of steel, cigarettes, firearms, textiles, pharmaceuticals, furniture, and paper and board. The city was the capital of the ancient Greco-Buddhist center of Gandhara and was named Peshawar by the Mogul emperor Akbar. The Kushan leader Kanishka (2nd cent. A.D.) made it his capital. For centuries, it was the target of successive Afghan, Persian, and Mongol invaders. It was taken by the Sikhs (early 19th cent.), from whom the British captured it in 1848. It became an important outpost of British India and was a base for British military operations against rebellious Pathan tribes.

Pe·tah Tiq·wa (pĕt'ə tĭk'və), town (1975 est. pop. 106,800), W central Israel. Its industries produce textiles, plastics, processed foods, tires and other rubber products, and soap. There are extensive citrus groves on the outskirts, and building stone is quarried nearby. Petah Tiqwa was founded in 1878 as the first modern Jewish agricultural settlement in Palestine.

Pet·a·lu·ma (pĕt'ə-lōō'mə), city (1978 est. pop. 31,500), Sonoma co.,

W Calif.; inc. 1858. It is a large poultry and dairy center. Cheese, twine, fishlines, canvas goods, and fabricated metal are also made.

Pe·tén (pĕ-tĕn'), region, c.15,000 sq mi (38,850 sq km), N Guatemala. It is a humid expanse of dense, tropical hardwood forests interrupted by savannas and crisscrossed by ranges of hills. There are large, permanent lakes, notably Lake Petén Itzá. The region is relatively inaccessible and has been only partly developed. It produces lumber, chicle, and some rubber and cacao. Petén was once a center of the Old Empire of the Maya. It is noted today chiefly as the scene of large-scale excavations of great archaeological ruins, particularly Tikal and Uaxactún. Although the Spanish nominally conquered the area and Cortés passed through it on his march to Honduras (1524–25), efforts at subjugation were sporadic until the Itzá tribe was driven out (1697) from their stronghold at Lake Petén Itzá.

Pe·ter·bor·ough (pē'tər-bûr'ō, -bər-ə), city (1976 pop. 59,683), SE Ont., Canada, NE of Toronto. It is at the falls of the Otonabee River, which connects, through the Trent Canal, with Lakes Ontario and Huron. Settled early in the 19th cent. as a lumber town, it is now a railroad and industrial center. Archaeologically valuable Indian sites are nearby.

Peterborough, municipal borough (1971 pop. 70,021), Cambridgeshire, E central England, on the Nene River. The city is a rail, engineering, and farm trade center; products include diesel engines, farm machinery, and processed foods. The Benedictine abbey was founded c.655. In 870 it was destroyed by the Danes, in the 10th cent. it was restored, in the 11th plundered, and in 1116 burned. The impressive cathedral, formerly the abbey church, was damaged by Cromwell's men in 1643. The bishop's palace and remains of the old abbey buildings and ancient gates are noteworthy. Queen Katharine of Aragón is buried here.

Peterborough, town (1970 pop. 3,807), Hillsborough co., S N.H., on the Contoocook River between Kenne and Nashua; granted 1738, inc. 1760. Both a resort and a mill town, it has had textile mills since the early 19th cent. The town is best known for its colony for composers, artists, and writers, planned by Edward MacDowell and founded and sustained by his widow.

Pe·ter·head (pē'tər-hĕd'), burgh (1974 est. pop. 14,994), Grampian region, NE Scotland, on a peninsula on the North Sea. It is the easternmost burgh of Scotland and it has a good harbor. Chiefly a center of herring fisheries, Peterhead has fish canneries, distilleries, and woolen mills. The town was founded in 1593.

Pe·ters·burg (pē'tərz-bûrg'). **1.** City (1970 pop. 2,632), seat of Menard co., central Ill., on the Sangamon River NNW of Springfield, in a farm and coal area; inc. as a town 1836, as a city 1841. The graves of Ann Rutledge and Edgar Lee Masters are here. **2.** Town (1970 pop. 2,697), seat of Pike co., SW Ind., SE of Vincennes; laid out 1817, inc. 1924. It has coal mines and ships farm produce. **3.** Independent city (1978 est. pop. 47,800), SE Va., on the Appomattox River; inc. 1850. A port of entry and an important tobacco market, it has industries producing cigarettes, luggage, and optical parts. Fort Henry was built here in 1646 on the site of an Indian village. A trading post was then established, and in 1784 three villages—Petersburg, Blandford (both laid out 1748), and Pocahontas (1752)—were combined as Petersburg town. In the American Revolution the area was taken (1781) by the British; from Petersburg Lord Cornwallis began the campaign that ended at his defeat at Yorktown. In the Civil War Petersburg, which guarded the southern approaches to Richmond, was under siege from June 15, 1864, to April 3, 1865. Union forces entered Richmond on the same day Petersburg fell, and Gen. Robert E. Lee surrendered the remnants of his army at Appomattox Courthouse one week later. Petersburg National Battlefield (est. 1926) encompasses much of the battle scene; many old earthworks and tunnels are preserved. Other points of interest include Blandford Cemetery, Blandford Church (1735–37), and Center Hill Mansion (1823; now a museum). Virginia State College is here. **4.** Town (1970 pop. 2,177), seat of Grant co., W.Va., in the Eastern Panhandle, on the South Branch of the Potomac River, in a farm area; settled c.1745.

Pe·tos·key (pə-tŏs'kē), resort city (1975 est. pop. 6,200), seat of Emmet co., N Mich., on Little Traverse Bay SW of Cheboygan, in a farm and lake region; settled c.1860, inc. as a village 1879, as a city 1895. Wood products and cement are made.

Pe·tra (pē'trə), ancient rock city, in present-day Jordan, known to the Arabs as Wadi Musa for the stream that flows through it. It was early occupied by the Edomites (descendants of Esau) and by the Nabataeans (an Arab tribe), who had their capital here from the 4th cent. B.C. until the Roman occupation in A.D. 106. It was for many centuries the focal point of a vast caravan trade. An early seat of Christianity, it was conquered by the Moslems in the 7th cent. and in the 12th cent. was captured by the Crusaders. Petra was unknown to the Western world until its ruins were visited by Johann Burckhardt in 1812. A narrow pass between towering walls leads to the plain upon which stood the ancient city. The plain is surrounded by hills in which tombs have been carved in the pink rock.

Pet·ri·fied Forest National Park (pĕt'rə-fīd'), 94,189 acres (38,146.5 hectares), E Ariz.; est. as a national monument 1906, as a national park 1962. A part of the Painted Desert, it contains the largest known display of petrified wood in the world. There are six separate "forests," with great logs of jasper and agate lying on the ground surrounded by the varied colors of endless fragments and small chips. Dating from the Triassic period, these "stone trees" were killed by natural processes, such as fire, insect attacks, and fungus (or rot). The trees were deeply buried in mud and sand that contained silica-rich volcanic ash. The logs became petrified as the mineral, carried into the wood by ground water, replaced the wood cells. As the surrounding material was eroded away, the petrified trees were exposed on the surface. Prehistoric Indians lived among the stone trees; ruins of their dwellings and their petroglyphs (ancient rock art) are present.

Pet·ro·dvo·rets (pĕt'rəd-və-rĕts', pyĕt'-), formerly **Pe·ter·hof** (pē'tər-hôf'), city, NW European USSR, on Neva Bay of the Gulf of Finland. Petrodvorets is a port, a rail terminus, and a resort center. The city grew up around the palaces and gardens built for Peter I, who founded it in 1711. It became the most lavish of the czar's summer residences, containing several palaces surrounded by vast parks famous for their fountains and cascades. Under the Soviet government, the palaces were converted to museums. Largely destroyed during World War II, Peterhof has since been restored.

Pet·ro·kre·post (pĕt'rə-krĕp'əst), town and fortress, NW European USSR, E of Leningrad. The town, the terminus of a railroad and of the lateral canals on Lake Ladoga, has shipbuilding and repair yards. On an island in Lake Ladoga stands the fortress, dominating the lake's access from the Neva River. Built in 1323 by the republic of Novgorod, the fortress fell to Sweden in 1611. Peter I captured it from the Swedes in 1702, during the Northern War, and named it Schlüsselburg, envisioning it as the major link in Russia's line of defense to the Baltic Sea. The fortress soon lost its military significance and was used until the 1917 Bolshevik Revolution as a prison. In 1928 it was converted into a museum. The town fell (1941) to the Germans during World War II; its recapture (1943) by Russian forces opened the land route to besieged Leningrad.

Pe·tro·le·um (pə-trō'lē-əm), county (1970 pop. 675), 1,655 sq mi (4,286.5 sq km), central Mont., in an agricultural area bounded on the N by the Missouri River and on the E by the Musselshell River; formed 1924; co. seat Winnett. Livestock, grain, petroleum, and natural gas are important to its economy.

Pet·ro·pav·lovsk (pĕt'rə-päv'lŏfsk), city (1976 est. pop. 196,000), capital of North Kazakhstan oblast, Central Asian USSR, in Kazakhstan, on the Ishim River and at the junction of the Trans-Kazakhstan and Trans-Siberian railroads. Small motors, agricultural machinery, leather, felt, and foodstuffs are produced. On the caravan route between Turkistan and China, Petropavlovsk was founded as a fort in 1752 and became a center for trade between Russia and the central Asian kingdoms. Its industrial development began in the late 19th cent. and intensified after World War II.

Petropavlovsk or **Pet·ro·pav·lovsk-Kam·chat·ski** (pĕt'rə-päv'lŏfsk-kəm-chät'skē), city (1976 est. pop. 202,000), capital of Kamchatka oblast, Far Eastern USSR. It is a major port and naval base on the Northern Sea Route. Free of ice seven months of the year, it is the base for a large fishing and whaling fleet.

Pe·tróp·o·lis (pə-trŏp'ə-les, pə-trô'po͞o-lēs), city (1970 pop. 116,080), Rio de Janeiro state, SE Brazil, picturesquely situated in hills just N of Rio de Janeiro. It is a fashionable resort, with a healthful climate, beautiful wooded estates, and tree-lined avenues. There are also industries producing textiles, coffee, and cereals. Colonized by German immigrants in 1845, it soon became the summer residence of the emperor Dom Pedro II. Among its many fine buildings is the old imperial palace, now a museum.

Pet·ro·za·vodsk (pĕt'rə-zä-vŏtsk'), city (1976 est. pop. 216,000), capital of Karelian Autonomous Republic, NW European USSR, a port on Lake Onega. It produces lumbering equipment and has shipyards, fish canneries, sawmills, and wood plants. Novgorodians worked the nearby iron deposits in the Middle Ages. Peter I founded a metal factory here in 1703.

Pet·tis (pĕt'əs), county (1970 pop. 34,137), 679 sq mi (1,758.6 sq km), central Mo.; formed 1833; co. seat Sedalia. Agricultural products include corn, wheat, oats, hay, potatoes, and livestock. There are barite and coal mines and limestone quarries in the area.

Pev·en·sey (pĕv'ən-zē), village, East Sussex, S England, on the English Channel. Modern Pevensey, called Pevensey Bay, is a shore resort. In the old town are remains of Roman walls and a Norman castle. The town, the landing place of William the Conqueror, was a member of the Cinque Ports.

Pforz·heim (pfôrts'hīm), city (1975 pop. 108,635), Baden-Württemberg, SW West Germany, on the Enz River, at the N end of the Black Forest. It is a major center of the West German jewelry and watchmaking industry. An important medieval trade center, Pforzheim often changed hands until it passed to the margraves of Baden in the 13th cent. Pforzheim was damaged in the Thirty Years' War (1618–48) and was devastated (1689) by the French in the War of the Grand Alliance; later, more than three quarters of the city was destroyed in World War II. Noteworthy buildings include an 11th cent. church and the Romanesque Church of St. Martin.

Pha·ros (fâr'ŏs), peninsula, extending into the Mediterranean Sea, N Egypt, NE Africa, forming two harbors at Alexandria. Originally an island, it was joined to the mainland by a mole, constructed by order of Alexander the Great. On Pharos stood the celebrated lighthouse completed (c.280 B.C.) by Ptolemy II, which is usually included among the Seven Wonders of the Ancient World. No precise description of it has survived. It was destroyed by an earthquake in the 14th cent.

Pharr (fär), city (1970 pop. 15,829), Hidalgo co., extreme S Texas; inc. 1916. There are natural-gas wells in the city, and a recycling plant manufactures propane, diesel fuel, and kerosene.

Phar·sa·la (fär'sä-lä) or **Phar·sa·lus** (fär-sā'ləs), ancient city, Thessaly, Greece. Near here in 48 B.C., Julius Caesar routed Pompey.

Phelps (fĕlps). **1.** County (1970 pop. 29,567), 677 sq mi (1,753.4 sq km), central Mo., in the Ozarks, drained by the Meramec and Gasconade rivers; formed 1857; co. seat Rolla. Grapes, apples, melons, and berries are grown. The county has stands of oak, hickory, cottonwood, and elm timber and pyrite and fire-clay mines. **2.** County (1970 pop. 9,553), 545 sq mi (1,411.6 sq km), S Nebr., in a farm area bounded in the N by the Platte River; formed 1873; co. seat Holdrege. Flour processing and livestock raising are done.

Phe·nix City (fē'nĭks'), city (1978 est. pop. 26,500), seat of Russell co., E Ala., on the Chattahoochee River opposite Columbus, Ga., in a cotton area; inc. 1883. In 1954 the state governor placed Phenix City under martial law for about five months—a result of the corruption that had long prevailed in the city.

Phi·ga·li·a (fĭ-gā'lē-ə, -lyə), ancient city of Greece, in SW Arcadia (now Arkadhia). It gives its name to the Phigalian Marbles, a frieze c.100 ft (30 m) long and 2 ft (61 cm) high, in high relief, representing battles between the Lapithae, a legendary people from Thessaly, and the Centaurs and between the Amazons and the Greeks. The frieze, dating from c.420 B.C., is now in the British Museum.

Phil·a·del·phi·a (fĭl'ə-dĕl'fē-ə), **1.** City (1970 pop. 6,274), seat of Neshoba co., E central Miss., NW of Meridian, in a farm area. Lumber and cotton are processed here. **2.** City (1970 pop. 1,950,098), coextensive with Philadelphia co., SE Pa., on the Delaware River c.100 mi (160 km) upstream at the influx of the Schuylkill River; chartered 1701. It is the fourth-largest city and port in the United States, and one of the largest freshwater ports in the world. It ranks high in the production of textiles, clothing, chemicals, electronic equipment, metal products (especially machinery), and a diversity of other manufactures. Its printing and publishing industry is important, and there are major oil refineries. Philadelphia is also an insurance and banking center. The site was first occupied by Indians, and in the 17th cent. there was a Swedish settlement. In 1681 Philadelphia, the "City of Brotherly Love," was founded as a Quaker colony by William Penn. Its commercial, industrial, and cultural growth was rapid, and by 1774 it was second only to London as the largest English-speaking city. It was the seat of the Continental Congress and served as the American capital from 1777 to 1788, except during the British occupation (Oct., 1777–June, 1778). It was the capital of the new republic from 1790 to 1800, as well as the state capital (to 1799). A nucleus of American culture in colonial times, Philadelphia is still the seat of many philosophical, artistic, dramatic, musical, and scientific societies. Among these are the Pennsylvania Academy of the Fine Arts (1805), the Academy of Natural Sciences, the American Philosophical Society (1743), and the Science Museum of the Franklin Institute (1824). Musical activities flourish in the city, which has an outstanding symphony orchestra. In Fairmount Park, the largest in the city, are the Philadelphia Museum of Art and zoological gardens. Among the many historical monuments and shrines are Independence Hall, where the Declaration of Independence was signed and the Liberty Bell is kept; the neighboring Congress Hall, where Congress met (1790–1800) and where Washington gave his farewell address; and Carpenters' Hall, where the First Continental Congress met. The historic 18th cent. houses in the Society Hill section and on Germantown Ave. are also tourist attractions. Philadelphia has over 30 educational institutions, including the Univ. of Pennsylvania and Temple Univ. A U.S. mint and a naval shipyard are in the city. In the 1950s an ambitious urban redevelopment program was instituted, and today Philadelphia has one of the nation's largest industrial parks.

Phi·lae (fī'lē'), former island, SE Egypt, NE Africa, in the Nile River above Aswan High Dam. Of its temples, all dating from late Egyptian and classical times (600 B.C.–A.D. 600), the most famous was the temple to Isis, built by the early Ptolemies and not closed to pagan worship until the reign of Justinian. The island is now covered by the waters of Lake Nasser.

Phil·ip (fĭl'ĭp), city (1970 pop. 983), seat of Haakon co., W central S.Dak., WSW of Pierre. It is in a dairy and livestock region.

Phi·lippe·ville (fē-lēp-vēl'): *see* Skikda, Algeria.

Phi·lip·pi (fĭ-lĭp'ī), ancient city, E Macedonia. Inhabited by Thracians and then Thasians, it was developed and fortified by Philip II of Macedon. Near the city was fought the decisive battle in which Octavian (Augustus) and Antony defeated (42 B.C.) Brutus and Cassius. It was the scene of St. Paul's first preaching in Europe.

Phil·ip·pi (fĭ'lə-pē), city (1970 pop. 3,002), seat of Barbour co., N W.Va., on the Tygart River S of Fairmont; settled c.1780.

Phil·ip·pines, Republic of the (fĭl'ə-pēnz'), republic (1976 est. pop. 43,751,000), 115,830 sq mi (300,000 sq km), SW Pacific, in Malay Archipelago off the SE Asia mainland. It comprises over 7,000 islands and rocks, of which only c.400 are permanently inhabited. The 11 largest islands contain about 95% of the total land area. Quezon City, on Luzon, is the capital and the second-largest city. Baguio is the summer capital. The northernmost point of land, the islet of Y'Ami in the Batan Islands, is separated from Formosa by the Bashi Channel (c.50 mi/80 km wide). The Philippines extend 1,152 mi (1,855 km) from north to south, between Formosa and Borneo, and 688 mi (1,108 km) from east to west, and are bounded by the Philippine Sea on the east, the Celebes Sea on the south, and the South

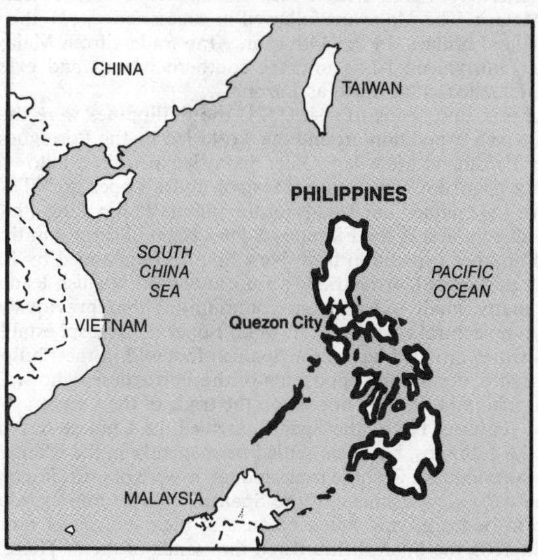

China Sea on the west. They comprise three natural divisions—the northern, which includes Luzon and nearby islands; the central, occupied by the Visayan islands and Palawan and Mindoro islands; and the southern, containing Mindanao Island and the Sulu Archipelago.

The Philippines are chiefly of volcanic origin. Most of the larger islands are traversed by mountain ranges, with Mt. Apo (9,690 ft/ 2,954 m), on Mindanao, the highest peak. Narrow coastal plains, wide valleys, volcanoes, dense forests, and mineral and hot springs further characterize the larger islands. Earthquakes are common. Of the many navigable rivers, Cagayan, on Luzon, is the largest; there are also large lakes on Luzon and Mindanao. The Philippines are entirely within the tropical zone.

The great majority of the people of the Philippines belong to the Malay group and are known as Filipinos. The non-Malayan inhabitants are the Negritos (negroid pygmies) and the Dumagats (similar to the Papuans of New Guinea). The Filipinos live mostly in the lowlands and constitute the largest Christian group in that part of the world. Roman Catholicism, a heritage from their Spanish conquerors, is professed by 84% of the population. The Christian Filipinos are divided into eight groups differing from one another in speech and other cultural elements. Largest of these groups are the Visayan, in the Visayan Islands; the Tagalog, native to the provinces adjoining Manila; and the Ilocano, of northwestern Luzon. Some 70 native languages are spoken in the Philippines; the nine most important languages are of the Malayo-Polynesian linguistic group. The government adopted (1946) Tagalog as the basis of the new national language, now known as Pilipino.

Economy. With their tropical climate, heavy rainfall, and naturally fertile volcanic soil, the Philippines are predominantly agricultural. Rice, corn, and coconuts take up about 80% of all cropland. Sugar cane, sweet potatoes, manioc, bananas, hemp, tobacco, and coffee are also important crops. Chief agricultural exports are coconut products, lumber and plywood, sugar, and hemp (the Philippines lead the world in its production). Carabao (water buffalo), pigs, chickens, goats, and ducks are widely raised, and there is dairy farming near the large cities. Fishing is a common occupation; the Sulu Archipelago is noted for its pearls and mother-of-pearl shell. The islands have one of the world's greatest stands of commercial timber, and they have abundant mineral resources, with copper, gold, iron, and chromite (the Philippines are fifth in its production) the most valuable. Nonmetallic minerals include rock asphalt, gypsum, asbestos, sulfur, and coal. Limestone, adobe, and marble are quarried. Manufacturing is concentrated in metropolitan Manila, near the nation's prime port, but there has been considerable industrial growth on Cebu, Negros, and Mindanao in recent years. Chief products are processed foods, cement, chemicals, metalware, textiles, beverages, tobacco products, wood and cork materials, wearing apparel, and electrical machinery. The production and repair of motor vehicles is also an important industry. Major exports are foods, lumber, and minerals, and the major imports are machinery, transportation equipment, fuels and lubricants, and base metals.

History. The Negritos are believed to have migrated some 30,000 years ago across land bridges then existing from Borneo, Sumatra, and Malaya. The Malayans followed in successive waves, the earliest by land bridges. In the 14th cent. Arab traders from Malay and Borneo introduced Islam into the southern islands and extended their influence as far north as Luzon.

The first Europeans to visit (1521) the Philippines were those in the Spanish expedition around the world led by the Portuguese explorer Ferdinand Magellan. Other Spanish expeditions followed, including one from New Spain (Mexico) under López de Villalobos, who in 1542 named the islands for the infante Philip, later Philip II. Spanish conquest of the Filipinos did not begin in earnest until 1564, when another expedition from New Spain, commanded by Miguel López de Legaspi, arrived. He soon established Spanish leadership over many small independent communities that previously had known no central rule. By 1571, when López de Legaspi established the Spanish city of Manila, the Spanish foothold in the Philippines was secure, despite the opposition of the Portuguese, who were eager to maintain their monopoly on the trade of the Orient.

For centuries before the Spanish arrived the Chinese had traded with the Filipinos, but none settled permanently in the islands until after the conquest. Chinese trade and labor were of great importance in the early development of the Spanish colony, but the Chinese came to be feared and hated because of their increasing numbers, and in 1603 the Spanish murdered thousands of them. There were frequent uprisings by the Filipinos, who resented the encomienda system. By the end of the 16th cent. Manila carried on a flourishing trade with China, India, and the East Indies. The Philippines supplied some wealth (including gold) to Spain, and the richly laden galleons plying between the islands and New Spain were often attacked by English freebooters. The period from 1600 to 1663 was marked by continual wars with the Dutch and with Moro pirates. The warlike Moros harassed the population; intermittent campaigns were conducted against them but without conclusive results until the mid-19th cent.

As the Spanish Empire waned, the Jesuit orders became more influential in the Philippines and acquired great amounts of property. It was the opposition to the clergy that in large measure brought about the rising sentiment for independence. Spanish injustice, bigotry, and economic oppression fed the movement. In 1896 revolution began in the province of Cavite and spread throughout the major islands. The Filipino leader, Emilio Aguinaldo, achieved considerable success before a peace was patched up with Spain. Neither side honored its agreements, however, and a new revolution was brewing when the Spanish-American War broke out in 1898. After the U.S. naval victory in Manila Bay on May 1, 1898, Com. George Dewey supplied Aguinaldo with arms and urged him to rally the Filipinos against the Spanish. By the time U.S. land forces had arrived, the Filipinos had taken the entire island of Luzon, except for the old walled city of Manila, which they were besieging. They had also declared their independence and established a republic under the first democratic constitution ever known in Asia. Their dreams of independence were crushed when the Philippines were transferred from Spain to the United States in the Treaty of Paris (1898), which closed the Spanish-American War; in Feb., 1899, Aguinaldo led a new revolt, this time against U.S. rule. Defeated on the battlefield, the Filipinos turned to guerrilla warfare. Their subjugation cost the United States far more money and took far more lives than the Spanish-American War. It also bitterly divided the American people, with sentiment polarizing along political lines. The insurrection was effectively ended with the capture (1901) of Aguinaldo, but the question of Philippine independence remained a burning issue in the politics of both the United States and the islands. The matter was complicated by growing economic ties between the two countries. Although comparatively little American capital was invested in island industries, U.S. trade bulked larger and larger until the Philippines became almost entirely dependent upon the American market. Free trade, established by an act of 1909, was expanded in 1913. In 1913 steps were taken to prepare the Filipino people for self-rule. Francis B. Harrison, who was appointed governor general, worked toward replacing many Americans in key positions of government with Filipinos. The Philippine assembly already had a popularly elected lower house, and the Jones Act, passed by the U.S. Congress in 1916, provided for a popularly elected upper house as well, with power to approve all appointments made by the governor general. It also gave the islands their first definite pledge of independence, although no specific date was set. Investigating commissions sent to the Philippines reported unfavorably on the islands' readiness for self-government, and in 1921 the trend toward bringing Filipinos into the government was reversed. Gov. Gen. Henry L. Stimson, who served from 1927 to 1929, somewhat allayed the turmoil raised among the Filipinos by the previous administration.

The advent of the Great Depression in the United States in the 1930s and the first aggressive moves by Japan in Asia (1931) shifted U.S. sentiment sharply toward granting independence to the Philippines. The Hare-Hawes-Cutting Act (1932) provided for complete independence of the islands in 1945 after 10 years of self-government under U.S. tutelage. Congress passed the bill over President Herbert Hoover's veto (1933). It had been drawn up with the aid of a commission from the Philippines, but Manuel L. Quezon, the leader of the dominant Nationalist party, opposed it, partially because of its threat of American tariffs against Philippine products but principally because of the provisions leaving naval bases in U.S. hands. Under his influence, the Philippine legislature rejected the bill. The Tydings-McDuffie Independence Act (1934) closely resembled the Hare-Howes-Cutting Act, but struck the provisions for American bases and carried a promise of further study to correct "imperfections or inequalities." The Philippine legislature ratified the bill; a constitution, approved by President Franklin D. Roosevelt (Mar., 1935) was accepted by the Philippine people in a plebiscite (May). Quezon was elected the first president (Sept.). When Quezon was inaugurated on Nov. 15, 1935, the Commonwealth of the Philippines was formally established. Quezon was re-elected in Nov., 1941.

To develop defensive forces against possible aggression, Gen. Douglas MacArthur was brought to the islands as military adviser in 1935. War came suddenly to the Philippines on Dec. 8, 1941, when Japan attacked without warning. Japanese troops invaded the islands in many places and launched a pincer drive on Manila. MacArthur's scattered defending forces (about 80,000 troops, four fifths of them Filipinos) were forced to withdraw to Bataan Peninsula and Corregidor Island, where they entrenched and tried to hold until the arrival of reinforcements, meanwhile guarding the entrance to Manila Bay and denying that important harbor to the Japanese. But no reinforcements were forthcoming. The Japanese occupied Manila on Jan. 2, 1942. MacArthur was ordered out by President Roosevelt and left for Australia on Mar. 11. The besieged U.S.-Filipino army on Bataan finally crumbled on Apr. 9, 1942. Many individual soldiers refused to surrender, however, and guerrilla resistance, organized and coordinated by U.S and Philippine army officers, continued throughout the Japanese occupation. Japan's efforts to win native loyalty found expression in the establishment (Oct. 14, 1943) of a "Philippine Republic," with José P. Laurel, former supreme court justice, as president. But the people suffered greatly from Japanese brutality, and the puppet government gained little support. The first liberation forces surprised the Japanese by landing (Oct. 20, 1944) at Leyte after months of U.S. air strikes against Mindanao. The Philippine government was established at Tacloban, Leyte, on Oct. 23. The landing was followed (Oct. 23-26) by the greatest naval engagement in history, called variously the Battle of Leyte Gulf and the Second Battle of the Philippine Sea. A great U.S. victory, it effectively destroyed the Japanese fleet and opened the way for the recovery of all the islands. The Japanese suffered over 425,000 dead in the Philippines. The Philippine congress met on June 9, 1945, for the first time since its election in 1941. It faced enormous problems. The land was devastated by war, the economy destroyed, the country torn by political warfare and guerrilla violence.

Manuel Roxas became the first president of the Republic of the Philippines when independence was granted, as scheduled, on July 4, 1946. Filipino ties with the United States were hardly severed, however. An act passed by the U.S. Congress provided for free trade between the two countries for eight years, and a Philippine plebiscite in Mar., 1947, approved granting the United States equal trading rights until 1974 in return for enormous rehabilitation funds. In Mar., 1947, the two countries signed a military assistance pact (since renewed) and the Philippines gave the United States a 99-year lease on designated military, naval, and air bases.

The enormous task of reconstructing the war-torn country was complicated in central Luzon by the Communist-dominated Hukbalahap guerrillas (Huks), who resorted to terror and violence in their efforts to achieve land reform and gain political power. They were finally brought under control after the minister of national defense, Ramón Magsaysay, combined strong military pressure with a conciliatory approach, offering homesteads on Mindanao to Huks who surrendered. Magsaysay was elected president in Nov., 1953. He had promised sweeping economic changes, and he did make progress in land reform, opening new settlements outside crowded Luzon island. His death in an airplane crash in Mar., 1957, was a serious blow to national morale.

In foreign affairs the Philippines maintained a firm anti-Communist policy; the Philippines joined the Southeast Asia Treaty Organization in 1954. There were difficulties with the United States over American military installations in the islands, and, despite formal recognition (1956) of full Philippine sovereignty over these bases, tensions increased until some of the bases were dismantled (1959) and the 99-year lease period was reduced. The early 1960s were marked by efforts to combat the mounting inflation that had plagued the republic since its birth; by attempted alliances with neighboring countries; and by a territorial dispute with Britain over North Borneo (later Sabah). Ferdinand E. Marcos, who succeeded to the presidency in the 1965 elections, inherited this dispute; in Sept., 1968, he approved a congressional bill annexing Sabah to the Philippines. Malaysia suspended diplomatic relations, and the matter was referred to the UN. The continuing need for land reform fostered a new Huk uprising in central Luzon, and in Aug., 1969, President Marcos began a major military campaign to subdue them. Civil war also threatened on Mindanao, where bands of Moros were attacking Christian settlers. In Nov., 1969, Marcos won an unprecedented reelection, but the election was accompanied by violence and charges of fraud, and Marcos's second term began with increasing civil disorder. When Pope Paul VI visited Manila in Nov., 1970, an attempt was made on his life. President Marcos declared martial law in Sept., 1972, charging that a Communist rebellion threatened. A new constitution was drafted (1973) providing direct powers for the president; a referendum gave Marcos the right to remain in office beyond the expiration of his term (1973). Meanwhile, the Moslem rebellion spread from Mindanao to the Sulu Archipelago. In 1977 a tentative accord was reached between Libya (which had given aid to the Moros), the Philippine government, and the rebels providing for the creation of 13 autonomous provinces to be governed by Moslems; however, there was strong opposition among Christians and by 1977 the agreement had not yet been implemented. In the wake of the Communist victories in Southeast Asia (1975), Marcos reoriented Philippine foreign policy away from dependence on the United States toward more cooperation with the nations of Asia. The People's Republic of China was recognized (1976), and the Philippines joined the Association of Southeast Asian Nations (ASEAN), the successor organization to SEATO.

Phil·ips·burg (fĭl'ĭps-bûrg'), city (1970 pop. 1,128), seat of Granite co., W Mont., NW of Butte; settled 1867, inc. 1890. Silver, manganese, and coal are mined here, and there are livestock and grain farms in the area.

Phi·lis·ti·a (fĭ-lĭs'tē-ə, -tyə), ancient region of SW Palestine, comprising a coastal strip along the Mediterranean and a portion of S Canaan. Strategically located on a commercial route from Egypt to Syria, the cities of the region formed a confederacy that was never completely conquered by the Jews.

Phil·lips (fĭl'ĭps). **1.** County (1970 pop. 40,046), 686 sq mi (1,776.7 sq km), E Ark., bounded on the E by the Mississippi River and drained by the White and St. Francis rivers; formed 1820; co. seat Helena. Its agriculture includes fruit, cotton, truck crops, soybeans, livestock, and dairy products. Cottonseed and soybean oil are produced. **2.** County (1970 pop. 4,131), 680 sq mi (1,761.2 sq km), NE Colo., bordering on Nebr. and drained by Frenchman Creek; formed 1889; co. seat Holyoke. It is in an agricultural area that produces wheat and small grains and has some manufacturing. **3.** County (1970 pop. 7,888), 897 sq mi (2,323.2 sq km), N Kansas, in a rolling prairie region bordering in the N on Nebr. and drained by the North Fork of the Solomon River; formed 1867; co. seat Phillipsburg. In an agricultural area yielding wheat, corn, milo, cattle, and pork, it has oil and gas wells and petroleum refineries. **4.** County (1970 pop. 5,386), 5,213 sq mi (13,501.7 sq km), N Mont., in an agricultural area bordering on Sask., Canada, and bounded on the S by the Missouri River; formed 1915; co. seat Malta. Grain, dairy products, livestock, and natural gas are important to its economy. Part of Lewis and Clark National Forest is in the southwest.

Phillips, city (1970 pop. 1,511), seat of Price co., N Wis., NW of Tomahawk, in a dairy region; settled 1874, inc. 1891.

Phil·lips·burg (fĭl'ĭps-bûrg'). **1.** City (1970 pop. 3,241), seat of Phillips co., N Kansas, NW of Salina; platted 1872, inc. 1880. It is a trade center in a corn-growing area and has an oil refinery. There is an annual rodeo. **2.** Town (1970 pop. 17,849), Warren co., NW N.J., on the Delaware River opposite Easton, Pa.; settled 1739, inc. 1861. The city's industrial growth began after the arrival of the railroad and the establishment of iron and steel works here in the mid-1800s.

Phnom Penh (pə-nôm' pĕn'), city (1970 est. pop. 470,000), capital of Cambodia, SW Cambodia, at the confluence of the Mekong and Tônlé Sap rivers. Immediately following the Communist victory (1975) in the Cambodian civil war, the entire population of Phnom Penh was forcibly removed from the city to the countryside to work the land. By 1977 the city was estimated to contain only 50,000 people, most of whom were believed to be government and military personnel. Prior to the onset of civil war (1970), Phnom Penh was the commercial center of the country, serving as the receiving, distribution, and collection point for the nation's exports and imports. The city had large markets and establishments that processed agricultural goods. Phnom Penh was also the transportation center of Cambodia. It was the focus of seven highways radiating out to the provinces and was the terminus of the country's two rail lines. The city was founded in the 14th cent. and was made the Khmer capital after the abandonment (c.1432) of Angkor. It became the capital of Cambodia in 1867. The city was occupied by the Japanese in World War II. During the civil war in Cambodia (1970-75) the population of Phnom Penh tripled as refugees poured in from the war-torn countryside. As the seat of government and the stronghold of the

Lon Nol regime, it was repeatedly under siege. By late 1974 the city was sustaining daily rocket attacks, and by early 1975 it was completely isolated by land, the only link with the outside being the international airport. Two weeks after Lon Nol's departure for exile in the United States, the Khmer Rouge (Communist) forces entered the city; the following day they ordered the population into the provinces to work in the rice fields. (The ostensible reason for this drastic action was to prevent imminent famine.) Since then little information has emerged from Phnom Penh; however, fragmentary reports indicate that life as it was previously known has ceased to exist. All of the city's cultural and educational institutions—the National Univ. of Phnom Penh, schools of art and engineering, a Buddhist university, and a national military academy—are believed to have been abolished. Phnom Penh remains, however, as capital of Cambodia. Whether the regime intends any substantial resettlement of the city in the immediate future is not known.

Pho·cae·a (fō-sē'ə), ancient city, W Asia Minor, N of Smyrna (Izmir), in present Turkey. It was northernmost of the Greek Ionian cities. In the 7th cent. B.C. it grew into a maritime state. In 540 B.C., after a siege by the Persians, most of the inhabitants left, and the city never recovered.

Pho·cis (fō'səs), ancient region of central Greece. It included Delphi, Mt. Parnassus, and Elatea. After the First Sacred War of c.590 B.C., Phocis lost control of Delphi to a council of states. With Athenian help Phocis regained (457 B.C.) hold of Delphi, thus precipitating the Second Sacred War. Early in the next century Phocis passed under Theban control. The Third Sacred War (355–346 B.C.) began with Phocis trying to re-establish itself and ended with the victory of Philip II of Macedon.

Phoe·ni·cia (fə-nē'shə, -nĭsh'ə), ancient territory on the E Mediterranean Sea occupied by Phoenicians. These people were Semitic-speaking Canaanites, and in the 9th cent. B.C. the Greeks gave the new appellation Phoenicians to those Canaanites who lived on the seacoast and traded with the Greeks. The name Phoenicia may be applied to all those places on the shores of the eastern Mediterranean where the Phoenicians established colonies. More often it refers to the great Phoenician cities, notably Tyre and Sidon on the coast of present-day Lebanon.

By 1250 B.C. the Phoenicians were well established as the navigators and traders of the Mediterranean world. They were organized into city-states; the greatest of these were Tyre and Sidon; others were Tripoli, Aradus, and Byblos. The most important Phoenician colonies that later became independent states were Utica and Carthage (founded in the 9th cent. B.C.). There is evidence that in Egyptian service Phoenicians sailed down the western coast of Africa, and possibly their ships even rounded Africa and reached the East Indies. Their carrying trade was enormous, and their wares included articles of wood, metal, ivory, and dyed cloth. In the 6th cent. B.C., they submitted to the tolerant empire of the Persians, keeping their own autonomy but gradually being absorbed into the Persian pattern. In Roman times the cities continued to exist, but Hellenistic culture absorbed the last traces of Phoenician civilization.

Phoe·nix (fē'nĭks), city (1978 est. pop. 669,000), state capital and seat of Maricopa co., S Ariz., on the Salt River; inc. 1881. It is the hub of the rich Salt River valley and an important center of data processing and electronics research and production. Aircraft, fabricated metals, machinery, food products, textiles, and apparel are also among its manufactures. The sunny, dry climate has made Phoenix a winter retreat and a noted health resort. The city was founded (c.1868) on the site of ancient Indian canals. The completion (1911) of the Roosevelt Dam on the Salt River brought power and abundant water to the community and opened a new era of farming. Phoenix boomed during World War II, when three airfields were opened, bringing thousands of servicemen into the area. The phenomenal growth continued after the war; veterans stationed in Phoenix returned to make it their home, and manufacturing concerns moved here to utilize the large labor supply. Among the city's many outstanding parks are the Desert Botanical Gardens, Japanese Flower Gardens, Southwestern Arboretum, and Camelback Mountain. Among its museums are Arizona Museum, with pioneer relics; Heard Museum, with Indian exhibitions; the Phoenix Art Museum; and the Pueblo Grande Museum, containing the excavations of Indian ruins c.800 years old.

Phoenix Islands, group of eight islands (1968 pop. 1,018), 11 sq mi (28.5 sq km), central Pacific, N of Samoa. The Phoenix Islands were discovered between 1823 and 1840 by British and American explorers, but most of them were annexed by Great Britain in the late 19th cent. After the United States took over Howland and Baker islands in 1935, Britain included (1937) the Phoenix group in the Gilbert and Ellice Islands colony. In 1938 the United States claimed sovereignty over Canton and Enderbury islands, and in 1939 Britain and the United States agreed to exercise joint control over the two islands for a period of 50 years.

Phoe·nix·ville (fē'nĭks-vĭl'), industrial borough (1970 pop. 14,823), Chester co., SE Pa., on the Schuylkill River; settled 1720, inc. 1849. Iron deposits in the region led to the early development of an iron industry and later (1886) to the manufacture of steel.

Phryg·i·a (frĭj'ē-ə), ancient region, central Asia Minor (now central Turkey). The Phrygians, who settled here c.1200 B.C., came from the Balkans and apparently spoke an Indo-European language. A kingdom flourished from the 8th to the 6th cent. B.C., when it became dominated by Lydia. Phrygia was best known to the Greeks as a source of slaves. Northern Phrygia became part of Galatia with the invasion of the Gauls (3rd cent. B.C.).

Phu·ket (po͝o'kĕt'), island, 206 sq mi (534 sq km), a province of Thailand, in the Andaman Sea, off the W coast of the Malay Peninsula. Flat, with isolated hills, the island is one of Thailand's chief tin-mining regions and also produces rubber, coconuts, and pepper. The Chinese have mined tin here since ancient times. Phuket town was founded in the 1st cent. B.C. by colonists from India. European merchants began trading here in the 16th cent. The island, contested by the Siamese and the Burmese during the 18th cent., was finally incorporated into Thailand in the 19th cent.

Pia·cen·za (pyä-chĕn'tsä), city (1971 pop. 106,461), capital of Piacenza prov., in Emilia-Romagna, on the Po River. It is an agricultural, commercial, and industrial center. Manufactures include agricultural machinery, chemicals, and food products. The city was a Roman stronghold (called Colonia Placentia) against the Gauls and was later occupied by the Goths, the Lombards, and the Franks. A free commune by the 12th cent., Piacenza joined the Lombard League. In 1545 it formed, with Parma and its territory, the duchy of Parma and Piacenza, ruled, until 1731, by the Farnese family. Noteworthy buildings include the Lombard-Gothic Palazzo del Comune (1281); the cathedral (1122–1233), with frescoes by Guercino; and the churches of San Savino and Madonna di Campagna.

Pi·at·ra-Ne·amt (pyä'trä-nyämt'), city (1974 est. pop. 66,568), NE Rumania, in the Bacău region. Oil refining, food processing, and the manufacture of chemicals are among the industries.

Pi·att (pī'ət), county (1970 pop. 15,509), 437 sq mi (1,131.8 sq km), central Ill., drained by the Sangamon River; formed 1841; co. seat Monticello. Its agriculture includes corn, oats, soybeans, wheat, livestock, poultry, and dairy products. Fiber tiles, patent medicines, and health foods are manufactured.

Piau·í (pyou-ē'), state (1975 est. pop. 1,988,200), 96,886 sq mi (250,935 sq km), NE Brazil, on the Atlantic Ocean; capital Teresina.

Pia·ve (pyä'vā), river, c.137 mi (220 km) long, rising in the Carnic Alps, Venetia, NE Italy, and flowing generally S, past Belluno, to the Gulf of Venice. Hydroelectric power is produced along the upper Piave; the lower river is used for irrigation.

Pic·ar·dy (pĭk'ər-dē), region and former province, N France, on the English Channel. It has three main regions: the plateau north of Paris, an important wheat and beet area; the Somme River valley, with manufacturing cities; and the coast, with fishing and commercial seaports and beach resorts. The name Picardy appeared about the 13th cent., designating the many small feudal holdings added to the crown by Philip II. During the Hundred Years' War the area was contested by France and England. Louis XI occupied it in 1477, securing it for France.

Pic·a·yune (pĭk'ē-o͞on', -ə-yo͞on'), city (1970 pop. 10,467), Pearl River co., S Miss., near the Pearl River and the La. border; inc. 1904. It is the trade, processing, and shipping center for a tung-tree, beef cattle, and dairy area. Tung oil, truck bodies, and blankets are among the city's products.

Pic du Mi·di d'Os·sau (pēk do͞o mē-dē' dô-sō'), double peak, 10,322 ft (3,148.2 m) high, SW France, in the Pyrenees, near the Spanish border. It is in a winter-sports area.

Pi·chin·cha (pə-chĭn'chə, pē-chēn'chä), volcano, 15,918 ft (4,855 m) high, N Ecuador, near Quito. It last erupted in 1881. On its lower slopes, in the battle of Pichincha on May 24, 1822, patriot forces routed the Spanish royalists and freed the territory that later became Ecuador.

Pick·a·way (pĭk'ə-wā'), county (1970 pop. 40,071), 504 sq mi (1,305.4 sq km), S central Ohio, intersected by the Scioto River; formed 1810; co. seat Circleville. In an agricultural area that produces soybeans, corn, livestock, and poultry, it has sand and gravel pits and a food-processing industry. Paper, plastics, glass, metal products, and machinery are made.

Pick·ens (pĭk'ənz). **1.** County (1970 pop. 20,326), 887 sq mi (2,297.3 sq km), W Ala., in the Black Belt, bordering on Miss. and drained by the Tombigbee and Sipsey rivers; formed 1820; co. seat Carrollton. It has cotton and poultry farming, lumber logging and sawmilling, some mining, and a textile industry. **2.** County (1970 pop. 9,620), 225 sq mi (582.8 sq km), N Ga., in the Blue Ridge; formed 1853; co. seat Jasper. It is in a resort and farming area (corn, cotton, hay, fruit, and livestock), with marble quarrying, lumbering, and industry (textile finishing and clothing manufacturing). **3.** County (1970 pop. 58,956), 492 sq mi (1,297.6 sq km), NW S.C., bounded in the E by the Saluda River, in the W by the Keowee and Seneca rivers, and bordering N.C. in the N; formed 1826; co. seat Pickens. Its agriculture includes cotton, corn, and poultry. It has granite quarries, a lumbering industry, and plants manufacturing textiles and garments, aluminum, and machinery. Part of the Blue Ridge is in the north.

Pickens, town (1970 pop. 2,954), seat of Pickens co., extreme NW S.C., W of Greenville; settled 1868, inc. 1908. Mainly a resort, it also makes textiles and lumber products. The site of Fort Prince George, the center of conflict in the Cherokee War of 1760–62, is nearby.

Pick·ett (pĭk'ət), county (1970 pop. 3,774), 158 sq mi (409.2 sq km), N Tenn., on the Cumberland Plateau, bounded in the N by Ky. and drained by the Obey and Wolf rivers; formed 1879; co. seat Byrdstown. Its agriculture includes livestock, tobacco, and corn. Bituminous-coal mining, lumbering, and clothing manufacturing are done.

Pi·co (pē'kō'), island (1974 est. pop. 18,014), 167 sq mi (433 sq km), in the N Atlantic, one of the central Azores. It takes its name from a volcanic mountain, Pico Alto, which rises to 7,711 ft (2,351.9 m).

Pi·co Ri·ve·ra (pē'kō rĭ-vâr'ə), city (1978 est. pop. 50,500), Los Angeles co., S Calif., on the San Gabriel and Rio Hondo rivers; inc. 1958.

Pic·tou (pĭk'to͞o), town (1976 pop. 4,588), N N.S., Canada, on Pictou Harbour, an inlet of Northumberland Strait. It is a lobster-fishing port. Pictou was settled (1763) by a group of colonists from Philadelphia and later received many settlers from the Scottish Highlands.

Pic·tured Rocks National Lakeshore (pĭk'chərd): *see* National Parks and Monuments Table.

pied·mont (pēd'mŏnt'), any area near the foot of a mountain, particularly the plateau (the Piedmont) extending from N.Y. to Ala. E of the Appalachian Mts. and W of the Atlantic coastal plain. In Md., Va., and N.C. it is east of the Blue Ridge Mts. The plateau is cut by numerous small rivers, whose fall line is along its eastern edge.

Piedmont, region (1975 est. pop. 4,541,271), 9,807 sq mi (25,400 sq km), NW Italy, bordering on France in the W and on Switzerland in the N; capital Turin. The mostly mountainous and hilly region has the Alps in the north and west and the Apennines in the south. In the more elevated parts of Piedmont, forest products and fruit are produced and cattle are raised. In the fertile valley of the upper Po River wheat, maize, rice, grapes, honey, and chestnuts are grown. Piedmont has considerable industry, well-developed hydroelectric facilities, and an extensive transportation network. There is a substantial tourist industry.

The area of Piedmont was incorporated by Rome in the 1st cent. B.C. It came to be known as Piedmont by the 13th cent., growing out of Turin and Ivrea, western marches of the Lombard kingdom of Italy. The marches passed by marriage (11th cent.) to the Savoy dynasty. In the 12th cent. free communes were instituted in many cities, while others remained under feudal lords. By the 15th cent. Savoy emerged as the chief power. The French often entered Piedmont, either as allies or as enemies, and greatly influenced Piedmontese history and culture. Piedmont was a major battlefield in the Italian Wars (15th–16th cent.), the wars of Louis XIV, and the French Revolutionary Wars. After 1814 the region became the nucleus of Italian unification.

Pie·dras Ne·gras (pyā'*th*räs nā'gräs), ruined city of the Classic era of the Maya, NW Petén, Guatemala, in the Usumacinta valley.

Piedras Negras, city (1970 pop. 41,033), Coahuila state, N Mexico, on the Rio Grande opposite Eagle Pass, Texas. Founded in 1849, it grew as an international shipping point.

Pierce (pîrs). **1.** County (1970 pop. 9,281), 342 sq mi (885.8 sq km), SE Ga., bounded on the S by the Satilla River, on the NE by the Little Satilla River, and drained by the Alabaha River; formed 1857; co. seat Blackshear. In a coastal-plain farming (tobacco, cotton, corn, sweet potatoes, and livestock) and timber area, it has clothing and shoe manufacturing. **2.** County (1970 pop. 8,493), 573 sq mi (1,484.1 sq km), NE Nebr., in an agricultural area drained by branches of the Elkhorn River; formed 1856; co. seat Pierce. Grain, livestock, dairy products, and poultry form the basis of its agriculture. **3.** County (1970 pop. 6,323), 1,038 sq mi (2,688.4 sq km), N central N.Dak., in an agricultural area watered by numerous small lakes; formed 1887; co. seat Rugby. Wheat, barley, and flax are grown here. **4.** County (1970 pop. 412,344), 1,676 sq mi (4,340.8 sq km), W central Wash., drained by the White and Puyallup rivers; formed 1852; co. seat Tacoma. In a rich agricultural area yielding fruit, nuts, bulbs, truck crops, livestock, and dairy products, it has sand and gravel pits and timber. Clothing, wood and metal products, furniture, paper, chemicals, plastics, concrete, machinery, and boats are manufactured. Parts of Mt. Rainier National Park and Columbia National Forest are here. **5.** County (1970 pop. 26,652), 591 sq mi (1,530.7 sq km), W Wis., bounded in the W by the St. Croix River and in the SW and S by the Mississippi River and Lake Pepin, which all form the Minn. border; formed 1853; co. seat Ellsworth. It is in a generally rugged terrain, which has dairying and poultry and stock raising. Its industries include food processing, lumbering, and some manufacturing.

Pierce, city (1970 pop. 1,360), seat of Pierce co., NE Nebr., NNW of Norfolk; settled 1870.

Pi·e·ri·a (pī-îr'ē-ə), region of ancient Macedonia, W of the Thermaic Gulf (the modern Gulf of Thessaloniki). It included Mt. Pierus, seat of the worship of Orpheus and the Muses, and Mt. Olympus.

Pierre (pîr), city (1978 est. pop. 11,700), state capital (since 1889) and seat of Hughes co., central S.Dak., on the E bank of the Missouri River, opposite Fort Pierre; inc. 1883. Its economy is centered around agriculture (chiefly grains and cattle) and the state government. Originally the fortified capital of the Aricara Indians, it served as the trade center of the middle Missouri River from 1822 to 1855. The city boomed with the arrival of the railroad (1880), becoming an important trading and shipping center for a farm and ranch area.

Pierre·latte (pyĕr-lät'), town (1968 pop. 9,873), Drôme dept., SE France. The center of France's nuclear industry, Pierrelatte has a large uranium-producing complex.

Pie·tar·saa·ri (pē-ĕt'ər-sä'rē), city (1970 pop. 18,725), Vaasa prov., W Finland, on the Gulf of Bothnia. It is an important port and industrial center. Most of the inhabitants speak Swedish. The city was founded in 1652.

Pie·ter·mar·itz·burg (pē'tər-măr'ĭts-bûrg'), city (1970 pop. 114,822), capital of Natal, E South Africa, in the foothills of the Drakensberg Range. The city is an administrative and industrial center. Its products include wattle-bark extract, furniture, footwear, chocolate, cloth, and diesel engines. Motor vehicles are assembled in the city. Pietermaritzburg was founded in 1838 and became capital of Natal when the province was annexed by Great Britain in 1843. Points of interest include the Church of the Vow (1839), built to commemorate the 1838 Boer victory over Zulu forces, and Fort Napier, erected by the British in 1843.

Pie·ters·burg (pē'tərz-bûrg'), city (1970 pop. 27,174), Transvaal, NE South Africa. It is a commercial center for the surrounding agricultural area. Pietersburg was founded in 1884. In 1900 during the South African War it was briefly the capital of both the Transvaal and the Orange Free State. It was occupied by British troops in 1901.

Pig·gott (pĭg'ət), city (1974 est. pop. 3,361), a seat of Clay co., extreme NE Ark., NE of Jonesboro; settled c.1850, inc. 1891. It produces shoes and has poultry, dairy, and hog farms.

Pike (pīk). **1.** County (1970 pop. 25,038), 673 sq mi (1,743.1 sq km), SE Ala., in the coastal plain, drained by the Conecuh and Pea rivers; formed 1821; co. seat Troy. It is in a cotton, peanut, and poultry farming area. Lumbering, food processing, and clothing manufacturing are done. **2.** County (1970 pop. 8,711), 600 sq mi (1,554 sq km), SW Ark., bounded on the S by the Little Missouri River; formed 1833; co. seat Murfreesboro. Its agriculture includes peaches, strawberries, cotton, truck crops, soybeans, poultry, and cattle. It has lumbering, gypsum quarrying, metal mining, and a poultry-dressing industry. Tourist attractions include Crater of Diamonds, Caddo Indian Burial Mounds, and Narrows Dam on Lake Greeson. **3.** County (1970 pop. 7,316), 230 sq mi (595.7 sq km), W Ga., bounded on the W by the Flint River; formed 1822; co. seat Zebulon. It is in a piedmont agricultural area that produces cotton, peaches, and truck crops. Its industry includes lumbering, food canning, and clothing manufacturing. **4.** County (1970 pop. 19,185), 829 sq mi (2,147.1 sq km), W Ill., bounded on the W and SW by the Mississippi River, on the E by the Illinois River, and drained by Bay and McCraney creeks; formed 1821; co. seat Pittsfield. It has agriculture (corn, wheat, hay, apples, livestock, and poultry), timber, and marble and granite quarries. Flour, cheese and other dairy products, shoes, and costume jewelry are manufactured. **5.** County (1970 pop. 12,281), 335 sq mi (867.7 sq km), SW Ind., bounded on the N by the White River and its East Fork and drained by the Patoka River; formed 1816; co. seat Petersburg. It is in a bituminous-coal and agricultural (grain and tobacco) area, with timber, oil wells, and clay pits. Its manufacturing includes concrete blocks, wood products, flour, and nursery stock. **6.** County (1970 pop. 61,059), 786 sq mi (2,035.7 sq km), E Ky., in the Cumberlands, bounded on the NE by Tug Fork and the W.Va. border, on the SE by the Va. border, and drained by Levisa and Russell forks; formed 1821; co. seat Pikeville. It is in a mountainous bituminous-coal region that also has some timber. Its agriculture includes dairy products, poultry, cattle, apples, Irish and sweet potatoes, soybeans, corn, and tobacco. **7.** County (1970 pop. 31,813), 410 sq mi (1,061.9 sq km), SW Miss., bordered on the S by La. and drained by the Bogue Chitto and Tangipahoa River; formed 1815; co. seat Magnolia. Farming, cotton, corn, and truck crops, dairying, and lumbering are major activities here. **8.** County (1970 pop. 16,928), 681 sq mi (1,763.8 sq km), E Mo., bounded in the E by the Mississippi River and crossed by the Salt River; formed 1818; co. seat Bowling Green. Major products include wheat, corn, oats, soybeans, apples, and livestock. There are also limestone quarries, and some manufacturing is done. **9.** County (1970 pop. 19,114), 443 sq mi (1,147.4 sq km), S Ohio, intersected by the Scioto River and by Sunfish and Beaver creeks; formed 1815; co. seat Waverly. Its agriculture includes corn, soybeans, wheat, oats, hay, livestock, and poultry. Chemical and metal products are made. **10.** County (1970 pop. 11,818), 545 sq mi (1,411.6 sq km), NE Pa., in a forested lake region bounded in the E by the Delaware River; formed 1814; co. seat Milford. The area is noted for textile manufacturing, timber, and recreational areas.

Pikes Peak (pīks), 14,110 ft (4,303.6 m) high, central Colo., in the Front Range of the Rocky Mts.; discovered by U.S. explorer Zebulon Pike in 1806. There are many higher peaks in the Rockies, but this is the best known and most conspicuous because of its location on the edge of the Great Plains. Its summit, generally snow-covered, is reached by a cog railroad and a highway.

Pike·ville (pīk'vĭl'). **1.** City (1975 est. pop. 5,300), seat of Pike co., E Ky., in the Cumberlands on Levisa Fork SW of Williamson, W.Va. It is a trade and shipping center for a bituminous-coal and timber region. Cement blocks, beverages, and meat and food products are made. **2.** Town (1970 pop. 1,454), seat of Bledsoe co., central Tenn., on the Sequatchie River NNE of Chattanooga, in a timber and farm region. Hosiery, wood products, cheese, and flour are made.

Pi·ła (pē'lə), town (1975 est. pop. 49,300), NW Poland. Once the capital of Grenzmark Posen-West Prussia, it is now chiefly a trade and industrial center. The city was chartered in 1380.

Pi·la·tus (pē-lä'to͞os), mountain, 6,800 ft (2,074 m) high, in the Alps of the Four Forest Cantons, central Switzerland. According to medieval legend, the corpse of Pontius Pilate was thrown into a small lake on the mountain.

Pil·co·ma·yo (pēl'kô-mä'yô), river, c.700 mi (1,125 km) long, rising in the Bolivian Andes E of Lake Poopó and flowing SE in a marshy course across the Gran Chaco to the Paraguay River near Asunción. It forms part of the Argentina-Paraguay border.

Pil·lars of Her·cu·les (pĭl'ərz; hûr'kyə-lēz'), ancient mythological name for promontories flanking the E entrance to the Strait of Gibraltar. They are usually identified with Gibraltar in Europe and with Mt. Acha at Ceuta in Africa.

Pi·ma (pē'mə), county (1970 pop. 351,667), 9,241 sq mi (23,934.2 sq km), S Ariz., in a mountainous region bordering on Mexico; formed 1864; co. seat Tucson. The Santa Catalina Mts. are in the northeast. Irrigated farming (cotton, alfalfa, citrus fruits, and truck crops) is found along the Santa Cruz River in the east. There are copper, gold, silver, and lead mines in the northwest. Saguaro National Monument and Papago Indian Reservation are in the county.

Pi·nal (pə-năl'), county (1970 pop. 68,579), 5,364 sq mi (13,892.7 sq km), S central Ariz., in a plateau and mesa region irrigated by water from the Gila and San Carlos rivers; formed 1875; co. seat Florence. Its agriculture includes cotton, alfalfa, citrus fruits, and truck crops. Copper is mined and processed.

Pi·nang or **Pe·nang** (both: pə-năng'), city (1971 est. pop. 270,000), capital of Pinang state, Malaysia, on the Strait of Malacca on Pinang Island. The city is Malaysia's leading port. The island is largely agricultural, with some rubber production. Pinang Island was the first British settlement on the Malay Peninsula.

Pi·nar del Ri·o (pē-när' dĕl rē'ō), city (1970 pop. 73,206), capital of Pinar del Rio prov., W Cuba. The city, founded in 1699, is famous for the tobacco grown in the Vuelta Abajo district.

Pinck·ney·ville (pĭngk'nē-vĭl'), city (1970 pop. 3,377), seat of Perry co., S Ill., SW of Mt. Vernon, in a farm and coal region; inc. 1861.

Pin·dus (pĭn'dəs), chief mountain range of Greece, extending c.100 mi (160 km) S from the Albanian border through NW Greece. Mt. Smólikas (8,650 ft/2,638.3 m) is the highest peak. The Pindus are a continuation of the Dinaric Alps but have a lower limestone content than the Dinarics. The sparsely populated range is rich in timber.

Pine (pīn), county (1970 pop. 16,821), 1,412 sq mi (3,657.1 sq km), E Minn., bounded on the E by the St. Croix River and Wis. and drained by the Kettle River; formed 1856; co. seat Pine City. In an agricultural area yielding livestock, dairy products, grain, and potatoes, it has timber and copper mines. Its industry includes food processing plants, steelworks, and paper mills.

Pine Bar·rens (băr'ənz), coastal-plain region, c.3,000 sq mi (7,770 sq km), S and SE N.J. It is composed chiefly of sandy soils, swamp-edged streams, pine stands, and tracts of cranberries and blueberries. Originally a well-forested area of pine, cedar, and oak, its trees were indiscriminately cut for shipbuilding and charcoal making until the 1860s, when they were nearly exhausted.

Pine Bluff (blŭf), city (1978 est. pop. 53,900), seat of Jefferson co., S central Ark., on the Arkansas River; inc. 1839. It is a port and trade center for an agricultural area and has industries producing electric transformers, wood and paper products, metal goods, and furniture. Pine Bluff Arsenal, established during World War II, is the center of U.S. army chemical, biological, and toxicological research.

Pine City, village (1970 pop. 2,143), seat of Pine co., E Minn., NE of St. Cloud near the Wis. border, in a farm and dairy region; platted 1869, inc. 1881.

Pine·dale (pīn'dāl'), town (1970 pop. 948), seat of Sublette co., W Wyo., on a branch of the Green River NNW of Rock Springs. It is a resort in a livestock, dairying, and timber area.

Pi·nel·las (pĭ-nĕl'əs), county (1970 pop. 522,329), 264 sq mi (683.8 sq km), W central Fla., largely a peninsula between the Gulf of Mexico and Tampa Bay, bordered on the W by a chain of barrier islands; formed 1911; co. seat Clearwater. It is in a citrus-fruit growing, dairying, and tourist area. Its industry includes canning, some clothing manufacturing, and wood processing. Furniture, leather goods, and concrete are made.

Pi·ne·ro·lo (pē'nā-rô'lō), city (1976 est. pop. 36,864), Piedmont, NW Italy, at the foot of the Alps. It is an agricultural and industrial center. Manufactures include paper, textiles, machinery, and processed food. First mentioned in the 10th cent., Pinerolo was a strongly fortified citadel that passed to the house of Savoy in the 13th cent. It was often in French hands from 1536 to 1814. The Man with the Iron Mask was held here for some years after his seizure in 1679.

Pines, Isle of (pīnz), island (1970 pop. 30,103), 1,180 sq mi (3,056 sq km), off SW Cuba. Pine forests cover much of the island and there are numerous mineral springs. Marble is quarried from low ridges in the northern part; the southern quarter of the island is an elevated plain. The economy is based on fishing and agriculture (citrus fruits and vegetables). Until the early 1960s much of the land was owned by American citizens, and the mild, healthful climate and excellent fishing waters made the island an important resort. Discovered by Columbus in 1494, the Isle of Pines was later used as a penal colony and was a rendezvous for buccaneers. During the colonial period it was a summer resort for the Spanish military. The island was ceded to the United States after the Spanish-American War (1898); and because its name was omitted from the Platt Amendment, which defined Cuba's boundaries, it was claimed by both countries. In 1907 the U.S. Supreme Court declared that the island did not belong to the United States; a treaty later (1925) confirmed Cuba's claim.

Pines, Isle of, island (1969 pop. 978), c.58 sq mi (150 sq km), South Pacific, a part of New Caledonia. Formerly a penal colony, the island is now a popular tourist resort.

Pine·ville (pīn'vĭl'). **1.** Town (1970 pop. 2,817), seat of Bell co., S Ky., on the Cumberland River N of Middlesboro. It is a resort and tourist center in a coal and timber area. **2.** City (1970 pop. 8,951), Rapides parish, central La., on the Red River opposite Alexandria; settled in the early 18th cent., inc. 1878. In a timber area, it makes lumber products. **3.** Town (1970 pop. 444), seat of McDonald co., extreme SW Mo., in the Ozarks on Elk River S of Neosho. It is a resort in an area yielding grain and dairy products. **4.** Town (1970 pop. 1,187), seat of Wyoming co., S W.Va., NW of Bluefield near the Guyandot River. It is a trade center in a coal-mining region.

P'ing-tung or **Ping·tung** (both: pĭng'dōōng'), town (1973 est. pop. 169,657), S Taiwan. It is a major sugar-refining center. Other industries produce metals, machinery, chemicals, and alcoholic beverages.

Pi·ni·ós (pē'nē-ôs') or **Pe·ne·us** (pə-nē'əs), river, 134 mi (215.6 km) long, rising in the Pindus Mts., NW Greece, and flowing generally E past Lárisa and through the Vale of Tempe into the Aegean Sea.

Pin·na·cles National Monument (pĭn'ə-kəlz): *see* National Parks and Monuments Table.

Pinsk (pĭnsk, pēnsk), city (1976 est. pop. 84,000), W European USSR, in Belorussia, in the Pripyat Marshes at the confluence of the Pina and Pripyat rivers. It has long been a noted water transport junction; timber is now the chief export. Industries include shipbuilding and repair and the manufacture of metal products, building materials, and clothing. Mentioned in the chronicles in 1097 as part of the Kievan state, the city became the capital of Pinsk duchy in the 13th cent. It passed to Lithuania in 1320 and to Poland in 1569. Pinsk was transferred to Russia in 1793 with the second partition of Poland; it reverted to Poland in 1921 but was ceded to the USSR in 1945.

Pi·oche (pē-ōch'), village (1970 pop. 600), seat of Lincoln co., E Nev., NNE of Las Vegas, at an altitude of 6,100 ft (1,860.5 m). Zinc, lead, silver, and gold are mined, and livestock and truck crops raised.

Pio·tr·ków Try·bu·nal·ski (pyô'tər-kōōf' trĭb'ōō-näl'skē), city (1975 est. pop. 64,200), central Poland. A textile center, it also manufactures mining equipment, agricultural machinery, glass, chemicals, bricks, and leather goods. It was first mentioned in 1217 and became the seat of several Polish diets (1347–1578) and tribunals (1578–1792). The city passed (1815) to Russia and was (1867–1915) the capital of Petrokov province. It reverted to Poland in 1919. Piotrków Trybunalski has several old churches and the ruins of a castle built by Casimir the Great.

Pipe Spring National Monument (pīp' sprĭng'): *see* National Parks and Monuments Table.

Pipe·stone (pīp'stōn'), county (1970 pop. 12,791), 464 sq mi (1,201.8 sq km), SW Minn., bordering on S.Dak. and drained by headwaters of the Rock River; formed 1857; co. seat Pipestone. It is in a livestock and grain (corn, oats, and barley) area. Its industry includes food processing and boatbuilding.

Pipestone, city (1970 pop. 5,328), seat of Pipestone co., SW Minn., near the S.Dak. border NE of Sioux Falls, S.Dak.; settled 1874, platted 1876, inc. 1881. It is a trade center of a farm and dairy region. Nearby are pipestone quarries, once used by Indians and now part of Pipestone National Monument.

Piq·ua (pĭk'wā', -wə), city (1970 pop. 20,741), Miami co., W Ohio, on the Miami River; settled 1797, chartered 1929. It is an industrial city with manufactures including airplane and automobile parts, steel and iron, paper, aluminum, wood, and metal products.

Pi·ra·ci·ca·ba (pîr'ə-sə-kä'bə), city (1970 pop. 125,490), São Paulo state, SE Brazil, on the Tietê River. The city houses a noted agricultural institute.

Pi·rai·évs (pē-rā-ĕfs') or **Pi·rae·us** (pī-rē'əs), city (1971 pop. 187,458), E central Greece, in Attica, on the Saronic Gulf. It is the port of Athens and the chief port in Greece. A commercial center, Piraiévs has shipyards and industries that manufacture chemicals, textiles, and machinery. The construction of Piraiévs was planned by Themistocles and executed (c.450 B.C.) by the architect Hippodamus of Miletus. The Long Walls, two parallel walls about 600 ft (183 m) apart, connected Athens with Piraiévs. Modern development of Piraiévs began after Greece achieved independence in the 19th cent.

Pír·gos or **Pyr·gos** (both: pîr'gôs), town (1971 pop. 20,599), capital of Ilía prefecture, SW Greece, in the Peloponnesus, near the mouth of the Alfiós River. It is a commercial center and has industries that manufacture cigarettes and alcoholic beverages.

Pir·ma·sens (pĭr'mä-zĕns'), city (1974 est. pop. 54,631), Rhineland-Palatinate, SW West Germany, near the French border. It is a leading manufacturer of shoes and boots. Other products include chemicals, plastics, and machines. Founded in the 8th cent., Pirmasens belonged to the counts of Hanau-Lichtenberg until 1736. It later passed to Hesse-Darmstadt and in 1816 to Bavaria.

Pir·na (pĭr'nä), city (1974 est. pop. 49,771), Dresden district, SE East Germany, on the Elbe River. Manufactures include rayon, paper, steel, furniture, and ceramics. Known in 1233, Pirna passed to Bohemia in 1298 and to Meissen in the early 15th cent. The Saxonians surrendered (1756) here to Prussia in the Seven Years' War.

Pi·sa (pē'zə, -sä), city (1976 est. pop. 103,479), capital of Pisa prov., Tuscany, N central Italy, on the Arno River near the Tyrrhenian Sea. Pisa is a commercial and industrial center; manufactures include glass, textiles, pharmaceuticals, and processed food. Probably a Greek colony, later certainly an Etruscan town, it became a Roman colony (180 B.C.) and prospered. During the 9th to 11th cent. A.D. it developed into a powerful maritime republic, fighting the Arabs throughout the Mediterranean and rivaling Genoa and Venice. While competing with Genoa for the possession of Corsica and Sardinia, Pisa was crushed by the Genoese in the naval battle of Meloria (1284). As a Ghibelline center in the 13th and 14th cent., the city was continually at war with Florence, to which it fell in 1406. At the same time, a school of sculpture founded by Nicola Pisano flourished in Pisa. The university (founded in the 14th cent.) enjoyed a great reputation during the Renaissance; Galileo, who was born in Pisa in 1564, was a student and later a teacher here. Pisa was badly damaged in World War II but was extensively reconstructed after 1945; the characteristic Pisan style, a variation of the Romanesque, was largely retained. The city's noteworthy structures include the Romanesque cathedral (1068–1118), the marble baptistery (1153–1278), the marble Leaning Tower (180 ft/55 m high and 16 ft/4.9 m out of the perpendicular), and the churches of Santa Maria della Spina (early 14th cent.) and Santa Caterina.

Pis·cat·a·quis (pĭs-kăt'ə-kwĭs), county (1970 pop. 16,285), 3,892 sq mi (10,080.3 sq km), N central Maine, drained by the Piscataquis and Pleasant rivers and the West Branch of the Penobscot River; formed 1838; co. seat Dover-Foxcroft. It is a potato-growing and lumbering area, with some industry (textile mills and plants manufacturing wood products, furniture, and shoes). The county has hundreds of lakes with good recreational facilities.

Pis·cat·a·way Park (pĭs-kăt'ə-wā'): *see* National Parks and Monuments Table.

Pis·co (pĭs'kō, pē'skō), city (1969 est. pop. 26,700), capital of Ica dept., SW Peru, a port on the Pacific Ocean. The major industries are the production of the famous Pisco brandy and the cultivation and processing of cotton.

Pí·sek (pē'sĕk'), city (1975 est. pop. 25,477), SW Czechoslovakia, in Bohemia, on the Otava River. It has woodworking, tobacco, and textile industries. Písek was founded in the 13th cent. and later suffered heavily in the Thirty Years' War (1618–48). The city has a 13th cent. palace and several Gothic churches.

Pi·sid·i·a (pĭ-sĭd'ē-ə, pī-), ancient country of S Asia Minor, S of Phrygia and N of Pamphylia. It was a mountainous country, traversed by the Taurus range. Its warlike tribes were independent until the country was incorporated into a Roman province in the early 1st cent. A.D.

Pis·to·ia (pē-stô'yä), city (1976 est. pop. 84,700), capital of Pistoia prov., Tuscany, central Italy, at the foot of the Apennines. It is an agricultural and industrial center. Manufactures include leather and metal goods, glass, textiles, canned foods, and footwear. Pistoia was under Roman rule from the 6th cent. B.C. Hampered by wars and internal strife, it fell under the hegemony of Florence in the 14th cent.

Pit·cairn Island (pĭt'kârn'), volcanic island (1975 est. pop. 67), 2.5 sq mi (6.5 sq km), South Pacific, SE of Tuamotu Archipelago. A British possession since 1839, the island is officially administered by the British High Commissioner to New Zealand. The island was discovered in 1767 by Philip Carteret, a British admiral, but was named after Robert Pitcairn, the midshipman who first sighted it. It was colonized in 1790 by mutineers from the *Bounty* and Tahitian women. Their descendants still inhabit the island. In 1957 the remains of the *Bounty* were discovered here.

Pitch Lake (pĭch), c.114 acres (46 hectares), SW Trinidad Island, Trinidad and Tobago. The lake is believed to be formed and supplied by the seepage of natural pitch, a form of petroleum, from the surrounding oil-rich region. The pitch, hard around the edges of the pool, becomes more viscous toward the center. The seemingly inexhaustible supply has yielded millions of tons of pitch since the 16th cent. Fossils of prehistoric animals have been found here.

Pi·teş·ti (pē-tĕsht', -tĕsh'tē), city (1977 est. pop. 123,943), S central Rumania, in Walachia, on the Argesul River. It is the commercial center of the Argeş region and an important rail junction.

Pit·kin (pĭt'kən), county (1970 pop. 6,185), 974 sq mi (2,522.7 sq km), W central Colo., drained by branches of the Colorado River and including part of the Sawatch Mts. in the E; formed 1881; co. seat Aspen. Livestock grazing, mining (silver, lead, and bituminous coal), and tourism are important to its economy.

Pitt (pĭt), county (1970 pop. 73,900), 685 sq mi (1,774.2 sq km), E central N.C., drained by the Tar River; formed 1760; co. seat Greenville. It is in a coastal-plain agricultural (tobacco, cotton, corn, soybeans, and peanuts) and lumbering area.

Pitt Island, 537 sq mi (1,390.8 sq km), W British Columbia, Canada, in Hecate Strait, between Banks Island and the mainland.

Pitts·bor·o (pĭts'bĕr'ō, -bə-rə). **1.** Village (1970 pop. 188), seat of Calhoun co., N central Miss., ENE of Grenada, in an agricultural and timber area. **2.** Town (1970 pop. 1,447), seat of Chatham co., central N.C., W of Raleigh; founded c.1787. Wood products and textiles are made here, and poultry is processed.

Pitts·burg (pĭts'bûrg'), county (1970 pop. 37,521), 1,241 sq mi (3,214.2 sq km), SE Okla., bounded in the N by the Canadian River and drained by small Gaines Creek, with the Ouachita Mts. in the SE corner; formed 1907; co. seat McAlester. Its agriculture includes peanuts, soybeans, grain, cotton, vegetables, dairy products, and cattle. It has oil and natural-gas wells, coal mines, rock quarries, and a food-processing industry.

Pittsburg. 1. Industrial city (1978 est. pop. 27,800), Contra Costa co., W Calif., on the edge of the San Francisco Bay area, at the junction of the Sacramento and the San Joaquin rivers; laid out 1849, inc. 1903. Manufactures include steel, chemicals, roofing materials, and cans. Coal was discovered here in 1855 and was mined until 1902. **2.** City (1978 est. pop. 18,000), Crawford co., SE Kansas, near the Mo. line; founded 1876 as a mining town, inc. 1880. It is a mining center near large coal deposits. Clay, limestone, zinc, lead, and oil are also found in the area. The city's manufactures include coal-mining equipment, aircraft, livestock food supplements, building supplies, and clay pipe. **3.** City (1970 pop. 3,844), seat of Camp co., E Texas, SW of Texarkana; settled 1854. It processes sweet potatoes and pine timber.

Pitts·burgh (pĭts'bərg), city (1978 est. pop. 430,000), seat of Allegheny co., SW Pa., at the confluence of the Allegheny and the Monongahela rivers, here forming the Ohio River; inc. 1816. A port of entry, it is a leading industrial center and has access to large reserves of raw material. Steel is the city's chief manufacture. Glass, machinery, mine-safety equipment, chemicals, petroleum products, paper goods, and electrical equipment are also produced. Printing and publishing are important, and extensive industrial research and testing are done. The downtown area is now known as the Golden Triangle, and the new Gateway Center is the headquarters for a number of large corporations.

Pittsburgh was founded on the site of the Indian town of Shannopin, a late-17th cent. fur-trading post with many canoe routes and trails. Fort Duquesne, built by the French in the middle of the 18th cent., later fell to the English and was renamed Fort Pitt. The village surrounding the fort was settled in 1760, prospering thereafter with the opening of the Northwest Territory. Pittsburgh is the seat of the Carnegie-Mellon Univ., the Univ. of Pittsburgh, Duquesne Univ., and other schools. The city has a fine park system, of which Schenley Park is the principal unit. The blockhouse of old Fort Pitt is preserved in Point State Park. Two botanical conservatories, a planetarium, an observatory, a civic arena (with a retractable dome), an aviary, the Flag Plaza, the Three Rivers Stadium, and a zoo are among the city's other features.

Pitts·field (pĭts'fēld'). **1.** City (1970 pop. 4,244), seat of Pike co., W Ill., SE of Quincy, in a farm area; laid out 1833, inc. 1869. Shoes are made here. **2.** City (1978 est. pop. 53,600), seat of Berkshire co., W Mass., between mountain ranges, on branches of the Housatonic River; inc. as a town 1761, as a city 1889. The city is the metropolis of the Berkshire resort area. Electrical products are produced.

Pitts·ton (pĭts'tən), industrial city (1978 est. pop. 11,600), Luzerne co., NE Pa., on the Susquehanna; settled c.1770 by the Susquehanna Company of Connecticut, inc. as a city 1894. It is a mining center for anthracite coal.

Pitt·syl·va·ni·a (pĭt'səl-vān'yə, -vā'nē-ə), county (1970 pop. 58,789), 1,001 sq mi (2,592.6 sq km), S Va., in the Piedmont, bounded in the S by N.C., in the N by the Roanoke River, and drained by the Dan and Banister rivers; formed 1766; co. seat Chatham. It is in a rich tobacco-growing area. Clothing, textiles, and tires are manufactured.

Piu·ra (pyo͞o'rä), city (1972 pop. 81,683), capital of Piura dept., NW Peru, in the irrigated Piura valley of the coastal desert. It is the commercial center for the cotton, sugar cane, rice, and corn raised in the region. San Miguel de Piura, the first settlement in Peru, was founded by Francisco Pizarro in 1532, but the site was unhealthful, and the settlement was moved to the present Piura.

Pi·ute (pī-o͞ot'), county (1970 pop. 1,164), 753 sq mi (1,950.3 sq km), SW central Utah, in a mountain and plateau area crossed by the Sevier River; formed 1865; co. seat Junction. It is in a rich mining area yielding uranium, gold, silver, lead, zinc, talc, mercury, perlinte, and manganese. Its agriculture includes livestock, dairy products, and potatoes. Lumbering and some manufacturing are done.

Pla·cen·tia (plə-sĕn'shə, -shē-ə), city (1978 est. pop. 33,900), Orange co., S Calif.; inc. 1926. Once a rural farming community, it became a residential city with shopping centers and light industries.

Placentia Bay, c.100 mi (160 km) long and up to 80 mi (129 km) wide, SE N.F., Canada. Offshore on Aug. 14, 1941, Franklin D. Roosevelt and Winston Churchill signed the Atlantic Charter on board the British battleship *Prince of Wales.*

Plac·er (plăs'ər), county (1970 pop. 77,632), 1,431 sq mi (3,706.3 sq km), E Calif., a narrow strip extending E from the Sacramento Valley, across the Sierra Nevada, to Lake Tahoe, on the Nev. border, and drained by the Bear and Rubicon rivers and the Middle Fork of the American River; formed 1851; co. seat Auburn. Its agriculture includes fruit, nuts, olives, livestock, poultry, and dairy products. Mining and quarrying and sawmilling are done. Clay and metal products are manufactured. There are winter sports and hunting, fishing, and camping areas in the mountains, and parts of Tahoe and El Dorado national forests are in the east.

Plac·er·ville (plăs'ər-vĭl'), city (1971 est. pop. 5,376), seat of El Dorado co., N central Calif., ENE of Sacramento; inc. as a town 1854, as a city 1903. It grew with the discovery of gold (1848) at nearby Coloma. Gold mining is still carried on in the region.

Plac·id, Lake (plăs'ĭd), 4 mi (6.4 km) long and c.1.5 mi (2.4 km) wide, NE N.Y., in the Adirondack Mts., near Mt. Marcy. The lake is a noted winter-sports center.

plain (plān), large area of level or nearly level land. Elevated plains are called plateaus, or tablelands, and very low, wet plains are called swamps. Plains have different names in different climates and countries. They include the tundras, steppe, prairies, pampas, savanna, llanos, flood plains of rivers, coastal plains, loess plains, arid plains, and lacustrine plains. The erosive action of water, glaciation, the draining of a lake, deposition of sediment, and the uplift of a continental shelf are some causes of the formation of plains.

Plain·field (plān'fēld'). **1.** Town (1970 pop. 11,957), Windham co., E Conn., on the Quinebaug River; settled 1689, inc. 1699. Textiles have been made here since the early 19th cent. **2.** Residential town (1977 est. pop. 8,650), Hendricks co., central Ind., WSW of Indianapolis. Grain is grown in the area. **3.** City (1978 est. pop. 42,500), Union co., NE N.J.; settled 1684 by Friends, inc. as a city 1869. Formerly a residential city in the New York metropolitan area, it has become the urban center of 10 closely allied municipalities, with diversified industries, including printing, construction, and the manufacture of packaging machinery, starch and chemicals, processed metals, housewares, and tools and gauges.

Plains (plānz). **1.** Town (1970 pop. 683), Sumter co., SW central Ga., WSW of Americus. **2.** City (1970 pop. 1,087), seat of Yoakum co., NW Texas, near the N.Mex. border SW of Lubbock, in a ranch and oil-field area.

Plain·view (plān'vyōō'). **1.** Uninc. city (1978 est. pop. 32,200), Nassau co., SE N.Y., on Long Island. It is chiefly residential. **2.** City (1978 est. pop. 20,600), seat of Hale co., NW Texas, on the Llano Estacado; inc. 1907. The city has large meat-packing and meat-processing industries. Manufactures include farm machinery, hybrid seeds, and irrigation equipment. Major archaeological remains of a late Ice Age civilization that hunted prehistoric giant bison were found here in 1944–45.

Plain·ville (plān'vĭl'), town (1970 pop. 16,733), Hartford co., central Conn., in the sandy headwaters of the Quinnipiac River; settled 1657, inc. 1869. Electrical products, ball bearings, tools, and other metal products are manufactured. During the Civil War the town was a manufacturing center, and from the late 1890s through the 1920s it had a major railroad yard.

Plan·kin·ton (plăng'kən-tən), city (1970 pop. 613), seat of Aurora co., SE central S.Dak., W of Mitchell, in a livestock, dairying, and grain region. Poultry is raised, and flour milled.

Pla·no (plā'nō), city (1978 est. pop. 44,500), Collin co., N Texas; inc. 1873. It is in a farm and livestock area on the blackland prairie. There are cotton gins and feed and flour mills. Among the manufactures are brass fittings, cast-iron products, and boats and trailers.

Plant City (plănt), city (1970 pop. 15,451), Hillsborough co., W central Fla.; inc. 1885. It is a processing, trade, and shipping center in a farm region. The city is known for its strawberries. It was settled on the site of an Indian village and developed with the coming of the railroad in 1884.

Plaque·mine (plăk'mĭn, -ə-mĭn), town (1970 pop. 7,739), seat of Iberville parish, SE La., on the Mississippi River S of Baton Rouge; inc. 1832. It is in an area yielding rice, sugar cane, timber, and oil. The Intracoastal Waterway and Bayou Plaquemine are connected here with the Mississippi by a series of locks.

Plaque·mines (plăk'mĭnz, -ə-mĭnz), parish (1970 pop. 25,225), 1,030 sq mi (2,667.7 sq km), extreme SE La., bounded on the W and S by Barataria Bay and the Gulf of Mexico, on the E by Breton Sound and the Gulf of Mexico; formed 1807; parish seat Pointe a la Hache. In the south there are sea swamps, and at the tip are islands separated by the mouths, called "passes," through which the Mississippi enters the Gulf. There is some agriculture, including citrus fruit, rice, sugar cane, truck crops, and lily bulbs. Natural gas, oil, and sulfur are found. Among its industries are seafood and vegetable canning and boatbuilding. The parish has areas for hunting, fishing, and fur trapping.

Plas·sey (plăs'ē), village, West Bengal state, NE India. In Plassey, Robert Clive decisively defeated (1757) the Nawab of Bengal, preparing the way for British dominion over northeast India.

Pla·ta, Rí·o de la (rē'ō dā lä plä'tä), estuary, c.170 mi (275 km) long, SE South America, formed by the Paraná and Uruguay rivers. Between Argentina and Uruguay, the estuary is c.120 mi (195 km) wide at its mouth on the Atlantic Ocean and decreases to c.20 mi (30 km) near its head. Dredged channels permit navigation by large vessels. Discovered (1516) by Juan Díaz de Solís, it was explored by Ferdinand Magellan in 1520 and by Sebastian Cabot from 1526 to 1529. The first settlement on its banks was made (1536) at Buenos Aires.

Pla·tae·a (plə-tē'ə), ancient city of Greece, in S Boeotia, on the slope of Mt. Cithaeron (Kithairón). In 479 B.C. Plataea was the scene of the decisive defeat of the Persians by the Greeks. At the beginning of the Peloponnesian War, Thebes besieged the city for two years (429–427 B.C.), and then captured and sacked it. Subsequently rebuilt, it was razed (c.373 B.C.) by the Thebans and reconstructed by Alexander the Great.

pla·teau (plă-tō'), elevated, level or nearly level portion of the earth's surface, larger in summit area than a mountain and bounded on at least one side by steep slopes. The origin of plateaus is assumed to be similar to that of mountains, the earth movements involved being distributed more uniformly over a wider area.

Platte (plăt). **1.** County (1970 pop. 32,081), 427 sq mi (1,105.9 sq km), W Mo., bounded in the S and W by the Missouri River; formed 1838; co. seat Platte City. An agricultural region yielding corn, oats, and tobacco, it is drained by the Little Platte River. **2.** County (1970 pop. 26,544), 667 sq mi (1,727.5 sq km), E central Nebr., bounded in the S by the Platte River and drained by the Loup River; formed 1855; co. seat Columbus. It is in an agricultural area producing grain, livestock, dairy foods, and poultry. **3.** County (1970 pop. 6,486), 2,086 sq mi (5,402.7 sq km), SE Wyo., watered by the North Platte and Laramie rivers; formed 1911; co. seat Wheatland. Its agriculture includes sugar beets, grain, livestock, and beans. It has iron-ore mines, stone quarries, and some manufacturing.

Platte, river, c.310 mi (500 km) long, formed by the confluence of the North Platte (680 mi/1,095 km long) and South Platte (430 mi/690 km long) rivers at North Platte, Nebr. It flows generally east across southern Nebr. to join the Missouri River at Plattsmouth. The river is too flood-prone in spring and too shallow and braided the rest of the year for navigation. Much of its water is diverted for irrigation, municipal uses, and hydroelectric power production. In the 19th cent., the Mormon Trail followed the north bank, and the Oregon Trail followed the south bank.

Platte City, city (1970 pop. 2,022), seat of Platte co., W Mo., on the Platte River NNW of Kansas City, in a farm and tobacco region; laid out 1828, inc. 1842. American troops camped here in the Mexican War, and the city was burned twice in the Civil War.

Platt National Park (plăt): *see* National Parks and Monuments Table.

Platts·burg (plăts'bûrg'), city (1970 pop. 1,832), seat of Clinton co., NW Mo., NNE of Kansas City, in a farm region; founded c.1835, inc. 1850.

Platts·burgh (plăts'bûrg'), city (1978 est. pop. 21,800), seat of Clinton co., NE N.Y., on Lake Champlain; settled 1767, inc. 1902. It is a trade and distribution point, with plants that make paper and plastics. A major source of employment is the adjoining Plattsburgh Air Force Base, a strategic air command installation. The city is also a summer-vacation center. During the War of 1812 a makeshift American fleet decisively defeated the British in a pitched battle on Lake Champlain near Plattsburgh.

Platts·mouth (plăts'məth, -mouth'), city (1970 pop. 6,371), seat of Cass co., SE Nebr., S of Omaha, near the junction of the Platte and Missouri rivers; founded 1854–55. It is a railroad and trade center. Stone and clay deposits are nearby.

Plau·en (plou'ən), city (1974 est. pop. 80,353), Karl-Marx-Stadt dist., S East Germany, on the White Elster River at the NW foot of the Erzgebirge. It has been a textile-milling center since the 15th cent. Other manufactures include machinery, machine tools, electrical equipment, and motor vehicles. Originally founded by the Slavs, Plauen became (c.1224) the seat of a branch of the Teutonic Knights. It passed to Bohemia in 1327 and to Saxony in 1466.

Plea·sant Grove (plĕz'ənt grōv'). **1.** Town (1977 est. pop. 5,774), Jefferson co., N central Ala., a suburb W of Birmingham; inc. 1933. **2.** City (1970 pop. 5,327), Utah co., N central Utah, in the valley of Utah Lake NNW of Provo; settled 1849 by Mormons, inc. 1855. It is a processing center and shipping point in an irrigated fruit and livestock region served by the Provo River project.

Pleas·an·ton (plĕz'ən-tən). **1.** City (1970 pop. 18,328), Alameda co., W Calif., a suburb of the San Francisco-Oakland area, in a vineyard and dairy region; inc. 1894. Wine and cheese are produced, and there are also publishing and research enterprises. **2.** City (1970 pop. 5,407), Atascosa co., SW Texas, S of San Antonio on the Atascosa River; settled in the 1850s, inc. 1928. It is a market and shipping center for a region of cattle ranches, truck farms, and oil fields.

Pleas·ants (plĕz'ənts), county (1970 pop. 7,274), 130 sq mi (336.7 sq km), NW W.Va., bounded on the NW by the Ohio River and the Ohio border, on the N by Middle Island Creek; formed 1851; co. seat St. Mary's. Its agriculture includes livestock, tobacco, and truck crops. It has oil and natural-gas wells and refineries, as well as some coal mines and timber. There is some chemical manufacturing.

Pleas·ant·ville (plĕz'ənt-vĭl'). **1.** Residential and resort city (1978 est. pop. 14,700), Atlantic co., SE N.J., just W of Atlantic City; settled 1702, inc. 1888. Tourism, shellfishing, deep-sea fishing, and boatbuilding are important activities. **2.** Residential village (1970 pop. 7,110), Westchester co., SE N.Y., NNE of New York City. It has a publishing industry.

Plen·ty·wood (plĕn'tē-wŏŏd'), city (1970 pop. 2,381), seat of Sheridan co., extreme NE Mont., NE of Fort Peck Reservoir; inc. 1912. It grew as a wheat-raising center.

Ple·ven (plĕv'ən) or **Plev·na** (plĕv'nə), city (1973 est. pop. 110,500),

N Bulgaria. A commercial center for a fertile agricultural region, it has food-processing industries and manufactures cotton textiles, ceramics, agricultural machinery, cement, and rubber goods. An old Thracian settlement, Pleven was later occupied by the Romans. It became a trade center under Turkish rule (15th–19th cent.). The city is famous for its defense by the Turks against Russian and Rumanian troops in the Russo-Turkish War of 1877–78.

Płock (plôtsk), city (1975 est. pop. 87,800), E central Poland, a port on the Vistula River. Located on a pipeline from the USSR to Poland, Płock is a major oil-refining and petrochemical center. Other industries include metalworking, sawmilling, and the manufacture of farm machinery and barges. Known in the 10th cent., it became a bishopric in 1075. The city passed in 1793 to Prussia and in 1815 to Russia. It reverted to Poland in 1921. Płock's 12th cent. cathedral, containing tombs of Polish kings and dukes, was badly damaged during World War II.

Plo·ieş·ti (plô-yĕsht′, -yĕsh′tē), city (1977 est. pop. 199,269), S central Rumania, in Walachia. It is the center of a rich oil region. The city has large refineries and oil storage installations and is an industrial center with varied manufactures. Founded in 1596, Ploieşti grew in the 19th cent. into the largest oil-producing center of southeast Europe. After Rumania signed (1940) a mutual cooperation pact with the Axis powers that provided substantial Rumanian oil to Germany, the Allies heavily bombed the city.

Plom·bières (plôN-byĕr′), village (1968 pop. 1,183), Vosges dept., NE France, in the Vosges Mts. It is a fashionable spa, with radioactive springs used since Roman times for medicinal purposes.

Plov·div (plôv′dĭf′, -dĭv′), city (1973 est. pop. 274,700), S central Bulgaria, on the Maritsa River. It is a transportation hub and the chief market for a fertile area. Plovdiv's major industries are food processing, lead and zinc smelting, brewing, and the manufacture of textiles, metal products, and shoes. There are also motor repair works in the city. Originally built by the Thracians, the city was captured in 341 B.C. by Philip II of Macedon, who established a military post here. It was the capital of Thracia under the Romans. Destroyed (early 13th cent.) by the Bulgarians, Plovdiv later became the center of the Bogomils. The city passed to Russia in 1877 and was united with Bulgaria in 1885. Plovdiv has several Orthodox churches and Turkish mosques. The ancient town walls and gate still stand.

Plu·mas (plo͞o′məs), county (1970 pop. 11,707), 2,566 sq mi (6,645.9 sq km), NE Calif., in the Sierra Nevada, drained by the Feather River, here formed by forks flowing in scenic canyons; formed 1854; co. seat Quincy. It has stock grazing, hay farming, and dairying in the mountain basins, the largest of which is the Sierra Valley. Copper, silver, and gold mining, sand and gravel quarrying, and lumbering of pine, fir, and cedar are done. It has hunting, fishing, hot springs, and winter sports. It contains parts of Plumas and Tahoe national forests and part of Lassen Volcanic National Park.

Plym·outh (plĭm′əth), county borough (1976 est. pop. 259,100), Devonshire, SW England, on Plymouth Sound, an inlet of the English Channel. Situated on a peninsula between the estuaries of the Plym and Tamar rivers, Plymouth is an important port and naval base. The southern waterfront and adjacent promenade are called the Hoe. Foodstuffs and raw materials are imported and manufactures of many kinds exported. Other items traded are granite, marble, kaolin, and fish. The Royal Marine Barracks and Naval Dockyard (1691) are here. In 1588 the port was the rendezvous of the anti-Armada fleet. From here Drake, Hawkins, Raleigh, and several later explorers set forth. Plymouth was held by the Parliamentarians for four years during the civil war, while the rest of Devonshire and Cornwall were Royalist. The first English factory to make Chinese porcelain was established in Plymouth in 1768. Among the principal points of interest on the Hoe are the old Royal Citadel (17th cent.), the upper part of Smeaton's lighthouse brought from Eddystone, an Armada memorial, and a naval war memorial.

Plymouth. 1. County (1970 pop. 24,322), 863 sq mi (2,235.2 sq km), NW Iowa, on the S.Dak. border in the W (formed here by the Big Sioux River), and drained by the Floyd and Little Sioux rivers; formed 1851; co. seat Le Mars. It is in a prairie agricultural area (cattle, hogs, poultry, corn, oats, and barley), with sand and gravel pits. **2.** County (1970 pop. 333,314), 654 sq mi (1,693.9 sq km), SE Mass., bounded on the E by Massachusetts and Cape Cod bays, on the S by Buzzards Bay; formed 1685; co. seat Plymouth. It has agriculture (cranberries, truck crops, and poultry) and some industry. Communities along its shoreline are popular summer resorts.

Plymouth. 1. Town (1970 pop. 10,321), Litchfield co., W Conn.; settled 1725, inc. 1795. Locks, tools and dies, and other metal products are manufactured. **2.** City (1975 est. pop. 7,900), seat of Marshall co., N Ind., S of South Bend; settled 1834 on the site of an Indian village, inc. 1872. In a farm area, it also processes food and makes automobile parts. **3.** Uninc. town (1970 pop. 18,606), seat of Plymouth co., SE Mass., on Plymouth Bay; founded 1620. Rope and twine, wire, metal products, and textiles are manufactured. The town, with summer resort facilities and major historic attractions, has a large tourist industry. Its harbor, now used by fishing boats and pleasure craft, was the scene of the landing by the Pilgrims in 1620, and the town was the first permanent white settlement in New England. Most famous of its many monuments is Plymouth Rock, returned to its original site in 1880; according to legend, the Pilgrims stepped on this boulder when disembarking from the *Mayflower.* The *Mayflower II,* a replica of the original ship, is moored here. The sites of the first houses are marked by tablets on Leyden St., the first street laid out by the Pilgrims. A number of 17th cent. houses on nearby streets are maintained as museums. Cole's Hill and Burial Hill contain graves of many of the first settlers, and Pilgrim Hall has numerous valuable relics. Near the site of the original village is the 80-ft (24.4-m) granite National Monument to the Forefathers (1889). Of great interest is nearby Pilgrim Village, a re-creation of the early settlement. The town also has a marine museum and aquarium. **4.** Industrial city (1970 pop. 11,758), Wayne co., SE Mich.; inc. 1867. Telephones, automotive parts, business machines, heating equipment, paper products, steel tanks, and packaging materials are made. **5.** Village (1978 est. pop. 26,500), Hennepin co., SE Minn., a suburb of Minneapolis–St. Paul; inc. 1955. Computer systems, heating units, games, and aluminum and steel products are manufactured. **6.** Resort town (1970 pop. 4,225), Grafton co., central N.H., on the Pemigewasset River S of Franconia Notch; inc. 1763. Wood products and electrical equipment are made. The town's old buildings include the courthouse (now a library) where Daniel Webster pleaded his first case. **7.** Town (1970 pop. 4,774), seat of Washington co., NE N.C., on the Roanoke River near Albemarle Sound; founded in the late 18th cent., inc. 1807. It is a processing point in a farm and timber area. **8.** Borough (1970 pop. 9,536), Luzerne co., NE Pa., on the Susquehanna River WSW of Wilkes-Barre, in an anthracite-coal area; inc. 1866. Textile products are made. **9.** Village (1970 pop. 283), Windsor co., S central Vt., SE of Rutland. President Calvin Coolidge was born here, and the farmhouse where he took the oath of office is nearby. He is buried in the local cemetery. **10.** City (1970 pop. 5,810), Sheboygan co., E Wis., between Sheboygan and Fond du Lac; inc. 1877. Cheese is made, and the Wisconsin Cheesemakers' Association and the Wisconsin Cheese Exchange are here.

Pl·zeň (pŭl′zĕn-yə), city (1975 est. pop. 156,461), W Czechoslovakia, in Bohemia, at the confluence of several rivers. One of Czechoslovakia's largest cities, Plzeň is famous for its beer (Pilsner), exported worldwide, and for the huge Skoda works (nationalized and renamed Lenin Works), which produce heavy machinery, machine tools, locomotives, automobiles, and armaments. Other industries in the city include distilling, sugar refining, papermaking, and the production of cement and pottery. Founded in 1290 by King Wenceslaus II of Bohemia, the city was an important Bohemian trade center. It remained a stronghold of Catholicism in the Hussite Wars (15th cent.). One of the earliest printing presses was established in Plzeň in 1468. Rapid industrialization dates from the late 19th cent. Plzeň was part of the Austro-Hungarian monarchy until 1918, when it was included in newly independent Czechoslovakia. It was taken by German forces in 1939 and became a leading producer of German armaments during World War II. In 1945 the city was liberated and returned to Czechoslovakia. Plzeň's historic buildings include the 13th cent. Gothic Church of St. Bartholomew and a 16th cent. Renaissance town hall.

Po (pō), longest river of Italy, c.405 mi (650 km) long, rising in the Cottian Alps of Piedmont, NW Italy. It winds generally east in a wide valley to enter the Adriatic Sea through several mouths. Its marshy delta is constantly expanding eastward. The Po River is navigable for small craft c.300 mi (480 km) upstream, but seasonal variations in flow hamper navigation. It is extensively used for irrigation. The Po valley is the most important industrial and agricultural region of Italy.

Po·ca·hon·tas (pō′kə-hŏn′təs). **1.** County (1970 pop. 12,793), 580 sq mi (1,502.2 sq km), N central Iowa, in a prairie agricultural area; formed 1851; co. seat Pocahontas. Cattle, hogs, and poultry are raised, and corn, oats, and soybeans are grown. There are coal deposits here. **2.** County (1970 pop. 8,870), 943 sq mi (2,442.4 sq km), E central W.Va., in the Alleghenies, drained by the Greenbrier and Gauley rivers and a fork of Cheat River; formed 1821; co. seat Marlinton. Its agriculture includes livestock, grain, dairy products, and fruit. Lumbering and leather tanning are done. Most of the county is in Monongahela National Forest.

Pocahontas. 1. City (1974 est. pop. 5,448), seat of Randolph co., NE Ark., on the Black River N of Walnut Ridge, in a farm area. Shoes and truck parts are made. **2.** City (1970 pop. 2,338), seat of Pocahontas co., NW Iowa, NW of Fort Dodge; platted 1870, inc. 1892.

Po·ca·tel·lo (pō'kə-tĕl'ō), city (1978 est. pop. 42,500), seat of Bannock co., SE Idaho, between mountains on the Portneuf River near its junction with the Snake River; inc. 1889. A railroad center since 1882, Pocatello is a major shipping and processing point for a livestock and farm area. It has an important mining industry, canneries, and plants making prefabricated homes, electronic equipment, and steel fabricators. Tourism is significant. Pocatello is the seat of Idaho State Univ. and the headquarters for Caribou National Forest.

Po·co·no Mountains (pō'kə-nō'), range of the Appalachian system, c.2,000 ft (610 m) high, NE Pa. Forested and having many lakes and streams, the Poconos are a major year-round resort area.

Po·do·lia (pō-dō'lyə), region, W central European USSR, in the Ukraine, separated in the S from the Moldavian SSR by the Dnestr and in the W from Galicia by the Southern Bug. It borders on Volhynia in the north. A fertile hilly plain drained by the Dnestr and the Southern Bug, Podolia is one of the richest and most densely populated agricultural regions of the Ukraine. The principal crops are sugar beets, wheat, tobacco, and sunflowers. Dairy farming and beekeeping are also important, and phosphate is mined. Food processing, especially sugar milling, is the major industry. Podolia was part of Kievan Russia from the 10th cent. In the 14th cent. Polish colonists began to convert the region from steppe into arable farmland. Western Podolia was annexed to Poland in 1430; the eastern section was part of Lithuania until 1569 and passed to Russia in 1793. The western portion was transferred to Austria in 1772, belonged to Poland from 1918 to 1939, and was then annexed by the USSR.

Po·dolsk (pŏ-dôlsk', pə-dôl'yəsk), city (1976 est. pop. 191,000), central European USSR, on the Pakhra River, a tributary of the Moskva. The center of a fertile agricultural region, Podolsk has electrotechnical industries and factories that produce heavy machinery, oil-refining equipment, and cables. Podolsk was a fief of the Danilov monastery in Moscow until 1764 and received a city charter in 1781. Prior to the 1917 Bolshevik Revolution, it was a frequent meeting place for Lenin and other revolutionaries.

Po Hai or **P'o-hai** (both: bō' hī'), arm of the Yellow Sea, indenting the coast of N China; the Liaotung Gulf is its northeast extension. The Huang Ho empties into the Po Hai.

Po·hang (pō'hăng'), city (1975 pop. 110,000), SE South Korea, on Yongil Bay of the Sea of Japan. The chief economic activities are fish canning and the production of iron and steel.

Poin·sett (poin'sĕt', -sĭt), county (1970 pop. 26,843), 762 sq mi (1,973.6 sq km), NE Ark., intersected by Crowley's Ridge and drained by the St. Francis and L'Anguille rivers; formed 1838; co. seat Harrisburg. Its agriculture includes cotton, rice, corn, soybeans, alfalfa, truck crops, livestock, and dairy products. It has hardwood timber, some mining, and manufacturing (clothing, shoes, and furniture).

Point Bar·row (băr'ō), northernmost point of Alaska, on the Arctic Ocean, at lat. 71°23′ N and long. 156°30′ W. Discovered in 1826 by Frederick W. Beechey, a British explorer, it was named for the British geographer Sir John Barrow. Navigation is open for only two or three months a year.

Pointe a la Hache (point' ăl'ə-hăsh'), village (1972 est. pop. 500), seat of Plaquemines parish, extreme SE La., SE of New Orleans on the E bank of the Mississippi. It is in a delta region yielding sugar cane, rice, and fruit. Seafood is canned, and wood products made. Fur trapping and fishing are also important.

Pointe-à-Pi·tre (pwăNt-ə-pē'trə), city (1967 pop. 33,107), Guadeloupe dept., French West Indies. It is on Grande-Terre Island at the southern entrance of the Rivière Salée. Pointe-à-Pitre exports sugar, rum, coffee, and bananas.

Pointe Cou·pee (point' ko͞o-pē'), parish (1970 pop. 22,002), 564 sq mi (1,460.8 sq km), SE central La., bounded on the E by the Mississippi River, on the N by tributaries of the Red River, and on the W by the Atchafalaya River; formed 1805; parish seat New Roads. Cotton, sugar cane, corn, sweet and white potatoes, and hay are grown, and livestock is raised. There are fisheries, timber, and some industry, including the processing of farm products.

Pointe-Noire (pwăNt-nwär'), city (1970 pop. 135,000), SW Congo Republic, Africa, a port on the Atlantic Ocean. It exports tropical timber, cotton, palm products, peanuts, and coffee. Plywood, aluminum ware, and soap are manufactured, and the city has shipbuilding and food-processing industries. It is also a noted center for sport fishing. Founded in 1883, it gained importance only after the building (1948) of the railroad to Brazzaville and the construction (1934-39) of an artificial harbor.

Point Pe·lee (pē'lē'), peninsula, c.10 mi (16 km) long, extending into W Lake Erie, S Ont., Canada, near Leamington. It is the southernmost part of the Canadian mainland.

Point Pleas·ant (plĕz'ənt). **1.** Residential and resort borough (1970 pop. 15,968), Ocean co., E N.J., near the Manasquan Inlet; settled 1850, inc. 1920. **2.** City (1970 pop. 6,122), seat of Mason co., W W.Va., on the Ohio River at the mouth of the Kanawha River; settled around Fort Blair (built 1774), inc. 1833. The surrounding area has coal mines and diversified agriculture; the city produces chemicals, iron products, and furniture. In the Battle of Point Pleasant (Oct. 10, 1774), a large force of Indians was defeated by Andrew Lewis and frontiersmen.

Point Rey·es National Seashore (rā'ĭs), 64,546 acres (26,141 hectares), W Calif.; est. 1962. Included in the area are steep bluffs overlooking the Pacific Ocean, lagoons, and esteros enclosed by sand dunes, rolling hills, and forests. On offshore rocks are bird rookeries and sea-lion herds. The San Andreas Fault passes through the park; there is a 15 to 20 ft (4.6-6.1 m) horizontal displacement of rock (a result of the 1906 earthquake).

Poi·tiers (pwä-tyā'), city (1975 pop. 81,313), capital of Vienne dept., W central France, on the Clain River. Poitiers's industries include metallurgy, machine building, printing, and the manufacture of chemicals and electrical equipment. The city was the capital of the Pictons, a Gallic people. Christianized early in Roman times, it was a stronghold of orthodoxy under its first bishop, St. Hilary of Poitiers (4th cent.), and, because of its important monasteries, was a great religious center of Gaul. A residence of Visigoth kings, the city was captured (507) by the Franks under Clovis I. It was the location of the brilliant court of Eleanor of Aquitaine. At Poitiers in 1356 Edward the Black Prince defeated and captured John II of France and his son, Philip the Bold of Burgundy. Charles VII founded a university here in 1432. In the Wars of Religion (1562-98) the city was unsuccessfully besieged (1568) by the Huguenots; in 1577 the Peace of Bergerac (also known as the Edict of Poitiers) was signed here granting religious freedom. Architecturally, Poitiers is one of the most interesting cities in Europe. There are Roman amphitheaters and baths, the baptistery of St. John (4th-12th cent.), the Cathedral of St. Pierre (12th-14th cent.), the courthouse (12th-15th cent., formerly a royal residence), as well as numerous other churches and late medieval and Renaissance residences.

Poi·tou (pwä-to͞o'), region and former province, W France, stretching from the Atlantic coast eastward beyond the Vienne River. The Vendée region, or Lower Poitou, is mostly a pastoral hedgerow country, with swamps in the west and the south. A narrow strip, the Vendean plain, is an intensive wheat-growing region. Upper Poitou is a rich agricultural area; it also has a large dairy industry. A part of the Roman province of Aquitaine, Poitou fell to the Visigoths (5th cent.) and to the Franks (507). The counts of Poitiers, who originated in the 9th cent., assumed the title duke of Aquitaine. The area was frequently contested by England and France until the end of the Hundred Years' War, when Charles VII incorporated it in the French crown lands.

Po·land (pō'lənd), republic (1970 pop. 32,589,209), 120,725 sq mi (312,677 sq km), central Europe, bordering on East Germany in the W, on the Baltic Sea in the N, on the USSR in the E, and on Czechoslovakia in the S. Warsaw is the capital.

The country is largely low-lying, except in the south, which includes the Carpathians, the Sudeten Mts., and the Małopolska Hills. The highest point is Rysy Mt. (c.8,200 ft/2,500 m), located in the High Tatra Mts. near the Czechoslovak border. Poland's main rivers (including the Vistula, the Oder, the Warta, and the Western Bug) are connected to the Baltic Sea and are important traffic lanes. The

country has three important Baltic ports (Gdańsk, Gdynia, and Szczecin) and a dense rail network. There are many lakes, especially in the north. About 50% of Poland's land area is arable (with the best soil in the south) and about 25% is forested.

Poland is divided into 22 provinces (including 5 city provinces):

NAME	CAPITAL	NAME	CAPITAL
City Provinces		Koszalin	Koszalin
Kraków		Kraków	Kraków
Łódź		Łódź	Łódź
Poznań		Lublin	Lublin
Warsaw		Olsztyn	Olsztyn
Wrocław		Opole	Opole
Provinces		Poznań	Poznań
Białystok	Białystok	Rzeszów	Rzeszów
Bydgoszcz	Bydgoszcz	Szczecin	Szczecin
Gdańsk	Gdańsk	Warszawa	Warsaw
Katowice	Katowice	Wrocław	Wrocław
Kielce	Kielce	Zielona Góra	Zielona Góra

Economy. Industry is largely controlled and planned by the state, but farming, which contributed a little less than half the national product in the 1970s, is mainly privately run. The chief agricultural products are potatoes, sugar beets, rye, wheat, and barley. The country's leading manufactures include iron and steel, machinery, electronic equipment, cement, chemicals, textiles, forest products, and processed food. The chief minerals produced are coal, iron ore, zinc-lead ores, sulfur, and petroleum. Much natural gas is also produced. Poland's leading exports are coal, processed meat, ships, railroad freight cars, and metal products; the main imports are iron ore, petroleum, metal products, fertilizer, and cotton.

History. The territorial dimensions of Poland have varied considerably during its history. In the 9th and 10th cent. the Poles gained hegemony over the other Slavic groups that occupied what is roughly present-day Poland. Under duke Mieszko I (reigned 960–92) of the Piast dynasty the conversion of Poland to Christianity began (966). The Piasts expanded their domains, and in 1025 Boleslaus I (reigned 992–1025) took the title of king. In the 14th cent. Lithuanians gained influence in the area. After a period of disunity, Lithuanian-Polish forces defeated the Teutonic Knights (1410) and reunified the kingdom. Ladislaus Jagiello, grand duke of Lithuania, became king of Poland as Ladislaus II (reigned 1386–1434). The Jagiello dynasty ruled a Polish-Lithuanian empire reaching from the Baltic Sea to the Black Sea until 1572; this period is considered the golden age of Poland.

After the death (1572) of the last Jagiello, kings were elected by the nobility, and factional rivalry prevented the establishment of a strong dynasty. Wars with Russia, Sweden, and Turkey weakened the kingdom. In 1655, Charles X of Sweden overran Poland, while Czar Alexis of Russia attacked from the east. The Poles managed to regroup and to save the country from complete dismemberment, but the ensuing peace treaties cost them much territory and independence. Russia, Prussia, and Austria vied for control of the Polish throne. These powers agreed to three successive partitions (1772, 1793, 1795), which resulted in the disappearance of Poland from the map of Europe.

Napoleon I created a Polish buffer state, the grand duchy of Warsaw, in 1807; after his defeat, the Congress of Vienna (1814–15) established a nominally independent Polish kingdom (Congress Poland) in personal union with the czar of Russia. When a Polish nationalist insurrection erupted in 1830, Russian forces entered Warsaw and annexed the kingdom. Other rebellions, all suppressed, broke out in Galicia in 1846, in Prussian and Austrian Poland in 1848, and in Russian Poland in 1863.

The defeat of the partitioning powers in World War I allowed Poland to regain its independence, which was proclaimed on Nov. 9, 1918. Nationalist leader Joseph Pilsudski was declared chief of state, and a republican constitution was adopted. The Treaty of Versailles (1919) gave Poland access to the Baltic Sea and forced Germany to return Prussian Poland to Poland. Gdańsk became a free city and parts of Silesia were awarded to Poland as a result of plebiscites. The Polish-Russian border became a subject of dispute, and war broke out between Poland and Russia. In the Treaty of Riga (1921) Poland secured parts of its claims to the eastern provinces. In 1926 a parliamentary government was suspended by a military coup d'état that made Pilsudski virtual dictator. After his death (1935) a powerful governing clique assumed control. In the 1930s Poland attempted to steer a precarious course among the powers of Europe. On Sept. 1, 1939, Germany invaded Poland and thus precipitated World War II. Polish resistance was crushed, and the country was partitioned between Germany and the USSR. After the German attack (1941) on the USSR, all Poland passed under German rule. The country suffered tremendous losses in life and property in the war. About 6 million Poles, including 3 million Jews, were killed. When Soviet troops entered Poland, a provisional Polish government was established (July, 1944) under Soviet auspices at Lublin. The last German troops were expelled in early 1945.

At the Potsdam Conference (July–Aug., 1945), the sections of Prussia east of the Oder and Neisse rivers, including Gdańsk and the southern part of East Prussia, were assigned to Poland. The present Polish-Soviet border was also fixed by treaty. A unicameral parliament was established (1946), pro-Soviet Communists took control, and legal opposition to the government was limited. The constitution of 1952 made Poland a people's republic on the Soviet model. Relations with the Vatican were severed; the church became a chief target of government persecution. In 1956 workers and students in Poznań rioted in a mass demonstration against Communist and Soviet control of Poland. Władysław Gomułka was elected leader of the Polish Communist party and became the symbol of revolt against Moscow. Collectivization of agriculture was halted, and the Poles were given far more freedom than under the previous regime. Relations with the church improved, and economic and cultural ties with the West were broadened. By the early 1960s Gomułka was tightening the party's hold on Poland; intellectual freedom was again curbed. In 1970 increasing prices of food led to riots by workers. Gomułka was ousted and replaced by Edward Gierek, who sought to ease the living conditions of the average citizen and further improve relations with the church.

Government. The 1952 constitution vests legislative power in the unicameral Sejm (parliament), whose members are elected to four-year terms. The Sejm appoints the council of ministers, which is the state's main executive and administrative body, and the council of state, which nominates high officials, passes on international agreements and has judicial functions. In practice, power is controlled by the politburo of the Polish United Workers' Party, whose first secretary is the most powerful individual in the country.

Poland, town (1970 pop. 2,015), Androscoggin co., SW Maine, W of Auburn; settled c.1768, inc. 1795. It includes the resort Poland Spring, long known for its mineral water. The Mansion House, Poland's first inn, dates from 1797.

Pol·dhu (pôl'dyōō), village, Cornwall, SW England. Guglielmo Marconi sent the first transatlantic radio transmission (1901) from Poldhu to Newfoundland.

Po·lish Cor·ri·dor (pō'lĭsh kôr'ĭ-dər), strip of German territory awarded to newly independent Poland by the Treaty of Versailles in 1919. The strip, 20 to 70 mi (32–112.6 km) wide, gave Poland access to the Baltic Sea. It contained the lower course of the Vistula, except the area constituting the Free City of Danzig. Gdynia was developed as Poland's chief port and came to rival the port of Danzig. Free German transit was permitted across the corridor, which separated East Prussia from the rest of Germany. Although the territory had once formed part of Polish Pomerania, a large minority of the population was German-speaking. The arrangement caused chronic friction between Poland and Germany. In Mar., 1939, Germany demanded the cession of Danzig and the creation of an extraterritorial German corridor across the Polish Corridor. Poland rejected these demands and obtained a French and British guarantee against aggression. On Sept. 1, 1939, the Polish-German crisis culminated in the German invasion of Poland and World War II.

Polk (pōk). **1.** County (1970 pop. 13,297), 860 sq mi (2,227.4 sq km), W Ark., bordered on the W by Okla. and drained by the Ouachita and Saline rivers; formed 1844; co. seat Mena. Its agriculture includes cotton, corn, truck crops, potatoes, fruit, livestock, poultry, and dairy products. It has lumber (logging and sawmills) and industry (poultry dressing, clothing manufacturing, and aluminum foundries). The county contains part of Ouachita National Forest. **2.** County (1970 pop. 228,515), 1,858 sq mi (4,812.2 sq km), central Fla., in a lake region bounded on the E by the Kissimmee River and partly on the N by the Withlacoochee River, and drained by the Peace River; formed 1861; co. seat Bartow. It is a leading producer of citrus fruit and phosphate. Truck crops, corn, cattle, and poultry are also important. Among its manufactures are clothing and textiles, furniture, fertilizers and insecticides, and metal products. **3.** County (1970 pop. 29,656), 312 sq mi (808.1 sq km), NW Ga., in a valley and ridge area on the Ala. border; formed 1851; co. seat Cedartown. Its agriculture includes cotton, corn, fruit, sweet potatoes, hay, livestock, and dairy products. It has timber, iron mines, and stone quarries. Prepared meats, frozen foods, yarn and nylon cord, furniture, wood products, paper, cement, and farm machinery are manufactured. **4.** County (1970 pop. 286,130), 578 sq mi (1,497 sq km), central Iowa, drained by the Des Moines, Raccoon, and Skunk rivers and by Beaver Creek; formed 1846; co. seat Des Moines. Its prairie agriculture includes hogs, cattle, poultry, corn, soybeans, and oats. Bituminous coal is mined in the south. **5.** County (1970 pop. 34,435), 2,012 sq mi (5,211.1 sq km), NW Minn., bounded on the W by N.Dak. and the Red River of the North, and drained by the Poplar, Sandhill, and Red Lake rivers; formed 1858; co. seat Crookston. In an agricultural area yielding sugar beets, potatoes, small grains, sunflowers, livestock, poultry, and dairy products, it has a food-processing industry and some mining and manufacturing. **6.** County (1970 pop. 15,415), 637 sq mi (1,649.8 sq km), SW central Mo., in the Ozarks, drained by the Pomme de Terre and Little Sac rivers; formed 1835; co. seat Bolivar. Corn, wheat, oats, hay, dairy cattle, and poultry are its major products. **7.** County (1970 pop. 6,468), 433 sq mi (1,121.5 sq km), E central Nebr., bounded in the N by the Platte River and drained by the Big Blue River; formed 1870; co. seat Osceola. Its agriculture includes livestock, grain, dairy products, and poultry. **8.** County (1970 pop. 11,735), 239 sq mi (619 sq km), W N.C., bounded in the S by the S.C. border; formed 1855; co. seat Columbus. This forest and piedmont area has farming (peaches, cotton, corn, and sweet potatoes), livestock raising (horses and cattle), and a tourist industry. **9.** County (1970 pop. 35,349), 739 sq mi (1,914 sq km), NW Oregon, bounded in the E by the Willamette River, with the Coast Range extending along the W border; formed 1845; co. seat Dallas. Its agriculture includes fruit, truck crops, grain, dairy products, and poultry. Among its industries are food processing, lumbering, and the manufacture of concrete products and construction machinery. **10.** County (1970 pop. 11,669), 434 sq mi (1,124.1 sq km), extreme SE Tenn., bordered in the E by N.C. and in the S by Ga., with the Unicoi Mts. in the NE, and drained by the Hiwassee and Ocoee rivers; formed 1839; co. seat Benton. In the southeast is a copper-producing region, with mines and smelters. In an agricultural area yielding soybeans, corn, cattle, and dairy products, it has textile mills, clothing factories, steelworks, and pine and oak timber. The county is largely included in Cherokee National Forest. **11.** County (1970 pop. 14,457), 1,100 sq mi (2,849 sq km), E Texas, bounded on the W and SW by the Trinity River, on the NE by the Neches River; formed 1846; co. seat Livingston. Lumbering, farming (cotton, corn, vegetables, pecans, and sorghums), livestock raising, and dairying are important. There are deposits of oil, silica, chalk, and sandstone. **12.** County (1970 pop. 26,666), 934 sq mi (2,419.1 sq km), NW Wis., bounded in the W by the St. Croix River and the Minn. border; formed 1853; co. seat Balsam Lake. It is in a dairying region. Its industry includes food processing and the manufacture of clothing, metal products, and aircraft.

Po·lotsk (pô'lətsk), city (1976 est. pop. 75,000), W European USSR, on the Western Dvina River at its confluence with the Polota. It is a large rail junction and agricultural trade center and has lumber mills, plants for processing food and flax, motor vehicle repair shops, and oil refineries. Manufactures include building materials, farm implements, metal goods, and glass filaments. Polotsk was the capital of a principality of the same name from the 10th to 13th cent., when it passed to Lithuania. A flourishing center for trade, it was transferred to Russia in 1772.

Pol·son (pōl'sən), city (1970 pop. 2,464), seat of Lake co., NW Mont., on Flathead Lake, in a farm region; settled 1899, inc. 1909.

Pol·ta·va (pŏl-tä'və), city (1976 est. pop. 270,000), capital of Poltava oblast, S European USSR, in the Ukraine, on the Kiev-Kharkov highway and on the Vorskla River, a tributary of the Dnepr. It is an industrial center and important rail junction in the rich black-earth agricultural region. The city has railroad shops, food- and tobacco-processing plants, and factories that produce machinery, railroad equipment, automobiles, tractors, building materials, footwear, leather goods, textiles, and wood products. Poltava was the site of a Slavic settlement in the 8th and 9th cent. It became part of Lithuania in 1430. In the 17th cent. it was the chief town of a Ukrainian Cossack regiment. Poltava was a flourishing commercial center in the 18th and 19th cent., a principal focus of the Ukrainian literary and national movement, and, under Czar Nicholas I, a place of exile.

Pol·y·ne·sia (pŏl'ə-nē'zhə, -shə), one of the three main divisions of Oceania, in the central and S Pacific Ocean. The larger islands are volcanic; the smaller ones are generally coral formations. The principal groups are the Hawaiian Islands, Samoa, Tonga, and the islands of French Polynesia.

Pom·e·ra·ni·a (pŏm'ə-rā'nē-ə), region of N central Europe, extending along the Baltic Sea from a line W of Stralsund, East Germany, to the Vistula River in Poland. From 1919 to 1939 Pomerania was divided among Germany, Poland, and the Free City of Danzig (Gdańsk). The German part constituted the Prussian province of Pomerania (14,830 sq mi/38,410 sq km), with Stettin (Szczecin) as its capital. The Polish part formed the province of Pomerelia (6,335 sq mi/16,408 sq km), with Bydgoszcz as its capital. After the Potsdam Conference in 1945, all that part (c.2,800 sq mi/7,250 sq km) of former Prussian Pomerania west of the Oder (but excluding Stettin) was incorporated into the Soviet-occupied German state of Mecklenburg; the remaining and much larger part was transferred to Polish administration and organized ultimately into the provinces of Koszalin, Szczecin, and Gdańsk.

A part of the North European plain, Pomerania is a primarily agricultural lowland, with generally poor, often sandy or marshy soil. It is dotted with numerous lakes and forests and is drained by many rivers, including the Oder, Ina, and Rega. Cereals, sugar beets, and potatoes are the main crops; livestock raising and forestry are important occupations.

History. By the 10th cent. A.D., when its recorded history began, Pomerania was inhabited by Slavic tribes. It was conquered by Boleslaus I (992–1025) of Poland but became an independent duchy early in the 11th cent. Poland regained control in the 12th cent. and introduced Christianity. The country was split into two principalities: West Pomerania became a part of the Holy Roman Empire (1181), and Pomerelia, as East Pomerania came to be known, became independent in 1227, was annexed to Poland in 1294, and was taken in 1308–9 by the Teutonic Knights. Pomerelia, including Danzig, was formally restored to Poland (1466); it remained an integral part of Poland until 1772, when it passed to Prussia. In 1919 part of it was given back to Poland. After the outbreak (1939) of World War II all parts of Pomerelia were reannexed to Germany, which lost them again to Poland in 1945.

In the meantime, Pomerania continued as a duchy of the Holy Roman Empire until the death (1637) of Bogislav XIV. The Peace of Westphalia (1648) gave Hither Pomerania (the western part) to Sweden, while Farther Pomerania (the eastern part) went to the electorate of Brandenburg (after 1701, the kingdom of Prussia). Sweden lost about half of its part of Pomerania to Prussia in 1720; the rest of Swedish Pomerania, nominally part of the Holy Roman Empire until its dissolution in 1806, passed to Denmark (1814) and then to Prussia (1815). Thus, from 1815 to 1919, all Pomerania and all Pomerelia were in Prussian hands. Most of Pomerania remained within German boundaries until 1945. After the transfer in 1945 of the larger part of Pomerania to Polish administration, the German-speaking population was largely expelled.

Pom·er·oy (pŏm'ə-roi). **1.** Village (1970 pop. 2,672), seat of Meigs co., SE Ohio, on the Ohio River SW of Parkersburg, W.Va., in an agricultural area; settled in the early 19th cent., inc. 1840. **2.** City (1970 pop. 1,823), seat of Garfield co., SE Wash., NE of Walla Walla, in a grain, cattle, and irrigated farm region; settled 1864, platted 1878, inc. 1886.

Po·mo·na (pə-mō'nə), city (1978 est. pop. 83,400), Los Angeles co., S Calif., a residential and industrial suburb of Los Angeles; inc. 1888. Citrus fruits and vegetables are canned and shipped. Pomona is the seat of California State Polytechnic Univ.

Pom·pa·no Beach (pŏm'pə-nō'), city (1978 est. pop. 53,000), Broward co., SE Fla., on the Atlantic coast and the Intracoastal Waterway; inc. 1908. It is a resort city with ocean beaches, excellent fishing, and a harness-racing track. Among the city's manufactures are pleasure boats, plastic and metal products, and electronic equipment. The raising of winter vegetables has long been an important industry. The city has many miles of small canals.

Pom·pe·ii (pŏm-pā', -pā'ē), ancient city of S Italy, a port near Naples at the foot of Mt. Vesuvius. Possibly an old Oscan settlement, it was a Samnite city for centuries before it passed under Roman rule (1st cent. B.C.). Pompeii then became a flourishing port and a prosperous resort with many villas. An eruption of Mt. Vesuvius in A.D. 79 buried Pompeii under cinders and ashes that preserved the ruins of the city with magnificent completeness. The long-forgotten site of the city was rediscovered in 1748. The habits and manners of life in Roman times have been revealed in great detail by the plan of the streets and footpaths, the statue-decorated public buildings, and the shops and homes of the artisans. The houses and villas have yielded rare and beautiful examples of Roman art.

Po·na·pe (pō'nä-pā'), volcanic island, 129 sq mi (334 sq km), W Pacific, in the E Caroline Islands. It is a flat dome of black basaltic rock, rising to c.2,100 ft (640 m), with a rim of fertile coastal land. There are deposits of bauxite, iron, and iron sulfate; copra, dried bonito, and handicrafts are the chief products. Ruins of ancient stone walls, dikes, and basaltic columns dot the island.

Pon·ca (pŏng'kə), city (1970 pop. 984), seat of Dixon co., NE Nebr., WNW of Sioux City, Iowa, on the Missouri River, in a livestock and grain area.

Ponca City, city (1978 est. pop. 27,700), Kay co., N Okla., on the Arkansas River; founded 1893 with the opening of the Cherokee Strip, inc. 1899. It is a trade, processing, and shipping hub in a grain, livestock, and oil area. There are oil refineries and plants that make oil-well equipment. The city has many parks, a pioneer museum, and an Indian museum.

Pon·ce (pôn'sā), city (1970 pop. 128,233), S Puerto Rico. It is the island's chief Caribbean port and an agricultural trade and distribution center. Industries include tourism, the processing of agricultural products, rum distilling, and varied manufacturing. It was founded in the early 16th cent.

Pon·cha Pass (pŏn'chə), 9,012 ft (2,748.7 m) high, central Colo., in the N tip of the Sangre de Cristo Mts. It was much used in the 19th cent. by Indians, overland immigrants, and mountain men.

Pon·de·ra (pŏn'də-rā'), county (1970 pop. 6,611), 1,643 sq mi (4,255.4 sq km), N Mont., in an agricultural area drained by Lake Frances and branches of the Marias River; formed 1919; co. seat Conrad. Its economy depends on grain, sugar beets, and livestock. Part of Lewis and Clark National Forest is in the west.

Pon·di·cher·ry (pŏn'də-chĕr'ē, -shĕr'ē), union territory (1971 pop. 471,347), 183 sq mi (474 sq km), of India. It comprises four noncontiguous enclaves. Under an agreement with France, India took over administration of the four enclaves on Nov. 1, 1954; the formal transfer occurred in Aug., 1962.

Pon·ta Del·ga·da (pôn'tə dĕl-gä'də), city (1974 est. pop. 20,195), capital of Ponta Delgada dist., in the Azores, Portugal. An important port on São Miguel island, it is the chief commercial center of the Azores. It is also a winter resort.

Pont·char·train, Lake (pŏn'chər-trān'), shallow lake, c.630 sq mi (1,630 sq km), N of New Orleans. It is linked with Lake Maurepas at its western end and with the Gulf of Mexico and at its eastern end through Lake Borgne. The lake is tidal and has brackish water.

Pon·te·fract (pŏn'tə-frăkt', pŭm'frĭt), municipal borough (1973 est. pop. 31,340), West Yorkshire, N England. Furniture, iron products, and textiles are made. Situated on the edge of coal fields, Pontefract grew around a great castle built in the 11th cent. on the site of a Saxon fort. Richard II died here.

Pon·te·ve·dra (pŏn'tə-vā'drə), city (1975 est. pop. 31,200), capital of Pontevedra prov., NW Spain, in Galicia, on the Atlantic Ocean at the mouth of the Lérez River. It is a fishing port. Clothing, leather goods, and fertilizers are made. Among its old structures are the Gothic Church of Santa María, the picturesque ruins of a 14th cent. convent, and a Roman bridge.

Pon·ti·ac (pŏn'tē-ăk'). **1.** City (1975 est. pop. 10,200), seat of Livingston co., N central Ill., on the Vermilion River NE of Bloomington; settled 1833, platted 1837, inc. 1857. It is a trade and shipping center in a farm area. **2.** Industrial city (1978 est. pop. 76,000), seat of Oakland co., SE Mich., on the Clinton River; founded 1818, inc. as a city 1861. Carriage making, important in the 1880s, gave way to the automobile industry.

Pon·ti·a·nak (pôn'tē-ä'näk), city (1971 pop. 217,555), capital of West Kalimantan prov., W Borneo, Indonesia, at the mouth of a small stream in the Kapuas delta. An important port, it serves an area producing rubber, palm oil, sugar, coconuts, pepper, rice, tobacco, and gold.

Pon·tics (pŏn'tĭks), mountain system, N Turkey, extending c.700 mi (1,125 km) along the S coast of the Black Sea. The Pontics, which generally lack porous rock, have been greatly dissected by the large amount of surface drainage. The northern slopes receive an average annual precipitation of c.95 in. (241 cm) and have lush vegetation. The southern slopes are much drier. Coal, antimony, and copper are mined in the mountains.

Pon·tine Marshes (pŏn'tēn', -tīn'), low-lying region, c.300 sq mi (780 sq km), in S Latium, central Italy, between the Tyrrhenian Sea and the Apennine foothills. It is crossed by drainage canals and the Appian Way, a Roman-built road. In pre-Roman and early Roman times the area was populated and fertile, but it was later abandoned because of the malaria in its marshlands. The Roman emperors Trajan and Theodoric and several popes started reclamation works, but a drainage system was not completed until the 1930s under Mussolini. During World War II the drainage works were damaged and the region was flooded. Wheat and cotton are now produced, and livestock is raised.

Pon·toise (pôɴ-twäz'), city (1975 pop. 27,240), capital of Val-d'Oise dept., N central France. It is the site of a technical school.

Pon·to·toc (pŏn'tə-tŏk'). **1.** County (1970 pop. 17,363), 501 sq mi (1,297.6 sq km), N Miss., drained by the Skuna and Yocona rivers; formed 1836; co. seat Pontotoc. Farming (cotton and corn) and dairying are major occupations. It has timber and deposits of clay and bauxite. **2.** County (1970 pop. 27,867), 719 sq mi (1,862.2 sq km), S central Okla., bounded in the N by the Canadian River and drained by Clear Boggy Creek and the Blue River; formed 1907; co. seat Ada. Its agriculture includes cotton, corn, oats, hay, sorghums, livestock, and dairy products. It also has oil and natural-gas fields and a food-processing industry. Rubber, stone, clay, and glass products and machinery are made.

Pontotoc, city (1970 pop. 3,453), seat of Pontotoc co., NE Miss., W of Tupelo, in a farm area; inc. 1837. Cottonseed oil and lumber are made here.

Pon·tus (pŏn'təs), ancient country, NE Asia Minor, on the Black Sea coast. On its inland side were Cappadocia and western Armenia. The greatest Pontic ruler was Mithradates VI, who conquered Asia Minor, gained control of the Crimea, and threatened Rome in Greece. After his defeat, the Romans joined Pontus to the province of Galatia-Cappadocia.

Pon·ty·pool (pŏn'tē-po͞ol'), urban district (1976 est. pop. 88,700), Gwent, SE Wales. There are coal-mining, steel, nylon, glass, and toy industries.

Pon·ty·pridd (pôn'tə-prē*th*'), urban district (1973 est. pop. 34,180), Mid Glamorgan, S Wales. It is a railroad junction and coal port. Electrical equipment, cables, and chains are made.

Poole (po͞ol), municipal borough (1976 est. pop. 110,600), Dorset, S England, on the N side of Poole Harbour. Poole has shipbuilding, pottery-making, and other industries. It is a naval supply station and has a considerable coastal trade.

Poo·na (po͞o'nə), city (1971 pop. 853,226), Maharashtra state, W central India. It is a commercial center with metalworks. There are several palaces and temples from the 17th and 18th cent., when Poona was the capital of the Mahrattas. Under British rule it was an important military center.

Po·o·pó (pō'ə-pō'), salt lake, 965 sq mi (2,499.4 sq km), on the high plateau of W Bolivia. It is more than 11,000 ft (3,355 m) above sea level and averages only 10 ft (3 m) in depth.

Po·pa·yán (pō'pə-yän'), city (1973 pop. 77,669), capital of Cauca dept., SW Colombia, on a volcanic terrace high above the Cauca River. Textiles and foodstuffs are produced in the city. Popayán was founded in 1536 and during colonial times was a wealthy and aristocratic trade center. After Colombia gained independence in the 19th cent., the city lost its commercial pre-eminence but remained a major cultural center.

Pope (pōp). **1.** County (1970 pop. 28,607), 816 sq mi (2,113.4 sq km), NW central Ark., in the Ozarks, bounded on the S by the Arkansas River, formed 1829; co. seat Russellville. It has diversified farming (fruit, cotton, truck crops, potatoes, corn, soybeans, livestock, poultry, and dairy products), a food-processing industry, sawmills, coal mines, and natural-gas wells. **2.** County (1970 pop. 3,857), 381 sq mi (986.8 sq km), extreme SE Ill., bounded on the E and SE by the Ohio River and drained by Bay and Lusk creeks; formed 1816; co. seat Golconda. Its agriculture includes fruit, corn, wheat, and livestock. Wood products are made. **3.** County (1970 pop. 11,107), 669 sq mi (1,732.7 sq km), W central Minn., in an agricultural area drained by the Chippewa River and watered by Lake Minnewaska and several smaller lakes; formed 1862; co. seat Glenwood. In an agricultural area yielding livestock, grain, dairy products, and poultry, it has a food-processing industry and manufactures machinery. There are many resorts in the county.

Pop·lar Bluff (pŏp'lər blŭf'), city (1978 est. pop. 17,400), seat of Butler co., SE Mo., in the Ozark foothills, on the low bluffs of the Black River near the Ark. line; inc. 1870. It is a trade, shipping, and medical center in a rich farming area. Plastics, wood and paper products, electric transformers, shoes, and concrete items are manufactured.

Pop·lar·ville (pŏp'lər-vĭl'), town (1970 pop. 2,312), seat of Pearl River co., S Miss., SSW of Hattiesburg, in an area producing tung nuts, pecans, and naval stores; inc. 1884.

Po·po·cat·é·petl (pō-pō'kä-tā'pət-əl), volcano, 17,887 ft (5,455.5 m) high, in the Cordillera de Anáhuac, central Mexico. The perpetually snow-capped cone is symmetrical, and the large crater has practically pure sulfur deposits only partially exploited. The volcano has been dormant since 1702. Occasionally it emits vast clouds of smoke.

Por·ban·der (pôr-bŭn'dər), town (1971 pop. 96,756), Gujurat state, W central India, on the Arabian Sea. Fishing is an important industry. Porbander produces ghee, textiles, vegetable oil, salt, cement, and chemicals.

Por·cu·pine (pôr'kyə-pīn'), river, 448 mi (720.8 km) long, rising in the Ogilvie Mts., NW Yukon Territory, Canada. It flows in a great arc northeast through the Eagle Plain, then west into Alaska and to the Yukon River at Fort Yukon. The river was discovered in 1842.

Po·ri (pô'rē), city (1975 est. pop. 80,343), capital of Turku-and-Pori prov., SW Finland, near the mouth of the Kokemaënjoki River. Timber and metals are exported, and wood products are manufactured. Pori was chartered in 1564 and was initially dominated by the Hanseatic League.

Pork·ka·la (pôr'kə-lə), small strategic peninsula, in the Gulf of Finland, near Helsinki. In accord with the Soviet-Finnish armistice of 1944, Finland leased it to the USSR for 50 years, for use as a naval base. The USSR returned it to Finland in 1956.

Pó·ros (pô'rôs), island (1971 pop. 4,051), c.8 sq mi (20 sq km), SE Greece, in the Aegean Sea near the Argolis peninsula of the Peloponnesus. It is famous for its fine marble. There are remains of a temple of Poseidon.

Pors·grunn (pôrs'grōōn'), town (1977 est. pop. 31,712), Telemark co., SE Norway, a port on the Frierfjord (an arm of the Skagerrak); chartered 1842. Manufactures include chemicals, fertilizers, metal goods, and electrical appliances. The town is noted for its porcelain.

port (pôrt), a harbor and its terminal facilities for the transfer of goods and passengers to or from waterborne means of transport. Port cities are located on oceans, bays, lakes, rivers, and canals in places where access to the hinterland provides a large volume of commerce.

Port Ad·e·laide (ăd'l-ād'), city (1976 pop. 36,020), South Australia state, S Australia, a suburb of Adelaide, on an inlet of Gulf St. Vincent. It is the principal port and wool-trading center of the state. The chief exports are wheat, flour, and wool. Sulfuric acid, processed foods, and automobile parts are made.

Port·a·down (pôrt'ə-doun'), municipal borough (1971 pop. 22,207), Co. Armagh, central Northern Ireland, on the Bann River. It is an important railroad and industrial center.

Por·tage (pôr'tĭj, pōr'-). **1.** County (1970 pop. 125,868), 495 sq mi (1,282.1 sq km), NE Ohio, intersected by the Cuyahoga River and by tributaries of the Mahoning River; formed 1807; co. seat Ravenna. Its agriculture includes potatoes, fruit, dairy products, truck crops, and livestock. It has sand and gravel pits, and its industries include lumbering, food processing, and textile milling. Paper, plastics, concrete, metals and metal products, motor vehicles, electrical equipment, and toy balloons are manufactured. There are many small lakes in the county. **2.** County (1970 pop. 47,541), 810 sq mi (2,097.9 sq km), central Wis., intersected by the Wisconsin River, and also drained by the Plover and Waupaca rivers; formed 1836; co. seat Stevens Point. In an irrigated agricultural area yielding dairy products, potatoes, beans, cucumbers, and other vegetables, it has sand and gravel pits, timber, and a food-processing industry.

Portage. **1.** Town (1978 est. pop. 24,100), Porter co., NW Ind., a suburb of Gary, on Lake Michigan; inc. 1959. A new port, accommodating ocean vessels, began operating here in the early 1970s. The town, located in a highly industrialized portion of the state, has a steel industry. **2.** City (1978 est. pop. 36,900), Kalamazoo co., SW Mich.; inc. 1963. Its manufactures include pharmaceuticals, industrial and medical gases, machinery, and paper and plastic products. **3.** City (1970 pop. 7,821), seat of Columbia co., central Wis.; inc. 1854. It is on the Wisconsin at the point where that river is close enough to the Fox River to make canoe portage feasible. In 1673 Louis Jolliet and Father Marquette were the first white men to use this important portage link in the water route from the Great Lakes to the Mississippi. The path is now a ship canal, and the present city is a farm trade center with some light manufacturing industry.

Por·tage la Prai·rie (pôr'tĭj lə prâr'ē), city (1976 pop. 12,555), S Man., Canada. It is the center of a farming region and has some industries.

Port Al·ber·ni (ăl-bûr'nē), city (1976 pop. 19,585), SW British Columbia, Canada, on Vancouver Island. It is a fishing port with boatbuilding and wood-products industries.

Por·ta·les (pôr-tăl'ĭs), city (1970 pop. 10,554), seat of Roosevelt co., E N.Mex., near the Texas line; inc. 1910. It is the trade center of an agricultural and livestock area. There are oil wells in the vicinity.

Port Al·len (ăl'ən), city (1970 pop. 5,728), seat of West Baton Rouge parish, SE La., on the Mississippi River opposite Baton Rouge, in a sugar-cane region; laid out 1854, inc. 1923.

Port An·ge·les (ăn'jə-ləs), city (1978 est. pop. 16,900), seat of Clallam co., NW Wash., on Juan de Fuca Strait opposite Victoria, British Columbia; inc. 1890. A port of entry with a fine harbor, Port Angeles is a boating and fishing center and has pulp and paper mills. The city is also a resort.

Port Ap·ra (äp'rə): *see* Apra Harbor, Guam.

Port Ar·thur (är'thər): *see* Lü-shun, China.

Port Arthur, city (1978 est. pop. 52,200), Jefferson co., SE Texas, on Sabine Lake; inc. 1898. A deepwater port of entry on the Sabine-Neches Canal, it is a giant oil port, with many large refineries, tank farms, chemical plants, and shipyards. Oil-drilling equipment and metal, steel, and aluminum products are also manufactured. A ship channel was completed in 1899. Port Arthur boomed after the discovery (1901) of oil at Spindletop.

Port Au·gus·ta (ô-gŭs'tə), city (1971 pop. 12,095), South Australia, S Australia, at the head of Spencer Gulf. It is a port and railroad center.

Port-au-Prince (pôrt'ə-prĭns'), city (1971 pop. 458,675), capital of Haiti, SW Haiti, on a bay at the end of the Gulf of Gonaïves. The country's chief seaport, it exports mainly coffee and sugar. The city has food- and tobacco-processing plants, rum distilleries, and textile and cement industries. Founded in 1749 by French sugar planters, it replaced Cap-Haïtien as capital of the French colony in 1770 and in 1804 it became the capital of newly independent Haiti. Port-au-Prince has suffered frequently from earthquakes, fires, and civil warfare. The city is like an amphitheater, with business and commercial quarters along the water and residences on the hills above. Landmarks include the French-built quay (1780), the Univ. of Haiti (est.

1944), the National Palace, and the Basilica of Notre Dame.

Port Char·lotte (shär'lət), uninc. town (1970 pop. 10,769), Charlotte co., SW Fla., on Charlotte Harbor (an inlet of the Gulf of Mexico) and the Peace and Myakka rivers. It is a planned residential community—one of several on a peninsula once owned by the Vanderbilt family. The area, formerly cattle pasture land, has been rapidly developed since the 1950s. Port Charlotte has 145 mi (233.3 km) of manmade waterways, many with access to the Gulf of Mexico.

Port Ches·ter (chĕs'tər), village (1978 est. pop. 23,300), Westchester co., SE N.Y., an industrial suburb of New York City, on Long Island Sound at the mouth of the Byram River, and on the Conn. border; settled after 1660, inc. 1868. Candy, nuts and bolts, brushes, toys, and lamps are among its many manufactures.

Port Clin·ton (klĭn'tən), village (1970 pop. 7,202), seat of Ottawa co., N Ohio, on Lake Erie WNW of Sandusky at the mouth of the Portage River; founded c.1828. A commercial fishing village in a farm and fruit-growing region, it also has a cannery and makes rubber and gypsum products.

Port Col·borne (kōl'bərn), town (1976 pop. 20,536), S Ont., Canada, on Lake Erie, at the S end of the Welland Ship Canal. An important transshipment center, it has a nickel refinery, grain elevators, and a cement plant.

Port E·liz·a·beth (ē-lĭz'ə-bĕth), city (1970 pop. 392,231), Cape Prov., SE South Africa, on Algoa Bay, an arm of the Indian Ocean. It is a tourist center and a major seaport that ships diamonds, wool, fruit, and other items. Automobile assembly is the chief industry; shoe manufacturing, metal and timber processing, and electrical engineering are also important. Port Elizabeth was founded by British settlers in 1820. The city grew rapidly after 1873, when a railroad to Kimberley was constructed.

Por·ter (pôr'tər, pōr'-), county (1970 pop. 87,114), 425 sq mi (1,100.8 sq km), NW Ind., bounded on the N by Lake Michigan, on the S by the Kankakee River, and drained by the Little Calumet and Grand Calumet rivers; formed 1835; co. seat Valparaiso. It is in a farming (corn, grain, hogs, cattle, poultry, and dairy products) and anthracite-coal mining area. Its industry includes food processing and the manufacture of textiles, fertilizers, plastics, metal products, and machinery.

Por·ter·ville (pôr'tər-vĭl'), city (1978 est. pop. 16,300), Tulare co., S central Calif., on the Tule River; founded 1859 on the old Los Angeles–San Francisco stage route, inc. 1902. The city is chiefly residential, with some agriculture and light manufacturing.

Port-Gen·til (pôr-zhäN-tēl'), city (1970 est. pop. 31,000), W Gabon, a seaport on Cape Lopez Bay, an arm of the Atlantic Ocean. Timber and locally manufactured plywood are exported.

Port Gib·son (gĭb'sən), town (1970 pop. 2,589), seat of Claiborne co., SW Miss., near the Mississippi River S of Vicksburg, in a cotton area; founded in the late 18th cent.

Port Glas·gow (glăs'gō, -kō), burgh (1974 est. pop. 22,278), Strathclyde region, W Scotland, on the Firth of Clyde. Its dry dock, built in 1762, was one of the first of its kind in Scotland. There are shipbuilding plants and textile, rope, and canvas factories.

Port Har·court (här'kərt, -kôrt'), city (1971 est. pop. 217,000), SE Nigeria, a deepwater port on the Bonny River in the Niger delta. Steel and aluminum products, pressed concrete, glass, tires, paint, footwear, furniture, and cigarettes are manufactured and bicycles and motor vehicles are assembled. Port Harcourt was founded by the British in 1912.

Port Hu·ron (hyŏŏr'ən, -ŏn'), city (1978 est. pop. 34,200), seat of St. Clair co., S Mich., a port of entry at the junction of the St. Clair River with Lake Huron; inc. 1857. It is a shipping center with railroad shops and plants making automobile wires and cables, and copper, brass, and paper products. The earliest settlement began (1686) with a French fort. The town grew after the building (1826) of Fort Gratiot Turnpike (between Port Huron and Detroit), ushering in a lumbering era. After the lumber industry declined (in the 1880s), local deposits of salt, oil, and natural gas were developed, and sawmilling, boatbuilding, and papermaking became important. The Fort Gratiot lighthouse, which marks the St. Clair straits off Port Huron, is the oldest on the Great Lakes.

Port Jef·fer·son (jĕf'ər-sən), uninc. residential town (1975 est. pop. 5,800), Suffolk co., SE N.Y., on N Long Island, on Long Island Sound opposite Bridgeport, Conn. It is a resort and yachting center.

Port Jer·vis (jĕr'vəs), resort city (1970 pop. 8,852), Orange co., SE N.Y., on the Delaware River near the point where N.Y., N.J., and Pa. meet; settled before 1700, inc. 1907. Textiles, metal products, and plated silverware are made.

Port·land (pôrt'lənd), urban district (1971 pop. 12,306), Dorset, S England. It is on the Isle of Portland, a small rocky peninsula. Portland stone has been used in St. Paul's Cathedral and other important London buildings. Lobsters and crabs are harvested.

Portland. **1.** Town (1970 pop. 8,812), Middlesex co., central Conn., on the Connecticut River; settled c.1690, inc. 1841. The town has boatyards. Machinery, corrugated boxes, and metal products are among its manufactures. **2.** City (1970 pop. 7,115), seat of Jay co., E Ind., on the Salamonie River NE of Muncie; founded 1835, inc. as a town 1843, as a city 1883. It is a processing point in a farm and livestock area. **3.** City (1978 est. pop. 56,700), seat of Cumberland co., SW Maine, situated on a small peninsula and adjacent land, with a large, deepwater harbor on Casco Bay; settled c.1632, set off from Falmouth and inc. 1786. It is a port of entry, the commercial center of the state, and the rail, highway, shipping, and processing center for a vast farming, lumbering, and resort area. Portland has shipyards, canneries, printing and publishing firms, foundries, and important lumbering, paper-milling, fishing, chemical, and textile industries. It was settled c.1632. The settlement known as Falmouth developed and, in spite of Indian raids in the late 17th cent., became a commercial center. It was almost completely destroyed by the British in 1775 but was rebuilt shortly afterward. Portland served as state capital from 1820 to 1832. In 1866 a fire destroyed much of the city. The lighthouse (est. 1791) is still in use. **4.** City (1978 est. pop. 348,000), seat of Multnomah co., NW Oregon, on the Willamette River near its junction with the Columbia; inc. 1851. It is a port of entry, a leading financial and industrial center, and an important deepwater port. Manufactures include lumber, wood products, paper, metals, machinery, foodstuffs, woolen textiles, clothing, and furniture. The city was founded in 1845. Its growth was rapid after 1850, when it served as a supply point for the Calif. gold fields, and continued with the coming of the railroad (1883) and the Alaska gold rush (1897–1900). Portland has an art museum, a museum of science and industry, a planetarium, a forestry center, a zoo, a Japanese garden, and a symphony orchestra.

Port La·vac·a (lə-văk'ə), city (1970 pop. 10,491), seat of Calhoun co., S Texas, on Lavaca Bay; inc. 1907. It is a port of entry and a resort city with fishing. There are shrimp- and oyster-processing industries.

Port Lou·is (lōō'ĭs, lōō'ē), city (1975 est. pop. 139,592), capital of Mauritius, NW Mauritius, a port on the Indian Ocean. The city's economy is dominated by its well-sheltered port, which handles Mauritius's international trade; there are extensive facilities for processing and storing sugar, the main export. Manufactures include cigarettes, rum and wine, food products, and aloe fiber. Port Louis was founded in 1735.

Port Mores·by (môrz'bē), town (1971 pop. 76,507), capital of Papua New Guinea, on New Guinea Island and the Gulf of Papua. Rubber, gold, and copra are exported. Port Moresby was founded by Capt. John Moresby, who landed here in 1873. The British occupied it in 1883.

Por·to (pōr'tōō): *see* Oporto, Portugal.

Pôr·to A·le·gre (pôr'tō ə-lĕg'rə), city (1975 est. pop. 1,043,964), capital of Rio Grande do Sul state, SE Brazil, on the Guaíba River. It is Brazil's major river port, exporting the products of the rich agricultural and pastoral hinterland. It has a shipyard, meat-packing plants, foundries, and varied processing industries. The city was founded (c.1742) by immigrants from the Azores. Since the 19th cent. its development has been aided by numerous German and Italian settlers. Modern business and government buildings coexist with many old, narrow streets and colonial buildings. The city has two large universities.

Por·to·fer·rai·o (pôr'tō-fĕr'ä-yō), town (1976 est. pop. 11,086), Tuscany, Italy, on the N coast of Elba Island. It handles most of the iron shipped from the island. It is also a seaside resort. Napoleon I resided here (1814–15) during his exile.

Port of Spain (spān), city (1973 est. pop. 60,450), capital of Trinidad and Tobago, on the Gulf of Paria. It is the industrial and commercial center of the country. From 1958 to 1962 Port of Spain was the capital of the now dissolved Federation of the West Indies. It exports agricultural products and asphalt. Bauxite from the Guianas and iron ore from Venezuela are transferred here for overseas shipment. The city has attractive public buildings and botanical gardens.

Por·to-No·vo (pôr'tō nō'vō), city (1975 est. pop. 104,000), capital of Benin, on Porto-Novo lagoon, an arm of the Gulf of Guinea. Porto-Novo is the trade center for an agricultural region whose chief product is palm oil; the city's exports include palm oil, cotton, and kapok. The Portuguese built a trading post here in the 17th cent. Africans were shipped as slaves from Porto-Novo to the Americas. Porto-Novo was incorporated into Dahomey colony in 1883 and in 1900 was made its capital.

Port Or·chard (ôr'chərd), town (1977 est. pop. 4,280), seat of Kitsap co., NW Wash., on an arm of Puget Sound opposite Bremerton; settled 1854, platted 1886, inc. 1893. It is a farm trade center.

Pôr·to Vel·ho (pōr'tōō vĕ'lyōō), city (1970 pop. 41,146), capital of Rondônia federal territory, NW Brazil, on the Madeira River. It is the last point of navigation on the river. The city's economy is based on the exploitation and shipment of the rubber and Brazil nuts found in surrounding forests.

Port Phil·lip Bay (fĭl'əp), large deepwater inlet of Bass Strait, 30 mi (48.3 km) long and 25 mi (40.2 km) wide, Victoria, SE Australia. Hobson's Bay is its northern arm.

Port Pir·ie (pĭr'ē), city (1971 pop. 13,269), South Australia state, S Australia, on an inlet of Spencer Gulf. It is a railroad center and has uranium refineries and smelting works for the silver-lead mines at Broken Hill. Silver-lead ore and refined lead are exported.

Port Ra·di·um (rā'dē-əm), mining village, N central Northwest Territories, Canada, on Great Bear Lake. The mines were discovered in 1930 and yielded deposits of pitchblende. During World War II the mines were expropriated by the Canadian government when scientists found that the ore contained a rich store of uranium oxide. They were exhausted and closed in 1960.

Port Roy·al Sound (roi'əl), arm of the Atlantic Ocean, between St. Helena and Parris islands to the N and Hilton Head Island to the S, in S S.C. The sound was named in 1562 by French explorer Jean Ribaut, founder of a short-lived Huguenot settlement on Parris Island.

Port Sa·id (sīd, sä-ēd'), city (1976 pop. 262,620), NE Egypt, a port on the Mediterranean Sea at the entrance to the Suez Canal. Salt is produced in Port Said by evaporating sea water, and there is a fishing industry. The city is a principal port for steamer service on the Nile River. Situated on a narrow peninsula between Lake Manzala and the sea, Port Said was founded in 1859 by the builders of the Suez Canal and named for Said Pasha, then khedive of Egypt.

Ports·mouth (pôrts'məth), county borough (1976 est. pop. 198,500), Hampshire, S England, on Spithead Channel. Since Henry VII built stone fortifications and docks here, Portsmouth has almost continuously been the foremost naval base of Great Britain. There are also aircraft-engineering and other industries. The Cathedral of St. Thomas of Canterbury dates partly from the 12th cent. Southsea Castle was built by Henry VIII. The house in which Charles Dickens was born has been converted into a museum, as has H.M.S. *Victory,* Adm. Nelson's flagship at Trafalgar in 1805.

Portsmouth. **1.** City (1978 est. pop. 24,100), Rockingham co., SE N.H., a port of entry with a good harbor and a state-owned port terminal at the mouth of the Piscataqua River opposite Kittery, Maine; settled c.1623, inc. 1653. It has a fishing industry and plants making shoes, wire and cable, and electric products. Tourism is also important. Portsmouth was a point for exporting lumber and fish and served as colonial capital until the Revolution. Shipbuilding was an early industry; men-of-war and privateers were made here during the Revolution and the War of 1812. The Portsmouth Naval Shipyard (est. 1800) is located on two islands (now joined together) in the Piscataqua River. It is an important submarine base and repair yard. The Treaty of Portsmouth, ending the Russo-Japanese War, was signed (1905) at the base. Portsmouth is the site of the U.S. Naval Disciplinary Command, and Pease Air Force Base, a large strategic air command installation. Many fine old houses are in Strawberry Banke, a restored colonial community. **2.** City (1978 est. pop. 24,600), seat of Scioto co., S Ohio, in a hilly area on the Ohio River at the mouth of the Scioto, across from South Portsmouth, Ky.; inc. 1814. It is an important industrial and rail center. Completion of the Ohio Canal (1832), linking Portsmouth with Cleveland, and the discovery of iron ore in the area started the city's industrial growth. Of interest are Mound Park, with ancient Indian burial grounds (now in the heart of the city), and traces of the old Ohio River Canal. **3.** Town (1971 pop. 12,521), Newport co., SE R.I., on Rhode Island; founded by William Coddington, John Clarke, Anne Hutchinson, and others 1638, inc. 1644. It is a summer resort and farming center. It was an early fishing, shipping, and shipbuilding center, with some farming. The first general assembly of the new colony met at Portsmouth in 1647. Coal mining was important in the 19th cent. **4.** City (1978 est. pop. 105,300), SE Va., on the Elizabeth River and Hampton Roads, adjacent to and opposite Norfolk, with which it is connected by two bridges and two tunnels; founded 1752 on the site of an Indian village, inc. 1858. Portsmouth has one of the world's largest shipyards, with more drydocks than any other yard in the nation, a naval ammunitions dump, and the headquarters of the Fifth U.S. coast guard district. Portsmouth is also a busy commercial seaport and a rail center, with railroad shops and terminals. A private shipyard was built in 1767; it served as a British base in the Revolutionary War, after which it became a U.S. base. In the Civil War the navy yard was burned and evacuated by the Federals in 1861 and then retaken in 1862. The nation's first battleship (*Texas*) was built here in 1892 and the first aircraft carrier (*Langley*) in 1922. Of interest in the city are Trinity Church (1762), the Shipyard Museum, and the Old Towne Historic District, with many old homes.

Port Su·dan (sōō-dăn'), city (1973 pop. 132,631), NE Sudan, on the Red Sea. The city serves a rich cotton-growing area of the Nile Valley. Construction of a railroad linking the Nile River and the Red Sea coast in 1905 led to the founding of Port Sudan as a harbor for the region.

Port Tal·bot (tăl'bət, tôl'-), municipal borough (1973 est. pop. 50,200), West Glamorgan, S Wales, at the mouth of the Avon River on Swansea Bay. The borough has large steelworks and is an export point for the coal and mineral industries of the Avon valley.

Port Town·send (toun'sənd), city (1977 est. pop. 5,655), seat of Jefferson co., NW Wash., on the Olympic Peninsula at the entrance to Puget Sound; settled 1851, inc. 1860. It is a resort center and a port of entry with a fine harbor. Seafood and lumber are processed.

Por·tu·gal (pôr'chə-gəl), republic (1970 pop. 8,668,267), 35,553 sq mi (92,082 sq km), SW Europe, on the W side of the Iberian Peninsula and including the Madeira Islands and the Azores in the Atlantic Ocean. The capital is Lisbon. Portugal is bordered by Spain on the east and north and by the Atlantic Ocean on the west and south. The country is crossed by rivers rising in Spain and flowing to the Atlantic; among them are the Douro, the Tagus, the Sado, and the Guadiana. The river valleys support agriculture, and in the Douro and Tagus valleys extensive vineyards are maintained. On the lower slopes there are olive groves; on the flatter uplands, as well as on the plains near the coast, grains are grown and livestock are raised. There are great variations in terrain and climate among the historic provinces. Trás-os-Montes in the extreme northeast has a rigorous mountain climate, as have parts of Entre-Minho-e-Douro (officially Douro). Beira has the highest mountains of the country, the scenic Serra de Estrela, dotted with resorts. Estremadura, in western Portugal, has broad alluvial plains, rising to cool and rocky uplands; along the Atlantic coast is a celebrated resort region. Most of Alentejo has a continental climate; although much of its soil is poor, together with Estremadura it is the granary of Portugal. In the southernmost of the

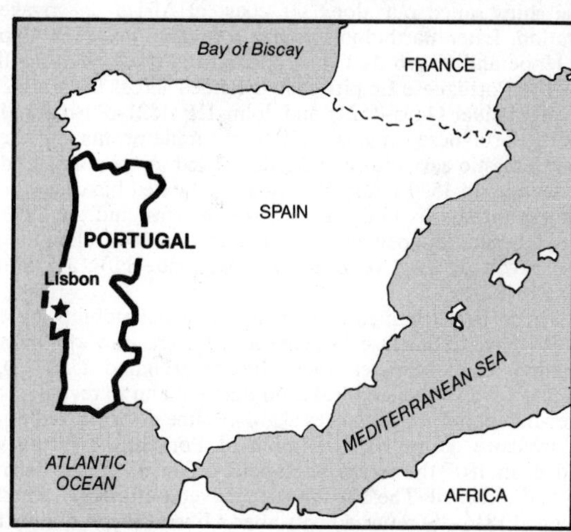

old provinces, Algarve, mountains curve across the north of the province down to Cape St. Vincent, the southwestern tip of Europe; citrus crops thrive in the mild Mediterranean climate, and there are acres of almond trees.

Portugal is divided into 18 districts and the autonomous districts of the Azores and the Madeira Islands:

NAME	CAPITAL	NAME	CAPITAL
Aveiro	Aveiro	Leiria	Leiria
Azores	Angra do Heroísmo	Lisboa	Lisbon
Beja	Beja	Madeira Is.	Funchal
Braga	Braga	Portalegre	Portalegre
Bragança	Bragança	Pôrto	Oporto
Castelo Branco	Castelo Branco	Santarém	Santarém
Coimbra	Coimbra	Setúbal	Setúbal
Évora	Évora	Viana do Castelo	Viana do Castelo
Faro	Faro	Vila Real	Vila Real
Guarda	Guarda	Viseu	Viseu

Economy. Portuguese agriculture is backward and inefficient; cereals must be imported. The country's fishing fleets bring in vital cargoes of sardine and tuna. Slightly more than half of the labor force is employed in various industries, of which food processing and the manufacture of textiles and chemicals are the largest; there is also a sizable cement industry, and low-grade coal, copper pyrites, iron ore, wolfram, and other minerals are mined. Most of the mines are in the northern mountains and in Beira. Portugal's forests provide a major portion of the world's supply of cork. The country trades heavily with West Germany, Great Britain, France, the United States, and Italy. Machinery and motor vehicles, textile fibers, petroleum, and cereals are major imports, and cotton textiles and wine as well as cork and other wood products are major exports. Portugal is one of the charter members of the European Free Trade Association (EFTA).

*History.*The Portuguese long considered themselves descendants of the Lusitanians, a Celtic people who came to the area after 1000 B.C. and stoutly resisted the Romans (2nd-1st cent. B.C.). Julius Caesar and Augustus completed the Roman conquest of the area, and the province of Lusitania thrived. Roman ways were adopted, and it is from Latin that the Portuguese language is derived. At the beginning of the 5th cent. A.D. the whole Iberian Peninsula was overrun by Germanic invaders; the Visigoths eventually established their rule. Present-day Algarve was part of the Byzantine empire during the 6th and 7th cent. In 711 the Visigoths were defeated by the Moors. It was during the long period of the Christian reconquest that the Portuguese nation was created. Alfonso VI of Castile obtained French aid in his wars against the Moors, in return for which he made Henry of Burgundy (c.1095) count of Coimbra. Henry's son Alfonso Henriques, after a victory over the Moors in 1139, began to style himself Alfonso I, king of Portugal. Spain recognized Portugal's independence in 1143, and by 1249 the Moors had been driven from Algarve.

The reign of John I (1385-1433) commenced with the decisive defeat of the Castilians and inaugurated the most glorious period of Portuguese history. Portugal entered an era of colonial and maritime expansion. Under the aegis of John's son, Henry the Navigator, Portuguese ships sailed out along the coast of Africa on voyages of exploration. Later Bartholomew Diaz rounded (1488) the Cape of Good Hope and Vasco da Gama opened the trade route to India (1498). The Portuguese Empire soon extended across the world. The reigns of Manuel (1495-1521) and John III (1521-57) marked the climax of Portuguese expansion. Reduced trade profits and the neglect of domestic agriculture and industry led to economic and political decline. In 1580 Philip II of Spain validated his claims to the Portuguese throne (as Philip I) by force of arms, and the "Spanish captivity" began. Independence was regained in 1640 when the Portuguese revolted; John of Braganza was made king as John IV (reigned 1640-56).

Gold from Brazil helped to re-create financial stability by 1730, but it also acted to increase royal absolutism. The powerful marquês de Pombal, chief minister under Joseph (reigned 1750-77), attempted to achieve monarchical centralization and to revitalize agriculture and commerce. Finances again became disorganized as Brazilian treasure dwindled, and most of Pombal's reforms were rescinded. In 1807 the forces of Napoleon I invaded and the royal family fled to Brazil. The French were driven out in 1811, but John VI (reigned 1816-26) returned only after a liberal revolution in 1820. After John's death (1826) his two sons, the absolutist Dom Miguel and the liberal Dom Pedro, struggled for power. Dom Pedro's forces prevailed, and under the reign of his daughter Maria (1826-53), a groundwork was laid for later commercial and legal reforms.

In 1910 a revolution overthrew the monarchy and established a republic. The change of rule did not cure Portugal's chronic economic problems. In World War I Portugal was at first neutral, then joined (1916) the Allies. The economy deteriorated, and in 1926 a military coup overthrew the government. António de Oliviera Salazar, the new finance minister, successfully reorganized the national accounts. He became prime minister in 1932 and dominated the government as dictator until he suffered a stroke in 1968. Portugal was neutral in World War II but allowed the Allies to establish naval and air bases. It became a member of the North Atlantic Treaty Organization in 1949 but was not admitted to the United Nations until 1955. Portugal's possessions Goa, Daman, and Diu were seized by India in 1961. In Africa armed resistance to Portuguese rule developed in Angola, Mozambique, and Portuguese Guinea in the early 1960s.

In 1968 Salazar was replaced by Marcello Caetano as prime minister. Under Caetano repression was somewhat eased. By early 1974 dissatisfaction with the debilitating war in Africa, together with political suppression and a deteriorating economy, resulted in growing unrest. On April 25, a group of army officers toppled the government, encountering a minimum of resistance from loyal forces and widespread acceptance from the people. The ruling military organization, the Armed Forces Movement (MFA), then took over direction of the country. Gen. António de Spínola, an officer who had been critical of the policies of the old regime but who did not take an active part in the coup, was appointed provisional president. Following the coup, the secret police was abolished and full civil liberties were restored. Overtures were made to guerrilla groups in Portugal's African territories for a peaceful transition to independence. Within the Portuguese government, however, conflicts soon developed between rightists and leftists. By the fall Spínola had resigned and a leftist, Gen. Francisco da Costa Gomes, was president. In 1975 the MFA formed a Revolutionary Council with supreme legislative and administrative powers; Gen. Vasco dos Santos Gonçalves, a leftist officer, was appointed premier. Elections for a constituent assembly were held on April 25, 1975, and the results, contrary to expectations on the part of the regime, indicated a strong endorsement of the moderate political parties, particularly the Socialists and the Popular Democrats (Social Democratic party after 1976), who together polled 65% of the vote. The apparent resistance of the government to accede to the outcome led to considerable strife between Socialists and Communists. In Sept., however, Costa Gomes yielded and dismissed Gonçalves. An attempted coup by leftist paratroopers was thwarted in 1975. Parliamentary elections were held in April, 1976, and the results of the 1975 election were essentially duplicated: the moderate parties won nearly two thirds of the vote, while the Communists came in a weak fourth. On June 27, 1976, Gen. António Ramahlo Eanes, a moderate who had helped foil the 1975 coup attempt, was overwhelmingly elected president, and the following month he invited Mário Soares, the Socialist leader, to become premier of Portugal's first elected government since 1926. By 1977 the country seemed to have overcome the worst of the post-coup turbulence and to have chosen a moderate form of democracy.

The republic, including the island groups, is divided into administrative districts, although the names of the six historic provinces are still used. Portuguese overseas territories in Africa—Angola, Cape Verde Islands, Guinea-Bissau, Mozambique, and São Tomé and Principe—all became independent in 1974 and 1975. Timor, in the Western Pacific, was annexed by Indonesia in 1975. Only Macao, in Asia, remains under Portuguese control.

Por·tu·guese In·di·a (pôr'chə-gēz', -gēs', ĭn'dē-ə), the former Portuguese possessions on the Indian subcontinent comprising the colonies of Goa, Daman, and Diu, which were annexed by India in Dec., 1961.

Port Wash·ing·ton (wŏsh'ĭng-tən, wô'shĭng-). **1.** Uninc. town (1970 pop. 15,923), Nassau co., SE N.Y., a suburb of New York City, on the N shore of Long Island and Manhasset Bay. It is a resort and yachting center. **2.** City (1970 pop. 8,752), seat of Ozaukee co., E Wis., on Lake Michigan N of Milwaukee; settled before 1835, inc. 1882. It is a fishing and shipping center. Tractors and prefabricated houses are manufactured.

Por·voo (pôr'vō): *see* Borgå, Finland.

Porz am Rhein (pôrts' äm rīn'), city (1974 est. pop. 83,831), North Rhine-Westphalia, W West Germany, on the Rhine River; chartered 1951. Manufactures of this industrial city include glass and electrical goods. Motor vehicles are assembled here. It was annexed to Cologne in 1975.

Po·sa·das (pə-sä'dəs, pō-sä'*th*äs), city (1970 pop. 97,514), capital of Misiones prov., NE Argentina, a port on the upper Paraná River. Posadas is a center of the maté industry; tobacco and cereals are also grown. The city was settled in 1849. Iguaçu Falls is nearby.

Po·sey (pō'zē), county (1970 pop. 21,740), 414 sq mi (1,072.3 sq km), extreme SW Ind., bounded on the W by the Wabash River and Ill. border, on the S by the Ohio River and Ky. border; formed 1814; co. seat Mount Vernon. In an agricultural area yielding grain, hogs, poultry, and fruit, it has petroleum fields. Clothing, plastics, asphalt, and metal products are manufactured.

Post (pōst), city (1970 pop. 3,854), seat of Garza co., NW Texas, SE of Lubbock; founded 1907, inc. 1916. It has a cotton-textile mill and ships cattle. Oil fields are nearby.

Pos·toj·na (pô'stoi-nä), town (1971 pop. 18,835), NW Yugoslavia, in Slovenia, on the Karst Plateau. A summer resort, it is the site of Europe's largest stalactite caverns, which are traversed by a subterranean river. Formerly Austrian, the town passed to Italy in 1919 and to Yugoslavia in 1947.

Potch·ef·stroom (pŏch'əf-strōōm'), town (1970 pop. 57,443), Transvaal, NE South Africa. Located in a fertile farming region, Potchefstroom is the center of one of the world's richest gold-mining districts. Uranium is also mined. The town has malt factories and timber and metal industries. The oldest town in the Transvaal, Potchefstroom was founded in 1838 and served as its capital until 1860. The fort where British forces were defeated during the Transvaal rebellion of 1880–81 is now a national monument. The town's main growth dates from 1933, when gold was discovered.

Po·teau (pō'tō'), city (1970 pop. 5,500), seat of Le Flore co., E Okla., on Poteau River, SSW of Fort Smith, Ark., in a timber, farm, and coal area; inc. 1898. There are oil and gas wells in the area.

Po·ten·za (pō-tĕn'tsä), city (1976 est. pop. 62,440), capital of Basilicata and of Potenza prov., S Italy, in the Apennines. It is an agricultural, commercial, and industrial center. Manufactures include machinery, chemicals, and processed food. The city was founded in the 2nd cent. B.C. by the Romans. Of note is the Church of San Francesco (begun 1274).

Po·ti (pô'tē), city (1976 est. pop. 54,000), SE European USSR, in Georgia, on the Black Sea at the mouth of the Rion River. It is a port that ships manganese, corn, lumber, and wine. The city was known as Phasis in the 5th cent. B.C., when it was a Greek colony. Later a Turkish fortress, it was taken by the Russians in 1828.

Pot·i·dae·a (pŏt'ĭ-dē'ə), ancient city, NE Greece, at the narrowest point of the Pallene peninsula in Chalcidice (now Khalkidhikí). It was a Corinthian colony (c.600 B.C.) but joined the Athenian-dominated Delian League. Potidaea revolted (432 B.C.) against Athens with Corinthian help. Athens recaptured (430 or 429 B.C.) the city. Philip II of Macedon took (356 B.C.) Potidaea and may have destroyed it in the ensuing war.

Po·to·mac (pə-tō'mək), river, 285 mi (458.6 km) long, formed SE of Cumberland, Md., by the confluence of its North and South branches and flowing generally SE to Chesapeake Bay. It forms part of the boundary between Md. and W.Va. and then separates Va. from both Md. and the District of Columbia. The upper course of the Potomac has cut several gaps across the parallel ridges of the Appalachian Mts.; the largest is at Harpers Ferry, W.Va. The river passes over the Great Falls above Washington, D.C. and enters a tidal estuary below the city. It is navigable for large ships to Washington, D.C. The river is noted for its beauty and historical associations.

Po·to·si (pə-tō'sē), city (1970 pop. 2,761), seat of Washington co., E Mo., SW of St. Louis, in a mineral, farm, and timber region. Lead was discovered here in the 1770s. There are also zinc, iron, and limestone deposits. Potosi is the center of a large barite industry.

Po·to·sí (pō-tō-sē'), city (1976 pop. 77,233), capital of Potosí dept., S Bolivia, at the foot of one of the world's richest ore mountains. In the cold, bleak high Andes at an altitude of c.13,780 ft (4,205 m), Potosí is one of the highest cities in the world. There is no agriculture in the region and scarcely any fuel. Potosí was founded in 1545 and during its first 50 years was a fabulous source of silver. Because of isolation and a series of disasters, such as the flood of 1626, the mines were unable to compete with those of Peru and Mexico. Improved technology and communications, however, have made possible the exploitation of silver, as well as tin, wolfram, and copper, and the revival of commercial life. The city's colonial landmarks include the Mint House, a replica of Spain's Escorial.

Pots·dam (pŏts'dăm'), city (1974 est. pop. 117,236), capital of Potsdam dist., central East Germany, on the Havel River, near Berlin. Manufactures include processed food, textiles, pharmaceuticals, electrical equipment, and locomotives. First mentioned in the late 10th cent. and chartered in the 14th cent., Potsdam was insignificant until Elector Frederick William of Brandenburg made it a residence (1660). The city's chief development came under Frederick II of Prussia (ruled 1740–86), who made Potsdam his chief residence and who built the palace and park of Sans Souci (1745–47) and the New Palace (1763–69). The royal family of Prussia continued to favor Potsdam as a residence, and numerous palaces were added by them. During World War II Potsdam was severely damaged, and in 1945 it was the scene of the Potsdam Conference. In addition to the numerous palaces, the city's notable structures include the Garrison Church (1731–35), where Frederick William I and Frederick II were buried until 1945, when their remains were transferred to Marburg, West Germany.

Potsdam, village (1970 pop. 10,303), St. Lawrence co., N N.Y., E of Ogdensburg, in a dairy and farm region; inc. 1831. Paper is made.

Pot·taw·at·tam·ie (pŏt'ə-wä'tə-mē). **1.** County (1970 pop. 86,991), 963 sq mi (2,494.2 sq km), SW Iowa, bounded in the W by the Missouri River, which here forms the Nebr. border; formed 1847, part of Sarpy co., Nebr., added 1943; co. seat Council Bluffs. In a prairie agricultural area (corn, cattle, and hogs), it has bituminous-coal mines. **2.** County (1970 pop. 11,755), 820 sq mi (2,123.8 sq km), NE Kansas, in a gently rolling to hilly area bounded on the S by the Kansas River, on the W by the Big Blue River; formed 1857; co. seat Westmoreland. It is in a livestock-ranching and grain-farming area. Clothing and construction machinery are made. **3.** County (1970 pop. 43,134), 794 sq mi (2,056.5 sq km), central Okla., intersected by the North Canadian, Canadian, and Little rivers; formed 1907; co. seat Shawnee. In an agricultural area yielding corn, cotton, livestock, poultry, truck crops, pecans, grain, and dairy products, it has oil and natural-gas wells. Its industry includes food processing and the manufacture of clothing, plastics, and machinery.

Pot·ter (pŏt'ər). **1.** County (1970 pop. 16,395), 1,092 sq mi (2,828.3 sq km), N Pa., in an agricultural and forested region that is the source of numerous streams; formed 1804; co. seat Coudersport. Its agriculture includes cattle, dairy products, and potatoes. Paper, leather and rubber products, and cheese are processed. There are oil and natural-gas wells in the area. **2.** County (1970 pop. 4,449), 869 sq mi (2,250.7 sq km), N central S.Dak., bounded in the W by the Missouri River; formed 1875; co. seat Gettysburg. It is in an irrigated agricultural area that has cattle ranching and dairying and produces wheat, corn, and small grains. It has some industry. **3.** County (1970 pop. 90,511), 901 sq mi (2,333.6 sq km), extreme N Texas, in the high plains of the Panhandle, drained by the Canadian River; formed 1876; co. seat Amarillo. This wheat and cattle-ranching region also has a helium field producing much of the world's supply, oil and natural-gas wells and pipelines, deposits of clay, caliche, dolomite, and gypsum, farming (oats, barley, grain sorghums, and forage), dairying, and hog and poultry raising.

Pot·ter·ies, the (pŏt'ə-rēz), district, c.9 mi (15 km) long and 3 mi (4.8 km) wide, Staffordshire, W central England, extending NW to SE in the upper Trent valley. The Potteries has been a center for the manufacture of china and earthenware since the 16th cent.; Josiah Wedgwood, Josiah Spode, and Thomas and Herbert Minton worked here. Most raw materials are now brought in from other districts.

Pot·ter's Bar (pŏt'ərz bär), urban district (1971 pop. 24,583), Hertfordshire, central England, in the Midlands. Most of Potter's Bar is within the Green Belt Zone, which is barred to industry.

Potts·town (pŏts'toun'), borough (1978 est. pop. 26,000), Montgomery co., SE Pa., on the Schuylkill River; settled c.1700, inc. 1815. The borough's manufactures include rubber tires and tubes, fabricated steel, aluminum castings, motor-vehicle parts, and baked goods. The state's first ironworks were established here in 1715.

Potts·ville (pŏts'vĭl'), industrial city (1978 est. pop. 18,400), seat of

Schuylkill co., E Pa., on the Schuylkill River; inc. 1847. Once a coal-mining center, it now manufactures varied products, including textiles, extrusions, trailers, and plastics. Pottsville was a rallying place for the Molly Maguires, who were tried here in 1877.

Pough·keep·sie (pə-kĭp'sē), city (1978 est. pop. 31,000), seat of Dutchess co., SE N.Y., on the Hudson River; settled 1687 by the Dutch, inc. as a city 1854. It is a trade center with a great variety of industries, including printing, electronics research, and the manufacture of business and milking machines, precision instruments, elevators, and hardware items. It became the temporary state capital in 1777. Poughkeepsie is the seat of Vassar College. An annual intercollegiate regatta is held on the river in June.

Pour·ri, Mont (môɴ po͝o-rē'), Alpine peak, 12,428 ft (3,790.5 m) high, Savoie dept., SE France, near the Italian border.

Pow·der (pou'dər). **1.** River, 150 mi (241.4 km) long, NE Oregon, rising S of La Grande in Elkhorn Ridge and flowing N and SE to the Snake River on the Idaho border. **2.** River, c.486 mi (780 km) long, rising in several branches in the S foothills of the Bighorn Mts., central Wyo., and flowing generally NE into Mont. to join the Yellowstone River near Terry.

Powder River, county (1970 pop. 2,862), 3,285 sq mi (8,508.2 sq km), SE Mont., in an agricultural area bordering on Wyo. and drained by the Powder and Little Powder rivers; formed 1919; co. seat Broadus. Livestock are raised. Part of Custer National Forest is in the west.

Pow·ell (pou'əl). **1.** County (1970 pop. 7,704), 173 sq mi (448.1 sq km), E Ky., drained by the Red River and several creeks; formed 1852; co. seat Stanton. In a hilly agricultural area yielding livestock, corn, hay, burley tobacco, and dairy products, it has oil wells, bituminous-coal mines, limestone quarries, and timber. It includes part of Cumberland National Forest. **2.** County (1970 pop. 6,611), 2,337 sq mi (6,052.8 sq km), W central Mont., in an agricultural and mining area drained by the Blackfoot River; formed 1901; co. seat Deer Lodge. Livestock, lead, and silver are its major products.

Powell Lake, manmade lake, 252 sq mi (652.7 sq km), S Utah, in Glen Canyon National Recreation Area. Its altitude is 3,700 ft (1,128.5 m).

Pow·er (pou'ər), county (1970 pop. 4,864), 1,411 sq mi (3,654.5 sq km), SE Idaho, in an irrigated agricultural and stock-raising region, drained by the Snake River; formed 1913; co. seat American Falls. Wheat, rye, potatoes, and sugar beets are grown. Part of Fort Hill Indian Reservation is in the mountainous region in the east.

Pow·e·shiek (pou'ə-shēk'), county (1970 pop. 18,803), 589 sq mi (1,525.5 sq km), central Iowa, in a rolling prairie agricultural area drained by forks of the English River; formed 1843; co. seat Montezuma. Farming yields hogs, cattle, poultry, corn, oats, and hay. Bituminous-coal deposits are in the south and west.

Pow·ha·tan (pou'ə-tăn'), county (1970 pop. 7,696), 269 sq mi (696.7 sq km), E central Va., bounded in the N by the James River and in the S by the Appomattox River; formed 1777; co. seat Powhatan. It has agriculture (tobacco, corn, and livestock) and lumbering.

Powhatan, village (1970 pop. 300), seat of Powhatan co., central Va., W of Richmond. It is a trade center in an agricultural area.

Pow·nal (pou'nəl), town (1970 pop. 2,441), Bennington co., extreme SW Vt., NW of Williamstown, Mass.; settled c.1760. Dutch squatters came to this area in the 1720s, but there was no permanent settlement until after 1760. Pownal was the scene of strife between holders of N.Y. and N.H. grants to the same land.

Pow·ys (pō'ĭs), nonmetropolitan county (1976 est. pop. 101,500), central Wales, created under the Local Government Act of 1972 (effective 1974); administrative center Llandrindod Wells. It comprises the former counties of Montgomeryshire and Radnorshire and portions of the former county of Breconshire.

P'o-yang or **Po·yang** (both: pō'yäng'), shallow lake, c.1,000 sq mi (2,590 sq km), N Kiangsi prov., SE China. One of China's largest lakes, it serves as a natural overflow reservoir for the Yangtze River, with which it is connected by canal. The lake basin is a rich rice-producing region.

Poz·nań (pôz'nän'yə), city (1975 est. pop. 516,000), W central Poland, a port on the Warta River. It is an important industrial and railway center. Manufactures include machinery (chiefly engines, freight cars, and machine tools), metals, and chemicals. Founded before the advent of Christianity in Poland, it became (968) the first Polish episcopal see and a nucleus of the Polish state. It remained in Poland until the second partition (1793), when it passed to Prussia. Poznań was included in the grand duchy of Warsaw in 1807, again passed to Prussia in 1815, and reverted to Poland in 1919. In World War II it was annexed to Germany, and thousands of Poles were expelled. The city has many old churches and museums with important art objects. Its most notable buildings are a Gothic cathedral (badly damaged in World War II) and a 16th cent. city hall.

Poz·zuo·li (pŏt-swō'lē), city (1976 est. pop. 58,800), Campania, S Italy, on the Bay of Naples. It is a port and an industrial and tourist center. Manufactures include iron and steel, machinery, and textiles. Pozzuoli was founded (6th cent. B.C.) by Greek exiles and was later a wealthy Roman seaport. Among the Roman remains are a large amphitheater and the ruins of the temple of Serapis (now partly under water), which once was a marketplace.

Prague (präg, prāg), city (1970 pop. 1,078,096), capital and largest city of Czechoslovakia, on both banks of the Vltava River. Prague is a leading European commercial and industrial center and is Czechoslovakia's most important industrial city. There are large engineering plants, machine-building and machine-tool enterprises, printing and publishing houses, and factories producing automobiles, rolling stock, airplanes, iron and steel, construction materials, chemicals, and a wide variety of consumer goods. Prague is also the see of a Roman Catholic archbishop, an Eastern Orthodox archbishop, and the archbishop of the Czechoslovak church.

The old section of Prague, which occupies the center of the city, is an architectural treasure enhanced by the beauty of its location on the hilly banks of the Vltava River. Hradčany Castle (14th cent.), the seat of the Czechoslovak presidents and the former royal residence, dominates the city. Next to it stands the largely Gothic Cathedral of St. Vitus, first built in the 10th cent.; it contains the tombs of St. Wenceslaus, St. John of Nepomuk, and many kings and emperors. The Hradčany quarter also contains many other fine churches and palaces. On the slope extending from the castle to the Vltava is the quaint Mala Strana (lesser town) quarter, the best-preserved part of old Prague. Mala Strana is connected with the Old Town, on the eastern bank of the river, by the 14th cent. Charles Bridge, the most beautiful of Prague's 13 bridges. The Old Town contains the Carolinum, the oldest part of Charles Univ. (1348); the adjacent Stavovske Theater; the vast Clementinum Library; the Gothic Old Town Hall (13th cent.; burned in May, 1945); the ancient clock of the seasons; the Gothic Tyn Cathedral (14th cent.); and the Powder Tower (15th cent., last of the city gates). Situated in the adjacent former Jewish quarter is the Old Synagogue (13th cent.). In the heart of modern Prague is Wenceslaus Square, with its statue of St. Wenceslaus. Other educational and cultural facilities in the city include a technical university (1707); the Czechoslovak Academy of Sciences; the National Gallery; and the National Museum.

History. The earliest settlements date from at least the 9th cent. Already an important trading center by the 10th cent., Prague achieved real prominence after King Wenceslaus I of Bohemia established (1232) a German settlement here. It grew rapidly in size and prosperity as Bohemia's capital and became under Emperor Charles IV (14th cent.) one of the most splendid cities of Europe. From the 14th to the early 17th cent. the emperors of the Holy Roman Empire resided at Prague as well as at Vienna. Hapsburg rule began in 1526. When in 1618 the Protestant Czech nobles felt the liberties of Bohemia threatened by Emperor Matthias, they vented their dissatisfaction by throwing two royal councilors and the secretary of the royal council of Bohemia out of the windows of Hradčany Castle (May 23, 1618). Although none of the victims of the so-called Defenestration of Prague were hurt, the event opened the Thirty Years' War. The Battle of the White Mountain (1620), fought near Prague, resulted in Bohemia's subjugation to Austrian rule. In the War of the Austrian Succession Prague was occupied by the French (1742) and the Prussians (1744); and in the Seven Years' War it was (1757) the scene of a major victory of Frederick II of Prussia. The center of the Czech national revival in the 19th cent., Prague played an important part in the Revolution of 1848 until its capture by the Austrians. In 1918 Prague became the capital of the newly created Czechoslovak republic. Occupied (1939–45) by the Germans, the city was liberated by Soviet troops after a Czech rebellion against the Germans.

Prai·rie (prâr'ē). **1.** County (1970 pop. 10,249), 661 sq mi (1,712 sq km), E central Ark., drained by the White River and Bayou des Arc; formed 1846; co. seats Des Arc and De Valls Bluff. In an agricultural area yielding cotton, rice, soybeans, and cattle, it has lumber, commercial fisheries, and clothing factories. **2.** County (1970 pop. 1,752), 1,727 sq mi (4,472.9 sq km), E Mont., in an agricultural area drained by the Yellowstone River; formed 1915; co. seat Terry. Livestock are raised.

Prairie du Chien (də shēn'), city (1970 pop. 5,540), seat of Crawford co., SW Wis., on the Mississippi River just above the mouth of the Wisconsin River and NW of Dubuque, Iowa; settled 1781, inc. 1872. The site of the city, at the western end of the Fox-Wisconsin water route from the Great Lakes to the Mississippi, was successively occupied by the French (1686), the British (c.1815), and the Americans (1814–16). In the early 19th cent. it was a center of fur trading. It later became a river port. Today it is a farm trade center. Aluminum products and automobile parts are manufactured.

prai·ries (prâr'ēz), generally level, originally grass-covered and treeless plains. Many of the prairies of the world were formerly used for grazing purposes, but more and more are now coming under cultivation; hence they are often referred to today as the "vanishing grasslands." Because they have the favorable climate and soil fertility characteristic of prairies, the wheat belts in the United States, the Ukraine, and the Pampas of Argentina are among the world's most productive agricultural regions.

Pra·tas Island (prä'täs), in the South China Sea, administered by Kwangtung prov., SE China. The island has guano deposits. Occupied by the Japanese in 1907–9 and again in 1939–45, Pratas Island passed to the Chinese government in 1950.

Pra·to (prä'tō), city (1976 est. pop. 154,362), Tuscany, central Italy. It is a major textile-making center, known for its wool industry since the 13th cent. Weaving machinery and leather goods are also manufactured. Prato was an Etruscan settlement. It came under Florence in the 14th cent. Among the city's noteworthy structures are the cathedral (12th–15th cent.), which has frescoes by Filippo Lippi and works by Donatello, Giovanni Pisano, and Andrea della Robbia, and a 13th cent. town hall and fortress.

Pratt (prăt), county (1970 pop. 10,056), 729 sq mi (1,888.1 sq km), S central Kansas, in a gently rolling plain drained by the South Fork of the Ninnescah River; formed 1867; co. seat Pratt. In a wheat and cattle area, it has oil and natural-gas fields and some manufacturing (including farm machinery).

Pratt, city (1970 pop. 6,736), seat of Pratt co., S Kansas, W of Wichita; inc. 1884. It is a shipping point for a wheat and cattle area.

Pratt·ville (prăt'vĭl'), city (1970 pop. 13,116), seat of Autauga co., central Ala.; inc. 1872. It has textile and related industries and a paper mill. Cotton gins have been manufactured here since 1838.

Prav·dinsk (präv'dyĭnsk), formerly **Fried·land** (frēt'länt'), town, NW European USSR, formerly in East Prussia. In 1807 the city was the scene of a battle in which Napoleon I defeated the Russians, thus precipitating the Treaty of Tilsit.

Pre·ble (prĕb'əl), county (1970 pop. 34,719), 427 sq mi (1,105.9 sq km), SW Ohio, bounded in the W by the Ind. border, and drained by Twin, Seven Mile, and Four Mile creeks and by the East Fork of the White River; formed 1808; co. seat Eaton. In an agricultural area that produces tobacco, grain, fruit, and livestock, it has limestone quarries, lumbering, and nurseries. Its industry includes some food processing and the manufacture of textiles, wood products, paper, chemicals, plastics, metal products, and machinery.

Pren·tiss (prĕn'təs), county (1970 pop. 20,133), 418 sq mi (1,082.6 sq km), NE Miss., drained by the East Fork of the Tombigbee River; formed 1870; co. seat Booneville. Agriculture (cotton, corn, and hay) and lumbering are major industries. Clay deposits are also found.

Prentiss, city (1970 pop. 1,789), seat of Jefferson Davis co., S Miss., NW of Hattiesburg, in a farm, timber, and livestock area.

Pře·rov (pər-zhĕr'ôf), city (1975 est. pop. 44,275), central Czechoslovakia, in Moravia. A railway center, Přerov also has iron and machinery works and manufactures optical and precision instruments and electrical equipment.

Pres·cott (prĕs'kət). **1.** City (1970 pop. 13,283), alt. 5,389 ft (1,643.6 m), seat of Yavapai co., central Ariz.; inc. 1883. It is a mining and ranching center, as well as a health and summer resort. Lumber and wood products, motors, molds and dies, electronic products, and clothing are manufactured. Gold was discovered in 1863, and Prescott was built in 1864 near Fort Whipple. It was twice territorial capital (1864–67, 1877–89). Annual events are a Frontier Days rodeo and a Smoki Indian ceremonial, featuring historic rituals and dances. **2.** City (1970 pop. 3,921), seat of Nevada co., SW Ark., NE of Hope, in a cotton, corn, and timber area; settled 1873. It is a market center for a peach-growing region.

Pres·i·den·tial Range (prĕz'ĭ-dĕn'shəl), group of the White Mts., N N.H., so called from the names of its peaks. Mt. Washington (6,288 ft/1,917.8 m) is the highest peak; a hotel and meteorological station are at the summit. A year-round resort center, it was developed for tourists in the mid-1800s.

Pre·sid·i·o (prə-sĭd'ē-ō), county (1970 pop. 4,842), 3,892 sq mi (10,008.3 sq km), extreme W Texas, in the Big Bend, with the Rio Grande (here forming the Mexican border) on the W and S; formed 1850; co. seat Marfa. It has large-scale ranching (cattle, sheep, and goats), some irrigated agriculture in the Rio Grande valley (cotton, grain, and truck crops), and silver mines.

Pres·pa, Lake (prĕs'pä), 112 sq mi (290 sq km), SW Yugoslavia, NW Greece, and E Albania. It is the highest lake (alt. 2,798 ft/853.4 m) of the Balkans.

Presque Isle (prĕsk' īl'), county (1970 pop. 12,836), 648 sq mi (1,678.3 sq km), NE Mich., bounded on the NE by Lake Huron, drained by the Black and Rainy rivers and by the North Branch of the Thunder Bay River; formed 1875; co. seat Rogers City. It is in a dairying and agricultural area yielding livestock, potatoes, and fruit. There are limestone quarries, fisheries, and resorts, as well as fishing in the Grand, Long, and Black lakes. Some manufacturing is done.

Presque Isle, city (1978 est. pop. 12,000), Aroostook co., NE Maine, inc. 1859. It is the trade, tourist, and shipping center of the Aroostook valley. During World War II an important air base served as a ferry point for planes to Britain.

Pres·ton (prĕs'tən), county borough (1976 est. pop. 131,200), Lancashire, N England, on the Ribble River. It has an active port. Preston is a center of cotton and rayon manufacturing; aircraft, motor vehicles, industrial machinery, and electrical appliances are also produced. A guild-merchant festival has been held in Preston every 20 years for more than four centuries. Preston was the scene of a Cromwellian victory in 1648 and of the surrender of the Jacobites after the rising of 1715.

Preston, county (1970 pop. 25,455), 645 sq mi (1,670.6 sq km), N W.Va., on the Allegheny Plateau, bordered in the N by Pa., in the E by Md., and drained by the Cheat River; formed 1818; co. seat Kingwood. There is bituminous-coal mining, lumbering, and limestone quarrying in the county. Its agriculture includes dairy products, poultry, and truck crops.

Preston. 1. Town (1970 pop. 226), seat of Webster co., W Ga., on the Kinchafoonee River W of Americus. Sawmilling is done. **2.** City (1970 pop. 3,310), seat of Franklin co., SE Idaho, near the Bear River SE of Malad City; settled 1866 by Mormons, inc. as a city 1913. It is a processing center of an irrigated farm and livestock region. **3.** Village (1970 pop. 1,413), seat of Fillmore co., SE Minn., on a branch of the Root River near the Iowa border SE of Rochester, in a farm and dairy region.

Pres·tons·burg (prĕs'tənz-bûrg'), town (1970 pop. 3,422), seat of Floyd co., E Ky., on the Levisa Fork W of Williamsson, in a truck-farming region yielding coal, oil, and natural gas; settled 1791.

Prest·wich (prĕst'wĭch'), municipal borough (1971 pop. 32,838), Greater Manchester, N England. It is a residential suburb and a cotton- and rayon-manufacturing center.

Prest·wick (prĕst'wĭk'), burgh (1974 est. pop. 13,138), Strathclyde region, SW Scotland, on the Firth of Clyde. It is a seaside resort. There are aircraft manufacturing, coal mining, and other industries.

Pre·to·ri·a (prĭ-tôr'ē-ə), city (1970 pop. 545,450), administrative capital of the Republic of South Africa and capital of its Transvaal prov. Although it is primarily an administrative center, there are important industries, especially iron and steel. The city has automobile assembly plants, railroad and machine shops, and flour mills. Founded in 1855, the city was named for Andries Pretorius, a Boer leader. Pretoria became the capital of the South African Republic (the Transvaal) in 1860. The Peace of Vereeninging, which ended the Boer War, was signed in Pretoria. When the Union of South Africa was founded in 1910, Pretoria became its administrative capital and Cape Town its parliamentary capital. Pretoria is the seat of the Univ. of South Africa. The Transvaal Museum, the National Historical Cultural Museum, and the National Zoological Gardens are also in the city.

Prib·i·lof Islands (prĭb'ə-lôf'), group of four volcanic islands, off SW Alaska in the Bering Sea, c.230 mi (370 km) N of the Aleutian Islands; discovered and named in 1786 by Gerasim Pribilof, a Russian navigator. The larger islands, St. Paul and St. George, are famous as the breeding place of the Alaska fur seal. The islands, part of the 1867 U.S. purchase of Alaska, became a seal reservation in 1868; they are presently administered by the U.S. Bureau of Fisheries. Prior to 1911 competition and ruthless hunting methods threatened extinction of the seals. Blue and white foxes are native to the islands. The Aleuts, brought to the islands in the late 1700s by the Russians, make a living by processing the seal and fox furs.

Pří·bram (pər-zhĭb'răm'), town (1975 est. pop. 32,345), W Czechoslovakia, in Bohemia. It is a center of gold and silver mining, with mine shafts more than 3,000 ft (915 m) deep.

Price (prīs), county (1970 pop. 14,520), 1,260 sq mi (3,263.4 sq km), N Wis., drained by the Flambeau and Jump rivers; formed 1879; co. seat Phillips. In a dairying area, with some potato farming, it has lumbering and paper mills and manufactures plastics and construction machinery. It contains a section of Chequamegon National Forest and several resort lakes.

Price, city (1970 pop. 6,218), seat of Carbon co., E central Utah, on the Price River SE of Provo; settled 1879, inc. 1911. It is a trade center in a coal, farm, and livestock area.

Prich·ard (prĭch'ərd), city (1978 est. pop. 39,800), Mobile co., SW Ala., an industrial suburb of Mobile; settled 1900, inc. 1925. Meat and seafood are packed, cotton is processed, and chemicals, fertilizer, naval stores, lumber, and paper products are manufactured.

Pri·e·ne (prī-ē'nē), ancient Ionian city of W Asia Minor, near the mouth of the Maeander (now Menderes) River. It was rebuilt in the 4th cent. B.C. and was the site of a temple of Athena Polias. It is an extremely well-preserved Greek city of this period.

Pri·lep (prē'lĕp'), city (1971 pop. 48,242), S Yugoslavia, in Macedonia. It is the trade center of an agricultural region; manufactures include tobacco, textiles, and leather products. Prilep has several medieval churches and monasteries.

prime me·rid·i·an (prīm' mə-rĭd'ē-ən), meridian that is designated zero degree (0°) longitude, from which all other longitudes are measured. By international convention, it passes through the original site of the Royal Observatory in Greenwich, England; for this reason, it is sometimes called the Greenwich meridian. Greenwich Mean Time, the standard basis for determining time throughout the world, is civil time measured at the prime meridian.

Prim·ghar (prĭm'gär'), town (1974 est. pop. 1,016), seat of O'Brien co., NW Iowa, NE of Sioux City, in a livestock and grain area.

Pri·mor·sky Kray (prē-môr'skē krī), administrative division (1970 pop. 1,721,000), c.64,900 sq mi (168,090 sq km), Far Eastern USSR, between Manchuria in the W and the Sea of Japan in the E; capital Vladivostok. The territory's coastal mountain range contains coal, iron ore, lead, zinc, lignite, tin, and silver. Fisheries (salmon and sardines) are located along the shore. An agricultural plain with millet and rice crops extends along the Manchurian border.

Prince Al·bert (prĭns ăl'bərt), city (1971 pop. 28,464), central Sask., Canada, on the North Saskatchewan River. It is a commercial and distribution center for a lumbering, fur-trapping, and farming area. There are wood-products and meat-packing industries. It was founded in 1866 as a Presbyterian mission to the Cree Indians.

Prince Charles (chärlz), island, 3,639 sq mi (9,425 sq km), part of the Canadian Arctic Archipelago, E Franklin dist., Northwest Territories, Canada. It was discovered in 1948.

Prince Ed·ward (ĕd'wərd), county (1970 pop. 14,379), 357 sq mi (924.6 sq km), S central Va., bounded in the N by the Appomattox River and drained by the Nottoway River; formed 1752; co. seat Farmville. It is in a rolling agricultural area that produces tobacco, grain, hay, cattle, and dairy products. Its manufactures include clothing, lumbering, wood products, shoes, electrical components, and golf balls.

Prince Edward Island, province (1975 pop. 119,000), 2,184 sq mi (5,657 sq km), E Canada, off N.B. and N.S. The capital is Charlottetown. One of the Maritime Provinces, it is separated on the south from N.S. and N.B. by the Northumberland Strait. The generally low, level land is c.140 mi (225 km) long and 5 to 35 mi (8-56 km) wide. White sandy beaches line the deeply indented north shore, and much of this favorite resort spot is now Prince Edward Island National Park (est. 1937). Low, red sandstone cliffs rim the south shores. The tide reaches back into the headwaters of the island's short rivers. Since earliest settlement, fishing has been an important industry, yielding an abundance of lobsters, oysters, codfish, halibut, mackerel, and herring. Livestock, fruit, and vegetables are produced, and potatoes are exported. Because of the lack of raw materials and cheap sources of power, manufacturing is largely limited to food processing.

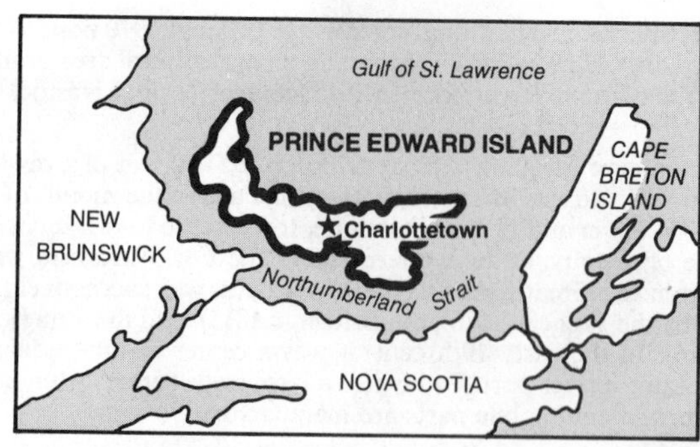

Micmac Indians lived on the island before white men arrived. It was discovered by Jacques Cartier in 1534. Samuel de Champlain named it Île St. Jean in 1603, and it was known by that name, or Isle St. John, until 1799, when it was renamed after Edward, duke of Kent. The first permanent settlement was made by the French in 1719 near present-day Charlottetown. The British gained permanent control under the Treaty of Paris in 1763, Prince Edward Island was annexed to N.S., but it became a separate colony in 1769. Responsible, or cabinet, government was granted in 1851. Prince Edward Island did not join the Canadian confederation until 1873.

Prince Fred·er·ick (frĕd'rĭk, frĕd'ə-rĭk), town (1970 pop. 400), seat of Calvert co., S Md., on Chesapeake Bay, SSW of Annapolis, in an agricultural and summer-resort area.

Prince George (jôrj), city (1976 pop. 59,929), central British Columbia, Canada, at the confluence of the Fraser and Nechako rivers. It is a distributing center for a lumber region. There are sawmills, pulp mills, chemical plants, and an oil refinery. In 1807 Simon Fraser of the North West Company established here the fur-trading post of Fort George, which was taken over (1821) by the Hudson's Bay Company. Settlement began c.1910 with the building of a railroad via Fort George to Prince Rupert.

Prince George, county (1970 pop. 29,092), 276 sq mi (714.8 sq km), SE Va., in a tidewater region bounded in the NW by the Appomattox River, in the N by the James River, and drained by the Blackwater and Nottoway rivers; formed 1700; co. seat Prince George. Its agriculture includes peanuts, truck crops, some tobacco, cotton, grain, poultry, livestock, and dairy products.

Prince George, village (1970 pop. 100), seat of Prince George co. (since 1785), E Va., E of Petersburg. Merchant's Hope Church (built in the 1650s) is nearby.

Prince Georg·es (jôr'jəz), county (1970 pop. 661,082), 485 sq mi (1,256.2 sq km), S central Md., bounded on the W by the Potomac River and the Va. line, on the E and NE by the Patuxent River, and drained by the Anacostia River; formed 1695; co. seat Upper Marlboro. Its agricultural areas produce tobacco, truck crops, dairy products, corn, and wheat. It has sand and gravel pits, and its industry includes food processing, lumber millwork, and the manufacture of furniture, chemicals, plastics, clay products, concrete, and machinery.

Prince of Wales, Cape (wālz), at the tip of the Seward Peninsula, NW Alaska, on the Bering Strait. It is the westernmost point of North America.

Prince of Wales Island, c.12,800 sq mi (33,150 sq km), S central Franklin dist., Northwest Territories, Canada, between Victoria and Somerset islands. The tundra-covered island has an irregular coastline and is deeply indented by Ommanney and Browne bays.

Prince of Wales Island, 2,231 sq mi (5,778.3 sq km), off SE Alaska, in the Alexander Archipelago. The island is heavily forested. Lumbering, fishing, and canning are the main industries.

Prince Ru·pert (rōō'pərt), city (1976 pop. 14,754), W British Columbia, Canada, on Kaien Island, in Chatham Sound near the mouth of the Skeena River, S of the Alaska border. A railroad and highway terminus and an ice-free port, it serves the mining, lumber, and agricultural areas of central and western British Columbia. It is a major fish-processing center, and there are wood-processing plants.

Princ·es Islands (prĭn′səz), Turkey: *see* Kizil Adalar.

Prin·cess Anne (prĭn′səs ăn′), town (1970 pop. 975), seat of Somerset co., SE Md., Eastern Shore, on Manokin River, SW of Salisbury, in an agricultural region; founded 1733, inc. 1867. It is a shipping center for a truck-farming area. Clothing factories and lumber and flour mills are here.

Prince·ton (prĭns′tən). **1.** City (1970 pop. 6,959), seat of Bureau co., N Ill., near the Illinois River WNW of La Salle, in a nursery and farm area; laid out 1833, inc. 1849. It has some manufacturing. **2.** City (1970 pop. 7,431), seat of Gibson co., SW Ind., N of Evansville, in a farm area yielding coal and oil; settled c.1812, inc. as a town 1818, as a city 1884. **3.** City (1970 pop. 6,292), seat of Caldwell co., W Ky., E of Paducah. A railroad center, the city ships tobacco, farm produce, livestock, hosiery, and fluorspar. **4.** City (1970 pop. 1,328), seat of Mercer co., N Mo., near the Iowa border NW of Kirksville, in a farm region; settled c.1840, platted 1846, inc. 1853. **5.** Borough (1978 est. pop. 13,600), Mercer co., W N.J.; settled late 1600s, inc. 1813. It is the seat of Princeton Univ., the Institute for Advanced Study, and numerous corporate research centers. Shortly after the Battle of Trenton, Princeton was the scene of a battle (Jan. 3, 1777) in which George Washington surprised and defeated a superior British force. In 1869 the first intercollegiate football game (between Rutgers and Princeton) took place here. **6.** City (1970 pop. 7,253), seat of Mercer co., S W.Va., NE of Bluefield, in a coal-mine and farm area; settled 1826, inc. 1837. A railroad center, it makes textiles, soap, missile components, and concrete products. It was the scene of engagements in the Civil War (May 16, 1862) and was burned by retreating Confederates.

Prince Wil·liam (wĭl′yəm), county (1970 pop. 111,102), 347 sq mi (898.7 sq km), NE Va., bounded in the E by the Potomac, and in the NE and N by Occoquan Creek and the Bull Run; formed 1727; co. seat Manassas. Its agriculture includes grain, tobacco, fruit, truck crops, livestock, poultry, and dairy products. It has granite quarries and heavy industries (concrete, metal products, machinery, and electrical and transportation equipment).

Prince William Forest Park: *see* National Parks and Monuments Table.

Prince William Sound, large, irregular, islanded inlet of the Gulf of Alaska, S Alaska. There are good harbors, and access to the interior is by highway and railroad. Extensive fishing and some mining are carried on in the area.

Prine·ville (prīn′vĭl′), city (1970 pop. 4,101), seat of Crook co., central Oregon, NE of Bend, in a lumbering and stock-raising region; settled 1868, laid out 1870, inc. 1880. The Crooked River project irrigates wheat and potato farms in the vicinity.

Pri·pyat (prĭp′yət) or **Pri·pet** (prĭp′ĕt), river, c.440 mi (710 km) long, rising NW of Kovel, NW Ukraine, W USSR, near the Polish border, and flowing generally E through the Pripyat Marshes, S Belorussia, into the Dnepr River in NE Ukraine. Navigable below Pinsk, it is connected by canals with the Western Bug River and the Neman River. The Pripyat Marshes are a forested, swampy area (c.38,000 sq mi/98,420 sq km) extending along the Pripyat River and its tributaries from Brest in the west to Mogilev in the northeast and Kiev in the southeast. With a dense network of rivers, lakes, and canals, the marshes are largely coextensive with the Polesye lowland. Drainage of the swamps was begun c.1870; the eastern part is now used for pasturage and cultivation.

Priš·ti·na or **Prish·ti·na** (both: prĭsh′tĭ-nä′), city (1971 pop. 69,524), S Yugoslavia, in Serbia. Priština is a commercial center where jewelry and textile products are made. It was a capital of the Serbian empire in the 14th cent.

Priz·ren (prēz′rĕn), city (1971 pop. 41,661), S Yugoslavia, in the Kossovo-Metohija region of Serbia. Its industries produce filigree silver jewelry, carpets, and embroideries. An important medieval trade center, the city reached its height as capital (1376–89) of Serbia.

Pro·ko·pyevsk (prə-kôp′yəfsk), city (1976 est. pop. 267,000), E Siberian USSR. A major coal producer of the Kuznetsk basin, it also manufactures mining machinery, chemicals, and food products. It was founded after the Russian Revolution of 1917.

Prome (prōm): *see* Pye, Burma.

Pros·ser (prŏs′ər), city (1977 est. pop. 3,335), seat of Benton co., S Wash., W of Richland at the falls of the Yakima River; settled 1880, inc. 1889. It is a marketing and processing center for irrigated farms served by the Yakima project.

Pro·stě·jov (prô′styĕ-yôf), city (1975 est. pop. 45,249), central Czechoslovakia, in Moravia. A trade center of the fertile Haná agricultural region, Prostějov has breweries, distilleries, and industries manufacturing farm machinery, textiles, electrical equipment, clothing, and iron goods. A 16th cent. castle dominates the city skyline.

Pro·vence (prô-väɴs′), region and former province, SE France. The fertile valley of the Rhône and the French Riviera produce fruits and vegetables (citrus fruits, olive oil, mulberry trees). Cattle are raised in the Camargue. The startling scenery has inspired such painters as Paul Cézanne. The coastal strip was settled c.600 B.C. by Greeks and later by Phoenician merchants. In the 2nd cent. B.C. the Romans established colonies. A part of Narbonensis, Provence was the oldest of the Roman possessions beyond the Alps. It was invaded by the Visigoths (5th cent.), the Franks (6th cent.), and the Arabs (8th cent.). But Roman institutions continued to have a profound cultural influence. In 879 the count of Arles established the kingdom of Cisjurane Burgundy, or Provence, which in 933 was united with Transjurane Burgundy to form the Kingdom of Arles. The major part of Provence, held by the house of Aragón, passed (1246) to the Angevin dynasty of Naples through marriage, and under the Angevins the towns became virtually independent republics. King René left Provence to his nephew, Charles of Maine, who left it to the French crown (1486).

Prov·i·dence (prŏv′ĭ-dəns), county (1970 pop. 581,470), 416 sq mi (1,077.4 sq km), N R.I., on the Conn. and Mass. borders, drained by the Blackstone and numerous other rivers; formed 1703 as Providence Plantations co., renamed 1729; co. seat Providence. It has agriculture (dairy products, poultry, and truck crops), granite quarries, sand and gravel pits, and fisheries. Its manufactures include textiles, furniture, paper, chemicals, plastics, leather, concrete, metal products, and machinery.

Providence, city (1978 est. pop. 158,600), state capital and seat of Providence co., NE R.I., a port at the head of Providence Bay; founded by Roger Williams 1636, inc. as a city 1832. It is a port of entry and a major trading center. The bay forms an excellent harbor from which oil and coal are shipped. Providence is widely known as a silverware- and jewelry-manufacturing center. Textiles, machinery, metal products, electronic equipment, rubber goods, and machine tools are also made, and there are printing and publishing enterprises. Roger Williams chose this site in 1636 after he was exiled from the colony of Massachusetts. He secured title to the land from Narragansett Indian chiefs and named the place in gratitude for "God's merciful providence." The settlement grew as a refuge for religious dissenters. Prosperity came in the 18th cent. with foreign commerce, especially trade in slaves and rum, and after the Revolution industrial development was rapid. The Brown brothers, John, Nicholas, and Moses, played leading roles in the growth of the town, prospering in foreign trade and fostering the textile and other industries. The city became sole capital of the state in 1900 (Newport had been joint capital until then). Providence is the seat of the noted Rhode Island School of Design, Brown Univ., Providence College, and other schools. It has several fine libraries, including the John Carter Brown Library of Brown Univ. and the Atheneum (1753), one of the oldest libraries in the United States. Among the city's many historic structures are the old statehouse (where the general assembly met 1762–1900; now a courthouse), the old market building (1773), and the First Baptist Meetinghouse (1775; the congregation was organized in 1638). On Prospect Terrace is Leo Friedlander's heroic statue of Roger Williams (1939). Another memorial to the founder is in Roger Williams Park, which contains a museum of natural history and a natural amphitheater.

Prov·ince·town (prŏv′ĭns-toun′), resort town (1970 pop. 2,911), Barnstable co., SE Mass., on the tip of Cape Cod; inc. 1727. The principal industries are tourism and fishing. The Pilgrims landed here in 1620 and stayed about a month before moving on to Plymouth. Permanent settlement was not made until c.1700. In the 20th cent. the town gained fame as a resort favored by artists.

Pro·vins (prô-văɴ′), town (1968 pop. 11,869), Seine-et-Marne dept., N central France. It is a tourist and commercial center. Built by the Romans on a rocky height, it was (11th–13th cent.) a prosperous trade hub and the site of one of the great fairs of Champagne. The picturesque upper town has preserved its ramparts that date from the 12th to the 14th cent.

Pro·vo (prō'vō'), city (1978 est. pop. 55,200), seat of Utah co., N central Utah, on the Provo River near Utah Lake; inc. 1851. It is a distribution, processing, and manufacturing center in an extensive mining and irrigated farm and fruit area. A major source of employment is a large steel mill nearby. Provo was settled by Mormons in 1849 and successfully defended against Indians in a war from 1865 to 1868. Railroad connections from Salt Lake City (1873) and Scofield (1878) made it a shipping point for the region's mines. Brigham Young Univ. is here.

Prow·ers (prou'ərz), county (1970 pop. 13,258), 1,626 sq mi (4,211.3 sq km), SE Colo., bordering on Kansas and drained by the Arkansas River; formed 1889; co. seat Lamar. It is in an irrigated agricultural area that produces alfalfa, sugar beets, corn, wheat, broomcorn, and sorghums, and cattle. Farm machinery is manufactured.

Prud·hoe Bay (prōō'dō'), inlet of the Arctic Ocean, N Alaska, in the Alaska North Slope region, E of the Colville River delta. In 1968 one of the largest oil reserves in North America was discovered here. Harsh climate and terrain make the extraction and shipment of the oil difficult.

Prus·sia (prŭsh'ə), former state, the largest and most important of the German states. Berlin was the capital. The chief member of the German Empire (1871–1918) and the Weimar Republic (1919–33), Prussia occupied more than half of all Germany and the major part of northern Germany. It consisted of 13 provinces: Berlin, Brandenburg, East Prussia (separated after 1919 from the rest of Prussia by the Polish Corridor), Hanover, Hesse-Nassau, Hohenzollern (a Prussian enclave between Württemberg and Baden in southwestern Germany), Pomerania, Rhine Province, Saxony, Schleswig-Holstein, Upper Silesia and Lower Silesia, and Westphalia. Grenzmark Posen; West Prussia was sometimes considered a 14th province. Prussia stretched from the borders of the Netherlands, Belgium, and Luxembourg in the west to those of Lithuania and Poland in the east. Prussia was mainly made up of low-lying land, drained by several rivers, notably the Rhine; the Weser; the Oder; and the Elbe, which divided the state into roughly equal eastern and western parts.

History. Before 1701 Prussia meant only the flat, sandy region later known as East Prussia that was separated from Brandenburg by a part of Poland (later known as West Prussia) and bordering on the Baltic Sea. The original inhabitants, the Borussi (or Prussians), were of Baltic stock. They were conquered by the Teutonic Knights in the 13th cent. The domain became (1525) a hereditary duchy under Polish suzerainty, and passed to the elector of Brandenburg in 1618. In the course of the 17th cent. the electors of Brandenburg, independent from Poland after 1660, increased their holdings eastward and westward. In 1701 Elector Frederick III had himself crowned "king in Prussia" and styled himself king Frederick I. King Frederick William I (reigned 1713–40) built up a strong army and developed a government-controlled economy and an obedient central bureaucracy. The landed aristocrats, the Junkers, were brought into military and state service and in turn were left free to enserf their peasants. Frederick II, or Frederick the Great (reigned 1740–86), entered upon a period of conquest, gaining Silesia from Austria (1745) and Pomerelia and Ermeland in the 1772 partition of Poland. Defeated by the French in the French Revolutionary Wars and the wars of Napoleon I, Prussia lost (1807) much territory, which it regained at the Congress of Vienna (1815) along with the entire Rhine Prov. and Westphalia, the northern half of Saxony, the remainder of Swedish Pomerania, and a large part of western Poland.

By 1834 Prussia had taken the lead in the economic unification of Germany. The March Revolution of 1848 was put down by force. In 1861 William I (regent since 1858) became king, and in 1862 he appointed as premier Otto von Bismarck, who effected the union of Germany under Prussian hegemony by means of deliberately planned wars against Denmark, Austria, and France. In 1871 William I of Prussia was proclaimed emperor of Germany.

In its main features the subsequent history of Prussia was that of Germany. The Prussian constitution was liberalized after Prussia became a republic in 1918, and the Junkers lost many of their estates through the cession of Prussian territory to Poland. Prussia was formally abolished as a state and partitioned among the four Allied occupation zones after 1945. Most of the former Prussian provinces became part of the new states of the Federal Republic of Germany and of the German Democratic Republic. The USSR annexed the northern part of East Prussia; Poland acquired the rest, as well as all Prussian territory east of the Oder and Neisse rivers.

Prut or **Pruth** (both: prōōt), river, c.530 mi (855 km) long, rising in the Carpathian Mts., W Ukraine, USSR, and flowing generally SE to the Danube River at Reni. It forms the border between Rumania and Moldavia, USSR.

Prze·myśl (pə-shĕm'ĭshl), city (1975 est. pop. 57,400), extreme SE Poland, on the San River in the Carpathian foothills. It is a trade center and has metalworking, clothing, food-processing, electrical engineering, and timber-working industries. The city was founded in the 8th cent. Between 981 and 1340 it was ruled by Kiev and Vladimir-Volynski. Przemyśl passed to Poland in the late 14th cent. Austria took the city in 1772; it reverted to Poland after World War I.

Pskov (pə-skôf'), city (1976 est. pop. 155,000), capital of Pskov oblast, NW European USSR, on the Velikaya River. It is an important rail junction in the heart of a flax-growing area. Industries include food processing and the manufacture of metals, machinery, building materials, and linen. Known in antiquity as Pleskov, it became (903) an outpost of Novgorod. Pskov became (1347) a flourishing democratic city-state that traded with the Hanseatic League. It was capital of Pskov Republic from 1348 to 1510. With its annexation (1510) by Moscow, Pskov lost its democratic institutions. Its importance, except as a strategic fortress, soon declined. The railroad station at Pskov was the scene (1917) of the abdication of Nicholas II. The historic core of Pskov is the inner walled city, containing a kremlin (12th–16th cent.), with towers in the Byzantine style, a cathedral, and numerous medieval churches and monasteries.

Ptol·e·ma·ïs (tôl'ə-mā'əs), town (1971 pop. 16,588), N Greece, in Macedonia. It was a small market town until 1958, when it began to be developed as an industrial center. Lignite, mined in vast quantities, is used to power thermoelectric plants, which produce electricity for iron and steel mills, aluminum factories, and chemical plants.

Pud·sey (pŭd'zē, -sē), municipal borough (1971 pop. 36,187), West Yorkshire, N England. It is a center of woolen- and worsted-textile industries.

Pueb·la (pwĕb'lə, pwā'blä), city (1974 est. pop. 466,000), capital of Puebla state (13,126 sq mi/33,996 sq km), E central Mexico. Located in a highland valley, it is an important agricultural, commercial, and manufacturing center, as well as a popular tourist spot. The site of Mexico's first textile-producing factory, Puebla now has cotton mills, onyx quarries, and pottery and food industries. The city is noted for the fine, colored tiles that decorate its buildings and numerous churches. The cathedral, built between 1552 and 1649, is one of the finest in Mexico; the theater, constructed in 1790, is said to be the oldest on the continent.

Pueb·lo (pwĕb'lō), county (1970 pop. 118,238), 2,401 sq mi (6,218.6 sq km), SE central Colo., drained by the Arkansas River and Fountain Creek; formed 1861; co. seat Pueblo. It is in an irrigated agricultural (livestock, sugar beets, and feed grains) and industrial (food processing, clay products, concrete, iron and steel products, and machinery) area. There is some mineral mining in the county. It includes part of San Isabel National Forest.

Pueblo, city (1978 est. pop. 103,800), seat of Pueblo co., S central Colo., on the Arkansas River in the foothills of the Rockies; inc. 1885. It is the center of shipping, trade, and industry for an extensive timber, coal, livestock, and irrigated-farm area. It has a huge steel industry. Wire, concrete, and lumber are also produced. A trading post, called Pueblo, was established here in 1842. After a severe flood in 1921, levees were constructed to control the Arkansas River.

Puer·to Bar·rios (pwĕr'tō bär'yōs'), city (1973 pop. 19,696), E Guatemala, capital of Izabal dept., on the Bay of Amatique, an arm of the Caribbean Sea. Bananas and coffee are the leading exports.

Puerto Ca·bel·lo (kä-bĕl'yō), city (1971 pop. 72,103), N Venezuela, a port on the Caribbean Sea. It ships meat, coffee, cacao, dyewoods, and copper ores. Strategically located, Puerto Cabello was subject to attacks by buccaneers and was a favorite market for Dutch smugglers during the colonial era.

Puerto Cor·tés (kōr-tās'), town (1974 pop. 25,817), NW Honduras, on the Caribbean Sea; founded c.1525. It is a principal banana port and a processing center.

Puerto de San·ta Ma·ri·a (də sän'tə mə-rē'ə), town (1970 pop. 42,111), Cádiz prov., S Spain, in Andalusia, on the Bay of Cádiz. It is a commercial center, exporting sherry.

Puerto Montt (mônt), city (1970 pop. 62,726), capital of Llanquihue prov., S central Chile, a port on Ancud Gulf, an inlet of the Pacific Ocean. It is a railroad terminus and the starting point for navigation through the inland waterways and among the islands to the south. The scenery—forested hills, lakes, narrow fjords, and peaks—makes Puerto Montt a popular resort. Founded in 1853, the city was settled largely by Germans. Sheep farming and fishing are important.

Puerto Ri·co (rē′kō), formerly **Por·to Ri·co** (pôr′tō rē′kō), island (1975 pop. 2,987,000), 3,425 sq mi (8,871 sq km), West Indies, c.1,000 mi (1,610 km) SE of Miami, Fla. Officially known as the Commonwealth of Puerto Rico (a self-governing entity in association with the United States), it includes the offshore islands of Mona, Vieques, and Culebra. The capital is San Juan. Smallest and easternmost of the Greater Antilles, Puerto Rico is bounded by the Atlantic Ocean on the north and the Caribbean Sea on the south. Mona Passage to the northwest separates the island from the Dominican Republic, and the Virgin Islands lie to the east.

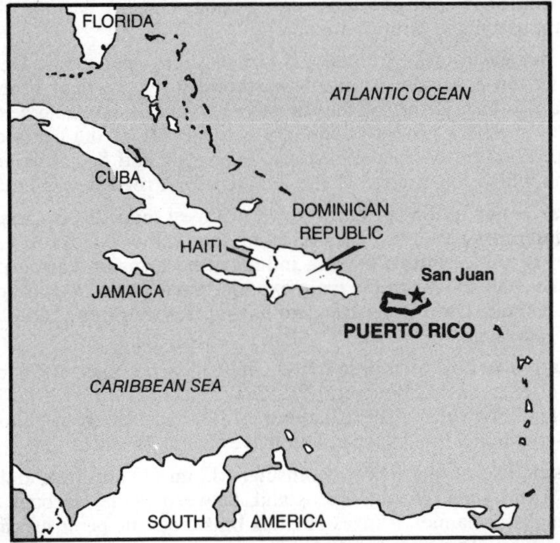

Puerto Rico is crossed by mountain ranges, notably the Cordillera Central, which rises to 4,389 ft (1,388 m) in the Cerro de Punta. The climate is mildly tropical, with little seasonal change. Rainfall is plentiful, despite some arid regions in the south. Hurricanes are likely to occur between Aug. and Oct. Puerto Rico's fertile soil supports one of the densest populations in the world. Overpopulation and insufficient jobs have contributed to social and economic problems and to heavy migration (mainly to New York City). The Puerto Ricans are descended from Spanish colonists, with an admixture of Indian and African strains. Spanish is the official language, and Roman Catholicism the main religion.

Economy. Sugar cane has long been the chief product; livestock raising (for meat and dairy production) ranks second among agricultural pursuits. Coffee, pineapples, and tobacco are other leading crops, and vegetable growing and canning are increasingly important. Heavy industry and manufacturing have come to replace agriculture as the greatest contributors to Puerto Rico's national income. The metallurgical and chemical industries are among the most important, along with oil refining (using crude oil from Venezuela) and the production of petrochemicals. Tourism is also a major industry. Puerto Rico's mineral resources are limited and not fully exploited. The United States is by far Puerto Rico's chief trading partner. The leading exports are raw and refined sugar and sugar products (rum, candy, and molasses).

History. Before the Spanish arrived, the island was inhabited by Arawak Indians. Christopher Columbus discovered the island in 1493 and named it San Juan Bautista, but he sailed on to Hispaniola to found a settlement. Juan Ponce de León began the actual conquest in 1508, landing at San Juan harbor, which he called Puerto Rico ("rich port"). A settlement was founded here in 1521. As hardship, disease, and Spanish reprisals eliminated the Arawak Indians altogether, they were replaced as plantation workers by African slaves, first introduced in 1513.

San Juan became a leading outpost of the Spanish Empire. Treasure-filled Spanish galleons anchored here on their long trip to Spain. Coffee was introduced in the 18th cent. to supplement sugar. Beginning in the 1820s there were some uprisings against Spanish rule, but they were quelled. As part of a reform movement that extended to Puerto Rico, slavery was abolished in 1873, and the new Spanish constitution of 1876 granted Puerto Rican representation in Spain's parliament. In Feb., 1898, largely through the efforts of the Puerto Rican statesman Luis Muñoz Rivera, Spain granted the island some autonomy. A few months later the Spanish-American War erupted. By the Treaty of Paris (Dec. 10, 1898), which ended the war, Puerto Rico was ceded to the United States. It remained under direct military rule until 1900, when the U.S. Congress set up an administration with an appointed U.S. governor.

During World War I U.S. holdings in Puerto Rico increased, and the change to a one-crop sugar economy was completed. Absentee ownership and one-crop culture aggravated the ills of overpopulation. Sanitary and health improvements under the U.S. occupation further accelerated population growth. Many Puerto Ricans criticized the American regime for its menace to the Hispanic roots of Puerto Rican culture.

In 1946 the U.S. government granted Puerto Rico increased local autonomy. The right of popular election of the governor followed, and Luis Muñoz Marín won the 1948 election. His administration undertook a program of agricultural reform and industrial expansion. On July 25, 1952, the Commonwealth of Puerto Rico was proclaimed. In the face of increasingly active movements both for statehood and for independence, a plebiscite was held in 1967 in which Puerto Ricans overwhelmingly voted in favor of the status quo.

Government. Puerto Rico's governor and both legislative houses are popularly elected for four-year terms. There are 29 senators and 54 representatives. An elected resident commissioner sits in the U.S. House of Representatives but cannot vote. The governor may serve an unlimited number of terms. On the local level Puerto Rico is divided into municipalities, each with its own mayor and assembly. Puerto Ricans share all the rights and obligations of U.S. citizenship, including service in the armed forces; however, they do not pay Federal taxes and cannot vote in national elections.

Educational Institutions. The island's rural vocational schools are exemplary. Institutions of higher learning include the Univ. of Puerto Rico (with its main branch at Río Piedras), the Inter-American Univ. at San Germán, the Catholic Univ. at Ponce, and a Catholic college for women at San Juan.

Pu·get Sound (pyo͞o′jĭt), arm of the Pacific Ocean, NW Wash., connected with the Pacific by Juan de Fuca Strait, entered through the Admiralty Inlet and extending in two arms c.100 mi (160 km) S to Olympia. The sound, which receives many streams from the Cascade Range, has numerous islands and is navigable for large ships. Discovered in 1787, the sound was explored and named by English Capt. George Vancouver for his aide, Peter Puget, in 1792.

Pu·la (po͞o′lä), city (1971 pop. 47,414), NW Yugoslavia, in Croatia, on the Adriatic at the S tip of the Istrian peninsula. A major seaport and an industrial center, it has shipyards, docks, and varied manufactures. Captured (178 B.C.) by the Romans, it was destroyed by Augustus, but was rebuilt by him and named Pietas Julia. It passed to Venice in 1148. The Treaty of Campo Formio (1797) transferred it to Austria. Under Austrian rule Pula became the chief naval base and arsenal of the Hapsburg empire. The city was ceded to Italy after World War I and to Yugoslavia after World War II. Pula has many well-preserved Roman ruins, notably a large amphitheater, the Porta Aurea (a triumphal arch of the 1st cent. B.C.), and the temple of Augustus and Roma (1st cent. A.D.).

Pu·lang·i (po͞o-läng′gē), river, c.200 mi (320 km) long, rising in the mountains of N Mindanao, the Philippines, and flowing SW to Liguasan Marsh then NW (as the Mindanao River) to Illana Bay at Cotabato. It drains a rich rice-growing area.

Pu·las·ki (po͞o-lăs′kē, pə-). **1.** County (1970 pop. 287,189), 765 sq mi (1,981.4 sq km), central Ark., intersected by the Arkansas River and drained by Bayou Meto and the Maumelle River; formed 1818; co. seat Little Rock. Its agriculture includes cotton, corn, hay, truck crops, livestock, and dairy products. It has timber, granite and stone quarries, and bauxite mines. Clothing, furniture, paper, chemicals, tires, concrete, and miscellaneous metal products and machinery are manufactured. **2.** County (1970 pop. 8,066), 254 sq mi (657.9 sq km), S central Ga., in a coastal plain area intersected by the Ocmulgee River; formed 1808; co. seat Hawkinsville. Cotton, peanuts, corn, and fruit are grown. Race horses are trained in the county. **3.** County (1970 pop. 8,741), 204 sq mi (528.4 sq km), extreme S Ill., bounded on the S by the Ohio River and drained by the Cache River; formed 1843; co. seat Mound City. Its agriculture includes livestock, wheat, fruit, poultry, cotton, truck crops, and dairy products. Lumbering, flour milling, and shipbuilding are done. **4.** County (1970 pop. 12,534), 433 sq mi (1,121.5 sq km), NW Ind., drained by Big Monon Creek and the Tippecanoe River; formed 1835; co. seat Winamac. It is in an agricultural area yielding livestock, dairy products, grain, poultry, and soybeans. Its industry includes clothing manufacturing. **5.** County (1970 pop. 35,234), 654 sq mi (1,693.9 sq km), SE central Ky., drained by the Cumberland River and small Fishing Creek; formed 1798; co. seat Somerset. In a hilly agricultural area that produces burley tobacco, corn, cattle, hogs, poultry, dairy products, and lespedeza, it has bituminous-coal mines, timber, oil fields, and stone quarries. Its industry includes food processing and the manufacture of textiles, wood products, furniture, glass, pottery, concrete, and machinery. It includes part of Cumberland National Forest. **6.** County (1970 pop. 53,967), 550 sq mi (1,424.5 sq km), central Mo., in the Ozarks, drained by the Gasconade River; formed 1818; co. seat Waynesville. Agricultural products include corn, wheat, fruit, and livestock. Part of Mark Twain National Forest is here. **7.** County (1970 pop. 29,564), 328 sq mi (849.5 sq km), SW Va., in the Great Appalachian Valley, drained by the New River; formed 1839; co. seat Pulaski. It is in an agricultural (stock raising and dairying) and mining (anthracite coal, iron, zinc, and limestone) area. Clothing and textiles, wood products, and furniture are made. Part of Jefferson National Forest is here.

Pulaski. **1.** City (1970 pop. 6,989), seat of Giles co., S Tenn., SSW of Nashville; settled 1807, inc. 1849. The first Ku Klux Klan was organized here in 1865. Clothing and shoes are manufactured. **2.** Town (1970 pop. 10,279), seat of Pulaski co., SW Va.; inc. 1886. It has textile, garment, furniture, and chemical industries.

Pull·man (pŏŏl'mən). **1.** Former city, since 1889 part of Chicago, Ill. It was founded in 1880 by G. M. Pullman as a model community for workers of his sleeping-car company; all property was company owned, and administration policies were paternalistic. The residents voted for incorporation with Chicago, and Illinois courts later required the company to sell all property not used for industrial purposes. In 1894 Pullman was the site of one of the most memorable strikes in U.S. history. **2.** City (1978 est. pop. 23,700), Whitman co., SE Wash., at the junction of the three forks of the Palouse River, near the Idaho border; inc. 1888. It is an agricultural center. Washington State Univ. is here.

Pu·log, Mount (pōō'lôg'), peak, 9,612 ft (2,931.7 m) high, NW Luzon, the Philippines, in the Cordillera Central.

pu·na (pōō'nä), high plateau region, 12,000 to 16,000 ft (3,660–4,880 m) high, between ridges of the Andes in Peru and Bolivia. Arid, cold, and, in general, covered by short coarse grass, the puna has long supported an Indian population. The icy wind sweeping the mineral-rich plateaus is also called puna.

Pu·na·ka or **Pu·na·kha** (both: pōō'nə-kə), town (1970 est. pop. 12,000), traditional capital of Bhutan, NW Bhutan. Founded in 1577, it is a fortress town with an important Buddhist monastery.

Punch·bowl (pŭnch'bōl'), hill, 500 ft (152.5 m) high, in the city of Honolulu, SE Oahu island, Hawaii. In the extinct volcanic crater at the summit is the National Memorial Cemetery of the Pacific, for men killed in World War II.

Pun·jab (pŭn'jäb', pŭn'jäb), historic region, NW Indian subcontinent. The Indus River bounds the region in part of the west and the Jumna River in part of the east. The five rivers that give Punjab its name, the Jhelum, the Chenab, the Ravi, the Sutlej, and the Beas, merge to form the Panjnad, which flows into the Indus. Except in the north, where there are forested mountains yielding salt and coal, the Punjab is a level alluvial plain. Rainfall is scant and irregular, but an extensive irrigation system has made possible enormous agricultural productivity. Wheat, millet, barley, cotton, and sugar cane are grown, and there are extensive fruit orchards. The Punjab has a large textile industry and much flour milling. The region formed one of the centers of the prehistoric Indus Valley Civilization, and after c.1500 B.C. it was the site of the earliest Aryan settlements. The Punjab was occupied by Alexander the Great and then by the Maurya empire. Moslems occupied western Punjab by the 8th cent. and firmly implanted Islam. Not until the late 12th cent. did they conquer eastern Punjab, which even afterward remained predominantly Hindu. Under the Mogul empire the Punjab reached its cultural height. When the empire declined in the late 18th cent., the Sikhs rose to dominance. The British, who emerged victorious in the two Sikh Wars (1846, 1849), annexed most of the Punjab and made it a province. In 1947 the Punjab was partitioned between India and Pakistan, approximately along the line between the main concentrations of the Moslem and the Hindu populations.

Pu·no (pōō'nō), city (1972 pop. 41,166), alt. 12,648 ft (3,857.6 m), capital of Puno dept., SE Peru, on Lake Titicaca. It is a commercial and transportation center. Wool, alcohol, and cacao are exported. Historical landmarks include a cathedral and some Indian ruins. Puno is also an archaeological center for the exploration of ancient Indian villages.

Pun·ta A·re·nas (pōōn'tə ə-rā'nəs), city (1970 pop. 61,813), capital of Magallanes prov., in Tierra del Fuego, S Chile, the only city on the Strait of Magellan and the southernmost city in the world. Punta Arenas was founded in 1847 to maintain Chile's claim to the strait. It was formerly a busy coaling station and is now an important center for export of Patagonian wool and mutton. Punta Arenas is popular with tourists in spite of a long rainy season. The city has one of the finest museums in South America.

Punta del Es·te (dĕl ĕs'tē), city, E Uruguay, on the Atlantic Ocean. Located on a narrow peninsula surrounded by excellent beaches, Punta del Este is one of the largest and most fashionable seaside resorts of South America. The city was the site of the conference (Aug., 1961) of the Inter-American Economic and Social Council, during which the charter of the Alliance for Progress was drafted.

Pun·ta·re·nas (pōōn'tä-rā'näs), town (1976 est. pop. 29,000), capital of Puntarenas prov., W Costa Rica, on the Gulf of Nicoya. It is the center of the country's banana industry and a major Pacific port. Bananas and coffee are the main exports. There is also a substantial coastal trade. Other industries are fishing (for shark and tuna) and fish processing.

Punx·su·taw·ney (pŭngk'sə-tô'nē), industrial borough (1970 pop. 7,792), Jefferson co., W central Pa., SSW of Du Bois; settled 1772 on the site of Shawnee villages, laid out c.1818, inc. 1850. It has bituminous-coal mines, packs meat, and makes metal products.

Pur·beck, Isle of (pər-bĕk'), peninsula, c.12 mi (20 km) long and c.8 mi (13 km) wide, Dorset, S England, between Poole Harbour and the English Channel. Ranges of chalk hills cross the peninsula from east to west. The region is noted for the production of marble and china clay. Purbeck was a favorite hunting ground of Saxon and Norman kings.

Pur·cell (pûr-sĕl'), city (1970 pop. 4,076), seat of McClain co., central Okla., SSE of Oklahoma City on the Canadian River; settled 1887, inc. 1946.

Pur·ga·toire (pûr'gə-twär'), river, 186 mi (299.3 km) long, SE Colo., rising at the meeting of the North Fork and Middle Fork rivers, it flows NE to the Arkansas River. The rock bed of Purgatoire has proven valuable for its records of dinosaur footprints.

Pu·ri (pŏŏr'ē), town (1971 pop. 72,712), Orissa state, E central India, on the Bay of Bengal. The life of the town centers around the Hindu cult of Juggernaut (Jagannath), a form of the Krishna incarnation of Vishnu. The images of Juggernaut and his sister and brother repose within a vast temple compound. Every summer each statue is mounted on an enormous temple cart and dragged by hundreds of pilgrims to a summer home 1 mi (1.6 km) distant. Limestone carving and toy making are the town's main industries.

Pu·rus (pōō-rōōs'), river, c.2,100 mi (3,380 km) long, rising in the Andes Mts., E Peru. It flows northeast in a meandering course across northwestern Brazil to the Amazon River.

Pur·vis (pûr'vəs), town (1976 est. pop. 2,062), seat of Lamar co., SE Miss., SSW of Hattiesburg, in a farm area.

Pu·san (pōō'sän'), city (1975 pop. 2,454,051), extreme SE South Korea, on the Korea Strait. It is the nation's largest port. Lying at the head of the Naktong River basin, it has served as a main southern gateway to Korea from Japan, which, during its rule over Korea (1910–45), developed Pusan's excellent natural harbor. Pusan's manufactures include woolen, cotton, and silk textiles, iron and steel, tires, plywood, frozen seafood, fishing nets, and wigs. There are also important shipbuilding and ship repair facilities. Nearby hot springs have made Pusan a popular resort city. The city became a major port under the Chinese Empire. It was invaded in 1592 by the Japanese, who had maintained a trading post here; however, the Japanese forces were recalled in 1598. In 1876 the Koreans were compelled to sign a treaty opening Pusan to Japanese trade and immigration. In 1883 the port was opened to general foreign commerce. Historic landmarks include the Kyongbok Palace (1394), the Changdok Palace, containing many valuable relics, and the Toksu Palace (1593), which houses the National Museum and Art Gallery.

Push·kin (pŏŏsh′kĭn), city (1976 est. pop. 86,000), NW European USSR, a residential and resort suburb of Leningrad. It produces road-building equipment and has an important botanical institute. Founded in 1708 on the site of a Finnish village. Pushkin served as a royal residence from 1725, with the huge baroque-style summer palace of Catherine II (built 1748–62) and that of Alexander I (built 1792–96) in the classical mode. The vast park at Pushkin has grottoes, pavilions, canals, lakes, and bridges.

Push·ma·ta·ha (pŏŏsh′mə-tä′hä), county (1970 pop. 9,385), 1,420 sq mi (3,677.8 sq km), SE Okla., drained by the Kiamichi and Little rivers; formed 1907; co. seat Antlers. It has agriculture (corn, cotton, potatoes, fruit, and livestock) and pine and oak timber. It manufactures clothing and wood products.

Pu·teaux (pü-tō′), suburb W of Paris (1975 pop. 35,514), Hauts-de-Seine dept., N central France, on the Seine River. An important industrial center, Puteaux is the birthplace of the French automobile industry. Other manufactures include electric locomotives, electrical equipment, aircraft, rubber goods, and perfume.

Pu·te·o·li (pyōō-tē′ə-lī), ancient city of Campania, S Italy, W of Naples. Founded c.520 B.C. by Samian Greeks, it came under Roman control by the end of the 4th cent. B.C. It became famous as a port, handling notably mosaics, pottery, and perfumes. The city was surrounded by handsome villas. Puteoli was destroyed by a series of Germanic invasions in the 5th cent. A.D.

Put in Bay (pŏŏt′ ĭn bā′), resort village (1970 pop. 135), N Ohio, on South Bass Island in Lake Erie, W of Toledo. Nearby, Oliver Hazard Perry decisively defeated the British naval forces in the Battle of Lake Erie (Sept. 10, 1813), forcing the British to evacuate Detroit. In the village is Perry's Victory and International Peace Memorial.

Put·nam (pŭt′nəm). **1.** County (1970 pop. 36,424), 779 sq mi (2,017.6 sq km), NE Fla., in a lake and swamp area drained by the St. Johns River; formed 1849; co. seat Palatka. Its agriculture includes corn, vegetables, peanuts, citrus fruit, poultry, and livestock. It has timber and deposits of clay, peat, and sand. **2.** County (1970 pop. 8,394), 339 sq mi (878 sq km), central Ga., bounded on the E by the Oconee River and drained by the Little River; formed 1807; co. seat Eatonton. It is in a piedmont stock-raising, dairying, and farming (cotton, corn, truck crops, and peaches) area, with a textile industry. **3.** County (1970 pop. 5,007), 160 sq mi (414.4 sq km), N central Ill., bounded on the N and W by the bend of the Illinois River; formed 1825; co. seat Hennepin. It has agriculture (corn, oats, wheat, livestock, poultry, and dairy products), bituminous-coal mining, and manufacturing of steel and wood products. It includes Senachwine and Sawmill lakes and fishing resorts. **4.** County (1970 pop. 26,932), 490 sq mi (1,269.1 sq km), W central Ind., drained by the Eel River and Raccoon and Mill creeks; formed 1821; co. seat Greencastle. In an agricultural area yielding grain, poultry, livestock, and dairy products, it has timber sawmilling, limestone quarrying, and textile manufacturing. **5.** County (1970 pop. 5,916), 518 sq mi (1,341.6 sq km), N Mo., bounded in the E by the Chariton River; formed 1845; co. seat Unionville. Farming (corn, wheat, and oats), livestock raising, and coal mining are major occupations. **6.** County (1970 pop. 56,696), 231 sq mi (598.3 sq km), SE N.Y., bounded on the W by the Hudson River, on the E by the Conn. border; formed 1812; co. seat Carmel. This hilly summer-resort region, with many lakes, has dairying, poultry raising, and truck farming. **7.** County (1970 pop. 31,134), 486 sq mi (1,258.7 sq km), NW Ohio, intersected by the Auglaize and Blanchard rivers; formed 1820; co. seat Ottawa. It is in an agricultural area that produces grain, sugar beets, livestock, and poultry. Its industry includes food processing, millwork, and the manufacture of chemical products and farm machinery. **8.** County (1970 pop. 35,487), 408 sq mi (1,056.7 sq km), central Tenn., in a hilly region on the Cumberland Plateau, drained by affluents of the Cumberland River; formed 1842; co. seat Cookeville. Its agriculture includes poultry, small grains, tobacco, corn, hay, cattle, and hogs. It has bituminous-coal mines and granite quarries. Its industry includes food processing and lumbering. Clothing, furniture, machinery, and household appliances are made. **9.** County (1970 pop. 27,625), 349 sq mi (903.9 sq km), W W.Va., drained by the Kanawha and Pocatalico rivers; formed 1848; co. seat Winfield. It has some bituminous-coal mines and oil and natural-gas wells. Livestock, fruit, truck crops, and tobacco are produced.

Putnam, city (1970 pop. 6,918), Windham co., NE Conn., at the falls of the Quinebaug River; settled 1693, inc. 1895, set off from Killingly 1855. Optical goods, textiles, and metal products are made.

Pu·tu·ma·yo (pōō′tōō-mä′yō) or **I·çá** (ē-sä′), river, c.1,000 mi (1,600 km) long, rising in the Andes, S Colombia, and flowing SE to the Amazon in NW Brazil. Mostly navigable, it marks part of Colombia's boundary with Ecuador and most of Colombia's frontier with Peru. The river valley, once a major source of rubber, has declined somewhat in economic importance.

Puy, Le (lə pwē′), city (1975 pop. 26,594), capital of Haute-Loire dept., S central France. A busy industrial city, Le Puy is the center of an old lace industry. The city grew after its shrine to the Virgin Mary became (10th cent.) a major place of pilgrimage. The modern section of Le Puy lies below a bare rock that towers almost 500 ft (152.5 m) above the city. At the foot of the rock lies the old city, with a cathedral (12th cent.) of extraordinarily daring construction and numerous Gothic buildings.

Puy·al·lup (pyōō-ăl′əp), city (1978 est. pop. 16,000), Pierce co., W Wash., on the Puyallup River; inc. 1890. It is located in a fertile farm valley noted for its berries and daffodil bulbs. Of interest is the mansion home (1890) of Ezra Meeker, the city's founder. A daffodil festival and a nine-day fair are held annually.

Puy-de-Dôme (pwē-də-dōm′), department (1975 pop. 580,033), 3,071 sq mi (7,953.9 sq km), S central France, in Auvergne; capital Clermont-Ferrand.

Puy de Dôme (pwē′ də dōm′), extinct volcano of the Massif Central and the second-highest peak (4,806 ft/1,465.8 m) of the Auvergne Mts., central France, W of Clermont-Ferrand. Crops are raised on the lower slopes; the highlands are used as pasturage. On its level summit is a meteorological observatory.

Pya·ti·gorsk (pyä′tĭ-gôrsk′), city (1976 est. pop. 103,000), Stavropol Kray, SE European USSR, on the Podkumok River in the N Caucasus. The city has an electrotechnical industry, shops for the repair of agricultural equipment, and factories that produce food, clothing, footwear, and other items for resort visitors. Founded in 1780, Pyatigorsk has been a spa since 1803.

Pyd·na (pĭd′nə), ancient town of Pieria, S Macedonia, near the Gulf of Salonica (now Thessaloníki). Nearby in 168 B.C. the Romans defeated the Macedonians and ended the kingdom of Macedon.

Pye (pyā) or **Prome** (prōm), city (1970 pop. 65,000), Pegu, S central Burma, on the Irrawaddy River. It is a commercial town and port. Pye was founded in the 8th cent. by the Pyus, who were conquered and absorbed by the Mon kingdom of Pegu, probably in the 9th cent. It was incorporated into British Burma in 1852.

Py·los (pī′lŏs, pē′lôs), ancient harbor, Messenia, SW Greece, on a bay of the Ionian Sea. Excavations have revealed a great Mycenaean palace of the 13th cent. B.C., perhaps the dwelling of King Nestor, and six hundred clay tablets that were important in the decipherment of the late Minoan script. The Bay of Pylos was the scene of an Athenian naval victory over Sparta in 425 B.C. and of the Battle of Navarino (1827) during the Greek War of Independence.

Pyong·yang (pyŭng′yăng′), city (1966 est. pop. 1,364,000), capital of North Korea, NW North Korea, on a high bluff above the Taedong River. Pyongyang, located near large iron and coal deposits, is a major industrial center; products include iron and steel, machinery, armaments, aircraft, textiles, sugar, and various light manufactures. Korea's oldest city, Pyongyang was founded, according to legend, in 1122 B.C. by remnants of the Chinese Shang dynasty. The city served as capital of the Choson kingdom (300–200 B.C.) and later became (108 B.C.) a Chinese colony and an important cultural center. Pyongyang fell c.1594 to the Japanese, who hoped to use it as a base for an invasion of China, but who then destroyed the city. Japanese invaders again devastated Pyongyang in 1894 and 1904. It became the capital of North Korea in 1948. After being ravaged in the Korean War, the city was rebuilt along modern lines. Only six gates remain of Pyongyang's former great walls. Other landmarks include three tombs (1st cent. B.C.) with remarkable murals, several old Buddhist temples, and a museum.

Pyr·a·mid Lake (pĭr′ə-mĭd′), 188 sq mi (487 sq km), W Nev. The lake is a remnant of ancient Lake Lahontan. Discovered in 1844 by U.S. explorer John Frémont, the lake was named for its large pyramidal rocks. It is located in Pyramid Lake Indian Reservation.

Pyr·e·nees (pĭr′ə-nēz′), mountain chain of SW Europe, 21,380 sq mi (55,374 sq km), between France and Spain, a formidable barrier between the Iberian Peninsula and the European mainland. It extends from the Bay of Biscay on the west to the Mediterranean Sea on the east. Of the three main ranges of the Pyrenees, the central section is the highest. The Pico de Aneto (11,168 ft/3,406.2 m) is the tallest peak. The Cantabrian Mts. are a western extension of the Pyrenees. The Pyrenees were formed during the Tertiary period. Exposed crystalline rock is found in the uplands, while folded limestone composes the lower slopes. Characteristic of the French Pyrenees, which are steeper than the southern slopes, are the torrents (*gaves*), often falling in cascades, and the natural amphitheaters (*cirques*), notably the famous Cirque de Gavarnie. The Pyrenees are a climatic divide. The northern slopes receive abundant rainfall while the southern slopes have a steppelike climate. The Franco-Spanish border, unchanged since the Peace of the Pyrenees (1659), generally follows the watershed. The mountain passes are high and difficult, but were often crossed by invading armies and barbarian hordes and by innumerable medieval pilgrims on their way to Santiago de Compostela. The Pyrenees are rich in timber and in pastures, and the many streams are utilized by hydroelectric power stations. Bauxite, talc, and zinc are mined. The population, partly of Basque and Bearnese stock, engages mostly in stock raising and agriculture.

Py·ré·nées-At·lan·tique (pē-rā-nā′zät-läɴ-tēk′), department (1975 pop. 534,748), 2,946 sq mi (7,630.1 sq km), SW France; capital Pau.

Py·ré·nées-O·ri·en·tales (pē-rā-nā′zô-rē-äɴ-täl′), department (1975 pop. 299,506), 1,578 sq mi (4,087 sq km), S France, in Roussillon, on the Mediterranean Sea; capital Perpignan.

Pyr·gos (pĭr′gôs): *see* Pirgos, Greece.

Qa·bis (kä′bĕs), or **Ga·bès** (gä′bĕs, gä-bĕs′), city (1966 pop. 32,330), E central Tunisia, on the Gulf of Gabès, an arm of the Mediterranean Sea. It is a fishing port and the center of an oasis noted for its date palms. The city was founded by the Romans.

Qaf·sah (kăf′sə, käf′sä) or **Gaf·sa** (găf′sə), town (1966 pop. 32,408), W central Tunisia, in an oasis. The town is a trade center for phosphates, dates, olives, and woolen goods. It is on the site of Capsa, a Numidian and later a Roman town.

Qal·yub (kŏl-yōōb′), town (1966 pop. 49,300), N Egypt, on the Nile River, near Cairo.

Qan·ta·rah, Al (ăl kăn′tər-ə), town, NE Egypt, on the E bank of the Suez Canal. It is on the ancient military road between Egypt and Syria. Al Qantarah is the terminus of a railroad to Palestine constructed during World War I, when the British Expeditionary Force in Egypt was based here.

Qa·tar (kŏt′ər, kä′tär), sheikdom (1973 est. pop. 160,000), c.4,400 sq mi (11,395 sq km), E Arabia, coextensive with the Qatar peninsula,

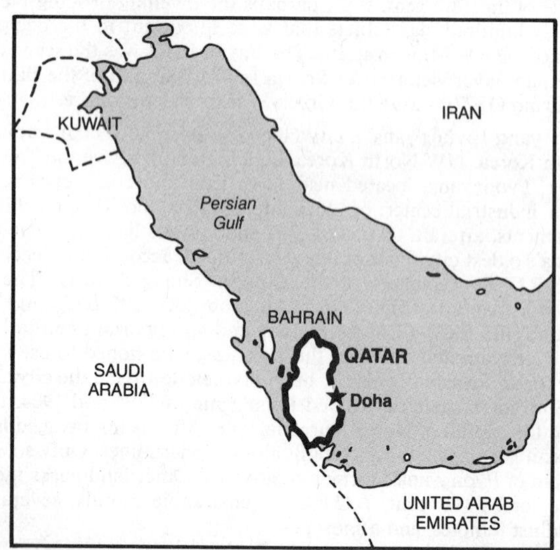

which projects into the Persian Gulf. The capital is Doha, or Bida (1971 est. pop. 95,000). Qatar is largely barren. Fishing and work in the oil industry support most of the population; oil began to be produced commercially in 1949. The sheikdom had close ties with Great Britain until Sept. 1, 1971, when it became independent.

Qat·ta·ra Depression (kŏt-tä′rŏ, kä-tä′rə), desert basin, c.7,000 sq mi (18,130 sq km), NW Egypt, NE Africa, in the Libyan Desert. It contains the lowest point (436 ft/133 m below sea level) in Africa.

Qaz·vin (käz-vēn′), city (1966 pop. 88,106), Tehran prov., NW Iran. A road and rail-transport center, the city has textile and flour mills and wineries. Qazvin was probably founded by Shapur II, king of Persia, in the 4th cent. A.D. It was captured by the Arabs in 644 and was the capital of Persia from 1548 to 1598.

Qeshm (kĕsh′əm), largest island of Iran, c.500 sq mi (1,295 sq km), S Iran, in the Strait of Hormoz. It is mostly rocky and has little vegetation. Dates and fruits are raised.

Qi·na (kĭn′ə), town (1970 est. pop. 77,600), capital of Qina governorate, E central Egypt, on the Nile River. Sugar cane and grains are grown nearby, and pottery is made in the town. Qina was built on the site of ancient Caene, of which many ruins remain. During World War II Qina was the terminus of a road through the mountains from Al Quayr on the Red Sea coast.

Qom (kôm), city (1966 pop. 134,292), Tehran prov., W central Iran. Located in a semiarid region, it is an industrial and transportation center. Its manufactures include textiles, glass, pottery, and shoes. Large deposits of petroleum have been found in the area. Qom has been a center of the Shiite Moslems since early Islamic times and became a pilgrimage center of in the 17th cent. Qom was pillaged by the Afghans in 1722, but in the 19th cent. its great shrine was lavishly restored and embellished.

Qua·nah (kwä′nə), city (1970 pop. 3,948), seat of Hardeman co., N Texas, near the Prairie Dog Town Fork of the Red River, WNW of Wichita Falls; settled 1886, inc. 1887.

Quan·ti·co (kwŏn′tĭ-kō), town (1970 pop. 719), Prince William co., N Va., on the Potomac River below Washington, D.C.; inc. 1927, reinc. 1934. A Marine Corps base and airfield are here.

Quay (kwā), county (1970 pop. 10,903), 2,875 sq mi (7,446.3 sq km), E N.Mex., bordered on the E by Texas and drained by the Canadian River and Tucumcari and Plaza Larga creeks; formed 1903; co. seat Tucumcari. Its economy depends on grain and livestock.

Que. Abbreviation for Quebec.

Que·bec (kwĭ-bĕk′) or **Qué·bec** (kə-bĕk′), province (1975 pop.

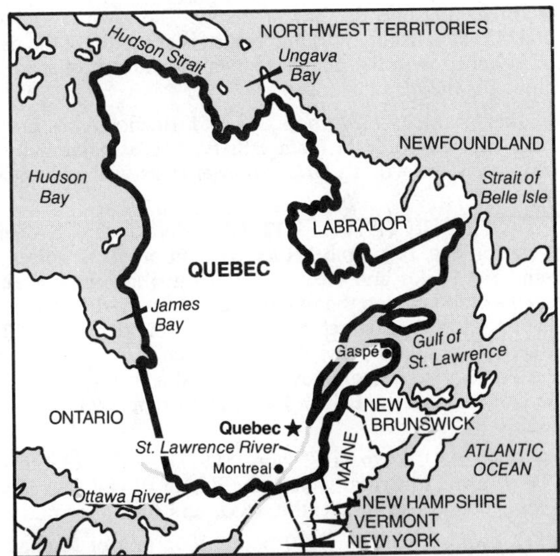

6,188,000), 594,860 sq mi (1,540,687 sq km), E Canada. The city of Quebec is the capital. Quebec is bounded on the north by Hudson Strait and Ungava Bay, on the east by Labrador and the Gulf of St. Lawrence, on the south by New Brunswick and the United States, and on the west by Ontario, James Bay, and Hudson Bay.

The Canadian Shield comprises the northern nine tenths of the province, which is relatively unexplored and uninhabited; the region has been planed by glacial action into a pattern of rounded hills, swiftly flowing rivers, and numerous lakes and bogs. Dense forests cover much of the surface, and the area is rich in minerals. South of the Canadian Shield lies the great St. Lawrence River. On both sides of the river south of Quebec city are the lowlands that are centers of agriculture, commerce, and industry. In the southeast are the Appalachian Highlands, which run parallel to the St. Lawrence. The Gaspé Peninsula, on the south bank of the St. Lawrence, borders on the Gulf of St. Lawrence. Quebec's climate is generally temperate, with variations among the regions. Tourism is important throughout the province during the summer season, and in the winter the Laurentian Mts. are a skiing attraction.

Economy. Quebec has vast resources of water power. The forests of the north yield wood for the province's pulp, paper, and lumber industries, and throughout the north country copper, iron, zinc, silver, and gold are mined. The iron-ore deposits in the Ungava Bay region have been exploited in recent decades. Asbestos is found in the far north and in the Thetford Mines region of the Appalachian Highlands. At Arvida, in the Saguenay valley, is one of the world's largest aluminum plants. The small farms of the lowlands yield dairy products, sugar beets, and tobacco. Quebec is second to Ontario among the Canadian provinces in industrial production. Its main manufactures are food products, beverages, transportation equipment, chemicals, and metal and paper products.

History. Since many continental explorations began in the region, Quebec has been called the cradle of Canada. In 1534 Jacques Cartier planted a cross on the Gaspé and the following year he sailed up the St. Lawrence. In 1608 Samuel de Champlain built a trading post on the site of the present-day Quebec city, and from this and subsequent settlements Catholic missionaries, explorers, and fur traders penetrated the American continent. The activities of private fur-trading companies ended in 1663, when Louis XIV made the region, known as New France, a royal colony. The long struggle to protect the colony and the fur trade against the hostile Iroquois Indians and the British was effectively lost in 1759, when the British defeated the French on the Plains of Abraham. By the Treaty of Paris of 1763 Great Britain acquired New France. By the Constitutional Act of 1791 the British detached the area west of the Ottawa River and made it the colony of Upper Canada (now Ontario). Quebec became known as Lower Canada, and in 1791 the first elective assembly was introduced. The resentment of the French community against the British led ultimately to the reunion of Upper and Lower Canada in 1841, and Quebec became known as Canada East. Responsible government was granted in 1849. With the formation of the Confederation of Canada in 1867, Canada East became the province of Quebec. English and French were made the official languages of both Quebec and the Canadian Parliament, and a dual school system was established within Quebec.

During the 20th cent. economic growth in Quebec was coupled with increasing determination to broaden provincial rights. Since the 1960s separatist groups, advocating an independent Quebec, have gained attention. In 1974 French was made the sole official language of the province, and all children were required to attend French language school. The coexistence of majority-French and minority-English cultures within the province and the reverse situation within Canada as a whole have remained sources of tension.

Quebec or **Québec**, city (1971 pop. 186,088; metropolitan pop. 480,502), provincial capital, S Que., Canada, at the confluence of the St. Lawrence and St. Charles rivers. The population is largely French speaking. Quebec is an important port and is an industrial, cultural, and tourist center. Part of the city is built on the waterfront and is called Lower Town; that part called Upper Town is on Cape Diamond, a bluff rising c.300 ft (90 m) above the St. Lawrence. Winding, narrow streets link the two sections of the city. The chief industries are shipbuilding and the manufacture of pulp and paper, leather products, textiles, clothing, machinery, and foods and beverages.

The site of Quebec was visited by Cartier in 1535, and in 1608 Champlain established a French colony in the present Lower Town; this was captured (1629) by the English, who held it until 1632. In 1663 Quebec was made the capital of New France and became the center of the fur trade. The city was unsuccessfully attacked by the English in 1690 and 1711. Finally in 1759 English forces under Wolfe defeated the French under Montcalm on the Plains of Abraham and captured Quebec. During the American Revolution, Americans under Richard Montgomery and Benedict Arnold failed (1775-76) to capture the city, although Arnold briefly held the Lower Town. Quebec became the capital of Lower Canada in 1791. After the union (1841) of Upper and Lower Canada, it was twice the capital of the United Provinces of Canada (1851-55 and 1859-65).

There are many notable old structures, including the Ursuline Convent (1639); the Basilica of Notre Dame (1647); Quebec Seminary (1663); and parts of the fortifications enclosing Old Quebec.

Qued·lin·burg (kvād'lən-bûrg', -lēn-bŏŏrKH'), city (1974 est. pop. 29,796), Halle district, W East Germany, at the foot of the lower Harz Mts. It is an agricultural market. Manufactures include food products, paper goods, and precision instruments. Quedlinburg was fortified in 922 by Henry I (Henry the Fowler). Henry I and his wife, St. Matilda (who with her son, Emperor Otto I, founded the celebrated convent in 936), are buried in the castle church.

Queen Annes (kwēn ănz'), county (1970 pop. 18,422), 373 sq mi (966.1 sq km), E Md., on the Eastern Shore, bounded on the E by the Del. border, on the W by Chesapeake Bay; formed 1706; co. seat Centreville. It is in a tidewater agricultural area that produces vegetables, fruit, dairy products, poultry, wheat, and corn.

Queen Char·lotte Islands (shär'lət), archipelago of several large and many small islands, off the coast of W British Columbia, Canada. There are valuable timber and fishing resources and several good harbors. The archipelago was visited in 1774 by Juan Pérez and in 1778 by Capt. James Cook; in 1787 it was surveyed by Capt. George Dixon. Hecate Strait separates it from the mainland.

Queen E·liz·a·beth Islands (ĭ-lĭz'ə-bəth), N part of the Arctic Archipelago, Franklin dist., Northwest Territories, N Canada. The islands are underlain by oil-bearing rock; extensive drilling has been under way since the early 1960s. The British explorer Sir William Parry discovered (1819-20) many of the islands.

Queens (kwēnz), borough of New York City (1970 pop. 1,973,708), land area c.113 sq mi (293 sq km), at the W end of Long Island, SE N.Y., coextensive with Queens co.; settled by the Dutch 1635, chartered as a borough of New York City 1898. It extends from the junction of the East River and Long Island Sound in the north across Long Island to Jamaica Bay and the Atlantic Ocean in the south. The borough is heavily industrialized, with extensive railroad yards. The first settlements in the area were made by the Dutch in 1635. Queens co. was organized in 1683. Several buildings of the 17th and 18th cent. remain. One of the first commercial nurseries in the country was established c.1737, and the community's collection of trees still includes several rare species. In the American Revolution British troops held the area after the Battle of Long Island (1776). In the 20th cent. growth was spurred with the opening of the Queensboro Bridge (1909) and a railroad tunnel (1910).

Queens·bor·ough-in-Shep·pey (kwēnz'bûr'ō-ĭn-shĕp'ē, -bər-ə-), municipal borough (1973 est. pop. 31,550), on the Isle of Sheppey, Kent, SE England, at the confluence of the Medway and Thames rivers. The borough was created in 1968 by the amalgamation of the municipal borough of Queensborough, the urban district of Sheerness, and the rural district of Sheppey.

Queens·land (kwēnz'lănd', -lənd), state (1976 pop. 2,037,032), 667,000 sq mi (1,727,530 sq km), NE Australia, bounded on the NE and E by the Coral Sea and the Pacific Ocean and on the NW by the Gulf of Carpentaria and Torres Strait; capital Brisbane. The major part of its coastline is sheltered by the Great Barrier Reef. Roughly half the state is in the tropical zone, with rain forests on Cape York Peninsula in the extreme north. The Great Dividing Range separates the fertile coastal strip from vast interior plains. The Great Artesian Basin in the interior provides water for a large livestock-raising area. Mainly an agricultural state, Queensland produces sugar cane, cotton, wheat, and tropical fruits. Mining is also important, especially copper, coal, lead, zinc, and bauxite. Oil and natural gas were discovered in the 1960s. In 1770 Capt. James Cook explored the coast of Queensland (then called Moreton Bay). Originally under the authority of New South Wales, Queensland served as a penal colony from 1824 to 1843. The area was separated from New South Wales and made a British colony in 1859. Queensland was federated as a state of the Commonwealth of Australia in 1901.

Quel·i·ma·ne (kĕl'ə-mä'nē), town (1970 pop. 71,289), capital of Zambézia district, E central Mozambique, a seaport on the Rio dos Bons Sinais near its mouth in the Indian Ocean. Exports include palm products, sisal, and tea. The Portuguese founded a trading station at Quelimane in 1544, and the town was an important slave market in the 18th and 19th cent.

Que·moy (kĭ-moi', kwĭ-, kē-), island group (1972 pop. 61,305), Formosa Strait, just off Fukien prov., China. It is a Nationalist outpost c.150 mi (240 km) west of Taiwan. The group consists of the islands of Quemoy and Little Quemoy and 12 islets in the mouth of Amoy Bay. Crops include sweet potatoes, peanuts, sorghum, barley, wheat, soybeans, vegetables, and rice. Fishing is also important. Quemoy island is heavily fortified; since 1949 it has been subjected to periodic bombardment from the Communist mainland.

Que Que (kwā kwā), city (1975 est. pop., 17,000), central Rhodesia, founded 1900. It is a gold-mining center and the focal point of Rhodesia's iron and steel industry.

Quer·cy (kĕr-sē'), region and former county, SW France. It consists of arid limestone plateaus cut by fertile valleys. Sheep raising is the chief activity. Of Gallo-Roman origin, Quercy became (9th cent.) a fief of the counts of Toulouse. It was savagely contested during the Hundred Years' War, after which it was united (1472) with the French crown.

Que·ré·ta·ro (kā-rā'tä-rō), city (1974 est. pop. 142,400), capital of Querétaro state (4,432 sq mi/11,478.9 sq km), central Mexico. It is a distribution center with industries producing machinery and farm implements; the city's cotton mills are among the most important in Mexico. Querétaro is also a popular tourist center. An Aztec city, Querétaro was conquered by the Spanish in 1531. The conspiracy (1810) that led to the revolution against Spain was planned here. In 1867 Emperor Maximilian and two of his generals were forced to surrender and then shot on a hill outside the city.

Quet·ta (kwĕt'ə), city (1972 pop. 120,000), capital of Baluchistan prov., W central Pakistan, at an altitude of c.5,500 ft (1,680 m), ringed by mountains. It commands the entrance through the strategic Bolan Pass into Afghanistan and is a trade center for Afghanistan, Iran, and much of central Asia. Fruits, vegetables, hides, and wool are the chief items traded.

Que·zal·te·nan·go (kā-säl'tā-näng'gō), city (1973 pop. 45,977), SW Guatemala, in the W highlands at an altitude of 7,500 ft (2,287.5 m). Its native market offers excellent textiles and handicrafts. The development of hydroelectric power has helped make it a leading industrial city of Central America.

Que·zon City (kā-sôn'), city (1975 est. pop. 994,700), official capital of the Republic of the Philippines, central Luzon, adjacent to Manila. Chiefly residential, it has a large textile industry. It is named for President Manuel Quezon, who in 1937 selected this site as the new capital of a free Philippines. It officially replaced Manila as capital in July, 1948. The Univ. of the Philippines is here.

Qui·be·ron (kē-brôɴ'), peninsula, Morbihan dept., NW France, in Brittany, projecting into the Bay of Biscay. The town of Quiberon (1968 pop. 4,595), a fishing port and resort, is at the tip of the peninsula and is linked with the mainland by a thin stretch of sand. Sardine canning is the major industry.

Quil·mes (kēl'mās), city (1970 pop. 355,265), Buenos Aires prov., E Argentina, on the Río de la Plata estuary. It is a major industrial city, with one of the world's largest breweries as well as numerous distilleries.

Qui·lon (kwē'lŏn), city (1971 pop. 124,072), Kerala state, SW India, on the Arabian Sea. It is a market for coconut products, spices, tea, coffee, and rice. Tiles and electrical apparatus are manufactured. Quilon is the oldest city on the Malabar Coast. In the 7th cent. it was noted by a Nestorian patriarch as the southernmost point of Christian influence in India. By the time the Dutch occupied Quilon in 1662, the Portuguese had already established a factory here. Soon after the Dutch came, the British East India Company took control of Quilon.

Quim·per (kăɴ-pâr'), town (1975 pop. 55,977), capital of Finistère dept., NW France, in Brittany, near the Bay of Biscay. It is famous for its pottery and also has textile, food, and furniture industries.

Quin·cy 1. (kwĭn'sē) Village (1970 pop. 2,500), seat of Plumas co., NE Calif., NW of Reno, Nev. The center of a timber and livestock area, it is also a resort and the headquarters for Plumas National Forest. **2.** (kwĭn'sē) City (1970 pop. 8,334), seat of Gadsden co., NW Fla., NW of Tallahassee near the Ga. border; founded c.1825, inc. as a town 1828, as a city 1923. The market center for an important tobacco area, it is also in a region producing large quantities of fuller's earth. **3.** (kwĭn'sē) City (1978 est. pop. 43,000), seat of Adams co., W Ill., on a bluff above the Mississippi; inc. 1839. It is a trade, industrial, and distributing center in a grain and livestock area. Quincy has a good harbor and was an important river port in the mid-19th cent. The sixth Lincoln-Douglas debate was held here on Oct. 13, 1858. **4.** (kwĭn'zē) City (1978 est. pop. 92,700), Norfolk co., E Mass., an industrial suburb of Boston, on Boston Bay; settled 1634, set off from Braintree 1792, inc. as a city 1888. It has a large shipbuilding industry and plants that make power transmissions, packaging machinery, soaps and detergents, and television tubes. The Plymouth Colony broke up (1627) a trading post established (1625) in the area by Thomas Morton, but a new settlement began in 1634. Ironworks began operation in 1644. Granite quarrying, started in 1750, became a large industry. The first railroad tracks in the United States were laid in Quincy in 1826. The city's large shipyards were of great importance in both world wars. John Adams and John Quincy Adams were born in Quincy. Their homes and places of birth are national historic sites.

Quin·ta·na Ro·o (kēn-tä'nä rô'ō), state (1970 pop. 88,150), 19,630 sq mi (50,842 sq km), SE Mexico, on the Caribbean; capital Chetumal. Occupying most of the eastern half of the Yucatán peninsula, it is wild, sparsely settled, and populated almost entirely by Indians. Some lumber, coconuts, and chicle are exported, and there is sponge and turtle fishing along the coast.

Quir·i·nal (kwĭr'ə-nəl), one of the seven hills of Rome, NE of Capitoline Hill. In the 16th cent. a papal palace was built here; it was the residence of the kings of Italy from 1870 to 1946 and is now the home of the president of Italy.

Quit·man (kwĭt'mən). **1.** County (1970 pop. 2,180), 156 sq mi (404 sq km), SW Ga., bounded on the W by the Chattahoochee River and the Ala. border; formed 1858; co. seat Georgetown. In a coastal-plain agricultural area yielding cotton, corn, peanuts, and pecans, it has lumbering, some mining, and manufacturing. **2.** County (1970 pop. 15,888), 411 sq mi (1,064.5 sq km), NW Miss., in a rich lowland cotton-growing area drained by the Coldwater and Tallahatchie rivers; formed 1877; co. seat Marks. The county's economy depends on lumbering and food processing.

Quitman. 1. City (1970 pop. 4,818), seat of Brooks co., extreme S Ga., WSW of Valdosta; inc. 1859. A trade and shipping center for a farm and livestock area, it has lumber and cotton mills. **2.** Town (1970 pop. 2,702), seat of Clarke co., SE Miss., on the Chickasawhay River S of Meridian, in a farm and timber area. **3.** City (1970 pop. 1,494), seat of Wood co., NE Texas, NNW of Tyler. It is a trading center in a cotton, oil, and livestock area.

Qui·to (kē'tō), city (1974 pop. 597,133), N central Ecuador, capital of Ecuador and of Pichincha prov. It is the educational, cultural, and political center of Ecuador. Quito lies at the foot of the Pichincha volcano in the hollow of a gently sloping, fertile valley. Only a short distance below the equator but at an elevation of 9,350 ft (2,851.8 m), Quito has a pleasant, balmy climate; however, it is subject to earthquakes and has been damaged several times. The city has textile mills and handicraft and other minor industries. Originally settled by the Quito Indians, the site was captured by the Incas and became the capital of the Inca Kingdom of Quito a few decades before the Spanish arrived and captured it (1534). In 1663 Quito became the seat of an *audiencia*, with boundaries foreshadowing that of present-day Ecuador. There was an abortive uprising against Spain in 1809, and not until 1822 was the city liberated. Quito has a Spanish colonial atmosphere, with many examples of fine early architecture, notably the great Church of San Francisco.

Qum·ran (kōōm-rän′), ancient village on the northwest shore of the Dead Sea, in present Jordan. It is famous for its caves, in some of which the Dead Sea Scrolls were found. In Israelite times it was the site of a small settlement. Between c.130 B.C. and c.110 B.C. Qumran was rebuilt by the Jewish sect whose library is represented by the Dead Sea Scrolls. It was destroyed (31 B.C.) by an earthquake and was rebuilt c.4 B.C. The Romans destroyed it (A.D. 68) and made use of the site as a military fortress.

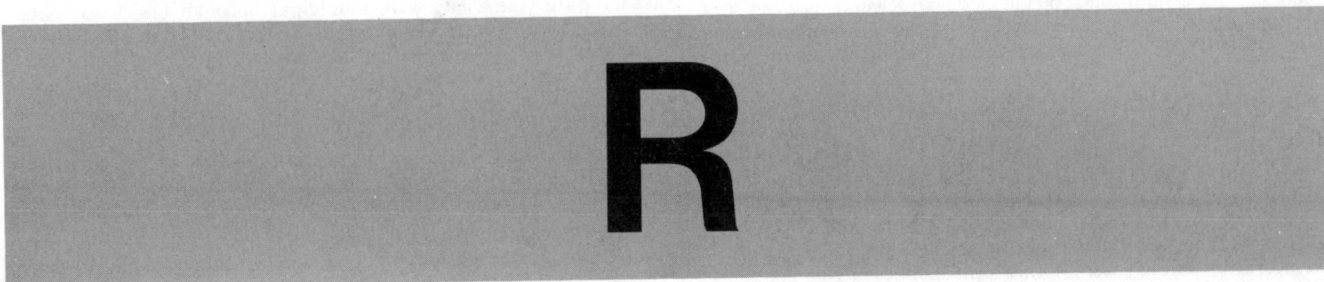

R

Rab (räb), island (1971 pop. 8,515), 40 sq mi (103.6 sq km) off Croatia, W Yugoslavia, in the Adriatic Sea. One of the Dalmatian islands, it is a popular seaside resort. The island was under Venetian rule from the 10th cent. until 1797, and it retains its ancient walls, the ruins of the palace of the Venetian governors, and the medieval palace of the former archbishops.

Ra·bat (rä-bät′), city (1971 pop. 367,620), capital of Morocco, on the Atlantic Ocean at the mouth of the Bou Regreg estuary, opposite Salé. The city is a minor port and has textile industries. There have been settlements here since ancient times. It became a Moslem fortress A.D. c.700. Prior to independence (1956), it was capital of the French protectorate of Morocco. Points of interest in Rabat are the old walls and the ruins of a large, unfinished mosque , that was built during the reign of Yakub (1184-99). Rabat was a stronghold of corsairs in the 17th and 18th cent.

Ra·baul (rə-boul′), town (1971 pop. 26,619), on New Britain island, Bismarck Archipelago, a part of Papua New Guinea. Rabaul has one of the finest harbors in the world. Copra is the chief export. The town is surrounded by active volcanoes and was nearly destroyed by eruptions in 1937. Totally destroyed by Allied bombing during World War II, it was rebuilt after the war.

Ra·bun (rā′bən), county (1970 pop. 8,327), 369 sq mi (955.7 sq km), extreme NE Ga., in the Blue Ridge on the N.C. and S.C. borders, drained by the Tallulah and the Little Tennessee rivers; formed 1819; co. seat Clayton. It has farming (hay, potatoes, fruit, and livestock), lumbering, and textile manufacturing.

Rac·coon (rə-kōōn′), river, rising in NW Iowa and flowing c.200 mi (320 km) SSE past Sac City and Jefferson to the Des Moines River.

Ra·ci·bórz (rät-sē′bōōsh), town (1975 est. pop. 45,900), S Poland, on the Oder River. A river port and rail junction, it also has industries producing machinery, machine tools, and electrical equipment. Chartered in 1217, it became (1288) the capital of a free imperial principality. It passed with Silesia to the house of Hapsburg in 1526 and to Prussia in 1745. It was incorporated into Poland in 1945.

Ra·cine (rə-sēn′), county (1970 pop. 170,838), 337 sq mi (872.8 sq km), SE Wis., bordered in the E by Lake Michigan and drained by the Fox and Root rivers; formed 1836; co. seat Racine. It is in a predominantly agricultural area yielding vegetables, corn, oats, dairy products, and livestock. It has some mining, and its industry includes food processing, lumbering, and iron and steel foundries.

Racine, industrial city (1978 est. pop. 91,800), seat of Racine co., SE Wis., on Lake Michigan, at the mouth of the Root River; inc. 1848. Its manufactures include farm machinery, heavy construction equipment, automobile parts, machine tools, floor wax, furniture, and electrical equipment. The first permanent settlement was established in 1834. Improvement of the harbor (c.1844) and the coming of the railroad (1855) brought industrial growth.

Rad·cliffe (răd′klĭf′), municipal borough (1971 pop. 29,320), Greater Manchester, N England. There are cotton and rayon mills and chemical, engineering, and paper plants. The parish church, founded in Norman times, has been restored.

Rad·nor·shire (răd′nər-shîr′), former county, E Wales. The county town was Presteigne. In 1974 Radnorshire became part of the new nonmetropolitan county of Powys.

Ra·dom (rä′dôm), city (1975 est. pop. 175,300), SE Poland. It has engineering and machine-building industries. There are also tanneries, food and tobacco processing plants, and various manufacturing industries. Radom probably originated as an assembly place for local diets. Casimir the Great of Poland founded the town of New Radom on the site in 1364. Radom passed to Austria in 1795 and to Russia in 1815, reverting to Poland after World War I.

Rae·ford (rā′fərd), town (1970 pop. 3,180), seat of Hoke co., S N.C. SW of Fayetteville. Cotton and rayon mills are here, and cottonseed oil is produced.

Ra·ges (rā′jēz) or **Rha·gae** (rā′jē), ancient and medieval city of Persia, located on the site of modern-day Ray, N Iran, a suburb of Tehran. It controlled the northeastern Persian trade route. It was occupied by the Parthians and the Arabs and flourished under the Seljuk Turks. Religious conflict between Sunni and Shiite Moslems resulted in the destruction of much of the city in 1186; further damage was done by the Mongols in 1220. By 1400 the city was deserted.

Ra·gu·sa (rä-gōō′zä), city (1976 est. pop. 53,700), capital of Ragusa prov., SE Sicily, Italy. Petroleum, asphalt, cheese, and plastics are produced in the city.

Rah·way (rô′wā′), industrial city (1978 est. pop. 27,700), Union co., NE N.J., on the Rahway River; settled c.1720 as part of Elizabethtown, inc. 1858. Chemicals, pharmaceuticals, and vacuum cleaners are among the city's manufactures.

Ra·ia·te·a (rä′yä-tā′ä), volcanic island (1967 est. pop. 6,200), 92 sq mi (238.3 sq km), South Pacific, largest and most important of the Leeward group of the Society Islands, French Polynesia. Raiatea's chief products are copra, oranges, tobacco, kapok, and vanilla. The island is believed to be the ancient Polynesian Maraiki, the religious and cultural center from which migrations to Hawaii, the Cook Islands, and New Zealand began c.600 years ago.

Rain·bow Bridge National Monument (rān′bō′ brĭj′), 160 acres (64.8 hectares), S Utah; est. 1910. Rainbow Bridge, the largest natural bridge in the world, is a symmetrical, pink, sandstone arch. Located in one of the most rugged and remote regions of the United States, it was discovered in 1909.

Rai·nier, Mount (rā-nîr′, rə-, rā′nîr): *see* Mount Rainier National Park.

Rains (rānz), county (1970 pop. 3,752), 210 sq mi (543.9 sq km), NE

Texas, bounded on the SW by the Sabine River; formed 1870; co. seat Emory. It has agriculture (cotton, fruit, truck crops, peanuts, and grains), livestock, and clay deposits.

Rain·y Lake (rān'ē), c.345 sq mi (895 sq km), on the U.S.-Canada border in N Minn. and W Ont. The lake, irregular in shape and dotted with islands, is located in rough woodlands. Its outlet, Rainy River (c.85 mi/135 km long), flows westerly to Lake of the Woods, passes International Falls and Baudette, Minn., and is used for logging.

Rai·pur (rī'po͝or), town (1971 pop. 205,909), Madhya Pradesh state, E central India, on the Kharun River. It is an agricultural-processing town.

Ra·jah·mun·dry (rä'jə-mŭn'drē), city (1971 pop. 165,900), Andhra Pradesh state, central India, on the Godavari River. It is a center of the tobacco industry and headquarters of the Godavari irrigation works. The Pushkaram religious festival, attracting thousands of pilgrims, is held here every 12 years.

Ra·ja·sthan (rä'jə-stän'), state (1971 pop. 25,724,142), 132,150 sq mi (342,269 sq km), NW India, bordered on the W by Pakistan; capital Jaipur. In the west of the state is the Thar (Indian) Desert; in the east is part of the upland region of the Deccan, where, with the aid of irrigation, millet, wheat, and cotton are grown. The Aravalli Range, which produces salt, sandstone, marble, coal, mica, and gypsum, crosses the state from the northeast to the southwest. The state was formed in 1948 from several former principalities of Rajputana.

Raj·kot (räj'kōt), city (1971 pop. 300,152), Gujarat state, W central India, on the Aji River. Formerly the capital of the Rajkot princely state and Saurashtra state, Rajkot is now an educational and cultural center. Textiles, hosiery, machine tools, plastics and electrical goods, and vegetable oil are produced.

Raj·pu·ta·na (räj'po͞o-tä'nə), historic region, NW India. Rajput tribal power rose here between the 7th and 13th cent., and the princes resisted the early Moslem incursions that began in the 11th cent. Rajput power reached its peak in the early 16th cent., but the area fell to the Moguls in 1568. The Moguls ruled Rajputana until the early 18th cent. The region passed to Great Britain in 1818. Under the British, Rajputana included more than 20 princely states. Most of these states were incorporated into Rajasthan after India gained independence in 1947.

Ra·leigh (rô'lē), county (1970 pop. 70,080), 604 sq mi (1,564.4 sq km), S W.Va., on the Allegheny Plateau, bounded on the E by New River and drained by the Coal and Buyandot rivers; formed 1850; co. seat Beckley. It has extensive bituminous-coal mines, some natural-gas fields, and sawmilling. Its agriculture includes livestock, strawberries, corn, potatoes, poultry, truck crops, and tobacco.

Raleigh. 1. Town (1970 pop. 1,018), seat of Smith co., S central Miss., NW of Laurel; inc. 1935. It is a trade center in an agricultural and timber area. **2.** City (1978 est. pop. 139,600), state capital, and seat of Wake co., central N.C. The site was selected for the capital in 1788, and the city was laid out and inc. 1792. It is a political, cultural, trade, and industrial center, with food, textile, and electrical manufactures. The first capitol (built 1792–94) burned in 1831 and was replaced by the present building, completed in 1840.

Ralls (rôlz), county (1970 pop. 7,764), 478 sq mi (1,238 sq km), NE Mo., on the Mississippi River and drained by the Salt River; formed 1820; co. seat New London. Wheat, corn, oats, livestock, and coal are major products here.

Ra·ma's Bridge (rä'məz): *see* Adam's Bridge.

Ra·mat Gan (rä'mät gän'), city (1975 est. pop. 121,100), W central Israel, adjacent to Tel Aviv. Founded in 1921, it is an important industrial center and a health resort. Food processing is the chief industry; construction materials are also made. Diamonds are processed on the city's outskirts.

Ram·bouil·let (räɴ-bo͞o-yā'), town (1968 pop. 15,918), Yvelines dept., N France. It is a summer resort in the heart of a magnificent forest. Sheep are raised, and radio equipment and plastics are made. The nearby chateau (14th–18th cent.), set in a beautiful park, is the official summer residence of French presidents.

Ram·la or **Ram·leh** (both: räm'lĕ), town, central Israel, in a farming area. Ramla was probably founded (c.716) by Arabs. It later became the capital of Palestine and was fought over constantly during the Crusades. Israeli forces took it in 1948.

Ram·pur (räm-po͝or'), city (1971 pop. 161,802), N central India. Pottery, fabrics, and cutlery are produced. Its library has many rare manuscripts and a fine collection of Mogul miniature paintings.

Ram·sey (răm'zē). **1.** County (1970 pop. 476,255), 155 sq mi (401.5 sq km), E Minn., crossed in the S by the Mississippi River; formed 1849; co. seat St. Paul. It is in an industrial and residential area, with oil and gas wells, a food-processing industry, and varied manufactures, including textiles, wood products, furniture, chemicals, plastics, concrete, and machinery. There are small lake resorts in the north. **2.** County (1970 pop. 12,915), 1,248 sq mi (3,232.3 sq km), NE central N.Dak., in a rich agricultural area watered by numerous lakes; formed 1873; co. seat Devils Lake. Its agriculture includes grain, livestock, dairy products, and poultry.

Rams·gate (rămz'gāt, -gĭt), municipal borough (1973 est. pop. 40,090), in the Isle of Thanet, Kent, SE England. It is a resort and yachting harbor. Ramsgate began as a fishing settlement, and extensive trade with Baltic ports developed early in the 18th cent. There are fine examples of Regency architecture here. Caves in the cliffs provided bombproof shelters during World War II.

Ran·ca·gua (räng-kä'gwä), city (1970 pop. 86,404), capital of O'Higgins prov., central Chile, in a fertile valley among the Andean foothills. Though one of Chile's largest copper mines is nearby, Rancagua is primarily an agricultural center.

Ran·chi (rän'chē), city (1971 pop. 176,225), Bihar state, E central India. It is a health resort, 2,128 ft (649 m) above sea level. The city houses one of the largest machine-tool factories in India.

Ran·dall (răn'dəl), county (1970 pop. 53,885), 916 sq mi (2,372.4 sq km), extreme N Texas, on the high plains of the Panhandle, drained by Tierra Blanca and Palo Duro creeks, here forming the Prairie Dog Town Fork of the Red River; formed 1876; co. seat Canyon. This cattle and wheat area also produces grain sorghums, barley, oats, hay, poultry, and dairy products.

Ran·ders (rä'nərs), city (1976 est. pop. 64,193), Århus co., N central Denmark, a seaport at the mouth of the Gudenå River in the Randers Fjord (an arm of the Kattegat). Manufactures include machinery, iron products, and gloves. The city is noted for its salmon fishing. Founded in the 11th cent., Randers was an important trade center in the Middle Ages.

Rand·fon·tein (rănt'fôn-tān'), city (1970 pop. 50,481), Transvaal, NE South Africa, near Johannesburg. Randfontein is a gold-mining center, and uranium is refined.

Ran·dolph (răn'dŏlf'). **1.** County (1970 pop. 18,331), 581 sq mi (1,504.8 sq km), E Ala., in a piedmont region drained by the Tallapoosa River; formed 1832; co. seat Wedowee. It has agriculture (cotton and livestock), timber, quarried products, and a textile industry. **2.** County (1970 pop. 12,645), 647 sq mi (1,675.7 sq km), NE Ark., bounded on the N by the Mo. line and drained by the Black, Current, Spring, and Elevan Point rivers; formed 1835; co. seat Pocahontas. In an agricultural area yielding cotton, grain, poultry, livestock, and dairy products, it has timber and some manufacturing. **3.** County (1970 pop. 8,734), 436 sq mi (1,129.2 sq km), SW Ga., in a coastal-plain agricultural area; formed 1828; co. seat Cuthbert. Cotton, corn, peanuts, and fruit are grown. It has lumbering and textile industries. **4.** County (1970 pop. 31,379), 594 sq mi (1,538.5 sq km), SW Ill., bounded on the W and S by the Mississippi River and drained by the Kaskaskia River; formed 1795; co. seat Chester. It has agriculture (corn, wheat, dairy products, and poultry), bituminous-coal mining, lumbering, and manufacturing of shoes and leather goods, flour, farm machinery, and wood products. **5.** County (1970 pop. 28,915), 457 sq mi (1,183.6 sq km), E Ind., bordered on the E by Ohio and drained by the Mississinewa and Whitewater rivers and by the West Fork of White River; formed 1818; co. seat Winchester. In an agricultural area yielding grain, poultry, dairy products, and livestock, it has lumbering, stone quarrying, and diversified manufacturing. **6.** County (1970 pop. 22,434), 473 sq mi (1,225.1 sq km), N central Mo., drained by tributaries of the Chariton and Salt rivers; formed 1829; co. seat Huntsville. Agriculture (corn, wheat, and oats), manufacturing, and bituminous-coal mining are the major occupations. **7.** County (1970 pop. 76,358), 798 sq mi (2,066.8 sq km), central N.C., in a piedmont area, drained by Deep River; formed 1779; co. seat Asheboro. It has farming (tobacco, hay, corn, wheat, dairy products, and poultry), textile manufacturing, and sawmilling. **8.** County (1970 pop. 24,596), 1,036 sq mi (2,683 sq km), NE central W.Va., at the base of the Eastern Panhandle on the Allegheny Plateau, drained by the Elk and Tygart Rivers; formed 1787; co. seat Elkins. Its agriculture includes livestock, fruit, and tobacco. There are bituminous-coal mines, and limestone quarries. It has furniture and factories and railroad shops. Much of the county is within Monongahela National Forest.

Randolph. **1.** Community (1970 pop. 18,233), Montgomery co., W central Md., a suburb of Washington, D.C. **2.** Town (1978 est. pop. 30,100), Norfolk co., E Mass.; settled c.1710, set off from Braintree and inc. 1793. It is chiefly residential. **3.** Town (1970 pop. 500), seat of Rich co., N Utah, E of Logan near the Bear River, in a livestock and grain area. Dairy products are processed.

Rand·wick (rănd'wĭk'), municipality (1976 pop. 119,460), New South Wales, SE Australia, part of the Sydney urban agglomeration. It has a large race track.

Ran·goon (răng-go͞on', răng'go͞on'), city (1973 est. pop. 2,055,365), capital of Burma, S central Burma, on the Rangoon River near its entrance into the Gulf of Martaban. Rangoon is the transportation hub of the country and its commercial and industrial center. The major exports are rice, teak, bran, petroleum, cotton, rubber, and copper; there are rice mills, sawmills, oil refineries, and shipyards. Probably founded in the 6th cent., it was until the 18th cent. a small fishing village, dominated—as is the modern city—by the the golden-spired Shwe Dagon Pagoda. Alaungapaya, the founder of the last line of Burmese kings, made the town his capital in 1753. Under his rule Rangoon was built up as the chief port of Burma. After it came under British rule in 1852, it was transformed into a modern city.

Ran·kin (răng'kən), county (1970 pop. 43,933), 755 sq mi (1,955.5 sq km), central Miss., bounded on the W and NW by the Pearl River; formed 1828; co. seat Brandon. Agriculture (cotton, corn, and truck crops) and lumbering are major industries. Natural-gas fields were discovered here in the 1930s.

Rankin, city (1970 pop. 1,105), seat of Upton co., W Texas, S of Midland, in a sheep and oil area; est. 1879.

Ran·noch, Loch (răn'əKH), lake, 9.5 mi (15.3 km) long and 1 mi (1.6 km) wide, Perthshire, central Scotland, in the Grampians. Part of a hydroelectric project, there is a power station at the west end. The lake is known for trout and salmon fishing.

Ran·som (răn'səm), county (1970 pop. 7,102), 863 sq mi (1,388.6 sq km), SE N.Dak., in a rich prairie area drained by the Sheyenne River; formed 1873; co. seat Lisbon. Farming and livestock raising are of major importance.

Ran·te·kom·bo·la (rän'tə-kŏm'bəl-ə), mountain, 11,286 ft (3,442.2 m) high, on the SW peninsula of Celebes island, Indonesia. It is the island's highest point.

Ran·toul (răn-to͞ol'), village (1978 est. pop. 22,500), Champaign co., E Ill., in a blackland farm area that yields corn and soybeans; inc. 1868.

Ra·pal·lo (rä-päl'lō), town (1976 est. pop. 29,226), in Liguria, NW Italy, on the Ligurian Sea and on the Italian Riviera. It is a seaport and seaside resort.

rape (rāp), name applied to each of the six obsolete territorial divisions (Hastings, Pevensey, Lewes, Bramber, Arundel, and Chichester) into which Sussex, England, is divided.

Rap·i·dan (răp'ə-dăn'), river, c.90 mi (144.8 km) long, rising in N Va. in the Blue Ridge E of Shenandoah and flowing SE and then NE to the Rappahannock River.

Rap·id City (răp'ĭd), city (1978 est. pop. 49,500), seat of Pennington co., SW S.Dak., on Rapid Creek, in an irrigated farm region served by the Bureau of Reclamation's Rapid Valley project; founded 1876 after the discovery of gold nearby, inc. 1882. It is the trade and transportation center of an extensive lumbering, ranching, and mining area. The city has meat-packing houses, flour mills, and firms that make jewelry, cement, and lumber. Nearby Ellsworth Air Force Base, a strategic air command installation, is a major source of employment. The city is also the tourist center of the Black Hills and the gateway to Mount Rushmore National Memorial, the Badlands National Monument, and Wind Cave National Park.

Ra·pides (rə-pēd'), parish (1970 pop. 118,078), 1,318 sq mi (3,413.6 sq km), central La., bounded on the NE by Big Saline Bayou and Catahoula Lake and drained by the Calcasieu and Red rivers; formed 1807; parish seat Alexandria. It is an agricultural and dairying area, growing cotton, hay, corn, sugar cane, and sweet and white potatoes. Some manufacturing is done, including the processing of farm products and timber. Part of Kisatchie National Forest is here.

Rap·pa·han·nock (răp'ə-hăn'ək), county (1970 pop. 5,199), 267 sq mi (691.5 sq km), N Va., in a piedmont region, rising in the N and NW to the Blue Ridge, and bounded on the N and NE by the Rappahannock River; formed 1833 when it was set off from Culpeper co.; co. seat Washington. It is in an agricultural area that produces corn, wheat, tobacco, hay, apples, truck farming, livestock, and dairy products. Clothing is manufactured. It includes part of the Shenandoah National Park.

Rappahannock, river, 212 mi (341.1 km) long, rising in the Blue Ridge Mts., N Va., and flowing generally SE to Chesapeake Bay. It is navigable to Fredericksburg. In the Civil War much fighting took place in the vicinity of the Rappahannock and the Rapidan, its largest tributary.

Rar·i·tan (răr'ĭ-tən), borough (1970 pop. 6,691), Somerset co., N central N.J., on the Raritan River just W of Somerville; inc. 1848. Dairying and manufacturing are done.

Raritan, river, 85 mi (136.8 km) long, rising in N central N.J., and flowing SE to Raritan Bay, an arm of Lower New York Bay, at Perth Amboy. Coal, brick products, refined metals, petroleum, clay, and sand are transported on the river.

Rar·o·ton·ga (răr-ə-tŏng'gə), formerly **Good·e·nough's Island** (go͝od'ĭ-nŭfs'), volcanic island (1968 est. pop. 10,900), 26 sq mi (67.3 sq km), South Pacific, capital of the Cook Islands. It is the largest, most important, and most southwesterly of the group. Citrus fruit, copra, and pearl shell are exported. Rarotonga was discovered in 1823 by the English missionary John Williams.

Ras al-Khai·mah (räs äl-khī'mä), sheikdom (1968 pop. 24,482), c.650 sq mi (1,685 sq km), part of the federation of United Arab Emirates, E Arabia, on the Persian Gulf. Previously affiliated with Sharjah, Ras al-Khaimah became a separate sheikdom under British protection in 1921. Oil production began in 1969. The sheikdom joined the United Arab Emirates in 1972.

Ras Da·shan (dä-shän'), highest peak of Ethiopia, 15,158 ft (4,623.2 m) high, N Amhara Plateau, E Africa. It is of volcanic origin and has many craters.

Rasht (răsht), city (1966 pop. 143,557), capital of Gilan prov., NW Iran, near the Caspian Sea. It is the trade center for an agricultural region where rice, cotton, silk, and peanuts are produced. Manufactures include textiles, food products, and soap. Rasht was reached by British trading expeditions from Russia in the 16th cent.

Ra·statt (rä'shtät, -stät), city (1970 pop. 29,850), Baden-Württemberg, SW West Germany, on the Murg River, near the French border. Manufactures include railroad cars, machinery, precision instruments, and furniture. Rastatt was first mentioned in 1247. It was destroyed (1689) by the French, but was rebuilt soon thereafter and served as the residence of the margraves of Baden-Baden from 1705 to 1771. The Treaty of Rastatt (Mar., 1714) complemented the treaties signed at Utrecht and Baden in 1713–14; together they ended the War of the Spanish Succession. As a result of the Treaty of Campo Formio (1797), a congress of the states of the Holy Roman Empire (attended by France) was held (1797-99) at Rastatt in order to determine compensation for the member states that had lost territory near the Rhine River to France during the French Revolutionary Wars; the congress was prematurely adjourned after the resumption of hostilities against France. Noteworthy buildings of the city include a baroque palace (17th–18th cent.) and several 18th cent. churches. The city's name is sometimes spelled Rastadt.

Rath·lin Island (răth'lĭn), 5 sq mi (13 sq km), N Northern Ireland; part of Co. Antrim. Its cliffs, of limestone and basalt, rise at Slieveacarn to 449 ft (137 m). Farming and fishing are important. St. Columba is said to have founded a church here in the 6th cent., and there are ruins of a castle in which Robert the Bruce is reputed to have hidden.

Rat·lam (rət-läm'), town (1971 pop. 118,625), Madhya Pradesh state, E central India, on the Bombay-Delhi RR. The town is a market for food grains, cotton, and oilseed. Manufactures include textiles, paper, straw products, and umbrellas.

Rat·na·pu·ra (rŭt'nə-po͝or'ə), town (1972 est. pop. 26,000), SW Sri Lanka (Ceylon). Located in a rubber- and rice-producing area, Ratnapura is Sri Lanka's major precious-stone center and has ruby and sapphire mines and an important gem-cutting industry. A hill topped by a Portuguese fort dominates the town.

Ra·ton (rə-tōn'), city (1970 pop. 6,962), seat of Colfax co., NE N.Mex., near the Colo. border NE of Santa Fe, at an altitude of 6,680 ft (2,037.4 m); settled in the 1870s on the Santa Fe Trail, laid out 1880, inc. 1891. It is a rail and trade center in a livestock and coal region. Electrical equipment and building blocks are made.

Rau·ma (rou'mä), city (1975 est. pop. 29,081), Turku ja Pori prov., SW Finland, on the Gulf of Bothnia. It is a port and has wood, paper, chemical, and lace and embroidery industries. Rauma was an early medieval trade center. In the late 19th cent. it had one of the largest sailing fleets in Finland.

Ra·val·li (rə-văl'ē), county (1970 pop. 14,409), 2,384 sq mi (6,174.6 sq km), W Mont., in an agricultural region including the fertile valley of the Bitterroot River and bordering on Idaho; formed 1893; co. seat Hamilton. Livestock, sugar beets, grain, and fruit are raised.

Ra·ven·na (rə-vĕn'ə, rä-vĕn'nä), city (1976 est. pop. 101,400), capital of Ravenna prov., in Emilia-Romagna, N central Italy, near the Adriatic Sea. It is an agricultural market and an industrial center. Manufactures include petroleum, furniture, cement, and processed food. Ravenna rose to importance under the Romans, who made Classis, its port, the station for their fleet in the northern Adriatic. In A.D. 402 Ravenna became capital of the Western Empire, and it was also the capital (5th–6th cent.) of the Ostrogoth kings Odoacer and Theodoric, who were responsible for some of the city's best buildings. Ravenna was the seat of the exarchs (governors of Byzantine Italy) from the late 6th cent. to 751, when its capture by the Lombards broke Byzantine power in Italy. Pepin the Short donated the lands of the exarchate to the pope in 756; this donation, confirmed by Charlemagne in 774, marked the beginning of the temporal power of the popes. Ravenna is famous for its colorful mosaics of the 5th and 6th cent., which show a strong Middle Eastern influence, and for its Roman and Byzantine buildings. Mosaics ornament the mausoleum of Galla Placidia (5th cent.), the octagonal baptistery (formerly a Roman bath), the churches of Sant' Apollinare Nuovo and Sant' Apollinare in Classe, and, richest of all, the Byzantine Church of San Vitale (consecrated 547). Also of note are the tombs of Theodoric and Dante, the Archbishop's Palace (with a museum), and the Academy of Fine Arts.

Ra·ven·na (rə-vĕn'ə, rĭ-văn'ə), city (1978 est. pop. 12,200), seat of Portage co., NE Ohio, in a lake and farm area; settled 1799, inc. 1852. Its manufactures include rubber, electric, and plastic products.

Ra·vens·berg (rä'vəns-bĕrKH), former county, W West Germany, now in North Rhine-Westphalia. It passed to the counts of Berg in 1346, to the dukes of Jülich in 1348 and, with Jülich, to the duchy of Cleves in 1521. In 1614 it was taken by Brandenburg.

Ra·vens·burg (rā'vənz-bûrg', rä'vəns-bŏŏrKH), city (1974 est. pop. 43,204), Baden-Württemberg, S West Germany. Its manufactures include machinery, pharmaceuticals, textiles, clothing, and processed foods. Ravensburg was founded in the late 11th cent. under the protection of the Guelphs. It became a free imperial city in the 13th cent. and flourished in the 15th and 16th cent. The city passed to Württemberg in 1810.

Ra·vi (rä'vē), one of the five rivers of the Punjab, 475 mi (764.3 km) long, rising in the Himalayas, NW India, and flowing generally W to join the Chenab River, NE Pakistan. Its waters are used extensively for irrigation and were contested by India and Pakistan until 1960, when a treaty was signed.

Ra·wal·pin·di (rä'wəl-pĭn'dē), city (1972 pop. 375,000), NE Pakistan. It occupies the site of an old village inhabited by the Rawals, a tribe of Yogis. The city has an oil refinery, gasworks, an iron foundry, railroad yards, and factories making tents, textiles, hosiery, chemicals, ordnance, finished steel, furniture, plywood, slate, and flour. Sikhs settled the area in 1765 and invited nearby traders to live in Rawalpindi. After the British occupied the Punjab (1849), it became a major British military outpost. From 1959 to 1970 it was the interim capital of Pakistan.

Raw·lins (rô'lĭnz), county (1970 pop. 4,393), 1,078 sq mi (2,792 sq km), NW Kansas, in a gently rolling plain area bordering in the N on Nebr. and watered by headstreams of Sappa and Beaver creeks; formed 1873; co. seat Atwood. It is in a grain and livestock area, with oil and gas fields and some manufacturing.

Rawlins, city (1970 pop. 7,855), seat of Carbon co., S Wyo., WNW of Laramie, at an altitude of 6,755 ft (2,060.3 m); founded 1867. In a tourist area, it is also a trade and distribution center for a ranch, oil, and uranium region.

Ray (rā), city (1966 pop. 102,835), Tehran prov., N Iran, a suburb of Tehran. The mausoleum of Hazrat Abd al-Azim, a major Shiite Moslem holy person, is here. The city was the site of the ancient and medieval city of Rages.

Ray, county (1970 pop. 17,599), 574 sq mi (1,486.7 sq km), NW Mo., bounded in the S by the Missouri River and drained by the Crooked River; formed 1820; co. seat Richmond. Agriculture (corn, wheat, oats, and potatoes) and coal mining dominate its economy.

Ray·mond (rā'mənd), town (1970 pop. 1,620), seat of Hinds co., SW Miss., WSW of Jackson, in a farm and timber area. It was the scene of a Civil War battle (May 12, 1863), resulting in a victory for Gen. U.S. Grant's forces.

Ray·mond·ville (rā'mənd-vĭl'), city (1970 pop. 7,987), seat of Willacy co., extreme S Texas, NNW of Brownsville; platted 1904, inc. 1921. In the irrigated section of the Rio Grande valley, it handles cotton, citrus fruit, vegetables, livestock, and oil.

Rayn·ham (rān'hăm), town (1970 pop. 6,705), Bristol co., SE Mass., NNE of Taunton; set off 1731. Ironworks were established in 1652.

Ray·town (rā'toun'), city (1978 est. pop. 32,600), Jackson co., W central Mo., a residential suburb of Kansas City; inc. 1950. It was the first stop on the Santa Fe Trail out of Independence, Mo.

Ray·ville (rā'vĭl'), town (1970 pop. 3,962), seat of Richland parish, NE La., near the Boeuf River E of Monroe, in a cotton and timber area. Veneer is manufactured.

Ré (rā), island, 33 sq mi (85.5 sq km), Charente-Maritime dept., off La Rochelle, W France, in the Bay of Biscay. The island is largely agricultural; it has oyster beds, some fishing, and a salt-extraction industry. The citadel, built (1681) by Vauban at Saint-Martin-de-Ré, is now a penitentiary.

Read·ing (rĕd'ĭng), county borough (1976 est. pop. 131,200), administrative center of Berkshire, S central England, on the Kennet River near its influx to the Thames. Reading is a market center with iron, engineering, malting, brewing, and biscuit and seed industries. It was occupied in 871 by the Danes, who burned it in 1006. A gateway and ruins of buildings, surrounded by a public park, remain of a Benedictine abbey founded in 1121 by Henry I, who is buried here. There are a 15th cent. grammar school, the Reading College of Technology, and the Univ. of Reading (1926; formerly a college, founded 1892, of Oxford Univ.), with agriculture and dairying departments.

Reading. 1. Town (1978 est. pop. 24,200), Middlesex co., NE Mass., a suburb of Boston; settled 1639, set off from Lynn and inc. 1644. Printing is the major industry. A 17th cent. tavern is in the town. **2.** City (1970 pop. 14,617), Hamilton co., SW Ohio, a suburb of Cincinnati; platted 1798, inc. 1851. It has diverse light manufacturing industries. **3.** City (1978 est. pop. 78,200), seat of Berks co., SE Pa., on the Schuylkill River, in the Pennsylvania Dutch region; laid out 1748, inc. as a city 1847. It is an important commercial, industrial, and railroad center. Its many manufactures include textiles, clothing, leather goods, and iron and steel products. Reading was an early iron-producing town; cannons were made here during the Revolution, and it was a Union ordnance center during the Civil War.

Rea·gan (rā'gən), county (1970 pop. 3,239), 1,133 sq mi (2,934.5 sq km), W Texas, at the N edge of Edwards Plateau, drained by Concho River; formed 1903; co. seat Big Lake.

Re·al (rē'ôl), county (1970 pop. 2,013), 625 sq mi (1,618.8 sq km), SW Texas, on Edwards Plateau, drained by the Nueces and Frio rivers; formed 1913; co. seat Leakey. It has ranching (goats, sheep, and cattle), timber (mainly cedar), and some agriculture.

Re·ci·fe (rə-sē'fĭ), city (1975 est. pop. 1,249,821), capital of Pernambuco state, NE Brazil, a port on the Atlantic Ocean. It lies partly on the mainland and partly on an island and is dissected by numerous waterways. Recife exports great quantities of the hinterland's products, including sugar, cotton, and coffee. Sugar refining and cotton milling are the chief industries. Founded by the Portuguese in 1548 as the port for nearby Olinda, Recife was settled by fishermen and sailors. The city was plundered by the British in 1595, and was occupied by the Dutch (1630–54), prospering under Maurice of Nassau. The city is the seat of two universities and has long been famed for its intellectual groups and political ferment. In addition to its modern buildings, Recife has a 17th cent. cathedral, an old Dutch fort, and an elaborate government palace.

Reck·ling·hau·sen (rĕk'lĭng-hou'zən), city (1974 est. pop. 123,229), North Rhine-Westphalia, W West Germany, an industrial center and transshipment point for the Ruhr district. Manufactures include iron and steel, machinery, chemicals, and textiles. Recklinghausen was chartered c.1230 and was held until 1803 by the archbishopric of Cologne. In 1815 it came under Prussian sovereignty.

Red (rĕd), river, 730 mi (1,174.6 km) long, rising in Yünnan prov., S China, and flowing SE, in deep, narrow gorges, through N Vietnam to form a great delta before entering the Gulf of Tonkin. The river carries a large quantity of silt, rich in iron oxide, that gives it a red color. The Red River has an irregular flow and is subject to flooding, especially during the June-Oct. high-water period; dikes and canals protect the delta from flood waters. Rice is the principal crop.

Red. 1. River, 1,222 mi (1,966.2 km) long, southernmost of the large tributaries of the Mississippi River. It rises in two branches in the Texas Panhandle and flows southeast between Texas and Okla. and between Texas and Ark. to Fulton, Ark. It then turns southward, enters La., and crosses southeast to the Atchafalaya and the Mississippi rivers. In Texas it flows rapidly through a canyon in semiarid plains, but later in its course it waters rich red-clay farm lands. For many years navigation was difficult on the lower course of the Red River due to fallen trees that floated downstream. The Great Raft, a 160-mi (257.4-km) logjam built through the centuries, was cleared from the river in the mid-1800s. The river is now navigable for small ships above Shreveport, La. **2.** River, often called the Red River of the North, c.310 mi (500 km) long, formed N of Lake Traverse, NE S.Dak., by the confluence of the Bois de Sioux and the Otter Tail rivers. It flows north between Minn. and N.Dak. and crosses the Canadian border into Man., emptying into Lake Winnipeg. The river drains the rich Red River valley region—the principal spring wheat- and flax-growing area of the United States and Canada.

Red Bank (băngk), borough (1978 est. pop. 11,900), Monmouth co., E N.J., on the Navesink estuary, in a fertile farm area; inc. 1908. An early shipping center, it is now a summer and winter resort with some light industry. Landmarks include Old Christ Church (1769) and the Allen House (1667).

Red Bluff (blŭf), city (1975 est. pop. 8,400), seat of Tehama co., N Calif., on the Sacramento River SSE of Redding; inc. 1876. It is a trade center in a farm and timber area.

Red·bridge (rĕd'brĭj'), borough (1971 pop. 238,614) of Greater London, SE England. Redbridge was created in 1965 by the merger of the municipal boroughs of Ilford and of Wanstead and Woodford, part of the municipal borough of Dagenham, and part of the urban district of Chigwell. Although primarily residential, Redbridge is an important shopping and commercial center. Its industries include light engineering and the manufacture of electrical components, photographic materials, and chemicals. Much of Epping Forest lies within the borough.

Red Cloud (kloud), city (1970 pop. 1,531), seat of Webster co., S Nebr., SSW of Grand Island, near the Kansas line and on the Republican River; inc. 1872. It is in a livestock, dairy, and poultry area.

Red Deer (dîr), city (1976 pop. 32,184), S central Alta., Canada, on the Red Deer River. It is the trade center for a region of dairying, mixed farming, and oil and gas production.

Red Deer, river, 385 mi (619.5 km) long, rising in the Rocky Mts. in Banff National Park, SW Alta., Canada, and flowing NE past Red Deer city, then SE and E across the plains to the South Saskatchewan River just over the Sask. border.

Red·ding (rĕd'ĭng). **1.** City (1978 est. pop. 40,400), seat of Shasta co., N central Calif., on the Sacramento River; inc. 1872. A tourist center for a mountain and lake region, it also has lumbering and food-processing industries. **2.** Town (1970 pop. 5,590), Fairfield co., SW Conn., NW of Bridgeport; settled c.1711, inc. 1767. Mark Twain's summer home was here, and he gave the town a library. The Israel Putnam Memorial Campground north of Redding Ridge commemorates the winter of 1778-79, when Putnam's troops camped here.

Red·ditch (rĕd'ĭch), new town and urban district (1976 est. pop. 51,600), Hereford and Worcester, central England. Redditch was designated one of the New Towns in 1964 to alleviate overpopulation in Birmingham and the Black Country. Its manufactures include needles, fishing tackle, springs, bicycles, and motorcycles.

Red·field (rĕd'fēld'), city (1970 pop. 2,943), seat of Spink co., NE central S.Dak., S of Aberdeen. It is a resort and farm trade center. Flour, dairy products, and grain are processed.

Red Lake (lāk), county (1970 pop. 5,388), 432 sq mi (1,119 sq km), NW Minn., drained by the Red Lake River; formed 1896; co. seat Red Lake Falls. It is in an agricultural area (corn, oats, barley, potatoes, dairy products, livestock) and has food-processing and lumbering industries. Mobile homes and machinery are made.

Red Lake Falls, city (1970 pop. 1,740), seat of Red Lake co., NW Minn., NE of Crookston; inc. as a village 1881, as a city 1898. A fur-trading post before 1800, it is now a dairy and farm trade center.

Red·lands (rĕd'ləndz), city (1978 est. pop. 35,100), San Bernardino co., S Calif., in the San Bernardino Valley; inc. 1888. Aircraft propulsion systems, furniture, and electrical vehicles are made.

Red Lodge (lŏj), city (1970 pop. 1,844), seat of Carbon co., S Mont., SW of Billings and NE of Yellowstone National Park; founded 1885, inc. 1892. It is a trade center of a ranch, farm, and coal-mine area.

Red·mond (rĕd'mənd), city (1970 pop. 11,020), King co., W Wash., a suburb of Seattle, on Lake Sammamish; inc. 1912. Its economy centers around research and development industries and the manufacture of electronic components and building products. Of interest are the preserved portions of the old brick Stevens Pass highway, and Marymoor Park, site of archaeological excavations.

Red Oak (ōk), city (1970 pop. 6,210), seat of Montgomery co., SW Iowa, on the East Nishnabotna River SE of Council Bluffs; inc. 1869. A processing point in a farm area, it also makes dry batteries.

Re·don·do Beach (rĭ-dŏn'dō), city (1978 est. pop. 64,400), Los Angeles co., S Calif., on the Pacific Ocean; inc. 1892. It is a residential and resort city, with boating facilities and fine beaches.

Red Riv·er (rĭv'ər). **1.** Parish (1970 pop. 9,226), 406 sq mi (1,051.5 sq km), NW La., bounded on the W by Bayou Pierre and Bayou Pierre Lake, on the E by Black Lake Bayou, and intersected by the Red River; formed 1848; parish seat Coushatta. Cotton, corn, and hay are grown. Natural-gas and oil fields are found here. Industries include cotton ginning and lumber milling. **2.** County (1970 pop. 14,298), 1,033 sq mi (2,675.5 sq km), NE Texas, bounded on the N by the Red River (here the Okla. border), on the S by the Sulphur River; formed 1836; co. seat Clarksville. It has diversified farming (cotton, corn, grains, hay, legumes, fruit, soybeans, vegetables, and pecans), livestock raising, some dairying, lumbering, and manufacturing of aluminum products, furniture, and books.

Red Sea, narrow sea, c.170,000 sq mi (440,300 sq km), c.1,450 mi (2,335 km) long and up to 225 mi (362 km) wide, between Africa and the Arabian Peninsula. The Gulf of Aqaba and the Gulf of Suez are the sea's northern arms; between them is the Sinai peninsula. The Red Sea is linked with the Gulf of Aden and the Arabian Sea by the straits of Bab el Mandeb. The sea is dotted with islands and with dangerous coral reefs. It is surrounded by exceedingly hot and dry deserts and steppes; the summer water temperature exceeds 85°F (29°C), and the water has a high salt content. The Red Sea was probably named for the reddish algae that appear in it at certain times of the year. Its importance as a trade route declined with the discovery of an all-water route around Africa in 1498. The opening of the Suez Canal in 1869 made the Red Sea one of the chief shipping routes connecting Europe with the Far East and Australia; however, the closing of the canal after the 1967 Arab-Israeli War, the building of pipelines to the Mediterranean Sea, and the construction of supertankers too large for the canal have combined to diminish the sea's importance as a commercial artery.

Red Wil·low (wĭl'ō), county (1970 pop. 12,191), 686 sq mi (1,776.7 sq km), S Nebr., bounded in the S by Kansas and drained by Beaver Creek and the Republican River; formed 1873; co. seat McCook. It is a fertile agricultural area, with dairying, grain, and livestock.

Red Wing (wĭng), city (1978 est. pop. 15,100), seat of Goodhue co., SE Minn., on the Mississippi River at the head of Lake Pepin; inc. 1857. Shoe manufacturing, leather processing, and flour milling are carried on. The early explorers found an Indian village here.

Red·wood (rĕd'wŏŏd'), county (1970 pop. 20,024), 874 sq mi (2,263.7 sq km), SW Minn., bounded on the N by the Minnesota River and drained by the Redwood and Cottonwood rivers; formed 1862; co. seat Redwood Falls. Its agriculture includes corn, oats, barley, and livestock. It has some mining and a food-processing industry (fats and oils), and it manufactures metal products and machinery.

Redwood City, city (1978 est. pop. 54,900), seat of San Mateo co., W Calif., on San Francisco Bay; inc. 1868. Food is processed, and electronic products, wire and cables, cement, plastics, automotive equipment, and chemicals are made. The city's large chrysanthemum industry dates from 1900.

Redwood Falls, town (1970 pop. 4,774), seat of Redwood co., SW Minn., on Redwood River, NW of New Ulm; platted 1865, inc. as a village 1875, as a city 1891. A shipping point for a farm area, it processes meat. Adjoining the city is Alexander Ramsey State Park, a picturesque area with gorges, cliffs, and falls.

Redwood National Park, 56,201 acres (22,761 hectares), along the Pacific coast, NW Calif.; est. 1968. Backed by coastal bluffs, 40 mi (64 km) of beach, lagoon, and rocky coast are preserved in their natural state; seals, sea lions, and birds live on offshore rocks. Inland, numerous stands of virgin California redwood, many over 2,000 years old, are found; the world's tallest tree, 367 ft (112 m) high, is located in the park.

Reed City (rēd), city (1970 pop. 2,286), seat of Osceola co., N central Mich., S of Cadillac, in a farm and lake region; inc. as a village 1874, as a city 1932. Aluminum products, tools, and flooring materials are made.

Reel·foot Lake (rēl'fŏŏt'), 20 mi (32 km) long, NW Tenn., near the Mississippi River; designated a national natural landmark by the National Park Service. It was formed when a depression created by earthquakes in 1811–12 was filled with Mississippi River water.

Reeves (rēvz), county (1970 pop. 16,526), 2,608 sq mi (67,568 sq km), extreme W Texas, in a plains area sloping to the Pecos River; formed 1883; co. seat Pecos. It has cattle and sheep ranching, and farming in the irrigated sections. Oil and sulfur are produced.

Re·fu·gio (rə-fŏŏr'ē-ō, -fyŏŏr'-), county (1970 pop. 9,494), 771 sq mi (1,996.9 sq km), S coastal Texas, bounded on the N by the San Antonio River, on the SW by the Aransas River, on the SE by Copano Bay, on the NE by San Antonio Bay; formed 1836; co. seat Refugio. This area, a leading Texas oil-producing county, also has ranching (cattle and sheep), agriculture (mainly cotton, also peanuts, grain sorghums, vegetables, fruit, and flax), and oil refineries.

Refugio, town (1970 pop. 4,340), seat of Refugio co., S Texas, on the Mission River NNE of Corpus Christi. A Spanish mission, Nuestra Señora del Refugio (our Lady of Refuge), established earlier near San Antonio Bay, was moved here in 1795. The town itself grew after settlement began in 1829, and it was formally established in 1834. Taken (1836) by the Mexicans in the Texas Revolution, Refugio was again occupied briefly by the Mexicans in the invasion of 1842. The bells in the Catholic church are said to be those of the original mission, taken and later returned by the Mexicans. Oil fields add prosperity to the old farming and ranching area.

Re·gens·burg (rā'gəns-bûrg', -bŏŏrKH), city (1974 est. pop. 133,183), Bavaria, SE West Germany, a port at the confluence of the Danube and Regen rivers. Its manufactures include machines, precision instruments, chemicals, leather goods, and printed materials. There are shipyards in the city. Regensburg, an important Roman frontier station, was known as Regina Castra. An abbey was founded in the mid-7th cent., and St. Boniface established an episcopal see in 739. Regensburg was captured (788) by Charlemagne when he subjugated Bavaria. The city was one of the most prosperous commercial centers of medieval Germany, trading especially with India and the Middle East. The city proper accepted the Reformation in the 16th cent., but soon thereafter it was strongly influenced by the Roman Catholic Counter Reformation (late 16th cent.). Its commerce declined in the 15th and 16th cent., as a result of the shifting of international trade routes. Regensburg was frequently the meeting place of the imperial diet from 1532, and from 1663 to 1806 it was the permanent seat of the diet. The diet that met here from 1801 to 1803 under the influence of Napoleon Bonaparte completely reorganized the moribund Holy Roman Empire. In 1810 the city passed to Bavaria and became the capital of the Upper Palatinate. Regensburg was badly damaged by the Allies in World War II, largely because it was an airplane-manufacturing center. Noteworthy structures of the city include the Gothic cathedral (13th–16th cent.); parts of the Porta Praetoria, a Roman gate (built A.D. 179); the Schottenkirche St. Jakob, a 12th cent. church; an 11th cent. chapel (with later decoration in the rococo style); the old city hall (14th–18th cent.), where the imperial diet met; and St. Emmeram, the episcopal residence (a former Benedictine convent founded in the 7th cent.).

Reg·gio di Ca·la·bri·a (rĕd'jō dē kä-lä'brē-ä), city (1976 est. pop. 177,883), capital of Reggio di Calabria prov., Calabria, extreme S Italy, on the Strait of Messina opposite Sicily. It is a beach resort and an important agricultural market for fruits and tobacco. Bergamot essence (used in perfume) is produced here. Known as Rhegium in ancient times, the city became (12th cent.) part of the kingdom of Sicily and later (13th cent.) of Naples. Its strategic position has resulted in numerous foreign invasions and incursions. The city has suffered many earthquakes—the worst ones in 1783 and 1908.

Reggio nell' E·mi·lia (nĕl' ā-mē'lyä), city (1976 est. pop. 129,674), capital of Reggio nell' Emilia prov., in Emilia-Romagna, N central Italy, on the Aemilian Way. Manufactures include food products, chemicals, and electrical equipment. Founded by Rome in the 2nd cent. B.C., it later became a free commune and in 1289 came under the Este family.

Re·gi·na (rĭ-jī'nə), city (1976 pop. 149,593), provincial capital, S Sask., Canada, on Wascana Creek. The city is the distribution and service center for one of the world's largest wheat-growing plains. Industries include agricultural processing, meat-packing, printing, bookbinding, oil refining, and automobile manufacturing. Regina was founded in 1882 when a railroad line was constructed through the region. It was the capital of the Northwest Territories from 1883 to 1905, when it became the capital of the newly created Sask. prov. From 1892 to 1920 Regina was the headquarters of the Northwest Mounted Police, and it is now western headquarters of the Royal Canadian Mounted Police.

Re·ho·vot (rĭ-hō'vōt) or **Re·ho·both** (rĭ-hō'bəth), town (1975 est. pop. 50,100), central Israel. Its industries include fruit packing and the production of citrus concentrates. Plastic products, imitation leather, cereals, and pharmaceuticals are also manufactured. Rehovot was founded in 1890 by Jewish immigrants from Russia. Chaim Weizmann, Israel's first president, lived here during the British mandate period (1922–48). His house is preserved.

Rei·chen·bach (rī'KHən-bäKH) or **Reichenbach im Vogt·land** (ĭm fôkt'länt'), city (1974 est. pop. 27,440), Karl-Marx-Stadt dist., S East Germany, at the foot of the Erzgebirge; chartered in the late 13th cent. Manufactures of this industrial city include textiles, clothing, plastics, and machinery.

Reichenbach Falls, waterfalls, total drop 656 ft (200 m), S central Switzerland, where the Reichenbach River joins the Aare River. Upper Reichenbach Falls is one of the highest cataracts (c.300 ft/90 m) in the Alps.

Rei·chen·hall (rī'KHən-häl'): *see* Bad Reichenhall, West Germany.

Reids·ville (rēdz'vĭl'), city (1970 pop. 1,806), seat of Tattnall co., SE Ga., W of Savannah, in a tobacco, cotton, and sweet-potato area; inc. 1838.

Rei·gate (rī'gĭt -gāt), municipal borough (1971 pop. 56,088), Surrey, S England. It is largely residential. Numerous parks attract visitors from London.

Reims (răNS): *see* Rheims, France.

Rein·deer Lake (rān'dîr'), one of the largest lakes in Canada, 2,467 sq mi (6,390 sq km), NE Sask. and NW Man. The Reindeer River drains it south to the Churchill River. The lake has many islands and is noted for its commercial and sport fishing.

Re·ma·gen (rā'mä'gən), town (1974 est. pop. 14,720), Rhineland-Palatinate, W West Germany, on the Rhine River. It is a rail junction from which mineral water is shipped. U.S. troops used the Ludendorff bridge at Remagen when they first crossed (Mar., 1945) the Rhine in World War II.

Rem·scheid (rĕm'shīt'), city (1975 pop. 135,810), North Rhine–Westphalia, W West Germany, on the Wupper River. It is the leading center of the West German tool and hardware industry; other products include textiles, household appliances, and machinery. Remscheid was chartered in 1808.

Rends·burg (rĕnts'bŏŏrKH), city (1974 est. pop. 34,625), Schleswig-Holstein, N West Germany, a port on the Eider River and the Kiel Canal. The city's manufactures include machinery, and there are shipyards. Rendsburg passed to the counts of Holstein in 1252 and was chartered soon thereafter. When Schleswig-Holstein rose (1848–51) against Denmark, Rendsburg was made the provisional capital. The Danes, however, regained control of the city. In 1866 Rendsburg was annexed by Prussia.

Ren·frew (rĕn'frŏŏ'), burgh (1971 pop. 18,589), Strathclyde region, W central Scotland, on the Clyde River, near Glasgow. It has shipyards and manufactures rubber, paint, and soap. A busy Clydeside port since the 12th cent., it became a burgh in the 14th cent. Until 1975 Renfrew was county town of Renfrewshire.

Ren·frew·shire (rĕn′frōō-shîr′, -shər), former county, central Scotland, on the Clyde River estuary. The county town was Renfrew. Renfrewshire was separated from Lanarkshire in 1404. In 1975 it became part of the Strathclyde region.

Ren·nell (rĕn′əl), coral limestone island, NE Coral Sea, southernmost of the Solomon Islands, W Pacific Ocean, S of Guadalcanal. The largest quantity of enclosed fresh water in the Pacific Ocean, Lake Te-Nggano (50 sq mi/129.5 sq km), is found at its eastern end.

Rennes (rĕn), city (1975 pop. 198,305), capital of Ille-et-Vilaine dept., NW France, at the junction of the Vilaine and Ille rivers. Among its products are textiles, leather goods, automobiles, electronic equipment, and petroleum. Rennes was an important Gallo-Roman town. In the 10th cent. it became the capital of the Breton county of Rennes and in 1196 of Brittany. Under the ancien régime it was the seat of the provincial Breton estates and the powerful parlement of Rennes. The town was ravaged by the Norsemen and also during the Hundred Years' War.

Re·no (rē′nō), county (1970 pop. 60,765), 1,260 sq mi (3,263.4 sq km), central Kansas, drained by the Arkansas River; formed 1867; co. seat Hutchison. It is in a wheat-growing region, with grain mills, oil and gas wells, salt refineries, and diversified manufactures (textiles, metal and wood products, furniture, chemicals, and farm machinery).

Reno, city (1978 est. pop. 80,900), seat of Washoe co., W Nev., on the Truckee River; inc. 1903. Tourism is the major industry. A crisp climate, extensive resort facilities, and free port privileges, in a state that permits quick divorce and legalized gambling, have promoted Reno's prosperity. Mining and agriculture are also important; irrigation is aided by the Truckee storage project. The site was once a popular campsite beside a ford on the Donner Pass route to Calif. The name Lake's Crossing was changed to Reno when the Central Pacific RR arrived in 1868 and the town was laid out. A rodeo, a national air race, and the state fair are held here annually. Reno is the site of a branch of the state university and a school of mines. It has a planetarium and an extensive collection of antique cars.

Rens·se·laer (rĕn′sə-lîr′, rĕn′sə-lər), county (1970 pop. 152,510), 665 sq mi (1,722.4 sq km), E N.Y., bounded on the W by the Hudson River, on the E by the Mass. and Vt. borders, and drained by the Hoosic River; formed 1791; co. seat Troy. It has dairying, farming (corn, clover, fruit, and potatoes), and poultry raising. It includes part of the Taconic Mts. and many small resort lakes.

Rensselaer. 1. City (1970 pop. 4,688), seat of Jasper co., NW Ind., on the Iroquois River NNW of Lafayette; settled c.1836, inc. 1897. It is a trade center for a livestock and grain area. **2.** City (1970 pop. 10,136), Rensselaer co., E N.Y., on the E bank of the Hudson River opposite Albany; settled 1630 by Dutch, inc. 1897. Chemicals and concrete products are among its manufactures. The city was formed by the union of several villages within the tract granted to Kiliaen Van Rensselaer by the Dutch West India Company.

Ren·ton (rĕn′tən), city (1978 est. pop. 27,700), King co., W Wash., an industrial suburb of Seattle, on Lake Washington; inc. 1901. It is a freshwater port of entry via the Lake Washington Ship Canal. Its clayworks were established in 1901 and its iron foundry in 1905. A Boeing aircraft plant began operation here during World War II and has since greatly expanded.

Ren·ville (rĕn′vĭl′). **1.** County (1970 pop. 21,139), 980 sq mi (2,538.2 sq km), SW central Minn., bounded on the S by the Minnesota River; formed 1855; co. seat Olivia. In an agricultural area yielding corn, oats, barley, potatoes, and livestock, it has some mining and a food-processing industry. Concrete, tools, and electronic equipment are manufactured. **2.** County (1970 pop. 3,828), 886 sq mi (2,294.7 sq km), N N.Dak., in an agricultural area drained by the Souris River; formed 1908; co. seat Mohall. It has diversified farming (dairy products, livestock, poultry, and wheat) and oil deposits.

Rep·ton (rĕp′tən), village (1971 pop. 32,667), Derbyshire, central England. It was once a capital of the kingdom of Mercia. A monastery, the seat of the bishopric of Mercia, stood here in the 7th cent. but was later destroyed by the Danes. There are remains of a priory founded in 1172.

Re·pub·lic (rĭ-pŭb′lĭk), county (1970 pop. 8,498), 719 sq mi (1,862.2 sq km), N Kansas, in a plains region bordering in the N on Nebr. and drained by Republican River; formed 1860; co. seat Belleville. It is in a corn, wheat, and livestock region, with some manufacturing of plastic products.

Republic, town (1977 est. pop. 1,053), seat of Ferry co., NE Wash., on the Sanpoil River NNE of Grand Coulee Dam; settled 1896 as a gold-mining camp, inc. 1900. It is a tourist center in a farming and lumbering region.

Re·pub·li·can (rĭ-pŭb′lĭ-kən), river, c.420 mi (675 km) long, formed in S Nebr. by the junction of the North Fork and Arikaree rivers. It is joined by the South Fork at Benkelman and flows east across the rolling grasslands of Nebr. and southeast across Kansas to join the Smoky Hill and form the Kansas River at Junction City. Its broad channel traverses a rich agricultural region. The river is included in the Missouri River basin project. Many dams and reservoirs have been built for flood control, irrigation, and power.

Re·sa·ca de la Pal·ma (rĭ-săk′ə dā lə păl′mə, rā-sä′kä *th*ä lä päl′mä), valley, an abandoned bed of the Rio Grande, N of Brownsville, Texas, where Mexican troops were defeated (May 9, 1846) by American forces led by Gen. Zachary Taylor.

Re·serve (rĭ-zûrv′), village (1970 pop. 500), seat of Catron co., W N.Mex., on the San Francisco River near the Ariz. border NNW of Silver City. At an altitude of 5,770 ft (1,759.9 m) in the Apache National Forest, it is in an agricultural and lumbering region.

Re·sis·ten·cia (rā-sēs-tān′syä), city, (1970 pop. 142,848), capital of Chaco prov., NE Argentina. It is the nucleus of an area of frontier settlements. The city carries on a lively trade from its port, Barranqueras, on the Paraná River. Cotton, cattle, leather, and quebracho (for tannin extraction) are the city's chief products; the economy of the surrounding region is based on farming, cattle raising, and lumbering. It was originally the site of an 18th cent. Jesuit mission.

Re·şi·ta (rĕ′shē-tsä), city (1974 est. pop. 75,029), W Rumania, in the Banat, in the W foothills of the Transylvanian Alps. The production of iron and steel, machinery, metals, and chemicals are the leading industries. Reşita was known in Roman times as a mining center for precious metals. The modern city was founded in 1768, when the first foundry was established.

Res·o·lu·tion Island (rĕz′ō-lōō′shən), 387 sq mi (1,002.3 sq km), off SE Baffin Island, Canada, in SE Franklin dist., Northwest Territories, at the E entrance of Hudson Strait.

Ré·u·nion (rā-ü-nyôɴ′), overseas department of France (1975 est. pop. 475,700), c.970 sq mi (2,510 sq km), one of the Mascarene Islands, in the Indian Ocean c.430 mi (690 km) E of Madagascar; capital Saint-Denis. The island is composed mainly of one active and several extinct volcanoes; its highest point is Le Piton des Neiges (10,069 ft/3,071 m). Settlement and cultivation are concentrated in the coastal lowlands. Since the 19th cent. sugar has been the island's chief product and export. Réunion was known to the Arabs and was visited by the Portuguese in the early 16th cent. The island was uninhabited until settled by the French c.1642. At first a penal colony, Réunion became a post of the French East India Company in 1665. In the 18th cent. the island was an exporter of coffee. After 1815, when coffee no longer could be produced competitively, sugar cane became the main crop. In 1947 Réunion became an overseas department.

Reus (rĕ′ōōs), city (1970 pop. 59,095), Tarragona prov., NE Spain, on the Mediterranean Sea, in Catalonia. Since the introduction (18th cent.) by English manufacturers of a cotton-spinning industry, Reus has grown into an important industrial center.

Reut·ling·en (roit′lĭng-ən), city (1975 pop. 96,157), Baden-Württemberg, SW West Germany. Manufactures include textiles, paper, leather goods, and machinery. Reutlingen was a free imperial city from the mid-13th cent. until it passed (1802–3) to Württemberg. The Church of St. Mary (13th–14th cent.) is an outstanding example of late German Gothic architecture.

Re·vere (rĭ-vîr′), city (1978 est. pop. 39,700), Suffolk co., E Mass., a residential suburb of Boston, on Massachusetts Bay; settled c.1630, set off from Chelsea and named for Paul Revere 1871, inc. as a city 1914. It has printing industries and varied manufactures.

Re·vil·la·gi·ge·do Islands (rĭ-vĭl′ə-gə-gē′dō), archipelago in the Pacific Ocean, c.450 mi (725 km) W of Colima state, Mexico. Socorro (110 sq mi/285 sq km) is the largest island. A volcano on San Benedicto island rose suddenly in 1952.

Re·wa (rē′wə), walled city (1971 pop. 69,197), Madhya Pradesh state, central India. It is the center of a district rich in coal, corundum, and limestone and a market for timber, cement, and limestone. It has several temples set in a park.

Rex·burg (rĕks'bûrg'), city (1973 est. pop. 9,761), seat of Madison co., E Idaho, in the Snake River valley NNE of Idaho Falls; founded 1883 by Mormons. It is a trade center for an area of ranches, irrigated farms of the Minidoka project, and dry farms.

Rey·kja·vík (rā'kyä-vēk'), city (1976 est. pop. 84,493), capital of Iceland, SW Iceland, on the Faxaflói. It is the center of the cod-fishing industry and the chief commercial and industrial hub of Iceland. Publishing is an important industry. Reykjavík is the seat of the parliament (Althing), of the Lutheran bishop of Iceland, and of the supreme court; there is a university (founded 1911). The city's heating system utilizes nearby hot springs. The founding of Reykjavík by Ingolfur Arnarson, thought to be the first settler in Iceland, is traditionally dated 874. Chartered in 1786, its modern growth began after 1904, when it became the capital.

Rey·nolds (rĕn'əldz), county (1970 pop. 6,106), 822 sq mi (2,129 sq km), SE Mo., in the Ozarks, drained by the Black River; formed 1845; co. seat Centerville. Livestock raising and farming are major activities here. The area has pine timber and granite quarries.

Re·zai·yeh (rĭ-zī'yə), city (1966 pop. 110,749), capital of West Azerbaijan prov., NW Iran, near Lake Rezaiyeh. It is the trade center for a fertile agricultural region. An important town by the 9th cent., Rezaiyeh was occupied a number of times by the Ottoman Turks. Rezaiyeh was the seat of the first U.S. Christian mission in Iran (1835). Around 1900 Christians made up more than 40% of the city's population; however, most of the Christians fled in 1918, and many who remained were massacred by Kurds.

Rezaiyeh, Lake, shallow salt lake, 1,815 sq mi (4,701 sq km), c.90 mi (145 km) long and 50 mi (80 km) wide, NW Iran, at an altitide of 4,180 ft (1,275 m). The largest lake in Iran, it has no outlet.

Rhae·tian Alps (rē'shən), section of the central Alps along the border of Switzerland, Italy, and Austria. It rises to 13,304 ft (4,057.7 m).

Rha·gae (rā'jē): *see* Rages, ancient city.

Rhea (rā), county (1970 pop. 17,202), 312 sq mi (808.1 sq km), E central Tenn., drained by the Tennessee River; formed 1807; co. seat Dayton. It is in a fruit-growing and agricultural area, with bituminous-coal mines and hardwood and pine timber. Its industry includes knitting mills and the manufacture of clothing, metal products, and machinery.

Rhe·gi·um (rē'jē-əm), ancient city, S Italy, on the Strait of Messina. Founded (c.720 B.C.) as a colony of Chalcis, later many Messenians settled here. It was powerful until its defeat and destruction (386 B.C.) by Dionysius the Elder of Syracuse.

Rheims (rēmz) or **Reims** (răɴs), city (1975 pop. 178,381), Marne dept., NE France, in Champagne. Rheims is situated amid large vineyards. Before the champagne industry took on its present proportions in the 18th cent., the chief products of Rheims were woolen textiles. As Durocotorum, the city of Remi, it was one of the most important cities of Roman Gaul. Clovis I was baptized and crowned (496) king of all Franks in the cathedral by St. Remi, the bishop of Rheims, and it became customary after Louis VII (1137) for the kings of France to be crowned here. In the present cathedral (13th–14th cent.), Joan of Arc stood next to Charles VII when, at her instance, he was crowned in 1429. During World War I heavy bombing destroyed the interior of the French Gothic cathedral, including most of the irreplaceable stained-glass windows. The town hall (17th cent.) and the old Church of St. Remi (11th–16th cent.) were also gravely damaged. In World War II, on May 7, 1945, German emissaries signed the unconditional surrender at Allied headquarters in Rheims.

Rhein·fel·den (rīn'fĕl-dən), town (1970 pop. 6,866), Aargau canton, N Switzerland, on the Rhine River, opposite the German town of the same name. Although it has one of the largest hydroelectric plants in Europe, Rheinfelden is chiefly noted as a tourist center with saline baths.

Rhen·ish Slate Mountains (rĕn'ĭsh slāt), extensive mountainous plateau, W West Germany, lying between W Hesse state and the borders of Belgium and Luxembourg. The plateau is dissected by the Rhine River and its tributaries.

Rhine (rīn), principal river of Europe, c.820 mi (1,320 km) long. It rises in the Swiss Alps and flows generally north, passing through or bordering on Switzerland, Liechtenstein, Austria, West Germany, France, and the Netherlands before emptying into the North Sea. The river carries more traffic than any other waterway in the world. Canals link the river with the Maas, Rhône-Saône, Marne, and Danube (via the Main River) valleys. The Rhine's highest source, the Hinter Rhine, issues from the Rheinwaldhorn Glacier more than 11,000 ft (3,355 m) above sea level. It forms the Rhine proper at Reichenau, south of Chur, Switzerland. At Basel the Rhine becomes the Upper Rhine of the Germans and turns sharply north across the broad-floored Rhine rift valley, a large graben, or down-faulted block, between the Black Forest and the Vosges Mts. Below Mainz, at Bingen, West Germany, the Rhine leaves the rift valley and flows for c.80 mi (130 km) across the Rhenish Slate Mts. in a steep gorge that is famous for its scenery, vineyards and superb wines, castles surviving from times when tolls were levied on the river's traffic, and legendary landmarks such as the Lorelei and the Drachenfels. Beyond Bonn the river becomes the Lower Rhine of the Germans and emerges onto the North German Plain as a broad, sluggish, and increasingly polluted river flowing on a bed of ancient deltaic deposits left by ancestors of the modern river. Just below Emmerich, on the border with the Netherlands, the modern delta begins, and the Rhine breaks up into two major distributaries, the Lek and the Waal. The Rhine was declared free to international navigation in 1868, and in 1919 navigation of the river between Basel and Krimpen, on the Lek, and Gorinchem, on the Waal, was placed under the authority of the Central Rhine Commission, with headquarters at Strasbourg. Navigation above Basel is controlled jointly by Switzerland and West Germany.

Rhine·beck (rīn'bĕk'), village (1970 pop. 2,336), Dutchess co., SE N.Y., in the foothills of the Berkshire Mts. near the Hudson River; settled before 1700, inc. 1834. It is the site of Beekman Arms, said to be the oldest hotel in the United States, and of a pre-Revolutionary Dutch Reformed church and cemetery.

Rhine·land (rīn'lănd'), region of W West Germany, along the Rhine River. The term is sometimes used to designate only the former Rhine Prov. of Prussia, but in its general meaning it also includes the Rhenish Palatinate, Rhenish and southern Hesse, and western Baden. The Treaty of Versailles (1919) after World War I provided for the Allied occupation of most of the region; the Ruhr district was occupied by French and Belgian forces from 1923 to 1925. The last occupation troops (who were French) withdrew from the Rhineland in June, 1930, five years before the terminal date set by the treaty. In Mar., 1936, the Nazi government of Germany began to remilitarize the Rhineland. The German fortifications in the Rhineland—the so-called Siegfried Line—were an extensive system of defenses in depth, penetrated by the Allies in World War II only after very heavy fighting.

Rhine·land·er (rīn'lăn'dər), city (1975 est. pop. 8,100), seat of Oneida co., N Wis., NNE of Wausau, in a lake and dairy region; settled 1880, chartered 1882, inc. 1894. Long a lumbering town, it still makes paper and wood products.

Rhine·land-Pa·lat·i·nate (rīn'lănd-pə-lăt'ĭ-nĭt'), state (1974 est. pop. 3,688,000), 7,658 sq mi (19,834 sq km), W West Germany; capital Mainz. The state was formed in 1946 by the merger of the Rhenish Palatinate, Rhenish Hesse, the southern portion of the former Rhine Prov. of Prussia, and a small part of the former Prussian province of Hesse-Nassau. Rhineland-Palatinate borders on France and the Saarland in the south and on Luxembourg and Belgium in the west. The majority of the population is employed in industry. Grain, potatoes, sugar beets, fruit, and tobacco are grown in the Rhine plain. There are vineyards in the Moselle and Rhine valleys.

Rhine Province, former province of Prussia, W West Germany. The province was also known as Rhenish Prussia and as the Rhineland. The northern section of the former province is now included in the state of North Rhine-Westphalia, and the southern section is in Rhineland-Palatinate. The province bordered in the west on the Netherlands, Belgium, and Luxembourg and in the south on France.

After the breakup (11th cent.) of the duchy of Lower Lorraine, the region split into more than 100 ecclesiastic and secular fiefs. As a result of the French Revolutionary Wars, France annexed the territory west of the Rhine, while the territory east of the Rhine was constituted (1803) the duchy (after 1806, grand duchy) of Berg. The award of the entire territory to Prussia at the Congress of Vienna (1814–15) represented the greatest Prussian territorial gain since the partitions of Poland. The entire region was constituted the Rhine

Prov. in 1824. Under the Treaty of Versailles (1919), the border territories of Eupen and of Malmedy and Moresnet were ceded to Belgium, and the southernmost corner of the province was included in the Saar Territory. These were recovered by Germany after 1935, but the status quo as of 1920 (with minor changes) was restored in 1945 after World War II.

Rhode Island, state (1975 pop. 908,000), 1,214 sq mi (3,144 sq km), NE United States, in New England, one of the Thirteen Colonies. Providence is the capital and the largest city. Its official name is the State of Rhode Island and Providence Plantations. The smallest of the 50 states and the second most densely populated, Rhode Island is bounded on the north and the east by Mass., on the south by the Atlantic Ocean, and on the west by Conn.

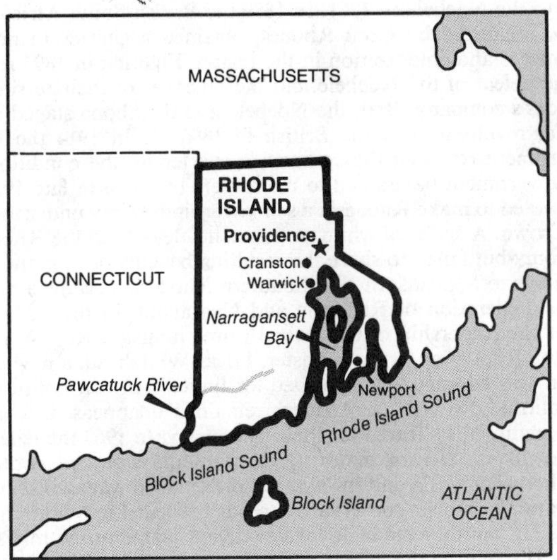

The dominant physiographic feature of the state is the Narragansett basin, a shallow lowland area of Carboniferous sediments, extending into southeastern Mass. and, in Rhode Island, being partly submerged in Narragansett Bay. The bay cuts inland c.30 mi (50 km) to Providence, where it receives the Blackstone River; it contains several islands, including Rhode Island, the largest (and the site of historic Newport); Conanicut Island, with the resort of Jamestown; and Prudence Island. The coastline between Point Judith and Watch Hill is marked by sandspits and barrier beaches, sheltering lagoons and salt marshes. Glaciation left many small lakes, and the rolling hilly surface of the state is cut by short, swift streams, with numerous falls.

Rhode Island is divided into 5 counties that have no political functions:

NAME	COUNTY SEAT	NAME	COUNTY SEAT
Bristol	Bristol	Providence	Providence
Kent	East Greenwich	Washington	West Kingston
Newport	Newport		

Economy. More than half the state is covered with forests, and agriculture is relatively unimportant to the economy. Most of the farm and pasture land is used for dairying and poultry raising, and the state is known for its Rhode Island Red chickens. Principal crops are potatoes, hay, apples, oats, and tomatoes. Commercial fishing is important. Narragansett Bay abounds in shellfish; flounder and porgy are also caught.

Eastern Rhode Island is intensively industrialized, and the concentration of the jewelry business in Providence is one of the largest in the world. Other products are silverware, textiles, primary and fabricated metals, machinery, electrical equipment, and rubber and plastic items. Rhode Island attracts many summer tourists. Its coast is lined with resorts noted for their fine swimming and boating facilities, and Block Island is a favorite vacation spot.

History. The region was probably visited (1524) by Giovanni da Verrazano, and in 1614 the area was explored by the Dutchman Adriaen Block. Roger Williams, banished (1635) from the Massachusetts Bay colony, established (1636) the first settlement in the area at Providence on land purchased from the Narragansett Indians. In 1638 other Puritan exiles bought, with Williams' aid, the island of Aquidneck (now Rhode Island) from the Narragansett Indians. There they established the settlements of Portsmouth (1638) and Newport (1639). The new colony organized a government in 1647 and was granted a royal charter in 1663. The early settlers were mostly of English stock. Many of them were attracted by the guarantee of religious freedom, an important principle with Williams. Jews settled in Newport in the first year of Williams' presidency (1654), and Quakers followed in large numbers. All the early settlers owned land that, following Williams' practice, was bought from the Indians. Fishing and trade supplemented the living won from the soil.

Until the American Revolution Newport was the commercial center of the colony, thriving especially on the triangular trade in rum, Negro slaves, and molasses. After the start of the Revolution Rhode Island militia under Nathanael Greene joined (1775) the Continental Army at Cambridge, and on May 4, 1776, the province renounced its allegiance to George III. The Revolution won, Rhode Island, jealous of its independence, refused to help strengthen the new confederation by approving a national import duty. It did not send delegates to the Constitutional Convention at Philadelphia and resisted ratifying the Constitution until the Federal government threatened to sever commercial relations with the state; even then, ratification passed (1790) by only two votes.

The post-Revolutionary era brought bankruptcy and currency difficulties, but it also marked the beginning of Rhode Island's industry. Samuel Slater built his first successful cotton-textile mill in the United States at Pawtucket in 1790. An abundance of water power led to the rapid development of manufacturing, in which merchants and shipping magnates invested their capital. The towns increased in population, and Providence surpassed Newport as the commercial center of the state. Since suffrage had long been restricted to freeholders, Rhode Island's increased urbanization resulted in the disenfranchisement of most of the townspeople. Agitation led by Thomas Wilson Dorr resulted in the adoption of a new constitution (1842) extending suffrage; however, the property qualification was not abolished until 1888.

Until well into the 20th cent. Rhode Island's political and economic life was dominated by mill owners presiding over small mill towns. English, Irish, and Scottish settlers had begun arriving in large numbers in the first half of the 19th cent., followed by French Canadians and, at the turn of the century, by Poles, Italians, and Portuguese. After World War I there was a long textile strike, centering in the Blackstone valley; this, together with the gradual removal of the mills to the South, led to a continuing decline in the cotton-textile industry. New industries such as electronics have been introduced. Measures have been taken to prevent future damage from hurricanes that have ravaged the coast in the past; the Hurricane Barrier Dam, designed to prevent damage from ocean waters, was built in the 1960s.

Government. Rhode Island's present constitution was adopted in 1842; it has been amended 36 times. The state's executive branch is headed by a governor elected for a two-year term and eligible for reelection. The bicameral legislature has a senate with 50 members and a house with 100, all elected for two-year terms. Local government is carried out on the city level. The state's leading educational institutions are Brown Univ. in Providence, and the Univ. of Rhode Island at Kingston.

Rhode Island, island, 15 mi (24 km) long and 5 mi (8 km) wide, S R.I., at the entrance to Narragansett Bay. The largest island in the state, it has fine beaches. Known to the Indians as Aquidneck, it was renamed Rhode Island (probably after the isle of Rhodes) in 1644.

Rhodes (rōdz) or **Ró·dhos** (rô'*th*ôs), island (1971 pop. 66,606), c.540 sq mi (1,400 sq km), SE Greece, in the Aegean Sea; largest of the Dodecanese, near Turkey. The island has fertile coastal strips where wheat, tobacco, cotton, olives, wine grapes, oranges, and vegetables are grown. The interior is mountainous. Shipbuilding, cattle raising, fishing, and sponge diving are important occupations. There is a large tourist industry. Rhodes was early influenced by the Minoan civilization of Crete and was colonized before 1000 B.C. by Dorians from Argos. By the 7th cent. B.C. it was dominated by the three city-states of Camirus, Lindus, and Ialysus, all commercial centers. In the early 7th cent. Rhodes established Gela, in Sicily, as its principal colony. Rhodes retained its independence until the Persian conquest in the late 6th cent. B.C. and joined (c.500 B.C.) the Ionian revolt that led to the Persian Wars. In 408 B.C. the three city-states of Rhodes united in a confederacy. The island was occupied by Macedon in 322 B.C., but it asserted its independence after the death of Alexander the Great (323 B.C.) and entered the period of its greatest prosperity, power, and cultural achievement. However, in the 2nd cent. B.C. its commerce—and hence its power—declined sharply. The island became involved in Rome's civil wars of the 1st cent. B.C., and in 43 B.C. it was seized and sacked by Caius Cassius, the Roman conspirator.

Rhodes remained in the Byzantine Empire until the capture of Constantinople (1204). It then passed under local lords, was held by Genoa (1248–50), was annexed (1256) by the emperor of Nicaea, and was conquered (c.1282) by the Knights Hospitalers. The knights defended the island against Ottoman attack until 1522–23, when it was captured by the forces of Sulayman I. Rhodes was taken by Italy in 1912 and was ceded to Greece in 1947. The modern city of Rhodes (1971 pop. 32,092), located at the northeastern tip of the island, is the capital of the Dodecanese prefecture. Its manufactures include cigarettes, soap, brandy, carpets, and processed foods. It is near the site of ancient Rhodes, planned in 408 B.C. by Hippodamus of Miletus. After repulsing a siege by Demetrius I of Macedon in 305 B.C., the citizens of ancient Rhodes erected (292–280 B.C.) in the harbor the Colossus of Rhodes, a bronze statue of Helios, which is one of the seven wonders of the ancient world. The colossus was destroyed in 224 B.C. by an earthquake. The present city was built largely by the Knights Hospitalers.

Rho·de·sia (rō-dē'zhə), republic (1977 est. pop. 6,625,000), 150,803 sq mi (390,580 sq km), S central Africa. Salisbury is the capital. Rhodesia is bordered on the north by Zambia, on the northeast and east by Mozambique, and on the southwest and west by Botswana. The terrain is mainly a plateau of four regions. The highveld, at elevations above 4,000 ft (1,219 m), crosses the country from southwest to northeast. On each side of it lies the middleveld, 3,000 to 4,000 ft (914–1,219 m) high, and beyond it the lowveld, at elevations below 3,000 ft (914 m). The fourth region, the Eastern Highlands, is a narrow, mountainous belt along the Mozambique border, where the highest point in Rhodesia, Inyangani (8,503 ft/2,592 m), stands. Rainfall varies from about 70 in. (178 cm) in the Highlands to less than 25 in. (64 cm) in the south. Rhodesia's official language is English, but the two major African groups, the Ndebele and Shona, speak their own languages. The population consists of about 250,000 whites, nearly 5 million Africans, and small minorities of Coloureds and Asians.

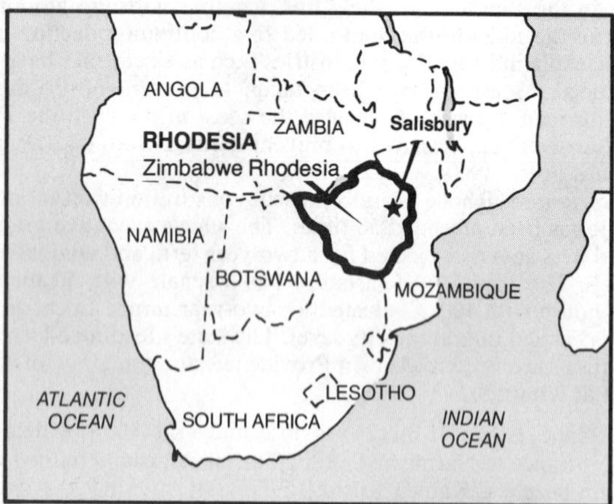

Economy. The Rhodesian economy is basically agricultural, with tobacco the principal cash crop and maize the chief food source. Farms owned by whites produce, besides these crops, cotton, sugar cane, vegetables, and fruits. The major products of African farms include sorghum, millet, rice, groundnuts, and cassava, as well as maize. There are also numerous tea plantations in the country, and in the middleveld there is excellent ranching land. Forests in southeastern Rhodesia yield valuable hardwoods. The country is endowed with a wide variety of mineral resources, including gold, tin, iron, and coal. Among Rhodesia's industrial products are iron and steel, cement, food products, textiles, and consumer goods. Most of Rhodesia's power is generated by a hydroelectric station on Kariba Lake. The country has good road and rail networks and internal air service. After Rhodesia unilaterally declared independence from Great Britain in 1965, UN economic sanctions were imposed. Agricultural exports, particularly tobacco and sugar, suffered; but minerals were less affected as new markets were found. Industrial production was reoriented from Zambian to domestic markets. Oil continued to reach Rhodesia via Mozambique and South Africa.

History. There are a number of Iron Age sites in Rhodesia, with artifacts dating from A.D. c.180. These early cultures were supplanted by migrating Bantu-speaking peoples after the 5th cent.; the ruins at Zimbabwe date from these migrations. In the early 16th cent. the Portuguese and Shona developed a trade in gold and other items. During the 1830s the Shona-speaking people were subjected to Ndebele invaders. British and Boer traders and hunters moved into the area, and the London Missionary Society established a mission to the Ndebele in 1861. In 1889 the British South Africa Company, organized by Cecil Rhodes, obtained a charter to promote commerce and colonization in the region. Fighting in 1893 resulted in the defeat of the Ndebele and the takeover of their territory by Rhodes's company. Both the Ndebele and the Shona staged unsuccessful revolts against the British in 1896–97. In 1914 the British government renewed the company's charter on the condition that self-government be granted to the settlers by 1924. In late 1922 settlers voted to make Rhodesia a self-governing colony under the British Crown. A series of white governments developed the Rhodesian economy but failed to share the resulting benefits or to grant political rights to Africans. In 1953 Southern Rhodesia became a member of the Federation of Rhodesia and Nyasaland. In the early 1960s under the leadership of the federal prime minister, Roy Welensky, and the Rhodesian prime minister, Edgar Whitehead, a new constitution was adopted that provided for limited African political participation; however, the Africans remained unappeased, while the white nationalists feared for their supremacy. In 1963 the federation broke up as African majority governments assumed control in Northern Rhodesia and in Nyasaland (renamed Malawi). After the federation's demise, conservative trends hardened in Southern Rhodesia. Ian Smith, a staunch conservative, became prime minister in 1964. The Smith government proclaimed a unilateral declaration of independence on Nov. 11, 1965. Britain called the proclamation an act of rebellion but refused to re-establish control by force. Britain requested UN economic sanctions against Rhodesia. In Nov., 1969, Rhodesia voted to become a republic as of Mar. 2, 1970. The new republican constitution provided for complete separation of the franchise along racial lines. In 1971 Britain and Rhodesia reached an accord that provided for limited African political participation. However, after a British commission's hearings in Rhodesia revealed widespread African opposition to the terms, Britain refused to recognize Rhodesian independence. The white Rhodesian government has retained control, although the banned African nationalist organizations have posed a serious challenge to the regime. However, a dramatic increase in black nationalist guerrilla attacks by the mid-1970s, and the combined pressure of Great Britain, South Africa, and the United States, forced the government to agree to discuss the eventual transfer of power to majority rule. In Sept., 1976, Smith offered a proposal for majority rule within two years. Black nationalist leaders, however, called for an immediate transfer and were suspicious of Smith's interpretation of the plan; the abortive Geneva conference (Oct., 1976) failed to resolve the differences and by 1977 Smith had withdrawn the proposal. Efforts by the British and American governments to reconvene the Geneva conference were unsuccessful. In the meantime guerrilla attacks from neighboring black African states increased; by 1977 there were infiltrations in the west and northwest (from Botswana and Zambia) and in the east (from Mozambique).

Government. The Rhodesian constitution provides for a cabinet, headed by a prime minister. Legislative authority is vested in a bicameral parliament. In the election of July, 1974, the Rhodesian Front won a sweeping victory.

Rhodesia and Ny·as·a·land, Federation of (nī-ăs'ə-lănd'), former federation (1953–63), SE Africa, composed of the self-governing British colony of Southern Rhodesia and the British protectorates of Northern Rhodesia and Nyasaland.

Rhod·o·pe (rŏd'ə-pē), mountain range, extending c.200 mi (320 km) from the Struma River, SE Bulgaria, to the lower Maritsa River, NE Greece. It consists of three sections—the Rhodope, Pirin Planina, and Rila Planina—and its highest peak, Musala, rises to 9,596 ft (2,926.8 m). The rugged range has few passes.

Rhon·dda (rŏn′də, rôn′*th*ə), municipal borough (1971 pop. 88,924), Mid Glamorgan, S Wales, on tributaries of the Taff River. Coal mining has declined in importance since the 1920s and 1930s.

Rhône (rōn), department (1975 pop. 1,429,647), 1,241 sq mi (3,214.2 sq km), E central France, in parts of Beaujolais and Lyonnais; capital Lyons.

Rhône, river, 505 mi (812.5 km) long, rising in the Rhône glacier, NE Valais, Switzerland. It flows west through a narrow, flat valley that separates the Bernese Alps from the Pennine Alps and enters Lake Geneva near Montreux. Leaving the lake at Geneva, it enters eastern France and is joined by the Saône River at Lyons. Navigable from this point, it flows south, separating the Massif Central from the French Alps. At Arles the river separates into the Grand Rhône and the Petit Rhône. Both branches are silted, and a canal has been built connecting the Rhône with the port of Marseilles. The Rhône valley south of Lyons is covered with excellent vineyards and fruit and vegetable gardens; in the extreme south silkworms are cultivated, and olives and flowers are important products.

Rhyl (rĭl), urban district (1973 est. pop. 22,150), Clwyd, N Wales, at the mouth of the Clwyd River. It is a seaside resort.

Ri·ad (rē-äd′): *see* Riyadh, Saudi Arabia.

Ri·al·to Island (rē-äl′tō), in the Grand Canal, Venice, NE Italy. A bridge connects Rialto and San Marco islands. Built between 1588 and 1591, it consists of a single marble arch and has arcades lined with shops.

Rib·ble (rĭb′l), river, c.75 mi (120 km) long, rising in the Pennines, North Yorkshire, N England, and flowing SW across Lancashire to the Irish Sea through a long, narrow estuary.

Ri·be (rē′bə), city (1970 com. pop. 17,187), capital of Ribe co., SW Denmark, on the Ribe River. One of the oldest cities of Denmark, Ribe was mentioned in the 9th cent. and became an episcopal see in 948. Its cathedral (built c.1130; restored 1884–1904) is a fine example of Danish Romanesque architecture. Other buildings of note are the Black Friars abbey (begun 1228), St. Catherine's Church (c.1230), and the city hall (c.1500).

Ri·bei·rão Prê·to (rē-bā-rouɴ′ prā′tō), city (1970 pop. 190,897), São Paulo state, SE Brazil. The city grew during the late 19th cent., with railroad construction and a large influx of Italian immigrants. During the 1930s coffee cultivation declined greatly, and the city became a processing center for a rich agricultural region producing chiefly sugar cane. Cotton and cereals are also grown, and cloth, cottonseed oil, food products, and alcoholic beverages are produced. Cattle breeding is also important.

Rice (rīs). **1.** County (1970 pop. 12,320), 721 sq mi (1,867.4 sq km), central Kansas, in a rolling plain region drained by the Little Arkansas and Arkansas rivers; formed 1867; co. seat Lyons. It is in a winter-wheat and livestock region, with oil and gas fields and salt mines. It has some chemical production and manufactures aircraft. **2.** County (1970 pop. 41,582), 495 sq mi (1,285 sq km), S Minn., drained by the Cannon and Straight rivers and watered by small lakes; formed 1853; co. seat Faribault. In an agricultural area yielding dairy products, livestock, corn, oats, and barley, it has some mining and dairy processing. Textile-mill products, plastics, and machinery are made.

Rich (rĭch), county (1970 pop. 1,615), 1,022 sq mi (2,649.6 sq km), N Utah, bordering on Idaho and Wyo. and drained in the E by the Bear River; formed 1864 as Richland co., renamed 1868; co. seat Randolph. It is in an agricultural area, with cattle and sheep raising, gas and oil exploration, and some manufacturing. Part of Cache National Forest and the Wasatch Range are in the west.

Rich·ard·son (rĭch′ərd-sən), county (1970 pop. 12,277), 548 sq mi (1,419.3 sq km), extreme SE Nebr., bordering on Kansas and Mo., bounded in the E by the Missouri River, and drained by the Nemaha River; formed 1854; co. seat Falls City. Products include feed, flour, grain, livestock, dairy goods, and poultry.

Rich·e·lieu (rĭsh′ə-lo͞o′), river, c.75 mi (120 km) long, issuing from the N end of Lake Champlain, near the N.Y.-Que. border, and flowing N across S Que. to the St. Lawrence River at Sorel. It is a link in the waterway connecting the Hudson and St. Lawrence rivers. Discovered (1609) by Samuel de Champlain, the French explorer, the river was the route of early explorers. It was an important military corridor in the French and Indian Wars, in the American Revolution, and in the War of 1812.

Rich·field (rĭch′fēld′). **1.** City (1978 est. pop. 41,900), Hennepin co., SE Minn., a residential suburb of Minneapolis; settled c.1851, inc. 1964. **2.** City (1970 pop. 4,471), seat of Sevier co., S central Utah, SE of Fillmore, in the Sevier River valley. Settled in 1863 by Mormons, it was abandoned during the Black Hawk War of 1865–68. It is now the processing center for a farm and dairy region. There are uranium, coal, and gypsum mines nearby.

Rich·land (rĭch′lənd). **1.** County (1970 pop. 16,829), 364 sq mi (942.8 sq km), SE Ill., bounded partly on the W by the Little Wabash River and drained by the Fox River and Bonpas Creek; formed 1841; co. seat Olney. It has agriculture (corn, wheat, livestock, apples, and poultry), timber, and manufacturing of shoes and wood and metal products. **2.** Parish (1970 pop. 21,774), 576 sq mi (1,491.8 sq km), NE La., bounded on the W by Bayou La Fourche and intersected by Boeuf River and many bayous; formed 1852; parish seat Rayville. It is chiefly an agricultural area yielding cotton, soybeans, corn, oats, sweet potatoes, and beef cattle. Industries include cotton processing and lumber milling. There is a large production of natural gas and some oil. The county contains hunting and fishing areas. **3.** County (1970 pop. 9,837), 2,079 sq mi (5,384.6 sq km), NE Mont., in an agricultural area bordering on N.Dak., bounded on the N by the Missouri River, and drained in the SE by the Yellowstone River; formed 1914; co. seat Sidney. Grain and livestock are major resources. **4.** County (1970 pop. 18,089), 1,450 sq mi (3,755.5 sq km), extreme SE N.Dak., in an agricultural area watered by the Sheyenne and Wild Rice rivers; formed 1873; co. seat Wahpeton. Grain, livestock, and dairy products are important. **5.** County (1970 pop. 129,997), 499 sq mi (1,292.4 sq km), N central Ohio, drained by forks of the Mohican River; formed 1808; co. seat Mansfield. Its agriculture includes livestock, grain, fruit, potatoes, and dairy products. It has sand and gravel pits, greenhouses, and diversified manufacturing (furniture, paper, stone, clay and glass products, metal products, and machinery). **6.** County (1970 pop. 233,868), 748 sq mi (1,937.3 sq km), central S.C., in the Sand Hills belt, partly bounded in the SW by the Congaree River, in the E by the Wateree River, and with parts of Lake Murray and Broad River in the W; formed 1785; co. seat Columbia. In an agricultural area yielding cotton, corn, poultry, livestock, and dairy products, it has granite quarries and its industries includes food processing and textile milling. Clothing, wood products, chemicals, metal products, and machinery are made. **7.** County (1970 pop. 17,079), 584 sq mi (1,512.6 sq km), SW Wis., bounded in the S by the Wisconsin River, and drained by the Pine and Kickapoo rivers; formed 1842; co. seat Richland Center. It is in a dairying and cattle-raising area. It has lumbering and food-processing industries and manufactures clothing and engines.

Richland, city (1978 est. pop. 31,800), Benton co., S Wash., at the confluence of the Columbia and Yakima rivers; inc. 1958. It is the headquarters of the Hanford Atomic Works (620 sq mi/1,605.8 sq km), on which the city's economy is based. The farming hamlet here was taken over (1942) by the U.S. government for an atomic bomb development plant. The city of Richland was built (1943–45) to house employees of the project. Federal ownership and management of the city were relinquished by 1958, and since 1961 the Hanford Works has been operated by an increasingly large number of private companies. With Kennewick and Pasco, Richland forms a tri-city community prospering in a farm and ranch area supported by the Columbia basin reclamation project.

Rich·mond (rĭch′mənd). **1.** County (1970 pop. 162,437), 325 sq mi (841.8 sq km), E Ga., bounded on the NE by the S.C. border, formed here by the Savannah River; formed 1777; co. seat Augusta. In a coastal-plain agricultural area yielding cotton, truck crops, fruit, corn, livestock, and dairy products, it has clay mining and lumber sawmilling. Its manufactures include textiles (yarn and clothing), furniture, paper, brick and tile, prepared meats, soybean oil, and cottonseed oil. **2.** County (1970 pop. 295,443), 57 sq mi (150.8 sq km), SE N.Y., in New York Bay, coextensive with Staten Island; formed 1683; co. seat St. George. It has beaches and amusement resorts on the east shore and a marshy, thinly inhabited region (drained by Richmond and Main creeks and Fresh Kills) in the west. Its communities are mainly residential. It has industries (shipbuilding and repairing, oil refining, lumber milling, and printing) and diverse manufacturing. **3.** County (1970 pop. 39,889), 477 sq mi (1,235.4 sq km), S N.C., bounded on the S by the S.C. border, on the W by the Pee Dee River; formed 1779; co. seat Rockingham. This piedmont region has farming (tobacco, cotton, and corn), textile manufacturing, and sawmilling. **4.** County (1970 pop. 6,504), 190 sq mi (492.1 sq km), E Va., along the S shore of the Northern Neck peninsula; formed 1692; co. seat Warsaw. Its agriculture includes truck crops, tobacco, grain, hay, corn, soybeans, livestock, and dairy products. It has fish and shellfish industries and pine lumbering.

Richmond. **1.** City (1978 est. pop. 69,700), Contra Costa co., W Calif., on San Pablo Bay, an inlet of San Francisco Bay; inc. 1905. It is a deepwater port and an industrial center with oil refineries and railroad repair shops. Manufactures include metal products, chemicals, canned foods, and electronics equipment. Originally part of a Spanish ranch founded in 1823, it was heavily settled with the coming of the railroad in 1899. **2.** City (1978 est. pop. 43,100), seat of Wayne co., E Ind., near the Ohio line; settled 1806 by Quakers from N.C., inc. as a city 1840. In the fertile Whitewater River valley, Richmond is primarily an industrial city. Buses and trucks, modular homes, and wire and cable are among its manufactures. **3.** City (1978 est. pop. 19,200), seat of Madison co., central Ky., in a bluegrass region; inc. 1800. It is a tobacco and livestock (cattle and thoroughbred horses) market. Eastern Kentucky Univ. is here. **4.** City (1970 pop. 4,948), seat of Ray co., NW Mo., near the Missouri River ENE of Kansas City, in a farm and coal region; laid out 1827, inc. 1835. **5.** Borough of New York City: *see* Staten Island. **6.** City (1970 pop. 5,777), seat of Fort Bend co., S Texas, on the Brazos River SW of Houston; founded 1882. Prominent at the time of the Texas Revolution, it was a busy plantation market until the Civil War and is now a town in a farm and livestock area with some oil. **7.** City (1978 est. pop. 222,600), state capital, E Va., at the head of navigation on the James River; settled 1637, inc. as a city 1782. It is a port of entry and a financial, commercial, shipping, and distribution center, with a deepwater port. It is also a major tobacco market. Textiles and apparel, chemicals, foodstuffs, metal items, paper and paper goods, and machinery are produced, and there are printing and publishing enterprises. The first permanent settlement was made in 1637. Fort Charles was built in 1645, and the site became a trading center. The city was laid out in 1737 under the patronage of William Byrd. It was made the capital in 1779 and was raided by the British in 1781. During the Civil War Richmond was the capital of the Confederacy and the constant objective of Federal forces. Much of the city was burned during the Confederate evacuation, Apr. 3, 1865. Places of interest include the state capitol (1785), which was designed by Thomas Jefferson; the Washington Monument; the Valentine Museum; the White House of the Confederacy, once the home of Jefferson Davis and now the Confederate Museum; St. John's Church (1741), where Patrick Henry made his famous "Give me liberty, or give me death" speech; the Edgar Allan Poe Shrine (the oldest building in the city, built c.1686); John Marshall's house (c.1790); the Robert E. Lee House (1844); and the Virginia Museum of Fine Arts.

Richmond National Battlefield Park: *see* National Parks and Monuments Table.

Richmond upon Thames (tĕmz), borough (1971 pop. 173,592) of Greater London, SE England. The borough was created in 1965 by the merger of the municipal boroughs of Barnes, Richmond, and Twickenham. Richmond upon Thames is mainly residential, with more than 5,000 acres (2,025 hectares) of public recreation grounds. Hampton Court Palace, the residence of Henry VIII, is here.

Rick·mans·worth (rĭk'mənz-wûrth'), urban district (1973 est. pop. 29,030), Hertfordshire, SE England, at the confluence of the Colne, Gade, and Chess rivers. Rickmansworth is an old market town. The district is largely residential and has many recreation spots, including woods, lakes, and Moor Park.

Ri·deau Canal (rē-dō', rē'dō), 126 mi (202.7 km) long, S Ont., Canada, connecting the Ottawa River at Ottawa with Lake Ontario at Kingston. The canal, which has 47 locks, follows the course of the Rideau River. It was built (1826–32) by army engineers to provide access from the St. Lawrence to Lake Ontario without exposure to attack by American forces on the U.S. shore of the St. Lawrence. The canal system is a recreation area and scenic attraction.

Ridge·field (rĭj'fēld'), residential town (1970 pop. 18,188), Fairfield co., SE Conn.; inc. 1709. The Battle of Ridgefield (Apr. 27, 1777) was fought here in an effort to stop William Tryon's men from retreating after a raid on Danbury. The town is noted for its quiet 18th cent. charm.

Ridge·land (rĭj'lənd), town (1970 pop. 1,165), seat of Jasper co., S S.C., W of Beaufort; inc. 1894. Sawmilling is done, and clothing is made here.

Ridge·wood (rĭj'wŏŏd'), residential village (1978 est. pop. 25,500), Bergen co., NE N.J.; inc. 1876. Various Revolutionary War historical sites are here.

Ridg·way (rĭj'wā'), borough (1970 pop. 6,022), seat of Elk co., NW Pa., on the Clarion River W of St. Marys; laid out 1833, inc. 1881. Electrical machinery is produced.

Rie·sa (rē'zä), city (1974 est. pop. 49,989), Dresden dist., SE East Germany, on the Elbe River. It is a river port and an industrial center. Manufactures include steel, furniture, and rubber goods.

Rif At·las (rĭf ăt'ləs), range of the Atlas Mts., NE Morocco, NW Africa, curving along the Mediterranean coast from Ceuta to Melilla. Composed of sedimentary rocks and uplifted during the Alpine orogeny, the range is a continuation of the Sierra Nevada of Spain and is separated from it by the Strait of Gibraltar. The region is inhabited by Berber tribes, who remained independent of any central authority until subdued (1925–26) by France and Spain.

Ri·ga (rē'gə), city (1970 pop. 733,000), capital of the Latvian SSR, NW European USSR, on the Daugava (Western Dvina) River near its entry into the Gulf of Riga. A major Baltic port, it is also a rail junction, a military base, and one of the USSR's leading industrial and cultural centers. Among Riga's industries are machine building, metalworking, shipbuilding, woodworking, food processing, and the manufacture of diesel engines, streetcars, chemicals, pharmaceuticals, radio and telephone equipment, meteorological instruments, textiles, building materials, and paper.

Riga is the site of a university (est. 1919), the Latvian Academy of Sciences (1946), and numerous other educational and cultural institutions. The old section, or Hansa town, is circled by a park-lined moat and includes the ancient castle of the Livonian Knights (rebuilt at various periods), the 13th cent. Lutheran cathedral (rebuilt 16th cent.), and the Parliament building (19th cent.). The old town, with its narrow, cobbled streets lined with gabled dwellings and warehouses, has retained much of its medieval character. Riga's new sections are spacious and modern.

History. The site had long been occupied by Baltic tribes when the monk Meinhard built a monastery c.1190 among a settlement of Livs. German merchants established a community at Riga in 1158. Bishop Albert of Livonia transferred his seat there in 1201 and founded the Livonian Knights, a German military religious order. The city, which became a member of the Hanseatic League in 1282, developed into a commercial and handicraft center. After the dissolution of the Livonian Order in 1561, Riga was independent and then passed to Poland (1581), Sweden (1621), and Russia (1710).

A leading Russian industrial center from the second half of the 19th cent., Riga had the third-largest number of industrial workers (after Moscow and St. Petersburg) by the 1890s. After World War I the independence of Latvia was proclaimed at Riga, which became the new country's capital. When Latvia was incorporated into the USSR in 1940, Riga was made the capital of the Latvian SSR. During World War II the city was again occupied (1941) by the Germans, from whom it was retaken (1944) by the Soviet army.

Riga, Gulf of, E arm of the Baltic Sea, W European USSR, bordering on Estonia and on Latvia. At its mouth it is nearly closed off by the Estonian island of Sarema. The gulf is frozen from Dec. to Apr.

Rig·by (rĭg'bē), city (1970 pop. 2,324), seat of Jefferson co., E Idaho, NNE of Idaho Falls; settled 1884 by Mormons, inc. as a village 1903, as a city 1915. It is a processing and shipping point for an irrigated farm and dairy region.

Ri·gi (rē'gē), mountain, in the Alps, N central Switzerland, between Lucerne, Zug, and Lauerz lakes, rising to 5,908 ft (1,801.9 m) at the Kulm, the highest peak. Ascended by rack-and-pinion railways it commands a panorama of nearly 180 mi (290 km).

Ri·je·ka (rē-ĕ'kä) or **Fi·u·me** (fē-ōō'mē, fyōō'mā), city (1971 pop. 132,933), NW Yugoslavia, in Croatia, on the Adriatic Sea and the Gulf of Quarnero. Yugoslavia's largest seaport, it exports ores, timber, and grain, and imports cotton, petroleum, and coal. The city's industries include shipbuilding and oil refining. Dating from Roman times, Rijeka was later held by the Franks. From the 9th to the 14th cent., Croatian dukes ruled the city, which passed to Austria in 1466. It was united with Croatia in 1776, but three years later Austria transferred it to Hungary. In 1814 it was restored to Austria, which transferred it to Hungary in 1822. The secret Treaty of London (1915) promised Rijeka to Yugoslavia, but Italy claimed it on the grounds that Italian-speaking inhabitants formed a majority of the population. While negotiations continued, an Italian free corps seized the city in Sept., 1919. By the Treaty of Rapallo (1920) Italy and Yugoslavia agreed to establish Rijeka as a free state. In 1922, however, Italian troops occupied Rijeka. The Treaty of Rome (1924) left Rijeka in Italian hands but awarded its eastern suburb, Susak, to Yugoslavia. In 1945 Rijeka passed under Yugoslav administration, and in 1947 the Allied peace treaty with Italy formally transferred it to Yugoslavia.

Rijs·wijk (rīs'vīk), city (1977 est. pop. 53,532), South Holland prov., W Netherlands, near The Hague. It has varied industries. The Treaty of Ryswick was signed here in 1697.

Ri·ley (rī'lē), county (1970 pop. 56,788), 597 sq mi (1,546.2 sq km), NE central Kansas, in a level to rolling plain, bounded on the E by the Big Blue River and drained in the S by the Kansas River; formed 1855; co. seat Manhattan. It is in a livestock and grain-farming area, with some mining. Wood products and concrete are made.

Rí·mac (rē'mäk), river, c.80 mi (130 km) long, rising in the Andes of W Peru, and flowing W through Lima to the Pacific Ocean near Callao. It is used extensively for irrigation. The scenic Rímac valley affords one of the chief lines of communication to the high, interior mountain basins.

Ri·mi·ni (rĭm'ə-nē, rē'mē-nē), city (1976 est. pop. 125,815), in Emilia-Romagna, N central Italy, on the Adriatic Sea. It is a commercial, industrial, and railroad center and a fashionable beach resort. Food products and wine are manufactured, and there are fisheries. Located at the junction of the Flaminian and Aemilian Ways, the city was a Roman colony of strategic importance (founded in the mid-3rd cent. B.C.). It later came under Byzantine rule. Rimini was included in Pepin the Short's donation to the popes (754). The Malatesta family seized power in Rimini in the 13th cent. In 1509 the city passed under papal control.

Ri·mous·ki (rĭ-mōō'skē), town (1976 pop. 27,897), S Que., Canada, on the S shore of the St. Lawrence River, NE of Quebec. It has lumber industries, dairy processing, and telecommunications-equipment plants.

Ring·gold (rĭng'gōld'), county (1970 pop. 6,373), 538 sq mi (1,393.4 sq km), S Iowa, on the Mo. border in the S, drained by the Little Platte and Grand rivers; formed 1847; co. seat Mount Ayr. Its rolling prairie agriculture includes hogs, cattle, poultry, corn, oats, and wheat. There are bituminous-coal deposits.

Ringgold, town (1970 pop. 1,381), seat of Catoosa co., NW Ga., SE of Chattanooga, Tenn.; inc. 1847. Textile products are made here, and stone is quarried nearby.

Ring·sted (rĭng'stĕ*th*), city (1976 est. pop. 17,000), Vestsjaelland co., E Denmark. It is the commercial and processing center of a rich agricultural region. A place of pagan worship in ancient Sjaelland, Ringsted later became a center of Christianity. The 12th cent. Benedictine monastery has a church containing the tombs of several Danish kings of the 12th–13th cent.

Ring·wood (rĭng'wŏŏd'), borough (1970 pop. 10,393), Passaic co., N N.J., in the Ramapo Mts. on the Wanaque River. Iron was found nearby in 1730; mines and works were developed after 1764 by Peter Hasenclever, who made Ringwood Manor his headquarters. Presented to the state in 1936, the estate (95 acres/38.5 hectares) became (1939) a park and the manor house was converted into a museum; Ringwood Manor is a national historic landmark.

Rí·o (rē'ō) and **Ri·o** (rē'ō, rē'ōō), respectively Spanish and Portuguese terms for river. For those not listed here, see under the second element of the name. Thus, for Río Lerma, *see* Lerma.

Rio Ar·ri·ba (ə-rē'bə), county (1970 pop. 25,170), 5,855 sq mi (15,164.5 sq km), NW N.Mex., bordered on the N by Colo. and drained by the Rio Grande, Rio Brazos, and Rio Chama; formed 1852; co. seat Tierra Amarilla. It has livestock grazing and farming (grain and chili peppers). There are national forest areas throughout the county.

Ri·o·bam·ba (rē'ō-bäm'bä), city (1974 pop. 58,029), capital of Chimborazo prov., central Ecuador, near Chimborazo volcano, in a high basin of the Andes. It was founded c.1530 and was completely destroyed by an earthquake in 1797. The convention that proclaimed Ecuador's independence from Greater Colombia met here in 1830.

Rio Blan·co (blăng'kō), county (1970 pop. 4,842), 3,263 sq mi (8,451.2 sq km), NW Colo., bordering on Utah and drained by the White River; formed 1889; co. seat Meeker. It is in a livestock-grazing region, with oil and natural-gas wells and coal mines. Part of White River National Forest and the Rocky Mts. are in the east.

Ri·o Bran·co (rē'ō brăng'kō, -kōō, rē'ōō), city (1970 pop. 34,531), capital of Acre state, NW Brazil, on the Acre River. Rubber and Brazil nuts are its chief products. Rio Branco was built on the site of a former rubber plantation.

Rio Branco, river, rising in the Guiana Highlands. It flows c.350 mi (565 km) south through Brazil to join the Rio Negro.

Ri·o de Ja·nei·ro (rē'ō də zhä-nâr'ō, rē'ōō dē zhä-nâr'ōō), state (1975 est. pop. 10,400,200), 16,568 sq mi (42,911 sq km), SE Brazil, on the Atlantic Ocean; capital Rio de Janeiro. In 1975 it absorbed the adjacent state of Guanabara, which contained the city of Rio de Janeiro.

Rio de Janeiro, city (1970 pop. 4,252,009), capital of Rio de Janeiro state, SE Brazil, on Guanabara Bay of the Atlantic Ocean. The second-largest city and former capital of Brazil, it is the cultural center of the country and a financial, commercial, and transportation hub. Rio, as it is popularly known, has one of the world's most beautiful natural harbors. The harbor is surrounded by low mountain ranges whose spurs extend almost to the waterside and thus divide the city. Among its natural landmarks are Sugar Loaf Mt. (1,296 ft/395 m) and Corcovado peak (2,310 ft/704 m), site of a colossal statue of Christ. The city acquired its modern outline in the early 1900s, and extensive public sanitation and remodeling are continuing. The longest underground urban highway, linking the northern and southern sections of the city, opened in 1968. The harbor is deep enough for even the largest vessels to come alongside the wharves. Through the port of Rio flows the major portion of Brazil's imports as well as its exports (iron ore, manganese, coffee, cotton, meat, and hides). Rio is also a distribution center for the coastal trade. The city's manufactures include textiles, foodstuffs, household appliances, cigarettes, chemicals, leather goods, metal products, and printed material. There are two major airports. Rio's climate is warm and humid, and the city has a worldwide reputation as a tourist center. Of particular attraction are the crescent-shaped beaches, especially the Copacabana, with its famous mosaic sidewalks. The most popular holiday is the pre-Lenten carnival, with its colorful street processions. Examples of Rio's famous modern architecture are the ministry of education, the Brazilian press association headquarters, and the museum of modern art. Older buildings house the national library, the municipal opera house, and several museums. The Itamarati Palace is also noteworthy. Notable churches include the ornate Candelária Church, the 18th cent. Church of Nossa Senhora da Glória, the 17th cent. Franciscan convent, and a 16th cent. Benedictine monastery. Rio has beautiful subtropical parks, including the Quinta da Boa Vista and the botanical garden (founded 1808).

According to tradition, the Rio de Janeiro area was visited in Jan., 1502, by Portuguese explorers who believed Guanabara Bay to be the mouth of a river; it was therefore named Rio de Janeiro. It is more likely that the region was discovered in 1504 by Gonçalo Coelho. In 1555 the French Huguenots established a colony, but they were driven out (1560–67) by Mem de Sá, governor general of the Portuguese colony of Brazil. At the same time the city of São Sebastião do Rio de Janeiro was founded by Mem de Sá's cousin. The settlement was captured and held for ransom by the French in 1711. In the 18th cent. Rio was designated the shipping point for all gold from the interior. It replaced Bahia (now Salvador) as the capital of Brazil in 1763 and subsequently became capital of the exiled royal court of Portugal (1808–21), the Brazilian empire (1822), and the federal republic (1889). It was superseded as capital by Brasília in 1960.

Ri·o Gran·de (rē'ōō grän'də), city (1970 pop. 98,863), Rio Grande do Sul state, S Brazil, on the Rio Grande River at the outlet of the Lagoa dos Patos (a tidal lagoon) to the Atlantic Ocean. Rio Grande has refrigeration plants, oil refineries, and factories producing foodstuffs, spices, leather, and tires. The city was founded in 1737.

Ri·o Grande (rē'ō grănd', grăn'dē), county (1970 pop. 10,494), 916 sq mi (2,372.4 sq km), S Colo., drained by the Rio Grande; formed 1874; co. seat Del Norte. It is in an irrigated agricultural (livestock, grain, hay, and potatoes) and mining (gold and silver) region. There is lumber millwork in the county. It includes part of Rio Grande and San Juan national forests.

Ri·o Gran·de (rē'ōō grän'də), name of several rivers of Brazil. The largest rises in south Minas Gerais state, SE Brazil, and flows c.650 mi (1,045 km) northwest to the Paranaíba River, with which it forms the Paraná River. Its lower course forms part of São Paulo's northern boundary.

Ri·o Grande (rē'ō grănd', grän'dē), river, c.1,885 mi (3,035 km) long, rising in SW Colo. in the San Juan Mts. and flowing S through the middle of N.Mex., past Albuquerque, then meandering SE as the border between Texas and Mexico, making a big bend and eventually emptying into the Gulf of Mexico at Brownsville, Texas, and Matamoros, Mexico. Other paired towns are Laredo, Texas, and Nuevo Laredo, Mexico, and El Paso, Texas, and Juárez, Mexico. The Rio Grande is unnavigable except near its mouth; Brownsville, Texas, is the river's chief port. Indian pueblos were thriving on its banks north of Las Cruces, N.Mex., when Francisco Vásquez de Coronado, the Spanish explorer, came (1540), and the Indians were then practicing irrigation of the arid country. Dams on the Rio Grande are used for irrigation, flood control, and regulation of flow. Elephant Butte Dam (completed 1916) and Caballo Dam (completed 1938) in N.Mex. create reservoirs that serve large areas. Further downstream north of Del Rio, Texas, is the Amisted Dam (completed 1969); it is 6 mi (9.7 km) long and impounds a huge reservoir. Amisted National Recreation Area is there. Below Laredo are Falcon Dam (completed 1954) and its large reservoir. Near the mouth of the Rio Grande is the citrus-fruit and truck-farm region commonly called the Rio Grande Valley and developed principally in the 1920s. An agreement between the United States and Mexico in 1945 provided for future projects to share the river's water. Shifts in the river's channel have led to border disputes between the United States and Mexico. Parts of its bed have been stabilized by canalization, and an international border commission mediates disputes. The 114-year controversy over the location of the border at El Paso was finally settled in 1968, when the water of the Rio Grande was diverted into a concrete channel. The river is known to the Mexicans as Río Bravo del Norte.

Ri·o Grande City (rē'ō grănd', grän'dē), uninc. town (1970 pop. 5,676), seat of Starr co., extreme S Texas, on the Rio Grande WNW of Brownsville. The region was settled c.1753 by the Spanish. Cattle, sheep, goats, and oil from the semiarid plains support the town, which was once a river port and is still a port of entry.

Ri·o Gran·de do Nor·te (rē'o͞o grän'də *tho͞o* nôr'tə), state (1970 pop. 1,855,700), 20,469 sq mi (53,015 sq km), NE Brazil, on the Atlantic Ocean; capital Natal.

Ri·o Gran·de do Sul (rē'o͞o grän'də *tho͞o* so͞ol'), state (1970 pop. 7,457,600), 108,951 sq mi (282,183 sq km), S Brazil, bordering on Argentina and Uruguay and on the Atlantic Ocean; capital Pôrto Alegre.

Ri·o·ja (rē-ō'hä), region, N Spain, in Old Castile, along the right bank of the Upper Ebro River. It is famous for its fine wines.

Riom (ryôɴ), town (1975 pop. 17,071), Puy-de-Dôme dept., S central France, in Auvergne. It has distilleries, tobacco plants, and factories making electrical appliances. Of Gallic origin, the Roman Ricomagus grew around the collegiate Church of St. Amable (1077; restored). It was the capital of the dukes of Auvergne.

Rí·o Tin·to (rē'ō tēn'tō), town (1970 pop. 7,903), Huelva prov., SW Spain, in the Sierra Morena. It is the center of the Río Tinto mining region, named for the river that crosses it. Known since Phoenician times, the area has some of the world's largest copper deposits and some iron and manganese.

Rip·ley (rĭp'lē). **1.** County (1970 pop. 21,138), 442 sq mi (1,144.8 sq km), SE Ind., drained by Laughery and Graham creeks; formed 1816; co. seat Versailles. It is in an agricultural area (grain, corn, tobacco, livestock, and dairying), with timber and limestone quarries. Its manufacturing includes food processing, wood and metal products, furniture, and shoes. **2.** County (1970 pop. 9,803), 639 sq mi (1,655 sq km), S Mo., in the Ozarks, drained by the Current and Little Black rivers; formed 1833; co. seat Doniphan. Agricultural products include cotton, corn, oats, and livestock.

Ripley. **1.** Town (1970 pop. 3,482), seat of Tippah co., NE Miss., SE of Memphis, Tenn., in a farm and dairy area; platted 1835, inc. 1837. Lumber milling is done. **2.** Town (1970 pop. 4,794), seat of Lauderdale co., W Tenn., near the Mississippi River NNE of Memphis, in a farm and timber area; inc. 1838. It has garment factories and a cannery. **3.** Town (1970 pop. 3,244), seat of Jackson co., NW W.Va., N of Charleston; settled in 1768, inc. 1832.

Rip·on (rĭp'ən), municipal borough (1973 est. pop. 12,580), North Yorkshire, N England, on the Ure River. It has foundries, varnish and paint factories, tanneries, and breweries and is a market town. There have been monasteries on the site since the 7th cent. The present cathedral dates from the 12th to the 15th cent. It has a Saxon crypt with a narrow passage called St. Wilfrid's Needle; the ability to pass through it was supposed to be an indication of chastity. The Wakeman's House (13th or 15th cent.), in the market place, was the residence of the mayor, or wakeman.

Ripon, city (1970 pop. 7,053), Fond du Lac co., central Wis., NW of Fond du Lac; settled 1844 as a Fourierist community, inc. 1858 as Ripon. Washing machines and canned foods are produced. It is claimed that the Republican Party was founded here (1854).

Ri·shon Le·ziy·yon (rē-shôn' lə-tsē-ōn'), town (1975 est. pop. 60,300), W central Israel. It has one of Israel's largest wineries. Rishon Leziyyon was founded in 1882.

Ris·ing Sun (rī'zĭng sŭn'), city (1970 pop. 2,305), seat of Ohio co., SE Ind., on the Ohio River SW of Cincinnati, Ohio; platted 1814, inc. as a town 1817, as a city 1849.

Ri·son (rī'zən), city (1970 pop. 1,214), seat of Cleveland co., S central Ark., SSW of Pine Bluff, in a cotton, corn, and hay region. Cotton ginning and sawmilling are done.

Ritch·ie (rĭch'ē), county (1970 pop. 10,145), 452 sq mi (1,170.7 sq km), NW W.Va., drained by the North and South forks of the Hughes River; formed 1843; co. seat Harrisville. The county has oil and natural-gas wells. Its major industry is hand-blown glass. It has livestock, dairy, grain, fruit, and tobacco farms.

Ritz·ville (rĭts'vĭl'), city (1977 est pop. 1,936) seat of Adams co., E central Wash., in the Columbia basin agricultural region SW of Spokane; settled 1878, platted 1881, inc. 1888. It is a processing and shipping center in a wheat and cattle region.

Ri·vas (rē'väs'), town (1971 pop. 10,007), SW Nicaragua. It is on the Isthmus of Rivas, a narrow land strip between Lake Nicaragua and the Pacific Ocean. Rivas was the seat of an Indian civilization at the time of the Spanish Conquest. During the California gold rush it controlled the transit route across Nicaragua.

riv·er (rĭv'ər), stream of water larger than a brook or creek. Land surfaces are never perfectly flat, and as a result the runoff after precipitation tends to flow downward by the shortest and steepest course in depressions formed by the intersection of slopes. Runoffs of sufficient volume and velocity join to form a stream that, by the erosion of underlying earth and rock, deepens its bed; it becomes perennial when it cuts deeply enough to be fed by ground water or when it has as its source an unlimited water reservoir. The lowest level to which a river can erode its bed is called base level. Sea level is the ultimate base level, but the floor of a lake or basin into which a river flows may become a local and temporary base level. Cliffs or escarpments and differences in the resistance of rocks create irregularities in the bed of a river and can thus cause rapids and waterfalls. A river tends to eliminate irregularities and to form a smooth gradient from its source to its base level. As it approaches base level, downward cutting is replaced by lateral cutting, and the river widens its bed and valley and develops a sinuous course that forms exaggerated loops and bends called meanders. A river may open up a new channel across the arc of a meander, thereby cutting off the arc and creating an oxbow lake.

Riv·er·head (rĭv'ər-hĕd'), uninc. town (1975 est. pop. 7,500), seat of Suffolk co., SE N.Y., on E Long Island, on the Peconic River E of Great Peconic Bay; settled c.1690. It is in a resort and farm area. Aircraft and trucks are manufactured.

Riv·er·i·na (rĭv'ə-rī'nə), region, 26,560 sq mi (68,790 sq km), New South Wales, SE Australia. Riverina is a rich agricultural area with associated processing industries. The Murrumbidgee River runs through the region.

River Rouge (ro͞ozh), city (1970 pop. 15,947), Wayne co., SE Mich., an industrial suburb of Detroit, on the Detroit and Rouge rivers; settled c.1817, inc. 1899. It is a port of entry, with automobile, shipbuilding, paper, and steel industries. The city grew in the 1920s with the expansion of the Ford Motor Company in the area.

Riv·er·side (rĭv'ər-sīd'), county (1970 pop. 459,074), 7,179 sq mi (18,586 sq km), SE Calif., with the Colorado River and the Ariz. border in the E and the Coast Ranges in the W; formed 1893; co. seat Riverside. The San Jacinto and San Bernardino ranges and the Little San Bernardino Mts. cross the county, which includes parts of the Colorado Desert, Coachella Valley, Salton Sea, and Palo Verde Valley. Its agriculture includes dairy products, eggs, poultry, livestock, citrus fruit, grapes, carrots, cereal grains, alfalfa, dates, potatoes, and cotton. There are cement-rock, gypsum, clay, sand, gravel, and iron deposits. Its industry includes extensive food processing and the manufacture of clothing, metal and wood products, furniture, chemicals, clay tile and pipe, concrete, and electronic equipment. It includes Joshua Tree National Monument and sections of Cleveland and San Bernardino national forests. It also includes desert resorts (notably Palm Springs), mountain recreational areas, and Lake Elsinore.

Riverside. 1. City (1978 est. pop. 152,600), seat of Riverside co., S Calif.; inc. 1883. It is famous for its orange industry. Other products include mobile homes, aluminum, aircraft and space components, and food machinery. The navel orange was introduced here in 1873; the original tree, still producing, is a tourist attraction. The first marketing cooperative, organized in Riverside in 1892, led to the founding of the California Fruit Growers Exchange. The city is the seat of the Univ. of California at Riverside (with a citrus research center, est. 1907). Mission Inn, a hotel in a unique mission setting, is in the city. **2.** Village (1970 pop. 10,432), Cook co., NE Ill., a residential suburb of Chicago, on the Des Plaines River; inc. 1875. It was planned as a model suburb by Frederick Law Olmsted and Calvert Vaux. The city has a number of buildings designed by Frank Lloyd Wright. The old water tower (late 19th cent.) is a national historic landmark.

Riv·er·ton (rĭv′ər-tən), city (1970 pop. 7,995), Fremont co., W central Wyo., NE of Lander on the Bighorn River at the mouth of the Wind River; founded and inc. 1906. It is the center of a livestock, timber, and irrigated farm area, with oil and natural-gas fields. Uranium is mined and processed nearby.

Riv·i·er·a (rĭv′ē-âr′ə), narrow coastal strip between the Alps and the Mediterranean, extending roughly from La Spezia, Italy, to Hyères, France. It is famous for its scenic beauty and for its mild winter climate and is dotted with fashionable resorts, hotels, and villas. Genoa is the center of the Italian Riviera and divides it into the Riviera di Levante (east) and the Riviera di Ponente (west). Also noteworthy is the rugged Cinqueterre coast near La Spezia. The French Riviera is also called the Côte d'Azur. Flowers for export and for use in the perfume industry are grown throughout the region. A panoramic highway runs along the Riviera from end to end; the Corniche du Littoral, which hugs the red cliffs of the coastline between Nice and Menton, France, is particularly famous.

Ri·vie·ra Beach (rĭ-vîr′ə), resort city (1978 est. pop. 24,600), Palm Beach co., SE Fla., on Lake Worth (a lagoon); inc. 1922. There are research and development firms and aerospace industries in the city.

Ri·vière du Loup (rē-vyâr′ dü lo͞o′), city (1976 pop. 13,103), E Que., Canada, on the S shore of the St. Lawrence River, NE of Quebec. It is a commercial and industrial center in a lumbering and agricultural region. It is also a resort and tourist center.

Riv·o·li Ve·ro·ne·se (rĭv′ə-lē vĕr′ə-nā′sē, -nā′zē), village (1975 est. pop. 50,003), Venetia, NE Italy, on the Adige River. It was the scene in Jan., 1797, of a decisive French victory over the Austrians.

Ri·yadh or **Ri·ad** (both: rē-äd′), city (1974 pop. 666,840), capital of Saudi Arabia, in the Nejd, central Saudi Arabia. It is situated in an oasis, c.240 mi (395 km) inland from the Persian Gulf. Riyadh is the focal point for desert travel and trade. Its architecture formerly represented the classic Arabic style, but in the oil boom of recent decades many buildings were torn down and replaced by large modern structures.

Roane (rōn). **1.** County (1970 pop. 38,881), 350 sq mi (906.5 sq km), E Tenn., in the Tennessee River valley; formed 1801; co. seat Kingston. Its agriculture includes fruit, tobacco, corn, wheat, livestock, and dairy products. It has coal and iron mines, limestone quarries, hardwood timber, and some industry (food processing and the manufacture of textiles, clothing, concrete, and machinery). **2.** County (1970 pop. 14,111), 486 sq mi (1,258.7 sq km), W central W.Va., drained by the Little Kanawha and Pocatalico rivers; formed 1856; co. seat Spencer. In a timber-producing region, it has some oil and natural-gas wells. Its agriculture includes livestock, fruit, and tobacco.

Ro·anne (rō-än′), town (1975 pop. 55,195), Loire dept., E central France, on the Loire River. Cotton and metals are the chief products; other industries include tanning and the spinning of artificial silk. Roanne (then Rodumna) was a crossroads in Gallo-Roman times. The Joseph-Dechelette Museum, noted for its ancient artifacts, is located in the town. Roanne also has several ruins from the Roman period.

Ro·a·noke (rō′ə-nōk′), county (1970 pop. 67,339), 262 sq mi (678.6 sq km), W central Va., in the Great Appalachian Valley, with the Blue Ridge in the SE; formed 1838; co. seat Salem (an independent city). It is in a rich agricultural and fruit-growing area, with dairying and stock and poultry raising. Textiles, furniture, metal products, and machinery are made. It contains part of the Jefferson National Forest and is traversed by the Appalachian Trail.

Roanoke, independent city (1978 est. pop. 100,300), SW Va., on the Roanoke River; settled c.1740, inc. 1882. Situated between the Blue Ridge and Allegheny Mts., Roanoke is the southern gateway to the Shenandoah valley. A tiny village until the coming of the railroad in 1882, Roanoke now has important railroad shops and factories manufacturing furniture, textile goods, electrical equipment, and metal products.

Roanoke, river, c.410 mi (660 km) long, rising in SW Va. and flowing generally SE across the Blue Ridge Mts. and into Albemarle Sound, NE N.C. The lower river is navigable for small craft. A comprehensive flood-control and hydroelectric-power scheme has been initiated on the river.

Roanoke Island, 12 mi (19.3 km) long and 3 mi (4.8 km) wide, NE N.C., off the Atlantic coast in Croatan Sound between Albemarle and Pamlico sounds. Tourism and fishing are the principal industries. English navigators brought back such glowing accounts of the island that Sir Walter Raleigh immediately dispatched a colonizing expedition. The colonists landed on Roanoke Island in Aug., 1585, but returned to England the next year. In 1587 Raleigh sent out another group under John White. Forced to return to England for supplies, White was unable to come back to Roanoke until 1591. Upon his return he found the colonists gone and the letters CROATOAN carved on a tree. This gave rise to a theory that the settlers had moved to Croatoan Island or had joined the Croatoan or Hatteras Indians. White was unable to search for the colonists because of bad weather. Another theory was later advanced with the discovery (1937–40) of some 40 stone tablets inscribed with what some believe to be the history of the "lost colony." The inscriptions tell of the death of many of the colonists from disease and Indian attacks and of the migration of others into the interior.

Roanoke Rap·ids (răp′ĭdz), industrial city (1978 est. pop. 14,000), Halifax co., N N.C., on the Roanoke River near the Virginia line; founded 1893, inc. 1931. Cotton textiles and paper products are manufactured.

Rob·bins·ville (rŏb′ĭnz-vĭl′), town and mountain resort (1970 pop. 777), seat of Graham co., W N.C., in Nantahala National Forest NNE of Murphy. There is sawmilling in the area.

Rob·ert Lee (rŏb′ərt lē′), city (1970 pop. 1,119), seat of Coke co., W Texas, N of San Angelo on the Colorado River. It is a market center for a farm and ranch area.

Rob·erts (rŏb′ərts). **1.** County (1970 pop. 11,678), 1,111 sq mi (2,877.5 sq km), extreme NE S.Dak., bordering on N.Dak. and Minn. and bounded in the E by Big Stone Lake and Lake Traverse; formed 1883; co. seat Sisseton. Its agriculture includes grain and hay farming, dairying, and cattle, sheep, and horse ranching. It has some mining and manufacturing. **2.** County (1970 pop. 967), 899 sq mi (2,328.4 sq km), extreme N Texas, on high plains of the Panhandle, drained by the Canadian River and its tributaries; formed 1876; co. seat Miami. This cattle-ranching area also produces wheat, maize, grain sorghums, hay, barley, oats, and some sheep and hogs.

Rob·ert·son (rŏb′ərt-sən). **1.** County (1970 pop. 2,163), 101 sq mi (261.6 sq km), NE Ky., in the Bluegrass, bounded on the W and SW by the Licking River and on the N by its North Fork; formed 1867; co. seat Mount Olivet. It is in a gently rolling upland agricultural area that produces livestock, poultry, burley tobacco, corn, and dairy products. **2.** County (1970 pop. 29,102), 474 sq mi (1,227.7 sq km), N Tenn., bounded in the N by Ky. and drained by the Red River and its affluents; formed 1796; co. seat Springfield. In a tobacco and livestock region, it has limestone quarries and some industry (food and tobacco processing and the manufacture of furniture, leather, and household appliances). **3.** County (1970 pop. 14,389), 874 sq mi (2,263.7 sq km), E central Texas, bounded on the

W by the Brazos River and on the E by the Navasota River; formed 1837; co. seat Franklin. It has diversified agriculture (cotton, corn, legumes, grain, hay, sorghums, watermelons, tomatoes, fruit, and truck crops), livestock (cattle, hogs, sheep, and poultry), and dairying. There are deposits of oil, natural gas, sand, gravel, stone, and lignite coal.

Rob·e·son (rŏb'ə-sən), county (1970 pop. 84,842), 949 sq mi (2,457.9 sq km), S N.C., in the coastal plain, bounded on the S by the S.C. border and drained by the Lumber River; formed 1786; co. seat Lumberton. This tobacco and timber area also has agriculture (corn, cotton, and soybeans) and textile mills.

Rob·in·son (rŏb'ĭn-sən), city (1970 pop. 7,178), seat of Crawford co., E Ill., near the Wabash River SSW of Terre Haute, Ind.; inc. 1886. The center of a farm and oil region, it has refineries and makes oil-field supplies, shoes, and ceramic products.

Ro·by (rō'bē), city (1970 pop. 784), seat of Fisher co., NW central Texas, N of Sweetwater. It is a trading center in an agricultural and cattle-raising area.

Ro·ca, Ca·bo da (kä'bo͞o *thə* rô'kə), cape, W Portugal, W of Lisbon. It is the western extremity of Europe.

Roch·dale (rŏch'dāl'), county borough (1976 est. pop. 210,200), Greater Manchester, NW England, on the Roch River. The chief industry is the spinning and weaving of cotton and woolen yarns. Rayon, rubber, leather, and asbestos are also produced.

Roche·fort (rôsh-fôr') or **Roche·fort-sur-Mer** (rôsh-fôr-sür-mâr'), city (1975 pop. 28,155), Charente-Maritime dept., W France, on the Charente River near the Bay of Biscay. It is a fishing port with shipyards and aircraft and machine industries. An important naval base in the days of sailing ships, it declined after the advent of steamships.

Ro·chelle, La (lä rô-shĕl'), city (1975 pop. 75,367), capital of Charente-Maritime dept., W France, on the Bay of Biscay. Industries include naval, aircraft, and automobile construction. La Rochelle is the principal French fishing port on the Atlantic coast. Chartered in the 12th cent., it was a Huguenot stronghold during the Wars of Religion and successfully resisted Catholic besiegers for half a year (1572–73). However, when Cardinal Richelieu resolved to crush the Huguenots, La Rochelle fell after a siege of 14 months (1627–28). La Rochelle prospered again as it became the chief center of trade with Canada, but it suffered from the loss of Canada by France and from the Continental System under Napoleon. The picturesque old fishing port in the heart of the city, the Renaissance town hall, and other old buildings make the city a favorite tourist center.

Roch·es·ter (rŏch'ĭ-stər), municipal borough (1973 est. pop. 56,030), Kent, SE England, on the Medway River. Cement, heavy machinery, electronic equipment, precision tools, and clothing are made. Rochester was the Roman Durobrivae and was important in Saxon times. St. Augustine founded a mission here in 604. A Norman wall 12 ft (3.7 m) thick surrounds ruins of a 12th cent. castle, which was besieged several times in the 13th and 14th cent. Charles Dickens's home at Gadshill is nearby.

Roch·es·ter (rŏch'ĕs'tər, -ĭ-stər). **1.** City (1970 pop. 4,631), seat of Fulton co., N Ind., E of Winamac; settled 1835, inc. 1853. Nearby Lake Manitou has resort facilities. **2.** Residential village (1970 pop. 7,054), Oakland co., SE Mich., ENE of Pontiac; settled 1817, inc. 1869. Tools, metal products, and machinery are made. The Parke Davis Biological Farm is nearby. **3.** City (1978 est. pop. 58,800), seat of Olmsted co., SE Minn.; inc. 1858. Its manufactures include electronic and electrical equipment, medical supplies, business machines, foodstuffs, and plastic, metal, wood, and fiber glass products. The city is famous as the home of the Mayo Clinic, founded (1889) by Dr. W. W. Mayo with his sons Charles Horace Mayo and William James Mayo. Rochester has museums of medical science, history, and antique vehicles. **4.** City (1978 est. pop. 19,400), Strafford co., SE N.H., on the Cocheco River near the Maine border; settled 1728, inc. as a city 1891. It has diverse industries. In the city are an art gallery and an antique aircraft museum. The White Mts. recreation area begins nearby. **5.** City (1978 est. pop. 258,100), seat of Monroe co., W N.Y., a port of entry on the Genesee River and Lake Ontario, in a rich fruit and truck farm region; inc. 1817. Its diverse manufactures include photographic, optical, dental, and gear-cutting equipment, process control and recording instruments, thermometers, office equipment, communications materials, automotive parts, wearing apparel, and jewelry. Permanent settlement began in 1812. The Erie Canal gave impetus to Rochester's growth, and flour milling became the first important industry. The city's cultural features include the Rochester Philharmonic and Eastman School of Music orchestras, a large choral group, the Rochester Museum of Art and Sciences, and the Memorial Art Gallery. Prominent residents have been Susan B. Anthony, Frederick Douglass, and George Eastman. Numerous parks and nurseries have earned Rochester the name Flower City.

Roche-sur-Yon, La (lä rôsh-sür-yôɴ), city (1975 pop. 44,713), capital of Vendée dept., W France, on the Yon River. An agricultural trade center, it also produces automobile and washing-machine parts, slippers, foundry products, and printed materials. It was founded in 1804 by Napoleon I as a town for non-Royalists.

Rock (rŏk). **1.** County (1970 pop. 11,346), 485 sq mi (1,256.2 sq km), extreme SW Minn., bordering on S.Dak. and Iowa, drained by the Rock River; formed 1857; co. seat Luverne. In an agricultural area yielding livestock, corn, oats, barley, and potatoes, it has some mining and meat packing. **2.** County (1970 pop. 2,231), 1,012 sq mi (2,621.1 sq km), N Nebr., bounded in the N by the Niobrara River; formed 1888; co. seat Bassett. It is predominately a grain and livestock area. **3.** County (1970 pop. 131,970), 721 sq mi (1,867.4 sq km), S Wis., on the Ill. border, drained by the Rock and Sugar rivers; formed 1836; co. seat Janesville. It is in a rich agricultural (corn, wheat, and tobacco), dairying, and livestock-raising area. Its industries include food processing and textile milling. Chemicals, plastics, cars, trucks, and fountain pens are manufactured. There are winter sports in the Beloit area.

Rock, river, c.285 mi (460 km) long, rising in SE Wis. and flowing SW through NW Ill. to the Mississippi River near Rock Island. It flows through a fertile farm area.

Rock·a·way (rŏk'ə-wā'), narrow peninsula, c.10 mi (16 km) long, SW Long Island, SE N.Y., in Queens borough of New York City. Separating Jamaica Bay from the Atlantic Ocean and isolated from the rest of New York City, the densely populated peninsula owes its growth to road and rail connections across the bay. There are fine sand beaches on Rockaway's southern side.

Rock·bridge (rŏk'brĭj'), county (1970 pop. 16,637), 601 sq mi (1,556.6 sq km), W central Va., at the S end of the Shenandoah Valley, with the Blue Ridge (here cut by the James River) in the E and SE; formed 1777; co. seat Lexington (an independent city). It has livestock raising, dairying, and fruit growing, with some lumbering and rock quarrying. Clothing is made. Its scenic mountain resort areas and medicinal springs attract tourists.

Rock·cas·tle (rŏk'kăs'əl), county (1970 pop. 12,305), 312 sq mi (808.1 sq km), SE central Ky., bounded on the SE by the Rockcastle River and drained by the Dix River and several creeks; formed 1810; co. seat Mount Vernon. In a hilly agricultural area that produces livestock, grain, and burley tobacco, it has bituminous-coal mines, timber, limestone and sandstone deposits, and oil wells. The county includes part of Daniel Boone National Forest and Great Saltpeter Caves.

Rock·dale (rŏk'dāl'), county (1970 pop. 18,152), 128 sq mi (331.5 sq km), N central Ga., in a piedmont region drained by the Yellow and South rivers; formed 1870; co. seat Conyers. Its agriculture includes cotton, corn, truck crops, grain, fruit, and livestock. Textiles, metal and wood products, furniture, and chemicals are manufactured.

Rock·ford (rŏk'fərd). **1.** Town (1970 pop. 603), seat of Coosa co., E central Ala., N of Montgomery, in a lumbering area. There are Indian relics nearby. **2.** Industrial city (1978 est. pop. 143,900), seat of Winnebago co., N Ill., on the Rock River near the Wis. line; inc. 1839. Rockford is the trade, processing, and shipping hub of an extensive agricultural region. Machine tools, screws and fasteners, and airplane and automobile parts are produced. Rockford was founded (1834) on the site of a battlefield of the Black Hawk War. The city has an extensive park and recreational system and a notable clock museum.

Rock·hamp·ton (rŏk-hămp'tən), city (1975 est. pop. 51,500), Queensland, E Australia, on the Fitzroy River. It is a rail center and trade center for the pastoral and mining regions of central Queensland. Rockhampton was founded in 1858.

Rock Hill. **1.** City (1970 pop. 6,815), St. Louis co., E Mo., a suburb W of St. Louis; inc. 1929. **2.** City (1978 est. pop. 36,400), York co., N S.C.; inc. 1870. An important textile center, it also has industries that produce fencing and plastics.

Rock·ing·ham (rŏk′ĭng-hăm′). **1.** County (1970 pop. 138,951), 691 sq mi (1,789.7 sq km), SE N.H., on the coast and drained by the Piscataqua, Exeter, and Lamprey rivers; formed 1769; co. seat Exeter. It has agriculture, shipping, and manufacturing (textiles, shoes, and wood products). There are resorts on its coastline and lakes. **2.** County (1970 pop. 72,402), 569 sq mi (1,473.7 sq km), N N.D., bounded on the N by the Va. border and drained by the Dan and Haw rivers; formed 1785; co. seat Wentworth. This piedmont tobacco and timber area also has textile mills, cigarette factories, and sawmills. **3.** County (1970 pop. 47,890), 865 sq mi (2,240.4 sq km), NW Va., partly in the central Shenandoah Valley, bordered in the W, NW, and N by W.Va., with the Alleghenies in the W and NW and the Blue Ridge along the SE border; formed 1777; co. seat Harrisonburg (an independent city). It has diversified farming (grain, apples, and peaches), dairying, and livestock raising, and is a leading poultry county. Limestone quarrying, lumbering, and food processing are done. It includes parts of George Washington National Forest and Shenandoah National Park and is traversed by the Appalachian Trail.

Rockingham, town (1975 est. pop. 6,200), seat of Richmond co., S N.C., SE of Charlotte near the Pee Dee River; settled c.1780, inc. 1887. Its manufactures include textiles and paper and metal products.

Rock Is·land (ī′lənd), county (1970 pop. 166,734), 420 sq mi (1,087.8 sq km), NW Ill., bounded on the N and W by the Mississippi River and drained by the Rock River; formed 1831; co. seat Rock Island. It has agriculture (corn, wheat, oats, poultry, and dairy products), bituminous-coal mines and clay pits, and manufacturing of heavy machinery. It includes Black Hawk and Campbell's Island state parks and part of a national wildlife refuge.

Rock Island, city (1970 pop. 50,166), seat of Rock Island co., NW Ill., on the Mississippi and Rock rivers, adjacent to Moline and opposite Davenport, Iowa; inc. 1841. These three cities, with East Moline, are called the Quad Cities. Farm equipment, machinery, metal and wood products, and rubber footwear are among Rock Island's manufactures. On a 1,000-acre (405-hectare) island in the Mississippi River is a U.S. arsenal, established in 1862. Arsenal Island was fortified by the British in the War of 1812 and by the Americans in 1816. During the Civil War it was the site of a Union military prison. A Confederate cemetery and a national cemetery are on the island.

Rock·land (rŏk′lənd), county (1970 pop. 229,903), 178 sq mi (461 sq km), SE N.Y., bounded on the E by the Hudson River (here widening into Tappan Zee), on the SW and S by the N.J. border, and drained by the Hackensack and Ramapo rivers; formed 1798; co. seat New City. Truck crops, fruit, and hothouse flowers are grown. There is some manufacturing. Several sections of Palisades Interstate Park, including Bear Mountain recreational area, are here.

Rockland. 1. City (1970 pop. 8,505), seat of Knox co., S Maine, ESE of Augusta; settled c.1770, inc. as a town 1848, as a city 1854. During the 1850s Rockland built a large number of clipper ships, and later in the century it shipped large quantities of lime from its quarries and kilns. The city today is a port of entry, a trading center, and a resort. Its main industry is fishing. **2.** Industrial town (1970 pop. 15,674), Plymouth co., E Mass.; settled 1673, set off from Abington and inc. 1874. Its products include abrasives, fiberglass boats, shoes, and sheet metal.

Rock·port (rŏk′pôrt′, -pōrt′). **1.** City (1970 pop. 2,565), seat of Spencer co., SW Ind., on the Ohio River SE of Evansville; settled 1807. Built on bluffs above the river, it is the site of Lincoln Pioneer Village, a memorial with reconstructed pioneer buildings. **2.** Town (1970 pop. 5,636), Essex co., NE Mass., on Cape Ann; settled 1690, set off from Gloucester 1840. It is a resort and artists' colony. **3.** City (1970 pop. 1,575), seat of Atchison co., extreme NW Mo., near the Missouri River, in a farm area; platted 1851. **4.** City (1970 pop. 3,879), seat of Aransas co., S Texas, on Aransas Bay NE of Corpus Christi. Built in the 1860s as a port to ship cattle, it is today a resort and fishing port on a channel leading to the Intracoastal Waterway.

Rock Rap·ids (răp′ĭdz), city (1975 est. pop. 2,735), seat of Lyon co., extreme NW Iowa, on the Rock River near the Minn. border; inc. 1886. It is in a farm and livestock area:

Rock·springs (rŏk-sprĭngz′), town (1970 pop. 1,221), seat of Edwards co., SW Texas, NE of Del Rio on the Edwards Plateau, near headwaters of the Colorado and Nueces rivers; settled 1887, inc. 1924. It is a center of goat and sheep ranching and an important mohair market. A tornado in 1927 almost destroyed the town. Nearby is Devil's Sinkhole, a large cavern remarkable for its bat colony.

Rock Springs, city (1970 pop. 11,657), alt. c.6,270 ft (1,910 m), Sweetwater co., SW Wyo., on Bitter Creek; inc. 1888. It is a cattle and sheep shipping point and the center of large natural trona mines that produce soda ash. Oil and gas production, electric-power distribution, a revived coal industry, and tourism are also important. Rock Springs was settled around a trading post and stage station established on the Oregon Trail in the 1860s.

Rock·ville (rŏk′vĭl′). **1.** Town (1970 pop. 2,820), seat of Parke co., W Ind., NNE of Terre Haute, in a farm area; settled 1823, inc. 1854. **2.** City (1970 pop. 41,821), seat of Montgomery co., W central Md.; inc. 1860. It has several scientific-research laboratories.

Rockville Cen·tre (sĕn′tər), residential village (1970 pop. 27,444), Nassau co., SE N.Y., on SW Long Island; inc. 1893. A state park is adjacent to the village.

Rock·wall (rŏk′wôl′), county (1970 pop. 7,046), 147 sq mi (380.7 sq km), NE Texas, drained by the East Fork of the Trinity River; formed 1873; co. seat Rockwall. It has agriculture (cotton, corn, oats, truck crops, cattle, poultry, and dairy products) and some timber.

Rockwall, city (1970 pop. 3,121), seat of Rockwall co., N Texas, near the East Fork of the Trinity River NE of Dallas. The blackland prairie yields cotton and grain, and the city processes aluminum and makes leather goods.

Rock·well City (rŏk′wĕl′), city (1970 pop. 2,396), seat of Calhoun co., W central Iowa, WSW of Fort Dodge, in a grain and livestock area; inc. 1882.

Rock·y Hill (rŏk′ē), town (1970 pop. 11,103), Hartford co., central Conn., a suburb of Hartford, on the Connecticut River; settled c.1650, inc. 1843. Firearms, chemical coatings, and metal products are made here. Rocky Hill was an important river port (1700–1820).

Rocky Mount. 1. City (1970 pop. 34,284), Edgecombe and Nash cos., E N.C., on the Tar River; settled by 1818, inc. 1867. The city is the processing and distribution center of a rich agricultural area (tobacco, cotton, and corn). Textiles, apparel, chemicals, and pharmaceuticals are manufactured. **2.** Town (1970 pop. 4,002), seat of Franklin co., SW Va., in the E of the foothills of the Blue Ridge Mts., S of Roanoke, in a tobacco area; combined with contiguous Mt. Pleasant and inc. 1873. An industrial town, it produces textiles, furniture, and plywood and has poultry and fruit farms.

Rocky Moun·tain House (moun′tən hous′), town (1971 pop. 2,968), S central Alta., Canada, at the foot of the Rocky Mts. and the confluence of the North Saskatchewan and Clearwater rivers. Founded in 1799 as a fortified post of the North West Company in Blackfoot Indian country, it was taken over (1821) and operated by the Hudson's Bay Company until 1875. It is now the gateway to a big-game hunting area.

Rocky Mountains, major mountain system of W North America and easternmost belt of the North American cordillera, extending more than 3,000 mi (4,800 km) from central N.Mex. to NW Alaska; Mt. Elbert (14,431 ft/4,399 m) in Colo. is the highest peak. The Rockies are located between the Great Plains on the east (from which they rise abruptly for most of their length) and a series of broad basins and plateaus on the west. The mountains form the Continental Divide, separating rivers draining to the Atlantic and Arctic oceans from those draining to the Pacific Ocean.

The Rockies were formed in the Mesozoic and Early Cenozoic eras during the Cordilleran orogeny. They are geologically complex, with remnants of an ancestral Rocky Mt. system and evidence that uplift, which involved almost all mountain-building processes, occurred as a series of pulses over millions of years. The mountains have since been eroded to expose ancient crystalline cores flanked by thick upturned layers of sedimentary rocks. Glaciers and snowfields, which today cover the northern ranges and the high peaks of the south, were at one time more extensive. Topographically, the Rockies are usually divided into five sections: the Southern Rockies, Middle Rockies, Northern Rockies (all in the United States), the Rocky Mountain system of Canada, and Brooks Range in Alaska. The Wyoming Basin, the system's principal topographic break, is sometimes considered a sixth section.

The Southern Rockies, in N.Mex., Colo., and southern Wyo., are dominated by two north-south belts of folded mountains that have been eroded to expose cores of Precambrian rocks rimmed by sedimentary rocks. Between the two belts are three basins known as the

North, South, and Middle "parks." To the southwest are the San Juan Mts., a nonlinear group of uplands composed mainly of volcanic rocks. The Southern Rockies are the system's highest section and include many peaks above 14,000 ft (4,267 m), among them Mt. Massive (14,418 ft/4,395 m) in the Sawatch Mts.

The Middle Rockies, chiefly in northeastern Utah and western Wyo., lie north of the Southern Rockies and are separated from them by the Wyoming Basin. The ranges of this section are lower and less continuous than those to the south. These ranges have been eroded down to their Precambrian cores and are rimmed by Paleozoic and Mesozoic sedimentary rocks. The highest peaks of the Middle Rockies are Gannet Peak (13,785 ft/4,202 m) in the Wind River Range and Grand Teton (13,766 ft/4,196 m) in the Teton Range.

The Northern Rockies, in northeastern Wash., northern and central Idaho, and northwestern Wyo., extend north from Yellowstone National Park to the U.S.-Canadian border. A series of north-south trending ranges separated by narrow trenches and valleys occupies most of northern Mont. and the Idaho Panhandle. Two especially distinctive trenches are the Rocky Mountain Trench, which extends northwest from Flathead Lake, and the Purcell Trench, which extends north from Coeur d'Alene Lake. The peaks of the Northern Rockies are generally lower than those to the south; among the highest are Borah Peak (12,655 ft/3,857 m) and Leatherman Peak (12,230 ft/3,728 m) in the Lost River Range.

The Rocky Mt. system of Canada is composed of two major sections: the high rugged peaks of the Canadian Rockies proper to the east, and the Columbia Mts. group on the west. The Canadian Rockies are located along the British Columbia–Alta. border and include Mt. Robson (12,972 ft/3,954 m; highest peak of the Rocky Mts. in Canada). The prominent Rocky Mountain Trench, west of the crest line, continues c.800 mi (1,290 km) into Canada from Mont. The Purcell Trench to the west also crosses into Canada and joins the Rocky Mountain Trench c.200 mi (320 km) north of the border.

The Rockies continue into the Yukon Territory and Northwest Territories as the Mackenzie, Richardson, and Franklin mts. In northern Alaska, the Brooks Range, a cold and treeless region rises to Mt. Chamberlin (9,020 ft/2,749 m) and forms the northernmost section of the Rocky Mts.

Mining is important throughout the entire system; gold, silver, lead, zinc, copper, and molybdenum are the chief minerals. The principal mining centers are Leadville and Cripple Creek, Colo.; the Butte-Anaconda district of Mont.; Coeur d'Alene, Idaho; and the Kootenay Trail region of British Columbia. Vast forests, largely under government control and supervision, are a major natural resource. Lumbering and other forestry activities are limited mainly to Mont., Idaho, and British Columbia, where commercially valuable stands are most abundant and accessible. The Rockies are also a year-round recreational attraction. The Rockies are a major barrier to overland transcontinental travel. The principal U.S. pass across the mountains is South Pass (alt. c.7,500 ft/2,300 m) at the southern end of the Wind River Range, southwestern Wyo. This pass was followed by the Oregon, Mormon, and California trails; the Sante Fe Trail skirted the southern end of the Rockies. In Canada the important passes are Kicking Horse (alt. 5,539 ft/1,688 m), which carries the Trans-Canada Highway, Crowsnest Pass, and Yellowhead Pass. Explorers of the U.S. Rockies have included Vasquez de Coronado (1540), Meriwether Lewis and William Clark (1804-6), Zebulon Pike (1806-7), John Frémont (1843-44), and Ferdinand Hayden (1871). Leading Canadian explorers were sieur de la Vérendrye (1738-39), Sir Alexander Mackenzie (1792-93), David Thompson (1799-1803), and Simon Fraser (1803-7).

Rocky Riv·er (rĭv′ər), city (1970 pop. 22,958), Cuyahoga co., NE Ohio, a suburb of Cleveland, on Lake Erie; settled 1815, inc. 1903. Chiefly residential, it has some light manufacturing.

Ro·court (rô-ko͝or′), town (1970 pop. 4,882), Liège prov., E Belgium. In the War of the Austrian Succession, the French under Maurice de Saxe defeated (Oct., 1746) the allied English and Austrians here.

Ro·dez (rô-dĕz′), city (1968 pop. 26,398), capital of Aveyron dept., S France. It is a farm trade center. An impressive cathedral (13th–16th cent.) of northern Gothic style is here.

Ro·dhos (rô′*th*ôs): *see* Rhodes, Greece.

Ro·dri·guez (rō-drē′gəs), island (1971 est. pop. 24,700), 42 sq mi (109 sq km), in the Indian Ocean, c.350 mi (565 km) E of Mauritius, of which it is an integral part. One of the Mascarene Islands, Rodriguez is surrounded by coral reefs. The main occupations are subsistence farming and fishing. Rodriguez was discovered in 1645 by the Portuguese, was briefly occupied (1691-93) by the Dutch, and was colonized (18th cent.) by the French from Mauritius. Britain took the island in 1810.

Roer·mond (ro͞or-mônt′), city (1977 est. pop. 36,533), Limburg prov., SE Netherlands, at the confluence of the Meuse and Roer rivers. Manufactures include chemicals and electrical equipment. Roermond was an important center of the cloth trade in medieval times.

Roe·se·la·re (ro͞o′sə-lä′rə), city (1977 pop. 51,247), West Flanders prov., W Belgium. It is an industrial center. At Roeselare, in 1794, the French defeated the Austrians.

Rog·er Mills (rŏj′ər mĭlz′), county (1970 pop. 4,452), 1,140 sq mi (2,952.6 sq km), W Okla., bounded in the W by the Texas border, in the N by the Canadian River, and intersected by the Washita River; formed 1907; co. seat Cheyenne. Its agriculture includes wheat, milo, cotton, broomcorn, rye, cattle, poultry and dairy products. It has oil and natural-gas wells.

Rog·ers (rŏj′ərz), county (1970 pop. 28,425), 685 sq mi (1,774.2 sq km), NE Okla., intersected by the Verdigris and Caney rivers; formed 1907; co. seat Claremore. It has stock raising and agriculture (grain, corn, barley, sweet potatoes, and dairy products), bituminous-coal mines, and oil and natural-gas wells. Its manufactures include chemicals, metal products, machinery, and auto parts. There are mineral springs in the county.

Rogers, city (1970 pop. 11,050), Benton co., extreme NW Ark., in the Ozarks; inc. 1881. The city is located in a resort area and has meat-processing plants.

Rogers City, city (1970 pop. 4,275), seat of Presque Isle co., NE Mich., on Lake Huron SE of Cheboygan; settled 1869, inc. as a village 1877, as a city 1944. It has limestone quarries and fisheries.

Rog·ers·ville (rŏj′ərz-vĭl′), town (1970 pop. 4,076), seat of Hawkins co., NE Tenn near the Holston River SW of Kingsport; founded 1786, inc. 1835. It is a processing center for a farm and timber region. A mineral spring is nearby.

Roger Wil·liams National Memorial (wĭl′yəmz): *see* National Parks and Monuments Table.

Rogue (rōg), river, c.200 mi (320 km) long, rising in SW Oregon, in the Cascade Range N of Crater Lake, and flowing SW and W through a fertile valley (noted for its orchards) and then across the Coast Range to the Pacific Ocean at Gold Beach.

Roh·tak (rō′tək), city (1971 pop. 124,072), Haryana state, N central India. Rohtak is a market for grain and sugar. Sculpture from the Buddhist period (c.600 B.C.) has been uncovered here.

Ro·lette (rō-lĕt′), county (1970 pop. 11,549), 913 sq mi (2,364.7 sq km), N N.Dak., in a prairie area bordering on Man.; formed 1873; co. seat Rolla. It is exclusively agricultural, yielding grain, livestock, dairy products, and poultry.

Rol·la (rŏl′ə). **1.** City (1978 est. pop. 12,700), seat of Phelps co., S central Mo.; inc. 1861. It is in a livestock and farm region of the Ozarks. Fan blades and dog food are produced. An annual Ozark festival is held here. Caves and springs abound in the area. **2.** City (1970 pop. 1,458), seat of Rolette co., N N.Dak., near the Canadian boundary NNW of Devils Lake, in a wheat and livestock region; inc. 1907. The city is the headquarters for the International Peace Garden.

Roll·ing Fork (rō′lĭng fôrk′), town (1970 pop. 2,034), seat of Sharkey co., W Miss., on Deer Creek N of Vicksburg, in a farm area; settled c.1827. There are Indian mounds nearby.

Ro·ma·gna (rō-mä′nyä), historic region, N central Italy, bordering on the Adriatic Sea in the E. The region was the center of Byzantine domination in Italy (540-751). Ravenna was the seat of the Byzantine exarchs; Rimini was a city of the Pentapolis. Despite the donations of Pepin the Short (754) and of Charlemagne (774), which gave the exarchate and the Pentapolis to the pope, later emperors continued to claim the territory. Otto IV recognized (1209) the papal rights, but effective papal rule was prevented at first by the free communes and later by the petty tyrants who ruled the cities. Cesare Borgia, made duke of Romagna (1501) by Pope Alexander VI, tried unsuccessfully to make the Romagna the nucleus of his own state. Shortly thereafter Pope Julius II effectively incorporated the Romagna into the Papal States. Papal rule, interrupted (1797-1814) by French occupation, ended in 1860, when the Romagna was annexed by the kingdom of Sardinia. Austrian troops had helped until 1859 to maintain the papal regime.

Ro·man (rô'män), town (1974 est. pop. 47,627), NE Rumania, in Moldavia, at the confluence of the Prut and Siretul rivers. Steel and chemicals are among the industrial products. The town was founded in the late 14th cent.

Ro·ma·ni·a (rō-mān'yə, -mā'nē-ə): *see* Rumania.

Rom·blon Islands (rôm-blôn'), island group (1970 pop. 167,665), 524 sq mi (1,357.2 sq km), Romblon prov., Philippines, in the Sibuyan Sea. The islands are volcanic, yet generally low-lying and oval-shaped. Rice and copra are grown.

Rome (rōm), city (1971 pop. 2,799,836), capital of Italy and see of the pope, whose residence, Vatican City, is a sovereign state within the city of Rome. Rome is also the capital of Latium, a region of central Italy, and of Rome prov. It lies on both banks of the Tiber River and its affluent, the Aniene River, between the Apennine Mts. and Tyrrhenian Sea. Called the Eternal City, it is one of the world's richest cities in history and art and one of its great cultural, religious, and intellectual centers. It is the focus of international traffic by road, rail, sea (at the port of Civitavecchia), and air. The economy of Rome depends to a very large extent on the tourist trade. The city is also a center of banking, insurance, printing, publishing, and fashion. Italy's movie industry (founded in 1936) is located at nearby Cinecitta. As an educational center Rome possesses, aside from its university (founded 1303), the colleges of the church, several academies of fine arts, and the Accademia di Santa Cecilia (founded 1584), the oldest academy of music in the world. The opera house of Rome is one of the grandest in Europe.

In the past half century Rome has expanded well beyond the walls started in the 3rd cent. by Emperor Aurelian. Long sections of the ancient walls have been preserved, however, and archaeology remains an essential element of modern city-planning in Rome. As in ancient times, the larger section of Rome lies on the left bank of the Tiber. Aside from modern residential quarters, the right-bank section contains Vatican City, including St. Peter's Church, the Castel Sant' Angelo, and the ancient quarter of Trastevere. In the left-bank section the Piazza Venezia, a central square, lies at the foot of the old Capitol and borders on the huge monument to King Victor Emmanuel II and on the Palazzo Venezia, a Renaissance palace. A broad avenue, the Via dei Fori Imperiali, runs from the Piazza Venezia southeast to the Colosseum, with the Emperors' Fora to the left, and the Capitol and the ancient Forum to the right. From the Colosseum the Via di San Gregorio continues south past the Arch of Constantine and the Baths of Caracalla to the Appian Way. From the Piazza Venezia another modern thoroughfare, the Via del Mare, leads southwestward to the Tiber River and then east past the Basilica of St. Paul's Without the Walls to the sea at Ostia. The narrow and busy Via del Corso leads north from the Piazza Venezia past the Piazza Colonna (now the heart of Rome) to the Piazza del Popolo at the gate of the old Flaminian Way. East of the Piazza del Popolo are the Pincian Hill and the famous Borghese Villa. In the widest westward bend of the Tiber River, west of the Via del Corso, is the Campo Marzio quarter, where most of the medieval buildings are located; there also are the Pantheon (now a church) and the parliament buildings. To the east of the Via del Corso the fashionable Via Condotti leads to the Piazza di Spagna; a flight of 132 steps ascends from that square to the Church of the Santa Trinitá dei Monti and the Villa Medici. The Quirinal palace is northeast of the Piazza Venezia.

Among the countless churches of Rome there are five patriarchal basilicas—St. Peter's, St. John Lateran, St. Mary Major (Santa Maria Maggiore), St. Lawrence outside the Walls, and St. Paul's outside the Walls. Among the many palaces and villas of Rome the Farnese palace (begun 1514) and the Farnesina (1508-11) are particularly famous; others, all dating from the 17th cent., are those of the great Roman families, the Colonna, Chigi, Torlonia, and Doria. Rome is celebrated for its beautiful Renaissance and Baroque fountains, such as the ornate Fontana di Trevi (18th cent.). The richest museums and libraries of Rome are in the Vatican. Others include the National (in the Villa Giulia), Capitoline and Torlonia museums, notable for their antiquities; and the Borghese, Corsini, Doria, and Colonna collections of paintings.

History. Ancient Rome was built on the east, or left, bank of the Tiber on seven hills; the Palatine, roughly in the center; the Capitoline, to the northwest; and the Quirinal, Viminal, Esquiline, Caelian, and Aventine in an outlying north-southwest curve. The hills of Rome were a meeting ground for Latins, Sabines, and Etruscans. In the 8th cent. B.C. the fortified elevation of the Palatine was probably taken by Etruscans. Tradition tells of the founding (753 B.C.) of Rome by Romulus and of the Tarquin family, the Etruscan royal house. It was probably Etruscan rule that civilized Rome. The Romans overthrew their foreign rulers c.500 B.C. and established the Roman republic, which lasted four centuries. The patrician class controlled the government, but the plebs (the major portion of the population) were allowed to elect the two patrician consuls, who held joint power. The ancient senate became more and more powerful until in the 3rd cent. B.C. it controlled the consuls completely.

In the 4th cent. B.C. Rome extended its influence over western Latium and southern Etruria. The Samnites were subdued and the inhabitants of Picenum, Umbria, Apulia, Lucania, and Etruria were pacified. The Punic Wars (264-146 B.C.) ended in the complete destruction of Carthage; Rome gained dominion over Spain, Sicily, Sardinia, Corsica, and the northern shores of Africa, indisputable hegemony in the Mediterranean, and an insatiable desire for conquest. Macedonia was made a Roman province (168 B.C.) and the Greeks and Egyptians became subject to Rome.

Senatorial corruption, class dissension, and slave revolts made the republic's last century turbulent. Pompey emerged as a popular champion in 79 B.C. His defeat of Mithradates VI brought Pontus, Syria, and Phoenicia under Roman dominion. On Pompey's return from the East (61 B.C.) he found an ally for his ambitions in Julius Caesar, a popular democratic leader of the best patrician blood. Having subdued Gaul in the Gallic Wars, Caesar returned to Italy, defeated Pompey (47 B.C.), and became master of Rome. He goverened through the old institutions, with wisdom and vigor. At his death (44 B.C.) the territories ruled by Rome included Spain (except part of the northwest), Gaul, Italy, part of Illyria, Macedonia, Greece, western Asia Minor, Bithynia, Pontus, Cilicia, Syria, Cyrenaica, Numidia, and the islands of the sea, and Rome completely controlled Egypt and Palestine.

Caesar's assassination brought anarchy, out of which emerged (31 B.C.) the rule of Octavian (later Augustus), usually considered the first Roman emperor. He organized provincial government and the army, rebuilt Rome, and patronized the arts and letters. His rule began a long period (200 years) of peace. During Augustus' reign (31 B.C.-A.D. 14) Roman roads were built and commerce and industry were greatly developed, particularly by sea, over which grain ships carried food from the ports of northern Africa. During the rule of Claudius I (A.D. 41-A.D. 54) half of Britain was conquered (A.D. 43) and Thrace, Lydia, and Judaea were made Roman provinces. His son Nero (A.D. 54-A.D. 68) brutally persecuted the new sect of Christians and rebuilt the city with broader streets after the great fire (A.D. 64). Domitian (A.D. 81-96), a despot and persecutor of Christians, virtually completed the conquest of Britain. Trajan (A.D. 98-117), one of the greatest of emperors, undertook great public works, defeated the Daci, and established Roman colonies there (in what is now modern Rumania), and pushed the eastern borders past Armenia and Mesopotamia. His successor, Hadrian (A.D. 117-138), built his wall (Hadrian's Wall) across Britain to hold back the barbarians. The successors of Hadrian were Antoninus Pius (138-161) and Marcus Aurelius (161-180), who ruled in what is commonly called the Golden Age of the empire.

With Commodus (180-192) the decline of the empire is usually said to have begun. The age of the praetorians was then at hand, when the rise and fall of emperors was determined by this elite corps of soldiers. Septimius Severus (193-211) was victorious against the Parthians and the Picts; Caracalla (211-17) built the famous baths named after him; Decius (249-51) was a violent persecutor of Christians; Valerian (253-59) was captured by the Persians; and Diocletian (284-305) enacted political and social reforms. By the Edict of Milan (313), Constantine I (306-37) granted universal religious tolerance and proclaimed his own conversion to Christianity. Constantine moved (323) his capital to Byzantium (renamed Constantinople). There was a brief resurgence of paganism under Julian the Apostate (361-63), but Christianity was securely established. After the death (395) of Theodosius I the empire was permanently divided into East and West, and Rome rapidly lost its political importance. The West sank into anarchy, and Italy was ravaged by invaders. Alaric I took Rome in 410, and Gaiseric conquered it in 455. In 476 the last emperor of the West, Romulus Augustulus, was deposed by the Goths under Odoacer.

The history of Rome in the Middle Ages is essentially that of two institutions, the papacy and the commune of Rome. In the 6th cent.

rule by the Goths yielded to that of Byzantium. Pope Gregory I (590–604), one of the greatest Roman leaders of all time, began to emancipate Rome from the Byzantine exarchs. Sustained by the people, the popes soon exercised greater power in Rome than did the imperial governors. The papal elections were, for the next 12 centuries, the main events in Roman history. During the 7th–9th cent. the Papal States emerged. The era after 800 was punctuated by visits to the city by the German kings to be crowned Holy Roman emperor or to secure the election of a pope to their liking. Papal elections, originally exercised by the citizens of Rome, came under the control of the great noble families.

Papal authority was challenged in the 12th cent. by the communal movement. A commune was set up (1144–55), but it was subdued by the intervention of Emperor Frederick I. Finally, a republic under papal patronage was established, headed by an elected senator. However, civil strife continued between popular and aristocratic factions and between Guelphs and Ghibellines. During the Babylonian Capitivity of the popes at Avignon (1309–78) Rome was desolate, economically ruined, and in constant turmoil. An attempt by Cola di Rienzi to revive ancient Roman institutions failed. After the Great Schism (1378–1417), Martin V returned to Rome, and with him began the true and effective dominion of the popes in Rome.

The successors of Martin V in the 15th cent. and the first half of the 16th cent. were chiefly interested in increasing the temporal power of the papacy, in patronizing the arts and letters, in beautifying the city, and in raising their own fortunes. Among the countless artists and architects who served the papal court, Donato Bramante, Michelangelo, Raphael, and Domenico Fontana were the chief creators of Rome as it is today. As a result of Clement VII's alliance with Francis I of France, Rome was stormed (1527) by the army of Emperor Charles V and subjected to a thorough plundering. In the 17th and 18th cent. the splendor of religious ceremonies, as well as the encouragement given by the popes to art, music, classical and archaeological studies, and the restoration of ancient monuments continued to make Rome a center of world culture.

In 1798 the French occupied Rome, deported the pope, and proclaimed Rome a republic. Pius VII re-entered Rome in 1800, but in 1808 Napoleon reoccupied the city and in 1809 annexed it to France. Papal rule was restored in 1814. Pope Pius IX, who ruled during a crucial period (1846–78), yielded to liberal demands and granted a constitution, but in 1848 Rome once more became a republic, under the leadership of Giuseppe Mazzini. French troops intervened and restored Pius IX, who made no further attempts at liberalism. The Italian kingdom, proclaimed in 1862, included most of the former Papal States but not Rome, which remained under papal rule as a protectorate of Napoleon III until his fall in 1870. In 1871 Rome became the capital of Italy. The popes refused to recognize their loss of temporal power until the Lateran Treaty (1929), which gave them sovereignty over Vatican City. In World War II Rome fell to the Allies on June 4, 1944. The postwar years were marked by a vigorous economic, artistic, and intellectual revival.

Rome. **1.** City (1978 est. pop. 29,100), seat of Floyd co., NW Ga., where the Etowah and Oostanaula rivers meet to form the Coosa River, in a farm, timber, and quarry area; inc. 1847. The city is a cotton market and has textile and lumber mills, clothing factories, and foundries. It was established (1834) on the site of a Cherokee village; the name was inspired by the seven hills upon which the city is built. During the Civil War Gen. N. B. Forrest captured (May, 1863) a Union cavalry force near here; Gen. W.T. Sherman burned the city in Nov., 1864. Limestone caves and mineral springs are in the area. **2.** Industrial city (1978 est. pop. 48,300), Oneida co., central N.Y., on the Mohawk River and the Barge Canal; laid out c.1786 on the site of Fort Stanwix, inc. as a city 1870. It is especially noted for its copper and brass manufactures. Iron-mill items, machine tools, road graders, and strip steel are among its many other products. Because of its location, the city was a busy portage point beginning with the Indians, and it had great strategic importance during the French and Indian Wars and in the American Revolution. Construction on the Erie Canal began (1817) in Rome. Nearby is Grissom Air Force Base.

Ro·me·o·ville (rō′mē-ō-vĭl′), residential village (1970 pop. 12,674), Will co., NE Ill., on the Des Plaines River, with access to the Illinois and Mississippi Canal and the Chicago Sanitary and Ship Canal; inc. 1901. It has oil refineries and manufactures medical supplies.

Rom·ney (rŏm′nē), trading town (1970 pop. 2,364), seat of Hampshire co., E W.Va., in the Eastern Panhandle, on the South Branch of the Potomac River S of Cumberland, Md.; founded 1762. It is a trade center for a fruit-growing area.

Rom·ney Marsh (rŏm′nē, rŭm′-), region, c.70 sq mi (180 sq km), Kent, SE England. A former coastal marsh, the region has been wholly reclaimed and now has good pasture land.

Røm·ø (rœm′œ′), island (1965 pop. 812), 39 sq mi (101 sq km), SW Denmark, in the North Sea, one of the North Frisian Islands. A whaling base in the 18th cent., it is now a resort and has fisheries. The island was held by Prussia from 1864 to 1920.

Roms·da·len (rōōms′dä′lĕn), valley, c.60 mi (95 km) long, SW Norway, flanked by the mountains of Dovrefjell. It is an ancient passage from the western coast to southern Norway. Many jagged peaks line the valley.

Ron·ces·val·les (rōn′thāz-vä′lyās), mountain pass, (alt. 3,468 ft/1,057.7 m), in the Pyrenees, between Pamplona, Spain, and Saint-Jean-Pied-de-Port, France. Tradition has made it the scene of the death of the hero Roland.

Ron·da (rôn′dä), town (1970 pop. 22,094), Málaga prov., S Spain, in Andalusia. One of the most colorful of Spanish towns, it is situated high in the mountains of Sierra de Ronda and is a popular resort. The old Moorish town, atop a hill, is separated from the lower new town by a deep gorge of the Guadalevín River.

Ron·dô·nia (rōōn-dô′nyə), federal territory (1975 est. pop. 141,300), 93,839 sq mi (243,043 sq km), NW Brazil, on the border with Bolivia; capital Pôrto Velho.

Ron·kon·ko·ma (rŏn-kŏn′kə-mə), uninc. town (1975 est. pop. 20,200), Suffolk co., SE N.Y., central Long Island, SE of Smithtown near Lake Ronkonkoma, in a residential and resort area.

Røn·ne (rœ′nə), city (1976 est. pop. 15,399), capital of Bornholm co., extreme E Denmark, on Bornholm Island, a port on the Baltic Sea; founded 1327. It is an industrial, fishing, and tourist center.

Roo·de·poort-Ma·rais·burg (rōō′də-pōōrt′mä-rä′bûrKH′), city (1970 pop. 115,366), Transvaal, NE South Africa; founded in 1887. It is a gold-mining center and a resort.

Rooks (rōōks), county (1970 pop. 7,628), 886 sq mi (2,294.7 sq km), N Kansas, in a rolling prairie region drained by the South Fork of the Solomon River; formed 1867; co. seat Stockton. It is in a wheat and livestock area, with oil fields and wood-products manufacturing.

Roo·sen·daal (rō′sən-däl), city (1977 est. pop. 52,652), North Brabant prov., SW Netherlands, near the Belgian border. It is a transportation and industrial center. Manufactures include food products and furniture.

Roo·se·velt (rō′zə-vĕlt′, rōō′-). **1.** County (1970 pop. 10,365), 2,385 sq mi (6,177.2 sq km), NE Mont., in a grain and livestock area bordered on N.Dak. and bounded on the S by the Missouri River; formed 1919; co. seat Poplar. Fort Peck Indian Reservation extends throughout much of the county. **2.** County (1970 pop. 16,479), 2,455 sq mi (6,358.5 sq km), E N.Mex., partly in the Llano Estacado and bounded on the E by the Texas border; formed 1903; co. seat Portales. Its agriculture includes cotton, wheat, milo, alfalfa, potatoes, broomcorn, peanuts, and livestock. It has some manufacturing and processing industries.

Roosevelt, river, c.400 mi (645 km) long, NW Brazil. It was called the Rio da Dúvida (River of Doubt) when it was explored by Theodore Roosevelt in 1913. Renamed in his honor, it is occasionally called Rio Teodoro.

Roo·se·velt-Cam·po·bel·lo International Park (rō′zə-vĕlt-kăm-pō-bĕl′ō, rōō′-): *see* National Parks and Monuments Table.

Roosevelt Island, c.90 mi (145 km) long, E Ross Ice Shelf, Ross Dependency, Antarctica, S of the Bay of Whales. Robert E. Byrd discovered the ice-covered island in 1934.

Ro·rai·ma (rô-rī′mə), federal territory (1975 est. pop. 48,200), 88,843 sq mi (230,103 sq km), NW Brazil, on the border of Venezuela and Guyana; capital Boa Vista.

Roraima, mountain, 9,094 ft (2,773.7 m) high, at the junction of the boundaries of Brazil, Guyana, and Venezuela. A giant table mountain, it is the highest point in the Guiana Highlands.

Rør·os (rœ′rōs), town (1970 pop. 5,147), Sør-Trøndelag co., E Norway, near the Swedish border. It is a winter sports center. Copper mines in the vicinity, opened in 1644, were for many years the most important in Norway.

Ror•schach (rôr'shäKH), town (1977 est. pop. 10,000), St. Gall canton, NE Switzerland, on the Lake of Constance. A prosperous commercial town in the Middle Ages, Rorschach is now a resort.

Ro•sa, Mon•te (mōn'tā rô'zä), massif, in the Pennine Alps, on the Swiss-Italian border. Its highest peak, the Dufourspitze, 15,200 ft (4,636 m), is the highest point in Switzerland. The Swiss side is covered by glaciers.

Ro•sa•rio (rô-sär'yō), city (1970 pop. 798,292), Santa Fe prov., E central Argentina, a port on the Paraná River on the E margin of the Pampa. Sugar refining and flour milling are the principal industries, and there are large meat-processing plants. Rosario was settled in the late 17th cent. but grew mainly after 1870 with the rapid development of the Pampa.

Ros•com•mon (rŏs-kŏm'ən), county (1971 pop. 53,519), 951 sq mi (2,463.1 sq km), central Republic of Ireland; co. town Roscommon. A part of the central plain of Ireland, the region is low-lying and contains many lakes and bogs. It is an agricultural county, specializing in cattle and sheep.

Roscommon, county town (1971 pop. 18,256) of Co. Roscommon, central Republic of Ireland. Noted for its Dominican priory and the remains of a castle, both dating from the 13th cent., Roscommon is a tourist center and market town.

Roscommon, county (1970 pop. 9,892), 521 sq mi (1,349.4 sq km), N central Mich., drained by the Muskegon River and branches of the Tittabawassee and Au Sable rivers; formed 1875; co. seat Roscommon. It is chiefly a forest area where lumber and wood products are manufactured. Some farming is done. The county contains several resorts and hunting and fishing areas.

Roscommon, village (1970 pop. 810), seat of Roscommon co., N central Mich., NE of Cadillac on the South Branch of the Au Sable River. It is a trade center for a resort and farm area.

Ro•seau (rō-zō'), county (1970 pop. 11,569), 1,676 sq mi (4,340.8 sq km), NW Minn., bounded on the NE by Lake of the Woods and drained by the Roseau River; formed 1894; co. seat Roseau. It is in an agricultural (grain, potatoes, dairy products, and livestock) and lumbering area, with a tourist industry.

Roseau, village (1970 pop. 2,552), seat of Roseau co., NW Minn., near the Canadian line WSW of Lake of the Woods. It is a port of entry and a trade and shipping center for a farm area.

Rose•bud (rōz'bŭd'), county (1970 pop. 6,032), 5,037 sq mi (13,045.8 sq km), E central Mont., in an agricultural area drained by the Tongue and Yellowstone rivers; formed 1901; co. seat Forsyth. Livestock, grain, and coal are its major products.

Rose•burg (rōz'bûrg'), city (1978 est. pop. 17,100), seat of Douglas co., SW Oregon; inc. 1872. It has an important lumbering industry and handles the produce of nearby ranches (sheep and cattle) and fruit orchards. Tourism is also important. The area is known for its good hunting and fishing.

Ro•selle (rō-zĕl'). **1.** City (1976 est. pop. 10,213), Cook and Du Page cos., NE Ill., WNW of Chicago, in a farm area; inc. 1922. Plastic products and furniture are made. **2.** Borough (1978 est. pop. 21,500), Union co., NE N.J.; set off from Linden 1890 and inc. 1894. It is chiefly residential, but there is some industry. Thomas Edison had a laboratory here.

Rose•mead (rōz'mēd'), city (1978 est. pop. 39,500), Los Angeles co., S Calif., a suburb of Los Angeles; founded 1867, inc. 1959.

Ro•sen•berg (rō'zən-bûrg'), city (1970 pop. 12,098), Fort Bend co., S Texas, on the Brazos River, in an oil and natural-gas area; inc. 1902. Its economic activities center around sulfur, salt, agricultural products (especially sugar), and oil. Rosenberg was founded with the coming of the railroad in 1883 and grew as a farm-marketing and shipping center. It attracted many German and Czech immigrants.

Ro•set•ta (rō-zĕt'ə), city (1966 pop. 36,700), N Egypt, in the Nile River delta. Rice is cultivated nearby, and rice milling and fish processing are the city's main industries. The Rosetta stone, a basalt slab inscribed by priests of Ptolemy V in hieroglyphic, demotic, and Greek, was found by Napoleon's troops near the city in 1799. It was taken by the British in 1801 and is now in the British Museum.

Rose•ville (rōz'vĭl'). **1.** City (1978 est. pop. 21,600), Placer co., N central Calif., in a fruit (especially grapes), grain, and livestock area, in the foothills of the Sierras; inc. 1909. Its products include wine, formica, and rocket parts. **2.** City (1978 est. pop. 55,600), Macomb co., SE Mich., a residential suburb of Detroit; inc. 1926. **3.** Village (1978 est. pop. 36,700), Ramsey co., SE Minn., a suburb of St. Paul; inc. 1948. It has steel, trucking, and computer industries.

Ros•kil•de (rôs'kĭl-ə), city (1976 est. pop. 49,338), capital of Roskilde co., E Denmark, a port on the Roskilde Fjord (an arm of the Isefjord). Manufactures include processed food, liquor, machines, leather goods, and pharmaceuticals. One of the oldest Danish cities, Roskilde was the capital of Denmark from the 10th cent. until 1443, when it was replaced by Copenhagen. The commercial prosperity of Roskilde revived in the 19th cent. By the Treaty of Roskilde (1658) Denmark ceded its lands in southern Sweden to Charles X of Sweden. Roskilde has a museum of Viking ships, and nearby is an atomic research center. The cathedral (late 12th cent.) contains some 40 royal tombs.

Ro•slavl (rə-slä'vəl), city (1976 est. pop. 55,000), W central European USSR, on the Oster River. It is a road and rail junction and a market town. Known from the 12th cent., Roslavl was chartered under Lithuanian rule in 1408 and ceded to Russia in 1667.

Ross (rôs), county (1970 pop. 61,211), 687 sq mi (1,779.8 sq km), S Ohio, intersected by the Scioto River and by Paint, Deer, Walnut, and Salt creeks; formed 1798; co. seat Chillicothe. It has corn and wheat farming, hog and cattle raising, and dairying. There are sand and gravel pits, and its industry includes lumbering and the manufacture of paper, chemicals, plastics, and concrete. Mound City Group National Monument is here.

Ross and Crom•ar•ty (krŏm'ər-tē), former county, N Scotland. The county town was Dingwall. Ross-shire, once independent, was incorporated with Cromarty in 1889. In 1975 Ross and Cromarty was divided between the Highland and Western Isles regions.

Ross•bach (rôs'bäKH'), village, Halle dist., S central East Germany. At Rossbach on Nov. 5, 1757, Frederick II of Prussia defeated the imperial army and the French in the Seven Years' War.

Ross Lake National Recreation Area: *see* National Parks and Monuments Table.

Ross Sea, arm of the Pacific Ocean, Antarctica, between Victoria Land and Marie Byrd Land. It was discovered in 1841 by Sir James Clark Ross, a British explorer. Ross Island with Mt. Erebus, an active volcano, is in the western part of the sea; Roosevelt Island is in the east. The Ross Sea's southern extension is the Ross Ice Shelf, a great frozen area whose 400-mi (643.6-km) seaward side is the source of huge icebergs. The Bay of Whales, the ice shelf's best-known inlet, lasted for c.50 years and was the site of Norwegian explorer Roald Amundsen's base for his trek to the South Pole in 1911; Little America, a U.S. base, was located nearby. McMurdo Sound, on the western side of Ross Sea, is usually free of pack ice in late summer; it has been the most important staging point for exploration and scientific investigation.

Ros•tock (rŏs'tŏk') or **Ros•tock-War•ne•mün•de** (rŏs'tŏk'vär'nə-mün'də), city (1974 est. pop. 210,167), capital of Rostock dist., N East Germany, on the Baltic Sea. It is an industrial center and a major seaport. Manufactures include agricultural machinery, chemicals, watches, processed food, and furniture. Originally a Slavic fortress, Rostock was chartered in the 13th cent. It became one of the chief members of the Hanseatic League. Its university (founded 1419) was an important center of learning. Historic structures include the 13th cent. Church of St. Mary and parts of the medieval city walls and gates.

Ros•tov (rŏs'tŏv', rə-stôf'), city (1974 est. pop. 31,000), E European USSR, on Lake Nero. It is a road and rail junction and has food-processing and flax-spinning plants. Rostov has been known since 862 and still retains its medieval aspect. It was annexed by the grand duchy of Moscow in 1474, was made the seat of an Orthodox metropolitan in 1587, and served as an important commercial center from the 16th to 19th cent. Rostov's ancient kremlin contains the Uspenski Cathedral and other splendid 13th cent. churches.

Ros•tov-na-Do•nu (rə-stôf'nä-dô'nōō), city (1970 pop. 921,000), capital of Rostov oblast, SE European USSR, on the Don River near its entrance into the Sea of Azov. It is a major port and rail hub and an important industrial, cultural, and scientific center. Rostov-na-Donu has ship and locomotive repair yards, plants processing food and tobacco, and factories that manufacture agricultural machinery, chemicals, building materials, electrical equipment, furniture, clothing, footwear, and leather goods. The city grew around a fortress erected in 1761. Chartered in 1797, it was a major grain-exporting center throughout the 19th cent. The city suffered much damage in World War II and was rebuilt after the war.

Ros·well (rŏz'wĕl', -wəl). **1.** City (1970 pop. 8,761), Fulton co., NW Ga., on the Chattahoochee River N of Atlanta; settled c.1837, inc. 1854. Its ante-bellum mansions attract many tourists. **2.** City (1978 est. pop. 40,600), seat of Chaves co., SE N.Mex., near the Pecos River; settled 1869 as a trading post, inc. 1903. It is the trade, marketing, and rail center of an irrigated farm area. The city grew rapidly after the discovery (1891) of artesian wells, with the coming (1894) of the railroad, and with the later discovery of oil. Nearby are Bitter Lake National Wildlife Refuge and Lincoln National Forest.

Ro·then·burg ob der Tau·ber (rō'tən-bo͝org' ôp dər tou'bər), town (1974 est. pop. 11,930), Bavaria, S West Germany, on the Tauber River. A picturesque medieval town, it is entirely walled (the walls dating from the 14th and 15th cent.) and contains numerous Gothic and Renaissance-style buildings. Noteworthy churches include St. Jakobskirche (1373–1436) and St. Johanniskirche (1393–1403). First mentioned in the 10th cent., Rothenburg was a free imperial city from the late 13th cent. until 1803, when it passed to Bavaria.

Roth·er·ham (rŏ*th*'ər-əm), borough (1976 est. pop. 249,400), S Yorkshire, N England, at the confluence of the Don and Rother rivers. It lies in a coal district. Manufactures include steel, brass, electrical equipment, glass, and chemicals.

Ro·to·ru·a (rō'tə-ro͞o'ə), city (1976 pop. 37,229), central North Island, New Zealand, in a region of lakes and hot springs. The area is important for timber and wood pulp.

Rot·ter·dam (rŏt'ər-dăm'), city (1971 pop. 679,032), South Holland prov., W Netherlands, on the Nieuwe Maas River near its mouth on the North Sea. Rotterdam is the major foreign-trade center of the Netherlands. The city's inner port is connected to Hoek van Holland, its outer port, by the New Waterway. The city also has shipyards, oil refineries, automobile assembly plants, and factories that manufacture clothing, paper, and electronic equipment.

Rotterdam was chartered in 1328 and experienced its greatest growth with the construction (1866–90) of the New Waterway, which made the port accessible to the largest oceangoing vessels; with the major expansion of industry in northwestern Germany from the late 19th cent.; and with the European economic boom after World War II. During World War II the entire center of the city was destroyed by German air bombardment (May 14, 1940) several hours after it had capitulated. Most of the old houses of Rotterdam were destroyed. Among the noteworthy buildings that survived the raid were the stock exchange (18th cent.), the city hall (1920), and the Boymans-Van Beuningen Museum, with its collection of paintings by Dutch masters.

Rou·baix (ro͞o-bā'), city (1975 pop. 109,553), Nord dept., N France, in French Flanders. It is one of the largest textile (chiefly wool) centers in France. A national textile school is here.

Rou·en (ro͞o-än', -äɴ'), city (1975 pop. 114,927), capital of Seine-Maritime dept., N France. Situated on the Seine near its mouth at the English Channel, Rouen functions as the port of Paris. Among its many manufactures are metal products, chemicals, drugs, textiles, paper, and leather goods. Of pre-Roman origin, Rouen was the victim of repeated raids (9th cent.) by Norsemen. By the 10th cent. it was the capital of Normandy and a leading European city. It was held (1419–49) in the Hundred Years' War by the English. Joan of Arc was tried and burned here in 1431. Rouen has been an archiepiscopal see since the 5th cent. and is particularly rich in ecclesiastical buildings. Damaged in World War II, but since restored, are the cathedral of Notre Dame (12th–15th cent.), the Church of St. Maclou and the palace of justice (both 15th–16th cent.), and the Grosse Horloge, a Renaissance clock tower. The houses where Pierre Corneille and Gustave Flaubert were born are both museums.

Rou·ergue (ro͞o-ârg'), region of S France, in the S Massif Central. One of the most mountainous areas of France, it is traversed by the Aveyron, Tarn, and other rivers, which form many deep gorges. Sheep are raised in great quantity and furnish milk for the Roquefort cheese industry. The county of Rouergue passed to the French crown in 1271, was ceded to England by the Treaty of Brétigny (1360), and reverted to France in 1368. The area passed eventually to the Bourbon family and was inherited in 1607 by Henry IV, who united it with the royal domain.

Round·up (round'ŭp'), city (1970 pop. 2,116), seat of Musselshell co., central Mont., on the Musselshell River N of Billings; inc. 1909. A cattle town in the late 19th cent., it is now the center of a ranch and farm region with coal and oil deposits.

Rour·ke·la (rôr-kā'lə), city (1971 pop. 125,427), Orissa state, E central India, at the confluence of the Koel and Lankh rivers. The city is built around large iron and steel plants. Other products are heavy machinery, fertilizers, and chemicals.

Rous·sil·lon (ro͞o-sē-yôɴ'), small region and former province, S France, bordering on Spain along the Pyrenees and on the Mediterranean. Wine, fruit, and olives are the chief products of this fertile and densely populated region, which also has a tourist industry. The area has changed hands many times, from the Romans, who arrived c.121 B.C., through the Visigoths, the Arabs, the Carolingians, the Spaniards, the counts of Barcelona, and the kings of Aragón, France, and Majorca. Louis XIII conquered it from Spain in 1642, and French possession was confirmed by the Treaty of the Pyrenees (1659).

Routt (rout), county (1970 pop. 6,592), 2,330 sq mi (6,034.7 sq km), NW Colo., bordering on Wyo. and drained by the Yampa River; formed 1877; co. seat Steamboat Springs. In an agricultural area yielding cattle, sheep, dairy products, wheat, and hay, it has bituminous-coal mines and oil and gas wells.

Rou·yn (ro͞o'ən, ro͞o-äɴ'), city (1976 pop. 17,678), extreme W Que., Canada. It is the commercial and service center of a copper, gold, zinc, and silver mining district.

Ro·va·ni·e·mi (rō'və-nē'ə-mē), city (1975 est. pop. 28,411), capital of Lappi prov. (Lapland), N Finland, at the confluence of the Ounas and Kemi rivers. Commercial and agricultural fairs and winter sports events are held in the city. It is the starting point of the Great Arctic Highway.

Ro·ve·re·to (rō'və-rā'tō), town (1976 est. pop. 32,069), in Trentino-Alto Adige, N Italy, on the Adige River. It is an agricultural and industrial center. Manufactures include machinery, silk, and forest products. Rovereto was taken (15th cent.) from the bishopric of Trent by Venice, which ceded it to Austria in 1517. It passed to Italy in 1919. Of note is the Church of San Marco (15th–17th cent.).

Ro·vi·go (rō-vē'gō), city (1976 est. pop. 51,784), capital of Rovigo prov., Venetia, N Italy, between the Adige and the Po rivers. It is an agricultural market and an industrial center. First mentioned in the 9th cent., Rovigo belonged to the Este family from 1194 to 1482, when it passed to Venice. In the city are the octagonal Church of the Madonna del Soccorso (late 16th cent.) and an excellent art gallery.

Ro·vinj (rō'vēn'yə), town (1971 pop. 16,402), NW Yugoslavia, in Croatia, on the Istrian coast of the Adriatic Sea. It is a seaport with shipbuilding and fishing industries. Rovinj belonged to Venice from 1283 until 1797, when it passed to Austria. Italy acquired it in 1918, and it was ceded to Yugoslavia in 1947.

Rov·no (rôv'nə), city (1976 est. pop. 162,000), capital of Rovno oblast, SW European USSR, in the Ukraine, on the Ustye River. It is an industrial center producing building materials, high-voltage apparatus, machinery, metal goods, clothing, soap, and foodstuffs. An old Ukrainian settlement, Rovno passed to Russia in 1793 and to Poland in 1921, reverting to the Ukraine in 1939.

Ro·vu·ma (ro͞o-vo͞o'mə), river, c.450 mi (724 km) long, rising in N Mozambique, near Lake Nyasa, and flowing E to the Indian Ocean. It forms most of the Tanzania-Mozambique border.

Row·an. **1.** (rou'ən) County (1970 pop. 17,010), 290 sq mi (751.1 sq km), NE Ky., bounded on the SW by the Licking River; formed 1856; co. seat Morehead. It is in a hilly clay-mining and agricultural (livestock, grain, and burley tobacco) area, with sawmills and sandstone quarries. Wood, stone, clay, and glass products are made. **2.** (rō'ăn') County (1970 pop. 90,035), 523 sq mi (1,354.6 sq km), W central N.C., bounded on the E by the Yadkin River and on the N by the South Yadkin River; formed 1753; co. seat Salisbury. This piedmont region has farming (cotton, corn, wheat, hay, dairy products, and poultry), textile manufacturing, granite and stone quarrying, and sawmilling.

Rox·bor·o (rŏks'bûr'ō), town (1970 pop. 5,370), seat of Person co., N N.C., N of Durham. It produces textiles and metal products and processes tobacco.

Rox·burgh·shire (rŏks'bûr'ə-shîr', -shər), former county, S Scotland. Jedburgh was the county town. Roxburghshire, once part of the ancient kingdom of Northumbria, suffered severely in the border wars between England and Scotland. Under the Local Government

Act of 1973 Roxburghshire became part of the Borders region.

Roy·al Gorge (roi'əl gôrj'), 10 mi (16 km) long, narrow canyon cut by the Arkansas River, S central Colo., often called the Grand Canyon of the Arkansas. The gorge was discovered in 1806 by an expedition led by U.S. explorer Zebulon Pike. Its near-vertical walls are more than 1,000 ft (305 m) high. One of the world's highest suspension bridges, 1,053 ft (321.2 m) above the river, crosses the canyon, and a cable railway ascends the canyon wall.

Royal Leam·ing·ton Spa (lĕm'ĭng-tən), municipal borough (1973 est. pop. 44,950), Warwickshire, central England, on the Leam River, a tributary of the Avon. The borough has ironworks and brick factories but is primarily a health resort with mineral springs.

Royal Oak (ōk), residential city (1978 est. pop. 74,800), Oakland co., SE Mich., a suburb of Detroit; settled c.1820, inc. as a city 1921.

Royal Tun·bridge Wells (tŭn'brĭj wĕlz'), municipal borough (1973 est. pop. 44,800), Kent, SE England. Mineral springs were discovered in 1606, and the town developed as a fashionable resort.

Ru·an·da-U·run·di (rōō-än'də-ōō-rōōn'dē), former colonial territory, central Africa. Germany gained rights to the region at the Conference of Berlin (1884-85). During World War I Belgium conquered (1916) the area, and in 1924 Ruanda-Urundi was formally constituted a mandate of the League of Nations under Belgian rule. When Ruanda-Urundi achieved independence on July 1, 1962, it was split into two territories, Rwanda and Burundi.

Ru·a·pe·hu (rōō'ə-pā'hōō), dormant volcanic peak, 9,175 ft (2,798.4 m) high, on North Island, New Zealand. A lake occupies its crater.

Rub al Kha·li (rōōb' äl kä'lē), desert of the Arabian Peninsula, c.225,000 sq mi (582,750 sq km). The desert slopes from an altitude of 3,300 ft (1,006.5 m) in the west to near sea level in the east. Sand dunes rise to c.660 ft (200 m) in the southwest; there are salt marshes and pans in the southeast. The desert is usually considered part of Saudi Arabia, but it is waterless and cannot support life. It was first explored by English explorer Henry Philby in 1932.

Ru·bi·con (rōō'bĭ-kŏn'), small stream that flows into the Adriatic and in Roman times marked the boundary between Cisalpine Gaul and ancient Italy. In 49 B.C. Julius Caesar crossed the Rubicon to march against Pompey in defiance of the senate's orders. He thus committed himself to conquer or to perish, and "to cross the Rubicon" now means to take an irrevocable step.

Rü·des·heim (rü'dəs-hīm'), town (1966 pop. c.7,500), Hesse, central West Germany, on the Rhine River. Vineyards in the area produce some of the most noted Rhine wines.

Ru·dolf, Lake (rōō'dŏlf'), c.2,500 sq mi (6,475 sq km), NW Kenya and SW Ethiopia, E Africa, in the Great Rift Valley. Surrounded by volcanic mountains, the lake has no outlet and is becoming saline.

Ru·dol·stadt (rōō'dôl-shtät'), city (1974 est. pop. 31,698), Gera dist., S East Germany, on the Thuringian Saale River. Manufactures include china, glass, machinery, and leather and metal goods. It is also a tourist resort. The Heidecksburg Palace, rebuilt in 1735, is now a museum.

Ru·eil-Mal·mai·son (rü-ā'yə-mäl-mā-zôɴ'), town (1975 pop. 67,727), Hauts-de-Seine dept., N central France. It is an industrial center where metals, armaments, photographic equipment, film, pharmaceuticals, and automobile accessories are produced. The town was originally a resort of the Merovingian kings (5th-7th cent.). It was bought by Cardinal Richelieu, who built an estate here. Napoleon lived here from 1800 to 1804, and Empress Josephine and her daughter are buried in the town. Napoleon's home, the famous Malmaison, is now a museum housing artifacts from the Napoleonic period.

Ru·fi·ji (rōō-fē'jē), river, c.375 mi (605 km) long, rising in the highlands of SW Tanzania, E Africa, and flowing NE, then E to the Indian Ocean opposite Mafia Island. There are irrigation and flood-control projects on the river.

Rug·by (rŭg'bē), municipal borough (1973 est. pop. 60,380), Warwickshire, central England. An important railroad junction and cattle market, it is known chiefly as the seat of Rugby School, founded in 1567. Rugby football originated at the school in 1823.

Rugby, city (1970 pop. 2,889), seat of Pierce co., N central N.Dak., ENE of Minot; platted 1885, inc. 1906. In a livestock and dairy area, it raises grain.

Rü·gen (rü'gən), island (1970 pop. 86,216), 358 sq mi (927.2 sq km), Rostock dist., N East Germany, in the Baltic Sea, separated from the mainland by the Strelasund. Agriculture and fishing are the main occupations. Famous chalk cliffs rise on the eastern shore. Rügen was conquered by Denmark in 1168, passed to Pomerania in 1325, and was taken by Prussia in 1815.

Ruhr (rŏŏr), region, c.1,300 sq mi (3,365 sq km), W West Germany. One of the world's greatest industrial complexes, it lies along and north of the Ruhr River (145 mi/233.3 km long), which rises in the hills of north-central West Germany and flows generally west to the Rhine River at Duisburg. Extensive high-quality coal deposits underlie the region in basins that are near the surface along the Ruhr River and at greater depths to the north along the Lippe River. The development of the Ruhr district began in the 19th cent. when the Krupp and Thyssen concerns built large integrated coal and steel empires. The Ruhr was occupied (1923-25) by French and Belgian forces, which greatly embittered German nationalist feeling. Some of the chief Ruhr industrialists helped Adolf Hitler to power in 1933. The Ruhr was a major bombing target for Allied forces during World War II. The International Authority for the Ruhr was set up in 1949 with responsibility for development of the region. Control passed to the European Coal and Steel Community in 1952 and to West Germany in 1954.

Ru·ma·ni·a (rōō-mā'nē-ə, -mān'yə) or **Ro·ma·ni·a** (rō-mā'nē-ə, -mān'yə), republic (1971 pop. 20,478,658), 91,699 sq mi (237,500 sq km), SE Europe, bordering on Hungary in the NW, on Yugoslavia in the SW, on Bulgaria in the S, on the Black Sea in the SE, and on the USSR in the E and N. Bucharest is the capital and largest city.

Rumania includes 7 historic and geographic regions: Walachia, Moldavia, Transylvania, and parts of Bukovina, Crişana-Maramureş, the Dobruja, and the Banat. The Danube River, which forms part of the border with Yugoslavia and almost all of the frontier with Bulgaria, traverses Rumania in the southeast; its tributary, the Prut, constitutes most of the border with the USSR. The Carpathian Mts., of which the Transylvanian Alps are a part, cut through Rumania in a wide arc from north to southwest; the highest peaks are Moldoveanu (8,343 ft/2,543 m) and Negoiu (8,317 ft/2,535 m). The country's climate is continental, with hot, dry summers and cold winters; severe droughts are common during the summer. The great majority of the inhabitants speak Rumanian, although there are also sizable minorities speaking Hungarian and German. By far the largest religious body is the Rumanian Orthodox Church.

Rumania is divided into 39 districts and the municipality of Bucharest:

NAME	CAPITAL	NAME	CAPITAL
Alba	Alba Iulia	Cluj	Cluj
Arad	Arad	Constanţa	Constanţa
Argeş	Piteşti	Covasna	Sfîntu-Gheorghe
Bacău	Bacău		
Bihor	Oradea	Dîmboviţa	Tîrgovişte
Bistriţa-Năsăud	Bristriţa	Dolj	Craiova
Botoşani	Botoşani	Galaţi	Galaţi
Brăila	Brăila	Gorj	Tîrgu-Jiu
Braşov	Braşov	Harghita	Miercurea-Ciuc
Bucharest (municipality)	Bucharest		
Buzău	Buzău	Hunedoara	Deva
Caraş-Severin	Reşiţa	Ialomiţa	Slobozia

NAME	CAPITAL	NAME	CAPITAL
Iaşi	Iaşi	Satu-Mare	Satu-Mare
Ilfov	Bucharest	Sibiu	Sibiu
Maramureş	Baia-Mare	Suceava	Suceava
Mehedinţi	Turnu-Severin	Teleorman	Alexandria
Mureş	Tîrgu Mureş	Timiş	Timişoara
Neamţ	Piatra-Neamţ	Tulcea	Tulcea
Olt	Slatina	Vaslui	Vaslui
Prahova	Ploieşti	Vîlcea	Rîmnicu-Vîlcea
Sălaj	Zalău	Vrancea	Focşani

Economy. Traditionally an agricultural country, Rumania greatly expanded its industrial base after World War II, so that in the early 1970s industry contributed more than half of the national product. The economy is almost entirely controlled and planned by the state. The chief farm products are wheat, maize, sugar beets, potatoes, sunflower seeds, and barley. There are extensive vineyards, and much wine is produced. Rumania is central Europe's largest producer of petroleum and natural gas (methane), with large oil fields around Ploieşti and on the Carpathian slopes in Moldavia. Other major resources include anthracite, lignite, iron and copper ores, salt, bauxite, gold, and chromium. The leading manufactures include refined petroleum, iron and steel, chemicals, textiles, cement, forest products, processed food, tires, and electronic and electric equipment. Rumania's leading exports are petroleum and petroleum products, farm produce, cement, and tractors; the main imports are iron ore, coked coal, metals, and electric equipment. The principal trading partners are the USSR, West Germany, Czechoslovakia, and East Germany.

History. Rumania occupies, roughly, ancient Dacia, which was a Roman province in the 2nd and 3rd cent. A.D. After the Romans left the region, the area was overrun successively by the Goths, the Huns, the Avars, the Bulgars, and the Magyars. After a period of Mongol rule (13th cent.), the history of the Rumanian people became in essence that of the two Rumanian principalities—Moldavia and Walachia—and of Transylvania, which for most of the time was a Hungarian dependency. The princes of Walachia (in 1417) and of Moldavia (mid-16th cent.) became vassals of the Ottoman Empire, but they retained considerable independence. Michael the Brave of Walachia defied both the Ottoman sultan and the Holy Roman emperor and at the time of his death (1601) controlled Moldavia, Walachia, and Transylvania, but his empire soon reverted to Turkish domination. Although the two principalities technically remained within the Ottoman Empire, after the Russo-Turkish War (1828–29) they actually became Russian protectorates. The Russians and Turks combined to suppress a Rumanian rebellion in 1848. The election (1859) of Alexander John Cuza as prince of both Moldavia and Walachia prepared the way for the official union (1861–62) of the two principalities as Rumania. Prince Karl of the house of Hohenzollern-Sigmaringen was chosen as Cuza's successor in 1866. Rumania gained full independence in 1878 but was obliged to restore south Bessarabia to Russia and to accept north Dobruja in its place. In 1881 Rumania was proclaimed a monarchy, and Prince Karl was crowned King Carol I.

After World War I, which it entered in 1916 on the Allied side, Rumania annexed Bessarabia from Russia, Bukovina from Austria, and Transylvania and the Banat from Hungary. A series of agrarian laws beginning in 1917 did much to break up the large estates and to redistribute the land to the peasants. The large Magyar population and other minority groups were a constant source of friction, however, and Rumanian politics in the 1920s became violent and unsettled. Fascist groups emerged, and in 1938 Carol II assumed dictatorial powers. The country remained neutral at the outbreak (1939) of World War II, but in 1940 it became a partner of the Axis. Marshal Ion Antonescu became dictator (1940) and replaced Carol with his son Michael. The Soviet Army invaded in 1944 and a Communist-led coalition goverment was set up in 1945. After the war Rumania recovered all its territories except Bessarabia, north Bukovina, and south Dobruja.

In Dec., 1947, Michael was forced to abdicate, and Rumania was proclaimed a people's republic. Nationalization of industry and natural resources was completed by a law of 1948, and there was also forced collectivization of agriculture. Beginning in 1963 Rumania's foreign policy became increasingly independent of that of the USSR. In 1969 Rumanian Communist Party leader Nicholas Ceauşescu and President Tito of Yugoslavia affirmed the sovereignty and equality of socialist nations. Rumania maintained close economic ties with the USSR in the 1970's but strengthened relations with the West, signing a trade agreement with the U.S. in 1975.

Government. Under the constitution adopted in 1965 the grand national assembly, consisting of 465 elected members, is the chief legislative body. In Mar., 1974, the government was reorganized; authority was concentrated in the new office of president of the republic. In practice, power in Rumania is controlled by the Rumanian Communist party, whose leading members are also the country's chief officials. The permanent presidium, led by the general secretary, is the party's highest body.

Ru·me·li·a (rōō-mē′lē-ə, -mēl′yə), region of S Bulgaria, between the Balkan and Rhodope mts. Historically, Rumelia denoted the Balkan possessions (particularly Thrace and Macedonia) of the Ottoman Empire.

Rum·ford (rŭm′fərd), town (1970 pop. 9,363), Oxford co., W Maine, at falls of the Androscoggin River NW of Augusta; settled 1780, inc. 1800. It has large paper mills and is in a ski resort area.

Run·corn (rŭng′kərn), urban district (1971 pop. 35,953), Cheshire, W England, on the Mersey River. It is on the Manchester Ship Canal and is the terminus of the Bridgewater Canal, which connects with the Mersey by a series of locks. Runcorn has shipyards, iron foundries, and tanning factories, and industries that manufacture pharmaceuticals and chemicals.

Run·nels (rŭn′əlz), county (1970 pop. 12,108), 1,060 sq mi (2,745.4 sq km), W central Texas, drained by the Colorado River and its tributaries; formed 1858; co. seat Ballinger. It is mainly agricultural, yielding cotton, grain sorghums, peanuts, truck crops, dairy products, poultry, cattle, and sheep. Clothing, metal products, and building materials are manufactured. There are oil wells and a meat-packing industry here.

Run·ny·mede (rŭn′ē-mēd′), meadow in Egham, Surrey, S England, on the S bank of the Thames River, W of London. Either on this meadow or on nearby Charter Island, King John accepted the Magna Carta (1215). There is a memorial to President John F. Kennedy on Runnymede.

Ru·pert (rōō′pərt), city (1970 pop. 4,563), seat of Minidoka co., S Idaho, in the Snake River valley E of Twin Falls; laid out 1905 as a model city by government engineers, inc. as a village 1906, as a city 1917. It is a shipping and processing point for farm products.

Rupert House, village and trading post, W Que., Canada, on the Rupert River E of its mouth on James Bay. Founded in 1668 as Fort Charles and later called Fort Rupert, then Rupert's House or Rupert House, it is the oldest fur-trading post of the Hudson's Bay Company. In the struggle between the English and French in Canada, the post was captured in 1686 by the French and alternately held by the French and British until the Peace of Utrecht (1713) restored it permanently to the Hudson's Bay Company. The post is the center of a large beaver sanctuary.

Ru·pert's Land (rōō′pərts), Canadian territory held (1670–1869) by the Hudson's Bay Company, named for Prince Rupert, first governor of the company. Under the charter granted (1670) to the company by Charles II, the region comprised the drainage basin of Hudson Bay, including large areas of Ont., Que., Man., Sask., Alta., and the Northwest Territories, and portions of Minn. and N.Dak. In 1869 the Hudson's Bay Company transferred Rupert's Land to Canada for £300,000 but retained certain blocks of land for trading and other purposes.

Ru·se (rōō′sā), city (1973 est. pop. 165,000), NE Bulgaria, a port on the Danube River. It has shipyards, a petroleum refinery, and varied manufactures. Founded (2nd cent. B.C.) as Prista, it became a Roman naval station. Under Turkish rule (15th–19th cent.) Ruse, known as Ruschuk, served as a military base. The city is noted for its old churches and mosques.

Rush (rŭsh). **1.** County (1970 pop. 20,352), 409 sq mi (1,059.3 sq km), E central Ind., drained by the Big Blue River and Flatrock Creek; formed 1821; co. seat Rushville. Its agriculture includes corn, soybeans, wheat, cattle, and hogs. It has stone quarrying and manufactures furniture, piston rings, and machinery. **2.** County (1970 pop. 5,117), 724 sq mi (1,875.2 sq km), central Kansas, watered by Walnut Creek; formed 1867; co. seat La Crosse. In an agricultural area yielding wheat, corn, and livestock, it has oil and gas fields, furniture and fiberglass manufacturing, and grain milling.

Rush·more, Mount (rŭsh'môr', -mōr'): *see* Mount Rushmore National Memorial.

Rush·ville (rŭsh'vĭl'). **1.** City (1970 pop. 3,300), seat of Schuyler co., W Ill., NE of Quincy; founded 1826, inc. 1831. A processing center in a grain and livestock region, it also has coal mines. **2.** City (1970 pop. 6,686), seat of Rush co., E Ind., ESE of Indianapolis; settled 1821. A trade and processing center of a farm area, it also makes furniture. **3.** City (1970 pop. 1,137), seat of Sheridan co., NW Nebr., ESE of Chadron, in a grain, livestock, and poultry area; founded c.1885.

Rusk (rŭsk). **1.** County (1970 pop. 34,102), 939 sq mi (2,432 sq km), E Texas, in a rolling wooded area drained by the Sabine and Angelina rivers; formed 1843; co. seat Henderson. A leading oil-producing county, it also has agriculture (cotton, corn, truck crops, fruit, peanuts, dairy products, cattle, and poultry) and deposits of clay, iron, lignite, and natural gas. Manufacturing, lumbering, and food processing are done. **2.** County (1970 pop. 14,238), 910 sq mi (2,356.9 sq km), NW Wis., drained by the Chippewa and Flambeau rivers; formed 1901 as Gates co., renamed 1905; co. seat Ladysmith. In a largely wooded area, with dairying and stock raising, it has lumbering, food-processing, and textile-milling industries.

Rusk, town (1970 pop. 4,914), seat of Cherokee co., E Texas, E of Palestine; settled 1846, inc. 1858. It is the center for a truck-farming, dairying, and lumbering section.

Rus·sell (rŭs'əl). **1.** County (1970 pop. 45,394), 627 sq mi (1,624 sq km), E Ala., in the coastal plain, bounded on the E by the Chattahoochee River and Ga.; formed 1832; co. seat Phenix City. Its agriculture includes cotton, livestock, and soybeans. It has timber and diversified manufactures (paper board, brick and tile, carpets, clothing, wood products, chemicals, and metal products). **2.** County (1970 pop. 9,428), 897 sq mi (2,323.2 sq km), central Kansas, in a sloping to gently rolling plain drained by the Saline and Smoky Hill rivers; formed 1867; co. seat Russell. It is in a livestock and wheat-farming area with oil fields. Wood and metal products are manufactured. **3.** County (1970 pop. 10,542), 238 sq mi (616.4 sq km), S Ky., in the Cumberland foothills, drained by the Cumberland River and Russell Creek; formed 1825; co. seat Jamestown. In a hilly agricultural area yielding corn, livestock, and burley tobacco, it has sawmills and factories producing cheese, clothing, and farm machinery. **4.** County (1970 pop. 24,533), 483 sq mi (1,251 sq km), SW Va., in the Alleghenies, drained by the Clinch River; formed 1785; co. seat Lebanon. It is in a hilly farming (tobacco, grain, clover, and fruit), dairying, and livestock-raising area with deposits of bituminous coal, lead, and zinc. Clothing and wood products are manufactured.

Russell, city (1970 pop. 5,371), seat of Russell co., central Kansas, W of Salina, in a farming and grazing area; inc. 1872. There are oil wells here, and oil-well equipment is made.

Russell Cave National Monument: *see* National Parks and Monuments Table.

Rus·sell·ville (rŭs'əl-vĭl'). **1.** City (1970 pop. 7,814), seat of Franklin co., NW Ala., S of Florence, in a cotton, grain, and mineral area; settled c.1815. **2.** City (1978 est. pop. 13,800), seat of Pope co., central Ark., in an area yielding coal, timber, and diverse agricultural products; settled 1835, inc. 1870. **3.** City (1970 pop. 6,456), seat of Logan co., S Ky., SW of Bowling Green, in an area yielding limestone, tobacco, livestock, and timber; settled 1788, inc. 1810.

Rüs·sels·heim (rüs'əls-hīm'), city (1974 est. pop. 60,221), Hesse, central West Germany, on the Main River. It is a center for the assembly of automobiles. Rüsselsheim was chartered in 1437 and passed to Hesse in 1479.

Rus·sia (rŭsh'ə). In its political meaning the term *Russia* applies to the Russian Empire until 1917 and to the Russian Soviet Federated Socialist Republic, largest republic of the USSR. The name is often loosely used to mean the whole Union of Soviet Socialist Republics. It is also used to designate the area inhabited by the Russian people as distinguished from other Eastern Slavs and from non-Slavic peoples. The following article deals with the formation and history of the Russian state and empire until 1917.

Numerous remains indicate that Russia was inhabited in the Paleolithic period. By the 7th cent. B.C. the northern shore of the Black Sea and the Crimea were controlled by the Scythians. Later the open steppes of Russia were invaded by numerous peoples, notably the Germanic Goths (3rd cent. A.D.), the Asian Huns (4th cent.), and the Turkic Avars (6th cent.). The Turkic Khazars built up (7th cent.) a powerful state in southern Russia, and the eastern Bulgars established (8th cent.) their empire in the Volga region. By the 9th cent. the eastern Slavs had settled in northern Ukraine, in Belorussia, and in the regions of Novgorod and Smolensk. The origin of the Russian state coincides with the arrival (9th cent.) of Scandinavian traders and warriors, the Varangians. Tradition has it that one of their leaders, Rurik, established himself peaceably at Novgorod by 862 and founded a dynasty. Rurik's successor, Oleg (reigned 879-912), transferred (882) his residence to Kiev, which remained the capital of Kievan Russia until 1169. Under Sviatoslav (reigned 964-72) the duchy reached the peak of its power. Christianity was made the state religion by Vladimir I (reigned 980-1015), who adopted (988-89) the Greek Orthodox rite. In 1169 Kiev was stormed by the Suzdal prince Andrei Bogolubski (reigned 1169-74), who made Vladimir the capital of the grand duchy. In 1237-40 the Mongols (commonly called Tatars) invaded southern and eastern Russia and established the empire of the Golden Horde, which lasted until 1480. Belorussia, most of the Ukraine, and part of western Russia were incorporated (14th cent.) into the grand duchy of Lithuania. Thus northeastern Russia became the main center of economic and political life. Under Ivan I (reigned 1328-41) Moscow took precedence over the other cities. The rulers of the grand duchy of Vladimir began to be called grand dukes of Moscow or Muscovy.

Under Ivan III (1462-1505) and his successor, Vasily III (1505-33), the Muscovite state expanded, and its rulers became more absolute. Ivan IV (Ivan the Terrible; reigned 1533-84) conquered the Tatar khanates of Kazan (1552) and Astrakhan (1556), establishing Russian rule over the huge area of the middle and lower Volga. At home Ivan crushed the opposition of the great feudal nobles—the boyars—and set up an autocratic government. With the death of Boris Godunov (reigned 1598-1605) began the "Time of Troubles." Polish troops invaded Russia in 1609 but were expelled in 1612. Michael Romanov was chosen czar in 1613, thus inaugurating the Romanov dynasty, which ruled Russia until 1917. Michael was succeeded by Alexis (reigned 1645-76), who gained eastern Ukraine from Poland. Russia in the 17th cent. was still medieval in culture and outlook. Serfdom, which became a legal institution in Russia in 1649, included growing numbers of persons and became increasingly oppressive.

Peter I (Peter the Great; reigned 1689-1725) created a regular conscript army and navy and Westernized the Russian administrative and fiscal systems. Securing a foothold on the Baltic Sea in the Northern War (1700-21), he founded (1703) St. Petersburg (now Leningrad) on the Gulf of Finland and transferred (1712) his capital there. The Russo-Turkish Wars of the next two centuries resulted in the expansion of Russia at the expense of the Ottoman Empire. Under Catherine II (Catherine the Great; reigned 1762-96) Russia became the chief power of continental Europe. Catherine continued Peter I's policies of absolute rule at home and of territorial expansion at the expense of neighboring states. In 1812 Napoleon began his great onslaught on Russia and took Moscow, but his army was repulsed and nearly annihilated in the winter of that year. Napoleon's downfall and the peace settlement made Russia and Austria the leading powers on the Continent.

Liberal ideas gained influence among the Russian aristocracy and educated bourgeoisie despite the growing intransigence of Alexander I (reigned 1801-25). They found an outlet in the unsuccessful Decembrist Conspiracy of 1825. Under Nicholas I (reigned 1825-55) Russia became the most reactionary European power. A clash of interests between Russia and the Western powers over the Ottoman Empire led to the Crimean War (1854-56), which revealed the inner weakness of Russia. Alexander II (reigned 1855-81) passed important liberal reforms, including the liberation (1861) of the serfs, before turning increasingly conservative. During the second half of the 19th cent. Russia continued its territorial expansion, reaching the frontiers of Afghanistan and China and the shores of the Pacific Ocean. The Trans-Siberian RR (constructed 1891-1905) opened much of Siberia to colonization and exploitation.

Nicholas II (reigned 1894-1917), the last Russian emperor, was a generally incompetent ruler surrounded by a reactionary entourage. The disastrous and unpopular Russo-Japanese War (1904-5) led to the Revolution of 1905. Nicholas was forced to grant a constitution, and a parliament was established. However, the new democratic freedoms were soon curtailed, as the government again became reactionary. The leftist Social Democratic party and its Bolshevik faction (formed in 1912) found adherents among the industrial workers and intellectuals.

Ill-prepared and cut off from its allies in the west, the country suffered serious reverses in World War I at the hands of the Germans and Austrians. Inflation, food shortages, and poor morale among the troops contributed to the outbreak of the Feb. Revolution of 1917. Nicholas abdicated in Mar., 1917 (he was executed in July, 1918). A provisional government first under Prince Lvov and then (July) under Alexander Kerensky was unable to enforce its authority. Finally, on Nov. 7, 1917, the Bolsheviks, led by V. T. Lenin, seized the government. Russia ended its involvement in World War I by signing the Treaty of Brest-Litovsk (Mar., 1918). Shortly thereafter civil war (complicated by foreign intervention) broke out in Russia; it continued until 1920, when the Soviet regime emerged victorious. Poland, Finland, and the Baltic countries emerged as independent states in the aftermath of the civil war; the Ukraine, Belorussia, and the Transcaucasian countries of Azerbaijan, Georgia, and Armenia proclaimed their independence but by 1921 were conquered by the Soviet armies. In 1917 Russia was officially proclaimed as the Russian Soviet Federated Socialist Republic, which in 1922 was united with the Ukrainian, Belorussian, and Transcaucasian republics to form the Union of Soviet Socialist Republics. For the history of Russia after 1917, *see* Union of Soviet Socialist Republics.

Rus·sian Soviet Federated Socialist Republic (rŭsh'ən), abbreviated **RSFSR**, constituent republic (1970 pop. 130,079,000), 6,591,100 sq mi (17,070,949 sq km), USSR. It is by far the largest, most populous, and economically most important of the 15 union republics that make up the USSR. Moscow, the capital of the USSR, is also the capital of the RSFSR. The republic was formed in 1917; for its history, *see* Russia; Union of Soviet Socialist Republics. The republic occupies most of eastern Europe and northern Asia, extending for c.5,000 mi (8,000 km) from the Baltic Sea in the west to the Pacific Ocean in the east and for 1,500 to 2,500 mi (2,400–4,000 km) from the Arctic Ocean in the north to the Black Sea, the Caucasus, the Altai and Sayan mts., and the Amur and Ussuri rivers in the south. The RSFSR is bounded by Norway and Finland in the northwest; by the Estonian, Latvian, Belorussian, and Ukrainian republics in the west; by the Georgian and Azerbaijan republics in the southwest; and by the Kazakh Republic, Mongolia, and China along the southern land border. The Kaliningrad oblast is an exclave separated by Lithuania and bordering on Poland.

The RSFSR occupies about 76% of the total land area and contains about 54% of the total population of the USSR. Its dominant relief features are (from west to east) the east European plain, the Urals, the west Siberian lowland, and the central Siberian plateau. Mt. Elbrus (18,481 ft/5,633 m), in the Caucasus, is the highest peak in the republic. The chief rivers draining the European RSFSR are the Don (into the Black Sea), the Volga (into the Caspian Sea), the Northern Dvina (into the White Sea), and the Pechora (into the Barents Sea). The climate, generally continental, varies from extreme cold in northern Russia and Siberia (where Verkhoyansk, the coldest place on earth, is situated) to subtropical along the Black Sea shore. The soil and vegetation zones include the entire tundra and taiga belts of the USSR, nearly the entire wooded steppe and the northern black-earth steppes, and isolated sections of the semidesert, desert, and subtropical zones.

The majority of the population are Russians (83%). There are also Ukrainians and such non-Slavic linguistic and ethnic groups as Tatars, Bashkirs, Chuvash, Komi, Komi Permyaks, Udmurts, Mari, Mordvinians, Jews, and various groups of the Far North and the Caucasus Mts. Administratively each area with a predominantly Russian population is constituted as a kray or oblast; non-Russian nationalities are constituted, in descending order of importance, as autonomous republics, autonomous oblasts, and national okrugs. The RSFSR has 16 autonomous republics: Bashkir, Buryat, Chechen-Ingush, Chuvash, Dagestan, Kabardino-Balkar, Kalmyk, Karelian, Komi, Mari, Mordvinian, North Ossetian, Tatar, Tuva, Udmurt, and Yakut; 6 krays: Altai, Khabarovsk, Krasnodar, Krasnoyarsk, Primorsky (Maritime), and Stavropol; 5 autonomous oblasts: Adyge, Gorno-Altai, Jewish (Birobidzhan), Karachay-Cherkess, and Khakass; 10 national okrugs: Agin-Buryat, Chukchi, Evenki, Khanty-Mansi, Komi-Permyak, Koryak, Nenets, Taymyr, Ust-Orda Buryat, and Yamalo-Nenets; and 49 oblasts.

The RSFSR may conveniently be divided into 9 physio-economic regions.

The central European area is flat, rolling country, with Moscow as its center, and forms a major industrial region. Trucks, ships, railway rolling stock, machine tools, electronic equipment, cotton and woolen textiles, and chemicals are the principal industrial products. The Volga and Oka rivers are the major water routes, and the Moscow-Volga and Don-Volga canals link Moscow with the Caspian and Baltic seas. Many rail lines serve the area.

The northern and northwestern European area, with Leningrad, the industrial center, has industries producing machine tools, electronic equipment, chemicals, ships, and precision instruments. The hills, marshy plains, lakes, and desolate plateaus contain rich deposits of coal, oil, iron ore, and bauxite, and the area is a prime source of lumber. The chief water routes are the Baltic-Belomorsk Canal and the Volga-Baltic Waterway.

The Volga area, stretching along the greatest river of European Russia, has highly developed hydroelectric power installations, including major dams at Volgograd (formerly Stalingrad), Kazan, Kuybyshev, and Balakovo. Farm machinery, ships, chemicals, and textiles are manufactured, and extensive oil and gas fields are worked. Agricultural products include wheat, vegetables, cotton, hemp, oil seeds, and fruit. Livestock raising and fishing are also important.

The north Caucasus area, descending northward from the principal chain of the Caucasus Mts. to a level plain, has rich deposits of oil, natural gas, and coal. Sochi is a popular resort. Farm machinery, coal, petroleum, and natural gas are the chief products. The Kuban River region, a fertile black-earth area, is one of the chief granaries of the RSFSR. Wheat, sugar beets, tobacco, and rice are grown, and cattle are raised. Other rivers include the Don, the Kuma, and the Terek, and the Volga-Don Canal is a major transportation route.

The southern half of the Ural area is a major center of Soviet petroleum and iron and steel production. Large deposits of iron ore, manganese, and aluminum ore are mined. Several trunk railroads serve the area, and rivers include the Kama and Belaya in the west and the Ural in the south.

Western Siberia, a vast plain—marshy and thinly populated in the north, hilly in the south—is of growing economic importance. At Novosibirsk and Kamen-na-Obi are large hydroelectric stations. The Kuznetsk Basin in the southwest is a center of coal mining, oil refining, and the production of iron, steel, machinery, and chemicals. The Ob-Irtysh drainage system crosses this area, which is also served by the Trans-Siberian and South Siberian rail lines. Agricultural products include wheat, rice, oats, and sugar beets, and livestock is raised.

Eastern Siberia is an area of plateaus, mountains, and river basins, and the major cities are located along the Trans-Siberian RR. A branch line links Ulan-Ude with Mongolia and Peking, China. There are hydroelectric stations at Bratsk, Krasnoyarsk, and Irkutsk. Coal, gold, graphite, iron ore, aluminum ore, zinc, and lead are mined in the area, and livestock is raised.

Northern and northeastern Siberia cover nearly half of Soviet territory; this is the least populated and least developed area of the USSR. The Ob, Yenisei, and Lena rivers flow to the Arctic. Through the use of atomic-powered icebreakers, the Northern Sea route has gained increasing economic importance. The Kolyma gold fields are the principal source of Soviet gold, and industrial diamonds are mined. Fur trapping and hunting are the chief activities in the taiga and tundra regions.

The far east region borders on the Pacific Ocean. Machinery is produced, and lumbering, fishing, hunting, and fur trapping are important.

Rus·ta·vi (ro͞os-tä'vē), city (1976 est. pop. 127,000), SE European USSR, in Georgia, on the Kura River. It has ironworks, steelworks, and chemical plants. The city, developed after 1948, is near the site of the ancient town of Rustavi, which was destroyed c.1400 by Tamerlane.

Rust·burg (rŭst'bûrg'), village (1970 pop. 500), seat of Campbell co., SW central Va., SSE of Lynchburg.

Rus·ton (rŭs'tən), city (1978 est. pop. 19,200), seat of Lincoln parish, N La.; settled 1884 as a railroad town and inc. the same year. It is the trading center of a rich farm, dairy, and natural-gas region.

Ruth·er·ford (rŭ*th*'ər-fərd, rŭth'-). **1.** County (1970 pop. 47,377), 562 sq mi (1,455.6 sq km), W N.C., bounded on the S by the S.C. border and drained by the Broad River; formed 1779; co. seat Rutherfordton. It has piedmont agriculture (cotton, corn, and sweet potatoes), timber and sawmills, varied manufactures (textiles, plastics, and furniture), and resorts. **2.** County (1970 pop. 59,428), 612 sq mi (1,585.1 sq km), central Tenn., in the central basin, drained by the Stones River; formed 1803; co. seat Murfreesboro. Its agriculture includes cotton, hay, grains, beans, cattle, and dairy products. It has lumbering and food-processing industries and manufactures clothing, wood products, furniture, paper, concrete, metal products, machinery, and transportation equipment.

Ruth·er·ford (rŭth'ər-fərd, rūth'-), borough (1978 est. pop. 19,400), Bergen co., NE N.J., a residential suburb of the New York City metropolitan area; inc. 1881. It is the seat of Fairleigh Dickinson Univ.

Ruth·er·ford·ton (rŭth'ər-fərd-tən), town (1970 pop. 3,245), seat of Rutherford co., W N.C., SE of Asheville, in a resort and farm area; founded 1779. Its manufactures include textile products and lumber.

Ruth·er·glen (rŭth'ər-glĕn'), burgh (1971 pop. 24,728), Strathclyde region, S central Scotland, on the Clyde River. Manufactures include chemicals, paper, textiles, and metal products. Rutherglen became a royal burgh in 1126 and was for some time one of the leading Clyde ports.

Rut·land (rŭt'lənd), former county (1971 pop. 27,463), central England. The county town was Oakham. In 1974 Rutland became part of the new nonmetropolitan county of Leicestershire.

Rutland, county (1970 pop. 52,637), 929 sq mi (2,406.1 sq km), W Vt., partly bounded in the W by Lake Champlain and N.Y. and rising to the Green Mts. in the E; formed 1781; co. seat Rutland. Its agriculture includes dairy products, fruit, poultry, and maple sugar. It has marble and slate quarrying, a food-processing industry, and manufacturing (clothing, lumbering, paper, chemicals, and machinery).

Rutland, city (1978 est. pop. 19,100), seat of Rutland co., W Vt., at the junction of Otter and East creeks; settled c.1770, inc. as a city 1892. Marble quarrying, which began c.1845, still flourishes in the area.

Rut·ledge (rŭt'lĭj), town (1970 pop. 863), seat of Grainger co., E Tenn., NE of Knoxville, in a timber and farm area. Hosiery and lumber products are made. There are marble quarries nearby.

Rüt·li (rüt'lē), meadows, central Switzerland, on the shore of the Lake of Lucerne. Here, according to the legend of William Tell, representatives of Uri, Schwyz, and Unterwalden met in 1307 to swear the Rütli Oath, on which Swiss freedom was founded. However, in the 19th cent. a written alliance of the three cantons, dated Aug. 1, 1291, was discovered.

Ru·wen·zo·ri (rōō'ən-zôr'ē, -zōr'ē), mountain range, E central Africa, on the Uganda-Zaire border, in the W arm of the Great Rift Valley between lakes Albert and Edward. The snow-capped summits are invariably shrouded in mist, and there are extensive glaciers and glacial lakes. The range may be the semifabulous Mountains of the Moon, erroneously supposed by the ancients to be the source of the Nile. Discovered in 1889 by Henry Stanley, the range was first climbed in 1906.

Rwan·da (rōō-än'də), republic (1977 est. pop. 4,330,000), 10,169 sq mi (26,338 sq km), E central Africa, bordering on Zaire in the W, on Uganda in the N, on Tanzania in the E, and on Burundi in the S. Kigali is the capital and largest town. Most of Rwanda is situated at elevations of 5,000 ft (1,520 m) or higher, and the country has a rugged relief made up of steep mountains and deep valleys. The principal geographic feature is the Virunga mountain range, which runs north of Lake Kivu and includes Rwanda's loftiest point, Volcan Karisimbi (14,787 ft/4,507 m). There is some lower land (at elevations below 3,000 ft/910 m) along the eastern shore of Lake Kivu and the Ruzizi River in the west and near the Tanzanian border in the east. The ethnic composition of the population is similar to that of Burundi. About 90% of the inhabitants are Hutu, agriculturalists who speak a Bantu language; 9% are Tutsi, who speak a Nilotic language; and 1% are Twa, who are a Pygmy group. Rwanda is the most densely populated country in Africa, and its population has a high annual growth rate (between 3% and 3.5%). About half of the population follows traditional religious beliefs, and most of the rest are Roman Catholic. A small number of Tutsi are Moslem. Kinyarwanda and French are the official languages. The economy of Rwanda is overwhelmingly agricultural, with most of the workers engaged in subsistence farming.

Economy. Economic development in Rwanda is hindered by its large and growing population and by poor access to the sea. The chief food crops are plantains, cassava, sweet potatoes, pulses, sorghum, and potatoes. The principal cash crops are coffee, tea, and pyrethrum. Large numbers of cattle, goats, and sheep are raised; most of the cattle are owned by the Tutsi; cattle are more a symbol of wealth and status than a source of food. Tin ore (cassiterite), wolframite, beryl, and colombo-tantalite are mined in significant quantities, and natural gas (methane) is produced at Lake Kivu. Rwanda's few manufactures include textiles, chemicals, and basic consumer goods such as processed food, beverages (especially beer), clothing, and footwear. The country has a good road network but no railroads. The annual value of Rwanda's imports is usually considerably higher than its earnings from exports. The main imports are foodstuffs, textiles, clothing, machinery, motor vehicles, and fuel; the principal exports are coffee, tin ore, wolframite, tea, and pyrethrum. The chief trading partners are Belgium and Luxembourg, Uganda, Japan, West Germany, and Kenya. Rwanda depends heavily on foreign aid.

History. By the late 18th cent. a single Tutsi-ruled state occupied most of present-day Rwanda. It was headed by a king (mwami), who controlled regionally based vassals (also Tutsi). They in turn dominated the Hutu, who, then as now, made up the vast majority of the population. Rwanda reached the height of its power under Mutara II (reigned early 19th cent.) and Kigeri IV (reigned 1853-95). Kigeri established a standing army and prohibited most foreigners from entering his kingdom. In 1890 Rwanda accepted German overrule without resistance and became part of German East Africa, but the Germans had virtually no influence over the affairs of the country and initiated no economic development. During World War I Belgian forces occupied (1916) Rwanda, and in 1919 it became Ruanda-Urundi (which in 1946 became a UN trust territory). Until the last years of Belgian rule the traditional social structure of Rwanda was not altered; however, considerable Christian missionary work was undertaken. In 1957 the Hutu demanded a change in Rwanda's power structure that would give them a voice in the country's affairs. Two Hutu political parties were formed, one by Joseph Gitera and the other by Grégoire Kayibanda. In 1959 Mutara III died and was succeeded by Kigeri V. The Hutu challenged the new mwami, and fighting between the Hutu and the Tutsi broke out. The Hutu emerged victorious, and some 100,000 Tutsi fled to neighboring countries. In 1960 Kayibanda became interim prime minister. In early 1961 a republic was proclaimed. Under pressure from the UN Belgium granted independence to Rwanda on July 1, 1962. Rwanda was headed by a president, who was also prime minister and whose considerable power included effective control of the 47-member national assembly. Kayibanda was elected as the first president and was re-elected in 1965 and 1969. In 1964, following an incursion from Tutsi-dominated Burundi, many Tutsi were killed in Rwanda, and numerous others left the country. In 1967-68 relations with Zaire were temporarily strained when some white mercenaries fighting in Zaire took refuge in Rwanda. In 1971-72 relations with Uganda were bitter after President Idi Amin of Uganda accused Rwanda of aiding groups trying to overthrow him. In Feb. and Mar., 1973, there was renewed fighting between the Hutu and Tutsi, and some 600 Tutsi fled to Uganda. On July 5, 1973, a military group toppled Kayibanda without violence and replaced him as head of state with Maj. Gen. Juvénal Habyalimana, the commander of the national guard. The Parmehutu party (Republican Democratic Movement-Parmehutu), the only legal party in the country, was suspended and the national assembly dissolved. Habyalimana, a Hutu from the northern part of the country, tried to improve relations between the Hutu and the Tutsi and to distribute power more evenly among the Hutu.

Ry·a·zan (rē'ə-zăn'), city (1976 est. pop. 432,000), capital of Ryazan oblast, E central European USSR, a port on the Oka River. Industries include oil refining, lignite processing, and the manufacture of machine tools and agricultural and transport equipment. One of Russia's oldest cities, Ryazan was founded in 1095 and was annexed by Moscow in 1521. Ryazan retains much medieval architecture and has picturesque churches with many-colored domes and gilded ornaments. A kremlin wall, dating from 1208, surrounds two former monasteries built in the 15th and the 17th cent.

Ry·binsk (rĭb'ĭnsk), city (1976 est. pop. 236,000), NE European USSR, on the upper Volga and the Rybinsk Reservoir, an artificial lake (c.2,000 sq mi/5,180 sq km). The site of a hydroelectric station, it is a major inland port with shipyards and factories producing road-building equipment, cables, and printing presses. Known since 1137, it has been a trade and shipping center for traffic between Moscow and Arkhangelsk since the 16th cent. The construction of the Volga-Neva canal system increased its importance as a port.

Ryb·nik (rĭb'nĭk), town (1975 est. pop. 103,000), S Poland. It is a railway junction and manufactures mining machinery, metal products, and chemicals. There are coal mines and coke plants nearby. Chartered in the 14th cent., Rybnik passed from Germany to Poland in 1921.

Ryde (rīd), municipal borough (1973 est. pop. 23,170), on the Isle of Wight, S England, on Spithead channel. It is one of the leading resorts of the island.

Rye (rī), municipal borough (1971 pop. 4,434), East Sussex, SE England, on the Rother River. It is a tourist resort and port with boatbuilding and netmaking industries. Rye had a thriving trade in the 17th cent. but decayed after the recession of the sea early in the 19th cent. There are remains of an ancient friary and a large Norman and Early English church.

Rye, city (1970 pop. 15,869), Westchester co., SE N.Y., a suburb of New York City, on Long Island Sound; settled 1660, inc. as a city 1942. It is chiefly residential, with a cancer-research center, a hardware and locks manufacturing company, and several corporate offices. The old Square House, an inn where many Revolutionary notables stayed, is now a museum.

Rye·gate (rī'gāt'), town (1970 pop. 261), seat of Golden Valley co., S central Mont., on the Musselshell River NW of Billings. Livestock, turkeys, and grain are raised. Dairy and poultry products are processed.

Ryu·kyu Islands (rē-ōō'kyōō), archipelago (1970 est. pop. 1,235,000), c.1,850 sq mi (4,790 sq km), SW Japan, in the W Pacific Ocean. The chain stretches about 650 mi (1,045 km) between Taiwan and Japan, separating the East China Sea from the Philippine Sea. The Ryukyus are composed of the Amami Islands, the Okinawa Islands, and the Sakishima Islands. The islands are the exposed tops of submarine mountains and are of volcanic or coral origin; there are several active volcanoes in the group. The chief agricultural products are sugar cane, sweet potatoes, pineapples, and rice. Fishing is also important.

The islands were the site of an ancient independent kingdom. The Chinese reached the islands in the 7th cent. In the 17th cent. the Japanese prince of Satsuma invaded the islands. Matthew Perry landed in the Ryukyus in 1853. The entire archipelago was incorporated into the Japanese empire in 1879. During World War II the Ryukyus were the scene of fierce fighting between U.S. and Japanese forces. After the war the islands south of lat. 30° N were placed (Aug., 1945) under a U.S. military governor. In 1951 the Japanese were given residual sovereignty over the islands, but the United States retained control. The archipelago was returned to Japan in May, 1972.

Rze·szów (zhĕs'ōōf'), city (1975 est. pop. 95,800), capital of Rzeszów prov., SE Poland. It is a railway junction and an important industrial center, whose major industries produce rolling stock, machinery, metals, building materials, foodstuffs, and clothing. An old commercial settlement, Rzeszów was chartered c.1340. It passed to Austria in 1772 and reverted to Poland in 1919.

Rzhev (ər-zhĕf'), city (1976 est. pop. 68,000), NW European USSR, on the Volga River and on a major rail line to Moscow. It has textile plants. Rzhev, an ancient trade center, was controlled by the Smolensk principality in the 12th cent. and taken by Novgorod in 1216.

S

Saa·le (zä'lə, sä'-), river, c.265 mi (425 km) long, rising in the Fichtelgebirge, E West Germany, and flowing N through SW East Germany, past Jena, Naumberg, and Halle, to the Elbe River SE of Magdeburg. The Saale's course is flanked by numerous medieval castles. Sugar beets are grown in the fertile lower valley.

Saal·feld (zäl'fĕlt'), city (1974 est. pop. 33,648), Gera dist., S East Germany, on the Thüringer Saale River. Manufactures include machine tools, electrical equipment, and dyes. Iron is mined and slate is quarried nearby. Saalfeld was founded c.1200 and in the 16th cent. was a silver-mining center. In 1806 the French defeated the Prussians here during the Napoleonic Wars. The city has a 14th cent. church, a 16th cent. city hall, a 13th cent. Franciscan monastery (now a museum), a 13th cent. castle, and an 18th cent. palace.

Saar (sär, zär), river, c.150 mi (240 km) long, rising in the Vosges Mts., NE France, and flowing N past Sarrebourg and Sarreguemines. It enters Saarland, western West Germany, and continues northwest past Saarbrücken into the Moselle River near Trier. The river flows through a heavily industrialized region.

Saar·brück·en (sär-brŏŏk'ən, zär-), city (1974 est. pop. 205,987), capital of Saarland, W West Germany, on the Saar River near the French border. It is the leading industrial center of the Saar coal basin. Manufactures include precision instruments, machinery, metal goods, printed materials, and beer. Located on the site of earlier Celtic, Roman, and Frankish settlements, Saarbrücken was chartered in 1321. It was the capital of the counts of Nassau-Saarbrücken from 1381 until its occupation (1793) by the French. The city passed to Prussia in 1815. From 1919 to 1935 and again from 1945 to 1957 Saarbrücken was included in the French-administered Saar Territory. The city retains the 15th cent. late Gothic Castle Church, the old city hall (1750), and a baroque church, the Ludwigskirche (1762–75).

Saa·re·maa (sär'ə-mä'), island: *see* Sarema.

Saar·land (sär'lănd', zär'länt'), formerly Saar or Saar Territory, state (1970 pop. 1,120,000), 991 sq mi (2,567 sq km), SW West Germany. Saarbrücken is the capital. The state is bounded by France in the south and west, by Luxembourg in the northwest, and by Rhineland-Palatinate in the north and east. A region of low, partly wooded hills, Saarland is drained by the Saar River. The state is highly industrialized, with a large iron and steel industry based on vast coal fields. Other manufactures include metal goods, glass, chemicals, and textiles. Agricultural production is limited. The state is served by a dense rail network and is connected with the Rhine-Marne Canal. The population is German-speaking and largely Roman Catholic. There is a university at Saarbrücken.

History. The Saarland possessed little unity before the 20th cent. In 1797, when it was ceded to France by the Treaty of Campo Formio, it was divided among France, the county of Saarbrücken (a dependency of Nassau), and the palatine duchy of Zweibrücken. The Treaty of Paris of 1815 divided the territory between Bavaria and Prussia. The Saar Territory came into existence as a political unit when the Treaty of Versailles (1919) made it an autonomous territory, administered by France under League of Nations supervision. When more than 90% of the votes cast in a 1935 plebiscite favored its reunion with Germany, the Saar was restored to Germany and constituted the Saarland prov. During World War II Hitler incorporated it (1940) with Lorraine (annexed from France) into the province of Westmark. The Saarland was placed under French military occupation in 1945 and in 1947 was given an autonomous government. In a referendum (1947) the population voted for economic union with France, and in 1948 a customs union went into effect. An agreement between France and West Germany in 1954 provided for an autonomous Saar; the economic union with France was to be maintained for 50 years. However, the agreement was rejected (Oct., 1955) by the Saarlanders in a popular referendum, and, in accordance with subsequent Franco-German agreements (1956), the Saar Territory became (Jan. 1, 1957) a state (Saarland) of the Federal Republic of Germany. The customs union with France was dissolved in July, 1959, whereupon the Saarland became economically integrated with West Germany.

Saar·lou·is (zär-lo͞o′ē, sär-, sär′lə-wē′), city (1974 est. pop. 40,057), Saarland, W West Germany, on the Saar River near the French border. Manufactures include steel, furniture, tobacco products, and bells. Coal is mined in the area. Founded (1680) by Louis XIV, Saarlouis became a major French frontier fortress. It was awarded (1815) to Prussia at the Congress of Vienna. It was administered by France from 1919 to 1935 and from 1945 to 1957.

Sa·ba (sä′bə), island (1969 est. pop. 972), 5 sq mi (13 sq km), Netherlands Antilles, one of the NW Leeward Islands. Saba is actually the cone of an extinct volcano rising to c.2,800 ft (855 m). Though it is a scenic island, there are no sheltered harbors, and landing is difficult. The chief settlement, called The Bottom, is in the crater of the volcano. Fishing and boatbuilding are the principal occupations.

Sa·ba·dell (sä′bə-dĕl′), city (1970 pop. 159,408), Barcelona prov., NE Spain, in Catalonia. Since medieval times it has been a leading textile center. Diesel engines, electrical equipment, and fertilizers are also made.

Sa·ba·e (sə-bä′ē), city (1975 pop. 45,700), Fukui prefecture, central Honshu, Japan. It is an agricultural market with textile and chemical industries.

Sa·ba·lan (sä′bə-län′) or **Sa·va·lan** (sä′və-län′), volcanic cone, 15,784 ft (4,814.1 m) high, NW Iran, near Ardabil. The prophet Zoroaster reputedly wrote the *Avesta* here.

Sa·bine (sə-bēn′). **1.** Parish (1970 pop. 18,638), 1,029 sq mi (2,655.1 sq km), W La., bounded on the W by the Sabine River, here forming the Texas border; formed 1843; parish seat Many. Chief farm crops are cotton, corn, hay, and peanuts. Natural-gas and oil fields and timber are found here. Industries include feed mills, sawmills, and garment factories. **2.** County (1970 pop. 7,187), 456 sq mi (1,181 sq km), E Texas, bounded on the E by the Sabine River (here the La. border); formed 1836; co. seat Hemphill. Most of the county is in Sabine National Forest, and lumbering and lumber milling are chief industries. It also has agriculture (cotton, corn, sweet potatoes, truck crops, and fruit), livestock (cattle, hogs, and poultry), and dairying. There is good hunting and fishing.

Sabine, river, c.575 mi (925 km) long, rising on the prairies NE of Dallas, Texas, and flowing SE across Texas, then S to mark the Texas-La. line. Near its mouth it broadens to form Sabine Lake (c.17 mi/27.4 km long; c.7 mi/11.3 km wide), then goes through Sabine Pass to the Gulf of Mexico.

Sabine Cross·roads (krôs′rōdz′, krŏs′-), locality, De Soto parish, NW La., near Mansfield. In the Civil War Union forces under Nathaniel P. Banks were defeated here on Apr. 8, 1864.

Sa·ble, Cape (sā′bəl), S Fla., southernmost extremity of the U.S. mainland. It is part of Everglades National Park.

Sable Island, low, sandy island, 25 mi (40.2 km) long and 1 mi (1.6 km) wide, off N.S., Canada, SE of Halifax. It is the exposed part of a sand shoal that stretches northeast-southwest for more than 100 mi (160 km). The island was known to mariners in the early 16th cent. Known as the "graveyard of the Atlantic," Sable Island is a major hazard to navigation; it now has a lighthouse, a lifesaving station, and a radio beacon. The island is also a breeding place for seals, which are protected by the government.

Sab·ze·var (săb′zə-vär′), city (1971 est. pop. 45,000), Khurasan prov., NE Iran. It is the trade center of a cotton-growing region. The city's manufactures include textiles and rugs.

Sac (sôk), county (1970 pop. 15,573), 578 sq mi (1,497 sq km), W Iowa, in a prairie agricultural area drained by the Raccoon and Boyer rivers; formed 1851; co. seat Sac City. Cattle and hogs are raised; corn, oats, and alfalfa are grown. There are gravel pits and coal deposits in the county.

Sac City, city (1970 pop. 3,268), seat of Sac co., W central Iowa, on the Raccoon River ESE of Sioux City; settled 1855, inc. 1856. It is a trade and processing center in a farm, livestock, and dairy area.

Sack·ets Harbor (săk′əts), village (1970 pop. 1,202), Jefferson co., N N.Y., on Lake Ontario. Settled 1801, it is now a summer resort. On May 27–29, 1813, American forces rebuffed British troops from Canada in a battle of the War of 1812.

Sack·ville (săk′vĭl′), town (1971 pop. 3,180), SE N.B., Canada, near the head of Chignecto Bay, an arm of the Bay of Fundy. The early French Acadian settlers reclaimed the nearby Tantramar marshes, creating fertile agricultural land. The first Baptist church in Canada was established here in the 1770s.

Sa·co (sô′kō), city (1970 pop. 11,678), York co., SW Maine, on the Saco River; settled 1631, inc. as Pepperellboro 1762, name changed 1805, inc. as a city 1867. Its manufactures include automotive parts, machinery, leather, shoes, and prefabricated homes.

Saco, river, c.105 mi (170 km) long, rising in the White Mts., N central N.H., and flowing SE through Maine to the Atlantic Ocean below Biddeford. The falls at Biddeford were an early source of water power for the textile industry.

Sac·ra·men·to (săk′rə-mĕn′tō), county (1970 pop. 634,373), 975 sq mi (2,525 sq km), N central Calif., in the Central Valley, bounded on the W by the Sacramento River and crossed by the American River; formed 1850; co. seat Sacramento. It is a rich agricultural area, yielding fruit, nuts, olives, vegetables, cattle, poultry, dairy products, sugar beets, grain, beans, alfalfa, hops, corn, and sorghums. It is a leading county in natural-gas production and gold dredging and also produces sand, gravel, stone, and silver. Clothing and textiles, furniture, chemical products, clay tile, concrete, metal products, machinery, and transportation equipment are manufactured.

Sacramento, city (1978 est. pop. 261,500), state capital and seat of Sacramento co., central Calif., on the Sacramento River at its confluence with the American River; settled 1839, inc. 1850. A deepwater port via a channel to Suisun Bay, it is the shipping, rail, processing, and marketing center for the truck and fruit farms of the Sacramento valley. Food processing is the city's major industry. The discovery of gold in 1848 at nearby Sutter's Mill (now Coloma) led to the platting of the town. Sacramento was made the state capital in 1854. Points of interest include the capitol building, Sutter's Fort, and the Crocker Art Gallery. Mather Air Force Base, McClellan Air Force Base, and an army depot are in the vicinity.

Sacramento, longest river of Calif., c.380 mi (610 km) long, rising near Mt. Shasta, N Calif., and flowing generally SW to Suisun Bay, an arm of San Francisco Bay, where it forms a large delta with the San Joaquin River. Many of the cities on or near the river sprang up in the 1848 gold rush. In recent years the Central Valley project has been developed to use the waters of the Sacramento with greater efficiency, particularly in the San Joaquin section.

Sa·do (sä′dō), island, 330 sq mi (855 sq km), in the Sea of Japan, off the W coast of N Honshu, Japan. The fertile central lowlands are an important rice-growing region; fishing and tourism are also important. Gold and silver mines have been worked since 1601.

Sa·do·vá (sä′dô-vä′), village, N Czechoslovakia, in Bohemia, near Hradec Králové. It was the site of a decisive Prussian victory over the Austrians in 1866, during the Austro-Prussian War.

Sa·fa·qis (sə-fä′kəs) or **Sfax** (sfăks), city (1975 pop. 171,297), E Tunisia, on the Gulf of Gabès, an arm of the Mediterranean Sea. It exports phosphates, olive oil, cereals, and sponges. Safaqis was the site of Phoenician and Roman colonies. Later it became a stronghold of the Barbary pirates.

Sa·fed Koh (sä-fĕd′ kō′), mountain range on the Pakistan-Afghanistan border, SE of Kabul, rising to 15,620 ft (4,764.1 m) at Mt. Sikaram. The range is crossed by the Khyber Pass and parallels the Kabul River. The northern slopes are nearly barren; pine and deodar grow on the main range, and the valleys support some agriculture.

Saf·ford (săf′ərd), city (1970 pop. 5,333), seat of Graham co., SE Ariz., on the Gila River, in an irrigated farm, livestock, and timber area; founded 1872, inc. 1901.

Sa·fi or **Saf·fi** (both: săf′ē), city (1971 pop. 129,113), W central Morocco, on the Atlantic Ocean. It is a center of the Moroccan fishing and canning industries. Phosphates are exported. Safi was a Portuguese base in the early 16th cent.; it then became (until 1660) the chief port of the Sadian dynasty. In World War II Allied forces landed here on Nov. 8, 1942.

Sa·fid Rud (sä-fēd′ ro͝od′), river, c.450 mi (725 km) long, rising in NW Iran and flowing E to the Caspian Sea at Rasht. A storage dam on the river was completed in 1962. The Safid Rud has cut a water gap through the Elburz Mts., which provides a major route between Tehran and the Caspian lowlands.

Sa·ga (sä′gä), city (1976 est. pop. 154,631), capital of Saga prefecture, W Kyushu, Japan. It is a railroad and coal-distributing center. Cotton textiles and ceramics are produced. A castle town in feudal times, Saga was the center of a rebellion in 1874.

Sag·a·da·hoc (săg′ə-də-hŏk′), county (1970 pop. 23,452), 257 sq mi (665.6 sq km), S Maine, on the coast, with the Androscoggin River joining the Kennebec at Merrymeeting Bay; formed 1854; co. seat Bath. It has farming and dairying, commercial fishing, and some industry (food processing, clothing manufacturing, and shipbuilding). There are summer colonies on the coast and offshore islands.

Sa·ga·mi·ha·ra (sä′gə-mə-hä′rə), city (1976 est. pop. 390,851), Kanagawa prefecture, central Honshu, Japan. It is a suburb of Tokyo, with chemical, food-processing, and metallurgical industries.

Sag·a·more Hill National Historic Site (săg′ə-môr′, -mōr′): *see* National Parks and Monuments Table.

Sa·gar (sä′gər), town (1971 pop. 127,458), Madhya Pradesh state, central India. Sagar is a market for wheat, cotton, and oilseed.

Sa·ghal·ien (sä′gəl-yĕn′), island: *see* Sakhalin.

Sag Harbor (săg), resort village (1970 pop. 2,363), Suffolk co., SE N.Y., on E Long Island, ENE of Southampton on Gardiners Bay; settled c.1707, inc. 1846. During the first half of the 19th cent. it was an important whaling port; today it is primarily a summer resort.

Sa·gi·naw (săg′ə-nô′), county (1970 pop. 219,743), 812 sq mi (2,103.1 sq km), central Mich., drained by the Saginaw River and its affluents; formed 1835; co. seat Saginaw. It is an agricultural area dealing in sugar beets, soybeans, corn, grain, livestock, and poultry. Automobile parts and equipment are manufactured. There are oil fields, salt deposits, and coal mines here.

Saginaw, city (1978 est. pop. 82,100), seat of Saginaw co., S Mich., on the Saginaw River, 15 mi (24 km) from Saginaw Bay (an inlet of Lake Huron); settled 1816, inc. 1857. Situated in an agricultural area, Saginaw is also a port of entry and an industrial center. Lewis Cass negotiated a treaty here (1819) with the Indians, who ceded much of Michigan to the United States. Fur trade was followed by a great pine-lumbering industry, which thrived until about 1890.

Sa·guache (sə-wäch′), county (1970 pop. 3,827), 3,144 sq mi (8,143 sq km), S Colo., including part of the fertile San Luis Valley in the E; formed 1866; co. seat Saguache. It is in an irrigated agricultural area (livestock, hay, and potatoes), with some mining and manufacturing. It includes ranges of the Rocky Mts. and parts of Gunnison, Rio Grande, Cochetopa, and San Isabel national forests.

Saguache, town (1970 pop. 642), seat of Saguache co., S central Colo., on Saguache Creek in the S foothills of the Sawatch Mts. W of the Sangre de Cristo Mts. and SSW of Salida, at an altitude of 7,800 ft (2,379 m); founded 1866, inc. 1891. It is a trading center in a livestock region. There are gold, silver, copper, and lead mines in the area.

Sa·gua la Gran·de (säg′wə lə grän′dē), city (1970 pop. 35,809), Las Villas prov., central Cuba, on the Sagua la Grande River. It is the commercial and processing center for an area raising sugar cane and cattle. The city's origins date back to the 17th cent.

Sa·gua·ro National Monument (sə-gwä′rō): *see* National Parks and Monuments Table.

Sag·ue·nay (săg′ə-nā′), river, c.125 mi (200 km) long, S Que., Canada. It issues from Lake St. John (c.375 sq mi/970 sq km) in two channels, the Grande Décharge and the Petite Décharge, separated by the Île d'Alma, and flows southeast to the St. Lawrence River at Tadoussac. Navigable below Chicoutimi, it flows through a picturesque gorge whose banks rise to more than 1,500 ft (457 m) at Eternity and Trinity capes. The Saguenay was first visited (1535) by Jacques Cartier. For more than three centuries it was a route traveled by explorers, missionaries, and fur traders; later it became a major lumber transportation route. In the 20th cent. pulp and paper mills and important hydroelectric stations were built.

Sa·gun·to (sə-go͞on′tō), town (1970 pop. 47,026), Valencia prov., E Spain, on the Palencia River, in Valencia. A seaport on the Mediterranean, it is an important metallurgical center. Saguntum was an ally of Rome when it was captured (219–218 B.C.) by the Carthaginians under Hannibal, precipitating the Second Punic War. It was conquered by the Romans in 214 B.C. and later fell to the Moors. In 1874 the restoration of the Spanish Bourbon dynasty was proclaimed here. On a ridge above the city are Roman remains.

Sa·hand (sä-händ′), peak, 12,140 ft (3,702.7 m) high, NW Iran, S of Tabriz. It is snow-covered most of the year and is traditionally associated with the prophet Zoroaster.

Sa·ha·ra (sə-hăr′ə, -hâr′ə, -hä′rə), world's largest desert, c.3,500,000 sq mi (9,065,000 sq km), N Africa, the W part of a great arid zone that continues into SW Asia. Extending more than 3,000 mi (4,830 km), from the Atlantic Ocean to the Red Sea, the Sahara is bounded on the north by the Atlas Mts., steppe, and the Mediterranean Sea; it stretches south c.1,200 mi (1,930 km) to the Sahel, a steppe in western and central Africa that forms its southern border. The desert includes most of Western Sahara, Mauritania, Algeria, Niger, Libya, and Egypt; the southern portions of Morocco and Tunisia; and the northern portions of Senegal, Mali, Chad, and Sudan. The eastern Sahara is usually divided into three regions—the Libyan Desert, which extends west from the Nile valley through western Egypt and eastern Libya; the Arabian Desert, or Eastern Desert, which lies between the Nile valley and the Red Sea in Egypt; and the Nubian Desert, which is in northeastern Sudan. Regions of sand dunes (erg) occupy only about 15% of the Sahara; "stone deserts," consisting of plateaus of denuded rock (hammada) or areas of coarse gravel (reg), cover about 70% of the region; mountains, oases, and transition zones account for the remainder. High mountain massifs rise in the central regions; they are the Ahagger (Hoggar) in southern Algeria, which rises to more than 9,000 ft (2,740 m); the Tibesti Massif in northern Chad, which rises to more than 11,000 ft (3,350 m); and the Aïr Mts. (Azbine) in northern Niger, which rise to more than 6,000 ft (1,830 m). The mountains are deeply dissected and were in the past infamous for the shelter they provided to marauders preying on desert traffic. The Sahara has one of the harshest climates in the world. Located in the trade winds belt, the region is subject to strong winds that blow constantly from the northeast between a subtropical high-pressure cell and an equatorial low-pressure cell. As air moves downward from the high-pressure into the low-pressure cell it becomes warmer and drier. The desiccating and dust-laden winds are sometimes felt north and south of the desert, where they are variously known as sirocco, khamsin, simoom, and harmattan. The northern slopes of the Atlas Mts. intercept most of the moisture from winds blowing inshore from the Mediterranean Sea. Border zones on the north and south, where the desert merges with the steppe, receive about 10 in. (25 cm) of rain a year with some seasonal regularity, but over most of the region rainfall is sparse, with an average annual total of less than 5 in. (12.7 cm); rainfall is usually torrential after long dry periods that sometimes last for years. The region's low relative humidity rarely exceeds 30% and is often in the 4% to 5% range. Daytime temperatures are high; Azizia, Libya, recorded the world's highest official temperature in the shade (136°F/58°C) in Sept., 1922. Heat loss is rapid at night and a diurnal range of 86°F (30°C) is common. Freezing temperatures are not uncommon at night from Dec. to Feb.

Sparse vegetation is found in most parts of the Sahara, with the exception of the sand-dune regions. The Nile and Niger rivers, both fed by rains outside of the desert, are the only permanent rivers in the region. Water is present at or just below the surface gravel in wadis (intermittent streams) that radiate from the mountain massifs, in scattered oases where the water table comes to the surface, and at greater depths in huge underground aquifers. The aquifers are believed to be filled with water dating from the Pleistocene epoch, when the Sahara was much wetter than it is today; the more than 20 lakes and areas of salt flats and boggy salt marshes are also considered relics from this pluvial period.

The principal ethnic groups of the Sahara are the Tuareg (of Berber origin), who dominate the mountains of the central Sahara; the peoples of mixed Berber and Arab origin in the western Sahara; and the Tibu (Tébu), of mostly Negroid origin, who dominate the Tibesti Massif.

Economy. Two thirds of the Sahara's estimated 2 million inhabitants (excluding those in the Nile valley) are concentrated in oases where date palms, fruits, vegetables, grains, and other crops are produced under irrigation. Nomads, with herds of sheep and goats and with camels for transportation, predominate in drier areas and continue to use oases, as in centuries past, for water, trade, and provisioning stops. Salt is still mined, as in the past, at Taoudenni, Mali, and at Bilma, Niger, and is transported by camel caravans across the desert. Extensive iron-ore deposits are worked in the Fort Gouraud area of Mauritania, and there are huge oil and gas deposits in Algeria and Libya.

History. At the time of the European Ice Age (c.50,000–100,000 years ago) the Sahara was a region of extensive shallow lakes and was well vegetated and occupied by a predominantly Negroid population. By Roman times most vegetation had disappeared. The camel was introduced probably in the 1st cent. A.D. and facilitated occupation by nomads (first the Berbers, later the Arabs), who lived in interdependence with the oasis dwellers. A profitable trans-Saharan trade in gold and slaves from western Africa, salt from the desert, and cloth and other products from the cities on the Mediterranean coast was carried on by the nomads from the 10th through the 19th cent. René Caillié was the first European explorer to cross (1822–24) the desert and emerge alive. Some areas of the Sahara remain unexplored, although a network of air and modern automobile routes now crosses the desert and links the major oases and mining areas. From west to east the four principal land routes across the desert are from Colomb-Bechar to Dakar; from Colomb-Bechar to Gao and Timbuktu by way of Reganne; from Touggourt to Agadès and Kano by way of In-Salah; and from Tripoli to Ghat.

Sa·ha·ran·pur (sə-hä'rən-po͝or'), city (1971 pop. 225,698), Uttar Pradesh state, N central India, on the Dharmaula River. Once a summer resort of the Mogul court, Saharanpur is now a district administrative center. Wood products and furniture are manufactured.

Sa·hel (sä-hĕl'), name applied to the semiarid region of Africa between the Sahara to the N and the savannas to the S, extending from Senegal, on the W, through Mauritania, Mali, Upper Volta, Niger, N Nigeria, Sudan, to Ethiopia on the E. Beginning in the late 1960s the Sahel was afflicted by a prolonged and devastating drought that further reduced the region's normally meager water supplies, shattered its agricultural economy, contributed to the starvation of hundreds of thousands of people, and forced the mass migration southward of many tribes. Many parts of the Sahel received heavy and prolonged periods of rainfall in mid-1974.

Sa·hi·wal (sä'ē-wäl', –väl'), city (1972 pop. 106,213), N Pakistan, on the Jhelum River. It is a market for food grains and cotton.

Sai·gon (sī-gŏn'), officially Ho Chi Minh City (since 1976), city (1971 est. pop. 1,804,900), S Vietnam, on the right bank of the Saigon River, a tributary of the Dong Nai. Saigon is a port and a commercial and industrial center. It has textile mills, canneries, glassworks, paper mills, shipyards, machine-assembling plants, and establishments processing the food and industrial crops of the country (rice, sugar cane, rubber) and manufacturing plastics, pharmaceuticals, building materials, and handicrafts.

An ancient Khmer settlement, Saigon passed (17th cent.) to the Annamese. It was captured by the French in 1859. Saigon grew from a small village to a large city under French rule. It was capital of Cochin China and from 1887 to 1902 was capital of the Union of Indochina. Saigon became the capital of the newly created state of South Vietnam in 1954. In the Vietnam War it served as military headquarters for U.S. and South Vietnamese forces and suffered considerable damage during the 1968 Tet offensive.

Sai·jo (sī-jō'), city (1975 pop. 39,100), Ehime prefecture, W Shikoku, Japan, on the Hiuchi Sea. It is an agricultural center.

Sa·i·ki (sä-ē'kē), city (1975 pop. 42,200), Oita prefecture, NE Honshu, Japan. It is a fishing port and agricultural market.

Sai·maa (sī'mä'), lake system, c.1,850 sq mi (4,790 sq km), occupying the heavily glaciated plateau of S central Finland. It comprises more than 120 connecting lakes; the large southern basin of the system constitutes Lake Saimaa proper (c.500 sq mi/1,295 sq km). There are numerous canals to facilitate steamship and lumber-raft traffic through the Saimaa Canal (c.37 mi/60 km long; completed 1856), which terminates at Vyborg, USSR, on the Gulf of Finland.

Saint Al·bans (sānt ôl'bənz), municipal borough (1976 est. pop. 123,800), Hertfordshire, E central England. A market town, it has printing and engineering industries and produces clothing and seeds. St. Albans is the site of the Roman Verulamium. King Offa of Mercia founded an abbey here in 793 to house the relics of Saint Alban, an early British martyr. The present cathedral was built mostly in the 11th cent. Part of nearby St. Michael's Church dates from the 10th cent. In the Wars of the Roses, St. Albans was the scene of a Yorkist victory in 1455 and a Lancastrian victory in 1461.

Saint Albans. 1. City (1970 pop. 8,082), seat of Franklin co., NW Vt., N of Burlington; inc. as a village 1856, as a city 1897. A port of entry and a trade center in a farming, dairying, and resort area, the city makes paper products and clothing. In the Civil War St. Albans was raided by Confederates from Canada. **2.** City (1970 pop. 14,356), Kanawha co., W W.Va., at the junction of the Coal and the Kanawha rivers; settled c.1790, inc. 1868. It is chiefly residential.

Saint An·drews (ăn'dro͞oz), burgh (1976 est. pop. 13,137), Fife, E Scotland, on the North Sea. A summer resort, it is famous for its golf courses. It was the ecclesiastical capital of Scotland until the Reformation. St. Andrews Cathedral, now a ruin, was founded in 1160 and vandalized by Protestants in 1559. The Univ. of St. Andrews, which dates from 1410, is the oldest in Scotland.

Saint Ann (ăn), city (1978 est. pop. 17,500), St. Louis co., E Mo., a suburb of St. Louis; inc. 1948.

Saint An·tho·ny (ăn'thə-nē), city (1970 pop. 2,877), seat of Fremont co., E Idaho, NNE of Idaho Falls; founded 1890, inc. 1905. It is a processing center for an irrigated dairy and farm area yielding potatoes, seed peas, and grain. Crystal Falls Cave is nearby.

Saint Au·gus·tine (ô'gə-stēn'), city (1970 pop. 12,352), seat of St. Johns co., NE Fla.; inc. 1824. Located on a peninsula between the Matanzas and San Sebastian rivers, it is separated from the Atlantic Ocean by Anastasia Island; the Intracoastal Waterway passes through the city. St. Augustine is a port of entry, a shrimping and shipping center, and a popular year-round resort; food-processing and aircraft-repair industries are also important. The oldest city in the United States, it was founded in 1565 by the Spanish explorer Pedro Menéndez de Avilés on the site of an ancient Indian village. St. Augustine passed to the English in 1763 at the end of the French and Indian Wars. In 1821 Spain ceded Florida to the United States, and St. Augustine grew rapidly until the Seminole War in the 1830s. Union troops held the city throughout the Civil War. Much of St. Augustine's colonial atmosphere remains. Among the old landmarks are Castillo de San Marcos, built 1672–96 and now a national monument, and Fort Matanzas, also a national monument, which was built by Spain in 1742. Other places of interest in the city are the old schoolhouse, the house reputed to be the oldest in the United States (said to date from the late 16th cent.), the slave market, and the cathedral (built 1793–97; partly restored).

Saint Aus·tell with Fow·ey (ô'stəl; foi, fō'ē), municipal borough (1973 est. pop. 32,710), Cornwall, SW England, at the mouth of the Fowey River on St. Austell Bay. The municipal borough of Fowey and the urban district of St. Austell were amalgamated in 1968. China clay is produced and exported, and tourism and fishing are also important. Fowey was an important port in the 14th cent. Castle Dore, to the north, is believed to be the castle of King Mark of Cornwall and thus is associated with the story of Tristram and Isolde.

Saint Ber·nard (bər-närd'), parish (1970 pop. 51,185), 510 sq mi (1,320.9 sq km), SE La., just SE of New Orleans, bounded on the N by Lake Borgne, on the E by Chandeleur Sound, on the S by Breton Sound, and on the W by the Mississippi River; formed 1807; parish seat Chalmette. Industrial communities near New Orleans have sugar refineries, oil refineries, stockyards, automobile assembly plants, and diversified manufacturing. The parish's agriculture consists of truck crops, dairy products, and livestock.

Saint Bernard, city (1970 pop. 6,131), Hamilton co., SW Ohio, a suburb N of Cincinnati; settled 1794, inc. as a village 1878. Soap is made here.

Saint Ber·nard (săɴ' bĕr-när'), two Alpine passes, both used since antiquity. The Great Saint Bernard (alt. 8,110 ft/2,473.6 m), on the Italian-Swiss border, links Valais canton, Switzerland, with Valle d'Aosta, Italy. Frequented by the Gauls and Romans, the pass also was crossed by Charlemagne, Emperor Henry IV, Frederick Barbarossa, and Napoleon I. The hospice, founded by St. Bernard of Menthon, is in the charge of Augustinian friars. The St. Bernard dogs bred by them were formerly used to search for lost travelers. The Great St. Bernard Road Tunnel, c.4 mi (6.4 km) long, linking Switzerland and Italy, was opened in 1964. The Little Saint Bernard (alt. 7,178 ft/2,189.3 m) connects Savoie dept., France, with Valle d'Aosta, Italy.

Saint Bon·i·face (bŏn'ə-fās'), city (1971 pop. 46,714), SE Man., Canada, on the Red River opposite Winnipeg. It is an industrial center, with large stockyards and meat-packing plants, oil refineries, flour mills, and breweries. St. Boniface was founded in 1818 as a Roman Catholic mission.

Saint-Bri·euc (săɴ'brē-œ'), town (1975 pop. 52,559), capital of Côtes-du-Nord dept., NW France, on the Gouet River near its mouth on the Bay of Saint-Brieuc, an arm of the English Channel. Metallurgy and textiles are the chief industries. Saint-Brieuc was probably founded in the 5th cent. and grew rapidly after a monastery was built (6th or 7th cent.). Many old houses remain, and there is a Gothic cathedral.

Saint Cath·a·rines (kăth'ər-ĭnz, kăth'rĭnz), city (1976 pop. 123,351), S Ont., Canada, on the Welland Ship Canal. An industrial center in a rich fruit-growing region, it has canneries and wineries as well as textile and paper mills; motor vehicle parts, machinery, electrical products, and farm implements are manufactured. St. Catharines was founded in 1790. The Royal Henley Regatta is held annually in Port Dalhousie, part of St. Catharines since 1961.

Saint-Cha·mond (săɴ'shä-môɴ'), city (1975 pop. 40,250), Loire dept., SE France, at the confluence of the Gier and Janon rivers. The city grew in the 19th cent. as a coal-mining center. Other products include weapons, textiles, bicycles, and toys.

Saint Charles (chärlz). **1.** Parish (1970 pop. 29,550), 288 sq mi (745.9 sq km), SE La., bounded on the N by Lake Pontchartrain, on the SE by Salvador and Cataouatche lakes, on the SW by Bayou Des Allemands, and intersected by the Mississippi River; formed 1807; parish seat Hahnville. Truck crops, rice, sugar cane, and corn are grown. Its industries include oil and natural-gas production and refining. **2.** County (1970 pop. 92,954), 551 sq mi (1,427.1 sq km), E Mo., bounded on the E by the Mississippi River and on the S by the Missouri River; formed 1812; co. seat St. Charles. It is in a rich agricultural area (corn, oats, wheat, hay, and potatoes), with coal deposits and some manufacturing.

Saint Charles. **1.** City (1970 pop. 12,945), Kane co., NE Ill., on the Fox River, in a farm and resort area; inc. 1850. Its manufactures include iron castings and furniture. **2.** City (1978 est. pop. 36,300), seat of St. Charles co., E Mo., on the low bluffs along the N bank of the Missouri River, in an industrial area; settled by French traders 1769, inc. as a city 1849. It is the trade and distribution center of a rich farm area. Shoes and metal products are manufactured. Coal mines are nearby. The earliest permanent white settlement on the Missouri River, St. Charles was an important trading post, a starting point on the westward Boone's Lick Trail, and the state capital from 1821 to 1826.

Saint Clair (klâr). **1.** County (1970 pop. 27,956), 641 sq mi (1,660.2 sq km), NE central Ala., in a hilly area bounded on the E by the Coosa River; formed 1818; co. seat Ashville. It has cotton and livestock farms, coal mines, and deposits of iron ore, clay, and limestone. Its manufactures include textiles, plastics, cement, metal products, and machinery. **2.** County (1970 pop. 285,199), 670 sq mi (1,735.3 sq km), SW Ill., bounded on the NW by the Mississippi River and drained by the Kaskaskia River and Silver Creek; formed 1790; co. seat Belleville. It includes the highly industrialized East Saint Louis area and Cahokia Mounds State Park. It has agriculture (corn, wheat, dairy products, poultry, and livestock), timber, bituminous-coal mines, and deposits of clay, sand, oil, and limestone. **3.** County (1970 pop. 120,175), 723 sq mi (1,872.6 sq km), SE Mich., bounded on the E by Lake Huron and the St. Clair River, on the S by Lake St. Clair, and drained by the Belle and Black rivers; formed 1821; co. seat Port Huron. Livestock, grain, and sugar beets are raised. Some manufacturing is done. There are salt mines, fisheries, and resorts here. **4.** County (1970 pop. 7,667), 699 sq mi (1,810.4 sq km), W Mo., drained by the Osage and Sac rivers; formed 1841; co. seat Osceola. It is in a farm (corn and hay), livestock, and coal region.

Saint Clair, Lake, c.490 sq mi (1,270 sq km), on the U.S.-Canadian border, between SW Ont. and SE Mich. The St. Clair River (41 mi/66 km long) flows into the lake from Lake Huron. The lake is one of the busiest sections of the St. Lawrence Seaway.

Saint Clair Shores (shôrz), city (1978 est. pop. 83,000), Macomb co., SE Mich., a residential suburb adjacent to Detroit, on Lake St. Clair; settled 18th cent. by the French, inc. 1925. It is a boating center.

Saint-Claude (săɴ-klōd'), town (1968 pop. 13,117), Jura dept., E France, in Franche-Comté, at the confluence of the Bienne and Tacon rivers. It is a resort and an industrial center where brier pipes, plastic products, glasses, and toys are manufactured. Serfdom survived in the town at the abbey of Saint-Claude until the abbey was suppressed by the French Revolution in 1789.

Saint-Cloud (săɴ-klōō'), town (1975 pop. 28,139), Hauts-de-Seine dept., N central France, a suburb W of Paris on the Seine River. It is a residential town and resort, with a famous racetrack. Aeronautic equipment, carburetors, radio equipment, motors, and cosmetics are produced. The palace of Saint-Cloud (built 1572; destroyed during the Franco-Prussian War in 1870) was a residence of many rulers of France. Napoleon I proclaimed the Empire at Saint-Cloud in 1804.

Saint Cloud (kloud), city (1978 est. pop. 40,200), seat of Stearns co., central Minn., on the Mississippi River, in a dairying region; inc. 1856. Granite has been quarried since 1868, and granite finishing is still a leading industry. Refrigeration equipment, lenses, paper products, valves, tanks, and generators are also made.

Saint Croix (kroi), county (1970 pop. 34,354), 734 sq mi (1,901.1 sq km), W Wis., bounded in the W by the St. Croix River and drained by the Eau Galle River; formed 1840; co. seat Hudson. It is in a dairying and stock-raising area. Its industry includes food processing and the manufacture of textiles, machinery, and equipment.

Saint Croix, island (1970 pop. 31,779), 80 sq mi (207.2 sq km), the largest of the U.S. Virgin Islands, in the West Indies.

Saint Croix. **1.** River, 75 mi (120.7 km) long, rising in the Chiputneticook Lakes and flowing SE to Passamaquoddy Bay, forming part of the U.S.-Canadian border. The river is used for power and to float logs downstream. In 1604 Samuel de Champlain established a colony on St. Croix Island near the river's mouth; it was abandoned in 1605. In 1798 the British insisted that the St. Croix was the Penobscot River, some miles to the west, and not the international boundary. However, the discovery of ruins of the French settlement on St. Croix Island verified the river as the St. Croix and therefore the border. **2.** River, 164 mi (263.9 km) long, rising in the lake district of NW Wis. and flowing S to the Mississippi River at Prescott, Wis. It forms part of the Wis.-Minn. border. A hydroelectric plant at St. Croix Falls supplies power to Minneapolis-St. Paul.

Saint Croix Island National Monument: *see* National Parks and Monuments Table.

Saint Croix National Scenic Riverway: *see* National Parks and Monuments Table.

Saint-Cyr-l'É·cole (săɴ-sîr'lā-kôl'), town (1975 pop. 16,537), Yvelines dept., N central France. A school for the daughters of impoverished noblemen was founded here in 1685 by Louis XIV and Mme. de Maintenon. The building later housed a military academy founded by Napoleon in 1808. It was destroyed in World War II, and the school was moved to Coëtquidan in Brittany. A new military school was opened in 1966.

Saint Da·vid's (dā'vĭdz), village, Dyfed, SW Wales. The village cathedral, one of the finest in Wales, is mainly Transitional Norman in style; after numerous additions and alterations it was restored in 1878. St. David's was for centuries one of the most important places of pilgrimage in Great Britain. Saint David's Head, northwest of the village, is the westernmost point of Wales.

Saint-De·nis (săɴ'də-nē'), city (1975 pop. 96,132), Seine-Saint-Denis dept., N central France, an industrial suburb N of Paris. Metals, chemicals, machinery, glass, paper, soap, and food products are the major manufactures. Saint-Denis was founded early in the Christian era and grew rapidly as a place of pilgrimage. In 626 King Dagobert I built a Benedictine abbey, which became the richest and most famous in France. Around 750 a new sanctuary was completed by Charlemagne. Joan of Arc blessed her weapons at the abbey. In the 12th cent. a basilica was built under the supervision of Abbé Suger, the abbot of Saint-Denis. Devastated during the French Revolution, the abbey was restored, with later work by the architect Eugène Viollet-le-Duc. Saint-Denis was the first cathedral considered essentially Gothic in construction. Within the cathedral are the tombs of many kings and leading personages of France, including Louis XVI and Marie Antoinette. The abbey is now a school.

Saint-Denis, city (1971 est. pop. 85,400), capital of the French overseas department of Réunion. It is a port on the Indian Ocean at the mouth of the St.-Denis River and exports sugar and rum. St.-Denis was founded in the late 17th cent. as a French way station to the Orient.

Saint-Dié (săN-dyā'), city (1975 pop. 25,423), Vosges dept., E France, in Lorraine, on the Meurthe River. It is an industrial center where wire, foundry products, chemical products, and machinery are manufactured. The city grew around a monastery founded in the 7th cent. The *Cosmographiae introductio* by Martin Waldseemüller, a geographic work that for the first time referred to the newly discovered continent as America, was printed in Saint-Dié in 1507.

Saint-Di·zier (săN'dē-zyā'), town (1975 pop. 37,226), Haute-Marne dept., NE France, on the Marne River. Its manufactures include machinery, musical instruments, and metals. Saint-Dizier has many structures dating from the 15th to the 18th cent. There is also a museum with Roman, Carthaginian, and early Christian artifacts.

Sainte Anne de Beau·pré (săN-tän' də bō-prā'), village (1976 pop. 3,284), S Que., Canada, on the St. Lawrence River NE of Quebec City. It is the site of a famous shrine established in 1620 by sailors who had been shipwrecked. A chapel was built in 1658 and a large church in 1876. Burned in 1922, the church was magnificently rebuilt; it houses relics and is one of Canada's foremost pilgrim resorts.

Sainte Gen·e·vieve (sānt jĕn'ə-vēv'), county (1970 pop. 12,867), 500 sq mi (1,295 sq km), E Mo., on the Mississippi River; formed 1812; co. seat Ste. Genevieve. It is in an agricultural region (corn, wheat, livestock, and hay), with lime, marble, and copper deposits. Part of Clark National Forest is here.

Sainte Genevieve, city (1970 pop. 4,468), seat of Ste. Genevieve co., E Mo., on the Mississippi River SSE of St. Louis; inc. 1893. The earliest permanent white settlement in the state, it was founded before 1750 by the French as a trading post. The salt springs, rich bottom lands, and nearby lead mines attracted settlers. The settlement was an important river port in the steamboat days. Today it is the trade center of a farm area and a shipping point for the lime and marble of the region.

Saint E·li·as, Mount (ĭ-lī'əs), 18,008 ft (5,492.4 m) high, in the St. Elias Mts., a section of the Coast Ranges, on the U.S.-Canadian border between SW Yukon Territory and SE Alaska. It was first seen by Vitus Bering, the Danish explorer, on July 16, 1741.

Saintes (săNt), town (1968 pop. 28,138), Charente-Maritime dept., W France, on the Charente River. It is a market for grains, brandy, and leather. The town was probably the capital of the Celtic Santones and was later occupied by the Romans. Louis IX defeated Henry III of England here in 1242. It has ruins of a Roman amphitheater and triumphal arch and two partially restored Romanesque churches (11th–12th cent.).

Sainte Thé·rèse (săNt' tā-rĕz'), city (1976 pop. 17,479), S Que., Canada, on the St. Lawrence River, NW of Montreal. It has factories producing automobiles, pianos, furniture, plywood, and clothing.

Saint-É·tienne (săN'tā-tyĕn'), city (1975 pop. 220,070), capital of Loire dept., SE France, in the Massif Central. It is a major steel and textile center. Other manufactures include ribbons (famous since the 15th cent.), silk, firearms, bicycles, automobile parts, and textile machinery. The textile and silk industry began in the 11th cent.; the first firearms were produced in the 16th cent. for Francis I. The first steel plant was built in 1815.

Saint Eu·sta·ti·us (yōō-stā'shē-əs, -shəs), island (1969 est. pop. 1,341), 8 sq mi (20.7 sq km), Netherlands Antilles, one of the Leeward Islands. The mountainous island thrives on agriculture and a growing tourist trade. Orangetown, the chief port, was settled by the Dutch in 1632; it became a center of contraband trade with the American colonies before and during the American Revolution. According to tradition, it was the first foreign port to salute (1776) the American flag. The island changed hands frequently. During the 18th cent. pirates and smugglers made it one of the leading trade centers of the West Indies.

Saint Fran·cis (frăn'sĭs), county (1970 pop. 30,799), 636 sq mi (1,647.2 sq km), E Ark., intersected by Crowley's Ridge and drained by the St. Francis and L'Anguille rivers; formed 1827; co. seat Forrest City. Its agriculture includes fruit, cotton, corn, sweet potatoes, and rice. It has lumbering, sand and gravel pits, and some industry (food processing and clothing manufacturing).

Saint Francis. 1. City (1970 pop. 1,725), seat of Cheyenne co., extreme NW Kansas, on the South Fork of the Republican River SW of McCook, Nebr.; founded 1885, inc. 1903. It is a trade and shipping center of a farm area. **2.** City (1970 pop. 10,489), Milwaukee co., SE Wis., a residential suburb of Milwaukee on Lake Michigan; inc. 1951. A power plant and a school for the deaf are here.

Saint Francis, river, c.470 mi (755 km) long, rising in the hills of SE Mo. and flowing S through NE Ark. to join the Mississippi River near Helena, Ark. The river forms part of the Ark.-Mo. border.

Saint Francis, Lake, an expansion of the St. Lawrence River, SE Ont. and S Que., Canada, SW of Montreal, extending between Cornwall and Valleyfield. It is part of the St. Lawrence Seaway.

Saint Fran·cis·ville (frăn'sĭs-vĭl'), town (1970 pop. 1,603), seat of West Feliciana parish, SE La., near the Mississippi River NNW of Baton Rouge, in a cotton and farm area. There are many ante-bellum houses in the region.

Saint Fran·cois (frăn'sĭs), county (1970 pop. 36,875), 457 sq mi (1,183.6 sq km), E Mo., partly in the St. Francois Mts. and drained by the Big and St. Francis rivers; formed 1821; co. seat Farmington. Agriculture (corn, wheat, hay, and livestock), mining (lead, zinc, cobalt, and nickel), and limestone quarrying are the major occupations.

Saint Fran·çois (săN' fräN-swä'). **1.** River, 165 mi (265.5 km) long, rising in Lac St. François, SE Que., Canada, and flowing SW through Lac Aylmer to Sherbrooke, then NW past Drummondville to Lac St. Pierre of the St. Lawrence River. There are several hydroelectric stations on its course. **2.** River, c.60 mi (95 km) long, rising in the Notre Dame Mts., SE Que., Canada, and flowing SE to the St. John River. It forms part of the boundary with Maine.

Saint Gall (sānt gôl', săN gäl'), canton (1977 est. pop. 383,200), 777 sq mi (2,012.4 sq km), NE Switzerland, bordering on the Lake of Constance in the N and on the Rhine River in the E. The canton and its capital, Saint Gall (1977 est. pop. 76,300), take their name from the Benedictine abbey erected (8th cent.) on the site of the hermitage of St. Gall, an Irish monk, around which the town grew. The city is a textile center. The abbots of St. Gall became princes of the Holy Roman Empire in the early 13th cent. Rebelling against the abbot, the city made an alliance with the Swiss Confederation (1454). In 1803 the town and the abbot's domains (secularized in 1798) were consolidated as a canton of the Swiss Confederation under Napoleon's Act of Mediation. One of the oldest scholastic centers north of the Alps, St. Gall has a library with a world-famous collection of medieval manuscripts.

Saint-Gau·dens National Historic Site (sānt-gô'dənz): *see* National Parks and Monuments Table.

Saint George (jôrj), town (1970 pop. 1,604), on St. George's Island, Bermuda. It was the capital of Bermuda until 1815, when it was replaced by Hamilton. During the American Civil War it harbored Confederate blockade-runners.

Saint George. 1. Town (1970 pop. 1,806), seat of Dorchester co., S S.C., near the Edisto River NW of Charleston; settled 1788, inc. 1889. Fertilizer and lumber products are made, and meat is processed. **2.** City (1970 pop. 7,097), seat of Washington co., extreme SW Utah, SW of Cedar City, in the Virgin River valley. Founded by Mormons (1861), it is noted for its livestock and fruit orchards.

Saint George's (jôr'jĭz) or **Saint George,** town (1970 pop., 7,303), capital of Grenada, in the West Indies. A port town on a deep and beautiful harbor, it exports cacao, nutmeg, and mace.

Saint George's Channel, strait, c.100 mi (160 km) long and 50 to 95 mi (80–155 km) wide, linking the Atlantic Ocean and the Irish Sea. It separates southeast Ireland from Wales.

Saint-Ger·main-en-Laye (săɴ'zhĕr-măɴ'äɴ-lā'), town (1975 pop. 37,509), Yvelines dept., N central France, on the Seine River, a residential suburb W of Paris. The town is known primarily for its 16th cent. Renaissance chateau, which was a royal residence until the French Revolution and now houses a museum of pre-Christian antiquities. Henry II and Louis XIV were among the kings born in the chateau. The magnificent chateau park was designed by André Lenôtre. Several important treaties (most notably the 1919 Treaty of Saint-Germain) were signed in the town.

Saint Gott·hard (sānt gŏt'ərd, săɴ' gô-tär'), mountain group of the Lepontine Alps, S central Switzerland, rising to Pizzo Rotondo (10,472 ft/3,194 m high). It is crossed by the St. Gotthard Pass, 6,935 ft (2,115.2 m) high, important since the 11th cent. The St. Gotthard Tunnel (9¼ mi/15 km long; maximum alt. 3,786 ft/1,154.7 m) is one of the longest Alpine tunnels.

Saint He·le·na (hə-lē'nə), island (1971 pop. 5,056), 47 sq mi (121.7 sq km), in the S Atlantic Ocean, 1,200 mi (1,930.8 km) W of Africa. Together with the islands of Ascension and Tristan da Cunha, it comprises the British dependency of St. Helena. The capital is Jamestown. Hemp, vegetables, sweet potatoes, and livestock are raised. Discovered by a Portuguese navigator in 1502, St. Helena was annexed by the Dutch in 1633. In 1659 it was occupied by the British East India Company, and in 1834 it became a British crown colony. St. Helena is best known as the place of exile of Napoleon I from 1815 until his death in 1821.

Saint Helena, parish (1970 pop. 9,937), 420 sq mi (1,087.8 sq km), SE La., bounded on the W by the Amite River, on the N by the Miss. border; formed 1810; parish seat Greensburg. Cotton, corn, sweet potatoes, and strawberries are grown. Lumber and wood products are manufactured.

Saint Hel·ens (hĕl'ənz), county borough (1976 est. pop. 194,400), Merseyside, NW England. It is the chief center of glass manufacture in England and also has iron and brass foundries, chemical and soap factories, and potteries.

Saint Helens, city (1970 pop. 6,212), seat of Columbia co., NW Oregon, on the Columbia River below Portland; founded 1847–48, inc. 1889. It is a shipping and trade center, with salmon fisheries and lumber, pulp, and paper mills.

Saint Helens, Mount, volcanic peak, 9,671 ft (2,949.7 m) high, SW Wash., in the Cascade Range.

Saint Hel·ier (hĕl'yər), town (1971 pop. 28,135), capital of Jersey, Channel Islands, Great Britain, on St. Aubin's Bay. It is a residential town, resort, and point of export for local produce. Royal Square was the scene of a battle (1781) in which the French unsuccessfully attempted to regain Jersey. On an adjacent island, protecting the harbor and connected with the mainland at low tide, is Elizabeth Castle, built in the late 16th cent. Near the castle is a rock that was supposedly the hermitage of St. Helier, or St. Helerius, the early missionary for whom the town was named.

Saint Hy·a·cinthe (sānt hī'ə-sĭnth', săɴt' yə-săɴt'), city (1976 pop. 37,500), S Que., Canada, on the Yamaska River, NE of Montreal. It has textile mills and plants manufacturing rubber and paper products, furniture, shoes, and leather goods.

Saint Ig·nace (ĭg'nəs), city (1970 pop. 2,892), seat of Mackinac co., N Mich., E Upper Peninsula, on the Straits of Mackinac, in a farm and dairy region; inc. as a village 1882, as a city 1883. Long the center of the Mackinac fur trade, it is now a fishing resort.

Saint James (jāmz), parish (1970 pop. 19,733), 249 sq mi (644.9 sq km), SE La., intersected by the Mississippi River; formed 1807; parish seat Convent. Its agriculture includes sugar cane, rice, corn, hay, truck crops, and tobacco. Sugar and rice milling and oil and gas extraction are among its industries.

Saint James. 1. City (1970 pop. 4,027), seat of Watonwan co., S Minn., SW of Mankato; inc. as a village 1871, as a city 1899. It is a shipping center for a farm region. Food products and tools are made here. **2.** Uninc. town (1970 pop. 10,500), Suffolk co., SE N.Y., on Long Island, in a farm and resort area.

Saint Jean (săɴ zhäɴ') or **Saint Johns** (sānt jŏnz'), city (1976 pop. 34,363), S Que., Canada, on the Richelieu River, SE of Montreal. It is an industrial center with textile and hosiery mills, food-canning plants, and varied light manufactures.

Saint-Jean-de-Luz (săɴ-zhäɴ'də-lüz'), town (1975 pop. 11,854), Pyrénées-Atlantiques dept., SW France, in the Basque Provinces, on the Bay of Biscay. It is a beach resort and a sardine- and tuna-fishing port. Saint-Jean-de-Luz has a 16th cent. Basque church.

Saint Jé·rôme (săɴ' zhā-rōm'), city (1976 pop. 25,175), S Que., Canada, on the North River, NW of Montreal. It is an industrial center with woolen and paper mills. Rubber and wood products are also manufactured.

Saint John (jŏn), city (1976 pop. 85,956), S N.B., Canada, at the mouth of the St. John River on the Bay of Fundy. A major year-round port, it has large dry docks and terminal facilities. The city has pulp and paper mills, oil and sugar refineries, and food-processing plants. The site was visited (1604) by Samuel de Champlain, and a fort and trading post was built (1631–35) by Charles de la Tour. In the struggle between France and England for possession of Acadia, the fort was captured and recaptured several times, finally becoming British in 1758. Loyalists from the United States established a settlement here called Parr Town. In 1785 it was incorporated with Carleton and named St. John, becoming the first incorporated city in Canada. Much of the old city was destroyed by fire in 1877.

Saint John, city (1970 pop. 1,477), seat of Stafford co., S central Kansas, WSW of Hutchinson; platted 1879, inc. 1885. It is the trade center of a grain and oil area.

Saint John, river, 418 mi (672.6 km) long, rising in N Maine and flowing NE to N.B., Canada, then SE below Edmundston and Fredericton to the Bay of Fundy at St. John. It forms part of the U.S.-Canadian border. At Grand Falls the river drops 75 ft (22.9 m) in a great cataract. At its mouth, within the city of St. John, are the Reversing Falls Rapids, caused by the strong tides of the Bay of Fundy, which force the river to reverse its flow at high tide. The river was discovered (1604) by Samuel de Champlain and Sieur de Monts. In the 17th and 18th cent. it was an important route for traders. It later became a major lumber transportation route.

Saint John's (jŏnz), city (1970 pop. 21,814), capital of Antigua, in the British West Indies, at the head of a harbor formed by an inlet. Tourism is important. In the 18th cent. St. John's served as a headquarters for the Royal Navy in the West Indies.

Saint John's, city (1976 pop. 86,576), provincial capital, SE N.F., Canada, on the NE coast of the Avalon Peninsula. Built on hills overlooking a fine harbor, it is the base of a large fishing fleet. The city's industries include shipbuilding, the manufacturing of fishing equipment and marine engines, and the storing, preserving, and processing of fish. St. John's is one of the oldest settlements in North America. In 1583 Sir Humphrey Gilbert took possession of the region for England. The settlement was captured and recaptured by France and England, becoming permanently British in 1762. It was at St. John's that Guglielmo Marconi heard (1901) the first transatlantic wireless message and from here that the first nonstop transatlantic flight was made in 1919.

Saint Johns (jŏnz), county (1970 pop. 31,035), 609 sq mi (1,567 sq km), NE Fla., in a partly swampy lowland area bounded on the W by the St. Johns River, on the E by the Atlantic Ocean; formed 1822; co. seat St. Augustine. Its agriculture includes vegetables, poultry, and dairy products. It has fishing and some forestry. Its industry includes food processing, clothing manufacturing, and shipbuilding.

Saint Johns. 1. City (1970 pop. 1,320), seat of Apache co., NE Ariz., ESE of Flagstaff near the N.Mex. border; founded 1874. Cattle and lumber are produced in this irrigated farm area. **2.** City (1970 pop. 6,672), seat of Clinton co., S Mich., N of Lansing, in a farm region; inc. as a village 1857, as a city 1904. Peppermint is an important crop. Bearings and other metal products are made.

Saint Johns, river, 285 mi (458.6 km) long, rising in the swampy region of SE Fla., N of Lake Okeechobee, and flowing N to Jacksonville, where it turns abruptly E and enters the Atlantic Ocean. The dredged river is navigable c.170 mi (275 km) upstream; there is a 30-ft (9-m) channel from Jacksonville to the ocean. The lower third of the river forms part of the Intracoastal Waterway.

Saint Johns·bur·y (jŏnz'bûr'ē, -bə-rē), town (1970 pop. 8,409), seat of Caledonia co., NE Vt., on the Passumpsic River NE of Montpelier; settled 1786, organized 1790. The town has a maple-sugar industry and granite works.

Saint John the Bap·tist (băp'tĭst), parish (1970 pop. 23,813), 227 sq mi (587.9 sq km), SE La., bounded on the N by Lake Maurepas, on the E by Lake Pontchartrain, on the S partly by Lake Des Allemands, and intersected by the Mississippi River; formed 1807; parish seat Edgard. Sugar cane, corn, truck crops, hay, and rice are grown. Among its industries are sugar refining and oil and gas extraction.

Saint Jo·seph (jō'zəf, -səf). **1.** County (1970 pop. 245,045), 467 sq mi (1,209.5 sq km), N Ind., bordered on the N by Mich. and drained by the St. Joseph, Yellow, and Kankakee rivers; formed 1830; co. seat South Bend. Its agriculture includes dairy products, corn, grain, fruit, mint, livestock, and truck crops. It has timber, a food-processing industry, and varied manufactures (clothing, furniture, paper, metal, rubber, and plastic products, chemicals, concrete, and miscellaneous machinery). There are lake resorts in the county. **2.** County (1970 pop. 47,392), 508 sq mi (1,315.7 sq km), SW Mich., bounded on the S by the Ind. border, drained by the St. Joseph River and its affluents; formed 1829; co. seat Centreville. It is chiefly a stock-raising and agricultural area where grain, potatoes, corn, and mint are grown and dairy products are made. There are several small lakes with resorts in the county.

Saint Joseph. 1. Town (1970 pop. 1,864), seat of Tensas parish, E La., near the Mississippi River SW of Vicksburg, Miss., in a cotton and petroleum region. There are some fine ante-bellum houses in the vicinity. **2.** City (1978 est. pop. 11,200), seat of Berrien co., SW Mich., on Lake Michigan at the mouth of the St. Joseph River; inc. 1834. A resort with beaches and mineral springs, it is also a port and a trade center for a fruit-growing region. Household appliances, refrigerators, automobile parts, and rubber goods are manufactured. Indian villages, a Jesuit mission, Fort Miami (a French trading fort built by Robert Cavelier, sieur de La Salle, in 1679), and a fur-trading post occupied this site before permanent American settlement began c.1830. **3.** City (1978 est. pop. 77,500), seat of Buchanan co., NW Mo., on the Missouri River; inc. 1845. A port of entry, it is a railroad center and the hub of a rich farming area. The city has meat-packing and food-processing plants. Among its manufactures are electrical products, wire rope, and wood products. The city was laid out c.1843 on the site of a trading post founded (1826) by Joseph Robidoux. In 1860 it became the eastern terminus of the pony express. Of interest are the pony-express stables and a museum of pioneer and Indian relics.

Saint Joseph, river, 210 mi (338 km) long, rising in S Mich. and flowing generally W in wide curves to Lake Michigan at Benton Harbor, Mich. The river was an important link to the Ohio River and Lake Erie for pioneer travelers.

Saint Kitts-Ne·vis (kĭts'nē'vĭs, -nĕv'ĭs), island state (1970 pop. 47,457), 120 sq mi (310.8 sq km), British West Indies, in the Leeward Islands. The state consists of the islands of St. Kitts or St. Christopher (68 sq mi/176.1 sq km), Nevis (50 sq mi/129.5 sq km), and Sombrero (2 sq mi/5.2 sq km). The capital is Basseterre on St. Kitts. A narrow strait separates the two larger islands, which are volcanic in origin, mountainous, and renowned for their scenery. Sugar, molasses, cotton, and coconuts are exported. St. Kitts and Nevis were discovered by Columbus in 1493, but settlement did not begin until the British arrived on St. Kitts in 1623. French settlers came to the island two years later. Nevis was first settled by the British in 1628. The Treaty of Paris of 1783 definitively awarded the islands to Britain. They were part of the colony of the Leeward Islands (1871–1956) and of the West Indies Federation (1958–62). In 1967, together with Anguilla, they became a self-governing state in association with Great Britain. Anguilla seceded later that year.

Saint Lam·bert (lăm'bərt), city (1976 pop. 20,318), S Que., Canada, on the St. Lawrence River. It is a residential suburb of Montreal.

Saint Lan·dry (lăn'drē), parish (1970 pop. 80,364), 930 sq mi (2,408.7 sq km), S central La., bounded on the E by the Atchafalaya River and drained by Bayou Teche; formed 1807; parish seat Opelousas. It is a fertile agricultural region growing cotton, corn, sugar cane, truck crops, pecans, rice, sweet potatoes, and hay. There is some industry, including the processing and shipping of farm products and lumber. The parish has natural-gas and oil fields.

Saint Law·rence (lôr'əns, lŏr'-), county (1970 pop. 112,309), 2,772 sq mi (7,179.5 sq km), N N.Y., bounded on the NW by the St. Lawrence River and drained by the St. Regis, Indian, Grass, Oswegatchie, and Raquette rivers; formed 1802; co. seat Canton. This plains area along the St. Lawrence, rising to the Adirondacks in the southeast, is a leading dairying county. It also has farming, maple-sugar production, lead, zinc, and pyrite mines, limestone and talc deposits, and diversified manufacturing. There are lake and river resorts.

Saint Lawrence, one of the principal rivers of North America, 744 mi (1,197 km) long. It issues from the northeastern end of Lake Ontario and flows northeast, first along the U.S.-Canadian border, then into S Que., Canada, past Montreal and Quebec city, to the Gulf of St. Lawrence, north of Cape Gaspé. It is the outlet of the Great Lakes and together with them forms a c.2,300-mi (3,700-km) waterway from the western end of Lake Superior to the Atlantic Ocean. In its upper course the river cuts through a part of the Canadian Shield. Below Cornwall, Ont., the river widens into Lake St. Francis. It widens again into Lake St. Louis, then descends through the Lachine Rapids to Montreal, head of navigation for oceangoing vessels. Between Sorel and Trois Rivières is Lake St. Peter, another widened section. Below the city of Quebec the river is tidal. It gradually increases in width to c.90 mi (145 km) at its mouth.

The St. Lawrence River is an important source of hydroelectric power. Canals have been constructed around the rapids, making the entire river navigable to all but the largest of vessels. The upper part is not navigable during the winter months because of ice accumulation. The St. Lawrence valley, long inhabited predominantly by French Canadians, is an agricultural region; potatoes, grains, hay, vegetables, and dairy cattle are raised.

The St. Lawrence River was visited in 1534 by Jacques Cartier; in 1535 he ascended it as far as Montreal. Quebec city was settled (1608) by Samuel de Champlain. The river system was long a highway for explorers, fur traders, and missionaries. The valley remained in the possession of the French until Canada was surrendered to the British at the close of the French and Indian Wars in 1763. The river became an international boundary after 1793, but British influence was generally dominant over the whole St. Lawrence valley and most of its southern tributaries until after the War of 1812.

Saint Lawrence, Gulf of, arm of the Atlantic Ocean, c.100,000 sq mi (259,000 sq km), SE Canada, extending c.250 mi (400 km) from the mouth of the St. Lawrence River to Newf. on the E. At its greatest width it is c.500 mi (805 km). In the gulf are Prince Edward Island, Anticosti Island, the Magdalen Islands, and numerous small islands. The Strait of Belle Isle, Cabot Strait, and the Strait of Canso lead to the Atlantic. The gulf is subject to frequent fog and is closed to navigation by ice from early Dec. to mid-Apr. It was visited by explorers before the 16th cent. It has important fishing grounds, especially for cod.

Saint Lawrence Island, c.90 mi (145 km) long and from 8 to 22 mi (13–35 km) wide, off W Alaska, in the Bering Sea. It is a barren island, inhabited by Eskimo engaged in whaling and fox trapping. It was discovered by Vitus Bering on St. Lawrence's Day, 1728.

Saint Lawrence Sea·way (sē'wā'), international waterway, 2,342 mi (3,768.3 km) long, consisting of a system of canals, dams, and locks in the St. Lawrence River and connecting channels between the Great Lakes; opened 1959. It provides passage for oceangoing vessels into central North America. The shipping season has been extended from mid-Apr. to mid-Dec. by increased use of icebreakers and air pumps to control ice formation in the locks. Iron ore, wheat, and coal are the principal cargoes carried on the seaway. Construction of the project was authorized jointly by Canada and the United States in the early 1950s. The seaway was included in the binational Water Quality Agreement of 1972 to prevent further pollution and improve water quality in the Great Lakes and the international section of the seaway.

Saint-Lô (săN-lō'), town (1968 pop. 19,613), capital of Manche dept., NW France, in Normandy. It is an agricultural center and has famous horse stables. Wood products, plaster, and clothing are manufactured. An old Gallo-Roman town, Saint-Lô was the scene of a massacre of Huguenots in the 16th cent. It has been rebuilt since its virtual destruction during the Allied invasion of Normandy in 1944.

Saint-Lou·is (săN-lo͞o-ē'), city (1973 est. pop. 99,000), NW Senegal, a port on an island in the Senegal River. It is a trade and export center for peanuts, hides, and skins. The oldest French colonial settlement in Africa, Saint-Louis was founded as a trade base in 1638. In 1659 a French fort was built here. Except for brief periods (1758–79 and 1809–15) of British ownership, Saint-Louis was long the capital of all French possessions in West Africa and was the capital of French West Africa from its inception in 1885 until 1902. From 1902 to 1958 Saint-Louis served as the capital of both Senegal and Mauritania.

Saint Lou·is (sānt lo͞o'ĭs, lo͞o'ē). **1.** County (1970 pop. 220,693), 6,092 sq mi (15,778 sq km), NE Minn., in an extensively watered area bordered on the N by Ont., Canada; formed 1855; co. seat Duluth.

The Vermilion and Mesabi iron ranges have immensely productive mines. Agriculture (dairy products and potatoes), lumbering, and food processing are also important. Clothing, chemicals, cement and concrete, metal products, and machinery are manufactured. **2.** County (1970 pop. 951,671), 497 sq mi (1,287.2 sq km), E Mo., bounded on the E by the Mississippi River, on the NW by the Missouri River; formed 1877; co. seat Clayton. It is in an agricultural (corn, wheat, hay, and potatoes) and horticultural area, with dairy farms and lime deposits. Wood products are made.

Saint Louis, city (1970 pop. 622,236), independent and in no county, E Mo., on the Mississippi River below the mouth of the Missouri River; inc. as a city 1822. It is the largest city of the state, a great river-freight handler, and a major rail center. A market for furs, livestock, grain, and other farm produce, St. Louis is also a wholesale, banking, and financial center. Its industries produce a wide variety of manufactures, including shoes and leather goods, beer, machinery, chemicals, aircraft, space capsules, and automotive vehicles and parts.

The site of the present city was chosen (1763) by Pierre Laclède for a fur-trading post. To honor Louis XV of France, it was named for his "name" saint, Louis IX of France. Transferred to the Spanish in 1770, it was retroceded to France in the time of Napoleon I and then sold to the United States as part of the Louisiana Purchase. However, the population and customs remained predominantly French until well into the 19th cent. St. Louis was the market and supply point for fur traders, mountain men, and explorers. The town grew rapidly after the War of 1812, when immigrants came in numbers to settle the West. St. Louis became one of the greatest U.S. river ports. Even after the railroads came, the river steamers were at their peak. After the Civil War—in which St. Louis was Unionist in sympathy—industry in the city expanded greatly. The oldest of the many bridges across the Mississippi was constructed by James B. Eads in 1874. The city has a symphony orchestra, a municipal opera, and many educational institutions, including Saint Louis Univ. and Washington Univ. New Cathedral is one of the largest Roman Catholic cathedrals in the country. One especially notable landmark is Gateway Arch, a giant stainless steel arch, 630 ft (192 m) high, designed by Eero Saarinen. The arch, which stands on the banks of the Mississippi, symbolizes St. Louis as the gateway to the West.

Saint Louis Park, city (1978 est. pop. 46,300), Hennepin co., SE Minn., a suburb of Minneapolis; settled 1854, inc. 1886. The manufactures of its industrial park include electronic equipment; machinery; metal, rubber, and plastic products; and processed foods.

Saint Lu·ci·a (lo͞o′shə, lo͞o-sē′ə), island (1970 pop. 101,100), 238 sq mi (616.4 sq km), British West Indies, one of the Windward Islands; capital Castries. Mountains, rising abruptly from the sea, create startlingly lovely scenic effects. The forests of the mountain slopes yield fine cabinet woods, and the volcanic soil is rich. Bananas and other tropical agricultural products are exported. Columbus probably discovered the island in 1502. The British, in the first attempt at colonization early in the 17th cent., were beaten back by the fierce Carib Indians. The island was later settled by the French, who signed a treaty with the Caribs in 1660. Thereafter St. Lucia was much contested by the two powers until the British regained it definitively in 1803. The island formed part of the British Windward Islands colony from 1958 to 1959, when the colony was dissolved. In 1967 St. Lucia became one of the six Associated States of the West Indies, with internal self-government.

Saint Lu·cie (lo͞o′sē), county (1970 pop. 50,836), 588 sq mi (1,522.9 sq km), SE Fla., on the Atlantic Ocean in a largely swampy lowland area, bordered on the E by a barrier beach enclosing Indian River lagoon; formed 1905; co. seat Fort Pierce. It is in a citrus-fruit, truck-farming, poultry, and livestock region, with forestry, fishing, and some mining. Canned fruit and vegetables, plastics, and concrete are produced. There is also a tourist industry.

Saint-Ma·lo (săɴ′mä-lō′), town (1975 pop. 45,030), Ille-et-Vilaine dept., NW France, on the English Channel. Built on a rocky promontory, Saint-Malo is a fishing port and tourist center. The major industries are deep-sea fishing, the drying of cod, and boatbuilding. A Welsh monk built a monastery nearby in the 6th cent., and in the 9th cent. refugees fleeing Norman raids on nearby Saint-Servan settled at the site of present-day Saint-Malo. It became a prosperous commercial seaport under the French in the 1500s. Between the 17th and 19th cent. French corsairs operated out of Saint-Malo, despite repeated English efforts to destroy the port and corsair fleet. Saint-Malo is famous for its ramparts and its 17th cent. architecture. Points of interest include the main gate to the city (15th cent.) and a chateau (15th cent.) that is now a municipal museum. In World War II German forces, retreating before the U.S. army, set the city ablaze. The city was the birthplace of Jacques Cartier.

Saint Ma·ries (sānt′ mə-rēz′), city (1970 pop. 2,571), seat of Benewah co., N Idaho, at the confluence of the St. Joe and St. Maries rivers SSE of Coeur d'Alene Lake; settled 1888, inc. 1913.

Saint Mar·tin (mär′tn), parish (1970 pop. 32,453), 738 sq mi (1,911.4 sq km), S La., divided into two sections separated by part of Iberia parish and traversed by the Atchafalaya and Grand rivers and Bayou Teche; formed 1807; parish seat St. Martinville. It is chiefly an agricultural area, growing sugar cane, cotton, corn, hay, peppers, sweet potatoes, and rice. Natural gas and oil are found. Fur trapping and lumbering are done, as well as the processing of farm products and lumber.

Saint Mar·tin (sānt mär′tn, săɴ′ mär-tăɴ′), island, 37 sq mi (95.8 sq km), West Indies, one of the Leeward Islands. Since its occupation in 1648 by the Dutch and the French, it has been divided; the northern part belongs to French Guadeloupe, and the southern part belongs to the Netherlands Antilles. A hilly, scenic island provided with good harbors, Saint Martin is a tourist resort. Cotton, sugar cane, and tropical fruits are raised.

Saint Mar·tin·ville (mär′tn-vĭl′), town (1970 pop. 7,153), seat of St. Martin parish, S La., on Bayou Teche N of New Iberia; settled by Frenchmen c.1760. An important resort during the early steamboat era, it declined after 1855. It is allegedly the site of the Evangeline romance, commemorated by nearby Longfellow-Evangeline State Park.

Saint Mar·y (mâr′ē), parish (1970 pop. 60,752), 624 sq mi (1,616.2 sq km), S La., on the Gulf Coast, bounded on the W and S by West Cote Blanche, East Cote Blanche, Atchafalaya bays, on the NE by Grand Lake, crossed by the Gulf Intracoastal Waterway, and drained by the Atchafalaya River and Bayou Teche; formed 1811; parish seat Franklin. Sugar cane, rice, corn, hay, and truck crops are grown. Industries include fisheries, natural-gas and oil extraction, seafood canning, and lumbering.

Saint Mar·ys (mâr′ēz), county (1970 pop. 47,388), 373 sq mi (966.1 sq km), S Md., bounded on the NE by the Patuxent River, on the E by Chesapeake Bay, on the S by the Potomac River and the Va. border; formed 1637. It is in a tidewater agricultural area that produces tobacco, corn, and wheat, with lumbering and commercial fishing.

Saint Marys. 1. City (1970 pop. 7,699), Auglaize co., W central Ohio, on the St. Marys River SW of Lima; inc. 1823. Rubber products are made. It is a resort center. **2.** Industrial borough (1970 pop. 7,470), Elk co., NW Pa., on Elk Creek E of Ridgway; settled 1842, inc. 1848. Carbon products are its chief manufacture. **3.** City (1970 pop. 2,348), seat of Pleasants co., NW W.Va., on the Ohio River ESE of Marietta, Ohio, in an oil and natural-gas area; settled c.1850.

Saint Marys. 1. River, 63 mi (101 km) long, flowing SE from Lake Superior to Lake Huron and forming part of the U.S.-Canadian line. Although frozen for about five months each year, the river and canal together form one of the world's busiest waterways. **2.** River, c.175 mi (280 km) long, rising in Okefenokee Swamp, SE Ga., and flowing E to the Atlantic Ocean. It forms part of the Ga.-Fla. border.

Saint Marys City, village (1970 pop. 540), St. Marys co., S Md., on the St. Marys River; est. 1634. English colonists, after purchasing a small village from the Indians, renamed it St. Marys and built Fort St. George. The first state assembly met here in 1635, and the village remained the provincial capital until Annapolis replaced it as capital in 1694.

Saint Mat·thews (măth′yo͞oz). **1.** City (1970 pop. 13,152), Jefferson co., N Ky., a residential suburb of Louisville; inc. 1950. **2.** Town (1970 pop. 2,403), seat of Calhoun co., central S.C., SE of Columbia; settled in the early 18th cent., inc. 1872. Clothing, fertilizer, and truck bodies are made.

Saint-Maur-des-Fos·sés (săɴ-môr′dā-fō-sā′), city (1975 pop. 80,920), Val-de-Marne dept., N central France, on the Marne River. It manufactures automobile parts, ball bearings, electrical equipment, asbestos and paper products, and furniture. St. Nicholas Church (12th–14th cent.) houses a statue of Our Lady of Miracles, an object of pilgrimages.

Saint Mau·rice (sānt môr′ĭs, săɴ′ mô-rēs′), river, c.325 mi (525 km) long, rising in the Laurentian Mts., S Que., Canada, and flowing SE and S to the St. Lawrence River at Trois Rivières. The river is important for the transportation of lumber.

Saint Mi·chael's Mount (mī'kəlz), pyramid-shaped rocky islet, 21 acres (8.5 hectares), Cornwall, SW England, in Mounts Bay. A natural causeway connects it at low tide with the mainland. A priory built in the 11th cent. is here.

Saint-Mi·hiel (săɴ' mē-yĕl'), town (1968 pop. 5,382), Meuse dept., NE France, in Lorraine, on the Meuse River. It grew around a Benedictine abbey founded in 709. Abbey buildings constructed in the 17th and 18th cent. are now used as a courthouse, library, and school.

Saint Mo·ritz (sānt' mə-rĭtz'), town (1977 est. pop. 7,400), Grisons canton, SE Switzerland, in the Upper Engadine, on the Lake of St. Moritz. One of the largest winter-sports centers in the world, it is surrounded by magnificent peaks. The Olympic winter games were held here in 1928 and in 1948.

Saint-Na·zaire (săɴ'nä-zâr'), city (1975 pop. 69,251), Loire-Atlantique dept., W France, at the mouth of the Loire River on the Bay of Biscay. Saint-Nazaire is an important seaport and a shipbuilding center, with aeronautical, metallurgical, chemical, and food industries. Built on the site of an ancient Gallo-Roman town, Saint-Nazaire belonged to the dukes of Brittany in the 14th and 15th cent.

Saint-O·mer (săɴ'tô-mâr'), city (1975 pop. 16,932), Pas-de-Calais dept., N France, in Flanders, on the Aa River. The chief manufactures are metals, textiles, telephone equipment, and beer. The city grew around a monastery founded in the 7th cent.

Sain·tonge (săɴ-tôɴzh'), region of W France, on the Bay of Biscay. Cattle and sheep raising, dairying, and the manufacture of cognac are the major occupations. The region was occupied by the Visigoths in 419 and by Clovis I in 507. It became part of England (1154) following the marriage of Eleanor of Aquitaine to Henry of Anjou. The region was incorporated into the French crown lands in 1372 and was a province of France until the Revolution in 1789.

Saint-Ou·en (săɴt'ōō-äɴ'), city (1975 pop. 43,588), Seine-Saint-Denis dept., N central France, on the Seine River. It is an industrial suburb and a terminal point for river shipping. Electrical equipment, metal products, pharmaceuticals, and perfumes are among the chief manufactures. In 1814 Louis XVIII signed the Declaration of Saint-Ouen, by which he became a constitutional monarch.

Saint Paul (sānt pôl'). **1.** City (1978 est. pop. 271,000), state capital and seat of Ramsey co., E Minn., on bluffs along the Mississippi River, contiguous with Minneapolis; inc. 1854. A port of entry and a major railroad hub, St. Paul is the industrial, commercial, and cultural center for a vast fertile region. Among the city's products are computers, electronic materials, automobiles, machinery, chemicals, paper, food products, beer, furniture, and steel and iron goods. A fur-trading post was established (early 1800s) at the confluence of the Mississippi and Minnesota rivers in what is now the historic village of Mendota. Traders, missionaries, and explorers were the first inhabitants; settlers followed from the East after securing treaties with the Indians. By 1823 the port was an important debarkation point and trading center. St. Paul became territorial capital in 1849 and state capital in 1858. Many lakes, public beaches, and nearby ski areas provide recreational facilities. The capitol, modeled after St. Peter's in Rome, has the largest unsupported marble dome in the world. **2.** City (1970 pop. 2,026), seat of Howard co., central Nebr., NNW of Grand Island; founded 1871. It is a trade and shipping point in a farm area.

Saint Paul's Church National Historic Site (pôlz): *see* National Parks and Monuments Table.

Saint Pe·ter (pē'tər), city (1970 pop. 8,339), seat of Nicollet co., S Minn., on the Minnesota River N of Mankato; settled 1853, inc. 1865. It is a farm trade center.

Saint Pe·ters·burg (pē'tərz-bûrg'), city (1978 est. pop. 238,200), Pinellas co., W Fla., on Tampa Bay and the Gulf of Mexico at the S end of the Pinellas peninsula; settled in the mid-1800s, inc. 1892. A port of entry with a large harbor, it is a popular winter resort. Manufactures include boats, trailers and campers, air conditioners, electronic equipment, and cement. The city also has citrus-fruit and commercial-fishing industries.

Saint-Pierre (sānt-pyâr', săɴ-pyâr'), town (1967 pop. 5,556), Martinique, French West Indies. Founded in 1635 and once the chief commercial city of the island, it was engulfed by a mass of flame, lava, and ash in the eruption (1902) of Pelée. Of the city's inhabitants (about 28,000), only one person survived.

Saint Pierre and Miq·ue·lon (săɴ pyâr'; mē-klôɴ'), French territory (1974 pop. 5,840), 93 sq mi (240.9 sq km), consisting of nine small islands, S of N.F., Canada, in the Gulf of St. Lawrence. The capital is St. Pierre on the island of the same name. Miquelon (83 sq mi/215 sq km) is the largest island. The islands are barren, rocky, and often fogbound, but their proximity to the Grand Banks makes them a valuable base for fishermen. Probably first settled by Basques, they were colonized by France in 1604. They were taken by the British (1713) but returned to France in 1763; twice retaken by the British, they were restored to France in 1814, with the provision that they be unfortified. They were granted local autonomy in 1935.

Saint-Quen·tin (săɴ'kän-tăɴ'), city (1975 pop. 67,243), Aisne dept., N France, on the Somme River. Foundry products, machinery, textiles, furniture, rubber, and food products are manufactured. Of Roman origin, the city was chartered in 1080. It became part of the royal domain in 1191 and was ceded briefly to Burgundy (1435–77). The city has a long history of sieges and captures, most notably by the Spanish (1557) during the Wars of Religion.

Saint-Ra·pha·ël (săɴ-rä'fə-ĕl'), town (1968 pop. 18,339), Var dept., SE France, on the French Riviera. It is an elegant resort and a small commercial port as well as an important naval air base and the site of a naval school.

Saint Re·gis (sānt rē'jĭs), settlement of Roman Catholic Iroquois on the S bank of the St. Lawrence River, on both sides of the U.S.-Canadian line, partly in Huntingdon co., Que., and partly in Franklin co., N.Y. The village was established c.1755 by a party of Catholic Iroquois from Caughnawaga, Que.

Saint Si·mons Island (sī'mənz), uninc. resort town (1970 pop. 5,346), Glynn co., SE Ga., on St. Simons Island in the Sea Islands NW of Brunswick.

Saint Ste·phen (stē'vən), town (1976 pop. 5,264), SW N.B., Canada, on the St. Croix River opposite Calais, Maine. The two towns, connected by an international bridge, form virtually a single community. St. Stephen was founded by Loyalists after the American Revolution.

Saint Tam·ma·ny (tăm'ə-nē), parish (1970 pop. 63,585), 886 sq mi (2,294.7 sq km), SE La., bounded on the S by Lake Pontchartrain, on the E by the Pearl River, here forming the Miss. line; formed 1810; parish seat Covington. It is in an agricultural area growing sugar cane, corn, sweet potatoes, and strawberries and raising livestock. There is some manufacturing, including boatbuilding. The parish also has hunting and fishing areas.

Saint Thom·as (tŏm'əs), city (1976 pop. 27,206), S Ont., Canada, S of London. The city is located in a rich agricultural area and has automobile plants and other factories.

Saint Thomas, island (1970 pop. 28,960), 32 sq mi (82.9 sq km), one of the U.S. Virgin Islands, West Indies. Tourism is the main economic activity.

Saint Thomas National Historic Site: *see* National Parks and Monuments Table.

Saint-Tro·pez (săɴ'trô-pā'), town (1975 pop. 4,523), Var dept., SE France, on the French Riviera. It is a popular beach resort and a picturesque small fishing port. From the 15th to the 17th cent. it was an independent republic.

Saint Vin·cent (sānt vĭn'sənt), island state (1970 pop. 89,129), 150 sq mi (388.5 sq km), British West Indies, in the Windward Islands. It comprises the island of St. Vincent (140 sq mi/362.6 sq km) and the small Grenadine islands to the north. The capital is Kingstown. St. Vincent island is mountainous, rising to 4,048 ft (1,234.6 m) at the now inactive Soufrière volcano, and well forested, with a healthful climate and abundant rainfall. Bananas, arrowroot, and copra are the chief exports, followed by other agricultural products including fine sea-island cotton. Tourism is also economically important. St. Vincent remained uncolonized until a British settlement was made in 1762. The French captured it in 1779, but it was restored to Britain in 1783. In 1902 much of the island was destroyed by an eruption of Soufrière. St. Vincent was part of the British colony of the Windward Islands (1880–1958) and of the West Indies Federation (1958–62). In 1969 it became a self-governing state in association with Great Britain.

Saint Vincent, Cape, high and rocky promontory at the SW extremity of Portugal. Several historic sea battles were fought nearby, the most notable in 1797, when the British defeated a large Spanish fleet. Prince Henry the Navigator, the Portuguese patron of exploration, lived nearby.

Saint Vincent, Gulf, inlet of the Indian Ocean, 90 mi (145 km) long and 45 mi (72.4 km) wide, SE South Australia state, Australia. Salt is obtained from the shores of the gulf by solar evaporation.

Saint-Vith (săɴ-vēt'), town (1970 pop. 3,001), Liège prov., E Belgium, near the West German border. An important road and rail junction in World War II, it was captured (Dec., 1944) by the Germans early in the Battle of the Bulge and was later taken (Jan. 23, 1945) by U.S. forces. The town was severely damaged in the fighting.

Sai·pan (sī-păn', -pän'), volcanic island (1970 pop. 7,967), 47 sq mi (121.7 sq km), W Pacific, in the Marianas Islands, U.S. Trust Territory of the Pacific Islands. Copra and scrap metal are the only exports, but sugar cane, coffee, and citrus fruits are grown, and the island has phosphate and manganese deposits. Saipan, with the other Marianas, was mandated to Japan in 1920 by the League of Nations. In World War II the island (site of a Japanese airbase) was taken by U.S. forces in 1944.

Sa·ïs (sā'ĭs), ancient city of Egypt, in the W central region of the Nile delta. It was the royal residence of the XXVI dynasty.

Sa·ja·ma (sä-hä'mä), mountain, 21,390 ft (6,524 m) high, in the Cordillera Occidental of the Andes, W Bolivia, near the border with Chile.

Sa·kai (sä'kī'), city (1976 est. pop. 768,702), Osaka prefecture, S Honshu, Japan, on Osaka Bay at the mouth of the Yamato River. An industrial center, it has engineering, iron, and steel works, chemical plants, machine factories, and textile mills. Sakai was a major port from the 15th to the 17th cent.

Sa·kar·ya (sə-kär'yə), river, c.490 mi (790 km) long, rising on the Anatolian plateau, NW Turkey. It flows generally north in a series of huge bends to the Black Sea at Karasu. There are hydroelectric-power plants on the river.

Sa·kha·lin (săk'ə-lēn'), formerly **Sa·gha·lien** (sä'gəl-yĕn'), island (c.29,500 sq mi/76,405 sq km), off the coast of the Soviet Far East, USSR, between the Sea of Okhotsk and the Sea of Japan. It is separated from the Soviet mainland on the west by the Tatar Strait and from Hokkaido, the northernmost island of Japan, by the Soya Strait. With the Kuril Islands it forms the Sakhalin oblast (1970 pop. 616,000) of the Soviet Far East. Two parallel mountain ranges, separated by a central valley, run the length of this elongated and heavily forested island. The climate is severe, but grains, beets, and potatoes are grown in the south. Lumbering, coal mining, herring fishing, and paper milling are the principal industries. There are oil fields in the northeast. Despite their small size, the coal and iron deposits are vital to the Soviet Far East region, where these minerals are scarce.

Sakhalin was explored by Russians in the 17th cent. and colonized by Russia and Japan in the 18th and 19th cent. It was under joint Russo-Japanese control until it passed entirely to Russia in 1875. Sakhalin became a czarist place of exile. By the Treaty of Portsmouth (1905) Russia retained the portion of Sakhalin north of lat. 50° N and Japan obtained the remainder. The Japanese territory was named Karafuto (a name sometimes applied to the whole island). Both countries colonized extensively and reduced the native population to a minority. After World War II the Japanese holdings were transferred to the USSR and nearly all the Japanese population was repatriated. In 1951 Japan renounced all claims to Sakhalin.

Sak·ka·ra (sə-kä'rə), necropolis (burial place) of ancient Memphis, Egypt, 3 mi (4.8 km) from the Nile on the border of the Libyan desert. Zoser built a step-pyramid here in the III dynasty, and on the grounds are many pyramids that date from the V and VI dynasties. The Serapeum, burial place of the Apis bulls, dates from the later period. The oldest dated papyrus (VI dynasty) was discovered here in 1893.

Sa·ku·ra-ji·ma (sä-kŏŏr'ə-jē'mə), peninsula, Kagoshima prefecture, S Kyushu, Japan, opposite Kagoshima. Formerly an island, Sakurajima became a peninsula in 1914 when lava from its three volcanic cones closed the channel. Fruits and turnips are raised.

Sa·la·do, Ri·o (rē'ō sə-lä'dō), name of several South American rivers, including more than 10 in Argentina. The most important is the Río Salado del Norte (c.1,250 mi/2,010 km long), rising in the Andes near Salta, north Argentina, and flowing southeast through a livestock-raising region to the Paraná River at Santa Fe. Salt and sulfur are mined along its upper course.

Sa·la·ir Ridge (sä'lə-ĭr'), range, c.200 mi (320 km) long, E Siberian USSR. Extending along the northern border of the Altai Kray, it rises to more than 2,000 ft (610 m) and forms the western edge of the Kuznetsk Basin; to the south it merges with the Kuznetsk Ala-Tau range. The Salair Ridge has iron, lead, silver, and zinc deposits.

Sal·a·man·ca (săl'ə-măng'kə, sä'lä-mäng'kä), city (1970 pop. 103,740), Guanajuato state, W central Mexico. Chiefly an oil center, it also serves as the commercial and distribution point for the surrounding agricultural region.

Salamanca, city (1975 est. pop. 131,374), capital of Salamanca prov., W central Spain, in León, on the Tormes River, c.2,600 ft (795 m) above sea level. There are food-processing and other industries. An ancient city, it was taken by Hannibal in 220 B.C. The Moors were driven out in 1085. Salamanca became world-famous after the foundation (c.1230) of its university by Alfonso IX. In the Peninsular War the city was in part demolished (1811) by the French. It was (1937–38) the capital of the Insurgents in the Spanish Civil War. There is a Roman bridge in the city, and the Plaza Mayor is among the finest colonnaded squares in Spain. Adjoining the old Gothic cathedral (12th cent.) is the imposing new cathedral (1513–1733), in which the Gothic, plateresque, and baroque styles are combined.

Sal·a·man·ca (săl'ə-măng'kə), city (1970 pop. 7,877), Cattaraugus co., W N.Y., on the Allegheny River S of Buffalo; settled in the 1860s, inc. 1913. Within the Allegany Indian Reservation and near Allegany State Park, it is a resort and a farm trade center. Manufactures include furniture and yarn.

Sal·a·mis (săl'ə-mĭs), ancient city on Cyprus, once the principal city. St. Paul visited it on his first missionary journey. Excavations revealed the ruins of a Greek theater; there are also many Roman ruins.

Salamis, island, E Greece, in the Saronic Gulf, W of Athens. It early belonged to Aegina but was later under Athenian control, except for a brief period after it was occupied (c.600 B.C.) by Megara. In the Persian Wars the allied Greek fleet, led by Themistocles, decisively defeated (480 B.C.) the Persians off Salamis.

Sale (sāl), municipal borough (1973 est. pop. 59,060), Greater Manchester, W England. It is a residential suburb of Manchester. Biscuits are made.

Sa·lé (sä-lā'), city (1970 pop. 155,557), NW Morocco, near Rabat. It has industries producing flour and fine carpets.

Sa·le·khard (sə-lĭ-ĸʜärt'), city (1970 pop. 22,000), capital of Yamalo-Nenets National Okrug, NW Siberian USSR, on the lower Ob River. It is a river port and has fish canneries, lumber mills, and shipyards. The city was founded as Obdorsk in 1595.

Sa·lem (sä'ləm), city (1971 pop. 308,303), Tamil Nadu state, SE India. Iron and manganese mining, mineral processing, and textile manufacturing are the major economic activities.

Salem, county (1970 pop. 60,346), 365 sq mi (945.4 sq km), SW N.J., bounded on the W by the Delaware River and drained by the Maurice and Salem rivers; formed 1694; co. seat Salem. Its agriculture includes dairy products, truck crops, poultry, and fruit. It has large marl deposits and manufacturing (clothing, chemicals, and glass).

Salem. 1. City (1970 pop. 1,277), seat of Fulton co., N Ark., NNW of Batesville near the Mo. border. It is a shipping point for a livestock, poultry, cotton, and feed area. **2.** City (1970 pop. 6,187), seat of Marion co., S central Ill., NE of Centralia, in an oil and farm region; inc. 1837. The birthplace of William Jennings Bryan is now a museum. **3.** City (1970 pop. 5,041), seat of Washington co., SE Ind., NW of Louisville, Ky.; settled c.1800, inc. as a town 1815, as a city 1933. The manufacture of wood products is its chief industry. **4.** City (1978 est. pop. 38,000), seat of Essex co., NE Mass., on an inlet of Massachusetts Bay; inc. 1629. The city has electrical, leather, and precision machine industries. Its many historical landmarks are tourist attractions. In 1626 Roger Conant led a group from Cape Ann to this site, called Naumkeag by the Indians. Salem's early history was darkened by the witchcraft trials of 1692. Its port was historically important as a center for the China trade and a privateering base in the American Revolution and in the War of 1812. Shipping declined after the War of 1812, and the city turned to manufacturing. Nathaniel Hawthorne's birthplace dates from the

17th cent., and the House of Seven Gables (1668) is preserved. The Peabody Museum, founded (1868) by George Peabody, contains exhibits of a museum organized in 1799 by the Salem East India Marine Society. **5.** City (1970 pop. 4,363), seat of Dent co., SE central Mo., SSE of Jefferson City, in a mining, timber, and farm region; settled c.1835, founded 1851, inc. 1881. **6.** Town (1970 pop. 20,142), Rockingham co., SE N.H.; settled 1652, inc. 1750. It is a marketing and distributing center, with steel-fabricating, printing, electronic, shoe, and wood-product industries. Of interest is Mystery Hill, site of large manmade structures believed to date from 2000 B.C. **7.** City (1970 pop. 7,648), seat of Salem co., SW N.J., on Salem Creek near the Delaware River SSE of Wilmington, Del. Salem was settled by Quakers in 1675 and was the first permanent English settlement in the Delaware valley. Incorporated in 1858, it grew as a port and farm trade center. **8.** City (1978 est. pop. 16,100), Columbiana co., NE Ohio, in a rich coal region; inc. 1806. Tools and dies, industrial machinery, pumps, and water systems are among its many and diverse manufactures. Settled (1803) by Quakers, Salem was an early abolitionist center and an important station on the Underground Railroad. **9.** City (1978 est. pop. 83,000), state capital and seat of Marion co., NW Oregon, on the Willamette River; inc. 1857. In a dairying, stock-raising, and farming area, it has numerous food-processing plants and a paper mill. Founded 1840–41 by Methodist missionaries, it became the capital of Oregon Territory in 1851 and remained the capital when Oregon became a state in 1859. Salem is the seat of Willamette Univ. **10.** City (1970 pop. 1,391), seat of McCook co., SE S.Dak., WNW of Sioux Falls; settled 1880, inc. 1885. A trade and distribution center in a grain, livestock, and dairy region, it has a poultry-packing plant. **11.** Independent city (1978 est. pop. 22,900), seat of Roanoke co., SW Va., on the Roanoke River, between the Blue Ridge and the Allegheny Mts.; first inc. 1806, inc. as a city 1967. Electrical equipment, rubber tires, fork lifts, steel items, locks, and tools and dies are made.

Salem Maritime National Historic Site: *see* National Parks and Monuments Table.

Sa·ler·no (sə-lûr′nō, sä-lâr′-), city (1976 est. pop. 161,645), capital of Salerno prov., Campania, S Italy, on the Gulf of Salerno, an inlet of the Tyrrhenian Sea. It is an agricultural, commercial, and industrial center. Manufactures include machinery, textiles, paper, and processed food. Originally a Greek settlement and later a Roman colony (founded 197 B.C.), Salerno became (6th cent.) a part of the duchy of Benevento and in the 9th cent. the seat of an independent principality, which fell to Robert Guiscard in 1076. In Sept., 1943, there was fierce fighting on the beaches near Salerno between the Allied landing forces and the Germans, who were pressed to retreat toward Naples. The famous medical school of Salerno (founded in the 9th cent., closed in the early 19th cent.) is believed to have been the first of its kind and reached its height in the 12th cent.

Sal·ford (sôl′fərd, săl′-), county borough (1976 est. pop. 261,000), Greater Manchester, NW England, on the Irwell River. It is a textile center and has an unusual number of parks and recreation grounds. Made a free town in 1230, Salford included Manchester in the Middle Ages. The Univ. of Salford specializes in science and technology.

Sal·gó·tar·ján (shôl′gō-tôr′yän), city (1977 est. pop. 44,000), N Hungary, near the Czechoslovakian border. It has ironworks and steelworks and manufactures agricultural machinery and glass. Salgótarján was a small settlement that grew rapidly after the discovery of coal in the early 19th cent. Nearby, on a basaltic hill, are the ruins of a medieval fortress.

Sa·li·da (sə-lī′də), city (1970 pop. 4,355), seat of Chaffee co., central Colo., on the Arkansas River between the Sangre de Cristo and Sawatch mts. WNW of Pueblo, at an altitude of 7,050 ft (2,150.3 m); founded c.1880 with the coming of the railroad, inc. 1891. It is a resort, trade center, and railroad center in a grain, mineral, and livestock area.

Sa·li·na (sə-lī′nə), city (1978 est. pop. 42,100), seat of Saline co., central Kansas, on the Smoky Hill River; founded 1858 by antislavery people, inc. 1870. It is the marketing and shipping center for an area that produces grain, livestock, oil, and natural gas, and a leading hub of the great hard-winter-wheat belt. The city has grain elevators, flour mills, and factories that make farm implements, lamps, and aircraft. Nearby is a noted Indian burial pit.

Sa·li·nas (sə-lē′nəs), city (1978 est. pop. 76,000), seat of Monterey co., W Calif.; inc. 1874. It is the shipping and processing center of a fertile valley. Lettuce, fruits, sugar beets, and dairy goods are produced, and spices, candy, and jams and jellies are made. The Alisal area (formerly called East Salinas), which was annexed by Salinas in 1964, was settled (1933) principally by migratory farm workers. John Steinbeck was born in Salinas, and his home is open to the public.

Salinas, river, c.150 mi (240 km) long, rising in the Santa Lucia Mts., S Calif., and flowing (partly underground) NW to Monterey Bay. The irrigated valley is the chief lettuce-producing region in the United States.

Sa·li·nas Gran·des (sä-lē′näs grän′dās), salt desert, c.3,200 sq mi (8,290 sq km), in Córdoba and Santiago del Estero provs., N Argentina. Sodium and potassium are mined here.

Sa·line (sə-lēn′). **1.** County (1970 pop. 36,107), 726 sq mi (1,880.3 sq km), central Ark., drained by the Saline River and its tributaries; formed 1835; co. seat Benton. Its agriculture includes cotton, truck crops, corn, and livestock. It has lumbering, aluminum-ore mining and processing, gravel and clay pits, and some manufacturing (furniture, chemicals, and clay products). **2.** County (1970 pop. 25,721), 384 sq mi (994.6 sq km), SE Ill., drained by the Saline River and partly in the Ozarks; formed 1847; co. seat Harrisburg. It has agriculture (corn, wheat, fruit, livestock, and dairy products), extensive bituminous-coal mining, and manufacturing (wood products, flour, and bricks). **3.** County (1970 pop. 46,592), 720 sq mi (1,864.8 sq km), central Kansas, in a rolling plains region intersected by the Smoky Hill, Saline, and Solomon rivers; formed 1860; co. seat Salina. In a winter-wheat and livestock area, it has some mining and diversified industry (food processing, grain mills, and the manufacture of textiles, furniture, concrete, metal products, and farm machinery). **4.** County (1970 pop. 24,837), 756 sq mi (1,958 sq km), central Mo., bounded on the N and the E by the Missouri River and drained by the Blackwater River; formed 1829; co. seat Marshall. It is in an agricultural area (corn, wheat, oats, cattle, hogs, and dairy products), with some manufacturing. **5.** County (1970 pop. 12,809), 575 sq mi (1,489.3 sq km), SE Nebr., drained by the Big Blue River; formed 1867; co. seat Wilber. Its agriculture includes livestock, grain, dairy products, and poultry.

Salis·bur·y (sôlz′bĕr′ē, -bə-rē) or **New Sar·um** (săr′əm, sâr′-), municipal borough (1973 est. pop. 35,460), administrative center of Wiltshire, S England. A market town with varied industries, Salisbury was founded in 1220, when the bishopric was moved here from Old Sarum. The great cathedral, a splendid example of Early English architecture with the highest spire in England (404 ft/123.2 m), was built mainly between 1220 and 1260. The 13th cent. palace of the bishops and numerous medieval churches are of interest.

Salisbury, city (1975 est. pop., with suburbs, 569,000), alt. 4,865 ft (1,483.8 m), capital of Rhodesia, NE Rhodesia. Salisbury is the trade center for an agricultural region whose main products are tobacco, maize, cotton, and citrus fruits. Manufactures include textiles, clothing, processed food and tobacco, beverages, steel, chemicals, furniture, fertilizers, and construction materials. Gold is mined in the area. Salisbury was founded (1890) by a mercenary force organized by Cecil J. Rhodes. It became a municipality in 1897 and a city in 1935. Salisbury was the capital of the Federation of Rhodesia and Nyasaland (1953–63). It is the site of the Univ. of Rhodesia, of Rhodes National Gallery, which has collections of African soapstone carvings, and of the National Museum, known for its archaeological holdings.

Salisbury. 1. City (1978 est. pop. 16,000), seat of Wicomico co., Md., on the Eastern Shore, at the head of the Wicomico River; settled 1732, inc. 1872. Poultry raising and processing is the city's major industry. The city is also a trade and service center for the Eastern Shore and has varied light manufacturing industries. **2.** City (1978 est. pop. 25,100), seat of Rowan co., W central N.C., in the Piedmont industrial region; inc. 1770. The production of textiles and garments is the major industry. Structural steel, brick, furniture, aluminum foil, and mobile homes are among the other manufactures. Granite quarries are nearby. The city has a great number of 18th and 19th cent. buildings, churches, and homes. The beautiful old county courthouse (1857) has been designated a historic site. The Old Stone House (1766) served as a fort in the French and Indian Wars. A nearby mill (built c.1816) is still in operation. The national cemetery in Salisbury was the site of one of the largest Confederate prison camps during the Civil War.

Salisbury Plain, undulating, mostly barren chalk plateau, c.300 sq mi (775 sq km), Wiltshire, S England. It is noted chiefly as the site of ancient monuments, of which Stonehenge is the most famous.

Sal·li·saw (săl'ə-sô'), city (1970 pop. 4,888), seat of Sequoyah co., E Okla., in a hilly region SE of Muskogee; founded c.1886. It is in a fertile farm area with limestone and coal deposits.

Salm·on (săm'ən), city (1970 pop. 2,910), seat of Lemhi co., E Idaho, at the confluence of the Salmon and Lemhi rivers near the Mont. border; inc. 1892. It was founded (1867) as a supply center after gold was discovered in the area. Later it became a distributing point for a mining, lumbering, and agricultural area. The Continental Divide is just to the east, and Salmon National Forest is nearby.

Salmon, river, c.425 mi (685 km) long, rising in many branches in the Sawtooth and the Salmon River mts., central Idaho. It flows northeast and is joined at Salmon by the Lemhi River, after which it flows west and is joined by the Middle Fork and the South Fork, then goes north to join the Snake River. The river's canyon, c.1 mi (1.6 km) deep and 10 mi (16.1 km) wide in some places, threads through a wilderness preserve. Though the swift waters and rapids are navigable downstream, it is impossible to return by the water route, thus giving the Salmon the name River of No Return. Salmon travel up the river to spawn.

Salmon River Mountains, central Idaho, between the Sawtooth Mts. and the Bitteroot Range, and bracketed by the Salmon River. The highest point is Twin Peaks (10,328 ft/3,150 m). The range includes sections of Challis and Salmon national forests.

Sa·lo·na (sə-lō'nə), ancient city of Dalmatia, NE of modern Split, Yugoslavia. A port on the Adriatic, it was used as a base for Roman conquest and was made a Roman colony and the capital of Illyricum in the 1st cent. B.C. The busy commercial city gained prestige when Diocletian, after retiring in A.D. 305, built a magnificent palace nearby. In the 7th cent. the people fled before invaders to Diocletian's palace, which they transformed into the city of Spalato (now Split, Yugoslavia). Salona was destroyed.

Sa·lon-de-Pro·vence (sä-lôN'də-prô-väNs'), town (1975 pop. 34,576), Bouches-du-Rhône dept., SE France. Its major manufactures are olive oil, electrical equipment, coffee, and soap. In the town are churches dating from the 12th to the 14th cent.

Sal·o·ni·ca (săl'ə-nē'kə, sə-lŏn'ĭ-kə): *see* Thessaloníki, Greece.

Sal·op (săl'əp), nonmetropolitan county (1976 est. pop. 359,000), W England; administrative center Shrewsbury. It was formed in 1974. The name Salop was also used to designate the former county of Shropshire.

Salt (sôlt), river, c.200 mi (320 km) long, rising near Queen City, NE Mo., and flowing generally S and E to the Mississippi River near the La. border.

Sal·ta (säl'tə), city (1970 pop. 176,216), capital of Salta prov., NW Argentina, in the Lerma valley. It is the center of a region rich in agricultural produce, minerals (chiefly oil), and forest products. Sugar, tobacco, wine grapes, and livestock are shipped. Salta's cathedral is well known, and ruins of many 17th cent. buildings dot the surrounding countryside. Founded in 1582, Salta became an important commercial and cultural center during the 17th cent.

Sal·ti·llo (säl-tē'yō), city (1974 est. pop. 200,700), capital of Coahuila state, N Mexico. It is located in an alluvial valley almost surrounded by mountains. The growing of cereal and cattle raising are the chief occupations, but textile manufacturing and food processing are also important. Founded in 1575, Saltillo was taken by Zachary Taylor's forces in the Mexican War and was occupied by French troops several times during the French intervention in Mexico.

Salt Lake, county (1970 pop. 458,607), 764 sq mi (1,978.8 sq km), N central Utah, in the Tableland area, drained by the Jordan River, bounded on the NW by Great Salt Lake, with the Wasatch Range in the E; formed 1852 as Great Salt Lake co., renamed 1868; co. seat Salt Lake City. Its agriculture includes sugar beets, alfalfa, grains, fruit, truck crops, and livestock. It has deposits of copper, coal, oil, gas, and barite. Its diversified industries include food processing, lumbering, and the manufacture of clothing and textiles, wood products, furniture, paper, chemicals, concrete, and machinery.

Salt Lake City, city (1978 est. pop. 164,500), alt. c.4,330 ft (1,320 m), state capital and seat of Salt Lake co., N central Utah, on the Jordan River near Great Salt Lake, at the foot of the Wasatch Range; inc. 1851. It is world headquarters of the Church of Jesus Christ of Latter Day Saints and a processing center for the products of an irrigated farm region that is also rich in minerals. Major industries include food processing, silver, lead, and copper smelting, the development and production of missiles and electronic equipment, oil refining, and printing and publishing. Founded in 1847 by Brigham Young as the capital of the Mormon community, the city achieved greatness as its economic hub. After 1849 it was a supply point for overland travel to Calif. Of interest are the Temple (built 1853-93), the state capitol (1914), Brigham Young's home, the Brigham Young Monument (1897), and a planetarium.

Sal·to (säl'tō), city (1975 pop. 71,881), capital of Salto dept., NW Uruguay, on the Uruguay River. Salto is a thriving cultural and commercial center for a farming and livestock region. There are boatbuilding and meat-packing industries. In the surrounding region are extensive orange and tangerine orchards as well as vineyards.

Sal·ton Sea (sôl'tən), saline lake, 370 sq mi (958.3 sq km), northern part of the Imperial Valley, SE Calif. It is 232 ft (70.8 m) below sea level. Salton Sea was formed as the Colorado River delta grew across the Gulf of California, severing its northern part. The area was a salt-covered depression known as Salton Sink until 1905, when a flood on the Colorado broke through an irrigation gap in its levee; the river flowed into the sink for two years before being checked. The Salton Sea's water level has gradually risen due to runoff from surrounding mountains and irrigation systems.

Salt River valley, irrigated region around the lower course of the Salt River, which rises in mountain streams near the Mogollon Rim of the Mogollon Plateau and flows SW to join the Gila River in S central Ariz. Indians used the Salt River for irrigation many centuries ago. In the 19th cent. American settlers began irrigated farming in the valley, and the Mormons used some of the old Indian canals at Mesa, Ariz. The Salt River project, the first large irrigation scheme undertaken under the Federal Reclamation Act of 1902, began in 1903, when construction started on Roosevelt Dam. Other dams supply water and power. The region is a rich producer of alfalfa, citrus fruits, lettuce, melons, and cotton.

Sa·lu·da (sə-lo͞o'də), county (1970 pop. 14,528), 458 sq mi (1,186.2 sq km), W central S.C., bounded in the N by the Saluda River and Lake Murray; formed 1896; co. seat Saluda. It is in a sparsely settled agricultural area that produces cotton and other farm products. Its industry includes food processing, textile mills, and lumbering.

Saluda. 1. Town (1970 pop. 2,442), seat of Saluda co., W S.C., W of Columbia; inc. 1897. Textile and lumber products are made, and poultry is processed. **2.** Village (1970 pop. 300), seat of Middlesex co., E Va., near the Rappahannock River ENE of West Point.

Saluda, river, c.200 mi (320 km) long, rising in the Blue Ridge, W S.C., and flowing SE across the Piedmont to the Broad River near Columbia.

Sa·luz·zo (sə-lo͞ot'sō), town (1971 pop. 17,828), Piedmont, NW Italy. It is an agricultural and industrial center. Manufactures include textiles, machinery, and processed food. It was the capital of the marquisate of Saluzzo from the 12th cent. to 1548, when it passed to France. The town came under the house of Savoy in 1601.

Sal·va·dor (săl'və-dôr') or **Ba·hi·a** (bä-ē'ə), city (1975 est. pop. 1,237,373), capital of Bahia state, E Brazil, a port on the Atlantic Ocean. It is the commercial center of a fertile crescent (the Recôncavo) and a shipping point for the cacao district to the south. Other exports include tobacco, sugar, hardwoods, industrial diamonds, and oil. Salvador is also a fashionable tourist center. Food processing is a leading industry. Founded in 1549, Salvador flourished with the development of sugar plantations and became the leading center of colonial Brazil. Because of an influx of black African slaves, the area is noted for its African heritage in music, dance, folk customs, and cuisine. The city was the capital of the Portuguese possessions in America until 1763. Many buildings and fortifications remain from the colonial period. In the early 19th cent. it was a center of the Brazilian independence movement and in 1912 was bombarded and heavily damaged by federal forces. The city, built on a peninsula, is divided into two sections connected by graded roads, elevators, and cable cars. It has many notable churches, including a 16th cent. cathedral.

Sal·ween (săl'wēn'), river of SE Asia, c.1,750 mi (2,815 km) long, rising in E Tibet, China, and flowing SE through Yünnan prov. in deep, narrow gorges parallel to the Mekong, Yangtze, and Irrawaddy rivers into Burma, where it cuts through the Shan Plateau and Karenni Hills and then empties into the Gulf of Martaban, E Burma, near Moulmein. Because of rapids, it is navigable only for c.75 mi (120 km) upstream.

Sal·yers·ville (săl′yərz-vĭl′), town (1970 pop. 1,196), seat of Magoffin co., E Ky., on the Licking River SSW of Ashland, in a mountain valley.

Salz·burg (sôlz′bûrg′, sälz′-, zälts′bŏŏrk′), province (1971 pop. 399,000), c.2,760 sq mi (7,150 sq km), W central Austria. The province borders on West Germany in the north and northwest. It is a predominately mountainous region, with parts of the Hohe Tauern Mts. and Salzburg Alps, and it is drained by the Salzach River. There are famous salt deposits that have long been worked, as well as gold, copper, and iron mines. Salzburg prov. is a scenic area, noted for its Alpine resorts and spas. The industry is varied; manufactures include machinery, aluminum, textiles, and forest products. Cattle and horses are raised. Kaprun dam, on the Salzach high in the mountains, includes one of the largest hydroelectric facilities in Europe.

The capital of Salzburg prov. and its chief city is Salzburg (1971 pop. 128,800), an industrial, commercial, and tourist center. Picturesquely situated on both banks of the Salzach River, the city is bounded by two steep hills, the Capuzinerberg (left bank) and the Mönchsberg, on the southern tip of which is the 11th cent. fortress of Hohensalzburg (right bank). The city's most noteworthy buildings are a late 7th cent. Benedictine abbey; the Franciscan church, consecrated in 1223; the early 17th cent. cathedral, modeled after St. Peter's in Rome; the Residenz (16th–18th cent.), formerly the archiepiscopal palace; Mirabell castle (early 18th cent.); and the Festspielhaus (1960), the city's chief concert hall. W. A. Mozart, Salzburg's most distinguished son, is honored by an annual summer music festival (est. 1925), which constitutes an important source of tourist revenue.

History. Originally inhabited by Celts, the territory was conquered by the Romans and became part of the province of Noricum. After the fall of the Roman Empire its history followed that of the city of Salzburg. An ancient Celtic settlement and later a Roman trading center named Juvavum, Salzburg developed in the early 8th cent. around the late 7th cent. monastery of St. Peter. By c.798 it was the seat of an archbishopric, and for almost 1,000 years it was the residence of the autocratic archbishops of Salzburg, the leading ecclesiastics of the German-speaking world. Secularized in 1802, Salzburg was transferred to Bavaria by the Peace of Schönbrunn (1809). The Congress of Vienna (1814–15) returned it to Austria.

Salz·git·ter (zälts′gĭt′ər), city (1974 est. pop. 120,090), Lower Saxony, E West Germany. Situated in one of the richest iron-ore producing regions in West Germany, it has blast furnaces, steel plants, and industries producing chemicals, textiles, machinery, and petroleum. The city was created in 1942 by the merger of 29 towns.

Salz·kam·mer·gut (zälts′käm′ər-gōōt′), resort area in Upper Austria, Styria, and Salzburg provs., W Austria. Known since antiquity for its salt mines, the region was banned to visitors until the early 19th cent. because the government wanted to prevent salt from being smuggled out. Salzkammergut is a summer and winter tourist center.

Sa·mar (sä′mär′), island (1970 pop. 1,024,336), 5,050 sq mi (13,079 sq km), Philippines, one of the Visayan Islands, NE of Leyte (from which it is separated by the narrow San Juanico Strait). It has commercial coconut plantations and is a leading banana producer. There are important lumbering, fishing, and mining industries; copper and iron ore are extracted.

Sa·ma·ra (sə-mä′rə): *see* Kuybyshev, USSR.

Samara, river, c.360 mi (580 km) long, rising in the foothills of the S Urals, E European USSR. It flows generally northwest and joins the Volga River at Kuybyshev, at the eastern extremity of the Samara Bend.

Sa·ma·rai (sä′mə-rī′), small island, 59 acres (23.9 hectares), at the SE tip of Papua New Guinea. It is a commercial and shipping center and a port of entry. An important European settlement before World War II, it was totally destroyed by Japanese bombing in 1942 and never regained its former prominence.

Sa·mar·i·a (sə-mâr′ē-ə), ancient city, central Palestine, on a hill NW of Nablus (Shechem). The site is now occupied by a village, Sabastiyah (Jordan). Samaria was built by King Omri as the capital of the northern kingdom of Israel in the early 9th cent. B.C. It fell in 721 B.C. to Sargon. The native population was deported, and the city was made the capital of an Assyrian province. It was destroyed in 120 B.C. by John Hyrcanus and was rebuilt by Herod the Great. According to tradition St. John the Baptist is buried here. Remains of a church of the Crusaders are in the city. Excavations (1908–10, 1931–35) uncovered fortifications and the palace of Omri.

Sam·ar·kand (săm′ər-kănd′), city (1970 pop. 267,000), capital of Samarkand oblast, Central Asian USSR, in Uzbekistan, on the Trans-Caspian RR. It is one of the oldest existing cities in the world and the oldest of central Asia. Modern Samarkand is a major cotton and silk center. Wine and tea are produced, and there are industries producing metal products, motor vehicle parts, leather goods, clothing, and footwear. The irrigated surrounding region has orchards, gardens, and wheat and cotton fields.

The old quarter of Samarkand, with its maze of narrow, winding streets, occupies the eastern part of the city and centers on the Registan, a great square. It contains Tamerlane's mausoleum; the Bibi Khan Mosque, with its turquoise cupola, erected by Tamerlane to the memory of his favorite wife; and the ruins of the observatory built by Ulugh-Beg, a grandson of Tamerlane.

History. Built on the site of Afrosiab, which dated from the 3rd or 4th millennium B.C., Samarkand was known to the ancient Greeks as Marakanda; ruins of the old settlement remain north of the present city. The chief city of Sogdiana, on the ancient trade route between the Middle East and China, Samarkand was conquered (329 B.C.) by Alexander the Great and became a meeting point of Western and Chinese culture. The Arabs took Samarkand in the 8th cent. A.D., and for five centuries it flourished as a trade center on the route between Baghdad and China. In 1220 Genghis Khan captured and devastated the city, but it revived in the 14th cent., when Tamerlane made it the capital of his empire. Under his rule the city reached its greatest splendor; sumptuous palaces and mosques were erected, and gardens laid out. The empire broke up in the late 15th cent. and was ruled by the Uzbeks for the following four centuries. Samarkand eventually became part of the emirate of Bukhara and fell to Russian troops in 1868. In 1925 it became the capital of the Uzbek SSR, but in 1930 it was replaced by Tashkent.

Sa·mar·ra (sə-mä′rə), town, N central Iraq, on the Tigris River. It is on the site of an ancient settlement and has given its name to a type of Neolithic pottery of the 5th millennium B.C. The present town was founded (836) by the Abbaside caliphs. The 17th cent. mosque with its golden dome is sacred to Shiite Moslems. There are notable ruins of many palaces, mosques, and other buildings.

Sam·bre (säN′brə), river, 120 mi (193 km) long, rising in N France and flowing NE to the Meuse River at Namur, SE Belgium. Canalized along most of its length, the river traverses the important Franco-Belgian coal basin and industrial district.

Sam·chok (säm′chŭk′), city (1966 pop. 26,000), E South Korea, a port on the Sea of Japan. It is a large industrial center in the heart of a rich coal and iron-ore mining area. Samchok has metallurgical, chemical, cement, carbide, and fertilizer plants.

Sam·chon·po (säm′chŭn′pô′), city (1975 pop. 35,900), S South Korea, on the Korea Strait. It is a fishing port and processing center.

Sam·ni·um (săm′nē-əm), ancient country of central and S Italy, mostly in the S Apennines, E of Campania and Latium and NE of Apulia. The desire of the Samnites to expand Samnium at the expense of Campania led to the Samnite Wars (343–290 B.C.).

Sa·mo·a (sə-mō′ə), chain of volcanic islands in the South Pacific, comprising the independent nation of Western Samoa and, E of long. 171° W, the islands of American Samoa, under U.S. control. The Samoan islands extend c.350 mi (565 km), with a total land area of c.1,200 sq mi (3,110 sq km), and lie midway between Honolulu, Hawaii, and Sydney, Australia. The major islands are volcanic and mountainous and are surrounded by coral reefs. The natives are Polynesians who may have arrived in the islands as early as 1000 B.C. From Samoa they swept out across the Pacific (c.1200 A.D.), carrying Polynesian civilization to innumerable other islands. European expansion into the islands in the early 18th cent. led to disorder and violence, compounded by tribal warfare. The first European missionaries arrived in 1830. Between 1847 and 1861 the United States, Great Britain, and Germany sent representatives to Samoa, and in 1878 the United States and the Samoan kingdom signed a treaty giving the United States certain trade privileges and the right to establish a naval station at Pago Pago. A tripartite treaty in 1899 between Great Britain, the United States, and Germany recognized U.S. interests east of long. 171° W; Germany was granted the western islands, and Great Britain withdrew from the area in consideration of rights in Tonga and the Solomon Islands. New Zealand seized the German islands in 1914 during World War I. In 1946 they became a UN trust territory held by New Zealand. In 1962 the independent nation of Western Samoa was created from the New Zealand territory; the eastern islands remained under U.S. control.

Sá·mos (sä'môs', sä'môs'), island (1971 pop. 32,664), c.181 sq mi (470 sq km), SE Greece, in the Aegean Sea; one of the Sporades, near Turkey. Largely mountainous, the island has much fertile soil; grapes, tobacco, citrus fruits, and currants are grown. Sámos was inhabited in the Bronze Age, and about the 11th cent. B.C. it was colonized by Ionian Greeks. By the 6th cent. B.C. the island was a commercial and maritime power and a cultural center. Sámos was conquered by the Persians toward the end of the 6th cent. B.C. but regained its independence after the Battle of Mycale (479 B.C.). The island declined after 322 B.C., when it fell out of Athenian hands. In the Middle Ages Sámos was held by a Genoese trading company from 1304 to 1329 and from 1346 to 1475, when it was captured by the Ottoman Empire. It was a semi-independent principality from 1832 until it passed to Greece in 1913.

Sa·mos·a·ta (sə-mŏs'ə-tə), ancient city of N Syria, on the Euphrates River. It was founded c.150 B.C. as the capital of the Commagene kingdom. Taken by the Romans in A.D. 72, it was of some importance in later Roman times. The Arabs took it in the 7th cent.

Sam·o·thrace (săm'ə-thrās') or **Sa·mo·thrá·ki** (sä'mə-thrä'kē), island (1971 pop. 3,012), c.71 sq mi (184 sq km), NE Greece, in the Aegean Sea. The island is largely mountainous. In ancient times Samothrace was an important center of worship. There are ruins of a religious sanctuary, some of which date to the 6th cent. B.C. The famous statue of the winged *Nike* (or *Victory*) *of Samothrace*, built c.200 B.C. to adorn a ship and later transferred to the island, was discovered on Samothrace in 1863 and is now in the Louvre in Paris. The island was ceded to Greece by the Ottoman Empire in 1913.

Samp·son (sămp'sən), county (1970 pop. 44,964), 945 sq mi (2,447.6 sq km), S central N.C., bounded on the W by the South River and drained by Black River; formed 1784; co. seat Clinton. It has coastal plain agriculture (tobacco, corn, cotton, peppers, soybeans, sweet potatoes, poultry, and hogs), some textile manufacturing, and a food-processing industry.

Sam·sun (säm-so͞on'), city (1975 pop. 169,060), capital of Samsun prov., N Turkey, a port on the Black Sea. It is a tobacco-processing center and an agricultural market. The ancient Amisus, it was founded (6th cent. B.C.) by Greek colonists. In the Middle Ages it was held by the Byzantines, the Seljuk Turks, and the Genoese before falling (14th cent.) to the Ottoman Turks.

San (sän), town (1972 pop. 18,000), central Mali, a port on the Bani River. It is the trade center for a region where peanuts, cotton, fruit, and vegetables are grown. The town manufactures bricks and has lime kilns and cotton gins.

San·a or **San'a** (both: sä-nä'), city (1975 est. pop. 135,000), capital of Yemen. The city lies inland on a high plain (alt. 7,250 ft/2,210 m). Sana is an Islamic cultural center and a commercial and marketing center noted for the grapes grown nearby. It has been settled from pre-Islamic times and has an ancient wall. It was under Ethiopian control in the 6th cent. In the 17th cent. and again from 1872 to 1918 it was occupied by Turkey.

San·an·daj (sän'ən-däj'), city (1971 est. pop. 58,000), capital of Kurdistan prov., W Iran. It is the trade center of a grain and sheep-raising region and is known for its rugs and fine woodwork.

San An·dre·as (săn' ăn-drā'əs), village (1970 pop. 1,564), seat of Calaveras co., central Calif., NE of Stockton; settled c.1848. Cement is manufactured.

San Andreas fault, great fracture of the earth's crust, the principal fault of an intricate network of faults extending more than 600 mi (965 km) from NW Calif. to the Gulf of California. It is located on the boundary between two sections of the earth's lithosphere—the North American plate and the Pacific plate—and separates southwest Calif. from the North American continent. The Pacific plate is moving northwest in relation to the North American plate, and it is believed that the total displacement along the fault since its formation more than 30 million years ago has been c.350 mi (565 km). Movement along the fault causes earthquakes; several thousand occur annually, although only a few are of moderate or great magnitude. The destructive San Francisco earthquake of 1906 was caused by a movement in which land surfaces on either side of the fault were displaced horizontally up to 21 ft (6.4 m).

San An·ge·lo (ăn'jə-lō), city (1978 est. pop. 69,600), seat of Tom Green co., W Texas, where two forks join to form the Concho River; laid out 1869, inc. 1903. It is an important wool and mohair market and a trade and shipping point for a wide area of sheep, goat, and cattle ranches, irrigated farms, and oil and natural-gas fields. Meat and dairy items, leather footwear, and oil field equipment are also produced. Founded beside a border military post, Fort Concho (1866; now restored as a museum), San Angelo was a rough frontier town in the 1870s; it grew after the coming of the railroad in 1888. Goodfellow Air Force Base adjoins the city.

San An·sel·mo (ăn-sĕl'mō), residential city (1970 pop. 13,031), Marin co., W Calif., near San Francisco; inc. 1907.

San An·to·ni·o (ăn-tō'nē-ō), city (1978 est. pop. 804,000), seat of Bexar co., S central Texas, at the source of the San Antonio River; inc. 1837. San Antonio is the industrial, trade, and financial center of a large agricultural area. Its manufactures include processed foods, aircraft, building materials, chemicals, wood products, clothing, and machinery. The city attracts thousands of tourists annually. Fort Sam Houston and Brooke Army Medical Center are in the city, and nearby are Lackland, Randolph Brooks, and Kelly air force bases.

The site had been visited by the Spanish long before the expedition under Martín de Alarcón founded a mission (San Antonio de Valero) and a presidio (San Antonio de Béjar, or Béxar) in 1718. San Antonio was the most important Texas settlement in Spanish and Mexican days. During the Texas Revolution it was captured by the Texans (Dec., 1835) and was the scene of the Mexican attack on the Alamo in Mar., 1836. After the Civil War and especially after the coming of the first railroad in 1877, San Antonio prospered as a cow town with a Spanish flavor, which it still retains. In addition to its many missions, points of interest include the Spanish governor's palace (c.1749), the Paseo del Río, a downtown river walk, and the Hertzberg Circus Collection. The Hemisfair Plaza contains the Institute of Texan Cultures and the 750-ft (228.8-m) Tower of the Americas. The Southwest Research Institute is notable for its research into the technical problems of the southwest region.

San Au·gus·tine (ô'gə-stēn'), county (1970 pop. 7,858), 473 sq mi (1,225.1 sq km), E Texas, bounded on the W by Attoyac Bayou, on the SW by the Angelina River; formed 1836; co. seat San Augustine. Agriculture (cotton, corn, sweet potatoes, truck crops, and fruit) and lumbering are important.

San Augustine, town (1970 pop. 2,539), seat of San Augustine co., E Texas, ESE of Nacogdoches. A Spanish mission, Nuestra Señora de los Dolores de los Ais (Our Lady of Sorrows of the Ais), was established in 1716, abandoned in 1719, and refounded in 1721. The town is today a center of a pine-timber and farm region.

San Be·ni·to (bə-nē'tō), county (1970 pop. 18,226), 1,396 sq mi (3,615.6 sq km), W Calif., with the San Benito River flowing through the S part of the Santa Clara Valley; formed 1874; co. seat Hollister. Its agriculture includes tomatoes, lettuce, bell peppers, asparagus, fruits, nuts, and cattle. It has quicksilver, granite, and sand and gravel deposits and a food-processing industry.

San Benito, city (1970 pop. 15,176), Cameron co., extreme S Texas; inc. 1911. San Benito is chiefly a processing center for citrus fruit and vegetables grown in the irrigated region of the lower Rio Grande valley. Truck trailers, electric equipment, and other products are manufactured. San Benito is also a retirement and winter tourist spot.

San Ber·nar·di·no (bûr'nər-dē'nō), county (1970 pop. 682,233), 20,117 sq mi (52,103 sq km), SE Calif., the largest county in the United States, bordered by Nev. and Ariz. and including parts of the Mojave and Colorado deserts and San Gabriel and San Bernardino mts.; formed 1853; co. seat San Bernardino. In the San Bernardino Valley fruit, wine grapes, dairy products, nuts, poultry, rabbits, and truck crops are produced. In the desert are cattle ranches and mines and quarries. Its industry includes food processing, winemaking, lumbering, and the manufacture of clothing, chemicals, plastics, concrete, metal products, machinery, and transportation equipment. San Bernardino National Forest, Joshua Tree National Monument, Death Valley National Monument, and Indian and military reservations occupy much of the county.

San Bernardino, city (1978 est. pop. 104,400), seat of San Bernardino co., S Calif., at the foot of the San Bernardino Mts.; inc. 1854. The city's many manufactures include steel, iron, and related products, propellants and rocket motors, electric golf carts, cement, and food items. The adjacent Norton Air Force Base is a major employer.

The area was explored (1772), named (1810), and first settled by Spanish explorers. A colony of Mormons arrived in the early 1850s and plotted the present city.

San Bernardino, Alpine pass, 6,770 ft (2,063 m) high, between Mesocco Valley and Rheinwald Valley, Grisons canton, SE Switzerland; used possibly since prehistoric times.

San Bernardino Mountains, part of the Coast Range, S Calif., extending c.60 mi (95 km) NW and SE through San Bernardino and Riverside cos. Notable peaks are San Bernardino Mt. (10,630 ft/3,242.2 m) and Mt. San Gorgonio (11,485 ft/3,503 m). This region embraces the mountain resort and recreational areas around Gregory, Arrowhead, and Big Bear lakes, in the San Bernardino National Forest.

San·born (săn′bərn), county (1970 pop. 3,697), 571 sq mi (1,478.9 sq km), SE central S.Dak., watered by the James River; formed 1883; co. seat Woonsocket. Its agriculture includes dairy products, livestock, poultry, and grain. It has some industry.

San Bru·no (brōō′nō), city (1978 est. pop. 39,800), San Mateo co., W Calif., a residential suburb on San Francisco Bay; inc. 1914. A Federal archives center and a U.S. Marine Corps reserve base are in the city.

San Car·los (kär′lōs), residential city (1978 est. pop. 26,900), San Mateo co., W Calif.; inc. 1925. The chief manufactures are electronic and communications equipment.

San Cle·men·te (klə-mĕn′tē), city (1978 est. pop. 25,300), Orange co., S Calif., on the Pacific coast; inc. 1928. San Clemente is a popular vacation spot, with several missions, a state park, and a national forest nearby.

San Cris·tó·bal (krĭ-stō′bəl), city (1970 pop. 25,829), S Dominican Republic, on a Caribbean coastal plain. The city was founded in the late 16th cent. The first Dominican constitution was signed here.

San Cristóbal, city (1971 pop. 151,717), capital of Táchira state, W Venezuela, in a mountainous region near the Colombian border. It is a commercial and industrial center. Textiles, leather products, ceramics, cement, and tobacco are produced, and coffee, sugar, and corn are exported. San Cristóbal was founded in 1561 and was severely damaged by an earthquake in 1875.

Sanc·ti-Spí·ri·tus (sängk′tē-spîr′ĭ-tōōs′), city (1970 pop. 57,703), Las Villas prov., central Cuba, on the Yayabo River. It is the commercial and processing center of an area that raises sugar cane, tobacco, and cattle. Founded in 1514, the city was moved to its present site in 1522. Sancti-Spíritus was the first important city to be captured by Fidel Castro's guerrilla forces (late 1958). Declared a historic monument on its 450th anniversary, the city retains some of its colonial atmosphere.

San·da·kan (sän′də-kän′), city (1970 pop. 42,413), Sabah, Malaysia, on N Borneo, on Sandakan Harbor, an inlet of the Sulu Sea. It is the trade hub for a rubber-producing and lumbering region. Sandakan was the capital of British North Borneo until 1947.

San·dal·wood Island (săn′dəl-wŏŏd′): *see* Vanua Levu.

San·de·fjord (sä′nə-fyôr′), town (1977 est. pop. 33,545), Vestfold co., SE Norway, near the mouth of the Oslofjord. An important shipping center since the 14th cent., it is also the base for a large whaling fleet operating in arctic waters. The town has shipyards, chemical works, and food-processing plants.

San·ders (săn′dərz), county (1970 pop. 7,093), 2,778 sq mi (7,195 sq km), NW Mont., in an agricultural region bordering on Idaho and drained by the Clark Fork and Flathead River; formed 1905; co. seat Thompson Falls. Its agriculture includes livestock, grain, and dairy products.

San·der·son (săn′dər-sən), uninc. town (1970 pop. 1,229), seat of Terrell co., W central Texas, N of the Rio Grande and W of the Pecos River. Wool, mohair, and cattle are shipped.

San·ders·ville (săn′dərz-vĭl′), city (1970 pop. 5,546), seat of Washington co., E central Ga., SW of Augusta; founded 1796, inc. 1812. It is a trade and processing center for a cotton, grain, livestock, and timber area.

Sand·hurst (sănd′hûrst′), village, Berkshire, S central England. It is the site of the Royal Military Academy.

San Di·e·go (dē-ā′gō), county (1970 pop. 1,357,854), 4,262 sq mi (11,038.6 sq km), SW Calif., bounded on the W by the Pacific Ocean, on the S by the Mexican border, with the Colorado Desert in the E; formed 1850; co. seat San Diego. Its agriculture includes cattle ranching, citrus fruit, truck crops, avocados, lima beans, hay, grain, grapes, olives, apples, walnuts, poultry, and dairy products. It has tuna fisheries, granite quarries, and sand and gravel pits. Clothing and textiles, furniture, chemicals, plastics, pottery, metal and wood products, and machinery are made. Its coastal and mountain resorts, hot springs, and scenery attract vacationers.

San Diego. 1. City (1978 est. pop. 820,000), seat of San Diego co., S Calif., on San Diego Bay; inc. 1850. It is an important port of entry and headquarters for the 11th U.S. naval district. San Diego has large aerospace, electronic, and shipbuilding industries and is a center for scientific research. It is also a distributing and processing point for a highly productive agricultural area. Sporting goods, clothing, rugs, furniture, and office equipment are among its manufactures. The city's delightful climate, ocean beaches, and many historic attractions, as well as its proximity to Mexico, draw visitors. Juan Rodríguez Cabrillo sailed into San Diego Bay in 1542 and claimed the land for Spain. Don Sebastian Viscaino encamped on Ballast Point in 1602 and established Mission San Diego de Alcalá, the first of a chain of missions. Inhabitants traded in cattle hides, and from 1850 to 1870 whaling was a major enterprise. Parts of Old Town are now a state historical park. San Diego is a cultural, educational, and medical center. Balboa Park contains a fine art gallery, several museums, and the San Diego Zoo. There is also a spectacular aquatic park. **2.** City (1970 pop. 4,490), seat of Duval co., S Texas, W of Alice; inc. 1935. It is the commercial center of an oil, ranch, and farm region.

San·do·mierz (sän-dô′myĕsh), town (1968 pop. 15,800), SE Poland, on the Vistula River. A river port and agricultural center, it also has industries producing glass, industrial porcelain, and wool products. Sandomierz was razed by the Tatars in 1241 and again in 1259 but was rebuilt (14th cent.) by Casimir III and became (16th cent.) a flourishing trade and cultural center and one of the most beautiful Polish towns. The town was heavily damaged by the Swedes in 1656 and lost its importance. It passed to Austria in 1772, to Russia in 1815, and reverted to Poland in 1919.

San·do·val (săn-dō′vəl), county (1970 pop. 17,492), 3,714 sq mi (9,619.3 sq km), NW central N.Mex., drained by the Rio Grande and Rio Puerco; formed 1903; co. seat Bernalillo. It has livestock grazing and farming (grain and chili peppers). Parts of Santa Fe National Forest and Bandelier National Monument are here.

Sand·point (sănd′point′), city (1970 pop. 4,144), seat of Bonner co., N Idaho, on Pend Oreille Lake; laid out 1898, inc. 1900. A lumber town, it is also a rail and highway center and a lake resort.

San·dring·ham (săn′drĭng-əm), village, Norfolk, E England, near the Wash River. Sandringham House, with its large estate, was purchased in 1861 by Edward VII, then prince of Wales. It has been used as a royal residence by Queen Alexandra, King George V, King George VI, and Queen Elizabeth II.

Sand Springs (sănd), city (1970 pop. 10,565), Tulsa co., NE Okla., an industrial suburb of Tulsa, on the Arkansas River; founded 1907. There are oil and natural-gas wells and textile and glass industries.

San·dus·ky (săn-dŭs′kē), county (1970 pop. 60,983), 410 sq mi (1,061.9 sq km), N Ohio, bounded in the NE by the Sandusky Bay of Lake Erie and intersected by the Sandusky and Portage rivers; formed 1820; co. seat Fremont. Its agriculture includes grain, fruit, sugar beets, and truck crops. It has limestone quarries and diversified manufacturing (textiles, paper, plastics, concrete, and metal products).

Sandusky. 1. City (1970 pop. 2,071), seat of Sanilac co., E Mich., E of Saginaw, in a farm and dairy region; inc. as a village 1885, as a city 1905. Rubber products are made. **2.** Industrial city (1978 est. pop. 31,700), seat of Erie co., N central Ohio, a port of entry on Sandusky Bay; inc. 1824. Sandusky has a fishing industry and many assorted manufactures and has been a tourist center since the 1880s.

Sand·vi·ken (sänd′vē′kən), city (1975 est. pop. 43,143), Gävleborg co., S central Sweden. A planned industrial city, it has ironworks and steelworks (founded 1862) that produce high-quality steel.

Sand·wich (sănd′wĭch, săn′wĭch), municipal borough (1973 est. pop. 4,420), Kent, SE England, on the Stour River. It is a resort and market center with some light industries. One of the Cinque Ports, Sandwich was the chief military port of the kingdom in the late 15th cent. Silting in the 16th cent. ruined the harbor. There are many medieval buildings.

Sandwich, resort town (1970 pop. 5,239), SE Mass., on W Cape Cod E of Bourne; settled c.1637, inc. 1639. Sandwich glass was made here from 1825 to 1888; the historical museum has a fine collection.

Sand·y Hook (săn'dē ho͝ok'), city (1970 pop. 192), seat of Elliott co., NE Ky., SW of Ashland, in a mountainous timber and agricultural area. Tobacco, corn, and hay are grown, and sawmilling is done. Daniel Boone National Forest is nearby.

Sandy Hook, low, sandy peninsula, NE N.J., separating Sandy Hook Bay from the Atlantic Ocean. At the northern end is Fort Hancock, which was built to protect New York harbor and was once used as a proving ground for heavy artillery. The Sandy Hook Lighthouse (85 ft/25.9 km high; built 1763) is the oldest in service in the United States. Henry Hudson's men explored this region in 1609. The British held the peninsula during the Revolution.

San Fer·nan·do (sän' fâr-nän'dō), city (1970 pop. 119,565), Buenos Aires prov., E Argentina. The city was established in 1806 to replace the port of Las Conchas, which had been destroyed by a storm.

San Fernando, city (1970 pop. 60,167), Cádiz prov., S Spain, in Andalusia. An Atlantic port, it has a naval academy and arsenal, naval workshops, and an observatory. Much salt is obtained from nearby marshes by evaporation.

San Fer·nan·do (săn' fər-năn'dō), city (1973 est. pop. 36,650), Trinidad and Tobago, on the Gulf of Paria. It is a commercial center.

San Fernando, city (1978 est. pop. 15,000), Los Angeles co., S Calif., in the San Fernando valley; inc. 1911. It has garment and electronic industries. The valley was first entered by white men in 1769, and from early days it was used for journeys to northern Calif. Gold was found here in 1842. San Fernando suffered extensive damage in the 1971 earthquake. San Fernando Mission (1797) is nearby.

San·ford (săn'fərd). **1.** City (1978 est. pop. 22,300), seat of Seminole co., central Fla., on Lake Monroe and the St. Johns River; inc. 1877. It is an agricultural center where citrus fruit and vegetables are processed. Electronic equipment, boats, clothing, and aluminum products are manufactured. The city was founded (1871) on the site of a trading post established (1837) near old Fort Mellon. **2.** Industrial town (1970 pop. 15,812), York co., SW Maine, on the Mousam River; inc. 1768. It was formerly a textile and garment manufacturing center, but there are now diversified industries. Nearby ocean beaches and recreational facilities attract summer vacationers. **3.** City (1970 pop. 11,716), seat of Lee co., central N.C.; inc. 1874. It is a processing center in a rich agricultural region. The city has a large tobacco market and is also one of the country's major brick-manufacturing cities.

San Fran·cis·co (săn' frən-sĭs'ko, frăn-), city (1970 pop. 715,674), coextensive with San Francisco co., W Calif., on the tip of a peninsula between the Pacific Ocean and San Francisco Bay, which are connected by the strait known as the Golden Gate; inc. 1850. It is the center of the web of industrial cities in the San Francisco Bay area, the marketplace for a large agricultural and mining region, the focus of many transportation routes, and the financial and insurance center of the West Coast. San Francisco and the bay area form the largest port on the West Coast and are a major center of trade with the Orient, Hawaii, and Alaska. Industries include food processing, shipbuilding, petroleum refining, and the manufacture of metal products and chemicals. The San Francisco area is a major cultural center of the West Coast and one of the greatest in the nation.

On his voyage around the world, Sir Francis Drake stopped (1579) in what is now the San Francisco Bay area. The city was founded in 1776, when a Spanish presidio and a mission were established at a location chosen by Juan Bautista de Anza. The little settlement, called Yerba Buena, although much visited by ships en route to the Far East, was still a village when the Mexican War broke out and a naval force under Commodore John D. Sloat took it (1846) in the name of the United States. It was then named San Francisco. When gold was discovered in California in 1848, San Francisco had a population of c.800; two years later it was incorporated with a population of c.25,000. The rush of gold seekers, adventurers, and settlers brought a period of lawlessness, when the Barbary Coast flourished. The city took on a cosmopolitan air, with newcomers arriving from all over the world. In this period the first Chinese settled in the city, and today San Francisco's Chinatown is the largest settlement of Chinese in the United States. In the years after the gold rush, San Francisco continued to grow as California became linked overland with the East, by the pony express in 1860 and by the transcontinental railroad in 1869. On the morning of April 18, 1906, the great San Andreas fault, which extends up and down the California coast, settled violently; and San Francisco was shaken by an earthquake that, together with the sweeping three-day fire that followed, all but destroyed the city. The San Francisco-Oakland Bay Bridge was opened in 1936, and the Golden Gate Bridge in 1937. By the time of the Golden Gate International Exposition (1939–40) the whole San Francisco Bay area was heavily industrialized, and it had become the leading commercial center of the West Coast. The United Nations Charter (1945) was drafted at San Francisco.

San Francisco is one of the most gracious and picturesque cities in the country. Its natural beauty and mild climate make it particularly attractive as a residential city. Of interest to the visitor are Mission Dolores (1782; at first called San Francisco de Asís); Golden Gate Park, where the California Academy of Sciences has two natural history museums, an aquarium, and a planetarium; the San Francisco Zoological Gardens (Fleishhacker Zoo); and the civic center, with a distinctive Renaissance-style city hall, a public library, and the municipally owned opera house. Art museums include the San Francisco Museum of Art, the M. H. De Young Memorial Museum, and the Palace of the Legion of Honor. Institutions of higher learning in the city include California State Univ., San Francisco, the Univ. of San Francisco, the Hastings College of Law, the Univ. of California, San Francisco, Lone Mountain College (formerly San Francisco College for Women), and several theological seminaries and junior colleges. The Presidio of San Francisco, the largest (1,542 acres/624 hectares) military encampment within the confines of an American city, is the headquarters of the Sixth U.S. Army. Continuously fortified since its establishment in 1776, the Presidio still contains remnants of the original Spanish adobe buildings, now incorporated into the officers' club.

San Francisco Bay, 50 mi (80 km) long and from 3 to 13 mi (4.8–21 km) wide, W Calif.; entered through the Golden Gate, a strait between two peninsulas. The bay is as deep as 100 ft (30 m) in spots, with a channel 50 ft (15 m) deep maintained through the sand bar off the Golden Gate. The Santa Clara Valley, part of a great depression paralleling the coast, is the landward extension of the bay. With San Pablo Bay and Suisun Bay, the natural harbor of San Francisco Bay is one of the best in the world. The English navigator Sir Francis Drake discovered the bay in 1579, but the Spanish explored it more fully in the late 18th cent.

San Francisco de Ma·co·rís (dā mä'kō-rēs'), city (1978 est. pop. 44,620), N Dominican Republic. It is the commercial and processing center for an agricultural region.

San Francisco Peaks, N Ariz., N of Flagstaff, consisting of Mt. Humphreys, 12,670 ft (3,864.4 m); Mt. Agassiz, 12,340 ft (3,763.7 m); and Mt. Fremont, 11,940 ft (3,641.7 m).

San Ga·bri·el (gā'brē-əl), residential city (1978 est. pop. 29,500), Los Angeles co., S Calif.; inc. 1913. Toys are manufactured. An annual three-day fiesta celebrates the founding (1771) of the San Gabriel mission, which was partly rebuilt after an earthquake in 1812.

San Gabriel Mountains, S Calif., E and NE of Los Angeles, running c.50 mi (80 km) W from Cajon Pass. San Antonio Peak (10,080 ft/3,074.4 m) is the highest of the range.

San·ga·mon (săng'gə-mən), county (1970 pop. 161,335), 880 sq mi (2,279.2 sq km), central Ill., drained by the Sangamon River and its tributaries; formed 1821; co. seat Springfield. It has agriculture (corn, wheat, oats, soybeans, livestock, poultry, and dairy products), bituminous-coal mines, limestone, clay, oil, sand, and gravel deposits, and manufacturing.

San·ger (săng'ər), city (1970 pop. 10,088), Fresno co., S central Calif., in the San Joaquin Valley; inc. 1911. It is a shipping and processing center for agricultural products.

San Ger·mán (sän' hĕr-män'), town (1970 pop. 11,613), SW Puerto Rico. It is the site of the Porta Coeli Convent (built 1511), one of the oldest churches in the Americas; it is now a museum.

Sang·i·he Islands (säng-gē'ə), volcanic group (314 sq mi/813.3 sq km), Indonesia, NE of Celebes. The islands are mountainous, forested, and fertile; tropical woods, rattan, copra, and nutmeg are produced. The area came under Dutch control in 1677.

San Gi·mi·gna·no (sän' jē-mē-nyä'nō), town (1976 est. pop. 2,900), Tuscany, central Italy. It is a tourist center that has preserved its medieval aspect. The city walls, the palaces, and the celebrated 14 towers (out of an original 72) still stand as they did in the 13th cent.

Also of note in the town are the cathedral (12th cent.; damaged in World War II), which is rich in works of art, and the Church of St. Augustine (13th cent.), with frescoes by Benozzo Gozzoli.

San·gli (säng'glē), town (1971 pop. 115,052), Maharashtra state, SE India, on the Krishna River. It is an agricultural market.

San·gre de Cris·to Mountains (săng'grē də krĭs'tō), part of the S Rocky Mts., extending c.220 mi (355 km) from S central Colo. into N central N.Mex. Most of the range is included in national forests.

San·i·lac (săn'ə-lăk'), county (1970 pop. 35,181), 961 sq mi (2,489 sq km), E Mich., bounded on the E by Lake Huron, drained by the Black and Cass rivers; formed 1849; co. seat Sandusky. Its crops include beans, sugar beets, fruit, and grain. Cattle are raised and dairy products are made. There is some manufacturing. The county contains fisheries and many resorts.

San Il·de·fon·so (săn ĭl'də-fŏn'sō) or **La Gran·ja** (lä gräng'hä), town (1970 pop. 4,164), Segovia prov., central Spain, in Old Castile. Near the town is the Spanish royal summer residence, built by Philip V (1721–23) in imitation of Versailles.

San I·si·dro (ĭ-sē'drō), city (1970 pop. 250,008), Buenos Aires prov., E Argentina. San Isidro grew around a chapel built in 1706. The city is known for its cathedral and historical museums.

San Ja·cin·to (jə-sĭn'tō), county (1970 pop. 6,702), 624 sq mi (1,616.2 sq km), E Texas, bounded on the N and E by the Trinity River and drained by headstreams of the San Jacinto River; formed 1869; co. seat Coldspring. Most of the county is in Sam Houston National Forest, and lumbering is the chief industry. It also has agriculture (corn, cotton, and truck crops), livestock, and oil wells.

San Jacinto, river, c.130 mi (210 km) long, rising in SE Texas as the West Fork and flowing S to Galveston Bay. Its chief tributary is Buffalo Bayou. In 1836 Texans under Sam Houston defeated a larger force of Mexicans in the final and decisive battle of the Texas Revolution on the San Jacinto near the mouth of Buffalo Bayou. The battlefield, a national historic landmark, is in San Jacinto State Park.

San Joa·quin (wŏ-kēn'), county (1970 pop. 291,073), 1,415 sq mi (3,664.9 sq km), central Calif., in the San Joaquin Valley, watered by the Mokelumne, Stanislaus, and Calaveras rivers; formed 1850; co. seat Stockton. It is in a richly irrigated agricultural area that produces grapes, asparagus, tomatoes, celery, nuts, rice, alfalfa, corn, sugar beets, dairy products, cattle, hogs, sheep, and poultry. It has sand and gravel quarrying and extensive industry.

San Joaquin, river, c.320 mi (515 km) long, rising in the Sierra Nevada, E Calif., and flowing W, then N through the Central Valley to form a large delta with the Sacramento River near Suisun Bay, an arm of San Francisco Bay. The San Joaquin is navigable c.40 mi (65 km) for oceangoing vessels to Stockton.

San Jo·se (hō-zā'), city (1978 est. pop. 590,100), seat of Santa Clara co., W Calif.; founded 1777, inc. 1850. It is in a rich fruit-growing area and has wineries and food-processing industries. Business machines, atomic-power equipment, and food machinery are among the manufactures. San Jose was the state capital from 1849 to 1851. To the north lies Mission San Jose de Guadalupe (1797) and to the west is Mission Santa Clara de Asís (1777).

San Jose Mission National Historic Site: *see* National Parks and Monuments Table.

San Jo·sé (săn' hō-zā') or **San Jo·sé de Ma·yo** (dā mä'yō), city (1975 pop. 28,427), capital of San José dept., S Uruguay, on the San José River. It is a commercial center for a large grain and livestock region. The city was founded in 1783 by settlers from Spain and was Uruguay's provisional capital in 1825–26. San José is noted for its architecture.

San José, city (1976 est. pop. 228,300), central Costa Rica, capital and largest city of Costa Rica. During colonial times the main industry of the region was tobacco raising; by the mid-19th cent. the city had become the center of a coffee-producing area. San José was founded (c.1738) at the beginning of the westward expansion from Cartago. In 1823 it became the country's capital. A modern city, it has a mixture of Spanish and North American architecture. In 1960 it was the site of two conferences of foreign ministers of the member states of the Organization of American States.

San José, town (1964 pop. 17,956), SW Guatemala, on the Pacific Ocean. It is a rail terminus and the major Pacific port of Guatemala.

San Juan (săn wän', săn hwän'), city (1970 pop. 112,500), capital of San Juan prov., W Argentina. Wine is the chief product, and vineyards dot the picturesque landscape. Fruits and grains are grown, cattle are raised, and the province is rich in minerals. Founded in 1562, San Juan figured prominently in the civil wars of the 19th cent.

San Juan, city (1970 pop. 452,749), capital and chief port of Puerto Rico, NE Puerto Rico. Sugar, tobacco, coffee, and fruit are exported, mainly to the United States. San Juan's industries include tourism, sugar refining, rum distilling, metalworking, publishing, and the manufacture of jewelry, clothing, shoes, textiles, furniture, pharmaceuticals, electronic equipment, machine tools, and plastics. The city's old section, situated on two rocky islets, is linked by bridges with the mainland. Ponce de León founded a settlement nearby in 1508. In 1521 the settlement was moved across the bay to the site of present-day San Juan. Strongly fortified, it withstood attacks by English buccaneers in 1595 but was sacked by the Dutch in 1625. San Juan gained increasing importance as a West Indian port during the 18th and 19th cent. U.S. troops occupied the city during the Spanish-American War in 1898. There are impressive historic buildings in the old city, including El Morro castle (begun 1539), San Cristóbal castle (begun 1631), and La Fortaleza (begun 1529), a former fort now used as the governor's official residence. Other San Juan landmarks include San José Church (founded c.1523), the oldest church in continuous use in the Western Hemisphere; Casa Blanca (1523); and the Cathedral of San Juan Bautista.

San Juan (săn wän'). **1.** County (1970 pop. 831), 391 sq mi (1,012.7 sq km), SW Colo., drained by the Animas River; formed 1876; co. seat Silverton. It is in a gold, silver, lead, zinc, and copper mining area, with livestock grazing. **2.** County (1970 pop. 52,517), 5,500 sq mi (14,245 sq km), extreme NW N.Mex., bordered on the W by Ariz., on the N by Colo., and drained by the Chaco and San Juan rivers; formed 1887; co. seat Aztec. Fruit, grain, and vegetables are grown. There are oil and natural-gas fields in the county. **3.** County (1970 pop. 9,606), 7,707 sq mi (19,961.1 sq km), extreme SE Utah, in a mountain and plateau area bordering on Colo. and Ariz., bounded on the W by the Colorado River and crossed in the S by the San Juan River; formed 1880; co. seat Monticello. In an irrigated farming area yielding hay, grain, sheep, and cattle, it has oil and gas wells and primary metal industries. **4.** County (1970 pop. 3,856), 179 sq mi (463.6 sq km), NW Wash., on the British Columbia border, and including most of the San Juan Islands; formed 1873; co. seat Friday Harbor. It has truck farming and salmon fishing.

San Juan (sän hwän'), river, c.110 mi (175 km) long, flowing from the SE corner of Lake Nicaragua E to the Caribbean Sea, near the port of San Juan del Norte. The lower course is the boundary between Nicaragua and Costa Rica.

San Juan (săn wän'), river, c.400 mi (645 km) long, rising in the San Juan Mts., SW Colo., and flowing W through N.Mex. and Utah to Lake Powell on the Colorado River. The San Juan is used extensively for irrigation; vegetables, fruits, and grains are grown in the river valley.

San Juan Cap·is·tra·no (săn wän' kăp'ĭ-strä'nō), city (1970 pop. 3,781), Orange co., S Calif.; inc. 1961. Although San Juan Capistrano has a small industrial park that manufactures sailboats, plastics, novelty clothing, and other items, the economy is based chiefly on tourism. Padre Junípero Serra founded a mission here in 1776. The mission church, completed in 1806, was ruined by an earthquake in 1812, but the chapel where Father Serra said Mass is still in daily use. Mexico nationalized mission lands in the early 1840s; President Abraham Lincoln returned the mission to the Roman Catholic Church in 1865. It is said that swallows come to the ruins of the church every Mar. 19, the Feast Day of St. Joseph, and depart on Oct. 23, the death date of St. John of Capistrano.

San Juan del Nor·te (sän hwän' dĕl nôr'tā), small town, SE Nicaragua, on the Caribbean Sea. It was occupied (1848) by the British to secure control of the Mosquito Coast and to check U.S. efforts to build an interoceanic canal. The port became the thriving eastern terminus of a transisthmian transport company operated by Cornelius Vanderbilt in the gold rush to California. In 1854 it was bombarded by the U.S. warship *Cyane* in retaliation for insults to the U.S. minister and damage to U.S. property in Nicaragua.

San Juan Hill (săn wän', săn hwän'), Oriente prov., E Cuba, near the city of Santiago de Cuba. It was the scene (July, 1898) of a battle in the Spanish-American War, in which Theodore Roosevelt and the Rough Riders took part.

San Juan Islands (săn wän'), archipelago of 172 islands, NW Wash., E of Vancouver Island. The islands were discovered and named c.1790 by Spanish explorers and were the subject of the San Juan boundary dispute between Great Britain and the United States; their ownership was decided in 1872. San Juan Island is a national historical park.

San Juan National Historic Site: *see* National Parks and Monuments Table.

Sankt Pöl·ten (zängkt pœl'tən), city (1973 pop. 50,144), Lower Austria prov., N central Austria. Manufactures include machinery, textiles, and paper. Chartered in the 12th cent., Sankt Pölten has a Romanesque cathedral and a town hall of the 16th–17th cent.

San Le·an·dro (lē-ăn'drō), residential and industrial city (1978 est. pop. 67,200), Alameda co., W Calif., on San Francisco Bay; inc. 1872. Food products, metal items, transportation equipment, electronic products, and furniture are among its manufactures.

San Lo·ren·zo (lə-rĕn'zō), uninc. city (1978 est. pop. 23,000), Alameda co., W Calif. It is chiefly residential.

San·lú·car de Bar·ra·me·da (sän-lo͞o'kär' dā bä'rä-mā'dä), city (1970 pop. 29,483), Cádiz prov., S Spain, on the Guadalquivir River estuary, in Andalusia. Manzanilla, a white wine, is a noted product. Sanlúcar flourished after the discovery of America, when all ships passed it to reach Seville. Columbus sailed from Sanlúcar in 1498, and Magellan in 1519.

San Lu·is (sän' lo͞o-ēs'), city (1970 pop. 59,113), capital of San Luis prov., W central Argentina. The city is the commercial center of an area producing cattle, grain, and wine. It is also a popular resort. Founded in 1594, San Luis was burned and sacked by Indians in 1712 and 1720.

San Lu·is (săn lo͞o'ĭs), city (1970 pop. 781), seat of Costilla co., S Colo., in the W foothills of the Sangre de Cristo Mts., on the E slope of the San Luis Valley SE of Alamosa, at an altitude of c.8,000 ft (2,440 m). Truck farming is done in the area.

San Luis O·bis·po (ō-bĭs'pō), county (1970 pop. 105,690), 3,184 sq mi (8,246.6 sq km), SW Calif., on the Pacific, with Santa Lucia Range in the NW and drained by the Salinas and Santa Maria rivers; formed 1850; co. seat San Luis Obispo. Its agriculture includes cattle, wheat, hay, truck crops, flower and vegetable seed, fruit and nuts, dairy goods, and poultry. It has oil and natural-gas fields, clay, sand and gravel quarries, and a food-processing industry. There are coast resorts, mineral springs, and old missions.

San Luis Obispo, city (1978 est. pop. 34,800), seat of San Luis Obispo co., S Calif., near San Luis Obispo Bay; inc. 1856. Furniture, building materials, and food items are among its products. In 1846 John G. Frémont seized the city for the United States. To escape torrential rains he quartered in the Franciscan mission, San Luis Obispo de Tolosa (1772), which is now a state landmark.

San Lu·is Po·to·sí (sän' lo͞o-ēs' pō'tō-sē'), city (1974 est. pop. 271,100), capital of San Luis Potosí state (24,417 sq mi/63,240 sq km), central Mexico. Situated on a plain almost entirely surrounded by low mountains, the city is a mining center and a rail junction. Industries include foundries, smelters, and factories producing clothing, leather goods, and beverages. San Luis Potosí was founded in 1576.

San Mar·cos (săn mär'kəs), city (1978 est. pop. 23,300), seat of Hays co., S central Texas, on the San Marcos River; inc. 1877. Meat is packed, and cattle feed, plastic products, heating equipment, and furniture are among the manufactures. The city is situated on the Balcones fault, where prehistoric earthquakes split the earth, releasing spring water to the surface and creating underground caves.

San Ma·ri·no (săn' mə-rē'nō), republic (1972 est. pop. 18,320), 24 sq mi (62 sq km), in the Apennines near the Adriatic Sea, SW of Rimini, N central Italy. It is the world's smallest republic. The capital is San Marino (1971 est. pop. 4,350); Serravalle is the only other town. Virtually all of the republic's inhabitants speak Italian and are Roman Catholic. Farming is the main occupation; cereals and fruit are grown, and cattle and hogs are raised. Major sources of revenue are tourism and the sale of postage stamps. Woolens, wine, and limestone are exported. Of note in San Marino are the Basilica of Santo Marino; towers (14th–16th cent.) built on each of the three peaks of Mt. Titano (2,300 ft/701 m), the chief geographic feature of present-day San Marino; the Gothic government house; and several art museums.

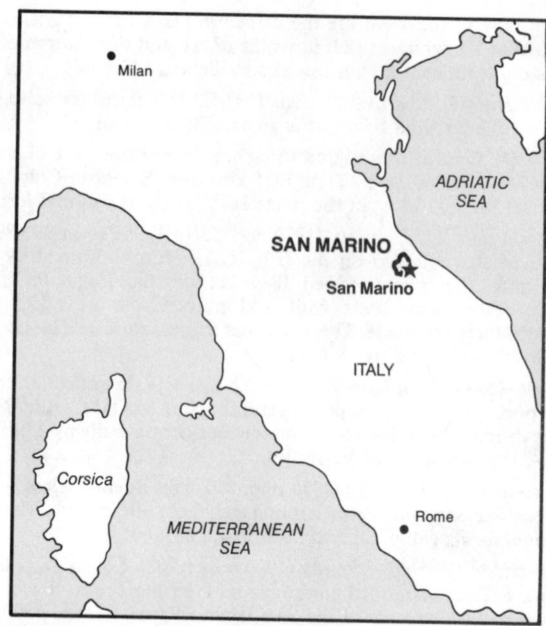

History. According to tradition Marino, a Christian stonecutter from Dalmatia, took refuge (early 4th cent.) on Mt. Titano. By the mid-5th cent. a community was formed; because of its relatively inaccessible location and its poverty, it has succeeded, with a few brief interruptions, in maintaining its independence. In 1631 its independence was recognized by the papacy. Italy and San Marino signed a treaty of friendship and economic cooperation in 1862 (renewed and expanded several times).

Government. Legislative power in San Marino is vested in the popularly elected grand council (made up of 60 members elected to five-year terms); every six months the council appoints two regents, who in conjunction with the 10-member council of state form the executive. The republic receives an annual subsidy from Italy. It mints its own coins, although Italian currency is in general use.

San Marino, residential city (1970 pop. 14,177), Los Angeles co., S Calif.; inc. 1913. The Henry E. Huntington Library and Art Gallery are here.

San Ma·te·o (mə-tā'ō), county (1970 pop. 557,361), 447 sq mi (1,157.7 sq km), W Calif., on San Francisco-San Mateo Peninsula; formed 1856 from San Francisco co.; co. seat Redwood City. The county produces artichokes, flowers, truck crops, hogs, and dairy products. There are stone, sand, and gravel quarries and redwood timber in the mountains. Its industry includes food processing and the manufacture of textiles, wood and paper products, furniture, chemicals, plastics, concrete, steel, and machinery.

San Mateo, city (1978 est. pop. 78,800), San Mateo co., W Calif., on San Francisco Bay; inc. 1894. It is a commercial and retail center. San Mateo was named by a Spanish expedition in 1776. The area was a Mexican colony from 1822 to 1846. San Mateo's main growth dates from the start of railroad service in 1863.

San Mi·guel (mē-gĕl'), city (1974 est. pop. 66,900), E El Salvador, at the foot of San Miguel volcano (6,996 ft/2,133.8 m). It has textile and dairy-products industries. The region produces cotton, henequen, and vegetable oil. San Miguel was founded in 1530.

San Mi·guel (mə-gĭl'). **1.** County (1970 pop. 1,949), 1,283 sq mi (3,323 sq km), SW Colo., drained by the Dolores and San Miguel rivers; formed 1861; co. seat Telluride. It is in a gold, silver, and lead mining area, with livestock grazing. It includes part of the San Miguel Mts. and of Montezuma and Uncompahgre national forests. **2.** County (1970 pop. 21,951), 4,741 sq mi (12,279.2 sq km), NE N.Mex., drained by the Canadian and Pecos rivers; formed 1862; co. seat Las Vegas. It has livestock, agriculture (wheat, beans, corn, and chili peppers), and gold mining. Parts of Santa Fe National Forest and the Sangre de Cristo Mts. are in the west.

San Ni·co·lás de los Gar·zas (sän' nē'kə-läs' dā lōs gär'səs), city (1970 pop. 111,502), Nuevo León state, N Mexico, in the Santa Catarina valley. It is situated on a major highway in an area where oranges are grown.

San Pab·lo (săn păb'lō), city (1978 est. pop. 18,600), Contra Costa co., W Calif., on San Pablo Bay, in a farm region; inc. 1948. One of the oldest Spanish settlements in the region, the city is now a commercial and medical center.

San Pa·tri·ci·o (pə-trĭsh'ē-ō), county (1970 pop. 47,288), 689 sq mi (1,784.5 sq km), S coastal Texas, bounded on the E by Aransas Bay, on the S by the Nueces River, on the N by the Aransas River; formed 1836; co. seat Sinton. This area is a leading oil and natural-gas producing county and also has agriculture (cotton, grain sorghums, corn, truck crops, peanuts, fruit, and livestock) and manufacturing.

San Pe·dro de Ma·co·rís (sän pā'drō dā mä'kō-rēs'), city (1970 pop. 42,473), SE Dominican Republic, on the Caribbean Sea at the mouth of the Higuamo River. It is the nation's leading sugar port. Textiles and alcohol are produced.

San Pe·dro Su·la (sän pā'drō sōō'lä), city (1974 pop. 150,991), capital of Cortés dept., NW Honduras. San Pedro Sula ships bananas and sugar.

San·pete (săn'pēt'), county (1970 pop. 10,976), 1,597 sq mi (4,136.2 sq km), central Utah, in a mountain area crossed by the Wasatch Plateau and watered by the Sevier and San Pitch rivers; formed 1852; co. seat Manti. It is in an agricultural (hay, sugar beets, fruit, and truck crops) and livestock (poultry, cattle, and sheep) area. It mines rock salt and gypsum, has a food-processing industry, and manufactures clothing and travel trailers.

San Quen·tin (kwĕn'tən), peninsula extending into San Francisco Bay, W Calif., N of San Francisco. The state prison here was begun in 1852.

San Ra·fael (rə-fĕl'), residential city (1978 est. pop. 45,900), seat of Marin co., W Calif., a suburb of San Francisco on the N portion of San Francisco Bay; inc. 1913. Electrical equipment and metal, plastic, and wood products are manufactured. It is the seat of the restored Mission San Rafael Arcángel (est. 1817).

San Re·mo (rā'mō), city (1976 est. pop. 53,200), in Liguria, NW Italy, on the Ligurian Sea and on the Italian Riviera. It is a fashionable resort and gaming center and a major flower market.

San Re·mo (rē'mō), uninc. town (1970 pop. 8,302), Suffolk co., SE N.Y., on the N shore of Long Island E of Kings Park.

San Sa·ba (să'bə, săb'ə), county (1970 pop. 5,540), 1,122 sq mi (2,906 sq km), on NE Edwards Plateau, Texas, bounded on the N and E by the Colorado River; formed 1856; co. seat San Saba. It has livestock ranching and agriculture (pecans, peanuts, grain sorghums, corn, oats, fruit, truck crops, barley, and some cotton).

San Saba, town (1970 pop. 2,555), seat of San Saba co., central Texas, on the San Saba River NW of Austin. Pecans, melons, sheep, cattle, turkeys, and wool and mohair are shipped.

San Sal·va·dor (săn săl'və-dôr'), city (1974 est. pop. 366,000), central El Salvador, capital and largest city of the country. Beer, tobacco products, clothing, textiles, and soap are produced. Built on the volcanic slope that parallels the Pacific coast, the city has suffered from recurrent and severe earthquakes and has been frequently rebuilt. San Salvador was founded early in the 16th cent. and for a time (1831–38) was the capital of the Central American Federation.

San Salvador, one of the Bahama Islands (1970 pop. 897), British West Indies. It was the first land discovered by Columbus in the New World in 1492. Indian inhabitants called it Guanahani, and it has also been named Watling or Watlings Island.

San Se·bas·tián (săn' sə-băs'chən, sə-bäs-tyän'), city (1975 est. pop. 166,250), capital of Guipúzcoa prov., N Spain, on the Bay of Biscay at the mouth of the Urumea River, in the Basque Provinces near the French frontier. Picturesquely situated at the foot of Mt. Urgull, it was a summer residence of the Spanish court in the 20th cent. and still is one of the most fashionable seaside resorts in Spain. There are fishing, steel, and paper industries. The city was rebuilt in the 19th cent. after its virtual destruction in the Peninsular War (1813). The San Sebastián pact, or republican manifesto, which precipitated the fall of the Spanish monarchy, was signed here in 1930.

San·ta An·a (săn'tə ăn'ə), city (1974 est. pop. 105,300), W El Salvador. It is the processing center for a sugar-cane, coffee, and cattle region. There are textile and foodstuffs industries. Nearby rises Santa Ana volcano (7,828 ft/2,387.5 m), the highest in El Salvador.

Santa Ana, city (1978 est. pop. 183,800), seat of Orange co., S Calif., in the fertile Santa Ana valley; inc. 1886. It is the governmental, business, medical, and industrial center of the large Anaheim–Santa Ana–Garden Grove metropolitan area. Among its many products are radios, electrical connectors, nuclear and aircraft components, and sporting goods.

Santa Bar·ba·ra (bär'bər-ə, bär'brə), county (1970 pop. 264,324), 2,738 sq mi (7,091.4 sq km), SW Calif., on the coast along the Santa Barbara Channel and drained by the intermittent Cuyama, Sisquoc, Santa Maria, and Santa Ynez rivers; formed 1850; co. seat Santa Barbara. It is in a diversified agricultural area that produces seed, sugar beets, truck crops, citrus and deciduous fruit, nuts, beans, alfalfa, and beef cattle. It has oil and natural-gas fields, sand and gravel quarries, and a food-processing industry. Textiles, wood products, porcelain, and machinery are made. It includes part of Los Padres National Forest.

Santa Barbara, city (1978 est. pop. 74,200), seat of Santa Barbara co., S Calif., on the Pacific Ocean; inc. 1850. A beautiful residential and resort city with many recreational facilities, it also has electronics and aerospace research and development firms, and an orchid industry. Oil fields are in the area and offshore. The region was discovered by Juan Cabrillo in 1542 and explored and named in 1602. A Spanish presidio, remnants of which remain, was founded in 1782. The Spanish mission (established in 1786) is considered one of the most beautiful of all of the California missions. Santa Barbara is known for its prevalent Spanish architecture. In Jan., 1969, an oil leak in an offshore drilling platform in the Santa Barbara Channel brought great destruction to the city's harbor and beaches.

Santa Barbara Islands, chain of eight rugged islands and many islets, extending c.150 mi (240 km) along the S Calif. coast from Point Conception to San Diego. The islands were discovered in 1542 by Juan Rodriguez Cabrillo, a Portuguese explorer in the service of Spain. Located from 13 to 68 mi (21–109 km) west of the mainland, they are divided into two groups. The Santa Barbara group, to the north, is separated from the mainland by the Santa Barbara Channel and the Santa Catalina group, to the south, by the San Pedro Channel and the Gulf of Santa Catalina. The islands are the exposed tops of low mountains.

Santa Cat·a·li·na (kăt'ə-lē'nə) or **Catalina Island,** S Calif., one of the Santa Barbara Islands, off Huntington Beach, Calif. It is a resort island, 22 mi (35.4 km) long and 1 to 8 mi (1.6–12.9 km) wide, with a picturesque, irregular coastline dotted with coves and beaches. It was discovered in 1542 and given its present name by Vizcaíno in 1602. In 1919 William Wrigley bought the island and constructed vacation and sports facilities. The casino on Sugar Loaf Mt. is the island's principal landmark.

Santa Cat·a·ri·na (kăt'ə-rē'nə), state (1975 est. pop. 3,351,400), 37,060 sq mi (95,985 sq km), S Brazil; capital Florianópolis.

Santa Cla·ra (klăr'ə, klä'rə), city (1970 pop. 131,504), capital of Las Villas prov., central Cuba. Cattle raising was the traditional industry until the 19th cent., when sugar cane became important. Tobacco processing and trading are carried on. Santa Clara was founded in 1689.

Santa Clara, county (1970 pop. 1,065,313), 1,305 sq mi (3,380 sq km), W Calif., extending SE from San Francisco Bay between Diablo Range and the Santa Cruz Mts.; formed 1850; co. seat San Jose. Its agriculture includes fruit, truck crops, livestock, poultry, dairy products, flowers, and seed. It has oil and gas wells and sand and gravel quarries, a food-processing industry, and diverse manufacturing.

Santa Clara, city (1978 est. pop. 84,100), Santa Clara co., W Calif.; inc. 1852. Electronic equipment, fiberglass, and plastic products are among the many manufactures. Points of interest include the Santa Clara de Asís Mission, founded in 1777.

San·ta Cruz (sän'tä krōōs'), city (1976 pop. 255,568), capital of Santa Cruz dept., central Bolivia, on the Piray River. A trade and processing center for sugar, coffee, rice, cattle, and lumber, it is of strategic and commercial importance because of its central location. A rail line completed in 1962 has given the city access to both the Atlantic and Pacific oceans. Santa Cruz was founded in the 16th cent. and was an early Jesuit missionary center.

San·ta Cruz (săn'tə krōōz'). **1.** County (1970 pop. 13,966), 1,246 sq mi (3,227.1 sq km), S Ariz., on the Mexican border; formed 1899; co. seat Nogales. Irrigated farming along the Santa Cruz River produces livestock, alfalfa, and cotton. Lead, zinc, silver, and copper are mined. Coronado National Forest extends throughout most of the county. **2.** County (1970 pop. 123,790), 439 sq mi (1,137 sq km), W Calif., in the Santa Cruz Mts., bounded on the W by the Pacific Ocean and Monterey Bay; formed 1850; co. seat Santa Cruz. The Pajaro River Valley is noted for its apples; the county also produces lettuce, vegetables, berries, nuts, sugar beets, dry beans, hay, dairy products, poultry, eggs, and livestock. It has granite quarries, sand and gravel pits, and light industry (food processing and canning, lumbering, and the manufacture of stone, clay and glass products). The county contains beach and mountain resorts.

Santa Cruz, city (1978 est. pop. 38,600), seat of Santa Cruz co., W Calif., on Monterey Bay; inc. 1866. It is a seaside city with many fine beaches and is surrounded by hills and redwoods. There are electronic and food-processing industries. Agriculture flourishes in the area.

Santa Cruz de Te·ne·ri·fe (dā tĕn'ə-rē'fā), city (1975 est. pop. 175,950), capital of Tenerife prov., Spain, a port on Tenerife island in the Canary Islands. Vegetables, sugar cane, tobacco, and bananas are exported. The city's splendid scenery and mild subtropical climate make it a favorite tourist resort.

San·ta Fe (săn'tə fā'), city (1970 pop. 312,427), capital of Santa Fe prov., NE Argentina, a river port near the Paraná, with which it is connected by canal. On the eastern margin of the Pampa, it is an important shipping point for agricultural products. The city also has some industry. Founded by the Spanish conquistador Juan de Garay (1573), Santa Fe has several notable churches.

Santa Fe, county (1970 pop. 54,774), 1,902 sq mi (4,926.2 sq km), N central N.Mex., drained in the NW by the Rio Grande; formed 1852; co. seat Santa Fe. It has livestock grazing, farming (grain and chili peppers), and mining (lead, zinc, coal, gold, and silver). It includes part of Santa Fe National Forest and the Sangre de Cristo Mts.

Santa Fe, city (1978 est. pop. 46,500), alt. c.7,000 ft (2,135 m), state capital and seat of Santa Fe co., N N.Mex., at the foot of the Sangre de Cristo Mts. It is an administrative, tourist, and resort center and a shipping point for Indian wares, minerals, and farm products. Founded c.1609 by the Spanish is on the site of ancient Indian ruins, it was a center of Spanish-Indian trade for over 200 years. The city is the oldest capital city in the United States. In the Pueblo revolt of 1680 the Spanish colonists were driven out; in 1692 they returned under Diego de Vargas. Shortly after Mexico gained independence from Spain (1821), extensive commerce with the United States developed by way of the Santa Fe Trail. In 1846 the region became a province of the United States. Among the city's churches are San Miguel Mission Church (c.1636), the Cathedral of St. Francis, and Cristo Rey Church, the largest adobe building in the United States. Among the many other points of interest are the Palace of the Governors, which now houses a state museum, the Laboratory of Anthropology, museums of international folk art and of Navaho ceremonial art, and a state-owned art gallery. There are many Indian pueblos in the region, and Bandelier National Monument is nearby.

Santa Fe Springs, city (1978 est. pop. 15,800), Los Angeles co., S Calif., in an oil and natural-gas region; inc. 1957.

Santa Fe Trail, important caravan route of the W United States, extending c.780 mi (1,255 km) from Independence, Mo., SW to Santa Fe, N.Mex. By the early 19th cent. small trapping parties had reached Santa Fe, then under Spanish rule. In Nov., 1821, William Becknell, a trader, returned with news that Mexico was free, and Santa Fe welcomed trade. From then on, annual wagon caravans made the 40- to 60-day trip over the trail and returned after a 4- to 5-week stay in Santa Fe. In 1850 a monthly stage line was started between Independence and Santa Fe. In 1880 the Santa Fe RR reached Santa Fe and marked the death of the trail.

Santa Is·a·bel (ĭs'ə-bĕl'): *see* Malabo, Equatorial Guinea.

Santa Ma·ri·a (mə-rē'ə), city (1970 pop. 156,929), Rio Grande do Sul state, S Brazil. It is a major railroad terminus and the site of an important military base. Leather goods, beer, and foodstuffs are produced. Santa Maria was established in 1797.

Santa Maria, city (1978 est. pop. 35,600), Santa Barbara co., S Calif., near San Luis Obispo Bay; inc. 1905. The economy is based largely on agriculture and oil. Other industries produce phonograph records, wire cables, marine equipment, and tire molds. Vandenberg Air Force Base is nearby.

Santa Mar·ta (mär'tə), city (1973 pop. 128,577), capital of Magdalena dept., N Colombia, a port on the Caribbean Sea. The city's banana industry is one of the most important in South America. Santa Marta also has fine beaches and is a tourist center. Founded by the Spanish explorer Rodrigo de Bastidas in 1525, it was often sacked by corsairs in the 16th cent. During colonial times the city was important as an outlet for the Magdalena River valley.

Santa Mon·i·ca (mŏn'ĭ-kə), city (1978 est. pop. 91,100), Los Angeles co., S Calif., on Santa Monica Bay; inc. 1886. Missiles, aircraft parts, electronic equipment, ceramics, chemicals, leather goods, furniture, and optical instruments are among the manufactures. The J. Paul Getty museum is here. The city has a 3-mi (4.8-km) oceanfront beach; nearby are several state parks.

San·tan·der (sän'tän-dĕr'), city (1975 est. pop. 164,994), capital of Santander prov., N Spain, in Old Castile, on the Bay of Biscay. It is a seaport and a popular resort. On the nearby peninsula of Magdalena is a former royal summer palace. An ancient port, Santander became, after the discovery of America, one of the busiest ports of northern Spain. The exploitation of nearby mines has favored the development of industries (ironworks and shipyards). The 13th cent. cathedral and the business district were destroyed by fire in 1941 but have been restored.

Santa Pau·la (pô'lə), city (1978 est. pop. 19,000), Ventura co., S Calif., on the Santa Clara River in a fertile valley that yields citrus fruits, avocados, and walnuts; laid out 1875, inc. 1902. Fruit packing and oil production are major industries, and there are plants that manufacture paper products, building materials, clothing, and ceramics.

San·ta·rém (săn'tə-rĕm'), town (1975 est. pop. 16,850), capital of Santarém dist. and Ribatejo, W central Portugal, above the right bank of the Tagus River. Agricultural produce is exported. The town has been important since Roman times because of its strategic location along the approaches to Lisbon. It was retaken from the Moors by the first king of Portugal, Alfonso I, in 1147.

Santa Ro·sa (rō'zə), city (1970 pop. 37,893), capital of La Pampa prov., central Argentina. It is surrounded by a rich agricultural and cattle-raising area. First settled in 1889, Santa Rosa attracted many Spanish, Italian, and French immigrants.

Santa Rosa, county (1970 pop. 37,741), 1,032 sq mi (2,672.9 sq km), NW Fla., between the Ga. border and the Gulf of Mexico, and bounded on the W by the Escambia River; formed 1842; co. seat Milton. It is in a rolling agricultural area that produces corn, peanuts, cotton, vegetables, and livestock. It also has oil and gas wells, forestry, and clothing and concrete manufacturing.

Santa Rosa. 1. City (1978 est. pop. 72,200), seat of Sonoma co., W Calif.; inc. 1868. It is an industrial city and a retail, financial, and medical center for the fertile Sonoma Valley. Luther Burbank lived here, and his gardens are preserved as a monument. Of interest also is the Church of One Tree, built (1874) from a single redwood and now housing the Robert L. Ripley Memorial Museum. In the vicinity are the Jack London Wolf House and memorial museum, Armstrong Redwoods State Park, and many other historic and natural attractions. **2.** Town (1970 pop. 1,813), seat of Guadalupe co., E central N.Mex., on the Pecos River ESE of Albuquerque; settled c.1865. It is a trade and shipping center for a livestock area.

Santa Rosa Island, narrow barrier beach between the Gulf of Mexico and Santa Rosa Sound, NW Fla., extending c.50 mi (80 km) parallel to the coast. It is the site of Fort Pickens and of a missile-launching station. The island is also a resort area.

Santa Tec·la (tĕk'lə), city (1970 est. pop. 38,000), central El Salvador. It was founded in 1854 after the capital, San Salvador, was destroyed in an earthquake. After San Salvador was rebuilt, Santa Tecla became a wealthy suburb. It is situated among coffee farms.

San·tee (săn-tē'), river, 143 mi (230.1 km) long, formed by the confluence of the Congaree and Wateree rivers, central S.C., and flowing SE to the Atlantic Ocean. A navigable canal (built 1792-1800) connects the Santee with the Cooper River. The Santee has been extensively developed for power and navigation.

San·ti·a·go (săn'tē-ä'gō, săn'-), city (1970 pop. 2,661,920), central Chile, capital of Chile and of Santiago prov., on the Mapocho River. The city was founded and named Santiago de Nueva Estremadura on Feb. 12, 1541, by Pedro de Valdivia. Santiago has spread over a broad valley plain and is today one of the largest cities in South America. Low foothills encompass the valley, and the snow-capped Andes, forming a superb backdrop, rise in the eastern distance. Textiles, foodstuffs, clothing, footwear, and other goods are produced.

There are also large iron and steel foundries in the city. Spacious parks, plazas, gardens, and wide avenues are characteristic features. Focal point of the intellectual and cultural development of Chile from colonial times to the present, Santiago has many national establishments, including a library, museum, theater, and university. The city has experienced several catastrophes. In Sept., 1541, the Araucanian Indians nearly wiped out the new settlement; it was completely leveled by an earthquake in 1647; and the Mapocho has frequently flooded the city. In the early 1970s Santiago was the scene of mass political demonstrations for and against the regime of Chilean President Salvador Allende, who died here during the coup d'état of Sept., 1973.

Santiago, city (1970 pop. 14,595), W central Panama. Santiago is a communications and commercial center in the Pacific lowlands.

Santiago de Com·po·ste·la (dā kŏm'pə-stĕl'ə) or **Santiago,** city (1970 pop. 70,893), La Coruña prov., NW Spain, in Galicia, on the Sar River. Here in the early 9th cent. the supposed tomb of the apostle St. James the Greater was reputedly discovered, and Alfonso II of Asturias had a sanctuary built on the site. The city grew around the shrine and became, after Jerusalem and Rome, the most famous Christian place of pilgrimage in the Middle Ages. Santiago de Compostela still thrives as a pilgrimage and tourist center. Its most remarkable building is the cathedral, which replaced the earlier sanctuary after its destruction (10th cent.) by the Moors. Originally built (11th–13th cent.) in Romanesque style, the cathedral has had baroque and plateresque additions and restorations. Other historic buildings include the Hospital Real (1501–11), built by Ferdinand and Isabella for the accommodation of poor pilgrims, and the Colegio Fonseca (16th cent.), a part of the university.

Santiago de Cu·ba (kyōō'bə), city (1970 pop. 275,970), capital of Oriente prov., SE Cuba. Santiago is situated on a cliff overlooking a bay. Minerals, agricultural produce, and woods are exported. Founded in 1514 by Diego de Velázquez and moved to its present site in 1588, Santiago served for some time as Cuba's capital. In its early days it was captured by French and English buccaneers and was a center of the smuggling trade with the British West Indies. Frenchmen fleeing the slave revolt in Haiti in the early 19th cent. settled in Santiago and heavily influenced the city's development. During the Spanish-American War of 1898, U.S. ships established a blockade in Santiago's harbor; when the Spanish admiral Pascual Cervera y Topete, bottled up in the harbor, made a desperate attempt to escape, his fleet was destroyed. The city retains many colonial landmarks, notably its cathedral (the largest in Cuba) and the crumbling forts that stand on high cliffs above the harbor.

Santiago del Es·te·ro (dĕl' ĕ-stĕr'ō), city (1970 pop. 119,127), capital of Santiago del Estero prov., N Argentina. It is a transportation hub of the Argentine Chaco and a commercial center for cattle. Nearby thermal springs have made the city a health resort.

Santiago de los Ca·ba·lle·ros (dā' lōs kä'bä-yâr'ōs), city (1970 pop. 155,151), N Dominican Republic, on the Yaque del Norte River. It is a rail and road junction in the center of the fertile region known as the Cibao lowland. The region produces subsistence crops, sugar cane, tobacco, coffee, and cotton. Tobacco products, beeswax, and honey are made in the city. Santiago was founded in 1495 and in 1844 was the site of a decisive battle in the Dominican Republic's war of independence.

San·to An·dré (sän'tōō än-drā'), city (1970 pop. 418,578), São Paulo state, S Brazil, a suburb of São Paulo. The city has industries that produce textiles, metal products, rubber goods, porcelain, foodstuffs, furniture, airplanes, munitions, and printed material.

San·to Do·min·go (săn'tō dō-mĭng'gō), former Spanish colony on the island of Hispaniola. The name is also given to the Dominican Republic, and in early days it applied to Haiti. Columbus discovered the island in 1492 and established a settlement on the northern coast, but when he returned in 1493 the settlers had vanished. He administered a new colony there until complaints against his rule caused him to be replaced (1500) by Francisco de Bobadilla. The colonists became farmers; the work was done for them under the encomienda system by Indians. Most of the natives perished under Spanish rule, and importation of black slaves was begun. Although Spain nominally owned the whole island, colonization had not been undertaken in the west. Later French planters were able to establish settlements in present-day Haiti. In the latter half of the 18th cent. sugar cane was introduced and sugar plantations became dominant. Spain ceded (1697) the western part of the island (then called Saint-Domingue) to France and in 1795 gave up the whole region. Spanish rule was restored in the east when the inhabitants, aided by the British, rebelled against the French in 1808–9. The Spanish themselves were ousted in 1821, but in 1822 the Haitians extended their rule over the entire island.

The Haitians were driven out in 1844 and the Dominican Republic was proclaimed. The capital is Santo Domingo (1970 pop. 671,402), on the Caribbean Sea, at the mouth of the Ozama River. Founded Aug. 4, 1496, by Bartholomew Columbus, brother of Christopher Columbus, it is the oldest continuously inhabited settlement in the Western Hemisphere. It became the base from which Diego de Velázquez set out to conquer Cuba. Prior to the conquest of Mexico and Peru, Santo Domingo was the seat of Spain's colonial administration in the New World. Santo Domingo was almost totally destroyed by a hurricane in 1930 but was rebuilt and renamed Ciudad Trujillo after dictator Rafael Leonidas Trujillo; the original name was restored in 1961 after his death. Although replete with historic sites, Santo Domingo today is a city of broad avenues and modern buildings. The cathedral, begun in 1514, is the oldest in the Western Hemisphere; it contains the reputed tomb of Christopher Columbus.

San·tos (săn'tōōs), city (1970 pop. 346,096), São Paulo state, SE Brazil, on the island of São Vicente in the Atlantic just off the mainland. It is southeast of the city of São Paulo, with which it is linked by rail and by the Via Anchieta highway. Santos is the world's greatest coffee port and the chief shipping point for the rich interior of São Paulo state. It handles the major share of Brazilian exports, including, besides coffee, oranges, bananas, cotton, and industrial products. Santos was founded c.1540 near the settlement of São Vicente. It was sacked by the English in 1591. Santos is a fashionable residential area and resort center with fine beaches.

San Vi·cen·te (săn' vĭ-sĕn'tē), city (1974 est. pop. 19,900), central El Salvador. Among its manufactures are shawls, hats, and tobacco products. San Vicente is the commercial center of a region that produces coffee, sugar cane, and indigo. From the nearby volcano San Vicente (7,360 ft/2,243 m) a wide vista of Central America may be seen.

São Cae·ta·no do Sul (souɴ' kī-tä'nōō dōō sōōl'), city (1970 pop. 150,171), São Paulo state, SE Brazil, an industrial suburb SE of the city of São Paulo. Because of their integration into São Paulo's industrial zone, São Caetano do Sul and nearby Santo André and São Bernardo do Campo are sometimes collectively referred to as "ABC." São Caetano do Sul assembles motor vehicles and produces metal products, electrical and communications equipment, chemicals, pharmaceuticals, textiles, clothing, and food products. The city was founded c.1631.

São Fran·cis·co (frə-sēsh'kōō), river, c.1,800 mi (2,900 km) long, rising in the Serra de Canastra, SW Minas Gerais state, Brazil, and flowing NE, then SE through the sertão region of E Brazil to the Atlantic Ocean. The river's flow varies with the season. The São Francisco, an ancient river that is embedded in the Brazilian Plateau, probably once entered the sea near Cape São Roque, northeast of its great bend. Paulo Afonso falls (275 ft/84 m high), east of the great bend, blocks navigation into the interior; a railroad circumvents the falls. The river is navigable along c.900 mi (1,450 km) of its middle course.

São João de Me·ri·ti (zhwouɴ *th*ĭ mə-rē'tĭ, də mĭr-ē-tē'), city (1970 pop. 163,934), Rio de Janeiro state, SE Brazil, a residential suburb NW of the city of Rio de Janeiro. Most of the labor force commutes to Rio.

São Lu·ís (lōō-ēsh'), city (1970 pop. 167,529), capital of Maranhão state, NE Brazil. It is a port city located on São Luís Island in São Marcos Bay of the Atlantic Ocean. São Luís is a trading and distribution center for agricultural products. Industries produce babassu oil, cacao, sugar, rum, and canned fruits. Founded in 1612 by the French and named in honor of Louis XIII, São Luís was captured in 1615 by the Portuguese and was occupied by the Dutch from 1641 to 1644. A noted cultural center in the 19th cent., it was also considered a bourgeois stronghold and merchant's center, in contrast to the aristocratic city of Alcantara across São Marcos Bay.

São Mi·guel (mē-gĕl'), island, 288 sq mi (746 sq km), one of the E Azores, in Ponta Delgada dist., Portugal. It is the largest island of the archipelago. The soil yields pineapples, oranges, tea, and other produce. The island's volcanic features, including a crater with a large lake, attract tourists.

Saône (sōn), river, 268 mi (431 km) long, rising in the Vosges Mts. near Épinal, E France, and flowing SW to join the Rhône River at Lyons. An important transportation link between Paris and Marseilles, it is connected by canals to the Moselle, Marne, Yonne, and Loire rivers. Because of its even and gentle flow, the Saône carries more traffic than the Rhône. There are famous vineyards along its course.

Saône-et-Loire (sōn'ā-lwär'), department (1975 pop. 569,810), 3,307 sq mi (8,565 sq km), E France, in Burgundy; capital Mâcon.

São Pau·lo (pou'lōō), state (1975 est. pop. 20,636,900), 95,713 sq mi (247,897 sq km), SE Brazil; capital São Paulo.

São Paulo, city (1970 pop. 5,186,752; metropolitan area 5,921,796), capital of São Paulo state, SE Brazil, on the Tietê River. The largest city of Brazil and of South America, São Paulo is an ultramodern metropolis with a tropical climate moderated by the city's altitude (2,700 ft/6,823 m). It dominates the vast hinterland of Brazil's wealthiest agricultural state and is the commercial, financial, and industrial center of Brazil. Through its Atlantic Ocean port of Santos it ships the farm produce of the interior. São Paulo's chief manufactures are textiles, processed foods, motor vehicles, heavy machinery, metal products, electrical equipment, pharmaceuticals, chemicals, furniture, clothing, shoes, paper, synthetic rubber, and tobacco products. Printing and publishing are also important. Abundant hydroelectric power has spurred industrial growth. The city is a major road, rail, and air transportation hub and has a modern subway system.

São Paulo was founded by Jesuit priests on Jan. 25, 1554, on the site of an old Indian village. In the 17th cent. it became a base for expeditions seeking mineral wealth and Indian slaves in the Brazilian interior. In 1681 São Paulo was made the administrative capital of the surrounding area, and in 1711 it achieved city status. However, it remained a minor commercial center for a sugar-cane and diversified agricultural region until the 1880s, when widespread coffee cultivation in São Paulo state brought sudden growth, prosperity, and an influx of European immigrants. The city has four universities, a medical school, a law school, two art museums, and the noted Butantan Institute, where snake serums are prepared.

São To·mé (tōō-mĕ'), town (1960 pop. 7,364), capital of São Tomé and Principe and a port on São Tomé island, in the Gulf of Guinea. It is the province's commercial center and main port. The chief exports are cocoa, coffee, copra, and palm products. The town has been the seat of a Roman Catholic bishop since 1534.

São Tomé and Prin·ci·pe (prēn'sē-pə), republic (1970 pop. 73,811), 372 sq mi (964 sq km), W Africa, in the Gulf of Guinea, c.150 mi (240 km) W of Gabon. São Tomé is the capital. The country, which consists of the islands of São Tomé, Principe, Pedras Tinhosas, and Rolas, was an overseas province of Portugal from the early 16th cent. until independence was achieved in July, 1975. Located just north of the equator, the islands are of volcanic origin. They have a tropical rain forest climate and a thick vegetation cover. Plantation-grown tropical produce, especially cocoa, is exported. The islands were discovered (1471) by Portuguese explorers, and in 1483 the São Tomé settlement was founded. They were proclaimed a colony of Portugal in 1522. The Dutch held the islands from 1641 to 1740, when they were recovered by the Portuguese. The plantation economy was established in the 18th cent. Following the proclamation of independence, Manuel Pinto da Costa, secretary-general of the Gabon-based Movement for the Liberation of São Tomé and Principe, became the first president of the republic. The first years of independence were accompanied by economic hardship caused by the departure of the Portuguese and the sizable population of contract and migrant workers.

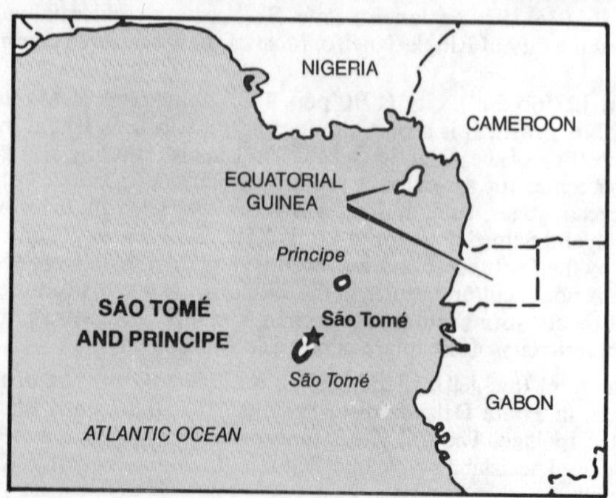

São Vi·cen·te (vē-sĕn'tĭ), city (1970 pop. 116,625), São Paulo state, SE Brazil, off the mainland on an island in the Atlantic. It was the first permanent Portuguese settlement (1532) in Brazil and during the 16th and parts of the 17th cent. was capital of the São Vicente captaincy. It was sacked in 1591 by the English pirate Thomas Cavendish. São Vicente is now a residential suburb of Santos (which is also on the island) and an ultramodern beach resort.

Sa·pe·le (sə-pā'lē), city (1969 est. pop. 71,000), S Nigeria, a port in the Niger delta. The center of the Nigerian timber industry, Sapele has sawmills and a large plywood and veneer factory; rubber is processed here, and shoes are manufactured. After the British established a vice consulate in the city in 1892, Sapele grew in importance as a port; in 1894 it came under British protection.

Sap·po·ro (säp-pō'rō), city (1976 est. pop. 1,276,547), capital of Hokkaido prefecture, SW Hokkaido, Japan. It is one of Japan's most rapidly growing urban centers. Food processing, lumbering, woodworking, and printing are the major industries. Sapporo is also a tourist center. It is famous for its annual snow festival and played host to the 1972 winter Olympics.

Sa·pul·pa (sə-pŭl'pə), city (1970 pop. 15,159), seat of Creek co., E central Okla.; inc. 1898. It is the trade center of a farm and oil region. Pottery and glass products are the chief manufactures; steel tanks and oil-drilling equipment are also made.

Sar·a·gos·sa (sâr'ə-gŏ'sə): *see* Zaragoza, Spain.

Sar·a·je·vo (sâr'ə-yā'vō), city (1971 pop. 292,241), capital of Bosnia and Hercegovina, S central Yugoslavia, on the Bosnia River. An important industrial and railway center, it has industries that manufacture metal products, electrical equipment, textiles, and tobacco. Lignite, iron ore, and manganese are mined nearby. The city is the seat of an Orthodox Eastern metropolitan, a Roman Catholic archbishop, and the chief ulema of the Yugoslav Moslems, who constitute a majority of the population. Founded in 1263, Sarajevo fell to the Turks in 1429. The town became an important Turkish military and commercial center and reached the peak of its prosperity in the 16th cent. The Congress of Berlin (1878) gave Sarajevo and the rest of Bosnia and Hercegovina to Austria-Hungary, to whom it belonged until its incorporation in 1918 into Yugoslavia. The city was a center of the Serbian nationalist movement. The assassination in Sarajevo of Archduke Francis Ferdinand and his wife on June 28, 1914, was an immediate cause of World War I. The city is noted for its Moslem architecture, including its Oriental marketplace and more than 100 mosques, the most important one dating from 1450.

Sar·a·nac Lake (sâr'ə-năk'), village (1970 pop. 6,086), Essex and Franklin cos., N N.Y., in the Adirondacks; settled c.1819 as a lumbering town, inc. 1892. It was developed as a health center after Edward L. Trudeau founded a tuberculosis sanatorium here in 1884. The sanatorium was closed in 1954, but its research center is still in operation.

Sa·ransk (sə-ränsk'), city (1976 est. pop. 241,000), capital of the Mordvinian Autonomous Republic, central European USSR. Machine building and food processing are the major industries. Saransk was founded as a fort in 1680.

Sa·ra·pul (sə-rä'pōōl), city (1976 est. pop. 107,000), E European USSR, in the Udmurt ASSR, on the Kama River. It has important dock facilities and is a rail junction. Industries include food processing, woodworking, tanning, and the production of oil field machinery, machine tools, radio equipment, steel, and aircraft parts. Founded in the late 16th cent., Sarapul was devastated in the Pugachev rebellion of 1773. During the early 19th cent. it served as a trade center on the route to Siberia.

Sar·a·so·ta (sâr'ə-sō'tə), county (1970 pop. 120,413), 587 sq mi (1,520 sq km), SW Fla., on the Gulf of Mexico, in a lowland area drained by the Myakka River and bordered by barrier beaches and Sarasota Bay; formed 1921; co. seat Sarasota. It is in a citrus-fruit, truck-farming, and cattle-raising area, with stone quarrying, lumbering, and manufacturing (textiles, furniture, concrete, metal products, and machinery). It is a year-round tourist region.

Sarasota, city (1978 est. pop. 47,100), seat of Sarasota co., SW Fla., on Sarasota Bay; settled c.1884, inc. 1914. It is a yachting and fishing resort with a construction industry, varied light manufacturing, and packing houses handling the citrus fruit, celery, and beef raised in the area. Sarasota is the former winter home of the Ringling Brothers and Barnum & Bailey Circus and is the site of the John and Mable Ringling Museum of Art, which is reputed to have the largest Rubens collection in the United States. Other attractions are the Circus Hall of Fame and the Cars of Yesterday Museum. Nearby, on the keys off the Gulf of Mexico, are many beautiful white-sand beaches.

Sar·a·to·ga (săr'ə-tō'gə), county (1970 pop. 121,764), 818 sq mi (2,126.8 sq km), E N.Y., partly in the S Adirondacks, bounded on the E by the Hudson River and on the S by the Mohawk River; formed 1791; co. seat Ballston Spa. It has dairying, farming (corn, vegetables, apples, and hay), poultry and stock raising, and manufacturing of paper, textiles, and chemicals.

Saratoga, residential city (1978 est. pop. 30,000), Santa Clara co., W Calif., in a vineyard and orchard area, in the foothills of the Santa Cruz Mts.; inc. 1956. Wine is produced in the city; local attractions include tours of the champagne cellars. The Villa Montalvo estate (1912), home of the late Sen. James Phelan, is a cultural center; its extensive facilities include art galleries, theaters, gardens, and an outdoor amphitheater.

Saratoga National Historical Park: *see* National Parks and Monuments Table.

Saratoga Springs, resort and residential city (1978 est. pop. 26,300), Saratoga co., E N.Y.; inc. as a village 1826, as a city 1915. Skidmore College is the largest source of employment. Lingerie, electronic and electrical supplies, farm equipment, and food products are among the city's manufactures. The last battle of the Saratoga campaign was fought near the city in 1777. The nearby Saratoga National Historical Park embraces the battlefield. After the Revolution, as the fame of its carbonated mineral waters spread, the village became a health and pleasure resort. In the 19th cent. Saratoga Springs was one of the most popular social and sporting centers in America. Horse racing, which continues to be one of its major attractions, was begun after 1863. An elaborate state-owned spa (1935) preserves and utilizes the waters and offers curative baths. Petrified gardens are to the west, and to the north is the cottage on Mt. McGregor where President Ulysses S. Grant completed his memoirs and spent the last weeks of his life.

Sa·ra·tov (sə-rä'təf), city (1977 est. pop. 856,000), capital of Saratov oblast, E European USSR, on the Volga River. Saratov's industries produce precision instruments, building materials, construction equipment, and electric generators. There are oil refineries, ship-repairing docks, gas plants, and chemical factories. The city was founded c.1590 as a Russian sentry post on the Volga. Although its military importance declined in the 18th cent., the city retained significance for its river trade.

Sar·celles (sär-sĕl'), city (1975 pop. 55,007), Val-d'Oise dept., N central France. Mostly residential, it has some light industry. A church dating partly from the 12th cent. and partly from the Renaissance is in Sarcelles.

Sar·din·i·a (sär-dĭn'ē-ə), region (1971 pop. 1,468,737), 9,302 sq mi (24,092 sq km), W Italy, mostly on the Mediterranean island of Sardinia, which is separated in the N from Corsica by the Strait of Bonifacio. The region also includes Asinara, Caprera, San Pietro, and La Maddalena islands. Cagliari is the capital of Sardinia, which is divided into the provinces of Cagliari, Nuoro, and Sassari. The highest point of the mostly mountainous island is Mt. Gennargentu (6,016 ft/1,834 m). The main agricultural area is the large Campidano Plain, located in the southwest and watered by the Manno and Tirso rivers. Natural pastures cover more than half the area of Sardinia; sheep and goats are widely raised. Wheat, barley, grapes, olives, cork, and tobacco are produced. Sardinia is rich in minerals, including zinc, lead, antimony, lignite, copper, iron, and salt. Fishing for tuna, lobster, and sardines is important. There is still little industry; manufactures include processed food, wine, refined petroleum, paper, cement, and textiles.

An early center of trade, Sardinia was mentioned in Egyptian sources in the 13th cent. B.C., and many traces of its prehistoric inhabitants remain. Phoenicians (c.800 B.C.) and Carthaginians (c.500 B.C.) settled here before Rome conquered the island (238 B.C.). After the fall of Rome, Sardinia passed to the Vandals (mid-5th cent. A.D.) and then to the Byzantines (early 6th cent.). The popes claimed suzerainty over it and helped repel Arab attacks (8th-11th cent.). Later Pisa and Genoa often fought (11th-14th cent.) for supremacy over the island. In 1297 Pope Boniface VIII bestowed the island on the house of Aragón, from which it passed to Spain (late 15th cent.), then Austria (1713), and finally the House of Savoy (1720). The Savoy kings of Sardinia tried to establish some order out of chaos on Sardinia with judicial, agrarian, and ecclesiastic reforms. Administrative autonomy was ended in 1847; however, the region received some autonomy under the Italian constitution of 1947.

Sardinia, kingdom of, name given to the possessions of the house of Savoy in 1720, when the island of Sardinia was awarded (by the Treaty of London) to Duke Victor Amadeus II of Savoy to compensate him for the loss of Sicily to Austria. Besides Sardinia, the kingdom included Savoy, Piedmont, and Nice; Liguria, including Genoa, was added by the Congress of Vienna in 1815. During the Risorgimento the kingdom expanded to include almost all Italy. Lombardy was added in 1859. In 1860 Parma, Modena, Bologna, Marche, and the Romagna were annexed by the kingdom. After the annexation (1861) of the Two Sicilies, Victor Emmanuel II of Sardinia was proclaimed king of Italy.

Sar·dis (sär'dĭs) or **Sar·des** (sär'dēz), ancient city of Lydia, W Asia Minor, at the foot of Mt. Tmolus, NE of the modern Izmir, Turkey. As capital of Lydia, it was the political and cultural center of Asia Minor from 650 B.C. until the death of Croesus (c.547 B.C.). The first coins were minted here in the 6th cent. B.C. An almost impregnable citadel, Sardis was nevertheless captured in 499 B.C. by the Ionians in the Persian Wars. In 133 B.C. it passed to the Romans. The city was destroyed by Tamerlane in the 14th cent. Excavations uncovered the temple of Artemis (dating from the 4th cent. B.C.) and inscriptions in old Lydian. The actual site of the city was not discovered until 1958.

Sardis, town (1970 pop. 2,391), seat of Panola co., NW Miss., S of Memphis, Tenn., near the Tallahatchie River, in a dairy and farm area; founded 1856.

Sa·re·ma or **Saa·re·maa** (both: sä'rĕ-mä), island, 1,048 sq mi (2,714 sq km), off the mainland of Estonia, W USSR, in the Baltic Sea, across the mouth of the Gulf of Riga. It is irregular in shape and has a level terrain. Dairy farming, stock raising, and fishing are the chief occupations. It is also a health resort. The island was ruled by the Livonian Knights until 1560, when it passed to Denmark, which in turn ceded (1645) it to Sweden. Sarema passed to Russia in 1710 and was incorporated into newly independent Estonia in 1917.

Sar·gas·so Sea (sär-găs'ō), part of the N Atlantic Ocean, lying roughly between the West Indies and the Azores and from about lat. 20° N to lat. 35° N, in the horse latitudes. The relatively still sea is the center of a great swirl of ocean currents and is a rich field for the marine biologist. It is noted for the abundance of gulfweed on its surface.

Sar·gent (sär'jənt), county (1970 pop. 5,937), 853 sq mi (2,209.3 sq km), SE N.Dak., in an agricultural area drained by the Wild Rice River; formed 1883; co. seat Forman.

Sar·go·dha (sär'gō-dä'), city (1972 pop. 140,000), E Pakistan, on the lower Jhelum Canal. It is a center for trade in cotton and other agricultural commodities. Sargodha has a steel-rolling mill and railroad repair shops.

Sa·ri (sä-rē'), city (1971 est. pop. 50,000), capital of Mazanderan prov., N Iran, near the Caspian Sea. It is the trade center for a farm region where citrus fruit, rice, and sugar cane are grown.

Sa·ri·ta (sə-rē'tə), village (1969 est. pop. 196), seat of Kenedy co., extreme S Texas, S of Kingsville. It is a rail center in a cattle-ranching area.

Sark (särk), island (1971 pop. 584), 2 sq mi (5 sq km), in the English Channel, E of Guernsey, one of the Channel Islands. It is divided into Great Sark and Little Sark, which are connected by a natural causeway, the Coupée. The interior is reached through tunnels from Creux Harbour, the landing place, on the east. The island belongs to Guernsey bailiwick; its local government is a survival of the feudal system. The economy is agricultural.

Sar·ma·tia (sär-mā'shē-ə, -shə), ancient district between the Vistula River and the Caspian Sea, occupied by the Sarmatians from the 3rd cent. B.C. through the 2nd cent. A.D. The term is vague and is also used to refer to the territory along the Danube and across the Carpathians where the Sarmatians were later driven by the Huns. They came into conflict with the Romans but later allied themselves with Rome, acting as buffers against the Germans. They were scattered or assimilated with the Germans by the 3rd cent. A.D.

Sar·ni·a (sär'nē-ə), city (1976 pop. 55,576), S Ont., Canada, on the St. Clair River, at the S end of Lake Huron opposite Port Huron, Mich. The city is a port and handles a large volume of freight. There are grain elevators, machinery plants, oil refineries, and chemical and synthetic-rubber industries.

Sa·ron·ic Gulf (sə-rŏn'ĭk), arm of the Aegean Sea, indenting SE Greece and separated from the Gulf of Corinth by the Isthmus of Corinth. The Saronic Gulf is the eastern terminus of the Corinth Canal, which cuts across the isthmus.

Sarps·borg (särps'bôr), city (1977 est. pop. 12,876), Ostfold co., SE Norway, a port on the Glåma River near its mouth in the Oslofjord. Manufactures include forest products, chemicals, and textiles. There is a large hydroelectric plant. Sarpsborg was founded in 1016 by Olaf II, was burned by the Swedes in 1567, and was rebuilt in 1839. Skjeberg Church, a medieval stone structure, is nearby.

Sar·py (sär'pē), county (1970 pop. 66,200), 239 sq mi (619 sq km), E Nebr., bounded in the E by the Missouri River and Iowa, in the W and S by the Platte River; formed 1857; co. seat Papillion. Feed, livestock, grain, fruit, dairy goods, and poultry are produced. The county lost territory to Pottawatamie co., Iowa, in 1943.

Sarthe (särt), department (1975 pop. 490,385), 2,397 sq mi (6,208.2 sq km), NW France; capital Le Mans.

Sa·se·bo (sä-sä'bō), city (1976 est. pop. 251,607), Nagasaki prefecture, W Kyushu, Japan. It is a port and naval base on the East China Sea. Shipbuilding, machine building, and food processing are the chief industries.

Sas·katch·e·wan (səs-kăch'ə-wən, -wän', săs-), province (1975 pop. 918,000), 251,700 sq mi (651,903 sq km), W Canada. Regina is the capital and largest city. Saskatchewan is bounded on the north by the Northwest Territories, on the east by Man., on the south by N.Dak. and Mont., and on the west by Alta. Its desolate northern third is part of the Laurentian Plateau. The principal rivers are the Churchill, the North and South Saskatchewan, and the Qu'Appelle. Between the Saskatchewan and Churchill rivers lies a mixed forest belt containing much marketable timber; a section is reserved as Prince Albert National Park. Only in south Saskatchewan has there been any substantial settlement or development.

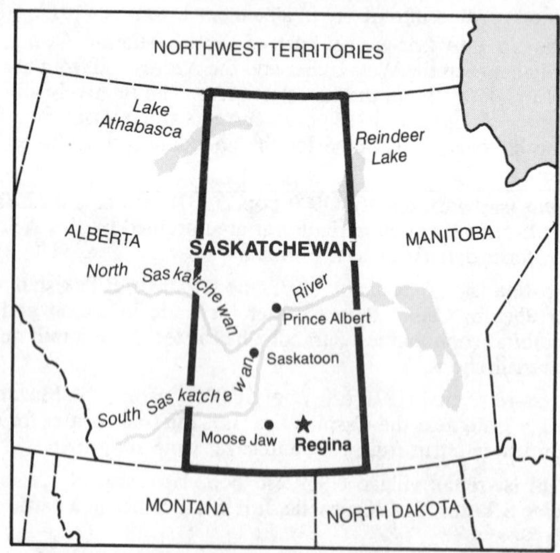

Except for a semiarid section in the southwest used for grazing and an area in the east and central portion given over to mixed farming and dairying, the land is devoted to the raising of hard wheat. The vast expanses of unbroken plain are well suited to large-scale mechanized farming. Oats, barley, rye, and flax are also grown throughout this region. Saskatchewan is rich in minerals. Uranium, copper, zinc, gold, coal, and potash are mined, and oil and natural gas have been discovered. Most of the province's industries process raw materials. A steel mill was opened in Regina in 1960.

The earliest trading posts in the area were established by the French (c.1750), but the first permanent settlement was made at Cumberland House in 1774 by the Hudson's Bay Company. Subsequently many other posts were set up by British fur traders along the region's waterways. In 1870 the Hudson's Bay Company ceded its rights to the Canadian government, and the area became part of the Northwest Territories. Saskatchewan became a province in 1905.

Saskatchewan, river, c.340 mi (550 km) long, formed by the confluence of the North Saskatchewan (c.760 mi/1,220 km long) and the South Saskatchewan (c.550 mi/890 km long) rivers near Prince Albert, central Sask., Canada; the system drains most of the Canadian prairie provinces. It flows generally east across the Man. border to Lake Winnipeg. The North Saskatchewan River rises in the Columbia ice field at the foot of Mt. Saskatchewan, southwest Alta., and flows generally east past Edmonton, into Saskatchewan prov., and then past North Battleford to Prince Albert. The South Saskatchewan River is formed in southern Alta. by the junction of the Bow and Oldman rivers. It flows east past Medicine Hat, then northeast into Saskatchewan prov., past Saskatoon, to Prince Albert. The Bow-South Saskatchewan-Saskatchewan system is c.1,200 mi (1,930 km) long. The Saskatchewan River and its branches were once important thoroughfares for explorers and trappers.

Sas·ka·toon (săs'kə-to͞on'), city (1976 pop. 133,750), S central Sask., Canada, on the South Saskatchewan River. It has grain elevators, grain and flour mills, stockyards, meat-packing plants, oil refineries, and potash-processing plants in the city. Saskatoon was settled in 1883 and grew rapidly after the coming of the railroad (1890).

Sas·sa·fras Mountain (săs'ə-frăs'), peak, 3,560 ft (1,085 m) high, NW S.C., in the Blue Ridge Mts., near the N.C. and Ga. lines. It is the highest point in the state.

Sas·sa·ri (säs'sä-rē), city (1976 est. pop. 115,990), capital of Sassari prov., NW Sardinia, Italy. It is an agricultural trade center, handling cheese, wine, fruit, and olive oil. Zinc and lead are mined nearby. Sassari was an important center in the Middle Ages and was held (13th cent.) by Genoa and (14th cent.) by the Aragonese. It passed to Piedmont in the early 18th cent.

Sa·tsu·ma (sä-tso͞o'mä), peninsula, Kagoshima prefecture, SW Kyushu, Japan. It gives its name to a famous porcelain, which was first manufactured here by Korean artisans in the 16th cent. Satsuma was a feudal province controlled by the powerful Shimazu clan, which exacted tribute from the Ryukyu Islands from the 17th to the 19th cent. In 1877 Takamori Saigo led the Satsuma clansmen in a rebellion against the imperial government. This rebellion, suppressed by the imperial army, was the last serious internal threat to the Meiji restoration.

Sa·tu-Ma·re (sä'to͞o-mä'rĕ), city (1977 est. pop. 103,612), NW Rumania, in Crişana-Maramureş, on the Someşul River, near the Hungarian border. The commercial and cultural center of a fertile agricultural region, it has industries that produce mining equipment, machine tools, metals, and textiles. The Peace of Szatmár, negotiated in 1711, ended the rebellion of Francis II Rakoczy.

Sa·u·di A·ra·bi·a (sä-o͞o'dē ə-rā'bē-ə, sou'-, sô'-), kingdom (1973 est. pop. 7,200,000), 829,995 sq mi (2,149,690 sq km), comprising most of the Arabian peninsula. Riyadh is the capital. Saudi Arabia is bounded on the west by the Gulf of Aqaba and the Red Sea, on the east by the Persian Gulf and Qatar, and on the north by Jordan, Iraq, a neutral zone, and Kuwait. To the southwest lies Yemen.

The south and southeast of the country are occupied entirely by the great Rub al Khali desert. Through the desert run largely undefined boundaries with the People's Republic of Yemen, Oman, and the United Arab Emirates. In addition to the Rub al Khali, Saudi Arabia has four major regions. The largest is the Nejd, a central plateau, which rises from c.2,000 ft (610 m) in the east to c.5,000 ft (1,520 m) in the west. Riyadh is located in the Nejd. The Hejaz stretches along the Red Sea from the Gulf of Aqaba south to Asir and is the site of the holy cities of Mecca and Medina. Asir, extending south to the Yemen Arab Republic border, has a fertile coastal plain. Inland mountains in the Asir region rise to more than 9,000 ft (2,743 m). The Eastern Province extends along the Persian Gulf and is the oil region of the country. The oasis of Al Hasa, located here, is probably the country's largest. Saudi Arabia's climate is generally hot and dry, although nights are cool, and there may be frosts in winter. The humidity along the coasts is high. The population is predominantly Moslem Arab of the Wahabi sect. The country is virtually an absolute monarchy, ruled under Moslem law.

Economy. Nomads and seminomads, who comprise about 60% of the population, raise camels, sheep, goats, and horses. Because of the

scarcity of water, agriculture has been restricted to Asir and to oases strung along the wadis, but recent irrigation projects have reclaimed many acres of desert. Mecca, Medina, and the port of Jidda have derived much income from religious pilgrims. The oil industry, located in the northeast along the Persian Gulf, dominates the economy. Oil is sent by pipeline (the Tapline) to Sidon, Lebanon, and by ship to Bahrain for refining. In Saudi Arabia there are refineries at Ras Tanura on the Persian Gulf. The oil boom after World War II led to the construction of the Ad-Dammam-Riyadh railroad, the development of Ad-Dammam as a deepwater port, and the bringing of electricity to the towns. Schools, hospitals, and new homes, particularly for the oil workers, have been built.

History. The origins of Saudi Arabia as a political unit lay with the puritanical Wahabi movement (18th cent.), which gained the allegiance of the powerful Saud family of the Nejd, in central Arabia. Supported by a large Bedouin following, the Sauds brought most of the peninsula under their control. The Wahabi movement was crushed (1811-18) by an Egyptian expedition, revived in the mid-19th cent., and was defeated again in 1891 by the Rashid dynasty. Beginning the Wahabi reconquest at the turn of the century, ibn Saud took Riyadh in 1902 and was master of the Nejd by 1906. On the eve of World War I he conquered the Al Hasa region from the Turks and extended his control over the Hejaz in 1924-25. The unified kingdom of Saudi Arabia was proclaimed in 1932.

Oil was discovered in 1936, and commercial production began in 1938. Saudi Arabia joined the Arab League in 1945, but it played only a minor role in the Arab wars with Israel in 1948, 1967, and 1973. Ibn Saud died in 1953 and was succeeded by his eldest son, Saud, who was deposed by his brother Faisal in 1964. The conservative Saudi leaders opposed the socialism and pan-Arabism of Egypt but agreed to send aid to Egypt and Jordan after the Arab defeat in the 1967 war with Israel. Saudi Arabia supported the royalists in the Yemeni civil war (1962-70). The government assumed a majority ownership of Aramco (a conglomerate of oil companies operating in the country) in 1974 and concluded an agreement for a full take-over in 1977. King Faisal was assassinated in 1975. His half-brother and successor, King Khalid pursued policies favoring limited oil price increases and a peaceful solution to the Arab-Israeli conflict.

Sauer (zou'ər), principal river of Luxembourg, c.100 mi (160 km) long, rising in the Ardennes, SE Belgium, and flowing E through Luxembourg, then S to join the Moselle River. With its tributaries, the Sauer drains all of Luxembourg.

Sau·gus (sô'gəs), town (1978 est. pop. 24,400), Essex co., NE Mass., a suburb of Boston on the Saugus River near the Atlantic Ocean; settled before 1637, set off from Lynn and inc. 1815. There are machine shops in the town. The Saugus ironworks (1646-c.1670; restored 1954) were the first successful enterprise of the kind in the colonies.

Saugus Iron Works National Historic Site: *see* National Parks and Monuments Table.

Sauk (sôk), county (1970 pop. 39,057), 841 sq mi (2,178.2 sq km), S central Wis., bounded in the NE and S by the Wisconsin River, and drained by the Baraboo River; formed 1840; co. seat Baraboo. It is in a dairying, corn-farming, and livestock-raising area, with sand and gravel pits, lumbering, and other industries (food processing and the manufacture of furniture, farm and garden machinery, and equipment). Several lake resorts are in the county.

Sauk Cen·tre (sĕn'tər), city (1970 pop. 3,750), Stearns co., central Minn., WNW of St. Cloud; settled 1856, platted 1863, inc. as a village 1876, as a city 1889. It is a trade center for a farm and dairy region. Sinclair Lewis laid the scene of *Main Street* in Sauk Centre, his birthplace.

Sauk Village, village (1970 pop. 7,479), Cook co., NE Ill., just SE of Chicago Heights; inc. 1957.

Sault Sainte Ma·rie (so͞o' sānt' mə-rē'), city (1976 pop. 81,048), S Ont., Canada, on the St. Marys River opposite Sault Ste Marie, Mich. Sault Ste Marie is an important port and manufacturing center. Iron and steel, lumber, pulp and paper products, and chemicals are made. It is a tourist center and the gateway to hunting and fishing resorts in nearby lake and forest regions. A fur-trading post was built on the site in 1783, and a canal and lock to bypass the St. Marys rapids was constructed by 1898. Americans destroyed the post and lock during the War of 1812; a new lock was opened in 1895.

Sault Sainte Marie, city (1978 est. pop. 16,300), seat of Chippewa co., N Mich., Upper Peninsula, on the St. Marys River opposite Sault Ste Marie, Ont.; inc. as a city 1887. Birch veneers, tools, dies, and stampings are made; there are also welding and ship-repair services. Because of the distance from markets, competition from other products, and the harsh winter weather, most of the city's larger industries closed during the 1960s. Tourism remains the economic mainstay. The famous Soo locks on the St. Marys River are a major attraction. Heavy-laden ocean vessels and Great Lakes freighters pass through the intricate system that links Lakes Superior and Huron. Excellent hunting and fishing in the area also draw sportsmen. The region at the rapids of the St. Marys River was discovered by Etienne Brule in 1615. Father Jacques Marquette established a Jesuit mission here in 1668, and trading posts followed. French occupation ended in 1763, and the British remained in control until the Americans arrived (1820). U.S. Fort Brady was built shortly thereafter. The Sault Ste Marie Canal was built in 1855 to facilitate the flow of ore. Other attractions in the city are the site of Father Marquette's mission and several historic homes. A giant international bridge connects Sault Ste Marie with its Canadian counterpart. Kincheloe Air Force Base, a strategic air command installation, is nearby.

Sault Sainte Marie Canals, two ship canals by-passing the rapids on the St. Marys River between Lake Superior and Lake Huron, at the cities of Sault Ste Marie, Mich., and Ont. The Canadian canal (1.4 mi/2.3 km long and 60 ft/18 m wide), which has one lock, was opened in 1895. It follows the route of the first canal constructed around the rapids (1797-98) by a fur company. The U.S. canal (1.6 mi/2.6 km long and 80 ft/24 m wide) was constructed (1853-55) by Mich. state and has since been reconstructed by the Federal government to accommodate larger vessels; it has four locks. Although closed by ice during the winter, the toll-free canals are among the busiest in the world and are a vital link in the Great Lakes Waterway. Most of the ships pass through the larger and deeper U.S. canal. The waterways are popularly called the Soo Canals.

Sau·mur (sō-mür'), town (1975 pop. 32,515), Maine-et-Loire dept., W France, on the Loire River. Saumur is noted for its religious-medal industry (dating from the 17th cent.) and for its sparkling white wines. Canned goods, clothing, toys, and liquors are also produced. Saumur, founded in Roman times, was seized from the counts of Blois in 1026 by Fulk Nerra, count of Anjou. As part of Anjou it was joined to the French crown in 1204 by Philip II. In the 16th cent. Saumur was given by Henry III to the then Protestant Henry of Navarre (later Henry IV). A famous Protestant academy was founded (1599), and the town became a bastion of the Huguenot movement. With the revocation of the Edict of Nantes in 1685, much of the population emigrated. Among the monuments in Saumur are the remarkable Romanesque Church of Notre-Dame-de-Nantilly (begun 12th cent.) and many Renaissance structures. Collections of art and tapestries are also preserved.

Saun·ders (sôn'dərz), county (1970 pop. 17,018), 759 sq mi (1,965.8 sq km), E Nebr., bounded in the E and N by the Platte River; formed 1867; co. seat Wahoo. It is known for its feed, grain, livestock, dairy goods, and poultry.

Sau·sa·li·to (sô'sə-lē'tō), residential city (1970 pop. 6,158), W Calif.,

N of San Francisco; inc. 1893. It is the northern terminus of the Golden Gate Bridge. There is an artists' colony here.

Sa·va (sä'vä), longest river of Yugoslavia, c.580 mi (930 km), rising in two headstreams in the Julian Alps, NW Yugoslavia, and flowing SE across N Yugoslavia to the Danube River at Belgrade. The Sava basin is a fertile agricultural region.

Sa·vai'i (sä-vī'ē), volcanic island (1971 pop. 40,572), Western Samoa. It is the largest (c.700 sq mi/1,810 sq km) and most westerly of the Samoan islands. Savai'i is fertile and mountainous. Bananas, copra, and cocoa are exported.

Sa·va·lan (sä'və-län'), mountain, Iran: *see* Sabalan.

sa·van·na or **sa·van·nah** (both: sə-văn'ə), tropical or subtropical grassland lying on the margin of the trade-wind belts. The climate of a savanna is characterized by a rainy period during the summer when the area is covered by grasses, and by a dry winter when the grasses wither. The most extensive savannas—all important pasture lands—are in Africa; others include the llanos and the campos of South America.

Sa·van·nah (sə-văn'ə). **1.** City (1978 est. pop. 112,500), seat of Chatham co., SE Ga., a port of entry on the Savannah River near its mouth; inc. 1789. It is a rail, fishing, and industrial center. Shipping is its major industry; manufactured goods are imported in great quantity, and exports include tobacco, cotton, sugar, clay, rosin, linerboard, and wood pulp. Chemicals, petroleum, rubber, lumber, plastics, and paper are among the city's many products. Savannah was founded by James Oglethorpe in 1733. During the American Revolution the British took Savannah on Dec. 29, 1778, and held it until July, 1782. With the growth of trade, the invention of the cotton gin, and the construction of a network of railroads, the city became a rival of Charleston as a commercial center. In the Civil War Fort Pulaski, on an island near the mouth of the Savannah River, was captured by Federals in 1862, but the city did not fall until Dec. 21, 1864, when Gen. W. T. Sherman entered it on completing his march to the sea. Savannah has beautiful, wide, shaded streets and many parks. Magnolias, pines, and ancient oaks are indigenous here. Despite devastating fires in 1796 and 1820, many old homes and buildings remain. The mansion birthplace of Juliette Gordon Low (built 1819–21) is owned and operated by the Girl Scouts of the U.S.A. as a memorial to their founder. The city's historic district was designated a national historic landmark in 1966. Savannah has numerous fine old churches, many schools, and several museums. An air force base and a U.S. coast guard station are here, and several beach and island resorts are nearby. **2.** City (1970 pop. 3,324), seat of Andrew co., NW Mo., N of St. Joseph; laid out 1842, inc. 1853. It ships farm products. **3.** Town (1970 pop. 5,576), seat of Hardin co., S Tenn., on the E bank of the Tennessee River SE of Jackson, in a farm area; laid out c.1823, inc. 1833. Clothing and wood products are made. Nearby are Shiloh National Military Park and Shiloh National Cemetery. There are Indian mounds here.

Savannah, river, 314 mi (505 km) long, formed by the confluence of the Tugaloo and Seneca rivers and flowing SE to the Atlantic Ocean; with the Tugaloo it forms the entire S.C.–Ga. boundary. Savannah, Ga., is the head of navigation for oceangoing ships.

Sa·voie (sä-vwä'), department (1975 pop. 305,118), SE France, bordering Italy; capital Chambéry.

Sa·vo·na (sä-vô'nä), city (1976 est. pop. 79,662), capital of Savona prov., in Liguria, NW Italy, on the Riviera. It is a major seaport. Manufactures include iron and steel, machinery, and chemicals. The seat of a marquisate in the Middle Ages, Savona was a busy commercial center until it was defeated and annexed by Génoa in 1528.

Sa·von·lin·na (sä'vōn-lĭn'nä), city (1975 est. pop. 28,336), Mikkeli prov., SE Finland. Situated in the Saimaa lake region, it is a resort, inland port, and road and railroad junction, with a large plywood industry, machine shops, and sawmills. It was built around the 15th cent. fortress of Olavinlinna and was chartered in 1639.

Sa·voy (sə-voi'), Alpine region of E France. The boundaries of old Savoy have changed with time, but presently the region is bounded on the north by Lake Geneva, on the west by the Rhône River, on the south by Dauphiné, and on the east by the Alpine crest on the Swiss and Italian borders. The region commands many important passes connecting France and Italy and includes the French portion of the highest Alpine peak, Mont Blanc. Agriculture and dairying are the region's chief occupations. Tourism is also important.

Savoy was inhabited by the Allobroges at the time Julius Caesar conquered the region. It became part of the first kingdom of Burgundy (5th cent.) and later of the kingdom of Arles (10th cent.), after which it was ceded to the Holy Roman Empire. In the 11th cent. Humbert the Whitehanded, a lord of Arles, consolidated the various feudal territories of the region. Under Amadeus VIII, Savoy became (early 15th cent.) a duchy extending far into France, Italy, and Switzerland. By the beginning of the 16th cent. Savoy fell under French and Swiss dominance. Emmanuel Philibert restored the territory and fortunes of the region and moved the ducal residence to Turin (1559), after which Savoy became essentially an Italian rather than a French state. When Victor Amadeus II became king of Sardinia in 1713, Savoy became a part of that new state. By the Treaty of Turin (1860), Piedmont, the ruling part of Savoy, ceded French Savoy to France. The region was annexed after a plebiscite.

Sa·watch Mountains (sə-wŏch'), high range of the Rocky Mts., W central Colo., extending c.110 mi (176 km) S from the Eagle River to the vicinity of Saguache. The Arkansas River bounds the range on the east; to the west lie the Elk Mts. and West Elk Mts. The highest summit in the range is Mt. Elbert (14,431 ft/4,398.6 m).

Sa·whaj (sô'häj), town (1970 est. pop. 85,300), capital of Sawhaj governorate, central Egypt, on the Nile River. It is located in a densely populated agricultural region. Products of the town include ginned cotton and woven textiles.

Saw·yer (sô'yər), county (1970 pop. 9,670), 1,259 sq mi (3,260.8 sq km), NW Wis., drained by the Chippewa River; formed 1883; co. seat Hayward. It is in a forested area, with dairying, potato farming, and cattle raising. Part of Chequamegon National Forest is here.

Saxe-Al·ten·burg (săks'äl'tən-bûrg'), former duchy, Thuringia, central Germany. Created a separate duchy in 1603, it was ruled by an Ernestine line of the house of Wettin. It passed (1672) to the dukes of Saxe-Gotha, but from 1826 to 1918 it was again a separate duchy. In 1920 it was incorporated into Thuringia.

Saxe-Co·burg (săks'kō'bûrg), former duchy, central Germany. A possession of the Ernestine branch of the house of Wettin, it was given by Ernest the Pious (d. 1675) of Saxe-Gotha to his son Albert. On Albert's death (1699) it passed to his younger brother, John Ernest, duke of Saxe-Saalfeld. The extinction (1825) of the related line of Saxe-Gotha-Altenburg resulted in a general redivision of the Ernestine possessions in 1826. The duchy of Saalfeld passed to the duke of Saxe-Meiningen, while Ernest III of Saxe-Coburg (or Ernest I) received the duchy of Gotha. Ernest I's brother was crowned (1831) as Leopold I, king of the Belgians, and his younger son was Prince Albert, consort of Queen Victoria of England. Thus the house of Saxe-Coburg-Gotha became the ruling dynasty of Belgium and of England. In 1920 Saxe-Gotha was incorporated into Thuringia, and Saxe-Coburg into Bavaria.

Saxe-Go·tha (săks'gō'thə), former duchy, Thuringia, central Germany. A possession of the Ernestine branch of the house of Wettin, it passed in the 16th cent. to the dukes of Saxe-Weimar. After the death (1605) of Duke John of Weimar, his territories were divided among his heirs. Saxe-Gotha gradually came under the control of Ernest the Pious, one of John's younger sons. On Ernest's death (1675), the succession was divided among his seven sons. Saxe-Gotha was awarded in 1826 to Ernest III of Saxe-Coburg (Ernest I of Saxe-Coburg-Gotha). Saxe-Altenburg became a separate duchy. In 1920 Saxe-Gotha was incorporated into Thuringia.

Saxe-Mei·ning·en (săks'mī'nĭng-ən), former duchy, Thuringia, central Germany. A possession of the Ernestine branch of the house of Wettin, it became a separate duchy in 1681 under Bernard, third son of Ernest the Pious of Saxe-Gotha. In the dynastic rearrangement that followed the extinction (1825) of the male line of Saxe-Gotha, the duke of Saxe-Meiningen received (1826) Saxe-Saalfeld from the duke of Saxe-Coburg (who obtained Gotha instead) and Saxe-Hildburghausen (whose duke was compensated with Saxe-Altenburg). Saxe-Meiningen sided (1866) with Austria in the Austro-Prussian War. The last duke abdicated in 1918, and in 1920 Saxe-Meiningen was incorporated into Thuringia.

Saxe-Wei·mar (săks'vī'mär'), former duchy, Thuringia, central Germany. The area passed in the division of 1485 to the Ernestine branch of the Wettin dynasty and remained with that branch after the redivision of the Wettin lands in 1547. The Ernestine lands were divided into the duchies of Weimar, Gotha, Coburg, Eisenach, and Altenburg. The male lines of Coburg, Gotha, and Eisenach having failed by 1640, their lands passed to the sons of Duke John of Weimar. Ernest the Pious, who had Gotha and Coburg, also inherited Altenburg in 1672; his possessions were again divided among his seven sons. An elder brother of Ernest the Pious, William, received Weimar and Eisenach; those duchies, however, were again separated under his heirs until the failure of the Eisenach line in 1741, when its territory (including Jena) reverted to duke Ernest Augustus I of Saxe-Weimar. Small as it was, the duchy of Saxe-Weimar-Eisenach, which resulted from the reunion in 1741, was the most important of the Thuringian principalities. It gained its greatest prosperity and cultural importance under duke Charles Augustus, who made Weimar an intellectual center of Europe. Charles Augustus sided against Napoleon I in the War of the Third Coalition, but was forced in 1806 to join the Confederation of the Rhine. Grand duke Charles Alexander sided (1866) with Prussia in the Austro-Prussian War. His grandson, William Ernest, abdicated in 1918, and in 1920 Saxe-Weimar-Eisenach was incorporated into Thuringia.

Sax·o·ny (săk'sə-nē). The geographic concept of Saxony has undergone great changes and acquired many meanings in the past 15 centuries. The land of the Saxons, Saxony was in Frankish times roughly the area in northwestern Germany between the Elbe and Ems rivers; it also included part of southern Jutland. After Charlemagne's conquest (772–804) of the Saxons, their land was incorporated into the Carolingian empire, and late in the 9th cent. the first duchy of Saxony, one of the five stem duchies of medieval Germany, was created. It occupied nearly all the territory between the Elbe and Saale rivers on the east and the Rhine on the west; it bordered on Franconia and Thuringia in the south.

The stem duchy was broken up into numerous fiefs late in the 12th cent. The ducal title of Saxony went to Bernard of Anhalt, founder of the Ascanian line of Saxon dukes. Widely separate territories continued after 1260 under separate branches of the Ascanians as Saxe-Lauenburg and Saxe-Wittenberg. The Golden Bull of 1356 raised the duke of Saxe-Wittenberg to the permanent rank of elector. Electoral Saxony, as his territory was called, was a relatively small area, and it lay east of the original stem duchy. In 1423 Frederick the Warlike, margrave of Meissen, added Electoral Saxony to his larger domains farther south; he became (1425) Elector Frederick I. Thus Saxony had shifted to east-central and eastern Germany from northwestern Germany.

In 1485 Saxony was partitioned between two sons of Elector Frederick II: Ernest received Electoral Saxony with Wittenberg and most of the Thuringian lands, while Albert received ducal rank and the Meissen territories, including Dresden and Leipzig. By the Capitulation of Wittenberg (1547) Ernest's nephew John Frederick was deprived of the electorate, which passed to duke Maurice of Saxony, a grandson of Albert. During the Thirty Years' War (1618–48) Electoral Saxony was occupied by Swedish troops and thoroughly devastated. In the Northern War (1700–1721) against Sweden, Elector Frederick Augustus I lost and regained the Polish throne. Unsuccessful wars with Prussia caused Saxony to lose prestige, and the union with Poland ended in 1763. The 18th cent. electors were lavish patrons of art and learning and greatly beautified their capital, Dresden. Leipzig led in the rise of German literature as well as in music, which reached its first peak with J. S. Bach.

Saxony sided with Prussia against France early in the French Revolutionary Wars but changed sides in 1806. For this act its elector was raised to royal rank, becoming King Frederick Augustus I. His failure to change sides again before the fall of Napoleon I cost him (1815) nearly half his kingdom at the Congress of Vienna. The principal remaining cities of the kingdom of Saxony were Dresden, Leipzig, Chemnitz (now Karl-Marx-Stadt), and Plauen. From 1871 until the abdication (1918) of Frederick Augustus III, Saxony was a member state of the German Empire. Meanwhile the larger part of the territories ceded in 1815 were incorporated with several other Prussian districts into the Prussian province of Saxony. This was united after 1945 with Anhalt to form the state of Saxony-Anhalt. The former kingdom of Saxony became after 1918 the state of Saxony and joined the Weimar Republic. After World War II the state of Saxony was reconstituted (1947) under Soviet occupation and joined (1949) the German Democratic Republic. It was abolished as an administrative unit in 1952 and was divided among the districts of Halle, Magdeburg, Leipzig, and Cottbus.

Sax·o·ny-An·halt (săk'sə-nē-än'hält), former state, c.10,000 sq mi (25,900 sq km), S East Germany. As constituted in 1947 under Soviet military occupation, Saxony-Anhalt consisted roughly of the former state of Anhalt, the former Prussian province of Saxony, and several small territories of the former state of Brunswick. Saxony-Anhalt was abolished as an administrative district in 1952.

Sa·ya·ma (sä-yä'mä), city (1976 est. pop. 109,069), Saitama prefecture, E central Honshu, Japan, on Lake Sayama. It is a resort and a center for various mechanical industries.

Sa·yan Mountains (sä-yän'), central Asia, chiefly in Central Asian USSR, in S Siberia. The Eastern Sayan Mts. extend c.680 mi (1,090 km) from the lower Yenisei River to the southwest end of Lake Baykal and rise to 11,686 ft (3,562 m) in the Munku-Sardyk; they form part of the USSR-Mongolian People's Republic border. The Western Sayan Mts., rising to 10,206 ft (3,111 m) in the Kyzyl-Tayga, extend c.400 mi (640 km) northeast from the Altai range to the central section of the Eastern Sayan Mts. There are a variety of mineral deposits in the Sayan Mts. Lumbering, agriculture, and hunting are the chief occupations.

Sayre (sâr). **1.** City (1970 pop. 2,712), seat of Beckham co., W Okla., on the North Fork of the Red River, in an oil, natural-gas, and cotton region; settled 1901, inc. as a town 1903, as a city 1910. It has railroad shops, makes carbon black, and refines oil. **2.** Borough (1970 pop. 7,434), Bradford co., NE Pa., SE of Elmira, N.Y.; laid out 1871, inc. 1891.

Sayre·ville (sâr'vĭl'), borough (1978 est. pop. 33,400), Middlesex co., E N.J., on the Raritan River; inc. 1919. Its manufactures include titanium, chemicals, plastics, photographic products, and ceramics.

Sca·fell (skô'fĕl'), mountain group, Cumberland, NW England, in the Lake District, in the Cumbrian Mts. It includes Scafell Pike (3,210 ft/978 m), the highest peak in England.

scale (skāl), in cartography, the ratio of the distance between two points on a map to the real distance between the two corresponding points portrayed. The scale may be expressed in three ways: numerically, as a ratio or a fraction, e.g., 1:100,000 or $^{1}/_{100,000}$; verbally, e.g., "one inch to one mile" (not "one inch equals one mile"); and graphically, by marking distances on a sample line.

Sca·man·der (skə-măn'dər), ancient name of the Küçük Menderes River, c.60 mi (95 km) long, NW Turkey. It flows west and northwest from the Kaz Dağı through the Troas into the Mediterranean Sea. It is frequently mentioned in the *Iliad*.

Scan·di·na·vi·a (skăn'də-nā'vē-ə), region of N Europe. It consists of the kingdoms of Sweden, Norway, and Denmark; Finland and Iceland are usually considered part of Scandinavia. Physiographically, Denmark belongs to the North European Plain rather than to the geologically distinct Scandinavian peninsula (which is part of the ancient Baltic Shield), occupied by Norway and Sweden. The peninsula (c.300,000 sq mi/777,000 sq km) is c.1,150 mi (1,850 km) long and from 230 to 500 mi (370–805 km) wide and is bordered by the Gulf of Bothnia, the Baltic Sea, the Kattegat and Skagerrak straits, the North Sea, the Atlantic Ocean, and the Arctic Ocean. It is mountainous in the west and slopes gently in the east and the south. The region was heavily glaciated during the Ice Age; Jostedalsbreen, the largest glacier of mainland Europe, is a remnant of the great ice sheet. The peninsula's western coast is deeply indented by fjords.

Nearly a quarter of the peninsula lies north of the Arctic Circle, reaching its northernmost point in Cape Nordkyn, Norway. The climate varies from tundra and subarctic in the north, to humid continental in the central portion, and to marine coastal in the south and southwest. The peninsula is rich in timber and minerals (notably iron and copper), and has a great hydroelectricity-generating capacity. Its coastal waters are important fishing grounds. Except for the Lapps and Finns in the north and east, the Scandinavian peoples speak a closely related group of Germanic languages.

Scap·a Flow (skăp'ə flō'), area of water, 15 mi (24 km) long and 8 mi (12.9 km) wide, in the Orkney Islands, off N Scotland. It was Britain's main naval base in both World Wars. The British vessel *Vanguard* was torpedoed in Scapa Flow in July, 1917, and the German fleet was scuttled here in 1919. In Oct., 1939, a German submarine slipped in and sank the *Royal Oak*, causing the British fleet to withdraw until 1940. The Churchill Barrier was begun the same year to block the eastern entrance to Scapa Flow by sinking 250,000 tons of rock in the sounds linking Mainland, Burray, South Ronaldsay, and two smaller eastern islands. The naval base was closed in 1957.

Scar·bor·o (skär'bûr'ō), town (1970 pop. 7,845), Cumberland co., SW Maine, between Saco and Portland; inc. 1658. It includes Scarboro village and the resorts Scarboro Beach and Prouts Neck. Thomas Cammock settled at Prouts Neck in the 1630s. The settlement suffered greatly in Indian raids and was practically destroyed in 1675.

Scar·bor·ough (skär'bûr'ō,-bər-ə), municipal borough (1973 est. pop. 43,300), North Yorkshire, NE England, on the North Sea. The town is primarily a resort. Sports tournaments are held annually. The site was recognized at an early time for its strategic importance. There are vestiges of a 4th cent. Roman signaling station and a 12th cent. castle. The ancient Church of St. Mary is on the site of a Cistercian priory.

Scars·dale (skärz'dāl'), village (1978 est. pop. 19,400), Westchester co., SE N.Y., a residential suburb of New York City; settled c.1701, inc. 1915.

Schaff·hau·sen (shäf-hou'zən), canton (1977 est. pop. 69,500), 115 sq mi (298 sq km), N Switzerland. Entirely on the right (northern) bank of the Rhine River, the canton consists of three noncontiguous agricultural and forested areas, which are largely surrounded by West German territory. The city of Schaffhausen (1970 pop. 37,035) is capital of the canton. It is an old city, picturesquely situated on the Rhine, and retains much of its medieval character. Iron and steel, metal goods, chemicals, woolen textiles, and watches are produced. Originally a Benedictine abbey (founded c.1050), Schaffhausen became (c.1208) a free city of the Holy Roman Empire, ruled first by abbots, then by the Hapsburgs, and, after c.1415, by local trade guilds. It joined the Swiss Confederation in 1501.

Schaum·burg-Lip·pe (shoum'bŏŏrkh-lĭp'ə), former state, N West Germany, E of the Weser River. In 1946 it was placed in Lower Saxony. The original region of Schauenburg included a considerable part of Westphalia in the 12th cent. In 1640 part of Schaumburg was divided between Brunswick-Lüneburg and Hesse-Kassel. The remainder of Schaumburg passed to Count Philip of Lippe, thus forming the county of Schaumburg-Lippe. The county became a principality in 1807. The last prince abdicated in 1918, and Schaumburg-Lippe joined the Weimar Republic.

Scheldt (skĕlt), river, c.270 mi (435 km) long, rising in N France and flowing generally NE across W Belgium and into the North Sea through the Western Scheldt estuary, SW Netherlands. Navigable for most of its length, the river is connected with a dense network of canals in northern France and Belgium. The upper Scheldt valley is an important steel-producing region. From the Peace of Westphalia in 1648 until 1863 (except during the Napoleonic period) the Netherlands possessed the right to close the Scheldt estuary and thus had a stranglehold over the port of Antwerp.

Sche·nec·ta·dy (skə-nĕk'tə-dē), county (1970 pop. 161,078), 207 sq mi (536.1 sq km), E N.Y., bounded on the N by Schoharie Creek and intersected by Mohawk River and the Barge Canal; formed 1809; co. seat Schenectady. It has dairying, fruit growing, and general farming.

Schenectady, city (1978 est. pop. 72,100), seat of Schenectady co., E N.Y., on the Mohawk River and the Barge Canal; founded 1661 by Arent Van Curler, inc. 1798. It is the home of the General Electric Company, established here in 1886. Several other companies also manufacture electrical equipment, and the production of chemicals is also important. The early inhabitants were victims of a bloody French and Indian massacre in 1690, but the community grew again, prospering as a stopping place for traders and settlers traveling west on the Mohawk River. The city's growth was particularly spurred by the opening (1820s) of the Erie Canal and the building (1830s) of the railroads. Notable among Schenectady's historic buildings are the homes in the old stockade area, which date from the early 1700s, and the pre-Revolutionary St. George's Episcopal Church. A science museum is maintained by the state in the former home and laboratory of Charles P. Steinmetz.

Schie·dam (skhē'däm'), city (1977 est. pop. 76,865), South Holland prov., W Netherlands, on the Nieuwe Maas River, near Rotterdam. It is famous for its gin. There are also shipyards and factories that manufacture chemicals, glass, and machinery.

Schie·hal·lion (shə-hăl'yən), mountain, 3,547 ft (1,081.8 m) high, Perthshire, central Scotland, near Loch Rannoch. In 1774 Nevil Maskelyne experimented here to determine the density of the earth.

Schil·ler Park (shĭl'ər), village (1970 pop. 12,712), Cook co., NE Ill., a residential suburb of Chicago; inc. 1914. It has some light industry.

Schlei·cher (shlī'kər), county (1970 pop. 2,277), 1,331 sq mi (3,447.3 sq km), W central Texas, on Edwards Plateau, drained by the San Sabo, South Concho, and Devils rivers; formed 1887; co. seat Eldorado. This sheep-ranching region also produces cattle, goats, poultry, and some agriculture (grain sorghums, oats, wheat, and cotton). There are oil, natural-gas, clay, and limestone deposits here.

Schles·wig (shlĕs'wĭg, shlās'vĭkh), former duchy, N Germany and S Denmark, occupying the S part of Jutland. The Eider River separates it from Holstein. German Schleswig forms part of Schleswig-Holstein. Danish Schleswig, known as North Schleswig, includes the cities of Åbenrå, Haderslev, Sønderborg, and Tønder.

The duchy of Schleswig, created in 1115, was a hereditary fief held from the kings of Denmark. In 1386 the count of Holstein received Schleswig as a hereditary fief. His descendant Christian I of Denmark inherited (1460) both Schleswig and Holstein. In the 16th cent. Schleswig and Holstein (which had also become a duchy) underwent complex subdivisions. The three main divisions were a ducal portion, ruled by the dukes of Holstein-Gottorp; a royal portion, ruled directly by the Danish kings; and a common portion, ruled jointly by the Danish kings and the dukes of Holstein-Gottorp. The Northern War (1700–1721) ended with the dispossession of Duke Charles Frederick of Holstein-Gottorp and the union of the ducal portion of Schleswig with the Danish crown. The ducal portion of Holstein passed to the Danish crown in 1773. Thus all Schleswig and Holstein were once more united under the Danish kings. Both duchies were annexed (1866) by Prussia. After World War I North Schleswig passed to Denmark by a plebiscite (1920) held in accordance with the Treaty of Versailles.

Schleswig, city (1970 pop. 32,518), Schleswig-Holstein, N West Germany, on the Schlei, an inlet of the Baltic Sea. The city's economy is based on the production of food products and leather and on fishing. One of the oldest cities in northern Germany, Schleswig was known by c.800. It was the residence of the dukes of Schleswig and (1514–1713) of the dukes of Holstein-Gottorp. It was the capital of Schleswig-Holstein from 1866 to 1917, when it was replaced as capital by Kiel. The fortified Gottorf, or Gottorp, castle (16th–18th cent.) in Schleswig now houses museums of art and early history.

Schles·wig-Hol·stein (shlĕs'wĭg-hōl'stīn', shlās'vĭkh-hôl'shtīn'), state (1970 pop. 2,494,000), c.6,050 sq mi (15,670 sq km), N West Germany. Kiel is the capital and chief port. Flanked on the west by the North Sea and on the east by the Baltic Sea, Schleswig-Holstein occupies the southern part of the Jutland peninsula and extends from the Elbe River northward to the Danish border. It includes some of the North Frisian Islands of the North Sea and the island of Fehmarn in the Baltic. The Kiel Canal links the North Sea and the Baltic. Schleswig-Holstein is drained by the Eider River, which forms the historic border between the former duchies of Schleswig (in the north) and Holstein (in the south). A low-lying region, the state has fertile agricultural land except in the center, where heaths and moors predominate. Farming (grain, potatoes, and vegetables) and cattle raising are widely pursued; shipping and fishing are important along the coasts. Manufactures include ships, processed food, textiles, clothing, and machinery. There are oil fields in the Dithmarschen region in the southwest.

History. From 1773 the kings of Denmark held both Schleswig and Holstein, but the duchies were in personal union with, and not part of, Denmark. Under the pressure of Danish nationalists King Frederick VII declared the complete union of Schleswig with Denmark in 1848. Revolution broke out among the predominantly German population of both duchies, and the German Confederation came to the aid of the rebels and occupied the duchies. After intermittent fighting, peace was made in 1850. The Treaty of London (1852), concluded by the major powers, guaranteed the continued status of the duchies in personal union with Denmark. In 1864 Prussia and Austria declared war on Denmark, which was easily defeated. Schleswig was placed under Prussian administration and Holstein under Austrian administration. The Austro-Prussian War of 1866 ended with a swift Prussian victory; Schleswig, Holstein, and the former Danish duchy of Lauenburg were annexed to Prussia and became the province of Schleswig-Holstein. After World War I the Danish majority of northern Schleswig determined by plebiscite (1920) the return of that part of the province to Denmark. The former free city of Lübeck and the Lübeck district of Oldenburg were incorporated into Schleswig-Holstein in 1937. After World War II Schleswig-Holstein was constituted (1946) as a state of West Germany.

Schley (slī), county (1970 pop. 3,097), 162 sq mi (419.6 sq km), SW central Ga., in the coastal plain; formed 1857; co. seat Ellaville. It is in an agricultural (cotton, corn, truck crops, peanuts, pecans, and peaches) and lumbering area.

Schmal·kal·den (shmäl'käl'dən), town (1974 est. pop. 15,017), Suhl dist., SW East Germany. It has been a metalworking center since the Middle Ages, and its manufactures include tools and machinery. Schmalkalden was chartered in the 13th cent., passed in 1583 to Hesse-Kassel, and, with it, passed to Prussia in 1866. In the town hall (built 1419) the Schmalkaldic League was founded in 1531. The inn where Martin Luther drew up (1537) the Schmalkaldic Articles, outlining the Protestant viewpoint, has been restored.

Schoen·brunn Village State Memorial (shān'brən, -bro͞on), E Ohio, S of New Philadelphia; site of the first town in Ohio, est. 1772 by Moravian missionary David Zeisberger and his Indian converts. During the American Revolution, the town was abandoned because of the British and Indian menace; later the town was burned by Indians. Restoration of the site began in 1923.

Scho·har·ie (skō-här'ē), county (1970 pop. 24,750), 624 sq mi (1,616.2 sq km), E central N.Y., partly in the Catskills and drained by Schoharie and Catskill creeks; formed 1795; co. seat Schoharie. It has dairying, farming (hay, fruit, truck crops, and potatoes), and poultry raising. There are resorts in the Catskills and several scenic caverns.

Schoharie, village (1970 pop. 1,125), seat of Schoharie co., E N.Y., W of Albany; inc. 1867. A stone fort (1772) is preserved as a museum.

School·craft (sko͞ol'krăft'), county (1970 pop. 8,226), 1,181 sq mi (3,058.8 sq km), S Upper Peninsula, Mich., bounded on the S by Lake Michigan and drained by the Indian and Manistique rivers; formed 1871; co. seat Manistique. It is a dairying and agricultural area, with some manufacturing. Lumbering and the making of wood products are the chief industries. There are many hunting and fishing resorts in the county.

Schreck·hör·ner (shrĕk'hör'nər), two peaks of the Bernese Alps, S central Switzerland. Gross Schreckhorn is 13,387 ft (4,083 m) high; Klein Schreckhorn reaches a height of 11,473 ft (3,499 m).

Schütt, Great (shüt), island, c.725 sq mi (1,880 sq km), SW Czechoslovakia, in the Danubian lowlands between the Danube River and its northern arm. It extends c.55 mi (90 km) from Bratislava to Komárno. The island's fertile soil produces a variety of crops. Opposite the Great Schütt lies the Little Schütt, an island c.30 mi (50 km) long and up to c.10 mi (16 km) wide, in northwestern Hungary between the Danube River and its southern arm. Wheat, rye, and dairy products are produced here.

Schuy·ler (skī'lər). **1.** County (1970 pop. 8,135), 434 sq mi (1,124.1 sq km), central Ill., partly bounded on the E by the Illinois River and drained by the La Moine River; formed 1825; co. seat Rushville. It has agriculture (livestock, poultry, corn, wheat, fruit, and dairy and meat products) and bituminous-coal mines. **2.** County (1970 pop. 4,665), 306 sq mi (792.5 sq km), N Mo., bounded in the W by the Chariton River and drained by the North Fabius River; formed 1845; co. seat Lancaster. It is in a farming (corn, wheat, oats, cattle, and poultry) area, with coal mining. **3.** County (1970 pop. 16,737), 330 sq mi (854.7 sq km), W central N.Y., in the Finger Lakes region, drained by Cayuta and Catherine creeks; formed 1854; co. seat Watkins Glen. This fruit-growing and general-farming area produces grain, poultry, livestock, hay, and dairy products. There is also diversified manufacturing. The county includes scenic Watkins Glen State Park and several waterfalls and gorges.

Schuyler, city (1970 pop. 3,597), seat of Colfax co., E Nebr., W of Fremont on the Platte River, in a prairie region; founded 1869. It is a rail and trade center for a grain and livestock area.

Schuyl·kill (sko͞ol'kĭl'), county (1970 pop. 160,089), 784 sq mi (2,030.6 sq km), E central Pa., in a mountainous region drained by the Schuylkill River; formed 1811; co. seat Pottsville. Anthracite coal is the major resource. Manufacturing (clothing and leather products), meat packing, and brewing are also done.

Schuylkill, river, c.130 mi (210 km) long, rising in Schuylkill co., E central Pa. and flowing generally SE to the Delaware River at Philadelphia. The Schuylkill is navigable by shallow-draft boats, and there are many industrial cities along its banks.

Schuylkill Ha·ven (hā'vən), industrial borough (1970 pop. 6,125), Schuylkill co., E Pa., on the Schuylkill River S of Pottsville, in an anthracite-coal area; settled 1748, laid out 1829, inc. 1840.

Schwa·bach (shvä'bäкн), city (1974 est. pop. 33,102), Bavaria, S West Germany. Manufactures include wire, chemicals, and processed foods. Schwabach was chartered in the late 14th cent., passed to Prussia in 1791, and to Bavaria in 1806. Noteworthy buildings include the Gothic church (1469–95), which contains carvings by Veit Stoss and a tabernacle by Adam Kraft, and the city hall (1528).

Schwä·bisch Gmünd (shvĕb'ĭsh gə-münt') or **Gmünd** (gə-münt'), city (1975 pop. 57,212), Baden-Württemberg, S West Germany, on the Rems River, at the N foot of the Swabian Jura Mts. It has long been known as a gold-working and silver-working center; other manufactures include machinery and glass. Founded by the mid-12th cent., Schwäbisch Gmünd was a free imperial city from 1268 until 1803, when it passed to Württemberg.

Schwäbisch Hall (häl) or **Hall** (häl), city (1975 pop. 32,415), Baden-Württemberg, S West Germany, on the Kocher River. It is a rail junction and has an iron and steel construction industry. The city is also a popular tourist center and, because of its saline baths, a health resort. Chartered in the 12th cent., Schwäbisch Hall was a free imperial city until 1803 and ruled considerable surrounding territory. Nearby is Comburg, a former fortified Benedictine abbey (founded 1075; now a teachers college).

Schwarz·wald (shvärts'vält'): *see* Black Forest.

Schwein·furt (shvīn'fo͝ort'), city (1974 est. pop. 56,976), Bavaria, E central West Germany, on the Main River. Manufactures include ball bearings, small motors, dyes, soap, and leather goods. The compound Paris green is also known as Schweinfurt green. Schweinfurt was known c.791. It was a free imperial city from 1282 to 1803, when it passed to Bavaria. The city was heavily bombed by the Allies during World War II.

Schwe·rin (shvā-rēn'), city (1974 est. pop. 104,984), capital of Schwerin dist., NW East Germany, on Schwerin Lake. It is the commercial, industrial, and transportation center of an agricultural and dairying region. Manufactures include chemicals, pharmaceuticals, ceramics, and tobacco products. Originally a Wendish settlement, Schwerin was chartered in 1161. In the early 17th cent. the city became the capital of Mecklenburg-Schwerin. It was occupied (1624–31) in the Thirty Years' War by imperial troops. The Peace of Westphalia (1648) gave its territories to the duke of Mecklenburg-Schwerin. Schwerin became the capital of the former state of Mecklenburg in 1934. Noteworthy buildings include the Gothic Protestant cathedral (14th–15th cent.) and the former grand ducal palace, built (19th cent.) on an island in Schwerin Lake.

Schwyz (shvēts), canton (1977 est. pop. 92,200), 351 sq mi (909 sq km), central Switzerland, one of the Four Forest Cantons, bordering on the Lake of Zürich in the N and the Lake of Lucerne in the SW. In 1240 Emperor Frederick II granted Schwyz (then under the Hapsburgs) a charter making it immediately subject to the Holy Roman Empire. The charter was revoked in 1274 by Rudolf I of Hapsburg, and in 1291 Schwyz concluded with Uri and Unterwalden the pact that became the basis of Swiss liberty. Its capital, Schwyz (1977 est. pop. 12,100), one of the oldest towns in Switzerland, is a summer resort. The town has a 16th cent. town hall with historic paintings, several baroque churches, and numerous patrician houses (17th–18th cent.).

Scil·ly Islands (sĭl'ē), archipelago of more than 150 isles and rocky islets (1970 est. pop. 2,000), off SW England, 28 mi (45 km) from Lands End. On the rocky coasts, now marked by lighthouses and lightships, scores of ships were wrecked, notably Sir Clowdisley Shovell's fleet in 1707. The growing of flowers for Bristol and London markets is one of the leading occupations. Tourism and vegetable growing are also important. On the largest island, St. Mary's, is the capital, Hugh Town, with the 16th cent. Star Castle at which Prince Charles (later Charles II) stopped in 1645 on his flight to Jersey.

Sci·o·to (sī-ō'tō), county (1970 pop. 76,951), 608 sq mi (1,574.7 sq km), S Ohio, bounded in the S by the Ohio River, here forming the Ky. border, and intersected by the Scioto and Little Scioto rivers; formed 1803; co. seat Portsmouth. It is in an agricultural area that produces grain, fruit, dairy products, and livestock. Its industry includes food processing, lumbering, and iron foundries.

Scioto, river, 237 mi (381.3 km) long, rising in W Ohio near Indian Lake and flowing E, then turning S to pass through Columbus and Chillicothe and enter the Ohio River at Portsmouth.

Sci·tu·ate (sĭch'o͞o-wāt, -wĭt). **1.** Resort town (1970 pop. 16,973), Plymouth co., SE Mass., on the Atlantic coast about midway between Boston and Plymouth, in an area of poultry, fruit, and truck farms; settled c.1630, inc. 1636. **2.** Town (1970 pop. 7,489), Providence co., W central R.I.; set off from Providence 1731. Scituate

Reservoir was formed by damming (1925) the Pawtuxet River. In the Revolution cannons were cast at ironworks here.

Sco·bey (skō'bē), city (1970 pop. 1,486), seat of Daniels co., NE Mont., NW of Williston, N.Dak., a port of entry on the Poplar River; settled c.1909, inc. 1919. It is a trading point in a grain and cattle region. Coal mines are in the area.

Scone (sko͞on), village, Perthshire, central Scotland. Old Scone, west of the modern village of New Scone, was the repository of the Coronation Stone and the coronation place of Scottish kings from Kenneth I to Charles II.

Sco·pus, Mount (skō'pəs), peak, 2,736 ft (834.5 m) high, NNE of Jerusalem. Dominating Jerusalem, it has always held strategic importance in the defense of the city. Roman legions camped here in A.D. 70 as did the Crusaders in 1099. After the Arab-Israeli War of 1948 Mt. Scopus was an Israeli-held enclave in Jordanian territory.

Scores·by Sound (skôrz'bē), arm of the Greenland Sea, E Greenland. It has numerous fjords that branch out generally westward to the ice cap. Some of the branches extend more than 180 mi (290 km) inland. At its mouth is the settlement of Scoresbysund (1969 pop. 339), a fishing and hunting base.

Sco·tia (skō'shə), Latin name for Scotland, used in the Middle Ages.

Scotia, village (1970 pop. 8,224), Schenectady co., E N.Y., on the Mohawk River opposite Schenectady; settled before 1660, inc. 1904.

Scot·land (skŏt'lənd), political division of Great Britain (1971 pop. 5,227,706), 30,414 sq mi (78,772 sq km), comprising the northern portion of the island of Great Britain and many surrounding islands. Scotland, England, and Wales have been united since 1707 under the name of the United Kingdom of Great Britain. They share one Parliament, but Scotland retains its own system of laws and education. Its governmental departments are under the direction of a secretary of state for Scotland, who is a member of the British cabinet. The Church of Scotland, which is Presbyterian, is the Established Church, but there are no restrictions on religious liberty. The eight universities are Edinburgh, Glasgow, Aberdeen, St. Andrews, Dundee, Stirling, Strathclyde, and Heriot-Watt. Fewer than 1,000 people, primarily in the far north, still speak only Gaelic, and fewer than 80,000 speak Gaelic in addition to English.

Scotland is separated from England by the Tweed River, the Cheviot Hills, the Liddell River, and Solway Firth. It is bounded on the north and west by the Atlantic Ocean and on the east by the North Sea. Because of Scotland's highly irregular outline and the deeply indented arms of the sea—usually called lochs when narrow and firths when broad—it has c.2,300 mi (3,700 km) of coastline. The Orkney and Shetland islands lie off the northern coast of the mainland and the Hebrides off the western. Scotland's principal rivers are the Clyde, the Forth, the Dee, the Tay, and the Tweed. The largest freshwater loch is Loch Lomond.

Scotland may be divided into three main geographic regions. The southern Uplands is a region of high, rolling moorland cut by numerous valleys. The central Lowlands is Scotland's most populous district and the locus of its commercial and industrial cities. Separated from the Lowlands by the Grampian Mts. are the Highlands of the north, a rough, mountainous region divided by the Great Glen and containing Ben Nevis (4,406 ft/1,343 m), the highest peak in Great Britain.

In 1975 the 33 historic counties of Scotland were abolished and replaced by 9 regional authorities and 3 island authorities:

REGIONAL AUTHORITIES	ISLAND AUTHORITIES
Borders	Orkney Islands
Central	Shetland Islands
Dumfries and Galloway	Western Isles
Fife	
Grampian	
Highland	
Lothian	
Strathclyde	
Tayside	

Economy. Most Scottish industry and commerce is concentrated in a few large cities on the waterways of the central Lowlands. Edinburgh, on the Firth of Forth, is the administrative and cultural capital of Scotland and a center of paper production and publishing. Glasgow is a thriving seaport and a center of shipbuilding, metalworks and engineering works, and varied light industry. The coal and iron deposits of the Strathclyde, Lothian, and Fife regions support the diverse industries of Glasgow and several other manufacturing cities. Other important industries are textile production, distilling, and fishing. Textiles, beer, and whisky, which are among Scotland's chief exports, are produced in many towns throughout the country. Salmon is taken from the Tay and the Dee rivers, and herring from the North Sea. Only about one fourth of the land is under cultivation (principally in cereals and vegetables), but sheep raising is important in the mountainous regions.

History. The Picts, of obscure origin, inhabited Scotland from prehistoric times. The Romans attempted vainly to penetrate Scotland, and their successive lines of forts and walls proved inadequate to contain the northern tribes of Picts and Celts. Although the Romans had little influence on Scottish life, Christianity had been introduced into Scotland before they left by St. Ninian and his disciples in the 5th cent. In the century and a half after the Roman evacuation (mid-5th cent.) four Scottish kingdoms came into being — that of the Picts in the north; that of the Scots who came from Ireland and founded Dalriada in what is now Argyllshire and the island of Iona; that of the Britains in Strathclyde; and that of Northumbria, founded by the Angles and settled largely by Germanic immigrants. After the decline of the Northumbrian power in Scotland began the raids of the Norsemen, who harried the country from the 8th to the 18th cent. In the mid-9th cent. the Scot king Kenneth I established his rule over nearly all the land north of the Firth of Forth. Under Malcolm III there began a reorganization of the Scottish church and a gradual anglicization of the Lowlands peoples. Although the clan system, based on blood relationships and personal loyalty to a chieftain, survived in the Highlands, feudal property laws were generally adopted in the Lowlands in the 11th and 12th cent.

In the reign of William the Lion Scotland became a fief of England by treaty (1174). In 1189 Richard I sold the Scots their freedom. The Norsemen were gradually pushed out of Scotland and finally defeated in 1263; only Orkney and Shetland remained in Norse hands until the 15th cent. When Alexander III died in 1286, a struggle for the throne ensued; Edward I of England installed his choice, John de Baliol (1292). Edward won Scottish submission, but Scotland rose in revolt, first under Sir William Wallace, then under Robert the Bruce (later Robert I). Robert was crowned king in 1306 and defeated Edward II at Bannockburn in 1314. The 14th cent. was marked by turbulence among the nobles and royal heirs and by repeated attacks from England. James I attempted to establish control over his nobles, but his murder in 1437 threw Scotland back into the old pattern of civil conflict over the next century.

The Reformation came to Scotland primarily through the efforts of John Knox (1505–1572). The religious issue was inextricably connected with opposition to the French Roman Catholic party of Mary of Guise (queen regent after 1542) and of her daughter Mary Queen of Scots, who lived in France as dauphine and then as queen. By the time Mary Queen of Scots arrived (1561) in Scotland, Catholicism had almost disappeared from the Lowlands. Mary's personal involvements and her struggle with Protestant nobles ended in her loss of the throne (1567), imprisonment in England, and execution (1587). Her son, James VI, succeeded to the English throne (1603) as James I of England. United under one crown, Scotland and England were finally at peace.

Presbyterianism and its maintenance became the great question in the 17th cent. The Covenanters declared their opposition to the liturgical forms imposed by Charles I and resisted his attempt to bring them to heel in the Bishops' Wars (1639–40). These wars led directly to the English Civil War. Charles's efforts to win the Scots by yielding rights to Presbyterianism in 1641 came too late. Scottish sympathies shifted to Charles, however, especially after his execution in 1649. Many Scots rallied to Charles II, and the Restoration (1660) was cause for rejoicing. The Stuarts, however, sought once more to restore episcopacy, and the Covenanters were, for many decades, subjected to severe persecution. With the Glorious Revolution (1688–89) Presbyterianism once more became the national church. But the Jacobites, supporters of the exiled Stuarts, caused great disruption, particularly in the Highlands. To assure the Hanoverian succession after Queen Anne's death, the union of England and Scotland was voted by both parliaments in 1707, providing for Scottish representatives in a Parliament of Great Britain.

The Jacobites attempted in 1715 and again in 1745 to destroy the union, but without success, and Scotland had peace at last. The economic results of the union eventually proved favorable to Scotland, and the people gradually enjoyed a higher standard of living. Thriving commerce within the British Empire led to expansion of shipping and shipbuilding, and Glasgow achieved eminence as a commercial center. Toward the end of the 18th cent. cotton spinning and weaving became Scotland's leading industries. By the end of the 19th cent., however, the metallurgical industry had come to dominate the economy. From Scotland emerged some of the first leaders of the British labor movement and the first labor representatives in Parliament came from Scottish mining areas. Concentration on heavy industry made Scotland an important arsenal in World War I. Scotland's participation in the benefits of the modern British welfare state has not lessened a persistent nationalist movement that urges greater autonomy for Scotland in the determination of local affairs.

Scotland. **1.** County (1970 pop. 5,499), 441 sq mi (1,142.2 sq km), NE Mo., drained by the North Fabius River and the North and South Wyaconda rivers; formed 1841; co. seat Memphis. It is in a grain, livestock, and dairying area. **2.** County (1970 pop. 26,929), 319 sq mi (826.2 sq km), S N.C., in a coastal plain and sandhills region, bounded on the SW by the S.C. border and drained by the Lumber River; formed 1899; co. seat Laurinburg. It has farming (cotton, tobacco, corn, and fruit), textile manufacturing, and sawmilling.

Scott (skŏt). **1.** County (1970 pop. 8,207), 898 sq mi (2,325.8 sq km), W Ark., in the Ouachita Mts., bordered on the W by Okla. and drained by the Poteau and Fourche La Fave rivers; formed 1833; co. seat Waldron. It is in a coal-mining, lumbering, and agricultural (livestock, cotton, corn, lespedeza, and dairy products) region. Its industry includes food processing, tobacco manufacturing, and furniture making. Part of Ouachita National Forest and a U.S. Wildlife preserve are in the county. **2.** County (1970 pop. 6,096), 251 sq mi (650.1 sq km), W central Ill., bounded on the W by the Illinois River and drained by Sandy and Mauvaise Terre creeks; formed 1839; co. seat Winchester. It is primarily agricultural, producing corn, wheat, oats, livestock, and poultry. **3.** County (1970 pop. 17,144), 193 sq mi (500 sq km), SE Ind., bounded on the N by the Muscatatuck River, and drained by its small tributaries; formed 1820; co. seat Scottsburg. Its agriculture includes grain, tobacco, truck crops, livestock, and poultry. It has timber, limestone deposits, and a food-processing industry. Wood products, furniture, and metal cans are made. **4.** County (1970 pop. 142,687), 454 sq mi (1,175.9 sq km), E Iowa, bounded in the E and S by the Mississippi River and in the N mainly by the Wapsipinicon River; formed 1837; co. seat Davenport. It is a prairie agricultural area (hogs, cattle, poultry, corn, wheat, and oats), with some industry. There are coal and limestone deposits here. **5.** County (1970 pop. 5,606), 724 sq mi (1,875.2 sq km), W Kansas, in a sloping to rolling plains area; formed 1873; co. seat Scott City. It is in a grain and livestock region, with some mining and manufacturing. **6.** County (1970 pop. 17,948), 284 sq mi (735.6 sq km), N central Ky., in the Bluegrass, bounded on the SW by Elkhorn Creek and drained by Eagle Creek; formed 1792; co. seat Georgetown. In a gently rolling agricultural area yielding burley tobacco, corn, and wheat, it has copper and aluminum industries, machinery manufacturing, and limestone quarrying. **7.** County (1970 pop. 32,423), 353 sq mi (914.3 sq km), SE Minn., bounded on the N and W by the Minnesota River; formed 1853; co. seat Shakopee. It has agriculture (dairy products, livestock, corn, oats, and barley), some mining, and industry (food processing, lumbering, and the manufacture of paper products, chemicals, plastics, concrete, metal products, and machinery). **8.** County (1970 pop. 21,369), 615 sq mi (1,592.9 sq km), central Miss., drained by the Strong and Leaf rivers and Tuscolameta Creek; formed 1833; co. seat Forest. Agriculture (cotton and corn) and lumbering are major occupations. **9.** County (1970 pop. 33,250), 421 sq mi (1,090.4 sq km), SE Mo., on the Mississippi River; formed 1821; co. seat Benton. This county grows and processes cotton. Corn, wheat, oats, hay, livestock, and lumber are also products of the area. **10.** County (1970 pop. 14,762), 544 sq mi (1,409 sq km), N Tenn., in a rugged region of the Cumberlands, bordered in the N by Ky. and drained by the South Fork of the Cumberland River; formed 1849; co. seat Huntsville. It is in a lumbering, coal and clay mining, and farming (corn, fruit, tobacco, hay, livestock, and vegetables) region. Clothing is manufactured. **11.** County (1970 pop. 24,376), 539 sq mi (1,396 sq km), SW Va., mainly in the Alleghenies, bounded in the S by Tenn., with parts of Clinch and Powell mts. in the W, and drained by the Clinch River and the North Fork of the Holston River; formed 1814; co. seat Gate City. Its agriculture includes fruit, tobacco, wheat, corn, dairy products, and livestock. It has some coal mining and lumbering and manufactures clothing and metal products.

Scott City, city (1970 pop. 4,001), seat of Scott co., W central Kansas, N of Garden City, in an irrigated region; settled 1884, inc. 1887. In an oil-producing area, the city has an oil refinery. Saddle horses are raised, and dairy products are made. A buffalo preserve and the remains of an Indian pueblo, El Cuartelejo, are in a state park north of the city.

Scott·dale (skŏt'dāl'), borough (1970 pop. 5,818), Westmoreland co., SW Pa., N of Connellsville, in a bituminous-coal area; laid out 1872, inc. 1874. It has some manufacturing.

Scotts·bluff (skŏts'blŭf'), city (1978 est. pop. 13,300), Scotts Bluff co., W Nebr., on the North Platte River near the Wyo. border; inc. 1900. It is the market, distribution, and processing point of an extensive irrigated farm region. There are beet-sugar refineries, canneries, meat-packing plants, and flour mills. The city is named for a nearby butte, Scotts Bluff (alt. 4,649 ft/1,418 m), a landmark to travelers on the Oregon and Mormon trails. In 1864 Fort Mitchell was established here as an outpost of Fort Laramie. Oregon Trail Museum is in Scotts Bluff National Monument, and Agate Fossil Beds National Monument is to the north.

Scotts Bluff (skŏts), county (1970 pop. 36,432), 726 sq mi (1,880.3 sq km), W Nebr., in an irrigated agricultural area bounded in the W by Wyo., and drained by the North Platte River; formed 1888; co. seat Gering. Beet sugar, flour, livestock, grain, beans, potatoes, dairy goods, and poultry are produced.

Scotts·bor·o (skŏts'bûr'ō, -bər-ə), city (1970 pop. 9,324), seat of Jackson co., NE Ala., E of Huntsville, near the Tennessee River, in a farm area; inc. 1870.

Scotts·burg (skŏts'bûrg'), town (1970 pop. 4,791), seat of Scott co., SE Ind., N of Louisville, Ky., in an agricultural area.

Scotts·dale (skŏts'dāl'), city (1978 est. pop. 79,300), Maricopa co., central Ariz.; inc. 1951. It is a resort and retirement center. Electronic equipment is manufactured and Arabian horses are bred.

Scotts·ville (skŏts'vĭl'), town (1970 pop. 3,584), seat of Allen co., S Ky., SE of Bowling Green, in a farming and dairying region.

Scran·ton (skrăn'tən), city (1978 est. pop. 92,600), seat of Lackawanna co., NE Pa., in a mountain region, on the Lackawanna River; settled in the 1700s, inc. 1866. It is the commercial and industrial center of an anthracite-coal section. Iron was first forged here in 1797. Mining decreased after World War II. Today the city's manufactures include plastics, heavy machinery, tanks, textiles, metal tools and parts, glass products, electronic equipment, and dental and medical supplies. Many lakes, state forests, and recreational sites are in the area.

Scre·ven (skrē'vən), county (1970 pop. 12,591), 651 sq mi (1,686.1 sq km), E Ga., bounded on the E by the S.C. border (formed here by the Savannah River) and on the SW by the Ogeechee River; formed 1793; co. seat Sylvania. It is in a coastal-plain agricultural (cotton, corn, potatoes, peanuts, and livestock) and forestry area. It also has textile industries.

Scun·thorpe (skŭn'thôrp'), municipal borough (1974 est. pop. 70,330), Humberside, E England. Situated on an ironstone field, it is a center of iron and steel manufacture. Light engineering and the production of tar and clothing are other industries.

Scur·ry (skûr'ē), county (1970 pop. 15,760), 904 sq mi (2,341.4 sq km), NW central Texas, in the rolling plains, drained by the Colorado River; formed 1876; co. seat Snyder. It has agriculture (cotton, grain sorghums, wheat, oats, peanuts, peaches, sheep, and poultry) and deposits of oil, natural gas, bentonite, coal, magnesium, limestone, and caliche.

Scu·ta·ri, Lake (sko͞o'tə-rē), c.25 mi (40 km) long and from 4 to 8 mi (6.4–12.9 km) wide, SE Europe, on the Yugoslav-Albanian border. It is the largest lake of the Balkan Peninsula and usually floods the surrounding area in the winter. Once an inlet of the Adriatic Sea, the lake is now separated from the sea by an alluvial isthmus. The lake is navigable by small steamers, and it abounds in fish.

Scy·ros (sī'rəs), island: *see* Skíros.

Scyth·i·a (sĭth'ē-ə), ancient region of Eurasia, extending from the Danube on the W to the borders of China on the E. The Scythians flourished from the 8th to the 4th cent. B.C. They spoke an Indo-Iranian language but had no system of writing. They were nomadic conquerors and skilled horsemen. The Scythians are traditionally associated with the area between the Danube and the Don, but modern excavations suggest that their origins were in western Siberia before they moved east into southern Russia in the early 1st millennium B.C. The Scythians traded (7th cent. B.C.) grain and their service as mercenaries for Greek wine and luxury items. They invaded (7th cent. B.C.) upper Mesopotamia and Syria and also made incursions into the Balkan Peninsula. They destroyed (c.325 B.C.) an expedition sent against them by Alexander the Great. After 300 B.C. they were driven out of the Balkans by the invading Celts. In southern Russia they were displaced (2nd or 1st cent. B.C.) by the related Sarmatians, and part of their empire became Sarmatia.

sea (sē), term used as synonymous with ocean, or a subdivision of an ocean (Caribbean Sea, Yellow Sea), or erroneously designating a large salt lake (Caspian Sea, Dead Sea, Aral Sea).

Sea·ford (sē'fərd). **1.** Town (1970 pop. 5,537), Sussex co., SW Del., on the Nanticoke River WSW of Georgetown, in a farm area; inc. 1865. It grew as an oyster-packing, shipping, and shipbuilding center. Today it processes food and has a large plant manufacturing nylon products. **2.** Uninc. hamlet (1970 pop. 17,379), Nassau co., SE N.Y., on the S shore of Long Island, on Great South Bay; settled 1643. It is a residential suburb of New York City and a resort village.

Sea·ham (sē'əm), urban district (1974 est. pop. 23,370), Durham, NE England, on the North Sea. Mining and shipping coal are major industries. Engineering and the manufacture of clothing and pottery are also important.

Sea Islands, chain of more than 100 low islands off the Atlantic coast of S.C., Ga., and N Fla. The ocean side of the islands is generally sandy; the side facing the mainland is marshy. Some islands remain uninhabited; others are resorts and wildlife sanctuaries. The Intracoastal Waterway passes through the Sea Islands. The Spanish discovered (16th cent.) and were the first to inhabit the islands, but abandoned them as the English steadily advanced in the area. James Oglethorpe, founder of the Georgia colony, built Fort Frederica on St. Simons Island between 1736 and 1754, during the English-Spanish struggle for control of the present southeastern United States. The ruins of the fort now constitute a national monument. The Sea Islands were the first important cotton-growing area in North America. The Union invasion in the Civil War and the distribution of land by the Federal government to newly freed slaves after the war ruined the wealth of the planters. Cotton culture gave way to diversified farming; poultry raising, oyster gathering, and fishing are also important today. Cumberland Island, largest of the Sea Islands, c.22 mi (35 km) long and from 1 to 5 mi (1.6–8 km) wide, has been designated a national seashore.

Seal Beach (sēl), city (1978 est. pop. 26,200), Orange co., S Calif., on the Pacific coast; inc. 1915. Aerospace items and salad dressings are produced. It is a beach city with an active art colony.

Sear·cy (sûr'sē), county (1970 pop. 7,731), 664 sq mi (1,720 sq km), N Ark., in the Ozarks, intersected by the Buffalo River and drained by the Middle Fork of the Little Red River; formed 1835; co. seat Marshall. It is in an agricultural (grain, cotton, hay, truck crops, and livestock) and lumbering area.

Searcy, city (1970 pop. 10,867), seat of White co., N central Ark., near the Little Red River NE of Little Rock, in a farm area. It packs food and manufactures shoes, office machines, and other products.

Sea·side (sē'sīd'). **1.** City (1978 est. pop. 34,200), Monterey co., W Calif., on Monterey Bay, in a fruit region; founded 1887, inc. 1954. Its economy is based on tourism and nearby U.S. Fort Ord. **2.** Resort city (1970 pop. 4,402), Clatsop co., NW Oregon, on the Pacific coast S of Astoria; inc. 1899. A monument marks the point considered the end of the Oregon Trail.

Seat Pleas·ant (sēt' plĕz'ənt), town (1970 pop. 7,217), Prince Georges co., W central Md., a suburb E of Washington, D.C.; inc. 1931.

Se·at·tle (sē-ăt'əl), city (1978 est. pop. 481,000), seat of King co., W Wash., built on seven hills, between Elliott Bay of Puget Sound and Lake Washington; inc. 1869. Seattle, the largest city in the Pacific Northwest, is the region's commercial, transportation, and industrial hub and a major port of entry, important in the Alaskan and Far Eastern trade. In addition to being a center of aircraft manufacturing and shipbuilding since World War II, the city has major food-processing and lumber industries; chemical products, metal goods, machinery, textiles and apparel, and paper, stone, clay, and glass items are also produced. Settled in 1851–52, Seattle remained a small lumber town until the coming of the railroad in 1884. The city became a boom town with the 1897 Alaska gold rush. The completion (1917) of a canal and locks made the city both a saltwater and a freshwater port. Aiding its industrial growth was the presence of coal in the area and the development of hydroelectric power. During the 1960s Seattle's port expanded enormously. The city, situated between the majestic Cascade and Olympic mountain ranges, with Mt. Rainier to the southeast and Mt. Baker to the northeast, is within easy reach of many national and state parks and recreation areas. It is a cultural center and the seat of the Univ. of Washington. Seattle was the site of the 1962 world's fair, the Century 21 Exposition. The symbol of that fair—a 600-ft (183-m) space needle—is a skyline landmark. Also remaining from the fair is the first publicly operated monorail in the United States.

Se·ba·go Lake (sĭ-bā'gō), c.12 mi (20 km) long and from 1 to 8 mi (1.6–12.9 km) wide, SW Maine, in a resort area. It is the source of Portland's water supply.

Se·bas·tian (sə-băs'chən), county (1970 pop. 79,237), 527 sq mi (1,364.9 sq km), W Ark., bounded on the W by the Okla. border, on the N by the Arkansas River, and drained by the Petit Jean River; formed 1851; co. seats Fort Smith and Greenwood. It has coal mines, oil and natural-gas wells, and timber. Its agriculture includes livestock, cotton, truck crops, fruit, poultry, and dairy products. Furniture, paper, clothing, glass, concrete, metal products, and machinery are manufactured.

Se·bring (sē'brĭng), city (1970 pop. 7,223), seat of Highlands co., S central Fla., NW of Lake Okeechobee; founded 1912, inc. 1913. There are citrus orchards in the region. An annual 12-hour endurance sports car race is held here.

Se·cau·cus (sē-kô'kəs), town (1970 pop. 13,228), Hudson co., NE N.J., on the Hackensack River, adjoining Jersey City; inc. 1917.

Se·cun·der·a·bad (sĭ-kŭn'dər-ə-băd'), town (1971 pop. 94,416), Andhra Pradesh state, S central India. A suburb of Hyderabad, the town is a major army base.

Se·cur·i·ty (sĭ-kyŏŏr'ə-tē), uninc. town (1970 pop. 8,700), El Paso co., central Colo.; near Colorado Springs.

Se·da·li·a (sĭ-dā'lē-ə, -dāl'yə), city (1978 est. pop. 21,800), seat of Pettis co., W central Mo.; inc. 1864. A rail center in a farm area, it has railroad shops, meat-packing plants, and a great variety of manufactures. Whiteman Air Force Base is west of the city.

Se·dan (sə-däɴ'), town (1968 pop. 23,037), Ardennes dept., NE France, on the Meuse River. A noted textile center since the 16th cent., Sedan also has metal and brewing industries. The town became part of French crown lands in 1642. It was a Protestant stronghold in the 16th and 17th cent. Sedan was the site of the decisive French defeat (1870) in the Franco-Prussian War and the surrender of Napoleon III. The town was the point of the first German breakthrough (1940) in the invasion of France in World War II.

Se·dan (sĭ-dăn'), city (1970 pop. 1,555), seat of Chautauqua co., SE Kansas, WNW of Coffeyville, near the Okla. border, in a farm and livestock area near oil fields; founded 1875, inc. 1876.

Sedge·moor (sĕj'mŏŏr'), marshy tract in Somerset, SW England. The forces of James II defeated the duke of Monmouth at Sedgemoor in 1685.

Sedg·wick (sĕj'wĭk'). **1.** County (1970 pop. 3,405), 544 sq mi (1,409 sq km), extreme NE Colo., bordering Nebr. and watered by the South Platte River; formed 1889; co. seat Julesburg. It is in an irrigated agricultural area that produces sugar beets, beans, and livestock. **2.** County (1970 pop. 350,694), 1,007 sq mi (2,608.1 sq km), S central Kansas, in a sloping to gently rolling plain drained by the Arkansas and Little Arkansas rivers; formed 1867; co. seat Wichita. It has wheat and livestock farms, scattered oil and gas fields, sand and gravel pits, and diversified industry. Furniture, chemicals, plastics, glassware, concrete, tools, and metal products are among its manufactures.

See·konk (sē'kŏngk), residential town (1970 pop. 11,116), Bristol co., SE Mass., at the R.I. line; settled 1636, set off from Rehoboth 1812.

Se·ges·ta (sĭ-jĕs'tə), ancient city of NW Sicily. Traditionally called a Trojan colony, it was the long-standing and bitter rival of Selinus. Athens undertook (415–413 B.C.) the disastrous expedition against Syracuse as an ally of Segesta. After this failure, Segesta got the help of Carthage, and Selinus was sacked (409 B.C.). Thereafter Segesta was a Carthaginian dependency with some interruptions until the First Punic War, when it surrendered to the Romans. It declined in the 1st cent. B.C.

Sé·gou or **Se·gu** (both: sā-gōō'), town (1971 est. pop. 36,400), SW

Mali, a port on the Niger River. It is the administrative and commercial center for an area where cotton, rice, millet, and peanuts are grown and cattle are raised. Cotton textiles are made in Ségou. In the late 17th cent. Ségou developed as the capital of a Bambara kingdom. In 1861 the town was captured by Al-hajj Umar, a militant Moslem reformer. In 1890 the town was occupied by the French. It is the headquarters of a large-scale agricultural development project on the Niger River that was begun in 1932 by the French.

Se·go·vi·a (sĕ-gō′vē-ä), city (1972 est. pop. 44,214), capital of Segovia prov., central Spain, in Old Castile, on the Eresma River. It stands on a rocky hill (3,297 ft/1,006 m high) crowned by the cathedral and the turreted alcazar (fortified palace). Under the Moors, it was a flourishing textile center but has since declined. There are light industries, and tourism is important. Segovia is of ancient origin and was favored by the Romans, who built (probably 1st cent.) the aqueduct (c.900 yd/820 m long) that still carries water to the city; it is one of the greatest Roman monuments in Spain. The city was repeatedly taken and lost by the Moors from 714 until Alfonso VI conquered it in 1079. It was a favorite residence of the kings of Castile.

Segovia, river, c.300 mi (480 km) long, rising in NW Nicaragua and flowing NE to the Caribbean Sea. Part of the Mosquito Coast region, once the object of dispute between Honduras and Nicaragua, the Segovia now forms the boundary between the two countries.

Se·guin (sə-gēn′), city (1970 pop. 15,934), seat of Guadalupe co., S central Texas, on the Guadalupe River; inc. 1853. Among its industrial products are bumpers, insulation, bricks, structural steel, fiber glass cloth, flour, packed meats, ditching machinery, furniture, and processed poultry. The city was founded (1831) by members of the Texas Rangers.

Se·gu·ra (sā-gŏŏ′rä), river, c.200 mi (320 km) long, rising on the NE slopes of the Sierra de Segura, SE Spain, and flowing generally E into the Mediterranean Sea. It is used for irrigation and hydroelectric-power generation.

Seine (sĕn), river, c.480 mi (770 km) long, rising in the Langres Plateau and flowing generally NW through N France. It passes Troyes, Melun, and Paris, meanders through Normandy, and empties into the English Channel in an estuary between Le Havre and Honfleur. With its tributaries and connecting canals, it drains the entire Paris basin. One of the most navigable rivers in France, it has been a great commercial artery since Roman times. The channel of the Seine is dredged and oceangoing vessels can dock at Rouen. Much of France's internal and foreign trade moves on the Seine.

Seine-et-Marne (sĕn-ā-märn′), department (1974 est. pop. 696,000), 2,290 sq mi (5,931.1 sq km), N central France, in Brie; capital Melun.

Seine-Mar·i·time (sĕn-mär-ē-tēm′), formerly **Seine-In·fé·ri·eure** (sĕn′ăN-fā-rē-ûr′), department (1974 est. pop. 1,187,000), 2,449 sq mi (6,342.9 sq km), N France, on the English Channel, mainly in Normandy; capital Rouen.

Seine-Saint-De·nis (sĕn-săNd′nē′), department (1974 est. pop. 1,392,-000), 91 sq mi (235.7 sq km), N central France, adjoining Paris; capital Bobigny.

Sei·stan (sā-stän′), border lowland region of SW Afghanistan and E Iran, c. 6,000 sq mi (15,540 sq km), fed mainly by the Helmand River and other streams. At low water, the region is reduced to two lagoons (Hamun-e Helmand and Gowd-e Zereh), and wheat, barley, and cotton are grown on the exposed land. Seistan corresponds roughly to ancient Drangiana. In the 2nd–3rd cent. A.D. it was held by the Scythians. From the 4th–7th cent. the region was the center of Zoroastrian worship. Seistan prospered under the Arabs from the 8th cent. A.D. until 1383, when Mongol conquerors destroyed the Helmand River control system. The area was disputed between Persia and Afghanistan from the 16th to early 20th cent.

Sek·on·di-Ta·ko·ra·di (sĕk′ən-dē-tä-kô-rä′dē), city (1970 pop. 161,071), capital of the Western Region, SW Ghana, on the Gulf of Guinea. An important seaport, Sekondi-Takoradi has shipbuilding, railroad-repair, and cigarette industries. The two parts of the city developed around Dutch and English forts built in the 17th cent. Sekondi, the older and larger of the two, prospered after the construction (1903) of a railroad to the mineral and timber resources of the hinterland. The two parts were amalgamated in 1946.

Sel·by (sĕl′bē), city (1970 pop. 957), seat of Walworth co., N S.Dak., N of Pierre. It is a trade and processing center for a farming and cattle-raising area.

Sel·den (sĕl′dən), uninc. village (1975 est. pop. 21,800), Suffolk co., SE N.Y., on Long Island.

Se·len·ga (sĕ′lĕng-gä′), river, 616 mi (991 km) long, rising in the Khangai Mts., NW Mongolian People's Republic, and flowing E, then N, across the Mongolia-USSR border to Lake Baykal. The Selenga, navigable from May to Oct., is Mongolia's chief river; its role as a transportation artery decreased with the advent of the Trans-Baykal RR.

Se·leu·ci·a (sə-lōō′shē-ə, -shə), ancient city of Mesopotamia, on the Tigris below modern Baghdad. Founded (c.312 B.C), it soon replaced Babylon as the main center for east-west commerce through the valley. The city was the eastern capital of the Seleucids until the Parthians conquered it. In a Parthian campaign Trajan burned the city, and in 164 B.C. it was destroyed by Romans. Another Seleucia was founded in Syria as the seaport for Antioch on the Orontes.

Se·li·nus (sə-lī′nəs), ancient city of Sicily. It was founded (c.628 B.C.) by Dorian Greeks. The constant rival of neighboring Segesta, Selinus persuaded Syracuse to interfere in a quarrel, which led to the unsuccessful Athenian expedition in Sicily (415–413 B.C.). Segesta invoked the aid of the Carthaginians, who sacked Selinus in 409 B.C. The city was rebuilt, but it did not prosper and was finally destroyed by Carthage in 250 B.C.

Sel·kirk Mountains (sĕl′kûrk′), rugged range of the Rocky Mts., SE British Columbia, Canada, near the Alta. border and extending NW c.200 mi (320 km) from the U.S. border. The range is almost encircled by the Columbia River.

Sel·kirk·shire (sĕl′kərk-shîr′, -shər), former county, SE Scotland. Selkirk was the county town. Once part of the Saxon kingdom of Northumbria, the county was annexed to Scotland in 1018. It suffered severely in the prolonged border wars between England and Scotland. Under the Local Government Act of 1973 Selkirkshire became part of the Borders region.

Sel·ma (sĕl′mə). **1.** City (1978 est, pop. 27,100), seat of Dallas co., S central Ala., on the Alabama River, in a fertile farm area; inc. 1820. Farm implements, foundry products, batteries, paper items, lumber, furniture, textiles, and clothing are among its manufactures. A Confederate arsenal and supply point, Selma was ravaged in 1865. Nearby is the site of Cahaba, capital of Ala. from 1819 to 1826. In 1965 Selma was the center of a black voter registration drive led by Dr. Martin Luther King. **2.** City (1970 pop. 7,459), Fresno co., S central Calif., SE of Fresno; founded 1872, inc. 1893. Food is processed here.

Sel·mer (sĕl′mər), town (1970 pop. 3,495), seat of McNairy co., SW Tenn., SSE of Jackson, in a farm, cotton, and timber region; inc. 1901. Footwear is made. Nearby is Shiloh National Military Park.

Se·ma·rang (sə-mä′räng), city (1971 pop. 646,590), capital of Central Java prov., N Java, Indonesia, on the Java Sea at the mouth of the Semarang River. An important port, it is one of the major commercial centers of Java. Tobacco, sugar, rubber, coffee, and kapok are exported. There are textile and shipbuilding industries.

Se·me·ru (sə-mĕr′ōō), volcanic peak, 12,060 ft (3,678.3 m) high, E Java, Indonesia. It is the island's highest point.

Sem·i·nole (sĕm′ə-nōl′). **1.** County (1970 pop. 83,692), 305 sq mi (790 sq km), E central Fla., bounded on the N and E by the St. Johns River; formed 1913; co. seat Sanford. Its agriculture includes citrus fruit, celery and other truck crops, corn, poultry, and livestock. It has lumbering, some mining, and a food-processing industry. **2.** County (1970 pop. 7,059), 246 sq mi (637.1 sq km), extreme SW Ga., bounded on the W by the Ala. and Fla. borders, and on the E by Spring Creek; formed 1920; co. seat Donalsonville. It is in a coastal plain agriculture (cotton, corn, truck crops, peanuts, and livestock) and timber area. There is clothing manufacturing in the county. **3.** County (1970 pop. 25,144), 630 sq mi (1,631.7 sq km), central Okla., bounded in the S by the Canadian River, in the N by the North Canadian River, and drained by the Little River; formed 1907; co. seat Wewoka. Its agriculture includes cattle, peanuts, alfalfa, cotton, corn, poultry, and dairy products. It has oil and gas wells, and some industry (food processing and the manufacture of clothing, metal products, and transportation equipment).

Seminole. 1. City (1970 pop. 7,878), Seminole co., central Okla., SE of Oklahoma City; settled 1890, inc. as a town 1908, as a city 1926.

Expanding with the oil boom in 1926, Seminole is now a residential and trade center for nearby oil fields. **2.** City (1970 pop. 5,007), seat of Gaines co., NW Texas, on the Llano Estacado SW of Lubbock, in a ranch, farm, and oil area; founded 1908, inc. 1936.

Sem·i·pa·la·tinsk (syĭ-mē'pə-lä'tyĭnsk), city (1973 est. pop. 259,000), capital of Semipalatinsk oblast, Central Asian USSR, in Kazakhstan, on the Irtysh River and the Turkistan-Siberia RR. It is a river port, rail terminus, and commercial center. Semipalatinsk has one of the USSR's biggest meat-packing plants; other industries include food processing, metal working, ship repairing, wool processing, and the manufacture of building materials. The city was founded as a fort in 1718.

Sem·li·ki (sĕm'lĭ-kē), river, c.130 mi (210 km) long, E Zaire, E central Africa, flowing N from Edward Nyanza to Albert Nyanza. It forms part of the Zaire-Uganda border.

Sem·mer·ing (zĕm'ər-ĭng), scenic resort region of the Eastern Alps, E Austria. The Alps here are crossed by the Semmering Pass, 3,215 ft (980.5 m) high and 275 ft (83.9 m) long. Beneath it runs the first mountain railroad in the world (built 1848–54).

Sem·nan (sĕm-nän'), city (1971 est. pop. 35,000), capital of Semnan governorate, Mazanderan prov., N Iran. It is the trade and transportation center of a fertile agricultural region. Manufactures include textiles and carpets. Semnan was destroyed by the Oghuz Turks (1036) and by the Mongols (1221).

Sen·a·to·bi·a (sĕn'ə-tō'bē-ə), town (1974 est. pop. 4,657), seat of Tate co., NW Miss., S of Memphis, Tenn., in a farm area; founded 1856. Underwear and lumber are produced. The area was burned by Union troops during the Civil War.

Sen·dai (sĕn'dī'), city (1970 pop. 545,065), capital of Miyagi prefecture, N Honshu, Japan, on Inshinomaki Bay. It has large industries that manufacture chemicals, metal goods, silk yarn, and machinery. Long an educational center, Sendai is the seat of Tohoku Univ. and the Industrial Art Institute.

Sen·e·ca (sĕn'ĭ-kə). **1.** County (1970 pop. 35,083), 330 sq mi (854.7 sq km), W central N.Y., in the Finger Lakes region, bounded on the E by Cayuga Lake and the Seneca River, partly on the W by Seneca Lake, and crossed by the Barge Canal in the NE; formed 1804; co. seats Ovid and Waterloo. It has dairying, farming (truck crops, fruit, wheat, potatoes, hay, and beans), and diversified manufacturing. **2.** County (1970 pop. 60,696), 551 sq mi (1,427.1 sq km), N Ohio, drained by the Sandusky River and its tributaries; formed 1820; co. seat Tiffin. In a rich agricultural area yielding grain, dairy products, livestock, and poultry, it has limestone quarries, clay pits, and diversified manufacturing. There are mineral springs here.

Seneca. 1. City (1970 pop. 2,182), seat of Nemaha co., NE Kansas, near the Nebr. border NNW of Topeka; laid out 1857, inc. 1870. It is the trade center of a farm area. A state park and a replica of a frontier post are nearby. **2.** Town (1970 pop. 6,573), Oconee co., extreme NW S.C., WSW of Greenville, in a cotton area; founded 1873, inc. 1874. A trade center, it manufactures textiles, clothing, and concrete products and processes poultry.

Seneca Falls, village (1970 pop. 7,794), Seneca co., W central N.Y., in the Finger Lakes region ENE of Geneva; settled c.1787, inc. 1831. Its products include machinery tools, power pumps, and knit goods. The *Lily*, a newspaper, was published here by Amelia Bloomer. Elizabeth Cady Stanton also lived here. Seneca Falls was the scene of the first women's rights convention (1848) in the United States.

Sen·e·gal (sĕn'ə-gôl'), republic (1977 est. pop. 4,375,000), 76,124 sq mi (197,161 sq km), W Africa, bordered by the Atlantic Ocean in the W, by Mauritania in the N, by Mali in the E, and by Guinea and Guinea-Bissau in the S. The Republic of the Gambia is an enclave in the southwest. The capital of Senegal is Dakar. Most of the country is low-lying, with a maximum altitude of c.200 ft (60 m). However, the southeast, which forms a small part of the Fouta Djallon region, rises to c.1,400 ft (430 m). Senegal's coast (c.250 mi/400 km long) is sandy from Saint-Louis to Dakar, situated near the tip of the Cape Verde peninsula, and is swampy or muddy south of Dakar. The country is mostly savanna, which becomes semidesert in the Sahel region of the north and northeast; the southwest is forested. The chief rivers of the country are the Senegal (which forms the boundary with Mauritania), the Falémé, the Gambia or Gambie, and the Casamance. Lake Guiers is located in the north.

Economy. Senegal is primarily an agricultural country, but industry in the cities is growing. Only about 10% of Senegal's land is

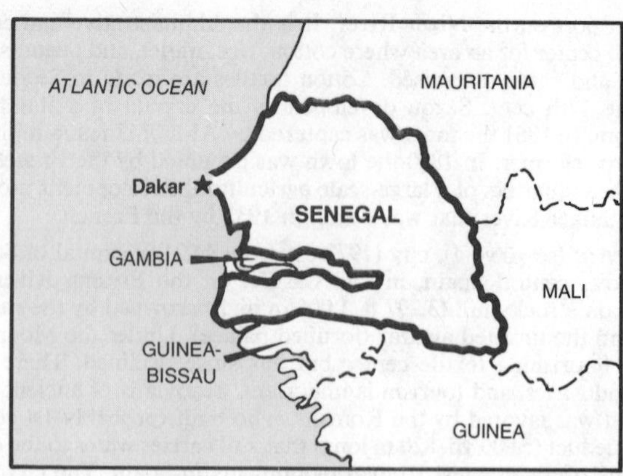

actually cultivated; much of the rest is used for pasturing livestock. The principal food crops are millet, manioc, sorghum, rice, maize, and pulses. Groundnuts are the chief cash crop; they are grown mainly on small farms. Some of Senegal's largest industries process groundnuts into oil and oilcake. Groundnuts, oil, and oilcake account for about 35% of Senegal's annual exports by value. Other manufactures include cement, chemicals, processed food, beverages, textiles, clothing, leather, footwear, and metal goods. Large numbers of cattle, sheep, and goats are raised. There is a sizable coastal fishing industry. The principal minerals extracted are phosphate rock, limestone, high-grade iron ore, and gold. The west-central part of Senegal is well served by railroads and major highways; a rail line runs from Dakar to Mali. Dakar is the country's leading port and also has a major international airport. The annual cost of Senegal's imports is usually considerably higher than its earnings from exports. The chief imports are foodstuffs (especially rice), machinery, textiles, transportation equipment, and petroleum products; the main exports (in addition to groundnuts and groundnut products) are calcium phosphate, processed fish, and hides and skins.

History. Archaeologists have found the remains of Paleolithic and Neolithic civilizations in the region now occupied by Senegal. The Wolof and Serer migrated into the area from the northeast around the middle of the first millennium A.D. The Tukolor settled in the Senegal River valley in the 9th cent., and from the 10th to 14th cent. their strong Moslem state of Tekrur dominated the valley. In the 14th cent. the Mali empire expanded westward from the region of the upper Niger River and conquered Tekrur. In the 15th cent. the Wolof established the Jolof empire in the region between the Sénégal River and the Siné wadi. Internal rivalries led to Jolof's breakup in the 17th cent. Beginning in the 15th cent., Portuguese traders used the Sénégal and Gambia rivers as routes to the interior, exchanging cloth and metal goods for gold dust, gum arabic, ivory, and small numbers of black African slaves. Trading stations were established at the mouths of the Sénégal and Casamance rivers and on Gorée Island and at Rufisque.

In the 17th cent. the Portuguese were displaced by the Dutch and the French. In 1677 the French captured Gorée from the Dutch, and it was for a time the main French naval base in west Africa. The chartered Royal Company of Senegal extended French influence far into the interior, increased the export of slaves, ivory, and gum, and encouraged with little success the cultivation of cotton and cacao. During the Seven Years' War (1756–63), Great Britain captured all the French posts in Senegal, returning only Gorée in 1763, and joined them with its holdings along the Gambia River to form the short-lived colony of Senegambia. By 1815 the French presence was limited to Saint-Louis, Gorée, and Rufisque, and during the first half of the 19th cent. there was little contact with the interior, whose trade was oriented to the north and east. As part of a French policy of assimilation, black African (along with the French-born) inhabitants of Senegal elected a deputy to the national assembly in Paris from the mid-19th cent. to independence in 1960. During the period from 1854 to 1865 (except for 1862) Capt. Louis Faidherbe was governor of Senegal, and he extended French influence and conquered Walo and Cayor. Faidherbe established schools for the black Africans and encouraged the cultivation of groundnuts. He halted the westward expansion of al-Hajj Umar, the Tukolor leader of the Tijaniyya Sufi brotherhood, who waged a large-scale holy war from a base in what is now Guinea beginning in the early 1850s.

In 1880, during the scramble for Africa, France held most of the

Senegal valley and the lower Casamance valley. In 1895 Senegal was made a French colony; it was part of French West Africa. In 1914 Blaise Diagne became the first black African elected from Senegal to the French national assembly, and he served until 1934. Under the French, Senegal's trade was reoriented toward the coast, its output of groundnuts increased dramatically, and railroads were built. During World War II, Senegal was aligned with the Vichy regime from 1940 to 1942, but then joined the Free French. In 1946 Senegal became part of the French Union and French citizenship was extended to all Senegalese. Politics in Senegal were led by its two French national assembly deputies, Lamine Gueye and Léopold Sédar Senghor, In 1948, Senghor founded the Senegalese Democratic Bloc, which dominated politics in Senegal in the 1950s.

In late 1958 Senegal became an autonomous republic within the French Community. In Jan., 1959, Senegal joined with Soudan (now Mali) to form the Mali Federation. On Aug. 20, 1960, Senegal withdrew from the federation, becoming an independent state within the French Community. At the time of independence, power was fairly evenly divided between the country's president, Senghor, and its prime minister, Mamadou Dia. In Dec., 1962, Dia staged an unsuccessful coup d'état and was arrested.

From the mid-1960s there was considerable unrest in the country caused by dissatisfaction with the growing concentration of power in Senghor's hands and by a declining economic situation resulting from lower world prices for groundnuts and reduced aid from France. The economic situation was worsened by a long-term drought in the Sahel region of northern Senegal. Major demonstrations and strikes became an almost annual occurrence and were particularly virulent in 1968, 1971, and 1973. In early 1967 there was an unsuccessful attempt on Senghor's life. In the early 1970s, about 70,000 black Africans from neighboring Portuguese Guinea fled to southern Senegal as a result of fighting in the colony between the Portuguese and black Africans. Also in the early 1970s, relations with Guinea were severed after President Sekou Touré of Guinea accused Senegal of supporting a mercenary raid on Conakry in 1970. Senghor was a leading force in establishing (1974) the West African Economic Community, which linked six former French territories.

Government. Under the 1963 constitution and later revisions, the president is directly elected to a 5-year term; he appoints the prime minister and other ministers, and they are responsible to him. Legislative power is vested in a 100-member national assembly.

Senegal, river, c.1,000 mi (1,610 km) long, formed in SW Mali, W Africa, by the confluence of the Bafing and Bakoy rivers, both of which rise in the Fouta Djallon, N Guinea. The river flows north, then generally west to form the Mauritania-Senegal border before entering the Atlantic Ocean. The river is tidal c.300 mi (480 km) upstream. It is an important source of irrigation water.

Sen·i·gal·lia (sā'nē-gäl'lyä), city (1972 est. pop. 32,800), in the Marche, central Italy, on the Adriatic Sea. It is a port, a seaside resort, and an industrial center. Manufactures include textiles, processed food, and construction materials. Made a Roman colony in the 3rd cent. B.C., it was later (6th cent. A.D.) one of the cities of the Byzantine Pentapolis. It became (12th cent.) a free commune and then was a papal fief under various rulers.

Sen·lis (säɴ-lēs'), town (1968 pop. 11,169), Oise dept., N central France, on the Nonette River. Wood products and mechanical and electrical equipment are the chief manufactures. Senlis has walls and towers from Gallo-Roman times and medieval ramparts and bastions. It also has the ruins of a château once inhabited by the first kings of France.

Sens (säɴs), town (1968 pop. 23,035), Yonne dept., N central France, on the Yonne River. Leather tanning and the manufacture of safes, electrical equipment, gears, and plastics are the chief industries. Sens was a Roman metropolis. The town was attacked by the Saracens in 731 and by the Normans in 886 and was annexed by the French crown in 1055. The town was a stronghold of the Holy League during the early 16th cent. A massacre of Huguenots took place at Sens in 1562. The Cathedral of Saint-Étienne (begun 1140), one of the oldest Gothic cathedrals, was largely built by William of Sens, who also reconstructed much of Canterbury Cathedral in England.

Sen·ta (sĕn'tä), city (1971 pop. 24,714), NE Yugoslavia, in the Vojvodina region of Serbia, on the Tisza River. A river port and an agricultural center, it has industries that produce foodstuffs, soap, and textiles. At Senta, in 1697, Prince Eugene of Savoy won a decisive victory over the Turks.

Sen·ti·num (sĕn-tī'nəm), ancient town of Umbria, E central Italy, near the modern town of Sassoferrato. In 295 B.C. the Romans defeated the Gauls and the Samnites here. The city wall and the remains of houses have been preserved.

Seoul (sōl), city (1970 pop. 5,536,377), capital of South Korea, NW South Korea, on the Han River. It has special status equivalent to that of a province and is the political, commercial, industrial, transportation, and cultural center of the nation. Since the partition of Korea in 1945 cut off access to raw materials in North Korea, the city has emphasized textile manufacturing, agricultural processing, and varied consumer industries. There are also tanneries, railroad repair shops, and large power plants. Founded in 1392, Seoul was the capital of the Yi dynasty, which ruled Korea until the country became (1910) a colony of Japan. It became the capital of South Korea in 1948. North Korean forces captured the city on June 28, 1950, only three days after the Korean War began; it then changed hands several times until UN troops took it in Mar., 1951. Heavily damaged during the war, the city was rebuilt along modern lines. Its population was greatly increased by refugees.

Se·pik (sā'pĭk'), river, c.700 mi (1,125 km) long, N New Guinea, rising near the border of West Irian and Papua New Guinea. It flows east to the Bismarck Sea. The river drains a vast mountainous region of central New Guinea.

Sept-Îles (sĕt-ēl'), city (1971 pop. 24,320), E Que., Canada, on the St. Lawrence River near its mouth. It is a major port exporting iron ore. The harbor was visited by Jacques Cartier in 1535, and a trading post was built on the site in 1650.

Se·quatch·ie (sĭ-kwŏch'ē), county (1970 pop. 6,331), 273 sq mi (707.1 sq km), SE central Tenn., partly in the Cumberlands, with Walden Ridge in the SE, and drained by the Sequatchie River; formed 1857; co. seat Dunlap. Its agriculture includes livestock, feed crops, fruit, and tobacco. It has coal mining and lumbering and manufactures clothing.

Se·quoi·a National Park (sĭ-kwoi'ə), 386,863 acres (156,680 hectares), E central Calif.; est. 1890. In the park are 35 groves of giant sequoias, spectacular granite mountains, and deep canyons. The General Sherman Tree, 272 ft (83 m) high and 37 ft (11.2 m) in diameter at its widest point, is the largest and is estimated to be more than 3,500 years old. Within the area are Mt. Whitney, the highest point in the conterminous United States, and the Great Western Divide, which separates westerly flowing waters from easterly flowing waters.

Se·quoy·ah (sĭ-kwoi'ə), county (1970 pop. 23,370), 696 sq mi (1,802.6 sq km), E Okla., in the hilly region of the Boston Mts., bounded in the E by the Ark. border and in the S by the Arkansas River; formed 1907; co. seat Sallisaw. Its agriculture includes vegetables, cotton, livestock, corn, and potatoes. It has coal mines, natural-gas wells, salt deposits, and timber. Wood products, furniture, chemicals, plastics, and auto parts are manufactured.

Se·raing (sə-rĕɴ'), city (1971 est. pop. 40,885), Liège prov., E Belgium, on the Meuse River, near Liège. A major center of heavy industry, it is the seat of the Cockerill steel and locomotive works. Nearby is Val Saint Lambert, one of the world's leading glassware-manufacturing centers.

Se·ram·pur or **Se·ram·pore** (both: sĕ'rəm-pôr'), town (1971 pop. 102,023), West Bengal state, E central India, on the Hooghly River, just N of Calcutta. Founded in 1799, Serampur was the center of Danish colonialism in India. Great Britain purchased the town from Denmark in 1845.

Ser·bi·a (sûr'bē-ə), constituent republic of Yugoslavia (1971 pop. 8,436,547), 34,107 sq mi (88,337 sq km), E Yugoslavia, the largest and most important of the republics. Belgrade is the capital of both Serbia and Yugoslavia. The republic consists of Serbia proper with the cities of Belgrade and Niš, the autonomous Vojvodina prov. with Subotica and Novi Sad, and the autonomous Kossovo-Metohija region with Priština. Serbia is largely mountainous in the west and south, but the northeast is part of the fertile Danubian plain, drained by the Danube, Sava, Tisza, and Morava rivers. Vojvodina and Serbia proper, the "breadbasket" of Yugoslavia, provide about half the country's total agricultural produce. Wheat, corn, hemp, sugar beets, and flax are the chief crops. Serbia proper has extensive vineyards and is one of Europe's major regions for fruit-growing (notably plums). It is also an important mining area.

The republic's population consists primarily of Serbs, with Slovenian, Croatian, Magyar, Montenegrin, Albanian, and Macedonian minorities. The Serbs distinguish themselves culturally from the closely related Croats and Slovenes through their membership in the Orthodox Eastern Church and use of the Cyrillic rather than the Roman alphabet.

History. Serbs settled in the Balkan Peninsula in the 6th and 7th cent. and accepted Christianity in the 9th cent. Their petty principalities were theoretically under a grand *zhupan,* who usually recognized Byzantine suzerainty. Rascia, the first organized Serbian state, was probably founded in the early 9th cent. in the Bosnian mountains; it steadily expanded from the 10th cent. Stephen Nemanja, whom the Byzantine emperor recognized as grand *zhupan* of Serbia in 1159, founded a dynasty that ruled for two centuries. Under Stephen Dushan, who became king in 1331 and czar in 1346, Serbia became the most powerful empire in the Balkan Peninsula, much of which it absorbed. After Stephen's death in 1355, however, the empire decayed and fell victim to the onslaught of the Ottoman Turks. The Serbs suffered defeat at the Maritsa River in 1371; in 1389, during the Battle of Kossovo Field, the cream of Serbian nobility was massacred and the fate of independent Serbia sealed. The Turks gradually absorbed the land, fully annexing it in 1459.

Serbia became a Turkish province, with its pashas residing at Belgrade. Turkish rule in Serbia was more oppressive than in most Turkish provinces. Although the Serbs were forbidden to possess weapons, frequent insurrections erupted. The liberation struggle began in 1804, when Karageorge ("Black George") led a rebellion that eventually freed the pashalik (province) of Belgrade from the Turks. In 1829 Russia forced the Treaty of Adrianople upon the sultan, who had to grant Serbian autonomy under Russian protection. In 1867 the last Turkish troops left Serbia. Prince Milan Obrenović supported the rebellion of Bosnia and Hercegovina against Turkish rule and in 1876 declared war on Turkey. The rout of the Serbs led Russia to enter the war on the Serbian side. The Congress of Berlin (1878) recognized Serbia's complete independence and substantially increased its territory.

Serbia's championship of Pan-Slavism in the Balkans engendered bitter rivalry with Bulgaria and Austria-Hungary. The strengthening of parliamentary government and expansion of the economy under King Peter I (acceded in 1903) greatly raised Serbia's prestige and exerted a powerful attraction on the South Slavs who remained under Austro-Hungarian rule. In 1912 Serbia and its Balkan allies declared war on and defeated Turkey. In 1913 Serbia turned against and defeated its former Bulgarian ally in the Second Balkan War. Serbia's victory made it the foremost Slavic power in the Balkans but greatly increased tensions with Austria-Hungary. When a Serbian nationalist assassinated Austrian archduke Francis Ferdinand in 1914, the empire declared war on Serbia, thus precipitating World War I. Serbia was overrun in 1915. In 1917 Serbian, Croatian, Slovenian, and Montenegrin representatives on Kérkira (Corfu) proclaimed the union of South Slavs. In 1918 the kingdom of the Serbs, Croats, and Slovenes, headed by Peter I of Serbia, officially came into existence. After that, the history of Serbia is essentially that of Yugoslavia.

Se·rem·ban (sə-rĕm-bän'), city (1971 pop. 79,915), capital of Negeri Sembilan, Malaysia, S Malay Peninsula, on the Linggi River. It is the commercial center of a rubber-growing and tin-mining area.

Ser·gi·pe (sər-zhē'pə), state (1970 pop. 901,618), 8,321 sq mi (21,551 sq km), NE Brazil, on the Atlantic Ocean; capital Aracaju.

Se·rov (syĕ'rəf), city (1973 est. pop. 99,000), E European USSR, in the E foothills of the Urals, on the Kakvy River. Serov produces cast iron and high-grade steel. The city was founded in 1894 in connection with the building of the Trans-Siberian RR.

Se·ro·we (sĕ-rō'wā), town (1971 pop. 15,723), E central Botswana. Located in a fertile, well-watered area, it is a trade center and the seat of the Ngwato tribe.

Ser·pu·khov (syĕr'pŏŏ-KHəf), city (1973 est. pop. 129,000), central European USSR, on the Oka River. It is an important textile center. A fortress town since 1339, it retains a stone kremlin (16th cent.).

Sér·rai (sâr'ā) or **Si·ris** (sī'rĭs), city (1971 pop. 39,897), capital of Sérrai prefecture, NE Greece, in Macedonia. It is a trade center for tobacco, grain, and livestock. Textiles and cigarettes are manufactured. The city was fortified under the Byzantine Empire and in the 14th cent. became a capital of Serbia. Sérrai was held by the Ottoman Empire from 1383 to 1913, when it passed to Greece. The city was occupied and damaged by Bulgaria in both world wars.

ser·tão (sər-touN'), semiarid hinterland of NE Brazil; c.250,000 sq mi (647,500 sq km). Its characteristic landscape is the caatinga, or thorny scrub forest. The chief occupation is stock raising. Periodic droughts have caused large-scale migrations to the Amazon basin and to the urban centers of southeastern Brazil. Reclamation activities were intensified in the 1950s and 1960s with the construction of numerous dams and hydroelectric projects, especially on the São Francisco River. In the 1960s a successful extensive regional economic development program was begun here. The area remains a focal point of social unrest. Peopled by leather-garbed cowboys, bandits, and religious fanatics, it has been a source of inspiration for numerous Brazilian writers.

Ses·tos (sĕs'təs), ancient town on the Thracian shore of the Hellespont opposite Abydos. It was here that Xerxes entered Thrace on his invasion of Greece, crossing the Hellespont on a bridge of boats. The city remained important in Roman times, but declined after the founding of Byzantium.

Ses·to San Gio·van·ni (sĕ'stō sän jō-vän'nē), city (1972 est. pop. 92,922), Lombardy, N Italy; an industrial suburb of Milan. Manufactures include iron, machinery, chemicals, and textiles.

Sète (sĕt), town (1968 pop. 40,476), Hérault dept., S France, in Languedoc, on the Mediterranean. It is one of the most important commercial and fishing ports of southern France, a wine-shipping center, with major gasoline refineries and plants making clothes, liqueurs, and chemical products.

Se·tes·dal (sā'təs-däl'), narrow valley, S Norway. It is drained by the Otra River and contains several lakes. Communication with the rest of the country has, until recently, been difficult; as a result the Setesdalers have retained their ancient dress, speech, customs, and handicrafts. Agriculture and fishing are the chief occupations.

Sé·tif (sā-tēf'), city (1974 est. pop. 157,065), capital of Sétif dept., NE Algeria. It is the commercial center of a region where native textiles and phosphates are manufactured and cereals are grown. Sétif was built by the French on the ruins of the Roman town of Sitifis, founded in the 1st cent. A.D.

Se·to (sā'tō), city (1970 pop. 92,681), Aichi prefecture, central Honshu, Japan. It has been an important porcelain center since the 13th cent.

Set·tsu (sāt'tsōō), city (1970 pop. 59,758), Osaka prefecture, SW Honshu, Japan. It is a suburb of Osaka.

Se·tú·bal (sə-tōō'bəl), city (1972 est. pop. 49,670), capital of Setúbal dist., S central Portugal, on the Bay of Setúbal at the mouth of the Sado River, in Estremadura. Its port handles wine, oranges, and cork. The city has a fishing fleet and shipyards and is a major sardine-canning center.

Se·van (sĕ-vän'), lake, c.540 sq mi (1,400 sq km), SE European USSR, in Armenia, at an altitude of 6,280 ft (1,915.4 m); it is 324 ft (99 m) deep. The largest lake of the Caucasus, it is fed by some 30 streams. Lake Sevan is free of ice in winter. Several hydroelectric stations have been built along the Razdan, an extensive system made possible by the steep gradient of the river and the great volume of the lake.

Se·vas·to·pol (sə-văs'tə-pōl', syĕ'və-stô'pəl), city (1973 est. pop. 252,000), SE European USSR, in the Ukraine, on the Crimean peninsula and the Bay of Sevastopol, an inlet of the Black Sea. The city is a port and a major naval base. Its industries include shipbuilding, lumber milling, food processing, and the production of bricks and furniture. Sevastopol stands near the site of the ancient Greek colony of Chersonesus (founded in 421 B.C.), the most important Greek colony in the Crimea until Scythian invasions forced it to become (179 B.C.–63 B.C.) a protectorate of King Mithridates VI. In the 1st cent. A.D. the cities of the Crimea became part of the Roman Empire. In the Middle Ages it remained a large trading, political, and cultural center. Sevastopol was founded as a city and port by Catherine II on the site of the Tatar village of Akhtiar after the Russian annexation (1783) of the Crimea. In the Crimean War Sevastopol resisted the besieging British, French, Turks, and Sardinians for 349 days (1854–55). The Russian fleet was sunk by the Russians themselves to block the entrance to the harbor. In Sept., 1855, the French successfully stormed the fortress and three days later the Russians were forced to abandon Sevastopol. In 1890 the city again became a chief naval base. The Sevastopol sailors mutinied during the 1905 revolution. The heroic resistance of Sevastopol in 1854–55 was, if possible, eclipsed by the stand the city made against the Germans in

World War II. During a siege lasting more than eight months, the city was virtually destroyed. For three weeks the defenders fought on in the rubble, until July 3, 1942, when German and Rumanian troops at last entered the city.

Sev·en Hills (sĕv′ən), city (1970 pop. 12,700), Cuyahoga co., N Ohio, a residential suburb of Cleveland, in a hilly area; inc. as a city 1961.

Seven Wonders of the World, in ancient classifications, were the Great Pyramid of Khufu or all the pyramids with or without the Sphinx; the Hanging Gardens of Babylon, with or without the walls; the Mausoleum at Halicarnassus; the Artemision at Ephesus; the Colossus of Rhodes; the Olympian *Zeus,* statue by Phidias; and the Pharos at Alexandria, or, instead, the walls of Babylon.

Sev·ern (sĕv′ərn), river, c.420 mi (675 km) long, rising in W Ont., Canada, and flowing NE through Severn Lake to Hudson Bay.

Severn, one of the principal rivers of Great Britain, c.200 mi (320 km) long, rising on Plinlimmon Mt., W Wales, and flowing NE and E to Shrewsbury, W England, and thence SE, S, and SW to the Bristol Channel. It is connected by canal with the Thames, Mersey, Trent, and other rivers. The Severn Road Bridge (opened 1966) is one of the world's longest (3,240 ft/988 m) suspension bridges. The river is important as a transportation route.

Se·ver·na Park (sə-vûr′nə), uninc. town (1970 pop. 16,358), Anne Arundel co., central Md., a suburb of Baltimore.

Se·ver·na·ya Zem·lya (sĕv′ər-nə-yä′ zĕm′lē-ä′), archipelago, c.14,300 sq mi (37,010 sq km), between the Kara and Laptev seas, Krasnoyarsk Kray, N Siberian USSR, off the Taymyr Peninsula. Extending north of lat. 80° N, it is composed of four major islands and several smaller ones. Glaciers are found on the larger islands.

Se·vier (sə-vîr′). **1.** County (1970 pop. 11,272), 585 sq mi (1,515.2 sq km), SW Ark., bounded on the W by the Okla. border, on the S by the Little River, on the E by the Saline River; formed 1828; co. seat De Queen. It is in a rich truck-farming area and produces cotton, livestock, poultry, and dairy products. It also has lumber and miscellaneous manufacturing. **2.** County (1970 pop. 28,241), 597 sq mi (1,546.2 sq km), E Tenn., bounded in the S and SE by N.C., with the Great Smoky Mts. along the S border, and drained by the French Broad River; formed 1794; co. seat Sevierville. Its agriculture includes livestock, fruit, tobacco, corn, and hay. It has lumbering and food-processing industries. Textiles, wood products, furniture, metal products, and machinery are made. **3.** County (1970 pop. 10,103), 1,929 sq mi (4,996.1 sq km), central Utah, in a mountain and plateau region crossed by the Sevier River; formed 1865; co. seat Richfield. In an agricultural area yielding livestock, hay, sugar beets, fruit, and truck crops, it has bituminous-coal mining and manufactures clothing and concrete products.

Sevier, river, c.280 mi (450 km) long, formed in SW Utah by the junction of Panguitch Creek and Assay Creek. It flows northward through canyons, then leaves the mountains and flows southwest through the Sevier Desert to Sevier Lake.

Se·vier·ville (sə-vîr′vĭl′), town (1970 pop. 2,661), seat of Sevier co., E Tenn., ESE of Knoxville, in a timber and farm area; laid out 1795, inc. 1887. Great Smoky Mountains National Park and Douglas Dam in the French Broad River are nearby.

Se·ville (sə-vĭl′, sĕ′-), city (1972 est. pop. 565,055), capital of Seville prov. and leading city of Andalusia, SW Spain, on the Guadalquivir River. Seville is a major port and cultural center. Wines, fruit, olives, cork, and minerals are exported. Its industries include the manufacture of tobacco, armaments, explosives, perfume, porcelain, pharmaceuticals, chemicals, textiles, and machinery. Seville was important in Phoenician times. It was favored by the Romans, who made it a judicial center of Baetica prov. Seville continued as the chief city of southern Spain under the Vandals and the Visigoths. Falling to the Moors in 712, it was a flourishing commercial and cultural center. Ferdinand III of Castile conquered it after a long siege and made it his residence (1248). With the discovery of the New World, Seville entered its greatest period of prosperity, being the chief port of trade with the new colonies until 1718, when it was superseded by Cádiz. Its economic recovery from the subsequent decline is only recent. In 1810 the French sacked the city. Seville was held by the insurgents throughout the civil war (1936–39).

Seville has kept much of its Moorish aspect. Its old quarters are crossed by tortuous, narrow streets, interrupted by fine squares, and lined with whitewashed houses with patios and balconies trimmed with iron filigree work. The Gothic cathedral (1401–1519), one of the world's largest, contains invaluable works of art and the tomb of Christopher Columbus. Adjoining the cathedral is the alcazar, built (14th cent.) in Moorish style. One of the world's most beautiful cities, Seville is the capital of bullfighting in Spain and a center of the Andalusian gypsies, famed for their songs and dances.

Sè·vres (sĕv′rə), town (1968 pop. 20,083), Hauts-de-Seine dept., N central France, on the Seine River; a residential suburb SW of Paris. The famous Sèvres ware porcelain is made in the town. Explosives, surgical supplies, and beer are also produced.

Sew·ard (sōō′ərd). **1.** County (1970 pop. 16,062), 646 sq mi (1,673.1 sq km), SW Kansas, in a plains region, bordered on the S by Okla., and drained by the Cimarron River; formed 1855; co. seat Liberal. It lies in a grain and livestock area, with a good supply of natural gas and water. It has a food-processing industry (especially meat-packing) and manufactures farm machinery. **2.** County (1970 pop. 14,460), 571 sq mi (1,478.9 sq km), SE Nebr., in an agricultural area drained by the Big Blue River; formed 1865; co. seat Seward. Products include grain, livestock, dairy goods, and poultry.

Seward. 1. City (1970 pop. 1,587), Kenai Peninsula borough, S Alaska, on Kenai Peninsula, at the head of Resurrection Bay; inc. 1912. It was founded in 1902 as the ocean terminus of the Alaska RR (built 1915–23). The airfield and ice-free harbor make it an important shipping and supply center for the Alaskan interior. It has a seafood cannery and freezer plant and a saw mill. Seward was almost completely devastated by an earthquake in 1964. **2.** City (1970 pop. 5,294), seat of Seward co., SE Nebr., WNW of Lincoln; founded 1868, inc. 1874. It is a farm trade center in a prairie region. China and fertilizers are manufactured.

Seward Peninsula, W Alaska, projecting c.200 mi (320 km) into the Bering Sea between Norton Sound and Kotzebue Sound, just below the Arctic Circle. The region is mostly bleak tundra, with long, cold winters. Placer-gold mining and trapping are the chief occupations of its sparse population.

Sey·chelles (sā-shĕlz′), independent republic (1978 est. pop. 61,000), c.110 sq mi (285 sq km), comprising approximately 85 islands in the Indian Ocean, c.600 mi (970 km) N of Madagascar and c.1,000 mi (1,600 km) E of Mombasa, Kenya. It was a British crown colony from 1903 until 1976. The capital and only urban center and port is Victoria, located on the largest island, Mahé (c.55 sq mi/140 sq km). Copra, coconuts, cinnamon, patchouli, vanilla, and tea are exported; fishing is an important local industry. Tourism is growing. The Seychelles were discovered by Vasco da Gama in 1502. In 1756 the French claimed the islands. Britain took possession of the Seychelles in 1794 and gained permanent control of them by the Treaty of Paris (1814). The islands were administered as part of Mauritius until 1903, when they were constituted a crown colony. The first elections to a legislative council were held in 1948. The Seychelles gained independence and were made a member of the Commonwealth in 1976.

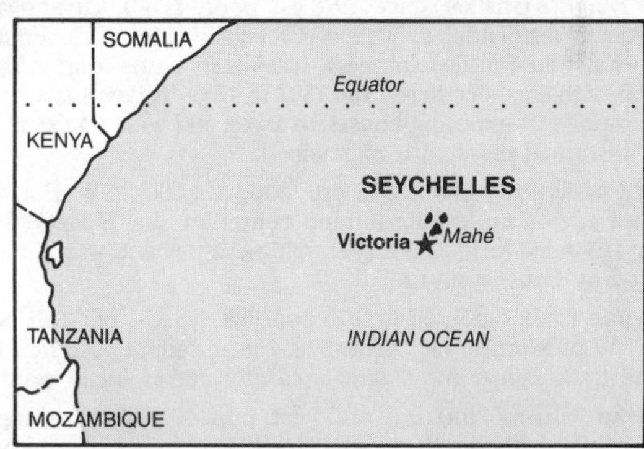

Sey·han (sā-hän′), river, c.320 mi (515 km) long, rising in the Anti-Taurus Mts., central Turkey, and flowing SW past Adana to the Mediterranean Sea. Cotton and grapes are grown in its valley. A dam on the river provides flood control, irrigation and municipal water, and electricity.

Sey·mour (sē′môr′). **1.** Town (1970 pop. 12,776), New Haven co., SW Conn., on the Naugatuck River; settled c.1678, inc. 1850. The industrial village of Seymour has metallurgical manufactures.

2. City (1975 est. pop. 13,100), Jackson co., SE Ind.; inc. 1864. A shipping center for a farm area, it is also a manufacturing city. **3.** City (1970 pop. 3,469), seat of Baylor co., N Texas, on the Brazos River SW of Wichita Falls; inc. 1906. The city, settled in 1878 as a ranch and farm community at the crossing of the old Dodge City and California trails, handles cotton, grain, oil, and cattle.

Sfax (sfäks): *see* Safaqis, Tunisia.

Sha·ba (shä′bä), formerly **Ka·tan·ga** (kä-täng′gə, kə-), region (1969 est. pop. 2,174,400), c.200,000 sq mi (518,000 sq km), SE Zaire. Shaba borders Angola on the southwest, Zambia on the southeast, and Lake Tanganyika on the east. The region encompasses the fertile Katanga Plateau (3,000–6,000 ft/915–1,830 m high), where profitable farming and ranching are carried on. In the eastern part of the region is an enormously rich mining area, which supplies most of the world's cobalt as well as extensive quantities of copper, tin, radium, uranium, and diamonds. Copper has been mined and exported by the region's inhabitants for centuries. The region's considerable industrial plant is largely concerned with the processing of minerals. In July, 1960, after the Democratic Republic of the Congo (now Zaire) became independent, Katanga proclaimed itself a republic and seceded from the central government. Under the leadership of its president, Moise Tshombe, and with Belgian aid, Katanga fought off repeated attempts by the central government to seize control. After two-and-a-half years of disorder and UN intervention, UN troops routed Tshombe's forces and ended the Katanga secession.

Shack·el·ford (shăk′əl-fərd), county (1970 pop. 3,323), 887 sq mi (2,297.3 sq km), N central Texas, drained by the Clear Fork of the Brazos River; formed 1856; co. seat Albany. This area, a leading Texas cattle-ranching county, also produces sheep, turkeys, wheat, cotton, and grain sorghums. It has oil and natural-gas wells.

Shad·ow Mountain National Recreation Area (shăd′ō): *see* National Parks and Monuments Table.

Shaf·ter (shăf′tər), city (1970 pop. 5,327), Kern co., S central Calif., NW of Bakersfield, in the San Joaquin Valley; founded 1913, inc. 1938.

Sha·hi (shä-hē′), city (1966 pop. 38,898), Mazanderan prov., N Iran. Manufactures include textiles and food products.

Shah·ja·han·pur (shä′jə-hän′po͝or), city (1971 pop. 135,604), Uttar Pradesh state, N central India, on the Garra River. Founded by the Mogul emperor Shah Jahan in 1647, it was named in his honor. It is now an agricultural market town.

Shak·er Heights (shā′kər), city (1978 est. pop. 34,200), Cuyahoga co., NE Ohio, a residential suburb of Cleveland; inc. 1912. Founded (1905) as a suburban development, it takes its name from a Shaker community that existed here from 1822 to 1889. Today it is a beautiful community of imposing houses, squares, and wide boulevards. A Shaker historical museum is maintained.

Shakh·ty (shäkʜ′tē), city (1973 est. pop. 214,000), SW European USSR; a major anthracite-mining center of the Donets Basin. Shakhty, founded in 1829 as a coal-mining settlement, was known as Aleksandrov-Grushevski until 1920.

Shak·o·pee (shăk′ō-pē), city (1970 pop. 6,876), seat of Scott co., E Minn., SW of Minneapolis; settled 1851 as a trading post, inc. 1870. It is the trade center for a farm area and makes metal products.

Sha·mo·kin (shə-mō′kĭn), city (1975 est. pop. 10,600), Northumberland co., E Pa.; settled c.1835, inc. 1864. Shamokin is a mining center for anthracite coal, and textiles, shoes, and trailers are produced here.

Shang·hai or **Shang-hai** (shăng′hī′), city (1970 est. pop. 11,000,000), in, but independent of, Kiangsu prov., E China, on the Whangpoo (Huang-p'u) River where it flows into the Yangtze estuary. It is administered directly by the central government. One of the world's great seaports, Shanghai is the largest city of China and the most populous on the continent of Asia. The only large port of central China not cut off from the interior by mountains, it is the natural seaward outlet of, and the gateway to, the productive Yangtze basin, one of China's richest regions. It handles the major share of the country's foreign shipping and a large coastwise trade. Despite a lack of fuel and raw materials, Shanghai is the leading industrial city of China, with large steelworks, textile mills, shipbuilding yards, oil-refining and gas-extracting operations, and plants making a great variety of light and heavy machinery, electrical and electronic equipment, machine tools, turbines, chemicals, pharmaceuticals, aircraft, tractors, motor vehicles, plastics, and assorted consumer goods. Shanghai is also a major publishing center. The commercial section of the city, the former International Settlement, is modern and Western in appearance, with broad streets and spacious boulevards lined with imposing buildings, many of skyscraper height. The Bund (which runs along the waterfront), Nanking Road, and Bubbling Well Road are the most noted thoroughfares. Typical Oriental buildings are found only in the original Chinese town (no longer walled), known as the Chapei quarter. Shanghai has annexed a large part of the surrounding rural area (over 2,000 sq mi/5,180 sq km); here peasant communities and collective farms produce the food crops (wheat, barley, vegetables, fruits, etc.) that support the city's population. Though water transport is of prime importance to Shanghai, new highways radiate to the countryside, and there are rail connections with Nanking and Hangchow. Shanghai has an international airport. A submarine base is in the harbor.

The name Shanghai dates from the Sung dynasty (11th cent.), but the town, which became a walled city in the 16th cent., was unimportant until it was opened to foreign trade by the Treaty of Nanking in 1842. The ensuing Western influence launched the city on its phenomenal growth. The greater part of the city was incorporated into the British concession (1843), just north of the old walled city, and into the U.S. concession of Hongkew (1862). In 1863 the United States and Great Britain consolidated their areas into the International Settlement. The French, who had obtained a concession in 1849, continued it as a separate entity. The foreign zones, which were under extraterritorial administration, maintained their own courts, police system, and armed forces. In 1927 Chiang Kai-shek, at the head of the Nationalist army and with the support of the Chinese Communists, captured Shanghai. Japan invaded and attacked the Chinese city in 1932 to force the government to break an unofficial boycott of Japanese goods. In Aug., 1937, the Japanese again attacked the Chinese city, and gallant resistance was overcome in Nov. The foreign zones were occupied by the Japanese after Dec. 7, 1941. In 1943 the United States and Great Britain renounced their claims in Shanghai, as did France in 1946. The city was restored to China at the end of World War II. In May, 1949, it fell to the Communists. Next to Peking, Shanghai is the country's foremost educational center. Shanghai has an astronomical observatory, several museums, an opera, a performing arts group, and many research institutes and learned societies.

Shan-hai-kuan or **Shan·hai·kwan** (both: shän′-hī′-kwän′), city, NE Hopeh prov., China, on the Po Hai. It has fertilizer and food-processing plants. Strategically situated where the Great Wall meets the sea and on the narrow coastal route to Manchuria, it has been the site of many battles and the route of many invasions.

Shan·non (shăn′ən). **1.** County (1970 pop. 7,196), 999 sq mi (2,587.4 sq km), S Mo., in the Ozarks and drained by the Current River; formed 1841; co. seat Eminence. Agriculture, timber, and copper and manganese mines are found in the area. **2.** County (1970 pop. 8,198), 2,100 sq mi (5,439 sq km), SW S.Dak., on the Nebr. border and drained by the White River; formed 1875 as a smaller county, and combined in 1943 with the adjacent Washington co.; it is unorganized and is attached to Fall River co. for judicial purposes. It raises livestock and manufactures leather footwear.

Shannon, principal river of the Republic of Ireland and longest (c.240 mi/385 km) in the British Isles. It rises near Cuilcagh Mt., in the northwest of Co. Cavan, and flows south through the Central Plain to Limerick, where it turns west in a broad estuary to the Atlantic Ocean between Loop Head and Kerry Head. Loughs Allen, Boderg, Ree, and Dreg are expansions of the river. The Shannon with its many tributaries drains a region of farmland and peat bogs. The river is connected with eastern Ireland by the Royal Canal and the Grand Canal. There is an important hydroelectric plant between Lough Derg and Limerick. The fisheries of the river are valuable.

Shan·non·town (shăn′ən-toun′), uninc. town (1970 pop. 7,491), Sumter co., central S.C., a residential suburb S of Sumter.

Shan·si (shän′sē′), province (1967 est. pop. 18,000,000), c.60,000 sq mi (155,400 sq km), NE China; capital T'ai-yüan. It is bounded on the west and the south by the Huang Ho and on the north by Inner Mongolia. Much of Shansi is a high plateau region. The soil is fertile, but scant rainfall and widespread erosion hamper the raising of sufficient food. Reforestation and irrigation projects have been instituted. The main food crops are winter wheat, corn, kaoliang, soybeans, millet, barley, and fruit. Cotton, tobacco, and grapes are grown as commercial crops. Livestock is raised in the northern grazing areas, and wool and hides are exported. Forestry is important in the mountainous regions. Shansi has rich and extensive coal and iron deposits. A salt lake in the southwestern part of the province is one of China's major inland sources of salt.

Shan·tar Islands (shən-tär'), archipelago, c.980 sq mi (2,540 sq km), Khabarovsk Kray, Far Eastern USSR, in the Sea of Okhotsk. Discovered in 1645, the islands are used as a fishing base and for fur trapping and lumbering.

Shan-t'ou (shän'tō') or **Swa·tow** (swä'tou'), city (1970 est. pop. 400,000), SE Kwangtung prov., China, a port on the South China Sea, in the Han River delta. When it was opened to foreign trade after the second Opium War, it was a minor fishing village. Today it is a trade and industrial center, with shipbuilding yards and machine shops. Fishing is still important.

Shan·tung or **Shan-tung** (both: shăn'tŭng', shän'dŏŏng'), province (1967 est. pop. 57,000,000), c.54,000 sq mi (139,860 sq km), NE China; capital Chi-nan. The eastern half of the province is a peninsula, situated between the Po Hai on the north and the Yellow Sea on the east and south. The mountain chain that forms the Liaotung peninsula in Manchuria continues into east and central Shantung. The western portion of the province, a level area, is part of the delta of the Huang Ho. Although the soil is fertile, rainfall is inadequate. Much of the land has been reclaimed, however, and half of the province is now under cultivation. Shantung is an important wheat-producing and cotton-producing province; kaoliang, millet, corn, soybeans, peanuts, sweet potatoes, fruits, and tobacco are also grown. Forest products are made, and pongee silk is produced by wild silkworms fed on oak leaves. Fishing is excellent along the rocky coast and offshore islands. Salt is also produced in the province. Oil is extracted near the mouth of the Huang Ho. Abundant coal and iron reserves are also exploited, and Shantung has deposits of gold, copper, and kaolin. Strategically located near Manchuria and with excellent harbors, Shantung has often been subjected to foreign encroachment. In recent years many Chinese have migrated from Shantung to Manchuria to escape extreme overcrowding.

Shao-hsing (shou'shĭng') or **Shao·hing** (shou'hĭng'), city (1970 est. pop. 225,000), N Chekiang prov., SE China, on the S shore of Hangchow Bay. It is a marketing center handling rice, silk, and tea. The most famous export is rice wine. Silk textiles, paper products, and handicrafts are made.

Shao-kuan (shou'gwän'), town (1970 est. pop. 125,000), N Kwangtung prov., SE China, on the North River. It is a trade center with coal and tungsten mining, a lead-zinc plant, and factories making chemicals, machine tools, rubber products, and cement.

Shao-yang or **Shao·yang** (both: shou'yäng'), city (1970 est. pop. 275,000), S Hunan prov., China. It is the trade center for the upper Tze valley and has coal and iron mines.

Sha·ri (shä-rē'), river, Chad: *see* Chari.

Shar·jah (shär-jä'), sheikdom (1968 pop. 31,480), c.1,000 sq mi (2,590 sq km), part of the federation of United Arab Emirates, E Arabia, on the Persian Gulf and the Gulf of Oman. The town of Sharjah (1968 pop. 19,198) is the second-largest town in the federation. Oil has been produced in Sharjah since 1961. Formerly a British protectorate, Sharjah joined the United Arab Emirates in 1971.

Shar·key (shär'kē), county (1970 pop. 8,937), 436 sq mi (1,129.2 sq km), W Miss., drained by the Sunflower River and other streams; formed 1876; co. seat Rolling Fork. Agriculture (cotton, corn, and oats) and timber are found here. Much of the county is in Delta National Forest.

Shar·on (shâr'ən). **1.** Resort town (1970 pop. 2,491), Litchfield co., NW Conn., in the Taconic Mts. WNW of Torrington; settled c.1738, inc. 1739. Noah Webster wrote his *Spelling Book* while he was teaching school here. **2.** Town (1970 pop. 12,367), Norfolk co., E Mass.; settled c.1650, inc. 1775. It is residential. **3.** City (1978 est. pop. 22,600), Mercer co., NW Pa., on the Shenango River, near the Ohio line; settled c.1800, inc. as a city 1920. An industrial city, its chief manufactures are steel, electrical appliances, and a variety of metal products. A campus of Pennsylvania State Univ. is here. **4.** Town (1970 pop. 200), Windsor co., E Vt., ENE of Rutland. The birthplace of Joseph Smith, founder of Mormonism, is marked by a granite monolith.

Sharon Springs, city (1970 pop. 1,012), seat of Wallace co., W Kansas, S of Goodland near the Smoky Hill River. It is a trading center for a livestock and grain region.

Shar·on·ville (shâr'ən-vĭl'), city (1970 pop. 11,393), Hamilton co., SW Ohio, a suburb of Cincinnati; surveyed 1796, inc. 1911.

Sharp (shärp), county (1970 pop. 8,233), 598 sq mi (1,548.8 sq km), N Ark., bounded on the N by the Mo. border and drained by the Strawberry and Spring rivers; formed 1868; co. seat Ash Flat. It is in an agricultural area that produces cotton, corn, hay, poultry, and livestock. Lumbering is done. There are fishing resorts in the county.

Sharps·burg (shärps'bûrg'). **1.** Town (1970 pop. 863), NW Md. The Civil War Battle of Antietam (or Sharpsburg) was fought here in 1862. Union forces repulsed Gen. R. E. Lee's troops, but the battle was one of the bloodiest of the war, with heavy losses on both sides. **2.** Borough (1970 pop. 5,453), Allegheny co., SW Pa., on the Allegheny River NE of Pittsburgh; inc. 1842. It has light industry.

Sharps·ville (shärps'vĭl'), borough (1970 pop. 6,126), Mercer co., NW Pa., just NE of Sharon; inc. 1874. Steel is made. Buhl Park, a large recreation center, is here.

Sha-shih (shä'shûr') or **Sha·si** (shä'sē'), city (1970 est. pop. 125,000), S Hupeh prov., China, on the Yangtze River. It is an important trade center for the northern Tung-t'ing lake basin and the site of a reservoir built to protect the central Hupeh plains from floods.

Shas·ta (shăs'tə), county (1970 pop. 77,640), 3,788 sq mi (9,811 sq km), N Calif., in a mountainous region, with the Klamath Mts. meeting the Cascade Range at the N end of the Central Valley; formed 1850; co. seat Redding. It is in a lumbering, stock-raising (beef and dairy cattle, hogs, sheep, and poultry), irrigated fruit-growing, and general farming (alfalfa, grain, vegetables, and berries) region. It has food processing, millwork, and manufacturing of cement and machinery. Hunting, fishing, and camping attract vacationers. It contains Lassen Volcanic National Park, and much of the county is included in Shasta and Lassen national forests.

Shasta, Mount, volcanic peak, 14,162 ft (4,319.4 m) high, N Calif., in the Cascade Range. Discovered c.1827, Mt. Shasta has long been extinct except for hot sulfurous springs near the top.

Shatt al Ar·ab (shăt' äl ăr'əb, shät' äl ä'räb), tidal river, 120 mi (193.1 km) long, formed by the confluence of the Tigris and Euphrates rivers, SE Iraq, and flowing SE to the Persian Gulf, forming part of the Iraq-Iran border. The Shatt al Arab flows through a broad, swampy delta that contains the world's largest date-palm groves. The river supplies fresh water to southern Iraq and Kuwait and is navigable for oceangoing vessels as far as Basra. Iraq and Iran have disputed navigation rights on the Shatt al Arab since 1935, when an international commission gave Iraq control of the Shatt al Arab.

Sha·wa·no (shô'nō'), county (1970 pop. 32,650), 919 sq mi (2,380.2 sq km), E central Wis., drained by the Wolf and Embarrass rivers; formed 1853; co. seat Shawano. It is in a generally forested area, with dairy farming. It has lumbering and food-processing industries and manufactures clothing and paper products. Shawano Lake is the largest of several lakes in the county.

Shawano, city (1975 est. pop. 6,700), seat of Shawano co., ENE Wis., on the Wolf River NW of Appleton; settled c.1840, inc. 1874. An early lumbering center, it now processes food and manufactures paper, wood products, and cheese.

Sha·win·i·gan (shə-wĭn'ĭ-gən), city (1971 pop. 27,792), S Que., Canada, on the St. Maurice River. Just north are the falls of the St. Maurice, 150 ft (45.8 m) high, with a hydroelectric station supplying power for the city's pulp and paper mills and plants that produce aluminum, abrasives, chemicals, cellulose, and textiles.

Shaw·nee (shô'nē', shô'nē'), county (1970 pop. 155,322), 548 sq mi (1,419.3 sq km), NE Kansas, on a dissected plain drained by the Kansas River; formed 1855; co. seat Topeka. Its agriculture includes stock raising and grain growing. It has limestone quarrying, sand and gravel pits, and a food-processing industry. Concrete, tires and rubber goods, metal products, and machinery are manufactured.

Shawnee. 1. City (1978 est. pop. 23,200), Johnson co., NE Kansas, a suburb of Kansas City; founded 1857, inc. 1922. Kitchenware is made. Shawnee was the original site of the Shawnee Indian Methodist Mission (1830). **2.** City (1978 est. pop. 26,600), seat of Pottawatomie co., central Okla., on the North Canadian River; inc. 1894. Shawnee boomed with the discovery of oil here in 1926 and is now the trade and rail center for a rich farm, dairy, and oil area.

Shaw·nee·town (shô'nē-toun'), city (1970 pop. 1,742), seat of Gallatin co., SE Ill., near the Ohio River SW of Evansville, Ind., in a farm and coal region. It was founded after the severe flood of 1937 forced part of Old Shawneetown, on the Ohio, to be moved to higher ground. Old Shawneetown, one of the state's oldest places, was settled after 1800, laid out in 1808, and incorporated in 1814. It was an important river port and commercial center in the early part of the 19th cent. Part of Old Shawneetown is now a state memorial.

She·ba (shē'bə), biblical name of a region of S Arabia. This region included present-day Yemen and the Hadhramaut. Its inhabitants, Sabaeans or Sabeans, migrated south at an early date from northwest Arabia. The Semitic colonization of Ethiopia was made in the 10th cent. B.C. from Sheba. In the same century the biblical queen of Sheba made her famous visit to Solomon. The country of Sheba was known as a region of great wealth. The culture of Sheba, which was at its height from the 6th to the 5th cent. B.C., is evidenced by the remains of a number of major engineering works. Sheba became (572) a Persian province and, with the rise of Mohammed, came under Islamic control and lost its separate identity.

She·be·li, We·bi (wĕ'bē shā-bĕl'ē), or **Webi Shi·be·li** (shē-bā'lē), river, c.1,000 mi (1,610 km) long, rising near Mt. Guramba, central Ethiopia, E Africa, and flowing SE into central Somali Republic. It comes within c.20 mi (30 km) of the Indian Ocean near Mogadisho, but turns southwest and parallels the coast for c.200 mi (320 km) before entering a swamp north of the Juba River. The river's lower valley is part of the chief agricultural region of Somali; sugar cane, cotton, and bananas are grown.

She·boy·gan (shĭ-boi'gən), county (1970 pop. 96,660), 505 sq mi (1,308 sq km), E Wis., bordered in the E by Lake Michigan and drained by the Sheboygan River; formed 1836; co. seat Sheboygan. It is a leading dairying and cheese-producing area. Its industry includes food processing, lumbering, iron and steel foundries, and the manufacture of furniture, textiles, paper, chemicals, plastics, metal products, and machinery.

Sheboygan, city (1978 est. pop. 48,500), seat of Sheboygan co., E Wis., a port of entry on Lake Michigan at the mouth of the Sheboygan River; inc. 1853. Plastics, stainless-steel products, leather goods, enamelware, furniture, knitted goods, and paper boxes are manufactured here. Dairying and brewing are also important industries. A fur-trading post was established here in 1795. Permanent settlement began c.1835, and Sheboygan grew into a shipping and industrial center. An Indian-mound park featuring a great number of excavated burial mounds is just south of the city.

Shef·field (shĕf'ēld), county borough (1974 est. pop. 513,310), South Yorkshire, N England, on the Don River. Sheffield is one of the leading industrial cities of England. It has been a center of cutlery manufacture since the 14th cent. Silver and electroplate goods, tools, and heavy steel goods, including plates for artillery and rails, are also made. The first Bessemer steelworks were built in Sheffield in 1859.

Sheffield, industrial city (1975 est. pop. 12,000), Colbert co., NW Ala., on the Tennessee River near Muscle Shoals, in an iron and coal area; inc. 1885. It is a railroad center and has plants that make agricultural chemicals, aluminum, stoves, and lingerie.

Sheffield Lake, residential city (1970 pop. 8,734), Lorain co., N central Ohio, near Lake Erie ENE of Lorain; inc. 1920.

She·ki (shĕk'ē), city (1970 pop. 43,000), S European USSR, in the Azerbaijan Republic, on the S slope of the Caucasus. It is a major center of silk production in a district that grows fruit and rice. Until its annexation (1805) by Russia it was the capital of a khanate under Persian sovereignty.

Shek·sna (shĕk-snä'), river, c.100 mi (160 km) long, flowing S between Lake Beloye and the Rybinsk Reservoir, NW European USSR. The Sheksna forms a link in the Volga-Baltic Waterway.

Shel·by (shĕl'bē). **1.** County (1970 pop. 38,037), 798 sq mi (2,066.8 sq km), central Ala., in a hilly region lying between the Coosa and Cahaba rivers; formed 1818; co. seat Columbiana. In an agricultural area yielding cotton, corn, and livestock, it has limestone quarries, some coal deposits, timber, and diversified industry. Textiles, wood products, concrete, metal products, and machinery are manufactured. **2.** County (1970 pop. 22,589), 772 sq mi (1,999.5 sq km), central Ill., drained by the South Fork of the Sangamon River and by the Kaskaskia and Little Wabash rivers; formed 1827; co. seat Shelbyville. It has agriculture (wheat, corn, hay, livestock, soybeans, dairy products, livestock, and poultry), bituminous-coal mining, and manufacturing of farm machinery. **3.** County (1970 pop. 37,797), 409 sq mi (1,059.3 sq km), central Ind., drained by the Big Blue River and Flatrock and Sugar creeks; formed 1821; co. seat Shelbyville. It is in a rich farming area that produces corn, hay, grain, livestock, and dairy goods. Its industry includes food processing, lumbering, and the manufacture of concrete, metal products, and machinery. **4.** County (1970 pop. 15,528), 587 sq mi (1,520.3 sq km), W Iowa, drained by the West Nishnabotna River and by Keg, Silver, and Mosquito creeks; formed 1851; co. seat Harlan. It is a prairie agricultural area, known for its cattle, hogs, poultry, corn, and alfalfa. It also has bituminous-coal deposits. **5.** County (1970 pop. 18,999), 383 sq mi (992 sq km), N central Ky., in the Bluegrass, bounded on the NW by Floyds Fork and drained by several creeks; formed 1792; co. seat Shelbyville. It is in a gently rolling upland agricultural area yielding burley tobacco, dairy products, livestock, and corn. Its industry includes food and tobacco processing. Chemical and metal products are made. **6.** County (1970 pop. 7,906), 501 sq mi (1,297.6 sq km), NE Mo., drained by the North and Salt rivers; formed 1835; co. seat Shelbyville. It is a farming (corn, wheat, oats, hay, and soybeans) and livestock-raising area, with coal deposits and stands of timber. **7.** County (1970 pop. 37,748), 408 sq mi (1,056.7 sq km), W Ohio, intersected by the Great Miami River and Loramie Creek; formed 1819; co. seat Sidney. Its agriculture includes soybeans, livestock, grain, and dairy products. It has gravel pits, iron foundries, lumbering, and manufacturing (plastics, metal products, and miscellaneous machinery and equipment). **8.** County (1970 pop. 722,111), 755 sq mi (1,955.5 sq km), extreme SW Tenn., bounded in the S by the Miss. border, in the W by the Mississippi River and Ark., and drained by the Loosahatchie and Wolf rivers; formed 1819; co. seat Memphis. It has cotton growing, dairying, and livestock raising. There are oil and gas wells and sand and gravel pits. Its industry includes food processing, textile milling, lumbering, and the manufacture of wood products, furniture, paper, chemicals, machinery, and electrical equipment. **9.** County (1970 pop. 19,672), 778 sq mi (2,015 sq km), E Texas, bounded on the E by the Sabine River (here the La. border), on the W by Attoyac Bayou; formed 1836; co. seat Center. This hilly wooded area includes part of Sabine National Forest and its chief industry is lumbering. It also has natural-gas wells, some oil, farming (truck crops, cotton, corn, peanuts, and pecans), beef and dairy cattle, and poultry.

Shelby. 1. City (1970 pop. 3,111), seat of Toole co., N Mont., NNW of Great Falls; inc. 1914. Once a cow town, then a center of dryland farming, it later benefited from the discovery (1921) of oil nearby. **2.** City (1975 est. pop. 15,300), seat of Cleveland co., W N.C., in a rich piedmont farming area; inc. 1843. Natural and synthetic textiles are made. To the south is Kings Mountain National Military Park, site of a battle fought on Oct. 7, 1780, between a party of backwoodsmen and a British and Tory force. **3.** City (1970 pop. 9,847), Richland co., N central Ohio, NW of Mansfield; founded 1828, inc. 1835. Steel tubing is made.

Shel·by·ville (shĕl'bē-vĭl'). **1.** City (1975 est. pop. 15,200), seat of Shelby co., central Ind., in a rich corn and livestock area; platted 1822, inc. as a city 1860. It is a farm-trade and manufacturing center. **2.** City (1970 pop. 4,182), seat of Shelby co., N Ky., W of Frankfort, in a bluegrass region; founded 1792. It is a large tobacco market and makes flour and chemicals. There are oil wells in the region. **3.** City (1970 pop. 601), seat of Shelby co., NE Mo., near the Salt River WNW of Hannibal, in a grain, livestock, and timber area. **4.** City (1975 est. pop. 11,900), seat of Bedford co., central Tenn., on the Duck River, in a farm and timber area; inc. 1819. Pencils, erasers, and pencil parts are its leading manufactures. The region is noted for the breeding of the Tennessee Walking Horse. Shelbyville was one of the country's early planned cities.

Shel·i·kof Strait (shĕl'ĭ-kôf'), c.150 mi (240 km) long, 30 mi (48 km) wide, S Alaska, between Alaska Peninsula and Kodiac and Afognak islands. It connects the Gulf of Alaska with the North Pacific Ocean. There are fish canneries on the eastern shore. Volcanic dust from the Katmai Volcano field in the west causes the sky to be constantly overcast. The western shore is uncharted.

Shell Lake (shĕl), city (1970 pop. 928), seat of Washburn co., NW Wis., S of Spooner, in a dairy and farm area; inc. 1908. Boats are made here.

Shel·ter Island (shĕl'tər), island (1970 pop. 1,644), 7 mi (11.3 km) long and 6 mi (9.7 km) wide, between the two peninsulas of E Long Island, SE N.Y. Settled in the 17th cent. by English colonists, the island has been a summer resort since the 1870s.

Shel·ton (shĕl'tən). **1.** City (1978 est. pop. 30,400), Fairfield co., SW Conn., on the Housatonic River opposite Derby; settled 1697, set off from Stratford 1789, inc. as a city 1915. Textiles, wire, pins and fasteners, and furniture are among the city's manufactures. **2.** City (1977 est. pop. 6,650), seat of Mason co., NW Wash., on an arm of Puget Sound NW of Olympia, in a lumbering and farming area; settled 1853, platted 1884, inc. 1889.

She·ma·kha (shĕ-mä-KHä'), city (1970 pop. 18,000), SE European USSR, in Azerbaijan, at the foot of the Caucasus. Its chief product is wine. Known since ancient times, Shemakha was an important silk center in the 16th cent. It passed to Russia in 1805. Its importance as a silk center declined after an earthquake in 1902.

Shen·an·do·ah (shĕn'ən-dō'ə), county (1970 pop. 22,852), 507 sq mi (1,313.1 sq km), NW Va., in the Shenandoah Valley, bounded in the W and NW by W.Va. and ridges of the Alleghenies, in the E and SE by Massanutten Mt., and drained by the North Fork of the Shenandoah River; formed 1772 as Dunmore co., renamed 1777; co. seat Woodstock. Its diversified agriculture includes livestock (especially poultry and dairy and beef cattle), wheat, fruit, and hay. It has some limestone quarrying, lumbering, and other industries (food processing, textile milling, and diversified manufactures). The county contains part of George Washington National Forest, mineral springs, and limestone caverns.

Shenandoah. **1.** City (1970 pop. 5,968), Page and Fremont cos., SW Iowa, SE of Council Bluffs; inc. 1871. It is a shipping center for a farm and livestock region and has varied small manufactures. **2.** Borough (1970 pop. 8,287), Schuylkill co., E Pa., N of Pottsville; settled 1835, laid out 1862, inc. 1866. It is an industrial city in an extensive anthracite-coal region. The first Greek Orthodox parish in the United States was organized here.

Shenandoah National Park, 193,537 acres (78,324 hectares), N Va., extending 80 mi (128.7 km) along the crest of the Blue Ridge; est. 1935. Elevations in the park range from 595 ft (181.5 m) at Front Royal to 4,049 ft (1,234.9 m) at the top of Hawksbill Mt. Heavily forested, the park contains a series of ridges and valleys, hollows, small hills, numerous streams, waterfalls, and trout-filled pools. The Appalachian Trail follows the crest.

Shenandoah valley, part of the Great Appalachian Valley, c.150 mi (240 km) long, N Va., located between the Blue Ridge and the Allegheny Mts. The valley is divided into two parts by Massanutten Mt., a ridge c.45 mi (70 km) long and c.3,000 ft (915 m) high. The Shenandoah River, c.150 mi (240 km) long, rises in two forks on either side of the ridge, uniting near Front Royal, Va., and flowing northeast to enter the Potomac at Harpers Ferry, W.Va. The Shenandoah valley was first explored in the early 1700s; the first white settlement was established in 1730. An important corridor in the westward pioneer movement, the valley became a rich agricultural area with fertile farm land, orchards, and pastures. It served as one of the Confederates' principal storehouses during the Civil War. Gen. R. E. Lee retreated through the valley after being checked in the Antietam campaign (1862) and the Gettysburg campaign (1863). By early 1865 the valley was completely lost to the South.

Shen·si (shĕn'sē'), province (1968 est. pop. 21,000,000), c.76,000 sq mi (196,840 sq km), N central China; capital Hsi-an. From north to south Shensi has four main regions—the loess plateau, fertile but dry; the Wei River valley, rich agriculturally and the center of population; the Tsingling divide, the highest range of the province; and the upper Han River valley. The valleys of the Wei and Han rivers and newly irrigated areas in the northwest are the main farming regions. Extensive reforestation, terracing, and irrigation have reclaimed much eroded land and increased agricultural output. Wheat, millet, cotton, soybeans, and corn are the chief crops. Rice, tea, and tung oil are produced in the south, and fruit orchards are cultivated in the upland areas. Livestock (notably sheep) are raised, and lumbering is important. Shensi has rich coal and iron deposits. Oil is extracted at Yen-ch'ang and salt is obtained from lakes. Since the 1960s Shensi has developed industrially. Shensi, especially the Wei River valley, was one of the early major political and cultural centers of northern China. The founders of the Chou and T'ang dynasties built their power here, and the Manchus gave the province its present boundaries. In 1935 the Communist army came to Shensi on its "long march," and from 1935 until the assumption of power in 1949 Shensi was the seat of the Chinese Communists.

Shen-yang (shŭn'yäng'), formerly **Muk·den** (mŏŏk'dĕn', mŏŏk'-), city (1970 est. pop. 3,750,000), capital of Liaoning prov., NE China, on the Hun River. It is the center of a highly industrialized area. Manufactures include heavy machinery, aircraft, tractors, automotive parts, cables, machine tools, transformers, textiles, chemicals, paper products, medicines, and cement. The city is connected by rail with all the major cities of Liaoning prov. and with Peking and North Korea. Farm communes have recently been established; the introduction of rice, improved agricultural techniques, and the intensive cultivation of wheat, corn, soybeans, and vegetables have made Shen-yang virtually self-sufficient in food. Following the establishment of the Chinese republic (1912), Shen-yang was the headquarters of several war lords. Here, in Sept., 1931, occurred the Mukden or Manchurian Incident, when the Japanese army used an explosion on the railroad north of Shen-yang as a pretext for occupying the city and beginning the occupation of all Manchuria. After 1931 the Japanese developed the city as an industrial center. Shen-yang fell to the Communists on Nov. 1, 1948, after a 10-month siege during which thousands starved; the defending Nationalist force was annihilated during a breakout attempt. The city has three sections—the old Chinese city, which is the administrative center; the new city, developed by the Japanese around the railroad; and a residential section beyond the railroad. The area doubled in population in the 1950s and 1960s.

Shep·herds·ville (shĕp'ərdz-vĭl'), city (1970 pop. 2,769), seat of Bullitt co., N Ky., S of Louisville, in a farm area.

Shep·pey, Isle of (shĕp'ē), c.30 sq mi (80 sq km), Kent, SE England, at the mouth of the Thames, separated from the mainland by The Swale, a narrow strait. It is largely flat, with wave-eroded cliffs to the north. Vegetables and grain are grown on the fertile soil, and sheep are raised.

Sher·brooke (shûr'brŏŏk'), city (1971 pop. 80,711), S Que., Canada, at the confluence of the Magog and the St. François rivers, E of Montreal. It is the commercial and market center for the surrounding farm region and has textile mills and plants producing mining machinery, rubber products, and leather goods.

Sher·burne (shûr'bərn), county (1970 pop. 18,344), 431 sq mi (1,116.3 sq km), central Minn., bounded on the W and S by the Mississippi River and drained by the Elk River; formed 1856; co. seat Elk River. In an agricultural area yielding potatoes, dairy products, livestock, grain, and poultry, it has some mining and manufacturing (processed food, concrete, and metal products).

Sher·i·dan (shĕr'ĭ-dən). **1.** County (1970 pop. 3,859), 893 sq mi (2,312.9 sq km), NW Kansas, in a rolling plain area drained by the South Fork of the Solomon River and the Saline River; formed 1873; co. seat Hoxie. It is in a grain and livestock-raising region, with some manufacturing. **2.** County (1970 pop. 5,779), 1,694 sq mi (4,387.5 sq km), extreme NE Mont., in a well-watered agricultural area bordering on N.Dak. and Sask. and drained by Medicine Lake and Big Muddy Creek; formed 1913; co. seat Plentywood. Grain is the chief crop. **3.** County (1970 pop. 7,285), 2,462 sq mi (6,376.6 sq km), NW Nebr., bordered in the N by S.Dak.; formed 1885; co. seat Rushville. It is drained by the Niobrara River and numerous small lakes. Grain, livestock, dairy goods, and poultry are produced. **4.** County (1970 pop. 3,232), 989 sq mi (2,651.5 sq km), central N.Dak., in an agricultural area watered by the Sheyenne River; formed 1908; co. seat McClusky. Livestock, poultry, dairy products, and wheat and other grains are important. **5.** County (1970 pop. 17,852), 2,532 sq mi (6,557.9 sq km), N Wyo., bordering on Mont. and watered by the Little Bighorn, Tongue, and Powder rivers; formed 1888; co. seat Sheridan. Its irrigated agriculture includes grain, livestock, sugar beets, and beans. It is in a bituminous-coal mining area, with lumbering and food-processing industries. Part of the Bighorn National Forest and the Bighorn Mts. are in the west.

Sheridan. **1.** City (1974 est. pop. 2,837), seat of Grant co., S central Ark., WNW of Pine Bluff. It makes lumber products and gins cotton. **2.** City (1975 est. pop. 10,300), seat of Sheridan co., N Wyo., on Goose Creek E of the Bighorn Mts., in a mineral, livestock, and irrigated farm region; inc. 1884. It is a railroad division point and a regional trade and market hub. It is the tourist center of an excellent hunting and fishing region. To the south is the site of the Fetterman massacre, where Indians wiped out a force of soldiers.

Sher·man (shûr'mən). **1.** County (1970 pop. 7,792), 1,055 sq mi (2,732.5 sq km), NW Kansas, in a gently rolling area bordering in the W on Colo. and watered by headstreams of Beaver and Sappa creeks; formed 1873; co. seat Goodland. It is in a grain and livestock area, with some mining and beet-sugar processing. **2.** County (1970 pop. 4,725), 571 sq mi (1,478.9 sq km), central Nebr., in a farming region drained by the Middle Loup River; formed 1873; co. seat Loup City. Grain, livestock, dairy goods, and poultry are important to its economy. **3.** County (1970 pop. 2,139), 830 sq mi (2,149.7 sq km), N Oregon, bounded and drained by the Columbia River in the N, by the Deschutes River in the W, by the John Day River in the E, and bordering on Wash.; formed 1889; co. seat Moro. Its agriculture includes wheat, barley, and oats. It has some manufacturing. **4.** County (1970 pop. 3,657), 916 sq mi (2,372.4 sq km), extreme N Texas, in high Panhandle plains, bounded on the N by the Okla. border and drained by the North Canadian River and its tributaries; formed 1876; co. seat Stratford. It has large-scale grain farming, beef, cattle, and poultry raising, and some dairying. There are natural-gas and some oil wells.

Sherman, city (1978 est. pop. 25,500), seat of Grayson co., N Texas, near the Red River; inc. 1858. Originally on a stagecoach route, it is now a highway and railroad junction. It has flour and feed mills, cotton gins, and textile factories. Business machines, instruments, and surgical dressings are also made.

's Her·to·gen·bosch (sĕr'tō-gən-bôs'), municipality (1973 est. pop. 84,914), capital of North Brabant prov., S central Netherlands, at the confluence of the Dommel and Aa rivers. It is an industrial and transportation center and has a large cattle market. Chartered in 1184, 's Hertogenbosch was a fortress city until 1876. Hieronymus Bosch was born here (c.1450).

Sher·wood (shûr'wŏŏd'), city (1970 pop. 6,744), Pulaski co., central Ark., NE of North Little Rock.

Sherwood Forest, formerly a large royal forest, mainly in Nottinghamshire, central England. Today remnants of the forest (c.150 sq mi/390 sq km) exist near Mansfield and Hucknall. The forest is most celebrated as the haunt of Robin Hood and his famous band.

Shet·land (shĕt'lənd), county (1978 est. pop. 19,000), 551 sq mi (1,427.1 sq km), extreme N Scotland, consisting of the Shetland Islands, an archipelago c.70 mi (110 km) long, NE of the Orkneys. The group consists of some 100 islands. About one fourth of them are inhabited. Mainland, Yell, Unst, Fetlar, Whalsey, and Bressay are the largest islands. Lerwick, on Mainland, is the administrative center. The surface of the islands is low and rocky, with few trees. In places cliffs rise above 1,000 ft (305 m). Oats and barley are the chief crops, but fishing and cattle and sheep raising are more important. The region is famous for its knitted woolen goods. The Shetland pony is bred here. Tourism is also significant. By the late 9th cent. the islands were occupied by the Norsemen, and traces of their speech and customs survive. Shetland was not annexed to Scotland until 1472, when the islands were taken over as an unredeemed pledge of King Christian I of Norway and Denmark for the dowry of his daughter, Margaret, who married James III of Scotland. Pictish forts are scattered throughout the islands, and a village from the Bronze Age has been unearthed at Jarlshof on Mainland.

Shey·enne (shī-ĕn', -ăn'), river, 325 mi (523 km) long, rising in central N.Dak. and flowing E and S, past Valley City and Lisbon, where it turns NE to join the Red River of the North near Fargo.

Shi·a·was·see (shī'ə-wô'sē), county (1970 pop. 63,075), 540 sq mi (1,398.6 sq km), S central Mich., drained by the Shiawassee, Maple, and Lookingglass rivers; formed 1837; co. seat Corunna. The area is chiefly agricultural, raising beans, grain, corn, potatoes, hay, and livestock. Dairy products are made. There is some light industry and manufacturing. Oil and gas fields and coal mines are found here.

Shi·ba·ta (shĭ-bä'tä), city (1970 pop. 49,000), Niigata prefecture, W central Honshu, Japan, on the Kaji River. It is a distribution point for rice and a center for iron, steel, and chemical industries.

Shi·bin al-Kom (shĭ-bēn' äl-kŏŏm'), city (1970 est. pop. 75,600), capital of Munufiyah governorate, N Egypt, in the Nile River delta. It is an agricultural market and a cotton-processing center.

Shick·shock Mountains (shĭk'shŏk'), range of the Appalachian system, E Que., Canada, a continuation of the Notre Dame Mts., extending c.100 mi (160 km) E to W near the N coast of the Gaspé Peninsula. Tabletop Mt., or Mt. Jacques Cartier (4,160 ft/1,268.8 m), is the highest point in southeast Canada.

Shiel, Loch (shēl), lake, 17 mi (27.4 km) long and 1 mi (1.6 km) wide, between Inverness-shire and Argyllshire, W Scotland. It is drained by a short stream into Loch Moidart.

Shih-chia-chuang (shûr'jē-ä'jwäng'), city (1970 est. pop. 1,500,000), capital of Hopeh prov., China, near the Shansi prov. border. It was a small village until the turn of the century, when it became a railroad junction. It has textile, fertilizer, pharmaceutical, automotive, and paper industries.

Shi·ko·ku (shĭ-kō'kōō), island (1973 est. pop. 3,948,000), 7,247 sq mi (18,770 sq km), S Japan, separated from Honshu and Kyushu by the Inland Sea. It has high mountains that limit agriculture and impede communication. Rice, grains, mulberry, palms, and camphor are the chief products. Industry is found on the northern side of the island.

Shil·ka (shĭl'kə), river, c.345 mi (555 km) long, formed E of Chita, Far Eastern USSR, by the confluence of the Onon and Ingoda rivers, both of which rise along the Mongolian-USSR border. It flows northeast, joining the Argun River to form the Amur River on the USSR-China border.

Shil·ling·ton (shĭl'ĭng-tən), borough (1970 pop. 6,249), Berks co., SE Pa., just SW of Reading; settled 1860. It has light industry.

Shil·long (shĭ-lông'), town (1971 pop. 87,659), capital of Assam and Meghalaya states, NE India. It is a summer resort c.5,000 ft (1,525 m) high in the Khasi Hills. Primitive tribes inhabit the surrounding district.

Shi·loh National Military Park (shī'lō): *see* National Parks and Monuments Table.

Shi·ma·da (shĭ-mä'dä), city (1970 pop. 66,489), Shizuoka prefecture, E central Honshu, Japan, on the Oi River. It is a distribution point for timber and rice and a center for chemical and mechanical industries.

Shi·mi·zu (shĭ-mē'zōō), city (1970 pop. 234,966), Shizuoka prefecture, E central Honshu, Japan, on Suruga Bay. A port and fishing center, it exports tea, oranges, and canned food.

Shi·mo·da (shĭ-mō'dä), town (1970 pop. 66,489), Shizuoka prefecture, E central Honshu, Japan, at the south extremity of Izu peninsula, on Shimoda Bay. It is a fishing base and an important port for the peninsula.

Shi·mo·ga (shĭ-mō'gə), town (1971 pop. 102,709), Karnataka state, SW India, on the Tunga River. Shimoga is district headquarters for a region that produces teak, sandalwood, rosewood, and rice.

Shi·mo·no·se·ki (shē'mō-nō-sä'kē), city (1970 pop. 258,425), Yamaguchi prefecture, extreme SW Honshu, Japan. An important port and fishing center on Shimonoseki Strait, Shimonoseki is a railroad and industrial center, with engineering works, shipyards, and metal and chemical plants. In the city is Akamagu, a 12th cent. shrine dedicated to Emperor Antoku and to the Taira clan. The Treaty of Shimonoseki, which ended the Sino-Japanese War, was negotiated and signed in 1895.

Shi·na·no (shĭ-nä'nō), river, longest of Honshu, Japan, c.230 mi (370 km) long. It rises in the mountains of central Honshu and flows generally northeast to the Sea of Japan.

Ship·ka (shĭp'kä), pass through the Balkans, alt. c.4,370 ft (1,335 m), central Bulgaria. Gabrovo, north of the pass, was the scene of a Russo-Bulgarian victory over the Turks in 1878.

Ship·ley (shĭp'lē), urban district (1974 est. pop. 28,550), West Yorkshire, N England, on the Aire River. Light engineering and the manufacture of woolens and worsteds are important industries.

Ship·pens·burg (shĭp'ənz-bûrg'), borough (1970 pop. 6,536), Cumberland and Franklin cos., S Pa., SW of Carlisle; settled 1730, inc. 1819. It has some manufacturing.

Shi·raz (shĭ-räz'), city (1966 pop. 269,865), capital of Fars prov., SW Iran, at an altitude of c.5,200 ft (1,575 m). It has long been known for its wines, carpets, and metalwork. Other manufactures include textiles, petrochemicals, cement, and sugar. An old settlement, Shiraz became an important commercial, military, and administrative center in the late 7th cent. From about the 10th cent. Shirazi traders were active along the East African coast. Tamerlane sacked the city in the late 14th cent. Under Karim Khan, the city served (1750–79) as capital of Persia; it declined after Karim's successor, Aga Mohammed Khan, moved the capital to Tehran.

Shir·bar·ghan (shîr-bär'gän), city (1967 pop. 50,440), N Afghanistan. It is a market for agricultural produce and Karakul lamb skins and is the site of an ancient citadel.

Shi·ré (shē'rā), river, c.250 mi (400 km) long, flowing from the S end of Lake Nyasa, Malawi, SE Africa, to the Zambezi River in central Mozambique. The upper Shiré is being developed for irrigation and power production. Cotton is raised in the valley.

Shir·ley (shûr'lē), town (1970 pop. 4,909), Middlesex co., N Mass., E of Fitchburg; settled c.1720, set off from Groton 1753. Twine is made, and apples are grown. A Shaker community was established here in 1793.

Shi·shal·din (shĭ-shôl'dĭn), volcano, 9,370 ft (2,857.9 m) high, on central Unimak Island, SW Alaska. It has been mildly active for about 150 years, with several eruptions recorded in recent years. It is locally known as Smoking Moses.

Shive·ly (shīv'lē), residential city (1975 est. pop. 17,700), Jefferson co., N Ky., adjacent to Louisville; settled c.1885, inc. 1938. Liquor distillation is the main industry.

Shi·zu·o·ka (shĭ-zōō'ô-kä), city (1970 pop. 416,378), capital of Shizuoka prefecture, E central Honshu, Japan, on Suruga Bay. It is a port and communications center and is known for its tea, oranges, and lacquer ware.

Shkha·ra (shŭk'hə-rä), peak, c.17,064 ft (5,200 m) high, in the Greater Caucasus, S European USSR, SW of Nalchik and on the border between the Georgian and Russian republics.

Shko·dër (shkō'dər), city (1971 est. pop. 56,500), capital of Shkodër prov., NW Albania, at the outlet of Lake Scutari. It is located in a fertile agricultural area that produces a variety of crops. Shkodër has industries that manufacture cement, textiles, tobacco products, foodstuffs, and leather goods. It is also an important fishing center. Shkodër became (168 B.C.) a Roman colony, passed to Byzantium, and was conquered by the Serbs in the 7th cent. A.D. It was captured by Sultan Mohammed II in 1479. Montenegrin troops occupied (1913) Shkodër in the Balkan Wars, but the European powers assigned the city to newly independent Albania.

Shoals (shōlz), town (1970 pop. 1,039), seat of Martin co., SW Ind., SW of Bedford; settled 1818, inc. c.1845.

Sho·la·pur (shō'lə-pōōr'), city (1971 pop. 398,361), Maharashtra state, W central India, on the Deccan plateau. Once a fortress town, Sholapur is now a textile-manufacturing city and a market for oilseed and tobacco.

Shore·view (shôr'vyōō'), residential village (1975 est. pop. 14,400), Ramsey co., SE Minn., a suburb of St. Paul; settled 1850, inc. 1957. It has a number of arsenals that produce arms and ammunition. The village is built around seven lakes.

Shore·wood (shôr'wŏŏd'), village (1970 pop. 15,576), Milwaukee co., SE Wis., between the Milwaukee River and Lake Michigan, a suburb of Milwaukee; settled c.1835, inc. 1900. It is mostly residential.

Sho·sho·ne (shō-shō'nē), county (1970 pop. 19,718), 2,609 sq mi (6,757.3 sq km), N Idaho, in a mining and lumbering area bounded on the E by Mont. and crossed by the Coeur d'Alene and St. Joe rivers; formed 1858; co. seat Wallace. It includes the Coeur D'Alene mining district and parts of Coeur d'Alene and St. Joe national forests. Lead, silver, zinc, and copper are mined.

Shoshone, city (1970 pop. 1,233), seat of Lincoln co., S Idaho, on the Little Wood River NNE of Twin Falls; founded 1882 with the coming of the railroad. It was first a cattle town, but became an irrigated farm and dairy center after 1905, particularly after Magic Dam was built in the Big Wood River in 1907. It is in the Minidoka project, ships wool, and is the gateway to Sun Valley and the Sawtooth Mts.

Shoshone Falls, 212 ft (65 m) high, flowing over a rim 900 ft (274.5 m) wide in the Snake River, S Idaho. Once a great spectacle, the falls have been reduced by irrigation projects upstream.

Shot·ter·y (shŏt'ə-rē), village, Warwickshire, central England, W of Stratford-on-Avon. The cottage in which Shakespeare's wife, Anne Hathaway, lived and its surrounding grounds are national property, preserved much as they were in Shakespeare's time.

Shreve·port (shrēv'pôrt'), city (1978 est. pop. 187,600), seat of Caddo parish, NW La., on the Red River near the Texas and Ark. borders; inc. 1839. It is an oil and natural-gas center, with important metal, cotton, and lumber manufactures. Dairy goods, feed and grain, machinery, telephones, defense products, and chemicals are also made. The city was founded after the Red River was laboriously cleared (1833–36) of logs and driftwood. It became the Confederate capital of the state in 1863. The discovery of oil in 1906 provided the greatest impetus for Shreveport's growth.

Shrews·bur·y (shrōōz'bûr'ē, -bər-ə), municipal borough (1974 est. pop. 56,670), Salop, W England, on the Severn River. Shrewsbury is a road and rail junction. There are varied manufactures, including diesel engines. It was an ancient Saxon and Norman stronghold. In 1403 Henry IV defeated Henry Percy (Hotspur) on a plain near Shrewsbury and had the body of the rebel displayed to the townspeople as proof.

Shrewsbury. 1. Town (1978 est. pop. 23,400), Worcester co., central Mass.; inc. 1727. Plastic goods are manufactured. **2.** City (1970 pop. 5,896), St. Louis co., E Mo., a suburb SW of St. Louis; inc. 1920.

Shrop·shire (shrŏp'shĭr', -shər), former county, W England. The county town was Shrewsbury. In Anglo-Saxon times Shropshire was a part of the kingdom of Mercia. In 1974 it became part of the new nonmetropolitan county of Salop.

Shu·men (shōō'mĕn), city (1969 est. pop. 69,600), NE Bulgaria. It is a railway junction and a market for grains and other agricultural products. Brewing, canning, flour milling, motor vehicle assembling, and the manufacture of spare parts for automobiles and tractors are the chief industries. Founded in 927, the city was fortified under Turkish rule (15th–19th cent.) and was strategically important in the Russo-Turkish Wars of the 18th and 19th cent.

Shush·tar (shōōsh-tär'), town (1966 pop. 24,000), Khuzistan prov., SW Iran, on the Karun River. It is an agricultural trade center and has long been known for its brocaded textiles and metalwork. Nearby are major petroleum fields.

Si or **Hsi** (both: sē) or **Si-kiang** (shē'jē-äng'), great river of S China, c.1,250 mi (2,010 km) long, rising in E Yunnan prov. and flowing generally E through Kwangsi and Kwangtung provs. to the South China Sea near Canton. At the junction with the Pei, east of Canton, the Si forms the vast Canton River delta (2,890 sq mi/7,485 sq km), consisting of a maze of channels and canals. The densely populated delta is one of China's chief economic areas.

Si·al·kot (sē-äl'kōt', syäl'-), city (1972 est. pop. 185,000), E Pakistan. It is a rail junction and a major trade and processing center. Manufactures include bicycles, surgical instruments, cutlery, rubber products, and ceramics. Textile weaving is also important.

Si·am (sī-ăm'): *see* Thailand.

Siam, Gulf of, shallow arm of the South China Sea, c.500 mi (800 km) long and up to 350 mi (560 km) wide, separating the Malay Peninsula from E Thailand, Cambodia, and Vietnam.

Si·an (sē'än', shē'-): *see* Hsi-an, China.

Si·ang (sē'äng, shē'-), river, China: *see* Hsiang.

Siang·tan (sē-äng'tän', shē'-): *see* Hsiang-t'an, China.

Siau·liai (shou'lā'), city (1973 est. pop. 103,000), W European USSR, in Lithuania. It is a rail hub and has railroad repair shops. Siauliai is also a major tanning, shoe-manufacturing, and flax-processing center. It belonged from 1589 until 1772 to the Polish crown and passed to Russia in 1795.

Ši·be·nik (shē-bĕn'ĭk), town (1971 pop. 30,090), W Yugoslavia, in Croatia, on the Adriatic Sea. It is a seaport and naval base and has aluminum, chemical, and textile industries. An early residence of the kings of Croatia, Šibenik was captured by Venice (1117), held by Hungary (1351–1412), and again passed to Venice. Austria held it from 1797 to 1918. It was incorporated into Yugoslavia in 1922.

Si·be·ri·a (sī-bîr'ē-ə), vast geographic region of the Asian USSR. It is generally understood to comprise the northern third of Asia, stretching from the Urals in the west to the mountain ranges of the Pacific Ocean watershed in the east and from the Laptev, Kara, and East Siberian seas (arms of the Arctic Ocean) in the north to the Kazakh steppes, the Altai and Sayan mountain systems, and the borders of the Mongolian People's Republic in the south. The Soviet Far East, which is commonly considered to be part of Siberia, is treated separately in Soviet regional schemes. Siberia, c.2,900,000 sq mi (7,511,000 sq km), has an estimated population (1970) of 29,820,000. Siberia's administrative units are the Yakut, Buryat, and Tuva autonomous republics, the Altai and Krasnoyarsk krays, and the Omsk, Novosibirsk, Tomsk, Kemerovo, Irkutsk, and Chita oblasts. Off Siberia in the Arctic Ocean are the New Siberian Islands, the Severnaya Zemlya Archipelago, and other islands.

Siberia may be divided, from north to south, into the zones of vegetation that run across the entire USSR—the tundras (extending c.200 mi/320 km inland along the entire Arctic coast), the taiga, the mixed forest belt, and the steppe zone. Forests occupy about 40% of Siberia's land. Siberia is drained, from south to north, by the mighty Ob, Yenisei, and Lena rivers (and their tributaries), which also provide the only means of north-south transportation. East-west transportation depends largely on the Trans-Siberian RR and to an increasing extent on the Arctic sea route.

The Siberian lowland occupies the western third of Siberia; it stretches from the Urals to the Yenisei River and is mainly a low-lying, often marshy, plain. Situated far from vulnerable frontiers, southwestern Siberia contains about 60% of Siberia's population, one of the USSR's major industrial complexes, and many important Soviet cities. The wooded steppe and fertile black earth of western Siberia favor agriculture and, especially in the Baraba Steppe, dairying. Wheat is the principal crop; rye, oats, potatoes, sunflowers, flax, and sugar beets are also important. The Kuznetsk Basin, in western Siberia, is one of the world's richest coal regions and also has large iron deposits. Rich oil and natural-gas fields have recently been discovered in the West Siberian lowlands, from which a network of pipelines now serves the European USSR.

Eastern Siberia extends from the Yenisei to a huge mountain chain, an offshoot of the mountains of Central Asia, comprising (from southwest to northeast) the Yablonovy, Stanovoy, Verkhoyansk, Kolyma, and Cherskogo ranges. In the center of eastern Siberia rise the Central Siberian uplands, which are separated from the northeastern mountains by the plateaus along the Vitim and Aldan rivers. South of the uplands lies Lake Baykal, the world's deepest lake, surrounded by mountains. Verkhoyansk, where the lowest temperatures on earth have been recorded, has summer heats rising above 90°F (32°C). Eastern Siberia is the USSR's leading producer of gold, diamonds, mica, and aluminum, and there are large reserves of iron ore, coal, graphite, and nonferrous metals. Exploitation of the region's rich waterpower resources began in the mid-1950s with the construction of two giant hydroelectric power stations on the Angara River at Irkutsk and Bratsk. Forestry and mining are major economic activities. Agriculture (wheat and oats) is practiced in the south, and animal husbandry is prevalent among the indigenous Siberian peoples. Reindeer breeding, fishing, sealing, hunting, and fur processing are important occupations in the Arctic north.

The great majority of the Siberian population is made up of Russians and Ukrainians. Non-Russian groups include Turkic-speaking nationalities, Buryat-Mongols, Finno-Ugric Ostyaks and Voguls, Samoyedes, and Tungus Evenki. The largely nomadic Mongol and Turkic herdsmen of southern Siberia have mostly settled down to agriculture under the Soviet government. The indigenous peoples of central and northern Siberia remain, for the most part, hunters and fishermen. The chief non-Christian religions in Siberia are Islam and Tibetan Buddhism.

History. In the historic period, southern Siberia frequently served as the point of departure for nomadic hordes, such as the Huns, the Mongols, and the Manchus, who conquered and lost immense empires. Among the political entities emerging after the breakup of the Mongol state of the Golden Horde in the mid-15th cent. was the Tatar khanate of Sibir, which was conquered by Russia in 1598. During the 17th cent. all of western Siberia was annexed to Russia. By 1640 Russian-sponsored Cossacks had reached the Sea of Okhotsk, and soon afterward they collided with Chinese troops. By the Treaty of Nerchinsk (1689) Russia abandoned to China the region later known as the Far Eastern Territory.

A colony of the Russian Empire, Siberia was administered by military governors who collected tribute but interfered little with native Siberian customs and religions. With the decline of the fur trade, previously an important source of wealth for Russia, in the early 18th cent., mining became the chief economic activity. From the early 17th cent. Siberia was used as a penal colony and a place of exile for political prisoners. Russian settlement of Siberia on a large scale began only with the construction (1892-1905) of the Trans-Siberian RR.

During the Russian civil war of 1918-20 Siberia was a a stronghold of the White forces, who established a regime at Omsk under Adm. A. V. Kolchak. Despite aid from a U.S., British, French, and Japanese expeditionary force, the anti-Bolshevik units were defeated and the Soviet government assumed control early in 1920. Under the Soviets, Siberia, especially the Ural-Kuznetsk complex, underwent dramatic economic development. Forced labor and population resettlement were instrumental in establishing mining, industrial, and agricultural installations. Post-World War II industrialization of Siberia continued at a rapid pace, with special concentration on southwestern Siberia and the Lake Baykal region. In the 1970s the Soviet government sought U.S., West European, and Japanese capital, credits, and technology for the development of Siberia's vast petroleum, natural-gas, timber, and other resources.

Si·be·ri·an Plat·form (sī-bîr'ē-ən plăt'fôrm') or **An·ga·ra Shield** (äng'gə-rä' shēld'), large, geologically stable area of Precambrian rocks, N Asia, comprising much of Siberia, USSR. It is bounded, in general, on the west by the Yenisei River, on the east by the Lena River, on the north by the Arctic Ocean, and on the south by the general latitude of Lake Baykal. Most of the region is covered by strata of lower Paleozoic sediments and it is thought that since the early Paleozoic era it has remained above the level of invading seas that flooded most continental land masses. Tundra and taiga cover the region, which also has a rich variety of minerals.

Si·bir (sə-bîr'), former city, southeast of present-day Tobolsk, W Siberian USSR. Founded in the 11th or 12th cent., it became (early 16th cent.) the capital of the Tatar khanate of Sibir. The Cossack Yermak took the city of Sibir in 1581, thus marking the start of Moscow's conquest of what is now Siberia.

Si·biu (sē-byōō'), city (1973 est. pop. 127,146), central Rumania. There are mechanical-engineering works and industries producing textiles, agricultural machinery, chemicals, and leather. Founded in the 12th cent. by German colonists, Sibiu was destroyed by the Tatars in 1241. In the 17th cent. it came under Austrian control. It is an Orthodox metropolitan see and has two cathedrals.

Sib·ley (sĭb'lē), county (1970 pop. 15,845), 583 sq mi (1,510 sq km), S central Minn., bounded on the E by the Minnesota River; formed 1853; co. seat Gaylord. In an agricultural area yielding livestock, dairy products, corn, oats, and barley, it has a food-processing industry and manufactures farm machinery.

Sibley, city (1975 est. pop. 2,955), seat of Osceola co., NW Iowa, near the Minn. border W of Spirit Lake; named 1872, inc. 1875. It is in a livestock and dairy area.

Si·bu·tu (sē'bōō-tōō', sē-bōō'tōō), island, 39 sq mi (101 sq km), Sulu prov., Philippines, the westernmost island in the Sulu Archipelago. Inadvertently omitted from the Treaty of 1898, it was ceded by Spain to the United States for $100,000 on Nov. 7, 1900.

Sic·i·ly (sĭs'ə-lē), region (1971 pop. 4,667,316), 9,925 sq mi (25,706 sq km), S Italy, mainly situated on the island of Sicily, which is in the Mediterranean Sea, separated from the Italian mainland by the narrow Strait of Messina. The region also includes the Egadi Islands, the Lipari Islands, the Pelagie Islands, Pantelleria Island, and Ustica Island. Palermo is the capital of Sicily, which is divided into the provinces of Agrigento, Caltanisetta, Catania, Enna, Messina, Palermo, Ragusa, Syracuse, and Trapani (named for their capitals).

The largest Mediterranean island, Sicily is triangular and formerly was sometimes called Trinacria; Capes Boeo (or Lilibeo), Passero, and Punta del Faro (or Peloro) are the vertices of the triangle. The island is almost entirely covered by hills and mountains (continuations of the Apennines); Mt. Etna (10,700 ft/3,261 m), in the east, is the highest point. The only wide valley is the fertile plain of Catania in the east, mostly located along the lower Simeto River. There are also narrow coastal strips in the south and west, and a small fertile plain (the Conca d'Oro) near Palermo in the northwest. Sicily has long been noted for the fertility of its soil, its pleasant climate, and its natural beauties.

Economy. Agriculture, the chief economic activity, has long been hampered by absentee ownership, primitive methods of cultivation, and inadequate irrigation. Wheat, barley, maize, olives, citrus fruit, almonds, wine grapes, and some cotton are produced; cattle, mules, donkeys, and sheep are raised. There are important tuna and sardine fisheries. Sicily's manufactures include processed food, chemicals, fertilizers, textiles, ships, leather goods, forest products, and refined petroleum. There are major petroleum fields in the southeast, and large quantities of natural gas and sulfur are also produced. The chief ports of the island are Palermo, Catania, and Messina.

History. The first known inhabitants of the island were the Elymi, Sicani, and Siculi. Phoenicians, Carthaginians, and Greeks later (9th-6th cent. B.C.) founded coastal settlements. The Greek cities flourished, and in the 5th cent. B.C., Syracuse gained hegemony over them. A century of antagonism between Greeks and Carthaginians was followed by strife between Romans and Carthaginians, which ended (c.215) with virtually all of Sicily under Roman rule. After the fall of Rome Sicily passed from the Vandals (mid-5th cent. A.D.), to the Goths (493), to the Byzantines (535), and then to the Arabs (9th cent.) The Arabs were displaced by the Norman conquest of Sicily (1060-91), led by Roger I. Roger II became (1130) the first king of Sicily; his descendant, Holy Roman Emperor Frederick II, reigned as king of Sicily from 1197 to 1250.

In 1266 Pope Clement IV crowned Charles I (Charles of Anjou) king of Naples and Sicily as his vassal. The unpopular government of the French brought on the Sicilian Vespers Revolt (1282) and the Sicilians chose Peter III of Aragón as king. With the accession of the house of Hapsburg to the Spanish throne (early 16th cent.), the island came under the control of a few powerful nobles and churchmen. In 1713 the Peace of Utrecht assigned Sicily to Savoy, which in 1720 exchanged it with Emperor Charles VI for Sardinia. The Spanish Bourbon kings ruled Sicily and Naples after 1735; popular revolts against them were mercilessly suppressed in 1820 and 1849. In 1860 Giuseppe Garibaldi conquered the island, which then voted to join the kingdom of Sardinia. Even after the unification of Italy was completed (1870), Sicily was neglected by the central government. Since the establishment (1950) of the Southern Italy Development Fund there have been land reforms and the island's economy has been generally developed.

Si·cy·on (sĭsh'ē-ŏn', sĭs'-), ancient city of Greece, in the Peloponnesus, NW of Corinth. Sicyon attained its greatest power under the Tyrant Cleisthenes in the 6th cent. B.C. Under the leadership of the general Aratus, Sicyon joined (3rd cent. B.C.) the Achaean League. With the destruction (146 B.C.) of Corinth by the Romans, Sicyon briefly regained power but subsequently declined. It was an important center of art.

Si·di-bel-Ab·bès (sē'dē-bĕl-ä-bĕs'), city (1974 est. pop. 151,148), W central Algeria, on the Mekerra River. It is the commercial center of an area of vineyards and market gardens. Until 1962 it was headquarters of the French Foreign Legion.

Sid·law Hills (sĭd'lô'), range, E Scotland. It extends c.30 mi (50 km) northeast from near Perth into Angus-shire. The highest hills, including Dunsinane, are more than 1,000 ft (305 m).

Sid·ley, Mount (sĭd'lē), mountain, 13,717 ft (4,183.7 m) high, Marie Byrd Land, Antarctica. It was discovered by Robert E. Byrd in 1934.

Sid·ney (sĭd'nē). **1.** Town (1970 pop. 1,061), seat of Fremont co., SW Iowa near the Missouri River SSE of Council Bluffs; inc. 1870. An annual state rodeo is held here. **2.** City (1970 pop. 4,543), seat of Richland co., NE Mont., in the Yellowstone valley, near the N.Dak. border NNE of Glendive; inc. 1911. Wheat and sugar beets are grown and livestock is raised in the area. There are oil and lignite fields nearby. **3.** City (1976 est. pop. 6,092), seat of Cheyenne co., W Nebr., in the Great Plains, region, on Lodgepole Creek between the North Platte and the South Platte rivers; founded 1867 by the Union Pacific RR. It grew up around Fort Sidney (established for the protection of railroad construction workers) and was an early supply point during the Black Hills gold rush (1876-77). Sidney is now a farm trade center in a rich wheat, oil, and natural-gas area. **4.** City (1975 est. pop. 16,700), seat of Shelby co., W central Ohio, on the Great Miami River; founded 1811, inc. 1834.

Si·don (sī'dən), ancient city, one of the great seaports of the Phoenicians, in present-day Lebanon. It was an important center for trade and was known for its purple dyes and for glassware. Although eclipsed by its own colony, Tyre, Sidon continued to be a port of prominence under the Persians, in the Hellenistic world, and in the later Roman Empire.

Sid·ra, Gulf of (sĭd'rə), arm of the Mediterranean Sea, between Misratah and Benghazi, Libya.

Sie·ben·ge·bir·ge (zē'bən-gə-bĭr'gə), wooded range of the Rhenish Slate Mts., W West Germany. It extends for c.10 mi (16 km) S of Bonn along the Rhine River and rises to 1,509 ft (460.2 m) in the Grosser Ölberg. It is particularly famous for the Drachenfels.

Sie·gen (zē'gən), city (1972 est. pop. 56,864), North Rhine-Westphalia, W West Germany, on the Sieg River. Iron ore is mined nearby, and the city has iron foundries and manufactures leather goods and machinery. Noteworthy buildings include the Nikolaikirche, a 13th cent. church. Peter Paul Rubens was born here.

Sie·mia·no·wi·ce Sla·skie (shĕ-myä-nô-vē'tsĕ shlăɴ'skyĕ), city (1973 est. pop. 70,400), S Poland. A center of the Katowice industrial region, it has ironworks, steelworks, and coal mines. Manufactures include machinery and metals.

Si·en·a (sē-ĕn'ə), city (1972 est. pop. 65,661), capital of Siena prov., Tuscany, central Italy. Rich in art treasures and historic architecture, it is one of the most popular tourist centers in Italy. It is also noted for its wine and its marble. According to tradition, Siena was founded at the beginning of Roman times by Senus, the son of Remus. It became a free commune in the 12th cent. and gradually developed into a wealthy republic. In the early 16th cent. the Spanish and French struggled for control of the city, which fell after a siege (1554-55) to Emperor Charles V. Shortly thereafter it passed to Cosimo I de' Medici, duke of Tuscany. The city's artistic fame is due mainly to the paintings of the Sienese school (13th-14th cent.). On the main square, the Piazza del Campo, are the imposing Gothic Palazzo Pubblico (1297-1310), the Mangia tower (334 ft/101.9 m high), a 14th cent. chapel, the Fonte Gaia, and several medieval palaces. The city's cathedral (11th-14th cent.) is a splendid example of Italian Gothic. Also of note are the Piccolomini library, the Baptistery of San Giovanni, the rich art gallery Pinacoteca, St. Dominic's Church, and Piccolomini palace. The city has a university founded in the 13th cent.

Si·er·ra (sē-ĕr'ə). **1.** County (1970 pop. 2,365), 958 sq mi (2,481.2 sq km), NE Calif., in the Sierra Nevada, bordered in the E by Nev. and drained by the Yuba River and tributaries; formed 1852; co. seat Downieville. It is in a mining (gold from rich lodes that have been mined since 1850, some lead, silver, and copper) and lumbering (pine, fir and cedar) region. It has cattle and sheep grazing, and some farming and dairying in the Sierra Valley. Old mining camps, hunting, fishing, and winter sports attract vacationers. It includes parts of Tahoe and Plumas national forests. **2.** County (1970 pop. 7,189) 4,166 sq mi (10,789.9 sq km), SW N.Mex., drained by the Rio Grande and Alamosa River; formed 1884; co. seat Truth or Consequences (formerly Hot Springs). It has livestock grazing and placer mining for gold. The county is a retirement and health center.

Si·er·ra Blan·ca (sē-ĕr'ə blăng'kə), village (1970 pop. 800), seat of Hudspeth co., extreme W Texas, SE of El Paso near the Mexican boundary. It is a railroad junction and shipping and trading point in a ranching, mining, and resort area.

Si·er·ra Le·one (sē-ĕr'ə lē-ōn'), republic (1977 est. pop. 3,150,000), 27,699 sq mi (71,740 sq km), W Africa, bordered by the Atlantic Ocean in the W, by Guinea in the N and E, and by Liberia in the S. Freetown is the capital. Sierra Leone's 350-mi (560-km) Atlantic coastline is made up of a belt (average width 30 mi/50 km) of low-lying mangrove swamps, except for the mountainous Sierra Leone Peninsula (on which Freetown is situated). The coastline is broken by numerous estuaries and has some wide, sandy beaches. Behind the coastal belt is a wooded plateau (average elevation: 1,000 ft/300 m). The eastern half of the country is mostly mountainous and includes Sierra Leone's loftiest point (6,390 ft/1,948 m), located near the Guinea border. Several rivers, including the Great Scarcies (which makes up a section of the boundary with Guinea) and the Mano (which forms part of the border with Liberia), flow through the country to the Atlantic Ocean. The headwaters of the Niger River are situated in the mountains of the northeast.

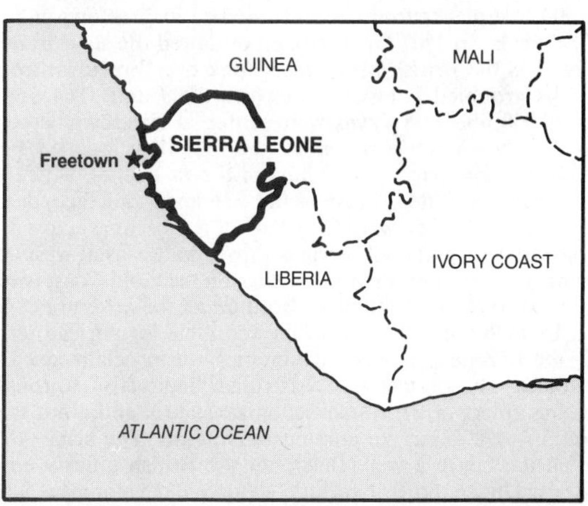

The great majority of the inhabitants of Sierra Leone are black Africans. The two main ethnic groups are the Mende (about 25% of the population), who speak a Mande language and live in the central and southern parts of the country, and the Temne (about 20% of the population), who speak a language closely related to Bantu and live in the north. Creoles, descendants of freed slaves from Nova Scotia, Canada, and the West Indies, make up about 25% of the population in Freetown. There are also small numbers of Indians and Lebanese. English is the country's official language. The majority of the population follows traditional religious beliefs, but there are significant numbers of Moslems and Christians.

Economy. Sierra Leone's economy is predominantly agricultural, with most of its workers engaged in subsistence farming. The principal food crops are rice, cassava, maize, pulses, plantains, and tomatoes. The leading cash crops are palm kernels, piassava, cacao, coffee, kola nuts, groundnuts, and ginger. Large numbers of poultry, cattle, sheep, and goats are raised. The country has an important but largely foreign-controlled mining industry. The main minerals extracted are diamonds, iron ore, bauxite, and titanium (ilmenite); together they account for about 80% of the country's exports by value. In 1972 the third-largest diamond ever found, the 969.8-carat "Star of Sierra Leone," was discovered in the country. In 1974 Sierra Leone joined the International Association of Producers of Bauxite, headquartered in Jamaica. The country's few manufactures include palm products, construction materials, refined petroleum, chemicals, lumber, and basic consumer goods such as processed food, beverages, clothing, footwear, furniture, and tobacco products. There is a growing fishing industry. Sierra Leone has limited rail and highway networks. Freetown has excellent port facilities; smaller ports are located at Bonthe (on Sherbro Island) and Pepel (near Freetown). The Great and Little Scarcies rivers are navigable by small craft for short distances. The cost of Sierra Leone's imports is usually slightly higher than its earnings from exports. The principal imports are machinery, manufactured consumer goods, foodstuffs, transportation equipment, and chemicals; the chief exports are minerals, palm kernels, cacao, and coffee. Sierra Leone's leading trade partners are Great Britain, Japan, the Netherlands, and United States.

History. The Temne were living along the northern coast of present-day Sierra Leone when the first Portuguese navigators reached the region in 1460. Beginning c.1500, European traders stopped regularly on the Sierra Leone peninsula, exchanging cloth and metal goods for ivory, timber, and black African slaves. Beginning in the mid-16th cent. Mande-speaking people migrated into Sierra Leone from present-day Liberia, and they eventually established the Mende states of Bullom, Loko, Boure, and Sherbro. In the early 17th cent. British traders became increasingly active along the Sierra Leone coast. Sierra Leone was a minor source of slaves during the 17th and 18th cent. In 1772 slavery was abolished in England, and after the American Revolutionary War (1775-83) attempts were made to resettle freed slaves in Africa. In 1787, 400 persons arrived at the Sierra Leone Peninsula, bought land from local leaders, and established the Province of Freedom near present-day Freetown. Most of the inhabitants died of disease in the first year. In 1790 a group of Temne (Moslems) destroyed what remained of the colony. A renewed attempt at settlement was made in 1792. The new colony named Freetown was controlled by the Sierra Leone Company, which forcefully held off the Temne, who resented the presence of foreigners. In 1800 about 500 free blacks from Jamaica arrived in Freetown via Nova Scotia, Canada. In 1807 Great Britain outlawed the slave trade, and in early 1808 the British government took over Freetown from the financially troubled company. Between 1808 and 1864 approximately 50,000 liberated slaves were settled at Freetown. Protestant missionaries were active there, and in 1827 they founded Fourah Bay College (now part of the Univ. of Sierra Leone), where black Africans were educated. Most of the freedmen and their descendants, known as Creoles, were Christians. They became active as missionaries, traders, and civil servants throughout coastal west Africa.

During the 19th cent. British holdings on the Gold Coast (present-day Ghana) were intermittently placed under the governor of Sierra Leone. In 1896 (near the end of the scramble for African territory among the European powers) the interior was proclaimed a British protectorate. The protectorate was ruled "indirectly" (through the indigenous rulers of the numerous small states), and a hut tax was imposed in 1898 to pay for administrative costs. The black Africans protested the tax in a war (1898), but the British quickly emerged victorious. Under the British little economic development was undertaken in the protectorate until the 1950s, although a railroad was built and the production for export of palm products and groundnuts was encouraged. In the 20th cent. the Creoles of the colony were largely excluded from higher government posts in favor of whites from Great Britain. After World War II the Creoles sought a larger voice in the affairs of Sierra Leone. In the elections of 1951 the protectorate-based Sierra Leone Peoples party (SLPP), led by Dr. Milton Margai (a Mende), emerged victorious. During the 1950s the black Africans were given more political responsibility, and educational opportunities were enlarged. In the economic sphere, mining (especially of diamonds and iron ore) increased greatly. On Apr. 27, 1961, Sierra Leone became independent, with Margai as prime minister; he died in 1964 and was succeeded by his brother, Albert M. Margai. The prime minister was appointed by the governor general and was responsible with his cabinet to the unicameral parliament. General elections were held in Mar., 1967, amid protests over Albert Margai's intention of creating a one-party state. The results of the election were never announced, but Siaka Stevens, leader of the opposition party (The All People's Congress-APC), formed (Mar. 21) a government at the request of the governor general. However, a military coup ousted Stevens a few minutes after he took the oath of office. Amid rioting by supporters of Stevens, the new government was toppled and replaced (Mar. 23-24) by a National Reformation Council. In Apr., 1968, lower-level army personnel led a revolt that overthrew the NRC and returned the nation to parliamentary government, with Stevens as prime minister. The following years were marked by considerable unrest. In Mar., 1971, after an abortive military coup, Stevens requested the presence of troops from Guinea to give his government support. On Apr. 19, 1971, Sierra Leone became a republic; Stevens began a five-year term as president and Sorie Ibrahim Koroma was appointed vice president and prime minister. The Guinean troops left Sierra Leone in early 1973, and parliamentary elections were held. The SLPP withdrew from the election and the APC won all but one of the elected parliamentary seats. Thus, Sierra Leone became in fact, if not in law, a one-party state.

Si·er·ra Ma·dre (sē-ĕr'ə mä'drā), residential city (1970 pop. 12,140), Los Angeles co., S Calif.; inc. 1907.

Sierra Madre, chief mountain system of Mexico, consisting of the Sierra Madre Oriental, the Sierra Madre Occidental, and the Sierra Madre del Sur and forming the dissected edges of the vast central Mexican plateau. Extending from northwest to southeast through Mexico from the U.S. border, the Sierra Madres, 6,000-12,000 ft (1,830-3,660 m) high, with deep, steep-sided canyons, have a great wealth of minerals including iron ore, lead, silver, and gold. Several hydroelectric-power stations have been built in the northern ranges. The Sierra Madre Oriental begins in barren hills south of the Rio Grande and runs for c.700 mi (1,125 km) roughly parallel to the coast of the Gulf of Mexico. It reaches an elevation of 18,700 ft (5,705 m) in Orizaba peak, which belongs also to the volcanic belt, Cordillera de Anáhuac. This belt, which divides Mexico in half at about lat. 19° N, on the other end joins the Sierra Madre Occidental This range, paralleling the Pacific coast for c.1,000 mi (1,610 km), extends southeast from Ariz. From c.5,000 ft (1,525 m) in the north, elevations reach over 10,000 ft (3,050 m) in the south. The Sierra Madre del Sur is a mass of uptilted mountains that touch the Pacific coast but form into no clearly defined range. It spreads over southern Mexico between the volcanic belt and the Isthmus of Tehuantepec and forms the natural harbor of Acapulco.

Sierra Ma·es·tra (mä-ĕs'trä), mountain range, Oriente prov., SE Cuba. It is the highest mountain system of Cuba and is rich in minerals, especially copper, manganese, chromium, and iron. Pico Turquino (6,560 ft/2,000.8 m) is the highest point.

Sierra Mo·re·na (mō-rā'nä), mountain range, SW Spain, extending c.375 mi (605 km) eastward from the Portuguese border to the Sierra de Alcaraz. Its highest peak is Bañuelo (c.4,340 ft/1,325 m). The range is rich in minerals, including copper, lead, and coal.

Sierra Ne·va·da (nə-vä'də), chief mountain range of S Spain, running from east to west for c.60 mi (95 km). The highest peak is Mulhacén (11,411 ft/3,480.4 m).

Sierra Ne·va·da (nə-vä'də, -văd'ə), mountain range, c.400 mi (645 km) long and from c.40 to 80 mi (65-130 km) wide, mostly in E Calif. It rises to 14,494 ft (4,420.7 m) in Mt. Whitney, the highest peak in the coterminous United States. The mountains extend northwest from Tehachapi Pass near Bakersfield, Calif., to the gap south of Lassen Peak. A tilted fault block in structure (the largest in the United States), the Sierra Nevada's eastern front rises sharply from the Great Basin, while its western slope descends gradually to the hills bordering the Central Valley of Calif. Snow-fed streams supply irrigation water to the Central Valley and to western Nev. and also generate hydroelectricity. Donner Pass (alt. 7,089 ft/2,162 m), the principal pass across the mountains, was used by thousands of immigrants in the middle and late 1800s. The Sierra Nevada are known for their magnificent scenery.

Sierra Ne·va·da de Mé·ri·da (nə-vä'də dā mā'rē-dä), mountain range, NW Venezuela, a spur of the Andes from 30 to 50 mi (48–80 km) and extending c.200 mi (320 km) NE from the Colombian border to the Caribbean coastal range. Pico Bolívar (16,411 ft/5,005.4 m high) is the highest point in Venezuela.

Sierra Vis·ta (vĭs'tə), city (1975 est. pop. 21,000), Cochise co., SE Ariz., SE of Tucson in the San Pedro valley. Cattle are raised, and there are copper, zinc, and lead mines.

Sierre (syĕr), town (1974 est. pop. 13,800), Valais canton, S Switzerland, on the Rhône River. It is the center of a rich horticultural region. There is a large aluminum plant nearby.

Sif·nos (sĭf'nôs') or **Siph·nos** (sĭf'nəs), island, c.32 sq mi (83 sq km), SE Greece, in the Aegean Sea; one of the Cyclades.

Sig·lu·fjör·dur (sĭk'lə-fyûr'*th*ər), town (1970 pop. 2,161), N Iceland, on the Greenland Sea. It is the capital of Iceland's herring industry.

Sig·our·ney (sĭg'ər-nē), city (1970 pop. 2,319), seat of Keokuk co., SE Iowa, NNE of Ottumwa; inc. 1868.

Sig·tu·na (sĭg'tü'nä), town (1970 pop. 3,648), Stockholm co., E Sweden. Founded c.1000, it was one of Sweden's earliest towns and its first capital. Sweden's first coin was minted here.

Si·gui·ri (sē-gîr'ē), town (1961 est. pop. 12,000), NE Guinea, on the Niger River. It is the commercial center for a region where rice, millet, and cotton are grown. Alluvial gold is found in the area.

Si·ha·nouk·ville (sē-hä'nək-vĭl'): *see* Kompong Som, Cambodia.

Si·kar (sē'kər), walled town (1971 pop. 70,987), Rajasthan state, NW India. It is a market for grain and tobacco.

Si·kas·so (sē-kä'sō), town (1971 est. pop. 26,600), S Mali. It is the trade center for a region where peanuts, cotton, rice, fruit, and vegetables are grown. The town has cotton ginning and rice milling.

Sike·ston (sīk'stən), city (1975 est. pop. 14,700), New Madrid and Scott cos., SE Mo.; inc. 1874. It is the shipping, marketing, and processing center of a cotton, soybean, and grain region.

Si·kho·te-A·lin (sē'kə-tā'ə-lēn'), mountain range, c.625 mi (1,005 km) long, S Far Eastern USSR. Its forests are a source of lumber, and there are deposits of coal, lead, zinc, silver, and tin.

Sik·kim (sĭk'ĭm), constitutional monarchy (1974 pop. 210,000), 2,745 sq mi (7,110 sq km), S central Asia, in the E Himalayas. The capital and only town is Gangtok. Sikkim is bordered on the west by Nepal, on the north by the Tibet region of China, on the east by Bhutan, and on the south by India. Most of Sikkim is mountainous, and rivers, including the Tista, flow through deep valleys hindering travel. In the mountains are extensive forests and grazing land for sheep, goats, cattle, and yaks. Corn is the major crop of the tropical

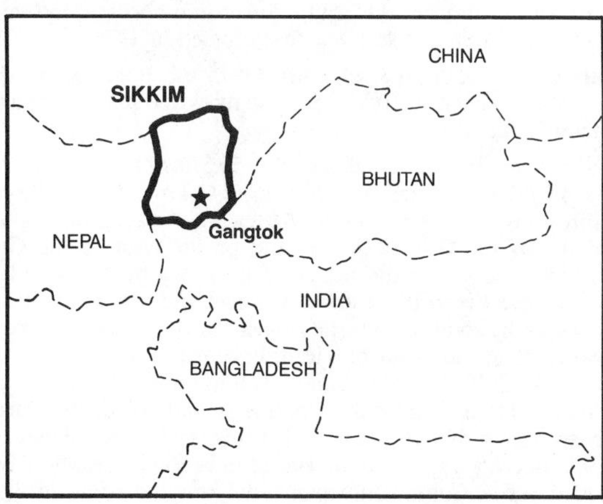

lowland valleys, and rice, millet, wheat, barley, legumes, fruits, and cardamom are also grown. Agriculture is chiefly for subsistence. Sikkim has some copper deposits, which are worked by primitive methods. There is a handicraft industry, and cotton weaving is common. Sikkim's people are predominantly of Nepalese extraction; the minority Bhotias (Tibetan in origin) and aboriginal Lepchas are mainly pastoral nomads. Although the Nepalese practice Hinduism, Buddhism is professed by the maharaja and the official class, and Sikkim is noted for its Buddhist monasteries. Tibeto-Burmese languages and dialects are spoken widely.

In the 16th cent. Tibetans began to settle Sikkim, whose native Lepchas were probably converted to Buddhism by Tibetan lamas. In 1642 a Tibetan king, from whom the last chogyal claimed descent, started a hereditary line of Sikkimese rulers. Gurkhas from Nepal invaded Sikkim several times in the 18th and 19th cent.; but the British forced the Gurkhas out of Sikkim (1814–16). Later (1835, 1849) the Sikkimese ceded territory to the British, who assumed a protectorate.

British protection ended when India won independence in 1947, but political and social unrest in led to a treaty (1950) by which the newly independent Sikkim became an Indian protectorate. India directed defense and foreign relations while Sikkim retained internal autonomy. The Indians financed construction of strategic roads that traverse the mountain passes into Tibet. These and other roads, in addition to other social and economic modernization, helped Sikkim to overcome its long isolation from the outside world. In 1974, under a new constitution, accepted under pressure by the kingdom's chogyal and passed by India's parliament, Sikkim was made an associated state of India. The chogyal was reduced to a titular position; in 1975 the office was abolished and replaced by a president (appointed by India) who headed a five-member cabinet.

Sil·ches·ter (sĭl'chĕs'tər, -chĭ-stər), village, Hampshire, S England. It is noted for the ruins of the Roman-British town Calleva Atrebatum. The outside walls, forum, amphitheater, and several temples were excavated beginning in the 1890s.

Si·le·sia (sī-lē'zhə, -shə, sī-), region of E central Europe, extending along both banks of the Oder River and bounded in the S by the mountain ranges of the Sudetes and the W Carpathians. Politically, almost all of Silesia is divided between Poland and Czechoslovakia. The Polish portion comprises most of the former Prussian provinces of Upper Silesia and Lower Silesia, both of which were transferred to Polish administration in 1945; the Polish portion also includes those parts of Upper Silesia that were ceded by Germany to Poland after World War I and part of the former Austrian principality of Teschen. A second, much smaller part of Silesia has belonged to Czechoslovakia since 1918.

Except in the south, Silesia is largely an agricultural and forested lowland, drained by the Oder and its tributaries. Along the slopes of the Sudetes there are numerous small industrial centers with traditional textile and glass industries. Czech Silesia comprises the rich Karvinna coal basin. The most important part of Silesia is, however, its southern tip—Upper Silesia, in Poland. One of the largest industrial concentrations of Europe, it has extensive coal and lignite deposits and zinc, lead, iron, and other ores; there are iron and steel mills, coke ovens, and chemical plants. Opole, the former capital of Upper Silesia, is an important trade center.

History. Slavic tribes settled here A.D. c.500, and Silesia was an integral part of Poland by the 11th cent. King Boleslaus III (reigned 1102–38), of the Piast dynasty, divided Poland into four hereditary duchies, of which Silesia was one. After 1200 the duchy of Silesia fell apart into numerous minor principalities. In the early 14th cent. the Silesian princes accepted the king of Bohemia as their suzerain. After a brief period under Hungarian rule in the 15th cent., Silesia passed to the house of Hapsburg in 1526. The Thirty Years' War (1618–48) brought untold misery to Silesia under successive Saxon, imperial, and Swedish occupation. It reverted to Austrian control at the Peace of Westphalia (1648) and was incorporated into the Bohemian crown domain in 1675. Prussia invaded Silesia in 1740 and was ceded virtually all of it at the end of the War of the Austrian Succession (1745).

During the Industrial Revolution of the late 18th and 19th cent. textile weaving and coal mining developed rapidly in Silesia, but industrialization brought great social tension between the predominantly German mill and mine owners and the largely Polish workers. After World War I the results of a plebiscite (1921) to determine the national status of Upper Silesia were favorable to Germany except in the easternmost part where the Polish population predominated. The larger part of the industrial district, including Katowice, passed to Poland. The contested city and district of Teschen were partitioned in 1920 between Poland and Czechoslovakia. As a result of the Munich Pact of 1938 most of Czech Silesia was partitioned between Germany and Poland, and after the German conquest of Poland in 1939 all Polish Silesia was annexed to Germany. After

World War II the pre-1938 boundaries were restored, but all formerly Prussian Silesia east of the Lusatian Neisse was placed under Polish administration. The Allies also allowed the expulsion of the German population from Czech Silesia, Polish Silesia, and Polish-administered Silesia. West Germany finally relinquished all claims to the area in a 1972 pact with Poland.

Si·lis·tra (sĭ-lĭs′trə), town (1969 est. pop. 37,400), NE Bulgaria, a port on the Danube River bordering Rumania. Products include foodstuffs, ceramics, and furniture. As the Roman Durostorum, it was founded in 29 B.C. Its importance continued under Byzantine and Bulgar rule. Conquered by the Turks (1388), the town was captured (1877) by the Russians, ceded to Bulgaria, transferred to Rumania in 1913, and returned to Bulgaria in 1940.

Sil·ke·borg (sĭl′kə-bôr′), city (1970 com. pop. 43,125), Århus co., central Denmark, on the Gudenå River. It is a tourist and health resort surrounded by beautiful lakes and woods. Nearby is a memorial to the playwright Kaj Munk, who was murdered (1944) in Silkeborg by Nazi terrorists.

Si·loam Springs (sĭ′lōm), city (1970 pop. 6,433), Benton co., NW Ark., in the Ozarks near the Okla. border NW of Fayetteville; inc. c.1880. Milk, fruit, and poultry are processed and shipped. Mineral springs and Baptist campgrounds are nearby. It is the seat of John Brown Univ.

Sils·bee (sĭlz′bē), city (1970 pop. 7,696), Hardin co., E Texas, near the Neches River NNW of Beaumont, in a truck-farm and timber area; inc. as a town 1906, as a city 1938. Laid out in 1894 on the site of a logging camp, it still processes native pine but also draws income from oil and cattle.

Sil·ver Bow (sĭl′vər bō′), county (1970 pop. 41,981), 715 sq mi (1,851.9 sq km), SW Mont., in a mountainous region crossed by the Continental Divide and bounded on the SW by the Big Hole River; formed 1881; co. seat Butte. Its natural resources include copper, zinc, lead, silver and manganese. Livestock and dairy products are also important.

Silver City, town (1970 pop. 8,557), seat of Grant co., SW N.Mex., near the Ariz. border NW of El Paso, Texas; founded 1870, inc. 1878. It is a trade and shipping point for nearby mines, ranches, and irrigated farms. It is also a health resort.

Silver Spring, uninc. city (1978 est. pop. 83,700), Montgomery co., W central Md., a residential suburb of Washington, D.C. It has a large naval ordnance laboratory, several large research laboratories, and a plant making precision instruments.

Silver Springs, mineral spring, N central Fla., source of the Silver River. The spring is one of the world's largest and most famous; a great variety of aquatic life may be seen through glass-bottomed boats. Hernando De Soto was probably the first white man to visit the spring (1539).

Sil·ver·ton (sĭl′vər-tən). **1.** Town (1970 pop. 797), seat of San Juan co., SW Colo., in the San Juan Mts. on the Animas River NNE of Durango, at an altitude of 9,302 ft (2,837.1 m); founded 1874, inc. 1885. It is a tourist and trade center, with lead, silver, copper, and gold mines in the vicinity. **2.** City (1970 pop. 6,538), Hamilton co., SW Ohio, a suburb NE of Cincinnati; inc. 1904. **3.** City (1970 pop. 1,026), seat of Briscoe co., NW Texas, on the Llano Estacado NE of Plainview, in a farm and ranch area.

Sil·vis (sĭl′vĭs), residential city (1970 pop. 5,907), Rock Island co., NW Ill., just E of Moline; inc. 1920.

Sil·vret·ta (sĭl-vrĕt′ə), mountain group of the Alps, in E Switzerland and SW Austria. Its highest peak, Piz Linard (11,185 ft/3,411.4 m), is in Switzerland; Piz Buin (10,869 ft/3,315 m) is the highest of the group in Austria.

Si·man·cas (sē-mäng′käs), village, Valladolid prov., NW Spain. The castle, an old fort rebuilt in the 15th cent., contains the Spanish national archives.

Sim·coe (sĭm′kō), town (1971 pop. 10,793), S Ont., Canada. It is a market, processing, and canning center for a region producing fruit, vegetables, and tobacco.

Simcoe, Lake, 539 sq mi (1,396 sq km), S Ont., Canada, between Georgian Bay and Lake Ontario. Lake Simcoe drains north through the Severn River to Georgian Bay and forms part of the Trent Canal system. There are several small resorts are on the lake.

Sim·fer·o·pol (sĭm′fə-rô′pəl), city (1973 est. pop. 269,000), capital of Crimean oblast, SE European USSR, in the Ukraine, on the Salgir River and on the Sevastopol-Kharkov RR. It is a land and water transport hub and a commercial center in a truck-farming and fruit-growing region. Industries include food processing, wine making, and the manufacture of machinery, machine tools, power station equipment, and knitwear. Tourism is economically important. Simferopol occupies the site of an ancient Scythian capital (founded 3rd cent. B.C.). Called Ak-Mechet under Tatar rule (15th–18th cent.), it was renamed after its annexation to Russia in 1784.

Sí·mi (sē′mē) or **Sy·me** (sē′mē, sī′-), island (1971 pop. 2,489), 22 sq mi (57 sq km), SE Greece, in the Aegean Sea, one of the Dodecanese Islands.

Si·mi Valley (sē′mē, sĭm′ē), city (1978 est. pop. 72,500), Ventura co., S Calif., in an oil, farm, and livestock region; laid out 1887, inc. 1969. Campers, toothpaste tubes, plastic containers, sportswear, sewage units, and liquid oxygen are manufactured.

Sim·la (sĭm′lə), town (1971 pop. 55,368), capital of Himachal Pradesh state, NW India. It is situated on a ridge (c.7,100 ft/2,165 m high) in the western Himalayas. Simla is a summer resort and the headquarters of the Indian army.

Sim·o·ïs (sĭm′ə-wəs), small river, NW Turkey, a tributary of the Scamander. It was the scene of many legendary events, particularly during the siege of Troy.

Si·mons·town (sī′mənz-toun′), town (1970 pop. 6,500), Cape Prov., SW South Africa, on False Bay, an arm of the Atlantic Ocean. It is a seaside resort and a station of the South African navy; industry centers around ship construction and repair. There is also a fishing industry. Simonstown was founded by the Dutch in 1741. In 1814 the town became the headquarters of the British South Atlantic squadron. In 1957 the base was turned over to South Africa, which allows Britain to continue using it.

Sim·plon (sĭm′plŏn′), pass, 6,590 ft (2,010 m) high, in the Lepontine Alps, Valais canton, S Switzerland. It is crossed by the Simplon Road, built (1800–1806) by Napoleon I. The Simplon RR passes through Simplon Tunnels I and II, the longest (both 12.3 mi/19.8 km) in the world. They cross the Swiss-Italian border from Brig to Isella. They have a maximum elevation of 2,313 ft (705.5 m).

Simp·son (sĭm′sən, sĭmp′-). **1.** County (1970 pop. 13,054), 239 sq mi (619 sq km), S Ky., bordered on the S by Tenn. and drained by West Fork of Drake Creek; formed 1819; co. seat Franklin. In a rolling agricultural area yielding dark tobacco, grain, strawberries, and livestock, it has timber, limestone quarries, and diversified manufacturing. **2.** County (1970 pop. 19,947), 587 sq mi (1,520.3 sq km), S central Miss., bounded on the W by the Pearl River and drained by the Strong River; formed 1824; co. seat Mendenhall. Agriculture and lumbering are the major occupations.

Simpson Desert. c.50,000 sq mi (129,500 sq km), SE Northern Territory, Australia. It runs west from the Queensland border c.200 mi (320 km) and north from the South Australian border c.150 mi (240 km). This uninhabited desert was first crossed in 1939.

Sims·bur·y (sĭmz′bĕr′ē, -bə-rē), town (1975 est. pop. 19,800), Hartford co., N Conn.; inc. 1670. Detonating fuses and machinery are the chief manufactures.

Si·nai (sī′nī′), peninsula, c.23,000 sq mi (59,570 sq km), NE Egypt. It is c.230 mi (370 km) long and 150 mi (240 km) wide and extends north into a broad isthmus linking Africa and Asia. Sinai is bounded on the east by the Gulf of Aqaba and on the west by the Gulf of Suez, which is linked to the Mediterranean Sea by the Suez Canal; the Negev desert is to the northeast. Level and sandy in the north, Sinai rises to the south in granitic ridges. Sharm al-Sheik, a promontory overlooking the Strait of Tiran, is near the tip of Sinai, at the mouth of the Gulf of Aqaba. Sinai has a hot and dry climate and is sparsely vegetated. Limestone quarrying and oil drilling are the main economic activities; nomadic herding is practiced. Jabal Musa, or Mt. Sinai, c.7,500 ft (2,290 m), is said to be the place where Moses received the Ten Commandments. Sinai was the scene of fighting during the Arab-Israeli Wars of 1956, 1967, and 1973.

Si·nai·a (sĭ-nī′ə), town (1966 pop. 11,976), S central Rumania, in Walachia, in the Transylvanian Alps. It is a resort and has heavy and light industries. Sinaia was the summer residence of the kings of Rumania. In the town are two former royal palaces, a 17th cent. monastery, and a castle.

Si·na·lo·a (sē′nə-lō′ə), state (1970 pop. 1,266,528), 22,582 sq mi

(58,487 sq km), W Mexico, on the Gulf of California and the Pacific Ocean; capital Culiacán. Sinaloa was joined with Sonora during the Spanish period; it became a separate state in 1830.

Sind (sĭnd), province (1969 est. pop. 11,100,000), c.50,000 sq mi (129,500 sq km), SE Pakistan, roughly coextensive with the lower Indus River valley and bounded by India on the E and S and by the Arabian Sea on the SW; capital Karachi. Hot and arid, Sind depends almost exclusively on irrigation for agriculture. It supports wheat, rice, millet, cotton, oilseed, sugar cane, fruits, and some tobacco. There are also sheep and cattle breeding and poultry farming. Other resources include gypsum and iron ore. The region is noted for handicrafts. Fishing is important in coastal areas. Sind may have been the site of the subcontinent's earliest civilization. The region was taken (5th cent. B.C.) by Darius I of Persia, invaded (325 B.C.) by Alexander the Great, annexed (c.3rd cent. B.C.) by the Maurya empire, overrun (165 B.C.) by the Huns, and ruled (1st–2nd cent. A.D.) by the Kushan dynasty. Arabs invaded Sind in 711, and it remained under direct or nominal Arab rule until the 11th cent., when it passed to the Moslem Turkic Ghaznivids. Arab religious, social, and cultural influences remain strong in present-day Sind. Although briefly incorporated into the Mogul empire by Akbar, the region remained for centuries under local Moslem dynasties. Sir Charles Napier, the British general, conquered the area in 1843. The British made Karachi the capital and administered Sind as part of the Bombay presidency until 1937, when it became an autonomous province. After Pakistan became independent in 1947, Karachi was made the national capital, and Sind's capital was shifted to Hyderabad. From 1955 to 1970, Sind was part of West Pakistan prov.; it became a separate province again in 1970.

Sin·ga·nal·lur (sĭng'gə-nə-lo͝or'), city (1971 pop. 112,206), Tamil Nadu state, S India. It is a suburb of Coimbatore.

Sin·ga·pore (sĭng'gə-pôr', sĭng'ə-, sĭng'gə-pôr'), republic (1976 pop. 2,300,000), 225 sq mi (583 sq km), consisting of the island of Singapore (210 sq mi/544 sq km) and about 60 small adjacent islands, SE Asia, at the S tip of the Malay Peninsula. Singapore city (1970 est. pop. 1,240,000), the capital, largest city, and chief port, is located on the southern shore of the island. The distinction between Singapore and Singapore city is disappearing, as the entire island becomes urbanized.

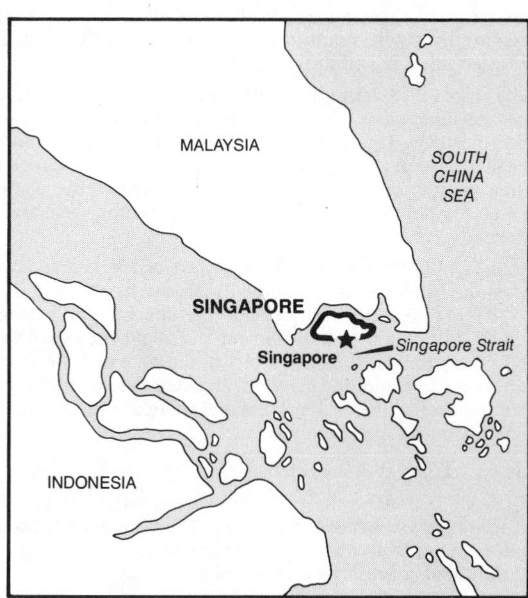

Lying just north of the equator and located between the Indian Ocean and the South China Sea, Singapore is situated at the convergence of some of the world's major sea-lanes. It is separated from Indonesia to the south by the Singapore Strait and from Malaysia to the north by the Johore Strait. Singapore island is low lying and is composed of a granitic core (rising to 580 ft/177 m at Bukit Timah, the country's highest point) surrounded by sedimentary lowlands. The coast is broken by many inlets, and there are extensive mangrove swamps. Keppel Harbor, site of the port of Singapore, is a natural deepwater anchorage between Singapore and the islands of Brani and Blakang Mati. Singapore has a tropical rain-forest climate with uniformly high temperatures and rainfall throughout the year. The island was once covered by rain forest, which is now limited to the central portion. Less than one fourth of Singapore's land is used for agriculture; tropical fruits and vegetables are intensively cultivated and poultry and hogs are raised. However, most of the cultivable land is given over to coconut and rubber plantations. Singapore is a major fishing center of southeast Asia. There are no exploitable natural resources in the country. Its power is produced by thermoelectric plants. Singapore has a good road system, a railroad crosses the island, and a causeway carrying road and rail traffic links Singapore to the mainland.

The population is about 76% Chinese; Malays and Indians constitute large minorities. Singapore is one of the world's most densely populated countries with about 9,200 people per sq mi (about 3,500 people per sq km). The country has a predominantly urban population and a massive urban renewal program was begun in the 1960s. As a result of family planning and a strict immigration policy, the birth rate had declined by the early 1970s to about 1.5%, down from 4.5% in the 1950s. Buddhism, Islam, Hinduism, and Christianity are the religions of Singapore. The country has four official languages—Malay, Chinese, Tamil, and English—and one of the highest literacy rates (a product of a fine uniform education system conducted in all the official languages). It also has one of the highest standards of living in Asia.

Economy. Singapore is one of the world's greatest commercial centers and has one of the world's largest and busiest ports. Commerce has historically been the chief source of income. For many years the largest importer in southeast Asia, Singapore is a free port and an entrepôt that re-exports more than half of what it imports, especially rubber, petroleum, textiles, timber, and tin. It also exports locally manufactured items such as electrical goods, petroleum products, processed food, rubber, and tin. The country imports most of its food requirements. Singapore's chief trading partners are Malaysia and Japan. With more than 300 factories and deepwater wharves, the Jurong Industrial Estate is Singapore's principal industrial complex. The country has a number of large petroleum storing and refining facilities; and Keppel Harbor has container handling facilities. Development of the former British naval base at Sembawang on the Johore Strait, now a commercial shipyard, has enhanced Singapore's status as a major center for shipbuilding and repairs.

History. Singapore was a trading center in the Srivijaya empire before it was destroyed in the 14th cent. by the Majapahit empire. It later became part of Johore in the Malacca Sultanate. The sparsely populated island was ceded (1819) to the British East India Company through the efforts of Sir T. Stamford Raffles; he founded the modern city of Singapore there that same year. In 1824 Singapore came under the complete control of the British and, although containing only a small fishing and trading village, quickly attracted Chinese and Malay merchants. The port grew rapidly, soon overshadowing Penang and Malacca in importance. With them Singapore became part of the Straits Settlements in 1826. The development of Malaya under British rule in the late 19th and early 20th cent. made Singapore one of the leading ports of the world for the export of tin and rubber. The construction of a railroad through the Malay Peninsula to Bangkok swelled Singapore's trade, and the building of airports made it more than ever a communication center. After the swift Japanese campaign in Malaya during World War II, Singapore was successfully attacked across the Johore Strait, and on Feb. 15, 1942, the British garrison surrendered; Singapore was reoccupied by the British in Sept., 1945. In 1946, Singapore, no longer a part of the Straits Settlements, was constituted, with Christmas Island and the Cocos-Keeling islands, a crown colony. Following a decade of Communist terrorism, Singapore, separated from Christmas Island and the Cocos-Keeling islands, became (June, 1959) a self-governing state. Under the policies of Prime Minister Lee Kuan Yew, Singapore's economic base was strengthened and a greater degree of social and cultural homogeneity was achieved. Singapore's industrial base was diversified, expanded, and modernized. Following the 1962 referendum, Singapore merged (Aug., 1963) with Malaya, Sarawak, and Sabah to form the Federation of Malaysia. However, because of deep divisions, especially racial antagonism between the Malays and the Chinese, Singapore was, by mutual agreement, separated from the federation in Aug., 1965, and became an independent republic. Singapore has remained in the Commonwealth of Nations and joined the United Nations in 1965.

Government. The country is governed by the constitution of 1966.

It has a parliamentary form of government, with a president as head of state and a prime minister as head of government. There is a 65-seat unicameral Parliament whose members are popularly elected.

Si·ning (shē'nĭng'): *see* Hsi-ning, China.

Sin·kao Shan (shĭn'gou'shän'), peak, Taiwan: *see* Hsin-Kao Shan.

Sin·kiang (sĭn'kyăng', shĭn'jē-äng'), autonomous region (1968 est. pop. 8,000,000), c.660,000 sq mi (1,709,400 sq km), NW China. It is officially known as the Sinkiang Uigur Autonomous Region and is also called Chinese Turkistan or Eastern Turkistan. The capital is Wu-lu-mu-ch'i. Sinkiang shares a 1,800-mile (2,900-km) border with the USSR on the north and west and is bordered by the Mongolian People's Republic on the east and by Kashmir and Tibet on the south. The great Altai, Tien Shan, and Kunlun mountain ranges enclose it on the north, west, and south, respectively; a barren plateau lies to the west. The rivers of Sinkiang, including the Tarim, the Yarkand, the Ili, the Ma-na-ssu, and the Ho-t'ien, rise in the mountains and flow from east to west. The level land, divided by the Tien Shan in central Sinkiang, comprises the Dzungaria, a grazing region to the north, and the Tarim basin, a vast desert to the south.

Economy. Sinkiang has a dry continental climate with great extremes of winter and summer temperature. Rainfall is scant, seldom exceeding 10 in. (30 cm) annually. The bulk of the population lives along the borders of the Dzungaria and the Tarim basin, where cotton and silk (both locally spun and woven) are produced, and wheat, rice, millet, potatoes, kaoliang, sugar beets, and fruit are grown. Turkic-speaking Uigurs (mainly Moslems), who comprise Sinkiang's majority ethnic group, have traditionally excelled in building the intricate systems of canals and wells that supply water to the fields. In recent years extensive areas of grazing land have been converted to raising wheat. Large-scale animal husbandry remains important, however, and the amount of livestock is increasing. Many of the Kazakh and Mongol stockherders are still at least seminomadic. Sinkiang is now linked to the Chinese rail network, but west and south of Wu-lu-mu-ch'i transportation is still concentrated along two ancient roads: the north road, which skirts the southern edge of the Dzungaria and connects Wu-lu-mu-ch'i with the Soviet Turkistan-Siberia RR, and the south road, which encircles the Tarim basin. The vast oil fields at Karamai are among the largest in China, and there are extensive deposits of coal, silver, copper, lead, nitrates, gold, and zinc. Large uranium reserves have been reported. New mines, as well as refineries, ironworks, steelworks, and chemical plants, have been established in recent years. Other industries include textile and cement production and sugar refining.

History. Sinkiang first passed under Chinese rule in the 1st cent. B.C., when the emperor Wu Ti defeated the Huns and occupied the region. In the 2nd cent. A.D. China lost Sinkiang to the Uzbek Confederation but reoccupied it in the mid-7th cent. It was conquered (8th cent.) by the Tibetans, overrun by the Uigurs, who established a kingdom here, and subsequently invaded (10th cent.) by the Arabs. Sinkiang passed to the Mongols in the 13th cent. An anarchic period followed until the Manchus established (1756) loose control. The subsequent relations between China and Sinkiang were marked by conflict and rebellion. Sinkiang became a Chinese province in 1881, but even as late as the establishment (1912) of the Chinese republic it remained more or less independent. Rebellions in 1936, 1937, and 1944 further eased Chinese rule. Late in 1949 Sinkiang capitulated to the Chinese Communists without a struggle. Because the 1953 census showed the Uigurs to comprise 74% of the population, Sinkiang prov. was reconstituted (1955) as the Sinkiang Uigur Autonomous Region. Autonomous districts were created as well for the Kazakh, Mongol, Hui, and Kirghiz minority groups. In the 1950s and 1960s the central government sent massive numbers of Chinese to Sinkiang to help develop water conservancy and mineral exploitation schemes. Consequently the Chinese are approaching numerical parity with the Uigurs. National defense has also been a consideration in the mass influx, because it is a strategic and sensitive region. In 1969 frontier incidents led to serious fighting between Soviet and Chinese forces along the Sinkiang-USSR border.

Si·nop (sə-nôp'), town (1970 pop. 15,096), capital of Sinop prov., N Turkey, on the Black Sea. It has an excellent harbor. Ancient Sinop was founded by colonists from Miletus in the 8th cent. B.C. and rose to great commercial and political importance. One of its chief exports was cinnabar, which derives its name from Sinop. The city fell (c.183 B.C.) to the kings of Pontus, whose capital it became. The Romans captured it (74–63 B.C.) and made it a free city. Under the Roman and Byzantine empires the city reached great prosperity. When the Byzantine Empire broke up in 1204, Sinop joined the Greek empire of Trebizond, but within a few years it was occupied by the Seljuk Turks. In 1853 a Russian naval squadron completely destroyed a Turkish flotilla here, hastening the onset of the Crimean War. Sinop was the birthplace of Diogenes.

Sint-Ni·klaas (sĭnt-nē'kläs'), city (1971 est. pop. 49,320), East Flanders prov., N Belgium. It is the commercial, industrial, and transportation center of the Waas region.

Sin·ton (sĭn'tən), town (1970 pop. 5,563), seat of San Patricio co., S Texas, N of Corpus Christi; settled c.1892, inc. 1916. A shipping point for cotton and cattle, it does aluminum smelting and oil refining.

Sin·tra (sēn'trə): *see* Cintra, Portugal.

Sint-Trui·den (sĭnt-troi'dən), town (1971 est. pop. 21,640), Limburg prov., E Belgium. It is an industrial center. Sint-Truiden developed around an abbey (7th cent.).

Sin·ui·ju (shĭn'ē-jōō'), city, (1967 est. pop. 165,000), W North Korea, on the Yellow Sea at the mouth of the Yalu River. It is a port and rail terminus and an industrial center, with manufactures of chemicals and aluminum. A bridge over the Yalu connects it with northeastern China.

Sin·yang (shĭn'yäng'): *see* Hsin-yang, China.

Sion (syôɴ), town (1974 est. pop. 22,900), capital of Valais canton, SW Switzerland, on the Rhône River. Sion is a wine and horticultural market center, with hydroelectric plants and coal mines nearby. An episcopal see since the 6th cent, Sion is rich in historic remains, among them a 13th cent. castle and a late Gothic cathedral.

Sioux (sōō). **1.** County (1970 pop. 27,996), 766 sq mi (1,983.9 sq km), NW Iowa, on the S.Dak. border (formed here by the Big Sioux River), drained by the Rock River and Floyd and West Branch Floyd rivers; formed 1851; co. seat Orange City. Its prairie agriculture includes hogs, cattle, poultry, corn, oats, and hay. **2.** County (1970 pop. 2,034), 2,063 sq mi (5,343.2 sq km), extreme NW Nebr., in an agricultural region bordering on S.Dak. and Wyo. and drained by branches of the White and Niobrara rivers; formed 1877; co. seat Harrison. Livestock, grain, dairy products, and poultry are important. Toadstool Park, an area of curious stone formation, is in the northeast. **3.** County (1970 pop. 3,632), 1,103 sq mi (2,856.8 sq km), S N.Dak.; formed 1914; co. seat Fort Yates. It is in an agricultural area yielding livestock, poultry, and grain. Part of Standing Rock Indian Reservation is within the county.

Sioux City, city (1978 est. pop. 85,100), seat of Woodbury co., NW Iowa, at the junction of the Big Sioux and Floyd rivers with the Missouri; inc. 1857. It is a shipping, trade, and industrial center for an extensive agricultural and livestock area and has a livestock market, a hog market, meat-packing houses, and processing plants for popcorn and honey. Fertilizers, electric tools, radios, feed, and seed are also produced.

Sioux Falls, city (1978 est. pop. 74,700), seat of Minnehaha co., SE S.Dak., on the Big Sioux River; settled 1856, inc. as a village 1877, as a city 1883. White settlers abandoned the site in 1862 because of Indian raids, but with the establishment (1865) of Fort Dakota it was resettled. Sioux Falls is the largest city in the state and the commercial, industrial, and shipping center of an extensive agricultural area. It has an important livestock market and there are meat-processing plants. Sandstone is quarried nearby.

Siph·nos (sĭf'nəs), island: *see* Sífnos.

Sip·par (sĭ-pär'), ancient city of N Babylonia, on the Euphrates in present Iraq near Baghdad. It was one of the capitals of Sargon and had a great temple to the sun god Shamash. Excavations have yielded thousands of inscribed clay tablets.

Si·qui·jor (sē'kĭ-hôr'), island, 130 sq mi (336.7 sq km), one of the Visayan Islands, the Philippines, just off the SE coast of Negros. The main town and chief seaport is also called Siquijor.

Si·re·tul (sîr'ə-tōōl'), river, c.450 mi (725 km) long, rising in the Carpathian Mts., W Ukraine, USSR, and flowing SE to the Danube River at Galaţi.

Si·ris (sĭ'rĭs): *see* Sérrai, Greece.

Sir·mi·um (sûr'mē-əm), ancient city of Pannonia. Near modern Sremska Motrovica, Yugoslavia, Sirmium was occupied in the 1st cent. B.C. by the Romans and was the chief city of Lower Pannonia.

Sí·ros (sîr'ôs'), city: *see* Hermoupolis, Greece.

Síros or **Sy·ros** (sī'rŏs'), island (1971 pop. 18,642), 33 sq mi (85.5 sq km), SE Greece, in the Aegean Sea, one of the Cyclades. Hermoupolis, or Síros, is the island's main town and port. Síros is the richest and most populous of the Cyclades.

Sis·ki·you (sĭs'kĭ-yōō), county (1970 pop. 33,225), 6,262 sq mi (16,218.6 sq km), N Calif., bounded on the N by the Oregon border, with the Klamath Mts. in the W, and part of the Cascade Range in the E; whose glacier-fed springs are sources of the Sacramento and McCloud rivers; formed 1852; co. seat Yreka. To the east of the Cascades lies a semiarid farming and lumbering region. Stock grazing, farming (potatoes, hay, barley, and wheat) and food processing are important. It has some mining and quarrying. The mountain scenery and hunting and fishing attract vacationers. Lava Beds National Monument and Klamath, Shasta, and Modoc national forests are in the county.

Sis·se·ton (sĭs'ə-tən), city (1970 pop. 3,094), seat of Roberts co., NE S.Dak., near the N.Dak border W of Lake Traverse; inc. 1892. A trade and processing center for a farm area and resort region, it is also the headquarters of the Sisseton Indian Reservation. Fort Sisseton, to the west, was established in 1864 for protection against Indians and was abandoned in 1888.

Si·ta·pur (sē'tə-pŏŏr'), town (1971 pop. 66,715), Uttar Pradesh state, N central India. It is a market for grain, oilseed, and jute; the leather and plywood industries are important.

Sit·ka (sĭt'kə), city (1978 est. pop. 7,400), Greater Sitka borough, SE Alaska, in the Alexander Archipelago, on Baranof Island; inc. 1971. Fishing is important; salmon, halibut, red snapper, crab, herring, abalone, and clams are caught. There are also canning, lumbering, and pulp-processing enterprises. Sitka was founded (1799) by Aleksandr Baranov and became the flourishing capital of Russian America. It remained the capital under U.S rule (1867–1900). Points of interest include Sitka National Historical Park, scene of a decisive battle (1804) between the Russians and the Indians, and the Russian Orthodox Cathedral of St. Michael (built 1844–48).

Sit·ting·bourne and Mil·ton (sĭt'ĭng-bôrn'; mĭl'tən), urban district (1974 est. pop. 31,960), Kent, SE England, on the old route of pilgrimage to Canterbury. It has paper, paint, and cement works.

Sit·twe (sĭt'wē), city (1969 est. pop. 81,000), capital of Arakan division, W Burma, at the mouth of the Kaladan River on the Bay of Bengal. It is an important port and rice-milling center. Originally a small fishing village, it became a port for the export of rice after the British occupied it in 1826.

Si·vas (sĭ-väs'), city (1970 pop. 133,979), capital of Sivas prov., central Turkey, on the Kızıl Irmak. A trade and manufacturing center, it has cement, textile, and rug factories. Iron ore is mined nearby. An important city of Asia Minor under the Romans, Byzantines, and Seljuk Turks and part of the Seljuk empire of Rum in the 12th cent., Sivas fell to the Mongols and later (15th cent.) to the Ottoman Turks.

Si·vash Sea (sē-väsh'), salt lagoon, c.1,000 sq mi (2,590 sq km), SW European USSR, in the Ukraine, extending along the NE coast of the Crimea. It is separated—except at the Genichesk Strait—from the Sea of Azov by the Tongue of Arabat, a narrow sandspit; the Perekop Isthmus separates it from the Black Sea in the north. The water has been the source of table salt and of several chemicals.

Si·wah or **Si·wa** (both: sē'wə), oasis, c.35 sq mi (90 sq km), NW Egypt, in the Libyan Desert. Dates and tea are grown in the oasis. It was the seat of the temple and oracle of Zeus Amon.

Si·wa·lik Hills (sĭ-wä'lĭk), southernmost range of the Himalayas, S central Asia, extending c.1,050 mi (1,690 km) from SW Kashmir through N India into S Nepal parallel to the main range. The Siwalik Hills are noted for their vast vertebrate fossil beds.

Sjael·land (shĕl'än') or **Zea·land** (zē'lənd), island (1965 pop. 2,055,-040), 2,709 sq mi (7,016.3 sq km), E Denmark, between the Kattegat and the Baltic Sea. Denmark's largest island, it is separated from Fyn by the Store Baelt and from Sweden by the Øresund. It includes most of Copenhagen, the Danish capital. Wheat growing, dairy farming, cattle breeding, and fishing are important occupations.

Ska·gen (skä'gən), city (1976 est. pop. 13,806), Nordjylland co., N Denmark, a port on Skagens Odde peninsula at the N end of Jylland and on the Kattegat. It has fisheries, shipyards, and fish canneries.

Skag·er·rak (skăg'ər-ăk'), strait, c.150 mi (240 km) long and 85 mi (135 km) wide, between Norway and Denmark, linking the North Sea and the Baltic Sea by way of the Kattegat.

Skag·it (skăj'ĭt), county (1970 pop. 52,381), 1,735 sq mi (4,493.7 sq km), NW Wash., bounded in the W by Rosario Strait, rising in the E to the Cascade Range, and drained by the Skagit River; formed 1883; co. seat Mount Vernon. Its agriculture includes fruit, vegetables, seed crops, bulbs, dairy products, cattle, and poultry. It has fishing, mining, lumbering, and diversified industry (food processing, boatbuilding, and oil refining). Part of Mt. Baker National Forest is here.

Skagit, river, c.150 mi (240 km) long, rising in the Cascade Range, British Columbia, and flowing SW through Wash. into Puget Sound. Gorge High Dam, Diablo Dam, and Ross Dam provide electricity for Seattle. The lakes formed behind the dams are in a 40-mi (64.4-km) stretch of river canyon that makes up Ross Lake National Recreation Area.

Skag·way (skăg'wā'), city (1970 pop. 675), Skagway-Yakutat census div., SE Alaska, at the head of Lynn Canal; founded 1897. It is the coastal terminus of the White Pass and Yukon Railway. During the gold rush (1897–98) it was a major disembarking point to the Klondike.

Ska·ma·ni·a (skə-măn'yə, -mā'nē-ə), county (1970 pop. 5,845), 1,672 sq mi (4,330.5 sq km), SW Wash., on the Oregon border in a mountain area watered by the Lewis and Columbia rivers; formed 1854; co. seat Stevenson. Its agriculture includes cattle, dairy products, fruit, and nuts. It has lumbering and manufactures wood products. Part of Columbia National Forest is here.

Ska·ra (skä'rä'), city (1970 pop. 10,284), Skaraborg co., S Sweden. It has industries that manufacture metal goods and footwear. Dating from at least the 9th cent., Skara is one of Sweden's oldest cities and has been an educational center since 1641, when Sweden's earliest institutions of higher learning were founded here.

Skar·a Brae (skâr'ə brā'), Stone Age village, in the Orkney Islands, N Scotland. Dating from c.2000 to 1500 B.C., the village was preserved under a sand dune until uncovered by a storm in 1851. It contains seven underground chambers furnished with stone dressers, tables, and beds.

Skee·na (skē'nə), river, c.360 mi (580 km) long, rising in the Stikine Mts., W British Columbia, Canada, and flowing S and SW to the Pacific Ocean near Prince Rupert. It is navigable for c.100 mi (160 km) upstream.

Skel·lef·te·å (shə-lĕf'tə-ō'), city (1972 est. pop. 41,262), Västerbotten co., NE Sweden, on the Gulf of Bothnia at the mouth of the Skellefte-älv River. The center of a rich mining region (copper, lead, gold, silver, and zinc), the city has several smelters and refineries. First chartered in 1621, it was rechartered in 1845 when its modern growth began.

Skel·lef·te älv (shə-lĕf'tə ĕlv'), river, c.255 mi (410 km) long, rising in Norrbotten co., N Sweden, and flowing SE to the Gulf of Bothnia at Skellefteå.

Skel·ligs (skĕl'ĭgz), rocky islands, off SW Republic of Ireland, in Co. Kerry, comprising Lemon Rock, Little Skellig, and Great Skellig. Great Skellig has a lighthouse and ruins of a monastery.

Skel·mers·dale (skĕl'mərz-dāl'), new town (1971 pop. 26,681), Lancashire, NW England. The area of the new town lies wholly within the area of the new urban district of Skelmersdale and Holland, created in 1968.

Skid·daw (skĭd'ô'), mountain, 3,054 ft (931.5 m) high, Cumbria, NW England, in the Lake District.

Ski·en (shē'ən, shā'-), city (1974 est. pop. 46,032), capital of Telemark co., SE Norway, a port on the Skienselva River. Manufactures include processed food, forest products, and footwear.

Skier·nie·wi·ce (skyĕr'nyə-vē'tsə), town (1970 pop. 25,600), E central Poland. It is a railway junction and manufacturing center where metals, glass, and ceramics are produced.

Skik·da (skēk'də), formerly **Phi·lippe·ville** (fē-lēp-vēl'), city (1974 est. pop. 127,968), NE Algeria, a port on the Gulf of Stora of the Mediterranean Sea. It was founded by the French in 1838 on the site of a Carthaginian colony.

Skí·ros (skîr'ôs') or **Scy·ros** (sī'rəs), island (1971 pop. 2,352), c.80 sq mi (205 sq km), E Greece, in the Aegean Sea, largest of the N Sporades. Wheat, figs, and olives are grown here, chromite and iron ore are mined, and fishing and sponge diving are important occupations. Skíros figures in many ancient legends. It was conquered c.469 B.C. by Athenians. The English poet Rupert Brooke, who died (1915) in World War I, is buried here.

Ski·ve (skē'və), city (1976 est. pop. 22,000), Viborg co., N Denmark, on the Limfjord at the mouth of the Skive River. It is a commercial center and a tourist resort.

Sko·kie (skō'kē), village (1978 est. pop. 67,400), Cook co., NE Ill., an industrial suburb adjacent to Chicago; inc. 1888. Its products include communications and electrical equipment and pharmaceuticals.

Skop·je (skôp'yə) or **Skop·lje** (skôp'lē-ə), city (1971 pop. 312,092), capital of Yugoslav Macedonia, S Yugoslavia, on the Vardar River. It is an important transportation and trade center as well as an industrial hub where metals, textiles, and glass are produced. Dating from Roman times, Skopje was captured by the Serbs in 1282 but fell to the Turks in 1392. It was taken by the Serbs in the Balkan Wars of 1912–13 and was included in Yugoslavia in 1918. Much of the city was rebuilt after a disastrous earthquake in 1963.

Sköv·de (skœv'də), city (1972 est. pop. 44,721), Skaraborg co., S Sweden, midway between Lakes Vänern and Vättern. Near the city is the Gothic church where several Swedish kings are buried.

Skow·he·gen (skou-hē'gən), town (1970 pop. 6,571), seat of Somerset co., central Maine, on the Kennebec River NNE of Waterville; settled 1771, set off from Canaan in 1823. It has varied manufactures and is a trade center for a resort and farming region. The Skowhegan State Fair is one of the oldest agricultural fairs in the country.

Skunk (skŭngk), river, 264 mi (424.8 km) long, central and SE Iowa, rising in NE Hamilton co. and flowing SE to the Mississippi River S of Burlington.

Skye (skī), island (1971 pop. 7,372), 670 sq mi (1,735 sq km), one of the Inner Hebrides, Inverness-shire, NW Scotland. It has an irregular coastline, and many of its lochs are rimmed by lofty, sheer precipices. Sheep and cattle raising, crofting, wool weaving, whiskey distilling, and fishing are the chief industries; diatomite is mined in the northeastern region. Gaelic is spoken.

Sla·gel·se (slā'yəl-sə), city (1976 est. pop. 27,900), Vestsjaelland co., S central Denmark. It is an industrial center and a rail junction.

Sla·ton (slăt'n), city (1970 pop. 6,583), Lubbock co., NW Texas, just SE of Lubbock in the Llano Estacado. It is a railroad division point and the trading, processing, and shipping center of a farm area.

Slave (slāv), river, c.310 mi (500 km) long, S Mackenzie dist., Northwest Territories, Canada. It comprises the middle sections of the Mackenzie River system and channels the waters of Lake Athabasca and the Peace River into Great Slave Lake at Fort Resolution. It is navigable for steamers except for the rapids between Fort Fitzgerald and Fort Smith.

Slave Coast, name given by European traders to the coast bordering the Bight of Benin on the Gulf of Guinea, W Africa. It was the principal source of slaves from West Africa from the 16th cent. to the mid-19th cent.

Sla·vo·ni·a (slə-vō'nē-ə), historic region, N Yugoslavia, now part of Croatia. It is between the Drava River in the north and the Sava River in the south. Wheat and corn are the major crops, and the leading industry is food processing. The region was originally part of the Roman province of Pannonia. In the 7th cent. a Slavic state was established here. With Croatia, Slavonia was united with Hungary in 1102, came under Turkish rule in the 16th cent., and was recovered by Hungary from the Turks (1699). As a result of the Revolution of 1848 Slavonia was made an Austrian crown land, but in 1868 it was restored to Hungary and was united with Croatia. It has been part of Yugoslavia since 1918.

Slav·yansk (sləv-yänsk'), city (1973 est. pop. 131,000), S European USSR, in the Ukraine, in the Donets Basin. Founded in 1676, the city is a railroad junction and has salt and soda works and machine and ceramic plants.

Slay·ton (slāt'n), village (1970 pop. 2,351), seat of Murray co., SW Minn., E of Pipestone, in a grain and livestock area; settled 1880, inc. 1887. A state park and fish hatchery are nearby.

Sleep·ing Bear Dunes National Lakeshore (slē'pĭng bâr'): *see* National Parks and Monuments Table.

Sli·dell (slī-dĕl'), city (1975 est. pop. 19,500), St. Tammany parish, SE La., near Lake Pontchartrain; inc. 1888. Originally a shipbuilding and brick-manufacturing town in a farm and timber region, it is now primarily residential.

Sli·go (slī'gō), county (1971 pop. 50,275), 694 sq mi (1,797 sq km), N Republic of Ireland; co. town Sligo. The interior of the county is mountainous. Cattle raising is the chief occupation.

Sligo, urban district (1971 pop. 14,080), county town of Co. Sligo, N Republic of Ireland, at the mouth of the Garavogue River on Sligo Bay. It is a seaport and fishing center, with a woolen trade and other industries. There are remains of a Dominican monastery (Sligo Abbey) built in the 13th cent.

Sli·ven (slē'vĕn), city (1969 est. pop. 81,100), E central Bulgaria, at the foot of the Balkan Mts. A textile center, it also produces textile machinery, glass, electrical goods, wood and metal products, foodstuffs, and wine. Coal is mined nearby.

Slope (slōp), county (1970 pop. 1,484), 1,225 sq mi (3,172.8 sq km), SW N.Dak., bordering on Mont. and drained by the Little Missouri River; formed 1915; co. seat Amidon. It is in a rich agricultural area yielding livestock, wheat, and hay; there are also extensive lignite deposits in the county.

Slough (slou), municipal borough (1974 est. pop. 88,420), Berkshire, central England. After World War I Slough underwent rapid industrial development.

Slo·vak·i·a (slō-văk'ē-ə, -vä'kē-ə), constituent land (18,917 sq mi/ 48,995 sq km) of Czechoslovakia. Bratislava is the capital. Slovakia is bordered by Moravia in the west, Austria in the southwest, Hungary in the south, the Ukraine in the east, and Poland in the north. Most of Slovakia is traversed by the Carpathian Mts., including the Tatra and the Beskids. Gerlachovka (8,737 ft/2,663 m), in the High Tatra, is the highest peak. Southern Slovakia is a part of the Little Alföld, a plain. Its fertile soil is drained by the Danube and its tributaries, notably the Váh. The mountainous part of Slovakia has vast forests and pastures, used for intensive sheep grazing, and is rich in mineral resources, including high-grade iron ore, copper, mercury, gold, silver, lead, and lignite. There are also numerous mineral springs, notably at Piešt'any, and many popular resorts. Farms, vineyards, orchards, and pastures for stock form the basis of southern Slovakia's economy. Mining, shipbuilding, and agricultural and metal processing have become important industries since World War II. Major Slovakian cities include Bratislava and Komarno, which are the major Danubian ports, and Košice, Trnava, and Nitra. The Slovaks comprise about 87% of the population; other groups include Hungarians (about 10%), Ukrainians, and Czechs. Roman Catholicism is the chief religion. Although the Slovaks and the Czechs are ethnically and linguistically related, they have been politically and culturally separate for 1,000 years.

History. The area now constituting Slovakia was settled by Slavic tribes in the 5th–6th cent. A.D. In the 9th cent. Slovakia formed part of the empire of Moravia. From the Magyar conquest of Slovakia early in the 10th cent. until 1918 Slovakia was generally under Hungarian rule. After the Ottoman Turkish victory at Mohács in 1526 over Louis II of Hungary and Bohemia, Slovakia, along with western Hungary, fell under Hapsburg rule. It thus escaped Turkish domination but became a stronghold of the great Hungarian nobles. In the 18th cent. Maria Theresa and Joseph II pursued religious freedom and social reform in Slovakia but greatly intensified Germanization. This policy spurred a Slovak national revival, which grew steadily in the 19th cent. During the anti-Hapsburg revolutions of 1848 the Slovaks formulated a set of demands for increased political and linguistic rights. After the establishment of the dual Austro-Hungarian Monarchy in 1867, Magyarization intensified, thus further heightening Slovak nationalism.

On Oct. 30, 1918, the Slovak National Council formally proclaimed independence from Hungary and incorporation into Czechoslovakia. Slovak nationalists were dissatisfied with the new republic, in which Slovakia held the status of a simple province; they demanded full autonomy on a basis of complete equality for both Czechs and Slovaks. After the Munich Pact of 1938 Slovakia became an autonomous state within reorganized Czechoslovakia. In 1939 Slovakia became a nominally independent state under German protection and entered World War II as Germany's ally. Soviet troops, with the powerful aid of Slovak partisans, drove the Germans out of Slovakia late in 1944.

The Allied victory in 1945 restored Slovakia to its territorial status before the Munich Pact, and the constitution of 1948 recognized Slovakia as one of the constituent states of a re-established Czechoslovakia. The accession in 1948 of a Communist government revived the old antagonism between Czechs and Slovaks. The liberal Communist regime of Alexander Dubček, which came into power in 1967, responded to Slovak discontent by promising federalization of the republic. The new Socialist Federal Republic came into being on Jan. 1, 1969; the constituent Czech and Slovak republics received autonomy over local affairs.

Slo·ve·ni·a (slō-vē′nē-ə), constituent republic of Yugoslavia (1971 pop. 1,725,088), 7,817 sq mi (20,246 sq km), NW Yugoslavia; capital Ljubljana. Most of Slovenia is situated in the Karst plateau and in the Julian Alps. Although farming and livestock raising are the chief occupations, Slovenia is the most industrialized and urbanized of all the Yugoslav republics. Iron, steel, and aluminum are produced, and there are mineral resources of oil, coal, and mercury. Until 1918 the region was largely comprised in the Austrian crown lands of Carinthia, Carniola, and Styria. In ancient times this region was inhabited by Illyrian and Celtic tribes. In the 1st cent. B.C. they fell under the Roman provinces of Pannonia and Noricum. The region was settled in the 6th cent. A.D. by the South Slavs, who set up an early Slav state that in 788 passed to the Franks. In 843 the region passed to the dukes of Bavaria. In 1335 Carinthia and Carniola passed to the Hapsburgs. From that time until 1918 Slovenia was part of Austria. In 1918 Slovenia was included in the kingdom of Serbs, Croats, and Slovenes (called Yugoslavia after 1929), and in 1919 Austria formally ceded the region. In World War II Slovenia was divided (1941) among Germany, Italy, and Hungary. After the war, Slovenia was made (1945) a constituent republic of Yugoslavia.

Sluis (slois), municipality, Zeeland prov., SW Netherlands, near the Belgian border. In 1340 Edward III of England defeated the fleet of Philip VI of France off Sluis in the first important engagement of the Hundred Years' War.

Słupsk (slo͞opsk), city (1973 est. pop. 73,500), NW Poland. It is a rail junction and commercial center, with food-processing plants, breweries, distilleries, and industries manufacturing metals, farm machinery, furniture, and chemicals. It was chartered in 1310.

Sme·de·re·vo (smĕ′dĕ-rĕ-vô), town (1971 pop. 40,289), E Yugoslavia, in Serbia, a port on the Danube River. Its industries include oil refining, steel manufacturing, and winemaking. Dating from Roman times, Smederevo was the capital of Serbia in the 15th cent.

Smeth·port (smĕth′pôrt′), borough (1970 pop. 1,883), seat of McKean co., NW Pa., SE of Bradford, in a wooded area; laid out 1807, inc. 1853. The region is agricultural and has coal mines and oil wells. Chemicals are manufactured.

Smith (smĭth). **1.** County (1970 pop. 6,757), 893 sq mi (2,312.9 sq km), N Kansas, bordering in the N on Nebr. and drained by North Fork of Solomon River, with the geographic center of the U.S. (39°50′ N 98°35′ W) 2 mi (3.2 km) NW of Lebanon; formed 1867; co. seat Smith Center. It is in the corn-belt region, with livestock raising and some manufacturing. **2.** County (1970 pop. 13,561), 642 sq mi (1,662.8 sq km), S central Miss., drained by the Leaf and Strong rivers; formed 1833; co. seat Raleigh. Farming (corn and cotton), lumbering, and bentonite mining are major occupations here. **3.** County (1970 pop. 12,509), 323 sq mi (836.6 sq km), N central Tenn., in the central basin, drained by the Cumberland River and its tributaries; formed 1799; co. seat Carthage. Its agriculture includes tobacco and corn farming, livestock raising, and dairying. It has lumbering and other industries (clothing, furniture, shoes, and aluminum). **4.** County (1970 pop. 97,096), 934 sq mi (2,419.1 sq km), E Texas, bounded on the N by the Sabine River, on the W by the Neches River; formed 1846; co. seat Tyler. It has diversified farming, rose growing, livestock raising, and dairying, with oil fields and deposits of iron ore, salt, silica, and lignite.

Smith Center, city (1970 pop. 2,389), seat of Smith co., N central Kansas, near the Nebr. line NW of Salina, in a corn and dairy area; founded 1871, inc. 1886.

Smith·field (smĭth′fēld′). **1.** Town (1970 pop. 6,677), seat of Johnston co., E central N.C., SE of Raleigh on the Neuse River; settled before 1746, inc. 1861. It processes tobacco and lumber and makes apparel. **2.** Town (1970 pop. 13,468), Providence co., N R.I.; set off from Providence and inc. 1731. Long a textile town, it now has diversified industries. It was settled early in the 18th cent., mainly by Friends. **3.** Town (1970 pop. 2,713), Isle of Wight co., SE Va., NW of Portsmouth; settled c.1633, inc. 1752. It is noted for its hams. St. Luke's Church nearby was built in either 1632 or 1682 and is one of the oldest Protestant churches in America. Also nearby is Bacon's Castle (1655), a house seized and fortified by Nathaniel Bacon's followers in Bacon's Rebellion.

Smith·land (smĭth′lənd), city (1970 pop. 514), seat of Livingston co., W Ky., on the left bank of the Ohio River at the mouth of the Cumberland River ENE of Paducah. In an agricultural area yielding burley tobacco and corn, it has fluorspar mines and stone quarries.

Smith·ville (smĭth′vĭl′), town (1970 pop. 2,997), seat of De Kalb co., central Tenn., N of McMinnville, in a farm, tobacco, and timber region; inc. 1843. It has lumber mills, gristmills, and a number of nurseries.

Smok·y (smō′kē), river, c.250 mi (400 km) long, rising in Jasper National Park, W Alta., Canada, and flowing generally NE to the Peace River. It was discovered (1792) by Alexander Mackenzie.

Smoky Hill, river, c.560 mi (900 km) long, rising on the Great Plains, E Colo., and flowing E across Kansas to join the Republican River and form the Kansas River at Junction City. The Smoky Hill basin is included in the Missouri River basin project.

Smo·lensk (smō-lĕnsk′, smô-), city (1973 est. pop. 234,000), capital of Smolensk oblast, W central European USSR, a port on the Dnepr River. It is an important rail junction, a distribution point for the region's agricultural products, and a commercial, cultural, and educational center. Smolensk has one of the USSR's largest flax-processing mills. Other industries include metalworking, machine building, flour milling, food processing, and the manufacture of textiles. One of Russia's oldest cities, Smolensk was already a commercial center in the late 9th cent. Its control of the key portages between the Dnepr and Western Dvina rivers and its connection with the Hanseatic cities of the Baltic Sea and with Moscow gave Smolensk its early strategic importance. The city declined in the 11th cent. but revived in the 12th cent. Smolensk was sacked by the Mongols in 1238–40. The westward expansion of the grand duchy of Moscow made Smolensk a target of prolonged struggle between Moscow and Poland-Lithuania; it was ultimately passed to the Russians by the Treaty of Andrusov (1667). Its location made Smolensk a target for Napoleon I, who seized the city in Aug., 1812, but was forced to retreat a month later. The city, scene of some of World War II's heaviest fighting, was captured by the Germans in 1941 and retaken by Soviet troops in 1943.

Smyr·na (smûr′nə): *see* İzmir, Turkey.

Smyrna. **1.** City (1978 est. pop. 17,600), Cobb co., NW Ga., a residential suburb of Atlanta; inc. 1872. The city grew with the coming of the railroad. Smyrna was almost totally destroyed during the Civil War. **2.** Town (1970 pop. 5,698), Rutherford co., central Tenn., SE of Nashville; inc. 1869. The Sam Davis home (restored as a Confederate shrine) and Stewart Air Force Base are nearby.

Smyth (smĭth), county (1970 pop. 31,349), 435 sq mi (1,126.7 sq km), SW Va., in the Great Appalachian Valley, drained by headstreams of the Holston River; formed 1832; co. seat Marion. Its agriculture includes cabbage, fruit, dairy products, and livestock. It has limestone quarries and some salt and gypsum mines. Its industry includes food processing, textile milling and the manufacture of clothing, furniture, wood and clay products, and machinery. It is traversed by the Appalachian Trail.

Snake (snāk), river, 1,038 mi (1,670 km) long, NW United States, the chief tributary of the Columbia. It rises in northwest Wyo., in Yellowstone National Park, flows through Jackson Lake in Grand Teton National Park, then south and west into Idaho and northwest to the Henrys Fork River. The combined stream runs southwest, then northwest, crossing Idaho through the Snake River plain. The Snake makes a bend into Oregon and turns north to form the Idaho-Oregon and Idaho-Wash. lines (receiving several tributaries), then turns at Lewiston, Idaho and flows generally west to join the Columbia River near Pasco, Wash. Hell's Canyon is the greatest of the Snake's many gorges and one of the deepest in the world. Extending c.125 mi (200 km) north along the Oregon-Idaho line, it reaches a maximum depth of c.7,900 ft (2,410 m). The Snake was discovered by the Lewis and Clark expedition (1803–6) and was of major importance in U.S. expansion into the Pacific Northwest. It is a major source of electricity, having numerous hydroelectric power plants. The upper and middle courses of the Snake and its tributaries are much used for irrigation.

Sneed·ville (snēd′vĭl′), village (1970 pop. 874), seat of Hancock co., NE Tenn., NE of Knoxville, in a mountainous farm area.

Snef·fels, Mount (snĕf'əlz), peak 14,143 ft (4,313.6 m) high, SW Colo. in the San Juan Mts. of the Rockies, W of Ouray.

Sno·ho·mish (snō-hō'mĭsh), county (1970 pop. 265,236), 2,098 sq mi (5,433.8 sq km), NW central Wash., drained by the Skykomish and Snoqualmie rivers; formed 1861; co. seat Everett. Its agriculture includes dairy products, peas, corn, fruit, and poultry. It has gold, silver, copper, and nickel mining. Its diversified industry includes lumbering, food processing, boatbuilding, and the manufacture of concrete, metal products, machinery, and aircraft equipment. The Cascade Range is in the east. Parts of Mt. Baker and Snoqualmie national forests are here.

Snow·don (snōd'n), highest mountain of Wales, 3,560 ft (1,085.8 m) high, Caernarvonshire, NW Wales. Its five peaks are separated by passes. There is a rack and pinion railway (opened 1896) from Llanberis to the summit.

Snow Hill (snō), town (1970 pop. 2,201), seat of Worcester co., Md., Eastern Shore, on the Pocomoke River SE of Salisbury; founded 1642, inc. 1812. It is a trade, processing, and shipping center of a farm region.

Snow·y Mountains (snō'ē), range of the Australian Alps, SE Australia. It is the site of the Snowy Mts. Hydroelectric Scheme, designed to double Australia's power output and to quadruple the irrigated area of the Murray basin by 1980.

Sny·der (snī'dər), county (1970 pop. 29,269), 327 sq mi (846.9 sq km), central Pa., bounded in the E by the Susquehanna River; formed 1855; co. seat Middleburg. It is in a fruit-growing and dairying area that also raises cattle and hogs. Its manufactures include clothing, lumber and wood products, furniture, concrete, electrical supplies, and metal ornaments.

Snyder, city (1970 pop. 11,171), seat of Scurry co., NW Texas, in a prairie and mesquite region; inc. 1907. Oil production is the city's main industry; natural gas is also refined and processed. Ranching and farming are important, and cotton, grain, sorghum, wheat, fruits, and vegetables are the chief crops.

So·bat (sō'bät), river, c.200 mi (320 km) long, on the Ethiopia-Sudan border, E Africa. It flows generally northwest through southeast Sudan to the White Nile at Taufikia. It is navigable in the flood season.

So·chi (sô'chē), city (1973 est. pop. 241,000), Krasnodar Kray, S European USSR, on the E shore of the Black Sea, in the foothills of the Caucasus. It was established as a spa in 1910.

Society Islands (sə-sī'ə-tē), island group (1970 pop. 105,328), South Pacific, a part of French Polynesia. The group comprises the Windward Islands and the Leeward Islands (total land area c.650 sq mi/1,680 sq km), two clusters of volcanic and coral islands lying in a 450-mi (724-km) chain. Only eight of the islands are inhabited. The islands are mountainous, and there are breadfruit, pandanus, and coconut trees; the limited fauna includes wild pigs, rats, and small lizards. The major products are copra, sugar, rum, mother-of-pearl, and vanilla. The Society Islands were discovered in 1767 by the English navigator Samuel Wallis, who claimed them for Great Britain. A year later, however, the French navigator Louis Antoine de Bougainville established a French claim. They were named the Society Islands in 1769 by Capt. James Cook. The group became a protectorate of France in 1843 and a colony in 1880. In 1946 the Society Islands were made an overseas territory of France.

So·cor·ro (sə-kôr'ō), county (1970 pop. 9,763), 6,603 sq mi (17,101.8 sq km), central N.Mex., drained by the Rio Puerco and Rio Grande; formed 1852; co. seat Socorro. It has grain farming and livestock grazing. It includes Gran Quivira National Monument, parts of Cibola National Forest, and several mountain ranges.

Socorro, city (1970 pop. 5,849), seat of Socorro co., W central N.Mex., on the Rio Grande S of Albuquerque, in a livestock, farm, and mine area. The New Mexico Institute of Mining and Technology is here.

So·co·tra (sə-kō'trə, sō-), island, 1,383 sq mi (3,582 sq km), Southern Yemen, at the mouth of the Gulf of Aden. The mountainous interior rises to c.5,000 ft (1,525 m). The island's inhabitants farm, fish, and herd; exports include dried fish, aloes, ghee, and pearls. Known to the ancient Greeks, Socotra was occupied by the East India Company in 1834, and in 1886 it became part of Britain's Aden protectorate. In 1967 Socotra was joined to the newly formed nation of Southern Yemen.

So·da Springs (sō'də), city (1974 est. pop. 3,487), seat of Caribou co., SE Idaho, near the Bear River ESE of Pocatello, in a grain and livestock region. Founded in 1863 by a dissenting Mormon sect, it was abandoned and refounded in 1870 by other Mormons. It is now a shipping point in a cattle and sheep area.

Sö·der·hamn (sœ'dər-hä'mən), city (1970 pop. 12,467), Gävleborg co., E Sweden, a seaport on the Gulf of Bothnia; chartered 1620. It is a commercial and industrial center. Iron ore, forest products, and fish are exported.

Sö·der·täl·je (sœ'dər-tĕl'yə), city (1972 est. pop. 77,875), Stockholm co., E Sweden, on a narrow bay of Lake Mälaren, near Stockholm. It is an industrial center and a health resort. Manufactures include motor vehicles, machinery, tobacco products, and pharmaceuticals. It was a trade center during the Viking era (9th–11th cent.).

Soest (zōst), city (1972 est. pop. 40,338), North Rhine-Westphalia, W West Germany. It is a manufacturing city and an agricultural trade center. Known in the 7th cent., Soest was the chief town of Westphalia in the Middle Ages and was a flourishing member of the Hanseatic League. The city passed to Brandenburg in 1614. Soest was badly damaged in World War II. It has many noteworthy medieval and Renaissance buildings.

So·fi·a (sō-fē'ə, sō'fē-ə), city (1972 est. pop. 920,000), capital of Bulgaria, W central Bulgaria, on a high plain surrounded by the Balkan Mts. Among its manufactures are engineering and metal products, machinery, textiles, rubber and leather goods, furniture, footwear, chemicals, and bricks and tiles. A Thracian settlement once occupied the site of Sofia. It was taken by the Romans in 29 A.D. and flourished under them. Destroyed by the Huns in 447, the city was rebuilt (6th cent.) by Byzantine emperor Justinian I. It formed part of the first (809–1018) and second (1186–1382) Bulgarian kingdoms. Sofia passed to the Ottomans in 1382. Taken by the Russians in the Russo-Turkish War of 1877–78, it became (1879) the capital of newly independent Bulgaria. During World War II the Russians captured Sofia from the Germans (1944) and installed a Communist government. The city has a university (founded 1889) and numerous other educational and cultural facilities.

Sog·na·fjord or **Sog·ne Fjord** (both: sông'nə-fyôr'), inlet of the Norwegian Sea, SW Norway, in Sogn og Fjordane co. Extending c.110 mi (180 km) inland and reaching a depth of c.4,000 ft (1,220 m), it is the longest and deepest fjord in Norway. In some places the mountains drop a sheer 3,000 ft (915 m) to the water on both sides of the fjord. It is a popular tourist area with many well-known resorts.

Sois·sons (swä-sôɴ'), city (1968 pop. 25,890), Aisne dept., N France, on the Aisne River. It is an agricultural and industrial center. Soissons was an old Roman town. Clovis I defeated the Roman legions at Soissons in 486, and the city was the capital of several Merovingian kings (5th–7th cent.). Pepin the Short dethroned Childeric III here in 751. Throughout the 19th and 20th cent. the city was the scene of warfare, culminating in the German invasion of 1940. Part of the Abbey of Saint-Jean-des-Vignes survives, as does the nearby Abbey of St. Médard, a burial place of Merovingian kings.

So·ka (sō'kä), city (1970 pop. 123,269), Saitama prefecture, E central Honshu, Japan. It is a suburb of Tokyo.

So·ko·to (sō-kō'tō, sō'kə-tō), city (1969 est. pop. 104,000), NW Nigeria, on the Sokoto River. It is the commercial center for a wide region and a collection place for hides, skins, and groundnuts. It also has cement, pottery, and leather tanning and dyeing industries. The city is a place of pilgrimage for Moslems.

So·la·no (sō-lä'nō), county (1970 pop. 171,989), 823 sq mi (2,131.6 sq km), central Calif., bounded on the SW by San Pablo Bay, on the S by Carquinez Strait and Suisun Bay, on the SE by Sacramento River, formed 1850; co. seat Fairfield. Its agriculture includes stock raising (sheep), dairying, and asparagus, truck, grain, sugar beets, alfalfa, fruit, and nut farming. It has natural-gas wells, lumber, extensive food processing, concrete manufacturing, and shipbuilding. There is a waterfowl hunting in the county.

Sol·dot·na (sŏl-dŏt'nə), city (1970 pop. 1,202), seat of Kenai Peninsula borough, S central Alaska.

Sol·e·dad (sŏl'ə-dăd'), city (1970 pop. 4,222), Monterey co., W Calif., SE of Salinas, in a farm and dairy area; inc. 1921. The ruins of Mission La Soledad (1791) and a state prison are here. Pinnacles National Monument is nearby.

So·lent, The (sō'lənt), channel, c.30 mi (50 km) long and .75 to 5 mi (1.2–8 km) wide, between the Isle of Wight and Hampshire, S England. It serves as an anchorage for ships entering Southampton.

Sol·fe·ri·no (sōl-fə-rē'nō), village (1971 pop. 1,811), Lombardy, N Italy, near Mantua. Here, on June 24, 1859, the French and Sardinians fought a bloody battle with the Austrians. Although the battle resulted in no clear decision, the Austrians withdrew to their strategic fortresses. Napoleon III, shocked by the huge losses and aware of the difficulties of continuing the war, soon afterward met Emperor Francis Joseph of Austria at Villafranca Di Verona, where a preliminary peace was arranged.

So·li (sō'lī), ancient city of Cilicia, SW of Tarsus, in present-day Turkey. It was founded c.700 B.C. by colonists from Rhodes. An important port at the time of Alexander the Great, Soli was destroyed in the 1st cent. B.C. by Tigranes of Armenia. It was rebuilt by Pompey.

So·li·hull (sō'lĭ-hŭl'), county borough (1974 est. pop. 107,800), West Midlands, central England, a mainly residential suburb of Birmingham. Automobiles, chemicals, and tools are made.

So·ling·en (zō'lĭng-ən), city (1972 est. pop. 176,734), North Rhine-Westphalia, W West Germany, on the Wupper River opposite Remscheid. It is a major center of the West German cutlery industry. Solingen steel, used in making knives, scissors, razors, and surgical instruments, is famous for its excellence.

Sol·na (sōl'nä), city (1972 est. pop. 55,430), Stockholm co., E Sweden, an industrial suburb of Stockholm. Manufactures include machinery, electrical goods, paper, and chocolate. It is the seat of the Swedish motion-picture industry and has numerous scientific institutes, including the Nobel Institute.

Sol·o·mon Islands (sŏl'ə-mən), independent volcanic group (1978 est. pop. 215,000), c.15,500 sq mi (40,150 sq km), SW Pacific, E of New Guinea. The 900-mi (1,448-km) chain formerly comprised the British Solomon Islands Protectorate (1970 pop. 161,524; c.11,500 sq mi/29,790 sq km) and the northernmost islands of Bougainville and Buka, which still belong to Papua New Guinea. The Solomons are sparsely populated and are mountainous and heavily wooded. The only significant export crops are copra and timber. There is very little manufacturing. A Spanish explorer, Álvaro de Mendeña de Neira, was the first European to visit the islands (1568). European settlers and missionaries arrived throughout the 18th and 19th cent.

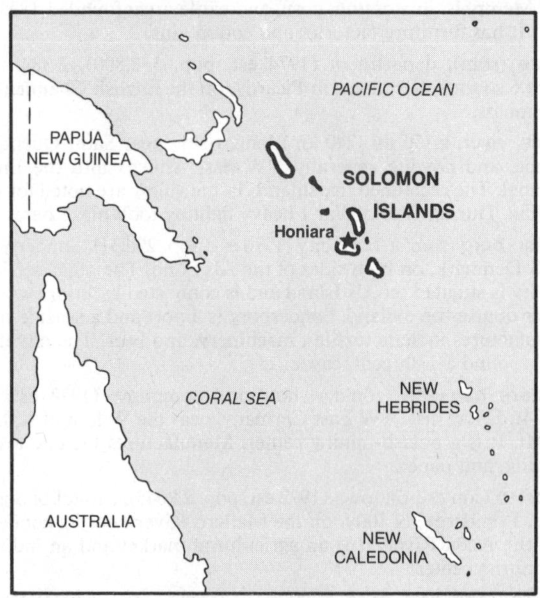

In 1885 the German New Guinea Company established control over the northern Solomons. The southern islands were placed under British protectorate in 1893; the eastern islands in 1898. In 1900 Germany transferred its islands (except Bougainville and Buka) to Great Britain in return for British withdrawal from Western Samoa. Bougainville and Buka were occupied by Australian forces during World War I and were placed under Australian mandate by the League of Nations in 1920. During World War II, Choiseul, New Georgia, Ysabel, and Guadalcanal were occupied by the Japanese (1942) but were liberated by U.S. forces (1943–44). The islands became independent in 1978.

So·lon (sō'lən), city (1970 pop. 11,147), Cuyahoga co., NE Ohio, a suburb of Cleveland; founded 1820, inc. as a city 1960. Its manufactures include metal products, machinery, electrical equipment, tools, containers, and electronic components.

So·lo·thurn (zō'lō-to͝orn'), canton (1974 est. pop. 227,000), 306 sq mi (793 sq km), NW Switzerland. Very irregular in shape, Solothurn lies mostly in the Jura Mts. Cereals are grown and cattle are raised. Industry is largely concentrated in Solothurn (1974 est. pop. 16,400), the capital. Situated on the Aare, Solothurn was a Roman settlement. It had been a free town of the Holy Roman Empire since 1218 and was admitted to the Swiss Confederation in 1481. The town retains much of its historic character. Electrical equipment, watches, and metal goods are made.

So·lo·vets·ki Islands (sŏl'ə-vĕt'skē), archipelago, c.150 sq mi (390 sq km), N European USSR, in the White Sea at the entrance of Onega Bay. It was used as a military fortress against Sweden in the 16th and 17th cent. From the reign of Ivan IV until 1956 the islands were a place of exile for criminals and for political and religious prisoners.

So·lu·tré-Pouil·ly (sô-lü-trā'po͞o-yē'), village (1968 pop. 378), Saône-et-Loire dept., E central France, in Burgundy. It is known for its white wines. It is the site of a rock shelter and a burial place of prehistoric man (discovered 1867).

Sol·vay (sŏl'vā'), village (1970 pop. 8,280), Onondaga co., central N.Y., WNW of Syracuse; inc. 1894. Castings, chemicals, and electrical fixtures are produced.

Sol·way Firth (sŏl'wā), arm of the Irish Sea, c.40 mi (65 km) long, separating NW England from SW Scotland. Its salmon fisheries are important. Hadrian's Wall terminated at Bowness, on the south shore.

So·ma·li Democratic Republic (sō-mä'lē) or **So·ma·li·a** (sō-mä'lē-ə, -mäl'yə), republic (1977 est. pop. 3,290,000), 246,200 sq mi (637,657 sq km), extreme E Africa, directly S of the Arabian Peninsula across the Gulf of Aden. Mogadisho is the capital. Somalia comprises almost the entire African coast of the Gulf of Aden and a longer stretch on the Indian Ocean. It is bounded on the northwest by the Djibouti Republic, on the west by Ethiopia, on the southwest by Kenya, and on the south and east by the Indian Ocean. Arid, semi-

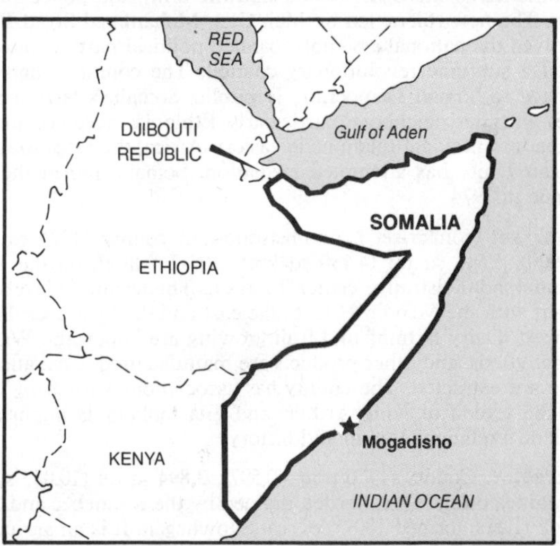

desert conditions make the country relatively unproductive. In most areas a barren coastal lowland (widest in the south) is abruptly succeeded by a rise to the great interior plateau, which is generally c.3,000 ft (910 m) high and stretches toward the northern and western highlands. The Juba and the Shebeli are the only important rivers. Somalia has no railroads. More than 80% of the republic's population is made up of Somali, who speak a Cushitic language and are Sunni Moslems by faith. Islam is the state religion. Although Somali is the national tongue, Arabic, Italian, and English are used officially. There are Bantu-speaking tribes in the southwest and numerous Arabs in the coastal towns. Somalia also has Italian, Indian, and Pakistani minorities.

Economy. Pastoralism is the dominant mode of life; both nomadic and sedentary herding of camels, sheep, goats, and cattle are carried on. Live animals, hides, skins, and clarified butter (ghee) are exported. The major cash crop is bananas. Other crops include citrus fruits, peanuts, cotton, sorghum, and millet. Somalia's most valuable mineral resource is uranium. Raw material processing constitutes the bulk of Somalian industry, which includes meat and fish (notably tuna) canning, sugar refining, oilseed processing, leather tanning, and the production of cotton textiles, iron products, soap, shoes, and cement.

History. Between the 7th and 10th cent. immigrant Moslem Arabs and Persians established trading posts along Somalia's Gulf of Aden and Indian Ocean coasts. During the 15th and 16th cent. Somali warriors regularly joined the armies of the Moslem sultanates in their battles with Christian Ethiopia. British, French, and Italian imperialism all played an active role in the region in the 19th cent. The British opportunity came when Egyptian forces, having occupied much of the region in the 1870s, withdrew in 1884 to fight the Mahdi in Sudan. British penetration led to the establishment of a protectorate. France first acquired a foothold in the area in the 1860s. An Anglo-French agreement of 1888 defined the boundary between the Somalian possessions of the two countries. Italy first asserted its authority in the area in 1889 by creating a small protectorate in the central zone. In 1925 Jubaland, or the Trans-Juba (east of the Juba River), was detached from Kenya to become the westernmost part of the Italian colony. In 1936 Italian Somaliland was combined with Somali-speaking districts of Ethiopia to form a province of Italian East Africa. During World War II the British conquered Italian Somaliland. Britain ruled the combined regions until 1950. In accordance with UN decisions, Italian Somaliland, renamed Somalia, was granted internal autonomy in 1956 and independence in 1960. Pan-Somali sentiment had been disappointed, however, by Britain's return (1954) to Ethiopia of the Ogaden region, with its large Somali population. In the early years of independence the government was faced with a severely underdeveloped economy and with a strong demand for a "Greater Somalia" that would include the estimated 350,000 Somalis in adjacent areas of Kenya, French Somaliland (now Djibouti), and Ethiopia. Hostilities between Somalia and Ethiopia erupted in 1964, and Kenya also became involved in the conflict; peace was restored in 1967. In 1967 the inhabitants of French Somaliland rejected independence in favor of continued association with France. In 1969 President Abd-i-rashid Ali Shermarke was assassinated, and the army and police staged a coup. The new rulers, led by Maj. Gen. Mohammed Siyad Barrah, dissolved the national assembly, banned political parties, and established a supreme revolutionary council. The country's name was changed to Somali Democratic Republic. Somalia's territorial disputes with her neighbors, particularly Ethiopia, have continued to plague international relations in eastern Africa; the Organization of African Unity has attempted mediation. Somalia joined the Arab League in 1974.

Som·er·set (sŏm'ər-sĕt'), nonmetropolitan county (1974 est. pop. 398,900), 1,613 sq mi (4,180 sq km), SW England, on the Bristol Channel; administrative center Taunton. The terrain is level in the center, with the Mendip Hills to the east and the Quantock Hills to the west. Dairy farming and fruit growing are important. Woolens, leather goods, and other products are manufactured. Coal and limestone are extracted. The county has associations with King Alfred and the legend of King Arthur, and Glastonbury is important in England's religious legend and history.

Somerset. 1. County (1970 pop. 40,597), 3,894 sq mi (10,085 sq km), W Maine., on the Que. border, drained by the Kennebec and Sebasticook rivers; formed 1809; co. seat Skowhegan. It is an agricultural (dairy products and poultry), lumbering, and manufacturing (wood products, paper, and shoes) area. There are resorts in the Belgrade Lakes region. **2.** County (1970 pop. 18,924), 339 sq mi (878 sq km), S Md., a peninsula on the Eastern Shore, bounded on the SE by Pocomoke Sound and the Va. line, on the W by Tangier Sound of the Chesapeake Bay; formed 1666; co. seat Princess Anne. It is in a sandy, partly marshy tidewater area with agricultural districts that produce fruit, strawberries, white potatoes, truck, poultry, and dairy products. It has timber, a large seafood industry, vegetable and seafood canneries, clothing manufacturing, and cutlery production. There are popular sport-fishing, muskrat-trapping, and wild-fowl hunting areas in the county. **3.** County (1970 pop. 198,372), 307 sq mi (795 sq km), N central N.J., bounded on the NE by the Passaic River, with Appalachian ridges in the NW, and part of Watchung Mts. in the NE, and drained by branches of the Raritan River; formed 1688; co. seat Somerville. Its agriculture includes dairying, cattle raising, horse breeding, and corn, wheat, soybeans, oats, and hay farming. It has stone quarrying and diversified industry. **4.** County (1970 pop. 76,037), 1,078 sq mi (2,792 sq km), S Pa., in a mountainous area, bordered in the S by Md., with the Allegheny Mts. in the E, and drained by the Casselman and Youghiogheny rivers; formed 1795; co. seat Somerset. It is in a mining (anthracite, bituminous coal, and limestone) and agricultural (oats, buckwheat, maple products, livestock, and poultry) area. Its industries include food processing, lumbering, and the manufacture of concrete, metal products, and machinery.

Somerset. 1. City (1975 est. pop. 11,000), seat of Pulaski co., S Ky., in a farm, coal, and limestone area of the Cumberland foothills; inc. 1810. It has railroad shops and yards and diversified manufactures. **2.** Town (1975 est. pop. 18,900), Bristol co., SE Mass., on the Taunton River; settled 1677, set off from Swansea and inc. 1790. It has varied industries. **3.** Borough (1970 pop. 6,269), seat of Somerset co., SW Pa., SSW of Johnstown; settled 1771, laid out 1795, inc. 1804. It is in a resort area and has some manufacturing. There are coal mines in the region.

Somers Point, city (1970 pop. 7,919), Atlantic co., SE N.J., on Great Egg Bay NW of Ocean City; settled c.1695, inc. 1886.

Som·ers·worth (sŏm'ərz-wûrth'), city (1970 pop. 9,026), Strafford co., SE N.H., on the Salmon Falls River opposite Berwick, Maine; settled 1670, inc. as a town 1754, as a city 1893. Textiles are made.

Som·er·vell (sŏm'ər-vĕl'), county (1970 pop. 2,793), 197 sq mi (510.2 sq km), N central Texas, in a wooded, hilly region drained by the Brazos River; formed 1875; co. seat Glen Rose. This scenic resort area has some agriculture (grains, peanuts, fruit, pecans, cattle, and poultry). There is a nuclear power plant.

Som·er·ville (sŏm'ər-vĭl'). **1.** City (1978 est. pop. 77,100), Middlesex co., E Mass., a residential and industrial suburb of Boston, on the Mystic River; settled 1630, set off from Charlestown 1842, inc. as a city 1872. Slaughtering and meat-packing are its leading industries. There are a number of historical attractions from the Revolution. **2.** Residential borough (1970 pop. 13,652), seat of Somerset co., N central N.J., on the Raritan River; settled 1683, inc. as a borough 1909. It is a farm trade center. Electronic parts and pharmaceuticals are made. **3.** Town (1970 pop. 1,816), seat of Fayette co., SW Tenn., E of Memphis, in a cotton, corn, and cattle area; founded 1825, inc. 1827. It has furniture factories and cotton gins.

Somme (sôm), department (1974 est. pop. 532,800), 2,384 sq mi (6,174.6 sq km), N France, in Picardy, on the English Channel; capital Amiens.

Somme, river, c.150 mi (240 km) long, rising near Saint-Quentin, N France, and flowing generally NW past Amiens into the English Channel. The reclaimed marshlands in the valley are noted for truck farming. During World War I heavy fighting took place here.

Søn·der·borg (sœn'ər-bôr), city (1976 est. pop. 29,231), Sønderjylland co., S Denmark, on both sides of the Als Sund. The older section of the city is situated on Als Island and is connected by bridge with the newer quarter on Jylland. Sønderborg is a port and a seaside resort. Manufactures include textiles, machinery, and beer. The city developed around a 13th cent. castle.

Son·ders·hau·sen (zôn'dərs-hou'zən), commune (1973 est. pop. 23,374), Erfurt dist., SW East Germany, near the W foot of Kyffhäuser Mt. It is a potash-mining center. Manufactures include textiles, clothing, and paper.

Son·dri·o (sôn'drē-ō), town (1972 est. pop. 23,045), capital of Sondrio prov., Lombardy, N Italy, on the Mallero River near its confluence with the Adda River. It is an agricultural market and an industrial and tourist center.

Song Ba (sông' bä'), river, c.200 mi (320 km) long, rising in the Annamese Cordillera, Vietnam, and flowing S past An Tuc to Hau Bon, then SE past Son Hoa to the South China Sea at Tuy Hoa. Diverse tropical crops are grown along its lower course.

Song·hai or **Song·hay** (both: sŏng'hī'), largest of the ancient native empires of W Africa. The state was founded (c.700) by Berbers on the Middle Niger. The rulers accepted Islam c.1000. Songhai reached its greatest extent under Askia Mohammed I (c.1493–1528). After he was deposed by his son the empire slowly began to decline.

Son·ne·berg (zôn'ə-bĕrgh'), city (1973 est. pop. 29,287), Suhl dist., S East Germany; first mentioned 1317. It is the center of the Thuringian toy industry. Other products include clothing and cartons.

So·no·ma (sə-nō'mə), county (1970 pop. 204,885), 1,604 sq mi (4,154.4 sq km), W Calif., on the Pacific, in the Coast Ranges, bounded on the S by San Pablo Bay and drained by the Russian River; formed 1850; co. seat Santa Rosa. It has a large poultry industry, as well as dairying and wine-grape growing, cattle raising, and general farming. It has some mining and quarrying, lumbering, and industry. There are many vacation resorts along the lower Russian River, mineral and hot springs, and redwood groves.

Sonoma, city (1970 pop. 4,112), Sonoma co., W Calif., N of San Francisco, in a farm and dairy region; founded 1835, inc. 1850. The California republic under J. C. Frémont was founded here (1846). San Francisco de Solano Mission (1823) and the first hotel (early 1840s) operated north of San Francisco are in the city.

So·no·ra (sə-nôr'ə, sō-nō'rä), state (1970 pop. 1,098,720), 70,484 sq mi (182,554 sq km), NW Mexico, on the Gulf of California, S of Ariz.; capital Hermosillo. Originally part of Nueva Viscaya, Sonora became a separate state in 1930. It played a key role in the Mexican Revolution against Spain (1910).

So·no·ra (sə-nôr'ə), city (1970 pop. 2,149), seat of Sutton co., W Texas, on an intermittent Dry Fork of the Devils River S of San Angelo; settled 1888, inc. 1917. It is a retail center for a ranching region. Livestock, wool, and mohair are shipped.

Son·so·na·te (sôn'sə-nä'tā), city (1971 pop. 33,562), SW El Salvador. It is the commercial center of one of the richest agricultural regions of El Salvador, where dairy products, coffee, sugar, tobacco, Peruvian balsam and subsistence crops are produced. The city was founded in 1524.

Soo Canals (sōō): *see* Saulte Sainte Marie Canals.

Soo·chow (sōō'chou', sōō'jō'): *see* Su-chou, China.

So·per·ton (sō'pər-tən), city (1970 pop. 2,596), seat of Treutlen co., central Ga., WNW of Savannah, in the coastal plain. Fertilizer, turpentine, and clothing are made.

So·pot (sô'pôt), city (1973 est. pop. 50,600), N Poland, on the Baltic Sea and the Gulf of Danzig. It is a seaside resort and tourist center. Sopot belonged to the city of Danzig (now Gdańsk) from 1283 to 1807. It passed to Prussia in 1814. Included in the Free City of Danzig in 1919, it was ceded to Poland in 1945.

Sop·ron (shô'prôn), city (1973 est. pop. 50,738), NW Hungary, near the Austrian border. It is a commercial center and produces cotton textiles, woolens, and wines. There are also fruit-preserving and sugar-refining industries. Originally a Celtic settlement, it became a military outpost under the Romans. Hungarians settled the area in the 10th and 11th cent. Sopron was the site of the coronation of King (later emperor) Ferdinand III of Hungary and Bohemia in 1625. It was transferred to Austria after World War I but was returned to Hungary after a plebiscite (1921). Sopron is one of the oldest cultural centers in Hungary.

So·rac·te (sə-răk'tē), isolated mountain, 2,267 ft (691.4 m) high, in Latium, central Italy, N of Rome. It was celebrated in the poetry of Vergil and Horace.

So·rel (sə-rĕl'), city (1971 pop. 19,347), S Que., Canada, at the confluence of the St. Lawrence and Richelieu rivers. It is a grain-shipping center with an important shipbuilding industry. Iron and steel, metal products, textiles, and clothing are made. The city is on the site of Fort Richelieu, built in 1665.

So·ri·a (sôr'ē-ə), town (1972 est. pop. 25,608), capital of Soria prov., N central Spain, in Old Castile, on the Duero River. It is the center of a pastoral region. Nearby are the ruins of Numantia.

Sor·ø (sô'rœ), town (1970 pop. 13,930), Vestsjaelland co., E Denmark. It is a cultural and resort center. There is an academy founded by Christian IV in 1623.

So·ro·ca·ba (sōōr'ōō-kä'bə), city (1970 pop. 165,990), São Paulo state, S Brazil, on the Sorocaba River. It is a transportation hub and a manufacturing center where textiles, cement, vegetable oils, agricultural machinery, explosives, and food products are made. Natural resources in the area include iron, gold, and silver.

Sor·ren·to (sōr-rĕn'tō), town (1972 est. pop. 11,300), Campania, S Italy, on the Sorrento Peninsula, which separates the Bay of Naples from the Gulf of Salerno. It is a tourist center and a summer resort.

So·sno·wiec (sô-snô'vyĕts), city (1973 est. pop. 148,400), S Poland. An industrial center of the Katowice region, it has coal mines and ironworks and steelworks as well as industries producing machinery, chemicals, and metals. Sosnowiec passed to Prussia in 1795, to Russia in 1815, and reverted to Poland in 1919.

Sou·der·ton (sou'dər-tən), borough (1970 pop. 6,366), Montgomery co., SE Pa., ENE of Pottstown; settled 1860, inc. 1887. It has light industry.

Sou·fri·ère (sōō'frē-ĕr'), dormant volcano, 4,813 ft (1,468 m) high, on Guadeloupe Island, French West Indies. It is the highest mountain in the Lesser Antilles.

Soufrière, volcano, 4,048 ft (1,234.6 m) high, on St. Vincent Island, British West Indies. On May 7, 1902, Soufrière erupted, laying waste a third of St. Vincent and killing more than 1,000 people.

Sou·li (sōō'lyē): *see* Suli, Greece.

Sou·ris (sōōr'ĭs), river, c.450 mi (725 km) long, rising in S Sask., Canada, and flowing SE with a great loop into N.Dak. (passing Minot), then N and NE to the Assiniboine River in SW Man.

Sousse (sōōs): *see* Susah, Tunisia.

South, the (south), region of the United States embracing the southeastern and south-central parts of the country. It includes, at the most, 14 states: Md., Va., N.C., S.C., Ga., Fla., Ky., Tenn., Ala., Miss., Ark., La., Okla., and Texas. Traditionally all states south of the Mason-Dixon line and the Ohio River (except W.Va.) make up the South, but to many the region is restricted to the 11 states south of the Potomac River that comprised the Confederacy. Mo. and Del. are two other states in which the Southern tradition is strong. The South has long been a section apart, even though it is not isolated by any formidable natural barriers. It is itself divided into many distinctive regions: the coastal plains along the Atlantic Ocean and the Gulf of Mexico; the Piedmont; the Great Smoky Mts. areas of bluegrass, black-soiled prairies, and clay hills; bluffs, flood plains, bayous, and delta lands and the interior plains, the Ozark Plateau, and large stretches of arid lands. The climate, however, is one unifying factor. Winters are mild, and no month averages below freezing; rainfall is heavy. The long, hot growing season (nine months along the Gulf) and the fertile soil (much of it overworked or ruined by erosion) make the South an agricultural region where such staples as tobacco, cotton, corn, and, to a lesser extent, rice and sugar cane have long flourished; citrus fruits, livestock, and timber are also important. Petroleum and natural gas are the region's chief mineral resources. It now has diversified industries, and the tourist trade is important here.

History. The basic agricultural economy of the Old South, determined by the climate and the soil, led to the introduction (1617) of the black as a source of cheap labor under the twin institutions of the plantation and slavery. Slavery might well have expired had not the invention of the cotton gin (1793) given it a firmer hold, but even so there would have remained the problem of racial tension.

Although Southerners had taken care to have their "peculiar institution" protected by the Constitution of the United States, it was not until the period beginning with the Missouri Compromise (1820-21) that the South definitely became a "conscious minority" in the nation. That event marked the rise of Southern sectionalism, rooted in the political doctrine of States' rights. When differences with the North, especially over the issue of the extension of slavery into the Federal territories, ultimately appeared insoluble, the South turned (1860-61) the doctrine of States' rights into secession (or independence), which in turn led inevitably to the Civil War. Most of its major battles and campaigns were fought in the South, and by the end of the war, with slavery abolished and most of the area in ruins, the Old South had died.

The period of Reconstruction following the war set the South's political and social attitude for years to come. During this difficult time radical Republicans, Carpetbaggers, blacks, and scalawags ruled the South with the support of Federal troops. White Southerners, objecting to this rule, resorted to terrorism and, with the aid of such organizations as the Ku Klux Klan, drove the Reconstruction governments from power. The breakdown of the plantation system gave rise to sharecropping, the tenant-farming system of agriculture that still exists in some areas. The last half of the 19th cent. saw the introduction of textile mills and various industries in the South. The troubled economic and political life of the region in the years between 1880 and World War II was marked by the rise of the Farmers' Alliance, Populism, and Jim Crow laws.

Since World War II the South has been experiencing profound political, economic, and social change. Southern reaction to the policies of the New Deal, the Fair Deal, the New Frontier, and the

Great Society led conservative Southern Democrats into a close congressional liaison with Northern conservative Republicans. The influx of new industries into the region has made the economic life of the South more diversified and more similar to that of other sections of the United States. Aided by advocates of full civil rights for all citizens and supported by the Federal courts, the Southern black has made great progress in ending political inferiority and social segregation.

South Af·ri·ca, Republic of (ăf'rĭ-kə), republic (1977 pop. 26,335,000), 471,442 sq mi (1,221,037 sq km), S Africa, bordering on the Atlantic Ocean in the west, on Namibia in the northwest, on Botswana and Rhodesia in the north, on Mozambique and Swaziland in the northeast, and on the Indian Ocean in the east and south. The administrative capital of South Africa is Pretoria, the legislative capital is Cape Town, and the judicial capital is Bloemfontein. The republic is divided into four provinces: Cape of Good Hope, Natal, Orange Free State, and Transvaal. In addition, under acts of parliament about 14% of the country's land area has been designated to be set aside for black Africans in ultimately independent territories ("Bantustans"). As of late 1974 nine such territories were planned; all but three had been given limited internal self-government.

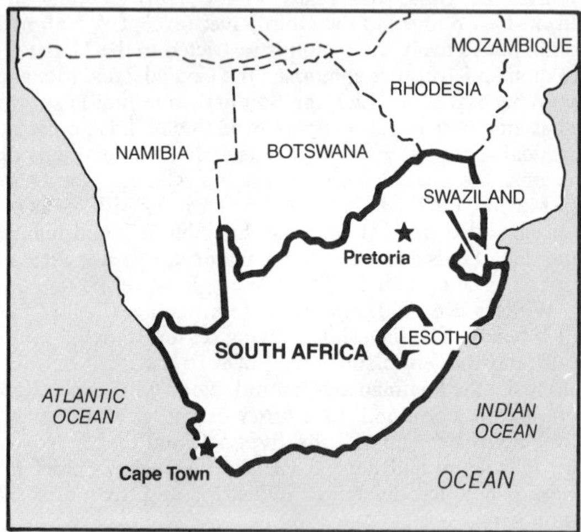

South Africa has three main geographic regions: a great interior plateau; an escarpment of mountain ranges that rims the plateau on the east, south, and west; and a marginal area lying between the escarpment and the sea. Most of the plateau consists of highveld, rolling grassland situated at 4,000-6,000 ft (1,220-1,830 m). In addition, in the northeast are the Witwatersrand (a ridge of rock where gold has been mined since 1886), the Bushveld Basin (a zone of savanna situated at 2,000-3,000 ft/610-910 m), and the Limpopo River basin; in the north are the southern fringes of the Kalahari desert; and in the west is the semiarid Cape middleveld, which includes part of the Orange River and is situated at 2,500-4,000 ft (760-1,220 m). The escarpment reaches its greatest heights (10,000-11,000 ft/3,050-3,350 m) in the Drakensberg Mts. in the east. The marginal area varies in width between 35 and 150 mi (60-240 km), and most of it is bordered by a narrow, low-lying coastal strip. The region also includes considerable stretches of grassland in the east; mountains and the semiarid Great and Little Karroo tablelands in the south; and desert (a southern extension of the Namib desert) in the west. Kruger National Park is in northeastern South Africa.

According to the official 1970 census the population of South Africa consisted of 14,741,000 black Africans ("Bantus"); 3,751,000 whites; 2,019,000 people of mixed white, Malayan, and black African descent ("Coloureds"); and 620,000 Asians, most of whom are Hindus of Indian descent. The black Africans fall into 10 main groups, based on their first language: Zulu, Xhosa, Tswana, Sepedi or North Sotho, Seshoeshoe or South Sotho, Shangaan, Swazi, Venda, South Ndebele, and North Ndebele. Afrikaans is the first language of about 60% of the whites and about 90% of the Coloureds; English is the first language of most of the rest of the whites and Coloureds and of most of the Asians. Many of the black Africans also speak English or Afrikaans, which are both official languages of the country.

Economy. Until about 1870 the economy of the region was almost entirely based on agriculture. With the discovery (1867) of diamonds and gold, mining became the foundation for rapid economic development. By 1945 manufacturing was the leading contributor to the annual national product. By 1970 manufacturing contributed about 23% of the annual national product; trade and tourism about 14%; mining and quarrying about 12%; finance, insurance, and real estate about 10%; and agriculture, forestry, and fishing about 9.5%. About 37% of the work force was employed in agriculture and only about 29% in manufacturing (including construction); in addition, many nonwhites worked as subsistence farmers. The economy is controlled by whites, but nonwhites make up more than 75% of the work force. Many black Africans work as contract laborers for specified periods in factories and mines, and they live (apart from their families) in special housing complexes. Roughly 500,000 black African migratory workers from Mozambique, Rhodesia, Lesotho, Malawi, and Botswana are employed in South Africa, mainly in the gold mines.

About 15% of the land in South Africa is arable. The chief crops grown are maize, wheat, oats, barley, sorghum, potatoes, groundnuts, deciduous and citrus fruit, cotton, tobacco, and sugar cane. In addition, large numbers of dairy and beef cattle, sheep, goats, and hogs are raised. The principal manufactures are processed food, beverages (including wine), textiles, clothing, forest products, chemicals, iron and steel, metal products, machinery, and motor vehicles. The leading minerals extracted (in terms of value) are gold, copper ore, coal, gem and industrial diamonds, asbestos, iron ore, platinum, chrome, antimony, and manganese. There is a large fishing industry, and much fish meal is produced. The country has good road and rail networks.

History. Little is known about the prehistory of South Africa, but remains of early man dating to the late Paleolithic period have been found in the country. By the first millennium A.D. the southern and central parts of the country were inhabited by Khoikhoi (Hottentots) and small bands of San (Bushmen). By about 1500, Bantu-speaking black Africans from east-central Africa had reached the region. In 1488 Bartolomeu Dias, a Portuguese navigator, became the first European to round the Cape of Good Hope. No permanent European settlement was made, however, until 1652, when Jan van Riebeeck and about 90 other persons set up a provisioning station for the Dutch East India Company on the Cape of Good Hope. Soon van Riebeeck began to trade with nearby Khoikhoi and brought in as slaves black Africans from western and eastern Africa and Malayans. By 1662 about 250 Europeans were living near the Cape, and gradually they moved inland, founding Stellenbosch in 1679. In 1689 about 200 Huguenot refugees from Europe arrived. By 1707 there were about 1,780 freeholders of European descent in South Africa, and they owned about 1,100 slaves. By the early 18th cent. most San and Khoikhoi had migrated into inaccessible parts of the country to avoid European domination. During the 18th cent. interbreeding between Khoikhoi, slaves, and Europeans began to create what later became known as the Coloured population. At the same time white farmers (known as Boers) began to journey increasingly farther from the Cape in search of pasture and cropland. By 1750 some farmers had migrated inland, where they encountered the Xhosa. In 1779 the first of a long series of so-called Kaffir Wars broke out between them, primarily over land and cattle ownership. The whites sought to establish the Great Fish River as the southern frontier of the Xhosa. In 1814 the Cape territory was assigned to Great Britain by the Congress of Vienna. In 1820, 5,000 British settlers were given small farms near the Great Fish. They were intended to form a barrier to the southern movement of the Xhosa. They were the first large body of Europeans not to be assimilated into the Afrikaner culture that had developed in the 17th and 18th cent. Great Britain disaffected the Boers by remodeling the administration, by tending to call for better treatment of the Coloured and black Africans who worked for the Boers as servants or slaves, by granting free nonwhites legal rights equal to those of the whites, and by restricting the acquisition of new land by the Boers. Seeking freedom from the restrictions of British rule as well as new land, about 12,000 Boers left the Cape between 1835 and 1843 in what is known as the Great Trek. The "Voortrekkers" migrated beyond the Orange River. Some remained in the highveld of the interior, forming isolated communities and small states. In the 1850s the Boer republics of the Orange Free State and the Transvaal were established.

In 1860 the first indentured laborers from India arrived in Natal to work on the sugar plantations, and by 1900 they outnumbered the whites there. Diamonds were discovered in 1867 along the Vaal and Orange rivers and in 1870 at what became (1871) Kimberley; in 1886 gold was discovered on the Witwatersrand. These discoveries (especially that of gold) spurred great economic development in South Africa; foreign trade increased dramatically, and the number of whites rose from about 300,000 in 1870 to about 1,000,000 in 1900. At the same time there were complex political developments. In 1871 the British annexed the diamond-mining region, despite the protests of the Orange Free State. In 1889 Cape Colony and the Orange Free State joined in a customs union, but the Transvaal (led by Paul Kruger, known as Oom Paul) adamantly refused to take part. In 1890 Cecil J. Rhodes, an ardent advocate of federation in South Africa, became prime minister of Cape Colony, and by 1894 he was encouraging the non-Afrikaner whites (known as the Uitlanders) in the Transvaal to overthrow Kruger. Tension mounted in the following years as the British supported the Uitlanders against the dominant Afrikaners. In 1899 the Transvaal and the Orange Free State declared war on Great Britain. The South African War (Boer War, 1899-1902) was won by the British. In 1910 the Union of South Africa, with dominion status, was established by the British; it included Cape of Good Hope, Natal, the Orange Free State, and the Transvaal as provinces. Under the Union's constitution power was centralized; the Dutch language (and in 1925 Afrikaans) was given equal status with English, and each province retained its existing franchise qualifications. After elections in 1910 Louis Botha became the first prime minister; he headed the South African party. In 1912 J. B. M. Hertzog founded the Afrikaner-oriented Nationalist party. By 1914, largely as a result of the efforts of Mohandas K. Gandhi, Indian immigration into South Africa had stopped. Botha led (1914) South Africa into World War I on the side of the Allies and quickly squashed a revolt by Afrikaners who opposed this alignment. In 1915 South African forces captured South West Africa (now Namibia) from the Germans, and after the war the territory was placed under the Union. In 1919 Botha was succeeded as prime minister by J. C. Smuts. In 1924 Hertzog became prime minister and remained in office until 1939; from 1934 to 1939 he was supported by Smuts, with whom he formed the United South African National party. Hertzog led an Afrikaner cultural and economic revival, was influential in gaining British recognition of South African independence, and curtailed the power of nonwhites. Winning a crucial vote in parliament (Sept., 1939), Smuts became prime minister again and brought South Africa into the war on the British (Allied) side; Hertzog, who was not alarmed by Nazi German aggression and had little affection for Great Britain, went into opposition. The Nationalist party won the 1948 elections. Nationalist Party leader D. F. Malan was prime minister from 1948 to 1954, and he was followed by J. G. Strijdom (1954-58), H. F. Verwoerd (1958-66), and B. J. Vorster (1966-78; president 1978-79)—all members of the Nationalist party. Their governments greatly strengthened white control of the country. The policy of segregating whites and nonwhites, known as apartheid, in almost all social relations was further implemented by laws that included additional curbs on free movement and the planned establishment of a number of independent homelands for black Africans. The black Africans had long protested their inferior treatment; in 1960 a peaceful protest ended when police opened fire, killing about 70 protesters and wounding about 190 others. In the 1960s most leaders of the opposition to apartheid (whites as well as blacks) were either in jail or were living in exile.

In 1961 South Africa left the British Commonwealth of Nations and became a republic. In the 1960s there were international attempts to wrest South West Africa from South Africa's control, but South Africa tenaciously maintained its hold on the territory. From the late 1960s the Vorster government began to try to start a dialogue on racial and other matters with independent black African nations; these attempts met with little success, except with countries such as Malawi, Lesotho, Batswana, and Swaziland, which were economically dependent on South Africa. South Africa was strongly opposed to the establishment of black rule in the white-dominated countries of Angola, Mozambique, and Rhodesia and gave military assistance to the whites there. However, by late 1974, with independence for Angola and Mozambique under black rule imminent, South Africa, as one of the few remaining white-ruled nations of Africa, faced the prospect of further isolation from the international community. In the early 1970s increasing numbers of whites protested apartheid, and the Nationalist party itself was divided largely on questions of race relations.

Government. Under the 1961 constitution as amended, South Africa has a bicameral parliament made up of a senate (54 members) and a house of assembly (171 members). The country's chief executive and head of state is the president, elected to a normally nonrenewable 7-year term by an electoral college consisting of members of parliament and presided over by the chief justice of South Africa or another judge designated by him. The president acts on the advice of the executive council or cabinet (led by the prime minister), which must have the support of a majority in the house. In practice, the prime minister is the country's leading executive figure.

South Am·boy (ăm'boi), city (1970 pop. 9,338), Middlesex co., E N.J., with a harbor at the mouth of the Raritan River opposite Perth Amboy; settled 1651, inc. as a borough 1888, as a city 1908. It became an important coal port in 1832 when it was made the terminal of the state's first railroad. The city continues to ship coal. In 1950 an explosion of munitions destined for Pakistan caused several deaths and damages worth millions of dollars.

South A·mer·i·ca (ə-mĕr'ĭ-kə), fourth-largest continent (1971 est. pop. 195,000,000), c.6,880,000 sq mi (17,819,000 sq km), the southern of the two continents of the Western Hemisphere. It is divided politically into 11 independent countries—Argentina, Bolivia, Brazil, Chile, Colombia, Ecuador, Guyana, Paraguay, Peru, Uruguay, and Venezuela—and the colonies of Surinam (Dutch) and French Guiana. The continent extends c.4,750 mi (7,640 km) from Punta Gallinas, Colombia, in the north to Cape Horn, Chile, in the south. At its broadest point, near where it is crossed by the equator, the continent extends c.3,300 mi (5,300 km) from east to west. South America is connected to North America by the Isthmus of Panama; it is washed on the north by the Caribbean Sea, on the east by the Atlantic Ocean, and on the west by the Pacific Ocean. Topographically the continent is divided into three sections: the South American cordillera, the interior lowlands, and the continental shield. The continental shield, in the east, which is separated into two unequal sections by the Amazon geosyncline, contains the continent's oldest rocks. Geologic studies in South America have supported the theory of continental drift and have shown that until 135 million years ago South America was joined to Africa. Extending down the middle of the continent is a series of lowlands running southward from the llanos of the north, through the selva of the great Amazon basin and the Gran Chaco, to the Pampa of Argentina. Paralleling the Pacific shore is the great cordillera composed of the Andes ranges and high intermontane valleys and plateaus. The Andes rise to numerous snow-capped peaks; Mt. Aconcagua (22,835 ft/6,960 m) in Argentina is the highest point in the Western Hemisphere. The Andes region is seismically active and prone to earthquakes. Volcanoes are present but currently inactive. Patagonia, a windy, semiarid plateau region, lies to the east of the Andes in southern Argentina. On the Pacific coast the land between the Andes and the sea widens northward from the islands of south Chile. In north Chile lies the barren Desert of Atacama. The continent's great river systems empty into the Atlantic Ocean and the Caribbean Sea; from north to south they are the Magdalena, Orinoco, Amazon, and Paraguay-Paraná systems. Excluding Lake Maracaibo, which is actually an arm of the Caribbean Sea, Lake Titicaca, on the Peru-Bolivia border, is the largest of the continent's lakes. South America embraces every climatic zone—tropical rainy, desert, high alpine—and vegetation varies accordingly.

Economy. Since the 17th cent. the exploitation of the continent's resources and the development of its industries have been the result of foreign investment and initiative, especially that of Spain, Great Britain, and the United States. Since World War II many of the nations of South America have sought greater economic independence. Foreign-owned companies have been nationalized, and raw materials, once almost exclusively exported, are now being used in local industries. An increasing number of South American industrial centers have developed heavy industries. An obstacle to industrial growth is the scarcity of coal. The continent therefore relies on its petroleum reserves, most notably in Venezuela and also in Argentina, Bolivia, Colombia, Ecuador, and Peru, as a source of fuel. Iron ore deposits are plentiful in the Guiana and Brazilian highlands, and copper is abundant in the central Andes mountain region of Chile and Peru. Other important mineral resources include tin in Bolivia, manganese in Brazil, and bauxite in Guyana.

A generally impoverished subsistence farming prevails in much of the continent, with about 50% of the people working only about 5% of the land. Dense forests, steep slopes, and unfavorable climatic conditions, along with crude agricultural methods, limit the amount of cultivable land. Commercial agriculture, especially the plantation type, fares better in terms of production because of the large scale and the opportunity to use modern mechanized farming methods. Among the agricultural exports are coffee, bananas, sugar cane, tobacco, cacao, and grains. Livestock grazing is important in the grassland regions; Argentina and Uruguay export meat. In the interior of the continent hunting and gathering of forest products are the chief economic activities of the indigenous peoples. In the more accessible areas forest products are removed for export. Fishing is especially important off the west coast of the continent.

History. European exploration of South America began at the beginning of the 16th cent. Under the Treaty of Tordesillas, Portugal claimed what is now Brazil, and Spanish claims were established throughout the rest of the continent with the exception of the Guianas. An Iberian culture and Roman Catholicism were early New World transplants—as were coffee, sugar cane, and wheat. Spain and Portugal maintained their colonies until the first quarter of the 19th cent., when successful revolutions resulted in the creation of independent states. The subjugation of the indigenous Indian civilizations was a ruthless accompaniment to settlement efforts, particularly those of Spain. The Inca Empire, centered at Cuzco, Peru, was conquered (1531–35) by Francisco Pizarro; other Indian cultures quickly declined or retreated in the face of conquest, conversion attempts, and subjugation. Today Indians constitute a significant portion of the continent's Andean population, especially in Bolivia, Ecuador, Peru, and Paraguay. Elsewhere in South America the population is generally mestizo, although Argentina, Uruguay, southern Brazil, and Chile have primarily European populations. There are sizable black populations in the Guianas, northeastern Brazil, Colombia, and Venezuela. Immigration since 1800 has brought European, Middle Eastern, and Asian (especially Japanese) peoples to the continent, particularly to Argentina and Brazil. The population of South America is growing at a faster rate than that of any other part of the world except Central America. Outside the cities the population density of the continent is very low, with vast portions of the interior virtually uninhabited; most of the people live within 200 mi (320 km) of the coast. With the post-World War II trend of rural-to-city migration, urban population growth is rapidly expanding.

South·amp·ton (south-ămp′tən, -hămp′-), county borough (1974 est. pop. 213,710), Hampshire, S England, at the head of Southampton Water. Southampton is England's chief port for passenger ships. It has large shipbuilding and yacht-building industries, aircraft and marine engine factories, and food-processing plants. Southampton is the site of the Roman Clausentum and of the Saxon Hamtune or Suth-Hamtun. The Crusaders under Richard I, Henry V on his expedition to France (1415), and the Pilgrims all embarked from Southampton. Until the 16th cent. Southampton had a lucrative trade with Venice in goods from the East. In the 18th cent. it was a fashionable spa. Growing trade, construction, and the coming of the steamboat all worked to convert the spa back into a commercial port. The city suffered considerable damage in World War II.

Southampton, county (1970 pop. 18,582), 602 sq mi (1,559.2 sq km), SE Va., in the tidewater region, bounded in the S by N.C., in the W by the Meherrin River, in the E by the Blackwater River, and drained by the Nottoway River; formed 1749 when it was set off from one of the original Virginia shires (1634); co. seat Courtland. It is in a diversified agricultural area that produces cotton, peanuts, fruit, grain, livestock, and truck crops. Its industry includes food processing, textile milling, and lumbering.

Southampton, village (1970 pop. 4,904), Suffolk co., SE N.Y., on SE Long Island E of Brookhaven; settled 1640, inc. 1894. It is a summer resort with many fine estates and a number of old homes. The Parrish Memorial Art Museum is here.

Southampton Island, c.15,700 sq mi (40,660 sq km), E Keewatin dist., Northwest Territories, Canada, at the entrance to Hudson Bay. It is separated from the mainland by Ross Welcome Sound and Frozen Strait. With lowlands in the west, the tundra-covered island rises to c.2,000 ft (610 m) in the east.

South Aus·tra·lia (ô-strāl′yə), state (1978 est. pop. 1,273,000), 380,070 sq mi (984,381 sq km), S central Australia, bounded on the S by the Indian Ocean; capital Adelaide. Many small islands off the south coast are included in the state. Much of South Australia is wasteland—deserts, mountains, salt lakes, and swampland. The only important river is the Murray, in the extreme southeast. Agriculture consists of the raising of barley and grapes and of wheat, oats, and rye. Livestock are grazed in the northern plains. Iron ore, salt, and gypsum are mined, and coal and natural gas are exploited. The chief manufactures are industrial metals and transportation equipment.

South Australia's coastal areas were visited by the Dutch in 1627. The British explorer Matthew Flinders noted likely settlement sites in 1802. The English Parliament passed the South Australian Colonization Act in 1834, and in Dec., 1836 the first colonists arrived. In 1901 South Australia was federated as a state of the commonwealth. Northern Territory, which had been included in the state in 1863, was transferred in 1911 to the commonwealth government.

South Bend. 1. City (1978 est. pop. 111,500), seat of St. Joseph co., N Ind., on the St. Joseph River; inc. as a city 1865. It lies in a farming and mint-growing region. Its manufactures include automotive parts, paints, nonelectrical machinery, rubber, plastics, metal products, and farm equipment. South Bend was settled c.1820 as a post of the American Fur Company on the site of a French mission and trading post. **2.** City (1977 est. pop. 1,805), seat of Pacific co., SW Wash., on the Willapa River near Willapa Bay; founded 1869, platted 1889, inc. 1890. Seafood and timber are processed.

South·bor·o (south′bûr′ō, -bər-ə), town (1970 pop. 5,798), Worcester co., E Mass., E of Worcester; settled 1660, set off from Marlboro 1727. Meat processing is done. St. Mark's School (1865) is here.

South Bos·ton (bŏs′tən), independent city (1975 est. pop. 6,800), S Va., on the Dan River ENE of Danville; chartered 1796, inc. as a town 1884, as a city 1960. It is a processing center in a tobacco area.

South·bridge (south′brĭj′), town (1975 est. pop. 17,600), Worcester co., S Mass., on the Quinebaug River; settled 1730, inc. 1816. Textiles are produced.

South Bur·ling·ton (bûr′lĭng-tən), city (1970 pop. 10,032), Chittenden co., NW Vt., on Lake Champlain; inc. 1971. It is a resort with some light industry.

South·bur·y (south′bĕr′ē, -bə-rē), town (1970 pop. 7,852), New Haven co., SW Conn., N of Bridgeport; settled 1673, set off from Woodbury 1787. The town has many country and summer houses. Steel traps and tacks are made here.

South Car·o·li·na (kăr′ə-lī′nə), state (1975 pop. 2,845,000), 31,055 sq mi (80,432 sq km), SE United States, one of the Thirteen Colonies. Columbia is the capital and the largest city. Roughly triangular in shape, South Carolina is bounded on the north by N.C., on the southwest by Ga. (with the Savannah River and its tributaries forming the state line), and on the southeast by the Atlantic Ocean.

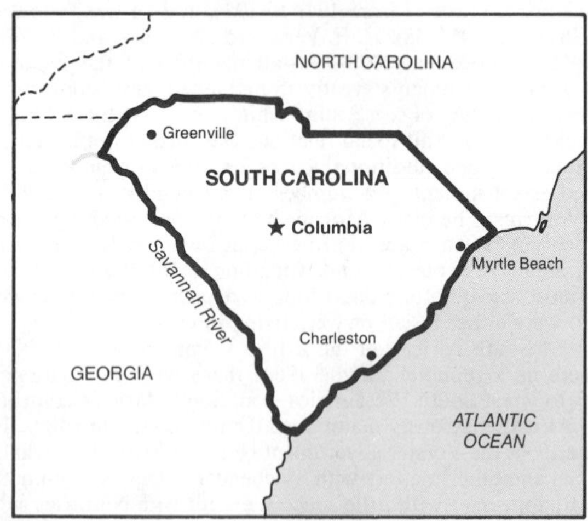

The long coast, firm and even above Georgetown, becomes generally marshy to the south and is sliced by a network of rivers and creeks, creating a maze of inlets and the famous Sea Islands. The coastal climate is humid and subtropical, with long, hot summers and short, mild winters. In this area are found cypress swamps, moss-hung oaks, beautiful flowering gardens, ante-bellum plantations, and the quaint historic seaports of Georgetown, Beaufort, and Charleston. The coastal plain is separated from the rolling Piedmont plateau of the up-country by the Fall Line, which runs generally parallel to the coast, passing through Columbia. Inland the climate is temperate, becoming progressively cooler as the altitude increases. In the extreme northwest are the Blue Ridge Mts.; they occupy only c.500 sq mi (1,290 sq km) in the state, with Sassafras Mt. (3,560 ft/1,085 m) the highest point. Rainfall is abundant and well distributed throughout South Carolina. The Pee Dee, Santee, Edisto, and Savannah river systems drain the state, flowing from the highlands to the sea with increasing speed, creating rapids and waterfalls. The older dams—the Saluda (with its reservoir, Lake Murray) and Buzzard Roost dams on the Saluda River, and the Santee and Pinoplis dams (with their reservoirs, Lake Marion and Lake Moultrie) on the Santee River—have been supplemented by the Clark Hill and Hartwell dams on the Savannah River, part of a planned 11-dam project.

Thousands of vacationers are attracted to famous Myrtle Beach and to the Sea Island resorts. The state's historical places of interest include Fort Sumter National Monument, Kings Mountain National Military Park, and Cowpens National Battlefield.

South Carolina is divided into 46 counties:

NAME	COUNTY SEAT	NAME	COUNTY SEAT
Abbeville	Abbeville	Greenwood	Greenwood
Aiken	Aiken	Hampton	Hampton
Allendale	Allendale	Horry	Conway
Anderson	Anderson	Jasper	Ridgeland
Bamberg	Bamberg	Kershaw	Camden
Barnwell	Barnwell	Lancaster	Lancaster
Beaufort	Beaufort	Laurens	Laurens
Berkeley	Moncks Corner	Lee	Bishopville
Calhoun	St. Matthews	Lexington	Lexington
Charleston	Charleston	McCormick	McCormick
Cherokee	Gaffney	Marion	Marion
Chester	Chester	Marlboro	Bennettsville
Chesterfield	Chesterfield	Newberry	Newberry
Clarendon	Manning	Oconee	Walhalla
Colleton	Walterboro	Orangeburg	Orangeburg
Darlington	Darlington	Pickens	Pickens
Dillon	Dillon	Richland	Columbia
Dorchester	St. George	Saluda	Saluda
Edgefield	Edgefield	Spartanburg	Spartanburg
Fairfield	Winnsboro	Sumter	Sumter
Florence	Florence	Union	Union
Georgetown	Georgetown	Williamsburg	Kingstree
Greenville	Greenville	York	York

Economy. The leading industries are based largely on the state's agricultural products—the huge textile and clothing industries, centered in the Piedmont, being based on that region's cotton crop, and lumbering and related enterprises on the c.12,500,000 acres (5,058,750 hectares) of forestland that cover the state. Other leading manufactures are chemicals, machinery, and foodstuffs. Perhaps the state's most impressive industrial installation is the Atomic Energy Commission's enormous Savannah River Atomic Energy Plant (mostly in Aiken co.), which produces plutonium and other nuclear materials. Principal minerals are nonmetallic—cement, stone, clays, and sand and gravel; South Carolina ranks second (1970) in the nation in vermiculite output and fourth in feldspar and mica. In agriculture, tobacco and soybeans have surpassed cotton as South Carolina's chief crops; the state ranks third (1972) in the nation in tobacco production. Cotton lint, corn, and cattle are economically important, and peanuts, pecans, sweet potatoes, and peaches (South Carolina is second in their production) are grown in abundance. Fishing is a major commercial enterprise; the chief catches are hard blue crabs and shrimps.

History. At an unknown coastal site in the Carolinas, what may have been the first white settlement in North America was founded (1526; not permanent) by an expedition under the Spanish explorer Lucas Vásquez de Ayllón. Later Hernando De Soto penetrated (1540) the Savannah River region. Spanish missions soon extended north from Florida almost to the site of present-day Charleston, and they remained until the arrival of the English. In 1663 Charles II awarded the territory from lat. 36° N to lat. 31° N to eight of his prominent supporters. The northern and southern sections of Carolina developed separately. The first permanent colony was established in 1670 at Albemarle Point under William Sayle. In 1680 the colony moved across the river to Oyster Point, which was better suited for defense. There the colonists established their capital, called Charles Town (later Charleston), which was to become the chief center of culture and wealth in the South. Wealthy colonists set up plantations worked by indentured servants and black and Indian slaves, while freemen cultivated the 50 acres (20 hectares) granted them by the proprietors.

The colony was divided into North and South Carolina in 1713. In 1715–16 the settlers were attacked by the once friendly Yamasee Indians, who had become resentful of exploitation by the Carolina traders. These attacks further revealed the lack of protection afforded by the proprietors, and in 1719 the colonists rebelled and received royal protection. South Carolina formally became a royal colony in 1729. The founding (1733) of Georgia to the south provided a buffer against the Spanish. To counterbalance the vast number of black slaves being imported for plantation labor, white immigration was encouraged. Regional antipathies developed between the small, self-sufficient farmers of the up-country and the powerful plantation lords of lower Carolina. Finally the legislature was impelled to grant certain up-country demands, including the establishment of courts in the region.

South Carolinians were leaders in the movement for independence, and in Mar., 1776, an independent government of South Carolina was set up with John Rutledge as president. In the American Revolution the British successfully besieged Charleston in 1780. In the ensuing Carolina campaign the British were ultimately forced to retreat, although they held Charleston until Dec., 1782. In 1786 the site of Columbia was chosen for the new capital. South Carolina ratified the Federal Constitution in May, 1788. Complete religious liberty was established in the state charter of 1790.

Eli Whitney's cotton gin enabled cotton plantations to spread far into the upper regions; thus the planters continued to dominate state policies. In the late 1820s cotton from the fertile western states glutted the market, and prosperity declined in South Carolina. Discontent was aggravated by national tariff policies that were unfavorable to South Carolina's agrarian economy. In 1832 the state passed its nullification act, declaring the tariff laws not binding upon South Carolina citizens. President Andrew Jackson acted firmly for the Union in this crisis, and in 1833 South Carolina repealed its act. John C. Calhoun, supported by Robert Y. Hayne, became the acknowledged leader of the whole South with his defense of the States' Rights doctrine; his political philosophy was later to form the intellectual basis for the Confederacy.

After Lincoln's election South Carolina was the first state to secede (Dec. 20, 1860) from the Union. Gov. Francis W. Pickens immediately demanded all Federal property within the state, including Fort Sumter, which was held by Union men under Maj. Robert Anderson. The firing on Sumter by Confederate batteries on Apr. 12, 1861, precipitated the Civil War. Charleston's forts withstood severe bombardments in 1863, and the state was saved from heavy military action until early in 1865. Then Gen. William T. Sherman, commanding the army that had marched through Georgia, advanced north through the state. The deliberate devastation, culminating in the burning of Columbia, was appalling. The constitution of 1868, which established universal male suffrage and ended property qualifications for office holding, gained the state readmittance (June, 1868) to the Union. During the period from 1868 to 1874 accomplishments such as the building of schools and railroads were offset by waste and corruption in the state government and by high taxation. By 1882 the vast majority of blacks had lost the vote; white political supremacy was assured.

Although some vigorous planters and merchants managed to recoup their fortunes after the war, farm tenancy (replacing the old plantation system) held most of the state's farmers in economic bondage. The Panic of 1873 was followed by two decades of agrarian hard times. Popular discontent was not ameliorated until the election (1890) of Benjamin Tillman, leader of the up-country farmers, as governor. Tillmanites reapportioned taxes and representation, expanded public education, and also initiated "Jim Crow laws" that excluded virtually all blacks from the crucial Democratic primaries.

Agriculture again suffered a setback in the 1920s, due largely to the boll weevil and soil depletion. Industry, especially the textile industry, also suffered in the Great Depression of the 1930s. New Deal legislation and the state road-building program provided South Carolina with some relief. Black migration began on a scale sufficient to bring the whites into the majority in the state by 1930. A state court decision in 1947 opened the Democratic primaries to the black vote. Under the governorship (1951–55) of James F. Byrnes, the poll tax was abolished as a voting requirement, steps were taken to curb Ku Klux Klan activities, and the educational system was greatly expanded. Integration of the schools after the 1954 U.S. Supreme Court decision met considerable opposition, but by 1970 all of the public school districts were technically in compliance with Federal desegregation requirements.

Government. The executive branch is headed by a governor elected for a four-year term and ineligible for re-election. South Carolina's bicameral legislature has a senate with 46 members elected for four-year terms and a house of representatives with 124 members elected for two years.

Educational Institutions. Among those prominent in the state are The Citadel (The Military College of South Carolina), at Charleston; Clemson Univ., at Clemson; South Carolina State College, at Orangeburg; and the Univ. of South Carolina, at Columbia.

South Charles·ton (chärlz'tən), city (1970 pop. 16,333), Kanawha co., W W.Va., on the Kanawha River, in a highly industrialized area; settled 1782, inc. 1917. It is an important chemical-manufacturing center. Adena Indian Mound (opened 1856–57) is nearby.

South Chi·na Sea (chī'nə), W arm of the Pacific Ocean, c.1,000,000 sq mi (2,590,000 sq km), between the SE Asian mainland and Taiwan, the Philippines, and Borneo. The Gulf of Tonkin and the Gulf of Siam are its chief embayments. The southwestern part of the sea from the Gulf of Siam to the Java Sea is a submerged plain in which the water is generally shallow (less than 200 ft/61 m). In contrast, the northeastern part of the sea is a deep basin, reaching depths of up to c.18,000 ft (5,490 m). Many islands dot the sea, which is a region subject to violent typhoons.

South Da·ko·ta (də-kō'tə), state (1975 pop. 683,000), 77,047 sq mi (199,552 sq km), N central United States, admitted to the Union in 1889 simultaneously with N.Dak. (they are the 40th and 39th states). Pierre is the capital. South Dakota is bounded on the north by N.Dak.; on the east by Minn. (with the Minnesota River forming part of the boundary) and Iowa (from which it is separated by the Big Sioux River); on the south by Nebr. (with the Missouri River forming part of the boundary); and on the west by Wyo. and Mont.

South Dakota shows some of the earliest geologic history of the continent in the rock formations of the ancient Black Hills and in the Badlands. At their extreme between the White River and the south fork of the Cheyenne, the Badlands display in their deeply eroded clay gullies not only colorful, fantastic shapes but also a wealth of easily accessible marine and land fossils. The whole of South Dakota has a continental climate; summer brings a succession of hot, cloudless days, and in the winter blizzards sweep across bare hillsides. The average annual rainfall is low. From east to west the state rises some 6,000 ft (1,829 m) to Harney Peak (7,242 ft/2,207 m) in the Black Hills, highest point in the United States east of the Rockies. Through the center of the state the Missouri River cuts a wide path southward; other principal rivers include the James and Big Sioux to the east, and the Cheyenne, the Belle Fourche, the Moreau, the Grand River, and the White River to the west. Almost one third of the region west of the Missouri River, a semiarid, treeless plain, belongs to the Indians, most of whom live on reservations such as Cheyenne River, Pine Ridge, Rosebud, and Standing Rock. Among the state's attractions are Wind Cave National Park, Jewel Cave National Monument, Badlands National Monument, and the famous mammoth carvings of the Mount Rushmore National Memorial.

South Dakota is divided into 67 counties:

NAME	COUNTY SEAT	NAME	COUNTY SEAT
Aurora	Plankinton	Brookings	Brookings
Beadle	Huron	Brown	Aberdeen
Bennett	Martin	Brule	Chamberlain
Bon Homme	Tyndall	Buffalo	Gannvalley
Butte	Belle Fourche	Lake	Madison
Campbell	Mound City	Lawrence	Deadwood
Charles Mix	Lake Andes	Lincoln	Canton
Clark	Clark	Lyman	Kennebec
Clay	Vermillion	McCook	Salem
Codington	Watertown	McPherson	Leola
Corson	McIntosh	Marshall	Britton
Custer	Custer	Meade	Sturgis
Davison	Mitchell	Mellette	White River
Day	Webster	Miner	Howard
Deuel	Clear Lake	Minnehaha	Sioux Falls
Dewey	Timber Lake	Moody	Flandreau
Douglas	Armour	Pennington	Rapid City
Edmunds	Ipswich	Perkins	Bison
Fall River	Hot Springs	Potter	Gettysburg
Faulk	Faulkton	Roberts	Sisseton
Grant	Milbank	Sanborn	Woonsocket
Gregory	Burke	Shannon	Hot Springs
Haakon	Philip	Spink	Redfield
Hamlin	Hayti	Stanley	Fort Pierre
Hand	Miller	Sully	Onida
Hanson	Alexandria	Todd	Winner
Harding	Buffalo	Tripp	Winner
Hughes	Pierre	Turner	Parker
Hutchinson	Olivet	Union	Elk Point
Hyde	Highmore	Walworth	Selby
Jackson	Kadoka	Washabaugh	Kadoka
Jerauld	Wessington Springs	Yankton	Yankton
Jones	Murdo	Ziebach	Dupree
Kingsbury	De Smet		

Economy. In the productive region east of the Missouri, livestock and livestock products comprise the primary source of income. Cattle and sheep ranching predominate in the west. Wheat and corn are South Dakota's chief cash crops, but oats, flaxseed, soybeans, and barley are also grown. Although there is diversified industry in the main cities of Sioux Falls and Rapid City, meat packing and food processing constitute by far the major industries of the state. Gold is South Dakota's most important mineral, and the town of Lead in the Black Hills is the country's leading gold-mining center. In 1972 the state ranked second in the nation in gold production, second in beryllium concentrate, and second in tin concentrate. Stone, sand and gravel, and cement are also important minerals.

History. At the time of the white man's arrival, South Dakota was inhabited by the agricultural Arikara and the nomadic Sioux (Dakota) Indians. The United States acquired the region as part of the Louisiana Purchase, and it was partially explored by Lewis and Clark (1804–6). Individual traders made the region their home, and the posts founded by Pierre Chouteau and the American Fur Company were the first bases for settlement. (Fort Pierre was established in 1817.) The introduction of the steamboat on the upper Missouri in 1831 brought renewed vigor to the fur trade, but it was not until the 1850s that any real settlement developed. Two land companies were established at Sioux Falls in 1856, and in 1859 Yankton, Bon Homme, and Vermillion were laid out. In 1861 the Dakota Territory was set up, embracing not only present-day N.Dak. and South Dakota but also eastern Wyo. and eastern Mont. Yankton was the capital.

Rumors of gold in the Black Hills in 1874 excited national interest, and Americans began to pour into the area. However, much of the Black Hills region had been granted (1868) to the Sioux by treaty, and when the Indians refused to sell, warfare again broke out. The defeat (1876) of George A. Custer and his men by Sitting Bull, Crazy Horse, and Gall in the battle of Little Bighorn (in what is now Mont.) did not prevent the white man from gradually acquiring more and more Indian land, including the gold-lined Black Hills. The extinction of the buffalo herds as well as Sitting Bull's death (1890) and the subsequent massacre of Big Foot's band at Wounded Knee Creek were factors leading to the end of Indian resistance. During the 1870s the gold fever mounted; the town of Lead began its long, productive career. Herds of cattle were first brought to the grasslands of western South Dakota partly to supply food for the miners. Settlement in the east also increased, and the period from 1878 to 1886 was the time of the great Dakota land boom, when population increased threefold. Agitation for statehood developed; in 1889 Congress passed an enabling act. The Dakotas were divided; South Dakota became a state with Pierre as capital.

The unusually severe winter of 1886–87 and recurrent droughts added to the difficulties of the farmers, who sought political action in the Populist Party, which won a resounding victory in 1896. Initiative and referendum (1898) and other progressive measures of the day were enacted, but prosperity quickly returned South Dakota to political conservatism. The extension of railroads encouraged further expansion of agriculture. New prosperity-depression cycles occurred after the boom of World War I. The combination of droughts and the Great Depression brought widespread calamities in the late 1920s and early 30s, and the state's population declined by 50,000 between 1930 and 1940. In the postwar period adoption of improved farming techniques resulted in a steady increase in agricultural production. This was accompanied, however, by the consolidation of small farms into large units and the displacement of many small farmers.

Government. South Dakota is governed under the 1889 constitution. The legislature consists of 35 senators and 75 representatives, all elected for two-year terms. The governor is also elected for two years.

Educational Institutions. Those prominent in South Dakota include Augustana College, at Sioux Falls; South Dakota School of Mines and Technology, at Rapid City; South Dakota State Univ., at Brookings; the Univ. of South Dakota, at Vermillion; and Northern State College, at Aberdeen.

Southeast A·sia (ā'zhə, ā'shə), region of Asia (est. pop. 240,000,000), c.1,740,000 sq mi (4,506,600 sq km), bounded roughly by the Indian subcontinent on the W, China on the N, and the Pacific Ocean on the E. Southeast Asia includes the Indochina Peninsula, which juts into the South China Sea, the Malay Peninsula, and the Malay Archipelago. The region has 9 independent countries: Burma, Thailand, Malaysia, Cambodia, Laos, Vietnam, the Philippines, Singapore, and Indonesia. It also includes Brunei, a British protectorate, and Portuguese Timor. Peninsular Southeast Asia is traversed by many mountains and drained by great rivers such as the Salween, Irrawaddy, Chao Phraya, and Mekong. Insular Southeast Asia is made up of numerous volcanic and coral islands. Southeast Asia has a generally tropical rainy climate except for the northwestern part, which has a humid subtropical climate. The wet monsoon winds are vital for the economic well-being of the region. Tropical forests cover most of the area. Rice is the chief crop; rubber, tea, spices, and coconuts are also important. The region has many minerals and produces most of the world's tin. Population is unevenly distributed, with the highest density in lowland areas. Most of the people live in small agrarian villages; the largest cities are Djakarta, Indonesia; Bangkok, Thailand; Singapore; Manila, Philippines; and Saigon, Vietnam. There is a great diversity in culture, history, religion, and ethnic composition. Many different languages are spoken, such as those of the Tibeto-Burman, Mon-Khmer, and Malayo-Polynesian families. Religions include Buddhism, Islam, Roman Catholicism, and Confucianism.

Most of the influences that molded the societies of Southeast Asia came from early Indian and Chinese sources. Several great civilizations, including those of the Khmers and Malays, flourished here. In the late 15th cent. strong Islamic influences were overshadowed by the arrival of the Europeans; only Thailand remained free of colonial occupation. Because of Southeast Asia's strategic location the region became a battleground between Allied and Japanese forces during World War II. Since the war the countries of Southeast Asia have become independent nations. They have been plagued by political turmoil, weak economies, ethnic strife, and social inequities. Throughout the 1960s and early 1970s there were open conflicts between Communist and non-Communist factions throughout most of the region, especially in Vietnam, Laos, and Cambodia.

South El Mon·te (ĕl mŏn'tē), city (1975 est. pop. 14,400), Los Angeles co., S Calif., in the San Gabriel Valley; inc. 1958. Its manufactures include home furnishings, household appliances, building materials, and various farm implements.

South·end-on-Sea (sou'thĕnd-ŏn-sē'), county borough (1974 est. pop. 163,380), Essex, E England, at the mouth of the Thames River. It is a popular seaside resort. Its manufactures include aluminum foil, paint, electronic equipment, and pharmaceuticals.

Southern Alps (ălps), mountain range, on South Island, New Zealand, paralleling the W coast. It rises to 12,349 ft (3,766.4 m) at Mt. Cook, New Zealand's highest peak. Extensively glaciated, the snow-capped range has many deep gorges.

South·ern Pines (sŭ*th*'ərn pīnz), winter resort town (1970 pop. 5,937), Moore co., central N.C., SW of Raleigh; settled 1885, inc. 1887. Noted for its mild climate, it is a center for riding, hunting, and other sports.

Southern Ye·men (yĕm'ən, yā'mən): *see* Yemen, Southern.

South Eu·clid (yōō'klĭd), city (1978 est. pop. 28,500), Cuyahoga co., NE Ohio, a suburb of Cleveland; inc. as a city 1940. It is mostly residential.

South Farm·ing·dale (fär'mĭng-dāl'), uninc. town (1978 est. pop. 20,700), Nassau co., SE N.Y., on Long Island.

South·field (south'fēld'), city (1978 est. pop. 77,700), Oakland co., SE Mich., a suburb of Detroit, on the Rouge River; laid out 1817, inc. as a city 1958. The city has varied light manufacturing and a warehousing industry.

South·gate (south'gāt'), city (1978 est. pop. 35,900), Wayne co., SE Mich., a residential suburb of Detroit; settled 1840–60, inc. 1958.

South Gate (gāt), city (1970 pop. 56,909), Los Angeles co., S Calif., an industrial suburb of Los Angeles; inc. 1923.

South Geor·gia (jôr'jə), island, c.1,450 sq mi (3,755 sq km), S Atlantic Ocean, c.1,200 mi (1,930 km) E of Cape Horn. It is a dependency of the British colony of the Falkland Islands and, with the Falklands, is claimed by Argentina. Capt. James Cook took possession of South Georgia in 1775. The island is located in the world's greatest whaling area.

South Gla·mor·gan (glə-môr'gən), nonmetropolitan county (1974 pop. 392,250), S Wales, created under the Local Government Act of 1972 (effective 1974). It comprises the county borough of Cardiff and portions of the former counties of Glamorganshire and Monmouthshire.

South Had·ley (hăd'lē), residential town (1975 est. pop. 17,000), Hampshire co., W Mass., on the Connecticut River near the Holyoke Range; settled 1684, inc. 1775. Its paper industry dates from the early 19th cent. Today electronic equipment, machinery, metal stampings, and concrete products are also made. The first navigable canal in the United States began operation here in 1795.

South Ha·ven (hā'vən), city (1970 pop. 6,471), Van Buren co., SW Mich., on Lake Michigan N of Benton Harbor; settled before 1840, inc. as a village 1869, as a city 1902. It is a port of entry and a trade and shipping center for a resort and fruit-growing area. Pianos, castings, and wood products are made.

South Hol·land (hŏl'ənd), province (1973 est. pop. 3,018,905), c.1,085 sq mi (2,810 sq km), W Netherlands, bounded by the North Sea in the W; capital The Hague. A fertile lowland, protected by dunes and dikes along the coast, its physical geography is similar to that of North Holland, with which it was united until 1840 as Holland.

South Holland, village (1978 est. pop. 27,100), Cook co., NE Ill., a suburb of Chicago; settled 1846 by Dutch, inc. 1894.

South Hous·ton (hyōō'stən), city (1970 pop. 11,527), Harris co., S Texas, an industrial suburb of Houston; inc. 1911.

South Hunt·ing·ton (hŭnt'ĭng-tən), uninc. town (1970 pop. 8,946), Suffolk co., SE N.Y., on Long Island S of Huntington.

South·ing·ton (sŭ*th*'ĭng-tən), town (1975 est. pop. 34,300), Hartford co., central Conn.; settled 1696, inc. 1779. Manufacturing began in Southington in the 1770s. Machine tools, chemicals, primary and fabricated metals, and electrical and transportation equipment are among the town's manufactures.

South Island (1974 est. pop. 842,300), 58,093 sq mi (150,461 sq km), New Zealand. It is the larger but less populous of the two principal islands of the country. It is separated from North Island by Cook Strait and from Stewart Island by Foveaux Strait. The Southern Alps extend almost the entire length of the island. Grain, fruit, timber, and sheep are the leading products; some coal, gold, and oil is found here.

South Kings·town (kĭng'stən, kĭngz'toun'), town (1975 est. pop. 17,300), seat of Washington co., S R.I.; settled 1641, inc. 1674 as Kingstown, divided into South Kingstown and North Kingstown 1723. Textiles are the chief product. The Narragansett Indians made their last stand in King Philip's War at nearby Great Swamp, now a historic site called Great Swamp Reservation.

South Mi·am·i (mī-ăm'ē, -ăm'ə), city (1970 pop. 11,780), Dade co., SE Fla., a suburb of Miami; settled 1899, inc. 1926.

South Mil·wau·kee (mĭl-wô'kē), industrial city (1978 est. pop. 23,200), Milwaukee co., SE Wis., on Lake Michigan; settled 1835, inc. 1897. Excavating machinery, electrical transmission equipment, and leather products are among the many manufactures.

South Og·den (ôg′dən, ŏg′-), city (1970 pop. 9,991), Weber co., N Utah, just S of Ogden; settled 1848 by Mormons, inc. 1936.

South·old (south′ōld′), resort town (1976 est.pop. 18,300), Suffolk co., SE N.Y., on NE Long Island, on Long Island Sound ENE of Riverhead; settled 1640.

South Or·ange (ôr′ĭnj, ŏr′-), village (1975 est. pop. 15,900), Essex co., NE N.J., inc. 1869. It is mostly residential and is the home of Seton Hall Univ.

South Ork·ney Islands (ôrk′nē), group in the South Atlantic, c.850 mi (1,370 km) SE of Cape Horn. Discovered in 1821, they were claimed by the British and are included as dependencies of the colony of the Falkland Islands. The Argentine government, which also claims them, maintains meteorological and wireless stations on Laurie Island.

South Par·is (păr′ĭs), town (1970 pop. 2,315), seat of Oxford co., W Maine, NW of Auburn; settled 1779, inc. 1793. Leather goods and food and wood products are made.

South Pas·a·de·na (păs′ə-dē′nə), city (1978 est. pop. 23,600), Los Angeles co., S Calif., a suburb of Los Angeles; inc. 1888. It is chiefly residential.

South Pass, broad, level valley (alt. c.7,550 ft/2,305 m), SW Wyo., cutting across the Rocky Mts. On the Oregon Trail, it served as a gateway for immigration to the Far West. It is now a national historic landmark.

South Plain·field (plān′fēld′), borough (1978 est. pop. 21,100), Middlesex co., NE N.J.; inc. 1926. Its businesses and industries include research and consulting firms, chemicals, electrical machinery, structural steel, gypsum products, and toys.

South Platte (plăt), river, c.450 mi (725 km) long, rising in the Rocky Mts. in many branches, which then join in central Colo. It flows in a narrow canyon east and northeast to Denver, then is broad and shallow as it crosses the Great Plains to join the North Platte in central Nebr., where it forms the Platte River. On the Great Plains portion, grazing and irrigated agriculture are important. The upper course of the South Platte is part of the Bureau of Reclamation's Colorado-Big Thompson project.

South Pole, southern end of the earth's axis, lat. 90° S and long. 0°. It is distinguished from the south magnetic pole. The South Pole was reached by Roald Amundsen, a Norwegian explorer, in 1911.

South·port (south′pôrt′, -pōrt′), county borough (1974 est. pop. 85,250), Merseyside, NW England. A seaside resort with light industries, it has several art and technical schools.

Southport. 1. Uninc. town (1970 pop. 11,976), Chemung co., S N.Y., SSW of Elmira. **2.** City (1970 pop. 2,220), seat of Brunswick co., SE N.C., on the Atlantic coast S of Wilmington and at the mouth of the Cape Fear River; founded 1792. Fort Johnston was begun here in 1745 and completed in 1764. The city has resort activities and is a fish-processing point noted for its shrimp fisheries.

South Port·land (pôrt′lənd, pōrt′-), city (1978 est. pop. 22,200), Cumberland co., SW Maine, on the Fore River and Casco Bay, part of the Portland metropolitan area; inc. 1898. Ships have been built here since the 17th cent. The city also has an industrial center with many varied light manufactures. Fort Preble was built here before the War of 1812. Portland Head Light, near the fort, is the oldest lighthouse (1791) on the Maine coast.

South Riv·er (rĭv′ər), borough (1970 pop. 15,428), Middlesex co., E N.J.; settled 1720, inc. 1898. Dress manufacturing is its major industry.

South Saint Paul (sānt pôl′), city (1978 est. pop. 20,800), Dakota co., SE Minn., a suburb of St. Paul, on the Mississippi River; inc. 1887. It is known for its stockyards and meat-packing industries. Its 75-acre (30-hectare) public livestock market is one of the nation's largest, serving some 20 states. Building materials and sheet-metal and wood products are also made. The city was settled in 1853 on the site of an Indian village.

South Salt Lake (sôlt lāk′), residential city (1970 pop. 7,810), Salt Lake co., central Utah, S of Salt Lake City; inc. 1938.

South San Fran·cis·co (săn′ frən-sĭs′kō), city (1978 est. pop. 49,300), San Mateo co., W Calif.; inc. 1908. South San Francisco has several industrial parks; its manufactures include steel, chemicals, metal products, paints, and processed foods.

South Seas, name given by early explorers to the whole of the Pacific Ocean. In recent times the name has been used to mean only the central Pacific, the south Pacific, and the southwest Pacific. More particularly it is applied to the South Sea Islands and the waters about them.

South Shet·land Islands (shĕt′lənd), barren, snow-covered archipelago off N Antarctic Peninsula, W Antarctica; Livingston and King George islands are the largest. Formerly bases for sealers and whalers and also for antarctic exploration, they now have scientific bases. Discovered by the British mariner William Smith in 1819, the islands are claimed by Great Britain and are part of the British Antarctic Territories. Argentina and Chile challenge this claim.

South Shields (shēldz), county borough (1974 est. pop. 98,610), Tyne and Wear, NE England, at the mouth of the Tyne River. It is a significant port that exports coal and iron. Shipbuilding and marine engineering are the main industries, and chemicals and paints are made.

South Sioux City (so͞o), city (1976 est. pop. 8,504), Dakota co., NE Nebr., on the Missouri River opposite Sioux City, Iowa, in a rich agricultural area; founded 1887.

South Su·bur·ban (sə-bûr′bən), city (1971 pop. 273,762), West Bengal state, NE India. It is a suburb of Calcutta.

South Vi·et·nam (vē-ĕt′näm′, -năm′): *see* Vietnam.

South·wark (sŭ*th*′ərk, south′wûrk′), borough (1971 pop. 259,982) of Greater London, SE England, on the Thames River. The borough was created in 1965 by the amalgamation of the metropolitan boroughs of Bermondsey, Camberwell, and Southwark. Printing, engineering, and furniture manufacture are the main industries. Camberwell is mainly residential. The old Southwark area is situated at the convergence of roads to London. It had a number of famous inns. The Bankside district of Southwark contains the Globe Theater and other places associated with Shakespeare. Dulwich College, in Camberwell, is a public school that opened in 1619.

South·well (south′wĕl′, -wəl), rural district (1971 pop. 57,365), Nottinghamshire, central England. It includes the small civil parish of Southwell, which since 1884 has been the cathedral town of Nottinghamshire. The present cathedral, begun c.1110, is on the site of a church said to have been founded in the 7th cent. by Paulinus.

South West Af·ri·ca (ăf′rĭ-kə): *see* Namibia.

South West·bur·y (wĕst′bĕr′ē), uninc. town (1970 pop. 10,978), Nassau co., SE N.Y., on Long Island.

South Wil·liams·port (wĭl′yəmz-pôrt′, -pōrt′), borough (1970 pop. 7,153), Lycoming co., N central Pa., on the West Branch of the Susquehanna River opposite Williamsport; inc. 1886.

South Wind·sor (wĭn′zər), town (1975 est. pop. 15,700), Hartford co., N Conn.; set off from Windsor 1845. Oliver Wolcott, a signer of the Declaration of Independence, was born here.

South·wick (south′wĭk′), rural town (1970 pop. 6,330), Hampden co., SW Mass., SW of Springfield; inc. 1770.

South York·shire (yôrk′shĭr′, -shər), metropolitan county (1974 est. pop. 1,319,180), N central England, created under the Local Government Act of 1972 (effective 1974). It is subdivided into four metropolitan districts. South Yorkshire comprises the county boroughs of Barnsley, Doncaster, Rotherham, and Sheffield, and parts of the former counties of Nottinghamshire and Yorkshire (West Riding).

So·vetsk (səv-yĕtsk′), formerly **Til·sit** (tĭl′zĭt, -sĭt), town (1970 pop. 38,456), NW European USSR, on the Neman River at the mouth of the Tilse. It is a rail junction, a river port, and an industrial and commercial center in an agricultural area. Lumbering and woodworking are the chief industries; others include the production of machines, iron, cotton cloth, and Tilsit cheese. The town grew around a castle built in 1288 by the Teutonic Knights and was chartered in 1552. Tilsit, an East Prussian town, was the site of treaties negotiated in 1807 between Napoleon I, Emperor Alexander I of Russia, and King Frederick William III of Prussia. The Russo-French alliance collapsed in 1812. Prussia was a virtual vassal of France under the treaties, freeing itself in 1813. Tilsit was occupied by Soviet forces in World War II and was transferred, along with other sections of East Prussia, to the USSR at the Potsdam Conference of 1945.

Soviet Far East, region (1970 pop. 5,800,000), c.2,400,000 sq mi (6,216,000 sq km), encompassing the entire northeast coast of Asia and including the Primorsky Kray (Maritime Territory), Khabarovsk Kray, and Amur, Magadan, Kamchatka, and Sakhalin oblasts of the USSR. Although often considered a part of Siberia, the Soviet Far East is treated separately in USSR regional schemes. The Soviet Far East is bounded on the northwest by the Yakut Autonomous SSR, on the north by the East Siberian Sea, on the northeast by the Bering Sea, on the southeast by the Sea of Japan, on the south by China (Manchuria), and on the southwest by the Yablonovy Mts. Other ranges in this mountainous area include the Stanovoy, Dzhugdzhur, and Kolyma. Arctic tundra covers the far north of the region, and forest taiga occupies the central section. In the south are the fertile Amur and Ussuri river valleys, which support crops of wheat, oats, soybeans, and sugar beets.

The Soviet Far East is virtually self-sufficient economically. Iron and steel manufacturing, oil refining, lumbering, and machine building are among the many industries that have developed in such important urban centers as Vladivostok, Komsomolsk, Khabarovsk, Ussuriysk, and Nikolayevsk. Large thermoelectric stations furnish industrial power. Coal is mined in the Buryea River basin and on Sakhalin, whose northern half also contains major oil fields. The Kolyma gold fields constitute the chief source of Soviet gold, and there are rich deposits of iron ore, lignite, lead, zinc, and silver in the Soviet Far East. Fishing, fur hunting, and trapping are important occupations. Major means of transport in the region include the Trans-Siberian RR and the Amur River. More than 25 ethnic groups inhabit the Soviet Far East, among them Russians, Jews, Koryaks, Tungus, Chukchi, and Kamchatkans.

History. Russian colonization of the area began in the late 16th cent. In 1856–57 the Russians took advantage of a weak Chinese empire to occupy all of the territory north of the Amur, and in 1860 they seized the land east of the Ussuri. With completion of the Trans-Siberian RR, Russian settlement of the Far East accelerated. Russia retained northern Sakhalin under the Treaty of Portsmouth (1905), but Japan was awarded the rest of the island.

After the Russian Revolution (1917), Japanese forces landed at Vladivostok and occupied large parts of the Russian territory. In 1920 the Far Eastern Republic was formed as a buffer state between Japan and the Soviet Union. In 1922 the Japanese forces withdrew, the republic was dissolved, and the area was incorporated into the USSR as an oblast. In the settlement following World War II the USSR acquired the southern half of Sakhalin and the Kuril Islands. In 1969 Sino-Soviet clashes erupted along the Amur and Ussuri frontiers between the countries. Subsequent negotiations bogged down, and both sides reinforced their military capabilities on the long borders.

So·ya Strait (sō′yä): *see* La Pérouse Strait, Japan-USSR.

Spa (spä, spô), town (1971 est. pop. 9,557), Liège prov., E Belgium, in the Ardennes. Its therapeutic mineral springs and baths, frequented since the 16th cent., made it so well known that the word "spa" now designates any similar health resort. The town had its greatest vogue in the 18th and 19th cent.

Spain (spān), country (1970 pop. 33,823,918), 194,884 sq mi (504,750 sq km), SW Europe. It consists of the Spanish mainland (190,190 sq mi/492,592 sq km), which occupies the major part of the Iberian Peninsula; of the Balearic Islands in the Mediterranean Sea; and of the Canary Islands in the Atlantic Ocean. Madrid is the capital. The Spanish territories in Africa have been reduced to two enclaves in Morocco. Continental Spain extends from the Pyrenees, which separate it from France, and from the Bay of Biscay, an arm of the Atlantic Ocean, southward to the Strait of Gibraltar, which separates it from Africa. (Gibraltar itself is a British possession.) The eastern and southeastern coast of Spain, from the French border to the Strait of Gibraltar, is washed by the Mediterranean. In the west Spain borders on the Atlantic Ocean both north and south of its frontier with Portugal.

The center of Spain forms a vast plateau extending from the Cantabrian Mts. in the north to the Sierra Morena in the south and from the Portuguese border in the west to the low ranges that separate the plateau from the Mediterranean coast in the east. It is traversed from west to east by mountain chains—notably the Sierra de Guadarrama—and the valleys of the Douro (Duero), Tagus, and Guadiana rivers. Except for some fertile valleys and irrigated lands, the central plateau is arid and thinly populated. It comprises the historic region of Castile, the heart of Spain, and in the west León and Estremadura. The chief cities are Madrid, Burgos, Valladolid, and Toledo in Castile; León, Zamora, and Salamanca in León; and Badajoz in Es-

tremadura. To the northeast of the central plateau is the broad valley of the Ebro, which traverses Aragón and flows into the Mediterranean. Aragón has Zaragoza as its chief city; it is historically and geographically connected with Catalonia, which occupies the Mediterranean coast from the French border to the mouth of the Ebro. Barcelona, the chief Catalan city, is the largest port and the second-largest city of Spain. The western Pyrenees and the northern coast, paralleled by the Cantabrian Mts., are occupied by Navarre, with the city of Pamplona; the Basque Provinces, with the ports of Bilbao and San Sebastián; Santander; and Asturias, with Oviedo and the port of Gijón. The extreme northwestern section, occupied by Galicia, has a deeply indented coast and excellent ports. Along the eastern coast, south of Catalonia, extend the regions of Valencia and Murcia, named after their chief cities. The Balearic Islands, with Palma as their capital, are off the coast of Valencia. The southernmost part of Spain, south of the Sierra Morena, is Andalusia; it is crossed by the fertile Guadalquivir valley. The chief cities of Andalusia are Seville, Córdoba, and Granada, the Mediterranean port of Málaga, and the Atlantic port of Cádiz. The Sierra Nevada, rising from the Mediterranean coast, has the highest peak (Mulhacén, 11,411 ft/3,478 m) in continental Spain.

The Spanish people, despite a strongly centralized government, display great regional diversity. Separatist tendencies remain particularly strong among the Catalans and the Basques. The Castilian dialect has become the standard Spanish language, but Catalan (akin to Provençal), Galician (akin to Portuguese), and Basque, unrelated to any other language, are still spoken and written extensively in their respective districts. The Roman Catholic Church is the established church in Spain. Illiteracy is high among the lower classes despite laws compelling elementary education.

Economy. Spain has made great economic progress in recent years, but it is still not among the most prosperous nations in Western Europe. Tourism is its greatest source of income. Primarily an agricultural country, it produces large crops of wheat, potatoes, sugar beets, barley, tomatoes, olives, citrus fruit, grapes, and cork. The best-known wine regions are those of Rioja, in the upper Ebro valley, and of Málaga and Jerez de la Frontera, in Andalusia. Agriculture is handicapped in many regions by an obsolete system of land tenure, by lack of mechanization, by insufficient irrigation, and by soil exhaustion and erosion. The major industries produce textiles, iron and steel, and chemicals. Industries are concentrated chiefly in the Madrid region; in Valladolid; in Catalonia, which has large textile manufactures; in Valencia; and in Asturias and the Basque Provinces, where the rich mineral resources of the Cantabrian Mts. (iron, coal, and zinc) are exploited. Copper is mined extensively at Río Tinto; other mineral resources include lead, silver, tin, and mercury. Petroleum is found near Burgos. Fishing, notably for sardines, tunny, cod, and anchovies, is an important source of livelihood for the coastal population, and fish canning is a major industry.

Spain's greatest trade is with the United States, West Germany, France, and Great Britain. Among leading exports are fruit, wine, and other food products, ships and other transportation equipment, and chemicals; major imports include machinery, petroleum, and iron and steel. Overland communications are generally poor. Most Spanish railroads, unlike those of the rest of Western Europe, use broad-gauged tracks.

Spain is divided into 50 provinces:

NAME	CAPITAL	NAME	CAPITAL
Álava	Vitoria	Lérida	Lérida
Albacete	Albacete	Logroño	Logroño
Alicante	Alicante	Lugo	Lugo
Almería	Almería	Madrid	Madrid
Ávila	Ávila	Málaga	Málaga
Badajoz	Badajoz	Murcia	Murcia
Baleares	Palma	Navarra	Pamplona
Barcelona	Barcelona	Orense	Orense
Burgos	Burgos	Oviedo	Oviedo
Cáceres	Cáceres	Palencia	Palencia
Cádiz	Cádiz	Pontevedra	Pontevedra
Castellón de la Plana	Castellón de la Plana	Salamanca	Salamanca
Cuidad Real	Cuidad Real	Santa Cruz de Tenerife	Santa Cruz de Tenerife
Córdoba	Córdoba	Santander	Santander
Cuenca	Cuenca	Segovia	Segovia
Gerona	Gerona	Sevilla	Seville
Granada	Granada	Soria	Soria
Guadalajara	Guadalajara	Tarragona	Tarragona
Guipúzcoa	San Sebastián	Teruel	Teruel
Huelva	Huelva	Toledo	Toledo
Huesca	Huesca	Valencia	Valencia
Jaén	Jaén	Valladolid	Valladolid
La Coruña	La Coruña	Vizcaya	Bilbao
Las Palmas	Las Palmas de Gran Canaria	Zamora	Zamora
León	León	Zaragoza	Saragossa

History. Civilization in Spain dates back to the Stone Age. The Basques antedated the Iberians, who mixed with Celtic invaders at an early period. The Phoenicians (9th cent. B.C.) and later the Carthaginians and Greeks established colonies on the coast and in the Balearic Islands. In the 3rd cent. B.C. the Carthaginians began to conquer most of the Iberian Peninsula, but the Romans expelled them in the second of the Punic Wars (218-201 B.C.). The Romans met strong resistance in the north, but by the 1st cent. A.D. their control was virtually complete. Except for the Basques, the Iberian population became thoroughly romanized, a process that brought political unity, law, and economic prosperity. In the 5th cent. A.D. Spain was overrun by Germanic invaders, and the Visigoths imposed (419 A.D.) a weak rule under which the Roman Catholic clergy acquired power. In 711 a Moslem Berber army crossed the Strait of Gibraltar and soon conquered the entire peninsula.

The Moors, as the Berber conquerors were called, established a Moslem state that reached its greatest splendor under Abd ar-Rahman III (reigned 912-61). The Moorish capital, Córdoba, became the chief center of learning in Europe. Commerce, craftsmanship, and agriculture also thrived. The Basques (in Navarre) and the Franks (in Catalonia) held out against the Moors. In succeeding centuries Christian kingdoms, notably Aragón, Léon, and Castile, gradually expanded through royal inheritance and warfare at the expense of the Moorish emirates. Under Castilian leadership the Christian reconquest was completed, and the fall of Granada (1492), the last Moorish outpost, made Ferdinand V of Aragón and Isabella I rulers of all Spain; the Moriscos (Christian Moors) were persecuted and finally expelled in 1609.

An economic revolution followed the discovery (1492) of America by Christopher Columbus. Almost all of South America, Central America, southern North America, and the Philippines were added to the Spanish world empire in the 16th cent. Gold and silver, the primary objective of the conquistadores, flowed into Spain in fabulous quantities. Spain in the 16th cent. (the Golden Century) was the first power of the world, with a brilliant cultural, artistic, and intellectual life. In the Italian Wars (1494-1559), Spain defeated France and added Naples and the duchy of Milan to its dependencies. At the end (1556) of the reign of Charles I, first of the Hapsburg kings (who ruled from 1516 to 1700), Spain was on its way to becoming a centralized and absolute monarchy. During the 16th cent. the church enlarged its already dominant position in Spanish life; the inquisition reached its greatest power, while at the same time the Catholic Reformation was advanced.

Spanish military might began to wane with the loss of parts of the Netherlands (1579), the defeat of the Spanish Armada (1588) by England, and losses to France in the 17th cent. Portugal, united with Spain by Philip II in 1580, rebelled and regained its independence in 1640. Charles II (reigned 1665-1700) lacked a direct heir, and his choice of Louis XIV's grandson, Philip, as successor provoked the War of the Spanish Succession (1701-14). The Peace of Utrecht confirmed Philip V as Spain's first Bourbon king. The Bourbons tried ineffectively to check the steady economic decline brought on by costly wars, the loss of foreign holdings, impoverishment of the soil, and corruption among government officials and clergy. Spain suffered its greatest humiliation in 1808 when Napoleon's French forces occupied the country and installed Joseph Bonaparte on the Spanish throne.

By 1814 Spanish resistance forces and the British had expelled the French, and Ferdinand VII (reigned 1814-33) was restored in 1812 at Cádiz by the first national Cortes (parliament).

A nationalist and liberal upsurge swept over Spain and its overseas empire, with the result that by 1825 most of Latin America had gained independence and Spain itself was bitterly contested by reactionary, moderate, and republican factions. Uprisings and coups d'état were frequent, and two Carlist Wars (1833-39 and 1872-76) were fought on behalf of the royal pretender Don Carlos (brother of Ferdinand VII) and his heirs. By the end of the 19th cent. socialism and anti-clericalism had gained wide followings.

Under Alfonso XIII (reigned 1886-1931) Spain lost the remainder of its empire in the Spanish-American War (1898) and remained neutral in World War I. Great social and economic unrest marked the postwar period. In 1931, after a republican victory in municipal elections, Alfonso XIII was deposed and a republic established. The Popular Front (republicans, Socialists, Communists, and syndicalists) was victorious in the national elections of 1936. Before the government had time to carry out its program, a military rebellion precipitated the Spanish civil war of 1936-39. The Insurgents, led by Gen. Francisco Franco, supported by conservative groups, and aided by Fascist Germany and Italy, defeated the Loyalists, who received meager aid from the USSR. A dictatorship was set up under Franco. Although it gave aid to the Axis, Spain remained a nonbelligerent in World War II. It entered the UN in 1955. An agreement with the United States in 1953 provided for U.S. bases in Spain and for economic and military aid. In 1956 Spanish Morocco became part of the independent state of Morocco; in 1968 Spanish Equatorial Guinea became independent; and in 1969 Ifni was ceded to Morocco.

Widespread strikes, student demonstrations, and Basque and Catalan separatist agitation gave evidence of unrest in the 1960s. Church leaders began to criticize aspects of the dictatorship. In 1966 Franco imposed certain limits on his own power and granted a somewhat liberalized constitution. Political violence and impatience for change intensified. Franco died in 1975; Prince Juan Carlos de Bourbon, his designated heir, became chief of state and was sworn in as king of a restored Spanish monarchy. In an atmosphere of political ferment verging on civil war, Juan Carlos I and his ministers advanced a program of reforms aimed at making Spain a parliamentary democracy. Amnesty was granted to political prisoners (1976), political parties, including the Communist, were legalized (1977), and the Cortes was reorganized and democratized (1977).

Government. Spain is a monarchy headed by King Juan Carlos I, who became chief of state three weeks before the death in 1975 of dictator Francisco Franco. The king appoints a cabinet, the prime minister of which is the head of government. In accordance with reforms approved in a 1976 plebiscite and enacted in 1977, legislative power is invested in a parliament (Cortes) consisting of a congress with 350 deputies serving four-year terms and a senate with 207 members serving six-year terms. All of the deputies (proportioned one for each 175,000 inhabitants) and about half of the senators (two for each of the 50 provinces and the Moroccan enclaves of Ceuta and Melilla) are elected by popular vote; the remaining senators are appointed by the king. Congress initiates legislation, which must be approved by a majority of both chambers.

Spal·ding (spôl′dĭng), county (1970 pop. 39,514), 201 sq mi (520.6 sq km), W central Ga., bounded on the W by the Flint River; formed 1851; co. seat Griffin. It is in a piedmont agricultural (cotton, corn, peppers, fruit, and livestock) and timber area. Its industry includes canned fruit and vegetables, weaving and knitting mills, textile finishing, and clothing manufacturing.

Span·dau (spän′dou′, shpän′-), district of West Berlin, Germany, at the confluence of the Havel and Spree rivers. It is a canal port and industrial center. Manufactures include steel and electrical equipment. Chartered as a town in 1232, it is the site of a major fortress built on the Havel River in 1560-94. The fortress was occupied in the Thirty Years' War by the Swedes (1631-34) and in the French Revolutionary Wars by the French (1806-13). It later became a dread political prison; several major Nazi war criminals were imprisoned here after the Nuremberg trials (1945-46). Spandau was incorporated into Berlin in 1920.

Span·ish (spăn'ĭsh), river, c.150 mi (240 km) long, issuing from Spanish Lake, S Ont., Canada, NW of Sudbury, and flowing generally S through Biskotasi and Agnew lakes to Lake Huron opposite Manitoulin island. There are several hydroelectric stations on the river.

Spanish Fork (fôrk), city (1970 pop. 7,284), Utah co., N central Utah, S of Provo near Spanish Fork River and Utah Lake; settled 1850 by Mormons, inc. 1855. It is a processing center of an irrigated farm and livestock area served by the Strawberry valley project.

Spanish Main (mān), mainland of Spanish America, particularly the coast of South America from the isthmus of Panama to the mouth of the Orinoco River. Spanish treasure fleets, sailing home from the New World, passed through the Caribbean north of the Main and were attacked by English buccaneers, raiding from the islands and coast. Pirates congregated here until the 19th cent.

Spanish Sa·ha·ra (sə-hâr'ə, -hăr'ə, -hä'rə): *see* Western Sahara.

Spanish Town, city (1970 pop. 41,600), SE Jamaica, on the Cobre River. It is the commercial and processing center of a rich agricultural region, as well as the main rail and highway communications hub for traffic to and from Kingston (the capital) and other parts of Jamaica.

Sparks (spärks), city (1978 est. pop. 36,200), Washoe co., W Nev., just E of Reno; inc. 1905. The city has important railroad, distributing, and warehousing activities. Tourism is a major industry, and there is some light manufacturing.

Spar·ta (spär'tə), city of ancient Greece, capital of Laconia, on the Eurotas (Evrótas) River in the Peloponnesus. The narrow, mountain-walled valley attracted invaders by its fertility, and the city-state of Sparta was created by invading Dorian Greeks, who later conquered the countryside of Laconia and Messenia (c.735-715 B.C.). In the 7th cent. B.C. Sparta enjoyed a period of wealth and culture, but after 600 B.C. it cultivated only the military arts. The city became an armed camp established (according to the official legend) by Lycurgus. The ruling class, the Spartiates, gave themselves wholly to rigorous military training and war. By the 6th cent. B.C. Sparta was the strongest Greek city. In the Persian Wars, Sparta fought beside Athens, but rivalry between the two city-states increased in the mid-5th cent. B.C. Sparta triumphed in the great Peloponnesian War (431-404 B.C.), which wrecked the Athenian empire. Soon afterward the Spartans were weakened by a war with Persia and lost their ascendancy in Greece to the Thebans c.371 B.C. Sparta fell an easy prey to Macedonia and declined. Under the Romans the city prospered. It was devastated by the Goths in A.D. 395. The modern city of Sparta (now Spárti; 1971 pop. 10,549) dates only from the 19th cent. The ruins of old Sparta, including sanctuaries and a theater, are nearby.

Sparta. 1. City (1970 pop. 2,172), seat of Hancock co., N central Ga., NE of Macon; inc. 1805. It is a trade, processing, and shipping center for a farm, granite, and timber area. **2.** Resort town (1970 pop. 1,304), seat of Alleghany co., NW N.C., near the Va. border W of Mount Airy. Lumber is manufactured here, and there are grain farms in the region. **3.** Town (1970 pop. 5,038), seat of White co., central Tenn., in the Cumberland Plateau NNW of Chattanooga, in a coal, limestone, and timber area; founded 1809, inc. 1833. Textiles and metal products are manufactured. **4.** City (1970 pop. 6,258), seat of Monroe co., W central Wis., on the La Crosse River ENE of La Crosse, in a farm and dairy area; settled c.1850, inc. as a village 1857, as a city 1883.

Spar·tan·burg (spärt'n-bûrg'), county (1970 pop. 173,724), 831 sq mi (2,152.3 sq km), NW S.C., bounded on the N by the N.C. border, on the SW by the Enoree River, and drained by the Tyger River; formed 1785; co. seat Spartanburg. It is in a peach-farming area that also produces cotton and corn. Its industries include textile milling, food processing, and the manufacture of wood products, chemicals, concrete, and machinery.

Spartanburg, city (1978 est. pop. 46,900), seat of Spartanburg co., NW S.C., in the Piedmont near the N.C. line; inc. 1831. The city is a noted textile center and an important commercial, transportation, and trade focus in an extensive cotton and farm-produce region. Textiles, machinery, ceramics, chemicals, plastic, rubber, and products of wood, metal, and paper are manufactured. A huge textile-research center is here. In the Revolutionary War 11 major battles were fought in the area. City and county were named for the "Spartan" regiment of Revolutionary troops recruited in the area. In the Civil War the city was a busy supply-manufacturing point. Spartanburg is the seat of several colleges, a branch of the Univ. of South Carolina, and a state school for the deaf and blind (est. 1849).

Spar·ti·ven·to, Cape (spär'tĭ-vĕn'tō), SE extremity of the "toe" of Italy, in Calabria, extending into the Ionian Sea.

Spear·man (spîr'mən), town (1970 pop. 3,435), seat of Hansford co., extreme N Texas, NNE of Amarillo, in a wheat and cattle section of the Panhandle; inc. 1921. A natural-gas plant is here.

Speed·way (spēd'wā'), town (1975 est. pop. 14,600), Marion co., central Ind., just W of Indianapolis; inc. 1926. The Indianapolis Speedway is here.

Spe·nard (spə-närd'), uninc. town (1970 pop. 18,089), Anchorage dist., S central Alaska, a suburb of Anchorage.

Spen·cer (spĕn'sər). **1.** County (1970 pop. 17,134), 396 sq mi (1,025.6 sq km), SW Ind., bounded on the S by the Ohio River and the Ky. line and drained by the Anderson River and Little Pigeon Creek; formed 1818; co. seat Rockport. In an agricultural area yielding grain, livestock, and poultry, it has clay pits, and its industry includes food processing and lumbering. **2.** County (1970 pop. 5,488), 193 sq mi (500 sq km), central Ky., in the Bluegrass, drained by the Salt River and several creeks; formed 1824; co. seat Taylorsville. It is in a rolling upland agricultural area that produces livestock, burley tobacco, and corn. It has some manufacturing.

Spencer. 1. Town (1970 pop. 2,553), seat of Owen co., SW Ind., on the White River NW of Bloomington, in a farm area; settled c.1815. **2.** City (1975 est. pop. 10,200), seat of Clay co., NW Iowa, on the Little Sioux River; inc. 1880. The city lies in a rich farm area. Beef is processed here. Work clothes, tools, prefabricated buildings, and beauty aids are among Spencer's manufactures. **3.** Town (1970 pop. 8,779), Worcester co., central Mass., W of Worcester; settled 1721, inc. 1775. Its footwear industry dates from the early 19th cent. **4.** Town (1970 pop. 1,179), seat of Van Buren co., central Tenn., S of Sparta, in the Cumberlands. Livestock is raised. Great Falls Dam is nearby. **5.** City (1970 pop. 2,271), seat of Roane co., W central W.Va., NNE of Charleston; settled 1812, inc. 1858. Clothing and rubber products are made.

Spencer Gulf, inlet of the Indian Ocean, 200 mi (322 km) long and 80 mi (129 km) wide, SE South Australia state, Australia, between Eyre and Yorke peninsulas. The gulf is the major outlet for iron ore from Middleback Range.

Spey (spā), river, c.105 mi (170 km) long, rising in the Mondhliath Mts., NE Scotland, and flowing generally NE through the Moray Firth to the North Sea. The river is rapid and unnavigable. There are important salmon fisheries on the lower Spey.

Spey·er (shpī'ər), city (1972 est. pop. 43,415), Rhineland-Palatinate, SW West Germany, on the Rhine River. The city, sometimes called Spires in English, is a river port and industrial center; manufactures include paper, chemicals, and textiles. There are also shipyards in the city. Speyer is a noted cultural and historical center of the Rhine plain. Its site was originally settled by the Celts. The city was destroyed (c.450) by the Huns but was later rebuilt and became (7th cent.) an episcopal see; in 1146 the Second Crusade was preached at Speyer by St. Bernard of Clairvaux. Speyer was made a free imperial city in 1294, but its bishops ruled substantial territories on both sides of the Rhine as princes of the Holy Roman Empire. Speyer was occupied by the French during the French Revolutionary Wars and formally ceded to France by the Treaty of Campo Formio (1797). Speyer and the episcopal lands west of the Rhine were subsequently given to Bavaria at the Congress of Vienna (1815); they were incorporated into the Rhenish Palatinate, of which Speyer was the capital until 1945. The city has retained parts of its medieval wall and gates. Its four-towered Imperial Cathedral (begun c.1030) is one of the greatest Romanesque buildings in Germany and contains the tombs of eight emperors. The Historical Museum of the Palatinate has large collections of pre-Roman and Roman materials and includes a wine museum.

Spe·zia, La (lä spĕt'syä), city (1972 est. pop. 124,027), capital of La Spezia prov., in Liguria, NW Italy, on the Gulf of La Spezia (an arm of the Ligurian Sea) at the E end of the Riviera. It is the chief Italian naval station and arsenal and the seat of a school of navigation. The city is also a commercial port, with shipyards and industries producing steel and petroleum. There is a notable cathedral (14th-16th cent.); nearby are the ruins of the important Roman town of Luna, which was destroyed in the Middle Ages.

Spice Islands (spīs): *see* Moluccas, Indonesia.

Spink (spĭnk), county (1970 pop. 10,595), 1,505 sq mi (3,898 sq km), NE central S.Dak., drained by the James River; formed 1873; co seat Redfield. It is in an agricultural area that produces grain, and has livestock and poultry raising and dairying. It has some manufacturing.

Spir·it Lake (spĭr'ĭt), city (1975 est. pop. 3,393), seat of Dickinson co., NW Iowa, near the Minn. border; settled 1856, inc. 1879. It is a farm trade and resort center.

Spits·bergen (spĭts'bûr'gən), largest island (15,075 sq mi/39,044 sq km) of Svalbard, a Norwegian possession in the Arctic Ocean. It is indented by large bays. The island served as the starting point for numerous polar expeditions.

Split (splĭt, splēt), city (1971 pop. 151,875), W Yugoslavia, in Croatia, on the Dalmatian coast of the Adriatic Sea. It is a major seaport and an important tourist and seaside resort. Shipbuilding and the production of cement, chemicals, and textiles are the leading industries. Split grew around the palace of Diocletian (who died here), built between 295 and 305. At various times part of Venice and Austria, it was included in Yugoslavia in 1818. The palace of Diocletian is the most remarkable among the Roman remains in Split. The city's other ancient buildings include the cathedral and the baptistery, both originally Roman temples; parts of its ancient walls and gates; and the town hall.

Spo·kane (spō-kăn'), county (1970 pop. 287,487), 1,758 sq mi (4,553.2 sq km), E Wash., drained by the Spokane River; formed 1858; co. seat Spokane. Its agriculture includes wheat and hay, truck farming, dairying, and livestock raising. It also has mineral mining, lumbering, and diversified manufacturing (clothing, concrete, metal products, construction machinery, and transportation equipment).

Spokane, city (1978 est. pop. 174,000), seat of Spokane co., E Wash., inc. 1881. It is a port of entry and commercial, transportation, and industrial center. The irrigated farms of the Columbia basin project contribute to the city's prosperity. The area has mineral deposits and cattle ranches and yields wheat, fruit, and other farm products. Spokane's industries include lumbering, food processing and packing, aluminum smelting, metal refining, and the manufacture of paper, clay, and cement products. A trading fort was established here in 1810; settlement began in 1871. Spokane has an art center and is the seat of several colleges and a university. Numerous lakes in the vicinity provide recreational facilities. Fairchild Air Force Base is to the west.

Spo·le·to (spō-lā'tō), city (1972 est. pop. 30,900), Umbria, central Italy. It is an industrial and tourist center. Manufactures include processed food, ceramics, and textiles. An Umbrian and later an Etruscan town, the city flourished after being taken (242 B.C.) by the Romans. It later became (c.A.D. 571) the seat of an important Lombard duchy. After centuries of control by emperors, the duchy in 1213 came under direct papal rule (to 1860). In the city are ruins of a bridge, the arch of Drusus (A.D. 21), a theater, and an amphitheater, all dating from the Roman era. The 4th cent. basilica of San Salvatore is a remarkable example of early Christian architecture.

Spor·a·des (spôr'ə-dēz, spŏr'-), islands, E and SE Greece, in the Aegean Sea. The Northern Sporades include Skíros, Skiathos, Skópelos, and some smaller islands off the coast of Évvoia and western Turkey. Límnos and Lesbos are sometimes also included. The Southern Sporades include Ikaría, Sámos, the Dodecanese, and sometimes Khíos. The main products of the Sporades are olive oil, wine, and citrus fruit.

Spots·wood (spŏts'wo͝od'), borough (1970 pop. 7,891), Middlesex co., central N.J., SSE of New Brunswick; inc. 1908. It was the site of a Revolutionary War ironworks.

Spot·syl·va·ni·a (spŏt'sĭl-vān'yə, -vā'nē-ə), county (1970 pop. 16,424), 409 sq mi (1,059.3 sq km), NE Va., bounded in the N by the Rapidan and Rappahannock rivers, and in the SW and W by the North Anna River; formed 1720; co. seat Spotsylvania. Its agriculture includes corn, wheat, tobacco, truck crops, dairy products, poultry, and livestock. It has lumbering and manufactures metal products and machinery. Major Civil War battles are commemorated by the Fredericksburg and Spotsylvania County Battlefields Memorial.

Spree (shprā), river, c.250 mi (400 km) long, rising in the Lausitz Mts., SE East Germany, near the Czechoslovak border. It flows north past Cottbus, then northwest through the Spree Forest, and from there it meanders east, north, and west before passing through Berlin to join the Havel River at Spandau.

Spring·dale (sprĭng'dāl'), city (1977 pop. 20,875), Benton and Washington cos., NW Ark.; inc. 1878. The economy centers on poultry, eggs, and dairy products. There is also varied light manufacturing. The surrounding Ozark Mts. draw many tourists.

Spring·er Mountain (sprĭng'ər), 3,820 ft (1,165.1 m) high, N Ga. It is the southernmost peak of the Blue Ridge Mts. and the southern terminus of the Appalachian Trail.

Spring·field (sprĭng'fēld'). **1.** Town (1970 pop. 1,660), seat of Baca co., SE Colo., in a grain, livestock, and dairy area; inc. 1889. **2.** City (1970 pop. 5,949), Bay co., NW Fla., E of Panama City; inc. 1935. **3.** City (1970 pop. 1,001), seat of Effingham co., E Ga., NNW of Savannah. **4.** City (1978 est. pop. 86,300), state capital and seat of Sangamon co., central Ill., on the Sangamon River; settled 1818, inc. as a city 1840. In a rich agricultural and coal region, it is a governmental, commercial, medical, and insurance center, with varied manufactures. Abraham Lincoln, who was instrumental in having Springfield made the state capital in 1839, lived and practiced law here from 1837 to 1861. He is buried nearby, with his wife and three of their children. Lincoln's home is preserved as a shrine. Other places of interest include the capitol (1867–87); the old capitol (1837), where Lincoln made his "House Divided" speech; and several Lincoln museums. **5.** City (1970 pop. 2,961), seat of Washington co., central Ky., WNW of Danville, in a bluegrass region; founded 1793. **6.** Industrial city (1978 est. pop. 168,000), seat of Hampden co., SW Mass., on the Connecticut River; inc. 1641. A port of entry, the city has insurance, chemical, plastic, metallurgical, paper, and printing industries. It was settled (1636) by Puritans under William Pynchon and was a station on the Underground Railroad. The U.S. Armory, which operated here from 1794 to 1966, was famous for the development of the Springfield and the Garand army rifles; it now contains an arms museum. The city is the seat of Springfield College, whose basketball hall of fame traces game's development since its invention here in 1891 by Dr. James Naismith. **7.** City (1977 est. pop. 5,576), Calhoun co., S Mich., W of Battle Creek; inc. 1952. A state recreational area is nearby. **8.** City (1978 est. pop. 132,600), seat of Greene co., SW Mo., in a resort area of the Ozarks; inc. 1846. It is the industrial, trade, and shipping center of a rich area producing dairy products, livestock, poultry, grains, and fruits. The city has railroad shops, flour mills, food-processing plants, and factories making clothing, furniture, typewriters, adhesives, truck trailers, and rubber and paper products. **9.** Township (1975 est. pop. 14,400), Union co., NE N.J., W of Newark; settled c.1717. In 1780 the British, marching on Morristown, razed Springfield village. **10.** City (1978 est. pop. 73,700), seat of Clark co., W central Ohio, on the Mad River; settled 1799, inc. as a city 1850. A manufacturing center in a rich farm area, it is known for its production of farm machinery. The city grew with the building of the National Road (1838), the arrival of the railroads (mid-1800s), and the establishment of farm-machinery plants (late 1800s). **11.** City (1978 est. pop. 36,900), Lane co., W central Oregon, between the McKenzie and Willamette rivers; inc. 1885. In a rich dairy, livestock, and farm region, and near the forested foothills of the Cascade Range, it has important lumbering and forest-product industries. Foods and chemicals are also produced. **12.** Industrial town (1970 pop. 10,063), Windsor co., SE Vt., on the Cascades of the Black River, in a fruit- and dairy-farming area; settled c.1772. The machine-tool industry is the town's largest industry. **13.** City (1970 pop. 9,720), seat of Robertson co., N Tenn., N of Nashville, in a limestone, farm, and timber area; founded c.1796, inc. 1819. It is an important tobacco market. **14.** Uninc. town (1970 pop. 12,500), Fairfax co., NE Va., a suburb of Washington, D.C.

Spring·hill (sprĭng'hĭl'), city (1970 pop. 6,496), Webster parish, NW La., NE of Shreveport at the Ark. border; inc. 1922. It is in a farm area and has paper mills and sawmills.

Spring Lake Park, village (1970 pop. 6,417), Anoka and Ramsey cos., SE Minn., a suburb N of Minneapolis; inc. 1953.

Springs (sprĭngz), city (1970 pop. 142,812), Transvaal, NE South Africa. It is the industrial center of a gold- and uranium-mining area. Manufactures include processed metals, chemicals, paper, glass, machine tools, bicycles, and printed materials.

Spring Valley. **1.** Uninc. community (1978 est. pop. 36,000), within the confines of San Diego, San Diego co., SW Calif., on Sweetwater Lake. It is residential, with some light industry. The Bancroft Ranch House Museum (1856) is a national historic landmark. **2.** City (1970 pop. 5,605), Bureau co., N Ill., on the Illinois River W of La Salle; inc. 1886. It is in an agricultural and mining area. **3.** Residential village (1975 est. pop. 21,400), Rockland co., SE N.Y., near the N.J. line; inc. 1902. It is a summer resort.

Spring·view (sprĭng'vyo͞o'), village (1970 pop. 260), seat of Keya Paha co., N Nebr., N of Ainsworth near the S.Dak. border. It is a trade center in an area yielding feed, grain, livestock, dairy and poultry products, and flour.

Spring·ville (sprĭng'vĭl'), city (1970 pop. 8,790), Utah co., N central Utah, S of Provo; settled 1850 by Mormons, inc. 1953. In an irrigated area served by the Strawberry valley project, the city is a canning and dairying center and a fruit-shipping point. Steel and cast-iron pipe are made. There is a notable art gallery here.

Spuy·ten Duy·vil Creek (spī'tən dī'vəl), tidal channel, now a ship canal, c.1 mi (1.6 km) long, SE N.Y., in New York City. It separates the northern tip of Manhattan Island from the mainland and connects the Hudson and the Harlem rivers.

Squaw Valley (skwô), valley, NE Calif., in the Sierra Nevada Mts., NW of Lake Tahoe. A well-known ski resort, it was the site of the 1960 Winter Olympics.

Sri Lan·ka (srē lăng'kə), formerly **Cey·lon** (sĭ-lŏn'), island (1976 pop. 14,000,000), 25,332 sq mi (65,610 sq km), in the Indian Ocean, just SE of India. The capital is Colombo. The pear-shaped island is 140 mi (225 km) across at its widest point and 270 mi (435 km) long. The narrow northern end is almost linked to southeastern India by Adam's Bridge, a chain of shoals. About four fifths of the island is flat or gently rolling; mountains in the south-central area rise to Pidurutalagal (8,291 ft/2,527 m), the highest point on the island. Sri Lanka has a generally uniform subtropical climate; the average lowland temperature is 80°F (27°C), but humidity is high. Rainfall, largely carried by monsoons, is adequate for agriculture, except in the north.

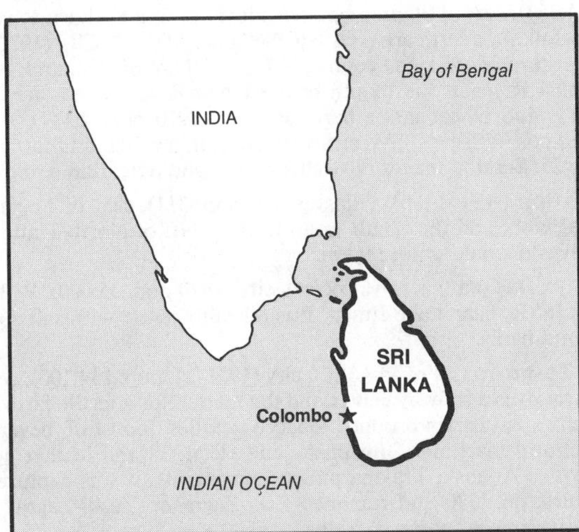

Economy. The country's economy is agricultural, emphasis being on export crops; tea, rubber, and coconut (all plantation-grown) comprise 90% of the island's exports. Cocoa, coffee, cinnamon, cardamom, pepper, cloves, nutmeg, citronella, and tobacco are also exported. Rice, fruit, and vegetables are grown for local consumption. Sri Lanka leads the world (1970) in the production of amorphous graphite, its principal mineral industry. Also mined are precious and semiprecious gems. Substantial deposits of iron ore have not yet been exploited. Industry is centered chiefly around the processing of agricultural products, especially the money crops—tea, rubber, and coconut. A great variety of consumer goods are also manufactured. The port of Colombo, on which most of the country's railroads and, to some extent, its road system converge, handles most of the foreign trade. The population of Sri Lanka is composed mainly (about 69%) of Sinhalese, who are Hinayana Buddhists; Hindu Tamils make up a large minority (about 23%), and there are smaller groups of Moslem Moors, Burghers (descendants of Dutch and Portuguese colonists), and Eurasians. The official language is Sinhala; Tamil is a secondary language, and English is widely spoken. Education is free, through the university level; the literacy rate is about 80%.

History. The most ancient of the inhabitants were probably the ancestors of the Veddas, an aboriginal people (numbering about 3,000) now living in remote mountain areas. They were conquered in the 6th cent. B.C. by the Sinhalese, who were originally from northern India; *Ramaya,* the ancient Hindu epic, probably reflects this conquest. The Sinhalese settled in the north and developed an elaborate irrigation system. They founded their capital at Anuradhapura, which, after the introduction of Buddhism from India in the 3rd cent. B.C., became one of the chief world centers of that religion. Buddhism stimulated the fine arts in Sri Lanka, its classical period being from the 4th to the 6th cent. The proximity of Sri Lanka to southern India resulted in many Tamil invasions. The Chola of southern India conquered Anuradhapura in the early 11th cent. and made Pollonarrua their capital. The Sinhalese soon regained power, but in the 12th cent. a Tamil kingdom arose in the north, and the Sinhalese were driven to the southwest. Arab traders, drawn by the island's spices, arrived in the 12th and 13th cent.; their descendants are the Moslem Moors. The Portuguese conquered the coastal areas in the early 16th cent. and introduced the Roman Catholic religion. By the mid-17th cent. the Dutch had taken over the Portuguese possessions and the rich spice trade. In 1795 the Dutch possessions were occupied by the British, who made the island a crown colony in 1798. In 1815 the island was brought under one rule for the first time when the central area, previously under the rule of Kandy, was conquered.

Under the British, tea, coffee, and rubber plantations were developed, and schools, including a university, were opened. A movement for independence arose during World War I. The constitution of 1931 granted universal adult suffrage to the inhabitants; but demands for independence continued, and in 1946 a more liberal constitution was enacted. Full independence was finally granted to the island on Feb. 4, 1948, with dominion status in the British Commonwealth. Riots in 1958 between Sinhalese and the Tamil minority over demands by the Tamils for official recognition of their language and the establishment of a separate Tamil state under a federal system resulted in severe loss of life. The Federal party of the Tamils was outlawed in 1961, following new disorders. Certain Western business facilities were nationalized (1962), and the country became involved in disputes with the United States and Great Britain over compensation. The elections in 1965 gave a parliamentary plurality to the more moderate socialist party. Closer relations with the West were established and compromise arrangements were made for recompensing nationalized companies. However, economic problems and severe inflation continued, aggravated by a burgeoning population (between 1946 and 1970 the population almost doubled). In 1970 the three-party anticapitalist coalition won a landslide victory following considerable pre-election violence. Social welfare programs, including rice subsidies and free hospitalization, were launched but failed to satisfy the extreme left, which, under the Marxist People's Liberation Front, attempted to overthrow the government (Apr.–May, 1971). With Soviet, British, and Indian aid, the rebellion was quelled after heavy fighting. In 1972 the country adopted a new constitution, declared itself a republic while retaining membership in the Commonwealth of Nations, and changed its name from Ceylon to Sri Lanka. In the early 1970s the government was confronted with a severe economic crisis as the country's food supplies and foreign exchange reserves dwindled in the face of rising inflation, high unemployment, a huge trade deficit, and the traditional policy of extensive social-welfare programs.

Sri·na·gar (srē-nŭg'ər), city (1971 pop. 403,413), historic capital of Kashmir, on the Jhelum River. Situated in the Vale of Kashmir, Srinagar is one of the most famed and beautiful summer resorts of the East. There are many canals, and transportation is chiefly by boat. In place of the hand-woven shawls (cashmeres) for which the city was famed, machine-made silks, woolens, and carpets are now manufactured. Other products include plywood and cement. The city was founded in the 6th cent. Extensive Buddhist ruins are near the city. In 1948 Srinagar became the capital of the Indian sector of the disputed state of Jammu and Kashmir.

Sri·ran·gam (srē-rŭng'gəm), town (1971 pop. 51,069), Tamil Nadu state, SE India, on an island in the Cauvery River. A group of temples to Vishnu (built c.1600) attracts many Hindu pilgrims. Srirangam is also a market for grain, sugar cane, and pepper.

Staf·ford (stăf'ərd), municipal borough (1974 est. pop. 54,530), administrative center of Staffordshire, W central England, on the Sow River, above its junction with the Trent River. The chief industry is the manufacture of electrical goods, locomotives, and engines; other products are concrete, shoes and shoe-repairing machinery, and salt. Izaak Walton was born in Stafford, and his cottage at Shallowford nearby is now a museum.

Stafford. **1.** County (1970 pop. 5,943), 795 sq mi (2,059.1 sq km), central Kansas, in a gently rolling area watered by Rattlesnake Creek; formed 1867; co. seat Saint John. It is in a wheat, corn, and livestock region, with scattered oil fields and some manufacturing. **2.** County (1970 pop. 24,587), 270 sq mi (699.3 sq km), NE Va., bounded in the E by the Potomac River, and in the SW and S by the Rappahannock River; formed 1666; co. seat Stafford. Its agriculture includes grain, tobacco, legumes, truck crops, dairy products, livestock, and poultry. It has river fisheries and manufactures pens and mechanical pencils.

Stafford. **1.** Textile town (1970 pop. 8,680), Tolland co., NE Conn., at the Mass. border NNW of Storrs; settled and inc. 1719. **2.** Village (1970 pop. 650), seat of Stafford co. (since 1715), NE Va., near the Potomac River NNE of Fredericksburg, in an agricultural area.

Staf·ford·shire (stăf'ərd-shîr', -shər), county (1974 est. pop. 984,620), 1,157 sq mi (2,996.6 sq km), W central England; administrative center Stafford. The terrain is gently undulating except for a district of rugged moorlands in the north. Much of the land is devoted to cattle pasturage, but the county is primarily industrial. The region was once a part of the Anglo-Saxon kingdom of Mercia.

Staines (stānz), urban district (1974 est. pop. 56,730), Surrey, SE England, on the Thames River. There is some industry here, including the manufacture of diesel engines.

Staked Plain (stākt): *see* Llano Estacado.

Stal·in·grad (stä'lĭn-grăd', stăl'ĭn-): *see* Volgograd, USSR.

Stam·ford (stăm'fərd), municipal borough (1974 pop. 14,670), Lincolnshire, E central England, on the Welland River. It is a market town. Products include diesel engines and electrical equipment, bricks, and tiles. It is the supposed site of a defeat of the Picts and Scots by the Saxons in 449 and was one of the Five Boroughs of the Danes. Stamford is noted for its architecture, including many 17th and 18th cent. buildings of Lincolnshire limestone.

Stamford, city (1978 est. pop. 105,000), Fairfield co., SW Conn., on Long Island Sound; settled 1641, inc. 1893 as a city within the town of Stamford (the two were consolidated in 1949). Office equipment, postage meters, bearings, chemicals, and cosmetics are among the manufactures. It is a residential community for many New York City commuters. A branch of the Univ. of Connecticut is here.

Stan·ards·ville (stăn'ərdz-vĭl'), town (1970 pop. 296), seat of Greene co., N central Va., N of Charlottesville.

Stan·dish (stăn'dĭsh), city (1970 pop. 1,184), seat of Arenac co., E Mich., near Saginaw Bay NNW of Bay City; inc. as a village 1893, as a city 1904. It is a trade and processing center of a farm and dairy region.

Stan·ford (stăn'fərd). **1.** City (1970 pop. 2,474), seat of Lincoln co., central Ky., SSW of Lexington; founded 1786. A fort was built here in 1775. **2.** Town (1970 pop. 505), seat of Judith Basin co., central Mont., SE of Great Falls. It is a shipping point for a grain, livestock, coal, and timber region. Dairy and poultry products, flour, and wool are produced.

Stan·is·laus (stăn'ĭs-lô, -lôs), county (1970 pop. 194,506), 1,511 sq mi (3,913.5 sq km), central Calif., a level land area in the San Joaquin Valley, bordered by the Coast Range and watered by the San Joaquin, Tuolumne, and Stanislaus rivers; formed 1854; co. seat Modesto. Its agriculture includes dairy and beef cattle, dairy products, poultry, fruit, nuts, alfalfa, beans, grain, and truck crops. There are sand and gravel pits and some mining. Its industry includes food processing, lumbering, and diverse manufacturing.

Sta·ni·slav (stə-nyĭ-släf'): *see* Ivano-Frankovsk, USSR.

Stan·ley (stăn'lē), county (1970 pop. 2,457), 1,414 sq mi (3,662.3 sq km), central S.Dak., bounded in the N by the Cheyenne River, in the E by the Missouri River, and drained by the Bad River; formed 1873; co. seat Fort Pierre. It is in a farming and cattle-raising area that produces wheat, barley, oats, and alfalfa. It has some manufacturing.

Stanley, city (1970 pop. 1,581) seat of Mountrail co., NW N.Dak., W of Minot; inc. 1910. In a dairy region, it raises wheat and corn. There are coal mines nearby.

Stanley Falls, seven cataracts on the Lualaba River, extending c.60 mi (95 km) between Kisangani and Ubundu, N central Zaire, central Africa. The falls have a drop of c.200 ft (60 m) and are circumvented by a short railroad.

Stanley Pool, lakelike expansion of the Congo River, c.320 sq mi (830 sq km), along the Zaire-Congo Republic border, W central Africa, c.350 mi (565 km) from the Congo's mouth. It is 22 mi (35 km) long and 14 mi (23 km) wide.

Stan·ley·ville (stăn'lē-vĭl'): *see* Kisangani, Zaire.

Stan·ly (stăn'lē), county (1970 pop. 42, 822), 398 sq mi (1,030.8 sq km), S central N.C., in a piedmont region bounded on the E by the Yadkin River (here becoming the Pee Dee River); formed 1841; co. seat Albemarle. It has farming (cotton, lespedeza, corn, and wheat), dairying, poultry raising, lumbering, and textile manufacturing.

Stan·o·voy Range (stăn'ō-voi), mountain range, c.450 mi (725 km) long, Far Eastern USSR, extending E from the Olekma River and forming part of the border between the Yakut Autonomous SSR and the Amur oblast.

Stan·ton (stăn'tn). **1.** County (1970 pop. 2,287), 676 sq mi (1,750.8 sq km), SW Kansas, in a sloping to rolling plain bordering in the W on Colo.; formed 1873; co. seat Johnson. It is in an agricultural area that raises both dryland and irrigated corn, milo, wheat, and sugar beets. It has natural-gas wells and some manufacturing. **2.** County (1970 pop. 5,758) 431 sq mi (1,116.3 sq km), NE Nebr., in a farming region drained by the Elkhorn River; formed 1867; co. seat Stanton. Grain, livestock, dairy goods, and poultry are produced.

Stanton. **1.** Residential city (1975 est. pop. 23,300), Orange co., SW Calif.; inc. 1956. **2.** City (1970 pop. 2,037), seat of Powell co., E central Ky., near the Red River ESE of Lexington, in an agricultural area. There are oil wells, coal mines, clay pits, stone quarries, and stands of timber in the area. Cumberland National Forest is nearby. **3.** City (1970 pop. 1,089), seat of Montcalm co., central Mich., NE of Grand Rapids; inc. as a village 1869, as a city 1881. **4.** City (1970 pop. 1,363), seat of Stanton co., NE Nebr., on the Elkhorn River SE of Norfolk, in a farm area; settled 1869, inc. 1871. **5.** City (1970 pop. 517), seat of Mercer co., central N.Dak., NNW of Bismarck on the Missouri River at the mouth of the Knife River. There are dairy, poultry, and wheat farms here. **6.** City (1970 pop. 2,117), seat of Martin co., W Texas, SW of Sweetwater, in a ranch and farm area; inc. 1925. Stanton makes oil-well supplies and irrigation equipment.

Sta·ple·ton (stā'pəl-tən), village (1970 pop. 311), seat of Logan co., central Nebr., on the South Loup River NNE of North Platte. It is in a livestock and grain region.

Sta·ra·ya Rus·sa (stä'rə-yə rōō'sə), city (1970 pop. 35,000), W European USSR, near Lake Ilmen. It is a health resort with salt springs and mud baths.

Sta·ra Za·go·ra (stä'rä zä-gô'rä), city (1971 est. pop. 114,100), central Bulgaria. It is a railway center and the market for a fertile farm area. The city's industries produce tobacco, textiles, foodstuff, beverages, agricultural machinery, furniture, and electrical and leather goods. Known as Augusta Trajana under Roman rule, it was captured by the Turks in 1370 and renamed Eski-Zagra or Yeski-Zagra, from which its present name is derived.

Star City (stär), city (1970 pop. 2,032), seat of Lincoln co., SE Ark., SE of Pine Bluff, in an agricultural area; inc. 1876.

Stark (stärk). **1.** County (1970 pop. 7,510), 291 sq mi (753.7 sq km), N central Ill., drained by the Spoon River and Indian Creek; formed 1839; co. seat Toulon. It has agriculture (corn, oats, wheat, livestock, and poultry), bituminous-coal mining, and manufacturing of horse collars, dairy products, cement blocks, and monuments. **2.** County (1970 pop. 19,613), 1,316 sq mi (3,408.4 sq km), W N.Dak., in an agricultural area drained by the Heart River; formed 1879; co. seat Dickinson. Livestock, dairy products, poultry, and wheat are important. It is also rich in lignite and clay deposits. **3.** County (1970 pop. 372,210), 576 sq mi (1,492.2 sq km), NE Ohio, intersected by the Tuscarawas River; formed 1808; co. seat Canton. It has agriculture (livestock, grain, fruit, and dairy products), coal mines, oil and gas wells, limestone quarries, and sand and gravel pits. Its industry includes food processing, lumbering, iron and steel foundries, and varied manufacturers.

Starke (stärk), county (1970 pop. 19,280), 310 sq mi (802.9 sq km),

NW Ind., bounded on the NW by the Kankakee River and drained by the Yellow River; formed 1835; co. seat Knox. It is in a rich agricultural area yielding large mint and onion crops, livestock, and poultry. Its manufacturing includes textiles, metal products, and machinery.

Starke, town (1970 pop. 4,848), seat of Bradford co., N Fla., SW of Jacksonville, in a mineral region. Clothing is made here.

Stark·ville (stärk'vĭl'), city (1970 pop. 11,369), seat of Oktibbeha co., E Miss., in a livestock, dairy, and farm area; inc. 1837. Textiles and clothing, milk products, and clocks are made. Mississippi State Univ. is nearby.

Starr (stär), county (1970 pop. 17,707), 1,211 sq mi (3,136.5 sq km), extreme S Texas, bounded on the SW and S by the Rio Grande and Mexican border; formed 1848; co. seat Rio Grande City. In the irrigated valley truck crops, citrus fruit, cotton, corn, grain, and peanuts are grown. The uplands have large ranches. There are deposits of oil, natural gas, and clay.

Starved Rock (stärvd), cliff, 140 ft (42.7 m) high, overlooking the Illinois River between La Salle and Ottawa, N Ill. It was visited by the French explorers Louis Jolliet and Father Marquette in 1673 and by Robert La Salle and Henri de Tonti in 1679. Legend says that in the 18th cent. the Ottawa Indians drove a band of Illinois Indians onto the cliff, where they died of thirst and starvation. Starved Rock has been designated a national historic landmark.

Stass·furt (shtäs'fo͝ort'), city (1973 est. pop. 26,272), Magdeburg dist., W East Germany. It is a center of a potash-mining region. Manufactures of the city include chemicals, foodstuffs, furniture, and electronic equipment.

State Col·lege (stāt' kŏl'ĭj), borough (1970 pop. 33,778), Centre co., central Pa.; settled 1859, inc. 1896. State College is mostly residential; nearby is the Pennsylvania State Univ.

Sta·ten Island (stăt'n), (1975 est. pop. 330,000), c.60 sq mi (160 sq km), SE N.Y., in New York Bay, SW of Manhattan, forming Richmond co. of New York state and Staten Island borough of New York City. It is separated from N.J. by Kill Van Kull and Arthur Kill, which are crossed by bridges. Ferries connect the island with Manhattan, and the Verrazano-Narrows Bridge links it with Brooklyn. Since the completion of the Verrazano-Narrows Bridge (1964), Staten Island has had an influx of new residents and industries and has lost much of its semirural character. The industrial area of Staten Island is located in the north, where docks line the northern and eastern shores. Beaches and parks, including part of Gateway National Recreation Area, are found along the southeastern coast. The island was visited by Henry Hudson in 1609 and was called Staaten Eylandet by the Dutch. Hostile Indians drove off the first white settlers, but by 1661 a permanent settlement had been founded.

Stat·en·ville (stăt'n-vĭl'), town (1970 pop. 650), seat of Echols co., S Ga., ESE of Valdosta on the Alapaha River near the Fla. border. Lumber and naval stores are processed.

States·bor·o (stāts'bûr'ō), city (1975 est. pop. 15,200), seat of Bulloch co., E Ga.; founded 1803, inc. 1902. It has a large tobacco market, an iron foundry, and textile, meat-packing, and lumbering industries.

States·ville (stāts'vĭl'), city (1975 est. pop. 22,200), seat of Iredell co., W central N.C., on a plateau in the Blue Ridge foothills; founded 1789, inc. 1847. It is a commercial and industrial center with several furniture factories.

Staun·ton (stăn'tən), city (1978 est. pop. 21,400), seat of Augusta co., W central Va., in the Shenandoah Valley; settled 1732, inc. as a city 1871. It is a trade and industrial center in a farm area known for its poultry, livestock, and apples. Other products include clothing, furniture, safety razors, and soft drinks. The city manager form of government originated in Staunton in 1908. In the city is Woodrow Wilson's birthplace, dedicated as a national shrine in 1941.

Sta·vang·er (stä-väng'ər), city (1974 est. pop. 84,334), capital of Rogaland co., SW Norway, a port on the Stavangerfjord (an arm of the Boknfjord). It is an important commercial and industrial center where ships are built and fish processed. Founded in the 8th cent., Stavanger was an episcopal see from c.1125 to 1682.

Stav·ro·pol (stăv'rō-pôl), city (1973 est. pop. 219,000), capital of Stavropol Kray, S European USSR, on the Stavropol Plateau. It has machine-tool, wool, leather, and food-processing industries. There are natural-gas fields in the area.

Stavropol Kray or **Stavropol Territory,** administrative division (1970 pop. 2,306,000), 31,120 sq mi (80,601 sq km), S European USSR, in the North Caucasus, the N foothills of the main Caucasian range, and the dry steppes to the NE. The central part of the territory occupies the Stavropol Plateau, a hilly region drained by the Kuma and Kuban rivers. There are oil and natural-gas deposits. Winter wheat, corn, sunflowers, cotton, grapes, and vegetables are cultivated. Sheep raising is an important occupation. The territory was first organized in 1924 as the North Caucasus Territory. Renamed Ordzhonikidze Territory in 1937, it was given its present name in 1943.

Steam·boat Springs (stēm'bōt'), town (1970 pop. 2,340), seat of Routt co., NW Colo., on the Yampa River NW of Denver at an altitude of c.6,762 ft (2,062.4 m), in a resort, timber, and farm area; founded 1875, inc. 1907. It has mineral springs.

Stearns (stûrnz), county (1970 pop. 95,400), 1,342 sq mi (3,475.8 sq km), central Minn., bounded on the E by the Mississippi River and watered by the Sauk River, with numerous small lakes; formed 1855; co. seat St. Cloud. It is in an agricultural area (dairy products, livestock, corn, oats, and barley), with granite quarrying and diverse industry.

Steele (stēl). **1.** County (1970 pop. 26,931), 425 sq mi (1,100.8 sq km), S Minn., drained by the Straight River; formed 1855; co. seat Owatonna. In an agricultural area yielding dairy products, livestock, corn, oats, barley, and poultry, it has some mining. Wood products, concrete, metal products, and machinery are manufactured. **2.** County (1970 pop. 3,749), 710 sq mi (1,838.9 sq km), E N.Dak., in a farming area watered by the Goose River; formed 1883; co. seat Finley. Its main crop is wheat. Livestock, dairy products, and potatoes are also produced.

Steele, city (1970 pop. 696), seat of Kidder co., central N.Dak., E of Bismarck, in a grain-growing area. It has grain elevators and is a shipping center for livestock, poultry, wheat, barley, and rye.

Steele, Mount, 16,644 ft (5,076.4 m) high, in the St. Elias Mts., SW Yukon Territory, Canada, in Kluane National Park.

Steel·ton (stēl'tən), industrial borough (1970 pop. 8,556), Dauphin co., SE Pa., on the Susquehanna River below Harrisburg; settled 1865, inc. 1880. It has a large steel mill. The first practical production of Bessemer steel in America was begun here (1867).

Steel·ville (stēl'vĭl'), city (1970 pop. 1,392), seat of Crawford co., E central Mo., E of Rolla, in a timber and farm area; settled 1833, inc. 1872. It has iron mines and makes shoes.

Ste·ger (stā'gər), village (1973 est. pop. 9,285), Cook and Will cos., NE Ill., S of Chicago Heights; inc. 1896. Furniture is made.

Stel·len·bosch (stĕl'ən-bôs', -bo͝osh'), city (1970 pop. 29,955), Cape Prov., SW South Africa, in the Eerste River valley. It is a wine-making and fruit-growing center. Other industries include sawmilling and the manufacture of bricks and tiles.

Stel·vi·o Pass (stĕl'vē-ō), alt. 9,048 ft (2,759.6 m), in the central Alps, N Italy, near the Swiss and Austrian borders. It is crossed by the highest road in the Alps, connecting the Valtellina with the upper Adige River valley.

Sten·dal (stĕn'däl', shtĕn'-), city (1973 est. pop. 38,927), Magdeburg dist., W East Germany, on the Uchte River. It is a major rail junction and has sugar refineries, metalworks, food canneries, and chemical factories. Stendal was founded in 1151 by Albert the Bear. The city joined the Hanseatic League c.1350. It contains noteworthy medieval structures. Marie Henri Beyle (1783–1842), the French author, took his pen name (Stendhal) from the city.

Sten·ness, Loch of (stĕn'əs), lake on Mainland island, off N Scotland. An isthmus between Harray and Stenness lochs holds the Standing Stones of Stenness, two rings of flat tablets dating from c.2000 B.C.

Ste·pa·na·kert (stĕp'ə-nə-kĕrt'), city (1973 est. pop. 32,000), capital of Nagorno-Karabakh Autonomous oblast, Azerbaijan Republic, S European USSR. Silk, wine, and food are processed.

Ste·phens (stē'vənz). **1.** County (1970 pop. 20,331), 173 sq mi (448.1 sq km), NE Ga., bordered on the E by S.C. and drained by headstreams of Broad River; formed 1905; co. seat Toccoa. It is in a piedmont agricultural (cotton, corn, hay, sweet potatoes, and livestock) and manufacturing (knitting mills, thread, clothing, furniture, and machinery) area. The western part of the county is in Chattahoochee National Forest. **2.** County (1970 pop. 35,902), 891 sq mi

(2,307.7 sq km), S Okla., drained by Wildhorse Creek; formed 1907; co. seat Duncan. Its agriculture includes cotton, corn, grain, broomcorn, oats, and watermelons, and it has cattle ranching and dairying. There are oil and gas wells. Its industry includes food processing, petroleum refining, and oil-field machinery manufacturing. **3.** County (1970 pop. 8,414), 899 sq mi (2,328.4 sq km), N central Texas, drained by the Clear Fork of the Brazos River and its tributaries; formed 1858; co. seat Breckenridge. It has diversified agriculture (corn, grain, grain sorghums, peanuts, fruit, and truck crops), livestock raising, and oil and natural-gas wells.

Ste·phen·son (stē'vən-sən), county (1970 pop. 48,861), 568 sq mi (1,471.1 sq km), bounded on the N by the Wis. border and drained by the Pecatonica River; formed 1837; co. seat Freeport. Its agriculture includes livestock, corn, wheat, oats, barley, nursery products, poultry, and dairy products.

Ste·phen·ville (stē'vən-vĭl'), city (1970 pop. 9,277), seat of Erath co., N central Texas, on the Bosque River SW of Fort Worth; founded 1856, inc. 1888. It processes the produce of the region's poultry and cotton farms, dairies, orchards, and pecan groves.

steppe (stĕp), temperate grassland of Eurasia, consisting of level, generally treeless plains extending over the lower regions of the Danube, in a broad belt over S and SE European USSR and Central Asian USSR, and stretching E to the Altai and S to the Transbaykal and Manchurian plains. The term is sometimes applied to the corresponding temperate grasslands of Hungary, the prairies of the United States, the pampas of South America, and the highveld of South Africa; it is sometimes also applied to the semiarid regions on the fringe of the hot deserts.

Ster·ling (stûr'lĭng), county (1970 pop. 1,056), 914 sq mi (2,367.3 sq km), W Texas, in rolling prairies drained by the North Concho River; formed 1891; co. seat Sterling City. This ranching region also has some poultry raising and grain farming. There are oil and natural-gas wells.

Sterling. 1. City (1975 est. pop. 10,500), seat of Logan co., NE Colo., on the South Platte River; inc. 1884. It is the trading center of an agricultural area. An oil boom occurred here in the 1950s. **2.** City (1975 est. pop. 16,400), Whiteside co., NW Ill., on the Rock River opposite Rock Falls; inc. 1841. It is an industrial center in a farm region.

Sterling City, city (1970 pop. 780), seat of Sterling co., W Texas, NW of San Angelo on the North Conch River, in a natural-gas area. Carbon black, chemicals, and fertilizer are made.

Ster·li·ta·mak (stûr'lĭ-tə-măk'), city (1973 est. pop. 203,000), Bashkir Autonomous Republic, E European USSR, on the Belaya River. It is a port and the center of a chemical complex. Milling and construction equipment and food products are also made.

Steu·ben (stōō'bən, styōō'-). **1.** County (1970 pop. 20,159), 309 sq mi (800.3 sq km), extreme NE Ind., bordered on the N by Mich. and on the E by Ohio; formed 1835; co. seat Angola. In an agricultural area with stock raising and dairying, it has timber, some mining, and manufacturing. It is a resort area containing 100 lakes. **2.** County (1970 pop. 99,546), 1,410 sq mi (3,651.9 sq km), SW N.Y., bounded on the S by the Pa. border and drained by the Canisteo, Cohocton, Tioga, and Chemung rivers; formed 1796; co. seat Bath. This dairying, grape-growing, and truck-farming area also has timber, sand and gravel pits, and diversified manufacturing.

Steu·ben·ville (stōō'bən-vĭl', styōō'-), city (1978 est. pop. 26,900), seat of Jefferson co., E central Ohio, on the Ohio River; laid out c.1797, inc. as a city 1851. The city's major industry is the production of steel.

Ste·ven·age (stē'və-nĭj), urban district (1974 est. pop. 69,590), Hertfordshire, E central England. Stevenage was the first new town to be designated under the New Towns Act of 1946, a program to decentralize population and industry. Manufactures include photographic apparatus, aircraft equipment, and electronic equipment.

Ste·vens (stē'vənz). **1.** County (1970 pop. 4,198), 731 sq mi (1,893.3 sq km), SW Kansas, in a gently rolling plain bordered on the S by Okla. and crossed by the Cimarron River; formed 1886. It is in a farming area, with natural-gas fields and some manufacturing. **2.** County (1970 pop. 11,218), 558 sq mi (1,445.2 sq km), W Minn., drained by the Pomme de Terre River; formed 1862; co. seat Morris. It is in a grain and livestock area, with some mining and plastics and concrete manufacturing. **3.** County (1970 pop. 17,405), 2,481 sq mi (6,425.8 sq km), NE Wash., on the British Columbia border, watered by the Columbia River, which is its W boundary; formed 1863; co. seat Colville. Its agriculture includes grain, fruit, dairy products, and livestock. It has copper, gold, silver, zinc, and magnesite mining, lumbering, and aluminum foundries. It includes parts of Colville National Forest and Spokane Indian Reservation.

Ste·ven·son (stē'vən-sən), town (1970 pop. 916), seat of Skamania co., SW Wash., ENE of Vancouver on the Columbia River near the foothills of the Cascades. It is in a lumbering, fruit-raising, and dairying region.

Stevens Point, city (1978 est. pop. 23,500), seat of Portage co., central Wis., on the Wisconsin and Plover rivers; inc. 1858. The major industries are insurance and the manufacture of paper and furniture.

Stew·art (stōō'ərt, styōō'-). **1.** County (1970 pop. 6,511), 452 sq mi (1,170.7 sq km), W Ga., bounded on the W by the Ala. line, formed here by the Chattahoochee River; formed 1830; co. seat Lumpkin. It is in a coastal-plain agricultural (cotton, corn, peanuts, fruit, and livestock) and timber area. **2.** County (1970 pop. 7,319), 470 sq mi (1,217.3 sq km), NW Tenn., bounded on the N by Ky., on the W by the Tennessee River, and drained by the Cumberland River; formed 1803; co. seat Dover. Its agriculture includes corn, hay, tobacco, livestock, and dairying. It manufactures clothing and metal products. Fort Donelson National Military Park and Fort Henry are here.

Stewart, river, 331 mi (533 km) long, rising in the Mackenzie Mts., central Yukon Territory, Canada, and flowing generally W to the Yukon River S of Dawson. Navigable for most of its length, it is a transportation route for lead ore.

Stewart Island, volcanic island (1966 pop. 329), 670 sq mi (1,735.3 sq km), S New Zealand, S of South Island across Foveaux Strait. Mountainous and scenic, it is a summer resort. Frozen fish and granite are exported. It was discovered in 1808 by the British, who bought it in 1864 from the Maori natives.

Steyr (stī'ər, shtī'-), city (1971 pop. 40,587), Upper Austria prov., central Austria, on the Enns and Steyr rivers. It has been an ironworking center since the Middle Ages.

Stig·ler (stĭg'lər), city (1970 pop. 2,347), seat of Haskell co., E Okla., SSE of Muskogee, in an agricultural and coal region.

Sti·kine (stĭ-kēn'), river, 335 mi (539 km) long, rising in the Stikine Mts., NW British Columbia, Canada. It flows in an arc west and southwest, crossing southeast Alaska, to the Pacific Ocean north of Wrangell Island. The Stikine was one of the routes during the Klondike gold rush (1897–98). The river is a noted salmon stream.

Still·wa·ter (stĭl'wô'tər, -wŏt'ər), county (1970 pop. 4,632), 1,794 sq mi (4,646.5 sq km), S Mont., in an irrigated agricultural region drained by the Yellowstone River; formed 1913; co. seat Columbus. Sugar beets and beans are grown, and livestock is raised.

Stillwater. 1. City (1970 pop. 10,191), seat of Washington co., E Minn., on the St. Croix River; inc. 1854. Shoes and clothing are among its manufactures. Stillwater was an early lumber center. **2.** City (1970 pop. 31,126), seat of Payne co., N central Okla.; inc. 1899. Seat of Oklahoma State Univ., it is also the market and processing center of a farm and livestock area.

Stil·well (stĭl'wĕl', -wəl), city (pop. 1,916), seat of Adair co., E Okla., near the Ark. border ENE of Muskogee, in a hilly farm, fruit, and timber region.

Stin·nett (stĭ-nĕt'), town (1970 pop. 2,014), seat of Hutchinson co., extreme N Texas, on the high plains of the Panhandle, near the Canadian River NNE of Amarillo. It is a trade center in an oil and farm region and ships cattle and grain.

Štip (shtēp), town (1971 pop. 27,289), SE Yugoslavia, in Macedonia. It is a processing center for opium poppies and has mineral waters. Štip was an important center of the medieval Serbian and Bulgarian empires.

Stir·ling (stûr'lĭng), burgh (1973 est. pop. 29,799), central Scotland, on the Forth River. The center of a large farm district, it has stock markets and light industries. There are coal fields nearby. Stirling Castle long rivaled Edinburgh as a royal residence. The castle was the birthplace of James II and (probably) James III and James IV.

Stir·ling·shire (stûr'lĭng-shîr', -shər), former county, central Scotland. The county town was Stirling. Under the Local Government Act of 1973 Stirlingshire was divided between the Central and Strathclyde regions.

Stock·bridge (stŏk′brĭj′), town (1970 pop. 2,312), Berkshire co., W Mass., on the Housatonic River, in the Berkshire mts.; inc. 1739. It is a summer resort center, and its proximity to ski areas attracts winter visitors. Stockbridge was founded in 1734 as a mission for the Muhhekanuk Indians. A leading summer theater, an art colony, several galleries, and a museum that contains many Norman Rockwell paintings are here.

Stock·holm (stŏk′hōlm′, -hōm′), city (1972 est. pop. 699,238), capital of Sweden and of Stockholm co., E Sweden. It is Sweden's economic, transportation, administrative, and cultural center. Manufactures include machinery, textiles, motor vehicles, rubber, processed food, printed materials, porcelain, and liquor. The city also has a large port and an important shipbuilding industry. Founded in the mid-13th cent., Stockholm became an important trade center dominated by the Hanseatic League. In 1520 Christian II of Denmark and Norway proclaimed himself king of Sweden at Stockholm; a large number of Swedish nobles had gathered to attend the coronation, and Christian had about 100 of the anti-Danish nobility murdered. The Stockholm massacre led to the successful uprising of Swedes under Gustavus Vasa, who became king of Sweden as Gustavus I (1523–60). Gustavus made Stockholm the center of his kingdom and ended the privileges there of the Hanseatic merchants. Stockholm was made the official capital of Sweden in 1634, about the same time that it became a European intellectual center under Queen Christina, who attracted men like the philosopher Descartes to her court.

Modern Stockholm has broad streets, many parks, and well-planned housing projects that have made it virtually a slumless city. Often called the Venice of the North, it is built on several peninsulas and islands. On Städsholmen Island, which has retained much of its medieval character, are churches dating from the 13th cent. and several old Hanseatic houses. Also on the island are the Great Square, site of the Stockholm massacre; the Riddarhuset, a 17th cent. structure in the Dutch Renaissance style; and the Royal Palace, built (1754) in Italian Renaissance style. Stockholm is the seat of a university (founded 1877), a technical university, a school of economics, and royal academies of music, science, art, and medicine. Each year the Nobel prizes (except the Nobel Peace Prize) are awarded here.

Stock·port (stŏk′pôrt′), county borough (1974 est. pop. 138,750), Greater Manchester, W central England, on the slopes of a narrow valley at the head of the Mersey River. Engineering and cotton textiles are the largest industries. There is a 14th cent. church and a grammar school founded in the 15th cent.

Stock·ton (stŏk′tən). **1.** City (1978 est. pop. 124,500), seat of San Joaquin co., central Calif., on the San Joaquin River; inc. 1850. It is an inland seaport located at the head of the San Joaquin delta. It is also a railroad center and a processing and distributing point for farm products from the San Joaquin valley. It has many canneries. Farm machinery, building materials, and boats are also made. The city was an outfitting center in the gold-rush days. **2.** City (1970 pop. 1,818), seat of Rooks co., N central Kansas, on the South Fork of the Solomon River NW of Salina, in a grain, livestock, and oil region; settled 1871, inc. 1880. **3.** City (1970 pop. 1,063), seat of Cedar co., W Mo., in an Ozark region near the Sac River NW of Springfield. Grain and livestock are raised.

Stock·ville (stŏk′vĭl′), village (1970 pop. 61), seat of Frontier co., S Nebr., NNE of McCook on Medicine Creek. Nearby are the sites of prehistoric Indian villages.

Stod·dard (stŏd′ərd), county (1970 pop. 25,771), 823 sq mi (2,131.6 sq km), SE Mo., on the St. Francis River in the W and drained by the Castor River and drainage canals; formed 1833; co. seat Bloomfield. Grain, livestock, and lumber are major products. Cotton is grown and processed.

Stoke-on-Trent (stōk′ŏn-trĕnt′), county borough (1974 est. pop. 262,120), Staffordshire, W central England. Situated in a coal field, it is the center of the Staffordshire pottery-making industry. Coal is mined, and brick, tile, chemicals, and tires are manufactured.

Stokes (stōks), county (1970 pop. 23,782), 457 sq mi (1,183.6 sq km), N N.C., bordered on the N by Va. and drained by the Dan River; formed 1798; co. seat Danbury. It has piedmont agriculture (tobacco and corn) and timber.

Stol·berg (shtôl′bĕrĸʜ′), city (1972 est. pop. 56,910), North Rhine-Westphalia, W West Germany; chartered 1856. It is a center of the West German brass industry, started (c.1600) by Protestant settlers from Aachen. Other manufactures include chemicals, pharmaceuticals, and glass.

Stone (stōn). **1.** County (1970 pop. 6,838), 608 sq mi (1,574.7 sq km), N Ark., in the Ozarks, bounded on the NE by the White River; formed 1873; co. seat Mountain View. It lies in a resort, pine and hardwood timber, and agricultural (poultry, livestock, grain, hay, dairy products, and cotton) region. Clothing is manufactured. **2.** County (1970 pop. 8,101), 448 sq mi (1,160.3 sq km), SE Miss., drained by the Biloxi River and the Red and Black creeks; formed 1916; co. seat Wiggins. Farming (cotton, corn, and pecans), dairying, and lumbering are done. **3.** County (1970 pop. 9,921), 449 sq mi (1,162.9 sq km), in the Ozarks, drained by the White and James rivers; formed 1851; co. seat Galena. It is an agricultural region, known for livestock, tomatoes, strawberries, and corn. Lead and zinc are mined.

Stone·ham (stōn′əm), town (1978 est. pop. 22,000), Middlesex co., NE Mass., a suburb of Boston; settled 1645, inc. 1725. It has a shoe industry and other manufactures.

Stone·ha·ven (stōn′hā′vən), burgh (1971 pop. 4,729), E Scotland, on the North Sea. A resort town, its products include whiskey and leather and woolen goods. Nearby are the ruins of Dunnottar Castle built in the 7th cent.

Stone·henge (stōn′hĕnj′), group of standing stones on Salisbury Plain, Wiltshire, S England. Pre-eminent among Megalithic monuments in the British Isles, the great prehistoric structure is enclosed within a circular ditch 300 ft (91 m) in diameter. Within the circular trench the stones are arranged in four series: the outermost is a circle of sandstones about 13.5 ft (4.1 m) high connected by lintels; the second is a circle of bluestone menhirs; the third is horseshoe shaped; the innermost is ovoid. Within the ovoid lies the Altar Stone. It was once widely believed to be a druid temple, but the druids probably did not arrive in Britain until c.250 B.C. Most archaeologists agree that Stonehenge served some sort of religious function, and one astronomer has theorized that Stonehenge was used as a huge astronomical instrument that could accurately measure solar and lunar movements as well as eclipses.

Stones River National Battlefield (stōnz): *see* National Parks and Monuments Table.

Stone·wall (stōn′wôl′), county (1970 pop. 2,397), 926 sq mi (2,398.3 sq km), in rolling plains with hilly areas and mesquite woodlands, drained by the Salt and Double Mountain forks of the Brazos River, formed 1876; co. seat Aspermont. This cattle-ranching area also has general agriculture (cotton, corn, grain, dairy products, and poultry), oil wells, and some minerals (gypsum, clay, stone, salt, and copper).

Ston·ing·ton (stōn′ĭng-tən), town (1975 est. pop. 16,300), New London co., SE Conn., on a peninsula in Long Island Sound; settled 1649 from Plymouth, inc. 1662. Fishing, boatbuilding, and the manufacture of precision tools and textiles are the leading industries. Stonington was once an important shipbuilding and whaling center, and many houses built by sea captains still remain.

Ston·y Brook (stō′nē), uninc. town (1970 pop. 6,600), Suffolk co., SE N.Y., on the N shore of Long Island ENE of Northport. It is a resort and residential community restored (1941) in the style of an 18th cent. village. A museum houses old carriages, sleds, and wagons.

Stony Point, uninc. town (1970 pop. 8,270), SE N.Y., on the Hudson River and N of Nyack. It has an orthopedic hospital for children. Nearby is the Stony Point Battle Reservation, memorializing the storming of Stony Point by Anthony Wayne's forces in the Revolution (1779).

Sto·re Baelt (stô′rə bĕlt′) and **Lil·le Baelt** (lĭl′ə bĕlt′), two shallow straits, S Denmark, connecting the Kattegat with the Baltic Sea. The Store Baelt, c.40 mi (65 km) long and from 10 to 20 mi (16–32 km) wide, separates Sjaelland and Fyn islands. The Lille Baelt, c.30 mi (50 km) long and from .5 to 18 mi (.8–29 km) wide, between Fyn and Jutland, is crossed by a road and railroad bridge.

Sto·rey (stôr′ē), county (1970 pop. 695), 262 sq mi (678.6 sq km), W Nev., bounded on the N by the Truckee River; formed 1861; co. seat Virginia City. Mt. Davidson is the site of the famous Comstock Lode. Some mining is still done today, but the county's main source of revenue is tourism.

Stor·fjord (stôr′fyôr′), deep inlet of the Norwegian Sea, c.70 mi (110 km) long, Møre og Romsdal co., SW Norway. A scenic area with sheer cliffs and waterfalls, Storfjord branches into several fjords.

Storm Lake (stôrm), city (1970 pop. 8,591), seat of Buena Vista co., NW Iowa, on Storm Lake WNW of Fort Dodge; inc. 1873.

Storrs (stôrz), community (1970 pop. 10,691), a part of the town of Mansfield, Tolland co., NE Conn. It is the seat of the Univ. of Connecticut.

Sto·ry (stôr'ē), county (1970 pop. 62,783), 568 sq mi (1,471.1 sq km), central Iowa, drained by the Skunk River; formed 1846; co. seat Nevada. It is a prairie agricultural area, raising hogs, cattle, poultry, corn, oats, and soybeans. There are limestone quarries, sand and gravel pits, and bituminous-coal deposits in the county.

Stough·ton (stōt'n). **1.** Town (1978 est. pop. 26,900), Norfolk co., E Mass.; founded 1637, inc. 1726. Shoes, woolen textiles, and electrical equipment are among its manufactures. **2.** City (1970 pop. 6,096), Dane co., S Wis., on the Yahara River SSE of Madison, in a farm and dairy area; settled 1847; inc. as a village 1868, as a city 1882. Truck bodies and parts are made.

Stour·bridge (sto͝or'brĭj, stour'-), municipal borough (1974 est. pop. 55,660), Hereford and Worcester, W central England. In the 16th cent. Stourbridge's famous glassmaking industry was established. Other products are chains, tools, and heating apparatus.

Stow (stō), city (1978 est. pop. 26,000), Summit co., NE Ohio, a suburb of Akron; settled 1802, inc. as a city 1960. Chiefly residential, it has some industry, including automobile assembly.

Stowe (stō), resort town (1970 pop. 2,388), Lamoille co., N central Vt.; settled 1794, inc. 1896. It is surrounded by mountains and is one of New England's largest ski resort areas.

Straf·ford (străf'ərd), county (1970 pop. 70,431), 376 sq mi (973.8 sq km), SE N.H., on the Maine border and drained by the Salmon Falls and Cocheco rivers; formed 1769; co. seat Dover. Wood products, textiles, shoes, machinery, leather goods, and brick are manufactured.

Stral·sund (shträl'zo͝ont'), city (1973 est. pop. 72,244), Rostock dist., N East Germany, on the Strelasund opposite Rügen Island. It is an industrial center and seaport. Manufactures include metal goods, furniture, refined sugar and other processed food, and beer. Founded in 1209, Stralsund became (late 13th cent.) a leading member of the Hanseatic League. In the Thirty Years' War, Stralsund was aided by Danish, then by Swedish, troops, and at the Peace of Westphalia (1648) it passed to Sweden. The city was taken by the French in 1807 and passed to Denmark by the Treaty of Kiel (1814) and to Prussia at the Congress of Vienna (1814–15). It was heavily damaged in World War II. Noteworthy are its medieval buildings and gates.

Stran·raer (strən-rär', străn-), burgh (1973 est. pop. 10,174), Wigtownshire, SW Scotland, at the head of Loch Ryan. A fishing port, it has a prosperous trade with Northern Ireland. Food-processing industries are there.

Stras·bourg (sträz-bo͝or'), city (1968 pop. 254,038), capital of Bas-Rhin dept., NE France, on the Ill River near its junction with the Rhine River. It is the intellectual and commercial capital of Alsace. The city's chief industries are metal casting, machine and tool construction, oil and gas refining, and boatbuilding. Iron, potassium, gasoline, and numerous industrial products are shipped through Strasbourg's great port on the Rhine. Chief among the city's historical monuments is the Roman Catholic cathedral, begun in 1015 and completed in 1439, a masterpiece of Rhenish architecture, with a famous astronomical clock installed in 1574.

History. In Roman times Strasbourg was called Argentoratum and was an important city in the province of Upper Germany. After becoming part of the Holy Roman Empire in 923, Strasbourg came under the temporal rule of its bishops. In 1262, after some struggles with the bishops, the burghers secured the status of a free imperial city for the city proper. Strasbourg accepted the Reformation in the 1520s and the city became an important Protestant center. The city's prosperity began to decline in the early 17th cent. and was severely damaged by the Thirty Years' War (1618–48). In 1681 Louis XIV seized Strasbourg, which increasingly adopted French customs and speech after the French Revolution. Ceded to Germany by the Treaty of Frankfurt (1871), it was recovered by France in 1919, following World War I. The city was occupied by the Germans and severely damaged in World War II. In 1949 Strasbourg became the seat of the Council of Europe. In 1967 some 30 neighboring towns were absorbed into a new Community of Strasbourg, thus giving the city's metropolitan area a population in excess of 400,000.

Strat·ford (străt'fərd), city (1971 pop. 24,508), S Ont., Canada, on the Avon River, SW of Toronto. It is an industrial center, with plants manufacturing textiles, furniture, automobile parts, and rubber and leather products. The city is the home of the noted Stratford Shakespearean Festival (started 1953).

Stratford. **1.** Town (1978 est. pop. 51,300), Fairfield co., SW Conn., at the mouth of the Housatonic River on Long Island Sound; inc. 1639. Aircraft engines, helicopters, machinery, hardware items, and asbestos products are among its many manufactures. The American Shakespeare Festival Theater and Academy opened here in 1955. **2.** Town (1970 pop. 2,139), seat of Sherman co., extreme N Texas, on the high plains of the Panhandle, NE of Dalhart. It is a market and shipping point for a wheat, cattle, and oil region.

Strat·ford-up·on-A·von (străt'fərd-ə-pŏn-ā'vŏn), municipal borough (1974 est. pop. 19,760), Warwickshire, central England, on the Avon River. A market town with light industries including brewing, canning, and the manufacture of aluminum goods, Stratford owes its great fame to its associations with William Shakespeare. A gabled building on Henley St., believed to be the poet's birthplace, is open to the public. The site of the home he purchased in 1597, and where he died in 1616, is marked (the building having been torn down in 1759). The town's principal memorial is a theater, where annual Shakespeare festivals are held. The first theater, built in the late 19th cent., was destroyed by fire in 1926, but the attached gallery, library, and museum were saved. The new theater was dedicated in 1932.

Strath·more (străth-môr'), valley, c.55 mi (90 km) long and 5 to 10 mi (8–16 km) wide, in Tayside region, E central Scotland, running from NE to SW between the Grampians and the Sidlaw Hills. It has some of Scotland's best farmland, producing oats, barley, and hay.

Strau·bing (shtrou'bĭng), city (1972 est. pop. 44,546), Bavaria, SE West Germany, on the Danube River. It is an agricultural market of Lower Bavaria and an industrial center. Manufactures include machinery, precision instruments, textiles, and beer.

stream (strēm), general term applied to all bodies of water flowing in channels regardless of their size.

Strea·tor (strē'tər), city (1978 est. pop. 14,800), La Salle and Livingston cos. N central Ill., on the Vermillion River; inc. 1882. It is an industrial center in an area of clay and shale deposits.

Stre·sa (strā'zä), town (1971 pop. 5,168), Piedmont, N Italy, on the W shore of Lake Maggiore. Its lovely gardens and villas and the scenic Borromean Islands nearby have made it one of the most popular resorts in the Italian lake country.

Stret·ford (strĕt'fərd), municipal borough (1978 est. pop. 53,470), Greater Manchester, NW England. Contiguous with Manchester and Salford, it has a large dock area and varied manufactures.

Strey·moy (strā'moi') or **Strøm·ø** (strœm'œ'), island (1966 pop. 14,078), 144 sq mi (373 sq km), Denmark, the largest of the Faeroe Islands.

Strongs·ville (strôngz'vĭl'), city (1975 est. pop. 19,900), Cuyahoga co., NE Ohio, a residential suburb of Cleveland; settled 1816, inc. 1927. It has a textbook-publishing company, a research laboratory, and some light manufacturing.

Strouds·burg (stroudz'bûrg'), borough (1970 pop. 5,451), seat of Monroe co., E Pa., near the Delaware Water Gap N of Easton; settled 1738, inc. 1815. It is a popular resort in the Pocono Mts. and has some manufacturing.

Stru·ma (stro͞o'mä), river, 216 mi (347.5 km) long, rising in the mountains of W Bulgaria and flowing S through NE Greece to the Aegean Sea.

Stru·mi·ca (stro͞o'mĭt-sä), town (1971 pop. 76,964), SE Yugoslavia, in Macedonia. It is an agricultural center. Strumica, an ancient town, was long under Turkish rule; it was ceded to Bulgaria in 1913 and to Yugoslavia in 1919.

Struth·ers (strŭ*th*'ərz), city (1970 pop. 15,343), Mahoning co., NE Ohio, an industrial suburb of Youngstown, on the Mahoning River; founded 1800, inc. 1922. It is an iron and steel center.

Stry (strē), city (1973 est. pop. 52,000), SW European USSR, in the Ukraine, on the Stry River (a tributary of the Dnestr) and in the Carpathian foothills. It is a major rail junction and a center of the Drogobych oil region. Industries include lumbering and woodworking, railroad-car repairing, flour milling, and the production of machinery and machine tools. An old Ukrainian settlement, Stry was chartered in 1431.

Stu·art (sto͞o'ərt, styo͞o-). **1.** City (1970 pop. 4,820), seat of Martin co., SE Fla., NNW of West Palm Beach. The eastern terminus of the St.

Lucie Canal to Lake Okeechobee, Stuart has yachting and fishing facilities. **2.** Town (1970 pop. 947), seat of Patrick co., S Va., on the Mayo River W of Martinsville in the foothills of the Blue Ridge. It is a trade center in an agricultural and timber area. Knit goods and canned foods are made.

Stur·bridge (stûr'brĭj'), town (1970 pop. 4,878), Worcester co., S Mass.; inc. 1738. Tourism is its major industry; Old Sturbridge Village, a model of an early American village, draws year-round visitors.

Stur·geon Bay (stûr'jən), city (1970 pop. 6,776), seat of Door co., NE Wis., at the head of Sturgeon Bay (an inlet of Green Bay); inc. 1883. A ship canal here cuts across Door Peninsula to Lake Michigan. A summer resort, the city also has cherry orchards and engages in shipping, shipbuilding, and canning.

Stur·gis (stûr'jĭs). **1.** City (1970 pop. 9,295), St. Joseph co., SE Mich., SSE of Kalamazoo; settled 1828, inc. as a village 1855, as a city 1896. It is a trade and industrial center. **2.** City (1970 pop. 4,536), seat of Meade co., W S.Dak., near the Black Hills NNW of Rapid City; laid out 1878 near Fort Meade, inc. 1888. A trade and distributing center of a livestock, wheat, and timber region, it has sawmills and a large stock market.

Stuts·man (stŭts'mən), county (1970 pop. 23,550), 2,264 sq mi (5,863.8 sq km), central N.Dak., in an agricultural area drained by the James River; formed 1873; co. seat Jamestown. Dairying, livestock ranching, and poultry raising are important.

Stutt·gart (shtŏŏt'gärt'), city (1972 est. pop. 630,390), capital of Baden-Württemberg, SW West Germany, on the Neckar River. It is a major transportation point. Manufactures include electrical and photographic equipment, optical goods, textiles, clothing, printed materials, beverages, pianos, and motor vehicles. It is also a tourist center and the site of industrial fairs. Stuttgart was chartered in the 13th cent. The city expanded rapidly in the 19th and 20th cent. as its industrial plant grew. After World War I it became famous for the innovative architecture of its numerous modern buildings. The center of the city was almost totally destroyed in World War II. After 1945 many old buildings were restored and striking modern structures (such as the city hall and the concert hall) were erected. The city has several museums, a university, and an academy of fine arts.

Stutt·gart (stŭt'gärt', -gərt), city (1970 pop. 10,477), a seat of Arkansas co., E central Ark.; inc. 1889. It is a trade and processing center of a rice-growing area that is also noted for its duck hunting. Shoes are manufactured in the city.

Styr·i·a (stîr'ē-ə), province (1971 pop. 1,192,442), 6,324 sq mi (16,379 sq km), central and SE Austria; capital Graz. Bordering on Yugoslavia in the south, Styria is predominately mountainous, with many forests, pastures, and meadowlands. It is the chief Austrian mining district (iron ore, lignite, and magnesite) and has a well-developed metals industry. Cattle, horses, and poultry are raised, and forestry is an important occupation. There are many Alpine resorts.

Sub·lette (sə-blĕt'), county (1970 pop. 3,755), 4,851 sq mi (12,564.1 sq km), W Wyo., watered by the Green River; formed 1921; co. seat Pinedale. In an agricultural region that raises hay and livestock, it has oil and gas wells. Part of Bridger National Forest and Wind River Range are in the north.

Sublette, city (1970 pop. 1,208) seat of Haskell co., SW Kansas, SW of Dodge City, in an oil, natural-gas, and farm area; inc. 1923.

Su·bo·ti·ca (sōō'bə-tēt'sə), city (1971 pop. 88,787), N Yugoslavia, in the Vojvodina region of Serbia. An important railway junction and an industrial center, it has factories that produce foodstuffs, textiles, chemicals, electrical products, and agricultural machinery.

Su·cea·va (sōō-chä'vä), town (1973 est. pop. 49,019), NE Rumania, in Bukovina, on the Suceava River. It is a commercial center and has industries that manufacture food products, leather, textiles, and cellulose.

Su-chou or **Soo·chow** (both: sōō'chou', sōō'jō'), or **Wuh·si·en** (wōō'-shē-ĕn'), city (1970 est. pop. 1,300,000), SE Kiangsu prov., E central China, on the Grand Canal near Tai Lake. Su-chou, famous for its silks since the Sung dynasty, is still a silk center; it also has cotton and embroidery manufactures and an important food-processing industry. On the city's outskirts are a small integrated steel complex and plants making chemicals, paper, machine tools, and motor vehicles. Su-chou was capital of the Wu kingdom in the 5th cent. B.C. A nine-storied pagoda here (c.250 ft/80 m high) may be the tallest in China.

Sü·chow (shü'jō'): *see* Hsü-chou, China.

Su·cre (sōō'krā), city (1969 est. pop. 48,000), S central Bolivia, constitutional capital of Bolivia and capital of Chuquisaca dept. Since 1898 La Paz has been the administrative capital of Bolivia. The city lies in a mountain valley on the eastern slope of the Andes at an altitude of c.8,500 ft (2,595 m). Sucre is a major agricultural center and supplies the mining communities of the barren altiplano. It also has an oil refinery.

Su·dak (sōō-däk'), town, SE European USSR, in the Ukraine, on the Crimean peninsula. It is a resort on the Black Sea. Its major industries are rose-oil processing and the production of fine quality wines and champagnes. Russia acquired Sudak in 1783 with the rest of the Crimea.

Su·dan (sōō-dăn'), officially Democratic Republic of the Sudan, republic (1977 est. pop. 18,415,000), 967,494 sq mi (2,505,813 sq km), NE Africa, bordering on Egypt in the north, on the Red Sea in the northeast, on Ethiopia in the east, on Kenya, Uganda, and Zaire in the south, on the Central African Empire and Chad in the west, and on Libya in the northwest. Khartoum is the capital. The main geographical feature of Sudan, Africa's largest country, is the Nile River, which, with its tributaries (including the Atbara, Blue Nile, and White Nile rivers), traverses the eastern part of the country from south to north. The Nile system provides irrigation for strips of agricultural settlement for much of its course in Sudan. In the extreme north, the Nile broadens into Lake Nasser, formed by the Aswan High Dam in Egypt. Much of the rest of the country is made up of an undulating plateau (1,000–2,000 ft/305–610 m high), which rises to higher levels in the mountains located in the northeast near the Red Sea, as well as in the central, western, and extreme southern portions of the country. The highest point in the Sudan is Kinyeti (10,456 ft/3,187 m), in the southeast. Rainfall diminishes from south to north in Sudan; thus, the south is characterized by swampland and rain forest, the center by savanna and grassland, and the north by desert and semidesert.

The inhabitants of Sudan are divided into three main groups: the northerners, about 39% of the population; the southerners, about 30%; and the westerners, about 13%. The northerners, who inhabit the country roughly north of 12° N lat. and mainly near the Nile, are Moslem (mostly of the Sunni branch), speak Arabic (the country's official language), and follow Arab cultural patterns. The westerners, so called because they immigrated from W Africa, are also Moslem, live mostly in the central Sudan, and work as farmers or agricultural laborers. The southerners largely follow traditional religious beliefs, although some are Christian; they are farmers or pastoralists, and most speak Nilotic languages. The leading ethnic groups in the south are the Dinka, Nuer, Shilluk, Bari, and Azande. The great majority of the country's population lives in villages or small towns;

the only sizable cities are Port Sudan, Wad Madani, Al Ubayyid, and the conurbation of Khartoum, Omdurman, and Khartoum North. The desert and semidesert of the north are largely uninhabited.

Economy. Sudan is an overwhelmingly agricultural country; much of the farming is of a subsistence kind and is carried out largely outside the money economy. Commercial agriculture in the later 1960s contributed about 34% of the annual national product, industry only about 11%. The government plays a major role in planning the economy. The leading farm crops are cotton, durra and other millets, groundnuts, sesame, dates, plantains, sugar cane, and coffee. Large numbers of cattle, sheep, goats, and camels are raised. A variety of forest products are produced, the most important being gum arabic. The leading products of the country's small mining industry are chromite; copper, manganese, and iron ores; salt; and gold. Manufacturing is mostly related to basic consumer needs. The country has a very limited transportation network.

Educational Institutions. In the early 1970s only about 19% of the population 10 years or older was literate. The main institutions of higher education are the Univ. of Khartoum and the Khartoum branch of the Univ. of Cairo.

History. Northeast Sudan, called Nubia in ancient times, was colonized (c.2000 B.C.) by Egypt. From the 8th cent. B.C. to the 4th cent. A.D. this region was ruled by the Cush kingdom, centered first at Napata and after c.600 B.C. at Meroë. From c.750 to c.650 B.C., Cush conquered and ruled Egypt. Meroë was a center of trade and ironworking, and from there iron technology probably spread to other parts of Africa. Most of the inhabitants of Nubia were converted to Coptic Christianity in the 6th cent. A.D., and by the 8th cent. two states flourished in the area. These states long resisted invasions from Islamic Egypt; however, by the 15th cent. the states collapsed, and Nubia was converted to Islam. The southern part of modern Sudan remained pagan. Much of the north was ruled by the Moslem state of Funj from the 16th cent. until 1821, when it was conquered by armies sent by Mohammed Ali of Egypt. The Egyptians founded (1823) Khartoum as their headquarters and developed Sudan's trade in ivory and slaves. Ismail Pasha (in office 1863–79) tried to extend Egyptian influence farther south in Sudan. This campaign, supported by the British, provoked a complex revolt (1881) by the Mahdi, who sought to end Egyptian influence and to purify Islam in Sudan. The Mahdists prevailed and Britain and Egypt decided to abandon Sudan. The Mahdi died in 1885, but his successor, the Khalifa Abdallahi, continued to build up the theocratic Mahdist state. In the 1890s the British intervened and destroyed the power of the Mahdists. Agreements in 1899 established the condominium government of the Anglo-Egyptian Sudan. Sudan was administered by a governor general, appointed by Egypt; however, the British controlled the government of Sudan. The Sudanese and Egyptians continued to oppose British rule. The British administered southern Sudan separately from the north. In the 1948 elections the Independence Front, which favored the creation of an independent republic, gained a majority over the National Front, which sought union with Egypt. After the 1952 revolution in Egypt, Britain and Egypt agreed to prepare Sudan for independence in 1956. In 1955 southerners, fearing that the new nation would be dominated by the Moslem north, began a revolt that lasted 17 years. Nevertheless, Sudan achieved independence as a parliamentary republic in 1956, as planned. In 1958 Gen. Ibrahim Abboud led a military coup that ended the parliamentary system. Abboud in 1964 agreed to the reestablishment of civilian government. The new regime also had little success in coping with the country's problems, and in 1969 Col. Jaafar al-Numeiry staged a successful coup. He established a leftist government, banned all political parties, and nationalized numerous banks and industries. In July, 1971, a Communist-led coup attempt was defeated. A bloody civil war, which resulted in the death of about 1.5 million southerners, was ended by an agreement between the government and the Southern-Sudan Liberation Front. Southern Sudan gained considerable autonomy. Also in 1972 the Sudanese Socialist Union drew up a new constitution for the country. In 1973 a constitution calling for one-party parliamentary government and granting much autonomy to the south was adopted. In Mar., 1973, Palestinian guerrillas belonging to the black September group kidnaped three foreign diplomats in Khartoum; the Palestinians unsuccessfully demanded the release of fellow guerrillas imprisoned in Jordan and Israel. They killed the diplomats shortly before surrendering (Mar. 4) to Sudanese authorities.

Sud·bur·y (sŭd'bĕr'ē, -bə-rē), city (1973 est. pop. 100,446), central Ont., Canada. It is the center of Canada's largest mining region, which produces much of the world's nickel and large quantities of copper, platinum, gold, silver, and sulfur.

Sudbury, town (1970 pop. 13,506), Middlesex co., E Mass.; inc. 1639. Electrical and electronic equipment is manufactured. The tavern (built 1686; restored) that was the scene of Henry Wadsworth Longfellow's *Tales of a Wayside Inn* is here.

Su·de·tes (so͞o-dā'tēz), mountain range, along the border of Czechoslovakia and Poland, extending c.185 mi (300 km) between the Elbe and Oder rivers. It is continued on the west by the Erzgebirge and on the east by the Carpathians. The mineral deposits of the Sudetes include coal, lignite, iron, nickel, copper, silver, graphite, kaolin, and pyrites. Industry flourishes on both slopes of the Sudetes; lumber products, glass and porcelain, paper, and textiles are the chief products. There are also numerous mineral springs and resorts. The region was largely German-speaking until 1945.

Su·ez (so͞o-ĕz'), city (1970 est. pop. 315,000), NE Egypt, at the N end of the Gulf of Suez and at the S terminus of the Suez Canal. An important port with extensive facilities, it is also a refueling station, a holding area for ships entering the canal, and a center for the storage and refining of oil. Petroleum products, paper, and fertilizers are major manufactures. Although the site of the city was occupied in antiquity, Suez was little more than a small village throughout most of its history. After the completion (1869) of the Suez Canal the city became a major port. Its economy suffered during the periods that the canal was closed following the Arab-Iraeli Wars. During the 1973 War the city was damaged, and parts of it were occupied by Israeli forces.

Suez Canal, waterway of Egypt extending from Port Said to Port Tawfiq (near Suez) and connecting the Mediterranean Sea with the Gulf of Suez and thence with the Red Sea. The canal is somewhat more than 100 mi (160 km) long. Proceeding south from Port Said, it runs in an almost undeviating straight line to Lake Timsah. From there a cutting leads to the Bitter Lakes (now one body of water), and a final cutting then reaches the Gulf of Suez. The canal has no locks and can accommodate ships of almost any draft.

The canal was planned by the French engineer Ferdinand de Lesseps, who also supervised construction (1859–69). Great Britain, which had opposed the construction of the canal, became the largest shareholder in 1875. After 1888 management was placed in the hands of the Suez Canal Company. Under the Anglo-Egyptian treaty of 1936, which made Egypt virtually independent, Britain reserved rights for the protection of the canal, but after World War II Egypt pressed for evacuation of British troops from the area. In 1954 Britain agreed to withdraw, and in June, 1956, the British completed their evacuation of armed forces from Egypt and the canal zone.

Egyptian President Gamal Abdal Nasser nationalized (July, 1956) the Suez Canal and set up the Egyptian Canal Authority to replace the existing privately owned company. On Oct. 29, 1956, Israel, having been denied passage through the canal since 1950 and having suffered repeated border raids from Egypt, invaded Egyptian territory. Within a few days France and Great Britain sent armed forces to retake the Suez Canal. Intervention by the United Nations brought an armistice in early Nov. The canal, blocked for more than six months because of damage and sunken ships, was cleared with UN help and reopened in Apr., 1957. Despite UN efforts to guarantee the free passage of vessels through the canal, Egypt prevented Israeli ships from using the waterway. The canal was closed by Egypt during the Arab-Israeli War of 1967, after which it formed part of the boundary between Egypt and the Israeli-occupied Sinai peninsula. Egypt lost considerable revenue as a result of the closing of the canal, but friendly Arab countries agreed to subsidize the Egyptian economy with contributions roughly equaling the former income from the canal. In Oct., 1973, Egyptian troops crossed the canal and breached the Israeli lines. In the ensuing conflict a complex military situation evolved, with Egyptians retrieving land on the east bank of the canal while Israeli units established a salient on the west bank. In early 1974 Egypt and Israel signed an agreement providing for a disengagement of military forces and for a withdrawal of Israeli troops into the Sinai. With both banks of the canal again in Egyptian hands, the government of Egypt, with the assistance of the U.S. navy, began clearing the canal of the mines and wreckage left from the 1967 War.

Suf·field (sŭf'ēld), town (1970 pop. 8,634), Hartford co., N Conn., on the Connecticut River at the Mass. border N of Hartford, in a tobacco-growing region; settled c.1670, inc. 1674. Cigar making began in 1810.

Suf·folk (sŭf'ək), nonmetropolitan county (1974 est. pop. 561,560), E central England; administrative center Ipswich. The terrain is low and undulating, and the region is mainly agricultural. Along the coast fishing is important. In Anglo-Saxon times Suffolk was part of the kingdom of East Anglia. Until 1974 it was divided into the administrative units of East Suffolk and West Suffolk.

Suffolk. 1. County (1970 pop. 735,190), 56 sq mi (145 sq km), E Mass., on Massachusetts and Boston bays; formed 1643; co. seat Boston. It comprises the city of Boston, several islands in Boston Harbor and the bay, and a number of outlying residential communities. **2.** County (1970 pop. 1,127,030), 929 sq mi (2,406.1 sq km), SE N.Y., on central and E Long Island, bounded on the S by the Atlantic, on the E by Block Island Sound, on the N by Long Island Sound; formed 1683; co. seat Riverhead. This residential and summer-resort region has areas of large estates (notably Southampton). There is a huge duck-raising industry along the south shore. It also has large oyster beds, fisheries, potato and truck farming, dairying, and some manufacturing.

Suffolk, independent city (1975 est. pop. 47,400), SE Va., on the Nansemond River near the Dismal Swamp; settled 1720, inc. as a town 1808, as a city 1910. The rail, commercial, and industrial center for a farm and timber area, it is an important peanut market and processing center. There are also tea-processing plants. Suffolk was burned by the British in 1779, occupied by Union troops on May 12, 1862, and besieged by Confederate Gen. James Longstreet in Apr., 1863.

Suhl (zo͞ol), city (1973 est. pop. 36,187), capital of Suhl district, SW East Germany. It is an industrial city manufacturing precision instruments, chemicals, toys, porcelain, and motor vehicles. Suhl was a noted center of the German arms industry in the Thirty Years' War (1618–48).

Suir (sho͝or), river, 85 mi (137 km) long, rising on Devilsbit Mt., central Republic of Ireland. It flows south through a fertile agricultural region, then east past Clonmel and Waterford to the Barrow River, with which it forms Waterford Harbour.

Suit·land (so͞ot'lənd), uninc. city (1978 est. pop. 26,700), Prince Georges co., central Md., a suburb of Washington, D.C. It is the seat of the U.S. Bureau of the Census and of the U.S. navy hydrographic office.

Suitland Parkway: *see* National Parks and Monuments Table.

Sui·yuan (swē'yü-än'), former province (c.126,000 sq mi/326,340 sq km), N China. The region of Suiyuan, part of Inner Mongolia, is chiefly a high arid plateau. Livestock raising and the growing of grains, chiefly wheat, support most of the people. Suiyuan was overrun (1937) by the Japanese, who included it in Menchiang (Mongol Border Land). In 1954 it was made part of the Inner Mongolian Autonomous Region.

Su·kho·na (so͞o-khô'nə), river, c.350 mi (565 km) long, N European USSR, flowing from Kubeno Lake NE into the Yug River at Veliki Ustyug to form the Northern Dvina River.

Su·khu·mi (so͞o-kho͞o'mē), city (1973 est. pop. 109,000), capital of the Abkhaz Autonomous Republic, SE European USSR, in Georgia, on the Black Sea. It is a port, a rail junction, and a major subtropical resort, whose sulfur baths have been frequented since Roman times.

Suk·kur (so͝ok'ər), city (1972 pop. 159,000), SE Pakistan, on the Indus River. It is an important commercial and industrial city and a center for trade with Afghanistan. Its industries produce cotton and silk textiles, hosiery, and foodstuffs. Handloom weaving is also important. Sukkur Barrage, a dam across the Indus, waters more than 5 million acres (2,023,000 hectares).

Su·lai·ma·ni·yah (so͞o'lā-mä'nē-ə), town (1970 est. pop. 98,100), NE Iraq. The town is a trade center and tourist resort. Inhabited by Kurds, it is known as a center of Kurdish nationalism. It was founded in 1789.

Su·lai·man Mountains (so͞o'lī-män'), range, extending c.250 mi (400 km) from N to S along the W edge of the Indus River valley, central Pakistan.

Su·la·we·si (so͞o'lə-wā'sē): *see* Celebes, Indonesia.

Su·li or **Sou·li** (both: so͞o'lyē), small mountainous district, N Greece, in Epirus. Its inhabitants, who lived in fortlike villages in the mountains, remained independent during most of the occupation of Greece by the Ottoman Turks. In 1803, however, Ali Pasha massacred many of them after concluding a false truce.

Sul·li·van (sŭl'ə-vən). **1.** County (1970 pop. 19,889), 457 sq mi (1,183.6 sq km), SW Ind., bounded on the W by the Wabash River and the Ind. border; formed 1816; co. seat Sullivan. Its agriculture includes grain, fruit, livestock, poultry, and dairy products. It has bituminous-coal mines, limestone quarries, oil and gas wells, and timber. Clothing, wood products, plastics, and instruments are made. **2.** County (1970 pop. 7,572), 654 sq mi (1,693.9 sq km), N Mo., in an agricultural area; formed 1845; co. seat Milan. Corn, oats, and livestock are the major products. **3.** County (1970 pop. 30,949), 539 sq mi (1,396 sq km), SW N.H., on the Vt. border and drained by the Connecticut, Cold, and Sugar rivers; formed 1827; co. seat Newport. Its agriculture includes fruit, poultry, and truck crops. Textiles, shoes, machinery, tools, and paper are manufactured, and there are resorts in the Sunapee Lake area. **4.** County (1970 pop. 52,580), 980 sq mi (2,538.2 sq km), SE N.Y., bounded on the W and SW by the Delaware River and the Pa. border; formed 1809; co. seat Monticello. This vacation region, with many mountain and lake resorts, has poultry raising, dairying, fruit growing, lumbering, and some manufacturing. **5.** County (1970 pop. 5,961), 478 sq mi (1,238 sq km), NE Pa., drained by Loyalsock and Muncy creeks; formed 1847; co. seat Laporte. It has some agriculture, anthracite-coal mining, lumbering, and manufacturing of clothing and shoes. It is in a scenic resort region of mountain lakes. **6.** County (1970 pop. 127,329), 413 sq mi (1,069.7 sq km), NE Tenn., in the Great Appalachian Valley, bounded in the N by Va., and drained by the South Fork of the Holston River; formed 1799; co. seat Blountville. Its agriculture includes tobacco, corn, truck crops, fruit, hay, dairy products, and livestock. It also has bituminous-coal mining, lumbering, and diversified industries. Part of Cherokee National Forest is here.

Sullivan. 1. City (1970 pop. 5,059), seat of Moultrie co., central Ill, SE of Decatur, in a grain area; inc. 1869. Shoes are made. **2.** City (1970 pop. 4,683), seat of Sullivan co., SW Ind. near the Wabash River S of Terre Haute; platted 1842. Coal is mined.

Sul·ly (sŭl'ē), county (1970 pop. 2,362), 1,004 sq mi (2,600.4 sq km), central S.Dak., bounded in the W by the Missouri River; formed 1873; co. seat Onida. It is in an agricultural (wheat, poultry, and dairying) and livestock-ranching (cattle and sheep) area, with some manufacturing.

Sul·mo·na (so͞ol-mô'nä), town (1971 pop. 20,548), Abruzzi, central Italy, on the Gizio River. A commercial and industrial center, it lies in a small, fertile plain enclosed by Apennine peaks.

Sul·phur (sŭl'fər). **1.** City (1970 pop. 15,106), Calcasieu parish, SW La.; inc. 1914. It is a trade center for an area producing rice, natural gas, timber, and oil. Its industry centers chiefly around petroleum products, chemicals, and related enterprises. In 1924 oil was discovered nearby. **2.** Resort city (1970 pop. 5,158), seat of Murray co., S Okla., SW of Ada near the Washita River and the Arbuckle Mts., in a ranch, farm, and quarry region; settled 1895. Platt National Park, with mineral springs, is nearby.

Sulphur Springs, city (1970 pop. 10,642), seat of Hopkins co., NE Texas, in a dairy area; inc. 1859. Milk and milk products are produced here. A trading post was established on the site in 1845. Sulphur Springs grew as a cotton and farm-produce market.

Su·lu Archipelago (so͞o'lo͞o'), island group (1970 pop. 427,386), 1,086 sq mi (2,812.7 sq km), the Philippines, SW of Mindanao. Lying between the Celebes and Sulu seas, it includes over 400 volcanic islands and coral islets extending almost to Borneo. Fishing is the major source of livelihood. The archipelago is also the prime source for pearls, marine turtles, seashells, and sea cucumbers. The islands are heavily forested, but local farming is nonetheless carried on and meets the needs of the people. The inhabitants are Moros, a Malayan people who were converted when Islam spread from Malaya and Borneo in the 14th and 15th cent. Formerly notorious as pirates, the Moslem Moros resisted Spanish rule until the 19th cent. The Moro sultanate (est. in the 16th cent.) passed to U.S. control in 1899. In 1940 the sultanate was abolished and Sulu became part of the Philippine Commonwealth.

Sulz·bach (zo͞olts'bäkh'), city (1972 est. pop. 19,233), Saarland, SW West Germany; chartered 1946. It is an industrial center of the Saar coal basin.

Su·ma·tra (so͞o-mä'trə), island (1971 est. pop. 20,800,000), c.183,000 sq mi (473,970 sq km), Indonesia, in the Indian Ocean along the equator, S and W of the Malay Peninsula and NW of Java. The westernmost and second-largest island of Indonesia, Sumatra is c.1,110 mi (1,790 km) long and c.270 mi (435 km) wide and is fringed with smaller islands off its western and eastern coasts. The Barisan, a volcanic mountain range, traverses its length, reaching 12,467 ft (3,800 m) at Mt. Kerintji. Rising in the Barisan range are several large rivers, including the Hari, Indragiri, and Musi. In the north is Lake Toba, a great salt lake. Because of the hot, moist climate and the heavy rainfall, the vegetation is luxuriant. Much of the eastern half of the island is swampland. The interior is covered largely by impenetrable rain forests. Among the animals found here are elephants, clouded leopards, tapirs, tigers, Malayan bears, and snakes. About 70% of the country's income is produced on the island, which has some of Indonesia's richest oil fields, its finest coal fields, and deposits of gold and silver. Its offshore islands are known for their tin and bauxite. Most of the country's rubber is grown in Sumatra; pepper, coffee, tea, sugar cane, and oil palms are also grown. The Deli region is famous for its tobacco. Rice, corn, and root crops are raised for local consumption. Timber cut includes camphor and ebony.

Sumatra is sparsely settled, with principal centers at Medan and Palembang; also important are Djambi and Padang. The four largest ethnic groups are the Atjehnese, Batak, Menangkabu, and coastal Malays. In the interior highlands are found the Gajo-Alas and the Rejang-Lampoeng groups. Islam is the predominant religion, though there are many Christians among the Batak and the Gajo-Alas. Chinese, Arabs, and Indians live on the coasts, and some 15 different languages are spoken on the island.

Sumatra had early contact with the Hindu civilization, and by the 7th cent. A.D. the powerful Hindu-Sumatran kingdom of Sri Vijaya flourished here. The kingdom extended its control over much of Indonesia and also over the Malay Peninsula. By the 14th cent. the island fell under the Javanese kingdom of Majapahit. The Arabs, who may have first arrived in the 10th cent., established the sultanate of Achin, which reached its height in the 17th cent. The first European to visit Sumatra was Marco Polo (c.1292). Following the Portuguese, who came in 1509, the Dutch arrived in 1596 and gradually gained control of all the native states. The British had brief control over parts of the island in the late 18th and early 19th cent. The Achinese (now more commonly the Atjehnese) launched a rebellion in 1873 and were not subdued by the Dutch until 1904. In World War II Japanese troops landed (Feb., 1942) in Sumatra and occupied it throughout the war. After independence was granted (1949), all of Sumatra was included in the new republic of Indonesia. Since then there has been much tribal agitation and repeated demands for local autonomy. The Atjehnese have waged occasional guerrilla warfare against the government, and in 1958 a full-scale rebellion was launched by dissident army officers. It spread to other islands before being quelled by the government. Sumatra now comprises eight provinces.

Sum·ba (so͞om'bä), island (1961 pop. 251,126), 4,305 sq mi (11,150 sq km), Indonesia, one of the Lesser Sundas, in the Indian Ocean, S of Flores across Sumba Strait. The island is noted for horse breeding. It was first visited by Europeans in 1522 and passed to the direct control of the Dutch in 1866.

Sum·ba·wa (so͞om-bä'wä), island (1961 pop. 407,596), 5,964 sq mi (15,447 sq km), Indonesia, one of the Lesser Sundas, between the Flores Sea and the Indian Ocean. Sumbawa has many volcanic peaks. The soil is fertile, and tropical fruit and rice are produced. Cattle raising is important.

Su·mer (so͞o'mər), term used today to designate the S part of ancient Mesopotamia. From the earliest date of record, southern Mesopotamia was occupied by a people speaking a non-Semitic language, known as Sumerians. Their origin is uncertain. Some evidence suggests that they may have come as conquerors from the East (possibly from Iran or India). Modern excavations have shown there was in the 5th millennium B.C. a prehistoric village culture in the area. By 3000 B.C. a flourishing urban civilization existed. Sumerian civilization was predominantly agricultural. The Sumerians were adept at building canals and effective systems of irrigation. They were also skilled in the use of such metals as copper, gold, and silver. The Sumerians are credited with inventing the Cuneiform system of writing. Between the years 3000 and 2340 the kings of important Sumerian cities were able to extend their control over large areas, forming various dynasties. However, Mesopotamia was also the home of a group of people speaking Semitic languages and with a different culture. The increasing Semitic strength culminated in the establishment (c.2340) of the Akkadian dynasty by Sargon, who for the first time imposed a wide imperial organization over the whole of Mesopotamia. This conquest gave impetus to the blending, already long in progress, of Sumerian and Semitic cultures. After the collapse of Akkad (c.2180) under the pressure of invading barbarians from the northeast, the Sumerians had a final revival under the third dynasty of Ur (c.2060). After this dynasty fell (c.1950) to the western Amorities and the Guti, the Sumerians were never again able to gain a political hegemony. The control of the country passed to Babylonia, and the Sumerians, as a nation, disappeared.

Sum·ga·it (so͞om'gä-ēt'), city (1973 est. pop. 145,000), Azerbaijan Republic, S European USSR, on the Caspian Sea at the mouth of the Sumgait River. It has a pipe-rolling mill and aluminum and synthetic rubber factories.

Sum·mers (sŭm'ərz), county (1970 pop. 13,213), 350 sq mi (906.5 sq km), S W.Va., on the Allegheny Plateau, drained by the New River; formed 1871; co. seat Hinton. It is in a bituminous-coal and natural-gas region. Its agriculture includes livestock, dairy products, fruit, and tobacco.

Sum·mers·ville (sŭm'ərz-vĭl'), town (1970 pop. 2,429), seat of Nicholas co., central W.Va., ESE of Charleston; founded c.1820, inc. 1897. In 1861 Nancy Hart, a Confederate spy, led an attack on the town in which a Union force was captured and most of the town burned.

Sum·mer·ville (sŭm'ər-vĭl'), town (1970 pop. 5,043), seat of Chattooga co., extreme NW Ga., on the Chattooga River NNW of Rome; inc. 1839. Textiles are manufactured. Marble deposits and a state fish hatchery are nearby.

Sum·mit (sŭm'ĭt). **1.** County (1970 pop. 2,665), 604 sq mi (1,564.4 sq km), central Colo., drained by the Blue River; formed 1861; co. seat Breckenridge. It has zinc, gold, and silver mining and livestock grazing. Part of Gore Range Eagles Nest Wilderness Area, Arapaho National Forest, two recreational reservoirs, and four major Colorado ski areas are in the county. **2.** County (1970 pop. 553,371), 408 sq mi (1,056.7 sq km), NE Ohio, drained by the Cuyahoga and Tuscarawas rivers; formed 1840; co. seat Akron. Its agriculture includes corn, clover, poultry, livestock, and dairy products. It has mining (sand and gravel pits, limestone and sandstone quarries, and salt deposits) and diversified industry. Textiles, furniture, paper, chemicals, rubber and plastics, concrete, metal products, and machinery are manufactured. **3.** County (1970 pop. 5,879), 1,849 sq mi (4,788.9 sq km), NE Utah, bordering on Wyo., and drained in the W by the Weber River; formed 1854; co. seat Coalville. It has agriculture (truck crops, hay, grain, dairying, cattle, and sheep) and silver, copper, lead, and zinc mines. Wasatch National Forest and the Uinta Mts. are here.

Summit. 1. Village (1970 pop. 11,569), Cook co., NE Ill., a suburb of Chicago; inc. 1890. It has one of the world's largest grain-milling plants. **2.** City (1978 est. pop. 22,000), Union co., NE N.J., a residential suburb of the New York City metropolitan area; settled c.1720, inc. 1869. Pharmaceuticals are made, and several major companies have research facilities here. It was the site of an important American lookout post during the Revolutionary War.

Sum·ner (sŭm'nər). **1.** County (1970 pop. 23,553), 1,186 sq mi (3,071.7 sq km), S Kansas, in a level to gently rolling plain bordering in the S on Okla., and drained by the Chikaskia, Ninnescah, and Arkansas rivers; formed 1867; co. seat Wellington. It is in a wheat, corn, and livestock area, with oil and gas fields and timber. Wood and metal products, machinery, and aircraft equipment are made. **2.** County (1970 pop. 56,266), 534 sq mi (1,383.1 sq km), N Tenn., bounded in the N by Ky., in the S by the Cumberland River, and drained by headstreams of the Red River and by Drake Creek; formed 1786; co. seat Gallatin. Its agriculture includes corn, tobacco, grains, fruit, vegetables, cattle, poultry, and dairy products. Clothing, furniture, concrete, metal products, and machinery are among its manufactures.

Sumner, town (1970 pop. 533), a seat of Tallahatchie co., NW central Miss., SE of Clarksdale. Cottonseed products are made.

Sum·ter (sŭm'tər). **1.** County (1970 pop. 16,974), 915 sq mi (2,370 sq km), W Ala., in the Black Belt, on the Miss. line and bounded E by Tombigbee River; formed 1832; co. seat Livingston. It has cotton, livestock, corn, and bee farming, lumber logging and sawmills, and small industries (clothing, paper, and concrete). **2.** County (1970 pop. 14,839), 555 sq mi (1,437.5 sq km), central Fla., bounded partly on the W and S by the Withlacoochee River, with swamps in the W and many scattered lakes; formed 1853; co. seat Bushnell. In an agricultural area yielding citrus fruit and truck crops, as well as corn, cattle, and poultry, it has limestone deposits, some timber, and industry (concrete and steel products). **3.** County (1970 pop. 26,931), 488 sq mi (1,263.9 sq km), SW central Ga., bounded on the E by the Flint River; formed 1831; co. seat Americus. It is in a coastal-plain agricultural (cotton, corn, watermelons, peaches, pecans, peanuts, and livestock) and timber area. Its manufactures include clothing, furniture, pottery, and metal products. **4.** County (1970 pop. 79,425), 672 sq mi (1,740.9 sq km), central S.C., bounded in the W by the Wateree River, in the NE by the Lynches River, and drained by the Black River; formed 1785; co. seat Sumter. In an agricultural (tobacco, corn, cotton, livestock, and poultry) and lumbering area, it has sand and gravel pits, and its industry includes food processing, textile mills, and clothing, furniture, tools, and machinery manufacturing.

Sumter, city (1978 est. pop. 24,300), seat of Sumter co., central S.C.; founded 1785, inc. 1845. It is the trade, processing, and shipping center of an important lumber, livestock, and farm region. Textile products, furniture, electric storage batteries, frozen foods, medical supplies, fabricated steel, and paints and varnishes are made here.

Sun (sŭn), river, c.130 mi (210 km) long, rising in the Rocky Mts., NW Mont., and flowing generally E to the Missouri River at Great Falls. The Sun River project irrigates c.92,000 acres (37,260 hectares).

Sun·bur·y (sŭn'bĕr'ē, -bə-rē), city (1975 est. pop. 12,000), seat of Northumberland co., E central Pa., on the Susquehanna River at the confluence of its N and W branches; laid out 1772, inc. 1921. Textile products are manufactured. It was the site of an Indian village in the early 18th cent. In 1742 a mission was established, and in 1756 Fort Augusta was built.

Sun·bur·y-on-Thames (sŭn'bĕr'ē-ŏn-tĕmz'), urban district (1974 est. pop. 40,210), Surrey, SE England, on the Thames. Sand and gravel are excavated in the district. There are motion-picture studios in Sunbury-on-Thames.

Sun·chon (so͞on'chŭn'), city (1970 pop. 61,000), SW South Korea. It is a railroad junction and an agricultural center.

Sun·da Islands (sŭn'də), Indonesia, between the South China Sea and the Indian Ocean, comprising the W part of the Malay Archipelago. It includes: the Greater Sunda Islands, to which belong the largest islands of Borneo, Sumatra, Java, and Celebes, and the Lesser Sundas, which lie east of Java and include Sumbawa, Flores, Timor, and Sumba.

Sun·dance (sŭn'dăns'), town (1970 pop. 1,056), seat of Crook co., NE Wyo., just NW of the Black Hills NW of Rapid City, S.Dak., at an altitude of 4,750 ft (1,448.8 m). It is a trade point in a livestock, grain, coal, and oil region. Devils Tower National Monument and part of Black Hills National Forest are nearby.

Sund·by·berg (sŭnd'bü-bĕr'yə), city (1970 pop. 28,016), Stockholm co., E Sweden, an industrial suburb of Stockholm; founded 1877. Manufactures include chemicals, paper, chocolate, and cables.

Sun·der·land (sŭn'dər-lənd), county borough (1974 est. pop. 215,280), Tyne and Wear, NE England, at the mouth of the Wear River. Established as a shipbuilding center and coal-shipping port in the 14th cent., Sunderland today exports iron and steel and manufactured goods. Shipbuilding is still an important industry, as are engineering, coal mining, and the manufacture of aircraft components, electrical goods, glass, clothes, chemicals, and pottery.

Sunds·vall (sŭnts'väl'), city (1972 est. pop. 65,712), Västernorrland co., E Sweden, a major seaport on the Sundsvallfjärden, an arm of the Gulf of Bothnia. Manufactures include cellulose, metal goods, aluminum, and dairy products. Sundsvall was chartered in 1621.

Sun·flow·er (sŭn'flou'ər), county (1970 pop. 37,047), 694 sq mi (1,797.5 sq km), W Miss., in a rich lowland cotton-growing region drained by the Sunflower River; formed 1844; co. seat Indianola. It produces corn, oats, hay, and alfalfa. Some lumbering is done.

Sun·ga·ri (so͝ong'gə-rē', so͞on'gä'rē'), river of NE China, c.1,150 mi (1,850 km) long, rising in the Ch'ang-pai Mts. and flowing generally N to the Amur River on the China-USSR border. It is the northernmost river system in China and forms a main gateway to the southern Manchurian plain.

Sung·kiang (so͝ong'gyäng', -jē-äng'), former province (c.32,000 sq mi/82,880 sq km), NE China. It was one of nine provinces created in Manchuria by the Chinese Nationalist government after World War II. In 1954 it became part of Heilungkiang prov.

Sun·ny·vale (sŭn'ē-vāl'), city (1978 est. pop. 105,600), Santa Clara co., W Calif., near San Francisco; settled 1849, inc. 1912. Its manufactures include electronic and electrical equipment, food products, pharmaceuticals, and paper products.

Sun Prai·rie (prâr'ē), city (1975 est. pop. 13,300), Dane co., S Wis., NE of Madison, in a farm area; settled after 1837, inc. as a village 1868, as a city 1958. Georgia O'Keeffe was born here.

Sun·set Crater National Monument (sŭn'sĕt'): *see* National Parks and Monuments Table.

Sun Valley, mountain resort city (1970 pop. 180), alt. c.6,000 ft (1,830 m), Blaine co., S central Idaho; inc. 1967. It was founded as a ski resort in 1936 by the Union Pacific RR. The railroad purchased 4,300 acres (1,740 hectares) near the declining mining village of Ketchum and built the resort as a means of attracting more passenger traffic to the West.

Sun·wui (so͞on'wē): *see* Hsin-hui, China.

Su·per·i·or (sə-pîr'ē-ər). **1.** Uninc. town (1970 pop. 4,975), Pinal co., SE Ariz., SE of Phoenix. It was settled after the Silver King and Silver Queen mines were discovered (c.1875). Today copper, gold, and silver are mined, and there is a copper smelter. **2.** Town (1970 pop. 993), seat of Mineral co., W Mont., WNW of Missoula, in a mine, timber, and grain area; settled c.1889, inc. 1948. **3.** City (1978 est. pop. 29,600), seat of Douglas co., NW Wis., on Superior Bay of Lake Superior, at the mouths of the St. Louis and the Nemadji rivers; inc. 1883. Its superb natural harbor, shared with Duluth, Minn., has some of the nation's largest coal and ore docks. Superior has shipyards, huge grain elevators, an oil refinery, and a large dairy products industry. The area was visited by French explorers (1661 and 1679) and grew after iron ore was discovered (1880s).

Superior, Lake, largest freshwater lake in the world, 31,820 sq mi (82,414 sq km), 350 mi (563 km) long and 160 mi (257 km) at its greatest width, bordered on the W by NE Minn., on the N and E by Ont., Canada, and on the S by NW Mich. and NW Wis.; largest, highest, and deepest of the Great Lakes, having a surface elevation of 602 ft (183.6 m) and a maximum depth of 1,302 ft (397 m). Lake Superior drains into Lake Huron through the St. Marys River and receives the waters of many short, swift-flowing streams. Lake Superior is part of the Great Lakes–St. Lawrence Seaway system, and can be reached by oceangoing and lake vessels. The principal cargoes are grain and iron ore. Ice impedes navigation from mid-Dec. to the end of Mar. at the lake's outlet and from early Dec. to the end of Apr. in harbors on the south shore. Etienne Brulé, the French explorer, probably discovered the lake in 1616.

Su·ra (so͞o-rä'), river, c.540 mi (870 km) long, rising E of Penza, S central European USSR, and flowing N to empty into the Volga River. It is navigable for about 100 mi (160 km) upstream.

Su·ra·ba·ja (so͝or'ə-bī'ə), city (1971 est. pop. 1,273,000), capital of East Java prov., NE Java, Indonesia, on the Kali Mas River just above its mouth at the W end of Madura Strait. Surabaja is a major naval base, with a huge shipyard. An industrial center, it has railroad shops, an automobile assembly plant, and an oil refinery. Manufactures include textiles, glass, fertilizer, shoes, tobacco products, machinery, metal products, processed foods, tools, and cement.

Su·ra·kar·ta (so͝or'ə-kär'tä), city (1961 pop. 367,626), on central Java, Indonesia, on the Solo River. It is a trade center for an area producing tobacco, rice, and sugar. Manufactures include textiles, leather work, machinery, metal products, furniture, and cigarettes, but Surakarta is particularly noted for its batik cloth and goldwork. Surakarta's outstanding feature is the vast, walled palace of the sultan, virtually a city in itself. The European section of the city resembles an old Dutch town.

Su·rat (so͝or'ət, so͞o-răt'), city (1971 pop. 471,656), Gujarat state, W central India, on the Gulf of Cambay. Surat became one of India's most populous cities and busiest ports during the 17th cent.; but in 1664, at the height of its opulence, it was sacked by the Mahrattas. Today the city is a small port, and a railroad junction. There are textile mills, cotton gins, and engineering works.

Su·resnes (sü-rĕn'), city (1968 pop. 40,606), Hauts-de-Seine dept., N central France, a residential and industrial suburb of Paris. Its manufactures include automobiles, aeronautic equipment, pharmaceuticals, and metal pipes. In 1593 a conference in Suresnes between Catholics and Protestants led to the adoption of Catholicism by Henry IV.

Su·ri·nam (so͝or'ə-năm'), also known as **Su·ri·na·me** (so͝or'ə-nä'-mə), country (1971 pop. 384,900), 63,037 sq mi (163,266 sq km), formerly a part of the kingdom of the Netherlands, NE South America, on the Atlantic Ocean. Part of the Guiana region, it is separated from Brazil on the south by the Tumuc-Humac Mts., from Guyana on the west by the Courantyne River, and from French Guiana on the east by the Maroni River. The capital is Paramaribo, situated on the Surinam River. Creoles, Asian Indians, and Indonesians form the largest population groups. Dutch is the official language, although many others are spoken. Surinam is internally self-governing and has a parliamentary form of government. It is a leading producer of bauxite, most of which is shipped to the United States. Other exports are rice, citrus fruits, bananas, and shrimp. Rice is the chief subsistence crop; sugar cane, coffee, and coconuts are also cultivated. The leading industries process bauxite, foodstuffs, and timber.

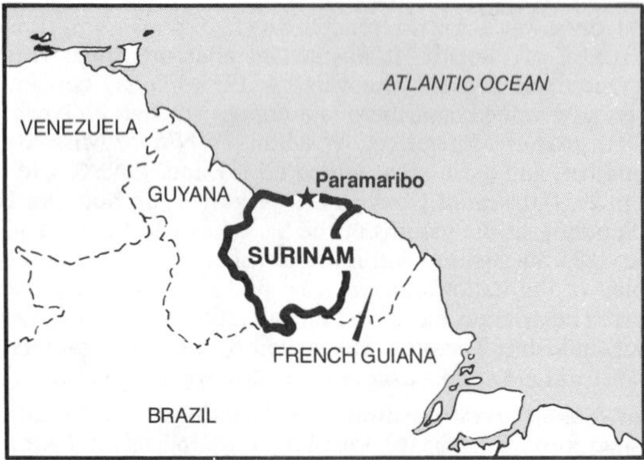

Although the first Dutch expeditions took place in 1597–98, the first colony, on Essequibo Island in present-day Guyana, was not founded until 1616. To exploit the territory the Dutch West India Company was founded in 1621. Dutch control was interrupted by English and French attacks and by a slave insurrection (1762–63). The Treaty of Breda gave all the English territory in Guiana to the Dutch, but in 1815 the Congress of Vienna awarded the area of Guyana to Britain while reaffirming the Dutch hold on Surinam. The Dutch granted Surinam a parliament in 1866. In 1954 Surinam officially became an internally autonomous part of the kingdom of the Netherlands. In May, 1974, the Dutch agreed to grant Surinam full independence by the end of 1975.

Sur·ma (so͝or'mä), river, 320 mi (515 km) long, rising in the Manipur hills, NE India, and flowing generally SW to join Meghna River in NE Bangladesh. The alluvial valley yields rice, tea, and oilseed.

Sur·rey (sûr'ē), nonmetropolitan county (1974 est. pop. 993,820), SE England; administrative center Guildford. The North Downs cross the county from east to west. There is dairy farming, market gardening, and wheat and oats cultivation. Manufactures include radio and radar equipment and aircraft. In Anglo-Saxon times Surrey was variously under the dominion of Mercia and Wessex and was overrun in the 9th cent. by the Danes. In 1974 a small area in the southeast was assigned to the new nonmetropolitan county of West Sussex.

Sur·ry (sûr'ē). **1.** County (1970 pop. 51,415), 536 sq mi (1,388.2 sq km), NW N.C., in a piedmont region bounded on the N by the Va. border, on the S by the Yadkin River; formed 1770; co. seat Dobson. It has farming (tobacco, corn, and dairy products), sawmilling, granite quarrying, and some manufacturing. **2.** County (1970 pop. 5,882), 277 sq mi (717.4 sq km), SE Va., in a tidewater region bounded in the N and NE by the James River, and in the S by the Blackwater River; formed 1652; co. seat Surry. Its agriculture includes peanuts, corn, cotton, dairy products, and livestock. It has pine lumbering and some manufacturing.

Surry, town (1970 pop. 269), seat of Surry co., SE Va., near the James River ESE of Petersburg. Thomas Rolfe's home (built c.1650) and Bacon's Castle are nearby.

Su·sa (so͞o'zə, -sə), ancient city, capital of Elam, SW of modern Dizful, Iran. From the 4th millennium B.C., Elam was under the cultural influence of Mesopotamia. Destroyed in the 7th cent. B.C., Susa was revived in the empire of the Achaemenid rulers of Persia.

Su·sah (so͞o'sä) or **Sousse** (so͞os), city (1966 pop. 58,161), NE Tunisia, on the Gulf of Hammamet, an arm of the Mediterranean Sea. It is a fishing port and export point for olive oil. The city was founded (c.9th cent. B.C.) by the Phoenicians, destroyed (A.D. 434) by the Vandals, and rebuilt by Justinian.

Su·san·ville (so͞o'zən-vĭl'), city (1970 pop. 6,608), seat of Lassen co., NE Calif., E of Redding, in a timber, dairy, and grain area; inc. 1900. Lassen Volcanic National Park is nearby.

Sus·que·han·na (sŭs'kwə-hăn'ə), county (1970 pop. 34,344), 833 sq mi (2,157.5 sq km), NE Pa., in a hilly lake region drained by the Susquehanna River, with the Lackawanna River rising in the E part of the county; formed 1810; co. seat Montrose. Its agriculture includes dairying and cherry-tree farming. It has lumbering, iron foundries, and other industries (clothing, shoes, and electrical equipment).

Susquehanna, river, 444 mi (715 km) long, rising in Otsego Lake, central N.Y., and zigzagging SE and SW through E central Pa. to Chesapeake Bay near Havre de Grace, Md. The bay is the drowned lower course of the river. The Susquehanna River traverses an anthracite-coal region, and there are many mining and industrial cities on its banks. The shallow, swift-flowing river is unsuited for navigation, but has several hydroelectric power plants. In June, 1972, the river, swollen by the heavy rains of Hurricane Agnes, breached 40-ft (12-m) dikes in places and flooded most of the basin, causing one of the greatest flood disasters in the history of the United States.

Sus·sex (sŭs'ĭks), former county, SE England. The old kingdom of the South Saxons (Sussex) was founded by King Aelle in the late 5th cent. Later the region was incorporated into Wessex. In 1974 the former administrative units of East Sussex and West Sussex were reorganized as nonmetropolitan counties.

Sussex. 1. County (1970 pop. 80,356), 950 sq mi (2,460.5 sq km), S Del., bounded on the S and W by the Md. line, on the N by the Mispillion River, on the E by Delaware Bay and the Atlantic Ocean; formed 1683; co. seat Georgetown. It has agriculture (fruit, poultry, dairy products, and truck crops), some manufacturing, canning, fishing, and shipbuilding. There are many resorts along the coast. **2.** County (1970 pop. 77,528), 527 sq mi (1,365 sq km), extreme NW N.J., mountain and lake area, bounded on the W by the Delaware River, on the N by the N.Y. border; N by N.Y. line, formed 1753; co. seat Newton. Its agriculture includes poultry, fruit, dairy products, and livestock. It has lead and zinc-ore mines, limestone quarries, sand and gravel pits, and timber. Its industry includes clothing, wood products, and machinery manufacturing. **3.** County (1970 pop. 11,464), 494 sq mi (1,279.5 sq km), SE Va., in a tidewater region bounded in the NE by the Blackwater River and drained by the Nottoway River; formed 1752; co. seat Sussex. Its agriculture includes peanuts, soybeans, cotton, corn, hogs, and cattle. It has lumbering and food-processing industries and manufactures wood and paper products.

Sussex, village (1970 pop. 40), seat of Sussex co., SE Va., SSE of Petersburg. The county courthouse dates from 1828.

Suth·er·land Falls (sŭ*th*'ər-lənd), waterfall, 1,904 ft (580.7 m) high, between Lake Quill and Arthur River, SW South Island, New Zealand. It is the world's fifth-highest waterfall.

Suth·er·land·shire (sŭ*th*'ər-lənd-shîr', -shər) or **Sutherland,** former county, N Scotland. Dornoch was the county town. Under the Local Government Act of 1973, Sutherlandshire became part of the Highland region.

Sut·lej (sŭt'lĕj), longest of the five rivers of the Punjab, c.900 mi (1,450 km) long, rising in the Kailas Range, SW Tibet (China), and flowing generally W, meandering through the Himalayas in India, then onto the Punjab plain where it forms part of the India-Pakistan border; continuing into Pakistan, it is joined by the Chenab River. The Sutlej is extensively used for irrigation.

Sut·ter (sŭt'ər), county (1970 pop. 41,935), 603 sq mi (1,561.8 sq km), N central Calif., a low-lying region in the Sacramento Valley, bounded on the W by the Sacramento River, on the E by the Feather River; formed 1850; co. seat Yuba City. It is a leading peach-growing county, also producing rice, beans, prunes and other fruit, nuts, sugar beets, grain, hay vegetables, dairy products, sheep, and poultry. Lumbering, mining and quarrying, fruit and vegetable processing, and machinery manufacturing are also done.

Sut·ton (sŭt'n), borough (1971 pop. 168,775) of Greater London, SE England. Sutton was created in 1965 by the merger of the municipal boroughs of Sutton and Cheam and of Beddington and Wallington with the urban district of Carshalton. It is mainly residential, but plastics, chemicals, and paper goods are produced.

Sutton, county (1970 pop. 3,175), 1,493 sq mi (3,866.9 sq km), W Texas, in broken uplands of Edwards Plateau, drained by the Devils and North Llano rivers; formed 1887; co. seat Sonora. This sheep- and goat-ranching region ships wool and mohair, and raises beef cattle, horses, hogs, and poultry. Tourism is also important.

Sutton, town (1970 pop. 1,031), seat of Braxton co., central W.Va., on the Elk River NE of Charleston, in a gas, oil, and agricultural area; inc. 1826. It is a trading point for an area yielding livestock, fruit, tobacco, lumber, flour, and feed. Marble and granite are quarried.

Sutton Cold·field (kōld'fēld', kōl'fēld'), municipal borough (1974 est. pop. 84,160), West Midlands, central England, residential suburb of Birmingham. There is a metal products industry. The moated New Hall dates partly from the 13th cent.

Sutton Hoo (hōō), archaeological site near Woodbridge, East Suffolk, E England, containing 11 barrows. Excavations here in 1938-39 revealed remains of a Saxon ship (c.660), which with its gold and silver treasures is now in the British Museum.

Sut·ton-in-Ash·field (sŭt'n-ĭn-ăsh'fēld'), urban district (1974 est. pop. 40,240), Nottinghamshire, central England. There are lace and hosiery factories, coal mines, and other industries.

Su·va (sōō'vä), city (1971 est. pop. 63,200), capital of Fiji, on the SE coast of Viti Levu island, S Pacific. It is a major shipping and commercial center. Coconut oil and soap are manufactured; sugar, copra, gold, and tropical fruits are exported.

Su·wan·nee (sə-wä'nē), county (1970 pop. 15,559), 686 sq mi (1,776.7 sq km), N Fla., in a flatwoods area with small lakes, bounded by the Suwannee and Santa Fe rivers; formed 1858; co. seat Live Oak. It is in a farming (corn, peanuts, cotton, tobacco, and vegetables) and stock-raising (cattle and hogs) area, with some lumbering and limestone deposits. Its industry includes food and wood processing and clothing manufacturing.

Suwannee, river, c.240 mi (390 km) long, rising in Okefenokee Swamp, SE Ga., and winding generally S through N Fla. to the Gulf of Mexico; it is dredged to accommodate shallow-draft vessels for 135 mi (217 km) upstream.

Su·won (sōō'wŭn'), city (1970 est. pop. 140,000), capital of Kyonggi prov., NW South Korea. Suwon has large silk and rayon textile mills.

Su·zu·ka (sōō-zōō'kä), city (1973 est. pop. 85,000), Mie prefecture, central Honshu, Japan. It is a manufacturing center with chemical, mechanical, and textile industries.

Sval·bard (sväl'bärd), archipelago (1978 est. pop. 3,000), 23,958 sq mi (62,051 sq km), a possession of Norway, located in the Arctic Ocean, c.400 mi (640 km) N of the Norwegian mainland and between lat. 74° N and 81° N. The islands form plateaus intersected by deep fjords. Spitsbergen, the largest island, contains the highest mountain of the group (Newtontoppen, c.5,650 ft/1,725 m) and the principal settlements. The warm North Atlantic Drift makes navigation possible for more than half the year along the western coasts. Ice fields and glaciers cover more than 60% of the area. Although there is some sealing, whaling, and fishing, the chief wealth of the islands is derived from their mineral resources, most notably coal; deposits of asbestos, copper, gypsum, iron, marble, zinc, and phosphate also exist.

Discovered (1194) by the Vikings, the islands were forgotten until their rediscovery (1596) by William Barentz, the Dutch navigator. English and Dutch whalers quarreled for a decade over the territory; in 1618 they compromised, the Dutch limiting their operations to the northern part, leaving the rest to the English, the French, and the Hanseatic League. After the decline of whaling, the group became (18th cent.) a hunting ground for Russian and Scandinavian fur traders. For a half century after the discovery of coal (19th cent.), Norway, Russia, and Sweden negotiated for the islands. By a treaty signed at Paris in 1920 they were awarded to Norway who took formal possession in 1925. In World War II Svalbard was raided (Aug., 1941) by an Allied party that evacuated the civilian population to England and rendered the mines inoperable. In Sept., 1943, German battleships and destroyers completed the devastation of the mines and mining installations by bombarding the islands. After the war the mining settlements were rebuilt. Coal mining concessions operated by the USSR since the 1920s account for about one third of the coal shipped from Svalbard.

Svend·borg (svĕn'bôr), city (1976 est. pop. 32,200), Fyn co., S Denmark, a seaport on the Svendborg Sund (an arm of the Lille Baelt); founded c.1200.

Sverd·lovsk (svĕrd-lôfsk'), city (1973 est. pop. 69,000), capital of Sverdlovsk oblast, E European USSR, in the E foothills of the central Urals, on the Iset River. It is a leading industrial, scientific, and cultural center. Its industries include metallurgy, gem cutting, and the manufacture of heavy machinery, chemicals, pharmaceuticals, building materials, and electrical apparatus. Nearby are gold and copper mines. The first ironworks were established in 1726. Its importance was enhanced by the building of the Great Siberian Highway through the city in 1783 and by the construction of the Trans-Siberian RR in the 19th cent.

Svish·tov (svēsh-tôf'), town (1969 est. pop. 22,900), N Bulgaria, a port on the Danube River. It is an agricultural center. With a history dating to Roman times, it became, under Turkish rule (15th-19th cent.), an important commercial and military center.

Swa·bi·a (swā'bē-ə), historic region, mainly in S Baden-Württemberg and SW Bavaria, SW West Germany. It is bounded in the east by Upper Bavaria, in the west by France, and in the south by Switzerland and Austria. It includes the former Prussian province of Hohenzollern. The main physical features of Swabia are the Black Forest; the valley of the upper Danube River, which rises here; the Swabian Jura, a mountain range that extends parallel to and north of the Danube; and the valley of the upper Neckar River. The Rhine and the Lake of Constance form the western and southern borders. The easternmost section of Swabia is part of the Danubian plateau of Bavaria and is a Bavarian province (c.3,940 sq mi/10,205 sq km), with Augsburg as capital. Stuttgart, the capital of Baden-Württemberg, is the chief city of western Swabia. Predominantly made up of agricultural or forested country, Swabia is famous for the loveliness of its landscape. Farming, forestry, and livestock raising are major occupations. Industrial products include textiles, machinery, chemicals, and metal goods.

Settled in the 3rd cent. by the Germanic Suebi and Alemanni during the great migrations, the region became one of the five basic or stem duchies of medieval Germany in the 9th cent., when it far exceeded its present boundaries. In 1079 the duchy was bestowed on the house of Hohenstaufen, which in 1138 also obtained the imperial dignity. On the extinction (1268) of the dynasty, Swabia broke up into small temporal and ecclesiastic lordships and lost its political identity. The Swiss part became independent in 1291 and the Hapsburg territories in Alsace passed to France in 1648, but Breisgau and the other Hapsburg domains in southern Baden remained Austrian until c.1806, except from 1469 to 1477, when they were ruled by Charles the Bold of Burgundy. The rest of Swabia was held by a multitude of princes, counts, bishops, and knights. The prosperous Swabian towns often became free imperial cities and formed defensive leagues. In the 19th cent. some (especially Stuttgart) revived as industrial centers. At the diet of Regensburg (1801-3), many of the small ecclesiastic and feudal holdings were taken over by Baden, Württemberg, and Bavaria.

Swain (swān), county (1970 pop. 8,835), 524 sq mi (1,357.2 sq km), W N.C., in a heavily forested mountain area, largely included in Great Smoky Mountains National Park, bounded on the N by Tenn., on the W by the Little Tennessee River; formed 1871; co. seat Bryson City. It has farming (corn and potatoes), cattle and poultry raising, and lumbering.

Swains·bo·ro (swānz'bûr'ō), city (1970 pop. 7,325), seat of Emanuel co., E Ga., NW of Savannah; founded 1814, inc. 1853. The area's timber and cotton are processed, and naval stores are produced.

Swamp·scott (swŏmp'skət), town (1970 pop. 13,578), Essex co., E Mass., a resort on Massachusetts Bay; settled 1629, set off from Lynn and inc. 1852. It has a fishing industry.

Swan·quar·ter (swän'kwôr'tər) or **Swan Quar·ter** (swän' kwôr'tər), town (1970 pop. 350), seat of Hyde co., E N.C., SE of Belhaven on an inlet of Pamlico Sound.

Swan·sea (swŏn'zē, -sē) or **A·ber·ta·we** (ăb-ər-tou'ē), county borough (1972 est. pop. 171,520), West Glamorgan, S Wales, on Swansea Bay at the mouth of the Tawe River. It is a metallurgical center; steel and tinplate are the most important products. Other industries are engineering, shipbuilding, oil refining, and the export of anthracite. Crude oil, metals, timber, grain, and rubber are imported. Swansea ware, of rich blue coloring with decorative painting, was made at the Swansea potteries in the first half of the 19th cent.

Swansea, town (1970 pop. 12,640), Bristol co., SE Mass., a suburb of Fall River, on an inlet of Mount Hope Bay; founded 1667, inc. 1785. Once a vast farmland, it is now chiefly residential.

Swarth·more (swôrth'môr'), residential borough (1970 pop. 6,156), Delaware co., SE Pa., NNE of Chester; inc. 1893. Swarthmore College (1869) is here.

Swa·tow (swä'tou'): *see* Shan-t'ou, China.

Swa·zi·land (swä'zē-lănd'), kingdom (1972 est. pop. 442,000), 6,705 sq mi (17,366 sq km), SE Africa. The capital is Mbabane. Swaziland is bordered on the south, west, and north by the Republic of South Africa and on the east by Mozambique. The country is mountainous, with steplike plateaus descending from the highveld (3,500-5,000 ft/1,067-1,524 m) in the west through the middleveld (1,500-3,000 ft/457-914 m) and the lowveld (500-1,500 ft/152-457 m), then rising to the rolling plateau of the Lebombo Mts. Swaziland is cut by four major river systems. English is Swaziland's official language, but the bulk of the population speaks SiSwati, a Zulu dialect. About 40% of the people practice traditional religions; most of the rest are Christians.

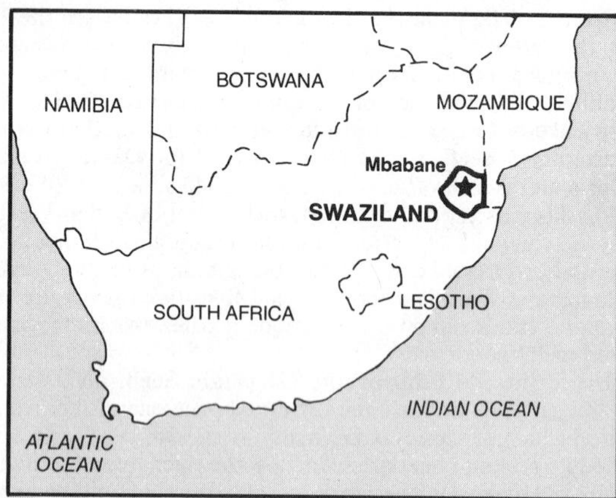

Economy. The country has excellent agricultural and ranching land. Sugar cane, citrus fruits, rice, cotton, maize, sorghum, tobacco, and peanuts are the principal crops. Cattle, a major export, and goats are raised in large numbers. About 56% of the land is held by the Swazi and the rest mainly by South Africans and Europeans. The forests of the highveld yield wood pulp and timber. Swaziland's mineral resources include iron ore, asbestos, coal, tin, and barite. Industry in Swaziland consists chiefly of food processing and the manufacture of light consumer goods. Swaziland's major trading partners are Great Britain and South Africa. Swaziland, Lesotho, and Botswana form a customs and monetary unit with South Africa.

History. The Swazi probably moved southward into the Mozambique area in the 16th cent. Fleeing Zulu attacks (19th cent.), they arrived in present-day Swaziland. During the 1800s, Europeans entered the area, and in 1888 the Swazi king granted them self-government. In 1903 Swaziland became a British High Commission Territory. On Sept. 6, 1968, Swaziland achieved complete independence. The Swazi people continue to find a common cause in resistance to Afrikaners' demands for incorporation into South Africa.

Government. The king is the head of state. Until 1973, when King Sobhuza II (reigned 1921-) assumed personal power, the monarch administered through a cabinet and a prime minister chosen by parliament. Parliament consisted of a senate and a house of assembly. The king's party, the Imbokodvo National Movement, was the country's dominant political force. Parliament could not legislate on questions regulated by Swazi law or custom unless authorized by the Swazi National Council, an advisory body consisting of the king, the queen mother, and all adult male Swazis.

Swe·den (swēd'n), constitutional monarchy (1973 est. pop. 8,200,000), 173,648 sq mi (449,750 sq km), N Europe, occupying the E part of the Scandinavian peninsula and bordering on Norway in the W, on Finland in the NE, on the Gulf of Bothnia in the E, on the Baltic Sea in the S, and on the Øresund (The Sound), the Kattegat, and the Skagerrak in the SW. The country includes several islands, notably Gotland and Öland, in the Baltic. Stockholm is the capital and largest city.

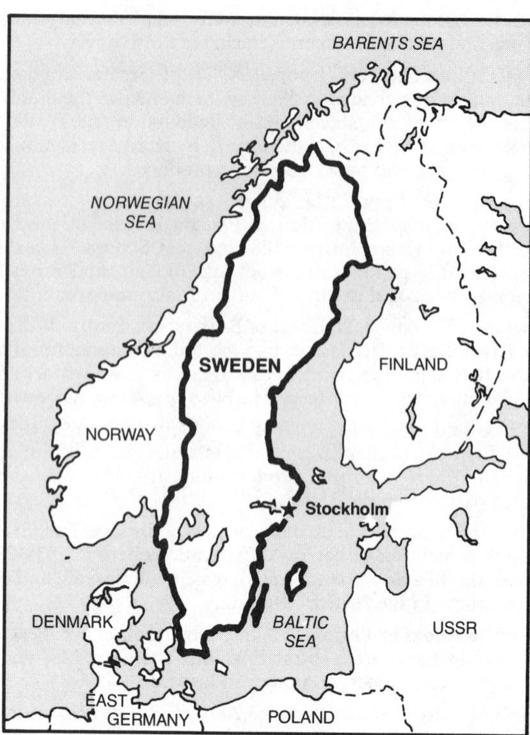

Sweden falls into two main geographic regions: the north (Norrland), comprising about two thirds of the country, which is mountainous (except for a narrow strip of lowland along the Gulf of Bothnia); and the south (Svealand and Götaland), which is low-lying and where most of the population lives. About 55% of Sweden's land area is forested and only about 8% is arable. The country has several large rivers, which generally flow in a southeastward direction; these include the Götaälv, the Dalälven, the Indalsälven, the Angermanälven, the Umeälv, the Skellefteälv, the Luleälv, and the Torneälv. There are also a number of large lakes, including lakes Vänern, Vättern, Mälaren, Storsjön, Hjälmaren, Siljan, and Uddjaur. The highest point in Sweden is Kebnekaise (6,965 ft/2,123 m), located in the Kölen (Kjölen) Mts. in Lapland. The great majority of the population speaks Swedish and is descended from Scandinavian tribes; there are small Finnish and Lapp-speaking minorities. Virtually all Swedes belong to the Lutheran church.

Sweden is divided into 24 counties:

NAME	CAPITAL	NAME	CAPITAL
Älvsborg	Vänersborg	Malmöhus	Malmö
Blekinge	Karlskrona	Norrbotten	Luleå
Gävleborg	Gävle	Örebro	Örebro
Göteborg and Bohus	Göteborg	Östergötland	Linköping
Gotland	Visby	Skaraborg	Mariestad
Halland	Halmstad	Södermanland	Nyköping
Jämtland	Östersund	Stockholm	Stockholm
Jönköping	Jönköping	Uppsala	Uppsala
Kalmar	Kalmar	Värmland	Karlstad
Kopparberg	Falun	Västerbotten	Umeå
Kristianstad	Kristianstad	Västermorrland	Härnösand
Kronoberg	Växjö	Västmanland	Västerås

Economy. Sweden is a highly industrialized country and has one of the highest living standards in the world. At the start of the 1970s manufacturing contributed about 32% of the annual national income and agriculture about 4%. Farming is concentrated in the southern part of the country; the leading commodities produced are dairy goods, grain, sugar beets, and potatoes. Poultry, hogs, and cattle are raised. Sweden is one of the world's leading producers of iron ore;

important mines are at Kiruna, Gällivare, and Grängesberg. Copper, lead, and zinc ores and pyrite are also extracted in sizable quantities.

The country's chief industrial centers are Stockholm, Göteborg, Malmö, Uppsala, Västerås, Hälsingborg, and Norrköping. The leading manufactures include iron and steel, metal goods, machinery, forest products, construction materials, textiles, clothing, processed food, refined petroleum, motor vehicles, and ships. Sweden is known for its fine glassware and its high-quality steel cutlery and blades. Much hydroelectric power is generated. Sweden carries on a large foreign trade, and the value of imports usually slightly exceeds the value of exports. The main imports are machinery, transport equipment, food, petroleum and petroleum products, and chemicals; the chief exports are machinery, ships, forest products, iron ore, and iron and steel. The principal trade partners are West Germany, Great Britain, Denmark, Norway, and the United States. Sweden is a member of the European Free Trade Association, and in 1972 it signed an industrial free trade agreement with the European Common Market.

History. In early historic times, Svealand was inhabited by the Svear, who conquered and merged with their southern neighbors the Gotar by the 6th cent. A.D. The early Swedes extended (10th cent.) their influence to the Black Sea and warred for centuries with their Danish and Norwegian neighbors. St. Ansgar introduced Christianity c.829, but paganism was fully eradicated only in the 12th cent. by Eric IX, who also conquered Finland. In 1319 Sweden and Norway were united under Magnus VII, and in 1397 Queen Margaret effected the personal union of Sweden, Norway, and Denmark through the Kalmar Union. In 1523 the Swedes rebelled and made Gustavus Vasa their king as Gustavus I.

Gustavus (reigned 1523–60) eliminated the influence of the Hanseatic League in Sweden, strengthened the central authority, and made Lutheranism the state religion. Gustavus II (reigned 1611–32) made Sweden the dominant Protestant power of continental Europe by acquiring Ingermanland, Karelia, and Livonia in wars with Russia and Poland. Western Pomerania, Wismar, and Bremen were added to the realm in 1648. The southern provinces of Sweden were definitively recovered from Denmark in 1660. Under Charles XI (reigned 1660–97), Sweden became an absolute monarchy, and the great nobles lost their independence. In the Northern War (1700–21) Sweden was crushed and had to yield most of its earlier conquests.

Internally, Sweden was torn in the 18th cent. by political intrigue and civil discord. The constitution of 1720 gave increased powers to the Riksdag (diet), but Gustavus III (reigned 1771–92), restored absolutism, which prevailed until a constitutional monarchy was established in the midst of wars against Napoleon in 1809. The Congress of Vienna compensated (1814) Sweden for its loss of Pomerania and Finland with Norway, which remained a separate kingdom in personal union with Sweden until 1905.

The 19th century brought industrial development and progressive liberalization in government. Under Gustavus V (reigned 1907–50), Sweden averted involvement in World Wars I and II. Universal taxpayer suffrage was introduced in 1907, and in 1910 a workmen's compensation insurance law began the long series of Swedish welfare legislation. Sweden entered the United Nations in 1946 and refused to join the North Atlantic Treaty Organization (NATO) in 1949 in order not to compromise its neutrality. The Social Democrats controlled the government from 1945 to 1976, when a nonsocialist coalition assumed power. Charles XVI Gustavus succeeded Gustavus VI in 1973.

Government. Under the constitution of 1975 the king is head of state but has little power. Legislative power is vested in the unicameral Riksdag, made up of 349 members elected by a system of proportional representation to three-year terms. The country's executive is the cabinet, headed by the prime minister, which must have the confidence of the Riksdag. Public administration is to a large extent decentralized, so that elected county and municipal governments play a major role in running the country.

The educational and cultural level in Sweden is high, and the school system is outstanding. There are universities at Göteborg, Karlstad, Linköping, Lund, Örebro, Stockholm, Umeå, Uppsala, and Växjö. The Nobel Prizes (except the Peace Prize) are awarded annually in Sweden. Social welfare legislation has long been advanced and comprehensive; it provides for universal pension systems (for old age, disability, and survivors) and also for maternity benefits, workmen's compensation insurance, sickness insurance, and allowances for all children.

Sweet Grass (swēt' grăs'), county (1970 pop. 2,980), 1,840 sq mi (4,765.6 sq km), S Mont., in an agricultural area drained by the Yellowstone River; formed 1895; co. seat Big Timber. Livestock and grain are its major products. Part of Absaroka National Forest is in the south.

Sweet·wa·ter (swēt'wô'tər, -wŏt'ər), county (1970 pop. 18,391), 10,429 sq mi (27,011.1 sq km), SW Wyo., bordering on Utah and Colo. and watered by the Green River; formed 1867 as Carter co., renamed 1869; co. seat Green River. It is in an oil- and gas-mining and sheep-raising area. Chemicals and fertilizers are manufactured.

Sweetwater, city (1970 pop. 12,020), seat of Nolan co., W Texas; inc. 1884. It is a rail shipping point for cattle ranches in a mesquite and scrub oak region.

Świd·ni·ca (shvēd-nē'tsä), town (1973 est. pop. 49,400), SW Poland. It has textile mills, sugar refineries, and various manufactures. Świdnica and the surrounding principality came to the Bohemian crown in 1368 and were ceded to Prussia in 1745.

Swift (swĭft), county (1970 pop. 13,177), 739 sq mi (1,914 sq km), W Minn., drained by the Pomme de Terre and Chippewa rivers; formed 1870; co. seat Benson. It is in a grain and livestock area with some mining and farm-machinery manufacturing.

Swift Cur·rent (kûr'ənt), city (1971 pop. 15,415), SW Sask., Canada, on Swift Current Creek. It is a distributing and processing center for a farm and oil region. Other industries are helium extraction and the manufacture of farm machinery and plastic goods.

Swin·don (swĭn'dən), municipal borough (1974 est. pop. 90,330), Wiltshire, S central England. Swindon was a small village until 1841, when the Great Western RR opened its locomotive and car works here.

Świ·no·ujś·cie (shfē'nô-ōō'ēsh-chĕ), town (1973 est. pop. 39,500), NW Poland, on Usedom Island, at the mouth of the Świna River. It is a fishing center and seaside resort. Chartered in 1765, the town became part of the Prussian province of Pomerania; it passed to Poland in 1945.

Swin·ton and Pen·dle·bur·y (swĭn'tən; pĕn'dəl-bĕr'ē, -bə-rē), municipal borough (1974 est. pop. 39,960), Greater Manchester, NW England. The borough has coal mines, cotton mills, and factories for pottery and storage batteries.

Swish·er (swĭsh'ər), county (1970 pop. 10,373), 896 sq mi (2,320.6 sq km), Texas, on the Llano Estacado; formed 1876; co. seat Tulia. Wheat growing, cattle ranching, dairying, and farming (sorghums, oats, barley, alfalfa, fruit, truck crops, and livestock) are important.

Swit·zer·land (swĭt'sər-lənd), also called the **Swiss Confederation,** republic (1970 pop. 6,269,783), 15,941 sq mi (41,287 sq km), central Europe. The federal capital is Bern. Switzerland borders on France in the west and southwest, with the Jura Mts. and the Lake of Geneva forming the frontier; in the north it is separated from West Germany by the Rhine River and the Lake of Constance; its eastern neighbors are Austria and Liechtenstein; in the southeast and south it is divided from Italy by the Alpine crests, the Lake of Lugano, and Lago Maggiore. Between the Jura Mts. and the Central Alps, which occupy the southern section (more than half) of the country, there is a long, narrow plateau, crossed by the Aare River and containing the lakes of Neuchâtel and Zürich.

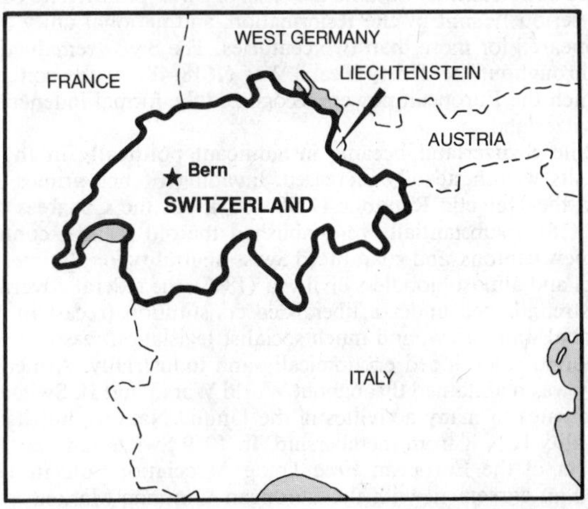

German, French, Italian, and Romansh (a Rhaeto-Roman dialect) are the national languages of Switzerland. German dialects (Schwyzerdütsch) are spoken by about 65% of the inhabitants; French, spoken by about 18%, predominates in the southwest; Italian, spoken by about 12%, is the language of Ticino, in the south. The few Romansh-speakers (less than 1%) are in the southeast. Approximately 48% of the population is Protestant and 49% Roman Catholic.

Switzerland is divided into 19 cantons and 6 half cantons:

NAME	CAPITAL	NAME	CAPITAL
Aargau	Aarau	Nidwalden (half canton)	Stans
Ausser-Rhoden (half canton)	Herisan	Obwalden (half canton)	Sarnen
Basel-Land (half canton)	Liestal	Saint Gall	Saint Gall
Basel-Stadt (half canton)	Basel	Schaffhausen	Schaffhausen
Bern	Bern	Schwyz	Schwyz
Fribourg	Fribourg	Solothurn	Solothurn
Geneva	Geneva	Thurgau	Frauenfeld
Glarus	Glarus	Ticino	Bellinzona
Grisons	Chur	Uri	Altdorf
Inner-Rhoden (half canton)	Appenzell	Valais	Sion
Lucerne	Lucerne	Vaud	Lausanne
Neuchâtel	Neuchâtel	Zug	Zug
		Zürich	Zürich

Economy. Although poor in natural resources except water power, Switzerland has attained prosperity through the export of its manufactures and through its technological achievements. The industries, employing almost half of the labor force, are mainly in the north; the chief manufactures are textiles, chemicals, machinery, instruments, watches and watch movements (in which Switzerland leads the world), jewelry, and foodstuffs (notably chocolate). The major source of power is hydroelectricity; the railroads, more than half of which are government-owned, use electric power on all main lines. In the alpine regions there is much dairying, with cheese the principal export. The agricultural yield of fruits, vegetables, and grains is supplemented by imports; other main imports are machinery, motor vehicles, and iron and steel. The main trading partners are West Germany, France, Italy, the United States, and Great Britain. Imports exceed exports, but Switzerland is favored by the huge profits of its tourist industry; each year several million people visit its alpine resorts. The stability of Swiss politics and of the Swiss currency has, moreover, made Switzerland a world banking center.

History. In 58 B.C. the local Helvetii were conquered by the Romans. The area passed to the Franks in the 6th cent. Divided (9th cent.) between Swabia and Transjurane Burgundy, it was united (1033) under the Holy Roman Empire. Encroachments by the house of Hapsburg on the privileges of the three localities of Uri, Schwyz, and Unterwalden resulted in the conclusion (1291) of a defensive league among them. The legendary hero of this event is William Tell. Joined by Lucerne, Zürich, Zug, Glarus, and Bern, the league decisively defeated the Hapsburgs by 1388 and expanded further in the 15th cent. Emperor Maximilian I granted Switzerland virtual independence in 1499. By 1513 the confederation had 13 members, and this number was maintained until 1798. The French defeated the Swiss at Marignano in 1515, after which a "perpetual alliance" with France and neutrality became the basis of Swiss policy. The cantons were seriously split by the Reformation, and national unity almost disappeared for more than two centuries. The Swiss remained neutral throughout the Thirty Years' War (1618–48), at the conclusion of which the European powers recognized the formal independence of Switzerland.

While Switzerland became insignificant politically in the 18th cent., its wealth steadily increased. Invading French armies established the Helvetic Republic (1798–1803), but the Congress of Vienna (1815) substantially re-established the old regime, confirmed nine new cantons, and guaranteed Swiss neutrality for all time. After a brief and almost bloodless civil war (1847), the federal government was strengthened under a liberalized constitution (recast in 1874). National unity grew, and much socialist legislation was enacted as the country developed economically and industrially. Armed neutrality was maintained throughout World Wars I and II. Switzerland participates in many activities of the United Nations, but its strict neutrality bars it from membership. In 1959 Switzerland became a member of the European Free Trade Association, and in 1972 it signed an agreement with the European Common Market.

Government. The federal constitution (1874) assigns specified functions, notably communications, foreign relations, and tariffs, to the confederation, leaving the cantons sovereign in other respects. There is universal male suffrage; women were granted (1971) the vote in federal elections and may vote in most cantonal and local elections. A council of states (two members from each canton, one from each half canton) and a 200-member national council (whose members are directly elected every four years) together form the federal assembly. The chief executive, or federal council, is composed of seven members (elected for four years by the federal assembly) and includes the president of the confederation (elected by the federal assembly annually). Switzerland frequently employs the referendum as well as the popular initiative to achieve political change. Cantonal constitutions differ widely. In Unterwalden, Clarus, and Appenzell the entire male electorate legislates directly in yearly outdoor meetings called *Landsgemeinden;* elsewhere a unicameral legislative council and an elected executive council are common.

Switzerland, county (1970 pop. 6,306), 221 sq mi (572.4 sq km), SE Ind., bounded on the E and S by the Ohio River and the Ky. line; formed 1814; co. seat Vevay. Its agriculture includes grain, tobacco, livestock, and truck crops. It has nonmetallic mineral mining and manufactures flour, plastics, shoes, and metal stampings.

Syb·a·ris (sĭb′ər-ĭs), ancient city of Magna Graecia, S Italy, in Bruttium, on the Gulf of Tarentum (now Taranto). It was founded in 720 B.C. by Achaeans and people from Argolis, the Troezenians. It became a wealthy Greek city, and its inhabitants were reputed to live voluptuous lives. The Troezenians destroyed the city in 510 B.C.

Syc·a·more (sĭk′ə-môr′), city (1970 pop. 7,843), seat of De Kalb co., N Ill., SW of Rockford; founded 1840, inc. 1859. It is a trade and processing point in a farm area.

Syd·ney (sĭd′nē), city (1973 est. pop. 57,770; urban agglomeration pop. 2,874,380), capital of New South Wales, SE Australia, surrounding Port Jackson inlet on the Pacific Ocean. Sydney is Australia's chief port and main cultural and industrial center. Its main exports are wool, wheat, flour, sheepskins, and meat; the chief imports are petroleum, coal, timber, and sugar. Sydney has shipyards, oil refineries, textile mills, brass foundries, and automobile, electronics, and chemical plants. The city was founded in 1788 as the first penal settlement of Australia.

Sydney, city (1971 pop. 33,230), Cape Breton Island, N.S., Canada, on the NE coast at the head of an arm of Sydney Harbour. It is the port and the commercial, trade, and industrial center of an important coal-mining area. The city has steel mills and plants manufacturing wood, food products, and chemicals.

Syk·tyv·kar (sĭk′tĭf-kär′), city (1973 est. pop. 142,000), capital of Komi ASSR, NW European USSR, a port on the Sysola River near its entry into the Vychegda. Lumbering and the manufacture of wood products are the chief industries. A settlement existed on the site of Syktyvkar by the late 16th cent.

Syl·a·cau·ga (sĭl′ə-kôg′ə), city (1970 pop. 12,255), Talladega co., central Ala.; inc. 1839. It is a processing center for a cotton, livestock, and timber area. In addition to marble, which has been produced here since 1840, textiles, clothing, metal goods, and dairy products are made. Iron ore is also found in the area.

Syl·het (sĭl-hĕt′), town (1961 est. pop. 37,740), E Bangladesh, on the Surma River. It is the administrative center for a district of rice and tea cultivation and extensive limestone quarrying. Sylhet is also a center of Islamic culture.

Syl·va (sĭl′və), town (1970 pop. 1,561), seat of Jackson co., W N.C., in the mountains WSW of Asheville. Wood and paper products and textiles are made.

Syl·va·ni·a (sĭl-vān′yə). **1.** City (1970 pop. 3,199), seat of Screven co., E Ga., NNW of Savannah near the Savannah River; founded 1847, inc. 1854. It is a processing center for a productive cotton and lumber area. **2.** City (1975 est. pop. 14,800), Lucas co., NW Ohio, a suburb of Toledo at the Mich. line; inc. 1867. It is chiefly residential, but building materials are made.

Syl·ves·ter (sĭl-vĕs′tər), city (1970 pop. 4,226), seat of Worth co., SW Ga., E of Albany; inc. 1898.

Sy·me (sē′mē, sī′-): *see* Sími, Greece.

Syr·a·cuse (sîr′ə-kyo͞oz′, -kyo͞os′), city (1972 est. pop. 111,344), capital of Syracuse prov., SE Sicily, Italy, on the Ionian Sea. It is a port and a food-processing and tourist center. The old town, on the small island of Ortygia, is connected by a bridge with the mainland, where the more modern districts are situated. Founded (734 B.C.) by Greek colonists from Corinth, Syracuse grew rapidly and soon founded colonies of its own. Its democratic government was suppressed by Gelon, tyrant of Gela, who took possession of the city in 485 B.C. Hiero I made it one of the great centers of Greek culture. Soon after Hiero's death a democracy was again established; it lasted from 466 B.C. to 406 B.C. In 406 B.C. Dionysius the Elder became tyrant. Under his long rule Syracuse reached the high point of its power and territorial expansion. After the death of Dionysius there were several decades of democratic government until tyranny was re-established by Agathocles and Hiero II (4th-3rd cent. B.C.). Syracuse suffered catastrophically when it abandoned its traditional ally Rome in favor of Carthage, in the second of the Punic Wars. The city fell in 212 B.C. and was sacked; Syracuse thence was reduced to the status of a provincial town. The period from Dionysius the Elder to 212 B.C. was brilliant in terms of culture. Plato visited Syracuse several times, and Theocritus and Archimedes lived here. Syracuse suffered another major setback in the late 9th cent. A.D., when it was badly damaged by Arab conquerors. It was captured by the Normans in 1085. Numerous remains testify to the city's past greatness.

Syracuse. 1. City (1970 pop. 1,720), seat of Hamilton co., SW Kansas, near the Colo. border on the Arkansas River W of Garden city; inc. 1887. **2.** City (1978 est. pop. 177,000), seat of Onondaga co., central N.Y., on Onondaga Lake and the Barge Canal; settled c.1788, inc. 1848. It is a port of entry, and its many manufactures include air conditioners, electrical and electronic equipment, automobile and aircraft parts, soda ash, chinaware, shoes, and typewriters. Saltmaking was the city's chief industry until after the Civil War. However, favorable location on the Erie Canal (opened there in 1819) and on railroads stimulated industrial development.

Syr Dar·ya (sĭr' där'yə), river, c.1,380 mi (2,220 km) long, Central Asian USSR. One of the principal rivers of central Asia, it is formed in the Fergana Valley of eastern Uzbekistan. It flows west through Tadzhikistan, then northwest through Uzbekistan and Kazakhstan, past Kzyl-Orda, and into the northern end of the Aral Sea. Unfit for navigation, its waters are used for irrigating the important cotton-growing areas along its course and for hydroelectric power. The Syr Darya forms the northern and eastern limits of the Kyzyl Kum desert. Alexander the Great in his conquest of Persia reached the river c.329 B.C.

Syr·i·a (sîr'ē-ə), officially **Syrian Arab Republic**, republic (1970 pop. 6,292,000), 71,467 sq mi (185,100 sq km), W Asia, bordering on Lebanon and the Mediterranean Sea in the W, on Turkey in the NW and N, on Iraq in the E and S, and on Jordan and Israel in the SW. Damascus is the country's capital and its largest city.

Syria falls into two main geographic regions, a western region and a much larger eastern region. The western region, which includes

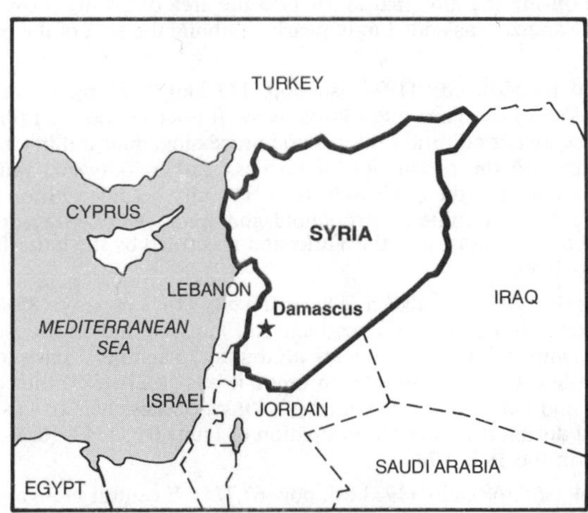

about two thirds of the country's population, can be subdivided into four parallel north-south zones. In the far west is a narrow, discontinuous lowland strip along the Mediterranean. It is bordered, and partly cut, by the Jabal an Nusayriyah, a mountain range (average elevation: 4,000 ft/1,220 m that is crossed by deep valleys. In the east the Jabal an Nusayriyah drops sharply to the Great Rift Valley, which in Syria contains the Orontes River. East of the rift are mountain ranges, including the Anti-Lebanon Mts. (which include Mt. Hermon, 9,232 ft/2,814 m, Syria's loftiest point) and scattered ranges in northwest Syria. Within these ranges are several fertile basins, including ones occupied by Damascus and Aleppo. The eastern region of Syria is made up of a plateau (average elevation: 2,000 ft/610 m), which is in large part bisected by a series of ranges that fan out northeastward from the Anti-Lebanon Mts. In the south are the Jabal ad Duruz Mts., from which the plain of Hawran extends westward to the Sea of Galilee. Other mountains are located in the north. Much of the southern section of the plateau forms part of the Syrian Desert; otherwise, the plateau is largely covered with steppe. There are irrigated, cultivated areas along the Euphrates River in the east.

Most Syrians are of Arab descent and speak Arabic, the country's official language. The chief minorities are the Kurds (numbering about 250,000) and the Turkomans (30,000), most of whom live in the north; the Armenians (130,000), many of whom live in Aleppo; and the Circassians (25,000), most of whom live in and near Qunaytirah in the southwest. About 75% of the country's inhabitants are Moslem, mostly of the Sunni sect. There are also significant numbers of Shiite Moslems. The Druses (120,000), whose religion combines Moslem and Christian elements, live in the south, principally in the Jabal ad Duruz. Approximately 15% of the people are Christian; the largest Christian groups are the Greek Orthodox, the Armenian Orthodox, and the Syrian Orthodox.

Economy. In the 1970s agriculture was still the mainstay of the Syrian economy, employing about 50% of the work force and contributing about 27% of the annual national product. However, the steadily increasing role of industry contributed about 20% of the annual national product. The state plays a major role in the country's economy, particularly in industry and commerce. About one third of Syria's land area is estimated to be arable. The best farmland is located along the coast and in the Jabal an Nusayriyah, around Aleppo, in the region between Hama and Hims, in the Damascus area, and in the land between the Euphrates and Khabur rivers, which is known as Al Jazirah. The Tabqa Dam on the Euphrates, completed in 1975, provides irrigation water for c.1,600,000 acres (c.650,000 hectares) of farmland. The principal crops are wheat, barley, cotton, grapes, tomatoes, sugar beets, olives, citrus fruit, lentils, onions, tobacco, and potatoes. Poultry, sheep, goats, and cattle are raised.

Damascus, Aleppo, and Hims are the chief industrial centers. The main manufactures are refined petroleum, cement, textiles, processed food, beverages, chemicals, cigarettes, soap, and glass. Handicraftsmen make articles of silk, leather, and glass. The principal minerals extracted are petroleum, found mainly at Qarah Shuk (Karachuk) in the extreme northeast; natural gas, found mainly in the Al Jazirah region; phosphates; and salt. Syria's limited transportation network serves mainly the western part of the country. Latakia is the leading seaport. The annual value of Syria's imports is usually considerably greater than the value of its exports. The principal imports are foodstuffs, machinery, iron and steel, textiles, petroleum and petroleum products, and motor vehicles; the chief exports are raw cotton and cotton textiles, wool, cereals, and live animals. The leading trade partners are Italy, the USSR, Lebanon, and West Germany.

History. Until the 20th cent. the term Syria generally denoted those lands of the Levant, or eastern littoral of the Mediterranean, that correspond to modern Syria and Lebanon, most of Israel and Jordan, western Iraq, and northern Saudi Arabia. Syria has always been an object of conquest, and it has been held by foreign powers during much of its history. The Amorites, coming c.2100 B.C. from the Arabian peninsula, were the first important Semitic people to settle in the region. From the 15th to the 13th cent. B.C. the area probably was part of the empire of the Hittites, although it came under Egyptian rule for long periods during that time. The first great indigenous culture was that of Phoenicia (located mostly in present-day Lebanon), which flourished after 1250 B.C. In the 10th cent. B.C. two Hebrew kingdoms were organized in Palestine. Syria suffered (11th-6th cent. B.C.) long invasions and intermittent control by the Assyrians and Babylonians. Under the Persian Empire, with its efficient administrative system, Syria's standard of living improved (6th-4th cent. B.C.). Alexander the Great conquered Syria between 333 and 331 B.C., and his short-lived empire was followed by that of the Seleucids. Their control of Syria was constantly threatened by Egypt, which usually held the south until Antiochus III conquered (early 2nd cent. B.C.) the region. The Romans conquered Syria by 63 B.C., but they continued to fight the Parthians here.

More significant for the future, Christianity was started in Palestine and soon exerted some influence over all of Syria.

After the division of Rome into the Eastern and Western empires in the 4th cent., Syria came under Byzantine rule. Between 633 and 640, Moslem Arabs conquered the region, and most Syrians were converted to Islam. Damascus was the usual capital of the Umayyad caliph (661–750) and enjoyed a period of great splendor. Groups of Christians remained in the Moslem areas, and they generally rendered aid to the Christians who came to Syria on Crusades (11th–14th cent.). By the late 11th cent. the Seljuk Turks had captured most of Syria, and the Christians fought against them as well as against Saladin, who triumphed (late 12th cent.) over both the Christians and his fellow Moslems. In the mid-13th cent. Syria was overrun by the Mongols, who were routed (1260) by Baybars, the Mameluke ruler of Egypt. The Mamelukes held control for most of the time until 1516, when the Ottoman Empire annexed the area. For most of the four centuries of Ottoman control, Syria's economy continued to be weak, and its government was in the hands of several Syrian families who often fought each other.

In 1832–33 Ibrahim Pasha, the son of Mohammed Ali of Egypt, annexed Syria to Egypt. Egypt held Syria until 1840, when the European powers (particularly Great Britain) forced its return to the Ottomans. During the rest of the 19th cent. the Syrian economy was modernized somewhat, but growing resentment of Ottoman rule developed. After bloody fighting between Christians and Druses, Lebanon (largely inhabited by Christians) was given considerable autonomy in 1860. The peace settlement after World War I gave (1920) France a League of Nations mandate over the Levant States (roughly present-day Syria and Lebanon). From this time the term Syria referred approximately to its present territorial extent. Agitation by Arab nationalists induced France to make Lebanon a completely separate state in 1926, but nationalist agitation continued for the independence of Syria. British and Free French forces invaded and occupied Syria in June, 1941. In accordance with previous promises, the French proclaimed the creation of an independent Syrian republic in Sept., 1941, and the country achieved complete independence in 1944.

Independent Syria has been characterized chiefly by economic growth and diversification, by political instability, and by conflict with Israel. A member of the Arab League, Syria joined other Arab states in the unsuccessful war (1948–49) against Israel. Lt. Col. Adib al-Shishakli ruled the country as a dictator from 1949 to 1954. In the following years the Ba'ath party, which combined Arab nationalism with a socialist program, emerged as the most influential political party. The Ba'athists led in joining Syria with Egypt to form (Feb., 1958) the United Arab Republic (UAR). By late 1959, Egypt had become dominant in the UAR, and it governed Syria almost as if it were a province. In 1961 a group of Syrian army officers seized power, withdrew the country from the UAR, and established the independent Syrian Arab Republic. In 1963 a joint Ba'ath-military government undertook to nationalize much of the economy and redistribute land to the peasants.

During the Arab-Israeli War of 1967 Israel captured the Golan Heights (stretching about 12 mi/19 km into Syria northeast of the Sea of Galilee), and it held onto this territory after a cease-fire went into effect. Gen. Hafez al-Assad, a moderate nationalist who emphasized the need to lessen Syria's economic and military dependence on the USSR, came to power in 1970 and was elected president in 1971. In the Arab-Israeli War of 1973 Syria failed to recover the Golan Heights. Border skirmishes with Israel and Palestinian guerrilla incursions into Israel from Syria were frequent through the 1960s and 70s. In 1976–77 a Syrian military force intervened in Lebanon to halt the civil war there between Moslems and Christians.

Government. Syria is ruled by an elected president who is assisted by a People's Assembly, partially elected and partially appointed. The National Progressive Front selects candidates to run for the Assembly. Independent candidates are also allowed to run.

Syr·i·an Desert (sîr'ē-ən), arid wasteland, SW Asia, between the cultivated lands along the E Mediterranean coast and the fertile Euphrates River valley. It extends north from the Arabian Desert in Saudi Arabia and comprises western Iraq, eastern Jordan, and southeast Syria. Several nomadic tribes inhabit the desert. Palmyra and other oases served as staging posts on ancient Mediterranean-Mesopotamian trade routes.

Sy·ros (sī'rŏs'), island: see Síros.

Syz·ran (sĭz'rən-yə), city (1973 est. pop. 180,000), S central European USSR, on the Volga River near its junction with the Syzran. The city is a major river port and rail center. Manufactures include hydroturbines and combines, and there are large oil refineries and tanneries. Oil, asphalt, limestone, and slate are extracted in the area.

Szcze·cin (shchĕ'tsēn), city (1973 est. pop. 358,000), NW Poland, formerly capital of the Prussian province of Pomerania, on the Oder near its influx into the Zalew Szczeciński. A major Baltic port, Szczecin is also an industrial center with shipyards, ironworks, coke works, and industries producing chemicals, metals, and foodstuffs. Until 1637 it was an important member of the Hanseatic League. At the Peace of Westphalia (1648) it passed to Sweden, but at the end of the Northern War Sweden ceded it (1720) to Prussia. During World War II the city suffered heavy damage from repeated bombings. Although four fifths of Szczecin is on the left bank of the Oder, the Potsdam agreement of 1945 transferring Pomerania east of the Oder to Polish administration was interpreted to include the city in the transfer. The German population was expelled and replaced by Poles.

Szcze-ciń-ski (shchĕ-tsēn'skē), town (1973 est. pop. 48,600), NW Poland. It is a rail junction and has metalworking and chemical industries. Chartered in the 13th cent., the town later joined the Hanseatic League. The town, which was virtually obliterated during World War II, was incorporated into Poland in 1945.

Sze·chwan (sĕ'chwän', sŭ'-), province (1968 est. pop. 70,000,000), c.220,000 sq mi (569,800 sq km), SW China; capital Ch'eng-tu. A naturally isolated region completely surrounded by mountains, Szechwan is accessible to the rest of China by the Yangtze River. It has been greatly expanded in the 1960s and 1970s; railroads now connect Ch'eng-tu with Chungking and Szechwan with Shensi and Yünnan provs. Central Szechwan is generally a rough plateau with a red sandstone formation. The basin includes the fertile, densely populated Chengtu Plain, the only large, level area in the province; however, extensive terracing adds much cultivated land. Szechwan is the country's leading rice producer and ranks second in the production of sugar cane. Potatoes, citrus fruits, wheat, corn, sugar beets, sweet potatoes, and beans are also grown. Szechwan is a major cotton producer; other economic crops include ramie hemp, medicinal herbs, tea, and oilseed. About 20% of the province is forested, and tung oil is a major export. Silk, grown on both mulberry and oak trees, is still produced. In the western areas (formerly Sikang prov.), there is much grazing land, and the province's cattle population is said to be the largest in the country. Salt has been mined since ancient times; other mineral resources include oil, natural gas, coal, iron, copper, lead, zinc, asbestos, and mercury.

Szechwan has often been an independent kingdom. It was early a center of Thai culture; its Indian influence came in via the Burma-Yünnan trade route. The Chinese Communists controlled much of northern Szechwan in the early 1930s, and the province served as a refuge during the long march. In 1955 the area of Sikang prov. east of the Yangtze was added to it, nearly doubling the area of the province.

Sze·ged (sĕ'gĕd), city (1973 est. pop. 133,206), S Hungary, at the confluence of the Tisza and Maros rivers. It is a river port, a railroad hub, and an agricultural center. Food processing, flour milling, boatbuilding, and the production of textiles, leather footwear, and tobacco are among the city's industries. The city became (9th or 10th cent.) a Magyar military stronghold and trade center. Szeged was sacked by the Tatars and the Turks and was ruled by the latter from 1542 to 1686.

Szé·kes·fe·hér·vár (sā'kĕsh-fĕ'hâr-vär), city (1973 est. pop. 85,405), W central Hungary. It is a road and rail junction and an industrial center, with industries producing aluminum, machinery, chemicals, and leather. Dating from Roman times, it was (1027–1527) the coronation and burial place of Hungary's kings. Székesfehérvár was destroyed during the Turkish occupation of Hungary (1543–1688) and rebuilt in the 18th cent.

Szol·nok (sôl'nôk), city (1973 est. pop. 67,774), E central Hungary, at the confluence of the Tisza and Zagyva rivers. It is a river port and a road and rail junction. Manufactures include chemicals, cellulose, footwear, and furniture.

Szom·bat·hely (sôm'bŏt-hā'), city (1973 est. pop. 70,608), W Hungary, near the Austrian border. An important railway junction, it produces agricultural machinery, textiles, and shoes and is also a market for local farm products. The city was founded in 48 A.D. by the Roman emperor Claudius.

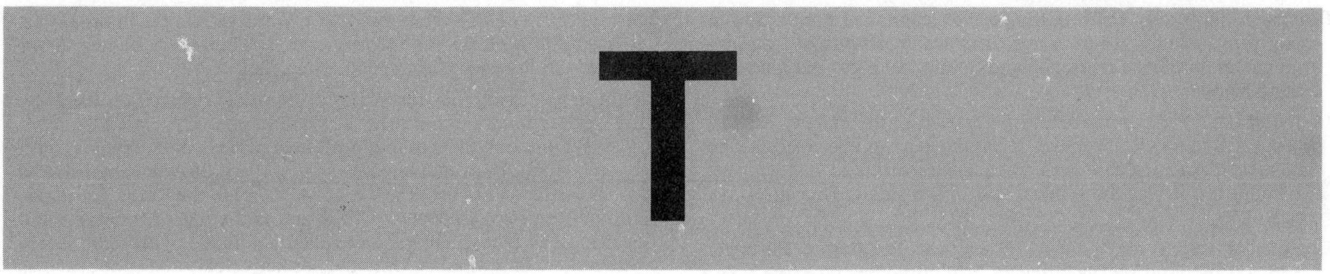

Ta·al, Lake (tä-äl'), 94 sq mi (243.5 sq km), SW Luzon, the Philippines, S of Manila. It contains Volcano Island, with Mt. Taal, an active volcano rising to 984 ft (300 m). Its last major eruption occurred in 1968.

Ta·bas·co (tə-băs'kō), state (1970 pop. 768,327), 9,783 sq mi (25,338 sq km), E Mexico, on the Gulf of Campeche; capital Villahermosa. Tabasco is predominantly a densely forested tropical plain. The climate is sultry, and rainfall in some areas exceeds 200 in. (508 cm) annually. Tropical agriculture (bananas, cacao, sugar cane, hardwoods, and fruits) and cattle raising are the leading economic activities, but rich oil fields discovered along the coast have brought sweeping economic and social changes to Tabasco.

Ta·ble Bay (tā'bəl), inlet of the Atlantic Ocean, 6 mi (9.7 km) wide, lying off W Cape Prov., Republic of South Africa. Table Mt. (3,567 ft/1,088 m) overlooks the bay, which was discovered in the late 15th cent. by Portuguese voyagers to India. The shore was settled by the Dutch in 1652, with the founding of Cape Town.

Tá·bor (tä'bôr'), city (1971 est. pop. 22,050), W Czechoslovakia, in Bohemia. Machinery, machine tools, electrical equipment, and textiles are the chief manufactures. Founded in 1420 by John Žižka, the city became the stronghold of the Taborites, the extreme wing of the Hussites. Tábor retains the tower of a 13th cent. castle and a town hall with a collection of Hussite relics.

Ta·bor, Mount (tā'bər), 1,929 ft (588.3 m) high, N Israel, in Galilee. Ruins of an ancient stronghold crown its summit. Many Christians believe it was the scene of Jesus' transfiguration.

Ta·bo·ra (tə-bôr'ə, -bōr'ə), city (1967 pop. 21,012), capital of Tabora region, W central Tanzania. It is a trade and transportation center; peanuts, cotton, cattle, and other agricultural commodities are shipped. Tabora, founded in 1852 by Arab traders, was captured in 1891 by Germany and became a center of administration of German East Africa.

Ta·briz (tə-brēz'), city (1971 est. pop. 420,000), capital of East Azerbaijan prov., NW Iran, on the Aji Chai (Talkheh) River, in the foothills of Mt. Sahand, at an elevation of c.4,600 ft (1,400 m). The fourth-largest city in Iran, it is a commercial, industrial, and transportation center. Its manufactures include carpets, textiles, food products, shoes, and soap. Tabriz, then known as Tauris, was (3rd cent. A.D.) the capital of Armenia. It was sacked by the Oghuz Turks c.1029, but by 1054, when it was captured by the Seljuk Turks, Tabriz was a provincial capital. In 1295 Ghazan Khan, the Mongol ruler of Persia, made it the chief administrative center of an empire stretching from Egypt to the Oxus River and from the Caucasus to the Indian Ocean. Tabriz was captured by Tamerlane in the late 14th cent., and Shah Ismail made it the capital of his empire from 1501 until his defeat (1514) by the Ottoman Turks. The Ottomans occupied Tabriz from 1585 to 1603. Nevertheless, by the 17th cent. it was a major commercial center. The city was again occupied (1724–30) by the Ottomans and was held by Russia (1827–28). The city has often been devastated by earthquakes and has few historical remains; of these the most important are the Blue Mosque (15th cent.) and the Ark, or Ali Shah, Mosque (14th cent.).

Ta·ca·ná (tä'kə-nä'), volcano, 13,333 ft (4,066.6 m) high, on the Mexico-Guatemala boundary; second-highest peak in Central America.

T'a-ch'eng (dä'chĕng') or **Chu·gu·chak** (cho͞o'go͞o'chäk'), town, N Sinkiang Uigur Autonomous Region, China, in the Dzungarian basin, bordering on the USSR. It is an agricultural hub and has a lumbering industry. Iron deposits are in the area.

Ta·chi·ka·wa (tä'chē-kä'wə), city (1970 pop. 117,057), Tokyo Metropolis, E central Honshu, Japan. It is an industrial suburb.

Ta·clo·ban (tä-klō'bän'), city (1970 pop. 49,908), capital of Leyte prov., NE Leyte, the Philippines, on an inlet of Leyte Gulf. It is a port and trade center and was the first landing place (Oct., 1944) for U.S. troops in the World War II campaign that liberated the Philippines from the Japanese.

Ta·co·ma (tə-kō'mə), city (1978 est. pop. 150,000), seat of Pierce co., W Wash., on Commencement Bay and Puget Sound at the mouth of the Puyallup River; inc. 1884. It is a major seaport and railroad terminus and one of the chief industrial cities in the Northwest. It is an important center for forest-products industries and has railroad shops and plants that manufacture chemicals and electrochemical products, explosives, paints, bleaches, fertilizers, heavy cranes and machines, adhesives, minerals, metals and alloys, furniture, boats, clothing, and food products. It is the gateway to several national parks and scenic wonders. McChord Air Force Base, Fort Lewis (a major army training center), and the state national guard headquarters are nearby.

Ta·con·ic Mountains (tə-kŏn'ĭk), range of the Appalachian Mts., extending c.150 mi (240 km) between the Green Mts. and the Hudson Valley along parts of New York's border with Vt., Mass., and Conn. Mt. Equinox (3,816 ft/1,163.9 m) is the highest point. The Berkshire Hills, western Mass., are part of the range.

Tad·ous·sac (tăd'ə-săk'), village (1971 est. pop. 1,000), S Que., Canada, at the confluence of the Saguenay and St. Lawrence rivers. It is a summer resort. Jacques Cartier visited here in 1535. An attempt (1600–1601) to establish a French colony failed, but Tadoussac later became the site of the oldest Christian mission in Canada and an important French fur-trading post.

Ta·dzhik Soviet Socialist Republic (tä-jĭk', -jēk') or **Ta·dzhik·i·stan** (tə-jĭk'ĭ-stăn', -stän', tä-), constituent republic (1978 est. pop. 3,538,000), 55,251 sq mi (143,100 sq km), Central Asian USSR. Dushanbe is the capital. Tadzhikistan borders on China in the east, Afghanistan in the south, the Kirghiz SSR in the north, and the Uzbek SSR in the west and northwest. Parts of the Pamir and Trans-Alai mt. systems are in the east. The Amu Darya, Syr Darya, and Zeravshan are the chief rivers and are used for irrigation. The easternmost section of the republic constitutes the Gorno-Badakhshan Autonomous Oblast.

Tadzhikistan's economy is based mostly on agriculture, livestock raising, mining, and raw-material processing. Long-staple cotton, wheat, barley, fruit (including wine grapes), and mulberry trees (for silk) are cultivated. Karakul sheep, dairy cattle, and yaks are raised. The republic's mountains yield coal, antimony, gold, salt, fluorspar, and numerous other minerals. Cotton ginning, silk spinning, fruit canning, winemaking, carpet weaving, metalworking, machine building, and the manufacture of cotton, silk, and woolen textiles and leather goods are the leading industries.

By the 9th and 10th cent. the Tadzhiks, an Iranian people of the Sunni Moslem religion, had achieved much success in agriculture and trade, but the territory was conquered by the Mongols in the 13th cent. In the 16th cent. it became part of Bukhara. Russia took control of it in the 1880s and 1890s, but the Tadzhiks remained split among several administrative-political entities, and their territories were exploited for their raw materials. After the 1917 Russian Revolution the Tadzhiks rebelled; the Red Army did not establish control over them until 1921. Tadzhikistan became an autonomous republic within Uzbekistan in 1924; in 1929 it became a constituent republic of the USSR.

Tae·gu (tī'go͞o'), city (1970 pop. 1,082,750), S South Korea, on the Kum River. It is a railroad junction, an industrial center, and a collection and distribution point for an extensive agricultural and mining region. Taegu has important textile industries. In Aug., 1950, it became the temporary capital of Korea.

Tae·jon (tī'jŭn'), city (1970 pop. 414,598), capital of South Chungchong prov., central South Korea. It is a railroad hub and agricultural center, with rice mills, silk and textile factories, and food-processing plants.

Ta·fi·lelt (tăfī-lĕlt') or **Ta·fi·let** (-lĕt'), oasis in the Sahara, SE Morocco, c.530 sq mi (1,375 sq km). It has date groves and small trading settlements. After c.760 it was an independent kingdom for nearly two centuries. It was the original seat of the ruling dynasty of Morocco.

Tag·an·rog (tăg'ən-rŏg'), city (1973 est. pop. 268,000), S European USSR, on the Gulf of Taganrog, an arm of the Sea of Azov. It is a port, exporting mainly grains and coal. Iron and steel milling, metallurgy, ship repairing, leather working, commercial fishing, agricultural processing, and the manufacture of heavy machinery and furniture are the city's major industries. A Pisan colony on the site was destroyed by the Mongols in the 13th cent.; Turks later settled here. In 1698 Peter the Great founded Taganrog as a fortress and naval base. The Turks recaptured it twice (1712 and 1739), but it was ceded by Turkey to Russia in 1774. Taganrog is a military and naval base. The writer Anton Chekhov was born here.

Ta·gus (tā'gəs), river, c.585 mi (940 km) long, rising in the Sierra de Gúdar, E Spain, and draining the central part of the Iberian Peninsula. The Tagus flows northwest through the mountains, past Teruel, then north across the Meseta of central Spain, past Toledo, to form part of the Spanish-Portuguese border. Entering Portugal, it flows southwest, past Santarém and into the Atlantic Ocean at Lisbon. The estuary of the Tagus (12 mi/19 km long) is one of Europe's finest harbors. The river is navigable for c.80 mi (130 km) upstream. The middle Tagus is used for irrigation.

Ta·hi·ti (tə-hē'tē), island (1970 est. pop. 84,552), South Pacific, in the Windward group of the Society Islands, French Polynesia. The capital is Papeete. Tahiti is the largest (402 sq mi/1,041.2 sq km) and most important of the French islands. The peninsula of Taiarapu, which forms eastern Tahiti, is joined to the western part of the island by the Isthmus of Taravao. Tahiti is mountainous, with Mt. Orohena (7,618 ft/2,322.5 m) the highest peak. The chief products are tropical fruits, copra, vanilla, and sugar cane; there are pearl fisheries off the coast. The island was settled by Polynesians in the 14th cent.; the first European to discover it was the English navigator Samuel Wallis, and later visits (1769, 1773, 1777) were made by Capt. James Cook. In 1843 Tahiti became a French protectorate and later (1880) a colony. During World War II the Tahitians supported the Free French; in 1946 all the indigenous inhabitants became French citizens. Paul Gauguin did many paintings in Tahiti, and Robert Louis Stevenson spent some time here.

Tah·le·quah (tăl'ə-kwô'), city (1970 pop. 9,254), seat of Cherokee co., NE Okla., NE of Muskogee, in a farm, strawberry, and livestock region. It was settled by Cherokees in 1839 and made their capital. A museum containing Indian relics is here.

Ta·hoe, Lake (tä'hō, tā'-), 193 sq mi (500 sq km), on the Calif.-Nev. line. The lake occupies a basin in the Sierra Nevada and is drained by the Truckee River. It lies 6,228 ft (1,900 m) above sea level and is 1,645 ft (501.7 m) deep; it is a resort.

Ta·ho·ka (tə-hō'kə), city (1970 pop. 2,956), seat of Lynn co., NW Texas, S of Lubbock on the Llano Estacado; settled 1903, inc. 1915. It ships cattle, cotton, black-eyed peas, and dairy products. To the northeast is Tahoka Lake, a landmark for early travelers.

Ta·houa (tou'ə), town (1970 est. pop. 20,335), SW Niger. It is a farming community and trade center; gypsum and phosphates are mined.

Tah·pan·hes (tä'pən-hēz'), ancient city, NE Egypt, on Lake Manzala. The site is now on the Suez Canal.

Tah·qua·me·non (tə-kwä'mə-nən, -mə-nŏn'), river, c.80 mi (130 km) long, rising in the E Upper Peninsula, N Mich., and flowing E and NE to Whitefish Bay of Lake Superior. It is celebrated in Longfellow's *Hiawatha.*

T'ai or **Tai** (tī), lake, c.1,300 sq mi (3,365 sq km), on the border between Kiangsu prov. and Chekiang prov., E China. The lake basin is one of the richest agricultural regions in China; rice, wheat, and cotton are grown.

T'ai-chung or **Tai·chung** (both: tī'jŏŏng'), city (1970 est. pop. 337,000), W central Taiwan. T'ai-chung is a central distributing and processing center for rice, sugar cane, and bananas. The city has textile, machine-building, food-processing, and chemical industries.

Ta·if (tä'īf), city (1963 est. pop. 54,000), W Saudi Arabia, in the Hejaz. It is c.5,000 ft (1,525 m) above sea level. The graves of two infant sons of the Prophet Mohammed are in Taif.

tai·ga (tī'gə), coniferous-forest belt of Eurasia, bordered on the N by the treeless tundra and on the S by the steppe. This vast belt, comprising about one third of the forest land of the world, extends south from the tundra to about lat. 62° N in Norway, Sweden, and Finland but dips still farther south to about lat. 53° N in the Urals. It extends through northern European USSR across the Ural Mountains and over most of Siberia. It has a continental climate, with long, severe winters of 6 or 7 months. Only the hardier cereals and roots, such as barley, oats, and potatoes, can be cultivated. The principal species of trees are cedar, pine, spruce, larch, birch, and aspen.

Tai·myr Peninsula (tī-mîr'): *see* Taymyr Peninsula.

T'ai-nan or **Tai·nan** (both: tī'nän'), city (1970 est. pop. 468,300), W central Taiwan, on the Taiwan (Formosa) Strait. The third-largest city of Taiwan, it produces metals, textiles, machinery, and processed foods. It is also a center for the marketing and processing of sugar cane, rice, peanuts, and salt, and there is an important fishing industry. Settled in 1590, T'ai-nan is the oldest city of Taiwan. Once called Taiwan or Taiwanfu, it was the political center of the island until the transfer of government to Taipei in 1885.

Taí·na·ron, Cape (tā'nə-rôn'): *see* Matapan, Cape.

Tai·pei or **Tai·peh** (both: tī'pā'), city (1970 est. pop. 1,740,800), N Taiwan, capital of Taiwan and provisional capital of the Republic of China. Taiwan's largest city, it is the administrative, cultural, and industrial center of the island. The major industries produce wood and paper products, textiles, metals, machinery, chemicals, food products, and fertilizers. Founded in the 18th cent., Taipei replaced T'ai-nan as the capital of Taiwan prov. in 1885. It underwent considerable enlargement and modernization under Japanese rule (1895-1945). In 1949, when the Communists forced Chiang Kai-shek to flee from the mainland of China, Taipei became the headquarters of the Nationalists. In 1967 the city became a special municipality with a status equal to that of a province.

Tai·ping (tī'pĭng'), city (1970 pop. 54,645), Perak, Malaysia, central Malay Peninsula. Once the leading tin-mining center of Malaya, the city is situated at the foot of Maxwell's Hill (alt. c.4,000 ft/1,220 m).

Tai·sha (tī'shə), town, Shimane prefecture, SW Honshu, Japan, on the Sea of Japan. It is a religious center, famous as the site of the ancient Izumo shrine. Traditionally all the Shinto gods convene here each Oct. The town also has a 6th cent. Buddhist temple.

T'ai-shan (tī'shän') or **Tai** (tī), peak, 5,069 ft (1,546 m) high, W Shantung prov., E China. The peak is revered by Buddhists and Taoists and has long been the goal of pilgrimages; it has many temples and shrines.

T'ai-tung or **Tai·tung** (both: tī'dŏŏng'), city (1970 est. pop. 59,500), W Taiwan. Rice, sugar cane, and peanuts are the major crops marketed and processed in T'ai-tung.

Tai·wan (tī'wän'), island (1976 est. pop. 16,426,386), 13,885 sq mi (35,962 sq km), in the Pacific Ocean, separated from the mainland of S China by the 100-mi (161-km) wide Formosa Strait. Together with many nearby islets, including the Pescadores and the island groups of Quemoy and Ma-tsu, it forms the seat of the Republic of China. The provisional capital is Taipei; Nanking, on mainland China, is regarded as the official capital of the republic. The heavily forested hills and mountains of central and eastern Taiwan reach their summit at Hsin-kao Shan (13,113 ft/3,999.5 m high); there are about 70 peaks exceeding 10,000 ft (3,050 m). This mountainous area produces some minerals (gold, silver, copper, and coal), but its main resources are forest products, including valuable hardwoods and natural camphor. The broad coastal plain in the west supports most of the island's population and is the chief agricultural zone. Typhoons are common.

Economy. Taiwan, with a semitropical climate and rainfall ranging from moderate to heavy, produces abundant food crops. Rice is the chief crop, followed by sugar cane; sweet potatoes, potatoes, bananas, peanuts, citrus fruits, pineapples, and tea are also important. The island has a sizable fishing fleet. Industry, once concerned mainly with rice and sugar milling, has diversified to include food processing and the production of textiles and chemicals. The manufacture of consumer goods, especially electrical appliances, has also become increasingly significant. Most industries are privately run, but the government operates those considered essential to national defense, such as steel. Railroad and bus lines are also government-operated. Major exports are textiles, clothing, fruits, and vegetables; imports include nonelectrical and electrical machinery and transportation equipment.

History. The earliest Chinese settlements on Taiwan began in the 7th cent., chiefly from the mainland provinces of Fukien and Kwangtung. The island was reached in 1590 by the Portuguese, who named it Formosa. In 1624 the Dutch founded forts in the south, while the Spanish established bases in the north. The Dutch expelled the Spaniards in 1641 and assumed control of the entire island. They in turn were forced out in 1662, when Koxinga, a general of the Ming dynasty of China, seized the island and established an independent kingdom. The island fell to the Manchus in 1683. Chinese immigration increased, and the aboriginal population was gradually pushed into the interior. Japan, attracted by the island's strategic and economic importance, acquired Taiwan by the Treaty of Shimonoseki (1895) after the First Sino-Japanese War. Japan exploited the island for the benefit of the Japanese home economy but not for colonization. Under Japan Taiwan's economy was modernized and industrialized, railroads were built, and the large cities expanded.

In accordance with the Cairo declaration of 1943 and the Potsdam Conference of 1945, Taiwan was returned to China as a province after World War II. In 1949, as the Chinese Communists gained complete control of the mainland, the Nationalist government of Chiang Kai-shek and the remnants of his army took refuge on the island. The Chinese Communists planned an invasion of Taiwan in 1950, but it was thwarted when President Truman ordered the U.S. 7th Fleet to patrol Formosa Strait. Taiwan's territorial status remained a major issue among the great powers. In 1955, following repeated attacks by the People's Republic of China against the Nationalist-held islands of Quemoy and Ma-tsu, the United States entered into a mutual security treaty with the Nationalists. In 1958 there was continuous, intensive shelling of Quemoy and Ma-tsu, and an invasion was again threatened. The United States, however, reasserted its determination to defend Taiwan, although it stressed that there was no commitment to help the Nationalist government return to the mainland. With U.S. economic aid, Taiwan enjoyed spectacular economic growth after 1950.

Caught between a native Taiwanese movement for independence and the continuing threat from Communist China, the Nationalist government was far from secure in the 1960s and 1970s. China's seat in the United Nations was taken away from the Republic of China and given to the People's Republic in 1971. Taiwan's international position continued to weaken in the early 1970s as the United States sought to improve relations with the People's Republic of China and as more large countries, such as Canada and Japan, moved to recognize the mainland government.

On Dec. 15, 1978, President Jimmy Carter announced that the United States was establishing diplomatic relations with the mainland government. Consequently, relations with Taiwan were severed on Dec. 31, 1978. The U.S.-Taiwanese defense treaty was to remain in effect until Dec. 31, 1979. At the end of 1978 Peking made overtures to Taiwan to initiate trade relations; these were rejected. The situation between the two nations remained unsettled.

Government. Taiwan's government is based on the constitution of 1946, which was drawn up to govern all of China. An elected national assembly of 1,488 members chooses the president and vice-president and is empowered to amend the constitution. The government is made up of five yuan, or branches. The executive yuan is headed by the president; the legislative yuan, which is elected, has 493 members and handles all legislation; the judicial yuan is appointed by the president and serves as the highest judicial authority; the control yuan is in charge of censorship; and the examination yuan supervises examinations for government positions.

T'ai-yüan or **Tai·yüan** (both: tī'yōō-än'), city (1970 est. pop. 2,725,000), capital of Shansi prov., N China, on the Fen River, in one of the world's richest coal and iron areas. It is a mining and smelting center with a large iron and steel complex and plants making heavy machinery, chemicals, plastics, fertilizer, cement, paper products, and processed foods. An ancient city, it fell to the Communists in 1949 after a siege in which thousands starved.

Ta·izz (tă-ĭz'), city (1971 est. pop. 80,000), S Yemen. It is an agricultural marketing center and was the administrative capital of Yemen from 1948 to 1962.

Ta·ji·mi (tə-jē'mē), city (1970 pop. 63,522), Gifu prefecture, central Honshu, Japan. It is the country's leading producer of ceramics.

Taj·rish (tăj-rēsh'), city (1966 pop. 157,486), Tehran prov., N Iran, a suburb of Tehran. It is a summer resort.

Ta·ju·mul·co (tä'hōō-mōōl'kō), inactive volcano, 13,816 ft (4,213.9 m) high, W Guatemala. It is the highest mountain in Central America.

Ta·ka·da (tə-kä'də), city (1970 pop. 56,000), Niigata prefecture, W central Honshu, Japan. It is an agriculture market and center for chemical industries.

Ta·ka·mat·su (tä'kə-mät'sōō), city (1970 pop. 274,367), capital of Kagawa prefecture, NE Shikoku, Japan, a port on the Inland Sea. Lacquer ware and paper products are manufactured here.

Ta·ka·o·ka (tä'kə-ō'kə), city (1970 pop. 159,664), Toyama prefecture, W central Honshu, Japan, on the Sho River. It has mechanical, textile, and paper industries.

Ta·ka·ra·zu·ka (tə-kä'rə-zōō'kə), city (1970 pop. 127,179), Hyogo prefecture, SW Honshu, Japan. It is a suburb of the Osaka-Kobe area.

Ta·ka·sa·go (tä'kə-sä'gō), city (1970 pop. 68,900), Hyogo prefecture, SW Honshu, Japan. It is an industrial center.

Ta·ka·sa·ki (tä'kə-sä'kē), city (1970 pop. 193,072), Gumma prefecture, central Honshu, Japan. A transportation and industrial center with flour mills, silk textile factories, and food-processing plants, it is known chiefly for its statue (130 ft/40 m high) of Kannon, goddess of mercy.

Ta·ka·ya·ma (tä'kə-yä'mə), city (1970 pop. 56,459), Gifu prefecture, W central Honshu, Japan, on the Jinzu River. It is an agricultural market and handicrafts center.

Ta·kla Ma·kan (tä'klə mə-kän'), vast sandy desert, c.125,000 sq mi (323,750 sq km), central Sinkiang Uigur Autonomous Region, NW China, between the Kunlun Mts. on the S and the Tien Shan Mts. and Tarim River on the N. It occupies most of the Tarim basin. The earliest Chinese contacts with the West were made along the Takla Makan oases.

Ta·ko·ma Park (tə-kō'mə), city (1975 est. pop. 17,600), Montgomery and Prince Georges cos., W central Md., a residential suburb of Washington, D.C.; inc. 1890.

Ta·la·ve·ra de la Rei·na (tä'lə-vâr'ə dā' lä rā'nə), town (1970 pop. 45,327), Toledo prov., central Spain, in New Castile, on the Tagus River. It is in an agricultural region and is known for its fine ceramics industry. Here, in 1809, the English and Spanish under Wellesley (the duke of Wellington) defeated the French under Joseph Bonaparte.

Tal·bot (tôl'bət, tăl'-). **1.** County (1970 pop. 6,625), 390 sq mi (1,010 sq km), W Ga., bounded on the NE by the Flint River and intersected by the Fall Line; formed 1827; co. seat Talbotton. It is in an agricultural (cotton, corn, grain, peaches, and livestock) and lumbering area. **2.** County (1970 pop. 23,682), 261 sq mi (676 sq km), E Md., a peninsula on the Eastern Shore, bounded on the W by Chesapeake Bay; formed 1661; co. seat Easton. In a tidewater agricultural area yielding truck crops, grain, dairy products, poultry, and cattle, it has a large seafood industry and other food processing. There is some clothing manufacturing in the county. It is a resort area (yachting, hunting, and fishing), with colonial mansions and gardens.

Tal·bot·ton (tôl'bə-tən, tăl'-), town (1970 pop. 1,045), seat of Talbot co., W Ga., NE of Columbus, in an agricultural area; founded 1828. Franklin D. Roosevelt State Park is nearby.

Tal·ca (täl'kə), city (1970 pop. 102,522), capital of Talca prov., S central Chile, in the central valley of Chile between Santiago and Concepción. Agriculture is important in the region, which is also Chile's greatest wine-producing area. Talca has distilleries, foundries, a tannery, and factories making matches, shoes, tobacco products, paper, and flour. Talca was founded in 1692; here (Feb. 12, 1818) Bernardo O'Higgins formally proclaimed Chile's independence.

Tal·ca·hua·no (täl'kə-wä'nō), city (1970 est. pop. 115,600), S central Chile, a port on the Pacific Ocean. Talcahuano is an important naval base. It has a large fishing industry, and fish are canned and exported. It also has extensive dry-dock facilities, metallurgical plants, and petroleum refineries, and it handles the exporting of the agricultural products of the interior.

Tal·ia·ferro (tä'lə-vər), county (1970 pop. 2,423), 195 sq mi (505 sq km), NE central Ga., drained by the Little and Ogeechee rivers; formed 1825; co. seat Crawfordville. It is in a piedmont agricultural (cotton, corn, grain, fruit, and livestock) and sawmilling area.

Ta-lien or **Ta·lien** (both: däl'yĕn'), city (1970 est. pop. 4,000,000), S Liaoning prov., China, on the Liao-tung peninsula in the Bay of Korea. It has been combined with Lü-shun (Port Arthur) into the joint municipality of Lü-ta. With a huge, well-protected harbor, modern freight-handling facilities, and fine rail connections, Ta-lien is the chief commercial port of Manchuria. It is also a major industrial center with large shipyards, fisheries, an oil refinery, textile mills, chemical and fertilizer plants, and factories making locomotives, rolling stock, and electrical equipment. The city first became important when Russia occupied it (1898) and developed it as the southern terminus of the South Manchurian RR. The Japanese acquired the territory in 1905, and it was thereafter known as the Kwantung leasehold. In 1945 Russia occupied it and received a free lease from Nationalist China on half the city's port facilities, an arrangement that continued under the Communist government. Russian troops remained here until 1955.

Tal·la·de·ga (tăl'ə-dē'gə), county (1970 pop. 65,280), 750 sq mi (1,942.5 sq km), E central Ala., bounded on the W by the Coosa River; formed 1832; co. seat Talladega. It has cotton and corn farming, limestone and stone quarrying, iron mining, and timber sawmilling. Its industry includes food processing, textile mill products, clothing, wood products, fertilizers, mineral products, iron and steel foundries, metal products, and machinery manufacturing. Parts of Talladega Mts. and Talladega National Forest are in the east.

Talladega, city (1978 est. pop. 17,800), seat of Talladega co., NE central Ala., in the Blue Ridge foothills; inc. 1835. Textiles and clothing are its chief products; sawmill machinery and iron castings are also made. In Nov., 1813, Andrew Jackson defeated the Creek Indians here.

Tal·la·has·see (tăl'ə-hăs'ē), city (1978 est. pop. 86,400), state capital and seat of Leon co., NW Fla.; inc. 1825. The state government, Florida State Univ., and Florida Agricultural and Mechanical Univ. are major sources of employment. Lumber and wood products are manufactured, and food is processed. When De Soto arrived here in 1539, he found a flourishing settlement of Apalachee Indians. Spanish missionaries and settlers followed, but the Indian village remained the major settlement until Tallahassee was founded (1824) as the capital of the Florida Territory. The ordinance of secession was adopted here in 1861.

Tal·la·hatch·ie (tăl'ə-hăch'ē), county (1970 pop. 19,338), 644 sq mi (1,668 sq km), NW central Miss., drained by the Tallahatchie River; formed 1833; co. seats Sumner and Charleston. Cotton, corn, soybeans, livestock, and timber are the major products.

Tal·la·poo·sa (tăl'ə-pōō'sə), county (1970 pop. 33,840), 705 sq mi (1,826 sq km), E Ala., in a piedmont region drained by the Tallapoosa River and Martin Lake; formed 1832; co. seat Dadeville. It has cotton and potato farming and timber logging and sawmills. Its industry includes textiles (weaving, knitting, and yarn mills and clothing manufacturing) and iron production.

Tallapoosa, river, 268 mi (431.2 km) long, rising in NW Ga. and flowing SW through E Ala. It joins the Coosa River near Montgomery, Ala., to form the Alabama River.

Tal·linn (tăl'ĭn, tä'lĭn), city (1974 est. pop. 392,000), capital of the Estonian SSR, NW European USSR, on the Gulf of Finland, opposite Helsinki. It is a major Baltic port, a rail and highway junction, and an industrial center. Tallinn has extensive military and naval installations. Industries include shipbuilding, metalworking, food and fish processing, and the manufacture of machinery, electrical and radio apparatus, oil-field equipment, cables, and building materials.

Tallinn was first mentioned by the Arabian geographer Idrisi in 1154. It was destroyed in 1219 by Waldemar II of Denmark, and it was sold (1346) with the rest of Estonia by Waldemar IV to the Livonian Knights. Upon the dissolution of the Livonian Order in 1561, it passed to Sweden. Captured by Peter I in 1710, Tallinn was formally ceded to Russia in 1721. In 1870 it was linked by rail with St. Petersburg (now Leningrad). Tallinn became the capital of independent Estonia in 1919 and of the Estonian SSR in 1940. It suffered considerable damage during the German occupation in World War II. The historical center of Tallinn consists of an upper town, on a steep hill topped by a medieval cathedral, and an adjoining lower town dating from Hanseatic times. The lower town is surrounded by a medieval wall with massive round towers.

Tall·madge (tăl'mĭj), city (1975 est. pop. 14,900), Summit co., NE Ohio, an industrial suburb of Akron; settled 1807, inc. 1950.

Tal·lu·lah (tə-lōō'lə), village (1970 pop. 9,643), seat of Madison parish, NE La., near the Mississippi River WNW of Vicksburg, Miss., in a cotton and timber area.

Ta·lo·ga (tə-lō'gə), town (1970 pop. 363), seat of Dewey co., W Okla., on the Canadian River SE of Woodward, in a stock-raising and agricultural area. Cotton and broomcorn are grown, and fuller's earth is mined.

Ta·ma (tā'mə), county (1970 pop. 20,147), 720 sq mi (1,864.8 sq km), E central Iowa, drained by the Iowa River and Wolf Creek; formed 1843; co. seat Toledo. It is in a prairie agricultural area yielding hogs, poultry, livestock, soybeans, corn, and oats. There is a small meat-packing plant as well as a seed-corn plant.

Ta·ma·le (tə-mä'lē), town (1970 pop. 81,612), capital of the Northern Region, N Ghana. It is a road junction and agricultural trade center.

Tam·al·pa·is, Mount (tăm'əl-pī'əs), peak, 2,604 ft (794.2 m) high, W Calif., across the Golden Gate from San Francisco. The mountain is a game preserve and a resort.

Ta·man Peninsula (tə-män'), c.20 mi (30 km) long and 8 mi (12.9 km) wide, Krasnodar Kray, SE European USSR, projecting westward between the Sea of Azov and the Black Sea. It is separated from the Crimea by the Kerch Strait. There are gas and petroleum deposits.

Ta·ma·qua (tə-mä'kwə), borough (1970 pop. 9,246), Schuylkill co., E Pa., NE of Pottsville; settled 1799, laid out 1829, inc. 1832. In an anthracite-coal area, it has some manufacturing.

Ta·ma·tave (tăm'ə-täv', tä'mə-), city (1972 est. pop. 59,503), NE Malagasy Republic, on Madagascar. Situated on the Indian Ocean, it is the nation's chief port and is connected by rail with Tananarive. Tamatave exports sugar, coffee, cloves, and rice. Food processing is the chief industry. The town was founded in the 18th cent. around a European trading post; it was occupied repeatedly by the French. Severely damaged (1927) by a storm, the city was subsequently rebuilt.

Ta·mau·li·pas (tä'mou-lē'pəs), state (1970 pop. 1,456,858), 30,734 sq mi (79,601 sq km), NE Mexico, on the Gulf of Mexico. Victoria is the capital. The state's greatest source of wealth is petroleum and its by-products, but agriculture and cattle raising are also important. Tamaulipas is a leading national producer of sugar cane and cotton; cereals, tobacco, and corn are other major crops, and citrus fruits are cultivated around Victoria. The state's industries include vegetable-oil extraction, flour milling, and the manufacture of chemicals and soap. Tourism is economically important. The territory was first explored by the Spanish in 1519; colonization began in 1747, and independence was won from Spain in 1824.

Tam·bov (täm-bôf', -bôv'), city (1973 est. pop. 245,000), capital of Tambov oblast, S central European USSR. A rail junction and manufacturing center, it produces machine tools, instruments, and chemicals. Founded in 1636 as an outpost against the Crimean Tatars, Tambov became (18th cent.) an administrative center.

Tam·il Na·du (tăm'əl nä'dōō), state (1971 pop. 41,199,168), 50,180 sq mi (129,966 sq km), SE India, on the Bay of Bengal. Rice, cotton, tea, tobacco, groundnuts, and millet are the principal crops. The main industries are food processing and the manufacture of cloth. Building materials and vehicles are also manufactured. There are irrigation canals and hydroelectric stations along the Cauvery River. An extensive rail network links the coastal cities with inland areas. Tamil Nadu was the seat of the Chola empire (10th–13th cent.). Moslems controlled the area for about a century. The Portuguese established trading posts (16th cent.), followed by the Dutch, French, and British (early 17th cent.). After a sharp struggle (1741–63) the British gained control, and the state (then called Madras) was considerably enlarged, but in 1953 Telugu-speaking areas were transferred to Andhra Pradesh, and in 1956 Kannada-speaking areas were transferred to Mysore (now Karnataka) and Malayalam areas to Kerala. This reduced the state's area by half.

Tam·pa (tăm'pə), city (1978 est. pop. 278,000), seat of Hillsborough co., W Fla., a port of entry with a harbor on Tampa Bay (an inlet of the Gulf of Mexico); inc. 1855. Tampa is a resort, a processing and shipping hub for the products of the area, a phosphate-mining center, and a large port with phosphate docks and elevators. It has a shrimp fleet, citrus-packing houses, huge breweries, and a noted cigar industry. The bay was visited by Pánfilo de Narváez in 1528, and in 1539 De Soto rescued the sole survivor of that expedition. The first white settlement began in 1823, and U.S. Fort Brooke was built in 1824. In the Civil War the town was taken (May, 1864) by Union troops.

Tam·pe·re (tăm'pə-rā', tăm'-), city (1971 est. pop. 161,257), Häme prov., SW Finland, on the banks of the rapids between lakes Näsijärvi and Pyhäjärvi. It is the second-largest city in Finland and a leading textile center. There are also locomotive works and other industries. The city, an important trade center since the 11th cent., was chartered in 1775.

Tam·pi·co (tăm-pē'kō), city (1970 pop. 179,584), Tamaulipas state, E Mexico, on the Pánuco River, a few miles inland from the Gulf of Mexico. One of Mexico's most important seaports, Tampico has one of the nation's largest fishing industries (oyster and shrimp) and a considerable export trade in cattle, hides, and other agricultural products. In pre-Columbian times Tampico was the site of the Huastec kingdom. Spanish settlement dates back to the 1530s. With the discovery of oil (c.1900), rapid development of petroleum industries began.

Tam·worth (tăm'wûrth'), city (1973 est. pop. 24,790), New South Wales, E Australia. It is an agricultural center and a transportation junction.

Tamworth, municipal borough (1974 est. pop. 43,830), Staffordshire, W central England. Its products include clothing, textiles, aluminum ware, paper, bricks, tiles, and agricultural machinery. Tamworth was burned by the Danes in the 9th cent. and rebuilt by Queen Æthelflæd in the 10th cent. The Church of St. Editha, built in the 8th cent. and rebuilt in 1345, is here. The town gives its name to a widely known breed of hog.

Ta·na (tä'nə), river, c.500 mi (805 km) long, rising near Mt. Kenya, central Kenya, E Africa, and flowing E, then S across Kenya to the Indian Ocean. There are hydroelectric plants and irrigation projects in the Tana basin.

Tana or **Tsa·na** (tsä'nə), largest lake of Ethiopia, c.1,400 sq mi (3,625 sq km), S of Gondar. One of the streams feeding it is regarded as the source of the Blue Nile.

Ta·na·be (tə-nä'bē), city (1970 pop. 49,000), Wakayama prefecture, SW Honshu, Japan, on Tanabe Bay. It is a commercial and fishing port with a processing industry.

Ta·na·elv (tä'nä-ĕlv'), river, 205 mi (330 km) long, rising in Finnmark co., N Norway, and flowing NE to Tanafjord, NE Norway. It forms part of the Norway-Finland border and is noted for its salmon.

Tan·a·na (tăn'ə-nô'), river, 600 mi (966 km) long, rising in W Yukon Territory near the Alaskan border and flowing NW across Alaska to the Yukon River. The Tanana valley, near Fairbanks, is central Alaska's chief farming area; grains and vegetables are grown. The Tanana, discovered c.1860, became an important route to the Yukon goldfields in 1898.

Ta·nan·a·rive (tə-năn'ə-rēv'), city (1972 est. pop. 366,530), capital of the Malagasy Republic and its Tananarive prov., on the island of Madagascar. Tananarive is the largest city in the Malagasy Republic and is its administrative, communications, and economic center. It is the trade center for a productive agricultural region whose main crop is rice. Its manufactures include food products (especially meat), beverages, cigarettes, and textiles. Tananarive was founded c.1625 as a walled citadel. In 1797 it was made the fixed residence of the Merina rulers, whose conquests (1810-28) made it the capital of almost all Madagascar. The city was captured by the French in 1895 and incorporated into their Madagascar protectorate. Today Tananarive is a modern city, built on the slopes of a ridge that rises to c.4,700 ft (1,435 m). At the top of the ridge is the former Merina royal residence.

Tan·dil (tän-dēl'), city (1970 pop. 54,000), Buenos Aires prov., E Argentina. Founded in 1823, it was a military outpost against raiding Indians. Tandil is today an attractive resort city. Some of Argentina's best granite is quarried in the surrounding hills.

Ta·ne·ga·shi·ma (tə-nā'gə-shē'mə), island, 176 sq mi (455.8 sq km), off S Kyushu, Japan. Fishing and farming are important here. Mendez Pinto, a Portuguese voyager on his way to China, landed (1543) here and introduced firearms to Japan.

Ta·ney (tā'nē), county (1970 pop. 13,023), 615 sq mi (1,592.9 sq km), S Mo., in the Ozarks, drained by the White River, which here forms Lake Taneycomo; formed 1837; co. seat Forsyth. It is a resort and agricultural area, known especially for livestock.

Tan·ga (tăng'gə), city (1967 pop. 61,058), capital of Tanga prov., NE Tanzania, a port on the Indian Ocean. It is a commercial, industrial, and transportation center. Exports include sisal, tea, and coffee. Among its manufactures are rolled steel, plywood, clothing, and twine.

Tan·gan·yi·ka (tăn'gən-yē'kə, tăng'-): *see* Tanzania, United Republic of.

Tanganyika, Lake, second-largest lake of Africa, c.12,700 sq mi (32,895 sq km), E central Africa, on the borders of Tanzania, Zaire, Zambia, and Burundi. It is c.420 mi (675 km) long and up to 45 mi (72 km) wide. The lake lies in the Great Rift Valley (alt. 2,534 ft/772 m) and is the world's second deepest (c.4,700 ft/1,435 m) freshwater lake. Lake Tanganyika has important fisheries. John Speke and Sir Richard Burton, the British explorers, were the first (1858) Europeans to see the lake.

Tan·gier (tăn-jîr'), city (1971 pop. 187,894), N Morocco, on the Strait of Gibraltar. The city is almost wholly without manufacturing industries, but the port is active. The walled Moorish town adjoins a modern European garden suburb. Tangier was probably founded by the Phoenicians. The chief port and commercial center of Morocco until the founding (808) of Fez, it had previously been under the rule of the Romans, and later of the Portuguese and English. By the mid-19th cent. it had become the diplomatic center of Morocco. When the rest of the country was divided between Spanish and French protectorates in 1912, the status of Tangier remained vague. Finally, in 1923-24, the city was included in an international zone administered by France, Spain, and Britain. Tangier remained under international control until 1956, when it was returned to Morocco.

Tan·gi·pa·ho·a (tăn'jə-pə-hō'ə), parish (1970 pop. 65,875), 808 sq mi (2,092.7 sq km), SE La., bounded on the E partly by the Tchefuncta River, on the S by Lake Maurepas and Lake Pontchartrain, on the N by the Miss. line, on the W partly by the Natalbany River, and drained by the Tangipahoa River; formed 1869; parish seat Amite City. The parish is chiefly agricultural, the north part being involved in the dairy industry and the south portion devoted to farming, especially strawberries, and to fishing and crabbing. Industries include oyster processing, cotton ginning, grain milling, and the manufacture of lumber. There are resorts and hunting and fishing opportunities.

T'ang-ku-la Shan-Mo (täng'ko͞o'lä' shän'mō') or **Tang·la** (täng'lä', däng'lä'), mountain range, SE extension of the Karakorum range, in the central Tibetan plateau, on the Tibet-Tsinghai prov. border, W China, rising to c.20,000 ft (6,100 m).

T'ang-shan or **Tang·shan** (both: täng'shän'), city (1970 est. pop. 1,200,000), NE Hopeh prov., China. In one of the greatest natural disasters in human history, T'ang-shan was totally destroyed by an earthquake on the night of July 28, 1976. More than 750,000 people were killed, and the city was reduced to a vast heap of rubble.

Ta·nim·bar Islands (tä-nĭm'bär', tə-) group of about 30 islands (1965 est. pop. 50,000), c.2,100 sq mi (5,440 sq km), E Indonesia, in the Banda Sea, between the Aru Islands and Timor, in the Moluccas. Important products are copra, tortoise shell, and trepang.

Ta·nis (tā'nĭs), ancient city of Egypt, in the E delta of the Nile. It was a significant city in the XIX and XXI dynasties. Threatened inundation by Lake Manzala caused it to be abandoned after the 6th cent. A.D.

Tan·jore (tăn-jôr', -jōr'): *see* Thanjavur, India.

Tan·ta (tän'tä), city (1970 est. pop. 254,000), capital of Gharbiyah governorate, N Egypt, in the Nile River delta. It is a cotton-ginning center and the main railroad hub of the delta.

Tan-tung (tän'to͝ong') or **An·tung** (än'to͝ong'), city (1970 est. pop. 450,000), SE Liaoning prov., China, at the mouth of the Yalu River, opposite Korea. A port city, its manufactures include paper, oakleaf silk, medicines, machinery, precision instruments, cement, and processed fish and other marine products.

Tan·za·ni·a, United Republic of (tăn'zə-nē'ə, tăn-zā'nē-ə), republic (1977 est. pop. 15,755,000), 364,898 sq mi (945,087 km), E Africa, formed in 1964 by the union of the republics of Tanganyika and Zanzibar. Mainland Tanzania is bordered on the south by Mozambique, Malawi, and Zambia; on the west by Zaire, Burundi, and Rwanda; on the north by Uganda and Kenya; and on the east by the Indian Ocean. Lake Nyasa forms part of the southern boundary,

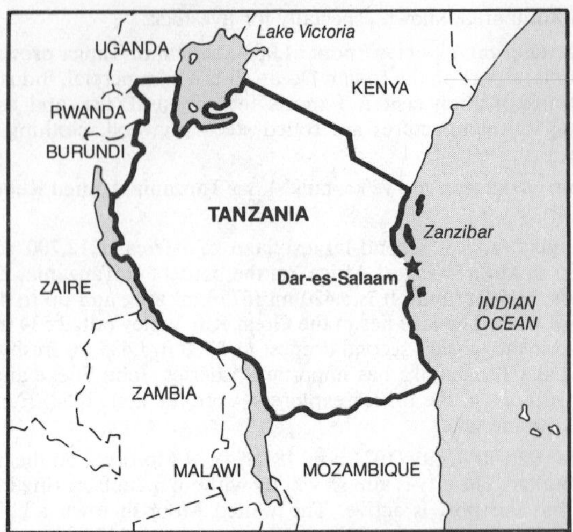

Lake Tanganyika part of the western boundary, and Victoria Nyanza part of the northern boundary. Dar-es-Salaam is the capital.

Tanzania falls into three major geographic zones—a narrow lowland coastal strip along the Indian Ocean, a vast interior plateau, and a number of scattered mountainous regions. The coastal zone (10-40 mi/16-60 km wide) receives considerable rainfall and has much fertile soil. The plateau (average elevation 3,500-4,500 ft/1,070-1,370 m) extends over most of the interior and is cut in two places by branches of the Great Rift Valley. The western branch contains Lake Tanganyika, and the eastern branch runs through central Tanzania about 500 ft (150 m) below the level of the plateau. The mountainous regions include Mt. Meru (14,979 ft/4,566 m) and Mt. Kilimanjaro (19,340 ft/5,895 m, the highest point in Africa) in the northeast; the Usambara, Nguru, and Uluguru mts. in the east; the Livingstone Mts. and the Kipengere Range near Lake Nyasa in the south; and the Ufipi Highlands in the southwest.

Economy. Tanzania is overwhelmingly agricultural, with most workers engaged in subsistence farming. The principal crops are cassava, millet, sorghum, wheat, yams, rice, maize, plantains, and pulses. The chief cash crops are sisal, cotton, coffee, cashew nuts, sugar cane, tea, tobacco, and pyrethrum; these commodities account for the bulk of the exports originating on the mainland. In addition, cattle, sheep, and goats are raised. Manufactures are largely limited to processed agricultural goods, beverages, and basic consumer items. Refined petroleum, aluminum goods, and construction materials are also produced, and motor vehicles are assembled. Minerals extracted in significant quantities include diamonds, gold, salt, tin, and mica. Immense deposits of coal and iron ore have been found near Lake Nyasa, but because of their inaccessibility they have not been exploited as yet on a large scale. Tanzania's road and rail networks serve mainly the coastal, central, and north-central parts of the country.

History. In 1959 Dr. L. S. B. Leakey, a British anthropologist, discovered at Olduvai Gorge in northeastern Tanzania the fossilized remains of *Homo habilis,* who lived about 1.75 million years ago and was a direct ancestor of modern man (*Homo sapiens*). Tanzania was later the site of Paleolithic cultures. By the beginning of the first millennium A.D. scattered parts of the country were thinly populated. By about A.D. 900 traders from southwestern Asia and India had settled on the coast. They exported small numbers of black Africans as slaves. By this time there were also commercial contacts with China and India. By about 1200 Kilwa Kisiwani (situated on an island) was a major trade center, handling gold and other goods from the near interior of Tanzania. By about 1000 the migration of Bantu-speaking black Africans into the interior of Tanzania from the west and the south was well under way. The Bantu were organized in relatively small political units.

In 1498 Vasco da Gama became the first European to visit the Tanzanian coast. By 1506 Portugal controlled most of the coast of eastern Africa. The Portuguese impact was mostly negative, but Kilwa's trade seems to have grown as a result of the contact. In 1587 the Zimba, a warlike black African group from southeastern Africa, sacked Kilwa and killed about 40% of its inhabitants. In 1698 the Portuguese were expelled from the East African coast with the help of Arabs from Oman. In the early 18th cent. the commerce of East Africa increased after the Omani Bu Said dynasty replaced the Yarubi rulers in 1741. Oman's commercial activity was centered on Zanzibar, from which it controlled the overseas trade. By the early 19th cent. numerous towns on the Tanzanian coast had been founded or revived. Sayyid Said, the great Bu Saidi ruler, took a great interest in East Africa. He brought with him many Arabs, who settled in the mainland towns as well as on Zanzibar. As a result the Swahili language (a blend of Bantu and Arabic) and culture gained new adherents. About the same time new caravan routes into the far interior were opened up. The caravans following the southern route obtained mainly slaves and ivory; along the more northerly routes ivory was the chief commodity purchased. In the middle third of the 19th cent. several European missionaries and explorers visited various parts of Tanzania. From the 1860s to the early 1880s Mirambo, a Nyamwezi, headed a large state that controlled much of the caravan trade of central and northern Tanzania. About the same time Tippu Tib, a Zanzibari, organized large caravans that passed through Tanzania to present-day Zambia and Zaire, where ivory and slaves were obtained.

In 1886 Germany established a vague sphere of influence over mainland Tanzania, except for a narrow strip of land along the coast that remained under the suzerainty of the sultan of Zanzibar, who leased it to the Germans. The German East Africa Company (founded 1887) governed the territory, called German East Africa. The company's aggressive conduct resulted in a major but unsuccessful resistance movement along the coast by black Africans, Arabs, and Swahilis. An Anglo-German agreement (1890) added Rwanda, Burundi, and other regions to German East Africa. The German government (1891) took over the country (which by then included the coast) and declared it a protectorate. However, it was not until 1898, with the death of the Hehe ruler, Mkwawa, who strongly opposed European rule, that the Germans succeeded in controlling the country. During the period 1905 to 1907 the trans-tribal, quasi-religious Maji Maji revolt against German rule engulfed most of southeastern Tanzania. Under the Germans several new crops (including sisal, cotton, and plantation-grown rubber) were introduced. In addition, many new Christian missions were established.

During World War I British and South African troops occupied (1916) most of German East Africa. In the postwar period the League of Nations made Tanganyika a British mandate. The British, especially during the administration (1925-31) of Gov. Sir Donald Cameron, attempted to rule "indirectly" through existing African leaders. Tanganyika, however, had few indigenous large-scale political units. Therefore African leaders had to be established in newly defined constituencies. As a result the British policy considerably altered the patterns of African life in Tanganyika. The British developed the territory's economy largely along the lines established by the Germans. Increasing numbers of Africans worked for a wage on plantations. Also after 1945 black Africans gradually gained more seats on the territory's legislative council.

In 1954 Julius Nyerere and Oscar Kambona transformed the Tanganyika African Association (founded in 1929) into the more politically oriented Tanganyika African National Union (TANU). TANU easily won the general elections of 1958-60. When Tanganyika became independent on Dec. 9, 1961, Nyerere became its first prime minister. In Dec., 1962, Tanganyika became a republic within the British Commonwealth of Nations. On Apr. 27, 1964, after a leftist revolution in Zanzibar, Tanganyika and Zanzibar merged to form one republic, which was named Tanzania in Oct., 1964. Nyerere became Tanzania's first president. Nyerere was overwhelmingly re-elected president in 1965 and 1970. Although formally united with the mainland, Zanzibar followed an independent course in many respects. In Feb., 1967, Nyerere issued the Arusha Declaration, a major policy statement that called for "egalitarianism" and self-reliance. The government was to be decentralized, and government officials were not to receive unnecessary privileges or large salaries. Major economic institutions such as banks and large industries were to be nationalized; the development of the rural areas was to be emphasized over urban development; and the country was to develop principally by using its own resources and not to become dependent on loans and other aid from foreign countries or industrial firms.

In the early 1970s there was tension between Tanzania and Uganda (headed by Idi Amin) that occasionally led to border clashes. Relations with Kenya also deteriorated in 1977 to the point where the border between the two countries was closed. Tanzania has supported various movements against white minority rule in S Africa; in 1976 Nyerere participated in discussions with U.S. Secretary of State Henry A. Kissinger aimed at finding a solution to the crisis in Rhodesia. In 1979, in response to a Ugandan border incursion, Tanzania invaded Uganda. Accompanied by dissident Ugandan rebels, Tanzania overthrew the despotic regime of Idi Amin Dada in a rare example of one black African nation interfering in the internal affairs of another. Perhaps out of concern with the sensitivity of this issue, Nyerere did not restore Obote to power in Uganda.

Government. Tanzania's head of state and chief executive is the president, who is nominated to a five-year term by a conference of delegates. The president is assisted by a cabinet, which includes the country's first vice president (a Zanzibari) and a prime minister (who is also second vice president). There is a national assembly made up of 193 members. All bills passed by the assembly must be approved by the president before they become law.

Ta·or·mi·na (tä'ôr-mē'nä), town (1972 est. pop. 6,600), E Sicily, Italy, overlooking the Ionian Sea and at the foot of Mt. Etna. It is a world-famous winter resort. Known in the 8th cent. B.C. and refounded by Carthaginians in the 4th cent. B.C., Taormina later flourished under the Greeks and the Romans. It was taken by the Arabs (early 10th cent.), fell to the Normans (late 11th cent.), and declined after the 15th cent.

Taos (tous), county (1970 pop. 17,516), 2,256 sq mi (5,843 sq km), N N.Mex., in a mountainous area bordered on the N by Colo. and drained by the Rio Grande; formed 1852; co. seat Taos. Grain farming and livestock raising are the major occupations. There are Pueblo Indian settlements in the south. Part of Carson National Forest is in the east.

Taos, town (1970 pop. 2,475), alt. c.7,000 ft (2,135 m), seat of Taos co., N N.Mex., between the Rio Grande and the Sangre de Cristo Mts.; founded c.1615, inc. 1934. Taos developed as an art colony after 1898 and attracted many painters and writers, notably John Marin and D. H. Lawrence. Taos was the center of the Pueblo revolt (1680) and of an anti-American revolt (1847).

T'ao-yuan or **Tao·yuan** (both: tou'yōō-än'), city (1970 est. pop. 87,500), N Taiwan. Situated in a rich agricultural area, it is a market center for local produce.

Ta·pa·chu·la (tä'pə-chōō'lə), city (1970 pop. 60,620) Chiapas state, SE Mexico, at the foot of the Chiapas highlands near the Guatemala border. It is the commercial center of a coffee-growing region and an important transportation link between Mexico and Central America. Ceramics and leather goods are produced by local artisans.

Ta·pa·jós (tä'pä-zhôs'), river, c.600 mi (965 km) long, formed at the border of Mato Grosso, Pará, and Amazonas states, central Brazil, by the confluence of the Juruena and Teles Pirez rivers. It flows northeast across western Pará into the Amazon River at Santarém.

Tap·pa·han·nock (tăp'ə-hăn'ək), town (1970 pop. 1,111), seat of Essex co., E Va., on the Rappahannock River NE of Richmond; founded 1680, inc. 1914.

Tap·pan (tăp'ən), uninc. residential village (1970 pop. 7,424), Rockland co., SE N.Y., near the N.J. border SSW of Nyack. The De Wint Mansion, George Washington's headquarters in 1780 and 1783, is here. Maj. John André was tried and hanged at Tappan. The Tappan Zee, a widening of the Hudson River, is nearby.

Tar·a (tăr'ə), village, Co. Meath, E Republic of Ireland. The Hill of Tara (507 ft/154.6 m high) was the seat of the kings of Ireland from ancient times until the 6th cent. The hill was the scene of the defeat of the Danes in 980 and the Irish insurgents in 1798.

Ta·ran·to (tä'rən-tō'), city (1972 est. pop. 229,195), capital of Taranto prov., Apulia, S Italy, on the Gulf of Taranto, an arm of the Ionian Sea. Taranto is an important military port and an agricultural, industrial, and fishing center. Manufactures include chemicals, machinery, and ships. Founded by colonists from Sparta in the 8th cent. B.C., Taranto was destroyed (927) by the Arabs but was later rebuilt by the Byzantines. Its harbor, protected by the Italian fleet, was bombed several times in World War II.

Tarbes (tärb), city (1968 pop. 55,375), capital of Hautes-Pyrénées dept., SW France, on the Adour River. It is an industrial, commercial, and tourist center in a cattle- and horse-raising area. In addition to the traditional forging and leather industries, there are machinery and electrical-equipment manufactures. Invaded and destroyed many times in the course of its history, Tarbes was joined to the French crown in the 16th cent.

Tar·bor·o (tär'bûr'ō, -bə-rə), town (1970 pop. 9,425), seat of Edgecombe co., E N.C., E of Rocky Mount on the Tar River; laid out 1760. It is the market center of a tobacco, cotton, and corn area and manufactures textiles, wood products, and machinery.

Ta·ri·fa (tä-rē'fä), town (1970 pop. 15,833), Cádiz prov., S Spain, in Andalusia. A minor seaport on the Strait of Gibraltar, it is the southernmost city of the European mainland. It was founded by the Greeks and later became the first Roman colony in Spain.

Ta·ri·ja (tä-rē'hä), city (1969 est. pop. 23,000), alt. 6,421 ft (1,958.4 m), capital of Tarija dept., S Bolivia. Tarija lies in a fertile valley in the eastern watershed of the Andes near the oil fields of the Chaco. It is famous for vineyards and orchards that grow olives, pears, peaches, and apples.

Ta·rim (dä'rēm', tä'-), chief river of Sinkiang Uigur Autonomous Region, NW China, c.1,300 mi (2,090 km) long, flowing generally E, along the N edge of the Takla Makan desert, to Lo-pu po (Lop Nor), a salt lake. The river gives its name to the arid Tarim basin, a great depression surrounded by the lofty Tien Shan, Kunlun, and Pamir mts. The Silk Road between China and Europe passed through the basin.

Tar·kwa (tär-kwä'), town (1970 pop. 11,001), S Ghana. It is an administrative and commercial center. One of the world's largest manganese mines is located nearby.

Tarn (tärn), department (1974 est. pop. 334,000), 2,231 sq mi (5,778.3 sq km), S France, in Languedoc; capital Albi.

Tarn, river, c.235 mi (380 km) long, rising in the Cévennes mts., S France, and flowing SW before emptying into the Garonne River. Deep gorges and canyons formed by the river are tourist attractions.

Tarn-et-Ga·ronne (tärn'ā-gä-rôn'), department (1974 est. pop. 182,000), 1,440 sq mi (3,729.6 sq km), SW France; capital Montauban.

Tar·nów (tär'nōōf'), city (1973 est. pop. 92,100), SE Poland. It is a railway junction and a leading center for the manufacture of basic chemicals and fertilizers. Settled by the 12th cent., Tarnów passed to Austria in 1772 and reverted to Poland after World War I.

Tar·now·skie Gó·ry (tär-nôf'skē-ə gŏŏr'ē), town (1973 est. pop. 42,700), S Poland. It is an industrial center where machinery, mining and railway equipment, and chemicals are produced. Nearby are coal, zinc, and lead mines. Chartered in 1526, the town passed from Germany to Poland after World War I.

Tar·pon Springs (tär'pən), resort city (1970 pop. 7,118), Pinellas co., W Fla., on the Gulf coast NW of Tampa near the mouth of the Anclote River; founded 1882. It is a sponge center. Greek fishermen take out the sponging fleet, and colorful Greek religious festivals draw many visitors. Apparel and chemicals are produced.

Tar·quin·i·i (tär-kwĭn'ē-ī'), ancient city of Etruria, central Italy, NW of Rome. The head of the Etruscan League, it was defeated in wars with Rome in the 4th cent. B.C. In the 3rd cent. B.C. it lost its independence. Much knowledge of Etruscan life has been gained from paintings on the walls of tombs in the necropolis of Tarquinii.

Tar·ra·go·na (tăr'ə-gō'nə), city (1972 est. pop. 83,827), capital of Tarragona prov., NE Spain, in Catalonia, on the Mediterranean Sea at the mouth of the Francolí River. A port and commercial center, it has a large wine export. An Iberian town, ancient Tarraco was captured (218 B.C.) by the Romans in the Second Punic War and became a flourishing commercial center. It fell to the Visigoths (5th cent.) and to the Moors (8th cent.) and was recovered in the early 12th cent. for Christian Spain.

Tar·rant (tăr'ənt), county (1970 pop. 716,317), 861 sq mi (2,230 sq km), N Texas, drained by the West and Clear forks of the Trinity River; formed 1849; co. seat Fort Worth. Its economy is based on diversified industry (including oil refining, meat packing, grain milling, and the manufacture of automobiles and aerospace equipment), dairying, general farming, and livestock raising.

Tar·ra·sa (tä-rä'sä), town (1970 pop. 138,697), Barcelona prov., NE Spain, in Catalonia. It is an industrial town, long famous for its woolen textiles. The town dates back to Roman days.

Tar·ry·town (tăr'ē-toun'), village (1970 pop. 11,115), Westchester co., SE N.Y., a residential suburb of New York City, on the E bank of the Hudson River opposite Nyack; settled in the 17th cent. by Dutch, inc. 1870. Precision instruments are manufactured. Of interest are Sunnyside, the home of Washington Irving; Philipsburg

Manor, an estate including a Dutch farmhouse (c.1683) and a restored operating gristmill; and Lyndhurst (1838), a Gothic Revival mansion overlooking the Hudson.

Tar·sus (tär'səs), city (1970 pop. 74,510), S Turkey, in Cilicia, on the Tarsus River near the Mediterranean Sea. It is an agricultural trade center; copper, zinc, chromium, and coal are mined in the region. Ancient Tarsus was one of the most important cities of Asia Minor. It reached the height of its prosperity and cultural achievement under Roman rule. The apostle Paul was born here.

Tar·tu (tär'tōō), city (1973 est. pop. 95,000), W European USSR, in Estonia, a port on the Ema River. It is an important industrial and cultural center and a rail junction. Food processing, metalworking, printing and publishing, lumbering, and the production of textiles, leather footwear, and agricultural machinery are the leading industries. The city was founded in 1030 and later developed as a trade center of the Hanseatic League. Captured by Peter I in 1704, during the Northern War, it was ceded to Russia by the Treaty of Nystad in 1721. Tartu is built around a hill topped by an old fortified castle and a restored 13th cent. cathedral.

Tash·kent (tăsh-kĕnt'), city (1974 est. pop. 1,552,000), capital of Tashkent oblast and of the Uzbek SSR, Central Asian USSR, in the foothills of the Tien Shan Mts. The largest and one of the oldest cities of Central Asia, it is the economic heart of the region. It is also a major cultural center, a rail and highway junction, and an important air terminal. The city lies in a great oasis along the Chirchik River and on the Trans-Caspian RR. There is extensive trade in grain and raw cotton. Tashkent has one of the largest cotton textile mills in Asia. Other industries include railroad workshops, food- and tobacco-processing plants, and factories that manufacture machinery, electrical equipment, chemicals, pharmaceuticals, paper, furniture, pottery, hosiery, and perfume. The modern Russian section of the city coexists with an old Oriental quarter with narrow, twisting streets, numerous mosques, and a bazaar. First mentioned in the 1st cent. B.C., Tashkent came under Arabic rule in the 7th cent. A.D. and passed to the Turks in the 12th cent. The city was captured in the 13th cent. by Genghis Khan and in the 14th cent. by Tamerlane. It was taken by Russian forces in 1865.

Tas·ma·ni·a (tăz-mā'nē-ə), island state (1970 est. pop. 410,800), 26,383 sq mi (68,332 sq km), SE Commonwealth of Australia. It is separated from Australia by the Bass Strait and lies 150 mi (240 km) south of the state of Victoria. Tasmania includes many offshore islands. The Indian Ocean is to the west and the Pacific Ocean to the east. Hobart is the state capital. Tasmania is geologically similar to the Australian continent. The island is mountainous and partly forested. Great Lake in the interior is the largest lake and the reservoir of an important hydroelectric plant. The state's major manufactures are metals and metal products and textiles. The raising of sheep for wool in the east and dairy farming in the northwest are also important. The mining of copper, zinc, tin, lead, and iron has increased in recent years. Tasmania was federated as a state in the Commonwealth of Australia in 1901.

Ta·tar Autonomous Soviet Socialist Republic (tä'tər), autonomous republic (1970 pop. 3,131,000), 26,255 sq mi (68,000 sq km), E European USSR, in the middle Volga and lower Kama river valleys. Kazan is the capital. The republic is a leading Soviet oil and natural-gas producer. There are also important deposits of brown coal, limestone, gypsum, dolomite, and marl. Lumbering and food, leather, and fur processing are major Tatar industries. Manufactures include machinery, chemicals, and pharmaceuticals. The low, rolling plain that makes up most of the republic's territory yields fodder crops, wheat and other cereals, sugar beets, sunflowers, and flax. The region was conquered (13th cent.) by the Mongols of the Golden Horde. Russian colonization followed its capture (1552) by Czar Ivan IV. The Tatar ASSR was organized in 1920.

Tatar Strait, narrow body of water, c.350 mi (565 km) long and from 5 to 80 mi (8–129 km) wide, S Far Eastern USSR, between the island of Sakhalin and the Asian mainland. It connects the Sea of Japan, in the south, with the Sea of Okhotsk, in the north.

Tate (tāt), county (1970 pop. 18,544), 405 sq mi (1,049 sq km), NW Miss., bounded on the W and NW by the Coldwater River; formed 1873; co. seat Senatobia. Farming (cotton, corn, and hay), dairying, and lumbering are the major occupations.

Ta·te·ba·ya·shi (tä'tə-bə-yä'shē), city (1970 pop. 61,130), Gumma prefecture, central Honshu, Japan, on the Tone River. It is a manufacturing center with mechanical and textile industries.

Ta·tra (tä'trə) or **Ta·tras** (tä'trəz), highest group of the Carpathian mountain system, in E central Europe. The High Tatra extends c.40 mi (65 km) along the Polish-Czechoslovakian border; its highest peak, Gerlachovka (8,711 ft/2,656.9 m), is in northeastern Czechoslovakia. The Low Tatra lies entirely in Czechoslovakia; it rises to 6,702 ft (2,044 m) in the Ďumbier. The extensively glaciated mountains have numerous lakes, moraines, and hanging valleys that have made the region a year-round resort area.

Tatt·nall (tăt'nəl), county (1970 pop. 16,557), 490 sq mi (1,269.1 sq km), SE central Ga., bounded on the SW by the Altamaha River; formed 1801; co. seat Reidsville. It is in a coastal-plain agricultural (cotton, tobacco, pecans, peanuts, and livestock) and timber area. There is clothing manufacturing in the county.

Ta-t'ung or **Ta·tung** (both: dä'tōōng'), city (1970 est. pop. 300,000), N Shansi prov., China. It is an important industrial and railway center in a region of great coal deposits. Manufactures include locomotives, motor vehicles, textiles, and cement.

Taun·ton (tôn'tən, tŏn'-), municipal borough (1974 est. pop. 38,300), administrative center of Somerset, SW England, on the Trove River. Its manufactures include textiles, shirts, gloves, and precision instruments. Taunton is also a market and railroad junction. There are three well-known schools for boys: King's College (1293), Queen's College (1843), and Taunton School (1847).

Taunton, industrial city (1978 est. pop. 40,500), seat of Bristol co., SE Mass., on the Taunton River; settled 1638, inc. as a city 1864. Silverware, jewelry, and clothing are among its manufactures.

Tau·nus (tou'nəs), range of the Rhenish Slate Mts., W West Germany, extending NE from the Rhine River, N of Mainz. The Taunus is covered by forests. Its southern slopes, the Rheingau region, have famous vineyards. There are many well-known mineral spas in the Taunus.

Tau·po, Lake (tou'pō), largest lake of New Zealand, 234 sq mi (606 sq km), 552 ft (168.4 m) deep, on central North Island. Lake Taupo, located in the Hot Springs dist., is surrounded by volcanoes. It is known for rainbow trout.

Tau·rang·a (tou-räng'ə), city (1974 est. pop. 32,500), N central North Island, New Zealand, on the Bay of Plenty. It is the leading New Zealand port for overseas trade. Wood products are the largest exports and fertilizers the largest imports.

Tau·rus (tôr'əs), mountain chain, S Turkey, extending c.350 mi (565 km) roughly parallel to the Mediterranean coast of S Asia Minor. It forms the southern border of the Anatolian plateau. Its northeastern extension across the Seyhan River is called the Anti-Taurus. The highest peak of the Taurus is the Ala Dağ (12,251 ft/3,736.6 m), at its eastern end. The range has important chromium deposits and other minerals.

Ta·var·es (tə-vâr'ēz), resort town (1970 pop. 3,261), seat of Lake co., central Fla., between Lakes Eustis and Dora, in a citrus-fruit area.

Ta·was City (tô'wəs), city (1970 pop. 1,666), seat of Iosco co., NE Mich., NNE of Bay City; settled c.1853, inc. as a village 1885, as a city 1895. It is a resort and fishing center. A state park and ski area are nearby.

Tax·co (täs'kō), town (1970 pop. 27,089), Guerrero state, S Mexico. It was founded in 1529 as a silver-mining community. Clinging to the side of a mountain, Taxco has steep, cobbled streets, overhanging grilled balconies, red-tile roofs, and fine glazed tiles set in white or pastel adobe walls. Modern buildings are prohibited by the government.

Tay (tā), longest river of Scotland, 118 mi (190 km), rising on Ben Lui in the Grampians and flowing NE, then SE to enter the North Sea through the Firth of Tay (25 mi/40.2 km long). The river has important salmon fisheries.

Tay·lor (tā'lər). **1.** County (1970 pop. 13,641), 1,051 sq mi (2,722 sq km), N Fla., on the Gulf of Mexico, in a flatwoods area with large swamps and many small lakes; formed 1856; co. seat Perry. It has forestry, agriculture (corn, peanuts, and livestock), limestone deposits, and fishing. Clothing and paper and metal products are made. **2.** County (1970 pop. 7,865), 403 sq mi (1,043.8 sq km), W central Ga., bounded on the N and E by the Flint River and intersected by the Fall Line; formed 1852; co. seat Butler. It is in an agricultural (cotton, corn, peanuts, pecans, peaches, and livestock) and lumbering area. Its manufactures include clothing and wood products. **3.** County (1970 pop. 8,790), 528 sq mi (1,367.5 sq km), SW Iowa, on the Mo. border and drained by the One Hundred and Two, Little Platte, and East Nodaway rivers; formed 1847; co. seat Bedford. It is a prairie agricultural area where corn, hogs, cattle, poultry, and hay are raised. Coal deposits are mined in the west. **4.** County (1970 pop. 17,138), 277 sq mi (717.4 sq km), central Ky., drained by the Green River and several creeks; formed 1848; co. seat Campbellsville. In a rolling agricultural area yielding burley tobacco, corn, oats, hay, wheat, and livestock, it has timber, limestone quarries, and industry (food processing, textile milling, and the manufacture of furniture and metal products). **5.** County (1970 pop. 97,853), 912 sq mi (2,362.1 sq km), W central Texas, drained to the N by tributaries of Brazos River, to the S by Colorado River tributaries; formed 1858; co. seat Abilene. It has diversified farming (cotton, grain sorghums, wheat, peanuts, fruit, and truck crops), livestock raising, and dairying. There are deposits of oil and natural gas here. **6.** County (1970 pop. 13,878), 174 sq mi (450.7 sq km), N W.Va., on the Allegheny Plateau, drained by the Tygart River; formed 1844; co. seat Grafton. Its agriculture includes livestock, fruit, and tobacco. It has some coal mines and natural-gas fields, and there is timber in the region. **7.** County (1970 pop. 16,958), 975 sq mi (2,525.3 sq km), N central Wis., drained by the Black, Yellow, Jump, and Rib rivers; formed 1875; co. seat Medford. It is in a dairying, farming (potatoes and truck crops), and stock-raising area, with lumbering, cheese processing, and manufacturing (plastics, shoes, glass, and engines). It contains a section of the Chequamegon National Forest and several small resort lakes.

Taylor. 1. City (1978 est. pop. 78,500), Wayne co., SE Mich., a suburb of Detroit adjacent to Dearborn; founded 1847 as a township, inc. as a city 1968. Its manufactures include tools and machines. **2.** Village (1970 pop. 240), seat of Loup co., central Nebr., NW of Ord on the North Loup River, in an area yielding dairy products, poultry, livestock, and grain. Cheesebrough Canyon is nearby. **3.** Borough (1970 pop. 6,977), Lackawanna co., NE Pa., on the Lackawanna River SW of Scranton; settled c.1800. Anthracite coal is mined. **4.** City (1970 pop. 9,616), Williamson co., central Texas, NE of Austin; platted 1876, inc. 1877. The blackland prairie here was long famous for its vast cotton yield. The prosperity of the city has increased by oil production, dairying, poultry raising, and manufacturing and food processing.

Tay·lors·ville (tā′lərz-vĭl′). **1.** City (1970 pop. 897), seat of Spencer co., NW central Ky., on the Salt River SE of Louisville, in a bluegrass agricultural area. Corn, burley tobacco, and livestock are raised. Dairy and animal-fat products are manufactured. The city also has feed and flour mills. **2.** Town (1970 pop. 1,231), seat of Alexander co., W central N.C., WSW of Winston-Salem, in a farm area; inc. 1887. Wood and textile products are manufactured.

Tay·lor·ville (tā′lər-vĭl′), city (1975 est. pop. 10,700), seat of Christian co., central Ill., in a farm and coal area; inc. 1882. Soybeans are processed, coal is mined, and farm and coal equipment and paper, plastic, and metal products are made.

Tay·my·ra (tī-mîr′ə), river, c.400 mi (645 km) long, rising in the center of the Taymyr Peninsula, N Siberian USSR, and flowing NE through Lake Taymyr (c.2,700 sq mi/6,995 sq km) into the Taymyr Gulf of the Kara Sea.

Tay·myr Peninsula or **Tai·myr Peninsula** (both: tī-mîr′), northernmost projection of Siberia, N central Siberian USSR, between the estuaries of the Yenisei and Khatanga rivers and extending into the Arctic Ocean. Cape Chelyuskin at the tip of the peninsula is the northernmost point of the Asian mainland. The peninsula is covered mostly with tundra.

Taze·well (tāz′wĕl′, -wəl). **1.** County (1970 pop. 118,649), 652 sq mi (1,688.7 sq km), central Ill., bounded on the NW by the Illinois River and drained by the Mackinaw River; formed 1827; co. seat Pekin. It has agriculture (corn, oats, soybeans, wheat, livestock, poultry, and hay), bituminous-coal mines, sand and gravel deposits, timber, and diversified manufacturing (farm machinery, clothing, washing machines, and clay, metal, wood, and leather products). **2.** County (1970 pop. 39,816), 522 sq mi (1,352 sq km), SW Va., in the Alleghenies, bounded in the N by W.Va., drained by the Clinch River and a headstream of the Holston River; formed 1799; co. seat Tazewell. It is in a livestock (cattle and sheep), dairying, and farming (fruit, clover, and corn) area. Its mining includes coal, limestone, shale, clay, and natural gas, and it has diversified industry (food processing, lumbering, and clothing, furniture, and machinery manufacturing). It includes part of the Jefferson National Forest.

Tazewell. 1. Town (1970 pop. 1,860), seat of Claiborne co., NE Tenn., NNE of Knoxville; inc. 1830. **2.** Town (1970 pop. 4,168), seat of Tazewell co., extreme SW Va., SW of Bluefield, W.Va., in a mountainous coal and farm region; settled 1769, inc. 1866.

Ta·zoult (tä-zo͞olt′), formerly **Lam·bèse** (läm-bĕz′), town (1966 pop. 6,000), NE Algeria. It is noted for the ruins of a Roman town (Lambaesis) founded in the 2nd cent. under Emperor Hadrian as the encampment of the third Augustan Legion and destroyed in the 5th cent.

Tbi·li·si (tə-bē-lē′sē) or **Tif·lis** (tĭf′lĭs), city (1974 est. pop. 984,000), capital of the Georgian SSR, SE European USSR, on the Kura River and the Transcaucasian RR. Located in a mountain-ringed basin, Tbilisi is the economic, administrative, and cultural heartland of Transcaucasia. It is also a major transportation center. Industries include filmmaking, printing and publishing, machine building, food processing, tanning, silk weaving, and the production of machine tools, electrical equipment, locomotives, and plastics. Orchards and vineyards surround the city. The region's mineral springs provide the basis for numerous health resorts. Tbilisi is one of the USSR's oldest cities, settled as early as the 4th cent. B.C. Until the 6th cent. it was under Persian, Georgian, and Iberian rule. The city lay along the natural trade route between the Caspian and Black seas but was also astride one of the world's great crossroads of invasion and migration. Tbilisi was a stronghold of Moslem power and a commercial center from the 8th to 11th cent.; during this period Arabs, Khazars, Seljuks, and Ottoman Turks successively ruled the city. From 1096 to 1225 it flourished as the capital of an independent Georgian state. It was ruled from the 13th to 18th cent. by Mongols, Iranians, and Turks before coming under Russian control in 1800–1801. It developed as a revolutionary center from the second half of the 19th cent. and played a leading role in the Revolution of 1905. Tbilisi rises in terraces from both banks of the Kura and has an old medieval section and a modernized section.

Tczew (chĕf), town (1973 est. pop. 44,600), N Poland, a port on the Vistula River. It is a rail junction, with railroad workshops and industries producing riverboats, farm machinery, construction materials, and enamelware. Chartered in 1260, Tczew became part of Poland in the late 13th cent. It passed to Russia in 1772 and was not reincorporated into Poland until 1919.

Tea·neck (tē′nĕk), residential suburban township (1978 est. pop. 40,600), Bergen co., NE N.J., near the Hackensack River; settled in the early 1600s, inc. 1895. Porcelain, bulbs and fuses, food seasonings, and textile looms are among its manufactures.

Te·bes·sa (tə-bĕs′ə), town (1974 est. pop. 58,008), NE Algeria, in the Atlas Mts. The town is an important agricultural market and is noted for its silk embroidery and carpets. The surrounding area is an important phosphate- and iron-mining region.

Te·cum·seh (tə-kŭm′sə, -sē), city (1970 pop. 2,058), seat of Johnson co., SE Nebr., SE of Lincoln on the North Fork of the Nemaha River; founded 1859. Grain, livestock, poultry produce, and fruit are processed.

Tees (tēz), river, c.70 mi (110 km) long, rising on Cross Fell in the Pennines, N England, and flowing generally E between Durham and North Yorkshire and through Cleveland to the North Sea. Its upper valley is picturesque; the lower reaches pass through Teeside, a major industrial region.

Tee·side (tē′sīd′), county borough (1974 est. pop. 393,960), Cleveland, NE England at the mouth of the Tees River. Boulby Cliff, the highest cliff on the English coastline, is within its borders. Iron ore and coal are mined in the surrounding area. There are iron and steel works, shipbuilding and repair yards, and chemical plants that are among the largest in England and produce items ranging from explosives to synthetic fabrics. It is also a market for cattle and agricultural produce.

Te·gu·ci·gal·pa (tā-go͞o′sē-gäl′pä), city (1970 est. pop. 232,300), capital and largest city of Honduras, in a small valley in the mountains of S central Honduras. Textiles, sugar, and cigarettes are produced in the city. Old Tegucigalpa, built on a steep hill, has narrow streets and sidewalks, overhanging balconies, and stair-stepped streets. Across the Choluteca River lies Comayagüela, the more modern section. Founded late in the 16th cent., Tegucigalpa was a colonial center of silver and gold mining.

Te·ha·ma (tə-hā'mə), county (1970 pop. 29,517), 2,984 sq mi (7,729 sq km), N Calif., in the N part of the Central Valley (here drained by the Sacramento River), with extensions of the Klamath Mts. and the Coast Range in the W and the Sierra Nevada in the E; formed 1856; co. seat Red Bluff. It is a leading producer of sheep and wool, with poultry and livestock raising, dairying, and fruit and alfalfa farming. It has lumbering, some mining, and food processing.

Teh·ran (tĕ-răn', -rän, tā-) or **Te·he·ran** (tĕ'ə-răn', -rän', tā'-), city (1966 pop. 2,719,730), capital of Iran and Tehran prov., N Iran, near Mt. Damavand. It is Iran's largest city and its administrative, commercial, and industrial center. Manufactures include textiles, sugar, and cement; motor vehicles are assembled. The city has a large bazaar and is a leading center for the sale and export of carpets. It became the capital of Persia in 1788. Under Reza Shah Pahlevi (reigned 1925–41) the city was much modernized. Tehran's importance and population grew greatly in the 20th cent., and today it is one of the major cities of the Middle East.

Te·huan·te·pec (tə-wän'tə-pĕk'), town (1970 pop. 67,520), Oaxaca state, S Mexico, on a wide bend of the Tehuantepec River not far from the Gulf of Tehuantepec, an arm of the Pacific. The town is on the Isthmus of Tehuantepec, a rolling, tropical lowland.

Teign·mouth (tĭn'məth), urban district (1971 pop. 12,554), Devonshire, SW England, at the mouth of the Teign River on the English Channel. Teignmouth is a seaport and resort. The harbor, important in the Middle Ages, is now used chiefly by yachtsmen and fishermen.

Te·ka·mah (tə-kā'mə), city (1970 pop. 1,848), seat of Burt co., E Nebr., NNW of Omaha; settled 1854, inc. 1855. It is a farm trade center. Reservoir Hill nearby gives a broad view of the Missouri valley.

Te·kir·dağ (tĕ-kîr'dä, -däg), city (1970 pop. 35,387), capital of Tekirdağ prov., NW Turkey, on the Sea of Marmara. It is a small port and an agricultural trade center.

Tel A·viv-Jaf·fa (tĕl' ə-vēv'jä'fə, -yä'fä), city (1972 pop. 362,200), W central Israel, on the Mediterranean Sea. It is Israel's largest city and its commercial, financial, communications, and cultural center. Construction is the main industry; textiles, clothing, and processed food are the chief manufactures, and pharmaceuticals, electrical appliances, printed materials, and chemicals are also produced. The city is a tourist resort with wide, attractive beaches. Tel Aviv was founded in 1909 by Jews from Jaffa who wished to build a modern suburb. The population grew dramatically in the late 1920s, again after Hitler came to power (1933) in Germany, and after World War II. When the state of Israel was proclaimed on May 14, 1948, Tel Aviv was briefly the capital; in 1949 the government was transferred to Jerusalem. In 1950 Tel Aviv and Jaffa were merged.

Tel el A·mar·na or **Tell el A·mar·na** (both: tĕl' ĕl ə-mär'nə), ancient locality, Egypt, near the Nile N of Asyut. Ikhnaton's capital, Akhetaton, was in Tel El Amarna. About 400 tablets with inscriptions in Akkadian cuneiform were found here in 1887. Constituting correspondence between Amenhotep III and Ikhnaton and the governors of the cities in Palestine and Syria, they shed much light on ancient Egypt and the Middle East.

Tel·fair (tĕl'fâr'), county (1970 pop. 11,394), 440 sq mi (1,140.6 sq km), S central Ga., bounded on the S by the Ocmulgee River; formed 1807; co. seat McRae. It has coastal-plain agriculture (cotton, corn, melons, pecans, fruit, and livestock) and forestry (lumber and naval stores).

Tel·ford (tĕl'fərd), new town (1971 pop. 75,579), Salop, W England. Its industries include engineering, quarrying, brickmaking, coal mining, iron founding, and brewing.

Tell City (tĕl), city (1970 pop. 7,933), Perry co., S Ind., on the Ohio River NE of Owensboro, Ky.; settled 1857 by Swiss. Boats, electronic tubes, and furniture are made.

Tel·ler (tĕl'ər), county (1970 pop. 3,316), 553 sq mi (1,432.3 sq km), central Colo.; formed 1899; co. seat Cripple Creek. It is in a mining (gold and silver), ranching, and livestock-grazing area, and includes part of Front Range and Pike National Forest. Cripple Creek district, a famous gold-producing region, is in the south.

Tel·lu·ride (tĕl'yə-rīd), city (1970 pop. 553), seat of San Miguel co., SW Colo., in the San Juan Mts. SSE of Grand Junction, at an altitude of 8,500 ft (2,592.5 m); inc. 1887. It is a trade and tourist center in a livestock, dairying, and mining region. Gold, silver, and lead are mined.

Te·ma (tā'mə), city (1970 pop. 58,815), SE Ghana, on the Gulf of Guinea. With the opening of an artificial harbor in 1961, Tema developed from a small fishing village to become Ghana's leading seaport and an industrial center. Manufactures include aluminum, steel, refined petroleum, soap, processed fish, chocolate, textiles, cement, and chemicals.

Te·mir·tau or **Te·mir-Tau** (both: tā'mîr-tou'), city (1973 est. pop. 187,000), Central Asian USSR, in Kazakhstan, on the Nura River. It is a major industrial center, with large iron and steel plants and synthetic rubber and metal factories.

Tem·pe (tĕm'pē), city (1978 est. pop. 92,400), Maricopa co., S Ariz., in the Salt River valley; inc. 1894. It is a health resort, an agricultural center, and the seat of Arizona State Univ. Indian ruins are nearby.

Tempe, Vale of (vāl), valley, c.5 mi (8 km) long, E central Greece, NE Thessaly, between Mt. Olympus and Mt. Óssa. Traversed by the Piniós River, the valley is famous for its rugged grandeur. Strategically important as a route into central Greece, the valley was fortified by the Romans and the Byzantines.

Tem·pel·hof (tĕm'pəl-hôf'), district, E West Berlin. A workers' residential quarter and a film-production center, it became part of the U.S. occupation sector after 1945. The district includes Tempelhof Field, West Berlin's chief airport and the main terminal of the U.S. airlift during the Soviet blockade (June, 1948–May, 1949).

Tem·ple (tĕm'pəl), city (1978 est. pop. 41,000), Bell co., central Texas; inc. 1882. In a rich blackland region, Temple has grain and textile mills, railroad shops, and plants making a wide variety of products. U.S. Fort Hood is in the area.

Temple City, residential suburban city (1978 est. pop. 30,400), Los Angeles co., S Calif.; settled 1827, inc. 1960. It has light manufacturing and service businesses.

Te·mu·co (tā-mōō'kō), city (1970 pop. 104,400), capital of Cautin prov., S central Chile, on the Cautin River. It is a commercial city dealing in livestock and in the agricultural produce of the region. Temuco, founded in 1881, was the point from which the colonization of southern Chile was begun, chiefly by German immigrants.

Ten·a·fly (tĕn'ə-flī), residential suburban borough (1970 pop. 14,827), Bergen co., NE N.J., on the Hudson River; settled 1640, inc. 1894.

Te·na·li (tā-nä'lē), city (1971 pop. 102,937), Andhra Pradesh state, SE India, on the Krishna River delta. It is a market for rice.

Ten·e·dos (tĕn'ə-dŏs), island: *see* Bozcaada.

Ten·er·iffe or **Ten·e·rife** (both: tĕn'ə-rĭf, tĕn-ə-rēf', tā'nā-rē'fā), island (1970 pop. 500,381), 795 sq mi (2,059 sq km), in the Atlantic off NW Africa, the largest of the Canary Islands, Spain. A scenic island, it is dominated by Mt. Teide, a snow-capped volcanic peak (12,198 ft/3,720.4 m).

Ten·nes·see (tĕn'ə-sē', tĕn'ə-sē'), state (1975 pop. 4,163,000), 42,244 sq mi (109,412 sq km), S central United States, admitted 1796 as the 16th state (slaveholding). Nashville is the capital and the second-largest city. The largest city is Memphis. The state is bounded on the north by Ky. and Va.; on the east by N.C.; on the south by Ga., Ala., and Miss.; and on the west by the Mississippi River, which separates it from Mo. and Ark.

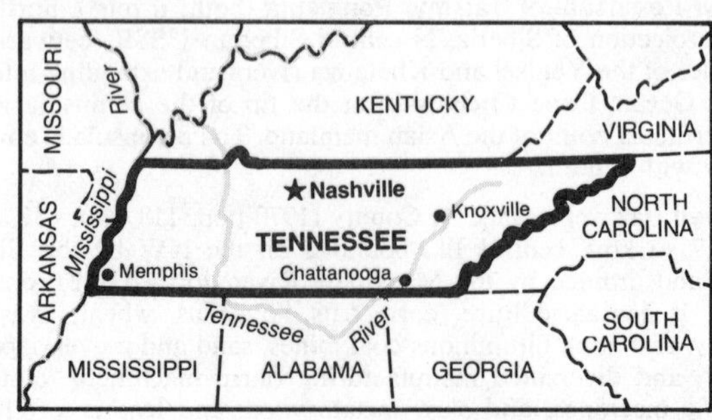

Although Tennessee is now primarily industrial, with 54% of its population in urban areas, many Tennesseans still derive their livelihood from the land of the state's three sharply defined regions: eastern Tennessee, middle Tennessee, and western Tennessee. In eastern Tennessee the Great Smoky Mts., Cumberland Plateau, and the narrow river valleys and heavily forested foothills generally restrict farming there to the subsistence level; but this region has two of the state's most industrialized cities, Chattanooga (fourth-largest) and Knoxville (third-largest). Middle Tennessee is hemmed in by the Tennessee River, which flows southwest through eastern Tennessee into northern Ala., looping back up into western Tennessee in its circuitous route to the Ohio. Gently rolling, fertile, bluegrass country, it is ideal for livestock raising and dairy farming. Western Tennessee, with its rich river bottomlands on which most of the state's cotton is grown, lies between the Tennessee and the Mississippi rivers. The average rainfall across the state ranges from 40 to 50 in. (101.6–127 cm), and the climate is generally mild.

Many beautiful lakes have been built by the Tennessee Valley Authority (TVA) and the army corps of engineers; 26 large ones are publicly owned. The TVA has also developed the Land Between the Lakes, an enormous Kentucky-Tennessee recreation area. Twenty-three state parks, covering some 132,000 acres (53,420 hectares) and parts of the Great Smoky Mountains National Park, Cherokee National Forest, and Cumberland Gap National Historical Park are in Tennessee. Sportsmen and visitors are attracted to Reelfoot Lake. Tennessee also has many sites of historic interest, including the Hermitage, home of Andrew Jackson; the Andrew Johnson National Historic Site; the Fort Donelson and Shiloh national military parks; and Stones River National Battlefield. Part of the Chickamauga and Chattanooga National Military Park is also in Tennessee.

Tennessee is divided into 95 counties:

NAME	COUNTY SEAT	NAME	COUNTY SEAT
Anderson	Clinton	Lauderdale	Ripley
Bedford	Shelbyville	Lawrence	Lawrenceburg
Benton	Camden	Lewis	Hohenwald
Bledsoe	Pikeville	Lincoln	Fayetteville
Blount	Maryville	Loudon	Loudon
Bradley	Cleveland	McMinn	Athens
Campbell	Jacksboro	McNairy	Selmer
Cannon	Woodbury	Macon	Lafayette
Carroll	Huntingdon	Madison	Jackson
Carter	Elizabethton	Marion	Jasper
Cheatham	Ashland City	Marshall	Lewisburg
Chester	Henderson	Maury	Columbia
Claiborne	Tazewell	Meigs	Decatur
Clay	Celina	Monroe	Madisonville
Cocke	Newport	Montgomery	Clarksville
Coffee	Manchester	Moore	Lynchburg
Crockett	Alamo	Morgan	Wartburg
Cumberland	Crossville	Obion	Union City
Davidson	Nashville	Overton	Livingston
Decatur	Decaturville	Perry	Linden
De Kalb	Smithville	Pickett	Byrdstown
Dickson	Charlotte	Polk	Benton
Dyer	Dyersburg	Putnam	Cookeville
Fayette	Somerville	Rhea	Dayton
Fentress	Jamestown	Roane	Kingston
Franklin	Winchester	Robertson	Springfield
Gibson	Trenton	Rutherford	Murfreesboro
Giles	Pulaski	Scott	Huntsville
Grainger	Rutledge	Sequatchie	Dunlap
Greene	Greeneville	Sevier	Sevierville
Grundy	Altamont	Shelby	Memphis
Hamblen	Morristown	Smith	Carthage
Hamilton	Chattanooga	Stewart	Dover
Hancock	Sneedville	Sullivan	Blountville
Hardeman	Bolivar	Sumner	Gallatin
Hardin	Savannah	Tipton	Covington
Hawkins	Rogersville	Trousdale	Hartsville
Haywood	Brownsville	Unicoi	Erwin
Henderson	Lexington	Union	Maynardville
Henry	Paris	Van Buren	Spencer
Hickman	Centerville	Warren	McMinnville
Houston	Erin	Washington	Jonesboro
Humphreys	Waverly	Wayne	Waynesboro
Jackson	Gainesboro	Weakley	Dresden
Jefferson	Dandridge	White	Sparta
Johnson	Mountain City	Williamson	Franklin
Knox	Knoxville	Wilson	Lebanon
Lake	Tiptonville		

Economy. The state's leading crops are tobacco, soybeans, hay, and cotton; cattle and dairy products are also principal farm commodities. Tennessee's leading mineral, in dollar value, is stone; zinc ranks second (Tennessee leads the nation in its production), followed by cement and coal. Tennessee is also the nation's largest producer of pyrites and is third in the production of phosphate rock. The state's leading manufactures are chemicals and related products, foods, textiles and apparel, electrical machinery, primary metals, and stone, clay, and glass items. Aluminum production has been important since World War I, and more recently a missile industry has been developed in Bristol.

History. Cherokee, Chickasaw, Shawnee, and Creek Indians were in the region when it was first visited by a European expedition under Hernando de Soto in 1540. French explorers came down the Mississippi River, claiming both sides for France, and c.1682 sieur de La Salle built Fort Prudhomme, possibly on the site of present-day Memphis. The French established additional trading posts in the area. English fur traders and hunters came over the mountains from the Carolinas and Virginia, prevailed over the Cherokee Indians, and made ineffectual the French claims to the area, which in any event was lost (1763) by the French as a result of the French and Indian Wars. The first permanent settlement was made (1769) in the Watauga River valley of eastern Tennessee by Virginians. In 1772 these hardy settlers living beyond the frontier formed the Watauga Association, the first attempt at government in Tennessee, and in 1777, at their request, North Carolina organized those settlements into Washington co.; Jonesboro, the county seat and oldest town in Tennessee, was founded two years later.

When, after the Revolution, North Carolina ceded its western lands to the Federal government, the eastern Tennessee settlers formed a short-lived independent government (1784–88), the state of Franklin. In 1790 the Federal government created the Territory of the United States South of the River Ohio (Southwest Territory). In 1796 Tennessee, with substantially its present boundaries, was admitted to the Union, with its capital at Knoxville. Settlers, numbering over 100,000 by 1800, swarmed in via such overland routes as the Wilderness Road and Cumberland Gap. Others poled keelboats from the Ohio River up the Cumberland and Tennessee rivers. The virtues and vices of their strongly egalitarian society were exemplified by Andrew Jackson, who was prominent in the faction-ridden politics of Tennessee. By the time Jackson became (1829) President, the state was prospering. The first steamboat had reached Nashville in 1819, the year in which Memphis was platted.

Although black slaves were numerous in western Tennessee, the state was pro-Union. Secession was rejected in a popular referendum on Feb. 9, 1861. However, after the firing on Fort Sumter and Lincoln's call for troops, a second referendum (June 8) approved secession by a two-thirds majority, and the state seceded. In the Civil War Tennessee was, after Virginia, the biggest battleground. In Apr., 1862, one of the bloodiest battles of the war was fought at Shiloh near the Mississippi state line, and Memphis fell to a Union fleet in June. The Confederates managed to hold on to Knoxville until Sept., 1863. The Union military government which had been set up under Andrew Johnson in 1862 was succeeded in Apr., 1865, by a civil government. The state was readmitted to the Union in Mar., 1866.

That effective instrument for re-establishing "white supremacy" in the South, the Ku Klux Klan, was founded (1866) in Tennessee, at Pulaski. A new constitution drafted in 1870 rejected the reforms of the radical Republicans; black suffrage was limited by means of the poll tax, and former Confederates were re-enfranchised. Economically, the farm-tenancy system, which had replaced the plantation system, brought much misery; industry, however, made advances after the war.

A statewide prohibition bill (not repealed until 1939) was passed over a governor's veto in 1909. In 1925 the state attracted international attention with the famous Scopes Trial at Dayton, and the fact that the state law banning the teaching of evolution was not repealed until 1967 indicates the strong hold that Protestant fundamentalism had on the people. Although opposed by private power companies, the TVA (est. in 1933) succeeded in providing hydroelectric power cheaply and in abundance, bringing modern comforts to thousands. TVA was chiefly responsible for the significant growth and diversification of industry, especially during and after World War II. The 1954 Supreme Court decision outlawing racial segregation in the public schools had far-reaching consequences for Tennessee. Despite rioting, considerable progress in integration has been made.

Government. The present constitution dates from 1870. Tennessee's executive branch is headed by a governor, elected for a four-year term and not permitted to succeed himself. The state's bicameral legislature has a senate, with 33 members elected for four-year terms, and a house, with 99 members elected for two-year terms.

Educational Institutions. Among those prominent in the state are the Univ. of Tennessee, chiefly at Knoxville; East Tennessee State Univ., at Johnson City; Fisk Univ. and Vanderbilt Univ., at Nashville; Memphis State Univ., at Memphis; Tennessee Technological Univ., at Cookeville; and the Univ. of the South, at Sewanee.

Tennessee, river, c.650 mi (1,045 km) long, the principal tributary of the Ohio River. It is formed by the confluence of the Holston and French Broad rivers near Knoxville, Tenn., and follows a U-shaped course to enter the Ohio River at Paducah, Ky. Its drainage basin covers c.41,000 sq mi (106,190 sq km) and includes parts of seven states. Navigation was long impeded by variations in channel depths and by rapids. However, the Tennessee Valley Authority (TVA) has converted the river into a chain of lakes held back by nine major dams. As a result of these improvements river traffic has increased; flooding has been controlled; a water-oriented recreation industry has been established; and electric power generated at the dams has attracted new industries to the region. A canal, scheduled for completion in the 1980s, will link the Tennessee River with the Gulf of Mexico by way of the Tombigee River.

Te·noch·ti·tlán (tā-nōch'tēt-län'), ancient city in the central valley of Mexico. The capital of the Aztecs, it was founded (A.D. c.1345) on a marshy island in Lake Texcoco. It was a flourishing city (est. pop. between 200,000 and 300,000), connected with the mainland by three great causeways. Cortés took the city in 1521 after a three-month siege, razed it, and captured the ruler, Cuauhtémoc, successor to Montezuma.The Spaniards founded present-day Mexico City on the ruins.

Te·nos (tē'nôs), island: *see* Tínos.

Ten·sas (tĕn'sô), parish (1970 pop. 9,732), 626 sq mi (1,621.3 sq km), NE La., bounded on the E and SE by the Mississippi River, on the W by the Tensas River; formed 1843; parish seat St. Joseph. It is agricultural, growing corn, cotton, oats, hay, and soybeans and raising livestock. There are cotton gins, sawmills, and oil wells here. The parish, which includes St. Joseph and Bruin lakes, has a fish hatchery and many fishing areas.

Te·o·ti·hua·cán (tā'ō-tē'wä-kän'), ancient commercial and religious center in the central valley of Mexico, NE of Mexico City. Once thought to be the religious center of the Toltec, it is now held to be the relic of an earlier civilization. Teotihuacán is the largest and most impressive urban site of ancient America. The Pyramid of the Sun, the tallest in Mexico, is 216 ft (65 m) high and covers approximately 10 acres (4 hectares) at the base; it dominates the symmetrical ground plan laid out in grid fashion along major thoroughfares, including the city's central axis—the Street of the Dead. Other buildings along this axis include the Pyramid of the Moon; the Citadel containing the Temple of Quetzalcoatl, so called because of its carvings of feathered serpents; the Temple of Agriculture; and the Quetzalpapalotl Palace.

Te·pic (tā-pēk'), city (1970 pop. 87,540), capital of Nayarit state, W Mexico, on the Tepic River. A commercial center on the coastal line of the Mexican National Railways and on a major highway, Tepic lies in a prosperous maize, sugar-cane, and cattle-raising area. The city has sugar mills and textile factories.

Te·pli·ce (tĕ'plĭ-tsĕ) or **Te·pli·ce-Ša·nov** (tĕ'plĭ-tse-shä'nôf), city (1971 est. pop. 51,124), NW Czechoslovakia, in Bohemia, in the Erzgebirge Mts. near the East German border. It is a road and rail hub and an industrial center in the heart of a lignite-mining area. In addition to coal the city produces machinery, glassware, ceramics, metal goods, chemicals, textiles, and foodstuffs. Teplice is also a famous resort and spa, whose hot mineral springs were known prior to Roman times.

Te·ra·mo (tâ'rä-mō), city (1972 est. pop. 38,300), capital of Teramo prov., Abruzzi, central Italy. It is an agricultural and industrial center. Manufactures include textiles and processed food.

Ter·cei·ra (tər-sā'rə), island (1960 pop. 72,485), 153 sq mi (396 sq km), in the N Atlantic, one of the central Azores, Portugal. Grains, cattle, and embroidery are exported. There is a U.S. air base here.

Te·rek (tyä'rĭk), river, c.370 mi (595 km) long, S European USSR, rising in the Caucasus, in Georgia, in glaciers W of Mt. Kazbek, and flowing N through the Daryal Gorge, then E and NE into the Caspian Sea. In its lower course the Terek is used for irrigation.

Te·re·si·na (tä'rä-zē'nä), city (1970 pop. 181,071), capital of Piauí state, NE Brazil, on the Parnaíba River. It is the main commercial and agricultural distribution center in the Parnaíba valley; cattle, hides and skins, rice, maize, cotton, and manioc are shipped through the city. Its industries produce textiles, sugar, soap, and lumber.

Ter·mez (tĕr-mĕz'), city (1973 est. pop. 50,000), capital of Surkhan-Darya oblast, Central Asian USSR, in Uzbekistan, a port on the Amu Darya River, near the Afghanistan border. It is the center of an agricultural region and has cotton and food processing and brick and tile industries.

Ter·na·te (tər-nä'tā), volcanic island (c.40 sq mi/105 sq km), E Indonesia, in the Molucca Sea, one of the Moluccas. It is heavily forested and mountainous. Exports include spices and copra. Despite its relatively small size, the island was for centuries a major spice center and one of the most important islands of the Moluccas.

Ter·ni (tĕr'nē), city (1972 est. pop. 108,379), capital of Terni prov., Umbria region, central Italy, on the Nera River. Manufactures include iron and steel, munitions, textiles, machinery, and chemicals. Hydroelectric power is generated at nearby waterfalls, formed in 272 B.C., when the Romans connected the Velino River with the Nera.

Ter·no·pol (tər-nō'pəl, tyĭr-nô'pəl), city (1973 est. pop. 104,000), capital of Ternopol oblast, SW European USSR, in the Ukraine, on the Seret River, a tributary of the Dnestr. It is an important rail junction and highway hub. Industries include food processing and the manufacture of machinery, building materials, electrical apparatus, clothing, leather products, footwear, and porcelain.

Terre·bonne (tĕr'bŏn', -bôn'), parish (1970 pop. 76,049), 1,368 sq mi (3,543.1 sq km), SE La., on the Gulf of Mexico and bounded partly on the E by small Bayou Pointe au Chien, on the W by the Atchafalaya River, and crossed by the Gulf Intracoastal Waterway; formed 1822; parish seat Houma. It has a swampy coast indented by Atchafalaya, Caillou, and Terrebonne bays and includes several lakes. There is some agriculture, fur trapping, hunting, and fishing done here. Seafood and vegetables are canned, sugar and lumber are milled, and oil is found in the area.

Ter·re Haute (tĕr'ə hōt', hŭt'), city (1978 est. pop. 61,800), seat of Vigo co., W Ind., on the Wabash River; inc. 1816. The commercial and trade center of a farming and coal-mining region, its many products include food items, phonograph records, aluminum and steel products, and farm and communications equipment. Founded (1811) as Fort Harrison, it grew as a river town. Terre Haute is the seat of Indiana State Univ.

Ter·rell (tĕr'əl). **1.** County (1970 pop. 11,416), 329 sq mi (852.1 sq km), SW Ga., bounded on the NE by the Kinchafoonee River, on the SW by Ichawaynochaway Creek; formed 1856; co. seat Dawson. It has coastal-plain agriculture (cotton, corn, fruit, and peanuts), timber, and garment and rubber-products manufacturing. **2.** County (1970 pop. 1,940), 2,391 sq mi (6,192.7 sq km), extreme W Texas, bounded on the S by the Rio Grande and the Mexican border, partly on the E by the Pecos River; formed 1905; co. seat Sanderson. Ranching (sheep, goats, some cattle, and horses) and hunting are the major occupations.

Terrell, city (1975 est. pop. 14,800), Kaufman co., N Texas; inc. 1883. Cattle are raised on the surrounding rich blackland prairies. The state mental hospital in Terrell is the largest source of employment in the city. There are also small manufacturing plants.

Ter·ry (tĕr'ē), county (1970 pop. 14,118), 899 sq mi (2,328.4 sq km), NW Texas, on the Llano Estacado, here crossed by intermittent Sulphur Springs Creek and Sulphur Draw; formed 1876; co. seat Brownfield. It has farming (especially grain sorghums, also corn, peanuts, cotton, peas, alfalfa, fruit, and truck crops), dairying, poultry and livestock raising, and oil and natural-gas wells.

Terry, city (1970 pop. 870), seat of Prairie co., E Mont., on the Yellowstone River NE of Miles City, in an irrigated farm and livestock area; settled 1883, inc. 1910.

Te·ruel (tĕ-rwĕl'), town (1970 pop. 21,638), capital of Teruel prov., E central Spain, in Aragón, at the confluence of the Guadalaviar and Alfambra rivers. The city is an agricultural trade center; the province has iron and coal mines and sulfur, zinc, and manganese deposits. The center of bitter fighting in the Spanish Civil War of 1936–39, it was largely destroyed, but has since been rebuilt.

Te·schen (tĕsh'ən), former principality (c.850 sq mi/2,200 sq km), now divided between Czechoslovakia and Poland. Teschen was its chief town. After World War I Teschen was divided, giving the western section, including the Karviná basin, to Czechoslovakia and the eastern agricultural section to Poland. During World War II the entire region was annexed to Germany, but in 1945 the status quo as of 1920 was restored despite Polish claims.

Te·te (tā'tə, tā'tā), town (1970 pop. 53,232), capital of Tete district, W central Mozambique, on the Zambezi River. It is a trade center.

Te·ton (tēt'n). **1.** County (1970 pop. 2,351), 457 sq mi (1,183.6 sq km), E Idaho, in a plateau area bordering on Wyo. and including the irrigated valley of the Teton River; formed 1915; co. seat Driggs. Dairying and agriculture (potatoes, dry beans, and sugar beets) are found. Part of Targhee National Forest is in the southwest. **2.** County (1970 pop. 6,116), 2,294 sq mi (5,941.5 sq km), N central Mont., in an irrigated agricultural area drained by the Teton River and crossed in the W by Lewis and Clark National Forest; formed 1893; co. seat Choteau. Grain growing and livestock raising are the major occupations. **3.** County (1970 pop. 4,823), 4,000 sq mi (10,360 sq km), NW Wyo., bordering on Idaho and Yellowstone National Park, and watered by the Snake River and Jackson Lake; formed 1921; co. seat Jackson. It is in a grain and livestock area, with coal deposits and clothing factories. Jackson Hole, Grand Teton National Park, and parts of Teton National Forest and Teton Range are here.

Teton. 1. River, 143 mi (230 km) long, rising in several branches in the Rocky Mts., NW Mont., and flowing E to the Marias River. **2.** River, c.60 mi (100 km) long, rising in W Wyo., in forks that unite in SE Idaho. The Teton flows north and west to Henrys Fork River, near Rexburg, Idaho.

Teton Range, part of the Rocky Mts., NW Wyo. and SE Idaho, just S of Yellowstone National Park. The highest peaks are within Grand Teton National Park, with Grand Teton (13,747 ft/4,192.8 m) the highest peak in the range. Teton Pass (8,431 ft/2,571.5 m) and Phillips Pass (10,700 ft/3,263.5 m) are just south of the park. The Teton Range includes part of Targhee National Forest.

Te·tuán (tā-twän'), city (1971 pop. 139,105), N Morocco. It has some light industry and is an export point for livestock and agricultural products. Its old casbah and mosques are tourist attractions.

Teu·to·burg Forest (tōō'tō-bûrg', toi'tō-bōōrkh'), hilly range, in N central West Germany. Near Detmold is a monument commemorating the victory (A.D. 9) of the Germans under Arminius over the Roman legions under Varus.

Tewkes·bur·y (tyōōks'bĕr'ē, -bə-rē), municipal borough (1974 est. pop. 9,010), Gloucestershire, W central England, on the Avon River near its junction with the Severn. Once noted for mustard production, it now has minor manufactures. At Bloody Meadow, south of the town, Edward IV in 1471 defeated the Lancastrians in the Wars of the Roses.

Tewksbury, town (1975 est. pop. 24,500), Middlesex co., NE Mass.; settled 1637, set off from Billerica and inc. 1734. It was the site of an Indian colony. Now a residential area, it has light manufacturing.

Tex·ar·kan·a (tĕk'sär-kăn'ə), city (1978 est. pop.: in Texas, 34,200; in Ark., 21,000), Bowie co. (Texas) and seat of Miller co. (Ark.), on the Texas-Ark. line; inc. 1880. Texarkana is a transportation and trade center for a large agricultural and pine forest area. Its many industries include cotton processing, lumbering and woodworking, and the manufacture of tires, clay products, rock wool, mobile homes, furniture, and garments.

Tex·as (tĕk'səs), state (1975 pop. 12,276,000), 267,339 sq mi (692,408 sq km), including 4,369 sq mi (11,316 sq km) of water surface, SW United States, admitted to the Union in 1845 as the 28th state. Austin is the capital. The second-largest state in the Union, Texas is roughly spade-shaped. Its Panhandle projects north into Okla., with N.Mex. on the west. Below the Panhandle, Texas continues its western border with N.M. but has a southwest shoulder that thrusts west between N.Mex. and Mexico. The state is bounded on the southwest and south by Mexico (the Rio Grande marks the entire international boundary), on the southeast by the Gulf of Mexico, and on the east by La. (the Sabine River forms much of the line). In the extreme northeast, Texas has a short border with Ark., and to the north, the Red River marks the boundary with Okla. west to the Panhandle.

The vast expanse of Texas contains great regional differences. East Texas is Southern in character, with pine-covered hills, cypress swamps, and remnants of the great cotton plantations founded before the Civil War. Cotton farming has been supplemented by diversified agriculture, including rice cultivation; almost all of the state's huge rice crop comes from east Texas. The real wealth of east Texas, however, comes from the immense, rich oil fields discovered in the 1920s. Oil is the principal economic support of Beaumont and Port Arthur and the basis for much of the heavy industry that crowds the Gulf Coast. The industrial heart of the coastal area is Houston, the largest city in the South and Southwest and the sixth-largest in the United States.

The low Gulf coastal plains grow drier in the south and become semiarid as they near the Rio Grande. The southern Gulf Coast is a popular tourist area, and some of the ports, such as Galveston and Corpus Christi, have economies dependent on both heavy industry and tourism. Brownsville is the shipping center for the intensively farmed and irrigated section along the lower Rio Grande, where citrus fruits and winter vegetables are grown. The long stretch of plain along the Rio Grande valley is largely given over to cattle ranching; in this area ranches average c.9,000 acres (3,640 hectares) in size. Texas has c.1,000 mi (1,610 km) of border with Mexico, and many southern and western Texas towns have a colorful Mexican flavor. Some towns are bilingual, and in some areas persons of Mexican descent make up as much as 80% of the population. Laredo is the most important gateway to Mexico, with an excellent highway to Mexico City and a thriving border commerce.

The first region to be farmed when Americans came to Texas in the 1820s was the bottoms of the lower Brazos and the Colorado rivers, but not until settlers moved into the rolling blackland prairies of central and north-central Texas was the agricultural wealth of the area realized. The heart of this region is the trading and shipping center of Waco; at the southwest extremity is San Antonio, the commercial center of a wide cotton, grain, and cattle country. To the north, Dallas and its neighboring city of Fort Worth are the focus of one of the most rapidly developing industrial sections of the country. Their oil refining, grain milling, and cotton and food processing have been supplemented since World War II by huge aircraft-manufacturing and electronics industries.

The Balcones Escarpment marks the western margin of the Gulf coastal plain; in central Texas the line is visible in a series of waterfalls and rough, tree-covered hills. To the west lie the south central plains and the Edwards Plateau. No traces of the subtropical lushness of the Gulf coastal plain are found in these regions; the climate is strictly continental, with occasional blizzards blowing across the flat land in winter. The Red River area, including the farming and oil center of Wichita Falls, can have extreme cold in winter, though without the severity that is intermittently experienced in the commercial center of the Panhandle, Amarillo, or in the dry-farming area around Lubbock. Cattle ranching, which began in the late 1870s, still persists, and huge ranches vie with extensive wheat and cotton farms for domination of the treeless land. Oil and grain, however, have revolutionized the economy of this section of the state, and industry has grown.

All of west Texas (that part of the state west of long. 100° W) is semiarid. South of the Panhandle lie the rolling plains around Abilene, a region cultivated in cotton, sorghum, and wheat and the site of newer oil fields discovered in the 1940s. The dry fields of west Texas are still given over to ranching, except for small irrigated areas that can be farmed. San Angelo serves as the commercial center of this area. The land beyond the Pecos River, rising to the mountains, offers the finest scenery of Texas. There are found the Davis Mts. and Guadalupe Peak, the highest point (8,751 ft/2,667 m) in the state. The wilderness of the Big Bend of the Rio Grande is typical of the barrenness of most of this area, where water and people are almost equally scarce.

Texas is divided into 254 counties:

NAME	COUNTY SEAT	NAME	COUNTY SEAT
Anderson	Palestine	Hansford	Spearman
Andrews	Andrews	Hardeman	Quanah
Angelina	Lufkin	Hardin	Kountze
Aransas	Rockport	Harris	Houston
Archer	Archer City	Harrison	Marshall
Armstrong	Claude	Hartley	Channing
Atascosa	Jourdanton	Haskell	Haskell
Austin	Bellville	Hays	San Marcos
Bailey	Muleshoe	Hemphill	Canadian
Bandera	Bandera	Henderson	Athens
Bastrop	Bastrop	Hidalgo	Edinburg
Baylor	Seymour	Hill	Hillsboro
Bee	Beeville	Hockley	Levelland
Bell	Belton	Hood	Granbury
Bexar	San Antonio	Hopkins	Sulphur Springs
Blanco	Johnson City	Houston	Crockett
Borden	Gail	Howard	Big Spring
Bosque	Meridian	Hudspeth	Sierra Blanca
Bowie	Boston	Hunt	Greenville
Brazoria	Angleton	Hutchinson	Stinnett
Brazos	Bryan	Irion	Mertzon
Brewster	Alpine	Jack	Jacksboro
Briscoe	Silverton	Jackson	Edna
Brooks	Falfurrias	Jasper	Jasper
Brown	Brownwood	Jeff Davis	Fort Davis
Burleson	Caldwell	Jefferson	Beaumont
Burnet	Burnet	Jim Hogg	Hebbronville
Caldwell	Lockhart	Jim Wells	Alice
Calhoun	Port Lavaca	Johnson	Cleburne
Callahan	Baird	Jones	Anson
Cameron	Brownsville	Karnes	Karnes City
Camp	Pittsburg	Kaufman	Kaufman
Carson	Panhandle	Kendall	Boerne
Cass	Linden	Kenedy	Sarita
Castro	Dimmitt	Kent	Jayton
Chambers	Anahuac	Kerr	Kerrville
Cherokee	Rusk	Kimble	Junction
Childress	Childress	King	Guthrie
Clay	Henrietta	Kinney	Brackettville
Cochran	Morton	Kleberg	Kingsville
Coke	Robert Lee	Knox	Benjamin
Coleman	Coleman	Lamar	Paris
Collin	McKinney	Lamb	Littlefield
Collingsworth	Wellington	Lampasas	Lampasas
Colorado	Columbus	La Salle	Cotulla
Comal	Bew Braunfels	Lavaca	Hallettsville
Comanche	Comanche	Lee	Giddings
Concho	Paint Rock	Leon	Centerville
Cooke	Gainesville	Liberty	Liberty
Coryell	Gatesville	Limestone	Groesbeck
Cottle	Paducah	Lipscomb	Lipscomb
Crane	Crane	Live Oak	George West
Crockett	Ozona	Llano	Llano
Crosby	Crosbyton	Loving	Mentone
Culberson	Van Horn	Lubbock	Lubbock
Dallam	Dalhart	Lynn	Tahoka
Dallas	Dallas	McCulloch	Brady
Dawson	Lamesa	McLennan	Waco
Deaf Smith	Hereford	McMullen	Tilden
Delta	Cooper	Madison	Madisonville
Denton	Denton	Marion	Jefferson
De Witt	Cuero	Martin	Stanton
Dickens	Dickens	Mason	Mason
Dimmit	Carrizo Springs	Matagorda	Bay City
Donley	Clarendon	Maverick	Eagle Pass
Duval	San Diego	Medina	Hondo
Eastland	Eastland	Menard	Menard
Ector	Odessa	Midland	Midland
Edwards	Rocksprings	Milam	Cameron
Ellis	Waxahachie	Mills	Goldthwaite
El Paso	El Paso	Mitchell	Colorado City
Erath	Stephenville	Montague	Montague
Falls	Marlin	Montgomery	Conroe
Fannin	Bonham	Moore	Dumas
Fayette	La Grange	Morris	Daingerfield
Fisher	Roby	Motley	Matador
Floyd	Floydada	Nacogdoches	Nacogdoches
Foard	Crowell	Navarro	Corsicana
Fort Bend	Richmond	Newton	Newton
Franklin	Mount Vernon	Nolan	Sweetwater
Freestone	Fairfield	Nueces	Corpus Christi
Frio	Pearsall	Ochiltree	Perryton
Gaines	Seminole	Oldham	Vega
Galveston	Galveston	Orange	Orange
Garza	Post	Palo Pinto	Palo Pinto
Gillespie	Fredericksburg	Panola	Carthage
Glasscock	Garden City	Parker	Weatherford
Goliad	Goliad	Parmer	Farwell
Gonzales	Gonzales	Pecos	Fort Stockton
Gray	Pampa	Polk	Livingston
Grayson	Sherman	Potter	Amarillo
Gregg	Longview	Presidio	Marfa
Grimes	Anderson	Rains	Emory
Guadalupe	Seguin	Randall	Canyon
Hale	Plainview	Reagan	Big Lake
Hall	Memphis	Real	Leakey
Hamilton	Hamilton	Red River	Clarksville
Reeves	Pecos	Titus	Mount Pleasant
Refugio	Refugio	Tom Green	San Angelo
Roberts	Miami	Travis	Austin
Robertson	Franklin	Trinity	Groveton
Rockwall	Rockwall	Tyler	Woodville
Runnels	Ballinger	Upshur	Gilmer
Rusk	Henderson	Upton	Rankin
Sabine	Hemphill	Uvalde	Uvalde
San Augustine	San Augustine	Val Verde	Del Rio
San Jacinto	Coldspring	Van Zandt	Canton
San Patricio	Sinton	Victoria	Victoria
San Saba	San Saba	Walker	Huntsville
Schleicher	Eldorado	Waller	Hempstead
Scurry	Snyder	Ward	Monahans
Shackelford	Albany	Washington	Brenham
Shelby	Center	Webb	Laredo
Sherman	Stratford	Wharton	Wharton
Smith	Tyler	Wheeler	Wheeler
Somervell	Glen Rose	Wichita	Wichita Falls
Starr	Rio Grande City	Wilbarger	Vernon
Stephens	Breckenridge	Willacy	Raymondville
Sterling	Sterling City	Williamson	Georgetown
Stonewall	Aspermont	Wilson	Floresville
Sutton	Sonora	Winkler	Kermit
Swisher	Tulia	Wise	Decatur
Tarrant	Fort Worth	Wood	Quitman
Taylor	Abilene	Yoakum	Plains
Terrell	Sanderson	Young	Graham
Terry	Brownfield	Zapata	Zapata
Throckmorton	Throckmorton	Zavala	Crystal City

Economy. In the state as a whole, mineral resources compete with industry for primary economic importance. Texas is the wealthiest mineral producer in the nation. It normally produces one third of the country's petroleum and contains one half of its known reserves. Texas also ranks first in the production of natural gas, natural-gas liquids, asphalt, and pyrites. It is second in helium, magnesium compounds (from salt water), salt, sulfur, sodium sulfate, clays, and gypsum; third in cement; and fourth in lime and talc. Chemicals and chemical products are the state's chief manufactures (19 out of the 20 largest U.S. chemical companies are based here), followed by petroleum, food and food products, transportation equipment, machinery, and primary and fabricated metals.

Agriculturally, Texas is one of the most important states in the country. It has more farms, farmland, cattle, sheep, and lambs than any other state, and it is the nation's leading producer of cotton and cottonseed. Principal crops are cotton lint, sorghum grain, hay, and rice. Texas is second among the states in the production of carrots, onions, spinach, watermelons, and honeydew melons. It is a leading commercial fishing state; the value of its fish catch in 1972 was exceeded only by that of California. Principal catches are shrimp, oysters, and menhaden.

History. The region that is now Texas was known early to the Spanish, who were, however, slow to settle there. Cabeza de Vaca, shipwrecked off the coast in 1528, wandered through the area in the 1530s, and Francisco Vásquez de Coronado probably crossed the northwest section in 1541. The first Spanish settlement was made (1682) at Ysleta on the site of El Paso by refugees from present N.M. after the Pueblo revolt of 1680. Several missions were established in the area; but the Comanche, the Apache, and other Indian tribes were unfriendly, and the missionaries withdrew. New attempts at settlement were prompted by the French threat from Louisiana, and missions, sometimes protected by presidios, were established at San Antonio (1718), Goliad (1749), Nacogdoches (permanently, 1779), and elsewhere. In general Spanish attempts to gain wealth from the wild region and to convert the Indians were unsuccessful, and in most places occupation was desultory.

By the early 19th cent. Americans were covetously eyeing Texas, especially after the Louisiana Purchase (1803) had extended the U.S. border to that fertile wilderness. In 1821 Stephen F. Austin led 300 families across the Sabine to the region between the Brazos and Colorado rivers, where they established, under a Spanish grant, the first American settlement in Texas. The newly independent government of Mexico, pleased with Austin's prospering colony, readily offered grants to other American settlers. Americans from all over the Union, but particularly from the South, poured into Texas. By 1830 the Americans outnumbered the Mexican settlers by more than three to one and had formed their own compact society. The Mexican government became understandably alarmed. The Texans, hoping to achieve a greater measure of self-government, petitioned Mexico for separate statehood. When Stephen Austin presented the petition in Mexico City, he was imprisoned, and Texas was regarrisoned by Mexican troops. The Texas Revolution broke out (1835) in Gonzales when the Mexicans attempted to disarm the Americans and were routed. The American settlers then drove all the Mexican troops from Texas. At a convention called at Washington-on-the-Brazos, Texas declared its independence (Mar. 2, 1836). A constitution was adopted and David Burnet was named interim president. The arrival of Antonio López de Santa Ana with a large army to crush the rebellion resulted in the heroic and tragic defense of the Alamo. The small Texas army, commanded by Samuel Houston, protected their rear, retreating strategically until Houston finally manuevered Santa Ana into a cul-de-sac near the site of present-day Houston. In the battle of San Jacinto (Apr. 21, 1836), Houston surprised the larger Mexican force during its afternoon siesta and scored a resounding victory. Santa Ana was captured and compelled to recognize the independence of Texas.

Texans sought annexation to the United States, but antislavery forces in the United States vehemently opposed the admission of another slave state, and Texas remained an independent republic under its Lone Star flag for almost 10 years. Sam Houston, the hero of the revolution, was the leading figure of the republic, serving twice as president. A combination of factors—confusion in the land system, low credit abroad, and the expense of maintaining the Texas Rangers and protecting Texas from marauding Mexican forces—contributed to impoverishing the republic and increasing the urgency for its annexation to the United States. President John Tyler narrowly pushed the admission of Texas through Congress shortly before the expiration of his term; Texas formally accepted annexation in July, 1845. This act was the immediate cause of the Mexican War. After Gen. Zachary Taylor defeated the Mexicans at Palo Alto and Resaca de la Palma, the Mexican forces retreated back across the Rio Grande.

During the Civil War Texas was the only Confederate state not overrun by Union troops. Remaining relatively prosperous, it liberally contributed men and provisions to the Southern cause. Texas was readmitted to the Union in Mar., 1870. Reconstruction ended in 1874 when the Democrats took control of the government In the decades following the war the Western element in Texas was strengthened as stock raising became dominant. From the open range and then from great fenced ranches Texas cowboys drove herds of longhorn cattle north over trails such as the Chisholm Trail. As railroads advanced across the state during the 1870s, farmlands were increasingly settled, and the small farmers came into violent conflict with the ranchers. During the 1880s demands for economic reform and limitation of the railroads' vast land domains were championed by the Farmers' Alliance and Gov. James S. Hogg.
James S. Hogg.

The transformation of Texas into a partly urban and industrial society was greatly hastened by the uncovering of the state's tremendous oil deposits beginning in 1901. Texas industry developed rapidly during the first years of the 20th cent., but conditions worsened for the tenant farmers, who by 1910 made up the majority of cultivators. World War I had a somewhat liberating effect on Texas blacks, but the reappearance of the Ku Klux Klan after the war helped to enforce "white supremacy." The economic boom of the 1920s was accompanied by further industrialization. The Great Depression of the 1930s, while severe, was less serious than in most states; the chemical and oil industries in particular continued to grow. The significance of the petrochemical and natural-gas industries increased during World War II, when the aircraft industry also rose to prominence and the establishment of military bases throughout Texas greatly contributed to the state's economy. Postwar years brought continued prosperity and industrial expansion, abetted by numerous flood-control, irrigation, and hydro-electric power projects. Texas has made considerable progress in desegregating its schools since the 1954 Supreme Court decision on school integration, and the number of black voters has increased by more than 25% since the 1965 voting rights law.

Government. The state's present constitution was adopted in 1876. The executive branch is headed by a governor elected for a four-year term (before 1974 the governor's term was two years). The legislature has a senate with 31 members elected for four-year terms and a house with 150 representatives elected for two-year terms.

Educational Institutions. Among the many institutions of higher learning in Texas are the Univ. of Texas, mainly at Austin, but with large branches at Arlington and El Paso; Baylor Univ., at Waco; East Texas State Univ., at Commerce; North Texas State Univ., at Denton; Rice Univ., at Houston; Southern Methodist Univ., at Dallas; Texas Arts and Industries Univ., at Kingsville; Texas Agricultural and Mechanical Univ., at College Station; Texas Christian Univ., at Fort Worth; and Texas Southern Univ. and the Univ. of Houston, both at Houston.

Texas. 1. County (1970 pop. 18,320), 1,183 sq mi (3,064 sq km), S central Mo., in the Ozarks; formed 1845; co. seat Houston. It is a dairy and poultry region, with corn, wheat, and hay. There are also barite deposits. Part of Mark Twain National Forest is here. **2.** County (1970 pop. 16,352), 2,062 sq mi (5,340.6 sq km), extreme NW Okla., in the high plains of the Panhandle, bounded in the N by Kansas, in the S by Texas, and intersected by the North Canadian River and small Coldwater and Golf creeks; formed 1907; co. seat Guymon. Its agriculture includes wheat, corn, soybeans, milo, alfalfa, livestock, and poultry. It has oil and gas wells, and some manufacturing (food products and farm machinery).

Texas City, city (1978 est. pop. 41,200), Galveston co., S Texas, on Galveston Bay, opposite the city of Galveston; inc. 1911. It is an industrial city and port with giant oil refineries, petrochemical plants, a huge tin smelter, and factories making fertilizer, valves, and pipes.

Tex·el (tĕk'səl, tĕs'əl), island (1970 pop. 11,394), 71 sq mi (184 sq km), North Holland prov., NW Netherlands, in the North Sea, the largest and southernmost of the West Frisian Islands. It is a popular summer resort.

Thad·de·us Kos·ci·usz·ko National Memorial (thăd'ē-əs kŏs'ē-ŭs'-kō): *see* National Parks and Monuments Table.

Thai·land (tī'lănd, -lənd), formerly **Si·am** (sī-ăm'), constitutional monarchy (1976 pop. 42,700,000), 198,455 sq mi (514,000 sq km), SE Asia. Bangkok is the capital. Occupying a central position on the Southeast Asia peninsula, Thailand is bordered by Burma on the west and northwest, by Laos on the north and east (the Mekong River forms much of the line), by Cambodia on the southeast, and by the Gulf of Siam and Malaysia on the south. A southward extension into the Malay Peninsula gives Thailand a long coastline on the Gulf of Siam and on the Andaman Sea. The heart of the country, the fertile and thickly populated central plain, is virtually one vast rice paddy, entirely flat and rarely more than a few feet above sea level. It is watered by the Chao Phraya and lesser rivers and is veined by a system of canals for irrigation and drainage. Bangkok and Ayutthaya, the old capital, are in that basin. The north is mountainous, with peaks rising to c.8,500 ft (2,590 m); mountains stretch south along the boundary with Burma on the west. Extensive forests in the north yield teak, which is cut, hauled to the rivers by elephants, and floated to market. Although the population in the north is relatively sparse, rice is intensively cultivated in the river valleys. Most of northeastern and eastern Thailand is occupied by the Korat (Khorat) plateau, which is cut off from the rest of the country by highlands and the Phetchabun Mts. It is a hilly, dry, and generally poor

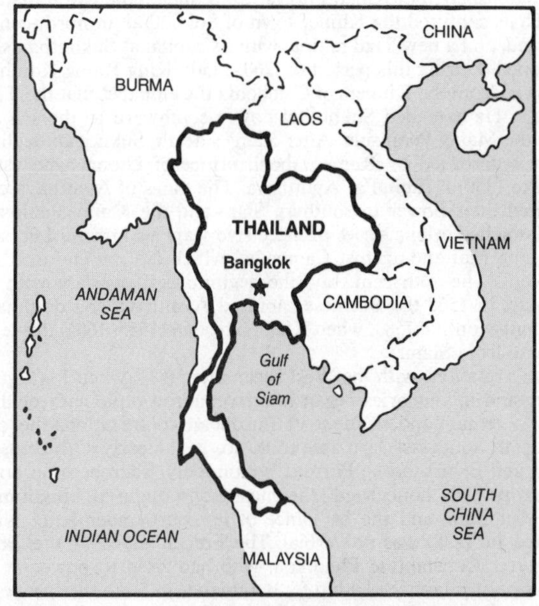

region, where livestock raising is dominant. Peninsular Thailand in the south (which includes Phuket and other offshore islands) is largely mountainous and covered with jungles. It is the principal source of the rubber and tin that rank Thailand third (1970) in world production of both. Thailand has a tropical and monsoonal climate.

While the ethnic minorities generally speak their own languages, Thai (linguistically related to Chinese) is the official tongue; English predominates among the Western languages. Hinayana Buddhism is the state religion; more than 93% of the people are Buddhists.

Economy. Thailand is heavily agricultural, with rice the leading crop and the major factor in a normally favorable trade balance; Thailand is second only to the United States in the amount of rice exported. Other commercial crops include rubber, corn, kenaf, jute, tapioca, cotton, tobacco, kapok, and sugar cane. Thailand's teak, once a major export, still supplies a large share of the world market. Marine and freshwater fisheries are important; fish provide most of the protein in the diet, and some of the deep-sea catches (mackerel, shark, shrimp, crab) are now being exported. Tin, by far the most valuable mineral, is a major export item. Tungsten, lead, zinc, and antimony are also mined for export. Iron ore, gold, precious and semiprecious stones, salt, lignite, petroleum, asphaltic sand, and glass sand are exploited on a smaller scale.

Industry is minor and is centered chiefly in the processing of agricultural products; rice milling is by far the most important, followed by sugar producing, textile spinning and weaving, and the processing of rubber, tobacco, and forest products. Lumbering is concentrated in the north. A major tin smelter is on Phuket Island. Small factories manufacture building materials, glass, pharmaceuticals, and various consumer goods. Handicraft production exceeds total factory output and has a ready market in the tourist trade. Tourism is an important source of foreign exchange; Bangkok is now a key point on round-the-world air routes. It is the political, commercial, cultural, and transportation center of the country, with the only port that can accommodate oceangoing vessels. Thailand's railroads originate in Bangkok and extend to Chiangmai, the Korat plateau, and to Cambodia, Laos, and Malaysia. A corresponding network of paved highways has recently been constructed. Thailand's inland waterways—a complex, interconnected system of rivers, streams, and canals—have been important arteries since ancient times. Local trade is chiefly in the hands of the large Chinese minority (about 3,000,000), and as a consequence there is tension between Thais and Chinese.

History. Like other countries of Southeast Asia, Thailand in prehistoric times was peopled through successive migrations from central Asia into territory already inhabited by primitive Negrito tribes. The main body of Thais remained in Yünnan, China, where by A.D. 650 they had organized the independent kingdom of Nanchao. By 1000, however, the Chinese had made it a tributary state. With the destruction of the kingdom of Nanchao by the Mongols under Kublai Khan in 1253, the slow infiltration of Thailand from the north turned into a mass migration. By that time the Khmer Empire was well established in the Chao Phraya valley and on the Korat plateau. The Thais captured the Khmer town of Sukhothai, in north-central Thailand, and a new Thai nation, with its capital at Sukhothai, soon developed. During this period (c.1260–1350), King Rama Kamheng borrowed from the Khmers of Cambodia the alphabet that the Thais still use. He extended Sukhothai power southward to the sea and down the Malay Peninsula. After Rama's death, Sukhothai declined and was absorbed by Rama Tibodi, prince of Utong, who established (c.1350) a capital at Ayutthaya. The kings of Ayutthaya consolidated their power in southern Siam and the Malay Peninsula, then launched a long series of indecisive wars against the Lao state of Chiang Mai and against Cambodia, which did not end until the 19th cent. The 16th cent. saw the beginnings of warfare with the Burmese; in 1568 the Burmese captured Ayutthaya and dominated the country until c.1583, when King Naresuan (1555–1605) drove the Burmese from Siam.

Siam's relations with the West began after 1511, when Portuguese traders and missionaries began to arrive; adroit diplomacy enabled Siam to remain independent of European colonization, the only country in Southeast Asia able to do so. In the early 17th cent. the Dutch and British broke Portugal's monopoly. Siam became, so far as Europe was concerned, the most consequential kingdom in Southeast Asia, and the brilliance of its court under King Narai (reigned 1657–88) was proverbial. The French, aided by the Greek adventurer Constantine Phaulkon, who had risen to power at the Siamese court, launched a bid for dominance in Siam that provoked an antiforeign coup d'état (1688). Phaulkon was executed, and Siam was closed to most foreigners for over a century. Gen. Chakkri (reigned 1782–1809), later known as Rama I, moved the capital from Thon Buri across the river to Bangkok and founded the Chakkri dynasty, thereafter the ruling house of Siam.

In the 19th cent. the authority of Bangkok was at last established over northern Siam, and relations with the West were resumed; Siam signed commercial treaties with Great Britain (1826) and the United States (1833). The Independence was threatened, however, when Great Britain extended her sway to Malaya and Burma, and France carved out an empire in Indochina. By opening their posts to European trade, by bringing in Western advisers, by strengthening the central administration as against the hereditary provincial chieftains, and by playing off British against French interests, the Siamese managed to stay free. Even so, the establishment of Siam's boundaries meant the surrender of its claims to Laos (1893) and parts of Cambodia (1907) and of its suzerainty over Kedah, Perlis, Kelantan, and Terengganu (1909), on the Malay Peninsula. The Westernization of Siam took place under an absolute monarchy and was chiefly the work of Mongkut (reigned 1851–68), or Rama IV, and his son Chulalongkorn (reigned 1868–1910), or Rama V.

Siam became a constitutional monarchy in 1932, when a bloodless coup d'état forced Prajadhipok (reigned 1925–35), Rama VII, to grant a constitution. The two young leaders of the coup, Pibul Songgram and Pridi Phanomyang, both educated in Europe and influenced by Western ideas, came to dominate Thai politics in the ensuing years. Songgram, a militarist, became premier in 1938. He changed the country's name to Thailand and instituted a program of expansion. Taking advantage of the French defeat (1940) in World War II, he renewed Thai claims in Cambodia and Laos. Japanese "mediation" resulted (1941) in territorial concessions to Thailand. In Dec., 1941, Pibul, despite the objections of Pridi Phanomyang, permitted the Japanese to enter Thailand, and in 1942 the government, under Japanese pressure, declared war on Great Britain and the United States. With the help of the United States, Pridi formed a militant anti-Japanese underground. After the war Thailand was forced to return territories in northern Malaya and Burma "granted" it by Japan in 1943 and those acquired in 1941 to French and British control. Pridi Phanomyang became premier and restored the name Siam as a repudiation of Pibul's policies. Inflation, corruption in government, and the mysterious death (1946) of King Ananda all contributed to the overthrow (1947) of Pridi's government by Pibul.

Under Pibul's military dictatorship, the name Thailand was again adopted; Thailand signed (1950) a technical and economic aid agreement with the United States and sent troops in support of the United Nations action in Korea. Thailand has received huge military grants from the United States and is the seat (since 1954) of the Southeast Asia Treaty Organization. The country, increasingly apprehensive over its proximity to Communist China, has remained consistently pro-Western in international outlook. The present king is Bhumibol Adulyadej (Rama IX; crowned in 1950 after a four-year regency).

In 1957 a military coup led by Field Marshal Sarit Thanarat finally overthrew Pibul Songgram, making Gen. Thanom Kittikachorn premier. In 1958, however, with the stated purpose of combating Communism, Sarit deposed his own premier, suspended the constitution, and declared martial law. When Sarit died in 1963, Thanom Kittikachorn was returned to power. A new constitution was finally promulgated in 1968. Under Sarit and Thanom the country's economy in the 1960s continued to boom; spurred by a favorable export market and considerable U.S. aid, it expanded at a rate of 7.5% per year. Thailand strongly supported the U.S. policy in South Vietnam. The nation's foreign policy was closely geared to the U.S. presence in Southeast Asia and its economy became increasingly dependent upon U.S. military spending and subsidies.

Economic reversals came in 1970 when the international demand for rice dropped substantially and the prices of tin and rubber fell. In addition, the security of the country appeared threatened by the spread of the Vietnam War into Cambodia and Laos and by growing insurgencies, chiefly Communist-led, in three separate areas within Thailand itself. In Nov., 1971, Premier Thanom Kittikachorn and three military aides abolished the constitution and the parliament and imposed military rule. In 1972 this new military junta launched a major four-month operation against the insurgents, but achieved few concrete results. In 1973 the military regime of Thanom was toppled after a week of student demonstrations and violence. King Bhumibol Adulyadej appointed a civilian as premier (the first in 20 years); in June, 1974, an all-civilian cabinet was formed, and the following Oct. a new constitution was promulgated. The civilian governments that ruled through 1975 and 1976 continued to be confronted with Communist insurgency in border areas, with left-wing student agitation, and with the threat of right-wing military coups. Relations with the United States deteriorated sharply following the Communist takeovers in Vietnam, Cambodia, and Laos (1975). By June, 1976, the last U.S. base in Thailand was closed and most U.S. military personnel had left. Student demonstrations against the return of Thanom to Thailand from exile (Oct., 1976) precipitated a military coup (Oct. 6) led by Admiral Sa-ngad Chaloryu. Sa-ngad suspended the 1974 constitution and installed a civilian premier and an Administrative Reform Committee to govern the country.

Thames (tĕmz), river, c.160 mi (260 km) long, rising NW of Woodstock, S Ont., Canada, and flowing SW past London and Chatham to Lake St. Clair. It is navigable to Chatham, near which was fought (1813) the Battle of the Thames in the War of 1812.

Thames, principal river of England, c.210 mi (340 km) long. It rises in four headstreams and flows generally eastward across southern England and through London to the North Sea at The Nore. In its upper course—around and above Oxford—it is often called Isis. The Thames drains c.5,250 sq mi (13,600 sq km). It is joined by canals (including the Oxford, Thames and Severn, and Grand Junction) that cover a wide area. The river is navigable by barges to Lechlade, below which there are a number of locks. The Thames is tidal to Teddington; there is a 23-ft (7-m) difference between low and high tide at London Bridge. The upper valley of the Thames is a broad, flat basin of alluvial clay soil, through which the river winds and turns constantly in all directions. At Goring Gap the valley narrows, separating the Chiltern Hills from the Berkshire Downs. The lower valley forms a second broad basin through which the Thames also meanders. Between Oxford and London, the valley is predominantly agricultural, with scattered villages. The Greater London conurbation along the river's lower course is one of the most important industrial regions of Great Britain.

Thames (thāmz, tĕmz), river, c.15 mi (25 km) long, formed by the confluence of the Yantic and Shetucket rivers at Norwich, E Conn., and flowing S to Long Island Sound at New London. Primarily a tidal estuary, it is the site of the U.S. coast guard academy and a U.S. navy submarine base.

Thanh Hoa (tän' wä'), city (1960 pop. 31,211), NE central Vietnam, near the mouth of the Song Ma River, in a cotton-growing area. Iron and phosphate are mined, and building materials are manufactured. Antimony and chromite deposits have been found in the vicinity.

Than·ja·vur (tən-jä'vŏŏr), formerly **Tan·jore** (tăn-jôr'), city (1971 pop. 140,547), Tamil Nadu state, SE India. It is a rice-milling center on the Cauvery River delta, known as the "rice bowl" of India. Thanjavur is also a center of craftsmanship noted especially for its silks and bronze ware and is a leading southern Indian music and dance center.

Thap·sus (thăp'səs), ancient N African seaport, c.100 mi (161 km) SE of Carthage in what is now Tunisia. The last stronghold of Pompey's party, the town was besieged in 46 B.C. by Julius Caesar. The defeat of Scipio here marked the end of opposition to Caesar in Africa.

Thar Desert (tär) or **Great In·di·an Desert** (ĭn'dē-ən), extensive arid region, c.500 mi (805 km) long and c.250 mi (400 km) wide, S Asia, in NW India and E Pakistan, between the Indus and Sutlej river valleys on the W and the Aravalli Range on the E.

Thá·sos (thā'sŏs), island (1971 pop. 13,316), c.170 sq mi (440 sq km), NE Greece, in the Aegean Sea. Timber, olive oil, honey, wine, and lead-zinc ores are its chief products; stock raising and fishing are important occupations.

Thay·er (thā'ər), county (1970 pop. 7,779), 577 sq mi (1,494.4 sq km), SE Nebr., in an agricultural area bounded in the S by Kansas and drained by the Little Blue River; formed 1871; co. seat Hebron. Its agriculture includes grain, livestock, dairy products, and poultry.

Thebes (thēbz), city of ancient Egypt. Al Uqsur and Al Karnak now occupy parts of its site. The temples and tombs that have survived, including the tomb of Tutankhamen, are among the most splendid in the world, and the site of the ancient city has been the scene of much important archaeological work.

Thebes, chief city of Boeotia, in ancient Greece. It was originally a Mycenaean city. Thebes is rich in associations with Greek legend and religion. The modern Thivai occupies the site of the Theban acropolis, part of which still survives. There are also remains of the prehistoric city and the temple of Ismenian Apollo.

The Dalles (dălz), city (1971 est. pop. 10,927), seat of Wasco co., N Oregon, on the Columbia River; inc. 1857. It is a busy inland port; ships passing through the locks at Bonneville Dam (c.50 mi/80 km downstream) can tie up at The Dalles and proceed upstream through the locks of The Dalles Dam. A processing and shipping point in an area producing sweet cherries, wheat, and beef, the city has a flour mill, fruit-processing plants, and an aluminum reduction plant. Mobile homes are also made. The Lewis and Clark expedition camped here in 1805, and later the site became the terminus of the Oregon Trail.

Thed·ford (thĕd'fərd), village (1970 pop. 303), seat of Thomas co., central Nebr., N of North Platte on the Middle Loup River, in a livestock and grain area. The Halsey Division of Nebraska National Forest is nearby.

The Mar-A-La·go National Historic Site (mâr'ə-lä'gō): *see* National Parks and Monuments Table.

Theodore Roo·se·velt Birthplace National Historic Site, Theodore Roosevelt Inaugural National Historic Site, Theodore Roosevelt Island (rō'zə-vĕlt, rōz'vĕlt): *see* National Parks and Monuments Table.

Theodore Roosevelt National Memorial Park, 70,436 acres (28,525 hectares), W N.Dak., in the Badlands and on the Little Missouri River; est. 1947. Roosevelt first came to the area in 1883 to hunt bison and other big game. In 1884 he established the Elkhorn Ranch, and for several years after that he returned to the ranch for short periods. The landscape of the park is marked by tablelands, buttes, canyons, and rugged hills; animal life is diverse.

Ther·mop·o·lis (thər-mŏp'ə-lĭs), resort town (1970 pop. 3,063), seat of Hot Springs co., N central Wyo., on the Bighorn River; founded 1897. A trade center for a livestock, farm, oil, and coal region, it also has an oil refinery. Wind River Canyon and Hot Spring State Park are nearby.

Ther·mop·y·lae (thər-mŏp'ə-lē), pass, E central Greece, SE of Lamía, between the cliffs of Mt. Oeta and the Malic Gulf. Silt accumulation has gradually widened the pass. In ancient times it was used as an entrance into Greece from the north. It is the site of several famous battles.

Thes·sa·lo·ní·ki (thĕ'sä-lō-nē'kē) or **Sa·lo·ni·ca** (să-lə-nē'kə, sə-lŏn'ĭ-kə), city (1971 pop. 345,799), capital of Thessaloníki prefecture, N Greece, in Macedonia; on the Gulf of Thessaloníki, an inlet of the Aegean Sea, at the neck of the Khalkidhikí Peninsula. It is the second-largest city in Greece, a major modern port, and an industrial and commercial center. Exports include grain, food products, tobacco, manganese and chrome ores, and hides. Industries produce textiles, machinery, metal goods, flour, cement, and explosives.

Thessaloníki was founded (c.315 B.C.) by Cassander, king of Macedon, who named it for his wife. It flourished after 146 B.C. as the capital of the Roman province of Macedon. Thessaloníki had from early times a sizable Jewish colony, and it was an early Christian diocese. Under the Byzantine Empire Thessaloníki was second only to Constantinople. The city was occupied by the Saracens in A.D. 904 and by the Normans of Sicily in 1185. When in 1204 the leaders of the Fourth Crusade created a Latin empire the kingdom of Thessaloníki, comprising most of northern and central Greece, was its largest fief. In 1246 the city fell to the Nicaeans, who in 1261 restored it to the Byzantine Empire. Thessaloníki was conquered by the Ottoman Turks in 1430 and remained in Ottoman hands until it was conquered by Greece in 1912 during the Balkan Wars. A great fire in 1917 destroyed much of the city. Thessaloníki suffered considerable damage in World War II, and its large (c.50,000) Jewish population was nearly liquidated by the Germans.

Thes·sa·ly (thĕs'ə-lē), largest ancient region of Greece, in N central Greece. It corresponded roughly to the present-day nomes (provs.) of Lárisa and Tríkkala, which form part of the modern region known as Thessaly. Ancient Thessaly was almost completely walled in by mountains, including Pindus, Óssa, and Othrys (now Othrís), and the plains were extremely fertile. Civilization dates from prehistoric times. Before 1000 B.C. a tribe called the Thessalians entered the area from the northwest. The Thessalians were powerful in the 6th cent. B.C. and again in the 4th cent. B.C., when Jason, the tyrant of Pherae, succeeded (374 B.C.) in uniting the region. Under the Roman emperors Thessaly was first joined to Macedonia and then

(A.D. 4th cent.) became a separate province. It passed (1355) to the Turks and was ceded to Greece in 1881.

Thet·ford Mines (thĕt'fərd mīnz'), city (1971 pop. 22,003), S Que., Canada, NE of Sherbrooke and S of Quebec. The city, developed after the discovery (1876) of large asbestos deposits, is located in one of the world's largest asbestos-producing regions. Chromium and feldspar are also mined.

Thi·bo·daux (tĭb'ə-dō'), city (1975 est. pop. 15,900), seat of Lafourche parish, SE La., on Bayou Lafourche; inc. 1838. It is the commercial center of an oil, sugar-cane, and farm area in the bayou country. Petrochemicals are manufactured, and the area to the south attracts many fishermen and shellfish gatherers.

Thief River Falls (thēf), city (1970 pop. 8,618), seat of Pennington co., NW Minn., N of Red Lake Falls at the confluence of the Thief and the Red Lake rivers. It is a trade center of a farm region, and poultry is processed. A Federal wildlife refuge is nearby.

Thiès (tyĕs), city (1970 est. pop. 90,700), W Senegal, on the Dakar-Niger RR. It is the trade center for a farming region where groundnuts, cassava, and livestock are raised. Manufactures include construction materials, wood furniture and plywood, and textiles. Aluminum phosphate is mined nearby and processed in Thiès.

Thion·ville (tyôN-vēl'), town (1970 est. pop. 40,254), Moselle dept., NE France, in Lorraine. It is a center for metallurgical and chemical industries. The town was a favorite of Charlemagne.

Thí·ra or **The·ra** (both: thîr'ə), volcanic island (1971 pop. 6,196), c.30 sq mi (80 sq km), SE Greece, in the Aegean Sea; one of the Cyclades. It is noted for its wine. Pumice stone and powdered tufa are exported. According to tradition the island was first settled by Phoenicians.

Thjórs·á (thyōrs'ou'), longest river of Iceland, c.150 mi (240 km) long. It rises on the eastern slopes of the Hofsjökull and flows southwest to the Atlantic Ocean.

Thom·as (tŏm'əs). **1.** County (1970 pop. 34,562), 541 sq mi (1,401.2 sq km), S Ga., bounded on the S by the Fla. border and drained by the Ochlockonee River; formed 1825; co. seat Thomasville. It has coastal-plain agriculture (cotton, tobacco, corn, melons, pecans, peanuts, and livestock) and lumbering. **2.** County (1970 pop. 7,501), 1,070 sq mi (2,771.3 sq km), NW Kansas, in a rolling plain drained by headstreams of the Solomon and Saline rivers; formed 1873; co. seat Colby. It is in a grain and livestock area, with some machinery manufacturing. **3.** County (1970 pop. 954), 716 sq mi (1,854.4 sq km), central Nebr., in an agricultural area drained by the Middle Loup and Dismal rivers; formed 1825; co. seat Thedford. Livestock and grain are farmed. Much of the Halsey Division of Nebraska National Forest is in the east.

Thomas Jef·fer·son Memorial (jĕf'ər-sən), monument, 18 acres (7 hectares), in East Potomac Park, on the Tidal Basin, Washington, D.C.; authorized by Congress 1934, built 1938–43, dedicated 1943. The white marble building, designed by John Russell Pope, is a circular structure with a domed ceiling, surrounded by 26 columns. Inside is a 19-ft (5.8-m) statue of Jefferson by the sculptor Rudulph Evans.

Thom·as·ton (tŏm'ə-stən). **1.** Town (1970 pop. 6,223), Litchfield co., W Conn., N of Waterbury; settled 1728, set off from Plymouth 1875. Seth Thomas established a clock factory here in 1812. Clocks and hardware are made today. **2.** City (1975 est. pop. 18,200), seat of Upson co., W central Ga., near the Flint River; inc. 1857. It is a textile center with textile mills (the first was established in 1833). Of interest are an old covered bridge and a number of historic homes.

Thom·as·ville (tŏm'əs-vĭl'). **1.** City (1975 est. pop. 18,400), seat of Thomas co., SW Ga., near the Fla. line; inc. 1831. It is a farm trade center, with a large fresh-vegetable market. Industries include meat-packing, baking, printing, and lumbering. The city is a winter resort, with excellent hunting and fishing in the area. An annual rose festival is held here. **2.** Industrial city (1975 est. pop. 15,500), Davidson co., central N.C., in the Piedmont; inc. 1854. It has cotton mills and textile, garment, and furniture industries.

Thomp·son (tŏmp'sən), city (1971 pop. 19,001), central Man., Canada, on the Burntwood River. A mining town, it developed after large nickel deposits were discovered in the area in 1956.

Thompson, river, 304 mi (489 km) long, Canada, formed by the North Thompson and the South Thompson rivers at Kamloops, S British Columbia, and flowing W and S to the Fraser River at Lytton. The river was discovered (1808) by Simon Fraser and named by him for David Thompson, a fellow explorer.

Thompson Falls, town (1970 pop. 1,356), seat of Sanders co., NW Mont., NW of Missoula; inc. 1910. In mining and timber country, it has several sawmills.

Thom·son (tŏm'sən), industrial city (1970 pop. 6,503), seat of McDuffie co., E Ga., W of Augusta; inc. 1854. Thomson is on the Fall Line and manufactures lumber and cotton products.

Thon Bu·ri (tŭn bŏŏr'ē), city (1970 pop. 627,989), central Thailand, on the W bank of the Chao Phraya River across from Bangkok. Part of metropolitan Bangkok, Thon Buri is a center of rice milling, saw-milling, and light manufacturing industries. It was capital of Siam from 1767 to 1782.

Thou·sand Islands (thou'zənd), a group of more than 1,800 islands and 3,000 shoals in the St. Lawrence River, E of Lake Ontario, N N.Y. and S Ont., Canada, stretching c.50 mi (80 km) along the U.S.-Canada line. Most of the islands are in Canada; Wolfe Island, Ont. (48 sq mi/124 sq km), is the largest. There are numerous parks on the islands, including Canada's St. Lawrence Islands National Park. The five-span Thousand Islands Bridge and highway (7 mi/11.3 km long; opened 1938) between the New York and Ontario mainlands crosses several islands and channels.

Thousand Oaks (ōks), residential city (1978 est. pop. 65,500), Ventura co., S Calif., in a farm area; inc. 1964. It has some light manufacturing. Oxnard Air Force Base is nearby.

Thrace (thrās), region, SE Europe, occupying the SE tip of the Balkan Peninsula and comprising NE Greece, S Bulgaria, and European Turkey. It is washed by the Black Sea in the northeast and by the Sea of Marmara and the Aegean Sea in the south. The Rhodope Mts. separate Greek from Bulgarian Thrace, and the Maritsa River (called the Évros in Greece) separates Greek from Turkish Thrace. The chief cities are İstanbul, Edirne (formerly Adrianople), and Gallipoli (all in Turkey); İstanbul (Constantinople) is generally considered a separate entity. With the exception of the mountainous Bulgarian section, Thrace is mainly agricultural, producing tobacco, wheat, silk, cotton, olive oil, and fruit. Coal, chromium, and wolfram are mined.

At the dawn of history the ancient Thracians extended as far west as the Adriatic Sea, but they were pushed eastward (c.1300 B.C.) by the Illyrians and later (5th cent. B.C.) by the Macedonians. In the north, however, Thrace at that period still extended to the Danube. The Thracians did not absorb Greek culture, and their tribes formed separate petty kingdoms. Many Greek colonies were founded in Thrace by c.600 B.C. Thrace was united as a kingdom under the chieftain Sitalces but after his death (428 B.C.) the state again broke up. By 342 B.C. all Thrace was held by Philip II of Macedon, and after 323 B.C. most of the country was in the hands of Lysimachus. It fell apart once more after Lysimachus' death (281 B.C.), and it was conquered by the Romans late in the 1st cent. B.C.

The region benefited greatly from Roman rule, but from the barbarian invasions of the 3rd cent. A.D. until modern times it was almost continuously a battleground. The northern section passed (7th cent.) to the Bulgarians; the southern section remained in the Byzantine Empire, but it was largely conquered (13th cent.) by the second Bulgarian empire. In 1361 the Ottoman Turks took Adrianople, and in 1453, after the fall of Constantinople, all of Thrace fell to the Turks.

After the annexation (1885) of northern Thrace by Bulgaria, the political meaning of the term Thrace became restricted to its southernmost part. In the first of the Balkan Wars (1912–13) Turkey ceded to Bulgaria all of Thrace west of the Maritsa River. After World War I, Bulgaria ceded (1919) the southern part of its share of Thrace to Greece. By the Treaty of Sèvres (1920) Greece also obtained most of eastern Thrace; the treaty, however, was superseded by the Treaty of Lausanne (1923), which restored to Turkey all Thrace east of the Maritsa. The Greek-Bulgarian boundary remains a point of dispute between the two countries.

Three Pa·go·das Pass (thrē' pə-gō'dəz), mountain pass, alt. 925 ft (282 m), at the S end of the Dawna Range, on the Burma-Thailand border. It is the chief route between southeast Burma and the Chao Phraya valley of Thailand.

Three Rivers, city (1970 pop. 7,355), St. Joseph co., SW Mich., S of Kalamazoo on the St. Joseph River at the confluence of the Portage and Rocky rivers; settled 1828, inc. as a village 1855, as a city 1895.

It is a farm trade center and industrial city, manufacturing boxes, paper products, and heating equipment. A French trading post and a Jesuit mission were here in the 17th cent.

Throck·mor·ton (thrŏk′môr′tən), county (1970 pop. 2,205), 920 sq mi (2,382.8 sq km), N Texas, drained by the Brazos River and its Clear Fork; formed 1858; co. seat Throckmorton. It has farming (wheat, cotton, grain sorghums, oats, fruit, and truck crops), livestock raising, some dairying, and some manufacturing.

Throckmorton, town (1970 pop. 1,105), seat of Throckmorton co., N Texas, SW of Wichita Falls; settled before 1880, inc. 1917. The rolling hills yield grain, cotton, cattle, poultry, and some oil.

Thu·le (thōō′lē), name given by the ancients to the most northerly land of Europe, an island discovered and described (c.310 B.C.) by the Greek navigator Pytheas and variously identified with Iceland, Norway, and the Shetland Islands.

Thu·le (thōō′lē, tōō′-), town in Thule dist. (1969 pop. 718), NW Greenland, NW of Cape York. In World War II the United States built a military base that is now the most important U.S. defense area in Greenland. It is also a base for Danish and U.S. scientific operations on the ice cap and on Peary Land.

Thun (tōōn), city (1974 est. pop. 36,800), Bern canton, central Switzerland, on the Aare River. Metal products, watches, and clothing are made. The Lake of Thun, c.20 sq mi (52 sq km), is at the foot of the Bernese Alps.

Thun·der Bay (thŭn′dər), city (1971 pop. 108,411), SW Ont., Canada, on Thunder Bay inlet of Lake Superior. The city was created in 1970 by the amalgamation of the twin cities of Fort William and Port Arthur and two adjoining townships. It is one of Canada's major ports, shipping wheat, lumber, coal, and iron ore. The city has shipyards, grain elevators, lumber and pulp and paper mills, and an oil refinery. Manufactures include structural steel, transportation equipment, and chemical products.

Thur·gau (tōōr′gou), canton (1974 est. pop. 185,000), 388 sq mi (1,005 sq km), NE Switzerland, bordered in the N by the Lake of Constance and watered by the Thur River; capital Frauenfeld. Cereals and fruit are grown, cattle are raised, and wine is produced. Thurgau was acquired (1264) by the Hapsburgs and was conquered (1460) and ruled by the Swiss cantons until 1798, when the French invaded Switzerland. In 1803 it became a canton of Switzerland.

Thu·ri·i (thyōō′rē-ī′), ancient city of Magna Graecia, S Italy, in Bruttium, on the Gulf of Tarentum. It was founded by Pericles in 443 B.C. to replace ruined Sybaris. Thurii became an ally of Rome and was pillaged (204 B.C.) by Hannibal.

Thu·rin·gi·a (thōō-rĭn′jē-ə), former state, central Germany, located in what is now SW East Germany. As constituted in 1946 under Soviet military occupation, Thuringia consisted of the prewar state of Thuringia (c.4,540 sq mi/11,760 sq km), with the addition of former Prussian enclaves and border areas, notably Erfurt and Mühlhausen. In 1952 the state was abolished as an administrative unit. The region of Thuringia extends to the foot of the Harz Mts. in the north and is crossed by the Thuringian Forest, which stretches from the Werra River in the west to the Thüringer Saale River in the southeast and rises to an altitude of 3,222 ft (982.7 m) in the Grosser Beerberg. The ancient Thuringians were conquered by the Franks during the 6th cent. A.D. and were converted (8th cent.) to Christianity by St. Boniface. In the 11th cent. the landgraves of Thuringia emerged as princes of the Holy Roman Empire and ruled over much of the territory that is modern Thuringia. The succession to Thuringia was long contested after 1247; the major part eventually fell to the house of Wettin, who in 1423 became electors of Saxony. After the division (1485) of the Wettin lands, Thuringia was split into several duchies. All the Thuringian territories except Saxe-Meiningen sided with Prussia in the Austro-Prussian War of 1866. The Thuringian states had been members of the German Confederation from 1815; they joined the North German Confederation in 1866 and the German Empire in 1871. Their rulers were expelled in 1918, and in 1920 the state of Thuringia was founded under the Weimar Republic.

Thur·rock (thûr′ək), urban district (1974 est. pop. 127,090), Essex, SE England, on the Thames River. It includes Tilbury, which has large docks that are part of the Port of London. Among Thurrock's industries are oil refining and the manufacture of soap, margarine, paper board, cement, and shoes. Queen Elizabeth I reviewed the troops at Tilbury Fort in 1588 at the time of the Spanish Armada.

Thurs·ton (thûrs′tən). **1.** County (1970 pop. 6,942), 388 sq mi (1,004.9 sq km), NE Nebr., in a farm area bounded in the E by the Missouri River and Iowa; drained by Logan Creek; formed 1889; co. seat Pender. It is in a feed, livestock, and grain area. Winnebago Indian Reservation is in the east; Omaha Indian Reservation is in the southeast. **2.** County (1970 pop. 76,894), 714 sq mi (1,849.3 sq km), W Wash., bounded in the NE by the Nisqually River and in the N by Puget Sound; formed 1852; co. seat Olympia. Its agriculture includes livestock, fruit, nuts, poultry, and dairy products. It has oystering, food processing, lumbering, and manufacturing (paper, plastics, concrete, and metal products). Part of Snoqualmie National Forest is in the county.

Tia·hua·na·co (tyä′wä-nä′kō), ancient Indian ruin, W Bolivia, S of Lake Titicaca, near the Peruvian border. Nearly 13,000 ft (3,965 m) above sea level, Tiahuanaco was probably the center of a pre-Incan Indian empire. Building was begun at some time before A.D. 500, and there is evidence of additional construction (c.1100–1300). Its structures were built of massive blocks weighing up to 100 tons. The stones, fitted together without mortar, were cut, squared, dressed, and notched with a precision equaled in no other aboriginal South American civilization.

Tia·ret (tyä-rā′), city (1974 est. pop. 63,039), NW Algeria, capital of Tiaret dept. Since Roman times it has been the center of a prosperous agricultural area.

Ti·ber (tī′bər), river, 251 mi (404 km) long, rising in the Etruscan Apennines, central Italy. It flows generally south across Tuscany, Umbria, and north Latium, then southwest through Rome to empty into the Tyrrhenian Sea. It is connected with the Arno River by the Chiana Canal. The upper Tiber and its tributaries are used to generate electricity. Subject to floods, the banks of the Tiber, especially in Rome, are diked.

Ti·be·ri·as (tī-bîr′ē-əs), town (1972 pop. 23,800), NE Israel, on the Sea of Galilee, 682 ft (208 m) below sea level. It is a trade center for agricultural settlements, a resort and spa noted for its thermal springs, and a lake port. There are machine shops, fisheries, and sausage, candy, and box factories in Tiberias. Named for Emperor Tiberius, the town was built A.D. c.20. After the destruction of Jerusalem, Tiberias became (2nd cent.) a center of Jewish learning; parts of the Mishna and Jerusalem Talmud were edited here. Tiberias was captured by the Arabs in 637, taken by the Crusaders in the 11th cent., recaptured by Saladin in 1187, and occupied by Egypt in 1247. It became part of the Ottoman Empire in the 16th cent. In 1922 it joined Palestine.

Ti·bet (tĭ-bĕt′), autonomous region (1970 est. pop. 1,700,000), c.471,700 sq mi (1,221,700 sq km), SW China. The capital is Lhasa. A Chinese autonomous region since 1951, Tibet is bordered on the southeast by Burma, on the south by India, Bhutan, Sikkim, and Nepal, on the west by India and Kashmir, and on the north and east by Chinese provinces. Almost completely surrounded by mountain ranges (including the Himalayas in the south and the Kunlun in the north), Tibet is largely a plateau averaging c.16,000 ft (4,880 m) in height. Many of the mightiest rivers of eastern Asia, especially the Yangtze, the Mekong, and the Salween, rise in Tibet; the most im-

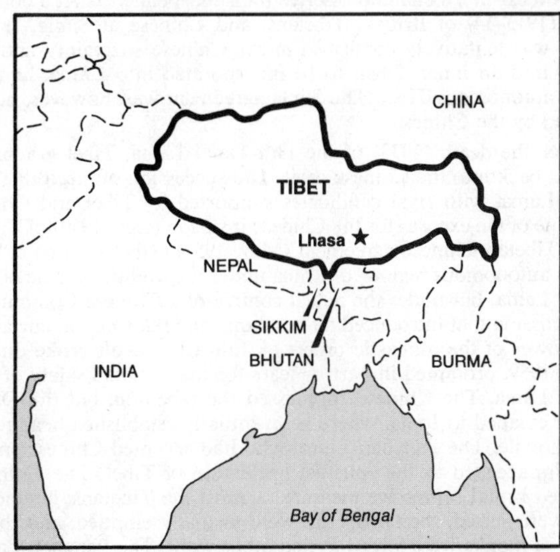

portant is the navigable Tsangpo (the Brahmaputra), which follows an easterly course through southern Tibet. North of the Tsangpo are many salt lakes, the largest being Na-mu Hu (Tengri Nor) in the east.

The inhabitants of Tibet are of Mongolian stock and speak a Tibeto-Burman language. Before the unsuccessful revolt of 1959, many of the dwellers of the cities were Lamaist monks, who may have comprised as much as one sixth of the country's male population. The chief figures of Lamaism, the Dalai Lama and the Panchen Lama (or Tashi Lama, for the lamastery at Tashi Lumpo), were at least the nominal heads of the Tibetan government. In general administration was equally divided between lamas and laymen belonging to the feudal aristocracy.

Economy. In this land of scant rainfall and a short growing season, the only extensive agricultural region is the Tsangpo valley, where barley, wheat, potatoes, millet, and turnips are grown. In this valley are nearly all the large cities, including Lhasa. Most other areas of Tibet are suited only for grazing; yaks, which can withstand the intense cold, are the principal domestic animals, and there are also large herds of goats and sheep. Much of the population is engaged in a pastoral life, but the advances made by irrigation and the growing of forage crops is decreasing nomadism. Stockbreeding cooperatives have been organized. In addition to vast salt reserves, Tibet has large deposits of gold, copper, and radioactive ores, but mining was long prohibited for religious reasons. Motor roads now connect divergent sections of the country.

History. Over the centuries the Tsangpo valley was the focus of ancient trade routes from India, China, and central Asia. Tibet emerged from obscurity to flourish in the 7th cent. A.D. as an independent kingdom with its capital at Lhasa. The Chinese first established relations with Tibet during the T'ang dynasty (618–906), and there were frequent wars of conquest. The Tibetan kingdom was associated with early Mahayana Buddhism, which the scholar Padmasambhava fashioned (8th cent.) into Lamaism. Toward the end of the 12th cent. many Indian Buddhists, fleeing before the Moslem invasion, went to Tibet. From the 13th cent.–18th cent. Tibet fell under Mongol influence. In 1270 Kublai Khan, emperor of China, was converted to Lamaism by the abbot of the Sakya lamasery; the abbot returned to Tibet to found the Sakya dynasty (1270–1340) and to become the first priest-king of Tibet. In 1720 the Manchu dynasty replaced Mongol rule in Tibet. China thereafter claimed suzerainty, often merely nominal.

During the 18th cent. British authorities in India attempted to establish relations with Lhasa, but the Gurkha invasion of 1788 and the subsequent Gurkha war (1792) with Tibet brought an abrupt end to the rapprochement. Throughout the 19th cent. Tibet maintained its traditional seclusion. Meanwhile, Ladakh, long part of Tibet, was lost to the rulers of Kashmir, and Sikkim was detached (1890) by Britain. In 1893 Britain succeeded in obtaining a trading post at Yatung, but continued Tibetan interference led to a military expedition (1904) to Lhasa, which enforced the granting of trade posts at Yatung, Chiang-tzu, and Ka-erh. Subsequently, Britain recognized (1906, 1907) China's suzerainty over Tibet. However, with the overthrow of the Manchu dynasty in China, the Tibetans expelled (1912) the Chinese in Tibet and reasserted their independence. At a conference (1913–14) of British, Tibetans, and Chinese at Simla, India, Tibet was tentatively confirmed under Chinese suzerainty and divided into an inner Tibet, to be incorporated into China, and an outer autonomous Tibet. The Simla agreement was, however, never ratified by the Chinese.

After the death (1933) of the 13th Dalai Lama, Tibet gradually drifted back into the Chinese orbit. The succession of the 10th Panchen Lama, with rival candidates supported by Tibet and China, was one of the excuses for the Chinese invasion (Oct., 1950) of Tibet. By a Tibetan-Chinese agreement (May, 1951), Tibet became a "national autonomous region" of China under the traditional rule of the Dalai Lama, but under the actual control of a Chinese Communist Commission that introduced far-reaching land reforms and curtailed the power of the monastic orders. A full-scale revolt broke out in Mar., 1959, prompted in part by fears for the personal safety of the Dalai Lama. The Chinese suppressed the rebellion, but the Dalai Lama escaped to India, where he eventually established headquarters in exile. The Panchen Lama, who had accepted Chinese sponsorship, acceded to the spiritual leadership of Tibet. The Chinese adopted brutal repressive measures against the Tibetans; landholdings were seized, the lamaseries were virtually emptied, and thousands of monks were forced to find other work. The Panchen Lama was deposed in 1964 after making statements supporting the Dalai Lama; he was replaced by a secular Tibetan leader. In 1962 China launched attacks along the Indian-Tibetan border. Following a cease-fire, Chinese troops withdrew behind the disputed line in the east, but continued to occupy part of Ladakh in Kashmir. Some of the border areas are still in dispute. In 1965 the Tibetan Autonomous Region was formally established.

Ti·ci·no (tē-chē'nō), canton (1970 pop. 245,458), 1,086 sq mi (2,813 sq km), S Switzerland, on the S slope of the central Alps, bordering on Italy; capital Bellinzona. Largely a mountainous region, Ticino embraces the Ticino River valley and part of Lago Maggiore and of the Lake of Lugano. Although it has a pastoral economy, wine is widely produced in the valleys and corn and tobacco are cultivated. Industry, mainly in the south, produces metal goods and chemicals; there is an extensive hydroelectric system along the Ticino River. Ticino is noted for its resorts. A part of Transpadane Gaul under the Roman Empire, Ticino later shared the history of Lombardy until the Swiss confederates captured it (15th–16th cent.) from the duchy of Milan. It was ruled until 1798 by Schwyz and Uri cantons and became a Swiss canton in 1803.

Ticino, river, 154 mi (248 km) long, rising in Ticino canton, S Switzerland, and flowing generally S through Lago Maggiore into N Italy, joining the Po River below Pavia. In Switzerland the Ticino is used to generate electricity. It provides irrigation in Italy. It was the scene (218 B.C.) of Hannibal's victory over Scipio.

Ti·con·der·o·ga (tī'kŏn-də-rō'gə), resort village (1970 pop. 3,268), Essex co., NE N.Y., on a neck of land between Lakes George and Champlain; settled in the 17th cent., inc. 1889. The falls of Lake George furnish power for its paper mill. Fort Carillon, built here by the French in 1755, was successfully defended by Montcalm against James Abercromby in 1758, but it fell to Jeffrey Amherst in 1759 and was renamed Fort Ticonderoga. It was captured (May 10, 1775) by a detachment of Green Mountain Boys under Ethan Allen and troops commanded by Benedict Arnold. In the Saratoga campaign it was abandoned (1777) without a fight by Arthur St. Clair to John Burgoyne. The fort was restored as a museum in 1909.

Tie·nen (tē'nən), town (1971 est. pop. 24,013), Brabant prov., central Belgium. It is a commercial and industrial center, with a major beet sugar refining industry. Tienen has suffered numerous sieges in its history. In 1831 the last fighting in the struggle for Belgian independence took place here.

Tien Shan (tē-ĕn' shän'), mountain system of central Asia, extending c.1,500 mi (2,415 km) from the Pamir Mts., USSR, NE through Sinkiang Uigur Autonomous Region, NW China, to the China-Mongolia border; Pobeda Peak (24,406 ft/7,443.8 m), on the USSR-China line, is the highest point. Because of the dry climate, the Tien Shan's snow line is generally above 11,000 ft (3,355 m). Coal, iron, lead, and zinc are mined in the region; grains are the predominant crop in the valleys. China and the USSR are linked by several passes, notably the Terek Pass (12,730 ft/3,882.7 m) on the route connecting Kashgar and Samarkand.

Tien·tsin or **T'ien-ching** (both: tĭn'tsĭn', tē-ĕn'-), city (1970 est. pop. 4,500,000), NE China. It is in east central Hopeh prov. but is administered by the central government. Tientsin is a port at the confluence of the Hai River (c.30 mi/50 km from its mouth) with the Grand Canal. Tientsin is a leading international port and is connected by rail with much of China. The city is an important manufacturing center, with iron and steel works, textile mills, a chemical industry based on salt, paper mills, and plants making automobiles, precision instruments, cement, fertilizer, rubber products, and woolen carpets. Strategically located on the overland route to Manchuria, Tientsin has been a frequent military objective. Agreements (1860) between China, Britain, and France made Tientsin a treaty port and conceded parts of the city for foreign settlements and garrisons. In the Boxer Rebellion (1900) there was a joint foreign occupation. With the abolition of the last foreign concessions in 1946 Tientsin was completely restored to Chinese sovereignty.

Ti·er·ra Am·a·ril·la (tĭ-ĕr'ə ăm'ə-rĭl'ə), village (1970 pop. 800), seat of Rio Arriba co., N N.Mex., near the San Juan Mts. and the Colo. border NNW of Santa Fe, at an altitude of c.7,460 ft (2,275.3 m). Livestock, grain, poultry, and potatoes are produced. There are coal mines in the vicinity.

Ti·er·ra del Fu·e·go (tĭ-ĕr'ə dĕl fo͞o-ā'gō, fyo͞o-, tyĕ'rä dĕl fwā'gō), archipelago, 28,476 sq mi (73,753 sq km), off S South America, separated from the mainland by the Strait of Magellan. The Andes extend through the western part, and the plateau of Patagonia continues into the eastern section. Tierra del Fuego is divided into two sections, the eastern part belonging to Argentina (the territory of Tierra del Fuego) and the larger western part to Chile (a part of Magallanes prov.). The economy is based on the raising of sheep and the exploitation of petroleum. Tierra del Fuego was discovered by Magellan in 1520. The introduction of sheep farming and the discovery of gold in the 1880s led to European, Argentine, and Chilean immigration. The aboriginal peoples were gradually killed off by disease.

Tif·fin (tĭf'ĭn), city (1978 est. pop. 19,300), seat of Seneca co., N central Ohio, on the Sandusky River in a farm area; inc. 1835. Radiators, china, glassware, heavy machinery, machine parts, wire and cable, and electrical equipment are made.

Tif·lis (tĭf'lĭs): *see* Tbilisi, USSR.

Tift (tĭft), county (1970 pop. 27,288), 266 sq mi (688.9 sq km), S central Ga., bounded on the NE by the Alapaha River and drained by the Withlacoochee River; formed 1905; co. seat Tifton. It has coastal-plain agriculture (cotton, corn, peanuts, livestock, and tobacco) and some manufacturing.

Tif·ton (tĭf'tən), city (1975 est. pop. 12,700), seat of Tift co., S central Ga.; inc. 1890. Tobacco, cotton, and truck crops are marketed and textiles and clothing, lumber, and plastics are produced.

Ti·gre (tē'grā), city (1970 pop. 152,335), Buenos Aires prov., E Argentina. A railroad terminus and river port, Tigre is a fruit market and has sawmills and shipyards.

Ti·gris (tī'grĭs), river of SW Asia, c.1,150 mi (1,850 km) long, rising in the Taurus Mts., E Turkey, and flowing SE through Iraq to join the Euphrates River, with which it forms the Shatt al Arab. It flows swiftly and receives many tributaries, including the Diyala and the Great and Little Zab. Dams across the river divert water for irrigation. The Tigris is subject to sudden, devastating floods, and the Wadi Ath Tharthar flood-control project protects Baghdad and vicinity from floods in addition to irrigating c.770,000 acres (311,850 hectares) of land. The Tigris is navigable to Baghdad for shallow-draft vessels; above Baghdad rafts carry much of the trade to Mosul. In antiquity some of the great cities of Mesopotamia stood on the banks of the Tigris, and the river served as an important transportation route. The Tigris flood plain was cultivated by irrigation from the earliest times.

Ti·jua·na (tē-hwä'nä), city (1970 pop. 277,306), Baja California state, NW Mexico, just S of the U.S. border. It is noted for its racetracks and bullfights. An irrigated agricultural area surrounds the city.

Ti·kal (tē-käl'), ruined city of the Classic Period of the Maya, N central Petén, Guatemala. The largest and possibly the oldest of the Maya cities, Tikal consists of nine groups of courts and plazas built on hilly land above surrounding swamps and interconnected by bridges and causeways. Temples and palaces rise above the plazas. The tallest structure, a temple, is 229 ft (70 m) high.

Til·burg (tĭl'bûrg'), city (1973 est. pop. 154,069), North Brabant prov., S Netherlands, near the Belgian border. Manufactures include textiles, textile machinery, dyes, and leather. The town hall was used in the first half of the 19th cent. as a royal residence by King William II of the Netherlands.

Til·den (tĭl'dən), village (1970 pop. 450), seat of McMullen co., S Texas, on the Frio River E of Cotulla.

Til·la·mook (tĭl'ə-mo͝ok'), county (1970 pop. 18,034), 1,115 sq mi (2,888.6 sq km), NW Oregon, bounded in the W by the Pacific Ocean, and drained in the N by the Nehalem River; formed 1853; co. seat Tillamook. It is in a dairying and cheese-producing area, with fisheries and seafood processing. It has lumbering and millwork.

Tillamook, city (1970 pop. 3,968), seat of Tillamook co., NW Oregon, near the head of Tillamook Bay W of Portland; inc. 1891. The trade and processing center of a fishing, lumbering, and dairying region, it is noted for its cheese.

Till·man (tĭl'mən), county (1970 pop. 12,901), 901 sq mi (2,333.6 sq km), SW Okla., bounded in the S by the Red River and the Texas border, in the N by the North Fork of the Red River, and drained by the Deep Red Run; formed 1907; co. seat Frederick. Its agriculture includes cotton, alfalfa, grain, livestock, poultry, and dairy products. It manufactures clothing, leather, and aircraft engines.

Til·sit (tĭl'zĭt): *see* Sovetsk, USSR.

Tim·a·ru (tĭm'ə-ro͞o), city (1971 pop. 28,326), central South Island, New Zealand, on the Pacific Ocean. Frozen meats and other products are exported.

Tim·ber Lake (tĭm'bər), city (1970 pop. 625), seat of Dewey co., N central S.Dak., NNW of Pierre. It is a trading point for a farming region yielding grain, dairy products, livestock, and poultry.

Tim·buk·tu (tĭm'bŭk-to͞o', tĭm-bŭk'to͞o), city (1971 est. pop. 11,900), central Mali, near the Niger River. Connected with the Niger by a series of canals, Timbuktu is served by the small river port of Kabara. Its salt trade and handicraft industries make it an important meeting place for the nomadic people of the Sahara. Timbuktu was founded (11th cent.) by the Tuareg. By the 14th cent., it was famous for its gold trade. Under the Songhai empire (15th and 16th cent.) the city was a great Moslem educational center. Timbuktu was sacked in 1593 by invaders from Morocco.

Tim·gad (tĭm'găd'), ruined city, Algeria, S of Constantine. It is sometimes called the Pompeii of North Africa because of the extensive remains of the Roman city founded here by Trajan in A.D. 100.

Ti·mi·şoa·ra (tē'mē-shwä'rä), city (1973 est. pop. 204,687), W Rumania, in the Banat, on the Beja Canal. It is a railroad hub and an industrial center, with engineering works, plants processing food and tobacco, and factories manufacturing textiles, machinery, and chemicals.

Tim·mins (tĭm'ĭnz), town (1973 est. pop. 42,970), central Ont., Canada, on the Mattagami River. Timmins is the commercial center of the rich Porcupine gold-mining district, where gold was first discovered in 1909. Silver, copper, lead, and zinc are also mined.

Ti·mor (tē'môr, tē-môr'), island (c.13,200 sq mi/34,190 sq km), largest and easternmost of the Lesser Sundas; a province of Indonesia. The island is long, narrow, and almost wholly mountainous. Rice, coconuts, and coffee are grown, sandalwood is cut, and stretches of grassland support cattle. The inhabitants, who are of Malay and Papuan stock, are predominantly Christian. The Portuguese were the first Europeans to establish themselves in Timor; their claim to the island was disputed by the Dutch, who arrived in 1613. By a treaty of 1859 (effective in 1914), the border between the Dutch and the Portuguese was finally settled. In World War II Timor was occupied (early 1942) by the Japanese. With the creation of the Republic of Indonesia in 1950 Dutch Timor, comprising the western half of the island, became Indonesian territory; the Portuguese retained the eastern half as a colony. In the wake of the overthrow of the Portuguese dictatorship (1974) and the new government's policy of decolonization, civil war erupted in eastern Timor (1975). Fearing that an independent eastern Timor would be susceptible to Chinese influence, Indonesia invaded the eastern half of the island and incorporated it as the 27th province of the country (1976).

Tim·pa·no·gos Cave National Monument (tĭm'pə-nō'gəs): *see* National Parks and Monuments Table.

Ti·ni·an (tĭn'ē-ăn', tē-nē-än'), island (1970 pop. 710), 39 sq mi (101 sq km), W Pacific, one of the Mariana Islands in the U.S. Trust Territory of the Pacific Islands. The island lies immediately southwest of Saipan. The planes that dropped atomic bombs on Hiroshima and Nagasaki were flown from Tinian.

Tin·ley Park (tĭn'lē), village (1977 pop. 23,207), Cook and Will counties, NE Ill., a residential suburb of Chicago; inc. 1892. A state mental hospital is here.

Tí·nos or **Te·nos** (both: tē'nôs), island (1971 pop. 8,232), 79 sq mi (204.6 sq km), SE Greece, in the Aegean Sea; one of the Cyclades. Wine, figs, wheat, and silk are produced on Tínos, and green marble is quarried. The island was a colony of Venice from 1390 to 1715, when it was captured by the Ottoman Turks.

Tin·tag·el (tĭn-tăj'əl), village, Cornwall, SW England. It is south of Tintagel Head, a promontory connected to the mainland by a narrow, rocky neck of land. Nearby is the reputed birthplace of King Arthur.

Ti·o·ga (tī-ō'gə). **1.** County (1970 pop. 46,513), 524 sq mi (1,357.2 sq km), S N.Y., bounded on the S by the Pa. border, intersected by the Susquehanna River, and drained by Cayuta, Catatonk, and Owego creeks; formed 1791; co. seat Owego. This dairying area, with diversified manufacturing, also has poultry raising and grain growing. **2.** County (1970 pop. 39,691), 1,146 sq mi (2,968.1 sq km), N Pa., bounded in the N by N.Y., drained in the E by the Tioga River and in the W by Pine Creek; formed 1804; co. seat Wellsboro. Its agriculture includes maple sugar, buckwheat, hay, and dairy products. It has bituminous-coal mining and diversified manufacturing (clothing, leather goods, metal products, and machinery).

Ti·o·nes·ta (tī'ō-nĕs'tə), borough (1970 pop. 711), seat of Forest co., NW Pa., ENE of Oil City on the Allegheny River at the mouth of Tionesta Creek. The area yields lumber, flour, dairy products, gas, and oil.

Tip·pah (tĭp'ə), county (1970 pop. 15,852), 464 sq mi (1,201.7 sq km), N Miss., bordering on Tenn. in the N; formed 1836; co. seat Ripley. It is drained by the Tallahatchie River. Cotton and pine timber are important to its economy.

Tip·pe·ca·noe (tĭp'ē-kə-nōō'), county (1970 pop. 109,378), 500 sq mi (1,295 sq km), W central Ind., intersected by the Wabash River; formed 1826; co. seat Lafayette. In a rich agricultural area yielding grain and livestock, it has bituminous-coal mining and diversified industry (food processing, lumbering, and the manufacture of wood products, plastics, and concrete).

Tippecanoe, river, c.170 mi (275 km) long, rising in the lake district of NE Ind. and flowing SW to the Wabash River, near Lafayette. U.S. Gen. William Henry Harrison fought the Shawnee Indians in the Battle of Tippecanoe, Nov. 7, 1811, on the site of the present-day town of Battle Ground, Ind.

Tip·per·ar·y (tĭp'ə-râr'ē), county (1971 pop. 123,565), 1,643 sq mi (4,255.4 sq km), S central Republic of Ireland; co. town Tipperary. Administratively, the county is divided into North Riding (its administrative center at Nenagh) and South Riding (its administrative center at Clonmel). The region is part of the central plain of Ireland, but the terrain is diversified by several mountain ranges. It is one of the richest agricultural areas in Ireland. Dairy farming and cattle raising are most important. Other industries are slate quarrying and the manufacture of meal and flour.

Tip·ton (tĭp'tən). **1.** County (1970 pop. 16,650), 261 sq mi (676 sq km), central Ind., drained by Cicero Creek, small Turkey Creek, and the South Fork of Wildcat Creek; formed 1844; co. seat Tipton. In an agricultural area yielding corn, wheat, hogs and cattle, it has timber and some manufacturing (canned fruit and vegetables, hardware, and auto parts). **2.** County (1970 pop. 28,001), 459 sq mi (1,188.8 sq km), W Tenn., bounded in the W by the Mississippi River and in the N by the Hatchie River; formed 1823; co. seat Covington. It is in a cotton-growing and livestock-raising area, which also produces corn and soybeans. Its industry includes food processing, textile milling, lumbering, and the manufacture of furniture, paper, pottery, and radio and television equipment.

Tipton. 1. City (1970 pop. 5,313), seat of Tipton co., central Ind., SSE of Kokomo; laid out 1845. It is a food-processing center and makes automobile parts. **2.** City (1970 pop. 2,877), seat of Cedar co., E Iowa, ENE of Iowa City; inc. 1857. Cheese is made here.

Tip·ton·ville (tĭp'tən-vĭl'), town (1970 pop. 2,407), seat of Lake co., NW Tenn., at the S end of Reelfoot Lake in a cotton and resort region; inc. 1900.

Ti·ra·në (tē-rä'nə) or **Ti·ra·na** (tē-rä'nä), city (1971 est. pop. 174,800), capital of Albania and of Tiranë prov., central Albania, on the Ishm River. It is the largest city and the chief industrial and cultural center of the country. Tiranë is located on a fertile plain that yields a variety of agricultural products. Its manufactures include textiles, metal products, footwear, agricultural machinery, and foodstuffs. Lignite is mined nearby.

Ti·ras·pol (tĭ-răs'pəl), city (1973 est. pop. 122,000), SW European USSR, in Moldavia on the Dnestr River. It is a major agricultural processing center. Tiraspol was founded (1792) as a Russian fortress on the site of a Moldavian settlement.

Tîr·go·viş·te (tûr'gô-vēsh'tĕ), town (1973 est. pop. 37,249), S central Rumania, in Walachia, in a petroleum-producing region. Oil refining and the manufacture of oil-field equipment are the chief industries. Many tourists are attracted by Tîrgovişte's historic buildings, including a remarkable 16th cent. cathedral with nine towers, a 15th cent. monastery, and the ruins of a 14th cent. castle.

Tîr·gu-Mu·reş (tûr'gōō-mōō'rĕsh), city (1973 est. pop. 109,873), central Rumania, chief city of the Magyar Autonomous Region, in Transylvania, on the Mureşul River. It is a major industrial center, with sugar refineries, distilleries, and industries manufacturing food products, chemicals, fertilizers, machinery, and furniture. Tîrgu-Mureş is also a market for agricultural products. There are pedagogical and medical-pharmaceutical institutes in the city. Dating from the 12th cent., Tîrgu-Mureş was the scene (1704) of the proclamation of Francis II Rakoczy as "ruling prince" of Hungary. The city remained part of Hungary until 1918, when Rumania acquired Transylvania.

Tir·no·vo (tûr'nə-vō'): *see* Trnovo, Bulgaria.

Ti·ruch·i·rap·al·li (tĭ-rōōch'ĭ-rä'pə-lē) or **Trich·i·nop·o·ly** (trĭch'ə-nŏp'ō-lē), city (1971 pop. 307,400), Tamil Nadu state, SE India, on the Cauvery River. It is an important educational, religious, and commercial city known for its gold and silver filigree work and brassware. The city's chief landmark is the shrine of Srirangam, an elaborately carved monument to the Hindu god Siva, at the base of a 270-ft (82.4-m) rock.

Ti·ru·nel·ve·li (tə-rōō-nĕl'və-lē), town (1971 pop. 108,498), Tamil Nadu state, SE India. Now an agricultural trading center with a sugar refinery, it was once an imperial city of the Chola kingdom (c.900–1200). St. Francis Xavier conducted missionary activity in the area (c.1545).

Ti·rup·pur (tĭ'rōō-pōōr), city (1971 pop. 113,302), Tamil Nadu state, S India. Located in the Eastern Ghats valley, Tiruppur is a commercial center with cotton spinning and weaving industries.

Ti·ryns (tĭ'rĭnz), ancient city of Greece, in the NE Peloponnesus, near Argos. Excavations have revealed not only extensive pre-Homeric palaces of the Mycenaean period but also remains going far back in prehistory. The old city was prominent in Greek legend.

Tish·o·ming·o (tĭsh'ə-mĭng'gō), county (1970 pop. 14,940), 443 sq mi (1,147.4 sq km), extreme NE Miss., bordering on Ala. in the E and on Tenn. in the N; formed 1836; co. seat Iuka. This hilly county is drained by tributaries of the Tennessee and Tombigbee rivers. It is a farming area (cotton, corn, sweet potatoes, and hogs), with some lumbering. Clay, sandstone, limestone, phosphates, and bauxite are its natural resources.

Tishomingo, city (1970 pop. 2,663), seat of Johnston co., S Okla., on the Washita River and Lake Texoma, ENE of Ardmore; settled 1850. Long the chief city of the Chickasaw Indians, Tishomingo is now the trade center of a farm and resort area.

Ti·sza (tĭs'ə), river, c.600 mi (965 km) long, formed by two headstreams in the Carpathians. It flows generally south across eastern Hungary into northern Yugoslavia, where it enters the Danube River east of Novi Sad. There are hydroelectric facilities on the river in Hungary. The Tisza is navigable for small craft to Szolnok and is also used to float timber.

Ti·ti·ca·ca (tē'tē-kä'kä), lake, c.3,200 sq mi (8,290 sq km), 110 mi (177 km) long, and c.900 ft (275 m) deep, in the Andes, on the Bolivia-Peru border; largest freshwater lake in South America and the world's highest large lake (c.12,500 ft/3,815 m above sea level). The lake is divided into two basins by the Strait of Tiquina. Fed by many short mountain streams, the lake is drained by the Desaguadero River to Lake Poopó. A center of Indian life from pre-Incan times, the shores of Titicaca are presently crowded with Indian villages and terraced fields. The almost constant temperature of the water (51°F/11°C) modifies the climate and makes possible the growing of maize and wheat at so high an altitude. In the lake are the islands of Titicaca and Coati, the legendary birthplace of the Incas, that contain ruins of past civilizations.

Ti·to·grad (tē'tō-gräd'), town (1971 pop. 54,509), capital of Montenegro, S Yugoslavia, at the confluence of the Ribnica and Morača rivers. A commercial center, it has industries producing furniture, tobacco, and foodstuffs.

Ti·tov Ve·les (tē'tôf vĕ'lĕs), town (1971 pop. 36,026), SE Yugoslavia, in Macedonia, on the Vardar River. It is a road and rail junction and the market center for an agricultural and silk-producing region.

Ti·tus (tī'təs), county (1970 pop. 16,702), 418 sq mi (1,082.6 sq km), NE Texas, bounded on the N by the Sulphur River, on the S by Cypress Bayou; formed 1846; co. seat Mount Pleasant. This forested area also has farming (cotton, sweet potatoes, peanuts, corn, fruit, and truck crops), dairying, livestock-raising, and deposits of oil, coal, natural gas, asphalt, and clay.

Ti·tus·ville (tī'təs-vĭl'). **1.** City (1978 est. pop. 28,800), seat of Brevard co., E Fla., on Indian River (a lagoon); inc. 1886. It is a regional trade center. The construction in the 1950s of the space center on nearby Cape Canaveral brought much activity to the area and caused the city's population to increase tenfold in less than a decade.

2. City (1970 pop. 7,331), Crawford co., NW Pa., NNE of Oil City, in a natural-gas and oil area; settled 1796, laid out 1809, inc. 1866. It has a steel mill and makes tools and electrical equipment. Nearby is a state park where the drilling of the first oil well in the United States was successfully completed in 1859.

Tiv·er·ton (tĭv′ər-tən), rural town (1970 pop. 12,559), Newport co., SE R.I., between the Sakonnet River and the Mass. line; settled 1680, included in Massachusetts until 1746, inc. 1747. Tiverton is a summer resort center in a farm area, and there are oyster fisheries.

Ti·vo·li (tĭv′ə-lē, tē′vō-lē), city (1972 est. pop. 42,228), in Latium, central Italy, on the Aniene River. The city is situated on a terrace dominating nearby Rome and the plain to the sea. It is celebrated for the waterfalls formed by the Aniene River and for the Villa d'Este. There are ruins of several Roman villas, notably that of Emperor Hadrian.

Ti·zi Ou·zou (tē-zē′ o͞o-zo͞o′), city (1974 est. pop. 108,000), capital of Tizi Ouzou dept., N Algeria. It is the administrative and commercial center of an agricultural region where figs and olives are grown.

Tji·re·bon (chĕr′ĭ-bŏn′), city (1971 pop. 178,529) and seaport, N Java, Indonesia, on the Java Sea. Crops grown in the fertile coastal plain include sugar and rice. The city has diversified manufactures and is the seat of a private university.

Tlal·ne·pan·tla (tläl′nā-pän′tlä), city (1970 pop. 45,575), Mexico state, S central Mexico, on the Tlalnepantla River. It is a communications and industrial center that owes its importance largely to its proximity to Mexico City. Smelting, metalworking, machine-building, and chemical manufacturing are the chief industries. There are important archaeological ruins nearby.

Tla·que·pa·que (tlä′kā-pä′kā), city (1970 pop. 59,760), Jalisco state, SW Mexico, in the Guadalajara valley. Its folklore and handicrafts, as well as its proximity to Guadalajara, make Tlaquepaque a popular tourist spot.

Tlax·ca·la (tläs-kä′lä), state (1970 pop. 420,638), 1,555 sq mi (4,027.5 sq km), E central Mexico; capital Tlaxcala. It is the smallest and one of the most densely populated Mexican states. The western part lies within Mexico's central plateau; the remainder, however, is extremely mountainous, with a temperate to cold climate. Maguey, cereals, and subsistence crops are grown in the valleys. In the mountains are the sources of the Río Balsas.

Tlem·cen (tlĕm-sĕn′), city (1974 est. pop. 115,054), NW Algeria, capital of Tlemcen dept. Its location on a crossroads between the Mediterranean coast and the Sahara and between Algeria and Morocco has made it a commercial center since ancient times. Tlemcen retains the atmosphere of medieval Moslem life. It still exports the carpets, woolens, and leather goods for which it has long been noted.

To·ba (tō′bä), largest lake of Indonesia, 448 sq mi (1,160.3 sq km), N Sumatra. It is drained by the Asahan River. In the lake is Samosir, a large island (205 sq mi/531 sq km) that is linked to the mainland by an isthmus.

To·ba·go (tə-bā′gō): *see* Trinidad and Tobago.

To·bol (tə-bôl′), river, c.1,050 mi (1,690 km) long, rising in the Mugodzhar Hills, NE Kazakhstan, Central Asian USSR, and flowing NE past Kustanay, into the Russian Republic, and past Kurgan to join the Irtysh River at Tobolsk. It is navigable in its lower course.

To·bolsk (tə-bôlsk′), city (1970 pop. 49,260), W Siberian USSR, a port on the Irtysh River near its confluence with the Tobol. Industries include shipbuilding; woodworking; fish, fur, leather, and flax processing; meat packing; and furniture and carpet making. Founded in 1587 by Cossacks on the site of a Tatar village, Tobolsk was one of Russian Siberia's first towns.

To·bruk (tō′bro͝ok′), city (1970 est. pop. 28,000), NE Libya, a port on the Mediterranean Sea. It was a fiercely contested objective in World War II.

To·can·tins (tō′kăɴ-tēɴs′), river, 1,640 mi (2,639 km) long, formed in S central Goiás state, Brazil, by the confluence of two headstreams. It flows north to the Pará River, the southern distributary of the Amazon, southwest of Belém. It is only partly navigable because of rapids. There are diamond washes near Carolina.

Toc·co·a (tə-kō′ə), city (1970 pop. 6,971), seat of Stephens co., NE Ga., N of Athens, in the Appalachian foothills near the Tugaloo River; inc. 1875. It is an industrial center in a farm area.

Todd (tŏd). **1.** County (1970 pop. 10,823), 376 sq mi (973.8 sq km), SW Ky., bounded on the S by Tenn., and drained by the Elk and West forks of the Red River; formed 1819; co. seat Elkton. In a rolling agricultural area yielding dark tobacco, corn, wheat, hogs, cattle, and fruit, it has timber, stone quarries, and some industry (clothing manufacturing, aluminum foundries, and machinery production). It includes Blue and Gray State Park and Jefferson State Monument. **2.** County (1970 pop. 22,114), 942 sq mi (2,440 sq km), central Minn., watered by Long Prairie River, with Lake Osakis in the SW; formed 1855; co. seat Long Prairie. It is in an agricultural area that produces dairy goods, livestock, and grain. There is some mining, dairy processing, and clothing manufacturing. **3.** County (1970 pop. 6,606), 1,388 sq mi (3,594.9 sq km), S S.Dak., on the Nebr. line, drained by the South Fork of the White River, and including all of the closed area of the Rosebud Indian Reservation; formed 1909; it is unorganized and is attached to Tripp co. for judicial purposes. It manufactures twine.

To·dos os San·tos Bay (tô′*th*o͞o-zo͞o sän′to͞os), inlet of the Atlantic Ocean, 25 mi (40 km) long and 20 mi (32 km) wide, E Bahia, Brazil. It receives the Paraguaçu River. Brazil's first oilfield (1939) is located north of Salvador, the bay's chief city. Itaparica, a large island at the mouth of the bay, has saltworks and oilfields. The fertile Reconcavo lowland surrounds the bay; subsistence crops, sugar cane, cotton, and tobacco are raised there. Todos os Santos Bay was discovered in 1501 by Amerigo Vespucci.

To·go (tō′gō), republic (1977 est. pop. 2,300,000), 21,622 sq mi (56,000 sq km), W Africa, bordering on the Gulf of Guinea in the S, on Ghana in the W, on Upper Volta in the N, and on Benin in the E. Lomé is the country's capital and its largest city. Togo is made up of five parallel geographic regions running from east to west. In the extreme south is a narrow sandy coastal strip (c.30 mi/50 km long), which is fringed by lagoons and creeks. A region (c.50 mi/80 km wide) of fertile clay soils lies north of the coast. The third region is made up of the clay-covered Mono Tableland, which reaches an altitude of c.1,500 ft (460 m) and is drained by the Mono River. North of the tableland is a mountainous area comprising the Togo mts. and the Atakora mts. and including Mt. Agou (c.3,940 ft/1,200 m), Togo's loftiest point. The fifth region, in the extreme north, is the rolling sandstone Oti Plateau. The country is almost entirely covered with savanna, which has somewhat thicker vegetation in the south and somewhat thinner vegetation in the far north.

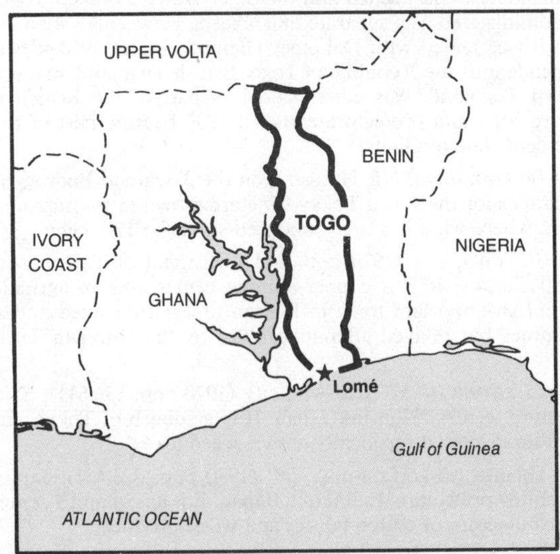

Economy. The majority of Togo's workers are engaged in agriculture, but since the early 1960s mining has played an increasingly important role in the economy. The principal food crops are manioc, millet, maize, rice, pulses, and sweet potatoes. The leading cash crops are coffee, cacao, and palm crops, which are raised mainly in the south; in addition, cotton and groundnuts are grown in the north. Large numbers of sheep, goats, hogs, and cattle are raised. Major deposits of phosphates at Akoumapé (in the southeast) began to be worked on a large scale in 1963; in 1971, Togo was the world's seventh leading producer of phosphate ores. Small quantities of chromite, bauxite, limestone, and iron ore are also mined in Togo, and marble is quarried. The country's few manufactures consist of basic consumer goods such as foodstuffs, beverages, clothing, footwear, and furniture. Processed phosphates and handicrafts are also important. Togo's limited road and rail transportation facilities are concentrated in the central and southern parts of the country; Lomé is the main port. The cost of Togo's imports is usually much higher than its earnings from exports. The main imports are machinery, manufactured consumer goods, foodstuffs, and petroleum products; the leading exports are phosphates, cacao, coffee, palm products, and groundnuts.

History. Togo became independent on April 27, 1960. Sylvanus Olympio, the head of the Committee of Togolese Unity, became its first president in 1961. Until 1966 there were tense relations with neighboring Ghana who sought to merge Togo with Ghana—a plan that Togo strongly resisted. A successful coup d'état took place on Jan. 13, 1963, during which Olympio was assassinated. Nicolas Grunitzky, an important political figure in the 1950s who had gone into exile (1958) in Dahomey (Benin), returned to Togo and became president. Grunitzky, leader of the Democratic Union of Togolese Populations, unsuccessfully attempted to unify the country. On Jan. 13, 1967, he was toppled in a bloodless army coup led by Lt. Col. Ghansimgbe Eyadema, who became president in Apr., 1967. Eyadema was confirmed overwhelmingly as president in elections in 1972. Togo has maintained close relations with France and has received considerable economic aid from that country and from West Germany.

To·go·land (tō'gō-lănd') or **Togo,** historic region (c.33,500 sq mi/86,765 sq km), W Africa, bordering on the Gulf of Guinea in the S. The western section of Togoland is now part of Ghana, and the eastern portion constitutes the Republic of Togo. From the 17th cent. until the early 19th cent. the Ashanti (situated in present-day Ghana) raided Togoland for slaves. European penetration of the region began in the 1840s with the arrival of German missionaries and merchants. A German protectorate over southern Togoland was recognized by the Conference of Berlin (1884–85). German military expeditions gained control of northern Togoland during the 1890s, and the protectorate's boundaries were further delimited in treaties with France (1897) and Great Britain (1904). Germany instituted much economic development; however, German levies of direct taxes and forced labor aroused resentment among the Togolese. In Aug., 1914, British and French forces easily captured Togoland from the Germans. In 1922 the League of Nations divided the region into two mandates, one French and the other British. French Togoland was administered as a separate unit except between 1934 and 1937, when it was joined with Dahomey (Benin), and in 1960 it became independent as the Republic of Togo. British Togoland, made up of western Togoland, was administered as part of the British Gold Coast colony and protectorate and in 1957 became part of the independent state of Ghana.

To·kaj (tô'koi), town, NE Hungary, on the Tisza and Bodrog rivers. The grapes for the noted Tokay wine are grown in the surrounding region, where wine has been produced since the 12th cent.

To·kat (tō-kät'), city (1970 pop. 44,110), capital of Tokat prov., N central Turkey. It is a copper-refining center and an agricultural market. An important town in Roman times, it declined under the Byzantines but revived after its capture by the Ottoman Turks in 1402.

To·ko·ro·za·wa (tō-kō'rō'zä-wä), city (1970 pop. 136,611), Saitama prefecture, central Honshu, Japan. It is a suburb of Tokyo and an agricultural market for locally grown green tea.

To·ku·shi·ma (tō-kōō'shĭ-mä), city (1970 pop. 223,451), capital of Tokushima prefecture, E Shikoku, Japan. It is a port and a center for the manufacture of cotton fabrics and wood products.

To·ky·o (tō'kē-ō'), city (1973 est. pop. 8,738,997), capital of Japan and of Tokyo prefecture, E central Honshu, at the head of Tokyo Bay. Greater Tokyo, the world's most populous city, consists of an urban area divided into 23 wards, a county area with farms and mountain villages, and the Izu Islands stretching to the south of Tokyo Bay. It is the administrative, financial, educational, and cultural center of Japan and a major industrial hub surrounded by numerous suburban manufacturing complexes. The city, which lies on the Kanto plain, is intersected by the Sumida River and has an extensive network of canals. Yokohama is its seaport, but there is a large man-made port at the mouth of the Sumida, through which such items as iron, steel, machinery, and chemicals are exported. The world's first public monorail line runs between downtown Tokyo and nearby Haneda international airport. The city's transportation system also includes "bullet trains" that travel at more than 300 mi (480 km) per hour between Tokyo and Osaka. Among the diverse industries of Tokyo are machine building, metalworking, printing and publishing, food processing, oil refining, and the manufacture of electronic apparatus, transport equipment, automobiles, steel, chemicals, cameras and optical goods, furniture, leather products, textiles, and a wide variety of consumer items.

The present city was founded in the 12th cent. as the village of Edo (also Yedo or Yeddo). In 1456–57 Ota Dokan, ruler of the Kanto region under the Japanese shogunate, constructed a castle at Edo. The castle passed in 1590 to Ieyasu Tokugawa, founder of the Tokugawa line of shoguns, who made Edo the capital of a province and, after formally assuming the title of shogun in 1603, the capital of the shogunate. The imperial capital, however, remained at Kyoto. On April 11, 1868, the last Tokugawa shogun surrendered Edo Castle to the imperial forces. The emperor, restored to power, made Edo his capital, renaming the city Tokyo. Tokyo's famed landmarks include the Meiji and Hie shrines; the temples of Sengakuji, Gokokuji, and Sensoji; and the Korakuen, a 17th cent. landscape garden. The Ginza is Tokyo's shopping and entertainment center; the Marunouchi quarter is the business center. One of the world's foremost educational centers, Tokyo has about 100 universities and colleges. There are numerous museums and more than 200 parks and gardens. Frequent rebuilding in the wake of an earthquake and fire in 1923 and heavy bombing during World War II has made Tokyo one of the most modern cities on the globe.

Tol·bu·khin (tôl-bōō'kĭn), city (1969 est. pop. 64,100), NE Bulgaria, a commercial and cultural center of the Dobruja region. Foodstuffs, cotton textiles, metal goods, and farm machinery are produced.

To·le·do (tə-lē'dō, tō-lā'*th*ō), city (1972 est. pop. 46,023), capital of Toledo prov., central Spain, in New Castile, on a granite hill surrounded on three sides by a gorge of the Tagus River. Toledo is of pre-Roman origin; known in ancient times as Toletum, it fell to the Romans in 193 B.C. In the 6th cent. Toledo prospered as a capital of the Visigothic kingdom. Its greatest prosperity began under Moorish rule (712–1085), first as the seat of an emir and after 1031 as the capital of an independent kingdom. Under the Moors and later under the kings of Castile, who made it their chief residence, Toledo was a center of the Moorish, Spanish, and Jewish cultures. Its commercial decline began in the 16th cent., but at the same time Toledo gained increased importance as the spiritual capital of Spanish Catholicism. The somber seat of the Grand Inquisitors, it was the center of the mysticism symbolized by its adopted citizen, El Greco, whose name has become inseparable from that of Toledo. The city's general aspect has changed little since El Greco painted his famous *View of Toledo.* Its chief landmark, the alcázar (fortified palace), was originally a Moorish structure, restored in the 13th cent. and transformed (1535, 1576) to serve as residence for Charles V and Philip II. Toledo is surrounded by partly Moorish, partly Gothic walls and gates. Of Moorish origin also is the Alcántara bridge across the Tagus. The Gothic cathedral, begun in 1226, is one of the finest in Spain and houses El Greco's *Espolio* and other paintings by him in its lovely baroque chapels.

To·le·do (tə-lē'dō). **1.** Village (1970 pop. 1,068), seat of Cumberland co., SE central Ill., SSE of Mattoon, in an agricultural area. Brooms and shoes are manufactured. **2.** City (1970 pop. 2,361), seat of Tama co., central Iowa, ESE of Marshalltown, in a farming area; inc. 1866. **3.** City (1978 est. pop. 366,000), seat of Lucas co., NW Ohio, on the Maumee River at its junction with Lake Erie; inc. 1837. With a fine natural harbor and a vast system of railroads and highways, Toledo is a port of entry and one of the chief shipping centers on the Great Lakes. Oil, coal, farm products, and numerous manufactures are exported; iron ore is the principal import. Toledo is also an industrial and commercial center, with large oil refineries, a glassmaking industry, shipyards, and plants that manufacture jeeps, automobile parts, machinery, scales, and chemicals. Other products are paints, metal stampings, tools, die castings, plastics, and cosmetics. The city was settled (1817) as Port Lawrence. In 1835–36 occurred the "Toledo War," an Ohio-Mich. boundary dispute, which was settled by Congress in favor of Ohio when Mich. became a state. The city is the seat of the Univ. of Toledo and several colleges. Points of interest include the Toledo Museum of Art and a large zoological park. The site of the Battle of Fallen Timbers, a national historic landmark, is in a nearby state park.

To·len·ti·no (tŏl'ən-tē'nō), town (1971 pop. 16,780), in the Marche, central Italy, on the Chienti River. In 1797 Pope Pius VI signed at Tolentino a humiliating treaty with Napoleon Bonaparte, under which the pope gave up considerable territory and numerous works of art.

Tol·land (tŏl'ənd), county (1970 pop. 103,440), 416 sq mi (1,077.4 sq km), NE Conn., on the Mass. line, drained by the Willimantic, Hockanum, Scantic, and Hop rivers; constituted 1785; former co. seat Rockville. Manufacturing (textiles, thread, wood products, buttons, and paper goods) and agriculture (dairy products, poultry, truck crops, potatoes, tobacco, and fruit) are the major industries here. It contains several state forests and lake resorts.

To·lu·ca (tə-lōō'kə), city (1970 pop. 114,079), capital of Mexico state, central Mexico. Located on the central plateau, Toluca (alt. c.8,760 ft/2,670 m) has a year-round cool climate. It was established as a settlement in 1530 by Hernán Cortés. The surrounding plain is fertile, producing grain, fruits, and vegetables. Cattle raising is important. The city has flour, cotton, and woolen mills and a brewery. Toluca is known for its basket weaving, pottery, and embroidery.

Tom (tŏm, tôm), river, c.525 mi (845 km) long, rising in the Ala-Tau range, S Siberian USSR. It flows north through the Kuznetsk Basin into the Ob River.

To·ma·ko·mai (tō'mə-kō'mī), city (1970 pop. 101,573), Hokkaido prefecture, S Hokkaido, Japan, on the Pacific Ocean. It is a commercial port and the site of Japan's largest paper and newsprint industry.

To·mar (tŏŏ-mär'), town (1960 pop. 14,118), central Portugal, in Ribatejo. It has paper and textile mills and other industries but is noted chiefly as the center of the Knights Templars and later of the Military Order of Christ.

To·ma·szów Ma·zo·wiec·ki (tô-mä'shŏŏf mä'zô-vyĕt'skē), city (1973 est. pop. 56,900), E central Poland. It is a railroad junction and has industries manufacturing woolens, synthetic textiles and fibers, and farm implements.

Tom·big·bee (tŏm-bĭg'bē), river, c.400 mi (645 km) long, rising in NE Miss. and flowing SE into W Ala., then generally S to join the Alabama River and form the Mobile River before entering Mobile Bay at Mobile. The Tombigbee is an important artery for manufactured goods. Dams and locks improve navigation on the river.

Tomb·stone (tōōm'stōn'), city (1970 pop. 1,241), Cochise co., SE Ariz.; inc. 1881. With its pleasant climate and legendary past, it is a well-known tourist attraction. The city became a national historic landmark in 1962. Silver was discovered there in 1877 by Ed Schieffelin, a prospector, who two years later laid out and named the city. Tombstone quickly became one of the richest and most lawless mining towns in the Southwest. Its newspaper, *Epitaph*, was first published in 1880. The city was county seat from 1881 to 1929. Large-scale mining ended by 1890. Among Tombstone's many picturesque landmarks are Boot Hill Graveyard, where many desperados are buried; Bird Cage Theater, now a museum; and O.K. Corral, scene of a climactic gun battle between the Clanton gang and Wyatt Earp, his brother Virgil, and Doc Holliday.

Tom Green (tŏm' grēn'), county (1970 pop. 71,047), 1,500 sq mi (3,885 sq km), W Texas, on Edwards Plateau, drained by the North, Middle, and South Concho rivers, here joining to form the Concho River; formed 1874; co. seat San Angelo. This mainly ranching region also has farming (cotton, oats, wheat, grain sorghums, fruit, pecans, and truck crops), dairying, and poultry raising. There are oil wells, deposits of clay, sand, and limestone, and railroad foundries.

Tomp·kins (tŏmp'kənz, tŏmp'-), county (1970 pop. 77,064), 482 sq mi (1,248.4 sq km), W central N.Y.; formed 1817; co. seat Ithaca. It has dairying, farming (grain, poultry, and fruit), and manufacturing. It includes the end of Cayuga Lake.

Tomp·kins·ville (tŏm'kənz-vĭl', tŏmp'-), town (1970 pop. 2,207), seat of Monroe co., S Ky., SSE of Glasgow near the Tenn. border, in a farm area. The old Mulkey meetinghouse (1798), now a state shrine, is nearby.

Tomsk (tŏmsk, tômsk), city (1973 est. pop. 374,000), capital of Tomsk oblast, W central Siberian USSR, on the Tom River. It is a major river port and freight transit point. Machine tools, electric motors, ball bearings, instruments, and chemicals are made. Founded in 1604 around a fort built by Boris Godunov, Tomsk was a major Siberian trade center until bypassed by the construction of the Trans-Siberian RR in the 1890s.

Toms River (tŏmz), uninc. residential town (1970 pop. 7,303), seat of Ocean co., E N.J., on an inlet of Barnegat Bay and SE of Lakehurst, in a poultry area; settled 1664. It is a fishing resort. The courthouse (1850) contains a county museum. The village, a privateering center in the American Revolution, was burned in 1782 by Loyalists and British.

Ton·a·wan·da (tŏn'ə-wŏn'də), city (1978 est. pop. 21,100), Erie co., NW N.Y., on the Niagara River at the terminus of the Barge Canal; inc. as a village 1854, as a city 1903. An industrial suburb of Buffalo and a lake port, Tonawanda is a commercial center and a transshipment point; its manufactures include steel, office equipment, and plastics.

Ton·bridge (tŭn'brĭj'), urban district (1974 est. pop. 31,520), Kent, SE England. Tonbridge is mainly residential, with light industry including printing and sawmilling. It is a railroad junction. The public school was founded in 1553.

Tøn·der (tœ'nər), town (1970 pop. 11,631), Sønderjylland co., SW Denmark, near the West German border. It has long been famous for its lace industry and also has breweries and meat-packing plants.

Ton·ga (tŏng'gə), island kingdom (1978 est. pop. 92,000), 270 sq mi (699 sq km), South Pacific, c.2,000 mi (3,220 km) NE of Sydney, Australia. Tonga is the only surviving independent kingdom in the South Pacific. The more than 150 islands constitute three main groups: Tongatabu (the seat of Nukualofa, the capital) in the south, Vavau in the north, and Haapai in the center. Most of the islands are volcanic, with active craters, but several are coral atolls. The native

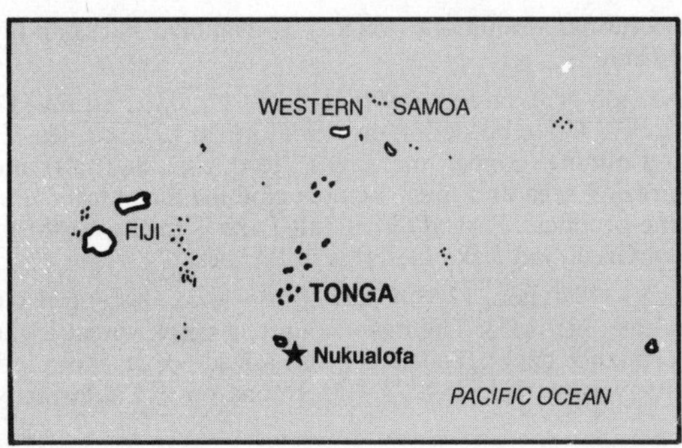

Polynesians grow subsistence crops and export copra and bananas. Because of compulsory primary education the literacy rate is relatively high. Every male Tongan over the age of 16 is entitled to an allotment of land, but the shortage of land precludes a holding for many. Under British influence since the 18th cent., Tonga was a self-governing protectorate from 1900 and has been an independent member of the Commonwealth since 1970.

Tong·e·ren (tông'ər-ən), town (1970 pop. 17,028), Limburg prov., E Belgium. It is the trade center of the productive Hesbaye farm region. Of note are the Church of Notre Dame (9th–15th cent.), with Romanesque cloisters, as well as the town's Roman walls and its many houses dating from the 16th and 17th cent.

Ton·kin (tŏn'kĭn', tŏng'-), historic region (c.40,000 sq mi/103,600 sq km), SE Asia, now forming the heartland of northern Vietnam. Tonkin was bordered on the north by China, on the east by the Gulf of Tonkin, on the south by the historic region of Annam, and on the south and west by Laos. The region of Tonkin was conquered in 111 B.C. by the Chinese, who ruled until they were ousted in A.D. 939, at which time the area became independent. The inhabitants began a southward expansion. After the division of the Vietnamese lands between two dynasties in 1558, the northern half was ruled from the city of Tonkin (modern Hanoi); thus the name Tonkin came to be applied by Europeans to the whole area. To open the Red River to French trade, French expeditions were sent into Tonkin in 1873 and 1882. In 1884 Annam accepted a French protectorate, conceding France a separate protectorate over Tonkin with control more direct than over Annam. After World War II Tonkinese and Annamese nationalist leaders joined in demanding independence for the state of Vietnam, and Tonkin was torn by guerrilla warfare between the

French and the Viet Minh nationalists led by Ho Chi Minh.

Tonkin, Gulf of, NW arm of the South China Sea, c.300 mi (480 km) long and 150 mi (240 km) wide, between N Vietnam and China. The shallow gulf (less than 200 ft/61 m deep) receives the Red River. An alleged attack (Aug., 1964) by North Vietnamese gunboats against U.S. naval forces stationed in the gulf led to increased U.S. involvement in the Vietnam War.

Tôn·lé Sap (tŏn'lā săp'), lake, central Cambodia; largest lake of SE Asia. It occupies the depression of the Cambodian plain and is fed by many streams; the Tônlé Sap River, c.70 mi (115 km) long, drains the lake southward into the Mekong River near Phnom Pénh. At low water in the dry season (Nov.-May), the lake covers c.1,100 sq mi (2,850 sq km). During the summer floods, however, the waters of the Mekong back up into Tônlé Sap (which forms a natural reservoir), raising the lake's level c.30 ft (9 m) and more than tripling its area. Approximately 2,500 sq mi (6,475 sq km) of surrounding forest are inundated by the floodwaters and provide a breeding ground for fish. The fisheries of the lake are one of Cambodia's major natural resources. The Tônlé Sap lake and river are also part of an important inland waterway system.

To·no·pah (tō'nə-pä'), uninc. town (1970 pop. 1,650), seat of Nye co., S central Nev. Settled in 1900 after the discovery of rich silver deposits, it is today a supply point and trading center for a mining and ranching area.

Tøns·berg (tœns'bâr'), city (1974 est. pop. 10,370), capital of Vestfold co., SE Norway, a port on the Skagerrak near the mouth of the Oslofjord. It is a shipping and whaling center and has industries that manufacture forest products and processed food. Tønsberg was founded c.870 and is Norway's oldest city.

Ton·to National Monument (tŏn'tō): *see* National Parks and Monuments Table.

Too·el·e (tōō-ĕl'ə), county (1970 pop. 21,545), 6,923 sq mi (17,930.6 sq km), NW Utah, bordering on Nev.; formed 1852; co. seat Tooele. It is in a mining (copper, manganese, lead, zinc, and salt) and livestock-grazing area and manufactures clothing, chemicals, concrete, and lime products. Part of Great Salt Lake is in the northeast, and much of Great Salt Lake Desert is in the west.

Tooele, city (1970 pop. 12,539), seat of Tooele co., N central Utah, in a farm area; inc. 1853. The major source of employment is the U.S. army ordnance depot. Tooele also has a lead-ore smelting industry. The city was settled (1849) by Mormons on a California wagon route.

Toole (tōōl), county (1970 pop. 5,839), 1,950 sq mi (5,050.5 sq km), N Mont., in an agricultural area bordering on Alta. and drained by the Marias River; formed 1914; co. seat Shelby. Grain, livestock, petroleum, and natural gas are found here.

Toombs (tōōmz), county (1970 pop. 19,151), 368 sq mi (953 sq km), E central Ga., bounded on the S by the Altamaha River, on the NE by the Ohoopee River; formed 1905; co. seat Lyons. It has coastal-plain agriculture (cotton, corn, peanuts, tobacco, cattle, and hogs) and some manufacturing.

Too·woom·ba (tə-wōōm'bə), city (1973 est. pop. 61,000), Queensland, E Australia, in the Eastern Highlands, at the edge of the Darling Downs, c.2000 ft (610 m) above sea level. The city is an agricultural market center with food-processing and farm-machinery industries and railroad workshops. It is also a summer resort.

To·pe·ka (tə-pē'kə), city (1978 est. pop. 121,300), state capital and seat of Shawnee co., NE Kansas, on the Kansas River; inc. 1857. In a rich agricultural region, it is an important shipping point for cattle and wheat and a wholesaling, marketing, and processing center for farm products. There are railroad shops, grain mills, meat-packing houses, and plants that make tires and rubber products, shoes, and cellophane. The city is an important center for psychiatric research and therapy. A ferry was established here in 1842 on the Oregon Trail. The city was laid out in 1854 by Free State settlers from Lawrence and New England. Topeka was selected state capital when Kansas was admitted to the Union in 1861.

to·pog·ra·phy (tə-pŏg'rə-fē), description or representation of the features and configuration of land surfaces. Topographic maps use symbols and coloring, with particular attention given to the shape and elevations of terrain. Relief is portrayed by means of contour lines, hachures, shading, or coloring to represent elevations, depressions, and depths of water; natural and man-made features, such as rivers, sand dunes, forests, urbanized areas, bridges, tunnels, roads, and power lines, are indicated by means of symbols and color overlays.

Tops·field (tŏps'fēld'), town (1970 pop. 5,225), Essex co., NE Mass., NNW of Salem; settled c.1635, inc. 1650. The kitchen of the Parson Capon House (1683) is reproduced in the Metropolitan Museum, New York. The town has an annual fair.

Tor·bay (tôr'bā'), county borough (1974 est. pop. 106,400), Devon, SW England. Torbay, created in 1968, comprises the former municipal borough of Torquay and the urban districts of Paignton and Brixham. On Tor Bay, it is a noted resort area. William of Orange landed at Brixham in 1688.

Tor·gau (tôr'gou'), city (1973 est. pop. 21,637), Leipzig dist., S central East Germany, a port on the Elbe River. Manufactures include chemicals, glass, pottery, and agricultural machinery. Long a strategic crossing point on the Elbe, Torgau was chartered in the 13th cent. In the Seven Years' War, Prussia defeated (1760) the Austrians near the city. Torgau passed to Prussia in 1815. On Apr. 27, 1945, near the end of World War II, advance elements of the U.S. and Soviet armies made contact for the first time here.

Tor·ne·älv (tôr'nə-ĕlv'), river, c.320 mi (515 km) long, rising in N Sweden at the Norwegian border; it drains Torneträsk lake (126 sq mi/326 sq km), flows SW into the Gulf of Bothnia, and forms the Swedish-Finnish frontier below its junction with the Muonioälv. The Torneälv is rich in salmon.

To·ron·to (tə-rŏn'tō), city (1971 pop. 712,786; metropolitan pop. 2,628,043), provincial capital, S Ont., Canada, on Lake Ontario. The second-largest city in Canada, it is a port of entry and an important commercial, financial, and industrial center as well as the banking and stock-exchange center of the country. Its importance as a port increased with the opening (1959) of the St. Lawrence Seaway. Toronto's industries include slaughtering and meat packing, printing and publishing, and the manufacture of aircraft, farm implements, electrical machinery, and metal products. The site was an early fur-trading center. The French built (1749) Fort Rouille here, but it was destroyed (1759) to prevent its occupation by the British. The British purchased the site from Indians in 1787. It was chosen (1793) as the capital of Upper Canada and was named York. In 1834 it was incorporated as Toronto.

Toronto, city (1970 pop. 7,705), Jefferson co., E central Ohio, on the Ohio River N of Steubenville; laid out 1818, successively called Newburg, Sloan's Station, and Toronto (1881). Steel products are manufactured, and coal is mined.

Tor·rance (tôr'əns, tŏr'-), county (1970 pop. 5,290), 3,346 sq mi (8,666 sq km), central N.Mex., formed 1903; co. seat Estancia. Its agriculture includes grain, beans, alfalfa, potatoes, and livestock. It includes parts of Cibola and Lincoln national forests.

Torrance, industrial city (1978 est. pop. 134,900), Los Angeles co., S Calif.; inc. 1921. It has large aircraft, electronics, and oil industries. Among its manufactures are aluminum products, steel, chemicals, and oil-field equipment.

Tor·re An·nun·zi·a·ta (tôr'ā ə-nōōn'sē-ä'tə), city (1972 est. pop. 57,096), Campania, S Italy, on the Bay of Naples and at the foot of Mt. Vesuvius. It is a port and seaside resort. Founded in the 14th cent., the city was destroyed by the eruption of Vesuvius in 1631.

Torre del Gre·co (dĕl grĕk'ō), city (1972 est. pop. 93,192), Campania, S Italy, on the Bay of Naples, near Mt. Vesuvius. It is a fishing port and a popular seaside resort. The coral industry has been a specialty of the town since the 16th cent.

Tor·rens, Lake (tôr'ənz, tŏr'-), shallow salt lake, 2,230 sq mi (5,775.7 sq km), central South Australia state, Australia. In a rift valley, it is 120 mi (193 km) long and is Australia's second-largest lake.

Tor·re·ón (tôr'ē-ōn'), city (1970 pop. 223,104), Coahuila state, N Mexico, on the Nazas River. It is the center of an extensively irrigated district. Cotton and wheat are the principal crops, and cattle raising is important. Torreón's industries include a rubber factory, foundries, cotton and flour mills, a brewery, and a large smelter.

Tor·res Strait (tôr'əs), channel, c.95 mi (153 km) wide, between New Guinea and Cape York Peninsula of Australia. It connects the Arafura and Coral seas. The strait is shallow and hazardous for navigation. Pearl fishing is the main activity in the strait.

Tor·ring·ton (tôr'ĭng-tən, tŏr'-). **1.** City (1978 est. pop. 30,900), Litchfield co., NW Conn., on the Naugatuck River; inc. 1740. It is an industrial and commercial hub and is known for its metal and machinery manufactures. The first machine-made brass goods in the country were produced in Torrington in 1834. The city was also the site of the world's first condensed-milk plant, and the process of homogenization was invented in Torrington. **2.** Town (1970 pop. 4,237), seat of Goshen co., SE Wyo., on the North Platte River NNE of Cheyenne near the Nebr. border, in an irrigated farm area served by the North Platte project. Stock ranches, oil wells, and coal and uranium mines are in the region. The town has a sugar refinery. Fossil beds are nearby.

Tórs·havn (tôrs-houn'), city (1966 pop. 9,738), capital of the Faeroe Islands, on SE Streymoy Island, Denmark. It is a shipping center and a major fishing port. The city has fish-processing plants and a shipyard.

Tor·tu·ga (tôr-to͞o'gə), island, c.70 sq mi (180 sq km), off N Haiti. It was a notorious rendezvous of pirates in the 17th cent. It is called Île de la Tortue by the Haitians.

To·ruń (tôr'o͞on', -o͞on'yə), city (1973 est. pop. 140,440), N central Poland, on the Vistula River. It is a river port and a railway junction. The major industries produce machinery, precision instruments, electrical equipment, textiles, and chemicals. It grew around a castle founded in 1231 by the Teutonic Knights. A flourishing trade center, it was a member of the Hanseatic League (14th–16th cent.). Toruń's importance made it an object of repeated disputes between Poland and the Teutonic Knights. A religious riot here (1724) caused Russia and Prussia to guarantee the rights of religious minorities in Poland. The city passed to Prussia in 1793 and again in 1815. It reverted to Poland in 1919. It was the birthplace of Copernicus; its university (founded 1945) bears his name.

To·to·wa (tō'tə-wə), borough (1975 est. pop. 11,500), Passaic co., NE N.J., a suburb of Paterson on the Passaic River; inc. 1898. Perfumes, furniture, plastics, detergents, spices, and food products are made.

Tot·to·ri (tə-tôr'ē, -tōr'ē), city (1970 pop. 113,151), capital of Tottori prefecture, S Honshu, Japan, a port on the Sea of Japan. Lumber, raw silk, and fruit are exported. The city also produces wood and paper items and is a market for rice.

Toug·gourt (to͞o-go͝ort'), town and oasis (1974 est. pop. 34,800), E Algeria, in the Sahara. An important administrative, commercial, and tourist center that was once famous for its abundant date crops, it is now a center of the Saharan oil industry.

Toul (to͞ol), town (1968 pop. 14,780), Meurthe-et-Moselle dept., NE France, on the Moselle River. It is largely an agricultural center but has clothing and glass industries. During the Middle Ages Toul was one of the bishoprics vital to the defense of France's eastern border. Confirmed as a French possession by the Peace of Westphalia in 1648, Toul played a significant role during the Franco-German conflicts of succeeding centuries.

Tou·lon (to͞o-lôN'), city (1968 pop. 174,746), Var dept., SE France, in Provence, on the Mediterranean Sea. Toulon is the principal naval center of France; shipbuilding and ship repairing are major industries. Chemicals, machinery, furniture, and cork are also produced. Toulon first achieved eminence as a hostel for errant Crusaders during the Middle Ages. The city was the scene of many historic naval battles. The young Napoleon gained distinction (1793) by retaking the city from the English. During World War II much of the French fleet was scuttled (1942) to avoid its capture by the Germans.

Tou·lon (to͞o'lŏn'), city (1970 pop. 1,207), seat of Stark co., N central Ill., NNW of Peoria, in a farm area; inc. 1859. Cheese is made.

Tou·louse (to͞o-lo͞oz'), city (1968 pop. 370,796), capital of Haute-Garonne dept., S France, on the Garonne River. One of the great cultural and commercial centers of France, it is also the center of the French aeronautics industry. Originally part of Roman Gaul, it was the capital of Aquitaine from 781 until 843. Toulouse was an artistic and literary center of medieval Europe. In the late 12th cent. the counts of Toulouse were suzerains of practically the entire region of Languedoc. Ruling with great wisdom and tolerance, the counts held a brilliant court that attracted the best troubadours and was the center of southern French literature. However, between 1208 and 1229 the area was laid waste by northern lords. The counts fell from power, and in 1271 the county passed to the French crown and from that time on formed much of Languedoc prov.

Tou·raine (to͞o-rān'), region and former province, W central France, centering around Tours and drained by the Loire, Cher, and Vienne rivers. Touraine, with its fertile valleys, orchards, and vineyards, is known as the "garden of France." Its numerous chateaus, built mainly in the 15th and 16th cent., are noted tourist attractions. Descartes, Rabelais, and Balzac were born in Touraine. Touraine passed (1152) under English domination and was retaken (1204) by Philip II of France and united with the French crown.

Tou·rane (to͞o-rän'): *see* Da Nang, Vietnam.

Tour·coing (to͝or-kwăN'), city (1968 pop. 98,755), Nord dept., N France, in French Flanders. With the adjacent city of Roubaix, it forms one of the most important textile centers of France.

Tour·nai (to͝or-nā'), city (1971 est. pop. 32,404), Hainaut prov., SW Belgium, on the Scheldt River. The manufactures of this commercial and industrial center include textiles, carpets, cement, and processed food. One of Belgium's oldest cities, Tournai was the fortified capital of a Roman province. The city was destroyed by the Normans in 881. It belonged to France after 1187 but was captured and attached to the Spanish Netherlands in 1521. It was taken several times by the French in the wars of the 17th–18th cent. Tournai has been a cultural center since the 12th cent.

Tou·ro Synagogue National Historic Site (to͝or'ō): *see* National Parks and Monuments Table.

Tours (to͝or), city (1968 pop. 128,120), capital of Indre-et-Loire dept., W central France, in Touraine, on the Loire River. It is a wine market and a tourist center, with metallurgical, chemical, electrical, clothing, and printing industries. An old Gallo-Roman town, the city was a center of medieval Christian learning. It was here that Charles Martel halted (732) the Moorish conquest of Europe. The history of Tours is essentially that of Touraine, of which it was the capital. The city has produced great painters, sculptors, goldsmiths, and tapestry weavers. In World War II it was briefly (June, 1940) the seat of the French government.

To·wan·da (tə-wŏn'də), borough (1970 pop. 4,224), seat of Bradford co., NE Pa., in the Susquehanna valley NW of Scranton; settled 1794, laid out 1812, inc. 1828. A trade center in a farm and dairy area, it also has some manufacturing.

Tow·er Ham·lets (tou'ər hăm'lĭts), borough (1971 pop. 164,948), of Greater London, SE England. Tower Hamlets was formed in 1965 by the merger of the metropolitan boroughs of Bethnal Green, Poplar, and Stepney. The southern boundary of Tower Hamlets fronts on the Thames and includes some of London's busiest docks. The borough's industries include furniture making, dressmaking, and brewing. Both the Tower of London and the Royal Mint are in the borough.

town (toun), in the United States. In the New England states the town is the basic unit of local government. Elsewhere in the United States the term signifies either a place incorporated as a town or simply a population center. A township is a geographic division of a county, usually made up of 36 sections, each with an area of 1 sq mi (2.6 sq km). Except in the Middle Atlantic states, townships are seldom units of local government.

Tow·ner (tou'nər), county (1970 pop. 4,645), 1,043 sq mi (2,701.4 sq km), N N.Dak., in a prairie area bordering on Man. and watered by numerous creeks and streams; formed 1883; co. seat Cando. Wheat, barley, oats, flax, potatoes, and some sunflowers are grown. Dairy products, livestock, and poultry are also important.

Towner, city (1970 pop. 870), seat of McHenry co., N central N.Dak., E of Minot on the Souris River, in a diversified farming area. Dairy products, poultry, and grain are processed, and hay and cattle are shipped.

Towns (tounz), county (1970 pop. 4,565), 166 sq mi (430 sq km), N Ga., bounded on the N by Chatuge Lake and the N.C. border and drained by the Hiawassee River; formed 1856; co. seat Hiawassee. It has agriculture (corn, hay, potatoes, fruit, and livestock), lumbering, and a resort area that is part of Chattahoochee National Forest.

Town·send (toun'zənd), town (1970 pop. 1,371), seat of Broadwater co., W central Mont., on the Missouri River NE of Butte; settled 1883, inc. 1895. Formerly a mining town, it still has some mines but also has ranching and irrigated farming.

Townsend, Mount, 7,260 ft (2,214.3 m) high, SE New South Wales, in the Australian Alps. It is Australia's second-highest peak.

Towns·ville (tounz'vĭl'), city (1973 est. pop. 76,500), NE Queensland, Australia, on Cleveland Bay. It is a major port. Wool, hides, meat, copper, and sugar are the chief exports. Copper and sugar refining, meat packing and freezing, and cement making are other industries.

Tow·son (tou'sən), uninc. city (1978 est. pop. 83,600), seat of Baltimore co., N Md., a residential and industrial suburb of Baltimore; settled c.1750. It has varied manufactures.

To·ya·ma (tō-yä'mə), city (1970 pop. 269,276), capital of Toyama prefecture, E central Honshu, Japan, on Tokyo Bay. It is the main center of Japan's patent medicine industry and also has industries that produce cotton and rayon yarn and pulp.

To·yo·ha·shi (tō'yə-hä'shē), city (1970 pop. 258,547), Aichi prefecture, central Honshu, Japan. It is a leading silk and cotton production center.

To·yo·ta (tō-yō'tə), city (1970 pop. 197,193), Aichi prefecture, central Honshu, Japan. It is a major industrial center dominated by the Toyota Automatic Loom Works.

Trab·zon (trăb-zŏn') or **Treb·i·zond** (trĕb'ĭ-zŏnd'), city (1970 pop. 80,795), capital of Trabzon prov., NE Turkey, a port on the Black Sea. A commercial center, it exports food products and tobacco. The city was founded in the 8th cent. B.C. by Greek colonists. It grew in importance after its incorporation (1st cent. A.D.) into the Roman Empire and became a prosperous port under the Byzantine Empire. It reached its greatest splendor after the establishment (1204) of the empire of Trebizond, which endured until 1461, when it was annexed by the Ottoman Empire. Trabzon was included (1920) in the short-lived independent state of Armenia. The city's large Greek population was deported in 1922-23.

Tra·cy (trā'sē), city (1975 pop. 16,055), San Joaquin co., central Calif., in the San Joaquin Valley; inc. 1910. Food products and glass are made. A pumping plant here is part of the Central Valley project.

Tra·fal·gar, Cape (trə-făl'gər), on the SW coast of Spain, near the NW shore of the Strait of Gibraltar. The Battle of Trafalgar, which took place off the cape on Oct. 21, 1805, was the famous naval victory of Adm. Horatio Nelson over allied French and Spanish fleets.

Trail (trāl), city (1971 pop. 11,149), SE British Columbia, Canada, on the Columbia River just N of the Wash. border. It is a metal-smelting center for a mining area that produces lead, zinc, silver, and gold. Sulfuric acid and fertilizers are manufactured.

Traill (trāl), county (1970 pop. 9,571), 861 sq mi (2,230 sq km), E N.Dak., in an agricultural area bounded on the E by the Red River of the North and drained by the Goose River; formed 1875; co. seat Hillsboro. Its main crops include wheat, soybeans, sugar beets, corn, barley, and flax.

Tra·lee (trə-lē'), urban district (1971 pop. 12,287), county town of Co. Kerry, SW Republic of Ireland, on the Lee River. It is a seaport linked with Blennerville on Tralee Bay by a 1-mi-long (1.6-km) canal. Boots, shoes, knitwear, and plastics are produced, and there is a tannery.

Tra·ni (trä'nē), town (1972 est. pop. 40,552), in Apulia, S Italy, on the Adriatic Sea. It is a seaport, a beach resort, and an agricultural center famous for its wine. Trani enjoyed great prosperity at the time of the Crusades (11th–13th cent.) and again in the 15th–16th cent. Its *ordinamenta maris* of 1063 probably constitute the first medieval code of maritime law.

Trans-A·lai (trăns'ä-lī', trănz'-), mountain range, central Asia, a part of the Pamir-Alai mountain system. The Trans-Alai extends c.125 mi (200 km) west from the China-USSR border into the USSR along the Kirghiz SSR-Tadzhik SSR border. It rises to 23,382 ft (7,131.5 m) in Lenin Peak.

Trans·ant·arc·tic Mountains (trăns'ănt-ärk'tĭk, trănz'-), mountain chain stretching across Antarctica from Victoria Land to Coats Land, separating the E Antarctic and W Antarctic subcontinents. Mt. Markham (14,275 ft/4,353.9 m high), near the Ross Ice Shelf, is the highest peak.

Trans·bay·ka·li·a or **Trans·bai·ka·li·a** (both: trăns'bī-kô'lē-ə, -lyə, -käl'ē-ə, -käl'yə, trănz'-), region, SE Siberian USSR, extending from Lake Baykal to the Amur River. It consists of plateaus and mountain ranges separated by wide, deep river valleys.

Trans·car·pa·thi·an Oblast (trăns'kär-pā'thē-ən, trănz'-): *see* Zakarpatskaya Oblast, USSR.

Trans·cau·ca·sia (trăns'kô-kā'zhə, -shə, trănz'-), region of the USSR (1978 est. pop. 13,733,000), extending from the Greater Caucasus to the Turkish and Iranian borders, between the Black and Caspian seas. Between the Greater Caucasus in the north and the Lesser Caucasus in the south is the Colchis lowland. The Kura, Rion, Inguri, and Alazan rivers are important for both hydroelectricity and irrigation. The region's natural resources are oil, manganese, copper, clays, and building stones. Manufactures include oil-industry machinery, mining equipment, metal products, automobiles, chemicals, plastics, cotton and silk cloth, and leather footwear. The area's chief crops are cotton, grain, sugar beets, sunflowers, tobacco, citrus fruits, tea, and plants for essential oils. Transcaucasia's mineral springs have given rise to numerous health resorts; seaside resorts also abound. The region was broken up (1918) into the republics of Armenia, Azerbaijan, and Georgia. After the three republics were conquered by the Red Army, the Transcaucasian SFSR was formed; it joined the USSR in Dec., 1922, becoming one of the four original federated republics. In 1936 Armenia, Azerbaijan, and Georgia were re-established as separate union republics.

Trans·co·na (trăns-kō'nə, trănz-), city (1971 pop. 22,490), SE Man., Canada. It is a suburb of Winnipeg.

Trans·kei, The (trăns-kā', -kī'), semiautonomous black African homeland, or Bantustan, c.16,500 sq mi (42,735 sq km), E Republic of South Africa; capital Umtata. The Transkei is bounded by the Great Kei River in the south, by the Indian Ocean in the east, by Natal in the north, and by Lesotho in the northwest. Much of the territory is hilly or mountainous, and there is little good farmland. Cattle and sheep are raised. Most of the Transkei's income is provided by citizens who work as migrant laborers in the mines and factories of the Orange Free State and the Transvaal. The Transkei's Indian Ocean coastline has good fishing, especially for sardines. Much of the territory was annexed in 1848 by Britain and in 1865 was joined to Cape Province. The Transkei was separated from Cape Province in 1963 to become the first of a projected nine internally self-governing black African areas within South Africa. In 1976 South Africa granted The Transkei its independence. However, both the Organization of African Unity (OAU) and the United Nations refused to recognize the new state, preferring instead a future undivided, black-ruled South Africa.

Trans·vaal (trăns-väl', trănz-), province (1970 pop. 8,901,054), 110,450 sq mi (286,065 sq km), NE Republic of South Africa; capital Pretoria. The Transvaal is bounded on the north and west by the Limpopo River, which forms the border with Rhodesia and Botswana, on the east by Mozambique and Swaziland, and on the south by the Vaal River, the border with the Orange Free State. It is mainly situated in the highveld, at an altitude of 3,000 to 6,000 ft (915–1,830 m). Cattle and sheep are raised, and maize, wheat, tobacco, citrus fruits, cotton, groundnuts, and temperate-zone crops are cultivated. The Transvaal's wealth, however, lies mainly in its minerals. Since 1886 mines in the province have supplied much of the world's gold. In addition, the province produces most of the country's diamonds, coal, asbestos, and uranium and all of its platinum and chromium. Its industries produce iron and steel, explosives, mining equipment, and varied consumer goods.

The Transvaal, inhabited by Bantu-speaking black Africans in the early 19th cent., was settled in the mid-1830s by Boer farmers. Having forced out most of the Africans, the Boers scattered over the huge territory. In 1852 Great Britain, which at the time also held Cape Colony and Natal, recognized the right of the Boers beyond the Vaal River to administer their own affairs. In 1857 the South African Republic was inaugurated in the southwestern Transvaal but claimed sovereignty over the whole territory.

In 1877 Britain, seeking to unify South Africa, annexed the South African Republic. In late 1880, however, the Boers began an armed revolt against the British and proclaimed a new republic; Britain then granted the South African Republic internal self-government. In 1886 large gold deposits were discovered on what later came to be called the Witwatersrand, and many foreigners, especially Britons and Germans, entered the republic. The foreigners threatened to overwhelm the Boers, whom they soon outnumbered by more than two to one. The Boers denied political rights to the foreigners and taxed them heavily. Tension between Boers and Britons in South Africa increased, and in 1899 the South African War broke out. The Transvaal was annexed by Britain in 1900, but guerrilla fighting continued. The Treaty of Vereeniging (1902) ended the war and made the Transvaal (as well as the Orange Free State) a crown colony of the British Empire. The Transvaal was granted self-government in 1907 and in 1910 became a founding province of the Union of South Africa.

Tran·syl·va·ni·a (trăn'səl-vā'nē-ə, -vān'yə), historic region and province (21,292 sq mi/55,146 sq km), central Rumania. A high plateau, Transylvania is separated in the south from Walachia by the Transylvanian Alps and in the east from Moldavia and Bukovina by the Carpathian Mts. (of which the Transylvanian Alps are a continuation). In the north and west Transylvania borders on Crisana-Maramureş and in the southwest on the Banat. The Transylvanian plateau, 1,000 to 1,600 ft (305–488 m), is drained by the Mureşul River and other tributaries of the Danube.

One of the most advanced regions of Rumania, Transylvania is rich in mineral resources, notably lignite, iron, lead, manganese, gold, copper, natural gas, salt, and sulfur. There are large iron and steel, chemical, and textile industries. Stock raising, agriculture, wine production, and fruit growing are important occupations. Timber is another valuable resource. Sizable Hungarian and German minorities, as well as some Jews and Gypsies, live in Transylvania. The area now constituting Transylvania was the nucleus of the Dacian (Getic) kingdom, which in A.D. 107 became part of the Roman province of Dacia. After the withdrawal (A.D. 271) of the Romans the region was overrun (3rd–10th cent.) by the Visigoths, the Huns, the Gepidae, the Avars, and the Slavs. The Magyar tribes first entered the region in the 10th cent. King Stephen I placed it under the Hungarian crown in 1003. The valleys in the east and southeast were settled by the Székely, a Turkic people akin to the Magyars. In the 12th and 13th cent. the areas in the south and northeast were settled by German colonists, and large numbers of Rumanians, called Vlachs or Walachians, were in the region by 1222.

After the Turkish conquest (mid-16th cent.) of central Hungary,Transylvania became semi-independent, with Austrian and Turkish influences vying for supremacy for nearly two centuries. In 1604 Stephen Bocskay led a rebellion against Austrian rule, and in 1606 he was recognized by the emperor as prince of Transylvania. Under Bocskay's successors the principality was the chief center of Hungarian culture and humanism and the main bulwark of Protestantism in eastern Europe. In 1711 Austrian control was definitely established over all Transylvania, whose princes were replaced by Austrian governors.

The Rumanians of the region began to demand greater recognition c.1790. When the Hungarians declared their independence in 1848, the Rumanians sided with the Austro-Russian forces that suppressed the rebellion. In 1867 Transylvania became an integral part of Hungary. After World War I the Rumanians of Transylvania proclaimed their union with Rumania. Transylvania was then seized by Rumania and was formally ceded by Hungary in the Treaty of Trianon (1920). It was now the turn of the Magyar and German nationalists to complain of Rumanian oppression. During World War II Hungary annexed (1940) northern Transylvania, which was, however, returned to Rumania after the war.

Transylvania, county (1970 pop. 19,713), 382 sq mi (989.4 sq km), W N.C., in the Blue Ridge, bounded on the S by the S.C. border and drained by the French Broad River; formed 1861; co. seat Brevard. It has farming (vegetables, hay, and corn), dairying, poultry raising, lumbering, manufacturing, and resorts.

Tran·syl·va·ni·an Alps (trăn'səl-vā'nē-ən, -vān'yən), southern branch of the Carpathian Mts., extending c.225 mi (360 km) E across central Rumania from the Danube River at the Iron Gate. Moldoveanu (8,343 ft/2,544.6 m) and Negoiu are the highest peaks. The range, composed of crystalline massifs, is densely forested and is a famous hunting ground. There are coal, iron, and lignite deposits.

Tra·pa·ni (trä'pä-nē), city (1972 est. pop. 60,100), capital of Trapani prov., W Sicily, Italy, a seaport on a promontory in the Mediterranean Sea. The city's exports include marsala wine, macaroni, olives, and tuna fish. The city was an important Carthaginian naval base and later fell to Rome (241 B.C.).

Trav·an·core (trăv'ən-kôr', -kōr'), former princely state, 7,622 sq mi (19,741 sq km), SW India, on the Arabian Sea. It is now in Kerala state. The region of Travancore has coastal lowlands and a hilly interior. Rainfall is heavy, and rice, sugar cane, coconuts, and cotton are important lowland crops. The hill region provides half of India's cardamom and much coffee, tea, rubber, and timber. There is a large Christian minority.

Trav·erse (trăv'ərs), county (1970 pop. 6,254), 568 sq mi (1,471 sq km), W Minn., bordering on S.Dak. and N.Dak. and bounded on the W by Lake Traverse and the Bois de Sioux River; formed 1862; co. seat Wheaton. It is in an agricultural area that produces small grains, cattle, sugar beets, dairy goods, and poultry. There is some manufacturing in the county.

Traverse City, city (1978 est. pop. 20,100), seat of Grand Traverse co., N Mich., at the head of the West Arm of Grand Traverse Bay, in a resort and cherry-growing region; inc. 1881. Tourism and food processing are major industries. Wall decorations, gear sets, tools and dies, and cranes are among the manufactures. The production of lumber was the major economic activity until c.1915, when the supply was depleted and farming begun.

Trav·is (trăv'ĭs), county (1970 pop. 295,516), 1,012 sq mi (2,621.1 sq km), central Texas, crossed SW to NE by Balcones Escarpment and drained by the Colorado River; formed 1840; co. seat Austin. It has farming (cotton, grain sorghums, corn, oats, forage, potatoes, peaches, and pecans), dairying, poultry raising, and ranching. There are clay and limestone deposits. Manufacturing and processing are done. There are recreational areas in the west.

Treas·ure (trĕzh'ər), county (1970 pop. 1,069), 985 sq mi (2,551.1 sq km), S central Mont., in an agricultural area drained by the Yellowstone River; formed 1919; co. seat Hysham. It is a grain and livestock area.

Treasure Island, artificial island, 400 acres (162 hectares), San Francisco Bay, Calif., between Alcatraz Island and Berkeley, and N of Yerba Buena Island, with which it is connected by a causeway. It was constructed as the site of the Golden Gate International Exposition (1939–40) and became a U.S. naval base in 1941.

Treb·i·zond (trĕb'ĭ-zŏnd'): *see* Trabzon, Turkey.

Tre·go (trē'gō), county (1970 pop. 4,436), 901 sq mi (2,333.6 sq km), W central Kansas, in a prairie region drained by the Saline and Smoky Hill rivers; formed 1867; co. seat WaKeeney. It is in a cattle and grain area, with oil and gas wells, crude-oil production, and some manufacturing.

Trel·le·borg (trĕl'ə-bôr'), city (1972 est. pop. 29,038), Malmöhus co., extreme S Sweden, a port on the Baltic Sea. Manufactures include machinery, rubber, cement, and refined sugar. Although Trelleborg was founded in the Middle Ages, its main growth dates from the late 19th cent.

Trem·pea·leau (trĕm'pə-lō'), county (1970 pop. 23,344), 735 sq mi (1,903.7 sq km), W Wis., bounded partly on the W by the Trempealeau River, on the SW by wooded bluffs along the Mississippi River, on the SE by the Black River; formed 1854; co. seat Whitehall. It is in a dairying and agricultural area that produces corn, hay, oats, soybeans, truck crops, and livestock. It processes butter and cheese, has lumbering industries, and manufactures furniture and metal products.

Trent (trĕnt), city (1972 est. pop. 93,415), capital of Trentino-Alto Adige and of Trent prov., N Italy, on the Adige River. It is an industrial and tourist center. Manufactures include leather goods, textiles, printed materials, and food products. The city was probably founded in the 4th cent. B.C. A succession of prince-bishops ruled, except for a few short intervals, until 1802, when the bishopric was secularized and became a part of Tyrol in Austria. Because Trent had always been Italian in language and culture, there developed a strong movement for union with Italy. Union was achieved in 1919 by the Treaty of Saint-Germain.

Trent, river, c.170 mi (275 km) long, rising on Biddulph Moor, Staffordshire, W England, and flowing generally NE through central England before joining the Ouse River. The Trent is the third-longest river of England. Water from the Trent is used as coolant in thermal power plants along its course.

Trent Canal, waterway system, 240 mi (386.2 km) long, S Ont., Canada, connecting Lake Ontario, from the Bay of Quinte, with Lake Huron at Georgian Bay; built 1833–48. It utilizes the Trent River to Rice Lake, the Otonabee River through Peterborough, the Kawartha Lakes and artificial channels to Lake Simcoe and Lake Couchiching, and the Severn River to Georgian Bay. Designed to shorten the shipping route between Lakes Ontario and Huron, it has proved more valuable as a source of water power.

Tren·ti·no-Al·to A·di·ge (trĕn-tē'nō-äl'tō ä'dē-jā'), region (1972 est. pop. 848,615), 5,256 sq mi (13,613 sq km), N Italy, bordering on Switzerland in the NW and on Austria in the N; capital Trent. The region is divided into Trento and Bolzano provs. The terrain is almost entirely mountainous, except for a narrow strip along the upper Adige River, where most of the population is concentrated. The region includes the Tyrolean Alps south of the Brenner Pass and, in the east, part of the Dolomites. Agriculture forms the backbone of the regional economy, with cereals, fruit, and dairy cattle the principal items. The chief manufactures are aluminum, forest products, processed food, and chemicals. There is also a large tourist industry.

Most of the region was included from the 11th cent. to 1802-3 in the episcopal principalities of Trent and Bressanone; in 1815 it was put under direct Austrian administration and incorporated into the Tyrol. Trent passed to Italy in 1866. After World War I the Treaty of Saint-Germain (1919) gave Bolzano to Italy, which resulted in agitation by its German-speaking population. Following an agreement (1946) between the Italian and Austrian governments, Italy (1947) granted the region considerable autonomy. Both German and Italian were made official languages. However, the German-speaking population in Bolzano prov. continued to demand greater autonomy and received the backing of Austria. This led to serious tension between the two countries. In 1960 Italy and Austria entered into direct negotiations. It was only in 1971 that a treaty was signed and ratified; this agreement stipulated that disputes in Bolzano would be submitted for settlement to the International Court of Justice in The Hague, that the province would receive increased legislative and administrative autonomy from Italy, and that Austria would not interfere in Bolzano's internal affairs.

Tren·ton (trĕn'tən), town (1971 pop. 14,589), SE Ont., Canada, on the Bay of Quinte at the mouth of the Trent River and the S end of the Trent Canal. Its manufactures include textiles, electronic components, and paper and steel products.

Trenton. **1.** City (1970 pop. 1,074), seat of Gilchrist co., N Fla., W of Gainesville, in a lumbering and farming area. **2.** Town (1970 pop. 1,523), seat of Dade co., extreme NW Ga., SW of Chattanooga, Tenn., at the foot of Lookout Mt. Lumber and hosiery are produced. **3.** City (1978 est. pop. 26,100), Wayne co., SE Mich., on the Detroit River opposite Grosse Ile, in a farm area; settled 1816, inc. as a city 1957. An early river port, it has oil refineries and plants making steel and iron molds, chemicals, automobile engines, and building materials. **4.** City (1970 pop. 6,063), seat of Grundy co., N Mo., on the Thompson River NE of Kansas City, in a farm region; laid out 1841, inc. 1857. A group of socialists conducted a cooperative venture in business and education here from 1897 to 1905. **5.** Village (1970 pop. 770), seat of Hitchcock co., S Nebr., W of McCook on the Republican River, in an area yielding grain, livestock, and dairy and poultry produce. **6.** City (1978 est. pop. 98,100), state capital (since 1790) and seat (since 1719) of Mercer co., W N.J., at the head of navigation on the Delaware River; settled 1679, inc. as a city 1792. Situated between Philadelphia and New York City, it is an important transportation hub and industrial center. Trenton's leading manufactures are steel cables, rubber goods, automobile parts, textiles, plastics, and a great variety of metal products. In the American Revolution Trenton was the scene of a battle when Washington crossed (Dec. 25, 1776) the ice-clogged Delaware and surprised and captured (Dec. 26) 918 Hessians. **7.** Town (1970 pop. 539), seat of Jones co., SE N.C., SE of Kinston on the Trent River. Sawmilling is done. **8.** City (1970 pop. 4,226), seat of Gibson co., W Tenn., on the North Fork of the Forked Deer NNW of Jackson, in a cotton and farm area; inc. 1826. The site of Davy Crockett's home is nearby.

Tres Ma·rí·as, Las (läs träs' mə-rē'əs), archipelago in the Pacific Ocean, c.60 mi (100 km) W of Nayarit state, Mexico. Of the four islands two—María Madre, the largest, and María Magdalena—produce maguey, salt, and lumber.

Treut·len (tro͞ot'lĭn), county (1970 pop. 5,647), 194 sq mi (502.5 sq km), E central Ga., bounded on the SW by the Ocoee River; formed 1917; co. seat Soperton. It is in a coastal plain agriculture (tobacco, corn, peanuts, and livestock) and naval-stores area.

Tre·vi·so (trə-vē'zō), city (1972 est. pop. 91,092), capital of Treviso prov., Venetia, NE Italy. It is an agricultural and industrial center. Manufactures include machinery, chemicals, and food products. In the early Middle Ages, Treviso was the seat of a Lombard duchy, then of a Frankish march. It later became a free commune and in 1339 fell to Venice. Severely damaged in the two world wars, Treviso remains picturesque, with canals and fortifications of the 16th and 17th cent.

Trib·une (trĭb'yo͞on'), city (1970 pop. 1,047), seat of Greeley co., W central Kansas, near the Colo. line NW of Garden City, in a wheat area; founded 1886, inc. 1888. A state university experimental farm is nearby.

Trich·i·nop·o·ly (trĭch'ə-nŏp'ō-lē): *see* Tiruchirapalli, India.

Tri·chur (trĭ-cho͝or'), city (1971 pop. 76,241), Kerala state, SE India. It is a market for betel and cashew nuts and has a wood-carving industry. Trichur is known for its ancient temples and churches.

Trier (trîr), city (1972 est. pop. 102,752), Rhineland-Palatinate, W West Germany, a port on the Moselle River, near the Luxembourg border. Trier is an industrial city and the main center of the Moselle wine region. Manufactures include textiles, rolled steel, metal products, and precision instruments. Founded by Augustus c.15 B.C., Trier was made (1st cent.) the capital of the Roman province of Belgica and later became (3rd cent.) the capital of the prefecture of Gaul; it was named after the Treveri, a people of eastern Gaul. It was captured (early 5th cent.) by the Franks. The city was made an episcopal see in the 4th cent. and an archiepiscopal see c.815. Its archbishops became powerful temporal princes. It was occupied by the French in 1797. At the Congress of Vienna the city and most of the archbishopric were awarded (1815) to Prussia. Trier suffered considerable damage in World War II.

Among the city's Roman monuments are the Porta Nigra (early 4th cent.), a fortified gate; an amphitheater (c.100), which can seat about 25,000 persons; ruins of the imperial baths (4th cent.); and the basilica. Trier also has a Romanesque cathedral, built (11th-12th cent.) around a 4th cent. nucleus and containing the Holy Coat of Treves (supposed to be the seamless coat of Jesus). Other noteworthy buildings include the Gothic Church of Our Lady (13th cent.). The remains of St. Matthew are preserved in a shrine in the pilgrimage church of St. Matthew (built in the 12th cent. around an earlier Benedictine monastery). Karl Marx was born in Trier (1818).

Tri·este (trē-ĕst', -ĕs'tē), city (1972 est. pop. 272,412), capital of Friuli-Venezia Giulia and of Trieste prov., extreme NE Italy, on the Gulf of Trieste, at the head of the Adriatic Sea. A major seaport with large shipyards, it is also a commercial and industrial center. Manufactures include steel, petroleum, and textiles. An ancient settlement, it was made a Roman colony (2nd cent. B.C.), was later held by the Lombards, and was taken by Charlemagne in the late 8th cent. In the 12th cent. it became a free commune. After two centuries of struggle with its rival Venice, Trieste placed itself (1382) under the control of the duke of Austria. In 1719 it was made a free port, which was the sole Austrian port and a natural outlet for central Europe. In 1867 the crown land of Trieste was made the capital of Küstenland prov. Despite its Austrian status, Trieste preserved linguistic and cultural ties with Italy. After World War I Trieste and its province were annexed (1919) by Italy. After World War II the area was claimed by Yugoslavia, but the Western powers opposed Yugoslavia's claim. As a compromise a new state, the Free Territory of Trieste, was created (1947) under the protection of the UN Security Council. The Free Territory included the city of Trieste and a coastal zone of Istria. When the Security Council could not agree on a governor for the territory, Anglo-American forces occupied Zone A, consisting of Italian-speaking Trieste and its environs, while the Yugoslavs occupied Zone B, the remainder of the Free Territory. Tension between Italy and Yugoslavia continued until 1954, when Zone A was placed under Italian administration and Zone B under Yugoslav civil administration.

Trieste has some Roman ruins, including those of an amphitheater. On a hill are the Romanesque Cathedral of San Giusto (part of which dates from the 5th cent.) and an imposing castle (14th-17th cent.).

Trigg (trĭg), county (1970 pop. 8,620), 408 sq mi (1,056.7 sq km), SW Ky., bounded on the S by Tenn., on the W by the Tennessee River, and drained by the Cumberland and Little rivers; formed 1820; co. seat Cadiz. In a gently rolling agricultural area yielding dark tobacco, wheat, corn, and livestock, it has timber, limestone deposits, and industry (clothing, metal products, and machinery manufacturing). It includes part of Kentucky Woodlands Wildlife Refuge.

Tri·glav (trē'gläv), peak, 9,392 ft (2,864.6 m) high, NW Yugoslavia, in the Julian Alps, near the Italian and Austrian borders. It is the highest peak in the Julian Alps and in Yugoslavia.

Trík·ka·la (trē'kə-lə), town (1971 pop. 34,794), capital of Tríkkala prefecture, N central Greece, in Thessaly. It is the commercial center of an agricultural and pastoral region. The town claims to be the birthplace of Asclepius, the legendary physician.

Tri·ko·ra Peak (trĭ-kôr'ə), 15,518 ft (4,733 m) high, in the Djajawidjaja Mts., Irian Barat, the second-highest peak in Indonesia.

Trim·ble (trĭm′bəl), county (1970 pop. 5,349), 146 sq mi (378 sq km), N Ky., in the outer Bluegrass, bounded on the W and N by the Ohio River and the Ind. border and drained by the Little Kentucky River; formed 1837; co. seat Bedford. In a gently rolling upland agricultural area that produces burley tobacco, cattle, fruit, and grain, it has timber and limestone deposits.

Trin·co·ma·lee (trĭng′kə-mə-lē′), town (1972 est. pop. 44,000), capital of Eastern prov., NE Sri Lanka (Ceylon), on the Bay of Bengal. Trincomalee has one of the world's finest natural harbors and can accommodate the largest vessels. Tea is the chief export; hides and dried fish are also shipped. Trincomalee is a railroad terminus and an important road junction and is noted for its rice and coconut plantations. There is some pearl fishing. The Hindu Temple of a Thousand Columns, built by early Tamil settlers from southern India, was destroyed (1622) by the Portuguese; on its site is a fort built (1676) by the Dutch. Britain and France sought (18th cent.) to wrest the city from the Dutch; it was captured (1795) by the British. During World War II Trincomalee was the British naval headquarters in the Pacific theater, and a British naval base remained at Trincomalee until 1957, when Ceylon took it over.

Trin·i·dad (trĭn′ĭ-dăd′), town (1970 pop. 31,500), Las Villas prov., central Cuba. Tobacco processing is the chief industry. Trinidad is a living relic of the colonial period and has been declared a national monument.

Trinidad, city (1970 pop. 9,901), seat of Las Animas co., S Colo., on the Purgatoire River near the N.Mex. border, at an altitude of c.6,000 ft (1,830 m); settled 1859 on the old Santa Fe Trail near Raton Pass; inc. 1879. It is the trade, shipping, and industrial center of a farm, timber, livestock, and coal area.

Trinidad and To·ba·go (tə-bā′gō), country (1970 pop. 945,200), 1,980 sq mi (5,129 sq km), West Indies, a member of the Commonwealth of Nations. The capital is Port of Spain. The country consists of two islands: Trinidad (1,864 sq mi/4,828 sq km) and Tobago (116 sq mi/300 sq km). Lying just north of the Orinoco River delta in Venezuela, Trinidad is largely flat or undulating except for a range of low mountains (the highest point is Mt. Aripo, 3,085 ft/940 m) in the north. Pitch Lake, in the southwest, is the world's largest (114 acres/46 hectares) basin of natural asphalt. Tobago, just northeast of Trinidad, is the exposed top of a mountain ridge (maximum height 2,000 ft/610 m) that is densely forested. The climate of both islands is warm and humid, and rainfall (from June to Dec.) is abundant. The population of the islands is predominantly of black African descent; about one third of the people are East Indian, and the remainder are of European, Middle Eastern, or Chinese origin. The main exports are petroleum and petroleum products, sugar, cocoa, asphalt, and chemicals. The islands have a large tourist industry.

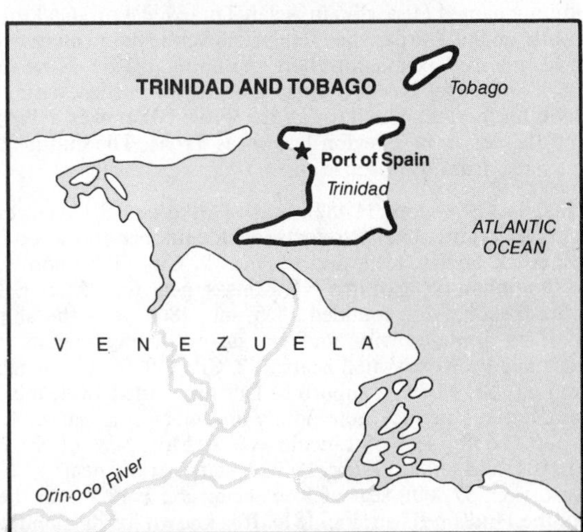

Trinidad was discovered by Christopher Columbus in 1498 but was not colonized because of the lack of precious metals. It was raided by the Dutch (1640) and the French (1677, 1690) and by British sailors. Britain captured it in 1797 and received formal title in 1802. Tobago, briefly settled by the English in the 17th cent., was held by the Dutch and the French before being acquired by the British in 1803. The islands were joined politically in 1888. Before becoming an independent nation in 1962, the islands were part of the short-lived West Indies Federation from 1958 to 1962. The country has a parliamentary form of government.

Trin·i·ty (trĭn′ĭ-tē). **1.** County (1970 pop. 7,615), 3,173 sq mi (8,218 sq km), NW Calif., mostly within the Klamath Mts., and drained by the Trinity, Eel, and Mad rivers; formed 1850; co. seat Weaverville. It is in a gold-mining area, with deposits of silver and manganese. There is timber sawmilling, livestock raising, dairying, and some farming. Mountain scenery, camping, and hiking attract vacationers. Much of the county is within Trinity, Shasta, and Mendocino national forests. **2.** County (1970 pop. 7,628), 707 sq mi (1,831.1 sq km), E Texas, bounded on the SW by the Trinity River, on the SE by the Neches River; formed 1850; co. seat Groveton. Lumbering is the chief industry in this wooded area, which includes part of Davy Crockett National Forest. It also has livestock (cattle, hogs, poultry, and sheep) and some agriculture (cotton, corn, peanuts, sugar cane, and forage crops).

Trinity, river rising in N Texas in three forks; the Clear Fork runs into the West Fork at Fort Worth, and the Elm Fork joins the West Fork at Dallas. The Trinity then flows c.510 mi (820 km) southeast to Trinity Bay, an arm of Galveston Bay. The waters of upper tributaries and the main stream are impounded in reservoirs that provide water for the Dallas-Fort Worth metropolitan area, flood control, and water for irrigation.

Trinity Bay, inlet of the Atlantic Ocean, 80 mi (129 km) long, SE N.F., Canada, between the Avalon Peninsula and the mainland. Trinity, a small port on the west shore, was the terminal of the first transatlantic cable, laid (1866) by Cyrus Field.

Trip·o·li (trĭp′ə-lē), city (1971 est. pop. 175,000), NW Lebanon, on the Mediterranean Sea. Citrus fruits, cotton, and other goods are exported from Tripoli. It has an oil refinery and is the terminus of an oil pipeline from Iraq. It was probably founded after 700 B.C.; in Persian times it was the capital of the Phoenician federation of Tyre, Sidon, and Aradus. The city flourished under the Seleucid and Roman empires. In A.D. 638 it was captured by the Arabs. After a long siege, in which its great library was destroyed, it was taken (1109) by the Crusaders. It was sacked by the sultan of Egypt in 1289 and was later rebuilt. The British conquered it from the Turks in 1918, and it became part of Lebanon in 1920.

Tripoli, city (1971 est. pop. 162,200), capital of Libya and of Tripoli dist., NW Libya, a port on the Mediterranean Sea. It is a commercial, industrial, administrative, and transportation center. Manufactures include processed food, textiles, tobacco products, and woven goods. The city was founded (probably in the 7th cent. B.C.) as Oea by Phoenicians from Tyre. The main city of the historic region of Tripolitania, it was later captured by the Romans (1st cent. B.C.), the Vandals (5th cent. A.D.), and the Arabs (7th cent.). Taken in 1510 by the Spanish, it was granted (1528) to the Knights of St. John, who held it until 1551, when it was taken by the Ottoman Turks. From 1711 to 1835 Tripoli was the seat of the Karamanlı dynasty. The city was a major base of the Barbary pirates, whom the United States fought (1801-5) in the Tripolitan War. In 1911 Tripoli passed to Italy, and later it was made the capital of the Italian colony of Libya. During World War II the city was captured (1943) by the British. Parts of the Roman walls and an arch of Marcus Aurelius remain.

Trip·o·lis (trĭp′ə-lĭs), town (1971 pop. 20,209), capital of Arcadia prefecture, S Greece, in the Peloponnesus. It is a transportation and agricultural center, and textiles and leather are produced. It was the seat of the Ottoman governors of Morea (the Peloponnesus). The town was devastated (1825) in the Greek War of Independence.

Trip·o·li·ta·ni·a (trĭp′ə-lĭ-tā′nē-ə), historic region, W Libya, bordering on the Mediterranean Sea. Tripoli is the chief city. In the 7th cent. B.C. the Phoenicians established colonies on the coast at Leptis, Oea (later Tripoli), and Sabratha. The coastal zone was later held by Carthage and was taken by Numidia in 146 B.C. Rome captured Tripolitania in 46 B.C. In A.D. 435 it fell to the Vandals, and it was captured by the Byzantines a century later. In the 7th cent. the Arabs gained control. The Normans briefly held the region in the mid-12th cent., and from the mid-13th to the mid-15th cent. Tripolitania was ruled from Tunisia. The Ottoman Turks captured the region in 1553, and it became a stronghold of Barbary pirates.

Tripp (trĭp), county (1970 pop. 8,171), 1,620 sq mi (4,195.8 sq km), S S.Dak., on the Nebr. border, bounded in the N by the White River and watered by several creeks and artificial lakes; formed 1873; co.

seat Winner. Its agriculture includes dairy products, livestock, poultry, and grain. It has some mining and manufacturing.

Trip·u·ra (trĭp′ər-ə), state (1971 pop. 1,556,342), 4,036 sq mi (10,453 sq km), NE India, bordered by Bangladesh on the N, W, and S; capital Agartala. Tripura's population is mainly engaged in agriculture. The region was annexed by the Mogul empire in 1733, passed under British rule in the 19th cent., and was joined to India in 1949. Tripura became a union territory in 1956 and a state in 1972.

Tris·tan da Cu·nha (trĭs′tăn də ko͞o′nə), group of volcanic islands (1978 est. pop. 300), in the S Atlantic, about midway between S Africa and S South America. The only habitable island of the group is Tristan da Cunha, formed by a volcano rising to c.6,760 ft (2,060 m); the other islands are Gough, Nightingale, and Inaccessible, the last being the home of the flightless rail, an almost extinct bird. Fishing is the chief industry. The island group was discovered by the Portuguese in 1506. In 1816 it was annexed by Great Britain, and in 1938 it became a dependency of the colony of St. Helena. The volcano, long dormant, erupted in 1961; the population was evacuated but returned in 1963.

Tri·van·drum (trĭ-văn′drəm), city (1971 pop. 409,627), capital of Kerala state, SW India, a port on the Arabian Sea. Manufactures include tires, tile, plywood, and titanium products.

Tr·na·va (tûr′nə-və), city (1971 est. pop. 39,253), S Czechoslovakia, in Slovakia. The market for a fertile agricultural region, it has steelworks and sugar refineries and manufactures agricultural machinery and railroad cars. Founded in the 6th or 7th cent., Trnava has many churches and monasteries, notably the fine Gothic cathedral.

Tr·no·vo or **Tir·no·vo** (both: tûr′nə-vō′), city (1969 est. pop. 43,700), N central Bulgaria, on the Yantra River. It is a commercial center and produces foodstuffs, textiles, and leather. The site was probably a Roman fortress. The second Bulgarian kingdom came into existence at Trnovo when Ivan I was proclaimed czar in 1186. It was the capital of Bulgaria under Ivan II. The city fell to the Turks in 1393. The independence of Bulgaria was proclaimed here in 1908.

Tro·as (trō′ăs′) or **the Tro·ad** (trō′ăd′), region about ancient Troy, on the NW coast of Asia Minor, in present NW Turkey. Traversed by Mt. Ida (Kaz Dağı), Troas was the scene of the events of the *Iliad.*

Tro·gir (trō′gîr′), town (1971 pop. 18,424), W Yugoslavia, partly on the Adriatic island of Čiovo and partly on the mainland, separated by a channel. Founded by the Greeks in the 3rd cent. B.C., Trogir passed to Venice in 1420 and to Austria in 1797. It was included in Yugoslavia in 1920. The town has a 9th cent. church, a 13th cent. cathedral, a 15th cent. town hall, and several medieval and Renaissance palaces.

Trois Ri·vières (trwä′ rē-vyĕr′), city (1971 pop. 55,869), S Que., Canada, at the confluence of the St. Lawrence and St. Maurice rivers. It is a port and an industrial center, with the pulp and paper industry dominant. The city was founded (1634) by Champlain.

Troll·hät·tan (trôl′hĕt′ən), city (1972 est. pop. 48,840), Älvsborg co., S Sweden, on the Götaälv River near Lake Vänern. The Götaälv River, which falls 108 ft (33 m) in about 1 mi (1.6 km) at Trollhättan, is used to generate hydroelectricity, which has made the city a major industrial center. Manufactures include locomotives, metal goods, airplane engines, footwear, clothing, and motor vehicles.

Trom·sø (trŏm′sō′), city (1974 est. pop. 42,246), capital of Troms co., NW Norway, on the island of Tromsøy; chartered 1794. The chief city of arctic Norway, it has large herring fisheries and is a base for seal hunters. Manufactures include ships and rope.

Trond·heim (trŏn′hām′), city (1974 est. pop. 133,205), capital of Sør-Trøndelag co., central Norway, a port on the Trondheimsfjord (an arm of the Atlantic Ocean). The second-largest city of Norway, it is a commercial, industrial, and shipping center. Manufactures include metal goods, construction materials, processed food, and forest products.

Founded in 997 by Olaf I, the first Christian king of Norway, the city was the political and religious capital of medieval Norway. In 1681 it was severely damaged by a fire. In the mid-19th cent. it reemerged as an important economic center. Haakon VII was crowned (1906) in Nidaros Cathedral as the first king of modern, independent Norway; subsequent rulers have also been crowned here. In World War II Trondheim was occupied by the Germans on the first day (April 9, 1940) of their invasion of Norway. It became a major German naval base and as such was frequently bombed by the Allies. Trondheim's celebrated cathedral, originally a church erected over the tomb of Olaf II (St. Olaf) in the 11th cent., was built in the 12th and 13th cent., but it was later ravaged by several fires; reconstruction was begun in 1869.

Trond·heims·fjord (trŏn′hāmz-fyôr′), inlet of the Norwegian Sea, c.80 mi (130 km) long, W central Norway. The valleys draining into the fjord comprise one of the most fertile agricultural regions of Norway.

trop·ics (trŏp′ĭks), also called tropical zone or torrid zone, all the land and water of the earth situated between the Tropic of Cancer at lat. 23½° N and the Tropic of Capricorn at lat. 23½° S. Every point within the tropics receives the perpendicular rays of the sun at noon on at least one day of the year. The sun is directly overhead at lat. 23½° N on June 21 or 22, the summer solstice, and at lat. 23½° S on Dec. 21 or 22, the winter solstice. The average annual temperature of the tropics is higher and the seasonal change of temperature is less than in other zones. The seasons in the tropics are marked by changes in wind or rainfall rather than temperature. Several different climatic types can be distinguished within the tropical belt, depending on distance from the ocean, prevailing wind conditions, and elevation. The tropics contain the world's largest regions of tropical rain-forest climate (Amazon and Congo basins). Toward the northern and southern limits are low-latitude savanna, steppe, and desert climates. Tropical highland climates (having the characteristics of temperate climates) also occur where high mountain ranges lie in the zone. Rubber, tea, coffee, cocoa, spices, bananas, pineapples, oils and nuts, and lumber are the leading agricultural exports of the countries in the tropical zone.

Troup (tro͞op), county (1970 pop. 44,466), 447 sq mi (1,157.7 sq km), W Ga., bounded on the W by the Ala. border and drained by the Chattahoochee River; formed 1826; co. seat La Grange. It has piedmont agriculture (cotton, corn, truck crops, fruit, and livestock), timber, and textile manufacturing.

Trous·dale (trouz′dāl), county (1970 pop. 5,155), 114 sq mi (295.3 sq km), N Tenn., bounded in the S by the Cumberland River; formed 1870; co. seat Hartsville. It has livestock, tobacco, corn, and grain farming and manufactures clothing and shoes.

Trou·ville-sur-Mer (tro͞o′vēl-sür-mâr′), town (1968 pop. 6,429), Calvados dept., N France, on the English Channel. It is a popular beach resort.

Trow·bridge (trō′brĭj), urban district (1974 est. pop. 19,790), Wiltshire, S England. It is a market town and a center for the manufacture of woolen goods.

Troy (troi), ancient city made famous by Homer's account of the Trojan War. It is also called Ilion or, in Latin, Ilium. Its site is the mound now named Hissarlik, in Asian Turkey, c.4 mi (6.4 km) from the mouth of the Dardanelles. Heinrich Schliemann identified the site and conducted excavations here beginning in 1871. Nine successive cities or villages have occupied the site, the earliest dating from the Neolithic period. The Troy of the Trojan War was a Phrygian city and the center of a region known as Troas. The culture of the Trojans dates from the Bronze Age.

Troy. 1. City (1970 pop. 11,482), seat of Pike co., SE Ala., on the Conecuh River; inc. 1843. Products include lumber and wood items, textiles, truck bodies, feed, and pecans. **2.** City (1970 pop. 1,047), seat of Doniphan co., extreme NE Kansas, near the Missouri River W of St. Joseph, Mo.; founded 1855, inc. 1860. It is the shipping center of an apple-growing and vinegar-producing region. Indian mounds have been excavated nearby. **3.** City (1978 est. pop. 62,000), Oakland co., SE Mich., a suburb of Detroit; settled 1821, inc. 1955. Its manufactures include automobile and electronic parts. **4.** City (1970 pop. 2,538), seat of Lincoln co., E Mo., NW of St. Louis; settled 1801, laid out 1819, inc. 1852. A state park is nearby. **5.** City (1978 est. pop. 57,400), seat of Rensselaer co., E N.Y., on the east bank of the Hudson River; inc. 1816. It is known for its manufacture of collars and shirts. Other important products are abrasives, auto parts, instruments, railroad supplies, and apparel. Henry Hudson explored (1609) the area near present Troy, and the site was included in the patroonship given to Kiliaen Van Rensselaer by the Dutch West India Company. The town was laid out in 1786. **6.** Town (1970 pop. 2,429), seat of Montgomery co., central N.C., E of Albermarle. It is a trade center and manufactures textiles and wood products. **7.** City (1978 est. pop. 18,500), seat of Miami co., W central Ohio, on the Great Miami River, in a farm area; inc. 1814. Food-processing machinery, motor generators, gummed paper, and tools are manufactured. Growth and industrialization came with the arrival of the Miami and Erie Canal in 1837.

Troyes (trwä), city (1968 pop. 74,898), capital of Aube dept., NE France, on the Seine River. It is an industrial town. Hosiery is the main product. Troyes became the capital of Champagne in the 11th cent. Its commercial importance was reflected in its annual fairs, which set standards of weights and measures for the whole of Europe, the troy weight having survived to this day. Troyes was the first town taken by Joan of Arc on her march to Rheims. The city has some fine Gothic structures, including the Cathedral of St. Peter and St. Paul (13th–16th cent.) and the Church of St. Urbain (begun 1262).

Tru·ji·llo (tro͞o-hē'yō), city (1972 pop. 127,535), capital of La Libertad dept., NW Peru. Trujillo processes sugar cane and rice and produces textiles, leather goods, food products, and cocaine. Founded in 1534, the city declared its independence from Spain in 1820, served as provisional capital of Peru in 1825, and was the main headquarters for Simón Bolívar. The pre-Incan ruins of Chanchan are nearby.

Trujillo, town (1971 pop. 25,921), capital of Trujillo state, W Venezuela. It is an agricultural market. Trujillo was founded in 1578 and was sacked by French pirates in 1678.

Truk (trŭk, tro͝ok), island group (1970 pop. 15,153), c.39 sq mi (100 sq km), W Pacific, in the E Caroline Islands. Truk consists of c.55 volcanic islands surrounded by an atoll reef and many islets. The chief products are copra and dried fish. During World War II Truk was the site of an important Japanese naval base.

Trum·bull (trŭm'bəl), county (1970 pop. 232,579), 608 sq mi (1,593.3 sq km), NE Ohio, bounded in the E by the Pa. border and drained by the Mahoning and Grand rivers; formed 1800; co. seat Warren. Its agriculture includes livestock, dairy products, clover, and grain, and it has sand and gravel pits. Steel milling, food processing, lumbering, and manufacturing are done.

Trumbull, town (1978 est. pop. 34,600), Fairfield co., SW Conn.; settled in the 1660s, inc. 1797. It has some light industry.

Tru·ro (tro͝or'ō'), town (1971 pop. 13,047), central N.S., Canada, near the head of Cobequid Bay, an arm of the Bay of Fundy. It is a railroad and industrial center, with lumber mills, printing plants, and other factories. An early Acadian settlement called Cobequid, the town was destroyed (1755) when the Acadians were expelled. After 1759 it received settlers from New England and Northern Ireland.

Truth or Con·se·quenc·es (tro͞oth' ôr kŏn'sə-kwĕn-səz), town (1970 pop. 4,656), seat of Sierra co., SW N.Mex., on the Rio Grande SW of Albuquerque, in a farm and mine region. Formerly known as Hot Springs (for its mineral springs), the city attracts health seekers and has a hospital for crippled children. Elephant Butte Dam is nearby.

Trut·nov (tro͞ot'nôf'), city (1971 est. pop. 25,241), N Czechoslovakia, in Bohemia, near the Polish border. It is a center of the Czech linen industry.

Try·on (trī'ən), village (1970 est. pop. 150), seat of McPherson co., W central Nebr., NNW of North Platte, in a grain, livestock, and poultry area.

Tsai·dam (tsī'däm'), arid basin, c.350 mi (565 km) long and c.100 mi (160 km) wide, between two branches of the Kunlun range, central Tsinghai prov., W China. A salt marsh occupies most of the area. Oil fields and refineries are found here, and iron ore is mined.

Tsa·na (tsä'nə), lake, Ethiopia: *see* Tana.

Tse·lin·ny Kray (tsā'lĭ-nē krī') or **Vir·gin Lands Territory** (vûr'jĭn), former administrative division, c.231,000 sq mi (598,290 sq km), Central Asian USSR, in Kazakhstan. Created in 1960 by the merger of Kokchetav, Kustanay, North Kazakhstan, Pavlodar, and Akmolinsk oblasts, it was abolished in the late 1960s.

Tse·lin·o·grad (tsĕ-lĭn'ə-gräd'), city (1973 est. pop. 201,000), capital of Tselinny Kray, Central Asian USSR, in Kazakhstan, on the Ishim River. It is a railroad junction and a center for the production of agricultural machinery and chemicals.

Tsi·nan (tsĭ'nän'): *see* Chi-nan, China.

Tsing·hai (chĭng'hī'), province (1967 est. pop. 2,000,000), c.250,000 sq mi (647,500 sq km), W China. Hsi-ning is the capital. Tsinghai lies in the Tibetan highlands at an average elevation of 9,800 ft (3,000 m) and is mainly a high, desolate plateau. The central region has the vast, swampy Tsaidam basin, and in the northeast there is the large Koko Nor or Tsinghai salt lake, the largest lake in China. In the mountain gorges of the south rise the Huang Ho, Yangtze, and Mekong rivers. The chief economic and most densely settled area is the northeast around Hsi-ning; here coal is mined and grain and potatoes are grown. Stock breeding is also important. From the 1950s to the 1970s there was an influx of Chinese to work in the mineral-extraction industries in the Tsaidam basin (oil, iron ore, salt, borax, and potash). Salt is so abundant that it is used for building blocks and for road pavement. Thousands of miles of highways link Hsi-ning and the Tsaidam basin with adjoining provinces; a railroad links Hsi-ning with Lan-chou, in Kansu prov.

Historically a part of Tibet, the Tsinghai region passed to the Mongol overlords of China in the 14th cent. It came under Chinese control after 1724 and was administered from Hsi-ning as the Koko Nor territory. In 1928 Tsinghai became a province of China. The Communist government established autonomous districts for the Tibetan, Chinese Moslem, Kazakh, and Mongol minorities.

Tsing·tao (tsĭng'tou', chĭng'dou'): *see* Ch'ing-tao, China.

Tsin·kiang (jĭn'jē-äng'): *see* Ch'üan-chou, China.

Tsin·ling or **Ch'in Ling** (both: chĭn'lĭng'), mountain range, outlier of the Kunlun Mts., between the Wei and Han rivers, Shensi prov., central China; T'ai-pai Shan (13,494 ft/4,115.7 m) is the highest peak. Coal is mined in the central region.

Tsi·tsi·har (tsē'tsē'här'): *see* Ch'i-ch'i-ha-erh, China.

Tskhin·va·li (tskĭn'və-lē), city (1973 est. pop. 32,000), capital of South Ossetian Autonomous Oblast, Georgian Republic, S European USSR. The city has lumber mills and electrical-products plants.

Tsu (tso͞o), city (1970 pop. 125,203), capital of Mie prefecture, S Honshu, Japan, on Ise Bay. It is a commercial and manufacturing center, with glass and food-processing factories.

Tsu·chi·u·ra (tso͞o'chē-o͞or'ə), city (1970 pop. 89,958), Ibaraki prefecture, central Honshu, Japan. It is an agricultural and commercial center.

Tsu·ga·ru Strait (tso͞o-gä'ro͞o), c.100 mi (160 km) long and 15–25 mi (24–40 km) wide, separating Honshu and Hokkaido, N Japan, and connecting the Sea of Japan with the Pacific Ocean.

Tsu·meb (tso͞o'mĕb'), town (1970 pop. 12,338), N Namibia. It is the commercial and distribution center for a region where copper, lead, and zinc are mined.

Tsun-i or **Tsun·yi** (both: dzo͞o'nē'), town (1970 est. pop. 275,000), N Kweichow prov., SW China. It is the commercial and agricultural distribution center of northern Kweichow. Iron and manganese ore are mined, food is processed, and textiles (cotton and silk), chemicals, and machine tools are manufactured.

Tsu·ru·ga (tso͞o-ro͞o'gə), city (1970 pop. 56,445), Fukui prefecture, central Honshu, Japan, a port on the Sea of Japan. Among the principal products are rayon and cement.

Tsu·ru·o·ka (tso͞o'ro͞o-ō'kə), city (1970 pop. 74,000), Yamagata prefecture, NE Honshu, Japan. It is an agricultural center.

Tsu·shi·ma (tso͞o-shē'mə), two Japanese islands in Korea Strait. The islands are rocky, and fishing is the main occupation. Nearby, in May, 1905, occurred the major naval battle of the Russo-Japanese War, in which the Russian Baltic fleet, under Admiral Rozhdestvenski, suffered nearly total disaster in its encounter with the Japanese fleet under Count Togo.

Tu·a·mo·tu Islands (to͞o'ə-mō'to͞o) or **Low Archipelago** (lō), coral group (1971 pop. 8,226), South Pacific, part of French Polynesia. They comprise c.80 atolls in a 1,300-mi (2,092-km) chain, with a total land area of c.330 sq mi (855 sq km). Rangiroa is the largest island; Fakarava is the most important commercially. The islands have coconut, pandanus, and breadfruit trees and produce pearl shell and copra. They were discovered by the Spanish in 1606, came under a French protectorate in 1844, and were annexed by France in 1881.

Tu·ap·se (to͞o'əp-sä'), city (1973 est. pop. 56,000), Krasnodar Kray, SE European USSR, on the Black Sea. It is a major petroleum port and the terminal of the pipeline from the Grozny oil fields. The city refines oil and manufactures equipment for the oil industry.

Tü·bing·en (tü'bĭng-ən), city (1972 est. pop. 69,261), Baden-Württemberg, SW West Germany, on the Neckar River. Manufactures include textiles, machinery, metal goods, precision instruments, and printed materials. Tübingen was chartered c.1200, passed to Württemberg in the mid-14th cent., and became the second capital of Württemberg in the mid-15th cent. Noteworthy buildings include the city hall (1435), the late-Gothic Church of St. George (15th cent.), and Hohentübingen, a castle first mentioned in the 11th cent. Tübingen is famous for its university (founded 1477).

Tuck·er (tŭk′ər), county (1970 pop. 7,447), 421 sq mi (1,090.4 sq km), NE W.Va., at the base of the Eastern Panhandle, on the Allegheny Plateau, with Laurel Ridge along the W border; formed 1856; co. seat Parsons. The county has timber (logging and sawmills) and limestone quarries. Its agriculture includes livestock, dairy products, and fruit. There is some chemical, charcoal, and shoe manufacturing. Most of the county lies in Monongahela National Forest.

Tuck·a·hoe (tŭk′ə-hō′), suburban village (1970 pop. 6,236), Westchester co., SE N.Y., SW of White Plains; settled 1684, inc. 1903. Electronic equipment is made.

Tuc·son (tōō′sŏn′), city (1978 est. pop. 301,600), seat of Pima co., SE Ariz.; inc. 1877. Situated in a desert valley surrounded by mountains, Tucson is an important transportation and tourist center; its dry, sunny climate attracts vacationers and health seekers. The city also has large electronic, optic, and research industries and serves as the processing and distributing center for cotton and livestock and for the many mining (chiefly copper) operations. The first Spanish settlers arrived in the late 17th cent., and in 1700 Mission San Xavier del Bac was founded. The present city was established (1776) as a walled presidio, and Tucson became a military border post of New Spain, of Mexico, and, after its transfer under the Gadsden Purchase, of the United States. Tucson served as territorial capital from 1867 to 1877.

Tu·cu·mán (tōō-kə-män′), city (1970 pop. 765,962), capital of Tucuman prov. NW Argentina. It is the commercial center of an area that produces sugar, cereals, fruit, and lumber. The city was founded in 1565 and was moved to its present site in 1685.

Tu·cum·car·i (tōō′kəm-kâr′ē), city (1970 pop. 7,189), seat of Quay co., E N.Mex., near the Canadian River NW of Clovis; settled c.1900, inc. 1908. It is a railroad division point and a trade center in a farm and cattle area.

Tu·de·la (tōō-dā′lä), town (1970 pop. 20,942), Navarre, N Spain, on the Ebro River. The surrounding region produces vegetables, fruit, grapes, and olives. There are sugar refineries and varied manufactures. Tudela flourished under the Moors and was later the second city of the kingdom of Navarre. In a battle nearby the French won a major victory (1808) in the Peninsular War. There is a fine 12th cent. cathedral and an old Roman bridge.

Tu·la (tōō′lə), ancient city in the present state of Hidalgo, central Mexico. It was one of the chief urban centers of the Toltec. The city is believed to be Tollán, the legendary Toltec capital. Archaeological investigations have revealed impressive architectural remains, including pyramidal structures and ball courts.

Tula, city (1973 est. pop. 486,000), capital of Tula oblast, N central European USSR, on the Upa River, a tributary of the Oka. It is an important rail and highway hub and a manufacturing city; it produces mining and transport equipment, deepwater pumps, boilers, ventilators, and armaments. First mentioned in 1146, in the 16th cent. the city became a fortress of the grand duchy of Moscow. Peter I built Russia's first arms factory here in 1712, based on the discovery of iron and coal deposits. Tula became a center of the Russian ironworking industry. During World War II the city withstood heavy German assaults. The 16th cent. kremlin, with turreted walls, has been preserved.

Tu·lare (tōō-lâr′, -lâr′ē), county (1970 pop. 188,322), 4,812 sq mi (12,463 sq km), S central Calif., extending E from the San Joaquin Valley to the crest of the Sierra Nevada and drained by the Kaweah, St. Johns, Tule, and Kern rivers; formed 1852; co. seat Visalia. It is in an agricultural area that produces cattle, dairy goods, citrus fruit, deciduous fruit, nuts, grapes, hay, alfalfa, grain, cotton, poultry, eggs, potatoes, and bees. It has lumbering, quarrying and mining, and diversified industry. Sequoia National Park, parts of Kings Canyon National Park, and Inyo and Sequoia national forests are in the county.

Tulare, city (1975 est. pop. 18,200), Tulare co., S central Calif., in the San Joaquin Valley; inc. 1888. It is a processing and shipping center for a farm, cotton, and dairy region. Truck bodies, chalkboard, and aluminum and concrete products are made.

Tul·chin (tōōl′chĭn), city, SW European USSR, in the Ukraine, on the Selnitsa River. It is the center of an agricultural district and has food-processing, clothing, and shoe industries. Probably founded by Hungarians, it later became a Polish fortress. It was assigned (1649) to the Ukraine. It reverted to Polish rule in 1654 but passed to Russia in 1793.

Tu·lia (tōōl′yə), city (1970 pop. 5,294), seat of Swisher co., extreme N Texas, S of Amarillo on the plains of the Panhandle; settled 1890, inc. 1909. The town produces cattle, milk, grain, and some cotton.

Tul·la·ho·ma (tŭl′ə-hō′mə), city (1975 pop. 15,577), Coffee and Franklin cos., central Tenn.; settled c.1850 as a railroad labor camp, inc. 1903. It is an industrial center in a timber and farm area; manufactures include sporting goods and wood products.

Tul·la·more (tŭl′ə-môr′), urban district (1971 pop. 6,809), county town of Co. Offaly, central Republic of Ireland. It is a marketing and processing center for a farm area.

Tulle (tōōl, tül), town (1968 pop. 20,016), capital of Corrèze dept., S central France. Firearms and other goods are made here. Tulle was built around a 7th cent. monastery. Tulle cloth was first manufactured in the town.

Tul·sa (tŭl′sə), county (1970 pop. 399,982), 573 sq mi (1,484.1 sq km), NE Okla., intersected by the Arkansas River and Bird Creek; formed 1907; co. seat Tulsa. It has stock raising and agriculture (dairy products, poultry, corn, grain, truck crops, pecans, fruit, and cotton), with gas production, limestone quarries, and sand and gravel pits. Clothing, wood products, paper, chemicals, metal products, machinery, and transportation equipment are manufactured.

Tulsa, city (1978 est. pop. 334,700), seat of Tulsa co., NE Okla., on the Arkansas River E of its junction with the Cimarron; inc. 1898. It became an inland port with the opening (1971) of the McClellan-Kerr Waterway, a 440-mi (708-km) system linking it with the Gulf of Mexico. It is an important center of the nation's petroleum industry with large refineries and plants that produce petroleum products and related equipment. Mining, metal processing, machinery manufacturing, and the aerospace industry are also important.

T'u-lu-fan (tōō′lōō′fän′) or **Tur·fan** (tōōr′fän′), town and oasis, in the T'u-lu-fan depression (c.5,000 sq mi/12,950 sq km), E Sinkiang Uigur Autonomous Region, China. It is an agricultural center. The T'u-lu-fan depression, the lowest point (505 ft/154 m below sea level) in China, was the center (A.D. 200–400) of a flourishing civilization in which Indian and Persian elements were combined.

Tu·ma·ca·co·ri National Monument (tōō′mə-kä′kə-rē): *see* National Parks and Monuments Table.

Tu·ma·co (tōō-mä′kō), city (1968 est. pop. 80,300), SW Colombia, a port on the Pacific Ocean. It is located on a small island just off the coast and has a very hot climate. Tumaco is a commercial center. Coffee, cacao, tobacco, vegetables, and other products of the interior are exported.

Tum·kur (tōōm-kōōr′), town (1971 pop. 70,476), Karnataka state, S central India. It is a health center and a market for vegetable oil, tobacco, chilies, and coconuts. Brick, tile, and iron and steel products are manufactured.

Tum·wa·ter (tŭm′wô′tər, -wŏt′ər), historic town (1977 est. pop. 5,960), Thurston co., W Wash., near Puget Sound just S of Olympia; founded 1845, inc. 1875. It was the first American settlement in the area and is sometimes considered the end of the Oregon Trail. The town has a large brewery.

tun·dra (tŭn′drə), treeless plains of N North America and N Eurasia, lying principally along the Arctic Circle, on the coasts and islands of the Arctic Ocean, and to the N of the coniferous forest belt. The tundra area is widest in northern Siberia on the Kara Sea and reaches as far south as 60° N at the neck of the Kamchatka peninsula. Although sometimes called the Arctic steppe and situated mainly within the Arctic Circle, it reaches southward into the Scandinavian, Timan, and Ural mts.

T'ung-hua or **Tung·hwa** (both: tōōng′hwä′), city (1970 est. pop. 275,000), SW Kirin prov., China, in a mountainous region, and on a railroad to Korea. Abundant coal and iron reserves are in the area, and the city has iron and steel works. Motor vehicles, machinery, and paper products are also manufactured.

Tung-t'ing or **Tung·ting** (both: dōōng′tĭng′), shallow lake, Hunan prov., SE China. Depending on the season, the Tung-t'ing varies in size from 1,400 to 4,000 sq mi (3,626–10,360 sq km). It attains its maximum extent during the period of heavy summer rains. The heavily populated lake basin is one of China's leading rice-producing regions.

Tun·gus·ka Basin (to͝on-go͞os'kə), c.400,000 sq mi (1,036,000 sq km), E Siberian USSR, between the Yenisei and Lena rivers. It has a huge untapped coal reserve.

Tu·ni·ca (to͞o'nĭ-kə, tyo͞o'-), county (1970 pop. 11,854), 458 sq mi (1,186.2 sq km), NW Miss., bounded in the NW and W by the Mississippi River, here forming the Ark. border, and partly bounded in the E by the Coldwater River; formed 1836; co. seat Tunica. It is in a rich agricultural area (cotton and corn), with cotton ginning, cottonseed processing, and lumber milling.

Tunica, town (1970 pop. 1,685), seat of Tunica co., NW Miss., near the Mississippi River below Memphis, Tenn., in a farm area. Cottonseed and lumber products are made.

Tu·nis (to͞o'nĭs), city (1966 pop. 468,997), capital of Tunisia, NE Tunisia, on the Lake of Tunis. Access to the Gulf of Tunis (an arm of the Mediterranean) is by a canal terminating at a subsidiary port, Halq al Wadi (La Goulette). Products include textiles, carpets, and olive oil. There are railroad workshops and a lead smelter. Tunis has notable mosques, the Univ. of Tunis (1960), and a national museum. The ruins of Carthage are nearby, to the northeast. Tunis is probably pre-Carthaginian. Surviving from the Middle Ages are walls, an aqueduct, and a mosque.

Tu·ni·sia (to͞o-nē'zhə, tyo͞o-), republic (1977 est. pop. 5,995,000), 63,378 sq mi (164,150 sq km), NW Africa. The capital is Tunis. Tunisia, occupying the eastern portion of the great bulge of North Africa, is bounded on the west by Algeria, on the north and east by the Mediterranean Sea, and on the southeast by Libya. It has a highly irregular coastline that affords many bays and several fine harbors, notably Bizerte, Qabis, Safaqis, and Susah. Part of the Atlas Mts. runs through northern Tunisia; but the mountains in Tunisia rarely exceed 4,000 ft (1,219 m) in elevation. In the south, below the Chott Djerid (a great salt lake), stretches the Sahara Desert. The population is largely Berber and Moslem, Arabic being the official language; there is a Jewish community dating back to ancient times.

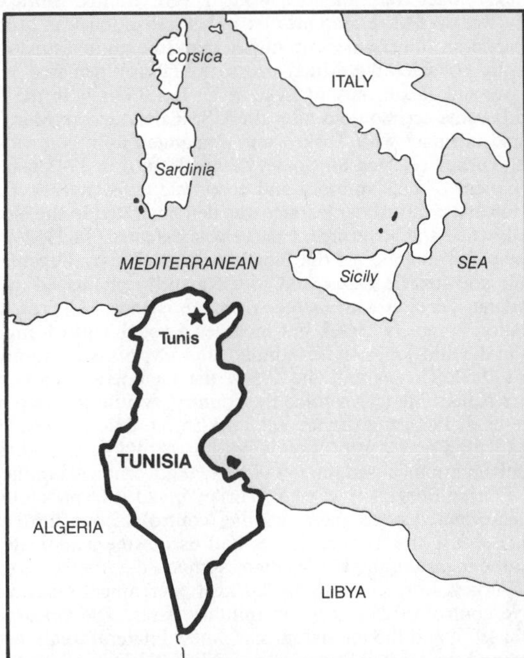

Economy. Tunisia's economy is based on agriculture; the leading crops are wheat, barley, grapes, olives, citrus fruits, and dates. Mineral production is the second most important sector of the economy. Phosphates and iron are found in quantity and some zinc and lead are also mined. Petroleum was found in 1964, and production began in 1966. Subsequently other oil fields were discovered, and production has increased substantially. Tunisia's manufacturing industries include steelworks, textile factories, food-processing plants, and sugar refineries. Petroleum, phosphates, and olive oil are the country's leading exports; its imports, which exceed exports, are headed by machinery, metal products, and transportation equipment.

History. The coast of Tunisia was settled in the 12th cent. B.C. by Phoenicians. It was then ruled by Carthage (6th cent. B.C.) and by Rome (2nd cent. B.C.). Held by Vandals (5th cent. A.D.) and Byzantines (6th cent.), it was conquered by Arabs in the 7th cent., who founded Al Qayrawan; and the Berber population was converted to Islam. Successive Moslem dynasties ruled, interrupted by Berber rebellions. When the Zirids (reigned from 972), Berber followers of the Fatimids, provoked the Fatimids in Cairo (1050), the latter sent thousands of Arab tribesmen to ravage Tunisia. In 1159 Tunisia was conquered by the Almohad caliphs of Morocco, who were succeeded by the Berber Hafsids (c.1230–1574), under whom Tunisia prospered. Spain seized many of the coastal cities (14th cent.), but they were recovered for Islam by the Ottoman Turks. Under its Turkish governors, the beys, Tunisia attained virtual independence. In the late 16th cent. the coast became a pirate stronghold.

In the 19th cent. Tunisia's heavy debts gave European powers cause for intervention. France, Great Britain, and Italy took over Tunisia's finances in 1869. A number of incidents, including attacks on Algeria (a French possession) by Tunisians, led to a French invasion of Tunisia. The bey was forced to sign treaties that provided for the organization of a French protectorate. The protectorate was opposed by Italy, which had economic interests and a sizable group of nationals in Tunisia. In the years immediately preceding World War I, threats of annexation were made. In 1920 the Tunisian nationalist Destour (Constitutional) party was organized. In 1934 a more radical faction, led by Habib Bourguiba, formed the Neo-Destour party. In World War II Tunisia came under Vichy rule after the fall of France (June, 1940). After the war nationalist agitation intensified. In 1950 France granted Tunisia a large degree of autonomy. The French population in Tunisia, however, opposed further reforms, and negotiations broke down. Bourguiba was arrested (1952), and his imprisonment precipitated a wave of violence. Full independence was negotiated in 1956, and Bourguiba became prime minister. The country became a republic in 1957 when the bey, Sidi Lamine, was deposed by a vote of the constituent assembly. Bourguiba followed a generally pro-Western foreign policy, but relations with France were strained over the French presence in northern Africa. France finally agreed to evacuate naval installations at Bizerte in 1963. Relations between Tunisia and Algeria deteriorated over the issue of border disputes after the latter gained (1962) independence from France. Bourguiba's support for a negotiated settlement with Israel of the Middle East problem caused strains in Tunisia's relations with other Arab countries. Bourguiba also accused Egypt (1958) and Algeria (1962) of complicity in attempts by Tunisian exiles to assassinate him. Domestically, Bourguiba's policies emphasized modernization and planned economic growth. An agrarian reform plan, involving the formation of cooperatives, was begun in 1962, but it aroused widespread opposition and was halted in 1969. In the early 1970s there was increasing conflict within the ruling Destour party. With evidence of general unrest, the succession to Bourguiba, whose health was poor, was a question of increasing importance.

Government. Tunisia is governed under the 1959 constitution; the president and national assembly are elected every five years. An amendment to the constitution in 1975 made Bourguiba president for life. A cabinet is appointed by the president, and a council, made up of the cabinet and leaders of the Destour party, advises the president.

Tun·ja (to͞on'hä), city (1968 est. pop. 72,700), capital of Boyacá dept., central Colombia, on the Pan-American Highway. It is a commercial center and distribution point for the products of the region (coal, emeralds, mineral water, and agricultural products) and for the cattle of the eastern llanos.

Tunk·han·nock (tŭngk-hăn'ək), residential borough (1970 pop. 2,251), seat of Wyoming co., NE Pa., on the Susquehanna River NW of Scranton; settled 1775, laid out 1790, inc. 1841.

Tu·ol·um·ne (to͞o-ŏl'ə-mē), county (1970 pop. 22,169), 2,252 sq mi (5,832.7 sq km), central Calif., in the Sierra Nevada, drained by the Tuolumne and Stanislaus rivers; formed 1850; co. seat Sonora. It has lumbering, mining and quarrying, cattle grazing, some farming, and poultry raising. Many campgrounds, lakes and streams, and winter-sports facilities attract visitors. The east part of the county is within Yosemite National Park, and it includes Stanislaus National Forest.

Tu·pe·lo (to͞o′pə-lō, tyo͞o′-), city (1978 est. pop. 23,500), seat of Lee co., NE Miss.; founded 1859, inc. 1870. It is the trade, processing, and shipping center for a cotton and livestock area. On the Civil War battlefield of Tupelo, now a national battlefield, Union troops repulsed an attack by Gen. N. B. Forrest (July 14, 1864) but nevertheless retreated.

Tupelo National Battlefield: *see* National Parks and Monuments Table.

Tur·fan (to͝or′făn′): *see* T'u-lu-fan, China.

Tu·rin (to͝or′ĭn, tyo͝or′-, tyo͞o-rĭn′), city (1972 est. pop. 1,172,476), capital of Piedmont and of Turin prov., NW Italy, at the confluence of the Po and Dora Riparia rivers. It is a major industrial center and a transportation hub. Manufactures include motor vehicles, tires, textiles, clothing, machinery, electronic equipment, leather goods, furniture, chemicals, and vermouth. It is an international center of clothing fashions. The most important Roman town of the western Po valley, Turin was later a Lombard duchy and then a Frankish county. It passed c.1280 to the house of Savoy. From 1720 to 1861 it was the capital of the kingdom of Sardinia. During the War of the Spanish Succession it suffered a long siege, which ended with the victory of Eugene of Savoy over the French. In 1798 Charles Emmanuel IV of Savoy was obliged by the French to abdicate and to abandon Turin, but Victor Emmanuel I returned in 1814, and the city soon became the center of Italian national aspirations. From 1861 to 1864 it was the capital of the new Italian kingdom. Because of its industrial importance, Turin suffered heavy damage in World War II; most of the important remaining buildings date from the 17th-19th cent. Of note are the Academy of Science, which contains the rich Egyptian Museum; the Cathedral of San Giovanni (late 15th cent.), which has an urn containing a shroud in which, it is said, Jesus was wrapped after the descent from the Cross. On a hill overlooking the city is the basilica of Superga (1717-31), containing the tombs of many of the dukes of Savoy and kings of Sardinia.

Tur·key (tûr′kē), republic (1970 pop. 35,666,549), 301,380 sq mi (780,574 sq km), SW Asia and SE Europe, bordering on Iraq in the SE, on Syria and the Mediterranean Sea in the S, on the Aegean Sea in the W, on Greece and Bulgaria in the NW, on the Black Sea in the N, and on the USSR and Iran in the E. Asian Turkey (made up largely of Asia Minor), which includes 97% of the country, is separated from European Turkey (made up of eastern Thrace) by the Bosporus, the Sea of Marmara, and the Dardanelles. Northeast Asian Turkey includes part of Armenia, and southeast Asian Turkey includes part of Kurdistan. Ankara is the capital.

European Turkey, which includes Edirne and most of Istanbul, is largely rolling agricultural land and is drained by the Ergene River. Asian Turkey is mostly made up of highland and mountains, with some narrow strips of lowland in the west on the coasts of the Aegean Sea and the Sea of Marmara and along the Simav, Gediz, and Menderes rivers; in the north on the Black Sea coast and along the Sakarya and Kızıl Irmak rivers; and in the south on the Mediterranean coast and along the Aksu, Göksu, Seyhan, and Ceyhan rivers. The center of western Asian Turkey is made up of the vast semiarid Plateau of Anatolia (average height c.3,000 ft/914 m), which includes Lakes Tuz and Beyşehir and which is fringed in the north by the Köroğlu Mts. and in the south by the Taurus Mts. In northeast Turkey are the Pontic Mts. and in eastern Turkey are the Eastern Taurus Mts. Great Ararat Mt. (16,945 ft/5,165 m), the highest point in Turkey, and Lake Van are in the extreme eastern part of the country. Southeastern Turkey is drained by the upper courses of the Tigris and Euphrates rivers.

Although the Turks regard the Osmanlis, or Ottomans, as their ancestors, they are a composite ethnic mixture. Ninety percent of the population speaks Turkish, the official language, as its first language, and 6% speaks Kurdish; there are also small Arabic, Circassian, Greek, Armenian, Georgian, and Laze-speaking minorities. Almost 99% of the people are Moslem, mostly of the Sunni branch; there are also small groups of Orthodox Christians (Istanbul is the seat of the Ecumenical Patriarch), Gregorians, Roman Catholics, Protestants, and Spanish-speaking Jews.

Economy. The Turkish economy is basically agricultural, but since the late 1940s the pace of industrialization has accelerated. The most productive farmland is in western Turkey, and there is extensive pastureland in most parts of the country. The chief crops are wheat, barley, maize, rye, oats, rice, cotton, fruit, and tobacco. Large numbers of sheep, goats, and cattle are raised. The principal minerals extracted are coal, lignite, copper and iron ores, chromite, antimony, and mercury. Some petroleum is produced. The country's chief manufactures include iron and steel, construction materials (especially cement), forest products, cotton and woolen textiles, processed food (especially refined sugar and raisins), wine, refined petroleum, and chemical fertilizer. Turkey is also noted for the manufacture of carpets; Meerschaum pipes and artifacts; and pottery. The country has a very limited rail network.

History. Although Anatolia (the western portion of Asian Turkey) is one of the oldest inhabited regions of the world, the history of Turkey as a national state began only with the collapse of the Ottoman Empire in 1918. After World War I the victorious Allies reduced the once mighty empire to a small state comprising the northern half of the Anatolian peninsula. Sultan Mohammed VI accepted the treaty, but Turkish nationalists resisted under the leadership of Mustafa Kemal (from 1934 known as Kemal Atatürk), who set up a Turkish national government at Ankara. The Turks defeated the Allied-supported Greeks (1922), consolidated their rule of the whole peninsula, and achieved international recognition for virtually the present boundaries of Turkey in 1923. Sovereignty over the Zone of the Straits was restored to Turkey in1936. Turkey was formally proclaimed a republic in Oct., 1923, but Kemal governed as a dictator until his death in 1938. Between these dates the nationalist regime transformed the country: Islam ceased to be the state religion, European legal codes were adopted, women were granted suffrage, and literacy was spread. Economically, industrialization was promoted by reliance on state ownership rather than foreign investment.

Despite considerable Allied pressure, Turkey declared war on Germany and Japan only in Feb., 1945. Relations with the Soviet Union became acrimonious after the USSR denounced (Mar., 1945) its friendship pact with Turkey and demanded joint control of the Straits. Turkey rejected all Soviet demands, and in 1947 it became the recipient of U.S. military and economic assistance. A Turkish move toward political democracy was demonstrated in the elections of 1950, when the government party was defeated. In 1952 Turkey became a full member of the North Atlantic Treaty Organization; U.S. air and missile bases were subsequently established at İzmir and Adana. Tension with Greece over the island of Cyprus, whose population is mostly Greek but includes a vocal Turkish minority, began in the mid-1950s and continued after Cyprus became independent in 1960. Throughout the 1950s, the Democratic party under Premier Adnan Menderes ruled the country. A military coup ousted Menderes in 1960, and the second Turkish republic was established in 1962 with a new constitution. Civil unrest and leftist and rightist political agitation caused the fall of several governments in the early 1970s. During Ecevit's brief term, Turkey invaded Cyprus following a Greek-oriented coup there, gaining control of one third of the island. Because the Turkish forces had used American-made arms during the invasion, the U.S. Congress imposed a total arms embargo on Turkey. In retaliation, the Turkish government seized administrative control of 25 American military bases. The embargo was lifted in 1975, and the following year a new bilateral treaty between Turkey and the United States, under which Turkey retained partial control of the bases, was signed. Ecevit's coalition collapsed in 1974 and, after lengthy negotiations, Demirel formed a coalition. Ecevit returned to power briefly in 1977, but his government fell after 10 days in office, and Demirel once again became premier.

Government. Under the 1961 constitution as amended, legislative power is vested in the bicameral grand national assembly, made up of the national assembly and the senate. The national assembly is composed of 450 members elected to four-year terms; the senate includes 150 members elected to six-year terms, 15 members appointed by the president of Turkey, and a small number of life members. The president, who is head of state, is elected by the grand national assembly to one nonrenewable seven-year term. The country's executive is the prime minister, who presides over the council of ministers and who must have the confidence of the national assembly.

Tur·ki·stan or **Tur·ke·stan** (both: tûr′kə-stăn′, -stän′), historic region of central Asia. Western or Russian Turkistan extended from the Caspian Sea in the west to the Chinese frontier in the east and from the Aral-Irtysh watershed in the north to the borders of Iran and Afghanistan in the south. Eastern, or Chinese, Turkistan comprised the western provinces of China, now constituting the Sinkiang Uigur Autonomous Region. Southern, or Afghan, Turkistan referred to a small area of northern Afghanistan. Politically, Russian Turkistan (now known officially as Soviet Central Asia) includes the Turkmen, Uzbek, Tadzhik, and Kirghiz republics and the southern portion of the Kazakh Republic. Much of the western part of the Soviet territory is composed of two deserts, the Kara-Kum and the Kyzyl-Kum. The eastern part, rough and hilly, rises to include the mountains of part of the Pamir highland and of the Tien Shan system. Athwart the eastern section extends the Fergana Valley, one of Asia's most fertile regions. Turkistan is regarded as a single region because a combination of geographic and historical factors made it the bridge linking the Eastern and Western worlds and the route taken by many of the great conquerors and migrating peoples. Following the Russian Revolution of 1917, the Turkistan autonomous soviet republic (1918) and the Bukhara and Khorezm soviet republics (1920) were set up in the region. However, in 1924 the southern part of Russian Turkistan was divided along geographic and ethnic lines into new divisions—the Uzbek SSR, the Turkmen SSR, the Tadzhik SSR (a union republic as of 1929), the Kirghiz autonomous oblast (made an autonomous republic in 1926 and a union republic in 1936), and the Kara-Kalpak autonomous oblast (which became an autonomous republic in 1932); the northern part of Turkistan was included in the Kazakh SSR.

Turk·men Soviet Socialist Republic (tûrk′mĕn′, -mən) or **Turk·men·i·stan** (tûrk′mĕn-ĭ-stăn′, -stän′), constituent republic (1978 est. pop. 2,627,000), 188,455 sq mi (488,100 sq km), Central Asian USSR. It borders on Afghanistan and Iran in the south, the Uzbek SSR and the Kazakh SSR in the east and northeast, and the Caspian Sea in the west. Ashkhabad is the capital. The desert lands of Kara-Kum occupy 90% of the total area; the population is concentrated in oases at the foot of the Kopet Dagh Mts. in the south and along the Amu Darya, Murgab, and Tedzhen rivers. The republic's numerous mineral resources include oil, natural gas, salt, phosphate, mirabilite, sulfur, ozokerite, iodine, bromine, witherite, bentonite, lignite, barites, clays, and such building stones as limestone and gypsum. More than 90% of the cultivated land is irrigated. Cotton is the chief crop; wheat, barley, maize, millet, sesame, vegetables, melons, wine grapes, kenaf, jute, and alfalfa are also cultivated. Karakul sheep (which provide wool for the region's famous carpets), horned cattle, horses, and camels are raised, and silkworms are bred. Turkmenistan's industries include cotton ginning, silk spinning, metalworking, ship and railroad car repairing, fish canning (along the Caspian), meat processing, oil refining, and the production of chemicals, textiles, and building materials. The republic has numerous hydroelectric stations. The Turkomans (or Turcomans) make up about 60% of the population; the remainder are Russians, Uzbek, Kazakhs, Tatars, Ukrainians, and Armenians. The Turkomans are a Turkic-speaking people of the Sunnite Moslem religion. Unlike other Central Asian groups, they still retain tribal and clan divisions. Turkmenistan formally became a constituent Soviet republic in 1925.

Turks and Cai·cos Islands (tûrks; kā′kəs), dependency of Great Britain (1970 est. pop. 5,675), 166 sq mi (430 sq km), British West Indies. There are more than 30 cays and islands, of which only 6 are inhabited. The capital is on Grand Turk. The islands are geographically a southeastern continuation of the Bahama Islands. Salt is the main export. The islands were discovered (1512) by Ponce de León.

Tur·ku (tŏŏr′kōō), city (1971 est. pop. 155,497), capital of Turku-and-Pori prov., SW Finland, at the mouth of the Aurajoki River on the Baltic Sea. The center of the fertile agricultural region of southwest Finland, it is also the country's largest port and an important industrial city. There are shipyards, steel mills, machine shops, textile mills, and clothing factories. Known as the "cradle of Finnish culture," Turku is among Finland's oldest cities. It was the capital of Finland until 1812.

Tur·lock (tûr′lŏk), city (1975 est. pop. 16,400), Stanislaus co., central Calif.; inc. 1908. It is the center of the Turlock irrigation district, which uses the waters of the Tuolumne River for a rich farm area. The city has canneries and poultry-processing plants.

Tur·ner (tûr′nər). **1.** County (1970 pop. 8,790), 293 sq mi (758.9 sq km), S central Ga., drained by the Withlacoochee and Alapaha rivers; formed 1905; co. seat Ashburn. It has coastal-plain agriculture (peanuts, pecans, soybeans, cotton, corn, and melons) and forestry (lumber and naval stores). **2.** County (1970 pop. 9,872), 612 sq mi (1,585.1 sq km), SE S.Dak., drained by branches of the Vermilion River; formed 1871; co. seat Parker. Its agriculture includes corn, oats, barley, wheat, soybeans, hay, dairy products, livestock, and poultry. It manufactures clothing.

Turn·hout (türn′hout), city (1971 est. pop. 38,194), Antwerp prov., N Belgium, near the Dutch border. Manufactures of this industrial city include paper products, textiles, lace, and electrical equipment.

Tur·nu-Se·ve·rin (tŏŏr′nōō-sĕ′vĕ-rēn′), city (1973 est. pop. 63,406), SW Rumania, in Walachia, on the Danube River opposite Yugoslavia. It is a river port and has large shipyards, a plywood factory, and several food-processing plants. The surrounding area is known for its extensive rose gardens and its white wine. Turnu-Severin was founded on the site of Drobeta, an ancient town believed to be the oldest Roman settlement in Rumania.

Tur·qui·no (tŏŏr-kē′nō), peak, 6,560 ft (2,000.8 m) high, SE Cuba, in the Sierra Maestra range. It is the highest point on the island. The mountain was the scene of intense guerrilla activity during the revolution led by Fidel Castro.

Tus·ca·loo·sa (tŭs′kə-lōō′sə), county (1970 pop. 116,029), 1,333 sq mi (3,452.5 sq km), W central Ala., in the coastal plain, drained by the Black Warrior, Sipsey, and North rivers; formed 1818; co. seat Tuscaloosa. It has cotton, corn, and bee farming, bituminous-coal mining, and lumbering. Its industry includes food processing, textile milling, petroleum refining, and iron and steel production.

Tuscaloosa, city (1978 est. pop. 71,000), seat of Tuscaloosa co., W central Ala., on the Black Warrior River; inc. 1819. It is a transportation, manufacturing, and medical center, with industries centered on the region's coal, iron, cotton, and timber. Food is processed, and rubber tires, chemicals, paper, adhesives, petroleum products, plastics, and textiles are manufactured. The city is primarily known as the seat of the Univ. of Alabama. Tuscaloosa was settled (1816) on the site of an Indian village after the Creek revolt of 1813. It was state capital from 1826 to 1846.

Tus·ca·ny (tŭs′kə-nē), region (1971 pop. 3,470,915), 8,876 sq mi (22,989 sq km), N central Italy, bordering on the Tyrrhenian Sea in the W and including the Tuscan Archipelago; capital Florence. The region is mostly hilly and mountainous. There is much fertile soil, especially in the Arno River valley and in the Maremma, a coastal strip. The Apennines are in northern and eastern Tuscany; in the northwest are the Alpi Apuane, where the famous Carrara marble is quarried; and there are also mountains in the south, where iron, magnesium, and quicksilver are produced. In addition, borax is produced in the Maremma, and iron is mined on Elba island. Farm products of the region include cereals, olives, tobacco, and grapes; sheep, goats, and hogs are widely raised. Manufactures include cotton and woolen textiles, chemicals, machinery, motor vehicles, precision instruments, glass, refined petroleum, and fertilizer. The wine produced in the Chianti district near Siena is world famous.

Modern Tuscany corresponds to the larger part of ancient Etruria. The Romans conquered the region in the mid-4th cent. B.C. It was a Lombard duchy (6th-8th cent. A.D.) and later a powerful march under the Franks (8th-12th cent.). Matilda (d.1115), the last Frankish ruler, bequeathed her lands to the papacy. Most cities became (11th-12th cent.) free communes; some of them (Pisa, Lucca, Siena, and Florence) developed into strong republics. Guelph (pro-papal) and Ghibelline (pro-imperial) strife was particularly violent in Tuscany. In the late Middle Ages and throughout the Renaissance, Tuscany was a center of the arts and of learning. The Tuscan spoken language became the literary language of Italy. Florence gained control over most Tuscan cities in the 14th-15th cent. Under the Medici, the ruling family of Florence, Tuscany became (1569) a grand duchy.

The French Revolutionary armies invaded Tuscany in 1799, and it was briefly included in the kingdom of Etruria (1801-7) before it was annexed to France by Napoleon I. In 1814 Tuscany again became a grand duchy, under the returning Ferdinand III, Leopold II (1824–59), and Ferdinand IV (1859–60). In 1860 Tuscany voted to unite with the kingdom of Sardinia.

Tus·ca·ra·was (tŭs′kə-rô′wəs), county (1970 pop. 77,211), 569 sq mi (1,471.7 sq km), E Ohio, intersected by the Tuscarawas River, Stillwater Creek, and small Sugar, Sandy, and Conotton creeks; formed 1808; co. seat New Philadelphia. It has agriculture (livestock, dairy products, and grain, bituminous-coal mines, oil and gas wells, and sand and gravel pits. Its industry includes food processing, lumbering, and the manufacture of furniture, chemicals, plastics, concrete, metal products, and machinery.

Tus·co·la (tŭs-kō′lə), county (1970 pop. 48,603), 815 sq mi (2,110.9 sq km), E Mich., bounded on the NW by Saginaw Bay, drained by the Cass River and its affluents; formed 1850; co. seat Caro. Sugar beets, beans, grain, potatoes, and fruit are grown, and livestock are raised in this agricultural area. There is some coal mining done here.

Tuscola, city (1970 pop. 3,917), seat of Douglas co., E central Ill., E of Decatur; platted 1857, inc. 1861. Industrial chemicals are made.

Tus·cu·lum (tŭs′kyŏŏ-ləm), city of ancient Latium. The ruins of this city are near modern Frascati, SE of Rome, Italy. According to legend, Tusculum was founded by Telegonus, son of Ulysses. It was a favorite summer residence of Roman nobles; Pliny the Younger, Cicero, and the emperors Nero and Titus were among those who built villas here.

Tus·cum·bi·a (tŭs-kŭm′bē-ə). **1.** City (1970 pop. 8,828), seat of Colbert co., NW Ala., on the Tennessee River near Muscle Shoals S of Florence; settled c.1815 on the site of a Chickasaw village. Helen Keller's birthplace is here. **2.** Town (1970 pop. 256), seat of Miller co., central Mo., on the Osage River SE of Eldon. It is a resort in an agricultural and barite-mining area.

Tus·ke·gee (tŭs-kē′gē), city (1970 pop. 11,028), seat of Macon co., SE Ala., in a cotton, corn, and dairy region; settled before 1763, inc. 1843. It has gristmills and plants making cottonseed oil and fertilizer. Tuskegee is best known as the seat of the Tuskegee Institute, chartered and opened in 1881 by Booker T. Washington as Tuskegee Normal and Industrial Institute.

Tus·tin (tŭs′tĭn), residential city (1978 est. pop. 33,700), Orange co., S Calif.; founded 1868, inc. 1927. Lumber, plumbing and piping, plastics, office equipment, and food products are made here. This rapidly growing city is part of the greater Los Angeles area.

Tu·ti·co·rin (tōō′tĭ-kô-rĭn′), city (1971 pop. 155,310), Tamil Nadu state, SE India. An important fishing center, it has a Fishing Technological Institute and is also famous for its pearl oysters. Other products include cotton cloth, embroidery, boats, and salt. Tuticorin was founded c.1540 by the Portuguese.

Tu·tu·i·la (tōō′tōō-ē′lä), island (1970 pop. 24,548), 52 sq mi (135 sq km), largest island of American Samoa. The island has a rugged eastern area, with a fertile plain in the southwest. Copra, canned fish, and handicrafts are the island's chief products.

Tu·va Autonomous Soviet Socialist Republic (tōō′və), administrative division (1970 pop. 231,000), 65,830 sq mi (170,500 sq km), extreme S Siberian USSR, on the Mongolian border. Kyzyl is the capital. The area is a mountain basin, c.2,000 ft (610 m) high. The eastern part is forested and elevated, and the west is a drier lowland. The area includes the upper course of the Yenisei River. Cattle, horses, sheep, goats, reindeer, and camels are raised in the elevated steppe areas, and grain is cultivated in the irrigated lowlands. Lumbering is carried on extensively. The fur trade remains important in the northeast. Among the republic's industries are food processing, leather making, woodworking, auto repairing, and the manufacture of building materials. The Tuvinians are a Turkic-speaking people; their religion is Tibetan Buddhism. They have a rich folklore and are skilled artisans in silver, bronze, wood, and stone. In 1921 the Bolsheviks established a Tuvinian People's Republic, popularly called Tannu-Tuva. It was annexed by the USSR in 1944 as an autonomous oblast and became an autonomous republic in 1961.

Tux·e·do Park (tŭk-sē′dō), residential and resort village (1970 pop. 861), Orange co., SE N.Y., SSW of Newburgh on Tuxedo Lake in the Ramapos near the N.J. border; inc. 1952. Pierre Lorillard, a wealthy sportsman, developed an exclusive colony (1886) known for its sports and social functions. The tailless dress coat, named for the colony, is said to have originated here. Sterling Forest Gardens and Palisades Interstate Park are nearby.

Tux·tla (tōōst′lə) or **Tuxtla Gu·tiér·rez** (gōō-tyâr′əs), city (1970 pop. 66,851), capital of Chiapas state, SE Mexico, in the fertile Grijalva valley and at the foot of the Chiapas highlands. Agriculture and cattle raising are the chief occupations, and there is trade in timber. Tuxtla is the focal distribution point for the products of the region.

Tuz, Lake (tōōz), shallow salt lake, c.625 sq mi (1,620 sq km), central Turkey. Salt is mined.

Tu·zi·goot National Monument (tōō′zĭ-gōōt′): *see* National Parks and Monuments Table.

Tuz·la (tōōz′lä), city (1971 pop. 53,852), central Yugoslavia, in Bosnia and Hercegovina. It has chemical, coking, and textile industries. Plums are grown in the vicinity, lignite and salt are mined, and some oil is extracted. The city's salt springs were known in Roman times.

Tver (tə-vâr′): *see* Kalinin, USSR.

Tweed (twēd), river, 97 mi (156 km) long, rising in the Southern Uplands of Scotland. It flows east through southern Scotland and then northeast, forming part of the Scotland-England border before entering the North Sea at Berwick in northeastern England. The Tweed system drains most of southeastern Scotland. In Scotland the Tweed waters a sheep-farming region. It also has rich salmon fisheries.

Twiggs (twĭgz), county (1970 pop. 8,222), 364 sq mi (942.8 sq km), central Ga., bounded on the W by the Ocmulgee River; formed 1809; co. seat Jeffersonville. It has coastal-plain agriculture (cotton, corn, grain, peanuts, and peaches) and a kaolin-mining area.

Twin Falls (twĭn), county (1970 pop. 41,807), 1,947 sq mi (5,042.7 sq km), S Idaho, in a livestock and dairying area bordering on Nev. and bounded in the N by the Snake River; formed 1907; co. seat Twin Falls. Irrigated regions are in the north, along the Snake River, and in the southwest, along Salmon Falls Creek. Potatoes, dry beans, sugar beets, onions, flax, and apples are grown.

Twin Falls, city (1978 est. pop. 24,300), seat of Twin Falls co., S Idaho, in the Snake River valley; inc. 1905. The city was begun as a center of a private irrigation project, which is now supplemented by the Minidoka project of the U.S. Bureau of Reclamation. One of the falls of Twin Falls in the nearby gorge is harnessed for hydroelectric power. Sugar beets, potatoes, corn, beans, and grains are processed, as well as livestock and dairy products. Several trout farms are in the area. Craters of the Moon National Monument is nearby.

Two Harbors, city (1970 pop. 4,437), seat of Lake co., NE Minn., on Lake Superior NE of Duluth; settled 1882, inc. as a village 1888, as a city 1907. It was an early shipping point for lumber, and since 1884 it has shipped much iron ore.

Two Rivers, city (1975 est. pop. 13,100), Manitowoc co., E Wis., on Lake Michigan at the mouth of the Twin River; inc. 1878. Two Rivers is closely associated with its twin city, Manitowoc, both of which are highly industrialized.

Ty·gart (tī′gərt), river, c.160 mi (260 km) long, rising in E W.Va. and flowing N to join the West Fork and form the Monongahela at Fairmont. Tygart River Dam (completed 1938), near Grafton, forms a large reservoir in Tygart Lake State Park.

Ty·ler (tī′lər). **1.** County (1970 pop. 12,417), 919 sq mi (2,380.2 sq km), E Texas, bounded on the N and E by the Neches River; formed 1846; co. seat Woodville. It has lumbering, farming (corn, cotton, vegetables, and sugar cane), some livestock raising, dairying, and some oil drilling. **2.** County (1970 pop. 9,929), 256 sq mi (663 sq km), NW W.Va., bounded on the NW by the Ohio River and the Ohio border, drained by Middle Island Creek; formed 1814; co. seat Middlebourne. It is in an oil and natural-gas region, including "Big Moses" gas well, and has sawmills. Its agriculture includes livestock, dairy, tobacco, grain, and truck crops.

Tyler, city (1978 est. pop. 62,400), seat of Smith co., E Texas; inc. 1850. In the heart of the rich East Texas oil field, Tyler has refineries and other oil-based industries. It also has a foundry and plants manufacturing pipes, tires, and electrical equipment. The city's rose-growing industry is one of the nation's largest. There is a huge municipal rose garden, and in season every street and yard is colored with blooms.

Ty·ler·town (tī′lər-toun′), town (1970 pop. 1,736), seat of Walthall co., S Miss., near the La. border SE of McComb, in a farm and timber area.

Tyn·dall (tĭn′dəl), city (1970 pop. 1,245), seat of Bon Homme co., SE S.Dak., near the Nebr. border NW of Yankton; founded 1879. It is a trade center of a grain, livestock, and dairy region.

Tyne (tīn), river, c.30 mi (50 km) long, NE England, formed near Hexham, Northumberland, by the confluence of the North Tyne (33 mi/53 km long; rising in SW Cheviot Hills) and the South Tyne (32 mi/52 km long; rising in the N Pennines). The Tyne flows eastward through the Tyneside conurbation to the North Sea at Tynemouth. The lower Tyne is lined with docks, shipbuilding yards, a variety of industrial plants, and coal-mining and ironworking towns.

Tyne and Wear (wâr), metropolitan county (1974 est. pop. 1,198,390), NE England, created under the Local Government Act of 1972 (effective 1974). It is subdivided into five metropolitan districts. Tyne and Wear comprises the county boroughs of Gateshead, Newcastle upon Tyne, South Shields, Sunderland, Tynemouth, and parts of the former counties of Durham and Northumberland.

Tyne·mouth (tīn'məth), county borough (1974 est. pop. 67,880), Tyne and Wear, NE England, on the Tyne River. Tynemouth is a shipbuilding center and a coal and fishing port. Its manufactures include furniture, textiles, glassware, machine tools, and die castings. It is also a resort.

Tyre (tīr), ancient city of Phoenicia, S of Sidon. It is the present-day Sur in Lebanon, a small town on a peninsula jutting into the Mediterranean from the mainland of Syria south of Beirut. The date of the founding of the city is extremely uncertain, but by 1400 B.C. it was a flourishing city. The maritime supremacy of Tyre was established by 1100 B.C. Throughout its long history Tyre frequently came under foreign rule. It was besieged by the Assyrians and the Chaldeans and fell to the Persians. The city was sacked by Alexander the Great but recovered quickly. In 64 B.C. it became a part of the Roman Empire. Christianity was introduced early into Tyre, and a splendid cathedral, of which there are remains, was built in the 4th cent. After the rise of Islam, Tyre came under Moslem rule and later under that of the Crusaders. It was destroyed by the Moslems in 1291 and never recovered its former greatness. The principal ruins of the city today are those of buildings erected by the Crusaders.

Ty·rol (tĭ-rōl', tī'rōl', tĭr'ōl'), province (1971 pop. 540,771), 4,882 sq mi (12,644 sq km), W Austria; capital Innsbruck. Bordering on West Germany in the north and on Italy and Switzerland in the south, it is an almost wholly Alpine region, traversed by the Inn River. The main part of the province is separated from the fertile East Tyrol by a corridor belonging partly to Italy and partly to Salzburg prov., Austria. The Tyrolean Alps, which culminate in the Ötztal Alps, are famed for their beauty and attract many tourists. Subsistence farming, cattle raising, forestry, and viticulture are the main occupations. Some industry is located at Innsbruck, Landeck, and Kufstein. The saltworks near Solbad Hall are an important source of revenue. The now little-worked silver and copper mines of Tyrol, known since antiquity, and its strategic position commanding the Brenner Pass across the Alps gave the region a fairly important role in European history.

The Tyrol was inhabited by Rhaetic tribes when it was conquered (15 B.C.) by the Romans. It was invaded (6th cent. A.D.) by Teutonic tribes and later by the Franks, who held all Tyrol by the 8th cent. Large parts of southern Tyrol (now in Italy) were ruled from the 11th cent. to 1802–3 by the bishops of Trent and Brixen. The two bishoprics fell to Austria after the Peace of Lunéville (1801) between France and Austria. The northern section (constituting the present Tyrol) passed (1363) to Austria. In 1805 the Treaty of Pressburg awarded all Tyrol to Napoleon's ally, Bavaria, but when war broke out (1809) the Tyrolean peasants stubbornly defied the French and Bavarian troops. In 1810 Napoleon attached most of southern Tyrol to Italy. Both parts were restored (1815) to Austria by the Congress of Vienna. The Treaty of Saint-Germain (1919) awarded southern Tyrol to Italy. The ruthless Italianization policy of the Fascist government created much unrest and friction in the period between the two world wars. The Italian constitution of 1947, however, gave southern Tyrol the status of an autonomous region, with full protection of minority rights.

Ty·rone (tĭ-rōn'), county (1971 pop. 138,975), 1,261 sq mi (3,266 sq km), Northern Ireland; co. town Omagh. There are uplands in the north and center; in the east and west are lowlands. Dairy and beef cattle are raised and barley, potatoes, and oats are grown. Manufactured products include linens, woolens, and other textiles, whiskey, and processed foods. Tourism is significant. Tyrone was organized as a shire in the beginning of the 17th cent.

Ty·rone (tī'rōn'), industrial borough (1970 pop. 7,072), Blair co., central Pa., NNE of Altoona in the Alleghenies; laid out c.1850, inc. 1857. Paper and metal products are made. It is an important outlet for bituminous-coal products.

Tyr·rell (tĭr'əl), county (1970 pop. 3,806), 390 sq mi (1,010 sq km), NE N.C., in a forested and swampy tidewater area bounded on the N by Albemarle Sound, on the E by the Alligator River, on the W by Phelps Lake; formed 1729; co. seat Columbia. It has farming (corn and soybeans), sawmilling, and fishing.

Tyr·rhe·ni·an Sea (tĭ-rē'nē-ən), part of the Mediterranean Sea, c.475 mi (765 km) long and from 60 to 300 mi (97–483 km) wide, between the Ligurian Sea, the Italian peninsula, Sicily, Sardinia, and Corsica. The Strait of Messina connects it with the Ionian Sea.

Tyu·men (tyōō-mĕn'), city (1973 est. pop. 299,000), SW Siberian USSR, on the Tura River. On the Trans-Siberian RR, Tyumen is a major transfer point for river and rail freight. It has shipyards and machine plants. The surrounding area is rich in petroleum and natural gas. Tyumen was founded in 1585 and is the oldest city in Siberia. It was formerly an important center of trade with China.

U

U·ban·gi (yōō-băng'ē, ōō-bäng'-ē), river, c.700 mi (1,125 km) long, formed on the Zaire–Central African Empire border, central Africa, by the confluence of the Uele and Bomu rivers and flowing W and S to the Congo River, of which it is the chief northern tributary.

U·be (ōō'bē), city (1970 pop. 152,935), Yamaguchi prefecture, SW Honshu, Japan, on the Inland Sea. It has a modern harbor and an important chemical industry. Coal is mined under the sea near Ube.

U·ca·ya·li (ōō'kä-yä'lē), river, c.1,000 mi (1,610 km) long, formed by the confluence of the Apurímac and Urubamba rivers, E Peru, and flowing N through a mountain and jungle wilderness to the Marañón River, SW of Iquitos. The Ucayali is navigable for its entire course by small craft.

U·dai·pur (ōō'dī-pŏŏr', ōō-dī'pŏŏr') or **Me·war** (mĕ-wär'), former princely state, now part of Rajasthan state, NW India. The Udaipur region, thickly wooded in the south and west, is mostly an alluvial plain watered by many intermittent streams. Grains and cotton are grown. Udaipur was probably founded in the early 8th cent. It accepted British overlordship in 1818 and joined Rajasthan in 1948.

Udaipur, city (1971 pop. 161,278), Rajasthan state, NW India, founded c.1560. It is an agricultural market and a weaving and embroidery center. The city is especially noted for its maharaja's palace, which overlooks scenic Pichola Lake.

Ud·de·val·la (ŭd'ə-väl'ə), city (1972 est. pop. 47,306), Göteborg och Bohus co., SW Sweden, a port on the Byfjorden, an arm of the

Skagerrak. Manufactures include textiles, clothing, furniture, and machinery. There are shipyards in the city.

U·di·ne (o͞o'dĕ-nā), city (1972 est. pop. 101,500), capital of Udine prov. (1972 est. pop. 519,793), Friuli-Venezia Giulia, NE Italy. Manufactures include machinery, textiles, clothing, and chemicals. In the 10th cent. Emperor Otto II gave the city to the patriarchs of Aquileia, who made it their capital (13th cent.). Udine passed to Venice in 1420 and to Austria in 1797 and 1814. It was annexed by Italy in 1866. In World War I the city was the headquarters of the Italian army (1915-17) and was occupied by Austria (1917-18).

Ud·murt Autonomous Soviet Socialist Republic (o͝od'mo͝ort'), autonomous republic (1970 pop. 1,417,000), 16,255 sq mi (42,100 sq km), E European USSR, in the forested foothills of the Urals, between the Kama and Vyatka rivers; capital Izhevsk. The terrain is mostly low and hilly, with wide river valleys. Grain (especially rye), flax, hemp, sugar beets, peas, and potatoes are cultivated. The republic's extensive timber, peat, and oil shale resources are only partially exploited. There are also deposits of quartz sand, clays, limestone, coal, and other minerals. Engineering, steel milling, metallurgy, lumbering, machine building, and food and flax processing are important industries. The predecessors of the Udmurts inhabited the region in Neolithic times. They were controlled by the Bulgar state from the 8th to 13th cent. The Russians gradually brought the Udmurts under their rule in the 16th cent. The area became the Votyak Autonomous oblast in 1920, the Udmurt Autonomous oblast in 1932, and an autonomous republic in 1934.

Ue·le (wĕl'ā), river, c.700 mi (1,125 km) long, rising in NE Zaire, central Africa, and flowing W to merge with the Bomu River and thus form the Ubangi River. It has many rapids.

U·fa (o͞o-fä'), city (1974 est. pop. 871,000), capital of Bashkir Autonomous Republic, E European USSR, in the Urals at the confluence of the Belaya and Ufa rivers. Ufa produces electrical and mining equipment and has oil refineries and a major chemical industry.

U·gan·da (yo͞o-găn'də, o͞o-gän'dä), republic (1977 est. pop. 12,115,000), 91,133 sq mi (236,036 sq km), E central Africa, bordering on Tanzania and Rwanda in the S, on Zaire in the W, on Sudan in the N, and on Kenya in the E; capital Kampala. Most of Uganda, which lies astride the equator, is made up of a fertile plateau (average elevation 4,000 ft/1,220 m), in the center of which is Lake Kyoga. The plateau is bounded on the west by the western branch of the Great Rift Valley, which includes lakes Albert and Edward and the Albert Nile River; on the southwest by the Ruwenzori Range, which includes Margherita Peak (16,794 ft/5,119 m), Uganda's highest point, and by the Virunga Mts.; on the south by Victoria Nyanza (Lake Victoria); and on the east and north by several mountain ranges. Altogether, about 18% of Uganda is made up of water surface and about 7% comprises highland situated at more than 5,000 ft (1,520 m). Virtually all of Uganda's inhabitants are black Africans, and about 90% of the people live in rural areas. Approximately 66% of the people speak one of the Bantu languages; the main Bantu ethnic groups, all of which live in the southern half of the country, are the Ganda (who make up about 15% of the country's total population), the Soga, the Ankole, the Nyoro, and the Toro. Other language groups in Uganda are the Nilotic, whose speakers live in the north and make up about 15% of the population; the Nilo-Hamitic, whose members live in the northeast and make up about 13% of the population; and the Sudanic, whose speakers live in the northwest and make up about 5% of the population. English is the country's official language; Swahili is widely spoken in commercial centers. About 30% are Christian and about 5% are Moslem.

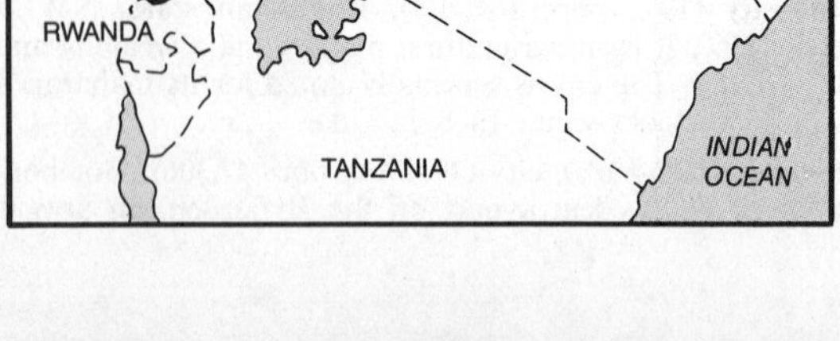

Economy. The economy of Uganda is overwhelmingly agricultural, and farming employs about 90% of the work force. The chief food crops are cassava, sweet potatoes, plantains, millet, sorghum, maize, and pulses. The principal cash crops are cotton, coffee, tea, tobacco, sugar cane, and groundnuts. Large numbers of poultry, cattle, goats, and sheep are raised. There is a sizable fishing industry, and much hardwood (especially mahogany) is cut. Copper ore is the leading mineral produced; other minerals extracted include tin and iron ores, manganese, beryl, chromite, and wolfram. Uganda's few manufactures are limited mainly to processed agricultural goods (especially cured coffee and ginned cotton) and basic consumer items, but they also include textiles, cement, chemical fertilizer and insecticides, and metal goods. There is a large hydroelectric plant at Owen Falls, located on the Victoria Nile River where it leaves Victoria Nyanza. Uganda has two main rail lines; one traverses the southern part of the country, the other connects Tororo on the Kenya border with Gulu in the north. The country is linked by rail with Mombasa, Kenya, on the Indian Ocean. Uganda's main trade partners are Great Britain, the United States, Japan, and West Germany. Uganda is a member of the East African Community.

History. Archaeological remains indicate that Uganda was the site of Paleolithic and Neolithic civilizations. Around A.D. 1100 Bantu-speaking people migrated into southwest Uganda from the west. By the 14th cent. they were organized in several kingdoms (known as the Cwezi states), which had been established by Hima (also called Tutsi) migrants from present-day southwest Ethiopia. Around 1500, Nilotic-speaking Lwo people from present-day southeast Sudan conquered the Cwezi states and established the Bito dynasties of Buganda, Bunyoro, and Ankole. Later in the 16th cent. other Lwo-speakers conquered northern Uganda, forming the Alur and Acholi ethnic groups. In the 17th cent. the Lango and Teso migrated into Uganda. During the 16th and 17th cent. Bunyoro was the leading state of southern Uganda. From about 1700 Buganda began to expand and by 1800 it controlled a large territory bordering Victoria Nyanza from the Victoria Nile to the Kagera River. Buganda was centrally organized under the autocratic kabaka (king). The Ganda raided widely for cattle, ivory, and slaves. In the 1840s Moslem traders from the Indian Ocean coast reached Buganda, and they exchanged firearms, cloth, and beads for ivory and slaves. Beginning in 1869 Bunyoro, ruled by Kabarega, challenged Buganda's ascendency. By the mid-1880s, however, Buganda again dominated southern Uganda.

In 1862 John Hanning Speke, a British explorer, became the first European to visit Buganda. He met with Mutesa I. Members of the British Protestant Church Missionary Society arrived in 1877, and they were followed in 1879 by representatives of the French Roman Catholic White Fathers. Each of the missions gathered a group of converts, which in the 1880s became fiercely mutually antagonistic. At the same time the number of Ganda converts to Islam was growing. In 1884 Mutesa died and his successor, Kabaska Mwanga, soon began to persecute the Christians out of fear for his own position. In 1888 Mwanga was deposed by the Christians and Moslems and replaced by his brother Kiwewa. A month later the Moslems ousted Kiwewa and replaced him with another brother, Kalema. In early 1890 Mwanga permanently regained his throne but lost much of his power to Christian chiefs. In 1890 Great Britain and Germany signed a treaty that, in part, gave the British rights to what was to become Uganda. Later that year Frederick Lugard established British authority in southern Uganda. In 1894 Great Britain officially made Uganda a protectorate. The British at first ruled Uganda through Buganda, but when Mwanga opposed their growing power, they deposed him and replaced him with his infant son Daudi Chwa. From the late 1890s to 1918 the British established their authority in the rest of Uganda by negotiating treaties and by using force where necessary. In 1900 an agreement was signed with Buganda that transformed the kingdom into a constitutional monarchy controlled largely by Protestant chiefs. In 1904 the commercial cultivation of cotton was begun, and cotton became the major export crop; coffee and sugar production accelerated in the 1920s. The country attracted few permanent settlers, and the cash crops were mostly produced by African smallholders. Many Asians settled in Uganda, where they played a leading role in the country's commerce. During the 1920s and 1930s the British considerably reduced Buganda's independence. In 1921 a legislative council for the protectorate was established; it

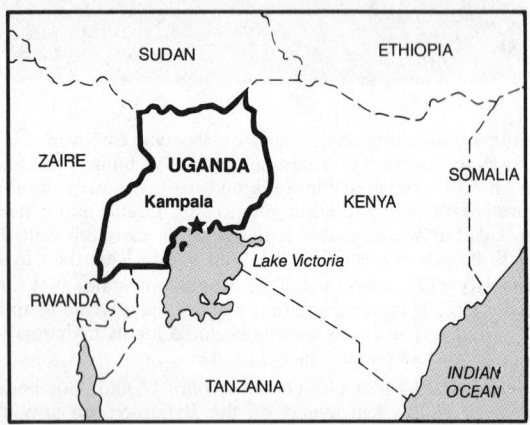

was not until the mid-1950s that a substantial number of seats was allocated to the Africans. In subsequent years, as Great Britain prepared Uganda for independence, Buganda intermittently demanded to be set up as a separate state. In 1961 there were three main political parties in Uganda—the Uganda People's Congress (UPC), whose members were mostly non-Ganda; the Democratic party, made up chiefly of Roman Catholic Ganda; and the Kabaka Yekka (Kabaka only) party, comprising only Ganda.

On Oct. 9, 1962, Uganda became independent, with A. Milton Obote, a Lango leader of the UPC, as prime minister. Buganda was given considerable autonomy. In 1963 Uganda became a republic, and Mutesa II was elected president. The first years of independence were dominated by a struggle between the central government and Buganda. In 1966 Obote curbed Buganda's autonomy. In 1967 a new constitution was introduced, giving the central government much power. In 1969 Obote decided to follow a leftist course in the hope of bridging the country's ethnic and regional differences. However, in Jan., 1971, Obote was deposed in a coup d'état by Maj. Gen. Idi Amin. By the end of 1971 Amin was in firm control. In 1972 Amin ordered Asians who were not citizens of Uganda to leave the country. Relations with Tanzania and Kenya were strained after Amin came to power, partly because many exiled Ugandans loyal to Obote fled to these countries and plotted against Amin. The main targets of Amin's purges were the Lango and Acholi tribes, but he also moved against the army. By 1977, according to Amnesty International, as many as 300,000 Ugandans may have been killed. Amin was implicated in the hijacking to Uganda (1976) of a French airliner containing many Jewish passengers. The hostages were subsequently rescued by an Israeli commando raid on Entebbe.

In 1979, with internal unrest spreading to the military, Amin attacked a border province of neighboring Tanzania to stabilize his hold on the army. Under the pretext of revenging this most recent of many border disputes between the two countries, Tanzanian President Julius Nyerere, an implacable foe of the despot Amin, launched a full-scale invasion of Uganda. Nyere's troops, joined by various Ugandan rebel groups, were welcomed as liberators by the populace. Amin's army quickly collapsed and their leader fled to a remote village in the north. On Apr. 13, 1979, Yusufu Lule was sworn in as the new president of Uganda. He pledged to restore democracy and rebuild the country's ravaged economy.

U·ga·rit (o͝o′gə-rēt′), ancient city, capital of Ugarit kingdom, W Syria. Although the name of this city was known from Egyptian and Hittite sources, its location and history were a mystery until the accidental discovery (1928) of an ancient tomb at the small Arab village of Ras Shamrah, near modern Latakia. This led to excavations (begun in 1929 and still continuing) that have established the identity of the mound as the site of ancient Ugarit. The city was probably occupied from the Neolithic period. By the 4th millennium it had reached a high stage of development and was part of the general civilization of ancient Syria. Early in the 2nd millenium Ugarit formed an alliance with Egypt, and from this period Egyptian influence was strong in the city. The city was also the most important center of Minoan trade in Syria. The 15th and 14th cent. B.C. were the period of highest prosperity for Ugarit. The art of this period shows that an important Mycenaean colony existed in the city. Foreign invasions and economic change in the 12th cent. B.C. caused Ugarit to decline. Among the more important discoveries at Ugarit are tablets from the 14th cent. B.C. Written in a cuneiform script, in a hitherto unknown language, Ugaritic, they record the poetic works and myths of the ancient Canaanites.

U·i·jong·bu (o͞o′ē-jŭng′bo͞o′), city (1970 pop. 94,518), NW South Korea. It is an agricultural center.

U·in·ta (yo͞o-ĭn′tə), county (1970 pop. 7,100), 2,086 sq mi (5,402.7 sq km), SW Wyo., bounded in the S and W by Utah and watered by branches of the Green River; formed 1869; co. seat Evanston. It is in a livestock and grain-farming region, with coal and oil deposits.

U·in·tah (yo͞o-ĭn′tə), county (1970 pop. 12,684), 4,487 sq mi (11,621.3 sq km), NE Utah, in a mountain and plateau area crossed by the Green River and bordering on Colo.; formed 1880; co. seat Vernal. It has livestock raising, oil and gas wells, and limestone and phosphate-rock quarries. Clothing and leather goods are made.

Uinta Mountains, range of the Rocky Mts. extending c.120 mi (195 km) E from NE Utah to SW Wyo. It rises to Kings Peak (13,528 ft/4,126 m), the highest point in Utah. Phosphates are mined.

U·ist (yo͞o′ĭst, o͞o′-), islands of the Outer Hebrides, NW Scotland. North Uist (1971 pop. 1,732) is 18 mi (29 km) long and 13 mi (21 km) wide, with a much indented coast. The east is hilly and boggy, but the west has some fertile land. South Uist (1971 pop. 3,781) is c.22 mi (35 km) long and 7 mi (11 km) wide. A testing range for rockets was erected here in 1954.

Ui·ten·hage (yo͞o′tən-hāg′, oi′tən-hä′gə), town (1970 pop. 70,517), Cape Prov., S South Africa, on the Zwartkops River. It has railroad workshops, textile mills, and motor-vehicle assembly plants.

U·ji (o͞o′jē), town (1970 pop. 103,497), Kyoto prefecture, S Honshu, Japan. It is a resort and is noted for its tea and for cormorant fishing. Uji has an 11th cent. monastery with a beautiful pavilion.

Uj·jain (o͞o′jīn′), city (1971 pop. 203,278), Madhya Pradesh state, central India, on the Sipra River. Many pilgrims visit Ujjain, which Hindus consider one of the holiest places in India. Nearby is a ruined ancient city, which may have been inhabited in the late 2nd millennium B.C. This city has been identified as the capital of the semilegendary kingdom of Avanti. Later it was the central city of the Malwa kingdoms, and in the 8th cent. A.D. it became the center of Sanskrit learning.

U·ki·ah (yo͞o-kī′ə), city (1975 est. pop. 11,100), seat of Mendocino co., W Calif., in a lumber and fruit-growing region; inc. 1876. Masonite is manufactured.

U·kraine (ū′krān′, ū-krān′) or **Ukrainian Soviet Socialist Republic,** constituent republic (1970 pop. 47,136,000), 232,046 sq mi (601,000 sq km), SW European USSR. Kiev is the capital. The Ukraine borders on Poland in the northwest; on Czechoslovakia, Hungary, Rumania, and the Moldavian SSR in the southwest; on the Black Sea and the Sea of Azov in the south; on the Russian Soviet Federated Socialist Republic (RSFSR) in the east and northeast; and on the Belorussian SSR in the north. In terms of population, economic value, and historic importance, the Ukraine ranks second after the RSFSR among the Soviet Union's 15 constituent republics. Drained by the Dnepr, the Dnestr, the Southern Bug, and the Donets rivers, the Ukraine consists largely of fertile steppes, extending from the Carpathians and the Volhynian-Podolian uplands in the west to the Donets Ridge in the southeast. In the north and northwest is the wooded area of the Pripyat Marshes, with gray podsol soil and numerous swamps; wooded steppes extend across the central Ukraine, and a fertile, treeless, black-earth steppe covers the south. The continental climate of the republic is greatly modified by proximity to the Black Sea.

Economy. The Ukraine supplies approximately 25% of the USSR's foodstuffs. The steppe is one of the chief wheat-producing regions of Europe. Other major crops include corn, rye, barley, potatoes, sugar beets, melons, and flax. The Ukraine possesses numerous raw materials and power resources, and its central and eastern regions form one of the world's densest industrial concentrations. About 30% of the Soviet Union's heavy industrial output comes from the Ukraine. The heavy metallurgical, machine-building, and chemical industries are based on the iron mines of Krivoy Rog, the manganese ores of Nikopol, and the coking coal and anthracite of the Donets Basin. The Dneproges dam powers one of the world's greatest hydroelectric stations and has made the Dnepr navigable for nearly its entire length. The region also produces aluminum, zinc, mercury titanium, nickel, oil, natural gas, and bauxite. Odessa is the principal port on the Black Sea. The western Ukraine, although mainly agricultural, has large petroleum and natural-gas centers, coal industries, and rich salt deposits. The republic's leading industrial products include machinery, cast iron, steel, rolled metals, tractors, cement and other building materials, mineral fertilizers, glass, paper, plywood, pottery, china, furniture, textiles, clothing, and leather footwear. Food processing is also a major industry.

History. In ancient times a major part of the present-day Ukraine was inhabited by the Scythians, who were later displaced by the Sarmatians. Early in the Christian era a series of invaders (Goths, Huns, Avars) overran the Ukrainian steppes, and in the 7th cent. the Khazars included much of the Ukraine in their empire. The Ukrainians themselves can be traced to Neolithic agricultural tribes in the Dnepr and Dnestr valleys. In the 9th cent. a Varangian dynasty from Scandinavia established itself at Kiev. Having freed the Slavs from Khazar domination, the Varangians united them in powerful Kievan Russia. Following Yaroslav's reign (1019–54), which marked the zenith of Kiev's power, Kievan Russia split into principalities. The Mongols (Tatars) conquered the entire region in the 13th cent. In the mid-14th cent. Lithuania began to expand eastward and southward, liberating the Ukraine from the Tatars. The Ukraine flourished under Lithuanian rule, but after the organic union of Poland and Lithuania in 1569 the Ukraine came under Polish rule, enserfment of the Ukrainian peasants proceeded apace, and the Ukrainian Orthodox Church suffered persecution. Meanwhile the Black Sea shore, ruled by the khans of Crimea, was absorbed into the Ottoman Empire in 1478. The term "Ukraine," which may be translated as "borderland," came into general usage in the 16th cent. The harsh conditions of Polish rule led many Ukrainians to flee serfdom and religious persecution by escaping beyond the area of the lower Dnepr rapids. These fugitives, known as Cossacks or Kozaks, waged revolutions against both Polish (1648) and Russian (1658) domination. After the Russo-Polish War the Ukraine was partitioned (1667). Russia obtained left-bank Ukraine, east of the Dnepr River and including Kiev; Poland retained right-bank Ukraine. The Cossacks again rose against Russian domination, but were defeated (1709). The Ukraine, its political autonomy terminated, was divided into three provinces in the 1770s. In 1783 Russia annexed the khanate of Crimea. The Polish partition treaties of 1772, 1793, and 1795 awarded Podolia and Volhynia to Russia, thus reuniting left-bank and right-bank Ukraine; eastern Galicia went to Austria.

A movement for Ukrainian national and cultural revival blossomed in the late 19th cent. despite czarist repression. Following the overthrow of the czarist regime in 1917 a Ukrainian central council was set up, and the Ukraine proclaimed complete independence in Jan., 1918. Soviet troops were sent in, but the Central Powers then overran the territory with their own soldiers and forced the Red Army to withdraw. The World War I armistice of Nov., 1918, in turn forced the withdrawal from the Ukraine of the Central Powers. Meanwhile an independent republic in western Ukraine had been proclaimed in Lvov. In Jan., 1919, the union of the two Ukraines was proclaimed; however, Soviet troops eventually regained control of the Ukraine, which in 1922 became one of the original constituent republics of the USSR. Lenin's attempts to assuage Ukrainian nationalism were abandoned by Stalin, who also imposed agricultural collectivization on the Ukraine and requisitioned all grain for export. The republic suffered severe wartime devastation during the German occupation (1941–44). Several major territorial changes occurred during and after World War II. South Bessarabia, recovered from Rumania in 1940, was incorporated into the Ukraine, while the former Moldavian ASSR was detached from the republic and merged with central Bessarabia as the Moldavian SSR. The northern parts of Bukovina and Bessarabia were added to the Ukraine, as was east Galicia, including Lvov, formally ceded by Poland in 1945. Zakarpatskaya oblast, which had been part of Czechoslovakia since 1919, was also ceded in 1945, thus completing the process by which all Ukrainian lands were united into a single republic. Crimea was annexed to the Ukraine in 1954.

U·lan Ba·tor (o͞o'län bä'tôr'), city (1971 est. pop. 282,000), capital of the Mongolian People's Republic, E central Outer Mongolia, on the Tola River. Manufactures include woolen textiles and related goods, leather and footwear, soap, paper, iron castings, matches, glassware, beer and spirits, and processed foods. Coal mined nearby provides power. The city has the only university (founded 1942) in the country and a library with ancient Mongolian, Chinese, and Tibetan manuscripts. Founded in 1649 as a monastery town, Ulan Bator still preserves the monastery section, the former center of the city, and the residence of the Living Buddha, once Mongolia's spiritual leader.

U·lan-U·de (o͞o-län'o͞o-dā'), city (1973 est. pop. 279,000), capital of the Buryat Autonomous Soviet Socialist Republic, SE Siberian USSR, on the Selenga River. It is a major transportation hub. Industries include railroad maintenance, boatbuilding, ship repairing, sawmilling, food and wool processing, meat canning, and the manufacture of machinery and locomotives. Founded in 1649 as a Cossack winter encampment, Ulan-Ude became a fortress in 1689 and a city in 1775. The city became the capital of the Far Eastern republic in 1920 and of the Buryat-Mongol Autonomous SSR in 1923.

Ul·has·na·gar (o͞ol'həs-nä'gər), city (1971 pop. 168,462), Maharashtra state, W central India. It is a suburb of Bombay.

U·li·thi (o͞o-lē'thē), atoll comprising 40 islets, 1.75 sq mi (4.5 sq km), W Pacific, in the W Caroline Islands. The atoll became (1920) part of the Japanese mandate in the Pacific.

Ulls·wa·ter (ŭlz'wô'tər, -wŏt'ər), lake, 7.5 mi (12.1 km) long, NW England. It is divided into three reaches. The waterfall of Aira Force (65 ft/19.8 m high) and Helvellyn Mt. are nearby.

Ulm (o͞olm), city (1972 est. pop. 93,407), Baden-Württemberg, S West Germany, on the Danube River. It is an active river port and rail junction. Manufactures include textiles, clothing, processed food, beer, and foundry products. A canal links Ulm with the Neckar River. Known in 854, Ulm became (14th cent.) a free imperial city in Swabia and ruled a considerable territory north of the Danube, reaching its zenith in the 15th cent. Changes in international trade routes during the 15th and 16th cent. and the religious wars in Germany caused its decline. Ulm accepted the Reformation c.1530 and was a member of the Schmalkaldic League. The city and its territory were awarded to Bavaria in 1803 but were transferred to Württemberg in 1810. In World War II more than half of the city, including many old and historic buildings, was destroyed. The famous Gothic minster, begun in 1377, is the second-largest Gothic church in Germany. Albert Einstein was born (1879) in Ulm.

Ul·san (o͞ol'sän'), city (1970 pop. 108,000), SE South Korea, a port on the Korea Strait. It is an industrial center, with oil refineries, chemical plants, and a large sugar refinery.

Ul·ster (ŭl'stər), county (1970 pop. 141,241), 1,143 sq mi (2,960.4 sq km), SE N.Y., bounded on the E by the Hudson River; formed 1683; co. seat Kingston. It has dairying, farming (fruit, truck crops, and potatoes), poultry raising, and manufacturing. The area lies mainly in the Catskills and has numerous small resort lakes.

Ulster, northernmost of the historic provinces of Ireland. Modern Ulster consists of nine counties. Six (Antrim, Armagh, Down, Fermanagh, Londonderry, and Tyrone) now make up Northern Ireland, which is often referred to as Ulster; the remaining three (Cavan, Donegal, and Monaghan) are in the Republic of Ireland.

U·lú·a (o͞o-lo͞o'ä), river, c.200 mi (320 km) long, rising in the Sierra de Gujiquiro, W Honduras, and flowing north to the Caribbean Sea. The Ulúa, with its tributaries, drains almost the entire western half of the country.

Ul·ya·novsk (o͞ol-yä'nəfsk), city (1973 est. pop. 395,000), capital of Ulyanovsk oblast, W central European USSR, a port on the Volga and Svigaya rivers. It is a major rail and water transport center and trades in grain, wool, and potash. Industries include food processing, flour milling, brewing, vodka distilling, and the manufacture of motor vehicles, machine tools, and metal and milling equipment. Ulyanovsk was founded in 1648 on the site of a Tatar village as a strongpoint to defend Russia's southern frontier. It was taken by the Cossacks in 1670, was the scene of fighting during the Pugachev insurrection of 1773–74, and was virtually destroyed by fire in 1864. The city was the birthplace of V. I. Lenin (whose original name was Vladimir I. Ulyanov).

U·lys·ses (yo͞o-lĭs'ēz), city (1970 pop. 3,779), seat of Grant co., SW Kansas, SW of Garden City, in a melon and grain area; inc. 1921.

U·man (o͞o-män'), city (1973 est. pop. 73,000), SW European USSR, in the Ukraine, at the confluence of the Kamenka and Umanka rivers. Its industries include poultry processing, fruit preserving, and the production of machinery, bricks, clothing, and vitamins. In the late 17th cent. Uman was an important fortress for protection against Crimean Tatar attacks on the Ukraine. In 1768 the city was the scene of a Ukrainian peasant and Cossack uprising that resulted in a general massacre.

U·ma·nak (o͞o'mə-năk'), town (1969 pop. 1,006), W Greenland, on an inlet of Baffin Bay. A hunting and fishing base, it has a canning factory and is a center for sealing operations.

U·ma·til·la (yo͞o'mə-tĭl'ə), county (1970 pop. 44,923), 3,231 sq mi (8,368.3 sq km), NE Oregon, bordering on Wash., bounded in the NW by the Columbia River, and drained by the Umatilla River; formed 1862; co. seat Pendleton. Its agriculture includes wheat and livestock. It has food-processing and lumbering industries and manufactures furniture, paper, concrete, machinery, and trailers.

Um·bri·a (ŭm'brē-ə), region (1972 est. pop. 780,598), 3,265 sq mi (8,456.4 sq km), central Italy; capital Perugia. Crossed by the Apennines in the east, Umbria is almost entirely mountainous or hilly. Farming, mostly on a small scale, is the chief occupation. Manufactures of the region include chemicals, iron and steel, processed food, and cotton and woolen textiles. The Umbri were among the first inhabitants of the region, settling here by 600 B.C. Knowledge of them is derived mainly from inscriptions found in Umbria. There are also many Etruscan remains from a later period. Umbria was conquered by the Romans in the 3rd cent. B.C., and after the fall of Rome it passed to the Goths and then to the Byzantines. From the

6th to the 11th cent. it was usually included in the powerful Lombard duchy of Spoleto. In the 12th cent. free communes developed in most cities. Local autonomy and petty tyrannies prevailed until the 16th cent., when the popes conquered Umbria. Umbria was held by France from 1798 to 1800 and from 1808 to 1814, when it was restored to the papacy. There were several revolts (1831, 1848, 1859) against papal rule, and in 1860 the region voted to join the kingdom of Sardinia.

U·me·å (o͞o'mə-ō'), city (1972 est. pop. 57,898), capital of Västerbotten co., NE Sweden, on an inlet of the Gulf of Bothnia at the mouth of the Umeälv River; founded 1622. Manufactures include forest products, furniture, and machinery.

U·me·älv (o͞o'mə-ĕlv'), river, c.285 mi (460 km) long, rising in N Sweden in a lake on the Norwegian border and flowing SE to the Gulf of Bothnia at Umeå.

Umm al-Qai·wain (o͞om' ăl-kī-wīn'), sheikdom (1968 pop. 2,900), c.300 sq mi (780 sq km), part of the federation of United Arab Emirates, E Arabia, on the Persian Gulf. Fishing and agriculture are the main economic activities, although oil production began in 1964. The sheikdom, formerly a British protectorate, became part of the United Arab Emirates in 1971.

Um·nak (o͞om'năk'), island, c.83 mi (134 km) long, off SW Alaska, one of the largest of the Aleutian Islands. A volcanic peak, Mt. Vsevidof, 7,236 ft (2,207 m) high, is here.

Um·ta·li (o͞om-tä'lē), city (1973 est. pop. 18,500), E Rhodesia, near the Mozambique border. Umtali is the commercial center for a rich agricultural and gold-mining region. Its industries include automobile assembly, petroleum refining, and the manufacture of textiles.

Um·ta·ta (o͞om-tä'tə), town (1970 pop. 25,216), capital of The Transkei, SE South Africa, on the Umtata River. Founded in 1860 as a military post, Umtata became capital of the newly created Transkei African homeland in 1963.

Un·a·las·ka (ŭn'ə-lăs'kə), rugged island, 30 mi (48.3 km) long, off SW Alaska, one of the largest of the Aleutian Islands. Discovered (c.1759) by Russian explorers, the island was a center of Russian fur trade until it was superseded by Kodiak. Spruces planted on Unalaska by the Russians remain among the few trees in the Aleutians.

Un·com·pah·gre (ŭn'kəm-pä'grē), river, c.75 mi (120 km) long, rising in the San Juan Mts., SW Colo., and flowing NW past Montrose to the Gunnison River at Delta. Its waters are used for irrigation.

Uncompahgre Peak, 14,301 ft (4,361.8 m) high, SW Colo., in the San Juan Mts. of the Rockies and NE of Ouray.

Un·ga·va Bay (ŭng-gä'və, -gā'və), inlet of the Atlantic Ocean, N Que., Canada, extending c.200 mi (320 km) S from Hudson Strait between the N Que. mainland and the N tip of the Labrador peninsula. It is 160 mi (257.4 km) wide at its mouth.

U·ni·coi (yo͞o'nə-koi'), county (1970 pop. 15,254), 185 sq mi (479.2 sq km), NE Tenn., with the Bald Mts. partly along the N.C. border in the S and SW and drained by the Nolichunky River; formed 1875; co. seat Erwin. Its agriculture includes tobacco, fruit, corn, hay, and livestock. It manufactures clothing, chemicals, and machinery.

U·ni·mak (yo͞o'nə-măk'), volcanic island, 70 mi (113 km) long, off W Alaska, one of the Aleutian Islands, nearest of the chain to the Alaska Peninsula.

Un·ion (yo͞on'yən). **1.** County (1970 pop. 45,428), 1,052 sq mi (2,724.7 sq km), S Ark., bounded on the S by the La. border, on the NE and E by the Ouachita River; formed 1829; co. seat El Dorado. It is in an important oil region, with drilling, refining, and manufacturing of oil products. Its agriculture includes fruit, peanuts, sweet potatoes, corn, cotton, and livestock. It has timber (logging and sawmills) and industry (poultry dressing and shipbuilding). **2.** County (1970 pop. 8,112), 240 sq mi (621.6 sq km), NE Fla., in a flatwoods area with several small lakes; formed 1921; co. seat Lake Butler. It is in a farming area (corn, vegetables, peanuts, and cattle), with lumbering and some clothing manufacturing. **3.** County (1970 pop. 6,811), 309 sq mi (800.3 sq km), N Ga., bounded on the N by the N.C. border and drained by the Nottely River; formed 1832; co. seat Blairsville. It has agriculture (hay, corn, potatoes, and livestock), lumbering, and resort areas. **4.** County (1970 pop. 16,071), 416 sq mi (1,072.3 sq km), SW Ill., bounded on the W by the Mississippi River, on the NW corner by the Big Muddy River, and drained by the Cache River; formed 1818; co. seat Jonesboro. It has agriculture (wheat, corn, fruit, truck crops, poultry, and dairy products), manufacturing of wood products, machinery, flour, shoes, and monuments, and limestone, granite, and marble quarries. It includes part of Shawnee National Forest. **5.** County (1970 pop. 6,582), 168 sq mi (435.1 sq km), E Ind., bounded on the E by the Ohio border and drained by East Fork of Whitewater River; formed 1821; co. seat Liberty. In a farming (grain and hogs) and dairying region, it has some industry (fertilizers, paint, and tractors). **6.** County (1970 pop. 13,557), 426 sq mi (1,103.3 sq km), S Iowa, drained by the Little Platte, Grand, and Thompson rivers; formed 1851; co. seat Creston. It is in a rolling prairie agricultural region. Hogs, cattle, and poultry are raised, and corn and oats are the chief crops. Bituminous-coal deposits are found here. **7.** County (1970 pop. 15,882), 343 sq mi (888.4 sq km), W Ky., bounded on the W by the Ohio River and the Ill. and Ind. borders, on the SW by the Tradewater River; formed 1811; co. seat Morganfield. In a rolling upland agricultural area yielding livestock, corn, wheat, hay, and tobacco, it has bituminous-coal mines, sand and gravel pits, and some manufacturing (textiles and metal products). **8.** Parish (1970 pop. 18,447), 906 sq mi (2,346.5 sq km), N La., bounded on the N by Ark., on the E by the Ouachita River, and drained by Bayou D'Arbonne; formed 1839; parish seat Farmerville. It contains a large natural-gas field and does cotton ginning and lumber milling. Corn, cotton, hay, peanuts, sweet potatoes, and soybeans are farmed. **9.** County (1970 pop. 19,096), 422 sq mi (1,093 sq km), N Miss., drained by the Tallahatchie River and its tributaries; formed 1870; co. seat New Albany. Agriculture (cotton, corn, sorghum, fruit, and poultry), dairying, and lumbering are important. The county includes part of Holly Springs National Forest. **10.** County (1970 pop. 543,116), 103 sq mi (266.8 sq km), NE N.J., bounded on the NW by the Passaic River, on the E by Newark Bay and Arthur Kill, and drained by the Rahway River; formed 1857; co. seat Elizabeth. It is in an industrial and residential area with many commuter communities. Its industry includes food processing and textile mills. Clothing, furniture, paper, chemicals, metal products, and machinery are made. There is some truck farming and dairying in the county. **11.** County (1970 pop. 4,925), 3,817 sq mi (9,886 sq km), extreme NE N.Mex., bordered on the N by Colo., on the E by Okla. and Texas, and drained by headwaters of North Canadian and Cimarron rivers; formed 1893; co. seat Clayton. It has grain farming and livestock grazing. Capulin Mountain National Monument is in the northwest. **12.** County (1970 pop. 54,714), 639 sq mi (1,655 sq km), SW N.C., in a piedmont area bounded on the S and SW by the S.C. border, on the NE by the Rocky River; formed 1842; co. seat Monroe. Farming (cotton, corn, and hay), dairying, and sawmilling are done. **13.** County (1970 pop. 23,786), 434 sq mi (1,124.1 sq km), central Ohio, drained by Darby Creek; formed 1820; co. seat Marysville. It is in an agricultural area that has livestock, dairying, grain, and poultry. Its industry includes food processing and the manufacture of chemicals, plastics, metal products, machinery, electrical equipment, and auto parts. **14.** County (1970 pop. 19,377), 2,032 sq mi (5,262.9 sq km), NE Oregon, in a mountain area crossed by the Grande Ronde River; formed 1864; co. seat La Grande. Its agriculture includes livestock and wheat. It also has lumbering and manufactures travel trailers. The Blue Mts. extend through most of the county. Parts of Whitman and Umatilla national forests are here. **15.** County (1970 pop. 28,603), 318 sq mi (823.6 sq km), central Pa., bounded in the E by the West Branch of the Susquehanna River; formed 1813; co. seat Lewisburg. It has dairying and grist-mill products. Its industry includes textile mills and lumbering. There is a forested recreational area in the western part of the county. **16.** County (1970 pop. 29,230), 515 sq mi (1,333.9 sq km), N S.C., bounded in the N by the Pacolet River, in the S by the Enoree River, in the E by the Broad River, and drained by the Tyger River; formed 1785; co. seat Union. Its agriculture includes cotton, corn, and sweet potatoes. Clothing, chemicals, plastics, and machinery are manufactured. Part of Sumter National Forest is here. **17.** County (1970 pop. 9,643), 452 sq mi (1,170.7 sq km), extreme SE S.Dak., bordering on Iowa and Nebr., bounded in the E by the Big Sioux River and in the S by the Missouri River; formed 1862 as Cole co., renamed 1864; co. seat Elk Point. In an agricultural area yielding livestock, poultry, grain, and dairy products, it has a food-processing industry. Clothing, machinery, transformers, and truck trailers are made. **18.** County (1970 pop. 9,072), 212 sq mi (549.1 sq km), NE Tenn., in a partly mountainous region traversed by ridges of the Appalachians, bounded in the NW by the Powell River and crossed by the Clinch River; formed 1797 as Cocke co., renamed 1846; co. seat Maynardville. Its agriculture includes tobacco, fruit, and livestock. It manufactures household furniture.

Union. 1. City (1970 pop. 5,183), seat of Franklin co., E Mo., near the Missouri River WSW of St. Louis, in a farm and livestock area with lead deposits; laid out 1826, inc. 1888. Shoes are made. **2.** Township (1978 est. pop. 50,100), Union co., NE N.J.; settled 1749 by colonists from Conn., set off from Elizabethtown 1808. Steel and metal products and paints are manufactured. Union was the site of a Revolutionary battle (1780). **3.** City (1975 est. pop. 10,600), seat of Union co., N S.C.; settled 1791, inc. 1837. Textiles, fertilizer, and metal products are made. **4.** Town (1970 pop. 566), seat of Monroe co., SE W.Va., ESE of Hinton, in a lumbering and flour-milling area.

Un·ión, La (lä' o͞on-yôn'), city (1971 pop. 17,207), SE El Salvador, on the Gulf of Fonseca, an arm of the Pacific Ocean. La Unión is the southern terminus of an important railroad system and a major port. It is on the Inter-American Highway and at the foot of the Conchagua volcano. The city was severely damaged in 1947 when the volcano erupted simultaneously with an earthquake.

Unión, La, town (1970 pop. 13,145), Murcia prov., SE Spain. It is a center for the rich lead, silver, iron, and zinc mines of the vicinity, which have been worked since Carthaginian times.

Un·ion Beach (yo͞on'yən), borough (1970 pop. 6,472), Monmouth co., E N.J., on Raritan Bay SE of Perth Amboy; inc. 1925.

Union City. 1. Residential city (1975 est. pop. 27,100), Alameda co., W Calif., in a farm region; inc. 1959. Steel and iron products, aluminum, and sugar are manufactured. **2.** City (1978 est. pop. 50,700), Hudson co., NE N.J., on the Palisades overlooking the Hudson River, directly opposite New York City; inc. 1925. The city has many small firms, most of them in the embroidery field. **3.** City (1975 est. pop. 12,200), seat of Obion co., W Tenn., near the Ky. line; inc. 1867. It is a trade, processing, and shipping center in a rich livestock, grain, and cotton region. Automobile parts, shoes, aluminum and sheet-metal products, and textiles are made. Three Civil War battles were fought in the vicinity in 1862–63.

Un·ion·dale (yo͞on'yən-dāl'), uninc. residential city (1978 est. pop. 22,500), Nassau co., SE N.Y., on Long Island.

Union of So·vi·et Socialist Republics (sō'vē-ĕt', -vē-ət), republic (1976 est. pop. 255,524,000), 8,649,489 sq mi (22,402,200 sq km), E Europe and N Asia. It is also known as the Soviet Union. The USSR borders on Rumania, Hungary, Czechoslovakia, and Poland in the west; on the Baltic Sea, Finland, and Norway in the northwest; on the Barents, Kara, Laptev, East Siberian, and Chukchi seas (arms of the Arctic Ocean) in the north; on the Bering Sea, Sea of Okhotsk, and Sea of Japan (arms of the Pacific Ocean) in the east; on China, Mongolia, and Afghanistan in the south; and on Iran, the Caspian Sea, Turkey, and the Black Sea in the southwest. Moscow is the capital.

Covering about one seventh of the earth's land area, the USSR is the world's largest nation and ranks third (after China and India) in population. Its maximum extent from east to west is about 6,800 mi (10,940 km) and from north to south about 2,800 mi (4,510 km). It can be divided into four regions. European USSR stretches from the country's western boundary to the Urals and the Caspian Sea. Central Asian USSR includes that part of the southern Soviet Union from the Caspian Sea to just beyond the Irtysh River. Siberian USSR comprises the vast region of the RSFSR between the Urals and the western boundaries of Chukchi National okrug, Magadan oblast, Khabarovsk kray, and Amur oblast. The Soviet Far East, the fourth region, is made up of the extreme eastern part of the RSFSR.

European USSR lies generally below 1,000 ft (305 m), except for the Urals, the Caucasus, and scattered highlands. In the north is a narrow belt of tundra extending along the entire length of the country and characterized by permafrost and a light cover of vegetation. The tundra zone in the European USSR includes the Kola Peninsula, the islands of Kolguyev and Novaya Zemlya, Franz Josef Land, and the mouths of the Onega, Northern Dvina, and Pechora rivers. South of the tundra is a somewhat wider belt of taiga, or evergreen forest, that fans out east of the Urals to include most of Siberian USSR and the Soviet Far East. The taiga zone in European USSR includes lakes Ladoga and Onega, the largest lakes in Europe. South of the taiga is a vast region known as the Russian Plain or Fertile Triangle, which stretches from the country's western boundary to the Yenisei River (in Central Siberian USSR) and takes in the rest of European USSR, except for the arid region between the Black and Caspian seas, which is made up largely of the Caucasus and has a climate and soil cover similar to Central Asian USSR. The Russian Plain includes the country's most fertile and productive farmland. It is drained by several large rivers, including the Dnepr, Don, Volga, Oka, Western Dvina, Neman, Dnestr, Southern Bug, and Kama in European USSR and the Irtysh and Ob in Siberian USSR. The chief rivers of European USSR are connected with one another by a system of artificial waterways; among them are the Baltic-White Sea canal, the Moscow-Volga canal, the Volga-Baltic waterway, and the Volga-Don canal. The vast Pripyat Marshes are in western European USSR. There is a narrow subtropical strip along the Black Sea littoral of the southern Crimea and of the Caucasus. Central Asian USSR is an arid region of poor soils. Most of the western part is low-lying, but there are highlands in the east and lofty mountains in the south and southeast. Mt. Communism (24,590 ft/7,495 m), the highest point in the USSR, is located in the Pamir Range in the extreme south.

Central Asian USSR also includes the Aral Sea; lakes Tengiz, Balkhash, and Issyk Kul; the Syr Darya River and parts of the Amu Darya, Ural, Irtysh, and Ili rivers; and the Kyzyl-Kum and Kara-Kum deserts. The tundra zone of Siberian USSR includes the Severnaya Zemlya archipelago, the New Siberian Islands, and Bolshoy Lyakhov Island in the Arctic Ocean, and the mouths of the Ob, Yenisei, Khatanga, Lena, Yana, and Indigirka rivers. The rest of the region can be divided into three areas—the West Siberian Plain, which lies between the Urals and the Yenisei River and is part of the Russian Plain; the Central Siberian Plateau, which is situated between the Yenisei and Lena rivers; and a mountainous area, which lies east of the Lena and continues into the Soviet Far East. Much of the West Siberian Plain is marshland; the area is drained chiefly by the Ob and Irtysh rivers. Bordering the plain in the southeast is a mountainous area that includes the Sayan Mts.

The Central Siberian Plateau averages about 2,000 ft (610 m) in height and is covered with taiga. Notable rivers here include the Lower Tunguska, the Stony Tunguska, the Angara, the Olenek, and the Vilyuy. The mountainous area in eastern Siberian USSR includes the Verkhoyansk and Cherskogo (Cherski) ranges in the center and the Patom Plateau, the Yablonovy Range, and Lake Baykal in the southwest. Most of the Soviet Far East, including Sakhalin Island in the Pacific Ocean, is covered with taiga; however, the north and northeast (extending to the center of Kamchatka Peninsula) is covered with tundra. The tundra zone also includes Wrangel Island in the Arctic Ocean and Cape Dezhnev on the Bering Strait.

The Soviet Far East is largely mountainous, although there are fertile lowlands in the northeast and southeast. In the northeast are the Chukotsk Mts. (situated on the Chukchi Peninsula), the Kolyma Range, and Koryaksky Range; in the east is the Dzhugdzhur Range; and in the southeast are the Sikhote-Alin and Bureya ranges. The Amur and Ussuri rivers (which form parts of the boundary with China) are located in the southeast, and the Anadyr and Penzhina rivers are in the northeast.

The USSR includes more than 100 ethnic groups. The major ones according to the 1970 census were the Russians (129 million), Ukrainians (41 million), and Belorussians (9 million), who all speak Slavic languages and together make up 74% of the country's population; the Uzbek (9 million), Tatars (6 million), Kazakhs (5 million), and Azerbaijan (4.4 million), who all speak a Turkic language; the Armenians (3.6 million); the Georgians (3.2 million); the Lithuanians (2.7 million); and the Moldavians (2.7 million). Between the censuses of 1959 and 1970 the population of the USSR grew by 33 million, or 16% (from 209 million to 242 million). In 1970 56% of the country's inhabitants were classified as urban. About 75% of the population lives in European USSR. Russian is the country's official language. Formal religion plays a relatively small role in the nation's affairs. The majority of those professing religious faith adhere to the Russian Orthodox Church, whose head is the patriarch of Moscow and All Russia. In the late 1960s it was estimated that the Russian Orthodox Church had 30 million regular worshippers. Also important are the Armenian Orthodox Church and the Georgian Orthodox Church. Roman Catholics are numerous in Lithuania, with substantial minorities in Latvia, Belorussia, and western Ukraine. Protestantism is represented chiefly by Lutherans, who are concentrated in the Baltic republics, and by Evangelical Christian Baptists. After Christianity the largest religious following is that of Islam, whose members (about 30 million in the late 1960s) are mostly Turkic-speaking persons living in the southwestern part of the country. The majority of the Moslems belong to the Sunni branch. There are about 2.2 million Jews (virtually all of whom live in European USSR), and approximately 500,000 Buddhists.

The Soviet Union includes 15 constituent (Union) republics:

NAME	CAPITAL	NAME	CAPITAL
Armenian SSR	Yerevan	Lithuanian SSR	Vilnius
Azerbaijan SSR	Baku	Moldavian SSR	Kishinev
Belorussian SSR	Minsk	Russian SFSR	Moscow
Estonian SSR	Tallinn	Tadzhik SSR	Dushanbe
Georgian SSR	Tbilisi	Turkmen SSR	Ashkhabad
Kazakh SSR	Alma-Ata	Ukrainian SSR	Kiev
Kirgiz SSR	Frunze	Uzbek SSR	Tashkent
Latvian SSR	Riga		

Economy. In 1917 the Soviet Union was an overwhelmingly agricultural country, although there had been considerable industrial development since the late 19th cent. The economy was devastated by the destruction incurred during World War I and the civil war that followed the Communist takeover. Between 1918 and 1921 the state took control of the whole economy. This led to inefficiency and confusion, and in 1921 there was a partial return to the market economy with the adoption of the New Economic Policy (NEP). In 1928 the NEP was abandoned, and the first Five-Year Plan (1928–32) was drawn up by Gosplan (the state planning commission), setting goals and priorities for virtually the entire economy. Under the Five-Year Plans the production of capital goods was emphasized, and consumer goods were largely neglected. The plan for the period from 1971 to 1975 was the first to set a higher growth rate for the manufacture of consumer goods, but the larger capital-goods sector of the economy still predominated.

In the 1970s the state owned and operated the industrial and service sectors of the economy. State farms (sovkhozy) and collective farms (kolkhozy), whose land is rented by a group of farmers at no cost and in perpetuity from the state and which are run by elected officials, embraced more than 95% of the country's cultivated land. However, about 30% of the country's total agricultural output was produced on the small amount of privately held cultivated land (owned principally by collective farmers). Western economists estimated that Soviet output in the early 1970s was equal roughly to half the U.S. gross national product (GNP). This level of production indicates the high level of growth achieved by the USSR in its short history despite the costly changeover from a market economy to a socialist economy and the great destruction and loss of manpower suffered in World War II. In 1971 industry and transport accounted for 78% of output (compared to 42% in 1913), and agriculture contributed 16% (compared to 58% in 1913). In the late 1960s the civilian work force numbered about 110 million, of whom approximately 50% were women. About 34% of the workers were employed in manufacturing and construction, 30% in agriculture and forestry, 11% in education and health, 8% in transport and communications, and 6% in commerce. The amount of cultivated land increased from 321 million acres (130 million hectares) in 1933 to 511 million acres (207 million hectares) in 1971, mostly as a result of the campaign (begun in 1954) to open Central Asian USSR and western Siberian USSR to the cultivation of grain. Although 30% more land is farmed in the USSR than in the United States, Soviet farm output is only about 80% of that of the United States. The main reasons for the lower yield per acre are a shorter growing season, more frequent natural impediments, and lesser amounts of modern farm machinery and fertilizer. The following are some of the main agricultural commodities produced in the Soviet Union (with the USSR's world rank in 1972 in terms of quantity produced): wheat (1st); maize (4th); barley (1st); rye (1st); oats (1st); rice (11th); butter (1st); cheese (3rd); beef and veal (2nd); pork (2nd); mutton, lamb, and goat meat (2nd); lard (1st); raw sugar (1st); honey (1st); tea (5th); tobacco (3rd); cotton (2nd); and wool (2nd). The main wheat-growing regions are in the Ukraine, in the Kuban steppe, on the steppes east of the middle Volga River, in northern Kazakh SSR, and in southwestern Siberian USSR. Large numbers of poultry, cattle, hogs, sheep, goats, horses, camels, and reindeer are raised.

In 1971 the Soviet Union was the world's third leading fishing nation in terms of the total weight of fish caught. The catch is principally made up of marine fish such as cod, haddock, and herring. The leading fishing port is Arkhangelsk, on the White Sea; Okhotsk, on the Sea of Okhotsk, also has important fisheries. In 1971 the USSR ranked 4th in the world in the number of whales caught.

Approximately one third of the land area in the Soviet Union is covered with forest, about three quarters of which is made up of softwood coniferous trees. The USSR is the world's leading producer of timber. The Soviet Union is extremely rich in mineral resources. The following are the main minerals produced (with the country's rank in 1972 in quantity extracted): antimony (4th); asbestos (2nd); bauxite (4th); chromite (1st); coal and lignite (1st); cobalt (3rd); copper ore (5th); fluorite (3rd); gold (2nd); iron ore (1st); lead (2nd); manganese ore (1st); mercury (2nd); molybdenum (3rd); nickel (2nd); crude petroleum (2nd); phosphate rock (2nd); platinum (1st); potash (1st); pyrite (1st); salt (3rd); silver (3rd); sulfur (4th); tin (3rd); tungsten (1st); and zinc ore (2nd). Coal and lignite are found mainly in the Donets Basin (Ukraine), around Vorkuta (northeastern European USSR), in the Moscow Basin, in the Kuznetsk Basin (east of Novosibirsk), in the Karaganda Basin (Kazakh SSR), on the Taymyr Peninsula, and in the Tunguska Basin and Lena Basin (both in eastern Siberian USSR). Petroleum is produced principally in western Siberian USSR (notably at Agansk and around Tyumen) and also at Baku, Grozny, and Maikop in the Caucasus and in the Ural-Volga region at Kuybyshev, in the Bashkir ASSR, and around Perm. Iron ore is mined in the Urals, at Krivoy Rog, around Kursk, in Kazakh SSR, and in many other areas; manganese ore is found at Nikopol (Ukraine) and Chiatura (Georgian SSR); chromite is mined in Kazakh SSR; gold is produced in eastern Siberian USSR, notably in the Kolyma Gold Fields and at Aldan; diamonds are found in the Vilyui River Basin in the Yakut ASSR; and bauxite is mined in western Siberian USSR, near Leningrad, and in Kazakh SSR.

The Soviet Union is a great industrial nation whose factories produce a wide range of manufactures. The country's leading industrial centers are also its largest cities, and most of them are situated in European USSR—Moscow, Leningrad, Gorky, Kuybyshev, Rostov-na-Donu, and Volgograd in the RSFSR; Kiev, Kharkov, Odessa, Dnepropetrovsk, and Donetsk in the Ukraine; Baku and Tbilisi in the Transcaucasus; and Sverdlovsk in the Urals. Increasingly, however, cities in other parts of the country are becoming major industrial centers. In 1971 and 1972 the Soviet Union was the world's leading producer of crude steel. The Soviet Union has great amounts of energy resources, notably coal, petroleum, natural gas, and water power. Between 1940 and 1971 the production of electricity increased 16-fold, reaching 800 billion kilowatt hours in 1971 (about half the U.S. output and second in the world). Large hydroelectric facilities are located on the Dnepr, Volga, Kama, Angara, and Yenisei rivers. Nuclear power is also being developed. European USSR has a dense rail network, as does Siberian USSR between the Urals and Novosibirsk. The Trans-Siberian RR runs along southern Siberia to Vladivostok on the Pacific Ocean; it is linked with the Turkistan-Siberia RR, which in turn is connected with the Trans-Caspian RR. The country's all-weather road system is relatively poor, and most freight transportation is by train rather than by truck. Most of the USSR is linked by airplane, and there is service to numerous foreign cities. The Soviet Union has access to several seas, but most of its ports are icebound in winter. Its only ice-free ports are those on the Black Sea (notably Odessa, Nikolayev, and Zhdanov); Kaliningrad, on the Baltic; and Murmansk, on the Arctic Ocean. Other large ports include Arkhangelsk, Leningrad, Riga, and Vladivostok. The country's great rivers provide important inland traffic lanes.

The Soviet Union carries on a substantial foreign trade. In 1972 the total value of Soviet international trade was equivalent to $38 billion, whereas that of the United States was $108 billion and that of West Germany $103 billion. In the period from 1964 to 1972 the USSR had a small annual trade deficit five times and an annual trade surplus four times. The principal exports are machinery, crude petroleum, iron and steel, and timber; the chief imports are machinery, manufactured consumer goods, and foodstuffs. About two thirds of Soviet trade is with the Communist nations of Eastern Europe. International trade is a monopoly of the state, and it is conducted in accordance with the government's overall goals for the economy. A noteworthy feature of Soviet trade relations in the late 1960s and early 70s were the agreements with private Western firms to build major installations in the USSR.

Government. Soviet Russia was the first state to be based on Marxist socialism. As officially created by the treaty of union of 1922, which joined the RSFSR, the Ukraine, Belorussia, and Transcaucasia (divided in 1936 into the Georgian, Armenian, and Azerbaijan republics), the USSR comprised Russia and the remainder of the Russian Empire as it had emerged from the Russian Revolution of 1917 and the ensuing civil war. The civil war had been complicated by Allied intervention and by war (1920) with Poland. The peace treaty (1921) with Poland, the declarations of independence of Finland, Estonia, Latvia, and Lithuania, and the seizure by Rumania of Bessarabia had greatly reduced the size of the former Russian Empire, establishing what the governments of Western Europe called a cordon sanitaire (quarantine belt) separating Communist Russia from the rest of Europe. Temporarily accepting this quarantine, Vladimir I. Lenin and the other leaders of the Soviet Union set about repairing the damage caused by the revolution and the civil war. The constitution of 1924 was based theoretically on the dictatorship of the proletariat and founded economically on the public ownership of the land and the means of production according to the revolutionary proclamation of 1917. In 1936 a new constitution (called the Stalin constitution) was promulgated. It had a veneer of radical innovation, but in practice the state operated much as it had before, with the Communist party of the Soviet Union (CPSU) continuing to be the most important political organization in the country. In 1977 a draft constitution was presented to the Supreme Soviet. Differing only slightly from the three previous constitutions, the 1977 document reiterated the economic guarantees on which the state was based and reasserted the supremacy of the party. The CPSU indirectly controls all levels of government, and the general secretary of the Communist party is usually the most powerful person in the USSR.

The chief administrative divisions of the USSR were set up so that each included one preponderant ethnic group that was given formal autonomy but little actual power. The USSR is made up of 15 constituent republics, each of which has its own government and is entitled, under the constitution, to maintain a separate army and to have its own foreign representation. However, no separate military forces are in fact maintained, and the only republics to have diplomats abroad are the Ukrainian SSR and the Belorussian SSR, which in 1945 were admitted into the United Nations. In practice, the governments of the constituent republics (and of the other administrative divisions) have authority only in the field of cultural affairs and are mostly concerned with implementing national policy. The constituent republics have the constitutional right to secede from the USSR, but it is highly unlikely that they will ever exercise this right. The next most important administrative divisions, after the constituent republics, are the autonomous republics, of which there are 20. Each has a unicameral supreme soviet and a council of ministers. Each of the other administrative divisions has only a unicameral soviet. These divisions include autonomous oblasts; oblasts; national okrugs; krays; and urban and rural districts.

Under the 1936 constitution national legislative power is vested in the Supreme Soviet. It is divided into the Soviet of the Union, elected by universal suffrage on the basis of one deputy for every 300,000 citizens, and the Soviet of the Nationalities, also elected by universal suffrage and containing 32 representatives from each of the constituent republics, 11 representatives from each of the autonomous oblasts, and 1 representative from each of the national okrugs. Members of the Supreme Soviet must be either members of the Communist party or so-called nonparty persons (the only legal political party being the Communist party); they serve four-year terms. Laws are passed by a simple majority in both houses; constitutional amendments require a two-thirds majority. The Supreme Soviet usually meets for a short session twice a year. In a joint session, it elects a presidium, or permanent committee, headed by a chairman, or president (who is the official head of state of the USSR), and also including one deputy chairman from each constituent republic, a secretary, and 20 members. Among the presidium's various functions are convening (or dissolving) the Supreme Soviet, carrying out the functions of the Supreme Soviet when it is not in session, interpreting existing laws, issuing edicts, and conducting national referenda. The Supreme Soviet appoints the council of ministers headed by a chairman (known as the country's premier), which is the chief executive and administrative body of the state.

History. The fundamental policy of the CPSU from its beginning has been complete socialization. In 1922 Germany recognized the Soviet Union and most other Western nations except the United States followed suit in 1924. A struggle for leadership followed Lenin's death in early 1924; Joseph V. Stalin and Leon Trotsky were the two main protagonists, with Stalin emerging victorious by the late 1920s. Stalin's program called for a more gradual transformation of Soviet society than did Trotsky's and had as its primary objective the consolidation of Communism in the USSR rather than Trotsky's ideal of immediate world revolution. Nevertheless, the Soviet Union continued to guide the Communist parties abroad through the Third International, or Comintern, and at home the New Economic Policy instituted in 1921 was replaced by full government planning with the adoption of the first Five-Year Plan (1928–32). A system of collective and state farms was imposed over widespread peasant opposition. Industrialization was accelerated, and the production of desperately needed industrial raw materials and capital equipment was stressed at the expense of consumer goods. One of the major results of the successive Five-Year Plans was the spectacular industrial and agricultural development of the Urals, Siberian USSR, and Central Asian USSR. The level of literacy, very low in 1917, was steadily raised in all parts of the country, and free medical and social services were extended to the population. At the same time, the state increased its hold over all political, social, and cultural aspects of life. The secret police became a major instrument of state control, and much power was given to the civil service. The system of controls gave rise to a large and powerful bureaucracy.

Religious bodies were severely persecuted in the early years of the Soviet Union, but in the mid-1930s there was a measure of relaxation in official policy. However, relations with the Roman Catholic Church and with the Jewish community remained hostile. The mid-1930s also saw a conservative trend in official attitudes toward culture: Family life was emphasized again, and divorces and abortions were made difficult to obtain; great men and events in pre-1917 Russian history were extolled in literature; and experimentation in education gave way to a return to structure and discipline. Following the murder (1934) of Sergei M. Kirov, one of Stalin's closest associates, and the announcement of the discovery of an alleged plot against Stalin's regime headed by the exiled Trotsky, there began a series of purges that culminated in the great purge from 1936 to 1938; the victims were generally sentenced to death or to long terms of hard labor. Much of the purge was carried out in secret. Independent influence in society was thus ended, and monolithic unity under Stalin was achieved by 1939.

In 1933 the United States recognized the USSR, and in 1934 the Soviet Union was admitted into the League of Nations. In the mid-1930s the USSR sought friendly relations with its neighbors, declared its renunciation of imperialistic expansion, and advocated total disarmament. Soviet-controlled Communist parties in other countries became friendlier to more moderate socialists and to liberals and in 1936 joined leftist Popular Front coalitions in France and Spain. The Western nations did not invite the USSR to take part in the negotiations with Germany leading to the Munich Pact (1938), and a radical shift in Soviet foreign policy ensued. On Aug. 23, 1939, the USSR concluded a nonaggression pact with Nazi Germany. Shortly afterward Hitler invaded Poland, precipitating World War II. Lithuania, Latvia, and Estonia were occupied (Sept.–Oct., 1939) by the Soviet Union, and in mid-1940 were transformed into constituent republics of the USSR. Finland opposed Soviet demands, and the Finnish-Russian War of 1939–40 resulted; it ended in a hard-earned Soviet victory. Finland ceded territory, which was organized into the Karelo-Finnish SSR (which in 1956 became part of the RSFSR as the Karelian ASSR). Rumania was forced (1940) to cede Bessarabia and northern Bukovina, and the Moldavian SSR was created. In Apr., 1941 a nonaggression treaty with Japan was signed.

Although defense preparations were accelerated, when Germany attacked on June 22, 1941, the Soviet Union was caught by surprise.

By the end of 1941 the Germans had overrun Belorussia and most of the Ukraine, had surrounded Leningrad, and were converging on Moscow. A Soviet counter-offensive saved Moscow, but in June, 1942, the Germans launched a new drive directed against Stalingrad (now called Volgograd). The Soviets drove the invaders back in an almost uninterrupted offensive and in 1944 entered Poland and the Balkan Peninsula. Early in 1945 Soviet troops marched into east Prussia. The converging Soviet armies then closed in on Berlin in a climactic drive. On May 2, 1945, Berlin fell; on May 7 the USSR together with the Western Allies accepted the surrender of Germany. The Soviet victory was obtained at the great price of at least 20 million lives and staggering material losses. The United States contributed much aid to the USSR through Lend-Lease. Understandings concerning the conduct of war and postwar policies had been reached by the USSR, the United States, and Great Britain at the Moscow Conferences (1941–47), the Teheran Conference (1943), the Yalta Conference (1945), and the Potsdam Conference (1945). In accordance with a previous agreement, the Soviet Union declared war on Japan on Aug. 8, 1945. A swift campaign brought Soviet forces deep into Manchuria and Korea by the date (Sept. 2, 1945) Japan surrendered. As a direct result of the war, the USSR received the southern half of Sakhalin Island and the Kuril Islands from Japan; the northern part of east Prussia from Germany; and some additional territory from Finland. By agreements in 1945 with Poland and Czechoslovakia the USSR also vastly increased the area of the Belorussian and Ukrainian republics.

Cooperation between the USSR and the Western powers ceased soon after the armistice, and relations between the Soviet Union and the United States became increasingly strained, leading to the Cold War. Friction became particularly acute in the jointly occupied countries of Germany, Austria, and Korea and in the United Nations (of which the USSR was a charter member), preventing the conclusion of joint peace treaties with Germany, Austria, and Korea and agreements over reparations and the control of nuclear weapons. Increasing Soviet influence in Poland, Czechoslovakia, Hungary, Rumania, Bulgaria and Albania and the continued tight control of East Germany created fears in the Western world of unlimited Soviet expansion. The USSR justified its policies by its fears of encirclement by hostile capitalist nations. In 1948 Yugoslavia declared its independence from the Soviet bloc. In 1949 the USSR recognized the newly established Communist government of China, and a 30-year alliance was signed in early 1950. Relations with the Western powers worsened considerably after the outbreak of the Korean War (1950–53), which the West ascribed to Soviet instigation, and Stalin's dictatorship took on an increasingly tyrannical character.

The death of Stalin on Mar. 5, 1953, ushered in a new era in Soviet history. "Collective leadership" at first replaced one-man rule, and the power of the secret police was curtailed. Soviet citizens began to gain a greater degree of personal freedom and civil security. Georgi Malenkov succeeded Stalin as premier, while Nikita S. Khrushchev, as first secretary of the central committee of the CPSU, played an increasingly important role in policy planning. In 1955 Malenkov was replaced as premier by Nikolai Bulganin. At the 20th All Union Congress (Feb. 1956), Khrushchev bitterly denounced the dictatorial rule and personality of Stalin in a secret speech that was later obtained by foreigners. Khrushchev replaced Bulganin as premier in 1958, thus becoming leader of both the government and the CPSU; he modified some of the more dictatorial aspects of Stalin's rule, but the CPSU continued to dominate all facets of Soviet life. Khrushchev retained many of Stalin's basic economic policies, but there were important changes. Management of the economy (especially industry) was decentralized (1957) in an attempt to reduce inefficiency and delays. In agriculture, vast tracts of virgin land (especially in Central Asian USSR and western Siberian USSR) were opened to the cultivation of grain, notably maize; taxation of collective farmers' private plots was reduced; and the Machine Tractor Stations, established in the late 1920s and 1930s as a means of supervising the collective farms by controlling their use of farm machinery, were abolished in 1958 and their equipment sold to the collectives. Somewhat larger amounts of consumer goods were manufactured. Foreign policy became more flexible; the Soviet Union negotiated a peace treaty with Austria (1955), established diplomatic relations with West Germany (1955), dissolved the Cominform (1956), allowed foreigners to travel in the USSR, and set up cultural exchanges with Western nations. In addition, in 1955 the Soviet Union began to form alliances with, and give aid to, the non-Communist nations of the Middle East, especially Egypt and Syria, and other non-Communist underdeveloped countries. Relations with the Communist countries of Eastern Europe were formalized and strengthened by the establishment of the Council for Mutual Economic Assistance and the Warsaw Treaty Organization. In June, 1956, a revolt against Soviet influence in Poland was defeated by the Polish army, but the Poles managed to gain some concessions from Moscow; an uprising in Hungary in Oct., 1956, was crushed ruthlessly by Soviet troops.

In the technological race between the Soviet Union and the West (principally the United States), the USSR exploded (1953) a hydrogen bomb; announced (1957) the development of intercontinental ballistic missiles; orbited (1957) the first artificial earth satellite (called Sputnik); and in 1961 sent Yuri Gagarin in the first manned orbital flight. The USSR participated in the international negotiations on nuclear disarmament and agreed (1958) to a voluntary moratorium on nuclear tests, but resumed testing in 1961. In 1963 the USSR signed a milestone treaty with the United States and Great Britain banning atmospheric nuclear tests. The question of divided Berlin (a focal point of the cold war) remained unresolved through several rounds of negotiations and a number of "Berlin crises." In June, 1964, the Soviet Union signed a separate peace treaty with East Germany. At the 22nd CPSU congress in 1961 the attack on Stalin was continued, and the reputations of many purge victims of the 1930s were rehabilitated. Stalin's body was removed from its place of honor in the Kremlin next to Lenin's; his name was erased from the geography of the USSR (for example, Stalingrad was renamed Volgograd). At the 22nd congress the Sino-Soviet conflict (which had begun in the late 1950s) emerged, stated at first in terms of a dispute with Albania (a close ally of China). In Oct., 1962, despite seemingly improved relations with the West, the USSR came into sharp conflict with the United States over the presence of Soviet missiles in Cuba. Some analysts maintain that the Cuban missile crisis marked a turning point in U.S.-Soviet relations because the USSR realized the extent to which the United States would protect what it considered its vital interests. In 1963 a "hot line" (direct and instantaneous teletype communications) was set up between the heads of government of the USSR and the United States.

In a well-prepared and bloodless move by CPSU leaders, Khrushchev was ousted from his positions of power Oct. 14–15, 1964. He was replaced as first secretary of the CPSU by Leonid I. Brezhnev and as premier by Alexei N. Kosygin. The official reasons given for Khrushchev's ouster were his advanced age (70) and his declining health. The real reason was dissatisfaction with the policies and style of his government. Specifically, Khrushchev was criticized for the inadequate performance of the economy, especially the agricultural sector; for the humiliation of the USSR in the Cuban missile crisis; for the widening rift with China; and for his flamboyant personal style. The new leaders stressed collective leadership (as opposed to Khrushchev's one-man rule), but because of his position at the head of the CPSU Brezhnev held an advantage and by 1970 was the most powerful person in the country.

In 1973 a number of close associates of Brezhnev were made full members of the politburo in the first major shake-up of that important body since 1964. In the later 1960s the official attitude toward Stalin became somewhat less hostile. In internal affairs the new leaders stressed economic development, and in foreign affairs they generally pursued peaceful coexistence with the West. A major program to decentralize decision-making in industry was begun. Under the new economic system individual firms made their own decisions on levels of production based on prevailing prices, and their efficiency was judged individually on the amount of profit they made. These firms still had to operate within the constraints of the overall Five-Year Plans, which established the basic course of the Soviet economy, and of the annual national government budget. Industrial production (and the productivity of individual workers) increased slowly after 1964, largely because available technology was not sufficiently advanced. Agricultural production increased dramatically. In 1973, after a slump the previous year, the grain harvest reached the record level of 222 million metric tons.

Beginning in the mid-1960s leading writers, scientists, and intellectuals protested certain aspects of Soviet life, especially curbs on the free flow of ideas, corruption in government, and inefficiency. Although the dissidents were small in number and had little popular support, they were often sentenced to terms in prison or exiled. From the later 1960s many Jews asked to leave the country. For a time the government made emigration for them exceptionally difficult but in the early 1970s considerable numbers of Jews were able to emigrate. Between 1971 and 1973 about 75,000 Jews left the USSR, compared to approximately 2,000 annually in preceding years. In 1974 the USSR agreed to ease its emigration policy in return for favored-nation trade status with the United States. Also contributing to disquiet in the country were the spokesmen of several ethnic groups (notably the Lithuanians, Latvians, and Tatars) who demanded increased autonomy for their people.

Formal Soviet-U.S. relations during the 1960s and 1970s were businesslike and at times good. There were serious differences over U.S. participation in the Vietnam War, during the India-Pakistan War of 1971, during the Arab-Israeli Wars of 1967 and 1973, and over Soviet support of Marxist guerrilla movements in Africa in the late 1970s. In 1969 the USSR, the United States, and some 100 other nations signed a treaty banning the spread of nuclear weapons to countries not already possessing them. Strategic arms limitation talks (SALT) between the United States and the Soviet Union were begun in 1969. When U.S. President Richard M. Nixon visited Moscow in 1972 an agreement partially limiting nuclear arms was signed, along with accords on cooperation in space exploration, environmental matters, and trade. Under Nixon's successor, Gerald R. Ford, another interim SALT agreement was reached at Vladivostok. The Nixon and Ford presidencies, during which Henry A. Kissinger was Secretary of State, were the highwater marks of the "era of détente." During the administration of U.S. President Jimmy Carter, however, Soviet-U.S. relations appeared to take a sudden turn for the worse. The prime reason was Carter's outspoken support for human rights around the world, a position the Soviet government considered to be a major interference in its internal affairs. Despite the surface hostility, negotiations were continuing in 1977 on a new SALT treaty, as were contacts on lower levels.

In 1968 Soviet relations with the Communist nations of Eastern Europe reached a critical stage when Soviet troops (and forces of some of the other Warsaw Treaty Organization members) invaded (Aug. 21) Czechoslovakia in a successful effort to curb the trend toward liberalization there. Rumania and Yugoslavia explicitly denounced the invasion but relations between the USSR and Eastern Europe soon improved. A major objective of Soviet foreign policy in the early 1970s was to gain official recognition of the post-World War II settlement in Europe. In 1970 a landmark treaty with West Germany was signed (ratified in 1972) confirming existing boundaries in Europe (notably the eastern border of East Germany) and also renouncing the use of force to settle disputes. In 1972 meetings were held to set up a European security conference, which the USSR hoped would also help make permanent the status quo in Europe, and the conference formally opened in 1973. Also in 1973 the North Atlantic Treaty Organization and the Warsaw Treaty Organization began negotiations on the mutual and balanced reduction of military forces in Europe.

The Sino-Soviet conflict worsened after 1964. In 1969 there were numerous border clashes. Both countries enlarged their border forces and maintained them in the early 1970s despite somewhat less tense relations. In the 1967 Arab-Israeli War, the Soviet Union backed the Arabs but gave them little material assistance. In 1970–71 the USSR equipped Egypt's army with modern weapons and also sent about 16,000 military advisers and soldiers. However, in mid-1972 Egypt forced virtually all Soviet military personnel to leave the country, apparently feeling that the USSR was delivering insufficient quantities of sophisticated arms. Soviet-Egyptian relations improved during the period of the 1973 Arab-Israeli War, when the USSR supplied arms to Egypt and Syria, but they soon worsened again, and by 1977, Soviet influence in the Middle East was negligible. In the 1970s there was a notable increase in the size and quality of the Soviet military, particularly the navy and the substantial Soviet tank force in Eastern Europe. In 1977, in a move that confirmed him as the most powerful figure in the USSR, Brezhnev was elevated to the presidency of the presidium of the Supreme Soviet, thus becoming chief of state

Union Springs, city (1970 pop. 4,324), seat of Bullock co., SE Ala., ESE of Montgomery, in the Black Belt; settled 1836. It is a farm trade center.

Un·ion·town (yo͞on'yən-toun'), city (1978 est. pop. 15,400), seat of Fayette co., SW Pa., near the W.Va. line; settled c.1767, inc. as a city 1916. It is a farm-trade center and an industrial city. Formerly noted for its production of coal and coke, the city now has industries with diversified manufacturing. Products include meters, trailers, aircraft parts, steel scaffolding, stone crushers, and tires. The city is the seat of the Fayette campus of Pennsylvania State Univ. Each autumn a foliage festival is held in Uniontown. Gen. George C. Marshall was born here. Nearby, on the old National Road, is Fort Necessity, built by George Washington. Gen. Edward Braddock is buried near the fort.

Un·ion·ville (yo͞on'yən-vĭl'), city (1970 pop, 2,075), seat of Putnam co., N Mo., NW of Kirksville, in a farm and livestock area; inc. 1855.

United Ar·ab E·mir·ates (ăr'əb; ĕ-mîr'ĭts, -āts'), federation (1968 pop. 179,138), c.30,000 sq mi (77,700 sq km), E Arabia, on the Persian Gulf and the Gulf of Oman. The federation consists of seven emirates: Abu Dhabi, Ajman, Dubai, Fujairah, Ras al-Khaimah, Sharjah, and Umm al-Qaiwain. The states were formerly known as the Trucial States, Trucial Coast, or Trucial Oman. The term *trucial* refers to the fact that the sheiks ruling the seven constituent states were bound by truces concluded with Great Britain in 1820 and by an agreement made in 1892 accepting British protection. Before British intervention, the area was notorious for its pirates and was called the Pirate Coast. After World War II the British granted internal autonomy to the sheikdoms. Discussion of federation began in 1968. Ras al-Khaimah at first opted for independence but reversed its decision in Feb., 1972.

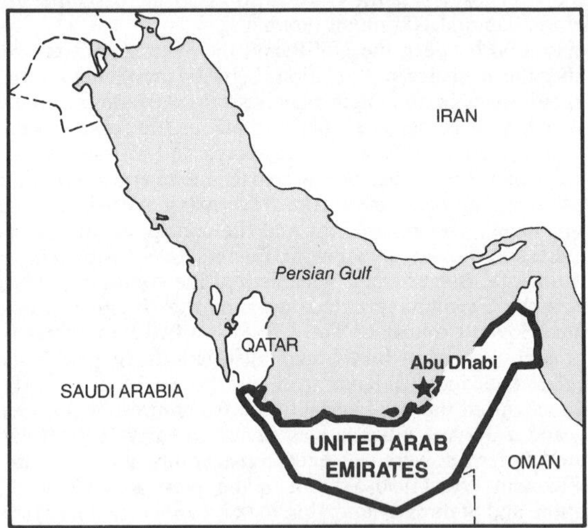

The highest legislative body in the federation is the supreme council, and a representative federal council holds executive power. Local matters are dealt with by the sheiks. Abu Dhabi is the temporary capital, pending the building of a new one on the border between Abu Dhabi and Dubai. Most of the inhabitants are Arabs, but there are also Persians, Baluchis, and Indians. The majority are Sunni Moslems, although some Shi'a Moslems live in Dubai. The area is rich in oil deposits, which have been exploited since the early 1960s. Fishing and pearling are also important occupations.

United Arab Republic, political union (1958–61) of Egypt and Syria. The capital was Cairo. The two countries were merged (1958) into a single unit comprising the Southern (Egypt) and the Northern (Syria) Regions, with Gamal Abdal Nasser as president. As an initial step toward creating a pan-Arab union, the republic abolished Syrian and Egyptian citizenship, termed its inhabitants Arabs, and called the country "Arab territory." It considered the Arab homeland to be the entire area between the Persian Gulf and the Atlantic coast. With Yemen, it formed (1958) a loose federation called the United Arab States. In 1961 Syria withdrew from the union after a military coup, and Yemen soon followed, thus ending the union.

United King·dom of Great Britain and Northern Ireland (kĭng'dəm): *see* Great Britain.

United Prov·inc·es (prŏv'ĭn-səs), former state, N India, now almost coextensive with the modern state of Uttar Pradesh. The United Provinces embraced the plain of the Ganges, the heartland of India. This region was the scene of the ancient Hindu epics, the *Ramayana* and the *Mahabharata*. Many Hindu pilgrims flock to the holy cities along the Ganges, notably Allahabad and Varanasi. The east, as the scene of Buddha's life, remains sacred to Buddhism. Though the region was overwhelmingly Hindu in population, it was under Moslem rule from the 12th to the 18th cent. Great Britain first acquired territory here in 1764. The United Provinces of Agra and Oudh was formed in 1877 by merging the presidency of Agra and the kingdom of Oudh. The provincial area was occupied by rebels in the Indian Mutiny. After the partition of India in 1947 many Moslems migrated from the United Provinces to Pakistan. In 1950 the new state of Uttar Pradesh was formed.

United States, republic (1975 pop. 213,121,000), 3,615,191 sq mi (9,363,353 sq km), North America, consisting of 50 states and a federal district; Washington, D.C., is the capital. The full name is the United States of America. The outlying territories and areas of the United States include: in the West Indies, Puerto Rico (since 1952 a commonwealth associated with the United States) and the Virgin Islands of the United States (purchased from Denmark in 1917); in the Pacific Ocean, Guam (ceded by Spain after the Spanish-American War), American Samoa, Wake Island, Canton Island, Enderbury Island, and several other islands. The United States also has trusteeship under the United Nations of the Caroline, Marshall, and part of the Marianas island chains. The Panama Canal Zone is held under lease from the government of Panama.

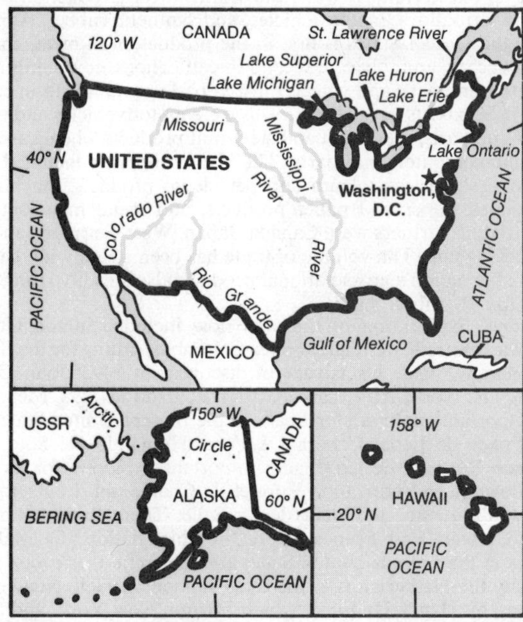

Excluding Alaska and Hawaii, the conterminous United States stretches across central North America from the Atlantic Ocean on the east to the Pacific Ocean on the west, and from Canada on the north to Mexico and the Gulf of Mexico on the south. The state of Alaska is located in extreme northwestern North America between the Arctic and Pacific oceans and is bordered by Canada on the east. The state of Hawaii, an island chain, is situated in the central Pacific Ocean c.2,100 mi (3,400 km) southwest of San Francisco.

The conterminous United States may be divided into seven broad physiographic divisions: from east to west, the Atlantic-Gulf Coastal Plain; the Appalachian Highlands; the Interior Plains; the Interior Highlands; the Rocky Mt. system; the Intermontane Region; and the Pacific Mt. system. An eighth division, the Laurentian Uplands, a part of the Canadian Shield, dips into the United States from Canada in the Great Lakes region. It is an area of little local relief, with an irregular drainage system and many lakes, as well as some of the oldest exposed rocks in the United States.

The Atlantic-Gulf Coastal Plain extends along the east and southeast coasts of the United States from eastern Long Island to the Rio Grande; Cape Cod and the islands off southeastern Mass. are also part of this region. Although narrow in the north, the Atlantic Coastal Plain widens in the south, merging with the Gulf Coastal Plain in Ga. The Atlantic and Gulf coasts are essentially coastlines of submergence, with numerous estuaries, embayments, islands, sandspits, and barrier beaches backed by lagoons. The northeast coast has many fine natural harbors, but south of the great capes of the N.C. coast there are few large bays. A principal feature of the lagoon-lined Gulf Coast is the great delta of the Mississippi River.

The Atlantic Coastal Plain rises in the west to the rolling Piedmont at the foot of the Appalachian Mts. These ancient mountains, a once-towering system now worn low by erosion, extend southwest from southeastern Canada to the Gulf Coastal Plain in Ala. In eastern New England, the Appalachians extend to the Atlantic Ocean, forming a rocky, irregular coastline. The Appalachians and the Adirondack Mts. of N.Y. include all the chief highlands of eastern United States; Mt. Mitchell (6,684 ft/2,037 m high), in the Black Mts. of N.C., is the highest point of eastern North America.

Extending more than 1,000 mi (1,610 km) from the Appalachians to the Rocky Mts. and lying between the Great Lakes in the north and the Gulf Coastal Plain in the south are the undulating Interior Plains. Once covered by a great inland sea, the Interior Plains are underlain by sedimentary rock. Almost all of the region is drained by one of the world's greatest river systems—the Mississippi-Missouri-Ohio. The Interior Plains may be divided into two sections: the fertile central lowlands, the agricultural heartland of the United States; and the Great Plains, a treeless plateau that gently rises from the central lowlands to the foothills of the Rocky Mts. The Black Hills of S.Dak. form the region's only upland area. The Interior Highlands are located just west of the Mississippi River between the Interior Plains and the Gulf Coastal Plain. This region consists of the rolling Ozark Plateau to the north and the Ouachita Mts. to the south.

West of the Great Plains are the lofty Rocky Mts. This system extends into northwestern United States from Canada and runs south into N.Mex. The highest of the numerous high peaks in the Rockies is Mt. Elbert (14,433 ft/4,399 m). Along the crest of the Rockies is the continental divide, separating Atlantic-bound drainage from that heading for the Pacific Ocean. Between the Rocky Mts. and the ranges to the west is the Intermontane region, an arid expanse of plateaus, basins, and ranges. The Columbia Plateau, in the north of the region, was formed by volcanic lava and is drained by the Columbia River and its tributary the Snake River, both of which have cut deep canyons into the plateau. The enormous Colorado Plateau, an area of sedimentary rock, is drained by the Colorado River and its tributaries; here the Colorado River has entrenched itself to form the Grand Canyon. West of the plateaus is an area of extensive desert and semidesert. The lowest point in North America, in Death Valley National Monument (282 ft/86 m below sea level), is here. The largest basin in the region is the Great Basin, an area of interior drainage and of numerous salt lakes, including the Great Salt Lake.

Between the Intermontane region and the Pacific Ocean is the Pacific Mt. system, a series of ranges generally paralleling the coast, formed by faulting and volcanism. The Cascade Range extends south from Canada into northern Calif., and from there is continued south by the Sierra Nevada. Mt. Whitney (14,495 ft/4,418 m), in the Sierra Nevada, is the highest peak in the conterminous United States. West of the Cascades and the Sierra Nevada are the Coast Ranges, which extend along the length of the U.S. Pacific coast. The San Andreas Fault, a crack in the earth's crust, parallels the trend of the Coast Ranges from San Francisco Bay to northwestern Mexico; earthquakes are common along its entire length. The Pacific Coastal Plain is narrow, and in many cases the mountains plunge directly into the sea.

Alaska may be divided into four physiographic regions; they are, from north to south, the Arctic Lowlands; the Rocky Mt. system; the Central Basins and Highlands Region; and the Pacific Mt. system, which parallels Alaska's southern coast and which rises to Mt. McKinley (20,320 ft/6,194 m), the highest peak of North America. The Hawaiian Islands may also be considered part of the Pacific Mt. system. These islands are the tops of volcanoes that rise from the floor of the Pacific Ocean; Mauna Kea and Mauna Loa are active volcanoes.

The terrain of northern United States was formed by the great continental ice sheets that covered northern North America during the Pleistocene epoch. The southern edge of the ice sheet is roughly traced by a line of moraines extending west from eastern Long Island and then along the course of the Ohio and Missouri rivers to the Rocky Mts.; land north of this line is covered by glacial material. Alaska and the mountains of northwestern United States had extensive mountain glaciers and were heavily eroded. Large glacial lakes occupied sections of the Great Basin; the Great Salt Lake and the other lakes of this region are remnants of the glacial lakes.

The United States has an extensive inland waterways system, much of which has been improved for navigation and flood control and developed to produce hydroelectricity and irrigation water. The

Mississippi-Missouri-Red Rock river system (c.3,890 mi/6,300 km long), is the longest in the United States and the second longest in the world. With its hundreds of tributaries the Mississippi basin drains more than half of the nation. The Yukon, Columbia, Colorado, and Rio Grande also have huge drainage basins. Other notable river systems include the Connecticut, Hudson, Delaware, Susquehanna, Potomac, James, Alabama, Brazos, San Joaquin, and Sacramento. The Great Salt Lake and Iliamna are the largest U.S. lakes outside of the Great Lakes and Lake of the Woods, which are shared with Canada. Since the opening of the St. Lawrence Seaway (1959) the Great Lakes have become a fourth U.S. seacoast, enabling oceangoing vessels to reach Duluth, Minn. The Illinois Waterway connects the Great Lakes with the Mississippi River, and the New York State Barge Canal links them with the Hudson. The Intracoastal Waterway provides sheltered passage for shallow draft vessels along the Atlantic and Gulf Coasts.

The United States has a broad range of climates, varying from the tropical rain-forest of Hawaii and the tropical savanna of southern Fla. to the subarctic and tundra climates of Alaska. East of the 100th meridian are the humid subtropical climate of southeastern United States and the humid continental climate of northeastern United States. Extensive forests are found in both these regions. West of the 100th meridian are the steppe climate and the grasslands of the Great Plains; trees are found along the water courses. In southwestern United States are the deserts, with the hottest and driest spots in the country. Along the Pacific coast are the Mediterranean-type climate of southern Calif. and, extending north into southeastern Alaska, the marine West Coast climate. The Pacific Northwest is one of the wettest parts of the United States and is densely forested. The Rocky Mts., Cascades, and Sierra Nevada have typical highland climates and are also heavily forested.

More than half of the population is urban, and the great majority of the inhabitants are of European descent. About 11% of the total population is black; there are smaller groups of Japanese and Chinese. Living for the most part on reservations are remnants of the aboriginal inhabitants, the heterogeneous Indian tribes. In addition to the original group of British settlers in the colonies of the Atlantic coast, numerous other national groups were introduced by immigration. Large numbers of black slaves were imported chiefly to work on the plantations of the South. When the United States was developing rapidly with the settlement of the West (where some earlier groups of French and Spanish settlers were absorbed), immigrants from Europe poured into the land. An important early group was the Scotch-Irish. Just before the middle of the 19th cent., Irish and German immigrants were predominant. A little later the Scandinavian nations supplied many settlers. After the Civil War, the immigrants came mainly from Italy, Greece, Russia, Poland, and from Austria-Hungary and the Balkans. There were lesser strains from Portugal, Spain, and the Levant, and to some extent the earlier sources continued to supply immigrants. A number of immigrants came, too, from the Far East. There is complete religious freedom in the United States, and the overwhelming majority of Americans are Christians. In turn, the majority of Christians are Protestants; the second-largest Christian group embraces the Roman Catholic Church; the Orthodox Eastern Church is also represented. In addition, a considerable group of Americans adhere to Judaism.

The United States is divided into 50 states and the District of Columbia:

NAME	CAPITAL	NAME	CAPITAL
Alabama	Montgomery	Maryland	Annapolis
Alaska	Juneau	Massachusetts	Boston
Arizona	Phoenix	Michigan	Lansing
Arkansas	Little Rock	Minnesota	St. Paul
California	Sacramento	Mississippi	Jackson
Colorado	Denver	Missouri	Jefferson City
Connecticut	Hartford	Montana	Helena
Delaware	Dover	Nebraska	Lincoln
Florida	Tallahassee	Nevada	Carson City
Georgia	Atlanta	New Hampshire	Concord
Hawaii	Honolulu	New Jersey	Trenton
Idaho	Boise	New Mexico	Santa Fe
Illinois	Springfield	New York	Albany
Indiana	Indianapolis	North Carolina	Raleigh
Iowa	Des Moines	North Dakota	Bismarck
Kansas	Topeka	Ohio	Columbus
Kentucky	Frankfort	Oklahoma	Oklahoma City
Louisiana	Baton Rouge	Oregon	Salem
Maine	Augusta	Pennsylvania	Harrisburg

NAME	CAPITAL	NAME	CAPITAL
Rhode Island	Providence	Vermont	Montpelier
South Carolina	Columbia	Virginia	Richmond
South Dakota	Pierre	Washington	Olympia
Tennessee	Nashville	West Virginia	Charleston
Texas	Austin	Wisconsin	Madison
Utah	Salt Lake City	Wyoming	Cheyenne

Economy. The United States is the greatest industrial nation in the world. Its mineral and agricultural resources are tremendous. Although it has been virtually self-sufficient in the past, an enormous increase in consumption, especially of energy, is making it increasingly dependent on certain imports. It is, nevertheless, the world's largest producer of both electrical and nuclear energy. It leads all nations in the production of natural gas, lead, copper, aluminum, sulfur, and salt. With the Soviet Union it is the leading producer of coal and steel. It ranks second in the production of crude oil, iron ore, silver, and zinc. The United States produces 69% of the world's mica, 68% of its molybdenum, 54% of its uranium, and 46% of its magnesium. It leads the world in the production of pig iron and ferroalloys, motor vehicles, and synthetic rubber. Agriculturally, the United States is first in the production of meat, cheese, corn, soybeans, and tobacco; second in cattle, hogs, cow's milk, butter, cotton lint, oats, and wheat; third in barley; and fourth in sugar. Major U.S. exports in the early 1970s were motor vehicles and parts, aircraft and parts, food, iron and steel-mill products, chemicals, and electronic computers and parts. The leading imports included ores and metal scraps, petroleum and petroleum products, machinery, automobiles, paper and paper products, and metal manufactures. Major trading partners were Canada, Japan, West Germany, and the United Kingdom. The volume of trade has been steadily increasing. By 1972 the nation's gross national product, also steadily rising, had passed the $1-trillion mark.

History. Exploration of the area now included in the United States was spurred after Christopher Columbus, sailing for the Spanish monarchy, made his voyage of discovery in 1492. John Cabot explored the North American coast for England in 1498. Men who were important explorers for Spain in the present United States include Ponce de León, Cabeza De Vaca, Hernando De Soto, and Francisco Vásquez de Coronado; important explorers for France were Giovanni da Verrazano, Samuel de Champlain, Louis Jolliet, Jacques Marquette, and sieur de La Salle. There three nations—England, Spain, and France—were the chief nations to establish colonies in the present United States although others also took part, especially the Netherlands in the establishment of New Netherland (explored by Henry Hudson), which became New York, and Sweden in a colony on the Delaware River. The first permanent settlement in the present United States was Saint Augustine (Florida), founded in 1565 by the Spaniard Pedro Menéndez De Avilés. Spanish control came to be exercised over Florida, West Florida, Texas, and a large part of the West, including California. For the purposes of finding precious metals and of converting heathens to Catholicism, the Spanish colonies in the present United States were relatively unfruitful and thus were never fully developed. The French established strongholds on the St. Lawrence River (Quebec and Montreal) and spread their influence over the Great Lakes country and along the Mississippi; the colony of Louisiana was a flourishing French settlement.

The French government, like the Spanish, tolerated only the Catholic faith, and it implanted the rigid and feudalistic seignorial system of France in its North American possessions. Partly for these reasons the French settlements attracted few colonists. The English settlements, which were on the Atlantic seaboard, developed in patterns more suitable to the New World, with greater religious freedom and economic opportunity. The first permanent English settlement was made at Jamestown (Virginia) in 1607. The first English settlements in Virginia were managed by a chartered commercial company, the Virginia Company; economic motives were paramount to the company in founding the settlements. The Virginia colony early passed to control by the crown and became a characteristic type of English colony—the royal colony. Another type—the corporate colony—was initiated by the settlement of the Pilgrims at Plymouth Colony in 1620 and by the establishment of the more important Massachusetts Bay colony by the Puritans in 1630. Religious motives were important in the founding of these colonies. The colonists of Massachusetts Bay brought with them from England the charter and the governing corporation of the colony, which thus became a corporate one, that is, one controlled by its own resident corporation. The corporate status of the Plymouth Colony, evinced in the Mayflower Compact, was established by the purchase (1626) of company and charter from the holders in England. Connecticut and Rhode Island, which were offshoots of Massachusetts, owed allegiance to no English company; their corporate character was confirmed by royal charters, granted to Connecticut in 1662 and to Rhode Island in 1663. A third type of colony was the proprietary, founded by lords proprietors under quasi-feudal grants from the king; prime examples are Maryland (under the Calvert family) and Pennsylvania (under William Penn). The religious and political turmoil of the Puritan Revolution in England, as well as the repression of the Huguenots in France, helped to stimulate emigration to the English colonies. Hopes of economic betterment brought thousands from England as well as a number from Germany and other continental countries. To obtain passage across the Atlantic, the poor often indentured themselves to masters in the colonies for a specified number of years. The colonial population was also swelled by criminals transported from England as a means of punishment. Once established as freedmen, former bondsmen and transportees were frequently allotted land with which to make their way in the New World.

The colonies were subject to English Mercantilism in the form of Navigation Acts, begun under Cromwell and developed more fully after the Stuart Restoration. The colonies at first benefited by these acts, which established a monopoly of the English market for certain colonial products. Distinct colonial economies emerged, reflecting the regional differences of climate and topography. Agriculture was of primary importance in all the sections. In New England many crops were grown, maize being the closest to a staple, and agricultural holdings were usually of moderate size. Fur trade was at first important, but it died out when the New England Confederation defeated Philip in King Philip's War and the Indians were dispersed. Fishing and commerce gained in importance, and the economic expansion of Massachusetts encouraged the founding of other New England colonies. In the middle colonies small farms abounded, interspersed with occasional great estates, and diverse crops were grown, wheat being most important. Land there was almost universally held through some form of feudal grant, as it was also in the South. Commerce grew quickly in the middle colonies, and large towns flourished, notably Philadelphia and New York. By the late 17th cent. small farms in the coastal areas of the South were beginning to give way to large plantations; these were profitably developed with the slave labor of blacks, who were imported in ever-increasing numbers. Plantations were almost exclusively devoted to cultivation of the great Southern staples—tobacco, rice, and, later, indigo. Fur trade and lumbering were long important. Although some towns developed, the Southern economy remained the least diversified and the most rural in colonial America. In religion, too, the colonies developed in varied patterns. In Massachusetts the religious theocracy of the Puritan oligarchy flourished. By contrast, Rhode Island allowed full religious freedom; there Baptists were in the majority, but other sects were soon in evidence. New Jersey and South Carolina also allowed complete religious liberty, and such colonies as Maryland and Pennsylvania established large measures of toleration. Maryland was at first a haven for Catholics, and Pennsylvania similarly a haven for Quakers, but within a few decades numerous Anglicans had settled in those colonies. Anglicans were also much in evidence further south, as were Presbyterians, most of them Scotch-Irish.

Politically, the colonies developed representative institutions, the most important being the vigorous colonial assemblies. Popular participation was somewhat limited by property qualifications. In the proprietary colonies, particularly, the settlers came into conflict with the executive authority. Important points of difference arose over the granting of large estates to a few, over the great power of the proprietors, over the failure of the proprietors (who generally lived in England) to cope with problems of defense, and over religious grievances, frequently stemming from a struggle for dominance between Anglicans and other groups. In corporate Massachusetts religious grievances were created by the zealous Puritan demand for conformity. These conflicts, together with England's desire to coordinate empire defenses against France and to gain closer control of the colonies' thriving economic life, stimulated England to convert corporate and proprietary colonies into royal ones. By 1702 New Hampshire, Massachusetts, and New Jersey had been made royal colonies, and later the Carolinas and Georgia were transferred to royal control. Pennsylvania and Maryland remained nominally proprietary, but modifications of their charters in effect brought them under royal sway. Only Connecticut and Rhode Island remained substantially free of royal interference and, as corporate colonies, continued to select their own governors. In general, royal control brought more orderly government and greater religious toleration, but it also focused the colonists' grievances on the mother country. The policies of the governors, who were the chief instruments of English will in the colonies, frequently met serious opposition. The colonial assemblies clashed with the governors—notably with Edmund Andros and Francis Nicholson—especially over matters of taxation. The assemblies successfully resisted royal demands for permanent income to support royal policies and used their powers over finance to expand their own jurisdiction. As the 18th cent. progressed, colonial grievances were exacerbated. The British mercantile regulations, beneficial to agriculture, impeded the colonies' commercial and industrial development, and the pattern of trade enforced by the Navigation Acts kept the colonies denuded of specie. However, economic and social growth continued, and by the mid-18th cent. there had been created a greater sense of a separate, thriving, and distinctly American, albeit varied, civilization. In New England, Puritan values were modified by the impact of commerce and by the influence of the Enlightenment, while in the South the planter aristocracy developed a lavish and gentlemanly mode of life. Enlightenment ideals also gained influential adherents in the South. Higher education flourished in such institutions as Harvard, William and Mary, and King's College (now Columbia Univ.). The varied accomplishments of Benjamin Franklin epitomized colonial common sense at its most enlightened and productive level. A religious movement of importance emerged in the revivals of the Great Awakening, stimulated by Jonathan Edwards; the movement ultimately led to a strengthening of Methodism. Also inherent in this movement was egalitarian sentiment, which progressed but was not to triumph in the colonial era. One manifestation of egalitarianism was the long-continued conflict between the men of the frontiers and the wealthy Eastern oligarchs who dominated the assemblies, a conflict exemplified in the Regulator Movement. Colonial particularism, still stronger than national feeling, caused the failure of the Albany Congress to achieve permanent union. However, internal strife and disunity remained a less urgent issue than the controversy with Great Britain. After the British and colonial forces had combined to drive the French from Canada and the Great Lakes region in the French and Indian War (1754-60). The colonists felt less need of British protection; but at this very time the British began colonial reorganization in an effort to impose on the colonists the costs of their own defense. Thus was set off the complex chain of events that united colonial sentiment against Great Britain and culminated in the American Revolution (1775-83).

The Revolution resulted in the independence of the Thirteen Colonies: Massachusetts, New Hampshire, Connecticut, Rhode Island, New York, New Jersey, Pennsylvania, Delaware, Maryland, Virginia, North Carolina, South Carolina, and Georgia; their territories were recognized as extending north to Canada and west to the Mississippi River. The revolution also broadened representation in government, advanced the movement for separation of church and state in America, increased opportunities for westward expansion, and brought the abolition of the remnants of feudal land tenure. The view that the Revolution had been fought for local liberty against strong central control reinforced the particularism of the states and was reflected in the weak union established under the Articles of Confederation. Before ratification of the Articles (1781), conflicting claims of states to Western territories had been settled by the cession of Western land rights to the Federal government. The Ordinance of 1787 established a form of government for territories and a method of admitting them as states to the Union. But the national government floundered. It could not obtain commercial treaties or enforce its will in international relations, and, largely because it could not raise adequate revenue and had no executive authority, it was weak domestically. Local economic depressions bred discontent that erupted in Shays' Rebellion, further revealing the weakness of the Federal government. Advocates of strong central government bitterly attacked the Articles of Confederation; supported particularly by professional and propertied groups, they had a profound influence on the Constitution drawn up by the Federal Constitutional Convention of 1787. The Constitution created a national government with ample powers for effective rule, which were limited by "checks and balances" to forestall tyranny or radicalism. Its concept of a strong, orderly Union was popularized by the *Federalist* papers of Alexander Hamilton, James Madison, and John Jay, which played an important part in winning ratification of the Constitution by the separate states.

The first man to be elected President was the hero of the Revolution, George Washington. Washington introduced many government practices and institutions, including the cabinet. Alexander Hamilton, as Washington's Secretary of the Treasury, promulgated a strong state and attempted to advance the economic development of the young country by a neomercantilist program; this included the establishment of a protective tariff, a mint, and the first Bank of the United States as well as assumption of state and private Revolutionary debts. The controversy raised by these policies bred divisions along factional and, ultimately, party lines. Hamilton and his followers, who eventually formed the Federalist Party, favored wide activity by the Federal government under a broad interpretation of the Constitution. Their opponents, who adhered to principles laid down by Thomas Jefferson and who became the Democratic Republican or Democratic Party, favored limited Federal jurisdiction and activities. To an extent these divisions were supported by economic differences, as the Democrats largely spoke for the agrarian point of view and the Federalists represented propertied and mercantile interests. A significant divergence between the factions concerned the French Revolution, which seemed heroically egalitarian to the Democrats and dangerously radical to the Federalists; the latter favored friendship with England, while the Democrats sought bonhomie with France. Political differences reflected divergent visions of American civilization. Although Americans generally were proud of the absence of a privileged aristocracy and feudal encumbrances on their land, they differed on more subtle questions of egalitarianism and progress.

The Federalists were victorious electing John Adams to the presidency in 1796. Federalist conservatism and anti-French sentiment were given vent in the Alien and Sedition Acts of 1798 and in other acts. The so-called Revolution of 1800 swept the Federalists from power and brought Jefferson to the presidency. Jefferson did bring a plainer and more republican style to government, and under him the Alien and Sedition Acts and other Federalist laws were allowed to lapse or were repealed. He moved toward stronger use of Federal powers, however, in negotiating the Louisiana Purchase (1803). In foreign policy Jefferson steered an officially neutral course between Great Britain and France. Under Jefferson's successor, James Madison, the continued British depredations of American shipping, combined with the clamor of American "war hawks" who coveted Canada and Fla., led to the War of 1812. The Treaty of Ghent (1814) settled no specific issues of the war, but did confirm the independent standing of the young republic.

Politically, the period that followed was the so-called era of good feeling. Democrats of all sections had by now adopted a Federalist approach to national development and were temporarily in agreement on a nationalist, expansionist economic policy. These policies were continued under James Monroe. The Monroe Doctrine (1823), proclaimed U.S. opposition to European intervention or colonization in the American hemisphere. Domestically, the strength of the Federal government was increased by the judicial decisions of John Marshall, who had already helped establish the power of the U.S. Supreme Court. By 1820, however, sectional differences were arousing political discord. In the North, merchants, manufacturers, inventors, farmers, and factory hands were busy with commerce, agricultural improvements, and the beginnings of the Industrial Revolution. In the South, cotton was king, and the new states of Ala., La., and Miss. were the pride of the cotton kingdom. The accession of Fla. (1819) further swelled the domain of the South. The American West was expanding as the frontier rapidly advanced. Around the turn of the century settlement of territory west of the Appalachians had given rise to the new states of Ken., Tenn., and Ohio. Settlers continued to move farther west, and the frontier remained a molding force in American life. The Missouri Compromise (1820) temporarily resolved the issue of slavery in new states, but under the presidency of John Quincy Adams sectional differences were aggravated. Particular friction was created by the tariff of 1828, which was highly favorable to Northern manufacturing but a "Tariff of Abominations" to the agrarian South. In the 1820s and 1830s the advance of democracy brought manhood suffrage to many states and virtual direct election of the President, and party nominating conventions replaced the caucus.

An era of political vigor was begun with the election (1828) of Andrew Jackson to the presidency. He provided powerful executive leadership, attuned to popular support, committing himself to a strong foreign policy and to internal improvements for the West. His stand for economic individualism and his attacks on such bastions of the moneyed interests as the Bank of the United States won the approval of the growing middle class. But the South became increasingly dissident, and John C. Calhoun emerged as its chief spokesman with his states' rights doctrine. Opponents of Jackson's policies, including both Northern and Southern conservative propertied interests, amalgamated to form the Whig party.

The West was winning greater attention in American life, and in the 1840s expansion to the Pacific was fervently proclaimed as the "manifest destiny" of the United States. Annexation of the republic of Texas (1845) precipitated the Mexican War; by the Treaty of Guadalupe Hidalgo (1848) the United States acquired two fifths of the territory then belonging to Mexico, including Calif. and the present American Southwest. In 1853 these territories were rounded out by the Gadsden Purchase. Although in the dispute with Great Britain over the Columbia River country, Americans demanded "Fifty-four forty or fight," under President James K. Polk a peaceful if more modest settlement was reached. Thus the United States gained its Pacific Northwest, and "manifest destiny" was virtually fulfilled. In Calif. the discovery of gold in 1848 brought the rush of forty-niners. The westward movement was also stimulated by many other factors, such as the great profits from open-range cattle ranching. Also, the American farmer, with his abundant land, was often profligate in its cultivation, and as the soil depleted he continued to move farther west, settling the virgin territory. Occupation of the West was also sped by European immigrants hungry for land.

By the mid-19th cent. the territorial gains and westward movement of the United States were focusing legislative argument on the extension of slavery. Only with great effort was the Compromise of 1850 achieved, and it was to be the last great compromise between the sections. The new Western states, linked in outlook to the North, had long since caused the South to lose hold of the House of Representatives, and Southern parity in the Senate was threatened by the prospective addition of more free states than slaveholding ones. The passage of the Kansas-Nebraska Bill (1854), which repealed the Missouri Compromise, led to violence between factions in "bleeding Kansas" and spurred the founding of the new Republican party. Southerners, unable to accept the end of slavery and fearful of slave insurrection (especially after the revolt led by Nat Turner in 1831), felt threatened by the abolitionists, who regarded themselves as leaders in a moral crusade. As the conflict became more embittered it rent the older parties. The Whig party was shattered, and its Northern wing was largely absorbed in the new antislavery Republican party. The climax came in 1860 when the Republican Abraham Lincoln defeated three opponents to win the presidency. Southern leaders, feeling there was no possibility of fair treatment under a Republican administration, resorted to secession from the Union and formed the Confederacy. The attempts of the seceding states to take over Federal property within their borders (notably Fort Sumter in Charleston, S.C.) precipitated the Civil War (1861-65), which resulted in a complete victory for the North and the end of all slavery. Military rule in parts of the South continued through the administrations of Ulysses S. Grant, which were also notable for their outrageous corruption. A result of the disputed election of 1876, in which the decision was given to Rutherford B. Hayes over Samuel J. Tilden, was the end of Reconstruction and the re-entry of the South into national politics.

The remainder of the 19th cent. was marked by railroad building and the disappearance of the American frontier. Great mineral wealth was discovered and exploited, and important technological innovations sped industrialization. Thus developed an economy based on steel, oil, railroads, and machines. Mammoth corporations such as the Standard Oil trust were formed, and "captains of industry" like John D. Rockefeller and financiers like J. P. Morgan controlled huge resources. Into the "land of promise" poured new waves of immigrants; some acquired dazzling riches, but many others suffered in a competitive and unregulated economic age. Behind the façade of the "Gilded Age," with its aura of peace and general prosperity, a whole range of new problems was created, forcing varied groups to promulgate new solutions.

Labor began to combine against grueling factory conditions, but the opposition of business to unions was frequently overpowering, and the bulk of labor remained unorganized. Belief in laissez faire and the influence of big business in both national parties, especially in the Republican party, delayed any widespread reform. Under President Benjamin Harrison the Sherman Antitrust Act was passed (1890).

By the 1890s a new wave of expansionist sentiment was affecting U.S. foreign policy. With the purchase of Alaska (1867) and the rapid settlement of the last Western territory, Oklahoma, American capital and attention were directed toward the Pacific and the Caribbean. The United States annexed the Hawaiian Islands in 1898. In that year expansionist energy found release in the Spanish-American War, which resulted in U.S. acquisition of Puerto Rico, the Philippine Islands, and Guam and in a U.S. quasi-protectorate over Cuba. American ownership of the Philippines involved military subjugation of the people; the Philippine Insurrection (1899–1901) cost more American lives and dollars than the Spanish-American War. Widening its horizons, the United States formulated the Open Door policy (1900), which expressed its interest in China. Established as a world power with interests in two oceans, the United States intervened in the Panama revolution (1903) to facilitate construction of the Panama Canal; this was but one of its many involvements in Latin American affairs under Theodore Roosevelt and later Presidents.

By the time of Roosevelt's administration (1901–9), the progressive reform movement had taken definite shape in the country. In its politics as shaped by R. M. La Follette and others, progressivism adopted many populist planks but promoted them from a more urban and forward-looking viewpoint. Roosevelt made some attacks on trusts, and he promoted regulation of interstate commerce as well as passage of the Pure Food and Drug Act (1906). In 1912 the presidency was won by the Democratic reform candidate, Woodrow Wilson. Wilson's "New Freedom" brought many progressive ideas to legislative fruition. The Federal Reserve System and the Federal Trade Commission were established, and the Adamson Act and the Clayton Antitrust Act were passed. Perhaps more than on the national level, progressivism triumphed in the states in legislation beneficial to labor, in the furthering of education, and in the democratization of electoral procedures.

When World War I burst upon Europe, Wilson made efforts to keep the United States neutral; in 1916 he was re-elected on a peace platform. However, American sympathies and interests were actively with the Allies, and on Apr. 6, 1917, the United States entered the war and provided crucial manpower and supplies for the Allied victory. After the war, isolationist sentiment against participation in the League of Nations, an integral part of the Treaty of Versailles (1919), caused the U.S. Senate to reject membership.

The country voted for a return to "normalcy" when it elected Warren G. Harding President in 1920; but the ensuing period was a time of rapid change. The nation became increasingly urban, and everyday life was transformed as the "consumer revolution" brought the spreading use of automobiles, telephones, radios, and other appliances. The pace of living quickened, and mores became less restrained, while fortunes were rapidly accumulated on the skyrocketing stock market, in real estate speculation, and elsewhere. But agriculture was not prosperous, and industry and finance became dangerously overextended. In 1929 there began the Great Depression, which reached worldwide proportions.

In the 1932 election President Herbert Hoover was overwhelmingly defeated by the Democrat Franklin D. Roosevelt. The new President immediately instituted his New Deal with vigorous measures. Congress, called into special session, enacted a succession of laws, some of them to meet the economic crisis with relief measures, others to put into operation long-range social and economic reforms. The program created a vast machinery by which the state could promote social welfare, but economic recovery from the depression proved elusive.

The ominous situation abroad was chiefly responsible for Roosevelt's unprecedented election to a third term in 1940. As Axis aggression led to the outbreak of the European war in Sept., 1939, the United States still strove to stay out of it, despite increasing sympathy for the Allies. In Mar., 1941, Lend-Lease aid was extended to the British and, in November, to the Russians. The threat of war had already caused the adoption of selective service to build the armed strength of the nation. On Dec. 7, 1941, Japanese bombs fell on Pearl Harbor. The United States promptly declared war, and four days later Germany and Italy declared war on the United States. The country efficiently mobilized its vast resources, transforming factories to war plants and building a mighty military force. After Roosevelt's sudden death in Apr., 1945, Harry S. Truman became President. A month later the European war ended when Germany surrendered. Before the defeat of Japan, the United States developed and used the atomic bomb, which precipitated the Japanese surrender, on Aug. 14, 1945.

Peacetime readjustment was successfully effected. The government's G.I. Bill enabled many former servicemen to obtain free schooling, and millions of other veterans were absorbed by the economy, which boomed in fulfilling the demands for long-unobtainable consumer goods. The shortening of the postwar factory work week and the proportionate reduction of wages precipitated a rash of strikes, causing the government to pass the Taft-Hartley Labor Act (1947). Some inflation occurred by 1947 as wartime economic controls were abandoned. Congress passed a host of President Truman's measures relating to minimum wages, public housing, farm surpluses, and credit regulation; thus was instituted acceptance of comprehensive government intervention in times of prosperity. The nation's support of Truman's policies was signified when it returned him to the presidency in 1948 in an upset victory over Thomas E. Dewey.

The most striking postwar development was America's new peacetime involvement in international affairs. The United States was a founding member of the United Nations (1945). Relations between the United States and the Soviet Union worsened during the late 1940s. The Truman Doctrine attempted to thwart Soviet expansion in Europe; massive loans, culminating in the Marshall Plan, were vital in reviving European economies and thus in diminishing the appeal of Communism. As the Cold War intensified, the United States took steps (1948) to nullify the Soviet blockade of Berlin and played the leading role in forming (1949) a new alliance of Western nations, the North Atlantic Treaty Organization (NATO). In the Korean War (1950–53), U.S. forces played the chief part in combating the North Korean and Chinese attack on South Korea. At home, the fear of domestic Communism and subversion almost became a national obsession. Security measures and loyalty checks in the government and elsewhere were tightened, alleged Communists were prosecuted and employees in varied fields were dismissed for questionable political affiliations. The most notorious prosecutor of alleged Communists was Senator Joseph McCarthy, whose extreme methods were later recognized as threats to freedom of speech and democratic principles.

By the mid-1950s America was in the midst of a great industrial boom, and stock prices were skyrocketing. In foreign affairs the Eisenhower administration was internationalist in outlook, although it sternly opposed Communist power and threatened "massive retaliation" for Communist aggression. Some antagonism came from the neutral nations of Asia and Africa, partly because of the U.S. association with former colonial powers and partly because U.S. foreign aid more often than not had the effect of strengthening ruling oligarchies abroad. In the race for technological superiority the United States exploded (1952) the first hydrogen bomb, but was second to the USSR in launching (Jan. 31, 1958) an artificial satellite and in testing an intercontinental guided missile. However, spurred by Soviet advances, the United States made rapid progress in space exploration and missile research. In the crucial domestic issue of racial integration, the U.S. Supreme Court in a series of decisions supported the efforts of black citizens to achieve full civil rights. In contrast to the increased liberalism of the Supreme Court was the appearance of such ultraconservative political groups as the John Birch Society. In 1959 Alaska and Hawaii became the 49th and 50th states of the Union. Despite hopes for "peaceful coexistence," negotiations with the USSR for nuclear disarmament failed to achieve accord, and Berlin remained a serious source of conflict. In 1961 Dwight D. Eisenhower, the oldest President ever to hold office, gave way to the youngest President ever elected, John F. Kennedy, who defeated the Republican candidate, Richard M. Nixon. President Kennedy called for "new frontiers" of American endeavor, but had difficulty securing Congressional support for his domestic programs (integration, tax reform, medical benefits for the aged). Kennedy's foreign policy combined such humanitarian innovations as the Peace Corps and the Alliance for Progress with the traditional opposition to Communist aggrandizement.

After breaking relations with Cuba, which, under Fidel Castro, had clearly moved within the Communist orbit, the United States supported (1961) an ill-fated invasion of Cuba by anti-Castro forces. In 1962, in reaction to the presence of Soviet missiles in Cuba, the United States risked war by blockading Soviet military shipments to Cuba and demanded the dismantling of Soviet bases there. The tensions of the Cold War eased when in 1963 the United States and the Soviet Union reached an accord on a limited ban of nuclear testing. On Nov. 22, 1963, President John F. Kennedy was assassinated while riding in a motorcade in Dallas, Texas. His successor, Lyndon B. Johnson, furthered progress toward racial equality with a momentous Civil Rights Act (1964), a Voting Rights Act (1965), and the 24th Amendment to the Constitution, which abolished the poll tax. But, international problems dominated Johnson's second term, and Johnson himself pursued an aggressive course, escalating American participation in the Vietnam War. He increased the number of U.S. troops in Vietnam from 16,000 to more than 500,000, provoking increasing opposition at home, manifested in antiwar demonstrations.

Serious race riots that erupted in cities across the nation, most devastatingly in the Watts district of Los Angeles (1965) and in Detroit and Newark (1967), and by various racial and political assassinations, notably those of Martin Luther King, Jr., and Senator Robert F. Kennedy (1968). Other manifestations of social upheaval were the increase of drug use, especially among youths, and the rising rate of crime, most noticeable in the cities. Opposition to American involvement in the Vietnam War so eroded President Johnson's popularity that he chose not to run again for President in 1968; his position as leader of the Democratic party had been seriously challenged by Senator Eugene McCarthy, who ran as a peace candidate in the primary elections. Antiwar forces in the Democratic party received a setback with the assassination of Senator Kennedy, also a peace candidate, and the way was opened for the nomination of Vice President Hubert H. Humphrey, a supporter of Johnson's policies, as the Democratic candidate for President. Violence broke out during the Democratic national convention in Chicago when police and national guardsmen battled some 3,000 demonstrators in what a national investigating committee later characterized as "a police riot."

The Republican candidate, Richard M. Nixon, ran on a platform promising an end to the Vietnam War and stressing the need for domestic "law and order"; he won a narrow victory, receiving 43.4% of the popular vote to Humphrey's 42.7%. A third-party candidate, Gov. George C. Wallace of Alabama, carried five Southern states. The Congress remained Democratic. Pronouncing the "Nixon doctrine"—that thenceforth other countries would have to carry more of the burden of fighting Communist domination, albeit with substantial American economic aid—Nixon began a slow withdrawal of American troops from Vietnam. Criticism that he was not moving fast enough in ending the war increased after the revelation (1969) of a long hushed-up massacre of Vietnamese civilians by American soldiers in the hamlet of My Lai and the publication (1971) of excerpts from the Pentagon Papers, a study of U.S. involvement in Vietnam that suggested serious policy errors as well as deception on the part of the government.

Massive antiwar demonstrations continued, and when Nixon in the spring of 1970 ordered U.S. troops into neutral Cambodia to destroy Communist bases and supply routes there, a wave of demonstrations, some of them violent, swept American campuses. Four students were killed by national guardsmen at Kent State Univ. in Ohio, and 448 colleges and universities temporarily closed down. Antiwar activity declined, however, when American troops were removed from Cambodia after 60 days. The institution of draft reform, the continued withdrawal of U.S. soldiers from Vietnam, and a sharp decrease in U.S. casualties all contributed toward dampening antiwar sentiment and removing the war as an issue from public debate. Racial flare-ups abated after the 1960s (although the issue of the bussing of children to achieve integration continued to arouse controversy); a measure of domestic unrest was still seen in sporadic terrorist bombings (a wing of the Capitol was one target), serious prison riots, and increased revolutionary activity by militant blacks. The growing movement of woman demanding social, economic, and political equality with men also reflected the changing times. A dramatic milestone in the country's space program was reached in July, 1969, with the landing of two men on the moon, the first of several such manned flights. Significant unmanned probes of several of the planets followed, and in 1973 the first space station was orbited.

In domestic policy Nixon appeared to favor an end to the many reforms of the '60s. He was accused by civil rights proponents of wooing Southern support by seeking delays in the implementation of school integration. Such actions by his administration were overruled by the Supreme Court. Nixon twice attempted to appoint conservative Southern judges to the U.S. Supreme Court and was twice frustrated by the Senate, which rejected both nominations. In an attempt to control the spiraling inflation inherited from the previous administration, Nixon concentrated on reducing federal spending. He vetoed numerous appropriations bills passed by Congress, especially those in the social service and public works areas, although he continued to stress defense measures, such as the establishment of an antiballistic missiles (ABM) system, and foreign aid. Important legislative accomplishments included the Postal Reorganization Act (1970), which converted the Post Office into an independent government agency, and a "revenue-sharing" bill (1972), which provided direct Federal payments to the states for local needs, a significant step toward the decentralization of government. Federal budget cuts contributed to a general economic slowdown but failed to halt inflation, so that the country experienced the unprecedented misfortune of both rising prices and rising unemployment; the steady drain of gold reserves after almost three decades of enormous foreign aid programs, a new balance-of-trade deficit, and the instability of the dollar in the international market also affected the economy.

In Aug., 1971, Nixon resorted to the freezing of prices, wages, and rents in an effort to control inflation. Nixon made a dramatic visit to the People's Republic of China in Feb., 1972, ending more than 20 years of hostility between the two countries. A trip to Moscow followed in the spring, culminating in the signing of numerous agreements between the United States and the Soviet Union. Nixon was re-elected (Nov., 1972) in a landslide, but his second term, however, was marred and finally destroyed by the Watergate affair, which began when five men employed by Nixon's re-election committee were arested after breaking into the Democratic party's national headquarters in Washington, D.C. A massive coverup of the burglary involving the President himself was gradually revealed in investigations, special Senate hearings, and the publication of taped White House conversations. Under the threat of impeachment, Nixon resigned on Aug. 9, 1974. His successor, Gerald R. Ford, weathered a severe economic recession but was faced with persistently high unemployment and inflation rates.

Jimmy Carter, the Democratic candidate, was elected president in 1976. The spiraling energy crisis aggravated inflation, which continued to be the country's most severe problem. Carter's greatest achievements were in foreign affairs, in which he officially recognized the People's Republic of China, signed a treaty that gave control of the Panama Canal to Panama, and negotiated the first peace treaty between Israel and Egypt, ending thirty years of war between those two countries. Discussions with the Soviets for a second treaty regulating nuclear armaments continued.

Government. The government of the United States is that of a federal republic. There is a division of powers between the Federal government and the state governments. The Federal government consists of three branches: the executive, the legislative, and the judicial. The executive power is vested in the President and, in the event of his incapacity, the Vice President. The executive conducts the administrative business of the nation with the aid of a cabinet. The Congress, the legislative branch, is bicameral and consists of the Senate and the House of Representatives. The judicial branch is formed by the Federal courts and headed by the U.S. Supreme Court. The members of the Congress are elected by universal suffrage as are the members of the electoral college, which formally chooses the President and the Vice President.

PRESIDENTS OF THE UNITED STATES

(with Vice Presidents, political parties, and dates in office)

George Washington, 1789-97
John Adams

John Adams [Federalist] 1797-1801
Thomas Jefferson

Thomas Jefferson [Democratic-Republican] 1801-9
Aaron Burr, 1801-5
George Clinton, 1805-9

James Madison [Democratic-Republican] 1809-17
George Clinton, 1809-12
(no Vice President, April, 1812-March, 1813)
Elbridge Gerry, 1813-14
(no Vice President, Nov., 1814-March, 1817)

PRESIDENTS OF THE UNITED STATES *(with Vice Presidents, political parties, and dates in office)*

James Monroe [Democratic-Republican] 1817-25
Daniel D. Tompkins

John Quincy Adams [Democratic-Republican] 1825-29
John C. Calhoun

Andrew Jackson [Democratic] 1829-37
John C. Calhoun, 1829-32
(no Vice President, Dec., 1832-March, 1833)
Martin Van Buren, 1833-37

Martin Van Buren [Democratic] 1837-41
Richard M. Johnson

William Henry Harrison [Whig] 1841
John Tyler

John Tyler [Whig] 1841-45
(no Vice President)

James Knox Polk [Democratic] 1845-49
George M. Dallas

Zachary Taylor [Whig] 1849-50
Millard Fillmore

Millard Fillmore [Whig] 1850-53
(no Vice President)

Franklin Pierce [Democratic] 1853-57
William R. King, 1853
(no Vice President, April, 1853-March, 1857)

James Buchanan [Democratic] 1857-61
John C. Breckinridge

Abraham Lincoln [Republican] 1861-65
Hannibal Hamlin, 1861-65
Andrew Johnson, 1865

Andrew Johnson [Democratic/National Union] 1865-69
(no Vice President)

Ulysses Simpson Grant [Republican] 1869-77
Schuyler Colfax, 1869-73
Henry Wilson, 1873-75
(no Vice President, Nov., 1875-March, 1877)

Rutherford Birchard Hayes [Republican] 1877-81
William A. Wheeler

James Abram Garfield [Republican] 1881
Chester A. Arthur

Chester Alan Arthur [Republican] 1881-85
(no Vice President)

Grover Cleveland [Democratic] 1885-89
Thomas A. Hendricks, 1885
(no Vice President, Nov., 1885-March, 1889)

Benjamin Harrison [Republican] 1889-93
Levi P. Morton

Grover Cleveland [Democratic] 1893-97
Adlai E. Stevenson

William McKinley [Republican] 1897-1901
Garret A. Hobart, 1897-99
(no Vice President, Nov., 1899-March, 1901)
Theodore Roosevelt, 1901

Theodore Roosevelt [Republican] 1901-9
(no Vice President, Sept., 1901-March, 1905)
Charles W. Fairbanks, 1905-9

William Howard Taft [Republican] 1909-13
James S. Sherman, 1909-12
(no Vice President, Oct., 1912-March, 1913)

Woodrow Wilson [Democratic] 1913-21
Thomas R. Marshall

Warren Gamaliel Harding [Republican] 1921-23
Calvin Coolidge

Calvin Coolidge [Republican] 1923-29
(no Vice President, 1923-25)
Charles G. Dawes, 1925-29

Herbert Clark Hoover [Republican] 1929-33
Charles Curtis

Franklin Delano Roosevelt [Democratic] 1933-45
John N. Garner, 1933-41
Henry A. Wallace, 1941-45
Harry S. Truman, 1945

Harry S. Truman [Democratic] 1945-53
(no Vice President, 1945-49)
Alben W. Barkley, 1949-53

Dwight David Eisenhower [Republican] 1953-61
Richard M. Nixon

John Fitzgerald Kennedy [Democratic] 1961-63
Lyndon B. Johnson

Lyndon Baines Johnson [Democratic] 1963-69
(no Vice President, 1963-65)
Hubert H. Humphrey, 1965-69

Richard Milhous Nixon [Republican] 1969-74
Spiro T. Agnew, 1969-73
(no Vice President, Oct. 10, 1973-Dec. 6, 1973)
Gerald R. Ford, 1973-74

Gerald Rudolph Ford [Republican] 1974-77
(no Vice President, Aug. 9, 1974-Dec. 19, 1974)
Nelson A. Rockefeller, 1974-77

James Earl Carter [Democratic] 1977-
Walter F. Mondale, 1977-

U·ni·ver·si·ty City (yōō′nə-vûr′sĭ-tē), city (1978 est. pop. 45,000), St. Louis co., E Mo., a residential suburb of St. Louis; inc. 1906.

University Heights, city (1978 est. pop. 18,100), Cuyahoga co., NE Ohio, a residential suburb of Cleveland; inc. 1925. It is the seat of John Carroll Univ.

University Park, city (1978 est. pop. 24,200), Dallas co., N Texas, surrounded by Dallas on three sides; inc. 1924. A residential suburb, the city is the seat of Southern Methodist Univ.

Un·ter·wal·den (o͞on'tər-väl'dən), canton, central Switzerland, one of the Four Forest Cantons. A mountainous, forested, and chiefly pastoral region, Unterwalden is divided into the half cantons of Obwalden (1974 est. pop. 26,000), 190 sq mi (492 sq km), in the west, with its capital at Sarnen, and Nidwalden (1974 est. pop. 26,300), 106 sq mi (274.5 sq km), in the east, with its capital at Stans. Dairying and woodworking are the main occupations of Obwalden, while Nidwalden has orchards, cement works, and glassworks.

Up·land (ŭp'lənd), city (1978 est. pop. 43,000), San Bernardino co., S Calif., at the foot of the San Gabriel Mts.; inc. 1906. Citrus fruits and grapes are packed and processed, and paint, orchard heaters, auto parts, and feed products are made.

U·po·lu (o͞o-pō'lo͞o), volcanic island, Western Samoa, S Pacific, the most populous of the Samoan islands. Upolu's land area is c.430 sq mi (1,115 sq km). The island is well watered, and its fertile soil yields cacao, rubber, bananas, and coconuts. Robert Louis Stevenson spent his last years on Upolu, residing at his home, Vailima.

Up·per Ar·ling·ton (ŭp'ər är'lĭng-tən), city (1978 est. pop. 34,500), Franklin co., central Ohio, a residential suburb of Columbus; inc. 1918.

Upper Aus·tri·a (ôs'trē-ə), province (1971 pop. 1,223,444), 4,625 sq mi (11,979 sq km), NW Austria; capital Linz. Bordering on West Germany in the west and Czechoslovakia in the north, the province is predominantly hilly and forested. Agriculture is the main occupation. The area of Upper Austria was included in the Roman province of Noricum. In 1156 it was made a duchy by Frederick I and given to the Babenberg dukes of Austria.

Upper Hutt (hŭt), city (1971 pop. 20,001), S North Island, New Zealand, on the Hutt River. It is largely residential but has some light industries.

Upper Marl·bor·o (märl'bûr'ō), town (1970 pop. 646), seat of Prince Georges co. (since 1732), central Md., ESE of Washington. It is a tobacco market. Northampton estate, with gardens laid out in 1788, is nearby.

Upper Sad·dle River (săd'l), borough (1970 pop. 7,949), Bergen co., NE N.J., near the N.Y. border NNE of Paterson; inc. 1894.

Upper San·dus·ky (săn-dŭs'kē), village (1970 pop. 5,645), seat of Wyandot co., N central Ohio, on the Sandusky River NNW of Marion, in a farm area; laid out 1843, inc. 1848. The first Methodist Episcopal mission in Ohio was built here c.1823.

Upper Volta (vŏl'tə), republic (1977 est. pop. 6,230,000), 105,869 sq mi (274,200 sq km), W Africa, bordering on Mali in the W and N, on Niger in the NE, on Benin in the SE, and on Togo, Ghana, and Ivory Coast in the S. Ouagadougou is the capital. The country is made up mainly of vast monotonous plains and of low hills that rise to c.2,300 ft (700 m) in the southwest. Precipitation is low and the soil is of poor quality. Rainfall is heaviest in the southwest, which is covered largely with savanna; the rest of the country is semidesert. Upper Volta has several rivers, none of which are navigable. In the southwest is the Komoé (Comoé) River, which flows through Ivory Coast to the Gulf of Guinea; in the center are the Black, Red, and White Volta rivers, which join in Ghana to form the Volta River; and in the northeast are several small tributaries of the Niger. The population of Upper Volta is made up of black Africans. The main ethnic groups are the Mossi (about 48% of the total population), Lobi (about 7%), Bobi (about 7%), and Gurunsi (about 6%), all of whose members speak a Voltaic language; Fulani (about 11%); Mande (about 7%); and Senufo (about 6%). French is the country's official language. Most of the inhabitants follow traditional religious beliefs; there are also about 1 million Moslems and approximately 220,000 Roman Catholics.

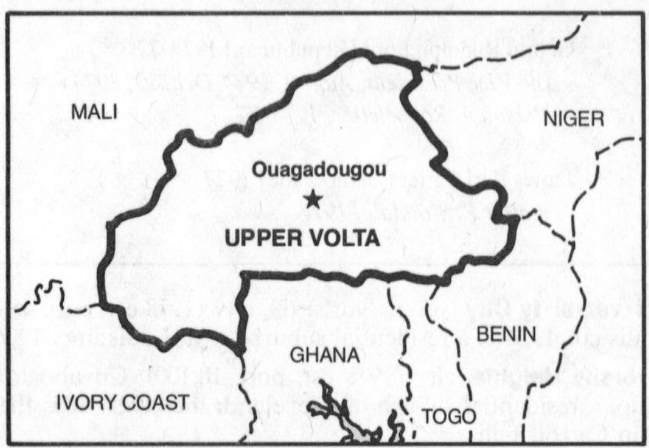

Economy. Upper Volta is a poor agricultural country, with the great majority of its workers engaged in subsistence farming. Less than 10% of the country's land area is cultivable. The principal agricultural commodities are sorghum, millet, maize, groundnuts, karite nuts, rice, cotton, yams, and sesame. Large numbers of cattle, sheep, and goats are raised. The country's manufactures, limited largely to basic consumer goods, include processed foods, beverages, ginned cotton, textiles, bicycles, construction materials, leather, cigarettes, and matches. Upper Volta has a small mining industry that produces manganese and limestone; there are also small deposits of copper ore, bauxite, and uranium. The country has a good road network; a railroad runs from Ouagadougou to the seaport of Abidjan, Ivory Coast, via Bobo-Dioulasso. The annual cost of Upper Volta's imports is usually much higher than its earnings from exports. The principal imports are foodstuffs, machinery, motor vehicles, textiles and clothing, and metals; the leading exports are live animals, cotton, groundnuts and groundnut oil, and hides and skins.

History. Neolithic remains have been found in northern Upper Volta. By about 1100 A. D. the principal inhabitants of the western part of the country were the Bobo, Lobi, and Gurunsi. Around 1,400 invaders on horseback from present-day Ghana established the Mossi states of Ouagadougou, Yatenga, Tengkodogo, and the state of Gourma. By using religion and a complex administrative system the conquerors created powerful states that endured for more than 500 years. Ouagadougou was headed by the Morho Naba. The Mossi states had strong armies and were able to repel most attacks by the Mali and Songhai empires during the period from the 14th to 16th cent. Near the end of the 19th cent. scramble for African territory among the European powers, France gained control over the region of Upper Volta. An Anglo-French agreement in 1898 established the boundary with the Gold Coast (now Ghana). The region of Upper Volta was administered as part of Soudan (then called Upper Senegal-Niger and now mostly part of Mali) until 1919, when it was made a separate protectorate. In 1947 Upper Volta was re-established as a separate territory within the French Union. By the mid-1950s the Voltaic Democratic Union (UDV) was the leading political party. On Aug. 5, 1960, Upper Volta achieved full independence. The constitution of 1960 established a strong presidential government, and Maurice Yaméogo became the first president. He managed to reduce the traditional power of the Mossi states, but his authority was weakened by ethnic conflicts and by a weak economy. In late 1965 Yaméogo was overwhelmingly re-elected president, but in Jan., 1966, he was ousted in a bloodless coup d'état headed by Lt. Col. Sangoulé Lamizana, who became head of state. Lamizana dissolved the national assembly and temporarily prohibited political activity. In 1970 a new constitution was approved in a national referendum. The chief executive officer was the president and legislative power was vested in a 57-member unicameral national assembly. In elections in Dec., Gérard Kango Ouedraogo became prime minister. However, in Feb., 1974, the army, headed by Lamizana, again intervened in the political process, dissolving the national assembly, ousting Ouedraogo, and suspending the 1970 constitution. During the 1960s and early 1970s Upper Volta received much financial aid from France. The country (especially the north) was severely affected by the long-term drought that began in the late 1960s.

Upp·sa·la (ŭp'sə-lə, -sä'lə), city (1972 est. pop. 132,560), capital of Uppsala co., E Sweden, on the Fyrisån River. Manufactures include metal goods, clothing, footwear, processed food, and printed materials. The city developed near Gamla Uppsala, now a small village, which became the pagan capital of Sweden in the 6th cent. An archiepiscopal see was established at present-day Uppsala in 1270, and the cathedral of Uppsala (13th cent.), the finest Gothic church in Sweden, became the coronation place of Swedish kings and is the burial place of Gustavus I, the botanist Linnaeus, and the scientist and religious teacher Swedenborg. The Univ. of Uppsala, founded in 1477, is the oldest university of northern Europe and has ranked among the world's great universities since its reorganization in 1593. The university's library contains more than 1 million volumes and about 20,000 manuscripts.

Up·shur (ŭp'shər). **1.** County (1970 pop. 20,976), 589 sq mi (1,525.5

sq km), NE Texas, partly bounded on the S by the Sabine River and drained by Cypress and Little Cypress bayous; formed 1846; co. seat Gilmer. This partly wooded area also has oil fields and deposits of natural gas, clay, lignite, and iron ore. Farming (sweet potatoes, truck crops, fruit, peanuts, and cotton), livestock raising, and some dairying are done. **2.** County (1970 pop. 19,092), 352 sq mi (911.7 sq km), NE central W.Va., on the Allegheny Plateau, drained by the Buckhannon River; formed 1851; co. seat Buckhannon. The county has bituminous-coal mines, natural-gas wells, and timber. Its agriculture includes livestock, dairy products, fruit, and grain. There is some industry in the county.

Up·son (ŭp'sən), county (1970 pop. 23,505), 333 sq mi (862.5 sq km), W central Ga., bounded on the W and S by the Flint River; formed 1824; co. seat Thomaston. It has piedmont agriculture (peaches, cotton, corn, truck crops, and livestock) and textile manufacturing.

Up·ton (ŭp'tən), county (1970 pop. 4,697), 1,312 sq mi (3,398.1 sq km), W Texas, with high prairies in the N and E and Castle and King mts. in the SW; formed 1887; co. seat Rankin. This ranching area (chiefly sheep, with some goats and cattle) also has oil and natural-gas wells and deposits of clay, potash, and salt.

Ur (ûr, o͝or), ancient city of Sumer, S Mesopotamia. The city is also known as Ur of the Chaldees. It was an important center of Sumerian culture and is identified in the Bible as the home of Abraham. The site was discovered in the 19th cent., but it was not until the excavations of C. Leonard Woolley in the 1920s and 1930s that a partial account of its history could be constructed. Remains found at the site seem to indicate that Ur existed as far back as the late Al' Ubaid period and that the city was an important commercial center even before the first dynasty was established (c.2500 B.C.). Among the most important remains of the first dynasty, which has revealed a luxurious material culture, are the royal cemetery, where the standard of Ur was found, and the Temple of Ninhursag at Ubaid, bearing the inscriptions of the kings of the first dynasty.

Ur was captured c.2340 by Sargon, and this era, called the Akkadian period, marks an important step in the blending of Sumerian and Semitic cultures. After this dynasty came a long period of which practically nothing is known except that a second dynasty rose and fell. The third dynasty was established c.2060 B.C. under King Ur-Nammu, who built the great ziggurat that has stood, although crumbled and covered with sand, throughout the centuries. An inscription in the Museum of the Ancient Orient in Istanbul was identified (1952) as a fragment of the code of Ur-Nammu. It predates the code of Hammurabi by 300 years and is the oldest known law code yet discovered. The third dynasty of Ur fell (c.1950 B.C.) to the Elamites and later to Babylon. The city was destroyed and rebuilt throughout the years by various kinds and conquerors, including Nebuchadnezzar and Nabonidus in the 6th cent. About the middle of the 6th cent., Ur went into a decline from which it never recovered.

U·ral (yo͝or'əl), river, c.1,580 mi (2,540 km) long, rising in the S Urals, SE Russian Republic, USSR, and flowing S through NW Kazakhstan and into the Caspian Sea at Guryev. The Ural River is a transport route to the north for oil, fish, and lumber; grain and cattle are generally shipped south on the river. It is a source of water for the towns and agricultural areas in the steppe area of western Kazakhstan.

U·rals or **Ural Mountains** (yo͝or'əlz), E European USSR, forming, together with the Ural River, the traditional boundary between Europe and Asia and separating the Russian plain from the W Siberian lowlands. The Urals extend c.1,500 mi (2,400 km) north and south from the Arctic tundra to the deserts north of the Caspian Sea. The polar section (north of lat. 64° N) is covered by tundra. The northern section (between lat. 64° N and lat. 61° N), a rocky treeless range, has the highest peaks, Naroda and Telpos-Iz. The central Urals (between lat. 61° N and lat. 55° N) are also known as the Ore Urals and have many low passes. The southern section (between lat. 55° N and lat. 51° N), known as the Mugodzhar Hills, consists of several high, parallel ridges that rise to 5,377 ft (1,639 m) in the Yaman-Tau. The southern Urals are drained by the Ural River into the Caspian Sea. The waterways in the west are the Kama and Belaya rivers, tributaries of the Volga, and, in the east, the Ob-Irtysh drainage system.

Except in the polar and northern sections, the mountains are densely forested, and lumbering is an important industry. The great mineral resources of the USSR are in the Urals. Iron ore is mined in the south, and there are rich deposits of coal, copper, manganese, gold, aluminum, and potash. Oil fields and refineries along the Kama and Belaya rivers in the western Urals produce much of the USSR's oil. Emeralds, chrysoberyl, topaz, and amethyst are mined, as are deposits of bauxite, asbestos, zinc, lead, silver, platinum, nickel, chrome, and tungsten. The Urals industrial area is in the central and southern Urals and the adjacent lowlands. The population consists primarily of Russians, with some Bashkirs, Tatars, Udmurts, and Komi-Permyaks.

U·ralsk (yo͞o-rälsk'), city (1973 est. pop. 144,000), capital of Uralsk oblast, Kazakhstan, W Central Asian USSR, on the Ural River. Among its industries are the repair of agricultural equipment, food processing, tanning, and the manufacture of engine and motor parts. Uralsk was founded in 1622 by Cossacks, who fought against the Bolsheviks in 1918–19.

U·ra·ni·um City (yo͞o-rā'nē-əm), town (1971 pop. 1,867), NW Sask., Canada, on Lake Athabasca near the Northwest Territories line. It is the center of a large uranium-mining area developed in the 1950s.

U·ra·wa (o͞o-rä'wä), city (1970 pop. 269,397), capital of Saitama prefecture, central Honshu, Japan. It is a commercial center and has a university.

Ur·ban·a (ûr-băn'ə). **1.** City (1978 est. pop. 35,600), seat of Champaign co., E central Ill., adjoining Champaign; inc. 1833. Urbana is a trade, medical, and educational center in a rich farm area. Electronic systems are manufactured. The city is best known as a seat of the Univ. of Illinois at Urbana-Champaign. Chanute Air Force Base is near the city. **2.** City (1975 est. pop. 11,000), seat of Champaign co., W central Ohio, in a rich farm and livestock area; inc. 1814. It has hatcheries and plants that make airplane lights, plastics, paper, fans, and welded products. During the War of 1812 Urbana was an outfitting point for the Great Lakes area. The Ohio Caverns are nearby, and several lakes are in the area.

Ur·ban·dale (ûr'bən-dāl'), city (1975 est. pop. 17,200), Polk co., central Iowa, a residential suburb of Des Moines; inc. 1917. It has light industry and warehousing.

Ur·bi·no (o͞or-bē'nō), town (1972 est. pop. 12,900), in the Marche, central Italy. It is an agricultural and tourist center, located on the site of a former Roman community. The town flourished under the Montefeltro family (12th–16th cent.) and then under the Della Rovere family (1508–1631), before coming under the papacy. The court of Federigo da Montefeltro, 2nd duke of Urbino (1444–82), was a great artistic center during the Renaissance. Urbino was particularly noted for its school of painting (15th–17th cent.) and for the manufacture of majolica ware. The Palazzo Ducale (1444–82) today houses a major museum, with paintings by Raphael (born in the town), Titian, Piero della Francesca, and others.

Ur·fa (o͞or-fä'), city (1970 pop. 100,654), capital of Urfa prov., SE Turkey. It is the trade center for a productive agricultural region. The city was called Edessa until its incorporation in the Ottoman Empire in 1637.

Ur·gench (o͞or-gĕnch'), ancient city of central Asia, on the site of present-day Kunya-Urgench, in the Turkmen SSR. A major trade and craft center from the 10th to 13th cent., Urgench became the capital of the khanate of Khorezm in the 12th cent. The city was destroyed by the Mongols in the early 13th cent., partially rebuilt, and finally abandoned in the 16th cent.

Urgench, city (1973 est. pop. 84,000), capital of Khorezm oblast, Central Asian USSR, in Uzbekistan, on the Amu Darya River, in the Khiva oasis. It is a large port and has cotton and food-processing industries.

U·ri (o͞or'ē), canton (1974 est. pop. 33,500), 415 sq mi (1,075 sq km), central Switzerland, one of the Four Forest Cantons; capital Altdorf. The region was incorporated into the Holy Roman imperial bailiwick of Zürich after 1098. The scene of the events of the William Tell legend, Uri in 1291 formed with Schwyz and Unterwalden the league that became the nucleus of Switzerland.

U·rua·pan (o͞or-wä'pän'), city (1970 pop. 82,677), Michoacán state, W Mexico. An attractive city with fine gardens and parks, it is in a semitropical, mountainous agricultural region. The city, founded in 1540, is the center of the manufacture of gourd lacquerware by the Tarascan Indians. Local craftsmen also produce glassware, woodwork, and embroideries.

U·ru·bam·ba (o͞or'ə-bäm'bə), river, c.450 mi (725 km) long, rising in the Andes Mts., S Peru, and flowing N to join the Apurímac River and form the Ucayali. The Urubamba is extensively used for irrigation. High above the Urubamba's gorge are the ruins of the terraced Inca city of Machu Picchu.

U·ru·guay (yŏŏr'ə-gwī', -gwā'), republic (1971 est. pop. 2,900,000), 68,536 sq mi (177,508 sq km), SE South America. The capital is Montevideo. The second-smallest country in South America, Uruguay extends from a short Atlantic coastline along the north bank of the Rió de la Plata to the Uruguay River, which separates it from Argentina. To the north is Brazil. The land is an area of topographical transition from the humid Argentine Pampa to the uplands of southern Brazil. North of the alluvial plain, known as the Banda Oriental, Uruguay generally has long, sweeping slopes and grasslands, wooded valleys with slow-moving rivers, and long ranges of low hills, with some huge granite blocks that stand out against the horizon. The land has a faint purplish hue. Although Uruguay is within the temperate zone, climatic variations are moderate; generally the climate is warm, with rainfall evenly distributed through the seasons, but in some years there are severe droughts.

Economy. Uruguay's greatest natural resource is its rich agricultural land, more than 70% of which is devoted to livestock raising. Sheep and cattle are so important that public rights of way have been built for driving the herds. Grains make up the bulk of the harvested crops. Wheat is the major food crop, followed by rice, an important export. Corn is the principal feed concentrate. Oats, barley, and grain sorghums are also grown, and oil crops (flaxseed and sunflower seed) and sugar beets and sugar cane are important. The country has abundant fisheries.

Despite Uruguay's basically agricultural economy, its dependence upon imports for most raw materials, and its lack of fuel resources, there is considerable industrialization. Hydroelectric power has been developed. The processing of agricultural and animal products accounts for about half of the manufacturing activity. Meat, wool, and hides and skins constitute about 80% of Uruguay's exports. Other manufactures include beverages, textiles, construction and building materials, chemicals, metallurgical goods, and petroleum and coal derivatives. Marble, stone, and granite were long thought to be the only important mineral resources, but in the late 1960s iron ore and some uranium were found. Uruguay's magnificent beaches, such as those at Punta del Este, are great economic assets. The country's transportation facilities are extensively developed. The state owns the railroad, as well as the power, telephone, oil refining, and other industries.

History. Although the Rió de la Plata was explored as early as 1515, not until 1624 did the Spanish establish the first permanent settlement, at Soriano in southwest Uruguay. The Portuguese founded (1680) a short-lived settlement at Colonia, and in 1717 they fortified a hill on the site of Montevideo. The Spanish drove them out (1724) and from then until the wars of independence controlled the Banda Oriental. On the pampas stock raising spread; gradually the unbounded range gave way to huge estancias and small settlements concentrated about the ranch buildings.

When the revolutionary banner was raised in the Argentine in 1810, the leaders of the Banda Oriental, notably Artigas, accepted the cause, but in 1814 Artigas broke with the military junta of Buenos Aires and began a struggle for Uruguayan independence that lasted until the Brazilian occupation of Montevideo in 1820. Five years later a small group, known as the Thirty-three Immortals, under the guidance of Lavalleja, declared Uruguay independent; in 1827 at Ituzaingó Brazil was defeated. Great Britain, opposing Brazilian expansion south to the Río de la Plata, helped ultimately to create an independent Uruguay as a buffer state between Argentina and Brazil. The peace (1828) stipulated that the new Uruguayan constitution should be acceptable to both the larger nations. It was the rough and hardy gaucho who fought for independence, and the traditions, personal loyalties, and rivalries of the gauchos helped to keep the nation in almost continual strife for three quarters of a century after independence was won. In the long struggles that ensued, the two dominant political parties of Uruguay emerged, the Colorados and the Blancos. When in 1864 Brazil presented a claim for damages to property and nationals during the civil wars, Uruguay refused to accept it. Brazil invaded and, aided by the Uruguayan general Vanancio Flores (a Colorado), overthrew the Blanco president. Paraguay came to the assistance of the Blancos, whereupon Argentina, Brazil, and Uruguay formed a tripartite alliance against Paraguay.

During the 19th and 20th cent. waves of immigration, chiefly from Europe, augmented the Uruguayan population. In Batlle y Ordóñez's second term as president (1911–15) began the social and material progress that made Uruguay one of the more stable and prosperous nations of Latin America. By a coup d'état in 1933, Gabriel Terra suspended the constitution of 1919. Yet, under Terra's rule, which ended in 1938, the socialistic measures for public welfare were not reversed but forwarded; the labor code was broadened, social benefits increased, and industry further nationalized.

Batlle's influence on Uruguayan political practice outlasted his death; concerned lest the country again fall prey to dictatorial caudillos, he had advocated the creation of an executive governing council. This reform, inspired by the Swiss multiple-executive system, was adopted in 1951; the office of president was replaced by a nine-man council with a president, chosen from the majority party, to act as titular head of state. Factionalism and apathy within the council, however, hindered action on social and economic problems, which became pressing in the mid-1950s and acute during the 60s. The increasing use of synthetic fibers and the steadily declining price of wool cut deeply into Uruguay's exports of wool and leather, and the country suffered an increasingly unfavorable balance of trade. Runaway inflation and high unemployment ensued. The vast and inefficient bureaucracy became a burden to the economy. In 1958 the Colorados, who had been in power for over 93 years, were overwhelmingly defeated by the conservative Blancos, who won again in 1962 by a narrower margin. Throughout the 1960s and early 1970s the economic decline continued relentlessly, intensified by droughts and floods and accompanied by massive social unrest—riots, strikes, and the emergence of a terrorist Marxist guerrilla group, the Tupamaro National Liberation Front. In 1967 a new constitution abolished the plural executive and reinstated a powerful president. That same year the Colorado party returned to power. Successive presidents ruled with increasingly dictatorial powers. As the Tupamaros increased terrorist activities, the army assumed great power, moving against the Tupamaros with arrests and torture.

Uruguay, river, c.1,000 mi (1,610 km) long, rising in S Brazil and flowing in an arc W, SW, and S to the Río de la Plata, an estuary; it forms part of the Brazil-Argentina border and the entire Argentina-Uruguay line. The lower river is navigable for oceangoing vessels for 130 mi (209.2 km) upstream. Its upper course is broken by waterfalls.

U·ruk (ōō'rŏŏk'), ancient city: *see* Erech.

U·rum·chi (ōō-rŏŏm'chē): *see* Wu-lu-mu-ch'i, China.

U·sam·ba·ra (ōō'səm-bä'rə), mountains, c.70 mi (115 km) long and from 20 to 40 mi (30–65 km) wide, NE Tanzania. On its slopes, which rise to c.8,000 ft (2,440 m), coffee, sisal, tea, and cinchona are produced; rice is grown in the swampy foothills.

U·se·dom (ōō'zə-dôm'), island, 164 sq mi (425 sq km), in the Baltic Sea. Formerly in Pomerania prov., Germany, it was divided in 1945 between East Germany (which received most of the island) and Poland. Usedom is separated from the mainland by Stettin Lagoon. It is generally lowland, with forests and several lakes. Grain and potatoes are the principal agricultural products; the main sources of income are tourism and fishing.

Ush·ant (ŭsh'ənt), island, 10 sq mi (25.9 sq km), Finistère dept., off Brittany, in the Atlantic. The chief occupations are sheep raising and fishing. In 1778 and 1794 naval battles occurred near here between the French and English.

Us·hua·ia (o͞o-swä'yə), city (1970 pop. 5,500), Tierra del Fuego Territory, S Argentina, a port on the Beagle Canal, on Ushuaia Bay. Settled in the 1870s by English missionaries, it was taken over by Argentine naval forces in 1884.

Usk (ŭsk), river, c.60 mi (95 km) long, rising in the Black Mts., S Wales and flowing generally SE to Bristol Channel near Newport. The upper Usk is noted for its beauty and its excellent fishing.

Üs·kü·dar (o͞os'kə-där'), urban district (1970 pop. 143,938), part of İstanbul, Turkey, on the Asian side of the Bosporus. It is a commercial and industrial center. Known as Chrysopolis in ancient times, it enjoyed its greatest prosperity after the Ottoman conquest (15th cent.). During the Crimean War Üsküdar was a base (1854–56) of the British army and the site of the military hospital made famous by the work of Florence Nightingale.

Us·pal·la·ta Pass (o͞os'pə-yä'tə), c.12,500 ft (3,815 m) high, over the Andes between Mendoza, Argentina, and Santiago, Chile. A trail for men and pack animals was used before the Transandine Railway was built. The Pan-American Highway now runs through the pass. The Christ of the Andes stands in the pass.

Us·su·ri (o͞o-so͝or'ē), river, c.365 mi (590 km) long, formed by the confluence of the Ulukhe and Daubikhe rivers, S Primorsky Kray, Far Eastern USSR. It flows north to the Amur River at Khabarovsk, forming part of the USSR-China border. The Ussuri abounds in fish and is used to transport timber.

Us·su·riysk (o͞o'so͞o-rēsk'), city (1973 est. pop. 138,000), Primorsky Kray, Far Eastern USSR, on the Suyfun River. It is a coal-mining center. Whaling, sugar refining, and the manufacture of motors and agricultural machinery are also important.

U·ster (o͞os'tər), town (1974 est. pop. 22,300), Zürich canton, NE Switzerland, between Pfäffiker and Greifen lakes. Textiles and machinery are made.

Ú·stí nad La·bem (o͞o'styē näd'lə-bĕm'), city (1971 est. pop. 72,661), NW Czechoslovakia, in Bohemia, on the Elbe and Bilina rivers and near the East German border. The city has ironworks and chemical, machine-building, and food-processing industries. Founded in the 13th cent., Ústí nad Labem was ceded to Germany in 1938 by the Munich Pact but reverted to Czechoslovakia in 1945.

Ust-Ka·me·no·gorsk (o͞ost'kə-mŭn'ə-gôrsk'), city (1973 est. pop. 247,000), capital of East Kazakhstan oblast, Central Asian USSR, in Kazakhstan, on the Irtysh River and in the foothills of the W Altai Mts. It is a river port and an industrial center with zinc, lead, and titanium-magnesium smelters. The city was founded in 1720 as the Russian military outpost.

Ust·yurt (o͞ost-yo͞ort'), desert plateau, c.62,000 sq mi (160,580 sq km), Central Asian USSR, between the Caspian and Aral seas. It rises to between c.490 and 980 ft (150–300 m). Its seminomadic population raises sheep, goats, and camels.

U·su·lu·tán (o͞o'so͞o-lo͞o-tän'), city (1971 pop. 19,616), S El Salvador. Near the volcano of the same name, the city is the commercial center of an agricultural region.

U·su·ma·cin·ta (o͞o'so͞o-mä-sēn'tä), river, c.600 mi (965 km) long, formed at the Guatemala-Mexico border by the Chixoy and Pasion rivers and flowing NE through Tabasco state, Mexico, to the Bay of Campeche. It is navigable for c.300 mi (485 km) upstream by small boats and is used to move logs and chicle downstream.

U·tah (yo͞o'tô', -tä'), state (1975 pop. 1,196,000), 84,916 sq mi (219,932 sq km), including 2,577 sq mi (6,674 sq km) of inland water surface, W United States, one of the Rocky Mt. states; admitted 1896 as the 45th state of the Union. Salt Lake City is the capital and largest city; it is also the headquarters of the Church of Jesus Christ of Latter-Day Saints, which founded the state and to a large extent still dominates it. Utah is bounded on the north by Idaho, on the northeast by Wyo., on the east by Colo., on the south by Ariz., and on the west by Nev.

The state has two dissimilar regions abruptly separated by the Wasatch Range (part of the Rocky Mts.), which runs generally south from the Idaho border. To the east of the Wasatch rise massive mountains and irregular plateaus; along its western foothills lie the major cities of Utah, while further west is the Great Basin. In the northeast the snow-capped Uinta Mts. reach the state's highest ele-

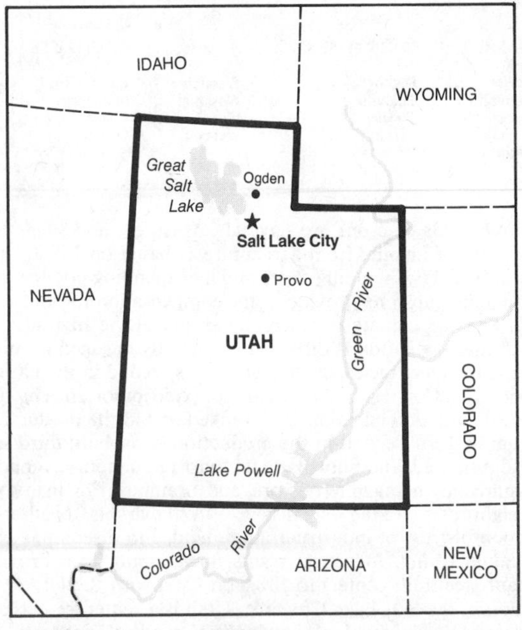

vation in Kings Peak (13,528 ft/4,123 m). The dissected Colorado Plateau stretches southward, rugged and largely uninhabitable except in isolated river valleys. Deep, tortuous canyons cut by the Colorado River and its tributaries impede travel but create vistas of remarkable grandeur. Western Utah, part of the Great Basin, was once submerged beneath an extensive Pleistocene lake, Lake Bonneville. During many thousands of years the amount of water in the lake fluctuated, then subsided, leaving behind a salt-strewn desert, wide expanses of arid but nonalkaline soil, and a series of lakes. Great Salt Lake, the largest of these, has through evaporation reached a concentration of mineral salts several times that of the ocean. The haze-covered Oquirrh Mts., rising south of the lake, dip to form pleasant beaches at the water's edge, then emerge as islands within the lake and rise again in the Promontory Mts. on the northern lake shore. Utah Lake, to the south, is the largest natural body of fresh water in the state and drains into Great Salt Lake through the Jordan River. Between Great Salt Lake and the Wasatch Range and curving southwest toward the Ariz. line is a river-crossed strip, an agricultural oasis and the center of the life of Utah.

Irrigation of the rich but arid land has long been a vital problem in Utah's agricultural development. Construction on the $325 million Central Utah project began in 1967; when completed, a vast complex of dams, canals, and aqueducts will carry water across the Wasatch Range to the Salt Lake valley. The arduous task of converting deserts to productive soil has confined the tilled land to a small percentage of the state's total area.

Thousands of visitors come annually to view the state's many natural wonders, most notably Great Salt Lake and the spectacular Bryce Canyon and Zion national parks. Other attractions are Canyonlands and Arches national parks; Natural Bridges, Cedar Breaks, Dinosaur, Hovenweep, Rainbow Bridge, and Timpanogos Cave national monuments; Glen Canyon National Recreation Area; and Golden Spike National Historic Site. The Bonneville Salt Flats are famous as an automotive speedway. There are many national forests and a number of Indian reservations. Capitol Reef National Park contains ancient cliff dwellings, caves with interesting glyphs, and numerous artifacts of prehistoric man.

Utah is divided into 29 counties:

NAME	COUNTY SEAT	NAME	COUNTY SEAT
Beaver	Beaver	Iron	Parowan
Box Elder	Bringham City	Juab	Nephi
Cache	Logan	Kane	Kanab
Carbon	Price	Millard	Fillmore
Daggett	Manila	Morgan	Morgan
Davis	Farmington	Piute	Junction
Duchesne	Duchesne	Rich	Randolph
Emery	Castle Dale	Salt Lake	Salt Lake City
Garfield	Panguitch	San Juan	Monticello
Grand	Moab	Sanpete	Manti

NAME	COUNTY SEAT	NAME	COUNTY SEAT
Sevier	Richfield	Wasatch	Heber City
Summit	Coalville	Washington	St. George
Tooele	Tooele	Wayne	Loa
Uintah	Vernal	Weber	Ogden
Utah	Provo		

Economy. Major crops are hay, wheat, barley, and sugar beets, but the bulk of income from agriculture is based on livestock and livestock products, including sheep and an expanding poultry industry. Abundant sunshine provides some compensation for inadequate rainfall, and the climate is moderate, except in the high altitudes. Much of the population is directly or indirectly engaged in mining. Copper is the chief metal; in 1970 Utah was second in the nation in its production. Copper is followed in economic importance by petroleum, coal, and molybdenum. Utah ranked first in the production of beryllium in 1970; second in the production of asphalt; third in silver, lead, tin, gold, and fluorspar; and fourth in mercury, vanadium, potassium salts, manganiferous ore, and uranium. For many years high freight rates and the long distances from markets, together with a Mormon distrust of industrialization, tended to discourage extensive manufacturing. Industrial plants now extend from Provo (an important steel mill center) to Brigham City, with the largest concentration in the Salt Lake City area. Utah is a center for aerospace research and the production of all kinds of missiles, spacecraft, electronic systems, and related items. Other major manufactures are processed foods, machinery, fabricated metalware, and petroleum products.

History. Although some of Coronado's men may have entered southern Utah in 1540, the first definite penetration by Europeans did not occur until 1776, when the Spanish missionaries Silvestre Vélez de Escalante and Francisco Atanasio Domínguez opened the route for the Old Spanish Trail between Santa Fe and Utah Lake. The large area of which Utah was a part was officially recognized as a Spanish possession (it passed to the United States in 1848 with the Treaty of Guadalupe Hidalgo after the Mexican War). In the 1820s the mountain men, in search of rich beaver streams, thoroughly explored the region. The discovery of Great Salt Lake is generally credited to James Bridger. In 1841 the first California-bound emigrant train, usually called the Bidwell party, made its way across the Great Salt Lake Desert. Several years later Miles Goodyear became Utah's first settler when he set up a trading post at the site of present-day Ogden.

Permanent settlement began in 1847 with the arrival of the first of the Mormons. Under Brigham Young's direction the ground was plowed and planted, the Temple foundation was laid, and Salt Lake City was built. In 1850 a large area, of which the present state was a part, was constituted Utah Territory and Young was appointed governor. The Indians, dispossessed of their lands, became embittered, and the Mormons were threatened by the powerful Ute Indians, eventually leading to the Walker War (1853-54) and the Black Hawk War (1865-68). Numerous petitions for statehood were denied because of the practice of polygamy, publicly avowed by the Mormons in 1852. Clashes between church and Federal interpretation of the law were frequent. Antagonism, much of it based on misunderstanding, grew out of proportion. In 1857 a "state of substantial rebellion" was declared by the Federal government; Young was removed from his post, and President James Buchanan directed U.S. army troops to proceed against the Mormons. The affair, known as the "Utah War" or the "Mormon campaign," was finally settled peacefully, but great ill feeling had developed, and several difficult decades followed. In 1890 a church edict advising members to abstain from the practice of polygamy was ratified.

Long before Utah became a state in 1896, its area had been reduced to its present size by the creation of the territories of Nevada and Colorado in 1861 and Wyoming Territory in 1868. The influx of settlers included many non-Mormon groups, and cultural and economic isolation had been broken by the development of mining as well as by the completion of the Union Pacific RR (1869). In political life the violent conflict between Mormons and non-Mormons has disappeared. Urbanization has proceeded rapidly. World War II spurred industrial growth, and the development of hydroelectric power during the 1950s attracted new industries.

Government. Utah still operates under its first constitution, adopted in 1895. The state's executive branch is headed by a governor elected for a four-year term. Utah's bicameral legislature has a senate with 30 members elected for four-year terms and a house of representatives with 69 members elected for two years.

Educational Institutions. Those prominent in the state include Brigham Young Univ., at Provo; the Univ. of Utah, at Salt Lake City; and Utah State Univ., at Logan.

Utah, county (1970 pop. 137,776), 2,014 sq mi (5,216.3 sq km), N central Utah, drained by Utah Lake and its tributaries; formed 1852; co. seat Provo. It is in an irrigated agricultural area that produces livestock, hay, sugar beets, and truck crops. Lead and zinc mining and food processing are done. Among its maufacturers are clothing, explosives, concrete, iron and steel, metal products, machinery, and equipment. Part of the Wasatch Range is in the east.

Utah Lake, c.145 sq mi (375 sq km), N central Utah; largest freshwater lake in the state. It drains through the Jordan River to the Great Salt Lake. The waters of Utah Lake were formerly much used for irrigation and in the 1930s showed signs of exhaustion. The Provo River project and the Strawberry Valley project now irrigate the region.

U·ti·ca (yo͞o'tĭ-kə), ancient N African city, NW of Carthage. According to tradition, it was founded by Phoenicians from Tyre c.1100 B.C. Utica usually allied itself with Carthage, but in the Third Punic War it sided with Rome. Upon the destruction of Carthage (146 B.C.), Utica was made the capital of the Roman province of Africa. It fell (A.D. 439) to the Vandals, was recaptured (534) by the Byzantines, and was finally destroyed (c.700) by the Arabs. Excavations at the site have yielded two Punic cemeteries and Roman ruins, including baths and a villa with mosaics.

Utica, city (1978 est. pop. 79,100), seat of Oneida co., central N.Y., on the Mohawk River and the Barge Canal, in a large dairy region; inc. 1862. It is a port of entry, and its manufactures include textiles, electronic and aviation equipment, tools, and firearms. Settled in 1773 on the site of old Fort Schuyler (1758), it was destroyed (1776) in an Indian and Tory attack and resettled after the Revolution. Its location on the Erie and other canals and on the railroads stimulated its industrial development.

U·trecht (yo͞o'trĕkt'), province (1973 est. pop. 838,435), c.500 sq mi (1,295 sq km), central Netherlands, bounded by the IJsselmeer in the N; capital Utrecht.

Utrecht, city (1978 est. pop. 269,574), capital of Utrecht prov., central Netherlands, on a branch of the Lower Rhine River. It is a transportation, financial, and industrial center. Manufactures include machinery, cement, processed minerals, food products, and chemicals. It is the site of a major trade fair. Utrecht was founded by the Romans. In the late 7th cent. it was made an episcopal see for St. Willibrord, the Apostle to the Frisians. The bishops of Utrecht, as princes of the Holy Roman Empire, later ruled the area around the city. Utrecht received a liberal charter in 1122, but difficulties with the bishops continued sporadically until 1527, when the bishop was forced to transfer his territorial rights to Emperor Charles V. Utrecht joined (1577) in the rebellion against Philip II of Spain, and on Jan. 23, 1579, the seven provinces of the northern Netherlands, from then on known as the United Provinces, drew together for their common defense in the Union of Utrecht. In 1713 several treaties forming part of the Peace of Utrecht were signed here. Today Utrecht is a picturesque city, crossed by numerous canals.

U·tre·ra (o͞o-trâr'ə), city (1970 pop. 22,000), Seville prov., S Spain, in Andalusia, on a branch of the Guadalquivir River. It is a rail junction and processing center of a fertile agricultural region. Horses and bulls are bred.

U·tsu·no·mi·ya (o͞o'tso͞o-nō'mē-ä), city (1970 pop. 301,231), capital of Tochigi prefecture, central Honshu, Japan. It is a tobacco-processing center and a tourist resort. Landmarks include the 9th cent. Oyaji temple.

Ut·tar Pra·desh (o͝ot'ər prə-dāsh', -dĕsh'), state (1971 pop. 88,341,-144), 113,454 sq mi (293,846 sq km), N central India, bordered on the N by Nepal and Tibet; capital Lucknow. The state was formed in 1950 by merging the United Provinces and the former princely states of Benares, Rampur, and Tehri. The northern area falls within the Himalaya zone, with many peaks higher than 20,000 ft (6,100 m), and there is a hilly region along the southern border; most of the state, however, is a low-lying fertile plain formed by the Jumna, Ganges, and Gogra rivers. The economy is predominantly agricultural, and industry is centered on processing sugar and cotton.

U·val·de (yo͞o-văl'dē), county (1970 pop. 17,348), 1,312 sq mi (3,398.1 sq km), SW Texas, crossed E to W by Balcones Escarpment and drained by the Nueces, Frio, and Leona rivers; formed 1850; co. seat Uvalde. It has ranching (goats, sheep, and also cattle), mohair and wool shipping, extensive beekeeping, some irrigated agriculture (truck crops and fruit), and asphalt mines.

Uvalde, city (1970 pop. 10,764), seat of Uvalde co., SW Texas; founded c.1854, inc. as a city 1921. A large plaza and many pecan and oak trees lend grace to the town, which draws wealth from surrounding ranches and an irrigated farm area. It has cattle-feed lots, meat-packing houses, grain-storage facilities, and great asphalt mines. Tourism is increasing in importance; there is excellent hunting and fishing in the area, which has brushy hills and deep canyons, with underground caverns and springs. A research and extension center of Texas A & M Univ. is here.

Ux·bridge (ŭks'brĭj), industrial town (1970 pop. 8,253), Worcester co., S Mass., on the Blackstone River SE of Worcester; settled 1662, set off from Mendon 1727. Textiles are manufactured.

Ux·mal (o͞oz-mäl'), ancity city, northern Yucatán peninsula, Mexico. A Late Classic period Maya center situated in the Puuc hills, Uxmal flourished between 600 and 900. It is one of the finest expressions of Maya architecture known as the Puuc style. Its impressive structures include the unique Pyramid of the Magician; the Nunnery, with elaborately decorated façades of stone mosaic friezes; and the Governor's Palace, with some 20,000 carved stone elements in its façade. The site was abandoned shortly after 950 but was reoccupied briefly in the 15th cent. by the Xiu, a Mexican group who soon abandoned the site after wresting power from the Cocom Itzá at Mayapán.

Uz·bek Soviet Socialist Republic (o͞oz'bĕk', o͞oz-bĕk', o͞oz-byĕk') or **Uz·bek·i·stan** (o͞oz-bĕk'ĭ-stăn', -stän'), constituent republic (1970 pop. 11,963,000), 173,552 sq mi (449,500 sq km), Central Asian USSR. Tashkent is the capital. The republic borders on Afghanistan in the south, on the Turkmen SSR in the southwest, on the Kazakh SSR in the west and north, and on the Kirghiz SSR and the Tadzhik SSR in the east. The Kara-Kalpak ASSR is included in Uzbekistan. The terrain of the republic encompasses two unequal sections: the larger northwest area, which is part of the Kyzyl-Kum desert; and the smaller southeast area, which has fertile loess soil and touches on the Tien Shan Mts. The Aral Sea lies on the northwest frontier. Central Asia's two major rivers—the Amu Darya and Syr Darya—pass through Uzbek territory. The Khiva oasis is irrigated by the Amu Darya, the fertile Fergana Valley by the Syr Darya and its tributaries, the Tashkent oasis by the Chirchik and Angren rivers, and the Samarkand and Bukhara oases by the Zeravshan.

The republic grows about 85% of the USSR's cotton crop and half of its rice. Other crops include cereals, alfalfa, fruits, wine grapes, kenaf, sesame, tobacco, and sugar cane. The republic ranks first in the USSR as a region of irrigated agriculture. Livestock are raised in the more arid western areas. Cotton, silk, and wool provide the basis for Uzbekistan's extensive textile industry. Machine building, metallurgy, food processing, hydroelectric power production, and the manufacture of iron and steel, machine tools, chemicals, fertilizer, building materials, and clothing are leading industries. Uzbekistan is rich in mineral resources. The Fergana Valley is the site of oil fields. Western Uzbekistan has large natural-gas deposits. Coal, zinc, copper, tungsten, molybdenum, lead, fluorspar, wolfram, ozokerite, sulfur, limestone, marl, and clays are also found.

Uzbekistan was the site of one of the world's oldest civilized regions: the ancient Persian province of Sogdiana, it was conquered in the 4th cent. B.C. by Alexander the Great. Turkic nomads entered the area in the 6th cent. A.D. It passed in the 8th cent. to the Arabs, who introduced Islam, and in the 12th cent. to the Seljuk Turks. Genghis Khan captured the region in the 13th cent., and in the 14th cent. Tamerlane made his native Samarkand the center of his huge empire. The realm was much reduced under his successors, the Timurids, and began to disintegrate by the end of the 15th cent. In the early 16th cent., the Uzbek invaded the region from the northwest. A remnant of the empire of the Golden Horde, they took their name from Uzbeg Khan (d. 1340), from whom their dynasty claimed descent. The Uzbek empire broke up into separate principalities in the 17th cent. Weakened by internecine warfare, these states were conquered by Russian forces, who took Tashkent in 1865, Samarkand and Bukhara in 1868, and Khiva in 1873. In 1918 the Turkistan ASSR was organized on Uzbek territory, in 1920 the Khorezm and Bukharan People's Republics were established, and finally, in 1924, the Uzbek-populated areas were united in the Uzbek SSR. Tadzhikistan was part of the Uzbek SSR until 1929, when it became a separate republic. In 1936 the Kara-Kalpak ASSR was joined with Uzbekistan.

Uzh·go·rod (o͞ozh'gə-rŏd'), city (1973 est. pop. 70,000), capital of Zakarpatskaya oblast, SW European USSR, in the Ukraine, in the SW Carpathian foothills and on the Uzh River. It is a rail and highway junction and the economic and cultural heart of Transcarpathian Ukraine. Industries include metalworking, automobile repairing, meat packing, winemaking, brandy distilling, and the manufacture of plywood, furniture, bricks, tiles, clothing, and footwear. The city has long been important militarily because of its position guarding the southern approach to the Uzhok Pass over the Carpathians. Uzhgorod was founded in the 8th or 9th cent. and belonged to Kievan Russia in the 10th and 11th cent. Conquered by the Magyars at the end of the 11th cent., the city remained under Hungarian rule until it passed to Austria-Hungary in 1867. Uzhgorod was the center of the Ukrainian national and Russophile movements in the 19th and early 20th cent. The city passed to Czechoslovakia in 1919, was under Hungarian occupation from 1938 to 1944, and was included in the Ukraine after World War II.

Vaal (väl), river, c.750 mi (1,205 km) long, rising in SE Transvaal, NE Republic of South Africa, S Africa, and flowing SW to the Orange River. It forms most of the Transvaal-Orange Free State border. The river's flow is almost totally regulated and provides water power for industries on the Witwatersrand. Vaal Dam, one of the country's largest, is located southeast of Vereeniging; it stores water for use by the mines. The Vaalhartz Dam, near the junction of the Vaal and the Hartz rivers, is an important part of the Vaal River Development Project.

Vaa·sa (vä'sə), city (1971 est. pop. 49,442), capital of Vaasa prov., W Finland, on the Gulf of Bothnia. It is a port and agricultural market. Chartered in 1606, Vaasa was rebuilt closer to the sea after a devastating fire in 1852. It was the capital of White Finland during the civil war of 1918.

Vác (väts), town (1973 est. pop. 32,619), N central Hungary, on the Danube River. A commercial center producing textiles, cement, and photographic articles, it is also a summer resort. Dating from Roman times, Vác was made (1008) a bishopric by St. Stephen. It has an 18th cent. cathedral, an episcopal palace, and an 18th cent. triumphal arch. The name was formerly spelled Vacz or Vacs.

Vac·a·ville (văk'ə-vĭl', vä'kə-), city (1979 est. pop. 38,200), Solano co., central Calif., in a farm area; inc. 1892. Food products are made. A state prison medical facility is here, and Travis Air Force Base and Hospital are to the south.

Va·duz (vä-dōōts'), town (1974 est. pop. 4,382), capital of Liechtenstein, W Liechtenstein, on the Rhine River. It is a tourist center. A beautiful medieval castle (now an art museum) dominates the town. Vaduz was destroyed (1499) in the war between the Swiss and the Holy Roman Empire and was rebuilt in the early 16th cent.

Váh (väk), river, c.245 mi (390 km) long, E Czechoslovakia. It is formed by the union of the Biély Váh, rising in the High Tatra, and the Cierny Váh, rising in the Low Tatra, and flows SW into the Danube at Komárno. The Orava and Nitra rivers are its chief tributaries.

Vai·den (vād'n), town (1970 pop. 716), a seat of Carroll co., central Miss., ESE of Greenwood, in a cotton and timber region.

Vai·gach (vī-gäch'), island: *see* Vaygach.

Vaiont Dam, 858 ft (261.7 m) high, on the Vaiont River, a tributary of the Piave River, in Venetia, NE Italy, near Belluno. Vaiont Dam, one of the highest in the world, was completed in 1961 and is used to generate electricity. After heavy rains in 1963, landslides into the Vaiont reservoir caused the stored water to spill over the dam, sweeping away the village of Longarone and flooding nearby hamlets; some 2,000 people drowned.

Va·lais (vä-lā'), canton (1974 est. pop. 211,000), 2,021 sq mi (5,234.4 sq km), S Switzerland; capital Sion. Bordering on France and Italy, it has some of the highest peaks (Matterhorn, Dufourspitze, Dom, and Weisshorn) in Switzerland. Mainly a livestock-raising and agricultural canton, it is also known for its fine wines. The Valais has a well-developed hydroelectric system, and its industries produce metal products and chemicals. Zermatt is the largest of its numerous resorts and winter sports centers. Most of the population is French-speaking and Roman Catholic. Taken by the Romans in 57 B.C., the region later passed to the Burgundians and to the Franks. In 999 Rudolf III of Burgundy made the bishop of Sion lord of Valais, but the country later split, with the Lower Valais passing to Savoy. In 1475 the bishop of Sion and the communes of the Upper Valais, which had gained considerable autonomy, defeated the duke of Savoy, and from then until 1798 the Lower Valais was held in subjection by the Upper Valais. Made a canton of the Helvetic Republic in 1798, an independent republic in 1802, and a French department in 1810, the Valais became a canton of the Swiss Confederation in 1815.

Val·cour Island (văl'kōōr'), c.2 mi (3.2 km) long, NE N.Y., in Lake Champlain, S of Plattsburgh. Here on Oct. 11, 1776, an American fleet under Benedict Arnold was attacked by the British and forced to retreat to Crown Point.

Val·day Hills or **Val·dai Hills** (both: väl-dī'), upland region, NW European USSR, composed of a series of glacial moraines that rise to c.1,100 ft (340 m). The region forms the watershed of the upper Volga, the Western Dvina, and the Dnepr rivers and also of the rivers that flow into Lake Ilmen. Numerous glacial lakes are found here; Lake Seliger is the largest.

Val-de-Marne (väl'də-märn'), department (1974 est. pop. 1,264,000), 94 sq mi (243.5 sq km), N central France, adjoining Paris on the SE; capital Créteil.

Val·dez (văl-dēz'), city (1970 pop. 1,005), S Alaska, at the head of Valdez Arm inside Prince William Sound; inc. 1901. It has tourist and fishing industries. Salmon spawning grounds are here. The city's excellent ice-free harbor was explored by the Spaniards in 1790. Valdez was established in 1898 as a debarkation point for men seeking a route to the Yukon gold fields that would obviate the necessity of paying duty to Canada. The city was devastated by a 1964 earthquake and was rebuilt 5 mi (8 km) west of the old site. Carefully planned, many of its new buildings are of Swiss design. Valdez is the southern terminus of the oil pipeline that originates in Prudhoe Bay. Its port facilities were greatly enlarged in the mid-1970s.

Val·di·vi·a (väl-dē'vē-ə), city (1970 pop. 90,942), capital of Valdivia prov., S central Chile, on the Valdivia River. It is a leading commercial and industrial center. Founded in 1552, the city did not grow until the arrival in the mid-19th cent. of German immigrants who founded the first industries (beer and shoes).

Val-d'Oise (väl-dwäz'), department (1974 est. pop. 821,000), 482 sq mi (1,248.4 sq km), N central France, N of Paris; capital Pontoise.

Val d'Or (väl dôr'), town (1971 pop. 17,421), SW Que., Canada, SE of Rouyn. It is a mining center. Gold was discovered in the region in 1909.

Val·dos·ta (văl-dŏs'tə), city (1979 est. pop. 37,200), seat of Lowndes co., S Ga., near the Fla. line, in a lake region; inc. 1860. Valdosta is a large naval-stores market and a processing, distributing, and commercial center for a tobacco, cotton, and livestock area. Manufactures include turpentine, pine lumber, and wood and paper products.

Vale (vāl), town (1970 pop. 1,448), seat of Malheur co., E Oregon, on the Malheur River near the Idaho border; settled 1864, inc. 1912. It is a shipping center for a livestock and farm region.

Va·lence (vä-läNs'), city (1968 pop. 62,358), capital of Drôme dept., SE France, in Dauphiné, on the Rhône River. Its many manufactures include metallurgical products, textiles, leather goods, jewelry, and munitions. An old Roman town, it was taken by the Visigoths (413) and the Arabs (c.730), then changed hands many times. It was actually ruled by its own bishops from 1150 until the 15th cent., when its citizens put themselves under the protection of the dauphin. Its university was founded in 1452.

Va·len·ci·a (və-lĕn'shē-ə, -shə, vä-lĕn'thyä), region (1972 est. pop. 3,162,224) and former kingdom, E Spain, on the Mediterranean. The country is chiefly mountainous, with a fertile coastal plain. Irrigation and an intensive system of cultivation were started by the Moors. Citrus and other fruits, rice, vegetables, cereals, olive oil, and wine are now produced and exported. Many prehistoric remains have been found in Valencia. Inhabited by the Iberians in early times, it was later colonized by Greek and Carthaginian traders. It passed to the Moors in the 8th cent. At the fall of the caliphate of Córdoba it became (1022) an independent emirate. The Cid briefly ruled the city and district of Valencia (1094–99). After a brief period of independence Valencia joined the Aragonese confederation and later was incorporated into the Spanish state. The region has had an economic revival in the 20th cent.; increased tourism has been an important factor.

Valencia, city (1972 est. pop. 662,557), capital of Valencia prov., E Spain, on the Turia River. It is in a fertile region a short distance from its busy Mediterranean port, El Grao, on the Gulf of Valencia. It is an active industrial and commercial center producing textiles, metal products, chemicals, furniture, and *azulejos,* or colored tiles. There also are important shipyards. First mentioned in the 2nd cent. B.C., Valencia was a Roman colony. Under the Moors, from the 8th to the 13th cent., it was twice the seat of an independent state. From 1094 to 1099 it was ruled by the Cid. Its university was founded in 1501. During the civil war Valencia served (1936–37) as the seat of the Loyalist government. Among its chief landmarks are the cathedral (13th–15th cent.), called La Seo, with a Gothic bell tower (the Miguelete); the Torres de Serranos, 14th cent. fortified towers built on Roman foundations; and the Renaissance palace of justice. The city also has a fine art gallery. The *Tribunal de las Aguas,* which settles disputes over the irrigation of the outlying garden region, has met regularly in the city since the 10th cent.

Valencia, city (1971 pop. 366,154), capital of Carabobo state, N Venezuela. It is Venezuela's fourth-largest city and one of its major industrial centers. Products include motor vehicles, chemicals, textiles, cattle feed, and consumer goods. The city is a market for sugar cane and cotton and for cattle driven from the Orinoco llanos. Valencia was founded in 1555. It was briefly the national capital in 1812 and again in 1830, when a convention held here proclaimed Venezuela's secession from Greater Colombia. Valencia's industrialization dates from the 1950s.

Va·len·ci·a (və-lĕn'chē-ə, -chə), county (1970 pop. 40,576), 5,656 sq mi (14,649 sq km), W N.Mex., bounded on the W by the Ariz. line and drained by the Rio Puerco and San Jose River; formed 1852; co. seat Los Lunas. It has livestock grazing and farming.

Va·len·ci·ennes (vä-läN-syĕn'), city (1968 pop. 46,626), Nord dept., N France, on the Escaut (Scheldt) River. It is located in a rich coal basin, which fuels its textile, machinery, and metallurgical industries. An important place in medieval Hainaut, it became famous (15th cent.) for its lace industry.

Val·en·tine (văl'ən-tīn'), city (1970 pop. 2,662), seat of Cherry co., N Nebr., on the Niobrara River near the S.Dak. border, in a cattle-grazing area; settled 1882. Nearby are a state fish hatchery and a wildlife refuge.

Va·lla·do·lid (vä'lyä-*thō*-lē*th*'), city (1972 est. pop. 247,160), capital of Valladolid prov., N central Spain, in León, at the confluence of the Pisuerga and Esgueva rivers and on the Canal de Castilla. A communications and industrial center, Valladolid is also an important grain market. Of obscure origin (its name is derived from Arabic), it was conquered by the Christians from the Moors in the 10th cent. and largely replaced Toledo as the chief residence of the kings of Castile in the 15th cent. Valladolid was the scene of the marriage of Ferdinand and Isabella in 1469. Christopher Columbus died (1506) in the city, Emperor Charles V often resided here, and Philip II was born here in 1527. It declined greatly after 1561, when Philip II made Madrid the Spanish capital (the capital was returned to Valladolid 1600–1606). Today Valladolid remains an important cultural center and archiepiscopal see. Its university, founded in 1346, has a rich library with valuable manuscripts. The house of Columbus and the house where Cervantes wrote part of *Don Quixote* have been preserved. Other landmarks include the late Renaissance cathedral, the Colegio de San Gregorio (15th cent.), with a lavishly adorned façade, and the former royal palace.

Val·le·jo (və-lā'ō), city (1979 est. pop. 70,500), Solano co., W Calif., on San Pablo Bay at the mouth of the Napa River; inc. 1866. It is a port and a trade and processing center for farm products. It has a U.S. naval shipyard on Mare Island, just west of the city. Founded by Adm. David Farragut in 1854, it covers 1,500 acres (607 hectares) and has four drydocks and eight shipbuilding ways. Submarines and destroyers are built and repaired here. Vallejo was created on the property of Gen. Mariano G. Vallejo to be the state capital; it was the nominal capital from 1852 to 1853. A junior college and the California Maritime Academy are in Vallejo. A state park is nearby.

Val·let·ta (və-lĕt'ə), city (1970 est. pop. 15,600), capital of Malta, NE Malta. It is strategically located on a rocky promontory between two deep harbors. Dockyards employ more workers than any other industry. The 16th cent. town contains many relics of the Knights of St. John of Jerusalem (the Knights Hospitalers, or Knights of Malta). Valletta contains a 16th cent. cathedral, the old governor's palace, the Royal Univ. (1769), and a library with a museum of antiquities. The city was severely damaged by air raids in World War II.

Val·ley (văl'ē). **1.** County (1970 pop. 3,609), 3,676 sq mi (9,520.8 sq km), central Idaho, in a mountainous and plateau area bounded in the E by the Middle Fork of the Salmon River and drained by the North Fork of the Payette River; formed 1917; co. seat Cascade. Stock raising, lumbering, dairying, agriculture (timothy, clover, and rye), and quicksilver and tungsten mining are major occupations. **2.** County (1970 pop. 11,471), 4,974 sq mi (12,882.7 sq km), NE Mont., in an irrigated agricultural area bordering on Sask., bounded on the S by the Missouri River, and drained by the Milk River; formed 1893; co. seat Glasgow. Grain and livestock are raised. **3.** County (1970 pop. 5,783), 569 sq mi (1,473.7 sq km), central Nebr., in an agricultural region drained by the North Loup and Middle Loup rivers; formed 1873; co. seat Ord. It yields livestock, grain, dairy products, and poultry.

Valley City, city (1970 pop. 7,843), seat of Barnes co., SE N.Dak., on the Sheyenne River W of Fargo. Settled in 1872 with the coming of the railroad, it was incorporated in 1883. The trade center of a grain, livestock, and dairy region, the city has flour mills and dairy-processing plants.

Val·ley·field (văl'ē-fēld'), city (1971 pop. 30,173), S Que., Canada, on the Beauharnois canal, at the NE end of Lake St. Francis, SW of Montreal. It is a port of entry and industrial center.

Valley Forge (fôrj, fōrj), on the Schuylkill River, NE Pa., NW of Philadelphia. Here, during the American Revolution, the main camp of the Continental Army was established (Dec., 1777–June, 1778) under the command of Gen. George Washington. The winter was severe and there was much illness and suffering. The site is now a state park.

Valley Stream, village (1979 est. pop. 38,800), Nassau co., SE N.Y., on Long Island, a residential suburb of New York City; inc. 1925.

Val·my (väl-mē'), village (1968 pop. 291), Marne dept., NE France, in the Argonne region. The "cannonade of Valmy," a Franco-Prussian artillery skirmish, was fought near here on Sept. 20, 1792. This was the first important engagement in the French Revolutionary Wars.

Va·lois (väl-wä'), historic region, now comprised in Aisne and Oise depts., N France. It is a rich agricultural area. A county and later a duchy, Valois was the appanage of the royal house of Valois.

Va·lo·na (və-lō'nə): *see* Vlorë, Albania.

Val·pa·rai·so (văl'pə-rī'zō, -rā'zō), city (1970 est. pop. 292,800), capital of Valparaiso dept., central Chile. It is the chief port of Chile and the terminus of a trans-Andean railroad. An important industrial center, it manufactures textiles, sugar, paint, enamelware, cottonseed oil, shoes and leather goods, chemicals, and metal products. From a narrow waterfront terrace, steep hills rise to make Valparaiso an amphitheater, with wharves and business quarters at the base and residential sections above. So steep is the ascent that funicular railways are used. The city faces a wide bay, which, although partly protected by breakwaters, often carries severe northern gales in winter. Valparaiso was founded in 1536 by the Spanish conquistador Juan de Saavedra but was not permanently established until 1544 by Pedro de Valdivia. It was frequently raided by English and Dutch pirates throughout the 16th and 17th cent. Relatively unimportant in colonial times, the city grew in the late 19th cent. It has a technical school, a Catholic university, and a Chilean naval academy.

Val·pa·rai·so. 1. (văl'pə-rī'zō) Resort city (1970 pop. 6,504), Okaloosa co., NW Fla., on Choctawhatchee Bay ENE of Pensacola; settled 1919, inc. 1921. Eglin Air Force Base is nearby. **2.** (văl'pə-rā'zō) City (1979 est. pop. 21,600), seat of Porter co., NW Ind.; inc. 1850. Ball and roller bearings, magnets, electric switches, and automobile parts are among its manufactures. The city is the seat of Valparaiso Univ. (noted for its huge chapel) and a technical institute (est. 1874).

Val·tel·li·na (väl-tā-lē'nä), Alpine valley of the upper Adda River, c.75 mi (120 km) long, in Lombardy, N Italy, extending from Lake Como to the Stelvio Pass. The valley is a fertile agricultural region, known for its wine. With the adjoining counties of Bormio and Chiavenna, the Valtellina was seized (1512) from Milan by the Grisons, which subsequently ruled the district—its richest and most populous possession—as a subject territory. By the start of the Thirty Years' War (1618–48), the stoutly Roman Catholic inhabitants of the Valtellina were ready for revolt against the Grisons, the majority of whose population was Protestant; in 1620 they rose and massacred their Protestant masters. These internal troubles quickly assumed European proportions, because the valley commanded the passages between Austria and the Grisons and Venice and Spanish-held Milan. The Valtellina became the pawn of the participants in the Thirty Years' War and the victim of their complicated intrigues. The massacre of 1620 led to a series of military interventions by Spain, Austria, the pope, the Catholic party of the Grisons, France, and the Protestant majority of the Grisons (largely financed by Venice). The valley was sacked in turn by these armies and in 1627 passed under Spanish control; transportation of Spanish reinforcements through the Valtellina into Germany contributed to several victories by the imperial party, notably at Nördlingen (1634). When France fully entered the war on the Protestant side, a French army was again dispatched (1635) to the Valtellina. Henri de Rohan conquered the valley but failed to restore it to the full control of his Grisons allies. Incensed, the Grisons Protestants, led by the preacher-soldier George Jenatsch, secretly negotiated with the Catholic powers, who promised to restore the Valtellina to the Grisons if the French were expelled. However, Rohan, ill and weakly supported by the French government, had to evacuate the Grisons in 1637. By the Peace of Milan (1639) the Grisons fully recovered the Valtellina; it remained in the Grisons until 1797, when it was incorporated into the Cisalpine Republic. The Valtellina passed (1815) to the Lombardo-Venetian kingdom (held by Austria) and later (1859) to Italy.

Val Ver·de (văl vûr'dē), county (1970 pop. 27,471), 3,241 sq mi (8,394.2 sq km), SW Texas, partly on Edwards Plateau, bounded on the SW and S by the Rio Grande and the Mexican border; formed 1885; co. seat Del Rio. It is a leading sheep-raising county and also has cattle and horse ranches and some irrigated agriculture (alfalfa, fruit, and truck crops). It includes Amistad Recreation Area.

Van (văn, vän), city (1970 pop. 46,751), capital of Van prov., E Turkey, near the E shore of Lake Van, at an altitude of 5,659 ft (1,726 m). It is the trade center for a fruit- and grain-growing region. Van was the cradle of an ancient Armenian civilization. It was the capital of the old Vannic kingdom of Urartu or Ararat. Near the city is the mound of Toprakkale, where excavations in the 19th cent. uncovered the remains of the town of Urartu. In 1939 archaeologists discovered fortifications and various materials dating from the 8th cent. B.C. Many of the Armenians living in the region were massacred by the Turks in 1895. Lake Van, c.1,450 sq mi (3,755 sq km), is the largest lake in Turkey. It is alkaline and has no apparent outlet.

Van Bu·ren (văn byo͝or'ən). **1.** County (1970 pop. 8,275), 699 sq mi (1,810.4 sq km), N central Ark., drained by the Middle Fork of the Little Red River; formed 1833; co. seat Clinton. Its agriculture includes strawberries, truck crops, potatoes, corn, cotton, hay, and livestock. There is some industry. Part of Ozark National Forest is in the southwest. **2.** County (1970 pop. 8,643), 487 sq mi (1,261.3 sq km), SE Iowa, on the Mo. border, drained by the Des Moines and Fox rivers; formed 1836; co. seat Keosauqua. It is primarily an agricultural county in a prairie region, growing corn, soybeans, and hay and raising cattle, poultry, and hogs. There are bituminous-coal mines and limestone quarries here. **3.** County (1970 pop. 56,173), 603 sq mi (1,561.8 sq km), SW Mich., bounded on the W by Lake Michigan and drained by the Paw Paw and Black rivers; formed 1837; co. seat Paw Paw. It is a fruit-growing region, specializing in apples, peaches, cherries, strawberries, and grapes. There are also grain, livestock, and truck farms, and dairy products are made. Some manufacturing is done. The county has oil wells, fisheries, and nurseries, as well as resorts. **4.** County (1970 pop. 3,758), 254 sq mi (657.9 sq km), central Tenn., in a hilly region in the Cumberlands, bounded in the N by the Caney Fork of the Cumberland River; formed 1840; co. seat Spencer. Its agriculture includes tobacco, corn, cattle, and hogs. It has lumbering and manufactures yarn, thread, and clothing.

Van Buren. 1. City (1970 pop. 8,373), seat of Crawford co., NW Ark., on the Arkansas River opposite Fort Smith; laid out c.1838, inc. 1843. It processes and ships products from nearby farms and manufactures wood products. **2.** Town (1970 pop. 714), seat of Carter co., S Mo., in the Ozarks on the Current River NW of Poplar Bluff. Lumber products are made.

Vance (văns), county (1970 pop. 32,691), 249 sq mi (644.9 sq km), N N.C., bounded on the N by Va., on the SW by the Tar River, formed 1881; co. seat Henderson. This piedmont tobacco and timber area has important tungsten deposits, textile manufacturing, sawmilling, and agriculture (corn, cotton, and tobacco).

Vance·burg (văns'bûrg'), city (1970 pop. 1,773), seat of Lewis co., NE Ky., on the Ohio River WNW of Ashland, in a farm and tobacco area; settled c.1796, inc. 1827. Shoes and wood products are made.

Van·cou·ver (văn-ko͞o'vər), city (1971 pop. 426,256), metropolitan pop. 1,082,352), SW British Columbia, Canada, on Burrard Inlet of the Strait of Georgia, opposite Vancouver Island and just N of the Wash. border. It is the largest city in western Canada and the chief Canadian Pacific port. It is also the major western terminus of trans-Canadian railroads, highways, and airways. Its location on hills with views of the harbor and the mountains of the Coast Range as well as its mild winter climate make it a year-round tourist center. The city's industries include shipbuilding, fish processing, and sugar and oil refining. It has textile and knitting mills and plants making metal, wood, paper, and mineral products. Vancouver is the western terminus of the tramontane pipeline bringing oil to the west coast from Edmonton. At Point Grey in metropolitan Vancouver is the Univ. of British Columbia. Stanley Park (900 acres/364 hectares), one of the city's many parks, has a zoo and famous gardens and specimens of native trees. The city was settled before 1875 and called Granville; it was incorporated in 1886, after a rail link was built, and named in honor of Capt. George Vancouver.

Vancouver, city (1979 est. pop. 49,000), seat of Clark co., SW Wash., on the Columbia River opposite Portland, Oregon, with which it is connected by a bridge; inc. 1857. An important deepwater port, it has shipyards, lumber mills, and an enormous grain elevator. Founded by the Hudson's Bay Company as Fort Vancouver in 1825-26, it was an important fur center and an early focus for settlement in the state. After the area was ceded to the United States in 1846, the U.S. Army established (1849) a fort that is still in operation and contributed to Vancouver's importance during World War II, when the city's shipping and manufacturing boomed. Vancouver has a junior college, a veterans hospital, and state schools for the blind and deaf. It is the headquarters for Gifford Pinchot National Forest. Historic attractions include Fort Vancouver National Historic Site; Covington House (1845), one of the oldest houses in the state; and the Ulysses S. Grant house and museum, where Grant lived when he was stationed here in 1852-53. A state trout hatchery is nearby.

Vancouver Island (1971 est. pop. 360,000), 12,408 sq mi (32,137 sq km), SW British Columbia, Canada, in the Pacific Ocean; largest island off W North America. It is c.285 mi (460 km) long and c.30 to 80 mi (50-130 km) wide and is separated from the mainland by Queen Charlotte, Georgia, and Juan de Fuca straits. The rugged island, a partially submerged portion of the Coast Mts., rises to 7,219 ft (2,201.8 m) at Golden Hinde Mt. The Pacific coastline is deeply indented by numerous fjords and inlets. The island has a mild, humid climate; western Vancouver Island receives the greatest amount of precipitation in North America. The island is heavily forested, and lumbering and wood processing are major industries. Vancouver Island is underlaid by a mineral-rich batholith, from which iron, copper, and gold are mined. Coal is extracted from a depression at the edge of the batholith; the mines at Nanaimo provide most of the coal for British Columbia. Fishing, agriculture, and tourism are other important economic activities. Population is concentrated along the east coast. The island was sighted (1774) by Juan Pérez, the Spanish explorer, and Capt. James Cook was the first (1778) to land here. In 1788 John Meares, an English trader, built a fort on Nootka Sound, which was later occupied by Spanish forces. In 1792 the island was circumnavigated by Capt. George Vancouver. British sovereignty over Vancouver was confirmed (1846) when the U.S.-Canada line was drawn through Juan de Fuca Strait. Vancouver Island was made a crown colony in 1849 and in 1866 became part of British Columbia.

Van·da·lia (văn-dāl'yə), city (1970 pop. 5,160), seat of Fayette co., S central Ill., on the Kaskaskia River SSE of Springfield, in a farm and oil region; inc. 1821. Shoes and packing material are made. It was the second capital of the state, 1820-39, and was on the National Road. Points of interest include the old capitol (1836; now a state memorial) and the Lincoln collection in the library.

Van·der·bilt Mansion National Historic Site (văn'dər-bĭlt'): *see* National Parks and Monuments Table.

Van·der·burgh (văn'dər-bûrg'), county (1970 pop. 168,772), 241 sq mi (624.2 sq km), SW Ind., bounded on the S by the Ohio River and the Ky. border; formed 1818; co. seat Evansville. Its agriculture includes winter wheat, hogs, soybeans, and corn. It has bituminous-coal mines, oil and gas wells, sand and gravel pits, and lumber.

Van·der·grift (văn'dər-grĭft'), industrial borough (1970 pop. 7,889), Westmoreland co., SW Pa., on the Kiskiminetas River NE of New Kensington; laid out 1895, inc. 1897. Steel products are its chief manufactures.

Vä·nern (vä'nərn), lake, c.2,145 sq mi (5,555 sq km), SW Sweden, fed by the Klarälven and drained by the Götaälv SW into the Kattegat. It is the largest lake in Sweden and the third largest in Europe. The deep lake, traversed by the Göta Canal, accommodates small ocean-going vessels. There are vast stands of forest north of the lake; pulp and paper mills line the shore.

Vä·ners·borg (vä'nərs-bôrg'), city (1970 pop. 19,465), capital of Älvsborg co., SW Sweden, at the SW tip of Lake Vänern; founded in the mid-17th cent. It is an industrial center and a port on the Göta Canal.

Van Horn (văn hôrn'), uninc. village (1970 pop. 2,240), seat of Culberson co., W Texas, in the mountains SE of El Paso. Near the site of an old stagecoach stop, Van Horn is a shipping point on the railroad and a highway junction in a country of cattle and sheep ranches.

Vannes (vän), town (1968 pop. 36,576), capital of Morbihan dept., NW France, in Brittany, on the Gulf of Morbihan. It is an important agricultural and tourist center with food-processing and textile factories and flour mills. The surrounding region is noted for its pre-Christian megalithic monuments. Vannes was the capital of the kingdom (later duchy) of Brittany (9th-16th cent.). Points of interest include the Cathedral of St. Peter (13th-19th cent.), which contains the tomb of St. Vincent Ferrer, and ramparts built during the 13th cent. Francis I (reigned 1515-47) was born in Vannes.

Va·nu·a Le·vu (və-no͞o'ə lĕv'o͞o) or **San·dal·wood Island** (săn'dəl-wo͝od'), volcanic island, 2,137 sq mi (5,534.8 sq km), S Pacific, second largest of the Fiji Islands. Mt. Thruston (3,139 ft/960 m) is the highest peak. The Ndreketi is the principal river. The large east peninsula is connected with the rest of the island by a narrow isthmus. There are gold mines and sugar-cane plantations on Vanua Levu.

Van Wert (văn wûrt'), county (1970 pop. 29,194), 409 sq mi (1,059.3 sq km), W Ohio, bounded in the W by the Ind. border and drained by the Little Auglaize River and headwaters of the Auglaize River;

formed 1837; co. seat Van Wert. It has diversified farming (corn, wheat, oats, livestock, and poultry) and limestone quarries. Its manufactures include clothing, wood products, paper, concrete, metal products, farm machinery, electrical equipment, and auto parts.

Van Wert, city (1975 est. pop. 10,800), seat of Van Wert co., NW Ohio, near the Ind. line, in a rich grain-farming area; inc. 1848. Fabricated metal products, electronic equipment, cheeses, cigars, and woodworking machinery are made. The city is known for its peonies, which blossom during the first two weeks in June. An annual peony festival is held.

Van Zandt (văn zănt'), county (1970 pop. 22,155), 845 sq mi (2,188.6 sq km), NE Texas, bounded on the NE by the Sabine River, partly on the E by the Neches River; formed 1848; co. seat Canton. It has diversified agriculture (cotton, corn, grains, legumes, sweet potatoes, fruit, truck crops, and pecans), nursery stock (especially roses), extensive dairying, and poultry and livestock raising. There are large deposits of oil, salt, and natural gas.

Var (vär), department (1974 est. pop. 574,000), 2,316 sq mi (5,998.4 sq km), SE France, in Provence; capital Draguignan.

Va·ra·na·si (və-rä'nə-sē), formerly **Be·na·res** (bə-när'əs, -ēz'), city (1971 pop. 583,856), Uttar Pradesh state, N central India, on the Ganges River. Although a rail hub and trade center, Varanasi is chiefly important as a holy city. Thought to be one of the world's oldest cities, it is the holiest city of the Hindus, who call it Kasi. There are about 1,500 temples, palaces, and shrines. Few of these, however, date back further than the 17th cent., since Moslem invasions destroyed many Hindu religious sites. The most famous Hindu temples are the Golden temple, dedicated to Siva, and the Durga temple with its swarms of sacred monkeys. The banks of the Ganges in the city are bordered by ghats, or flights of steps, that Hindus descend in order to bathe in the sacred river. Hindus believe that to die in Varanasi releases them from the cycle of rebirths and enables them to enter heaven. About one million religious pilgrims visit the city annually. Varanasi is of importance to other religions also. Buddha is said to have begun preaching at Sarnath, 4 mi (6.4 km) outside the city. The mosque of the emperor Aurangzeb stands on the city's highest ground and is the most notable building of the Moslem period in India. Varanasi is also famous for its silk brocades and brassware.

Var·dar (vär'där'), river, c.240 mi (385 km) long, rising in the Šar Planina, S Yugoslavia, and flowing NE, then SE through NE Greece to the Aegean Sea near Thessaloníki.

Va·re·se (vä-rā'zā), city (1972 est. pop. 84,523), capital of Varese prov., Lombardy, N Italy, near the Swiss border. Situated near several Italian lakes, it is a popular tourist center. Manufactures include silk, machinery, and leather goods. The Este palace in Varese (1768–80) has lovely gardens. On a nearby hill is a church originally founded (late 4th cent.) by St. Ambrose.

Var·kaus (vär'kous'), town (1971 est. pop. 24,004), Kuopio prov., S central Finland, on Lake Saimaa. In an abundant forest region, it is a major timber, pulp, and paper-manufacturing center.

Var·na (vär'nə), city (1971 est. pop. 235,200), E Bulgaria, on the Black Sea. It is a major port and an industrial center, with shipyards and one of Bulgaria's largest cotton textile industries. Other products include foodstuffs, machinery, metal goods, soap, chemicals, ceramics, and household appliances. Varna is also an international summer resort. Founded in 580 B.C. as the Greek colony of Odessus, it passed to the Roman Empire in the 1st cent. A.D. The city passed to the second Bulgarian kingdom in 1201, was captured by the Turks in 1391, and became an active seaport under their rule. In 1444 the Turks under Murad II won a decisive victory over Crusader forces led by Ladislaus III of Poland and Hungary, who was killed. The Battle of Varna was the last major attempt by Christian Europe to stem the Ottoman tide. Varna was (1854) the chief naval base of the British and French forces in the Crimean War. The city was liberated from Turkish rule in 1878 and ceded to newly independent Bulgaria. It now has a university (founded 1920), a polytechnic institute, a naval academy, an archaeological museum, and ruins of a 5th cent. basilica and a 6th cent. Byzantine fortress.

Varns·dorf (värns'dôrf), city (1970 pop. 13,927), NW Czechoslovakia, in Bohemia, on the East German border. It is a railway junction and has industries that produce hosiery, cotton, velvet, linen textiles, dyes, and wooden articles. The village was known in the 14th cent.; the city was founded in 1849 through the union of six settlements.

Väs·ter·ås (vĕs'tə-rōs'), city (1972 est. pop. 117,946), capital of Västmanland co., E Sweden, a port on Lake Mälaren at the mouth of the Svartån River. It is the main center of the Swedish electrical-goods industry; other manufactures include metal goods, textiles, and glass. Founded by 1100, Västerås was one of Sweden's great medieval cities, with a cathedral (13th cent.) and a fortified castle (14th cent.). Eleven important diets convened here, notably the Västerås Recess (1527), which created the Lutheran state church, and the diet of 1544, which made the Swedish throne hereditary.

Väs·ter·vik (vĕs'tər-vēk'), city (1972 est. pop. 22,123), Kalmar co., SE Sweden, on an inlet of the Baltic Sea. Manufactures of this industrial center include paper, furniture, matches, and prefabricated houses. There are also shipyards, iron foundries, and fisheries.

Vat·i·can (văt'ĭ-kən), residence of the pope at Rome. Since the so-called Roman Question was ended by the Lateran Treaty of 1929 between Pope Pius XI and King Victor Emmanuel III (negotiated by Cardinal Gasparri and Mussolini), the Vatican City has been an independent state (108.7 acres/44 hectares), with the pope as its absolute ruler. The Vatican City is a roughly triangular tract of land within Rome, on the west bank of the Tiber River and west of the Castel Sant' Angelo. In its southeast corner is the piazza of St. Peter's Church. North of the piazza is a quadrangular area containing administrative buildings and the Belvedere Park. West of Belvedere Park are the pontifical palaces, and beyond the palaces lie the Vatican Gardens, which make up half the area of the little state. The Leonine Wall forms the western and southern boundaries. The political freedom of the Vatican is guaranteed and protected by Italy.

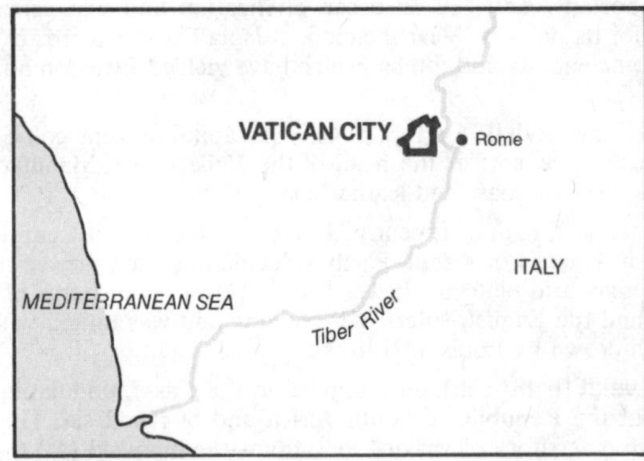

The state has its own citizenship, issues its own currency and postage stamps, and has its own flag and a large diplomatic corps. The civil government of the Vatican City is run by a lay governor and a council, all appointed by and responsible to the pope. The law is the canon law and the courts are part of the judicial system of the church. The Vatican is above all the seat of the central government of the Roman Catholic Church. The principal chapel in the Vatican is the Sistine Chapel, the ceiling of which was painted (1508–12) by Michelangelo. The history of the Vatican as a papal residence dates from the 5th cent., when, after Emperor Constantine I had built the basilica of St. Peter's, Pope Symmachus built a palace nearby. The Vatican City has its own newspaper (*Osservatore Romano*), railroad station, and broadcasting facility (first established by Marconi under Pius XI).

Vat·na·jö·kull (vät'nä-yœ'kool), glacier, c.3,150 sq mi (8,160 sq km), SE Iceland; largest glacier in Europe. At an elevation of from 4,200 to 6,100 ft (1,280–1,860 m), it covers a huge volcanic plateau. It descends in some 40 branches into adjoining valleys. Some peaks rise above the ice, notably Öraefajökull.

Vät·tern (vĕt'ərn), lake, 733 sq mi (1,898.5 sq km), c.80 mi (130 km) long and up to 20 mi (32 km) wide, S central Sweden, drained by the Motala Ström E into the Baltic Sea. It is the second-largest lake in Sweden. The Göta Canal crosses the northern part of the deep lake. Visingsö (c.10 sq mi/25 sq km) is the largest island in the Vättern; a prehistoric burial mound is there.

Vau·cluse (vō-klüz'), department (1974 est. pop. 380,600), 1,377 sq mi (3,566.4 sq km), SE France, in Provence; capital Avignon.

Vaud (vō), canton (1974 est. pop. 520,000), 1,239 sq mi (3,209 sq km), W Switzerland; capital Lausanne. Bordering on France in the west, it lies roughly between the Lake of Geneva, the Lake of Neuchâtel, the Jura Mts., and the Bernese Alps. Cereals, tobacco, and other crops are grown, livestock is raised, and wine is produced in its large, fertile central region. There are watchmaking towns in the west. Montreux and Vevey are among its numerous resorts. The population is French-speaking and mainly Protestant. Originally occupied by Celts, the region was conquered by the Romans in 58 B.C. In 1536 it was conquered by Bern and forced to accept the Reformation. In 1798, having revolted against its Bernese rulers, it became the canton of Leman in the Helvetic Republic. In 1803 it joined the Swiss Confederation under its present name.

Väx·jö (vĕk'shœ'), city (1972 est. pop. 43,411), capital of Kronoberg co., S Sweden. Manufactures include textiles, machines, sporting goods, and furniture. An old city, Växjö became the seat of a bishop c.1170 and was chartered in 1342. It has an impressive cathedral (12th cent.) and the Småland Museum, which contains a notable collection of glassware.

Vay·gach or **Vai·gach** (both: vī-gäch'), island, 1,312 sq mi (3,398 sq km), Nenets National Okrug, off NE European USSR, in the Arctic Ocean, between the mainland and Novaya Zemlya. The island is covered with tundra and has zinc, copper, and lead deposits.

Ve·ga (vā'gə), town (1970 pop. 839), seat of Oldham co., extreme N Texas, in the plains of the Panhandle W of Amarillo. It is a trade and shipping center for a cattle and grain region.

Ve·ii (vē'ī'), ancient city of Etruria, 11 mi (18 km) NW of Rome, Italy. One of the most powerful member cities of the Etruscan League, it was constantly at war with Rome. It fell (c.396 B.C.) to the Roman army under Camillus after a 10-year siege. The city was an important center of Etruscan civilization and was especially noted for its statuary. Near the modern Isola Farnese there are ruins of the ancient city and tombs, which have yielded Etruscan antiquities.

Vej·le (vī'lə), city (1976 est. pop. 48,848), capital of Vejle co., central Denmark, a seaport at the head of the Vejle Fjord. Manufactures include textiles, soap, and leather goods.

Ve·lay (və-lā'), region, E central France, in the Massif Central and partly in Haute-Loire dept. Partly volcanic in origin, the region has many high, arid plateaus. It was held in turn by the counts of Toulouse and the English rulers of Aquitaine and was united with the French crown by Louis VIII in the 13th cent.

veld or **veldt** (both: vĕlt), term applied to the grassy, undulating plateaus of the Republic of South Africa and of Rhodesia. The veld comprises territory of varying elevation—the highveld (4,000–6,000 ft/1,220–1,830 m), the middleveld (2,000–4,000 ft/610–1,220 m), and the lowveld (500–2,000 ft/150–610 m). The highveld, the largest of the plateaus, is in the Republic of South Africa. Abundant crops of potatoes and maize are grown, large cattle herds are grazed, and industrial and mining centers are found.

Ve·li·ki U·styug (və-lē'kē o͞o-styo͞ok'), city (1970 pop. 37,000), N central European USSR, a port on the Northern Dvina River. Industries include shipbuilding and the production of textiles, bristles and brushes, and silver handicrafts. First mentioned as Ustyug in the 12th cent., the city later became (16th–17th cent.) a wealthy and important trade and transport center on the road from Moscow to Arkhangelsk and from European Russia to Siberia. Veliki Ustyug also became a noted center of painting and artistic handicrafts (particularly in silver) during the 17th and 18th cent. Its commercial role declined after the founding of St. Petersburg in 1703. City landmarks include the Voznesensky Church and several monasteries.

Ve·li·ki·ye Lu·ki (və-lē'kē-ə lo͞o'kē), city (1973 est. pop. 95,000), W central European USSR, on the Lovat River. A railroad junction, it has industries producing foodstuffs, bricks, radios, and furniture. Dating from the 12th cent., the city was controlled first by Novgorod, then by Lithuania, and, from 1448, by Moscow.

Vel·lore (vĕ-lôr'), town (1971 pop. 139,082), Tamil Nadu state, SE India, on the Palar River. It is a district administrative center and an agricultural market town. In the 18th cent. Vellore was the stronghold of the Indian leaders Haidar Ali and his son Tippoo Sahib. An ancient fort, a 14th cent. temple of the Hindu god Siva, and a medical college are in the town.

Vel·sen (vĕl'sən), municipality (1973 est. pop. 66,989), North Holland prov., W Netherlands, near the mouth of the North Sea Canal. It is a center of the Dutch steel industry.

Ve·nan·go (və-năng'gō), county (1970 pop. 62,353), 678 sq mi (1,756 sq km), NW Pa., drained by the Allegheny River; formed 1800; co. seat Franklin. It is in an oil-producing region, with coal mines and natural-gas wells. Its industry includes dairy processing, clothing, lumbering, and the manufacture of chemicals, plastics, metal products, and miscellaneous machinery.

Ven·dée (väɴ-dā'), department (1974 est. pop. 430,200), 2,594 sq mi (6,718.5 sq km), W France, on the Bay of Biscay, in Poitou; capital La Roche-sur-Yon. The offshore islands of Noirmoutier and Yeu are included in the department. Largely an agricultural (wheat, cattle raising) and forested region, the Vendée has many beach resorts and fishing ports. Canned fish, leather, textiles, fishing boats, and uranium are the chief products. The department gave its name to the insurrection of 1793 to 1796, which began here. The peasants of the Vendée, who had lived amiably with the local nobility, began violently to oppose the French Revolution when it turned against the Roman Catholic Church. Under Henri La Rochejaquelein and others, an army of more than 50,000 men was raised to clear the region of Revolutionary authorities. The army occupied Saumur and planned to continue through Brittany, Maine, and Normandy to join the Chouans, the anti-Revolutionary peasants of those regions. However, the important city of Nantes held out against the Vendeans, who marched as far north as Granville but were then forced by lack of discipline to return south late in 1793. Overtaken at Le Mans and Savenay by the republican army, they were totally defeated and suffered terrible reprisals. Robespierre's overthrow led to the peace of La Jaunaie (1795), by which the government granted an amnesty and freedom of worship to the Vendeans. Renewed conflict began in 1796, when royalist émigrés, backed by Great Britain, tried to land at Quiberon in Brittany; they were routed by government forces under Gen. Lazare Hoche. The comte d'Artois (later Charles X), who had landed on the isle of Yeu, took fright and abandoned the Vendean leaders to capture and execution. Smaller royalist uprisings occurred in 1815 (against Napoleon I) and in 1832, when the duchess de Berry tried to stir up the Vendée for the Bourbon cause against Louis Philippe.

Ven·dôme (väɴ-dōm'), town (1968 pop. 16,728), Loir-et-Cher dept., N central France, in Orléanais. It is a manufacturing town with food processing and metal and electrical industries. During the Middle Ages the town was the prosperous center of the county (duchy after 1515) of Vendôme. Henry IV inherited the duchy as part of the Bourbon lands, united it with the royal domain in 1589, and gave the title of the duke of Vendôme to his illegitimate son César in 1598. Among the numerous monuments are the ruins of a chateau built during the 12th, 14th, and 15th cent.

Ve·ne·ti·a (və-nē'shē-ə, -shə), region (1972 est. pop. 4,168,445), 7,095 sq mi (18,376 sq km), NE Italy, bordering on the Gulf of Venice (an arm of the Adriatic Sea) in the E and on Austria in the N; capital Venice. Venetia falls into two geographic zones, a mountainous and hilly area in the north, which includes parts of the Dolomites and Carnic Alps (Alpi Carniche), and the fertile Venetian Plain in the south, which is partly marshy near the Adriatic. Venetia's main rivers are the Po, the Adige, and the Piave. The region borders on Lake Garda in the west. Most of Venetia's workers are engaged in agriculture; the leading crops are cereals, fruit, sugar beets, and hemp. Mulberry trees are grown for use in sericulture. Manufactures include textiles, chemicals, paper, processed food, wine, and ships. Tourism is important. Venetia derives its name from the Veneti, a people who settled the region c.1000 B.C. and who came under Roman rule in the 2nd cent. B.C. Emperor Augustus joined Venetia and Istria to form a separate province, whose capital was Aquileia. Venetia was devastated by Attila, king of the Huns, in the mid-5th cent. A.D. About the 10th cent. the towns of the region began to reacquire importance. They were ruled at first by their bishops and later developed into free communes. Some towns (including Verona and Padua) grew powerful under the rule of noble families, but the republic of Venice gradually became dominant, and by the early 15th cent. its territories included virtually all of present-day Venetia. By the Treaty of Campo Formio (1797) Venetia passed to Austria, and by the Treaty of Pressburg (1805) it was made part of the Napoleonic kingdom of Italy. In 1814 Venetia was restored to Austria, which held it to the end of the Austro-Prussian War (1866), when it was ceded to Italy. After World War II Udine prov. in the east was detached from Venetia and was combined with that part of Venezia Giulia not ceded by Italy to Yugoslavia to form the region of Friuli-Venezia Giulia. Venetia suffered severe flooding in 1966.

Ve·ne·zi·a Giu·lia (və-nĕt'sē-ə jo͞ol'yä), former region, NE Italy, on the Adriatic Sea. It was formed after World War I from part of the territories ceded by Austria to Italy in 1919, and included eastern Friuli, Trieste, Istria, and part of Carniola. Fiume was added in 1921. After World War II most of Venezia Giulia was ceded by Italy to Yugoslavia. A section (c.285 sq mi/740 sq km) formed the Free Territory of Trieste (divided in 1954 between Italy and Yugoslavia). The rest of Venezia Giulia remained Italian and was merged with Udine prov.

Ven·e·zue·la (vĕn'ə-zwā'lə, -zwē'lə), republic (1973 est. pop. 11,300,000), 352,143 sq mi (912,050 sq km), N South America. The capital is Caracas. Venezuela has a coastline 1,750 mi (2,816 km) long on the Caribbean Sea in the north. It is bordered on the south by Brazil, on the west and southwest by Colombia, and on the east by Guyana. Dependencies include Margarita Island, Tortuga Island, and many smaller island groups in the Caribbean.

Geographically Venezuela is a land of vivid contrasts, with four major divisions: the Venezuelan highlands, the coastal lowlands, the basin of the Orinoco River, and the Guiana Highlands. An almost inaccessible and largely unexplored wilderness south of the Orinoco, the Guiana Highlands occupy more than half of the national territory and are noted for scenic wonders such as Angel Falls. Iron ore, gold, diamonds, and other minerals are found near Ciudad Bolívar. The dense forests of the region yield rubber, tropical hardwoods, and other forest products. The boundary with Brazil is mostly mountainous. The Orinoco, one of the great rivers of South America, has its source in this region. The Orinoco basin is a great pastoral area. North of the Orinoco are the llanos, the vast, hot Orinoco plains, where there is a great cattle industry. Oil is found north of the Orinoco. The most vital region economically, however, is an area in the coastal plains, the lowlands around Lake Maracaibo. There, since 1918, Dutch, British, and American interests have developed the astonishingly rich oil fields. The coastal lowlands are exceedingly hot, but coastal ranges rise abruptly from the Caribbean to cool altitudes of 6,000 to 7,000 ft (1,830–2,130 m). These ranges soon become a region of hills, intermontane basins, and plateaus known as the Venezuelan highlands and are a spur of the Andes. Farther to the southwest the mountains rise to their greatest height at Pico Bolívar (16,427 ft/5,010 m) in the Sierra Nevada de Mérida. Densely populated, the highland region is the political and commercial hub of the nation. Coffee, the keystone of the economy before the oil boom, comes from the slopes and cocoa from the lower foothills.

Economy. Because of its oil riches Venezuela has the highest per capita national income in Latin America. Oil accounts for about 90% of the export income, 70% of government revenues, and 20% of the gross national product. Other exports are iron ore, coffee, cocoa, rice, and cotton. The main imports are manufactured goods—especially machinery, vehicles, and chemicals—and food. A large amount of oil is exported to the Netherlands Antilles for refining. The government has used oil revenues to stimulate manufacturing industries. Food processing and the manufacture of textiles, shoes, chemicals, and automobiles have become well established. Heavy metal works are being established on the Orinoco near Santo Tomé de Guayana (Ciudad Guayana). Despite recent government reform programs, Venezuela's great wealth remains in the hands of a small minority.

History. Columbus discovered the mouths of the Orinoco in 1498. Amerigo Vespucci, coming upon an island off the Paraguaná peninsula in 1499, nicknamed it Venezuela (little Venice) because of Indian villages built above the water on stilts; the name held and was soon applied to the mainland. Spanish settlements were established on the coast, but the major task of the conquest was accomplished by German adventurers—Ambrosio de Alfinger, George of Speyer, and especially Nikolaus Federmann.

After Napoleon took control of Spain a revolution began (1810) in Venezuela, under Francisco de Miranda. In 1811 complete independence was declared. Later Simón Bolívar and his lieutenants, working from Colombia, were able to liberate Venezuela. The victory of Carabobo (1821) secured independence from Spain.

Venezuelan politics have been restive, with control passing from landholding caudillos to a series of dictators, elected presidents, military juntas, and finally to a popularly elected president. There have been several uprisings and outbreaks of terrorism since then, along with uneasy relationships with some of its neighbors over border disputes and control of offshore oil reserves.

Government. Venezuela is governed under a 1961 constitution. The president and members of congress are elected for five-year terms. The president may not succeed himself.

Ven·ice (vĕn'ĭs), city (1976 est. pop. 362,494), capital of Venetia and of Venice prov., NE Italy, built on 118 alluvial islets within a lagoon in the Gulf of Venice, an arm of the Adriatic Sea. The city is connected with the mainland, 2.5 mi (4 km) away, by a rail and highway bridge. Between the islands run about 150 canals crossed by some 400 bridges. The Grand Canal is the main traffic artery; its chief bridge is the Rialto. Gondolas and other boats are the only means of conveyance. Houses are built on piles. Venice is a tourist, commercial, and industrial center. Manufactures include lace, jewelry, flour, and glass, and the city is a center for shipbuilding.

History. With Istria, Venice formed a province of the Roman Empire. In the late 5th cent. refugees fleeing the Lombard invaders of northern Italy sought safety on the largely uninhabited islands. The communities organized themselves (697) under a doge and by the 9th cent. had formed the city of Venice. The city secured (10th cent.) most of the coast of Dalmatia, thus gaining control of the Adriatic, and began to build up its eastern empire, obtaining trade and other privileges in the ports of the eastern Mediterranean. (The influence of the Middle East, particularly Byzantium, is most clearly expressed in St. Mark's Church—rebuilt 1063–73—located on the city's principal square.) Strategic points in the Ionian, the Aegean, and the eastern Mediterranean were taken, notably Crete (1216). The great traveler Marco Polo represented the enterprising spirit of Venice in the 13th and 14th cent. After defeating (1380) its rival Genoa, Venice was indisputably the leading European sea power. In the 15th cent. Venice reached the height of its power. The city engaged in a rich trade between Europe and Asia, all Venetia on the mainland was conquered, and Venetian ambassadors made the power of the city felt at every court of the known world.

The Renaissance marked the height of Venice's artistic glory. The Venetian school of painting, which besides its giants—Titian and Tintoretto—also included Giovanni Bellini, Giorgione, and Veronese, gave Venice its present aspect of a city of churches and palaces, floating on water, blazing with color and light, and filled with art treasures.

The fall of Cyprus (1571), Crete (1669), and the Peloponnesus (1715) to the Turks ended Venetian dominance in the eastern Mediterranean. When, in 1797, Napoleon I delivered Venice to Austria, the republic fell without fighting. Having expelled the Austrians in 1848, however, Venice under the leadership of Daniele Manin heroically resisted siege until 1849. In 1866 Venice and Venetia were united with the kingdom of Italy.

The center of animation in Venice is St. Mark's Square and the Piazzetta, which leads from the square to the sea. On the square are St. Mark's Church; the Gothic Doges' Palace (14th–15th cent.), from which the Bridge of Sighs (c.1600) leads to the former prisons; the Old and New Law Courts (16th–17th cent.); the campanile (325 ft/99 m high; built in the 10th cent. and rebuilt after it collapsed in 1902); the Moors' Clocktower (late 15th cent.); the elegant Old Library (1553); St. Moses' Church; and the twin columns supporting the statues of St. Theodore stepping on a crocodile and of a winged lion of St. Mark, the emblem of Venice. On an island facing the Piazzetta is the Church of San Giorgio Maggiore (1566–1610).

Venice, resort city (1970 pop. 7,015), Sarasota co., SW Fla., on the Gulf coast S of Sarasota, in a citrus-fruit area. It was developed in 1924 as a retirement community.

Ven·lo (vĕn′lō′), municipality (1973 est. pop. 62,675), Limburg prov., SE Netherlands, on the Maas (Meuse) River, near the West German border. It is a trade center for fruit and vegetables. Manufactures include lumber and chemicals. Venlo was formerly a fortress city; it frequently changed hands during the wars of the 16th, 17th, and 18th cent. Noteworthy buildings include a 15th cent. Gothic church and the 16th cent. town hall.

Ven·ti·mi·glia (vĕn′tə-mēl′yə), town (1972 est. pop. 21,000), Liguria, NW Italy, on the Ligurian Sea and the Italian Riviera, near the French border. It is a seaport, a popular beach resort, and a major flower market. Landmarks include the 11th cent. Church of San Michele and the cathedral (11th–12th cent.), with a 5th cent. piscina. In nearby Grimaldi are grottoes in which prehistoric remains have been found.

Vents·pils (vĕnt′spīls), city (1970 pop. 40,467), W European USSR, in Latvia, on the Baltic Sea, at the mouth of the Venta River. An ice-free seaport, it also has shipyards, fisheries, and varied manufactures.

Ven·tu·ra (vĕn-to͝or′ə, -tyo͝or′ə), county (1970 pop. 378,497), 1,863 sq mi (4,825.2 sq km), SW Calif., in a mountainous area, with a wide delta of the Santa Clara River entering the Pacific, including small islands in the central part of the Santa Barbara chain; formed 1872; co. seat Ventura. It is a leading Calif. county in lemon growing and also produces other citrus fruit, lima beans, walnuts, sugar beets, and truck crops. It has oil and natural-gas fields and refineries. Its industry includes food processing, clothing, lumber millwork, and diverse manufacturing.

Ventura, city (1979 est. pop. 69,500), seat of Ventura co., SW Calif., on the Pacific coast in a farm and oil region; inc. 1866. Fruit and vegetable packing, petroleum production, and research (especially in electronics and missiles) are the major industries. A mission called San Buenaventura (still the official name of the city), founded by Junípero Serra in 1782, has been restored.

Ve·ra·cruz (vĕr′ə-kro͞oz′, vâr′ä-kro͞os′), state (1970 pop. 3,815,419), 27,759 sq mi (71,896 sq km), E central Mexico; capital Jalapa de Enriquez. Stretching c.430 mi (690 km) along the Gulf of Mexico and reaching from 30 to 100 mi (48–161 km) inland, Veracruz rises from a tropical coastal plain into the temperate valleys and highlands of the Sierra Madre Oriental. In ancient times the area was a hub of pre-Columbian civilizations, including the Olmec, the Huastec, and the Totonac-speaking tribes.

Veracruz, city (1970 pop. 214,072), Veracruz state, E central Mexico, on the Gulf of Mexico. Rivaling Tampico as the country's main port, it is also the commercial and industrial center of an important oil region, as well as a major tourist resort with beautiful scenery, fine beaches, and excellent accommodations. The city stands on a low, sandy plain surrounded by dunes and swamps, some of which have been reclaimed and are very fertile. In 1519 the Spanish explorer Hernán Cortés landed near the site later chosen (1599) for the present city. Veracruz was easy prey for the buccaneers of the 17th and 18th cent. The harbor is guarded by the fortress of San Juan de Ulúa, which was begun in the 17th cent. and was the last stronghold of the Spanish before their expulsion in 1821. Veracruz was blockaded in 1838 by the French, who sought compensation for damages suffered by French nationals. In 1847 U.S. troops under Gen. Winfield Scott landed at Veracruz to begin the major campaign of the Mexican War. The War of the Reform involved foreign intervention in Veracruz; in Dec., 1861, Spanish troops landed here as the first contingent of a joint European force. French and British forces arrived the following month. When it became apparent that France was bent on actual conquest, the Spanish and British withdrew from the joint force. The adventure of the empire of Maximilian ensued. In 1914 an incident involving U.S. sailors in Tampico led President Woodrow Wilson to land troops in Veracruz, where they briefly besieged the customhouse. Mexico responded by severing diplomatic relations.

Ver·cel·li (vĕr-chĕl′ē), city (1972 est. pop. 56,832), capital of Vercelli prov., Piedmont, N Italy, on the Sesia River. It is an important rice market and has food-processing and textile industries. A Roman town and later a free commune, it passed to the Visconti of Milan in 1335 and was ceded by them to the house of Savoy in 1427.

Ver·de (vûr′dē), river, c.190 mi (310 km) long, rising in central Ariz. and flowing S to the Salt River. The valley supported early Indian civilizations and is dotted with ruins, such as those at Tuzigoot National Monument. Bartlett and Horseshoe dams are on the river.

Verde, Cape (vûrd), peninsula, extending into the Atlantic Ocean, W Senegal; the westernmost point of Africa. The cape was discovered by the Portuguese in 1445. The Cape Verde Islands are c.350 mi (560 km) to the west.

Ver·dun (vər-dŭn′), city (1971 pop. 74,718), S Que., Canada, on the S shore of Montreal Island, on the St. Lawrence River. It is a residential suburb of Montreal.

Ver·dun (vər-dŭn′, vĕr-dœɴ′), town (1968 pop. 22,013), Meuse dept., NE France, in Lorraine, on the Meuse River. A strategic transportation center, Verdun has varied industries, including textiles, candy making, and printing. The town was a prosperous commercial center in Roman times and also during the Carolingian period in the 800s. An episcopal see since the 4th cent., Verdun, with its surrounding area, was one of the three bishoprics (with Metz and Toul) seized (1552) by Henry II of France from the Holy Roman Empire. The town itself was a free imperial city before it passed to France. The Peace of Westphalia (1648), ending the Thirty Years' War, confirmed Verdun in French possession. Fortified by Sébastien Vauban during the reign of Louis XIV, Verdun thereafter became important strategically. After 1871 the town became the principal French fortress facing Germany and was surrounded by a ring of defenses. The longest battle of World War I was fought at Verdun in 1916. In 1918 the Americans and French were victorious in the Verdun sector and at Saint-Mihiel. Almost totally destroyed, Verdun was rebuilt after the war. The town and the battlefield of Verdun, with their huge military cemeteries and numerous impressive monuments, form a national sanctuary.

Ve·ree·ni·ging (fə-rē′nĭ-kĭng′), city (1970 pop. 172,549), Transvaal, NE South Africa, on the Vaal River. An industrial center, its chief products are iron, steel pipes, bricks and tiles, and processed lime and coal. Thermal power plants supply electricity to nearby goldfields. The city, founded in 1892, owed its early growth to nearby coal mines. The Treaty of Vereeniging (1902), which ended the South African War, was negotiated here and signed in Pretoria.

Ver·kho·yansk (vyĕr′khə-yänsk′), town, Yakut Autonomous Republic, NE Siberian USSR, on the Yana River, near the Arctic Circle. The lowest temperature recorded here was −90°F (−68°C). Founded in 1638, it was a place of political exile until 1917.

Verkhoyansk Range, mountain chain, c.600 mi (970 km) long, E Siberian USSR, in the Yakut Autonomous Republic. The world's lowest temperatures for inhabited places have been recorded in this region.

Ver·man·dois (vĕr′mäɴ-dwä′), region, N France, now in Somme and Aisne depts. The region is largely agricultural (wheat, beets) but has some industry. Vermandois became an earldom under Charlemagne (9th cent.). The region was annexed to the French crown (c.1200), ceded to Burgundy (1435), recovered by Louis XI (1477), and incorporated into the province of Picardy.

Ver·mil·ion (vər-mĭl′yən). **1.** County (1970 pop. 97,047), 899 sq mi (2,328.4 sq km), E Ill., bounded on the E by the Ind. border and drained by the Vermilion and Little Vermilion rivers; formed 1826; co. seat Danville. It has agriculture (corn, oats, soybeans, wheat, truck crops, dairy products, livestock, poultry), bituminous-coal mining, and diversified manufacturing. **2.** Parish (1970 pop. 43,071), 1,205 sq mi (3,121 sq km), S La., on the Gulf of Mexico, bounded partly on the N and NW by Bayou Queue de Tortue, on the SE by Vermilion Bay, drained by the Vermilion River, and crossed by the Gulf Intracoastal Waterway; formed 1844; parish seat Abbeville. Its agriculture includes cotton, corn, sugar cane, hay, sweet potatoes, and rice. Among its industries are sugar and rice milling, cotton ginning, canning, lumber milling, and oil and natural-gas extraction. The region is also noted for fur trapping, cattle raising, duck hunting, and fishing.

Vermilion, city (1970 pop. 9,872), Erie and Lorain cos., N central Ohio, near Lake Erie on the Vermilion River ESE of Sandusky, in a fishing and resort area; settled 1808, inc. 1838.

Ver·mil·lion (vər-mĭl′yən), county (1970 pop. 16,793), 263 sq mi (681.2 sq km), W Ind., bounded on the W by the Ill. border, on the E by the Wabash River, and drained by the Vermilion River; formed 1824; co. seat Newport. It has agriculture (grain, fruit, and livestock), some timber, and bituminous-coal mines. Clothing and chemicals are made.

Vermillion, city (1970 pop. 9,128), seat of Clay co., extreme SE S.Dak., on the Vermilion River SSW of Sioux Falls; settled 1859, inc. 1873. It is a trade center of a rich farm and livestock region. Fort Vermillion, a fur-trading post, was built (1835) nearby. The city was the site of an encampment of Mormons (1845-46) on their way to Utah.

Ver·mont (vər-mŏnt'), state (1975 pop. 474,000), 9,609 sq mi (24,887 sq km), NE United States, in New England, admitted to the Union in 1791 as the 14th state. Montpelier is the capital. Vermont is bounded on the north by the Canadian province of Que., on the east by N.H. (the Connecticut River forms the line), on the south by Mass., and on the west by N.Y. (Lake Champlain constitutes more than half this boundary).

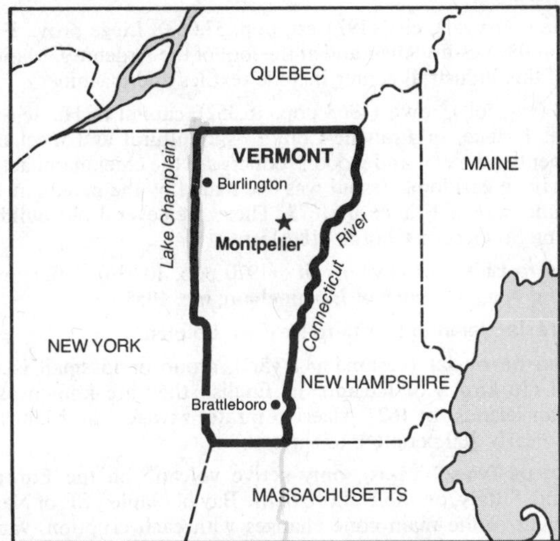

The forested Green Mts. constitute the dominant topographical feature of the state. They consist of at least four distinct groups, all traversing the state generally in a north-south direction. Largest and most important are the Green Mts. proper, which extend down the center of the state from the Canadian border to the Mass. line, rising to Vermont's highest peak in Mt. Mansfield (4,393 ft/1,339 m). The Taconic Mts., occupying the southwestern portion of the state, contain Vermont's important marble deposits. East of the Green Mts. and extending from the Canadian border to somewhat below the middle of the state are the Granite Hills, so called because of their valuable stone. The fourth group, sometimes called the Red Sandrock Hills, extends along the Vermont shore of Lake Champlain. In eastern Vermont there are also isolated peaks, or monadnocks, not connected with the principal ranges.

The rivers of Vermont flow either into the Connecticut River or into Lake Champlain. The Winooski rises east of the Green Mts. and cuts directly through them to Lake Champlain. Grand Isle co., comprising several islands and a peninsula jutting down into Lake Champlain from Canada, is connected to Vermont proper by causeways. Vermont has the variable climate of the northeast temperate zone, with abundant rainfall and a growing season that varies from 120 days in the Connecticut valley to 150 in the Lake Champlain region. Winter brings heavy snows, which usually cover the ground for at least three full months.

Vermont is divided into 14 counties:

NAME	COUNTY SEAT	NAME	COUNTY SEAT
Addison	Middlebury	Grand Isle	North Hero
Bennington	Bennington and Manchester	Lamoille	Hyde Park
		Orange	Chelsea
		Orleans	Newport
Caledonia	St. Johnsbury	Rutland	Rutland
Chittenden	Burlington	Washington	Montpelier
Essex	Guildhall	Windham	Newfane
Franklin	St. Albans	Windsor	Woodstock

Economy. With its rugged terrain, much of it still heavily wooded, Vermont has limited areas of arable land, but the state is well suited to grazing. Dairy farming has long been dominant in agriculture, and the state ships milk in huge quantities. Hay is the state's chief crop, and Vermont is famous for its maple syrup; apples and potatoes are other important products. The state's most valuable mineral resources are stone, asbestos, sand and gravel, and talc. In Rutland and Proctor industry is based on the quarrying and finishing of marble, and at Barre the famous Vermont granite is quarried and processed. The manufacture of nonelectric machinery is a major industry. The manufacture of computer components, food products, pulp and paper, and plastics has helped to compensate for the declining textile industry. Tourism is also vitally important to the state economy. Every summer thousands of vacationers are drawn by the scenic mountains and the picturesque New England villages, while climbers attack the many accessible peaks and hikers take on the Long Trail that runs the length of the state along the Green Mt. ridge. In the winter skiers flock to the slopes at Mad River Glen, Bromley, Stowe, and elsewhere.

History. The first white man known to have entered the area that is now Vermont was Samuel de Champlain, who, after beginning the colonization of Quebec, journeyed south with a Huron war party in 1609 to the beautiful lake to which he gave his name. In 1666 the French built a fort on the Isle La Motte in Lake Champlain. However, this and later French settlements were abandoned, and until well into the 18th cent. the region was something of a no man's land. Fort Dummer, built (1724) by the English near the site of Brattleboro to protect settlers from Indians, is considered the first permanent settlement in what is now Vermont.

In 1741 Benning Wentworth became royal governor of New Hampshire. The boundary between New Hampshire and New York was not precisely defined in his commission, and both Wentworth and Gov. George Clinton of New York claimed the land between Lake Champlain and the Connecticut River. In 1749 Wentworth made the first of the New Hampshire Grants—the township called Bennington—to a group that included his relatives and friends. The New Yorkers contested the validity of these grants, and a territorial dispute lasting 40 years ensued. Regional pride among the New England settlers played a large part in creating resistance to New York authority, which was upheld by the British crown in 1764. Chief among the leaders of this resistance was Ethan Allen, who organized the Green Mountain Boys. At the beginning of the Revolution the Green Mountain Boys captured Ticonderoga, and Seth Warner took Crown Point. In Jan., 1777, Vermont (as its citizens were soon calling the region) proclaimed itself an independent state at Westminster. Chiefly because of the opposition of New York, the Continental Congress refused to recognize Vermont as the 14th colony or state. For ten years Vermont remained an independent state, performing all the offices of a sovereign government and gradually becoming more and more independent. Not until 1791, after the dispute with New York was finally adjusted (1790) by payment of $30,000, did Vermont enter the Union. It was the first state to be admitted after the adoption of the Constitution by the 13 original states.

In the next two decades Vermont had the greatest population increase in its history, from 85,425 in 1790 to 217,895 in 1810. The War of 1812 was unpopular in Vermont, as in the rest of New England, and during the war extensive smuggling across the Canadian border was carried on. At this early period in its history Vermont, lacking an aristocracy of wealth, was the most democratic state in New England. Jeffersonian Democrats held control for most of the first quarter of the 19th cent. In the Mexican War, which it viewed as an undertaking solely to increase slave territory, Vermont was very apathetic, but no Northern state was more energetic in support of the Union cause in the Civil War.

The economy of the state, meanwhile, was in the midst of a series of sharp dislocations. The rise of manufacturing in towns and villages during the early 19th cent. had created a demand for foodstuffs for the nonfarming population. Consequently commercial farming began to crowd out the subsistence farming that had predominated since the mid-18th cent. Grain and beef cattle became the chief market produce, then sheep for the wool textile mills of southern New England. After the Civil War the rural population declined as many farmers migrated westward or to the cities, and abandoned farms became a common sight. The transition to dairy farming in the 20 years following the war staved off a permanent decline in Vermont's agriculture.

Government. Vermont is governed under a constitution adopted in 1793. The state legislature, called the general assembly, consists of a senate with 30 members and a house of representatives with 150 members, all elected to serve two-year terms. The governor is elected for a two-year term.

Educational Institutions. Among Vermont's institutions of higher education are Bennington College, Middlebury College, and the Univ. of Vermont, at Burlington.

Ver·nal (vûr'nəl), city (1970 pop. 3,908), seat of Uintah co., NE Utah, in the Uintah Basin S of the Uinta Mts. and N of the Green River; inc. 1879. It is the trade and processing center of a grain, livestock, and mine region. Oil fields are in the vicinity.

Ver·non (vûr'nən), city (1971 pop. 13,283), S British Columbia, Canada, near the north end of Okanagan Lake. There are lumber mills and sawmills in the district.

Vernon. 1. Parish (1970 pop. 53,794), 1,357 sq mi (3,514.6 sq km), W La., bounded on the W by the Sabine River, here forming the Texas border, and drained by the Calcasieu River; formed 1871; parish seat Leesville. Corn, cotton, peanuts, hay, and sweet potatoes are grown, and livestock is raised. There are sand and gravel pits. **2.** County (1970 pop. 19,065), 838 sq mi (2,170.4 sq km), W Mo., on the Osage River and drained by the Little Osage and Marmaton rivers; formed 1851; co. seat Nevada. It is an agricultural area (corn, wheat, and oats), with livestock and poultry. Coal and asphalt pits and oil wells are also found here. **3.** County (1970 pop. 24,557), 802 sq mi (2,077.2 sq km), SW Wis., bounded in the W by the Mississippi, which here forms the Iowa and Minn. borders, and drained by the Kickapoo, Bad Axe, and Baraboo rivers; formed 1851 as Bad Axe co., renamed 1862; co. seat Viroqua. It is in a dairying, tobacco-farming, and livestock-raising area. Its industry includes cheese processing, lumbering, and engine manufacturing.

Vernon. 1. Town (1979 est. pop. 29,000), Tolland co., N Conn.; settled c.1726, inc. 1808. Manufactures include electronic components, silk screens, and textiles. Vernon merged with Rockville in 1965 and is closely associated with the nearby towns of Ellington and Tolland in the greater Hartford area. **2.** Town (1970 pop. 440), seat of Jennings co., SE Ind., on Vernon Creek NW of Madison, in an agricultural area. **3.** City (1970 pop. 11,454), seat of Wilbarger co., N Texas, near the Okla. line; inc. 1890.

Ve·ro Beach (vîr'ō), city (1975 est. pop. 16,500), seat of Indian River co., E Fla., on Indian River (a lagoon and part of the Intracoastal Waterway), in a citrus-fruit region; founded c.1888, inc. 1919.

Vé·roi·a or **Ve·ri·a** (both: vĕr'ē-ä), town (1971 pop. 29,528), capital of Imathía prefecture, N Greece, in Macedonia. In ancient times the town was known as Berea or Beroea.

Ve·ro·na (və-rō'nə), city (1976 est. pop. 271,381), capital of Verona prov., Venetia, NE Italy, on the Adige River. It is an industrial and agricultural center. Manufactures include food products, textiles, and chemicals. It was an important settlement even before its conquest by Rome in 89 B.C. During the barbarian invasions of Rome (5th–6th cent. A.D.) Odoacer made it his fortress, and Theodoric made it his favorite residence. Verona later became the seat of Frankish counts. In the 12th cent. it was made a free commune. Along with other communes of Venetia, Verona formed (1164) the Veronese League, which opposed Emperor Frederick I. The Ghibelline della Scala, or Scaligeri, family became lords of Verona in the 1260s, and under them the city reached its greatest power. In 1387 it fell to Milan. Venice conquered Verona in 1405 and ruled it until 1797. In the 19th cent. Austria, which then ruled Venetia, made Verona one of its chief fortresses in northern Italy. The Congress of Verona was held here in 1822. After Austrian rule of Venetia was ended (1866), Verona joined the kingdom of Italy. Among the numerous points of interest in Verona are the Romanesque Church of San Zeno Maggiore (9th–15th cent.), the large Roman amphitheater (1st cent. A.D.), and a Roman theater.

Ver·sailles (vər-sī', vĕr-), city (1968 pop. 90,829), capital of Yvelines dept., N central France. An insignificant village, it was made famous by Louis XIV, who built (mid-17th cent.) the palace and grounds that have become almost synonymous with the name Versailles. The huge structure, representing French classical style at its height, was the work of Louis Le Vau, J. H. Mansart, and Charles Le Brun. André Le Nôtre laid out the park and gardens, which are decorated with fountains, reservoirs, and sculptures by such artists as Antoine Coysevox. A huge machine was built at Marly-le-Roi to supply water for the fountains. The park contains two smaller palaces, the Grand Trianon and the Petit Trianon, as well as numerous temples, grottoes, and other decorative structures. The scene of the beginnings of the French Revolution, Versailles never again became a royal residence; under Louis Philippe it became a national monument and museum. The palace was the scene of the proclamation of the German Empire (1871) and of the Third French Republic. Several important treaties were signed at Versailles, most notably the 1919 treaty ending World War I and establishing the League of Nations. Versailles is today one of the greatest tourist centers in France. The city has some industry, such as the manufacture of trailers, insulators, chemical products, and watches.

Ver·sailles (vər-sālz'). **1.** Town (1970 pop. 1,020), seat of Ripley co., SE Ind., W of Aurora, in a farm area. **2.** City (1970 pop. 5,679), seat of Woodford co., N central Ky., W of Lexington, in a fertile bluegrass region; founded 1792. Horses are bred here, and electrical equipment is made. **3.** City (1970 pop. 2,244), seat of Morgan co., central Mo., SW of Jefferson City, in a resort and farm area; founded 1835, inc. 1878.

Ver·viers (vĕr-vyā'), city (1971 est. pop. 33,063), Liège prov., E Belgium, on the Vesdre River and at the foot of the Ardennes. Manufactures of this industrial center include textiles and machinery.

Ve·soul (və-zo͞ol'), town (1968 pop. 16,352), capital of Haute-Saône dept., E France, in Franche-Comté. Agricultural and mechanical equipment and metal and wood products are the chief manufactures. Formerly an earldom, Vesoul was decimated by the plague in 1586. It became part of France in 1678. There are several old buildings, including St. George Church (18th cent.).

Ves·ta·vi·a Hills (vĕs-tā'vē-ə), city (1970 pop. 10,836), Jefferson co., N central Ala., a suburb of Birmingham; inc. 1955.

Ves·te·rå·len Islands (vĕs'tə-rô'lən): *see* Lofoten.

Vest·man·na·ey·jar (vĕst'mä'nə-ā'yär'), group of 15 small islands, c.10 mi (16 km) S of Iceland. In English they are known as the Westman Islands. In 1627 Algerian pirates ravaged the islands and carried nearly 300 people into slavery.

Ve·su·vi·us (və-so͞o'vē-əs), only active volcano on the European mainland, S Italy, on the E shore of the Bay of Naples, SE of Naples. The height of the main cone changes with each eruption, varying within a few hundred feet of the 4,000-ft (1,219-m) level; in 1969 the height was 4,190 ft (1,278 m). The second summit, Monte Somma (3,770 ft/1,149 m), is a ridge that half encircles the cone and is separated from it by a valley (c.3 mi/5 km long). Its lower slopes are extremely fertile, dotted with villages, and covered with vineyards that produce the famous Lachryma Christi wine. The base of the mountain (circumference c.45 mi/70 km) is encircled by a railroad, and a chairlift reaches almost to the rim of the crater (diameter c.2,300 ft/700 m). On the western slope, at 1,995 ft (608 m), there is a seismological observatory (built 1840–45). The outline of Vesuvius forms part of the backdrop of Naples; it is often surmounted by a faint plume of smoke. The earliest recorded eruption (A.D. 79) was described by Pliny the Younger in two letters to Tacitus; the eruption buried Pompeii, Herculaneum, and Stabiae under cinders, ashes, and mud. Pliny the Elder was killed by the eruption, which he had come to investigate. Frequent eruptions have been recorded since then, notably in 512, in 1631, six times in the 18th cent., eight times in the 19th cent. (notably in 1872), and in 1906, 1929, and 1944. The eruptions vary greatly in severity.

Vesz·prém (vĕs'prām'), town (1973 est. pop. 42,924), W Hungary, near the Lake of Balaton. It is a commercial center producing chemicals, knitted goods, and foodstuffs. Made a bishopric by St. Stephen in 1001, Veszprém has an 18th cent. episcopal palace, a cathedral (rebuilt many times), a former citadel, and a museum containing Roman remains. The tall Turkish minaret is now a fire tower.

Ve·vay (vē'vē), city (1970 pop. 1,463), seat of Switzerland co., SE Ind., on the Ohio River E of Madison, in a farming region; settled c.1800 by Swiss, laid out 1813. Shoes are made.

Ve·vey (və-vā'), town (1974 est. pop. 17,500), Vaud canton, W Switzerland, on the Lake of Geneva. It is a resort. Various goods are manufactured, including chocolate, pharmaceuticals, and metal products. About every 25 years a great wine festival is held.

Vex·in (vĕk-săN'), region of N France. It is mainly agricultural, with some industry in the valleys. By the Treaty of Saint-Clair-sur-Epte (911) the northernmost part (Vexin Normand) was assigned to Rollo of Normandy; the rest (Vexin Français) remained with the French crown.

Vi·a·reg·gio (vē'ə-rĕj'ō), city (1972 est. pop. 56,405), Tuscany, N central Italy, on the Tyrrhenian Sea. It is a fishing center and a fashionable beach resort.

Vi·borg (vē'bôr'), city (1976 est. pop. 32,100), capital of Viborg co., N central Denmark. It is a commercial and industrial center and a rail junction. Manufactures include tobacco and textiles. It is one of the oldest Danish cities and became an episcopal see in 1065. The city's cathedral (c.1130), the largest granite church in Denmark, was restored (1862-76) in its original Romanesque style. There is also a 13th cent. Dominican abbey church with a remarkable Flemish altarpiece.

Vi·cen·za (vē-chĕn'zä), city (1972 est. pop. 118,057), capital of Vicenza prov., Venetia, NE Italy. It is an agricultural, commercial, and industrial center. Manufactures include machinery, chemicals, and processed food. Originally a Roman town, Vicenza became a free commune and joined (12th cent.) the Lombard League. It was stormed by Emperor Frederick II in 1236 and later fell to various powers (including Verona and Milan) before being annexed (1404) by Venice. Andrea Palladio (1508-80) made Vicenza famous for his interpretation of classical architecture. The basilica, the Loggia del Capitano, the Teatro Olimpico, the Villa Capra (called La Rotonda), and the Palazzo Chiericato (now housing a museum), all designed by Palladio, inspired the Georgian style in England and the Colonial style in the United States. Vicenza also has a noted Gothic cathedral, with a polyptych (1356) by Lorenzo Veneziano. Bartolomeo Montagna was the founder, in the late 15th cent., of the Vicenza school of painting.

Vi·chy (vĭsh'ē, vē-shē'), city (1968 pop. 33,506), Allier dept., central France, on the Allier River. Vichy's hot mineral springs have made it one of the foremost spas in Europe. The Vichy government was the regime set up here by Marshal Henri Pétain in July, 1940, subsequent to the Franco-German armistice of June 22. Its effective control extended only to unoccupied France and its colonies. The Vichy government, which was never recognized by the Allies, became a German tool in the hands of such men as Pierre Laval, Jean François Darlan, and Jacques Doriot, although German expectations were never completely satisfied. When the Allies invaded North Africa in Nov., 1942, Hitler annulled the armistice of 1940 and occupied all France.

Vicks·burg (vĭks'bûrg'), city (1979 est. pop. 31,400), seat of Warren co., W Miss., on bluffs above the Mississippi River at the mouth of the Yazoo; inc. 1825. An important port, it is the commercial, processing, and shipping center for a cotton, timber, and livestock area. Its many manufactures include lumber and related products, lighting fixtures, machinery, mobile homes, chemicals, feed, fertilizers, and food products. There was a French fort near here in the early 18th cent., and the Spanish established Fort Nogales in 1791. The area, known to the English as Walnut Hills, came into U.S. possession in 1798. Rev. Newitt Vick founded a mission nearby c.1812 and in 1819 began laying out lots for a town. Vicksburg became a busy river port, and in the Civil War it was a major objective in Grant's Vicksburg campaign. The city fell July 4, 1863, after 14 months of naval shelling, 7 months of land assault, and 47 days of total siege. River traffic, which fell off greatly in the late 19th and early 20th cent., has been aided by the U.S. Mississippi River Commission, whose headquarters are at Vicksburg. Nearby is the U.S. Waterways Experiment Station. Ante-bellum homes are in the city and the surrounding area. In Apr., 1973, sections of the city were inundated as the level of the Mississippi and Yazoo rivers rose in flood. In Vicksburg National Military Park are preserved trenches and fortifications of the Civil War siege. North of the city is a national cemetery containing 18,113 Civil War dead, including c.13,000 unknown Union soldiers brought from temporary burial places all over the South. Most of the Confederate soldiers killed in the campaign were buried in the Vicksburg city cemetery.

Vic·to·ri·a (vĭk-tôr'ē-ə, -tōr'ē-ə), state (1978 est. pop. 3,730,000), 87,884 sq mi (227,620 sq km), SE Australia, bounded on the S and E by the Indian Ocean, Bass Strait, and the Tasman Sea; capital Melbourne. The Australian Alps and other mountains of the Eastern Highlands traverse the state; the highest point is Mt. Bogong (6,508 ft/1,985 m). The climate is generally temperate and pleasant. The large but frequently dry rivers such as the Campaspe and the Mitta Mitta are important for irrigation purposes.

Despite its small size, Victoria is one of Australia's leading agricultural states. Wheat, grown largely in the northeast, is the most important crop, followed by oats, barley, fruits, and vegetables. Livestock and dairying are also important. Sheep are raised in the southwest and dairy cattle in the south. Victoria was the first state in Australia to develop industry. The manufacture of automobiles is the state's largest industry, followed by textiles, clothing, and food processing. Gold mining has declined sharply; however, the mining of rich brown coal.

Settlement began in the 1830s when sheep ranchers from Tasmania came looking for pasture. In 1851 Victoria was made a separate British colony, which was granted full constitutional self-government in 1855. The discovery of gold in 1851 led to a rapid population increase. Victoria was federated as a state of the Commonwealth of Australia in 1901.

Victoria, city (1971 pop. 61,761; metropolitan pop. 195,800), capital of British Columbia, SW Canada, on Vancouver Island and Juan de Fuca Strait. It is the largest city on the island and its major port and business center. Victoria is noted as a residential city because of its mild climate, beautiful scenery, and many parks (including Beacon Hill Park) and drives. It is also a popular center for American and Canadian tourists. It has sawmills and woodworking plants, fish-processing factories, grain elevators, and cold-storage plants. The city is the base of a deep-sea fishing fleet; a large naval installation is nearby. Founded (1843) as Fort Camosun, a Hudson's Bay Company post, the city was later called Fort Victoria. With the discovery (1858) of gold on the British Columbia mainland, Victoria became the port, supply base, and outfitting center for miners on their way to the gold fields. In 1866, when the island was administratively united with the mainland, Victoria remained the capital of the colony and became the provincial capital in 1871.

Victoria, city (1970 pop. 83,897), capital of Tamaulipas state, NE Mexico, on the San Marcos River at the foot of the Sierra Madre Oriental. Henequen and citrus fruits are the principal products, and pine forests are exploited. Victoria was founded in 1750.

Victoria, county (1970 pop. 53,766), 892 sq mi (2,310.3 sq km), S Texas, drained by the Guadalupe and San Antonio rivers, formed 1836; co. seat Victoria. This cattle-ranching area also has dairying, agriculture (corn, cotton, hay, and grains), poultry and livestock raising, oil and natural-gas wells, and clay, sand, and gravel deposits.

Victoria, city (1979 est. pop. 48,800), seat of Victoria co., S Texas, on the Guadalupe River, in a prosperous farm, cattle, and oil area. The Victoria Barge Canal (completed in 1962) connects the city with the Intracoastal Waterway. Victoria has food-processing plants, aircraft shops, and factories manufacturing petrochemicals, concrete, metal, machinery, clothing, and boats.

Victoria de las Tu·nas (də läs to͞o'näs), city (1970 pop. 53,739), Oriente prov., E Cuba. It is the marketing center for an important livestock-raising area.

Victoria Falls, waterfall, c.1 mi (1.6 km) wide, with a maximum drop of 420 ft (128 m), in the Zambezi River, S central Africa, on the Zambia-Rhodesia border. The falls are formed as the Zambezi plummets into a narrow chasm (c.400 ft/120 m wide) carved by its waters along a fracture zone in the earth's crust. Numerous islets at the crest of the falls divide the water to form a series of falls. The thick mist and loud roar produced there are perceptible from a distance of about 25 mi (40 km). The Boiling Pot, the beginning of a winding gorge (c.50 mi/80 km long) through which the river flows below the falls, is spanned by a 650-ft (198-m) long bridge that is 310 ft (94 m) above the river. The discovery (1855) of the falls is generally credited to David Livingstone, the British explorer, who named them for Queen Victoria.

Victoria Island, c.81,930 sq mi (212,200 sq km), part of the Arctic Archipelago, Franklin dist., Northwest Territories, N Canada. It is the third-largest island of Canada. The island was discovered by the British explorers Thomas Simpson and Peter W. Dease in the late 1830s; John Rae explored it in 1851.

Victoria Land, part of E Antarctica, S of New Zealand; Cape Adare is to the NE. Bounded on the east by the Ross Sea and on the west by Wilkes Land, it consists of ranges of the Transantarctic Mts., with a high plateau in the interior. The region was discovered during the expedition (1839-43) of Sir James Clark Ross.

Victoria Nile, river, section of the White Nile, c.260 mi (420 km) long, central Uganda, E central Africa. It drains from the northern end of Lake Victoria at Jinja and flows generally north and west, over Ripon Falls and Owen Falls (both now submerged), through shallow Lake Kyoga, and thence over Murchison Falls to Lake Albert. Hydroelectric plants are located at Owen and Murchison falls. The river is navigable from Lake Albert to Murchison Falls.

Victoria Ny·an·za (nē-ăn'zə, nī-), largest lake of Africa and the world's second-largest freshwater lake, c.26,830 sq mi (69,490 sq km), E central Africa, on the Uganda-Tanzania-Kenya border. Victoria Nyanza (c.255 mi/410 km long and c.155 mi/250 km wide) occupies a shallow depression (c.250 ft/75 m deep) on the Equatorial Plateau (alt. 3,725 ft/1,135 m) between two arms of the Great Rift Valley. It has an irregular shoreline and many small islands. Numerous streams feed Victoria Nyanza, which is one of the chief headwater reservoirs of the Nile; the Victoria Nile drains the lake to the north. At Owen Falls Dam on the Victoria Nile the lake's waters are used to generate hydroelectricity. The lake basin is densely populated and intensely cultivated, and the lake is an important fishery. The first European to see Victoria Nyanza (originally called Ukerewe) was John Speke, the British explorer, in 1858; Henry Stanley explored the region in 1875. It is also called Lake Victoria.

Vi·da·lia (vĭ-dāl'yə). **1.** City (1970 pop. 9,507), Toombs and Montgomery cos., SE Ga., W of Savannah, in a farm area. It is an important tobacco market. **2.** Town (1970 pop. 5,538), seat of Concordia parish, E La., on the Mississippi River opposite Natchez, Miss.; settled c.1800, est. 1811. In 1939 the town was moved back from the river when new levees were built.

Vi·din (vē'dĭn), city (1969 est. pop. 42,600), extreme NW Bulgaria, a port on the Danube River. Metal goods, ceramics, and gold and silver filigree work are produced. Vidin is the seat of the Bulgarian metropolitan. Founded in the 1st cent. A.D. as the Roman fortress of Bononia, Vidin became (14th cent.) the capital of the independent West Bulgarian kingdom. It was captured by the Turks in 1396. Under Turkish rule it served (1794-1807) as the residence of the pasha Osman Pazvantoğlu. Vidin has several mosques, old churches, synagogues, a bazaar, and ruins of a medieval fortress.

Vi·dor (vĭ'dôr'), city (1970 pop. 9,738), Orange co., SE Texas, E of Beaumont. Established c.1900 as a farming and ranching community and developed later as a lumber camp, it is today primarily a residential and commercial center in the Beaumont-Port Arthur industrial area.

Vi·en·na (vē-ĕn'ə), city and province (1971 pop. 1,614,800), 160 sq mi (414 sq km), administrative seat of Lower Austria and capital of Austria, NE Austria, on the Danube River. The former residence of the Holy Roman emperors and, after 1806, of the emperors of Austria, Vienna is one of the great historic cities of the world and a melting pot of the Germanic, Slavic, Italian, and Hungarian peoples and cultures. Located on a plain surrounded by the Wienerwald (Vienna Woods) and the Carpathian foothills, it is a cultural, industrial, commercial, and transportation center. The city is divided into 23 districts grouped roughly in two semicircles around the Innere Stadt, or Inner City. Vienna's industries, mainly concentrated on the left bank of the Danube and in the southern districts, produce machinery, textiles, chemicals, and furniture. There are also large oil refineries, breweries, and distilleries. The production of handicrafts (fashion articles, ceramics, and leather) occupies a large portion of the population. Vienna's musical and theatrical life, its parks, coffee houses, and museums, make it a great tourist attraction.

History. Originally a Celtic settlement, Vienna, then called Vindobona, became an important Roman military and commercial center. After the Romans withdrew (late 4th cent.), it rapidly changed hands among various invaders. The Magyars, who gained possession of Vienna early in the 10th cent., were driven out by Leopold I of Babenberg. Construction on Vienna's noted Cathedral of St. Stephen began c.1135. Bohemia conquered Austria in 1251. In 1282 Vienna became the official residence of the house of Hapsburg. The city was occupied (1485-90) by Matthias Corvinus of Hungary and was besieged by the Turks (1529 and 1683). Early in the 18th cent. a new circle of fortifications was built around the city and many magnificent buildings were erected: the Hofburg (the imperial residence), the Karlskirche, St. Peter's Church, the Belvedere (summer residence of Prince Eugene of Savoy), the Kinsky Palace, the Schwarzenburg Palace, and the winter residence of Prince Eugene. Empress Maria Theresa (reigned 1740-80) enlarged the old university, founded in 1365, and completed the royal summer palace, Schönbrunn. Joseph II (1765-90) opened the Prater, a large imperial garden that now contains an amusement park, to the public. Haydn, Mozart, Beethoven, and Schubert lived in Vienna and gave it lasting glory. In 1805 and 1809 Vienna was occupied by Napoleon. The era of political reaction under Prince Metternich, which followed the Congress of Vienna (1814-15), was also famous for the waltzes of Joseph Lanner and the Strauss family, and for the farces of Nestroy, the comedies of Raimund, and the tragic dramas of Grillparzer.

The modern city dates from Francis Joseph's reign (1848-1916). By 1860 the old ramparts about the inner city had been replaced by the famous boulevard the Ringstrasse. The principal edifices on or near the Ringstrasse are the neo-Gothic Rathaus, with many statues and a tower 320 ft (98 m) high; the domed museums of natural history and of art; the Votivkirche; the parliament buildings; the palace of justice; the famous opera house and the Burgtheater; the Künstlerhaus, with painting exhibitions; the Musikverein, containing the conservatory of music; and the Academy of Art. In the late 19th and early 20th cent. Vienna flourished again as a cultural and scientific center. Rokitansky, Wagner-Jauregg, and Billroth (to whom Brahms dedicated the string quartets Op. 51) worked at the General Hospital; at the same time Freud was developing his theory of psychoanalysis. Vienna attracted Brahms, Mahler, Richard Strauss, and Arnold Schönberg and his disciples, who gave it a further period of musical greatness. Krauss, Werfel, Hofmannsthal, Schnitzler, and Wassermann dominated the literary scene. Vienna suffered hardships during World War I. At the end of the war it became the capital of the small republic of Austria. In 1922 Vienna became an autonomous province *(Bundesland)* of Austria. On Mar. 15, 1938, Adolf Hitler entered Vienna, and Austria was annexed to Germany. During World War II the city suffered considerable damage. The Jewish population (115,000 in 1938), residing mainly in the Leopoldstadt district, was reduced to 6,000 by the war's end. The Russian army entered the city in Apr., 1945. Vienna and Austria were divided into four occupation zones by the victorious Allies. The occupation lasted until 1955, when, by treaty, the four powers reunited Austria as a neutral state.

Vienna. 1. City (1970 pop. 2,341), seat of Dooly co., S central Ga., S of Macon; inc. 1841. It is a trade and processing center for a farm area producing pecans, fruit, and truck crops. **2.** City (1970 pop. 1,325), seat of Johnson co., S Ill., NW of Paducah, Ky., in a farm and timber area; inc. 1837. **3.** Town (1970 pop. 505), seat of Maries co., central Mo., near the Gasconade River SSE of Jefferson City, in an agricultural area. **4.** Town (1979 est. pop. 21,200), Fairfax co., N Va., a residential suburb of Washington, D.C.; inc. 1890. Originally called Springfield, it became the site of Fairfax county's first courthouse in 1742. It grew mainly after World War II. **5.** Industrial city (1970 pop. 11,549), Wood co., NW W.Va., on the Ohio River; laid out 1794, inc. 1935. Glass is manufactured here.

Vienne (vyĕn), department (1974 est. pop. 347,400), 2,697 sq mi (6,985.2 sq km), W central France, in Poitou; capital Poitiers.

Vienne, town (1968 pop. 29,057), Isère dept., SE France, on the Rhône River. It is a farm trade center with textile, metallurgical, and pharmaceutical industries. The capital of the Allobroges, Vienne (then Vienna) became one of the chief cities of Roman Gaul. Rich in Roman remains, Vienne has the temple of Augustus and Livia (c.25 B.C.), which rivals the Maison Carrée of Nîmes. Recent excavations have unearthed a Roman cultural center.

Vien·tiane (vyĕn-tyän', vyäɴ-tyän'), city (1973 est. pop. 174,229), capital of Laos, in the north-central part of the country, on the Mekong River, c.130 mi (210 km) SE of the former royal capital of Luang Prabang. A rail line links the two cities. Vientiane is a trading center for forest products, lac, textiles, and hides. Prior to the Communist takeover of Laos (1975) there was much commerce between Vientiane and Thailand (directly across the Mekong). Commerce and retailing in the capital were dominated by Chinese and Vietnamese minorities. Vientiane is a city noted for its canals, its houses on stilts, and its numerous pagodas, one of which now houses an architectural museum. The capital of a Lao kingdom from 1707, the city was sacked by the Siamese in 1827. It passed under French rule in 1893 and became capital of the French protectorate of Laos in 1899.

Vier·sen (fĭr'zən), city (1972 est. pop. 86,030), North Rhine-Westphalia, W West Germany. It is a textile-manufacturing center; other products include machinery, leather goods, and processed food.

Viet·nam (vē-ĕt'näm', -năm', vyĕt'-), state, Southeast Asia. It was formed (1949-50) by the union of the historic areas of Tonkin (northern Vietnam), Annam (central Vietnam), and Cochin China (southern Vietnam). The region is covered largely by forested mountains and plateaus. In the north is the delta of the Red River and in the south that of the Mekong River. The central section of Vietnam lies between the Annamese range and the South China Sea. The region has a tropical monsoon climate, modified by local conditions. Agriculture is the principal occupation of the population, and rice is by far the leading crop. The Vietnamese (more than 80% of the population) are a basically Mongoloid people. Most consider themselves Buddhists, but religious practice typically combines elements

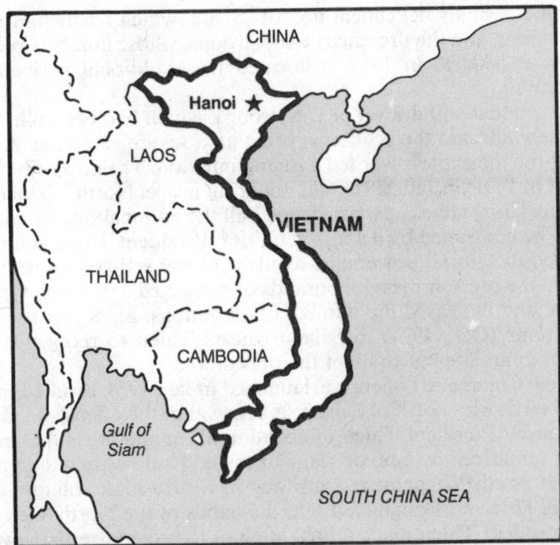

of Taoism, Confucianism, ancestor worship, and animism as well. Religious activity has been somewhat curtailed, although a sizable number of Roman Catholics still worship here. Principal highland tribes are the Tahi, Muong, Nung, Meo, and Man in the north and the Montagnards in the south. There are large numbers of Chinese in the urban centers, notably in the Cholon area of Saigon.

History. The early history of Vietnam is that of Tonkin, Annam, and Cochin China. The first Europeans to arrive were the Portuguese in 1535. Dutch, French, and English traders came in the 17th cent., at which time missionaries entered the area, winning many converts to Roman Catholicism. The persecution of missionaries and of their Vietnamese converts by the ruler of Vietnam was a factor prompting French conquest in the 19th cent. The French captured Saigon in 1859 and after a period of warfare organized (1867) the colony of Cochin China. In 1884 France declared protectorates over Tonkin and Annam; in 1887 it merged Tonkin, Annam, and Cochin China with Cambodia to form a union of Indochina, to which Laos was added in 1893.

A nationalist movement arose in Vietnam in the early 20th cent. and gained momentum during the Japanese occupation in World War II. The Japanese allowed the French Vichy administration to continue as a figurehead power until Mar., 1945, when they ousted it and established the autonomous state of Vietnam, comprising Tonkin, Annam, and Cochin China. At the end of World War II the Viet Minh party (the League for the Independence of Vietnam, a coalition of nationalist and Communist groups), headed by Ho Chi Minh, established a republic with its capital at Hanoi. The Chinese Nationalists, who occupied North Vietnam for 7 months after the war (in accordance with a decision made at the Potsdam Conference), did not challenge Ho's power.

The French attempted to reassert their authority in Vietnam following the war. In Mar., 1946, France signed an agreement with Ho Chi Minh, recognizing Vietnam as a free state within the Indochina federation and the French Union. French troops were then permitted to replace the Chinese in the north. However, differences immediately arose over whether Cochin China was included in the independent state of Vietnam; in June, 1946, France supported the establishment of a separate republic of Cochin China. Fighting broke out (Nov., 1946) between Vietnamese and French troops in Haiphong, and French ships shelled the city, killing some 6,000 civilians. The next month the Viet Minh attacked the French at Hanoi, ushering in the prolonged and bloody guerrilla conflict that became known as the French Indochina War (1946-54). In an attempt to win popular support France concluded a treaty (ratified Feb., 1950) granting Vietnam, now including Cochin China, independence within the French Union. The new state was promptly recognized by the United States, Great Britain, and other states; meanwhile the Ho regime was recognized by the USSR, Communist China, and other Soviet allies. The government failed to win the general support of the Vietnamese, many of whom saw it as a French puppet. Thousands of non-Communists joined the Viet Minh, and the war reached an eventual stalemate, with the French controlling the cities and a few isolated outposts and the Viet Minh occupying most of the countryside. France formally asked U.S. aid for the regime in Feb., 1950. By 1954 the United States was paying about 80% of the French war costs in Vietnam. In early 1954 Viet Minh forces closed in on Dienbienphu, upon which the French had staked the defense of the Red River delta. Dienbienphu fell in May, and at the Geneva Conference of 1954 France had to accept disadvantageous terms for an armistice. As a temporary expedient Vietnam was divided into two parts along a line approximating the 17th parallel (lat. 17° N). North Vietnam, where the Viet Minh were the strongest, went to the Communist government of Ho Chi Minh, while South Vietnam was placed under the control of the French-backed government. Freedom of movement between the two areas was to be permitted for a period of 300 days, thereby facilitating the regroupment of Communist forces in the north and non-Communist forces in the south. The unification of the country under one government was to be effected through general elections, later scheduled for July, 1956. These elections, which were considered likely to favor the Communists, were never held; the South Vietnamese government refused to participate on the grounds that it had not signed the Geneva agreements and was therefore not bound by them. The region was divided then into two new political entities: North Vietnam, or the Democratic Republic of Vietnam, and South Vietnam, or the Republic of Vietnam.

North Vietnam, the former Democratic Republic of Vietnam (1973 est. pop. 22,300,000), 61,293 sq mi (158,750 sq km), includes all of Tonkin and a portion of Annam. The capital is Hanoi. It is bordered by China on the north, the Gulf of Tonkin on the east, South Vietnam on the south, and Laos on the west. The northern and western sections have high mountain masses that are continuations of the mountains of the Chinese provinces of Yunnan and Kwangsi to the north. The mountains give way in the east to an alluvial plain—the delta of the Red River and other streams that empty into the Gulf of Tonkin. To the south is a narrow coastal strip backed by mountains that extend almost to the gulf. Population is concentrated in the fertile Red River delta. In the delta are the great rice fields and the major cities of Hanoi and Haiphong, the chief harbor.

Economy. Although only 15% of the total land is arable, the economy is basically agricultural. Farming is largely unmechanized; water buffalo and oxen are used, and irrigation pumps are manually controlled. Rice fields comprise almost all of the cultivated land, but corn, sweet potatoes, cassava, taro, and beans are also grown, as are some industrial crops (cotton, jute, hemp, sugar cane, oilseeds); now considerable wheat and other grains are imported from China and the USSR. The mountainous areas contain dense rain forests, and timber is the country's most valuable economic crop. Fishing is also important. North Vietnam has virtually all of Vietnam's mineral deposits; coal is by far the most abundant resource, and tin, iron, zinc, lead, gold, chromite, manganese, uranium, and phosphate are also mined. Industrial progress has been hampered by the many years of war; the massive U.S. bombings (1965-68, 1972) virtually destroyed the country's limited industrial plant, but much has been reconstructed.

After the partition of Vietnam (1954) some 900,000 people fled south during the period permitting free movement between the two zones. Many of the emigrants were Catholics, but many were also fleeing the land-reform program initiated even before the defeat of the French. Ho Chi Minh maintained good relations with both China and the USSR, receiving enormous aid from both countries while skillfully protecting independence. A three-year economic rehabilitation program (1958-60) and a five-year plan (1961-66), financed with Soviet and Chinese aid, were aimed at improving both industry and agriculture. Electric power production was increased fifteenfold, new mineral deposits were located, mining operations were expanded, and many new industries were established, especially in Hanoi and Haiphong.

The War. Much national effort was devoted to the support of Communist insurgents in South Vietnam (the Viet Cong). (The North Vietnamese also supported Communist insurgency movements in Laos and, later, in Cambodia.) North Vietnamese troops and supplies were soon infiltrating South Vietnam by way of what came to be known as the Ho Chi Minh Trail in Laos. Open warfare inevitably resulted, and the United States, deeply committed to the support of the non-Communist government of South Vietnam, became increasingly involved. The U.S. bombing of North Vietnam began after two U.S. destroyers were reportedly attacked (Aug., 1964) by North Vietnamese torpedo boats in the Gulf of Tonkin. In Mar., 1968, raids north of lat. 19° N were halted to promote peace negotiations. In Nov., 1968, all bombing ceased. U.S. "protective reaction" air strikes against military installations south of lat. 19° N were resumed by 1970, however. In Apr., 1972, in response to a major Communist drive from North Vietnam, the United States re-instituted mass bombings throughout the country; Haiphong harbor and six other North Vietnamese ports, as well as rivers and canals, were mined and effectively closed to shipping. The country's industrial plant was destroyed, transportation lines were cut, and many nonmilitary targets—including the extensive system of dikes in the Red River delta and numerous residential areas—were hit. Morale nevertheless remained high; damaged transportation facilities were constantly repaired, and "ant tactics" kept supplies laboriously moving from Communist China. After the cease-fire (Jan., 1973) reconstruction began with foreign aid.

South Vietnam, the former Republic of Vietnam (1973 est. pop. 19,700,000), 67,108 sq mi (173,809 sq km), includes Cochin China and the major part of Annam. The capital is Saigon, now renamed Ho Chi Minh City. It is bounded by North Vietnam on the north, the South China Sea on the east and south, and Cambodia and Laos on the west. Population is concentrated in the fertile Mekong delta, which occupies the southern two fifths of the country. With heavy rainfall and rich alluvial soil, it is one of the world's greatest rice-growing regions. To the north the Annamese range—the southern spur of a mountain range that originates in China—rises to more than 8,000 ft (2,440 m). It contains a notable plateau known as the Central Highlands (alt. 600-1,600 ft/180-490 m), which, although sparsely populated, contains rubber, coffee, and tea plantations. East of the Annamese range are the Central Lowlands (central to all of Vietnam), a narrow coastal strip where short, often torrential rivers, flowing west to east, form fertile deltas given over to the growing of rice.

Economy. South Vietnam is predominantly agricultural, and rice is the chief crop; fiber plants, peanuts, tea, coffee, and sugar cane are secondary crops. Massive U.S. aid has rebuilt war-torn areas and financed public works and irrigation projects. Fishing is widespread, especially along the coast, and forestry is important. The south is poor in mineral resources. Silica, limestone, and clay support small glass, cement, and ceramic enterprises. In the Ho Chi Minh area textile mills, assorted manufacturing plants, and establishments processing the food and forest products of the country (for example, rice, timber, fish) are found.

The War. A few months after the partition of Vietnam in 1954 South Vietnam withdrew from the French Union and thus attained complete sovereignty. In a referendum held in Oct., 1955, the electorate approved the establishment of a republic with Ngo Dinh Diem as president. The republic, proclaimed on Oct. 26, 1955, was recognized as the legal government of Vietnam by the United States, France, Great Britain, and other Western powers. Diem suppressed the Cao Dai, a religious sect with its own private army (the Binh Xuyen), and the Hoa Hao, an occultist religious group, both of which opposed him. But his authoritarian policies—rigid press censorship, interference with elections, restriction of opposition parties, and mass arrests—drew increasing criticism. His land-reform programs were ineffective, and by late 1961 the Viet Cong, supported by North Vietnam, had won control of virtually half the country. The United States increased its military and economic aid to combat the Communist threat and at the same time put pressure on President Diem for democratic reforms. Resentment against the government was dramatized by the Buddhist crisis, which erupted in May, 1963, as a result of government persecution. A number of self-immolations by Buddhist monks followed. Large antigovernment demonstrations provoked police shootings, mass arrests, and more repressive government measures. These actions, along with the increasing loss of territory to the Viet Cong, prompted Diem's own military commanders to resort to a coup (Nov. 1, 1963), in which Diem and his brother, Ngo Dinh Nhu (who headed the secret police), were murdered. A period of great political instability followed. In 1964 the North Vietnamese army began infiltrating into South Vietnam to assist the Viet Cong, and the guerrilla conflict expanded into the Vietnam War. In June, 1965, a military junta came to power with Gen. Nguyen Van Thieu as chief of state and Air Vice Marshal Nguyen Cao Ky as prime minister. A new constitution (approved Mar., 1967) provided for a strong executive and a bicameral legislature. In Sept., 1967, Thieu and Ky were elected president and vice president, respectively. Heavy fighting in the rural areas forced thousands of people to seek refuge in the cities, where serious overcrowding ensued. Heavy damage was sustained in the Tet offensive of 1968, especially in Hue and in the Saigon area. The country became increasingly dependent upon U.S. aid, which reached massive proportions, and the presence of U.S. troops, whose numbers peaked at almost 550,000 in 1969, dislocated the traditional agricultural economy.

The gradual withdrawal of U.S. troops, which had begun in 1969, adversely affected the economy, bringing a severe recession. At the same time the endless war fed a raging inflation. The cease-fire, concluded in Paris in Jan., 1973, did not bring peace; North Vietnamese and Viet Cong attacks persisted, and half the national budget continued to be consumed by defense activities. President Thieu gradually assumed dictatorial powers; he abolished local self-government, restricted the press, arrested thousands of suspected Viet Cong sympathizers, and increased the number of executions. Mass protest demonstrations (Oct., 1974) in Saigon caused Thieu to reorganize his cabinet in an attempt to quiet the opposition.

A new Communist offensive, launched in late 1974, resulted in the loss of territories north of Saigon. A renewal of the offensive led to a decision by President Thieu to abandon all inland areas that could not be reinforced by land or sea. Chaos and panic were widespread, and the South Vietnamese army was in nearly total collapse. The cities of Hue and Danang fell into the hands of the North Vietnamese. President Thieu made a final attempt to reorganize the government (Mar., 1975), but by early Apr. demands for his resignation were voiced. On Apr. 21 the Communist forces were perilously close to Saigon, and the governmental structure collapsed. President Thieu resigned, and there were several futile attempts to reorganize a government. On Apr. 30 Communist forces entered Saigon to receive the surrender of the South Vietnamese government.

The Aftermath. A provisional revolutionary government was established. Negotiations followed between northern representatives and the defeated South Vietnamese to decide on the appointment of one government for all of Vietnam.

A new national assembly was elected on Apr. 25, 1976, and convened in Hanoi for the first time on June 24. On July 2 it proclaimed the reunification of the country, which was to be called the Socialist Republic of Vietnam. It also named Ton Duc Thang, the incumbent president of North Vietnam, as president of the new republic. A cabinet composed largely of former North Vietnamese ministers, with the addition of six South Vietnamese, was appointed. Recent developments in Cambodia have strained relations among all countries in the area. The USSR, a major supporter of Vietnam, and China are engaged in a struggle for supremacy in the area. In late 1978 China invaded Vietnam but retreated behind its own boundaries after a series of inconclusive engagements between Chinese and Vietnamese troops (1979).

Government. Following the reunification of the country the national assembly named a committee to draft a constitution for the Socialist Republic of Vietnam. Pending its promulgation the Democratic Republic of Vietnam's constitution of 1969 was to be in effect throughout the country. Full political control rests with the Communist Party and its supporting organizations. All power theoretically belongs to the people and is exercised through their elected representatives. The highest organ of state authority is the national assembly, a unicameral body elected by universal suffrage. Members are elected for four-year terms that may be prolonged in case of war or other extraordinary circumstances. The national assembly elects the president and two vice presidents for a four-year term of office. The national assembly also chooses a premier on the recommendation of the president; it names the other members of the council of ministers on the recommendation of the premier. For administrative purposes the country is divided into provinces, autonomous zones, and municipalities. Although representative councils are elected at all these levels, power is tightly centralized through the overall control exercised by the Politburo of the Vietnamese Communist Party, whose members in fact hold all the key positions in the government.

The government also embarked on a major reorganization of Vietnam's provincial boundaries, reducing the number of provinces in both the northern and southern parts of the country. A large-scale redistribution of the population was also implemented, including resettlement of many residents of the southern cities into rural areas and the shifting of surplus labor from the south to the north. This redistribution was apparently less drastic than the one implemented by neighboring Cambodia in mid-1969.

Vigne·male (vēn'yə-mäl'), mountain, 10,820 ft (3,300 m) high, S France, on the Franco-Spanish border. It is the highest peak in the French Pyrenees.

Vi·go (vē'gō), city (1970 pop. 197,144), Pontevedra prov., NW Spain, in Galicia, on an inlet of the Atlantic Ocean. It is a naval base and one of the most active ports of Spain and a center of tuna and sardine fishing. It also has shipyards, canneries, petroleum and sugar refineries, and various light industries.

Vi·go (vē′gō, vī′-), county (1970 pop. 114,528), 415 sq mi (1,074.9 sq km), W Ind., bounded on the W by the Ill. line, intersected by the Wabash River, and drained by Honey Creek; formed 1818; co. seat Terre Haute. It has stock raising, wheat farming, and bituminous-coal mining. Its industry includes food processing and diverse manufacturing (clothing, wood products, furniture, chemicals, plastics, glass, concrete, and machinery).

Vi·ja·ya·wa·da (vĭj′ə-yə-wä′də), formerly **Bez·wa·da** (bĕz-wä′də), city (1971 pop. 317,258), Andhra Pradesh state, SE India, near the Krishna River delta. It is a transportation and the trade center for the Eastern Ghats.

Vi·las (vī′ləs), county (1970 pop. 10,958), 867 sq mi (2,245.5 sq km), N Wis., bordered in the N by Mich. and drained by the Wisconsin and Manitowish rivers; formed 1893; co. seat Eagle River. Lumbering and some manufacturing are done. It is mostly a resort area, with woods and numerous lakes. Northern State Forest is here.

Vil·jan·di (vĭl′yän′dē), town (1967 est. pop. 19,000), NW European USSR, in Estonia. It is a rail terminus and has fur-processing, canning, textile, and match factories. Founded in 1283, Viljandi was an important medieval trade center and a member of the Hanseatic League.

Vil·lach (fĭl′äKH), city (1973 est. pop. 50,993), Carinthia prov., S Austria, on the Drava River. It is an industrial and rail center. Manufactures include forest products, machinery, and chemicals. Nearby is a mineral spa. Originally a Roman settlement, Villach belonged to the bishopric of Bamberg from the early 11th cent. to 1759, when it passed to Austria. The city has Gothic and baroque churches and a museum with prehistoric artifacts.

Vil·la·fran·ca di Ve·ro·na (vē′lə-fräng′kə dē və-rō′nə), town (1971 pop. 22,366), Venetia, NE Italy. In 1859 Napoleon III and Emperor Francis Joseph of Austria met here after the Austrian defeats at Magenta and Solferino and signed a preliminary peace treaty, which was formalized the same year by the Treaty of Zürich. Sardinia, Napoleon's ally, was not represented. Austria ceded Lombardy, which was added to Sardinia; Venetia remained Austrian. The rulers of Tuscany were to be reinstated, and the Italian states were to form a confederation under the presidency of the pope. Sardinia ignored the last two clauses; to obtain Napoleon's consent for this course, Victor Emmanuel II of Sardinia ceded Nice and Menton to France (1860). The exclusion of Sardinia from the Treaty of Villafranca, an act that nearly deprived Victor Emmanuel of his leading role in the Risorgimento, was deeply resented throughout Italy and greatly harmed Franco-Italian relations.

Vi·lla·her·mo·sa (vē′yä-ĕr-mō′sä), city (1970 pop. 99,565), capital of Tabasco state, SE Mexico, on the Grijalva River. The city, which has good communications facilities, is the commercial and distribution center for the surrounding region. Oil is the chief product. Villahermosa was founded in the 16th cent. On the city's outskirts is a large collection of Olmec stone sculptures.

Vi·lla·vi·cen·ci·o (vē′yə-vē-sĕn′sē-ō′), city (1968 est. pop. 62,500), capital of Meta dept., central Colombia, on the Meta River. Cattle, coffee, bananas, rice, and rubber are the main products of the region.

Vi·lla·vi·ci·o·sa (vē′yə-vē′sē-ō′sə), town (1970 pop. 17,213), Oviedo prov., NW Spain, in Asturias, on the Bay of Biscay. It is a major fishing port with cider distilleries.

Ville·franche-sur-Saône (vēl-fräNsh′sür-sōn′), town (1968 pop. 26,338), Rhône dept., E central France, on the Morgon River. Its industries include weaving, the dyeing of cotton fabric, and the manufacture of clothing and agricultural tools. The town is also a river port and the trade center for Beaujolais wine, which is made in the region. Villefranche-sur-Saône was founded in 1212 and became capital of Beaujolais in the 14th cent. Points of interest include the Church of Notre-Dame-de-Marais (12th-16th cent.) and several Renaissance houses.

Ville Platte (vēl plăt′), town (1970 pop. 9,692), seat of Evangeline parish, S central La., NW of Opelousas; settled by the French in the early 19th cent. It is the marketing and processing center for a rice, cotton, timber, and petroleum region.

Ville·ur·banne (vēl′ür-bän′), city (1968 pop. 119,879), Rhône dept., E central France, a suburb E of Lyons. It is a major industrial center.

Vil·ni·us (vĭl′nē-əs), city (1974 est. pop. 420,000), capital of the Lithuanian SSR, W European USSR, on the Neris River. It is a rail and highway junction, a commercial and industrial city, and a center of education and the arts. Industries include machine building, metallurgy, food and tobacco processing, and the manufacture of electrical equipment, machine tools, and construction materials. Vilnius was officially founded in 1323, when the Lithuanian prince Gediminas made it his capital and built his castle here. The city also became (1415) the metropolis of the Lithuanian Orthodox Eastern Church. Vilnius gained importance both as a strategic point and as a trade and transportation center. The city declined after the merger of Lithuania and Poland, and its Lithuanian-Belorussian culture was replaced by Polish institutions. In the third partition of Poland (1795) Vilnius passed to Russia, where it became a provincial capital (1801-1815). After World War I the city was disputed between Poland and the newly independent Lithuania, which claimed it as its capital. The Paris Peace Conference assigned the city to the Lithuanians, to whom the Russians gave it (1920) after capturing it from the Poles. In the same year, however, Poland retook Vilnius, which became part of Poland (1922) after a plebiscite of doubtful validity. A theoretical state of war between Poland and Lithuania continued until 1927, and diplomatic relations were resumed only in 1938, when Lithuania abandoned its claim to Vilnius. In 1939 Soviet troops occupied the city, and it was transferred to Lithuania, which in 1940 was incorporated into the USSR. Vilnius was occupied by the Germans in World War II and was heavily damaged. The city's university, founded by Stephen Báthory as a Jesuit academy in 1579, is one of Europe's oldest. Vilnius is also the seat of the Lithuanian Academy of Sciences (founded 1941). The city was a leading East European center of Jewish culture from the 16th cent. until the virtual extermination of its large Jewish population by the Germans in World War II. The city's historic nucleus contains numerous old churches and synagogues. The old town hall is now a museum. Also of interest is the Ausros Vartai or Pointed Gate, the sole remnant of the city walls built (1552) by Sigismund I Jagiello. Above the gate is a shrine containing an image of the Virgin, long an object of pilgrimage.

Vil·yu·i (vĭl-yo͞o′ē), river, c.1,520 mi (2,445 km) long, rising in the central Siberian uplands, E Siberian USSR, in the Yakut Autonomous Republic, and flowing E through an agricultural area into the Lena River. It is navigable for c.750 mi (1,205 km) upstream and abounds in fish. In the Vilyui basin there are deposits of diamonds, iron ore, coal, and gold. The Vilyui Range, which rises to 3,300 ft (1,006 m), separates the Lena and Olenek river basins.

Vi·ña del Mar (vēn′yə dĕl mär′), city (1970 est. pop. 153,100), central Chile. Practically a suburb of Valparaiso, Viña del Mar is one of the most popular resort cities in South America. There are luxurious hotels, villas, clubs, and gardens and fine beaches. The city has several industries—textile mills, an oil refinery, and a large sugar refinery. It is also a market for the agricultural products of the region.

Vin·cennes (văN-sĕn′), town (1968 pop. 49,143), Val-de-Marne dept., N central France, an industrial and residential suburb E of Paris. Radio, electrical, and photographic equipment, typewriters, bicycle accessories, machinery, and machine tools are produced. A royal residence since the 12th cent., Vincennes was a favorite of Louis IX. The huge castle dates in part from the 14th cent. Its interior fortress, the keep, in which many famous prisoners were held, has been converted into a museum. Among the many kings of France who lived at Vincennes were Charles V, Charles IX, and Francis I; Henry V of England and Cardinal Mazarin died in the castle. Vincennes also has one of the most famous zoos in Europe.

Vin·cennes (vĭn-sĕnz′), city (1979 est. pop. 19,400), seat of Knox co., SW Ind., on the Wabash River; inc. 1814. The city is the center of an extensive farm area. Its many manufactures include storage batteries, spring assemblies, fabricated steel, glass, wood, paper, and plastic products, fertilizer, seed, farm implements, and food products. Although 1702 is a traditional date for its founding, French fur traders had almost certainly come long before that time. By 1732 it had been fortified by the younger sieur de Vincennes and was an important French settlement. It was occupied by the British in 1763. Vincennes was capital (1800-13) of Indiana Territory until succeeded by Corydon. The old French settlement changed its character as German immigrants, Yankees, and others made it into an American river port.

Vin·dhya Range (vĭn'dyə), chain of hills, c.600 mi (965 km) long, rising to c.3,000 ft (915 m), Madhya Pradesh state, central India. The Vindhya Range has been the historic dividing line between north and south India, separating the Sanskrit-speaking Aryan invaders from the Dravidian peoples of the Deccan. The massive sandstone of the range, long an important building material, was used for the famous group of Buddhist stupas at Sanchi (built 3rd cent. B.C.–11th cent. A.D.), the 11th cent. Jain and Brahman temples at Khajraho, and the 15th cent. palaces of Gwalior.

Vine·land (vīn'lənd), city (1979 est. pop. 54,900), Cumberland co., S N.J., in a poultry and fruit area; settled 1861, inc. 1952 when combined with Landis township. It has cooperative markets, large glassworks, and food-processing and clothing industries.

Vi·ni·ta (vĭ-nē'tə), city (1970 pop. 5,847), seat of Craig co., NE Okla., NE of Tulsa; founded c.1870. An agricultural and livestock center, it also has some manufacturing. A rodeo is held annually.

Vin·land (vĭn'lənd), section of North America discovered by Leif Ericsson in the 11th cent. The sources for the knowledge of Leif Ericsson's exploration differ as to whether it was planned or accidental, but it is definitely known that he found a land containing grapes and self-sown wheat, which he called Vinland. Later expeditions, particularly that of Thorfinn Karlsefni, attempted to rediscover that land. There has been much speculation as to the identification of Vinland, but from the examination of the evidence that is available or may be deduced pertaining to directions, distances, topography, climate, and flora and fauna, the southern coast of New England has been generally accepted.

Vin·ni·tsa (vĭn'ĭt-sə), city (1973 est. pop. 251,000), capital of Vinnitsa oblast, in Podolia, the Ukraine, SW European USSR, on the Southern Bug River. A railroad junction in a sugar-beet district, the city has food-processing and other industries.

Vin·son Mas·sif (vĭn'sən mə-sēf'), peak, 16,860 ft (5,142.3 m) high, W Antarctica, in the Ellsworth Mts. It is the highest peak in Antarctica.

Vin·ton (vĭn'tən), county (1970 pop. 9,420), 411 sq mi (1,064.5 sq km), S Ohio, drained by Raccoon Creek and Salt and Little Raccoon creeks; formed 1850. It has bituminous-coal mining and agriculture (livestock, grain, and fruit). Its industry includes lumbering and the manufacture of explosives and hardware products.

Vinton. 1. City (1970 pop. 4,845), seat of Benton co., E central Iowa, on the Cedar River between Cedar Rapids and Waterloo; settled 1839, inc. 1869. It is a trade and processing center in a farm area. **2.** Town (1970 pop. 6,347), Roanoke co., SW Va., E of Roanoke; settled 1797, inc. 1884. Textiles and feed are made.

Vir·gin·ia (vûr-jĭn'yə, -jĭn'ē-ə), state (1975 pop. 4,973,000), 40,815 sq mi (105,711 sq km), E United States, most northerly of the Southern states, the first of the Thirteen Colonies. It is officially styled a commonwealth. Richmond is the capital.

Virginia is roughly triangular in shape. Its base rests on N.C. and Tenn. to the south, and in the north it touches on W.Va. and Md.; its eastern side is bounded by Md. and the Atlantic Ocean and its western side by Ky. and W.Va. The small section of Virginia that with Md. and Del. occupies the Delmarva peninsula is separated from the main part of Virginia and is called the Eastern Shore. The coastal plain or Tidewater region of eastern Virginia, generally flat and partly swampy, is cut by four great tidal rivers—the Potomac (forming most of the line with Md.), the Rappahannock, the York, and the James—all of which empty into Chesapeake Bay. At its western extension the tidewater rises to c.300 ft. (90 m) at the fall line (passing through Richmond) and gives way to the Piedmont—rolling, generally fertile country that broadens gradually as it extends south to the N.C. line. Rising rather abruptly in the western Piedmont is the Blue Ridge range, topped by the state's highest peak, Mt. Rogers (5,720 ft/1,743 m). Between the Blue Ridge and the westward-lying Allegheny Mts., both part of the Appalachian range, lies the valley and ridge province, commonly called the Valley of Virginia. Chief of the valleys, which are joined by narrow gaps, is the rich, beautiful, and historic valley of the Shenandoah.

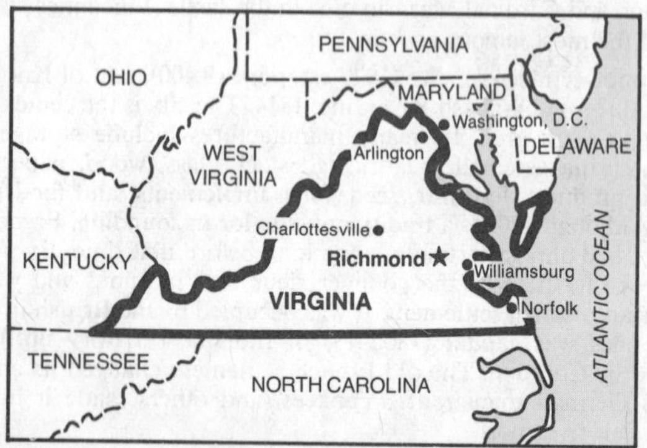

Virginia's shores, mountains, mineral springs, natural wonders, and numerous historic sites draw thousands of visitors annually. Major tourist attractions are Shenandoah National Park, Colonial National Historical Park, and Arlington House National Memorial. Other historic points of interest include Appomattox Court House National Historical Park; Manassas and Richmond national battlefield parks; Booker T. Washington and George Washington Birthplace national monuments; Jamestown National Historic Site; National Capital Parks (shared with Washington, D.C., and Md.); and several national cemeteries and military parks.

Virginia is divided into 96 counties and 38 independent cities:

NAME	COUNTY SEAT	NAME	COUNTY SEAT
Accomac	Accomac	King George	King George
Albemarle	Charlottesville	King William	King William
Alleghany	Covington	Lancaster	Lancaster
Amelia	Amelia Courthouse	Lee	Jonesville
Amherst	Amherst	Loudoun	Leesburg
Appomattox	Appomattox	Louisa	Louisa
Arlington	Arlington	Lunenburg	Lunenburg
Augusta	Staunton	Madison	Madison
Bath	Warm Springs	Mathews	Mathews
Bedford	Bedford	Mecklenburg	Boydton
Bland	Bland	Middlesex	Saluda
Botetourt	Fincastle	Montgomery	Christianburg
Brunswick	Lawrenceville	Nansemond	Suffolk
Buchanan	Grundy	Nelson	Lovingston
Buckingham	Buckingham	New Kent	New Kent
Campbell	Rustburg	Northampton	Eastville
Caroline	Bowling Green	Northumberland	Heathsville
Carroll	Hillsville	Nottoway	Nottoway
Charles City	Charles City	Orange	Orange
Charlotte	Charlotte Courthouse	Page	Luray
Chesterfield	Chesterfield	Patrick	Stuart
Clarke	Berryville	Pittsylvania	Chatham
Craig	New Castle	Powhatan	Powhatan
Culpeper	Culpeper	Prince Edward	Farmville
Cumberland	Cumberland	Prince George	Prince George
Dickenson	Clintwood	Prince William	Manassas
Dinwiddie	Dinwiddie	Pulaski	Pulaski
Essex	Tappahannock	Rappahannock	Washington
Fairfax	Fairfax	Richmond	Warsaw
Fauquier	Warrenton	Roanoke	Salem
Floyd	Floyd	Rockbridge	Lexington
Fluvanna	Palmyra	Rockingham	Harrisonburg
Franklin	Rocky Mount	Russell	Lebanon
Frederick	Winchester	Scott	Gate City
Giles	Pearisburg	Shenandoah	Woodstock
Gloucester	Gloucester	Smyth	Marion
Goochland	Goochland	Southampton	Courtland
Grayson	Independence	Spotsylvania	Spotsylvania
Greene	Standardsville	Stafford	Stafford
Greensville	Emporia	Surry	Surry
Halifax	Halifax	Sussex	Sussex
Hanover	Hanover	Tazewell	Tazewell
Henrico	Richmond	Warren	Front Royal
Henry	Martinsville	Washington	Abingdon
Highland	Monterey	Westmoreland	Montross
Isle of Wight	Isle of Wight	Wise	Wise
James City	Williamsburg	Wythe	Wytheville
King and Queen	King and Queen Courthouse	York	Yorktown

INDEPENDENT CITIES

Alexandria	Emporia	Lynchburg	Salem
Bedford	Fairfax	Martinsville	South Boston
Bristol	Falls Church	Newport News	Staunton
Buena Vista	Franklin	Norfolk	Suffolk
Charlottesville	Fredericksburg	Norton	Virginia Beach
Chesapeake	Galax	Petersburg	Waynesboro
Clifton Forge	Hampton	Portsmouth	Williamsburg
Colonial Heights	Harrisonburg	Radford	Winchester
Covington	Hopewell	Richmond	
Danville	Lexington	Roanoke	

Economy. Virginia's economy is among the most diversified in the South, but agriculture is still the basic element. Tobacco leads among crops, but hay, corn, peanuts, sweet potatoes, and apples are also important. Livestock and livestock products are a major source of agricultural income; the Valley of Virginia is well known as a cattle area, and dairy and poultry farms are also important there. The coastal fisheries are large, and Virginia shellfish—especially oysters and crabs—yield a large annual catch. Coal is Virginia's chief mineral; stone, cement, sand, and gravel are also important.

The manufacture of chemical products is Virginia's leading industry. The nation's tobacco manufacture has long been focused in Richmond. Roanoke is a center for the rail transport equipment industry, and a high proportion of the nation's shipyards are concentrated on the shores of Hampton Roads. Norfolk serves as a major U.S. naval base, and Portsmouth as a U.S. naval shipyard; Hampton is an important center for aeronautical research. Other leading industries include the manufacture of food, textile, and paper products.

History. Virginia (named for Elizabeth I) at first designated the whole vast area of North America not held by the Spanish or French. The colony on Roanoke Island, organized by Sir Walter Raleigh, failed, but the English soon made another attempt slightly farther north. The London Company, a group of merchants, sent three ships and 144 men to establish a base, and the tiny force founded (May 13, 1607) Jamestown on the James River, the first permanent English settlement in America. By 1608, despite the firm and resourceful leadership of John Smith, hunger and disease had reduced their numbers to 38. Although sickness and starvation continued to take a heavy toll, the settlement at last began to make headway. Tobacco, first cultivated in 1612, gave the company new hope of a profitable return on its investment. As incentives to immigration the company instructed Gov. George Yeardley in 1619 to form a house of burgesses—the first representative assembly in the New World—and in 1620 began to import women. In 1622 an Indian attack that left 350 settlers dead delivered a fatal blow to the company, and in 1624, beset by internal dissension, it surrendered its charter to the crown. Virginia became a royal colony, the first in English history.

By 1641 the colony was well established and extended on both sides of the James up to its falls. Most of the settlers (about 7,500 in 1641) were Anglicans, and during the civil war in England many well-to-do Englishmen came to Virginia. The colony was loyal to the crown until 1652, when an expedition sent by Oliver Cromwell forced it to adhere to the Puritan Commonwealth. A prosperous era of expanding tobacco and fur trade came to an end with the Restoration in 1660. The Navigation Acts forced the tobacco trade to use only English ships and English ports, which were at first insufficient to handle it; tobacco piled up in Virginia and in England, and prices plummeted. The 1690s brought renewed prosperity as the Navigation Acts began to be applied more beneficially and Virginia tobacco was profitably marketed in Europe.

Expansion of the plantation system was made possible only with the use of black slave labor (first recognized by law in 1662), and tens of thousands of blacks were being imported every year by the end of the century. Small independent cultivators, unable to compete with the plantation-slave system, formed the nucleus of a poor white class that drifted southward or pioneered to the west. Settlement of the Piedmont was begun by the end of the century, and the Shenandoah valley was opened c.1730. Soil exhaustion from continuous tobacco cultivation speeded the westward march, as did the settlement activities of land speculators. The last of the French and Indian Wars, in which Virginians—notably Col. George Washington—were prominent, ended the French obstacle to westward migration.

With Massachusetts, Virginia was a leader in the movement that culminated in the American Revolution. In 1773 the burgesses at Williamsburg (the capital since 1699) formed an intercolonial committee of correspondence. The Virginia leaders proposed (May, 1774) a congress of all the colonies; delegates were chosen at the First Virginia Convention (Aug.) and in Sept., Virginia's Peyton Randolph was elected president of the First Continental Congress. The Fifth Virginia Convention (May 6–June 29, 1776) declared the colony's independence and adopted a declaration of rights and the first constitution of a free American state. Patrick Henry was elected the first governor. Although the British had burned Norfolk in Jan., 1776, they did not invade the state in full force until 1779, when they took Portsmouth and Suffolk. The British cause was lost as American land forces and a French fleet combined to bring about Cornwallis's surrender (Oct. 19, 1781) in the Yorktown Campaign. During the Revolution a degree of religious freedom had been instituted in Virginia under the lead of Thomas Jefferson. A liberal law for formal emancipation of slaves was passed in 1782 and remained in force for more than 20 years. In 1786 a statute for religious freedom, championed by James Madison, established complete religious equality for all Virginians. In replacing the unsatisfactory Articles of Confederation with the Constitution of the United States, Virginians, especially James Madison, again played leading roles.

The Old Dominion yielded (1784) its claim to the Northwest Territory to the federal government and ceded (1789) a portion of its Potomac lands for the creation of the District of Columbia. In 1792 Kentucky, a Virginia county since 1776, was admitted to the Union as a separate state. Virginia itself, stimulated by western complaints, embarked on a vigorous policy of internal improvements in the second and third decades of the 19th cent. Economically the whole state benefited from transportation improvements, from the growth of scientific agriculture and the spread of wheat cultivation, and from the growth of such industries as tobacco processing and iron manufacture. In the east cotton replaced tobacco as the staple, and as the newer Southern states eclipsed Virginia in cotton production the tidewater became a breeding ground for the slaves they needed. Elsewhere in the state, especially in the west, antislavery sentiment was strong in the early 19th cent., and following the slave insurrection (1831) of Nat Turner the house of delegates voted down a bill to abolish slavery by the narrow margin of seven votes. The insurrection did result in harsher laws and more conservative policies regarding blacks. A liberal political spirit triumphed in the constitution of 1851, which granted suffrage to "every white male citizen."

For the most part Virginians labored to avert conflict between North and South. Secession came (Apr. 17, 1861) only after all attempts to keep peace had failed. Virginia joined the Confederacy, and Richmond became the Confederate capital. Most Virginians west of the Appalachians opposed secession, and on June 20, 1863, this section was admitted to the Union as the new state of West Virginia. Virginia was the chief battleground of the Civil War. Gen. Robert E. Lee surrendered the remnants of his Army of Northern Virginia at Appomattox Courthouse on Apr. 9, 1865. President Jefferson Davis had already fled Richmond, and the Confederacy soon collapsed.

In 1870, after the Virginia assembly had ratified the 14th and 15th amendments to the Constitution, the state was readmitted to the Union. The end of slavery and the hard agricultural times of postwar decades also ended the plantation system in Virginia and brought some increase in farm tenancy, but the economy benefited from diversification as fruit farming and the cigarette industry became important. In 1902 a new state constitution invoked rigorous literacy tests for voting, completing the long process of reducing the black electorate. During the years preceding World War I Virginia's prosperity grew as dairy farming in particular gained importance, and during the war agriculture boomed, as did industry, especially the important shipbuilding works at Hampton Roads.

In the mid-1920s Harry Flood Byrd assumed direction of the state's powerful Democratic organization. Byrd, governor from 1926 to 1930 and U.S. Senator from 1933 until 1965, put through a sound reorganization of the state government, brought about the passage of the first antilynching law of any state, and improved the highway system. Liberals criticized his austere financial policy for scrimping on public education and welfare. In the Great Depression of the 1930s the state's economy fared relatively well, being built around consumer goods—foods, textiles, and tobacco—which remained in demand.

In World War II Virginia was the scene of much military training, and the shipyards at Hampton Roads and other industries again aided the war effort. In the prosperous postwar period the conservative Byrd organization maintained its power. After the 1954 Supreme Court decision on integration, attempts at desegregating Virginia's schools proceeded slowly.

Government. The Virginia constitution was revised extensively in the late 1960s. The legislature (called the general assembly) consists of a house of delegates of 100 members, elected to two-year terms, and a senate with 30 members, elected to four-year terms. The governor serves a four-year term and is ineligible for re-election.

Educational Institutions. Among Virginia's institutions of higher learning are the Univ. of Virginia, mainly at Charlottesville, with Mary Washington College at Fredericksburg; the College of William and Mary in Virginia, mainly at Williamsburg; Hampton Institute, at Hampton; Randolph-Macon College, at Ashland; Randolph-Macon Woman's College, at Lynchburg; Sweet Briar College, at Sweet Briar; Virginia Military Institute and Washington and Lee Univ., at Lexington; Virginia Polytechnic Institute and State Univ., at Blacksburg; and Virginia State College, at Petersburg.

Virginia. **1.** City (1970 pop. 1,814), seat of Cass co., central Ill., WNW of Springfield, in a farm area; platted 1836, inc. 1857. **2.** City (1975 est. pop. 11,400), St. Louis co., NE Minn., on the Mesabi range; inc. 1892. In addition to its iron mines—both open-pit (these are tourist attractions) and underground—the city has foundries and plants that manufacture garments and chipboard. It is also the trade center of a resort area. A junior college and the Minnesota Museum of Mining are here. The Laurentian Divide is to the north, and a state park, with an abandoned mine, is nearby.

Virginia Beach, independent city (1979 est. pop. 256,500), SE Va., on the Atlantic coast; inc. 1906. In 1963 Princess Anne co. and the former small town of Virginia Beach were merged, giving the present city an area of 302 sq mi (782 sq km). Its economy centers on tourism, agriculture (truck crops and livestock), and four large military bases now within the city limits. Virginia Beach has long been a popular resort, with beautiful beaches and excellent sportfishing. The huge area now added to the city offers additional fishing as well as game hunting. One section is famous for its thoroughbred horses and another for its Guernsey cows — it sends breeding stock abroad and is the nation's largest producer-distributor of Guernsey milk. Among the many points of interest in the city are the Cape Henry memorial cross, site of the landing of the first colonists in 1607; the Cape Henry lighthouse (1791; restored); the nation's oldest brick residence (1636; restored); and the Alan B. Shepard civic center, a geodesic aluminum-domed structure. Seashore State Park is within the city limits, and Virginia Wesleyan College is on the Norfolk-Virginia Beach border. The Chesapeake Bay Bridge-Tunnel (opened 1964) links Virginia Beach with the Eastern Shore of Va. and Md.

Virginia City. **1.** Town (1970 pop. 149), seat of Madison co., SW Mont., SE of Butte. Founded in 1863, when gold was discovered in Alder Gulch, Virginia City became a lusty, brawling mining town. It was the first town in Montana Territory to be incorporated (1864), and the first newspaper in the territory was printed here (1864). Today the town is a tourist attraction; many of its old buildings are restored. **2.** Village (1970 pop. 695), seat of Storey co., W Nev., SE of Reno, at an altitude of 6,500 ft (1,982.5 m). It was founded in 1859, when rich deposits of gold and silver, notably the famous Comstock Lode, were discovered nearby. During the 1870s Virginia City was the mining metropolis of the West, with a population of c.35,000. It had fine residences, clubs, restaurants, an opera house, several churches, and 110 saloons. The first newspaper printed in the state, the *Territorial Enterprise*, was published here after 1861. The town is now a tourist center.

Vir·gin Islands (vûr'jĭn), group of about 100 small islands, West Indies, E of Puerto Rico. The islands, divided politically between the United States and Great Britain, are subject to occasional hurricanes between Aug. and Oct. and suffer from light earthquakes. The water supply is almost completely dependent on rainfall and is preserved in cisterns. The tropical climate, with its cooling northeast trade winds, and the picturesque quality of the islands, enhanced by their Old World architecture, have encouraged a large tourist trade.

The islands were discovered and named by Columbus in 1493. The Virgin Islands of the United States (1978 est. pop. 102,000), 133 sq mi (344.5 sq km), were purchased from Denmark in 1917 for $25 million because of their strategic position alongside the approach to the Panama Canal. Under a law passed in 1954 the islands are administered by the U.S. Dept. of the Interior; a governor and senate are locally elected. There are 68 islands in this group. The capital is Charlotte Amalie on St. Thomas. Food crops are raised; sugar cane is no longer grown but rum is still distilled. Meat packing is a growing industry.

Immediately to the northeast are the British Virgin Islands, a colony (1978 est. pop. 10,000), 59 sq mi (152.8 sq km). They are ruled by a governor (appointed by the crown) and an executive and a legislative council. There are more than 30 islands; 16 are inhabited. Road Town, the capital, is on Tortula. Livestock raising, farming, and fishing are the principal economic activities, in addition to tourism. Britain acquired the islands from the Dutch in 1666.

Virgin Islands National Park, 14,419 acres (5,840 hectares), St. John, Virgin Islands; est. 1956. The park, with beaches, coves, and headlands, is rich in tropical-plant, animal, and marine life. Prehistoric Carib Indian relics and the remains of Danish colonial sugar plantations are in the park.

Virgin Lands Territory: *see* Tselinny Kray, USSR.

Vi·run·ga (vĭ-rōōng'gə), volcanic range of mountains, E Zaire, N Rwanda, and SW Uganda, central Africa, NE of Lake Kivu. It separates the basins of the Nile and Congo rivers. Karisimbi (14,786 ft/4,509.7 m) is the highest peak.

Vis (vēs), island (1971 pop. 5,054), 35 sq mi (91 sq km), W Yugoslavia, off the Dalmatian coast in the Adriatic. A popular resort, it is the center of the Yugoslav fish-canning industry and is noted for its wine. Its chief town, also named Vis, is a picturesque village on the north coast. Ancient Issa was a Greek colony from the 4th cent. B.C. and later prospered under the Romans. From 996 to 1797 it was a Venetian possession, and in the Napoleonic Wars (1803–15) it changed hands among the Austrians, the French, and the British. From the Congress of Vienna (1815) until 1918 the island belonged to Hungary. Two important naval battles occurred off Vis: in 1811 the British won a victory over the French, and in the Austro-Prussian War of 1866 the Austrians under Admiral Wilhelm von Tegetthoff thoroughly defeated the Italian fleet. There are many ancient remains on the island, notably Roman baths, mosaic pavements, and several old Roman Catholic and Orthodox Eastern churches.

Vi·sa·kha·pat·nam (vĭ-sä'kə-pŭt'nəm), city (1971 pop. 352,504), Andhra Pradesh state, SE India, a port on the Bay of Bengal. Established by the British in the 17th cent., Visakhapatnam was famous for cloth and for local handicrafts, especially ornamental objects of ivory, horn, and silver. Today it is a district administrative center and a health resort and has a naval base, a shipyard, and an oil refinery. Andhra Univ. is on the outskirts of the city.

Vi·sa·lia (vĭ-sāl'yə), city (1979 est. pop. 43,600), seat of Tulare co., S central Calif., in the San Joaquin Valley; founded 1852, inc. 1874. Its economy is centered chiefly around agriculture. Electronic products are also manufactured.

Vi·sa·yan Islands (vĭ-sī'ən), large island group (1970 est. pop. 10,000,000), c.24,000 sq mi (62,160 sq km), in and around the Visayan Sea, central Philippines. The group includes Bohol, Cebu, Leyte, Masbate, Negros, Panay, Samar, and hundreds of smaller islands. The coastal plains of Samar and Leyte are densely populated. Cebu, Negros, and Panay are the commercial heart of the Visayan Islands.

Vis·by (vĭz'bē), city (1972 est. pop. 22,946), capital of Gotland co., SE Sweden, on Gotland Island and on the Baltic Sea. It is an industrial center and a popular resort and has a modern ice-free port. Manufactures include cement and refined sugar. The city is a Lutheran episcopal see. Visby was a pagan religious center. In the 11th cent. it became a prominent international trade center of the Hanseatic League. An independent republic, the city was the commercial center of northern Europe, minting its own coins and developing an influential international maritime code. The ruins of 10 fine churches and the restored Cathedral of St. Mary, all from the 12th and 13th cent., testify to its former greatness. Visby's decline began in 1280, when Sweden conquered Gotland while the city was suffering civil strife. It was sacked (1361) and captured (1362) by the Danes, who returned (1370) it to the Hanseatic League. Pirates made it their stronghold for the next two centuries, and its trade was taken over by Lübeck. Visby passed to Denmark in 1570, and again came under Sweden in 1645. The city only began to recover its trade in the 19th cent.

Vis·count Mel·ville Sound (vī'kount mĕl'vĭl'), 250 mi (402 km) long and 100 mi (161 km) wide, arm of the Arctic Ocean, N Northwest Territories, Canada, between Victoria and Prince of Wales islands on the S and Melville and Bathurst islands on the N. It is a section of the Northwest Passage. Through McClure Strait on the west it is linked with the Beaufort Sea; Barrow Strait and McClintock Channel lead east and southeast. A crossroads of arctic waterways, it is navigable only under favorable weather conditions. The western part was discovered (1850–53) by Sir Robert McClure.

Vi·se·u (vē-zā'ōō), town (1970 pop. 16,140), capital of Viseu dist. and Beira Alta, N central Portugal. The town has agricultural-processing and textile industries. It was founded by the Romans.

Vi·so, Mon·te (môn'tā vē'zō), peak, 12,602 ft (3,843.6 m) high, on the French-Italian border; highest of the Cottian Alps. The Po River rises here.

Vis·ta (vĭs'tə), uninc. town (1979 est. pop. 33,000), San Diego co., S Calif., near the Pacific coast, in an agricultural and resort area; inc. 1963.

Vis·tu·la (vĭs'chōō-lə), longest river and principal waterway of Poland, c.665 mi (1,070 km) long. It rises in the West Beskid range of the Carpathians, southern Poland, and flows northeast past Kraków, northwest past Warsaw, and north to the Gulf of Danzig on the Baltic Sea. Navigable for small craft for almost its entire length, the Vistula is connected by canals with the Oder, Dnepr, Neman, and Pregel rivers.

Vi·tebsk (vē'tĕpsk', -tĕbsk'), city (1973 est. pop. 258,000), capital of Vitebsk oblast, Belorussia, W European USSR, on the Western Dvina River. It is a river port and large railroad junction in an agricultural district. Manufactures include processed food, textiles, and building materials. Vitebsk dates from the 11th cent. and was the capital of a Russian principality that came under (14th cent.) Lithuanian rule. It passed to Russia again in 1772.

Vi·ter·bo (vē-tûr'bō), city (1972 est. pop. 39,200), capital of Viterbo prov., Latium, central Italy, near Lake Bolsena. It is an agricultural center with food-processing industries. A Roman colony called Vicus Elbii, the city later (11th cent.) passed to the papacy. It became a favorite residence of the popes, and several conclaves were held here. The city has a picturesque medieval quarter, with palaces and houses built in the 13th–14th cent. and with numerous fountains. Landmarks include the pinnacled palace of the popes and the Gothic loggia (both 13th cent.); the Romanesque Cathedral of San Lorenzo (12th cent.), with a fine campanile; and the former Convent of St. Mary (11th and 14th–15th cent.), which now houses the municipal museum.

Vi·ti Le·vu (vē'tē lĕv'o͞o), volcanic island, 4,010 sq mi (10,386 sq km), S Pacific, largest and most important of the Fiji Islands. On Viti Levu are Suva, the capital and chief port of the Fijis, and Lautoka, an important town. Mt. Victoria (c.4,340 ft/1,325 m), the highest peak in the Fijis, is on the island. Sugar cane, pineapples, rice, coconuts, and cotton are the major products; dairying and gold mining are also important industries. Sugar and copra are the chief exports.

Vi·tim (vĭ-tēm'), river, c.1,140 mi (1,835 km) long, rising in the Transbaykalian Mts., E Siberian USSR, in the Buryat Autonomous Republic, and flowing S, NE, then N into the Lena River at Vitim. It is navigable for five months of the year.

Vi·to·ri·a (vĭ-tôr'ē-ə, -tōr'ē-ə), city (1972 est. pop. 143,963), capital of Álava prov., N Spain, in the Basque Provinces. It is a manufacturing and administrative center. It was probably founded in the 6th cent. by the Visigoths.

Vi·tó·ri·a (vĭ-tôr'ē-ə, -tōr'ē-ə), city (1970 pop. 121,978), capital of Espírito Santo state, E Brazil, on an island in Espírito Santo Bay of the Atlantic Ocean. It is one of Brazil's chief ore ports and is linked by rail with rich iron deposits at Itabira, in Minas Gerais state. Besides processing ore, the city produces sugar, leather goods, coffee, furniture, and paper bags. Vitória was founded in 1535 as a defensive position against the Indians. The city repelled several French attacks and the English pirate Thomas Cavendish (late 16th cent.), as well as an attempted Dutch invasion in 1625.

Vi·try-le-Fran·çois (vē-trē'lə-fräN-swä'), town (1968 pop. 16,879), Marne dept., NE France, on the Marne River. Woolens, earthenware, metal products, containers, and lumber are the chief manufactures. The town was founded by Francis I in 1545. During World War I it was the headquarters of Gen. Joseph Joffre; in World War II the town was almost completely destroyed. Monuments include the Notre Dame Church (17th–18th cent.).

Vi·try-sur-Seine (vē-trē'sür-sĕn'), city (1968 pop. 77,846), Val-de-Marne dept., N central France, on the Seine River; an industrial suburb SE of Paris. The chief products are lighting and heating equipment, paper and cardboard, and chemicals.

Vit·to·ri·a (vĭ-tôr'ē-ə, -tōr'ē-ə), city (1972 est. pop. 46,757), SE Sicily, Italy; founded 1607. It is an important center of wine and olive-oil production and export.

Vit·to·rio Ve·ne·to (vĭ-tôr'ē-ō vĕn'ə-tō', -tōr'ē-ō, vā'nə-), town (1972 est. pop. 30,947), Venetia, NE Italy, in the Alpine foothills. It is an industrial and commercial center and a spa.

Vi·tu Islands (vē'to͞o), volcanic group, 37 sq mi (95.8 sq km), in the Bismarck Archipelago, part of Papua New Guinea. The group is the chief copra center of Papua New Guinea.

Vi·va·rais (vē-vä-rā'), region, roughly coextensive with Ardèche dept., SE France. Its mountainous terrain rises to 5,753 ft (1,754.7 m) in the Mézenc. Cattle raising and silk manufacturing are the chief occupations. The medieval county of Viviers or Vivarais, a part of the kingdom of Arles, was held in fief by the counts of Toulouse, who lost it to the French crown in 1229.

Viz·i·a·nag·ram (vĭz'ē-ə-nŭg'rəm), city (1971 pop. 86,608), Andhra Pradesh state, E India, near the Bay of Bengal. It is a market for grain, peanuts, and sugar. There are leather-tanning and jute-processing industries. On the city's outskirts is the largest ferro-manganese plant in India.

Vlaar·ding·en (vlär'dĭng-ən), city (1973 est. pop. 81,785), South Holland prov., SW Netherlands, on the Nieuwe Maas (New Meuse) River, near Rotterdam. It is an industrial city and a major port.

Vla·di·mir (vlăd'ə-mĭr', vlə-dē'mĭr'), city (1973 est. pop. 256,000), capital of Vladimir oblast, W central European USSR, on the Klyazma River. A rail junction, it has industries producing machinery, tractors, chemicals, cotton textiles, and plastics. It was founded in the early 12th cent. by Vladimir II of Kiev. The dukes of Moscow emerged as the most powerful Russian princes, and in 1364 they acquired Vladimir; they assumed the title of grand dukes and for a time afterward had themselves crowned here. The city's landmarks include the Uspensky (Assumption) Cathedral (1158–61) with a museum of religious art and tombs of the early princes of Vladimir.

Vla·di·mir-Vo·lyn·ski (vlə-dē'mĭr-və-lĭn'skē), city (1967 est. pop. 23,000), W European USSR, in the Ukraine. One of the oldest Ukrainian settlements, it was founded in the 9th cent. and supposedly refounded in 988 by the Grand Duke Vladimir I of Kiev. It became an Eastern Orthodox bishopric and the capital of the grand duchy of Vladimir or Lodomeria. The settlement was fortified and became a large trading center between the 10th and 13th cent. Originally dependent on Kiev, the duchy became independent in 1154 and for some time included all of Volhynia. It was united with the duchy of Galich in 1188 to form the Galich-Volhynian duchy, of which it was the capital from 1300. The city passed to Lithuania in the late 14th cent. It changed hands often but finally went to Russia in 1795. The Treaty of Riga (1921) awarded the city to Poland, but it was included in the Ukraine in 1939. Notable architectural monuments are the Mstyslavsky or Uspensky Cathedral (1157–60) and remains of old fortress walls (12th–13th cent.).

Vla·di·vos·tok (vlăd'ə-və-stŏk', -vŏs'tŏk'), city (1973 est. pop. 481,000), capital of Primorsky Kray (Maritime Territory), Far Eastern USSR, on a peninsula that extends between two bays of the Sea of Japan. It is the chief Soviet port on the Pacific (kept open in winter by icebreakers), the terminus of the Trans-Siberian RR and the Northern Sea Route, the chief base of the Soviet navy in the Pacific, and a base for fishing and whaling fleets. The city has large shipyards, chemical and engineering factories, fish canneries, and food plants. Founded in 1860, Vladivostok grew rapidly after the completion (1903) of the Trans-Siberian RR. It developed as a naval base after the loss (1905) of Port Arthur to Japan. In World War I the Allies used the city as a major supply depot, and after the Russian Revolution of 1917 they occupied it. Most of the occupying forces were Japanese, but there were also about 7,500 Americans and contingents of British, Italian, and French troops. In 1920, when Vladivostok was included in the newly proclaimed Far Eastern Republic, the Japanese continued to occupy the region and installed a counterrevolutionary Russian puppet government. By 1922 all the interventionist forces had withdrawn, and the city came under Soviet control. In World War II Vladivostok was a major port for lend-lease supplies. The city is the chief Soviet cultural center in the Far East.

Vlis·sing·en (vlĭs'ĭng-ən) or **Flush·ing** (flŭsh'ĭng), city (1973 est. pop. 42,639), Zeeland prov., SW Netherlands, on the S coast of the former island of Walcheren. It has oil refineries, shipyards, machinery factories, and iron and steel plants and is an important port for traffic with England. Chartered in 1247, Vlissingen was one of the first Dutch towns to rebel (1572) against Spain. Because it dominates the approach (via the Western Scheldt) to Antwerp, Vlissingen has been the scene of several battles. During World War II the Allies captured (1944) the city from the Germans after bitter fighting.

Vlo·rë (vlôr'ə, vlōr'ə) or **Va·lo·na** (və-lō'nə), city (1971 est. pop. 51,400), capital of Vlorë prov., SW Albania, on Vlorë Bay of the Adriatic Sea. Vlorë is a major seaport and a commercial center. Its industries produce foodstuffs, leather products, cement, and tobacco products. A commercial fishing fleet is based at Vlorë. Petroleum, natural gas, bitumen, and salt deposits are found nearby. The Stalin (Kuçovë) oil field near Berat is linked to Vlorë by pipeline. Vlorë Bay, strategically located at the mouth of the Adriatic Sea, has long been the site of military installations. In the 5th cent. Vlorë became an episcopal see. The city was prominent in the struggle (11th–12th cent.) between the Normans of Sicily and the Byzantines. It passed to Serbia in 1345 and to the Ottoman Empire in 1464; it was held by the Turks until 1912, when Albanian independence was proclaimed here. Vlorë was occupied by the Italians from 1914 to 1920 and from 1939 to 1944; it was bombed during World War II.

Vl·ta·va (vŭl'tə-və), longest river of Czechoslovakia, c.270 mi (435 km) long, rising in the Bohemian Forest, SW Czechoslovakia, and flowing SE, then N, past Prague, to the Elbe River at Mělník. There are several large hydroelectric stations on the river.

Vo·de·na (vō-*th*ā-nä'): *see* Edhessa, Greece.

Voj·vo·di·na (voi'və-dē'nə), autonomous province (1971 pop. 1,950,-268), 8,301 sq mi (21,500 sq km), NE Yugoslavia, in Serbia. Novi Sad is the chief city. A part of the great Danubian plain, it is watered by the Danube, the Tisza, and the Sava rivers and is one of the most densely populated and most prosperous parts of Yugoslavia. It is the breadbasket of Yugoslavia; fruit (notably plums, used for brandy) and vegetables are extensively cultivated. Cattle raising is also important, and food processing is the most significant industry. The region was part of Hungary and Croatia before its conquest by the Turks in the 16th cent. and was restored to the Hungarian crown by the Treaty of Passarowitz (1699). Parts of the region were included in the military frontier of southern Hungary in the 18th cent., and the whole region was settled with Serbian and Croatian fugitives from the Ottoman Empire, as well as by German colonists. The region was ceded (1920) to Yugoslavia by the Treaty of Trianon, and it received autonomy in 1946.

vol·ca·no (vŏl-kā'nō), aperture in the earth's crust through which gases, molten rock, or lava, and solid fragments are discharged. The term is commonly applied both to the vent and to the conical mountain (cone) built up around the vent by the ejected rock materials. Usually the mountain has as its apex a cavity, or crater, in which is the mouth of the vent. Volcanoes are described as active, dormant, or extinct. About 500 are known to be active. Belts of volcanoes are found in the ocean along the crest of the mid-ocean ridge system. Belts of volcanoes also occur where a crustal plate is being subducted into the earth's interior along converging crustal plate boundaries. One such belt is associated with volcanic island arcs and ocean trenches surrounding the Pacific Ocean; another occurs along the north shore of the Mediterranean Sea and extends eastward through Asia Minor and the Himalayan Mts. Isolated volcanoes in the mid-ocean area of the Pacific have given rise to abundant sea mounts and volcanic island chains, such as the Hawaiian chain.

Volcanic eruptions may take one or more of four chief forms, or phases, known as Hawaiian, Strombolian, Vulcanian, and Peleean. In the Hawaiian phase there is a relatively quiet effusion of basaltic lava unaccompanied by explosions or the ejection of fragments; the eruptions of Mauna Loa on the island of Hawaii are typical. The Strombolian phase derives its name from the volcano Stromboli in the Lipari, or Aeolian, Islands, north of Sicily. It applies to continuous but comparatively mild discharges in which viscous lava is emitted in recurring explosions; the ejection of incandescent material produces luminous clouds. A more explosive volcanic eruption is the Vulcanian. In this phase the magma (lava before emission) accumulates in the upper level of the vent but is blocked by a plug of hardened lava that forms at the orifice between consecutive explosions. When the explosive gases have reached a critical pressure within the volcano, masses of solid and liquid rock erupt into the air and clouds of vapor form over the crater; unlike the Strombolian phase, the clouds are not incandescent. The Peleean is the most violent phase of volcanic action, manifesting emission of fine ash, hot, gas-charged fragments of lava, and superheated steam in an incandescent "cloud" that travels downhill at great speed. This phase derived its name from Mont Pelée, on Martinique, which erupted in this fashion in 1902; a series of violent explosions formed a cloud that annihilated all life in its path; finally the mountain itself was blown apart. Eruptions also occur under the sea. The soil resulting from decomposition of volcanic materials is extremely fertile.

Volcano Islands, island group, c.11 sq mi (30 sq km), W Pacific. The group consists of three islands, of which Iwo Jima is the most important. There are sugar-cane plantations and sulfur mines on the islands. The inhabitants are Japanese, Koreans, and Formosans. Japanese fishermen and sulfur miners arrived in 1887, and Japan annexed the islands in 1891. Captured by U.S. forces in World War II, the islands were placed under U.S. administration from 1945 until 1968, when they were restored to Japan.

Vo·len·dam (vō'lən-däm'), town (1970 pop. c.12,000), North Holland prov., N central Netherlands, on the IJsselmeer, near Amsterdam. A picturesque town largely unchanged since the 17th cent., it is a famous tourist spot and is much frequented by artists.

Vol·ga (vŏl'gə, vôl'-, vōl'-), river, c.2,300 mi (3,700 km) long, central and E European USSR. It is the longest river of Europe and the principal waterway of the USSR, being navigable (with locks bypassing the dams) almost throughout its course. Its basin forms about one third of European USSR. Rising at an altitude of only 742 ft (226 m) in the Valday Hills, it winds east through the Rybinsk Reservoir and past Gorky to Kazan, where it turns south and continues its broad, majestic course past Ulyanovsk and Volgograd. From Volgograd (c.300 mi/480 km upstream) the Volga River flows in a course below sea level through the Caspian lowland. There a distributary parallels the mainstream and is connected with the Volga by numerous channels. The Volga enters the Caspian Sea through a wide delta below Astrakhan. The Volga-Baltic Waterway links the Volga with the Baltic Sea and with the Baltic-White Sea Canal; the Volga-Don Canal links the Volga with the Azov and Black seas; the Moscow Canal connects it directly with Moscow. In its upper course the Volga traverses numerous lakes. Below Gorky it broadens considerably. The Zhiguli Mts. cause the river to make a sharp bend (the Samara Bend), which reaches its easternmost point at Kuybyshev. The Volga is navigable from late Apr. to late Nov. at Shcherbakov and from early Mar. to mid-Dec. at Astrakhan. Dams and reservoirs have been constructed in the Volga basin for flood control, improved navigation, irrigation, and hydroelectric power. There are many important hydroelectric stations along the upper Volga. At Ivankovo, northwest of Moscow, a dam creates the vast Volga Reservoir or Moscow Sea, covering an area of c.125 sq mi (320 sq km). The Volga has played an important part in the life of the Russian people, for centuries serving as the chief thoroughfare of Russia and as the lifeline of Russian colonization to the east. It carries one half of the total river freight of the USSR and irrigates the vast steppes of the lower Volga region. Grain, oil (chiefly from Baku, shipped across the Caspian Sea), salt, fish, and caviar (from the Volga delta and the Caspian Sea) are shipped upstream; lumber is the main commodity shipped downstream.

The Volga was known to the ancient Greeks as the Rha, but little was known about the river until the early Middle Ages, when Slavic tribes settled along its upper course, the Bulgars along its middle course, and the Khazars in the south. Its importance as a trade route dates from that time. The Russians soon extended their control as far as Nizhny Novgorod (Gorky), founded in 1221. The Mongol invasion of the 13th cent. resulted in the direct control by the Golden Horde of the Volga below Nizhny Novgorod and in the creation (15th cent.) of the Tatar khanates of Kazan and Astrakhan, which fell to Moscow only in the 16th cent. Sarai, on the Volga River near modern Volgograd, was the capital of the empire of the Golden Horde. The conquest of these territories was largely the work of the Cossacks, who used the Volga and its tributaries for their epic forays into Siberia (under Yermak in the 16th cent.) and into the Caspian Sea (under Stenka Razin, in the 17th cent.). Many of the Finnic and Turco-Tatar nationalities still live in the middle and lower Volga regions, notably in the Chuvash ASSR, the Mari ASSR, the Mordvinian ASSR, the Tatar ASSR, and the Udmurt ASSR. The Kalmyrs settled in the lower Volga region in the early 17th cent. The lower Volga was the center of the great peasant rebellions under Stenka Razin and Pugachev. After their suppression Catherine II settled many German colonists in the region around Saratov.

Vol·ga-Bal·tic Waterway (vŏl'gə-bôl'tĭk, vôl'-, vōl'-), canal and river system, c.685 mi (1,100 km) long, N European USSR. It links the Volga River and the Leningrad industrial area. It consists of the Moscow-Volga Canal, the Volga River, the Rybinsk Reservoir, the Mariinsk system (composed of the Sheksna River, the White Lake Canal, the Kovzha River, the Mariinsk Canal, and the Vytegra River), the Onega Canal, the Svir River, the Ladoga Canals, and the Neva River to Leningrad. The waterway was begun in 1709 to connect St. Petersburg with the interior. The major canals were built in the 1930s. The waterway has been reconstructed and modernized (reopened May, 1964), the principal addition being a dam across the Sheksna River near Cherepovets, which deepened the waterway as far as the Kovzha River, facilitating the use of larger vessels. Although more extensive, this waterway follows the historic Baltic-Volga trade route, in use since the 9th cent.

Vol·go·grad (vŏl'gō-grăd', vôl'-, vōl'-), formerly **Sta·lin·grad** (stä'-

lĭn-grăd′, stăl′ĭn-), city (1974 est. pop. 885,000), capital of Volgograd oblast, SE European USSR, a port on the Volga River and the eastern terminus of the Volga-Don Canal. Volgograd is also a major rail center, with connections to Moscow, the Donets Basin, the Caucasus, and southwestern Siberia. One of the world's largest hydroelectric power dams stands on the Volga just above the city. A center of heavy industry, Volgograd has shipyards, oil refineries, iron, steel, and aluminum mills, and tank, tractor, cable, machinery, and chemical factories. Other industries include food processing, flour milling, distilling, sawmilling, tanning, and the manufacture of farm and oil-field equipment. Founded in 1589 as a stronghold to defend Russia's newly acquired land along the Volga, the city was originally called Tsaritsyn. It fell to the Cossack rebels under Stenka Razin in 1670 and Yemelyan Pugachev in 1774. In the 19th cent. it became an important commercial center.

The city was named Stalingrad in 1925. It was virtually destroyed during World War II in a battle that marked a major turning point in the war and a landmark in military history. In Sept., 1942, a German army exceeding 500,000 men (including Italians, Hungarians, and Rumanians) and commanded by Gen. Friedrich von Paulus began an all-out attack on Stalingrad, which was defended by 16 Soviet divisions under Gen. Vasily I. Chuikov. Stalin ordered that the city be held at all costs. After two months of house-to-house fighting the Germans had taken most of the city, but the Soviet garrison, receiving supplies from the east bank of the Volga, continued to hold out, thus giving Gen. Grigori Zhukov time to prepare a counteroffensive. Adolf Hitler refused, against the advice of his general staff, to allow Paulus to withdraw. In Nov., 1942, two Soviet forces, advancing from the north and south in a pincers movement, encircled the Germans. In Dec. a German relief force was routed. Paulus surrendered the remnants of his army on Feb. 2, 1943. The combined German and Soviet losses during the battle were staggering—the Germans alone suffered approximately 300,000 casualties. The Soviets followed up their victory with a mighty westward drive. Rebuilding of Stalingrad began immediately after the city's liberation. It was renamed Volgograd in 1961, following Nikita Khrushchev's renewed denunciations of Stalin's dictatorship.

Vol·hyn·i·a (vŏl-hĭn′ē-ə, vō-lĭn′yə), historic region, SW European USSR, in the Ukraine, around the headstreams of the Pripyat and Western Bug rivers in an area of forests, lakes, and marshlands. One of the oldest Slavic settlements in Europe, it derived its name from the extinct city of Volyn or Velyn, said to have stood on the Western Bug. Volhynia was divided (c.1388) between Poland (western part) and Lithuania (eastern part). With the Polish-Lithuanian union of 1569, Volhynia became a quasi-autonomous province of Poland. During the second and third partitions of Poland (1793, 1795) Volhynia passed to Russia and was made (1797) a province. In 1921 the Treaty of Riga returned western Volhynia to Poland, but the rest passed to the Ukraine. Poland ceded its section of Volhynia to the USSR in 1939, and the Soviet-Polish border agreement of 1945 confirmed it as a Soviet possession. This section now constitutes the Volyn oblast, a rich agricultural lowland and coal-mining area.

Vo·log·da (vô′ləg-də), city (1973 est. pop. 198,000), N central European USSR, on the Vologda River. It is a major river and rail junction in a dairying region. There are shipyards, machine factories, lumber mills, and flax-processing plants. Vologda is famed for its lace. Founded in 1147 by merchants from Novgorod, it passed to Moscow in the 15th cent. In Vologda's old kremlin is the Cathedral of St. Sophia, built (1568–70) by Ivan IV.

Vó·los (vô′lôs′), city (1971 pop. 51,290), capital of Magnisia prefecture, E Greece, in Thessaly, on the Gulf of Vólos, an inlet of the Aegean Sea. The principal port of Thessaly, Vólos is a transportation, commercial, and industrial center. Its leading exports are tobacco, wheat, wine, and olives. Vólos is located near the sites of the ancient towns of Pagasae and Demetrias. It was known as Gholos under Ottoman rule and was modernized after it passed to Greece in 1881. An earthquake in 1955 damaged much of the city.

Vol·sin·i·i (vŏl-sĭn′ē-ī′), ancient city of Etruria, Italy, probably on the site of modern Orvieto. It was a powerful member of the Etruscan League, and the spirit of the league was broken when Romans conquered and thoroughly sacked Volsinii in 280 B.C. A new Volsinii was founded near Lacus Volsiniensis (Lake Bolsena).

Volsk (vôlsk), city (1973 est. pop. 71,000), S central European USSR, a port on the Volga River. It has food and metal processing and cement industries.

Vol·ta (vŏl′tə, vōl′-, vôl′-), river, c.290 mi (465 km) long, formed in central Ghana, W Africa, by the confluence of the Black Volta (c.840 mi/1,350 km long) and the White Volta (c.450 mi/725 km long), both of which rise in Upper Volta. The river flows generally south, through a large delta, to the Gulf of Guinea at Ada. Lake Volta (c.3,275 sq mi/8,480 sq km), one of the world's largest man-made lakes, extends c.280 mi (450 km) upstream behind Akosombo Dam, southeast Ghana, in the Ajena Gorge. The dam (370 ft/113 m high; completed 1965) regulates the flow of the Volta River, stores water for irrigation, and generates hydroelectricity that supports a large aluminum industry.

Vol·ta Re·don·da (vŏl′tə rĭ-dŏn′də, vōl′-, vôl′-), city (1970 pop. 120,645), Rio de Janeiro state, E Brazil, on the Paraíba River. Its proximity to sources of hydroelectricity and basic raw materials and to the industrial centers of Rio de Janeiro and São Paulo favored its selection as the site of a huge government-run steel mill. The city was founded in 1941, and construction of the steel mill began the following year, with production starting in 1946. Within a decade Volta Redonda, one of the most ambitious industrial projects in South America, was producing more than half of Brazil's ingots and rolled steel.

Vol·ter·ra (vōl-tĕr′ä), town (1972 est. pop. 13,400), Tuscany, central Italy. A powerful Etruscan town, it later (12th–13th cent.) was a free commune and passed to Florence in the 14th cent. Of note are well-preserved Etruscan gates and tombs, medieval walls, and the Palazzo dei Priori (13th cent.). The powerful fortress (built 14th–15th cent.) is now a prison. There is an Etruscan museum in the town.

Vol·tur·no (vŏl-tŏŏr′nō, vōl′-, vôl′-), chief river of S Italy, 109 mi (175.4 km) long, rising in the Apennines of Molise and flowing SE, then SW through Campania to the Tyrrhenian Sea.

Vo·lu·sia (vō-lōō′shə), county (1970 pop. 169,487), 1,062 sq mi (2,750.6 sq km), E Fla., in a lowlands area bordered on the W and partly on the S by the St. Johns River and by the Atlantic on the E, with many lakes and scattered swamps; formed 1854; co. seat De Land. It is in a farm area that produces citrus fruit, vegetables, dairy products, and poultry. Canned fruit and vegetables, clothing, chemicals, plastics, concrete, metal products, and machinery are made. There are tourist resorts along the coast.

Vor·arl·berg (fôr′ärl′bûrg, fŏr′-), province (1971 pop. 271,473), 1,004 sq mi (2,600 sq km), extreme W Austria, bordering on Switzerland, Liechtenstein, and West Germany; capital Bregenz. The province is a cattle-raising and dairy-farming region noted for its Alpine scenery. It has considerable industry. Hydroelectric works dot the Bregenzer Ache and Ill rivers, and textile mills are found in almost every town. Beautiful embroidery and lace are produced by artisans. Vorarlberg has numerous popular winter sports resorts. The province is bounded on the west by the Rhine River. Vorarlberg was part of the Roman province of Rhaetia and was acquired by the powerful counts of Montfort in the Middle Ages. The Hapsburgs gained possession of it piecemeal in the 14th, 15th, and 16th cent., and in 1523 it became a crown land, administered by the Tyrol. Culturally related to the Swiss, the inhabitants of Vorarlberg voted for unification with Switzerland after World War I. The attempt failed because of Allied and Swiss opposition.

Vor·ku·ta (vôr-kōō′tə), city (1973 est. pop. 92,000), Komi Autonomous Republic, NE European USSR, above the Arctic Circle. Founded in 1932 as the site of large Soviet forced-labor camps, Vorkuta became a city in 1942. Some of the camps may still exist, though they were reportedly dissolved after Stalin's death in 1953.

Vo·ro·nezh (və-rô′nĭsh), city (1974 est. pop. 729,000), capital of Voronezh oblast, E central European USSR, on the Voronezh River. A river port and a major industrial center in a black-earth agricultural region, it has industries producing machinery, locomotives, synthetic rubber, oil, and food products. A nuclear power station operates at Voronezh. Founded in 1586 as a frontier fortress against Crimean and Nogai Tatar attacks from the southern steppe, it became a shipbuilding center in the Azov campaign (1695–96) of Peter I. During World War II it was largely destroyed (1942-43) when a German advance was stopped here; it was rebuilt completely after the war. The architectural monuments, the Nikolsk church (early 18th cent.), and the Potemkin palace (18th cent.) were restored. The Univ. of Voronezh, originally the Univ. of Tartu, was transferred here in 1918. There are Scythian burial mounds outside the city.

Vo·ro·shi·lov·grad (vôr′ə-shē′ləf-grăd′), city (1973 est. pop. 414,000), capital of Voroshilovgrad oblast, S central European

USSR, in the Ukraine, at the confluence of the Lugan and Olkhov rivers, in the Donets Basin. Its products include locomotives, steel pipes, mining equipment, machine tools, textiles, and processed food. It was founded in 1796 around a cannon foundry and is the oldest center of the Donets Basin. Named Lugansk in 1889, it was called Voroshilovgrad from 1935 to 1958, Lugansk from 1959 to 1969, and renamed Voroshilovgrad in 1970.

Vosges (vōzh), department (1974 est. pop. 405,800), 2,303 sq mi (5,964.8 sq km), E France, largely in Lorraine; capital Épinal.

Vosges, mountain range, E France, between the Alsatian plain in the E and the plateau of S Lorraine in the W. It extends generally north and parallel to the Rhine River for c.120 mi (190 km) from the Belfort Gap. The Vosges, old crystalline mountains flanked by sandstone, have gently rounded or nearly flat summits. The highest point is the Ballon de Guebwiller (4,672 ft/1,425 m). The slopes (steep in Alsace, gentle in Lorraine) are forested (chiefly by pines) up to c.3,500 ft (1,070 m). Vineyards producing Riesling and other wines grow on the Alsatian slopes. Lumbering, dairying, and paper and textile manufacturing are the chief occupations. There are resorts, notably Plombières-les-Bains. The Moselle, Meurthe, Sarre, and Ill rivers rise in the Vosges.

Vot·kinsk (vôt'kĭnsk), city (1973 est. pop. 79,000), Udmurt Autonomous Republic, E European USSR, on a tributary of the Kama River. It has machine plants, sawmills, and brickyards. The home of Tchaikovsky, who was born in Votkinsk, is now a museum.

Voy·a·geurs National Park (voi'ĭ-jərz): *see* National Parks and Monuments Table.

Vra·tsa (vrä'tsə), city (1969 est. pop. 47,600), NW Bulgaria, in the foothills of the Balkan Mts. It is a commercial and crafts center and a railway junction. Vratsa has textile, metal processing, and ceramics industries. It was an administrative and garrison town under Ottoman Turkish rule (15th–19th cent.).

Vrin·da·ban (vrĭn'də-bŭn'), formerly **Brin·da·ban** (brĭn'də-bŭn'), town (1971 pop. 29,475), Uttar Pradesh state, central India, on the Jumna River. Legends about the youth of the Hindu god Krishna center about the town, which is a popular pilgrim site. There are c.1,000 shrines, chiefly from the 16th cent., including the elegant Red Temple of Krishna.

Vyat·ka (vyät'kə): *see* Kirov, USSR.

Vyatka, river, c.850 mi (1,370 km) long, rising in the foothills of the central Urals, E European USSR, and flowing first N, then NW past Kirov, and finally SE into the Kama River near Mamadysh.

Vyaz·ma (vyäz'mə), city (1970 pop. 44,145), N central European USSR, on the Vyazma River, a tributary of the Dnepr. It is a rail junction and has machine building, auto repair, food- and flax-processing, and tanning industries. Founded in the 9th cent., Vyazma became an important trade and military center that was an object of contention among Russia, Lithuania, and Poland. During World War II it was held by the Germans, who destroyed it prior to withdrawal.

Vy·borg (vē'bôrg'), city (1973 est. pop. 68,000), NW European USSR, NW of Leningrad and near the Finnish border, on Vyborg Bay and the Gulf of Finland. A Baltic port and railroad junction, it is an export center for lumber. It also has shipyards and industries producing machinery, electrical equipment, food products, furniture, and paper. Vyborg was a trading point for Novgorod in the 12th cent. but actually grew around a Swedish castle built here in 1293. Vyborg became a port for the Hanseatic League and was chartered in the 15th cent. In 1710 Peter the Great seized Vyborg, and it was incorporated with Finland (then under Russian sovereignty) in 1812. Vyborg remained Finnish until 1940, when it was occupied by the Soviet Union. It was recaptured by Finnish forces in 1941 and was finally seized by the Soviets in 1944 and awarded to them by the Finnish-Soviet peace treaty (1947). In the city are a tower (1550), several towers of the town hall (15th–17th cent.), and a fort (18th cent.).

Vy·cheg·da (vĭch'əg-də), river, c.700 mi (1,125 km) long, rising in several headstreams in the Urals, NE European USSR, and flowing generally W into the Northern Dvina River at Kotlas. It is navigable (Apr.–Nov.) c.595 mi (960 km) to Voldino and is used for timber flotation. Solvychegodsk and Syktyvkar are two of its chief ports. In the 16th cent. the Vychegda was an important water route to Siberia.

Vyg (vĭg), lake, c.300 sq mi (780 sq km), NW European USSR, in Karelia, between Lake Onega and the White Sea. The lake is a part of the Baltic–White Sea Canal.

Vy·te·gra (vĭt'ĭ-grə), river, c.45 mi (70 km) long, rising near Lake Kovzhas, NW European USSR, and flowing generally NW into Lake Onega. It is one of the most important parts of the Volga-Baltic Waterway. At its mouth is the city of Vytegra, a center for the lumber industry and for ship and farm-implement repairs.

Waal (väl), main arm of the Rhine River, 52 mi (84 km) long, branching off the Rhine near the West German border and flowing W through central Netherlands, past Nijmegen, to join the Maas (Meuse) near Gorinchem. The joined rivers form the upper Merwede.

Wa·bash (wô'băsh). **1.** County (1970 pop. 12,841), 222 sq mi (575 sq km), SE Ill., bounded on the E and S by the Wabash River and drained by Bonpas Creek; formed 1824; co. seat Mount Carmel. It has agriculture (corn, wheat, soybeans, and livestock), manufacturing of electrical and sports equipment, clothing, flour, and paper products, and deposits of petroleum, natural gas, and bituminous coal. **2.** County (1970 pop. 35,553), 398 sq mi (1,030.8 sq km), N Ind., drained by the Wabash, Eel, Salamonie, and Mississinewa rivers; formed 1832; co. seat Wabash. It is in an agricultural area that produces livestock, soybeans, wheat, and corn. Its manufactures includes furniture, abrasive products, metal products, and farm machinery. There is timber sawmilling in the county.

Wabash, city (1975 est. pop. 13,300), seat of Wabash co., N central Ind., on the Wabash River; inc. 1849. Wabash is in a fertile area that yields grain, vegetables, and fruit. Rubber products, paperboard, temperature controls, insulation, and electrical parts are manufactured. The U.S. government built (1820) a mill here for the Miami Indians. A treaty concluded (1826) with the Indians opened the way for the first permanent white settlers, who arrived in 1827. Wabash was the world's first electrically lighted city.

Wabash, river, c.475 mi (765 km) long, rising in Grand Lake, W Ohio, and flowing NW into Ind., then generally SW through Ind., becoming the Ind.-Ill. border before emptying into the Ohio River; largest northern tributary of the Ohio. The Wabash's major tributaries are the Tippecanoe and White rivers. Dams on the Wabash control floods, produce hydroelectricity, and regulate navigation; sand and gravel barges constitute the chief traffic on the river. In the fertile Wabash basin corn and livestock are raised.

Wa·ba·sha (wô'bə-shô), county (1970 pop. 17,224), 522 sq mi (1,352 sq km), SE Minn., bounded on the E by the Mississippi River and Wis. border and drained by the Zumbro River; formed 1849; co. seat Wabasha. In an agricultural area yielding livestock, dairy products, corn, oats, and barley, it has some mining, a food-processing industry, and clothing and machinery manufacturing.

Wabasha, city (1970 pop. 2,371), seat of Wabasha co., SE Minn., on the Mississippi River at the foot of Lake Pepin; platted 1843, inc. 1858. An early fur-trading post was here. Wabasha is now a farm trade center.

Wa·baun·see (wə-bôn'sē), county (1970 pop. 6,397), 792 sq mi (2,051.3 sq km), E Kansas, in a dissected plain bounded on the N by the Kansas River; formed 1855; co. seat Alma. It is in a grain and livestock-farming area, with some mining and manufacturing.

Wa·co (wā'kō), city (1979 est. pop. 99,600), seat of McLennan co., E central Texas, on the Brazos River, just below the mouth of the Bosque; inc. 1856. It is a trading, shipping, and industrial center. Tires, glass, and paper are among the manufactures. Rich blacklands supported cotton plantations and cattle ranches before the Civil War, but after the war the city suffered a severe decline. Prosperity returned when its suspension bridge (still a tourist attraction) was built across the Brazos (1870) and the railroad arrived (1881). Waco is the seat of Baylor Univ.

Wa·de·na (wŏ-dē'nə), county (1970 pop. 12,412), 536 sq mi (1,388.2 sq km), central Minn., drained by the Crow Wing River; formed 1858; co. seat Wadena. It is in a livestock-farming and dairying region, with timber and furniture manufacturing.

Wadena, village (1970 pop. 4,640), seat of Wadena co., W Minn., WNW of Brainerd, in a farm and lake area; settled 1871, inc. 1881.

Wades·bor·o (wādz'bûr'ō), town (1970 pop. 3,977), seat of Anson co., S N.C., near the S.C. border ESE of Charlotte; settled c.1785. It has textile and lumber mills.

Wa·di Hal·fa (wä'dē hăl'fə), town, N Sudan, on the Nile. It is the terminus of a railroad from Khartoum and the point at which cotton, wheat, livestock, and other goods are transferred to steamers going down the Nile into Egypt. Archaeological expeditions have worked to excavate and preserve the area's numerous Egyptian antiquities, which faced inundation from the reservoir of the Aswan High Dam. Founded in the 19th cent., Wadi Halfa became (1885–98) the headquarters of the Anglo-Egyptian army as it prepared to reconquer Sudanese territory from the Mahdi. The railroad up the Nile to Wadi Halfa was built to support Lord Kitchener's forces during the reconquest. During World War II the town served as a staging post on the Allied communication line with Egypt via central Africa.

Wad Ma·da·ni (wäd mä-dä'nē), city (1969 est. pop. 71,000), capital of Blue Nile prov., E Sudan, on the Blue Nile River. It is linked by rail with Khartoum and is the chief center of a cotton-growing region. Wheat, barley, and livestock are other products of the city. Wad Madani has an agricultural research station. A small Turco-Egyptian administrative post in the 19th cent., the city grew rapidly after the implementation (1925) of a program to develop the Al Jazirah region.

Wads·worth (wôdz'wûrth'), city (1970 pop. 13,142), Medina co., NE Ohio, an industrial suburb of Akron; settled c.1816, inc. 1866.

Wag·ga Wag·ga (wŏg'ə wŏg'ə), city (1973 est. pop. 30,310), New South Wales, SE Australia, on the Murrumbidgee River. It is the center of an agricultural district with food-processing and rubber-goods plants and foundries. There is a Roman Catholic cathedral in the city. Wagga Wagga also has teacher-training, technical, and agricultural colleges.

Wag·on·er (wăg'ən-ər), county (1970 pop. 22,163), 563 sq mi (1,458.2 sq km), NE Okla., bounded in the S by the Arkansas River and drained by the Neosho and Verdigris rivers; formed 1907; co. seat Wagoner. Its agriculture includes grain, cotton, corn, livestock, soybeans, and dairy products. It has some oil and gas wells and manufactures air and gas compressors, motors, and generators.

Wagoner, city (1970 pop. 4,959), seat of Wagoner co., E Okla., SE of Tulsa, in an agricultural and resort area; founded c.1887. A museum with Indian and Civil War relics is here.

Wa·gram (vä'gräm), town, Lower Austria prov., NE Austria, in the Marchfeld, near Vienna. On July 5–6, 1809, Napoleon I gained one of his most brilliant victories here. Despite their heroic conduct and the able leadership of Archduke Charles, the Austrians were forced to fall back by French field artillery fire. Napoleon's "grand battery" of 100 guns was the largest concentration of artillery that had until then been used for massed fire. Six days later Austria was forced to conclude an armistice.

Wa·hi·a·wa (wä'hē-ə-wä'), residential city (1970 pop. 17,598), Honolulu co., Hawaii, on central Oahu, in a pineapple region.

Wah·ki·a·kum (wô-kī'ə-kəm), county (1970 pop. 3,592), 261 sq mi (676 sq km), SW Wash., bounded in the S by the Columbia River; formed 1854; co. seat Cathlamet. Its agriculture includes dairy products, livestock, hay, and potatoes. It also has fishing and lumbering.

Wa·hoo (wä'ho͞o), city (1970 pop. 3,835), seat of Saunders co., E Nebr., NNE of Lincoln, in a farm area; founded 1865.

Wah·pe·ton (wô'pə-tən), city (1970 pop. 7,076), seat of Richland co., extreme SE N.Dak., on the Red River; inc. 1883. A trade and processing center of a livestock, dairy, and grain area, it is also the seat of a junior college and a Federal Indian school. Pottery is made.

Wai·a·na·e Mountains (wī'ä-nä'ā), volcanic range, W Oahu island, Hawaii. It rises to Mt. Kaala (4,025 ft/1,227.6 m), the highest point on the island.

Wai·ka·to (wī'kä'tō), river, 270 mi (434.4 km) long, rising in Lake Taupo, central North Island, New Zealand, and flowing NW into the Tasman Sea. It is New Zealand's longest river. The Waikato's hydroelectric stations are the main source of power for North Island. Coal is mined in the reclaimed swamplands of the Waikato basin. Dairy farming is an important activity in the basin. The river is navigable for 80 mi (129 km) up to Cambridge.

Wai·ki·ki (wī'kē-kē'), famous beach and resort center SE of Honolulu on SE Oahu island, Hawaii. Surfboard riding is the major sport at Waikiki, which is noted for its huge waves. The Honolulu Zoo is in nearby Kapiolani Park.

Wai·lu·ku (wī-lo͞o'ko͞o), city (1970 pop. 7,979), seat of Maui co., Hawaii, on the N central coast of Maui W of Haiku-Pauwela, in an agricultural area. It is a tourist center.

Wai·me·a (wī-mā'ə), town (1970 pop. 1,569), Hawaii, on the S coast of Kauai island. James Cook landed on the site in 1778, and the first missionaries to come to Hawaii settled here. Nearby are the remains of a fort built by the Russians in 1815, when they tried to gain influence in Hawaii.

Wa·ka·ya·ma (wä-kä'yä-mä), city (1970 pop. 365,267), capital of Wakayama prefecture, SW Honshu, Japan, on the Inland Sea. It is a railroad hub and a manufacturing center where cotton and flannel textiles are produced.

Wake (wāk), county (1970 pop. 229,006), 858 sq mi (2,222.2 sq km), central N.C., in a piedmont region drained by the Neuse River; formed 1770; co. seat Raleigh. Farming (tobacco, cotton, soybeans, corn, poultry, dairy products, and hogs), food processing, metal and machinery manufacturing, printing, and sawmilling are important.

Wa·Kee·ney (wô'kē'nē), city (1970 pop. 2,334), seat of Trego co., NW Kansas, between the Saline and Big Creek rivers WNW of Hays; platted 1878, inc. 1880. Early settlers sold fertilizer made from buffalo bones left on the prairie after the animals had been killed. There are wheat fields to the west, and fossils have been found here.

Wake·field (wāk'fēld'), county borough (1972 est. pop. 59,840), West Yorkshire, N central England, on the Calder River. It has been a center of the cloth industry from the 14th cent. Modern manufactures include woolen goods, chemicals, machine tools, and power presses. The site was occupied by the Danes and Saxons; Richard, duke of York, was slain in the Battle of Wakefield in 1460.

Wakefield, town (1979 est. pop. 25,600), Middlesex co., NE Mass., a suburb of Boston; settled 1639, inc. 1812.

Wake Island (wāk), atoll with three islets (Wake, Wilkes, and Peale), (1978 est. pop. 1,000), 3 sq mi (7.8 sq km), central Pacific, between Hawaii and Guam. It is a U.S. commercial and military base under the jurisdiction of the Federal Aviation Agency. There is no indigenous population. Wake Island was discovered by the Spanish in 1568, visited by the British in 1796 and named after Capt. William Wake, and annexed by the United States in 1898. The island became (1935) a commercial air base on the route to the Orient and later served as a U.S. military base. In Dec., 1941, Wake Island was seized by the Japanese. U.S. forces bombed the island from 1942 until Japan's surrender in 1945.

Wa·kul·la (wŏ-kŭl'ə), county (1970 pop. 6,308), 601 sq mi (1,556.6 sq km), NW Fla., in a lowland area bounded on the S by Apalachee Bay, on the W by the Ochlockonee River, and drained by the St. Marks and Wakulla rivers; formed 1843; co. seat Crawfordville. It is in a tourist, forestry, and fishing area, with agriculture (cattle, hogs, corn, peanuts, and vegetables), food processing, and petroleum refining. The coast is a national wildlife refuge, and the western interior is part of Apalachicola National Forest.

Wa·la·chi·a or **Wal·la·chi·a** (both: wə-lā'kē-ə), historic region (29,568 sq mi/76,581 sq km), S Rumania. The Transylvanian Alps separate it in the northwest from Transylvania and the Banat; the Danube separates it from Yugoslavia in the west, Bulgaria in the south, and northern Dobruja in the east; in the northeast it adjoins Moldavia. Bucharest, the Rumanian capital, is its chief city. The Oltul River, a tributary of the Danube, divides Walachia into Muntenia or Greater Walachia (20,265 sq mi/52,486 sq km) in the east and Oltenia or Lesser Walachia (9,303 sq mi/24,095 sq km) in the west. With the rich Ploieşti oil fields and the industrialized area near Bucharest, Walachia is economically the most developed region of Rumania. Its industries (chemicals, heavy machinery, and shipbuilding) provide employment for about half of the country's labor force. Walachia is also a rich agricultural area and the "bread basket" of Rumania.

The region was part of the Roman province of Dacia and has retained its Romanic speech despite centuries of invasion and foreign rule. Although theoretically part of the Byzantine Empire, Walachia was successively occupied (6th-11th cent.) by the Lombards, the Avars, and the Bulgarians. By the 12th cent. it had passed under the Cumans, who in turn succumbed (1240) to the Mongols. When the Mongol wave receded, the native inhabitants descended from their mountain refuges, and the principality of Walachia was founded (c.1290) by their leader Radu Negru, or Rudolf the Black. The name *Vlachs* (or *Walachs* or *Wallachs*) was given them by their Slavic neighbors. Although some claim that the Vlachs are direct descendants of the Dacians (mainly on the ground that they preserved their Latin speech), it is more than likely that they represent a composite ethnic mixture. The sister principality, Moldavia, came into existence about the same time as Walachia. Cîmpulung, the earliest capital of Walachia, was later replaced by Curtea-de-Arges. Mircea the Great of Walachia (reigned 1386-1418) shared in the defeats of Kossovo (1389) and Nikopol (1396) at the hands of the Turks and was obliged to pay tribute to the sultan.

Walachia continued to be governed by its own princes under Turkish suzerainty. Like Moldavia, it was torn by strife among the great landowners (or boyars) and among rival claimants to the throne. Prince Vlad the Impaler (reigned 1456-62) restored some order by putting 20,000 persons to death within six years. He refused tribute to the sultan, defeated the Turks, and impaled the Turkish prisoners. His rivalry with Stephen the Great of Joldavia cost him his throne. A last attempt to free all Rumanians from foreign domination was made (1593-1601) by Michael the Brave, who massacred the Turks in Walachia and conquered Transylvania and Moldavia. His death delivered Walachia back into the hands of the Turks.

In the Russo-Turkish Wars of the 18th cent. Walachia was repeatedly occupied by Russian and Austrian troops. The oppressive rule of the Phanariots lasted until 1822, when the Rumanians rebelled against the Greeks, who at the same time began their war of independence against Turkey. Native governors were again appointed, and the Treaty of Adrianople in 1829 made Walachia an almost autonomous state, tributary to Turkey but under Russian protection. A Rumanian national uprising (1848-49) in Walachia was suppressed by Russian intervention. Russian troops occupied (1853) Walachia and Moldavia early in the Crimean War; however, to purchase Austrian neutrality, they evacuated the lands in 1854, and the two Danubian Principalities (as Walachia and Moldavia were called) passed under Austrian occupation. The Congress of Paris (1856), which ended the Crimean War, guaranteed the principalities virtual independence under the nominal suzerainty of Turkey. With the accession (1859) of Alexander John Cuza as prince of both Moldavia and Walachia, the history of modern Rumania began.

Wał·brzych (välb'zhĭk), city (1973 est. pop. 127,400), SW Poland. Coal mining is the chief economic activity. The city's importance dates from the 19th cent. industrialization of the Lower Silesian coal basin. Wałbrzych sustained great damage during World War II, after which it passed from Germany to Poland.

Wal·che·ren (väl'kə-rən), region, Zeeland prov., SW Netherlands, on the North Sea at the entrance to the Scheldt estuary. Dunes line the North Sea coast, and diked lowlands predominate elsewhere. Agriculture and cattle raising are the mainstays of Walcheren's economy; the main crops are wheat, vegetables, fruit, and sugar beets. Walcheren also has a considerable tourist trade, attracted largely by the medieval buildings in Middelburg. Walcheren was occupied by the Germans and suffered heavy bombardment during World War II. The dikes were bombed and the region largely flooded in order to force German evacuation. In the autumn of 1944 British troops landed on Walcheren, drove out the remaining Germans, and went on to capture the Belgian port of Antwerp, which became a major Allied base.

Wal·den (wôl'dən), town (1970 pop. 907), seat of Jackson co., N Colo., on the headstream of the North Platte River between Park Range and the Medicine Bow Mts. and SW of Laramie, Wyo., near the Wyo. border, at an altitude of 8,300 ft (2,531.5 m). It is a supply point in a livestock and grain region. There are coal mines nearby.

Wal·do (wôl'dō), county (1970 pop. 23,328), 737 sq mi (1,908.8 sq km), S Me., on Penobscot Bay, drained by the Penobscot, Sebasticook, and Sheepscot rivers; formed 1827; co. seat Belfast. Its agriculture in the inland lake region includes apples, potatoes, truck crops, dairy products, and poultry. It has lumber sawmilling and some industry (food processing and the manufacture of furniture, chemical products, and shoes).

Wal·dron (wôl'drən), city (1970 pop. 2,132), seat of Scott co., W Ark., SE of Fort Smith, in a timber, dairy, and farm area of the Ouachita Mts.; inc. 1852.

Wales (wālz), western peninsula (1971 pop. 2,723,596) of Great Britain, 8,016 sq mi (20,761 sq km), W of England; politically united with England since 1536. Wales is bounded on the north by the Irish Sea, on the south by the Bristol Channel, and on the west by Cardigan Bay and St. George's Channel. Across the Menai Strait is the Welsh island of Anglesey. The Cambrian Mts. cover most of Wales, with the high point at Snowdon (3,560 ft/1,085 m). The eastern rivers—the Dee, the Severn, and the Wye—drain into England. The Usk flows into the Bristol Channel.

The eastern boundary, drawn in 1536, united England and Wales politically but disregarded cultural and linguistic distribution. Welsh-speaking areas were added to the English border counties. In 1971 more than 41,000 inhabitants of Wales spoke Celtic Welsh only, and more than 673,000 spoke Welsh in addition to English.

In 1974 the 13 historic counties of Wales were abolished and replaced by 8 new counties:

NAME	ADMINISTRATIVE CENTER	NAME	ADMINISTRATIVE CENTER
Clwyd	Mold	Mid Glamorgan	Cardiff
Dyfed	Carmathen	Powys	Llandrindod Wells
Gwent	Cwmbran	South Glamorgan	Cardiff
Gwynedd	Caernarvon	West Glamorgan	Swansea

Economy. The northern counties of Wales are characterized by farms and pastoral highlands. The coastal towns of the Lleyn Peninsula are tourist and vacation centers for northern England's industrial cities. The industrial wealth of Wales is concentrated in the southern counties bordering on the Bristol Channel. This area has large steelworks, oil refineries, tinplate and copper foundries, and the southern Wales coal fields, which have been revitalized since World War II.

History. Welsh tradition stretches back into prehistory. The Roman impress upon Wales was light, and Welsh clans of shepherds and farmers continued to dominate large areas of Great Britain. Border wars were chronic between the Welsh and the English kingdoms. The disparate clans of pastoral people gradually coalesced. Hywel Dda, king of Wales in the mid-10th cent., collected Welsh law and custom into a unified code. William I of England tried to deal with the Welsh by setting up border earldoms to protect his newly won kingdom from their incursions. Wales was increasingly threatened with English conquest, but dissension within England in the early 12th cent. relaxed pressure on the Welsh princes, and medieval Welsh culture approached its full blossom. English conquest of Wales was finally accomplished by Edward I in 1282. The Statute of Rhuddlan (1284) established English rule. To placate Welsh sentiment, Edward had his son made prince of Wales in 1301; thus originated the English custom of entitling the king's eldest son prince of Wales. Changes in Welsh life, although few, included a gradual cultural decline and the growth of market towns through trade with England. Henry VII, the first Tudor king, who ascended the English throne in 1485, was the grandson of Owen Tudor, a Welshman. The Welsh feudal aristocracy became versed in English manners and were received at the English court. The Act of Union (1536) and supplementary legislation completed the process of administrative assimilation by abolishing all Welsh customary law at variance with the English and by establishing English as the language of all legal proceedings.

The Reformation came belatedly to Wales. Catholic tradition died slowly. In the 17th–18th cents. evangelical Protestantism took a strong hold. The Industrial Revolution brought exploitation of the mineral wealth of Wales, at first in the north and then in the rich coal fields of the south. By the early 19th cent. the effects of industrialization threatened both cottage industry and agriculture. Poverty and unemployment intensified in the years of economic decline following World War I. After World War II the Labour government, which drew substantial support from the socialist stronghold of southern Wales, undertook a full-scale program of industrial redevelopment. This included reorganization of the coal mines and tinplate manufacture under government control, introduction of diversified industry, and improvement of communications, housing, and technical education. Welsh nationalism has revived, both in the arts and politically.

Wal·hal·la (wôl-hăl'ə), town (1970 pop. 3,662), seat of Oconee co., NW S.C., near the Blue Ridge W of Greenville, in a resort area; founded c.1850, inc. 1855. A U.S. trout hatchery is nearby.

Walk·er (wô'kər). **1.** County (1970 pop. 56,246), 805 sq mi (2,085 sq km), NW central Ala., in a hilly area crossed by Mulberry Fork; formed 1823; co. seat Jasper. It has cotton and livestock farming, coal mining, and timber. Its industry includes food processing, petroleum refining, and the manufacture of concrete, clothing, furniture, metal products, and machinery. **2.** County (1970 pop. 50,691), 445 sq mi (1,152.6 sq km), extreme NW Ga., bordered on the N by Tenn., on the W by Ala.; formed 1833; co. seat La Fayette. It has agriculture (cotton, corn, soybeans, hay, dairy products, and poultry), textile manufacturing, and coal mining. Part of Chicamauga and Chattanooga National Military Park and Chattahoochee National Forest are here. **3.** County (1970 pop. 27,680), 790 sq mi (2,046.1 sq km), E central Texas, bounded on the NE by the Trinity River and drained by tributaries of the Trinity and San Jacinto rivers; formed 1846; co. seat Huntsville. This rolling wooded area, which includes part of Sam Houston National Forest, has deposits of fuller's earth, clay, and oil; livestock raising, dairying, and farming (cotton, corn, fruit, forage and truck crops) are done.

Walker, resort village (1970 pop. 1,073), seat of Cass co., N central Minn., on Leech Lake NNE of Wadena, in a dairy region. Greater Leech Lake Indian Reservation is nearby.

Walker Lake, salt lake, c.105 sq mi (270 sq km), W Nev., SE of Carson City. Fed by the Walker River, it is a remnant of prehistoric Lake Lahontan and has no outlet.

Wal·lace (wŏl'ĭs), county (1970 pop. 2,215), 911 sq mi (2,359.5 sq km), W Kansas, in a gently sloping to rolling plains region, bordering in the W on Colo. and drained by headstreams of the Smoky Hill River: formed 1868; co. seat Sharon Springs. It is in an irrigated agricultural area that produces wheat, corn, feed grains, sugar beets, cattle, hogs, and sheep.

Wallace, city (1970 pop. 2,206), seat of Shoshone co., N Idaho, near the Mont. border ESE of Coeur d'Alene; settled 1883, inc. as a village 1888, as a city 1893. The city was founded after the discovery of gold in the region and is now the smelting and distributing center for an area with lead and zinc mines.

Wal·la·chi·a (wə-lā'kē-ə), region: *see* Walachia.

Wal·la·sey (wŏl'ə-sē), county borough (1972 est. pop. 96,070), Merseyside, W central England, on the tip of Wirral peninsula at the mouth of the Mersey River. There is some industry, including flour milling and ship repairing, but Wallasey is largely residential.

Wal·la Wal·la (wŏl'ə wŏl'ə), county (1970 pop. 42,176), 1,262 sq mi (3,268.6 sq km), SE Wash., on the Oregon border, watered by the Snake, Columbia, Touchet, and Walla Walla rivers; formed 1854; co. seat Walla Walla. It is in a rich agricultural area that produces wheat, peas, onions, potatoes, alfalfa, asparagus, grapes, and sugar beets. Its industry includes food processing, lumbering, and paper milling. The Whitman National Monument is here.

Walla Walla, city (1979 est. pop. 24,400), seat of Walla Walla co., SE Wash., at the junction of the Walla Walla River and Mill Creek, near the Oregon border; inc. 1862. It is a trade, processing, and distributing center for a rich farm area. Fruits and vegetables (especially green peas) are canned and frozen in numerous plants here, grain is processed for animal feeds, and cans are manufactured. The city also has a pulp and paper mill. The old fur-trading Fort Walla Walla was established near that site in 1818, and in 1836 the mission of Marcus Whitman was also built nearby. Wagon trains began bringing settlers in the 1840s, and Steptoeville (later Walla Walla) grew around the U.S. military Fort Walla Walla (est. 1856; now a veterans hospital). The name was changed when the settlement became county seat in 1859. Walla Walla is the seat of Whitman College, the state's first institution of higher learning. Walla Walla College, a junior college, and the state penitentiary are also in the city. The Whitman mission nearby has been restored as a national historic site.

Wal·ler (wŏl'ər), county (1970 pop. 14,285), 509 sq mi (1,318.3 sq km), SE Texas, bounded on the W by the Brazos River and drained by tributaries of the San Jacinto River; formed 1873; co. seat Hempstead. It has agriculture (especially watermelons and truck crops, also corn, rice, peanuts, hay, soybeans, livestock, and poultry), natural-gas and oil wells, and clothing factories.

Wal·ling·ford (wŏl'ĭng-fərd), town (1979 est. pop. 38,400), New Haven co., S Conn.; inc. 1670. Its silverware industry dates from c.1835. Fruit growing and the manufacture of plastics, steel, precision instruments, and hardware are among the town's other industries. Several old buildings remain. Located in Wallingford is the Choate preparatory school for boys, now coordinated with Rosemary Hall, a girls' preparatory school formerly located at Greenwich, Conn. A branch of the Oneida Community was founded in the town in 1851.

Wal·lis and Fu·tu·na Islands (wŏl'ĭs, fo͞o-to͞o'nä), French overseas territory (1969 pop. 8,546), South Pacific, W of Samoa and NE of Fiji. Comprising two small groups, the Wallis Islands and the Hoorn Islands, which are c.120 mi (190 km) apart, it is sometimes called Wallis Archipelago. Copra was an important export until the mid 1960s, when an attack of rhinoceros beetles ravaged the islands' palm trees. Timber is still exported. The islands came under French control in 1842 and became an overseas territory in 1961.

Wal·low·a (wŏ-lou'ə), county (1970 pop. 6,247), 3,178 sq mi (8,231 sq km), extreme NE Oregon, bordering on Wash. and Idaho, and drained by the Wallowa and Grande Ronde rivers, with the Grand Canyon of the Snake River extending N-S along the E boundary; formed 1887; co. seat Enterprise. Its agriculture includes livestock, grain, and hay, and it has a lumbering industry. Part of Umatilla National Forest is in the northwest.

Walls·end (wôlz'ĕnd'), municipal borough (1972 est. pop. 46,070), Tyne and Wear, NE England, on the Tyne River. In a coal-mining region, Wallsend has shipbuilding and engineering industries. It is the eastern terminus of Hadrian's Wall.

Wal·nut Canyon National Monument (wôl'nŭt', -nət): *see* National Parks and Monuments Table.

Walnut Creek, residential city (1979 est. pop. 51,900), Contra Costa co., W Calif., in the San Francisco Bay area; inc. 1914. Numerous industrial research firms, corporate headquarters, and a naval weapons station are here.

Walnut Ridge, city (1970 pop. 3,800), a seat of Lawrence co., NE Ark., W of Paragould, in a farm region.

Wal·pi (wäl'pē), pueblo, NE Ariz., on a mesa NE of Flagstaff; founded c.1700. Its inhabitants are Pueblo Indians who speak the Hopi language (Uto-Aztecan linguistic stock). One of the most picturesque pueblos of the Southwest, it is a major tourist attraction. The pueblo was founded as a refuge in anticipation of Spanish retaliation for the revolt of the Pueblo Indians (1680). The antelope ceremony is held here in Aug. on even years, and the famous Hopi snake dance in Aug. on odd years. The pueblo, however, is gradually being deserted for the new village of Polacca at the foot of the mesa.

Wal·pole (wôl'pōl', wŏl'-), town (1975 pop. 18,600), Norfolk co., E Mass., SW of Boston; settled 1659, inc. 1724. Textiles and paper products are the chief manufactures.

Wal·sall (wôl'sôl), county borough (1972 est. pop. 183,780), West Midlands, W central England, in the Black Country. Coal mining, iron and brass founding, limestone quarrying, and the manufacturing of leather goods and aircraft parts are among its industries.

Wal·sen·burg (wôl'sən-bûrg'), city (1970 pop. 4,329), seat of Huerfano co., S Colo., S of Pueblo, in a grain, livestock, and coal area at an altitude of 5,071 ft (1,546.7 m); laid out 1873 on the site of a Spanish village.

Walsh (wôlsh), county (1970 pop. 16,251), 1,286 sq mi (3,330.7 sq km), NE N.Dak., in a rich agricultural area bounded on the E by the Red River of the North and drained by the Forest and Park rivers; formed 1881; co. seat Grafton. Its agriculture includes dairy products, livestock, grain, potatoes, and sugar beets.

Wal·sing·ham (wôl'sĭng-əm), rural district (1971 pop. 17,416), Norfolk, E central England. It is the site of Walsingham Abbey, one of the great shrines of medieval England.

Wal·ter·bor·o (wôl'tər-bûr'ō), town (1970 pop. 6,257), seat of Colleton co., S S.C., W of Charleston, in a fertile farm region noted for its hunting and fishing; settled 1784, inc. 1826. Textile, lumber, and food products are manufactured.

Wal·ters (wôl'tərz), city (1970 pop. 2,611), seat of Cotton co., SW Okla., near the Red River, in a farm, cattle, and oil area.

Wal·thall (wôl'thôl), county (1970 pop. 12,500), 430 sq mi (1,113.7 sq km), S Miss., bordered in the S by La. and drained by the Bogue Chitto; formed 1910; co. seat Tylertown. Farming (cotton and corn), dairying, poultry raising, and lumbering are the major occupations.

Walthall, village (1970 pop. 161), seat of Webster co., central Miss., E of Greenwood.

Wal·tham (wôl'thăm, -thəm), city (1979 est. pop. 53,600), Middlesex co., E Mass., a suburb of Boston, on the Charles River; settled c.1634, set off from Watertown 1738, inc. as a city 1884. Electronic equipment and parts, precision instruments, and small tools are among its varied manufactures. Brandeis Univ. is in the city.

Wal·tham Forest (wôl'təm, -thəm), borough (1971 pop. 233,528) of Greater London, SE England. The borough was formed in 1965 by the merger of the municipal boroughs of Chingford, Leyton, and Walthamstow. Waltham Forest is primarily residential. William Morris lived in Water House. The hunting lodge of Queen Elizabeth I is now a museum.

Waltham Holy Cross, urban district (1971 pop. 14,585), Essex, SE England. The great abbey, the Norman nave of which is now used as a parish church, was built in 1030 to contain a cross found in Somerset and believed to have miraculous powers. The abbey was enlarged in 1060 by King Harold, who is thought to have been buried here.

Wal·ton (wôl'tən). **1.** County (1970 pop. 16,087), 1,053 sq mi (2,727.3 sq km), NW Fla., bounded on the E by the Choctawhatchee River, on the N by the Ala. border, and on the S by the Gulf of Mexico; formed 1824; co. seat De Funiak Springs. Its coastal-plain agriculture includes corn, peanuts, cotton, poultry, cattle, and hogs. It has lumbering, food-processing, and clothing industries. Part of Choctawhatchee National Forest is in the southwest. **2.** County (1970 pop. 23,404), 330 sq mi (854.7 sq km), N central Ga., bounded on the NE by the Apalachee River and drained by the Alcovy River; formed 1818; co. seat Monroe. It has piedmont agriculture (cotton, corn, grain sorghum, and peaches), timber, and textile manufacturing.

Walton and Wey·bridge (wā'brĭj), urban district (1972 est. pop. 51,820), Surrey, SE England. It is a residential suburb of London.

Wal·ton-le-Dale (wôl'tən-lə-dāl'), urban district (1971 pop. 26,841), Lancashire, N England. There are engineering works and textile and paper industries.

Wal·vis Bay (wôl'vĭs), city (1970 pop. 21,725), W central Namibia, on Walvis Bay, an arm of the Atlantic Ocean. With the surrounding area, c.430 sq mi (1,110 sq km), it is an exclave of South Africa, although it is administered by Namibia. Walvis Bay is the principal port of Namibia and is the terminus of a railroad from the hinterland. Whaling and fishing fleets are stationed here, and the city has fish canneries. The town of Walvis Bay and its surrounding region were annexed by Great Britain in 1878 and incorporated into the Cape Colony. When Namibia (South West Africa) was annexed (1884) by Germany, Walvis Bay became a British exclave administered by the Cape Colony.

Wal·worth (wôl'wûrth', -wərth). **1.** County (1970 pop. 7,842), 718 sq mi (1,859.6 sq km), N central S.Dak., bounded in the W by the Missouri River; formed 1873; co. seat Selby. It is in a dairying, grain-farming, and livestock-raising area, with some mining and manufacturing and a tourist industry. **2.** County (1970 pop. 63,444), 557 sq mi (1,442.6 sq km), SE Wis., bounded in the S by Ill., drained by Turtle Creek and several other small streams; formed 1836; co. seat Elkhorn. It comprises an extensive dairying region and has other agriculture. Its industry includes food processing, iron foundries, textile mills, and the manufacture of fertilizers, plastics, metal products, farm machinery, and electrical equipment. It is in a resort area with several lakes, notably Lake Geneva.

Wamps·ville (wŏmps'vĭl'), village (1970 pop. 586), seat of Madison co., central N.Y., E of Syracuse, in a dairying area.

Wan·a·que (wŏn'ə-kyo͞o), borough (1970 pop. 8,636), Passaic co., NE N.J., in the Ramapos NNW of Paterson; inc. 1918. It is near the Wanaque Reservoir, the largest in the state.

Wands·worth (wŏndz'wûrth), borough (1971 pop. 298,931) of Greater London, SE London, on the Thames River. The borough was created in 1965 by the merger of the metropolitan borough of Battersea with most of the metropolitan borough of Wandsworth. In the 18th cent. there were cloth industries worked by Huguenots who had come to England after the revocation of the Edict of Nantes in 1685. Battersea Park, along the Thames between the Albert and Chelsea bridges, contains a lake, amphitheater, and theater. Wandsworth Technical College and Battersea Technical College are in the borough.

Wang·a·nu·i (wŏng'ə-no͞o'ē), city (1974 est. pop. 35,800), SW North Island, New Zealand, near the mouth of the Wanganui River. Wanganui is a distribution center and port for coastal trade.

Wan-hsien (wän'shē-ĕn'), city (1970 est. pop. 175,000), E Szechwan prov., China, an important port on the Yangtze River at the beginning of the Yangtze gorges. It is a major tung-oil exporting point.

Wan·ne-Eick·el (vä'nə-ī'kəl), city (1972 est. pop. 95,494), North Rhine-Westphalia, W West Germany, a port on the Rhine-Herne Canal. It is a coal-mining and industrial center of the Ruhr district. Its manufactures include chemicals, processed food, and electrical goods. The city was formed in 1926 by the merger of the twin cities of Wanne and Eickel.

Wan·tagh (wŏn'tô'), uninc. residential city (1979 est. pop. 22,300), Nassau co., SE N.Y., on the S shore of Long Island. A causeway leads to Jones Beach State Park.

Wa·pa·ko·net·a (wô'pô-kō-nĕt'ə), city (1970 pop. 7,324), seat of Auglaize co., W central Ohio, SSW of Lima, in a farm and livestock area; inc. 1833.

Wap·el·lo (wŏp'ə-lō), county (1970 pop. 42,149), 436 sq mi (1,129.2 sq km), SE Iowa, drained by the Des Moines River and Cedar Creek; formed 1843; co. seat Ottumwa. It is in a prairie-agricultural area where hogs, cattle, poultry, corn, soybeans, and wheat are raised. There are bituminous-coal mines.

Wapello, city (1970 pop. 1,873), seat of Louisa co., SE Iowa, on the Iowa River near the Mississippi River NNW of Burlington, in a farm area; settled 1837, inc. 1856. Ancient Indian mounds are nearby.

Wa·ran·gal (və-rŭng'gəl), city (1971 pop. 207,520), Andhra Pradesh state, SE India. It is a market town for grain and oilseed.

Ward (wôrd). **1.** County (1970 pop. 58,560), 2,044 sq mi (5,294 sq km), N central N.Dak., in an agricultural area drained by the Souris and Des Lacs rivers; formed 1885; co. seat Minot. Wheat and other grains, livestock, poultry, and dairy produce are important. It also has lignite deposits, sand and gravel pits, and some manufacturing. **2.** County (1970 pop. 13,019), 827 sq mi (2,141.9 sq km), extreme W Texas, bounded on the W and S by Pecos River; formed 1887; co. seat Monahans. It has oil and natural-gas fields, livestock ranches, and some irrigated agriculture (cotton, alfalfa, grain sorghums, fruit, and truck crops).

War·dha (wär'də, vŭr'də), town (1971 pop. 69,037), Maharashtra state, central India. It is a market for cotton.

Ware (wâr), county (1970 pop. 33,525), 912 sq mi (2,362.1 sq km), SE Ga., drained by the Satilla River; formed 1824; co. seat Waycross. It has coastal-plain agriculture (tobacco, corn, dairy products, and livestock), textile manufacturing, and forestry. Okefenokee Swamp occupies the southern part of the county.

Ware, industrial town (1970 pop. 8,187), Hampshire co., central Mass., W of Worcester; settled c.1717, inc. 1761. Its chief products are textiles and shoes.

Ware·ham (wâr'əm, -hăm), town (1975 pop. 12,700), Plymouth co., SE Mass., on an inlet of Buzzards Bay; settled 1678, inc. 1739. It is a resort and a shipping point for cranberries and shellfish.

War·ley (wôr'lē), county borough (1971 pop. 163,388), West Midlands, central England. Constituted in 1966, Warley comprises substantially the former county borough of Smethwick and the municipal boroughs of Oldbury and Rowley Regis. The borough is highly industrialized. Its manufactures include scales and weighing machines, steel tubes, chemicals, tar, nails and screws, and electrical appliances. There are also glass, iron, brass, and engineering works.

Warm Springs (wôrm). **1.** Watering place, Meriwether co., W Ga., famous in treating and studying the aftereffects of poliomyelitis. The salutary properties of the water springing from Pine Mt. were known to the Indians, and white men learned of them in the late 18th cent. By the 1830s a resort was established. It was destroyed by fire in 1865 but was rebuilt and was fashionable at the end of the 19th cent. Franklin D. Roosevelt, who found the water beneficial after his attack of poliomyelitis, established (1927) the Georgia Warm Springs Foundation to help other victims of the disease, and he gave the foundation his 2,600-acre (1,052-hectare) farm here. He retained the cottage known as the Little White House (now a national shrine), in which he died in 1945. **2.** Village (1970 pop. 1,101), seat of Bath co., W Va., in the Alleghenies W of Staunton, in a recreation area. It has medicinal hot springs.

War·ner Rob·ins (wôr'nər rŏb'ĭnz), city (1979 est. pop. 42,700), Houston co., central Ga., in an agricultural region; inc. 1943. The surrounding area yields peanuts, grain, fruit, and livestock. The city grew with the construction of the adjacent Robins Air Force Base, and its economy is centered around that vast military complex, now one of the largest air force installations in the South and headquarters of the Continental Air Command. Before World War II a small country hamlet called Wellston (pop. c.50) existed on that site. After the air force base and a major air force supply depot were established, a boom city grew around them. The city and base were named for Gen. Warner Robins (1882–1940), considered the originator of the air force's present systems of supply and maintenance.

War·ren (wôr'ən, wŏr'-). **1.** County (1970 pop. 6,669), 284 sq mi (735.6 sq km), central E Ga., bounded on the W by the Ogeechee River; formed 1793; co. seat Warrenton. Farming (cotton, corn, peas, and grain), sawmilling, and granite quarrying are done. **2.** County (1970 pop. 21,595), 541 sq mi (1,401.2 sq km), W Ill., drained by Henderson and Swan creeks; formed 1825; co. seat Monmouth. Its agriculture includes livestock, corn, wheat, soybeans, oats, poultry, clover, and dairy products. There are deposits of bituminous coal and clay and factories manufacturing farm machinery, pottery, and sheet-metal products. **3.** County (1970 pop. 8,705), 368 sq mi (953.1 sq km), W Ind., bounded on the W by the Ill. border, on the SE by the Wabash River; formed 1827; co. seat Williamsport. It is in an agricultural area (cotton, corn, and livestock), with timber and some mining and manufacturing. **4.** County (1970 pop. 27,432), 572 sq mi (1,481.5 sq km), S central Iowa, drained by the North, South, and Middle rivers; formed 1846; co. seat Indianola. Cattle, hogs, poultry, corn, oats, and wheat are raised in this prairie agricultural area. Bituminous-coal deposits are found here. **5.** County (1970 pop. 57,884), 546 sq mi (1,414.1 sq km), S Ky., bounded on the N by the Green River and drained by the Barren River; formed 1796; co. seat Bowling Green. It is in a rolling agricultural area that yields dark tobacco, corn, dairy products, livestock, strawberries, and poultry. It has timber, limestone and asphalt quarries, oil and gas wells, and diversified industry. **6.** County (1970 pop. 44,981), 581 sq mi (1,504.8 sq km), W Miss., bounded in the W by the Mississippi River (here forming the La. line), in the E and S by the Big Black River, and intersected by the Yazoo River; formed 1809; co. seat Vicksburg. Cotton, corn, livestock, and timber are its major products. **7.** County (1970 pop. 9,699), 426 sq mi (1,103.3 sq km), E central Mo., bounded in the S by the Missouri River; formed 1833; co. seat Warrenton. Agricultural products include wheat, corn, oats, and livestock. There are also coal mines and clay pits. **8.** County (1970 pop. 73,960), 362 sq mi (937.6 sq km), NW N.J., in a hilly region bounded on the W by the Delaware River, on the SE and E by the Musconetcong River; formed 1824; co. seat Belvidere. It has agriculture (dairy products, truck crops, and grain) and manufacturing (clothing, lumber products, paper, plastics, metal products, and machinery). **9.** County (1970 pop. 49,402), 887 sq mi (2,297.3 sq km), E N.Y., in the S Adirondacks, bounded on the E by Lake George and drained by the Hudson and Schroon rivers; formed 1813; co. seat Lake George. This year-round resort region has many lakes, state parks, dude ranches, and hiking and skiing trails. Dairying, poultry and stock raising, some farming, and garnet mining are done. **10.** County (1970 pop. 15,810), 424 sq mi (1,098.2 sq km), N N.C., bounded on the N by the Va. border and Lake Gaston, drained by the Roanoke River; formed 1779; co. seat Warrenton. It has piedmont agriculture (tobacco, cotton, and corn) and timber. **11.** County (1970 pop. 85,505), 408 sq mi (1,056.7 sq km), SW Ohio, intersected by the Little Miami River, and also drained by Todd, Caesar, and Turtle creeks; formed 1803; co. seat Lebanon. It is in an agricultural area (livestock, grain, fruit, and tobacco), and has sand and gravel pits. Its industry includes lumbering and the manufacture of furniture, paper, chemicals, concrete, metal products, machinery, and transportation equipment. **12.** County (1970 pop. 47,682), 905 sq mi (2,344 sq km), NW Pa., in a plateau area drained by the Allegheny River; formed 1800; co. seat Warren. Its agriculture includes dairying and potato farming. It has oil and gas wells, food-processing and lumbering industries, and manufacturing (chemicals, plastics, concrete, metal products, machinery, and electrical equipment). **13.** County (1970 pop. 26,972), 439 sq mi (1,137 sq km), central Tenn., with its E portion in the Cumberlands, drained by affluents of the Caney Fork of the Cumberland River and Center Hill Reservoir; formed 1807; co. seat McMinnville. It is in a livestock-raising and agricultural area. Textiles, hosiery, clothing, lumbering, metal products, machinery, and electrical equipment are manufactured. **14.** County (1970 pop. 15,301), 219 sq mi (567.2 sq km), N Va., in the Shenandoah Valley, with the Blue Ridge in the E, and the North and South Forks joining here to form the Shenandoah River; formed 1836; co. seat Front Royal. Its agriculture includes livestock raising, dairying, and farming (fruit, corn, wheat, and hay). It has diversified manufacturing (hosiery, textiles, clothing, furniture, plastics, chemicals, and cement). Parts of Shenandoah National Forest and Shenandoah National Park are in the county.

Warren. 1. City (1970 pop. 6,433), seat of Bradley co., S Ark., S of Pine Bluff, in a farm area. Wood products are made, and meat is processed. **2.** City (1979 est. pop. 166,300), Macomb co., SE Mich., a suburb of Detroit; est. 1837, inc. as a city 1957. It is an important metalworking center. Steel is processed, and tools and dies, and automobile engines, bodies, and parts are made. **3.** City (1970 pop. 1,999), seat of Marshall co., NW Minn., NNW of Crookston, in a wheat region; settled 1878, inc. 1891. **4.** City (1979 est. pop. 58,900), seat of Trumbull co., NE Ohio, in the fertile Mahoning valley; settled 1799, inc. as a city 1905. Steel, metal-forming machinery, electrical equipment, and automobile and truck parts are the principal manufactures. **5.** Borough (1979 est. pop. 12,200), seat of Warren co., NE Pa., on the Allegheny River; inc. 1832. Warren is located in beautiful wooded country near oil and natural-gas reserves. Oil is refined, and electrical equipment and metal products are made. A Seneca Indian village here was burned by Colonel Brodhead in 1781. Laid out c.1795, Warren was an early lumbering center. It had an oil boom in 1860. **6.** Town (1970 pop. 10,523), Bristol co., E R.I., a suburb of Providence on the Kickemuit River and Narragansett Bay; inc. 1747. An early whaling and shipbuilding center, it still has an active boatbuilding industry. Other manufactures are luggage, electrical items, and jewelry. Brown Univ. was first chartered here (1764) as Rhode Island College. During the American Revolution Lafayette made his headquarters in Warren after taking command (1778) of the American forces in the area.

War·rens·burg (wôr'ənz-bûrg', wŏr'-), city (1970 pop. 13,125), seat of Johnson co., W Mo.; inc. as a city 1855. It is situated in a dairy and farm region. Local manufactures include clothing, lawnmowers, and electronic components.

War·ren·ton (wôr'ən-tən, wŏr'-). **1.** City (1970 pop. 2,073), seat of Warren co., E Ga., W of Augusta; inc. as a town 1810, as a city 1908. It is in an agricultural and timber area. **2.** City (1970 pop. 2,057), seat of Warren co., E Mo., WNW of St. Louis, in a fire-clay region; est. in the early 19th cent., inc. 1864. The southeastern part of Warren co. was settled by pioneers who followed Daniel Boone. **3.** Town (1970 pop. 1,035), seat of Warren co., N N.C., near the Va. border NNE of Raleigh; inc. 1779. It is a logging and lumber-milling center. **4.** Town (1970 pop. 4,027), seat of Fauquier co., N Va., NNW of Fredericksburg, in a rich farm area of the Piedmont; settled in the 18th cent., inc. 1810. Warrenton is a horse-breeding, racing, and fox-hunting center. Purebred cattle are also raised.

War·ri (wôr′ē), city (1969 est. pop. 64,000), S Nigeria, a port on the Warri River. It is a transshipment point where oceangoing vessels meet Niger River boats. The main items shipped from Warri are rubber, palm products, cacao, groundnuts, and hides and skins. The city's industries include bicycle assembly, rubber processing, and ship repair. Traditional artisans make canoes, rope, and mats. According to legend, Warri was founded in the 15th cent. by a Benin prince. By the 17th cent. it was independent of Benin, and in the 19th cent. it became wealthy in the palm oil trade. Warri came under British protection in 1884.

War·rick (wôr′ĭk, wŏr′-), county (1970 pop. 27,972), 391 sq mi (1,012.7 sq km), SW Ind., bounded on the S by the Ohio River and the Ky. border; formed 1813; co. seat Boonville. Its agriculture includes livestock, fruit, grain, truck crops, and tobacco. It has bituminous-coal mines, limestone deposits, sawmills, and manufactures (clothing, concrete, and machinery).

War·ring·ton (wôr′ĭng-tən, wŏr′-), county borough (1972 est. pop. 66,760), Cheshire, NW England, on the Mersey River and on the Manchester Ship Canal. Manufactures include wire and other metal products, chemicals, soap, leather goods, and beer. The site was occupied very early; the Church of St. Elphin has a Saxon crypt. The grammar school dates from 1526, and there are several half-timbered houses. Warrington academy for religious dissenters (1757–83) included Joseph Priestly among its faculty.

Warrington, uninc. residential town (1970 pop. 15,848), Escambia co., extreme NW Fla., a suburb of Pensacola, on Pensacola Bay. Although chiefly residential, it has shipyards and waterfront industries.

War·saw (wôr′sô), city (1975 est. pop. 1,436,000), capital of Poland and of Warsaw prov., central Poland, on both banks of the Vistula River. It is a political, cultural, and industrial center, a major transportation hub, and one of Europe's great historic cities. The city has industries producing steel machinery, motor vehicles, chemicals, electrical equipment, and textiles. The city probably grew around a castle built in the 13th cent. by a duke of Masovia. In 1413 Warsaw became the capital of the duchy of Masovia, which was incorporated with Great Poland in 1526. After Kraków burned, Warsaw replaced it (1596) as Poland's capital. In 1807 it became the capital of the grand duchy of Warsaw and was the scene in 1812 of a diet that proclaimed the re-establishment of Poland. In 1813, however, the city fell to the Russians, and in 1815 it became the capital of the nominally independent kingdom of central Poland, awarded by the Congress of Vienna to the Russian crown. German forces took the city in 1915, during World War I. In Nov., 1918, it was liberated by Polish troops and proclaimed capital of the restored Polish state. The city was the scene in 1926 of a military coup that established Marshal Joseph Pilsudski's dictatorship. During World War II the city was occupied (1939–45) by German troops and subjected to systematic destruction. In 1940 the Germans isolated the Jewish ghetto, which in 1942 contained about 500,000 persons. When Warsaw was liberated (Jan., 1945) by Soviet troops, only about 200 Jews remained. The postwar decision to retain Warsaw as the national capital resulted in a large-scale reconstruction. The medieval Stare Miasto (old town), with its marketplace and 14th cent. cathedral, was rebuilt according to the prewar pattern. Warsaw has many educational and cultural institutions, including the Univ. of Warsaw (founded in 1818) and the Polish Academy of Sciences.

Warsaw. 1. City (1970 pop. 7,506), seat of Kosciusko co., NE Ind., WNW of Fort Wayne, in a lake region; settled c.1836, inc. 1854. A resort center, it also manufactures automobile and boat parts, furniture, and ordnance. **2.** City (1970 pop. 1,232), seat of Gallatin co., N Ky., on the Ohio River SW of Covington. It is a trade center in the outer Bluegrass agricultural region. Furniture is made, and there are lumber mills and nurseries here. **3.** Resort city (1970 pop. 1,423), seat of Benton co., central Mo., on the Osage River SSW of Sedalia, in a hunting and fishing region; founded 1837, inc. 1839. **4.** Village (1970 pop. 3,619), seat of Wyoming co., W central N.Y., SSW of Batavia, in a farm region; settled 1803, inc. 1843. Elevators and electrical equipment are made. **5.** Town (1970 pop. 511), seat of Richmond co., E Va., SE of Fredericksburg near the Rappahannock River. Seafood, dairy products, and lumber are processed.

War·ta (vär′tä), river, c.475 mi (765 km) long, rising in the Jura Krakowska, S Poland, and flowing N and W to the Oder River. The Warta is navigable for large craft to Poznan and for small craft for about half its course. The lower course of the river, formerly in Germany, became (1945) part of Poland as a result of the Potsdam Conference.

Wart·burg (wôrt′bûrg′), town (1970 pop. 541), seat of Morgan co., NE central Tenn., WNW of Knoxville, in the Cumberlands.

War·wick (wôr′ĭk, wŏr′-), municipal borough (1972 est. pop. 17,850), administrative center of Warwickshire, central England, on the Avon River. Although there is some commerce and manufacturing, Warwick is best known for Warwick Castle, located on the site of a fortress built in 915 by Æthelflæd, the daughter of King Alfred. The castle was begun in the 14th cent. and was converted into a mansion in the 17th. St. Mary's Church dates partly from the 12th cent. In the church are a Norman crypt and monuments to Richard de Beauchamp, earl of Warwick, to his countess, and to Robert Dudley, earl of Leicester.

War·wick (wôr′wĭk, wŏr′-), city (1979 est. pop. 87,300), Kent co., central R.I., at the head of Narragansett Bay; settled by Samuel Gorton 1642, inc. as a city 1931. Its textile industry dates from 1794. Other manufactures include machinery, pipes and tubing, and fluorescent-lighting fixtures. Warwick village was nearly destroyed (1676) in King Philip's War. Gaspee Point, south of Pawtuxet, was the scene of the burning of the British revenue cutter *Gaspee* in 1772; annual "Gaspee Days" commemorate the event. Warwick is the seat of the Seminary of Our Lady of Providence.

War·wick·shire (wôr′ĭk-shîr′, -shər, wŏr′-), nonmetropolitan county (1974 est. pop. 468,270), 975 sq mi (2,525 sq km), central England; administrative center Warwick. The region is a varied one, largely given to agriculture (wheat and other grains, dairying, sheep and cattle grazing). There are many industries including the manufacture of motor vehicles, aircraft, and textiles.

Wa·satch (wô′săch), county (1970 pop. 5,863), 1,191 sq mi (3,084.7 sq km), N central Utah, in a mountain area largely within Uinta National Forest and irrigated by the Strawberry River in the S and the Provo River in the N; formed 1862; co. seat Heber City. Its agriculture includes grain, alfalfa, sugar beets, truck crops, fruit, cattle, and sheep. It also has lead and zinc mines and manufactures clothing.

Wasatch Range, part of the Rocky Mts., extending c.250 mi (400 km) south from SE Idaho to central Utah. Mt. Timpanogos, the highest peak (12,008 ft/3,662.4 m), is the site of Timpanogos Cave National Monument. Many streams on the western flank of the Wasatch carry water into the fertile Salt Lake oasis, which stretches along the foothills; all of Utah's principal cities and most of the state's population are found here. Emigrant Canyon, a site of early Mormon activities, is now a major winter resort area.

Was·co (wŏs′kō), county (1970 pop. 20,133), 2,381 sq mi (6,166.8 sq km), N Oregon, bounded in the W by the Cascade Range, in the N by the Columbia River and Wash., and drained by the Deschutes River; formed 1854; co. seat The Dalles. It has wheat, fruit, and livestock farming. Its industry includes lumbering and aluminum products.

Wasco, city (1970 pop. 8,269), Kern co., S central Calif., NW of Bakersfield, in an agricultural area; settled 1907, inc. 1945. The Friant-Kern Canal nearby is part of the Central Valley project.

Wa·se·ca (wŏ-sē′kə), county (1970 pop. 16,663), 415 sq mi (1,074.9 sq km), S Minn., drained by the Le Sueur River, with Lake Elysian in the NW; formed 1857; co. seat Waseca. In an agricultural area yielding livestock, dairy products, corn, oats, and barley, it has a food-processing industry.

Waseca, city (1970 pop. 6,789), seat of Waseca co., SE Minn., N of Albert Lea, in a farm and dairy region; settled 1854, inc. 1868. It produces frozen foods, radio equipment, and sporting goods. A state university agricultural school and experiment station is here.

Wash, The (wŏsh, wôsh), inlet of the North Sea, 20 mi (32 km) long and 15 mi (24 km) wide, between Lincolnshire and Norfolk, E England. It is mostly shallow with sandbars and low, marshy shores. Dredged ship channels lead to King's Lynn and Boston.

Wash·a·baugh (wŏsh′ə-bô, wôsh′-), county (1970 pop. 1,389), 1,061 sq mi (2,748 sq km), SW central S.Dak., bounded in the N by the White River, and lying in the Pine Ridge Indian Reservation; formed 1883; it is unorganized and is attached to Jackson co. for judicial purposes. Cattle are raised.

Wash·a·kie (wŏsh′ə-kē′, wôsh′-), county (1970 pop. 7,569), 2,262 sq mi (5,858.6 sq km), N central Wyo., watered by the Bighorn River;

formed 1911; co. seat Worland. It is in an irrigated agricultural (sugar beets, beans, and livestock) area, with oil and natural-gas fields. Its industry includes beet-sugar processing.

Wash·burn (wŏsh'bərn, -bûrn', wôsh'-), county (1970 pop. 10,601), 817 sq mi (2,116 sq km), NW Wis., in a wooded area drained by the Namekagon and Yellow rivers; formed 1883; co. seat Shell Lake. Its agriculture includes dairy products, livestock, potatoes, cranberries, and poultry. Meat processing, lumbering, and boatbuilding are done. The county has many lakes and a summer resort area.

Washburn. 1. City (1970 pop. 804), seat of McLean co., central N.Dak., N of Bismarck on the Missouri River. Wheat, rye, and flax are grown. **2.** City (1970 pop. 1,957), seat of Bayfield co., N Wis., on Chequamegon Bay of Lake Superior N of Ashland; founded 1884, inc. 1904. It is a trade center for a dairy, lumber, and fruit region.

Wash·ing·ton (wŏsh'ĭng-tən, wôsh'-), state (1975 pop. 3,522,000), 68,192 sq mi (176,617 sq km), including 1,483 sq mi (3,841 sq km) of inland water surface, extreme NW United States, in the Pacific Northwest, admitted 1889 as the 42nd state. Olympia is the capital; Seattle, Spokane, and Tacoma are the largest cities. Washington is bounded on the north by the Canadian province of British Columbia, on the east by Idaho, on the south by Oregon (with the Columbia River marking much of the boundary), and on the west by the Pacific Ocean, with Puget Sound in the northwest and two inlets, Grays Harbor and Willapa Bay, farther south.

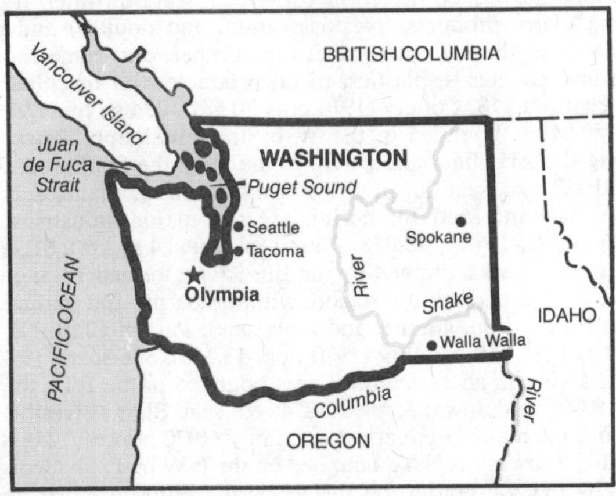

The deep cut of Puget Sound creates the Olympic Peninsula—a wet region with dense rain forests of spruce, fir, cedar, and hemlock—much of it included in Olympic National Park, where the Olympic Mts. rise 7,965 ft (2,428 m) in Mt. Olympus. Puget Sound is navigable and has many beautiful bays, on which are situated such important commercial and industrial cities as Seattle, Tacoma, and Everett. The lofty Cascade Range runs north and south, rising to 14,410 ft (4,392 m) in Mt. Rainier, and divides the state into western and eastern sections. The Cascades block the eastward movement of warm ocean air, creating abundant rainfall to the west and almost semiarid conditions to the east. Thus, the coastal region is one of the wettest areas in the country, averaging in some places as much as 150 in. (381 cm) of rainfall per year and containing some of the heaviest stands of timber in the world. In contrast, the dry eastern section is comparatively treeless with just sufficient rainfall for dry farming and for cattle and sheep raising. Spokane is the commercial and transportation hub of the entire region between the Cascades and the Rockies. However, irrigation has converted many of the river valleys east of the Cascades into garden areas.

Washington's great water resources provide not only irrigation but enormous hydroelectric power. Because of the rapid drop from its origin to its mouth, the Columbia is one of the greatest sources of hydroelectric power in the world. Grand Coulee Dam, one of the world's largest concrete dams and greatest potential power-producing structures, and Bonneville Dam have been supplemented, on the river's upper course, by Chief Joseph and Rocky Reach dams (completed 1961), Priest Rapids Dam (1962), and Wanapum Dam (1963), and, on its lower course, by The Dalles Dam (1957), John Day Dam (1968), and McNary Dam (1953), all shared with Oregon. The Snake River in the east and the Yakima River in south-central Washington also have important irrigation projects, and dams on the Skagit River supply power to Seattle and the surrounding area.

Thousands of visitors are annually attracted to Mount Rainier National Park, Olympic National Park, North Cascades National Park (created 1968), Fort Vancouver and Whitman Mission national historic sites, and Coulee Dam National Recreation Area. Rugged mountain slopes and the simple grandeur of the scenery draw many climbers during the summer months, and in winter excellent snowfields near Seattle and Tacoma attract crowds of skiers.

Washington is divided into 39 counties:

NAME	COUNTY SEAT	NAME	COUNTY SEAT
Adams	Ritzville	Lewis	Chehalis
Asotin	Asotin	Lincoln	Davenport
Benton	Prosser	Mason	Shelton
Chelan	Wenatchee	Okanogan	Okanogan
Clallam	Port Angeles	Pacific	South Bend
Clark	Vancouver	Pend Oreille	Newport
Columbia	Dayton	Pierce	Tacoma
Cowlitz	Kelso	San Juan	Friday Harbor
Douglas	Waterville	Skagit	Mount Vernon
Ferry	Republic	Skamania	Stevenson
Franklin	Pasco	Snohomish	Everett
Garfield	Pomeroy	Spokane	Spokane
Grant	Ephrata	Stevens	Colville
Grays Harbor	Montesano	Thurston	Olympia
Island	Coupeville	Wahkiakum	Cathlamet
Jefferson	Port Townsend	Walla Walla	Walla Walla
King	Seattle	Whatcom	Bellingham
Kitsap	Port Orchard	Whitman	Colfax
Kittitas	Ellensburg	Yakima	Yakima
Klickitat	Goldendale		

Economy. The state's leading industry, centered in Seattle, is the manufacture of jet airplanes, missiles, and spacecraft. Washington's second largest industry is food processing. Washington leads the country in the production of apples and is also a major wheat-producing state. It is the nation's second largest producer of sweet cherries, as well as of asparagus and green peas. Other major crops are hay and potatoes. Cattle, dairy goods, and poultry are also economically important. Salmon is the biggest fishing catch, but halibut, bottomfish, oysters, and crabs are also caught in significant numbers.

More than one half of the state's area is forested, and the lumber and wood-products industry is the state's third largest industry. Many of Washington's cities (Tacoma, Bellingham, Everett, Anacortes) began as sawmill centers, and lumber, pulp, paper, and related items are still their major products. Other important manufactures in the state are chemicals and primary metals. The abundant water power and the rich aluminum and magnesium ores found in the Okanogan Highlands in the northeast have made Washington the nation's leading aluminum producer. Washington's chief minerals are sand and gravel, cement, stone, uranium, and diatomite. Gold, silver, lead, and zinc are also found in the Okanogan Highlands.

History. Washington's early history is shared with that of the whole Oregon Territory. Capt. James Cook's English expedition (1778) first opened up the area to the maritime fur trade with China, and British fur companies were soon exploring the West. In 1792 the British explorer George Vancouver sailed into Puget Sound and mapped the area, and the American fur trader Robert Gray sailed up the Columbia River, establishing U.S. claim to the areas that it drained. In 1818 a treaty provided for joint rights for the United States and Great Britain in the Columbia River country. Fort Vancouver, on the site of present-day Vancouver, sheltered American overland traders and later the American missionaries who were the first real settlers in the area north of the Columbia. Despite American demands for the ouster of the British from the whole of the Columbia River country up to a lat. of 54°40′N, the boundary was set at lat. 49°N in 1846.

Partly as a means of increasing protection against Indian attacks, the Oregon Territory, embracing the Washington area, was created in 1848; but in 1853 the region was divided, and Washington Territory (containing a part of what is now Idaho) was set up. (The Idaho section was cut away when Idaho Territory was formed in 1863.) Meanwhile, a small settlement sprang up at New Market (near present Olympia). Settlers recognized the potential of the Puget Sound country and poured into the area in ever-increasing numbers. Lumber and fishing industries arose to supply the demand to the south, and other towns, including Seattle, were founded. Treaties with the coast Indians were quickly concluded, but the more warlike inland tribes revolted, and hostilities with the Cayuse, the Yakima, and the Nez Perce tribes continued for many years. (Today about half of

Washington's Indians, who number about 33,000, live in or near reservations.) In 1860 gold strikes brought a rush of prospectors to the Walla Walla area. The great influx of settlers was delayed, however, until the 1880s, when transport by rail became possible. The population almost quadrupled between 1880 and 1890.

By the time Washington became a state in 1889, the eastern plains had been given over to cattle and sheep, agriculture was flourishing in the fertile valleys, and the lumber industry had been founded. Seattle, the primary departure point for the Klondike, became a boom town after the discovery of gold in Alaska in 1897. At the same time came labor clashes that gave Washington a reputation as a radical state.

Washington was an important center of the defense industry during World War II, particularly with the immense aircraft industry in Seattle. The large Japanese-American population in the state (more than 15,000 persons) was moved eastward, suffering great physical and emotional hardship. In the postwar period labor-management relations eased, owing in part to the political power of organized labor, and also to the continuing widespread prosperity brought by booming aircraft production and expanding uranium- and aluminum-processing industries.

Government. Washington still operates under its first constitution, adopted in 1889. Its executive branch is headed by a governor elected for a four-year term. The bicameral legislature has a senate with 49 members elected for four-year terms and a house of representatives with 98 members elected for two-year terms.

Educational Institutions. Among the state's most prominent are the Univ. of Washington and Seattle Univ., at Seattle; Washington State Univ., at Pullman; Gonzaga Univ., at Spokane; Pacific Lutheran Univ., at Tacoma; and Whitman College, at Walla Walla.

Washington. **1.** County (1970 pop. 16,241), 1,066 sq mi (2,761 sq km), SW Ala., in the coastal plain on the Miss. border, bounded on the E by the Tombigbee River; formed 1800; co. seat Chatom. Its agriculture includes corn, soybeans, cattle, hogs, and cotton. Naval stores and agricultural chemicals are produced. **2.** County (1970 pop. 77,370), 962 sq mi (2,491.6 sq km), NW Ark., in the Ozarks, bordered on the W by the Okla. line and drained by the Illinois and White rivers; formed 1828; co. seat Fayetteville. It has agriculture (livestock, poultry, dairy products, truck crops, fruit, and grain), lumber, limestone deposits, and diversified industry. **3.** County (1970 pop. 5,550), 2,526 sq mi (6,542.3 sq km), NE Colo., drained by small branches of the South Platte River; formed 1887; co. seat Akron. Wheat farming, cattle ranching, and oil refining are important. There is some manufacturing in the county. **4.** County (1970 pop. 11,453), 585 sq mi (1,515.2 sq km), NW Fla., bounded on the W by the Choctawhatchee River, with many small lakes; formed 1825; co. seat Chipley. In a rolling agricultural area yielding corn, peanuts, cotton, vegetables, and livestock, it has a forestry industry. **5.** County (1970 pop. 17,480), 674 sq mi (1,745.6 sq km), E central Ga., bounded on the E by the Ogeechee River, on the W by the Oconee River; formed 1784; co. seat Sandersville. It has coastal-plain agriculture (cotton, corn, pecans, and truck crops), sawmilling, and kaolin mining. **6.** County (1970 pop. 7,633), 1,462 sq mi (3,786.6 sq km), W Idaho, in a mountainous region bounded in the W by the Snake River and Oregon; formed 1879; co. seat Weiser. Stock raising, farming (hay, sugar beets, fruit, and truck crops), and quicksilver mining are the major occupations. It includes part of Weiser National Forest. **7.** County (1970 pop. 13,780), 564 sq mi (1,460.8 sq km), SW Ill., bounded on the N by the Kaskaskia River and drained by Little Muddy River and Beaucoup Creek; formed 1818; co. seat Nashville. It has agriculture (corn, wheat, livestock, fruit, poultry, and dairy products), bituminous-coal mining, and manufacturing of machinery and flour. **8.** County (1970 pop. 19,278), 516 sq mi (1,336.4 sq km), S Ind., bounded on the N by the Muscatatuck River and East Fork of the White River and drained by the Blue River; formed 1813; co. seat Salem. It is in a grain-growing and poultry area, with stone quarries, timber, and some industry. **9.** County (1970 pop. 18,967), 568 sq mi (1,471.1 sq km), SE Iowa, in a prairie agricultural area drained by the Skunk and English rivers; formed 1838; co. seat Washington. Cattle, poultry, and hogs are raised, and corn, oats, and wheat are grown. **10.** County (1970 pop. 9,249), 891 sq mi (2,307.7 sq km), N Kansas, in a gently rolling plain bordering in the N on Nebr. and drained by the Little Blue River; formed 1855; co. seat Washington. It is in a grain and livestock area, with some mining and manufacturing. **11.** County (1970 pop. 10,728), 307 sq mi (795 sq km), central Ky., in the Bluegrass, bounded on the NW by Beech Fork and drained by several creeks; formed 1792; co. seat Springfield. In a rolling upland agricultural area yielding burley tobacco, livestock, dairy products, poultry, corn, and wheat, it has limestone quarries, timber, and industry (creamery and tobacco processing and clothing and wood-products manufacturing). **12.** Parish (1970 pop. 41,987), 665 sq mi (1,722.4 sq km), E La., bounded on the N by the Miss. border, on the E by the Pearl River, partly on the W by the Tchefuncta River, and drained by the Bogue Chitto; formed 1819; parish seat Franklin. Cotton, corn, hay, sweet potatoes, and peanuts are grown, and livestock is raised. Industries include cotton and cottonseed processing, lumber and paper milling, and other manufacturing. The parish has extensive forests. **13.** County (1970 pop. 29,859), 2,554 sq mi (6,614.9 sq km), the most easterly co. in Maine, on the U.S. - Canadian line, drained by the St. Croix, Machias, East Machias and Dennys rivers; formed 1789; co. seat Machias. It has agriculture and an extensive food-processing industry, lumbering, and paper milling. There are many resorts and hunting and fishing areas in the county. **14.** County (1970 pop. 103,829), 459 sq mi (1,189 sq km), N Md., bounded on the N by the Pa. line, on the S and SW by the Potomac River; formed 1776; co. seat Hagerstown. Its agriculture includes corn, wheat, peaches, apples, berries, truck crops, poultry, and dairy products. Ornamental fish and aquatic plants are raised here. It has limestone quarrying, lumber (millwork), and diversified industry. Tourism is also important. **15.** County (1970 pop. 83,003), 386 sq mi (999.7 sq km), E Minn., bounded on the S by the Mississippi River, on the E by the St. Croix River and the Wis. border; formed 1849; co. seat Stillwater. It is in a farming (dairy products, livestock, grain, and poultry) and recreation area, with sand and gravel pits, timber, and some industry. Furniture, chemicals, plastics, metal products, and machinery are manufactured. **16.** County (1970 pop. 70,581), 734 sq mi (1,901.1 sq km), W Miss., bounded in the W by the Mississippi River (here forming the Ark. line) and partly in the E by the Sunflower River; formed 1827; co. seat Greenville. Farming (cotton, alfalfa, and truck crops), dairying, and lumbering are the major industries here. **17.** County (1970 pop. 15,086), 760 sq mi (1,968.4 sq km), SE central Mo., in the Ozarks, drained by the Big River; formed 1813; co. seat Potosi. It is an agricultural region, with lumbering and mining (barite, lead, zinc, iron, granite, and limestone). Part of Clark National Forest is here. **18.** County (1970 pop. 13,310), 386 sq mi (999.7 sq km), E Nebr., in an agricultural area bounded in the E by the Missouri River and Iowa; formed 1854; co. seat Blair. Livestock and grain are the major products. **19.** County (1970 pop. 52,725), 836 sq mi (2,165.2 sq km), E N.Y., bounded on the NW by Lake George, on the E by the Va. border, on the W by the Hudson River; formed 1772; co. seat Hudson Falls. It has farming (potatoes, fruit, corn, oats, and poultry), dairying, timber, slate and limestone quarrying, and manufacturing. **20.** County (1970 pop. 14,038), 343 sq mi (888.4 sq km), E N.C., bounded on the N by Albemarle Sound; formed 1799; co. seat Plymouth. This tidewater area has timber, farming (peanuts, tobacco, and corn), a paper industry, fishing, and hunting. **21.** County (1970 pop. 57,160), 641 sq mi (1,660.2 sq km), SE Ohio, bounded in the SE by the Ohio River and the W.Va. border and intersected by the Muskingum and Little Muskingum rivers; formed 1788; co. seat Marietta. Its agriculture includes livestock, dairy products, fruit, and truck crops. It has oil and gas wells and lumbering and food-processing industries. Furniture, plastics, concrete, metal products, machinery, and instruments are made. **22.** County (1970 pop. 42,302), 424 sq mi (1,098.2 sq km), NE Okla., bounded in the N by the Kansas border and drained by the Caney River; formed 1907; co. seat Bartlesville. It is an agricultural area (livestock, truck crops, grain, poultry, and dairy products), with oil and gas fields and lead and zinc mines. Its industry includes food processing, iron and steel foundries, and lead and zinc smelting. **23.** County (1970 pop. 157,920), 716 sq mi (1,854.4 sq km), NW Oregon, in a mountain area in the Coast Range; formed 1843 as Twality or Falatine co., renamed 1849; co. seat Hillsboro. Its agriculture includes fruit, truck crops, grain, poultry, and dairy products. Food processing and lumbering are done. Wood products, paper, plastics, concrete, metal products, machinery, electronic instruments, and transportation equipment are manufactured. **24.** County (1970 pop. 210,876), 857 sq mi (2,219.6 sq km), SW Pa., bounded in the W by W.Va. and in the E by the Monongahela River; formed 1781; co. seat Washington. It is in a bituminous-coal mining and livestock-raising area, which also has dairying and grain farming. Its manufactures include clothing, paper, chemicals, glass, concrete, metal products, machinery, and electrical equipment. The area was in dispute between Va. and Pa. until 1784. **25.** County (1970 pop. 85,706), 321 sq mi (831.4 sq km), S R.I., on the Conn. border and Block Island Sound, drained by the Wood, Hunt, Queen, and Pawcatuck rivers; formed 1729 as Kings co., renamed 1781; co. seat West Kingston. Its agriculture includes potatoes, corn, fruit, truck crops, tobacco, poultry, and dairy products. It has granite quarries and diversified industries. There are resorts on Block Island Sound and Narragansett Bay. **26.** County (1970 pop. 73,924), 323 sq mi (836.6 sq km), NE Tenn., in the Great Appalachian Valley, with mountain ridges in the S and SE, bounded in the NE by the Watauga River and drained by the Nolichunky River; formed 1777 as the first co. in Tenn. at the request of the Watauga Association of settlers; co. seat Jonesboro. It is in an agricultural area that produces fruit, tobacco, corn, hay, livestock, and dairy goods. Its industry includes food processing, textile mills, lumbering, and the manufacture of clothing, furniture, concrete, metal products, machinery, housewares, and radio and television equipment. **27.** County (1970 pop. 18,842), 594 sq mi (1,538.5 sq km), S central Texas, bounded on the E by the Brazos River, on the N by Yegua Creek; formed 1836; co. seat Brenham. This rich agricultural area yields cotton, corn, grains, hay, peanuts, fruit, truck crops, pecans, dairy products, poultry, beef cattle, hogs, and sheep. There are oil and natural-gas wells here. Some lumbering, manufacturing, and processing are done. **28.** County (1970 pop. 13,669), 2,427 sq mi (6,285.9 sq km), extreme SW Utah, in a mountain area bordering on Ariz. and Nev. and drained by the Virgin River; formed 1852; co. seat St. George. It is in a livestock-grazing area and manufactures clothing. Zion National Park, Zion National Monument, and Dixie National Forest are in the county. **29.** County (1970 pop. 47,659), 707 sq mi (1,831.1 sq km), central Vt., drained by the Winooski and Mad rivers; formed 1810 as Jefferson co., renamed 1814; co. seat Montpelier. Its agriculture includes dairy products, corn, and maple sugar. It also has stone quarries and diversified industry, including food processing and lumbering. **30.** County (1970 pop. 40,835), 574 sq mi (1,486.7 sq km), SW Va., in the Great Appalachian Valley, bounded in the S by Tenn. and drained by the North, Middle, and South forks of the Holston River; formed 1776; co. seat Abingdon. It is in a cattle-raising and farming (burley tobacco, poultry, fruit, clover, corn, and dairy products) area. Clothing, concrete, steel products, and machinery are made. It contains part of Jefferson National Forest and a section of the Appalachian Trail. **31.** County (1970 pop. 63,839), 429 sq mi (1,111.1 sq km), E Wis., in a hilly region drained by the Milwaukee and Menomonee rivers; formed 1836; co. seat West Bend. It is in a dairying, stock-raising, and fruit and vegetable farming area. It has lumbering, food processing, and other industries producing paper, chemicals, plastics, concrete, machinery, outboard motors, and travel trailers.

Washington. **1.** City (1970 pop. 4,094), seat of Wilkes co., NE Ga., NW of Augusta; settled 1773, laid out 1780. It was the colony's temporary capital during the American Revolution. Eli Whitney's workshop is nearby. The last cabinet meeting of the Confederacy was held here in May, 1865. Many ante-bellum houses remain. **2.** City (1975 est. pop. 11,200), seat of Daviess co., SW Ind.; settled 1805, inc. as a city 1871. It has railroad shops and a rubber industry. **3.** City (1970 pop. 6,317), seat of Washington co., SE Iowa, SSW of Iowa City, in a farm area; inc. 1864. **4.** City (1970 pop. 1,584), seat of Washington co., NE Kansas, near the Nebr. border NW of Topeka, in a farm area; inc. 1875. **5.** City (1970 pop. 8,499), Franklin co., E Mo., on the Missouri River WSW of St. Louis; settled c.1822, platted 1837, inc. 1841. It is a farm trade center. Shoes, metal and dairy products, and corncob pipes are made. **6.** Borough (1970 pop. 5,943), Warren co., NW N.J., NE of Phillipsburg on the old Morris Canal; settled 1741, inc. 1868. Plastic containers and furniture are made. **7.** City (1970 pop. 8,961), seat of Beaufort co., E N.C., at the head of the Pamlico estuary SSE of Greenville; founded 1771. It is a market, processing, and shipping center in a tobacco, grain, and livestock area and has clothing and lumber mills. **8.** City (1979 est. pop. 20,400), seat of Washington co., SW Pa., in a bituminous-coal region; settled 1769, laid out 1781, inc. as a city 1924. Chief among its many manufactures are glass, steel, and electronic products. David Bradford House, erected in 1788, was a meeting place in the Whiskey Rebellion (1794). Le Moyne House (1812) was the home of Dr. Francis Le Moyne, an abolitionist leader. Washington and Jefferson College (1781; oldest college west of the Alleghenies) is here. Also in the city are a race track and a trolley museum. **9.** Town (1970 pop. 189), seat of Rappahannock co., N Va., in the E foothills of the Blue Ridge W of Warrenton; platted 1749 by George Washington. Apples are shipped.

Washington, D.C., capital of the United States (1970 pop. 756,510), coextensive (since 1878, when Georgetown became a part of Washington) with the District of Columbia, on the Potomac River; inc. 1802. It is the center of a metropolitan area (1970 pop. 2,861,123) extending into Md. and Va. and is the legislative, administrative, and judicial center of the United States. The Federal government is virtually the only source of employment. Washington is also a major tourist attraction, drawing millions of visitors every year.

In 1790 the rivalry of Northern and Southern states for the capital's location ended when Thomas Jefferson's followers supported Alexander Hamilton's program for Federal assumption of state debts in return for an agreement to situate the national capital on the banks of the Potomac River. George Washington selected the exact spot. The "Federal City" was designed by Pierre L'Enfant and laid out by Andrew Ellicott. Construction began on the White House in 1792 and on the Capitol the following year. Congress held its first session in Washington in 1800. In the War of 1812 the British captured and sacked (1814) Washington, burning most of the public buildings, including the Capitol and the White House. The city grew slowly. Even after 1850 it was still "a sea of mud," and not until the 20th cent. did it assume its present gracious and urban aspect. After 1901 Washington was developed on the basis of the resurrected L'Enfant plan—a gridiron arrangement of streets cut by diagonal avenues radiating from the Capitol and White House, with an elaborate system of parks.

The stately city spreads out over 69 sq mi (179 sq km), including 8 sq mi (20.7 sq km) of water surface, with broad tree-shaded thoroughfares and open vistas at frequent intervals. The numerous impressive government buildings are built of white or gray stone in the classical style. Among the city's fine parks is West Potomac Park, which extends south from the Lincoln Memorial and includes the Tidal Basin, flanked by the famous Japanese cherry trees. Besides the Capitol and the White House, some other important government buildings are the Senate and the House of Representatives office buildings, the Supreme Court Building, the Pentagon (in Virginia), the Federal Bureau of Investigation Building, the Library of Congress, the National Archives Building, and Constitution Hall. Best known of the city's many statues and monuments are the Washington Monument, the Lincoln Memorial, and the Thomas Jefferson Memorial. The Arlington Memorial Bridge across the Potomac connects the capital with Arlington National Cemetery. The city's many institutions of higher education include Georgetown Univ., George Washington Univ., and Howard Univ. The many cultural attractions of the capital are the National Gallery of Art, the various centers under the auspices of the Smithsonian Institution, and the John F. Kennedy Center for the Performing Arts. The Smithsonian Institution is one of many public and private institutions dedicated to scientific research and education.

In 1871 Washington lost its charter as a city and a territorial government was inaugurated to govern the entire District of Columbia. Congress took direct control of the District's government in 1874, providing for a mayor appointed by the President and a commission chosen by Congress; the residents were disfranchised. The Twenty-third Amendment (1961) to the Constitution gave inhabitants the right to vote in presidential elections; the District of Columbia was accorded three electoral votes, the minimum number. In 1970 legislation was enacted authorizing election of a nonvoting delegate to the House of Representatives. The present system of government, approved in a referendum in May, 1974, provides for an elected mayor and a 13-member city council but reserves for Congress the right to review the budget and legislation passed by the council and to retain direct control over an enclave containing most of the Federal buildings and monuments. The first elections were held in Nov., 1974.

Washington Court House, city (1975 est. pop. 12,500), seat of Fayette co., SW Ohio, on Paint Creek, in a productive farm, dairy, and poultry area; laid out and founded c.1810, inc. 1831. Its many manufactures include shoes, gloves, and automobile and aircraft parts.

Washington Island, atoll (1968 pop. 437), 3 sq mi (7.8 sq km), central Pacific, one of the Line Islands and part of the British colony of the Gilbert and Ellice Islands. Discovered by the American explorer Edmund Fanning in 1798, it was annexed by Great Britain in 1889 and became part of the colony in 1916. Copra is the only export.

Washington Island, c.20 sq mi (50 sq km), NE Wis., in NW Lake Michigan, just off the N tip of the Door Peninsula. It includes the town of Washington (1970 pop. 446). The island was visited by the

French explorers Radisson (1657) and La Salle (1679). It has a large Icelandic settlement and is a popular resort area.

Washington Monument, hollow shaft, 555 ft 5⅛ in. (169.3 m) high, located on a 106-acre (43-hectare) site at the west end of the Mall, Washington, D.C.; dedicated 1885. In 1783 Congress passed a resolution approving an equestrian statue of George Washington, and in 1791 architect Pierre L'Enfant included a site for the statue near the present location of the monument. However, Washington objected to the idea. After Washington's death in 1799 plans for a memorial were discussed but none were adopted until 1832, when the Washington National Monument Society, a private group, was formed. Its activity brought gifts of money, as well as blocks of stone from each state, some foreign governments, and private individuals. These "tribute blocks" carry inscriptions on the inside walls of the monument. Architect Robert Mills's elaborate Greek temple design was accepted for the monument, and on July 4, 1848, the cornerstone was laid. Work on the project was interrupted by political quarreling in the 1850s; by the time of the Civil War funds became scarce. It was not until 1876 that Congress took over the project and appropriated money for the monument. The base, entirely different from Mills's design, was completed in 1880; the aluminum top was positioned in 1884; and the monument was opened to the public in 1888. The top may be reached by stairs or elevator.

Washington Park, village (1970 pop. 9,524), St. Clair co., SW Ill., just E of East St. Louis; inc. 1917. Electronic equipment is made.

Wash·i·ta (wŏsh'ĭ-tô, wôsh'-), county (1970 pop. 12,141), 1,009 sq mi (2,613.3 sq km), W Okla., intersected by the Washita River and Elk Creek; formed 1907; co. seat Cordell. Its agriculture includes wheat, cotton, livestock, dairy products, and alfalfa. It has oil and gas wells and manufactures textiles and motor vehicle parts.

Washita, river, c.450 mi (725 km) long, rising in the Texas Panhandle near the Okla. line and flowing generally SE across Okla. to the Red River. The Battle of the Washita (1868), in which Gen. George A. Custer defeated the Cheyenne Indians, took place on the river, near Cheyenne. Fort Cobb Dam on Pond Creek and Foss Dam on the Washita both serve the Washita basin project.

Wash·oe (wŏsh'ō, wôsh'ō), county (1970 pop. 121,068), 6,375 sq mi (16,511.3 sq km), NW Nev., in a mountainous area bordering on Calif. and Oregon, and irrigated in the S by the Truckee storage project; formed 1861; co. seat Reno. It has agriculture, cattle raising, mining, and some manufacturing, but its main industries are tourism and gambling.

Wash·te·naw (wŏsh'tə-nô, wôsh'-), county (1970 pop. 234,103), 711 sq mi (1,841.5 sq km), SE Mich., drained by the Huron and Raisin rivers; formed 1826; co. seat Ann Arbor. Its agriculture includes livestock, poultry, grain, corn, beans, sugar beets, and alfalfa. Dairy products are made. The county has several small lakes and resorts.

Wa·tau·ga (wŏ-tô'gə), county (1970 pop. 23,404), 317 sq mi (821 sq km), NW N.C., in the Blue Ridge, bordered on the NW by Tenn. and drained by the Watauga River and South Fork of the New River; formed 1849; co. seat Boone. It has farming (tobacco, vegetables, and corn), poultry and livestock raising, and sawmilling.

Wa·ter·bur·y (wô'tər-bĕr'ē, -bə-rē, wŏt'ər-), city (1979 est. pop. 106,100), New Haven co., W Conn., on the Naugatuck River; settled 1651, inc. as a city 1853. Its brass industry dates from the mid-18th cent. Clocks and watches, tools, instruments, plastics, and electronic parts are among the many other manufactures of Waterbury.

wa·ter·fall (wô'tər-fôl', wŏt'ər-), sudden drop in a stream formed where the stream passes over a layer of harder rock—often igneous—to an area of softer, and therefore more easily eroded, rock. Normally, as a stream grows older, the waterfall, by undercutting and by the erosion of the brink and of the stream bed above the fall, moves upstream and loses height until it eventually becomes a series of rapids and finally disappears. Because of the waterpower that waterfalls and rapids can provide, their existence is often a determinant of the location of cities.

Wa·ter·ford (wô'tər-fərd, wŏt'ər-), county (1971 pop. 77,315), 710 sq mi (1,839 sq km), S Republic of Ireland; co. town Waterford. Although the terrain is largely hilly, there are lowlands in the east. Principal rivers are the Blackwater, the Bride, and the Suir, which forms most of the northern boundary. The coastline, on the south, is indented by Dungarvan Harbour and Waterford Harbour and by Youghal Bay and Tramore Bay. The county has much farming and grazing land; dairy and beef cattle and sheep are important. Fishing, food processing, tanning, and glassmaking are other industries. Waterford was rebellious under English domination, notably in the latter part of the 16th cent., when it suffered severely during the revolt of the Desmonds.

Waterford, county borough (1971 pop. 31,968), county town of Co. Waterford, S Republic of Ireland, on the Suir River near the head of Waterford Harbor. The city is a port for the produce, especially dairy products and meat, of southern Ireland. Other industries are fishing, food processing, and the manufacture of footwear and fertilizers. The making of Waterford glass, famous in the 18th cent., died out in the mid-19th cent. but has been revived. Waterford was established very early as a walled Danish settlement. King John granted the first charter in the 13th cent. In 1618 the charter was withdrawn because the people refused to accept the religious supremacy of the king of England. Waterford was besieged by Oliver Cromwell in 1649 and taken by Henry Ireton in 1650. The borough contains remains of 13th cent. Franciscan and Dominican foundations that were suppressed in the 16th cent.; there are also Protestant and Roman Catholic cathedrals.

Waterford, town (1975 pop. 18,200), New London co., SE Conn., on Long Island Sound; settled c.1653, inc. as a separate town from New London, 1801. It is mainly residential, with some publishing and other light industry. Both commercial and sport fishing are carried on. The Millstone atomic power plant produces the town's electricity.

Wa·ter·loo (wô'tər-lo͞o', wŏt'ər-), town (1971 est. pop. 18,540), Brabant prov., central Belgium, near Brussels. The Battle of Waterloo was fought just south of here on June 18, 1815. The battle is commemorated by a large monument (built 1823–27).

Waterloo, city (1971 pop. 36,677), SE Ont., Canada. It is a suburb of Kitchener. The district was settled (1800–1805) by Mennonites from Pennsylvania.

Waterloo. **1.** City (1970 pop. 4,546), seat of Monroe co., SW Ill., S of East St. Louis, in a farm area; platted 1818, inc. 1849. **2.** City (1979 est. pop. 77,700), seat of Black Hawk co., NE Iowa, on the Cedar River; inc. 1868. Originally a center for sawmills and flour mills, Waterloo is now a trade and industrial center in a farm and livestock area. The city's chief industries are meat-packing and the manufacture of farm machinery, plastics, and heating and air-conditioning equipment. **3.** Industrial village (1970 pop. 5,418), seat of Seneca co., W central N.Y., NE of Geneva, in the Finger Lakes region between Seneca and Cayuga lakes; inc. 1824. Cosmetics are made here, and there are canneries.

Wa·ter·town (wô'tər-toun', wŏt'ər-). **1.** Town (1975 est. pop. 19,200), Litchfield co., W Conn.; set off from Waterbury and inc. 1780. Synthetic textiles, thread, plastics, chemicals, mattresses, and metal goods are among its manufactures. A method for processing silk thread developed here (1849) and led to the foundation of a major silk industry in the 19th cent. Taft preparatory school and portions of a state park and a state forest are in the town. **2.** Town (1979 est. pop. 33,200), Middlesex co., E Mass., on the Charles River; settled 1630, inc. 1785. It is an industrial suburb of Boston, and its manufactures include machinery, automotive parts, rubber-coated fabrics, clothing, and food products. A Federal arsenal, built in 1816, was greatly enlarged during the two world wars; most of it is now owned by the town, but the U.S. government has retained a section for research. The Perkins School for the Blind, est. in Boston 1829, moved to Watertown 1912. **3.** City (1979 est. pop. 28,700), seat of Jefferson co., N N.Y., on the Black River, in a dairy region; settled c.1800, inc. as a city 1869. The falls on the river (more than 100 ft/30 m high) provide power for its many small industries. Talc, lead, zinc, and iron are mined in the area. Watertown also has a tourist business as a result of its proximity to Canada, the Adirondacks, and the Thousand Islands resort area. A junior college is in the city, and the huge Camp Drum military reservation (national guard) and an air force station are nearby. **4.** City (1979 est. pop. 15,300), seat of Codington co., NE S.Dak., on the Big Sioux River; inc. 1885. It is the distributing, shipping, and trading center for a large agricultural area, and it has food-processing industries. **5.** Industrial city (1979 est. pop. 17,900), Dodge and Jefferson counties, SE Wis., at the falls of the Rock River, in a farm and dairy area; inc. 1853. Carl Schurz lived here from 1855 to 1857. His wife, Margarethe, established (1856) the first kindergarten in the United States here; it has been restored and moved to the grounds of the Octagon House (c.1849), the city's historical museum.

Water Valley, city (1970 pop. 3,285), a seat of Yalobusha co., N central Miss., SSE of Memphis, Tenn., in a farm area; inc. 1858. It ships watermelons and produces cheese and clothing.

Wa·ter·ville (wô'tər-vĭl', wŏt'ər-). **1.** City (1979 est. pop. 16,800), Kennebec co., S Maine, at the falls of the Kennebec River; settled 1754, inc. as a city 1888. It is the trade and medical center of a lake resort area, with railroad shops as well as textile and paper and pulp mills. During the early 1900s the town had five shipyards that built many ocean and river vessels. Present-day Waterville is the seat of Colby College and of Thomas College. **2.** Town (1974 est. pop. 973), seat of Douglas co., central Wash., NE of Wenatchee, in a wheat, cattle, and timber region; founded 1886.

Wa·ter·vliet (wô'tər-vlēt', wŏt'ər-), industrial city (1970 pop. 12,404), Albany co., E N.Y., on the Hudson River, opposite Troy, near the terminus of the New York State Barge Canal; founded by the Dutch 1735, inc. as a city 1896. The U.S. arsenal, which specializes in the production of heavy ordnance, was established in 1813. In 1776 Ann Lee founded in Watervliet the first American community of Shakers.

Wat·ford (wŏt'fərd), municipal borough (1972 est. pop. 77,690), Hertfordshire, SE England. Watford's great publishing industry produces many of the nation's periodicals. Other industries are engineering, the manufacture of woolen and silk goods, paper, and chemicals, and brewing. There is a technical college and a teacher-training school.

Watford City, village (1970 pop. 1,768), seat of McKenzie co., W N.Dak., SE of Williston. Settled in 1914 with the coming of the railroad, it was incorporated in 1934. The village is a railhead and a trade center of a grain, livestock, and dairy region. Oil fields and lignite mines are nearby.

Wat·kins Glen (wŏt'kĭnz), resort village (1970 pop. 2,716), seat of Schuyler co., W central N.Y., in the Finger Lakes region, at the S end of Seneca Lake; inc. 1842. It is in a grape and wine area and has extensive saltworks. Its setting of cliffs, waterfalls, and unusual rock formations made by an interwinding stream attracts many visitors. The resort hotel here is famed for its mineral spring water. An international Grand Prix sports car race is held here annually. Watkins Glen State Park adjoins the village.

Wat·kins·ville (wŏt'kĭnz-vĭl'), town (1970 pop. 986), seat of Oconee co., NE central Ga., SSW of Athens, in an agricultural area.

Wat·ling Street (wŏt'lĭng), important ancient road in England, built by the Romans. It ran north from London until it met the Fosse Way and then west to Wroxeter in Salop, a distance of more than 100 mi (161 km). It passed through St. Albans. The ancient road is still in use in certain places.

Wa·ton·ga (wə-tŏng'gə), city (1970 pop. 3,696), seat of Blaine co., W central Okla., NW of Oklahoma City; settled 1892. It is the center of an agricultural area growing cotton and wheat.

Wat·on·wan (wŏt'n-wŏn'), county (1970 pop. 13,298), 433 sq mi (1,121.5 sq km), S Minn., drained by the Watonwan River; formed 1860; co. seat St. James. It is in an agricultural area that yields corn, oats, barley, livestock, dairy products, and poultry. Its food-processing industry includes poultry, eggs, and frozen vegetables.

Wat·se·ka (wŏt-sē'kə), city (1970 pop. 5,294), seat of Iroquois co., E Ill., SSE of Kankakee, in a farm area; platted 1860 on the site of a trading post (1821), inc. 1867. Electrical equipment is made.

Wat·son·ville (wŏt'sən-vĭl'), city (1975 est. pop. 16,500), Santa Cruz co., W Calif., on the Pajaro River near Monterey Bay; inc. 1868. It is a trade and processing center for vegetables, fruits (especially apples), and flowers. Granite is quarried, and bricks, aluminum parts, and pajamas are made. The site was discovered by the Portola expedition in 1769. A yearly rally of operable antique aircraft is held here. Nearby are beach and mountain resorts and several state parks.

Wat·ten·scheid (văt'ən-shĭt'), city (1972 est. pop. 81,356), North Rhine-Westphalia, W West Germany, an industrial center of the Ruhr district. Its manufactures include metal goods, electrical equipment, and textiles.

Wau·chu·la (wô-cho͞o'lə), city (1970 pop. 3,007), seat of Hardee co., central Fla., E of Bradenton. Settled around a fort built during the Seminole War, it produces vegetables and fruit.

Wau·ke·gan (wô-kē'gən), residential and industrial city (1979 est. pop. 64,400), seat of Lake co., NE Ill., on Lake Michigan; inc. 1859. It has a good harbor and is a port of call on the St. Lawrence Seaway route. Its industries are closely allied with those of neighboring North Chicago. Pharmaceuticals, chemicals, and iron, steel, and wood products are among the area's many manufactures. Waukegan was settled (1835) as Little Fort near an old French stockade established on the site of an Indian village.

Wau·ke·sha (wô'kə-shô), county (1970 pop. 231,335), 554 sq mi (1,434.9 sq km), SE Wis., in a hilly region drained by the Fox and Bark rivers; formed 1846; co. seat Waukesha. It is in a dairying and farming area, with limestone quarries. Its industry includes food processing, lumbering, and the manufacture of furniture, paper, chemicals, plastics, concrete, metal products, machinery, and electrical equipment. It has many resort lakes.

Waukesha, city (1979 est. pop. 49,000), seat of Waukesha co., SE Wis., on the Fox River; inc. 1896. An industrial center in a dairy area, it became a health resort after the Civil War. Its bottled waters are shipped widely. Manufactures include engines, bearings, castings, aluminum products, and electrical and electronic equipment. Carroll College, a county technical institute, and an extension center of the Univ. of Wisconsin are here. Old Indian mounds are preserved in the city's Cutler Park.

Wau·kon (wô-kŏn'), city (1970 pop. 3,883), seat of Allamakee co., NE Iowa, near the Mississippi River ESE of Decorah, in a game area; settled 1851, inc. 1883. Effigy Mounds National Monument is in the area.

Wau·pac·a (wô-păk'ə), county (1970 pop. 37,780), 751 sq mi (1,945.1 sq km), E central Wis., drained by the Wolf and Embarrass rivers; formed 1851; co. seat Waupaca. It is in a dairying and farming (livestock, potatoes, and vegetables) area. Its industry includes cheese processing, textile milling, lumbering, iron foundries, and varied manufactures.

Waupaca, city (1970 pop. 4,342), seat of Waupaca co., central Wis., on the Waupaca River NW of Oshkosh; settled 1859, inc. 1875. It is the center of a lake resort area.

Wau·ri·ka (wô-rē'kə), city (1970 pop. 1,833), seat of Jefferson co., S Okla., SE of Lawton, in a farm area; settled c.1890, inc. 1903.

Wau·sau (wô'sô), city (1979 est. pop. 32,500), seat of Marathon co., central Wis., on the Wisconsin River; settled 1839, inc. 1872. It is an industrial and commercial city in a dairy region. Its industries include paper, paper products, machinery, and insurance. A branch of the Univ. of Wisconsin is in Wausau. Rib Mountain State Park is nearby.

Wau·se·on (wô'sē-ŏn), village (1970 pop. 4,932), seat of Fulton co., NW Ohio, WSW of Toledo, in a farm and dairy area; settled 1835, inc. 1857.

Wau·sha·ra (wô-shăr'ə), county (1970 pop. 14,795), 627 sq mi (1,623.9 sq km), central Wis., drained by the small White and Pine rivers and by Willow Creek; formed 1851; co. seat Wautoma. It is in a dairying and farming (rye, poultry, and potatoes) area. Its industry includes food processing and lumbering. There are many small resort lakes in the county.

Wau·to·ma (wô-tō'mə), city (1970 pop. 1,624), seat of Waushara co., central Wis., W of Oshkosh, in a dairy area; inc. 1901 as a village, 1940 as a city. It is a summer resort.

Wau·wa·to·sa (wô'wə-tō'sə), city (1979 est. pop. 54,900), Milwaukee co., SE Wis., an industrial suburb adjacent to Milwaukee, on the Menominee River; settled 1835, inc. as a city 1897. Metal products are among its manufactures.

Wa·ver·ly (wā'vər-lē). **1.** City (1975 est. pop. 7,351), seat of Bremer co., NE Iowa, on the Cedar River NNW of Waterloo; inc. 1859. Farm produce is processed, and excavating equipment is made. **2.** Village (1970 pop. 5,261), Tioga co., S central N.Y., on the Chemung River ESE of Elmira at the Pa. border, in a farm area; inc. 1853. Poultry feed is produced. **3.** Resort village (1970 pop. 4,858), seat of Pike co., S Ohio, on the Scioto River S of Chillicothe, in a farm area; founded 1829 as Uniontown, renamed 1830, inc. 1842. Lake White State Park is nearby. **4.** Town (1970 pop. 3,794), seat of Humphreys co., central Tenn., W of Nashville near the Tennessee River, in a farm and timber area; laid out 1836, inc. 1837.

Wax·a·hach·ie (wŏk'sə-hăch'ē), city (1970 pop. 13,452), seat of Ellis co., N Texas; inc. 1861. A market center (especially for cattle) in the rich blackland prairie, it also has small manufacturing industries.

Way·cross (wā'krôs'), city (1979 est. pop. 19,600), seat of Ware co.,

SE Ga.; settled 1818, inc. 1874. Waycross is a rail and highway center in a productive pine lumber, livestock, tobacco, and pecan area. It has railroad shops, a tobacco auction market, and a great variety of manufactures. Waycross is the gateway to Okefenokee Swamp Park (a national wildlife refuge). A state park is also nearby.

Way·land (wā'lənd), town (1970 pop. 13,461), Middlesex co., E Mass., W of Boston; settled c.1638, inc. 1835. Electronic and chemical research is carried on.

Wayne (wān). **1.** County (1970 pop. 17,858), 645 sq mi (1,670.6 sq km), SE Ga., bounded on the NE by the Altamaha River, on the SW by the Little Satilla River; formed 1803; co. seat Jesup. It has coastal plain agriculture (tobacco, cotton, corn, and livestock), forestry, pulp and paper mills, and textile manufacturing. **2.** County (1970 pop. 17,004), 716 sq mi (1,854.4 sq km), SE Ill., drained by the Little Wabash River and Elm Creek; formed 1819; co. seat Fairfield. It has agriculture (livestock, poultry, fruit, corn, wheat, hay, and redtop seed), oil wells, and factories manufacturing clothing, auto parts, and paper boxes. **3.** County (1970 pop. 79,109), 405 sq mi (1,049 sq km), E Ind., bordered by Ohio and drained by the Whitewater River; formed 1810; co. seat Richmond. Its agriculture includes grain, poultry, livestock, dairy products, and flowers. Lumbering and food processing are done. Textiles, wood and metal products, chemicals, plastics, and machinery are made. **4.** County (1970 pop. 8,405), 532 sq mi (1,377.9 sq km), S Iowa, on the Mo. border, drained by a branch of the Chariton River; formed 1846; co. seat Corydon. It is primarily agricultural, yielding mainly corn, soybeans, and livestock. There is some diversified light industry, and bituminous-coal deposits are mined in the east and south. **5.** County (1970 pop. 14,268), 440 sq mi (1,139.6 sq km), S Ky., in the Cumberland foothills, bordered in the S by Tenn. and crossed by the Cumberland River; formed 1800; co. seat Monticello. In a hilly agricultural area that produces burley tobacco, grain, and livestock, it has lumber, bituminous-coal mines, oil wells, and rock quarries. **6.** County (1970 pop. 2,669,604), 605 sq mi (1,567 sq km), SE Mich., bounded on the E by the Detroit River and Lakes Erie and St. Clair and drained by the Huron River and the River Rouge; formed 1796; co. seat Detroit. It has agriculture (truck crops, livestock, and dairy products), nurseries, commercial fisheries, and salt mines. Detroit and the manufacturing cities of its metropolitan area produce most of the U.S. automobiles and auto parts. **7.** County (1970 pop. 16,650), 827 sq mi (2,141.9 sq km), SE Miss., bordered in the E by Ala. and drained by the Chickasawhay River; formed 1809; co. seat Waynesboro. Farming (cotton and corn), cattle raising, and lumbering are the major occupations. **8.** County (1970 pop. 8,546), 766 sq mi (1,983.9 sq km), SE Mo., in the Ozarks, drained by the St. Francis and Black rivers; formed 1818; co. seat Greenville. Its products include corn, oats, hay, livestock, and oak, pine, and hickory timber. There are granite deposits in the county. **9.** County (1970 pop. 10,400), 443 sq mi (1,147.4 sq km), NE Nebr., in a farm area drained by Logan Creek; formed 1870; co. seat Wayne. Livestock, grain, dairy products, and poultry are important. **10.** County (1975 pop. 82,194), 606 sq mi (1,569.5 sq km), W N.Y., bounded on the N by Lake Ontario and crossed by the Barge Canal; formed 1823; co. seat Lyons. This rich fruit-, truck-, and nut-growing region also has dairy farming and diversified manufacturing. **11.** County (1970 pop. 85,408), 567 sq mi (1,468.5 sq km), E central N.C., in a coastal plain area drained by the Neuse River; formed 1779; co. seat Goldsboro. It has farming (tobacco, cotton, and corn), sawmilling, and some manufacturing. **12.** County (1970 pop. 87,123), 561 sq mi (1,453 sq km), N central Ohio, intersected by Killbuck Creek and the Lake Fork of the Mohican River; formed 1812; co. seat Wooster. Its agriculture includes livestock, poultry, grain, fruit, and dairy products. There are bituminous-coal mines, oil and gas wells, and sand and gravel pits. Food processing and lumbering are among its diverse industries. Chemicals, plastics, concrete, metal products, machinery, and miscellaneous transportation equipment are made. **13.** County (1970 pop. 29,581), 741 sq mi (1,919.2 sq km), extreme NE Pa., in a lake region bounded in the E by the Delaware River and N.J. border; formed 1798; co. seat Honesdale. In a fruit-growing region, it has diversified manufacturing (clothing, wood products, furniture, and electronic equipment) and recreation facilities. **14.** County (1970 pop. 12,365), 739 sq mi (1,914 sq km), S Tenn., bounded in the S by Ala. and drained by the Buffalo River and tributaries of the Tennessee River; formed 1817; co. seat Waynesboro. Its agriculture includes corn, cotton, soybeans, and livestock. It has limestone and iron-ore deposits and factories manufacturing clothing, shoes, metal products, and electronic equipment. **15.** County (1970 pop. 1,483), 2,486 sq mi (6,438.7 sq km), S central Utah, in a mountain and plateau area crossed by the Dirty Devil River and bounded on the E by the Green River; formed 1892; co. seat Loa. It is in a livestock and farming area, with some mining and manufacturing. Capital Reef National Monument is here. **16.** County (1970 pop. 37,581), 513 sq mi (1,328.7 sq km), SW W.Va., on the Allegheny Plateau, bounded on the W by the Tug Fork and Big Sandy rivers and the Ky. border; formed 1842; co. seat Wayne. The county has bituminous-coal mines, oil and natural-gas wells, timber, and sand and gravel pits. Glassware and concrete are made.

Wayne 1. City (1979 est. pop. 20,100), Wayne co., SE Mich., a suburb of Detroit, on the Lower Rouge River; inc. as a village 1869, and with surrounding areas as a city 1960. It has automobile and aircraft industries and other varied manufactures. **2.** City (1970 pop. 5,379), seat of Wayne co., NE Nebr., SW of Sioux City, Iowa, in a farm area; laid out 1881. **3.** Town (1970 pop. 1,385), seat of Wayne co., W W.Va., S of Huntington; settled 1842, inc. 1882. It is in a bituminous-coal region.

Waynes·bor·o (wānz'bûr'ō). **1.** City (1970 pop. 5,530), seat of Burke co., E Ga., S of Augusta; laid out 1783, inc. 1812. A cotton center, it also makes wood and metal products and processes vegetables. **2.** Town (1974 est. pop. 4,671), seat of Wayne co., SE Miss., near the Chickasawhay River S of Meridian, in a timber, livestock, poultry, and farm area; inc. 1876. Among its products are lumber, electric blankets, and gloves. **3.** Borough (1979 est. pop. 9,600), Franklin co., S Pa., near the Md. line; laid out 1797, inc. 1818. A trade, processing, and shipping center in an agricultural area (dairy farms and fruit orchards), it is also a manufacturing city. Pottery, prefabricated homes, and machine tools are produced. **4.** City (1970 pop. 1,983), seat of Wayne co., S Tenn., WNW of Lawrenceburg, in a farm and timber area; inc. 1827. **5.** Independent city (1979 est. pop. 16,500), central Va., in the Shenandoah valley; settled c.1736, inc. as a city 1948. A manufacturing center in a farm area, it has plants making a wide variety of products. A boy's military school and a girls' preparatory school are in the city. Nearby are Sherando Lake Recreation Area (in George Washington National Forest) and a game refuge.

Waynes·burg (wānz'bûrg'), borough (1970 pop. 5,152), seat of Greene co., SW Pa., S of Washington; laid out 1796, inc. 1816. It is an important coal-producing region.

Waynes·ville (wānz'vĭl'). **1.** City (1970 pop. 3,375), seat of Pulaski co., central Mo., in the Ozarks WSW of Rolla, in a farm area; founded 1831, inc. 1901. Fort Leonard Wood is nearby. **2.** Resort town (1970 pop. 6,488), seat of Haywood co., W N.C., in the Blue Ridge WSW of Asheville; settled before 1800, inc. 1871. In a farm, timber, and mining (kaolin and mica) area, Waynesville produces rubber goods and shoes. Nearby is Great Smoky Mountains National Park.

Wa·zir·i·stan (wä-zîr'ĭ-stän'), region (1961 pop. 395,000), 4,473 sq mi (11,585 sq km), North-West Frontier Province, Pakistan, on the Afghanistan border. An extremely arid and mountainous region, it is divided into North Waziristan, inhabited by farming Wazir tribes, and South Waziristan, populated by seminomad Mahsuds. The two tribes, both of Pathan descent, have constant blood feuds and supplement their meager incomes by brigandage. They live in fortress-like mountain villages or tent camps and export some timber and firewood, hides, ghee (clarified butter), and iron to other parts of Pakistan for cash income. Fertile valleys in parts of northern Waziristan support wheat, maize, barley, and millet; livestock are also raised. In southern Waziristan the hills are used for grazing, and forests on the higher slopes provide timber. Protected by the mountain fastness, the Wazirs and Mahsuds historically resisted British authority. When the Durand Line was established as the border between Afghanistan and British India in 1893, Waziristan became an independent territory outside the bounds of effective British rule. Since Waziristan became part of Pakistan in 1947, the government has continued the British practice of pacification through payment of subsidies to tribal chieftains. The tribes, led by the Faqir of Ipi, have reportedly received arms from Afghanistan, which has agitated for an independent Pushtunistan composed of all border Pathan tribal lands.

Waz·zan (wä-zän'), town (1960 pop. 26,203), N Morocco. It is a sacred city.

Weak·ley (wēk'lē), county (1970 pop. 28,827), 576 sq mi (1,491.8 sq km), NW Tenn., bounded in the N by Ky. and drained by headstreams of the Obion River; formed 1823; co. seat Dresden. Its agriculture includes tobacco, corn, cotton, livestock, sweet potatoes, and fruit. It has kaolin deposits and timber and manufactures clothing, shoes, and miscellaneous equipment.

Weald, the (wēld), area between the North Downs and the South Downs, SE England, forming part of the counties of East Sussex, West Sussex, Surrey, Hampshire, and Kent. The Weald is now largely agricultural.

Wear (wîr), river, c.65 mi (100 km) long, rising in the Pennines, W Durham, NE England, and flowing to the North Sea at Sunderland. Navigable for barges to Durham, the river waters a rich agricultural area. The lower Wear passes through an industrial region.

Weath·er·ford (wĕ*th*'ər-fərd), city (1970 pop. 11,750), seat of Parker co., N central Texas; inc. 1856. It is in a fertile region that yields peanuts, watermelons, and peaches. Oilfield equipment is manufactured. A junior college and a railroad museum are here. Fort Wolters, a major helicopter center, is nearby.

Wea·ver·ville (wē'vər-vĭl'), village (1970 pop. 1,489), seat of Trinity co., NW Calif., NW of Redding; settled 1850. Trinity National Forest is nearby.

Webb (wĕb), county (1970 pop. 78,859), 3,306 sq mi (8,562.5 sq km), SW Texas, bounded on the W and SW by the Rio Grande; formed 1848; co. seat Laredo. This irrigated agricultural area produces onions, other vegetables, melons, corn, hay, peanuts, livestock, oil and natural gas, clay, and coal.

We·ber (wē'bər), county (1970 pop. 126,278), 581 sq mi (1,504.8 sq km), N Utah, drained by the Weber and Ogden Rivers; formed 1852; co. seat Ogden. It is in an irrigated agricultural area that produces livestock, alfalfa, grain, sugar beets, fruit, and truck crops. It has potash, soda, and borate deposits and manufactures clothing, wood products, concrete, machinery, and transportation equipment.

Weber, river, c.125 mi (200 km) long, rising in the Uinta Mts., N central Utah, and flowing N and NW to join the Ogden River at Ogden. The combined stream flows to the Great Salt Lake. The Weber has long been used for irrigation and is now part of the U.S. Bureau of Reclamation's Weber basin project, which irrigates more than 80,000 acres (32,380 hectares) and provides water for industrial and municipal use. Among the dams on the Weber are Wanship Dam (completed 1957) and Echo Dam.

Web·ster (wĕb'stər). **1.** County (1970 pop. 2,362), 195 sq mi (505 sq km), W central Ga.; formed 1853; co. seat Preston. It has farming (corn, truck crops, fruit, peanuts, and pecans), livestock raising, and sawmilling. **2.** County (1970 pop. 48,391), 718 sq mi (1,859.6 sq km), NW central Iowa, drained by the Des Moines River; formed 1851; co. seat Fort Dodge. It is a rich farming area in a prairie area where cattle, hogs, and poultry are raised; the principal crops are corn, soybeans, and hay. There are coal, clay, gypsum, and sand and gravel deposits. Its industries include meat packing, grain milling, and the manufacture of chemicals, machinery, and petroleum products. **3.** County (1970 pop. 13,282), 339 sq mi (878 sq km), W Ky., bounded on the NE by the Green River, on the SW by the Tradewater River; formed 1860; co. seat Dixon. It is in a rolling agricultural area that produces livestock, grain, hay, poultry, tobacco, and fruit. It also has bituminous-coal mines and natural-gas wells. **4.** Parish (1970 pop. 39,939), 615 sq mi (1,592.9 sq km), NW La., on the Ark. border, partly bounded on the E by Black Lake Bayou and drained by Bayou Dorcheat; formed 1871; parish seat Minden. Its agriculture consists of corn, cotton, hay, peanuts, and sweet potatoes. The parish has natural-gas and oil wells and refineries, sand and gravel pits, cotton gins, and lumber and paper mills. **5.** County (1970 pop. 10,047), 416 sq mi (1,077.4 sq km), central Miss., drained by the Big Black River and tributaries of the Yalobusha River; formed 1874; co. seat Walthall. Agriculture (cotton, lespedeza, corn, and poultry) and lumbering are the primary occupations. **6.** County (1970 pop. 15,562), 590 sq mi (1,528.1 sq km), S central Mo., in the Ozarks, drained by the James River; formed 1855; co. seat Marshfield. Corn, wheat, fruit, and tomatoes are grown in this region. Livestock and oak timber are also important. **7.** County (1970 pop. 5,396), 575 sq mi (1,489.3 sq km), S Nebr., bounded in the S by Kansas and drained by the Republican River; formed 1871; co. seat Red Cloud. Its agriculture includes livestock, grain, dairy products, and poultry. **8.** County (1970 pop. 9,809), 551 sq mi (1,427.1 sq km), E central W.Va., on the Allegheny Plateau, drained by the Elk and Gauley rivers; formed 1860; co. seat Webster Springs. It has bituminous-coal mines, some oil and natural-gas fields, timber, and hunting and fishing areas. Its agriculture includes livestock, fruit, and tobacco.

Webster. 1. Town (1970 pop. 14,917), Worcester co., S Mass., near the Conn. line; settled c.1713, set off from Dudley and Oxford and inc. 1832. The chief manufactures are footwear, fabrics, and textiles. Webster was named for Daniel Webster and became a textile center in the early 19th cent. through the efforts of Samuel Slater, a pioneer in the U.S. textile industry. **2.** City (1970 pop. 2,252), seat of Day co., NE S.Dak., NNW of Watertown, in a livestock and wheat area; platted 1880. It is a resort for a lake region and a trade and shipping center.

Webster City, city (1970 pop. 8,488), seat of Hamilton co., central Iowa, on the Boone River ESE of Fort Dodge; settled 1851, inc. 1874. It is an industrial city and trade center. Farm and laundry machinery is manufactured, and food is processed.

Webster Groves, city (1979 est. pop. 24,400), St. Louis co., E Mo., a residential suburb of St. Louis; inc. 1896. It is the seat of Webster College and Eden Theological Seminary.

Webster Springs, town (1970 pop. 1,038), seat of Webster co., E central W.Va., ENE of Charleston, in a farm, mine, and timber area; settled 1860, inc. 1892. It is a health resort, with mineral springs, and a center for sportsmen.

Wed·dell Sea (wĕd'əl), arm of the Atlantic Ocean, W Antarctica, SE of South America, bordered by the Antarctic Peninsula and Coats Land. The vast Ronne and Filchner ice shelves are at the head of the sea. It is named for James Weddell, a British navigator who claimed to have discovered the sea in 1823. It was investigated by the British explorer William Bruce from 1902 to 1904 and was studied most fully during the International Geophysical Year (1957-58).

We·dow·ee (wē-dou'ē), town (1970 pop. 842), seat of Randolph co., E Ala., near the confluence of the Little Tallapoosa and Tallapoosa rivers ESE of Talladega. It has cotton gins.

Wee·haw·ken (wē'hô-kən, wē-hô'kən), township (1970 pop. 12,958), Hudson co., NE N.J., on the Hudson River opposite New York City, with which it is connected by the Lincoln Tunnel; inc. 1859. It has railroad shops and varied industries. Highwood, the James Gore King estate, was the scene in 1804 of the duel between Aaron Burr and Alexander Hamilton. A bronze bust commemorates Hamilton, who was fatally wounded.

Wei (wā), river, c.450 mi (725 km) rising in W Kansu prov., China, and flowing E through Kansu and Shensi provs. to the Huang Ho. Its wide, alluvial valley was the site of some of the earliest centers of Chinese civilization (such as the Chou dynasty, 1122-255 B.C.) and is now the agricultural and population center of Shensi prov. Cotton, wheat, millet, and fruits are extensively grown. The Wei is navigable below Hsi-an, the capital of Shansi; Pao-chi is also on the river.

Wei-hai-wei or **Wei·hai·wei** (both: wā'hī'wā'), city, NE Shantung prov., China, a seaport on the Po Hai. The harbor is protected by Liu-kung Island. The city was part of a territory (c.285 sq mi/740 sq km), also called Wei-hai-wei, which was leased by Great Britain from 1898 to 1930. The British leasehold comprised the city, a strip of land along the coast, and nearby islands. Under the British, the city was developed from a small village into a major port and coaling station, and a naval base (Port Edward) was established on Liu-kung Island.

Wei·mar (vī'mär), city (1973 est. pop. 63,265), Erfurt district, S East Germany, on the Ilm River. It is an industrial, transportation, and cultural center. Manufactures include textiles, agricultural machinery, electrical equipment, chemicals, pharmaceuticals, printed materials, and processed food. Weimar became important in the 16th cent. when it was made the capital of the duchy (after 1815 the grand duchy) of Saxe-Weimar. It developed as a cultural center of international importance. Lucas Cranach the Elder worked here (16th cent.), and from 1708 to 1717 Johann Sebastian Bach was court organist and concertmaster at Weimar. Under Dowager Duchess Amalia (1739-1807) and her son, Charles Augustus (1775-1828), Weimar reached the peak of its fame as a cultural center. With the arrival (1775) of Goethe at the court, Weimar and Goethe became virtually synonymous. Goethe not only made Weimar the literary capital of Europe during his lifetime, but he also attracted such men as Herder and Schiller, established and directed the Weimar theater, and as chief minister of Charles Augustus was active in the physical improvement of the city. The Weimar state theater was the site of the first performances of most of Goethe's and many of Schiller's plays. After Goethe's death (1832) Weimar's active cultural life continued. Franz Liszt was musical director here in the mid-19th cent., and Richard Wagner's opera *Lohengrin* was first performed (1850) in Weimar. The fact that Friedrich Nietzsche (1844–1900) lived and died at Weimar resulted in the foundation here of the important Nietzsche Archives by his sister. In 1919 Weimar was the scene of the German national assembly that established the republican government known as the Weimar Republic. The Bauhaus art school was first established (1919) in Weimar. Among the landmarks of the city are the former grand ducal palace (built 1789–1803) and the ducal crypt with the graves of Goethe and Schiller; the residences of Goethe, Schiller, and Liszt; Goethe's garden cottage; the Goethe National Museum; and the nearby ducal castle of Tiefurt. The city has a state college of music and an academy of art and architecture, and it is the seat of the Goethe and Schiller archives. Buchenwald, the Nazi concentration camp (1937–45), was located nearby.

Weir·ton (wîr'tən), city (1979 est. pop. 24,400), Brooke and Hancock cos., NW W.Va., in the Northern Panhandle, on the Ohio River; settled 1790s, inc. 1947. It is a steel-manufacturing center.

Wei·ser (wē'sər), city (1970 pop. 4,108), seat of Washington co., W Idaho, at the confluence of the Weiser and Snake rivers N of Payette. Laid out in 1877, it was moved in 1890 a mile west after a fire destroyed the original city. It is in an irrigated farm, dairy, and livestock region.

Weis·sen·fels (vīs'ən-fĕls), city (1973 est. pop. 43,994), Halle district, S East Germany, on the Saale River. It is an industrial city and lignite-mining center. Manufactures include shoes, paper, and machinery. Chartered in the 12th cent., Weissenfels passed to Prussia in 1815. The baroque palace (17th cent.) served as the residence of the dukes of Saxe-Weissenfels from 1680 to 1746.

Weiss·horn (vīs'hôrn'), peak, 14,782 ft (4,508.5 m) high, Valais canton, S Switzerland, one of the highest in the Pennine Alps.

Welch (wĕlch), coal-mining city (1975 est. pop. 3,800), seat of McDowell co., S W.Va., on the Tug Fork River NW of Bluefield; settled 1885, inc. 1894.

Weld (wĕld), county (1970 pop. 89,297), 4,002 sq mi (10,365.2 sq km), N Colo., bordering on Wyo. and Nebr. and watered by the South Platte and Cache La Poudre rivers; formed 1861; co. seat Greeley. It has coal mines, oil and gas fields, and irrigated agriculture (sugar beets, grain, beans, and livestock). Clothing, wood products, concrete, and farm machinery are made.

Wel·kom (vĕl'kəm), city (1970 pop. 67,472), Orange Free State, central South Africa. It is a commercial center, and there are mines in the vicinity. Founded in 1947, Welkom is a planned city.

Wel·land (wĕl'ənd), city (1971 pop. 44,397), SE Ont., Canada, on the Welland Ship Canal. It is a canal port and an industrial center. Cotton, iron, steel, and many other goods are made in Welland. The city is also a distributing center for a fruit-growing area.

Welland Ship Canal, 27.6 mi (44.4 km) long, SE Ont., Canada, connecting Lake Ontario with Lake Erie and bypassing Niagara Falls. Built between 1914 and 1932 by Canada to replace a canal opened in 1829, it can accommodate the largest lake ships. Its eight locks overcome a 326-ft (99.4-m) difference in level between the lakes. It is part of the St. Lawrence Seaway system.

Welles·ley (wĕlz'lē), town (1979 est. pop. 25,800), Norfolk co., E Mass., a residential suburb SW of Boston; settled 1660, inc. 1881. Its many educational institutions include several private preparatory schools, Babson College, and Wellesley College.

Wel·ling·bor·ough (wĕl'ĭng-bə-rə), urban district (1972 est. pop. 38,440), Northamptonshire, central England. It is a very old market town. Formerly known for its chalybeate spring, Wellingborough is now a rail center with leather factories, iron foundries, breweries, flour mills, and chemical works. It has a public school that was founded in 1595.

Wel·ling·ton (wĕl'ĭng-tən), city (1974 est. pop. 141,800), capital of New Zealand, extreme S North Island, on Port Nicholson, an inlet of Cook Strait. It is a great communications and transportation center and is an important port for coastal trade. Wellington has garment, transportation-equipment, food-processing, and textile industries. It was founded in 1840 and supplanted Auckland as the capital in 1865. Notable are the governor-general's residence, the Parliament building, the National Art Gallery, and the Dominion Museum. Victoria Univ. of Wellington was founded in 1962. Wellington has a symphony orchestra and ballet as well as opera companies. It is the seat of a Roman Catholic archbishopric.

Wellington. 1. City (1970 pop. 8,072), seat of Sumner co., S central Kansas, S of Wichita, in an agricultural region; founded and laid out 1871, inc. 1872. It was on the old Chisholm Trail, and oil fields in the vicinity were developed in the 1930s. The city has railroad shops and flour mills and manufactures aircraft parts and furniture. **2.** City (1970 pop. 2,884), seat of Collingsworth co., extreme N Texas, ESE of Amarillo; settled 1888, inc. 1909. Long a prairie cow town, Wellington now handles cotton, grain, turkeys, truck crops, and apples, as well as cattle.

Wells (wĕlz), municipal borough (1972 est. pop. 8,750), Somerset, SW England. Primarily a cathedral town, it has changed little since medieval times. The first church was erected by King Ine of Wessex in the early 8th cent. The earliest part of the present cathedral dates from 1176 and much of the woodwork dates from the 15th cent. There are more than 300 13th cent. sculptured figures. The grounds of the present bishop's palace include ruins of the original 13th cent. structure and the complete 14th cent. moat and wall. There is a theological college in Wells.

Wells. 1. County (1970 pop. 23,821), 368 sq mi (953 sq km), NE Ind., drained by the Wabash and Salamonie rivers; formed 1835; co. seat Bluffton. In an agricultural area yielding livestock, dairy products, soybeans, and grain, it has limestone deposits and a food-processing industry. Glassware, metal products, and machinery are made. **2.** County (1970 pop. 7,847), 1,299 sq mi (3,364.4 sq km), central N.Dak., in an agricultural area drained by the Sheyenne and James rivers; formed 1873 as Gingras co.; co. seat Fessenden. Wheat, barley, oats, and sunflowers are grown.

Wells·bor·o (wĕlz'bûr'ō), borough (1970 pop. 4,003), seat of Tioga co., N Pa., SW of Elmira, N.Y.; settled c.1800, laid out 1806, inc. 1830. Glass and wood products are made.

Wells·burg (wĕlz'bûrg'), city (1970 pop. 4,600), seat of Brooke co., NW W.Va., in the industrial Northern Panhandle, on the Ohio River NNE of Wheeling; platted 1790, chartered 1797.

Wells·ville (wĕlz'vĭl'). **1.** Village (1970 pop. 5,815), Allegany co., SW N.Y., at the confluence of the Genesee and Dyke rivers WNW of Elmira; settled c.1795, inc. 1871. An oil-refining center, it makes oil-well supplies, heating equipment, steam turbines, and generators. **2.** City (1970 pop. 5,891), Columbiana co., E central Ohio, on the Ohio River just WSW of East Liverpool; founded 1797. Clay products are made.

Wels (vĕls), city (1971 pop. 47,081), Upper Austria province, W Austria, on the Traun River. It is an industrial and rail center and an agricultural market. Manufactures include machinery, paper, and textiles. Nearby are natural-gas wells. A town in Roman times, Wels later became a stronghold against the Avars and the Magyars. Noteworthy buildings include the parish church and the castle where Emperor Maximilian I died in 1519.

Welsh Marches (wĕlsh), lands in Wales along the English border. After the Norman conquest of England in the 11th cent., William I established the border earldoms of Chester, Shrewsbury, and Hereford to protect his English kingdom. Norman barons were encouraged by William's successors to conquer and hold other earldoms in the east of Wales. These nobles ruled as petty feudal princes, owing allegiance only to the king. The Act of Union (1536) abolished the more than 100 marcher lordships, providing for their division into Welsh shires or their incorporation into English counties.

Wel·wyn Garden City (wĕl'ĭn), urban district (1972 est. pop. 40,570), Hertfordshire, E central England. It was founded by Sir Ebenezer Howard in 1920. After World War II the district's growth was planned to relieve London of overpopulation.

We·natch·ee (wə-năch'ē), city (1979 est. pop. 18,200), seat of Chelan co., central Wash., on the Columbia River in the foothills of the Cascade Range; inc. 1892. It is a resort and a commercial center in a fertile fruit-growing valley famous for its apples. An apple-blossom festival is held annually in the spring. Wenatchee's major industries are food processing and the production of aluminum. In the city are Wenatchee Valley College, an agriculture experiment and research station, and a museum containing prehistoric Indian artifacts.

Wen·chow or **Wen-chou** (wŭn'jō'), city (1970 est. pop. 250,000), SE Chekiang prov., SE China. It is a small deep-sea port on the Ou River 12 mi (19.3 km) from the East China Sea and a major trade center for an area producing tea, cotton, and oranges. Manufactures include condensed milk, paper and bamboo products, fertilizer, and handicrafts. Founded in the 4th cent. A.D., Wenchow retains many ancient buildings.

Went·worth (wĕnt'wûrth'), village (1970 pop. c.125), seat of Rockingham co., N N.C., NW of Reidsville.

Wer·ni·ge·rode (vĕr'nē-gə-rō'də), city (1973 est. pop. 33,479), Magdeburg district, W East Germany, at the N foot of the Harz Mts. It is an industrial city and a tourist center. Manufactures include machinery, paper, processed food, and pharmaceuticals. Noteworthy buildings in the picturesque city include a medieval castle (rebuilt in the 19th cent.), formerly the seat of the princes of Stolberg-Wernigerode; two Romanesque-early Gothic churches; and a 15th cent. city hall.

We·sel (vā'zəl), city (1972 est. pop. 46,387), North Rhine–Westphalia, W West Germany, on the Rhine River near the mouth of the Lippe River. It is a river port, a transshipment point, and an industrial center in the Ruhr district. Manufactures include precision instruments, processed food, glass, and iron goods. First mentioned in the 8th cent., Wesel passed to the counts of Cleves in the early 13th cent. and in 1407 joined the Hanseatic League. The city came under the control of Brandenburg in 1666. Wesel was almost totally destroyed in World War II. In Mar., 1945, the Allies crossed the Rhine here in a major amphibious and airborne operation. The city contains a Gothic church, the Willibrordikirche (1424–1506).

We·ser (vā'zər), river, c.300 mi (480 km) long in E central West Germany, flowing generally N to the North Sea through a long estuary. Navigable to Kassel on the Fulda River, the Weser is connected by the Midland canal system with the Rhine, the Ems, and the Elbe rivers.

Wes·la·co (wĕs'lĭ-kō), city (1975 est. pop. 16,700), Hidalgo co., extreme S Texas, in the irrigated region of the lower Rio Grande valley; inc. 1921. It has a giant citrus-canning plant, fruit- and vegetable-processing companies, a garment factory, and other agriculture-related industries. Agricultural research is also conducted. The city's name was derived from the initials of the W. E. Stewart Land Company, which promoted the townsite in 1917. Its Spanish-style architecture and palm-lined streets abloom with flowers give it a leisurely air. It is linked to Mexico by an international bridge. A wildlife preserve is nearby.

Wes·sing·ton Springs (wĕs'ĭng-tən), city (1970 pop. 1,300), seat of Jerauld co., SE S.Dak., in the Wessington Hills NW of Mitchell; founded 1880. It is a trade and shipping center for a dairy, livestock, and grain region.

West Al·lis (wĕst ăl'ĭs), city (1979 est. pop. 67,000), Milwaukee co., SE Wis., a suburb of Milwaukee; inc. 1902. Mobile equipment, electronic products, generators, and heavy machinery are among its manufactures. The Wis. state fair and annual international skating and auto races are held here. A veterans' hospital is nearby.

West Ba·ton Rouge (băt'n rōōzh'), parish (1970 pop. 16,864), 203 sq mi (525.8 sq km), SE central La., bounded on the E by the Mississippi River; formed 1807; parish seat Port Allen. Sugar cane, corn, hay, and cotton are grown. The area is also noted for oil, lumber, and sugar milling.

West Bend (bĕnd), industrial city (1975 est. pop. 20,000), seat of Washington co., E Wis., on the Milwaukee River; inc. 1885, consolidated with Barton in 1961. Tools and dies, machine tools, washers, dairy items, and leather products are made here. A two-year branch of the Univ. of Wisconsin is located in West Bend, and a state park is nearby.

West Ber·lin (bûr-lĭn'): *see* Berlin.

West·bor·ough (wĕst'bûr'ō), town (1970 pop. 12,594), Worcester co., E central Mass., on the Assabet River; inc. 1717. The town, which is largely residential, produces abrasives, electronic components, tools, dyes, and other products. The birthplace of Eli Whitney, the inventor of the cotton gin, is preserved.

West Branch (brănch). **1.** Town (1977 est. pop. 1,612), Cedar co., SE Iowa , E of Iowa City, in a farm region; settled c.1850 by Quakers, inc. 1875. It is the birthplace of Herbert Hoover (his house is preserved) and the seat of the Herbert Hoover Presidential Library. **2.** City (1970 pop. 1,912), seat of Ogemaw co., N Mich., E of Cadillac, in a hunting and fishing area; inc. as a village 1885, as a city 1905.

West Bridge·wa·ter (brĭj'wô'tər, -wŏt'ər), town (1970 pop. 6,429), Plymouth co., E Mass., S of Brockton; settled 1651, set off from Bridgewater and inc. 1822. It is a residential community, with light manufacturing.

West Brom·wich (brŭm'ĭj, -ĭch, brŏm'wĭch), county borough (1972 est. pop. 165,440), West Midlands, W central England. The borough's area was enlarged in 1966 by the addition of most of the borough of Wednesbury and other areas. On the site of a 12th cent. Benedictine priory, the town has coal mines, foundries, and electrical engineering and chemical works.

West·brook (wĕst'brŏŏk'), city (1975 est. pop. 15,000), Cumberland co., SW Maine, an industrial suburb W of Portland; founded 1657, inc. as a city 1891. Its manufactures include shoes and paper and wood products. An industrial park opened in the city in 1969.

West·bur·y (wĕst'bĕr'ē, -bə-rē), residential village (1975 est. pop. 15,100), Nassau co., SE N.Y., on Long Island; settled 1650, inc. 1932. Harness races are held at Roosevelt Raceway.

West Cald·well (kôl'dwĕl'), borough (1970 pop. 11,913), Essex co., NE N.J., a residential suburb of Newark and New York City; inc. 1904. It has some light manufacturing.

West Car·roll (kăr'əl), parish (1970 pop. 13,028), 356 sq mi (922 sq km), NE La., bounded on the E by Bayou Macon, on the W by the Boeuf River, and on the N by the Ark. line; formed 1877; parish seat Oak Grove. It is an agricultural area, growing cotton, corn, hay, sweet potatoes, and seed oats. Also found here are oil and natural-gas wells, lumber mills, and cotton gins.

West Car·roll·ton (kâr'əl-tən), city (1975 est. pop. 13,700), Montgomery co., SW Ohio, a suburb of Dayton on the Miami River.

West·ches·ter (wĕst'chĕs'tər), county (1970 pop. 894,406), 443 sq mi (1,147.4 sq km), SE N.Y., bounded on the W by the Hudson River, on the SE by Long Island Sound, on the SE and E by the Conn. border; formed 1683; co. seat White Plains. It is chiefly a residential surburban region of metropolitan New York City, with horticulture, farming, dairying, and poultry raising.

West Ches·ter (chĕs'tər), borough (1979 est. pop. 21,800), seat of Chester co., SE Pa., W of Philadelphia; inc. 1799. Primarily residential, West Chester is also the trade and processing center of a fertile agricultural region. The borough's chief manufactures are pharmaceuticals, canned mushrooms, and fire-fighting equipment. West Chester State College is here, and Brandywine State Historical Park, on the site of the Battle of Brandywine (1777), is nearby. The Turk's Head Inn (1747), an early stagecoach stop, still stands.

West·cliffe (wĕst'klĭf'), town (1970 pop. 243), seat of Custer co., S central Colo., on a branch of the Arkansas River WSW of Pueblo, at an altitude of 7,800 ft (2,379 m). Grain, livestock, and dairy products are processed. There are silver and lead mines in the vicinity.

West Co·lum·bi·a (kə-lŭm'bē-ə). **1.** City (1970 pop. 7,838), Lexington co., central S.C., a suburb W of Columbia; inc. 1894. Textile, metal, rubber, and plastic products are made. **2.** City (1970 pop. 3,335), Brazoria co., S Texas, near the Brazos River SSW of Houston; founded 1826. With neighboring East Columbia, it was a center of plantations in Stephen F. Austin's colony and was briefly (1836) capital of the Texas republic. Today it is a farm and livestock center in a rich oil area.

West Co·vi·na (kō-vē'nə), city (1979 est. pop. 76,100), Los Angeles co., S Calif., in the San Gabriel valley; inc. 1923. Before World War II it was a small rural community.

West Des Moines (də moin'), city (1979 est. pop. 23,800), Polk co., S central Iowa, a suburb W of Des Moines; inc. 1893 as Valley Junction, renamed 1938.

Wes·ter·ly (wĕs'tər-lē), town (1979 est. pop. 13,900), Washington co., extreme SW R.I., between the Pawcatuck River and Block Island Sound; inc. 1669. Its large textile industry dates from 1814, and granite has been quarried here since c.1850. Westerly has other varied manufactures and a substantial trade market. It is also famous as a summer resort. The town actually embraces 11 villages, among them Westerly, on the Pawcatuck River; Watch Hill, a resort severely damaged in the hurricanes of 1938 and 1954; and Avondale. Points of interest include many old buildings. A bridge built in 1932 connects the village of Westerly with Conn. A lighthouse and a U.S. coast guard station are maintained at Watch Hill.

West·ern Aus·tra·lia (wĕs'tərn ô-strāl'yə), state (1978 est. pop. 1,171,000), 975,920 sq mi (2,527,633 sq km), Australia, comprising the entire W part of the continent, and bounded on the N, W, and S by the Indian Ocean. Perth is the capital. Western Australia is the largest state of the commonwealth, but only its southwest corner is fertile and substantially settled; the rest is arid and scarcely habitable. Half the population lives in the Perth metropolitan area; there are also 10,000 aborigines living on reservations throughout the state. State-owned goldfields cover much of Western Australia, and there is a vast central desert. The large lakes in the interior are usually dry, and the northern rivers are intermittent; the only important river is the Swan in the southwest. The climate is tropical in the north and temperate in the southwest. Agriculture is confined primarily to the southwest and around Perth. About one half of the cultivated land is in wheat. Sheep graze in the north and southwest, and wool is a major product. Meat, dairy products, and timber are also important. The mining of gold, coal, iron, and other minerals is steadily increasing. Industry expanded significantly during the 1960s. Dirck Hartog, a Dutchman who arrived in 1616, was the first white man known to have visited the coast. A penal colony was founded at Albany in 1826, and the first free settlement was established in the Perth-Fremantle area in 1829. During the 1850s Britain sent some 10,000 convicts to aid the settlers. Gold was discovered in the 1880s. In 1901 Western Australia became a state of the Commonwealth of Australia.

Western Islands: *see* Hebrides, the.

Western Re·serve (rĭ-zûrv′), tract of land in NE Ohio, on the S shore of Lake Erie, retained by Conn. in 1786 when it ceded its claims to its western lands. In 1792 Conn. gave 500,000 acres (202,500 hectares), called firelands, to its citizens whose property was burned during the American Revolution. The Connecticut Land Company bought the remaining land in 1795; the next year, one of its directors, Moses Cleaveland, established the first permanent settlement in the reserve, Cleveland. The reserve was included in the Northwest Territory as Trumbull co. Later this region was divided into ten counties and parts of four others.

Western Sahara, region (1977 est. pop. 130,000), 102,703 sq mi (266,000 sq km), NW Africa, bordering on the Atlantic Ocean in the W, on Morocco in the N, on Algeria in the NE, and on Mauritania in the E and S. The main towns are El Aaiún, Villa Cisneros, and Semara. The region is extremely arid and is almost entirely covered with stones, gravel, or sand. Rocky highlands in the east reach c.1,500 ft (460 m). The permanent population is made up of Arabs and Berbers (together about 55% of the total) and Spaniards.

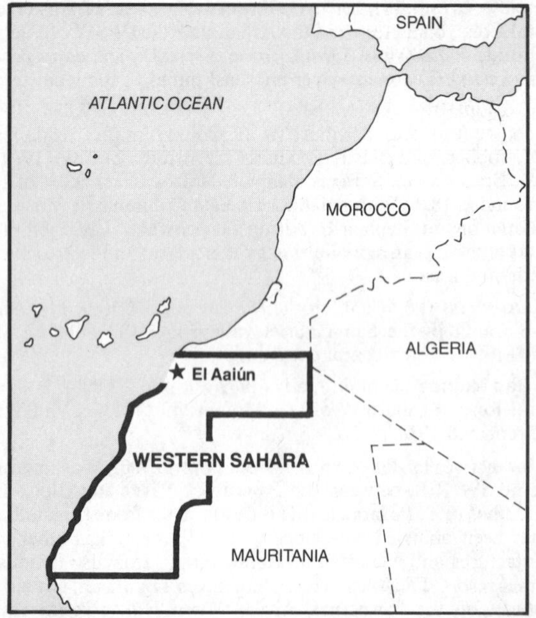

Economy. The traditional economy is limited to the raising of goats, camels, and sheep and the cultivation of date palms. There are several coastal fisheries. Large deposits of phosphates were first exploited by a Spanish-controlled firm in the early 1970s. Potash and iron deposits have also been found but are not yet worked. The province has a limited transportation network.

History. Portuguese navigators reached this region in 1434. However, there was little European contact with the region until the 19th cent. In 1884 Spain claimed a protectorate over the coast from Cape Bojador to Cape Blanc. The boundaries of the protectorate were extended by Franco-Spanish agreements in 1900, 1904, and 1920. The Spanish had only slight contact with the interior until the 1950s. In 1957 a rebel movement ousted the Spanish, who regained control of the region with French help in Feb., 1958. In Apr., 1958, Spain joined the previously separate districts of Saguia el Hamra and Rio de Oro to form the province of Spanish Sahara. Throughout the 1970s dissident inhabitants of Spanish Sahara formed organizations seeking independence for the province. At the same time neighboring nations (notably Mauritania, Morocco, and Algeria) pressured Spain to call a referendum on the area's future in accordance with UN resolutions. In Feb., 1976, Mauritania and Morocco proceeded with the annexation of this territory. Relations between Algeria and Libya have been strained because of this issue.

Western Sa·mo·a (sə-mō′ə), independent state (1978 est. pop. 170,000), South Pacific, comprising the W half of the Samoa island chain. There are nine major islands with a total land area of 1,097 sq mi (2,842 sq km). Apia, the capital, is on Upolu island. All the islands are mountainous, fertile, and surrounded by coral reefs. The population, which is Polynesian, is engaged largely in subsistence agriculture. Tourism is important. All of the Samoan islands west of

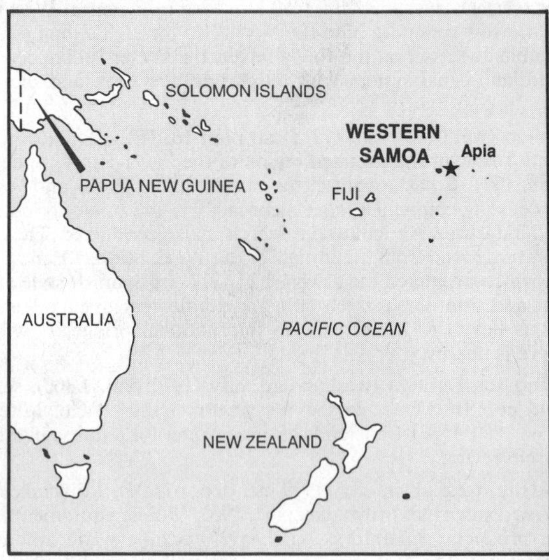

long. 171° W were awarded to Germany under the terms of an 1899 treaty between Germany, the United States, and Great Britain. New Zealand seized the islands from Germany in 1914 and obtained a mandate over them from the League of Nations in 1921. The United Nations made Western Samoa a trusteeship of New Zealand in 1946. In 1961 a UN-supervised plebiscite was held, and on Jan. 1, 1962, independence was proclaimed.

Western Springs, village (1974 pop. 13,728), Cook co., NE Ill., a suburb of Chicago; inc. 1886.

Wes·ter·ville (wĕs′tər-vĭl′), city (1975 est. pop. 16,500), Delaware and Franklin cos., central Ohio; inc. 1858. Seed, grain cleaners, and fabricated steel are made.

West Fe·li·ci·a·na (fə-lĭsh′ē-ăn′ə), parish (1970 pop. 11,376), 405 sq mi (1,049 sq km), E central La., bounded on the W by the Mississippi River and on the N by the Miss. line; formed 1824; parish seat St. Francisville. Corn, cotton, hay, and sugar cane are grown. Industries include canneries, cotton gins, lumber mills, and sand and gravel pits.

West·field (wĕst′fēld′). **1.** City (1979 est. pop. 35,000), Hampden co., SW Mass., a residential and industrial suburb of Springfield, on the Westfield River; settled c.1660, inc. as a city 1920. Bicycles, machinery, and paper and metal products are made. **2.** Town (1979 est. pop. 31,900), Union co., NE N.J.; settled late 17th cent. as part of Elizabethtown, inc. 1903. It is completely residential.

West Flan·ders (flăn′dərz), province (1971 est. pop. 1,059,011), W Belgium, bordering on the North Sea in the W, on the Netherlands in the NE, and on France in the S; capital Bruges. It has considerable fertile soil; grain, flax, and dairy cattle are raised. Fishing is pursued in the North Sea. The province's varied manufactures include textiles and linen, both long-standing industries.

West·ford (wĕst'fərd), town (1970 pop. 10,368), Middlesex co., NE Mass., a suburb of the greater Boston area; settled 1653, set off from Chelmsford and inc. 1729. There are apple orchards and granite quarries (which have long been in operation). The major manufacture is textile machinery. Although chiefly a residential community today, it was once a busy industrial town, attracting many French and Irish immigrants. Its many colonial structures include two 17th cent. saltbox houses and the Old Fletcher Tavern (1713).

West Frank·fort (frăngk'fərt), city (1970 pop. 8,854), Franklin co., S Ill., N of Marion; inc. 1905. It is the trade and industrial center of a coal-mining area. Oil fields are nearby.

West Gla·mor·gan (glə-môr'gən), nonmetropolitan county (1974 est. pop. 372,560), S Wales, created under the Local Government Act of 1972 (effective 1974). It comprises the county borough of Swansea and portions of the former county of Glamorganshire.

West Hart·ford (härt'fərd), town (1979 est. pop. 65,200), Hartford co., central Conn., a suburb of Hartford; settled c.1679, inc. 1854. Tobacco is sorted and packed, and machine tools and parts, aircraft accessories, air conditioners, electrical equipment, vacuum cleaners, and typewriter ribbons are among the many manufactures. The town has numerous commercial and professional offices. It is the seat of St. Joseph College, the Univ. of Hartford, and the American School for the Deaf (1817). Of interest is Noah Webster's birthplace.

West Ha·ven (hā'vən), town (1979 est. pop. 53,000), New Haven co., S Conn., a suburb across the West River from New Haven; settled 1638, inc. as a separate borough 1873. Although chiefly residential, there is some manufacturing industry. The Univ. of New Haven is here.

West Hel·e·na (hĕl'ə-nə), city (1973 pop. 10,838), Phillips co., E Ark.; inc. 1917. Originally a suburb of Helena, its expanding industry has caused it to grow larger than Helena. Wood products are its chief manufactures. St. Francis National Forest is nearby.

West Hemp·stead (hĕmp'stĕd', hĕm'-), uninc. city (1979 est. pop. 26,500), Nassau co., SE N.Y., on Long Island. It is residential.

West In·dies (ĭn'dēz), archipelago, between North America and South America, curving c.2,500 mi (4,020 km) from Fla. to the coast of Venezuela and separating the Caribbean Sea and the Gulf of Mexico from the Atlantic Ocean. The archipelago, sometimes called the Antilles, is divided into three groups, the Bahama Islands, the Greater and Lesser Antilles, and the Dutch and Venezuelan islands off the northern coast of Venezuela.

Many of the islands are mountainous, and some have partly active volcanoes. Hurricanes occur frequently, but the warm climate (tempered by northeast trade winds) and the clear tropical seas have made the West Indies a very popular resort area. Some 25 million people live on the islands, and the majority of inhabitants are of black African descent.

Several of the islands were discovered (1492) by Christopher Columbus. In 1496 the first permanent European settlement was made by the Spanish on Hispaniola. By the middle 1600s the English, French, and Dutch had established settlements in the area, and in the following century there was constant warfare among them for control of the islands. Large numbers of Africans were imported to provide slave labor for the sugar-cane plantations.

The political status of the islands varies: Barbados, Cuba, Haiti, the Dominican Republic, Jamaica, the Bahama Islands, and Trinidad and Tobago are independent. The Netherlands Antilles officially have equal status with Holland in the Kingdom of the Netherlands. Guadeloupe and Martinique are overseas departments of France, Puerto Rico is a self-governing commonwealth associated with the United States, and the U.S. Virgin Islands have territorial status. In 1958 ten British territories joined to form the West Indies Federation. It was slated for independence in 1962. Trinidad and Tobago, Jamaica, and Barbados were the principal members. However, Jamaica, the most populous and prosperous member, voted (1961) to leave the federation, fearing that it would have to support the economically underdeveloped members; Trinidad and Tobago followed suit, and the federation was dissolved in May, 1962.

West King·ston (kĭng'stən), village (1970 pop. 700), seat of Washington co., S R.I., in South Kingstown.

West La·fay·ette (lăf'ē-ĕt', lä'fē-), city (1979 est. pop. 21,100), Tippecanoe co., W Ind., on the Wabash River; inc. 1924. Purdue Univ. is here. Nearby is the Tippecanoe battlesite, where William Henry Harrison fought (1811) the Indian chief Tecumseh.

West·lake (wĕst'lāk'), city (1975 est. pop. 17,400), Cuyahoga co., NE Ohio, a suburb of Cleveland; inc. as a city 1956. Among its manufactures are ink and plastics.

West Long Branch (lŏng brănch), borough (1970 pop. 6,845), Monmouth co., E N.J., SW of Long Branch; settled 1711, inc. 1908. Shadow Lawn, Woodrow Wilson's summer White House, and Monmouth College are here.

West Lo·thi·an (lō'*th*ē-ən), former county, S central Scotland, on the Firth of Forth. The county town was Linlithgow. Under the Local Government Act of 1973, West Lothian was divided between the Lothian and Central regions.

West·meath (wĕst-mēth', wĕst'mē*th*), county (1971 pop. 53,570), 681 sq mi (1,764 sq km), central Republic of Ireland; co. town Mullingar. A part of the central plain of Ireland, the county is mostly level and fertile, with many lakes and bogs. The principal river is the Inny, a tributary of the Shannon. Cattle raising is the chief occupation. There is some manufacture of textiles. Westmeath was separated from Meath as an independent county in 1543.

West Mem·phis (mĕm'fĭs), city (1979 est. pop. 27,500), Crittenden co., NE Ark., near the Mississippi River; founded c.1910, inc. as a city 1935. It is a timber and cotton center.

West Mid·lands (mĭd'ləndz), metropolitan county (1974 est. pop. 2,785,460), central England, created under the Local Government Act of 1972 (effective 1974). West Midlands comprises the county boroughs of Birmingham, Coventry, Dudley, Solihull, Walsall, Warley, West Bromwich, and Wolverhampton, and portions of the former counties of Staffordshire, Worcestershire, and Warwickshire.

West Mif·flin (mĭf'lən), borough (1979 est. pop. 25,900), Allegheny co., SW Pa., a suburb of Pittsburgh, on the Monongahela River. There are steelworks and a household appliances plant in West Mifflin.

West·min·ster (wĕst'mĭn'stər). **1.** Residential city (1979 est. pop. 70,100), Orange co., S Calif.; founded 1870 as a temperance colony for Presbyterians, inc. 1957. It has several industrial parks. **2.** City (1979 est. pop. 27,200), Adams and Jefferson cos., N central Colo., a residential suburb of Denver; inc. 1911. Telephone-switching equipment and electro-mechanical products are manufactured in the city. Standley Lake provides the city with both water and recreation. **3.** City (1970 pop. 7,207), seat of Carroll co., N Md., NW of Baltimore, in a farm region; settled c.1764, inc. 1819. The city was an important Union supply base during the Gettysburg campaign. **4.** Town (1970 pop. 1,875), Windham co., SE Vt., on the Connecticut River S of Bellows Falls, in a farm and timber area; granted 1735 or 1736 by Massachusetts colony to the inhabitants of Taunton, regranted 1752, chartered 1772. Here, on Jan. 15, 1777, Vermont was declared (under the name New Connecticut) an independent state.

Westminster, City of, borough (1971 pop. 225,632) of Greater London, SE England, on the Thames River. The borough was created in 1965 by the merger of the metropolitan boroughs of the City of Westminster, Paddington, and St. Marylebone. Westminster is the location of the principal offices and residences of Great Britain's national government. Important offices and departments are in Whitehall and Downing streets. The monarch lives in Buckingham Palace. Parliament meets in Westminster Palace. Paddington has an important railroad terminal. In St. Marylebone are the administrative offices of the British Broadcasting Corp., London's chief shopping district, and Harley St., a center of medical practice. There is also a clothing industry in St. Marylebone. Westminster School is a leading public school, founded in the 14th cent. and re-established by Queen Elizabeth I in 1560. Other notable features of the borough are Westminster Cathedral, Westminster Abbey, Saint James's Palace, the National Gallery, the Imperial College of Science and Technology, St. James's Park, Hyde Park, parts of Regent's Park and Kensington Gardens, Mme Tussaud's waxworks, and Kensal Green Cemetery, resting place of several literary figures. Westminster Bridge is the second-oldest bridge in London.

West Mon·roe (mŭn-rō'), city (1970 pop. 14,868), Ouachita parish, N La., on the Ouachita River, opposite Monroe, in a forest and lake area; inc. 1851. Its chief industries are lumber and paper milling.

West·more·land (wěst-môr′lənd). **1.** County (1970 pop. 376,935), 1,024 sq mi (2,652.2 sq km), SW Pa., bounded in the N by the Kiskiminetas and Conemaugh rivers, in the SW by the Monongahela River, and partly in the NW by the Allegheny River, formed 1773; co. seat Greensburg. It is in a pasturing, fruit, and grain-farming area with anthracite and bituminous coal mines, oil and gas wells, and limestone quarries. Its industry includes food processing and the manufacture of clothing, paper, chemicals, plastics, glass, concrete, steel, metal products, machinery, electrical equipment, and automobiles. **2.** County (1970 pop. 12,142), 229 sq mi (593.1 sq km), E Va., on the N shore of the Northern Neck peninsula, bounded in the SW by the Rappahannock River; formed 1653; co. seat Montross. Its agriculture includes truck crops, especially tomatoes, tobacco, hay, livestock, and poultry. It has fishing and seafood processing and manufactures wood products and plastics. There are many resorts.

Westmoreland, city (1970 pop. 485), seat of Pottawatomie co., NE Kansas, NW of Topeka, in a cattle and grain region.

West·mor·land (wěst′mər-lənd), former county, N England. The county town was Appleby. In 1974 Westmorland became part of the new nonmetropolitan county of Cumbria.

West·mount (wěst′mount), city (1971 pop. 23,606), S Que., Canada, on Montreal Island. A western residential suburb of Montreal, it became a city in 1908.

West New York (nōō yôrk, nyōō), town (1979 est. pop. 34,900), Hudson co., NE N.J., atop the Palisades across the Hudson River from New York City; settled 1790, inc. 1898. It is a residential town with some light industry. West New York is the leading embroidery center in the United States. The waterfront on the Hudson is 1 mi (1.6 km) long and can accommodate oceangoing vessels.

Wes·ton (wěs′tən), county (1970 pop. 6,307), 2,407 sq mi (6,234.1 sq km), NE Wyo., bordering on S.Dak.; formed 1890; co. seat Newcastle. It is in a livestock and grain farming area, with oil and gas wells, bentonite mines, and a petroleum-refining industry. It contains part of the Black Hills and Black Hills National Forest in the northeast.

Weston. 1. Residential town (1970 pop. 7,417), Fairfield co., SW Conn., NW of Bridgeport; settled c.1670, inc. 1787. **2.** Town (1970 pop. 10,870), Middlesex co., E Mass., W of Boston; settled c.1642, set off from Watertown and inc. 1713. The town is mainly residential. Regis College and the Weston College Geophysical Observatory are here. **3.** Trading town (1970 pop. 7,323), seat of Lewis co., N W.Va., on the West Fork of the Monongahela River SSW of Clarksburg, in a farm and natural-gas area; founded 1818. Glassmaking is the main industry.

Wes·ton-su·per-Mare (wěst′ən-sōō′pər-mâr′), municipal borough (1972 est. pop. 50,730), Avon, SW England, on the Bristol Channel. It is a seaside resort with attractions that include Worlebury Hill, with its Iron Age hill fort and a fine view of the opposite coast of Wales; a long esplanade; and Brean Down, a bird sanctuary. There are light industries with products that include shoes, aluminum window frames, and scientific instruments.

West Or·ange (ôr′ĭnj), town (1979 est. pop. 42,900), Essex co., NE N.J., a residential suburb of Newark; set off from Orange 1862, inc. 1900. Glenmont, Thomas Edison's home in Llewellyn Park, and his laboratory (now a museum) are included in the Edison National Historic Site. The Edison plant manufactures electrical equipment.

West Palm Beach (päm), city (1970 pop. 57,375), seat of Palm Beach co., SE Fla., on Lake Worth (a lagoon) opposite Palm Beach, with which it is connected by bridges; inc. 1894. It is a winter resort and a center for the research and production of aeronautical and electronic equipment. The city was developed by Henry M. Flagler in 1893 as a commercial center for Palm Beach.

West·pha·lia (wěst-fāl′yə), region and former province of Prussia, W West Germany. After 1945 the province was incorporated into the West German state of North Rhine–Westphalia. The region of Westphalia occupies, roughly, a triangle formed by a line drawn eastward from the Rhine River at the Dutch border to the Weser River at Minden, a line drawn from Minden southwestward to Siegen, and a line drawn to the northwest from Siegen and parallel to the Rhine. The region is drained by the Ems, Weser, Ruhr, and Lippe rivers; it is hilly in the east and south and forms a low plain in the northwest. The land consists partly of fertile soil and partly of sandy tracts, moors, and heaths. The Ruhr valley, in the west, is part of the great Westphalian coal basin and of the Ruhr district.

Westphalia first appears as the name of the western third of the duchy of Saxony in the 10th cent. Westphalia survived the breakup (1180) of the Saxon duchy as a regional concept, although it lost political unity. The larger part of Westphalia came under the rule of ecclesiastical princes. All of the region was later included as part of the Westphalian Circle of the Holy Roman Empire (formed c.1500). The anarchy caused by numerous local feudal lords and robber barons led, in the 12th cent., to the creation of the secret Vehmgericht, a criminal tribunal, which had its center at Arnsberg. In the later Middle Ages most of the important Westphalian towns prospered as members of the Hanseatic League. As a result of the demise of the house of Cleves and the Thirty Years' War (terminated in 1648 by the Peace of Westphalia), the elector of Brandenburg obtained Ravensberg, Mark, and the bishopric of Minden; thus Prussia obtained a foothold in western Germany.

The bishoprics of Münster, Paderborn, and Osnabrück and the duchy of Westphalia were secularized only in 1803 by the Diet of Regensburg as a result of the French Revolutionary Wars. In 1807, after the signing of the Treaty of Tilset, Napoleon seized all Prussian possessions west of the Elbe, as well as the electorates of Hesse-Kassel and Hanover and the duchy of Brunswick; the northern section of these territories was annexed by France, and the southern section was constituted as the kingdom of Westphalia, with Napoleon's brother Jérôme Bonaparte as king. However, only a small part of Westphalia was included in that kingdom, which collapsed in 1813. At the Congress of Vienna the major part of Westphalia proper was awarded (1815) to Prussia; and Hanover, Hesse-Kassel, and Brunswick were restored. Westphalia continued as a Prussian province until 1945.

West Pitts·ton (pĭts′tən), borough (1970 pop. 7,074), Luzerne co., NE Pa., on the Susquehanna River NE of Wilkes-Barre, in an anthracite-coal region; inc. 1857. It was the site of Fort Jenkins, burned by the British in 1778. Cigars and machinery are manufactured.

West Plains (plānz), city (1975 est. pop. 7,100), seat of Howell co., S central Mo., ESE of Springfield and near the Ark. line, in a farm and timber region; settled 1840, platted 1858, inc. 1912.

West Point. 1. City (1970 pop. 8,714), seat of Clay co., NE Miss., NW of Columbus; inc. 1858. It is the rail, trade, and processing center for a cotton, dairy, livestock, and timber area. **2.** City (1970 pop. 3,385), seat of Cuming co., NE Nebr., SE of Norfolk on the Elkhorn River; inc. 1858. It is a farm trade center.

West·port (wěst′pôrt′). **1.** Residential town (1979 est. pop. 27,700), Fairfield co., SW Conn., on Long Island Sound at the mouth of the Saugatuck River; settled 1645–50, inc. 1835. It is a summer resort and a residence for New York City commuters, especially artists and writers. The town has a summer theater. William Tryon landed at Compo Beach before his raid on Danbury in 1777. A number of 18th cent. houses remain. Nearby are a state park and a fish hatchery. **2.** Resort and industrial town (1970 pop. 9,791), Bristol co., SE Mass., between Fall River and New Bedford; settled 1670, set off from Dartmouth 1787. The town was devastated in King Philip's War and later rebuilt.

West Prus·sia (prŭsh′ə), former province of Prussia, 9,867 sq mi (25,556 sq km), NE Germany, extending S from the Baltic Sea, between Pomerania on the W and East Prussia on the E. Danzig was the capital. The larger part of the region belonged to Poland until the Polish partitions of 1772 and 1793. The Treaty of Versailles (1919) gave most of West Prussia to Poland and made Danzig and its environs a free city. The remainder of West Prussia was divided between the Prussian province of Grenzmark Posen–West Prussia and the district of West Prussia, incorporated with the province of East Prussia. The whole territory was again annexed to Germany at the outbreak (1939) of World War II, but in 1945 the Potsdam Conference placed it under Polish administration.

West Saint Paul (sānt pôl′), city (1979 est. 18,300), Dakota co., SE Minn., a suburb of St. Paul; inc. 1889. Meat is processed, and plastics, textiles, and clothing are manufactured.

West Spring·field (sprĭng′fēld′), industrial town (1979 est. pop. 28,000), Hampden co., SW Mass., on the Connecticut River opposite Springfield; settled 1654, set off from Springfield and inc. 1774. Paper, ignition systems, and hair-care preparations are manufactured. Storrowton, a reconstructed colonial village, is on the grounds of the annually held Eastern States Exposition.

West Sus·sex (sŭs′ĭks), nonmetropolitan county (1974 est. pop.

629,890), SE England, created under the Local Government Act of 1972 (effective 1974). It is composed of the former county of West Sussex and parts of the former counties of East Sussex and Surrey.

West Un·ion (yo͞on'yən). **1.** City (1970 pop. 2,624), seat of Fayette co., NE Iowa, NE of Waterloo, in a farm and dairy area; inc. 1857. Echo Valley State Park is nearby. **2.** Village (1970 pop. 1,951), seat of Adams co., SW Ohio, near the Ohio River WNW of Portsmouth, in a tobacco area; laid out 1804, inc. 1859. **3.** Town (1970 pop. 1,141), seat of Doddridge co., N W.Va., W of Clarksburg; inc. 1850.

West Vir·gin·ia (vûr-jĭn'yə), state (1975 pop. 1,794,000), 24,181 sq mi (62,629 sq km), E central United States, admitted (1863) as the 35th state of the Union. Charleston is the capital. Extremely irregular in both outline and terrain, West Virginia has two narrow projections—the Northern Panhandle and the Eastern Panhandle. The jagged Virginia-West Virginia line continues southwest from the Eastern Panhandle. In the southwest the state is bounded by Ky. (the Tug Fork forms the state line) and in the west by Ohio (from which it is separated by the Ohio River).

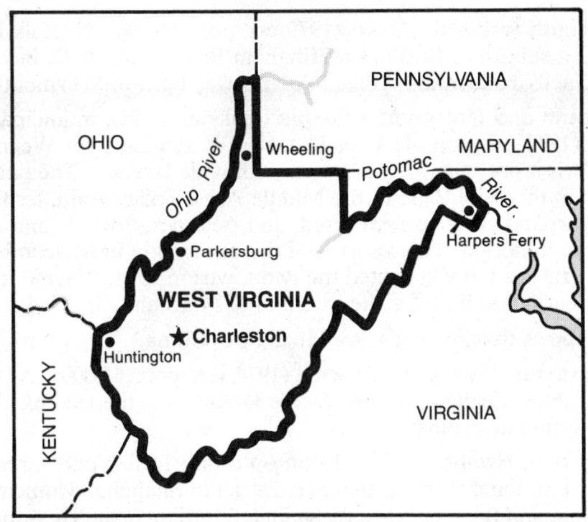

Nicknamed the Mountain State, West Virginia is hilly and rugged, with the highest mean altitude (1,500 ft/457 m) of any state east of the Mississippi. Nearly all of West Virginia is in the Allegheny Plateau. The Eastern Panhandle, a part of the Appalachian ridge and valley country, contains the state's lowest point (240 ft/73 m) near Harpers Ferry, as well as its highest point, Spruce Knob (4,860 ft/1,481 m). West Virginia is well drained; its important rivers include the Tug Fork, the Big Sandy River, the New River, the Kanawha, the Little Kanawha, the Cheat, and the Monongahela. West Virginia's climate is of the humid continental type, with hot summers (except in the highest areas) and cool to cold winters.

West Virginia's natural beauty is spectacular, and the excellent hunting, fishing, hiking, camping, and skiing offered here attract a growing tourist industry. Numerous mineral springs are scattered throughout the state. Other tourist attractions include Harpers Ferry National Historical Park and various prehistoric Indian mounds.

West Virginia is divided into 55 counties:

NAME	COUNTY SEAT	NAME	COUNTY SEAT
Barbour	Philippi	Lewis	Weston
Berkeley	Martinsburg	Lincoln	Hamlin
Boone	Madison	Logan	Logan
Braxton	Sutton	McDowell	Welch
Brooke	Wellsburg	Marion	Fairmont
Cabell	Huntington	Marshall	Moundsville
Calhoun	Grantsville	Mason	Point Pleasant
Clay	Clay	Mercer	Princeton
Doddridge	West Union	Mineral	Keyser
Fayette	Fayetteville	Mingo	Williamson
Gilmer	Glenville	Monongalia	Morgantown
Grant	Petersburg	Monroe	Union
Greenbrier	Lewisburg	Morgan	Berkeley Springs
Hampshire	Romney	Nicholas	Summersville
Hancock	New Cumberland	Ohio	Wheeling
Hardy	Moorefield	Pendleton	Franklin
Harrison	Clarksburg	Pleasants	St. Marys
Jackson	Ripley	Pocahontas	Marlinton
Jefferson	Charles Town	Preston	Kingwood
Kanawha	Charlestown	Putnam	Winfield

NAME	COUNTY SEAT	NAME	COUNTY SEAT
Raleigh	Beckley	Upshur	Buckhannon
Randolph	Elkins	Wayne	Wayne
Ritchie	Harrisville	Webster	Webster Springs
Roane	Spencer	Wetzel	New Martinsville
Summers	Hinton	Wirt	Elizabeth
Taylor	Grafton	Wood	Parkersburg
Tucker	Parsons	Wyoming	Pineville
Tyler	Middlebourne		

Economy. Except for the river-bottom lands, a few small plateaus, and the northern end of the rolling, fertile Valley of Virginia in the Eastern Panhandle, farming is not extensive. Hay, apples, corn, and tobacco are the principal crops, while cattle, dairy products, apples, and eggs lead in market receipts. West Virginia has extensive natural resources; it is the nation's leading producer of bituminous coal (its reserves total more than 60 million tons), and its production of natural gas is the highest of any state east of the Mississippi. Stone, cement, salt, and oil are also important. The state has major glass and chemical industries concentrated in the highly industrialized Ohio and Kanawha river valleys. Other manufactures include primary and fabricated metals and machinery. Steel mills extend south from Pittsburgh into the Northern Panhandle; Wheeling is a major manufacturing hub in that area. Lumber has long been an important resource in the state; about 65% of the land is still forested, most of it in valuable hardwoods.

History. This part of Virginia, which later became West Virginia, was cut off from the eastern regions by rugged mountains and remained uninhabited for more than a century after Virginia had thriving settlements. German families established (c.1730) a settlement on the Potomac and named it Mecklenburg; now called Shepherdstown, it is the oldest town in the state. Settlers began making their way over the Allegheny Mts., and they eventually came into conflict with the French, who claimed the Ohio valley; this conflict was the direct cause of the last French and Indian War (1754–63). After the war, great numbers of settlers poured back over the mountains, ignoring the British proclamation of 1763, which, in the hopes of avoiding Indian troubles, forbade settlement west of the Alleghenies. The Indians resented this encroachment on their hunting grounds, and their hostility was increased by unjust treatment from white settlers. During the American Revolution the area was disturbed by three major Indian invasions led by the British.

Population growth and prosperity were spurred by the opening of the Mississippi River with the Louisiana Purchase in 1803. The area became an increasingly important part of Virginia, but the predominance of small farms and the almost total absence of slavery contributed to a sense of estrangement from the eastern part of the state. Separate statehood for the northwestern counties came about during the Civil War. Creation of a new state was overwhelmingly approved in the referendum of Oct. 24, 1861, and President Lincoln proclaimed admission of a new state, West Virginia, effective June 20, 1863. In the Civil War the Confederates failed to control the region militarily. The strategically important Eastern Panhandle was the scene of continual fighting; not originally a part of West Virginia, it had been quickly annexed (1863) because it contained the Baltimore and Ohio RR.

In 1885 the capital, which had been shuttled back and forth between Wheeling and Charleston, became fixed at Charleston. The now famous Hatfield-McCoy feud began in 1882 and lasted until 1896. Of great significance was the state's industrial expansion in the late 19th cent., based on rich resources and supported by the immigration of southern blacks and northern laborers.

West Virginia's great chemical industry was founded during World War I and greatly expanded during World War II. Both wars also brought unprecedented boom periods to the mines and the steel mills. The state's rapid industrialization was long accompanied by serious labor problems. Unionization was bitterly resisted by mine owners, and strikes were often marked by serious and extended violence. The Great Depression in 1930 intensified difficulties, but reform measures under the New Deal finally assured the miners their right to organize; membership in the United Mine Workers of America soared. During the 1950s, economic weakness in the coal industry helped bring about the highest unemployment rate in the country and a major exodus of population. Economic conditions improved during the 1960s, as Federal aid poured into the state and massive efforts were made to attract new industry.

Government. West Virginia's present constitution dates from 1872. The executive branch is headed by a governor elected for a four-year term. The state's bicameral legislature has a senate with 34 members elected for four-year terms and a house of delegates with 100 members elected for two-year terms.

Educational Institutions. The most prominent in the state is West Virginia Univ., which has two main campuses at Morgantown.

West War·wick (wôr'wĭk, -ĭk), town (1979 est. pop. 25,500), Kent co., central R.I., on the Pawtuxet River; set off from Warwick and inc. 1913. Textile manufacturing is a leading industry. West Warwick includes the village of River Point.

West·we·go (wĕst-wē'gō), city (1970 pop. 11,402), Jefferson parish, SE La., a suburb of New Orleans.

West·wood (wĕst'wŏŏd'). **1.** Residential town (1970 pop. 12,750), Norfolk co., E Mass., in the greater Boston area; settled 1640, inc. 1897. It has several early 18th cent. buildings. **2.** Residential borough (1970 pop. 11,105), Bergen co., NE N.J., a suburb in the New York-northern New Jersey metropolitan area; inc. 1894. Some light manufacturing is carried on.

Westwood Lakes, uninc. village (1970 pop. 12,811), Dade co., SE Fla., a residential suburb of Miami.

West York·shire (yôrk'shĭr', -shər), metropolitan county (1974 est. pop. 2,079,530), N central England, created under the Local Government Act of 1972 (effective 1974). West Yorkshire comprises the county boroughs of Bradford, Leeds, Halifax, Dewsbury, Huddersfield, and Wakefield, and portions of the former county of Yorkshire (West Riding).

Weth·ers·field (wĕ*th*'ərz-fēld'), town (1979 est. pop. 27,500), Hartford co., central Conn., on the Connecticut River, adjoining Hartford on the N; settled 1634 by colonists from Watertown, Mass.; inc. 1637. Wethersfield is the oldest permanent English settlement in the state. Its many colonial buildings include the Joseph Webb House, where Gen. George Washington and the Comte de Rochambeau met secretly in 1781 to coordinate the efforts of French forces with the American army in the Revolutionary War.

Wet·ter·horn (vĕt'ər-hôrn'), peak, c.12,150 ft (3,700 m) high, Bern canton, S central Switzerland, in the Bernese Alps N of the Finsteraarhorn.

Wet·ting·en (vĕt'ĭng-ən), town (1974 est. pop. 19,500), Aargau canton, N Switzerland. It is the site of the Zürich power station and of industries that produce textiles and metal goods. In the town is a former Cistercian monastery, founded in 1227 and now a school, which contains outstanding stained glass and the tomb of the Holy Roman Emperor Albert I.

We·tump·ka (wē-tŭmp'kə), city (1970 pop. 3,912), seat of Elmore co., central Ala., on the Coosa River below Jordan Dam NNE of Montgomery, on the fall line; settled 1820, inc. 1834. It is a farm trade center with cotton gins and cloth and lumber mills. Indian mounds and the site of Fort Toulouse (18th cent.) are nearby.

Wet·zel (wĕt'səl), county (1970 pop. 20,314), 363 sq mi (940.2 sq km), NW W.Va., bounded on the W by the Ohio River and the Ohio border; formed 1846; co. seat New Martinsville. It has oil and natural-gas wells, lumber, and sand and gravel pits. Its agriculture includes livestock, dairy, poultry, grain, tobacco, and truck crops.

Wetz·lar (vĕts'lär), city (1972 est. pop. 37,721), Hesse, central West Germany, on the Lahn River. Situated in a region where iron ore is mined, the city has a metallurgical industry. Other manufactures include optical equipment (cameras, microscopes, and binoculars), machinery, and textiles. Wetzlar was a free imperial city from 1180 to 1803. The supreme court of the Holy Roman Empire was located in the city from 1693 to 1806. The city passed to Prussia in 1815 and formed an enclave between Nassau and Upper Hesse. It suffered considerable damage in World War II. Noteworthy structures include the cathedral (9th cent.) and the ruins of Kalsmunt castle (13th cent.).

We·wa·hitch·ka (wē'wô-hĭch'kă), town (1970 pop. 1,733), seat of Gulf co., NW Fla., SW of Tallahassee; inc. 1925. Tupelo honey is shipped.

We·wo·ka (wē-wō'kə), city (1970 pop. 4,673), seat of Seminole co., central Okla., SE of Oklahoma City, in an oil and farm area. Settled in 1902 on the site of an Indian village, it was incorporated in 1925.

Wex·ford (wĕks'fərd), county (1971 pop. 86,351), 910 sq mi (2,357 sq km), SE Republic of Ireland; co. town Wexford. Wexford is chiefly an agricultural region; wheat is the chief crop, and cattle and pigs are raised. The name of the county is Danish in origin.

Wexford, urban district (1971 pop. 11,744), county town of Co. Wexford, SE Republic of Ireland, on Wexford Harbour. Wexford is a business center. Products include woolens, agricultural machinery, furniture, mineral water, and cured bacon. English invaders landed here and signed a treaty with the Irish in 1169; Oliver Cromwell sacked the town in 1649; and the United Irishmen made it their headquarters in 1798. Selskar, or St. Sepulchre, Abbey dates from the 12th cent. Of interest are the Church of St. Patrick and the old Bull Ring (scene of bullfights). Wexford was once noted for its fairs and tourneys.

Wexford, county (1970 pop. 19,717), 559 sq mi (1,447.8 sq km), NW Mich., intersected by the Manistee River and drained by the Clam River; formed 1869; co. seat Cadillac. There is some farming (corn, grain, and hay) and livestock, and poultry raising. The county contains summer and winter resorts and part of Manistee National Forest.

Wey·mouth (wā'məth), town (1979 est. pop. 56,400), Norfolk co., E Mass., a suburb of Boston on Hingham Bay; settled 1622, inc. 1635. It is chiefly residential. Abigail Adams was born in Weymouth.

Weymouth and Mel·combe Re·gis (mĕl'kəm rē'jĭs), municipal borough (1972 est. pop. 41,410), Dorset, SW England, on Weymouth Bay. It is a port and a resort town with wide beaches. The port was active in the wool trade in the Middle Ages. Today grain, fertilizers, and Portland stone are exported, and potatoes, flowers, and tomatoes are imported. The resort facilities are mostly in Melcombe Regis; Elizabeth I amalgamated the two towns in 1571. Weymouth was an embarkation base for the invasion of Normandy in 1944.

Wham·po·a (hwäm'pō'ä'): *see* Huang-pu, China.

Whang·a·rei (hwäng'gə-rā'), city (1974 est. pop. 33,000), N North Island, New Zealand, on the Pacific Ocean. It is the leading city on the Northland Peninsula.

Whang·poo, Hwang·poo, or **Huang-p'u** (all: hwäng'pōō'), river, 60 mi (97 km) long, rising in the lake district of Shanghai Municipality, E China, and flowing NE past Shanghai into the Yangtze estuary at Wu-sung. It is a major navigational route. Its dredged channel, lined with wharves, warehouses, and industrial plants, provides access to Shanghai for oceangoing vessels.

Whar·ton (hwôr'tn, wôr'-), county (1970 pop. 36,729), 1,076 sq mi (2,786.8 sq km), S Texas, on the Gulf coastal plains, drained by the Colorado and San Bernard rivers; formed 1846; co. seat Wharton. A leading sulfur-producing area, it also has oil and natural-gas wells, farming (rice, cotton, corn, wheat, flax, hay, sorghums, soybeans, pecans, truck crops, and fruit), cattle ranching, dairying, poultry and livestock raising, and varied manufacturing.

Wharton, city (1970 pop. 7,881), seat of Wharton co., S Texas, on the Colorado River SW of Houston; founded 1847, inc. 1902. It has small industries related to the oil, cattle, rice, cotton, and sulfur produced in the area.

What·com (hwŏt'kəm), county (1970 pop. 81,983), 2,126 sq mi (5,506.3 sq km), NW Wash., on the British Columbia border and Puget Sound, and drained by the Baker, Nooksack, and Skagit rivers; formed 1854; co. seat Bellingham. It is in a rich agricultural region yielding dairy products, fruit, truck crops, bulbs, and sugar beets. It has salmon-fishing, lumbering, and diversified industry. It includes Mt. Baker and Mt. Shuksan, in Mt. Baker National Forest.

Wheat·land (hwēt'lənd, wēt'-), county (1970 pop. 2,529), 1,420 sq mi (3,677.8 sq km), central Mont., in an agricultural area drained by the Musselshell River; formed 1917; co. seat Harlowton. Livestock and grain are the major products.

Wheatland, town (1970 pop. 2,498), seat of Platte co., SE Wyo., NNW of Cheyenne, in a livestock and irrigated farm area; settled c.1885. Eagle's Nest Gap, a pass on the Oregon Trail, is nearby.

Whea·ton (hwēt'n, wēt'n). **1.** City (1979 est. pop. 41,900), seat of Du Page co., NE Ill., a residential suburb of Chicago; inc. 1859. **2.** Uninc. city (1979 est. pop. 73,800), Montgomery co., central Md., a residential suburb of Washington, D.C. It was named for Union Gen. Frank Wheaton, who defended nearby Fort Stevens in the Civil War. **3.** Village (1970 pop. 2,029), seat of Traverse co., W Minn., near the S.Dak. border E of Lake Traverse, in a farm area; settled 1884, inc. 1887.

Wheat Ridge, city (1979 est. pop. 29,000), Jefferson co., N central Colo., a residential suburb of Denver; inc. 1969. An annual carnation festival is held here.

Whee·ler (hwē'lər, wē'-). **1.** County (1970 pop. 4,596), 306 sq mi (792.5 sq km), SE central Ga., bounded on the NE and E by the Oconee River, on the S by the Altamaha River; formed 1912; co. seat Alamo. It has coastal-plain agriculture (corn, peanuts, melons, and fruit) and forestry. **2.** County (1970 pop. 1,051), 576 sq mi (1,491.8 sq km), NE central Nebr., in an agricultural area drained by the Cedar River; formed 1877; co. seat Bartlett. Stock and grain are the major products. **3.** County (1970 pop. 1,849), 1,707 sq mi (4,421.1 sq km), N central Oregon, in a mountain area crossed by the John Day River; formed 1899; co. seat Fossil. It is in a lumbering and livestock-raising area. **4.** County (1970 pop. 6,434), 914 sq mi (2,367.3 sq km), extreme N Texas, in the E Panhandle, bounded on the E by the Okla. border and drained by North Fork of the Red River; formed 1876; co. seat Wheeler. This area, underlaid by part of a huge natural-gas and oil field, has gas refineries, carbon-black plants, deposits of clay, caliche, silica, and gypsum, and agriculture (grain sorghums, cotton, fruit, truck crops, dairy products, cattle, hogs, poultry, and horses).

Wheeler, town (1970 pop. 1,116), seat of Wheeler co., extreme N Texas, ENE of Amarillo; founded c.1880, inc. 1924. It is the trade center of a farm, livestock, and oil area.

Whee·ling (hwē'lĭng, wē'-). **1.** Village (1975 est. pop. 18,500), Cook co., NE Ill., a suburb of Chicago; founded c.1830, inc. 1894. **2.** City (1979 est. pop. 42,700), seat of Ohio co., W.Va., in the N Panhandle, on the Ohio River; settled 1769, inc. as a city 1836. It is an important manufacturing and commercial center in an area rich in coal and natural gas. Its many industrial products include steel, iron, chemicals, ceramics, glass, tobacco, plastics, and textiles. Fort Fincastle, renamed Fort Henry, was built (1774) to protect the settlement from Indian raids; in 1782 it was the scene of one of the last skirmishes of the American Revolution, in which a party of British and Indian attackers was driven off. Wheeling became the western terminus of the National Road in 1818, a port of entry in 1831, and a railhead in 1852. Wheeling became the first capital of the state in 1863. The city has a symphony orchestra, an opera workshop, and a racetrack. It is the seat of Wheeling College. Points of interest include the site of Fort Henry, St. Joseph's Cathedral, and Oglebay Park, with museums, a nature center, and an outdoor theater.

Whis·key·town-Shas·ta-Trin·i·ty National Recreation Area (hwĭs'kē-toun'shăs'tə-trĭn'ĭ-tē): *see* National Parks and Monuments Table.

Whit·by (hwĭt'bē, wĭt'-), town (1971 pop. 25,324), SE Ont., Canada, NE of Toronto, on Lake Ontario. The town's manufactures include tires and electronic equipment.

White (hwīt, wīt). **1.** County (1970 pop. 39,253), 1,041 sq mi (2,696 sq km), NE central Ark., bounded on the E by the White River, drained by the Bayou des Arc and intersected by the Little Red River; formed 1835; co. seat Searcy. It is a major producer of strawberries; also raised are livestock, cotton, potatoes, vegetables, and pecans. It has timber and industry (frozen fruit and vegetables, steel products, and calculating machines). There are mineral springs in the county. **2.** County (1970 pop. 7,742), 243 sq mi (629.4 sq km), NE Ga., drained by the Chattahoochee River; formed 1857; co. seat Cleveland. It has agriculture (cotton, corn, hay, potatoes, and poultry), lumber, and a resort area. Chattahoochee National Forest occupies the northern part. **3.** County (1970 pop. 17,312), 502 sq mi (1,300.2 sq km), SE Ill., bounded on the E by the Wabash River and drained by the Skillet Fork and Little Wabash rivers; formed 1815; co. seat Carmi. It has agriculture (wheat, corn, livestock, and poultry), manufacturing of clothing, wood products, buttons, and dairy products, a food-processing industry, and oil and natural-gas wells. **4.** County (1970 pop. 20,995), 497 sq mi (1,287.2 sq km), NW Ind., partly bounded on the E by the Tippecanoe River; formed 1834; co. seat Monticello. An agricultural area yielding corn, oats, soybeans, and dairy products, it also has stone quarries. Its manufactures include furniture, food processing, and machinery. **5.** County (1970 pop. 16,329), 382 sq mi (989.4 sq km), central Tenn., in the Cumberlands, bounded in the S and W by the Caney Fork; formed 1806; co. seat Sparta. It has livestock-raising and farming (tobacco, corn, hay, and vegetables). Clothing, furniture, plastics, and electrical equipment are manufactured.

White. 1. River, c.690 mi (1,110 km) long, rising in the Boston Mts., NW Ark., and flowing first N into SW Mo., then generally SE through NE Ark. to the Mississippi River. The White is navigable for shallow-draft vessels c.300 mi (480 km) upstream. There are three major dam projects on the river. **2.** River, 307 mi (494 km) long, rising near Muncie, E Ind., and flowing SW through Indianapolis to the Wabash River. With the White East Fork (282 mi/454 km long), its chief tributary, the White drains much of southern Ind. **3.** River, 507 mi (816 km) long, rising in NW Nebr. and flowing N then E through S S.Dak. to the Missouri River near Chamberlain. It drains much of the Badlands.

White Bear Lake, city (1979 est. pop. 24,100), Ramsey and Washington cos., SE Minn., on White Bear Lake; inc. 1922. It is a residential and resort suburb of Minneapolis-St. Paul. A junior college is here, and Bald Eagle Lake is nearby.

White Cloud (kloud), city (1970 pop. 1,044), seat of Newaygo co., W Mich., NE of Muskegon; inc. as a village 1879, as a city 1950. Furniture and wood products are made. Manistee National Forest is nearby.

White·fish Bay (hwīt'fĭsh', wīt'-), village (1970 pop. 17,402), Milwaukee co., SW Wis., a residential suburb of Milwaukee on Lake Michigan; inc. 1892.

White·hall (hwīt'hôl', wīt'-). **1.** City (1979 est. pop. 25,300), Franklin co., central Ohio, a suburb of Columbus; inc. 1948. Manufactures include water coolers and packaged meats. A defense-industry supply center is here. **2.** Borough (1970 pop. 16,450), Allegheny co., SW Pa., a residential suburb of Pittsburgh; inc. 1948. **3.** City (1970 pop. 1,486), seat of Trempealeau co., W Wis., on the Trempealeau River SSE of Eau Claire; settled 1855, inc. as a village 1887, as a city 1941.

White·ha·ven (hwīt'hā'vən, wīt'-), municipal borough (1972 est. pop. 26,460), Cumbria, NW England, at the mouth of Solway Firth. Whitehaven is a seaport and industrial town. There are coal mines (some of which extend under the sea), iron foundries, and other industries. Whitehaven was attacked by John Paul Jones in 1778.

White·horse (hwīt'hôrs', wīt'-), city (1971 pop. 11,217), S Yukon, Canada, on the Yukon River. Since 1952 it has been the territorial capital. Whitehorse is on the Alaska Highway and is the terminal of the White Pass and Yukon Railway from Skagway, Alaska. The city is the center of a copper-mining, hunting, and fur-trapping region that attracts tourists. It was an important supply and stagecoach center during the Klondike gold rush (1897–98).

White Horse, Vale of the, Oxfordshire, S central England. The vale is the valley of the Ock River. The region is rich in associations with Alfred the Great, who was born in Wantage, the central town of the vale. According to tradition, his victory at Ashdown in 871 was commemorated by the White Horse on White Horse Hill, although it is probably of a much earlier date. The figure of the horse, over 350 ft (107 m) long, is at Uffington, near Wantage, and its outline is visible for miles. It was formed by cutting away the turf to expose the white chalk of the hillside beneath. An Iron Age fort, Uffington Castle is on the hilltop. There are other "white horses" of various ages in Wiltshire, Berkshire, Yorkshire, and elsewhere, but that at Uffington is the most famous.

White House, official name of the executive mansion of the President of the United States, on Pennsylvania Ave., Washington, D.C., facing Lafayette Square. The building, constructed of Virginia freestone, is of simple and stately design fronted by a portico of high Ionic columns reaching from the ground to the roof pediment. The White House, designated "the Palace" in the original plans, was designed by James Hoban on a site chosen by George Washington. It is the oldest public building in Washington, its cornerstone having been laid in 1792. John Adams was the first President to live here (1800). The building was restored after being burned (1814) by British troops, and the smoke-stained gray stone walls were painted white. Despite popular myth the name White House was applied to the building some time before it was painted. The name became official when President Theodore Roosevelt had it engraved upon his stationery.

White Mountains, part of the Appalachian system, N N.H. and SW Maine, rising to 6,288 ft (1,917.8 m) at Mt. Washington in the Presidential Range and to 5,249 ft (1,600.9 m) at Mt. Lafayette in the Franconia Mountains. Crawford Notch separates these two main groups. Formed in the latter part of the Paleozoic era, the White Mts. are composed chiefly of granite and have been extensively glaciated.

White Nile (nīl), river, one of the chief tributaries of the Nile, E Africa. The name is sometimes used for the 600 mi (970 km) long section of the river known as the Bahr el Abiad which extends upstream from Khartoum to the junction of the Bahr el Jebel and the Bahr el Ghazal at Lake No, c.100 mi (160 km) above Malakal. In a wider sense it is applied to the entire c.2,300 mi (3,700 km) long stem of the Nile draining from the headwaters of Victoria Nyanza. In this wider sense, its remotest headstream is the Luvironza River in Burundi, which flows into the Ruvuvu River and which, in turn, is a tributary of the Kagera River, one of the principal headstreams feeding into Victoria Nyanza. Known as the Victoria Nile for approximately the next 260 mi (430 km), it flows north and west through Uganda into Lake Albert. It leaves Lake Albert as the Albert Nile and flows north c.100 mi (160 km) to Nimule, where it enters the Sudan and becomes the Bahr el Jebel. From Nimule to Rejaf is a zone of rapids. At Juba it leaves the highlands of central Africa and enters the broad Sudan plain; downstream at Bor, it flows through the Sudd, a vast swampy area. At Khartoum it joins with the Blue Nile to form the Nile proper.

White Oak, uninc. community (1970 pop. 19,769), Montgomery and Prince Georges cos., central Md., in the suburbs of Washington, D.C. It is the site of a naval ordnance laboratory.

White Pass, 2,888 ft (880.8 m) high, in the Coast Mts., on the Alaska-British Columbia border, NE of Skagway. A hazardous trail through the pass was made (1897) by prospectors going to the Klondike. Between 1898 and 1900 the White Pass and Yukon Railway was built from Skagway to White Horse, Yukon, to provide transportation from the Pacific tidewater to the Yukon valley.

White Pine (pīn), county (1970 pop. 10,150), 8,904 sq mi (23,061.4 sq km), E Nev., in a mountainous region bordering on Utah; formed 1869; co. seat Ely. Its many ranges have copper, lead, gold, and silver mines.

White Plains (plānz), city (1979 est. pop. 47,200), seat of Westchester co., SE N.Y., N of New York City; settled 1683; inc. as a village 1866, as a city 1916. Mainly residential, the city has some light industries and serves as the headquarters for several corporations and laboratories. The Battle of White Plains (1776), a principal engagement of the Revolutionary War, followed Gen. George Washington's retreat from New York City. Washington briefly made his headquarters in White Plains at the Elijah Miller House, which still stands.

White River, town (1970 pop. 617), seat of Mellette co., S S.Dak., SSW of Pierre on the South Fork of the White River. It is a trading center for a farming and livestock-raising area yielding cattle and hogs. A hydroelectric plant is here.

White Rock, city (1971 pop. 10,349), SE British Columbia, Canada, on Georgia Strait and on the U.S. border. The city is a customs port and resort center with a residential area.

White Sands (săndz), uninhabited desert area, S central N.Mex. It is a center for U.S. military-weapons research and testing. On July 16, 1945, the first atomic bomb was exploded at Holloman Air Force Base (formerly Alamogordo Air Base). Each military branch maintains facilities at White Sands Missile Range, the busiest U.S. missile range. The area encloses White Sands National Monument and San Andres National Wildlife Refuge.

Whites·burg (hwīts'bûrg', wīts'-), town (1970 pop. 1,137), seat of Letcher co., E Ky., SE of Hazard near the Va. border, in a coal and limestone area; settled 1840, inc. 1872.

White Sea, c.36,680 sq mi (95,000 sq km), NW European USSR, an inlet of the Barents Sea. Its northern section, opening into the Barents Sea between the Kola and Kanin peninsulas, is connected with the southern body of the sea by a narrow strait c.100 mi (160 km) long and 30 to 35 mi (48-56 km) wide. Kandalashka Bay, in the southern section, is the deepest part of the sea (1,115 ft/340 m). A canal system (140 mi/225 km long) connects the White Sea, at Belomorsk, with the Baltic Sea, at Leningrad. Icebreakers keep the major sections of the sea open in winter. There are lumber exports, fisheries (herring and cod), and seal herds. The White Sea was significant in the 16th cent. as the only sea outlet for Muscovite trade.

White·side (hwīt'sīd', wīt'-), county (1970 pop. 62,877), 687 sq mi (1,779.3 sq km), NW Ill., bounded on the NW by the Mississippi River and drained by the Rock River and by Rock and Elkhorn creeks; formed 1836; co. seat Morrison. It has agriculture (livestock, corn, wheat, oats, hay, truck crops, and poultry), limestone quarries, a food-processing industry, and diverse manufacturing (machinery, home appliances, metal and wire products, and gas engines). It has hunting areas along the Mississippi.

White Sul·phur Springs (sŭl'fər), city (1970 pop. 1,200), seat of Meagher co., central Mont., on the Smith River E of Helena, in a ranch, timber, and mine area; settled 1878, inc. 1888. Nearby mineral springs make this a health resort.

White·ville (hwīt'vĭl', wīt'-), town (1970 pop. 5,292), seat of Columbus co., SE N.C., W of Wilmington; founded 1810. It is a tobacco market and trade center and manufactures wood products and apparel.

White·wa·ter (hwīt'wô'tər, -wŏt'ər, wīt'-), city (1977 pop. 10,942), Jefferson and Walworth counties, SE Wis., in a dairy and farm area; inc. 1885. It has a foundry and plants that make eyeglasses, television parts, and dairy products.

Whit·field (hwĭt'fēld', wĭt'-), county (1970 pop. 55,108), 281 sq mi (727.8 sq km), NW Ga., bounded on the N by the Tenn. line, on the E by the Conasauga River; formed 1851; co. seat Dalton. It has agriculture (cotton, corn, hay, fruit, and livestock) and some manufacturing. Part of Chattahoochee National Forest is in the south.

Whi·ting (hwī'tĭng, wī'-), city (1970 pop. 7,054), Lake co., NW Ind., on Lake Michigan SE of Chicago, Ill., in the Calumet industrial region; settled 1885, inc. 1903. It has large oil refineries. Soap and chemicals are made.

Whit·ley (hwĭt'lē, wĭt'-). **1.** County (1970 pop. 23,395), 337 sq mi (872.8 sq km), NE Ind., drained by the Eel River; formed 1835; co. seat Columbia River. It is in an agricultural area yielding livestock, grain, truck crops, poultry, soybeans, and dairy products. Textiles, plastics, and metal products are made. **2.** County (1970 pop. 24,145), 459 sq mi (1,188.8 sq km), SE Ky., in the Cumberlands, bordered on the S by Tenn. and drained by the Cumberland and Laurel rivers; formed 1818; co. seat Williamsburg. Its agriculture includes tobacco, corn, oats, Irish potatoes, apples, poultry, cattle, dairy products, and lespedeza. It also has bituminous-coal mines, gas wells, hardwood timber, and industry (clothing, metal products, and machinery manufacturing). Part of Cumberland National Forest is here.

Whitley Bay, municipal borough (1972 est. pop. 37,590), Tyne and Wear, NE England, on the North Sea. Chartered as a municipal borough in 1954, it is the main resort on the Northumberland coast. It is also a residential area for the industrial concentration on the northern bank of the Tyne River.

Whitley City, uninc. village (1970 pop. 1,060), seat of McCreary co., S Ky., WNW of Middlesboro, in a coal and timber area of the Cumberlands.

Whit·man (hwĭt'mən, wĭt'-), county (1970 pop. 37,900), 2,153 sq mi (5,576.3 sq km), SE Wash., on the Idaho border, in a plateau region drained by the Snake and Palouse rivers; formed 1871; co. seat Colfax. Its agriculture includes wheat, barley, oats, hogs, and vegetables. It has lumbering and manufactures farm machinery.

Whitman, town (1970 pop. 13,059), Plymouth co., SE Mass., S of Boston; settled c.1670, inc. 1875. It is an industrial town that manufactures shoe polish, plastics, foundry products, burial vaults, and textile machinery.

Whitman Mission National Historic Site: *see* National Parks and Monuments Table.

Whit·ney, Mount (hwĭt'nē, wĭt'-), peak, 14,494 ft (4,420.7 m) high, E Calif., in the Sierra Nevada at the E border of Sequoia National Park. It is the second-highest peak in the United States.

Whit·sta·ble (hwĭt'stə-bəl, wĭt'-), urban district (1972 est. pop. 25,550), Kent, SE England. Formerly the port for Canterbury pilgrims, today it is a resort and residential area. Whitstable oysters have long been famous.

Whit·ti·er (hwĭt'ē-ər, wĭt'-), city (1979 est. pop. 69,500), Los Angeles co., S Calif., in a fruit and oil area; inc. 1898. The city's manufactures include automobile, aircraft, and missile parts, oil pumps, and clay and steel products. It was founded by Quakers in 1887.

Why·al·la (hwī-ăl'ə, wī-), city (1973 est. pop. 33,250), South Australia state, S Australia, on Spencer Gulf. The city has shipbuilding and iron and steel industries.

Wi·baux (wē'bō'), county (1970 pop. 1,465), 890 sq mi (2,305.1 sq km), E Mont., in an agricultural area bordering on N.Dak. and drained by Beaver Creek; formed 1914; co. seat Wibaux. Grain and livestock are its major products.

Wibaux, town (1970 pop. 644), seat of Wibaux co., E Mont., ESE of Glendive on Beaver Creek near the N.Dak. border. It is a shipping point in a sheep, cattle, and grain region. Machine parts are made.

Wich·i·ta (wĭch'ə-tô'). **1.** County (1970 pop. 3,274), 724 sq mi (1,875.2 sq km), W Kansas, in a rolling plain area; formed 1873; co. seat Leoti. It is in a wheat, sorghum, corn, oats, cattle, hogs, and sheep farming region, which is partly under irrigation. There are oil and gas wells and some manufacturing plants. **2.** County (1970 pop. 120,563), 611 sq mi (1,582.5 sq km), N Texas, bounded on the N by the Red River and drained by the Wichita River and small creeks; formed 1858; co. seat Wichita Falls. A leading petroleum-producing county, this area also has farming (corn, oats, wheat, cotton, corn, hay, fruit, and truck crops), livestock raising, and dairying.

Wichita, city (1979 est. pop. 269,100), seat of Sedgwick co., S central Kansas, at the confluence of the Arkansas and the Little Arkansas rivers; inc. 1870. It is the chief commercial and industrial center of southern Kansas and has railroad shops, flour mills, meat-packing plants, grain elevators, oil refineries, and a huge aircraft industry. Other manufactures are heaters, air conditioners, and trenching equipment. Wichita is located on the site of a village (1863–65) inhabited by Wichita Indians who had been driven out of Oklahoma and Texas for their Union sympathies during the Civil War. The city was founded in 1868 by settlers serving the Chisholm Trail. In 1872 the railroad was extended to Wichita and the city boomed as a cow town. After 1880 it became the trade center of an agricultural and livestock region. Oil was discovered just east of Wichita in 1915. Today the city has many civic and cultural facilities, including art museums, a symphony orchestra, a new modernistic convention and cultural complex (Century II), and a large speech-and-hearing rehabilitation center. It has fine parks, a zoo, a "cow-town" restoration, and two large stadiums. It is the seat of Wichita State Univ., Friends Univ., and Kansas Newman College. McConnell Air Force Base is nearby.

Wichita Falls, city (1979 est. pop. 97,000), seat of Wichita co., N Texas, on the Wichita River; inc. 1889. Settled in the 1870s, the town did not grow until the coming of the railroad in 1882. The city achieved tremendous prosperity with the oil booms of 1919 and 1937 in the vicinity. Electrical supplies, medical products, and electronic products are manufactured here. Agriculture and ranching are also important to the city's economy.

Wick (wĭk), burgh (1973 est. pop. 7,804), N Scotland, on Wick Bay at the mouth of the Wick River. It is an important herring port. Manufactures include knitwear, whiskey, herring oil, and hosiery.

Wick·liffe (wĭk'ləf, -lĭf'). **1.** City (1970 pop. 1,211), seat of Ballard co., SW Ky., on the Mississippi River just below the influx of the Ohio River below Cairo, Ill., in an agricultural and timber area. It has clay pits, and pottery and lamps are made. A buried Indian village nearby has yielded valuable archaeological material. **2.** City (1979 est. pop. 17,000), Lake co., NE Ohio; inc. 1916. Chemicals, machinery, and meters are manufactured.

Wick·low (wĭk'lō), county (1971 pop. 66,295), 782 sq mi (2,025 sq km), E Republic of Ireland; co. town Wicklow. The Wicklow Mts. and their foothills occupy almost the entire area of the county. Sheep and cattle are raised, and grains are cultivated. Wicklow also has copper mines and granite and slate quarries. The people of the mountainous district were long able to maintain their independence of English control, and Wicklow was not organized as a shire until 1606.

Wicklow, urban district (1971 pop. 3,908), county town of Co. Wicklow, E Republic of Ireland, on the Irish Sea. It is a seaport and market for an area of potato growing and cattle and sheep raising. Chemicals, fertilizers, and farm implements are the chief manufactures.

Wi·com·i·co (wī-kŏm'ĭ-kō'), county (1970 pop. 54,236), 381 sq mi (986.8 sq km), SE Md., on the Eastern Shore, bounded on the N by the Del. border, on the E by the Pocomoke River, on the SW by the Wicomico River, on the W and NW by the Nanticoke River; formed 1867; co. seat Salisbury. It is in a tidewater agricultural (fruit, strawberries, truck crops, sweet potatoes, poultry, and dairy products) and timber area. Its industry includes food processing, clothing manufacturing, wood products, plastics, concrete, and metal products. There is fishing and muskrat trapping in the county.

Wid·nes (wĭd'nəs), municipal borough (1972 est. pop. 57,420), Cheshire, NW England, on the Mersey River. It is an important alkali-processing center and timber market.

Wie·ner Neu·stadt (vē'nər noi'shtät), city (1971 pop. 34,800), Lower Austria prov., E Austria. It is an industrial and rail center. Manufactures include locomotives, heavy machinery, and textiles. Founded in 1192, Wiener Neustadt was the birthplace of Emperor Maximilian I (1459–1519). The city was severely damaged in World War II. The 12th cent. castle of the Babenbergs, dukes of Austria, became a military academy in 1752.

Wie·ner·wald (vē'nər-vält), forested range, NE Austria, just W of Vienna. An outlier of the Eastern Alps, it rises to 2,930 ft (893.6 m) in the Schöpfl. The best-known summit, however, is the Kahlenberg (1,585 ft/483 m) near Vienna. The beautiful forests, streams, and hills of the Wienerwald have made it a favorite excursion and resort area for the Viennese.

Wies·ba·den (vēs'bä'dən, vĭs'-), city (1972 est. pop. 252,232), capital of Hesse, central West Germany, on the Rhine River, at the S foot of the Taunus Mts. The city, an industrial center and a market for Rhine wines, is one of the most famous spas of Europe. Manufactures include metal goods, chemicals, plastics, pharmaceuticals, machinery, and textiles. There are also motion-picture and television studios. Wiesbaden was founded as a Celtic settlement in the 3rd cent. B.C. In the 1st and 2nd cent. A.D. it was a popular Roman spa known as Aquae Mattiacorum; there are remains of the Roman water conduits and walls. It later became a free imperial city and passed to the county (later duchy) of Nassau in 1281. In 1806 the city was made the capital of Nassau and with it passed to Prussia in 1866. After World War I Wiesbaden was the seat (1918–29) of the Allied Rhineland Commission. Noteworthy buildings in the city include the castle (1837–41), the Kurhaus (1905–7), and the State Theater of Hesse (1892–94).

Wig·an (wĭg'ən), county borough (1972 est. pop. 81,420), Greater Manchester, N England, on the Douglas River. There are coal mines in the vicinity. The borough has a wide variety of industries. In the Middle Ages Wigan was an important market town and was noted for its pottery and pewter and for bell founding. There were ironworks in the 19th cent. Wigan is thought to have been the site of the Roman station Coccium. The fine Church of All Saints has a Norman tower.

Wig·gins (wĭg'ĭnz), town (1970 pop. 2,995), seat of Stone co., SE Miss., N of Gulfport, in a farm area. Lumber, pine products, and pickles are made here.

Wight, Isle of (wīt), island county (1974 est. pop. 109,680), 147 sq mi (381 sq km), S England, across the Solent and Spithead channels from Hampshire; administrative center Newport. The island is 23 mi (37 km) long from the eastern Foreland to the Needles and 13 mi (21 km) wide. The Medina, which almost bisects the island, and the East Yar and the West Yar are the chief rivers. The mild climate, the scenery, pleasant villages, and a beautiful coastline make the island popular as a winter resort. There are shipbuilding, aircraft, and other industries. Quarrying is also done. The island was conquered by the Romans in A.D. 43 and probably settled later by the Jutes. It was annexed to the kingdom of Wessex in 661 and Christianized c.700. The Isle of Wight was the headquarters of the Danes at the end of the 10th cent. William I bestowed the lordship of the island upon William Fitz-Osbern. In 1293 it returned permanently to the crown. At Carisbrooke Castle, now in ruins, King Charles I was imprisoned (1647–48). In 1890 the island was established as a separate administrative county. Queen Victoria's seaside home, Osborne House, is near the famous yachting center at Cowes.

Wig·town·shire (wĭg'tən-shîr', -shər), former county, SW Scotland. In 1973 Wigtownshire became part of the Strathclyde region.

Wil·bar·ger (wĭl'bär'gər), county (1970 pop. 15,355), 952 sq mi (2,465.7 sq km), N Texas, bounded on the N by the Red River and drained by the Pease River and small creeks; formed 1858; co. seat Vernon. In the north it is a rich agricultural area (cotton, wheat, oats, grain sorghums, corn, fruit, and truck crops); in the south it has large-scale cattle ranching and some dairying and poultry raising. There are oil wells and clay mines here.

Wil·ber (wĭl'bər), city (1970 pop. 1,483), seat of Saline co., SE Nebr., SSW of Lincoln; platted 1873, inc. 1879. It is located in a grain, livestock, and dairy region.

Wil·bra·ham (wĭl'brə-hăm), town (1970 pop. 11,984), Hampden co., S central Mass.; settled 1730, inc. 1763. It is mainly residential. Ice cream is made here.

Wil·bur·ton (wĭl'bər-tən), city (1970 pop. 2,504), seat of Latimer co.,

SE Okla., E of McAlester, in a cattle and coal region; settled 1890, inc. 1902.

Wil·cox (wĭl'kŏks'). **1.** County (1970 pop. 16,303), 899 sq mi (2,328.4 sq km), SW central Ala., in the Black Belt, drained by the Alabama River; formed 1819; co. seat Camden. It has cotton and cattle farming, lumber logging, and some industry (clothing, paper, and millwork). **2.** County (1970 pop. 6,998), 383 sq mi (992 sq km), S central Ga., bounded on the E by the Ocmulgee River and drained by the Alapaha River; formed 1857; co. seat Abbeville. It has coastal-plain agriculture (cotton, corn, melons, and peanuts) and forestry.

Wil·der·ness Road (wĭl'dər-nəs), principal avenue of westward migration for U.S. pioneers from c.1790 to 1840, blazed in 1775 by the American frontiersman Daniel Boone and an advance party of the Transylvania Company. Feeders from the east (Richmond, Va.) and the north (Harpers Ferry, W.Va.) converged at Fort Chiswell in the Shenandoah valley. Boone's road ran southwest from here through the valley, then west across the Appalachian Mts. and through Cumberland Gap into the Ky. Bluegrass region and to the Ohio River. The road followed old buffalo traces and Indian paths, but much of it had to be cut through the wilderness. In the early years, many travelers fell victim to hostile Indians. After Ky. became a state in 1792, the road was widened to accommodate wagons. Private contractors, authorized to keep up sections of the road, charged tolls for its use. With the building of the National Road, the Wilderness Road was neglected and finally abandoned in the 1840s. Since 1926 the Wilderness Road has been a section of U.S. Route 35, the Dixie Highway.

Wil·helms·ha·ven (vĭl'hĕlms-hä'fən), city (1972 est. pop. 104,333), Lower Saxony, NW West Germany, on Jade Bay, an inlet of the North Sea. It is a major oil port and an industrial center. Manufactures include heavy machinery, automobile chassis, electrical equipment, typewriters, and clothing. The city is also a summer resort. It is connected by a canal with Emden and by an oil pipeline with the Ruhr district. Founded in 1869 on territory purchased from Oldenburg in 1853, it was the chief German naval base on the North Sea until the end of World War II, after which its naval installations were dismantled. In 1956 it again was made a naval base. The city has marine biological and geological institutes and an ornithological station.

Wilkes (wĭlks). **1.** County (1970 pop. 10,184), 468 sq mi (1,212.1 sq km), NE Ga., bounded on the NE by the Clark Hill Reservoir of the Savannah River; formed 1777; co. seat Washington. It has piedmont agriculture (cotton, corn, hay, sweet potatoes, peaches, and livestock), lumbering, and some manufacturing. **2.** County (1970 pop. 49,524), 757 sq mi (1,960.6 sq km), NW N.C., mostly in the Blue Ridge, and drained by the Yadkin River; formed 1777; co. seat Wilkesboro. Agriculture (poultry, tobacco, and corn), lumbering, and the manufacture of glass, textiles, and furniture are important.

Wilkes-Bar·re (wĭlks'băr'ē), city (1979 est. pop. 55,000), seat of Luzerne co., E Pa., on the E bank of the Susquehanna River; settled 1769, inc. as a city 1871. Once a major anthracite-coal center, Wilkes-Barre has factories manufacturing pencils, radios, tires, and oil-refining equipment. The city was named for John Wilkes and Isaac Barre, defenders of the colonies before Parliament. The settlement was burned in 1778 by the British and Indians, just after the Wyoming Valley massacre, and was again burned in the second Pennamite Wars. Wilkes-Barre is the seat of Wilkes College, King's College, and a branch of Pennsylvania State Univ. The city has a symphony orchestra and a racetrack; the Swetland Homestead (early 1800s) is of historical interest. Much of Wilkes-Barre was severely damaged by the flooding of the Susquehanna in June, 1972.

Wilkes·bor·o (wĭlks'bûr'ō, -bə-rə), town (1970 pop. 2,417), seat of Wilkes co., NW N.C., on the Yadkin River W of Winston-Salem; founded c.1777, inc. 1889. It is in an area of small farms and has a large poultry-packing plant. Wood products are manufactured.

Wil·kin (wĭl'kən), county (1970 pop. 9,389), 752 sq mi (1,947.7 sq km), W Minn., drained by the Otter Tail River and bounded on the W by the Bois de Sioux River and the Red River of the North; formed 1858; co. seat Breckenridge. Its agriculture includes wheat, grain, potatoes, and livestock. It has some clothing manufacturing.

Wil·kins·burg (wĭl'kĭnz-bûrg'), residential borough (1979 est. pop. 22,900), Allegheny co., SW Pa., a suburb of Pittsburgh; settled c.1800, inc. 1887.

Wil·kin·son (wĭl'kən-sən). **1.** County (1970 pop. 9,393), 458 sq mi (1,186.2 sq km), central Ga., bounded on the E by the Oconee River; formed 1803; co. seat Irwinton. It has coastal-plain agriculture (cotton, corn, potatoes, grain, melons, and peanuts), kaolin mining, and sawmilling. **2.** County (1970 pop. 11,099), 674 sq mi (1,745.7 sq km), extreme SW Miss., bordering on La. in the S and W, with the Mississippi River forming the W boundary; formed 1802; co. seat Woodville. Farming (cotton and corn), lumbering, and cattle raising are primary activities here. The county includes part of Homochitto National Forest.

Will (wĭl), county (1970 pop. 247,825), 847 sq mi (2,193.7 sq km), NE Ill., bounded on the E by the Ind. border and drained by the Des Plains, Du Page, and Kankakee rivers; formed 1836; co. seat Joliet. It has agriculture (corn, oats, soybeans, fruit, truck crops, livestock, poultry, and dairy products), bituminous-coal mines, limestone deposits, and diversified manufacturing.

Wil·la·cy (wĭl'ə-sē), county (1970 pop. 15,570), 591 sq mi (1,530.7 sq km), extreme S Texas, bounded on the E by Laguna Madre, which is separated from the Gulf of Mexico and crossed by the Gulf Intracoastal Waterway; formed 1911, reorganized 1921; co. seat Raymondville. This rich agricultural region irrigated by the Rio Grande yields citrus fruit, truck crops, cotton, beef and dairy cattle, sheep, hogs, and poultry. It also has deposits of oil, salt clay, and sulfur.

Wil·lam·ette (wə-lăm'ĭt), river, 294 mi (473 km) long, rising in several headstreams in the Cascade Range, W Oregon, and flowing N past Eugene, Salem, and Portland to the Columbia River just NW of Portland. The river is navigable for most of its course. Its wide, fertile valley is a major fruit-growing and dairying region. There is also diversified agriculture, manufacturing, and an important lumber industry. The river and its tributaries have been harnessed for flood control, navigation, and power production with numerous dams and reservoirs in the Willamette basin. First settled in the 1830s, the valley was the goal of many pioneers traveling west to the Oregon Country on the Oregon Trail. The region quickly became the chief source of food products on the West Coast, especially with the Calif. gold rush in the mid-1800s. Rapidly settled after 1846, the valley has long been the most densely populated part of Oregon.

Wil·lem·stad (vĭl'əm-stät'), city (1960 pop. 43,547), Curaçao, capital of the Netherlands Antilles. The city is the commercial and industrial center of the Netherlands Antilles as well as a free port and tourist center. It is especially important as a transshipment point and qxrefining center for petroleum sent across the Gulf of Venezuela from Maracaibo. Neat and attractive, its streets lined by narrow, gabled houses, Willemstad has a distinctive Dutch character.

Wil·liam How·ard Taft National Historic Site (wĭl'yəm hou'ərd tăft): *see* National Parks and Monuments Table.

Wil·liams (wĭl'yəmz). **1.** County (1970 pop. 19,031), 2,064 sq mi (5,345.8 sq km), NW N.Dak., in an agricultural area bounded on the S by the Missouri River; formed 1873; co. seat Williston. Its agriculture includes grain, livestock, dairy products, and turkeys. There are also coal mines and oil wells here. **2.** County (1970 pop. 33,669), 421 sq mi (1,090.4 sq km), extreme NW Ohio, bounded in the N by Mich. and in the W by Ind., and intersected by the St. Joseph and Tiffin rivers; formed 1824; co. seat Bryan. Its agriculture includes livestock, poultry, and grain. It has food-processing and lumbering industries and manufactures furniture, chemicals, plastics, metal products, machinery, electrical equipment, and automobile parts.

Wil·liams·burg (wĭl'yəmz-bûrg'), county (1970 pop. 34,243), 935 sq mi (2,421.7 sq km), E central S.C., bounded in the S by the Santee River and drained by the Black River; formed 1785; co. seat Kingstree. It is in an agricultural area that produces tobacco, corn, soybeans, hay, truck crops, hogs, cattle, and poultry. Its industry includes textile milling and lumbering. Hunting and fishing attract tourists.

Williamsburg. **1.** City (1970 pop. 3,687), seat of Whitley co., SE Ky., on the Cumberland River WNW of Middlesboro, in a coal, gas, farm, and timber area. Many Indian artifacts have been found in this area. **2.** Historic city (1975 est. pop. 9,800), seat of James City co., SE Va., on a peninsula between the James and York rivers; settled 1632, laid out and renamed 1699, inc. 1722. It is a great tourist attraction and is also important as the seat (since 1693) of the College of William and Mary in Virginia. Williamsburg became the temporary capital after the burning of Jamestown (1676), then served as capital of Va. from 1699 to 1779. It was the scene of important conventions during the movement for American independence, but it declined after the capital was moved (1779) to Richmond. In 1926, with the financial supportof John D. Rockefeller, Jr., a large-scale restoration of the city was begun; 700 buildings were removed, 83 were renovated, and 413 were rebuilt on their original sites. Today Williamsburg has its colonial appearance, with green formal gardens and many craft shops where revived trades are practiced. Among the historic structures are the colonial capitol (reconstructed); Raleigh Tavern (reconstructed), rendezvous of Revolutionary patriots; the courthouse of 1770; the Bruton Parish Church (1710–15); the governor's palace (reconstructed); the public gaol; the magazine; and many old homes. The Abby Aldrich Rockefeller Museum here houses one of the finest folk art collections in the country. The Colonial Parkway passes through Williamsburg, connecting it with the Jamestown and Yorktown sections of Colonial National Historical Park.

Wil·liam·son (wĭl'yəm-sən). **1.** County (1970 pop. 49,021), 429 sq mi (1,111.1 sq km), S Ill., drained by the Big Muddy River and South Fork of the Saline River; formed 1839; co. seat Marion. It has agriculture (corn, wheat, livestock, fruit, and dairy products) and bituminous-coal mines. **2.** County (1970 pop. 34,423), 593 sq mi (1,535.9 sq km), central Tenn., drained by the Harpeth River; formed 1799; co. seat Franklin. Its agriculture includes livestock, tobacco, and dairy products. Furniture, paper, shoes, concrete, metal products, and household appliances are manufactured. **3.** County (1970 pop. 37,305), 1,104 sq mi (2,859.4 sq km), central Texas, drained by the San Gabriel River and its forks; formed 1848; co. seat Georgetown. It has ranching (sheep, goats, and cattle) in the hilly western section, and farming (cotton, corn, oats, grain sorghums, grain, hay, peanuts, fruit, and truck crops) in the rich blackland prairies of the eastern section. Poultry raising, some dairying, and beekeeping are also important.

Williamson, city (1970 pop. 5,831), seat of Mingo co., SW W.Va., on the Tug Fork SW of Charleston, in a rich coal area; inc. as a town 1892, as a city 1905. Furniture is made. Natural-gas and oil wells are found in the vicinity.

Wil·liams·port (wĭl'yəmz-pôrt'). **1.** Town (1970 pop. 1,661), seat of Warren co., W Ind., on the Wabash River SW of Lafayette; settled 1829. It is in a grain-growing area. **2.** City (1979 est. pop. 34,500), seat of Lycoming co., central Pa., on the Susquehanna River; settled 1772, inc. as a borough 1806, as a city 1866. The scene of several Indian massacres in colonial times, Williamsport grew with the development of the lumber industry in the 19th cent. Today it is a tourist center, a manufacturing city, and the trade and distribution point for an agricultural area. The chief manufactures are aircraft engines and parts, wire and rope products, valves, machinery, piping, electronic and photographic equipment, and clothing. Williamsport is the home of Little League baseball; the Little League world series is held here each year. Also in the city are Lycoming College and Williamsport Area Community College.

Wil·liams·ton (wĭl'yəm-stən), town (1970 pop. 6,570), seat of Martin co., NE N.C., on the Roanoke River ESE of Rocky Mount; inc. 1779. It is a tobacco market and a processing center in a farm area. Wood products and fertilizer are made.

Wil·liams·town (wĭl'yəmz-toun'), city (1971 pop. 29,983), Victoria, SE Australia, part of the Melbourne urban agglomeration, on Hobson's Bay. The city has oil refineries and other industries.

Williamstown. **1.** City (1970 pop. 2,063), seat of Grant co., N Ky., S of Covington, in a fertile farm area; settled before 1792. It is a trade and shipping center. **2.** Town (1970 pop. 8,454), Berkshire co., extreme NW Mass., in the Berkshires on the Hoosic River W of North Adams; settled 1749, inc. 1765. Wire and film are made. The town is in a resort area and has several pre-Revolutionary houses. Williams College is here.

Wil·liams·ville (wĭl'yəmz-vĭl'), residential village (1970 pop. 6,878), Erie co., NW N.Y., a suburb NE of Buffalo; settled c.1800, inc. 1869. Gelatin is produced.

Wil·li·man·tic (wĭl'ə-măn'tĭk), town (1975 est. pop. 14,000), former seat of Windham co., E Conn., ESE of Hartford; inc. 1893. Its cotton spinning industry dates from 1822. Although known as the Thread City, it also manufactures machinery, electrical equipment, and hardware.

Wil·lis·ton (wĭl'ĭs-tən), city (1975 est. pop. 10,800), seat of Williams co., NW N.Dak., on the Missouri River; inc. 1904. It was an early riverboating town, and its importance increased with the discovery (1951) of rich oil reserves in the Williston Basin. It is the trade, processing, and shipping center of an agricultural region, with an oil refinery, stockyards, grain elevators, dairy-processing plants, and railroad shops. Huge reserves of lignite, as well as natural gas, salt, and leonardite are in the area. Of interest are a frontier museum and nearby Fort Union Trading Post National Historic Site and Fort Buford State Historic Site. A junior college branch of the Univ. of North Dakota and a state agricultural experiment station are here.

Williston Park, residential village (1970 pop. 9,154), Nassau co., SE N.Y., on W Long Island N of Mineola; inc. 1926. Paint is made.

Will·mar (wĭl'mär), city (1979 est. pop. 14,800), seat of Kandiyohi co., central Minn., in a dairy and farm region; settled 1856, inc. as a city 1901. It is a medical center and a railroad division point. It has a turkey hatchery and a turkey-processing plant. Food products, plastics, machinery, and clothing are also made. A state junior college, a vocational institute, and a state mental hospital are here.

Wil·lough·by (wĭl'ə-bē), city (1975 est. pop. 19,000), Lake co., NE Ohio, on the Chagrin River, near Lake Erie; settled c.1800, inc. as a city 1951. Manufactures include rubber products, electronic components, clothing, and fused quartz. Nearby is Kirtland Temple (1833–36), a Mormon church.

Wil·lo·wick (wĭl'ō-wĭk'), city (1975 est. pop. 22,000), Lake co., NE Ohio, a residential suburb of Cleveland on Lake Erie; inc. 1924.

Wil·lows (wĭl'ōz), city (1971 est. pop. 4,369), seat of Glenn co., N central Calif., NNW of Sacramento; inc. 1886. It is the headquarters of Mendocino National Forest. Rice is grown and dairy products are made.

Wil·mette (wĭl-mĕt'), village (1979 est. pop. 32,200), Cook co., NE Ill., a residential suburb of Chicago, on Lake Michigan; inc. 1872.

Wil·ming·ton (wĭl'mĭng-tən). **1.** City (1979 est. pop. 72,400), seat of New Castle co., NE Del., on the Delaware River and tributary streams, the Christina and the Brandywine; settled 1638, inc. as a city 1832. A port of entry handling considerable domestic and foreign shipping, it has large shipyards and railroad shops and is a great chemical center, with manufacturing plants and company headquarters. Other industries include copper smelting, automobile assembling, and the manufacture of explosives and munitions, and rubber, leather, and petroleum products. Fort Christina, built here by the Swedes in 1638, was taken by the Dutch (1655) and then by the British (1664). In 1682 William Penn came into possession of the region. Shipping and manufacturing grew early, and industry was well developed when E. I. du Pont established a powder mill on the Brandywine here in 1802. **2.** Town (1975 est. pop. 17,500), Middlesex co., NE Mass., a suburb of Boston, on the Ipswich River; settled 1639, inc. 1730. Economic enterprises include space research and plastic and electronics manufacturing. **3.** City (1979 est. pop. 56,700), seat of New Hanover co., SE N.C., a port of entry on the Cape Fear River, c.30 mi (50 km) from its mouth; settled 1732, inc. as a city 1866. It has a fine harbor and is the state's largest port, receiving domestic petroleum products and shipping tobacco, wood products, and scrap metal. Wilmington is also a rail hub, a resort, and a sports fishing center. Among its manufactures are metal and wood products, textiles and clothing, boilers, and fertilizers. British Gen. Cornwallis held the town in 1781. During the Civil War it was the last Confederate port to close. **4.** City (1975 pop. 10,300), seat of Clinton co., SW Ohio, in a farm area; settled 1810, inc. 1828. Wood-boring tools, air compressors, castings, and automobile parts are made.

Wil·son (wĭl'sən). **1.** County (1970 pop. 11,317), 574 sq mi (1,486.7 sq km), SE Kansas, in a dissected plain watered by the Verdigris and Fall rivers; formed 1855; co. seat Fredonia. Its agriculture includes wheat, soybeans, corn, maize, cattle, hogs, and poultry. It has oil and gas fields, soybean-oil mills, and plants manufacturing cement and concrete, farm machinery, and boats. **2.** County (1970 pop. 57,486), 375 sq mi (971.3 sq km), E central N.C., in a coastal-plain tobacco and timber area drained by Contentnea Creek; formed 1855; co. seat Wilson. Some processing and manufacturing are done. **3.** County (1970 pop. 36,999), 567 sq mi (1,468.5 sq km), N central Tenn., bounded in the N by the Cumberland River; formed 1799; co. seat Lebanon. It has livestock raising and farming (corn, tobacco, and dairy products). Its diversified industry includes food processing, textile milling, and the manufacture of clothing, furniture, metal products, and machinery. **4.** County (1970 pop. 13,401), 802 sq mi (2,077.2 sq km), S Texas, drained by the San Antonio River; formed

1860; co. seat Floresville. It has agriculture (peanuts, watermelons, grain sorghums, corn, cotton, livestock, and poultry) and deposits of oil and clay.

Wilson. **1.** City (1979 est. pop. 32,400), seat of Wilson co., E N.C., in a rich agricultural region; inc. 1849. It is a commercial and industrial center producing textile goods, metal products, and processed foods. **2.** Borough (1970 pop. 8,406), Northampton co., E Pa., just W of Easton; inc. 1920. Woolens and paper products are manufactured.

Wilson, Mount, peak, 5,710 ft (1,741.6 m) high, S Calif., in the San Gabriel Mts., NE of Pasadena. It is the site of Mt. Wilson Observatory (est. 1904), which is operated jointly by the Carnegie Institution and the California Institute of Technology and houses a 100-in. (254-cm) telescope.

Wil·son's Creek National Battlefield (wĭl'sənz): *see* National Parks and Monuments Table.

Wil·ton (wĭl'tən), municipal borough (1971 pop. 3,815), Wiltshire, S central England. Carpets have been made in Wilton for centuries. Felt and farm machinery are other important products. Three sheep fairs are held annually. Wilton was an ancient capital of Wessex and the residence of Saxon kings. In the 9th cent. Wilton was the site of a battle between King Alfred and the Danes. It was the seat of a bishopric until 1050. Wilton House, in which Sir Philip Sidney wrote *Arcadia,* was partly designed by Inigo Jones. It is the seat of the earl of Pembroke.

Wilton, town (1975 est. pop. 16,700), Fairfield co., SW Conn.; settled c.1701, inc. 1802.

Wil·ton Man·ors (wĭl'tən măn'ərz), city (1975 est. pop. 13,700), Broward co., SE Fla.; inc. 1947. It is a residential community in the greater Fort Lauderdale area. An electronic research firm is here.

Wilt·shire (wĭlt'shîr', -shər), county (1974 est. pop. 501,200), 1,345 sq mi (3,484 sq km), S central England; administrative center Salisbury. More than half of Wiltshire is occupied by the great chalk Salisbury Plain and by the Marlborough Downs. Primarily an agricultural county, Wiltshire affords large areas for sheep grazing in the uplands, and the fertile valleys of the Lower Avon, the East Avon, and the Kennet rivers have extensive dairy farming. Pigs are also raised and grains cultivated. Textiles, metal products, processed foods, farm machinery, and electrical goods are manufactured. The county is rich in historic associations. At Stonehenge, Avebury, and Silbury Hill are the largest and oldest monuments of the early British, dating back 4,000 years. Old Sarum was the seat of a bishopric until the 13th cent., when the office was transferred to Salisbury, famous since then for its cathedral. Wilton, known for its carpets, was once the capital of the powerful Saxon kingdom of Wessex, where in the 9th cent. many of King Alfred's battles against the Danes were fought. His grandson, Athelstan, is buried at Malmesbury Abbey, and according to legend, Queen Guinevere spent her last days in the nunnery at Amesbury. Among Wiltshire's famous sons or residents are Joseph Addison, John Dryden, John Gay, George Herbert, and Christopher Wren.

Win·a·mac (wĭn'ə-măk'), town (1970 pop. 2,341), seat of Pulaski co., N central Ind., on the Tippecanoe River and W of Rochester; settled 1837, inc. 1868. Clothing and steel springs are made.

Win·chen·don (wĭn'chən-dən), industrial town (1970 pop. 6,635), Worcester co., N central Mass., near the N.H. border NNW of Fitchburg; settled 1753, inc. 1764. Furniture, wood products, toys, apparel, and machinery are made.

Win·ches·ter (wĭn'chĕs'tər, -chə-stər), municipal borough (1972 est. pop. 31,620), administrative center of Hampshire, S central England. It was called Caer Gwent by the Britons, Venta Belgarum by the Romans, and Wintanceastre by the Saxons. Winchester was the capital of the Anglo-Saxon kingdom of Wessex. Even after the Norman Conquest, when London gradually gained political ascendancy, Winchester remained England's center of learning and attracted many religious scholars. At this time it was also a wool center. The city has held a position of great ecclesiastical influence, reflected in its magnificent cathedral. This Norman structure, which replaced a Saxon church, was consecrated in 1093. In the 14th cent. it was enlarged and transformed into the present Gothic cathedral. It is the burial place of Saxon kings and queens and of William of Wykeham, Samuel Wilberforce, Izaak Walton, and Jane Austen. In Winchester are remains of Wolvesey Castle, where Queen Mary I lived in 1554. The Norman castle, where several parliaments met, was damaged by Oliver Cromwell's soldiers; a round table, supposedly of King Arthur, hangs in the Great Hall. Winchester is still a historic cathedral town, virtually untouched by modern industry and construction. Winchester College, one of the great English public schools, was founded (1382; opened 1394) by William of Wykeham, bishop of Winchester, and is still partly housed in 14th cent. buildings.

Winchester. **1.** Town (1970 pop. 11,106), Litchfield co., NW Conn., in the Litchfield Hills; settled 1732, inc. 1771. It includes Winsted (1970 pop. 8,954), an industrial center where electrical appliances, machine products, pipe organs, and fishing tackle are manufactured. Winchester lies at the gateway to the Berkshire Hills, in a region of lakes and mountain laurel. **2.** City (1970 pop. 1,788), seat of Scott co., W central Ill., WSW of Springfield, in a farm area; platted 1830, inc. 1843. Here Stephen A. Douglas taught school, and Lincoln made his first speech on the Kansas-Nebraska issue. **3.** City (1970 pop. 5,493), seat of Randolph co., E Ind., E of Muncie; settled 1819. It is a shipping center for livestock, grain, poultry, and dairy products. Glass containers are made. Nearby are ancient Indian earthworks. **4.** City (1979 est. pop. 16,700), seat of Clark co., N central Ky.; inc. 1793. The center of a tobacco and livestock area on the edge of the Bluegrass country, it has plants making steel tubing, bed springs, synthetic rubber, truck axles, and lamps. The city is the headquarters of Cumberland National Forest. Two state parks are in the vicinity. **5.** Town (1979 est. pop. 22,600), Middlesex co., E Mass., a residential suburb of Boston; settled 1640, inc. 1850. **6.** Town (1970 pop. 5,256), seat of Franklin co., S Tenn., WNW of Chattanooga, in a clover-growing, dairy, and livestock area; founded c.1814, inc. 1821. Straw hats and medicines are produced. **7.** City (1979 est. pop. 23,200), seat of Frederick co., N Va., in the Shenandoah Valley; settled 1732 near an Indian village in Lord Fairfax's domain, inc. as a city 1874. It is the trade, processing, and shipping center for an orchard district noted for its apples. Its manufactures include brake linings, cans, furniture, clothing, plastics, and travel trailers. George Washington began his career as a surveyor here in 1748. During the French and Indian War Winchester was a center for defense against Indian raids, and Washington, who commanded the Virginia troops, had his headquarters here. Gen. Daniel Morgan lived in Winchester and is buried in Mt. Hebron Cemetery. During the Civil War the city suffered severely, changing hands many times. Stonewall Jackson headquartered here during the winter of 1861–62, and Gen. Philip Sheridan during the winter of 1864–65. Both headquarters are preserved. Also of interest are Washington's office, the old Presbyterian Church (1790; now an armory), and an annual apple festival. Willa Cather and Richard E. Byrd were born in Winchester.

Wind Cave National Park (wĭnd), 28,059 acres (11,355 hectares), in the Black Hills, SW S.Dak.; est. 1903. Wind Cave, discovered in 1881, was named for the strong air currents that blow alternately in and out of it depending on whether the atmospheric pressure is higher or lower than the air pressure inside the cave. There are 10.5 mi (16.9 km) of explored passageways. The cave's temperature remains constant at 47°F (8°C). Unusual calcite-crystal and boxwork formations make Wind Cave unique. The park's surface is characterized by rolling grasslands with herds of bison, elk, deer, and pronghorn antelope; several prairie dog towns are in the park.

Win·der (wīn'dər), city (1970 pop. 6,605), seat of Barrow co., N Ga., ENE of Atlanta; settled as Jug Tavern, inc. 1893 as Winder. It is a trade and processing center in a farm area, and there are textile mills.

Win·der·mere (wĭn'dər-mîr), lake, 10.5 mi (17 km) long and 1 mi (1.6 km) wide, in the Lake District, between Lancashire and Westmorland, NW England. It is c.210 ft (60 m) deep and lies among wooded hills near Scafell and other mountains. The largest lake in England, it is fed by many streams and is drained by the Leven River to Morecambe Bay. The lake's shores are indented to form little bays, and it has several islands. It attracts many tourists.

Wind·ham (wĭn'dəm). **1.** County (1970 pop. 84,515), 514 sq mi (1,331.1 sq km), NE Conn., on the Mass. and R.I. lines, drained by the Quinebaug, Natchaug, Shetucker, Little, Moosup, and French rivers; constituted 1726; former co. seats Putnam and Willimantic. It is an agricultural area (dairy products, poultry, and truck crops), with manufacturing centers producing textiles, thread, machinery, cutlery, metal products, clothing, paper and rubber goods, shoes, chemicals, furniture, optical goods, and wood products. It has several state parks and forests and resorts on small lakes. **2.** County (1970 pop. 33,476), 784 sq mi (2,030.6 sq km), SE Vt., bounded in the S by Mass., in the E by the Connecticut River, and rising to the

Green Mts. in the W, drained by the West and Deerfield rivers; formed 1781; co. seat Newfane. It is in a dairying and lumbering region. Its manufactures include wood products, furniture, paper, wire, and machinery.

Windham, town (1975 est. pop. 22,000), Windham co., E Conn.; inc. 1692. It includes the industrial city of Willimantic.

Wind·hoek (vĭnt'ho͝ok'), city (1970 pop. 61,260), capital of Namibia. It is an administrative, communications, and economic center. Windhoek is one of the world's major trade centers for Karakul sheepskins. Clothing is manufactured, and meat and bone meal are processed in the city. Windhoek was originally the headquarters of a Nama chief. Occupied by German forces in 1885, it became the seat of administration and was later (1892) made the capital of the German colony of South West Africa. During World War I Windhoek was captured by South African troops. Today the city retains a German flavor, and many of its residents are of German background. Windhoek stands 5,428 ft (1,655.5 m) above sea level and is surrounded by hills, three of which have castles built in German medieval style.

Win·disch (vĭn'dĭsh), town (1970 pop. 7,444), Aargau canton, N Switzerland, on the Reuss River near its confluence with the Aare. Textiles and cables are made here. Originally a Helvetian settlement, it later became a Roman camp called Vindonissa. A Roman amphitheater, which accommodated over 10,000 spectators, has been excavated. Nearby are the remains of the once powerful monastery of Königsfelden, erected in 1310 on the site where Holy Roman Emperor Albert I had been murdered. The monastery church is a splendid example of Swiss Gothic architecture, with fine stained glass.

Win·dom (wĭn'dəm), city (1970 pop. 3,952), seat of Cottonwood co., SW Minn., WSW of Mankato; platted 1870, inc. as a village 1875, as a city 1920. It is a farm trade center, and lawn mowers and snowplows are made.

Wind River Range, part of the Rocky Mts., W Wyo., running SE c.120 mi (190 km) and constituting part of the Continental Divide. Gannett Peak (13,785 ft/4,204.4 m) is the highest point in Wyo. South Pass (alt. 7,550 ft/2,302.8 m), at the southern end of the range, was the most important Rocky Mts. pass on the Oregon Trail.

Wind·sor (wĭn'zər). **1.** Town (1971 pop. 3,775), central N.S., Canada, at the mouth of the Avon River on an arm of Minas Basin. It is the center of a gypsum and limestone-quarrying area. Manufactures include fertilizers, building materials, and lumber products. Windsor was settled by Acadians (1703) and called Pisiquid. After their expulsion it was settled by New Englanders and renamed in 1764. It is the site of Fort Edward, built (1750) by the British. King's College, the first English university in Canada, was founded in Windsor in 1789 but moved in 1923 to Halifax as part of Dalhousie Univ. **2.** City (1971 pop. 203,300), S Ont., Canada, on the Detroit River opposite Detroit, Mich. It is a port of entry and an important industrial center producing automobiles, salt, and chemicals. The city was settled by the French in 1749. After the American Revolution many Loyalists settled in the area. The city is the seat of Windsor Univ.

Windsor, officially **New Windsor,** municipal borough (1972 est. pop. 30,360), Berkshire, S central England, on the Thames River. There are a few light industries. In Elizabethan times about 70 inns enlivened the town. Sir Christopher Wren built the town hall, and Grinling Gibbons did much of the wood carving in the Church of St. John the Baptist. The real importance of the town has always been derived from Windsor Castle, the chief residence of English rulers since William I. The castle was improved and rebuilt by successive sovereigns. Henry II erected the Round Tower, and Edward IV began the construction of St. George's Chapel, one of the most splendid churches in England, where the Knights of the Garter are installed with medieval ceremony. In the chapel several of England's kings are buried. Some of the vaults are now used to store art treasures, national archives, and museum collections. In Frogmore, the royal mausoleum, Queen Victoria and Prince Albert are buried.

Windsor, county (1970 pop. 44,082), 962 sq mi (2,491.6 sq km), E Vt., bounded in the E by the Connecticut River and drained by the Ottauquechee, White, Black, and Williams rivers; formed 1781; co. seat Woodstock. Its agriculture includes dairy products, poultry, and maple sugar. It has talc, soapstone, and pyrophyllite mining, and diversified industry (food processing, textile mills, lumbering and wood products, and miscellaneous machinery manufacturing). There are summer and winter resorts.

Windsor. 1. Town (1975 est. pop. 23,500), Hartford co., N Conn., at the confluence of the Farmington and Connecticut rivers, just N of Hartford. Settled by Plymouth Colony in 1633, Windsor was the first English settlement in Connecticut and is the state's oldest town. Although primarily residential, the town produces nuclear installations for submarines and generators, tool and machine parts, computer equipment, metal products, paper, and soft drinks. Windsor's once renowned tobacco industry has declined. **2.** Town (1970 pop. 2,199), seat of Bertie co., NE N.C., between Albemarle Sound and the Roanoke River; settled 1721, inc. 1776. It is a lumber-processing center. Cotton and peanuts are grown. **3.** Town (1970 pop. 4,158), Windsor co., SE Vt., on the Connecticut River SSW of White River Junction; chartered 1761 by Gov. Benning Wentworth of New Hampshire, settled 1764, rechartered 1772 by New York Colony. The convention that adopted a constitution and organized Vermont as an independent state (under that name) was held here in July, 1777, and in 1778 the first legislature met in the town. Machinery and rubber products are made. Old Constitution House has been restored as an inn.

Windsor Locks, town (1975 est. pop. 15,000), Hartford co., N Conn., on the Connecticut River; settled 1663, set off from Windsor and inc. 1854. A tobacco-farming center since the 17th cent., it also has a large aircraft plant and paper industries. The town developed industrially after a canal, with locks, was built (1829) around the rapids.

Wind·ward Channel (wĭnd'wərd), strait, c.50 mi (80 km) wide, between Cuba and Haiti, connecting the Atlantic Ocean and Caribbean Sea. It provides a direct route from the eastern United States to the Panama Canal.

Wind·ward Islands (wĭnd'wərd), S group of the Lesser Antilles in the West Indies, curving generally S for c.300 mi (480 km) from the Leeward Islands toward NE Venezuela. The Windward Islands consist of French Martinique, Grenada, and the British Windward Islands (c.700 sq mi/1,810 sq km). The British islands consist of Dominica, St. Lucia, and St. Vincent. Since 1967 the islands have been associated states of the United Kingdom. Of volcanic origin, the islands are generally rugged, mountainous, and well forested, and they have many streams and lakes. They produce a variety of tropical agricultural crops for export including bananas, spices, limes, and cacao. The islands are subject to hurricanes in the autumn. The most encouraging development has been the growth of the tourist trade. The culture varies from island to island, but the French influence is strong and many inhabitants speak a French patois. The long struggle for dominance in the islands was a good part of the worldwide Anglo-French conflict. It was only after the close of the Napoleonic Wars that the Congress of Vienna established the present ownership.

Win·field (wĭn'fēld'). **1.** City (1975 est. pop. 11,000), seat of Cowley co., S central Kansas, on the Walnut River; inc. 1873. The surrounding area is basically agricultural, with some petroleum production. Among the city's manufactures are crayons, gas burners, steel drums, industrial boilers, and aircraft. **2.** Town (1970 pop. 328), seat of Putnam co., W W.Va., on the Kanawha River NW of Charleston, in an agricultural and coal-mining area.

Wink·ler (wĭngk'lər), county (1970 pop. 9,640), 887 sq mi (2,297.3 sq km), W Texas, bounded on the N by the N.Mex. border, with Cap Rock escarpment in the E; formed 1887; co. seat Kermit. It has oil and natural-gas fields and large-scale cattle ranching.

Winn (wĭn), parish (1970 pop. 16,369), 950 sq mi (2,460.5 sq km), N central La., bounded on the W by Saline Bayou, on the SE by Bayou Castor, and intersected by the Dugdemona River; formed 1852; parish seat Winnfield. Its agriculture consists of corn, cotton, hay, sweet potatoes, truck crops, and peanuts. There are salt mines, limestone quarries, lumber mills, and some manufacturing.

Win·ne·ba (wĭ-nā'bə), town (1970 pop. 30,800), S Ghana, a fishing port on the Gulf of Guinea. Coconuts, cassava, and corn are raised. The town was an important trading and export center, but after the opening of the harbor at Tema in 1961 its commercial traffic declined considerably.

Win·ne·ba·go (wĭn'ə-bā'gō). **1.** County (1970 pop. 246,623), 519 sq mi (1,344.2 sq km), N Ill., bounded on the N by the Wis. border and drained by the Rock, Pecatonica, and Kishwaukee rivers; formed 1836; co. seat Rockford. It has agriculture (dairy products, grain, and livestock) and extensive manufacturing (including machine tools and fasteners). **2.** County (1970 pop. 12,990), 401 sq mi (1,038.6 sq km), N Iowa, on the Minn. border and drained by Lime Creek; formed 1851; co. seat Forest City. It is a prairie agricultural area, specializing in feed grain and livestock. Corn, oats, and sugar beets are grown. Sand and gravel pits are found here, and recreational vehicles are made. **3.** County (1970 pop. 129,946), 448 sq mi (1,160.3 sq km), E central Wis., bounded in the E by Lake Winnebago and drained by the Wolf and Fox rivers; formed 1840; co. seat Oshkosh. It is in a dairying and farming region, with limestone quarries. Its industry includes food processing, textile milling, lumbering, iron foundries, and diverse manufacturing.

Winnebago, Lake, 215 sq mi (557 sq km), E Wis. Fed and drained by the Fox River, the lake is part of an all-water route between the Great Lakes and the Mississippi River. Oshkosh and Fond du Lac are on the lake, which is a major recreation area.

Win·ne·muc·ca (wĭn'ə-mŭk'ə), city (1970 pop. 3,587), seat of Humboldt co., N Nev., on the Humboldt River NE of Reno; inc. 1917. Founded in 1850 as a trading post called French Ford, it was renamed when the Central Pacific RR reached here in 1868. The city is the center of a ranching and mining area.

Win·ner (wĭn'ər), city (1970 pop. 3,789), seat of Tripp co., S S.Dak., SSE of Pierre. It was founded after 1908 when the eastern part of the Rosebud Indian Reservation was opened to white settlement. Winner is a shipping and trade center for a dairy, grain, and livestock area.

Win·ne·shiek (wĭn'ə-shēk'), county (1970 pop. 21,758), 688 sq mi (1,781.9 sq km), NE Iowa, on the Minn. border and drained by the Upper Iowa and Turkey rivers; formed 1847; co. seat Decorah. It is a prairie agricultural area where cattle, hogs, and poultry are raised, and corn, oats, and hay are grown. Farm products are made here, and there are many limestone quarries.

Win·net·ka (wĭ-nĕt'kə), village (1979 est. pop. 14,800), Cook co., NE Ill., a residential suburb of Chicago, on Lake Michigan; inc. 1869. A correspondence school for the blind is here.

Win·nett (wĭn'ĕt, -ĭt), town (1970 pop. 271), seat of Petroleum co., central Mont., on a branch of the Musselshell River E of Lewistown. It is a shipping point in a wheat region. Livestock is raised, and there are oil wells in the area.

Winn·field (wĭn'fēld'), city (1970 pop. 7,142), seat of Winn parish, N central La., NNW of Alexandria, in a pine-woods region; inc. 1855. Salt mines and limestone quarries are nearby. Kisatchie National Forest is to the west.

Win·ni·bi·go·shish, Lake (wĭn'ĭ-bĭ-gō'shĭsh), 179 sq mi (464 sq km), N central Minn., in Chippewa National Forest, E of Bemidji. The outlet of the lake, one of the largest reservoirs of the Mississippi headwaters, is dammed.

Win·ni·peg (wĭn'ə-pĕg), city (1971 pop. 246,246; 1972 est. metropolitan area pop. 548,573), provincial capital, SE Man., Canada, at the confluence of the Red and Assiniboine rivers. It is the largest city of the Prairie Provinces and one of the world's largest wheat markets. A railroad, commercial, industrial, and distribution center, it has an international airport, railroad shops, grain elevators, stockyards, meat-packing plants, flour mills, and varied manufacturing industries. The city's history reflects the history of early French and British explorers and fur traders. Several posts had been built in the region, including those of fierce rivals, the Hudson's Bay Company and the North West Company. The two companies were merged in 1821. Fort Gibraltar, a post of the North West Company on the site of present-day Winnipeg, was renamed Fort Garry and became the leading post in the region. In 1835 its name was changed to Winnipeg. Settlement was spurred by the construction of a rail line in 1881. The Univ. of Manitoba is here.

Winnipeg, river, c.200 mi (320 km) long, issuing from the N end of Lake of the Woods, SW Ont., Canada, and flowing in a winding course generally NW to the SE end of Lake Winnipeg, SE Man. There are six hydroelectric stations on its course. The river was first traveled by the sons of Vérendrye, the Canadian explorer, and was much used by explorers and fur traders.

Winnipeg, Lake, third-largest lake of Canada, 9,465 sq mi (24,514 sq km), 264 mi (425 km) long and from 25 to 68 mi (40–109 km) wide, S central Man., Canada, N of Winnipeg. It is a remnant of glacial Lake Agassiz. It receives the Red, Winnipeg, and Saskatchewan rivers and is drained northeast by the Nelson River to Hudson Bay. It is surrounded by valuable timber land; there are several summer resorts on its shores. The lake has extensive fishing resources. It was discovered in 1733 and was an important route of early explorers and fur traders.

Win·ni·pe·go·sis, Lake (wĭn'ə-pə-gō'sĭs), 2,086 sq mi (5,403 sq km), 125 mi (201 km) long and 25 mi (40 km) wide, W Man., Canada. It is a remnant of glacial Lake Agassiz. It drains southeast into Lake Manitoba and thence into Lake Winnipeg. Lake Winnipegosis has important pike fisheries.

Win·ni·pe·sau·kee, Lake (wĭn'ə-pə-sô'kē), 71 sq mi (184 sq km), E central N.H. The lake is irregular in shape, with many indentations. Lake Winnipesaukee drains into the Merrimack River. The region around the lake is a popular summer resort.

Winns·bor·o (wĭnz'bûr'ō). **1.** Town (1970 pop. 5,349), seat of Franklin parish, NE La., SE of Monroe; founded c.1844, inc. 1902. It is a trading center and has cotton gins, sawmills, and oil wells. **2.** City (1970 pop. 3,411), seat of Fairfield co., N central S.C., N of Columbia; settled in the mid-18th cent., inc. 1816. Textile, meat, and lumber products are made.

Wi·no·na (wĭ-nō'nə), county (1970 pop. 44,409), 620 sq mi (1,605.8 sq km), SE Minn., bounded on the E by the Mississippi River and Wis.; formed 1854; co. seat Winona. In an agricultural area yielding dairy products, livestock, poultry, corn, oats, barley, and potatoes, it has some mining and a food-processing industry. Textiles, furniture, paper products, plastics, concrete, stone and metal products, and machinery are manufactured.

Winona. 1. City (1979 est. pop. 25,800), seat of Winona co., SE Minn., on the Mississippi River; inc. 1857. Automotive products, flour, metal, and heavy road equipment are made here. An early trading and lumber center, Winona grew as river traffic increased, and the city developed as a manufacturing and commercial center. **2.** City (1970 pop. 5,521), seat of Montgomery co., N central Miss., E of Greenwood, in a farm and livestock area. Cotton and dairy products are made.

Wi·noo·ski (wĭ-nōōs'kē), city (1970 pop. 7,309), Chittenden co., NW Vt., on the Winooski River NE of Burlington; set off from Colchester 1922. Its development began shortly after the Revolution, with the building of mills at the falls here. Dresses, furniture, and wire are among its manufactures.

Wi·noos·ki (wə-nōōs'kē), river, 90 mi (145 km) long, rising in NE Vt. and flowing SW, then NW, across the state, passing Montpelier and Waterbury and entering Lake Champlain near Burlington and Winooski. There are flood-control and hydroelectric power projects on the river.

Wins·low (wĭnz'lō). **1.** City (1970 pop. 8,066), Navajo co., NE Ariz., near the Little Colorado River ESE of Flagstaff, in a cattle-raising and lumbering region; founded 1880, inc. 1900. Nearby is Meteor Crater, a depression nearly a mile (1.6 km) in diameter and c.600 ft (185 m) deep. **2.** Town (1970 pop. 7,299), Kennebec co., S Maine, on the Kennebec River opposite Waterville; settled 1764, inc. 1771. The town manufactures paper products, raises poultry, and processes milk. A blockhouse, all that remains of Fort Halifax (1754), is nearby.

Win·ston (wĭn'stən). **1.** County (1970 pop. 16,654), 615 sq mi (1,592.9 sq km), NW Ala., drained by branches of Sipsey Fork; formed 1850, name changed from Hancock in 1858; co. seat Double Springs. It has cotton, melon, and poultry farming, lumber sawmilling, and some manufacturing (clothing, furniture, and metal products). William B. Bankhead National Forest extends throughout the county. **2.** County (1970 pop. 18,406), 606 sq mi (1,569.5 sq km), E central Miss., drained by the Noxubee River and the headwaters of the Pearl River; formed 1833; co. seat Louisville. Farming (cotton and corn), dairying, and lumbering are the major occupations.

Win·ston-Sa·lem (wĭn'stən-sā'ləm), city (1979 est. pop. 144,400), seat of Forsyth co., central N.C., in the Piedmont; inc. 1913. Winston-Salem is the nation's chief tobacco manufacturer. The village of Bethabara, the first Moravian settlement in the state, was established nearby in 1753. In 1766 the Moravians built their central town, Salem, a few miles away, and most of the industries and residents of Bethabara moved there. Winston was established in 1849 as the county seat. The two communities were united in 1913. Moravian culture has been sustained in the city through long-range efforts to restore the 18th cent. village of Old Salem (some 40 buildings erected between 1767 and 1811 are extant). Also of interest is historic Bethabara Park. Winston-Salem is the seat of Wake Forest Univ., Winston-Salem State Univ., Salem College, North Carolina School of the Arts, a technical institute, a school of medicine, a Bible college, and Salem Academy (est. 1772).

Win·ter Ha·ven (wĭn'tər hā'vən), resort city (1970 pop. 16,136), Polk co., central Fla; settled 1883. It is a marketing, processing, and shipping center for one of the state's chief citrus-fruit regions. There are 97 lakes within a radius of 5 mi (8 km) and a boat course of some 17 connected lakes. Tourist attractions include an annual citrus festival and the nearby Cypress Gardens, a famed water-skiing center.

Winter Park, residential and resort city (1970 pop. 21,895), Orange co., central Fla., just N of Orlando in a citrus area; settled in the 1850s, inc. 1887. Within the city are 12 lakes, some of which are connected by navigable canals. The city is known for its large oak trees and parks.

Win·ter·set (wĭn'tər-sĕt'), city (1970 pop. 3,654), seat of Madison co., SW Iowa, SW of Des Moines; founded 1846, inc. 1876. It is a shipping center for grain and livestock.

Win·ter·thur (vĭn'tər-to͝or'), city (1972 est. pop. 91,400), Zürich canton, N Switzerland. An industrial center, it is an important rail junction and has manufactures of railroad equipment (including locomotives and diesel engines), cotton textiles, clothes, and other goods. It is also a cultural center with an old music festival and two excellent art collections. Winterthur was ruled by the counts of Kyburg (whose castle stands south of the city) until 1264, when it passed to the Hapsburgs. It became a free city of the Holy Roman Empire in 1415 and in 1467 was bought by Zürich.

Win·throp (wĭn'thrəp), residential and resort town (1979 est. pop. 19,800), Suffolk co., E Mass.; settled 1635, inc. 1852. Several houses of historical interest (17th–18th cent.) still stand, including Gov. John Winthrop's house.

Win·ton (wĭn'tən), town (1970 pop. 917), seat of Hertford co., NE N.C., SE of Murfreesboro on the Chowan River, in a sawmilling area.

Wir·ral (wîr'əl), urban district (1972 est. pop. 27,170), Merseyside, NW England, on a peninsula between the Mersey and Dee estuaries.

Wirt (wûrt), county (1970 pop. 4,154), 235 sq mi (608.7 sq km), W W.Va., drained by the Little Kanawha River; formed 1848; co. seat Elizabeth. Its agriculture includes livestock, poultry, dairy, grain, fruit, truck crops, and tobacco. It has some oil and lumber and manufactures transportation equipment.

Wis·cas·set (wĭs-kăs'ĕt, -ĭt), town (1970 pop. 2,244), seat of Lincoln co., S Maine, NE of Bath; settled 1663, inc. 1802. A resort town, it has colonial homes and a brick courthouse built in 1824. The town flourished in the sailing-ship days.

Wis·con·sin (wĭs-kŏn'sən, -sĭn), state (1975 pop. 4,593,000), 56,154 sq mi (145,439 sq km), including 1,449 sq mi (3,753 sq km) of inland water surface, N central United States, admitted as the 30th state of the Union in 1848. Madison is the capital. Water marks most of Wisconsin's boundaries; to the east lies Lake Michigan and to the north, Lake Superior; between the two, in the northeast corner, Wisconsin is divided from the Upper Peninsula of Michigan partly by the Menominee River. To the west the St. Croix and the Mississippi mark most of the boundary with Minn. and Iowa, and to the south lies Ill.

Despite the tempering effects of the water, winters in the north are cold and summers in the south are warm; however, most of Wisconsin is known for its pleasantly cool summer temperatures. The most notable feature of the state is its profusion of lakes, over 8,500, ranging in size from Lake Winnebago (215 sq mi/557 sq km) to relatively tiny glacial lakes of surprising beauty. The Wisconsin River, with its extensive dam system, runs generally southward through the middle of the state until it turns west (just northwest of Madison) to flow into the Mississippi. Running a parallel course just to the east, Wisconsin's major watershed extends in a broad arch from north to south.

Wisconsin's frontage on both Lakes Superior and Michigan, as well as its many beautiful lakes, streams, and its northern woodlands, has made it a haven for hunters, fishermen, and water and winter sports enthusiasts. Tourism is a growing industry.

Wisconsin is divided into 72 counties:

NAME	COUNTY SEAT	NAME	COUNTY SEAT
Adams	Friendship	Burnett	Grantsburg
Ashland	Ashland	Calumet	Chilton
Barron	Barron	Chippewa	Chippewa Falls
Bayfield	Washburn	Clark	Neillsville
Brown	Green Bay	Columbia	Portage
Buffalo	Alma	Crawford	Prairie du Chien
Dane	Madison	Oconto	Oconto
Dodge	Juneau	Oneida	Rhinelander
Door	Sturgeon Bay	Outagamie	Appleton
Douglas	Superior	Ozaukee	Port Washington
Dunn	Menomonie	Pepin	Durand
Eau Claire	Eau Claire	Pierce	Ellsworth
Florence	Florence	Polk	Balsam Lake
Fond du Lac	Fond du Lac	Portage	Stevens Point
Forest	Crandon	Price	Phillips
Grant	Lancaster	Racine	Racine
Green	Monroe	Richland	Richland Center
Green Lake	Green Lake	Rock	Janesville
Iowa	Dodgeville	Rusk	Ladysmith
Iron	Hurley	St. Croix	Hudson
Jackson	Black River Falls	Sauk	Baraboo
Jefferson	Jefferson	Sawyer	Hayward
Juneau	Mauston	Shawano	Shawano
Kenosha	Kenosha	Sheboygan	Sheboygan
Kewaunee	Kewaunee	Taylor	Medford
La Crosse	La Crosse	Trempealeau	Whitehall
Lafayette	Darlington	Vernon	Viroqua
Langlade	Antigo	Vilas	Eagle River
Lincoln	Merrill	Walworth	Elkhorn
Manitowoc	Manitowoc	Washburn	Shell Lake
Marathon	Wausau	Washington	West Bend
Marinette	Marinette	Waukesha	Waukesha
Marquette	Montello	Waupaca	Waupaca
Menominee	Keshena	Waushara	Wautoma
Milwaukee	Milwaukee	Winnebago	Oshkosh
Monroe	Sparta	Wood	Wisconsin Rapids

Economy. The nation's largest dairy herds graze in Wisconsin, and the state leads in the production of milk and cheese. After dairy products and cattle, its most valuable farm commodities are hogs and corn. Wisconsin also ranks first in the production of hay and alfalfa; other important crops are oats, potatoes, cranberries, and a great variety of fruits and vegetables. Food processing, predictably, is one of the state's foremost industries, surpassed only by the manufacture of machinery, which is centered in Milwaukee, Madison, and Racine. The pulp, paper, and paper-products industrial complex in Green Bay and Appleton is the largest (1970) in the nation. Other important manufactures are transportation equipment, metal products, farm implements, and lumber. Almost all Wisconsin's major industries are to be found within Milwaukee, where the traditional industries of beer brewing and meat-packing are rivaled by the manufacture of heavy machinery and diesel and gasoline engines.

Iron ore, sand and gravel, stone, and zinc are the chief mineral resources. Important copper deposits were discovered in the north in the early 1970s. The state's greatest natural resource since its earliest days has been lumber. Giant forests (white pines in the north, hardwoods elsewhere) once covered all except the southern prairie. Reforestation measures have saved the valuable lumber industry, and today c.45% of Wisconsin's land area is forested. Wisconsin has 14 ports on the Great Lakes, all capable of accommodating ocean vessels. The superb harbor at Superior has sizable shipyards and among the nation's largest coal and ore docks.

History. The Frenchman Jean Nicolet arrived at the site of Green Bay in 1634 in search of fur pelts and the Northwest Passage. He was followed by other traders and missionaries. Meanwhile the growth of white civilization in the East was bringing the Ottawa, the Huron, and other Indian tribes into Wisconsin, where they in turn unsettled the older Indian inhabitants. Only the Menominee remained relatively settled. Nicolas Perrot helped (1667) establish Green Bay as the center of the Wisconsin fur trade, and in 1686 he formally claimed all the region for France. The fur trade flourished despite the 50-year war between the Fox Indians and the French. Like all of New France, Wisconsin fell to the British with the end of the French and Indian Wars (1763). The British hold continued even after the end of the American Revolution. After Jay's Treaty (1794) northwestern strongholds were turned over to the Americans, but the British continued to dominate the fur trade from the Canadian border. In the War of 1812 Wisconsin again fell into British hands. It was only with the Treaty of Ghent (1814) that effective U.S. territorial control began.

Present-day Wisconsin was transferred from Illinois Territory to Michigan Territory in 1818. After 1825 a considerable number of easterners began arriving via the new Erie Canal and the Great Lakes to work the lead mines and clear farmland. They settled in the Milwaukee area and along the waterways. The incursions of aggressive white settlers brought trouble with the Indians, culminating in the Black Hawk War (1832); this revolt, brutally crushed, was the last major Indian uprising in the area. In 1836 Wisconsin was made a territory, and the capital city of Madison was founded. Wisconsin achieved statehood in 1848. The influx of German immigrants after 1848 was especially heavy, and some parts of Wisconsin assumed the tidy semi-German look that has persisted.

The meat-packing and brewing industries of Milwaukee began to assume importance in the 1860s. The state's great dairy industry developed, spurred by an influx of skilled dairy farmers from New York and Scandinavia. In the 1870s lumbering became the state's most important industry. Oshkosh and La Crosse flourished. With lumbering came large paper and wood products industries.

A trend toward liberal political views was stimulated in Wisconsin by socialist thought, which was introduced early. In the early 20th cent., reform sentiment blossomed in the Progressive movement, under the tutelage of the Republican leader, Robert M. La Follette. The movement resulted in a direct primary law (1903), in legislation to regulate railroads and industry, in pure food acts, in high civil service standards, and in efforts toward cooperative nonpartisan action to solve labor problems. Wisconsin was generally prosperous in the 1920s; industrialization made rapid strides. The Great Depression of the 1930s struck particularly hard in industrialized Milwaukee. During World War II Wisconsin's shipbuilding industry flourished. In the prosperous postwar era, urbanization and industrial growth continued.

Government. Wisconsin still operates under its first constitution, adopted in 1848. Its executive branch is headed by a governor elected for a four-year term (before 1970 it was for a two-year term). Wisconsin's bicameral legislature has a senate with 33 members elected for four-year terms and an assembly with 99 members elected for two years.

Educational Institutions. The extensive Univ. of Wisconsin has campuses at Madison, Eau Claire, Green Bay, La Crosse, Milwaukee, Oshkosh, Kenosha, Platteville, River Falls, Stevens Point, Menomonie, Superior, and Whitewater. Other notable institutions of higher learning are Beloit College, at Beloit; Lawrence Univ., at Appleton; and Marquette Univ., at Milwaukee.

Wisconsin, river, c.430 mi (690 km) long, rising in the lake district, NE Wis., and flowing generally SW across central Wis. to the Mississippi River near Prairie du Chien. At Portage it is connected by a short canal with the Fox River, and thus with Lake Michigan. There are many hydroelectric power facilities on the river. The scenic Dells of the Wisconsin are a famous gorge.

Wisconsin Rap·ids (răp'ĭdz), city (1979 est. pop. 18,000), seat of Wood co., central Wis., on the Wisconsin River; inc. 1869. Paper and pulp, plastics, chemicals, and iron and steel are produced. Dairy farms, agriculture, and a large cranberry industry also contribute to the city's economy. Two towns on the river here, Grand Rapids (east bank) and Centralia (west bank), were consolidated in 1900, and the name was changed in 1920 to Wisconsin Rapids. A state-owned nursery is nearby.

Wise (wīz). **1.** County (1970 pop. 19,687), 922 sq mi (2,388 sq km), N Texas, drained by the West Fork of the Trinity River; formed 1856; co. seat Decatur. It has diversified farming (grain, peanuts, corn, cotton, hay, fruit, and truck crops), livestock raising, extensive dairying, clay mining, and limestone quarrying. Some manufacturing is done. **2.** County (1970 pop. 35,947), 412 sq mi (1,067.1 sq km), SW Va., in the Alleghenies, with the Cumberlands in the NW along the Ky. border, bounded in the SE by the Clinch River and drained by the Powell and Pound rivers; formed 1856; co. seat Wise. Its agriculture includes grain, livestock, fruit, tobacco, and dairy products. It has coal mining and food-processing, clothing, and lumbering industries. It includes mountain resorts and part of the Jefferson National Forest.

Wise, town (1970 pop. 2,891), seat of Wise co., extreme SW Va., in the Cumberlands NW of Bristol, near the Ky. border. It is the center of an area of coal mines and apple orchards.

Wis·mar (vĭs'mär), city (1973 est. pop. 56,724), Rostock district, NW East Germany, on the Baltic Sea. It is an industrial center and an oil and fishing port. Manufactures include machinery and processed food, and there are shipyards in the city. Wismar was (1256–1306) the residence of the princes of Mecklenburg and later became one of the most flourishing members of the Hanseatic League. Under the Peace of Westphalia (1648) the city passed to Sweden, but in 1803 Sweden pledged it to Mecklenburg-Schwerin with the privilege of recall within 100 years. In 1903 Sweden renounced all rights to the city. Wismar was badly damaged in World War II. There are several Gothic churches and many medieval houses.

Wit·bank (wĭt'băngk), town (1972 est. pop. 37,456), Transvaal, NE South Africa. It is the industrial center of South Africa's most important coalfield. Aside from coal-processing plants, the town has plants producing carbide, cyanide of potassium, steel, and vanadium. A miners' training college is in Witbank.

Wit·ten (vĭt'ən), city (1972 est. pop. 96,566), North Rhine-Westphalia, W West Germany, on the Ruhr River. It is a modern industrial city, manufacturing iron and steel, glass, and chemicals. Witten was first mentioned in the 13th cent. and was chartered in 1825.

Wit·ten·berg (wĭt'n-bûrg', vĭt'ən-bĕrKH'), city (1973 est. pop. 48,614), Halle district, central East Germany, on the Elbe River. A city with a noted history, it is today an industrial and mining center and a rail junction. Manufactures include machinery, chemicals, processed food, and rubber products. Wittenberg was (1273–1422) the seat of the Ascanian dukes of Saxe-Wittenberg, who in 1356 became electors of Saxony. In 1423 Saxe-Wittenberg passed to the margraves of Meissen, who in 1425 were given electoral rank. Elector Frederick III founded (1502) the Univ. of Wittenberg, which became the center of the Protestant Reformation when Martin Luther and Philip Melanchthon taught here. In 1517 Luther nailed his 95 theses on the door of the Schlosskirche, and in 1520 he burned the papal bull against him outside the Elster gate. The first complete Lutheran Bible was printed (1534) at Wittenberg. In 1547 Emperor Charles V captured Wittenberg after the Battle of Mühlberg, where elector John Frederick I of Saxony was captured. By the Capitulation of Wittenberg, in the same year, John Frederick, representing the Ernestine line of the house of Wettin, ceded the electoral dignity and the duchy of Saxony to Maurice, of the collateral Albertine line. Primarily the focus of the Lutheran Reformation, 16th cent. Wittenberg was also a center of German art. Lucas Cranach, the elder, founded a school of painting here. The city declined after 1547, when Dresden, residence of the Albertine dukes, replaced it as Saxon capital. In 1815 Wittenberg passed to Prussia, and in 1817 the Univ. of Wittenberg was absorbed by the Univ. of Halle. Among Wittenberg's most notable structures are the Schlosskirche (15th cent.), where Luther and Melanchthon are buried; the town church (14th-15th cent.), where Luther preached; the houses where Luther, Melanchthon, and Lucas Cranach, the elder, lived; and the city hall (16th cent.).

Witt·stock (vĭt'shtôk'), town (1970 pop. 10,606), Potsdam dist., N central East Germany. Manufactures include woolen textiles, machinery, and forest products. At Wittstock in 1636 the Swedes under Baner defeated the Saxon and Imperial forces under Melchior Hatzfeldt in an important battle of the Thirty Years' War.

Wit·wa·ters·rand (wĭt-wô'tərz-rănd', -wŏt'ərz-) or **the Rand** (rănd), region, Transvaal, Republic of South Africa. The area, which forms the watershed between the Vaal and Olifants rivers, is c.25 mi (40 km) wide and extends more than 60 mi (100 km) from west to east in a series of parallel ranges 5,000 to 6,000 ft (1,525–1,830 m) above sea level. The Rand is one of the world's richest gold-mining regions. The gold occurs in reefs, or thin bands, that are mined at depths of up to 10,000 ft (3,050 m). Development of the Rand dates from the 1880s. Although many of the older mines are now nearly exhausted, the Rand still produces more than two thirds of South Africa's gold and almost one third of total world output. Silver and iridium are recovered as gold-refining by-products, and the region also has coal mines. The Rand also has such industries as engineering, steelmilling, metallurgy, machine building, diamond cutting, food processing, and the manufacture of chemicals, cement, furniture, and clothing.

Wło·cła·wek (vlô-tslä'vĕk), city (1973 est. pop. 85,000), central Poland, a port on the Vistula (Wisła) River. It is an agricultural market center and has industries producing farm machinery, chemicals, metalware, paper, cellulose, and pottery. Nearby are salt domes, lignite deposits, and sulfur springs. The city was founded in the 12th cent., passed to Russia in 1815, and reverted to Poland after World War I. Landmarks include a 14th cent. Gothic cathedral and a 17th cent. episcopal palace.

Wo·burn (wo͝o'bərn), village, Bedfordshire, S central England. It is famous for Woburn Abbey (seat of the dukes of Bedford), an 18th cent. mansion constructed on the site of a Cistercian Abbey founded in 1145. It contains a noteworthy art collection with many classical works brought from Rome in the 18th cent.

Wo·burn (wō'bərn, wo͝o'-), city (1979 est. pop. 33,300), Middlesex co., NE Mass.; settled 1640, inc. as a city 1888. It has electrical, chemical, and leather industries.

Wo·king (wō'kĭng), urban district (1972 est. pop. 77,290), Surrey, SE England. Woking is primarily a residential suburb of London. There are printing, rubber, and packing industries. Of interest is a mosque built in 1889, headquarters of the Moslem Society of Great Britain.

Wol·cott (wo͝ol'kət), town (1970 pop. 12,495), New Haven co., central Conn.; inc. 1796. Tools and novelties are made. Bronson Alcott was born nearby.

Wolfe (wo͝olf), county (1970 pop. 5,669), 227 sq mi (588 sq km), E Ky., in the Cumberland foothills, drained by the Red River and the North Fork of the Kentucky River; formed 1860; co. seat Campton. It is in a mountain agricultural area that produces livestock, fruit, and tobacco and has some mining and manufacturing. It includes Natural Bridge State Park and part of Cumberland National Forest.

Wol·fen·büt·tel (vôl'fən-bü'təl), city (1972 est. pop. 40,291), Lower Saxony, E West Germany, on the Oker River, near the East German border. It is an agricultural market and an industrial center. Manufactures include machinery, chemicals, and liquor. Wolfenbüttel developed around an 11th cent. castle that became (c.1280) a favorite Guelphic residence. The famous library, founded in the 17th cent., contains about 350,000 volumes, some 3,000 incunabula, and more than 8,000 manuscripts.

Wolf National Scenic Riverway (wo͝olf): *see* National Parks and Monuments Table.

Wolf Point, city (1970 pop. 3,095), seat of Roosevelt co., NE Mont., on the Missouri River N of Miles City; settled 1878, inc. 1915. A shipping center for a wheat and cattle region, it is also a servicing point for nearby oil fields. Fort Peck Indian Reservation is to the north.

Wolfs·burg (vôlfs'bo͝orKH), city (1972 est. pop. 130,156), Lower Saxony, NE West Germany, on the Midland Canal. A small village in 1937, Wolfsburg grew and prospered as the headquarters of the Volkswagen automobile company.

Wolf Trap Farm Park: *see* National Parks and Monuments Table.

Wol·las·ton Lake (wo͝ol'ə-stən), 796 sq mi (2,062 sq km), NE Sask., Canada, NW of Reindeer Lake. It drains into both the Churchill and the Mackenzie river systems.

Wol·lin or **Wo·lin** (both: vô'lēn'), island, 95 sq mi (246 sq km), off the coast of Pomerania, in the Baltic Sea, and belonging to Poland. Wollin is separated from the mainland by the Zalew Szczeciński (Stettiner Haff). It is generally a lowland, with forests and several lakes. Fishing and livestock raising are the chief industries. There are numerous bathing resorts. The principal town, Wollin, is a fishing port. A fortress and Slavic settlement once occupied the site of the town, whose history dates from the 9th cent. The island passed to Sweden in 1648, to Prussia in 1721, and to Poland after World War II. It is administratively part of Szczecin prov.

Wol·lon·gong (wo͝ol'ən-gông'), city (1973 est. pop. 165,240; urban agglomeration pop. 205,780), New South Wales, SE Australia. It is an important iron and steel center. There are other industries, including copper refining and textile and chemical manufacturing. Port Kembla, which was absorbed by Wollongong in 1947, is a major port. A branch of the Univ. of New South Wales is in Wollongong.

Wol·ver·hamp·ton (wo͝ol'vər-hămp'tən), county borough (1972 est. pop. 269,460), West Midlands, W central England, in the Black Country. The area of the borough was greatly enlarged in 1966. It is highly industralized; its products include automobiles, hardware, rayon, tires, and chemicals. St. Peter's Church in Wolverhampton dates mostly from the 13th and 15th cent. Its grammar school was established in 1512. There are two teacher-training schools and a technical school.

Won·ju (wŭn'jo͞o'), city (1970 pop. 111,972), N South Korea. It is an agricultural center.

Won·san (wŭn'sän), city (1967 est. pop. 215,000), capital of Kangwon prov., SE North Korea, on the Sea of Japan. It is a major port and naval base, with a natural harbor protected by a line of islands. The city has important fish and fish-processing industries. Oil refining, rice processing, and the manufacture of locomotives, textiles, and leather goods are also important. Opened to foreign trade in 1883, Wonsan became a Japanese naval base in World War II. It suffered heavy damage during the Korean War.

Wood (wo͝od). **1.** County (1970 pop. 89,722), 619 sq mi (1,603.2 sq km), NW Ohio, bounded in the NW by the Maumee River and intersected by the Portage River; formed 1820; co. seat Bowling Green. It has diversified farming (corn, wheat, livestock, poultry, oats, and fruit) and limestone quarries. Its industry includes food processing, lumbering, iron and steel foundries, and the manufacture of paper, chemicals, plastics, concrete, metal products, and machinery. **2.** County (1970 pop. 18,589), 721 sq mi (1,867.4 sq km), NE Texas, bounded on the SW and S by the Sabine River; formed 1850; co. seat Quitman. Its agriculture includes fruit, truck crops, sweet potatoes, corn, legumes, cotton, hay, dairy products, poultry, cattle, hogs, and horses. There is extensive lumbering, and the county has deposits of oil, natural gas, lignite, and clay. **3.** County (1970 pop. 86,818), 368 sq mi (953.1 sq km), W W.Va., bounded on the W by the Ohio River and the Ohio border and drained by the Little Kanawha River; formed 1798; co. seat Parkersburg. It has oil and natural-gas wells, deposits of clay, and some coal mining. Its agriculture includes livestock, tobacco, and truck crops. There is extensive manufacturing (boxes, paper products, chemicals, plastics, rubber, glassware, pottery, concrete, and metal products). **4.** County (1970 pop. 65,362), 807 sq mi (2,090.1 sq km), central Wis., intersected by the Wisconsin River and drained by the Yellow River; formed 1856; co. seat Wisconsin Rapids. It is in a dairying, cranberry-growing, and general farming area. Its industry includes cheese processing, lumbering, and the manufacture of wood products, furniture, paper, concrete, metal products, and farm machinery.

Wood·bine (wo͝od'bīn'), city (1970 pop. 1,002), seat of Camden co., extreme SE Ga., SW of Brunswick on the Satilla River, in a fishing and sawmilling area. Boxes are manufactured.

Wood·burn (wo͝od'bərn), city (1970 pop. 7,495), Marion co., NW Oregon, NNE of Salem in the Willamette River valley; platted 1871, inc. 1889. It is the processing center of a farm region.

Wood·bur·y (wo͝od'bĕr'ē, -bə-rē), county (1970 pop. 103,052), 871 sq mi (2,255.9 sq km), W Iowa, bounded on the W by the Big Sioux and Missouri rivers and drained by the Little Sioux and the West Fork Little Sioux rivers; formed 1851; co. seat Sioux City. Livestock are raised, and corn, oats, and wheat are grown. Industries include meat-packing, grain milling, and the manufacture of lumber, chemicals, beverages, textiles, and paper products.

Woodbury. **1.** Town (1970 pop. 5,869), Litchfield co., W Conn., W of Waterbury, in a dairy region; settled 1672, chartered 1674. Electrical equipment is made. Among its old buildings are several churches and Globe House (c.1745), where Samuel Seabury was elected first presiding bishop of the Protestant Episcopal Church. **2.** Residential city (1975 est. pop. 11,700), seat of Gloucester co., SW N.J., in the Philadelphia-Camden metropolitan area; settled 1683, inc. as a city 1871. Originally a Quaker settlement, Woodbury attempted to stay aloof from the Revolutionary War, but the armies of both sides occupied the town and many battles were fought in the vicinity. The city's 18th cent. buildings include the Cooper House, where Cornwallis stopped in 1777, and a Friends meetinghouse. **3.** Town (1970 pop. 1,725), seat of Cannon co., central Tenn., E of Murfreesboro, in a livestock and dairy region; inc. 1843.

Wood·ford (wŏŏd'fərd). **1.** County (1970 pop. 28,012), 528 sq mi (1,367.5 sq km), central Ill., bounded on the W by Lake Peoria (a widening of the Illinois River) and drained by the Mackinaw River and Crow Creek; formed 1841; co. seat Eureka. It has agriculture (corn, oats, soybeans, wheat, truck crops, livestock, and poultry), bituminous-coal mining, manufacturing of tile and concrete blocks, and a food-processing industry. **2.** County (1970 pop. 14,434), 193 sq mi (500 sq km), E central Ky., in the Bluegrass, bounded on the W by the Kentucky River, on the N and NE by Elkhorn Creek; formed 1788; co. seat Versailles. In a gently rolling agricultural area yielding dairy products, livestock, poultry, burley tobacco, hemp, and bluegrass seed, it has fluorspar mines and industry (distilled liquor, clothing, glassware, and equipment manufacturing).

Wood·land (wŏŏd'lənd), city (1979 est. pop. 27,600), seat of Yolo co., N central Calif., in a rich farm area yielding tomatoes, rice, sugar beets, and alfalfa; inc. 1871. It is a center for the manufacture of mobile homes, with numerous plants and related warehousing operations. The city has many historic homes and is the site of a state historical farm.

Woodlawn, uninc. town (1970 pop. 28,821 including Woodmoor), Baltimore co., N Md., a residential suburb of Baltimore.

Wood·mere (wŏŏd'mîr'), uninc. residential town (1979 est. pop. 19,700), Nassau co., SE N.Y., on Long Island.

Wood·ridge (wŏŏd'rĭj'), village (1979 est. pop. 20,900), Du Page co., NE Ill.; inc. 1959. It is a residential community W of Chicago in a wooded and farm area.

Wood River (wŏŏd), city (1970 pop. 13,186), Madison co., SW Ill., on the Mississippi River just above its junction with the Missouri; inc. 1923. It has oil refineries and pipeline terminals.

Wood·ruff (wŏŏd'rəf), county (1970 pop. 11,566), 591 sq mi (1,530.7 sq km), N central Ark., bounded on the W by the White River and drained by the Cache River; formed 1862; co. seat Augusta. Its agriculture includes cotton, rice, soybeans, fruit, and livestock. It has timber and commercial fisheries. Its industry includes clothing manufacturing, wood products, and some food processing.

Woods (wŏŏdz), county (1970 pop. 11,920), 1,298 sq mi (3,361.8 sq km), NW Okla., bounded in the N by Kansas, in the W and S by the Cimarron River, and drained by the Salt Fork of the Arkansas River; formed 1907; co. seat Alva. In an agricultural area yielding wheat, sorghums, alfalfa, cattle, and dairy products, it has oil and gas wells and some manufacturing.

Woods·field (wŏŏdz'fēld'), village (1970 pop. 3,239), seat of Monroe co., SE Ohio NE of Marietta in an agricultural area; founded 1815, inc. 1835. Coal, vaults, and tools are produced.

Woods Hole, village and seaport in the town of Falmouth, Barnstable co., SE Mass., at the SW extremity of Cape Cod. It is the site of an important marine biology laboratory and of the Oceanographic Institution.

Wood·son (wŏŏd'sən), county (1970 pop. 4,789), 497 sq mi (1,287.2 sq km), SE Kansas, in a dissected plain crossed in the NE by the Neosho River, in the SW by the Verdigris River; formed 1855; co. seat Yates Center. It is in a livestock and grain-growing region, with scattered oil fields and clothing manufacturing.

Wood·stock (wŏŏd'stŏk'). **1.** City (1979 est. pop. 11,600), seat of McHenry co., NE Ill.; inc. 1845. Its manufactures include typewriters and business machines, metal products, and dairy items. **2.** Town (1970 pop. 5,714), Ulster co., SE N.Y., in the foothills of the Catskill Mts. There is an artists' colony (founded 1902), which sponsors a gallery. In 1969 the town was overwhelmed by thousands of youths attending a summer rock music festival. **3.** Town (1970 pop. 2,608), seat of Windsor co., E Vt., on the Ottauquechee River W of White River Junction; chartered 1761, settled 1765. An early publishing center, it is now a resort with many fine old houses and a famous village green. **4.** Town (1970 pop. 2,338), seat of Shenandoah co., N Va., SSW of Winchester in the Shenandoah Valley; settled 1752, inc. 1872. It is a farm trade center with processing plants. The courthouse here dates from 1791. On top of nearby Massanutten Mt. is an observation tower from which the seven horseshoe bends of the Shenandoah River can be seen.

Woods·ville (wŏŏdz'vĭl'), town (1970 pop. 1,500), seat of Grafton co., W N.H., on the Connecticut River NNE of Hanover. Wood products are made.

Wood·ville (wŏŏd'vĭl'). **1.** Town (1970 pop. 1,734), seat of Wilkinson co., extreme SW Miss., near the La. border SSE of Natchez, in a timber area; inc. 1811. **2.** Town (1970 pop. 2,662), seat of Tyler co., E Texas, NNW of Beaumont, in a pine-woods region; settled c.1847, inc. 1929.

Wood·ward (wŏŏd'wərd), county (1970 pop. 15,537), 1,251 sq mi (3,240.1 sq km), NW Okla., intersected by the North Canadian and Cimarron rivers; formed 1907; co. seat Woodward. Its agriculture includes wheat, rye, broomcorn, alfalfa, vegetables, cattle, and dairy products. It has oil and gas wells and manufactures clothing.

Woodward, city (1970 pop. 9,412), seat of Woodward co., NW Okla., on the North Canadian River W of Enid; settled 1893. A market for a livestock, farm, and dairy region, it also has some manufacturing and is the headquarters for many gas and oil companies.

Woo·mer·a-Mar·a·ling·a (wŏŏ'mər-ə-mâr'ə-lĭng'gə), town (1971 pop. 4,069), in the state of South Australia, S Australia, near Lake Torrens. It is the site of a missile-testing range used by Australia and its allies. Australia's first earth satellite was launched here in 1967. Nearby is a U.S. space-tracking station.

Woon·sock·et (wŏŏn-sŏk'ĭt, wŏŏn-). **1.** City (1979 est. pop. 46,800), Providence co., N R.I., on both sides of the Blackstone River; settled c.1666, set off from Cumberland 1867, inc. as a city 1888. The demise of the textile industry, which long shaped the city, has hurt its economy. Worsted weaving and package dyeing are still carried on; new manufactures are electronic equipment, plastics, and sporting goods. The inhabitants are primarily Franco-Americans whose forebears came from Canada. Of interest are the river falls in the center of the city and the unusual potholes worn by swirling stones in the riverbed. Also in Woonsocket are a library in which Abraham Lincoln spoke (1860) and the John Arnold House (1712). **2.** City (1970 pop. 852), seat of Sanborn co., SE S.Dak., S of Huron, in a grain, dairy, and livestock area; settled 1883, inc. 1888.

Woos·ter (wŏŏs'tər), city (1979 est. pop. 20,400), seat of Wayne co., N central Ohio, in a farm area; inc. 1817. Paper, brass, food and rubber products, and camera equipment are made. A state agricultural research and develdpment station is nearby. The city is the seat of The College of Wooster, which was chartered in 1866 and opened in 1870.

Worces·ter (wŏŏs'tər), county borough (1972 est. pop. 74,170), administrative center of Hereford and Worcester, W central England, on the Severn River. The making of porcelain, gloves, and Worcestershire sauce are long-established industries; recently metal goods and machines have also been manufactured. The site became the seat of a bishopric c.680. The present cathedral is chiefly 14th cent., with a Norman crypt and tombs; in it are held, alternately with Hereford and Gloucester, the Festivals of the Three Choirs. There are several old parish churches and timbered houses. The Commandery, restored in 1954, was a hospital in the 11th cent. In the English Civil War Worcester was the scene of Cromwell's final victory with the complete rout of Charles II and the Scots in 1651. There are two very old public schools: Royal Grammar School (13th cent.) and King's School (1541).

Worcester, town (1970 pop. 41,198), Cape Prov., SW South Africa. It produces wine and liquor and processes the fruits and vegetables of the surrounding area. There are textile and metal industries. Worcester's large thermoelectric station powers the electrified railroad that runs through the nearby Hex River Mts. The town was founded in 1820 and was named for the marquess of Worcester, governor of Cape of Good Hope Colony. A technical college and the Drostdy (1825), which is a national monument and the home of the Afrikaner Museum, are in Worcester.

Worcester. **1.** County (1970 pop. 24,442), 479 sq mi (1,241 sq km), SE coastal Md., on the Eastern Shore, bounded on the S by the Va. border, on the SW and NW by the Pocomoke River, on the N by the Del. border; formed 1742; co. seat Snow Hill. It is in a tidewater agricultural area (truck crops, potatoes, fruit, dairy products, and poultry), with softwood timber, lumber mills, and vegetable canneries. Its shores have small fishing towns and resorts. Cypress swamps, including part of Pocomoke State Forest, lie along the Pocomoke River. **2.** County (1970 pop. 637,079), 1,513 sq mi (3,918.7 sq km), central Mass., bordering on N.H., R.I., and Conn., and drained by the Blackstone, Nashua, Assabet, Millers, and Ware rivers; formed 1731; co. seat Worcester. Its rivers yield water power for a number of industrial cities. It also has dairy and truck farms. There are a number of state forests and resort lakes in the county.

Worcester, industrial city (1979 est. pop. 166,800), seat of Worcester co., central Mass., a port of entry on the Blackstone River; inc. 1722. The canalization (1828) of the Blackstone River marked the beginning of Worcester's rapid industrial development. Its manufactures now include machinery, metal goods, chemicals, plastics, pharmaceuticals, electrical equipment, textiles, shoes, and abrasives. There is also a printing and publishing industry. Settled in 1673, the city suffered Indian attacks in 1675 and 1683. In Shays' Rebellion the courthouse was besieged (1786) by insurgents. The first woman's suffrage national convention was held (1850) in Worcester. Edward Everett Hale was pastor here from 1842 to 1856. Worcester is the seat of Clark Univ., Worcester Polytechnic Institute, Worcester State College, College of the Holy Cross, and several junior colleges. It has a number of notable museums, two zoos, and an annual music festival (inaugurated in 1858). Also of interest is a huge three-level shopping center with a Plexiglas dome. In 1953 a severe tornado struck the city, causing the loss of many lives and extensive property damage. Lake Quinsigamond and two state parks are in the vicinity.

Worces·ter·shire (wo͝os'tər-shîr', -shər), former county, W central England. In 1974 Worcestershire became part of the new nonmetropolitan county of Hereford and Worcester.

Wor·king·ton (wûr'kĭng-tən), municipal borough (1972 est. pop. 28,700), Cumbria, NW England, at the mouth of the Derwent River. Workington has a good harbor. Coal mines are in the vicinity, and there are clothing and carpet industries, engineering works, steel mills, and tanneries. In 1974 Workington became part of the new nonmetropolitan county of Cumbria.

Work·sop (wûrk'səp, wûr'-), municipal borough (1972 est. pop. 36,320), Nottinghamshire, central England. The borough contains a portion of Sherwood Forest. It is a coal-mining center with many industries, including foundries, chemical works, glassworks, and breweries. There are remains of an Augustinian priory founded in 1103.

Wor·land (wûr'lənd), city (1970 pop. 5,055), seat of Washakie co., N central Wyo., on the Bighorn River; laid out 1905. In an oil, natural-gas, and irrigated farm area, it has sugar-beet and oil refineries.

Worms (vôrms), city (1972 est. pop. 76,658), Rhineland-Palatinate, S central West Germany, on the Rhine River. It is an industrial city and a leading wine trade center. Manufactures include leather goods, chemicals, and machinery. One of the most venerable historic centers of Europe, Worms was originally a Celtic settlement called Borbetomagus. It was captured and fortified by the Romans under Drusus in 14 B.C. and was known as Civitas Vangionum. It became the capital of the first kingdom of Burgundy in the 5th cent.; much of the *Nibelungenlied* is set in Worms at the Burgundian court. The city was an early episcopal see, and its bishops ruled some territory on the right bank of the Rhine as princes of the Holy Roman Empire until 1803, when the bishopric was secularized and passed to Hesse-Darmstadt. The city itself, however, early escaped episcopal control; in 1156 it was created a free imperial city. Numerous important meetings, including about 100 imperial diets, were held here. The best known of these meetings were the episcopal synod of 1076, which declared Pope Gregory VII deposed; the conference that led in 1122 to the Concordat of Worms; the diet of 1495; and the diet of 1521. The city suffered heavy damage in the Thirty Years' War (1618–48). It was annexed by France in 1797 and passed to Hesse-Darmstadt at the Congress of Vienna (1814–15). Worms was occupied (1918–30) by French troops after World War I. The city was more than half destroyed in World War II, but was reconstructed after 1945. Worms had one of the oldest Jewish settlements in Germany. Its Romanesque-Gothic synagogue, founded in 1034, was destroyed by the Nazis in 1938 but was rebuilt after the war and reopened in 1961. Of note is the city's Romanesque cathedral (11th–12th cent.). Near Worms is the Liebfrauenkirche (13th–15th cent.), a church surrounded by vineyards, which gave its name to the area's noted white wine, Liebfraumilch.

Worth (wûrth). **1.** County (1970 pop. 14,770), 579 sq mi (1,499.6 sq km), bounded on the NW by the Flint River, on the N by Lake Blackshear, and drained by the Ochlockonee River; formed 1853; co. seat Sylvester. It has coastal plain agriculture (peanuts, cotton, tobacco, corn, and melons) and forestry. **2.** County (1970 pop. 8,968), 400 sq mi (1,036 sq km), N Iowa, on the Minn. border, drained by the Shell Rock River; formed 1851; co. seat Northwood. Hogs, cattle, and poultry are raised, and corn, oats, and soybeans are grown in this prairie agricultural region. **3.** County (1970 pop. 3,359), 267 sq mi (691.5 sq km), NW Mo., bordered by Iowa in the N and drained by the Grand River; formed 1861; co. seat Grant City. It is an agricultural region, specializing in livestock.

Worth, village (1971 pop. 12,200), Cook co., NE Ill., a suburb of Chicago; inc. 1914.

Wor·thing (wûr'*th*ĭng), municipal borough (1972 est. pop. 89,090), West Sussex, S England. It is a seaside resort. Protected by the South Downs, the area has an unusually warm climate. Fruits, vegetables, and flowers are cultivated. There are many prehistoric and Roman ruins (including the largest earthwork in England), relics of which are found in the Worthing museum.

Wor·thing·ton (wûr'*th*ĭng-tən). **1.** City (1975 est. pop. 9,900), seat of Nobles co., SW Minn., near the Iowa border ENE of Sioux Falls, S.Dak., in a lake region. It is a farm trade center. **2.** City (1970 pop. 15,326), Franklin co., central Ohio, a suburb of Columbus; settled 1803, inc. 1835. It is the seat of the Pontifical College Josephinum.

Wound·ed Knee (wo͞on'dĭd nē'), creek, rising in SW S.Dak. and flowing NW to the White River; site of the last major battle of the Indian Wars. After the death of Sitting Bull, a band of Sioux Indians, led by Big Foot, fled into the badlands, where they were captured by the 7th Cavalry on Dec. 28, 1890, and brought to the creek. These Indians, adherents of the Ghost Dance religion of the prophet Wovoka, wore "bullet-proof" ghost shirts. On Dec. 29, the Indians were ordered disarmed; but when a medicine man threw dust into the air, a warrior pulled a gun and wounded an officer. The U.S. troops opened fire, and within minutes almost 200 men, women, and children were shot. The soldiers later claimed that it was difficult to distinguish the Sioux women from the men.

Wran·gel Island or **Wran·gell Island** (both: răng'gəl), island, 1,740 sq mi (4,507 sq km), in the Arctic Ocean, between the East Siberian Sea and the Chukchi Sea, off NE USSR. It is separated from the mainland by Long Strait. Generally barren, frozen, and rocky, it has an arctic station and a permanent settlement. The island is a breeding ground for polar bears, polar foxes, seals, and lemmings. During the summer it is visited by numerous varieties of birds. The island was sought by Russian Baron Ferdinand von Wrangel during his arctic expedition of 1820–24; he had heard of it from Siberian natives, but he did not succeed in finding it. It was finally discovered by Thomas Long, captain of an American whaling ship, who named it for Wrangel. Later George W. De Long, an American explorer, discovered that it was a small island and not a part of the mainland, as at first believed. In 1911 a group of Russians made a landing on the island, and in 1921 Vilhjalmur Stefansson, the Canadian explorer, sent a small party to Wrangel with a view to claiming it for Great Britain. In 1926 the Soviet government established the first permanent colony here, ousting the few of Stefansson's Eskimo settlers who had remained. The Soviet freighter *Chelyuskin*, trying to discover (1933) whether an ordinary cargo ship could navigate the Northeast Passage, was crushed in the ice off Wrangel Island. The party was marooned on the island but was later rescued.

Wran·gell Island (răng'gəl), island, 30 mi (48 km) long and 5 to 14 mi (8.1–22.5 km) wide, off SE Alaska in the Alexander Archipelago, S of the mouth of the Stikine River. It was occupied in 1834 by Russians, who named it for Ferdinand von Wrangel. The city of Wrangell, on the northern coast, grew around a fort built to prevent encroachment by the Hudson's Bay Company traders. From 1867 to 1877 it was a U.S. military post, and later it became an outfitting point for hunters and explorers as well as for miners using the Stikine River route to the Yukon. Lumbering, fishing, mining, and fur farming are pursued.

Wrangell Mountains, S Alaska, extending c.100 mi (160 km) SE from the Copper River to the Canadian border, where they meet the St. Elias Mts. Mt. Blackburn (16,523 ft/5,039.5 m) is the highest peak. There is a cosmic radiation observatory on Mt. Wrangell (14,006 ft/4,269 m).

Wrath, Cape (răth), promontory, Sutherlandshire, NW extremity of the Scottish mainland. The headland, 368 ft (112 m) high, has a lighthouse.

Wray (rā), town (1970 pop. 1,953), seat of Yuma co., NE Colo., near the Nebr. border, in a wheat, corn, and grain area; inc. 1906. Beecher Island battleground, with a memorial to an Indian engagement here in 1868, is nearby.

Wren·tham (rĕn'thəm), town (1970 pop. 7,315), Norfolk co., SE Mass., SW of Boston; settled 1669, inc. 1673. Metal products are made. Helen Keller and Anne Sullivan lived here.

Wrex·ham (rĕk'səm), municipal borough (1972 est. pop. 39,140), Clwyd, NE Wales, in the coalfield of N Wales. Besides mining, Wrexham's industries include the manufacture of steel, rayon, plastics, and leather goods. St. Giles's Church was rebuilt in 1472 after a fire; most of the present structure dates from the 16th cent. Elihu Yale is buried here.

Wright (rīt). **1.** County (1970 pop. 17,294), 577 sq mi (1,494.4 sq km), N central Iowa, drained by the Iowa and Boone rivers; formed 1851; co. seat Clarion. It is a rolling prairie agricultural area dealing primarily in livestock and grain. There are bituminous-coal deposits, as well as sand and gravel pits, and the county contains several small lakes. **2.** County (1970 pop. 38,933), 674 sq mi (1,745.7 sq km), S central Minn., bounded on the N by the Mississippi River, on the E by the Crow River; formed 1855; co. seat Buffalo. In an agricultural area yielding dairy products, livestock, poultry, corn, oats, and barley, it has a food-processing industry and manufactures furniture, plastics, and stone and metal products. **3.** County (1970 pop. 13,667), 684 sq mi (1,771.6 sq km), S central Mo., in the Ozarks, drained by the Gasconade River; formed 1841; co. seat Hartville. Agriculture (fruit, poultry, and dairy products), lumbering, and lead and zinc mining are the major occupations. Part of Mark Twain National Forest is here.

Wright Brothers National Memorial: *see* National Parks and Monuments Table.

Wrights·ville (rīts'vĭl'), city (1970 pop. 2,106), seat of Johnson co., E central Ga., NE of Dublin; inc. 1866. Vegetables are canned, and clothing manufactured.

Wro·cław (vrôt'släf') or **Bres·lau** (brĕs'lou, brĕz'-), city (1973 est. pop. 560,300), capital of Wrocław prov., SW Poland, on the Oder (Odra) River. A railway center and river port, the city is also an industrial center with manufactures of machinery, iron goods, textiles, railroad equipment, and food products. Wrocław probably was a Slavic settlement when it was made (c.1000) an episcopal see subordinate to the archbishop of Gniezno. It became (1163) the capital of the duchy of Silesia, ruled by a branch of the Polish Piast dynasty. Sacked by the Mongols in 1241, the city was rebuilt by German settlers and developed as a trade center. Passing (1335) to Bohemia, it became a member (1368–1474) of the Hanseatic League. It was ceded to the Hapsburgs in 1526 and to Prussia in 1742. The city grew considerably in the 19th cent. in commercial and industrial importance and was the site of two large semiannual trade fairs. Its university was founded in 1811, when it absorbed the university formerly at Frankfurt-an-der-Oder. Wrocław was badly damaged during a Soviet siege in World War II. After 1945 the German inhabitants were expelled and replaced by Poles. Historic buildings include a 13th cent. cathedral, several Gothic churches, and a Gothic town hall that houses a historical museum.

Wu-ch'ang or **Wu·chang** (both: wōō'chäng'), former city, since 1950 part of Wu-han, E Hupeh prov., China, on the right bank of the Yangtze River at the mouth of the Han. The oldest of the three Wu-han cities, it dates from the Han dynasty (200 B.C.–A.D. 200). The first outbreak of the Revolution of 1911, which led to the formation of the Chinese republic, occurred here on Oct. 10. The day is celebrated as the Double Tenth, the tenth day of the tenth month. The city's numerous institutions of higher learning include Wu-han Univ.

Wu-chou (wōō'jō'), city (1970 est. pop. 150,000), E Kwangsi Chuang Autonomous Region, S China, a port on the Hsi River at the mouth of the Kwei River. It is an important trade and shipping center on the Kwangsi-Kwangtung border. Industries include sugar refining, food processing, and the manufacture of chemicals. It was the site of a major U.S. air force base in World War II.

Wu-han or **Wu·han** (both: wōō'hän'), city (1970 est. pop. 4,250,000), capital of Hupeh prov., central China, at the junction of the Han and Yangtze rivers. The great industrial, commercial, and transportation center of central China, Wu-han comprises (since 1950) the former cities of Han-k'ou, Han-yang, and Wu-ch'ang. Situated in the heart of China, virtually equidistant from Peking, Shanghai, Canton, and Ch'eng-tu, it is an air, river, and rail hub dominating the middle Yangtze plain; China's main north-south railroad runs through the city. The Yangtze is here spanned by a mile-long bridge that accommodates both trains and motor vehicles. The busy port on the Yangtze, although about 600 mi (970 km) from the sea, handles large oceangoing vessels. Wu-han is one of the most important industrial centers in China; it has the country's largest integrated iron-steel complex. Also in the city are railroad shops, automotive works, textile mills (Wu-han has the country's largest cotton mill), food-processing establishments, and plants making heavy machinery, glass, cement, fertilizer, and paper products. The many institutions of higher learning include Hupeh Univ., Wu-han Univ., and a medical college. A bridge across the Han River links Hank'ou and Han-yang.

Wu-hsi (wōō'shē'), city (1970 est. pop. 900,000), S Kiangsu prov., China, on the Grand Canal and the N bank of T'ai Lake. It is a silk-producing center. Foods (especially grains) are processed, and machine tools, paper products, fertilizer, and motor vehicles are also made. Wu-hsi has long been famous for its little figurines of opera and drama characters, still being produced today. A small walled city in the early 19th cent., Wu-hsi rapidly replaced Su-chou as the economic center of the T'ai Lake basin. It is on a railroad connecting Shanghai with much of northern China. The city's name, which is translated as "without tin," refers to the tin mines in the area that were exhausted during the Han dynasty.

Wuh·sien (wōō'shē-ĕn'): *see* Su-chou, China.

Wu-hu or **Wu·hu** (both: wōō'hōō'), city (1970 est. pop. 300,000), E central Anhwei prov., China. It is a deepwater port on the Yangtze River, linked by rail with Nanking and Shanghai. It is a commercial center, a major rice market, and the distribution and processing point for the agricultural products of the region.

Wu-lu-mu-ch'i (wōō'lōō'mōō'chē') or **U·rum·chi** (ōō-rōōm'chē), city (1970 est. pop. 500,000), capital of Sinkiang Uigur Autonomous Region, NW China, in the Dzungarian basin. Wu-lu-mu-ch'i is an administrative and commercial center at the junction of several caravan routes from the USSR, Lan-chou (Kansu prov.), and Kashgar. Since 1963 it has been linked to the Chinese rail network via Lan-chou. The main industrial center of Sinkiang, Wu-lu-mu-ch'i has iron and steel works, tanneries, cotton mills, food-processing establishments, and plants manufacturing motor vehicles, agricultural machinery, chemicals, machine tools, and cement. Coal, tin, and silver mines are nearby. The population is mostly Uigur with minorities of Mongols, Chinese, and Kazakhs. Chinese influence began as early as about 122 B.C. when Emperor Wu Ti of the Han dynasty conquered eastern Sinkiang. Wu-lu-mu-ch'i is the seat of Sinkiang Univ., a medical college, and several technical institutes. The city was officially called Tihwa until 1954.

Wu·pat·ki National Monument (wōō-păt'kē): *see* National Parks and Monuments Table.

Wup·per (vōōp'ər), river, c.65 mi (100 km) long, W central West Germany. It is formed by several headstreams and winds in a tortuous course north and southwest past Wuppertal, Remscheid, and Solingen into the Rhine River. Its middle course is heavily industrialized. The river is used to generate power.

Wup·per·tal (vōōp'ər-täl'), city (1972 est. pop. 413,153), North Rhine-Westphalia, W West Germany, on the Wupper River. It is an industrial center, formed in 1929 by the merger of Barmen, Elberfeld, Vohwinkel, and several smaller towns. Manufactures include textiles, metal goods, chemicals, pharmaceuticals, and paper. Barmen was first mentioned in the 11th cent. and Elberfeld in the 12th cent. Elberfeld pioneered in legislation for relief of the poor by a system that it adopted in the mid-19th cent. and that was widely imitated. As a major production center of ball bearings and chemicals in World War II, the city was heavily damaged by Allied bombing raids. Noteworthy buildings include the city hall (1912–22) and the opera house (1956). There is a museum of the history of clocks and watches.

Würt·tem·berg (wûrt'əm-bûrg', wōōrt'-, vür'təm-bĕrg'), former state, SW Germany. In 1952 it was incorporated into the new state of Baden-Württemberg. Stuttgart was the capital. The former state bordered on Baden in the northwest, west, and southwest, on Hohenzollern and Switzerland (from which it was separated by the Lake of Constance) in the south, and on Bavaria in the east and northeast. It included the Swabian Jura in the south, part of the Black Forest in the west, and the cities of Ulm, Esslingen, Heilbronn, Tübingen, and Friedrichshafen. The southern part of Württemberg was the core of the medieval duchy of Swabia; Württembert north of Stuttgart was part of Franconia. Franconia broke up into numerous fiefs in the 10th cent., and Swabia suffered the same fate in the 13th cent. when the house of Hohenstaufen became extinct. Among the local lords who obtained immediacy under the Holy Roman Empire were the counts of Württemberg, whose original domains, established by the 11th cent., were centered around Esslingen. In the following centuries the counts considerably expanded their territory, but aside from Stuttgart they held no important towns, most of the Swabian cities (for example, Ulm, Hall, Gmünd, Esslingen, and Rottweil) being independent as free imperial cities. They also acquired (1397) the principality of Montbéliard in France and several places in Alsace. The various territories were subdivided among the branches of the family, but in 1482 count Eberhard V declared the indivisibility of the holdings. Württemberg was raised to ducal rank in 1495. In 1519, however, the Swabian league of cities, fearing the rising power of Württemberg, expelled duke Ulrich I from his domains, and in 1520 it sold the duchy to the newly elected emperor Charles V. Ulrich, a turbulent individual, never ceased in his attempts to recover his lands. In 1524 he helped the rebelling peasants in the great Peasants' War, and in 1525 he invaded Württemberg with an army of Swiss mercenaries. The Swiss cantons, however, soon summoned their troops home, and Ulrich had to flee again. A Protestant convert, Ulrich secured (1534) the help of Philip of Hesse, a leading defender of the Reformation, and, through Philip, of Francis I of France; at the same time the peasants of Württemberg were rising against the unpopular government of King (later Emperor) Ferdinand I. At the Battle of Lauffen (1534), Ulrich and Philip routed Ferdinand's troops. Ferdinand was obliged to restore Württemberg to Ulrich, although nominally Ulrich was to hold the duchy as a fief from Austria. Immediacy under the empire was restored only in 1599. With Ulrich's return Lutheranism was introduced. However, large parts of southern Württemberg remained in the hands of the house of Hapsburg and of a number of powerful abbeys; these territories were incorporated into Württemberg only later. As a result, a large minority of the present population is Roman Catholic. Württemberg was repeatedly the scene of fighting in the wars of the 17th and 18th cent. Duke Frederick II, through his alliance with Napoleon I, obtained the rank of elector in 1803 and became king of Württemberg as Frederick I in 1806, after joining the Confederation of the Rhine. Between 1802 and 1810 the territories of Württemberg were more than doubled and reached their final frontiers. Frederick retained both his royal title and his lands at the Congress of Vienna, after having passed (1813) from the French to the Allied camp. William I, his successor, granted a liberal constitution in 1819. During the reign (1864-91) of King Charles Württemberg sided against Prussia in the Austro-Prussian War of 1866, joined Prussia's side in the Franco-Prussian War of 1870-71, and became (1871) a member of the German Empire. Charles's successor,William II, abdicated in 1918, and Württemberg joined (1919) the Weimar Republic. After World War II northern Württemberg was a part of the temporary state of Württemberg-Baden, and southern Württemberg was a part of the temporary state of Württemberg-Hohenzollern until the state of Baden-Württemberg was formed in 1952.

Würt·tem·berg-Ba·den (vür'təm-bĕrg-bä'dən), former state, c.6,060 sq mi (15,700 sq km), SW West Germany. The state was formed after 1945 and comprised the northern parts of the former states of Baden and Württemberg. In 1952 it became part of Baden-Württemberg.

Würt·tem·berg-Ho·hen·zol·lern (vür'təm-bĕrg-hō'ən-tsôl'ərn), former state, c.4,020 sq mi (10,410 sq km), SW West Germany. Tübingen was the capital. Formed after 1945, the state comprised southern Württemberg, the former Prussian province of Hohenzollern, and the former district of Lindau, Bavaria. In 1952 it became part of Baden-Württemberg.

Würz·burg (vürts'bŏŏrk), city (1972 est. pop. 114,160), capital of Lower Franconia, Bavaria, S central West Germany, on the Main River. It is an industrial city and the center of a wine-producing region. Manufactures include printing presses, machine tools, chemicals, textiles, and beer. Würzburg was originally a Celtic settlement and was made an episcopal see by Saint Boniface in 741. After the breakup (10th cent.) of the duchy of Franconia, its bishops ruled a vast territory on both sides of the Main as princes of the Holy Roman Empire. In 1168 the bishops assumed the title of dukes of Eastern Franconia, of which they held a major part. During the Peasants' War the bishop of Würzburg temporarily lost (1524-25) his territory to the rebels, but he held out at his fortress of Marienberg against Götz von Berlichingen. Later, the splendor-loving prince-bishops transformed (17th-18th cent.) the city into one of the finest residences of Europe and founded (1582) the Univ. of Würzburg, where the anthropologist and pathologist Rudolf Virchow and the physicist Wilhelm Roentgen taught in the 19th cent. Secularized after the Treaty of Lunéville (1801), Würzburg passed (1803) to Bavaria; was made (1805) a separate electorate in favor of Ferdinand, the dispossessed grand duke of Tuscany; and reverted (1815) to Bavaria. The city was severely damaged during World War II. Noteworthy landmarks include the baroque former episcopal residence (1720-44); the Romanesque cathedral (11th-13th cent.), containing works by the sculptor Tilman Riemenschneider; the Marienkapelle (1377-1479), a late Gothic chapel; the Old Main Bridge; and Marienberg fortress (the episcopal residence from the mid-13th to the 18th cent.).

Wu-su or **Wu·su** (both: wōō'sōō'), town and oasis, N Sinkiang Uigur Autonomous Region, China, in the Dzungarian basin. It is an oil-producing center in the Karamai oil fields.

Wu-t'ai Shan (wōō'tī' shän'), mountain range, extending c.150 mi (240 km) across NE Shansi and NW Hopeh provs., NE China. The mountains, rising to c.11,500 ft (3,510 m), are sacred to the Mongols and contain lamaseries frequented by pilgrims.

Wy·an·dot (wī'ən-dŏt'), county (1970 pop. 21,826), 406 sq mi (1,051.5 sq km), N central Ohio, drained by the Sandusky River; formed 1845; co. seat Upper Sandusky. In an agricultural area yielding livestock, grain, poultry, fruit, and dairy products, it has limestone quarries and some industry.

Wy·an·dotte (wī'ən-dŏt'), county (1970 pop. 186,845), 152 sq mi (393.7 sq km), NE Kansas, in a rolling to hilly area bounded on the N by the Missouri River and drained by the Kansas River; formed 1859; co. seat Kansas City. It has dairying and general farming, gas fields, timber, and sand and gravel pits. Its manufactures include clothing, wood products, furniture, paper, chemicals, cement, concrete, metal products, and machinery.

Wyandotte, industrial city (1979 est. pop. 36,300), Wayne co., SE Mich., a suburb of Detroit on the Detroit River; inc. as a city 1867. Manufactures include automobile parts and barrels. Bessemer steel was first commercially produced in the city in 1864 by W. F. Durfee. A Wyandot Indian village was here in the 19th cent.; of interest is a totem pole depicting Wyandot history.

Wyandotte Cave, one of the largest natural caverns in the United States, S Ind., W of New Albany; discovered in 1798. There are 23 mi (37 km) of passages and several large and beautiful chambers on five levels. Saltpeter was mined here until the middle of the 19th cent.

Wye (wī), river, c.130 mi (210 km) long, rising on Plynlimon Mt., W Wales, and flowing generally SE past Builth Wells, Wales, and Hereford and Monmouth, in England, to the estuary of the Severn River. It is noted for its beautiful valley, especially the part that forms the Gloucestershire-Monmouthshire boundary. Reservoirs on the Elan River, a tributary, provide water for Birmingham.

Wynne (wĭn), city (1974 est. pop. 7,292), seat of Cross co., E Ark. N of Forrest City; founded 1863. A shipping point for a farm area, it also makes shoes and lumber products.

Wy·o·ming (wī-ō'mĭng), state (1975 pop. 365,000), 97,914 sq mi (253,597 sq km), W United States, admitted as the 44th state of the Union in 1890. Cheyenne is the capital and largest city. The state, exactly rectangular in shape, is bounded on the north by Mont., on the east by S.Dak. and Nebr., on the south by Colo. and Utah, and on the west by Utah, Idaho, and Mont.

Wyoming is traversed by the Rocky Mts., which angle across the state from the northwest. East of the mountains is the rolling country of the Great Plains, actually a mile-high region covered with grasses, cactus, and sage, and interrupted by the upward thrust of mountain ranges. Only in the central section of the state is the sweep of plain unbroken, and across this stretch the wagon trains rolled westward over the Oregon Trail. In the extreme northeast the low, wooded Black Hills give way to eroded badlands extending west to the banks of the Powder River, which wanders through some of the most famous cattle country in the United States. West beyond the Powder is the tall grass country that was the hunting ground of the Crow Indians until the Sioux, following the buffalo, pushed the Crow into the

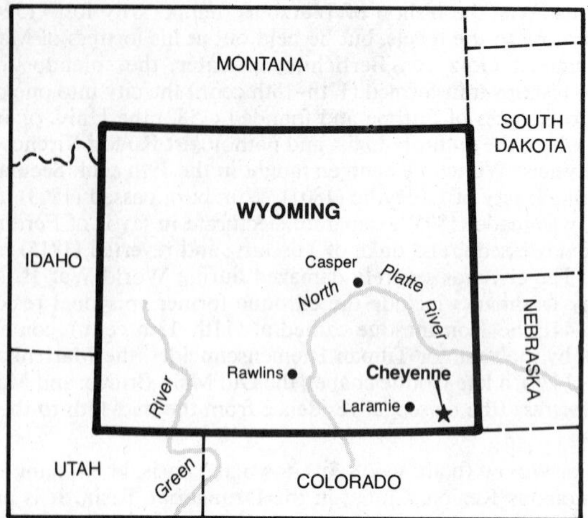

mountains. The Sioux fell in turn before the relentless advance of the white man, and today farms and ranches occupy this fertile and beautiful plains area.

In southeastern Wyoming the higher tablelands are interrupted by the Laramie and Medicine Bow ranges. Across this region travelers to the Pacific coast made their way when the Indian wars of the 1860s endangered treks on the regular Oregon Trail. The railroad followed the wagons and coaches as the Union Pacific laid its tracks along this southerly course. In southwestern Wyoming is the natural gateway through the Rockies: the broad, grassy South Pass. Immediately north of the pass is the Wind River Range, reaching the highest altitude in the state at Gannett Peak (13,785 ft/4,202 m). Still farther north rise the Gros Ventre and Absaroka ranges, and to the west, near the Idaho line, the glorious Tetons loom above a lake and valley country of incomparable beauty. From the mountain heights snows melt to feed a number of rivers; the Snake begins its long, winding journey into Idaho and on to the Columbia, the Yellowstone travels north and east into the Missouri, and the Green River flows south to join the Colorado. This wealth of surface water supplements the scant rainfall, and water is impounded for irrigation, flood control, and in some cases hydroelectric power.

The state's natural beauty draws hundreds of thousands of visitors annually, making tourism a major source of revenue. Wyoming has two spectacular national parks: Grand Teton, which embraces the most stunning portion of the Teton Range, and Yellowstone, which occupies the entire northwest corner of the state and is the oldest and largest of all the national parks. Its geysers and hot springs are world famous, as is the breathtaking Grand Canyon of the Yellowstone. Wyoming is also prime hunting and fishing country. The nation's largest herds of elk and antelope are here; deer, moose, and bear are plentiful, and the rivers, lakes, and streams teem with fish. Also in the state are Devils Tower and Fossil Butte national monuments and two national recreational areas, Bighorn Canyon and Flaming Gorge. In addition, the multitude of rodeos, annual roundups, frontier celebrations, and dude ranches are drawing an increasingly large number of vacationers every year.

Wyoming is divided into 23 counties:

NAME	COUNTY SEAT	NAME	COUNTY SEAT
Albany	Laramie	Natrona	Casper
Big Horn	Basin	Niobrara	Lusk
Campbell	Gillette	Park	Cody
Carbon	Rawlins	Platte	Wheatland
Converse	Douglas	Sheridan	Sheridan
Crook	Sundance	Sublette	Pinedale
Fremont	Lander	Sweetwater	Green River
Goshen	Torrington	Teton	Jackson
Hot Springs	Thermopolis	Uinta	Evanston
Johnson	Buffalo	Washakie	Worland
Laramie	Cheyenne	Weston	Newcastle
Lincoln	Kemmerer		

Economy. Dry farming, producing hay, wheat, and barley, is supplemented by the more diversified yield (especially sugar beets and dry beans) of the irrigated fields. Most of the inhabitants of the state derive their livelihood directly or indirectly from farming or ranching. The most valuable farm commodities, in terms of cash receipts, are cattle, sheep, sugar beets, dairy products, and wool. Sparse grasses over much of the region necessitate a large grazing area for each animal; sheep graze in places unfit for cattle, and both sheep and cattle range by permit in the national forests. Cooperative grazing tracts are on the increase. Horses, a prized essential of ranching, are carefully raised and trained. Oil wells were first drilled in the 1860s, and today petroleum is the state's most important mineral, followed by natural gas, sodium salts, and uranium. In 1970 Wyoming ranked first in the production of sodium carbonate and second in uranium. Low-grade coal is also abundant, and considerable amounts of gold, iron, copper, and various clays are mined. The production of petroleum and petroleum products, centered in Casper, comprises the state's leading industry. Other important manufactures are processed foods and clay, glass, and wood products. Wyoming has millions of acres of forest.

History. Portions of present Wyoming were at one time claimed by Spain, France, and England; the acquisition of the territory by the United States was completed through five major annexations—the Louisiana Purchase in 1803, the Treaty of 1819 with Spain, cession by the republic of Texas in 1836 and partition from Texas after it was annexed in 1845, the Treaty of Guadalupe Hidalgo (1848) after the Mexican War, and the international agreement (1846) with Great Britain concerning the Columbia River country. The early development of Wyoming was closely linked with the fur trade and the great westward migrations. French trappers and explorers may have reached the area in the middle to late 18th cent., but the first authentic accounts of the region were provided by John Colter, who, trapped in the Wyoming mountains for several years, returned to St. Louis in 1810 with fantastic accounts of the steaming geysers and great canyons of the Yellowstone. Colter returned west, and other fur traders made their way into Wyoming. The overland party on its way to found Astoria on the Columbia River went through Teton Pass in 1811. The following year Robert Stuart, returning from Astoria, crossed South Pass and followed much of the route that was to be the Oregon Trail. Only the hardiest and most self-sufficient could survive the Indian attacks and the rugged isolation of the country.

With the expeditions of William H. Ashley, the Mountain Men entered the country, and some of the most famous of those early explorers—Thomas Fitzpatrick, James Bridger, and Jedediah S. Smith—crossed and recrossed the land. Attracted by the fur trade, Capt. B. L. E. de Bonneville organized a sizable expedition, and his were the first wagons to go (1832) through South Pass. The first permanent trading post was Fort William (1834), famous under its later name, Fort Laramie. In 1843 Fort Bridger (now in a state park) was built. The area also aroused the interest of John C. Frémont, who made an expedition in 1842. By the 1840s the route west through Wyoming was in steady use by caravans headed toward Oregon, and fur-trading posts became stations on the Oregon Trail. As the fur trade declined, many former trappers and mountaineers settled along the trail, furnishing horses and other supplies to the migrants and purchasing debilitated stock to be put to pasture and sold the following year.

Mormons trekking to Utah (Brigham Young led the first party in 1847) and forty-niners rushing to the gold fields of California joined the many thousands traversing the mountain passes of Wyoming. A number of Mormons settled for a time in western Wyoming. The death of Mormon pioneers in a blizzard (1856) and the thousands of graves along the Oregon Trail give an indication of the toll taken by disease, starvation, Indian attacks, and winter snows. Despite the hardships, telegraph stations (1861) and stagecoach and freight lines were established, and in 1860-61 pony express riders heroically crossed Wyoming on their route between St. Joseph, Missouri, and Sacramento, California. Indian trouble in the early 1860s forced the rerouting of stagecoaches to the south, along the Overland Trail.

The Indians, displaced from their former homes in the east and west, and waging internecine warfare for control of the rich buffalo ranges, feared encroachment by the whites on their hunting grounds, especially after the opening (1864) of the Bozeman Trail. Treaties were made and broken by both sides, and wars with the Sioux persisted, particularly in the Powder River valley. Meanwhile, southern Wyoming was relatively free of Indian attacks, and a gold rush, stimulated by the discoveries at South Pass (1867), brought the first heavy influx of settlers to that region; the flow was increased by the uncovering of vast coal deposits in southwestern Wyoming. Probably the greatest stimulus to settlement was the completion (1868) of the Wyoming sector of the Union Pacific RR. Towns, including

Cheyenne, sprang up beside the tracks, and trade thrived on the demands of the road crews and the new settlers.

In 1868 the region became the Territory of Wyoming, with Cheyenne as its capital. Wyoming pioneered in political equality when, in 1869, the first territorial legislature granted the vote to women. The territory continued to advance economically as huge herds of cattle were driven up over the Texas, or Long, Trail. Indian rebellions had been quelled by the late 1870s. The Arapaho Indians were placed on the Wind River Reservation with their former enemies, the Shoshone Indians, and cattlemen safely moved their herds to grasslands throughout Wyoming. But the days of dangerous living were not over. Cattle rustling became so common that the authorities could not control it, and juries grew fearful of returning just verdicts against criminals. The Wyoming Stock Growers Association was organized in 1873 to protect cattle owners, and members frequently formed vigilante groups to administer their own justice. The struggle reached its height in the Johnson county cattle war of 1892. Lawlessness was also exemplified by the Hole-in-the-Wall gang, which broadened its activities to include bank and train robberies as well as cattle theft.

Gradually, vast areas were fenced in and winter pastures were established. The influx of sheep in the late 1890s, however, brought new violence. Cattlemen made frantic efforts to exclude the sheep from close grazing on the precious grasslands. Homesteaders were also unwelcome, and many left when they realized that the country was unsuited for small acreage cultivation. However, population increase was steady, advancing from about 9,000 in 1870 to more than 90,000 in 1900. With expanding population came development in other ways: eager frontiersmen rapidly (and somewhat chaotically) established schools, and in 1887 the Univ. of Wyoming was founded.

Statehood was achieved in 1890, and in keeping with its frontier ideals, Wyoming adopted a liberal state constitution that included the secret ballot. The Carey Act of 1894, providing for the reclamation and settlement of land, stimulated further agrarian development and, in addition, pointed out the need for conservation and efficient use of water. The establishment of national parks protected the timberlands and extensive grazing areas, and water power was harnessed to furnish electricity for farms and industries.

In politics, the progressive movement found numerous adherents in Wyoming; in 1915, after one of the most bitter fights in the state's history, progressive forces triumphed over the railroad and related interests with the establishment of a state utilities commission. A workmen's compensation law was passed in 1915, and in that year also the legislature authorized the Univ. of Wyoming to accept Federal grants for agricultural experiments and demonstrations. In 1924 Wyoming became the first state to elect a woman governor, Nellie Taylor Ross. By then the state ranked fourth in the nation in the production of crude oil, but the valuable finds at Teapot Dome are probably remembered best as the symbol of corruption in the administration of President Warren Harding.

Under the New Deal Wyoming was well served by the national soil conservation programs, which benefited dry farmers who had extended operations into semiarid regions and had suffered severely in the drought years beginning in the late 1920s. The cooperative movement in agriculture also gained ground in this period and has since continued to grow. One of the most important events in the state since World War II has been the discovery of uranium. New oil finds have also helped to offset economic losses resulting from a disastrous four-year-long drought in the 1950s.

Government. The state still operates under its first constitution, adopted in 1890. The executive branch is headed by a governor elected for a four-year term. Wyoming's bicameral legislature has a senate with 30 members elected for four-year terms and a house of representatives with 61 members elected for two years. The state sends two Senators and one Representative to the U.S. Congress and has three electoral votes.

Wyoming. 1. County (1970 pop. 37,688), 598 sq mi (1,548.8 sq km), W N.Y., drained by the Genessee River and Tonawanda and Cattaraugus creeks; formed 1841; co. seat Warsaw. This dairying and farming area (vegetables, grain, and fruit) also has timber, salt works, and diversified manufacturing. **2.** County (1970 pop. 19,082), 398 sq mi (1,030.8 sq km), NE Pa., in a hilly region drained by the Susquehanna River; formed 1842; co. seat Tunkhannock. It is in a dairying region with some quarrying. Its industry includes food processing, lumbering, and the manufacture of clothing, paper, shoes, and campers. **3.** County (1970 pop. 30,095), 504 sq mi (1,305.4 sq km), S W.Va., on the Allegheny Plateau, drained by the Guyandot River and its tributaries; formed 1850; co. seat Pineville. It is an important coal-mining county and also has timber and some natural gas. Its agriculture includes livestock, poultry, truck crops, fruit, and tobacco.

Wyoming, city (1979 est. pop. 58,600), Kent co., W Mich., in the greater Grand Rapids metropolitan area, on the Grand River; settled 1832, inc. 1959. Aircraft and automobile parts, home appliances, and aluminum are produced in the city.

Wyoming Valley, c.20 mi (30 km) long and 3 to 4 mi (4.8–6.4 km) wide, in Luzerne co., NE Pa., on the Susquehanna River. Wilkes-Barre is the metropolis of this once rich anthracite region, which is now a major manufacturing area. The valley was the scene of a long contest between Connecticut and Pennsylvania over conflicting land claims based on 17th cent. charters. After the Susquehanna Company purchased (1754) land here from the Indians at the Albany Congress, a temporary settlement of the region in 1762–63 led to the first permanent settlement in 1769.

Wy·o·mis·sing (wī'ō-mĭs'ĭng), industrial borough (1970 pop. 7,136), Berks co., SE Pa., on the Schuylkill River opposite Reading; inc. 1906. Textile machinery and hosiery are made.

Wythe (wĭth), county (1970 pop. 22,139), 460 sq mi (1,191.4 sq km), partly in the Great Appalachian Valley, with the Alleghenies in the N and NW, and drained by the New River; formed 1790; co. seat Wytheville. It has farming (cabbage, grain, and potatoes) and cattle grazing on its bluegrass pastures. Its mining includes lead and zinc ores, and it has diversified industry (food processing, yarn mills, clothing, lumbering, metal products, and machinery.

Wythe·ville (wĭth'vĭl'), town (1970 pop. 6,069), seat of Wythe co., SW Va., SW of Roanoke on a high plateau in a bluegrass section; founded 1792, inc. 1839. It is the processing center of a livestock, timber, and fertile farm area. Textile products are made.

Xán·thi (ksän'thē) or **Xan·the** (zăn'thē), city (1972 pop. 24,867), capital of Xánthi prefecture, NE Greece, in Thrace. Tobacco is produced.

Xan·thus (zăn'thəs), ancient city of Lycia, W Asia Minor, in present Turkey. On the Xanthus River, it was besieged and taken by the Persians (c.546 B.C.) and centuries later (c.42 B.C.) by the Romans. Both times the inhabitants destroyed their city before surrendering. Many works of art from Xanthus, such as the archaic sculptured reliefs and the Nereid monument, are now in the British Museum.

Xau·en (hou'ān'): *see* Chechaouèn, Morocco.

Xe·nia (zēn'yə, zē'nē-ə), city (1979 est. pop. 30,900), seat of Greene co., SW Ohio; inc. 1814. It is a trade and industrial center in a farm area. Rope, twine, monuments and markers, plastics, potato chips, valves, and hydraulic lifts are among its manufactures. A tornado destroyed about half of the city on April 3, 1974.

Xin·gu (shēng-gōō'), river, 1,230 mi (1,979.1 km) long, rising in central Mato Grosso state, Brazil, and winding N across Pará state into the Amazon River at the head of the Amazon delta. The Xingu, with many rapids and falls, passes through wild, partly unexplored country, and only in its lower course is it navigable.

Xo·chi·mil·co (sō'chē-mēl'kō), town (1970 pop. 117,083), a suburb of Mexico City. Mainly a commercial and tourist center, it is famous for its canals lined with poplars and flowers. The Indians established soil-covered rafts in Lake Xochimilco; the rafts became islands rooted to the lake bottom and continued to supply vegetables and flowers to Mexico City.

Ya-an or **Ya·an** (both: yä'än'), city (1970 est. pop. 100,000), SW Szechwan prov., China, on a tributary of the Min River. It is a tea center and a highway hub.

Ya·blo·no·vy Range (yä'blə-nə-vē'), mountain chain, in Transbaykalia, SE Siberian USSR. Forming part of the watershed between the Arctic and Pacific oceans, it extends from the Mongolian border in the southwest to the Olekma River in the northeast. It is crossed by the Trans-Siberian RR at Chita. Sokhondo, c.8,230 ft (2,510 m) is the highest point.

Yad·kin (yăd'kən), county (1970 pop. 24,599), 335 sq mi (867.7 sq km), NW N.C., in a piedmont region bounded on the N and E by the Yadkin River; formed 1850; co. seat Yadkinville. It has tobacco growing, dairying, and lumbering.

Yad·kin·ville (yăd'kən-vĭl'), town (1970 pop. 2,232), seat of Yadkin co., central N.C., W of Winston-Salem. Wood products are made.

Yak·i·ma (yăk'ə-mô', -mə), county (1970 pop. 145,212), 4,268 sq mi (11,054.1 sq km), S Wash., in a mountain area in the Cascade Range, divided by the Yakima valley; formed 1865; co. seat Yakima. It is in a fertile irrigated agricultural area that produces fruit, truck crops, dairy products, grain, hay, hops, potatoes, and livestock. Its industry includes food processing, lumbering, and the manufacture of clothing, paper, plastics, concrete, metal products, farm and food-processing machinery, and transportation equipment. It includes part of Snoqualmie National Forest and Yakima Indian Reservation.

Yakima, city (1979 est. pop. 57,000), seat of Yakima co., S central Wash., on the Yakima River just below its confluence with the Naches; inc. 1886. It is the trade and shipping center of an irrigated agricultural valley noted for its fruit, hops, and mint. It has several fruit canneries and plants that manufacture lumber products, blue jeans, and aircraft equipment and parts. Located in Yakima are a junior college, a school of nursing, a state fish hatchery, the central Washington fairgrounds, and a race track. A state park is to the east, and an Indian reservation is in the vicinity.

Yakima, river, 203 mi (326.6 km) long, rising in the Cascade Range, central Wash., and flowing SE past Yakima to the Columbia River near Kennewick. The U.S. Bureau of Reclamation's Yakima project (begun in 1906) utilizes the Yakima and its tributaries to irrigate c.460,000 acres (186,300 hectares) and has helped make the river valley an important farming and fruit-growing region.

Ya·kut Autonomous Soviet Socialist Republic (yə-kōōt'), autonomous division (1970 pop. 664,000), c.1,200,000 sq mi (3,108,000 sq km), NE Siberian USSR; capital Yakutsk. Yakut is bounded in the north by the Laptev and East Siberian seas of the Arctic Ocean, in the south by the Stanovoy Range, in the west by the Central Siberian Uplands, and in the east by the Verkhoyansk Range. It also encompasses the New Siberian and Lyakhov islands in the Arctic Ocean. The terrain is largely plain, with tundra in the north and taiga elsewhere. More than 40% of the territory lies inside the Arctic Circle. The Lena, Yana, Indigirka, and Kilyma rivers cross the republic, which includes virtually the entire Lena basin. There are many lakes in the lowlands. The rivers are used for navigation (during the summer) and for flotage and have great hydroelectric potential. The Yakutsk-Skovorodino highway is the chief overland route. Yakutia has no railroads. Air transport and winter sledge are widespread.

One of the world's coldest inhabited regions, the republic has extremely severe winters and short summers; temperatures of -103°F (-75°C) have been recorded in some cities. Agriculture is possible only in the south, along the Lena and its tributaries. Wheat, barley, rye, and leaf and root vegetables are grown. Diamond mining is the republic's main industry. There is extensive gold mining in the Aldan district. Lumbering, fishing, hunting, fur breeding and trapping, livestock raising (especially horses), and reindeer herding are also important economic activities. Much lumber is exported. Yakutia has printing works, food-processing plants, and factories that produce cement and other building materials, clothing, and leather footwear. Bone carving has long been a noted art among the Yakut, who make up about half of the republic's population; most of the remainder are Russians, and there are small Evenki, Eveny, and Chukchi minorities. About half the people are urban dwellers. The Yakut, who speak a Turkic language with Mongolian influence, settled around the Lena River between the 13th and 15th cent. Russian colonization began after the establishment of a fort at Yakutsk in 1632. The autonomous republic was organized in 1922.

Ya·kutsk (yə-kōōtsk'), city (1973 est. pop. 126,000), capital of Yakut Autonomous Republic, E Siberian USSR, a port on the Lena River. It is also a highway center and has tanneries, sawmills, and brickworks. Yakutsk was founded in 1632.

Yal·o·bush·a (yăl'ə-bŏŏsh'ə), county (1970 pop. 11,915), 488 sq mi (1,263.9 sq km), N central Miss., drained by the Yocona and Skuna rivers; formed 1833; co. seats Water Valley and Coffeeville. Cotton, corn, livestock, dairy products, and timber are major resources.

Yal·ta (yôl'tə, yäl'-), city (1973 est. pop. 66,000), SW European USSR, in S Crimea, on the Black Sea. Picturesquely situated near the seashore, Yalta is the largest resort in the Crimea. There are many hotels, sanatoriums, and tourist and rest homes—many of which were built as villas by the nobility before the Russian Revolution. Churchill, Roosevelt, and Stalin met in Yalta in Feb., 1945. Yalta is on the site of an ancient Greek colony.

Ya·lu (yä'lōō), river, c.500 mi (805 km) long, rising in the Ch'ang-pai Mts. in Kirin prov., NE China, and flowing SW to the Bay of Korea at Tan-tung. It forms part of the China–North Korea border. In places it is navigable for shallow-draft vessels, but its chief commercial use is for floating timber to sawmills. The Supung Dam above Sinuiju, North Korea, one of the largest dams in Asia, supplies hydroelectric power to China and North Korea. Chinese troops entered the Korean War by crossing the Yalu.

Ya-lung (yä'lŏŏng'), river, c.800 mi (1,290 km) long, rising in the Kunlun Mts., S Tsinghai prov., W China, and flowing S across W Szechwan prov. to the Yangtze River at the Yünnan line. It flows through deep gorges and is one of the Yangtze's longest tributaries.

Ya·ma·ga·ta (yä'mə-gä'tə), city (1970 pop. 204,127), capital of Yamagata prefecture, N Honshu, Japan. It is a silk-reeling center. Agricultural production is also important.

Ya·ma·gu·chi (yä'mə-gōō'chē), city (1970 pop. 73,000), capital of Yamaguchi prefecture, SW Honshu, Japan. It was a castle city from the 14th to 16th cent. and the site of many Buddhist temples and a mission established (1550) by St. Francis Xavier.

Ya·ma·lo-Ne·nets National Okrug (yə-mä'lə-nə-nĕts'), administrative division (1970 pop. 80,000), c.290,000 sq mi (751,100 sq km), NW Siberian USSR, on both sides of the Gulf of Ob and including the Yamal peninsula; capital Salekhard. The area has frozen ground (permafrost) and tundra, forest tundra, and taiga vegetation. Reindeer raising, fishing, and fur trapping are the chief occupations. There are deposits of iron ore, coal, natural gas, and peat in the region. The population consists of Russians, Nentsy, Khanty, and Komi. The area was organized as an okrug in 1930. The Soviet government maintained forced labor camps in the area.

Yam·bol or **Jam·bol** (both: yäm'bōl'), city (1969 pop. 70,300), SE Bulgaria. It is a commercial center and produces cotton textiles, machinery, and cement. There are mineral springs nearby. Dating from Roman times, Yambol was a residence of Turkish beys in the 15th to 18th cent.

Yam·hill (yăm'hĭl'), county (1970 pop. 40,213), 709 sq mi (1,836.3 sq km), NW Oregon, bounded in the E by the Willamette River and drained by the Yamhill River; formed 1843; co. seat McMinnville. Its agriculture includes grain, hay, legume seeds, fruit, truck crops, poultry, and dairy products. It has food-processing and lumbering industries, and manufactures machines and mobile homes.

Yam·pa (yăm'pə), river, c.250 mi (400 km) long, NW Colo., rising in the Rocky Mts. and flowing N then W to the Green River just E of the Utah border.

Ya·mu·na (yä'mə-nə), river, India: *see* Jumna.

Yan·cey (yăn'sē), county (1970 pop. 12,629), 311 sq mi (805.5 sq km), W N.C., bounded on the NW by the Tenn. border and the Bald Mts. and crossed by ranges of the Blue Ridge and the Black Mts.; formed 1833; co. seat Burnsville. Its economy depends on farming (tobacco, hay, potatoes, and corn), stock raising, lumbering, and mining (mica, kaolin, and feldspar). The county includes part of Pisgah National Forest and has many resort areas.

Yan·cey·ville (yăn'sē-vĭl'), uninc. village (1970 pop. 1,274), seat of Caswell co., N N.C., NE of Greensboro; founded 1833. It has lumber and textile mills.

Yang-chou or **Yang·chow** (both: yăng'chou', yäng'jō'), city (1970 est. pop. 210,000), N Kiangsu prov., China, on the Grand Canal. It is an agricultural market center with fertilizer and machine-tool industries. An ancient walled city, Yang-chou was in the 6th cent. one of the three capitals of the Sui dynasty and was later an important cultural center under the T'ang dynasty. It was a center of Nestorian Christianity and was governed by Marco Polo from 1282 to 1285. It is famous for its storytellers (who still perform today) as well as for its historic buildings and former palaces.

Yang-ch'üan (yäng'chü-än'), city, E central Shansi prov., China, on the highway and railroad linking T'ai-yüan with Hopeh prov. The center of an important coal-producing area, it is a growing industrial city. Iron is also mined nearby.

Yang·tze (yăng'sē', yăngt'-), longest river of China and of Asia, c.3,450 mi (5,550 km) long, rising in the Tibetan Highlands, SW Tsinghai prov., W China, and flowing generally E through central China into the East China Sea at Shanghai. The Yangtze and its tributaries drain more than 750,000 sq mi (1,942,500 sq km). The river passes through one of the world's most populated regions and has long been used as a major east-west trade and transportation route in China. The Yangtze's turbulent upper course (1,600 mi/2,575 km long) flows southeast through forested, steep-walled gorges 2,000–4,000 ft (610–1,220 m) deep. After receiving the Ya-lung River at the Szechwan-Yünnan prov. border, the Yangtze turns northeast toward the Szechwan basin. At I-pin, on the western edge of the Szechwan basin, the river becomes the Yangtze proper. Leaving the Szechwan basin, the Yangtze receives the Wu River and flows through the spectacular Yangtze gorges that extend from Feng-chieh to I-ch'ang; there the river is a serious hazard and at times navigation is impossible. Temples and pagodas are perched on prominent hills along the gorges.

East of I-ch'ang, the Yangtze enters the lake-studded middle basin of Hupeh, Hunan, and Kiangsi provs., a rich agricultural and industrial region. Dikes protect large areas of the river's middle basin from flood waters. Although the Yangtze does not often experience the devastating floods that characterize the Huang Ho, it has occasionally caused wide damage; great floods occurred in 1931 and 1954. The fertile middle basin is China's most productive agricultural region; rice is the main crop. The river enters the East China Sea through the extensive, ever-expanding delta region of Anhwei and Kiangsu provs. Dikes have been built to reclaim coastal marshes and create additional farmland. The Yangtze carries its greatest volume during the summer rainy season. It is navigable for ocean liners to Wu-han, c.600 mi (970 km) upstream; during the summer high-water period, I-ch'ang, c.1,000 mi (1,610 km) upstream, is the head of navigation. Smaller craft can sail to I-pin, c.1,500 mi (2,410 km) from the sea.

Yank·ton (yăngk'tən), county (1970 pop. 19,039), 518 sq mi (1,341.6 sq km), SE S.Dak., on the Nebr. border, drained by the James River and bounded in the S by the Missouri River; formed 1862; co. seat Yankton. Its agriculture includes corn, soybeans, alfalfa, wheat, oats, barley, livestock, and dairy products. Among its diversified manufactures are metal products, machinery, radio and TV components, and aircraft equipment.

Yankton, city (1975 est. pop. 12,300), seat of Yankton co., extreme SE S.Dak., on the Missouri River; inc. 1869. A railroad and trade center in a farm and livestock region, it has grain elevators, creameries, and plants manufacturing a great variety of products, including elevators, trailers, and aircraft and electronic components. Settled 1858 as a fur-trading post, Yankton was the Dakota territorial capital from 1861 to 1879; the old capitol building still stands. Nearby Lewis and Clark Lake, formed by Gavins Point Dam (completed 1956), is part of the Missouri River basin project.

Ya·oun·dé (yä'ōōn-dā'), city (1970 est. pop. 178,000), capital of the United Republic of Cameroon. Manufactures include cigarettes, dairy products, clay and glass goods, and lumber. Yaoundé is a regional trade center for coffee, cacao, copra, sugar cane, and Para rubber. Yaoundé was founded in 1888 by German traders as a base for tapping the ivory trade. It was occupied by Belgian troops during World War I and after the war was (except for 1940–46) the capital of French Cameroon. Yaoundé is the site of the Univ. of Cameroon (1962), which includes schools of teaching and agriculture. The city has many other educational and research institutes, including a school of administration and law (1960) and a school of journalism (1970).

Yap (yăp, yäp), island group (1970 pop. 2,856), c.25 sq mi (65 sq km), in the W Caroline Islands, W Pacific. Yap is the principal cable station of the Pacific and an important radio transmitting point. It consists of 4 large and 10 small islands surrounded by a coral reef. Discovered by the Spanish in 1791, Yap was sold to Germany in 1899. It became part of the Japanese-mandated area of the Pacific under the League of Nations in 1920 and fell to U.S. forces in 1945 during World War II. Yap is known for the stone disks used as money by the Micronesian natives.

Yar·kand (yär-kănd'), river, c.500 mi (805 km) long, rising in the Karakorum range, Sinkiang Uigur Autonomous Region, NW China, and flowing NE to join with the Kashgar to form the Tarim River. The city of Yarkand, the largest settlement in the Tarim Basin, forms an oasis at the western end of the Takla Makan desert; it is the main trade center for goods traveling from Sinkiang to the USSR and India. Marco Polo visited Yarkand in 1271 and in 1275. The city was on the important Silk Road between China and Europe. A treaty signed here in 1874 opened Yarkand to trade with India.

Yar·mouth (yär'məth), city (1971 pop. 8,516), SW N.S., Canada, on the Atlantic Ocean. Yarmouth is a summer resort and tourist center. The region was visited (1604) by Champlain and became a French fishing settlement. In 1759 a few settlers came to the site of the city from Yarmouth, Mass., and called it after their former home. The city was founded in 1761, when a larger group of settlers came from Sandwich, Mass. They were followed by Acadians (1767) from the Grand Pré district and by United Empire Loyalists (1785).

Yarmouth, county borough (1972 est. pop. 49,830), Norfolk, E England, a port on a long, narrow peninsula between the North Sea and

Breydon Water. It is a resort and fishing port, with engineering and textile industries. The Church of St. Nicholas was founded early in the 12th cent. The Tolhouse (14th cent.), now a museum, is one of the oldest municipal buildings in Great Britain. Although heavily bombed during World War II, the older part of the town retains many of the "rows," narrow lanes from 29 in. to 6 ft (73.7 cm-1.8 m) wide.

Yarmouth. 1. Town (1970 pop. 4,854), Cumberland co., SW Maine, NNE of Portland; settled 1636, inc. 1849. The settlement was raided several times in the Indian wars of the 17th and 18th cent. The town makes wood products and cans fish. **2.** Resort town (1970 pop. 12,033), Barnstable co., SE Mass., on the S shore of Cape Cod; settled and inc. 1639. The main street of this picturesque town is lined with well-preserved old houses. Of special interest is the Thacher House (1680). The Yarmouth port is a historic district.

Yar·muk (yär-mo͝ok'), river, c.50 mi (80 km) long, rising near the Jordan-Syria border and flowing generally W to the Jordan River, S of the Sea of Galilee. One of the larger rivers of Jordan, it is extensively used for hydroelectricity and irrigation. The East Ghor Canal branches from the Yarmuk and irrigates c.30,000 acres (12,140 hectares) in the eastern Ghor region of the Jordan valley.

Ya·ro·slavl (yä'rə-slä'vəl), city (1974 est. pop. 558,000), capital of Yaroslavl oblast, E European USSR, on the upper Volga River. It is a river port, a major rail junction, and an industrial and commercial center. Yaroslavl has linen and leather factories dating from the 17th cent. and textile mills dating from the 18th. Other industries include oil refining, food and tobacco processing, printing, shipbuilding, and the manufacture of machinery and diesel engines. According to tradition, the city was founded by Yaroslav the Wise of Kievan Russia in 1010, although it was not mentioned in the chronicles until 1071. In 1218 it became the capital of the independent Yaroslavl principality, which was absorbed by Moscow in 1463. The city flourished during the 16th and 17th cent. as a commercial center on the Moscow-Arkhangelsk route from the White Sea to the Middle East. In 1564 the first modern Russian ships were built at Yaroslavl, and in 1722 it became the site of Russia's first cloth factory. It was a major Russian manufacturing city by the 18th cent., notably for textiles. Until the construction of the Moscow-Volga Canal in 1937, the city served as Moscow's Volga port. Yaroslavl's landmarks include the 12th cent. Spaso-Preobrazhenski Monastery, several 17th cent. churches, and the Volkov theater (1911).

Yar·ra (yăr'ə), river, 115 mi (185 km) long, rising in the Great Dividing Range, S Victoria, Australia, and flowing generally W through Melbourne to Port Phillip Bay. There are recreational facilities along the river.

Yas·na·ya Po·lya·na (yäs'nə-yə pəl-yä'nə), village, central European USSR, just S of Tula. It was the birthplace and residence of Leo Tolstoy, who is buried at his estate here. The writer's home was looted during German occupation of the village in 1941, but has since been restored and is a museum and literary shrine.

Yates (yāts), county (1970 pop. 19,831), 344 sq mi (891 sq km), W central N.Y., in the Finger Lakes region bounded on the E by Seneca Lake and drained by Flint Creek; formed 1823; co. seat Penn Yan. This grape-growing region also yields other fruit, truck crops, wheat, potatoes, hay, and dry beans.

Yates Center, city (1970 pop. 1,967), seat of Woodson co., SE Kansas, W of Fort Scott; inc. 1884. It is a trade and shipping center for a hay-producing area. There are oil wells in the area.

Ya·tsu·shi·ro (yät'so͞o-shîr'ō), city (1970 pop. 76,000), Kumamoto prefecture, W Kyushu, Japan, on Yatsushiro Bay and the Kuma River estuary. It is a major commercial and fishing port and industrial center with chemical and foodstuffs industries.

Yau·co (you'kō'), town (1970 pop. 12,922), SW Puerto Rico, on the Yauco River. It is the trade and processing center of a sugar, coffee, tobacco, and cotton area.

Yav·a·pai (yăv'ə-pī'), county (1970 pop. 37,005), 8,091 sq mi (20,955.7 sq km), W central Ariz., in a plateau region with Black Hills in the NE and Black Mesa in the N; formed 1864; co. seat Prescott. The Verde, Santa Maria, and Agua Fria rivers cross the county. Gold, silver, copper, lead, and zinc are found here. Livestock and fruit are also produced. The county includes Montezuma Castle National Monument in the northeast and Prescott National Forest.

Yazd (yäzd), city (1971 est. pop. 98,000), Esfahan prov., central Iran, in a desert region. The city is a trade center for cotton, carpets, and pistachios. Textiles and felt goods are the chief manufactures. Yazd was an important Zoroastrian center in Sassanid times. It was conquered by the Arabs in 642, and in the 13th cent., when Marco Polo visited Yazd, it was a large, flourishing city. Shah Ismail annexed it to Persia in the 16th cent. Yazd is a picturesque city, with narrow, winding streets and several fine medieval mosques, religious schools, and tombs. Its Zoroastrian community, the largest in Iran, erected a modern fire temple in 1942. There is an elaborate underground water system.

Ya·zoo (yă-zo͞o'), county (1970 pop. 27,314), 938 sq mi (2,429.4 sq km), W central Miss., bounded in the E and SE by the Big Black River; formed 1823; co. seat Yazoo City. It is intersected by the Yazoo and other streams in the Yazoo system. Agriculture (cotton, corn, and hay) and lumbering are important. Oil fields are found here.

Yazoo, river, 188 mi (302.5 km) long, formed in W central Miss. by the confluence of the Tallahatchie and Yalobusha rivers. Prevented by natural levees from joining the Mississippi River sooner, the Yazoo parallels the Mississippi for c.175 mi (280 km), meandering southwest along the eastern edge of the Mississippi's flood plain before entering it near Vicksburg. The Yazoo is navigable for shallow-draft vessels. Although subject to flooding, the fertile plain between the two rivers, called the Delta, is a major cotton growing region. In the spring of 1973 about 2,800 sq mi (7,250 sq km) of the Yazoo basin were inundated by water backed up because of floods on the Mississippi River; parts of the Delta region were saturated for as long as four months, causing a delay in spring plantings.

Yazoo City, city (1970 pop. 11,688), seat of Yazoo co., W central Miss., on the Yazoo River; inc. 1830. It is a trade, processing, and industrial center in a cotton, cattle, and soybean area. Oil is refined, and clothing and fertilizer are the leading manufactures. Union troops occupied the city in May, 1864, and burned many buildings.

Yea·don (yăd'n), borough (1975 est. pop. 12,058), Delaware co., SE Pa., a suburb of Philadelphia; inc. 1894.

Ye·gor·evsk (yə-gôr'yəfsk), city (1973 est. pop. 70,000), W central European USSR. It is a cotton-milling and textile center and also produces machine tools and asbestos.

Ye·lets (yə-lĕts'), city (1973 est. pop. 107,000), E central USSR, on the Sosna River, a tributary of the Don. A rail junction in a black-earth agricultural district, the city exports livestock and grain. Yelets has been famed for its lace since the 19th cent. Other industries include food processing, leather tanning, and the manufacture of machinery and hydroelectric equipment. First mentioned in 1146, Yelets was taken by Tamerlane in 1395. Virtually abandoned after a Tatar raid in the 15th cent., the city revived in the 17th cent. and became an important commercial center.

Yel·ga·va (yĕl'gə-və), city (1973 est. pop. 59,000), W European USSR, in Latvia, on the Lielupe River. It is a major rail hub and a trade center for grain and timber. The city has textile plants and sugar mills, as well as industries that manufacture machinery, linen, leather, and metal products. The city grew around a fortress established by the Livonian Knights in the 13th cent. but was destroyed by the Lithuanians in 1345. In 1561 Yelgava became the residence of the dukes of Courland; it passed to Russia with the duchy in 1795. German troops held Yelgava during World War I. In 1919, during the struggle for Latvian independence, the city was occupied in turn by Soviet forces, by German free corps, and by the Latvians. Part of independent Latvia from 1920 to 1940, Yelgava was then seized by the USSR, held by the Germans from 1941 to 1944, and liberated by Soviet troops.

Yell (yĕl), county (1970 pop. 14,208), 929 sq mi (2,406.1 sq km), W central Ark., bounded on the NE by the Arkansas River, and drained by the Petit Jean and Fourche La Fave rivers; formed 1840; co. seats Danville and Dardanelle. In an agricultural area yielding cotton, corn, hay, livestock, poultry, and dairy products, it has timber (logging, sawmills), sand and gravel pits, and some industry (poultry dressing and prepared feeds).

Yel·low (yĕl'ō), river, China: *see* Huang Ho.

Yel·low·knife (yĕl'ō-nīf'), town (1971 pop. 6,122), capital of the Northwest Territories, S Mackenzie dist., Northwest Territories, Canada, on the N shore of Great Slave Lake at the mouth of the Yellowknife River. The town was founded (1935) after the discovery of rich deposits of gold. It is a mining, supply, and transportation center, with an airport, radio and meteorological stations, a public school, a post of the Royal Canadian Mounted Police, and regional offices of other federal agencies. Another mine was discovered in 1944 and a new townsite was established the next year. Yellowknife, named after an Indian tribe, became capital of the Northwest Territories in 1967.

Yellow Med·i·cine (mĕd'ĭ-sən), county (1970 pop. 14,523), 758 sq mi (1,963.2 sq km), SW Minn., bordering S.Dak., bounded on the NE by the Minnesota River, and drained by the Lac qui Parle and Yellow Medicine rivers; formed 1871; co. seat Granite Falls. It is in an agricultural area that produces corn, oats, barley, and livestock. It has food processing, some mining, and machinery manufacturing. A Sioux Indian reservation is in the northeast.

Yellow Sea, arm of the Pacific Ocean, between China and Korea. It has a maximum depth of 500 ft (152.5 m). Po Hai, Korea Bay, and the Liao-tung Gulf are its major inlets. South of the Korean peninsula, it becomes the East China Sea.

Yel·low·stone (yĕl'ō-stōn'), county (1970 pop. 87,367), 2,642 sq mi (6,842.8 sq km), S Mont., in an agricultural area drained by the Yellowstone River; formed 1883; co. seat Billings. Sugar beets, beans, livestock, dairy products, and wool are processed. There is some manufacturing.

Yellowstone, river, 671 mi (1,079.6 km) long, rising in NW Wyo., and flowing NE through Mont. to enter the Missouri River near the N.Dak. line. It drains c.70,400 sq mi (182,335 sq km). The most scenic aspects of the river are found in Yellowstone National Park in northwest Wyo. There, the river feeds and drains Yellowstone Lake, 139 sq mi (360 sq km), the largest high-altitude (alt. 7,331 ft/2,236 m) lake in North America. After leaving the lake, the river drops 109 ft (33.2 m) at Upper Falls, then 308 ft (94 m) at Lower Falls, before entering the deep and spectacular Grand Canyon of the Yellowstone (19 mi/30.6 km long); Tower Falls, 132 ft (40.3 m) high, is at the northern end of the canyon. The river's waters have been used for irrigation since the late 1860s. The U.S. Bureau of Reclamation operates several projects on the Yellowstone that are used for irrigation, flood control, power production, and recreation. These include the Huntley project near Billings, Mont., the Buffalo Rapids project near Glendive, Mont., and the Savage unit of the Missouri River basin project.

Yellowstone National Park, first (est. 1872) and largest (2,221,773 acres/899,818 hectares) national park in the United States, NW Wyo., extending into Mont. and Idaho. It lies mainly on a broad plateau in the Rocky Mts., on the Continental Divide, c.8,000 ft (2,440 m) above sea level, surrounded by mountains from 10,000 to 14,000 ft (3,050–4,270 m) high. Most of the plateau is formed from once molten lava; volcanic activity is still evidenced by nearly 10,000 hot springs, 200 geysers, and many vents and hot-mud pots. The more prominent geysers are unequaled in size, power, and variety of action. Old Faithful, the best known although not the largest geyser, erupts at an average interval of 64.5 min and shoots c.11,000 gal (41,640 liters) of water some 150 ft (45.8 m) high. Mammoth Hot Springs, a series of five terraces with reflecting pools, continues to grow, as residue from the mineral-rich steaming water is deposited. The park's other natural wonders include petrified forests, lava formations, and the "black glass" Obsidian Cliff. Eagle Peak, 11,370 ft (3,466 m), is the highest point in the park. Yellowstone Lake, the Grand Canyon of the Yellowstone, and waterfalls are notable features on the Yellowstone River, which crosses the park. Evergreen forests cover 90% of the park, and a great variety of flowers and other plant life are found. Bears, mountain sheep, elk, bison, moose, and more than 200 kinds of birds make their homes in Yellowstone Park, which is one of the world's largest wildlife sanctuaries.

Yem·en (yĕm'ən, yă'mən), also **North Yemen,** officially Yemen Arab Republic, republic (1975 pop. 5,237,893), 75,290 sq mi (195,000 sq km), SW Asia, at the SW tip of the Arabian Peninsula. Sana is the capital. Yemen is bordered on the north and northeast by Saudi Arabia, on the southeast by Southern Yemen, and on the west by the Red Sea. The Arabic name, Al-Yaman, probably means the right hand and describes the country's position as one stands before the Kaaba in Mecca facing east. Because it was confused with another Arabic word meaning happiness, the name was rendered Arabia Felix (happy or fortunate Arabia) by European geographers. In classical times, however, the name was applied to a much larger area than the present Yemen.

The country consists of a narrow coastal plain, the interior high-

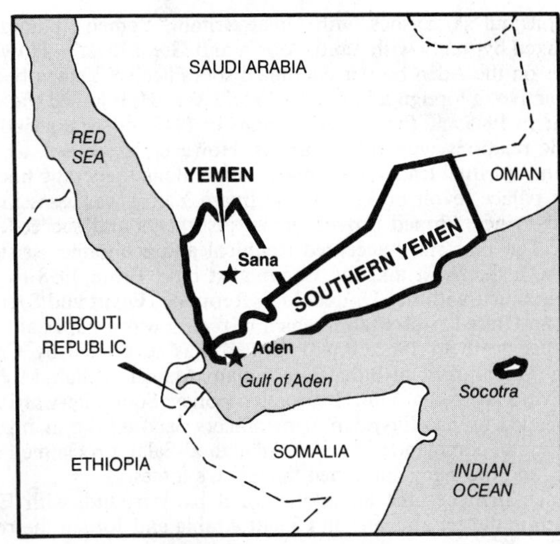

lands, and the eastern desert. The coastal plain, or Tihamah, which is about 20 to 50 mi (30–80 km) wide, extends along the length of western Yemen. It is a hot and virtually rainless region with high humidity and is composed of alluvial and talus material carried down from the highlands; there is little vegetation. The interior highlands, which are actually a section of the upturned Arabian plateau, are the highest part (rising to more than 12,000 ft/3,660 m) of the Arabian Peninsula. The highlands receive an annual average rainfall of c.20 in. (50 cm), making them also the wettest part of the peninsula; most of the precipitation occurs during the summer rainy season. Vegetation is of the subtropical variety and varies with altitude. Numerous wadis radiate from the highlands, but Yemen has no permanent streams; oases and springs provide local water needs. The eastern side of the highlands, which is in a rain shadow, slopes eastward into the great sandy expanses of the Rub al Khali desert. The great majority of the people of Yemen are Arabs; most of those living on the Tihamah are of mixed Arab and African stock. About 60% of the population are Sunni Moslems, while the remainder are Zaidi Moslems. Between 1948 and 1950 about 50,000 Yemeni Jews emigrated to Israel.

Economy. Except for salt, Yemen has no commercially exploitable minerals. The moist and fertile highlands are the country's chief agricultural region. A variety of grains, as well as vegetables, fruits, cotton, coffee, and qat (a narcotic shrub) are raised. In the Tihamah irrigated cotton and dates are grown; however, this area is primarily a livestock-raising region in which sheep, goats, and camels are raised. Manufacturing, which is largely based on agricultural products, has been generally increasing but provides the country with little revenue. Handicrafts play an important role in the economy. Agricultural products provide about 70% of Yemen's exports, with cotton, coffee, qat, salt, and hides and skins the leading commodities.

History. The earliest recorded civilizations of southern Arabia were the Minaean and Sabaean. The Sabaean kingdom flourished from c.750 B.C. to c.115 B.C., with Marib the capital after c.600 B.C. Sabaean society was highly developed technically, as witnessed by the remains of a great dam at Marib that was the center of a large irrigation system. The Himyarites, who followed the Sabaeans, were invaded by the Romans (1st cent. B.C.) and were occupied by the Ethiopians (A.D. c.340–378). During the second Himyarite kingdom Christianity and Judaism took root in Yemen. Ethiopia again conquered the country in 525. After a Persian period (575–628), Islam came to Yemen, which was soon reduced to a province of the Moslem caliphate. After the breakup of the caliphate, Yemen came under the control of the rising Rassite dynasty, imams of the Zaidi sect who built the theocratic political structure of Yemen that lasted until 1962. The Fatamid caliphs of Egypt occupied most of Yemen from c.1000 until c.1175, when it fell to the Ayyubids, who ruled until c.1250. By 1520 Yemen formed part of the Ottoman Empire, which exercised nominal sovereignty until the end of World War I.

A turbulent wave of Wahabism, a puritanical sect of Islam, swept across the Arabian Peninsula at the opening of the 19th cent. and drove out the Zaidi imams. Ibrahim Pasha of Egypt drove out the Wahabis in 1818, and the Egyptians remained until 1840. The Turks then replaced the Egyptians. In 1934 after a brief Saudi Arabian invasion and skirmishes with Great Britain, Yemen's boundaries were fixed by treaty with Saudi Arabia and Great Britain. However, clashes on the Aden border continued sporadically. Yemen became more active in foreign affairs after World War II; it joined the Arab League in 1945 and the United Nations in 1947 and established diplomatic relations with other nations. However, the imam, as both king and spiritual leader, continued to rule along theocratic lines. In 1948 a palace revolt broke out, and Imam Yahya was assassinated. Crown Prince Ahmad drove out the insurgents and succeeded as imam. The new ruler accepted technical and economic assistance from both the West and the Communist bloc. From 1958 to 1961 Yemen joined with the United Arab Republic (Egypt and Syria) to form the United Arab States, which in reality was a paper alliance. After his death in 1962, Imam Ahmad was succeeded by Crown Prince Mohammed al-Bahr (later Imam Mansur Billah Mohammed), who favored a neutralist foreign policy. Soon afterward a revolt headed by pro-Egyptian army officers deposed the imam. The ruling junta, commanded by Col. Adallah al-Salal, proclaimed a republic, and the army contained the imam's forces.

Yemen then became an international battleground, with Egypt supporting the republicans and Saudi Arabia and Jordan the royalists. The Yemeni republicans split into opposing factions on the issue of Egyptian support. In an administrative reorganization in 1966, the independent government of Premier Hassan al-Amri was ousted by a strongly pro-Egyptian regime, with al-Salal assuming the office of premier. In Nov., 1967, al-Salal's government was overthrown while he was abroad, and a three-man republican council was formed with Qadi Abd al-Rahman al-Iryani (one of the anti-Egyptian leaders) as chairman; al-Amri resumed the premiership. Fighting between the republicans and the royalists continued until 1970, when Saudi Arabia formally recognized the republican regime and stopped all aid to the royalists. Between 1967 and 1974 there were frequent border clashes between Yemen and Southern Yemen. On June 12, 1974, Chairman al-Iryani resigned after a period of internal political tension, and the next day a group of army officers led by Col. Ibrahim al-Hamidi staged a nonviolent coup d'état.

Government. Yemen is governed by the constitution of 1970 and has a republican form of government. The chairman (president) of the presidential council, which has three members, is the head of state and has the most power. The consultative council, the legislative body, is composed of 179 members, of which 20 are appointed by the president.

Yemen, Southern, officially People's Democratic Republic of Yemen, republic (1973 est. pop. 1,555,000), 111,074 sq mi (287,682 sq km), SW Asia, at the S edge of the Arabian Peninsula. Aden is the capital. Southern Yemen is bordered on the north by Saudi Arabia, on the east by Oman, on the south by the Gulf of Aden and the Arabian Sea, and on the west by Yemen. The islands of Karaman, in the Red Sea, Perim, in the Bab el Mandeb, and Socotra and the Kuria Maria islands, in the Arabian Sea, are part of the republic.

The country consists of a coastal plain, a belt of mountains, and a portion of the Arabian plateau. The coastal plain, which is very narrow and stretches more than 700 mi (1,125 km) along the southern edge of the Arabian Peninsula, is a hot and virtually rainless region with high humidity and sparse vegetation (mainly thorn bushes). In the western part of the country are low mountains that are actually part of the edge of the Arabian plateau. The ranges increase in height from east to west, with the western portion also being the wettest part of the country. The mountains lead up to the level of the plateau, which, in the east, rises from the coastal plain. In the north the plateau slopes down into the barren, sandy expanses of the Rub al Khali desert. The mountains and the plateau, both of which have an arid, rocky, and rugged environment, are cut by numerous wadis that usually carry water only during the summer rainy season, and then mainly in the form of floods. Wadi Hadhramaut, in the central part of the country, is the largest wadi. Its wide upper and middle portions have alluvial soil and moisture from intermittent streams and are the country's best farmlands; the lower portion is dry and uninhabited. The majority of the people of Southern Yemen are Arabs; those on Socotra are of mixed European, African, and Arab descent. The tribal social structure is still prevalent in the country, although its importance diminishes along the coast, where there has been long contact with foreigners. Only the tribes in the north are entirely nomadic.

Economy. Southern Yemen is one of the world's poorest nations and relies heavily on foreign aid. Its economy was greatly damaged during the time that the Suez Canal was closed following the 1967 Arab-Israeli war. Subsistence agriculture is the mainstay of the economy. Modern methods of irrigation have increased the amount of tilled land. The principal farming areas are Wadi Hadhramaut, which produces coffee and tobacco; Wadis Bana and Tibban, which produce cotton; and the moist highland region in the west, which produces grains and dates. Pastoralism is important throughout the country, and fishing is a major economic activity along the coast. Nearly all of Southern Yemen's industry is located in or near Aden. Petroleum refining is the most important industry, and petroleum products and ships' fuel account for nearly three quarters of the country's exports. Several light industries produce domestically consumed goods such as soap, cigarettes, and textiles. Cotton, fish products, and hides and skins are major exports. The region has also been a major supplier of incense. Southern Yemen's chief imports are food, textiles, and crude petroleum.

History. Southern Yemen was once part of a larger region called Al-Yaman, the Arabia Felix of early European geographers. A number of empires, including the Minaean, Sabaean, and Himyarite, flourished there. The region came under Moslem influence in the 7th cent. In the 16th cent. it became part of the Ottoman Empire and came under the suzerainty of the imams of Yemen. The British presence in Southern Yemen began in 1839, when forces of the British East India Company occupied Aden. In 1854 and 1857 the Kuria Maria and Perim islands were ceded to the British and other mainland areas were purchased by them. Between 1886 and 1914, Britain signed a number of protectorate treaties with local rulers. In 1937 the area was divided for administrative purposes into the East Aden protectorate and the West Aden protectorate.

In 1959, 6 small states of the West Aden protectorate formed the Federation of the Emirates of the South; it was later enlarged to 10 members. Despite considerable opposition from its population, the Aden colony proper was made part of the federation (1963), which was then renamed the Federation of South Arabia. By 1965, 16 tribal states had joined the federation. However, nationalist groups opposed to the federation began a terrorist campaign against the British. Two rival nationalist groups emerged: the National Liberation Front (NLF) and the Front for the Liberation of Occupied South Yemen (FLOSY). The NLF, which had emerged as the dominant group by 1967, forced the collapse of the federation after taking control of the governments of all the component states. Britain accelerated its withdrawal, and Southern Yemen became independent in Nov., 1967, with Qahtan al-Shaabi of the NLF the first president. In 1970 the country received a new constitution and was renamed the People's Democratic Republic of Yemen. Since independence there have been border disputes and armed clashes with Oman and the Yemen Arab Republic. An accord was signed with the Yemen Arab Republic in 1972 calling for the end of fighting and the merger of the two countries.

Government. The country is governed under the constitution of 1970, which vests power in a three-man presidential council. The president is the head of state and the prime minister the head of government, and there is a 101-member supreme people's council. The National Liberation Front is the only legal political party.

Yen·a·ki·ye·vo (yĕn'ə-kē'yə-və), city (1970 pop. 92,000), SE European USSR, in the Ukraine, on the Bulavin River. It is a large center for coal, iron and steel, and chemicals in the Donets Basin. Yenakiyevo was founded in 1883 as a coal-mining settlement.

Yen-an or **Yen·an** (both: yĕn'än'), city (1971 est. pop. 45,000), N Shensi prov., China, on the Yen River. Now a market and tourist center, it is famed as the terminus of the long march and the de facto capital (1936–47, 1948–49) of the Chinese Communists, who established arsenals, several colleges, and a military academy (now a museum) here. The city's many loess caves served as homes and air raid shelters during World War II. Points of interest include the former homes of Mao Tse-tung and Chou En-lai, and a nine-story pagoda built during the Sung dynasty (960–1279) and now made into a monument to the revolution.

Ye·ni·sei (yĕn'ĭ-sā'), chief river of Siberia, c.2,500 mi (4,025 km) long, central Siberian USSR. It is formed at Kyzyl, Tuva Autonomous Oblast, by the junction of the Bolshoi Yenisei and Maly Yenisei rivers, which rise in the east Sayan Mts. along the USSR–Mongolian border. It flows westward, then generally north to enter the Kara Sea through a c.250-mi (400-km) long estuary composed of Yenisei Bay and Yenisei Gulf. The river is frozen during the winter months. In the spring ice in the upper Yenisei melts before that in the lower river, causing extensive flooding as water backs up behind the frozen portion of the river. The Yenisei's upper course is turbulent, with many rapids, and has a great hydroelectric generating potential. The river's middle course widens and is navigable for steamers. There is fishing for sturgeon and salmon in the river's lower reaches.

Yen·ping (yĕn'pĭng'): *see* Nan-p'ing, China.

Yen-t'ai (yĕn'tī') or **Che·foo** (jŭ'fōō'), city (1970 est. pop. 180,000), N Shantung prov., China. It is Shantung's largest fishing port. The city also has fruit orchards, and wine and brandy are produced. Yen-t'ai was opened to foreign trade in 1862. By the Chefoo Convention, signed here in 1876, many new treaty ports were established. A rail line, built in 1955, links Yen-t'ai with Ch'ing'-tao and Chi-nan.

Ye·ot·mal (yā-ōt'mäl'), town (1971 pop. 64,829), Maharashtra state, central India. Yeotmal is a district administrative center, a cattle-breeding town, and a market for peanuts, cotton, and timber.

Yeo·vil (yō'vĭl'), municipal borough (1972 est. pop. 25,960), Somerset, SW England, on the Yeo River. It is a market town and a leather-making center. Glove making has been a local specialty since the 16th cent.; helicopters and processed foods are also produced.

Yer·ba Bue·na Island (yûr'bə bwā'nə), 300 acres (121.5 hectares), W Calif., in San Francisco Bay. It is the midpoint of the San Francisco-Oakland Bay Bridge, which crosses the island through a tunnel. On the island are several government installations, including a lighthouse service and a naval training station.

Ye·re·van (yĕr'ə-vän'), city (1974 est. pop. 870,000), capital of the Armenian Soviet Socialist Republic, SE European USSR, on the Razdan River. One of the USSR's oldest towns and a leading industrial, cultural, and scientific center, Yerevan is also a rail junction and carries on a brisk trade in agricultural products. The city's industries produce metals, machine tools, electrical equipment, chemicals, textiles, and food products. Archaeological evidence indicates that the fortress of Yerbuni stood on Yerevan's site in the 8th cent. B.C. The city, known in the 7th cent. A.D., was the capital of Armenia under Persian rule and became historically and strategically important as a crossroads of the caravan routes between Transcaucasia and India. After the downfall (15th cent.) of Tamerlane's empire, to which Yerevan belonged, the city passed back and forth between Persia and Turkey. In 1440 it became the center of East Armenia. During the 17th cent. Yerevan was a frontier fort and a caravan trading point. It became the capital of the Yerevan khanate of Persia in 1725. Taken by Russia in 1827, the city was formally ceded by the Treaty of Turkmanchai (1828). Yerevan was the center of independent Armenia from 1918 to 1920, when it became the capital of the newly formed Armenian SSR. Educational and cultural facilities include a university, the Armenian Academy of Sciences, a state museum, and several libraries. There are ruins of a 16th cent. Turkish fortress.

Yer·ing·ton (yĕr'ĭng-tən), city (1970 pop. 2,010), seat of Lyon co., W Nev., on the Walker River SE of Carson City; settled 1860, inc. 1907. It is a trading center in a ranching and mining region.

Ye·şil Ir·mak (yĕ-shēl' ər-mäk'), river, c.260 mi (420 km) long, rising NE of Sivas, N Turkey, flowing NW, then NE, past Tokat and Amasya into the Black Sea near Samsun.

Yev·pa·to·ri·ya (yĕf'pə-tôr'ē-ə, -tōr'ē-ə), city (1973 est. pop. 86,000), S European USSR, in the Ukraine, in the Crimea. It is a Black Sea port, a rail hub, and a vacation and health resort. Fishing, food processing, wine making, limestone quarrying, weaving, and the manufacture of building materials, machinery, and furniture are the chief industries. Yevpatoriya stands on the site of the ancient Greek colony of Kerkinitida, founded in the 6th cent. B.C. In the 1st cent. B.C. the area was captured by the Pontian king Mitridat Evpator. Yevpatoriya came under the control of the Turko-Tatars in the 13th cent.; they later became vassals of Turkey, which took the city in 1478. Russia annexed Yevpatoriya along with the rest of the Crimea in 1783, and during the Crimean War it was occupied (1854) by British, French, and Turkish troops. Historic landmarks include a 16th cent. mosque and the ruins of the Tatar fortress (15th cent.).

Yin-ch'uan or **Yin·chwan** (both: yĭn'chwän'), city, capital of Ninghsia Hui Autonomous Region, China, on the Huang Ho. It is a shipping point for the products of the fertile Ninghsia plain. Textiles are manufactured, and coal is mined in the area. Marco Polo visited Yin-ch'uan in the 13th cent.

Ying-k'ou or **Ying·kow** (both: yĭng'kou', -kō'), city (1970 est. pop. 215,000), S Liaoning prov., China, on the Liao River near its mouth on the Po Hai. It is on the South Manchurian RR in an area producing rice, cotton, and oakleaf silk. It has fishing and lumbering industries and plants making textiles, petrochemicals, machinery, automotive parts, and paper products.

Yoa·kum (yō'kəm), county (1970 pop. 7,344), 830 sq mi (2,149.7 sq km), NW Texas, on the Llano Estacado bounded on the W by the N.Mex. border; formed 1876; co. seat Plains. Its agriculture includes cattle, sheep, goats, horses, hogs, mules, some dairy products, grain sorghums, corn, peanuts, cotton, alfalfa, watermelons, and sunflowers. It also has oil and natural-gas wells and refineries.

Yok·kai·chi (yōk-kī'chē), city (1970 pop. 229,234), Mie prefecture, W Honshu, Japan, a port on Ise Bay. It is a manufacturing center that produces banko ware (a kind of porcelain), cotton textiles, tea, vegetable oil, and cement.

Yo·ko·ha·ma (yō'kə-hä'mə), city (1973 est. pop. 2,494,975), capital of Kanagawa prefecture, SE Honshu, Japan, on the W shore of Tokyo Bay. Yokohama is a leading port and belongs to the extensive urban-industrial belt around Tokyo. Among its industries are shipyards, steel mills, oil refineries, chemical plants, and factories that produce transportation equipment, electrical apparatus, automobiles, machinery, primary metals, and textiles. In 1854 Matthew C. Perry visited Yokohama, which was then a small fishing village. In 1859 it became a port for foreign trade and the site of a foreign settlement that enjoyed extraterritorial rights. Known especially for its exports of raw silk, Yokohama also handled canned fish and other local products. Foreign trade led to the rapid growth of Yokohama, which served during the last half of the 19th cent. as Tokyo's outer port. The capital has since expanded the facilities and operations of its own port; but Yokohama is still important in the export of machinery, iron, and steel, as well as in its traditional staple of raw silk. Yokohama formally became a city in 1889. Extraterritoriality was abolished in 1899. Virtually destroyed by an earthquake and fires in 1923, Yokohama was quickly rebuilt; the city was modernized, and extensive improvements were made in its harbor. Yokohama suffered heavy bombardment during World War II, but it revived and prospered. The city has four universities, a variety of Christian churches, Shinto shrines, and temples, and numerous parks and gardens, notably Nogeyama Park, which was created after the earthquake. The filling in of shallow areas of the bay for port facilities and industrial use has continued.

Yo·ko·su·ka (yō'kə-sōō'kə), city (1970 pop. 347,576), Kanagawa prefecture, E central Honshu, Japan. It has an important naval base (founded 1868), shipyards, arsenals, and ironworks.

Yo·lo (yō'lō'), county (1970 pop. 91,788), 1,028 sq mi (2,662.5 sq km), N central Calif., in the Sacramento Valley, bounded on the E by the Sacramento River and drained by Cache and Putah creeks; formed 1850; co. seat Woodland. It has extensive year-round agricultural production, which includes sugar beets, tomatoes, asparagus, alfalfa, rice, fruit, olives, beans, grain, stock raising, and dairying. There is mineral mining in the county, and its industry includes food processing, petroleum refining, and the manufacture of plastics, steel and wood products, and concrete.

Yon·kers (yŏng'kərz), city (1979 est. pop. 185,700), Westchester co., SE N.Y., on the E bank of the Hudson, in a hilly region just N of the Bronx (New York City); inc. 1855. Its elevator works date from 1852. Other manufactures are aerosol valves, sugar, seafood cocktails, chemicals, cable, wire, telephone parts, art supplies, and electronic duplicators. The area was included in the land grant given (1646) by the Dutch West India Company to the New Netherland lawyer Adriaen Van der Donck. It was a trading center in colonial days. Water power from the Nepperhan River attracted early industries, such as the elevator works and several carpet mills. Today Yonkers is the seat of Sarah Lawrence College and the Boyce Thompson Institute for Plant Research. Also in the city are Philipse Manor, built in the 17th cent. by Frederick Philipse, the Hudson River Museum and Space Planetarium, and Yonkers Raceway.

Yonne (yôn), department (1974 est. pop. 295,000), 2,867 sq mi (7,425.5 sq km), N central France; capital Auxerre.

York (yôrk), county borough (1972 est. pop. 104,780), North Yorkshire, N England, at the confluence of the Ouse and Foss rivers. York is especially noted for the manufacture of cocoa, chocolate, and confectionery. Instrument making, printing, and light engineering are among its other industries. York was a British settlement occupied by the ancient tribe of Brigantes. It was an important military post of the British province of the Roman Empire. Emperor

Hadrian visited York in 120 and built an earthen rampart to keep out the Picts and the Celts. The emperors Septimus Severus (211) and Constantius I (306) died here and Constantine I was proclaimed emperor at York in 306. York became an important center in the Kingdom of Northumbria. In the 7th cent. St. Paulinus, the first archbishop of York, was consecrated. The archbishopric of York is the ecclesiastical center of the north of England, second only to Canterbury in importance. In the 8th cent. York was one of the most famous centers of education in Europe. Alcuin was born here and became the headmaster of St. Peter's School, now one of the oldest public schools in England. In the Middle Ages York was a busy wool market. The Cathedral of St. Peter, commonly known as York Minster, occupies the site of the wooden church in which King Edwin was baptized by St. Paulinus on Easter Day in 627. The present edifice dates partly from the Norman period. There are many other notable medieval structures in York. The ancient city is enclosed by walls dating in part from Norman times. Four of the gates, including Micklegate and Monk Bar, remain. The Univ. of York was founded in 1963; there is also a teacher-training school. York has several important museums. The York Plays reached their height in the 15th cent. and were revived at the Yorkshire Festival of 1951.

York. 1. County (1970 pop. 111,576), 1,000 sq mi (2,590 sq km), SW corner of Maine, lying between N.H. and the coast, drained by Salmon Falls and the Piscataquis, Saco, Mousam, and Ossipee rivers; formed 1652 (the oldest county in the state), named Yorkshire in 1658, with Cumberland and Lincoln cos. set off in 1670; co. seat Alfred. It has truck gardens and dairying, lumbering, commercial fishing, and a food-processing industry. Textile mill products, clothing, paper products, plastics, leather goods, shoes, concrete, metal products, and machinery are made. There are summer resorts on the coast and lakes. **2.** County (1970 pop. 13,685), 577 sq mi (1,494.4 sq km), SE Nebr., in a farming region drained by branches of the Big Blue River; formed 1870; co. seat York. Agriculture yields livestock, grain, dairy, and poultry produce. **3.** County (1970 pop. 272,603), 909 sq mi (2,354.3 sq km), S Pa., bounded in the E by the Susquehanna River and in the S by Md., with part of South Mt. lying in the NW; formed 1749; co. seat York. Its agriculture includes apple orchards. It has limestone quarries and diversified manufactures (food products, cigars, textiles and clothing, lumber, furniture, paper, chemicals, leather, concrete, iron and steel, metal products, machinery, and transportation equipment). **4.** County (1970 pop. 85,216), 685 sq mi (1,774.2 sq km), N S.C., bounded in the W by the Broad River, in the E by the Catawba River, in the N by N.C., and containing the lower end of Catawba Lake, formed by a dam N of Rock Hill; formed 1785; co. seat York. It is in a fertile agricultural area that produces cotton, grain, peaches, truck farming, poultry, and dairy products. Its industry includes food processing and the manufacture of textiles and clothing, lumber, paper, chemicals, rubber, metal products, and machinery. It contains part of Kings Mt. National Military Park in the northwest. **5.** County (1970 pop. 33,203), 129 sq mi (334.1 sq km), SE Va., in the tidewater region, on the NE shore of the peninsula, bounded in the SW by the James River, in the S by Hampton Roads, and in the E by Chesapeake Bay; formed 1634; co. seat Yorktown. Its agriculture includes truck farming, tobacco growing, and stock raising. It has fishing, oystering, and a seafood industry, and petroleum refining is done.

York. 1. Town (1970 pop. 5,690), York co., SW Maine, NE of Kittery; settled 1624, chartered 1641, inc. 1652. The early settlement suffered much in the Indian wars and was nearly destroyed in 1692. York is said to have been the site of the first sawmill (c.1624) in what is now the United States. The stone jail (built 1653) is now a museum; a colonial church, a garrison house, and other old buildings are preserved. **2.** City (1970 pop. 6,778), seat of York co., SE Nebr., W of Lincoln; platted 1869, inc. 1872. The trade and distribution center of an irrigated farm region, it also processes and ships large quantities of eggs. **3.** City (1970 pop. 50,335), seat of York co., SE Pa., on Codorus Creek, in a fertile agricultural area; laid out 1741, inc. as a city 1887. It is a market, trade, processing, and distribution center in the Pennsylvania Dutch country. In addition to food and related products, its factories make air conditioners, turbines, chains, and textile and paper products. York was a meeting place (1777–78) of the Continental Congress. During the Civil War it was occupied briefly (1863) by Confederates. **4.** Town (1970 pop. 5,661), seat of York co., N S.C., SW of Charlotte, N.C.; settled c.1751, inc. 1841. It is a processing center for a farm area. Textiles, machinery, and hardware are manufactured.

York, Cape, NW Greenland, in N Baffin Bay, W of Melville Bay. The Cape York meteorites were discovered by U.S. explorer Robert E. Peary, who brought the largest (c.100 tons) to the American Museum of Natural History, New York City. In 1932 a monument to Peary was erected at Cape York.

Yorke Peninsula (yôrk), 160 mi (257.4 km) long and 35 mi (56.3 km) wide, SE South Australia state, Australia, between Spencer Gulf and Gulf St. Vincent. It is a farming area.

York Factory, fur-trading post, NE Man., Canada, on Hudson Bay, at the mouth of the Hayes River, just E of the mouth of the Nelson River. The name was used for several late-17th cent. posts in the area, which changed hands during the struggle between England and France for control of the rich fur trade. The British gained final control after the Peace of Utrecht (1713). The present post (built 1788–93) was a major warehouse for the Hudson's Bay Company. It was closed in 1957.

York·shire (yôrk'shîr', -shər), former county, N England. Largest of the English counties, it was divided into three administrative counties, or ridings: East Riding, with Beverley as its county town; North Riding, with Northallerton as its county town; West Riding, with Wakefield as its county town. There are numerous prehistoric remains, including barrows and huge monoliths. The city of York (Eboracum) was an important Roman post in northern Britain. After the Roman withdrawal in the 5th cent. and the Saxon invasions, the region became a part of the kingdom of Northumbria. In the Middle Ages Yorkshire was the center of British monastic life; by the early 16th cent. there were more than 100 abbeys, priories, nunneries, and friaries. In this period Yorkshire had a large woolen-cloth industry. The dissolution of the religious houses under Henry VIII was met with great resistance, notably in the Pilgrimage of Grace (1536). In 1974 Yorkshire was reorganized; most of East Riding became part of the new nonmetropolitan county of Humberside, most of North Riding and West Riding became part of the new nonmetropolitan county of North Yorkshire, an area of West Riding became part of the new metropolitan county of West Yorkshire, and an area of North Riding became part of the new nonmetropolitan county of Cleveland.

York·ton (yôrk'tən), city (1971 pop. 13,430), SE Sask., Canada, NE of Regina. It is a railroad center and has large stockyards, warehouses, a flour mill, brick and cement plants, and a farm-implement plant.

York·town (yôrk'toun'), historic town (1970 pop. 311), seat of York co., SE Va., on the York River 10 mi (16 km) from its mouth on Chesapeake Bay; settled 1631, laid out 1691. It is included in the Colonial National Historical Park. The town, once an important tobacco port, reached its zenith c.1750. The Yorktown campaign (1781) brought to a close the American Revolution; the battlefield surrounds the town. In the Civil War Yorktown was besieged (Apr.–May, 1862) in the Peninsular campaign, and the city was taken by Union troops on May 4. Places of interest in Yorktown include the customhouse (c.1706; restored 1929); Grace Church (1697); the Moore House (c.1725), in which the terms of Cornwallis's surrender were negotiated; and the Yorktown Monument (1881), commemorating the victory of 1781.

York·ville (yôrk'vĭl'), village (1974 est. pop. 2,453), seat of Kendall co., NE Ill., on the Fox River SW of Aurora; inc. 1873. It ships grain.

Yo·sem·i·te National Park (yō-sĕm'ĭ-tē), 761,320 acres (308,335 hectares), E central Calif.; est. 1890 as a result of the efforts of conservationist John Muir. Located in the Sierra Nevada, it is a glacier-scoured area of great beauty. Enclosed within the park is the famed Yosemite Valley (alt. c.4,000 ft/1,220 m), surrounded by cliffs and pinnacles; Half Dome, which reaches a height of c.4,800 ft (1,465 m) above the valley, and El Capitan, which rises perpendicularly c.3,600 ft (2,000 m) above the valley, are the highest of the surrounding peaks. The world's three largest monoliths of exposed granite are found in the park. There are also many beautiful lakes, rivers, streams, and waterfalls, the most noted of which is Yosemite Falls, the highest in North America, with a drop of 2,425 ft (739.6 m) in two segments; Ribbon Falls has a 1,612-ft (491.7-m) drop. Three groves of sequoias are within the park's limits, which also include fine growths of other trees and more than 1,000 varieties of flowering plants. In the scenic Hetch Hetchy Valley is the reservoir that supplies water to San Francisco.

Yosh·kar-O·la or **Iosh·kar-O·la** (both: yŏsh-kär'ō-lä'), city (1973 est. pop. 188,000), capital of the Mari ASSR, E central European USSR. Manufactures include pharmaceuticals and agricultural machinery. The city was founded in 1578 as a Russian outpost.

Yo·su (yŭ'sōō'), city (1970 pop. 113,651), S South Korea, on the Korea Strait. It is a trading port and fishing base and the site of South Korea's major oil refinery. Yosu also has an important thermal power plant. There is a large canning industry.

Young (yŭng), county (1970 pop. 15,400), 888 sq mi (2,300 sq km), N Texas, drained by the Brazos River; formed 1856; co. seat Graham. Its agriculture includes wheat, cotton, grain sorghums, oats, barley, fruit, truck crops, dairy products, livestock, and poultry. There are oil and natural-gas fields, clay and sand pits, and large deposits of coal. Some processing and manufacturing are done.

Youngs·town (yŭngz'toun'), city (1979 est. pop. 120,900), seat of Mahoning co., NE Ohio, near the Pa. line, in an extensive coal and iron region; founded 1797, inc. 1849. It is one of the largest iron and steel centers in the country. Rubber goods, electric lamps, machinery, plant equipment, aluminum extrusions, automobiles and parts, rolling mill equipment, and sprinkler systems are also produced. Discovery of iron ore, coal, and limestone led to the construction of the first iron furnace in 1803. The city's growth was spurred by the opening of the Pennsylvania and Ohio Canal (1839), the arrival of the railroad (1853), and the establishment of steel plants in the 1890s.

Y·pres (ē'prə), town (1970 pop. 20,825), West Flanders prov., SW Belgium, near the French border. It is an agricultural market and an industrial center. Manufactures include textiles, textile-making machinery, and processed food. During the Middle Ages Ypres was one of the most powerful towns of Flanders, with a flourishing cloth industry that rivaled those of Ghent and Bruges. However, political and social unrest and foreign wars led to the decline of this industry. A center of resistance to Spanish rule, the town was taken (1584) and sacked by Alessandro Farnese. It was held by France from 1678 to 1716 and from 1792 to 1814. In World War I Ypres was the scene of three great battles. Ypres was completely destroyed during the war and was later rebuilt. Among the city's restored buildings are the Gothic Cathedral of St. Martin and the magnificent cloth-workers hall (both originally built in the 13th cent.). On the ramparts of the fortifications built (late 17th cent.) by Vauban is a British memorial gate designed by Sir Reginald Blomfield, and outside the town's walls are some 40 military cemeteries.

Yp·si·lan·ti (ĭp'sə-lăn'tē), city (1979 est. pop. 25,700), Washtenaw co., SE Mich., on the Huron River; inc. 1832. It is a residential, commercial, and farm-trade center and an industrial city where automobiles and automobile parts are manufactured. Indian trails crossed this site and an Indian village and a French trading post (1809-c.1819) were here. Eastern Michigan Univ. is in the city.

Y·re·ka (wī-rē'kə) or **Yreka City,** town (1970 pop. 5,394), seat of Siskiyou co., N Calif., NE of Eureka; inc. 1857. It boomed as a mining camp after the discovery of gold (1851) at Yreka Flat.

Y·ser (ē-zĕr'), river, c.50 mi (80 km) long, rising in N France and flowing generally NE through NW Belgium and into the North Sea at Nieuwpoort. It connects a network of canals. It was the scene of heavy fighting in World War I.

Y·stad (ōō'stəd), city (1970 pop. 14,164), Malmöhus co., S Sweden, a seaport on the Baltic Sea. It is a commercial and industrial center and a popular summer resort. At Ystad in 1799 Gustavus IV issued his declaration of war against Napoleon I.

Yü·an (yōō-än'), river, 540 mi (868.9 km) long, rising in S Kweichow prov. and flowing generally NE to Tung-t'ing lake, Hunan prov., SE China. Navigation above Ch'ang-te is limited by rapids to small craft. The Yüan valley, a major north-south trade route, yields tungsten, iron ore, and tung yu (wood oil).

Yu·ba (yōō'bə), county (1970 pop. 44,736), 638 sq mi (1,652.4 sq km), N central Calif., extending NE from Feather River through the Sacramento Valley to the lower W slope of the Sierra Nevada, and drained by the Yuba and Bear rivers; formed 1850; co. seat Marysville. Its agriculture includes fruit growing (peaches, pears, grapes, figs, prunes, olives, and nuts), farming (hops, rice, barley and other grains, alfalfa, and truck crops), dairying, and stock raising. It has lumbering, some mining, and food processing. It includes parts of Plumas and Tahoe national forests, and there is hunting and fishing.

Yuba City, town (1975 est. pop. 15,200), seat of Sutter co., N central Calif., on the Feather River; founded 1849 during the gold rush; inc. 1908. It is a processing center for fruits, nuts, and vegetables.

Yu·ca·tán (yōō'kə-tăn', -tän'), state (1970 pop. 774,011), 14,868 sq mi (38,508 sq km), SE Mexico, occupying most of the N part of the Yucatán peninsula; capital Mérida. It became a state when Mexico won independence (1821) but seceded from 1839 to 1843.

Yucatán, peninsula, c.70,000 sq mi (181,300 sq km), mostly in SE Mexico, separating the Caribbean Sea from the Gulf of Mexico. The peninsula is largely a low, flat, limestone tableland rising to c.500 ft (155 m) in the south. To the north and west the plain continues as the Campeche Bank, stretching under shallow water c.150 mi (240 km) from the low, sandy shore line. The eastern coast rises in low cliffs in the north and is indented by bays and paralleled by islands and cays in the south. Short ranges of hills cross the peninsula at scattered intervals.

The land has tropical dry and rainy seasons, but generally in the north the climate is hot and dry and in the south hot and humid. Most of the northern half, although covered with only a few inches of subsoil, is one of the most important henequen-raising regions of the world; the uncultivated area is under a dense growth of scrub, cactus, sapote wood, and mangrove thickets. Subsistence crops, tobacco, and cotton also are grown. Magnificent forests of tropical hardwoods provide the basis for a lumber industry. This area teems with tropical life, including the jaguar, the armadillo, the iguana, and the Yucatán turkey. Fishing is important along the Yucatán coast. The peninsula's fine beaches are being developed as tourist resorts.

Yuc·ca House National Monument (yŭk'ə): *see* National Parks and Monuments Table.

Yu·go·sla·vi·a (yōō'gō-slä'vē-ə), federal republic (1971 pop. 20,504,516), 98,766 sq mi (255,804 sq km), SE Europe, largely in the Balkan Peninsula. Belgrade is the capital. Yugoslavia is bounded by Italy and the Adriatic Sea on the west, by Austria and Hungary on the north, by Rumania and Bulgaria on the east, and by Greece and Albania on the south.

The Adriatic coast, known as Dalmatia, is dotted with numerous ports, and there are many islands. About four fifths of Yugoslavia is mountainous, with an average altitude of 1,500 to 2,000 ft (457.5-610 m). The chief mountain chain, the Dinaric Alps, runs parallel to the Adriatic coast. In the northwest, the Julian Alps culminate at 9,396 ft (2,865.8 m) in the Triglav. Yugoslavia is traversed, in the northeast, by the Danube and its tributaries.

Yugoslavia is a federation of 6 people's republics:

NAME	CAPITAL	NAME	CAPITAL
Bosnia and Hercegovina	Sarajevo	Montenegro	Titograd
Croatia	Zagreb	Slovenia	Ljubljana
Macedonia	Skopje	Serbia	Belgrade

Economy. Since World War II, Yugoslavia has become the most heavily industrialized nation in the Balkans. It is rich in mineral resources, notably lignite, petroleum, iron, copper, lead, zinc, bauxite, antimony, chrome, and manganese. Metal processing is the most important manufacturing industry; others produce textiles, chemicals, and processed food. The most important industrial area is the Sava valley between Belgrade and Zagreb. Despite industrialization, nearly half the labor force is still engaged in agriculture. Livestock raising is important. Leading crops are corn, wheat, fruits, sugar beets, potatoes, and rye. The country has extensive vineyards. Metal products and foodstuffs are the leading exports, and machinery, transportation equipment, and other manufactured goods are the chief imports. Yugoslavia entered into a trade agreement with the Common Market in 1970. Tourism is increasing; the Adriatic coast has numerous resorts, and there are many mineral spas throughout the country. All major industrial enterprises have been nationalized. While there are state and cooperative farms, most productive land remains privately owned.

History. Yugoslavia came into existence as a result of World War I. (The earlier histories of its six component republics are treated separately, under their respective names.) In 1914 only Serbia (which included the present Yugoslav republic of Macedonia) and Montenegro were independent states; Croatia, Slovenia, and Bosnia and Hercegovina belonged to the Austro-Hungarian Monarchy, the defeat of which led the way to unification of the South Slavs. The "Kingdom of the Serbs, Croats, and Slovenes" was formally proclaimed in Dec., 1918, with the Serbian king, Peter I, on the new throne. King Alexander, who ascended the throne on Peter I's death (1921), proclaimed a dictatorship in 1929 during a crisis over Croatian separatism and changed the name of the kingdom to Yugoslavia. Troubles with Croatian and Macedonian nationalists culminated (1934) in Alexander's assassination. His son, Peter II, succeeded under the regency of Alexander's cousin, Prince Paul.

In Apr., 1941, German troops invaded and overran neutral Yugoslavia. While Peter II established a government in exile in London, two rival Yugoslav resistance groups emerged: the *chetniks* under Draja Mihajlović and an army under the Communist Tito (Josip Broz). Tito was supported by the USSR, and he won the support of Great Britain as well. By late Oct., 1944, the Germans had been driven from Yugoslavia. In 1945 Tito became premier in a coalition government from which non-Communists were soon excluded. The constituent assembly proclaimed a federal republic in Nov., 1945.

The constitution of 1946 gave wide autonomy to the six newly created republics, but actual power remained in the hands of Tito and the Communist party. Close ties were maintained with the USSR until 1948, when a breach between the Yugoslav and Soviet Communist parties occurred. Economic and military assistance was there after received from the West. More cordial relations with the USSR were resumed in 1955, but new rifts occurred because of Soviet intervention in Hungary (1956) and Czechoslovakia (1968). Domestically Yugoslavia's form of communism included the abandonment of agricultural collectivization (1953), the delegation of economic power to workers' councils, and comparative intellectual freedom.

Government. Yugoslavia is governed under a constitution adopted in 1974. The bicameral parliament is made up of the 220-member federal chamber and the 88-member chamber of the republics and provinces; members of both bodies are elected to four-year terms. The country's executive is composed of the nine-member collective presidency, made up of one representative from each republic and province and the president of the presidency (who is selected by parliament). The members of the presidency usually serve nonrenewable five-year terms; in May, 1974, Tito was elected president for life by parliament.

Yu·kon (yo͞o'kŏn'), city (1975 est. pop. 12,980), Canadian co., central Okla., W of Oklahoma City, in an agricultural region.

Yukon, river, c.2,000 mi (3,220 km) long, rising in Atlin Lake, NW British Columbia, Canada, and receiving numerous headwater streams; one of the longest rivers of North America. It flows generally northwest, past Dawson and across the Alaska border, to Fort Yukon, thence generally southwest through central Alaska until, in a wide swing north, it enters Norton Sound of the Bering Sea through a delta that is 60 mi (96.5 km) wide. The river is incised in the Yukon Plateau; marshy land borders much of its upper course. The Yukon is navigable for river boats three months of the year to Whitehorse, c.1,775 mi (2,860 km) upstream. The Yukon basin is one of the most sparsely populated and least developed regions of North America. Much of its history, exploration, and development centers on the river system. Its lower reaches were explored (1836-37, 1843) by Russians, and in 1843 Robert Campbell of the Hudson's Bay Company explored the upper course. During the Klondike gold rush (1897-98) the Yukon was a major route to the gold fields. Greater development of the basin occurred in the mid-1900s due to its strategic location, and several military installations were later built. The Yukon River is a major salmon-spawning ground, and salmon fishing is an important seasonal activity.

Yukon Territory, territory (1975 pop. 21,000), 207,076 sq mi (536,327 sq km), NW Canada. The capital and largest town is Whitehorse. The triangle-shaped territory is bordered on the north by the Beaufort Sea of the Arctic Ocean, on the east by the Mackenzie dist., Northwest Territories, on the south by British Columbia and Alaska, and on the west by Alaska.

Although most of the territory is a watershed for the Yukon River and its tributaries, the northern and southeastern regions drain into the Mackenzie River system. Immediately south of the desolate arctic coast the country is uninhabited and generally unknown. The other parts of the territory have great natural beauty, with snow-fed lakes backed by perpetually white-capped mountains, and forests and streams abounding with wildlife. Winters are long and cold, with low humidity. During the short summers longer days and surprisingly warm sun bring a profusion of wild flowers and enable the hardier grains and vegetables to mature. The few settlements are situated on the riverbanks. Transportation facilities are limited, and for many years the Yukon River system was the main artery. Air transportation now plays a vital role. The Alaska Highway and other all-weather roads have been built since World War II.

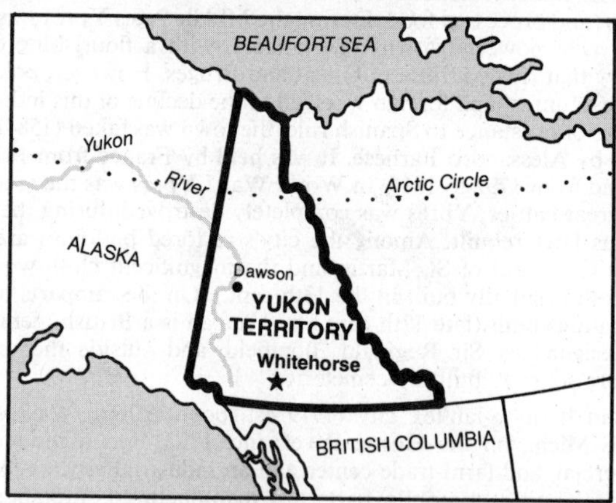

The leading industry in the territory is mining; asbestos, copper, silver, zinc, lead, and gold are the principal minerals. Manufacturing and fishing are relatively unimportant, but tourism is gaining in importance.

The territory's history began with the explorations in the 1840s of fur traders for the Hudson's Bay Company. Several trading posts were built on the Yukon River, and after the famous gold strikes in the Klondike River region in the 1890s more than 30,000 prospectors trekked across the icy barriers in search of gold. The Canadian government acquired the Yukon from the Hudson's Bay Company in 1870 and administered it as part of the Northwest Territories. The Yukon was made a separate district (1895) and then a separate territory (1898) with Dawson as capital. Whitehorse became the capital in 1952.

The government consists of a federally appointed governor and an elected legislative council of seven members. The territory elects one delegate to the Canadian Parliament. About 2,000 Indians and Eskimo are included in the population.

Yu·ma (yo͞o'mə). **1.** County (1970 pop. 60,827), 9,983 sq mi (25,856 sq km), SW Ariz., in an agricultural area drained by the Gila River and bounded on the N by the Bill Williams River, on the W by the Colorado River and the Calif. border, on the S by the Mexican border; formed 1864; co. seat Yuma. Cotton, alfalfa, citrus fruit, truck crops, and pecans are raised on irrigated farmland. Part of Colorado River Indian Reservation is here. **2.** County (1970 pop. 8,544), 2,383 sq mi (6,172 sq km), NE Colo., bordering on Kansas and Nebr., and drained by the Arikaree River; formed 1889; co. seat Wray. It is in an agricultural area that produces livestock, grain, dairy products, and poultry. There is some manufacturing.

Yuma, city (1979 est. pop. 32,900), seat of Yuma co., extreme SW

Ariz., on the E bank of the Colorado River near the confluence of the Gila River; founded 1854, inc. as a city 1914. It is a major trade center of an extensive farm area irrigated by the Yuma project. The project has turned more than 100,000 acres (40,500 hectares) of desert into a fertile farm region known for its cattle, citrus fruits, melons, winter vegetables, grains, and cotton. Two nearby military installations contribute to the city's economy—the sprawling Yuma Proving Grounds and a U.S. Marine Corps air station. Yuma is also a growing resort center because of its mild climate. Early missions were built in the area by Fathers Eusebio Kino and F. T. H. Garcés, but there was no white settlement until after Fort Yuma was built (1850) to protect overlanders on the route to California. After 1858 Yuma was a river port and the center of a gold-mining boom. Points of interest in the area include Fort Yuma (on the west bank of the river), the territorial prison (built 1875; now a museum), St. Thomas Mission (16th cent.), three dams on the Colorado River, and the California sand dunes.

Yü-men or **Yü·men** (both: yōō'mŭn'), city (1970 est. pop. 325,000), NW Kansu prov., China. It is China's leading petroleum center, with oil fields and refineries. Yü-men, on the Old Silk Road to Sinkiang, is named for an ancient gateway in the Great Wall.

yun·gas (yōōng'gəs), region of lowland valleys in the eastern piedmont of the Andes Mts., 5,000–8,000 ft (1,525–2,440 m) high, extending from the Peru-Bolivia border SE into central Bolivia. They receive excessive rainfall and are warm and humid. Although isolated and very difficult of access, the yungas assumed economic importance in the early 20th cent. as a major source of rubber and quinine. Coca, coffee, tobacco, and fruit are grown. With improved communications the region's economy has grown, especially in the more accessible valleys close to La Paz, Bolivia, which have been developed as resorts.

Yün·nan or **Yün-nan** (both: yōō-nän'), province (1968 est. pop. 23,000,000), c.162,000 sq mi (419,580 sq km), SW China, bordering on Burma in the W and Laos and Vietnam in the S; capital K'unming. The eastern half of the province is a limestone plateau with karst scenery and unnavigable rivers flowing through deep mountain gorges; the western half is characterized by mountain ranges and rivers running north and south. The rugged, vertical terrain produces a wide range of flora and fauna, and the province has been called a natural zoological and botanical garden. Agriculture is restricted to the few upland plains, open valleys, and terraced hillsides. Rice is the main crop; corn, wheat, sweet potatoes, soybeans (as a food crop), tea, sugar cane, and tobacco are also grown. Cotton is being developed. On the steep slopes in the west livestock is raised and timber is cut (teak in the southwest). However, Yünnan's chief source of wealth lies in its vast mineral resources. It is the country's leading tin producer; other deposits include iron, coal, lead, copper, zinc, gold, mercury, silver, antimony, and sulfur. Road and railroad traffic has been recently improved and K'un-ming is now a transportation center; an important railroad runs from K'un-ming to Hanoi, Vietnam, while transportation to Burma is maintained by the Burma Road. There are many minority groups in Yünnan. From ancient times the Chinese invaders gradully pushed the aboriginal tribes into mountain localities, where today, retaining their distinct languages and culture, they populate eight autonomous districts. The Miao, Yao, Lolo, Lao, Shan, Thai, and Lisu are some of the larger tribes; there is also a considerable Tibetan minority. Separated by rugged mountains from the central authority in northern China, Yünnan for centuries remained independent. In 1253 it was conquered by the Mongols of the Yüan dynasty, which destroyed the Thai kingdom of Nan Chao. Yünnan passed to the Manchus in 1659 and became a province of China under the control of the central government. It was a major center of Chinese resistance in World War II, and in 1950 it passed to Communist control.

Yve·lines (ēv-lēn'), department (1974 est. pop. 1,010,000), 877 sq mi (2,271.4 sq km), N central France, W of Paris; capital Versailles.

Y·ver·don (ē-vĕr-dōɴ'), town (1974 est. pop. 21,000), Vaud canton, W Switzerland, at the S end of the Lake of Neuchâtel. It is an old spa with Roman ruins. Typewriters are manufactured.

Zaan·dam (zän-däm'), municipality (1973 est. pop. 69,279), North Holland prov., W Netherlands, near Amsterdam. Manufactures include food products, chemicals, lumber, and machinery. Peter I of Russia stayed at Zaandam in 1697 to learn shipbuilding, which at the time was a flourishing industry of the city.

Zab (zăb), name applied to the two principal tributaries of the Tigris River. The Great Zab, 265 mi (426.4 km) long, rises in southeast Turkey and flows generally south through Iraq to the Tigris. The Little Zab, 250 mi (402.3 km) long, rises in northwest Iran and flows southwest through Iraq to the Tigris. Both rivers are extensively used for irrigation, flood control on the Tigris, and hydroelectricity.

Zab·rze (zäb'zhā), city (1973 est. pop. 201,200), S Poland. It is a railway junction in the Katowice mining and industrial region. Local coal deposits form the basis of Zabrze's coke and chemical industries. Founded in the 13th cent., Zabrze passed to Prussia in 1742. The city was renamed in 1915 in honor of German Field Marshal von Hindenburg; its old name was restored when it was ceded to Poland in 1945.

Za·ca·te·cas (zăk'ə-tā'kəs), city (1970 pop. 50,251), capital of Zacatecas state (28,125 sq mi/72,844 sq km), N central Mexico. With an altitude of more than 8,000 ft (2,440 m), it is situated in a deep ravine surrounded by arid hills. The city is characterized by old colonial buildings and narrow, steep cobbled streets, frequently broken by stone steps. Zacatecas is a distributing center for the mining country and has industries making fine serapes and other items. Founded in 1848, the strategically located city was a key point in the Mexican wars and revolutions of the 19th and early 20th cent. Its cathedral was heavily pillaged during these struggles.

Za·ca·te·co·lu·ca (zăk'ə-tĕk'ə-lōō'kə), city (1971 pop. 15,718), S central El Salvador. Baskets, cotton products, and lumber are made. The city was heavily damaged by an earthquake in 1932.

Za·dar (zä'där'), city (1971 pop. 43,817), W Yugoslavia, in Croatia, on the Dalmatian coast of the Adriatic Sea. A seaport and a tourist center, it has industries that produce liqueur, tobacco, and jute. Founded by the Illyrians in the 4th cent. B.C., Zadar became a Roman colony in the 2nd cent. B.C. It passed to the Byzantine Empire in 553 and was settled by the South Slavs in the 7th cent. Although disputed by Venice, Hungary, and Croatia, it remained under Byzantine protection until 1001, when Emperor Alexius I transferred it to Venice. At the end of the 11th cent. it was seized by Hungary, but the leaders of the Fourth Crusade, persuaded by the doge Enrico Dandolo, reconquered it for Venice in 1202. After a five-day siege the Crusaders sacked the city, an act for which they were condemned by Pope Innocent III. Hungary continued to dispute Zadar with Venice, which obtained permanent possession of the city only in 1409. The Treaty of Campo Formio (1797) gave it to Austria, where, from 1815 to 1918, it was the capital of the crownland of Dalmatia. Zadar passed to Italy by the Treaty of Saint-Germain

(1919), was occupied (1945) by Yugoslav forces at the end of World War II, and was formally ceded to Yugoslavia by the Italian peace treaty of 1947. The city has several Roman monuments and medieval churches and palaces.

Ża·gań (zhä'gän'), town (1970 est. pop. 21,500), W Poland, on the Bóbr River. It has lignite mines, textile mills, and glassmaking industries. Founded in the 12th cent., Żagań was the capital of a principality that passed to the Hapsburgs in the 16th cent. and to Prussia in 1745. The town was incorporated into Poland after World War II.

Za·gorsk (zə-gôrsk'), city (1973 est. pop. 97,000), central European USSR. It is a rail terminus and a handicraft center known for wood carvings and toys. Manufactures include farm machinery, lacquers and paints, concrete pipes, automobile components, textiles, and furniture. The city developed from a settlement around the Troitse-Sergiyeva Lavra, one of Russia's most famous monasteries (founded 1340). The original wooden church, built by the monk Sergius, was destroyed in a Tatar raid in 1391. The Lavra contains the Troitski Cathedral (15th cent.); the Uspenski Cathedral (16th cent.), with the tomb of Boris Godunov; and a treasure chamber with rich tapestries and many objects of liturgical art. The monastery, long a place of pilgrimage, was made into a museum in 1920.

Za·greb (zä'grĕb'), city (1971 pop. 566,084), capital of Croatia, NW Yugoslavia, on the Sava River. A major industrial and financial center, it has industries that produce machinery, machine tools, metal products, and chemicals. It is also a cultural center, with the Yugoslav Academy of Arts and Sciences (founded 1861), a university (founded 1669), an institute of nuclear physics, an observatory, and several fine museums and art galleries. Zagreb is the seat of a Roman Catholic archbishop, an Orthodox Eastern archbishop, a Protestant bishop, and a grand rabbi. The city was originally a suburb of the ancient Roman town of Andautonia. Invaded by Mongols in 1242, it became in the second half of the 13th cent. the chief city of Croatia and Slavonia, which were then provinces of Hungary. The city was merged in 1557 with nearby Gradec, which had been a free royal city under the Hungarian crown. The part of Croatia in which Zagreb is situated escaped Turkish domination. During the 19th cent. Zagreb was a center of the Yugoslav nationalist movement. With the formation of the dual Austro-Hungarian Monarchy in 1867, the city became the capital of an autonomous Croatia. A fine modern city, Zagreb has its historic center in the old Kaptol dist., with the Catholic cathedral (begun 1093) and the Catholic archiepiscopal palace (18th cent.).

Zag·ros (zăg'rəs), mountain system of W Iran, extending c.1,100 mi (1,770 km) from the Turkish-Soviet frontier SE to the Strait of Hormuz, forming the W and S border of the central Iranian plateau. The Zagros vary from the rugged, forested, and snowcapped mountains of the northwest, with numerous volcanic cones and large basins, to the parallel ridge and valley system of the central portion, with lowland salt marshes, and the low, irregular southwest region, characterized by bare rock and sand dunes. The northern half of the Zagros is heavily populated, and the fertile valleys support agriculture. In the uplands of the central range, tribal pastoralism predominates. In the southeast Zagros, dates and cereals are grown at oases. Iran's major oil fields lie along the western foothills of the central Zagros, where salt domes have trapped huge amounts of oil. In antiquity the Zagros formed the boundary between Assyria and Media.

Za·he·dan (zä'hĭ-dän'), city (1971 est. pop. 42,000), capital of Seistan and Baluchistan prov., SE Iran, near the borders with Pakistan and Afghanistan. It is a road junction and the terminus of a railroad that runs into Pakistan.

Zaire (zâr), formerly Democratic Republic of the Congo, republic (1977 est. pop. 25,915,000), c.905,000 sq mi (2,343,950 sq km), central Africa, bordering on Angola in the SW and W, on Cabinda and the Congo Republic in the W, on the Central African Empire and Sudan in the N, on Uganda, Rwanda, Burundi, and Tanzania in the E, and on Zambia in the SE. Kinshasa is its capital.

Zaire lies astride the equator, and virtually all of the country is part of the vast Congo River drainage basin. North-central Zaire is made up of a large plateau (average elevation, c.1,000 ft/305 m), which is covered with equatorial forest and has numerous swamps. The plateau is bordered on the east by mountains, which rise to the lofty Ruwenzori Mts. In south Zaire are highland plateaus (average elevation, c.3,000 ft/915 m), which are covered with savanna. The high Mitumba Mts. in the southeast include Lake Mweru (situated on the border with Zambia).

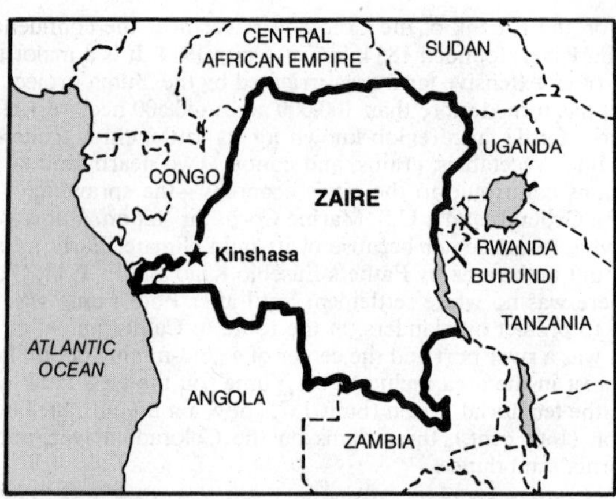

Economy. In the early 1970s Zaire's economy was still based on agriculture, with about half the workers engaged as subsistence farmers; but the country's considerable mineral and energy resources had helped to create substantial modern mining and manufacturing sectors. Less than 5% of Zaire's land area is cultivated or used as pasture. The sown land falls into three categories—that used for growing food crops, that used for growing cash crops on a small scale, and that used for growing cash crops on plantations. The principal food crops are cassava, yams, maize, millet, rice, groundnuts, plantains, and pulses. Rubber, coffee, cotton, and palm products are produced commercially in limited quantities on small farms. The main plantation commodities are palm products, coffee, tea, cacao, sugar cane, and rubber. Substantial numbers of goats, sheep, and cattle are raised.

Mining is centered in the Shaba, Kivu, and Kasai-Oriental regions. The products of Shaba include copper (the country's leading mineral product in terms of value), cobalt, zinc, manganese, uranium, cassiterite (tin ore), coal, gold, and silver. Tungsten, tantalum, silver, and gold are extracted in Kivu. Kasai-Oriental is by far the world's leading source of industrial diamonds (12.4 million carats in 1972), and some gem diamonds (1 million carats in 1972) are also found here. The extraction and processing of minerals is run largely by foreign-controlled firms. There are major deposits (as yet unexploited) of petroleum and natural gas (methane) in northeast Zaire; in addition, modest offshore deposits of petroleum were discovered near the mouth of the Congo River in 1973. In 1974 a state-owned company was created to take over the distribution of oil petroleum products. Approximately 40% of Zaire is covered with forest; considerable amounts of ebony and teak are produced annually as well as less valuable woods. The principal manufactures of Zaire are processed copper, zinc, and cassiterite; refined petroleum; basic consumer goods such as processed food, beverages, clothing, and footwear; iron and steel; cement; and chemicals. The numerous rivers of Zaire give it an immense potential for producing hydroelectricity.

History. The earliest inhabitants of the region of Zaire were probably Pygmies. By the end of the 1st millennium B.C. small numbers of black Africans had migrated into Zaire from the northwest. Scholars believe that the Bantu language family was started by some of these people in present-day southeast Shaba around the start of the 1st millennium A.D. Aided by their knowledge of iron technology and agriculture, the Bantu-speakers migrated to other parts of Zaire and Africa. From about 700 the copper deposits of south Shaba were worked by the Bantu and traded over wide areas. By about 1000 the Bantu had settled most of Zaire, forcing the Pygmies into small, scattered areas. By the early 2nd millennium the Bantu had increased in number and were coalescing into states, including Kongo, Luba, Lunda, Kuba, and (through intermarriage) Luba-Lunda states. Most of the states were ruled by a divine monarch, whose authority was checked by a council of high civil servants and elders. The Kuba kingdom was noted for its sculpture and decorative arts. In 1482 Diogo Cão, a Portuguese navigator, became the first European to visit Zaire. Soon thereafter the Portuguese established ties with the king of Kongo, and in the early 16th cent. they established themselves on parts of the coast of modern Angola.

The Portuguese had little influence on Zaire until the late 18th cent., when they backed black African and mulatto traders who traveled far inland. From the early 19th cent., Swahili and Nyamwezi traders from present-day Tanzania penetrated into southeast Zaire, where they exchanged cloth, beads, cowrie shells, and other goods for copper, salt, and ivory. Some of the traders established states with considerable power. Msiri (a Nyamwezi) established himself near Mwata Kazembe in 1856, and was a major force until 1891, when he was killed by the Belgians. From the 1860s to the early 1890s, Muhammad bin Hamad (known as Tippu Tib), a Swahili trader from Zanzibar, ruled a large portion of eastern Zaire northwest of Lake Tanganyika.

In the late 1870s the territory was unified by the efforts of Leopold II, king of the Belgians (reigned 1865–1909). Leopold believed that Belgium needed colonies to ensure its prosperity, and he privately set about to establish a colonial empire. In 1878 Leopold engaged the explorer Henry M. Stanley to establish the king's authority in the Congo basin. At the Conference of Berlin (1884–85) the European powers recognized Leopold's claim to the Congo basin, and in a ceremony (1885) at Banana, Leopold announced the establishment of the Independent State of the Congo, headed by himself. It was not until the mid-1890s, however, that Leopold's control was established in most parts of the state. In following years, Shaba was conquered, and eastern Zaire was wrested from the control of East African Arab and Swahili traders. Because he did not have sufficient funds to develop the Congo, Leopold sought and received loans from the parliament, and Belgium annexed the Congo in 1908. Much of the land was given to concessionaire companies, which in return were to build railroads or to effectively occupy a specified part of the country or merely to give the state a percentage of their profits. In addition, Leopold maintained a large estate in the region of Lake Leopold II. Private companies were also established to exploit the mineral wealth of Shaba and Kasai. Under Belgian rule the worst excesses (such as forced labor) of the Independent State were ended, but the Congo was still regarded almost exclusively as a field for European investment, and little was done to give the black Africans a significant role in the government or economy of the colony. European concerns established more large plantations, and vast mining operations were set up. Black Africans formed the labor pool for these European-managed operations. By the end of the 1920s mining was the mainstay of the economy, having far outdistanced agriculture. Some of the mining companies built cities for their workers, and there was a considerable movement of black Africans from the countryside to urban areas. Christian missionaries (the great majority of whom were Roman Catholic) were very active in the Congo. The literacy level was raised considerably by the missionaries, but virtually no black Africans were educated beyond the primary level until the mid-1950s, when two universities were opened. A noteworthy indigenous religious movement was that of Simaon Kimbangu, who, educated by Protestant missionaries, around 1920 established himself as a prophet and healer. He soon gathered a large following, and was jailed in 1921 by the colonial government. Kimbangu died in 1951 while still in jail. In the late 1930s Simon-Pierre Mpadi organized a major Kimbanguist revival and was forced by the Belgians to flee the Congo. In 1955, when demands by blacks in Africa for independence were mounting, Antoine van Bilsen, a Belgian professor, published a "30-Year Plan" for granting the Congo increased self-government. Congolese nationalists, notably Joseph Kasavubu (who headed ABAKO, a party based among the Kongo people) and Patrice Lumumba (who led the leftist Mouvement National Congolais), became increasingly strident. In Jan., 1959, there were serious nationalist riots in Kinshasa, and thereafter the Belgians steadily lost control of events in the Congo. At a round-table conference at Brussels in Jan.–Feb., 1960, it was decided that the Belgian Congo would become fully independent on June 30, 1960. Following elections, Lumumba in June became prime minister and Kasavubu head of state. However, shortly after independence began on June 30, the Republic of the Congo (as the nation was called) began to disintegrate as ethnic and political rivalries came to the fore. On July 4 the Congolese army mutinied, and on July 11 Moïse Tshombe declared Katanga (now called Shaba), of which he was provisional president, to be independent. Belgium sent troops to the country to protect its citizens and also its mining interests. On July 14 the UN Security Council voted to send a force to the Congo to help establish order, but under the UN Charter the force was not allowed to intervene in internal affairs. Therefore, Lumumba turned to the USSR for help against Katanga, but on Sept. 5 he was dismissed as prime minister by Kasavubu and replaced by Joseph Ileo. On Sept. 14 Col. Joseph Mobutu, the head of the army, seized power and dismissed both Kasavubu and Ileo. On Dec. 1 Lumumba, who probably had the largest national following of any Congo politician, was arrested by the army; he was murdered while allegedly trying to escape imprisonment in Katanga in mid-Feb., 1961. By the end of 1960 the Congo was divided into four quasi-independent parts. The secession of Katanga (under Tshombe), with its great mineral resources, particularly weakened the national government. In Feb., 1961, Ileo again became prime minister, with Kasavubu as head of state. In April, Tshombe was arrested. By July, however, Tshombe (again free) was proclaiming the independence of Katanga. In the same month Cyrille Adoula became prime minister. In August the UN forces began to take an active role in Congo affairs by disarming Katangese soldiers. In Nov., Mobutu launched an attack on Katanga aimed at ending the secession. However, the national scene remained confused; Adoula failed to command much support, the government's finances were weak, and there was considerable agitation by the followers of Lumumba. At the end of June, 1964, the last UN troops were withdrawn. In desperation, Kasavubu appointed Tshombe prime minister in July, 1964, but this move resulted in large-scale rebellions. With the help of U.S. arms, Belgian troops, and white mercenaries, the central government gradually regained control of the country. In mid-1965, Kasavubu appointed Evariste Kimba prime minister. In Nov., 1965, Mobutu again intervened, dismissing Kasavubu and proclaiming himself president. In late 1966 Mobutu abolished the office of prime minister, establishing a presidential form of government.

In 1971 the country was renamed Zaire, as was the Congo River; in 1972 Katanga was renamed Shaba—largely in an attempt to destroy the region's past association with secession—and Mobutu dropped his Christian names and called himself Mobutu Sese Seko. By the end of the 1960s, the country enjoyed political stability, although there was intermittent student unrest. In 1975, Mobutu was elected to a new seven-year term as president. In the early 1970s he centralized the administration of the nation. Zaire remained dependent on volatile world copper prices; when the copper market fell dramatically in 1975–76, the effect on Zaire's economy was catastrophic. By 1977 Mobutu's government had defaulted on several international loans and was in a state of great political instability. In Mar., 1977, a secessionist army invaded Shaba prov. from the neighboring state of Angola. Mobutu, charging Soviet and Cuban interference, was able to repel the invaders, but only with the help of 1,500 Moroccan troops (flown in on French planes), Egyptian pilots, and material aid from the United States and Europe.

Zaire, river, Africa: *see* Congo, river.

Za·kar·pat·ska·ya Oblast (zä'kər-pät'skə-yə) or **Trans·car·pa·thi·an Oblast** (trăns'kär-pā'thē-ən), administrative division (1970 pop. 1,057,000), 4,981 sq mi (12,900.8 sq km), SW European USSR, in the Ukraine, on the SW slopes of the Carpathian Mts.; capital Uzhgorod. It is thickly forested and largely agricultural. The plain in the southwest, which is drained by the Tisza River and its tributaries, supports crops of wheat, corn, tobacco, sugar beets, and potatoes. There are vineyards, fruit orchards, and walnut groves in the foothills. The oblast's mineral resources include brown coal, rock salt, fire clays, marble, limestone, and some iron, crude oil, and natural gas. Forests occupy nearly half the area of the oblast, and lumbering, along with the production of such items as wood chemicals, furniture, and cartons, is a leading industry. Among other industries are mining, food processing, brewing, wine making, tanning, and the manufacture of bricks, tiles, footwear, clothing, and textiles. The majority of the population is Ukrainian, with Hungarian, Russian, and Slovak minorities.

Inhabited by Slavic tribes from the 8th cent., the region was part of the Kievan state in the 10th and 11th cent. but was conquered by the Magyars, who ruled it until 1918. It has been variously known as Ruthenia or the Carpathian Ukraine. Its inhabitants were historically called Ruthenians. Until the early 20th cent. the region was among the most economically and culturally backward areas of the world. Hungarian absentee landlords owned virtually the entire country. After World War I the National Ruthenian councils of Uzhgorod, Khust, and Presov jointly called for the region's union with newly independent Czechoslovakia, which incorporated Transcarpathia in May, 1919. Although a guarantee of provincial autonomy embodied in the Treaty of St. Germain (Sept., 1919) did not materialize, the region fared better under Czechoslovak than under Magyar control. Economic modernization began, and the peasants were freed from their nearly servile status, but agrarian reform failed to break up all the large estates. In the wake of the Munich Pact (1938), the reorganized state of Czecho-Slovakia was pressured by

Germany to grant autonomy to Transcarpathia. After Czecho-Slovakia was dismembered in Mar., 1939, the region proclaimed its independence; but it was shortly occupied by and annexed to Hungary. Transcarpathia was reconquered late in 1944 by Soviet troops and local guerrillas, and in 1945 Czechoslovakia agreed to cede the area to the USSR. The oblast was formed in 1946.

Zá·kin·thos (zä'kĕn-thôs') or **Zan·te** (zăn'tē), island (1971 pop. 30,180), c.157 sq mi (407 sq km), W Greece, in the Ionian Sea, one of the Ionian Islands. Wine, currants, citrus fruits, and olive oil are produced on the fertile, intensively cultivated island.

Za·ko·pa·ne (zä'kə-pä'nē), town (1970 pop. 27,039), S Poland, at the foot of the Tatra Mts. A leading health resort and winter sports center, Zakopane was the site of the world skiing championship competitions in 1929 and 1939.

Za·ma (zā'mə), ancient town near the N coast of Africa, in present Tunisia. Although there was more than one town named Zama, tradition says that in 202 B.C. Scipio Africanus Major defeated Hannibal here in the decisive and final battle of the Second Punic War. There is good reason for believing the actual battle was fought at some nearby place.

Zam·be·zi (zăm-bē'zē), river, c.1,700 mi (2,735 km) long, rising in NW Zambia, S central Africa, and flowing in an S-shaped course generally E through E Angola, along the Zambia-Rhodesia border, and through central Mozambique to the Mozambique Channel of the Indian Ocean, near Chinde. The upper Zambezi flows over part of the great basalt plateau of Africa, the middle Zambezi is entrenched in the plateau, and the lower Zambezi flows through a wide valley. Many rapids interrupt the river's flow, making it unsuited for navigation; however, its navigable stretches are used for local traffic. Kariba Lake, impounded by Kariba Dam, is one of the world's largest man-made lakes. The Zambezi's banks are fertile and well populated. The river has a great hydroelectricity-generating potential; there is a small power plant at Victoria Falls, and the one at Kariba Dam is one of the largest in Africa.

Zam·bi·a (zăm'bē-ə), republic (1971 est. pop. 4,396,000), 290,584 sq mi (752,614 sq km), central Africa, bordering on Zaire in the N, on Tanzania in the NE, on Malawi and Mozambique in the E, on Rhodesia, Botswana, and Namibia in the S, and on Angola in the W. Lusaka is the capital.

Zambia is largely made up of a highland plateau, which rises in the east. The elevation here ranges from c.3,000 to 5,000 ft (915–1,520 m), and higher altitudes are attained in the Muchinga Mts. and on the Nyika plateau (adjacent to Malawi). Also in east Zambia are Lake Bangweulu, parts of lakes Mweru and Tanganyika, and the Luangwa and Chambeshi rivers. The Zambezi River drains much of the western part of the country (where the elevation is c.1,500–3,000 ft/460–915 m) and forms a large part of Zambia's southern boundary. The impressive Victoria Falls and the huge Lake Kariba (formed by Kariba Dam), both on the border with Rhodesia, are part of the Zambezi in the south. About 98% of the country's population is made up of Bantu-speaking black Africans. The remainder of the inhabitants are Europeans and Asians.

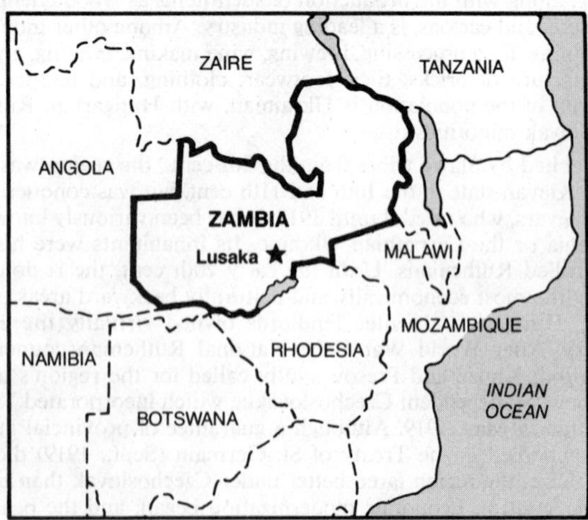

Economy. Zambia's economy is divided into a poor traditional sector and an affluent modern sector, dominated by the copper industry. Most Zambians work the country's relatively infertile soil as subsistence farmers; commercial agriculture is mostly confined to a small number of large farms. The leading crops are maize, sorghum, millet, beans, groundnuts, tobacco, and cotton. Large numbers of poultry, cattle, sheep, goats, and pigs are raised, and there is a small fishing industry. The mining and refining of copper constitutes by far the largest industry in the country and is concentrated in the cities of the Copperbelt. Zambia is among the world's five leading producers of copper; coal, zinc, lead, gold, silver, and cobalt are also mined. Other manufactures include food products, beverages, textiles and clothing, construction materials, iron and steel, forest products, and plastics. Most of Zambia's energy is supplied by hydroelectric plants, especially the one at Kariba Dam.

History. Parts of Zambia were probably inhabited by man about one million years ago, and subsequently Stone Age and Iron Age cultures flourished in the country. Some Bantu-speaking peoples (probably including the Tonga) reached the region by A.D. c.1200, but the ancestors of most of modern Zambia's ethnic groups arrived from present-day Angola and Zaire between the 16th and 18th cent. By the late 18th cent. traders (including Arabs, Swahilis, and black Africans) had penetrated the region from both the Atlantic and Indian Ocean coasts; they exported copper, wax, and slaves. In 1835 the Ngoni, a warlike group from southern Africa, entered Zambia. At about the same time the Kololo penetrated Zambia from the south, and they ruled the Lozi kingdom of Barotseland from 1838 to 1864. The Scottish explorer David Livingstone first visited Zambia in 1851; he discovered Victoria Falls in 1855, and in 1873 he died near Lake Bangweulu. In 1890 agents of Cecil Rhodes's British South Africa Company signed treaties with several African leaders, including Lewanika, the Lozi paramount chief, and proceeded to administer the region. It was divided into the protectorates of Northwestern and Northeastern Rhodesia until 1911, when the two were joined to form Northern Rhodesia. In the late 1890s company troops moved against several African peoples in the east (including the Ngoni and the Bemba) who had refused to recognize foreign rule. The mining of copper and lead began in the early 1900s, and by 1909 the central railroad from Livingstone to Ndola had been completed and about 1,500 Europeans had settled in the country. In 1924 the British took over the administration of the protectorate.

In the late 1920s massive copper deposits were discovered in what soon became known as the Copperbelt, and by 1937 about 4,000 skilled Europeans and about 20,000 unskilled black African laborers were working here. The Africans protested the discrimination and ill-treatment to which they were subjected by staging strikes in 1935, 1940, and 1956. They were not allowed to form unions but did organize self-help groups that brought together persons of diverse ethnic backgrounds. In 1946 delegates from these groups met in Lusaka and formed the Federation of African Welfare Societies, the first protectorate-wide black African movement, and in 1948 this organization was transformed into the Northern Rhodesia African Congress. Under the leadership of Harry Nkumbula in the early 1950s, it fought strenuously, if unsuccessfully, against the establishment of the Federation of Rhodesia and Nyasaland (1953–63), which joined Northern Rhodesia, Southern Rhodesia (now Rhodesia), and Nyasaland (now Malawi). The booming copper industry had attracted about 72,000 whites to Northern Rhodesia by 1958, and the blacks here experienced increasing European domination, similar to that in Southern Rhodesia. Kenneth Kaunda, a militant former schoolteacher, took over the leadership of the Africans from the more moderate Nkumbula and in 1959 formed a new party, the United National Independence Party (UNIP). Following a massive civil disobedience campaign in 1962, black Africans were given a larger voice in the affairs of the protectorate.

On Oct. 24, 1964, Northern Rhodesia became independent as the Republic of Zambia, with Kaunda as its first president; he was reelected in 1968 and 1973. Zambia remained a member of the Commonwealth of Nations. The main problems faced by Kaunda in the first decade of independence were uniting Zambia's diverse peoples, reducing European control of the economy, and coping with white-dominated Southern Rhodesia (which unilaterally declared its independence as Rhodesia in Nov., 1965). In the mid-1960s Kaunda overcame the resistance to central control of the Lozi (who had enjoyed considerable autonomy under the protectorate) and of the fanatical quasi-Christian Lumpa church (led by Alice Lenshina), centered in northern Zambia; separatist sentiment among the Bemba, however, continued into the 1970s. European economic control was reduced by increasing the number of trained Zambians, by diversifying the country's economy, and (from 1969) by the government's acquisition of a 51% interest in most major firms (especially mining and banking companies) operating in the country. Beginning in the late 1960s Kaunda faced formidable opposition from political and student groups protesting the growing concentration of power in his hands. Early in 1972 the influential United Progressive party, led by Simon Kapwepwe, was banned, and later in the year all political parties except the UNIP were outlawed, as Zambia became a one-party state. In the mid-1970s Kaunda attempted to promote a settlement that would lead to black majority rule in Rhodesia; at the same time, Zambia gave active support to guerrilla groups from Rhodesia and helped forge the African National Council from rival Zimbabwe (the black African name for Rhodesia) guerrilla organizations. The border with Rhodesia remained closed after 1973.

Zam·bo·an·ga (zäm'bō-äng'gə), city (1970 pop. 41,750), Zamboanga del Sur prov., SW Mindanao, the Philippines, at the tip of the Zamboanga peninsula, on Basilan Strait. One of the chief cities of Mindanao and a busy seaport, it is the hub of a major iron-producing and lumbering area, which has large-scale rubber cultivation, plywood mills, and resin plants. The city is situated at the foot of high mountains and is known for its delightful climate, beautiful parks, and nearby beaches. Tourist attractions include a 16th cent. Spanish fort and the mosque at Talangkusay, a famous Moro ceremonial center.

Za·mo·ra (zə-môr'ə, -mōr'ə), city (1972 est. pop. 50,311), capital of Zamora prov., NW Spain, in León, on the Duero River. It is a communications and agricultural marketing and processing center. Because of its strategic position, the city was contested during the Middle Ages, first between the Christians and Moors, then between León and Castile. Parts of the medieval fortifications are still preserved. There is a fine 12th cent. Romanesque cathedral.

Za·mość (zä'môsh'), town (1973 est. pop. 37,100), SE Poland, on the Wieprz River. It is a commercial center, trading mainly in agricultural products. The town's chief industries are metalworking and the manufacture of furniture and clothing. Zamość was founded in 1579. The town defended itself against a Cossack invasion in 1648 and against the Swedish king Charles X in 1656. The city passed to Austria in 1772 and to Russia in 1815; it reverted to Poland after World War I.

Zanes·ville (zānz'vĭl'), city (1979 est. pop. 36,400), seat of Muskingum co., central Ohio, on the Muskingum River at its junction with the Licking; inc. 1815. It is a trade and industrial center manufacturing pottery, glassware, and electrical equipment. The area has deposits of clay, oil, natural gas, sand, limestone, and iron ore. The site was selected by Ebenezer Zane, surveyor of Zane's Trace, the gateway to the Northwest Territory. Zane's great-great-grandson, the writer Zane Grey, was born in the city. A two-year interval as state capital (1810–12) and the city's location on waterways and the National Road spurred its growth. Of interest are the National Road-Zane Grey museum, several early homes of Federal design, and the nearby Ohio ceramics center.

Zan·jan (zän-jän'), city (1971 est. pop. 60,000), Gilan prov., NW Iran, on the Zanjanchal River. It is the trade center for an agricultural region where grain and fruit are grown. Manufactures include rugs and metalwork. It is served by roads and a railroad.

Zan·te (zăn'tē): *see* Zákinthos, Greece.

Zan·zi·bar (zăn'zĭ-bär', zăn'zĭ-bär'), region (1967 pop. 190,494), 950 sq mi (2,460.5 sq km), United Republic of Tanzania, E Africa, consisting of the coral islands of Zanzibar and Tumbatu, in the Indian Ocean. The economy of the region is almost exclusively agricultural; fertile soil is limited to the western half of the island. The chief commodities produced are cassava, sweet potatoes, rice, maize, plantains, citrus fruit, cloves, coconuts, and cacao. There is a sizable fishing industry. The few manufactures include clove oil and woven goods. Handicraftsmen make objects of wood, ivory, and metal. Lime is the only mineral resource. The main imports are foodstuffs and fuel; principal exports are cloves and copra.

History. The first permanent residents of Zanzibar seem to have been the ancestors of the Hadimu and Tumbatu, settling in A.D. c.1000. They belonged to various mainland ethnic groups and lived in small villages. Because they lacked central organization, they were easily subjected by outsiders. Traders from Arabia, the Persian Gulf region of modern Iran (especially Shiraz), and western India probably visited Zanzibar as early as the 1st cent. Traders from the Persian Gulf region began to settle in small numbers on Zanzibar in the late 11th or 12th cent.; they intermarried with the black Africans and eventually a hereditary ruler (known as the Mwenyi Mkuu or Jumbe), of mixed Shirazi-black African descent, emerged among the Hadimu. A similar ruler, called the Sheha, was set up among the Tumbatu. These two rulers helped solidify the ethnic identity of their respective peoples. By 1503 the Portuguese had gained control of Zanzibar and most of the East Africa coast. They established a trading station and a Roman Catholic mission on Zanzibar, but their cultural impact was minimal. In 1698 Arabs from Oman ousted the Portuguese from East Africa. The Omanis gained nominal control of the islands, but until the reign of Sayyid Said (1804–56) they took little interest in them. Said permanently moved his court to Zanzibar in 1841. Under Said, Arabs gained control of Zanzibar's fertile soil, forcing most of the Hadimu to migrate to the eastern part of the island. Zanzibar became the main center of the East Africa ivory and slave trade. Some of the slaves were used on the clove plantations, and others were exported to other parts of Africa and overseas. Zanzibar's trade was run by Omanis, who organized caravans into the interior of East Africa; the trade was largely financed by Indians resident on Zanzibar, many of whom were agents of Bombay firms. On Said's death in 1856 his African and Omani holdings were separated, with his son Majid becoming sultan of Zanzibar. Majid was succeeded as sultan by Barghash in 1870, by Khalifa in 1888, by Ali ibn Said in 1890, by Hamid ibn Thuwain in 1893, by Hamoud ibn Mohammed in 1896, by Ali in 1902, by Khalifa ibn Naroub in 1911, by Abdullah ibn Khalifa in 1960, and by Jamshid ibn Abdullah in 1963. From the 1820s British, German, and U.S. traders were active on Zanzibar. As early as 1841 the representative of the British government on Zanzibar was an influential adviser of the sultan, and this was especially the case under Sir John Kirk, the British consul from 1866 to 1887. In a treaty with Great Britain in 1873, Barghash agreed to halt the slave trade in his realm. During the scramble for African territory among European powers, Great Britain gained a protectorate over Zanzibar and Pemba by a treaty with Germany in 1890. The sultan's mainland holdings were incorporated in German East Africa (later Tanganyika), British East Africa (later Kenya), and Italian Somaliland. The British considered Zanzibar an essentially Arab country and maintained the prevailing power structure. The office of sultan was retained (although stripped of most of its power), and Arabs, almost to the exclusion of other groups, were given opportunities for higher education and were recruited for bureaucratic posts. The chief government official during the period 1890 to 1913 was the British consul general, and from 1913 to 1963 it was the British resident. From 1926 the resident was advised by a legislative assembly.

After World War II political activity in Zanzibar increased. In the 1950s three main political parties were established—the Zanzibar Nationalist party (ZNP) and its offshoot, the Zanzibar and Pemba People's party (ZPPP), both of which principally represented the Arabs, and the Afro-Shirazi party (ASP), whose followers were black Africans. In 1957 popularly elected representatives sat on the legislative council for the first time, and in 1961 they were given a majority of seats. In June, 1963, Zanzibar gained internal self-government, and a ZNP-ZPPP coalition emerged victorious in elections held in July.

On Dec. 10, 1963, Zanzibar (together with Pemba) became independent, with Sultan Jamshid ibn Abdullah as head of state and Prime Minister Mohammed Shamte Hamadi, also an Arab, as the leader of government. On Jan. 12, 1964, this arrangement was overthrown in a violent leftist revolt of the black Africans led by John Okello. A republic was declared, with Abeid Karume of the ASP as its president and as head of the Revolutionary Council (the country's chief governmental body). The sultan was forced into exile, all land was nationalized, the ZNP and ZPPP were banned, and numerous Arabs were imprisoned. Subsequently, many other Arabs and some Indians left the country. On Apr. 27, 1964, Zanzibar and Tanganyika agreed to merge, and the resulting republic was renamed Tanzania in Oct., 1964. Zanzibar has retained considerable independence in internal affairs, but its foreign relations and defense are handled by the central government in Dar es Salaam. Zanzibar's president serves as the first vice president of Tanzania.

Zanzibar, city (1967 pop. 68,490), capital of Zanzibar Mjini region, Tanzania, on the W coast of Zanzibar island, separated by a 22-mile (35.4-km) wide channel from the mainland of E Africa. It is the island's chief commercial center and seaport. Cloves and copra are the main exports. Founded in the 16th cent. as a Portuguese trade depot, the city remained insignificant until the 19th cent., when the sultan of Oman transferred (1841) his court here. It flourished as a major center of the East Africa ivory and slave trade and was regularly visited by U.S., British, and German trading vessels. In 1890 it became the capital of the British protectorate of Zanzibar, and in 1963 it was made the capital of the independent republic of Zanzibar. When Zanzibar merged with Tanganyika in 1964 to form Tanzania, the city of Zanzibar continued as the seat of the island's government. It is a picturesque, cosmopolitan city, with winding streets, colorful bazaars, and interesting architecture. Of note are several mosques, the former sultan's palace, and Anglican and Roman Catholic cathedrals.

Za·pa·ta (zə-pä'tə), county (1970 pop. 4,352), 957 sq mi (2,478.6 sq km), extreme S Texas, bounded on the W by the Rio Grande and the Mexican border; formed 1858; co. seat Zapata. In a ranching area, it has some dairying and farming (corn, cotton, grain sorghums, and feed) and deposits of oil and natural gas.

Zapata, uninc. town (1970 pop. 2,102), seat of Zapata co., S Texas, SSE of Laredo; settled 1770. Flooding from the completion of Falcon Dam on the Rio Grande caused Zapata to move to its present site in 1952.

Za·po·pan (zä'pō-pän'), city (1970 pop. 182,934), Jalisco state, SW Mexico; est. 1541. Its economy is rural; cattle raising is the chief occupation, and maize is grown. Manufactures include fertilizers, textiles, and the products of local artisans. Stone quarries are nearby. The sanctuary of Our Lady of Zapopan, who is said to have brought peace by a miraculous intervention in the Mixtón War (1539–42), makes the city a pilgrimage point as well as a tourist spot.

Za·po·ro·zhye (zä'pə-rô'zhə), city (1974 est. pop. 729,000), capital of Zaporozhye oblast, S European USSR, in the Ukraine, a port on the Dnepr River, opposite the island of Khortitsa. It is a major rail junction and industrial center and the site of the Dneproges dam and power station, one of the USSR's largest hydroelectric plants. Large quantities of grain are exported. The city has steel mills, coking plants, aluminum and magnesium works, and factories that produce farm machinery, transformers, automobiles, tractors, chemicals, soap, and other items. Well supplied with electricity, Zaporozhye forms, together with the adjoining Donets Basin and the Nikopol manganese and Krivoy Rog iron mines, one of the Ukraine's leading industrial complexes. The city, founded in 1770 on the site of the Zaporozhye Cossack camp, consists of old Zaporozhye (called Aleksandrovsk before 1921) and the new industrial Zaporozhye, which developed during the 1930s and adjoins the Dneproges installations and the port of Lenin. For nearly three centuries the Zaporozhye Cossacks served as the rallying point for the Ukrainians' struggle against social, national, and religious oppression. In the 17th cent. the Cossacks founded an independent state, which occupied most of southern Ukraine except the Black Sea littoral, a possession of the Crimean khans. After the union of Poland and Lithuania in 1569, the Ukraine came under Polish rule; but the Poles were too weak to defend Ukraine from frequent devastating Tatar raids. The need for self-defense led at the end of the 15th cent. to the rise of the Ukrainian Cossacks, who by the mid-16th cent. had formed a state along the lower and middle Dnepr. Although they formally recognized the sovereignty of the Polish kings, the Cossacks, for all practical purposes, enjoyed complete political independence. By the end of the 16th cent., however, Poland sought fuller control over the Ukraine and the Zaporozhye Cossacks. Persecution of the Ukrainian Orthodox Church after 1596 provoked repeated outbreaks among the Ukrainians, and the Cossacks, as staunch adherents of the Orthodox faith, participated actively in the rebellions.

In 1648 the Zaporozhye Cossacks, led by Hetman Bohdan Chmielnicki, began a series of campaigns that eventually defeated the Poles and freed the Ukraine from Polish domination. Chmielnicki's forces suffered defeat in 1651, however, and were forced at Belaya Tserkov to accept a treaty unfavorable to the Ukraine. In 1654 Chmielnicki persuaded the Cossacks to transfer their allegiance to the Russian czars. By the Treaty of Andrusov in 1667 the left bank of the Dnepr and Kiev were ceded to Russia. The Russians proceeded to encroach upon Cossack privileges much as the Poles had, thus engendering revolts in what was left of the Zaporozhye territory.

When Hetman Ivan Mazepa joined Charles XII of Sweden against Russia in the Northern War, he shared in the Swedish defeat at Poltava (1709). At that time, Czar Peter I virtually ended Cossack independence in Ukraine. Many Zaporozhye Cossacks fled to the khanate of Crimea, but in 1734 they were allowed to return to their old territory and to establish a new Cossack headquarters. Russia, however, continued to view the Cossacks with suspicion; and in 1775 the Russian army, on orders from Catherine II, destroyed the Zaporozhye camp, thus completely abolishing the last stronghold of Ukrainian independence. Most of the Zaporozhye Cossacks then moved to Turkish territory at the mouth of the Danube, where they founded a new community. In 1828–29, however, they returned to the Ukraine and settled along the shores of the Sea of Azov. When the Russians tried in the 19th cent. to settle them in the newly conquered northern Caucasus, the Cossacks rebelled and were disbanded (1865). Those Zaporozhye Cossacks who had remained in the Ukraine were allowed in 1787 to settle along the Black Sea shores between the Dnepr and Bug rivers; they became known as the Black Sea Cossacks. In 1792 they were resettled in the Kuban region and after 1860 became known as the Kuban Cossacks.

Za·qa·ziq, Az (äz' zə-kä-zēk'), city (1970 est. pop. 173,000), capital of Sharqiyah governorate, N Egypt, in the Nile River delta. It is Egypt's leading cotton market, as well as a trade center for grain and a rail and canal junction. Az Zaqaziq was established as a transportation and market center in the mid-19th cent., when cotton cultivation was spreading in the delta. Nearby are the ruins of the ancient city of Bubastis.

Za·ra·go·za (zä'rə-gō'zə) or **Sar·a·gos·sa** (sâr'ə-gŏs'ə), city (1972 est. pop. 501,140), capital of Zaragoza prov. and leading city of Aragón, NE Spain, on the Ebro River. An important commercial and communications center, it is situated in a fertile, irrigated agricultural region. Among the manufactures are wood products, foodstuffs, and paper. Of ancient origin, it was named Caesarea Augusta by Emperor Augustus. It fell to the Goths (5th cent.) and to the Moors (8th cent.), under whom it became (1017) the capital of an independent emirate. Charlemagne tried to take it but was defeated by the Moors (778). The Cid fought for a time in the service of the Moorish ruler of Zaragoza. The city was conquered (1118) by Alfonso I of Aragón, who made it the capital of his kingdom. The most notable event in the later history of Zaragoza was its heroic resistance against the French in the Peninsular War. The city resisted the first siege (1808), surrendering only after some 50,000 defenders had died in the second siege (1808–9). Zaragoza is a cultural center and is rich in works of art, many of which show Moorish influence. There are two cathedrals—La Seo (12th–16th cent.), formerly a mosque, and El Pilar (17th cent.), named after the sacred pillar near which the Virgin is said to have appeared in the vision of St. James the Greater. El Pilar contains frescoes by Velázquez and Goya. Also noteworthy are the Church of San Pablo, the Moorish castle of Aljafería (residence of the emirs and later the kings of Aragón), the *Ionja* (exchange building), and a 15th cent. stone bridge across the Ebro. The modern church of San Antonio de Padua contains the remains of Italian soldiers killed in the civil war (1936–39).

Za·ri·a (zä'rē-ə), city (1971 est. pop. 201,000), N Nigeria. It is a processing center for Nigeria's main cotton-growing region. Cottonseed, groundnut, and shea-nut oil are produced, bicycles are assembled, and cigarettes are manufactured. It was founded about 1000 and was one of the seven Hausa city-states. The city was captured by the Fulani in 1805 and included in the Sokoto caliphate. In 1901 British forces took the city. The old part of the city is surrounded by walls.

Ža·tec (zhä'tĕts'), city (1970 pop. 15,725), NW Czechoslovakia, in Bohemia. It is the center of a famous hop-growing region. Žatec was founded in the 11th cent. The city was captured and burned by the Swedes in 1639 during the Thirty Years' War.

Za·va·la (zə-vä'lə), county (1970 pop. 11,370), 1,292 sq mi (3,346.3 sq km), SW Texas, drained by the Nueces and Leona rivers; formed 1858; co. seat Crystal City. Partly in the irrigated Winter Garden region, it produces spinach, other vegetables, citrus fruit, grain sorghums, corn, oats, peanuts, hay, pecans, cattle, hogs, poultry, and dairy products. Asphalt mining is done.

Za·wier·cie (zä-vyĕr'chā), city (1973 est. pop. 40,900), S Poland, on the Warta River. Its industries produce iron and steel, machinery and machine tools, chemicals, metals, glass, and textiles. Lignite and iron ore are mined nearby. Zawiercie passed to Prussia in 1795, to Russia in 1815, and to Poland after World War I.

Za·yan·deh Rud (zī-yän-dĕ' rōōd'), river, c.250 mi (400 km) long, rising in the Zagros Mts., W central Iran, and flowing SE through an agricultural district to a swamp W of Yazd. It is used for irrigation along its entire length. The Kuh-Rang Dam (built 1953) diverts water from the upper course of the Karun River through a 2-mi (3.2-

km) tunnel into the Zayandeh Rud, where it is used to supplement irrigation in the Esfahan area.

Zay·san, Lake (zī-sän′), freshwater lake, c.700 sq mi (1,815 sq km), SE Kazakhstan, in the Altai Mts. It is crossed by the Irtysh River. The lake abounds in fish.

Zea·land (zē′lənd), island: *see* Sjaelland.

Zeb·u·lon (zĕb′yə-lən), city (1970 pop. 776), seat of Pike co., W central Ga., SSW of Griffin. Food canning is done.

Zee·land (zē′lənd, zā′länt′), province (1973 est. pop. 319,392), c.650 sq mi (1,685 sq km), SW Netherlands, bordering on Belgium in the S and the North Sea in the W; capital Middelburg. Much of the land is below sea level and is protected by dikes. Zeeland, a part of Holland from the 10th cent., later became a separate county, but it continued to be ruled by the counts of Holland, and its history was largely identical with that of Holland. In 1579 Zeeland joined the Union of Utrecht as one of the United Provinces of the Netherlands.

Zef·at (zĕf′ät), town (1972 pop. 13,600), NE Israel. It is a hot-weather resort and has a thriving artists' colony and many museums and old synagogues. Ceramics and handicrafts are produced in the town. It was founded A.D. c.70. Flavius Josephus, a Jewish historian and soldier, built fortifications that later formed the foundations of a 12th cent. Crusader castle built by the Knights Templars; its ruins still stand. After the expulsion of the Jews from Spain in 1492, many learned Jews moved to Zefat and made the town an important center of rabbinical and cabalistic studies, which it remained through the 17th cent. In 1563 the first Hebrew printing press in the Holy Land was established in Zefat. Largely destroyed by an earthquake in 1769, Zefat was repopulated by Russian Hasidim in 1776. The Arabs forced most Jews to leave Zefat in 1929, but Jews returned after the 1948 Arab-Israeli war.

Zeist (zīst), municipality (1973 est. pop. 57,155), Utrecht prov., central Netherlands, near Utrecht. It is largely residential; manufactures include pharmaceuticals and furniture. A settlement of the Moravian Brethren was established here in 1746.

Zeitz (tsits, zīts), city (1973 est. pop. 44,675), Halle dist., S East Germany, on the White Elster River. Manufactures include machinery, chemicals, furniture, and beer. Of note in the city are the late-Gothic Church of St. Michael and a castle (17th cent.), whose church contains the tomb of the 16th cent. scientist Georg Agricola.

Ze·la (zē′lə), ancient city of Pontus, NE Asia Minor. Here Mithridates VI defeated Triarius c.67 B.C., and in 47 B.C. Julius Caesar defeated Pharnaces, king of Pontus, recording the victory in his famous dispatch "*Veni, vidi, vici,*" "I came, I saw, I conquered."

Ze·rav·shan (zĕr′əf-shän′), river, c.460 mi (740 km) long, rising in the Turkestan Range of the Pamir-Alai mountain system, S Central Asian USSR, in Tadzhikstan. It flows westward through the agricultural Zeravshan valley, then into Uzbekistan, past Samarkand and Bukhara, and disappears in the desert near the Amu Darya. The valley, irrigated by the Katta-Kurgan reservoir, is one of the chief oases of Central Asia and is on the site of the ancient Sogdiana. The Zeravshan Mts., forming the southern watershed of the river, rise to c.18,480 ft (5,635 m). The range has coal and ore deposits.

Ze·ya (zā′ə), river, c.800 mi (1,290 km) long, rising in the Stanovoy Range, Far Eastern USSR, and flowing S to join the Amur River. It carries gold in its upper reaches, and its basin has gold, graphite, and lignite deposits. The lower course flows through the rich agricultural Zeya-Bureya Plain.

Zgierz (zə-gyĕsh′), city (1973 est. pop. 46,300), E central Poland. A textile center, it also manufactures chemicals, textile machinery, and metals. Chartered about 1300, Zgierz grew after the textile industry was established here in 1818.

Zhda·nov (zhdä′nəf), city (1973 est. pop. 442,000), S European USSR, in the Ukraine, on the Sea of Azov at the mouth of the Kalmius River. A seaport and railroad terminus, Zhdanov is also a large iron and steel center with machine plants, chemical works, and shipyards. Coal, salt, and grain are the chief exports.

Zhi·gu·li Mountains (zhē′go͝o-lyē′), wooded range, E European USSR, in the Samara bend of the Volga River at Kuybyshev. They rise to c.2,220 ft (675 m) and are rich in oil. The mountains supply water for the Kuybyshev hydroelectric station.

Zhi·to·mir (zhĭ-tô′mĭr′), city (1973 est. pop. 199,000), capital of Zhitomir oblast, SW European USSR, in the Ukraine, on the Teterev River, a tributary of the Dnepr. It is a road and rail junction in an agricultural area. Industries include lumber milling, food processing, granite quarrying, metalworking, and the manufacture of building materials. An old city on the trade route from Scandinavia to Constantinople, Zhitomir was known in 1240. It was part of the Kievan state and later passed to Lithuania (1320) and Poland (1569). Returned to Russia with the second partition of Poland (1793), it became an important provincial and trade center before the Bolshevik Revolution.

Zie·bach (zē′bäk′, -bä′, -bô′), county (1970 pop. 2,221), 1,982 sq mi (5,133.4 sq km), NW central S.Dak., bounded in the S by the Cheyenne River and drained by the Moreau River; formed 1911; co. seat Dupree. It is in a farming and cattle-ranching area that produces grain and dairy products. The county is within Cheyenne River Indian Reservation.

Zie·lo·na Go·ra (zhĕ-lô′nä go͝or′ə), city (1973 est. pop. 79,700), capital of Zielona Gora prov., W Poland. It is a railroad junction and has lignite mines. Famous for its wines, the city also produces railroad cars, woolen textiles, machinery, and foodstuffs. Founded in the 13th cent., Zielona Gora became a prominent town on the trade route from Berlin to Upper Silesia.

Zi·guin·chor (zē′găN-shôr′), city (1970 pop. 45,800), SW Senegal, a port on the Casamance River. Located in a rice-growing region, the city produces peanut oil, frozen fish, and orange juice and is an outlet to the sea for the shipment of peanuts and other products of the southern hinterlands. Ziguinchor was occupied by the French in 1888. The city has a mechanical training and crafts school and is the seat of a Roman Catholic bishopric.

Zil·ler·tal Alps (tsĭl′ər-täl′), range of the E Alps, astride the Austro-Italian border and extending 35 mi (56.3 km) NE into the Tyrol. The range rises to 11,555 ft (3,524.3 m) in the Hockfeiler. The Brenner Pass forms the divide between the Zillertal Alps and the Ötztal Alps. The Zillertal is noted for its magnificent scenery.

Zim·bab·we (zĭm-bäb′wē), ruined city, SE Rhodesia, near Fort Victoria. It was discovered by European explorers c.1870, and some believed it the biblical Ophir, where King Solomon had his mines. From 1890 to 1900 some 100,000 gold mining claims were staked out here, but all proved barren. Modern archaeological evidence has shown that Zimbabwe was first occupied by the earliest Iron Age people in the 3rd cent. It was abandoned sometime thereafter until it was reoccupied in the late 9th cent. or early 10th cent. The remaining ruins include a massive wall, constructed in the 11th cent., a strong fortress, nearby dwellings, and an elliptically shaped enclosure. The buildings were once richly decorated with stone carvings and gold and copper ornaments.

Zin·der (zĭn′dər), town (1970 est. pop. 24,000), S Niger. It is the trade center for an agricultural region where grains, manioc, and peanuts are grown, and cattle and sheep are raised. Manufactures include millet flour, beverages, and tanned goods. Zinder was situated on an old trans-Saharan caravan route that connected northern Nigeria with the African coast as early as the 11th cent. The walled town was the capital of a Moslem state controlled by Bornu from the 16th to the mid-19th cent. Zinder was conquered by the French in 1899 and during World War I was the scene of an unsuccessful Tuareg uprising against French control. The town grew after 1920, when nomads began settling here in large numbers, and from 1922 to 1926 it served as the capital of the French Niger colony. Parts of the old city wall and the 19th cent. palace of the ruler of Zinder still stand.

Zi·on (zī′ən), city (1975 est. pop. 18,000), Lake co., extreme NE Ill., on Lake Michigan; inc. 1902. Television sets, clothing, metal fittings, and food products are made, and there is a printing and publishing house. Zion was founded in 1901 by John Alexander Dowie of the Christian Catholic Church. Until 1935 it was a communal society with a theocratic government. Of interest are the Zion Hotel (1902) and Shiloh House, the mansion built (1902) for the Dowie family.

Zion National Park, 147,035 acres (59,550 hectares), SW Utah; est. 1919. The park is noted for its many scenic trails and its vividly colored cliffs, rock formations, and deep canyons. The fingerlike, box-shaped Koblob Canyons have sheer 1,500-ft (457.5-m) walls. Zion Canyon, the park's main attraction, is a multicolored gorge, 15 mi (24.1 km) long, .5 mi (.8 km) deep, cut by the Virgin River. Vegetation in the park ranges from desert type in the canyons to forests on the mesas. Small animals thrive in the area, and mule deer are common.

Zi·ro (zîr'ō), town (1973 pop. 99,982), Arunachal Pradesh union territory, extreme NE India, in the Himalayas near the Tibetan border.

Zit·tau (tsĭt'ou'), city (1973 est. pop. 42,428), Dresden district, SE East Germany, on the Lusatian Neisse River, near the Polish and Czechoslovak borders. Manufactures include textiles, chemicals, machinery, and motor vehicles. An old Slavic settlement, Zittau was chartered in the 13th cent. In 1346 it joined the Lusatian League. It passed to Saxony in the early 17th cent. There are several medieval churches in the city.

Zla·to·ust (zlä'tə-o͞ost'), city (1973 est. pop. 186,000), E European USSR, on the Ai River in the S Urals. It is a rail terminus and a metallurgical center. Besides steel mills, the city has metal-engraving works and factories that manufacture farm machinery, instruments, precision castings, small arms, abrasives, and clocks and watches. Zlatoust was founded in 1754.

Znoj·mo (znoi'mō), city (1971 est. pop. 27,085), S central Czechoslovakia, in Moravia. It is the center of a horticultural region; the city's industries produce ceramics, alcohol, textiles, and furniture. Chartered in 1226, it has several fine churches and a 15th cent. town hall. At Znojmo, in 1809, an armistice was signed by the Austrians and the French after the French victory at Wagram.

Zoar (zôr), village (pop. 228), Tuscarawas co., E central Ohio, on the Tuscarawas River; founded 1817, inc. 1884. It was founded by a group of Separatists from southern Germany who fled religious persecution and, under the leadership of Joseph Michael Bimeler, emigrated to America. The Quakers received them kindly in Philadelphia and assisted them in obtaining land in Ohio. The village of Zoar was laid out, a communistic system was adopted to solve economic difficulties, and a strict moral and religious life was maintained. After Bimeler's death (1853), the society declined, and in 1898 the communistic mode of life was abandoned.

Zom·ba (zŏm'bə), town (1971 est. pop. 20,000), S Malawi, in the Shire Highlands. Until 1966 it was the national capital. European planters founded the town c.1880. It is the commercial center for a region growing cotton, coffee, and tobacco; tung oil is produced, and there is a hydroelectric power plant nearby. Zomba Plateau (c.7,000 ft/ 2,135 m high) is a popular summer resort.

zone (zōn), in geography, an area with a certain physical unity that distinguishes it from other areas. The division of the earth into five climatic zones probably originated (5th cent. B.C.) with Parmenides, who recognized a torrid zone and north and south temperate zones and postulated north and south frigid (or arctic) zones; his classification was adopted by Aristotle and is still in use. The zones are based on latitude: the torrid zone lies between 23½° N and 23½° S, the temperate zones between these parallels and the polar circles, and the frigid zones from the polar circles to the poles. Later geographers, recognizing that climate is affected by such conditions as altitude, distance from water, prevailing winds, and ocean currents, have used other bases for zoning. Most geographers today recognize five major climatic groups, based mainly on the work of the German meteorologist Wladimir Köppen. Two of these groups—the rainy tropics and the dry tropics, which encompass four different climates—together correspond roughly to the former torrid zone. Two humid climate groups of the Köppen system, encompassing six climates, together correspond roughly to the former temperate zones. Köppen's two polar climates correspond roughly to the two former frigid zones. In addition to the five groups encompassing twelve climates, geographers also recognize a series of highland zones where many of the other climates of the world are duplicated.

Zon·gul·dak (zông'gəl-däk'), city (1970 pop. 77,135), capital of Zonguldak prov., N Turkey, a port on the Black Sea. It is the center of the major coal-producing region.

Zren·ja·nin (zrĕn'yə-nĭn), city (1971 pop. 129,846), NE Yugoslavia, in the Vojvodina region of Serbia, on the Begej River. A river port and a railway center, it has industries that produce foodstuffs, beer, textiles, and agricultural machinery. It was known as Nagybecskerek until its transfer (1918) from Hungary to Yugoslavia, and it was subsequently named Veliki Bečkerek (until the 1930s) and Petrovgrad (until 1947).

Zug (tso͝ok, zo͞og), canton (1974 est. pop. 72,400), 93 sq mi (240.9 sq km), N central Switzerland. It is a forested and mountainous region with orchards, meadows, and pastures in the valleys. Owned by the counts of Kyburg and later (after 1273) by the Hapsburg family, Zug joined the Swiss Confederation in 1352 and again in 1364, after a return to Hapsburg domination. Its capital, Zug (1974 est. pop. 22,700), is on the Lake of Zug (15 sq mi/38.9 sq km). It has manufactures of metalware, electrical equipment, textiles, and kirsch. Zug retains a medieval flavor. Its Church of St. Oswald is one of the most splendid late-Gothic churches in Switzerland.

Zug·spit·ze (tso͝ok'shpĭt'sə), mountain, 9,721 ft (2,965 m) high, in the Bavarian Alps on the West German-Austrian border. It is the highest peak of West Germany. A cog-and-pinion railroad connects the popular resort of Garmisch-Partenkirchen, at its foot, with the summit.

Zui·der Zee (zī'dər zē', zā'), former shallow inlet of the North Sea, c.80 mi (130 km) long, indenting NE Netherlands. Once a lake, it was joined to the North Sea by a great flood in the 13th cent. A vast drainage project, begun in 1920, split the old Zuider Zee into the IJsselmeer, south of the IJsselmeer dam, and the Waddenzee, between the dam and the West Frisian Islands.

Zu·lu·land (zo͞o'lo͞o-lănd'), historic region and home of the Zulu, c.10,000 sq mi (25,900 sq km), NE Natal prov., Republic of South Africa. Zululand is bordered by the Indian Ocean on the east, by Mozambique on the north, and by Swaziland on the west. The terrain rises from a low coastal plain to the foothills of the Drakensberg range. Although some maize is grown, the economy depends primarily on cattle raising. Zululand's two major commercial crops, sugar cane and cotton, are generally cultivated on European-owned coastal plantations. Sugar milling and some paper making are virtually the region's only industries. There is also considerable exploitation of wattle and eucalyptus. The Zulu, who belong to the southern branch of the Nguni-speaking peoples, constitute the majority of the population, and Zulu is the chief language. Most Zulu live as members of an extended family in a fenced compound (kraal), headed by the oldest man. Members of the family occupy beehive-shaped huts in the enclosure of the kraal, within which the cattle are kept penned. The prolonged absence of a majority of the men, many of whom are employed in the distant cities and mines of South Africa has, however, weakened Zulu society.

The name Zulu originally denoted a small tribe that, migrating southward, reached the area around the Tugela River in the late 17th cent. The Zulu became historically important in the early 19th cent. under Chaka, whose conquests reduced many tribes to vassalage and caused others to flee. His warlike successors soon encountered the Boer settlers migrating north into Natal as part of the Great Trek. The Zulu chief Dingaan ambushed and slaughtered about 500 Boers in 1838. In revenge about 3,000 Zulu were killed by the forces of Andries Pretorius in the Battle of Blood River. Subsequent Boer intervention in Zulu domestic affairs led in 1840 to the overthrow of Dingaan and the crowning of Mpanda, who became a vassal of the Boer republic of Natal. The British, who succeeded the Boers as rulers of Natal in 1843, encountered the hostility of Mpanda's son, Cetewayo. After he ignored an ultimatum that he submit to British rule, Great Britain launched an attack on Zululand in 1878 and, although suffering several grave defeats, finally triumphed in July, 1879. Faced with continuing Zulu rebellions, the British annexed Zululand in 1887; it became part of Natal in 1897.

The Bantustan (black African homeland) designated by the government of South Africa, in accordance with the Bantu Self-Government Act of 1959, to be the Zulu homeland has been named Kwazulu; it is 12,140 sq mi (31,442.6 sq km) in area and is made up of isolated tracts of land, only some of which were part of historical Zululand. It is, therefore, neither geographically unified nor territorially homogeneous. The area north of the Tugela River, where the largest tracts of Zulu territory lie, forms the hub of Kwazulu. Slightly more than half of South Africa's Zulu population lives in Kwazulu, which also has Xhosa, Sotho, and Swazi minorities.

Zu·ñi (zo͞o'nyē, -nē), pueblo (1970 pop. 3,958), McKinley co., W N.Mex., in the Zuñi Indian Reservation; built c.1695. Its inhabitants are Pueblo Indians of the Zuñian linguistic family. They are a sedentary people, who farm irrigated land and are noted for basketry, pottery, turquoise jewelry, and weaving, and for their ceremonial dances. The original seven Zuñi villages are usually identified with the mythical Seven Cities of Cibola, which were publicized by Marcos de Niza. In 1540 Francisco Vásquez de Coronado attacked the villages, thinking that they had vast stores of gold. The villages were abandoned in the Pueblo revolt of 1680. The present pueblo was built on the site of one of the original seven villages.

Zü·rich (zo͝or'ĭk, tsü'rĭkн), canton (1974 est. pop. 1,122,000), 668 sq mi

(1,730.1 sq km), N Switzerland, bounded in part by the Lake of Zürich in the S and West Germany in the N. It is a fertile agricultural region with orchards, meadows, and forests. Machinery and other metal goods as well as textiles are manufactured.

Zürich (1974 est. pop. 401,600), its capital, is the country's commercial and economic center as well as the intellectual center of German-speaking Switzerland. Its industrial products include machine tools, radios, clothes, paper, and textiles. It is the hub of a printing and publishing industry and has many banking and financial institutions. Occupied as early as the Neolithic period by lake dwellers, the site of Zürich was settled by the Helvetii. It was conquered (58 B.C.) by the Romans, and after the 5th cent. passed successively to the Alemanni, to the Franks, and to Swabia. It became a free imperial city after 1218, accepted a corporative constitution in 1336, and joined the Swiss Confederation in 1351. Its claim to the Toggenburg led to a ruinous war (1436–50) with the other confederates. In the 16th cent. Zürich, under the influence of Ulrich Zwingli, became the leading power of the Swiss Reformation and once more provoked a civil war. The Roman Catholic victory at Kappel (1531) ended Zürich's political leadership. In 1799 the city was the scene of two battles of the French Revolutionary Wars. Zürich developed as a cultural and scientific center in the 18th and 19th cent. It has the largest Swiss university (founded 1833), a world-famous polytechnic school (est. mid-19th cent.), and many museums. The Romanesque Grossmünster (11th–13th cent.), where Zwingli preached, the Fraumünster (12th and 15th cent.), the 17th cent. town hall, and numerous old residences contrast harmoniously with many fine modern structures. The educational reformer Heinrich Pestalozzi was born in the city, and James Joyce is buried here.

Zürich, Lake of, narrow, elongated lake, 34 sq mi (88.1 sq km), 25 mi (40.2 km) long, N Switzerland. It has a maximum depth of c.470 ft (145 m). The lake is connected to the Lake of Wallenstadt (Walensee) by the Linth Canal and also receives water from the Linth River. It is drained by the Limmat River. The gently sloping shores of the lake are covered with vineyards, orchards, and woods; houses and villas dot the slopes.

Zut·phen (zŭt'fən), city (1973 est. pop. 27,718), Gelderland prov., E central Netherlands, on the IJssel River. It is an administrative, industrial, and commercial center. Zutphen was chartered in 1191 and was an important Spanish stronghold during the Dutch struggle for independence (16th cent.). Sir Philip Sidney died (1586) in a skirmish at the city's walls while aiding his uncle, the earl of Leicester, in an unsuccessful attempt to take Zutphen from the Spanish. The city was captured by the Dutch under Maurice of Nassau in 1591.

Zwei·brück·en (tsfī-brük'ən), city (1972 est. pop. 37,522), Rhineland-Palatinate, W West Germany, near the Saarland border. Zweibrücken is a transportation center and has ironworks, steelworks, and factories that produce leather goods, machines, and textiles. It is also a noted horse-breeding center. Zweibrücken was chartered in 1352 and passed (1385) to the Palatinate branch of the Bavarian house of Wittelsbach. In 1410 it became the seat of the counts (later dukes) palatine of Zweibrücken under a cadet line of the Palatinate branch. Charles X of Sweden was the nephew of John II, duke palatine of Zweibrücken; his son, Charles XI of Sweden, inherited Zweibrücken in the late 17th cent., and the duchy remained in personal union with Sweden from 1697 until the death (1718) of Charles XII. The Zweibrücken line continued in Sweden until the death (1741) of Ulrica Leonora, sister of Charles XII, and in Zweibrücken until 1731, when the related Palatinate-Birkenfeld line acceded. On the death (1799) of Elector Charles Theodore of Bavaria, who had reunited all the Wittelsbach lands except Zweibrücken, the duke palatine of Zweibrücken, his next of kin, succeeded him as elector of Bavaria, thus completing the reunion of the family territories; as Maximilian I he became (1806) the first king of Bavaria. However, the duchy of Zweibrücken itself had been annexed (1797) to France. It was restored to Bavaria at the Congress of Vienna (1814–15) and since then has shared the history of the Rhenish Palatinate.

Zwick·au (tsfĭk'ou'), city (1973 est. pop. 123,830), Karl-Marx-Stadt district, S East Germany, on the Mulde River. It is an industrial city and the center of a coal mining region. Manufactures include machinery, tractors, textiles, mining equipment, and motor vehicles. Zwickau was chartered in the early 13th cent., and it was a free imperial city from 1290 to 1323, when it passed to the margraves of Meissen. The city was repeatedly plundered during the Thirty Years' War (1618–48). Noteworthy buildings include a basilica (12th–15th cent.), the Church of St. Catherine (14th cent.), and the city hall (15th cent.). Robert Schumann was born (1810) in Zwickau, and there is a Schumann museum here.

Zwol·le (zvô'lə), municipality (1973 est. pop. 77,122), capital of Overijssel prov., N central Netherlands, on the Zwartewater River. It is an administrative, transportation, and industrial center. Nearby was the monastery where Thomas à Kempis lived (15th cent.).

Ży·rar·dów (zhĭ-rär'do͞of'), city (1973 est. pop. 33,800), E central Poland. It is a textile center, known especially for its woolens. Leather goods are also manufactured in the city.

Appendix

In the past year the world has been in almost constant upheaval and change: revolutions in Africa, Central America, and Iran; stunning election upsets in Great Britain and Canada; new treaties and alliances between the United States and the People's Republic of China and between Egypt and Israel; escalating oil prices and a chronic fuel shortage in the major nations of the world; and rising inflation everywhere. Some of the most important events that have occurred during the past year are detailed in the following pages.

United States

In 1978 and 1979 the two greatest problems facing President Jimmy Carter were energy and inflation. Despite his achievements in foreign affairs, his inability to deal adequately with these issues created a lack of confidence among the American people.

In Mar. 1979, the worst accident in the history of nuclear power occured at the Three Mile Island plant near Harrisburg, Pa. While blame was shuffled between the plant construction firm, the power company, and the Nuclear Regulatory Commission, a widespread protest movement called for the immediate shutdown of all nuclear plants and the banning of construction of any new facilities. Such extreme actions were not taken, but the growth of nuclear power in the United States suffered a severe setback. With the loss of confidence in the safety of nuclear power, even among top government officials, the problem of developing alternative forms of energy became more immediately apparent.

Also in the first half of 1979, gasoline shortages and resulting spiraling prices became a national problem. The public was divided between those who believed the so-called "energy crisis" was genuine and those who thought it was a scheme by oil companies to increase their profits. President Carter's plan to decontrol oil prices and levy a windfall profits tax on the large corporations was met with furious debate in the U.S. Congress.

Adding further to the energy-related transportation problems of the United States was the worst air disaster in the country's history in June, 1979. The crash of a DC-10 at Chicago's O'Hare Airport killed 275 persons and led to the indefinite grounding of all DC-10s until the structural weakness that had caused the crash was explored and resolved.

The SALT II Treaty calling for a strategic balance in nuclear armaments was signed in Vienna at a conference held June 15-18 by Jimmy Carter and Soviet Communist Party Chief Leonid Brezhnev. Critics in the United States believed the treaty conceded too much weapons superiority to the Soviets. Others felt the signing would create a more positive climate for improved relations and negotiations between the United States and the Soviet Union.

Canada

Prime Minister Pierre Trudeau's Liberal government was voted out on June 4, 1979, when Joseph Clark, a conservative, was elected to replace him. The Liberal government had fallen out of favor because it had awarded a disproportionate amount of federal funds to Quebec during a period of rising unemployment and inflation in all provinces. Clark pledged to decentralize the federal government, cut taxes, and cooperate with Quebec in an effort to prevent its secession from the country.

Central America

In the fall of 1978 the relative stability of Central America was threatened by mounting revolutionary activities fo the Sandinist National Liberation Front against the Nicaraguan government of Anastasio Somoza Debayle. In Sept. 1978, a month-long civil war took thousands of lives. Although Somoza's National Guard successfully put down the uprising, Sandinist forces continued their attempt to overthrow his government. The capital city of Managua was the center of street warfare and a general business strike, although similar activity raged throughout the country and along the Costa Rican border. Economic exchange was disrupted throughout the surrounding regions. Somoza implemented a state of siege in June, 1979, and accused the governments of Costa Rica, Venezuela, Cuba, and Panama of aiding and arming the Sandinist guerrillas. Repressive regimes in El Salvador and Guatemala verbally supported Somoza's government.

The allegations against Panamanian gunrunners had far-reaching effects; conservative elements in the U.S. Congress opposed to enactment of the new Panama Canal Treaty, scheduled for late in 1979, attempted to use the charges as evidence of Panama's unreliability in maintaining neutral operation of the Canal Zone. Counter-lobbying efforts in Congress sought to discredit the Nicaraguan accusations; Panamanian businessmen made a public appeal to the American people not to allow the voiding of many of the original treaty's provisions before it went into effect.

South America

South America continued to be a target for human rights activists in 1978. The Carter Administration, the Roman Catholic Church, and organizations such as Amnesty International all brought pressure on South American military governments in response to reports of persistent and systematic violations of human rights in these countries.

In Nov., 1978, the attention of the world was drawn to Jonestown, Guyana, where 909 Americans, all members of a religious sect called the People's Temple, committed mass suicide.

Europe

In June, 1979, the first direct elections for a European Parliament were held. Each nation of the nine-member European Community was represented according to population. Its power and influence were felt to be questionable as the old Parliament, in existence since 1958, had proved little more than a debating society. The new Parliament's charter called it a consultative body, but observers hoped it could eventually lead to a stronger, united Europe with increased impact on world affairs.

In Great Britain Margaret Thatcher became the first woman Prime Minister and the first woman to lead a major Western nation in the 20th century. In elections held in May, 1979, her Conservative Party ousted James Callaghan's Labour Party. Thatcher's principal aims were a reduction of Britain's welfare state, a break with the socialist past, and a return to a strong market economy. She promised increased defense spending and stronger ties with the United States and France.

In Italy national elections in June, 1979, took a toll on Communist Party strength and set up new barriers to the party's bid for membership in a coalition government with the ruling Christian Democrats. It was the first setback in 33 years for the Italian Communists, who had been consistently growing in power. Several smaller parties, including the Radical Party, made surprising gains in the June elections leading to the speculation of future instability in the perennially shaky Italian government.

Pope John Paul II returned to his native Poland in June, 1979, and was greeted by an overwhelming throng of Catholic supporters, causing great embarrassment to the Communist government there and in Moscow. The Pope's call for greater religious freedom, more churches, the need for open national boundaries in Eastern Europe, and respect for human dignity and rights caused considerable concern in the Communist-bloc countries.

Middle East

Although the peace treaty between Israel and Egypt was signed on Mar. 27, 1979, there was still widespread unrest in the Middle East. In an effort to end Palestinian attacks, Israel continued to engage in border skirmishes with Lebanon. Egypt, boycotted by the United Arab Republic for its signing of the peace treaty, lost millions of dollars in trade and the political support of the other Arab nations.

In Iran the Ayatollah Khomeini was unable to consolidate mass support for his government. On Mar. 8, 1979, thousands of Iranian women demonstrated in protest of Khomeini's seeming opposition to women's rights, and in early June, 1979, Iranian government forces were sent into the southern coastal city of Khorramshahr, to quell an uprising of Arabs who were demanding autonomy. Throughout the world human-rights advocates denounced the executions of dozens of the Shah's former supporters that were ordered by the new government. In addition, the country was plagued by economic instability, caused in part by the temporary shutdown of the oil wells during Khomeini's takeover of the country.

Africa

From Jan., 1978, to the spring of 1979, five of Africa's more prominent leaders lost power. Kenya's Jomo Kenyatta, a symbol of African nationalism, died; South Africa's John Vorster resigned; Ian Smith's white minority government in Rhodesia was succeeded in Apr., 1979 by the black majority government of Bishop Abel Muzorewa (who renamed the country Zimbabwe Rhodesia); Uganda's Idi Amin was driven from his government's quarters in Kampala by a combined force of Tanzanian troops and Ugandan exiles; and Algeria's president, Houari Boumedienne, died in Dec., 1978.

Wars of independence and secession continue to be fought throughout the continent. The black-ruled nation of Namibia was granted independence from South Africa after prolonged negotiations. A bloody uprising in the copper-rich Shaba province of Zaire by Katangan rebels was put down after Zaire's President Mobutu called in the French Foreign Legion. In Zimbabwe, Rhodesia the Patriotic Front, led by Joshua Nkomo and Robert Mugabe, stepped up guerrilla attacks from bases in Mozambique.

Asia

After many years of negotiations with Chinese officials, including celebrated journeys made by Henry Kissinger and U.S. presidents Richard M. Nixon and Jimmy Carter, formal diplomatic relations were established between the United States and the People's Republic of China on Jan. 1, 1979. At the same time, diplomatic ties with Nationalist China were severed, although President Carter announced that the existing defense treaty with Taiwan would remain in effect until Dec. 31, 1979, Taiwan bitterly denounced the new policy and rejected an offer by Peking to establish trade relations.

In Dec., 1977, fighting broke out on the border between Vietnam and Cambodia, causing Cambodia to break off its diplomatic ties with Vietnam. The fighting escalated, resulting in the conquest of Cambodia by Vietnam some months later. This led to a build-up of tension between Vietnam and the leaders of China, who viewed the takeover as both an imperialistic action and as a threat to Chinese security because of the Soviet Union's military backing of Vietnam. The mistreatment by the Vietnamese government of ethnic Chinese living in Vietnam prompted China to send troops across the border into that country during May, 1979. After several weeks of battles, China withdrew its troops.

Meanwhile, hundreds of thousands of Chinese refugees fled Vietnam and Cambodia by sea. They paid Vietnamese officials bribes of up to $4,000 to be allowed to leave on boats that were often not permitted to land in any other country. Almost three fourths of the "boat people" died aboard ship from thirst, starvation, or disease, and many drowned. Those who were allowed to land frequently were forced to leave shortly after their arrival. Malaysia, claiming the refugees were causing major social and economic problems, developed a massive deportation plan that included the possible shooting of anyone who attempted to enter or return to Malaysia.

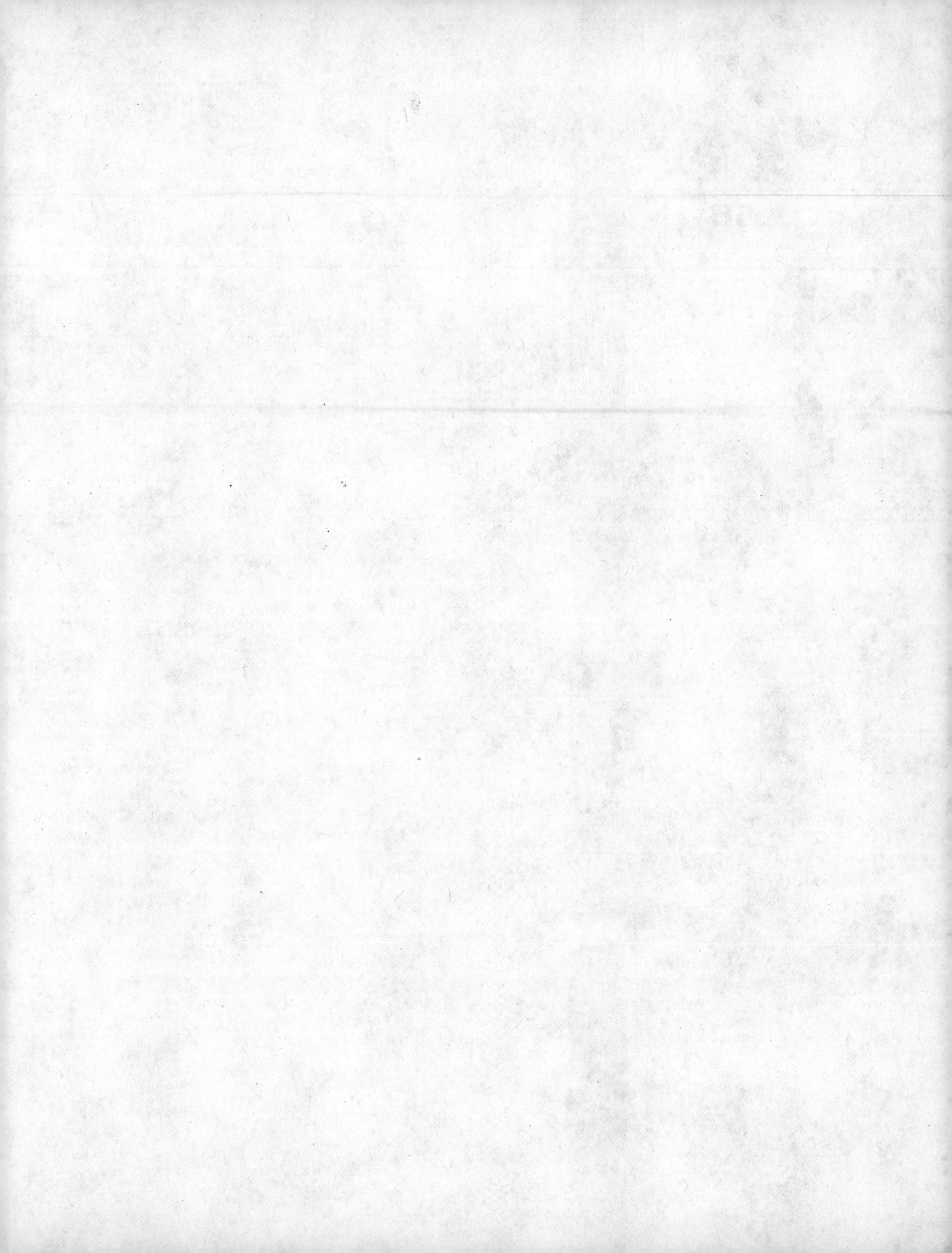

World Atlas

The maps in the following section were selected from *Goode's World Atlas,* published by Rand McNally & Company. They were chosen to give a quick overview of the world, showing physical features and the location of all countries, U.S. states, Canadian provinces, and major cities.

Contents

Map — Map Projection

map, conventionalized representation of spatial phenomena on a plane surface. Unlike photographs, maps are selective and may be prepared to show various quantitative and qualitative facts, including boundaries, physical features, patterns, and distribution. Each point on a map corresponds to a geographic position in accordance with a definite scale and projection.

Cartography, or mapmaking, antedates even the art of writing. Diagrams of areas familiar to them were made by Marshall Islanders, Eskimos, American Indians, and other preliterate peoples. Maps drawn by ancient Babylonians, Egyptians, and Chinese have been found. The oldest known map, now on exhibition in the Semitic Museum of Harvard, is a Babylonian clay tablet dating from c.2500 B.C. Our present system of cartography was established by the Greeks, who remained unexcelled until the 16th cent. Scientific measurements of earth distances by means of meridians and parallels were first made by Eratosthenes (3rd cent. B.C.). Of the ancient scholars, the mathematician and geographer Ptolemy (A.D. 2nd cent.), expounded on the principles of cartography; his system was followed for many centuries, although his basic error in underestimating the earth's size was not corrected until the age of Mercator. Only the Mediterranean world was represented with any accuracy in early maps. During the Middle Ages, while European cartographers produced artistic, idealized maps, Arabic mapmakers carried on the work of Ptolemy, and the Chinese produced the first printed maps.

Three major events contributed to the spectacular renaissance of cartography in Europe around 1500—the rediscovery and translation into Latin of Ptolemy's *Geographia*, the invention of printing and engraving, and the great voyages of discovery. This renaissance was manifested by the work of Gerardus Mercator in the first modern world atlas, published in 1570 by Abraham Ortelius, and by the decorative, painting-like maps of the French Sanson family (17th cent.). Improvements in the methods of surveying and increased emphasis on accuracy led to the noted work in the 18th cent. of the Frenchmen Guillaume Delisle and J. B. B. d'Anville, the founders of modern cartography.

After 1750 many European governments undertook the systematic mapping of their countries. The first important national survey was made in France (published 1756), followed by the Ordnance Survey of Great Britain (published 1801) and the topographic survey of Switzerland (organized 1832). In the United States the Geological Survey (established 1878) has mapped much of the country on varying scales. During the 19th cent. the demand for national maps was fulfilled, and famous world atlases were published.

With the advent of the 20th cent. the need arose for an international map of the world on a uniform scale. At several meetings of the International Geographical Congress (1891, 1909, 1913), the German Albrecht Penck presented and perfected plans for a world map on a scale of 1:1,000,000, to consist of about 1,500 sheets, each covering four degrees of latitude and six degrees of longitude in a modified conic projection. Uniformity of lettering and the use of layer tints to indicate relief were agreed upon. However, only part of the work has been completed. The greatest single contribution to the map of the world was made by the American Geographical Society of New York, which completed (1945) its 107-sheet *Map of Hispanic America.*

During World Wars I and II the science and art of mapping was greatly advanced. Aerial and satellite photography, radar, and sonar as the basis for mapmaking have made great technical advances since the end of World War II. Computer mapping was developed in the 1960s.

---

map projection, transfer of the features of the surface of the earth or another spherical body onto a flat sheet of paper. Only a globe can represent accurately the shape, orientation, and relative area of the earth's surface features; any projection produces distortion with regard to some of these characteristics. The particular projection chosen for a given map will depend on the use for which the map is intended. Some projections preserve correct relative distances in all directions from the center of the map (equidistant projection); some show areas equal to (equal-area projection) or shapes similar to (conformal projection) those on a globe of the same scale; some are useful in determining direction.

Many map projections can be constructed by the use of a light source to project the features of the globe onto a piece of paper (although in practice one performs the operation mathematically rather than with a light); other projections can be constructed only mathematically. Projections are classified as cylindrical, conic, or azimuthal according to the method of projection with a light source; many projections that can be constructed only mathematically are also classified according to this system. In a typical cylindrical projection, one imagines the paper to be wrapped as a cylinder around the globe, tangent to it along the equator. Light comes from a point source at the center of the globe or, in some cases, from a filament running from pole to pole along the globe's axis. In the former case the poles clearly cannot be shown on the map, as they would be projected along the axis of the cylinder out to infinity. In the latter case the poles become lines forming the top and bottom edges of the map. The Mercator projection, long popular but now less so, is a cylindrical projection of the latter type that can be constructed only mathematically. In all cylindrical projections the meridians of longitude, which on the globe converge at the poles, are parallel to one another; in the Mercator projection the parallels of latitude, which on the globe are equal distances apart, are drawn with increasing separation as their distance from the equator increases in order to preserve shapes. However, the price paid for preserving shapes is that areas are exaggerated with increasing distance from the equator. The poles themselves cannot be shown on the Mercator projection.

In a conic projection a paper cone is placed on a globe like a hat, tangent to it at some parallel, and a point source of light at the center of the globe projects the surface features onto the cone. The cone is then cut along a convenient meridian and unfolded into a flat surface in the shape of a circle with a sector missing. All parallels are arcs of circles with a pole (the apex of the original cone) as their common center, and meridians appear as straight lines converging toward this same point. Some conic projections are conformal (shape preserving); some are equal-area (size preserving). A polyconic projection uses various cones tangent to the globe at different parallels. Parallels on the map are arcs of circles but are not concentric.

In an azimuthal projection a flat sheet of paper is tangent to the globe at one point. The point light source may be located at the globe's center (gnomonic projection), on the globe's surface directly opposite the tangent point (stereographic projection), or at some other point along the line defined by the tangent point and the center of the globe, e.g., at a point infinitely distant (orthographic projection). In all azimuthal projections the tangent point is the central point of a circular map; all great circles passing through the central point are straight lines, and all directions from the central point are accurate. If the central point is a pole, then the meridians (great circles) radiate from that point and parallels are shown as concentric circles. The gnomonic projection has the useful property that all great circles (not just those that pass through the central point) appear as straight lines; conversely, all straight lines drawn on it are great circles. A navigator taking the shortest route between two points (always part of a great circle) can plot his course on a gnomonic projection by drawing a straight line between the two points.

Among the other commonly used map projections are the Mollweide homolographic and the sinusoidal, both of which are equal-area projections with horizontal parallels; they are especially useful for world maps. Goode's homolosine projection is a composite using the sinusoidal projection between latitudes 40° N and 40° S and the homolographic projection for the remaining parts. Interruptions, or splits, are often made in the ocean areas in order to show land areas with truer shapes.

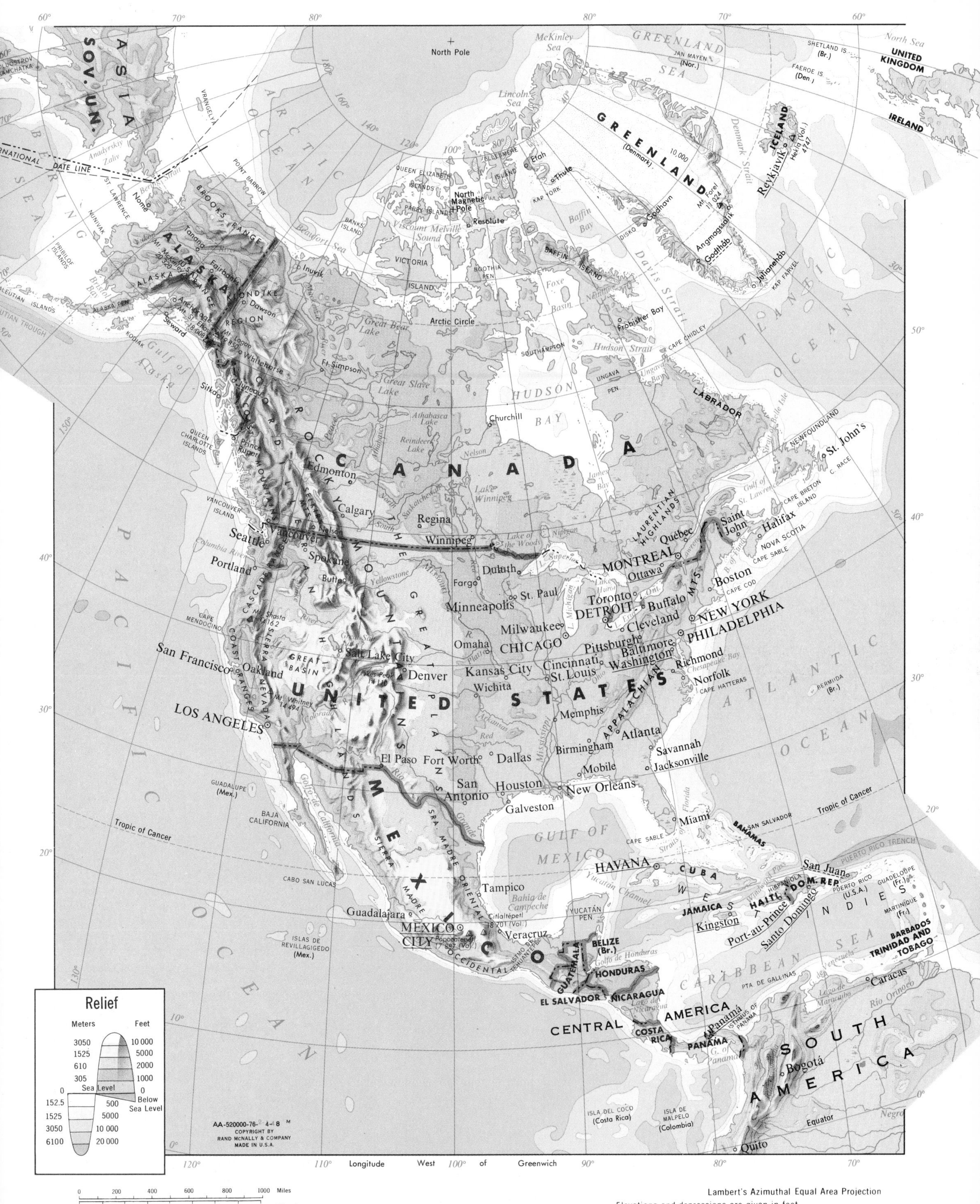

North Pole
GREENLAND SEA
GREENLAND (Denmark)
ICELAND
Reykjavík
UNITED KINGDOM
IRELAND
ASIA
SOV. UN.
ARCTIC OCEAN
BERING SEA
ALASKA
Anchorage
Fairbanks
Nome
Juneau
Sitka
Gulf of Alaska
CANADA
HUDSON BAY
Churchill
Edmonton
Calgary
Regina
Winnipeg
Vancouver
Seattle
Portland
Spokane
Butte
MONTREAL
Ottawa
Québec
Saint John
Halifax
St. John's
LABRADOR
NEWFOUNDLAND
NOVA SCOTIA
UNITED STATES
Duluth
Fargo
St. Paul
Minneapolis
Milwaukee
CHICAGO
DETROIT
Toronto
Buffalo
Cleveland
Boston
NEW YORK
PHILADELPHIA
Pittsburgh
Baltimore
Washington
Richmond
Norfolk
Omaha
Kansas City
St. Louis
Cincinnati
Wichita
Denver
Salt Lake City
San Francisco
Oakland
LOS ANGELES
Memphis
Atlanta
Birmingham
Savannah
Jacksonville
Mobile
New Orleans
Dallas
Fort Worth
El Paso
San Antonio
Houston
Galveston
Miami
GULF OF MEXICO
HAVANA
CUBA
MEXICO
Tampico
Guadalajara
MEXICO CITY
Veracruz
BELIZE (Br.)
GUATEMALA
HONDURAS
EL SALVADOR
NICARAGUA
COSTA RICA
PANAMA
CENTRAL AMERICA
JAMAICA
Kingston
HAITI
DOM. REP.
Port-au-Prince
Santo Domingo
San Juan
CARIBBEAN SEA
Caracas
Bogotá
Quito
SOUTH AMERICA
PACIFIC OCEAN
ATLANTIC OCEAN
Tropic of Cancer
Arctic Circle
Equator
Relief
Meters
Feet
3050
1525
610
305
0
152.5
1525
3050
6100
10 000
5000
2000
1000
Sea Level
0
500
5000
10 000
20 000
Below Sea Level
AA-520000-76- 4-8 M
COPYRIGHT BY RAND McNALLY & COMPANY
MADE IN U.S.A.
Longitude West of Greenwich
0 200 400 600 800 1000 Miles
0 400 800 1200 1600 Kilometers
Lambert's Azimuthal Equal Area Projection
Elevations and depressions are given in feet

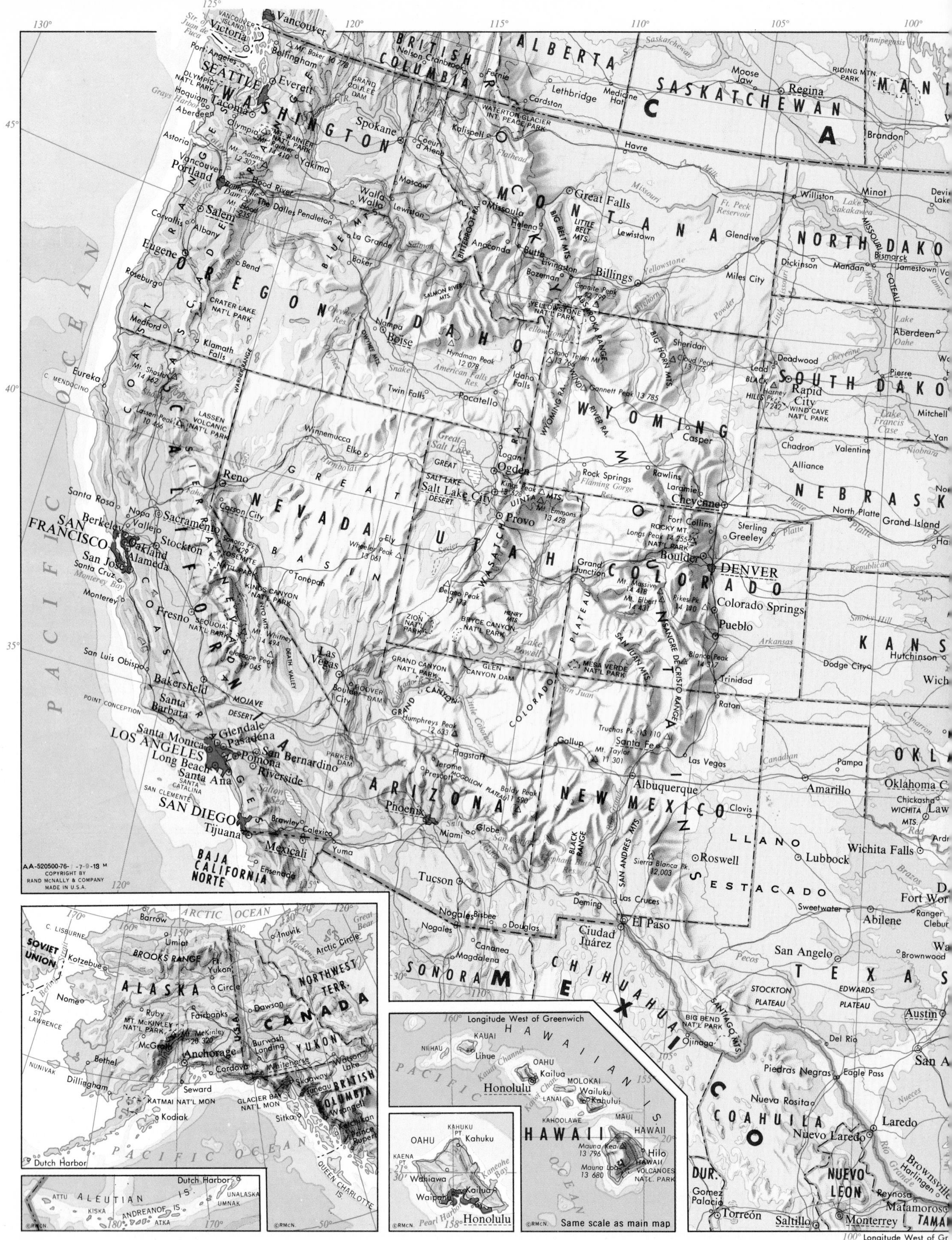

Polyconic Pro

Elevations and depressions are given in feet

ONTARIO
QUEBEC
NEW BRUNSWICK
PRINCE EDWARD ISLAND
NOVA SCOTIA
MAINE
WISCONSIN
MICHIGAN
ILLINOIS
INDIANA
OHIO
KENTUCKY
TENNESSEE
MISSOURI
ARKANSAS
MISSISSIPPI
ALABAMA
GEORGIA
FLORIDA
LOUISIANA
NORTH CAROLINA
SOUTH CAROLINA
VIRGINIA
WEST VIRGINIA
PENN.
NEW YORK
ATLANTIC OCEAN
GULF OF MEXICO
BAHAMAS
Relief
Meters
Feet
Sea Level
Below Sea Level

Conic Projection
Elevations and depressions are given in feet

QUEBEC
MONTREAL
Ottawa
Hull
Sherbrooke
ONTARIO
ALGONQUIN PROVINCIAL PARK
Peterborough
Kingston
Oshawa
TORONTO
LAKE ONTARIO
Surface 246 Feet above Sea Level
maximum depth 778 Feet
Hamilton
Niagara Falls
Rochester
BUFFALO
Syracuse
Utica
Schenectady
Albany
Troy
ADIRONDACK MTS.
NEW YORK
CATSKILL MTS.
Binghamton
Elmira
VERMONT
NEW HAMPSHIRE
MAINE
Portland
Concord
Manchester
Burlington
Montpelier
MASS.
BOSTON
Worcester
Springfield
Providence
Fall River
New Bedford
R.I.
CONN.
Hartford
New Haven
Bridgeport
LONG ISLAND SOUND
NEW YORK
Newark
Jersey City
Paterson
Trenton
NEW JERSEY
Atlantic City
PENNSYLVANIA
Scranton
Wilkes-Barre
Allentown
Bethlehem
Reading
Harrisburg
Lancaster
Altoona
Johnstown
PITTSBURGH
McKeesport
PHILADELPHIA
Camden
Wilmington
DEL.
Dover
MARYLAND
BALTIMORE
WASHINGTON D.C.
Arlington
Alexandria
VIRGINIA
Richmond
Charlottesville
Lynchburg
Newport News
Norfolk
Portsmouth
Virginia Beach
ATLANTIC OCEAN
APPALACHIAN MOUNTAINS
BLUE RIDGE
ALLEGHENY
CHESAPEAKE BAY
Relief
Meters
Feet
1525
610
305
152.5
0
Sea Level
5000
2000
1000
500
0
152.5
1525
3050
500
5000
10 000
AA-520596-76 5-7-10 M
COPYRIGHT BY
RAND McNALLY & COMPANY
MADE IN U.S.A.
0 20 40 60 80 100 120 Miles
0 20 40 60 80 100 120 140 160 180 200 Kilometers
80°
78°
76°
74°
72°
70°
44°
42°
40°
38°

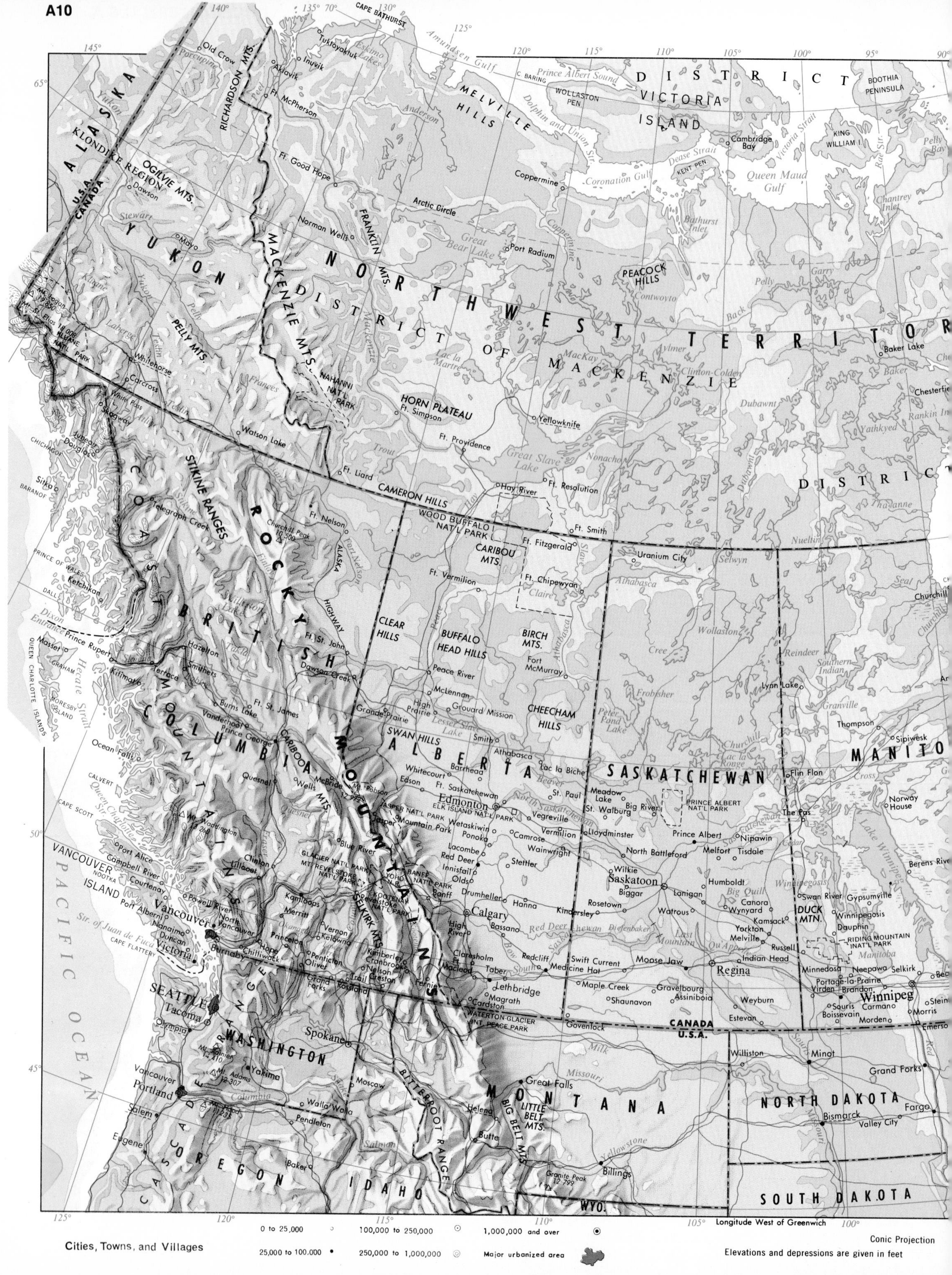
CAPE BATHURST
Amundsen Gulf
C. BARING
Prince Albert Sound
DISTRICT
VICTORIA
ISLAND
WOLLASTON PEN
BOOTHIA PENINSULA
MELVILLE HILLS
Dolphin and Union Str.
Cambridge Bay
Victoria Strait
KING WILLIAM I.
Dease Strait
KENT PEN
Coronation Gulf
Queen Maud Gulf
Coppermine
Chantrey Inlet
Bathurst Inlet
Tuktoyaktuk
Eskimo Lakes
Inuvik
Aklavik
Old Crow
Porcupine
Ft. McPherson
Peel
Anderson
RICHARDSON MTS.
ALASKA
U.S.A.
CANADA
KLONDIKE REGION
OGILVIE MTS.
Dawson
Ft. Good Hope
Arctic Circle
Stewart
YUKON
Mayo
Norman Wells
FRANKLIN MTS.
MACKENZIE MTS.
NORTHWEST TERRITORIES
Great Bear Lake
Port Radium
PEACOCK HILLS
Contwoyto
Pelly
Garry
Back
DISTRICT OF MACKENZIE
Mt. Logan 19,850
Mt. St. Elias 18,008
KLUANE NAT'L PARK
PELLY MTS.
Laberge
Whitehorse
Carcross
White Pass
Skagway
Teslin
Frances
NAHANNI NAT'L PARK
Lac la Martre
MacKay
Aylmer
Clinton-Colden
Baker Lake
Baker
Dubawnt
Chesterfield
Rankin Inlet
Yathkyed
HORN PLATEAU
Ft. Simpson
Yellowknife
Watson Lake
Juneau
Douglas
CHICHAGOF
Sitka
BARANOF
Ft. Providence
Great Slave Lake
Nonacho
Trout
Ft. Liard
Hay River
Ft. Resolution
CAMERON HILLS
WOOD BUFFALO NAT'L PARK
Ft. Smith
Thelon
Thlewiaza
Nueltin
STIKINE RANGES
Telegraph Creek
COAST
ROCKY
Ft. Nelson
Churchill Peak 10,500
ALASKA HIGHWAY
Ft. Nelson
Finlay
Ft. Fitzgerald
CARIBOU MTS.
Uranium City
Selwyn
Ft. Vermilion
Ft. Chipewyan
Claire
Athabasca
Seal
Churchill
PRINCE OF WALES I.
Ketchikan
DALL I.
Dixon Entrance
Prince Rupert
Masset
GRAHAM I.
QUEEN CHARLOTTE ISLANDS
MORESBY ISLAND
Hecate Strait
Williston Lake
BRITISH
CLEAR HILLS
BUFFALO HEAD HILLS
BIRCH MTS.
Fort McMurray
Wollaston
Cree
Reindeer
Southern Indian
Ft. St. John
Hazelton
Terrace
Smithers
Kitimat
Dawson Creek
Peace River
McLennan
Grouard Mission
High Prairie
Grande Prairie
Lesser Slave Lake
CHEECHAM HILLS
Frobisher
Peter Pond Lake
Lynn Lake
Granville
Thompson
Sipiwesk
Burns Lake
Ft. St. James
Vanderhoof
Prince George
COLUMBIA
SWAN HILLS
Smith
Athabasca
Lac la Biche
ALBERTA
Churchill
Lac la Ronge
SASKATCHEWAN
MANITOBA
Flin Flon
Ocean Falls
Quesnel
Wells
CARIBOO MTS.
McBride
Mt. Robson 12,972
Whitecourt
Barrhead
Edson
Ft. Saskatchewan
Beaver
St. Paul
Meadow Lake
St. Walburg
Big River
PRINCE ALBERT NAT'L PARK
Cross
Norway House
CALVERT
Queen Charlotte Sd.
CAPE SCOTT
Mt. Waddington 13,260
JASPER NAT'L PARK
Jasper
Edmonton
ELK ISLAND NAT'L PARK
Vegreville
North Saskatchewan
Mountain Park
Wetaskiwin
Ponoka
Camrose
Vermilion
Lloydminster
Wainwright
Prince Albert
Nipawin
The Pas
Lake Winnipeg
Blue River
Lacombe
Red Deer
Stettler
North Battleford
Melfort
Tisdale
Berens River
Port Alice
Campbell River
VANCOUVER ISLAND
NOOTKA
Courtenay
Clinton
Lillooet
GLACIER NAT'L PARK
MT. REVELSTOKE NAT'L PARK
YOHO
BANFF NAT'L PARK
Innisfail
Olds
Wilkie
Saskatoon
Biggar
Humboldt
Big Quill
Winnipegosis
Swan River
Gypsumville
Powell River
Kamloops
Revelstoke
KOOTENAY NAT'L PARK
Banff
Drumheller
Hanna
Rosetown
Lanigan
Canora
Wynyard
Port Alberni
Nanaimo
Duncan
Vancouver
North Vancouver
Merritt
SELKIRK MTS.
Calgary
Kindersley
Watrous
DUCK MTN.
Winnipegosis
Kamsack
Dauphin
Vernon
Kelowna
Okanagan
High River
Bassano
Red Deer
Diefenbaker
Last Mountain
Yorkton
Melville
RIDING MOUNTAIN NAT'L PARK
Str. of Juan de Fuca
CAPE FLATTERY
Victoria
Burnaby
Chilliwack
Hope
Princeton
Penticton
Oliver
Kimberley
Cranbrook
Nelson
Creston
Claresholm
Macleod
Taber
Redcliff
Medicine Hat
Swift Current
Moose Jaw
Qu'Appelle
Regina
Indian Head
Russell
Minnedosa
Neepawa
Selkirk
Manitoba
Grand Forks
Rossland
Trail
Fernie
Lethbridge
Magrath
Cardston
Maple Creek
Shaunavon
Gravelbourg
Assiniboia
Weyburn
Portage-la-Prairie
Virden
Brandon
Souris
Carmano
Boissevain
Morden
Winnipeg
Steinbach
Morris
Emerson
SEATTLE
Tacoma
Olympia
WATERTON-GLACIER INT. PEACE PARK
Govenlock
Estevan
CANADA
U.S.A.
PACIFIC OCEAN
Spokane
WASHINGTON
Mt. Rainier 14,410
Mt. Adams 12,307
Yakima
Milk
Williston
Minot
Vancouver
Portland
Moscow
Walla Walla
Great Falls
Missouri
Grand Forks
Columbia
BITTERROOT RANGE
MONTANA
NORTH DAKOTA
Salem
Pendleton
Helena
LITTLE BELT MTS.
BIG BELT MTS.
Bismarck
Valley City
Fargo
Eugene
Butte
Salmon
CASCADE RANGE
OREGON
Baker
IDAHO
Yellowstone
Billings
Granite Peak 12,799
WYO.
SOUTH DAKOTA
145°
140°
135°
70°
130°
125°
120°
115°
110°
105°
100°
95°
90°
65°
50°
45°
Longitude West of Greenwich
Cities, Towns, and Villages
0 to 25,000
25,000 to 100,000
100,000 to 250,000
250,000 to 1,000,000
1,000,000 and over
Major urbanized area
Conic Projection
Elevations and depressions are given in feet

Longitude West of Greenwich
Same scale as main map
QUEBEC
CAPE BAULD
Strait of Belle Isle
LONG RANGE MTS.
White Bay
Gulf of St. Lawrence
C. ST JOHN
GROS MORNE NAT'L PARK
Notre Dame Bay
Twillingate
Botwood
Deer Lake
Windsor
Gander
Corner Brook
Grand Falls
Bonavista Bay
Bonavista
Stephenville
TERRA NOVA NAT'L PARK
C. ST. GEORGE
St. George's Bay
St. George's
NEWFOUNDLAND
Trinity
Trinity Bay
CAPE RAY
Channel-Port-aux-Basques
St. John's
Cabot Strait
CAPE NORTH
Fortune Bay
Placentia Bay
Grand Bank
Burin
CAPE BRETON ISLAND
ST PIERRE AND MIQUELON (Fr.)
ATLANTIC OCEAN
Igloolik
BAFFIN ISLAND NAT'L PARK
BAFFIN ISLAND
Pangnirtung
CUMBERLAND PEN.
Cumberland Sound
PRINCE CHARLES ISLAND
FRANKLIN
Foxe Basin
MELVILLE PENINSULA
Arctic Circle
Frobisher Bay
HALL PEN.
FOXE PEN.
Lake Harbour
EVERETT MTS.
RESOLUTION
Foxe Channel
Hudson Strait
CAPE CHIDLEY
SOUTHAMPTON ISLAND
BELL PEN.
NOTTINGHAM ISLAND
C. HOPES ADVANCE
AKPATOK
Ivujivik
TORNGAT MTS.
Hebron
Ungava Bay
C. LOW
COATS
MANSEL
KEEWATIN
PENINSULE D'UNGAVA
Ft. Chimo
Nain
Hopedale
Makkovik
Povungnituk
OTTAWA ISLANDS
HUDSON BAY
Cartwright
MEALY MTS.
Battle Harbour
St. Anthony
Goose Bay
Churchill Falls
All islands within bays and straits lie within Northwest Territories.
Schefferville
LABRADOR
NEWFOUNDLAND
Lac Bienville
BELCHER ISLANDS
Natashquan
Mingan
ÎLE D'ANTICOSTI
Corner Brook
Stephenville
St. George
C. HENRIETTA MARIA
Ft. Severn
MTS. OTISH
QUEBEC
Sept-Îles
Clarke City
Gulf of St. Lawrence
Channel-Port-aux-Basques
PTE. LOUIS-XIV
James Bay
Ft. George
AKIMISKI
Gaspé
MTS. CHIC-CHOCS
Chandler
ÎLES DE LA MADELEINE
CAPE BRETON HIGHLANDS NAT'L PARK
Ft. Albany
St. Lawrence River
Matane
Rimouski
PEI
PRINCE EDWARD NAT'L PARK
New Waterford
Sydney
Chibougamau
Moosonee
NEW BRUNSWICK
Charlottetown
New Glasgow
NOVA SCOTIA
Dolbeau
Alma
Chicoutimi
Jonquière
St. Félicien
Roberval
Chambord
ONTARIO
Coral Rapids
Fraserdale
CANADA
U.S.A.
Fredericton
Moncton
Saint John
Truro
Windsor
Dartmouth
Halifax
Lunenburg
Bridgewater
Liverpool
Shelburne
Yarmouth
Digby
Kentville
Woodstock
La Tuque
Quebec
Lévis
MAINE
St. Andrews
St. Stephen
La Sarre
Amos
Senneterre
Hearst
Nakina
Armstrong Sta.
Kapuskasing
Cochrane
Rouyn
Val-d'Or
Parent
Shawinigan
Trois-Rivières
Victoriaville
Drummondville
Sherbrooke
Geraldton
Longlac
Iroquois Falls
Malartic
Timmins
Kirkland Lake
Joliette
Sorel
St. Hyacinthe
Granby
Dryden
Nipigon
Ville-Marie
Témiscaming
Chapleau
Cobalt
MONTREAL
Augusta
Marathon
PUKASKWA NAT'L PARK
Sturgeon Falls
Mattawa
Sudbury
North Bay
Hull
Valleyfield
Portland
Thunder Bay
Lake Superior
Pembroke
Renfrew
Ottawa
VERMONT
NEW HAMPSHIRE
Montpelier
Concord
ATLANTIC OCEAN
Sault Ste. Marie
Thessalon
Blind River
Espanola
Huntsville
Bancroft
Smiths Falls
Brockville
Kingston
Ogdensburg
BOSTON
CAPE COD
Parry Sound
Georgian Bay
MANITOULIN
Orillia
Peterborough
Alexandria Bay
MASS.
Hartford
Providence
R.I.
Marquette
Escanaba
MICHIGAN
Midland
Barrie
Lindsay
Trenton
Cobourg
Lake Ontario
NEW YORK
Duluth
Superior
Wiarton
Owen Sound
Oshawa
Whitby
Rochester
Albany
CONN.
Kincardine
TORONTO
Hamilton
Niagara Falls
BUFFALO
WISCONSIN
Lake Huron
Kitchener
St. Catharines
PENNSYLVANIA
NEW YORK
St. Paul
Green Bay
London
Port Huron
St. Thomas
Scranton
Newark
N.J.
Saginaw
Flint
Sarnia
Chatham
MINNEAPOLIS
Lake Michigan
Grand Rapids
Lansing
DETROIT
Windsor
Lake Erie
Madison
MILWAUKEE
Leamington
OHIO
CHICAGO
ILL.
Toledo
Relief
Meters
Feet
3050
1525
610
305
152.5
10 000
5000
2000
1000
500
Sea Level
0
AA-520200-76 8-8-14 M
COPYRIGHT BY RAND McNALLY & COMPANY MADE IN U.S.A.
Miles
Kilometers

Polyconic Projection
Elevations and depressions are given in feet

ATLANTIC OCEAN
PUERTO RICO
San Juan
Arecibo
Aguadilla
Mayagüez
Ponce
Caguas
Humacao
Guayama
Vieques
CARIBBEAN SEA
Christiansted
SAINT CROIX (U.S.A.)
ST. THOMAS
Charlotte Amalie
Crown Mt. 1558
St. Thomas Harbor
NORTH AMERICAN BASIN
BAHAMAS
Nassau
Miami
Norfolk
Richmond
Raleigh
Charlotte
Wilmington
Columbia
Charleston
Savannah
Jacksonville
St. Augustine
W. Palm Beach
Key West
CUBA
Santa Clara
Cienfuegos
Camagüey
Holguín
Guantánamo
Santiago de Cuba
Manzanillo
Cárdenas
Matanzas
Guanabacoa
JAMAICA
Kingston
Montego Bay
Spanish Town
Port Antonio
HAITI
Port-au-Prince
Cap-Haïtien
DOMINICAN REPUBLIC
Santo Domingo
HISPANIOLA
GREATER ANTILLES
WEST INDIES
PUERTO RICO TRENCH
LESSER ANTILLES
LEEWARD IS.
WINDWARD IS.
Pointe-à-Pitre
Basse-Terre
DOMINICA
Fort-de-France
ST. LUCIA
BARBADOS
Bridgetown
Kingstown
GRENADA
TRINIDAD AND TOBAGO
Port of Spain
CARIBBEAN SEA
Willemstad
Caracas
Maracaibo
Valencia
Barquisimeto
Maracay
Cumaná
Ciudad Bolívar
Ciudad Guayana
VENEZUELA
COLOMBIA
Bogotá
Medellín
Cali
Barranquilla
Cartagena
Santa Marta
Cúcuta
Bucaramanga
Panama
Colón
CANAL ZONE (U.S.A.)
GUYANA
BRAZIL
SERRA PACARAIMA
AMERICA
Longitude West of Greenwich
Relief
Meters
Feet
3050
10 000
1525
5000
610
2000
305
1000
152 5
500
0
Sea Level
0
152.5
500
1525
5000
3050
10 000
6100
20 000
50
100
200
300
400
500 Miles
100
200
400
600
800 Kilometers

Conic Projection

Elevations and depressions are given in feet

0 50 100 200 300 400 500 Miles
0 100 200 400 600 800 Kilometers

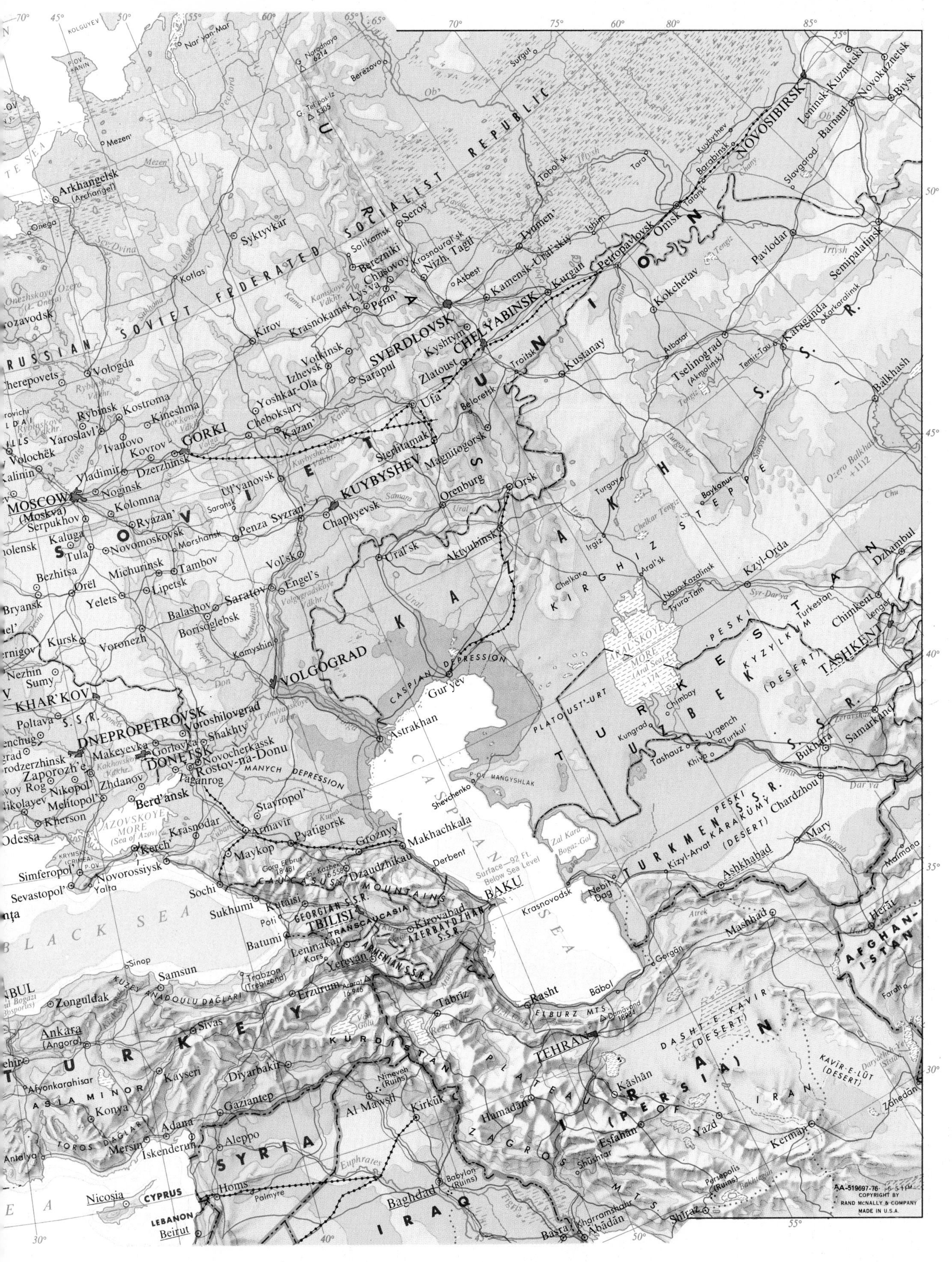

RUSSIAN SOVIET FEDERATED SOCIALIST REPUBLIC
U.S.S.R.
SOVIET UNION
KAZAKH S.S.R.
UZBEK S.S.R.
TURKMEN S.S.R.
GEORGIAN S.S.R.
AZERBAYDZHAN S.S.R.
ARMENIAN S.S.R.
TRANSCAUCASIA
TURKEY
SYRIA
IRAQ
IRAN (PERSIA)
AFGHANISTAN
LEBANON
CYPRUS
MOSCOW (Moskva)
Arkhangelsk (Archangel)
Nar'yan-Mar
Mezen'
Onega
Kotlas
Syktyvkar
Kirov
Vologda
Cherepovets
Rybinsk
Kostroma
Yaroslavl'
Ivanovo
Kineshma
Kovrov
GORKI
Dzerzhinsk
Vladimir
Kalinin
Noginsk
Kolomna
Serpukhov
Ryazan'
Kaluga
Tula
Novomoskovsk
Bezhitsa
Orël
Bryansk
Yelets
Lipetsk
Michurinsk
Tambov
Morshansk
Penza
Saransk
Syzran'
Ul'yanovsk
Kazan'
Cheboksary
Yoshkar-Ola
Izhevsk
Votkinsk
Sarapul
Perm'
Krasnokamsk
Lys'va
Chusovoy
Berezniki
Solikamsk
Serov
Krasnoural'sk
Nizh. Tagil
SVERDLOVSK
Asbest
Kamensk-Ural'skiy
CHELYABINSK
Kyshtym
Zlatoust
Ufa
Beloretsk
Sterlitamak
Magnitogorsk
Troitsk
Kurgan
Tyumen'
Tobol'sk
Ishim
Petropavlovsk
Omsk
Tara
Tatarsk
Barabinsk
Kuybyshev
NOVOSIBIRSK
Leninsk-Kuznetski
Novokuznetsk
Barnaul
Biysk
Slavgorod
Pavlodar
Semipalatinsk
Kokchetav
Tselinograd (Akmolinsk)
Karaganda
Karkaralinsk
Temir-Tau
Balkhash
Ozero Balkhash
Kustanay
Atbasar
Turgay
Baykonur
Irgiz
Chelkar
Aral'sk
Novo-Kazalinsk
Tyura-Tam
Kzyl-Orda
Turkestan
Chimkent
Dzhambul
TASHKENT
Lenger
KUYBYSHEV
Chapayevsk
Orenburg
Orsk
Aktyubinsk
Ural'sk
Vol'sk
Engel's
Saratov
Balashov
Borisoglebsk
Voronezh
Kursk
Kamyshin
VOLGOGRAD
Gur'yev
Astrakhan
CASPIAN DEPRESSION
MANYCH DEPRESSION
KAR'KOV
Sumy
Nezhin
Poltava
DNEPROPETROVSK
Dneprodzerzhinsk
Zaporozh'e
Krivoy Rog
Nikopol'
Makeyevka
Gorlovka
DONETSK
Voroshilovgrad
Shakhty
Novocherkassk
Rostov-na-Donu
Taganrog
Zhdanov
Melitopol'
Berd'ansk
Nikolayev
Kherson
Odessa
AZOVSKOYE MORE (Sea of Azov)
Kerch'
Simferopol'
Sevastopol'
Yalta
Novorossiysk
Krasnodar
Armavir
Maykop
Stavropol'
Pyatigorsk
Groznyy
Makhachkala
Derbent
Dzaudzhikau
Gora El'brus 18,481
G. Kazbek 16,558
CAUCASUS MOUNTAINS
Sochi
Sukhumi
Kutaisi
Poti
Batumi
TBILISI
Kirovabad
Leninakan
Kars
Yerevan
BAKU
Surface —92 Ft. Below Sea Level
CASPIAN SEA
Krasnovodsk
Nebit-Dag
Shevchenko
P-OV MANGYSHLAK
Zal Kara-Bogaz-Gol
PLATO UST'-URT
ARAL'SKOYE MORE (Aral Sea) +174
Chimbay
Kungrad
Urgench
Turtkul'
Tashauz
Khiva
Bukhara
Samarkand
KYZYLKUM (DESERT)
PESKI KYZYLKUM
PESKI KARAKUMY (DESERT)
KIRGHIZ STEPPE
Chardzhou
Mary
Kizyl-Arvat
Ashkhabad
Mashhad
Herat
Gorgān
Bābol
Rasht
ELBURZ MTS.
Damavand 18,934
TEHRAN
DASHT-E KAVIR (DESERT)
KAVIR-E-LUT (DESERT)
Kāshān
Esfahān
Yazd
Kerman
Zāhedān
Farah
Shiraz
Persepolis (Ruins)
Shūshtar
Hamadān
ZAGROS MTS.
Abādān
Khorramshahr
Basra
Baghdad
Babylon (Ruins)
Kirkūk
Al Mawsil
Nineveh (Ruins)
KURDISTAN
Tabrīz
Ararat 16,946
Erzurum
Trabzon (Trebizond)
Samsun
Sinop
Zonguldak
KUZEY ANADOLU DAĞLARI
Ankara (Angora)
Sivas
Kayseri
Diyarbakir
Gaziantep
Afyonkarahisar
ASIA MINOR
Konya
TOROS DAĞLARI
Adana
Mersin
Iskenderun
Aleppo
Antalya
Homs
Palmyre
Euphrates
Nicosia
Beirut
BLACK SEA
AA-519697-76
COPYRIGHT BY RAND McNALLY & COMPANY MADE IN U.S.A.

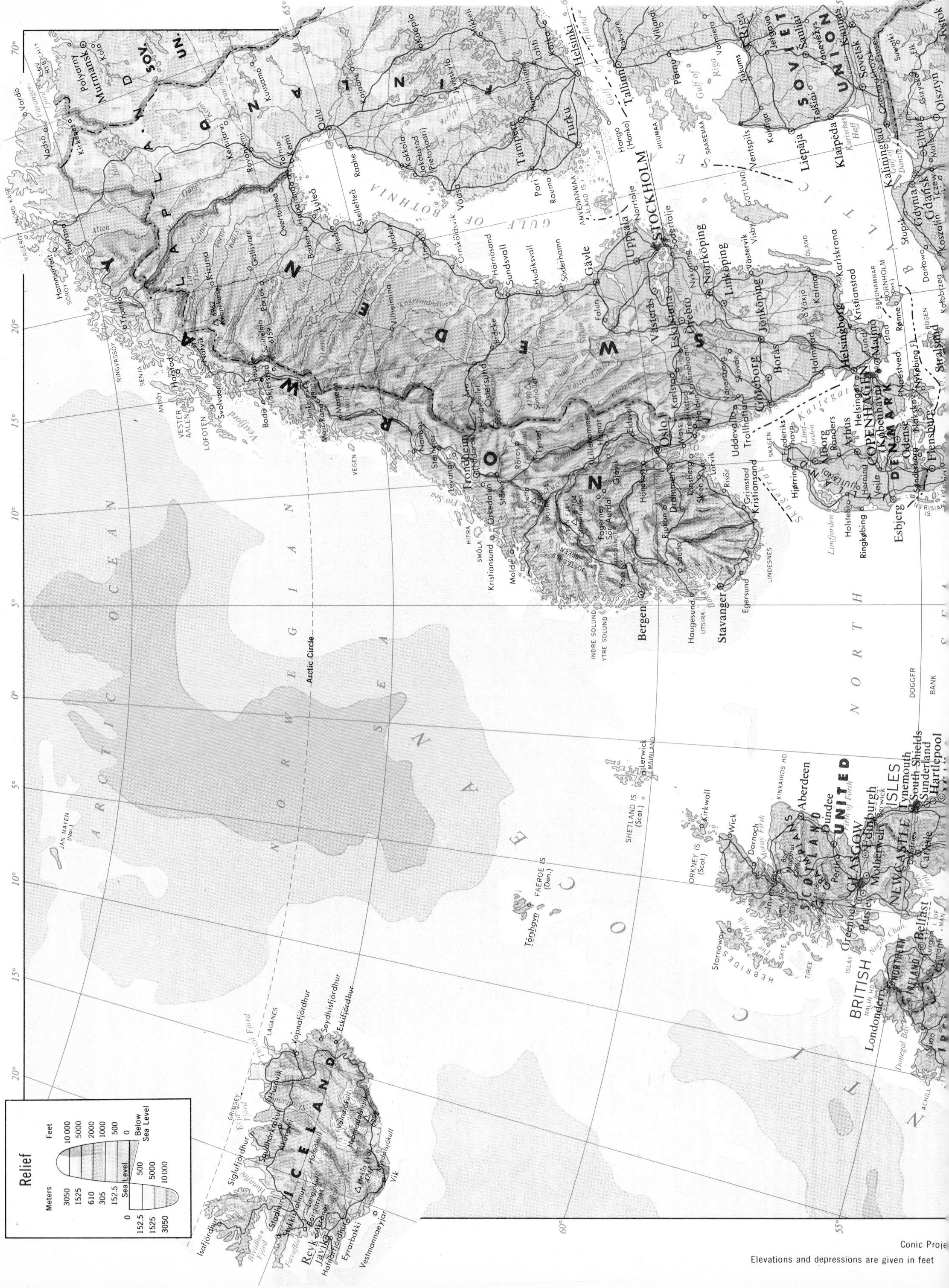

Conic Proje

Elevations and depressions are given in feet

50 100 150 200 250 300 Miles
100 200 300 400 500 Kilometers
Longitude West of Greenwich
Longitude East of Greenwich
AA-559400-76
COPYRIGHT BY RAND McNALLY & COMPANY MADE IN U.S.A.
FRANCE
SPAIN
PORTUGAL
ITALY
ALGERIA
TUNISIA
MOROCCO
GERMANY
YUGOSLAVIA
HUNGARY
AUSTRIA
ALBANIA
BAY OF BISCAY
MEDITERRANEAN SEA
TYRRHENIAN SEA
ADRIATIC SEA
IONIAN SEA
LIGURIAN SEA
ENGLISH CHANNEL
PARIS
MADRID
ROME
LISBON
BARCELONA
LONDON
AMSTERDAM
BRUSSELS
MUNICH
VIENNA
BUDAPEST
NAPLES
MILAN
TURIN
Marseille
Algiers
CORSICA (Fr.)
SARDINIA (It.)
MALTA

Bonne's Pro

Elevations and depressions are given in feet

POLAND
CZECHOSLOVAKIA
HUNGARY
ROMANIA
YUGOSLAVIA
BULGARIA
ALBANIA
GREECE
TURKEY
ASIA MINOR
U.S.S.R.
UKRAINIAN S.S.R.
MOLDAVIAN S.S.R.
R.S.F.S.R.
SYRIA
LEBANON
ISRAEL
JORDAN
SAUDI ARABIA
CYPRUS
EGYPT
LIBYA
BARQAH (CIRENAICA)
LIBYAN PLATEAU
LIBYAN DESERT
MUNKHAFAD AL QATTARAH
JABAL AL AKHDAR
SINAI PEN.
BLACK SEA
AZOVSKOYE MORE (Sea of Azov)
AEGEAN SEA
IONIAN SEA
MEDITERRANEAN SEA
RED SEA
DEAD SEA
CRETE
KATOWICE
BUDAPEST
BUCHAREST
DNEPROPETROVSK
DONETSK
ISTANBUL
ATHENS
ALEXANDRIA (Al Iskandariyah)
CAIRO (Al Qahirah)
Belgrade (Beograd)
Sofia (Sofiya)
Ankara (Angora)
Damascus (Dimashq)
Beirut
Jerusalem
Tel Aviv-Yafo
Amman
Nicosia
Tiranë
Odessa
Kishinev
Areas occupied by Israel since June 1967

0 50 100 150 200 250 300 Miles
0 100 200 300 400 500 Kilometers

SWITZERLAND
AUSTRIA
SLOVENIJA
HRVATSKA (CROATIA)
LIGURIAN SEA
TYRRHENIAN SEA
ADRIATIC
Golfo di Venezia
Golfo di Genova
Golfo di Salerno
CORSICA (Fr.)
SARDINIA (It.)
SICILY
AEGEAN SEA
MEDITERRANEAN SEA
CRETE
Same scale as main map
TURIN (Torino)
MILAN (Milano)
Genoa (Genova)
Venice (Venezia)
Bologna
Florence (Firenze)
ROME (Roma)
VATICAN CITY
SAN MARINO
NAPLES (Napoli)
Palermo
Catania
Cagliari
Sassari
Bastia
Ajaccio
Trieste
Ljubljana
Zagreb
Rijeka (Fiume)
Split
Pescara
Ancona
Bari
Messina
Reggio di Calabria
Iráklion (Candia)
Khania (Canea)
Conic Projection
Elevations and depressions are given in feet
©RMCN.

Relief
Meters
Feet
3050
10 000
1525
5000
610
2000
305
1000
152.5
500
0
Sea Level
0
152.5
500
1525
5000
3050
10 000
ROMANIA
CARPATII MERIDIONALI (TRANSYLVANIAN ALPS)
BANAT
VALACHIA
BUCHAREST (Bucureşti)
Braşov (Oraşul-Stalin)
Sibiu
Arad
Szeged
Timişoara
Craiova
Ploeşti
Buzău
Giurgiu
Ruse (Russe)
IRON GATE
Turnu-Severin
Belgrade (Beograd)
Zemun (Semlin)
Novi Sad
Subotica
Sarajevo
SRBIJA
JUGOSLAVIJA
CRNA GORA (MONTENEGRO)
Titograd
Skopje
ŠAR PLANINA
Niš
Priština
Shkodër (Scutari)
Tiranë
Durrës
ALBANIA
Vlorë (Valona)
Korçë
BULGARIA
Sofia (Sofiya)
Plovdiv (Philippopolis)
Stara Zagora
STARA PLANINA (BALKAN MTS.)
RHODOPE MTS.
Burgas
Varna (Stalin)
Pleven
Tolbukhin
Shumen
THRACE
Edirne (Adrianople)
Kirklareli
ISTRANCA DAG
Gelibolu (Gallipoli)
Çanakkale
Troy Ruins
Saros Körfezi
MACEDONIA
Thessaloniki
Kavalla
Alexandroúpolis (Dedeagats)
THÁSOS
SAMOTHRÁKI
IMROZ
ÁYION ÓROS (MOUNT ATHOS)
Áthos 6670
LIMNOS
Thermaikos Kólpos
Ólimbos 9550
THESSALÍA
Lárisa
Vólos
PINDHOS ÓROS
EPIRUS
Ioánnina (Yannina)
Kérkira
KÉRKIRA
GREECE
AEGEAN SEA
SPORÁDHES
Skíros
LÉSVOS
Mitilíni
KHÍOS
İzmir
DHODHEKÁNISOS (DODECANESE)
ATHENS (Athinai)
Piraiévs (Piraeus)
Thívai (Thebes)
DELPHI RUINS
Pátrai (Patras)
Korinthiakós Kólpos
Kórinthos (Corinth)
PELOPÓNNISOS
Tripolis
Spárti (Sparta)
Kalámai
Navplion
KIKLÁDHES
Sáronikós Kólpos
Argolikós Kólpos
Kiparissiakós Kólpos
Messiniakós Kólpos
ÍDHRA
Zákinthos
ZÁKINTHOS (ZANTE)
KEFALLINÍA
Argostólion
LEVKÁS
IONIAN ISLANDS
IONIAN SEA
Strait of Otranto
Brindisi
Lecce
Otranto
C. D OTRANTO
C. S. MARIA DI LEUCA
Gallipoli
AKR. AKRÍTAS
AKR. MALÉA
Longitude East of Greenwich
18°
20°
22°
24°
26°
28°
Miles
0 10 20 30 40 50 60 70 80 90 100 110 120
Kilometers
0 20 40 60 80 100 120 140 160 180 200

Conic Proj

Elevations and depressions are given in feet

NORWAY
Arendal
Grimstad
Lillesand
Egersund
Sogndal
Flekkefjord
Kristiansand
Farsund
Mandal
LINDESNES
Skagerrak
Kattegat
SWEDEN
Göteborg
Alingsås
Borås
Mölndal
Varberg
Falkenberg
Halmstad
Laholm
Ängelholm
Helsingborg
Landskrona
Malmö
Trelleborg
Skagen
Frederikshavn
Hjørring
Brønderslev
LAESØ
Ålborg
Limfjorden
Thisted
Løgstør
Nykøbing
Hobro
Mariager
Randers
Grenå
ANHOLT
Skive
Viborg
Struer
Holstebro
JYLLAND
Silkeborg
Århus
Herning
Skanderborg
Ringkøbing
Ringkøbing Fjord
Horsens
Vejle
Fredericia
Varde
Esbjerg
Kolding
Odense
Nyborg
FYN
Ribe
Haderslev
Faborg
Svendborg
Rudkøbing
Åbenrå
Tønder
Sønderborg
AERØ
LANGELAND
LOLLAND
FALSTER
Nakskov
Maribo
Nykøbing Fl.
FEHMARN
DENMARK
COPENHAGEN (København)
Hillerød
Helsingør
Roskilde
Køge
SJAELLAND
Holbaek
Kalundborg
Ringsted
Slagelse
Korsør
Naestved
Vordingborg
MØN
SAMSØ
BLÅVANDS HUK
FANØ
RØMØ
SYLT
FÖHR
NORTH FRISIAN IS.
Flensburg
SCHLESWIG
Schleswig
Husum
Eckernförde
Rendsburg
Kiel
Kiel Bay
Tönning
Heide
HOLSTEIN
Neustadt
Neumünster
Itzehoe
Lübeck
Bad Oldesloe
Elmshorn
BALTIC SEA
Rostock
Wismar
Güstrow
Teterow
Schwerin
GERMAN DEMOCRATIC REPUBLIC
MECKLENBURG
Parchim
Ludwigslust
Wittenberge
Perleberg
Salzwedel
Stendal
Gardelegen
Tangermünde
HELGOLAND
Cuxhaven
Bremerhaven
Stade
HAMBURG
Bergedorf
Lüneburg
LÜNEBURGER HEIDE
Uelzen
Soltau
Bremen
Verden
Delmenhorst
Oldenburg
Papenburg
Leer
Emden
Norden
Wilhelmshaven
FRISIAN ISLANDS
NORDERNEY
LANGEOOG
JUIST
BORKUM
AMELAND
TERSCHELLING
VLIELAND
TEXEL
Delfzijl
Groningen
Leeuwarden
Harlingen
Waddenzee
IJsselmeer
Den Helder
Emmen
Meppel
Meppen
Lingen
NIEDERSACHSEN
FED. REP. OF GERMANY
Nienburg
Celle
Hannover
Minden
Braunschweig
Magdeburg
Hildesheim
Wolfenbüttel
Hameln
Herford
Bielefeld
Detmold
Osnabrück
Rheine
Nordhorn
NETHERLANDS
Alkmaar
Zwolle
Almelo
Hengelo
Enschede
Deventer
Apeldoorn
AMSTERDAM
Zaandam
Haarlem
Leiden
Utrecht
The Hague ('s Gravenhage)
Delft
Vlaardingen
ROTTERDAM
Dordrecht
Nijmegen
Arnhem
Rheden
Münster
Coesfeld
Gütersloh
Paderborn
Hamm
Ahlen
Lippstadt
Soest
Göttingen
Northeim
Einbeck
Goslar
Quedlinburg
Halberstadt
Stassfurt
Bernburg
Blankenburg
Aschersleben
Eisleben
Halle
Merseburg
Sangerhausen
Nordhausen
Heiligenstadt
Sondershausen
Mühlhausen
THÜRINGEN
Eisenach
Erfurt
Gotha
Weimar
Jena
Arnstadt
Rudolstadt
Saalfeld
Ilmenau
Suhl
Zella-Mehlis
Schmalkalden
Meiningen
Hildburghausen
Sonneberg
Neustadt b.C.
Coburg
Kulmbach
Bayreuth
Bamberg
Schweinfurt
Bad Kissingen
Würzburg
Kitzingen
Forchheim
Erlangen
Aschaffenburg
Darmstadt
Hanau
Offenbach
FRANKFURT AM MAIN
TAUNUS
SPESSART
Bad Homburg
Wiesbaden
Mainz
Bingen
Bad Kreuznach
Kirn
HUNSRÜCK
Wittlich
Koblenz
Boppard
Mayen
Neuwied
Andernach
Limburg
Ahrweiler
Giessen
Marburg
Fulda
Bad Hersfeld
Kassel
Eschwege
HESSEN
WESTERWALD
Siegen
Siegburg
Bonn
COLOGNE (Köln)
Gummersbach
NORDRHEIN-WESTFALEN
Wuppertal
Solingen
Lüdenscheid
Hagen
Iserlohn
Arnsberg
Dortmund
Bochum
ESSEN
Mülheim
Gelsenkirchen
Oberhausen
Duisburg
Wesel
Kleve
Mönchengladbach
DÜSSELDORF
Roermond
Weert
Venlo
Eindhoven
Helmond
Tilburg
Breda
Hertogenbosch
Bergen op Zoom
Vlissingen
Zeebrugge
Oostende
Brugge
Gent
ANTWERP
Turnhout
Mechelen
Heerlen
Maastricht
Aachen
Düren
Eupen
Verviers
Spa
Liège
Herstal
Seraing
Huy
Namur
Charleroi
Dinant
Givet
Malmedy
EIFEL
ARDENNES
Bastogne
LUX.
BELGIUM
BRUSSELS
Leuven
Aalst
Anderlecht
Nivelles
Jumet
Mons
La Louvière
Valenciennes
Denain
Maubeuge
Hautmont
Fourmies
Cambrai
Douai
Arras
Lens
Béthune
Lille
Roubaix
Tourcoing
Armentières
Kortrijk
Roeselare
Torhout
Ieper
FLANDERS
Dunkerque
Calais
St. Omer
Boulogne-sur-Mer
Étaples
Hesdin
Crécy
Abbeville
St. Valéry
Le Tréport
FRANCE
Strait of Dover
Dover
Folkestone
Maidstone
Canterbury
Ramsgate
Margate
NORTH FORELAND
Sheerness
Southend-on-Sea
R. Thames
Harwich
Ipswich
Lowestoft
Great Yarmouth
Norwich
Thetford
Lynn
NORTH SEA
DOGGER BANK 60—120 Ft.
Longitude East of Greenwich
58°
56°
54°
52°
50°
2°
4°
6°
8°
10°
12°
14°
10 20 30 40 50 60 70 80 90 100 110 120 Miles
20 40 60 80 100 120 140 160 180 200 Kilometers

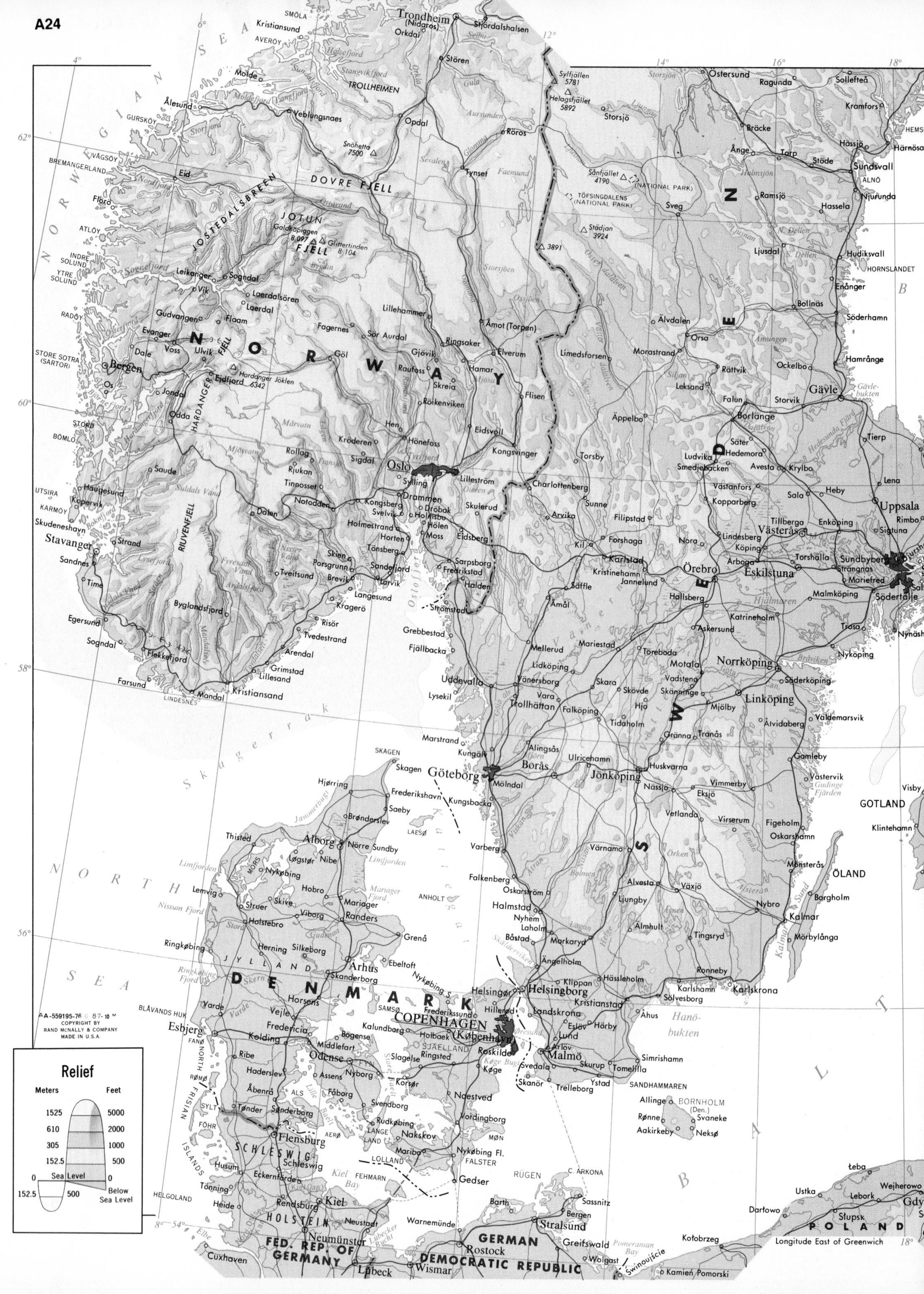
NORWAY
SWEDEN
DENMARK
NORWEGIAN SEA
NORTH SEA
Skagerrak
Kattegat
BALTIC
FED. REP. OF GERMANY
GERMAN DEMOCRATIC REPUBLIC
POLAND
Trondheim (Nidaros)
Stjørdalshalsen
Orkdal
Kristiansund
SMØLA
AVERØY
Støren
Molde
TROLLHEIMEN
Ålesund
GURSKØY
Veblungsnaes
Opdal
Røros
Snøhetta 7500
DOVRE FJELL
VÅGSØY
BREMANGERLAND
Eid
Florø
JOSTEDALSBREEN
JOTUN FJELL
Galdhøpiggen 8,097
Glittertinden 8,104
ATLØY
INDRE SOLUND
YTRE SOLUND
Leikanger
Sogndal
Vik
Laerdalsøren
Laerdal
RADØY
Gudvangen
Flaam
Fagernes
Sør Aurdal
Lillehammer
Evanger
Voss
Ulvik
Dale
STORE SOTRA (SARTOR)
Bergen
Os
Eidfjord
Hardanger Jøklen 6342
HARDANGER FJELL
Göl
Gjøvik
Ringsaker
Raufoss
Hamar
Skreia
Røikenviken
Jondal
Odda
STORD
BØMLO
Kröderen
Hen
Hønefoss
Eidsvoll
Rollag
Rjukan
Sigdal
Tinnosset
Oslo
Sylling
Lilleström
Kongsvinger
Saude
Haugesund
UTSIRA
Kopervik
KARMØY
Skudeneshavn
Stavanger
Sandnes
Strand
Time
RIUVENFJELL
Notodden
Kongsberg
Drammen
Svelvik
Drøbak
Holmsbu
Hölen
Moss
Skulerud
Holmestrand
Horten
Tönsberg
Dalen
Skien
Porsgrunn
Sandefjord
Tveitsund
Brevik
Larvik
Langesund
Sarpsborg
Fredrikstad
Halden
Eidsberg
Strömstad
Kragerö
Risör
Tvedestrand
Arendal
Grimstad
Lillesand
Kristiansand
Mandal
Byglandsfjord
Egersund
Sogndal
Flekkefjord
Farsund
LINDESNES
Tynset
Faemund
Åmot (Torpen)
Elverum
Flisen
Charlottenberg
Sylfjällen 5781
Helagsfjället 5892
Storsjö
Östersund
Ragunda
Sollefteå
Kramfors
Bräcke
Ånge
Torp
Stöde
Hässjö
Härnösa
HEMSÖ
Sundsvall
ALNÖ
Njurunda
Sånfjället 4190 (NATIONAL PARK)
TÖFSINGDALENS (NATIONAL PARK)
Sveg
Ramsjö
Hassela
Städjan 3924
3891
Ljusdal
Hudiksvall
HORNSLANDET
Enånger
Bollnäs
Söderhamn
Älvdalen
Orsa
Morastrand
Limedsforsen
Rättvik
Ockelbo
Hamrånge
Leksand
Gävle
Storvik
Falun
Äppelbo
Borlänge
Säter
Hedemora
Tierp
Ludvika
Smedjebacken
Avesta
Krylbo
Torsby
Västanfors
Kopparberg
Sala
Heby
Lena
Uppsala
Rimbo
Sigtuna
Sunne
Arvika
Filipstad
Tillberga
Enköping
Västerås
Lindesberg
Köping
Kil
Forshaga
Nora
Arboga
Torshälla
Sundbyberg
Strängnäs
Mariefred
Karlstad
Kristinehamn
Jannelund
Örebro
Eskilstuna
Säffle
Åmål
Hallsberg
Malmköping
Södertälje
Katrineholm
Askersund
Trosa
Nynäshamn
Grebbestad
Fjällbacka
Mellerud
Mariestad
Töreboda
Nyköping
Lidköping
Motala
Norrköping
Uddevalla
Vänersborg
Skara
Vadstena
Söderköping
Lysekil
Vara
Skövde
Trollhättan
Falköping
Hjo
Mjölby
Linköping
Valdemarsvik
Tidaholm
Åtvidaberg
Gränna
Tranås
Marstrand
Kungälv
Alingsås
Göteborg
Mölndal
Borås
Ulricehamn
Jönköping
Huskvarna
Gamleby
Västervik
Nässjö
Vimmerby
Kungsbacka
Eksjö
Visby
GOTLAND
Vetlanda
Virserum
Figeholm
Klintehamn
Oskarshamn
Varberg
Värnamo
Mönsterås
ÖLAND
Falkenberg
Oskarström
Alvesta
Växjö
Borgholm
Halmstad
Ljungby
Nybro
Kalmar
Nyhem
Laholm
Almhult
Tingsryd
Båstad
Markaryd
Mörbylånga
Ängelholm
Ronneby
Hässleholm
Karlshamn
Karlskrona
Klippan
Helsingborg
Kristianstad
Sölvesborg
Åhus
Hanö bukten
Landskrona
Eslöv
Hörby
Lund
Arlöv
Malmö
Simrishamn
Svedala
Skurup
Tomelilla
Skanör
Trelleborg
Ystad
SANDHAMMAREN
Allinge
BORNHOLM (Den.)
Rønne
Svaneke
Aakirkeby
Neksø
SKAGEN
Skagen
Hjørring
Frederikshavn
Saeby
Brønderslev
LAESØ
Thisted
Ålborg
Nörre Sundby
Løgstør
Nibe
Limfjorden
MORS
Nykøbing
Hobro
Mariager Fjord
Lemvig
Struer
Skive
Mariager
ANHOLT
Nissum Fjord
Viborg
Randers
Holstebro
Grenå
Ringkøbing
Herning
Silkeborg
JYLLAND
Århus
Ebeltoft
Ringkøbing Fjord
Skanderborg
Nykøbing S.
Helsingør
BLÅVANDS HUK
Varde
Horsens
SAMSØ
Frederikssund
Hillerød
Vejle
Kalundborg
COPENHAGEN (København)
Esbjerg
Fredericia
Bogense
Holbaek
Kolding
Middelfart
SJAELLAND
Roskilde
FANØ
Ribe
Odense
Slagelse
Ringsted
Køge
Køge Bugt
Haderslev
Assens
Nyborg
Korsør
RØMØ
NORTH FRISIAN ISLANDS
Åbenrå
ALS
Fåborg
Naestved
Svendborg
SYLT
Tønder
Sønderborg
Vordingborg
Rudkøbing
FÖHR
Flensburg
AERØ
LANGELAND
Nakskov
MØN
SCHLESWIG
Schleswig
Maribo
LOLLAND
Nykøbing Fl.
FALSTER
Husum
Eckernförde
FEHMARN
Kiel Bay
Gedser
RÜGEN
C. ARKONA
Tönning
HELGOLAND
Heide
Rendsburg
Kiel
Barth
Sassnitz
Bergen
HOLSTEIN
Neustadt
Warnemünde
Stralsund
Neumünster
Lübecker Bucht
Greifswald
Pomeranian Bay
Cuxhaven
Rostock
Wolgast
Swinoujście
Lübeck
Wismar
Kamień Pomorski
Kołobrzeg
Darłowo
Ustka
Słupsk
Lebork
Łeba
Wejherowo
Gdynia
Longitude East of Greenwich
AA-559195-76 87-10
COPYRIGHT BY RAND McNALLY & COMPANY MADE IN U.S.A.
Relief
Meters
Feet
1525
5000
610
2000
305
1000
152.5
500
0
Sea Level
0
152.5
500
Below Sea Level

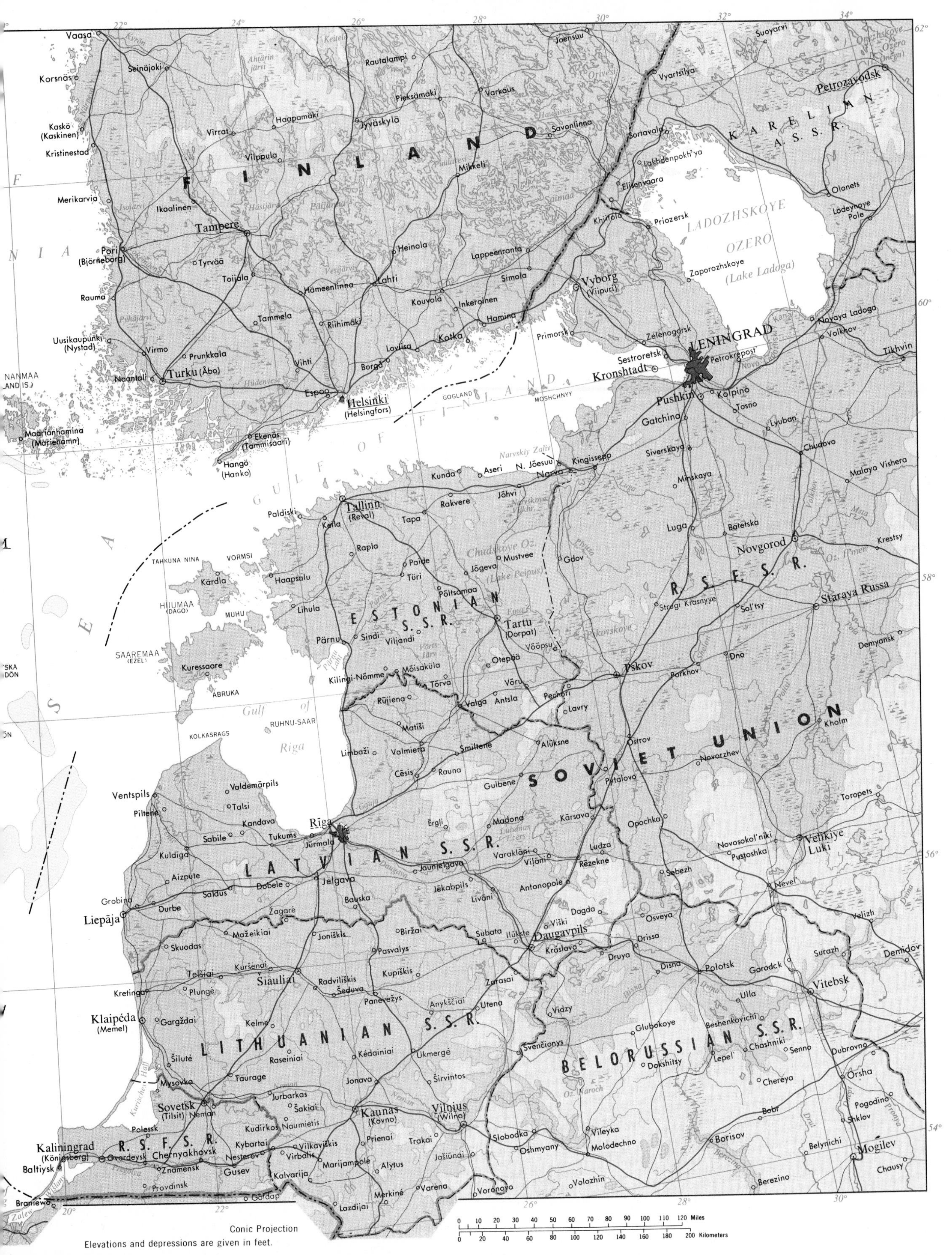

Conic Projection

Elevations and depressions are given in feet.

0 10 20 30 40 50 60 70 80 90 100 110 120 Miles

0 20 40 60 80 100 120 140 160 180 200 Kilometers

Conic Projection
Elevations and depressions are given in feet.

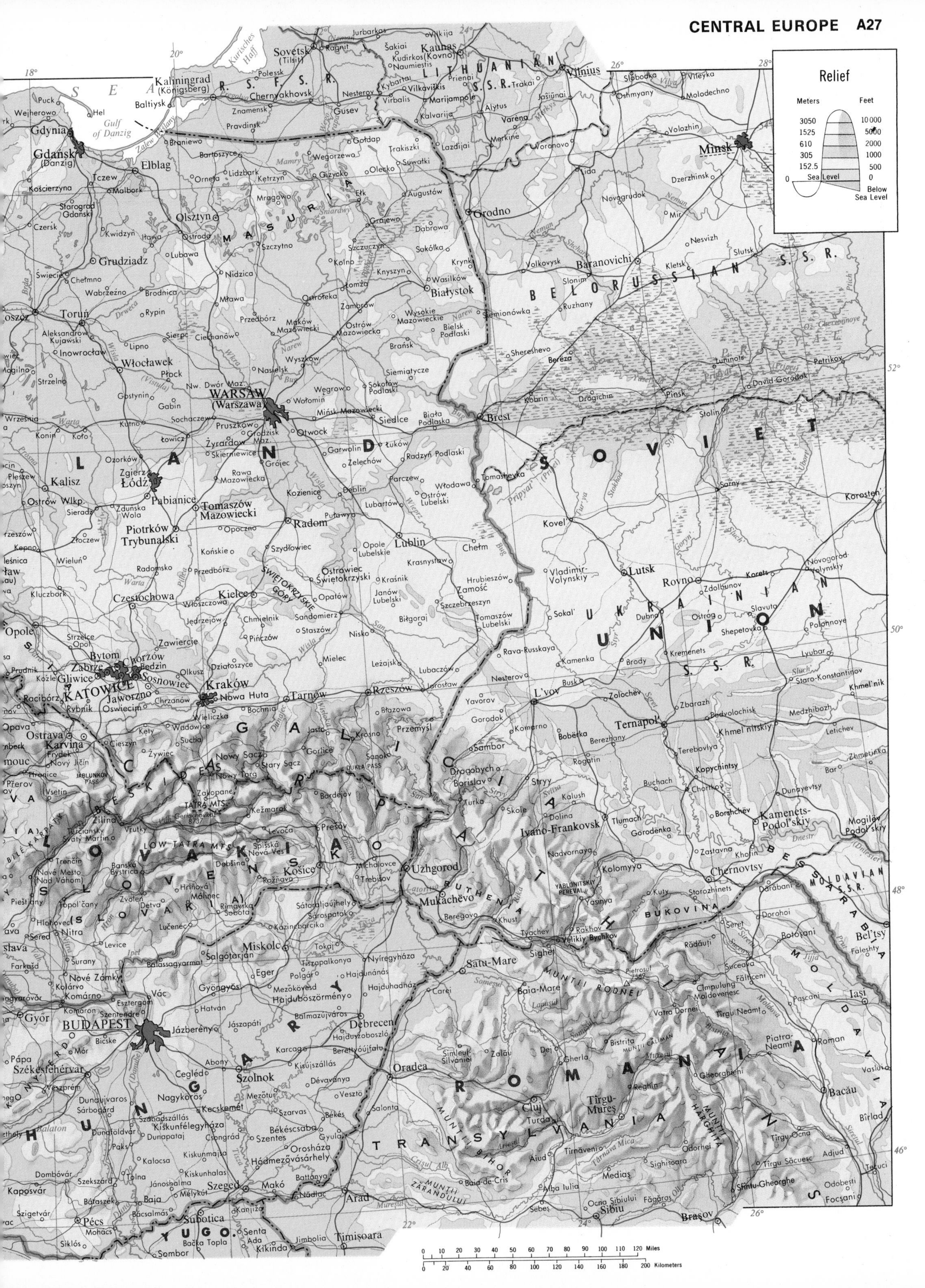
Relief
Meters
Feet
3050
10 000
1525
5000
610
2000
305
1000
152.5
500
0
Sea Level
0
Below Sea Level
R. S. F. S. R.
LITHUANIAN S.S.R.
BELORUSSIAN S.S.R.
SOVIET
UKRAINIAN
UNION
S. S. R.
POLAND
MASURIA
GALICIA
CARPATHIA
RUTHENIA
SLOVAKIA
HUNGARY
ROMANIA
TRANSYLVANIA
MOLDAVIA
BUKOVINA
BESSARABIA
MOLDAVIAN S.S.R.
YUGO.
Gulf of Danzig
Kurisches Haff
Zalew Wiślany
Kaliningrad (Königsberg)
Sovetsk (Tilsit)
Baltiysk
Chernyakhovsk
Gusev
Polessk
Znamensk
Pravdinsk
Nesterov
Kaunas
Vilnius
Jurbarkas
Šakiai
Kudirkos Naumiestis
Vilkaviškis
Marijampolė
Kalvarija
Alytus
Varėna
Trakai
Prienai
Kybartai
Virbalis
Lazdijai
Merkinė
Jašiūnai
Slobodka
Vileyka
Oshmyany
Molodechno
Volozhin
Minsk
Dzerzhinsk
Voronovo
Lida
Novogrudok
Mir
Grodno
Nesvizh
Baranovichi
Slutsk
Volkovysk
Slonim
Kletsk
Ruzhany
Semionówka
Shereshevo
Bereza
Kobrin
Drogichin
Pinsk
Luninets
David-Gorodok
Petrikov
Brest
Stolin
Pripyat' (Prypeć)
Sarny
Korosten'
Kovel'
Lutsk
Rovno
Zdolbunov
Korets
Novograd-Volynskiy
Vladimir-Volynskiy
Dubno
Ostrog
Slavuta
Shepetovka
Polonnoye
Kremenets
Lyubar
Staro-Konstantinov
Khmel'nik
Sokal'
Rava-Russkaya
Kamenka
Brody
Busk
Zolochev
L'vov
Nesterov
Yavorov
Gorodok
Komarno
Sambor
Drogobych
Borislav
Stryy
Skole
Turka
Rogatin
Kalush
Dolina
Ivano-Frankovsk
Tlumach
Gorodenka
Nadvornaya
Kolomyya
Ternopol'
Zbarazh
Podvolochisk
Terebovlya
Kopychintsy
Chortkov
Borshchev
Buchach
Kamenets-Podol'skiy
Dunayevtsy
Zastavna
Khotin
Chernovtsy
Storozhinets
Kuty
Mogilev-Podol'skiy
Medzhibozh
Letichev
Khmel'nitskiy
Zhmerinka
Bar
Uzhgorod
Mukachevo
Beregovo
Khust
Tyachev
Rakhov
Velikiy Bychkov
Yasinya
YABLONITSKIY PEREVAL
Gdynia
Gdańsk (Danzig)
Hel
Puck
Wejherowo
Elbląg
Tczew
Malbork
Kościerzyna
Starogard Gdański
Czersk
Kwidzyń
Grudziądz
Świecie
Chełmno
Wąbrzeźno
Brodnica
Toruń
Aleksandrów Kujawski
Inowrocław
Włocławek
Lipno
Rypin
Płock
Strzelno
Gostynin
Gabin
Kutno
Konin
Koło
Ozorków
Zgierz
Łódź
Pabianice
Kalisz
Ostrów Wlkp.
Sieradz
Zduńska Wola
Pleszew
Wieluń
Kluczbork
Kępno
Opole
Strzelce Opol.
Prudnik
Koźle
Gliwice
Bytom
Zabrze
Chorzów
Będzin
Sosnowiec
KATOWICE
Racibórz
Rybnik
Jaworzno
Chrzanów
Oświęcim
Olkusz
Zawiercie
Częstochowa
Kraków
Nowa Huta
Wieliczka
Bochnia
Tarnów
Piotrków Trybunalski
Tomaszów Mazowiecki
Opoczno
Radomsko
Końskie
Przedbórz
Włoszczowa
Kielce
Jędrzejów
Chmielnik
Pińczów
Działoszyce
Busko
Staszów
Sandomierz
Opatów
Ostrowiec Świętokrzyski
ŚWIĘTOKRZYSKIE GÓRY
Szydłowiec
Radom
Warka
Grójec
Rawa Mazowiecka
Skierniewice
Żyrardów
Sochaczew
Łowicz
Pruszków
Grodzisk Maz.
WARSAW (Warszawa)
Nw. Dwór Maz.
Wołomin
Otwock
Garwolin
Żelechów
Mińsk Mazowiecki
Siedlce
Węgrów
Sokołów Podlaski
Łuków
Radzyń Podlaski
Biała Podlaska
Kozienice
Dęblin
Puławy
Lubartów
Parczew
Włodawa
Ostrów Lubelski
Opole Lubelskie
Lublin
Chełm
Krasnystaw
Kraśnik
Janów Lubelski
Biłgoraj
Zamość
Szczebrzeszyn
Hrubieszów
Tomaszów Lubelski
Nisko
Mielec
Leżajsk
Lubaczów
Jarosław
Rzeszów
Przemyśl
Sanok
Krosno
Jasło
Gorlice
Nowy Sącz
Stary Sącz
Nowy Targ
Zakopane
Błażowa
DUKLA PASS
Olsztyn
Ostróda
Iława
Lubawa
Nidzica
Mława
Działdowo
Ciechanów
Sierpc
Przedbórz
Maków Mazowiecki
Pułtusk
Nasielsk
Wyszków
Ostrołęka
Ostrów Mazowiecka
Zambrów
Łomża
Wysokie Mazowieckie
Bielsk Podlaski
Brańsk
Siemiatycze
Białystok
Wasilków
Knyszyn
Krynki
Sokółka
Szczuczyn
Grajewo
Ełk
Augustów
Dąbrowa
Kolno
Szczytno
Mrągowo
Kętrzyn
Lidzbark
Orneta
Bartoszyce
Braniewo
Węgorzewo
Giżycko
Olecko
Gołdap
Suwałki
Trakiszki
Mamry
Śniardwy
Narew
Bug
Wisła
Warta
San
Wieprz
Drwęca
Pilica
Neman
Pripyat'
Horyn'
Styr
Sluch
Dnestr
Seret
Prut
Mureş
Someş
Tisza
Danube (Duna)
Balaton
Opava
Ostrava
Karviná
Frýdek
Nový Jičín
Přerov
Hranice
Vsetín
Cieszyn
Żywiec
Sucha
Wadowice
Kęty
JABLUNKOV PASS
BESKIDS
TATRA MTS.
Gerlachovka 8737
LOW TATRA MTS.
Žilina
Turčiansky Svätý Martin
Vrútky
Kežmarok
Levoča
Spišská Nová Ves
Prešov
Bardejov
Dobšiná
Rožňava
Košice
Michalovce
Trebišov
Trenčín
Nové Mesto (Nad Váhom)
Banská Bystrica
Zvolen
Detva
Hriňová
Málinec
Rimavská Sobota
Lučenec
Piešťany
Topoľčany
Hlohovec
Nitra
Sereď
Levice
Šurany
Nové Zámky
Kolárovo
Komárno
Farkašd
Sátoraljaújhely
Sárospatak
Kazincbarcika
Miskolc
Tokaj
Salgótarján
Balassagyarmat
Eger
Gyöngyös
Hatvan
Vác
Esztergom
Szentendre
Győr
BUDAPEST
Jászberény
Jászapáti
Mezőkövesd
Hajdúböszörmény
Tiszapalkonya
Polgár
Nyíregyháza
Hajdúnánás
Hajdúhadház
Balmazújváros
Debrecen
Hajdúszoboszló
Berettyóújfalu
Karcag
Kisújszállás
Abony
Szolnok
Cegléd
Nagykőrös
Kecskemét
Mezőtúr
Dévaványa
Vésztő
Szarvas
Békés
Békéscsaba
Gyula
Orosháza
Szentes
Csongrád
Kiskunfélegyháza
Szabadszállás
Dunapataj
Kalocsa
Kiskunmajsa
Kiskunhalas
Hódmezővásárhely
Makó
Szeged
Battonya
Mélykút
Jánoshalma
Baja
Bácsalmás
Bátaszék
Szekszárd
Tolna
Paks
Dunaföldvár
Dunaújváros
Sárbogárd
Székesfehérvár
Veszprém
Pápa
Bicske
Mór
Dombóvár
Kaposvár
Szigetvár
Pécs
Mohács
Siklós
Subotica
Senta
Ada
Bačka Topola
Sombor
Kikinda
Kanjiža
Jimbolia
Timişoara
Arad
Nădlac
Salonta
Oradea
Carei
Satu-Mare
Baia-Mare
Sighet
Pietrosul 7562
MUNTII RODNEI
Şimleul-Silvaniei
Zalău
Dej
Bistriţa
Gherla
MUNTII CALIMAN
Reghin
Cluj
Turda
Tîrgu-Mureş
Aiud
Tîrnăveni
Mediaş
Sighişoara
Alba Iulia
Sebeş
Ocna Sibiului
Sibiu
Făgăraş
Braşov
Sfîntu-Gheorghe
Tîrgu Săcuesc
Tîrgu Ocna
Adjud
Odobeşti
Focşani
Tecuci
Bîrlad
Vaslui
Bacău
Piatra-Neamţ
Roman
Tîrgu Neamţ
Paşcani
Iaşi
Botoşani
Dorohoi
Darabani
Fălticeni
Suceava
Rădăuţi
Seret
Cîmpulung Moldovenesc
Vatra Dornei
Gheorgheni
MUNTII HARGHITA
MUNTII BIHOR
MUNTII ZARANDULUI
Baia-de-Criş
Bel'tsy
Faleshty
Tîrnava Mică
Crişul Alb
18°
20°
22°
24°
26°
28°
46°
48°
50°
52°
0 10 20 30 40 50 60 70 80 90 100 110 120 Miles
0 20 40 60 80 100 120 140 160 180 200 Kilometers

ATLANTIC OCEAN
NORTH SEA
BALTIC SEA
Gulf of Bothnia
WHITE SEA
BARENTS SEA
BLACK SEA
CASPIAN SEA
UNITED KINGDOM
NORWAY
SWEDEN
FINLAND
DENMARK
POLAND
TURKEY
IRAQ
IRAN
SVALBARD (SPITSBERGEN) (Nor.)
ZEMLYA FRANTSA-IOSIFA (FRANZ JOSEF LAND)
NOVAYA ZEMLYA
KOLGUYEV
VAYGACH
P-OV YAMAL
P-OV GYDANSKIY
KARELIAN A.S.S.R.
ESTONIAN S.S.R.
LATVIAN S.S.R.
LITHUANIAN S.S.R.
BELORUSSIAN S.S.R.
UKRAINIAN S.S.R.
MOLD. S.S.R.
GEORGIAN S.S.R.
ARMENIAN S.S.R.
AZERBAIDZHAN S.S.R.
KOMI A.S.S.R.
RUSSIAN
URAL MTS.
WEST SIBERIAN LOWLAND
SOVIET
KAZAKH S.S.R.
KIRGHIZ STEPPE
UZBEK S.S.R.
TURKMEN S.S.R.
TADZHIK S.S.R.
KIRGHIZ S.S.R.
CASPIAN DEPRESSION
PLATO UST'-URT
ARAL'SKOYE MORE
PESKI KARAKUMY (DESERT)
PESKI KYZYL KUM (DESERT)
PESKI MUYUN-KUM
PECHORA BASIN
Arctic Circle
MOSCOW (Moskva)
LENINGRAD
KIEV
KHAR'KOV
DONETSK
DNEPROPETROVSK
GORKI
KUYBYSHEV
SVERDLOVSK
NOVOSIBIRSK
BAKU
TASHKENT
TEHRAN
STOCKHOLM
COPENHAGEN
BERLIN
WARSAW
HAMBURG
GLASGOW
Lambert's Azimuthal, Equal Area Project
Elevations and depressions are given in

Relief
Meters
Feet
3050
10 000
1525
5000
610
2000
305
1000
152.5
500
0
Sea Level
0
152.5
500
1525
5000
3050
10 000
Below Sea Level
SEVERNAYA ZEMLYA (NORTHERN LAND)
M. CHELYUSKIN
TAYMYR
P-OV
GORY BYRRANGA
Khatangskiy Zaliv
Nordvik
Khatanga
BOL'SHOY BEGICHEV
LAPTEV SEA
OCEAN
NOVOSIBIRSKIYE O-VA (NEW SIBERIAN ISLANDS)
KOTEL'NYY
FADDEYA
NOVAYA SIBIR
DE-LONGA
MALYY LYAKHOVSKIY
LYAKHOVSKIYE
EAST SIBERIAN SEA
M. SVYATOY NOS
M. BUOR-KHAYA
MEDVEZH'I
VRANGELYA (WRANGEL)
M. SHELAGSKIY
AYON
Ambarchik
CHUKOTSKIY P-OV
CHUKOTSKOYE NAGOR'YE
Anadyr'
Anadyrskiy Zaliv
Arctic Circle
Markovo
Penzhino
KORYAKSKIY KHREBET
M. OLYUTORSKIY
KARAGIN
Tilichiki
Gizhiga
M. TAYGONOS
Zaliv Shelekhova
Palana
P-OV KAMCHATKA
Klyuchevskaya (vol.) 15 584
Ust'-Kamchatsk
Verkhne-Kamchatsk
Petropavlovsk-Kamchatskiy
Ust'-Bol'sheretsk
KHREBET GYDAN (KOLYMSKIY)
Nizhne-Kolymsk
Sredne-Kolymsk
Allaykha
Zyryanka
Zashiversk
Kazach'ye
Tiksi
Bulun
Ust'-Olenëk
Yana
Abyy
Verkhoyansk
KHREBET CHERSKOGO
Gora Chen 10 171
VERKHOYANSKIY KHREBET
Oymyakon
Magadan
Yamsk
M. ALEVINA
Okhotsk
Zhigansk
Olenëk
Lena
Aldan
Aldanskaya
Ust'-Maya
Vilyuy
Vilyuysk
Yakutsk
Amga
YAKUT A.S.S.R.
YAKUT AUTONOMOUS SOVIET SOCIALIST REPUBLIC
GORY PUTORANA
Tura
Nizhnyaya Tunguska
Suntar
Olëkminsk
Tommot
Aldan
Nel'kan
DZHUGDZHUR KHREBET
Ayan
SEA OF OKHOTSK
M. YELIZAVETY
Okha
SHANTAR
Udskaya Guba
SAKHALIN (Sov. Union)
Aleksandrovsk
Poronaysk
M. TERPENIYA
Uglegorsk
Yuzhno-Sakhalinsk
Kholmsk
Korsakov
Baykit
Podkamennaya Tunguska
FEDERATED
Mukhtuya
Peleduy
Vitim
G. Golets-Purpula 6377
PATOM PLATEAU
Bodaybo
Golets-Skalistyy 9186
STANOVOY KHREBET
Chumikan
Nikolayevsk-na-Amure
Yartsevo
G. Polkan 3543
Yenisey
Kirensk
Tyndinskiy
Zeya
KHREBET BUREINSKIY
Komsomol'sk-na-Amure
Malmyzh
Sovetskaya Gavan'
Tatar Strait
Ilimsk
Angara
Nizhne-Angarsk
BURYAT
BAYKAL'SKIY KHREBET
Skovorodino
Svobodnyy
Belogorsk
Ust'-Tyrma
Bureya
Birobidzhan
Khabarovsk
SIKHOTE ALIN'
Krasnoyarsk
Kansk
Tayshet
Bratsk
Bratskoye Vdkhr.
Tulun
Balakhta
Nizhneudinsk
G. Piramida 10 801
Zhigalovo
Kachuga
Lake Baykal
Barguzin
A.S.S.R.
Surface elev. 1535 ft. above sea level
Vitim
YABLONOVYY KHREBET
Sretensk
Nerchinsk
NERCHINSKIY KHREBET
Nerchinskiy Zavod
Chita
Blagoveshchensk
Amur
LESSER KHINGAN RANGE
Minusinsk
Abakan
KHREBET
Cheremkhovo
Angarsk
Munku Sardyk 11 457
Irkutsk
Ulan-Ude
Kyren
Aginskoye
Borzya
Aksha
Argun
Nenchiang
Pok'ot'u
Ch'ich'iharh (Tsitsihar)
Hailun
Suihua
Sungari
Dalnerechensk
Spassk-Dal'niy
Ussuriysk
Artëm
Partizansk
Nakhodka
Vladivostok
Arsen'yev
Ol'ga
Kyzyl
SAYAN
TANNU-OLA
Petrovsk-Zabaykal'skiy
Gorodok
Kyakhta
Khöbsögöl Dalai
Uvs Nuur
Onon
Kerulen
Wench'üan
KHINGAN
GREATER
RANGE
HAERHPIN (Harbin)
P'oli
Mutanchiang
Fuyu
MANCHURIA
Chilin
Selenge Gol
Ulaan Baatar
Öndör Haan
Hara Usu
Jibhalanta
Jirgalanta
HANGAYN NURUU (KHANGAI MTS.)
MONGOLIA
Tsast Bogdo Ula 13865
ALTAI MTS.
Sayr Usa
GOBI OR SHAMO (DESERT)
T'aonan
Lupei
CH'ANGCH'UN
Shuangliao
Tunhua
Hunch'un
Najin
Chŏngjin
SEA OF JAPAN
Ch'ihfeng
FUSHUN
MUKDEN (Shenyang)
Weich'ang
Che'ngte
CHINA
Changchiak'ou
Fengchen
PEKING (Peiching)
TIENTSIN (T'ienching)
Paoting
Lüshun
Lüta
Korea Bay
Po Hai
SHANTUNG PANTAO
YELLOW SEA
P'yŏngyang
KOREA
Kaesŏng
SEOUL
Andong
Taegu
PUSAN
Korea Strait
Hami
Sakhalin
Sōya Kaikyō
Wakkanai
HOKKAIDŌ
Otaru
Sapporo
Esashi
HONSHŪ
JAPAN
Kanazawa
KYOTO
Tottori
Matsue
KŌBE
Okayama
ŌSAKA
Hiroshima
Kōchi
Longitude East of Greenwich
Miles
100
200
300
400
500
600
Kilometers
200
400
600
800
1000
AA-570000-76 -8-6-13 M
COPYRIGHT BY RAND McNALLY & COMPANY
MADE IN U.S.A.

Relief
Meters
3050
1525
610
305
152.5
0
Sea Level
152.5
1525
3050
Feet
10000
5000
2000
1000
500
0
500
5000
10000
Below Sea Level
0 50 100 150 200 250 300 Miles
0 100 200 300 400 500 Kilometers
ARCTIC OCEAN
BARENTS SEA
KARA SEA
White Sea
WESTERN SIBERIAN LOWLAND
PECHORA BASIN
TUNDRA
URAL
KOMI A.S.S.R.
RUSSIAN S.F.S.R.
KARELIAN A.S.S.R.
LAPLAND
NORWAY
SWEDEN
FINLAND
GULF OF BOTHNIA
Gulf of Finland
BALTIC SEA
ESTONIAN S.S.R.
LATVIAN S.S.R.
LITHUANIAN S.S.R.
MOSCOW (Moskva)
LENINGRAD
GORKI
SVERDLOVSK
BASHKIR A.S.S.R.
TATAR A.S.S.R.
UDMURT A.S.S.R.
CHUVASH A.S.S.R.
MARI A.S.S.R.
VALDAI HILLS
Arctic Circle
Arkhangelsk (Archangel)
Murmansk
Helsinki
Riga
Kazan'
Perm'
Kirov
Syktyvkar
Petrozavodsk
Vologda

Conic Projection

Elevations and depressions are given in feet.

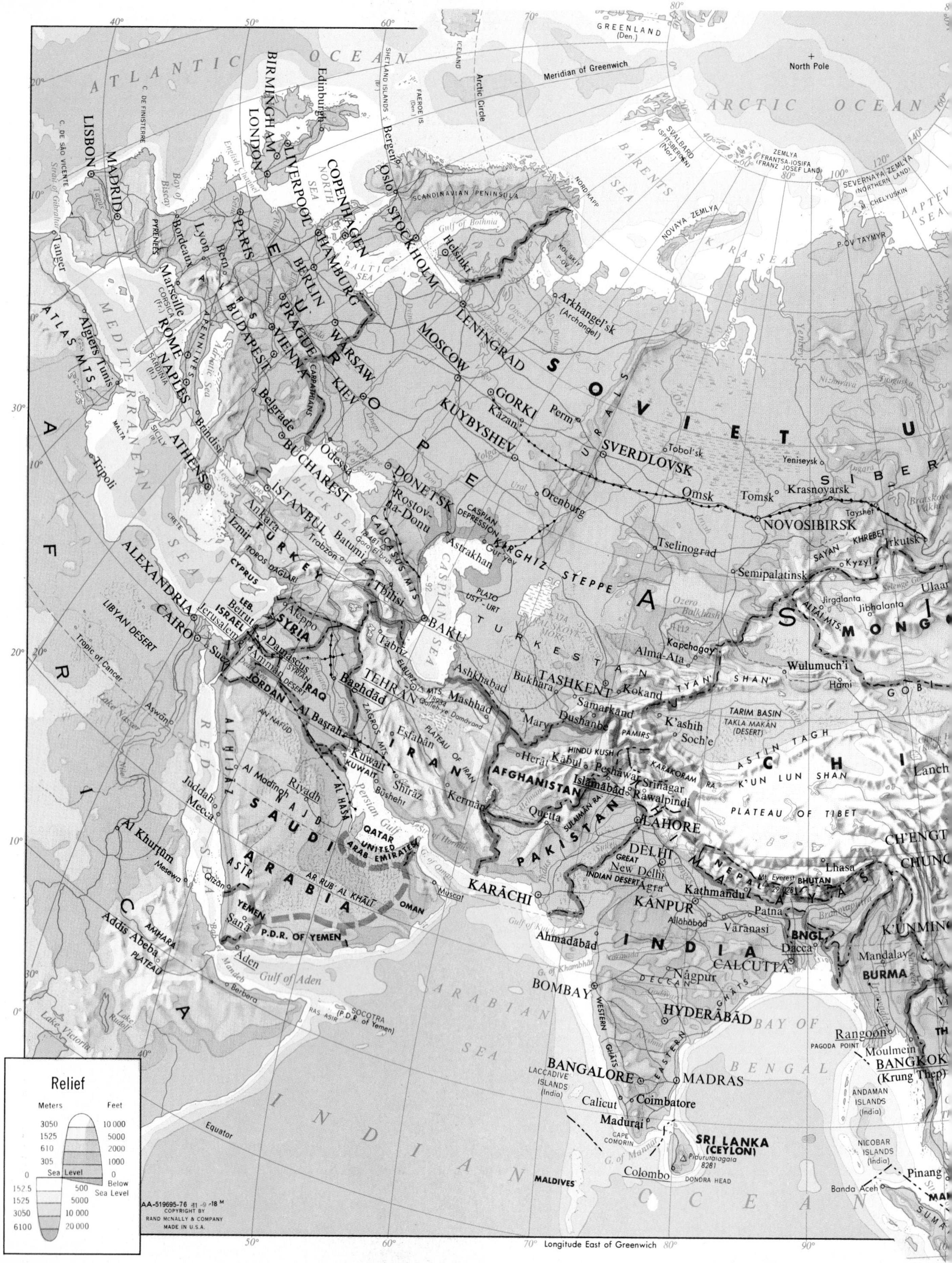

Lambert's Azimuthal, Equal Area Projectic

Elevations and depressions are given in feet

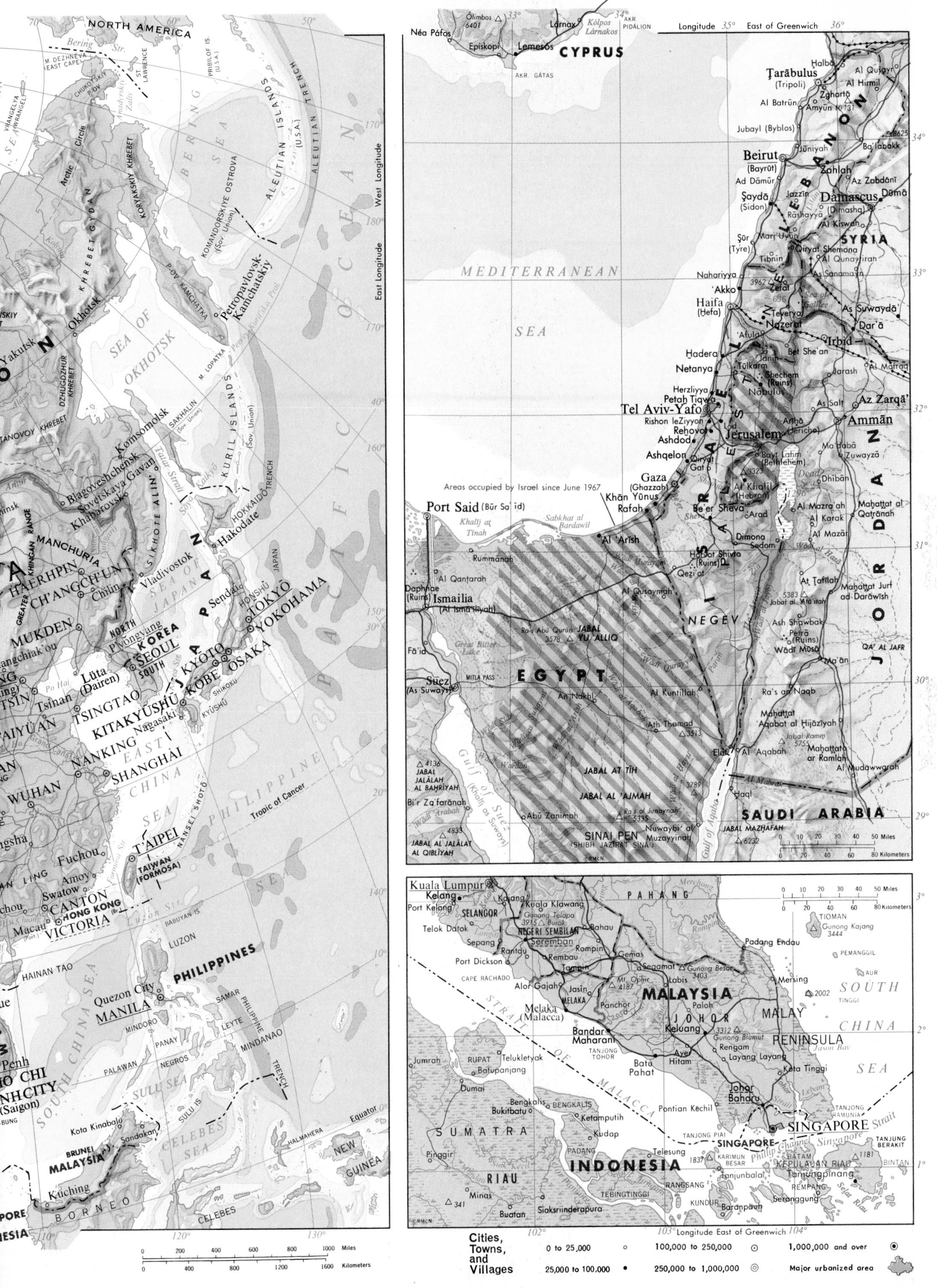

Cities, Towns, and Villages

0 to 25,000	○	100,000 to 250,000	⊙	1,000,000 and over	◉
25,000 to 100,000	•	250,000 to 1,000,000	◎	Major urbanized area	

Polyconic Project

Elevations and depressions are given in feet

INDIA · POLITICAL
AFGHANISTAN
PAKISTAN
JAMMU AND KASHMIR
HIMACHAL PRADESH
PUNJAB
HARYANA
RĀJASTHĀN
UTTAR PRADESH
NEPAL
SIKKIM
BHUTAN
ARUNACHAL PRADESH
ASSAM
NAGALAND
MEGHALAYA
MIZORAM
BANGLADESH
BIHAR
WEST BENGAL
GUJARAT
MADHYA PRADESH
MAHĀRĀSHTRA
ORISSA
ANDHRA PRADESH
KARNATAKA
KERALA
TAMIL NADU
SRI LANKA (CEYLON)
BURMA
THAILAND
TIBET
CHINA
IRAN
LAOS
Tropic of Cancer
ARABIAN SEA
BAY OF BENGAL
1-TRIPURA
2-MANIPUR
3-LAKSHADWEEP
4-DELHI
5-DĀDRA AND NAGAR HAVELI
6-PONDICHERRY
7-GOA, DAMĀN, AND DIU
Jalālābād
Dargai
Chārsadda
KHYBER PASS
MORGA RA.
Peshāwar
0 10 20 30 40 Miles
0 20 40 60 Kilometers
Ozero Balkhash
PESKI MUYUN-KUM
Dzhambul
KIRGIZSKIY KHREBET
Frunze
Chimkent
KIRGHIZ S.S.R.
Namangan
Kokand
Andizhan
Osh
Fergana
Leninabad
TADZHIK S.S.R.
Dushanbe
Kurgan-Tyube
PAMIRS
K'ashih (Kashgar)
TAKLA MAKAN
Soch'e (Yarkand)
SINKIANG UIGHUR AUT. REG.
Hotien
Feyzābād
Mazār-e Sharīf
HINDU KUSH
Gilgit
Chitrāl
KARAKORAM
KARAKORAM PASS
Mt. Godwin Austen (K-2) 28 250
JAMMU AND KASHMIR
Srīnagar
Islāmābād
Rāwalpindi
Kābul
Ghaznī
Jhelum
Jammu
Sialkot
Gujrānwāla
Amritsar
Lyallpur
LAHORE
Jullundur
Ludhiāna
Simla
Chandīgarh
Dehra Dūn
Hardwār
Nanda Devi 25 645
PAKISTAN
Quetta
Multān
Dera Ismāīl Khān
Dera Ghāzi Khān
Bahāwalpur
SULAIMAN RANGE
Patiāla
Ambāla
Sahāranpur
Meerut
Morādābād
DELHI
New Delhi
Rāmpur
Bareilly
Shāhjahānpur
Bīkaner
GREAT INDIAN DESERT
Alwar
Mathura
Bharatpur
Āgra
Āligarh
Farrukhābād
Lucknow
Faizābād
Gorakhpur
Darbhanga
Shikārpur
Sukkur
Mohenjo-Daro (Ruins)
KIRTHAR RANGE
Hyderābād
Jaipur
Ajmer
Jodhpur
Tonk
Gwalior
KĀNPUR
Allāhābād
Vārānasi (Benares)
Patna
Monghyr
Bhāgalpur
Jhānsi
Bānda
Mirzāpur
Gaya
Kota
ARAVALLI RA.
Udaipur
Abu Road
Pālanpur
Rann of Kutch
Bhuj
Māndvi
GUJARAT
AHMADĀBĀD
Rājkot
Jāmnagar
Porbandar
Junāgadh
Verāval
Diu
Bhāvnagar
Baroda
Surat
Damān
Ujjain
Bhopāl
Sāgar
Jabalpur
Indore
VINDHYA RA.
MADHYA PRADESH
Burhānpur
Amrāvati
Nāgpur
Raipur
Bilāspur
Dhule
Akola
Wardha
Chandrapur
Aurangābād
Nāsik
BOMBAY
MAHĀRĀSHTRA
Ahmadnagar
DECCAN
Nizāmābād
HYDERĀBĀD
Pune
Sholāpur
Sāngli
Gulbarga
Kolhāpur
Belgaum
Panaji (Panjim)
Hubli
Bellary
Kurnool
Warangal
Vijayawāda
Guntūr
Elūru
Machilīpatnam
Rājahmundry
Kākināda
Yanam
Vishākhapatnam
Vizianagaram
Berhampur
Cuttack
Bhubaneswar
Puri
Sambalpur
Raurkela
Jamshedpur
Rānchi
Asansol
Burdwān
Howrah
CALCUTTA
Kharagpur
Balasore
Mouths of the Ganges
Dacca
Chittagong
Khulna
Comilla
Rājshāhi
TIBETAN AUTONOMOUS REGION
TRANS-HIMALAYA
HIMALAYA
Lhasa
Gyangtse
Mt. Everest 29 028
Kāthmandu
Dhaulāgiri 26 810
Kānchenjunga 28 208
Gangtok
Thimbu
Punakha
Darjeeling
Cooch Behar
Gauhāti
Shillong
KHASI HILLS
Imphāl
Kohima
Jorhāt
Sibsāgar
Tinsukia
Sadiya
Myitkyina
Mogaung
Bhamo
Mandalay
Shwebo
Monywa
Myingyan
Yamethin
Yenangyaung
Magwe
Sittwe
Kyaukpyu
Sandoway
Henzada
Bassein
Rangoon
Pyinmana
PEGU YOMA
ARAKAN YOMA
Mouths of the Irrawaddy
PAGODA PT.
Nellore
Cuddapah
MADRAS
Kānchipuram
Vellore
Kolār
BANGALORE
Mangalore
Mysore
Pondicherry
Cuddalore
Salem
Kumbakonam
Nāgappattinam
Calicut
Mahe
Coimbatore
Ernākulam
Tiruchchirāppalli
Thanjāvūr
Madurai
LACCADIVE ISLANDS (India)
COROMANDEL COAST
EASTERN GHĀTS
WESTERN GHĀTS
Tiruchchirāppalli
Nāgappattinam
Ernākulam
Thanjāvūr
Madurai
Alleppey
Tuticorin
Quilon
Tirunelveli
Trivandrum
CAPE COMORIN
Jaffna
Trincomalee
Mannar
Puttalam
Anurādhapura
Kandy
Colombo
Pidurutalagala 8281
Galle
Matara
DONDRA HEAD
INDIAN OCEAN
Same scale as main map
Longitude East of Greenwich
Cities, Towns, and Villages
0 to 25,000
25,000 to 100,000
100,000 to 250,000
250,000 to 1,000,000
1,000,000 and over
Major urbanized area
50 100 200 300 400 500 Miles
100 200 400 600 800 Kilometers
©RMCN.

Polyconic Projection

Elevations and depressions are given in feet

Skovorodino
Moho
Chiuchichien
Sretensk
Chita
Nerchinsk
Nerchinski Zavod
Aginskoye
Aksha
Manchouli (Lupin)
Hailaerh (Hailar)
Choybalsan
Blagoveshchensk
Svobodnyy
Zavitinsk
Komsomol'sk-na-Amure
Sovetskaya Gavan'
Khabarovsk
Birobidzhan
HSIAOHSINGANLING SHANMO (LESSER KHINGAN RANGE)
SIKHOTE ALIN'
Ch'ich'ihaerh (Tsitsihar)
HEILUNGKIANG
HAERHPIN (Harbin)
Hailun
Fuchin
Spassk-Dal'niy
Ussuriysk
Vladivostok
Nakhodka
Vladimiro Aleksandrovskoye
Chilin (Kirin)
KIRIN
CH'ANGCH'UN
Shuangliao
Yenchi
Najin
Ch'ŏngjin
MUKDEN (Shenyang)
FUSHUN
LIAONING
Liaoyang
Yingk'ou
Chinchou
Hamhŭng
NORTH KOREA
Wŏnsan
P'yŏngyang
SEOUL (Sŏul)
Inch'ŏn
Kaesŏng
SOUTH KOREA
Andong
Kyŏngju
Taegu
Masan
PUSAN
Kunsan
Mokp'o
KOREAN ARCHIPELAGO
CHEJU I (QUELPART)
Korea Str.
Tsushima Str.
INNER MONGOLIAN AUT. REG.
CH'AHAERH
Huhohaot'e
Changchiak'ou (Kalgan)
PEKING (Peiching)
TIENTSIN (T'ienching)
HOPEH
Paoting
Chengting
Lüshun (Port Arthur)
Lüta (Dairen)
Po Hai
Yent'ai (Chefoo)
Weihai
Weifang
Tsinan (Chinan)
Chiaohsien
TSINGTAO (Ch'ingtao)
SHANTUNG
YELLOW SEA
Lienyünchiang
Chengchou
K'aifeng
HONAN
Hsüchou
KIANGSU
Huaiyin
Nanyang
Pangfou
Fouyang
NANKING (Nanching)
Chenchiang
Ch'angchou
Soochow
Wuhsi
SHANGHAI
SHANGHAI-SHIH
Sungchiang
Chiahsing
Hofei
ANHWEI
Anching
WUHAN
Hanyang
Wuch'ang
Shashih
Hangchou
Shaohsing
Ningpo
CHEKIANG
CHOUSHAN ARCHIPELAGO
Chinhua
Ch'üchou
Nanch'ang
KIANGSI
Ch'angsha
Hsiangt'an
Wenchou
Hengyang
Kanchou
Nanp'ing
Fuchou (Foochow)
FUKIEN
Ch'üanchou
Amoy (Hsiamen)
Changchou
Swatow (Shant'ou)
Meihsien
KWANGTUNG
CANTON (Kuangchou)
Kowloon
VICTORIA
HONG KONG (Br.)
Macau
Formosa Strait
Chilung
T'AIPEI
T'aichung
T'ainan
Kaohsiung
TAIWAN (FORMOSA)
Bashi Channel
SOUTH CHINA SEA
EAST CHINA SEA
SEA OF JAPAN
SEA OF OKHOTSK
PHILIPPINE SEA
PACIFIC OCEAN
NANSEI SHOTŌ
Tropic of Cancer
SAKHALIN (Sov. Un.)
Uglegorsk
Dolinsk
Yuzhno-Sakhalinsk
Kholmsk
Korsakov
Tatar Strait
KURIL ISLANDS (Sov. Union)
ITURUP
URUP
KUNASHIR
Sōya Kaikyō
Wakkanai
Asahikawa
Sapporo
Otaru
Muroran
Kushiro
Nemuro
HOKKAIDŌ
Hakodate
Aomori
Hirosaki
Kuji
Morioka
Akita
Ishinomaki
Sakata
Sendai
Yamagata
Niigata
Nagaoka
Nagano
Utsunomiya
Toyama
Kanazawa
Maebashi
TŌKYŌ
Chiba
YOKOHAMA
Fukui
Gifu
NAGOYA
HONSHŪ
KYOTO
KOBE
Nara
OSAKA
Wakayama
Tottori
Matsue
Okayama
Hiroshima
Kure
Shimonoseki
Tokushima
Takamatsu
Kōchi
Matsuyama
SHIKOKU
KITAKYŪSHŪ
Fukuoka
Kurume
Sasebo
Nagasaki
Kumamoto
KYŪSHŪ
Kagoshima
ŌSUMI GUNTŌ
TOKARA GUNTŌ
AMAMI GUNTŌ
OKINAWA GUNTŌ
Naha
SAKISHIMA GUNTŌ
IZU SHICHITO
JAPAN
Relief
Meters
Feet
3050
1525
610
305
152.5
0
Sea Level
10 000
5000
2000
1000
500
0
Below Sea Level
152.5
1525
3050
6100
500
5000
10 000
20 000
AA-569700-76
COPYRIGHT BY RAND McNALLY & COMPANY MADE IN U.S.A.
Longitude East of Greenwich
0 50 100 200 300 400 500 Miles
0 100 200 400 600 800 Kilometers

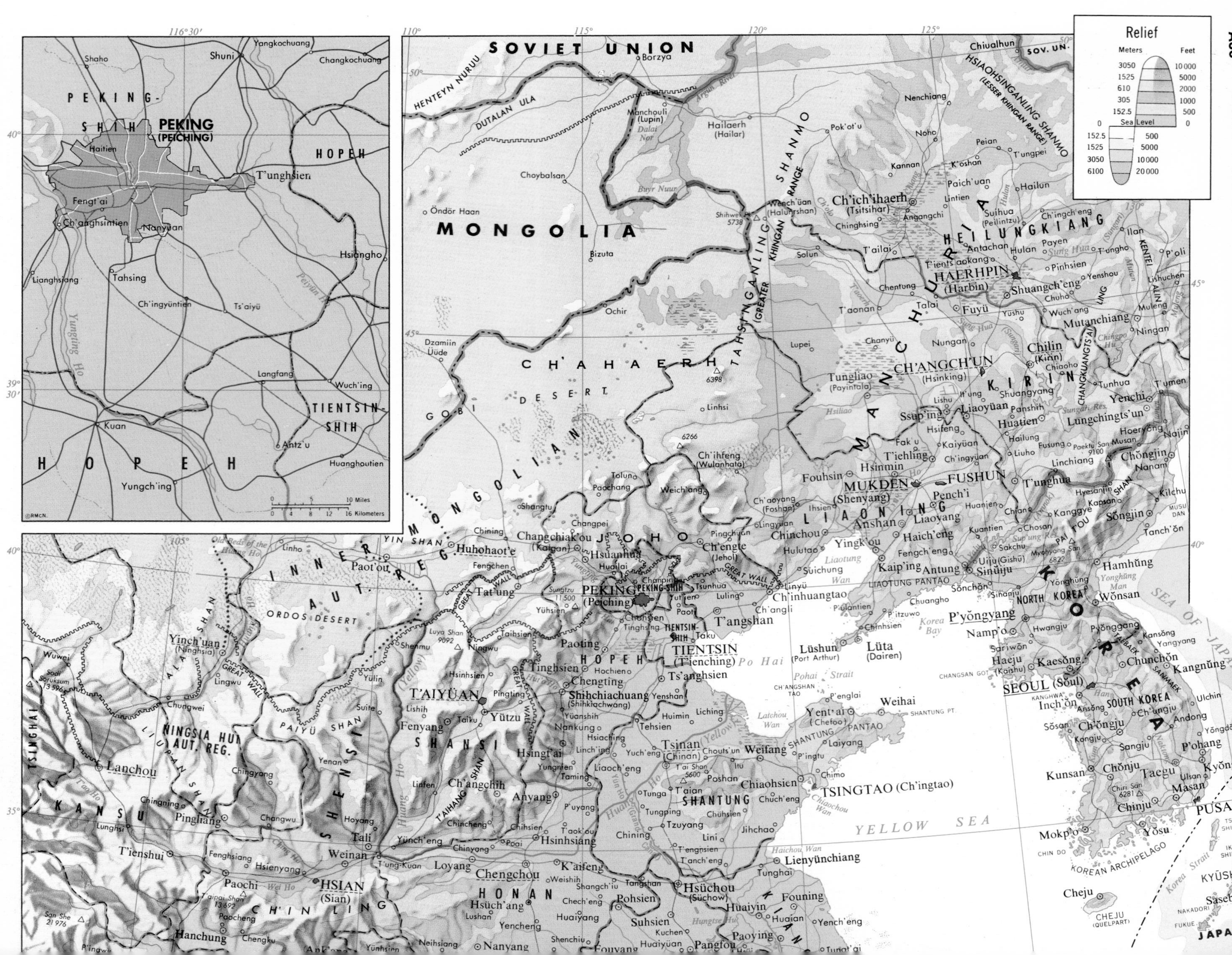

Lambert Conformal Conic Projecti
Elevations and depressions are given in feet

EAST CHINA SEA
SOUTH CHINA SEA
PHILIPPINE SEA
NANSEI SHOTO (RYUKYU ISLANDS)
SAKISHIMA-GUNTO
JAPAN
Tropic of Cancer
TAIWAN (FORMOSA)
T'AIPEI
Formosa Strait
PESCADORES
Bashi Channel
Luzon Strait
BATAN ISLANDS
BABUYAN ISLANDS
Balintang Channel
LUZON
PHILIPPINES
MANILA
Quezon City
PRATAS (China)
HSISHA CH'ÜNTAO (PARACEL IS.) (China)
HAINAN TAO
Gulf of Tonkin
LEICHOU PANTAO
VIETNAM
LAOS
CAMBODIA
THAILAND
ANNAMITIC CORDILLERA
Hanoi
Haiphong
Hue
Da Nang (Tourane)
HUPEH
HUNAN
KIANGSI
FUKIEN
CHEKIANG
KWANGTUNG
KWANGSI CHUANG AUT. REG.
KWEICHOW
YUNNAN
SZECHWAN
RED BASIN
WUHAN
CHUNGKING (Ch'ungch'ing)
K'UNMING (Yünnanfu)
CANTON (Kuangchou)
HONG KONG (Br.)
VICTORIA
Kowloon
Macau (Port.)
Ch'angsha
Nanch'ang
Fuchou (Foochow)
Amoy (Hsiamen)
Swatow (Shant'ou)
Hangchou
Ningpo
Wenchou
Nanning
Kueiyang
Kueilin
Liuchou
Haik'ou
Chanchiang (Ft. Bayard)
Longitude East of Greenwich
AA-560793-76 7-5-11 M
COPYRIGHT BY RAND McNALLY & COMPANY
MADE IN U.S.A.
0 50 100 150 200 250 300 Miles
0 100 200 300 400 500 Kilometers

Polyconic Projection
Elevations and depressions are given in feet

PHILIPPINE
SEA
LIPPINES
CATANDUANES
Catbalogan
SAMAR
Tacloban
LEYTE
DINAGATE
PHILIPPINE
GALATHEA DEEP
34 440
TRENCH
BOHOL
Butuan
Cagayan
MINDANAO
Davao
Apo
9692
Davao Gulf
PALAU IS.
(PELEW)
(P.I.T.T.)
(U. S.)
MIANGAS
SONSOROL
ISLANDS
KEPULAUAN
TALAUD
SANGIHE
SIAU
MOROTAI
Tondano
Ternate
HALMAHERA
Laut
Maluku
(Molucca Sea)
Laut
Halmahera
WAIGEO
Selat Dampier
BACAN
Labuha
KEPULAUAN OBI
OBI
TALIABU
MANGOLE
KEPULAUAN
SULA
SANANA
SALAWATI
Sorong
G. Kwoka
10 045
Manokwari
JAZIRAH
DOBERAI
NUMFOOR
BIAK
JAPEN
KEPULAUAN
SCHOUTEN
MAPIA
MISOL
Teluk Berau
Teluk
Cenderawasih
Fakfak
MALUKU
(MOLUCCAS)
SERAM
Piru
G. Binaiya
9931
Bula
BURU
Amboina
AMBON
Kaimana
S I A
BANDA
BESAR
KEPULAUAN KAI
KAI KECIL
KEPULAUAN
LUCIPARA
KEPULAUAN
TOKANGBESI
LAUT BANDA
(BANDA SEA)
GUNUNGAPI
Dobo
KEPULAUAN
ARU
TRANGAN
ADI
DAMAR
WETAR
BABAR
YAMDENA
KEPULAUAN
TANIMBAR
ALOR
I. DE ATAÚRO
MOA
SELARU
Dili
TIMOR
TIMOR
SEA
ARAFURA SEA
DOLAK
TANJUNG VALS
Merauke
MELVILLE
Dundas Str.
COBOURG
PEN.
CROKER
WESSEL IS.
Van
Diemen Gulf
BATHURST
Darwin
AUSTRALIA
Gulf of Carpentaria
C. ARNHEM
CAPE
YORK
PEN.
THURSDAY
C. YORK
Torres Strait
GREAT
BARRIER
REEF
DARU
Gulf
of Papua
CORAL SEA
TG. PERKAM
Jayapura
(Sukarnopura)
PEGUNUNGAN VAN REES
Mamberamo
PEGUNUNGAN
MAOKE
Puncak Trikora
15 518
Puncak Jaya
16 500
NEW GUINEA
Digul
Fly
Strickland
Sepik
Aitape
Wewak
Ramu
BISMARCK RA.
Madang
KARKAP
LONG
Mt. Giluwe 13 660
Mt. Wilhelm 15 400
Mt. Bangeta
13 454
PAPUA
NEW GUINEA
Lae
Salamaua
Huon Gulf
Morobe
13 100
Mt. Albert Edward
OWEN STANLEY RA.
Buna
Port Moresby
Mt. Victoria
13 363
SOUTH CAPE
Samarai
NINIGO IS.
HERMIT IS.
ADMIRALTY IS.
MANUS
MUSSAU
EMIRAU
NEW HANOVER
Kavieng
BISMARCK
ARCH.
NEW
IRELAND
Namatanai
Rabaul
Kokopo
VITU IS.
Talasea
The Father
7546
NEW BRITAIN
BOUGAINVILLE TRENCH
TROBRIAND IS.
WOODLARK
D'ENTRECASTEAUX IS.
Equator
125°
130°
135°
140°
145°
150°
20°
15°
10°
5°
0°
10°
0 50 100 200 300 400 500 Miles
0 100 200 400 600 800 Kilometers
Sinait
Cabugao
Lapog
Santo
Domingo
Vigan
La Paz
Dolores
Bangued
Bucay
Narvacan
Pilar
Abra
Candon
Santa Lucia
S. Cruz
Cervantes
Angaki
Bontoc
Mt. Amuyao
8865
CORDILLERA CENTRAL
Lubuagan
Chico
Tuao
Solana
Tuguegarao
Amulung
Iguig
S. Pablo
Cabagan
Centro
Ilagan
Gamu
Cauayan
Cagayan
Magat
S. Mariano
Divilacan Bay
Palanan Bay
PALANAN PT.
SIERRA MADRE
Bangar
Luna
Balaoan
Kiangan
Santiago
Angadanan
Echague
Jones
Bacnotan
San Fernando
S. Juan
Naguilian
Pulog
9612
Bagabag
Solano
Bauang
Aringay
Baguio
Bayombong
Bambang
Dupax
Casiguran
Diapitan Bay
DIJOHAN PT
SANTIAGO
Bolinao
CAPE BOLINAO
CABARRUYAN
Lingayen
Gulf
7388
Sto. Tomas
Consolacion
San
Fabian
Bani
Alaminos
Agno
Burgos
Dagupan
Binalonan
S. Nicolas
Tayug
S. Quintin
Urdaneta
Rosales
Casiguran Sd.
CAPE SAN ILDEFONSO
PHILIPPINE
16°
CAIMAN PT.
Dasol Bay
Lingayen
San Carlos
Infanta
Santa Cruz
Mangatarem
Bayambang
San Jose
Muños
Baler Bay
Baler
ENCANTO PT.
Camiling
Candelaria
High Pk.
6830
Gerona
Tarlac
Victoria
Cabanatuan
PALAUIG PT.
Palauig
LUZON
Iba
Concepcion
Penaranda
Gapan
Dingalan Bay
SEA
SOUTH CHINA SEA
Pinatubo
5709
Angeles
Arayat
S. Miguel
S. Fernando
Pampanga
S. Narciso
S. Antonio
Guagua
Subic
Malolos
Olongapo
Orani
Sto.
Maria
Montalban
Infanta
Polillo Str.
POLILLO IS.
POLILLO
PATNANONGAN
JOMALIG
Polillo
SAMPALOC PT.
Balanga
Orion
Malabon
Quezon
City
Manila
Bay
MANILA
Lamon Bay
Mariveles
Cavite
Pasig
San Juan
COCHINOS PT.
CORREGIDOR
Naic
Laguna
de Bay
Biñan
Sta. Rosa
Tauza
Indang
Silang
Sto. Tomas
Tanauan
Calamba
Sta. Cruz
Mauban
Nagcarlan
7177
Mt. Banahao
S. Pablo
Atimonan
Tiaong
BALESIN
CALAGUA IS.
Capalonga
Paracale
CABALETE
ALABAT
Mt. Cadig
2388
Labo
Talisay
Daet
Mt. Labo
4500
QUINABUCASAN PT.
14°
Nasugbu
Lopez B.
Calauag B.
Balayan
Lemery
S. Jose
Lipa
Taal
Lucena
Gumaca
Calauag
Ragay
San
Miguel
Bay
Libmanan
Lagonoy
CABRA
Lubang
AMBIL
LUBANG
IS.
LABUAN
GOLO
Balayan
Bay
MARICABAN
Rosario
Batangas
Unisan
Tayabas Bay
Macalelon
Catanauan
Ragay Gulf
S. Narciso
Naga
Vol. Isarog
6450
Pili
Verde I. Passage
VERDE
Lobo
Mompog Pass
BONDOC PEN.
Baao
Buhi
Tabaco
Mayon
8077
CAPE CALAVITE
Paluan
Calapan
Boac
Mt. Sto. Antonio
2859
Sta. Cruz
Torrijos
Gasan
MARINDUQUE
Polangui
Ligao
Camalig
Legaspi
Mt. Halcon
8491
Naujan
BONDOC PT.
San Pascual
Burias Pass
Mamburao
DUMALI PT.
Pinamalayan
MINDORO
BURIAS
Pilar
BANTON
Jones
SIBUYAN
Mt. Baco
8161
Sablayan
DONGON PT.
Mindoro Strait
Knob Pk.
3838
Bulalacao
Tablas Strait
Odiongan
ROMBLON
Romblon
TICAO
S. Jacinto
Aroroy
Masbate
S. Jose
TABLAS
SIBUYAN
Cajidiocan
MASBATE
SEA
Looc
ILIN
BURUNCAN PT.
BUSUANGA
TARA
0 10 20 30 40 Miles
0 10 20 30 40 50 60 Kilometers
©RMCN.
120°
122°

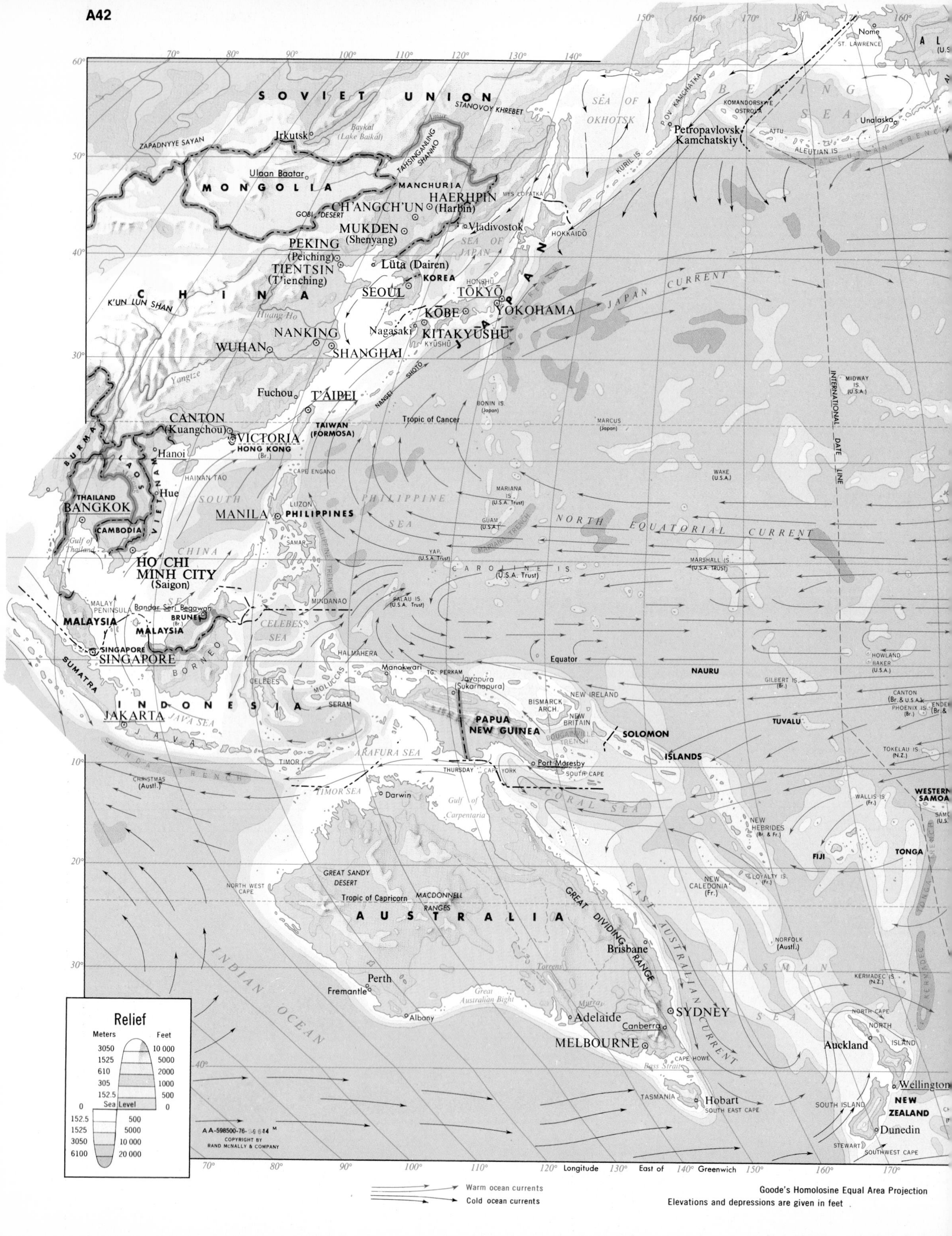

Relief

Meters	Feet
3050	10 000
1525	5000
610	2000
305	1000
152.5	500
0 Sea Level	0
152.5	500
1525	5000
3050	10 000
6100	20 000

AA-598500-76-6-6-14 M

70° 80° 90° 100° 110° 120° Longitude 130° East of 140° Greenwich 150° 160° 170°

Warm ocean currents

Cold ocean currents

Goode's Homolosine Equal Area Projection

Elevations and depressions are given in feet

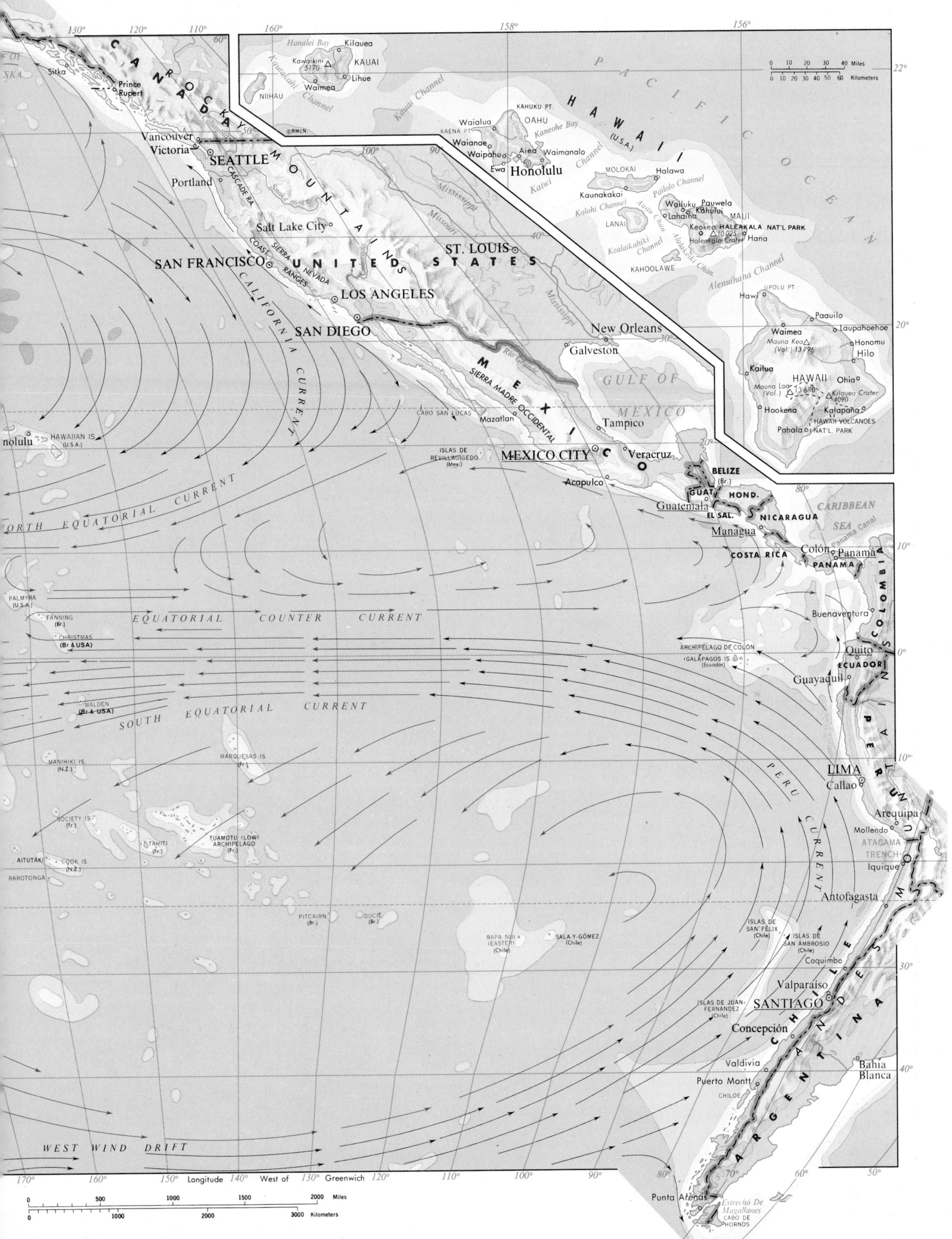
CANADA
ROCKY MOUNTAINS
Sitka
Prince Rupert
Vancouver
Victoria
SEATTLE
Portland
CASCADE RA.
Salt Lake City
SAN FRANCISCO
COAST RANGES
SIERRA NEVADA
UNITED STATES
ST. LOUIS
LOS ANGELES
SAN DIEGO
CALIFORNIA CURRENT
Mississippi
Missouri
New Orleans
Galveston
Rio Grande
MEXICO
SIERRA MADRE OCCIDENTAL
GULF OF MEXICO
Tampico
CABO SAN LUCAS
Mazatlan
MEXICO CITY
Veracruz
Acapulco
ISLAS DE REVILLAGIGEDO (Mex.)
BELIZE (Br.)
GUAT.
HOND.
Guatemala
EL SAL.
NICARAGUA
Managua
CARIBBEAN SEA
Panama Canal
COSTA RICA
Colón
Panama
PANAMA
COLOMBIA
Buenaventura
ARCHIPÉLAGO DE COLÓN (GALÁPAGOS IS.) (Ecuador)
Quito
ECUADOR
Guayaquil
PERU
LIMA
Callao
PERU CURRENT
Arequipa
Mollendo
ATACAMA TRENCH
Iquique
Antofagasta
ISLAS DE SAN FÉLIX (Chile)
ISLAS DE SAN AMBROSIO (Chile)
Coquimbo
Valparaíso
SANTIAGO
ISLAS DE JUAN FERNANDEZ (Chile)
Concepción
CHILE
ANDES MTS.
ARGENTINA
Valdivia
Puerto Montt
CHILOE
Bahía Blanca
Punta Arenas
Estrecho De Magallanes
CABO DE HORNOS
HAWAIIAN IS. (U.S.A.)
nolulu
NORTH EQUATORIAL CURRENT
PALMYRA (U.S.A.)
FANNING (Br.)
CHRISTMAS (Br. & USA)
EQUATORIAL COUNTER CURRENT
MALDEN (Br. & USA)
SOUTH EQUATORIAL CURRENT
MANIHIKI IS. (N.Z.)
MARQUESAS IS. (Fr.)
SOCIETY IS. (Fr.)
TAHITI (Fr.)
TUAMOTU (LOW) ARCHIPELAGO (Fr.)
AITUTAKI
COOK IS. (N.Z.)
RAROTONGA
PITCAIRN (Br.)
DUCIE (Br.)
RAPA NUI (EASTER) (Chile)
SALA-Y-GÓMEZ (Chile)
WEST WIND DRIFT
170° 160° 150° Longitude 140° West of 130° Greenwich 120° 110° 100° 90° 80° 70° 60° 50°
0 500 1000 1500 2000 Miles
0 1000 2000 3000 Kilometers
Hanalei Bay
Kilauea
Kawaikini 5170
KAUAI
Lihue
Waimea
NIIHAU
Kaulakahi Channel
Kauai Channel
©RMCN.
PACIFIC OCEAN
HAWAII (U.S.A.)
KAHUKU PT.
OAHU
Waialua
KAENA PT.
Kaneohe Bay
Waianae
Waipahu
Aiea
Waimanalo
Ewa
Honolulu
Kaiwi Channel
MOLOKAI
Halawa
Kaunakakai
Pailolo Channel
Kalohi Channel
Wailuku
Pauwela
Kahului
Lahaina
MAUI
LANAI
Auau Chan.
Keokea
HALEAKALA NAT'L PARK
10 025
Haleakala Crater
Hana
Kealaikahiki Channel
Alalakeiki Chan.
KAHOOLAWE
Alenuihaha Channel
UPOLU PT.
Hawi
Paauilo
Laupahoehoe
Waimea
Honomu
Mauna Kea (Vol.) 13 796
Hilo
Kailua
HAWAII
Mauna Loa (Vol.) 13 680
Ohia
Kilauea Crater 4090
Hookena
Kalapana
HAWAII VOLCANOES NAT'L PARK
Pahala
0 10 20 30 40 Miles
0 10 20 30 40 50 60 Kilometers
160° 158° 156° 22° 20°

Lambert's Azimuthal, Equal Area Projecti

Elevations and depressions are given in feet

PAPUA NEW GUINEA
Mt. Albert Edward 13 100
Buna
Mt. Victoria 13 363
Port Moresby
OWEN STANLEY RA.
TROBRIAND IS.
WOODLARK
D'ENTRECASTEAUX ISLANDS
Samarai
SOUTH CAPE
LOUISIADE ARCHIPELAGO
TAGULA
ROSSEL
Torres Strait
BANKS
HORN
CAPE YORK
CAPE YORK PENINSULA
Princess Charlotte Bay
OSPREY REEF
CAPE MELVILLE
CORAL SEA
Laura
Cooktown
Palmerville
GREAT BARRIER REEF
HOLMES REEFS
WILLIS IS.
ATHERTON
Cairns
Mungana
Mt. Bartle Frere 5287
Croydon
PLATEAU
FLINDERS REEFS
LIHOU REEFS
TREGROSSE IS.
Forsayth
Ingham
HINCHINBROOK
Halifax Bay
Townsville
GREGORY RANGE
MARION REEF
Charters Towers
Bowen
WHITSUNDAY
CUMBERLAND IS.
Richmond
Hughenden
CLARKE RA.
Repulse Bay
Mackay
Kynuna
Mt. Dalrymple 4190
CONNORS RANGE
NORTHUMBERLAND IS.
Winton
SWAIN REEFS
Capricorn Chan.
WRECK REEFS
Clermont
GREAT DIVIDING RANGE
Longreach
Barcaldine
Emerald
Rockhampton
Mount Morgan
CURTIS
Gladstone
Jericho
Dingo
Blackall
Yaraka
BUCKLAND TABLELAND
Tambo
Hervey Bay
SANDY CAPE
Bundaberg
FRASER
Maryborough
Quilpie
Charleville
Roma
Gympie
GREY RANGE
Thargomindah
Dalby
Toowoomba
DARLING DOWNS
Brisbane
N. STRADBROKE I.
Cunnamulla
St. George
Ipswich
Southport
Dirranbandi
Warwick
Hungerford
Mungindi
Mt. Roberts
Lismore
Moree
Tenterfield
Capoompeta
Glen Innes
Brewarrina
Walgett
Inverell
NEW ENGLAND RANGE
Grafton
Bourke
Narrabri
Armidale
Coonamble
The Round Mountain
Wilcannia
Cobar
Nyngan
Tamworth
WARRUMBUNGLE RA.
Kempsey
LIVERPOOL RA.
Port Macquarie
Nymagee
Dubbo
NEW SOUTH WALES
Cessnock
Maitland
Newcastle
Orange
Forbes
Bathurst
Lithgow
BLUE MTS.
SYDNEY
Botany Bay
Wentworth
West Wyalong
Hay
RIVERINA
Narrandera
Wollongong
Goulburn
Wagga Wagga
Deniliquin
Kerang
Echuca
Canberra
Jervis Bay
Albury
Mt. Kosciusko 7316
AUSTL. CAP. TER.
SNOWY MTS.
Bendigo
Benalla
Cooma
VICTORIA
Bega
Ballarat
Maryborough
Bombala
Geelong
MELBOURNE
CAPE HOWE
Port Phillip Bay
Bairnsdale
NINETY MILE BEACH
Wonthaggi
CAPE OTWAY
WILSON'S PROMONTORY
Bass Strait
KING
FLINDERS
FURNEAUX GROUP
HUNTER IS.
CAPE BARREN
TASMANIA
Ulverstone
Burnie
Devonport
Launceston
Mt. Ossa 5305
Strahan
Risdon
New Norfolk
Hobart
BRUNY
SOUTH EAST CAPE
TASMAN SEA
LORD HOWE (NEW S. WALES)
CHOISEUL
VELLA LAVELLA
SANTA ISABEL
NEW GEORGIA
RENDOVA
MALAITA
FLORIDA
SOLOMON ISLANDS
RUSSELL IS.
TULAGI
Honiara
GUADALCANAL
SAN CRISTÓBAL
RENNELL
SANTA CRUZ ISLANDS
TORRES IS.
BANKS ISLANDS
ESPÍRITU SANTO
MAEWO
PENTECOST
NEW HEBRIDES (British and French Condominium)
MALEKULA
AMBRIM
EPI
EFATE
Vila
ERROMANGA
TANA
ANEITYUM
PACIFIC OCEAN
ÎLES CHESTERFIELD (Fr.)
ÎLES BÉLEP
OUVÉA
LIFOU
ÎLES LOYAUTÉ (French)
MARÉ
NEW CALEDONIA (Fr.)
Nouméa
ÎLE DES PINS
Tropic of Capricorn
NORTH CAPE
Kaitaia
Russell
Hauraki Gulf
GREAT BARRIER
Devonport
Auckland
Bay of Plenty
EAST CAPE
NORTH ISLAND
Hamilton
North Taranaki Bight
Lake Taupo
New Plymouth
C. EGMONT
Mt. Ruapehu 9175
Gisborne
Hawke Bay
Napier
Hastings
South Taranaki Bight
Wanganui
Palmerston North
NEW ZEALAND
CAPE FAREWELL
Tasman Bay
Nelson
Lower Hutt
Wellington
Karamea Bight
Cook Strait
CAPE FOULWIND
Greymouth
Hokitika
SOUTH ISLAND
SOUTHERN ALPS
Mt. Cook 12 349
Pegasus Bay
Christchurch
CASCADE PT.
Canterbury Bight
Timaru
Dunedin
CAPE SAUNDERS
RESOLUTION ISLAND
Foveaux Strait
Invercargill
STEWART ISLAND
SOUTHWEST CAPE
Same scale as main map
©RMcN.
0 50 100 200 300 400 500 Miles
0 100 200 400 600 800 Kilometers

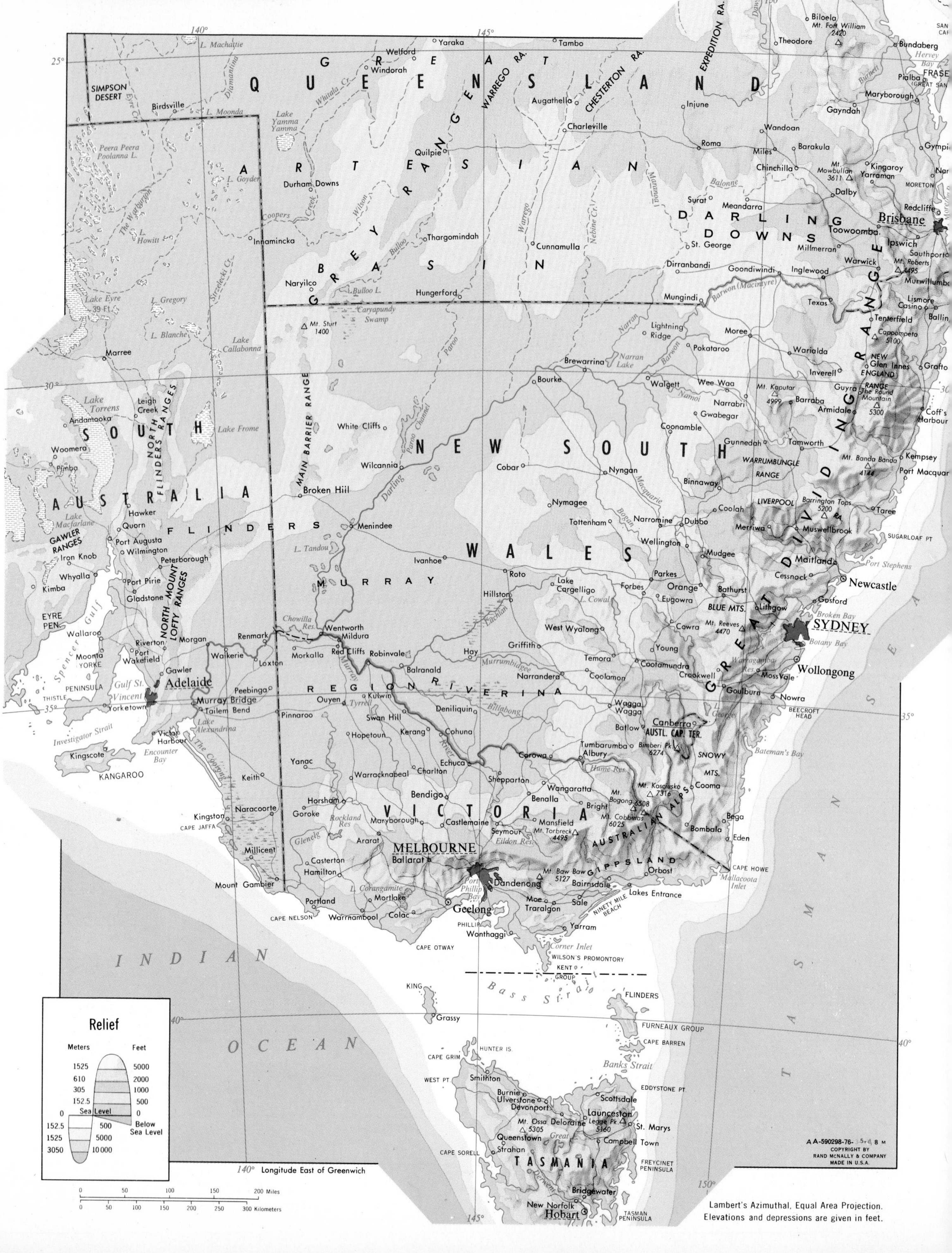

QUEENSLAND

GREAT ARTESIAN BASIN

GREY RANGE

DARLING DOWNS

SOUTH AUSTRALIA

NEW SOUTH WALES

VICTORIA

TASMANIA

MURRAY

RIVERINA REGION

FLINDERS

GREAT DIVIDING RANGE

AUSTRALIAN ALPS

GIPPSLAND

INDIAN OCEAN

TASMAN SEA

SIMPSON DESERT

Birdsville, Windorah, Welford, Yaraka, Tambo, Augathella, Charleville, Quilpie, Durham Downs, Innamincka, Thargomindah, Naryilco, Hungerford, Cunnamulla, St. George, Dirranbandi, Goondiwindi, Mungindi, Injune, Roma, Wandoan, Miles, Chinchilla, Barakula, Surat, Meandarra, Dalby, Toowoomba, Brisbane, Ipswich, Warwick, Inglewood, Millmerran, Gladstone, Biloela, Theodore, Bundaberg, Maryborough, Gayndah, Kingaroy, Yarraman, Redcliffe, Southport, Texas, Tenterfield, Casino, Moree, Inverell, Glen Innes, Armidale, Tamworth, Gunnedah, Narrabri, Wee Waa, Walgett, Lightning Ridge, Pokataroo, Brewarrina, Bourke, Cobar, Nyngan, Coonamble, Gwabegar, Barraba, Kempsey, Port Macquarie, Coff's Harbour, Grafton, Binnaway, Coolah, Dubbo, Narromine, Wellington, Mudgee, Merriwa, Muswellbrook, Maitland, Cessnock, Newcastle, Gosford, Sydney, Wollongong, Moss Vale, Nowra, Goulburn, Canberra, Bathurst, Lithgow, Orange, Parkes, Forbes, Eugowra, Cowra, Young, Cootamundra, Crookwell, Temora, Wagga Wagga, Griffith, West Wyalong, Lake Cargelligo, Hillston, Roto, Ivanhoe, Menindee, Broken Hill, Wilcannia, White Cliffs, Wentworth, Mildura, Red Cliffs, Robinvale, Balranald, Hay, Narrandera, Coolamon, Deniliquin, Swan Hill, Kerang, Cohuna, Echuca, Shepparton, Corowa, Albury, Tumbarumba, Batlow, Cooma, Bombala, Bega, Eden, Wangaratta, Benalla, Bright, Mansfield, Seymour, Bendigo, Castlemaine, Maryborough, Ballarat, Melbourne, Geelong, Dandenong, Moe, Traralgon, Sale, Bairnsdale, Orbost, Lakes Entrance, Yarram, Wonthaggi, Colac, Warrnambool, Portland, Mortlake, Hamilton, Casterton, Ararat, Horsham, Goroke, Charlton, Warracknabeal, Hopetoun, Yanac, Ouyen, Kulwin, Pinnaroo, Murray Bridge, Tailem Bend, Victor Harbour, Adelaide, Gawler, Kingston, Naracoorte, Millicent, Mount Gambier, Keith, Peebinga, Loxton, Waikerie, Renmark, Morgan, Riverton, Port Wakefield, Wallaroo, Moonta, Kadina, Yorketown, Kimba, Whyalla, Port Pirie, Port Augusta, Gladstone, Peterborough, Wilmington, Quorn, Hawker, Iron Knob, Woomera, Andamooka, Leigh Creek, Marree, Kingscote

Grassy, Smithton, Burnie, Ulverstone, Devonport, Deloraine, Launceston, Scottsdale, St. Marys, Campbell Town, Queenstown, Strahan, New Norfolk, Bridgewater, Hobart

Bass Strait, Banks Strait, Spencer Gulf, Gulf St. Vincent, Investigator Strait, Encounter Bay, Port Phillip Bay, Corner Inlet, Mallacoota Inlet, Bateman's Bay, Botany Bay, Broken Bay, Port Stephens, Hervey Bay

KANGAROO, KING, FLINDERS, FURNEAUX GROUP, CAPE BARREN, HUNTER IS., KENT GROUP, EYRE PEN., YORKE PENINSULA, WILSON'S PROMONTORY, CAPE HOWE, CAPE OTWAY, CAPE NELSON, CAPE JAFFA, CAPE GRIM, WEST PT, CAPE SORELL, EDDYSTONE PT, FREYCINET PENINSULA, TASMAN PENINSULA, BEECROFT HEAD, SUGARLOAF PT, MORETON, THISTLE

Lake Eyre −39 Ft., Lake Torrens, Lake Frome, Lake Gregory, Lake Blanche, Lake Callabonna, Lake Macfarlane, Lake Alexandrina, Lake Yamma Yamma, L. Machattie, L. Moonda, L. Goyder, L. Howitt, Peera Peera Poolanna L., L. Tandou, Narran Lake, L. Cowal, Hume Res., Eildon Res., L. Corangamite, Rockland Res, Caryapundy Swamp, Bulloo L., Lake George

Darling, Murray, Murrumbidgee, Lachlan, Paroo, Warrego, Barwon, Namoi, Macquarie, Bogan, Balonne, Maranoa, Nebine Cr., Wilson, Bulloo, Coopers Creek, Diamantina, Eyre Cr., Whitula Cr., Strzelecki Cr., The Warburton, Burnett, Dawson, Billabong, Wakool, Glenelg, Tyrrell, Derwent, Great, Warragamba, The Coorong, Narran, Paroo Channel

GAWLER RANGES, FLINDERS RANGES, NORTH FLINDERS RANGES, NORTH MOUNT LOFTY RANGES, MAIN BARRIER RANGE, WARREGO RA., CHESTERTON RA., EXPEDITION RA., NEW ENGLAND RANGE, WARRUMBUNGLE RANGE, LIVERPOOL RA., BLUE MTS., SNOWY MTS., AUSTL. CAP. TER.

Mt. Sturt 1400, Mt. Fort William 2420, Mt. Mowbullan 3611, Mt. Roberts 4495, Capoompeta 5100, The Round Mountain 5300, Mt. Kaputar 4999, Mt. Banda Banda 4144, Barrington Tops 5200, Mt. Reeves 4470, Bimberi Pk. 6274, Mt. Kosciusko 7316, Mt. Bogong 6508, Mt. Cobberas 6025, Mt. Torbreck 4495, Mt. Baw Baw 5127, Mt. Ossa 5305, Legge Pk. 5160

NINETY MILE BEACH

25°, 30°, 35°, 40°, 140°, 145°, 150°

140° Longitude East of Greenwich

Relief

Meters	Feet
1525	5000
610	2000
305	1000
152.5	500
0 Sea Level	0
152.5	500 Below Sea Level
1525	5000
3050	10 000

0 50 100 150 200 Miles

0 50 100 150 200 250 300 Kilometers

Lambert's Azimuthal, Equal Area Projection.
Elevations and depressions are given in feet.

A A-590298-76- 5-4-8 M

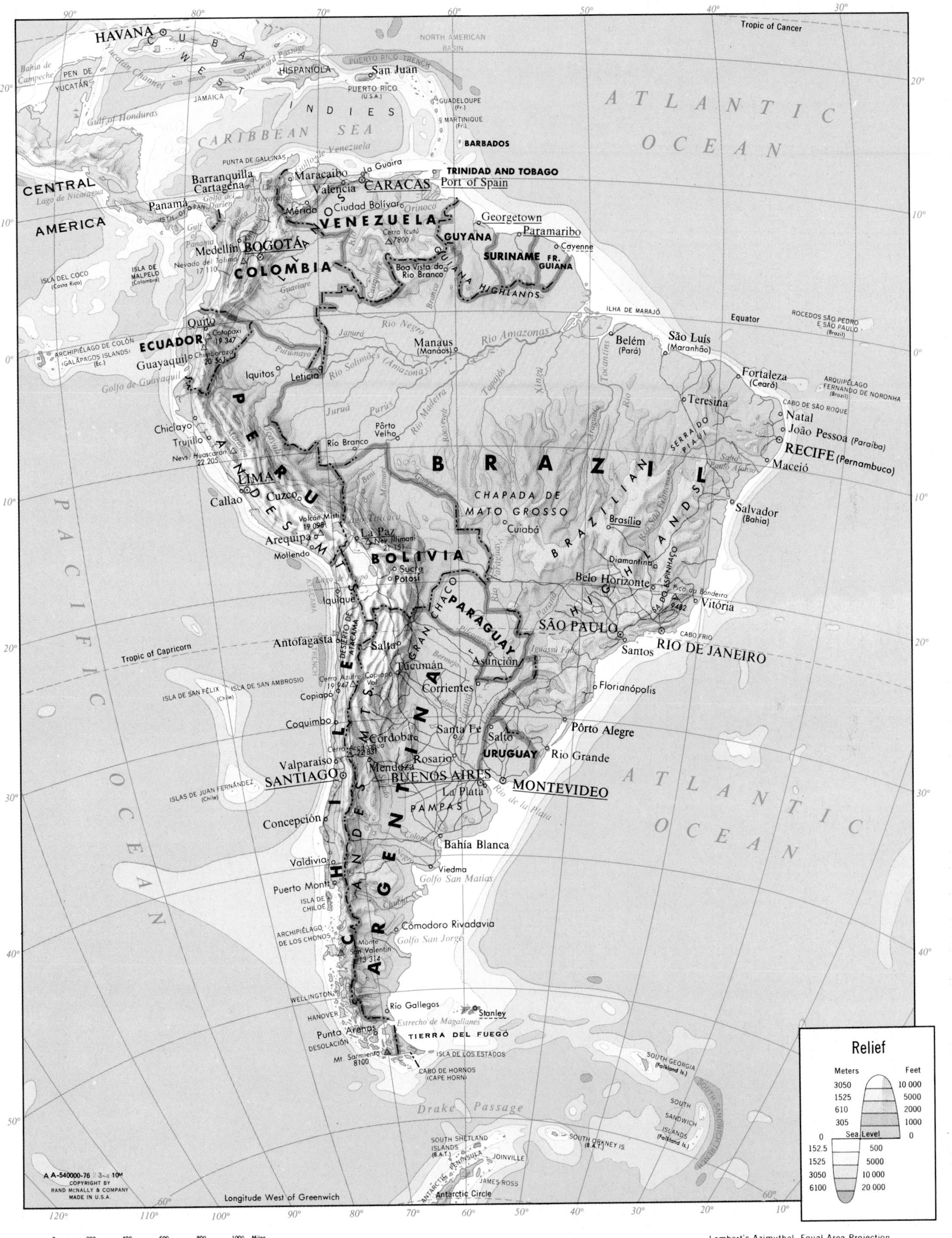
Tropic of Cancer
HAVANA
CUBA
WEST INDIES
HISPANIOLA
San Juan
PUERTO RICO (U.S.A.)
JAMAICA
NORTH AMERICAN BASIN
PUERTO RICO TRENCH
GUADELOUPE (Fr.)
MARTINIQUE (Fr.)
BARBADOS
CARIBBEAN SEA
ATLANTIC OCEAN
Bahía de Campeche
PEN. DE YUCATÁN
Yucatán Channel
Windward Passage
Gulf of Honduras
PUNTA DE GALLINAS
Golfo de Venezuela
CENTRAL AMERICA
Lago de Nicaragua
Barranquilla
Cartagena
Panama
Maracaibo
La Guaira
Valencia
CARACAS
TRINIDAD AND TOBAGO
Port of Spain
Mérida
Ciudad Bolívar
Orinoco
VENEZUELA
Georgetown
GUYANA
Paramaribo
SURINAME
Cayenne
FR. GUIANA
Cerro Icutú 7800
Medellín
BOGOTÁ
COLOMBIA
Nevado del Tolima 17 110
ISLA DEL COCO (Costa Rica)
ISLA DE MALPELO (Colombia)
Boa Vista do Rio Branco
GUIANA HIGHLANDS
ILHA DE MARAJÓ
Equator
ROCEDOS SÃO PEDRO E SÃO PAULO (Brazil)
Quito
Cotopaxi 19 347
ECUADOR
Chimborazo 20 561
Guayaquil
ARCHIPIÉLAGO DE COLÓN (GALÁPAGOS ISLANDS) (Ec.)
Golfo de Guayaquil
Manaus (Manáos)
Rio Negro
Rio Amazonas
Belém (Pará)
São Luís (Maranhão)
Iquitos
Leticia
Rio Solimões (Amazonas)
Fortaleza (Ceará)
ARQUIPÉLAGO FERNANDO DE NORONHA (Brazil)
Teresina
CABO DE SÃO ROQUE
Natal
João Pessoa (Paraíba)
RECIFE (Pernambuco)
Maceió
PERU
Chiclayo
Trujillo
Nevs. Huascarán 22 205
Pôrto Velho
Rio Branco
BRAZIL
SERRA DO PIAUÍ
CHAPADA DE MATO GROSSO
LIMA
Callao
Cuzco
ANDES MTS.
Volcán Misti 19 098
Arequipa
Mollendo
La Paz
Nev. Illimani 21 151
BOLIVIA
Sucre
Potosí
Cuiabá
Brasília
BRAZILIAN HIGHLANDS
Salvador (Bahia)
Diamantina
Belo Horizonte
Pico da Bandeira 9482
Vitória
Iquique
CHACO
PARAGUAY
GRAN CHACO
SÃO PAULO
Santos
RIO DE JANEIRO
CABO FRIO
Antofagasta
DESIERTO DE ATACAMA
Salta
Tropic of Capricorn
Asunción
Tucumán
Corrientes
Florianópolis
Cerro Azufre 19 947
Copiapó Vol.
ISLA DE SAN FÉLIX (Chile)
ISLA DE SAN AMBROSIO
Copiapó
Coquimbo
Córdoba
Santa Fe
Salto
Pôrto Alegre
Cerro Aconcagua 22 831
Rosario
URUGUAY
Rio Grande
Valparaíso
Mendoza
SANTIAGO
BUENOS AIRES
MONTEVIDEO
ISLAS DE JUAN FERNÁNDEZ (Chile)
La Plata
Río de la Plata
PACIFIC OCEAN
PAMPAS
Concepción
ARGENTINA
CHILE
Bahía Blanca
Valdivia
Viedma
Golfo San Matías
Puerto Montt
ISLA DE CHILOÉ
Comodoro Rivadavia
ARCHIPIÉLAGO DE LOS CHONOS
Golfo San Jorge
Monte San Valentín 13 314
WELLINGTON
Río Gallegos
Stanley
HANOVER
Estrecho de Magallanes
Punta Arenas
TIERRA DEL FUEGO
DESOLACIÓN
Mt. Sarmiento 8100
ISLA DE LOS ESTADOS
CABO DE HORNOS (CAPE HORN)
SOUTH GEORGIA (Falkland Is.)
SOUTH SANDWICH ISLANDS (Falkland Is.)
SOUTH SANDWICH TRENCH
Drake Passage
SOUTH SHETLAND ISLANDS (B.A.T.)
SOUTH ORKNEY IS. (B.A.T.)
ANTARCTIC PENINSULA
JOINVILLE
JAMES ROSS
Antarctic Circle
Longitude West of Greenwich
A A-540000-76 2-3-4 10M
COPYRIGHT BY RAND McNALLY & COMPANY MADE IN U.S.A.
Relief
Meters
Feet
3050 10 000
1525 5000
610 2000
305 1000
0 Sea Level 0
152.5 500
1525 5000
3050 10 000
6100 20 000
0 200 400 600 800 1000 Miles
0 400 800 1200 1600 Kilometers

Lambert's Azimuthal, Equal Area Projection
Elevations and depressions are given in feet

Longitude

Sinusoidal Projection

Elevations and depressions are given in feet

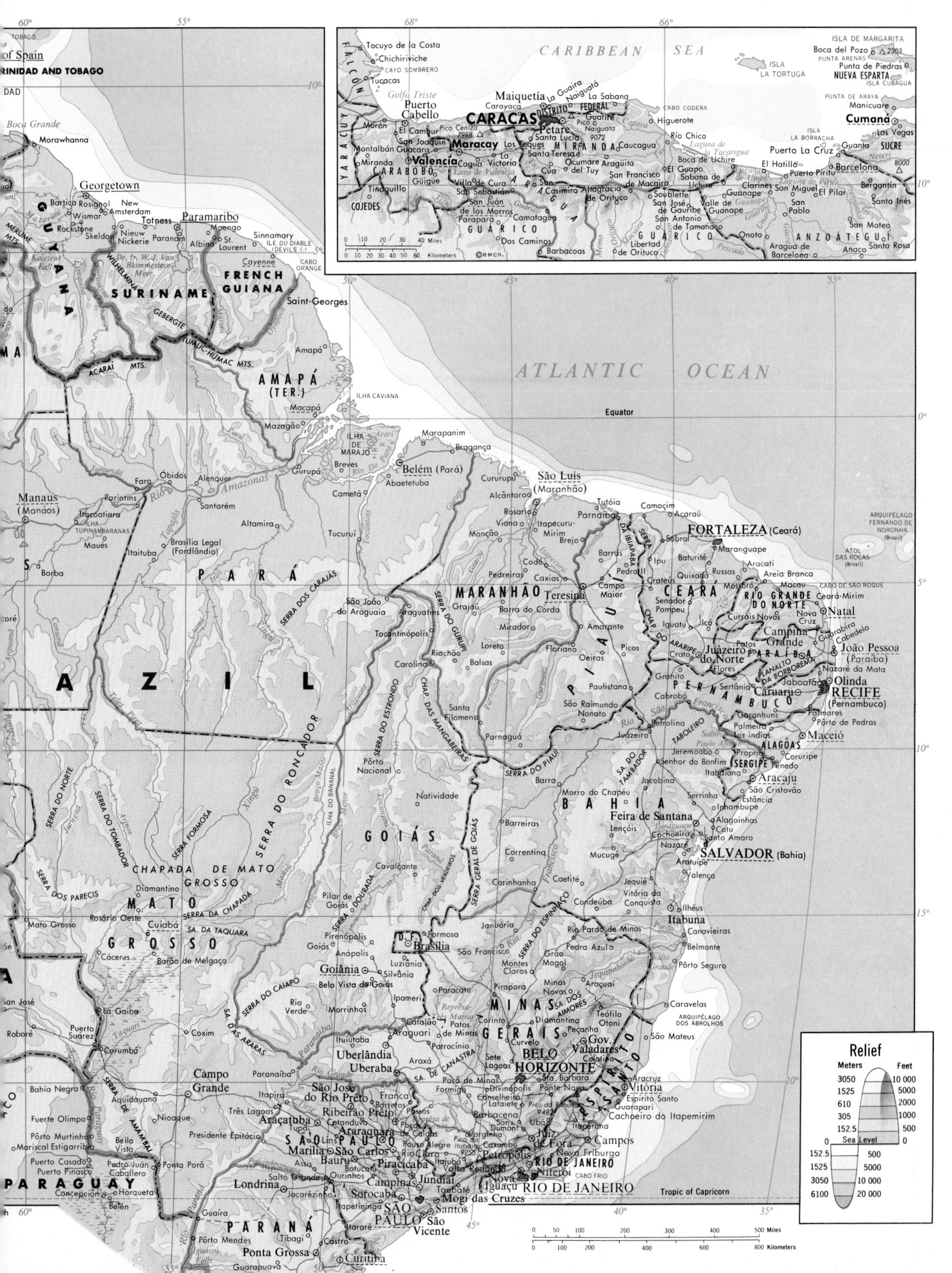
CARIBBEAN SEA
ISLA DE MARGARITA
Boca del Pozo
Punta de Piedras
NUEVA ESPARTA
ISLA CUBAGUA
ISLA LA TORTUGA
Tocuyo de la Costa
Chichiriviche
CAYO SOMBRERO
Tucacas
Golfo Triste
FALCÓN
YARACUY
Puerto Cabello
Maiquetía
La Guaira
Naiguatá
La Sabana
DISTRITO FEDERAL
CABO CODERA
Carayaca
CARACAS
Petare
Guatire
Higuerote
Morón
El Cambur
Pico Ceniza 7988
San Joaquín
Maracay
Los Teques
Santa Lucía
MIRANDA
Caucagua
Río Chico
Laguna de la Tacarigua
Montalbán
Guacara
Valencia
Cagua
La Victoria
Santa Teresa
Ocumare del Tuy
Aragüita
Boca de Uchire
Miranda
CARABOBO
Güigüe
Lago de Valencia
Villa de Cura
Cúa
El Guapo
Sabana de Uchire
San Francisco de Macaira
ARAGUA
Tinaquillo
San Sebastián
San Casimiro
Altagracia de Orituco
COJEDES
San Juan de los Morros
Parapara
Camatagua
GUÁRICO
Dos Caminos
Barbacoas
Libertad de Orituco
Soublette
San José de Guaribe
Valle de Guanape
San Antonio de Tamanaco
Guanape
Clarines
San Miguel
Onoto
Aragua de Barcelona
ANZOÁTEGUI
Anaco
Santa Rosa
San Mateo
San Pablo
El Pilar
Santa Inés
Bergantín
Barcelona
Puerto Píritu
El Hatillo
Puerto La Cruz
Guanta
SUCRE
Cumaná
Manicuare
PUNTA DE ARAYA
Las Vegas
LA BORRACHA
8000
0 10 20 30 40 Miles
0 10 20 30 40 50 60 Kilometers
©RMCN.
of Spain
TRINIDAD AND TOBAGO
Boca Grande
Morawhanna
Georgetown
Bartica
Rosignol
New Amsterdam
Wismar
Rockstone
Skeldon
Totness
Paramaribo
Moengo
Nieuw Nickerie
Paranam
Albina
St. Laurent
Sinnamary
ILE DU DIABLE (DEVILS I.)
Cayenne
CABO ORANGE
GUYANA
SURINAME
FRENCH GUIANA
Kaieteur Fall
Saint-Georges
TUMUC-HUMAC MTS.
ACARAÍ MTS.
Amapá
AMAPÁ (TER.)
Macapá
Mazagão
ILHA CAVIANA
ATLANTIC OCEAN
Equator
ILHA DE MARAJÓ
Marapanim
Bragança
Belém (Pará)
Breves
Abaetetuba
Gurupá
Cametá
Faro
Óbidos
Alenquer
Amazonas
Santarém
Parintins
Manaus (Manáos)
Itacoatiara
ILHA TUPINAMBARANAS
Maués
Altamira
Brasília Legal (Fordlândia)
Itaituba
Borba
Tucuruí
PARÁ
Cururupu
São Luís (Maranhão)
Alcântara
Rosário
Viana
Itapecuru-Mirim
Monção
Brejo
Tutóia
Parnaíba
Camocim
Acaraú
FORTALEZA (Ceará)
Sobral
Maranguape
Baturité
Aracati
Ipu
Quixadá
Russas
Areia Branca
Crateús
Macau
Mossoró
CABO DE SÃO ROQUE
Ceará-Mirim
Natal
CEARÁ
RIO GRANDE DO NORTE
Senador Pompeu
Currais Novos
Nova Cruz
Iguatu
Icó
Campina Grande
Cabedelo
João Pessoa (Paraíba)
PARAÍBA
Crato
Juazeiro do Norte
Patos
Nazaré da Mata
Olinda
RECIFE (Pernambuco)
Caruaru
Jaboatão
Sertânia
PERNAMBUCO
Flores
Garanhuns
Palmares
Pôrto de Pedras
Maceió
ALAGOAS
Palmeira dos Índios
Penedo
Coruripe
Propriá
SERGIPE
Aracaju
São Cristovão
Estância
ARQUIPÉLAGO FERNANDO DE NORONHA (Brazil)
ATOL DAS ROCAS (Brazil)
Codó
Caxias
Pedreiras
Teresina
Campo Maior
MARANHÃO
Barra do Corda
Grajaú
Mirador
Amarante
Floriano
Picos
Oeiras
PIAUÍ
Paulistana
São Raimundo Nonato
Loreto
Riachão
Balsas
Carolina
Tocantinópolis
Araguatins
São João do Araguaia
Imperatriz
SERRA DOS CARAJÁS
SERRA DO GURUPI
Santa Filomena
Parnaguá
CHAP. DAS MANGABEIRAS
SERRA DO ESTRONDO
Pôrto Nacional
Natividade
SERRA DO PIAUÍ
Barra
Petrolina
Juazeiro
Cabrobó
Jeremoabo
Senhor do Bonfim
Itabaiana
Jacobina
Morro do Chapéu
BAHIA
Serrinha
Inhambupe
Alagoinhas
Feira de Santana
Catu
Lençóis
Cachoeira
Santo Amaro
Nazaré
SALVADOR (Bahia)
Aratuípe
Valença
Barreiras
Correntina
Mucugê
Caetité
Jequié
Vitória da Conquista
Ilhéus
Itabuna
Canavieiras
Belmonte
Pôrto Seguro
Carinhanha
Condeúba
GOIÁS
Cavalcante
Pilar de Goiás
SERRA GERAL DE GOIÁS
SERRA DO RONCADOR
ILHA DO BANANAL
A Z I L
SERRA DO NORTE
SERRA DO TOMBADOR
SERRA FORMOSA
SERRA DOS PARECIS
CHAPADA DE MATO GROSSO
Diamantino
MATO GROSSO
Rosário Oeste
Cuiabá
SA. DA TAQUARA
Mato Grosso
Cáceres
Barão de Melgaço
Pirenópolis
Goiás
Anápolis
Formosa
D.F. Brasília
Luziânia
Silvânia
Goiânia
Bela Vista de Goiás
SERRA DO CAIAPÓ
Rio Verde
Morrinhos
Ipameri
Paracatu
Januária
São Francisco
Montes Claros
Grão Mogol
Pirapora
Rio Pardo de Minas
Pedra Azul
Araçuaí
Minas Novas
MINAS GERAIS
Caravelas
ARQUIPÉLAGO DOS ABROLHOS
Teófilo Otoni
Diamantina
Peçanha
Gov. Valadares
São Mateus
Corinto
Curvelo
Catalão
Patos de Minas
Araguari
Patrocínio
Ituiutaba
Uberlândia
Uberaba
Araxá
SA. DE CANASTRA
Sete Lagoas
BELO HORIZONTE
Sta. Bárbara
Colatina
Aracruz
Vitória
ESPÍRITO SANTO
Guarapari
Cachoeiro do Itapemirim
Pará de Minas
Divinópolis
Formiga
Conselheiro Lafaiete
Ponte Nova
Barbacena
São João del Rei
Ubá
Juiz de Fora
Itaperuna
Campos
Petrópolis
Nova Friburgo
RIO DE JANEIRO
Niterói
CABO FRIO
Nova Iguaçu
Volta Redonda
Corumbá
La Gaiba
Puerto Suárez
Roboré
San José
Coxim
SA. DAS ARARAS
Campo Grande
Bahía Negra
Aquidauana
SERRA DE AMAMBAÍ
Fuerte Olimpo
Pôrto Murtinho
Mariscal Estigarribia
Bella Vista
Puerto Casado
Puerto Pinasco
Pedro Juán Caballero
Ponta Porã
PARAGUAY
Concepción
Horqueta
Belén
Nioaque
Paranaíba
Itapira
São José do Rio Prêto
Franca
Ribeirão Prêto
Três Lagoas
Araçatuba
Presidente Epitácio
Tupã
Araraquara
SÃO PAULO
Marília
São Carlos
Bauru
Piracicaba
Assis
Ourinhos
Botucatu
Campinas
Jundiaí
Londrina
Jacarèzinho
Sorocaba
Taubaté
Mogi das Cruzes
Itapetininga
SÃO PAULO
Santos
São Vicente
Itararé
Guaíra
PARANÁ
Pôrto Mendes
Tibagi
Castro
Ponta Grossa
Guarapuava
Curitiba
Iguassú Falls
Tropic of Capricorn
Relief
Meters
Feet
3050
1525
610
305
152.5
0
10 000
5000
2000
1000
500
0
Sea Level
152.5
1525
3050
6100
500
5000
10 000
20 000
0 50 100 200 300 400 500 Miles
0 100 200 400 600 800 Kilometers

0 50 100 200 300 400 500 Miles
0 100 200 400 600 800 Kilometers

Sinusoidal Projection
Elevations and depressions are given in feet

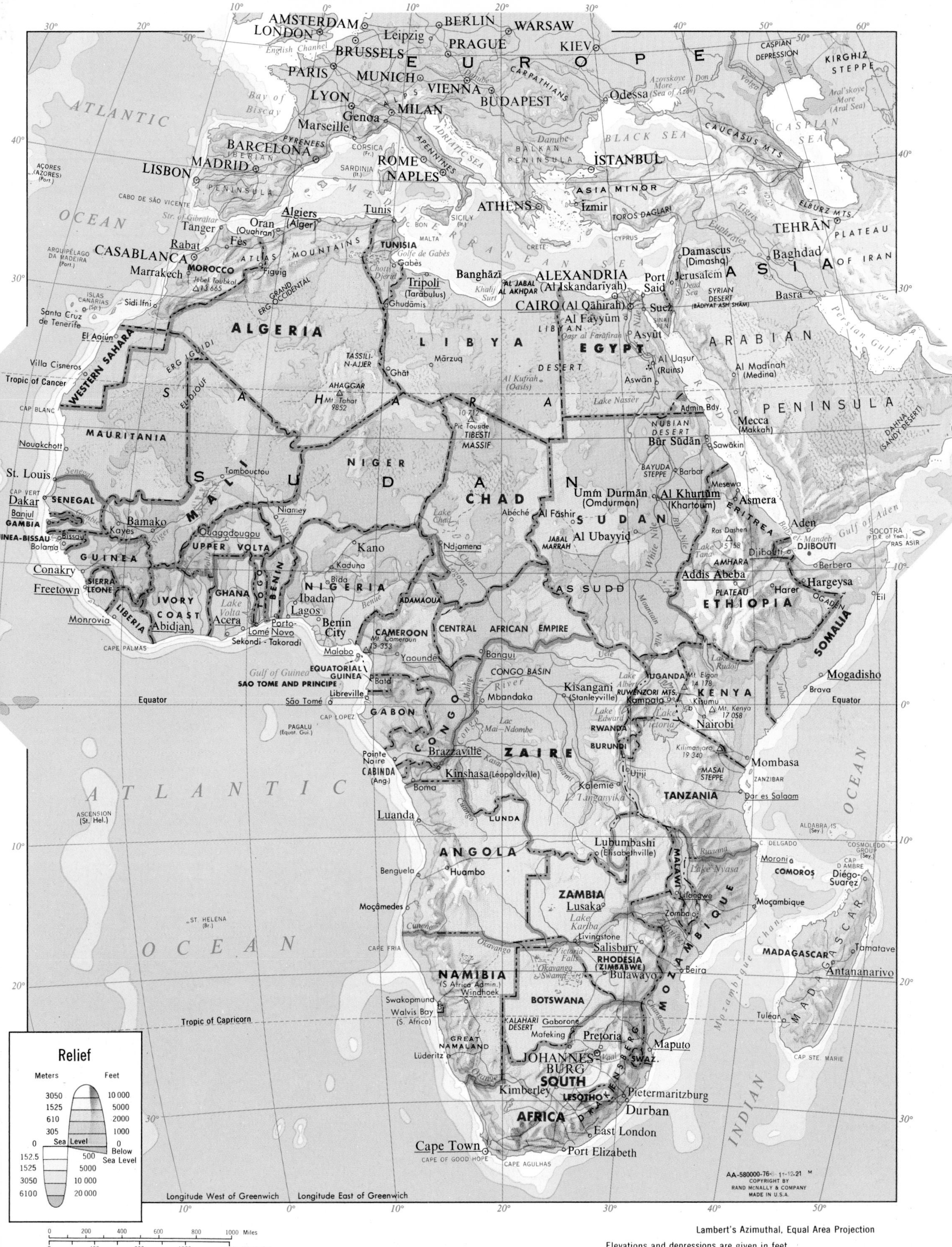
EUROPE
ASIA
AMSTERDAM
LONDON
BERLIN
WARSAW
Leipzig
PRAGUE
KIEV
BRUSSELS
English Channel
PARIS
MUNICH
VIENNA
CARPATHIANS
CASPIAN DEPRESSION
KIRGHIZ STEPPE
Bay of Biscay
LYON
ALPS
BUDAPEST
Odessa
Azovskoye More (Sea of Azov)
Aral'skoye More (Aral Sea)
ATLANTIC
Genoa
MILAN
Marseille
Danube
BLACK SEA
CAUCASUS MTS.
CASPIAN SEA
PYRENEES
BARCELONA
IBERIAN PENINSULA
CORSICA (Fr.)
ADRIATIC SEA
APENNINES
BALKAN PENINSULA
ISTANBUL
AÇORES (AZORES) (Port.)
LISBON
MADRID
SARDINIA (It.)
ROME
NAPLES
MEDITERRANEAN SEA
ASIA MINOR
CABO DE SÃO VICENTE
ATHENS
İzmir
TOROS DAGLARI
ELBURZ MTS.
OCEAN
Str. of Gibraltar
Oran (Ouahran)
Algiers (Alger)
Tunis
SICILY (It.)
C. BON
CYPRUS
Tigris
Euphrates
TEHRĀN
PLATEAU
Tanger
Rabat
Fès
MALTA
CRETE
Baghdad
OF IRAN
ARQUIPÉLAGO DA MADEIRA (Port.)
CASABLANCA
ATLAS MOUNTAINS
TUNISIA
Golfe de Gabès
Gabès
Damascus (Dimashq)
Marrakech
MOROCCO
Figuig
Chott Djerid
Banghāzī
AL JABAL AL AKHḌAR
ALEXANDRIA (Al Iskandarīyah)
Port Said
Jerusalem
Jebel Toubkal 13 665
GRAND ERG OCCIDENTAL
Tripoli (Tarābulus)
Khalīj Surt
Dead Sea
SYRIAN DESERT (BĀDIYAT ASH SHĀM)
Basra
ISLAS CANARIAS (Sp.)
Sidi Ifni
Ghudāmis
CAIRO (Al Qāhirah)
Suez
SINAI PEN.
Santa Cruz de Tenerife
ALGERIA
Al Fayyūm
Persian Gulf
El Aaiún
LIBYA
LIBYAN DESERT
Qaşr al Farāfirah
Asyūţ
ARABIAN
WESTERN SAHARA
ERG IGUIDI
TASSILI-N-AJJER
Mārzuq
EGYPT
Nile
Villa Cisneros
Ghāt
Al Uqşur (Ruins)
Al Madīnah (Medina)
Tropic of Cancer
EL DJOUF
AHAGGAR
Mt. Tahat 9852
SAHARA
Al Kufrah (Oasis)
Aswān
RED SEA
CAP BLANC
Lake Nasser
Admin. Bdy.
PENINSULA
DAHNA (SANDY DESERT)
MAURITANIA
Pic Toussidé 10 712
TIBESTI MASSIF
NUBIAN DESERT
Mecca (Makkah)
Nouakchott
Būr Sūdān
Sawākin
NIGER
BAYUDA STEPPE
Barbar
St. Louis
Senegal
Tombouctou
SUDAN
MALI
Mesewa
CAP VERT
Dakar
SENEGAL
CHAD
Umm Durmān (Omdurman)
Al Khurţūm (Khartoum)
Asmera
Banjul
GAMBIA
Gambia
Niamey
Abéché
Al Fāshir
ERITREA
Aden
Gulf of Aden
GUINEA-BISSAU
Bissau
Bolama
Bamako
Kayes
Ouagadougou
UPPER VOLTA
Lake Chad
JABAL MARRAH
Al Ubayyid
White Nile
Blue Nile
Ras Dashen 15 158
Bab el Mandeb
SOCOTRA (P.D.R. of Yem.)
RAS ASIR
DJIBOUTI
Djibouti
Kano
Ndjamena
Lake Tana
AMHARA
Berbera
GUINEA
Kaduna
Conakry
SIERRA LEONE
Bida
NIGERIA
Addis Abeba
Hargeysa
Freetown
IVORY COAST
GHANA
TOGO
BENIN
Ibadan
ADAMAOUA
AS SUDD
PLATEAU
Harer
ETHIOPIA
OGADEN
Eil
LIBERIA
Lake Volta
Lagos
Benue
Mountain Nile
Monrovia
Abidjan
Accra
Lomé
Porto-Novo
Benin City
CAMEROON
Mt. Cameroun 13 353
CENTRAL AFRICAN EMPIRE
SOMALIA
CAPE PALMAS
Sekondi-Takoradi
Malabo
Bangui
Uele
Lake Rudolf
Gulf of Guinea
EQUATORIAL GUINEA
Yaoundé
Mogadisho
SAO TOME AND PRINCIPE
Bata
CONGO BASIN
Congo River
Kisangani (Stanleyville)
Lake Albert
UGANDA
Mt. Elgon 14 178
RUWENZORI MTS.
Kampala
KENYA
Juba
Brava
Equator
Libreville
Ubangi
Mbandaka
Kisumu
Mt. Kenya 17 058
São Tomé
GABON
CAP LOPEZ
Lake Edward
RWANDA
Lake Victoria
Nairobi
PAGALU (Equat. Gui.)
CONGO
Lac Mai-Ndombe
BURUNDI
Brazzaville
ZAIRE
Kilimanjaro 19 340
Pointe Noire
CABINDA (Ang.)
Kasai
Lomami
Ujiji
Mombasa
Kinshasa (Léopoldville)
MASAI STEPPE
ZANZIBAR
ATLANTIC
Boma
Kalemie
Lake Tanganyika
TANZANIA
Dar es Salaam
ASCENSION (St. Hel.)
Cuango
Luanda
LUNDA
ALDABRA IS. (Sey.)
C. DELGADO
Lubumbashi (Elisabethville)
Ruvuma
COSMOLEDO GROUP (Sey.)
ANGOLA
MALAWI
Lake Nyasa
Moroni
COMOROS
CAP D'AMBRE
Benguela
Huambo
Diégo-Suarez
ZAMBIA
Lusaka
Lilongwe
Moçambique
Moçâmedes
Zomba
ST. HELENA (Br.)
Lake Kariba
Cunene
Livingstone
Salisbury
OCEAN
Okavango
Victoria Falls
RHODESIA (ZIMBABWE)
MADAGASCAR
Tamatave
CAPE FRIA
Okavango Swamp
Bulawayo
Beira
NAMIBIA (S. Africa Admin.)
MOZAMBIQUE
Antananarivo
Windhoek
Mozambique Channel
Swakopmund
Walvis Bay (S. Africa)
BOTSWANA
Tropic of Capricorn
KALAHARI DESERT
Gaborone
Changane
Tuléar
Mafeking
Pretoria
Maputo
GREAT NAMALAND
Lüderitz
JOHANNESBURG
Vaal
SWAZ.
CAP STE. MARIE
SOUTH AFRICA
Orange
Kimberley
LESOTHO
DRAKENSBERG
Pietermaritzburg
Durban
INDIAN
OCEAN
East London
Cape Town
CAPE OF GOOD HOPE
CAPE AGULHAS
Port Elizabeth
AA-580000-76- 11-12-21 M
COPYRIGHT BY RAND McNALLY & COMPANY MADE IN U.S.A.
Longitude West of Greenwich
Longitude East of Greenwich
Relief
Meters
Feet
3050
1525
610
305
0
Sea Level
152.5
1525
3050
6100
10 000
5000
2000
1000
0
Below Sea Level
500
5000
10 000
20 000
0 200 400 600 800 1000 Miles
0 400 800 1200 1600 Kilometers
Lambert's Azimuthal, Equal Area Projection
Elevations and depressions are given in feet.

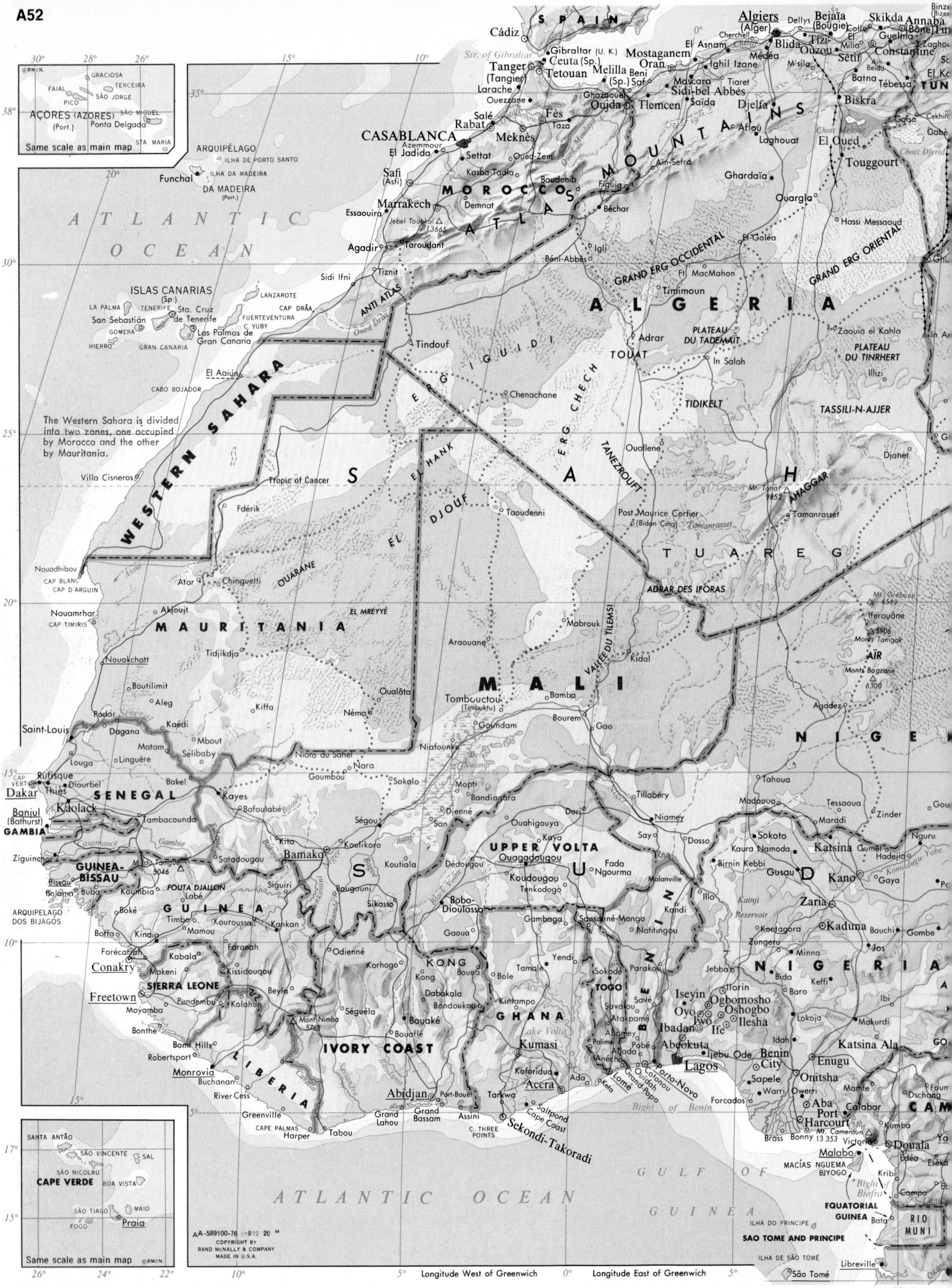

AÇORES (AZORES) (Port.)

Same scale as main map

ARQUIPÉLAGO DA MADEIRA (Port.)

ISLAS CANARIAS (Sp.)

The Western Sahara is divided into two zones, one occupied by Morocco and the other by Mauritania.

ATLANTIC OCEAN

CAPE VERDE

Same scale as main map

GULF OF GUINEA

Longitude West of Greenwich 0° Longitude East of Greenwich

Sinusoidal Projection

Elevations and depressions are given in feet

0 50 100 200 300 400 500 Miles

0 100 200 400 600 800 Kilometers

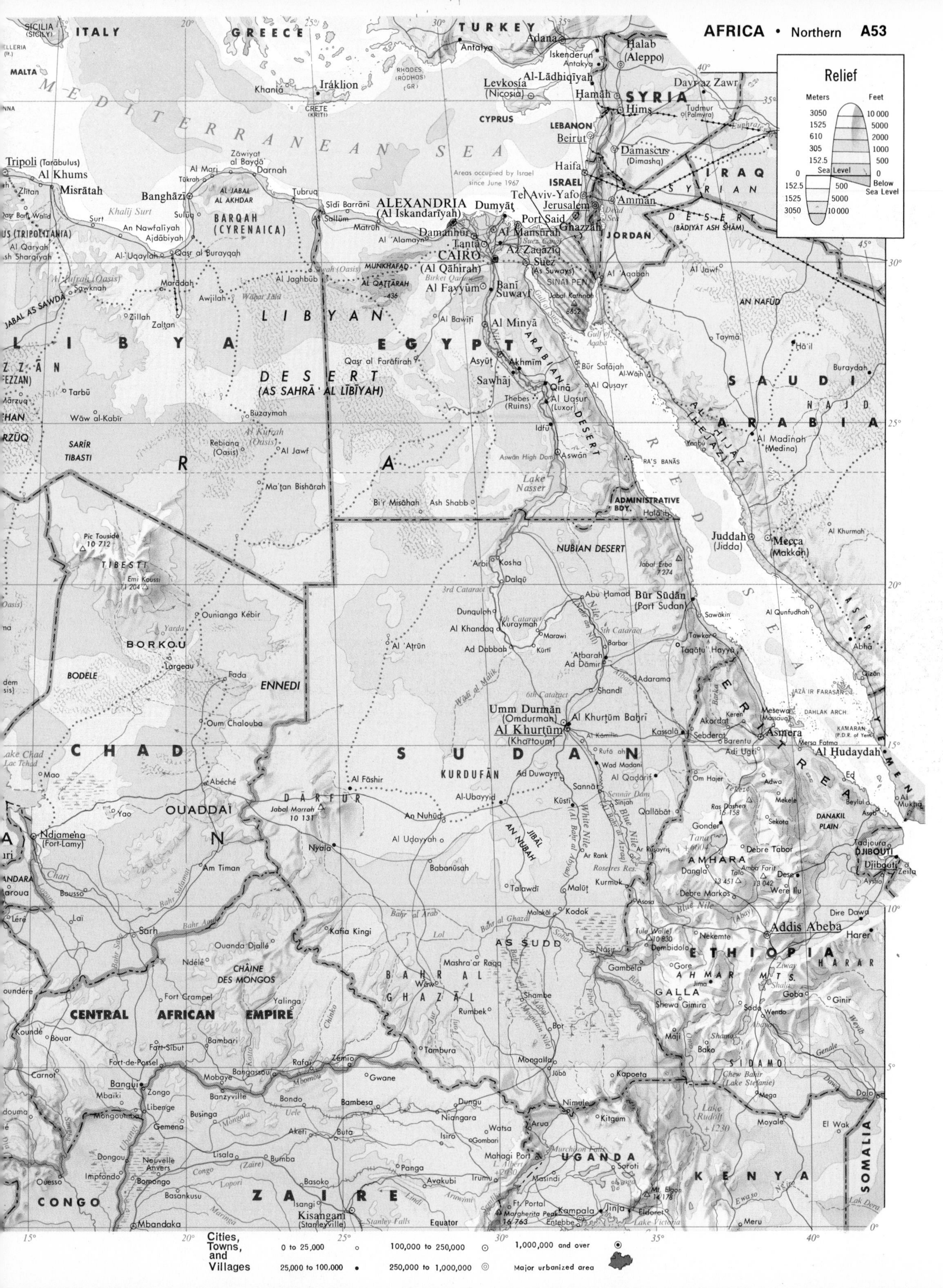
Relief
Meters
Feet
3050
1525
610
305
152.5
0
Sea Level
10 000
5000
2000
1000
500
0
152.5
1525
3050
500
5000
10 000
Below Sea Level
ITALY
GREECE
TURKEY
SICILIA (SICILY)
MALTA
MEDITERRANEAN SEA
CRETE (KRITI)
CYPRUS
RHODES (RÓDHOS) (GR)
Iráklion
Khaniá
Antalya
Adana
Iskenderun
Antakya
Halab (Aleppo)
Al-Lādhiqīyah
Levkosía (Nicosia)
Ḥamāh
Ḥimṣ
SYRIA
Dayr az Zawr
Tudmur (Palmyra)
LEBANON
Beirut
Damascus (Dimashq)
Haifa
ISRAEL
Tel Aviv-Yafo
Jerusalem
'Ammān
JORDAN
Ghazzah
IRAQ
SYRIAN DESERT (BĀDIYAT ASH SHĀM)
Areas occupied by Israel since June 1967
Tripoli (Tarābulus)
Al Khums
Misrātah
Banghāzī
Al Marj
Darnah
Zāwiyat al Bayḍā'
AL JABAL AL AKHDAR
BARQAH (CYRENAICA)
Ajdābiyah
Tubruq
Sīdī Barrānī
As Sallūm
ALEXANDRIA (Al Iskandarīyah)
Dumyāṭ
Port Said
Damanhūr
Ṭanṭā
Al Manṣūrah
Az Zaqāzīq
CAIRO (Al Qāhirah)
Suez (As Suways)
SINAI PEN.
Jabal Katrīnah 8652
Al Fayyūm
Banī Suwayf
Al Minyā
Asyūṭ
Akhmīm
Sawhāj
Qinā
Al Uqṣur (Luxor)
Thebes (Ruins)
Idfū
Aswān
Aswān High Dam
Lake Nasser
Būr Safājah
Al Quṣayr
ARABIAN DESERT
LIBYAN DESERT (AS SAHRĀ' AL LĪBĪYAH)
LIBYA
EGYPT
Qaṣr al Farāfirah
Al Jaghbūb
Siwah (Oasis)
MUNKHAFAḌ AL QAṬṬĀRAH -436
Al 'Alamayn
Maṭrūḥ
Marādah
Awjilah
Zillah
Zaltan
Buzaymah
Rebiana (Oasis)
Al Kufrah (Oasis)
Al Jawf
Ma'ṭan Bishārah
Bi'r Misāḥah
Ash Shabb
Tarbū
Wāw al-Kabīr
SARĪR TIBASTI
ADMINISTRATIVE BDY.
Ḥalā'ib
NUBIAN DESERT
Jabal Erba 7274
RA'S BANĀS
RED SEA
Juddah (Jidda)
Mecca (Makkah)
Al Madīnah (Medina)
Yanbu
AL HIJAZ (HEJAZ)
SAUDI ARABIA
NAJD
AN NAFŪD
Taymā'
Ḥā'il
Buraydah
Al Jawf
Al Khurmah
Al Qunfudhah
ASIR
Abhā
Qizān
YEMEN
Al Ḥudaydah
Al Mukhā
Pic Tousidé 10 712
TIBESTI
Emi Koussi 11 204
Ouninanga Kébir
BORKOU
BODELE
Largeau
Fada
ENNEDI
Oum Chalouba
CHAD
Lake Chad
Mao
Abéché
OUADDAÏ
Yao
Ndjamena (Fort-Lamy)
Am Timan
Bousso
Laï
Sarh
'Arbī
Kosha
Dalqū
3rd Cataract
Dunqulah
Abu Ḥamad
Būr Sūdān (Port Sudan)
Sawākin
Al Khandaq
Kuraymah
Marawi
Ad Dabbah
Kūrtī
Al 'Aṭrūn
'Aṭbarah
Ad Dāmir
Barbar
5th Cataract
Ṭawkar
Adarama
Shandī
6th Cataract
Umm Durmān (Omdurman)
Al Kharṭūm Baḥrī
Al Kharṭūm (Khartoum)
Kassalā
SUDAN
KURDUFĀN
Al Fāshir
DĀRFŪR
Jabal Marrah 10 131
Al-Ubayyiḍ
An Nuhūd
Al Uḍayyah
Nyala
Babanūsah
Ad Duwaym
Sannār
Kūstī
Sinjah
Al Qaḍārif
Qallābāt
White Nile
Blue Nile
Ar Ruṣayriṣ
Ar Rank
Roseires Res.
Kurmuk
Malūṭ
Talawdī
JIBĀL AN NUBAH
Malakāl
Kodok
Bahr al Ghazāl
Bahr al Arab
AS SUDD
Mashra' ar Raqq
BAHR AL GHAZĀL
Wāw
Rumbek
Shambe
Bor
Tambura
Mongalla
Jūbā
Kapoeta
Nimule
Kafia Kingi
Ouanda Djallé
Ndélé
CHAÎNE DES MONGOS
Fort Crampel
Yalinga
CENTRAL AFRICAN EMPIRE
Bambari
Fort-Sibut
Bouar
Fort-de-Possel
Bangassou
Mobaye
Rafai
Zémio
Gwane
Bangui
Zongo
Libenge
Banzyville
Bondo
Bambesa
Dungu
Niangara
Watsa
Gombari
Isiro
Buta
Aketi
Businga
Gemena
Mongoumba
Dongou
Nouvelle Anvers
Lisala
Bumba
Impfondo
Bomongo
Basoko
Panga
Avakubi
Irumu
Mahagi Port
Ft. Portal
Margherita Peak 16 763
Kisangani (Stanleyville)
Stanley Falls
Equator
Mbandaka
Basankusu
CONGO
ZAIRE
UGANDA
Arua
Kitgum
Soroti
Masindi
Kampala
Entebbe
Jinja
Mt. Elgon 14 178
Eldoret
Lake Victoria
KENYA
Lake Rudolf +1230
Moyale
El Wak
Meru
SOMALIA
ERITREA
Akordat
Keren
Mesewa (Massaua)
Asmera
Barentu
Adi Ugri
Sebderat
DAHLAK ARCH.
KAMARAN (P.D.R. of Yemen)
Om Hajer
Ras Dashen 15 158
Gonder
Tana
Debre Tabor
AMHARA
Dangla
Debre Markos
Dese
Were Ilu
Assab
DANAKIL PLAIN
Tadjoura
DJIBOUTI
Zeila
Dire Dawa
Harer
Addis Abeba
ETHIOPIA
HARAR
Nekemte
Dembidolo
Gore
Gambela
Nasir
AHMAR MTS.
GALLA
Jima
Shewa Gimira
Ziway
Goba
Ginir
Sodo
Maji
Bako
SIDAMO
Chew Bahir (Lake Stefanie)
Mega
Dolo
Genale
15°
20°
25°
30°
35°
40°
45°
0°
5°
10°
15°
20°
25°
30°
Cities, Towns, and Villages
0 to 25,000
25,000 to 100,000
100,000 to 250,000
250,000 to 1,000,000
1,000,000 and over
Major urbanized area

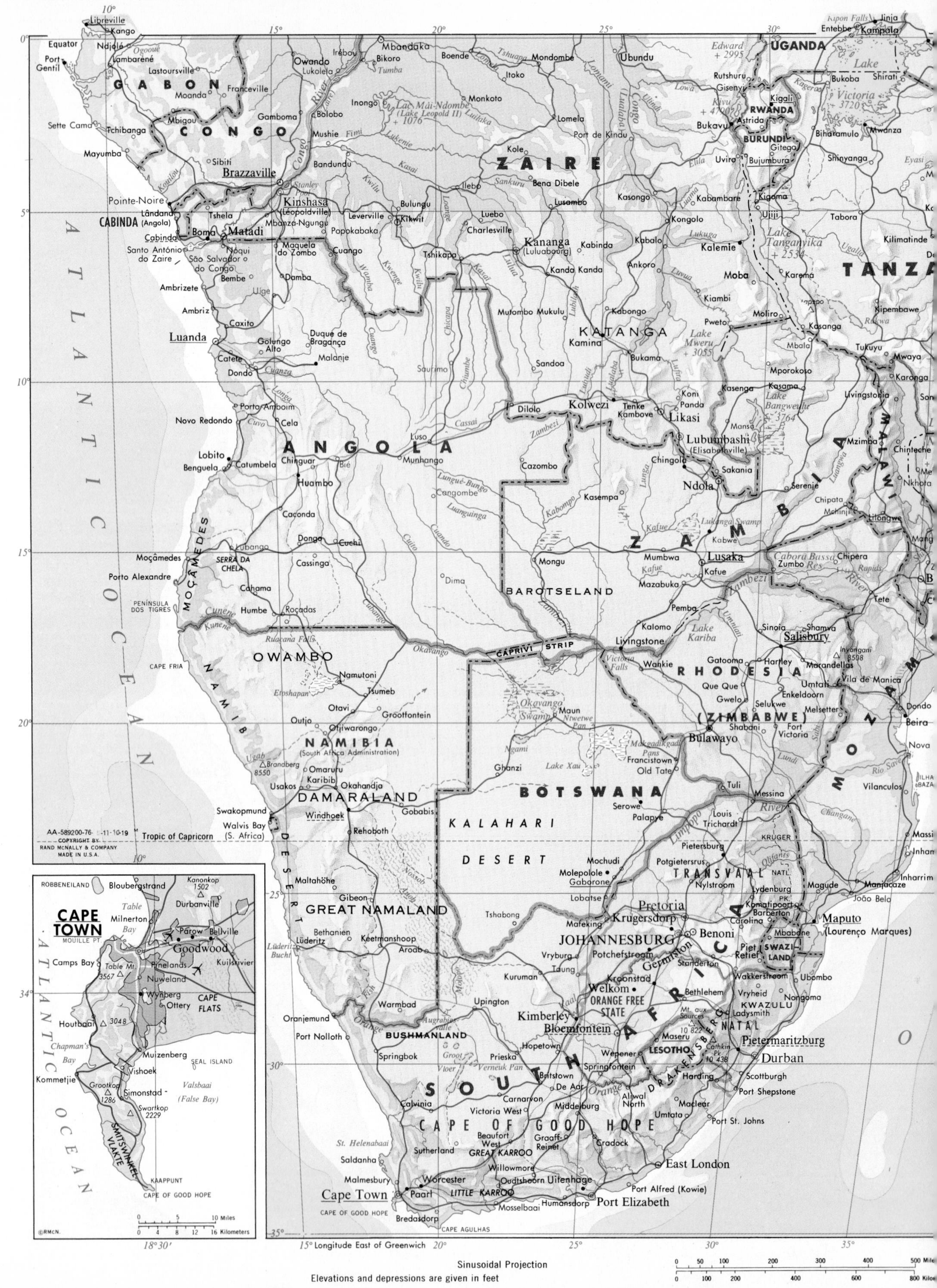
GABON
CONGO
ZAIRE
UGANDA
RWANDA
BURUNDI
TANZA
ANGOLA
CABINDA (Angola)
ZAMBIA
MALAWI
MOZAMBIQUE
RHODESIA (ZIMBABWE)
BOTSWANA
NAMIBIA
(South Africa Administration)
SOUTH AFRICA
SWAZILAND
LESOTHO
ATLANTIC OCEAN
KATANGA
BAROTSELAND
CAPRIVI STRIP
OWAMBO
DAMARALAND
GREAT NAMALAND
BUSHMANLAND
KALAHARI DESERT
NAMIB DESERT
MOÇÂMEDES
TRANSVAAL
ORANGE FREE STATE
NATAL
KWAZULU
CAPE OF GOOD HOPE
GREAT KARROO
LITTLE KARROO
DRAKENSBERG
Libreville
Kango
Equator
Port Gentil
Lambaréné
Lastoursville
Franceville
Moanda
Mbigou
Sette Cama
Tchibanga
Mayumba
Sibiti
Brazzaville
Pointe-Noire
Lândana
Cabinda
Santo António do Zaire
São Salvador do Congo
Boma
Matadi
Kinshasa
(Léopoldville)
Bandundu
Mbandaka
Boende
Inongo
Lac Mai-Ndombe
(Lake Leopold II)
Kikwit
Tshikapa
Ilebo
Luebo
Kananga
(Luluabourg)
Kabinda
Kolwezi
Likasi
Lubumbashi
(Elisabethville)
Kisangani
Ubundu
Bukavu
Uvira
Bujumbura
Kigali
Kigoma
Ujiji
Lake Tanganyika
Kalemie
Moba
Lake Mweru
Kasama
Lake Bangweulu
Ndola
Chingola
Kasempa
Mongu
Lusaka
Kafue
Livingstone
Victoria Falls
Lake Kariba
Salisbury
Bulawayo
Gwelo
Umtali
Beira
Luanda
Malanje
Lobito
Benguela
Huambo
Luso
Moçâmedes
Porto Alexandre
Novo Redondo
Cuanza
Kunene
Okavango
Zambezi
Okavango Swamp
Maun
Ghanzi
Francistown
Serowe
Gaborone
Tsumeb
Grootfontein
Otavi
Otjiwarongo
Omaruru
Karibib
Okahandja
Windhoek
Gobabis
Swakopmund
Walvis Bay
(S. Africa)
Tropic of Capricorn
Rehoboth
Maltahöhe
Gibeon
Bethanien
Keetmanshoop
Lüderitz
Warmbad
Oranjemund
Port Nolloth
Springbok
Upington
Kimberley
Bloemfontein
Welkom
Kroonstad
Johannesburg
Pretoria
Krugersdorp
Benoni
Germiston
Potchefstroom
Vryburg
Mafeking
Pietersburg
Louis Trichardt
Messina
Nylstroom
Lydenburg
Barberton
Maputo
(Lourenço Marques)
Pietermaritzburg
Durban
Ladysmith
Maseru
Port Shepstone
Umtata
Port St. Johns
East London
Port Elizabeth
Uitenhage
Port Alfred (Kowie)
Grahamstown
Cradock
Graaff-Reinet
Beaufort West
Oudtshoorn
Mosselbaai
Worcester
Paarl
Cape Town
Saldanha
Malmesbury
Calvinia
Victoria West
De Aar
Middelburg
Bredasdorp
CAPE AGULHAS
CAPE OF GOOD HOPE
AA-589200-76 -11-10-19
COPYRIGHT BY RAND McNALLY & COMPANY MADE IN U.S.A.
CAPE TOWN
ROBBENEILAND
Bloubergstrand
Table Bay
Milnerton
Durbanville
Parow
Bellville
Goodwood
Camps Bay
Table Mt. 3567
Pinelands
Nuweland
Wynberg
Ottery
CAPE FLATS
Kuilsrivier
Houtbaai
Chapman's Bay
Muizenberg
SEAL ISLAND
Vishoek
Kommetjie
Simonstad
Valsbaai (False Bay)
Swartkop 2229
SMITSWINKEL VLAKTE
KAAPPUNT
CAPE OF GOOD HOPE
0 5 10 Miles
0 4 8 12 16 Kilometers
18°30′
©RMcN.
15° Longitude East of Greenwich
Sinusoidal Projection
Elevations and depressions are given in feet
0 50 100 200 300 400 500 Miles
0 100 200 400 600 800 Kilometers

SOMALIA
Kismayu
Bur Gavo
Equator
Lamu
Witu
Malindi
Takaungu
Mombasa
Vanga
Tanga
Pangani
PEMBA ISLAND
ZANZIBAR
Zanzibar
Bagamoyo
Dar es Salaam
Morogoro
Kisaki
MAFIA
Utete
Kilwa Kivinje
Lindi
Mikindani
CABO DELGADO
Masasi
Mocímboa da Praia
Ibo
Porto Amélia
Lúrio
Memba
Nacala
Moçambique
António Enes
ILHA ANGOCHE
I N D I A N
ALDABRA IS. (Sey.)
COSMOLEDO GROUP (Sey.)
Moroni
GRANDE COMORE
COMOROS
MOHÉLI
ANJOUAN
ÎLES GLORIEUSES (Fr.)
CAP D'AMBRE
Diégo-Suarez
Dzaoudzi
MAYOTTE (Fr.)
NOSSI BÉ
Vohémar
Maromokotro 9436
Majunga
Mandritsara
Maroantsetra
Baie d'Antongil
ÎLE SAINTE-MARIE
CAP SAINT-ANDRÉ
Besalampy
ÎLE JUAN DE NOVA (Fr.)
Maintirano
ÎLES BARREN
Alaotra
Ambatondrazaka
Fénérive
Tamatave
Moramanga
Antananarivo
Vatomandry
Tsiafajavona 8671
Tsiribihina
Mahanoro
Antsirabe
Ambositra
Morondava
Fianarantsoa
Mananjary
BASSAS DA INDIA (Mad.)
Morombe
Mangoky
EUROPA (Mad.)
Manakara
Ivohibé
Farafangana
Tuléar
Betroka
Onilahy
Mananara
Mahaly
Trafonomby 6417
Fort-Dauphin
CAP STE. MARIE
MADAGASCAR
MOZAMBIQUE CHANNEL
Wolhuterskop
Jacksonstuin
MAGALIESBERG
Hartbeespoort
Kosmos
Hartbeespoortdam
Pretoria North
Pretoria
Swartspruit
Silverton
Rayton
Cullinan
Skeerpoort
Voortrekkerhoogte
Valhalla
Hennopsrivier
Lyttelton
Irene
Tierpoort
Foothills
WITWATERSBERG
Olievenhoutpoort
Halfway House
Bapsfontein
Kaalfontein
Tarlton
Krugersdorp
Modderfontein
Kempton Park
JOHANNESBURG
Alexandra
Putfontein
Randfontein
Roodepoort
Discovery
Edenvale
Florida
Maraisburg
Primrose
Boksburg
Benoni
Brakpan
Orlando
Turffontein
Rosettenville
Germiston
Alberton
Pimville
Springs
WITWATERSRAND
Arlington
Dannhauser
Dundee
Senekal
Paul Roux
Bethlehem
Harrismith
Glencoe
Wasbank
Nqutu
Mahlabatini
Kestell
ORANGE FREE STATE
ROYAL NATAL NAT'L. PK.
Clarens
Babanango
Fouriesburg
Ladysmith
Pomeroy
Nkandla
Melmoth
Ficksburg
Butha Buthe
Leribe
Mt. aux Sources
Bergville
Winterton
Colenso
Tugela Ferry
Weenen
Eshowe
Clocolan
Cathedral Pk. 9856
Estcourt
Kranskop
Greytown
Pitseng
MALOTI MTS.
Cathkin Pk. 10438
Mapumulo
Teyateyaneng
Mokhotlong
Mooirivier
Mt. Gilboa 5803
Stanger
Machache 9464
LESOTHO
Thabana Ntlenyana 11425
New Hanover
Dalton
Wartburg
Roma
Howick
Impendle
Nshoni
Pietermaritzburg
Camperdown
Verulam
Bulwer
Richmond
Pinetown
Durban
Underberg
Donnybrook
Creighton
Isipingo
Swartberg 7619
Mid Illovo
Mohale's Hoek
The Twins 8820
Qacha's Nek
Ixopo
Franklin
Umzimkulu
Umkomaas
Zastron
Matatiele
Umzinto
Scottburgh
Cedarville
Mt. Currie 7297
Park Rynie
Quthing
Sezela
Kokstad
Harding
Umtentweni
Herschel
Witberg 7853
Mount Fletcher
Ben Macdhui 9846
Mount Ayliff
Port Shepstone
Uvongo Beach
Margate
Lady Grey
Rhodes
Bizana
Barkly East
Mount Frere
Tabankulu
Port Edward
Jamestown
Rossouw 8430
Ugie
Maclear
Flagstaff
Qumbu
Elliot
Tsolo
Lusikisiki
Molteno
STORMBERG
Dordrecht
Sterkstroom
Indwe
Cala
Umtata
Libode
Ngqeleni
Port St. Johns
Lady Frere
Engcobo
RAME HEAD
Queenstown
Waverly
Mqanduli
Elliotdale
Tarkastad
Tylden
Cofimvaba
Tsomo
Idutywa
Whittlesea
Nqamakwe
Willowvale
Cradock
BANKBERG 6606
WINTERBERGE 7778
Cathcart
Butterworth
Kentani
Seymour
Stutterheim
Pearston
Frankfort
Komga
Adelaide
Keiskammahoek
Kei Mouth
Morgan's Bay
Somerset East
Bedford
Mgcleantown
Fort Beaufort
Alice
King William's Town
Berlin
Breidbach
Gonubie
East London
Kidd's Beach
SUURBERGE
Riebeek-Oos
Peddie
Alicedale
Grahamstown
Hamburg
Kirkwood
Salem
Addo
Bathurst
Alexandria
Port Alfred (Kowie)
Uitenhage
SAINT CROIX ISLAND
BIRD ISLAND
Port Elizabeth
KAAP RECIFE
SOUTH AFRICA
NATAL
CAPE OF GOOD HOPE
INDIAN OCEAN
Relief
Meters
Feet
3050
1525
610
305
152.5
Sea Level
10 000
5000
2000
1000
500
Longitude East of Greenwich
Cities, Towns, and Villages
0 to 25,000
25,000 to 100,000
100,000 to 250,000
250,000 to 1,000,000
1,000,000 and over
Major urbanized area
©RMCN.

POPULATION DENSITY
Arctic Circle
Tropic of Cancer
Equator
Longitude West of Greenwich
Tropic of Capricorn
Meridian of Greenwich
Antarctic Circle
Vancouver
Seattle
Portland
San Francisco
Los Angeles
San Diego
Phoenix
Denver
Minneapolis
Milwaukee
Chicago
Kansas City
St. Louis
Indianapolis
Toronto
Montreal
Boston
Hartford
New York
Philadelphia
Washington
Dallas
Atlanta
New Orleans
Miami
Havana
Mexico D. F.
Caracas
Medellin
Bogota
Lima
Fortaleza
Recife
Salvador
Belo Horizonte
Rio de Janeiro
São Paulo
Porto Alegre
Santiago
Buenos Aires
Montevideo
Glasgow
London
Paris
Hamburg
Leningrad
Stockholm
Minsk
Warsaw
Kiev
Vienna
Lyon
Milan
Madrid
Barcelona
Rome
Naples
Bucharest
Sofia
Istanbul
Athens
Lisbon
Algiers
Casablanca
Alexandria
Cairo
Lagos
Kinshasa
Johannesburg
Cape Town
0 500 1000 1500 Miles
0 500 1000 1500 2000 Kilometers
Population Density
per square kilometer (per square mile)
of Total Area
of Cultivated Land
9 (24) ARGENTINA 74 (192)
2 (5) AUSTRALIA 30 (79)
13 (33) BRAZIL 301 (780)
87 (229) CHINA 650 (1684)
97 (252) FRANCE 291 (728)
156 (653) GERMANY 775 (2008)
150 (429) INDIA & PAKISTAN 366 (949)
298 (777) JAPAN 1989 (5152)
11 (40) SOVIET UNION 183(473)
229 (611) UNITED KINGDOM 771 (1997)
23 (58) UNITED STATES 104 (269)

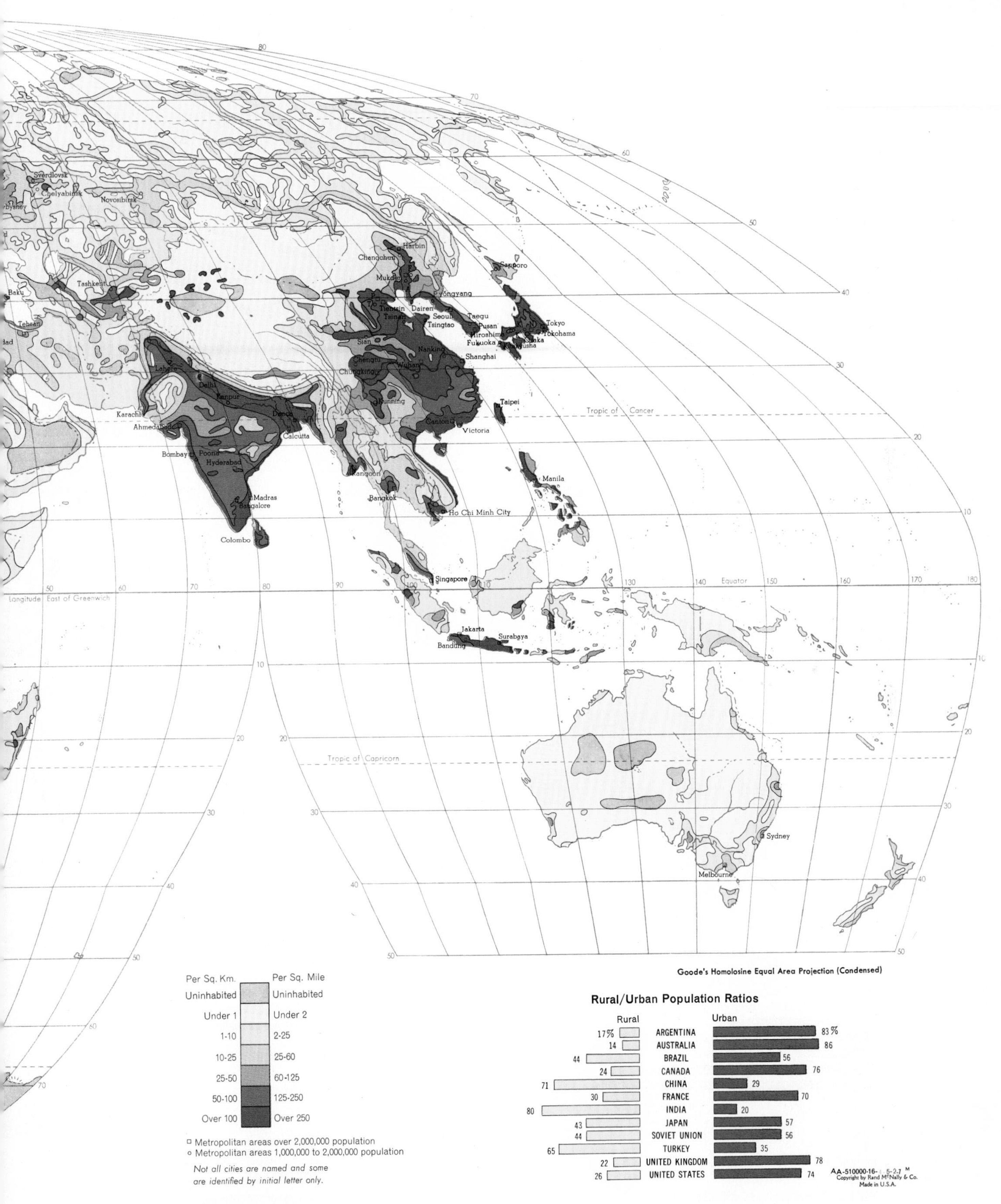

Sverdlovsk
Chelyabinsk
Novosibirsk
Tashkent
Baku
Tehran
Harbin
Changchun
Mukden
Sapporo
P'yongyang
Tientsin
Tsinan
Dairen
Seoul
Tsingtao
Taegu
Pusan
Hiroshima
Fukuoka
Tokyo
Yokohama
Sian
Nanking
Shanghai
Chengtu
Wuhan
Chungking
Lahore
Delhi
Kanpur
Kunming
Taipei
Karachi
Ahmedabad
Dacca
Calcutta
Canton
Victoria
Tropic of Cancer
Bombay
Poona
Hyderabad
Rangoon
Manila
Madras
Bangalore
Bangkok
Ho Chi Minh City
Colombo
Singapore
Equator
Longitude East of Greenwich
Jakarta
Bandung
Surabaya
Tropic of Capricorn
Sydney
Melbourne
Goode's Homolosine Equal Area Projection (Condensed)
Per Sq. Km.
Per Sq. Mile
Uninhabited
Uninhabited
Under 1
Under 2
1-10
2-25
10-25
25-60
25-50
60-125
50-100
125-250
Over 100
Over 250
□ Metropolitan areas over 2,000,000 population
○ Metropolitan areas 1,000,000 to 2,000,000 population
Not all cities are named and some are identified by initial letter only.
Rural/Urban Population Ratios
Rural
Urban
17% ARGENTINA 83%
14 AUSTRALIA 86
44 BRAZIL 56
24 CANADA 76
71 CHINA 29
30 FRANCE 70
80 INDIA 20
43 JAPAN 57
44 SOVIET UNION 56
65 TURKEY 35
22 UNITED KINGDOM 78
26 UNITED STATES 74
AA-510000-16-
Copyright by Rand McNally & Co.
Made in U.S.A.

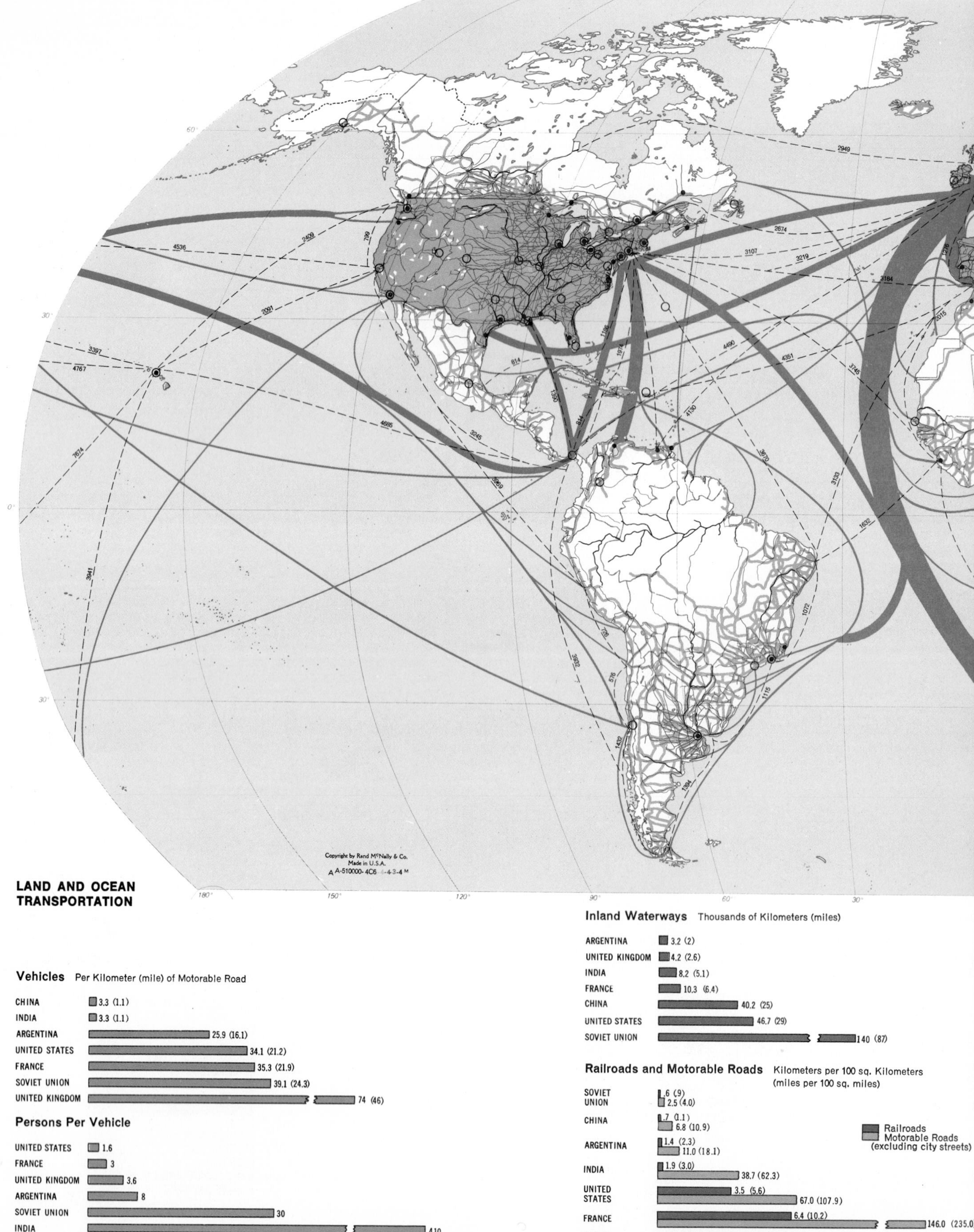

LAND AND OCEAN TRANSPORTATION

Vehicles Per Kilometer (mile) of Motorable Road

CHINA	3.3 (1.1)
INDIA	3.3 (1.1)
ARGENTINA	25.9 (16.1)
UNITED STATES	34.1 (21.2)
FRANCE	35.3 (21.9)
SOVIET UNION	39.1 (24.3)
UNITED KINGDOM	74 (46)

Persons Per Vehicle

UNITED STATES	1.6
FRANCE	3
UNITED KINGDOM	3.6
ARGENTINA	8
SOVIET UNION	30
INDIA	410
CHINA	1134

Inland Waterways Thousands of Kilometers (miles)

ARGENTINA	3.2 (2)
UNITED KINGDOM	4.2 (2.6)
INDIA	8.2 (5.1)
FRANCE	10.3 (6.4)
CHINA	40.2 (25)
UNITED STATES	46.7 (29)
SOVIET UNION	140 (87)

Railroads and Motorable Roads Kilometers per 100 sq. Kilometers (miles per 100 sq. miles)

	Railroads	Motorable Roads (excluding city streets)
SOVIET UNION	.6 (.9)	2.5 (4.0)
CHINA	.7 (1.1)	6.8 (10.9)
ARGENTINA	1.4 (2.3)	11.0 (18.1)
INDIA	1.9 (3.0)	38.7 (62.3)
UNITED STATES	3.5 (5.6)	67.0 (107.9)
FRANCE	6.4 (10.2)	146.0 (235.0)
UNITED KINGDOM	7.5 (12.0)	142.0 (227.7)

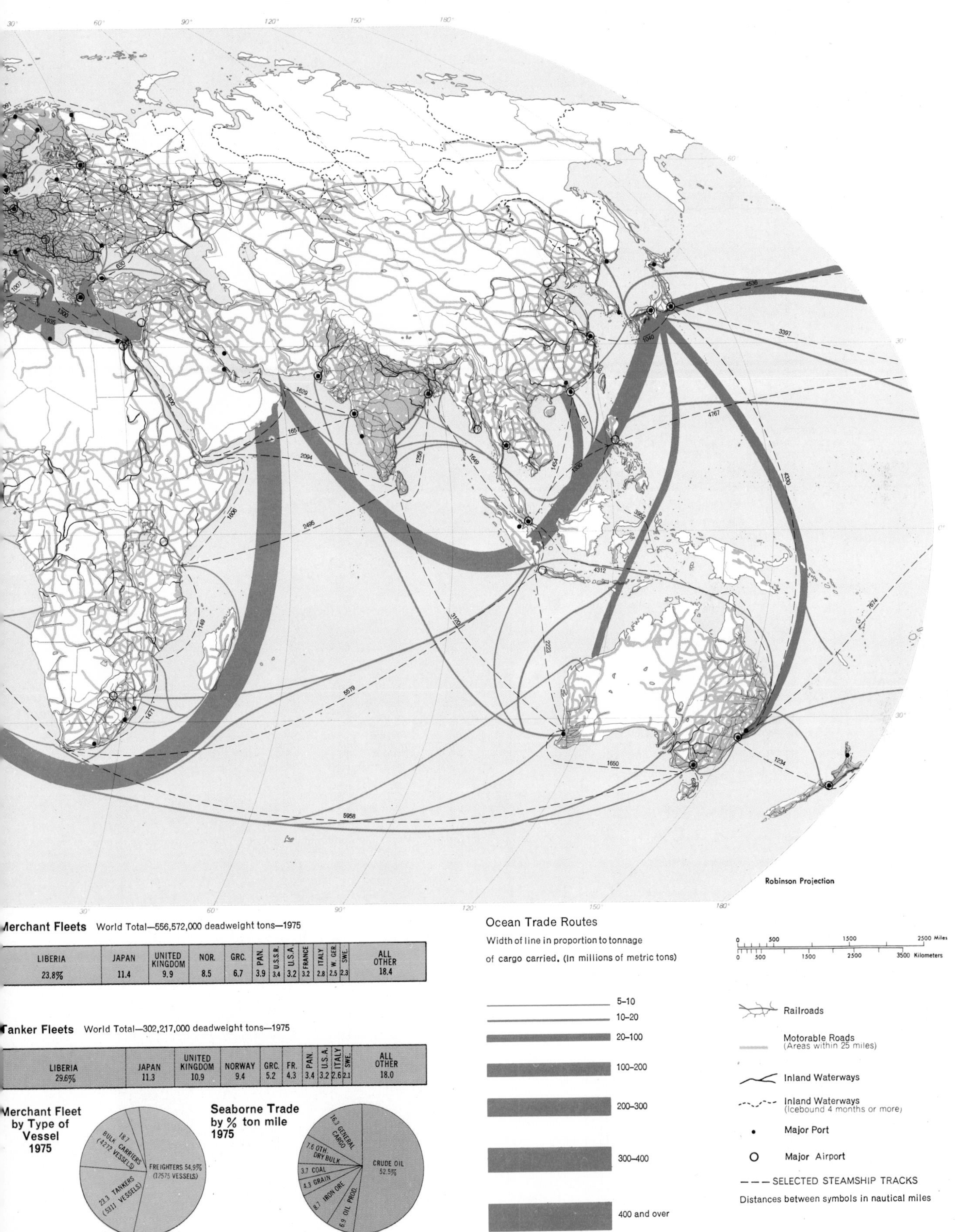

Merchant Fleets World Total—556,572,000 deadweight tons—1975

LIBERIA	JAPAN	UNITED KINGDOM	NOR.	GRC.	PAN.	U.S.S.R.	U.S.A.	FRANCE	ITALY	W. GER.	SWE.	ALL OTHER
23.8%	11.4	9.9	8.5	6.7	3.9	3.4	3.2	3.2	2.8	2.5	2.3	18.4

Tanker Fleets World Total—302,217,000 deadweight tons—1975

LIBERIA	JAPAN	UNITED KINGDOM	NORWAY	GRC.	FR.	PAN.	U.S.A.	ITALY	SWE.	ALL OTHER
29.6%	11.3	10.9	9.4	5.2	4.3	3.4	3.2	2.6	2.1	18.0

Merchant Fleet by Type of Vessel 1975

World Total—556,572,000 deadweight tons—1975

Seaborne Trade by % ton mile 1975

Ocean Trade Routes

Width of line in proportion to tonnage of cargo carried. (In millions of metric tons)

- 5–10
- 10–20
- 20–100
- 100–200
- 200–300
- 300–400
- 400 and over

- Railroads
- Motorable Roads (Areas within 25 miles)
- Inland Waterways
- Inland Waterways (Icebound 4 months or more)
- Major Port
- Major Airport
- SELECTED STEAMSHIP TRACKS

Distances between symbols in nautical miles

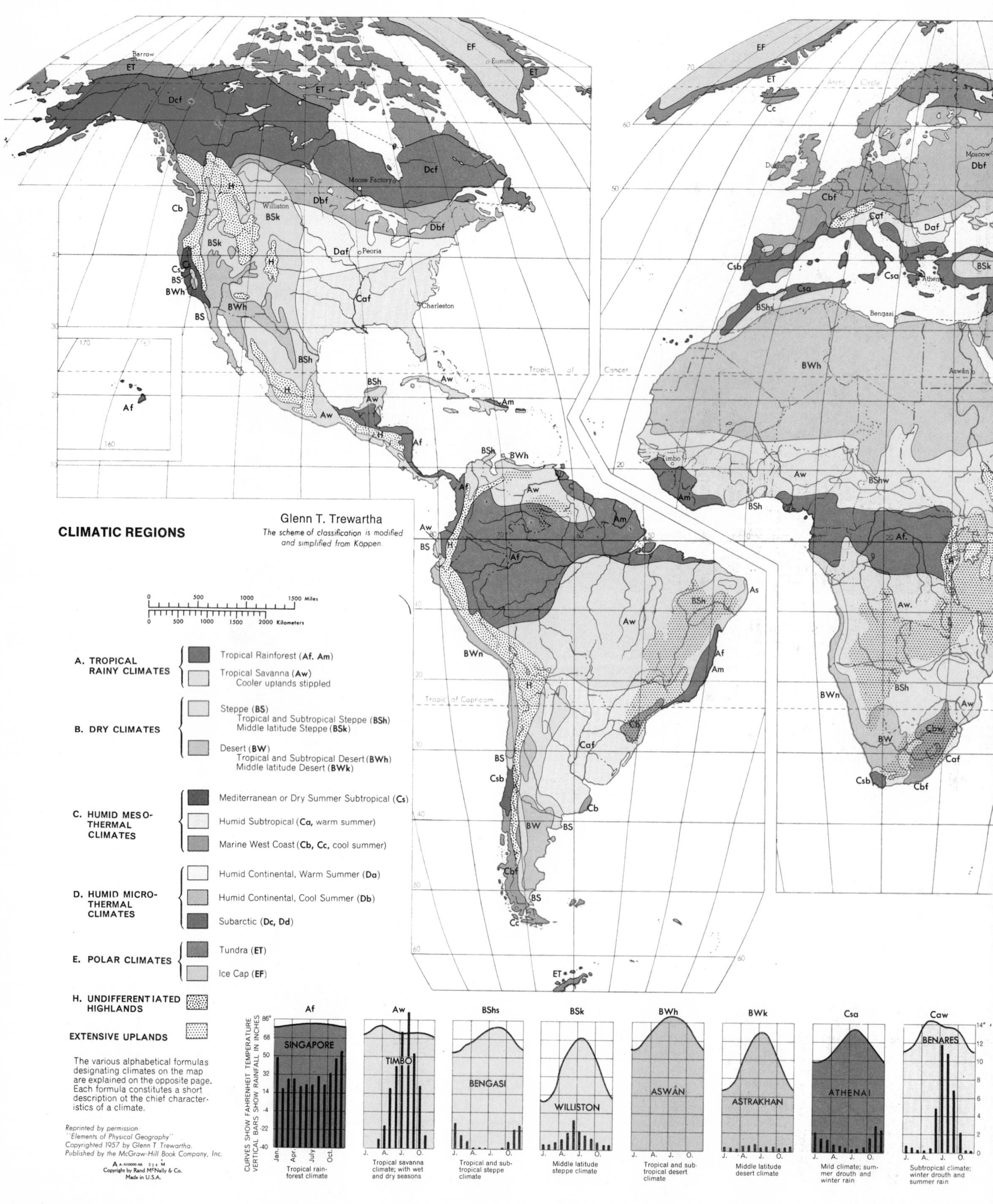
CLIMATIC REGIONS
Glenn T. Trewartha
The scheme of classification is modified and simplified from Köppen.
0 500 1000 1500 Miles
0 500 1000 1500 2000 Kilometers
A. TROPICAL RAINY CLIMATES
Tropical Rainforest (Af. Am)
Tropical Savanna (Aw)
Cooler uplands stippled
B. DRY CLIMATES
Steppe (BS)
Tropical and Subtropical Steppe (BSh)
Middle latitude Steppe (BSk)
Desert (BW)
Tropical and Subtropical Desert (BWh)
Middle latitude Desert (BWk)
C. HUMID MESO-THERMAL CLIMATES
Mediterranean or Dry Summer Subtropical (Cs)
Humid Subtropical (Ca, warm summer)
Marine West Coast (Cb, Cc, cool summer)
D. HUMID MICRO-THERMAL CLIMATES
Humid Continental, Warm Summer (Da)
Humid Continental, Cool Summer (Db)
Subarctic (Dc, Dd)
E. POLAR CLIMATES
Tundra (ET)
Ice Cap (EF)
H. UNDIFFERENTIATED HIGHLANDS
EXTENSIVE UPLANDS
The various alphabetical formulas designating climates on the map are explained on the opposite page. Each formula constitutes a short description of the chief characteristics of a climate.
Reprinted by permission
"Elements of Physical Geography"
Copyrighted 1957 by Glenn T. Trewartha.
Published by the McGraw-Hill Book Company, Inc.
Copyright by Rand McNally & Co.
Made in U.S.A.
Barrow
Eismitte
Moose Factory
Williston
Peoria
Charleston
Dublin
Moscow
Athenai
Bengasi
Aswân
Timbo
Tropic of Cancer
Tropic of Capricorn
Arctic Circle
CURVES SHOW FAHRENHEIT TEMPERATURE
VERTICAL BARS SHOW RAINFALL IN INCHES
Af
SINGAPORE
Tropical rain-forest climate
Aw
TIMBO
Tropical savanna climate; with wet and dry seasons
BShs
BENGASI
Tropical and sub-tropical steppe climate
BSk
WILLISTON
Middle latitude steppe climate
BWh
ASWÂN
Tropical and sub-tropical desert climate
BWk
ASTRAKHAN
Middle latitude desert climate
Csa
ATHENAI
Mild climate; summer drouth and winter rain
Caw
BENARES
Subtropical climate; winter drouth and summer rain

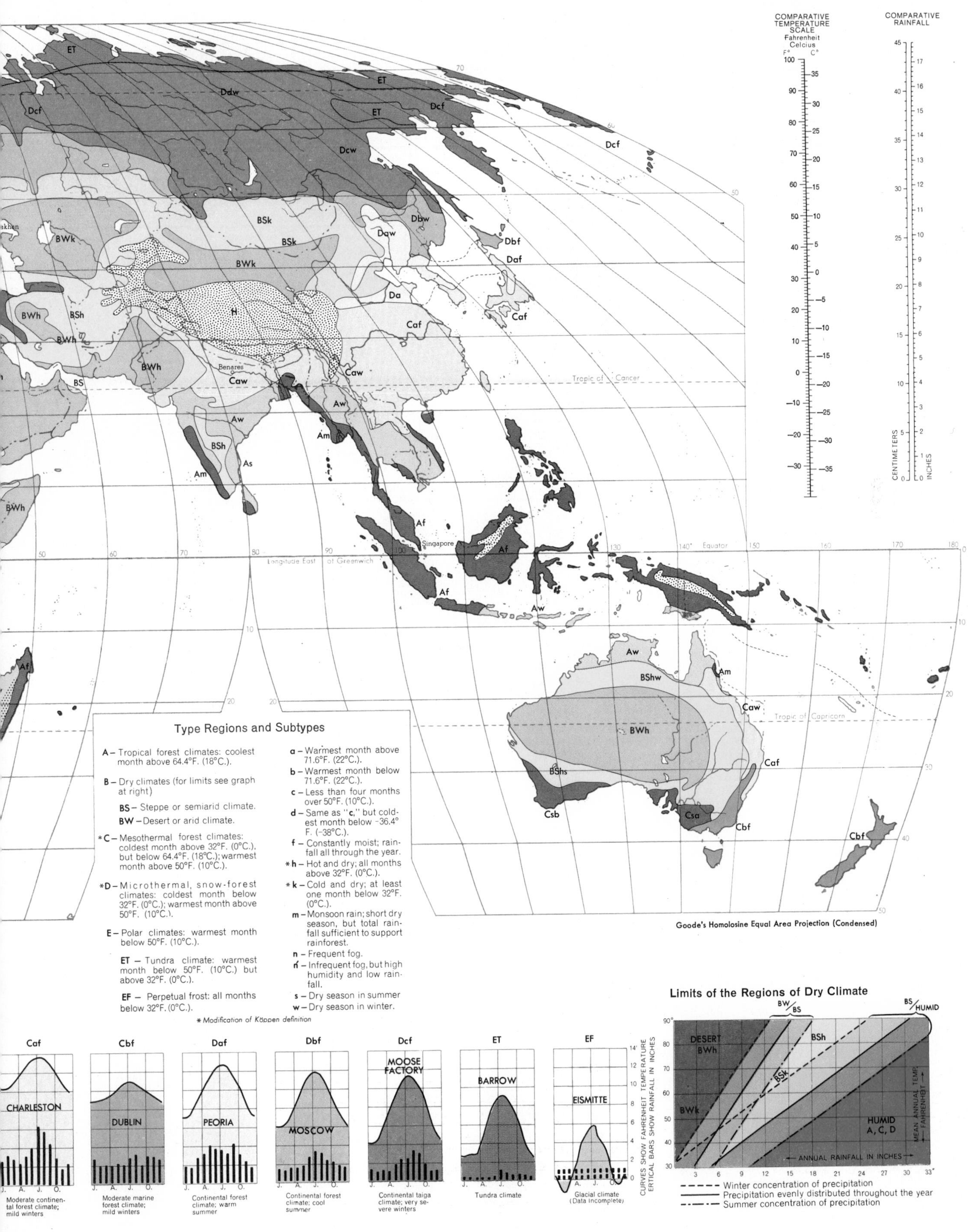

COMPARATIVE TEMPERATURE SCALE
Fahrenheit
Celcius
COMPARATIVE RAINFALL
CENTIMETERS
INCHES
Tropic of Cancer
Equator
Tropic of Capricorn
Longitude East of Greenwich
Singapore
Benares
Type Regions and Subtypes
A – Tropical forest climates: coolest month above 64.4°F. (18°C.).
B – Dry climates (for limits see graph at right)
BS – Steppe or semiarid climate.
BW – Desert or arid climate.
*C – Mesothermal forest climates: coldest month above 32°F. (0°C.), but below 64.4°F. (18°C.); warmest month above 50°F. (10°C.).
*D – Microthermal, snow-forest climates: coldest month below 32°F. (0°C.); warmest month above 50°F. (10°C.).
E – Polar climates: warmest month below 50°F. (10°C.).
ET – Tundra climate: warmest month below 50°F. (10°C.) but above 32°F. (0°C.).
EF – Perpetual frost: all months below 32°F. (0°C.).
a – Warmest month above 71.6°F. (22°C.).
b – Warmest month below 71.6°F. (22°C.).
c – Less than four months over 50°F. (10°C.).
d – Same as "c," but coldest month below -36.4° F. (-38°C.).
f – Constantly moist; rainfall all through the year.
*h – Hot and dry; all months above 32°F. (0°C.).
*k – Cold and dry; at least one month below 32°F. (0°C.).
m – Monsoon rain; short dry season, but total rainfall sufficient to support rainforest.
n – Frequent fog.
n' – Infrequent fog, but high humidity and low rainfall.
s – Dry season in summer
w – Dry season in winter.
* Modification of Köppen definition
Goode's Homolosine Equal Area Projection (Condensed)
Limits of the Regions of Dry Climate
BW/BS
BS/HUMID
DESERT BWh
BWk
BSh
BSk
HUMID A, C, D
MEAN ANNUAL TEMP. FAHRENHEIT
ANNUAL RAINFALL IN INCHES
– – – – Winter concentration of precipitation
——— Precipitation evenly distributed throughout the year
–·–·– Summer concentration of precipitation
Caf
CHARLESTON
Moderate continental forest climate; mild winters
Cbf
DUBLIN
Moderate marine forest climate; mild winters
Daf
PEORIA
Continental forest climate; warm summer
Dbf
MOSCOW
Continental forest climate; cool summer
Dcf
MOOSE FACTORY
Continental taiga climate; very severe winters
ET
BARROW
Tundra climate
EF
EISMITTE
Glacial climate (Data incomplete)
CURVES SHOW FAHRENHEIT TEMPERATURE
VERTICAL BARS SHOW RAINFALL IN INCHES

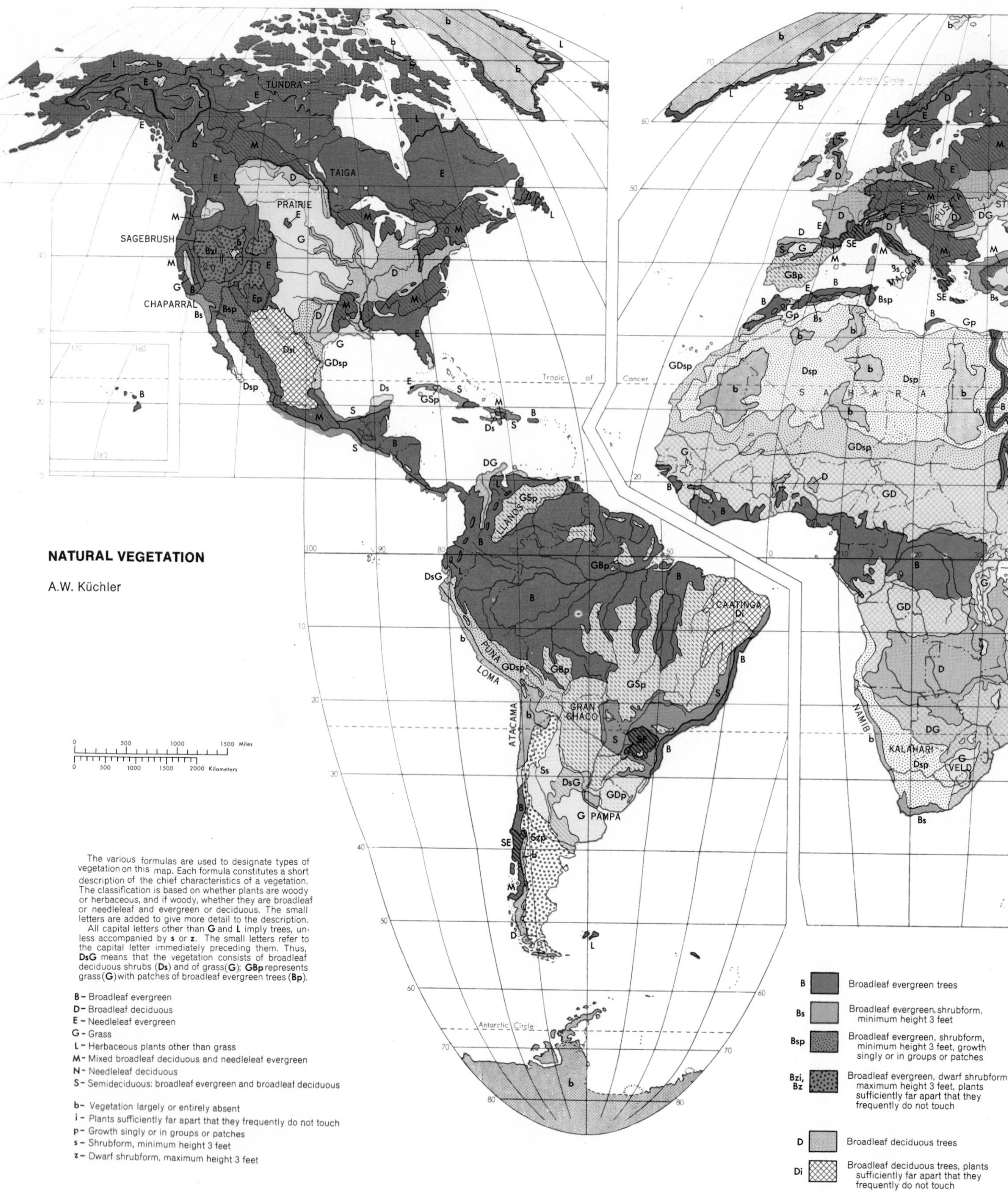

NATURAL VEGETATION

A.W. Küchler

The various formulas are used to designate types of vegetation on this map. Each formula constitutes a short description of the chief characteristics of a vegetation. The classification is based on whether plants are woody or herbaceous, and if woody, whether they are broadleaf or needleleaf and evergreen or deciduous. The small letters are added to give more detail to the description.

All capital letters other than **G** and **L** imply trees, unless accompanied by **s** or **z**. The small letters refer to the capital letter immediately preceding them. Thus, **DsG** means that the vegetation consists of broadleaf deciduous shrubs (**Ds**) and of grass(**G**); **GBp** represents grass(**G**) with patches of broadleaf evergreen trees (**Bp**).

- **B** - Broadleaf evergreen
- **D** - Broadleaf deciduous
- **E** - Needleleaf evergreen
- **G** - Grass
- **L** - Herbaceous plants other than grass
- **M** - Mixed broadleaf deciduous and needleleaf evergreen
- **N** - Needleleaf deciduous
- **S** - Semideciduous: broadleaf evergreen and broadleaf deciduous

- **b** - Vegetation largely or entirely absent
- **i** - Plants sufficiently far apart that they frequently do not touch
- **p** - Growth singly or in groups or patches
- **s** - Shrubform, minimum height 3 feet
- **z** - Dwarf shrubform, maximum height 3 feet

B	Broadleaf evergreen trees
Bs	Broadleaf evergreen, shrubform, minimum height 3 feet
Bsp	Broadleaf evergreen, shrubform, minimum height 3 feet, growth singly or in groups or patches
Bzi, Bz	Broadleaf evergreen, dwarf shrubform, maximum height 3 feet, plants sufficiently far apart that they frequently do not touch
D	Broadleaf deciduous trees
Di	Broadleaf deciduous trees, plants sufficiently far apart that they frequently do not touch

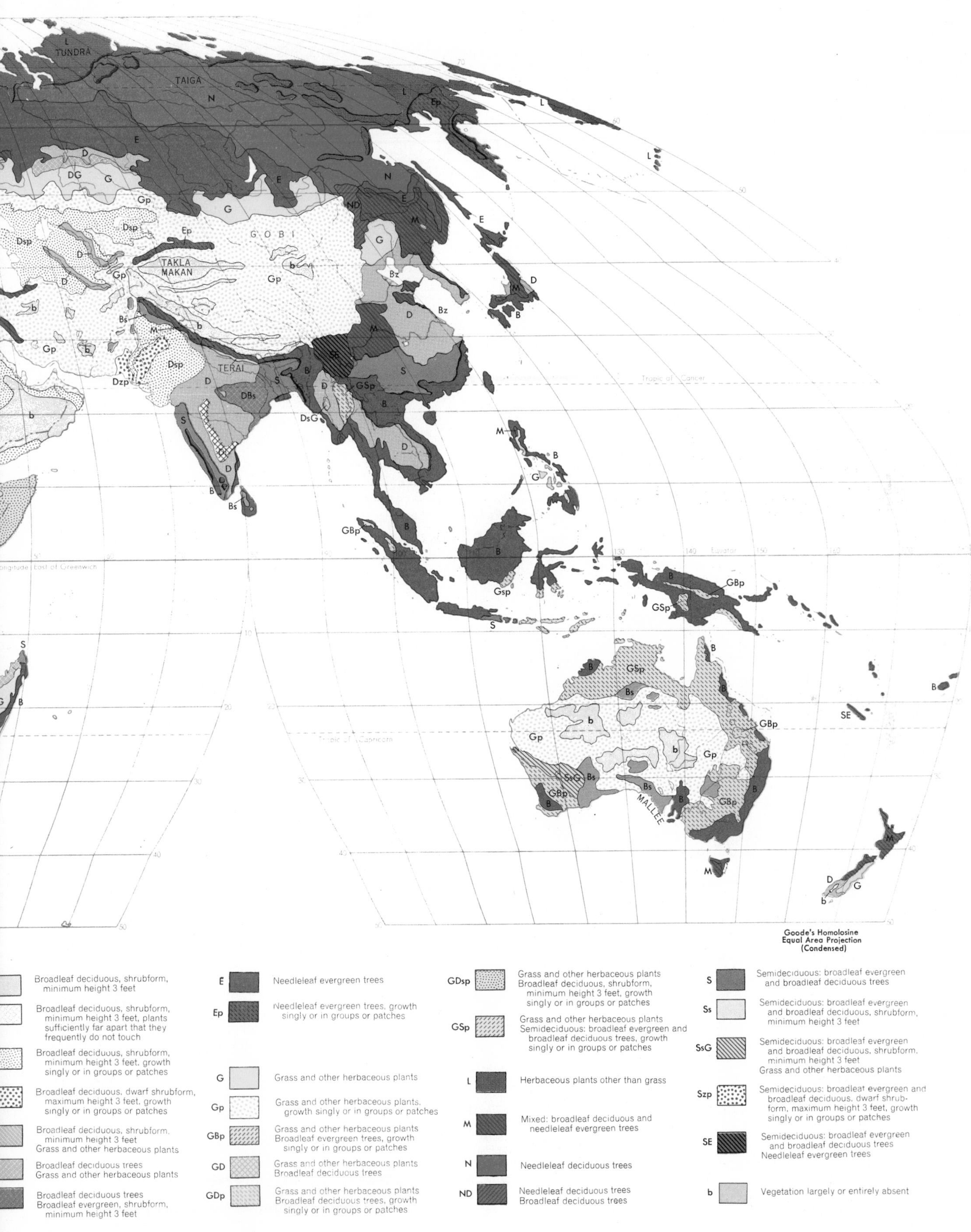
TUNDRA
TAIGA
GOBI
TAKLA MAKAN
TERAI
MALLEE
Tropic of Cancer
Tropic of Capricorn
Equator
Longitude East of Greenwich
Goode's Homolosine Equal Area Projection (Condensed)
Broadleaf deciduous, shrubform, minimum height 3 feet
Broadleaf deciduous, shrubform, minimum height 3 feet, plants sufficiently far apart that they frequently do not touch
Broadleaf deciduous, shrubform, minimum height 3 feet, growth singly or in groups or patches
Broadleaf deciduous, dwarf shrubform, maximum height 3 feet, growth singly or in groups or patches
Broadleaf deciduous, shrubform, minimum height 3 feet
Grass and other herbaceous plants
Broadleaf deciduous trees
Grass and other herbaceous plants
Broadleaf deciduous trees
Broadleaf evergreen, shrubform, minimum height 3 feet
E Needleleaf evergreen trees
Ep Needleleaf evergreen trees, growth singly or in groups or patches
G Grass and other herbaceous plants
Gp Grass and other herbaceous plants, growth singly or in groups or patches
GBp Grass and other herbaceous plants
Broadleaf evergreen trees, growth singly or in groups or patches
GD Grass and other herbaceous plants
Broadleaf deciduous trees
GDp Grass and other herbaceous plants
Broadleaf deciduous trees, growth singly or in groups or patches
GDsp Grass and other herbaceous plants
Broadleaf deciduous, shrubform, minimum height 3 feet, growth singly or in groups or patches
GSp Grass and other herbaceous plants
Semideciduous: broadleaf evergreen and broadleaf deciduous trees, growth singly or in groups or patches
L Herbaceous plants other than grass
M Mixed: broadleaf deciduous and needleleaf evergreen trees
N Needleleaf deciduous trees
ND Needleleaf deciduous trees
Broadleaf deciduous trees
S Semideciduous: broadleaf evergreen and broadleaf deciduous trees
Ss Semideciduous: broadleaf evergreen and broadleaf deciduous, shrubform, minimum height 3 feet
SsG Semideciduous: broadleaf evergreen and broadleaf deciduous, shrubform, minimum height 3 feet
Grass and other herbaceous plants
Szp Semideciduous: broadleaf evergreen and broadleaf deciduous, dwarf shrubform, maximum height 3 feet, growth singly or in groups or patches
SE Semideciduous: broadleaf evergreen and broadleaf deciduous trees
Needleleaf evergreen trees
b Vegetation largely or entirely absent

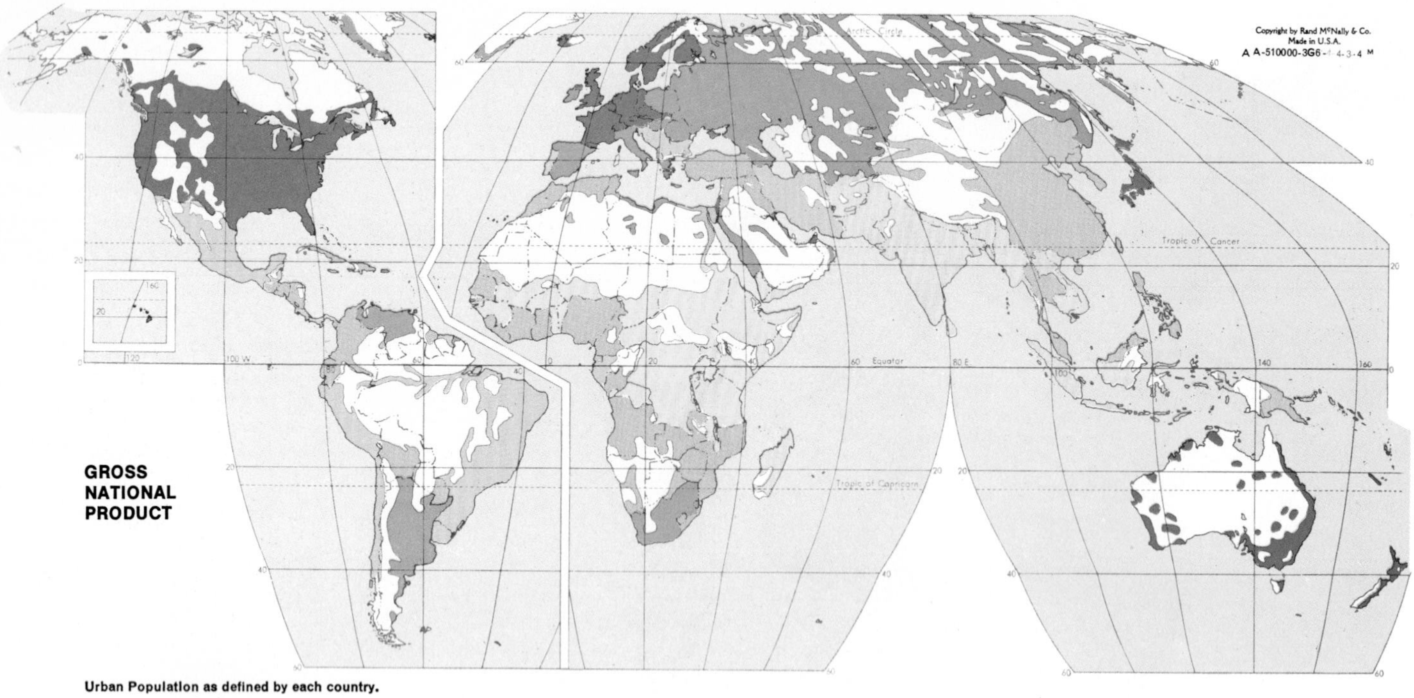

Copyright by Rand McNally & Co.
Made in U.S.A.
A A-510000-3G6-1-4-3-4 M
Arctic Circle
Tropic of Cancer
Equator
Tropic of Capricorn
GROSS NATIONAL PRODUCT

Urban Population as defined by each country.

Per Capita

World Av. $1,530 →

- >$3600 U.S. Dollars
- 1800–3600
- 900–1800
- 300–900
- <300
- Uninhabited or sparsely populated

Percentage of Population in Each Per Capita Category

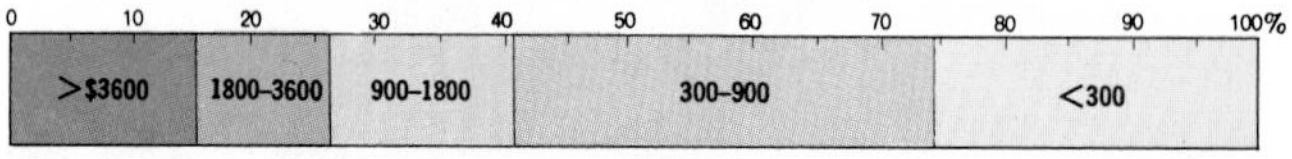

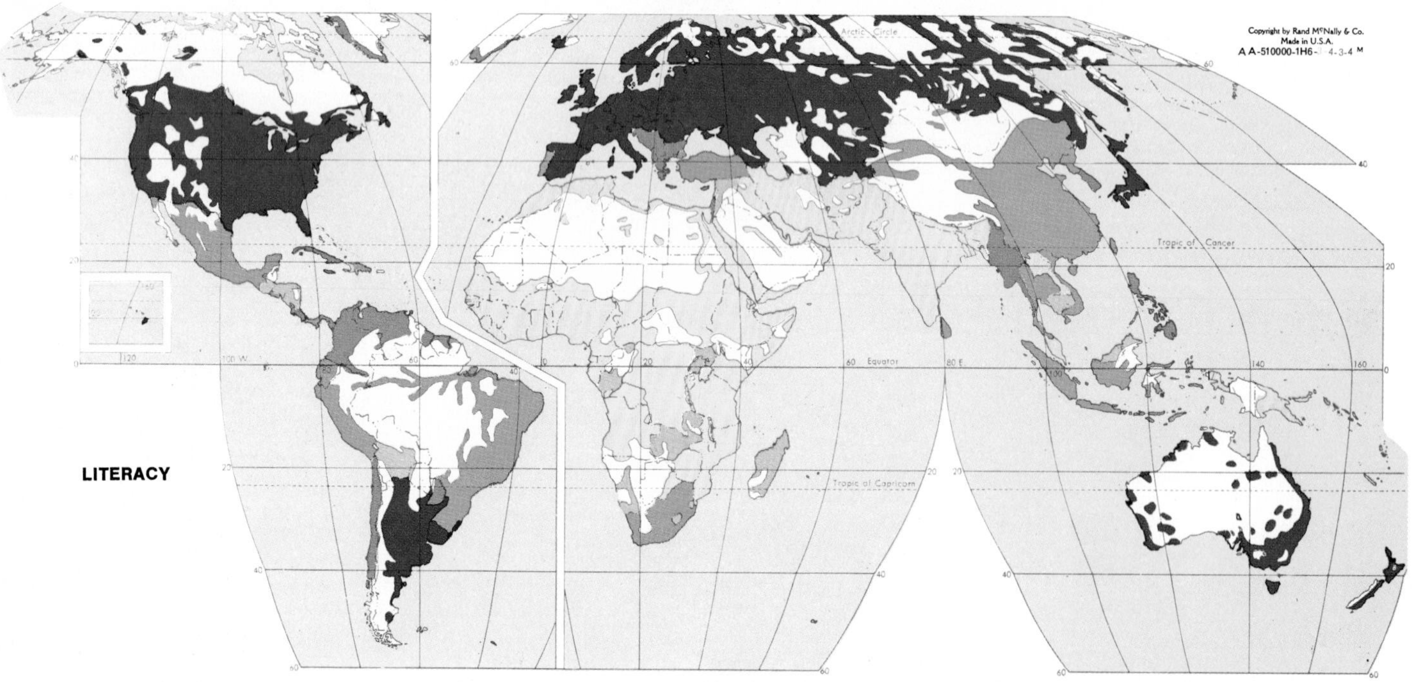

0
10
20
30
40
50
60
70
80
90
100%
>$3600
1800–3600
900–1800
300–900
<300

Copyright by Rand McNally & Co.
Made in U.S.A.
A A-510000-1H6-1-4-3-4 M
Arctic Circle
Tropic of Cancer
Equator
Tropic of Capricorn
LITERACY

World Av. 52% →

- >90%
- 70-90
- 50-70
- 30-50
- <30
- Uninhabited or sparsely populated

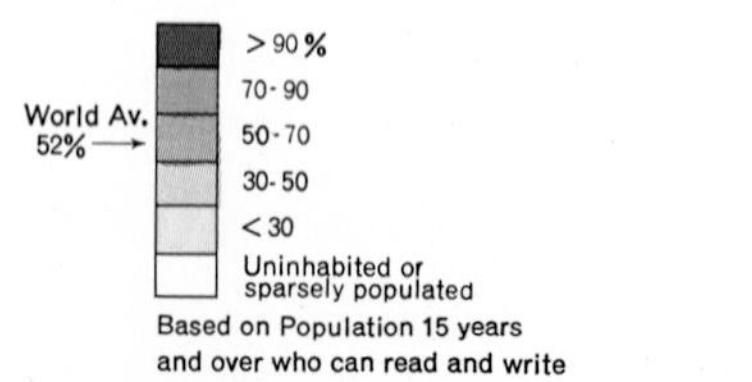

Based on Population 15 years and over who can read and write

Newspaper Circulation

per 1,000 population

INDIA 16
BRAZIL 40
FRANCE 231
SOVIET UNION 373
UNITED STATES 300
JAPAN 537

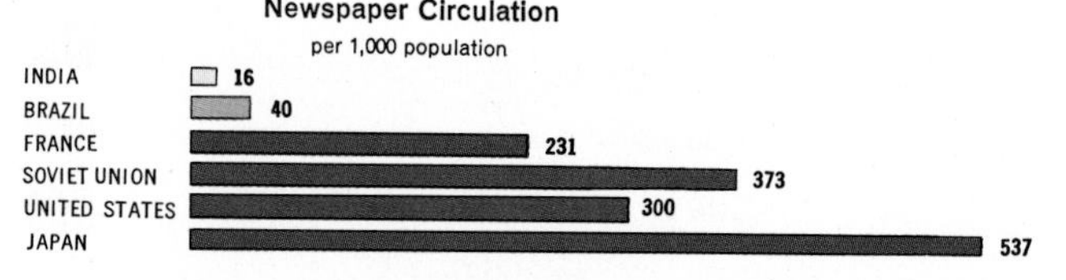

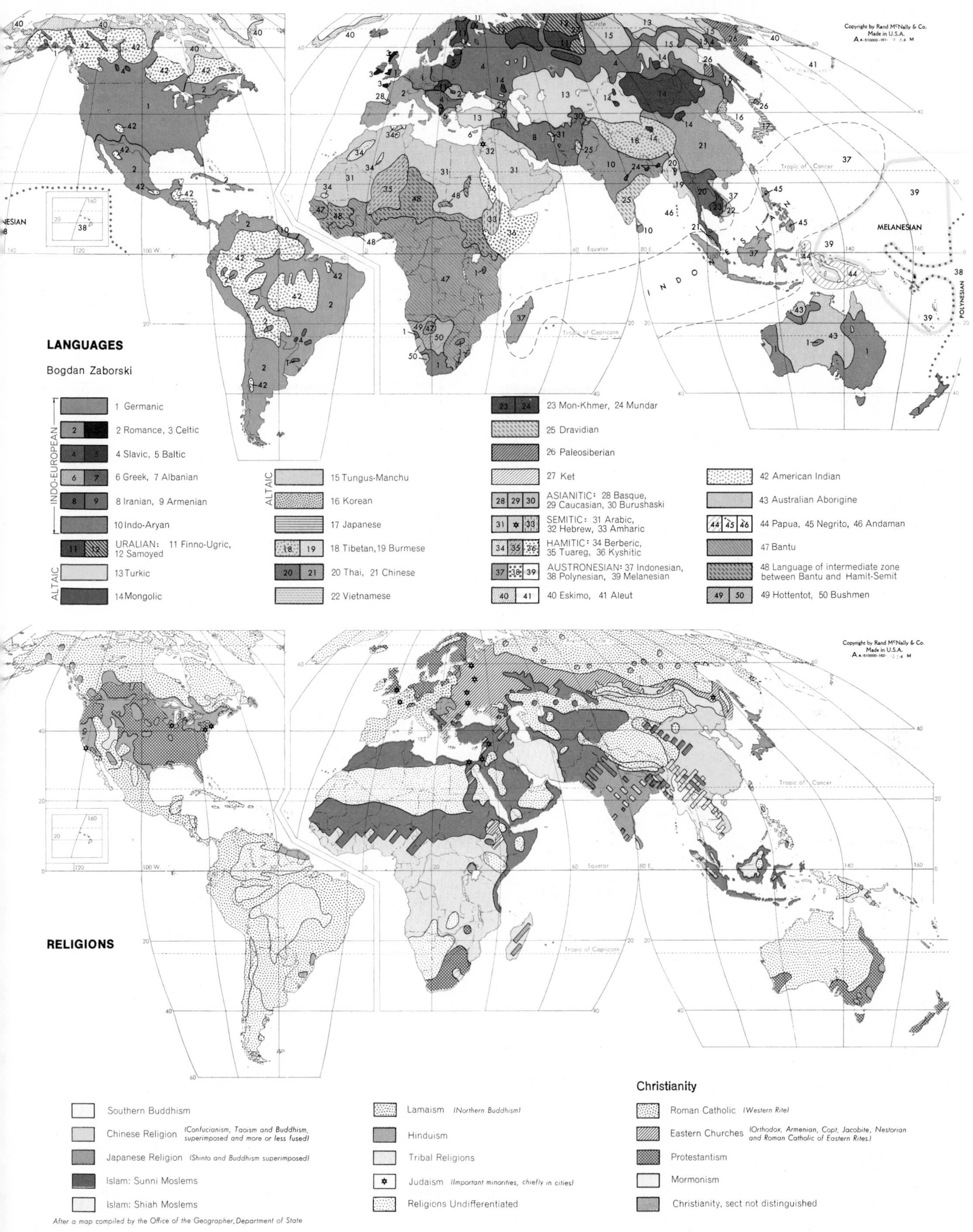
LANGUAGES
Bogdan Zaborski
INDO-EUROPEAN
1 Germanic
2 Romance, 3 Celtic
4 Slavic, 5 Baltic
6 Greek, 7 Albanian
8 Iranian, 9 Armenian
10 Indo-Aryan
URALIAN: 11 Finno-Ugric, 12 Samoyed
ALTAIC
13 Turkic
14 Mongolic
15 Tungus-Manchu
16 Korean
17 Japanese
18 Tibetan, 19 Burmese
20 Thai, 21 Chinese
22 Vietnamese
23 Mon-Khmer, 24 Mundar
25 Dravidian
26 Paleosiberian
27 Ket
ASIANITIC: 28 Basque, 29 Caucasian, 30 Burushaski
SEMITIC: 31 Arabic, 32 Hebrew, 33 Amharic
HAMITIC: 34 Berberic, 35 Tuareg, 36 Kyshitic
AUSTRONESIAN: 37 Indonesian, 38 Polynesian, 39 Melanesian
40 Eskimo, 41 Aleut
42 American Indian
43 Australian Aborigine
44 Papua, 45 Negrito, 46 Andaman
47 Bantu
48 Language of intermediate zone between Bantu and Hamit-Semit
49 Hottentot, 50 Bushmen
MELANESIAN
POLYNESIAN
INDONESIAN
Tropic of Cancer
Equator
Tropic of Capricorn
Copyright by Rand McNally & Co. Made in U.S.A.
RELIGIONS
Southern Buddhism
Chinese Religion (Confucianism, Taoism and Buddhism, superimposed and more or less fused)
Japanese Religion (Shinto and Buddhism superimposed)
Islam: Sunni Moslems
Islam: Shiah Moslems
Lamaism (Northern Buddhism)
Hinduism
Tribal Religions
Judaism (Important minorities, chiefly in cities)
Religions Undifferentiated
Christianity
Roman Catholic (Western Rite)
Eastern Churches (Orthodox, Armenian, Copt, Jacobite, Nestorian and Roman Catholic of Eastern Rites.)
Protestantism
Mormonism
Christianity, sect not distinguished

After a map compiled by the Office of the Geographer, Department of State

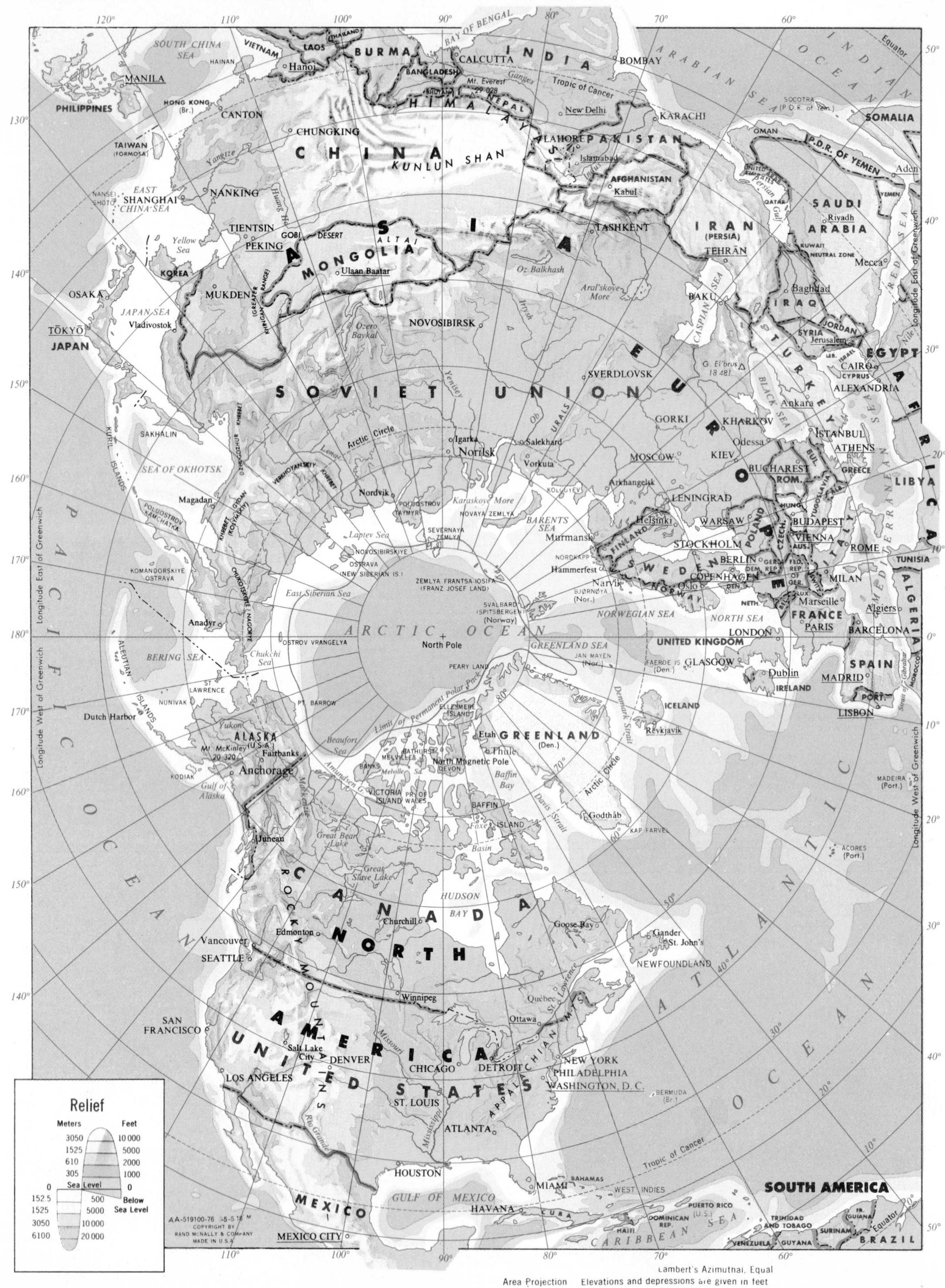

Lambert's Azimuthal Equal Area Projection Elevations and depressions are given in feet

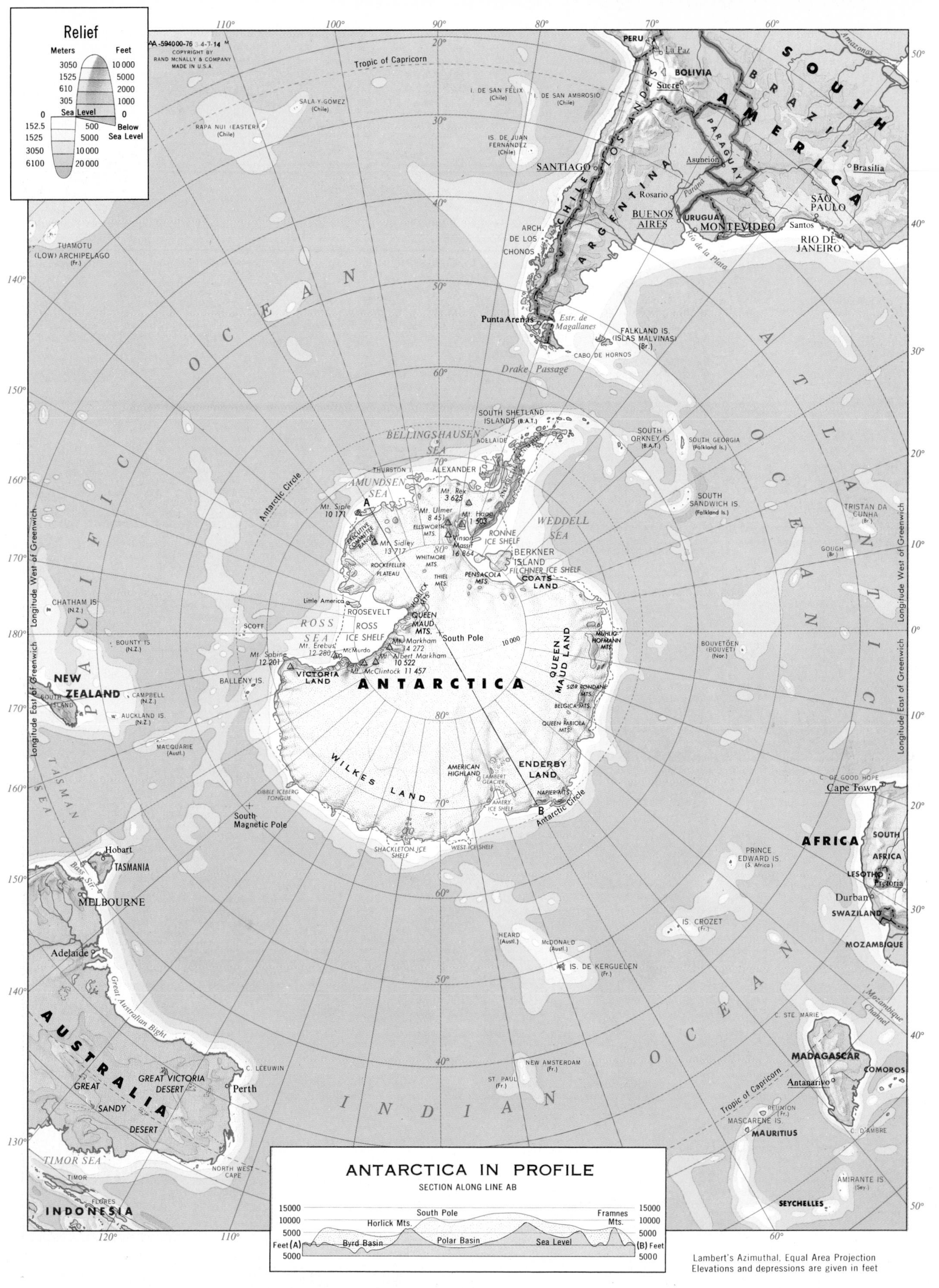

Lambert's Azimuthal, Equal Area Projection
Elevations and depressions are given in feet

glossary of foreign geographical terms

Annam...Annamese
Arab...Arabic
Bantu...Bantu
Bur...Burmese
Camb...Cambodian
Celt...Celtic
Chn...Chinese
Czech...Czech
Dan...Danish
Du...Dutch
Fin...Finnish
Fr...French
Ger...German
Gr...Greek
Hung...Hungarian
Ice...Icelandic
India...India
Indian...American Indian
Indon...Indonesian
It...Italian
Jap...Japanese
Kor...Korean
Mal...Malayan
Mong...Mongolian
Nor...Norwegian
Per...Persian
Pol...Polish
Port...Portuguese
Rom...Romanian
Rus...Russian
Siam...Siamese
So. Slav...Southern Slavonic
Sp...Spanish
Swe...Swedish
Tib...Tibetan
Tur...Turkish
Yugo...Yugoslav

å, *Nor., Swe*...brook, river
aa, *Dan., Nor*...brook
aas, *Dan., Nor*...ridge
āb, *Per*...water, river
abad, *India, Per*...town, city
ada, *Tur*...island
adrar, *Berber*...mountain
air, *Indon*...stream
akrotírion, *Gr*...cape
älf, *Swe*...river
alp, *Ger*...mountain
altipiano, *It*...plateau
alto, *Sp*...height
archipel, *Fr*...archipelago
archipiélago, *Sp*...archipelago
arquipélago, *Port*...archipelago
arroyo, *Sp*...brook, stream
ås, *Nor., Swe*...ridge
austral, *Sp*...southern
baai, *Du*...bay
bab, *Arab*...gate, port
bach, *Ger*...brook, stream
backe, *Swe*...hill
bad, *Ger*...bath, spa
bahía, *Sp*...bay, gulf
bahr, *Arab*...river, sea, lake
baia, *It*...bay, gulf
baía, *Port*...bay
baie, *Fr*...bay, gulf
bajo, *Sp*...depression
bak, *Indon*...stream
bakke, *Dan., Nor*...hill
balkan, *Tur*...mountain range
bana, *Jap*...point, cape
banco, *Sp*...bank
bandar, *Mal., Per.*
...town, port, harbor
bang, *Siam*...village
bassin, *Fr*...basin
batang, *Indon., Mal*...river
ben, *Celt*...mountain, summit
bender, *Arab*...harbor, port
bereg, *Rus*...coast, shore
berg, *Du., Ger., Nor., Swe.*
...mountain, hill
bir, *Arab*...well
birkat, *Arab*...lake, pond, pool
bit, *Arab*...house
bjaerg, *Dan., Nor*...mountain
bocche, *It*...mouth
boğazi, *Tur*...strait
bois, *Fr*...forest, wood
boloto, *Rus*...marsh
bolsón, *Sp* flat-floored desert valley
boreal, *Sp*...northern
borg, *Dan., Nor., Swe*...castle, town
borgo, *It*...town, suburb
bosch, *Du*...forest, wood
bouche, *Fr*...river mouth
bourg, *Fr*...town, borough
bro, *Dan., Nor., Swe*...bridge
brücke, *Ger*...bridge
bucht, *Ger*...bay, bight
bugt, *Dan., Nor., Swe*...bay, gulf
bulu, *Indon*...mountain
burg, *Du., Ger*...castle, town
buri, *Siam*...town
burun, burnu, *Tur*...cape
by, *Dan., Nor., Swe*...village
caatinga, *Port. (Brazil)*
...open brushland
cabezo, *Sp*...summit
cabo, *Port., Sp*...cape
campo, *It., Port., Sp*...plain, field
campos, *Port. (Brazil)*...plains
cañón, *Sp*...canyon
cap, *Fr*...cape
capo, *It*...cape
casa, *It., Port., Sp*...house
castello, *It., Port*...castle, fort
castillo, *Sp*...castle
càte, *Fr*...hill
çay, *Tur*...stream, river
cayo, *Sp*...rock, shoal, islet
cerro, *Sp*...mountain, hill
champ, *Fr*...field
chang, *Chn*...village, middle
château, *Fr*...castle
chen, *Chn*...market town
chiang, *Chn*...river
chott, *Arab*...salt lake
chou, *Chn*...capital of district; island
chu, *Tib*...water, stream
cidade, *Port*...town, city
cima, *Sp*...summit, peak
città, *It*...town, city
ciudad, *Sp*...town, city
cochilha, *Port*...ridge
col, *Fr*...pass
colina, *Sp*...hill
cordillera, *Sp*...mountain chain
costa, *It., Port., Sp*...coast
côte, *Fr*...coast
cuchilla, *Sp*...mountain ridge
dağ, *Tur*...mountain(s)
dake, *Jap*...peak, summit
dal, *Dan., Du., Nor., Swe*...valley
dan, *Kor*...point, cape
danau, *Indon*...lake
dar, *Arab*...house, abode, country
darya, *Per*...river, sea
dasht, *Per*...plain, desert
deniz, *Tur*...sea
désert, *Fr*...desert
deserto, *It*...desert
desierto, *Sp*...desert
détroit, *Fr*...strait
dijk, *Du*...dam, dike
djebel, *Arab*...mountain
do, *Kor*...island
dorf, *Ger*...village
dorp, *Du*...village
duin, *Du*...dune
dzong, *Tib.*
...fort, administrative capital
eau, *Fr*...water
ecuador, *Sp*...equator
eiland, *Du*...island
elv, *Dan., Nor*...river, stream
embalse, *Sp*...reservoir
erg, *Arab*...dune, sandy desert
est, *Fr., It*...east
estado, *Sp*...state
este, *Port., Sp*...east
estrecho, *Sp*...strait
étang, *Fr*...pond, lake
état, *Fr*...state
eyjar, *Ice*...islands
feld, *Ger*...field, plain
festung, *Ger*...fortress
fiume, *It*...river
fjäll, *Swe*...mountain
fjärd, *Swe*...bay, inlet
fjeld, *Nor*...mountain, hill
fjord, *Dan., Nor*...fiord, inlet
fjördur, *Ice*...fiord, inlet
fleuve, *Fr*...river
flod, *Dan., Swe*...river
flói, *Ice*...bay, marshland
fluss, *Ger*...river
foce, *It*...river mouth
fontein, *Du*...a spring
forêt, *Fr*...forest
fors, *Swe*...waterfall
forst, *Ger*...forest
fos, *Dan., Nor*...waterfall
fu, *Chn*...town, residence
fuente, *Sp*...spring, fountain
fuerte, *Sp*...fort
furt, *Ger*...ford
gang, *Kor*...stream, river
gangri, *Tib*...mountain
gat, *Dan., Nor*...channel
gàve, *Fr*...stream
gawa, *Jap*...river
gebergte, *Du*...mountain range
gebiet, *Ger*...district, territory
gebirge, *Ger*...mountains
ghat, *India*...pass, mountain range
gobi, *Mong*...desert
gol, *Mong*...river
göl, gölü, *Tur*...lake
golf, *Du., Ger*...gulf, bay
golfe, *Fr*...gulf, bay
golfo, *It., Port., Sp*...gulf, bay
gomba, gompa, *Tib*...monastery
gora, *Rus., So. Slav*...mountain
góra, *Pol*...mountain
gorod, *Rus*...town
grad, *Rus., So. Slav*...town
guba, *Rus*...bay, gulf
gundung, *Indon*...mountain
guntō, *Jap*...archipelago
gunung, *Mal*...mountain
haf, *Swe*...sea, ocean
hafen, *Ger*...port, harbor
haff, *Ger*...gulf, inland sea
hai, *Chn*...sea, lake
hama, *Jap*...beach, shore
hamada, *Arab*...rocky plateau
hamn, *Swe*...harbor
hāmūn, *Per*...swampy lake, plain
hantō, *Jap*...peninsula
hassi, *Arab*...well, spring
haus, *Ger*...house
haut, *Fr*...summit, top
hav, *Dan., Nor*...sea, ocean
havn, *Dan., Nor*...harbor, port
havre, *Fr*...harbor, port
háza, *Hung*...house, dwelling of
heim, *Ger*...hamlet, home
hem, *Swe*...hamlet, home
higashi, *Jap*...east
hisar, *Tur*...fortress
hissar, *Arab*...fort
ho, *Chn*...river
hoek, *Du*...cape
hof, *Ger*...court, farmhouse
höfn, *Ice*...harbor
hoku, *Jap*...north
holm, *Dan., Nor., Swe*...island
hora, *Czech*...mountain
horn, *Ger*...peak
hoved, *Dan., Nor*...cape
hsien, *Chn*...district, district capital
hu, *Chn*...lake
hügel, *Ger*...hill
huk, *Dan., Swe*...point
hus, *Dan., Nor., Swe*...house
île, *Fr*...island
ilha, *Port*...island
indsö, *Dan., Nor*...lake
insel, *Ger*...island
insjö, *Swe*...lake
irmak, irmagi, *Tur*...river
isla, *Sp*...island
isola, *It*...island
istmo, *It., Sp*...isthmus
järvi, jaur, *Fin*...lake
jebel, *Arab*...mountain
jima, *Jap*...island
jökel, *Nor*...glacier
joki, *Fin*...river
jökull, *Ice*...glacier
kaap, *Du*...cape
kai, *Jap*...bay, gulf, sea
kaikyō, *Jap*...channel, strait
kalat, *Per*...castle, fortress
kale, *Tur*...fort
kali, *Mal*...creek, river
kand, *Per*...village
kang, *Chn*...mountain ridge; village
kap, *Dan., Ger*...cape
kapp, *Nor., Swe*...cape
kasr, *Arab*...fort, castle
kawa, *Jap*...river
kefr, *Arab*...village
kei, *Jap*...creek, river
ken, *Jap*...prefecture
khor, *Arab*...bay, inlet
khrebet, *Rus*...mountain range
kiang, *Chn*...large river
king, *Chn*...capital city, town
kita, *Jap*...north
ko, *Jap*...lake
köbstad, *Dan*...market-town
kol, *Mong*...lake
kólpos, *Gr*...gulf
kong, *Chn*...river
kopf, *Ger*...head, summit, peak
köpstad, *Swe*...market-town
körfezi, *Tur*...gulf
kosa, *Rus*...spit
kou, *Chn*...river mouth
köy, *Tur*...village
kraal, *Du. (Africa)*...native village
ksar, *Arab*...fortified village
kuala, *Mal*...bay, river mouth
kuh, *Per*...mountain
kum, *Tur*...sand
kuppe, *Ger*...summit
küste, *Ger*...coast
kyo, *Jap*...town, capital
la, *Tib*...mountain pass
labuan, *Mal*...anchorage, port
lac, *Fr*...lake
lago, *It., Port., Sp*...lake
lagoa, *Port*...lake, marsh
laguna, *It., Port., Sp*...lagoon, lake
lahti, *Fin*...bay, gulf
län, *Swe*...county
landsby, *Dan., Nor*...village
liehtao, *Chn*...archipelago
liman, *Tur*...bay, port
ling, *Chn*...pass, ridge, mountain
llanos, *Sp*...plains
loch, *Celt. (Scotland)*...lake, bay
loma, *Sp*...long, low hill
lough, *Celt. (Ireland)*...lake, bay
machi, *Jap*...town
man, *Kor*...bay
mar, *Port., Sp*...sea
mare, *It., Rom*...sea
marisma, *Sp*...marsh, swamp
mark, *Ger*...boundary, limit
massif, *Fr*...block of mountains
mato, *Port*...forest, thicket
me, *Siam*...river
meer, *Du., Ger*...lake, sea
mer, *Fr*...sea
mesa, *Sp*...flat-topped mountain
meseta, *Sp*...plateau
mina, *Port., Sp*...mine
minami, *Jap*...south
minato, *Jap*...harbor, haven
misaki, *Jap*...cape, headland
mont, *Fr*...mount, mountain
montagna, *It*...mountain
montagne, *Fr*...mountain
montaña, *Sp*...mountain
monte, *It., Port., Sp.*
...mount, mountain
more, *Rus., So. Slav*...sea
morro, *Port., Sp*...hill, bluff
mühle, *Ger*...mill
mund, *Ger*...mouth, opening
mündung, *Ger*...river mouth
mura, *Jap*...township
myit, *Bur*...river
mys, *Rus*...cape
nada, *Jap*...sea
nadi, *India*...river, creek
naes, *Dan., Nor*...cape
nafud, *Arab*...desert of sand dunes
nagar, *India*...town, city
nahr, *Arab*...river
nam, *Siam*...river, water
nan, *Chn., Jap*...south
näs, *Nor., Swe*...cape
nez, *Fr*...point, cape
nishi, nisi, *Jap*...west
njarga, *Fin*...peninsula
nong, *Siam*...marsh
noord, *Du*...north
nor, *Mong*...lake
nord, *Dan., Fr., Ger., It.,*
Nor., Swe...north
norte, *Port., Sp*...north
nos, *Rus*...cape
nyasa, *Bantu*...lake
ö, *Dan., Nor., Swe*...island
occidental, *Sp*...western
ocna, *Rom*...salt mine
odde, *Dan., Nor*...point, cape
oedjoeng, *Mal*...point, cape
oeste, *Port., Sp*...west
oka, *Jap*...hill
oost, *Du*...east
oriental, *Sp*...eastern
óros, *Gr*...mountain
ost, *Ger., Swe*...east
öster, *Dan., Nor., Swe*...eastern
ostrov, *Rus*...island
oued, *Arab*...river, stream
ouest, *Fr*...west
ozero, *Rus*...lake
pää, *Fin*...mountain
padang, *Mal*...plain, field
pampas, *Sp. (Argentina)*
...grassy plains
pará, *Indian (Brazil)*...river
pas, *Fr*...channel, passage
paso, *Sp*...mountain pass, passage
passo, *It., Port.*
...mountain pass, passage, strait
patam, *India*...city, town
pei, *Chn*...north
pélagos, *Gr*...open sea
pegunungan, *Indon*...mountains
peña, *Sp*...rock
peresheyek, *Rus*...isthmus
pertuis, *Fr*...strait
peski, *Rus*...desert
pic, *Fr*...mountain peak
pico, *Port., Sp*...mountain peak
piedra, *Sp*...stone, rock
ping, *Chn*...plain, flat
planalto, *Port*...plateau
planina, *Yugo*...mountains
playa, *Sp*...shore, beach
pnom, *Camb*...mountain
pointe, *Fr*...point
polder, *Du., Ger*...reclaimed marsh
polje, *So. Slav*...plain, field
poluostrov, *Rus*...peninsula
pont, *Fr*...bridge
ponta, *Port*...point, headland
ponte, *It., Port*...bridge
pore, *India*...city, town
porthmós, *Gr*...strait
porto, *It., Port*...port, harbor
potamós, *Gr*...river
p'ov, *Rus*...peninsula
prado, *Sp*...field, meadow
presqu'île, *Fr*...peninsula
proliv, *Rus*...strait
pu, *Chn*...commercial village
pueblo, *Sp*...town, village
puerto, *Sp*...port, harbor
pulau, *Mal*...island
punkt, *Ger*...point
punt, *Du*...point
punta, *It., Sp*...point
pur, *India*...city, town
puy, *Fr*...peak
qal'a, qal'at, *Arab*...fort, village
qasr, *Arab*...fort, castle
rann, *India*...wasteland
ra's, *Arab*...cape, head
reka, *Rus., So. Slav*...river
reprêsa, *Port*...reservoir
rettō, *Jap*...island chain
ría, *Sp*...estuary
ribeira, *Port*...stream
riberão, *Port*...river
rio, *It., Port*...stream, river
río, *Sp*...river
rivière, *Fr*...river
roca, *Sp*...rock
rt, *Yugo*...cape
rūd, *Per*...river
saari, *Fin*...island
sable, *Fr*...sand
sahara, *Arab*...desert, plain
saki, *Jap*...cape
sal, *Sp*...salt
salar, *Sp*...salt flat, salt lake
salto, *Sp*...waterfall
san, *Jap., Kor*...mountain, hill
sat, satul, *Rom*...village
schloss, *Ger*...castle
sebkha, *Arab*...salt marsh
see, *Ger*...lake, sea
şehir, *Tur*...town, city
selat, *Indon*...stream
selvas, *Port. (Brazil)*
...tropical rain forests
seno, *Sp*...bay
serra, *Port*...mountain chain
serranía, *Sp*...mountain ridge
seto, *Jap*...strait
severnaya, *Rus*...northern
shahr, *Per*...town, city
shan, *Chn*...mountain, hill, island
shatt, *Arab*...river
shi, *Jap*...city
shima, *Jap*...island
shōtō, *Jap*...archipelago
si, *Chn*...west, western
sierra, *Sp*...mountain range
sjö, *Nor., Swe*...lake, sea
sö, *Dan., Nor*...lake, sea
söder, södra, *Swe*...south
song, *Annam*...river
sopka, *Rus*...peak, volcano
source, *Fr*...a spring
spitze, *Ger*...summit, point
staat, *Ger*...state
stad, *Dan., Du., Nor., Swe.*
...city, town
stadt, *Ger*...city, town
stato, *It*...state
step', *Rus*...treeless plain, steppe
straat, *Du*...strait
strand, *Dan., Du., Ger., Nor.,*
Swe...shore, beach
stretto, *It*...strait
strom, *Ger*...river, stream
ström, *Dan., Nor., Swe*...stream, river
stroom, *Du*...stream, river
su, suyu, *Tur*...water, river
sud, *Fr., Sp*...south
süd, *Ger*...south
suidō, *Jap*...channel
sul, *Port*...south
sund, *Dan., Nor., Swe*...sound
sungai, sungei, *Indon., Mal*...river
sur, *Sp*...south
syd, *Dan., Nor., Swe*...south
tafelland, *Ger*...plateau
take, *Jap*...peak, summit
tal, *Ger*...valley
tandjung, tanjong, *Mal*...cape
tao, *Chn*...island
târg, târgul, *Rom*...market, town
tell, *Arab*...hill
teluk, *Indon*...bay, gulf
terra, *It*...land
terre, *Fr*...earth, land
thal, *Ger*...valley
tierra, *Sp*...earth, land
tō, *Jap*...east; island
tonle, *Camb*...river, lake
top, *Du*...peak
torp, *Swe*...hamlet, cottage
tsangpo, *Tib*...river
tsi, *Chn*...village, borough
tso, *Tib*...lake
tsu, *Jap*...harbor, port
tundra, *Rus*...treeless arctic plains
tung, *Chn*...east
tuz, *Tur*...salt
udde, *Swe*...cape
ufer, *Ger*...shore, riverbank
umi, *Jap*...sea, gulf
ura, *Jap*...bay, coast, creek
ust'ye, *Rus*...river mouth
valle, *It., Port., Sp*...valley
vallée, *Fr*...valley
valli, *It*...lake
vár, *Hung*...fortress
város, *Hung*...town
varoš, *So. Slav*...town
veld, *Du*...open plain, field
verkh, *Rus*...top, summit
ves, *Czech*...village
vest, *Dan., Nor., Swe*...west
vik, *Swe*...cove, bay
vila, *Port*...town
villa, *Sp*...town
villar, *Sp*...village, hamlet
ville, *Fr*...town, city
vostok, *Rus*...east
wad, wādī, *Arab.*
...intermittent stream
wald, *Ger*...forest, woodland
wan, *Chn., Jap*...bay, gulf
weiler, *Ger*...hamlet, village
westersch, *Du*...western
wüste, *Ger*...desert
yama, *Jap*...mountain
yarimada, *Tur*...peninsula
yug, *Rus*...south
zaki, *Jap*...cape
zaliv, *Rus*...bay, gulf
zapad, *Rus*...west
zee, *Du*...sea
zemlya, *Rus*...land
zuid, *Du*...south

Index to Atlas

Abbreviations Used in This Index

Afg. .... Afghanistan
Afr. .... Africa
Ak. .... Alaska
Al. .... Alabama
Alb. .... Albania
Alg. .... Algeria
And. .... Andorra
Ang. .... Angola
Ant. .... Antarctica
Ar. .... Arkansas
Arch. .... Archipelago
Arc.O. .... Arctic Ocean
Arg. .... Argentina
A.S.S.R. .... Autonomous Soviet Socialist Republic
Atl.O. .... Atlantic Ocean
Aus. .... Austria
Austl. .... Australia
Az. .... Arizona

B. .... Bay, Bahia
Ba. .... Bahamas
Bngl. .... Bangladesh
Barb. .... Barbados
Bel. .... Belgium
Bhu. .... Bhutan
Bk. .... Bank
Bol. .... Bolivia
Bots. .... Botswana
Br. .... British
Braz. .... Brazil
Bru. .... Brunei
Bul. .... Bulgaria
Bur. .... Burma

C. .... Cape, Cerro
Ca. .... California
Cam. .... Cameroon
Camb. .... Cambodia
Can. .... Canal, Canada
Can.Is. .... Canary Is.
Cen.Afr.Emp. Central African Empire
Chan. .... Channel
Co. .... Colorado
Col. .... Colombia
Con. .... Congo
C.R. .... Costa Rica
Cr. .... Creek
Ct. .... Connecticut
C.V. .... Cape Verde
C.Z. .... Canal Zone
Czech. .... Czechoslovakia

De. .... Delaware
Den. .... Denmark

Dept. .... Department
Des. .... Desert
Dist. .... District
Div. .... Division
Dom.Rep. .... Dominican Republic

E. .... East
Ec. .... Ecuador
Eng. .... England
Equat. Gui. .... Equatorial Guinea
Eth. .... Ethiopia
Eur. .... Europe

Faer. .... Faeroe Is.
Falk. Is. .... Falkland Is.
Fd. .... Fjord
F.R.G. Federal Republic of Germany
Fin. .... Finland
Fk. .... Fork
Fl. .... Florida
For. .... Forest
Fr. .... France
Fr. Gu. .... French Guiana
Ft. .... Fort

G. .... Gulf
Ga. .... Georgia
Gam. .... Gambia
G.D.R. German Democratic Republic
Grc. .... Greece
Grnld. .... Greenland
Gt. Brit. .... Great Britain
Guad. .... Guadeloupe
Guat. .... Guatemala
Gui. .... Guinea
Guy. .... Guyana

Hai. .... Haiti
Hbr. .... Harbor
Hd. .... Head
Hi. .... Hawaii
Hond. .... Honduras
Hung. .... Hungary

I. .... Island
Ia. .... Iowa
Ice. .... Iceland
Id. .... Idaho
Il. .... Illinois
In. .... Indiana
Ind.O. .... Indian Ocean
Indon. .... Indonesia
Ire. .... Ireland
Is. .... Islands

Isr. .... Israel
Isth. .... Isthmus
It. .... Italy

Jam. .... Jamaica
Jap. .... Japan

Ken. .... Kenya
Kor. .... Korea
Ks. .... Kansas
Kuw. .... Kuwait
Ky. .... Kentucky

L. .... Lake, Loch, Lough
La. .... Louisiana
Lat. .... Latitude
Leb. .... Lebanon
Leso. .... Lesotho
Lib. .... Liberia
Liech. .... Liechtenstein
Lux. .... Luxembourg

Ma. .... Massachusetts
Mad. .... Madagascar
Mad.Is. .... Madeira Islands
Mala. .... Malaysia
Mand. .... Mandate
Mart. .... Martinique
Md. .... Maryland
Me. .... Maine
Medit. .... Mediterranean
Mex. .... Mexico
Mi. .... Michigan
Mn. .... Minnesota
Mo. .... Missouri
Mong. .... Mongolia
Mor. .... Morocco
Moz. .... Mozambique
Ms. .... Mississippi
Mt. .... Mount, Montana
Mtn. .... Mountain
Mts. .... Mountains

N.A. .... North America
Ne. .... Nebraska
NC. .... North Carolina
N.Cal. .... New Caledonia
ND. .... North Dakota
Nep. .... Nepal
Neth. .... Netherlands
New Hebr. .... New Hebrides
NH. .... New Hampshire
Nic. .... Nicaragua
Nig. .... Nigeria
N.Ire. .... Northern Ireland
NJ. .... New Jersey

NM. .... New Mexico
Nor. .... Norway
Nv. .... Nevada
NY. .... New York
N.Z. .... New Zealand

O. .... Ocean
Oh. .... Ohio
Ok. .... Oklahoma
Om. .... Oman
Or. .... Oregon

P. .... Pass
Pa. .... Pennsylvania
Pac.O. .... Pacific Ocean
Pak. .... Pakistan
Pan. .... Panama
Pap.N.Gui. .... Papua New Guinea
Par. .... Paraguay
Pass. .... Passage
P.D.R. of Yem. .... Yemen, People's Democratic Republic of
Pen. .... Peninsula
Phil. .... Philippines
Pk. .... Peak, Park
Plat. .... Plateau
Pln. .... Plain
Pol. .... Poland
Port. .... Portugal
P.R. .... Puerto Rico
Prov. .... Province
Pt. .... Point

R. .... River, Rio, Rivière
Ra. .... Range, Ranges
Reg. .... Region
Rep. .... Republic
Res. .... Reservation, Reservoir
Rf. .... Reef
Rh. .... Rhodesia
RI. .... Rhode Island
Rom. .... Romania
R.S.F.S.R. .... Russian Soviet Federated Socialist Republic
Rw. .... Rwanda

S. .... San, Santo, South
Sa. .... Serra, Sierra
S.A. .... South America
S.Afr. .... South Africa
Sal. .... El Salvador
Sau.Ar. .... Saudi Arabia
SC. .... South Carolina
Scot. .... Scotland
SD. .... South Dakota
Sd. .... Sound

S.L. .... Sierra Leone
Sol.Is. .... Solomon Is.
Som. .... Somalia
Sov.Un. .... Soviet Union
Sp. .... Spain
Spr., Sprs. .... Spring, Springs
S.S.R. .... Soviet Socialist Republic
St. .... Saint
Str. .... Strait
Sud. .... Sudan
Sur. .... Surinam
Swaz. .... Swaziland
Swe. .... Sweden
Switz. .... Switzerland
Swp. .... Swamp
Syr. .... Syria

Tan. .... Tanzania
Tas. .... Tasmania
Ter. .... Territory
Thai. .... Thailand
Tn. .... Tennessee
Trin. .... Trinidad and Tobago
Tun. .... Tunisia
Tur. .... Turkey
Tx. .... Texas

U.A.E. .... United Arab Emirates
Ug. .... Uganda
U.K. .... United Kingdom
UR. .... Uruguay
U.S., U.S.A. United States of America
Ut. .... Utah

Va. .... Virginia
Val. .... Valley
Ven. .... Venezuela
Viet. .... Vietnam
Vir.Is. .... Virgin Is.
Vol. .... Volcano
Vt. .... Vermont

Wa. .... Washington
Wi. .... Wisconsin
W.Sah. .... Western Sahara
W.Sam. .... Western Samoa
WV. .... West Virginia
Wy. .... Wyoming

Yugo. .... Yugoslavia

Page	Name	Region	Lat.	Long.
19	Aegean Sea	Asia-Eur.	39·04 N	24·56 E
23	Aerø (I.)	Den.	54·52 N	10·22 E
32	Afghanistan	Asia	33·00 N	63·00 E
52	Aflou	Alg.	33·59 N	2·04 E
51	Africa			
33	'Afula	Isr.	32·36 N	35·17 E
31	Afyonkarahisar	Tur.	38·45 N	30·20 E
53	Agadem	Niger	16·50 N	13·17 E
52	Agadez	Niger	16·58 N	7·59 E
52	Agadir	Mor.	30·30 N	9·37 W
31	Agdam	Sov.Un.	40·00 N	47·00 E
17	Agen	Fr.	44·13 N	0·31 E
41	Agno	Phil.	16·07 N	119·49 E
41	Agno (R.)	Phil.	15·42 N	120·28 E
20	Agnone	It.	41·49 N	14·23 E
35	Agra	India	27·18 N	78·00 E
20	Agri (R.)	It.	40·15 N	16·21 E
21	Agrínion	Grc.	38·38 N	21·06 E
48	Aguadas	Col.	5·37 N	75·27 W
13	Aguadilla	P.R.	18·27 N	67·10 W
12	Aguascalientes	Mex.	21·52 N	102·17 W
17	Aguilas	Sp.	37·26 N	1·35 W
48	Aguja, Pta. (Pt.)	Peru	6·00 S	81·15 W
54	Agulhas, C.	S.Afr.	34·47 S	20·00 E
40	Agung, Gunung (Mtn.)	Indon.	8·28 S	115·07 E
41	Agusan (R.)	Phil.	8·12 N	126·07 E
52	Ahaggar (Mts.)	Alg.	23·14 N	6·00 E
35	Ahmadābād	India	23·04 N	72·38 E
35	Ahmadnagar	India	19·09 N	74·45 E
53	Ahmar Mts.	Eth.	9·22 N	42·00 E
26	Ahrweiler	F.R.G.	50·34 N	7·05 E
25	Ahtärin-järvi (L.)	Fin.	62·46 N	24·25 E
24	Åhus	Swe.	55·56 N	14·19 E
34	Ahvāz	Iran	31·15 N	48·54 E
25	Ahvenanmaa (Åland Is.)	Fin.	60·36 N	19·55 E
43	Aiea	Hi.	21·18 N	157·52 W
49	Aimorès, Serra dos (Mts.)	Braz.	17·40 S	42·38 W
52	Aïn Beïda	Alg.	35·57 N	7·25 E
52	Aïn Salah	Alg.	27·13 N	2·22 E
18	Aïn Taïba	Alg.	30·20 N	5·30 E
17	Aïn-Témouchent	Alg.	35·20 N	1·23 W
48	Aipe	Col.	3·13 N	75·15 W
52	Aïr (Mts.)	Niger	18·00 N	8·30 E
33	Airhitam, Selat (Str.)	Indon.	0·58 N	102·38 E
17	Aisne (R.)	Fr.	49·28 N	3·32 E
41	Aitape	Pap.N.Gui.	3·00 S	142·10 E
21	Aitolikón	Grc.	38·27 N	21·21 E
21	Aitos	Bul.	42·42 N	27·17 E
43	Aitutaki (I.)	Cook Is.	19·00 S	162·00 W
27	Aiud	Rom.	46·19 N	23·40 E
18	Aix-en-Provence	Fr.	43·32 N	5·27 E
21	Aíyina	Grc.	37·37 N	22·12 E
21	Aíyina (I.)	Grc.	37·43 N	23·35 E
21	Aíyion	Grc.	38·13 N	22·04 E
25	Aizpute	Sov.Un.	56·44 N	21·37 E
18	Ajaccio	Fr.	41·55 N	8·42 E
44	Ajana	Austl.	28·00 S	114·45 E
53	Ajdābiyah	Libya	30·56 N	20·16 E
33	'Ajmah, Jabal al (Mts.)	Egypt	29·12 N	34·03 E
34	Ajman	U.A.E.	25·15 N	54·30 E
35	Ajmer	India	26·26 N	74·42 E
53	Aketi	Zaire	2·44 N	23·46 E
31	Akhaltsikhe	Sov.Un.	41·40 N	42·50 E
53	Akhdar, Al Jabal al (Mts.)	Libya	32·00 N	22·00 E
21	Akhelóös (R.)	Grc.	38·45 N	21·26 E
31	Akhisar	Tur.	38·58 N	27·58 E
21	Akhtopol	Bul	42·08 N	27·54 E
11	Akimiski (I.)	Can.	52·54 N	80·22 W
37	Akita	Jap.	39·40 N	140·12 E
52	Akjoujt	Mauritania	19·45 N	14·23 W
33	'Akko	Isr.	32·56 N	35·05 E
10	Aklavik	Can.	68·28 N	135·26 W
35	Akola	India	20·47 N	77·00 E
53	Akordat	Eth.	15·34 N	37·54 E
11	Akpatok (I.)	Can.	60·30 N	67·10 W
16	Akranes	Ice.	64·18 N	21·40 W
21	Akrítas, Akr. (C.)	Grc.	37·45 N	22·00 E
8	Akron	Oh.	41·05 N	81·30 W
31	Aksaray	Tur.	38·30 N	34·05 E
31	Aksehir	Tur.	38·20 N	31·20 E
31	Aksehir (L.)	Tur.	38·40 N	31·30 E
29	Aksha	Sov.Un.	50·28 N	113·00 E
36	Ak Su (R.)	China	40·34 N	77·15 E
31	Aktyubinsk	Sov.Un.	50·20 N	57·00 E
16	Akureyri	Ice.	65·39 N	18·01 W
7	Alabama (State)	U.S.	32·50 N	87·30 W
7	Alabama (R.)	Al.	31·20 N	87·39 W
41	Alabat (I.)	Phil.	14·14 N	122·05 E
31	Alacam	Tur.	41·30 N	35·40 E
34	Al Aflaj (Des.)	Sau.Ar.	24·00 N	44·47 E
49	Alagôas (State)	Braz.	9·50 S	36·33 W
49	Alagoinhas	Braz.	12·13 S	38·12 W
43	Alalakeiki Chan.	Hi.	20·40 N	156·30 W
53	Al 'Alamayn	Egypt	30·53 N	28·52 E
6	Alameda	Ca.	37·46 N	122·15 W
41	Alaminos	Phil.	16·09 N	119·58 E
19	Al 'Amirīyah	Egypt	31·01 N	29·52 E
31	Alanya	Tur.	36·40 N	32·10 E
55	Alaotra (L.)	Mad.	17·15 S	48·17 E
33	Al 'Aqabah	Jordan	29·32 N	35·00 E
33	Al 'Arīsh	Egypt	31·08 N	33·48 E
38	Ala Shan (Mts.)	China	38·02 N	105·20 E
5	Alaska (State)	U.S.	64·00 N	150·00 W
5	Alaska, G. of	Ak.	57·42 N	147·40 W
53	Al-'Aṭrūn	Sud.	18·13 N	26·44 E
30	Alatyr'	Sov.Un.	54·55 N	46·30 E
48	Alausí	Ec.	2·15 S	78·45 W
20	Alba	It.	44·41 N	8·02 E
17	Albacete	Sp.	39·00 N	1·49 W
27	Alba Iulia	Rom.	46·05 N	23·32 E
14	Albania	Eur.	41·45 N	20·00 E
44	Albany	Austl.	35·00 S	118·00 E
7	Albany	Ga.	31·35 N	84·10 W
9	Albany	NY	42·40 N	73·50 W
6	Albany	Or.	44·38 N	123·06 W
11	Albany (R.)	Can.	51·45 N	83·30 W
34	Al Başrah	Iraq	30·35 N	47·59 E
33	Al Batrūn	Leb.	34·16 N	35·39 E
53	Al Bawīṭī	Egypt	28·19 N	29·00 E
7	Albemarle Sd.	NC	36·00 N	76·17 W
20	Albenga	It.	44·04 N	8·13 E
44	Alberga, The (R.)	Austl.	27·15 S	135·00 E
53	Albert (L.)	Afr.	1·50 N	30·40 E
10	Alberta (Prov.)	Can.	54·33 N	117·10 W
41	Albert Edward, Mt.	Pap.N.Gui.	8·25 S	147·25 E
55	Alberton	S.Afr.	26·16 S	28·08 E
18	Albi	Fr.	43·54 N	2·07 E
49	Albina	Sur.	5·30 N	54·33 W
8	Albion	Mi.	42·15 N	84·50 W
9	Albion	NY	43·15 N	78·10 W
18	Alboran, Isla del (I.)	Sp.	35·58 N	3·02 W
17	Alboran Sea	Afr.-Eur.	35·54 N	4·26 W
24	Ålborg	Den.	57·02 N	9·55 E
17	Albox	Sp.	37·23 N	2·08 W
6	Albuquerque	NM	35·05 N	106·40 W
34	Al Buraymī	Om.	23·45 N	55·39 E
46	Albury	Austl.	36·00 S	147·00 E
17	Alcalá la Real	Sp.	37·27 N	3·57 W
20	Alcamo	It.	37·58 N	13·03 E
17	Alcañiz	Sp.	41·03 N	0·08 W
49	Alcântara	Braz.	2·17 S	44·29 W
17	Alcazar de San Juan	Sp.	39·22 N	3·12 W
17	Alcoy	Sp.	38·42 N	0·30 W
55	Aldabra Is.	Afr.	9·16 S	46·17 E
29	Aldan	Sov.Un.	58·46 N	125·19 E
29	Aldanskaya	Sov.Un.	61·52 N	135·29 E
22	Aldershot	Eng.	51·14 N	0·46 W
8	Alderson	WV	37·40 N	80·40 W
52	Aleg	Mauritania	17·03 N	13·55 W
50	Alegre (R.)	Braz.	22·22 S	43·34 W
50	Alegrete	Braz.	29·46 S	55·44 W
30	Aleksandrov	Sov.Un.	56·24 N	38·45 E
27	Aleksandrow Kujawski	Pol.	52·54 N	18·45 E
21	Aleksinac	Yugo.	43·33 N	21·42 E
17	Alençon	Fr.	48·26 N	0·08 E
49	Alenquer	Braz.	1·58 S	54·44 W
43	Alenuihaha Chan.	Hi.	20·20 N	156·05 W
19	Aleppo	Syr.	36·10 N	37·18 E
17	Alès	Fr.	44·07 N	4·06 E
20	Alessandria	It.	44·53 N	8·35 E
24	Ålesund	Nor.	62·28 N	6·14 E
6	Aleutian Is.	Ak.	52·40 N	177·30 W
29	Alevina, Mys (C.)	Sov.Un.	58·49 N	151·44 E
67	Alexander I	Ant.	71·00 S	71·00 W
55	Alexandra	S.Afr.	26·07 S	28·07 E
44	Alexandria	Austl.	19·00 S	136·56 E
9	Alexandria	Can.	45·50 N	74·35 W
8	Alexandria	In.	40·20 N	85·20 W
7	Alexandria	La.	31·18 N	92·28 W
21	Alexandria	Rom.	43·55 N	25·21 E
55	Alexandria	S.Afr.	33·40 S	26·26 E
9	Alexandria	Va.	38·50 N	77·05 W
	Alexandria, see Al Iskandarīyah			
9	Alexandria Bay	NY	44·20 N	75·55 W
21	Alexandroúpolis (Dedeagats)	Grc.	40·51 N	25·51 E
53	Al-Fāshir	Sud.	13·38 N	25·21 E
53	Al Fayyūm	Egypt	29·14 N	30·48 E
21	Alfiós (R.)	Grc.	37·33 N	21·50 E
52	Alger (Algiers)	Alg.	36·51 N	2·56 E
51	Algeria	Afr.	28·45 N	1·00 E
20	Alghero	It.	40·32 N	8·22 E
55	Algoabaai (B.)	S.Afr.	33·51 S	24·50 E
8	Algonac	Mi.	42·35 N	82·30 W
9	Algonquin Provincial Park	Can.	45·50 N	78·20 W
19	Al Hammām	Egypt	30·46 N	29·42 E
34	Al Hasā (Plain)	Sau.Ar.	27·00 N	47·48 E
34	Al Ḥijāz (Reg.)	Sau.Ar.	23·45 N	39·08 E
33	Al Hirmil	Leb.	34·23 N	36·22 E
34	Al Hudaydah	Yemen	14·43 N	43·03 E
34	Al Hufūf	Sau.Ar.	25·15 N	49·43 E
21	Aliákmon (R.)	Grc.	40·26 N	22·17 E
18	Alicante	Sp.	38·20 N	0·30 W
55	Alice	S. Afr.	32·47 S	26·51 E
55	Alicedale	S. Afr.	33·18 S	26·04 E
44	Alice Springs	Austl.	23·38 S	133·56 E
20	Alicudi (I.)	It.	38·34 N	14·21 E
35	Alīgarh	India	27·58 N	78·08 E
24	Alingsås	Swe.	57·57 N	12·30 E
9	Aliquippa	Pa.	40·37 N	80·15 W
53	Al Iskandarīyah (Alexandria)	Egypt	31·12 N	29·58 E
54	Aliwal North	S.Afr.	31·09 S	28·26 E
34	Al-Jabal Al-Akhḍar (Mts.)	Om.	23·30 N	56·43 W
33	Al Jafr, Qa'al (L.)	Jordan	30·15 N	36·24 E
53	Al Jaghbūb	Libya	29·46 N	24·32 E
53	Al Jawf	Libya	24·14 N	23·15 E
34	Al Jawf	Sau.Ar.	29·45 N	39·30 E
52	Al Jufrah (Oasis)	Libya	29·30 N	15·16 E
53	Al Kāmilīn	Sud.	15·09 N	33·06 E
33	Al Karak	Jordan	31·11 N	35·42 E
34	Al Khābūrah	Om.	23·45 N	57·30 E
33	Al Khalīl (Hebron)	Jordan	31·31 N	35·07 E
53	Al Khandaq	Sud.	18·38 N	30·29 E
53	Al Khums	Libya	32·35 N	14·10 E
34	Al Khurmah	Sau.Ar.	21·37 N	41·44 E
53	Al Khurṭum (Khartoum)	Sud.	15·34 N	32·36 E
53	Al-Khurṭum Baḥrī	Sud.	15·43 N	32·41 E
33	Al Kiswah	Syr.	3·31 N	36·13 E
23	Alkmaar	Neth.	52·39 N	4·42 E
53	Al Kufrah (Oasis)	Libya	24·45 N	22·45 E
33	Al Kuntillah	Egypt	29·59 N	34·42 E
34	Al Kuwayt (Kuwait)	Kuw.	29·04 N	47·59 E
19	Al Lādhiqīyah (Latakia)	Syr.	35·32 N	35·51 E
35	Allāhābād	India	25·32 N	81·53 E
29	Allaykha	Sov.Un.	70·32 N	148·53 E
8	Allegan	Mi.	42·30 N	85·55 W
9	Allegany Ind. Res.	NY	42·05 N	78·55 W
9	Allegheny (R.)	Pa.	41·10 N	79·20 W
9	Allegheny Front (Mts.)	U.S.	38·12 N	80·03 W
7	Allegheny Mts.	U.S.	37·35 N	81·55 W
8	Allegheny Plat	U.S.	39·00 N	81·15 W
9	Allegheny Res.	Pa.	41·50 N	78·55 W
22	Allen, Lough (B.)	Ire.	54·07 N	8·09 W
9	Allentown	Pa.	40·35 N	75·30 W
26	Aller R.	F.R.G.	52·43 N	9·50 E
6	Alliance	Ne.	42·06 N	102·53 W
8	Alliance	Oh.	40·55 N	81·10 W
34	Al Lidām	Sau.Ar.	20·45 N	44·12 E
17	Allier (R.)	Fr.	46·43 N	3·03 E
24	Allinge	Den.	55·16 N	14·48 E
34	Al Luḥayyah	Yemen	15·58 N	42·48 E
8	Alma	Mi.	43·25 N	84·40 W
28	Alma-Ata	Sov.Un.	43·19 N	77·08 E
33	Al Mabrak (R.)	Sau.Ar.	29·16 N	35·12 E
34	Al Madīnah (Medina)	Sau.Ar.	24·26 N	39·42 E
33	Al Mafraq	Jordan	32·21 N	36·13 E
34	Al Manāmah	Bahrain	26·01 N	50·33 E
53	Al Manṣūrah	Egypt	31·02 N	31·25 E
53	Al-Marj	Libya	32·44 N	21·08 E
34	Al Maṣīrah (I.)	Om.	20·43 N	58·58 E
34	Al Mawsil	Iraq	36·00 N	42·53 E
33	Al Mazār	Jordan	31·04 N	35·41 E
33	Al Mazra'ah	Jordan	31·17 N	35·33 E
23	Almelo	Neth.	52·20 N	6·42 E
17	Almería	Sp.	36·52 N	2·28 W
24	Älmhult	Swe.	56·35 N	14·08 E
53	Al Minyā	Egypt	28·04 N	30·45 E
21	Almirós	Grc.	39·13 N	22·47 E
9	Almonte	Can.	45·15 N	76·15 W
35	Almora	India	29·20 N	79·40 E
34	Al Mubarraz	Sau.Ar.	22·31 N	46·27 E
33	Al Mudawwarah	Jordan	29·20 N	36·01 E
34	Al Mukallā	P.D.R. of Yem.	14·27 N	49·05 E
34	Al Mukhā	Yemen	13·43 N	43·27 E
24	Alnö (I.)	Swe.	62·20 N	17·39 E
41	Alor (I.)	Indon.	8·07 S	125·00 E
33	Alor Gajah	Mala.	2·23 N	102·13 E
40	Alor Setar	Mala.	6·24 N	100·08 E
36	Alot'ai	China	47·52 N	86·50 E
8	Alpena	Mi.	45·05 N	83·30 W
18	Alps (Mts.)	Eur.	46·18 N	8·42 E
48	Alpujarra	Col.	3·23 N	74·56 W
53	Al Qaḍārif	Sud.	14·03 N	35·11 E
53	Al Qāhirah (Cairo)	Egypt	30·00 N	31·17 E
53	Al Qaryah ash Sharqiyah	Libra	30·36 N	13·13 E
34	Al Qaṭīf	Sau.Ar.	26·30 N	50·00 E
34	Al Qayṣūmah	Sau.Ar.	28·15 N	46·20 E
33	Al Qunaytirah	Syr.	33·09 N	35·49 E
34	Al Qunfudhah	Sau.Ar.	19·08 N	41·05 E
33	Al Quṣaymah	Egypt	30·40 N	34·23 E
53	Al Quṣayr	Egypt	26·14 N	34·11 E
33	Al Qusayr	Egypt	34·32 N	36·33 E
24	Als (I.)	Den.	55·06 N	9·40 E
26	Alsace (Reg.)	Fr.	48·25 N	7·24 E
24	Alsterån (R.)	Swe.	56·54 N	15·50 E
50	Alta Gracia	Arg.	31·41 S	64·19 W
48	Altagracia	Ven.	10·42 N	71·34 W
49	Altagracia de Orituco	Ven.	9·53 N	66·22 W
36	Altai Mts.	Asia	49·11 N	87·15 E
7	Altamaha (R.)	Ga.	31·50 N	82·00 W
49	Altamira	Braz.	3·13 S	52·14 W
50	Altamirano	Arg.	35·26 S	58·12 W
20	Altamura	It.	40·40 N	16·35 E
36	Altan Bulag	Mong.	50·18 N	106·31 E
16	Alten (R.)	Nor.	69·40 N	24·09 E
26	Altenburg	G.D.R.	50·59 N	12·27 E
48	Altiplano (Plat.)	Bol.	18·38 S	68·20 W
48	Alto Marañón, Rio (R.)	Peru	8·18 S	77·13 W
7	Alton	Il.	38·53 N	90·11 W
9	Altoona	Pa.	40·25 N	78·25 W
53	Al-Ubayyiḍ	Sud.	13·15 N	30·15 E
53	Al-Uḍayyah	Sud.	12·06 N	28·16 E
53	Al-'Ugaylah	Libya	30·15 N	19·07 E
9	Alumette I.	Can.	45·50 N	77·00 W
53	Al Uqṣur (Luxor)	Egypt	25·38 N	32·59 E
24	Älvdalen	Swe.	61·14 N	14·04 E
24	Alvesta	Swe.	56·55 N	14·29 E
34	Al Wajh	Sau.Ar.	26·15 N	36·32 E
35	Alwar	India	27·39 N	76·39 E
25	Alytus	Sov.Un.	54·25 N	24·05 E
44	Amadeus, (L.)	Austl.	24·30 S	131·25 E
11	Amadjuak (L.)	Can.	64·50 N	69·20 W
24	Åmål	Swe.	59·05 N	12·40 E
48	Amalfi	Col.	6·55 N	75·04 W
21	Amaliás	Grc.	37·48 N	21·23 E
49	Amambai, Serra de (Mts.)	Braz.	20·06 S	57·08 W
37	Amami Guntō (Is.)	Jap.	28·25 N	129·00 E
49	Amapá	Braz.	2·14 N	50·48 W
49	Amapá (Ter.)	Braz.	1·15 N	52·15 W
49	Amarante	Braz.	6·17 S	42·43 W
6	Amarillo	Tx.	35·14 N	101·49 W
20	Amaro, Mt.	It.	42·07 N	14·07 E
31	Amasya	Tur.	40·40 N	35·50 E
48	Amazonas (State)	Braz.	4·15 S	64·30 W
49	Amazonas, Rio (R.)	Braz.	2·03 S	53·18 W
35	Ambāla	India	30·31 N	76·48 E
48	Ambalema	Col.	4·47 N	74·45 W
29	Ambarchik	Sov. Un.	69·39 N	162·18 E
48	Ambato	Ec.	1·15 S	78·30 W
55	Ambatondrazaka	Mad.	17·58 S	48·43 E
26	Amberg	F.R.G.	49·26 N	11·51 E
41	Ambil (I.)	Phil.	13·51 N	120·25 E
41	Amboina	Indon.	3·45 S	128·17 E
41	Ambon (I.)	Indon.	4·50 S	128·45 E
55	Ambositra	Mad.	20·31 S	47·28 E
8	Amboy	Il.	41·41 N	89·15 W
55	Ambre, Cap d' (C.)	Mad.	12·06 S	49·15 E
45	Ambrim (I.)	New Heb.	16·25 S	168·15 E
54	Ambriz	Ang.	7·50 S	13·06 E
54	Ambrizete	Ang.	7·14 S	12·52 E
23	Ameland (I.)	Neth.	53·29 N	5·54 E
6	American Falls Res.	Id.	42·56 N	113·18 W
67	American Highland	Ant.	72·00 S	79·00 E
7	Americus	Ga.	32·04 N	84·15 W
10	Amery	Can.	56·34 N	94·03 W
21	Amfissa	Grc.	38·32 N	22·26 E
29	Amga	Sov.Un.	61·08 N	132·09 E
53	Amhara (Prov.)	Eth.	11·30 N	36·45 E
11	Amherst	Can.	45·49 N	64·14 W
8	Amherst	Oh.	41·24 N	82·13 W
17	Amiens	Fr.	49·54 N	2·18 E
67	Amirante Is.	Sey.	6·02 S	52·30 E
9	Amityville	NY	40·41 N	73·24 W
33	'Ammān	Jordan	31·57 N	35·57 E
21	Amorgós (I.)	Grc.	36·47 N	25·47 E
11	Amos	Can.	48·31 N	78·04 W
24	Åmot (Torpen)	Nor.	61·08 N	11·17 E

Page	Name	Region	Lat.	Long.
35	Amrāvati	India	20·58 N	77·47 E
35	Amritsar	India	31·43 N	74·52 E
23	Amsterdam	Neth.	52·21 N	4·52 E
9	Amsterdam	NY	42·55 N	74·10 W
26	Amstetten	Aus.	48·09 N	14·53 E
53	Am Timan	Chad	11·18 N	20·30 E
34	Amu Darya (R.)	Asia	40·40 N	62·00 E
41	Amulung	Phil.	17·51 N	121·43 E
10	Amundsen G.	Can.	70·17 N	123·28 W
67	Amundsen Sea	Ant.	72·00 S	110·00 W
24	Amungen (L.)	Swe.	61·07 N	16·00 E
38	Amur R.	China-Sov.Un.	49·38 N	127·25 E
41	Amuyao, Mt.	Phil.	17·04 N	121·09 E
21	Amvrakikos Kólpos (G.)	Grc.	39·00 N	21·00 E
33	Amyun	Leb.	34·18 N	35·48 E
49	Anaco	Ven.	9·29 N	64·27 W
6	Anaconda	Mt.	46·07 N	112·55 W
29	Anadyr'	Sov. Un.	64·47 N	177·01 E
33	Anadyrskiy Zaliv (B.)	Sov.Un.	64·10 N	178·00 W
40	Anambas, Kepulauan (Is.)	Indon.	2·41 N	106·38 E
49	Anápolis	Braz.	16·17 S	48·47 W
50	Añatuya	Arg.	28·22 S	62·45 W
50	Anchieta	Braz.	22·49 S	43·24 W
39	Anching	China	30·32 N	117·00 E
6	Anchorage	Ak.	61·12 N	149·48 W
12	Ancon	C.Z.	8·55 N	79·32 W
20	Ancona	It.	43·37 N	13·32 E
50	Ancud	Chile	41·52 S	73·45 W
50	Ancud, G. de	Chile	41·15 S	73·00 W
50	Andalgalá	Arg.	27·35 S	66·14 W
40	Andaman Is.	Andaman & Nicobar Is.	11·38 N	92·17 E
40	Andaman Sea	Asia	12·44 N	95·45 E
23	Anderlecht	Bel.	50·49 N	4·16 E
26	Andernach	F.R.G.	50·25 N	7·23 E
8	Anderson	In.	40·05 N	85·50 W
7	Anderson	SC	34·30 N	82·40 W
10	Anderson (R.)	Can.	68·32 N	125·12 W
47	Andes Mts.	S.A.	13·00 S	75·00 W
35	Andhra Pradesh (State)	India	16·00 N	79·00 E
19	Andikíthira (I.)	Grc.	35·50 N	23·20 E
28	Andizhan	Sov.Un.	40·51 N	72·39 E
37	Andong	Kor.	36·31 N	128·42 E
18	Andorra	And.	42·38 N	1·30 E
17	Andorra	Eur.	42·30 N	2·00 E
16	Andöy (I.)	Nor.	69·12 N	14·58 E
6	Andreanof Is.	Ak.	51·10 N	177·00 W
20	Andria	It.	41·17 N	15·55 E
21	Andros	Grc.	37·50 N	24·54 E
13	Andros I.	Ba.	24·30 N	78·00 W
21	Andrós (I.)	Grc.	37·59 N	24·55 E
9	Androscoggin (R.)	Me.	44·25 N	70·45 W
17	Andújar	Sp.	38·04 N	4·03 W
45	Aneityum (I.)	New Hebr.	20·15 S	169·49 E
41	Angadanan	Phil.	16·45 N	121·45 E
41	Angaki	Phil.	17·10 N	120·40 E
38	Angangchi	China	47·05 N	123·58 E
29	Angarsk	Sov.Un.	52·48 N	104·15 E
24	Ange	Swe.	62·31 N	15·39 E
48	Angel, Salto (Falls)	Ven.	5·44 N	62·27 W
12	Angel De La Guarda (I.)	Mex.	29·30 N	113·00 W
41	Angeles	Phil.	15·09 N	120·35 E
24	Ängelholm	Swe.	56·14 N	12·50 E
16	Angermanälvan (R.)	Swe.	64·02 N	17·15 E
26	Angermünde	G.D.R.	53·02 N	14·00 E
17	Angers	Fr.	47·29 N	0·36 W
40	Angkor (Ruins)	Camb.	13·52 N	103·50 E
22	Anglesey (I.)	Wales	53·35 N	4·28 W
5	Angmagssalik	Grnld.	65·40 N	37·40 W
50	Angol	Chile	37·47 S	72·43 W
8	Angola	In.	41·35 N	85·00 W
51	Angola	Afr.	14·15 S	16·00 E
17	Angoulême	Fr.	45·40 N	0·09 E
13	Anguilla (I.)	St. Kitts-Nevis-Anguilla	18·15 N	62·54 W
24	Anholt (I).	Den.	56·43 N	11·34 E
36	Anhsi	China	40·36 N	95·49 E
37	Anhwei	China	31·30 N	117·15 E
21	Anina	Rom.	45·03 N	21·50 E
9	Anita	Pa.	41·05 N	79·00 W
55	Anjouan (I.)	Comoros	12·14 S	44·47 E
38	Ank'ang	China	32·38 N	109·10 E
31	Ankara (Angora)	Tur.	39·55 N	32·50 E
26	Anklam	G.D.R.	53·52 N	13·43 E
39	Anlu	China	31·18 N	113·40 E
39	Anlung	China	25·01 N	105·32 E
9	Ann, C.	Ma.	42·40 N	70·40 W
8	Anna	Il.	37·28 N	89·15 W
52	Annaba (Bône)	Alg.	36·57 N	7·39 E
26	Annaberg-Bucholz	G.D.R.	50·35 N	13·02 E
34	An Nafūd (Des.)	Sau.Ar.	28·30 N	40·30 E
34	An Najaf	Iraq	32·00 N	44·25 E
33	An Nakhl	Egypt	29·55 N	33·45 E
40	Annamitic Cordillera (Mts.)	Laos-Viet.	17·34 N	105·38 E
9	Annapolis	Md.	39·00 N	76·25 W
8	Ann Arbor	Mi.	42·15 N	83·45 W
34	An Nāşirīyah	Iraq	31·08 N	46·15 E
53	An Nawfalīyah	Libya	30·57 N	17·38 E
7	Anniston	Al.	33·39 N	85·47 W
53	An-Nudūd	Sud.	12·39 N	28·18 E
48	Anori	Col.	7·01 N	75·09 W
21	Áno Theológos	Grc.	40·37 N	24·41 E
20	Áno Viánnos	Grc.	35·02 N	25·26 E
39	Anp'u	China	21·28 N	110·00 E
26	Ansbach	F.R.G.	49·18 N	10·35 E
48	Anserma	Col.	5·13 N	75·47 W
48	Ansermanuevo	Col.	4·47 N	75·59 W
38	Anshan	China	41·00 N	123·00 E
39	Anshun	China	26·12 N	105·50 E
44	Anson B.	Austl.	13·10 S	130·00 E
9	Ansonia	Ct.	41·20 N	73·05 W
31	Antakya	Tur.	36·20 N	36·10 E
31	Antalya (Adalia)	Tur.	37·00 N	30·50 E
31	Antalya Körfezi (G.)	Tur.	36·40 N	31·20 E
55	Antananarivo	Mad.	18·51 S	47·40 E
67	Antarctica		80·15 S	127·00 E
67	Antarctic Pen	Ant.	70·00 S	65·00 W
17	Antequera	Sp.	37·01 N	4·34 W
52	Anti Atlas (Mts.)	Mor.	28·45 N	9·30 W
11	Anticosti, Île d' (I.)	Can.	49·30 N	62·00 W
8	Antigo	Wi.	45·09 N	89·11 W
12	Antigua	Guat.	14·32 N	90·43 W
13	Antigua	N.A.	17·15 N	61·15 W
13	Antilles, Greater (Is.)	N.A.	20·30 N	79·15 W
13	Antilles, Lesser (Is.)	N.A.	12·15 N	65·00 W
48	Antioquia	Col.	6·34 N	75·49 W
48	Antioquia (Dept.)	Col.	6·48 N	75·42 W
50	Antofagasta	Chile	23·32 N	70·21 W
50	Antofalla, Salar de Des.)	Arg.	26·00 S	67·52 W
13	Antón	Pan.	8·24 N	80·15 W
55	Antongil, Baie d' (B.)	Mad.	16·15 S	50·15 E
55	António Enes	Moz.	16·14 S	39·58 E
22	Antrim Mts.	N.Ire.	55·00 N	6·10 W
55	Antsirabe	Mad.	19·49 S	47·16 E
50	Antuco (Vol.)	Chile	37·30 S	72·30 W
38	Antung	China	40·10 N	124·30 E
23	Antwerpen (Antwerp)	Bel.	51·13 N	4·24 E
38	Antz'u	China	39·23 N	116·44 E
35	Anuradhapura	Sri Lanka	8·24 N	80·25 E
25	Anykščiai	Sov.Un.	55·34 N	25·04 E
48	Anzá	Col.	6·19 N	75·51 W
28	Anzhero-Sudzhensk	Sov. Un.	56·08 N	86·08 E
20	Anzio	It.	41·28 N	12·39 E
49	Anzoátegui (State)	Ven.	9·38 N	64·45 W
20	Aosta	It.	45·45 N	7·20 E
53	Aouk, Bahr (R.)	Chad	9·30 N	20·45 E
7	Apalachicola (R.)	Fl.	29·43 N	84·59 W
48	Apaporis (R.)	Col.	0·48 N	72·32 W
40	Aparri	Phil.	18·15 N	121·40 E
21	Apatin	Yugo.	45·40 N	19·00 E
23	Apeldoorn	Neth.	52·14 N	5·55 E
48	Apía	Col.	5·07 N	75·58 W
21	Apíranthos	Grc.	37·07 N	25·32 E
41	Apo (Mtn.)	Phil.	6·56 N	125·05 E
7	Appalachian Mts.	U.S.	37·20 N	82·00 W
24	Äppelbo	Swe.	60·30 N	14·02 E
20	Appennino (Mts.)	It.	43·48 N	11·06 E
26	Appenzell	Switz.	47·19 N	9·22 E
8	Appleton	Wi.	44·14 N	88·27 W
20	Aprília	It.	41·36 N	12·40 E
31	Apsheronskiy, P-Ov. (Pen.)	Sov.Un.	40·20 N	50·30 E
48	Apure (R.)	Ven.	8·08 N	68·46 W
48	Apurimac (R.)	Peru	11·39 S	73·48 W
19	Aqaba, G. of	Asia	28·30 N	34·40 E
33	Aqabah, Wādī al (R.)	Egypt	29·48 N	34·05 E
49	Aquidauana	Braz.	20·24 S	55·46 W
53	Arab, Bahr al- (R.)	Sud.	9·46 N	26·52 E
53	Arabian Des.	Asia	27·06 N	32·49 E
51	Arabian Pen.	Asia	28·00 N	40·00 E
32	Arabian Sea	Asia	16·00 N	65·15 E
49	Aracaju	Braz.	11·00 S	37·01 W
49	Aracati	Braz.	4·31 S	37·41 W
49	Araçatuba	Braz.	21·14 S	50·19 W
49	Aracruz	Braz.	19·58 S	40·11 W
49	Araçuaí	Braz.	16·57 S	41·56 W
33	'Arad	Isr.	31·20 N	35·15 E
27	Arad	Rom.	46·10 N	21·18 E
42	Arafura Sea	Oceania	8·40 S	130·00 E
49	Aragua (State)	Ven.	10·00 N	67·05 W
49	Aragua de Barcelona	Ven.	9·29 N	64·48 W
49	Araguaía (R.)	Braz.	8·37 S	49·43 W
49	Araguari	Braz.	18·43 S	48·03 W
49	Araguatins	Braz.	5·41 S	48·04 W
49	Aragüita	Ven.	10·13 N	66·28 W
19	Araj (Oasis)	Egypt	29·05 N	26·51 E
34	Arāk	Iran	34·08 N	49·57 E
36	Arakan Yoma (Mts.)	Bur.	19·51 N	94·13 E
21	Arakhthos (R.)	Grc.	39·10 N	21·05 E
28	Aral'sk	Sov.Un.	46·47 N	62·00 E
15	Aral'skoye More (Aral Sea)	Sov.Un.	45·17 N	60·02 E
31	Aralsor (L.)	Sov. Un.	49·00 N	48·20 E
22	Aran (I.)	Ire.	54·58 N	8·33 W
22	Aran Is.	Ire.	53·04 N	9·59 W
17	Aranjuez	Sp.	40·02 N	3·24 W
52	Araouane	Mali	18·54 N	3·33 W
31	Arapkir	Tur.	39·00 N	38·10 E
49	Araraquara	Braz.	21·47 S	48·08 W
49	Araras, Serra das (Mts.)	Braz.	18·03 S	53·23 W
50	Araras, Serra das (Mts.)	Braz.	23·30 S	53·00 W
46	Ararat	Austl.	37·17 S	142·56 E
31	Ararat (Mtn.)	Tur.	39·50 N	44·20 E
49	Arari (L.)	Braz.	0·30 S	48·50 W
49	Araripe, Chapada do (Plain)	Braz.	5·55 S	40·42 W
31	Aras (R.)	Iran-Sov. Un.	39·15 N	47·10 E
49	Aratuípe	Braz.	13·12 S	38·58 W
48	Arauca	Col.	6·56 N	70·45 W
49	Arauca (R.)	Ven.	7·13 N	68·43 W
49	Araxá	Braz.	19·41 S	46·46 W
49	Araya, Punta de (Pt.)	Ven.	10·40 N	64·15 W
41	Arayat	Phil.	15·10 N	120·44 E
53	'Arbi	Sud.	20·36 N	29·57 E
24	Arboga	Swe.	59·26 N	15·50 E
20	Arborea	It.	39·50 N	8·36 E
22	Arbroath	Scot.	56·36 N	2·25 W
9	Archbald	Pa.	41·30 N	75·35 W
48	Archidona	Ec.	1·01 S	77·49 W
17	Arcos de la Frontera	Sp.	36·44 N	5·48 W
21	Arda (R.)	Bul.	41·36 N	25·18 E
34	Aradabīl	Iran	38·15 N	48·00 E
31	Ardahan	Tur.	41·10 N	42·40 E
24	Ardals Fd.	Nor.	58·53 N	7·55 E
30	Ardatov	Sov.Un.	54·58 N	46·10 E
23	Ardennes (Mts.)	Bel.	50·01 N	5·12 E
6	Ardmore	Ok.	34·10 N	97·08 W
9	Ardmore	Pa.	40·01 N	75·18 W
16	Åre	Swe.	63·12 N	13·12 E
13	Arecibo	P.R.	18·28 N	66·45 W
49	Areia Branca	Braz.	4·58 S	37·02 W
49	Arenas, Punta (Pt.)	Ven.	10·57 N	64·24 W
24	Arendal	Nor.	58·29 N	8·44 E
48	Arequipa	Peru	16·27 S	71·30 W
20	Arezzo	It.	43·28 N	11·54 E
47	Argentina	S.A.	35·30 S	67·00 W
50	Argentino (L.)	Arg.	50·15 S	72·45 W
21	Arges (R.)	Rom.	44·27 N	25·22 E
21	Argolikos Kólpos (G.)	Grc.	37·20 N	23·00 E
21	Argos	Grc.	37·38 N	22·45 E
21	Argostólion	Grc.	38·10 N	20·30 E
24	Århus	Den.	56·09 N	10·10 E
20	Ariano	It.	41·09 N	15·11 E
48	Ariari (R.)	Col.	3·34 N	73·42 W
48	Arica	Chile	18·34 S	70·14 W
27	Arieşul (R.)	Rom.	46·25 N	23·15 E
33	Arīḩā (Jericho)	Jordan	31·51 N	35·28 E
41	Aringay	Phil.	16·25 N	120·20 E
49	Arinos (R.)	Braz.	12·09 S	56·49 W
49	Aripuanã (R.)	Braz.	7·06 S	60·29 W
33	'Arīsh, Wādī al (R.)	Egypt	30·36 N	34·07 E
6	Arizona (State)	U.S.	34·00 N	113·00 W
7	Arkansas (State)	U.S.	34·50 N	93·40 W
6	Arkansas R.	Ok.	35·20 N	94·56 W
30	Arkhangelsk (Archangel)	Sov.Un.	64·30 N	40·25 E
22	Arklow	Ire.	52·47 N	6·10 W
24	Arkona, C.	G.D.R.	54·43 N	13·43 E
26	Arlberg Tun.	Aus.	47·05 N	10·15 E
17	Arles	Fr.	43·42 N	4·38 E
9	Arlington	Vt.	43·05 N	73·05 W
9	Arlington	Va.	38·55 N	77·10 W
24	Arlöv	Swe.	55·38 N	13·05 E
44	Arltunga	Austl.	23·19 S	134·45 W
22	Armagh	N.Ire.	54·21 N	6·25 W
48	Armaro	Col.	4·58 N	74·54 W
31	Armavir	Sov.Un.	45·00 N	41·00 E
48	Armenia	Col.	4·33 N	75·40 W
28	Armenian S.S.R.	Sov.Un.	41·00 N	44·39 E
23	Armentières	Fr.	50·43 N	2·53 E
46	Armidale	Austl.	30·27 S	151·50 E
11	Armstrong Station	Can.	50·21 N	89·00 W
23	Arnhem	Neth.	51·58 N	5·56 E
44	Arnhem, C.	Austl.	12·15 S	137·00 E
44	Arnhem Land, (Reg.)	Austl.	13·15 S	133·00 E
20	Arno (R.)	It.	43·45 N	10·42 E
9	Arnprior	Can.	45·25 N	76·20 W
23	Arnsberg	F.R.G.	51·25 N	8·02 E
26	Arnstadt	G.D.R.	50·51 N	10·57 E
54	Aroab	Namibia	25·40 S	19·45 E
41	Aroroy	Phil.	12·30 N	123·24 E
50	Arpoador, Ponta do (Pt.)	Braz.	22·59 S	43·11 W
34	Ar Ramādī	Iraq	33·30 N	43·12 E
22	Arran (I.)	Scot.	55·39 N	5·30 W
53	Ar Rank	Sud.	11·45 N	32·53 E
17	Arras	Fr.	50·21 N	2·40 E
34	Ar Rub Al Khālī (Des.)	Sau.Ar.	20·30 N	49·15 E
53	Ar-Ruşayriş	Sud.	11·38 N	34·42 E
21	Árta	Grc.	39·08 N	21·02 E
46	Artesian Basin, The	Austl.	26·45 S	141·40 E
41	Aru, Kepulauan (Is.)	Indon.	6·20 S	133·00 E
48	Aruba (I.)	Neth. Antilles	12·29 N	70·00 W
36	Arunachal Pradesh (Union Ter.)	India	27·35 N	92·56 E
55	Arusha	Tan.	3·22 S	36·41 E
11	Arvida	Can.	48·26 N	71·11 W
24	Arvika	Swe.	59·41 N	12·35 E
30	Arzamas	Sov. Un.	55·20 N	43·52 E
26	As	Czech.	50·12 N	12·13 E
35	Asansol	India	23·45 N	86·58 E
30	Asbest	Sov.Un.	57·02 N	61·28 E
9	Asbury Park	NJ	40·13 N	74·01 W
51	Ascension (I.)	Atl.O.	8·00 S	13·00 W
26	Aschaffenburg	F.R.G.	49·58 N	9·12 E
26	Aschersleben	G.D.R.	51·46 N	11·28 E
20	Ascoli Piceno	It.	42·50 N	13·55 E
21	Asenovgrad	Bul.	42·00 N	24·49 E
44	Ashburton (R.)	Austl.	22·30 S	115·30 E
33	Ashdod	Isr.	31·46 N	34·39 E
7	Asheville	NC	35·35 N	82·35 W
15	Ashkhabad	Sov.Un.	39·45 N	58·13 E
8	Ashland	Ky.	38·25 N	82·40 W
8	Ashland	Oh.	40·50 N	82·15 W
9	Ashland	Pa.	40·45 N	76·20 W
9	Ashley	Pa.	41·15 N	75·55 W
40	Ashmore Rf.	Indon.	12·08 S	122·45 E
33	Ashqelon	Isr.	31·40 N	34·36 E
53	Ash Shabb	Egypt	22·34 N	29·52 E
34	Ash Shaqrā'	Sau.Ar.	25·10 N	45·08 E
33	Ash Shawbak	Jordan	30·31 N	35·35 E
34	Ash Shiḩr	P.D.R. of Yem.	14·45 N	49·32 E
8	Ashtabula	Oh.	41·55 N	80·50 W
11	Ashuanipi (L.)	Can.	52·40 N	67·42 W
15	Asia Minor	Asia	38·18 N	31·18 E
20	Asinara, Golfo di (G.)	It.	40·58 N	8·28 E
20	Asinara (I.)	It.	41·02 N	8·22 E
34	Asīr (Reg.)	Sau.Ar.	19·30 N	42·00 E
51	Asir, Ras (C.)	Som.	11·55 N	51·30 E
24	Askersund	Swe.	58·43 N	14·53 E
53	Asmera	Eth.	15·17 N	38·56 E
53	Asosa	Eth.	10·13 N	34·28 E
53	As Sallūm	Egypt	31·34 N	25·09 E
33	As Salt	Jordan	32·02 N	35·44 E
35	Assam (State)	India	26·00 N	91·00 E
24	Assens	Den.	55·16 N	9·54 E
52	Assini	Ivory Coast	4·52 N	3·16 W
10	Assiniboia	Can.	49·38 N	105·59 W
49	Assis	Braz.	22·39 S	50·21 W
20	Assisi	It.	43·04 N	12·37 E
53	As-Sudd (Reg.)	Sud.	8·45 N	30·45 E
34	As Sulaymānīyah	Iraq	35·47 N	45·23 E
34	As Suwaydā'	Syr.	32·41 N	36·41 E
53	As Suways	Egypt	29·58 N	32·34 E
21	Astakós	Grc.	38·42 N	21·00 E
31	Astara	Sov.Un.	38·30 N	48·50 E
20	Asti	It.	44·54 N	8·12 E
36	Astin Tagh			

Page	Name	Region	Lat.	Long.
40	Banggi (I.)	Mala.	7·12 N	117·10 E
53	Banghāzī	Libya	32·08 N	20·06 E
40	Bangka (I.)	Indon.	2·24 S	106·55 E
40	Bangkalan	Indon.	6·07 S	112·50 E
35	Bangladesh	Asia	24·15 N	90·00 E
7	Bangor	Me.	44·47 N	68·47 W
8	Bangor	Mi.	42·20 N	86·05 W
9	Bangor	Pa.	40·55 N	75·10 W
22	Bangor	Wales	53·13 N	4·05 W
41	Bangued	Phil.	17·36 N	120·38 E
53	Bangui	Cen.Afr.Emp.	4·22 N	18·35 E
54	Bangweulu, L.	Zambia	10·55 S	30·10 E
41	Bani	Phil.	16·11 N	119·51 E
52	Bani (R.)	Mali	13·07 N	6·15 W
19	Banī Mazār	Egypt	28·29 N	30·48 E
53	Banī Suwayf	Egypt	29·05 N	31·06 E
40	Banjak, Kepulauan (I.)	Indon.	2·08 N	97·15 E
20	Banja Luka	Yugo.	44·45 N	17·11 E
40	Banjarmasin	Indon.	3·18 S	114·32 E
52	Banjul (Bathurst)	Gam.	13·28 N	16·39 W
40	Ban Kantang	Thai.	7·26 N	99·28 E
55	Bankberg (Mts.)	S.Afr.	32·18 S	25·15 E
45	Banks (Is.)	Austl.	10·10 S	143·08 E
5	Banks I.	Can.	73·00 N	123·00 W
45	Banks Is.	New Hebr.	13·38 S	168·23 E
46	Banks Str.	Austl.	40·45 S	148·00 E
40	Ban Kui Nua	Thai.	12·04 N	99·50 E
22	Bann (R.)	N.Ire.	54·50 N	6·29 W
48	Baños	Ec.	1·30 S	78·22 W
27	Banská Bystrica	Czech.	48·46 N	19·10 E
21	Bansko	Bul.	41·51 N	23·33 E
41	Banton	Phil.	12·54 N	121·55 E
22	Bantry	Ire.	51·39 N	9·30 W
22	Bantry B.	Ire.	51·25 N	10·09 W
40	Banyuwangi	Indon.	8·15 S	114·15 E
53	Banzyville	Zaire	4·14 N	21·11 E
55	Bapsfontein	S.Afr.	26·01 S	28·26 E
48	Baqueroncito	Col.	3·18 N	74·40 W
8	Baraboo	Wi.	43·29 N	89·44 W
27	Baranovichi	Sov.Un.	53·08 N	25·59 E
33	Baranpauh	Indon.	0·40 N	103·28 E
50	Barão de Juperanã	Braz.	22·21 S	43·41 W
49	Barão de Melgaço	Braz.	16·12 S	55·48 W
48	Baraya	Col.	3·10 N	75·04 W
48	Barbacoas	Col.	1·39 N	78·12 W
49	Barbacoas	Ven.	9·30 N	66·58 W
13	Barbados	N.A.	13·30 N	59·00 W
53	Barbar	Sud.	18·11 N	34·00 E
8	Barberton	Oh.	41·01 N	81·37 W
54	Barberton	S.Afr.	25·48 S	31·04 E
48	Barbosa	Col.	6·26 N	75·19 W
8	Barboursville	WV	38·20 N	82·20 W
13	Barbuda (I.)	Antigua	17·45 N	61·15 W
45	Barcaldine	Austl.	23·33 S	145·17 E
20	Barcellona	It.	38·07 N	15·15 E
18	Barcelona	Sp.	41·25 N	2·08 E
49	Barcelona	Ven.	10·09 N	64·41 W
48	Barcelos	Braz.	1·04 S	63·00 W
34	Bardar-e Pahlavī	Iran	37·16 N	49·15 E
33	Bardawīl, Sabkhat al (B.)	Egypt	31·20 N	33·24 E
27	Bardejov	Czech.	49·18 N	21·18 E
22	Bardsey (I.)	Wales	52·45 N	4·50 W
8	Bardstown	Ky.	37·50 N	85·30 W
28	Barents Sea	Sov.Un.	72·14 N	37·28 E
53	Barentu	Eth.	15·06 N	37·39 E
29	Barguzin	Sov.Un.	53·44 N	109·28 E
20	Bari	It.	41·08 N	16·53 E
48	Barinas	Ven.	8·36 N	70·14 W
10	Baring, C.	Can.	70·07 N	119·48 W
40	Barisan, Pegunungan (Mts.)	Indon.	2·38 S	101·45 E
40	Barito (Strm.)	Indon.	2·10 S	114·38 E
53	Barka (R.)	Eth.	16·44 N	37·34 E
55	Barkly East	S.Afr.	30·58 S	27·37 E
44	Barkly Tableland	Austl.	18·15 S	137·05 E
44	Barlee (L.)	Austl.	29·45 S	119·00 E
20	Barletta	It.	41·19 N	16·20 E
28	Barnaul	Sov.Un.	53·18 N	83·23 E
9	Barnesboro	Pa.	40·45 N	78·50 W
8	Barnesville	Oh.	39·55 N	81·10 W
9	Barnet	Vt.	44·20 N	72·00 W
22	Barnstaple	Eng.	51·06 N	4·05 W
53	Bāro (R.)	Eth.	7·40 N	34·17 E
38	Baroda	India	22·21 N	73·12 E
54	Barotse Pln.	Zambia	15·50 S	22·55 E
53	Barqah (Cyrenaica) (Prov.)	Libya	31·09 N	21·45 E
48	Barquisimeto	Ven.	10·04 N	39·16 W
49	Barra	Braz.	11·04 S	43·11 W
46	Barraba	Austl.	30·22 S	150·36 E
49	Barra do Corda	Braz.	5·33 S	45·13 W
22	Barra Is.	Scot.	57·00 N	7·30 W
48	Barrancabermeja	Col.	7·06 N	73·49 W
48	Barranquilla	Col.	10·57 N	75·00 W
49	Barras	Braz.	4·13 S	42·14 W
9	Barre	Vt.	44·15 N	72·30 W
49	Barreiras	Braz.	12·13 S	44·59 W
17	Barreiro	Port.	38·39 N	9·05 W
46	Barren, C.	Austl.	40·20 S	149·00 E
55	Barren, Îles	Mad.	18·18 S	43·57 E
49	Barretos	Braz.	20·40 S	48·36 W
10	Barrhead	Can.	54·08 N	114·24 W
9	Barrie	Can.	44·25 N	79·45 W
46	Barrington Tops (Mtn.)	Austl.	32·00 S	151·25 E
6	Barrow	Ak.	71·20 N	156·00 W
22	Barrow	Eng.	54·10 N	3·15 W
44	Barrow (I.)	Austl.	20·50 S	115·00 E
44	Barrow Creek	Austl.	21·23 S	133·55 E
22	Barrow R.	Ire.	52·35 N	7·05 W
26	Barth	G.D.R.	54·20 N	12·43 E
49	Bartica	Guy.	6·23 N	58·32 W
31	Bartin	Tur.	41·35 N	32·12 E
45	Bartle Frere, Mt.	Austl.	17·30 S	145·46 E
9	Barton	Vt.	44·45 N	72·05 W
27	Bartoszyce	Pol.	54·15 N	20·50 E
46	Barwon (R.)	Austl.	29·45 S	148·2 E
26	Barycz R.	Pol.	51·30 N	16·38 E
53	Basankusu	Zaire	1·14 N	19·45 E
26	Basel	Switz.	47·32 N	7·35 E
55	Bashee (R.)	S.Afr.	31·47 S	28·25 E
39	Bashi Chan	Phil.	21·20 N	120·22 E
30	Bashkir (A.S.S.R.)	Sov.Un.	54·12 N	57·15 E
40	Basilan (I.)	Phil.	6·37 N	122·07 E
20	Basilicata (Reg.)	It.	40·30 N	15·55 E
20	Baška	Yugo.	44·58 N	14·44 E
31	Baskale	Tur.	38·10 N	44·00 E
31	Baskunchak (L.)	Sov.Un.	48·20 N	46·40 E
53	Basoko	Zaire	0·52 N	23·50 E
10	Bassano	Can.	50·47 N	112·28 W
20	Bassano	It.	45·46 N	11·44 E
55	Bassas da India (I.)	Afr.	21·23 S	39·42 E
40	Bassein	Bur.	16·46 N	94·47 E
13	Basse Terre	Guad.	16·00 N	61·43 W
8	Bass Is.	Oh.	41·40 N	82·50 W
46	Bass Str.	Austl.	39·40 S	145·40 E
24	Båstad	Swe.	56·26 N	12·46 E
20	Bastia	Fr.	42·43 N	9·27 E
23	Bastogne	Bel.	50·02 N	5·45 E
52	Bata	Equat.Gui.	1·51 N	9·45 E
33	Batam I.	Indon.	1·03 N	104·00 E
39	Batan Is.	Phil.	20·58 N	122·20 E
39	Batangan, C.	Viet.	15·18 N	109·10 E
41	Batangas	Phil.	13·45 N	121·04 E
27	Bátaszék	Hung.	46·07 N	18·40 E
8	Batavia	Il.	41·51 N	88·18 W
9	Batavia	NY	43·00 N	78·15 W
8	Batesville	In.	39·15 N	85·15 W
22	Bath	Eng.	51·24 N	2·20 W
9	Bath	NY	42·25 N	77·20 W
45	Bathurst	Aust.	33·28 S	149·30 E
55	Bathurst	S.Afr.	33·26 S	26·53 E
44	Bathurst (I.)	Austl.	11·19 S	130·13 E
10	Bathurst Inlet	Can.	68·10 N	108·00 W
41	Batian (I.)	Indon.	1·07 S	127·52 E
34	Bāţlāq-E Gāvkhūnī (L.)	Iran	31·40 N	52·48 E
52	Batna	Alg.	35·41 N	6·12 E
7	Baton Rouge	La.	30·28 N	91·10 W
40	Battambang	Camb.	13·14 N	103·15 E
8	Battle Creek	Mi.	42·20 N	85·15 W
11	Battle Harbour	Can.	52·17 N	55·33 W
27	Battonya	Hung.	46·17 N	21·00 E
40	Batu Kepulauan (I.)	Indon.	0·10 S	99·55 E
31	Batumi	Sov.Un.	41·40 N	41·30 E
33	Batu Pahat	Mala.	1·51 N	102·56 E
33	Batupanjang	Indon.	1·42 N	101·35 E
49	Baturité	Braz.	4·16 S	38·47 W
41	Bauang	Phil.	16·31 N	120·19 E
52	Bauchi	Nig.	10·19 N	9·50 E
54	Baudouinville	Zaire	7·12 S	29·39 E
11	Bauld, C.	Can.	51·38 N	55·25 W
49	Bauru	Braz.	22·21 S	48·57 W
25	Bauska	Sov.Un.	56·24 N	24·12 E
26	Bautzen	G.D.R.	51·11 N	14·27 E
46	Baw Baw, Mt.	Austl.	37·50 S	146·17 E
40	Bawean (I.)	Indon.	5·50 S	112·40 E
18	Bay al Kabīr Wadi (R.)	Libya	29·52 N	14·28 E
41	Bayambang	Phil.	15·50 N	120·26 E
13	Bayamón	P.R.	18·27 N	66·13 W
28	Bayan-Aul	Sov.Un.	50·43 N	75·37 E
9	Bayard	WV	39·15 N	79·20 W
31	Bayburt	Tur.	40·15 N	40·10 E
8	Bay City	Mi.	43·35 N	83·55 W
36	Baydarag Gol (R.)	Mong.	46·09 N	98·52 E
26	Bayern (Bavaria) (State)	F.R.G.	49·00 N	11·16 E
17	Bayeux	Fr.	49·19 N	0·41 W
29	Baykal, Ozero (Baikal, L.)	Sov.Un.	53·00 N	109·28 E
29	Baykit	Sov.Un.	61·43 N	96·39 E
28	Baykonur	Sov.Un.	47·46 N	66·11 E
41	Bayombong	Phil.	16·28 N	121·09 E
17	Bayonne	Fr.	43·28 N	1·30 W
9	Bayonne	NJ	40·40 N	74·07 W
21	Bayramic	Tur.	39·48 N	26·35 E
26	Bayreuth	F.R.G.	49·56 N	11·35 E
9	Bays, L. of	Can.	45·15 N	79·00 W
33	Bayt Lahm (Bethlehem)	Jordan	31·42 N	35·13 E
17	Baza	Sp.	37·29 N	2·46 W
54	Bazaruto, Ilha do (I.)	Moz.	21·42 S	36·10 E
23	Beachy Head	Eng.	50·40 N	0·25 E
9	Beacon	NY	41·30 N	73·55 W
6	Beatrice	Ne.	40·16 N	96·45 W
8	Beattyville	Ky.	37·35 N	83·40 W
54	Beaufort West	S.Afr.	32·20 S	22·45 E
7	Beaumont	Tx.	30·05 N	94·06 W
8	Beaver (I.)	Mi.	45·40 N	85·30 W
10	Beaver (R.)	Can.	54·20 N	111·10 W
8	Beaver Dam	Wi.	43·29 N	88·50 W
8	Beaver Ind. Res.	Mi.	45·40 N	85·30 W
48	Bebará	Col.	6·07 N	76·39 W
21	Bečej	Yugo.	45·36 N	20·03 E
52	Béchar	Alg.	31·39 N	2·14 W
8	Beckley	WV	37·40 N	81·15 W
9	Bedford	Can.	45·10 N	73·00 W
22	Bedford	Eng.	52·10 N	0·25 W
8	Bedford	In.	38·50 N	86·30 W
8	Bedford	Oh.	41·23 N	81·32 W
9	Bedford	Pa.	40·05 N	78·20 W
55	Bedford	S.Afr.	32·43 S	26·19 E
27	Bedzin	Pol	50·19 N	19·10 E
8	Beech Grove	In.	39·43 N	86·05 W
46	Beecroft Hd.	Austl.	35·03 S	151·15 E
33	Be'er Sheva'	Isr.	31·15 N	34·48 E
33	Be'er Sheva' (R.)	Isr.	31·23 N	34·30 E
46	Bega	Austl.	36·50 S	149·49 E
54	Beira	Moz.	19·46 S	34·58 E
33	Beirut (Bayrūt)	Leb.	33·53 N	35·30 E
17	Beja	Port.	38·03 N	7·53 W
17	Béja	Tun.	36·52 N	9·20 E
52	Bejaïa (Bougie)	Alg.	36·46 N	5·00 E
34	Bejestān	Iran	34·30 N	58·22 E
27	Békés	Hung.	46·45 N	21·08 E
27	Békéscsaba	Hung.	46·39 N	21·06 E
37	Beketova	Sov.Un.	53·23 N	125·21 E
21	Bela Crkva	Yugo.	44·53 N	21·25 E
21	Bela-Slatina	Bul.	43·2 N	23·56 E
49	Bela Vista de Goia's	Braz.	16·57 S	48·47 W
40	Belawan	Indon.	3·43 N	98·43 E
30	Belaya (R.)	Sov.Un.	52·30 N	56·15 E
31	Belaya Tserkov'	Sov.Un.	49·48 N	30·09 E
11	Belcher Is.	Can.	56·20 N	80·40 W
8	Belding	Mi.	43·05 N	85·25 W
30	Belebey	Sov.Un.	54·00 N	54·10 E
49	Belém (Pará)	Braz.	1·18 S	48·27 W
50	Belén	Par.	23·30 S	57·09 W
45	Bélep, Îsles	N.Cal.	19·30 S	160·32 E
22	Belfast	N.Ire.	54·36 N	5·45 W
22	Belfast, Lough (B.)	Ire.	54·45 N	6·00 W
17	Belfort	Fr.	47·40 N	7·50 E
35	Belgaum	India	15·57 N	74·32 E
23	Belgium	Eur.	51·00 N	2·52 E
31	Belgorod	Sov.Un.	50·36 N	36·32 E
9	Belington	WV	39·00 N	79·55 W
21	Beli Timok (R.)	Yugo.	43·35 N	22·13 E
40	Belitung (I.)	Indon.	3·30 S	107·30 E
12	Belize	Belize	17·31 N	88·10 W
12	Belize	N.A.	17·00 N	88·40 W
8	Bellaire	Oh.	40·00 N	80·45 W
35	Bellary	India	15·15 N	76·56 E
50	Bella Union	Ur.	30·18 S	57·26 W
50	Bella Vista	Arg.	27·07 S	65·14 W
50	Bella Vista	Arg.	28·35 S	58·53 W
50	Bella Vista	Arg.	34·18 S	58·41 W
49	Bella Vista	Braz.	22·16 S	56·14 W
8	Bellefontaine	Oh.	40·25 N	83·50 W
17	Belle Île (I.)	Fr.	47·15 N	3·30 W
11	Belle Isle, Str. of	Can.	51·35 N	56·30 W
9	Belleville	Can.	44·15 N	77·25 W
8	Belleville	Il.	38·31 N	89·59 W
8	Bellevue	Mi.	42·30 N	85·00 W
8	Bellevue	Oh.	41·15 N	82·45 W
6	Bellingham	Wa.	48·46 N	122·29 W
67	Bellingshausen Sea	Ant.	72·00 S	80·30 W
20	Bellinzona	Switz.	46·10 N	9·09 E
48	Bello	Col.	6·20 N	75·33 W
9	Bellows Falls	Vt.	43·10 N	72·30 W
11	Bell Pen	Can.	63·50 N	81·16 W
20	Belluno	It.	46·08 N	12·14 E
50	Bell Ville	Arg.	32·33 S	62·36 W
54	Bellville	S.Afr.	33·54 S	18·38 E
49	Belmonte	Braz.	15·58 S	38·47 W
12	Belmopan	Belize	17·15 N	88·47 W
29	Belogorsk	Sov.Un.	51·09 N	128·32 E
50	Belo Horizonte	Braz.	19·54 S	43·56 W
8	Beloit	Wi.	42·31 N	89·04 W
30	Belomorsk	Sov.Un.	64·30 N	34·42 E
31	Belopol'ye	Sov.Un.	51·10 N	34·19 E
30	Beloretsk	Sov.Un.	53·58 N	58·25 E
28	Belorussian (S.S.R.)	Sov.Un.	53·30 N	25·33 E
30	Beloye (L.)	Sov.Un.	60·10 N	38·05 E
30	Belozersk	Sov.Un.	60·00 N	38·00 E
31	Bel'tsy	Sov.Un.	47·47 N	27·57 E
28	Belukha, Gol'tsy (Mtn.)	Sov.Un.	49·47 N	86·23 E
8	Belvidere	Il.	42·14 N	88·52 W
9	Belvidere	NJ	40·50 N	75·05 W
45	Belyando (R.)	Austl.	22·09 S	146·48 E
30	Belyy	Sov.Un.	55·52 N	32·58 E
28	Belyy (I.)	Sov.Un.	73·19 N	72·00 E
54	Bembe	Ang.	7·00 S	14·20 E
54	Bena Dibele	Zaire	4·00 S	22·49 E
46	Benalla	Austl.	36·30 S	146·00 E
17	Benavente	Sp.	42·01 N	5·43 W
6	Bend	Or.	44·04 N	121·17 W
31	Bendery	Sov.Un.	46·49 N	29·29 E
46	Bendigo	Austl.	36·39 S	144·20 E
26	Benešov	Czech.	49·48 N	14·40 E
20	Benevento	It.	41·08 N	14·46 E
32	Bengal, B. of	Asia	17·30 N	87·00 E
33	Bengkalis	Indon.	1·29 N	102·06 E
40	Bengkulu	Indon.	3·46 S	102·18 E
54	Benguela	Ang.	12·35 S	13·25 E
22	Ben Hope (Mtn.)	Scot.	58·25 N	4·25 W
48	Beni (R.)	Bol.	13·41 S	67·30 W
52	Beni-Abbès	Alg.	30·11 N	2·13 W
18	Benicarló	Sp.	40·26 N	0·25 E
51	Benin	Afr.	8·00 N	2·00 E
52	Benin City	Nig.	6·19 N	5·41 E
52	Beni Saf	Alg.	35·23 N	1·20 W
20	Benkovac	Yugo.	44·02 N	15·41 E
55	Ben Macdhui (Mtn.)	Leso-S.Afr.	30·38 S	27·54 E
9	Bennington	Vt.	42·55 N	73·15 W
55	Benoni	S.Afr.	26·11 S	28·19 E
26	Bensheim	F.R.G.	49·42 N	8·38 E
8	Benton	Il.	38·00 N	88·55 W
8	Benton Harbor	Mi.	42·05 N	86·30 W
52	Benue (R.)	Nig.	7·55 N	8·55 E
33	Benut (R.)	Mala.	1·43 N	103·20 E
8	Benwood	WV	39·55 N	80·45 W
21	Beograd (Belgrade)	Yugo.	44·48 N	20·32 E
33	Berakit, Tanjung (C.)	Indon.	1·16 N	104·44 E
21	Berat	Alb.	40·43 N	19·59 E
41	Berau, Teluk (B.)	Indon.	2·22 S	131·40 E
50	Berazategui	Arg.	34·46 S	58·14 W
51	Berbera	Som.	10·25 N	45·05 E
19	Berd'ansk	Sov.Un.	46·45 N	36·47 E
31	Berdichev	Sov.Un.	49·53 N	28·32 E
8	Berea	Ky.	37·30 N	84·19 W
8	Berea	Oh.	41·22 N	81·51 W
27	Beregovo	Sov.Un.	48·13 N	22·40 E
10	Berens River	Can.	52·22 N	97·02 W
27	Berettyóújfalu	Hung.	47·14 N	21·33 E
27	Berëza	Sov.Un.	52·29 N	24·59 E
27	Berezhany	Sov.Un.	49·25 N	24·58 E
30	Berezniki	Sov.Un.	59·25 N	56·46 E
30	Berëzovo	Sov.Un.	64·10 N	65·10 E
31	Bergama	Tur.	39·08 N	27·09 E
20	Bergamo	It.	45·43 N	9·41 E
49	Bergantīn	Ven.	10·04 N	64·23 W
23	Bergedorf	F.R.G.	53·29 N	10·12 E
26	Bergen	G.D.R.	54·26 N	13·26 E
24	Bergen	Nor.	60·24 N	5·20 E
23	Bergen op Zoom	Neth.	51·29 N	4·16 E
17	Bergerac	Fr.	44·49 N	0·28 E
55	Bergville	S.Afr.	28·46 S	29·22 E
35	Berhampur	India	19·19 N	84·48 E
5	Bering Sea	Asia-N.A.	58·00 N	175·00 W
6	Bering Str.	Ak.	64·50 N	169·50 W
6	Berkeley	Ca.	37·52 N	122·17 W
9	Berkeley Springs	WV	39·40 N	78·10 W
21	Berkovitsa	Bul.	43·14 N	23·08 E
26	Berlin, East	G.D.R.	52·31 N	13·28 E

Page	Name	Region	Lat.	Long.
7	Boston Mts.	Ar.	35·46 N	93·32 W
46	Botany B.	Austl.	33·58 S	151·11 E
21	Botevgrad	Bul.	42·54 N	23·41 E
16	Bothnia, G. of	Eur.	63·40 N	21·30 E
27	Botosani	Rom.	47·46 N	26·40 E
51	Botswana	Afr.	22·10 S	23·13 E
26	Bottrop	F.R.G.	51·31 N	6·56 E
49	Botucatú	Braz.	22·50 S	48·23 W
11	Botwood	Can.	49·08 N	55·21 W
52	Bouafle	Ivory Coast	6·59 N	5·45 W
52	Bouaké	Ivory Coast	7·41 N	5·00 W
53	Bouar	Cen.Afr.Emp.	5·57 N	15·36 E
52	Boudenib	Mor.	32·14 N	3·04 W
17	Bou Dia, C.	Tun.	35·18 N	11·17 E
18	Boufarik	Alg.	36·35 N	2·55 E
42	Bougainville Trench	Oceania	7·00 S	152·00 E
52	Bougouni	Mali	11·27 N	7·30 W
18	Bouira	Alg.	36·25 N	3·55 E
44	Boulder	Austl.	31·00 S	121·40 E
6	Boulder	Co.	40·02 N	105·19 W
6	Boulder City	Nv.	35·57 N	114·50 W
17	Boulogne-sur-Mer	Fr.	50·44 N	1·37 E
52	Bouna	Ivory Coast	9·16 N	3·00 W
67	Bounty Is	N.Z.	47·42 S	179·05 E
52	Bourem	Mali	16·43 N	0·15 W
17	Bourg-en-Bresse	Fr.	46·12 N	5·13 E
17	Bourges	Fr.	47·06 N	2·22 E
46	Bourke	Austl.	30·10 S	146·00 E
22	Bournemouth	Eng.	50·44 N	1·55 W
18	Bou Saada	Alg.	35·13 N	4·17 E
53	Bousso	Chad	10·33 N	16·45 E
52	Boutilimit	Mauritania	17·30 N	14·54 W
67	Bouvetöen	Atl.O.	54·26 S	3·24 E
20	Bovino	It.	41·14 N	15·21 E
10	Bow (R.)	Can.	50·35 N	112·15 W
45	Bowen	Austl.	20·02 S	148·14 E
7	Bowling Green	Ky.	37·00 N	86·26 W
8	Bowling Green	Oh.	41·25 N	83·40 W
41	Bowokan, Pulau-Pulau (Is.)	Indon.	2·20 S	123·45 E
22	Boyle	Ire.	53·59 N	8·15 W
8	Boyne City	Mi.	45·15 N	85·05 W
22	Boyne R.	Ire.	53·40 N	6·40 W
21	Bozcaada (Tenedos)	Tur.	39·50 N	26·05 E
21	Bozcaada (I.)	Tur.	39·50 N	26·00 E
6	Bozeman	Mt.	45·41 N	111·00 W
20	Bra	It.	44·41 N	7·52 E
20	Brač (I.)	Yugo.	43·18 N	16·36 E
9	Bracebridge	Can.	45·05 N	79·20 W
24	Bräcke	Swe.	62·44 N	15·28 E
49	Braço Maior (R.)	Braz.	11·00 S	51·00 W
49	Braço Menor (R.)	Braz.	11·38 S	50·00 W
20	Brádano (R.)	It.	40·43 N	16·22 E
9	Braddock	Pa.	40·24 N	79·52 W
22	Bradford	Eng.	53·47 N	1·44 W
8	Bradford	Oh.	40·10 N	84·30 W
9	Bradford	Pa.	42·00 N	78·40 W
8	Bradley	Il.	41·09 N	87·52 W
17	Braga	Port.	41·20 N	8·25 W
50	Bragado	Arg.	35·07 S	60·28 W
49	Bragança	Braz.	1·02 S	46·50 W
35	Brahmaputra (R.)	India	26·45 N	92·45 E
35	Brāhui (Mts.)	Pak.	28·32 N	66·15 E
31	Brăila	Rom.	45·15 N	27·58 E
55	Brakpan	S.Afr.	26·15 S	28·22 E
9	Brampton	Can.	43·41 N	79·46 W
50	Branca, Pedra (Mtn.)	Braz.	22·55 S	43·28 W
49	Branco (R.)	Braz.	2·21 N	60·38 W
54	Brandberg (Mtn.)	Namibia	21·15 S	14·15 E
26	Brandenburg	G.D.R.	52·25 N	12·33 E
26	Brandenburg (Reg.)	G.D.R.	52·12 N	13·31 E
10	Brandon	Can.	49·50 N	99·57 W
9	Brandon	Vt.	43·45 N	73·05 W
22	Brandon Hill	Ire.	52·15 N	10·12 W
9	Branford	Ct.	41·15 N	72·50 W
27	Braniewo	Pol.	54·23 N	19·50 E
27	Brańsk	Pol.	52·44 N	22·51 E
9	Brantford	Can.	43·09 N	80·17 W
49	Brasília	Braz.	15·49 S	47·39 W
49	Brasilia Legal (Fordlândia)	Braz.	3·45 S	55·46 W
21	Braşov (Oraşul-Stalin)	Rom.	45·39 N	25·35 E
52	Brass	Nig.	4·28 N	6·28 E
27	Bratislava	Czech.	48·09 N	17·07 E
29	Bratsk	Sov.Un.	56·10 N	102·04 E
9	Brattleboro	Vt.	42·50 N	72·35 W
26	Braunau	Aus.	48·15 N	13·05 E
26	Braunschweig	F.R.G.	52·16 N	10·32 E
51	Brava	Som.	1·20 N	44·00 E
24	Bråviken (R.)	Swe.	58·40 N	16·40 E
6	Brawley	Ca.	32·59 N	115·32 W
22	Bray	Ire.	53·10 N	6·05 W
8	Brazil	In.	39·30 N	87·00 W
47	Brazil	S.A.	9·00 S	53·00 W
47	Brazilian Highlands (Mts.)	Braz.	14·00 S	48·00 W
6	Brazos (R.)	U.S.	33·10 N	98·50 W
54	Brazzaville	Con.	4·16 S	15·17 E
21	Brčko	Yugo.	44·54 N	18·46 E
27	Brda R.	Pol.	53·18 N	17·55 E
26	Břeclav	Czech.	48·46 N	16·54 E
22	Brecon Beacons	Wales	52·00 N	3·55 W
23	Breda	Neth.	51·35 N	4·47 E
54	Bredasdorp	S.Afr.	34·15 S	20·00 E
26	Bregenz	Aus.	47·30 N	9·46 E
21	Bregovo	Bul.	44·07 N	22·45 E
55	Breidbach	S.Afr.	32·54 S	27·26 E
16	Breidha Fd.	Ice.	65·15 N	22·50 W
49	Brejo	Braz.	3·33 S	42·46 W
24	Bremangerland (I.)	Nor.	61·51 N	4·25 E
26	Bremen	F.R.G.	53·05 N	8·50 E
8	Bremen	In.	41·25 N	86·05 W
26	Bremerhaven	F.R.G.	53·33 N	8·38 E
26	Brenner P.	Aus.-It.	47·00 N	11·30 E
9	Brentwood	Md.	39·00 N	76·55 W
20	Brescia	It.	45·33 N	10·15 E
20	Bressanone	It.	46·42 N	11·40 E
17	Brest	Fr.	48·24 N	4·30 W
27	Brest	Sov.Un.	52·06 N	23·43 E
49	Breves	Braz.	1·32 S	50·13 W
24	Brevik	Nor.	59·04 N	9·39 E
46	Brewarrina	Austl.	29·54 S	146·50 E
20	Brežice	Yugo.	45·55 N	15·37 E
21	Breznik	Bul.	42·44 N	22·55 E
9	Bridgeport	Ct.	41·12 N	73·12 W
8	Bridgeport	Il.	38·40 N	87·45 W
8	Bridgeport	Oh.	40·00 N	80·45 W
9	Bridgeport	Pa.	40·06 N	75·21 W
9	Bridgeton	NJ	39·30 N	75·15 W
13	Bridgetown	Barb.	13·08 N	59·37 W
46	Bridgewater	Austl.	42·50 S	147·28 E
11	Bridgewater	Can.	44·23 N	64·31 W
9	Bridgton	Me.	44·04 N	70·45 W
22	Bridlington	Eng.	54·06 N	0·10 W
26	Brig	Switz.	46·17 N	7·59 E
46	Bright	Austl.	36·43 S	147·00 E
22	Brighton	Eng.	50·47 N	0·07 W
21	Brindisi	It.	40·38 N	17·57 E
20	Brinje	Yugo.	45·00 N	15·08 E
46	Brisbane	Austl.	27·30 S	153·10 E
9	Bristol	Ct.	41·40 N	72·55 W
22	Bristol	Eng.	51·29 N	2·39 W
9	Bristol	Pa.	40·06 N	74·51 W
9	Bristol	RI	41·41 N	71·14 W
7	Bristol	Tn.	36·35 N	82·10 W
9	Bristol	Vt.	44·10 N	73·00 W
7	Bristol	Va.	36·36 N	82·00 W
5	Bristol B.	Ak.	58·08 N	158·54 W
22	Bristol Chan.	Eng.	51·20 N	3·47 W
10	British Columbia (Prov.)	Can.	56·00 N	124·53 W
54	Britstown	S.Afr.	30·30 S	23·40 E
17	Brive-la-Gaillarde	Fr.	45·10 N	1·31 E
26	Brno	Czech.	49·18 N	16·37 E
9	Brockport	NY	43·15 N	77·55 W
11	Brockville	Can.	44·35 N	75·40 W
27	Brodnica	Pol.	53·16 N	19·26 E
27	Brody	Sov.Un.	50·05 N	25·10 E
46	Broken Hill	Austl.	31·55 S	141·35 E
9	Bromptonville	Can.	45·30 N	72·00 W
24	Brønderslev	Den.	57·15 N	9·56 E
8	Bronson	Mi.	41·55 N	85·15 W
6	Brooks Ra.	Ak.	68·20 N	159·00 W
8	Brookville	In.	39·20 N	85·00 W
9	Brookville	Pa.	41·10 N	79·00 W
22	Broom (B.)	Scot.	57·59 N	5·32 W
44	Broome	Austl.	18·00 S	122·15 E
26	Broumov	Czech.	50·33 N	15·55 E
8	Brownstown	In.	38·50 N	86·00 W
6	Brownsville	Tx.	25·55 N	97·30 W
6	Brownwood	Tx.	31·44 N	98·58 W
44	Bruce, Mt.	Austl.	22·35 S	118·15 E
8	Bruce Pen.	Can.	44·50 N	81·20 W
26	Bruchsal	F.R.G.	49·08 N	8·34 E
26	Bruck	Aus.	47·25 N	15·14 E
23	Brugge	Bel.	51·13 N	3·05 E
40	Brunei	Asia	4·52 N	113·38 E
7	Brunswick	Ga.	31·08 N	81·30 W
9	Brunswick	Me.	43·54 N	69·57 W
9	Brunswick	Md.	39·20 N	77·35 W
50	Brunswick, Pen. de	Chile	53·25 S	71·15 W
45	Bruny (I.)	Austl.	43·30 S	147·50 E
50	Brusque	Braz.	27·15 S	48·45 W
23	Bruxelles (Brussels)	Bel.	50·51 N	4·21 E
8	Bryan	Oh.	41·25 N	84·30 W
31	Bryansk	Sov.Un.	53·12 N	34·23 E
6	Bryce Canyon Natl. Park	Ut.	37·35 N	112·15 W
33	Buatan	Indon.	0·45 N	101·49 E
52	Buba	Guinea-Bissau	11·39 N	14·58 W
48	Bucaramanga	Col.	7·12 N	73·14 W
41	Bucay	Phil.	17·32 N	120·42 E
44	Buccaneer Arch.	Austl.	16·05 S	122·00 E
27	Buchach	Sov.Un.	49·04 N	25·25 E
52	Buchanan	Lib.	5·57 N	10·02 W
8	Buchanan	Mi.	41·50 N	86·25 W
45	Buchanan (L.)	Austl.	21·40 S	145·00 E
9	Buckhannon	WV	39·00 N	80·10 W
22	Buckhaven	Scot.	56·10 N	3·10 W
22	Buckie	Scot.	57·40 N	2·50 W
45	Buckland Tableland	Austl.	24·31 S	148·00 E
21	Bucureşti (Bucharest)	Rom.	44·23 N	26·10 E
8	Bucyrus	Oh.	40·50 N	82·55 W
27	Budapest	Hung.	47·30 N	19·05 E
48	Buenaventura	Col.	3·46 N	77·09 W
48	Buenaventura, Bahia de	Col.	3·45 N	79·23 W
9	Buena Vista	Va.	37·45 N	79·20 W
50	Buenos Aires	Arg.	34·20 S	58·30 W
48	Buenos Aires	Col.	3·01 N	76·34 W
50	Buenos Aires (Prov.)	Arg.	36·15 S	61·45 W
50	Buenos Aires (L.)	Arg.-Chile	46·30 S	72·15 W
9	Buffalo	NY	42·54 N	78·51 W
55	Buffalo (R.)	S.Afr.	28·35 S	30·27 E
10	Buffalo Head Hills	Can.	57·16 N	116·18 W
27	Bug (R.)	Pol.	52·29 N	21·20 E
31	Bug (R.)	Sov.Un.	48·12 N	30·13 E
48	Buga	Col.	3·54 N	76·17 W
7	Buggs Island L.	NC-Va.	36·30 N	78·38 W
20	Bugojno	Yugo.	44·03 N	17·28 E
30	Bugul'ma	Sov.Un.	54·40 N	52·40 E
30	Buguruslan	Sov.Un.	53·30 N	52·32 E
41	Buhi	Phil.	13·26 N	123·31 E
31	Buinaksk	Sov.Un.	42·40 N	47·20 E
54	Bujumbura	Burundi	3·23 S	29·22 E
54	Bukama	Zaire	9·08 S	26·00 E
54	Bukavu	Zaire	2·30 S	28·52 E
15	Bukhara	Sov.Un.	39·31 N	64·22 E
33	Bukitbatu	Indon.	1·25 N	101·58 E
40	Bukittingg	Indon.	0·25 S	100·28 E
54	Bukoba	Tan.	1·20 S	31·49 E
27	Bukovina (Reg.)	Sov.Un.	48·06 N	25·20 E
41	Bula	Indon.	3·00 S	130·30 E
41	Bulalacao	Phil.	12·32 N	121·25 E
54	Bulawayo	Rh.	20·12 S	28·43 E
14	Bulgaria	Eur.	42·12 N	24·13 E
45	Bulloo (R.)	Austl.	25·23 S	143·30 E
7	Bull Shoals Res.	Ar.-Mo.	36·35 N	92·57 W
29	Bulun	Sov.Un.	70·48 N	127·27 E
54	Bulungu	Zaire	6·04 S	21·54 E
55	Bulwer	S.Afr.	29·49 S	29·48 E
53	Bumba	Zaire	2·11 N	22·28 E
41	Buna	Pap.N.Gui.	8·58 S	148·38 E
44	Bunbury	Austl.	33·25 S	115·45 E
46	Bundaberg	Austl.	24·45 S	152·18 E
29	Buor-Khaya, Guba (B.)	Sov.Un.	71·45 N	131·00 E
34	Buraydah	Sau.Ar.	26·23 N	44·14 E
45	Burdekin (R.)	Austl.	19·22 S	145·07 E
31	Burdur	Tur.	37·50 N	30·15 E
35	Burdwān	India	23·29 N	87·53 E
29	Bureya	Sov.Un.	49·55 N	130·00 E
21	Burgas	Bul.	42·29 N	27·30 E
19	Burgaski Zaliv (G.)	Bul.	42·30 N	27·40 E
55	Bur Gavo	Som.	1·14 S	41·47 E
26	Burgdorf	Switz.	47·04 N	7·37 E
9	Burgess	Va.	37·53 N	76·21 W
41	Burgos	Phil.	16·03 N	119·52 E
18	Burgos	Sp.	42·20 N	3·44 W
24	Burgsvik	Swe.	57·04 N	18·18 E
35	Burhānpur	India	21·26 N	76·08 E
41	Burias I.	Phil.	12·56 N	122·56 E
41	Burias Pass.	Phil.	13·04 N	123·11 E
11	Burin	Can.	47·02 N	55·10 W
9	Burke	Vt.	44·40 N	72·00 W
44	Burketown	Austl.	17·50 S	139·30 E
7	Burlington	Ia.	40·48 N	91·05 W
9	Burlington	NJ	40·04 N	74·52 W
9	Burlington	Vt.	44·30 N	73·15 W
8	Burlington	Wi.	42·41 N	88·16 W
32	Burma	Asia	21·00 N	95·15 E
10	Burnaby	Can.	49·14 N	122·58 W
46	Burnie	Austl.	41·15 S	146·05 E
22	Burnley	Eng.	53·47 N	2·19 W
10	Burns Lake	Can.	54·14 N	125·46 W
31	Bursa	Tur.	40·10 N	28·10 E
53	Būr Safājah	Egypt	26·57 N	33·56 E
53	Būr Sa'īd	Egypt	31·15 N	32·19 E
53	Būr Sūdān (Port Sudan)	Sud.	19·30 N	37·10 E
8	Burt (L.)	Mi.	45·25 N	84·45 W
22	Burton-on-Trent	Eng.	52·48 N	1·37 W
41	Buru (I.)	Indon.	3·30 S	126·30 E
41	Buruncan Pt.	Phil.	12·11 N	121·23 E
51	Burundi	Afr.	3·00 S	29·30 E
29	Buryat A.S.S.R.	Sov.Un.	55·15 N	112·00 E
23	Bury St. Edmunds	Eng.	52·14 N	0·44 E
50	Burzaco	Arg.	34·35 S	58·23 W
34	Būshehr	Iran	28·48 N	50·53 E
54	Bushmanland (Reg.)	S.Afr.	29·15 S	18·45 E
53	Businga	Zaire	3·20 N	20·53 E
27	Busk	Sov.Un.	49·58 N	24·39 E
44	Busselton	Austl.	33·40 S	115·30 E
20	Busto Arsizio	It.	45·47 N	8·51 E
41	Busuanga (I.)	Phil.	12·20 N	119·43 E
53	Buta	Zaire	2·48 N	24·44 E
55	Butha Buthe	Leso.	28·49 S	28·16 E
18	Butler	In.	41·25 N	84·50 W
9	Butler	Pa.	40·50 N	79·55 W
6	Butte	Mt.	46·00 N	112·31 W
55	Butterworth	S.Afr.	32·20 S	28·09 E
22	Butt of Lewis (C.)	Scot.	58·34 N	6·15 W
41	Butuan	Phil.	8·40 N	125·33 E
40	Butung (I.)	Indon.	5·00 S	122·55 E
31	Buturlinovka	Sov.Un.	50·47 N	40·35 E
30	Buy	Sov.Un.	58·30 N	41·48 E
38	Buyr Nuur (L.)	China-Mong.	47·50 N	117·00 E
21	Buzău	Rom.	45·09 N	26·51 E
53	Buzaymah	Libya	25·14 N	22·13 E
31	Buzuluk	Sov.Un.	52·50 N	52·10 E
21	Byala	Bul.	43·26 N	25·44 E
27	Bydgoszcz	Pol.	53·07 N	18·00 E
8	Byesville	Oh.	39·55 N	81·35 W
24	Bygdin (L.)	Nor.	61·24 N	8·31 E
24	Byglandsfjord	Nor.	58·40 N	7·49 E
27	Bytom	Pol.	50·21 N	18·55 E
27	Bytow	Pol.	54·10 N	17·30 E

C

Page	Name	Region	Lat.	Long.
50	Caazapá	Par.	26·14 S	56·18 W
41	Cabagan	Phil.	17·27 N	121·50 E
41	Cabalete (I.)	Phil.	14·19 N	122·00 E
41	Cabanatuan	Phil.	15·30 N	120·56 E
49	Cabedelo	Braz.	6·58 S	34·49 W
48	Cabimas	Ven.	10·21 N	71·27 W
54	Cabinda	Ang.	5·10 S	10·00 E
54	Cabinda	Ang.	5·33 S	12·12 E
49	Cabo Frio	Braz.	22·53 S	42·02 W
8	Cabot Hd.	Can.	45·15 N	81·20 W
11	Cabot Str.	Can.	47·35 N	60·00 W
41	Cabra (I.)	Phil.	13·55 N	119·55 E
49	Cabrobó	Braz.	8·34 S	39·13 W
50	Cabuçu (R.)	Braz.	22·57 S	43·36 W
41	Cabugao	Phil.	17·48 N	120·28 E
21	Čačak	Yugo.	43·51 N	20·$2 E
49	Cáceres	Braz.	16·11 S	57·32 W
17	Cáceres	Sp.	39·28 N	6·20 W
50	Cachi, Nevados de (Pk.)	Arg.	25·05 S	66·40 W
50	Cachinal	Chile	24·57 S	69·33 W
49	Cachoeira	Braz.	12·32 S	38·47 W
50	Cachoeirá do Sul	Braz.	30·02 S	52·49 W
54	Caconda	Ang.	13·43 S	15·06 E
41	Cadig, Mt.	Phil.	14·11 N	122·26 E
8	Cadillac	Mi.	44·15 N	85·25 W
8	Cadiz	Oh.	40·15 N	81·00 W
17	Cádiz	Sp.	36·34 N	6·20 W
17	Cádiz, Golfo de	Sp.	36·50 N	7·00 W
17	Caen	Fr.	49·13 N	0·22 W
22	Caernarfon	Wales	53·08 N	4·17 W
22	Caernarfon B.	Wales	53·09 N	4·56 W
49	Caetité	Braz.	14·02 S	42·14 W
41	Cagayan	Phil.	8·13 N	124·30 E
40	Cagayan (R.)	Phil.	16·45 N	121·55 E
40	Cagayan Is.	Phil.	9·40 N	120·30 E
40	Cagayan Sulu (I.)	Phil.	7·00 N	118·30 E
20	Cagli	It.	43·35 N	12·40 E
20	Cagliari	It.	39·16 N	9·08 E
20	Cagliari, Golfo di (G.)	It.	39·08 N	9·12 E
49	Cagua	Ven.	10·12 N	67·27 W
13	Caguas	P.R.	18·12 N	66·01 W
54	Cahama	Ang.	16·17 S	14·19 E
17	Cahors	Fr.	44·27 N	1·27 E
49	Caiapó, Serra do (Mts.)	Braz.	17·52 S	52·37 W
48	Caicedonia	Col.	4·21 N	75·48 W
13	Caicos Is.	Turks & Caicos Is.	21·45 N	71·50 W
41	Caiman Pt.	Phil.	15·56 N	119·33 E
12	Caimito, (R.)	Pan.	8·50 N	79·45 W

Page	Name	Region	Lat.	Long.
49	Coruripe	Braz.	10•09 S	36•13 W
6	Corvallis	Or.	44•34 N	123•17 W
9	Corry	Pa.	41•55 N	79•40 W
8	Corydon	In.	38•10 N	86•05 W
8	Corydon	Ky.	37•45 N	87•40 W
20	Cosenza	It.	39•18 N	16•15 E
8	Coshocton	Oh.	40•15 N	81•55 W
8	Cosigüina (Vol.)	Nic.	12•59 N	83•35 W
55	Cosmoledo Group (Is.)	Afr.	9•42 S	47•45 E
13	Costa Rica	N.A.	10•30 N	84•30 W
48	Cotabambas	Peru	13•49 S	72•17 W
41	Cotabato	Phil.	7•06 N	124•13 E
52	Cotonou	Benin	6•21 N	2•26 E
48	Cotopaxi (Mtn.)	Ec.	0•40 S	78•26 W
22	Cotswold Hills	Eng.	51•35 N	2•16 W
26	Cottbus	G.D.R.	51•47 N	14•20 E
17	Cottian Alps (Mts.)	Fr.-It.	44•46 N	7•02 E
9	Coudersport	Pa.	41•45 N	78•00 W
50	Coulto, Serra do (Mts.)	Braz.	22•33 S	43•27 W
7	Council Bluffs	Ia.	41•16 N	95•53 W
49	Courantyne (R.)	Guy.-Sur.	4•28 N	57•42 W
10	Courtenay	Can.	49•41 N	125•00 W
22	Coventry	Eng.	52•25 N	1•29 W
17	Covilhã	Port.	40•18 N	7•29 W
8	Covington	In.	40•10 N	87•15 W
8	Covington	Ky.	39•05 N	84•31 W
8	Covington	Oh.	40•10 N	84•20 W
9	Covington	Va.	37•50 N	80•00 W
46	Cowal, L.	Austl.	33•30 S	147•10 E
44	Cowan, (L.)	Austl.	32•00 S	122•30 E
22	Cowes	Eng.	50•43 N	1•25 W
46	Cowra	Austl.	33•50 S	148•33 E
49	Coxim	Braz.	18•32 S	54•43 W
48	Coyaima	Col.	3•48 N	75•11 W
12	Cozumel, Isla de (I.)	Mex.	20•26 N	87•10 W
55	Cradock	S.Afr.	32•12 S	25•38 E
21	Craiova	Rom.	44•18 N	23•50 E
9	Cranberry (L.)	NY	44•10 N	74•50 W
10	Cranbrook	Can.	49•31 N	115•46 W
8	Crandon	Wi.	45•35 N	88•55 W
9	Cranston	RI	41•46 N	71•25 W
6	Crater Lake Natl. Park	Or.	42•58 N	122•40 W
49	Crateús	Braz.	5•09 S	40•35 W
49	Crato	Braz.	7•19 S	39•13 W
8	Crawfordsville	In.	40•00 N	86•55 W
23	Crécy	Fr.	50•13 N	1•48 E
10	Cree (L.)	Can.	57•35 N	107•52 W
55	Creighton	S.Afr.	30•02 S	28•52 E
20	Crema	It.	45•21 N	9•53 E
20	Cremona	It.	45•09 N	10•02 E
20	Cres (Tsrĕs)	Yugo.	44•58 N	14•21 E
20	Cres (I.)	Yugo.	44•50 N	14•31 E
8	Crestline	Oh.	40•50 N	82•40 W
10	Creston	Can.	49•06 N	116•31 W
20	Crete (I.)	Grc.	35•15 N	24•30 E
22	Crewe	Eng.	53•06 N	2•27 W
26	Crimmitschau	G.D.R.	50•49 N	12•22 E
9	Crisfield	Md.	38•00 N	75•50 W
48	Cristobal Colón, Pico (Pk.)	Col.	11•00 N	74•00 W
27	Crişul Alb (R.)	Rom.	46•20 N	22•15 E
21	Crna (R.)	Yugo.	41•03 N	21•46 E
21	Črna Gora (Montenegro) (Reg.)	Yugo.	42•55 N	18•52 E
20	Črnomelj	Yugo.	45•35 N	15•11 E
44	Croker (I.)	Austl.	10•45 S	132•25 E
8	Crooksville	Oh.	39•45 N	82•05 W
9	Cross (L.)	Can.	44•55 N	76•55 W
10	Cross Lake	Can.	54•37 N	97•47 W
8	Crosswell	Mi.	43•15 N	82•35 W
21	Crotone	It.	39•05 N	17•08 E
13	Crown Mtn.	Vir. Is. (U.S.A.)	18•22 N	64•58 W
8	Crown Point	In.	41•25 N	87•22 W
9	Crown Point	NY	44•00 N	73•25 W
45	Croydon	Austl.	18•15 S	142•15 E
22	Croydon	Eng.	51•22 N	0•06 W
67	Crozet Is.	Ind. O.	46•20 S	51•30 E
13	Cruz, Cabo (C.)	Cuba	19•50 N	77•45 W
50	Cruz Alta	Braz.	28•41 S	54•02 W
50	Cruz del Eje	Arg.	30•46 S	64•45 W
48	Cruzeiro do Sul	Braz.	7•34 S	72•40 W
8	Crystal Lake	Il.	42•15 N	88•18 W
27	Csongrád	Hung.	46•42 N	20•09 E
27	Csorna	Hung.	47•39 N	17•11 E
49	Cúa	Ven.	10•10 N	66•54 W
54	Cuando (R.)	Ang.	16•50 S	22•40 E
54	Cuango (Kwango) (R.)	Afr.	6•35 S	16•50 E
54	Cuanza (R.)	Ang.	9•05 S	13•15 E
50	Cuarto Saladillo (R.)	Arg.	33•00 S	63•25 W
13	Cuba	N.A.	22•00 N	79•00 W
49	Cubagua, Isla (I.)	Ven.	10•48 N	64•10 W
54	Cubango (Okavango) (R.)	Ang.-Namibia	17•10 S	18•20 E
54	Cuchi	Ang.	14•40 S	16•50 E
48	Cúcuta	Col.	7•56 N	72•30 W
8	Cudahy	Wi.	42•57 N	87•52 W
35	Cuddalore	India	11•49 N	79•46 E
35	Cuddapah	India	14•31 N	78•52 E
44	Cue	Austl.	27•30 S	118•10 E
48	Cuenca	Ec.	2•52 S	78•54 W
17	Cuenca	Sp.	40•05 N	2•07 W
12	Cuernavaca	Mex.	18•55 N	99•15 W
20	Cuglieri	It.	40•11 N	8•37 E
49	Cuiabá	Braz.	15•33 S	56•03 W
22	Cuillin Sd.	Scot.	57•09 N	6•20 W
54	Cuito (R.)	Ang.	14•15 S	19•00 E
13	Culebra (I.)	P.R.	18•19 N	65•32 W
45	Culgoa (R.)	Austl.	29•21 S	147•00 E
12	Culiacán	Mex.	24•45 N	107•30 W
40	Culion	Phil.	11•43 N	119•58 E
18	Cullera	Sp.	39•12 N	0•15 W
55	Cullinan	S.Afr.	25•41 S	28•32 E
9	Culpeper	Va.	38•30 N	77•55 W
8	Culver	In.	41•15 N	86•25 W
49	Cumaná	Ven.	10•28 N	64•10 W
9	Cumberland	Md.	39•40 N	78•40 W
7	Cumberland (R.)	U.S.	36•45 N	85•33 W
45	Cumberland Is.	Austl.	20•29 S	149•46 E
11	Cumberland Pen.	Can.	65•59 N	64•05 W
11	Cumberland Sd.	Can.	65•27 N	65•44 W
48	Cundinamarca (Dept.)	Col.	4•57 N	74•27 W
54	Cunene (Kunene) (R.)	Ang.-Namibia	17•05 S	12•35 E
20	Cuneo	It.	44•24 N	7•31 E
46	Cunnamulla	Austl.	28•00 S	145•55 E
12	Cupula, Pico (Mtn.)	Mex.	24•45 N	111•10 W
48	Curaçao (I.)	Neth. Antilles	12•12 N	68•58 W
50	Curacautín	Chile	38•25 S	71•53 W
50	Curicó	Chile	34•57 S	71•14 W
50	Curitiba	Braz.	25•20 S	49•15 W
49	Currais Novos	Braz.	6•02 S	36•39 W
55	Currie, Mt.	S.Afr.	30•28 S	29•23 E
21	Curtea de Argeş	Rom.	45•09 N	24•40 E
45	Curtis (I.)	Austl.	23•38 S	151•43 E
49	Curuá (R.)	Braz.	6•26 S	54•39 W
21	Ćurug	Yugo.	45•27 N	20•26 E
48	Curupira, Serra (Mts.)	Braz.-Ven.	1•00 N	65•30 W
49	Cururupu	Braz.	1•40 S	44•56 W
50	Curuzú Cuatiá	Arg.	29•45 S	57•58 W
49	Curvelo	Braz.	18•47 S	44•14 W
35	Cuttack	India	20•38 N	85•53 E
54	Cuvo (R.)	Ang.	10•55 S	14•00 E
26	Cuxhaven	F.R.G.	53•51 N	8•43 E
8	Cuyahoga Falls	Oh.	41•08 N	81•29 W
40	Cuyo Is.	Phil.	10•54 N	120•08 E
49	Cuyuni (R.)	Guy.-Ven.	6•40 N	60•44 W
48	Cuzco	Peru	13•36 S	71•52 W
8	Cynthiana	Ky.	38•20 N	84•20 W
32	Cyprus	Asia	35•00 N	31•00 E
14	Czechoslovakia	Eur.	49•28 N	16•00 E
27	Czersk	Pol.	53•47 N	17•58 E
27	Czestochowa	Pol.	50•49 N	19•10 E

D

Page	Name	Region	Lat.	Long.
52	Dabakala	Ivory Coast	8•16 N	4•36 W
48	Dabeiba	Col.	7•01 N	76•16 W
27	Dabrowa	Pol.	53•37 N	23•18 E
35	Dacca	Bngl.	23•45 N	90•29 E
26	Dachau	F.R.G.	48•16 N	11•26 E
35	Dādra & Nagar Haveli (Union Ter.)	India	20•00 N	73•00 E
41	Daet (Mtn.)	Phil.	14•07 N	122•59 E
52	Dagana	Senegal	16•31 N	15•30 W
31	Dagestan (Reg.)	Sov.Un.	43•40 N	46•10 E
41	Dagupan	Phil.	16•02 N	120•20 E
53	Dahlak Arch. (Is.)	Eth.	15•45 N	40•30 E
45	Dajarra	Austl.	21•45 S	139•30 E
52	Dakar	Senegal	14•40 N	17•26 W
38	Dalai Nor (L.)	China	48•50 N	116•45 E
24	Dalälven (R.)	Swe.	60•26 N	15•50 E
46	Dalby	Austl.	27•10 S	151•15 E
24	Dale	Nor.	60•34 N	5•46 E
7	Dale Hollow (L.)	Tn.	36•33 N	85•03 W
24	Dalen	Nor.	59•28 N	8•01 E
44	Daley (L.)	Austl.	14•15 S	131•15 E
44	Daley Waters	Austl.	16•15 S	133•30 E
6	Dallas	Tx.	32•45 N	96•48 W
20	Dalmacija (Reg.)	Yugo.	43•25 N	16•37 E
29	Dalnerechensk	Sov.Un.	46•07 N	133•21 E
53	Dalqu	Sud.	20•07 N	30•41 E
45	Dalrymple, Mt.	Austl.	21•14 S	148•46 E
55	Dalton	S.Afr.	29•21 S	30•41 E
49	Dam	Sur.	4•36 N	54•54 W
35	Damān	India	20•32 N	72•53 E
53	Damanhūr	Egypt	30•59 N	30•31 E
41	Damar (I.)	Indon.	7•15 S	129•15 E
54	Damaraland (Reg.)	Namibia	22•15 S	16•15 E
31	Damavand (Mtn.)	Iran	36•05 N	52•05 E
54	Damba	Ang.	6•41 S	15•08 E
34	Dāmghān	Iran	35•50 N	54•15 E
41	Dampier, Selat (Str.)	Indon.	0•40 N	131•15 E
44	Dampier Arch.	Austl.	20•15 S	116•25 E
44	Dampier Land (Pen.)	Austl.	17•30 S	122•25 E
53	Danakil Pln.	Eth.	12•45 N	41•01 E
39	Da Nang (Tourane)	Viet.	16•08 N	108•22 E
9	Danbury	Ct.	41•23 N	73•27 W
46	Dandenong	Austl.	37•59 S	145•13 E
53	Dānglā	Eth.	11•17 N	37•00 E
30	Danilov	Sov.Un.	58•12 N	40•08 E
21	Danilov Grad	Yugo.	42•31 N	19•08 E
31	Dankov	Sov.Un.	53•17 N	39•09 E
9	Dannemora	NY	44•45 N	73•45 W
55	Dannhauser	S.Afr.	28•07 S	30•04 E
9	Dansville	NY	42•30 N	77•40 W
19	Danube (R.)	Eur.	43•41 N	23•35 E
8	Danville	Il.	40•10 N	87•35 W
8	Danville	In.	39•45 N	86•30 W
8	Danville	Ky.	37•35 N	84•50 W
9	Danville	Pa.	41•00 N	76•35 W
7	Danville	Va.	36•35 N	79•24 W
16	Danzig, G. of	Pol.	54•41 N	19•01 E
33	Daphnae (Ruins)	Egypt	30•43 N	32•12 E
33	Dar'ā	Syria	32•37 N	36•07 E
27	Dărăbani	Rom.	48•13 N	26•38 E
52	Daraj	Libya	30•12 N	10•14 E
35	Darbhanga	India	26•03 N	85•09 E
55	Dar-es-Salaam	Tan.	6•48 S	39•17 E
53	Dārfūr (Prov.)	Sud.	13•21 N	23•46 E
35	Dargai	Pak.	34•35 N	72•00 E
52	D'Arguin, Cap (C.)	Mauritania	20•28 N	17•46 W
48	Darien	Col.	3•56 N	76•30 W
48	Darién, Golfo del (G.)	N.A.-S.A.	9•36 N	77•54 W
35	Darjeeling	India	27•05 N	88•16 E
46	Darling (R.)	Austl.	31•50 S	143•20 E
46	Darling Downs (Reg.)	Austl.	27•22 S	150•50 E
44	Darling Ra.	Austl.	30•30 S	115•45 E
22	Darlington	Eng.	54•32 N	1•35 W
8	Darlington	Wi.	42•41 N	90•06 W
26	Darłowo	Pol.	54•26 N	16•23 E
26	Darmstadt	F.R.G.	49•53 N	8•40 E
53	Darnah	Libya	32•44 N	22•41 E
22	Dartmoor	Eng.	50•35 N	4•05 W
11	Dartmouth	Can.	44•40 N	63•34 W
22	Dartmouth	Eng.	50•33 N	3•28 W
41	Daru (I.)	Pap.N.Gui.	9•17 S	143•13 E
20	Daruvar	Yugo.	45•37 N	17•16 E
40	Darvel B.	Mala.	4•50 N	118•40 E
44	Darwin	Austl.	12•25 S	131•00 E
50	Darwin, Cordillera (Mts.)	Chile-Arg.	54•40 S	69•30 W
34	Daryācheh-ye Rezā'īyeh (L.)	Iran	38•07 N	45•17 E
34	Dasht (R.)	Pak.	25•30 N	62•30 E
34	Dasht-e Kavīr Des.	Iran	34•43 N	53•30 E
34	Dasht-e-Lūt (Des.)	Iran	31•47 N	58•38 E
41	Dasol B.	Phil.	15•53 N	119•40 E
40	Datu, Tandjung (C.)	Indon.	2•08 N	110•15 E
25	Daugava (R.)	Sov.Un.	56•40 N	24•40 E
30	Daugavpils	Sov.Un.	55•52 N	25•32 E
10	Dauphin	Can.	51•09 N	100•00 W
41	Davao	Phil.	7•05 N	125•30 E
41	Davao G.	Phil.	6•30 N	125•45 E
7	Davenport	Ia.	41•34 N	90•38 W
45	Davenport	N.Z.	37•29 S	174•47 E
13	David	Pan.	8•27 N	82•27 W
27	David-Gorodok	Sov.Un.	52•02 N	27•14 E
9	Davis	WV	39•15 N	79•25 W
67	Davis Sea	Ant.	66•00 S	92•00 E
5	Davis Str.	Can.	66•00 N	60•00 W
26	Davos	Switz.	46•47 N	9•50 E
53	Dawa (R.)	Eth.	4•34 N	41•34 E
34	Dawāsir, Wādī ad (R.)	Sau.Ar.	20•48 N	44•07 E
40	Dawna Ra.	Bur.	17•02 N	98•01 E
10	Dawson	Can.	64•04 N	139•22 W
46	Dawson (R.)	Austl.	24•20 S	149•45 E
10	Dawson Creek	Can.	55•46 N	120•14 W
17	Dax	Fr.	43•42 N	1•06 W
34	Dayr az Zawr	Syr.	35•15 N	40•01 E
8	Dayton	Oh.	39•45 N	84•15 W
7	Daytona Beach	Fl.	29•11 N	81•02 W
9	Dayville	Ct.	41•50 N	71•55 W
54	De Aar	S.Afr.	30•45 S	24•05 E
33	Dead Sea	Isr.-Jordan	31•30 N	35•30 E
6	Deadwood	SD	44•23 N	103•43 W
9	Deal Island	Md.	38•10 N	75•55 W
50	Deán Funes	Arg.	30•26 S	64•12 W
8	Dearborn	Mi.	42•18 N	83•15 W
22	Dearg, Ben (Mtn.)	Scot.	57•48 N	4•59 W
10	Dease Str.	Can.	68•50 N	108•20 W
41	De Atauro (I.)	Indon.	8•20 S	126•15 E
6	Death Valley	Ca.-Nv.	36•55 N	117•12 W
21	Debar (Dibra)	Yugo.	41•31 N	20•32 E
18	Debdou	Mor.	34•01 N	2•50 W
27	Deblin	Pol.	51•34 N	21•49 E
27	Debno	Sov.Un.	50•24 N	25•44 E
27	Debrecen	Hung.	47•32 N	21•40 E
53	Debre Markos	Eth.	10•15 N	37•45 E
53	Debre Tabor	Eth.	11•57 N	38•09 E
8	Decatur	Il.	39•50 N	88•59 W
8	Decatur	In.	40•50 N	84•55 W
8	Decatur	Mi.	42•10 N	86•00 W
35	Deccan (Plat.)	India	19•05 N	76•40 E
26	Decin	Czech.	50•47 N	14•14 E
50	Dedo do Deus (Mt.)	Braz.	22•30 S	43•02 W
52	Dédougou	Upper Volta	12•38 N	3•28 W
22	Dee (R.)	Scot.	57•05 N	2•25 W
22	Dee (R.)	Wales	53•00 N	3•10 W
11	Deer Lake	Can.	49•10 N	57•25 W
8	Defiance	Oh.	41•15 N	84•20 W
26	Deggendorf	F.R.G.	48•50 N	12•59 E
44	DeGrey (R.)	Austl.	20•20 S	119•25 E
35	Dehra Dūn	India	30•09 N	78•07 E
27	Dej	Rom.	47•09 N	23•53 E
8	De Kalb	Il.	41•54 N	88•46 W
6	Delano Pk.	Ut.	38•25 N	112•25 W
8	Delaware	Oh.	40•15 N	83•05 W
7	Delaware (State)	U.S.	38•40 N	75•30 W
9	Delaware (R.)	U.S.	41•50 N	75•20 W
9	Delaware B.	De.-NJ	39•05 N	75•10 W
8	Delaware Res.	Oh.	40•30 N	83•05 E
26	Delemont	Switz.	47•21 N	7•18 E
23	Delft	Neth.	52•01 N	4•20 E
23	Delfzijl	Neth.	53•20 N	6•50 E
50	Delgada Pta. (Pt.)	Arg.	43•46 S	63•46 W
55	Delgado, Cabo (C.)	Moz.	10•40 S	40•35 E
35	Delhi	India	28•54 N	77•13 E
35	Delhi (State)	India	28•30 N	76•50 E
26	Delitzsch	G.D.R.	51•32 N	12•18 E
21	Dell Alice, Pt.	It.	39•23 N	17•10 E
53	Dellys	Alg.	36•59 N	3•40 E
26	Delmenhorst	F.R.G.	53•03 N	8•38 E
29	De-Longa (I.)	Sov.Un.	76•30 N	153•00 E
46	Deloraine	Austl.	41•30 S	146•40 E
8	Delphi	In.	40•35 N	86•40 W
8	Delphos	Oh.	40•50 N	84•20 W
6	Del Rio	Tx.	29•21 N	100•52 W
21	Delvine	Alb.	39•58 N	20•10 E
30	Dëma (R.)	Sov.Un.	53•40 N	54•30 E
53	Dembidolo	Eth.	8•46 N	34•46 E
6	Deming	NM	32•15 N	107•45 W
26	Demmin	G.D.R.	53•54 N	13•04 E
52	Demnat	Mor.	31•58 N	7•03 W
40	Dempo, Gunung (Vol.)	Indon.	4•04 S	103•11 E
22	Denbigh	Wales	53•15 N	3•25 W
13	Denham, Mt.	Jam.	18•20 N	77•30 W
23	Den Helder	Neth.	52•55 N	5•45 E
17	Denia	Sp.	38•48 N	0•06 E
46	Deniliquin	Austl.	35•20 S	144•52 E
6	Denison	Tx.	33•45 N	97•02 W
31	Denizli	Tur.	37•40 N	29•10 E
14	Denmark	Eur.	56•14 N	8•30 E
5	Denmark Str.	Grnld.	66•30 N	27•00 W
8	Dennison	Oh.	40•25 N	81•20 W
11	De Nouvelle-France (C.)	Can.	62•03 N	74•00 W
40	Denpasar	Indon.	8•35 S	115•10 E
9	Denton	Md.	38•55 N	75•50 W
44	D'entrecasteaux, Pt.	Austl.	34•50 S	114•45 E
41	D'entrecasteaux Is.	Pap.N.Gui.	9•45 S	152•00 E
6	Denver	Co.	39•44 N	104•59 W
8	De Pere	Wi.	44•25 N	88•04 W
9	Depew	NY	42•55 N	78•43 W
8	Depue	Il.	41•15 N	89•55 W
35	Dera Ghāzi Khān	Pak.	30•09 N	70•39 E
35	Dera Ismāīl Khān	Pak.	31•55 N	70•51 E
31	Derbent	Sov.Un.	42•00 N	48•10 E
44	Derby	Austl.	17•20 S	123•40 E
9	Derby	Ct.	41•20 N	73•05 W
22	Derby	Eng.	52•55 N	1•29 W
22	Derg, Lough (B.)	Ire.	53•00 N	8•09 W

Page	Name	Region	Lat.	Long.
21	Derventa	Yugo.	45•58 N	17•58 E
46	Derwent (R.)	Austl.	42•21 S	146•30 E
6	Deschutes R.	Or.	44•25 N	121•21 W
58	Dese	Eth.	11•00 N	39•51 E
50	Deseado, Rio (R.)	Arg.	46•50 S	67•45 W
7	Des Moines	Ia.	41•35 N	93•37 W
7	Des Moines (R.)	U.S.	43•45 N	94•20 W
31	Desna (R.)	Sov.Un.	51•05 N	31•03 E
50	Desolación (I.)	Chile	53•05 S	74•00 W
8	Des Plaines	Il.	42•02 N	87•54 W
26	Dessau	G.D.R.	51•50 N	12•15 E
26	Detmold	G.D.R.	51•57 N	8•55 E
8	Detroit	Mi.	42•22 N	83•10 W
27	Detva	Czech.	48•32 N	19•21 E
21	Deva	Rom.	45•52 N	22•52 E
27	Dévaványa	Hung.	47•01 N	20•58 E
31	Develi	Tur.	38•20 N	35•10 E
23	Deventer	Neth.	52•14 N	6•07 E
6	Devils Lake	ND	48•10 N	98•55 W
21	Devoll (R.)	Alb.	40•55 N	20•10 E
46	Devonport	Austl.	41•20 S	146•30 E
34	Dezfūl	Iran	32•14 N	48•37 E
33	Dezhneva, Mys (East Cape)	Sov.Un.	68•00 N	172•00 W
35	Dhaulāgiri (Mtn.)	Nep.	28•42 N	83•31 E
21	Dhenoúsa (I.)	Grc.	37•09 N	25•53 E
33	Dhibān	Jordan	31•30 N	35•46 E
21	Dhidhimótikhon	Grc.	41•20 N	26•27 E
21	Dhodhekánisos (Dodecanese) (Is.)	Grc.	38•00 N	26•10 E
35	Dhule	India	20•58 N	74•43 E
20	Dia (I.)	Grc.	35•27 N	25•17 E
49	Diable, Ile du (Devils I.)	Fr.Gu.	5•15 N	57•10 W
12	Diablo Heights	C.Z.	8•58 N	79•34 W
49	Diamantina	Braz.	18•14 S	43•32 W
44	Diamantina (R.)	Austl.	25•38 S	139•53 E
49	Diamantino	Braz.	14•22 S	56•23 W
41	Diapitan B.	Phil.	16•28 N	122•25 E
6	Dickinson	ND	46•52 N	102•49 W
9	Dickson City	Pa.	41•25 N	75•40 W
31	Dicle (R.)	Tur.	37•50 N	40•40 E
10	Diefenbaker (Res.)	Can.	51•20 N	108•10 W
50	Diego Ramirez, Islas (Is.)	Chile	56•15 S	70•15 W
55	Diego-Suarez	Mad.	12•18 S	49•16 E
36	Dien Bien Phan	Viet.	21•38 N	102•49 E
17	Dieppe	Fr.	49•54 N	1•05 E
11	Digby	Can.	44•37 N	65•46 W
41	Digul (R.)	Indon.	7•00 S	140•27 E
41	Dijohan Pt.	Phil.	16•24 N	122•25 E
17	Dijon	Fr.	47•21 N	5•02 E
28	Dikson	Sov.Un.	73•30 N	80•35 E
53	Dikwa	Nig.	12•06 N	13•53 E
41	Dili	Indon.	8•35 S	125•35 E
18	Di Linosa I.	It.	36•01 N	12•43 E
31	Dilizhan	Sov.Un.	40•45 N	45•00 E
8	Dillon Res.	Oh.	40•05 N	82•05 W
54	Dilolo	Zaire	10•19 S	22•23 E
54	Dima	Ang.	15•45 S	20•15 E
34	Dimashq (Damascus)	Syria	33•31 N	36•18 E
21	Dimbovita (R.)	Rom.	44•43 N	25•41 E
33	Dimona	Isr.	31•03 N	35•01 E
41	Dinagat I.	Phil.	10•15 N	126•15 E
23	Dinant	Bel.	50•17 N	4•50 E
20	Dinara Planina (Mts.)	Yugo.	43•50 N	16•15 E
41	Dingalan B.	Phil.	15•19 N	121•33 E
22	Dingle	Ire.	52•10 N	10•13 W
22	Dingle B	Ire.	52•02 N	10•15 W
45	Dingo	Austl.	23•45 S	149•26 E
22	Dingwall	Scot.	57•37 N	4•23 W
52	Diourbel	Senegal	14•40 N	16•15 W
35	Diphu Pass	China	28•15 N	96•45 E
53	Dire Dawal	Eth.	9•40 N	41•47 E
44	Dirk Hartog (I.)	Austl.	26•25 S	113•15 E
46	Dirranbandi	Austl.	28•24 S	148•29 E
44	Disappointment (L.)	Austl.	23•20 S	120•20 E
20	D'Ischia, I.	It.	40•26 N	13•55 E
55	Discovery	S.Afr.	26•10 S	27•53 E
5	Disko	Grnld.	70•00 N	54•00 W
30	Disna	Sov.Un.	55•34 N	28•15 E
7	District of Columbia	U.S.	38•50 N	77•00 W
49	Distrito Federal (Dist.)	Braz.	15•49 S	47•39 W
35	Diu	India	20•48 N	70•58 E
41	Divilacan B.	Phil.	17•26 N	122•25 E
49	Divinópolis	Braz.	20•10 S	44•53 W
8	Dixon	Il.	41•50 N	89•30 W
10	Dixon Ent.	Ak.-Can.	54•25 N	132•00 W
31	Diyarbakir	Tur.	38•00 N	40•10 E
53	Dja (R.)	Cam.	3•25 N	13•17 E
21	Djakovica	Yugo.	42•33 N	20•28 E
52	Djanet	Alg.	24•29 N	9•26 E
18	Djedi (R.)	Alg.	34•18 N	4•39 E
52	Djelfa	Alg.	34•40 N	3•17 E
18	Djerba, Ile de (I.)	Tun.	33•53 N	11•26 E
52	Djerid, Chott (L.)	Tun.	33•15 N	8•29 E
53	Djibouti	Djibouti	11•34 N	43•00 E
51	Djibouti	Afr.	11•35 N	48•08 E
17	Djidjelli	Alg.	36•49 N	5•47 E
24	Djursholm	Swe.	59•26 N	18•01 E
31	Dnepr (Dnieper) (R.)	Sov.Un.	46•47 N	32•57 E
31	Dneprodzerzhinsk	Sov.Un.	48•32 N	34•38 E
28	Dneprodzerzhinskoye Vdkhr	Sov.Un.	49•00 N	34•10 E
31	Dnepropetrovsk	Sov.Un.	48•23 N	34•10 E
31	Dnestr (Dniester) (R.)	Sov.Un.	48•21 N	28•10 E
44	Dobbyn	Austl.	19•45 S	140•02 E
25	Dobele	Sov.Un.	56•37 N	23•18 E
26	Döbeln	G.D.R.	51•08 N	13•07 E
41	Doberai Jazirah (Pen.)	Indon.	1•25 S	133•15 E
41	Dobo	Indon.	6•00 S	134•18 E
21	Doboj	Yugo.	44•42 N	18•04 E
27	Dobšina	Czech.	48•48 N	20•25 E
49	Doce (R.)	Braz.	19•01 S	42•14 W
49	Dr. Ir. W.J. van Blommestein Meer (Res.)	Sur.	4•45 N	55•05 W
6	Dodge City	Ks.	37•44 N	100•01 W
9	Dodgeville	NY	43•10 N	74•45 W
8	Dodgeville	Wi.	42•58 N	90•07 W
54	Dodoma	Tan.	6•11 S	35•45 E
23	Dogger Bk.	Eur.	55•07 N	2•25 E
31	Dogubayazit	Tur.	39•35 N	44•00 E
21	Doiran (L.)	Grc.	41•10 N	23•00 E
41	Dolak (I.)	Indon.	7•45 S	137•30 E
11	Dolbeau	Can.	48•52 N	72•16 W
17	Dôle	Fr.	47•07 N	5•28 E
30	Dolgiy (I.)	Sov.Un.	69•20 N	59•20 E
27	Dolina	Sov.Un.	48•57 N	24•01 E
53	Dolo	Som.	4•01 N	42•14 E
20	Dolomitiche, Alpi (Mts.)	It.	46•16 N	11•43 E
48	Dolores	Col.	3•33 N	74•54 W
43	Dolores	Phil.	17•40 N	120•43 E
10	Dolphin and Union Str.	Can.	69•22 N	117•10 W
26	Domažlice	Czech.	49•27 N	12•55 E
27	Dombóvár	Hung.	46•22 N	18•08 E
48	Domeyko, Cordillera (Mts.)	Chile	20•50 S	69•02 W
13	Dominica	N.A.	15•30 N	60•45 W
13	Dominican Republic	N.A.	19•00 N	70•45 W
22	Don (R.)	Scot.	57•19 N	2•39 W
26	Donawitz	Aus.	47•23 N	15•05 E
54	Dondo	Ang.	9•38 S	14•25 E
54	Dondo	Moz.	19•33 S	34•47 E
35	Dondra Hd.	Sri Lanka	5•52 N	80•52 E
22	Donegal	Ire.	54•44 N	8•05 W
22	Donegal, Mts. of	Ire.	54•44 N	8•10 W
22	Donegal Bay	Ire.	54•35 N	8•36 W
31	Donetsk (Stalino)	Sov.Un.	48•00 N	37•35 E
44	Dongara	Austl.	29•15 S	115•00 E
40	Donggala	Indon.	0•45 S	119•32 E
39	Dong Hoi	Viet.	17•25 N	106•42 E
54	Dongo	Ang.	14•45 S	15•30 E
41	Dongon Pt.	Phil.	12•43 N	120•35 E
53	Dongou	Con.	2•02 N	18•04 E
20	Donji Vakuf	Yugo.	44•08 N	17•25 E
55	Donnybrook	S.Afr.	29•56 S	29•54 E
8	Door Pen.	Wi.	44•40 N	87•36 W
20	Dora Baltea	It.	45•40 N	7•34 E
22	Dorchester	Eng.	50•45 N	2•34 W
17	Dordogne (R.)	Fr.	44•53 N	0•16 E
23	Dordrecht	Neth.	51•48 N	4•39 E
55	Dordrecht	S.Afr.	31•24 S	27•06 E
20	Dorgali	It.	40•18 N	9•37 E
26	Dornbirn	Aus.	47•24 N	9•45 E
22	Dornoch	Scot.	57•55 N	4•01 W
22	Dornoch Firth	Scot.	57•55 N	3•55 W
27	Dorohoi	Rom.	47•57 N	26•28 E
44	Dorre (I.)	Austl.	25•19 S	113•10 E
23	Dortmund	F.R.G.	51•31 N	7•28 E
31	Dörtyal	Tur.	36•50 N	36•20 E
49	Dos Caminos	Ven.	9•38 N	67•17 W
52	Dosso	Niger	13•03 N	3•12 E
7	Dothan	Al.	31•13 N	85•23 W
17	Douai	Fr.	50•23 N	3•04 E
52	Douala	Cam.	4•03 N	9•42 E
22	Douglas	Isle of Man	54•10 N	4•24 W
53	Doumé	Cam.	4•41 N	13•26 E
49	Dourada, Serra (Mts.)	Braz.	15•11 S	49•57 W
17	Douro, Rio (R.)	Port.	41•03 N	8•12 W
9	Dover	De.	39•10 N	75•30 W
23	Dover	Eng.	51•08 N	1•19 E
9	Dover	NH	43•15 N	71•00 W
9	Dover	NJ	40•53 N	74•33 W
8	Dover	Oh.	40•35 N	81•30 W
23	Dover, Str. of	Eur.	50•50 N	1•15 W
30	Dovlekanovo	Sov.Un.	54•15 N	55•05 E
24	Dovre Fjell (Plat.)	Nor.	62•03 N	8•36 E
54	Dow, L.	Bots.	21•22 S	24•52 E
8	Dowagiac	Mi.	42•00 N	86•05 W
52	Drâa, C.	Mor.	28•39 N	12•15 W
52	Drâa, Oued (R.)	Mor.	28•00 N	9•31 W
21	Draganovo	Bul.	43•13 N	25•45 E
21	Drăgăsani	Rom.	44•39 N	24•18 E
54	Drakensberg (Mts.)	Leso.-S.Afr.	29•15 S	29•07 E
47	Drake Passage	S.A.-Ant.	57•00 S	65•00 W
21	Dráma	Grc.	41•09 N	24•10 E
24	Drammen	Nor.	59•45 N	10•15 E
26	Drau (R.)	Aus.	46•44 N	13•45 E
20	Drava (R.)	Yugo.	46•37 N	15•17 E
20	Dravograd	Yugo.	46•37 N	15•01 E
26	Drawsko Pomorski	Pol.	53•31 N	15•50 E
36	Dre Chu (R.)	China	34•11 N	96•08 E
26	Dresden	G.D.R.	51•05 N	13•45 E
20	Drin (R.)	Alb.	42•13 N	20•13 E
20	Drina (R.)	Yugo.	44•09W	19•30 E
20	Drinit, Pellg I (Bght.)	Alb.	41•42 N	19•17 E
24	Dröbak	Nor.	59•40 N	10•35 E
22	Drogheda	Ire.	53•43 N	6•15 W
27	Drogichin	Sov.Un.	52•10 N	25•11 E
27	Drogobych	Sov.Un.	49•21 N	23•31 E
10	Drumheller	Can.	51•28 N	112•42 W
8	Drummond (I.)	Mi.	46•00 N	83•50 W
11	Drummondville	Can.	45•53 N	72•33 W
19	Druze, Jebel (Mts.)	Syr.	32•40 N	36•58 E
27	Drweca R.	Pol.	53•06 N	19•13 E
11	Dryden	Can.	49•47 N	92•50 W
52	Dschang	Cam.	5•34 N	10•09 E
13	Duarte, Pico (Mtn.)	Dom. Rep.	19•00 N	71•00 W
10	Dubawnt (L.)	Can.	63•27 N	103•30 W
10	Dubawnt (R.)	Can.	61•30 N	103•49 W
34	Dubayy	U.A.E.	25•18 N	55•26 E
46	Dubbo	Austl.	32•20 S	148•42 E
22	Dublin (Baile Atha Cliath)	Ire.	53•20 N	6•15 W
27	Dubno	Sov.Un.	50•24 N	25•44 E
9	Du Bois	Pa.	41•10 N	78•45 W
31	Dubovka	Sov.Un.	49•00 N	44•50 E
21	Dubrovnik (Ragusa)	Yugo.	42•40 N	18•10 E
7	Dubuque	Ia.	42•30 N	90•43 W
44	Duchess	Austl.	21•30 S	139•55 E
43	Ducie I.	Oceania	25•30 S	126•20 W
10	Duck Mtn.	Can.	51•35 N	101•00 W
48	Duda (R.)	Col.	3•25 N	74•23 W
17	Duero (R.)	Sp.	41•30 N	5•10 W
8	Dugger	In.	39•00 N	87•10 W
20	Dugi Otok (I.)	Yugo.	44•03 N	14•40 E
23	Duisburg	F.R.G.	51•26 N	6•46 E
48	Duitama	Col.	5•48 N	73•09 W
27	Dukla P.	Pol.	49•25 N	21•44 E
33	Dūmā	Syr.	33•34 N	36•17 E
41	Dumaguete City	Phil.	9•14 N	123•15 E
41	Dumai	Indon.	1•39 N	101•30 E
41	Dumali Pt.	Phil.	13•07 N	121•42 E
22	Dumbarton	Scot.	56•00 N	4•35 W
22	Dumfries	Scot.	54•05 N	3•40 W
53	Dumyāṭ	Egypt	31•22 N	31•50 E
27	Duna (R.)	Hung.	46•07 N	18•45 E
27	Dunaföldvar	Hung.	46•48 N	18•55 E
27	Dunajec (R.)	Pol.	49•52 N	20•53 E
27	Dunapataj	Hung.	46•42 N	19•03 E
27	Dunaujvaros	Hung.	46•57 N	18•55 E
22	Dunbar	Scot.	56•00 N	2•25 W
8	Dunbar	WV	38•20 N	81•45 W
10	Duncan	Can.	48•47 N	123•42 W
22	Duncansby Hd.	Scot.	58•40 N	3•01 W
22	Dundalk	Ire.	54•00 N	6•18 W
9	Dundalk	Md.	39•16 N	76•31 W
22	Dundalk B.	Ire.	53•55 N	6•15 W
8	Dundee	Il.	42•06 N	88•17 W
22	Dundee	Scot.	56•30 N	2•55 W
55	Dundee	S.Afr.	28•14 S	30•16 E
44	Dundras (L.)	Austl.	32•15 S	132•00 E
44	Dundras Str.	Austl.	10•35 S	131•15 E
22	Dundrum B.	Ire.	54•13 N	5•47 W
45	Dunedin	N.Z.	45•48 S	170•32 E
22	Dunfermline	Scot.	56•05 N	3•30 W
22	Dungarvin	Ire.	52•06 N	7•50 W
23	Dunkerque	Fr.	51•02 N	2•37 E
8	Dunkirk	In.	40•20 N	85•25 W
9	Dunkirk	NY	42•30 N	79•20 W
22	Dun Laoghalre	Ire.	53•16 N	6•09 W
9	Dunmore	Pa.	41•25 N	75•30 W
9	Dunnville	Can.	42•55 N	79•40 W
53	Dunqulah	Sud.	19•21 N	30•19 E
41	Dupax	Phil.	16•16 N	121•06 E
21	Dupnitsa	Bul.	42•15 N	23•07 E
54	Duque de Bragança	Ang.	9•06 S	15•57 E
50	Duque de Caxias	Braz.	22•46 S	43•18 W
9	Duquesne	Pa.	40•22 N	79•51 W
8	Du Quoin	Il.	38•01 N	89•14 W
8	Durand	Mi.	42•50 N	84•00 W
12	Durango	Mex.	24•02 N	104•42 W
12	Durango (State)	Mex.	25•00 N	106•00 W
50	Durazno	Ur.	33•21 S	56•31 W
55	Durban	S.Afr.	29•48 S	31•00 E
54	Durbanville	S.Afr.	33•50 S	18•39 E
25	Durbe	Sov.Un.	56•36 N	21•24 E
20	Durdevac	Yugo.	46•03 N	17•03 E
23	Düren	F.R.G.	50•48 N	6•30 E
22	Durham	Eng.	54•47 N	1•46 W
7	Durham	NC	36•00 N	78•55 W
46	Durham Downs	Austl.	27•30 S	141•55 E
21	Durrës	Alb.	41•19 N	19•27 E
9	Duryea	Pa.	41•20 N	75•50 W
35	Dushanbe	Sov.Un.	38•30 N	68•45 E
23	Düsseldorf	F.R.G.	51•14 N	6•47 E
38	Dutalan Ula (Mtn.)	Mong.	49•25 N	112•40 E
6	Dutch Harbor	Ak.	53•58 N	166•30 W
30	Dvinskaya Guba (G.)	Sov.Un.	65•10 N	38•40 E
26	Dvūr Králové nad Labem	Czech.	50•28 N	15•43 E
8	Dwight	Il.	41•00 N	88•20 W
36	Dzabhan Gol (R.)	Mong.	48•19 N	94•08 E
38	Dzamiin Üüde	Mong.	44•38 N	111•32 E
55	Dzaoudzi	Comoro Is.	12•44 S	45•15 E
15	Dzaudzhikau	Sov.Un.	48•00 N	44•52 E
30	Dzerzhinsk	Sov.Un.	56•20 N	43•50 E
29	Dzhugdzhur Khrebet (Mts.)	Sov.Un.	56•15 N	137•00 E
27	Dzialoszyce	Pol.	50•21 N	20•22 E
26	Dzierzoniów	Pol.	50•44 N	16•38 E
36	Dzungaria (Reg.)	China	44•39 N	86•13 E
	E			
8	Eagle	WV	38•10 N	81•20 W
6	Eagle Pass	Tx.	28•49 N	100•30 W
8	Earlington	Ky.	37•15 N	87•31 W
12	East, Mt.	C.Z.	9•09 N	79•16 W
9	East Aurora	NY	42•46 N	78•38 W
26	East Berlin	G.D.R.	52•31 N	13•28 E
23	Eastbourne	Eng.	50•48 N	0•16 E
45	East Cape (C.)	N.Z.	37•37 S	178•33 E
8	East Chicago	In.	41•39 N	87•29 W
37	East China Sea	Asia	30•28 N	125•52 E
8	East Cleveland	Oh.	41•33 N	81•35 W
26	Eastern Alps (Mts.)	Aus.-Switz.	47•03 N	10•55 E
35	Eastern Ghāts (Mts.)	India	13•50 N	78•45 E
36	Eastern Turkestan (Reg.)	China	38•23 N	80•41 E
9	Easthampton	Ma.	42•15 N	72•45 W
9	East Hartford	Ct.	41•45 N	72•35 W
8	East Jordan	Mi.	45•05 N	85•05 W
8	East Lansing	Mi.	42•45 N	84•30 W
8	East Liverpool	Oh.	40•40 N	80•35 W
55	East London	S.Afr.	33•02 S	27•54 E
11	Eastmain (R.)	Can.	52•12 N	73•19 W
9	Easton	Md.	72•45 N	76•05 W
9	Easton	Pa.	40•45 N	75•15 W
9	East Orange	NJ	40•46 N	74•12 W
8	East Peoria	Il.	40•40 N	89•30 W
9	East Providence	RI	41•49 N	71•22 W
9	East Rochester	NY	43•10 N	77•30 W
7	East St.Louis	Il.	38•38 N	90•10 W
28	East Siberian Sea	Sov.Un.	73•00 N	153•28 E
9	East Stroudsburg	Pa.	41•00 N	75•10 W
9	East Syracuse	NY	43•05 N	76•00 W
8	East Tawas	Mi.	44•15 N	83•30 W
8	Eaton	Oh.	39•45 N	84•40 W
8	Eaton Rapids	Mi.	42•30 N	84•40 W
7	Eau Claire	Wi.	44•47 N	91•32 W
24	Ebeltoft	Den.	56•11 N	10•39 E
9	Ebensburg	Pa.	40•29 N	78•44 W
26	Ebingen	F.R.G.	48•13 N	9•04 E
36	Ebi Nuur (L.)	China	45•09 N	83•15 E
20	Eboli	It.	40•38 N	15•04 E
52	Ebolowa	Cam.	2•54 N	11•09 E
17	Ebro, Río (R.)	Sp.	41•30 N	0•35 W
8	Eccles	WV	37•45 N	81•10 W
21	Eceabat (Maidos)	Tur.	40•10 N	26•21 E
41	Echague	Phil.	16•43 N	121•40 E

Page	Name	Region	Lat.	Long.
26	Echternach	Lux.	49•48 N	6•25 E
46	Echuca	Austl.	36•10 S	144•47 E
17	Écija	Sp.	37•20 N	5•07 W
26	Eckernförde	F.R.G.	54•27 N	9•51 E
47	Ecuador	S.A.	0•00 N	78•30 W
53	Ed	Eth.	13•57 N	41•37 E
52	Edéa	Cam.	3•48 N	10•08 E
22	Eden (R.)	Eng.	54•40 N	2•35 W
55	Edenvale	S.Afr.	29•06 S	28•10 E
26	Eder R.	F.R.G.	51•05 N	8•52 E
8	Edgerton	Wi.	42•49 N	89•06 W
21	Édhessa	Grc.	40•48 N	22•04 E
8	Edinburg	In.	39•20 N	85•55 W
22	Edinburgh	Scot.	55•57 N	3•10 W
21	Edirne (Adrianople)	Tur.	41•41 N	26•35 E
10	Edmonton	Can.	53•33 N	113•28 W
11	Edmundston	Can.	47•22 N	68•20 W
21	Edremit	Tur.	39•35 N	27•00 E
21	Edremit Körfezi (G.)	Tur.	39•28 N	26•35 E
54	Edward (L.)	Zaire	0•25 S	29•40 E
8	Edwardsville	Il.	38•49 N	89•58 W
8	Eel (R.)	In.	40•50 N	85•55 W
45	Efate (I.)	New Hebr.	18•02 S	168•29 E
8	Effingham	Il.	39•05 N	88•30 W
20	Egadi, Isole (Is.)	It.	38•01 N	12•00 E
27	Eger	Hung.	47•53 N	20•24 E
24	Egersund	Nor.	58•29 N	6•01 E
9	Egg Harbor	NJ	39•30 N	74•35 W
36	Egiin Gol (R.)	Mong.	49•41 N	100•40 E
45	Egmont, C.	N.Z.	39•18 S	173•49 E
31	Egridir Gölü (L.)	Tur.	38•10 N	30•00 E
51	Egypt	Afr.	26•58 N	27•01 E
26	Eichstätt	F.R.G.	48•54 N	11•14 E
24	Eid	Nor.	61•54 N	6•01 E
24	Eidsberg	Nor.	59•32 N	11•16 E
24	Eidsvoll	Nor.	60•19 N	11•15 E
26	Eifel (Plat.)	F.R.G.	50•08 N	6•30 E
44	Eighty Mile Beach	Austl.	20•45 S	121•00 E
51	Eil	Som.	7•53 N	49•45 E
26	Eilenburg	G.D.R.	51•27 N	12•38 E
55	Eilliot	S.Afr.	31•19 S	27•52 E
26	Einbeck	F.R.G.	51•49 N	9•52 E
23	Eindhoven	Neth.	51•29 N	5•20 E
48	Eirunepé	Braz.	6•37 S	69•58 W
26	Eisenach	G.D.R.	50•58 N	10•18 E
26	Eisenhüttenstadt	G.D.R.	52•08 N	14•40 E
26	Eisleben	G.D.R.	51•31 N	11•33 E
26	Ejdfjord	Nor.	60•28 N	7•04 E
25	Ekenäs (Tammisaari)	Fin.	59•59 N	23•25 E
24	Eksjö	Swe.	57•41 N	14•55 E
52	El Aaiún	W.Sah.	26•45 N	13•15 W
55	Elands (R.)	S.Afr.	31•48 S	26•09 E
18	El Asnam (Orléansville)	Alg.	36•14 N	1•32 E
33	Elat	Isr.	29•34 N	34•57 E
31	Elâzig	Tur.	38•40 N	39•00 E
20	Elba, Isola di (I.)	It.	42•42 N	10•25 E
48	El Banco	Col.	8•58 N	74•01 W
21	Elbasan	Alb.	41•08 N	20•05 E
18	El Bayadh	Alg.	33•42 N	1•06 E
26	Elbe (R.)	G.D.R.	53•47 N	9•20 E
6	Elbert, Mt.	Co.	39•05 N	106•25 W
31	Elbistan	Tur.	38•20 N	37•10 E
27	Elblag	Pol.	54•11 N	19•25 E
31	El'brus, Gora (Mt.)	Sov.Un.	43•20 N	42•25 E
31	Elburz Mts.	Iran	36•30 N	51•00 E
48	El Cajon	Col.	4•50 N	76•35 W
49	El Cambur	Ven.	10•24 N	68•06 W
48	El Carmen	Col.	9•54 N	75•12 W
26	Elde (R.)	G.D.R.	53•11 N	11•30 E
52	El Djouf (Des.)	Mauritania	21•45 N	7•05 W
7	El Dorado	Ar.	33•13 N	92•39 W
8	Eldorado	Il.	37•50 N	88•30 W
53	Eldoret	Ken.	0•31 N	35•17 E
6	Elephant Butte Res.	NM	33•25 N	107•10 W
13	Eleuthera (I.)	Ba.	25•05 N	76•10 W
17	El Ferrol	Sp.	43•30 N	8•12 W
8	Elgin	Il.	42•03 N	88•16 W
22	Elgin	Scot.	57•40 N	3•30 W
52	El Goléa	Alg.	30•39 N	2•52 E
53	Elgon, Mt.	Ken.	1•00 N	34•25 E
49	El Guapo	Ven.	10•07 N	66•00 W
18	El Hamada (Plat.)	Alg.	30•53 N	1•52 W
52	El Hank (Bluffs)	Mauritania-Mali	23•44 N	6•45 W
49	El Hatillo	Ven.	10•08 N	65•13 W
54	Elila (R.)	Zaire	3•00 S	26•50 E
25	Elisenvaara	Sov. Un.	61•25 N	29•46 E
9	Elizabeth	NJ	40•40 N	74•13 W
8	Elizabethtown	Ky.	37•40 N	85•55 W
52	El Jadida	Mor.	33•14 N	8•34 W
27	Elk	Pol.	53•53 N	22•23 E
8	Elk (R.)	WV	38•30 N	81•05 W
52	El Kairouan	Tun.	35•46 N	10•04 E
8	Elkhart	In.	41•40 N	86•00 W
8	Elkhorn	Wi.	42•39 N	88•32 W
9	Elkins	WV	38•55 N	79•50 W
10	Elk Island Natl. Park	Can.	53•37 N	112•45 W
6	Elko	Nv.	40•51 N	115•46 W
8	Elk Rapids	Mi.	44•55 N	85•25 W
9	Elkton	Md.	39•35 N	75•50 W
9	Ellenville	NY	41•40 N	74•25 W
5	Ellesmere I.	Can.	81•00 N	80•00 W
55	Elliotdale	S.Afr.	31•58 S	28•42 E
67	Ellsworth Highland	Ant.	77•00 S	90•00 W
26	Ellwangen	F.R.G.	48•47 N	10•08 E
8	Elm (R.)	WV	38•30 N	81•05 W
18	El Maadid	Mor.	31•32 N	4•30 W
8	Elmhurst	Il.	41•54 N	87•56 W
52	El Milia	Alg.	36•30 N	6•16 E
9	Elmira	NY	42•05 N	76•50 W
9	Elmira Heights	NY	42•10 N	76•50 W
48	El Misti (Vol.)	Peru	16•04 S	71•20 W
52	El Mreyyé (Des.)	Mauritania	19•15 N	7•50 W
23	Elmshorn	F.R.G.	53•45 N	9•39 E
52	El Oued	Alg.	33•23 N	6•49 E
48	El Pao	Ven.	8•08 N	62•37 W
6	El Paso	Tx.	31•47 N	106•27 W
49	El Pilar	Ven.	9•56 N	64•48 W
49	El Roboré	Bol.	18•23 S	59•43 W
8	Elroy	Wi.	43•44 N	90•17 W
12	El Salvador	N.A.	14•00 N	89•30 W
48	El Tigre	Ven.	8•49 N	64•15 W
31	El'ton (L.)	Sov.Un.	49•10 N	47•00 E
35	Elūru	India	16•44 N	80•09 E
17	Elvas	Port.	38•53 N	7•11 W
24	Elverum	Nor.	60•53 N	11•33 E
53	El Wak	Ken.	3•00 N	41•00 E
8	Elwood	In.	40•15 N	85•50 W
23	Ely	Eng.	52•25 N	0•17 E
6	Ely	Nv.	39•16 N	114•53 W
8	Elyria	Oh.	41•22 N	82•07 W
25	Ema (R.)	Sov.Un.	58•25 N	27•00 E
24	Emån (R.)	Swe.	57•15 N	15•46 E
31	Emba (R.)	Sov.Un.	46•50 N	54•10 E
48	Embalse Guri (L.)	Ven.	7•30 N	63•00 W
8	Embarrass (R.)	Il.	39•15 N	88•05 W
26	Emden	F.R.G.	53•21 N	7•15 E
45	Emerald	Austl.	28•34 S	148•00 E
53	Emi Koussi, (Mtn.)	Chad	19•50 N	18•30 E
20	Emilia-Romagna (Reg.)	It.	44•35 N	10•48 E
8	Eminence	Ky.	38•25 N	85•15 W
41	Emirau (I.)	Pap.N.Gui.	1•40 S	150•28 E
23	Emmen	Neth.	52•48 N	6•55 E
6	Emmons Mt.	Ut.	40•43 N	110•20 W
20	Empoli	It.	43•43 N	10•55 E
7	Emporia	Ks.	38•24 N	96•11 W
9	Emporium	Pa.	41•30 N	78•15 W
26	Ems R.	F.R.G.	52•52 N	7•16 E
26	Ems-Weser (Can.)	F.R.G.	52•23 N	8•11 E
26	Enånger	Swe.	61•36 N	16•55 E
12	Encantada, Cerro de la (Mtn.)	Mex.	31•58 N	115•15 W
41	Encanto Pt.	Phil.	15•44 N	121•46 E
50	Encarnación	Par.	27•26 S	55•52 W
48	Encontrados	Ven.	9•01 N	72•10 W
46	Encounter B.	Austl.	35•50 S	138•45 E
33	Endau (R.)	Mala.	2•29 N	103•40 E
42	Enderbury (I.)	Oceania	2•00 S	107•50 W
67	Enderby Land (Reg.)	Ant.	72•00 S	52•00 E
9	Endicott	NY	42•05 N	76•00 W
21	Enez	Tur.	40•42 N	26•05 E
9	Enfield	Ct.	41•55 N	72•35 W
13	Engano, Cabo (C.)	Dom.Rep.	18•40 N	68•30 W
40	Engaño, C.	Phil.	18•40 N	122•45 E
55	Engcobo	S.Afr.	31•41 S	27•59 E
31	Engel's	Sov.Un.	51•20 N	45•40 E
40	Enggano (I.)	Indon.	5•22 S	102•18 E
22	England (Reg.)	U.K.	51•35 N	1•40 W
8	English	In.	38•15 N	86•25 W
11	English (R.)	Can.	50•31 N	94•12 W
17	English Chan.	Eng.	49•45 N	3•06 W
6	Enid	Ok.	36•25 N	97•52 W
54	Enkeldoorn	Rh.	19•59 S	30•58 E
24	Enköping	Swe.	59•39 N	17•05 E
53	Ennedi (Plat.)	Chad	16•45 N	22•45 E
22	Ennis	Ire.	52•54 N	9•05 W
22	Enniscorthy	Ire.	52•33 N	6•27 W
22	Enniskillen	N.Ire.	54•20 N	7•25 W
26	Enns (R.)	Aus.	47•37 N	14•35 E
23	Enschede	Neth.	52•10 N	6•50 E
12	Ensenada	Mex.	32•00 N	116•30 W
39	Enshih	China	30•18 N	109•25 E
54	Entebbe	Ug.	0•04 N	32•28 E
50	Entre Ríos (Prov.)	Arg.	31•30 S	59•00 W
52	Enugu	Nig.	6•27 N	7•27 E
48	Envigado	Col.	6•10 N	75•34 W
20	Eolie, Isole (Is.)	It.	38•43 N	14•43 E
21	Epeirus (Reg.)	Grc.	39•35 N	20•45 E
17	Epernay	Fr.	49•02 N	3•54 E
45	Epi	New Hebr.	16•59 S	168•29 E
17	Épinal	Fr.	48•11 N	6•27 E
33	Episkopi	Cyprus	34•38 N	32•55 E
52	Equatorial Guinea	Afr.	2•00 N	7•15 E
53	Erba, Jabal (Mtn.)	Sud.	20•53 N	36•45 E
19	Erciyas (Mtn.)	Tur.	38•30 N	35•36 E
50	Erechim	Braz.	27•43 S	52•11 W
31	Ereğli	Tur.	37•40 N	34•00 E
31	Ereğli	Tur.	41•15 N	31•25 E
26	Erfurt	G.D.R.	50•59 N	11•04 E
21	Ergene (R.)	Tur.	41•17 N	26•50 E
25	Ērgli	Sov.Un.	56•54 N	25•38 E
9	Erie	Pa.	42•05 N	80•05 W
7	Erie, L.	U.S.-Can.	42•15 N	81•25 W
53	Eritrea (Reg.)	Eth.	16•15 N	38•30 E
26	Erlangen	F.R.G.	49•36 N	11•03 E
35	Ernākulam	India	9•58 N	76•23 E
22	Erne, Upper Lough (L.)	N.Ire.	54•20 N	7•24 W
22	Erne, Lough (L.)	N.Ire.	54•30 N	7•40 W
11	Ernest Sound	Ak.	55•52 N	132•10 W
45	Eromanga (I.)	New Hebr.	18•58 S	169•18 E
18	Er Ricani	Mor.	31•09 N	4•20 W
22	Errigal, Mt.	Ire.	55•02 N	8•07 W
26	Erzgebirge (Ore Mts.)	G.D.R.	50•29 N	12•40 E
31	Erzincan	Tur.	39•50 N	39•30 E
31	Erzurum	Tur.	39•55 N	41•10 E
24	Esbjerg	Den.	55•29 N	8•25 E
8	Escanaba	Mi.	45•44 N	87•05 W
26	Eschwege	F.R.G.	51•11 N	10•02 E
34	Esfahān	Iran	32•38 N	51•30 E
55	Eshowe	S.Afr.	28•54 S	31•28 E
8	Eskdale	WV	38•05 N	81•25 W
16	Eskifjördhur	Ice.	65•04 N	14•01 W
24	Eskilstuna	Swe.	59•23 N	16•28 E
10	Eskimo Lakes (L.)	Can.	69•40 N	130•10 W
31	Eskişehir	Tur.	39•40 N	30•20 E
17	Esla (R.)	Sp.	41•50 N	5•48 W
24	Eslöv	Swe.	55•50 N	13•17 E
48	Esmeraldas	Ec.	0•58 N	79•45 W
44	Esperance	Austl.	33•45 S	122•07 E
48	Espinal	Col.	4•10 N	74•53 W
49	Espinhaço, Serra do (Mts.)	Braz.	16•06 S	44•56 W
49	Espírito Santo	Braz.	20•27 S	40•18 W
49	Espírito Santo (State)	Braz.	19•57 S	40•58 W
45	Espiritu Santo (I.)	New Hebr.	15•45 S	166•50 E
25	Espoo	Fin.	60•13 N	24•41 E
50	Esquel	Arg.	42•47 S	71•22 W
52	Essaouira	Mor.	31•34 N	9•44 W
26	Essen	F.R.G.	51•26 N	6•59 E
49	Essequibo (R.)	Guy.	4•26 N	58•17 W
9	Essex	Md.	39•19 N	76•29 W
9	Essex	Vt.	44•30 N	73•05 W
8	Essexville	Mi.	43•35 N	83•50 W
26	Esslingen	F.R.G.	48•45 N	9•19 E
6	Estacado, Llano (Plain)	U.S.	33•50 N	103•20 W
50	Estados, Isla de los	S.A.	55•05 S	63•00 W
49	Estância	Braz.	11•17 S	37•18 W
52	Estcourt	S.Afr.	29•04 S	29•53 E
20	Este	It.	45•13 N	11•40 E
10	Estevan	Can.	49•07 N	103•05 W
28	Estonian S.S.R.	Sov.Un.	59•10 N	25•00 E
50	Estrêla (R.)	Braz.	22•39 S	43•16 W
49	Estrondo, Serra do (Mts.)	Braz.	9•52 S	48•56 W
27	Esztergom	Hung.	47•46 N	18•45 E
5	Etah	Grnld.	78•20 N	72•42 W
23	Étaples	Fr.	50•32 N	1•38 E
51	Ethiopa	Afr.	7•53 N	37•55 E
20	Etna, Mt. (Vol.)	It.	37•48 N	15•00 E
54	Etoshapan (L.)	Namibia	19•07 S	15•30 E
44	Eucla	Austl.	31•45 S	128•50 E
9	Euclid	Oh.	41•34 N	81•32 W
6	Eugene	Or.	44•02 N	123•06 W
23	Eupen	Bel.	50•39 N	6•05 E
34	Euphrates (R.)	Asia	36•00 N	39•30 E
6	Eureka	Ca.	40•45 N	124•10 W
34	Eurgun (Mtn.)	Iran	28•47 N	57•00 E
14	Europe			
24	Evanger	Nor.	60•40 N	6•06 E
8	Evanston	Il.	42•03 N	87•41 W
8	Evansville	In.	38•00 N	87•30 W
8	Evansville	Wi.	42•46 N	89•19 W
8	Evart	Mi.	43•55 N	85•10 W
44	Everard (L.)	Austl.	36•20 S	134•10 E
44	Everard Ra.	Austl.	27•15 S	132•00 E
35	Everest, Mt.	Nep.-China	28•00 N	86•57 E
6	Everett	Wa.	47•59 N	122•11 W
11	Everett Mts.	Can.	62•34 N	68•00 W
7	Everglades Natl. Park	Fl.	25•39 N	80•57 W
17	Évora	Port.	38•35 N	7•54 W
17	Évreux	Fr.	49•02 N	1•11 E
21	Evrotas (R.)	Grc.	37•15 N	22•17 E
21	Evvoia (I.)	Grc.	38•38 N	23•45 E
53	Ewaso Ng'iro (R.)	Ken.	0•59 N	37•47 E
22	Exe (R.)	Eng.	50•57 N	3•37 W
22	Exeter	Eng.	50•45 N	3•33 W
9	Exeter	NH	43•00 N	71•00 W
22	Exmoor	Eng.	51•10 N	3•55 W
22	Exmouth	Eng.	50•40 N	3•20 W
44	Exmouth, G.	Austl.	21•45 S	114•30 E
54	Eyasi, L.	Tan.	3•25 S	34•55 E
16	Eyja Fd.	Ice.	66•21 N	18•20 W
16	Eyrarbakki	Ice.	63•51 N	20•52 W
44	Eyre	Austl.	32•15 S	126•20 E
46	Eyre (L.)	Austl.	28•43 S	137•50 E
44	Eyre Pen.	Austl.	33•30 S	136•00 E
50	Ezeiza	Arg.	34•36 S	58•31 W
21	Ezine	Tur.	39•47 N	26•18 E

F

Page	Name	Region	Lat.	Long.
24	Fåborg	Den.	55•06 N	10•19 E
20	Fabriano	It.	43•20 N	12•55 E
48	Facatativá	Col.	4•49 N	74•09 W
53	Fada	Chad	17•06 N	21•18 E
52	Fada Ngourma	Upper Volta	12•04 N	0•21 E
29	Faddeya (I.)	Sov.Un.	76•12 N	145•00 E
24	Faemund (L.)	Nor.	62•17 N	11•40 E
20	Faenza	It.	44•16 N	11•53 E
14	Faeroe Is.	Eur.	62•00 N	5•45 W
21	Făgăras	Rom.	45•50 N	24•55 E
24	Fagerness	Nor.	61•00 N	9•10 E
50	Fagnano (L.)	Arg.-Chile	54•35 S	68•20 W
52	Faial I.	Açores	38•40 N	29•19 W
22	Fair (I.)	Scot.	59•34 N	1•41 W
6	Fairbanks	Ak.	64•50 N	147•48 W
8	Fairbury	Il.	40•45 N	88•25 W
8	Fairfield	Il.	38•25 N	88•20 W
9	Fairhaven	Ma.	41•35 N	70•55 W
9	Fair Haven	Vt.	43•35 N	73•15 W
9	Fairmont	WV	39•30 N	80•10 W
8	Fairmount	In.	40•25 N	85•45 W
9	Fairport	NY	43•05 N	77•30 W
8	Fairport Harbor	Oh.	41•45 N	81•15 W
35	Faizābād	India	26•50 N	82•17 E
13	Fajardo	P.R.	18•20 N	65•40 W
41	Fakfak	Indon.	2•56 S	132•25 E
38	Fak'u	China	42•28 N	123•20 E
39	Falalise, C.	Viet.	19•20 N	106•18 E
49	Falcón (State)	Ven.	11•00 N	68•28 W
9	Falconer	NY	42•10 N	79•10 W
52	Falemé (R.)	Afr.	13•40 N	12•00 W
24	Falkenberg	Swe.	56•54 N	12•25 E
22	Falkirk	Scot.	55•59 N	3•55 W
50	Falkland Is.	S.A.	50•45 S	61•00 W
24	Falköping	Swe.	58•09 N	13•30 E
9	Fall River	Ma.	41•42 N	71•07 W
22	Falmouth	Eng.	50•08 N	3•04 W
8	Falmouth	Ky.	38•40 N	84•20 W
33	False Divi Pt.	India	15•45 N	80•50 E
24	Falster (I.)	Den.	54•48 N	11•58 E
27	Fălticeni	Rom.	47•27 N	26•17 E
24	Falun	Swe.	60•38 N	15•35 E
19	Famagusta	Cyprus	35•08 N	33•59 E
50	Famatina, Sierra de (Mts.)	Arg.	29•00 S	67•50 W
39	Fan Ching Shan (Mts.)	China	26•46 N	107•42 E
39	Fanghsien	China	32•05 N	110•45 E
43	Fanning (I.)	Gilbert & Ellice Is.	4•20 N	159•00 W
20	Fano	It.	43•49 N	13•01 E
24	Fanø (I.)	Den.	55•24 N	8•10 E
55	Farafangana	Mad.	21•18 S	47•59 E
34	Farāh	Afg.	32•15 N	62•13 E
52	Faranah	Gui.	10•02 N	10•44 W
53	Farasān, Jaza'ir (Is.)	Eth.	16•45 N	41•08 E
19	Faras (R.)	Libya	30•18 N	17•19 E
19	Faregh, Wadi al (R.)	Libya	30•10 N	19•34 E
45	Farewell, C.	N.Z.	40•37 S	171•46 E
6	Fargo	ND	46•53 N	96•48 W
53	Farit, Amba (Mt.)	Eth.	10•51 N	37•52 E

Page	Name	Region	Lat.	Long.
27	Farkašd	Czech.	48·00 N	17·43 E
8	Farmersburg	In.	39·15 N	87·25 W
8	Farmington	Il.	40·42 N	90·01 W
9	Farmington	Me.	44·40 N	70·10 W
22	Farne (I.)	Eng.	55·40 N	1·32 W
9	Farnham	Can.	45·15 N	72·55 W
49	Faro	Braz.	2·05 S	56·32 W
17	Faro	Port.	37·01 N	7·57 W
25	Fåron (I.)	Swe.	57·57 N	19·10 E
44	Farquhar, C.	Austl.	23·50 S	112·55 E
8	Farrell	Pa.	41·10 N	80·30 W
35	Farrukhābād	India	27·29 N	79·35 E
21	Fársala (Pharsalus)	Grc.	39·18 N	22·25 E
24	Farsund	Nor.	58·05 N	6·47 E
50	Fartura, Serra da (Mts.)	Braz.	26·40 S	53·15 W
5	Farvel, Kap (C.)	Grnld.	60·00 N	44·00 W
21	Fasano	It.	40·50 N	17·22 E
31	Fatsa	Tur.	40·50 N	37·30 E
16	Fauske	Nor.	67·15 N	15·24 E
16	Faxaflói (B.)	Ice.	64·33 N	22·40 W
53	Fazzān (Fezzan) (Prov.)	Libya	26·45 N	13·01 E
52	Fdérik	Mauritania	22·45 N	12·38 W
17	Fécamp	Fr.	49·45 N	0·20 E
49	Federal, Distrito (Dist.)	Ven.	10·34 N	66·55 W
26	Fehmarn I.	F.R.G.	54·28 N	11·15 E
49	Feira de Santana	Braz.	12·16 S	38·46 W
26	Feldkirch	Aus.	47·15 N	9·36 E
20	Feltre	It.	46·02 N	11·56 E
55	Fénérive	Mad.	17·30 S	49·31 E
38	Fengchen	China	40·28 N	113·20 E
38	Fengch'eng	China	40·28 N	124·03 E
39	Fengchieh	China	31·02 N	109·30 E
38	Fenghsiang	China	34·25 N	107·20 E
38	Fengt'ai	China	39·51 N	116·19 E
39	Fengtu	China	29·58 N	107·50 E
38	Fengyang	China	32·55 N	117·32 E
8	Fenton	Mi.	42·50 N	83·40 W
38	Fenyang	China	37·20 N	111·48 E
31	Feodosiya (Kefe)	Sov.Un.	45·02 N	35·21 E
34	Ferdows	Iran	34·00 N	58·13 E
20	Ferentino	It.	41·42 N	13·18 E
28	Fergana	Sov.Un.	40·16 N	72·07 E
7	Fergus Falls	Mn.	46·17 N	96·03 W
20	Fermo	It.	43·10 N	13·43 E
22	Fermoy	Ire.	52·05 N	8·06 W
49	Fernando de Noronha, Arquipélago	Braz.	3·50 S	33·15 W
20	Ferrara	It.	44·50 N	11·37 E
48	Ferreñafe	Peru	6·38 S	79·48 W
17	Ferryville	Tun.	37·12 N	9·51 E
52	Fès	Mor.	34·08 N	5·00 W
22	Festiniog	Wales	52·59 N	3·58 W
31	Fethiye	Tur.	36·40 N	29·05 E
11	Feuilles, Rivière aux (R.)	Can.	58·30 N	70·50 W
55	Fianarantsoa	Mad.	21·21 S	47·15 E
21	Fier	Alb.	40·43 N	19·34 E
22	Fife Ness (C.)	Scot.	56·15 N	2·19 W
53	Fifth Cataract	Sud.	18·27 N	33·38 E
24	Figeholm	Swe.	57·24 N	16·33 E
52	Figuig	Mor.	32·20 N	1·30 W
42	Fiji	Oceania	18·50 S	175·00 E
67	Filchner Ice Shelf	Ant.	80·00 S	35·00 W
21	Filiatrá	Grc.	37·10 N	21·35 E
20	Filicudi (I.)	It.	38·34 N	14·39 E
19	Filigas (R.)	Tur.	41·10 N	32·53 E
24	Filipstad	Swe.	59·44 N	14·09 E
54	Fimi (R.)	Zaire	2·43 S	17·50 E
8	Findlay	Oh.	41·05 N	83·40 W
17	Finisterre, Cabo de (C.)	Sp.	42·52 N	9·48 W
44	Finke (R.)	Austl.	25·25 S	134·30 E
14	Finland	Eur.	62·45 N	26·13 E
25	Finland, G. of	Eur.	59·35 N	23·35 E
48	Finlandia	Col.	4·38 N	75·39 W
10	Finlay (R.)	Can.	57·45 N	125·30 W
26	Finsterwalde	G.D.R.	51·38 N	13·42 E
31	Firat (R.)	Tur.	39·40 N	38·30 E
20	Firenze (Florence)	It.	43·47 N	11·15 E
20	Firenzuola	It.	44·08 N	11·21 E
35	Fīrozpur	India	30·58 N	74·39 E
54	Fish (R.)	Namibia	27·30 S	17·45 E
11	Fisher Str.	Can.	62·43 N	84·28 W
9	Fitchburg	Ma.	42·35 N	71·48 W
44	Fitzroy (R.)	Austl.	18·00 S	124·05 E
45	Fitzroy (R.)	Austl.	23·45 S	150·02 E
44	Fitzroy Crossing	Austl.	18·08 S	126·00 E
8	Fitzwilliam (I.)	Can.	45·30 N	81·45 W
24	Fjällbacka	Swe.	58·37 N	11·17 E
24	Flaam	Nor.	60·15 N	7·01 E
6	Flagstaff	Az.	35·15 N	111·40 W
55	Flagstaff	S.Afr.	31·06 S	29·31 E
9	Flagstaff (L.)	Me.	45·05 N	70·30 W
6	Flaming Gorge Res.	Wy.	41·13 N	109·30 W
13	Flamingo Pt.	Vir.Is.	18·19 N	65·00 W
23	Flanders (Reg.)	Fr.	50·53 N	2·29 E
22	Flannan (Is.)	Scot.	58·13 N	8·14 W
6	Flathead L.	Mt.	47·57 N	114·20 W
10	Flattery C.	Wa.	48·22 N	125·10 W
24	Flekkefjord	Nor.	58·19 N	6·38 E
8	Flemingsburg	Ky.	38·25 N	83·45 W
26	Flensburg	F.R.G.	54·48 N	9·27 E
17	Flers-del-l'Orne	Fr.	48·43 N	0·37 W
44	Flinders (Reg.)	Austl.	32·15 S	138·45 E
46	Flinders (I.)	Austl.	39·35 S	148·10 E
45	Flinders (R.)	Austl.	18·48 S	141·07 E
45	Flinders Rfs.	Austl.	17·30 S	149·02 E
8	Flin Flon	Can.	54·46 N	101·53 W
8	Flint	Mi.	43·00 N	83·45 W
7	Flint (R.)	Ga.	31·25 N	84·15 W
24	Flisen	Nor.	60·35 N	12·03 E
8	Flora	Il.	38·40 N	88·25 W
8	Flora	In.	40·25 N	86·30 W
7	Florence	Al.	34·46 N	87·40 W
48	Florencia	Col.	1·31 N	75·13 W
50	Florencio Varela	Arg.	34·34 S	58·16 W
49	Flores	Braz.	7·57 S	37·48 W
40	Flores (I.)	Indon.	8·14 S	121·08 E
40	Flores Laut (Flores Sea)	Indon.	7·09 N	120·30 E
49	Floriano	Braz.	6·17 S	42·58 W
50	Florianópolis	Braz.	27·30 S	48·30 W
48	Florida	Col.	3·20 N	76·12 W
55	Florida	S.Afr.	26·11 S	27·56 E
50	Florida	Ur.	34·06 S	56·14 W
7	Florida (State)	U.S.	30·30 N	84·40 W
45	Florida (I.)	Sol.Is.	8·56 S	159·45 E
13	Florida, Strs. of	N.A.	24·10 N	81·00 W
7	Florida Keys (Is.)	Fl.	24·33 N	81·20 W
21	Florina	Grc.	40·48 N	21·24 E
24	Florö	Nor.	61·36 N	5·01 E
20	Flumendosa, R.	It.	39·45 N	9·18 E
8	Flushing	Mi.	43·05 N	83·50 W
41	Fly (R.)	Pap.N.Gui.	8·00 S	141·45 E
21	Foča	Yugo.	43·29 N	18·48 E
27	Focsani	Rom.	45·41 N	27·17 E
20	Foggia	It.	41·30 N	15·34 E
52	Fogo I.	C.V.	14·46 N	24·51 W
26	Fohnsdorf	Aus.	47·13 N	14·40 E
26	Föhr I.	F.R.G.	54·47 N	8·30 E
39	Fokang	China	23·50 N	113·35 E
20	Foligno	It.	42·58 N	12·41 E
23	Folkeston	Eng.	51·05 N	1·18 E
48	Fómeque	Col.	4·29 N	73·52 W
8	Fond du Lac	Wi.	43·47 N	88·29 W
20	Fondi	It.	41·23 N	13·25 E
12	Fonseca, Golfo de (G.)	Hond.	13·09 N	87·55 W
17	Fontainebleau	Fr.	48·24 N	2·42 E
48	Fonte Boa	Braz.	2·32 S	66·05 W
48	Fontibón	Col.	4·42 N	74·09 W
55	Foothills	S.Afr.	25·55 S	27·36 E
46	Forbes	Austl.	33·24 S	148·05 E
26	Forchheim	F.R.G.	49·43 N	11·05 E
52	Forecariah	Gui.	9·26 N	13·06 W
5	Forel, Mt.	Grnld.	65·50 N	37·41 W
9	Forest City	Pa.	41·35 N	75·30 W
22	Forfar	Scot.	57·10 N	2·55 W
20	Forli	It.	44·13 N	12·03 E
18	Formentera, Isla de (I.)	Sp.	38·43 N	1·25 E
49	Formiga	Braz.	20·27 S	45·25 W
50	Formosa	Arg.	27·25 S	58·12 W
49	Formosa	Braz.	15·32 S	47·10 W
50	Formosa (Prov.)	Arg.	24·30 S	60·45 W
49	Formosa, Serra (Mts.)	Braz.	12·59 S	55·11 W
33	Formosa Str.	Asia	24·30 N	120·00 E
45	Forsayth	Austl.	18·33 S	143·42 E
24	Forshaga	Swe.	59·34 N	13·25 E
26	Forst	G.D.R.	51·45 N	14·38 E
11	Fort Albany	Can.	52·20 N	81·20 W
49	Fortaleza (Ceará)	Braz.	3·35 S	38·31 W
8	Fort Atkinson	Wi.	42·55 N	88·46 W
55	Fort Beaufort	S.Afr.	32·47 S	26·39 E
8	Fort Branch	In.	38·15 N	87·35 W
10	Fort Chipewyan	Can.	58·46 N	111·15 W
6	Fort Collins	Co.	40·36 N	105·04 W
53	Fort Crampel	Cen.Afr.Emp.	6·59 N	19·11 E
55	Fort-Dauphin	Mad.	24·59 S	46·58 E
13	Fort-de-France	Mart.	14·37 N	61·06 W
53	Fort-de-Possel	Cen.Afr.Emp.	5·03 N	19·11 E
7	Fort Dodge	Ia.	42·31 N	94·10 W
9	Fort Edward	NY	43·15 N	73·30 W
44	Fortescue (R.)	Austl.	21·25 S	116·50 E
10	Fort Fitzgerald	Can.	59·48 N	111·50 W
11	Fort George	Can.	53·40 N	78·58 W
6	Fort Good Hope	Can.	66·19 N	128·52 W
22	Forth, Firth of	Scot.	56·04 N	3·03 W
53	Fort Hall	Ken.	0·47 S	37·13 E
54	Fort Jameson	Zambia	13·35 S	32·43 E
54	Fort Johnston	Malawi	14·16 S	35·14 E
10	Fort Liard	Can.	60·16 N	123·34 W
10	Fort Macleod	Can.	49·43 N	113·25 W
52	Fort MacMahon	Alg.	29·55 N	1·49 E
10	Fort McMurray	Can.	56·44 N	111·23 W
10	Fort McPherson	Can.	67·37 N	134·59 W
18	Fort Miribel	Alg.	28·50 N	2·51 E
18	Fort National	Alg.	36·45 N	4·15 E
10	Fort Nelson	Can.	58·57 N	122·30 W
10	Fort Nelson (R.)	Can.	58·44 N	122·20 W
6	Fort Peck Res.	Mt.	47·52 N	106·59 W
53	Fort Portal	Ug.	0·40 N	30·16 E
10	Fort Providence	Can.	61·27 N	117·59 W
6	Fort Randall Dam	U.S.	42·48 N	98·35 W
10	Fort Resolution	Can.	61·08 N	113·42 W
10	Fort St. James	Can.	54·26 N	124·15 W
10	Fort St. John	Can.	56·15 N	120·51 W
35	Fort Sandeman	Pak.	31·28 N	69·29 E
10	Fort Saskatchewan	Can.	53·43 N	113·13 W
11	Fort Severn	Can.	56·58 N	87·50 W
31	Fort Shevchenko	Sov.Un.	44·30 N	50·18 E
53	Fort Sibut	Cen.Afr.Emp.	5·44 N	19·05 E
10	Fort Simpson	Can.	61·52 N	121·48 W
7	Fort Smith	Ar.	35·23 N	94·24 W
10	Fort Smith	Can.	60·09 N	112·08 W
11	Fortune B.	Can.	47·25 N	55·25 W
10	Fort Vermilion	Can.	58·23 N	115·50 W
54	Fort Victoria	Rh.	20·07 S	30·47 E
8	Fortville	In.	40·00 N	85·50 W
8	Fort Wayne	In.	41·00 N	85·10 W
22	Fort William	Scot.	56·50 N	3·00 W
46	Fort William, Mt.	Austl.	24·45 S	151·15 E
6	Fort Worth	Tx.	32·45 N	97·20 W
6	Fort Yukon	Ak.	66·30 N	145·00 W
39	Foshan	China	23·02 N	113·07 E
20	Fossano	It.	44·34 N	7·42 E
20	Fossombrone	It.	43·41 N	12·48 E
8	Fostoria	Oh.	41·10 N	83·20 W
17	Fougéres	Fr.	48·23 N	1·14 W
38	Fouhsin	China	42·05 N	121·40 E
22	Foula (I.)	Scot.	60·08 N	2·04 W
39	Fouling	China	29·40 N	107·30 E
45	Foulwind, C.	N.Z.	41·45 S	171·37 E
52	Foumban	Cam.	5·43 N	10·55 E
38	Founing	China	33·55 N	119·54 E
53	Fourth Cataract	Sud.	18·52 N	32·07 E
52	Fouta Djallon (Mts.)	Gui.	11·37 N	12·29 W
38	Fouyang	China	32·53 N	115·48 E
45	Foveaux Str.	N.Z.	46·30 S	167·43 E
8	Fowler	In.	40·35 N	87·20 W
44	Fowler, Pt.	Austl.	32·05 S	132·30 E
8	Fox (R.)	Il.	41·35 N	88·43 W
8	Fox (R.)	Wi.	44·18 N	88·23 W
10	Foxe Basin	Can.	67·35 N	79·21 W
11	Foxe Chan.	Can.	64·30 N	79·23 W
11	Foxe Pen.	Can.	64·57 N	77·26 W
22	Foyle, Lough (B.)	Ire.	55·07 N	7·08 W
49	Franca	Braz.	20·28 S	47·20 W
21	Francavilla	It.	40·32 N	17·37 E
14	France	Eur.	46·39 N	0·47 E
10	Frances (L.)	Can.	61·27 N	128·28 W
54	Franceville	Gabon	1·38 S	13·35 E
54	Francistown	Bots.	21·17 S	27·28 E
8	Frankfort	In.	40·15 N	86·30 W
8	Frankfort	Ky.	38·10 N	84·55 W
8	Frankfort	Mi.	44·40 N	86·15 W
9	Frankfort	NY	43·05 N	75·05 W
55	Frankfort	S.Afr.	32·43 S	27·28 E
26	Frankfurt	G.D.R.	52·20 N	14·31 E
26	Frankfurt am Main	F.R.G.	50·07 N	8·40 E
8	Franklin	In.	39·25 N	86·00 W
9	Franklin	NH	43·25 N	71·40 W
8	Franklin	Oh.	39·30 N	84·20 W
9	Franklin	Pa.	41·25 N	79·50 W
55	Franklin	S.Afr.	30·19 S	29·28 E
10	Franklin, Dist. of	Can.	70·46 N	105·22 W
10	Franklin Mts.	Can.	65·36 N	125·55 W
46	Fraser (Great Sandy) (I.)	Austl.	25·12 S	153·00 E
10	Fraser (R.)	Can.	52·20 N	122·35 W
22	Fraserburgh	Scot.	57·40 N	2·01 W
50	Fray Bentos	Ur.	33·10 S	58·19 W
24	Fredericia	Den.	55·35 N	9·45 E
9	Frederick	Md.	39·25 N	77·25 W
9	Fredericksburg	Va.	38·20 N	77·30 W
11	Fredericton	Can.	45·48 N	66·39 W
24	Frederikshavn	Den.	57·27 N	10·31 E
24	Frederikssund	Den.	55·51 N	12·04 E
48	Fredonia	Col.	5·55 N	75·40 W
9	Fredonia	NY	42·25 N	79·20 W
24	Fredrikstad	Nor.	59·14 N	10·58 E
9	Freehold	NJ	40·15 N	74·16 W
9	Freeland	Pa.	41·00 N	75·50 W
8	Freeport	Il.	42·19 N	89·30 W
9	Freeport	NY	40·39 N	73·35 W
52	Freetown	S.L.	8·30 N	13·15 W
17	Fregenal de la Sierra	Sp.	38·09 N	6·40 W
26	Freiberg	G.D.R.	50·54 N	13·18 E
26	Freiburg	G.D.R.	48·00 N	7·50 E
44	Fremantle	Austl.	32·03 S	116·05 E
8	Fremont	Mi.	43·25 N	85·55 W
8	Fremont	Oh.	41·20 N	83·05 W
47	French Guiana	S.A.	4·20 N	53·00 W
8	French Lick	In.	38·35 N	86·35 W
12	Fresnillo	Mex.	23·10 N	102·52 W
10	Fresno	Ca.	36·43 N	119·47 W
48	Fresno	Col.	5·10 N	75·01 W
26	Freudenstadt	F.R.G.	48·28 N	8·26 E
46	Freycinet Pen.	Austl.	42·13 S	148·56 E
54	Fria, C.	Namibia	18·15 S	12·10 E
50	Frias	Arg.	28·43 S	65·03 W
26	Fribourg	Switz.	46·48 N	7·07 E
26	Frieburg	F.R.G.	47·59 N	7·50 E
26	Friedland	G.D.R.	53·39 N	13·34 E
26	Friedrichshafen	F.R.G.	47·39 N	9·28 E
49	Frio, Cabo (C.)	Braz.	22·58 S	42·08 W
18	Friol	Sp.	43·02 N	7·48 W
23	Frisian (Is.)	Neth.	53·30 N	5·20 E
20	Friuli-Venezia Giulia	It.	46·20 N	13·20 E
10	Frobisher L.	Can.	56·25 N	108·20 W
11	Frobisher Bay	Can.	63·48 N	68·31 W
11	Frobisher B.	Can.	62·49 N	66·41 W
46	Frome, L.	Austl.	30·40 S	140·13 E
9	Front Royal	Va.	38·55 N	78·10 W
16	Fro Sea	Nor.	63·49 N	9·12 E
20	Frosinone	It.	41·38 N	13·22 E
9	Frostburg	Md.	39·40 N	78·55 W
28	Frunze	Sov.Un.	42·49 N	74·42 E
27	Frýdek	Czech.	49·43 N	18·22 E
26	Frydlant	Czech.	50·56 N	15·05 E
37	Fuchin	China	47·13 N	132·11 E
39	Fuchou (Foochow)	China	26·02 N	119·18 E
39	Fuch'un (R.)	China	29·50 N	120·00 E
12	Fuerte, Rio del (R.)	Mex.	26·15 N	108·50 W
49	Fuerte Olimpo	Par.	21·10 S	57·49 W
52	Fuerteventura I.	Can.Is.	28·24 N	13·21 W
36	Fuhai	China	47·01 N	87·07 E
37	Fuji-san (Mtn.)	Jap.	35·23 N	138·44 E
37	Fuckien (Fuchien)	China	25·40 N	117·30 E
37	Fukui	Jap.	36·05 N	136·14 E
35	Fūlādī, Kūh-e (Mtn.)	Afg.	34·38 N	67·55 E
26	Fulda R.	F.R.G.	51·05 N	9·40 E
9	Fulton	NY	43·20 N	76·25 W
52	Funchal	Mad.Is.	32·41 N	16·15 W
48	Fundación	Col.	10·43 N	74·13 W
11	Fundy, B. of	Can.	45·00 N	66·00 W
11	Fundy Natl. Park	Can.	45·38 N	65·00 W
39	Funing Wan (B.)	China	26·48 N	120·35 E
50	Furnas, Reprêsa de	Braz.	21·00 S	46·00 W
45	Furneaux Group (Is.)	Austl.	40·15 S	146·27 E
26	Fürstenfeld	Aus.	47·02 N	16·03 E
26	Fürstenwalde	G.D.R.	52·21 N	14·04 E
26	Fürth	F.R.G.	49·28 N	11·03 E
48	Fusagasugá	Col.	4·22 N	74·22 W
38	Fushun	China	41·50 N	124·00 E
38	Fusung	China	42·12 N	127·12 E
39	Fuyang	China	30·10 N	119·58 E
38	Fuyü	China	45·20 N	125·00 E
24	Fyn (I.)	Den.	55·24 N	10·33 E
22	Fyne (L.)	Scot.	56·14 N	5·10 W
24	Fyresdal Vand (L.)	Nor.	59·04 N	7·55 E

G

Page	Name	Region	Lat.	Long.
54	Gaborone	Bots.	24·28 S	25·59 E
52	Gabès	Tun.	33·51 N	10·04 E
52	Gabès, Golfe de	Tun.	32·22 N	10·59 E
51	Gabon	Afr.	0·30 S	10·45 E
21	Gabrovo	Bul.	42·52 N	25·19 E
48	Gachetá	Col.	4·50 N	73·36 W
21	Gacko	Yugo.	43·10 N	18·34 E
7	Gadsden	Al.	34·00 N	86·00 W
21	Gaesti	Rom.	44·43 N	25·21 E
20	Gaeta	It.	41·18 N	13·34 E
52	Gafsa	Tun.	34·16 N	8·37 E
41	Gagrary (I.)	Phil.	13·23 N	123·58 E
20	Gaïdhouronísi (I.)	Grc.	34·53 N	25·58 E
14	Gaillac-sur-Tarn	Fr.	43·54 N	1·52 E
12	Gaillard Cut	C.Z.	9·03 N	79·42 W
7	Gainesville	Fl.	29·40 N	82·20 W
46	Gairdner, L.	Austl.	32·20 S	136·30 E

Page	Name	Region	Lat.	Long.
22	Galashiels	Scot.	55•40 N	2•57 W
46	Galati	Rom.	45•25 N	28•05 E
21	Galatina	It.	40•10 N	18•12 E
21	Galaxidhion	Grc.	38•26 N	22•22 E
24	Galdhöpiggen	Nor.	61•37 N	8•17 E
12	Galera, Cerro (Mtn.)	C.Z.	8•55 N	79•38 W
48	Galeras (Vol.)	Col.	0•57 N	77•27 W
7	Galesburg	Il.	40•56 N	90•21 W
9	Galeton	Pa.	41•45 N	77•40 W
21	Galibolu (Gallipoli)	Tur.	40•25 N	26•40 E
30	Galich	Sov.Un.	58•20 N	42•38 E
27	Galicia	Pol.-Sov.Un.	49•48 N	21•05 E
45	Galilee (L.)	Austl.	22•23 S	145•09 E
33	Galilee, Sea of	Isr.	32•53 N	35•45 E
13	Galion	Oh.	40•45 N	82•50 W
17	Galite, La I.	Alg.	37•36 N	8•03 E
53	Galla	Eth.	7•22 N	35•28 E
20	Gallarate	It.	45•37 N	8•48 E
35	Galle	Sri Lanka	6•13 N	80•10 E
48	Gallinas, Pta. de	Col.	12•10 N	72•10 W
21	Gallipoli	It.	40•03 N	17•58 E
8	Gallipolis	Oh.	38•50 N	82•10 W
16	Gällivare	Swe.	68•06 N	20•29 E
6	Gallup	NM	35•30 N	108•45 W
53	Galnale Doria R.	Eth.	5•35 N	40•26 E
8	Galt	Can.	43•22 N	80•19 W
22	Galty Mts.	Ire.	52•19 N	8•20 W
8	Galva	Il.	41•11 N	90•02 W
7	Galveston	Tx.	29•18 N	94•48 W
22	Galway	Ire.	53•16 N	9•05 W
22	Galway B.	Ire.	53•10 N	9•47 W
52	Gambaga	Ghana	10•32 N	0•26 W
53	Gambela	Eth.	8•15 N	34•33 E
52	Gambia	Afr.	13•38 N	19•38 W
52	Gambia (R.)	Afr.	13•20 N	15•55 W
54	Gamboma	Con.	1•53 S	15•51 E
24	Gamleby	Swe.	57•54 N	16•20 E
41	Gamu	Phil.	17•05 N	121•50 E
11	Gander	Can.	48•57 N	54•34 W
35	Ganges, Mouths of	India	21•18 N	88•40 E
35	Ganges (R.)	India	24•32 N	87•58 E
20	Gangi	It.	37•48 N	14•15 E
36	Gangtok	India	27•15 N	88•30 E
6	Gannett Pk.	Wy.	43•10 N	109•38 W
52	Gao	Mali	16•16 N	0•03 W
41	Gapan	Phil.	15•18 N	120•56 E
49	Garanhuns	Braz.	8•49 S	36•28 W
20	Garda, Lago di	It.	45•43 N	10•26 E
26	Gardelegen	G.D.R.	52•32 N	11•22 E
8	Garden (I.)	Mi.	45•50 N	85•50 W
9	Gardner	Ma.	42•35 N	72•00 W
21	Gargaliánoi	Grc.	37•07 N	21•50 E
25	Gargždai	Sov.Un.	55•43 N	20•09 E
50	Garin	Arg.	34•10 S	58•44 W
26	Garmisch-Partenkirchen	F.R.G.	47•38 N	11•10 E
17	Garonne (R.)	Fr.	44•43 N	0•25 W
8	Garrett	In.	41•20 N	85•10 W
10	Garry (L.)	Can.	66•16 N	99•23 W
27	Garwolin	Pol.	51•54 N	21•40 E
8	Gary	In.	41•35 N	87•21 W
48	Garzón	Col.	2•13 N	75•44 W
41	Gasan	Phil.	13•19 N	121•52 E
31	Gasan-Kuli	Sov.Un.	37•25 N	53•55 E
8	Gas City	In.	40•30 N	85•40 W
44	Gascoyne (R.)	Austl.	25•15 S	117•00 E
11	Gaspé	Can.	48•50 N	64•29 W
11	Gaspé, Péninsule de	Can.	48•23 N	65•42 W
8	Gassaway	WV	38•40 N	80•45 W
50	Gastre	Arg.	42•12 S	68•50 W
18	Gata, Cabo de	Sp.	36•42 N	2•00 W
18	Gata, Sierra de	Sp.	40•12 N	6•39 W
33	Gátes Akrotírion (C.)	Cyprus	34•30 N	33•15 E
22	Gateshead	Eng.	54•56 N	1•38 W
9	Gatineau (R.)	Can.	45•45 N	75•50 W
12	Gatun	C.Z.	9•16 N	79•25 W
12	Gatun, L.	Pan.-C.Z.	9•13 N	79•24 W
12	Gatun (R.)	Pan.	9•21 N	79•10 W
12	Gatun Locks	C.Z.	9•16 N	79•27 W
25	Gauja (R.)	Sov.Un.	57•10 N	24•30 E
41	Gauttier-Gebergte (Mts.)	Indon.	2•30 S	138•45 E
20	Gávdhos (I.)	Grc.	34•48 N	24•08 E
24	Gävle	Swe.	60•40 N	17•07 E
24	Gavle-bukten (B.)	Swe.	60•45 N	17•30 E
46	Gawler	Austl.	34•35 S	138•47 E
46	Gawler Ra.	Austl.	32•35 S	136•30 E
35	Gaya	India	24•53 N	85•00 E
52	Gaya	Nig.	11•58 N	9•05 E
8	Gaylord	Mi.	45•00 N	84•35 W
46	Gayndah	Austl.	25•43 S	151•33 E
31	Gaziantep	Tur.	37•10 N	37•30 E
27	Gdańsk (Danzig)	Pol.	54•20 N	18•40 E
27	Gdynia	Pol.	54•29 N	18•30 E
19	Gediz (R.)	Tur.	38•44 N	28•45 E
26	Gedser	Den.	54•35 N	12•08 E
46	Geelong	Austl.	38•06 S	144•13 E
41	Geelvink-baai (B.)	Indon.	2•20 S	135•30 E
44	Geikie Ra.	Austl.	17•35 S	125•32 E
26	Geislingen	F.R.G.	48•37 N	9•52 E
21	Gelibolu (Pen.)	Tur.	40•23 N	25•10 E
23	Gelsenkirchen	F.R.G.	51•31 N	7•05 E
33	Gemas	Mala.	2•35 N	102•37 E
53	Gemena	Zaire	3•15 N	19•46 E
31	Gemlik	Tur.	40•30 N	29•10 E
53	Genale (R.)	Eth.	5•00 N	41•15 E
50	General Madariaga	Arg.	36•59 S	57•14 W
50	General Pico	Arg.	36•46 S	63•44 W
50	General Roca	Arg.	39•01 S	67•31 W
50	General San Martín	Arg.	34•19 S	58•32 W
9	Genesee (R.)	NY	42•25 N	78•10 W
8	Geneseo	Il.	41•28 N	90•11 W
9	Geneva	NY	42•50 N	77•00 W
8	Geneva	Oh.	41•45 N	80•55 W
26	Geneva, L.	Switz.	46•28 N	6•30 E
26	Génève (Geneva)	Switz.	46•14 N	6•04 E
20	Genova (Genoa)	It.	44•23 N	9•52 E
20	Genova, Golfo di	It.	44•10 N	8•45 E
12	Genovesa (I.)	Ec.	0•08 N	90•15 W
23	Gent	Bel.	51•05 N	3•40 E
26	Genthin	G.D.R.	52•24 N	12•10 E
44	Geographe B.	Austl.	33•00 S	114•00 E
44	Geographic Chan.	Austl.	24•15 S	112•50 E
31	Geokchay	Sov.Un.	40•40 N	47•40 E
9	George (L.)	NY	43•40 N	73•30 W
49	Georgetown	Guy.	7•45 N	58•04 W
9	Georgetown	De.	38•40 N	75•20 W
8	Georgetown	Il.	40•00 N	87•40 W
8	Georgetown	Ky.	38•10 N	84•35 W
9	Georgetown	Md.	39•25 N	75•55 W
9	George Washington Birthplace Natl. Mon.	Va.	38•10 N	77•00 W
7	Georgia (State)	U.S.	32•40 N	83•50 W
28	Georgian S.S.R.	Sov.Un.	42•17 N	43•00 E
11	Georgian B.	Can.	45•15 N	80•50 W
44	Georgina (R.)	Austl.	22•00 S	138•15 E
31	Georgiyevsk	Sov.Un.	44•05 N	43•30 E
26	Gera	G.D.R.	50•52 N	12•06 E
50	Geral, Serra	Braz.	28•30 S	51•00 W
19	Geral de Goiás, Serra	Braz.	14•22 S	45•40 W
44	Geraldton	Austl.	28•40 S	114•35 E
11	Geraldton	Can.	49•43 N	87•00 W
27	Gerlachovka Pk.	Czech.	49•12 N	20•05 E
14	German Democratic Republic	Eur.	53•30 N	12•30 E
8	Germantown	Oh.	39•35 N	84•25 W
14	Germany, Federal Republic of	Eur.	51•45 N	8•30 E
55	Germiston	S.Afr.	26•19 S	28•11 E
41	Gerona	Phil.	15•36 N	120•36 E
18	Gerona	Sp.	41•55 N	2•48 E
9	Gettysburg	Pa.	39•50 N	77•15 W
35	Ghāghra (R.)	India	27•19 N	81•22 E
51	Ghana	Afr.	8•00 N	2•00 W
54	Ghanzi	Bots.	21•30 S	22•00 E
52	Ghardaïa	Alg.	32•29 N	3•38 E
52	Ghāt	Libya	24•52 N	10•16 E
53	Ghazāl, Bahr al- (R.)	Sud.	9•11 N	29•37 E
17	Ghazaouet	Alg.	35•19 N	1•09 W
33	Ghazzah (Gaza)	Gaza Strip	31•30 N	34•29
27	Gheorghieni	Rom.	46•48 N	25•30 E
27	Gherla	Rom.	47•01 N	23•55 E
52	Ghudāmis	Libya	30•07 N	9•26 E
20	Giannutri, I. di	It.	42•15 N	11•06 E
54	Gibeon	Namibia	24•45 S	16•40 E
17	Gibraltar	Eur.	36•08 N	5•22 W
18	Gibraltar, Str. of	Afr.-Eur.	35•55 N	5•45 W
8	Gibson City	Il.	40•25 N	88•20 W
44	Gibson Des.	Austl.	24•45 S	123•15 E
17	Gien	Fr.	47•43 N	2•37 E
26	Giessen	F.R.G.	50•35 N	8•40 E
20	Giglio, I. di	It.	42•23 N	10•55 E
18	Gijón	Sp.	43•33 N	5•37 W
6	Gila (R.)	Az.	32•41 N	113•50 W
45	Gilbert (R.)	Austl.	17•15 S	142•09 E
42	Gilbert Is.	Oceania	1•30 S	173•00 E
55	Gilboa, Mt.	S.Afr.	29•13 S	30•17 E
44	Gillen (I.)	Austl.	26•15 S	125•15 E
22	Gillingham	Eng.	51•23 N	0•33 E
8	Gilman	Il.	40•45 N	87•55 W
41	Giluwe, Mt.	Pap.N.Gui.	6•04 S	144•00 E
53	Ginir	Eth.	7•13 N	40•44 E
20	Ginosa	It.	40•35 N	16•48 E
20	Gioia del Colle	It.	40•48 N	16•55 E
49	Gi-Paraná (R.)	Braz.	9•33 S	61•35 W
48	Girardot	Col.	4•19 N	75•47 W
31	Giresun	Tur.	40•55 N	38•20 E
17	Gironde (Est.)	Fr.	45•31 N	1•00 W
22	Girvan	Scot.	55•15 N	5•01 W
45	Gisborne	N.Z.	38•40 S	178•08 E
54	Gisenyi	Rw.	1•43 S	29•15 E
54	Gitega	Burundi	3•39 S	30•05 E
21	Giurgui	Rom.	43•53 N	25•58 E
23	Givet	Fr.	50•80 N	4•47 E
29	Gizhiga	Sov.Un.	61•59 N	160•46 E
27	Gizycko	Pol.	54•03 N	21•48 E
21	Gjinokastër	Alb.	40•04 N	20•10 E
24	Gjövik	Nor.	60•47 N	10•36 E
6	Glacier Bay Natl. Mon.	Ak.	58•40 N	136•50 W
10	Glacier Natl. Park	Can.	51•45 N	117•35 W
46	Gladstone	Austl.	23•45 S	150•00 E
46	Gladstone	Austl.	33•15 S	138•20 E
6	Gladstone	Can.	50•15 N	98•50 W
8	Gladwin	Mi.	44•00 N	84•25 W
20	Glamoč	Yugo.	44•03 N	16•51 E
26	Glarus	Switz.	47•02 N	9•03 E
22	Glasgow	Scot.	55•54 N	4•25 W
26	Glauchau	G.D.R.	50•51 N	12•28 E
30	Glazov	Sov.Un.	58•05 N	52•52 E
26	Glda (R.)	Pol.	53•27 N	16•52 E
6	Glen Canyon Dam	Az.	36•57 N	111•25 W
55	Glencoe	S.Afr.	28•14 S	30•09 E
6	Glendale	Ca.	34•09 N	118•15 W
6	Glendive	Mt.	47•08 N	104•41 W
46	Glenelg (R.)	Austl.	37•20 S	141•30 E
46	Glen Innes	Austl.	29•45 S	152•02 E
9	Glens Falls	NY	43•20 N	73•40 W
24	Glittertinden (Mtn.)	Nor.	61•39 N	8•12 E
27	Gliwice	Pol.	50•18 N	18•40 E
6	Globe	Az.	33•20 N	110•50 W
26	Glogów	Pol.	51•40 N	16•04 E
24	Glomma (R.)	Nor.	61•22 N	11•02 E
24	Glommen (R.)	Nor.	60•03 N	11•15 E
55	Glorieuses, Îles	Afr.	11•28 S	47•50 E
22	Gloucester	Eng.	51•54 N	2•11 W
9	Gloucester	Ma.	42•37 N	70•40 W
9	Gloucester City	NJ	39•53 N	75•08 W
8	Glouster	Oh.	39•35 N	82•05 W
9	Gloversville	NY	43•05 N	74•20 W
30	Glubokoye	Sov.Un.	55•08 N	27•44 E
30	Glukhov	Sov.Un.	51•42 N	33•52 E
26	Gmünden	Aus.	47•57 N	13•47 E
27	Gniezno	Pol.	52•32 N	17•34 E
21	Gnjilane	Yugo.	42•28 N	21•27 E
35	Goa	India	15•45 N	74•00 E
53	Goba	Eth.	7•17 N	39•58 E
54	Gobabis	Namibia	22•25 S	18•50 E
36	Gobi or Shamo (Des.)	Mong.	43•29 N	103•15 E
35	Godāvari (R.)	India	17•42 N	81•15 E
44	Goddards Soak	Austl.	31•20 S	123•30 E
8	Goderich	Can.	43•45 N	81•45 W
5	Godhavn	Grnld.	69•15 N	53•30 W
10	Gods Lake	Can.	54•40 N	94•09 W
5	Godthåb	Grnld.	64•10 N	51•32 W
37	Godwin Austen, Mt.	Pak.	36•06 N	76•38 E
25	Gogland (I.)	Sov.Un.	60•04 N	26•55 E
49	Goiânia	Braz.	16•41 S	48•57 W
49	Goiás	Braz.	15•57 S	50•10 W
49	Goiás (State)	Braz.	12•35 S	48•38 W
31	Göksu (R.)	Tur.	36•40 N	33•30 E
24	Göl	Nor.	60•58 N	8•54 E
8	Golconda	Il.	37•21 N	88•32 W
27	Goldap	Pol.	54•17 N	22•17 E
12	Gold Hill (Mtn.)	C.Z.	9•03 N	79•08 W
26	Goleniów	Pol.	53•33 N	14•51 E
29	Golets-Purpula (Mtn.)	Sov.Un.	59•08 N	115•22 E
41	Golo	Phil.	13•38 N	120•17 E
20	Golo (R.)	Fr.	42•28 N	9•18 E
21	Golyamo Konare	Bul.	42•16 N	24•33 E
53	Gombari	Zaire	2•45 N	29•00 E
52	Gombe	Nig.	10•19 N	11•02 E
31	Gomel'	Sov.Un.	52•20 N	31•03 E
52	Gomera I.	Can.Is.	28•00 N	18•01 W
13	Gonaïves	Hai.	19•25 N	72•45 W
13	Gonâve, Ile de la	Hai.	18•50 N	73•30 W
53	Gonder	Eth.	12•39 N	37•30 E
55	Gonubie	S.Afr.	32•56 S	28•02 E
50	González Catán	Arg.	34•31 S	58•39 W
54	Good Hope, C. of	S.Afr.	34•21 S	18•29 E
8	Goodland	In.	40•50 N	87•15 W
54	Goodwood	S.Afr.	33•54 S	18•33 E
11	Goose Bay	Can.	53•19 N	60•33 W
35	Gorakhpur	India	26•45 N	82•39 E
53	Gore	Eth.	8•12 N	35•34 E
34	Gorgān	Iran	36•44 N	54•30 E
20	Gorgona (I.)	It.	43•27 N	9•55 E
31	Gori	Sov.Un.	42•00 N	44•08 E
20	Gorizia	It.	44•56 N	13•40 E
30	Gorki	Sov.Un.	56•15 N	44•05 E
30	Gor'kovskoye	Sov.Un.	56•38 N	43•40 E
30	Gor'kovskoye (Res.)	Sov.Un.	57•38 N	41•18 E
27	Gorlice	Pol.	49•38 N	21•11 E
26	Görlitz	G.D.R.	51•10 N	15•01 E
31	Gorlovka	Sov.Un.	48•17 N	38•03 E
21	Gorna-Oryakhovitsa	Bul.	43•08 N	25•40 E
21	Gornji Milanovac	Yugo.	44•02 N	20•29 E
28	Gorno-Altaysk	Sov.Un.	52•28 N	82•45 E
27	Gorodënka	Sov.Un.	48•40 N	25•30 E
30	Gorodets (Res.)	Sov.Un.	57•00 N	43•55 E
27	Gorodok	Sov.Un.	49•37 N	23•40 E
47	Gorodok	Sov.Un.	50•30 N	103•58 E
40	Gorontalo	Indon.	0•40 N	123•04 E
27	Goryn' R.	Sov.Un.	50•55 N	26•07 E
26	Gorzow Wielkopolski	Pol.	53•44 N	15•15 E
8	Goshen	Ind.	41•35 N	85•50 W
26	Goslar	F.R.G.	51•55 N	10•25 E
49	Gospa (R.)	Ven.	9•43 N	64•23 W
20	Gospić	Yugo.	44•31 N	15•03 E
21	Gostivar	Yugo.	41•46 N	20•58 E
27	Gostynin	Pol.	52•24 N	19•30 E
24	Göta alv (R.)	Swe.	58•11 N	12•03 E
24	Göta Can.	Swe.	58•35 N	15•24 E
24	Göteborg	Swe.	57•39 N	11•56 E
52	Gotel Mts.	Cam.-Nig.	7•05 N	11•20 E
26	Gotha	G.D.R.	50•57 N	10•43 E
24	Gotland (I.)	Swe.	57•35 N	17•35 E
25	Gotska Sandön (I.)	Swe.	58•24 N	19•15 E
26	Göttingen	F.R.G.	51•32 N	9•57 E
67	Gough (I.)	Atl.O.	40•00 S	10•00 W
11	Gouin, Rés	Can.	48•15 N	74•15 W
46	Goulburn	Austl.	34•47 S	149•40 E
52	Goumbou	Mali	14•59 N	7•27 W
52	Goundam	Mali	16•29 N	3•37 W
52	Gouré	Niger	13•53 N	10•44 E
9	Gouverneur	NY	44•20 N	75•25 W
10	Govenlock	Can.	49•15 N	109•48 W
50	Governador Ilhado (I.)	Braz.	22•48 S	43•13 W
50	Governador Portela	Braz.	22•28 S	43•30 W
49	Governador Valadares	Braz.	18•47 S	41•45 W
9	Gowanda	NY	42•30 N	78•55 W
50	Goya	Arg.	29•06 S	59•12 W
54	Graaff-Reinet	S.Afr.	32•10 S	24•40 E
20	Gracac	Yugo.	44•16 N	15•50 E
21	Gračanico	Yugo.	44•42 N	18•19 E
52	Graciosa I.	Açores	39•07 N	27•30 W
21	Gradačac	Yugo.	44•50 N	18•28 E
46	Grafton	Austl.	29•38 S	153•05 E
9	Grafton	WV	39•20 N	80•00 W
10	Graham (I.)	Can.	53•50 N	132•40 W
55	Grahamstown	S.Afr.	33•19 S	26•33 E
20	Graian Alps	Fr.-It.	45•17 N	6•52 E
49	Grajaú	Braz.	5•59 S	46•03 W
49	Grajaú (R.)	Braz.	4•24 S	46•04 W
27	Grajewo	Pol.	53•38 N	22•28 E
21	Gramada	Bul.	43•46 N	22•41 E
20	Grammichele	It.	37•15 N	14•40 E
22	Grampian Mts.	Scot.	56•30 N	4•55 W
12	Granada	Nic.	11•55 N	85•58 W
18	Granada	Sp.	37•13 N	3•37 W
50	Gran Bajo (Pln.)	Arg.	47•35 S	68•45 W
9	Granby	Can.	45•30 N	72•40 W
52	Gran Canaria I.	Can.Is.	27•39 N	15•39 W
50	Gran Chaco	Arg.-Par.	25•30 S	62•15 W
8	Grand (R.)	Mi.	42•58 N	85•13 W
13	Grand Bahama (I.)	Ba.	26•35 N	78•30 W
11	Grand Bank	Can.	47•06 N	55•47 W
52	Grand Bassam	Ivory Coast	5•12 N	3•44 W
22	Grand Canal	Ire.	53•21 N	7•15 W
6	Grand Canyon	Az.	36•05 N	112•10 W
6	Grand Canyon Natl. Park	Az.	36•15 N	112•20 W
13	Grand Cayman (I.)	Cayman Is.	19•15 N	81•15 W
6	Grand Coulee Dam	Wa.	47•58 N	119•28 W
49	Grande, Boca (Est.)	Ven.	8•46 N	60•17 W
12	Grande, Ciri (R.)	Pan.	8•55 N	80•04 W
48	Grande, Rio (R.)	Bol.	16•49 S	63•19 W
49	Grande, Rio			

Page	Name	Region	Lat.	Long.
48	Huascarán, Nevs.	Peru	9•05 S	77•50 W
50	Huasco	Chile	28•32 S	71•16 W
38	Huatien	China	42•38 N	126•45 E
8	Hubbard (L.)	Mi.	44•45 N	83•30 W
35	Hubli	India	15•25 N	75•09 E
22	Huddersfield	Eng.	53•39 N	1•47 W
24	Hudiksvall	Swe.	61•44 N	17•05 E
8	Hudson	Mi.	41•50 N	84•15 W
9	Hudson	NY	42•15 N	73•45 W
11	Hudson Bay	Can.	52•52 N	102•25 W
11	Hudson B.	Can.	60•15 N	85•30 W
9	Hudson Falls	NY	43•20 N	73•30 W
8	Hudson R.	NY	41•55 N	73•55 W
11	Hudson Str.	Can.	63•25 N	74•05 W
39	Hue	Viet.	16•28 N	107•42 E
18	Huelva	Sp.	37•16 N	6•58 W
17	Huésca	Sp.	42•07 N	0•25 W
17	Huéscar	Sp.	37•50 N	2•34 W
45	Hughenden	Austl.	20•58 S	144•13 E
44	Hughes	Austl.	30•45 S	129•30 E
38	Huhohaot'e	China	41•05 N	111•50 E
48	Huila	Col.	3•10 N	75•20 W
48	Huila, Nevado de	Col.	2•59 N	76•01 W
39	Huilai	China	23•02 N	116•18 E
39	Huili	China	26•48 N	102•20 E
38	Huimin	China	37•29 N	117•32 N
39	Huiyang	China	23•05 N	114•25 E
39	Huk'ou	China	29•58 N	116•20 E
38	Hulan	China	45•58 N	126•32 E
38	Hulan (R.)	China	42•20 N	126•30 E
9	Hull	Can.	45•26 N	75•43 W
22	Hull (R.)	Eng.	53•47 N	0•20 W
38	Hulutao	China	40•40 N	122•55 E
13	Humacao	P.R.	18•09 N	65•49 W
48	Humaitá	Braz.	7•37 S	62•58 W
48	Humaitá	Par.	27•08 S	58•18 W
54	Humansdorp	S.Afr.	33•57 S	24•45 E
54	Humbe	Ang.	16•50 S	14•55 E
22	Humber (L.)	Eng.	53•38 N	0•40 W
10	Humboldt	Can.	52•12 N	105•07 W
6	Humboldt (R.)	U.S.	40•30 N	116•50 W
6	Humphreys Pk.	Az.	35•20 N	111•40 W
26	Humpolec	Czech.	49•33 N	15•21 E
16	Hunaflói (B.)	Ice.	65•41 N	20•44 W
37	Hunan	China	28•08 N	111•25 E
37	Hunch'un	China	42•53 N	130•34 E
21	Hunedoara	Rom.	45•45 N	22•54 E
14	Hungary	Eur.	46•44 N	17•55 E
46	Hungerford	Austl.	28•50 S	144•32 E
39	Hung Shui Ho (R.)	China	25•00 N	107•22 E
38	Hungtse Hu (L.)	China	33•17 N	118•37 E
26	Hunsrück (Mts.)	F.R.G.	49•43 N	7•12 E
26	Hunte R.	F.R.G.	52•45 N	8•26 E
45	Hunter Is.	Austl.	40•33 S	143•36 E
8	Huntingburg	In.	38•15 N	86•55 W
9	Huntingdon	Can.	45•10 N	74•05 W
8	Huntington	In.	40•55 N	85•30 W
9	Huntington	Pa.	40•30 N	78•00 W
8	Huntington	WV	38•25 N	82•25 W
9	Huntsville	Can.	45•20 N	79•15 W
41	Huon G.	Pap.N.Gui.	7•15 S	147•45 E
37	Hupeh	China	31•20 N	111•58 E
33	Ḩurayḏin, Wādī (R.)	Egypt	30•55 N	34•12 E
8	Hurd, C.	Can.	45•15 N	81•45 W
50	Hurlingham	Arg.	34•20 S	58•38 W
8	Huron	Oh.	41•20 N	82•35 W
8	Huron	SD	44•22 N	98•15 W
7	Huron, L.	U.S.-Can.	45•15 N	82•40 W
16	Húsavik	Ice.	66•00 N	17•10 W
24	Huskvarna	Swe.	57•48 N	14•16 E
26	Husum	F.R.G.	54•29 N	9•04 E
6	Hutchinson	Ks.	38•02 N	97•56 W
38	Hut'o Ho (R.)	China	38•10 N	114•00 E
23	Huy	Bel.	50•33 N	5•14 E
16	Hvannadalshnukur (Mtn.)	Ice.	64•09 N	16•46 W
20	Hvar (I.)	Yugo.	43•08 N	16•28 E
38	Hwangju	Kor.	38•39 N	125•49 E
35	Hyderābād	India	17•29 N	79•28 E
35	Hyderābād	Pak.	25•29 N	68•28 E
35	Hyderābād (State)	India	23•29 N	76•50 E
17	Hyéres	Fr.	43•09 N	6•08 E
17	Hyéres, Iles d'	Fr.	42•57 N	6•17 E
38	Hyesanjin	Kor.	41•11 N	128•12 E
8	Hymera	In.	39•10 N	87•20 W
6	Hyndman Pk.	Id.	43•38 N	114•04 W

I

Page	Name	Region	Lat.	Long.
21	Ialomita (R.)	Rom.	44•37 N	26•42 E
27	Iasi	Rom.	47•10 N	27•40 E
41	Iba	Phil.	15•20 N	119•59 E
52	Ibadan	Nig.	7•17 N	3•30 E
48	Ibagué	Col.	4•27 N	75•13 W
21	Ibar (R.)	Yugo.	43•22 N	20•35 E
48	Ibarra	Ec.	0•19 N	78•08 W
51	Iberian Pen.	Port.-Sp.	41•00 N	0•07 W
9	Iberville	Can.	45•14 N	73•01 W
52	Ibi	Nig.	8•12 N	9•45 E
49	Ibiapaba, Serra da	Braz.	3•30 S	40•55 W
18	Ibiza, Isla de (Iviza I.)	Sp.	39•07 N	1•05 E
55	Ibo	Moz.	12•20 S	40•35 E
34	Ibrahim, Jabal (Mtn.)	Sau.Ar.	20•31 N	41•17 E
48	Ica	Peru	14•09 S	75•42 W
48	Icá (R.)	Braz.	2•56 S	69•12 W
48	Içana	Braz.	0•15 N	67•19 W
14	Iceland	Eur.	65•12 N	19•45 W
39	Ich'ang	China	30•38 N	111•22 E
49	Icó	Braz.	6•25 S	38•43 W
48	Icutú, Cerro	Ven.	7•07 N	65•30 W
53	Idah	Nig.	7•07 N	6•43 E
6	Idaho	U.S.	44•00 N	115•10 W
6	Idaho Falls	Id.	43•30 N	112•01 W
36	Ideriin Gol (R.)	Mong.	48•58 N	98•38 E
53	Idfū	Egypt	24•57 N	32•53 E
21	Idhra (I.)	Grc.	37•20 N	23•3 E
40	Idi	Indon.	4•58 N	97•47 E
20	Idriaj	Yugo.	46•01 N	14•01 E
55	Idutywa	S.Afr.	32•06 S	28•18 E
23	Ieper	Bel.	50•50 N	2•53 E
20	Ierápetra	Grc.(In.)	35•01 N	25•48 E
20	Iesi	It.	43•37 N	13•20 E
52	Ife	Nig.	7•30 N	4•30 E
52	Iferouâne	Niger	19•04 N	8•24 E
52	Iforas, Adrar des (Mts.)	Alg.-Mali	19•55 N	2•00 E
29	Igarka	Sov.Un.	67•22 N	86•16 E
21	Ighil Izane	Alg.	35•43 N	0•43 E
20	Iglesias	It.	39•20 N	8•34 E
52	Igli	Alg.	30•32 N	2•15 W
11	Igloolik	Can.	69•33 N	81•18 W
50	Iguaçu (R.)	Braz.	22•42 S	43•19 W
50	Iguassu (R.)	Braz.	25•45 S	52•30 W
50	Iguassu Falls	Braz.	25•40 S	54•16 W
49	Jiguatu	Braz.	6•22 S	39•17 W
52	Iguidi, Erg (Dune)	Alg.	26•22 N	6•53 W
41	Iguig	Phil.	17•46 N	121•44 E
38	Ihsien	China	41•30 N	121•15 E
30	Iijoki (R.)	Fin.	65•28 N	27•00 E
52	Ijebu-Ode	Nig.	6•50 N	3•56 E
23	IJsselmeer (L.)	Neth.	52•46 N	5•14 E
25	Ikaalinen	Fin.	61•47 N	22•55 E
21	Ikaría (I.)	Grc.	37•43 N	26•07 E
21	Ikhtiman	Bul.	42•26 N	23•49 E
54	Ikoma	Tan.	2•08 S	34•47 E
41	Ilagen	Phil.	17•09 N	121•52 E
38	Ilan	China	46•10 N	129•40 E
39	Ilan	Taiwan	24•50 N	121•42 E
27	Ilawa	Pol.	53•35 N	19•36 E
54	Ilebo (Port-Franqui)	Zaire	4•19 S	20•35 E
31	Ilek	Sov.Un.	51•30 N	53•10 E
31	Ilek (R.)	Sov.Un.	51•20 N	53•10 E
52	Ilesha	Nig.	7•38 N	4•45 E
22	Ilfracombe	Eng.	51•13 N	4•08 W
18	Ilhavo	Port.	40•36 N	8•41 W
49	Ilhéus	Braz.	14•52 S	39•00 W
29	Ilimsk	Sov.Un.	56•47 N	103•43 E
41	Ilin (I.)	Phil.	12•16 N	120•57 E
21	Iliodhrómia (I.)	Grc.	39•18 N	23•35 E
9	Ilion	NY	43•00 N	75•05 W
36	Ili R.	Sov.Un.	43•46 N	77•41 E
48	Illampu, Nevado	Bol.	15•50 S	68•15 W
41	Illana B.	Phil.	7•38 N	123•41 E
50	Illapel	Chile	31•37 S	71•10 W
26	Iller R.	F.R.G.	47•52 N	10•06 E
48	Illimani, Nevado (Pk.)	Bol.	16•50 S	67•38 W
7	Illinois	U.S.	40•25 N	90•40 W
8	Illinois (R.)	Il.	40•52 N	89•31 W
52	Illizi	Alg.	26•35 N	8•24 E
23	Ilmenau (R.)	F.R.G.	53•20 N	10•20 E
48	Ilo	Peru	17•46 S	71•13 W
40	Iloilo	Phil.	10•49 N	112•33 E
52	Ilorin	Nig.	8•30 N	4•32 E
30	Ilych (R.)	Sov.Un.	62•30 N	57•30 E
30	Imandra (L.)	Sov.Un.	67•40 N	32•30 E
50	Imbarié	Braz.	22•38 S	43•13 W
8	Imlay City	Mi.	43•00 N	83•15 W
26	Immenstadt	F.R.G.	47•34 N	10•12 E
20	Imola	It.	44•19 N	11•43 E
20	Imotski	Yugo.	43•25 N	17•15 E
49	Impameri	Braz.	17•44 S	48•03 W
55	Impendle	S.Afr.	29•38 S	29•54 E
20	Imperia	It.	43•52 N	8•00 E
53	Impfondo	Con.	1•37 N	18•04 E
35	Imphāl	India	24•42 N	94•00 E
21	Imroz (I.)	Tur.	40•10 N	25•27 E
16	Inari (L.)	Fin.	69•02 N	26•22 E
31	Ince Burun (C.)	Tur.	42•00 N	35•00 E
38	Inch'ŏn	Kor.	37•26 N	126•46 E
20	Incudine, Mt.	Fr.	41•53 N	9•17 E
24	Indalsälven (R.)	Swe.	62•50 N	16•50 E
41	Indang	Phil.	14•11 N	120•53 E
31	Inder (L.)	Sov.Un.	48•20 N	52•10 E
32	India	Asia	23•00 N	77•30 E
9	Indian (R.)	NY	44•05 N	75•45 W
9	Indiana	Pa.	40•40 N	79•10 W
7	Indiana	U.S.	39•50 N	86•45 W
8	Indianapolis	In.	39•45 N	86•08 W
10	Indian Head	Can.	50•29 N	103•44 W
12	Indio (R.)	Pan.	9•13 N	78•28 W
40	Indochina	Asia	17•22 N	105•18 E
40	Indonesia	Asia	4•38 S	118•45 E
35	Indore	India	22•48 N	76•51 E
40	Indragiri (R.)	Indon.	0•27 S	102•05 E
35	Indrāvati (R.)	India	19•15 N	80•54 E
24	Indre Solund (I.)	Nor.	61•09 N	4•37 E
35	Indus (R.)	Pak.	26•43 N	67•41 E
55	Indwe	S.Afr.	31•30 S	27•21 E
31	Inebolu	Tur.	41•50 N	33•40 E
31	Inego	Tur.	40•05 N	29•20 E
41	Infanta	Phil.	14•44 N	121•39 E
41	Infanta	Phil.	15•50 N	119•53 E
8	Ingersoll	Can.	43•05 N	81•00 W
45	Ingham	Austl.	18•45 S	146•14 E
26	Ingolstadt	F.R.G.	48•46 N	11•27 E
31	Ingur (R.)	Sov.Un.	42•30 N	42•00 E
54	Inhambane	Moz.	23•47 S	35•28 E
49	Inhambupe	Braz.	11•47 S	38•13 W
54	Inharrime	Moz.	24•17 S	35•07 E
50	Inhomirim	Braz.	22•34 S	43•11 W
36	Ining	China	43•58 N	80•49 E
48	Inirida (R.)	Col.	2•25 N	70•38 W
46	Injune	Austl.	25•52 S	148•30 E
25	Inkeroinem	Fin.	60•42 N	26•50 E
46	Innamincka	Austl.	27•50 S	140•48 E
13	Inner Brass (I.)	Vir.Is.	18•23 N	64•58 W
22	Inner Hebrides (Is.)	Scot.	57•20 N	6•20 W
36	Inner Mongolian Aut. Reg.	China	43•30 N	113•33 E
26	Inn R.	F.R.G.-Aus.	48•19 N	13•16 E
26	Innsbruck	Aus.	47•15 N	11•25 E
27	Inowroctaw	Pol.	52•48 N	18•16 E
52	In Salah	Alg.	27•13 N	2•22 E
7	International Falls	Mn.	48•34 N	93•26 W
10	Inuvik	Can.	68•40 N	134•10 W
45	Invercargil	N.Z.	47•18 S	168•27 E
46	Inverel	Austl.	29•50 S	151•32 E
22	Inverness	Scot.	57•30 N	4•07 W
46	Investigator Str.	Austl.	35•33 S	137•00 E
54	Inyangani, Mt.	Rh.	18•06 S	32•37 E
21	Ioánnina (Yannina)	Grc.	39•39 N	20•52 E
8	Ionia	Mi.	43•00 N	85•10 W
21	Ionian Is.	Grc.	39•10 N	20•05 E
19	Ionian Sea	Eur.	38•59 N	18•48 E
21	Ios (I.)	Grc.	36•48 N	25•25 E
7	Iowa	U.S.	42•05 N	94•20 W
7	Iowa City	Ia.	41•39 N	91•31 W
27	Ipel R.	Czech.-Hung.	48•08 N	19•00 E
48	Ipiales	Col.	0•48 N	77•45 W
39	Ipin (Süchow)	China	28•50 N	104•40 E
40	Ipoh	Mala.	4•45 N	101•05 E
46	Ipswich	Austl.	27•40 S	152•50 E
23	Ipswich	Eng.	52•05 N	1•05 E
49	Ipu	Braz.	4•11 S	40•45 W
31	Iput' (R.)	Sov.Un.	52•53 N	31•57 E
48	Iquique	Chile	20•16 S	70•07 W
48	Iquitos	Peru	3•39 S	73•18 W
20	Iráklion (Candia)	Grc.	35•20 N	25•10 E
32	Iran	Asia	31•15 N	53•30 E
34	Iran, Plat. of	Iran	32•28 N	58•00 E
40	Iran Mts.	Mala.	2•30 N	114•30 E
32	Iraq	Asia	32•00 N	42•30 E
33	Irbid	Jordan	32•33 N	35•51 E
31	Irbil	Iraq	36•10 N	44•00 E
30	Irbit	Sov.Un.	57•40 N	63•10 E
54	Irébou	Zaire	0•40 S	17•48 E
14	Ireland	Eur.	53•33 N	13•00 W
55	Irene	S.Afr.	25•53 S	28•13 E
28	Irgiz	Sov.Un.	48•30 N	61•17 E
30	Iriklinskoye Vdkhr (Res.)	Sov.Un.	52•20 N	58•50 E
54	Iringa	Tan.	7•46 S	35•42 E
39	Iriomote Jima (I.)	Jap.	24•20 N	123•30 E
22	Irish Sea	Eur.	53•55 N	5•25 W
29	Irkutsk	Sov.Un.	52•16 N	104•00 E
21	Iron Gate (Gorge)	Yugo.-Rom.	44•43 N	22•32 E
46	Iron Knob	Austl.	32•47 S	137•10 E
8	Iron Mountain	Mi.	45•49 N	88•04 W
8	Ironton	Oh.	38•30 N	82•45 W
8	Iroquois (R.)	Il.-n.	40•55 N	87•20 W
11	Iroquois Falls	Can.	48•41 N	80•39 W
35	Irrawaddy (R.)	Bur.	23•27 N	96•25 E
40	Irrawaddy, Mouths of the	Bur.	15•40 N	94•32 E
28	Irtysh (R.)	Sov.Un.	58•32 N	68•31 E
53	Irumu	Zaire	1•30 N	29•52 E
17	Irun	Sp.	43•20 N	1•47 W
22	Irvine	Scot.	55•39 N	4•40 W
8	Irvine	Ky.	37•40 N	84•00 W
12	Isaacs, Mt.	Pan.	9•22 N	79•01 W
48	Isabela (I.)	Ec.	0•47 S	91•35 W
8	Isabella Ind. Res.	Mi.	43•35 N	84•55 W
16	Isafjördhur	Ice.	66•09 N	22•39 W
53	Isangi	Zaire	0•46 N	24•15 E
26	Isar R.	F.R.G.	48•27 N	12•02 E
20	Isarco (R.)	It.	46•37 N	11•25 E
41	Isaroga (Vol.)	Phil.	13•40 N	123•23 E
20	Iseo, Lago di (L.)	It.	45•50 N	9•55 E
17	Isère (R.)	Fr.	45•24 N	6•04 E
23	Iserlohn	F.R.G.	51•22 N	7•42 E
20	Isernia	It.	41•35 N	14•14 E
52	Iseyin	Nig.	7•58 N	3•36 E
39	Ishan	China	24•32 N	108•42 E
28	Ishim	Sov.Un.	56•07 N	69•13 E
28	Ishim (R.)	Sov.Un.	53•17 N	67•45 E
37	Ishinomaki	Jap.	38•22 N	141•22 E
21	Ishm	Alb.	41•32 N	19•35 E
55	Isipingo	S.Afr.	29•59 S	30•58 E
53	Isiro	Zaire	2•47 N	27•37 E
31	Iskenderun	Tur.	36•45 N	36•15 E
19	Iskenderun Körfezi (G.)	Tur.	36•22 N	35•25 E
31	Iskilip	Tur.	40•40 N	34•30 E
21	Iskūr (R.)	Bul.	43•05 N	23•37 E
35	Islāmābād	Pak.	35•55 N	73•05 E
10	Island L.	Can.	53•47 N	94•25 W
22	Islay (I.)	Scot.	55•55 N	6•35 W
22	Isle of Man	Eur.	54•26 N	4•21 W
7	Isle Royale Natl. Park	U.S.	47•57 N	88•37 W
25	Isojärvi (L.)	Fin.	61•47 N	22•00 E
31	Isparta	Tur.	37•50 N	30•40 E
34	Israel	Asia	32•40 N	34•00 E
28	Issyk-Kul, Ozero (L.)	Sov.Un.	42•13 N	76•12 E
31	Istanbul	Tur.	41•02 N	29•00 E
31	Istanbul Boğazi (Bosporous) (Str.)	Tur.	41•10 N	29•10 E
21	Istiaía	Grc.	38•58 N	23•11 E
48	Istmina	Col.	5•10 N	76•40 W
20	Istra (Pen.)	Yugo.	45•18 N	13•48 E
21	Istranca Dağ (Mts.)	Bul.-Turk.	41•50 N	27•25 E
50	Itá	Par.	25•39 S	57•14 W
49	Itabaiana	Braz.	10•42 S	37•17 W
49	Itabuna	Braz.	14•47 S	39•17 W
49	Itacoatiara	Braz.	3•03 S	58•18 W
48	Itagüi	Col.	6•11 N	75•36 W
50	Itagui (R.)	Braz.	22•53 S	43•43 W
50	Itaipava	Braz.	22•23 S	43•09 W
50	Itaipu	Braz.	22•58 S	43•02 W
49	Itaituba	Braz.	4•12 S	56•00 W
50	Itajái	Braz.	26•52 S	48•39 W
49	Itajubá	Braz.	22•26 S	45•27 W
14	Italy	Eur.	43•58 N	11•14 E
50	Itambi	Braz.	22•44 S	42•57 W
49	Itapecurú (R.)	Braz.	4•05 S	43•49 W
49	Itapĕcuru-Mirim	Braz.	3•17 S	44•15 W
49	Itaperuna	Braz.	21•12 S	41•53 W
49	Itapetininga	Braz.	23•37 S	48•03 W
49	Itapira	Braz.	20•42 S	51•19 W
4	Itatiaia, Pico da (Pk.)	Braz.	22•18 S	44•41 W
8	Ithaca	Mi.	43•20 N	84•35 W
9	Ithaca	NY	42•25 N	76•30 W
21	Itháki (I.)	Grc.	38•27 N	20•48 E
54	Itoko	Zaire	1•13 S	22•07 E
48	Ituango	Col.	7•07 N	75•44 W
49	Ituiutaba	Braz.	18•56 S	49•17 W
37	It'ung	China	43•15 N	125•10 E
37	Iturup (I.)	Sov.Un.	45•35 N	147•15 E
50	Ituzaingo	Arg.	34•24 S	58•40 W
26	Itzehoe	F.R.G.	53•55 N	9•31 E
46	Ivanhoe	Austl.	32•53 S	144•10 E
27	Ivano-Frankovsk	Sov.Un.	48•53 N	24•46 E
30	Ivanovo	Sov.Un.	57•02 N	41•54 E
55	Ivohibé	Mad.	22•28 S	46•59 E
51	Ivory Coast	Afr.	7•43 N	6•30 W
20	Ivrea	It.	45•25 N	7•54 E
11	Ivujivik	Can.	62•17 N	77•52 W
52	Iwo	Nig.	7•38 N	4•11 E
55	Ixopo	S.Afr.	30•10 S	30•04 E
39	Iyang	China	28•52 N	112•12 E
30	Izhevsk	Sov.Un.	56•50 N	53•15 E
30	Izhma	Sov.Un.	65•00 N	54•05 E
30	Izhma (R.)	Sov.Un.	64•00 N	53•00 E
31	Izmail	Sov.Un.	45•00 N	28•49 E
31	Izmir	Tur.	38•25 N	27•05 E
21	Izmir Körfezi (G.)	Tur.	38•43 N	26•37 E
31	Izmit	Tur.	40•45 N	29•45 E

J

Page	Name	Region	Lat.	Long.
53	Jabal, Baḩr al (R.)	Sud.	7•02 N	30•45 E
35	Jabalpur	India	23•18 N	79•59 E
26	Jablonec (Nad			

Page	Name	Region	Lat.	Long.
	Nisou)	Czech.	50•43 N	15•12 E
27	Jablunkov P.	Czech.	49•31 N	18•35 E
49	Jaboatão	Braz.	8•14 S	35•08 W
50	Jacarepaguá	Braz.	22•55 S	43•22 W
49	Jacarézinho	Braz.	23•13 S	49•58 W
26	Jachymov	Czech.	50•22 N	12•51 E
8	Jackson	Ky.	37•32 N	83•17 W
8	Jackson	Mi.	42•15 N	84•25 W
7	Jackson	Ms.	32•17 N	90•10 W
8	Jackson	Oh.	39•00 N	82•40 W
7	Jackson	Tn.	35•37 N	88•49 W
7	Jacksonville	Fl.	30•20 N	81•40 W
7	Jacksonville	Il.	39•43 N	90•12 W
49	Jacobina	Braz.	11•13 S	40•30 W
26	Jade B.	F.R.G.	53•28 N	8•17 E
48	Jaén	Peru	5•38 S	78•49 W
18	Jaen	Sp.	37•45 N	3•48 W
46	Jaffa, C.	Austl.	36•58 S	139•29 E
35	Jaffna	Sri Lanka	9•44 N	80•09 E
33	Jahore Str.	Mala.	1•22 N	103•37 E
34	Jahrom	Iran	28•30 N	53•28 E
35	Jaipur	India	27•00 N	75•50 E
20	Jajce	Yugo.	44•20 N	17•19 E
35	Jajpur	India	20•49 N	86•37 E
40	Jakarta	Indon.	6•17 S	106•45 E
16	Jakobstad	Fin.	63•33 N	22•31 E
35	Jalālābād	Afg.	34•25 N	70•27 E
12	Jalapa Enríquez	Mex.	19•32 N	96•53 W
12	Jalisco (State)	Mex.	20•07 N	104•45 W
53	Jālū, Wāḥat (Oasis)	Libya	28•58 N	21•45 E
13	Jamaica	N.A.	17•45 N	78•00 W
40	Jambi	Indon.	1•45 S	103•28 E
21	Jambol	Bul.	42•28 N	26•31 E
40	Jambuair (C.)	Indon.	5•15 N	79•30 E
6	James (R.)	U.S.	46•25 N	98•55 W
9	James (R.)	Va.	37•35 N	77•50 W
11	James B.	Can.	53•53 N	80•40 W
44	James Ra	Austl.	24•15 S	133•30 E
47	James Ross (I.)	Ant.	64•20 S	58•20 W
9	Jamestown	NY	42•05 N	79•15 W
6	Jamestown	ND	46•54 N	98•42 W
55	Jamestown	S.Afr.	31•07 S	26•49 E
24	Jammerbugt (B.)	Den.	57•20 N	9•28 E
35	Jammu	India	32•50 N	74•52 E
35	Jammu and Kashmīr	India-Pak.	39•10 N	75•05 E
35	Jāmnagar	India	22•33 N	70•03 E
35	Jamshedpur	India	22•52 N	86•11 E
48	Jamundí	Col.	3•15 N	76•32 W
8	Janesville	Wi.	42•41 N	89•03 W
33	Janin	Jordan	32•27 N	35•19 E
16	Jan Mayen (I.)	Nor.	70•59 N	8•05 W
24	Jannelund	Swe.	59•14 N	14•24 E
27	Jánoshalma	Hung.	46•17 N	19•18 E
27	Janów Lubelski	Pol.	50•40 N	22•25 E
49	Januária	Braz.	15•31 S	44•17 W
33	Japan	Asia	36•30 N	133•30 E
37	Japan, Sea of	Asia	40•08 N	132•55 W
41	Japen (I.)	Indon.	1•30 S	136•15 E
50	Japeri	Braz.	22•38 S	43•40 W
48	Japurá (R.)	Braz.	1•30 S	67•54 W
33	Jarash	Jordan	32•17 N	35•53 E
49	Jari (R.)	Braz.	0•28 N	53•00 W
27	Jarocin	Pol.	51•58 N	17•31 E
27	Jaroslaw	Pol.	50•01 N	22•41 E
33	Jasin	Mala.	2•19 N	102•26 E
25	Jašiūnai	Sov.Un.	54•27 N	25•25 E
34	Jāsk	Iran	25•46 N	57•48 E
27	Jasło	Pol.	49•44 N	21•28 E
33	Jason B.	Mala.	1•53 N	104•14 E
8	Jasonville	In.	39•10 N	87•15 W
10	Jasper	Can.	52•53 N	118•05 W
8	Jasper	In.	38•20 N	86•55 W
10	Jasper Natl. Park	Can.	53•09 N	117•45 W
27	Jászapáti	Hung.	47•29 N	20•10 E
18	Játiva	Sp.	38•58 N	0•31 W
50	Jaú	Braz.	22•16 S	48•31 W
48	Jauja	Peru	11•43 S	75•32 W
25	Jaunjelgava	Sov.Un.	56•37 N	25•06 E
48	Javari (R.)	Col.-Peru	4•25 S	72•07 W
40	Jawa (Java) (I.)	Indon.	8•35 N	111•11 E
40	Jawa, Laut (Java Sea)	Indon.	5•10 S	110•30 E
26	Jawor	Pol.	51•04 N	16•12 E
27	Jaworzno	Pol.	50•11 N	19•18 E
41	Jaya (Pk.)	Indon.	4•00 S	137•15 E
41	Jayapura (Sukarnapura)	Indon.	2•30 S	140•45 W
27	Jászberény	Hung.	47•30 N	19•56 E
33	Jazzīn	Leb.	33•34 N	35•37 E
18	Jebal Aures (Mts.)	Alg.	35•16 N	5•53 E
52	Jebba	Nig.	9•07 N	4•46 E
27	Jędrzejów	Pol.	50•38 N	20•18 E
8	Jefferson	Wi.	42•59 N	88•45 W
7	Jefferson City	Mo.	38•34 N	92•10 W
8	Jeffersonville	In.	38•17 N	85•44 W
19	Jeib, Wadi el (R.)	Jordan-Isr.	30•30 N	35•20 E
25	Jēkabpils	Sov.Un.	56•29 N	25•50 E
26	Jelenia Góra	Pol.	50•53 N	15•43 E
25	Jelgava	Sov.Un.	56•39 N	23•40 E
17	Jemmapes	Alg.	36•43 N	7•21 E
26	Jena	G.D.R.	50•55 N	11•37 E
8	Jennings	Mi.	44•20 N	85•20 W
49	Jequié	Braz.	13•53 S	40•06 W
49	Jequitinhonha (R.)	Braz.	16•47 S	41•19 W
49	Jeremoabo	Braz.	10•03 S	38•13 W
16	Jerez de la Frontera	Sp.	36•42 N	6•09 W
45	Jericho	Austl.	28•38 S	146•24 E
6	Jerome	Az.	34•45 N	112•10 W
17	Jersey (I.)	Eur.	49•13 N	2•07 W
9	Jersey City	NJ	40•43 N	74•05 W
9	Jersey Shore	Pa.	41•10 N	77•15 W
33	Jerusalem	Isr.-Jordan	31•46 N	35•14 E
35	Jhānsi	India	25•29 N	78•32 E
35	Jhelum	Pak.	31•40 N	71•51 E
36	Jibhalanta	Mong.	47•49 N	97•00 E
27	Jiffa R.	Rom.	47•35 N	27•02 E
44	Jiggalong	Austl.	23•20 S	120•45 E
38	Jihchao	China	35•27 N	119•28 E
26	Jihlava	Czech.	49•23 N	15•33 E
53	Jilf al-Kabīr (Plat.)	Egypt	24•09 N	25•29 E
53	Jima	Eth.	7•41 N	36•52 E
21	Jimbolia	Rom.	45•45 N	20•44 E
12	Jiménez	Mex.	27•09 N	104•55 W
9	Jim Thorpe	Pa.	40•50 N	75•45 W
26	Jindřichov Hradec	Czech.	49•09 N	15•02 E
53	Jinja	Ug.	0•26 N	33•12 E
48	Jipijapa	Ec.	1•36 S	80•52 W
36	Jirgalanta	Mong.	48•08 N	91•40 E
21	Jiu (R.)	Rom.	44•45 N	23•17 E
54	João Belo	Moz.	25•00 S	33•45 E
49	João Pessoa (Paraíba)	Braz.	7•09 S	34•45 W
35	Jodhpur	India	26•23 N	73•00 E
25	Joensuu	Fin.	62•35 N	29•46 E
55	Johannesburg	S.Afr.	26•08 S	27•54 E
9	Johnsonburg	Pa.	41•30 N	78•40 W
8	Johnson City	Il.	37•50 N	88•55 W
9	Johnson City	NY	42•10 N	76•00 W
7	Johnson City	Tn.	36•17 N	82•23 W
42	Johnston (I.)	Oceania	17•00 N	168•00 W
9	Johnstown	NY	43•00 N	74•20 W
9	Johnstown	Pa.	40•20 N	78•50 W
37	Joho	China	42•31 N	118•12 E
40	Johor	Mala.	2•15 N	103•00 E
32	Johor (R.)	Mala.	1•39 N	103•52 E
32	Johor Bahru	Mala.	1•28 N	103•46 E
50	Joinville	Braz.	26•18 S	48•47 W
47	Joinville (I.)	Ant.	63•00 S	53•30 W
16	Jökullsá (R.)	Ice.	65•38 N	16•08 W
8	Joliet	Il.	41•37 N	88•05 W
11	Joliette	Can.	46•01 N	73•30 W
40	Jolo	Phil.	5•59 N	121•05 E
40	Jolo (I.)	Phil.	5•55 N	121•15 E
41	Jomalig (I.)	Phil.	14•44 N	122•34 E
25	Jonava	Sov.Un.	55•05 N	24•15 E
24	Jondal	Nor.	60•16 N	6•16 E
41	Jones	Phil.	13•56 N	122•05 E
41	Jones	Phil.	16•35 N	121•39 E
7	Jonesboro	Ar.	35•49 N	90•42 W
8	Jonesville	Mi.	42•00 N	84•45 W
25	Joniškis	Sov.Un.	56•14 N	23•36 E
26	Jönköping	Swe.	57•47 N	14•10 E
11	Jonquiere	Can.	48•25 N	71•15 W
7	Joplin	Mo.	37•05 N	94•31 W
32	Jordan	Asia	30•15 N	38•00 E
33	Jordan (R.)	Jordan	31•58 N	35•36 E
35	Jorhāt	India	26•43 N	94•16 E
52	Jos	Nig.	9•53 N	9•05 E
44	Joseph Bonaparte, G.	Austl.	13•30 S	128•40 E
24	Jostedalsbreen	Nor.	61•40 N	6•55 E
24	Jotun Fjell	Nor.	61•44 N	8•11 E
10	Juan de Fuca, Str. of	Wa.-Can.	48•25 N	124•37 W
55	Juan de Nova, Île	Afr.	17•18 S	43•07 E
12	Juan Diaz, (R.)	Pan.	9•05 N	79•30 W
47	Juan Fernández, Islas de	Chile	33•30 S	79•00 W
49	Juàzeiro	Braz.	9•27 S	40•28 W
49	Juazeiro do Norte	Braz.	7•16 S	38•57 W
50	Juárez	Arg.	37•42 S	59•46 W
53	Jūbā	Sud.	4•58 N	31•37 E
51	Juba R.	Som.	1•30 N	42•25 E
33	Jubayl (Byblos)	Leb.	34•07 N	35•38 E
17	Júcar (R.)	Sp.	39•10 N	1•22 W
12	Juchitán	Mex.	16•15 N	95•00 W
34	Juddah	Sau.Ar.	21•30 N	39•15 E
26	Judenburg	Aus.	47•10 N	14•40 E
39	Juian	China	27•48 N	120•40 E
23	Juist (I.)	F.R.G.	53•41 N	6•50 E
49	Juiz de Fora	Braz.	21•47 S	43•20 W
50	Jujuy	Arg.	24•14 S	65•15 W
50	Jujuy (Prov.)	Arg.	23•00 S	65•45 W
38	Jukao	China	32•24 N	120•33 E
55	Jukskei (R.)	S.Afr.	25•58 S	27•58 E
48	Juliaca	Peru	15•26 S	70•12 W
5	Julianehåb	Grnld.	60•07 N	46•20 W
20	Julijske Alpe (Mts.)	Yugo.	46•05 N	14•05 E
35	Jullundur	India	31•29 N	75•39 E
23	Jumet	Bel.	50•28 N	4•30 E
33	Jumrah	Indon.	1•48 N	101•04 E
49	Jumundá (R.)	Braz.	1•33 S	57•42 W
35	Junagādh	India	21•33 N	70•25 E
38	Junan	China	32•59 N	114•22 E
33	Junaynah, Ra's al (Mt.)	Egypt	29•02 N	33•58 E
49	Jundiaí	Braz.	23•12 S	46•52 W
6	Juneau	Ak.	58•25 N	134•30 W
39	Jungchiang	China	25•52 N	108•45 E
26	Jungfrau Pk.	Switz.	46•30 N	7•59 E
50	Junín	Arg.	34•35 S	60•56 W
48	Junín	Col.	4•47 N	73•39 W
33	Juniyah	Leb.	33•59 N	35•38 E
16	Junkeren (Mtn.)	Nor.	66•29 N	14•58 E
53	Jur (R.)	Sud.	6•38 N	27•52 E
22	Jura (I.)	Scot.	56•09 N	6•45 W
17	Jura (Mts.)	Switz.	46•55 N	6•49 E
22	Jura, Sd. of	Scot.	55•45 N	5•55 W
25	Jurbarkas	Sov.Un.	55•06 N	22•50 E
25	Jūrmala	Sov.Un.	56•57 N	23•37 E
48	Juruá (R.)	Braz.	5•27 S	67•39 W
49	Juruena (R.)	Braz.	12•22 S	58•34 W
48	Jutaí (R.)	Braz.	4•26 S	68•16 W
12	Juticalpa	Hond.	14•35 N	86•17 W
21	Južna Morava (R.)	Yugo.	42•30 N	22•00 E
24	Jylland	Den.	56•04 N	9•00 E
25	Jyväskylä	Fin.	62•14 N	25•46 E

K

Page	Name	Region	Lat.	Long.
55	Kaalfontein	S.Afr.	26•02 S	28•16 E
54	Kaappunt (C.)	S.Afr.	34•21 S	18•30 E
40	Kabaena (I.)	Indon.	5•35 S	121•07 E
52	Kabala	S.L.	9•43 N	11•39 W
54	Kabalo	Zaire	6•03 S	26•55 E
54	Kabambare	Zaire	4•47 S	27•45 E
54	Kabinda	Zaire	6•08 S	24•29 E
54	Kabompo (R.)	Zambia	14•00 S	23•40 E
54	Kabongo	Zaire	7•58 S	25•10 E
18	Kaboudia, Ras (C.)	Tun.	35•17 N	11•28 E
35	Kābul	Afg.	34•39 N	69•14 E
35	Kabul (R.)	Asia	34•44 N	69•43 E
54	Kabwe (Broken Hill)	Zambia	14•27 S	28•27 E
29	Kachuga	Sov.Un.	54•09 N	105•43 E
31	Kadiyevka	Sov.Un.	48•34 N	38•37 E
30	Kadnikov	Sov.Un.	59•30 N	40•10 E
52	Kaduna	Nig.	10•33 N	7•27 E
52	Kaédi	Mauritania	16•09 N	13•30 W
6	Kaena Pt.	Hi.	21•33 N	158•19 W
38	Kaesŏng (Kaijo)	Kor.	38•00 N	126•35 E
53	Kafia Kingi	Sud.	9•17 N	24•28 E
54	Kafue	Zambia	15•45 S	28•17 E
54	Kafue (R.)	Zambia	15•45 S	26•30 E
54	Kagera (R.)	Tan.	1•10 S	31•10 E
37	Kagoshima	Jap.	31•35 N	130•31 E
40	Kahayan (R.)	Indon.	1•45 S	113•40 E
6	Kahoolawe (I.)	Hi.	20•28 N	156•48 W
6	Kahuku Pt.	Hi.	21•50 N	157•50 W
6	Kahului	Hi.	20•53 N	156•28 W
41	Kai (Is.)	Indon.	5•35 S	132•45 E
33	Kaiang	Mala.	3•00 N	101•47 E
49	Kaieteur Fall	Guy.	4•48 N	59•24 W
38	K'aifeng	China	34•48 N	114•22 E
41	Kai Kecil (I.)	Indon.	5•45 S	132•40 E
29	Kaikyō (Str.)	Sov.Un.	45•45 N	141•20 E
6	Kailua	Hi.	21•18 N	157•43 W
41	Kaimana	Indon.	3•32 S	133•47 E
52	Kainji L.	Nig.	10•25 N	4•50 E
38	Kaip'ing	China	40•25 N	122•20 E
26	Kaiserslautern	F.R.G.	49•26 N	7•46 E
45	Kaitaia	N.Z.	35•30 S	173•28 E
6	Kaiwi Chan	Hi.	21•10 N	157•38 W
39	Kaiyüan	China	23•42 N	103•20 E
38	Kaiyuan	China	42•30 N	124•00 E
16	Kajaani	Fin.	64•15 N	27•16 E
33	Kajang (Mt.)	Mala.	2•47 N	104•05 E
31	Kakhovskoye (L.)	Sov.Un.	47•21 N	33•33 E
35	Kākināda	India	16•58 N	82•18 E
31	Kalach	Sov.Un.	50•15 N	40•55 E
36	Kaladan (R.)	Bur.	21•07 N	93•04 E
54	Kalahari Des.	Bots.	23•00 S	22•03 E
21	Kalámai	Grc.	37•04 N	22•08 E
8	Kalamazoo	Mi.	42•20 N	85•40 W
8	Kalamazoo (R.)	Mi.	42•35 N	86•00 W
34	Kalar (Mtn.)	Iran	31•43 N	51•41 E
35	Kalāt	Pak.	29•05 N	66•36 E
40	Kalatoa (I.)	Indon.	7•22 S	122•30 E
54	Kalemie (Albertville)	Zaire	5•56 S	29•12 E
38	Kalgan	China	40•45 N	114•58 E
44	Kalgoorlie	Austl.	30•45 S	121•35 E
19	Kaliakra, Nos (Pt.)	Rom.	43•25 N	28•42 E
30	Kalinin (Tver)	Sov.Un.	56•52 N	35•57 E
25	Kaliningrad (Königsberg)	Sov.Un.	54•42 N	20•32 E
6	Kalispell	Mt.	48•12 N	114•18 W
27	Kalisz	Pol.	51•45 N	18•05 E
16	Kalix (R.)	Swe.	67•12 N	21•41 E
24	Kalmar	Swe.	56•40 N	16•19 E
24	Kalmar Sund	Swe.	56•30 N	16•17 E
31	Kalmyk A.S.S.R.	Sov.Un.	46•56 N	46•00 E
27	Kalocsa	Hung.	46•32 N	19•00 E
54	Kalomo	Zambia	17•02 S	26•30 E
30	Kaluga	Sov.Un.	54•29 N	36•12 E
24	Kalundborg	Den.	55•42 N	11•07 E
27	Kalush	Sov.Un.	49•02 N	24•24 E
25	Kalvarija	Sov.Un.	54•24 N	23•17 E
29	Kalyma (R.)	Sov.Un.	66•32 N	152•46 E
30	Kama (L.)	Sov.Un.	55•28 N	51•00 E
30	Kama (R.)	Sov.Un.	56•10 N	53•50 E
34	Kamarān (I.)	P.D.R. of Yem.	15•19 N	41•47 E
54	Kambove	Zaire	10•58 S	26•43 E
29	Kamchatka (Pen.)	Sov.Un.	55•19 N	157•45 E
31	Kamenets-Podolskiy	Sov.Un.	48•41 N	26•34 E
20	Kamenjak, Rt (C.)	Yugo.	44•45 N	13•57 E
27	Kamenka	Sov.Un.	50•06 N	24•20 E
28	Kamen'-na-Obi	Sov.Un.	53•43 N	81•28 E
30	Kamensk-Ural'skiy	Sov.Un.	56•27 N	61•55 E
26	Kamenz	G.D.R.	51•16 N	14•05 E
26	Kamień Pomorski	Pol.	53•57 N	14•48 E
54	Kamina	Zaire	8•44 S	25•00 E
10	Kamloops	Can.	50•40 N	120•20 W
53	Kampala	Ug.	0•19 N	32•25 E
40	Kampar-Kiri (R.)	Indon.	0•30 N	101•30 E
40	Kampot	Camb.	10•41 N	104•07 E
26	Kamp R.	Aus.	48•30 N	15•45 E
10	Kamsack	Can.	51•34 N	101•54 W
30	Kamskoye (Res.)	Sov.Un.	59•03 N	56•48 E
31	Kamyshin	Sov.Un.	50•08 N	45•20 E
30	Kamyshlov	Sov.Un.	56•50 N	62•32 E
39	Kan (R.)	China	26•50 N	115•00 E
54	Kananga (Luluabourg)	Zaire	6•14 S	22•17 E
7	Kanawha (R.)	U.S.	37•55 N	81•50 W
19	Kanayis, Rasel (C.)	Egypt	31•14 N	28•08 E
37	Kanazawa	Jap.	36•34 N	136•38 E
35	Kānchenjunga (Mtn.)	India-Nep.	27•30 N	88•18 E
39	Kanchou	China	25•50 N	114•30 E
54	Kanda Kanda	Zaire	6•56 S	23•36 E
30	Kandalaksha	Sov.Un.	67•10 N	33•05 E
30	Kandalakshskiy (B.)	Sov.Un.	66•20 N	35•00 E
25	Kandava	Sov.Un.	57•03 N	22•45 E
52	Kandi	Benin	11•08 N	2•56 E
35	Kandy	Sri Lanka	7•18 N	80•42 E
9	Kane	Pa.	41•40 N	78•50 W
31	Kanevskoye (Res.)	Sov.Un.	50•10 N	30•40 E
46	Kangaroo (I.)	Austl.	36•05 S	137•05 E
34	Kangāvar	Iran	34•37 N	46•45 E
40	Kangean (I.)	Indon.	6•50 S	116•22 E
38	Kanggye	Kor.	40•55 N	126•40 E
38	Kanghwa (I.)	Kor.	37•38 N	126•00 E
54	Kango	Gabon	0•09 N	10•08 E
36	K'angting	China	30•15 N	101•58 E
30	Kanin, P-Ov. (Pen.)	Sov.Un.	68•00 N	45•00 E
30	Kanin Nos, Mys (G.)	Sov.Un.	68•40 N	44•00 E
21	Kanjiža	Yugo.	46•05 N	20•02 E
8	Kankakee	Il.	41•07 N	87•53 W
8	Kankakee (R.)	Il.	41•15 N	88•15 W
52	Kankan	Gui.	10•23 N	9•18 W
38	Kannan	China	47•50 N	123•30 E
52	Kano	Nig.	12•00 N	8•30 E
54	Kanonkop			

Page	Name	Region	Lat.	Long.
27	Kiskunhalas	Hung.	46•24 N	19•26 E
27	Kiskunmajsa	Hung.	46•29 N	19•42 E
55	Kismayu	Som.	0•18 S	42•30 E
52	Kissidougou	Gui.	9•11 N	10•06 W
16	Kistrand	Nor.	70•29 N	25•01 E
27	Kisujszallás	Hung.	47•12 N	20•47 E
27	Kisumu	Ken.	0•06 S	34•45 E
37	Kitakyūshū	Jap.	34•15 N	130•23 E
8	Kitchener	Can.	43•25 N	80•35 W
53	Kitgum	Ug.	3•29 N	33•04 E
19	Kíthira (I.)	Grc.	36•15 N	22•56 E
21	Kíthnos (I.)	Grc.	37•24 N	24•10 E
10	Kitimat	Can.	54•03 N	128•33 W
9	Kittanning	Pa.	40•50 N	79•30 W
9	Kittery	Me.	43•07 N	70•45 W
26	Kitzingen	F.R.G.	49•44 N	10•08 E
54	Kivu, Lac	Zaire	1•45 S	28•55 E
31	Kiyev (Kiev)	Sov.Un.	50•27 N	30•30 E
30	Kizel	Sov.Un.	59•05 N	57•42 E
31	Kızıl Irmak (R.)	Tur.	40•15 N	34•00 E
31	Kizlyar	Sov.Un.	44•00 N	46•50 E
15	Kizy-Arvat	Sov.Un.	38•55 N	56•33 E
55	Klaas Smits (R.)	S.Afr.	31•45 S	26•33 E
26	Kladno	Czech.	50•10 N	14•05 E
26	Klagenfurt	Aus.	46•38 N	14•19 E
25	Klaipéda (Memel)	Sov.Un.	55•43 N	21•10 E
6	Klamath Falls	Or.	42•13 N	121•49 W
24	Klarälven (R.)	Swe.	60•40 N	13•00 E
26	Klatovy	Czech.	49•23 N	13•18 E
26	Kleve	F.R.G.	51•47 N	6•09 E
24	Klintehamn	Swe.	57•24 N	18•14 E
31	Klintsy	Sov.Un.	52•46 N	32•14 E
24	Klippan	Swe.	56•08 N	13•09 E
20	Ključ	Yugo.	44•32 N	16•48 E
26	Klodzko	Pol.	50•26 N	16•38 E
10	Klondike Reg.	Ak.-Can.	64•12 N	142•38 W
10	Kluane (L.)	Can.	61•15 N	138•40 W
10	Kluane Natl. Pk.	Can.	60•25 N	137•53 W
27	Kluczbork	Pol.	50•59 N	18•15 E
47	Klyuchevskaya (Vol.)	Sov.Un.	56•13 N	160•00 E
21	Knezha	Bul.	43•27 N	24•03 E
8	Knightstown	In.	39•45 N	85•30 W
20	Knin	Yugo.	44•02 N	16•14 E
26	Knittelfeld	Aus.	47•13 N	14•50 E
41	Knob Pk.	Phil.	12•30 N	121•20 E
22	Knockmealdown Mts.	Ire.	52•13 N	8•09 W
8	Knox	In.	41•15 N	86•40 W
7	Knoxville	Tn.	35•58 N	83•55 W
27	Knyszyn	Pol.	53•16 N	22•59 E
37	Kōbe	Jap.	34•30 N	135•10 E
31	Kobelyaki	Sov.Un.	49•11 N	34•12 E
24	København (Copenhagen)	Den.	55•43 N	12•27 E
26	Koblenz	F.R.G.	50•18 N	7•36 E
27	Kobrin	Sov.Un.	52•13 N	24•23 E
31	Kobuleti	Sov.Un.	41•50 N	41•40 E
21	Kocani	Yugo.	41•54 N	22•25 E
20	Kočevje	Yugo.	45•38 N	14•51 E
26	Kocher R.	F.R.G.	49•00 N	9•52 E
37	Kōchi	Jap.	33•35 N	133•32 E
6	Kodiak	Ak.	57•50 N	152•30 W
5	Kodiak (I.)	Ak.	57•24 N	153•32 W
53	Kodok	Sud.	9•57 N	32•08 E
24	Køge	Den.	55•27 N	12•09 E
24	Køge Bugt (B.)	Den.	55•30 N	12•25 E
35	Kohīma	India	25•45 N	94•41 E
28	Kokand	Sov.Un.	40•27 N	71•07 E
25	Kokemäen (R.)	Fin.	61•23 N	22•03 E
16	Kokkola	Fin.	63•47 N	22•58 E
8	Kokomo	In.	40•30 N	86•20 W
41	Kokopo	Pap.N.Gui.	4•25 S	152•27 E
11	Koksoak (R.)	Can.	57•42 N	69•50 W
55	Kokstad	S.Afr.	30•33 S	29•27 E
35	Kolār	India	13•39 N	78•33 E
27	Kolárvo	Czech.	47•54 N	17•59 E
24	Kolding	Den.	55•29 N	9•24 E
54	Kole	Zaire	3•19 S	22•46 E
30	Kolguyev (I.)	Sov.Un.	69•00 N	49•00 E
26	Kolin	Czech.	50•01 N	15•11 E
25	Kolkasrags (Pt.)	Sov.Un.	57•46 N	22•39 E
26	Köln (Cologne)	F.R.G.	50•56 N	6•57 E
27	Kolno	Pol.	53•23 N	21•56 E
27	Kolo	Pol.	52•11 N	18•37 E
26	Kolobrzeg	Pol.	54•10 N	15•35 E
30	Kolomna	Sov.Un.	55•06 N	38•47 E
27	Kolomyya	Sov.Un.	48•32 N	25•04 E
28	Kolpashevo	Sov.Un.	58•16 N	82•43 E
30	Kolpino	Sov.Un.	59•45 N	30•37 E
30	Kol'skiy P-Ov. (Kola Pen.)	Sov.Un.	67•15 N	37•40 E
30	Kolva (R.)	Sov.Un.	61•00 N	57•00 E
54	Kolwezi	Zaire	10•43 S	25•28 E
29	Kolyma (R.)	Sov.Un.	66•30 N	151•45 E
66	Komadorskie Ostrova (Is.)	Sov.Un.	55•40 N	167•13 E
52	Komadougou Yobé (R.)	Niger-Nig.	13•20 N	12•45 E
27	Komárno	Czech.	47•46 N	18•08 E
27	Komarno	Sov.Un.	49•38 N	23•43 E
27	Komaron	Hung.	47•45 N	18•06 E
54	Komatipoort	S.Afr.	25•21 S	32•00 E
55	Komga	S.Afr.	32•36 S	27•54 E
28	Komi A.S.S.R.	Sov.Un.	61•31 N	53•15 E
54	Kommetjie	S.Afr.	34•09 S	18•19 E
36	Kommunizma, Pik (Pk.)	Sov.Un.	39•46 N	71•23 E
21	Komotiní	Grc.	41•07 N	25•22 E
40	Kompong Som (Sihanoukville)	Camb.	10•40 N	103•50 E
40	Kompong Thom	Camb.	12•41 N	104•39 E
31	Komrat	Sov.Un.	46•17 N	28•38 E
31	Komsomolets Zaliv (B.)	Sov.Un.	45•40 N	52•00 E
29	Komsomol'sk-na-Amure	Sov.Un.	50•46 N	137•14 E
30	Konda (R.)	Sov.Un.	60•50 N	64•00 E
54	Kondoa	Tan.	4•52 S	36•00 E
52	Kong	Ivory Coast	9•05 N	4•41 W
54	Kongolo	Zaire	5•23 S	27•00 E
24	Kongsberg	Nor.	59•40 N	9•36 E
24	Kongsvinger	Nor.	60•12 N	12•00 E
54	Koni	Zaire	10•32 S	27•27 E
27	Konin	Pol.	52•11 N	18•17 E
21	Kónitsa	Grc.	40•03 N	20•46 E
21	Konjic	Yugo.	43•38 N	17•59 E
38	Kongju	Kor.	36•21 N	127•05 E
31	Konotop	Sov.Un.	51•13 N	33•14 E
27	Końskie	Pol.	51•12 N	20•26 E
26	Konstanz	F.R.G.	47•39 N	9•10 E
52	Kontagora	Nig.	10•24 N	5•28 E
31	Konya	Tur.	36•55 N	32•25 E
10	Kootenay (R.)	Can.	49•45 N	117•05 W
10	Kootenay Natl. Park	Can.	51•06 N	117•02 W
24	Kopervik	Nor.	59•18 N	5•20 E
30	Kopeysk	Sov.Un.	55•07 N	61•36 E
24	Köping	Swe.	59•32 N	15•58 E
24	Kopparberg	Swe.	59•53 N	15•00 E
34	Koppeh Dāgh (Mts.)	Iran	37•28 N	58•29 E
20	Koprivnica	Yugo.	46•10 N	16•48 E
27	Kopychintsy	Sov.Un.	49•06 N	25•55 E
21	Korçë	Alb.	40•37 N	20•48 E
20	Korčula (I.)	Yugo.	42•50 N	17•05 E
38	Korea B.	China-Kor.	39•18 N	123•50 E
33	Korea	Asia	38•45 N	130•00 E
38	Korean Arch.	Kor.	34•05 N	125•35 E
37	Korea Str.	Kor.-Jap.	33•30 N	128•30 E
27	Korets	Sov.Un.	50•35 N	27•13 E
52	Korhogo	Ivory Coast	9•27 N	5•38 W
21	Korinthiakós Kólpos (G.)	Grc.	38•15 N	22•33 E
21	Kórinthos (Corinth)	Grc.	37•56 N	22•54 E
26	Körmend	Hung.	47•02 N	16•36 E
20	Kornat (I.)	Yugo.	43•46 N	15•10 E
31	Korosten'	Sov.Un.	50•51 N	28•39 E
37	Korsakov	Sov.Un.	46•42 N	143•16 E
25	Korsnäs	Fin.	62•51 N	21•17 E
17	Korsør	Den.	55•19 N	11•08 E
23	Kortrijk	Bel.	50•49 N	3•10 E
29	Koryakskiy Khrebet (Mts.)	Sov.Un.	62•00 N	168•45 E
26	Kościan	Pol.	52•05 N	16•38 E
27	Kościerzyna	Pol.	54•08 N	17•59 E
46	Kosciusko, Mt.	Austl.	36•26 S	148•20 E
53	Kosha	Sud.	20•49 N	30•27 E
38	K'oshan	China	48•00 N	126•30 E
27	Košice	Czech.	48•43 N	21•17 E
55	Kosmos	S.Afr.	25•45 S	27•51 E
21	Kosovska Mitrovica	Yugo.	42•51 N	20•50 E
20	Kostajnica	Yugo.	45•14 N	16•32 E
30	Kostroma	Sov.Un.	57•46 N	40•55 E
26	Kostrzyń	Pol.	52•35 N	14•38 E
26	Koszalin	Pol.	54•12 N	16•10 E
26	Kőszeg	Hung.	47•21 N	16•32 E
35	Kota	India	25•17 N	75•49 E
40	Kota Baharu	Mala.	6•15 N	102•23 E
40	Kotabaru	Indon.	3•22 S	116•15 E
40	Kota Kinabalu	Mala.	5•55 N	116•05 E
54	Kota Kota	Malawi	12•52 S	34•16 E
33	Kota Tinggi	Mala.	1•43 N	103•54 E
21	Kotel	Bul.	42•54 N	26•28 E
30	Kotel'nich	Sov.Un.	58•15 N	48•20 E
29	Kotel'nyy (I.)	Sov.Un.	74•51 N	134•09 E
25	Kotka	Fin.	60•28 N	26•56 E
30	Kotlas	Sov.Un.	61•10 N	46•50 E
21	Kotor	Yugo.	42•26 N	18•48 E
20	Kotor Varoš	Yugo.	44•37 N	17•23 E
53	Kotto (R.)	Cen.Afr.Emp.	5•17 N	22•04 E
6	Kotzebue	Ak.	66•48 N	162•42 W
52	Koudougou	Upper Volta	12•15 N	2•22 W
54	Kouilou (R.)	Con.	4•00 S	12•05 E
52	Koulikoro	Mali	12•53 N	7•33 W
53	Koundé	Cen.Afr.Emp.	6•08 N	14•32 E
28	Kounradskiy	Sov.Un.	47•25 N	75•10 E
52	Kouroussa	Gui.	10•39 N	9•53 W
52	Koutiala	Mali	12•29 N	5•29 W
25	Kouvola	Fin.	60•51 N	26•40 E
30	Kovda (L.)	Sov.Un.	66•45 N	32•00 E
27	Kovel'	Sov.Un.	51•13 N	24•45 E
30	Kovrov	Sov.Un.	56•23 N	41•21 E
39	Kowloon	Hong Kong	22•28 N	114•20 E
21	Koynare	Bul.	43•23 N	24•07 E
21	Kozáni	Grc.	40•16 N	21•51 E
27	Kozienice	Pol.	51•34 N	21•35 E
27	Koźle	Pol.	50•19 N	18•10 E
21	Kozloduy	Bul.	43•45 N	23•42 E
40	Kra, Isth. of	Thai.	9•30 S	99•45 E
55	Kraai (R.)	S.Afr.	30•50 S	27•03 E
24	Kragerö	Nor.	58•53 N	9•21 E
21	Kragujevac	Yugo.	44•01 N	20•55 E
27	Kraków	Pol.	50•05 N	20•00 E
17	Kraljevo	Yugo.	43•39 N	20•48 E
24	Kramfors	Swe.	62•54 N	17•49 E
20	Kranj	Yugo.	46•16 N	14•23 E
55	Kranskop	S.Afr.	28•57 S	30•54 E
26	Kraslice	Czech.	50•19 N	12•30 E
31	Krasnaya Sloboda	Sov.Un.	48•25 N	44•35 E
27	Kraśnik	Pol.	50•53 N	22•15 E
31	Krasnodar	Sov.Un.	45•03 N	38•55 E
30	Krasnokamsk	Sov.Un.	58•00 N	55•45 E
30	Krasnoslobodsk	Sov.Un.	54•20 N	43•50 E
30	Krasnotur'insk	Sov.Un.	59•47 N	60•15 E
30	Krasnoufimsk	Sov.Un.	56•38 N	57•46 E
30	Krasnoural'sk	Sov.Un.	58•21 N	60•05 E
30	Krasnovishersk	Sov.Un.	60•22 N	57•20 E
31	Krasnovodsk	Sov.Un.	40•00 N	52•50 E
29	Krasnoyarsk	Sov.Un.	56•13 N	93•12 E
27	Krasnystaw	Pol.	50•59 N	23•11 E
31	Krasnyy Kut	Sov.Un.	50•50 N	47•00 E
40	Kratie	Camb.	12•28 N	106•06 E
21	Kratovo	Yugo.	42•04 N	22•12 E
31	Kremenchug	Sov.Un.	49•04 N	33•26 E
31	Kremenchugskoye (Res.)	Sov.Un.	49•20 N	32•45 E
27	Kremenets	Sov.Un.	50•06 N	25•43 E
26	Krems	Aus.	48•25 N	15•36 E
25	Krestsy	Sov.Un.	58•18 N	32•26 E
25	Kretinga	Sov.Un.	55•55 N	21•17 E
52	Kribi	Cam.	2•57 N	9•55 E
24	Kristiansand	Nor.	58•09 N	7•59 E
24	Kristianstad	Swe.	56•02 N	14•09 E
24	Kristiansund	Nor.	63•07 N	7•49 E
24	Kristinehamn	Swe.	59•20 N	14•05 E
25	Kristinestad	Fin.	62•16 N	21•28 E
21	Kriva-Palanka	Yugo.	42•12 N	22•21 E
31	Krivoy Rog	Sov.Un.	47•54 N	33•22 E
20	Križevci	Yugo.	46•02 N	16•30 E
20	Krk (I.)	Yugo.	45•06 N	14•33 E
27	Krnov	Czech.	50•05 N	17•41 E
24	Kröderen	Nor.	60•07 N	9•49 E
31	Krolevets	Sov.Un.	51•33 N	33•21 E
27	Kroměříž	Czech.	49•18 N	17•23 E
29	Kronotskiy, Mys. (C.)	Sov.Un.	54•58 N	163•15 E
30	Kronshtadt	Sov.Un.	59•59 N	29•47 E
31	Kropotkin	Sov.Un.	45•25 N	40•30 E
27	Krosno	Pol.	49•41 N	21•46 E
27	Krotoszyn	Pol.	51•41 N	17•25 E
20	Krško	Yugo.	45•58 N	15•30 E
54	Kruger Natl. Park	S.Afr.	23•22 S	30•18 E
55	Krugersdorp	S.Afr.	26•06 S	27•46 E
21	Krujë	Alb.	41•32 N	19•49 E
40	Krung Thep (Bangkok)	Thai.	13•50 N	100•29 E
21	Kruševac	Yugo.	43•34 N	21•21 E
21	Kruševo	Yugo.	41•20 N	21•15 E
24	Krylbo	Swe.	60•07 N	16•14 E
31	Krymskiy P-Ov. (Crimea) (Pen.)	Sov.Un.	45•18 N	33•30 E
27	Krynki	Pol.	53•15 N	23•47 E
18	Ksar el Kebir	Mor.	35•01 N	5•48 W
33	Kuala Klawang	Mala.	2•57 N	102•04 E
33	Kuala Lumpur	Mala.	3•08 N	101•42 E
38	Kuan	China	39•25 N	116•18 E
39	Kuangchang	China	25•50 N	116•18 E
39	Kuangchou Wan (B.)	China	20•40 N	111•00 E
39	Kuangte	China	30•40 N	119•20 E
38	Kuantien	China	40•40 N	124•50 E
31	Kuba	Sov.Un.	41•05 N	48•30 E
31	Kuban' (R.)	Sov.Un.	45•10 N	37•55 E
31	Kuban (R.)	Sov.Un.	45•20 N	40•05 E
19	Kuban R.	Sov.Un.	45•14 N	38•20 E
30	Kubenskoye (L.)	Sov.Un.	59•40 N	39•40 E
36	Kuch'e	China	41•34 N	82•44 E
38	Kuchen	China	33•20 N	117•18 E
40	Kuching	Mala.	1•30 N	110•26 E
33	Kudap	Indon.	1•14 N	102•30 E
40	Kudat	Mala.	6•56 N	116•48 E
25	Kudirkos Naumiestis	Sov.Un.	54•51 N	23•00 E
39	Kueichih	China	30•35 N	117•28 E
39	Kueilin	China	25•18 N	110•22 E
39	Kueiyang	China	26•45 N	107•00 E
36	K'uerhlo	China	41•37 N	86•03 E
26	Kufstein	Aus.	47•34 N	12•11 E
54	Kuilsrivier	S.Afr.	33•56 S	18•41 E
21	Kukës	Alb.	42•03 N	20•25 E
21	Kula	Bul.	43•52 N	23•13 E
31	Kula	Tur.	38•32 N	28•30 E
25	Kuldīga	Sov.Un.	56•59 N	21•59 E
30	Kulebaki	Sov.Un.	55•22 N	42•30 E
26	Kulmbach	F.R.G.	50•07 N	11•28 E
28	Kulunda	Sov.Un.	52•38 N	74•00 E
38	Kum (R.)	Kor.	36•50 N	127•30 E
31	Kuma (R.)	Sov.Un.	44•50 N	45•10 E
37	Kumamoto	Jap.	32•49 N	130•40 E
21	Kumanovo	Yugo.	42•10 N	21•41 E
52	Kumasi	Ghana	6•41 N	1•35 W
52	Kumba	Cam.	4•38 N	9•25 E
35	Kumbakonam	India	10•59 N	79•25 E
21	Kumkale	Tur.	39•59 N	26•10 E
37	Kunashir (I.)	Sov.Un.	44•40 N	145•45 E
51	Kundelungu, Plateau des	Zaire	9•00 S	25•30 E
33	Kundur (I.)	Indon.	0•49 N	103•20 E
54	Kunene (Cunene) (R.)	Ang.-Namibia	17•05 S	12•35 E
24	Kungälv	Swe.	57•53 N	12•01 E
30	Kungur	Sov.Un.	57•27 N	56•53 E
35	Kungrad	Sov.Un.	42•59 N	59•00 E
24	Kungsbacka	Swe.	57•31 N	12•04 E
36	K'un Lun Shan (Mts.)	China	35•26 N	83•09 E
39	K'unming (Yünnanfu)	China	25•10 N	102•50 E
39	Kunsan	Kor.	35•54 N	126•46 E
16	Kuopio	Fin.	62•48 N	28•30 E
20	Kupa (R.)	Yugo.	45•32 N	14•50 E
41	Kupang	Indon.	10•14 S	123•37 E
25	Kupiškis	Sov.Un.	55•50 N	24•55 E
31	Kupyansk	Sov.Un.	49•44 N	37•38 E
31	Kura (R.)	Sov.Un.	41•10 N	45•40 E
36	Kurak Darya (R.)	China	41•09 N	87•46 E
53	Kuraymah	Sud.	18•34 N	31•49 E
31	Kurdistan (Reg.)	Tur.-Iran	37•40 N	43•30 E
53	Kurdufān (Prov.)	Sud.	14•08 N	28•39 E
21	Kŭrdzhali	Bul.	41•39 N	25•21 E
37	Kure	Jap.	34•17 N	132•35 E
25	Kuressaare	Sov.Un.	58•15 N	22•26 E
28	Kurgan	Sov.Un.	55•28 N	65•14 E
35	Kurgan Tyube	Sov.Un.	38•00 N	68•49 E
37	Kuril Is.	Sov.Un.	46•20 N	149•30 E
25	Kurisches Haff (Bay)	Sov.Un.	55•10 N	21•08 E
53	Kurmuk	Sud.	10•40 N	34•13 E
35	Kurnool	India	16•00 N	78•04 E
15	Kuršenai	Sov.Un.	56•01 N	22•56 E
31	Kursk	Sov.Un.	51•44 N	36•08 E
23	Kuršumlija	Yugo.	43•08 N	21•18 E
53	Kūrtī	Sud.	18•08 N	31•39 E
54	Kuruman	S.Afr.	27•25 S	23•30 E
37	Kurume	Jap.	33•10 N	130•30 E
37	Kushiro	Jap.	43•00 N	144•22 E
31	Kushum (R.)	Sov.Un.	50•30 N	50•40 E
28	Kushva	Sov.Un.	58•18 N	59•51 E
28	Kustanay	Sov.Un.	53•10 N	63•39 E
53	Kūstī	Sud.	13•09 N	32•39 E
31	Kütahya	Tur.	39•20 N	29•50 E
31	Kutaisi	Sov.Un.	42•15 N	42•40 E
40	Kutaradja	Indon.	5•30 N	95•20 E
35	Kutch, Gulf of	India	22•45 N	68•33 E
35	Kutch, Rann of	India	23•59 N	69•13 E
20	Kutina	Yugo.	45•29 N	16•48 E
27	Kutno	Pol.	52•14 N	19•22 E
30	Kutno (L.)	Sov.Un.	65•15 N	31•30 E
29	Kutulik	Sov.Un.	53•12 N	102•51 E
27	Kuty	Sov.Un.	48•16 N	25•12 E
16	Kuusamo	Fin.	65•59 N	29•10 E
32	Kuwait	Asia	29•00 N	48•45 E
30	Kuybyshev (Kuibyshev)	Sov.Un.	53•10 N	50•05 E
30	Kuybyshevskoye (Res.)	Sov.Un.	53•40 N	49•00 E
31	Kuznetsk	Sov.Un.	53•00 N	46•30 E
28	Kuznetsk Basin	Sov.Un.	57•15 N	86•15 E
20	Kvarnerski Zaliv (B.)	Yugo.	44•41 N	14•05 E
54	Kwango (Cuango) (R.)	Afr.	6•35 S	16•50 E
36	Kwangsi Chuang (Aut. Reg.)	China	24•00 N	108•30 E
37	Kwangtung	China	23•45 N	113•15 E
36	Kweichow (Kueichou)	China	27•00 N	106•10 E
54	Kwenge (R.)	Zaire	6•45 S	18•23 E
27	Kwidzyń	Pol.	53•45 N	18•56 E
54	Kwilu (R.)	Zaire	3•22 S	17•22 E
41	Kwoka, Gunung (Mtn.)	Indon.	0•45 S	132•26 E
36	Kyakhta	Sov.Un.	51•00 N	107•30 E
36	Kyaukpyu	Bur.	19•19 N	93•33 E
25	Kybartai	Sov.Un.	54•40 N	22•46 E
39	Ky Lam	Viet.	15•48 N	108•30 E
45	Kynuna	Austl.	21•30 S	142•12 E
53	Kyoga, L.	Ug.	1•30 N	32•45 E
38	Kyŏngju	Kor.	35•48 N	129•12 E
37	Kyōto	Jap.	35•00 N	135•46 E
29	Kyren	Sov.Un.	51•46 N	102•13 E
25	Kyrön (R.)	Fin.	63•03 N	22•20 E
30	Kyshtym	Sov.Un.	55•43 N	60•33 E
37	Kyūshū (I.)	Jap.	32•27 N	131•03 E
21	Kyustendil	Bul.	42•16 N	22•39 E
29	Kyzyl	Sov.Un.	51•37 N	93•38 E
15	Kyzyl Kum, Peski (Des.)	Sov.Un.	42•47 N	64•45 E
36	Kyzylsu (R.)	China	39•26 N	74•30 E
35	Kzyl-Orda	Sov.Un.	44•58 N	65•45 E

L

Page	Name	Region	Lat.	Long.
26	Laa	Aus.	48•42 N	16•23 E
48	La Asunción	Ven.	11•02 N	63•57 W
50	La Banda	Arg.	27•48 S	64•12 W
52	Labé	Gui.	11•19 N	12•17 W
26	Labe (Elbe) (R.)	Czech.	50•05 N	15•20 E
10	Laberge (L.)	Can.	61•08 N	136•42 W
31	Labinsk	Sov.Un.	44•30 N	40•40 E
33	Labis	Mala.	2•23 N	103•01 E
41	Labo	Phil.	13•39 N	121•14 E
41	Labo	Phil.	14•11 N	122•49 E
41	Labo, Mt.	Phil.	14•00 N	122•47 E
50	Laboulaye	Arg.	34•01 S	63•10 W
11	Labrador (Reg.)	Can.	53•05 N	63•30 W
48	Lábrea	Braz.	7•28 S	64•39 W
41	Labuan	Phil.	13•43 N	120•07 E
40	Labuan (I.)	Mala.	5•28 N	115•11 E
41	Labuha	Indon.	0•43 S	127•35 E
50	La Calera	Col.	4•43 N	75•58 W
17	La Calle	Alg.	36•52 N	8•23 E
35	Laccodive Is.	India	11•00 N	73•02 E
12	La Ceiba	Hond.	15•45 N	86•52 W
48	La Ceja	Col.	6•02 N	75•25 W
11	Lac-Frontière	Can.	46•42 N	70•00 W
30	Lacha (L.)	Sov.Un.	61•15 N	39•05 E
26	La Chaux de Fonds	Switz.	47•07 N	6•47 E
46	Lachlan (R.)	Austl.	33•54 S	145•15 E
12	La Chorrera	Pan.	8•54 N	79•47 W
9	Lackawanna	NY	42•49 N	78•50 W
10	Lac la Biche	Can.	54•46 N	112•58 W
10	Lacombe	Can.	52•28 N	113•44 W
9	Laconia	NH	43•30 N	71•30 W
18	La Coruña	Sp.	43•20 N	8•20 W
7	La Crosse	Wi.	43•48 N	91•14 W
48	La Cruz	Col.	1•37 N	77•00 W
8	Ladd	Il.	41•25 N	89•25 W
48	La Dorado	Col.	5•28 N	74•42 W
25	Ladozhskoye Ozero (Ladoga, L.)	Sov.Un.	60•59 N	31•30 E
55	Lady Frere	S.Afr.	31•48 S	27•16 E
55	Lady Grey	S.Afr.	30•44 S	27•17 E
55	Ladysmith	S.Afr.	28•38 S	29•48 E
41	Lae	Pap.N.Gui.	6•15 S	146•57 E
24	Laerdal	Nor.	61•03 N	7•24 E
24	Laerdalsören	Nor.	61•08 N	7•26 E
24	Laesø (I.)	Den.	57•17 N	10•57 E
8	Lafayette	In.	40•25 N	86•55 W
7	Lafayette	La.	30•15 N	92•02 W
49	La Gaiba	Braz.	17•54 S	57•32 W
22	Lagan	N.Ire.	54•30 N	6•00 W
24	Lagan (R.)	Swe.	56•34 N	13•25 E
16	Laganes (Pt.)	Ice.	66•21 N	14•02 W
12	Lagarto, R.	Pan.	9•08 N	80•05 W
24	Lågan (R.)	Nor.	59•15 N	9•47 E
52	Laghouat	Alg.	33•45 N	2•49 E
41	Lagonoy	Phil.	13•44 N	123•31 E
41	Lagonoy G.	Phil.	13•34 N	123•46 E
52	Lagos	Nig.	6•27 N	3•24 E
12	Lagos de Moreno	Mex.	21•21 N	101•55 W
6	La Grande	Or.	45•20 N	118•06 W
11	La Grande (R.)	Can.	53•55 N	77•30 W
44	La Grange	Austl.	18•40 S	122•00 E
7	La Grange	Ga.	33•01 N	85•00 W
8	Lagrange	In.	41•40 N	85•25 W
8	La Grange	Ky.	38•20 N	85•25 W
48	La Grita	Ven.	8•02 N	71•59 W
49	La Guaira	Ven.	10•36 N	66•54 W
50	Laguna	Braz.	28•19 S	48•42 W
41	Laguna de Bay (L.)	Phil.	14•24 N	121•13 E
48	Lagunillas	Bol.	19•42 S	63•38 W
13	La Habana (Havana)	Cuba	23•08 N	82•23 W
26	Lahn R.	F.R.G.	50•21 N	7•54 E
24	Laholm	Swe.	56•30 N	13•00 E
35	Lahore	Pak.	32•00 N	74•18 E
26	Lahr	F.R.G.	48•19 N	7•52 E
25	Lahti	Fin.	60•59 N	27•39 E
39	Lai, C.	Viet.	17•08 N	107•30 E
53	Lai	Chad	9•29 N	16•18 E
38	Laichou Wan (B.)	China	37•22 N	119•19 E
39	Laipin	China	23•42 N	109•20 E
38	Laiyang	China	36•59 N	120•42 E
50	Lajeado	Braz.	29•24 S	51•46 W
50	Lajes	Braz.	27•47 S	50•17 W
44	Lake Brown	Austl.	31•03 S	118•30 E
7	Lake Charles	La.	30•15 N	93•14 W
22	Lake Dist.	Eng.	54•25 N	3•20 W
8	Lake Forest	Il.	42•16 N	87•50 W
8	Lake Geneva	Wi.	42•36 N	88•28 W
11	Lake Harbour	Can.	62•43 N	69•40 W
7	Lakeland	Fl.	28•02 N	81•58 W
8	Lake Odessa	Mi.	42•50 N	85•15 W
9	Lake Placid	NY	44•17 N	73•59 W
8	Lakewood	Oh.	41•29 N	81•48 W
9	Lakewood	Pa.	40•05 N	74•10 W
25	Lakhdenpokh'ya	Sov.Un.	61•33 N	30•10 E
18	La Línea	Sp.	36•11 N	5•22 W
35	Lalitpur	Nep.	27•23 N	85•24 E
18	Lalla-Maghnia	Alg.	34•52 N	1•40 W
23	La Louviere	Bel.	50•30 N	4•10 E
11	La Malbaie	Can.	47•39 N	70•10 W
20	La Marmora, Pta. (Mtn.)	It.	40•00 N	9•28 E
48	Lamas	Peru	6•24 S	76•41 W
54	Lambaréné	Gabon	0•42 S	10•13 E
48	Lambayeque	Peru	6•41 S	79•58 W
9	Lambertville	NJ	40•20 N	75•00 W
48	La Mesa	Col.	4•38 N	74•27 W
21	Lamía	Grc.	38•54 N	22•25 E
41	Lamon B.	Phil.	14•35 N	121•52 E
12	Lampazos	Mex.	27•03 N	100•30 W
17	Lampedusa (I.)	It.	35•29 N	12•58 E
55	Lamu	Ken.	2•16 S	40•54 E
6	Lanai (I.)	Hi.	20•48 N	157•06 W
35	Lanak La (P.)	China	34•40 N	79•50 E
18	La Nao, Cabo de (C.)	Sp.	38•43 N	0•14 E
22	Lanark	Scot.	55•40 N	3•50 W
22	Lancaster	Eng.	54•04 N	2•55 W
8	Lancaster	Ky.	37•35 N	84•30 W
9	Lancaster	NH	44•25 N	71•30 W
9	Lancaster	NY	42•54 N	78•42 W
8	Lancaster	Oh.	39•40 N	82•35 W
9	Lancaster	Pa.	40•05 N	76•20 W
38	Lanchou	China	35•55 N	103•55 E
54	Lândana	Ang.	5•15 S	12•07 E
26	Landau	F.R.G.	49•13 N	8•07 E
26	Landsberg	F.R.G.	48•03 N	10•53 E
22	Lands End Pt.	Eng.	50•03 N	5•45 W
26	Landshut	F.R.G.	48•32 N	12•09 E
24	Landskrona	Swe.	55•51 N	12•47 E
21	Langadhás	Grc.	40•44 N	24•10 E
33	Langat (R.)	Mala.	2•46 N	101•33 E
39	Langchung	China	31•40 N	106•05 E
24	Langeland (I.)	Den.	54•52 N	10•46 E
24	Langesund	Nor.	58•59 N	9•38 E
24	Lang Fd.	Nor.	62•40 N	7•45 E
16	Langjökoll (Glacier)	Ice.	64•40 N	20•31 W
26	Langnau	Switz.	46•56 N	7•46 E
40	Langsa	Indon.	4•33 N	97•52 E
40	Lang Son	Viet.	21•52 N	106•42 E
10	Lanigan	Can.	51•52 N	105•02 W
36	Lanisung Chiang (Mekong)	China	24•45 N	100•31 E
9	Lansdale	Pa.	40•20 N	75•15 W
9	Lansford	Pa.	40•50 N	75•50 W
8	Lansing	Mi.	42•45 N	84•35 W
50	Lanús	Arg.	34•27 S	58•24 W
20	Lanusei	It.	39•51 N	9•34 E
52	Lanzarote I.	Can.Is.	29•04 N	13•03 W
40	Laoag	Phil.	18•13 N	120•38 E
40	Lao Kay	Viet.	22•30 N	102•32 E
48	La Oroya	Peru	11•30 S	76•00 W
40	Laos	Asia	20•15 N	102•00 E
52	La Palma I.	Can.Is.	28•42 N	19•03 W
50	La Pampa (Prov.)	Arg.	37•25 S	67•00 W
50	Lapa Rio Negro	Braz.	26•12 S	49•56 W
50	La Paz	Arg.	30•48 S	59•47 W
49	La Paz	Bol.	16•31 S	68•03 W
12	La Paz	Mex.	24•00 N	110•15 W
41	La Paz	Phil.	17•41 N	120•41 E
8	Lapeer	Mi.	43•05 N	83•15 W
16	Lapland (Reg.)	Eur.	68•20 N	22•00 E
50	La Plata	Arg.	34•54 S	57•57 W
41	Lapog	Phil.	17•44 N	120•28 E
8	La Porte	In.	41•35 N	86•45 W
25	Lappeenranta	Fin.	61•04 N	28•08 E
21	Lapseki	Tur.	40•20 N	26•41 E
28	aptev Sea	Sov.Un.	75•39 N	120•00 E
27	Lapusul (R.)	Rom.	47•29 N	23•46 E
50	La Quiaca	Arg.	22•15 S	65•44 W
20	L'Aquila	It.	42•22 N	13•24 E
34	Lār	Iran	27•31 N	54•12 E
52	Larache	Mor.	35•15 N	6•09 W
6	Laramie	Wy.	41•20 N	105•40 W
6	Laredo	Tx.	27•31 N	99•29 W
53	Largeau	Chad	17•55 N	19•07 E
20	Larino	It.	41•48 N	14•54 E
50	La Rioja	Arg.	29•18 S	67•42 W
50	La Rioja (Prov.)	Arg.	28•45 S	68•00 W
21	Lárisa	Grc.	39•38 N	22•25 E
33	Lárnakos, Kólpos (B.)	Cyprus	36•50 N	33•45 E
33	Lárnax	Cyprus	34•55 N	33•37 E
17	La Rochelle	Fr.	46•10 N	1•09 W
17	La Roche-sur-Yon	Fr.	46•39 N	1•27 W
44	Larrey Pt.	Austl.	19•15 S	118•15 E
24	Larvik	Nor.	59•06 N	10•03 E
49	La Sabana	Ven.	10•38 N	66•24 W
18	La Sagra (Mtn.)	Sp.	37•56 N	2•35 E
8	La Salle	Il.	41•20 N	89•05 W
11	La Sarre	Can.	48•43 N	79•12 W
6	Las Cruces	NM	32•20 N	106•50 W
50	La Serena	Chile	29•55 S	71•24 W
17	La Seyne-sur-Mer	Fr.	43•07 N	5•52 E
50	Las Flores	Arg.	36•01 S	59•07 W
36	Lashio	Bur.	22•58 N	98•03 E
52	Las Palmas de Gran Canaria	Can.Is.	28•07 N	15•28 W
20	La Spezia	It.	44•07 N	9•48 E
6	Lassen Pk.	Ca.	40•30 N	121•32 W
6	Lassen Volcanic Natl. Park	Ca.	40•43 N	121•35 W
10	Last Mountain (L.)	Can.	51•05 N	105•10 W
54	Lastoursville	Gabon	1•00 S	12•49 E
12	Las Tres Virgenes, Vol.	Mex.	26•00 N	111•45 W
6	Las Vegas	Nv.	36•12 N	115•10 W
6	Las Vegas	NM	35•36 N	105•13 W
49	Las Vegas	Ven.	10•26 N	64•08 W
50	Las Vizcachas, Meseta de (Plat.)	Arg.	49•35 S	71•00 W
48	Latacunga	Ec.	1•02 S	78•33 W
19	Latakia (Reg.)	Syr.	35•10 N	35•49 E
9	Latrobe	Pa.	40•25 N	79•15 W
11	La Tuque	Can.	47•27 N	72•49 W
28	Latvian (S.S.R.)	Sov.Un.	57•28 N	24•29 E
46	Launceston	Austl.	41•35 S	147•22 E
22	Launceston	Eng.	50•38 N	4•26 W
50	La Unión	Chile	40•15 S	73•04 W
18	La Unión	Sp.	37•38 N	0•50 W
45	Laura	Austl.	15•40 S	144•45 E
9	Laurel	De.	38•30 N	75•40 W
9	Laurel	Md.	39•06 N	76•51 W
7	Laurel	Ms.	31•42 N	89•07 W
5	Laurentian Highlands (Reg.)	Can.	49•00 N	74•50 W
20	Lauria	It.	40•03 N	15•02 E
26	Lausanne	Switz.	46•32 N	6•35 E
40	Laut (I.)	Indon.	3•39 S	116•07 E
50	Lautaro	Chile	38•40 S	72•24 W
40	Laut Kecil, Kepulauan (Is.)	Indon.	4•44 S	115•43 E
11	Laval	Can.	45•31 N	73•44 W
17	Laval	Fr.	48•05 N	0•47 W
45	Lavella (I.)	Sol.Is.	7•50 S	155•45 E
20	Lavello	It.	41•05 N	15•50 E
44	Laverton	Austl.	28•45 S	122•30 E
49	La Victoria	Ven.	10•14 N	67•20 W
21	Lávrion	Grc.	37•44 N	24•05 E
7	Lawrence	Ks.	38•57 N	95•13 W
9	Lawrence	Ma.	42•42 N	71•09 W
8	Lawrenceburg	In.	39•06 N	84•47 W
8	Lawrenceburg	Ky.	38•00 N	85•00 W
8	Lawrenceville	Il.	38•45 N	87•45 W
9	Lawsonia	Md.	38•00 N	75•50 W
6	Lawton	Ok.	34•36 N	98•25 W
34	Lawz, Jabal al (Mtn.)	Sau.Ar.	28•46 N	35•37 E
33	Layang Layang	Mala.	1•49 N	103•28 E
25	Laždijai	Sov.Un.	54•12 N	23•35 E
20	Lazio (Latium) (Reg.)	It.	42•05 N	12•25 E
6	Lead	SD	44•22 N	103•47 W
8	Leamington	Can.	42•05 N	82•35 W
22	Leamington	Eng.	52•17 N	1•25 W
7	Leavenworth	Ks.	39•19 N	94•54 W
27	Leba	Pol.	54•45 N	17•34 E
33	Lebam R.	Mala.	1•35 N	104•09 E
8	Lebanon	In.	40•00 N	86•30 W
8	Lebanon	Ky.	37•32 N	85•15 W
9	Lebanon	NH	43•40 N	72•15 W
8	Lebanon	Oh.	39•25 N	84•10 W
9	Lebanon	Pa.	40•20 N	76•20 W
34	Lebanon	Asia	34•00 N	34•00 E
19	Lebanon Mts.	Leb.	33•30 N	35•32 E
31	Lebedin	Sov.Un.	50•34 N	34•27 E
31	Lebedyan'	Sov.Un.	53•03 N	39•08 E
27	Lebork	Pol.	54•33 N	17•46 E
50	Lebú	Chile	37•35 S	73•37 W
21	Lecce	It.	40•22 N	18•11 E
20	Lecco	It.	45•52 N	9•28 E
26	Lech R.	F.R.G.	47•41 N	10•52 E
17	Le Creusot	Fr.	46•48 N	4•23 E
22	Leeds	Eng.	53•48 N	1•33 W
26	Leer	F.R.G.	53•14 N	7•27 E
22	Lee R.	Ire.	51•52 N	8•30 W
9	Leesburg	Va.	39•10 N	77•30 W
8	Leetonia	Oh.	40•50 N	80•45 W
23	Leeuwarden	Neth.	52•12 N	5•50 E
44	Leeuwin, C.	Austl.	34•15 S	114•30 E
44	Lefroy (L.)	Austl.	31•30 S	122•00 E
41	Legaspi	Phil.	13•09 N	123•44 E
46	Legge Pk.	Austl.	41•33 S	148•10 E
20	Legnano	It.	45•35 N	8•53 E
26	Legnica	Pol.	51•13 N	16•10 E
17	Le Havre	Fr.	49•31 N	0•07 E
22	Leicester	Eng.	52•37 N	1•08 W
44	Leichhardt, (R.)	Austl.	18•30 S	139•45 E
46	Leigh Creek	Austl.	30•33 S	138•30 E
24	Leikanger	Nor.	61•11 N	6•51 E
26	Leine R.	F.R.G.	51•58 N	9•56 E
22	Leinster	Ire.	52•45 N	7•19 W
8	Leipsic	Oh.	41•05 N	84•00 W
26	Leipzig	G.D.R.	51•20 N	12•24 E
8	Leitchfield	Ky.	37•28 N	86•20 W
23	Lek (R.)	Neth.	51•59 N	5•30 E
17	Lekef	Tun.	36•14 N	8•42 E
24	Leksand	Swe.	60•45 N	14•56 E
26	Le Locle	Switz.	47•03 N	6•43 E
50	Le Maire, Estrecho de (Str.)	Arg.	55•15 S	65•30 W
17	Le Mans	Fr.	48•01 N	0•12 E
41	Lemery	Phil.	13•51 S	120•55 E
33	Lemesós	Cyprus	34•39 N	33•02 E
24	Lemvig	Den.	56•33 N	8•16 E
24	Lena	Swe.	60•01 N	17•40 E
29	Lena (R.)	Sov.Un.	68•39 N	124•15 E
50	Lençóes Paulista	Braz.	22•30 S	48•45 W
49	Lençóis	Braz.	12•38 S	41•28 W
15	Lenger	Sov.Un.	41•38 N	70•00 E
33	Lenik (R.)	Mala.	1•59 N	102•51 E
30	Leningrad	Sov.Un.	59•57 N	30•20 E
28	Leninogorsk	Sov.Un.	50•29 N	83•25 E
31	Leninsk	Sov.Un.	48•40 N	45•10 E
28	Leninsk-Kuznetski	Sov.Un.	54•28 N	86•48 E
31	Lenkoran'	Sov.Un.	38•52 N	48•58 E
26	Leoben	Aus.	47•22 N	15•09 E
9	Leominster	Ma.	42•32 N	71•45 W
12	León	Mex.	21•08 N	101•41 W
12	León	Nic.	12•28 N	86•53 W
18	León	Sp.	42•38 N	5•33 W
20	Leonforte	It.	37•40 N	14•27 E
26	Lepontine Alpi (Mts.)	Switz.	46•28 N	8•38 E
31	Lepsinsk	Sov.Un.	45•32 N	80•47 E
16	Le Puy-en-Velay	Fr.	45•02 N	3•54 E
20	Lercara	It.	36•47 N	13•36 E
12	Lerdo	Mex.	25•31 N	103•30 W
53	Léré	Chad	9•42 N	14•14 E
55	Leribe	Leso.	28•53 S	28•02 E
18	Lérida	Sp.	41•38 N	0•37 E
9	Le Roy	NY	43•00 N	78•00 W
22	Lerwick	Scot.	60•08 N	1•27 W
21	Lesh (Alessio)	Alb.	41•47 N	19•40 E
20	Lésina, Lago di (L.)	It.	41•48 N	15•12 E
21	Leskovac	Yugo.	43•00 N	21•58 E
30	Lesnoy	Sov.Un.	66•45 N	34•45 E
54	Lesotho	Afr.	29•45 S	28•07 E
17	Les Sables-d'Olonne	Fr.	46•30 N	1•47 W
10	Lesser Slave L.	Can.	55•25 N	115•30 W
21	Lésvos (I.)	Grc.	39•15 N	25•40 E
26	Leszno	Pol.	51•51 N	16•35 E
10	Lethbridge	Can.	49•42 N	112•50 W
48	Leticia	Col.	4•04 S	69•57 W
23	Le Tréport	Fr.	50•03 N	1•21 E
40	Leuser, Gulung (Mtn.)	Indon.	3•36 N	97•17 E
23	Leuven	Bel.	50•53 N	4•42 E
21	Levádhia	Grc.	38•25 N	22•51 E
16	Levanger	Nor.	63•42 N	11•01 E
20	Levanna (Mtn.)	Fr.-It.	45•25 N	7•14 E
44	Leveque, C.	Austl.	16•26 S	123•08 E
54	Leverville	Zaire	5•13 S	18•43 E
27	Levice	Czech.	48•13 N	18•37 E
20	Levico	It.	46•02 N	11•20 E
11	Lévis	Can.	46•49 N	71•11 W
21	Levkás	Grc.	38•49 N	20•43 E
21	Levkás (I.)	Grc.	38•42 N	20•22 E
27	Levoča	Czech.	49•03 N	20•38 E
9	Lewes	De.	38•45 N	75•10 W
22	Lewes	Eng.	50•51 N	0•01 E
22	Lewis (I.)	Scot.	58•05 N	6•07 W
8	Lewisburg	WV	37•50 N	80•20 W
6	Lewiston	Id.	46•24 N	116•59 W
9	Lewiston	Me.	44•05 N	70•14 W
9	Lewiston	NY	43•11 N	79•02 W
6	Lewiston	Mt.	47•05 N	109•25 W
6	Lewistown	Pa.	40•35 N	77•30 W
8	Lexington	Ky.	38•05 N	84•30 W
9	Lexington	Va.	37•45 N	79•20 W
41	Leyte (I.)	Phil.	10•35 N	125•35 E
27	Lezajsk	Pol.	50•14 N	22•25 E
36	Lhasa	China	29•41 N	91•12 E
38	Lianghsiang	China	39•43 N	116•08 E

Page	Name	Region	Lat.	Long.
38	Liaoch'eng	China	36•27 N	115•56 E
38	Liao Ho (R.)	China	41•40 N	122•40 E
37	Liaoning (Prov.)	China	41•31 N	122•11 E
38	Liaotung Pantao (Pen.)	China	39•45 N	122•22 E
38	Liaotung Wan (B.)	China	40•25 N	121•15 E
38	Liaoyang	China	41•18 N	123•10 E
38	Liaoyüan	China	43•00 N	124•59 E
10	Liard (R.)	Can.	59•43 N	126•42 W
48	Libano	Col.	4•55 N	75•05 W
53	Libenge	Zaire	3•39 N	18•40 E
26	Liberec	Czech.	50•45 N	15•06 E
51	Liberia	Afr.	6•30 N	9•55 W
49	Libertad de Orituco	Ven.	9•32 N	66•24 W
8	Liberty	In.	39•35 N	84•55 W
8	Libertyville	Il.	42•17 N	87•57 W
41	Libmanan	Phil.	13•42 N	123•04 E
55	Libode	S.Afr.	31•33 S	29•03 E
17	Libourne	Fr.	44•55 N	0•12 W
52	Libreville	Gabon	0•23 N	9•27 E
51	Libya	Afr.	27•38 N	15•00 E
53	Libyan Des.	Libya	28•23 N	23•34 E
19	Libyan Plat.	Egypt	30•58 N	26•20 E
50	Licancábur, Cerro (Mtn.)	Chile	22•45 S	67•45 W
36	Lichiang	China	27•00 N	100•08 E
8	Licking (R.)	Ky.	38•30 N	84•10 W
20	Licosa, Pt.	It.	40•17 N	14•40 E
27	Lida	Sov.Un.	53•53 N	25•19 E
24	Lidköping	Swe.	58•31 N	13•06 E
27	Lidzbark	Pol.	54•07 N	20•36 E
39	Liechou Pan-Tao (Pen.)	China	20•40 N	109•25 E
17	Liechtenstein	Eur.	47•10 N	10•00 E
23	Liège	Bel.	50•40 N	5•30 E
39	Lienchiang	China	21•38 N	110•15 E
37	Lienyün	China	33•10 N	120•01 E
38	Lienyünchiang	China	34•43 N	119•27 E
26	Lienz	Aus.	46•49 N	12•45 E
25	Liepāja	Sov.Un.	56•31 N	20•59 E
27	Liesing	Aus.	48•09 N	16•17 E
26	Liestal	Switz.	47•28 N	7•44 E
9	Lievre, Riviére du (R.)	Can.	45•00 N	75•25 W
22	Liffey R.	Ire.	53•21 N	6•35 W
45	Lifou (I.)	N.Cal.	21•15 S	167•32 E
41	Ligao	Phil.	13•14 N	123•33 E
46	Lightning Ridge	Austl.	29•23 S	147•50 E
55	Ligonha (R.)	Moz.	16•14 S	39•00 E
8	Ligonier	In.	41•30 N	85•35 W
20	Liguria (Reg.)	It.	44•24 N	8•27 E
20	Ligurian Sea	Eur.	43•42 N	8•32 E
45	Lihou Rfs.	Austl.	17•23 S	152•43 E
39	Lihsien	China	29•42 N	111•40 E
6	Lihue	Hi.	21•59 N	159•23 W
25	Lihula	Sov.Un.	58•41 N	23•50 E
54	Likasi (Jadotville)	Zaire	10•59 S	26•44 E
17	Lille	Fr.	50•38 N	3•01 E
24	Lille Baelt (Str.)	Den.	55•09 N	9•53 E
24	Lillehammer	Nor.	61•07 N	10•25 E
24	Lillesand	Nor.	58•16 N	8•19 E
24	Lilleström	Nor.	59•56 N	11•04 E
10	Lillooet	Can.	50•30 N	121•55 W
54	Lilongwe	Malawi	13•59 S	33•44 E
8	Lima	Oh.	40•40 N	84•05 W
48	Lima	Peru	12•06 S	76•55 W
50	Limay	Arg.	39•50 S	69•15 W
25	Limbazi	Sov.Un.	57•32 N	24•44 E
26	Limburg	F.R.G.	50•22 N	8•03 E
24	Limedsforsen	Swe.	60•54 N	13•24 E
24	Limfjorden (Fd.)	Den.	56•14 N	7•55 E
24	Limfjorden (Fd.)	Den.	56•56 N	10•35 E
44	Limmen Bght.	Austl.	14•45 S	136•00 E
21	Limni	Grc.	38•47 N	23•22 E
21	Limnos (I.)	Grc.	39•58 N	24•48 E
17	Limoges	Fr.	45•50 N	1•15 E
48	Limón	C.R.	10•01 N	83•02 W
12	Limón B.	C.Z.	9•21 N	79•58 W
51	Limpopo R.	Afr.	23•15 S	27•46 E
50	Linares	Chile	35•51 S	71•35 W
12	Linares	Mex.	24•53 N	99•34 W
18	Linares	Sp.	38•07 N	3•38 W
50	Linares (Prov.)	Chile	35•53 S	71•30 W
20	Linaro, C.	It.	42•02 N	11•53 E
38	Linchiang	China	41•45 N	127•00 E
38	Linch'ing	China	36•49 N	115•42 E
39	Linch'uan	China	27•58 N	116•18 E
50	Lincoln	Arg.	34•51 S	61•29 W
22	Lincoln	Eng.	53•14 N	0•33 W
8	Lincoln	Il.	40•09 N	89•21 W
6	Lincoln	Ne.	40•49 N	96•43 W
22	Lincoln Wolds	Eng.	53•25 N	0•23 W
26	Lindau	F.R.G.	47•33 N	9•40 E
24	Lindesberg	Swe.	59•37 N	15•14 E
23	Lindesnes (C.)	Nor.	58•00 N	7•05 E
38	Lindho	China	40•45 N	107•30 E
55	Lindi	Tan.	10•00 S	39•43 E
53	Lindi R.	Zaire	1•00 N	27•13 E
9	Lindsay	Can.	44•20 N	78•45 W
38	Linfen	China	36•00 N	111•38 E
40	Linga (Is.)	Indon.	0•35 S	105•05 E
41	Lingayen	Phil.	16•01 N	120•13 E
41	Lingayen G.	Phil.	16•18 N	120•11 E
26	Lingen	F.R.G.	52•32 N	7•20 E
39	Lingling (Yungchow)	China	26•10 N	111•40 E
39	Lingting Yang (Can.)	China	22•00 N	114•00 E
52	Linguère	Senegal	15•24 N	15•07 W
38	Lingwu	China	38•05 N	106•18 E
38	Lingyüan	China	41•12 N	119•20 E
39	Linhai	China	28•52 N	121•08 E
38	Linhsi	China	43•30 N	118•02 E
38	Lini	China	35•04 N	118•21 E
39	Linkao	China	19•58 N	109•40 E
24	Linköping	Swe.	58•25 N	15•35 E
22	Linnhe (L.)	Scot.	56•35 N	4•30 W
49	Lins	Braz.	21•42 S	49•41 W
38	Lintien	China	42•08 N	124•59 E
8	Linton	In.	39•05 N	87•15 W
39	Linwu	China	25•20 N	112•30 E
38	Linyü	China	40•01 N	119•45 E
26	Linz	Aus.	48•18 N	14•18 E
41	Lipa	Phil.	13•55 N	121•10 E
20	Lipari	It.	38•29 N	15•00 E
20	Lipari (I.)	It.	38•32 N	15•04 E
31	Lipetsk	Sov.Un.	52•26 N	39•34 E
39	Lip'ing	China	26•18 N	109•00 E
27	Lipno	Pol.	52•50 N	19•12 E
23	Lippe (R.)	F.R.G.	51•36 N	6•45 E
26	Lippstadt	F.R.G.	51•39 N	8•20 E
39	Lip'u	China	24•38 N	110•35 E
20	Liri (R.)	It.	41•49 N	13•30 E
53	Lisala	Zaire	2•09 N	21•31 E
17	Lisbon	Port.	38•42 N	9•05 W
8	Lisbon	Oh.	40•45 N	80•50 W
9	Lisbon Falls	Me.	43•59 N	70•03 W
22	Lisburn	N. Ire.	54•35 N	6•05 W
6	Lisburne, C.	Ak.	68•20 N	165•40 W
38	Lishih	China	37•32 N	111•12 E
38	Lishu	China	43•12 N	124•18 E
38	Lishuchen	China	45•01 N	130•50 E
39	Lishui	China	28•28 N	120•00 E
46	Lismore	Austl.	28•48 S	153•18 E
67	Lister, Mt.	Ant.	78•05 S	163•00 E
33	Litani (R.)	Lib.	33•28 N	35•42 E
8	Litchfield	Il.	39•10 N	89•38 W
46	Lithgow	Austl.	33•23 S	149•31 E
20	Lithinon, Ark. (C.)	Grc.	34•59 N	24•35 E
28	Lithuanian S.S.R.	Sov.Un.	55•42 N	23•30 E
21	Litókhoron	Grc.	40•05 N	22•29 E
26	Litoměřice	Czech.	50•33 N	14•10 E
26	Litomyšl	Czech.	49•52 N	16•14 E
67	Little America	Ant.	78•30 S	161•30 W
40	Little Andaman I.	Andaman & Nicobar Is.	10•39 N	93•08 E
6	Little Belt Mts.	Mt.	47•00 N	110•50 W
6	Little Colorado (R.)	Az.	36•05 N	111•35 W
9	Little Falls	NY	43•05 N	74•55 W
13	Little Hans Lollick (I.)	Vir.Is.	18•25 N	64•54 W
8	Little Kanawha (R.)	WV	39•05 N	81•30 W
54	Little Karroo (Mts.)	S.Afr.	33•50 S	21•02 E
11	Little Mecatina (R.)	Can.	52•40 N	62•21 W
6	Little Missouri (R.)	SD	45•46 N	103•48 W
7	Little Rock	Ar.	34•42 N	92•16 W
9	Littleton	NH	44•15 N	71•45 W
8	Little Wabash (R.)	Il.	38•50 N	88•30 W
39	Liuan	China	31•45 N	116•29 E
39	Liuchou	China	24•25 N	109•30 E
38	Liuho	China	42•10 N	125•38 E
40	Liukang Tenggaya (Is.)	Indon.	6•56 S	118•10 E
38	Liup'an Shan (Mts.)	China	36•20 N	105•30 E
39	Liuyang	China	28•10 N	113•35 E
8	Livermore	Ky.	37•30 N	87•05 W
11	Liverpool	Can.	44•02 N	64•41 W
22	Liverpool	Eng.	53•25 N	2•52 W
45	Liverpool Ra.	Austl.	31•47 S	31•00 E
53	Livindo R.	Gabon	1•09 N	13•30 E
6	Livingston	Mt.	45•40 N	110•35 W
54	Livingstone	Zambia	17•50 S	25•53 E
54	Livingstonia	Malawi	10•36 S	34•07 E
20	Livno	Yugo.	43•50 N	17•03 E
31	Livny	Sov.Un.	52•28 N	37•36 E
20	Livorno (Leghorn)	It.	43•32 N	11•18 E
50	Livramento	Braz.	30•46 S	55•21 W
22	Lizard Pt.	Eng.	49•55 N	5•09 W
20	Ljubljana	Yugo.	46•04 N	14•29 E
20	Ljubuški	Yugo.	43•11 N	17•29 E
24	Ljungan (R.)	Swe.	62•50 N	13•45 E
24	Ljungby	Swe.	56•49 N	13•56 E
24	Ljusdal	Swe.	61•50 N	16•11 E
24	Ljusnan (R.)	Swe.	61•55 N	15•33 E
22	Llandudno	Wales	53•20 N	3•46 W
22	Llanelly	Wales	51•44 N	4•09 W
48	Llanos	Col.-Ven.	4•00 N	71•15 W
22	Lleyn Prom.	Wales	52•55 N	3•10 W
10	Lloydminster	Can.	53•17 N	110•00 W
50	Llullaillaco (Vol.)	Arg.	24•50 S	68•30 W
54	Loange (R.)	Zaire	6•10 S	19•40 E
54	Lobatsi	Bots.	25•13 S	25•35 E
50	Lobería	Arg.	38•13 S	58•48 W
54	Lobito	Ang.	12•30 S	13•34 E
48	Lobos de Tierra (I.)	Peru	6•29 S	80•55 W
26	Locarno	Switz.	46•10 N	8•43 E
39	Loching	China	28•02 N	120•40 E
22	Lochy (L.)	Scot.	56•57 N	4•45 W
9	Lock Haven	Pa.	41•05 N	77•30 W
8	Lockport	Il.	41•35 N	88•04 W
40	Loc Ninh	Viet.	12•00 N	106•30 E
33	Lod	Isr.	31•57 N	34•55 E
25	Lodeynoye Pole	Sov.Un.	60•43 N	33•24 E
20	Lodi	It.	45•18 N	9•30 E
27	Lódź	Pol.	51•46 N	19•13 E
16	Lofoten (Is.)	Nor.	68•26 N	13•42 E
8	Logan	Oh.	39•35 N	82•25 W
6	Logan	Ut.	41•46 N	111•51 W
8	Logan	WV	37•50 N	82•00 W
10	Logan, Mt.	Can.	60•54 N	140•33 W
8	Logansport	In.	40•45 N	86•25 W
51	Logone (R.)	Afr.	11•15 N	15•10 E
17	Logroño	Sp.	42•28 N	2•25 W
24	Løgstør	Den.	56•56 N	9•15 E
17	Loire (R.)	Fr.	47•19 N	1•11 W
48	Loja	Ec.	3•49 S	79•13 W
31	Lokhvitsa	Sov.Un.	50•21 N	33•16 E
52	Lokoja	Nig.	7•47 N	6•45 E
53	Lol R.	Sud.	9•06 N	28•09 E
24	Lolland	Den.	54•41 N	11•00 E
21	Lom	Bul.	43•48 N	23•15 E
54	Lomami (R.)	Zaire	0•50 S	24•40 E
50	Lomas de Zamora	Arg.	34•31 S	58•24 W
20	Lombardia (Reg.)	It.	45•20 N	9•30 E
41	Lomblen (I.)	Indon.	8•08 S	123•45 E
40	Lombok (I.)	Indon.	9•15 S	116•15 E
40	Lombok (Str.)	Indon.	9•00 S	115•28 E
52	Lomé	Togo.	6•08 N	1•13 E
54	Lomela	Zaire	2•19 S	23•33 E
54	Lomela (R.)	Zaire	0•35 S	21•20 E
53	Lomie	Cam.	3•10 N	13•37 E
22	Lomond, Loch	Scot.	56•15 N	4•40 W
27	Lomza	Pol.	53•11 N	22•04 E
9	Lonaconing	Md.	39•35 N	78•55 W
8	London	Can.	43•00 N	81•20 W
22	London	Eng.	51•30 N	0•07 W
8	London	Oh.	39•50 N	83•30 W
22	Londonderry	N.Ire.	55•00 N	7•19 W
44	Londonderry, C.	Austl.	13•30 S	127•00 E
49	Londrina	Braz.	21•53 S	51•17 W
8	Lonely (I.)	Can.	45•35 N	81•30 W
13	Long (I.)	Ba.	23•25 N	75•10 W
41	Long (I.)	Pap.N.Gui.	5•10 S	147•30 E
54	Longa (R.)	Ang.	10•20 S	13•50 E
6	Long Beach	Ca.	33•46 N	118•12 W
9	Long Branch	NJ	40•18 N	73•59 W
22	Longford	Ire.	53•43 N	7•40 W
9	Long I.	NY	40•50 N	72•50 W
9	Long Island Sd.	Ct.-NY	41•05 N	72•45 W
11	Longlac	Can.	49•41 N	86•28 W
9	Long Pt.	Can.	42•35 N	80•05 W
9	Long Point B.	Can.	42•40 N	80•10 W
11	Long Range Mts.	Can.	48•00 N	58•30 W
45	Longreach	Austl.	23•32 S	144•17 E
6	Longs Pk.	Co.	40•17 N	105•37 W
9	Longueuil	Can.	45•32 N	73•30 W
40	Long Xuyen	Viet.	10•31 N	105•28 E
41	Looc	Phil.	12•16 N	121•59 E
8	Loogootee	In.	38•40 N	86•55 W
22	Loop Head	Ire.	52•32 N	9•59 W
29	Lopatka, Mys (C.)	Sov.Un.	51•00 N	156•52 E
41	Lopez B.	Phil.	14•04 N	122•00 E
39	Lop'ing	China	29•02 N	117•12 E
8	Lorain	Oh.	41•28 N	82•10 W
35	Loralai	Pak.	30•31 N	68•35 E
17	Lorca	Sp.	37•39 N	1•40 W
45	Lord Howe (I.)	Austl.	31•44 S	157•56 E
49	Loreto	Braz.	7•09 S	45•10 W
48	Lorica	Col.	9•14 N	75•54 W
22	Lorne, Firth of	Scot.	56•10 N	6•09 W
26	Lörrach	F.R.G.	47•36 N	7•38 E
6	Los Angeles	Ca.	34•00 N	118•15 W
50	Los Angeles	Chile	37•27 S	72•15 W
50	Los Chonos, Archipielago de	Chile	44•35 N	76•15 W
50	Los Estados (I.)	Arg.	54•45 S	64•25 W
39	Loshan	China	29•40 N	103•40 E
20	Lošinj	Yugo.	44•30 N	14•29 E
20	Lošinj (I.)	Yugo.	44•35 N	14•34 E
12	Los Reyes	Mex.	19•35 N	102•29 W
49	Los Teques	Ven.	10•22 N	67•04 W
50	Los Vilos	Chile	31•56 S	71•29 W
17	Lot (R.)	Fr.	44•32 N	1•08 E
50	Lota	Chile	37•11 S	73•14 W
39	Loting	China	23•42 N	111•35 E
26	Lötschen Tun.	Switz.	46•26 N	7•54 E
40	Louangphrabang	Laos	19•47 N	102•15 E
8	Loudonville	Oh.	40•40 N	82•15 W
52	Louga	Senegal	15•37 N	16•13 W
8	Louisa	Ky.	38•05 N	82•40 W
45	Louisade Arch.	Pap.N.Gui.	10•44 S	153•58 E
11	Louis XIV, Pte.	Can.	54•35 N	79•51 W
7	Louisiana	U.S.	30•50 N	92•50 W
54	Louis Trichardt	S.Afr.	22•52 S	29•53 E
7	Louisville	Ky.	38•15 N	85•45 W
26	Louny	Czech.	50•20 N	13•47 E
17	Lourdes	Fr.	43•06 N	0•03 W
22	Louth	Eng.	53•27 N	0•02 W
30	Lovat (R.)	Sov.Un.	57•23 N	31•18 E
21	Lovech	Bul.	43•10 N	24•40 E
25	Loviisa	Fin.	60•28 N	26•10 E
11	Low, C.	Can.	62•58 N	86•50 W
54	Lowa (R.)	Zaire	1•30 S	27•18 E
8	Lowell	In.	41•17 N	87•26 W
9	Lowell	Ma.	42•38 N	71•18 W
8	Lowell	Mi.	42•55 N	85•20 W
10	Lower Arrow (L.)	Can.	49•40 N	118•08 W
45	Lower Hutt	N.Z.	41•08 S	175•00 E
23	Lowestoft	Eng.	52•31 N	1•45 E
27	Lowicz	Pol.	52•06 N	19•57 E
27	Low Tatra Mts.	Czech.	48•57 N	19•18 E
9	Lowville	NY	43•45 N	75•30 W
46	Loxton	Austl.	34•25 S	140•38 E
38	Loyang	China	34•45 N	112•32 E
45	Loyauté, Iles.	N.Cal.	21•17 S	168•16 E
21	Ložnica	Yugo.	44•31 N	19•16 E
31	Lozovaya	Sov.Un.	48•27 N	38•37 E
54	Lualaba (R.)	Zaire	1•00 S	25•45 E
54	Luama (R.)	Zaire	4•17 S	27•45 E
38	Luan (R.)	China	41•25 N	117•15 E
54	Luanda	Ang.	8•48 S	13•14 E
54	Luanguinga (R.)	Ang.	14•00 S	20•45 E
54	Luangwa (R.)	Zambia	11•25 S	32•55 E
29	Lubaczów	Pol.	50•08 N	23•10 E
26	Lubán	Pol.	51•08 N	15•17 E
25	Lubānas Ezers (L.)	Sov.Un.	56•48 N	26•30 E
41	Lubang	Phil.	13•49 N	120•07 E
41	Lubang (Is.)	Phil.	13•47 N	119•56 E
54	Lubango	Ang.	14•55 S	13•30 E
41	Lubao	Phil.	14•55 N	120•36 E
27	Lubartow	Pol.	51•27 N	22•37 E
27	Lubawa	Pol.	53•31 N	19•47 E
26	Lübben	G.D.R.	51•56 N	13•53 E
6	Lubbock	Tx.	33•35 N	101•50 W
26	Lübeck	F.R.G.	53•53 N	10•42 E
26	Lübecker Bucht (B.)	G.D.R.	54•10 N	11•20 E
54	Lubilash (R.)	Zaire	7•35 S	23•55 E
26	Lubin	Pol.	51•24 N	16•14 E
27	Lublin	Pol.	51•14 N	22•33 E
31	Lubny	Sov.Un.	50•01 N	33•02 E
41	Lubuagan	Phil.	17•24 N	121•11 E
54	Lubudi (R.)	Zaire	9•20 S	25•20 E
54	Lubumbashi (Élisabethville)	Zaire	11•40 S	27•28 E
20	Lucca	It.	43•51 N	10•29 E
22	Luce B.	Scot.	54•45 N	4•45 W
41	Lucena	Phil.	13•55 N	121•36 E
27	Lučenec	Czech.	48•19 N	19•41 E
20	Lucera	It.	41•31 N	15•22 E
39	Luchi	China	28•18 N	110•10 E
39	Luchih	China	31•17 N	120•54 E
39	Luchou	China	28•58 N	105•25 E
41	Lucipara (I.)	Indon.	5•45 S	128•15 E
26	Luckenwalde	G.D.R.	52•05 N	13•10 E
35	Lucknow	India	26•54 N	80•58 E
21	Luda Kamchiya (R.)	Bul.	42•46 N	27•13 E
54	Lüderitz	Namibia	26•35 S	15•15 E
54	Lüderitz Bucht (B.)	Namibia	26•35 S	14•30 E
35	Ludhiāna	India	31•00 N	75•52 E
8	Ludington	Mi.	44•00 N	86•25 W
24	Ludvika	Swe.	60•10 N	15•09 E
26	Ludwigsburg	F.R.G.	48•53 N	9•14 E
26	Ludwigshafen	F.R.G.	49•29 N	8•26 E
26	Ludwigslust	G.D.R.	53•18 N	11•31 E
54	Luebo	Zaire	5•15 S	21•22 E
54	Lufira (R.)	Zaire	9•32 S	27•15 E
30	Luga	Sov.Un.	58•43 N	29•52 E
26	Lugano	Switz.	46•01 N	8•52 E
55	Lugenda (R.)	Moz.	12•05 S	38•15 E
22	Lugnaquilla, Mt.	Ire.	52•56 N	6•30 W
20	Lugo	It.	44•28 N	11•57 E
17	Lugo	Sp.	43•01 N	7•32 W
21	Lugoj	Rom.	45•51 N	21•56 E
54	Luilaka (R.)	Zaire	2•18 S	21•15 E
22	Luimneach	Ire.	52•39 N	8•35 W
37	Lujchow Pen.	China	20•40 N	110•30 E
54	Lukanga Swp.	Zambia	14•30 S	27•25 E
54	Lukenie (R.)	Zaire	3•10 S	19•05 E
54	Lukolela	Zaire	1•03 S	17•01 E
21	Lukovit	Bul.	43•13 N	24•07 E
27	Luków	Pol.	51•57 N	22•25 E
54	Lukuga (R.)	Zaire	5•50 S	27•35 E

M

Page	Name	Region	Lat.	Long.
35	Mannar, G. of	India	8·47 N	78·33 E
26	Mannheim	F.R.G.	49·30 N	8·31 E
8	Mannington	WV	39·30 N	80·55 W
20	Mannu (R.)	It.	39·32 N	9·03 E
41	Manokwari	Indon.	0·56 S	134·10 E
17	Manresa	Sp.	41·44 N	1·52 E
54	Mansa	Zambia	11·12 S	28·53 E
11	Mansel (I.)	Can.	61·56 N	81·10 W
48	Manseriche, Pongo de	Peru	4·15 S	77·45 W
8	Mansfield	Oh.	40·45 N	82·30 W
9	Mansfield, Mt.	Vt.	44·30 N	72·45 W
49	Manso (R.)	Braz.	13·30 S	51·45 W
48	Manta	Ec.	1·03 S	80·16 W
20	Mantaova (Mantua)	It.	45·09 N	10·47 E
41	Manui (Is.)	Indon.	3·35 S	123·38 E
41	Manus (I.)	Pap.N.Gui.	2·22 S	146·22 E
31	Manych (R.)	Sov.Un.	47·00 N	41·10 E
15	Manych Dep.	Sov.Un.	46·32 N	42·44 E
31	Manych-Gudilo (Lake)	Sov.Un.	46·40 N	42·50 E
48	Manzanares	Col.	5·15 N	75·09 W
13	Manzanillo	Cuba	20·20 N	77·05 W
12	Manzanillo	Mex.	19·02 N	104·21 W
53	Mao	Chad	14·07 N	15·19 E
41	Maoke (Mtn.)	Indon.	4·00 S	138·00 E
39	Maoming	China	21·55 N	110·40 E
41	Mapia (I.)	Indon.	0·57 N	134·22 E
12	Mapimi, Bolsón de (Des.)	Mex.	27·27 N	103·20 W
10	Maple Creek	Can.	49·55 N	109·27 W
55	Mapumulo	S.Afr.	29·12 S	31·05 E
54	Maputo (Lourenço Marques)	Moz.	26·50 S	32·30 E
41	Maqueda Chan.	Phil.	13·40 N	123·52 E
54	Maquela do Zombo	Ang.	6·08 S	15·15 E
50	Mar, Serra do	Braz.	26·30 S	49·15 W
48	Maracaibo	Ven.	10·38 N	71·45 W
48	Maracaibo, Lago de	Ven.	9·55 N	72·13 W
49	Maracay	Ven.	10·15 N	67·35 W
53	Marādah	Libya	29·10 N	19·07 E
52	Maradi	Niger	13·29 N	7·06 E
31	Marāgheh	Iran	37·20 N	46·10 E
55	Maraisburg	S.Afr.	26·12 S	27·57 E
49	Marajó, Ilha de	Braz.	0·30 S	50·00 W
54	Marandelles	Rh.	18·10 S	31·36 E
49	Maranguape	Braz.	3·48 S	38·38 W
49	Maranhão	Braz.	5·15 S	45·52 W
46	Maranoa (R.)	Austl.	27·01 S	148·03 E
48	Marañón, Rio	Peru	4·26 S	75·08 W
49	Marapanim	Braz.	0·45 S	47·42 W
31	Maras	Tur.	37·40 N	36·50 W
11	Marathon	Can.	48·50 N	86·10 W
40	Maratua (I.)	Indon.	2·14 N	118·30 E
53	Marawi	Sud.	18·07 N	31·57 E
44	Marble Bar	Austl.	21·15 S	119·15 E
26	Marburg	F.R.G.	50·49 N	8·46 E
17	Marchena	Sp.	37·20 N	5·25 W
48	Marchena (I.)	Ec.	0·29 N	90·31 W
42	Marcus (I.)	Asia	24·00 N	155·00 E
9	Marcus Hook	Pa.	39·49 N	75·25 W
9	Marcy, Mt	NY	44·10 N	73·55 W
50	Mar del Plata	Arg.	37·59 S	57·35 W
31	Mardin	Tur.	37·25 N	40·40 E
45	Mare (I.)	N.Cal.	21·53 S	168·30 E
22	Maree (L.)	Scot.	57·40 N	5·44 W
12	Margarita	C.Z.	9·20 N	79·55 W
49	Margarita, Isla de	Ven.	11·00 N	64·15 W
22	Margate	Eng.	51·21 N	1·17 E
55	Margate	S.Afr.	30·52 S	30·21 E
53	Margherita Pk	Afr.	0·22 N	29·51 E
30	Maria A.S.S.R.	Sov.Un.	56·20 N	48·00 E
24	Mariager	Den.	56·38 N	10·00 E
24	Mariager Fd	Den.	56·44 N	10·32 E
42	Mariana Is.	Oceania	17·20 N	145·00 E
42	Mariana Trench	Oceania	12·00 N	144·00 E
13	Marianao	Cuba	23·05 N	82·26 W
50	Mariano Acosta	Arg.	34·28 S	58·48 W
12	Marias, Islas	Mex.	21·30 N	106·40 W
24	Maribo	Den.	54·46 N	11·29 E
20	Maribor	Yugo.	46·33 N	15·37 E
41	Maricaban (I.)	Phil.	13·40 N	120·44 E
67	Marie Byrd Land	Ant.	78·00 S	130·00 W
24	Mariefred	Swe.	59·17 N	17·09 E
24	Mariestad	Swe.	58·43 N	13·45 E
8	Marietta	Oh.	39·25 N	81·30 W
25	Marijampole	Sov.Un.	54·33 N	23·26 E
49	Marília	Braz.	22·02 S	49·48 W
41	Marinduque (I.)	Phil.	13·14 N	121·45 E
8	Marine City	Mi.	42·45 N	82·30 W
7	Marinette	Wi.	45·05 N	87·40 W
53	Maringa	Zaire	1·15 N	20·05 E
8	Marion	Il.	37·40 N	88·55 W
8	Marion	In.	40·35 N	85·45 W
8	Marion	Ky.	37·19 N	88·05 W
8	Marion	Oh.	40·35 N	83·10 W
45	Marion Rf.	Austl.	18·57 S	151·31 E
48	Mariquita	Col.	5·13 N	74·52 W
49	Mariscal Estigarribia	Par.	22·03 S	60·28 W
50	Marisco, Ponta do	Braz.	23·01 S	43·17 W
18	Maritime Alps	Fr.-It.	44·20 N	7·02 E
21	Maritsa (R.)	Grc.-Tur.	40·43 N	26·19 E
41	Mariveles	Phil.	14·27 N	120·29 E
33	Marj Uyan	Leb.	33·21 N	35·36 E
36	Marka Kul (L.)	Sov.Un.	49·15 N	85·48 E
24	Markaryd	Swe.	56·30 N	13·34 E
67	Markham, Mt.	Ant.	82·59 S	159·30 E
29	Markovo	Sov.Un.	64·46 N	170·48 E
31	Marks	Sov.Un.	51·40 N	46·40 E
26	Marktredwitz	F.R.G.	50·02 N	12·05 E
9	Marlborough	Ma.	42·21 N	71·33 W
8	Marlette	Mi.	43·25 N	83·05 W
8	Marlinton	WV	38·15 N	80·10 W
21	Marmara (I.)	Tur.	40·38 N	27·35 E
31	Marmara (Sea)	Tur.	40·40 N	28·00 E
17	Marne (R.)	Fr.	49·08 N	3·39 E
48	Maroa	Ven.	2·43 N	67·37 W
55	Maroantsetra	Mad.	15·18 S	49·48 E
48	Maro Jarapeto (Mtn.)	Col.	6·29 N	76·39 W
55	Maromokotro (Mtn.)	Mad.	14·00 S	49·11 E
49	Maroni (R.)	Fr.Gu.-Sur.	3·02 N	53·54 W
43	Marquesas Is.	Fr.Polynesia	8·50 S	141·00 W
7	Marquette	Mi.	46·32 N	87·25 W
53	Marra, Jabal (Mt.)	Sud.	13·00 N	23·47 E
52	Marrakech	Mor.	31·38 N	8·00 W
46	Marree	Austl.	29·38 S	137·55 E
20	Marsala	It.	37·48 N	12·28 E
17	Marseille	Fr.	43·18 N	5·25 E
8	Marseilles	Il.	41·20 N	88·40 W
8	Marshall	Il.	39·20 N	87·40 W
8	Marshall	Mi.	42·20 N	84·55 W
7	Marshall	Tx.	32·33 N	94·22 W
42	Marshall Is.	Pac.Is.Trust Ter.	10·00 N	165·00 E
8	Marshfield	Wi.	44·40 N	90·10 W
24	Marstrand	Swe.	57·54 N	11·33 E
40	Martaban, G. of	Bur.	16·34 N	96·58 E
40	Martapura	Indon.	3·19 S	114·45 E
9	Marthas Vineyard (I.)	Ma.	41·25 N	70·35 W
26	Martigny-Bourg	Switz.	46·06 N	7·00 E
21	Martina Franca	It.	40·43 N	17·21 E
13	Martinique	N.A.	14·50 N	60·40 W
9	Martinsburg	WV	39·30 N	78·00 W
8	Martins Ferry	Oh.	40·05 N	80·45 W
8	Martinsville	In.	39·25 N	86·25 W
10	Martre, Lac la	Can.	63·24 N	119·58 W
24	Mårvatn (L.)	Nor.	60·10 N	8·28 E
28	Mary	Sov.Un.	37·45 N	61·47 E
46	Maryborough	Austl.	25·35 S	152·40 E
46	Maryborough	Austl.	37·00 S	143·50 E
7	Maryland	U.S.	39·10 N	76·25 W
8	Marysville	Oh.	40·15 N	83·25 W
53	Mārzuq	Libya	26·00 N	14·09 E
53	Marzūq, Idehan (Dunes)	Libya	24·30 N	13·00 E
40	Masalembo-Besar (I.)	Indon.	5·40 S	114·28 E
38	Masan	Kor.	35·10 N	128·31 E
55	Masasi	Tan.	10·43 S	38·48 E
41	Masbate	Phil.	12·21 N	123·38 E
41	Masbate (I.)	Phil.	12·19 N	123·03 E
52	Mascara	Alg.	35·25 N	0·08 E
67	Mascarene Is	Mauritius	20·20 S	56·40 E
8	Mascoutah	Il.	38·29 N	89·48 W
54	Maseru	Leso.	29·09 S	27·11 E
34	Mashhad	Iran	36·17 N	59·30 E
53	Mashra'ar-Ragg	Sud.	8·28 N	29·15 E
53	Masindi	Ug.	1·44 N	31·43 E
34	Masjed Soleymān	Iran	31·45 N	49·17 E
22	Mask, Lough	Ire.	53·35 N	9·23 W
8	Mason	Mi.	42·35 N	84·25 W
7	Mason City	Ia.	43·08 N	93·14 W
20	Massa	It.	44·02 N	10·08 E
7	Massachusetts	U.S.	42·20 N	72·30 W
20	Massafra	It.	40·35 N	17·05 E
20	Massa Marittima	It.	43·03 N	10·55 E
9	Massena	NY	44·55 N	74·55 W
10	Masset	Can.	54·02 N	132·09 W
8	Massillon	Oh.	40·50 N	81·35 W
54	Massinga	Moz.	23·18 S	35·18 E
6	Massive, Mt.	Co.	39·05 N	106·30 W
27	Masuria (Reg.)	Pol.	53·40 N	21·10 E
54	Matadi	Zaire	5·49 S	13·27 E
12	Matagalpa	Nic.	12·52 N	85·57 W
11	Matagami (L.)	Can.	50·10 N	78·28 W
52	Matam	Senegal	15·40 N	13·15 W
12	Matamoros	Mex.	25·52 N	97·30 W
11	Matane	Can.	48·51 N	67·32 W
13	Matanzas	Cuba	23·05 N	81·35 W
35	Matara	Sri Lanka	5·59 N	80·35 E
40	Mataram	Indon.	8·45 S	116·15 E
55	Matatiele	S.Afr.	30·21 S	28·49 E
12	Matehuala	Mex.	23·38 N	100·39 W
20	Matera	It.	40·42 N	16·37 E
17	Mateur	Tun.	37·09 N	9·43 E
35	Mathura	India	27·39 N	77·39 E
25	Matiši	Sov.Un.	57·43 N	25·09 E
28	Matochkin Shar	Sov.Un.	73·57 N	56·16 E
49	Mato Grosso	Braz.	15·04 S	59·58 W
49	Mato Grosso (State)	Braz.	14·38 S	55·36 W
49	Mato Grosso, Chapada de (Plain)	Braz.	13·39 S	55·42 W
34	Matrah	Om.	23·36 N	58·27 E
53	Matrūh	Egypt	31·19 N	27·14 E
37	Matsue	Jap.	35·29 N	133·04 E
37	Matsuyama	Jap.	33·48 N	132·45 E
9	Mattaponi (R.)	Va.	37·45 N	77·00 W
26	Matterhorn Mt.	Switz.	45·57 N	7·36 E
8	Mattoon	Il.	39·30 N	88·20 W
48	Maturín	Ven.	9·48 N	63·16 W
41	Mauban	Phil.	14·11 N	121·44 E
49	Maués	Braz.	3·34 S	57·30 W
6	Maui (I.)	Hi.	20·52 N	156·02 W
8	Maumee	Oh.	41·30 N	83·40 W
8	Maumee (R.)	In.-Oh.	41·10 N	84·50 W
8	Maumee B	Oh.	41·50 N	83·20 W
54	Maun	Bots.	19·52 S	23·40 E
6	Mauna Kea (Vol.)	Hi.	19·52 N	155·30 W
6	Mauna Loa (Vol.)	Hi.	19·28 N	155·38 W
40	Maung Nakhon Sawan	Thai.	16·00 N	99·52 E
51	Mauritania	Afr.	19·38 N	13·30 W
67	Mauritius	Afr.	20·18 S	57·36 E
8	Mauston	Wi.	43·46 N	90·05 W
13	Mayagüez	P.R.	18·12 N	67·10 W
26	Mayen	F.R.G.	50·19 N	7·14 E
31	Maykop (Maikop)	Sov.Un.	44·35 N	40·10 E
36	Maymyo	Bur.	22·14 N	96·32 E
10	Mayo	Can.	63·40 N	135·51 W
22	Mayo, Mts. of	Ire.	54·01 N	9·01 W
41	Mayon (Vol.)	Phil.	13·21 N	123·43 E
55	Mayotte (I.)	Fr.	13·07 S	45·32 E
39	Mayraira Pt	Phil.	18·40 N	120·45 E
12	Mayran, Laguna de	Mex.	25·40 N	102·35 W
8	Maysville	Ky.	38·35 N	83·45 W
54	Mayumba	Gabon	3·25 S	10·39 E
9	Mayville	NY	42·15 N	79·30 W
8	Mayville	Wi.	43·30 N	88·45 W
54	Mazabuka	Zambia	15·51 S	27·46 E
49	Mazagão	Braz.	0·05 S	51·27 W
35	Mazār-i-Sharif	Afg.	36·48 N	67·12 E
49	Mazaruni (R.)	Guy.	5·58 N	59·37 W
12	Mazatenango	Guat.	14·30 N	91·30 W
12	Mazatlán	Mex.	23·14 N	106·27 W
25	Mažeikiai	Sov.Un.	56·19 N	22·24 E
33	Mazḫafah, Jabal (Mts.)	Sau.Ar.	28·56 N	35·05 E
20	Mazzara del Vallo	It.	37·40 N	12·37 E
20	Mazzarino	It.	37·16 N	14·15 E
54	Mbabane	Swaz.	26·18 S	31·14 E
53	Mbaïki	Cen.Afr.Emp.	3·53 N	18·00 E
54	Mbala (Abercorn)	Zambia	8·50 S	31·22 E
53	Mbandaka	Zaire	0·04 N	18·16 E
54	Mbanza-Ngungu	Zaire	5·20 S	14·55 E
54	Mbigou	Gabon	2·07 S	11·30 E
53	Mbomou (Bomu) (R.)	Cen.Afr. Emp.-Zaire	4·50 N	23·35 E
52	Mbout	Mauritania	16·03 N	12·31 W
54	Mchinji	Malawi	13·42 S	32·50 E
7	Mead, L.	Az.-Nv.	36·20 N	114·14 W
10	Meadow Lake	Can.	54·08 N	108·26 W
9	Meadville	Pa.	41·40 N	80·10 W
8	Meaford	Can.	44·35 N	80·40 W
11	Mealy Mts.	Can.	53·32 N	57·58 W
46	Meandarra	Austl.	27·47 S	149·40 E
9	Mechanicsburg	Pa.	40·15 N	77·00 W
9	Mechanicville	NY	42·55 N	73·45 W
18	Mecheria	Mor.	33·30 N	0·13 W
26	Mecklenburg	G.D.R.	53·34 N	12·18 E
40	Medan	Indon.	3·35 N	98·35 E
50	Medanosa, Punta	Arg.	47·50 S	65·53 W
48	Medellín	Col.	6·15 N	75·34 W
18	Medenine	Tun.	33·22 N	10·33 E
6	Medford	Or.	42·19 N	122·52 W
8	Medford	Wi.	45·09 N	90·22 W
27	Medias	Rom.	46·09 N	24·21 E
10	Medicine Hat	Can.	50·03 N	110·40 W
9	Medina	NY	43·15 N	78·20 W
8	Medina	Oh.	41·08 N	81·52 W
17	Medina del Campo	Sp.	41·18 N	4·54 W
18	Mediterranean Sea	Afr.-Asia-Eur.	36·22 N	13·25 E
17	Medjerda, Oued (R.)	Tun.	36·43 N	9·54 E
31	Medvedista (R.)	Sov.Un.	50·10 N	43·40 E
30	Medvezhegorsk	Sov.Un.	63·00 N	34·20 E
29	Medvezh'y (Is.)	Sov.Un.	71·00 N	161·25 E
44	Meekatharra	Austl.	26·30 S	118·38 E
26	Meerane	G.D.R.	50·51 N	12·27 E
35	Meerut	India	28·59 N	77·43 E
21	Megalópolis	Grc.	37·22 N	22·08 E
21	Mégara	Grc.	37·59 N	23·21 E
36	Meghelaya	India	25·30 N	91·30 E
39	Meihsien	China	24·20 N	116·10 E
39	Meiling Pass	China	25·22 N	115·00 E
26	Meiningen	G.D.R.	50·35 N	10·25 E
26	Meiringen	Switz.	46·45 N	8·11 E
26	Meissen	G.D.R.	51·11 N	13·28 E
50	Mejillones	Chile	23·07 S	70·31 W
53	Mekele	Eth.	13·31 N	39·19 E
52	Meknès	Mor.	33·56 N	5·44 W
40	Mekong R.	Thai.-Laos	17·53 N	103·57 E
33	Melaka (Malacca)	Mala.	2·11 N	102·15 E
46	Melbourne	Austl.	37·52 S	145·08 E
30	Melekess	Sov.Un.	54·20 N	49·30 E
30	Melenki	Sov.Un.	55·25 N	41·34 E
10	Melfort	Can.	52·52 N	104·36 W
53	Melik, Wadi al (R.)	Sud.	16·48 N	29·30 E
52	Melilla (Sp.)	Afr.	35·24 N	3·30 W
50	Melipilla	Chile	33·40 S	71·12 W
31	Melitopol	Sov.Un.	46·49 N	35·19 E
24	Mellerud	Swe.	58·43 N	12·25 E
55	Melmoth	S.Afr.	28·38 S	31·26 E
50	Melo	Ur.	32·18 S	54·07 W
52	Melrhir Chott (L.)	Alg.	33·52 N	5·22 E
54	Melsetter	Rh.	19·44 S	32·51 E
17	Melun	Fr.	48·32 N	2·40 E
10	Melville	Can.	50·55 N	102·48 W
45	Melville, C.	Austl.	14·15 S	145·50 E
44	Melville (I.)	Austl.	11·30 S	131·12 E
11	Melville (R.)	Can.	53·46 N	59·31 W
10	Melville Hills	Can.	69·18 N	124·57 W
11	Melville Pen	Can.	67·44 N	84·09 W
27	Mélykút	Hung.	46·14 N	19·21 E
55	Memba	Moz.	14·12 S	40·35 E
26	Memmingen	F.R.G.	47·59 N	10·10 E
49	Memo (R.)	Ven.	9·32 N	66·30 W
7	Memphis	Tn.	35·07 N	90·03 W
9	Memphremagog (L.)	Can.	45·05 N	72·10 W
8	Menasha	Wi.	44·12 N	88·29 W
31	Menderes (R.)	Tur.	37·50 N	28·20 E
50	Mendes	Braz.	22·32 S	43·44 W
6	Mendocino, C.	Ca.	40·25 N	124·22 W
8	Mendota	Il.	41·34 N	89·06 W
8	Mendota (L.)	Wi.	43·09 N	89·41 W
50	Mendoza	Arg.	32·48 S	68·45 W
50	Mendoza (Prov.)	Arg.	35·10 S	69·00 W
46	Menindee	Austl.	32·23 S	142·30 E
8	Menominee	Mi.	45·08 N	87·40 W
8	Menominee (R.)	Mi.-Wi.	45·37 N	87·54 W
18	Menorca, Isla de (Minorca)	Sp.	40·05 N	3·58 E
40	Mentawai (Is.)	Indon.	1·08 S	98·10 E
55	Mentz (R.)	S.Afr.	33·13 S	25·15 E
30	Menzelinsk	Sov.Un.	55·40 N	53·15 E
44	Menzies	Austl.	29·45 S	122·15 E
23	Meppel	Neth.	52·41 N	6·08 E
26	Meppen	F.R.G.	52·40 N	7·18 E
20	Merabéllou, Kólpos (G.)	Grc.	35·16 N	25·55 E
20	Merano	It.	46·39 N	11·10 E

Page	Name	Region	Lat.	Long.
41	Merauke	Indon.	8•32 S	140•17 E
50	Mercedes	Arg.	29•04 S	58•01 W
50	Mercedes	Ur.	33•17 S	58•04 W
33	Merchong (R.)	Mala.	3•08 N	103•13 E
11	Mercy, C	Can.	64•48 N	63•22 W
9	Meredith	NH	43•35 N	71•35 W
40	Mergui	Bur.	12•29 N	98•39 E
40	Mergui Arch	Asia	12•04 N	97•02 E
12	Mérida	Mex.	20•58 N	89•37 W
48	Mérida	Ven.	8•30 N	71•15 W
48	Mérida, Cordillera de	Ven.	8•30 N	70•45 W
9	Meriden	Ct.	41•30 N	72•50 W
7	Meridian	Ms.	32•21 N	88•41 W
25	Merikarvia	Fin.	61•51 N	21•30 E
25	Merkiné	Sov.Un.	54•09 N	24•10 E
27	Merkys R	Sov.Un.	54•23 N	25•00 E
50	Merlo	Arg.	34•25 S	58•44 W
8	Merrill	Wi.	45•11 N	89•42 W
9	Merrimack (R.)	Ma.-NH	43•10 N	71•30 W
10	Merritt	Can.	50•07 N	120•47 W
53	Mersa Fatma	Eth.	14•54 N	40•14 E
26	Merseburg	G.D.R.	51•21 N	11•59 E
22	Mersey (R.)	Eng.	53•15 N	2•51 W
31	Mersin	Tur.	37•00 N	34•40 E
33	Mersing	Mala.	2•25 N	103•51 E
22	Merthyr Tydfil	Wales	51•46 N	3•30 W
53	Meru	Ken.	0•01 N	37•45 E
49	Merume Mts.	Guy.	5•45 N	60•15 W
31	Merzifon	Tur.	40•50 N	35•30 E
21	Mesagne	It.	40•34 N	17•51 E
6	Mesa Verde Natl. Park	Co.	37•22 N	108•27 W
53	Mesewa (Massaua)	Eth.	15•40 N	39•19 E
21	Mesolóngion	Grc.	38•23 N	21•28 E
20	Messina	It.	38•11 N	15•34 E
54	Messina	S.Afr.	22•17 S	30•13 E
20	Messina (Str.)	It.	38•10 N	15•34 E
21	Messíni	Grc.	37•05 N	22•00 E
21	Méssiniakós Kólpos (G.)	Grc.	36•59 N	22•00 E
21	Mesta (R.)	Bul.	41•42 N	23•40 E
20	Mestre	It.	45•29 N	12•15 E
48	Meta (Dept.)	Col.	3•28 N	74•07 W
48	Meta (R.)	Col.	4•33 N	72•09 W
50	Metán	Arg.	25•32 S	64•51 W
54	Metangula	Moz.	12•42 S	34•48 E
21	Metkovic	Yugo.	43•02 N	17•40 E
8	Metropolis	Il.	37•09 N	88•46 W
17	Metz	Fr.	49•08 N	6•10 E
23	Meuse (R.)	Eur.	50•32 N	5•22 E
12	Mexicali	Mex.	32•28 N	115•29 W
9	Mexico	Me.	44•34 N	70•33 W
12	Mexico (State)	Mex.	19•50 N	99•50 W
5	Mexico	N.A.	23•45 N	104•00 W
12	Mexico, G. of	N.A.	25•15 N	93•45 W
12	Mexico City	Mex.	19•28 N	99•09 W
9	Meyersdale	Pa.	39•55 N	79•00 W
34	Meymaneh	Afg.	35•53 N	64•38 E
30	Mezen	Sov.Un.	65•50 N	44•05 E
30	Mezen (R.)	Sov.Un.	65•20 N	44•45 E
27	Mezökövesd	Hung.	47•49 N	20•36 E
27	Mezötur	Hung.	47•00 N	20•36 E
55	Mgeni (R.)	S.Afr.	29•38 S	30•53 E
6	Miami	Az.	33•20 N	110•55 W
7	Miami	Fl.	25•45 N	80•11 W
8	Miami (R.)	Oh.	39•20 N	84•45 W
8	Miamisburg	Oh.	39•40 N	84•20 W
34	Miāneh	Iran	37•15 N	47•13 E
41	Miangas (I.)	Phil.	5•30 N	127•00 E
39	Miaoli	Taiwan	24•30 N	120•48 E
26	Miastko	Pol.	54•01 N	17•00 E
27	Michalovce	Czech.	48•44 N	21•56 E
7	Michigan	U.S.	45•55 N	87•00 W
7	Michigan, L.	U.S.	43•20 N	87•10 W
8	Michigan City	In.	41•40 N	86•55 W
11	Michikamau (L.)	Can.	54•11 N	63•21 W
11	Michipicoten (I.)	Can.	47•49 N	85•50 W
12	Michoacán (State)	Mex.	19•15 N	101•30 W
31	Michurinsk	Sov.Un.	52•53 N	40•32 E
54	Middleburg	S.Afr.	31•30 S	25•00 E
40	Middle Andaman I.	Andaman & Nicobar Is.	12•44 N	93•21 E
9	Middlebury	Vt.	44•00 N	73•10 W
24	Middlefart	Den.	55•30 N	9•45 E
8	Middleport	Oh.	39•00 N	82•05 W
22	Middlesbrough (Teesside)	Eng.	54•35 N	1•18 W
9	Middletown	Ct.	41•35 N	72•40 W
9	Middletown	De.	39•30 N	75•40 W
9	Middletown	NY	41•26 N	74•25 W
8	Middletown	Oh.	39•30 N	84•25 W
17	Midi, Canal du	Fr.	43•22 N	1•35 E
55	Mid Illovo	S.Afr.	29•59 S	30•32 E
9	Midland	Can.	44•45 N	79•50 W
8	Midland	Mi.	43•40 N	84•20 W
42	Midway Is.	Pac.O.	28•00 N	179•00 W
31	Midye	Tur.	41•35 N	28•10 E
26	Miedzyrzecz	Pol.	52•26 N	15•35 E
27	Mielec	Pol.	50•17 N	21•27 E
31	Migorod	Sov.Un.	49•56 N	33•36 E
50	Miguel Pereira	Braz.	22•27 S	43•28 W
30	Mikhaylov	Sov.Un.	54•14 N	39•03 E
31	Mikhaylovka	Sov.Un.	50•05 N	43•10 E
55	Mikindani	Tan.	10•17 S	40•07 E
25	Mikkeli	Fin.	61•42 N	27•14 E
21	Míkonos (I.)	Grc.	37•26 N	25•30 E
26	Mikulov	Czech.	48•47 N	16•39 E
8	Milan	Mi.	42•05 N	83•40 W
20	Milan	It.	45•29 N	9•12 E
31	Milâs	Tur.	37•10 N	27•25 E
20	Milazzo	It.	38•13 N	15•17 E
46	Mildura	Austl.	34•10 S	142•18 E
6	Miles City	Mt.	46•24 N	105•50 W
9	Milford	Ct.	41•15 N	73•05 W
9	Milford	De.	38•55 N	75•25 W
9	Milford	Ma.	42•09 N	71•31 W
9	Milford	NH	42•50 N	71•40 W
22	Milford Haven	Wales	51•40 N	5•10 W
44	Miling	Austl.	30•30 S	116•25 E
10	Milk River	Can.	49•09 N	112•05 W
6	Milk R.	Can.-U.S.	48•25 N	108•45 W
31	Millerovo	Sov.Un.	48•58 N	40•27 E
8	Millersburg	Ky.	38•15 N	84•10 W
8	Millersburg	Oh.	40•35 N	81•55 W
9	Millersburg	Pa.	40•35 N	76•55 W
46	Millicent	Austl.	37•30 S	140•20 E
44	Millstream	Austl.	21•45 S	117•10 E
9	Millville	NJ	39•25 N	75•00 W
54	Milnerton	S.Afr.	33•52 S	18•30 E
21	Milos (Milo) (I.)	Grc.	36•45 N	24•35 E
9	Milton	Pa.	41•00 N	76•50 W
8	Milwaukee	Wi.	43•03 N	87•55 W
33	Minas	Indon.	0•52 N	101•29 E
50	Minas	Ur.	34•18 S	55•12 W
17	Minas de Ríontinto	Sp.	37•43 N	6•35 W
49	Minas Gerais (State)	Braz.	17•45 S	43•50 W
49	Minas Nova	Braz.	17•20 S	42•19 W
12	Minatitlan	Mex.	17•59 N	94•33 W
22	Minch, The (Chan.)	Scot.	58•04 N	6•04 W
22	Minch, The Little (Chan.)	Scot.	57•35 N	6•45 W
39	Min Chiang (R.)	China	26•30 N	118•30 E
39	Min Chiang (R.)	China	29•30 N	104•00 E
41	Mindanao (I.)	Phil.	7•30 N	125•10 E
41	Mindanao Sea	Phil.	8•55 N	124•00 E
52	Mindelo	C.V.Is.	16•53 N	25•00 W
26	Minden	F.R.G.	52•17 N	8•58 E
41	Mindoro (I.)	Phil.	13•04 N	121•06 E
41	Mindoro Str.	Phil.	12•28 N	120•33 E
9	Mineola	NY	40•43 N	73•38 W
31	Mineral'nyye Vody	Sov.Un.	44•10 N	43•15 E
8	Mineral Point	Wi.	42•50 N	90•10 W
8	Minerva	Oh.	40•45 N	81•10 W
20	Minervino	It.	41•07 N	16•05 E
11	Mingan	Can.	50•18 N	64•02 W
31	Mingechaur (R.)	Sov.Un.	41•00 N	47•20 E
44	Mingenew	Austl.	29•15 S	115•45 E
8	Mingo Junction	Oh.	40•15 N	80•40 W
17	Minho, Rio	Port.	41•28 N	9•05 W
52	Minna	Nig.	9•37 N	6•33 E
7	Minneapolis	Mn.	44•58 N	93•15 W
10	Minnedosa	Can.	50•14 N	99•51 W
7	Minnesota	U.S.	46•10 N	90•20 W
7	Minnesota (R.)	Mn.	45•04 N	96•03 W
8	Minonk	Il.	40•55 N	89•00 W
6	Minot	ND	48•13 N	101•16 W
30	Minsk	Sov.Un.	53•54 N	27•35 E
27	Mińsk Mazowiecki	Pol.	52•10 N	21•35 E
11	Minto (L.)	Can.	57•18 N	75•50 W
20	Minturno	It.	41•17 N	13•44 E
29	Minusinsk	Sov.Un.	53•47 N	91•45 E
36	Minya Konka (Mt.)	China	29•16 N	101•46 E
27	Mir	Sov.Un.	53•27 N	26•25 E
49	Mirador	Braz.	6•19 S	44•12 W
48	Miraflores	Col.	5•10 N	73•13 W
48	Miraflores	Peru	16•19 S	71•20 W
12	Miraflores Locks	C.Z.	9•00 N	79•35 W
48	Miranda	Col.	3•14 N	76•11 W
49	Miranda	Ven.	10•09 N	68•24 W
49	Miranda (State)	Ven.	10•17 N	66•41 W
34	Mirbāt	Om.	16•58 N	54•42 E
40	Miri	Mala.	4•13 N	113•56 E
50	Mirim, L.	Braz.-Ur.	33•00 S	53•15 W
35	Mirzāpur	India	25•12 N	82•38 E
8	Mishawaka	In.	41•45 N	86•15 W
50	Misiones (Prov.)	Arg.	27•00 S	54•30 W
27	Miskolc	Hung.	48•07 N	20•50 E
41	Misol (I.)	Indon.	2•00 S	130•05 E
53	Misrātah	Libya	32•23 N	14•58 E
11	Missinaibi (R.)	Can.	50•27 N	83•01 W
8	Mississinewa (R.)	In.	40•30 N	85•45 W
7	Mississippi	U.S.	32•30 N	89•45 W
9	Mississippi (L.)	Can.	45•05 N	76•15 W
7	Mississippi (R.)	U.S.	31•50 N	91•30 W
6	Missoula	Mt.	46•52 N	114•00 W
7	Missouri	U.S.	38•00 N	93•40 W
7	Missouri (R.)	U.S.	40•40 N	96•00 W
6	Missouri Coteau, (Plat.)	U.S.	47•30 N	101•00 W
11	Mistassibi (R.)	Can.	49•44 N	69•58 W
11	Mistassini (L.)	Can.	50•48 N	73•30 W
11	Mistassini (R.)	Can.	50•02 N	72•38 W
26	Mistelbach	Aus.	48•34 N	16•33 E
20	Mistretta	It.	37•54 N	14•22 E
8	Mitchell	In.	38•45 N	86•25 W
6	Mitchell	SD	43•42 N	98•01 W
45	Mitchell (R.)	Austl.	15•30 S	142•15 E
7	Mitchell, Mt.	NC	35•47 N	82•15 W
21	Mitilíni	Grc.	39•09 N	26•35 E
33	Mitla P.	Egypt	30•03 N	32•40 E
26	Mittelland (can.)	G.D.R.	52•18 N	10•42 E
26	Mittweida	G.D.R.	50•59 N	12•58 E
18	Mizdah	Libya	31•29 N	13•09 E
21	Mizil	Rom.	45•01 N	26•30 E
35	Mizoram (Union Ter.)	India	23•25 N	92•45 E
24	Mjölby	Swe.	58•20 N	15•09 E
24	Mjörn (L.)	Swe.	57•55 N	12•22 E
24	Mjösa	Nor.	60•41 N	11•25 E
24	Mjösvatn	Nor.	59•55 N	7•50 E
54	Mkalama	Tan.	4•07 S	34•38 E
26	Mladá Boleslav	Czech.	50•26 N	14•52 E
27	Mława	Pol.	53•07 N	20•25 E
55	Mlazi (R.)	S.Afr.	29•52 S	30•42 E
21	Mljet (I.)	Yugo.	42•40 N	17•45 E
41	Moa (I.)	Indon.	8•30 S	128•30 E
54	Moanda	Gabon	1•37 S	13•09 E
53	Mobaye	Cen.Afr.Emp.	4•19 N	21•11 E
7	Moberly	Mo.	39•24 N	92•25 W
7	Mobile	Al	30•42 N	88•03 W
7	Mobile B.	Al.	30•26 N	87•56 W
55	Moçambique	Moz.	15•03 S	40•42 E
54	Moçâmedes	Ang.	15•10 S	12•09 E
54	Moçâmedes (Reg.)	Ang.	16•00 S	12•15 E
34	Mocha	Yemen	13•11 N	43•20 E
54	Mochudi	Bots.	24•13 S	26•07 E
55	Mocímboa da Praia	Moz.	11•20 S	40•21 E
55	Modderfontein	S.Afr.	26•06 S	28•10 E
20	Modena	It.	44•38 N	10•54 E
17	Modica	It.	36•50 N	14•43 E
49	Moengo	Sur.	5•43 N	54•19 W
51	Mogadisho	Som.	2•08 N	45•22 E
36	Mogaung	Bur.	25•30 N	96•52 E
49	Mogi das Cruzes	Braz.	23•33 S	46•10 W
30	Mogilëv	Sov.Un.	53•53 N	30•22 E
31	Mogilëv-Podol'skiy	Sov.Un.	48•27 N	27•51 E
27	Mogilno	Pol.	52•38 N	17•58 E
36	Mogok	Bur.	23•14 N	96•38 E
6	Mogollon, Plat.	Az.	34•26 N	117•17 W
27	Mohács	Hung.	45•59 N	18•38 E
55	Mohale's Hoek	Leso.	30•09 S	27•28 E
9	Mohawk (R.)	NY	43•15 N	75•20 W
55	Moheli (I.)	Comoros	12•23 S	43•38 E
35	Mohenjo-Dero (Ruins)	Pak.	27•20 N	68•10 E
37	Moho	China	53•33 N	122•30 E
16	Mo-i-Rana	Nor.	65•54 N	13•15 E
25	Mõisaküla	Sov.Un.	58•07 N	25•12 E
11	Moisie (R.)	Can.	50•35 N	66•25 W
6	Mojave Desert	Ca.	35•05 N	117•30 W
55	Mokhotlong	Leso.	29•18 S	29•06 E
38	Mokp'o	Kor.	34•50 N	126•30 E
30	Moksha (R.)	Sov.Un.	54•50 N	43•20 E
20	Molat (I.)	Yugo.	44•15 N	14•40 E
27	Moldavia	Rom.	47•20 N	27•12 E
28	Moldavian S.S.R.	Sov.Un.	48•00 N	28•00 E
24	Molde	Nor.	62•44 N	7•15 E
24	Molde Fd.	Nor.	62•40 N	7•05 E
27	Moldova R.	Rom.	47•17 N	26•27 E
54	Molepolole	Bots.	24•15 S	25•33 E
20	Molfetta	It.	41•11 N	16•38 E
17	Molina de Aragón	Sp.	41•40 N	1•54 W
54	Moliro	Zaire	8•13 S	30•34 E
20	Moliterno	It.	40•13 N	15•54 E
48	Mollendo	Peru	17•02 S	71•59 W
24	Mölndal	Swe.	57•39 N	12•01 E
30	Molodechno	Sov.Un.	54•18 N	26•57 E
30	Mologa (R.)	Sov.Un.	58•05 N	35•43 E
6	Molokai (I.)	Hi.	21•15 N	157•05 E
54	Molopo (R.)	S.Afr.	27•45 S	20•45 E
55	Molteno	S.Afr.	31•24 S	26•23 E
55	Mombasa	Ken.	4•03 S	39•40 E
8	Momence	Il.	41•09 N	87•40 W
41	Mompog Pass	Phil.	13•35 N	122•09 E
48	Mompos	Col.	8•05 N	74•30 W
24	Møn (I.)	Den.	54•54 N	12•30 E
17	Monaco	Eur.	43•43 N	7•47 E
22	Monaghan	Ire.	54•16 N	7•20 W
13	Mona Pass.	N.A.	18•00 N	68•10 W
17	Monastir	Tun.	35•49 N	10•56 E
49	Monção	Braz.	3•39 S	45•23 W
30	Monchegorsk	Sov.Un.	69•00 N	33•35 E
26	Mönchengladbach	F.R.G.	51•12 N	6•28 E
12	Monclovra	Mex.	26•53 N	101•25 W
11	Moncton	Can.	46•06 N	64•47 W
54	Mondombe	Zaire	0•45 S	23•06 E
20	Mondoví	It.	44•23 N	7•53 E
53	Mongala R.	Zaire	3•20 N	21•30 E
53	Mongalla	Sud.	5•11 N	31•46 E
35	Monghyr	India	25•23 N	86•34 E
32	Mongolia	Asia	46•00 N	100•00 E
53	Mongos, Chaîne des (Mts.)	Cen.Afr.Emp.	8•04 N	21•59 E
54	Mongu	Zambia	15•15 S	23•09 E
54	Monkoto	Zaire	1•38 S	20•39 E
8	Monon	In.	40•55 N	86•55 W
9	Monongah	WV	39•25 N	80•10 W
9	Monongahela	Pa.	40•11 N	79•55 W
9	Monongahela (R.)	WV	39•30 N	80•10 W
21	Monopoli	It.	40•55 N	17•17 E
20	Monreale	It.	38•04 N	13•15 E
7	Monroe	La.	32•30 N	92•06 W
8	Monroe	Mi.	41•55 N	83•25 W
8	Monroe	Wi.	42•35 N	89•40 W
52	Monrovia	Lib.	6•18 N	10•47 W
23	Mons	Bel.	50•29 N	3•55 E
24	Mönsterås	Swe.	57•04 N	16•24 E
36	Montagh Ata (Mt.)	China	38•26 N	75•23 E
7	Montagne Tremblante Prov. Pk.	Can.	46•30 N	75•51 W
8	Montague	Mi.	43•30 N	86•25 W
41	Montalban	Phil.	14•47 N	12•11 E
49	Montalbán	Ven.	10•14 N	68•19 W
20	Montalcone	It.	45•49 N	13•30 E
6	Montana	U.S.	47•10 N	111•50 W
17	Montargis	Fr.	47•59 N	2•42 E
17	Montauban	Fr.	44•01 N	1•22 E
9	Montauk	NY	41•03 N	71•57 W
9	Montauk Pt.	NY	41•05 N	71•55 W
17	Montcalm, Pic de	Fr.	42•43 N	1•13 E
17	Mont-de-Marsan	Fr.	43•54 N	0•32 W
48	Monteagudo	Bol.	19•49 S	63•48 W
44	Monte Bello (Is.)	Austl.	20•30 S	114•10 E
50	Monte Caseros	Arg.	30•16 S	57•39 W
20	Montecristo, I. di	It.	42•20 N	10•19 E
13	Montego Bay	Jam.	18•30 N	77•55 W
50	Monte Grande	Arg.	34•34 S	58•28 W
17	Montélimar	Fr.	44•33 N	4•47 E
8	Montello	Wi.	43•47 N	89•20 W
12	Montemorelos	Mex.	25•14 N	99•50 W
20	Montepulciano	It.	43•05 N	11•48 E
6	Monterey	Ca.	36•36 N	121•53 W
6	Monterey B.	Ca.	36•48 N	122•01 W
48	Montería	Col.	8•47 N	75•57 W
50	Monteros	Arg.	27•14 S	65•29 W
12	Monterrey	Mex.	25•43 N	100•19 W
20	Monte Sant' Angelo	It.	41•43 N	15•59 E
49	Montes Claros	Braz.	16•44 S	43•41 W
20	Montevarchi	It.	43•30 N	11•45 E
50	Montevideo	Ur.	34•50 S	56•10 W
7	Montgomery	Al.	32•23 N	86•17 W
8	Montgomery	WV	38•10 N	81•25 W
8	Monticello	Il.	40•05 N	88•35 W
8	Monticello	In.	40•40 N	86•50 W
9	Monticello	NY	41•35 N	74•40 W
11	Mont-Joli	Can.	48•35 N	68•11 W
17	Montluçon	Fr.	46•20 N	2•35 E
20	Montone (R.)	It.	44•03 N	11•45 E
8	Montpelier	In.	40•35 N	85•20 W
8	Montpelier	Oh.	41•35 N	84•35 W
9	Montpelier	Vt.	44•20 N	72•35 W
17	Montpellier	Fr.	43•38 N	3•53 E
11	Montréal	Can.	45•30 N	73•35 W

Page	Name	Region	Lat.	Long.
26	Montreux	Switz.	46•26 N	6•52 E
9	Montrose	Pa.	41•50 N	75•50 W
22	Montrose	Scot.	56•45 N	2•25 W
13	Montserrat	N.A.	16•48 N	63•15 W
40	Monywa	Bur.	22•02 N	95•16 E
20	Monza	It.	45•34 N	9•17 E
55	Mooi (R.)	S.Afr.	29•00 S	30•15 E
55	Mooirivier	S.Afr.	29•14 S	29•59 E
46	Moonta	Austl.	34•05 S	137•42 E
44	Moora	Austl.	30•35 S	116•12 E
44	Moore (L.)	Austl.	29•50 S	128•12 E
9	Moore Res.	Vt.-NH	44•20 N	72•10 W
8	Mooresville	In.	39•37 N	86•22 W
10	Moose (L.)	Can.	54•14 N	99•28 W
11	Moose (R.)	Can.	51•01 N	80•42 W
10	Moose Jaw	Can.	50•23 N	105•32 W
9	Moosilauke (Mtn.)	NH	44•00 N	71•50 W
11	Moosonee	Can.	51•20 N	80•44 W
52	Mopti	Mali	14•30 N	4•12 W
48	Moquegua	Peru	17•15 S	70•54 W
27	Mór	Hung.	47•51 N	18•14 E
35	Moradabad	India	28•57 N	78•48 E
55	Moramanga	Mad.	18•48 S	48•09 E
24	Morastrand	Swe.	61•00 N	14•29 E
27	Morava (Moravia)	Czech.	49•21 N	16•57 E
26	Morava R.	Czech.	49•53 N	16•53 E
49	Morawhanna	Guy.	8•12 N	59•33 W
22	Moray Firth	Scot.	57•41 N	3•55 W
24	Mörbylånga	Swe.	56•32 N	16•23 E
10	Morden	Can.	49•11 N	98•05 W
30	Mordvin A.S.S.R.	Sov.Un.	54•18 N	43•50 E
22	More, Ben (Mtn.)	Scot.	58•09 N	5•01 W
22	Morecambe B.	Eng.	53•55 N	3•25 W
46	Moree	Austl.	29•20 S	149•50 E
8	Morehead	Ky.	38•10 N	83•25 W
12	Morelia	Mex.	19•43 N	101•12 W
17	Morena, Sierra	Sp.	38•15 N	5•45 W
8	Morenci	Mi.	41•50 N	84•50 W
50	Moreno	Arg.	34•25 S	58•47 W
10	Moresby I.	Can.	52•50 N	131•55 W
46	Moreton (I.)	Austl.	26•53 S	152•42 E
46	Moreton B.	Austl.	27•12 S	153•10 E
8	Morganfield	Ky.	37•40 N	87•55 W
55	Morgan's Bay	S.Afr.	32•42 S	28•19 E
9	Morgantown	WV	39•40 N	79•55 W
35	Morga Ra.	Afg.	34•02 N	70•38 E
37	Morioka	Jap.	39•40 N	141•21 E
41	Morobe	Pap.N.Gui.	8•03 S	147•45 E
51	Morocco	Afr.	32•00 N	7•00 W
55	Morombe	Mad.	21•39 S	43•34 E
50	Morón	Arg.	34•24 S	58•37 W
49	Morón	Ven.	10•29 N	68•11 W
55	Morondava	Mad.	20•17 S	44•18 E
55	Moroni	Comoros	11•41 S	43•16 E
41	Morotai (I.)	Indon.	2•12 N	128•30 E
31	Morozovsk	Sov.Un.	48•20 N	41•50 E
49	Morrinhos	Braz.	17•45 S	48•56 W
10	Morris	Can.	49•21 N	97•22 W
8	Morris	Il.	41•20 N	88•25 W
8	Morrison	Il.	41•48 N	89•58 W
49	Morro do Chapéu	Braz.	11•34 S	41•03 W
31	Morshansk	Sov.Un.	53•25 N	41•35 E
24	Mors (I.)	Den.	56•46 N	8•38 E
20	Mortara	It.	45•13 N	8•47 E
50	Morteros	Arg.	30•47 S	62•00 W
30	Morzhovets (I.)	Sov.Un.	66•40 N	42•30 E
6	Moscow	Id.	46•44 N	116•57 W
26	Mosel R.	F.R.G.	49•49 N	7•00 E
25	Moshchnyy (Is.)	Sov.Un.	59•56 N	28•07 E
55	Moshi	Tan.	3•21 S	37•20 E
16	Mosjøen	Nor.	65•50 N	13•10 E
30	Moskva (Moscow)	Sov.Un.	55•45 N	37•37 E
30	Moskva (R.)	Sov.Un.	55•50 N	37•05 E
27	Mosonmagyaróvár	Hung.	47•51 N	17•16 E
13	Mosquitos, Gulfo de los	Pan.	9•17 N	80•59 W
24	Moss	Nor	59•29 N	10•39 E
54	Mosselbaai	S.Afr.	34•06 S	22•23 E
49	Mossoró	Braz.	5•13 S	37•14 W
26	Most	Czech.	50•32 N	13•37 E
52	Mostaganem	Alg.	36•04 N	0•11 E
21	Mostar	Yugo.	43•20 N	17•51 E
24	Motala	Swe.	58•34 N	15•00 E
22	Motherwell	Scot.	55•45 N	4•05 W
17	Motril	Sp.	36•44 N	3•32 W
54	Mouille Pt.	S.Afr.	33•54 S	18•19 E
17	Moulins	Fr.	46•34 N	3•19 E
40	Moulmein	Bur.	16•30 N	97•39 E
18	Moulouya, Oued (R.)	Mor.	34•07 N	3•27 W
8	Mound City	Il.	37•06 N	89•13 W
8	Mound City Group Natl. Mon.	Oh.	39•25 N	83•00 W
8	Moundsville	WV	39•50 N	80•50 W
10	Mountain Park	Can.	52•55 N	117•14 W
55	Mount Ayliff	S.Afr.	30•48 S	29•24 E
8	Mount Carmel	Il.	38•25 N	87•45 W
8	Mount Carmel	Pa.	40•50 N	76•25 W
8	Mount Carroll	Il.	42•05 N	89•55 W
8	Mount Clemens	Mi.	42•36 N	82•52 W
55	Mount Fletcher	S.Afr.	30•42 S	28•32 E
8	Mount Forest	Can.	44•00 N	80•45 W
55	Mount Frere	S.Afr.	30•54 S	29•02 E
46	Mount Gambier	Austl.	37•30 S	140•53 E
8	Mount Gilead	Oh.	40•30 N	82•50 W
8	Mount Hope	WV	37•55 N	81•10 W
44	Mount Isa	Austl.	21•00 S	139•45 E
6	Mount McKinley Natl. Park	Ak.	63•48 N	153•02 W
44	Mount Magnet	Austl.	28•00 S	118•00 E
45	Mount Morgan	Austl.	23•42 S	150•45 E
8	Mount Morris	Mi.	43•10 N	83•45 W
9	Mount Morris	NY	42•45 N	77•50 W
8	Mount Pleasant	Mi.	43•35 N	84•45 W
6	Mount Rainier Natl. Park	Wa.	46•47 N	121•17 W
10	Mount Revelstoke Natl. Park	Can.	51•22 N	120•15 W
9	Mount Savage	Md.	39•45 N	78•55 W
8	Mount Sterling	Ky.	38•05 N	84•00 W
9	Mount Union	Pa.	40•25 N	77•50 W
8	Mount Vernon	Il.	38•20 N	88•50 W
8	Mount Vernon	In.	37•55 N	87•50 W
9	Mount Vernon	NY	40•55 N	73•51 W
8	Mount Vernon	Oh.	40•25 N	82•30 W
9	Mount Vernon	Va.	38•43 N	77•06 W
49	Moura	Braz.	1•33 S	61•38 W
22	Mourne, Mts.	N.Ire.	54•10 N	6•09 W
46	Mowbullan, Mt.	Austl.	26•50 S	151•34 E
53	Moyale	Ken.	3•28 N	39•04 E
52	Moyamba	S.L.	8•10 N	12•26 W
18	Moyen Atlas (Mts.)	Mor.	32•49 N	5•28 W
48	Moyobamba	Peru	6•12 S	76•56 W
51	Mozambique	Afr.	20•15 S	33•53 E
55	Mozambique Chan.	Afr.	24•00 S	38•00 E
31	Mozdok	Sov.Un.	43•45 N	44•35 E
31	Mozyr	Sov.Un.	52•03 N	29•14 E
54	Mporokoso	Zambia	9•23 S	30•05 E
55	Mpwapwa	Tan.	6•21 S	36•29 E
55	Mqanduli	S.Afr.	31•50 S	28•42 E
27	Mragowo	Pol.	53•52 N	21•18 E
52	M'sila	Alg.	35•47 N	4•34 E
55	Mtamvuna (R.)	S.Afr.	30•43 S	29•53 E
55	Mtata (R.)	S.Afr.	31•48 S	29•03 E
31	Mtsensk	Sov.Un.	53•17 N	36•33 E
40	Muang Kohn Kaen	Thai.	16•37 N	102•41 E
40	Muang Lamphum	Thai.	18•40 N	98•59 E
40	Muang Phitsanulok	Thai.	16•51 N	100•15 E
40	Muang Sakon	Thai.	17•00 N	104•06 E
33	Muar (R.)	Mala.	2•18 N	102•43 E
49	Mucugê	Braz.	13•02 S	41•19 W
46	Mudgee	Austl.	32•47 S	149•10 E
31	Muğla	Tur.	37•10 N	28•20 E
26	Mühldorf	F.R.G.	48•15 N	12•33 E
26	Mühlhausen	G.D.R.	51•13 N	10•25 E
25	Muhu (I.)	Sov.Un.	58•41 N	22•55 E
39	Mui Ron, C.	Viet.	18•05 N	106•45 E
54	Muizenberg	S.Afr.	34•07 S	18•28 E
27	Mukachëvo	Sov.Un.	48•25 N	22•43 E
29	Mukhtuya	Sov.Un.	61•00 N	113•00 E
26	Mulde R.	G.D.R.	50•30 N	12•30 E
38	Muleng	China	44•32 N	130•18 E
38	Muleng (R.)	China	44•40 N	130•30 E
45	Mulgrave (I.)	Austl.	10•08 S	142•14 E
23	Mülheim	F.R.G.	51•25 N	6•53 E
17	Mulhouse	Fr.	47•46 N	7•20 E
22	Mull (I.)	Scot.	56•40 N	6•19 W
40	Müller (Mts.)	Indon.	0•22 N	113•05 E
22	Mullet Pen	Ire.	54•15 N	10•12 W
22	Mullinger	Ire.	53•31 N	7•26 W
35	Multān	Pak.	30•17 N	71•13 E
40	Mulu (Mtn.)	Mala.	3•56 N	115•11 E
54	Mumbwa	Zambia	14•59 S	27•04 E
26	München (Munich)	F.R.G.	48•08 N	11•35 E
8	Muncie	In.	40•10 N	85•30 W
48	Mundonueva, Pico de	Col.	4•18 N	74•12 W
45	Mungana	Austl.	17•15 S	144•18 E
46	Mungindi	Austl.	32•00 S	148•45 E
54	Munhango	Ang.	12•15 S	18•55 E
29	Munku Sardyk (Mtn.)	Sov.Un.-Mong.	51•45 N	100•30 E
41	Muños	Phil.	15•44 N	120•53 E
22	Munster	Ire.	52•30 N	9•24 W
40	Muntok	Indon.	2•05 S	105•11 E
40	Muong Sing	Laos	21•06 N	101•17 E
16	Muonio (R.)	Fin.-Swe.	68•15 N	23•00 E
31	Muradiye	Tur.	39•00 N	43•40 E
31	Murat (R.)	Tur.	38•50 N	40•40 E
44	Murchison (R.)	Austl.	26•45 S	116•15 E
53	Murchison Falls	Ug.	2•15 N	31•41 E
17	Murcia	Sp.	38•00 N	1•10 W
27	Mureşul R.	Rom.	46•02 N	21•50 E
26	Müritz See (L.)	G.D.R.	53•20 N	12•33 E
36	Murku Sardyk (Pk.)	Sov.Un.-Mong.	51•56 N	100•21 E
30	Murmansk	Sov.Un.	69•00 N	33•20 E
30	Murom	Sov.Un.	55•30 N	42•00 E
37	Muroran	Jap.	42•21 N	141•05 E
46	Murray Bridge	Austl.	35•10 S	139•35 E
45	Murray Reg.	Austl.	33•20 S	142•30 E
46	Murray R.	Austl.	34•20 S	142•21 W
26	Mur R.	Aus.	47•10 N	14•08 E
46	Murrumbidgee (R.)	Austl.	34•30 S	145•20 E
20	Murska Sobota	Yugo.	46•40 N	16•14 E
35	Murwāra	India	23•54 N	80•23 E
46	Murwillumbah	Austl.	28•15 S	153•30 E
26	Mürz R.	Aus.	47•30 N	15•21 E
26	Murzzuschlag	Aus.	47•37 N	15•41 E
31	Mus	Tur.	38•55 N	41•30 E
21	Musala (Mtn.)	Bul.	42•05 N	23•24 E
34	Muscat	Om.	23•23 N	58•30 E
44	Musgrave Ra.	Austl.	26•15 S	131•15 E
54	Mushie	Zaire	3•04 S	16•50 E
40	Musi (Strm.)	Indon.	2•40 S	103•42 E
48	Musinga, Alto	Col.	6•40 N	76•13 W
8	Muskegon	Mi.	43•15 N	86•20 W
8	Muskegon (R.)	Mi.	43•20 N	85•55 W
8	Muskegon Heights	Mi.	43•10 N	86•20 W
8	Muskingum (R.)	Oh.	39•45 N	81•55 W
7	Muskogee	Ok.	35•44 N	95•21 W
9	Muskoka (L.)	Can.	45•00 N	79•30 W
41	Mussau (I.)	Pap.N.Gui.	1•30 S	149•32 E
22	Musselburgh	Scot.	55•55 N	3•08 W
31	Mustafakemalpasa	Tur.	40•05 N	28•30 E
37	Musu Dan (C.)	Kor.	40•51 N	130•00 E
46	Muswellbrook	Austl.	32•15 S	150•50 E
38	Mutan (R.)	China	45•30 N	129•40 E
38	Mutanchiang	China	44•28 N	129•38 E
54	Mutombo Mukulu	Zaire	8•12 S	23•56 E
28	Muyun-Kum (Des.)	Sov.Un.	44•30 N	70•00 E
55	Mvoti (R.)	S.Afr.	29•18 S	30•52 E
55	Mwanza	Tan.	2•31 S	32•54 E
54	Mwaya	Tan.	9•19 S	33•51 E
54	Mweru (L.)	Zaire-Zambia	8•50 S	28•50 E
18	Mya R.	Alg.	29•26 N	3•15 E
36	Myingyan	Bur.	21•37 N	95•26 E
40	Myinmoletkat (Pk.)	Bur.	13•58 N	98•34 E
36	Myitkyina	Bur.	25•33 N	97•25 E
27	Myjava	Czech.	48•45 N	17•33 E
38	Myohyang San (Mtn.)	Kor.	40•00 N	126•12 E
16	Mýrdalsjökull	Ice.	63•34 N	18•04 W
35	Mysore	India	12•31 N	76•42 E
25	Mysovka	Sov.Un.	55•11 N	21•17 E
54	Mzimba	Malawi	11•52 S	33•34 E
55	Mzimkulu (R.)	S.Afr.	30•12 S	29•57 E
55	Mzimvubu (R.)	S.Afr.	31•22 S	29•20 E

N

Page	Name	Region	Lat.	Long.
26	Naab R.	F.R.G.	49•38 N	12•15 E
25	Naantali	Fin.	60•29 N	22•03 E
44	Nabberu (L.)	Austl.	26•05 S	120•35 E
52	Nabeul	Tun.	36•34 N	10•45 E
33	Nābulus	Jordan	32•13 N	35•16 E
55	Nacala	Moz.	14•34 S	40•41 E
18	Naceur, Bou (Mt.)	Mor.	33•50 N	3•55 W
39	Na Cham	Viet.	22•02 N	106•30 E
26	Náchod	Czech.	50•25 N	16•08 E
13	Nadir	Vir.Is.	18•19 N	64•53 W
21	Nădlac	Rom.	46•09 N	20•52 E
27	Nadvornaya	Sov.Un.	48•37 N	24•35 E
24	Naestved	Den.	55•14 N	11•46 E
35	Nafūd ad Dahy (Des.)	Sau.Ar.	22•15 N	44•15 E
41	Naga	Phil.	13•37 N	123•12 E
36	Nagaland (State)	India	25•47 N	94•15 E
37	Nagano	Jap.	36•42 N	138•12 E
37	Nagaoka	Jap.	37•22 N	138•49 E
35	Nāgappattinam	India	10•48 N	79•51 E
37	Nagasaki	Jap.	32•48 N	129•53 E
41	Nagcarlan	Phil.	14•07 N	121•24 E
31	Nagornokarabakh	Sov.Un.	40•10 N	46•50 E
37	Nagoya	Jap.	35•09 N	136•53 E
35	Nāgpur	India	21•12 N	79•09 E
41	Naguilian	Phil.	16•33 N	120•23 E
26	Nagykanizsa	Hung.	46•27 N	17•00 E
27	Nagykörös	Hung.	47•02 N	19•46 E
37	Naha	Jap.	26•02 N	127•43 E
10	Nahanni Natl. Pk.	Can.	62•10 N	125•15 W
33	Nahariyya	Isr.	33•01 N	35•06 E
31	Nahr al Khābur (R.)	Syr.	35•50 N	41•00 E
50	Nahuel Huapi (L.)	Arg.	41•00 S	71•30 W
41	Naic	Phil.	14•20 N	120•46 E
49	Naiguatá	Ven.	10•37 N	66•44 W
49	Naiguata, Pico	Ven.	10•32 N	66•44 W
11	Nain	Can.	56•29 N	61•52 W
22	Nairn	Scot.	57•35 N	3•54 W
55	Nairobi	Ken.	1•17 S	36•49 E
55	Naivasha	Ken.	0•47 S	36•29 E
34	Najd (Des.)	Sau.Ar.	25•18 N	42•38 E
38	Najin	Kor.	42•04 N	130•35 E
34	Najran (Des.)	Sau.Ar.	17•29 N	45•30 E
38	Nakadorishima (I.)	Jap.	33•00 N	128•20 E
31	Nakhichevan	Sov.Un.	39•10 N	45•30 E
29	Nakhodka	Sov.Un.	43•03 N	133•08 E
40	Nakhon Ratchasima	Thai.	14•56 N	102•14 E
40	Nakhon Si Thammarat	Thai.	8•27 N	99•58 E
24	Nakskov	Den.	54•51 N	11•06 E
27	Nakto nad Notecia	Pol.	53•10 N	17•35 E
38	Naktong (R.)	Kor.	36•10 N	128•30 E
31	Nal'chik	Sov.Un.	43•30 N	43•35 E
52	Nālūt	Libya	31•51 N	10•49 E
34	Namak, Daryacheh-ye (L.)	Iran	34•58 N	51•33 E
34	Namakzār-e Shāhdād (L.)	Iran	31•00 N	58•30 E
41	Namatanai	Pap.N.Gui.	3•43 S	152•26 E
46	Nambour	Austl.	26•48 S	153•00 E
40	Nam Dinh	Viet.	20•30 N	106•10 E
54	Namib Des.	Namibia	18•45 S	12•45 E
51	Namibia	Afr.	19•30 S	16•13 E
46	Namoi (R.)	Austl.	30•10 S	148•43 E
18	Namous, Oued en (R.)	Alg.	31•48 N	00•19 W
6	Nampa	Id.	43•35 N	116•35 W
38	Namp'o	Kor.	38•47 N	125•28 E
16	Namsos	Nor.	64•28 N	11•14 E
36	Nam Tsho (L.)	China	30•30 N	91•10 E
23	Namur	Bel.	50•29 N	4•55 E
54	Namutoni	Namibia	18•45 S	17•00 E
40	Nan, Mae Nam (R.)	Thai.	18•11 N	100•29 E
10	Namaimo	Can.	49•10 N	123•56 W
38	Nanam	Kor.	41•38 N	129•37 E
39	Nanao Tao (I.)	China	23•30 N	117•30 E
39	Nanch'ang	China	28•38 N	115•48 E
39	Nancheng	China	26•50 N	116•40 E
37	Nanching (Nanking)	China	32•04 N	118•46 E
39	Nanch'ung	China	30•45 N	106•05 E
17	Nancy	Fr.	48•42 N	6•11 E
35	Nanda Devi (Mt.)	India	30•30 N	80•25 E
39	Nanhsiung	China	25•10 N	114•20 E
39	Nani Dinh	Viet.	20•25 N	106•08 E
38	Nankung	China	37•22 N	115•22 E
39	Nan Ling (Mts.)	China	25•15 N	111•40 E
44	Nannine	Austl.	26•50 S	118•30 E
39	Nanning	China	22•56 N	108•10 E
39	Nanp'an (R.)	China	24•50 N	105•30 E
39	Nanp'ing	China	26•40 N	118•05 E
37	Nansei-shotō (Ryukyu Islands)	Jap.	27•30 N	127•00 E
36	Nan Shan (Mts.)	China	38•43 N	98•00 E
17	Nantes	Fr.	47•13 N	1•37 W
9	Nanticoke	Pa.	41•10 N	76•00 W
9	Nantucket (I.)	Ma.	41•15 N	70•05 W
38	Nanyang	China	33•00 N	112•42 E
38	Nanyüan	China	39•48 N	116•24 E
39	Nao Chou (I.)	China	20•58 N	110•58 E
21	Náousa	Grc.	40•38 N	22•05 E
6	Napa	Ca.	38•20 N	122•17 W
9	Napanee	Can.	44•15 N	77•00 W
45	Napier	N.Z.	39•30 S	177•00 E
20	Naples	It.	40•37 N	14•12 E
48	Napo (R.)	Peru	1•49 S	74•20 W
8	Napoleon	Oh.	41•20 N	84•10 W
20	Napoli, Golfo di	It.	40•29 N	14•08 E
8	Nappanee	In.	41•30 N	86•00 W
37	Nara	Jap.	34•41 N	135•50 E
52	Nara	Mali	15•09 N	7•27 W
46	Naracoorte	Austl.	36•50 S	140•50 E
18	Narbonne	Fr.	43•12 N	3•00 E
21	Nardò	It.	40•11 N	18•02 E
48	Nare	Col.	6•12 N	74•37 W
27	Narew R.	Pol.	52•43 N	21•19 E
35	Narmada (R.)	India	22•17 N	74•45 E
30	Narodnaya (Mtn.)	Sov.Un.	65•10 N	60•10 E
30	Naro Fominsk	Sov.Un.	55•23 N	36•43 E
25	Närpeså (R.)	Fin.	62•35 N	21•24 E
9	Narragansett B.	RI	41•20 N	71•15 W
46	Narrandera	Austl.	34•40 S	146•40 E
44	Narrogin	Austl.	33•00 S	117•15 E
30	Narva	Sov.Un.	59•24 N	28•12 E
41	Narvacan	Phil.	17•27 N	120•29 E
16	Narvik	Nor.	68•21 N	17•18 E
25	Narvskiy Zaliv (B.)	Sov.Un.	59•35 N	27•25 E

Page	Name	Region	Lat.	Long.
30	Nar'yan-Mar	Sov.Un.	67•42 N	53•30 E
46	Naryilco	Austl.	28•40 S	141•50 E
28	Narym	Sov.Un.	58•47 N	82•05 E
35	Naryn (R.)	Sov.Un.	41•46 N	73•00 E
9	Nashua	NH	42•47 N	71•23 W
8	Nashville	Mi.	42•35 N	85•05 W
7	Nashville	Tn.	36•10 N	86•48 W
21	Našice	Yugo.	45•29 N	18•06 E
27	Nasielsk	Pol.	52•35 N	20•50 E
30	Näsijärvi (L.)	Fin.	61•42 N	24•05 E
35	Nasik	India	20•02 N	73•49 E
53	Naşir	Sud.	8•30 N	33•06 E
35	Nasirabad	Bngl.	24•48 N	90•28 E
11	Naskaupi (R.)	Can.	53•59 N	61•10 W
13	Nassau	Ba.	25•05 N	77•20 W
53	Nasser, L.	Egypt	23•50 N	32•50 E
24	Nässjö	Swe.	57•39 N	14•39 E
41	Nasugbu	Phil.	14•05 N	120•37 E
39	Nata	China	19•30 N	109•38 E
48	Natagaima	Col.	3•38 N	75•07 W
49	Natal	Braz.	6•00 S	35•13 W
54	Natal (Prov.)	S.Afr.	28•50 S	30•07 E
7	Natchez	Ms.	1•35 N	91•20 W
52	Natitingou	Benin	10•19 N	1•22 E
49	Natividade	Braz.	11•43 S	47•34 W
54	Natron, L.	Tan.	2•17 S	36•10 E
40	Natuna Besar (Is.)	Indon.	3•22 N	108•00 E
44	Naturaliste, C.	Austl.	33•30 S	115•10 E
9	Naugatuck	Ct.	41•25 N	73•05 W
41	Naujan	Phil.	13•19 N	121•17 E
26	Naumburg	G.D.R.	51•10 N	11•50 E
42	Nauru	Oceania	0•30 S	167•00 E
12	Nautla	Mex.	20•14 N	96•44
50	Navarino (I.)	Chile	55•30 S	68•15 W
12	Navojoa	Mex.	27•00 N	109•40 W
21	Návplion	Grc.	37•33 N	22•46 E
21	Náxos (I.)	Grc.	37•15 N	25•20 E
12	Nayarit	Mex.	22•00 N	105•15 W
49	Nazaré	Braz.	13•04 S	38•49 W
49	Nazaré da Mata	Braz.	7•46 S	35•13 W
12	Nazas, R.	Mex.	25•08 N	104•20 W
33	Nazerat	Isr.	32•43 N	35•19 E
31	Nazilli	Tur.	37•40 N	28•10 E
53	Ndélé	Cen.Afr.Emp.	8•21 N	20•43 E
53	Ndjamena (Fort-Lamy)	Chad	12•07 N	15•03 E
55	Ndjolé	Gabon	0•15 S	10•45 E
54	Ndola	Zambia	12•58 S	28•38 E
22	Neagh Lough	N.Ire.	54•40 N	6•47 W
33	Néa Páfos	Cyprus	34•46 N	32•27 E
21	Neápolis	Grc.	36•35 N	23•08 E
20	Neápolis	Grc.	35•17 N	25•37 E
22	Neath	Wales	51•41 N	3•50 W
46	Nebine Cr.	Austl.	27•50 S	147•00 E
31	Nebit-Dag	Sov.Un.	39•30 N	54•20 E
6	Nebraska	U.S.	41•45 N	101•30 W
26	Neckar R.	F.R.G.	49•16 N	9•06 E
50	Necochea	Arg.	38•30 S	58•45 W
8	Neenah	Wi.	44•10 N	88•30 W
10	Neepawa	Can.	50•13 N	99•29 W
23	Neetze (R.)	F.R.G.	53•04 N	11•00 E
33	Negeri Sembilan	Mala.	2•46 N	101•54 E
33	Negev (Des.)	Isr.	30•34 N	34•43 E
21	Negoi (Mtn.)	Rom.	45•33 N	24•38 E
21	Negotin	Yugo.	44•13 N	22•33 E
40	Negrais, C.	Bur.	16•08 N	93•34 E
50	Negro (R.)	Arg.	39•50 S	65•00 W
40	Negro, Rio	Braz.	0•18 S	63•21 W
40	Negros (I.)	Phil.	9•50 N	121•45 E
48	Neguá	Col.	5•51 N	76•36 W
39	Neichiang	China	29•38 N	105•01 E
8	Neillsville	Wi.	44•35 N	90•37 W
48	Neira	Col.	5•10 N	75•32 W
26	Neisse (R.)	Pol.	51•30 N	15•00 E
48	Neiva	Col.	2•55 N	75•16 W
53	Nekemte	Eth.	9•09 N	36•29 E
8	Nekoosa	Wi.	44•19 N	89•54 W
24	Neksø	Den.	55•05 N	15•05 E
29	Nel'kan	Sov.Un.	57•45 N	136•36 E
35	Nellore	India	14•28 N	79•59 E
10	Nelson	Can.	49•29 N	117•17 W
45	Nelson	N.Z.	41•15 S	173•22 E
46	Nelson, C.	Austl.	38•29 S	141•20 E
10	Nelson (R.)	Can.	56•50 N	93•40 W
8	Nelsonville	Oh.	39•30 N	82•15 W
52	Néma	Mauritania	16•37 N	7•15 W
25	Neman	Sov.Un.	55•02 N	22•01 E
30	Neman R.	Sov.Un.	53•28 N	24•45 E
27	Nembe	Nig.	4•35 N	6•26 E
37	Nemuro	Jap.	43•13 N	145•10 E
22	Nenagh	Ire.	52•50 N	8•05 W
38	Nenchiang	China	49•02 N	125•15 E
37	Nen Chiang (R.)	China	47•07 N	123•28 E
32	Nepal	Asia	28•45 N	83•00 E
20	Nera (R.)	It.	42•45 N	12•54 E
29	Nerchinsk	Sov.Un.	51•47 N	116•17 E
29	Nerchinskiy Khrebet (Mts.)	Sov.Un.	50•30 N	118•30 E
29	Nerchinskiy Zavod	Sov.Un.	51•35 N	119•46 E
21	Neretva (R.)	Yugo.	43•08 N	17•50 E
22	Ness, Loch	Scot.	57•23 N	4•20 W
27	Nesterov	Sov.Un.	50•03 N	23•58 E
25	Nesterov	Sov.Un.	54•39 N	22•38 E
21	Néstos (R.)	Grc.	41•25 N	24•12 E
33	Netanya	Isr.	32•19 N	34•52 E
14	Netherlands	Eur.	53•01 N	3•57 E
11	Nettilling (L.)	Can.	66•30 N	70•40 W
26	Neubrandenburg	G.D.R.	53•33 N	13•16 E
26	Neuburg	F.R.G.	48•43 N	11•12 E
26	Neuchâtel	Switz.	47•00 N	6•52 E
26	Neuchatel, Lac de (L.)	Switz.	46•48 N	6•53 E
26	Neuhaldensleben	G.D.R.	52•18 N	11•23 E
26	Neumarkt	F.R.G.	49•17 N	11•30 E
26	Neumünster	F.R.G.	54•04 N	10•00 E
26	Neunkirchen	Aus.	47•43 N	16•05 E
50	Neuquén	Arg.	38•52 S	68•12 W
50	Neuquen (Prov.)	Arg.	39•40 S	70•45 W
50	Neuquén (R.)	Arg.	38•45 S	69•00 W
26	Neusiedler See (L.)	Aus.	47•54 N	16•31 E
26	Neustadt	F.R.G.	49•21 N	8•08 E
26	Neustadt	F.R.G.	54•06 N	10•50 E
26	Neustadt bei Coburg	F.R.G.	50•20 N	11•09 E
26	Neustrelitz	G.D.R.	53•21 N	13•05 E
26	Neu Ulm	F.R.G.	48•23 N	10•01 E
26	Neuwied	F.R.G.	50•26 N	7•28 E
6	Nevada	U.S.	39•30 N	117•00 W
17	Nevada, Sierra	Sp.	37•01 N	3•28 W
6	Nevada, Sierra	U.S.	39•20 N	120•25 W
48	Nevado, Cerro el	Col.	4•02 N	74•08 W
12	Nevado de Colima (Mtn.)	Mex.	19•34 N	103•39 W
22	Neva Stantsiya	Sov.Un.	59•53 N	30•30 E
30	Nevel	Sov.Un.	56•03 N	29•57 E
49	Neveri	Ven.	10•13 N	64•18 W
17	Nevers	Fr.	46•59 N	3•10 E
21	Nevesinje	Yugo.	43•15 N	18•08 E
22	Nevis, Ben (Mtn.)	Scot.	56•47 N	5•00 W
13	Nevis I.	St. Kitts-Nevis-Anguilla.	17•05 N	62•38 W
21	Nevrokop	Bul.	41•35 N	23•46 E
31	Nevşehir	Tur.	38•40 N	34•35 E
30	Nev'yansk	Sov.Un.	57•29 N	60•14 E
8	New Albany	In.	38•17 N	85•49 W
49	New Amsterdam	Guy.	6•14 N	57•30 W
67	New Amsterdam (I.)	Ind.O.	37•52 S	77•32 E
9	Newark	De.	39•40 N	75•45 W
9	Newark	NJ	40•44 N	74•10 W
9	Newark	NY	43•05 N	77•10 W
8	Newark	Oh.	40•05 N	82•25 W
8	Newaygo	Mi.	43•25 N	85•50 W
9	New Bedford	Ma.	41•35 N	70•55 W
8	Newberg	Or.	45•17 N	122•58 W
7	New Bern	NC	35•05 N	77•05 W
8	New Boston	Oh.	38•45 N	82•55 W
9	New Brighton	Pa.	40•34 N	80•18 W
9	New Britain	Ct.	41•40 N	72•45 W
41	New Britain (I.)	Pap.N.Gui.	6•45 S	149•38 E
9	New Brunswick	NJ	40•29 N	74•27 W
11	New Brunswick	Can.	47•14 N	66•30 W
8	Newburg	In.	38•00 N	87•25 W
9	Newburgh	NY	41•30 N	74•00 W
22	Newbury	Eng.	51•24 N	1•26 W
9	Newburyport	Ma.	42•48 N	70•53 W
45	New Caledonia	Oceania	21•28 S	164•40 E
46	Newcastle	Austl.	33•00 S	151•55 E
9	New Castle	De.	39•40 N	75•35 W
22	Newcastle	Eng.	55•00 N	1•45 W
8	New Castle	In.	39•55 N	82•25 W
8	New Castle	Oh.	40•20 N	82•10 W
8	New Castle	Pa.	41•00 N	80•25 W
44	Newcastle Waters	Austl.	17•10 S	133•25 E
8	Newcomerstown	Oh.	40•15 N	81•40 W
35	New Delhi	India	28•43 N	77•18 E
45	New England Ra.	Austl.	29•32 S	152•30 E
11	Newfoundland	Can.	48•15 N	56•53 W
45	New Georgia (I.)	Sol.Is.	8•08 S	158•00 E
11	New Glasgow	Can.	45•35 N	62•36 W
41	New Guinea (I.)	Asia	5•45 S	140•00 E
7	New Hampshire	U.S.	43•55 N	71•40 W
55	New Hanover	S.Afr.	29•23 S	30•32 E
41	New Hanover (I.)	Pap.N.Gui.	2•37 S	150•15 E
8	New Harmony	In.	38•10 N	87•55 W
9	New Haven	Ct.	41•20 N	72•55 W
23	Newhaven	Eng.	50•45 N	0•10 E
8	New Haven	In.	41•05 N	85•00 W
45	New Hebrides	Oceania	16•02 S	169•15 E
41	New Ireland (I.)	Pap.N.Gui.	3•15 S	152•30 E
7	New Jersey	U.S.	40•30 N	74•50 W
8	New Lexington	Oh.	39•40 N	82•10 W
8	New Lisbon	Wi.	43•52 N	90•11 W
9	New London	Ct.	41•20 N	72•05 W
8	New London	Wi.	44•24 N	88•45 W
9	Newmarket	Can.	44•00 N	79•30 W
8	New Martinsville	WV	39•35 N	80•50 W
6	New Mexico	U.S.	34•30 N	107•10 W
46	New Norfolk	Austl.	42•50 S	147•17 E
7	New Orleans	La.	30•00 N	90•05 W
8	New Philadelphia	Oh.	40•30 N	81•30 W
45	New Plymouth	N.Z.	39•04 S	174•13 E
22	Newport	Eng.	50•41 N	1•25 W
22	Newport	Wales	51•36 N	3•05 W
9	Newport	NH	43•20 N	72•10 W
9	Newport	RI	41•29 N	71•16 W
9	Newport	Vt.	44•55 N	72•15 W
9	Newport News	Va.	36•59 N	76•24 W
8	New Richmond	Oh.	38•55 N	84•15 W
9	New Rochelle	NY	40•55 N	73•47 W
26	New Ross	Ire.	52•25 N	6•55 W
45	New South Wales	Austl.	32•45 S	146•14 E
8	Newton	Il.	39•00 N	88•10 W
9	Newton	Ma.	42•21 N	71•13 W
9	Newton	NJ	41•03 N	74•45 W
22	Newtownards	Ire.	54•35 N	5•39 W
11	New Waterford	Can.	46•15 N	60•05 W
9	New York	NY	40•40 N	73•58 W
7	New York (State)	U.S.	42•45 N	78•05 W
45	New Zealand	Oceania	39•14 S	169•30 E
34	Neyshābūr	Iran	36•06 N	58•45 E
31	Nezhin	Sov.Un.	50•03 N	31•52 E
54	Ngami (R.)	Bots.	20•56 S	22•31 E
53	Ngaoundéré	Cam.	7•19 N	13•35 E
55	Ngong	Ken.	1•27 S	36•39 E
55	Ngqeleni	S.Afr.	31•41 S	29•04 E
53	Nguigmi	Niger	14•15 N	13•07 E
52	Nguru	Nig.	12•53 N	10•26 E
40	Nha Trang	Viet.	12•08 N	108•56 E
52	Niafounke	Mali	16•03 N	4•17 W
8	Niagara	Wi.	45•45 N	88•05 W
9	Niagara Falls	NY-Can.	43•06 N	79•02 W
52	Niamey	Niger	13•31 N	2•07 E
53	Niangara	Zaire	3•42 N	27•52 E
40	Nias (I.)	Indon.	0•58 N	97•43 E
24	Nibe	Den.	56•57 N	9•36 E
12	Nicaragua	N.A.	12•45 N	86•15 W
13	Nicaragua, Lago de (L.)	Nic.	11•45 N	85•28 W
20	Nicastro	It.	38•39 N	16•15 E
18	Nice	Fr.	43•42 N	7•21 E
11	Nichicun (L.)	Can.	53•07 N	72•10 W
8	Nicholasville	Ky.	37•55 N	84•35 W
40	Nicobar Is.	Andaman & Nicobar Is.	8•28 N	94•04 E
19	Nicosia	Cyprus	35•10 N	33•22 E
48	Nicoya, Golfo de	C.R.	10•03 N	85•04 W
27	Nidzica	Pol.	53•21 N	20•30 E
26	Niedere Tauern (Mts.)	Aus.	47•15 N	13•41 E
26	Niedersachsen (Lower Saxony)	F.R.G.	52•52 N	8•27 E
26	Nienburg	F.R.G.	52•40 N	9•15 E
49	Nieuw Nickerie	Sur.	5•51 N	57•00 W
31	Niğde	Tur.	37•55 N	34•40 E
51	Niger	Afr.	18•02 N	8•30 E
51	Niger (R.)	Afr.	5•33 N	6•33 E
51	Nigeria	Afr.	8•57 N	6•30 E
37	Niigata	Jap.	37•47 N	139•04 E
6	Niihau (I.)	Hi.	21•50 N	160•05 E
23	Nijmegen	Neth.	51•50 N	5•52 E
31	Nikolayev	Sov.Un.	46•58 N	32•02 E
31	Nikolayevskiy	Sov.Un.	50•00 N	45•30 E
29	Nikolayevsk-na-Amure	Sov.Un.	53•18 N	140•49 E
30	Nikol'sk	Sov.Un.	59•30 N	45•40 E
21	Nikopol	Bul.	43•41 N	24•52 E
31	Nikopol	Sov.Un.	47•36 N	34•24 E
21	Nikšić	Yugo.	42•45 N	18•57 E
53	Nile (R.)	Afr.	19•15 N	32•30 E
8	Niles	Mi.	41•50 N	86•15 W
8	Niles	Oh.	41•15 N	80•45 W
50	Nilópolis	Braz.	22•48 S	43•25 W
52	Nimba, Mont	Ivory Coast	7•40 N	8•33 W
17	Nîmes	Fr.	43•49 N	4•22 E
53	Nimule	Sud.	3•38 N	32•12 E
46	Ninety Mile Bch.	Austl.	38•20 S	147•30 E
31	Nineveh (Ruins)	Iraq	36•30 N	43•10 E
38	Ningan	China	44•20 N	129•20 E
36	Ningerh	China	23•14 N	101•14 E
39	Ninghai	China	29•20 N	121•20 E
39	Ningming	China	22•22 N	107•06 E
39	Ningpo	China	29•56 N	121•30 E
36	Ningsia Hui Aut. Reg.	China	40•18 N	104•45 E
39	Ningte	China	26•38 N	119•33 E
38	Ningwu	China	39•00 N	112•12 E
39	Ninh Binh	Viet.	20•22 N	106•00 E
41	Ninigo Is.	Pap.N.Gui.	1•15 S	143•30 E
49	Nioaque	Braz.	21•14 S	55•41 W
6	Niobrara (R.)	Ne.	42•46 N	98•46 W
52	Nioro du Sahel	Mali	15•15 N	9•35 W
17	Niort	Fr.	46•17 N	0•28 W
10	Nipawin	Can.	53•22 N	104•00 W
8	Nipigon	Can.	48•58 N	88•17 W
11	Nipigon (L.)	Can.	49•37 N	89•55 W
11	Nipissing (L.)	Can.	45•59 N	80•19 W
21	Niš	Yugo.	43•18 N	21•55 E
21	Nišava (R.)	Yugo.	43•17 N	22•17 E
27	Nisko	Pol.	50•30 N	22•07 E
24	Nissan (R.)	Swe.	57•06 N	13•22 E
24	Nisser Vand (L.)	Nor.	59•14 N	8•35 E
24	Nissum Fd.	Den.	56•24 N	7•35 E
50	Niterói	Braz.	22•53 S	43•07 W
22	Nith (R.)	Scot.	55•13 N	3•55 W
27	Nitra	Czech.	48•18 N	18•04 E
27	Nitra R.	Czech.	48•13 N	18•14 E
8	Nitro	WV	38•25 N	81•50 W
43	Niue	Oceania	19•50 S	167•00 W
23	Nivelles	Bel.	50•33 N	4•17 E
35	Nizāmābād	India	18•48 N	78•07 E
29	Nizhne-Angarsk	Sov.Un.	55•49 N	108•46 E
31	Nizhne-Chirskaya	Sov.Un.	48•20 N	42•50 E
29	Nizhne-Kolymsk	Sov.Un.	68•32 N	160•56 E
29	Nizhneudinsk	Sov.Un.	54•58 N	99•15 E
30	Nizhniye Sergi	Sov.Un.	56•41 N	59•19 E
28	Nizhniy Tagil	Sov.Un.	57•54 N	59•59 E
29	Nizhnyaya (Lower) Tunguska (R.)	Sov.Un.	64•13 N	91•30 E
24	Njurunda	Swe.	62•15 N	17•24 E
55	Nkandla	S.Afr.	28•40 S	31•06 E
35	Noākhāli	Bngl.	22•52 N	91•08 E
8	Noblesville	In.	40•00 N	86•00 W
6	Nogales	Az.	31•20 N	110•55 W
12	Nogales	Mex.	18•49 N	97•09 W
12	Nogales	Mex.	31•15 N	111•00 W
30	Noginsk	Sov.Un.	55•52 N	38•28 E
38	Noho	China	48•23 N	124•58 E
17	Noirmoutier, Île de	Fr.	47•03 N	3•08 W
8	Nokomis	Il.	39•15 N	89•10 W
30	Nolinsk	Sov.Un.	57•32 N	49•50 E
6	Nome	Ak.	64•30 N	165•20 W
10	Nonacho (L.)	Can.	61•48 N	111•20 W
54	Nongoma	S.Afr.	27•48 S	31•45 E
10	Nootka (I.)	Can.	49•32 N	126•42 W
54	Nóqui	Ang.	5•51 S	13•25 E
24	Nora	Swe.	59•32 N	14•56 E
26	Norden	F.R.G.	53•35 N	7•14 E
26	Norderney I.	F.R.G.	53•45 N	6•58 E
24	Nord Fd.	Nor.	61•50 N	5•35 E
26	Nordhausen	G.D.R.	51•30 N	10•48 E
26	Nordhorn	F.R.G.	52•26 N	7•05 E
16	Nord Kapp	Nor.	71•07 N	25•57 E
26	Nördlingen	F.R.G.	48•51 N	10•30 E
26	Nord-Ostsee Kan. (Kiel Can.)	F.R.G.	54•03 N	9•23 E
26	Nordrhein-Westfalen (North Rhine-Westphalia)	F.R.G.	50•50 N	6•53 E
29	Nordvik	Sov.Un.	73•57 N	111•15 E
22	Nore R.	Ire.	52•34 N	7•15 W
6	Norfolk	Ne.	42•10 N	97•25 W
9	Norfolk	Va.	36•55 N	76•15 W
42	Norfolk	Oceania	27•10 S	166•50 E
29	Noril'sk	Sov.Un.	69•00 N	87•11 E
8	Normal	Il.	40•35 N	89•00 W
7	Norman, L.	NC	35•30 N	80•53 W
45	Norman (R.)	Austl.	18•27 S	141•29 E
45	Normanton	Austl.	17•45 S	141•10 E
10	Norman Wells	Can.	65•26 N	127•00 W
44	Nornalup	Austl.	35•00 S	117•00 E
24	Norra Dellen (L.)	Swe.	61•57 N	16•25 E
24	Norre Sundby	Den.	57•04 N	9•55 E
7	Norris (L.)	Tn.	36•17 N	84•10 W
9	Norristown	Pa.	40•07 N	75•21 W
24	Norrköping	Swe.	58•37 N	16•10 E
24	Norrtälje	Swe.	59•47 N	18•39 E
44	Norseman	Austl.	32•15 S	122•00 E
49	Norte, Serra do	Braz.	12•04 S	59•08 W
11	North, C.	Can.	47•02 N	60•25 W
45	North, C.	N.Z.	34•31 S	173•02 E
45	North, I.	N.Z.	37•34 S	171•12 E
9	North Adams	Ma.	42•40 N	73•05 W
44	Northam	Austl.	31•50 S	116•45 E
5	North America			
13	North American Basin	Atl.O.	23•45 N	62•45 W
44	Northampton	Austl.	28•22 S	114•45 E
22	Northampton	Eng.	52•14 N	0•56 W
9	Northampton	Ma.	42•20 N	72•45 W
9	Northampton	Pa.	40•45 N	75•30 W
40	North Andaman I.	Andaman & Nicobar Is.	13•15 N	93•30 E
8	North Baltimore	Oh.	41•10 N	83•40 W
9	North Berwick	Me.	43•18 N	70•46 W
9	Northbridge	Ma.	42•09 N	71•39 W
7	North Carolina	U.S.	35•40 N	81•30 W
8	North Chan (B.)	Can.	46•10 N	83•20 W
22	North Chan	N.Ire.-Scot.	55•15 N	7•56 W
8	North Chicago	Il.	42•19 N	87•51 W
6	North Dakota	U.S.	47•20 N	101•55 W
22	North Downs	Eng.	51•11 N	0•01 W
26	Northeim	F.R.G.	51•42 N	9•59 E
22	Northern Ireland	U.K.	54•48 N	7•00 W
44	Northern Territory	Austl.	18•15 S	133•00 E
46	North Flinders, Ra.	Austl.	31•55 S	138•45 E
23	North Foreland	Eng.	51•20 N	1•30 E
24	North Frisian Is.	Den.	55•16 N	8•15 E
12	North Gamboa	C.Z.	9•07 N	79•40 W
8	North Judson	In.	41•15 N	86•50 W
8	North Manchester	In.	41•00 N	85•45 W
46	North Mount Lofty Ranges	Austl.	33•50 S	138•30 E
6	North Platte	Ne.	41•08 N	100•45 W
6	North Platte (R.)	U.S.	41•20 N	102•40 W
8	North Pt.	Mi.	45•00 N	83•20 W
10	North Saskatchewan (R.)	Can.	52•40 N	106•45 W

Page	Name	Region	Lat.	Long.
16	North Sea	Eur.	56•09 N	3•16 E
45	North Stradbroke I.	Austl.	27•45 S	154•18 E
45	North Taranaki Bght.	N.Z.	38•23 S	172•03 E
9	North Tonawanda	NY	43•02 N	78•53 W
22	North Uist (I.)	Scot.	57•37 N	7•22 W
9	Northumberland	NH	44•30 N	71•30 W
45	Northumberland (Is.)	Austl.	21•42 S	151•30 E
10	North Vancouver	Can.	49•19 N	123•04 W
8	North Vernon	In.	39•05 N	85•45 W
44	North West C.	Austl.	21•50 S	112•25 E
22	Northwest Highlands	Scot.	56•50 N	5•20 W
10	Northwest Territories	Can.	64•42 N	119•09 W
23	Northwich	Eng.	53•15 N	2•31 W
22	North York Moors	Eng.	54•20 N	0•40 W
9	Norwalk	Ct.	41•06 N	73•25 W
8	Norwalk	Oh.	41•15 N	82•35 W
14	Norway	Eur.	63•48 N	11•17 E
9	Norway	Me.	44•11 N	70•35 W
8	Norway	Mi.	45•47 N	87•55 W
10	Norway House	Can.	53•59 N	97•50 W
16	Norwegian Sea	Eur.	66•54 N	1•43 E
9	Norwich	Ct.	41•20 N	72•00 W
23	Norwich	Eng.	52•40 N	1•15 E
9	Norwich	NY	42•35 N	75•30 W
9	Norwood	Ma.	42•11 N	71•13 W
8	Norwood	Oh.	39•10 N	84•27 W
55	Nossi Bé (B.)	Mad.	13•14 S	47•28 E
54	Nossob (R.)	Namibia	24•15 S	19•10 E
26	Noteć R.	Pol.	52•50 N	16•19 E
15	Noto	It.	36•49 N	15•08 E
24	Notodden	Nor.	59•35 N	9•15 E
37	Noto-Hantō (Pen.)	Jap.	37•18 N	137•03 E
11	Notre Dame B.	Can.	49•45 N	55•15 W
8	Nottawasaga B.	Can.	44•45 N	80•35 W
11	Nottaway (R.)	Can.	50•58 N	78•02 W
22	Nottingham	Eng.	52•58 N	1•09 W
11	Nottingham I.	Can.	62•58 N	78•53 W
52	Nouadhibou	Mauritania	21•02 N	17•09 W
52	Nouakchott	Mauritania	18•06 N	15•57 W
52	Nouamrhar	Mauritania	19•22 N	16•31 W
45	Noumea	N.Cal.	22•18 S	166•48 E
53	Nouvelle Anvers	Zaire	1•42 N	19•08 E
49	Nova Cruz	Braz.	6•22 S	35•20 W
50	Nova Iguaçu	Braz.	22•45 S	43•27 W
54	Nova Mambone	Moz.	21•04 S	35•13 E
20	Novara	It.	45•24 N	8•38 E
11	Nova Scotia	Can.	44•28 N	65•00 W
21	Nova Varoš	Yugo.	43•24 N	19•53 E
25	Novaya Ladogo	Sov.Un.	60•06 N	32•16 E
29	Novaya Sibir (I.)	Sov.Un.	75•42 N	150•00 E
28	Novaya Zemlya (I.)	Sov.Un.	72•00 N	54•46 E
21	Nova Zagora	Bul.	42•30 N	26•01 E
27	Nové Mesto (Nad Váhom)	Czech.	48•44 N	17•47 E
27	Nové Zámky	Czech.	47•58 N	18•10 E
30	Novgorod	Sov.Un.	58•32 N	31•16 E
20	Novi	It.	44•43 N	8•48 W
20	Novi Grad	Yugo.	44•09 N	15•34 E
21	Novi-Pazar	Bul.	43•22 N	27•26 E
21	Novi Pazar	Yugo.	43•08 N	20•30 E
21	Novi Sad	Yugo.	45•15 N	19•53 E
31	Novocherkassk	Sov.Un.	47•25 N	40•04 E
31	Novogorod-Severskiy	Sov.Un.	52•01 N	33•14 E
27	Novogrudok	Sov.Un.	53•35 N	25•51 E
28	Novo-Kazalinsk	Sov.Un.	45•47 N	62•00 E
28	Novokuznetsk (Stalinsk)	Sov.Un.	53•43 N	86•59 E
20	Novo Mesto	Yugo.	45•48 N	15•13 E
30	Novomoskossk	Sov.Un.	54•06 N	38•08 E
30	Novomoskovsk	Sov.Un.	48•37 N	35•12 E
54	Novo Redondo	Ang.	11•13 S	13•50 E
31	Novorossiyski	Sov.Un.	44•43 N	37•48 E
21	Novo-Selo	Bul.	44•09 N	22•46 E
28	Novosibirsk	Sov.Un.	55•09 N	82•58 E
29	Novosibirskiye O-va (New Siberian Is.)	Sov.Un.	76•45 N	140•30 E
31	Novoukrainka	Sov.Un.	48•18 N	31•33 E
31	Novouzensk	Sov.Un.	50•40 N	48•08 E
31	Novozybkov	Sov.Un.	52•31 N	31•54 E
27	Nový Jičín	Czech.	49•36 N	18•02 E
28	Novyy Port	Sov.Un.	67•19 N	72•28 E
27	Nowa Huta	Pol.	50•04 N	20•20 E
26	Nowa Sól	Pol.	51•49 N	15•41 E
46	Nowra	Austl.	34•55 S	150•45 E
27	Nowy Dwór Mazowiecki	Pol.	52•26 N	20•46 E
27	Nowy Sacz	Pol.	49•36 N	20•42 E
27	Nowy Targ	Pol.	49•29 N	20•02 E
55	Nqamakwe	S.Afr.	32•13 S	27•57 E
55	Nqutu	S.Afr.	28•17 S	30•41 E
55	Ntshoni (Mtn.)	S.Afr.	29•34 S	30•03 E
54	Ntwetwe Pan (Salt Flat)	Bots.	20•00 S	24•18 E
53	Nubah, Jibāl an- (Mts.)	Sud.	12•22 N	30•39 E
53	Nubian Des.	Sud.	21•13 N	33•09 E
48	Nudo Corpuna (Mt.)	Peru	15•53 S	72•04 W
48	Nudo de Pasco (Mt.)	Peru	10•34 S	76•12 W
6	Neuces R.	Tx.	28•20 N	98•08 W
10	Nueltin (L.)	Can.	60•14 N	101•00 W
49	Nueva Esparta	Ven.	10•50 N	64•35 W
6	Nueva Rosita	Mex.	27•55 N	101•10 W
13	Nuevitas	Cuba	21•35 N	77•15 W
12	Nuevo Laredo	Mex.	27•29 N	99•30 W
12	Nuevo Leon	Mex.	26•00 N	100•00 W
12	Nuevo San Juan	Pan.	9•14 N	79•43 W
44	Nullagine	Austl.	22•00 S	120•07 E
44	Nullarbor Plain	Austl.	31•45 S	126•30 E
41	Numfoor	Indon.	1•20 S	134•48 E
38	Nungan	China	44•25 N	125•10 E
6	Nunivak (I.)	Ak.	60•25 N	167•42 W
20	Nuoro	It.	40•29 N	9•20 E
28	Nurata	Sov.Un.	40•33 N	65•28 E
26	Nürnberg	F.R.G.	49•28 N	11•07 E
31	Nusabyin	Tur.	37•05 N	41•10 E
35	Nushki	Pak.	29•30 N	66•02 E
9	Nutter Fort	WV	39•15 N	80•15 W
33	Nuwaybi 'al Muzayyinah	Egypt	28•59 N	34•40 E
54	Nuweland	S.Afr	33•58 S	18•28 E
53	Nyala	Sud.	12•00 N	24•52 E
54	Nyasa, L.	Afr.	10•45 S	34•30 E
24	Nyborg	Den.	55•20 N	10•45 E
24	Nybro	Swe.	56•44 N	15•56 E
36	Nyenchhen Thanglha (Mts.)	China	29•55 N	88•08 E
24	Nyhem	Swe.	56•39 N	12•50 E
27	Nyíregyháza	Hung.	47•58 N	21•45 E
24	Nykøbing	Den.	56•46 N	8•47 E
24	Nykøbing Falster	Den.	54•45 N	11•54 E
24	Nykøbing Sjaelland	Den.	55•55 N	11•37 E
24	Nyköping	Swe.	58•46 N	16•58 E
54	Nylstroom	S.Afr.	24•42 S	28•25 E
46	Nymagee	Austl.	32•17 S	146•18 E
26	Nymburk	Czech.	50•12 N	15•03 E
22	Nymphe Bk.	Ire.	51•36 N	7•35 W
24	Nynäshamn	Swe.	58•53 N	17•55 E
46	Nyngan	Austl.	31•31 S	147•25 E
52	Nyong (R.)	Cam.	3•40 N	10•25 E
26	Nyrány	Czech.	49•43 N	13•13 E
27	Nysa	Pol.	50•29 N	17•20 E
30	Nytva	Sov.Un.	58•00 N	55•10 E

O

Page	Name	Region	Lat.	Long.
6	Oahu (I.)	Hi.	21•38 N	157•48 W
8	Oakdale	Ky.	38•15 N	85•50 W
8	Oakharbor	Oh.	41•30 N	83•05 W
6	Oakland	Ca.	37•48 N	122•16 W
8	Oakland City	In.	38•20 N	87•20 W
8	Oak Park	Il.	41•53 N	87•48 W
12	Oaxaca	Mex.	16•45 N	97•00 W
30	Ob (R.)	Sov.Un.	62•15 N	67•00 E
11	Oba	Can.	48•58 N	84•09 W
22	Oban	Scot	56•25 N	5•35 W
23	Oberhausen	F.R.G.	51•27 N	6•51 E
8	Oberlin	Oh.	41•15 N	82•15 W
26	Oberösterreich	Aus.	48•05 N	13•15 E
41	Obi	Indon.	1•25 S	128•15 E
49	Óbidos	Braz.	1•57 S	55•32 W
31	Oboyan'	Sov.Un.	51•14 N	36•16 E
30	Obskaya Guba (B.)	Sov.Un.	67•13 N	73•45 E
13	Ocala	Fl.	29•11 N	82•09 W
48	Ocaña	Col.	8•15 N	73•37 W
52	Occidental, Grand Erg (Dunes)	Alg.	29•30 N	00•45 W
48	Occidental, Cordillera	Col.	5•05 N	76•04 W
48	Occidental, Cordillera	Peru	10•12 S	76•58 W
12	Occidental, Sierra Madre	Mex.	29•30 N	107•30 W
9	Ocean City	Md.	38•20 N	75•10 W
9	Ocean City	NJ	39•15 N	74•35 W
10	Ocean Falls	Can.	52•21 N	127•40 W
9	Ocean Grove	NJ	40•10 N	74•00 W
21	Ocenele Mari	Rom.	45•05 N	24•17 E
36	Ochina Ho (R.)	China	41•15 N	100•46 E
38	Ochir	China	45•38 N	115•35 E
24	Ockelbo	Swe.	60•54 N	16•35 E
7	Ocmulgee (R.)	Ga.	32•35 N	83•30 W
21	Ocna-Sibiului	Rom.	45•52 N	24•04 E
7	Oconee (R.)	Ga.	32•45 N	83•00 W
8	Oconomowoc	Wi.	43•06 N	88•24 W
8	Oconto	Wi.	44•54 N	87•55 W
8	Oconto Falls	Wi.	44•53 N	88•11 W
49	Ocumare del Tuy	Ven.	10•07 N	66•47 W
24	Odda	Nor.	60•04 N	6•30 E
31	Ödemis	Tur.	38•12 N	28•00 E
24	Odense	Den.	55•24 N	10•20 E
26	Odenwald (For.)	F.R.G.	49•39 N	8•55 E
58	Oder R.	G.D.R.	52•40 N	14•19 E
31	Odessa	Sov.Un.	46•28 N	30•44 E
52	Odienné	Ivory Coast	9•30 N	7•34 W
41	Odioñgan	Phil.	12•24 N	121•59 E
27	Odobesti	Rom.	45•46 N	27•08 E
27	Odorhei	Rom.	46•18 N	25•17 E
27	Odra R.	Pol.	50•28 N	17•55 E
49	Oeiras	Braz.	7•05 S	42•01 W
26	Ofanto (R.)	It.	41•08 N	15•33 E
58	Offenbach	F.R.G.	50•06 N	8•50 E
58	Offenburg	F.R.G.	48•28 N	7•57 E
51	Ogaden (Plat.)	Eth.	6•45 N	44•53 E
52	Ogbomosho	Nig.	8•08 N	4•15 E
6	Ogden	Ut.	41•14 N	111•58 W
9	Ogdensburg	NY	44•40 N	75•30 W
10	Ogilvie Mts.	Can.	64•45 N	138•10 W
8	Oglesby	Il.	41•20 N	89•00 W
20	Oglio (R.)	It.	45•15 N	10•19 E
54	Ogooué (R.)	Gabon	0•50 S	9•20 E
20	Ogulin	Yugo.	45•17 N	15•11 E
7	Ohio	U.S.	40•30 N	83•15 W
8	Ohio R.	U.S.	37•25 N	88•05 W
26	Ohře (Eger) R.	Czech.	50•08 N	12•45 E
21	Ohrid	Yugo.	41•08 N	20•46 E
21	Ohrid (L.)	Alb.-Yugo.	40•58 N	20•35 E
24	Oieren (L.)	Nor.	59•50 N	11•25 E
9	Oil City	Pa.	41•25 N	79•40 W
17	Oise (R.)	Fr.	49•30 N	2•56 E
12	Ojinaga	Mex.	29•34 N	104•26 W
31	Oka (R.)	Sov.Un.	52•10 N	35•20 E
30	Oka (R.)	Sov.Un.	55•10 N	42•10 E
55	Okahandja	Namibia	21•50 S	16•45 E
10	Okanagan L.	Can.	50•00 N	119•28 W
52	Okano (R.)	Gabon	0•15 N	11•08 E
54	Okavango (Cubango) (R.)	Ang.-Namibia	17•10 S	18•20 E
54	Okavango Swp.	Bots.	19•30 S	23•02 E
37	Okayama	Jap.	34•39 N	133•54 E
7	Okeechobee, L.	Fl.	27•00 N	80•49 W
23	Oker (R.)	F.R.G.	52•23 N	10•00 E
29	Okha	Sov.Un.	53•44 N	143•12 E
29	Okhotsk	Sov.Un.	59•28 N	143•32 E
33	Okhotsk, Sea of	Asia	56•45 N	146•00 E
37	Oki Guntō (Arch.)	Jap.	36•17 N	133•05 E
37	Okinawa (I.)	Jap.	26•30 N	128•30 E
37	Okinawa Guntō (Is.)	Jap.	26•50 N	127•25 E
6	Oklahoma	U.S.	36•00 N	98•20 W
6	Oklahoma City	Ok.	35•27 N	97•32 W
24	Öland (I.)	Swe.	57•03 N	17•15 E
50	Olavarría	Arg.	36•49 N	60•15 W
27	Olawa	Pol.	50•57 N	17•18 E
20	Olbia	It.	40•55 N	9•28 E
10	Old Crow	Can.	67•51 N	139•58 W
26	Oldenburg	F.R.G.	53•09 N	8•13 E
9	Old Forge	Pa.	41•20 N	75•50 W
22	Oldham	Eng.	53•32 N	2•07 W
22	Old Head of Kinsale	Ire.	51•35 N	8•35 W
54	Old Tate	Bots.	21•18 S	27•43 E
9	Olean	NY	42•05 N	78•25 W
27	Olecko	Pol.	54•02 N	22•29 E
29	Olëkminsk	Sov.Un.	60•39 N	120•40 E
29	Olenëk (R.)	Sov.Un.	70•18 N	121•15 E
17	Oléron Île, d'	Fr.	45•52 N	1•58 W
27	Oleśnica	Pol.	51•13 N	17•24 E
29	Ol'ga	Sov.Un.	43•48 N	135•44 E
17	Olhão	Port.	37•02 N	7•54 W
55	Olievenhoutpoort	S.Afr.	25•58 S	27•55 E
54	Olifants (R.)	S.Afr.	23•58 S	31•00 E
21	Ólimbos	Grc.	40•03 N	22•22 E
33	Ólimbos (Mtn.)	Cyprus	34•56 N	32•52 E
49	Olinda	Braz.	8•00 S	34•58 W
8	Olive Hill	Ky.	38•15 N	83•10 W
17	Olivenza	Sp.	38•42 N	7•06 W
6	Oliver	Can.	49•11 N	119•33 W
50	Olivos	Arg.	34•15 S	58•29 W
26	Olkusz	Pol.	50•16 N	19•41 E
48	Ollagüe	Chile	21•17 S	68•17 W
8	Olney	Il.	38•45 N	88•05 W
27	Olomouc	Czech.	49•37 N	17•15 E
25	Olonets	Sov.Un.	60•58 N	32•54 E
41	Olongapo	Phil.	14•49 S	120•17 E
17	Olot	Sp.	42•09 N	2•30 E
26	Olsnitz	G.D.R.	50•25 N	12•11 E
27	Olsztyn	Pol.	53•47 N	20•28 E
26	Olten	Switz.	47•20 N	7•53 E
21	Oltenita	Rom.	44•05 N	26•39 E
19	Olt R.	Rom.	44•09 N	24•40 E
6	Olympia	Wa.	47•02 N	122•52 W
6	Olympic Natl. Park	Wa.	47•54 N	123•00 W
9	Olyphant	Pa.	41•30 N	75•40 W
29	Olyutorskiy, Mys (C.)	Sov.Un.	59•49 N	167•16 E
22	Omagh	N.Ire.	54•35 N	7•25 W
7	Omaha	Ne.	41•18 N	95•57 W
32	Oman	Asia	20•00 N	57•45 E
34	Oman, G. of	Asia	24•24 N	58•58 E
54	Omaruru	Namibia	21•25 S	16•50 E
20	Ombrone (R.)	It.	42•48 N	11•18 E
53	Om Hajer	Eth.	14•06 N	36•46 E
53	Omo R.	Eth.	5•54 N	36•09 E
8	Omro	Wi.	44•01 N	89•46 W
28	Omsk	Sov.Un.	55•12 N	73•19 E
30	Omutninsk	Sov.Un.	58•38 N	52•10 E
8	Onaway	Mi.	45•25 N	84•10 W
27	Ondava (R.)	Czech.	48•51 N	21•40 E
38	Öndör Haan	Mong.	47•20 N	110•40 E
30	Onega	Sov.Un.	63•50 N	38•08 E
30	Onega (R.)	Sov.Un.	63•20 N	39•20 E
9	Oneida	NY	43•05 N	75•40 W
9	Oneida (L.)	NY	43•10 N	76•00 W
9	Oneonta	NY	42•25 N	75•05 W
30	Onezhskaja Guba (B.)	Sov.Un.	64•30 N	36•00 E
30	Onezhskiy (Pen.)	Sov.Un.	64•30 N	37•40 E
30	Onezhskoye Ozero (Onega, L.)	Sov.Un.	62•02 N	34•35 E
36	Ongin	Mong.	46•00 N	102•46 E
55	Onilahy (R.)	Mad.	23•41 S	45•00 E
52	Onitsha	Nig.	6•09 N	6•47 W
29	Onon (R.)	Sov.Un.	50•33 N	114•18 E
49	Onoto	Ven.	9•38 N	65•03 W
44	Onslow	Austl.	21•53 S	115•00 E
11	Ontario	Can.	50•47 N	88•50 W
7	Ontario, L.	U.S.-Can.	43•35 N	79•05 W
44	Oodnadatta	Austl.	27•38 S	135•40 E
44	Ooldea Station	Austl.	30•35 S	132•08 E
7	Oologah Res.	Ok.	36•43 N	95•32 W
23	Oostende	Bel.	51•14 N	2•55 E
23	Ooster Schelde (R.)	Neth.	51•40 N	3•40 E
27	Opatow	Pol.	50•47 N	21•25 E
27	Opava	Czech.	49•56 N	17•52 E
50	Opdal	Nor.	62•37 N	9•41 E
9	Opeongo (L.)	Can.	45•40 N	78•20 W
39	Ophir, Mt.	Mala.	2•22 N	102•37 E
11	Opinaca (R.)	Can.	52•28 N	77•40 W
30	Opochka	Sov.Un.	56•43 N	28•39 E
27	Opoczno	Pol.	51•22 N	20•18 E
27	Opole	Pol.	50•42 N	17•55 E
27	Opole Lubelskie	Pol.	51•09 N	21•58 E
27	Oradea	Rom.	47•02 N	21•55 E
18	Oran	Alg.	35•46 N	0•45 W
50	Orán	Arg.	23•13 S	64•17 W
46	Orange	Austl.	33•15 S	149•08 E
9	Orange	Ct.	41•15 N	73•00 W
17	Orange	Fr.	44•08 N	4•48 E
49	Orange, Cabo	Braz.	4•25 N	51•30 W
54	Orange (R.)	Namibia-S.Afr.	29•15 S	17•30 E
54	Orange Free State	S.Afr.	28•15 S	26•00 E
9	Orangeville	Can.	43•55 N	80•06 W
41	Orani	Phil.	14•47 N	120•32 E
26	Oranienburg	G.D.R.	52•45 N	13•14 E
54	Oranjemund	Namibia	28•33 S	16•20 E
21	Orastie	Rom.	45•50 N	23•14 E
20	Orbetello	It.	42•27 N	11•15 E
46	Orbost	Austl.	37•43 S	148•20 E
48	Orchilla	Ven.	11•47 N	66•34 W
44	Ord (R.)	Austl.	17•30 S	128•40 E
38	Ordos Des.	China	39•12 N	108•10 E
31	Ordu	Tur.	41•00 N	37•50 E
31	Ordzhonikidze	Sov.Un.	43•05 N	44•35 E
24	Örebro	Swe.	59•16 N	15•11 E
8	Oregon	Il.	42•01 N	89•21 W
6	Oregon (State)	U.S.	43•40 N	121•50 W
24	Öregrund	Swe.	60•20 N	18•26 E
30	Orekhovo-Zuyevo	Sov.Un.	55•46 N	39•00 E
31	Orël	Sov.Un.	52•54 N	36•03 E
31	Orenburg	Sov.Un.	51•50 N	55•05 E
31	Orgeyev	Sov.Un.	47•27 N	28•49 E
36	Orhon Gol (R.)	Mong.	48•33 N	103•07 E
48	Oriental, Cordillera	Bol.	14•00 S	68•33 W
48	Oriental, Cordillera	Col.	3•30 N	74•27 W
12	Oriental, Sierra Madre	Mex.	25•30 N	100•45 W
9	Orillia	Can.	44•35 N	79•25 W
48	Orinoco, Rio	Ven.	8•32 N	63•13 W
41	Orion	Phil.	14•37 N	120•34 E
20	Oristano	It.	39•53 N	8•38 E
20	Oristano, Golfo di	It.	39•53 N	8•12 E
49	Orituco (R.)	Ven.	9•37 N	66•25 W
49	Oriuco (R.)	Ven.	9•36 N	66•25 W
25	Orivesi (L.)	Fin.	62•15 N	29•55 E
12	Orizaba	Mex.	18•52 N	97•05 E
24	Orkdal	Nor.	63•19 N	9•54 W
24	Örkedalen	Nor.	63•15 N	9•53 E
24	Örken (L.)	Swe.	57•11 N	14•45 E
24	Orkal (R.)	Nor.	62•55 N	9•50 E
22	Orkney (Is.)	Scot.	59•01 N	2•08 W
7	Orlando	Fl.	28•32 N	81•22 W
55	Orlando	S.Afr.	26•15 S	27•56 E
17	Orléans	Fr.	47•55 N	1•56 E
8	Orleans	In.	38•40 N	86•25 W
27	Orneta	Pol.	54•07 N	20•10 E
24	Ornö (I.)	Swe.	59•02 N	18•35 E
16	Örnsköldsvik	Swe.	63•10 N	18•32 E
20	Orobie, Alpi (Mts.)	It.	46•05 N	9•47 E
48	Orocué	Col.	4•48 N	71•26 W
22	Oronsay, Pass of	Scot.	55•55 N	6•25 W
20	Orosei, Golfo di	It.	40•12 N	9•45 E
27	Orosháza	Hung.	46•33 N	20•31 E
8	Orrville	Oh.	40•45 N	81•50 W
24	Orsa	Swe.	61•08 N	14•35 E
24	Örsdals Vand (L.)	Nor.	58•39 N	6•06 E
30	Orsha	Sov.Un.	54•29 N	30•28 E
31	Orsk	Sov.Un.	51•15 N	58•50 E
21	Orsova	Rom.	44•43 N	22•26 E

Page	Name	Region	Lat.	Long.
48	Ortega	Col.	3•56 N	75•12 W
17	Ortegal, Cabo	Sp.	43•46 N	8•15 W
17	Ortigueira	Sp.	43•40 N	7•50 W
20	Ortona	It.	42•22 N	14•22 E
48	Oruro	Bol.	17•57 S	66•59 W
20	Orvieto	It.	42•43 N	12•08 E
21	Oryakhovo	Bul.	43•43 N	23•59 E
24	Os	Nor.	60•24 N	5•22 E
30	Osa	Sov.Un.	57•18 N	55•25 E
37	Ōsaka	Jap.	34•40 N	135•27 E
8	Oscoda	Mi.	44•25 N	83•20 W
8	Osgood	In.	39•10 N	85•20 W
28	Osh	Sov.Un.	40•28 N	72•47 E
9	Oshawa	Can.	43•50 N	78•50 W
8	Oshkosh	Wi.	44•01 N	88•35 W
25	Oshmyany	Sov.Un.	54•27 N	25•55 E
52	Oshogbo	Nig.	7•47 N	4•34 E
21	Osijek	Yugo.	45•33 N	18•48 E
24	Oskarshamm	Swe.	57•16 N	16•24 E
24	Oskarström	Swe.	56•48 N	12•55 E
31	Oskol (R.)	Sov.Un.	51•00 N	37•41 E
24	Oslo	Nor.	59•56 N	10•41 E
24	Oslo Fd.	Nor.	59•03 N	10•35 E
31	Osmaniye	Tur.	37•10 N	36•30 E
26	Osnabrück	F.R.G.	52•16 N	8•05 E
50	Osorno	Chile	40•42 S	73•13 W
45	Osprey Reef	Austl.	14•00 S	146•45 E
46	Ossa, Mt.	Austl.	41•45 S	146•05 E
9	Ossining	NY	41•09 N	73•51 W
24	Ossjöen (L.)	Nor.	61•20 N	12•00 E
30	Ostashkov	Sov.Un.	57•07 N	33•04 E
23	Oste (R.)	F.R.G.	53•20 N	9•19 E
24	Oster-daläven (R.)	Swe.	61•40 N	13•00 E
24	Oster Fd.	Nor.	60•40 N	5•25 E
24	Östersund	Swe.	63•09 N	14•49 E
24	Östhammar	Swe.	60•16 N	18•21 E
27	Ostrava	Czech.	49•51 N	18•18 E
27	Ostróda	Pol.	53•41 N	19•58 E
31	Ostróg	Sov.Un.	50•21 N	26•40 E
31	Ostrogozhsk	Sov.Un.	50•53 N	39•03 E
27	Ostroleka	Pol.	53•04 N	21•35 E
30	Ostrov	Sov.Un.	57•21 N	28•22 E
27	Ostrowiec Swietokrzyski	Pol.	50•55 N	21•24 E
27	Ostrów Lubelski	Pol.	51•32 N	22•49 E
27	Ostrów Mazowiecka	Pol.	52•47 N	21•54 E
27	Ostrów Wielkopolski	Pol.	51•38 N	17•49 E
27	Ostrzeszów	Pol.	51•26 N	17•56 E
21	Ostuni	It.	40•44 N	17•35 E
21	Osum (R.)	Alb.	40•37 N	20•00 E
37	Ōsumi-Guntō (Arch.)	Jap.	30•34 N	130•30 E
9	Oswegatchie (R.)	NY	44•15 N	75•20 W
9	Oswego	NY	43•25 N	76•30 W
27	Oswiecim	Pol.	50•02 N	19•17 E
37	Otaru	Jap.	43•07 N	141•00 E
48	Otavalo	Ec.	0•14 N	78•16 W
54	Otavi	Namibia	19•35 S	17•20 E
21	Othonoí (I.)	Grc.	39•51 N	19•26 E
21	Óthris Óros (Mts.)	Grc.	39•00 N	22•15 E
11	Otish, Mts.	Can.	52•15 N	70•20 W
54	Otjiwarongo	Namibia	20•20 S	16•25 E
20	Otočac	Yugo.	44•53 N	15•15 E
21	Otranto	It.	40•07 N	18•30 E
21	Otranto, C. di	It.	40•06 N	18•32 E
21	Otranto, Strait of	It.-Alb.	40•30 N	18•45 E
8	Otsego	Mi.	42•25 N	85•45 W
24	Ottavand (L.)	Nor.	61•53 N	8•40 E
11	Ottawa	Can.	45•25 N	75•43 W
8	Ottawa	Il.	41•20 N	88•50 W
8	Ottawa	Oh.	41•00 N	84•00 W
11	Ottawa (R.)	Can.	46•05 N	77•20 W
11	Ottawa Is.	Can.	59•50 N	81•00 W
24	Otteråen	Nor.	59•13 N	7•20 E
9	Otter Cr.	Vt.	44•05 N	73•15 W
54	Ottery	S.Afr.	34•02 S	18•31 E
7	Ottumwa	Ia.	41•00 N	92•26 W
46	Otway, C.	Austl.	38•55 S	153•40 E
50	Otway, Seno (B.)	Chile	53•00 S	73•00 W
27	Otwock	Pol.	52•05 N	21•18 E
7	Ouachita (R.)	U.S.	33•25 N	92•30 W
7	Ouachita Mts.	Ok.-Ar.	34•29 N	95•01 W
53	Ouaddaï (Reg.)	Chad	13•04 N	20•00 E
52	Ouagadougou	Upper Volta	12•22 N	1•31 W
52	Ouahigouya	Upper Volta	13•35 N	2•25 W
52	Oualâta	Mauritania	17•11 N	6•50 W
52	Ouallene	Alg.	24•43 N	1•15 E
53	Ouanda Djallé	Cen.Afr.Emp.	8•56 N	22•46 E
52	Ouarane (Dunes)	Mauritania	20•44 N	10•27 W
52	Ouargla	Alg.	32•00 N	5•18 E
51	Oubangui (Ubangi) (R.)	Afr.	4•30 N	20•35 E
18	Oudrhes, L.	Mor.	32•33 N	4•49 W
54	Oudtshoorn	S.Afr.	33•33 S	23•36 E
52	Oued-Zem	Mor.	33•05 N	5•49 W
17	Ouessant, I. d'	Fr.	48•28 N	5•00 W
53	Ouesso	Con.	1•37 N	16•04 E
52	Ouezzane	Mor.	34•48 N	5•40 W
22	Oughter (L.)	Ire.	54•02 N	7•40 W
52	Ouidah	Benin	6•25 N	2•05 E
52	Oujda	Mor.	34•41 N	1•45 W
18	Ouled Nail, Montes des	Alg.	34•43 N	2•44 E
16	Oulu	Fin.	64•58 N	25•43 E
16	Oulu-jarvi (L.)	Fin.	64•20 N	25•48 E
53	Oum Chalouba	Chad	15•48 N	20•30 E
16	Ounas (R.)	Fin.	67•46 N	24•40 E
53	Ounianga Kébir	Chad	19•04 N	20•22 E
49	Ourinhos	Braz.	23•04 S	49•45 W
22	Ouse (R.)	Eng.	53•45 N	1•09 W
13	Outer Brass (I.)	Vir.Is.	18•24 N	64•58 W
22	Outer Hebrides (Is.)	Scot.	57•20 N	7•50 W
54	Outjo	Namibia	20•05 S	17•10 E
46	Ouyen	Austl.	35•05 S	142•10 E
50	Ovalle	Chile	30•43 S	71•16 W
54	Ovamboland	Namibia	18•10 S	15•00 E
16	Övertornea	Swe.	66•19 N	23•31 E
17	Oviedo	Sp.	43•22 N	5•50 W
54	Owando	Con.	0•29 S	15•55 E
9	Owasco (L.)	NY	42•50 N	76•30 W
9	Owego	NY	42•05 N	76•15 W
8	Owen	Wi.	44•56 N	90•35 W
8	Owensboro	Ky.	37•45 N	87•05 W
8	Owen Sound	Can.	44•30 N	80•55 W
41	Owen Stanley Ra.	Pap.N.Gui.	9•00 S	147•30 E
8	Owensville	In.	38•15 N	87•40 W
8	Owenton	Ky.	38•35 N	84•55 W
52	Owerri	Nig.	5•26 N	7•04 E
8	Owosso	Mi.	43•00 N	84•15 W
6	Owyhee Mts.	Id.	43•15 N	116•48 W
6	Owyhee Res.	Or.	43•27 N	117•30 W
6	Owyhee R.	Or.	43•04 N	117•45 W
22	Oxford	Eng.	51•43 N	1•16 W
8	Oxford	Mi.	42•50 N	83•15 W
8	Oxford	Oh.	39•30 N	84•45 W
22	Ox Mts.	Ire.	54•05 N	9•05 W
49	Oyapock (R.)	Braz.-Fr.Gu.	2•45 N	52•15 W
52	Oyem	Gabon	1•37 N	11•35 E
29	Oymyakon	Sov.Un.	63•14 N	142•58 E
52	Oyo	Nig.	7•51 N	3•56 E
41	Ozamiz	Phil.	8•06 N	123•43 E
7	Ozarks, L. of the	Mo.	38•06 N	93•26 W
7	Ozark Plat.	Mo.	36•37 N	93•56 W
20	Ozieri	It.	40•38 N	8•53 E
27	Ozorków	Pol.	51•58 N	19•20 E

P

Page	Name	Region	Lat.	Long.
36	Paan	China	30•08 N	99•00 E
54	Paarl	S.Afr.	33•45 S	18•55 E
27	Pabianice	Pol.	51•40 N	19•29 E
48	Pacaás Novos, Massiço de	Braz.	11•03 S	64•02 W
48	Pacaraima, Serra (Mts.)	Braz.-Ven.	3•45 N	62•30 W
48	Pacasmayo	Peru	7•24 S	79•30 W
36	Pach'u	China	39•50 N	78•23 E
12	Pachuca	Mex.	20•07 N	98•43 W
40	Padang	Indon.	1•01 S	100•28 E
33	Padang, Palau (I.)	Indon.	1•12 N	102•21 E
33	Padang Endau	Mala.	2•39 N	103•38 E
8	Paden City	WV	39•30 N	80•55 W
26	Paderborn	F.R.G.	51•43 N	8•46 E
20	Padova (Padua)	It.	45•24 N	11•53 E
7	Paducah	Ky.	37•05 N	88•36 W
38	Paektu San (Mt.)	China-Kor.	42•00 N	128•03 E
20	Pag (I.)	Yugo.	44•30 N	14•48 E
40	Pagai Selatan (I.)	Indon.	2•48 S	100•22 E
40	Pagai Utara (I.)	Indon.	2•45 S	100•02 E
51	Pagalu (I.)	Equat.Gui.	2•00 S	3•30 E
21	Pagasitikós Kólpos (G.)	Grc.	39•15 N	23•00 E
33	Pahang (State)	Mala.	3•02 N	102•57 E
40	Pahang R.	Mala.	3•39 N	102•41 E
38	Paich'uan	China	47•22 N	126•00 E
25	Paide	Sov.Un.	58•54 N	25•30 E
38	Paiho	China	32•30 N	110•15 E
25	Päijänna (L.)	Fin.	61•38 N	25•05 E
8	Painesville	Oh.	41•40 N	81•15 W
8	Paintsville	Ky.	37•50 N	82•50 W
39	Paise	China	24•00 N	106•38 E
22	Paisley	Scot.	55•50 N	4•30 W
48	Paita	Peru	5•11 S	81•12 W
38	Pai T'ou Shan (Mts.)	Korea	40•30 N	127•20 E
38	Paiyü Shan (Mtns.)	China	37•02 N	108•30 E
40	Pakanbaru	Indon.	0•43 N	101•15 E
32	Pakistan	Asia	28•00 N	67•30 E
40	Pakokku	Bur.	21•29 N	95•00 E
20	Pakrac	Yugo.	45•25 N	17•13 E
27	Paks	Hung.	46•38 N	18•53 E
20	Palagruža (Is.)	Yugo.	42•20 N	16•23 E
29	Palana	Sov.Un.	59•07 N	159•58 E
41	Palanan B.	Phil.	17•14 N	122•35 E
41	Palanan Pt.	Phil.	17•12 N	122•40 E
35	Pälanpur	India	24•08 N	73•29 E
54	Palapye	Bots.	22•34 S	27•28 E
41	Palau Is.	Pac.Is.Trust Ter.	7•15 N	134•30 E
41	Palauig	Phil.	15•27 N	119•54 E
41	Palauig Pt.	Phil.	15•28 N	119•41 E
40	Palawan (I.)	Phil.	9•50 N	117•38 E
25	Paldiski	Sov.Un.	59•22 N	24•04 E
40	Palembang	Indon.	2•57 S	104•40 E
18	Palencia	Sp.	42•02 N	4•32 W
48	Palermo	Col.	2•53 N	75•26 W
20	Palermo	It.	38•08 N	13•24 E
7	Palestine	Tx.	31•46 N	95•38 W
33	Palestine (Reg.)	Asia	31•33 N	35•00 E
36	Paletwa	Bur.	21•19 N	92•52 E
36	Palik'un	China	43•43 N	92•50 E
51	Palimé	Togo	6•54 N	0•38 E
35	Palk Str.	India	10•00 N	79•23 E
18	Palma de Mallorca	Sp.	39•35 N	2•38 E
49	Palmares	Braz.	8•46 S	35•28 W
50	Palmas	Braz.	26•20 S	51•56 W
52	Palmas, C.	Lib.	4•22 N	7•44 W
49	Palmeira dos Índios	Braz.	9•26 S	36•33 W
45	Palmerston North	N.Z.	40•21 S	175•43 E
45	Palmerville	Austl.	16•08 S	144•15 E
20	Palmi	It.	38•21 N	15•54 E
48	Palmira	Col.	3•33 N	76•17 W
43	Palmyra (I.)	Oceania	6•00 N	162•20 W
34	Palmyra (Ruins)	Syr.	34•25 N	38•28 E
15	Palmyre	Syr.	30•35 N	37•58 E
33	Paloh	Mala.	2•11 N	103•12 E
18	Palos, Cabo de (C.)	Sp.	39•38 N	0•43 W
31	Palu	Tur.	38•55 N	40•10 E
41	Paluan	Phil.	13•25 N	120•29 E
17	Pamiers	Fr.	43•07 N	1•34 E
35	Pamirs (Plat.)	Sov.Un.	38•14 N	72•27 E
7	Pamlico Sd.	NC	35•10 N	76•10 W
6	Pampa	Tx.	35•32 N	100•56 W
50	Pampa de Castillo (Plat.)	Arg.	45•30 S	67•30 W
41	Pampanga (R.)	Phil.	15•20 N	120•48 E
50	Pampas (Reg.)	Arg.	37•00 S	64•30 W
48	Pamplona	Col.	7•19 N	72•41 W
18	Pamplona	Sp.	42•49 N	1•39 W
9	Pamunkey (R.)	Va.	37•40 N	77•20 W
8	Pana	Il.	39•25 N	89•05 W
21	Panagyurishte	Bul.	42•30 N	24•11 E
35	Panaji (Panjim)	India	15•33 N	73•52 E
13	Panamá	N.A.	8•35 N	81•08 W
13	Panama, G. of	Pan.	7•45 N	79•20 W
13	Panama, Isth. of	Pan.	9•00 N	81•00 W
20	Panaria (Is.)	It.	38•37 N	15•05 E
20	Panaro (R.)	It.	44•47 N	11•06 E
40	Panay (I.)	Phil.	11•15 N	121•38 E
21	Pančevo	Yugo.	44•52 N	20•42 E
33	Panchor	Mala.	2•10 N	102•43 E
54	Panda	Zaire	10•59 S	27•24 E
31	Pandar-e Pahlavī	Iran	37•30 N	49•30 E
25	Panevėžys	Sov.Un.	55•44 N	24•21 E
28	Panfilov	Sov.Un.	44•12 N	79•58 E
53	Panga	Zaire	1•51 N	26•25 E
55	Pangani	Tan.	5•28 S	38•58 E
40	Pangkalpinang	Indon.	2•11 S	106•04 E
11	Pangnirtung	Can.	66•08 N	65•26 W
40	Panjang, Selat (Str.)	Indon.	1•00 N	102•00 E
38	Panshih	China	42•50 N	126•48 E
39	Pan Si Pan (Mtn.)	Viet.	22•25 N	103•50 E
41	Pantar (I.)	Indon.	8•40 S	123•45 E
17	Pantelleria (I.)	It.	36•43 N	11•59 E
12	Panuco (R.)	Mex.	21•59 N	98•20 W
35	Panvel	India	18•59 N	73•06 E
49	Pao (R.)	Ven.	9•52 N	67•57 W
38	Paochang	China	41•52 N	115•25 E
38	Paocheng	China	33•15 N	106•58 E
38	Paochi	China	34•10 N	106•58 E
8	Paoli	In.	38•35 N	86•30 W
36	Paoshan	China	25•14 N	99•03 E
37	Paoting	China	38•52 N	115•31 E
38	Paoting	China	42•04 N	125•00 E
38	Paot'ou	China	40•28 N	110•10 E
38	Paoying	China	33•14 N	119•20 E
27	Pápa	Hung.	47•18 N	17•27 E
12	Papantla de Olarte	Mex.	20•30 N	97•15 W
26	Papenburg	F.R.G.	53•05 N	7•23 E
41	Papua, Gulf of	Pap.N.Gui.	8•20 S	144•45 E
41	Papua New Guinea	Oceania	7•00 S	142•15 E
50	Paquequer Pequeno	Braz.	22•19 S	43•02 W
49	Pará, Rio do (R.)	Braz.	1•09 S	48•48 W
41	Paracale	Phil.	14•17 N	122•47 E
50	Paracambi	Braz.	22•36 S	43•43 W
49	Paracatu	Braz.	17•17 S	46•43 W
21	Paraćin	Yugo.	43•51 N	21•26 E
48	Parados, Cerro de los (Mtn.)	Col.	5•44 N	75•13 W
49	Paraguaçu (R.)	Braz.	12•25 S	39•46 W
48	Paraguaná, Pen. de	Ven.	12•00 N	69•55 W
47	Paraguay	S.A.	24•00 S	57•00 W
49	Paraguay, Rio	S.A.	21•12 S	57•31 W
12	Paraiso	C.Z.	9•02 N	79•38 W
52	Parakou	Benin	9•21 N	2•37 E
49	Paramaribo	Sur.	5•50 N	55•15 W
48	Paramillo (Mtn.)	Col.	7•06 N	75•55 W
33	Paran (R.)	Isr.	30•05 N	34•50 E
50	Paraná	Arg.	31•44 S	60•29 W
50	Paraná, Rio (R.)	Arg.	32•15 S	60•55 W
49	Paraná (R.)	Braz.	13•05 S	47•11 W
49	Paranaguá	Braz.	25•39 S	48•42 W
49	Paranaíba	Braz.	19•43 S	51•13 W
49	Paranaíba (R.)	Braz.	18•58 S	50•44 W
49	Paranam	Sur.	5•39 N	55•13 W
50	Paranápanema (R.)	Braz.	22•28 S	52•15 W
49	Parapara	Ven.	9•44 N	67•17 W
26	Parchim	G.D.R.	53•25 N	11•52 E
27	Parczew	Pol.	51•38 N	22•53 E
49	Pardo (R.)	Braz.	15•25 S	39•40 W
26	Pardubice	Czech.	50•02 N	15•47 E
49	Parecis, Serra dos (Mts.)	Braz.	13•45 S	59•28 W
11	Parent	Can.	47•59 N	74•30 W
48	Paria, Golfo de (G.)	Ven.	10•33 N	62•14 W
48	Parima, Serra (Mts.)	Braz.-Ven.	3•45 N	64•00 W
48	Pariñas, Punta (Pt.)	Peru	4•30 S	81•23 W
49	Parintins	Braz.	2•34 S	56•30 W
8	Paris	Can.	43•15 N	80•23 W
17	Paris	Fr.	48•51 N	2•20 E
8	Paris	Il.	39•35 N	87•40 W
8	Paris	Ky.	38•15 N	84•15 W
7	Paris	Tx.	33•39 N	95•33 W
6	Parker Dam	Az.-Ca.	34•20 N	114•00 W
8	Parkersburg	WV	39•15 N	81•35 W
46	Parkes	Austl.	33•10 S	148•10 E
55	Park Rynie	S.Afr.	30•22 S	30•43 E
20	Parma	It.	44•48 N	10•20 E
6	Parma	Oh.	41•23 N	81•44 W
49	Parnaguá	Braz.	9•52 S	44•27 W
49	Parnaíba	Braz.	3•00 S	41•42 W
49	Parnaiba (R.)	Braz.	3•57 S	42•30 W
21	Parnassós (Mtn.)	Grc.	38•36 N	22•35 E
25	Pärnu	Sov.Un.	58•24 N	24•29 E
25	Pärnu (R.)	Sov.Un.	58•40 N	25•05 E
25	Pärnu Laht (B.)	Sov.Un.	58•15 N	24•17 E
46	Paroo (R.)	Austl.	29•40 S	144•24 E
34	Paropamisus (Mts.)	Afg.	34•45 N	63•58 E
21	Páros	Grc.	37•05 N	25•14 E
21	Páros (I.)	Grc.	37•11 N	25•00 E
54	Parow	S.Afr.	33•54 S	18•36 E
50	Parral	Chile	36•07 S	71•47 W
8	Parry (I.)	Can.	45•15 N	80•00 W
5	Parry Is.	Can.	75•30 N	110•00 W
9	Parry Sound	Can.	45•20 N	80•00 W
7	Parsons	Ks.	37•20 N	95•16 W
9	Parsons	WV	39•05 N	79•40 W
20	Partinico	It.	38•02 N	13•11 E
29	Partizansk	Sov.Un.	43•15 N	133•19 E
6	Pasadena	Ca.	34•09 N	118•09 W
27	Paşcani	Rom.	47•46 N	26•42 E
26	Pasewalk	G.D.R.	53•31 N	14•01 E
41	Pasig	Phil.	14•34 N	121•05 E
50	Paso de los Libres	Arg.	29•33 S	57•05 W
7	Passaic	NJ	40•52 N	74•08 W
26	Passua	F.R.G.	48•34 N	13•27 E
17	Passero, C.	It.	36•34 N	15•13 E
50	Passo Fundo	Braz.	28•16 S	52•13 W
49	Passos	Braz.	20•45 S	46•37 W
48	Pastaza (R.)	Peru	3•05 S	76•18 W
48	Pasto	Col.	1•15 N	77•19 W
40	Pasuruan	Indon.	7•45 S	112•50 E
25	Pasvalys	Sov.Un.	56•04 N	24•23 E
50	Patagonia (Reg.)	Arg.	46•45 S	69•30 W
20	Paternò	It.	37•25 N	14•58 E
7	Paterson	NJ	40•55 N	74•10 W
35	Patiāla	India	30•25 N	76•28 E
50	Pati do Alferes	Braz.	22•25 S	43•25 W
35	Patna	India	25•33 N	85•18 E
41	Patnanongan	Phil.	14•50 N	122•25 E
8	Patoka (R.)	Ind.	38•25 N	87•25 W
29	Patom Plat.	Sov.Un.	59•30 N	115•00 E
49	Patos	Braz.	7•03 S	37•14 W
50	Patos, Lago dos	Braz.	31•15 S	51•30 W
49	Patos de Minas	Braz.	18•39 S	46•31 W
21	Pátrai (Patras)	Grc.	38•15 N	21•48 E
21	Patraïkós Kólpos (G.)	Grc.	38•16 N	21•19 E
49	Patrocínio	Braz.	18•48 S	46•47 W
40	Pattani	Thai.	6•56 N	101•13 E
9	Patton	Pa.	40•40 N	78•45 W
9	Patuxent (R.)	Md.	39•10 N	77•10 W
17	Pau	Fr.	43•18 N	0•23 W
8	Paulding	Oh.	41•05 N	84•35 W
49	Paulistana	Braz.	8•13 S	41•06 W
49	Paulo Afonso, Salto (falls)	Braz.	9•33 S	38•32 W
48	Pavarandocito	Col.	7•18 N	76•32 W
20	Pavia	It.	45•12 N	9•11 E
28	Pavlodar	Sov.Un.	52•17 N	77•23 E
31	Pavlograd	Sov.Un.	48•32 N	35•52 E
30	Pavlovskiy Posad	Sov.Un.	55•47 N	38•39 E
50	Pavuna	Braz.	22•48 S	43•21 W
8	Paw Paw	Mi.	42•15 N	85•55 W
9	Pawtucket	RI	41•53 N	71•23 W
21	Paxoi (I.)	Grc.	39•14 N	20•15 E
8	Paxton	Il.	40•35 N	88•00 W
38	Payen	China	46•00 N	127•20 E
30	Pay-Khoy, Khrebet (Mts.)	Sov.Un.	68•08 N	63•04 E
11	Payne (L.)	Can.	59•22 N	73•16 W
50	Paysandú	Ur.	32•16 S	57•55 W

Page	Name	Region	Lat.	Long.
21	Pazardzhik	Bul.	42•10 N	24•22 E
20	Pazin	Yugo.	45•14 N	13•57 E
9	Peabody	Ma.	42•32 N	70•56 W
10	Peace (R.)	Can.	55•40 N	118•30 W
10	Peace River	Can.	56•14 N	117•17 W
10	Peacock Hills	Can.	66•08 N	109•55 W
44	Peak Hill	Austl.	25•38 S	118•50 E
7	Pearl (R.)	La.-Ms.	31•06 N	89•44 W
6	Pearl Harbor	Hi.	21•20 N	157•53 W
55	Pearston	S.Afr.	32•36 S	25•09 E
66	Peary Land	Grnld.	82•00 N	40•00 W
55	Pebane	Moz.	17•10 S	38•08 E
21	Peć	Yugo.	42•39 N	20•18 E
49	Peçanha	Braz.	18•37 S	42•26 W
30	Pechenga	Sov.Un.	69•30 N	31•10 E
30	Pechora (R.)	Sov.Un.	66•00 N	52•30 E
30	Pechora Basin	Sov.Un.	67•55 N	58•37 E
30	Pechorskaya Guba (B.)	Sov.Un.	68•40 N	55•00 E
6	Pecos (R.)	U.S.	31•10 N	103•10 W
27	Pécs	Hung.	46•04 N	18•15 E
55	Peddie	S.Afr.	33•13 S	27•09 E
49	Pedra Azul	Braz.	16•03 S	41•13 W
49	Pedreiras	Braz.	4•30 S	44•31 W
50	Pedro de Valdivia	Chile	22•32 S	69•55 W
50	Pedro do Rio	Braz.	22•20 S	43•09 W
49	Pedro Juan Caballero	Par.	22•40 S	55•42 W
12	Pedro Miguel	C.Z.	9•01 N	79•36 W
12	Pedro Miguel Locks	C.Z.	9•01 N	79•36 W
49	Pedro II	Braz.	4•20 S	41•27 W
46	Peebinga	Austl.	34•43 S	140•55 E
22	Peebles	Scot.	55•40 N	3•15 W
7	Pee Dee (R.)	NC-SC	34•01 N	79•26 W
9	Peekskill	NY	41•17 N	73•55 W
45	Pegasus B.	N.Z.	43•18 S	173•37 E
26	Pegnitz R.	F.R.G.	49•38 N	11•40 E
40	Pegu	Bur.	17•17 N	96•29 E
36	Pegu Yoma (Mts.)	Bur.	19•16 N	95•59 E
21	Pehčevo	Yugo.	41•42 N	22•57 E
38	Peian	China	48•05 N	126•26 E
38	Peiching (Peking)	China	39•55 N	116•23 E
39	Peihai	China	21•30 N	109•10 E
39	Peili	China	19•08 N	108•42 E
38	Peiyün Ho	China	39•42 N	116•48 E
8	Pekin	Il.	40•35 N	89•30 W
38	Peking-Shih (Mun.)	China	40•07 N	116•00 E
18	Pelagie, Isole I.	It.	35•46 N	12•32 E
21	Pélagos (I.)	Grc.	39•17 N	24•05 E
17	Pelat, Mt.	Fr.	44•16 N	6•43 E
29	Peleduy	Sov.Un.	59•50 N	112•47 E
8	Pelee, Pt.	Can.	41•55 N	82•30 W
8	Pelee I.	Can.	41•45 N	82•30 W
26	Pell-Worm I.	F.R.G.	54•33 N	8•25 E
10	Pelly (L.)	Can.	66•08 N	102•57 W
10	Pelly (R.)	Can.	62•20 N	113•26 W
10	Pelly B.	Can.	68•57 N	91•05 W
10	Pelly Mts.	Can.	61•50 N	133•05 W
21	Peloponnisos	Grc.	37•28 N	22•14 E
50	Pelotas	Braz.	31•45 S	52•18 W
17	Pelvoux, Mt.	Fr.	44•56 N	6•24 E
30	Pelym (R.)	Sov.Un.	60•20 N	63•05 E
33	Pemanggil (I.)	Mala.	2•37 N	104•41 E
54	Pemba	Zambia	15•29 S	27•22 E
55	Pemba (I.)	Tan.	5•20 S	39•57 E
9	Pembroke	Can.	45•50 N	77•00 W
22	Pembroke	Wales	51•40 N	5•00 W
17	Peñalara (Mtn.)	Sp.	40•52 N	3•57 W
41	Penaranda	Phil.	15•20 N	120•59 E
17	Peñarroya-Peublonuevo	Sp.	38•18 N	5•18 W
17	Peñas, Cabo de	Sp.	43•42 N	6•12 W
50	Penas, Golfo de	Chile	47•15 S	77•30 W
38	Pench'i	China	41•25 N	123•50 E
52	Pendembu	S.L.	8•06 N	10•42 W
48	Penderisco (R.)	Col.	6•30 N	76•21 W
6	Pendleton	Or.	45•41 N	118•47 W
6	Pend Oreille L.	Id.	48•09 N	116•38 W
49	Penedo	Braz.	10•17 S	36•28 W
9	Penetanguishene	Can.	44•45 N	79•55 W
38	P'englai	China	37•49 N	120•45 E
20	Penne	It.	42•28 N	13•57 E
35	Penner (R.)	India	14•43 N	79•09 E
26	Pennine Alpi	Switz.	46•02 N	7•07 E
22	Pennine Chain (Mts.)	Eng.	53•44 N	1•59 W
8	Pennsboro	WV	39•10 N	81•00 W
7	Pennsylvania	U.S.	41•00 N	78•10 W
9	Penn Yan	NY	42•40 N	77•00 W
7	Penobscot (R.)	Me.	45•00 N	68•36 W
44	Penong	Austl.	32•00 S	133•00 E
7	Pensacola	Fl.	30•25 N	87•13 W
45	Pentecost (I.)	New Hebr.	16•05 S	168•28 E
10	Penticton	Can.	49•30 N	119•35 W
22	Pentland Firth	Scot.	58•44 N	3•25 W
31	Penza	Sov.Un.	53•10 N	45•00 E
22	Penzance	Eng.	50•07 N	5•40 W
26	Penzberg	F.R.G.	47•43 N	11•21 E
29	Penzhino	Sov.Un.	63•42 N	168•00 E
8	Peoria	Il.	40•45 N	89•35 W
9	Pepacton Res.	NY	42•05 N	74•40 W
21	Peqin	Alb.	41•03 N	19•48 E

Page	Name	Region	Lat.	Long.
17	Perdido, Mt.	Sp.	42•40 N	0•00
48	Pereira	Col.	4•49 N	75•42 W
31	Perekop	Sov.Un.	46•08 N	33•39 E
8	Pere Marquette	Mi.	43•55 N	86•10 W
30	Pereslavl'-Zalesskiy	Sov.Un.	56•43 N	38•52 E
30	Pereyaslav	Sov.Un.	50•05 N	31•25 E
50	Pergamino	Arg.	33•53 S	60•36 W
17	Périgueux	Fr.	45•12 N	0•43 E
48	Perija, Sierra de	Col.	9•25 N	73•30 W
41	Perkam, Tandjung (C.)	Indon.	1•20 S	138•45 E
26	Perleberg	G.D.R.	53•06 N	11•51 E
30	Perm	Sov.Un.	58•00 N	56•15 E
49	Pernambuco	Braz.	8•08 S	38•54 W
21	Pernik	Bul.	42•36 N	23•04 E
17	Perpignan	Fr.	42•42 N	2•48 E
9	Perry	NY	42•45 N	78•00 W
8	Perrysburg	Oh.	41•35 N	83•35 W
34	Persepolis (Ruins)	Iran	30•15 N	53•08 E
34	Persian G.	Asia	27•38 N	50•30 E
44	Perth	Austl.	31•50 S	116•10 E
9	Perth	Can.	44•40 N	76•15 W
22	Perth	Scot.	56•24 N	3•25 W
9	Perth Amboy	NJ	40•31 N	74•16 W
8	Peru	Il.	41•20 N	89•10 W
8	Peru	In.	40•45 N	86•00 W
47	Peru	S.A.	10•00 S	75•00 W
20	Perugia	It.	43•08 N	12•24 E
31	Pervomaysk	Sov.Un.	48•04 N	30•52 E
20	Pesaro	It.	43•54 N	12•55 E
49	Pescado (R.)	Ven.	9•33 N	65•32 W
20	Pescara	It.	42•26 N	14•15 E
20	Pescara (R.)	It.	42•18 N	13•22 E
31	Peschanyy, Mys (C.)	Sov.Un.	43•10 N	51•20 E
20	Pescia	It.	43•53 N	11•42 E
35	Peshāwar	Pak.	34•01 N	71•34 E
21	Peshtera	Bul.	42•03 N	24•19 E
8	Peshtigo	Wi.	45•03 N	87•46 W
33	Petah Tiqwa	Isr.	32•05 N	34•53 E
49	Petare	Ven.	10•28 N	66•48 W
8	Petenwell Res.	Wi.	44•10 N	89•55 W
9	Peterborough	Can.	44•20 N	78•20 W
46	Peterborough	Austl.	32•53 S	138•58 E
22	Peterborough	Eng.	52•35 N	0•14 W
22	Peterhead	Scot.	57•36 N	3•47 W
9	Peter Pt.	Can.	43•50 N	77•00 W
10	Peter Pond L.	Can.	55•55 N	108•44 W
8	Petersburg	Il.	40•01 N	89•51 W
8	Petersburg	In.	38•30 N	87•15 W
7	Petersburg	Va.	37•12 N	77•30 W
8	Petoskey	Mi.	45•25 N	84•55 W
33	Petra	Jordan	30•21 N	35•25 E
21	Petrich	Bul.	41•24 N	23•13 E
20	Petrinja	Yugo.	45•25 N	16•17 E
30	Petrokrepost	Sov.Un.	59•56 N	31•03 E
8	Petrolia	Can.	42•50 N	82•10 W
49	Petrolina	Braz.	9•18 S	40•28 W
28	Petropavlovsk	Sov.Un.	54•44 N	69•07 E
29	Petropavlovsk-Kamchatskiy	Sov.Un.	53•13 N	158•56 E
50	Petrópolis	Braz.	22•31 S	43•10 W
21	Petroseni	Rom.	45•24 N	23•24 E
31	Petrovsk	Sov.Un.	52•20 N	45•15 E
31	Petrovskoye	Sov.Un.	45•20 N	43•00 E
29	Petrovsk-Zabaykal'skiy	Sov.Un.	51•13 N	109•08 E
25	Petrozavodsk	Sov.Un.	61•46 N	34•25 E
30	Peza (R.)	Sov.Un.	65•35 N	46•50 E
26	Pforzheim	F.R.G.	48•52 N	8•43 E
40	Phan Rang	Viet.	11•30 N	108•43 E
40	Phet Buri	Thai.	13•07 N	99•53 E
9	Philadelphia	Pa.	40•00 N	75•13 W
33	Philippines	Asia	14•25 N	125•00 E
42	Philippine Sea	Asia	16•00 N	133•00 E
41	Philippine Trench	Phil.	10•30 N	127•15 E
9	Philipsburg	Pa.	40•55 N	78•10 W
46	Phillip (I.)	Austl.	38•32 S	145•10 E
33	Phillip Chan.	Indon.	1•04 N	103•40 E
9	Phillipi	WV	39•10 N	80•00 W
8	Phillips	Wi.	45•41 N	90•24 W
9	Phillipsburg	NJ	40•45 N	75•10 W
40	Phnom Penh	Camb.	11•39 N	104•53 E
6	Phoenix	Az.	33•30 N	112•00 W
42	Phoenix Is.	Oceania	4•00 S	174•00 W
9	Phoenixville	Pa.	40•08 N	75•31 W
40	Phu Bia (Pk.)	Laos	19•36 N	103•00 E
40	Phu-Quoc (I.)	Camb.	10•13 N	104•00 E
40	Phuket	Thai.	7•57 N	98•19 E
20	Piacenza	It.	45•02 N	9•42 E
20	Pianosa (I.)	It.	42•13 N	15•45 E
27	Piatra-Neamt	Rom.	46•54 N	26•24 E
49	Piauí	Braz.	7•40 S	42•25 W
49	Piauí, Serra do	Braz.	10•45 S	44•36 W
20	Piave (R.)	It.	45•45 N	12•15 E
20	Piazza Armerina	It.	37•23 N	14•26 E
53	Pibor R.	Sud.	7•21 N	32•54 E
13	Picara Pt.	Vir.Is.	18•23 N	64•57 W
20	Piccole Alpi Dolomitche (Mts.)	It.	46•05 N	12•17 E
39	Pichieh	China	27•20 N	105•18 E
17	Pico de Aneto (Mtn.)	Sp.	42•35 N	0•38 E
52	Pico I.	Açores	38•16 N	28•49 W
49	Picos	Braz.	7•13 S	41•23 W

Page	Name	Region	Lat.	Long.
33	Pidálion, Akrotírion (C.)	Cyprus	34•50 N	34•05 E
35	Pidurutalagala Mt.	Sri Lanka	12•27 N	80•45 E
9	Piedmont	WV	39•30 N	79•05 W
12	Piedras Negras	Mex.	28•41 N	100•33 W
25	Pieksämäki	Fin.	62•18 N	27•14 E
20	Piemonte	It.	44•30 N	7•42 E
9	Pierce	WV	39•15 N	79•30 W
6	Pierre	SD	44•22 N	100•20 W
27	Piešťany	Czech.	48•36 N	17•48 E
55	Pietermaritzburg	S.Afr.	29•36 S	30•23 E
54	Pietersburg	S.Afr.	23•56 S	29•30 E
54	Piet Retief	S.Afr.	27•00 S	30•58 E
27	Pietrosul Pk.	Rom.	47•35 N	24•49 E
20	Pieve di Cadore	It.	46•26 N	12•22 E
6	Pikes Pk.	Co.	38•49 N	105•03 W
26	Pila	Pol.	53•09 N	16•44 E
50	Pilar	Par.	27•00 S	58•15 W
41	Pilar	Phil.	12•55 N	123•41 E
41	Pilar	Phil.	17•24 N	120•36 E
49	Pilar de Goiás	Braz.	14•47 S	49•33 W
50	Pilcomayo (R.)	Par.	24•45 S	69•15 W
41	Pili	Phil.	13•34 N	123•17 E
27	Pilica R.	Pol.	51•00 N	19•48 E
25	Piltene	Sov.Un.	57•17 N	21•40 E
44	Pimba	Austl.	31•15 S	146•50 E
55	Pimville	S.Afr.	26•17 S	27•54 E
12	Pinacate, Cerro	Mex.	31•45 N	113•30 W
41	Pinamalayan	Phil.	13•04 N	121•31 E
40	Pinang	Mala.	5•21 N	100•09 E
31	Pinarbasi	Tur.	38•50 N	36•10 E
13	Pinar del Río	Cuba	22•25 N	83•35 W
41	Pinatubo (Mtn.)	Phil.	15•09 N	120•19 E
8	Pinckneyville	Il.	38•06 N	89•22 W
27	Pińczów	Pol.	50•32 N	20•33 E
21	Píndhos Oros (Mts.)	Grc.	39•48 N	21•19 E
7	Pine Bluff	Ar.	34•13 N	92•01 W
44	Pine Creek	Austl.	13•45 S	132•00 E
30	Pinega	Sov.Un.	64•40 N	43•30 E
30	Pinega (R.)	Sov.Un.	64•10 N	42•30 E
54	Pinelands	S.Afr.	33•57 S	18•30 E
20	Pinerolo	It.	44•47 N	7•18 E
55	Pinetown	S.Afr.	29•47 S	30•52 E
40	Ping, Mae Nam (R.)	Thai.	17•54 N	98•29 E
38	Pingchüan	China	40•58 N	118•40 E
33	Pinggir	Indon.	1•05 N	101•12 E
39	P'ingho	China	24•30 N	117•02 E
39	Pinghsiang	China	27•40 N	113•50 E
38	Pingliang	China	35•12 N	106•50 E
39	P'inglo	China	24•30 N	110•22 E
39	P'ingt'an	China	25•30 N	119•45 E
38	Pingting	China	37•50 N	113•30 E
39	P'ingtung	Taiwan	22•40 N	120•35 E
38	P'ingwu	China	32•20 N	104•40 E
38	Pinhsien	China	45•40 N	127•20 E
40	Pini (I.)	Indon.	0•07 N	98•38 E
21	Piniós (R.)	Grc.	40•33 N	21•40 E
13	Pinos, Isla de	Cuba	21•40 N	82•45 W
45	Pins, Ile des	N.Cal.	22•44 S	167•44 E
27	Pinsk	Sov.Un.	52•07 N	26•05 E
48	Pinta (I.)	Ec.	0•41 N	90•47 W
20	Piombino	It.	42•56 N	10•33 E
27	Piotrków Trybunalski	Pol.	51•23 N	19•44 E
21	Pipéri (I.)	Grc.	39•19 N	24•20 E
8	Piqua	Oh.	40•10 N	84•15 W
49	Piracicaba	Braz.	22•43 S	47•39 W
29	Piramida, Gol'tsy (Mtn.)	Sov.Un.	54•00 N	96•00 E
20	Piran	Yugo.	45•31 N	13•34 E
49	Pirapora	Braz.	17•39 S	44•54 W
49	Pirenópolis	Braz.	15•56 S	48•49 W
21	Pírgos	Grc.	37•51 N	21•28 E
49	Piritu, Laguna de	Ven.	10•00 N	64•57 W
26	Pirmasens	F.R.G.	49•12 N	7•34 E
26	Pirna	G.D.R.	50•57 N	13•56 E
21	Pirot	Yugo.	43•09 N	22•35 E
41	Piru	Indon.	3•15 S	128•25 E
31	Piryatin	Sov.Un.	50•13 N	32•31 E
20	Pisa	It.	43•52 N	10•24 E
48	Pisagua	Chile	18•43 S	70•12 W
48	Pisco	Peru	13•43 S	76•07 W
48	Pisco, Bahia de	Peru	13•43 S	77•48 W
9	Piseco (L.)	NY	43•25 N	74•35 W
26	Pisek	Czech.	49•18 N	14•08 E
20	Pisticci	It.	40•24 N	16•34 E
20	Pistoia	It.	43•57 N	11•54 E
48	Pitalito	Col.	1•45 N	75•09 W
43	Pitcairn	Oceania	24•30 S	133•00 W
16	Pite (R.)	Swe.	66•08 N	18•51 E
16	Piteå	Swe.	65•21 N	21•10 E
21	Pitesti	Rom.	44•51 N	24•51 E
44	Pithara	Austl.	30•27 S	116•45 E
55	Pitseng	Leso.	29•03 S	28•13 E
7	Pittsburg	Ks.	37•25 N	94•43 W
9	Pittsburgh	Pa.	40•26 N	80•01 W
9	Pittsfield	Ma.	42•25 N	73•15 W
9	Pittston	Pa.	41•20 N	75•50 W
38	P'itzuwo (Hsinchin)	China	39•25 N	122•19 E
48	Piura	Peru	5•13 S	80•46 W
11	Placentia B.	Can.	47•14 N	54•30 W
9	Placid (L.)	NY	44•20 N	74•00 W
9	Plainfield	NJ	40•38 N	74•25 W
8	Plainwell	Mi.	42•25 N	85•40 W

Page	Name	Region	Lat.	Long.
17	Plasencia	Sp.	40•02 N	6•07 W
30	Plast	Sov.Un.	54•22 N	60•48 E
50	Plata, R. de la	Arg.-Ur.	34•35 S	58•15 W
20	Platani (R.)	It.	37•26 N	13•28 E
48	Plato	Col.	9•49 N	74•48 W
6	Platte (R.)	U.S.	40•50 N	100•40 W
9	Plattsburgh	NY	44•40 N	73•30 W
26	Plauen	G.D.R.	50•30 N	12•08 E
9	Pleasant (L.)	NY	43•25 N	74•25 W
45	Plenty, B. of	N.Z.	37•23 S	177•10 E
27	Pleszew	Pol.	51•54 N	17•48 E
21	Pleven	Bul.	43•24 N	24•26 E
21	Pljevlja	Yugo.	43•20 N	19•21 E
27	Plock	Pol.	52•32 N	19•44 E
21	Ploeşti	Rom.	44•56 N	26•01 E
21	Plomárion	Grc.	38•51 N	26•24 E
17	Plomb du Cantal (Mt.)	Fr.	45•30 N	2•49 E
21	Plovdiv (Philippopolis)	Bul.	42•09 N	24•43 E
25	Plunge	Sov.Un.	55•56 N	21•45 E
22	Plymouth	Eng.	50•25 N	4•14 W
8	Plymouth	In.	41•20 N	86•20 W
9	Plymouth	Ma.	42•00 N	70•45 W
9	Plymouth	NH	43•50 N	71•40 W
9	Plymouth	Pa.	41•15 N	75•55 W
8	Plymouth	Wi.	43•45 N	87•59 W
26	Plzeň (Pilsen)	Czech.	49•46 N	13•25 E
20	Po (R.)	It.	45•00 N	11•23 E
38	Poar	China	35•10 N	113•08 E
52	Pobé	Benin	6•58 N	2•41 E
6	Pocatello	Id.	42•54 N	112•30 W
31	Pochëp	Sov.Un.	52•56 N	32•27 E
30	Pochinski	Sov.Un.	54•40 N	44•50 E
9	Pocomoke City	Md.	38•05 N	75•35 W
9	Pocono Mts.	Pa.	41•10 N	75•05 W
49	Poços de Caldas	Braz.	21•48 S	46•34 W
52	Poder	Senegal	16•35 N	15•04 W
29	Podkamennaya (Stony) Tunguska (R.)	Sov.Un.	61•43 N	93•45 E
30	Podol'sk	Sov.Un.	55•26 N	37•33 E
20	Poggibonsi	It.	43•27 N	11•12 E
38	Pohai Str.	China	38•05 N	121•40 E
38	P'ohang	Kor.	35•57 N	129•23 E
38	Pohsien	China	33•52 N	115•47 E
13	Pointe-à-Pitre	Guad.	16•15 N	61•32 W
54	Pointe Noire	Con.	4•48 S	11•51 E
8	Point Pleasant	WV	38•50 N	82•10 W
17	Poitiers	Fr.	46•35 N	0•18 E
38	Pok'ot'u	China	48•45 N	121•42 E
14	Poland	Eur.	52•37 N	17•01 E
41	Polangui	Phil.	13•18 N	123•29 E
25	Polessk	Sov.Un.	54•50 N	21•14 E
31	Poles'ye (Pripyat Marshes)	Sov.Un.	52•10 N	27•30 E
27	Polgár	Hung.	47•54 N	21•10 E
38	P'oli	China	45•40 N	130•38 E
20	Policastro, Golfo di	It.	41•00 N	13•23 E
21	Políkhnitos	Grc.	39•05 N	26•11 E
41	Polillo	Phil.	14•42 N	121•56 E
41	Polillo Is.	Phil.	15•05 N	122•15 E
41	Polillo Str.	Phil.	15•02 N	121•40 E
20	Polistena	It.	40•25 N	16•05 E
21	Poliyiros	Grc.	40•23 N	23•27 E
29	Polkan, Gol'tsy (Mtn.)	Sov.Un.	60•18 N	92•08 E
30	Polotsk	Sov.Un.	55•30 N	28•48 E
31	Poltava	Sov.Un.	49•35 N	34•33 E
30	Polyarnyy	Sov.Un.	69•10 N	33•30 E
26	Pomerania	Pol.	53•50 N	15•20 E
24	Pomeranian B.	G.D.R.	54•10 N	14•20 E
55	Pomeroy	S.Afr.	28•36 S	30•26 E
6	Pomona	Ca.	34•04 N	117•45 W
21	Pomorie	Bul.	42•24 N	27•41 E
13	Ponce	P.R.	18•01 N	66•43 W
35	Pondicherry	India	11•58 N	79•48 E
17	Ponferrada	Sp.	42•33 N	6•38 W
10	Ponoka	Can.	52•42 N	113•35 W
30	Ponoy	Sov.Un.	66•58 N	41•00 E
30	Ponoy (R.)	Sov.Un.	65•50 N	38•40 E
52	Ponta Delgada	Açores	37•40 N	25•45 W
50	Ponta Grossa	Braz.	25•09 S	50•05 W
49	Ponta Porã	Braz.	22•30 S	55•31 W
20	Pontedera	It.	43•37 N	10•37 E
49	Ponte Nova	Braz.	20•26 S	42•52 W
17	Pontevedra	Sp.	42•28 N	8•38 W
8	Pontiac	Il.	40•55 N	88•35 W
8	Pontiac	Mi.	42•37 N	83•17 W
40	Pontianak	Indon.	0•04 S	109•20 E
33	Pontian Kechil	Mala.	1•29 N	103•24 E
31	Pontic Mts.	Tur.	41•20 N	34•30 E
20	Pontremoli	It.	44•21 N	9•50 E
20	Ponza, Isole di	It.	40•55 N	12•58 E
22	Poole	Eng.	50•43 N	2•00 W
48	Poopó, Lago de	Bol.	18•16 S	67•57 W
48	Popayán	Col.	2•21 N	76•42 W
8	Poplar Plains	Ky.	38•20 N	83•40 W
12	Popocatépetl (Vol.)	Mex.	19•01 N	98•38 W
54	Popokabaka	Zaire	5•42 S	16•35 E
21	Popovo	Bul.	43•23 N	26•17 E
35	Porbandar	India	21•44 N	69•40 E
48	Porce (R.)	Col.	7•11 N	74•55 W
10	Porcupine (R.)	Can.	67•38 N	140•07 W
20	Pordenone	It.	45•58 N	12•38 E
20	Poreč	Yugo.	45•13 N	13•37 E
25	Pori (Björneborg)	Fin.	61•29 N	21•45 E

Q

R

Page	Name	Region	Lat.	Long.
24	Ragunda	Swe.	63•07 N	16•24 E
17	Ragusa	It.	36•58 N	14•41 E
35	Rāichūr	India	16•23 N	77•18 E
35	Raigarh	India	21•57 N	83•32 E
12	Rainbow City	C.Z.	9•20 N	79•23 W
6	Rainier, Mt.	Wa.	46•52 N	121•46 W
7	Rainy (L.)	Can.-Mn.	48•43 N	94•29 W
7	Rainy (R.)	Can.-Mn.	48•50 N	94•41 W
35	Raipur	India	21•25 N	81•37 E
8	Raisin (R.)	Mi.	42•00 N	83•35 W
35	Rājahmundry	India	17•03 N	81•51 E
40	Rajang, Balang (Strm.)	Mala.	2•10 N	113•30 E
35	Rājkot	India	22•20 N	70•48 E
35	Rājshāhi	Bngl.	24•26 S	88•39 E
27	Rakhov	Sov.Un.	48•02 N	24•13 E
31	Rakitnoye	Sov.Un.	50•51 N	35•53 E
26	Rakovnik	Czech.	50•07 N	13•45 E
30	Rakvere	Sov.Un.	59•22 N	26•14 E
7	Raleigh	NC	35•45 N	78•39 W
55	Rame Hd.	S.Afr.	31•48 S	29•22 E
34	Ramlat as Sab'atayn	Sau.Ar.	16•08 N	45•15 E
33	Ramm, Jabal (Mts.)	Jordan	29•37 N	35•32 E
35	Rāmpur	India	28•53 N	79•03 E
40	Ramree (I.)	Bur.	19•01 N	93•23 E
22	Ramsey	Isle of Man	54•20 N	4•25 W
23	Ramsgate	Eng.	51•19 N	1•20 E
24	Ramsjö	Swe.	62•11 N	15•44 E
41	Ramu (R.)	Pap.N.Gui.	5•35 S	145•16 E
40	Ranau, L.	Indon.	4•52 S	103•52 E
50	Rancagua	Chile	34•10 S	70•43 W
35	Rānchī	India	23•24 N	85•18 E
24	Randers	Den.	56•28 N	10•03 E
55	Randfontein	S.Afr.	26•10 S	27•42 E
9	Randolph	Vt.	43•55 N	72•40 W
24	Rands Fd.	Nor.	60•35 N	10•10 E
9	Rangeley	Me.	44•56 N	70•38 W
9	Rangeley (L.)	Me.	45•00 N	70•25 W
6	Ranger	Tx.	32•26 N	98•41 W
40	Rangoon	Bur.	16•46 N	96•09 E
35	Rangpur	Bngl.	25•48 N	89•19 E
33	Rangsang (I.)	Indon.	0•53 N	103•05 E
10	Rankin Inlet	Can.	62•45 N	94•27 W
33	Rantau	Mala.	2•35 N	101•58 E
40	Rantelkomboa (Mtn.)	Indon.	3•22 S	119•50 E
8	Rantoul	Il.	40•25 N	88•05 W
20	Rapallo	It.	44•21 N	9•14 E
43	Rapa Nui (Easter) (I.)	Chile	26•50 S	109•00 W
6	Rapid City	SD	44•06 N	103•14 W
25	Rapla	Sov.Un.	59•02 N	24•46 E
9	Rappahannock (R.)	Va.	38•20 N	75•25 W
9	Raquette (L.)	NY	43•50 N	74•35 W
27	Rara Mazowiecka	Pol.	51•46 N	20•17 E
43	Rarotonga	Cook Is.	20•40 S	163•00 W
33	Ra's an Naqb	Jordan	30•00 N	35•29 E
53	Ras Dashen (Mtn.)	Eth.	12•49 N	38•14 E
25	Raseiniai	Sov.Un.	55•23 N	23•04 E
34	Ra's Fartak (C.)	P.D.R. of Yem.	15•43 N	52•17 E
33	Rashayya	Leb.	33•30 N	35•50 E
19	Rashīd (Rosetta)	Egypt	31•22 N	30•25 E
34	Rasht	Iran	37•13 N	49•45 E
21	Raška	Yugo.	43•16 N	20•40 E
31	Rasskazovo	Sov.Un.	52•40 N	41•40 E
26	Rastatt	F.R.G.	48•51 N	8•12 E
17	Ras Uarc (C.)	Mor.	35•28 N	2•58 W
40	Rat Buri	Thai.	13•30 N	99•46 E
26	Rathenow	G.D.R.	52•36 N	12•20 E
22	Rathlin (I.)	Ire.	55•18 N	6•13 W
6	Raton	NM	36•52 N	104•26 W
24	Rättvik	Swe.	60•54 N	15•07 E
26	Ratzeburger See (L.)	G.D.R.	53•48 N	11•02 E
24	Raufoss	Nor.	60•44 N	10•30 E
25	Rauma	Fin.	61•07 N	21•31 E
25	Rauna	Sov.Un.	57•21 N	25•31 E
40	Raung (Mtn.)	Indon.	8•15 S	113•56 E
35	Raurkela	India	22•15 N	84•53 E
25	Rautalampi	Fin.	62•39 N	26•25 E
27	Rava-Russkaya	Sov.Un.	50•14 N	23•40 E
20	Ravenna	It.	44•27 N	12•13 E
8	Ravenna	Oh.	41•10 N	81•20 W
26	Ravensburg	F.R.G.	47•48 N	9•35 E
44	Ravensthorpe	Austl.	33•30 S	120•20 E
8	Ravenswood	WV	38•55 N	81•50 W
35	Rāwalpindi	Pak.	33•40 N	73•10 E
34	Rawāndūz	Iraq	36•37 N	44•30 E
26	Rawicz	Pol.	51•36 N	16•51 E
44	Rawlina	Austl.	31•13 S	125•45 E
6	Rawlins	Wy.	41•46 N	107•15 W
50	Rawson	Arg.	43•16 S	65•09 W
11	Ray, C.	Can.	47•40 N	59•18 W
40	Raya, Bukit (Mtn.)	Indon.	0•45 S	112•11 E
55	Rayton	S.Afr.	25•45 S	28•33 E
17	Raz, Pte. du	Fr.	48•02 N	4•43 W
21	Razgrad	Bul.	43•32 N	26•32 E
21	Razlog	Bul.	41•54 N	23•32 E
17	Ré, Île de	Fr.	46•10 N	1•53 W
22	Reading	Eng.	51•25 N	0•58 W
8	Reading	Mi.	41•45 N	84•45 W
8	Reading	Oh.	39•14 N	84•26 W
9	Reading	Pa.	40•20 N	75•55 W
50	Realengo	Braz.	23•50 S	43•25 W
53	Rebiana (Oasis)	Libya	24•10 N	22•03 E
20	Recanati	It.	43•25 N	13•35 E
44	Recherche, Arch. of the	Austl.	34•17 S	122•30 E
31	Rechitsa	Sov.Un.	52•22 N	30•24 E
49	Recife (Pernambuco)	Braz.	8•09 S	34•59 W
55	Recife, Kapp (C.)	S.Afr.	34•03 S	25•43 E
50	Reconquista	Arg.	29•01 S	59•41 W
6	Red (R.)	Can.-U.S.	49•11 N	97•18 W
7	Red (R.)	U.S.	31•40 N	92•55 W
10	Redcliff	Can.	50•05 N	110•47 W
10	Red Deer (R.)	Can.	52•05 N	113•00 W
11	Red Indian L.	Can.	48•40 N	56•50 W
11	Red Lake	Can.	51•03 N	93•49 W
9	Red Lion	Pa.	39•55 N	76•30 W
26	Rednitz R.	F.R.G.	49•10 N	11•00 E
50	Redonda, Isla	Braz.	23•05 S	43•11 W
36	Red R.	Viet.	22•25 N	103•50 E
52	Red Sea	Afr.-Asia	23•15 N	37•00 E
22	Ree, Lough	Ire.	53•30 N	7•45 W
8	Reed City	Mi.	43•50 N	85•35 W
8	Reedsburg	Wi.	43•32 N	90•01 W
46	Reeves, Mt.	Austl.	33•50 S	149•56 E
26	Rega (R.)	Pol.	53•48 N	15•30 E
26	Regen R.	F.R.G.	49•09 N	12•21 E
26	Regensburg	F.R.G.	49•02 N	12•06 E
52	Reggane	Alg.	27•08 N	0•06 E
20	Reggio	It.	44•43 N	10•34 E
20	Reggio di Calabria	It.	38•07 N	15•42 E
27	Reghin	Rom.	46•47 N	24•44 E
10	Regina	Can.	50•25 N	104•39 W
26	Regnitz (R.)	F.R.G.	49•50 N	10•55 E
54	Rehoboth	Namibia	23•10 S	17•15 E
33	Rehovot	Isr.	31•53 N	34•49 E
26	Reichenbach	G.D.R.	50•36 N	12•18 E
17	Reims	Fr.	49•16 N	4•00 E
50	Reina Adelaida, Arch.	Chile	52•00 S	74•15 W
10	Reindeer (L.)	Can.	57•36 N	101•23 W
34	Rema, Jabal (Mtn.)	Yemen	14•13 N	44•38 E
33	Rembau	Mala.	2•36 N	102•06 E
48	Remedios	Col.	7•03 N	74•42 W
33	Rempang I.	Indon.	0•51 N	104•04 E
45	Rendova (I.)	Sol.Is.	8•38 S	156•26 E
26	Rendsburg	F.R.G.	54•19 N	9•39 E
9	Renfrew	Can.	45•30 N	76•30 W
33	Rengam	Mala.	1•53 N	103•24 E
46	Renmark	Austl.	34•10 S	140•50 E
45	Rennel (I.)	Sol.Is.	11•50 S	160•38 E
17	Rennes	Fr.	48•07 N	1•02 W
9	Rensselaer	NY	42•40 N	73•45 W
6	Reno	Nv.	39•32 N	119•49 W
20	Reno (R.)	It.	44•10 N	10•55 E
9	Renovo	Pa.	41•20 N	77•50 W
8	Rensselaer	In.	41•00 N	87•10 W
6	Republican (R.)	Ne.-Ks.	39•40 N	97•40 W
45	Repulse B.	Austl.	20•56 S	149•22 E
17	Requena	Sp.	39•29 N	1•03 W
50	Resistencia	Arg.	27•24 S	58•54 W
21	Reşita	Rom.	45•18 N	21•56 E
5	Resolute	Can.	74•41 N	95•00 W
11	Resolution (I.)	Can.	61•30 N	63•58 W
45	Resolution (I.)	N.Z.	45•43 S	166•00 E
48	Restrepo	Col.	3•49 N	76•31 W
48	Restrepo	Col.	4•16 N	73•32 W
20	Réthimnon	Grc.	35•21 N	24•30 E
67	Reunion	Afr.	21•06 S	55•36 E
26	Reutlingen	F.R.G.	48•29 N	9•14 E
12	Revillagigedo, Islas	Mex.	18•45 N	111•00 W
35	Rewa	India	24•41 N	81•11 E
48	Reyes	Bol.	14•19 S	67•16 W
14	Reykjanes (C.)	Ice.	63•37 N	24•33 W
16	Reykjavik	Ice.	64•09 N	21•39 W
34	Rezā'īyeh	Iran	37•30 N	45•15 E
30	Rēzekne	Sov.Un.	56•31 N	27•19 E
20	Rhaetien Alps (Mts.)	It.	46•22 N	10•33 E
23	Rheden	Neth.	52•02 N	6•02 E
26	Rheine	F.R.G.	52•16 N	7•26 E
26	Rheinland-Pfalz (Rhineland-Palatinate)	F.R.G.	50•05 N	6•40 E
26	Rhein R.	F.R.G.	50•34 N	7•21 E
14	Rhine (R.)	Eur.	50•34 N	7•21 E
8	Rhinelander	Wi.	45•39 N	89•25 W
7	Rhode Island	U.S.	41•35 N	71•40 W
55	Rhodes	S.Afr.	30•48 S	27•56 E
51	Rhodesia	Afr.	17•50 S	29•30 E
21	Rhodope Mts.	Bul.	42•00 N	24•08 E
22	Rhondda	Wales	51•40 N	3•40 W
17	Rhône (R.)	Fr.	45•14 N	4•53 E
22	Rhum (I.)	Scot.	57•00 N	6•20 W
49	Riachão	Braz.	7•15 S	46•30 W
40	Riau	Indon.	0•30 N	104•55 E
33	Riau (Str.)	Indon.	0•04 N	104•27 E
22	Ribble, R.	Eng.	53•10 N	3•15 W
24	Ribe	Den.	55•20 N	8•45 E
49	Ribeirão Prêto	Braz.	21•11 S	47•47 W
48	Riberalta	Bol.	11•06 S	66•02 W
8	Rib Lake	Wi.	45•20 N	90•11 W
9	Rice (L.)	Can.	44•05 N	78•10 W
10	Richardson Mts.	Can.	66•58 N	136•19 W
9	Richardson Park	De.	39•45 N	75•35 W
9	Richelieu (R.)	Can.	45•05 N	73•25 W
9	Richford	Vt.	45•00 N	72•35 W
11	Richibucto	Can.	46•41 N	64•52 W
8	Richland Center	Wi.	43•20 N	90•25 W
45	Richmond	Austl.	20•47 S	143•14 E
9	Richmond	Can.	45•40 N	72•07 W
8	Richmond	In.	39•50 N	85•00 W
8	Richmond	Ky.	37•45 N	84•20 W
55	Richmond	S.Afr.	29•52 S	30•17 E
9	Richmond	Va.37rb135N		77•30 W
8	Richwood	WV	38•10 N	80•30 W
9	Rideau L.	Can.	44•40 N	76•20 W
9	Rigeley	WV	39•40 N	78•45 W
9	Ridgway	Pa.	41•25 N	78•40 W
55	Riebeek-Oos	S.Afr.	33•14 S	26•09 E
26	Ried	Aus.	48•13 N	13•30 E
26	Riesa	G.D.R.	51•17 N	13•17 E
20	Rieti	It.	42•25 N	12•51 E
55	Rievleidam (L.)	S.Afr.	25•52 S	28•18 E
25	Rīga	Sov.Un.	56•55 N	24•05 E
25	Riga, G. of	Sov.Un.	57•56 N	23•05 E
33	Rīgān	Iran	28•45 N	58•55 E
34	Rīgestān	Afr.	30•53 N	64•42 E
11	Rigolet	Can.	54•10 N	58•40 W
25	Riihimäki	Fin.	60•44 N	24•44 E
20	Rijeka (Fiume)	Yugo.	45•22 N	14•24 E
27	Rika R.	Sov.Un.	48•21 N	23•37 E
27	Rimavska Sobota	Czech.	48•25 N	20•01 E
24	Rimbo	Swe.	59•45 N	18•22 E
20	Rimini	It.	44•03 N	12•33 E
21	Rîmnicu Sărat	Rom.	45•24 N	27•06 E
21	Rîmnicu-Vîlcea	Rom.	45•07 N	24•22 E
11	Rimouski	Can.	48•27 N	68•32 W
24	Ringkøbing	Den.	56•06 N	8•14 E
24	Ringkøbing Fd.	Den.	55•55 N	8•04 E
24	Ringsaker	Nor.	60•55 N	10•40 E
24	Ringsted	Den.	55•27 N	11•49 E
16	Ringvassöy (I.)	Nor.	69•58 N	16•43 E
40	Rinjani (Mtn.)	Indon.	8•39 S	116•22 E
12	Rio Abajo	Pan.	9•01 N	78•30 W
48	Riobamba	Ec.	1•45 S	78•37 W
48	Rio Branco	Braz.	9•57 S	67•50 W
50	Río Branco	Ur.	32•33 S	53•29 W
49	Rio Branco	Braz.	2•35 N	61•25 W
49	Rio Chico	Ven.	10•20 N	65•58 W
49	Rio Claro	Braz.	21•25 S	47•33 W
50	Río Cuarto	Arg.	33•05 S	64•15 W
50	Rio de Janeiro	Braz.	22•50 S	43•20 W
49	Rio de Janeiro (State)	Braz.	22•27 S	42•43 W
50	Río Dercero	Arg.	32•12 S	63•59 W
50	Río Gallegos	Arg.	51•43 S	69•15 W
50	Rio Grande	Braz.	31•04 S	52•14 W
49	Rio Grande do Norte (State)	Braz.	5•26 S	37•20 W
50	Rio Grande do Sul (State)	Braz.	29•00 S	54•00 W
48	Ríohacha	Col.	11•30 N	72•54 W
51	Rio Muni	Equat.Gui.	1•47 N	8•33 E
48	Ríonegro	Col.	6•09 N	75•22 W
50	Río Negro (Prov.)	Arg.	40•15 S	68•15 W
50	Rio Negro, Embalse del (Res.)	Ur.	32•45 S	55•50 W
20	Rionero	It.	40•55 N	15•42 E
49	Rio Pardo de Minas	Braz.	15•43 S	42•24 W
48	Ríosucio	Col.	5•25 N	75•41 W
40	Riouw, Pulau-Pulau (Is.)	Indon.	0•30 N	104•55 E
49	Rio Verde	Braz.	17•47 S	50•49 W
8	Ripon	Wi.	43•49 N	88•50 W
44	Ripon (I.)	Austl.	20•05 S	118•10 E
53	Ripon Falls	Ug.	0•38 N	33•02 E
45	Risdon	Austl.	42•37 S	147•32 E
33	Rishon le Ziyyon	Isr.	31•57 N	34•48 E
8	Rising Sun	In.	38•55 N	84•55 W
24	Risor	Nor.	58•44 N	9•10 E
48	Ritacuva, Alto (Mtn.)	Col.	6•22 N	72•13 W
24	Riuvenfjell (Mts.)	Nor.	59•20 N	6•55 E
20	Riva	It.	45•54 N	10•49 E
50	Rivera	Ur.	30•52 S	55•32 W
52	River Cess	Lib.	5•46 N	9•52 W
9	Riverhead	NY	40•55 N	72•40 W
46	Riverina	Austl.	34•55 S	144•30 E
8	River Rouge	Mi.	42•16 N	83•09 W
9	Riverton	Va.	39•00 N	78•15 W
11	Rivière-du-Loup	Can.	47•50 N	69•32 W
34	Riyadh (Ar Riyāḑ)	Sau.Ar.	24•31 N	46•47 E
31	Rize	Tur.	41•00 N	40•30 E
21	Rizzuto, C.	It.	38•53 N	17•05 E
24	Rjukan	Nor.	59•53 N	8•30 E
17	Roanne	Fr.	46•02 N	4•04 E
7	Roanoke	Va.	37•16 N	79•55 W
7	Roanoke (R.)	NC-Va.	36•17 N	77•22 W
54	Robbeneiland (I.)	S.Afr.	33•48 S	18•22 E
45	Roberts, Mt.	Austl.	32•05 S	152•30 E
52	Robertsport	Lib.	6•45 N	11•22 W
11	Roberval	Can.	48•32 N	72•15 W
8	Robinson	Il.	39•00 N	87•45 W
46	Robinvale	Austl.	34•45 S	142•45 E
10	Robson, Mt.	Can.	53•07 N	119•09 W
54	Roçadas	Ang.	16•50 S	15•05 E
49	Rocas, Atol das	Braz.	3•50 S	33•46 W
47	Rocedos São Pedro E São Paulo (I.)	Braz.	1•50 N	30•00 W
50	Rocha	Ur.	34•26 S	54•14 W
22	Rochdale	Eng.	53•37 N	2•09 W
17	Rochefort	Fr.	45•55 N	0•57 W
8	Rochelle	Il.	41•53 N	89•06 W
8	Rochester	In.	41•05 N	86•20 W
7	Rochester	Mn.	44•01 N	92•30 W
9	Rochester	NH	43•20 N	71•00 W
9	Rochester	NY	43•15 N	77•35 W
8	Rock (R.)	Il.	41•40 N	89•52 W
8	Rock Falls	Il.	41•45 N	89•42 W
8	Rockford	Il.	42•16 N	89•07 W
45	Rockhampton	Austl.	23•26 S	150•29 E
7	Rock Hill	SC	34•55 N	81•01 W
46	Rockland Res.	Austl.	36•55 S	142•20 E
8	Rockport	In.	38•20 N	87•00 W
6	Rock Springs	Wy.	41•35 N	109•13 W
49	Rockstone	Guy.	5•55 N	57•27 W
8	Rockville	In.	39•45 N	87•15 W
9	Rockville Centre	NY	40•39 N	73•39 W
6	Rocky Mountain Natl. Park	Co.	40•29 N	106•06 W
5	Rocky Mts.	N.A.	50•00 N	114•00 W
17	Rodez	Fr.	44•22 N	2•34 E
19	Rōdhos	Grc.	36•24 N	28•15 E
19	Ródhos (I.)	Grc.	36•00 N	28•29 E
27	Rodnei, Muntii (Mts.)	Rom.	47•41 N	24•05 E
30	Rodniki	Sov.Un.	57•08 N	41•48 E
21	Rodonit, Kep I (C.)	Alb.	41•38 N	19•01 E
44	Roebourne	Austl.	20•50 S	117•15 E
44	Roebuck, B.	Austl.	18•15 S	121•10 E
23	Roermond	Neth.	51•11 N	6•00 E
23	Roeselare	Bel.	50•55 N	3•05 E
11	Roes Welcome Sd.	Can.	64•10 N	87•23 W
31	Rogachëv	Sov.Un.	53•07 N	30•04 E
21	Rogatica	Yugo.	43•46 N	19•00 E
27	Rogatin	Sov.Un.	49•22 N	24•37 E
8	Rogers City	Mi.	45•30 N	83•50 W
48	Rogoaguado (L.)	Bol.	12•42 S	66•46 W
26	Rogózno	Pol.	52•44 N	16•53 E
24	Röikenviken	Nor.	60•27 N	10•26 E
13	Rojo, Cabo	P.R.	17•55 N	67•14 W
26	Rokycany	Czech.	49•44 N	13•37 E
48	Roldanillo	Col.	4•24 N	76•09 W
24	Rollag	Nor.	59•55 N	8•48 E
46	Roma	Austl.	26•30 S	148•48 E
55	Roma	Leso.	29•28 S	27•43 E
11	Romaine (R.)	Can.	51•22 N	63•23 W
27	Roman	Rom.	46•56 N	26•57 E
14	Romania	Eur.	46•18 N	22•53 E
41	Romblon	Phil.	12•34 N	122•16 E
41	Romblon (I.)	Phil.	12•33 N	122•17 E
7	Rome	Ga.	34•14 N	85•10 W
9	Rome	NY	43•15 N	75•25 W
20	Rome	It.	41•52 N	12•37 E
8	Romeo	Mi.	42•50 N	83•00 W
31	Romny	Sov.Un.	50•46 N	33•31 E
24	Rømø (I.)	Den.	55•08 N	8•17 E
33	Rompin	Mala.	2•42 N	102•30 E
33	Rompin (R.)	Mala.	2•54 N	103•10 E
22	Ronaldsay, North (I.)	Scot.	59•21 N	2•23 W
22	Ronaldsay, South (I.)	Scot.	59•48 N	2•55 W
49	Roncador, Serra do	Braz.	12•44 S	52•19 W
8	Ronceverte	WV	37•45 N	80•30 W
17	Ronda	Sp.	37•45 N	5•10 W
48	Rondônia (Ter.)	Braz.	10•15 S	63•07 W
10	Ronge, Lac la	Can.	55•10 N	105•00 W
24	Rønne	Den.	55•08 N	14•46 E
24	Ronneby	Swe.	56•13 N	15•17 E
67	Ronne Ice Shelf	Ant.	77•30 S	38•00 W
55	Roodepoort	S.Afr.	26•10 S	27•52 E
49	Roosevelt (R.)	Braz.	9•22 S	60•28 W
67	Roosevelt I.	Ant.	79•30 S	168•00 W
44	Roper (R.)	Austl.	14•50 S	134•00 E
48	Roques, Islas los (Is.)	Ven.	21•25 N	67•40 W
48	Roraima (Ter.)	Braz.	2•00 N	62•15 W
49	Roraima (Mtn.)	Ven.-Guy.	5•12 N	60•52 W
24	Röros	Nor.	62•36 N	11•25 E
26	Rorschach	Switz.	47•27 N	9•28 E
26	Rosa, Monte	It.	45•56 N	7•51 E
41	Rosales	Phil.	15•54 N	120•38 E
50	Rosario	Arg.	32•58 S	60•42 W
49	Rosario	Braz.	2•49 S	44•15 W
41	Rosario	Phil.	13•49 N	121•13 W
50	Rosário do Sul	Braz.	30•17 S	54•52 W
49	Rosário Oeste	Braz.	14•47 S	56•20 W
6	Roseberg	Or.	43•13 N	123•30 W
52	Roseires Res.	Sud.	11•15 N	34•45 E
26	Rosenheim	F.R.G.	47•52 N	12•06 E
10	Rosetown	Can.	51•33 N	108•00 W
55	Rosettenville	S.Afr.	26•15 S	28•04 E
8	Rosiclare	Il.	37•30 N	88•15 W
49	Rosignol	Guy.	6•16 N	57•37 W
21	Rosiorii de Vede	Rom.	44•06 N	25•00 E
24	Roskilde	Den.	55•39 N	12•04 E
30	Roslavl'	Sov.Un.	53•56 N	32•52 E
20	Rossano	It.	39•34 N	16•38 E
9	Rosseau (L.)	Can.	45•15 N	79•30 W
45	Rossel (I.)	Pap.N.Gui.	11•31 S	154•00 E
10	Rossland	Can.	49•05 N	118•48 W

Page	Name	Region	Lat.	Long.
31	Rossosh	Sov.Un.	50•12 N	39•32 E
55	Rossouw	S.Afr.	31•12 S	27•18 E
67	Ross Sea	Ant.	76•00 S	178•00 W
67	Ross Shelf Ice	Ant.	81•30 S	175•00 W
26	Rostock	G.D.R.	54•04 N	12•06 E
30	Rostov	Sov.Un.	57•13 N	39•23 E
31	Rostov-na-Donu	Sov.Un.	47•16 N	39•47 E
16	Rösvatn (L.)	Nor.	65•36 N	13•08 E
6	Roswell	NM	33•23 N	104•32 W
26	Rothenburg	F.R.G.	49•20 N	10•10 E
22	Rothesay	Scot.	55•50 N	3•14 W
40	Roti (I.)	Indon.	10•30 S	122•52 E
46	Roto	Austl.	33•07 S	145•30 E
23	Rotterdam	Neth.	51•55 N	4•27 E
26	Rottweil	F.R.G.	48•10 N	8•36 E
17	Rouen	Fr.	49•25 N	1•05 E
8	Rough River Res.	Ky.	37•45 N	86•10 W
22	Rousay (I.)	Scot.	59•10 N	3•04 W
11	Rouyn	Can.	48•22 N	79•03 W
16	Rovaniemi	Fin.	66•29 N	25•45 E
20	Rovato	It.	45•33 N	10•00 E
20	Rovereto	It.	45•53 N	11•05 E
20	Rovigo	It.	45•05 N	11•48 E
20	Rovinj	Yugo.	45•05 N	13•40 E
48	Rovira	Col.	4•14 N	75•13 W
27	Rovno	Sov.Un.	50•37 N	26•17 E
55	Rovuma (Ruvuma) (R.)	Moz.-Tan.	10•50 S	39•50 E
40	Roxas	Phil.	11•30 N	122•47 E
22	Royal Can.	Ire.	53•28 N	6•45 W
55	Royal Natal Natl. Pk.	S.Afr.	28•35 S	28•54 E
8	Royal Oak	Mi.	42•29 N	83•09 W
8	Royalton	Mi.	42•00 N	86•25 W
27	Rožňava	Czech.	48•39 N	20•32 E
31	Rtishchevo	Sov.Un.	52•15 N	43•40 E
54	Ruacana Falls	Ang.-Namibia	17•15 S	14•45 E
45	Ruapehu (Mtn.)	N.Z.	39•15 S	175•37 E
28	Rubtsovsk	Sov.Un.	51•31 N	81•17 E
6	Ruby	Ak.	64•38 N	155•22 W
24	Rudkøbing	Den.	54•56 N	10•44 E
36	Rudog	China	33•42 N	79•56 E
53	Rudolf, L.	Ken.-Eth.	3•30 N	36•05 E
23	Rudolstadt	G.D.R.	50•46 N	13•30 E
53	Rufā'ah	Sud.	14•52 N	33•30 E
55	Rufiji (R.)	Tan.	8•00 S	39•20 E
52	Rufisque	Senegal	14•43 N	17•17 W
26	Rügen (Pen.)	G.D.R.	54•28 N	13•47 E
25	Ruhnu-Saar (I.)	Sov.Un.	57•46 N	23•15 E
26	Ruhr R.	F.R.G.	51•18 N	8•17 E
48	Ruiz, Nevado del	Col.	4•52 N	75•20 W
25	Rūjiena	Sov.Un.	57•54 N	25•19 E
54	Rukwa, L.	Tan.	8•00 S	32•25 E
21	Ruma	Yugo.	45•00 N	19•53 E
53	Rumbek	Sud.	6•52 N	29•43 E
9	Rumford	Me.	44•32 N	70•35 W
34	Rummah, Wādī ar (R.)	Sau.Ar.	26•17 N	41•45 E
33	Rummānah	Egypt	31•01 N	32•39 E
33	Rupat (I.)	Indon.	1•55 N	101•35 E
33	Rupat (Str.)	Indon.	1•55 N	101•17 E
11	Rupert (R.)	Can.	51•35 N	76•30 W
21	Ruse (Russe)	Bul.	43•50 N	25•59 E
8	Rushville	In.	39•35 N	85•30 W
49	Russas	Braz.	4•48 S	37•50 W
10	Russell	Can.	50•47 N	101•15 W
8	Russell	Ky.	38•30 N	82•45 W
45	Russell	N.Z.	35•38 S	174•13 E
45	Russell Is.	Sol.Is.	9•16 S	158•30 E
28	Russian S.F.S.R.	Sov.Un.	61•00 N	60•00 E
27	Ruthenia	Sov.Un.	48•25 N	23•00 E
9	Rutland	Vt.	43•35 N	72•55 W
54	Rutshuru	Zaire	1•11 S	29•27 E
20	Ruvo	It.	41•07 N	16•32 E
55	Ruvuma (Rovuma) (R.)	Moz.-Tan.	10•50 S	39•50 S
53	Ruwenzori Mts.	Afr.	0•53 N	30•00 E
27	Ruzhany	Sov.Un.	52•49 N	24•54 E
51	Rwanda	Afr.	2•10 S	29•37 E
30	Ryazan	Sov.Un.	54•37 N	39•43 E
30	Ryazhsk	Sov.Un.	53•43 N	40•04 E
30	Rybachiy, P-Ov. (Pen.)	Sov.Un.	69•50 N	33•20 E
30	Rybinsk	Sov.Un.	58•02 N	38•52 E
30	Rybinskoye Vdkhr. (Res.)	Sov.Un.	58•23 N	38•15 E
27	Rybnik	Pol.	50•06 N	18•37 E
22	Ryde	Eng.	50•43 N	1•16 W
31	Ryl'sk	Sov.Un.	51•33 N	34•42 E
27	Rypin	Pol.	53•04 N	19•25 E
27	Rzeszów	Pol.	50•02 N	22•00 E
30	Rzhev	Sov.Un.	56•16 N	34•17 E

S

Page	Name	Region	Lat.	Long.
26	Saale R.	G.D.R.	51•14 N	11•52 E
26	Saalfeld	G.D.R.	50•38 N	11•20 E
26	Saar	F.R.G.	49•25 N	6•50 E
26	Saarbrücken	F.R.G.	49•15 N	7•01 E
25	Saaremaa (Ezel) (I.)	Sov.Un.	58•28 N	21•30 E
50	Saavedra	Arg.	37•45 S	62•23 W
21	Šabac	Yugo.	44•45 N	19•49 E
40	Sabah (Reg.)	Mala.	5•10 N	116•25 E
49	Sabana de Uchire	Ven.	10•02 N	65•32 W
48	Sabanalarga	Col.	10•38 N	75•02 W
48	Sabanas Páramo (Mtn.)	Col.	6•28 N	76•08 W
40	Sabang	Indon.	5•52 N	95•26 E
54	Sabi (R.)	Rh.	20•18 S	32•07 E
25	Sabile	Sov.Un.	57•03 N	22•34 E
12	Sabinas	Mex.	28•05 N	102•30 W
67	Sabine, Mt.	Ant.	72•05 S	169•10 E
7	Sabine (R.)	U.S.	31•35 N	94•00 W
41	Sablayan	Phil.	15•49 N	120•47 E
11	Sable, C.	Can.	43•25 N	65•24 W
7	Sable, C.	Fl.	25•12 N	81•10 W
30	Sablya, Gora (Mtn.)	Sov.Un.	64•50 N	59•00 E
9	Sacandaga Res.	NY	43•10 N	74•15 W
26	Sachsen	G.D.R.	50•45 N	12•17 E
9	Sacketts Harbor	NY	43•55 N	76•05 W
50	Saco (R.)	Braz.	22•20 S	43•26 W
9	Saco (R.)	Me.	43•53 N	70•46 W
50	Sacra Famalia do Tinguá	Braz.	22•29 S	43•36 W
6	Sacramento	Ca.	38•35 N	121•30 W
34	Sa'dah	Yemen	16•50 N	43•45 E
35	Sadiya	India	27•53 N	95•35 E
37	Sado (I.)	Jap.	38•05 N	138•26 E
24	Saeby	Den.	57•21 N	10•29 E
52	Safi (Asfi)	Mor.	32•24 N	9•09 W
31	Safid Rud (R.)	Iran	36•50 N	49•40 E
35	Sāgar	India	23•55 N	78•45 E
8	Saginaw	Mi.	43•25 N	84•00 W
8	Saginaw B.	Mi.	43•50 N	83•40 W
31	Sagiz (R.)	Sov.Un.	48•30 N	56•10 E
9	Saguache	Co.	38•05 N	106•10 W
9	Saguache Cr.	Co.	38•05 N	106•40 W
11	Saguenay (R.)	Can.	48•20 N	70•15 W
18	Sagunto	Sp.	39•40 N	0•17 W
51	Sahara Des.	Afr.	23•44 N	1•40 W
18	Saharan Atlas (Mts.)	Mor.-Alg.	32•51 N	1•02 W
35	Sahāranpur	India	29•58 N	77•41 E
52	Saïda	Alg.	34•51 N	00•07 E
25	Saimaa	Fin.	61•24 N	28•45 E
22	St. Albans	Eng.	51•44 N	0•20 W
9	St. Albans	Vt.	44•50 N	73•05 W
8	St. Albans	WV	38•20 N	81•50 W
22	St. Albans Hd.	Eng.	50•34 N	2•00 W
55	St. André, Cap (C.)	Mad.	16•15 S	44•31 E
11	St. Andrews	Can.	45•05 N	67•03 W
22	St. Andrews	Scot.	56•20 N	2•40 W
11	St. Anthony	Can.	51•24 N	55•35 W
41	St. Antonio, Mt.	Phil.	13•23 N	122•00 E
7	St. Augustine	Fl.	29•53 N	81•21 W
22	St. Bees Hd.	Eng.	54•30 N	3•40 W
22	St. Brides B.	Wales	51•17 N	4•45 W
17	St. Brieuc	Fr.	48•32 N	2•47 W
9	St. Catharines	Can.	43•10 N	79•14 W
17	St. Chamond	Fr.	45•30 N	4•17 E
8	St. Charles	Il.	41•55 N	88•19 W
8	St. Charles	Mi.	43•20 N	84•10 W
8	St. Clair	Mi.	42•55 N	82•30 W
8	St. Clair (L.)	Can.-Mi.	42•25 N	82•30 W
8	St. Clair (R.)	Can.-Mi.	42•45 N	82•25 W
7	St. Cloud	Mn.	45•33 N	94•08 W
55	St. Croix I.	S.Afr.	33•48 S	25•45 E
13	Saint Croix (I.)	Vir.Is.	17•40 N	64•43 W
7	St. Croix (R.)	Can.-Me.	45•28 N	67•32 W
7	St. Croix R.	Mn.-Wi.	45•00 N	92•44 W
22	St. David's Hd.	Wales	51•54 N	5•25 W
17	St.-Denis	Fr.	48•26 N	2•22 E
10	St. Elias, Mt.	Can.	60•25 N	141•00 W
17	St. Étienne	Fr.	45•26 N	4•22 E
11	St. Félicíen	Can.	48•39 N	72•28 W
20	St. Florent, Golfe de	Fr.	42•55 N	9•08 E
9	St. Francis L.	Can.	45•00 N	74•20 W
46	St. George	Austl.	28•02 S	148•40 E
11	St. George, C.	Can.	48•28 N	59•15 W
11	St. George's	Can.	48•26 N	58•29 W
49	St. Georges	Fr.Gu.	3•48 N	51•47 W
11	St. George's B.	Can.	48•20 N	59•00 W
22	St. George's Chan.	Eng.-Ire.	51•45 N	6•30 W
26	St. Gotthard Tun.	Switz.	46•38 N	8•55 E
51	St. Helena	Atl.O.	16•01 S	5•16 W
54	St. Helenabaai (B.)	Afr.	32•25 S	17•15 E
22	St. Helens	Eng.	53•27 N	2•44 W
9	St. Hyacinthe	Can.	45•35 N	72•55 W
8	St. Ignace	Mi.	45•51 N	84•39 W
9	St. Jean	Can.	45•20 N	73•15 W
11	St. Jean, Lac	Can.	48•35 N	72•00 W
9	St. Jérôme	Can.	45•47 N	74•00 W
11	St. John	Can.	45•16 N	66•03 W
11	St. John (R.)	Can.	46•39 N	67•40 W
11	St. John, C.	Can.	50•00 N	55•32 W
13	St. John (I.)	Vir.Is.	18•16 N	64•48 W
11	St. John (R.)	N.A.	45•15 N	67•40 W
11	St. John's	Can.	47•34 N	52•43 W
8	St. Johns	Mi.	43•05 N	84•35 W
7	St. Johns (R.)	Fl.	29•54 N	81•32 W
9	St. Johnsbury	Vt.	44•25 N	72•00 W
8	St. Joseph	Mi.	42•05 N	86•30 W
7	St. Joseph	Mo.	39•44 N	94•49 W
8	St. Joseph (I.)	Can.	46•15 N	83•55 W
11	St. Joseph (L.)	Can.	51•31 N	90•40 W
8	St. Joseph (R.)	Mi.	41•45 N	85•50 W
22	St. Kilda (I.)	Scot.	57•10 N	8•32 W
13	St. Kitts (I.)	St. Kitts-Nevis-Anguilla	17•24 N	63•30 W
9	St. Lambert	Can.	45•29 N	73•29 W
49	St. Laurent	Fr.Gu.	5•27 N	53•56 W
6	St. Lawrence (I.)	Ak.	63•10 N	172•12 W
11	St. Lawrence, Gulf of	Can.	48•00 N	62•00 W
11	St. Lawrence R.	Can.-U.S.	48•24 N	69•30 W
8	St. Louis	Mi.	43•25 N	84•35 W
8	St. Louis	Mo.	38•39 N	90•15 W
52	St.-Louis	Senegal	16•02 N	16•30 W
13	St. Lucia	N.A.	13•54 N	60•40 W
22	St. Magnus B.	Scot.	60•25 N	2•09 W
17	St. Malo	Fr.	48•40 N	2•02 W
17	St. Malo, Golfe de	Fr.	48•50 N	2•49 W
55	Ste.-Marie, Île	Mad.	16•58 S	50•15 E
55	Ste. Marie, Cap	Mad.	25•31 S	45•00 E
46	St. Marys	Austl.	41•40 S	148•10 E
8	St. Marys	Can.	43•15 N	81•10 W
8	St. Mary's	Oh.	40•30 N	84•25 W
9	St. Marys	Pa.	41•25 N	78•30 W
8	St. Marys	WV	39•20 N	81•15 W
8	St. Matthews	Ky.	38•15 N	85•39 W
11	St. Maurice (R.)	Can.	47•20 N	72•55 W
26	St. Moritz	Switz.	46•31 N	9•50 E
17	St. Nazaire	Fr.	47•18 N	2•13 W
10	St. Paul	Can.	53•59 N	111•17 W
7	St. Paul	Mn.	44•57 N	93•05 W
67	St. Paul I.	Ind.O.	38•43 S	77•31 E
7	St. Petersburg	Fl.	27•47 N	82•38 W
11	St. Pierre & Miquelon	N.A.	46•53 N	56•40 W
26	St. Pölten	Aus.	48•12 N	15•38 E
17	St. Quentin	Fr.	49•52 N	3•16 E
11	St. Stephen	Can.	45•12 N	66•17 W
8	St. Thomas	Can.	42•45 N	81•15 W
13	St. Thomas (I.)	Vir.Is.	18•22 N	64•57 W
26	St. Veit	Aus.	46•46 N	14•20 E
13	St. Vincent	N.A.	13•20 N	60•50 W
46	St. Vincent, G.	Austl.	34•55 S	138•00 E
10	St. Walburg	Can.	53•39 N	109•12 W
48	Sajama, Nevada	Bol.	18•13 S	68•53 W
34	Sakākah	Sau.Ar.	29•58 N	40•03 E
6	Sakakawea, Lake	ND	47•49 N	101•58 W
54	Sakania	Zaire	12•45 S	28•34 E
31	Sakarya (R.)	Tur.	40•10 N	31•00 E
37	Sakata	Jap.	38•56 N	139•57 E
38	Sakchu	Kor.	40•29 N	125•09 E
29	Sakhalin (I.)	Sov.Un.	51•52 N	144•15 E
25	Sakiai	Sov.Un.	54•59 N	23•05 E
39	Sakishima-Gunto (Is.)	Jap.	24•25 N	125•00 E
31	Sakmara (R.)	Sov.Un.	52•00 N	56•10 E
31	Sal (R.)	Sov.Un.	47•20 N	42•10 E
24	Sala	Swe.	59•56 N	16•34 E
20	Sala Consilina	It.	40•24 N	15•38 E
50	Saladillo	Arg.	35•38 S	59•48 W
50	Salado (R.)	Arg.	26•05 S	63•35 W
12	Salado de los Nadadores Rio	Mex.	27•26 N	101•35 W
12	Salamanca	Mex.	20•36 N	101•10 W
9	Salamanca	NY	42•10 N	78•45 W
17	Salamanca	Sp.	40•54 N	5•42 W
53	Salamat, Bahr (R.)	Chad.	10•06 N	19•16 E
41	Salamaua	Pap.N.Gui.	6•50 S	146•55 E
48	Salamina	Col.	5•25 N	75•29 W
21	Salamis	Grc.	37•58 N	23•30 E
48	Salaverry	Peru	8•16 S	78•54 W
41	Salawati (I.)	Indon.	1•22 N	130•15 E
43	Sala-y-Gómez (I.)	Chile	26•50 S	105•50 W
48	Saldaña (R.)	Col.	3•42 N	75•16 W
54	Saldanha	S.Afr.	32•55 S	18•05 E
25	Saldus	Sov.Un.	56•39 N	22•30 E
46	Sale	Austl.	38•10 S	147•07 E
52	Salé	Mor.	34•09 N	6•42 W
30	Salekhard	Sov.Un.	66•35 N	66•50 E
8	Salem	Il.	38•40 N	89•00 W
35	Salem	India	11•39 N	78•11 E
8	Salem	In.	38•35 N	86•00 W
9	Salem	Ma.	42•31 N	70•54 W
9	Salem	NJ	39•35 N	75•30 W
8	Salem	Oh.	40•55 N	80•50 W
6	Salem	Or.	44•55 N	123•03 W
55	Salem	S.Afr.	33•29 S	26•30 E
8	Salem	WV	39•15 N	80•35 W
20	Salemi	It.	37•49 N	12•48 E
20	Salerno	It.	40•27 N	14•46 E
20	Salerno, Golfo di	It.	40•30 N	14•40 E
22	Salford	Eng.	53•26 N	2•19 W
27	Salgótarján	Hung.	48•06 N	19•50 E
6	Salina	Ks.	38•50 N	97•37 W
20	Salina (I.)	It.	38•35 N	14•48 E
12	Salina Cruz	Mex.	16•10 N	95•12 W
12	Salinas	Mex.	22•38 N	101•42 W
13	Salinas	P.R.	17•58 N	66•16 W
22	Salisbury	Eng.	50•35 N	1•51 W
9	Salisbury	Md.	38•20 N	75•40 W
54	Salisbury	Rh.	17•50 S	31•03 E
11	Salisbury (I.)	Can.	63•36 N	76•20 W
22	Salisbury Plain	Eng.	51•15 N	1•52 W
6	Salmon (R.)	Id.	45•30 N	115•45 W
9	Salmon (R.)	NY	44•35 N	74•15 W
44	Salmon Gums	Austl.	33•00 S	122•00 E
6	Salmon River Mts.	Id.	44•15 N	115•44 W
27	Salonta	Rom.	46•46 N	21•38 E
31	Sal'sk	Sov.Un.	46•30 N	41•20 E
6	Salt (R.)	Az.	33•28 N	111•35 W
50	Salta	Arg.	24•50 S	65•16 W
50	Salta (Prov.)	Arg.	25•15 S	65•00 W
12	Saltillo	Mex.	25•24 N	100•59 W
6	Salt Lake City	Ut.	40•45 N	111•52 W
50	Salto	Ur.	31•18 S	57•45 W
49	Salto Grande	Braz.	22•57 S	49•58 W
6	Salton Sea	Ca.	33•28 N	115•43 W
52	Saltpond	Ghana	5•16 N	1•07 W
24	Saltsjöbaden	Swe.	59•15 N	18•20 E
12	Salud, Mt.	Pan.	9•14 N	79•42 W
20	Saluzzo	It.	44•39 N	7•31 E
49	Salvador (Bahia)	Braz.	12•59 S	38•27 W
36	Salween R.	Bur.	26•46 N	98•19 E
31	Sal'yany	Sov.Un.	39•40 N	49•10 E
26	Salzburg	Aus.	47•48 N	13•04 E
26	Salzburg (State)	Aus.	47•30 N	13•18 E
26	Salzwedel	G.D.R.	52•51 N	11•10 E
13	Samana, Cabo	Dom.Rep.	19•20 N	69•00 W
41	Samar (I.)	Phil.	11•30 N	126•07 E
31	Samara (R.)	Sov.Un.	52•50 N	50•35 E
41	Samarai	Pap.N.Gui.	10•45 S	150•49 E
28	Samarkand	Sov.Un.	39•42 N	67•00 E
35	Sambalpur	India	21•30 N	84•05 E
27	Sambor	Sov.Un.	49•31 N	23•12 E
23	Sambre (R.)	Bel.	50•20 N	4•15 E
21	Samokov	Bul.	42•20 N	23•33 E
21	Sámos (I.)	Grc.	37•53 N	26•35 E
21	Samothráki (I.)	Grc.	40•23 N	25•10 E
41	Sampaloc Pt.	Phil.	14•43 N	119•56 E
24	Samsø (I.)	Den.	55•49 N	10•47 E
31	Samsun	Tur.	41•20 N	36•05 E
31	Samtredia	Sov.Un.	42•18 N	42•25 E
31	Samur (R.)	Sov.Un.	41•40 N	47•20 E
52	San	Mali	13•18 N	4•54 W
34	San'ā'	Yemen	15•17 N	44•05 E
52	Sanaga (R.)	Cam.	4•10 N	10•40 E
47	San Ambrosio, Isla de	Chile	26•40 S	80•00 W
41	Sanana (I.)	Indon.	2•15 S	126•38 E
34	Sanandaj	Iran	36•44 N	46•43 E
48	San Andrés	Col.	6•57 N	75•41 W
6	San Andres, Mts.	U.S.	33•00 N	106•40 W
6	San Andres Mts.	NM	33•45 N	106•40 W
12	San Andrés Tuxtla	Mex.	18•27 N	95•12 W
6	San Angelo	Tx.	31•28 N	100•22 W
20	San Antioco, I. di	It.	39•00 N	8•25 E
50	San Antonio	Chile	33•34 S	71•36 W
48	San Antonio	Col.	2•57 N	75•06 W
48	San Antonio	Col.	3•55 N	75•38 W
41	San Antonio	Phil.	14•57 N	120•05 E
6	San Antonio	Tx.	29•25 N	98•30 W
13	San Antonio, Cabo	Cuba	21•55 N	84•55 W
50	San Antonio de los Cobres	Arg.	24•15 S	66•29 W
49	San Antonio de Tamanaco	Ven.	9•42 N	66•03 W
50	San Antonio Oeste	Arg.	40•49 S	64•56 W
20	San Bartolomeo	It.	41•25 N	15•04 E
20	San Benedetto del Tronto	It.	42•58 N	13•54 E
6	San Bernardino	Ca.	34•07 N	117•19 W
50	San Bernardo	Chile	33•35 S	70•42 W
12	San Blas	Mex.	21•33 N	105•19 W
17	San Blas, C.	Fl.	29•38 N	85•38 W
50	San Carlos	Chile	36•23 S	71•58 W
48	San Carlos	Col.	6•11 N	74•58 W
41	San Carlos	Phil.	15•56 N	120•20 E
48	San Carlos	Ven.	9•36 N	68•35 W
50	San Carlos de Bariloche	Arg.	41•15 S	71•26 W
6	San Carlos Res.	Az.	33•05 N	110•29 W
49	San Casimiro	Ven.	10•01 N	67•02 W
20	San Cataldo	It.	37•30 N	13•59 E
13	Sánchez	Dom.Rep.	19•15 N	69•40 W
6	San Clemente (I.)	Ca.	33•02 N	118•36 W
48	San Cristóbal	Ven.	7•43 N	72•15 W
48	San Cristobal (I.)	Ec.	1•05 S	89•15 W
45	San Cristobal (I.)	Sol.Is.	10•47 S	162•17 E
20	San Croce, C.	It.	37•15 N	15•18 E
13	Sancti Spíritus	Cuba	21•55 N	79•25 W
17	Sancy, Puy de	Fr.	45•30 N	2•53 E
55	Sand (R.)	S.Afr.	28•30 S	29•30 E
40	Sandakan	Mala.	5•51 N	118•03 E
22	Sanday (I.)	Scot.	59•17 N	2•25 W
24	Sandefjord	Nor.	59•09 N	10•14 E
24	Sandhammar, C.	Swe.	55•24 N	14•37 E
6	San Diego	Ca.	32•43 N	117•10 W
24	Sandnes	Nor.	58•52 N	5•44 E
54	Sandoa	Zaire	9•39 S	23•00 E
27	Sandomierz	Pol.	50•39 N	21•45 E
20	San Donà di Piave	It.	45•38 N	12•34 E
36	Sandoway	Bur.	18•24 N	94•28 E
20	Sandrio	It.	46•11 N	9•53 E
44	Sandstone	Austl.	28•00 S	119•25 E
8	Sandusky	Mi.	43•25 N	82•50 W
8	Sandusky	Oh.	41•25 N	82•45 W
8	Sandusky (R.)	Oh.	41•10 N	83•20 W
8	Sandwich	Il.	42•35 N	88•53 W
46	Sandy C.	Austl.	24•25 S	153•10 E
50	San Estanislao	Par.	24•38 S	56•20 W
41	San Fabian	Phil.	16•14 N	120•28 E
50	San Felipe	Chile	32•45 S	70•43 W
48	San Felipe	Ven.	10•13 N	68•45 W

Page	Name	Region	Lat.	Long.
47	San Felix, Isla de	Chile	26•20 S	80•10 W
50	San Fernando	Arg.	34•11 S	58•34 W
50	San Fernando	Chile	36•36 S	70•58 W
41	San Fernando	Phil.	16•38 N	120•19 E
48	San Fernando de Apure	Ven.	7•46 N	67•29 W
48	San Fernando de Atabapo	Ven.	3•58 N	67•41 W
24	Sånfjället (Mtn.)	Swe.	62•19 N	13•30 E
7	Sanford	Fl.	28•46 N	80•18 W
50	San Francisco	Arg.	31•23 S	62•09 W
6	San Francisco	Ca.	37•45 N	122•26 W
12	San Francisco del Oro	Mex.	27•00 N	106•37 W
49	San Francisco de Macaira	Ven.	9•58 N	66•17 W
26	Sangerhausen	G.D.R.	51•28 N	11•17 E
53	Sangha (R.)	Afr.	2•40 N	16•10 E
41	Sangihe (I.)	Indon.	3•30 N	125•30 E
48	San Gil	Col.	6•32 N	73•13 W
20	San Giovanni in Fiore	It.	39•15 N	16•40 E
38	Sangju	Kor.	36•20 N	128•07 E
35	Sāngli	India	16•56 N	74•38 E
6	Sangre De Cristo, Range	U.S.	37•45 N	105•50 W
20	Sangro (R.)	It.	41•38 N	13•56 E
41	San Ildefonso, C.	Phil.	16•03 N	122•10 E
50	San Isidro	Arg.	34•13 S	58•31 W
41	San Jacinto	Phil.	12•33 N	123•43 E
49	San Joaquin	Ven.	10•16 N	67•47 W
50	San Jorge, Golfo	Arg.	46•15 S	66•45 W
49	San José	Bol.	17•54 S	60•42 W
6	San Jose	Ca.	37•20 N	121•54 W
13	San Jose	C.R.	9•57 N	84•05 W
41	San Jose	Phil.	12•22 N	121•04 E
41	San José	Phil.	14•49 N	120•47 E
41	San Jose	Phil.	15•49 N	120•57 E
12	San Jose (I.)	Mex.	25•00 N	110•35 W
50	San José de Feliciano	Arg.	30•26 S	58•44 W
49	San José de Gauribe	Ven.	9•51 N	65•49 W
50	San Juan	Arg.	31•36 S	68•29 W
48	San Juan	Col.	3•23 N	73•48 W
41	San Juan	Phil.	14•30 N	121•14 E
41	San Juan	Phil.	16•41 N	120•20 E
13	San Juan	P.R.	18•30 N	66•10 W
50	San Juan (Prov.)	Arg.	31•00 S	69•30 W
13	San Juan, Cabezas de	P.R.	18•29 N	65•30 W
6	San Juan (R.)	Ut.	37•10 N	110•30 W
50	San Juan Bautista	Par.	26•48 S	57•09 W
49	San Juan de los Morros	Ven.	9•54 N	67•22 W
12	San Juan del Sur	Nic.	11•15 N	85•53 W
6	San Juan Mts.	Co.	37•50 N	107•30 W
13	San Juan R.	Nic.	10•58 N	84•18 W
50	San Julián	Arg.	49•17 S	68•02 W
50	San Justo	Arg.	34•25 S	58•33 W
52	Sankarani R.	Gui.-Mali	11•10 N	8•35 W
26	Sankt Gallen	Switz.	47•25 N	9•22 E
54	Sankuru (R.)	Zaire	4•00 S	22•35 E
12	San Lazaro, C.	Mex.	24•58 N	113•30 W
50	San Lorenzo	Arg.	32•46 S	60•44 W
17	Sanlúcar	Sp.	36•46 N	6•21 W
48	San Lucas	Bol.	20•12 S	65•06 W
12	San Lucas, C.	Mex.	22•45 N	109•45 W
50	San Luis	Arg.	33•16 S	66•15 W
48	San Luis	Col.	6•03 N	74•57 W
50	San Luis (Prov.)	Arg.	32•45 S	66•00 W
12	San Luis (State)	Mex.	22•45 N	101•45 W
6	San Luis Obispo	Ca.	35•18 N	120•40 W
12	San Luis Potosi	Mex.	22•08 N	100•58 W
12	San Luis Potosí (State)	Mex.	22•45 N	101•45 W
20	San Marco	It.	41•53 N	15•50 E
21	San Maria di Léuca, C.	It.	39•47 N	18•20 E
41	San Mariano	Phil.	17•00 N	121•58 E
17	San Marino	Eur.	43•40 N	13•00 E
48	San Martín	Col.	3•42 N	73•44 W
50	San Martín (L.)	Arg.-Chile	48•15 S	72•30 W
49	San Mateo	Ven.	9•45 N	64•34 W
50	San Matías, Golfo	Arg.	41•30 S	63•45 W
39	Sanmen Wan (B.)	China	29•00 N	122•15 E
50	San Miguel	Arg.	34•17 S	58•43 W
41	San Miguel	Phil.	15•09 N	120•56 E
12	San Miguel	Sal.	13•28 N	88•11 W
49	San Miguel	Ven.	9•56 N	64•58 W
48	San Miguel (R.)	Bol.	13•34 S	63•58 W
41	San Miguel B.	Phil.	13•55 N	123•12 E
53	Sannār	Sud.	13•34 N	33•32 E
41	San Narciso	Phil.	15•01 N	120•05 E
41	San Narciso	Phil.	13•34 N	122•33 E
50	San Nicolás	Arg.	33•20 S	60•14 W
41	San Nicolas	Phil.	16•05 N	120•45 E
27	Sanok	Pol.	49•31 N	22•13 E
41	San Pablo	Phil.	14•05 N	121•20 E
49	San Pablo	Ven.	9•46 N	65•04 W
41	San Pascual	Phil.	13•08 N	122•59 E
50	San Pedro	Arg.	24•15 S	64•51 W
50	San Pedro	Par.	24•13 S	57•00 W
20	San Pietro, I. di	It.	39•09 N	8•15 E
41	San Quintin	Phil.	15•59 N	120•47 E
50	San Rafael	Arg.	34•30 S	68•13 W
48	San Rafael	Col.	6•18 N	75•02 W
20	San Remo	It.	43•48 N	7•46 E
27	San R.	Pol.	50•33 N	22•12 E
13	San Roman, C.	Ven.	12•00 N	69•45 W
48	San Roque	Col.	6•29 N	75•00 W
12	San Salvador	Sal.	13•45 N	89•11 W
48	San Salvador (I.)	Ec.	0•14 S	90•50 W
13	San Salvador (Watling) (I.)	Ba.	24•05 N	74•30 W
52	San Sebastian	Can.Is.	28•09 N	17•11 W
17	San Sebastián	Sp.	43•19 N	1•59 W
49	San Sebastián	Ven.	9•58 N	67•11 W
20	San Severo	It.	41•43 N	15•24 E
38	San She (Mtn.)	China	33•00 N	103•50 E
37	San Shui	China	23•14 N	112•51 E
6	Santa Ana	Ca.	33•45 N	117•52 W
12	Santa Ana	Sal.	14•02 N	89•35 W
50	Santa Anna, Cochilha de	Braz.	30•30 S	56•30 W
52	Santa Antão (I.)	C.V.Is.	17•20 N	26•05 W
6	Santa Barbara	Ca.	34•26 N	119•43 W
6	Santa Catalina (I.)	Ca.	33•29 N	118•37 W
50	Santa Catarina (State)	Braz.	27•15 S	50•30 W
13	Santa Clara	Cuba	22•25 N	80•00 W
50	Santa Clara	Ur.	32•46 S	54•51 W
12	Santa Clara, Sierra	Mex.	27•30 N	113•50 W
48	Santa Cruz	Bol.	17•45 S	63•03 W
50	Santa Cruz	Braz.	29•43 S	52•15 W
50	Santa Cruz	Braz.	22•55 S	43•41 W
6	Santa Cruz	Ca.	36•59 N	122•02 W
41	Santa Cruz	Phil.	13•28 N	122•02 E
41	Santa Cruz	Phil.	14•17 N	121•25 E
41	Santa Cruz	Phil.	15•46 N	119•53 E
41	Santa Cruz	Phil.	17•06 N	120•27 E
50	Santa Cruz (Prov.)	Arg.	48•00 S	70•00 W
48	Santa Cruz (I.)	Ec.	0•38 S	90•20 W
50	Santa Cruz (R.)	Arg.	50•05 S	66•30 W
52	Santa Cruz de Tenerife	Can.Is.	28•07 N	15•27 W
45	Santa Cruz Is.	Sol.Is.	10•58 S	166•47 E
20	Sant' Eufemia, Golfo di	It.	38•53 N	15•53 E
17	Santa Eugenia de Ribeira	Sp.	42•34 N	8•55 W
50	Santa Fe	Arg.	31•33 S	60•45 W
6	Santa Fe	NM	35•10 N	106•00 W
50	Santa Fe (Prov.)	Arg.	32•00 S	61•15 W
49	Santa Filomena	Braz.	9•09 S	44•45 W
12	Santa Genoveva, (Mtn.)	Mex.	23•30 N	110•00 W
39	Sant'ai	China	31•02 N	105•02 E
49	Santa Inés	Ven.	9•54 N	64•21 W
50	Santa Inés (I.)	Chile	53•45 S	74•15 W
45	Santa Isabel (I.)	Sol.Is.	7•57 S	159•28 E
49	Santa Lucia	Ven.	10•18 N	66•40 W
12	Santa Magarita (I.)	Mex.	24•15 N	112•00 W
50	Santa Maria	Braz.	29•40 S	54•00 W
20	Santa Maria	It.	41•05 N	14•15 E
41	Santa Maria	Phil.	14•48 N	120•57 E
52	Santa Maria I.	Açores	37•09 N	26•02 W
48	Santa Marta	Col.	11•15 N	74•13 W
6	Santa Monica	Ca.	34•01 N	118•29 W
50	Santana (R.)	Braz.	22•33 S	43•37 W
48	Santander	Col.	3•00 N	76•25 W
17	Santander	Sp.	43•27 N	3•50 W
49	Santarém	Braz.	2•28 S	54•37 W
17	Santarém	Port.	39•18 N	8•48 W
50	Santa Rosa	Arg.	36•45 S	64•10 W
6	Santa Rosa	Ca.	38•27 N	122•42 W
48	Santa Rosa	Col.	6•38 N	75•26 W
48	Santa Rosa	Ec.	3•29 S	78•55 W
41	Santa Rosa	Phil.	14•18 N	121•07 E
49	Santa Rosa	Ven.	9•37 N	64•10 W
48	Santa Rosa de Cabal	Col.	4•53 N	75•38 W
12	Santa Rosalía	Mex.	27•13 N	112•15 W
49	Santa Teresa	Ven.	10•14 N	66•40 W
50	Santa Vitória do Palmar	Braz.	33•30 S	53•16 W
7	Santee (R.)	SC	33•27 N	80•02 W
50	Santiago	Braz.	29•05 S	54•46 W
50	Santiago	Chile	33•26 S	70•40 W
41	Santiago	Phil.	16•42 N	121•33 E
12	Santiago, Rio Grande de	Mex.	21•15 N	104•05 W
41	Santiago (I.)	Phil.	16•29 N	120•03 E
13	Santiago de los Cabelleros	Dom.Rep.	19•30 N	70•45 W
13	Santiago de Cuba	Cuba	20•00 N	75•50 W
50	Santiago del Estero	Arg.	27•50 S	64•14 W
6	Santiago Mts.	Tx.	30•00 N	103•30 W
49	Santo Amaro	Braz.	12•32 S	38•33 W
50	Santo Angelo	Braz.	28•16 S	53•59 W
54	Santo Antonio do Zaire	Ang.	6•10 S	12•25 E
41	Santo Domingo	Phil.	17•39 N	120•24 E
13	Santo Domingo	Dom.Rep.	18•30 N	69•55 W
49	Santos	Braz.	23•58 S	46•20 W
49	Santos Dumont	Braz.	21•28 S	43•33 W
41	Santos Thomas	Phil.	14•07 N	121•09 E
41	Santo Tomas (Mtn.)	Phil.	16•23 N	120•32 E
50	Santo Tomé	Arg.	28•32 S	56•04 W
50	San Valentin (Mtn.)	Chile	46•41 S	73•30 W
20	San Vito	It.	45•53 N	12•52 E
39	Sanya	China	18•10 N	109•32 E
50	São Borja	Braz.	28•44 S	55•59 W
50	São Carlos	Braz.	22•02 S	47•54 W
49	São Cristovão	Braz.	11•04 S	37•11 W
49	São Francisco	Braz.	15•59 S	44•42 W
49	São Francisco, Rio	Braz.	8•56 S	40•20 W
50	São Francisco do Sul	Braz.	26•15 S	48•42 W
50	São Gabriel	Braz.	30•28 S	54•11 W
50	São Gonçalo	Braz.	22•55 S	43•04 W
50	São João da Barra	Braz.	21•40 S	41•03 W
50	São João del Rei	Braz.	21•08 S	44•14 W
50	São João de Meriti	Braz.	22•47 S	43•22 W
49	São João do Araguaia	Braz.	5•29 S	48•44 W
52	São Jorge I.	Açores	38•28 N	27•34 W
49	São José do Rio Pardo	Braz.	21•36 S	46•50 W
49	São José do Rio Prêto	Braz.	20•57 S	49•12 W
50	São Leopoldo	Braz.	29•46 S	51•09 W
49	São Luis (Maranhão)	Braz.	2•31 S	43•14 W
49	São Mateus	Braz.	18•44 S	39•45 W
52	São Miguel I.	Açores	37•59 N	26•38 W
17	Saône (R.)	Fr.	46•27 N	4•58 E
52	São Nicolau	C.V.	16•19 N	25•19 W
50	São Paulo	Braz.	23•34 S	46•38 W
49	São Paulo (State)	Braz.	21•45 S	50•47 W
48	São Paulo de Olivença	Braz.	3•32 S	68•46 W
49	São Raimundo Nonato	Braz.	9•09 S	42•32 W
49	São Roque, Cabo de	Braz.	5•06 S	35•11 W
54	São Salvador do Congo	Ang.	6•30 S	14•10 E
52	São Tiago I.	C.V.	15•09 N	24•45 W
52	São Tomé	São Tomé & Príncipe	0•20 N	6•44 E
51	Sao Tome & Principe	Afr.	1•00 N	6•00 E
18	Saoura, Oued (R.)	Alg.	29•39 N	1•42 W
49	São Vicente	Braz.	23•57 S	46•25 W
52	Sao Vincenti I.	C.V.	16•51 N	24•35 W
17	São Vinente, Cabo de	Port.	37•03 N	9•31 W
52	Sapele	Nig.	5•54 N	5•41 E
37	Sapporo	Jap.	43•02 N	141•29 E
53	Sara, Bahr (R.)	Chad-Cen.Afr.Empire	8•19 N	17•44 E
21	Sarajevo	Yugo.	43•15 N	18•26 E
9	Saranac Lake	NY	44•20 N	74•05 W
9	Saranac L.	NY	44•15 N	74•20 W
50	Sarandi	Arg.	34•26 S	58•21 W
30	Saransk	Sov.Un.	54•10 N	45•10 E
30	Sarapul	Sov.Un.	56•28 N	53•50 E
9	Saratoga Springs	NY	43•05 N	74•50 W
31	Saratov	Sov.Un.	51•30 N	45•30 E
39	Saravane	Laos	15•48 N	106•40 E
40	Sarawak	Mala.	2•30 N	112•45 E
27	Sárbogárd	Hung.	46•53 N	18•38 E
52	Sardalas	Libya	25•59 N	10•33 E
20	Sardinia (I.)	It.	40•08 N	9•05 E
53	Sarh (Fort-Archambault)	Chad	9•09 N	18•23 E
31	Sarikamis	Tur.	40•30 N	42•40 E
38	Sariwŏn	Korea	38•40 N	125•45 E
21	Şarkoy	Tur.	40•39 N	27•07 E
50	Sarmiento, Monte	Chile	54•28 S	70•40 W
8	Sarnia	Can.	43•00 N	82•25 W
27	Sarny	Sov.Un.	51•17 N	26•39 E
21	Saronikós Kólpos (G.)	Grc.	37•51 N	23•30 E
21	Saros Körfezi (G.)	Tur.	40•30 N	26•20 E
27	Sárospatak	Hung.	48•19 N	21•35 E
21	Šar Planina (Mts.)	Yugo.	42•07 N	21•54 E
24	Sarpsborg	Nor.	59•17 N	11•07 E
17	Sarreguemines	Fr.	49•06 N	7•05 E
17	Sarria	Sp.	42•14 N	7•17 W
20	Sartène	Fr.	41•36 N	8•59 E
17	Sarthe (R.)	Fr.	47•44 N	0•32 W
26	Sárvár	Hung.	47•14 N	16•55 E
31	Sarych, Mys (C.)	Sov.Un.	44•25 N	33•00 E
35	Sasarām	India	25•00 N	84•00 E
37	Sasebo	Jap.	33•12 N	129•43 E
26	Sašice	Czech.	49•14 N	13•31 E
10	Saskatchewan	Can.	54•46 N	107•40 W
10	Saskatchewan (R.)	Can.	53•45 N	103•20 W
10	Saskatoon	Can.	52•07 N	106•38 W
30	Sasovo	Sov.Un.	54•20 N	42•00 E
20	Sassari	It.	40•44 N	8•33 E
26	Sassnitz	G.D.R.	54•31 N	13•37 E
52	Satadougou	Mali	12•21 N	10•07 W
24	Säter	Swe.	60•21 N	15•50 E
30	Satka	Sov.Un.	55•03 N	59•02 E
27	Sátoraljaujhely	Hung.	48•24 N	21•40 E
27	Satu-Mare	Rom.	47•50 N	22•53 E
24	Saude	Nor.	59•40 N	6•21 E
16	Sáudharkrókur	Ice.	65•41 N	19•38 W
32	Saudi Arabia	Asia	22•40 N	46•00 E
8	Saugatuck	Mi.	42•40 N	86•10 W
8	Saugeer (R.)	Can.	44•20 N	81•20 W
9	Saugerties	NY	42•05 N	73•55 W
11	Sault Ste. Marie	Can.	46•31 N	84•20 W
11	Sault Ste. Marie	Mi.	46•29 N	84•21 W
45	Saunders, C.	N.Z.	45•55 S	170•50 E
54	Saurimo	Ang.	9•39 S	20•24 E
21	Sava (R.)	Yugo.	44•50 N	17•00 E
31	Savalan (Mtn.)	Iran	38•20 N	48•00 E
52	Savalou	Benin	7•56 N	1•58 E
7	Savannah	Ga.	32•04 N	81•07 W
7	Savannah (R.)	Ga.-SC	33•11 N	81•51 W
26	Sávava R.	Czech.	49•36 N	15•24 E
52	Savé	Benin	8•09 N	2•03 E
54	Save, Rio	Moz.	21•28 S	34•14 E
20	Savigliano	It.	44•38 N	7•42 E
20	Savona	It.	44•19 N	8•28 E
25	Savonlinna	Fin.	61•53 N	28•49 E
40	Sawahlunto	Indon.	0•37 S	100•50 E
53	Sawākin	Sud.	19•02 S	37•19 E
40	Sawankhalok	Thai.	17•16 N	99•48 E
53	Sawda, Jabal as (Mts.)	Libya	28•14 N	13•46 E
18	Sawfjjin, Wadi (R.)	Libya	31•18 N	13•16 E
53	Sawhāj	Egypt	26•34 N	31•40 E
53	Sawknah	Libya	29•04 N	15•53 E
40	Sawu, Laut (Savu Sea)	Indon.	9•15 S	122•15 E
40	Sawu (I.)	Indon.	10•15 S	122•00 E
52	Say	Niger	13•09 N	2•16 E
29	Sayan Khrebet (Mts.)	Sov.Un.	51•30 N	90•00 E
33	Şaydā (Sidon)	Leb.	33•34 N	35•23 E
34	Şayhūt	P.D.R. of Yem.	15•23 N	51•28 E
9	Sayre	Pa.	41•55 N	76•30 W
36	Sayr Usa	Mong.	44•51 N	107•00 E
34	Say'un	P.D.R. of Yem.	16•00 N	48•59 E
9	Sayville	NY	40•45 N	73•10 W
21	Sazan (Saseno) (I.)	Alb.	40•30 N	19•17 E
24	Scäffle	Swe.	59•10 N	12•55 E
32	Scandinavian Pen.	Eur.	62•00 N	14•00 E
22	Scarborough	Eng.	54•16 N	0•19 W
26	Schaffhausen	Switz.	47•42 N	8•38 E
11	Schefferville	Can.	54•52 N	67•01 W
23	Schelde, R.	Bel.	51•04 N	3•55 E
9	Schenectady	NY	42•50 N	73•55 W
20	Schio	It.	45•43 N	11•23 E
26	Schleswig	F.R.G.	54•32 N	9•32 E
26	Schleswig-Holstein	F.R.G.	54•40 N	9•10 E
26	Schmalkalden	G.D.R.	50•41 N	10•25 E
8	Schofield	Wi.	44•52 N	89•37 W
26	Schönebeck	G.D.R.	52•01 N	11•44 E
41	Schouten (I.)	Indon.	0•45 S	136•40 E
26	Schramberg	F.R.G.	48•14 N	8•24 E
9	Schroon (L.)	NY	43•50 N	73•50 W
9	Schuylkill-Haven	Pa.	40•35 N	76•10 W
26	Schwabach	F.R.G.	49•19 N	11•02 E
26	Schwäbische Alb (Mts.)	F.R.G.	48•11 N	9•09 E
26	Schwäbisch Gmünd	F.R.G.	48•47 N	9•49 E
26	Schwäbisch Hall	F.R.G.	49•08 N	9•44 E
26	Schwandorf	F.R.G.	49•19 N	12•08 E
40	Schwaner Mts.	Indon.	1•05 S	112•30 E
26	Schwarzwald (For.)	F.R.G.	47•54 N	7•57 E
26	Schwaz	Aus.	47•20 N	11•45 E
26	Schwechat	Aus.	48•09 N	16•29 E
26	Schwedt	G.D.R.	53•04 N	14•17 E
26	Schweinfurt	F.R.G.	50•03 N	10•14 E
26	Schwenningen	F.R.G.	48•04 N	8•33 E
26	Schwerin	G.D.R.	53•36 N	11•25 E
26	Schweriner See (L.)	G.D.R.	53•40 N	11•06 E
26	Schwyz	Switz.	47•01 N	8•38 E
20	Sciacca	It.	37•30 N	13•09 E
22	Scilly (Is.)	Eng.	49•56 N	6•50 W
8	Scioto (R.)	Oh.	39•10 N	82•55 W
22	Scotland	U.K.	57•05 N	5•10 W
9	Scottstown	Can.	45•35 N	71•15 W
10	Scott, C.	Can.	50•47 N	128•26 W
55	Scottburgh	S.Afr.	30•18 S	30•42 E
67	Scott Is.	Ant.	67•00 S	178•00 E
67	Scott Ra.	Ant.	68•00 S	55•00 E
8	Scottsburg	In.	38•40 N	85•50 W
46	Scottsdale	Austl.	41•12 S	147•37 E
8	Scottville	Mi.	44•00 N	86•20 W
9	Scranton	Pa.	41•45 N	75•45 W

Page	Name	Region	Lat.	Long.
9	Scugog (L.)	Can.	44•05 N	78•55 W
21	Scutari (R.)	Alb.	42•14 N	19•33 E
9	Seaford	De.	38•35 N	75•40 W
10	Seal (R.)	Can.	59•08 N	96•37 W
54	Seal I.	S.Afr.	34•07 S	18•36 E
6	Seattle	Wa	47•36 N	122•20 W
9	Sebago	Me.	43•52 N	70•20 W
12	Sebastion Vizcaino, Bahia	Mex.	28•45 N	115•15 W
40	Sebatik (I.)	Indon.	3•52 N	118•14 E
53	Sebderat	Eth.	15•30 N	36•45 E
21	Sebes	Rom.	45•58 N	23•34 E
8	Sebewaing	Mi.	43•45 N	83•25 W
31	Sebinkarahisar	Tur.	40•15 N	38•10 E
26	Sebnitz	G.D.R.	51•01 N	14•16 E
18	Sebou, Oued R.	Mor.	34•23 N	5•18 W
8	Sebree	Ky.	37•35 N	87•30 W
8	Sebring	Oh.	40•55 N	81•05 W
20	Secchia (R.)	It.	44•25 N	10•25 E
7	Sedalia	Mo.	38•42 N	93•12 W
17	Sedan	Fr.	49•49 N	4•55 E
33	Sedom	Isr.	31•04 N	35•24 E
25	Seduva	Sov.Un.	55•46 N	23•45 E
54	Seekoevlei (L.)	S.Afr.	34•04 S	18•33 E
18	Sefrou	Mor.	33•49 N	4•46 W
30	Seg (L.)	Sov.Un.	64•00 N	33•30 E
53	Segamat	Mala.	2•30 N	102•49 E
52	Ségou	Mali	13•27 N	6•16 W
48	Segovia	Col.	7•08 N	74•42 W
18	Segre (R.)	Sp.	41•54 N	1•10 E
52	Séguéla	Ivory Coast	7•57 N	6•40 W
18	Segura (R.)	Sp.	38•24 N	2•12 W
25	Seinäjoki	Fin.	62•47 N	22•50 E
17	Seine, Rivière (R.)	Fr.	49•21 N	1•17 E
50	Seio do Venus (Mtn.)	Braz.	22•28 S	43•12 W
52	Sekondi-Takoradi	Ghana	4•59 N	1•43 W
53	Sekota	Eth.	12•47 N	38•59 E
33	Selangor (State)	Mala.	2•53 N	101•29 E
21	Selanoutsi	Bul.	43•42 N	24•05 E
41	Selaru (I.)	Indon.	8•30 S	130•30 E
40	Selatan, Tandjung (C.)	Indon.	4•09 S	114•40 E
40	Selayar (I.)	Indon.	6•15 S	121•15 E
24	Selbu (L.)	Nor.	63•18 N	11•55 E
36	Selenge Gol. (R.)	Mong.	49•04 N	102•23 E
52	Selibaby	Mauritania	15•21 N	12•11 W
30	Seliger (L.)	Sov.Un.	57•14 N	33•18 E
10	Selkirk	Can.	50•09 N	96•52 W
10	Selkirk Mts.	Can.	51•00 N	117•40 W
7	Selma	Al.	32•25 N	87•00 W
54	Selukwe	Rh.	19•34 S	30•03 E
10	Selwyn (L.)	Can.	59•41 N	104•30 W
21	Seman (R.)	Alb.	40•48 N	19•53 E
40	Semarang	Indon.	7•03 S	110•27 E
40	Semarinda	Indon.	0•30 S	117•10 E
31	Semënovka	Sov.Un.	52•10 N	32•34 E
40	Semeru, Gunung (Mtn.)	Indon.	8•06 S	112•55 E
28	Semipalatinsk	Sov.Un.	50•28 N	80•29 E
28	Semiyarskoye	Sov.Un.	51•03 N	78•28 E
53	Semliki R.	Ug.-Zaire	0•45 N	29•36 E
26	Semmering	Aus.	47•39 N	15•50 E
31	Semnān	Iran	35•30 N	53•30 E
53	Senador Pompeu	Braz.	5•34 S	39•18 W
37	Sendai	Jap.	38•18 N	141•02 E
9	Seneca (L.)	NY	42•30 N	76•55 W
9	Seneca Falls	NY	42•55 N	76•55 W
51	Senegal	Afr.	14•53 N	14•58 W
52	Sénégal (R.)	Afr.	16•00 N	14•00 W
26	Senftenberg	G.D.R.	51•32 N	14•00 E
55	Sengunyane (R.)	Leso.	29•35 S	28•08 E
49	Senhor do Bonfim	Braz.	10•21 S	40•09 W
20	Senigallia	It.	43•42 N	13•16 E
20	Senj	Yugo.	44•58 N	14•55 E
16	Senja (I.)	Nor.	69•28 N	16•10 E
53	Sennar Dam	Sud.	13•38 N	33•38 E
11	Senneterre	Can.	48•20 N	77•22 W
21	Senta	Yugo.	45•54 N	20•05 E
33	Sepang	Mala.	2•43 N	101•45 E
50	Sepetiba, Baia de (B.)	Braz.	23•01 S	43•42 W
41	Sepik (R.)	Pap.N.Gui.	4•07 S	142•40 E
11	Sept-Îles	Can.	50•12 N	66•23 W
6	Sequoia Natl. Park	Ca.	36•34 N	118•37 W
23	Seraing	Bel.	50•38 N	5•28 E
41	Seram (I.)	Indon.	2•45 S	129•30 E
40	Serang	Indon.	6•13 S	106•10 E
33	Seranggung	Indon.	0•49 N	104•11 E
31	Serdobsk	Sov.Un.	52•30 N	44•20 E
27	Sered	Czech.	48•18 N	17•43 E
33	Seremban	Mala.	2•44 N	101•57 E
54	Serenje	Zambia	13•12 S	30•49 E
27	Seret	Czech.	48•17 N	17•43 E
27	Seret	Rom.	47•58 N	26•01 E
27	Seret R.	Sov.Un.	49•45 N	25•30 E
49	Sergipe (State)	Braz.	10•27 S	37•04 W
30	Sergiyevsk	Sov.Un.	53•58 N	51•00 E
21	Sérifos	Grc.	37•10 N	24•32 E
21	Sérifos (I.)	Grc.	37•42 N	24•17 E
50	Seropédica	Braz.	22•44 S	43•43 W
30	Serov	Sov.Un.	59•36 N	60•30 E
54	Serowe	Bots.	22•18 S	26•39 E
28	Serpukhov	Sov.Un.	54•53 N	37•27 E
21	Sérrai (Seres)	Grc.	41•06 N	23•36 E
49	Serrinha	Braz.	11•43 S	38•49 W
49	Sertânia	Braz.	8•28 S	37•13 W
33	Serting (R.)	Mala.	3•01 N	102•32 E
50	Seruí	Braz.	22•40 S	43•08 W
20	Sesia (R.)	It.	45•33 N	8•25 E
55	Sesmyl (R.)	S.Afr.	25•51 S	28•06 E
20	Sestri Levante	It.	44•15 N	9•24 E
30	Sestroretsk	Sov.Un.	60•06 N	29•58 E
17	Sète	Fr.	43•24 N	3•42 E
49	Sete Lagoas	Braz.	19•23 S	43•58 W
52	Setif	Alg.	36•18 N	5•21 E
52	Settat	Mor.	33•02 N	7•30 W
54	Sette-Cama	Gabon	2•29 S	9•40 E
18	Setúbal	Port.	30•32 N	8•54 W
11	Seul, Lac (L.)	Can.	50•20 N	92•30 W
24	Sevalen (L.)	Nor.	62•19 N	10•15 E
31	Sevan (L.)	Sov.Un.	40•10 N	45•20 E
28	Sevastopol' (Akhiar)	Sov.Un.	44•34 N	33•34 E
11	Severn (R.)	Can.	55•21 N	88•42 W
22	Severn (R.)	Eng.	51•42 N	2•25 W
30	Severnaya Dvina (Northern Dvina) (R.)	Sov.Un.	63•00 N	42•40 E
29	Severnaya Zemlya (Northern Land) (Is.)	Sov.Un.	79•33 N	101•15 E
6	Sevier R	Ut.	39•25 N	112•20 W
48	Sevilla	Col.	4•16 N	75•56 W
18	Sevilla	Sp.	37•29 N	5•58 W
21	Sevlievo	Bul.	41•02 N	25•05 E
6	Seward	Ak.	60•18 N	149•29 W
50	Sewell	Chile	34•01 S	70•18 W
67	Seychelles	Afr.	5•20 S	55•10 E
16	Seydhisfjördhur	Ice.	65•21 N	14•08 W
19	Seyhan (R.)	Tur.	37•28 N	35•40 E
31	Seym (R.)	Sov.Un.	51•23 N	33•22 E
8	Seymour	In.	38•55 N	85•55 W
55	Seymour	S.Afr.	32•33 S	26•48 E
55	Sezela	S.Afr.	30•33 S	30•37 W
20	Sezze	It.	41•32 N	13•30 E
52	Sfaz	Tun.	34•51 N	10•45 E
21	Sfîntu-Gheorghe	Rom.	45•53 N	25•49 E
23	's Gravenhage (The Hague)	Neth.	52•05 N	4•16 E
37	Sha (R.)	China	33•33 N	114•30 E
54	Shabani	Rh.	20•15 S	30•28 E
67	Shackleton Shelf Ice	Ant.	65•00 S	100•00 E
35	Shāhjahānpur	India	27•58 N	79•58 E
38	Shaho	China	40•08 N	116•16 E
34	Shahrezā	Iran	31•47 N	51•47 E
31	Shahsavār	Iran	36•40 N	51•00 E
31	Shakhty	Sov.Un.	47•41 N	40•11 E
53	Shala L.	Eth.	7•34 N	39•00 E
34	Shām, Jabal ash (Mtn.)	Om.	23•01 N	57•45 E
53	Shambe	Sud.	7•08 N	30•46 E
34	Shammar, Jabal (Mts.)	Sau.Ar.	27•13 N	40•16 E
53	Shamo (L.)	Eth.	5•58 N	37•00 E
9	Shamokin	Pa.	40•45 N	76•30 W
54	Shamva	Rh.	17•18 S	31•35 E
53	Shandī	Sud.	16•44 N	33•29 E
38	Shangch'iu	China	34•24 N	115•39 E
39	Shanghai	China	31•14 N	121•27 E
39	Shanghai Shih (Mun.)	China	31•30 N	121•45 E
39	Shangjao	China	28•25 N	117•58 E
38	Shangtu	China	41•38 N	113•22 E
22	Shannon R.	Ire.	52•30 N	9•58 W
38	Shanshan	China	42•51 N	89•53 E
37	Shansi (Shanhsi) (Prov.)	China	37•30 N	112•00 E
29	Shantar (I.)	Sov.Un.	55•13 N	138•42 E
39	Shant'ou (Swatow)	China	23•20 N	116•40 E
37	Shantung (Prov.)	China	36•08 N	117•09 E
39	Shantung Pantao (Pen.)	China	37•00 N	120•10 E
39	Shantung Pt.	China	37•28 N	122•40 E
39	Shaohsing	China	30•00 N	120•40 E
39	Shaokuan	China	24•58 N	113•42 E
38	Shaopo	China	32•33 N	119•30 E
44	Shark B.	Austl.	25•30 S	113•00 E
8	Sharon	Pa.	41•15 N	80•30 W
34	Sharr, Jabal (Mtn.)	Sau.Ar.	28•00 N	36•07 E
39	Shashih	China	30•20 N	112•18 E
6	Shasta, Mt.	Ca.	41•35 N	122•12 W
6	Shasta L.	Ca.	40•51 N	122•32 W
30	Shatsk	Sov.Un.	54•00 N	41•40 E
10	Shaunavon	Can.	49•40 N	108•25 W
8	Shawano	Wi.	44•41 N	88•13 W
11	Shawinigan	Can.	46•32 N	72•46 W
6	Shawnee	Ok.	35•20 N	96•54 W
8	Shawneetown	Il.	37•40 N	88•05 W
39	Shayang	China	31•00 N	112•38 E
27	Shchara (R.)	Sov.Un.	53•17 N	25•12 E
8	Sheboygan	Wi.	43•45 N	87•44 W
8	Sheboygan Falls	Wi.	43•43 N	87•51 W
33	Shechem (Ruins)	Jordan	32•15 N	35•22 E
22	Sheelin (L.)	Ire.	53•46 N	7•34 W
23	Sheerness	Eng.	51•26 N	0•46 E
22	Sheffield	Eng.	53•23 N	1•28 W
22	Shehy, Mts.	Ire.	51•46 N	9•45 W
30	Sheksna (R.)	Sov.Un.	59•50 N	38•40 E
29	Shelagskiy, Mys (C.)	Sov.Un.	70•08 N	170•52 E
8	Shelburn	In.	39•10 N	87•30 W
11	Shelburne	Can.	43•46 N	65•19 W
9	Shelburne	Can.	44•04 N	80•12 W
8	Shelby	Mi.	43•35 N	86•20 W
8	Shelby	Oh.	40•50 N	82•40 W
8	Shelbyville	Il.	39•20 N	88•45 W
8	Shelbyville	In.	39•30 N	85•45 W
8	Shelbyville	Ky.	38•10 N	85•15 W
34	Shelbyville Res.	Il.	39•30 N	88•45 W
29	Shelekhova, Zaliv (B.)	Sov.Un.	60•00 N	156•00 E
9	Shelton	Ct.	41•15 N	73•05 W
31	Shemakha	Sov.Un.	40•35 N	48•40 E
9	Shenandoah	Pa.	40•50 N	76•15 W
9	Shenandoah	Va.	38•30 N	78•30 W
9	Shenandoah Natl. Park	Va.	38•35 N	78•25 W
9	Shenandoah (R.)	Va.	38•55 N	78•05 W
38	Shenchiu	China	33•11 N	115•06 E
28	Shenkursk	Sov.Un.	62•10 N	43•08 E
38	Shenmu	China	38•55 N	110•35 E
36	Shensi (Shenhsi) (Prov.)	China	35•30 N	109•10 E
38	Shenyang (Mukden)	China	41•45 N	123•22 E
35	Sheopur	India	25•37 N	78•10 E
31	Shepetovka	Sov.Un.	50•10 N	27•01 E
46	Shepparton	Austl.	36•15 S	145•25 E
9	Sherbrooke	Can.	45•24 N	71•54 W
27	Shereshevo	Sov.Un.	52•31 N	24•08 E
6	Sheridan	Wy.	44•48 N	106•56 W
6	Sherman	Tx.	33•39 N	96•37 W
22	Shetland (Is.)	Scot.	60•35 N	2•10 W
34	Shevchenko	Sov.Un.	44•00 N	51•10 E
53	Shewa Gimira	Eth.	7•13 N	35•49 E
8	Shiawassee (R.)	Mi.	43•15 N	84•05 W
34	Shibām	P.D.R. of Yem.	16•02 N	48•40 E
37	Shichitō (Seven Is.)	Jap.	34•18 N	139•28 E
38	Shihchiachuang	China	38•04 N	114•31 E
39	Shihlung	China	23•05 N	113•58 E
39	Shihwanta Shan	China	22•10 N	107•30 E
38	Shihwei Pk.	China	47•11 N	119•59 E
35	Shikārpur	Pak.	27•51 N	68•52 E
37	Shikoku (I.)	Jap.	33•43 N	133•33 E
35	Shillong	India	25•39 N	91•58 E
37	Shimonoseki	Jap.	33•58 N	130•55 E
22	Shin, Loch	Scot.	58•08 N	4•02 W
53	Shinyanga	Tan.	3•40 S	33•26 E
9	Shippenburg	Pa.	40•00 N	77•30 W
33	Shiqma (R.)	Isr.	31•31 N	34•40 E
54	Shirati	Tan.	1•15 S	34•02 E
34	Shīrāz	Iran	29•32 N	52•27 E
54	Shire (R.)	Malawi	16•20 S	35•05 E
35	Shivpuri	India	25•31 N	77•46 E
33	Shivta (Ruins)	Isr.	30•54 N	34•36 E
21	Shkodër (Scutari)	Alb.	42•04 N	19•30 E
8	Shoals	In.	38•40 N	86•45 W
35	Sholāpur	India	17•42 N	75•51 E
8	Shorewood	Wi.	43•05 N	87•54 W
31	Shpola	Sov.Un.	49•01 N	31•36 E
7	Shreveport	La.	32•30 N	93•46 W
22	Shrewsbury	Eng.	52•43 N	2•44 W
38	Shuangch'eng	China	45•18 N	126•18 E
37	Shuangliao	China	43•37 N	123•30 E
38	Shuangyang	China	43•28 N	125•45 E
21	Shumen	Bul.	43•15 N	26•54 E
39	Shunan	China	29•38 N	119•00 E
38	Shuni	China	40•09 N	116•38 E
35	Shunning	China	24•34 N	99•49 E
34	Shuqrah	P.D.R. of Yem.	13•32 N	46•02 E
34	Shūrāb (R.)	Iran	31•08 N	55•30 E
31	Shur R.	Iran	35•40 N	50•10 E
34	Shūshtar	Iran	31•50 N	48•46 E
28	Shuya	Sov.Un.	56•52 N	41•23 E
39	Shweba	Bur.	22•23 N	96•13 E
33	Siak Ketjil (R.)	Indon.	1•01 N	101•45 E
33	Siakriinderapura	Indon.	0•48 N	102•05 E
35	Siālkot	Pak.	32•39 N	74•30 E
21	Siátista	Grc.	40•15 N	21•32 E
41	Siau (I.)	Indon.	2•40 N	126•00 E
25	Šiauliai (Shyaulyay)	Sov.Un.	55•57 N	23•19 E
20	Šibenik	Yugo.	43•44 N	15•55 E
32	Siberia	Asia	57•00 N	97•00 E
40	Siberut (I.)	Indon.	1•22 S	99•45 E
54	Sibiti	Con.	3•41 S	13•21 E
21	Sibiu	Rom.	45•47 N	24•09 E
40	Sibolga	Indon.	1•45 N	98•45 E
35	Sibsāgar	India	26•47 N	94•45 E
40	Sibuti	Phil.	4•40 N	119•30 E
41	Sibuyan (I.)	Phil.	12•19 N	122•25 E
40	Sibuyan Sea	Phil.	12•43 N	122•38 E
40	Sicapoo (Mtn.)	Phil.	18•05 N	121•03 E
17	Sicily (I.)	It.	37•38 N	13•30 E
48	Sicuaní	Peru	14•12 S	71•12 W
53	Sidamo	Eth.	5•08 N	37•45 E
20	Siderno Marina	It.	38•18 N	16•19 E
20	Sídheros, Akr. (C.)	Grc.	35•19 N	26•20 E
21	Sidhiró Kastron	Grc.	41•13 N	23•27 E
53	Sīdī Barrānī	Egypt	31•41 N	26•09 E
52	Sidi bel Abbès	Alg.	35•15 N	0•43 W
52	Sidi Ifni	Mor.	29•22 N	10•15 W
67	Sidley, Mt.	Ant.	77•25 S	129•00 W
8	Sidney	Oh.	40•20 N	84•10 W
33	Sidr, Wādī (R.)	Egypt	29•43 N	32•58 E
27	Siedlce	Pol.	52•09 N	22•20 E
26	Siegburg	F.R.G.	50•48 N	7•13 E
23	Siegen	F.R.G.	50•52 N	8•01 E
26	Sieg R.	F.R.G.	50•51 N	7•53 E
27	Siemiatycze	Pol.	52•26 N	22•52 E
27	Siemionówka	Pol.	52•53 N	23•50 E
40	Siem Reap	Camb.	13•32 N	103•54 E
20	Siena	It.	43•19 N	11•21 E
27	Sieradz	Pol.	51•35 N	18•45 E
27	Sierpc	Pol.	52•51 N	19•42 E
6	Sierra Blanca Pk.	NM	33•25 N	105•50 W
51	Sierra Leone	Afr.	8•48 N	12•30 W
21	Sífnos (I.)	Grc.	36•58 N	24•30 E
24	Sigdal	Nor.	60•01 N	9•35 E
27	Sighet	Rom.	47•57 N	23•55 E
27	Sighisoara	Rom.	46•11 N	24•48 E
16	Siglufjördhur	Ice.	66•06 N	18•45 W
31	Signakhi	Sov.Un.	41•45 N	45•50 E
48	Sigsig	Ec.	3•05 S	78•44 W
24	Sigtuna	Swe.	59•40 N	17•39 E
17	Sigüenza	Sp.	41•03 N	2•38 W
52	Siguiri	Gui.	11•25 N	9•10 W
31	Siirt	Tur.	38•00 N	42•00 E
29	Sikhote Alin (Mts.)	Sov.Un.	45•00 N	135•45 E
21	Síkinos (I.)	Grc.	36•45 N	24•55 E
35	Sikkim	Asia	27•42 N	88•25 E
27	Siklós	Hung.	45•51 N	18•18 E
41	Silang	Phil.	14•14 N	120•58 E
35	Silchar	India	24•52 N	92•50 E
27	Silesia	Pol.	50•58 N	16•53 E
31	Silifke	Tur.	36•20 N	34•00 E
19	Silistra	Bul.	44•01 N	27•13 E
24	Siljan (R.)	Swe.	60•48 N	14•28 E
24	Silkeborg	Den.	56•10 N	9•33 E
25	Šilutė	Sov.Un.	55•23 N	21•26 E
49	Silvânia	Braz.	16•43 S	48•33 W
9	Silver Creek	NY	42•35 N	79•10 W
55	Silverton	S.Afr.	25•45 S	28•13 E
17	Silves	Port.	37•15 N	8•24 W
9	Simcoe	Can.	42•50 N	80•20 W
9	Simcoe (L.)	Can.	44•30 N	79•20 W
40	Simeulue (I.)	Indon.	2•27 N	95•30 E
31	Simferopol' (Akmechet)	Sov.Un.	44•58 N	34•04 E
19	Simi (I.)	Grc.	36•27 N	27•41 E
35	Simla	India	31•09 N	77•15 E
27	Simleul-Silvaniei	Rom.	47•14 N	22•46 E
25	Simola	Fin.	60•55 N	28•06 E
54	Simonstad	S.Afr.	34•11 S	18•25 E
26	Simplon P.	Switz.	46•13 N	7•53 E
26	Simplon Tun.	It.-Switz.	46•16 N	8•20 E
44	Simpson Des.	Austl.	24•40 S	136•40 E
24	Simrishamn	Swe.	55•35 N	14•19 E
37	Simushir (I.)	Sov.Un.	47•15 N	150•47 E
21	Sinaia	Rom.	45•20 N	25•30 E
53	Sinai Pen.	Egypt	29•24 N	33•29 E
41	Sinait	Phil.	15•54 N	120•28 E
12	Sinaloa (State)	Mex.	25•15 N	107•45 W
38	Sinanju	Kor.	39•39 N	125•41 E
31	Sinap	Tur.	42•00 N	35•05 E
48	Sincé	Col.	9•15 N	75•14 W
48	Sincelejo	Col.	9•12 N	75•30 W
25	Sindi	Sov.Un.	58•20 N	24•40 E
31	Sinel'nikovo	Sov.Un.	49•19 N	35•33 E
33	Singapore	Asia	1•22 N	103•45 E
33	Singapore Str.	Indon.	1•14 N	104•20 E
21	Singitikós Kólpos (G.)	Grc.	40•15 N	24•00 E
36	Singu	Bur.	22•37 N	96•04 E
20	Sinj	Yugo.	43•42 N	16•39 E
53	Sinjah	Sud.	13•09 N	33•52 E
36	Sinkiang Uighur (Aut. Reg.)	China	40•15 N	82•15 E
49	Sinnamary	Fr.Gu.	5•15 N	57•52 W
20	Sinni (R.)	It.	40•05 N	16•15 E
50	Sino, Pedra de (Mtn.)	Braz.	22•27 S	43•02 W
54	Sinoia	Rh.	17•22 S	30•12 E
38	Sinŭiju	Kor.	40•04 N	124•33 E
26	Sion	Switz.	46•15 N	7•17 E
7	Sioux City	Ia.	42•30 N	96•25 W
6	Sioux Falls	SD	43•33 N	96•43 W
48	Sipí	Col.	4•39 N	76•38 W
10	Sipiwesk	Can.	55•27 N	97•24 W
40	Sipura (I.)	Indon.	2•15 S	99•33 E
17	Siracusa	It.	37•02 N	15•19 E
35	Sirājganj	Bngl.	24•23 N	89•43 E
44	Sir Edward Pellew Group (Is.)	Austl.	15•15 S	137•15 E
27	Siretul R.	Rom.	46•10 N	27•18 E
34	Sirhān, Wadi (R.)	Sau.Ar.	31•02 N	37•16 E
21	Síros (Ermoúpolis)	Grc.	37•30 N	24•56 E
21	Síros (I.)	Grc.	37•23 N	24•55 E
25	Sirvintos	Sov.Un.	55•02 N	24•59 E
20	Sisak	Yugo.	45•29 N	16•20 E
12	Sisal	Mex.	21•09 N	90•03 W
34	Sistān, Daryacheh-ye (L.)	Iran-Afg.	31•45 N	61•15 E
8	Sisterville	WV	39•30 N	81•00 W
20	Sitía	Grc.	35•09 N	26•10 E
6	Sitka	Ak.	57•08 N	135•18 W

Page	Name	Region	Lat.	Long.
40	Sittwe	Bur.	20•09 N	92•54 E
31	Sivas	Tur.	39•50 N	36•50 E
31	Siverek	Tur.	37•50 N	39•20 E
25	Siverskaya	Sov.Un.	59•17 N	30•03 E
53	Siwah (Oasis)	Egypt	29•33 N	25•11 E
53	Sixth Cataract	Sud.	16•26 N	32•44 E
24	Sjaelland (I.)	Den.	55•34 N	11•35 E
21	Sjenica	Yugo.	43•15 N	20•02 E
24	Skagen	Den.	57•43 N	10•32 E
24	Skagen (Pt.)	Den.	57•43 N	10•31 E
24	Skagerrak (Str.)	Eur.	57•43 N	8•28 E
6	Skagway	Ak.	59•30 N	135•28 W
24	Skälderviken (B.)	Swe.	56•20 N	12•25 E
29	Skalistyy, Golets (Mtn.)	Sov.Un.	57•28 N	119•48 E
24	Skanderborg	Den.	56•04 N	9•55 E
9	Skaneateles	NY	42•55 N	76•25 W
9	Skaneateles (L.)	NY	42•50 N	76•20 W
24	Skänninge	Swe.	58•24 N	15•02 E
24	Skanör	Swe.	55•24 N	12•49 E
21	Skantzoúra (Is.)	Grc.	39•03 N	24•05 E
24	Skara	Swe.	58•25 N	13•24 E
10	Skeena (R.)	Can.	54•10 N	129•40 W
55	Skeerpoort	S.Afr.	25•49 S	27•45 E
55	Skeerpoort (R.)	S.Afr.	25•58 S	27•41 E
49	Skeldon	Guy.	5•49 N	57•15 W
16	Skellefte (R.)	Swe.	65•18 N	19•08 E
16	Skellefteå	Swe.	64•47 N	20•48 E
24	Skern (R.)	Den.	55•56 N	8•52 E
22	Skerries (Is.)	Wales	53•30 N	4•59 W
21	Skíathos (I.)	Grc.	39•15 N	23•25 E
22	Skibbereen	Ire.	51•32 N	9•25 W
24	Skien	Nor.	59•13 N	9•35 E
27	Skierniewice	Pol.	51•58 N	20•13 E
18	Skikda (Philippeville)	Alg.	36•58 N	6•51 E
21	Skíros	Grc.	38•53 N	24•32 E
21	Skiros (I.)	Grc.	38•50 N	24•43 E
24	Skive	Den.	56•34 N	8•56 E
16	Skjalfandá (R.)	Ice.	65•24 N	16•40 W
16	Skjerstad	Nor.	67•12 N	15•37 E
20	Škofja Loka	Yugo.	46•10 N	14•20 E
27	Skole	Sov.Un.	49•03 N	23•32 E
21	Skópelos (I.)	Grc.	39•04 N	23•31 E
30	Skopin	Sov.Un.	53•49 N	39•35 E
21	Skopje	Yugo.	42•02 N	21•26 E
24	Skövde	Swe.	58•25 N	13•48 E
29	Skovorodino	Sov.Un.	53•53 N	123•56 E
20	Skradin	Yugo.	43•49 N	17•58 E
24	Skreia	Nor.	60•40 N	10•55 E
24	Skudeneshavn	Nor.	59•10 N	5•19 E
24	Skulerud	Nor.	59•40 N	11•30 E
25	Skuodas	Sov.Un.	56•16 N	21•32 E
24	Skurup	Swe.	55•29 N	13•27 E
31	Skvira	Sov.Un.	49•43 N	29•41 E
26	Skwierzyna	Pol.	52•35 N	15•30 E
22	Skye (I.)	Scot.	57•25 N	6•17 W
50	Skyring, Seno (B.)	Chile	52•35 S	72•30 W
24	Slagese	Den.	55•25 N	11•19 E
40	Slamet (Mtn.)	Indon.	7•15 S	109•15 E
21	Slănic	Rom.	45•13 N	25•56 E
21	Slatina	Rom.	44•26 N	24•21 E
10	Slave (R.)	Can.	59•40 N	111•21 W
28	Slavgorod	Sov.Un.	52•58 N	78•43 E
21	Slavonija (Reg.)	Yugo.	45•29 N	17•31 E
20	Slavonska Požega	Yugo.	45•18 N	17•42 E
21	Slavonski Brod	Yugo.	45•10 N	18•01 E
31	Slavyansk	Sov.Un.	48•52 N	37•34 E
22	Sligo	Ire.	54•17 N	8•19 W
24	Slite	Swe.	57•41 N	18•47 E
21	Sliven	Bul.	42•41 N	26•20 E
25	Slobodka	Sov.Un.	54•34 N	26•12 E
30	Slobodskoy	Sov.Un.	58•48 N	50•02 E
25	Sloka	Sov.Un.	56•57 N	23•37 E
27	Slonim	Sov.Un.	53•05 N	25•19 E
20	Slovenija (Reg.)	Yugo.	45•58 N	14•43 E
27	Slovensko (Slovakia)	Czech.	48•40 N	19•00 E
27	Sluch (R.)	Sov.Un.	50•56 N	26•48 E
20	Sluderno	It.	46•38 N	10•37 E
20	Slunj	Yugo.	45•08 N	15•46 E
27	Slupsk	Pol.	54•28 N	17•02 E
31	Slutsk	Sov.Un.	53•02 N	27•34 E
22	Slyne Head	Ire.	53•25 N	10•05 W
21	Smederevo (Semendria)	Yugo.	44•39 N	20•54 E
21	Smederevska Palanka	Yugo.	44•21 N	21•00 E
24	Smedjebacken	Swe.	60•09 N	15•19 E
31	Smela	Sov.Un.	49•14 N	31•52 E
9	Smethport	Pa.	41•50 N	78•25 W
10	Smith	Can.	55•10 N	114•02 W
10	Smithers	Can.	54•47 N	127•10 W
8	Smithland	Ky.	37•10 N	88•25 W
9	Smiths Falls	Can.	44•55 N	76•05 W
46	Smithton	Austl.	40•55 S	145•12 E
54	Smitswinkelvlakte	S.Afr.	34•16 S	18•25 E
6	Smoky Hill (R.)	Ks.	38•40 N	97•32 W
24	Smöla (I.)	Nor.	63•16 N	7•40 E
28	Smolensk	Sov.Un.	54•46 N	32•03 E
21	Smyadovo	Bul.	43•04 N	27•00 E
9	Smyrna	De.	39•20 N	75•35 W
6	Snake (R.)	Wa.	46•35 N	117•20 W
27	Sniardwy L.	Pol.	53•46 N	21•59 E
24	Snöhetta (Mtn.)	Nor.	62•18 N	9•12 E
22	Snowdon, Mt.	Wales	53•05 N	4•04 W
9	Snow Hill	Md.	38•15 N	75•20 W
45	Snowy Mts.	Austl.	36•17 S	148•30 E
53	Sobat R.	Sud.	9•04 N	32•02 E
49	Sobral	Braz.	3•39 S	40•16 W
27	Sochaczew	Pol.	52•14 N	20•18 E
36	Soch'e (Yarkand)	China	38•15 N	77•15 E
31	Sochi	Sov.Un.	43•35 N	39•50 E
43	Society Is.	Fr.Polynesia	15•00 S	157•30 W
48	Socorro	Col.	6•23 N	73•19 W
51	Socotra I.	P.D.R. of Yem.	13•00 N	52•30 E
24	Söderhamn	Swe.	61•20 N	17•00 E
24	Söderköping	Swe.	58•30 N	16•14 E
24	Södertälje	Swe.	59•12 N	17•35 E
38	Sodi Soruksum (Mtn.)	China	37•20 N	102•00 E
53	Sodo	Eth.	7•03 N	37•46 E
24	Södra Dellen (L.)	Swe.	61•45 N	16•30 E
26	Soest	F.R.G.	51•35 N	8•05 E
21	Sofiya (Sofia)	Bul.	42•43 N	23•20 E
48	Sogamoso	Col.	5•42 N	72•51 W
24	Sogndal	Nor.	58•20 N	6•17 E
24	Sogndal	Nor.	61•14 N	7•04 E
24	Sogne Fd.	Nor.	61•09 N	5•30 E
27	Sokal	Sov.Un.	50•28 N	24•20 E
31	Soke	Tur.	37•40 N	27•10 E
52	Sokodé	Togo	8•59 N	1•08 E
27	Sokolka	Pol.	53•23 N	23•30 E
52	Sokolo	Mali	14•51 N	6•09 W
52	Sokoto	Nig.	13•04 N	5•16 E
27	Sokotów Podlaski	Pol.	52•24 N	22•15 E
41	Solana	Phil.	17•40 N	121•41 E
41	Solano	Phil.	16•31 N	121•11 E
48	Soledad	Col.	10•47 N	75•00 W
30	Solikamsk	Sov.Un.	59•38 N	56•48 E
48	Solimões, Rio	Braz.	2•45 S	67•44 W
26	Solingen	F.R.G.	51•10 N	7•05 E
24	Sollefteå	Swe.	63•06 N	17•17 E
31	Sol'-Iletsk	Sov.Un.	51•10 N	55•05 E
42	Solomon Is.	Oceania	7•00 S	148•00 E
26	Solothurn	Switz.	47•13 N	7•30 E
30	Solov'etskiy (I.)	Sov.Un.	65•10 N	35•40 E
20	Šolta (I.)	Yugo.	43•20 N	16•15 E
26	Soltau	F.R.G.	53•00 N	9•50 E
38	Solun	China	47•32 N	121•18 E
9	Solvay	NY	43•05 N	76•10 W
24	Sölvesborg	Swe.	56•04 N	14•35 E
30	Sol'vychegodsk	Sov.Un.	61•18 N	46•58 E
22	Solway Firth	Eng.-Scot.	54•42 N	3•55 W
51	Somalia	Afr.	3•28 N	44•47 E
21	Sombor	Yugo.	45•45 N	19•10 E
49	Sombrero, Cayo	Ven.	10•52 N	68•12 W
9	Somerset	Pa.	40•00 N	79•05 W
55	Somerset East	S.Afr.	32•44 S	25•36 E
9	Somerville	Ma.	42•23 N	71•06 W
27	Somesul R.	Rom.	47•43 N	23•09 E
23	Somme (R.)	Fr.	50•02 N	2•04 E
50	Somuncurá, Meseta de (Plat.)	Arg.	41•15 S	68•00 W
35	Son (R.)	India	24•40 N	82•35 E
38	Sŏnchŏn	Kor.	39•49 N	124•56 E
55	Sondags (R.)	S.Afr.	33•17 S	25•14 E
24	Sonderborg	Den.	54•55 N	9•47 E
26	Sondershausen	G.D.R.	51•17 N	10•45 E
39	Song Ca (R.)	Viet.	19•15 N	105•00 E
54	Songea	Tan.	10•41 S	35•39 E
38	Sŏngjin	Kor.	40•38 N	129•10 E
40	Songkhla	Thai.	7•09 N	100•34 E
26	Sonneberg	G.D.R.	50•20 N	11•14 E
12	Sonora (State)	Mex.	29•45 N	111•15 W
12	Sonora (R.)	Mex.	28•45 N	111•35 W
6	Sonora Pk.	Ca.	38•22 N	119•39 W
48	Sonsón	Col.	5•42 N	75•28 W
41	Sonsorol Is.	Pac.Is.TrustTer.	5•03 N	132•33 E
39	Soochow (Wuhsien)	China	31•19 N	120•37 E
48	Sopetrán	Col.	6•30 N	75•44 W
24	Sopot	Pol.	54•26 N	18•25 E
26	Sopron	Hung.	47•41 N	16•36 E
20	Sora	It.	41•43 N	13•37 E
24	Sör Aurdal	Nor.	60•54 N	9•24 E
11	Sorel	Can.	46•01 N	73•07 W
46	Sorell, C.	Austl.	42•10 S	144•50 E
20	Soresina	It.	45•17 N	9•51 E
17	Soria	Sp.	41•46 N	2•28 W
49	Sorocaba	Braz.	23•29 S	47•27 W
31	Soroki	Sov.Un.	48•09 N	28•17 E
41	Sorong	Indon.	1•00 S	131•20 E
53	Soroti	Ug.	1•43 N	33•37 E
16	Söröy (I.)	Nor.	70•37 N	20•58 E
20	Sorrento	It.	40•23 N	14•23 E
41	Sorsogon	Phil.	12•51 N	124•02 E
25	Sortavala	Sov.Un.	61•43 N	30•40 E
38	Sŏsan	Kor.	36•40 N	126•25 E
30	Sosnogorsk	Sov.Un.	63•13 N	54•09 E
27	Sosnowiec	Pol.	50•17 N	19•10 E
30	Sos'va (R.)	Sov.Un.	63•10 N	63•30 E
49	Soublette	Ven.	9•55 N	66•06 W
20	Soúdhas, Kolpós (G.)	Grc.	35•33 N	24•22 E
21	Souflion	Grc.	41•12 N	26•17 E
17	Souk-Ahras	Alg.	36•18 N	8•19 E
38	Sŏul (Seoul)	Kor.	37•35 N	127•03 E
54	Sources, Mt. aux	Leso.-S.Afr.	28•47 S	29•04 E
10	Souris	Can.	49•38 N	100•15 W
52	Sousse	Tun.	36•00 N	10•39 E
51	South Africa	Afr.	28•00 S	24•50 E
22	Southampton	Eng.	50•54 N	1•30 W
9	Southampton	NY	40•53 N	72•24 W
11	Southampton I.	Can.	64•38 N	84•00 W
40	South Andaman I.	Andaman & Nicobar Is.	11•57 N	93•24 E
44	South Australia (State)	Austl.	29•45 S	132•00 E
8	South Bend	In.	41•40 N	86•20 W
9	Southbridge	Ma.	42•05 N	72•00 W
41	South C.	Pap.N.Gui.	10•40 S	149•00 E
7	South Carolina	U.S.	34•15 N	81•10 W
8	South Charleston	WV	38•20 N	81•40 W
40	South China Sea	Asia	15•23 N	114•12 E
6	South Dakota	U.S.	44•20 N	101•55 W
22	South Downs	Eng.	50•55 N	1•13 W
45	Southeast, C.	Austl.	43•47 S	146•03 E
23	Southend-on-Sea	Eng.	51•33 N	0•41 E
45	Southern Alps (Mts.)	N.Z.	44•08 S	169•18 E
44	Southern Cross	Austl.	31•13 S	119•30 E
11	Southern Indian (L.)	Can.	56•46 N	98•57 W
22	Southern Uplands	Scot.	55•15 N	4•28 W
8	South Fox (I.)	Mi.	45•25 N	85•55 W
47	South Georgia (I.)	Falk.Is.	54•00 S	37•00 W
8	South Haven	Mi.	42•25 N	86•15 W
9	Southington	Ct.	41•35 N	72•55 W
45	South I.	N.Z.	43•15 S	167•00 E
8	South Milwaukee	Wi.	42•55 N	87•52 W
6	South Platte (R.)	U.S.	40•40 N	102•40 W
8	South Pt.	Mi.	44•50 N	83•20 W
46	Southport	Austl.	27•57 S	153•27 E
22	Southport	Eng.	53•38 N	3•00 W
47	South Sandwich Is.	Falk.Is.	58•00 S	27•00 W
47	South Sandwich Trench	S.A.-Ant.	55•00 S	27•00 W
10	South Saskatchewan (R.)	Can.	53•15 N	105•05 W
22	South Shields	Eng.	55•00 N	1•22 W
22	South Shropshire Hills	Eng.	52•30 N	3•02 W
45	South Taranaki Bight	N.Z.	39•27 S	171•44 E
22	South Uist (I.)	Scot.	57•15 N	7•24 W
45	Southwest C.	N.Z.	47•17 S	167•12 E
25	Sovetsk (Tilsit)	Sov.Un.	55•04 N	21•54 E
29	Sovetskaya Gavan'	Sov.Un.	48•59 N	140•14 E
32	Soviet Union	Eur.-Asia	60•30 N	64•00 E
37	Sōya Misaki (C.)	Jap.	45•35 N	141•25 E
31	Sozh (R.)	Sov.Un.	52•17 N	31•00 E
21	Sozopol	Bul.	42•18 N	27•50 E
23	Spa	Bel.	50•30 N	5•50 E
14	Spain	Eur.	40•15 N	4•30 W
9	Spangler	Pa.	40•40 N	78•50 W
13	Spanish Town	Jam.	18•00 N	76•55 W
9	Sparrows Point	Md.	39•13 N	76•29 W
8	Sparta	Il.	38•07 N	89•42 W
8	Sparta	Mi.	43•10 N	85•45 W
7	Spartanburg	SC	34•57 N	82•13 W
21	Spárti (Sparta)	Grc.	37•07 N	22•28 E
20	Spartivento, C.	It.	37•55 N	16•09 E
20	Spartivento, C.	It.	38•54 N	8•52 E
29	Spassk-Dal'niy	Sov.Un.	44•30 N	133•00 E
20	Spátha, Akr. (C.)	Grc.	35•42 N	24•45 E
8	Spencer	In.	39•15 N	86•45 W
8	Spencer	WV	38•55 N	81•20 W
46	Spencer G.	Austl.	34•20 S	136•55 E
21	Sperkhiós (R.)	Grc.	38•54 N	22•02 E
22	Sperrin Mts.	N.Ire.	54•55 N	6•45 E
26	Spessart (Mts.)	F.R.G.	50•07 N	9•32 E
22	Spey (L.)	Scot.	57•25 N	3•29 W
26	Speyer	F.R.G.	49•18 N	8•26 E
20	Spinazzola	It.	40•58 N	16•05 E
27	Spišská Nová Ves	Czech.	48•56 N	20•35 E
26	Spittal	Aus.	46•48 N	13•28 E
20	Split	Yugo.	43•30 N	16•28 E
6	Spokane	Wa.	47•39 N	117•25 W
20	Spoleto	It.	42•44 N	12•44 E
21	Sporádhes (Is.)	Grc.	38•55 N	24•05 E
40	Spratly (I.)	China	8•38 N	11•54 E
26	Spree R.	G.D.R.	51•53 N	14•08 E
26	Spremberg	G.D.R.	51•35 N	14•23 E
54	Springbok	S.Afr.	29•35 S	17•55 E
8	Springfield	Il.	39•46 N	89•37 W
8	Springfield	Ky.	37•35 N	85•10 W
9	Springfield	Ma.	42•05 N	72•35 W
7	Springfield	Mo.	37•13 N	93•17 W
8	Springfield	Oh.	39•55 N	83•50 W
9	Springfield	Vt.	43•20 N	72•35 W
54	Springfontein	S.Afr.	30•16 S	25•45 E
11	Springhill	Can.	45•39 N	64•03 W
55	Springs	S.Afr.	26•16 S	28•27 E
8	Springvalley	Il.	41•20 N	89•15 W
9	Squam (L.)	NH	43•45 N	71•30 W
20	Squillace, Gulfo di (G.)	It.	38•44 N	16•47 E
21	Srbija (Serbia) (Reg.)	Yugo.	44•05 N	20•35 E
21	Srbobran	Yugo.	45•32 N	19•50 E
29	Sredne-Kolymsk	Sov.Un.	67•49 N	154•55 E
27	Śrem	Pol.	52•06 N	17•01 E
21	Sremska Karlovci	Yugo.	45•10 N	19•57 E
21	Sremska Mitrovica	Yugo.	44•59 N	19•39 E
29	Sretensk	Sov.Un.	52•13 N	117•39 E
32	Sri Lanka (Ceylon)	Asia	8•45 N	82•30 E
35	Srinagar	India	34•11 N	74•49 E
27	Sroda	Pol.	52•14 N	17•17 E
39	Ssuen	China	24•50 N	108•18 E
36	Ssumao	China	22•56 N	101•07 E
39	Ssŭnan	China	27•50 N	108•30 E
38	Ssup'ing	China	43•05 N	124•24 E
26	Stade	F.R.G.	53•36 N	9•28 E
16	Stadhur	Ice.	65•08 N	20•56 W
24	Städjan (Mtn.)	Swe.	61•53 N	12•50 E
00	Stafford	Eng.	52•48 N	2•06 W
9	Stamford	Ct.	41•03 N	73•32 W
54	Standerton	S.Afr.	26•57 S	29•17 E
8	Stanford	Ky.	37•29 N	84•40 W
55	Stanger	S.Afr.	29•22 S	31•18 E
24	Stangvik Fd.	Nor.	62•54 N	8•55 E
50	Stanley	Falk.Is.	51•46 S	57•59 W
53	Stanley Falls	Zaire	0•30 N	25•12 E
54	Stanley Pool (L.)	Zaire	4•07 S	15•40 E
29	Stanovoy Khrebet (Mts.)	Sov.Un.	56•12 N	127•12 E
14	Stara Planina (Balkan Mts.)	Bul.	42•50 N	24•45 E
30	Staraya Russa	Sov.Un.	57•58 N	31•21 E
21	Stara Zagora	Bul.	42•26 N	25•37 E
26	Stargard Szczeciński	Pol.	53•19 N	15•03 E
31	Starobel'sk	Sov.Un.	49•19 N	38•57 E
27	Starogard Gdański	Pol.	53•58 N	18•33 E
31	Staro-Konstantinov	Sov.Un.	49•45 N	27•12 E
31	Staro-Minskaya	Sov.Un.	46•19 N	38•51 E
22	Start Pt.	Eng.	50•14 N	3•34 W
27	Stary Sacz	Pol.	49•32 N	20•36 E
31	Staryy Oskol	Sov.Un.	51•18 N	37•51 E
26	Stassfurt	G.D.R.	51•52 N	11•35 E
27	Staszów	Pol.	50•32 N	21•13 E
9	State College	Pa.	40•50 N	77•55 W
9	Staunton	Va.	38•10 N	79•05 W
24	Stavanger	Nor.	58•59 N	5•44 E
31	Stavropol'	Sov.Un.	45•05 N	41•50 E
26	Stawno	Pol.	54•21 N	16•38 E
9	Steelton	Pa.	40•15 N	76•45 W
44	Steep Pt.	Austl.	26•15 N	112•05 E
8	Steger	Il.	41•28 N	87•38 W
26	Steiermark (Styria) (State)	Aus.	47•22 N	14•40 E
10	Steinbach	Can.	49•32 N	96•41 W
16	Steinkjer	Nor.	64•00 N	11•19 E
11	Stellarton	Can.	45•34 N	62•40 W
26	Stendal	G.D.R.	52•37 N	11•51 E
31	Stepanakert	Sov.Un.	39•50 N	46•40 E
46	Stephens, Port	Austl.	32•43 S	152•55 E
11	Stephenville	Can.	48•33 N	58•35 W
28	Stepnyak	Sov.Un.	52•37 N	70•43 E
55	Sterkstroom	S.Afr.	31•33 S	26•36 E
6	Sterling	Co.	40•38 N	103•14 W
8	Sterling	Il.	41•48 N	89•42 W
30	Sterlitamak	Sov.Un.	53•38 N	55•56 E
27	Šternberk	Czech.	49•44 N	17•18 E
26	Stettiner Haff (L.)	G.D.R.	53•47 N	14•02 E
10	Stettler	Can.	52•19 N	112•43 W
8	Steubenville	Oh.	40•20 N	80•40 W
8	Stevens Point	Wi.	44•30 N	89•35 W
10	Stewart (R.)	Can.	63•27 N	138•48 W
45	Stewart I.	N.Z.	46•50 S	168•06 E
26	Steyr	Aus.	48•03 N	14•24 E
10	Stikine (R.)	Can.	58•17 N	130•10 W
10	Stikine Ranges	Can.	59•05 N	130•00 W
21	Štip	Yugo.	41•43 N	22•07 E
22	Stirling	Scot.	56•05 N	3•59 W
24	Stjördalshalsen	Nor.	63•26 N	11•00 E
8	Stockbridge Munsee Ind. Res.	Wi.	44•49 N	89•00 W
26	Stockerau	Aus.	48•24 N	16•13 E
24	Stockholm	Swe.	59•23 N	18•00 E
22	Stockport	Eng.	53•24 N	2•09 W
6	Stockton	Ca.	37•56 N	121•16 W
22	Stockton	Eng.	54•35 N	1•25 W
24	Stöde	Swe.	62•26 N	16•35 E
22	Stoke-on-Trent	Eng.	53•01 N	2•12 W
27	Stokhod (R.)	Sov.Un.	51•24 N	25•20 E
21	Stolac	Yugo.	43•03 N	17•59 E
27	Stolin	Sov.Un.	51•54 N	26•52 E
24	Stömstad	Swe.	58•58 N	11•09 E
22	Stonehaven	Scot.	56•57 N	2•09 W
9	Stonington	Ct.	41•20 N	71•55 W
24	Stora (R.)	Den.	56•22 N	8•35 E
30	Stora Lule (R.)	Swe.	67•00 N	19•30 E
24	Stord (I.)	Nor.	59•54 N	5•15 E

Page	Name	Region	Lat.	Long.
24	Store Baelt (Str.)	Den.	55•25 N	10•50 E
16	Stören	Nor.	62•58 N	10•21 E
24	Store Sotra (Sartor)	Nor.	60•24 N	4•35 E
24	Stor Fd.	Nor.	62•17 N	6•19 E
55	Stormberg (Mts.)	S.Afr.	31•28 S	26•35 E
13	Stormy Pt.	Vir.Is.(U.S.A.)	18•22 N	65•01 W
22	Stornoway	Scot.	58•13 N	6•21 W
27	Storozhinets	Sov.Un.	48•10 N	25•44 E
24	Störsjo	Swe.	62•49 N	13•08 E
24	Störsjoen (L.)	Nor.	61•32 N	11•30 E
24	Störsjon (L.)	Swe.	63•06 N	14•00 E
24	Storvik	Swe.	60•37 N	16•31 E
8	Stoughton	Wi.	42•54 N	89•15 W
23	Stour (R.)	Eng.	52•09 N	0•29 E
22	Strabane	N.Ire.	54•59 N	7•27 W
45	Strahan	Austl.	42•08 S	145•28 E
26	Strakonice	Czech.	49•18 N	13•52 E
21	Straldzha	Bul.	42•37 N	26•44 E
26	Stralsund	G.D.R.	54•18 N	13•04 E
24	Strand	Nor.	59•05 N	5•59 E
22	Strangford, Lough (B.)	Ire.	54•30 N	5•34 W
24	Strängnas	Swe.	59•23 N	16•59 E
22	Stranraer	Scot.	54•55 N	5•05 W
17	Strasbourg	Fr.	48•36 N	7•49 E
8	Stratford	Can.	43•20 N	81•05 W
9	Stratford	Ct.	41•10 N	73•05 W
22	Stratford	Eng.	52•13 N	1•41 W
8	Stratford	Wi.	44•16 N	90•02 W
26	Straubing	F.R.G.	48•52 N	12•36 E
26	Strausberg	G.D.R.	52•35 N	13•50 E
8	Streator	Il.	41•05 N	88•50 W
21	Strehaia	Rom.	44•37 N	23•13 E
41	Strickland (R.)	Pap.N.Gui.	6•15 S	142•00 E
21	Strimonikós Kólpos (G.)	Grc.	40•40 N	23•55 E
20	Strómboli (Vol.)	It.	38•46 N	15•16 E
22	Stronsay (I.)	Scot.	59•09 N	2•35 W
9	Stroudsburg	Pa.	41•00 N	75•15 W
24	Struer	Den.	56•29 N	8•34 E
21	Struma (R.)	Bul.	41•55 N	23•05 E
21	Strumica	Yugo.	41•26 N	22•38 E
8	Struthers	Oh.	41•00 N	80•35 W
27	Stryy	Sov.Un.	49•16 N	23•51 E
27	Strzelce Opolskie	Pol.	50•31 N	18•20 E
27	Strzelin	Pol.	50•48 N	17•06 E
27	Strzelno	Pol.	52•37 N	18•10 E
44	Stuart Ra.	Austl.	29•00 S	134•30 E
40	Stung Treng	Camb.	13•36 N	106•00 E
27	Stupsk	Pol.	54•28 N	17•02 E
11	Sturgeon Falls	Can.	46•19 N	79•49 W
8	Sturgis	Ky.	37•35 N	88•00 W
8	Sturgis	Mi.	41•45 N	85•25 W
44	Sturt Cr.	Austl.	19•40 S	127•40 E
55	Stutterheim	S.Afr.	32•34 S	27•27 E
26	Stuttgart	F.R.G.	48•48 N	9•15 E
16	Stykkisholmur	Ice.	65•00 N	21•48 W
27	Styr' R.	Sov.Un.	51•44 N	26•07 E
39	Suao	Taiwan	24•35 N	121•45 E
25	Subata	Sov.Un.	56•02 N	25•54 E
41	Subic	Phil.	14•52 N	120•15 E
41	Subic B.	Phil.	14•41 N	120•11 E
21	Subotica	Yugo.	46•06 N	19•41 E
27	Suceava	Rom.	47•39 N	26•17 E
27	Suceava R.	Rom.	47•45 N	26•10 E
27	Sucha	Pol.	49•44 N	19•40 E
48	Sucio (R.)	Col.	6•55 N	76•15 W
22	Suck	Ire.	53•34 N	8•16 W
48	Sucre	Bol.	19•06 S	65•16 W
49	Sucre (State)	Ven.	10•18 N	64•12 W
34	Sudair	Sau.Ar.	25•48 N	46•28 E
51	Sudan	Afr.	14•00 N	28•00 E
52	Sudan (Reg.)	Afr.	15•00 N	7•00 E
11	Sudbury	Can.	46•28 N	81•00 W
26	Sudetes (Mts.)	Czech.	50•41 N	15•37 E
53	Suez, G. of	Egypt	29•53 N	32•33 E
8	Sugar (Cr.)	In.	39•55 N	87•10 W
46	Sugarloaf Pt.	Austl.	32•19 S	153•04 E
33	Suhaymī, Wādī as (R.)	Egypt	29•48 N	33•12 E
26	Suhl	G.D.R.	50•37 N	10•41 E
38	Suhsien	China	33•37 N	117•51 E
39	Suichuan (Mtn.)	China	26•25 N	114•10 E
38	Suichung	China	40•22 N	120•20 E
37	Suifenho	China	44•47 N	131•13 E
38	Suihua (Peilintzu)	China	46•38 N	126•50 E
22	Suir R.	Ire.	52•20 N	7•32 W
38	Suite	China	37•32 N	110•12 E
36	Suiyuan	China	41•31 N	107•04 E
40	Sukabumi	Indon.	6•52 S	106•56 E
40	Sukadana	Indon.	1•15 S	110•30 E
30	Sukhinichi	Sov.Un.	54•07 N	35•18 E
30	Sukhona (R.)	Sov.Un.	59•30 N	42•20 E
31	Sukhumi	Sov.Un.	43•00 N	41•00 E
35	Sukkur	Pak.	27•49 N	68•50 E
41	Sula (I.)	Indon.	2•20 S	125•20 E
35	Sulaimān Ra.	Pak.	29•47 N	69•10 E
31	Sulak (R.)	Sov.Un.	43•30 N	47•00 E
40	Sulawesi (Celebes) (Is.)	Indon.	2•15 S	120•30 E
24	Suldals Vand (L.)	Nor.	59•35 N	6•59 E
19	Sulina	Rom.	45•08 N	29•38 E
16	Sulitelma (Mtn.)	Nor.-Swe.	67•03 N	16•09 E
48	Sullana	Peru	4•57 N	80•47 W
8	Sullivan	Il.	41•35 N	88•35 W
8	Sullivan	In.	39•05 N	87•20 W
20	Sulmona	It.	42•02 N	13•58 E
36	Sulo	China	41•29 N	80•15 E
36	Sulo Ho	China	40•53 N	94•55 E
40	Sulu Arch.	Phil.	5•52 N	122•00 E
19	Suluntah	Libya	32•39 N	21•49 E
40	Sulu Sea	Phil.	8•25 N	119•00 E
40	Sumatra (I.)	Indon.	2•06 N	99•40 E
40	Sumba (I.)	Indon.	9•52 S	119•00 E
40	Sumbawa (I.)	Indon.	9•00 S	118•18 E
40	Sumbawa-Besar	Indon.	8•32 S	117•20 E
27	Sümeg	Hung.	46•59 N	17•19 E
11	Summerside	Can.	46•25 N	63•47 W
26	Šumperk	Czech.	49•57 N	17•02 E
28	Sumy	Sov.Un.	50•54 N	34•47 E
9	Sunbury	Pa.	40•50 N	76•45 W
40	Sunda Is.	Indon.	9•00 S	108•40 E
24	Sundals Fd.	Nor.	62•50 N	7•55 E
35	Sundarbans (Swp.)	Bngl.-India	21•50 N	89•00 E
40	Sunda Selat (Str.)	Indon.	5•45 S	106•15 E
40	Sunda Trench	Indon.	9•45 S	107•30 E
44	Sunday Str.	Austl.	15•50 S	122•45 E
24	Sundbyberg	Swe.	59•24 N	17•56 E
22	Sunderland	Eng.	54•55 N	1•25 W
24	Sundsvall	Swe.	62•24 N	19•19 E
38	Sungari Res.	China	42•55 N	127•50 E
39	Sungchiang	China	31•01 N	121•14 E
37	Sung Hua (R.) (Sungari)	China	46•09 N	127•53 E
38	Sungtzu (Mtn.)	China	39•40 N	114•50 E
31	Sungurlu	Tur.	40•08 N	34•20 E
24	Sunne	Swe.	59•51 N	13•07 E
29	Suntar	Sov.Un.	62•14 N	117•49 E
25	Suoyarvi	Sov.Un.	62•12 N	32•29 E
7	Superior	Wi.	46•44 N	92•06 W
11	Superior, L.	Can.-U.S.	47•38 N	89•20 W
38	Sup'ung Res.	Kor.-China	40•35 N	126•00 E
33	Şūr (Tyre)	Leb.	36•16 N	35•13 E
34	Sūr	Om.	22•23 N	59•28 E
40	Surabaya	Indon.	7•23 S	112•45 E
40	Surakarta	Indon.	7•35 S	110•45 E
27	Šurany	Czech.	48•05 N	18•11 E
46	Surat	Austl.	27•18 S	149•00 E
35	Surat	India	21•08 N	73•22 E
40	Surat Thani	Thai.	8•59 N	99•14 E
28	Surgut	Sov.Un.	61•18 N	73•38 E
40	Surin	Thai.	14•59 N	103•57 E
47	Surinam	S.A.	4•00 N	56•00 W
53	Surt	Libya	31•14 N	16•37 E
19	Surt, Khalij (G.)	Afr.	31•30 N	18•28 E
20	Susa	It.	45•01 N	7•09 E
20	Sušac	Yugo.	44•31 N	14•15 E
20	Sušak	Yugo.	45•20 N	14•24 E
20	Sušak (I.)	Yugo.	42•45 N	16•30 E
9	Susquehanna	Pa.	41•55 N	73•35 W
9	Susquehanna (R.)	Pa.	39•50 N	76•20 W
11	Sussex	Can.	45•43 N	65•31 W
39	Susung	China	30•18 N	116•08 E
54	Sutherland	S.Afr.	32•25 S	20•40 E
35	Sutlej (R.)	Pak.-India	30•15 N	72•25 E
55	Suurberge (Mts.)	S.Afr.	33•15 S	25•32 E
27	Suwatki	Pol.	54•05 N	22•58 E
7	Suwannee (R.)	Fl.-Ga.	29•42 N	83•00 W
28	Svalbard (Spitsbergen) (Is.)	Eur.	77•00 N	20•00 E
24	Svaneke	Den.	55•08 N	15•07 E
31	Svatovo	Sov.Un.	49•23 N	38•10 E
24	Svedala	Swe.	55•29 N	13•11 E
24	Sveg	Swe.	62•03 N	14•22 E
24	Svelvik	Nor.	59•37 N	10•18 E
25	Švenčionys	Sov.Un.	55•09 N	26•09 E
24	Svendborg	Den.	55•05 N	10•35 E
30	Sverdlovsk	Sov.Un.	56•50 N	61•10 E
21	Svilajnac	Yugo.	44•12 N	21•14 E
21	Svilengrad	Bul.	41•44 N	26•11 E
30	Svir' (R.)	Sov.Un.	60•55 N	33•40 E
25	Svir Kanal	Sov.Un.	60•10 N	32•40 E
21	Svishtov	Bul.	43•36 N	25•21 E
26	Svitavy	Czech.	49•46 N	16•28 E
27	Svitsa (R.)	Sov.Un.	49•09 N	24•10 E
28	Svobodnyy	Sov.Un.	51•28 N	128•28 E
16	Svolvaer	Nor.	68•15 N	14•29 E
29	Svyatoy Nos, Mys (C.)	Sov.Un.	72•18 N	139•28 E
45	Swain Rfs.	Austl.	22•12 S	152•08 E
54	Swakopmund	Namibia	22•40 S	14•30 E
22	Swale (R.)	Eng.	54•12 N	1•30 W
44	Swan (R.)	Austl.	31•30 S	126•30 E
46	Swan Hill	Austl.	35•20 S	143•30 E
44	Swanland	Austl.	31•45 S	119•15 E
22	Swansea	Wales	51•37 N	3•59 W
22	Swansea B.	Wales	51•25 N	4•12 W
55	Swartberg (Mtn.)	S.Afr.	30•08 S	29•34 E
54	Swartkop (Mtn.)	S.Afr.	34•13 S	18•27 E
55	Swartspruit	S.Afr.	25•44 S	28•01 E
55	Swaziland	Afr.	26•45 S	31•30 E
14	Sweden	Eur.	60•10 N	14•10 E
6	Sweetwater	Tx.	32•28 N	100•25 W
26	Świdnica	Pol.	50•50 N	16•30 E
26	Świdwin	Pol.	53•46 N	15•48 E
26	Śweibodziec	Pol.	52•16 N	15•36 E
26	Świebodzin	Pol.	50•51 N	16•17 E
27	Swiecie	Pol.	53•23 N	18•26 E
27	Swietokrzyskie Góry (Mts.)	Pol.	50•57 N	21•02 E
22	Swilly, Lough	Ire.	55•10 N	7•38 W
22	Swindon	Eng.	51•35 N	1•55 W
26	Świnoujście	Pol.	53•56 N	14•14 E
26	Switzerland	Eur.	46•30 N	7•43 E
8	Sycamore	Il.	42•00 N	88•42 W
46	Sydney	Austl.	33•55 S	151•17 E
11	Sydney	Can.	46•09 N	60•11 W
11	Sydney Mines	Can.	46•14 N	60•14 W
30	Syktyvkar	Sov.Un.	61•35 N	50•40 E
24	Sylfjällen (Mtn.)	Swe.	63•00 N	12•10 E
24	Sylling	Nor.	59•52 N	10•12 E
26	Sylt I.	F.R.G.	54•55 N	8•30 E
9	Syracuse	NY	43•05 N	76•10 W
19	Syra I.	Grc.	37•19 N	25•10 E
32	Syria	Asia	35•00 N	37•15 E
34	Syrian Des.	Asia	32•03 N	39•30 E
30	Syso'la (R.)	Sov.Un.	60•50 N	50•40 E
31	Syzran'	Sov.Un.	53•10 N	48•10 E
27	Szabadszallas	Hung.	46•52 N	19•15 E
26	Szamotuty	Pol.	52•36 N	16•34 E
27	Szarvas	Hung.	46•51 N	20•36 E
27	Szczebrzeszyn	Pol.	50•41 N	22•58 E
26	Szczecin (Stettin)	Pol.	53•25 N	14•35 E
26	Szczecinek	Pol.	53•42 N	16•42 E
27	Szczuczyn	Pol.	53•32 N	22•17 E
27	Szczytno	Pol.	53•33 N	21•00 E
36	Szechwan (Ssuch'uan) (Prov.)	China	31•20 N	103•00 E
27	Szeged	Hung.	46•15 N	20•12 E
27	Székesfehérvár	Hung.	47•12 N	18•26 E
27	Szekszárd	Hung.	46•19 N	18•42 E
36	Szengen	China	23•39 N	107•45 E
27	Szentendre	Hung.	47•40 N	19•07 E
27	Szentes	Hung.	46•38 N	20•18 E
27	Szigetvar	Hung.	46•05 N	17•50 E
27	Szolnok	Hung.	47•11 N	20•12 E
26	Szombathely	Hung.	47•13 N	16•35 E
26	Szprotawa	Pol.	51•34 N	15•29 E
27	Szydlowiec	Pol.	51•13 N	20•53 E
	T			
41	Taal (L.)	Phil.	13•58 N	121•06 E
41	Tabaco	Phil.	13•27 N	123•40 E
55	Tabankulu	S.Afr.	30•56 S	29•19 E
12	Tabasco (State)	Mex.	18•10 N	93•00 W
10	Taber	Can.	49•47 N	112•08 W
41	Tablas (I.)	Phil.	12•26 N	112•15 E
41	Tablas Str.	Phil.	12•17 N	121•41 E
54	Table B.	S.Afr.	33•41 S	18•27 E
54	Table Mt.	S.Afr.	33•58 S	18•26 E
12	Taboga (I.)	Pan.	8•48 N	79•35 W
12	Taboguilla (I.)	Pan.	8•48 N	79•31 W
49	Taboleiro (Plat.)	Braz.	9•34 S	39•22 W
26	Tábor	Czech.	49•25 N	14•40 E
54	Tabora	Tan.	5•01 S	32•48 E
52	Tabou	Ivory Coast	4•25 N	7•21 W
34	Tabrīz	Iran	38•00 N	46•13 E
12	Tacaná (Vol.)	Mex.-Guat.	15•09 N	92•07 W
49	Tacarigua, Laguna de la	Ven.	10•18 N	65•43 W
36	T'ach'eng (Chuguchak)	China	46•50 N	83•24 E
41	Tacloban	Phil.	11•06 N	124•58 E
48	Tacna	Peru	18•34 S	70•16 W
6	Tacoma	Wa.	47•14 N	122•27 W
9	Taconic Ra.	NY	41•55 N	73•40 W
50	Tacuarembó	Ur.	31•44 S	55•56 W
52	Tademaït, Plat. du	Alg.	28•00 N	2•15 E
48	Tadó	Col.	5•15 N	76•30 W
28	Tadzhik S.S.R.	Sov.Un.	39•22 N	69•30 E
38	Taebaek Sanmaek (Mts.)	Kor.	37•20 N	128•50 E
38	Taegu	Kor.	35•49 N	128•41 E
18	Tafilelt (Oasis)	Mor.	31•49 N	4•44 W
31	Taganrog	Sov.Un.	47•13 N	38•44 E
31	Taganrogskiy Zaliv (B.)	Sov.Un.	46•55 N	38•17 E
20	Tagliamento (R.)	It.	46•11 N	12•53 E
45	Tagula (I.)	Pap.N.Gui.	11•45 S	153•46 E
40	Tahan	Mala.	4•33 N	101•52 E
52	Tahat, Mt.	Alg.	23•22 N	5•21 E
43	Tahiti (I.)	Fr.Polynesia	17•30 S	149•30 W
25	Tahkuna Nina	Sov.Un.	59•08 N	22•03 E
6	Tahoe (L.)	Ca.-Nv.	39•09 N	120•18 W
52	Tahoua	Niger	14•54 N	5•16 E
39	Tahsien	China	31•12 N	107•30 E
39	Tahsing	China	39•44 N	116•19 E
38	Tahsinganling Shanmo (Greater Khingan Ra.)	China	46•30 N	120•00 E
38	T'aian	China	36•13 N	117•08 E
39	T'aichung	Taiwan	24•10 N	120•42 E
38	T'aihang Shan (Mts.)	China	35•45 N	112•00 E
39	T'ai Hu (L.)	China	31•13 N	120•00 E
38	Taiku	China	37•25 N	112•35 E
36	Tailagein Khara	Mong.	43•39 N	105•54 E
38	T'ailai	China	46•20 N	123•10 E
46	Tailem Bend	Austl.	35•15 S	139•30 E
39	T'ainan	Taiwan	23•08 N	120•18 E
19	Tainaron, Akra (C.)	Grc.	36•20 N	21•20 E
39	Taining	China	26•58 N	117•15 E
38	T'aipai Shan (Mtn.)	China	33•42 N	107•25 E
39	T'aipei	Taiwan	25•02 N	121•38 E
40	Taiping	Mala.	4•56 N	100•39 E
39	T'aishan	China	22•15 N	112•50 E
38	T'ai Shan (Mtn.)	China	36•16 N	117•05 E
50	Taitao, Pen. de	Chile	46•20 S	77•15 W
39	T'aitung	Taiwan	22•45 N	121•02 E
33	Taiwan (Formosa)	Asia	23•30 N	122•20 E
38	T'aiyüan	China	37•32 N	112•38 E
39	Taiyün (Mtn.)	China	25•40 N	118•08 E
17	Tajo, Río (Tagus)	Sp.	39•40 N	5•07 W
18	Tājūrā	Libya	32•56 N	13•24 W
40	Tak	Thai.	16•57 N	99•12 E
37	Takamatsu	Jap.	34•20 N	134•02 E
55	Takaungu	Ken.	3•41 S	39•48 E
10	Takla L.	Can.	55•25 N	125•53 W
36	Takla Makan (Des.)	China	39•22 N	82•34 E
38	Tal'ai	China	45•25 N	124•22 E
48	Talara	Peru	4•32 N	81•17 W
41	Talasea	Pap.N.Gui.	5•20 S	150•00 E
41	Talaud	Indon.	4•17 N	127•30 E
17	Talavera de la Reina	Sp.	39•58 N	4•51 W
53	Talawdī	Sud.	10•41 N	30•21 E
50	Talca	Chile	35•25 S	71•39 W
50	Talcahuano	Chile	36•41 S	73•05 W
29	Taldy-Kurgan	Sov.Un.	45•03 N	77•18 E
36	Tali	China	26•00 N	100•08 E
41	Taliabu	Indon.	1•30 S	125•00 E
41	Talim (I.)	Phil.	14•21 N	121•14 E
41	Talisay	Phil.	14•08 N	122•56 E
31	Talkheh Rūd (R.)	Iran	38•00 N	46•50 E
7	Tallahassee	Fl.	30•25 N	84•17 W
25	Tallinn (Reval)	Sov.Un.	59•26 N	24•44 E
53	Talo (Mt.)	Eth.	10•45 N	37•55 E
25	Talsi	Sov.Un.	57•16 N	22•35 E
50	Taltal	Chile	25•26 S	70•32 W
52	Tamale	Ghana	9•25 N	0•50 W
48	Tamaná, Cerro	Col.	5•06 N	76•10 W
49	Tamanaco (R.)	Ven.	9•32 N	66•00 W
52	Tamanrasset (R.)	Alg.	22•15 N	2•51 E
52	Tamanrasset	Alg.	22•45 N	5•34 E
9	Tamaqua	Pa.	40•45 N	75•50 W
22	Tamar (R.)	Eng.	50•35 N	4•15 W
55	Tamatave	Mad.	18•14 S	49•25 E
12	Tamaulipas (State)	Mex.	23•45 N	98•30 W
49	Tambador, Serra do	Braz.	10•33 S	41•16 W
40	Tambelan (Is.)	Indon.	0•38 N	107•38 E
46	Tambo	Austl.	24•50 S	146•15 E
31	Tambov	Sov.Un.	52•45 N	41•10 E
53	Tambura	Sud.	5•34 N	27•30 E
52	Tamgak, Monts	Niger	18•40 N	8•40 E
18	Tamgrout	Mor.	30•12 N	5•46 W
52	Tamgue, Massif du	Gui.	12•15 N	12•35 W
35	Tamil Nadu (State)	India	11•30 N	78•00 E
38	Taming	China	36•15 N	115•09 E
25	Tammela	Fin.	60•49 N	23•45 E
7	Tampa	Fl.	27•57 N	82•25 W
7	Tampa B.	Fl.	27•35 N	82•38 W
16	Tampere	Fin.	61•21 N	23•39 E
12	Tampico	Mex.	22•14 N	97•51 W
33	Tampin	Mala.	2•28 N	102•15 E
46	Tamworth	Austl.	31•01 S	151•00 E
55	Tana (R.)	Ken.	2•00 S	40•15 E
45	Tana (I.)	New Hebr.	19•32 S	169•27 E
16	Tana (R.)	Nor.-Fin.	69•20 N	24•54 E
40	Tanahbala (I.)	Indon.	0•30 S	98•22 E
40	Tanahmasa (I.)	Indon.	0•03 S	97•30 E
53	Tana L.	Eth.	12•09 N	36•41 E
44	Tanami	Austl.	19•45 S	129•50 E
20	Tanaro (R.)	It.	44•45 N	8•02 E
41	Tanauan	Phil.	14•04 N	121•10 E
38	T'anch'eng	China	34•37 N	118•22 E
38	Tanchŏn	Kor.	40•29 N	128•50 E
50	Tandil	Arg.	36•16 S	59•01 W
50	Tandil, Sierra del	Arg.	38•40 S	59•40 W
37	Tanega (I.)	Jap.	30•36 N	131•11 E
52	Tanezrouft	Alg.	24•17 N	0•30 W
55	Tanga	Tan.	5•04 S	39•06 E
54	Tanganyika, L.	Afr.	5•15 S	29•40 E
26	Tangermünde	G.D.R.	52•33 N	11•58 E
38	Tangho	China	32•40 N	112•50 E
52	Tangier	Mor.	35•52 N	5•55 W
39	Tangt'u	China	31•35 N	118•28 E
38	Tangshan	China	34•27 N	116•27 E
41	Tanimbar (Is.)	Indon.	8•00 S	132•00 E
33	Tanjong (C.)	Mala.	1•53 N	102•29 E
33	Tanjong Piai (I.)	Mala.	1•16 N	103•11 E
33	Tanjong Ramunia (C.)	Mala.	1•27 N	104•44 E
40	Tanjungbalai	Indon.	2•52 N	99•43 E
33	Tanjungbalai	Indon.	1•00 N	103•26 E
40	Tanjungkarang-Tebukbetung	Indon.	5•30 S	105•04 E
40	Tanjungpandan	Indon.	2•47 S	107•51 E
33	Tanjungpinang	Indon.	0•55 N	104•29 E

Page	Name	Region	Lat.	Long.
29	Tannu-Ola (Mts.)	Sov.Un.	51•00 N	94•00 E
34	Tannūrah, Ra's al (C.)	Sau.Ar.	26•45 N	49•59 E
51	Tanzania	Afr.	6•48 S	33•58 E
38	Taoerh (R.)	China	45•40 N	122•00 E
38	Táo Ho' (R.)	China	35•30 N	103•40 E
38	T'aok'ou	China	35•34 N	114•32 E
38	T'aonan	China	45•15 N	122•45 E
20	Taormina	It.	37•53 N	15•18 E
52	Taoudenni	Mali	22•57 N	3•37 W
39	Taoyüan	China	29•00 N	111•15 E
25	Tapa	Sov.Un.	59•16 N	25•56 E
38	Tapa Shan (Mts.)	China	32•25 N	108•20 E
49	Tapajós (R.)	Braz.	3•27 S	55•33 W
35	Tāpi (R.)	India	21•33 N	74•30 E
39	Tapieh Shan (Mts.)	China	31•40 N	114•50 E
53	Taqātu' Hayyā	Sud.	18•10 N	36•17 E
49	Taquara, Serra de	Braz.	15•28 S	54•33 W
49	Taquari (R.)	Braz.	18•35 S	56•50 W
29	Tara	Sov.Un.	56•58 N	74•13 E
41	Tara (I.)	Phil.	12•18 N	120•28 E
33	Ţarābulus (Tripoli)	Leb.	34•25 N	35•50 E
53	Tarābulus (Tripoli)	Libya	32•50 N	13•13 E
40	Tarakan	Indon.	3•17 N	118•04 E
20	Taranto	It.	40•30 N	17•15 E
20	Taranto, Golfo di	It.	40•03 N	17•10 E
48	Tarapoto	Peru	6•29 S	76•26 W
48	Tarata	Bol.	17•43 S	66•00 W
20	Taravo (R.)	Fr.	41•54 N	8•58 E
22	Tarbat Ness (Hd.)	Scot.	57•51 N	3•50 W
17	Tarbes	Fr.	43•04 N	0•05 E
53	Tarbū	Libya	26•07 N	15•49 E
46	Taree	Austl.	31•52 S	152•21 E
52	Tarfaya	Mor.	27•58 N	12•55 W
17	Tarifa	Sp.	36•02 N	5•35 W
48	Tarija	Bol.	21•42 S	64•52 W
34	Tarīm	P.D.R. of Yem.	16•13 N	49•08 E
36	Tarim (R.)	China	40•45 N	85•39 E
36	Tarim Basin	China	39•52 N	82•34 E
55	Tarka (R.)	S.Afr.	32•15 S	26•00 E
55	Tarkastad	S.Afr.	32•01 S	26•18 E
31	Tarkhankut, Mys (C.)	Sov.Un.	45•18 N	32•08 E
52	Tarkwa	Ghana	5•19 N	1•59 W
41	Tarlac	Phil.	15•29 N	120•36 E
55	Tarlton	S.Afr.	26•05 S	27•38 E
48	Tarma	Peru	11•26 S	75•40 W
17	Tarn (R.)	Fr.	44•03 N	2•41 E
27	Tarnów	Pol.	50•02 N	21•00 E
20	Taro (R.)	It.	44•41 N	10•03 E
52	Taroudant	Mor.	30•39 N	8•52 W
20	Tarquinia (Corneto)	It.	42•16 N	11•46 E
17	Tarragona	Sp.	41•05 N	1•15 E
31	Tarsus	Tur.	37•00 N	34•50 E
50	Tartagal	Arg.	23•31 S	63•47 W
19	Tartous	Egypt	34•54 N	35•59 E
30	Tartu (Dorpat)	Sov.Un.	58•23 N	26•44 E
15	Tashauz	Sov.Un.	41•50 N	59•45 E
28	Tashkent	Sov.Un.	41•23 N	69•04 E
45	Tasman B.	N.Z.	39•11 S	173•22 E
46	Tasmania (State)	Austl.	38•20 S	146•30 E
45	Tasmania (I.)	Austl.	41•28 S	142•30 E
46	Tasman Pen.	Austl.	43•00 S	148•30 E
42	Tasman Sea	Oceania	29•30 S	155•00 E
52	Tassili-n-Ajjer (Plat.)	Alg.	25•40 N	6•57 E
30	Tatar A.S.S.R.	Sov.Un.	55•30 N	51•00 E
28	Tatarsk	Sov.Un.	55•15 N	75•00 E
29	Tatar Str.	Sov.Un.	51•00 N	141•45 E
27	Tatra Mts.	Czech.-Pol.	49•15 N	19•40 E
39	Tattien Ting (Mtn.)	China	22•25 N	111•20 E
39	Tatu Ho (R.)	China	29•20 N	103•30 E
38	Tat'ung	China	40•00 N	113•30 E
49	Taubaté	Braz.	23•03 S	45•32 W
26	Tauern Tun.	Aus.	47•12 N	13•17 E
54	Taung	S.Afr.	27•25 S	29•45 E
9	Taunton	Ma.	41•54 N	71•03 W
28	Taunus (Mts.)	F.R.G.	50•15 N	8•33 E
45	Taupo, L.	N.Z.	38•38 S	175•27 E
25	Taurage	Sov.Un.	55•15 N	22•18 E
28	Tavda	Sov.Un.	58•00 N	64•44 E
30	Tavda (R.)	Sov.Un.	59•20 N	63•28 E
40	Tavoy	Bur.	14•04 N	98•19 E
31	Tavşanli	Tur.	39•30 N	29•30 E
8	Tawas City	Mi.	44•15 N	83•30 W
8	Tawas Pt.	Mi.	44•15 N	83•25 W
40	Tawitawi Group (Is.)	Phil.	4•52 N	120•35 E
53	Tawkar	Sud.	18•28 N	37•46 E
22	Tay, Firth of	Scot.	56•26 N	2•45 W
22	Tay (L.)	Scot.	56•25 N	5•07 W
22	Tay (R.)	Scot.	56•35 N	3•37 W
41	Tayabas B.	Phil.	13•44 N	121•40 E
29	Taygonos, Mys (C.)	Sov.Un.	60•37 N	160•17 E
8	Taylorville	Il.	39•30 N	89•20 W
34	Taymā	Sau.Ar.	27•45 N	38•55 E
29	Taymyr (Taimyr) (L.)	Sov.Un.	74•13 N	100•45 E
29	Taymyr, P-Ov (Taimyr Pen.)	Sov.Un.	75•15 N	95•00 E
29	Tayshet (Taishet)	Sov.Un.	56•09 N	97•49 E
39	Tayü	China	25•20 N	114•20 E
41	Tayung	Phil.	16•01 N	120•45 E
52	Taza	Mor.	34•08 N	4•00 W
28	Tazovskoye	Sov.Un.	66•58 N	78•28 E
31	Tbilisi	Sov.Un.	41•40 N	44•45 E
54	Tchibanga	Gabon	2•51 S	11•02 E
27	Tczew	Pol.	54•06 N	18•48 E
52	Tébessa	Alg.	35•27 N	8•13 E
33	Tebing Tinggi (I.)	Indon.	0•54 N	102•39 E
27	Tecuci	Rom.	45•51 N	27•30 E
8	Tecumseh	Mi.	42•00 N	84•00 W
22	Tees (R.)	Eng.	54•40 N	2•10 W
48	Tefé	Braz.	3•27 S	64•43 W
12	Tegucigalpa	Hond.	14•08 N	87•15 W
34	Tehrān	Iran	35•45 N	51•30 E
38	Tehsien	China	37•28 N	116•17 E
39	Tehua	China	25•30 N	118•15 E
12	Tehuacan	Mex.	18•27 N	97•23 W
12	Tehuantepec	Mex.	16•20 N	95•14 W
12	Tehuantepec, Golfo de	Mex.	15•45 N	95•00 W
17	Tejo, Rio	Port.	39•23 N	8•01 W
53	Tekeze (R.)	Eth.	13•38 N	38•00 E
21	Tekirdağ (Rodosto)	Tur.	41•00 N	27•28 E
12	Tela	Hond.	15•45 N	87•25 W
33	Telapa Burok	Mala.	2•51 N	102•04 E
31	Telavi	Sov.Un.	42•00 N	45•20 E
33	Tel Aviv-Yafo	Isr.	32•03 N	34•46 E
10	Telegraph Creek	Can.	57•59 N	131•22 W
6	Telescope Pk.	Ca.	36•12 N	117•05 W
49	Teles Pirez (R.)	Braz.	8•28 S	57•07 W
33	Telesung	Indon.	1•07 N	102•53 E
36	Telii Nuur (L.)	China	45•49 N	86•08 E
8	Tell City	In.	38•00 N	86•45 W
48	Tello	Col.	3•05 N	75•08 W
33	Telok Datok	Mala.	2•51 N	101•33 E
30	Tel'pos-Iz, Gora (Mtn.)	Sov.Un.	63•50 N	59•20 E
25	Telšiai	Sov.Un.	55•59 N	22•17 E
33	Telukletyak	Indon.	1•53 N	101•45 E
12	Temax	Mex.	21•10 N	88•51 W
31	Temir	Sov.Un.	49•10 N	57•15 E
28	Temir-Tau	Sov.Un.	50•08 N	73•13 E
50	Temperley	Arg.	34•32 S	58•24 W
20	Tempio Pausania	It.	40•55 N	9•05 E
26	Templin	G.D.R.	53•08 N	13•30 E
31	Temryuk	Sov.Un.	45•17 N	37•21 E
50	Temuco	Chile	38•46 S	72•38 W
40	Tenasserim	Bur.	12•09 N	99•01 E
52	Tenéré (Des.)	Niger	19•23 N	10•15 E
52	Tenerife I.	Can.Is.	28•41 N	17•02 W
17	Ténés	Alg.	36•28 N	1•22 E
38	T'enghsien	China	35•07 N	117•08 E
54	Tenke	Zaire	11•26 S	26•45 E
44	Tennant Creek	Austl.	19•45 S	134•00 E
7	Tennessee	U.S.	35•50 N	88•00 W
7	Tennessee (L.)	U.S.	35•35 N	88•20 W
7	Tennessee (R.)	U.S.	35•10 N	88•20 W
46	Tenora	Austl.	34•23 S	147•33 E
46	Tenterfield	Austl.	29•00 S	152•06 E
49	Teófilo Otoni	Braz.	17•49 S	41•18 W
12	Tepic	Mex.	21•32 N	104•53 W
26	Teplice Sanov	Czech.	50•39 N	13•50 E
48	Tequendama, Salto de (Falls)	Col.	4•34 N	74•18 W
20	Teramo	It.	42•40 N	13•41 E
31	Tercan	Tur.	39•40 N	40•12 E
52	Terceira I.	Açores	38•49 N	26•36 W
27	Terebovlya	Sov.Un.	49•18 N	25•43 E
31	Terek (R.)	Sov.Un.	43•30 N	45•10 E
49	Teresina	Braz.	5•04 S	42•42 W
50	Teresópolis	Braz.	22•25 S	42•59 W
30	Teriberka	Sov.Un.	69•00 N	35•15 E
31	Terme	Tur.	41•05 N	42•00 E
35	Termez	Sov.Un.	37•19 N	67•20 E
20	Termini	It.	37•58 N	13•39 E
12	Términos, Laguna de	Mex.	18•37 N	91•32 W
20	Termoli	It.	42•00 N	15•01 E
41	Ternate	Indon.	0•52 N	127•25 E
20	Terni	It.	42•38 N	12•41 E
27	Ternopol	Sov.Un.	49•32 N	25•36 E
29	Terpeniya, Mys (C.)	Sov.Un.	48•44 N	144•42 E
10	Terrace	Can.	54•31 N	128•35 W
20	Terracina	It.	41•18 N	13•14 E
9	Terrebonne	Can.	45•42 N	73•38 W
8	Terre Haute	In.	39•25 N	87•25 W
23	Terschelling (I.)	Neth.	53•25 N	5•12 E
17	Teruel	Sp.	40•20 N	1•05 W
21	Tešanj	Yugo.	44•36 N	17•59 E
36	Tesiin Gol (R.)	Mong.	50•14 N	94•30 E
10	Teslin (L.)	Can.	60•12 N	132•08 W
10	Teslin (R.)	Can.	61•18 N	134•14 W
52	Tessaoua	Niger	13•53 N	7•53 E
22	Test (R.)	Eng.	51•10 N	2•20 W
20	Testa del Gargano (Pt.)	It.	41-48 N	16•13 E
54	Tete	Moz.	16•13 S	33•35 E
31	Teterev (R.)	Sov.Un.	50•35 N	29•18 E
26	Teterow	G.D.R.	53•46 N	12•33 E
52	Tetouan	Mor.	35•42 N	5•34 W
21	Tetovo	Yugo.	42•01 N	21•00 E
30	Tetyushi	Sov.Un.	54•58 N	48•40 E
20	Tévere (Tiber) (R.)	It.	42•30 N	12•14 E
33	Teverya	Isr.	32•48 N	35•32 E
7	Texarkana	Ar.	33•26 N	94•02 W
7	Texarkana	Tx.	33•26 N	94•04 W
6	Texas	U.S.	31•00 N	101•00 W
23	Texel (I.)	Neth.	53•10 N	4•45 E
55	Teyateyaneng	Leso.	29•11 S	27•43 E
30	Teykovo	Sov.Un.	56•52 N	40•34 E
10	Tha-anne (R.)	Can.	60•50 N	96•56 W
55	Thabana Ntlenyana (Mtn.)	Leso.	29•28 S	29•17 E
32	Thailand	Asia	16•30 N	101•00 E
40	Thailand, G. of	Asia	11•37 N	100•46 E
40	Thaie Luang (L.)	Thai.	7•51 N	99•39 E
23	Thames (R.)	Eng.	51•26 N	0•54 E
19	Thamit R.	Libya	30•39 N	16•23 E
36	Thanglha Ri (Mts.)	China	33•15 N	89•07 E
39	Thanh-Hoa	Viet.	19•46 N	105•42 E
35	Thanjāvūr	India	10•51 N	79•11 E
46	Thargomindah	Austl.	27•58 S	143•57 E
21	Thásos (I.)	Grc.	40•41 N	24•53 E
33	Thatch Cay (I.)	Vir.Is.	18•22 N	64•53 W
26	Thaya R.	Aus.-Czech.	48•48 N	15•40 E
53	Thebes (Ruins)	Egypt	25•47 N	32•39 E
6	The Dalles	Or.	45•36 N	121•10 W
41	The Father (Mtn.)	Pap.N.Gui.	5•10 S	151•55 E
46	Theodore	Austl.	24•51 S	150•09 E
10	The Pas	Can.	53•50 N	101•15 W
46	The Round Mtn.	Austl.	30•17 S	152•19 E
21	Thessalía (Reg.)	Grc.	39•50 N	22•09 E
11	Thessalon	Can.	46•11 N	83•37 W
21	Thessaloníki	Grc.	40•38 N	22•59 E
55	The Twins (Mtn.)	Leso.-S.Afr.	30•09 S	28•29 E
52	Thiès	Senegal	14•48 N	16•56 W
35	Thimbu	Bhu.	27•33 N	89•42 E
16	Thingvallavatn (L.)	Ice.	64•12 N	20•22 W
17	Thionville	Fr.	49•23 N	6•31 E
53	Third Cataract	Sud.	19•53 N	30•11 E
24	Thisted	Den.	56•57 N	8•38 E
16	Thisti Fd.	Ice.	66•29 N	14•59 W
46	Thistle (I.)	Austl.	34•55 S	136•11 E
21	Thívai (Thebes)	Grc.	38•20 N	23•18 E
16	Thjórsá (R.)	Ice.	64•23 N	19•18 W
9	Thomas	WV	39•15 N	79•30 W
10	Thompson	Can.	55•48 N	97•59 W
45	Thomson (R.)	Austl.	29•30 S	143•07 E
16	Thórisvatn (L.)	Ice.	64•02 N	19•09 W
8	Thorntown	In.	40•05 N	86•35 W
9	Thorold	Can.	43•13 N	79•12 W
9	Thousand Is.	NY-Can.	44•15 N	76•10 W
21	Thrace	Grc.-Tur.	41•20 N	26•07 E
8	Three Oaks	Mi.	41•50 N	86•40 W
52	Three Points, C.	Ghana	4•45 N	2•06 W
8	Three Rivers	Mi.	42•00 N	83•40 W
5	Thule	Grnld.	76•34 N	68•47 W
26	Thun	Switz.	46•46 N	7•34 E
11	Thunder Bay	Can.	48•28 N	89•12 W
26	Thuner See (L.)	Switz.	46•40 N	7•30 E
26	Thüringen (Thuringia)	G.D.R.	51•07 N	10•45 E
22	Thurles	Ire.	52•44 N	7•45 W
45	Thursday (I.)	Austl.	10•17 S	142•23 E
22	Thurso	Scot.	58•35 N	3•40 W
67	Thurston Pen.	Ant.	71•20 S	98•00 W
54	Thysville	Zaire	5•08 S	14•58 E
41	Tiaong	Phil.	13•56 N	121•20 E
52	Tiaret	Alg.	35•28 N	1•15 E
50	Tibagi	Braz.	24•40 S	50•35 W
53	Tibasti, Sarir (Des.)	Chad	24•00 N	16•30 E
53	Tibesti Massif (Mts.)	Chad	20•40 N	17•48 E
36	Tibet, Plat. of	China	32•22 N	83•30 E
36	Tibetan Aut. Reg.	China	31•15 N	84•48 E
33	Tibnīn	Leb.	33•12 N	35•23 E
12	Tiburón (I.)	Mex.	28•45 N	113•10 W
41	Ticaco Pass	Phil.	12•38 N	123•50 E
41	Ticao (I.)	Phil.	12•40 N	123•30 E
9	Ticonderoga	NY	43•50 N	73•30 W
24	Tidaholm	Swe.	58•11 N	13•53 E
52	Tidikelt	Alg.	25•53 N	2•11 E
52	Tidjikdja	Mauritania	18•33 N	11•25 W
38	T'iehling	China	42•18 N	123•50 E
39	Tien Ch'ih (L.)	China	24•58 N	103•18 E
39	Tienmen	China	30•40 N	113•10 E
39	Tienpai	China	21•30 N	111•20 E
39	T'ienpao	China	23•18 N	106•40 E
38	T'ienshui	China	34•25 N	105•40 E
38	T'ientsaokang	China	45•58 N	126•00 E
38	Tietsin-Shih	China	39•30 N	117•13 E
39	T'ientung	China	23•32 N	107•10 E
24	Tierp	Swe.	60•21 N	17•28 E
55	Tierpoort	S.Afr.	25•53 S	28•26 E
50	Tierra del Fuego	Chile-Arg.	53•50 S	68•45 W
49	Tieté (R.)	Braz.	20•46 S	50•46 W
8	Tiffin	Oh.	41•10 N	83•15 W
50	Tigre	Arg.	34•09 S	58•35 W
48	Tigre (R.)	Peru	2•20 S	75•41 W
54	Tigres, Península dos	Ang.	16•30 S	11•45 E
34	Tigris (R.)	Asia	34•45 N	44•10 E
33	Tīh, Jabal at (Mts.)	Egypt	29•23 N	34•05 E
12	Tijuana	Mex.	32•32 N	117•02 W
50	Tijuca, Pico da	Braz.	22•56 S	43•17 W
31	Tikhoretsk	Sov.Un.	45•55 N	40.05 E
30	Tikhvin	Sov.Un.	59•36 N	33•38 E
34	Tikrīt	Iraq	34•36 N	43•31 E
29	Tiksi	Sov.Un.	71•42 N	128•32 E
23	Tilburg	Neth.	51•33 N	5•05 E
52	Tilemsi, Vallée du	Mali	17•50 N	0•25 E
29	Tilichiki	Sov.Un.	60•49 N	166•14 E
52	Tillabéry	Niger	14•14 N	1•30 E
24	Tillberga	Swe.	59•40 N	16•34 E
45	Timaru	N.Z.	44•26 S	171•17 E
31	Timashevskaya	Sov.Un.	45•47 N	38•57 E
52	Timbo	Gui.	10•41 N	11•51 W
24	Time	Nor.	58•45 N	5•39 E
52	Timimoun	Alg.	29•14 N	0•22 E
52	Timiris, Cap	Mauritania	19•23 N	16•32 W
21	Timis (R.)	Rom.	45•28 N	21•06 E
11	Timiskaming Station	Can.	46•41 N	79•01 W
11	Timmins	Can.	48•25 N	81•22 W
41	Timor (I.)	Indon.	10•08 S	125•00 E
42	Timor Sea	Asia	12•40 S	125•00 E
21	Timoşoara	Rom.	45•44 N	21•21 E
55	Tina (R.)	S.Afr.	30•50 S	28•44 E
49	Tinaguillo	Ven.	9•55 N	68•18 W
33	Tīnah, Khalīj at (G.)	Egypt	31•06 N	32•42 E
52	Tindouf	Alg.	27•43 N	7•44 W
33	Tinggi, Palau (I.)	Mala.	2•16 N	104•16 E
38	Tinghsien	China	38•30 N	115•00 E
38	Tinghsing	China	39•18 N	115•50 E
48	Tingo María	Peru	9•15 S	76•04 W
24	Tingsryd	Swe.	56•32 N	14•58 E
24	Tinnosset	Nor.	59•44 N	9•00 E
24	Tinnsjö	Nor.	59•55 N	8•49 E
50	Tinogasta	Arg.	28•07 S	67•30 W
21	Tínos (I.)	Grc.	37•45 N	25•12 E
52	Tinrhert, Plat. du	Alg.	27•30 N	7•30 E
35	Tinsukia	India	27•18 N	95•29 E
33	Tioman (I.)	Mala.	2•25 N	104•30 E
8	Tippecanoe (R.)	In.	40•55 N	86•45 W
22	Tipperary	Ire.	52•28 N	8•13 W
8	Tipton	In.	40•15 N	86•00 W
21	Tirane	Alb.	41•18 N	19•50 E
20	Tirano	It.	46•12 N	10•09 E
31	Tiraspol	Sov.Un.	46•52 N	29•38 E
31	Tire	Tur.	38•05 N	27•48 E
22	Tiree (I.)	Scot.	56•34 N	6•30 W
21	Tirgovişte	Rom.	44•54 N	25•29 E
21	Tirgu-Jiu	Rom.	45•02 N	23•17 E
27	Tirgu-Mureş	Rom.	46•33 N	24•35 E
27	Tirgu Neamt	Rom.	47•14 N	26•23 E
27	Tirgu-Ocna	Rom.	46•18 N	26•38 E
27	Tîrgu Săcuesc	Rom.	46•04 N	26•06 E
21	Tírnavos	Grc.	39•50 N	22•14 E
27	Tîrnăveni	Rom.	46•19 N	24•18 E
26	Tirol	Aus.	47•13 N	11•10 E
20	Tirso (R.)	It.	40•15 N	9•03 E
35	Tiruchchirāppalli	India	10•49 N	78•48 E
35	Tirunelveli	India	8•53 N	77•43 E
21	Tisza (R.)	Yugo.	45•50 N	20•13 E
27	Tisza R.	Hung.	46•30 N	20•08 E
48	Titicaca, Lago	Bol.-Peru	16•12 S	70•33 W
48	Titiribi	Col.	6•05 N	75•47 W
21	Titograd	Yugo.	42•25 N	20•42 E
21	Titovo Užice	Yugo.	43•51 N	19•53 E
21	Titov Veles	Yugo.	41•42 N	21•50 E
9	Titusville	Pa.	40•40 N	79•40 W
20	Tívoli	It.	41•58 N	12•48 E
40	Tizard Bk. and Rf.	China	10•51 N	113•20 E
52	Tizi-Ouzou	Alg.	36•44 N	4•04 E
49	Tiznados (R.)	Ven.	9•53 N	67•49 W
52	Tiznit	Mor.	29•52 N	9•39 W
12	Tlaxcala	Mex.	19•16 N	98•14 W
52	Tlemcen	Alg.	34•53 N	1•21 W
27	Tlumach	Sov.Un.	48•47 N	25•00 E
13	Tobago (I.)	N.A.	11•15 N	60•30 W
28	Tobol (R.)	Sov.Un.	56•02 N	65•30 E
48	Tocaima	Col.	4•28 N	74•38 W
49	Tocantinópolis	Braz.	6•27 S	47•18 W
49	Tocantins (R.)	Braz.	3•28 S	49•22 W
50	Tocopilla	Chile	22•03 S	70•08 W
49	Tocuyo de la Costa	Ven.	11•03 N	68•24 W
40	Togian (Is.)	Indon.	0•20 S	122•00 E
51	Togo	Afr.	8•00 N	0•52 E
25	Toijala	Fin.	61•11 N	21•46 E
27	Tokaj	Hung.	48•06 N	21•24 E
37	Tokara Guntō (Is.)	Jap.	29•45 N	129•15 E
31	Tokat	Tur.	40•20 N	36•30 E
42	Tokelau Is.	Oceania	8•00 S	176•00 W
28	Tokmak	Sov.Un.	42•44 N	75•41 E
37	Tokuno (I.)	Jap.	27•42 N	129•25 E
37	Tokushima	Jap.	34•06 N	134•31 E
37	Tokyo	Jap.	35•41 N	139•44 E
21	Tolbukhin	Bul.	43•33 N	27•52 E
8	Toledo	Oh.	41•40 N	83•35 W

Page	Name	Region	Lat.	Long.
17	Toledo	Sp.	39•53 N	4•02 W
7	Toledo Bend Res.	La.-Tx.	31•30 N	93•30 W
48	Tolima (Dept.)	Col.	4•07 N	75•20 W
48	Tolima, Nevado del	Col.	4•40 N	75•20 W
20	Tolmezzo	It.	46•25 N	13•03 E
20	Tolmin	Yugo.	46•12 N	13•45 E
27	Tolna	Hung.	46•25 N	18•47 E
40	Tolo, Teluk (B.)	Indon.	2•00 S	122•06 E
8	Toluca	Il.	41•00 N	89•10 W
12	Toluca	Mex.	19•17 N	99•40 W
12	Toluca, Nevado de	Mex.	19•09 N	99•42 W
38	Tolun	China	42•12 N	116•15 E
30	Tolyatti	Sov.Un.	53•30 N	49•10 E
8	Tomah	Wi.	43•58 N	90•31 W
8	Tomahawk	Wi.	45•27 N	89•44 W
27	Tomashevka	Sov.Un.	51•34 N	23•37 E
27	Tomaszow Lubelski	Pol.	50•20 N	23•27 E
27	Tomaszów Mazowiecki	Pol.	51•33 N	20•00 E
49	Tombador, Serra do	Braz.	11•31 S	57•33 W
7	Tombigbee (R.)	Al.	31•45 N	88•02 W
52	Tombouctou (Timbuktu)	Mali	16•46 N	3•01 W
24	Tomelilla	Swe.	55•34 N	13•55 E
17	Tomelloso	Sp.	39•09 N	3•02 W
40	Tomini, Teluk (B.)	Indon.	0•10 N	121•00 E
29	Tommot	Sov.Un.	59•13 N	126•22 E
28	Tomsk	Sov.Un.	56•29 N	84•57 E
9	Tonawanda	NY	43•01 N	78•53 W
41	Tondano	Indon.	1•15 N	124•50 E
24	Tønder	Den.	54•47 N	8•49 E
42	Tonga	Oceania	18•50 S	175•20 W
50	Tongoy	Chile	30•16 S	71•29 W
53	Tonj R.	Sud.	6•18 N	28•33 E
35	Tonk	India	26•13 N	75•45 E
39	Tonkin, Gulf of	Viet.	20•30 N	108•10 E
40	Tonle Sap (L.)	Camb.	13•03 N	102•49 E
26	Tönning	F.R.G.	54•20 N	8•55 E
24	Tönsberg	Nor.	59•19 N	10•25 E
39	Toohsien	China	25•30 N	111•32 E
46	Toowoomba	Aust.	27•32 S	152•10 E
7	Topeka	Ks.	39•02 N	95•41 W
27	Topol'čany	Czech.	48•38 N	18•10 E
12	Topolobampo	Mex.	25•45 N	109•00 W
21	Topolovgrad	Bul.	42•05 N	26•19 E
46	Torbreck, Mt.	Austl.	37•05 S	146•55 E
8	Torch (L.)	Mi.	45•00 N	85•30 W
24	Töreboda	Swe.	58•44 N	14•04 E
23	Torhout	Bel.	51•01 N	3•04 E
48	Toribío	Col.	2•58 N	76•14 W
20	Torino (Turin)	It.	45•05 N	7•44 E
16	Torino (R.)	Fin.-Swe.	67•00 N	23•50 E
16	Torne (R.)	Swe.	67•29 N	21•44 E
16	Torne Träsk (L.)	Swe.	68•10 N	20•36 E
11	Torngat Mts.	Can.	59•18 N	64•35 W
16	Tornio	Fin.	65•55 N	24•09 E
21	Toronaíos Kólpos (G.)	Grc.	40•10 N	23•35 E
11	Toronto	Can.	43•40 N	79•23 W
8	Toronto	Oh.	40•30 N	80•35 W
30	Toropets	Sov.Un.	56•31 N	31•37 E
31	Toros Dağlari (Taurus Mts.)	Tur.	37•00 N	32•40 E
24	Torp	Swe.	62•30 N	16•04 E
22	Torquay (Torbay)	Eng.	50•30 N	3•26 W
48	Torra, Cerro	Col.	4•41 N	76•22 W
17	Torre de Cerredo (Mtn.)	Sp.	43•10 N	4•47 W
20	Torre del Greco	It.	40•32 N	14•23 E
20	Torre Maggiore	It.	40•41 N	15•18 E
46	Torrens, L.	Austl.	30•07 S	137•40 E
12	Torreón	Mex.	25•32 N	103•26 W
45	Torres Is.	New Hebr.	13•18 N	165•59 E
17	Tôrres Novas	Port.	39•28 N	8•37 W
41	Torres Str.	Austl.	10•30 S	141•30 E
35	Torrijos	Phil.	13•19 N	122•06 E
9	Torrington	Ct.	41•50 N	73•10 W
24	Torsby	Swe.	60•07 N	12•56 E
24	Torshälla	Swe.	59•26 N	16•21 E
16	Tórshavn	Faer.	62•00 N	6•55 W
13	Tortola (I.)	Vir.Is.	18•34 N	64•40 W
20	Tortona	It.	44•52 N	8•52 W
17	Tortosa	Sp.	40•59 N	0•33 E
49	Tortuga, Isla la	Ven.	10•55 N	65•18 W
27	Toruń	Pol.	53•01 N	18•37 E
22	Tory (I.)	Ire.	55•17 N	8•10 W
30	Torzhok	Sov.Un.	57•03 N	34•53 E
20	Toscana	It.	43•23 N	11•08 E
50	Tostado	Arg.	29•10 S	61•43 W
31	Tosya	Tur.	41•00 N	34•00 E
30	Tot'ma	Sov.Un.	60•00 N	42•20 E
49	Totness	Sur.	5•51 N	56•17 W
22	Tottenham	Eng.	51•35 N	0•06 W
37	Tottori	Jap.	35•30 N	134•15 E
52	Touat (Oases)	Alg.	27•22 N	0•38 W
52	Touba	Ivory Coast	8•17 N	7•41 W
52	Toubkal Jebel (Mtn.)	Mor.	31•15 N	7•46 W
52	Touggourt	Alg.	33•09 N	6•07 E
18	Touil R.	Alg.	34•42 N	2•16 E
18	Toulon	Fr.	43•09 N	5•54 E
17	Toulouse	Fr.	43•37 N	1•27 E
40	Toungoo	Bur.	19•00 N	96•29 E
17	Tours	Fr.	47•23 N	0•39 E
53	Touside, Pic	Chad	21•10 N	16•30 E
24	Tovdalselv (R.)	Nor.	58•23 N	8•16 E
9	Towanda	Pa.	41•45 N	76•30 W
45	Townsville	Austl.	19•18 S	146•50 E
9	Towson	Md.	39•24 N	76•36 W
40	Towuti, Danau (L.)	Indon.	3•00 S	121•45 E
37	Toyama	Jap.	36•42 N	137•14 E
18	Tozeur	Tun.	33•59 N	8•11 E
31	Trabzon	Tur.	41•00 N	39•45 E
55	Trafonomby (Mtn.)	Mad.	24•32 S	46•35 E
25	Trakai	Sov.Un.	54•38 N	24•59 E
27	Trakiszki	Pol.	54•16 N	23•07 E
22	Tralee	Ire.	52•16 N	9•20 W
24	Tranas	Swe.	58•03 N	14•56 E
41	Trangan (I.)	Indon.	6•52 S	133•30 E
20	Trani	It.	41•15 N	16•25 E
36	Trans Himalaya Mts.	China	30•25 N	83•43 E
54	Transvaal	S.Afr.	24•21 S	28•18 E
27	Transylvania	Rom.	46•30 N	22•35 E
20	Trapani	It.	38•02 N	12•34 E
46	Traralgon	Austl.	38•15 S	146•33 E
20	Trasimeno, Lago	It.	43•00 N	12•12 E
17	Tras os Montes	Port.	41•33 N	7•13 W
26	Traun R.	Aus.	48•10 N	14•15 E
26	Traunstein	F.R.G.	47•52 N	12•38 E
8	Traverse City	Mi.	44•45 N	85•40 W
20	Travnik	Yugo.	44•13 N	17•43 E
26	Třebič	Czech.	49•13 N	15•53 E
21	Trebinje	Yugo.	42•43 N	18•21 E
27	Trebisov	Czech.	48•36 N	21•32 E
26	Třeboň	Czech.	49•00 N	14•48 E
45	Tregrosse Is.	Austl.	18•08 S	150•53 E
50	Treinta y Tres	Ur.	33•14 S	54•17 W
50	Trelew	Arg.	43•15 S	65•25 W
24	Trelleborg	Swe.	55•24 N	13•07 E
22	Tremadoc B.	Wales	52•43 N	4•27 W
20	Tremiti, Isole di	It.	42•07 N	16•33 E
27	Trenčín	Czech.	48•52 N	18•02 E
50	Trenque Lauquén	Arg.	35•50 S	62•44 W
22	Trent (R.)	Eng.	53•05 N	1•00 W
20	Trento	It.	46•04 N	11•07 E
20	Trento	It.	46•16 N	10•47 E
11	Trenton	Can.	44•05 N	77•35 W
9	Trenton	NJ	40•13 N	74•46 W
50	Tres Arroyos	Arg.	38•18 S	60•16 W
49	Três Lagoas	Braz.	20•48 S	51•42 W
49	Três Marias (Res.)	Braz.	18•15 S	45•30 W
48	Tres Morros, Alto de (Mtn.)	Col.	7•08 N	76•10 W
20	Treviglio	It.	45•30 N	9•34 E
20	Treviso	It.	45•39 N	12•15 E
36	Triangle, The (Reg.)	Asia	26•00 N	98•00 E
20	Trieste	It.	45•39 N	13•48 E
20	Trieste, G. of	It.	45•38 N	13•40 E
21	Tríkkala	Grc.	39•33 N	21•49 E
41	Trikora, Puncak (Pk.)	Indon.	4•15 S	138•45 E
35	Trincomalee	Sri Lanka	8•39 N	81•12 E
48	Trinidad	Bol.	14•48 S	64•43 W
6	Trinidad	Co.	37•11 N	104•31 W
13	Trinidad	Cuba	21•50 N	80•00 W
50	Trinidad	Ur.	33•29 S	56•55 W
49	Trinidad (I.)	Trin.	10•00 N	61•00 W
13	Trinidad and Tobago	N.A.	11•00 N	61•00 W
47	Trinidade, Ilha da	Braz.	21•00 S	32•00 W
12	Trinidad R.	Pan.	8•55 N	80•01 W
11	Trinity	Can.	48•59 N	53•55 W
11	Trinity B.	Can.	48•00 N	53•40 W
7	Trinity R.	Tx.	30•50 N	95•09 W
20	Trino	It.	45•11 N	8•16 E
21	Trípolis	Grc.	37•32 N	22•32 E
35	Tripura (State)	India	24•00 N	92•00 E
67	Tristan da Cunha Is.	Alt.O.	35•30 S	12•15 W
49	Triste, Golfo	Ven.	10•40 N	68•05 W
35	Trivandrum	India	8•34 N	76•58 E
27	Trnava	Czech.	48•22 N	17•34 E
41	Trobriand Is.	Pap.N.Gui.	8•25 S	151•45 E
20	Trogir	Yugo.	43•32 N	16•17 E
11	Trois-Rivières	Can.	46•21 N	72•35 W
30	Troitsk	Sov.Un.	54•06 N	61•34 E
30	Troitsko-Pechorsk	Sov.Un.	62•18 N	56•07 E
24	Trollhättan	Swe.	58•17 N	12•17 E
24	Trollheim (Mts.)	Nor.	62•48 N	9•05 E
16	Tromsö	Nor.	69•38 N	19•12 E
50	Tronador, Cerro	Arg.	41•7 S	71•56 W
24	Trondheim (Nidaros)	Nor.	63•25 N	11•35 E
24	Trosa	Swe.	58•54 N	17•25 E
11	Trout (L.)	Can.	51•16 N	92•46 W
10	Trout (L.)	Can.	61•10 N	121•30 W
11	Trout L.	Can.	51•13 N	93•20 W
9	Troy	NY	42•45 N	73•45 N
8	Troy	Oh.	40•00 N	84•10 W
21	Troy (Ruins)	Tur.	39•59 N	26•14 E
17	Troyes	Fr.	48•18 N	4•03 E
21	Trstenik	Yugo.	43•36 N	20•00 E
31	Trubchëvsk	Sov.Un.	52•36 N	32•46 E
48	Trujillo	Col.	4•10 N	76•20 W
12	Trujillo	Hond.	15•55 N	85•58 W
48	Trujillo	Peru	8•08 S	79•00 W
17	Trujillo	Sp.	39•27 N	5•50 W
48	Trujillo	Ven.	9•15 N	70•28 W
21	Trŭn	Bul.	42•49 N	22•39 E
11	Truro	Can.	45•22 N	63•16 W
22	Truro	Eng.	50•17 N	5•05 W
26	Trutnov	Czech.	50•36 N	15•36 E
26	Trzcianka	Pol.	53•02 N	16•27 E
26	Trzebiatow	Pol.	54•03 N	15•16 E
36	Tsaidam Swp.	China	37•19 N	94•08 E
38	Ts'aiyü	China	39•39 N	116•36 E
36	Tsast Bogda Ula (Mt.)	Mong.	46•44 N	92•34 E
28	Tselinograd	Sov.Un.	51•10 N	71•43 E
52	Tshela	Zaire	4•59 S	12•56 E
52	Tshikapa	Zaire	6•25 S	20•48 E
55	Tsiafajovona (Mtn.)	Mad.	19•17 S	47•27 E
31	Tsimlyanskiy (Res.)	Sov.Un.	47•50 N	43•40 E
38	Tsinan (Chinan)	China	36•40 N	117•01 E
36	Tsinghai (Prov.)	China	36•14 N	95•30 E
55	Tsiribihina (R.)	Mad.	19•45 S	43•30 E
55	Tsitsa (R.)	S.Afr.	31•28 S	28•53 E
55	Tsolo	S.Afr.	31•19 S	28•47 E
55	Tsomo	S.Afr.	32•03 S	27•49 E
55	Tsomo (R.)	S.Afr.	31•53 S	27•48 E
37	Tsugaru Kaikyō (Str.)	Jap.	41•25 N	140•20 E
54	Tsumeb	Namibia	19•10 S	17•45 E
39	Ts'unghua	China	23•30 N	113•40 E
38	Tsunhua	China	40•12 N	117•55 E
37	Tsushima Kaikyō (Str.)	Asia	33•52 N	129•30 E
43	Tuamotu (Low), Arch.	Fr.Polynesia	19•00 S	141•20 W
41	Tuao	Phil.	17•44 N	121•26 E
31	Tuapse	Sov.Un.	44•00 N	39•10 E
52	Tuareg (Reg.)	Alg.	21•26 N	2•51 E
50	Tubarão	Braz.	28•23 N	48•56 W
26	Tübingen	F.R.G.	48•33 N	9•05 E
53	Tubruq	Libya	32•03 N	24•04 E
49	Tucacas	Ven.	10•48 N	68•20 W
6	Tucson	Az.	32•15 N	111•00 W
50	Tucumán	Arg.	26•52 S	65•08 W
48	Tucupita	Ven.	9•00 N	62•09 W
49	Tucuruí	Braz.	3•34 S	49•44 W
17	Tudela	Sp.	42•03 N	1•37 W
55	Tugela (R.)	S.Afr.	28•50 S	30•52 E
55	Tugela Ferry	S.Afr.	28•44 S	30•27 E
8	Tug Fork (R.)	WV	37•50 N	82•30 W
41	Tuguegarao	Phil.	17•37 N	121•44 E
41	Tukangbesi (Is.)	Indon.	6•00 S	124•15 E
53	Tūkrah	Libya	32•34 N	20•47 E
10	Tuktoyaktuk	Can.	69•32 N	132•37 W
30	Tukum	Sov.Un.	57•00 N	22•50 E
25	Tukums	Sov.Un.	56•57 N	23•09 E
54	Tukuyu	Tan.	9•13 S	33•43 E
30	Tula	Sov.Un.	54•12 N	37•37 E
45	Tulagi (I.)	Sol.Is.	9•15 S	160•17 E
12	Tulancingo	Mex.	20•04 N	98•24 W
40	Tulangbawang (R.)	Indon.	4•17 S	105•00 E
48	Tulcán	Ec.	0•44 N	77•52 W
19	Tulcea	Rom.	45•10 N	28•47 E
31	Tul'chin	Sov.Un.	48•42 N	28•53 E
55	Tuléar	Mad.	20•16 S	43•44 E
54	Tuli	Rh.	20•58 S	29•12 E
33	Tūlkarm	Jordan	32•19 N	35•02 E
22	Tullamore	Ire.	53•15 N	7•29 W
26	Tulln	Aus.	48•21 N	16•04 E
7	Tulsa	Ok.	36•08 N	95•58 W
48	Tuluá	Col.	4•06 N	76•12 W
36	T'ulufan (Turfan)	China	43•06 N	88•41 E
29	Tulun	Sov.Un.	54•29 N	100•43 E
48	Tumaco	Col.	1•41 N	78•44 W
54	Tumba, Lac	Zaire	0•50 S	17•45 E
48	Tumbes	Peru	3•39 S	80•27 W
38	T'umen	China	43•00 N	129•50 E
38	Tumen (R.)	China	42•08 N	128•40 E
49	Tumeremo	Ven.	7•15 N	61•28 W
49	Tumuc-Humac Mts.	S.A.	2•15 N	54•50 W
22	Tunbridge Wells	Eng.	51•05 N	0•09 E
30	Tundra (Reg.)	Sov.Un.	70•45 N	84•00 E
37	Tung (R.)	China	24•13 N	115•08 E
38	Tunga	China	36•11 N	116•16 E
39	T'ungan	China	24•48 N	118•02 E
37	T'ungchiang	China	47•38 N	132•54 E
38	Tunghai	China	34•35 N	119•05 E
38	T'ungho	China	45•58 N	128•40 E
39	Tunghsiang	China	28•18 N	116•38 E
38	Tunghsien	China	39•55 N	116•40 E
39	Tungjen	China	27•45 N	109•12 E
38	T'ung-Kuan	China	34•48 N	110•25 E
39	T'ungku Chiao (Pt.)	China	19•40 N	111•15 E
38	Tungliao (Payintala)	China	43•30 N	122•15 E
38	T'ungpei	China	48•00 N	126•48 E
38	Tungping	China	35•50 N	116•24 E
39	Tungt'ing Hū (L.)	China	29•10 N	112•30 E
38	Tunhua	China	48•18 N	128•10 E
52	Tunis	Tun.	36•59 N	10•06 E
17	Tunis, Golfe de	Tun.	37•06 N	10•43 E
51	Tunisia	Afr.	35•00 N	10•11 E
48	Tunja	Col.	5•32 N	73•19 W
9	Tunkhannock	Pa.	41•35 N	75•55 W
49	Tupã	Braz.	21•47 S	50•33 W
49	Tupinambaranas, Ilha	Braz.	3•04 S	58•09 W
48	Tupiza	Bol.	21•26 S	65•43 W
9	Tupper Lake	NY	44•15 N	74•25 W
48	Tuquerres	Col.	1•12 N	77•44 W
29	Tura	Sov.Un.	64•08 N	99•58 E
30	Tura (R.)	Sov.Un.	57•15 N	64•23 E
48	Turbo	Col.	8•02 N	76•43 W
27	Turciansky Svätý Martin	Czech.	49•02 N	18•48 E
27	Turda	Rom.	46•35 N	23•47 E
36	Turfan Depression	China	42•16 N	90•00 E
55	Turffontein	S.Afr.	26•15 S	28•03 E
28	Turgay	Sov.Un.	49•42 N	63•39 E
15	Turgayka (R.)	Sov.Un.	49•44 N	66•15 E
21	Tŭrgovishte	Bul.	43•14 N	26•36 E
31	Turgutlu	Tur.	38•30 N	27•20 E
25	Türi	Sov.Un.	58•49 N	25•29 E
27	Turka	Sov.Un.	49•10 N	23•02 E
28	Turkestan	Sov.Un.	42•40 N	65•00 E
28	Turkestan (Reg.)	Sov.Un.	43•27 N	62•14 E
32	Turkey	Eur.-Asia	38•45 N	32•00 E
28	Turkmen S.S.R.	Sov.Un.	40•46 N	56•01 E
13	Turks (Is.)	Turks & Caicos Is.	21•40 N	71•45 W
25	Turku (Åbo)	Fin.	60•28 N	22•12 E
12	Turneffe (I.)	Belize	17•25 N	87•43 W
23	Turnhout	Bel.	51•19 N	4•58 E
26	Turnov	Czech.	50•36 N	15•12 E
21	Turnu Măgurele	Rom.	43•54 N	24•49 E
21	Turnu-Severin	Rom.	44•37 N	22•38 E
21	Turski Trstenik	Bul.	43•26 N	24•50 E
15	Turtkul'	Sov.Un.	41•28 N	61•02 E
29	Turukhansk	Sov.Un.	66•03 N	88•39 E
27	Turya R.	Sov.Un.	51•18 N	24•55 E
7	Tuscaloosa	Al.	33•10 N	87•35 W
8	Tuscola	Il.	39•50 N	88•20 W
39	Tushan	China	25•50 N	107•42 E
35	Tuticorin	India	8•51 N	78•09 E
49	Tutóia	Braz.	2•42 S	42•21 W
21	Tutrakan	Bul.	44•02 N	26•36 E
26	Tuttlingen	F.R.G.	47•58 N	8•50 E
42	Tuvalu	Oceania	5•20 S	174•00 E
34	Tuwayq, Jabal (Mts.)	Sau.Ar.	20•45 N	46•30 E
12	Tuxpan	Mex.	19•34 N	103•22 W
12	Tuxpan	Mex.	20•57 N	97•26 W
12	Tuxtla Gutiérrez	Mex.	16•44 N	93•08 W
16	Tuy	Sp.	42•07 N	8•49 W
49	Tuy (R.)	Ven.	10•15 N	66•03 W
39	Tuyün	China	26•18 N	107•40 E
31	Tuz Gölü (L.)	Tur.	39•00 N	33•30 E
21	Tuzla	Yugo.	44•33 N	18•46 E
24	Tvedestrand	Nor.	58•39 N	8•54 E
24	Tveitsund	Nor.	59•03 N	8•29 E
14	Tvertsa (L.)	Sov.Un.	56•58 N	35•22 E
22	Tweed (R.)	Scot.	55•32 N	2•35 W
27	Tyachev	Sov.Un.	48•01 N	23•42 E
36	Tyan' Shan' (Tien-Shan) (Mts.)	Sov.Un.-China	42•00 N	78•46 E
55	Tylden	S.Afr.	32•08 S	27•06 E
7	Tyler	Tx.	32•21 N	95•19 W
29	Tyndinskiy	Sov.Un.	55•22 N	124•45 E
22	Tyne (R.)	Eng.	54•59 N	1•56 W
22	Tynemouth	Eng.	55•04 N	1•39 W
24	Tynest	Nor.	62•17 N	10•45 E
24	Tyri Fd.	Nor.	60•03 N	10•25 E
9	Tyrone	Pa.	40•40 N	78•15 W
46	Tyrrell, L.	Austl.	35•12 S	143•00 E
17	Tyrrhenian Sea	It.	40•10 N	12•15 E
25	Tyrvää	Fin.	61•19 N	22•51 E
31	Tyub-Karagan, Mys (C.)	Sov.Un.	44•30 N	50•10 E
28	Tyukalinsk	Sov.Un.	56•03 N	71•43 E
31	Tyuleniy (I.)	Sov.Un.	44•30 N	48•00 E
28	Tyumen'	Sov.Un.	57•02 N	65•28 E
28	Tyura-Tam	Sov.Un.	46•00 N	63•15 E
39	Tzu Shui (R.)	China	26•50 N	111•00 E
38	Tzuyang	China	35•35 N	116•50 E

U

Page	Name	Region	Lat.	Long.
18	Uarc, Ras (C.)	Mor.	35•31 N	2•45 W
48	Uaupés	Braz.	0•02 S	67•03 W
49	Ubá	Braz.	21•08 S	42•55 W
53	Ubangi (Oubangui) (R.)	Afr.	4•30 N	20•35 E
49	Uberaba	Braz.	19•47 S	47•47 W
49	Uberlândia	Braz.	18•54 S	48•11 W
54	Ubombo	S.Afr.	27•33 S	32•13 E
40	Ubon Ratchathani	Thai.	15•15 N	104•52 E
36	Ubsa Nuur (L.)	Mong.	50•29 N	93•32 E
48	Ucayali (R.)	Peru	8•58 S	74•13 W
28	Uch-Aral	Sov.Un.	46•14 N	80•58 E
35	Udaipur	India	24•41 N	73•41 E
24	Uddevalla	Swe.	58•21 N	11•55 E
20	Udine	It.	46•05 N	13•14 E

Page	Name	Region	Lat.	Long.
30	Udmurt A.S.S.R.	Sov.Un.	57•00 N	53•00 E
40	Udon Thani	Thai.	17•31 N	102•51 E
49	Udskaya Guba (B.)	Sov.Un.	55•00 N	136•30 E
26	Uekermünde	G.D.R.	53•43 N	14•01 E
53	Uele R.	Zaire	3•55 N	23•30 E
30	Ufa	Sov.Un.	54•45 N	55•57 E
30	Ufa (R.)	Sov.Un.	56•00 N	57•05 E
54	Ugab (R.)	Namibia	21•10 S	14•00 E
54	Ugalla (R.)	Tan.	6•15 S	32•30 E
51	Uganda	Afr.	2•00 N	32•28 E
55	Ugie	S.Afr.	31•13 S	28•14 E
29	Uglegorsk	Sov.Un.	49•00 N	142•31 E
30	Ugra (R.)	Sov.Un.	54•43 N	34•20 E
21	Ugŭrchin	Bul.	43•06 N	24•23 E
8	Uhrichsville	Oh.	40•25 N	81•20 W
54	Uíge	Ang.	7•37 S	15•03 E
38	Uiju	Kor.	40•09 N	124•33 E
31	Uil (R.)	Sov.Un.	49•30 N	55•10 E
54	Uitenhage	S.Afr.	33•46 S	25•26 E
54	Ujiji	Tan.	4•55 S	29•41 E
35	Ujjain	India	23•18 N	75•37 E
40	Ujung Pandang (Makasar)	Indon.	5•08 S	119•28 E
30	Ukhta	Sov.Un.	63•08 N	53•42 E
30	Ukhta	Sov.Un.	65•22 N	31•30 E
25	Ukmerge	Sov.Un.	55•16 N	24•45 E
28	Ukrainian S.S.R.	Sov.Un.	49•15 N	30•15 E
36	Ulaan Baatar	Mong.	47•56 N	107•00 E
36	Ulaan Goom	Mong.	50•23 N	92•14 E
29	Ulan-Ude	Sov.Un.	51•59 N	107•41 E
38	Ulchin	Kor.	36•57 N	129•26 E
21	Ulcinj (Dulcigno)	Yugo.	41•56 N	19•15 E
54	Ulindi (R.)	Zaire	1•55 S	26•17 E
26	Ulm	F.R.G.	48•24 N	9•59 E
67	Ulmer, Mt.	Ant.	77•30 S	86•00 W
24	Ulricehamn	Swe.	57•49 N	13•23 E
38	Ulsan	Kor.	35•35 N	129•22 E
22	Ulster	Ire.-N.Ire.	54•41 N	7•10 W
31	Ulukişla	Tur.	36•40 N	34•30 E
46	Ulverstone	Austl.	41•20 S	146•22 E
24	Ulvik	Nor.	60•35 N	6•53 E
30	Ul'yanovsk	Sov.Un.	54•20 N	48•05 E
26	Ülzen	F.R.G.	52•58 N	10•34 E
31	Uman'	Sov.Un.	48•44 N	30•13 E
20	Umbria	It.	42•53 N	12•22 E
16	Ume (R.)	Swe.	64•57 N	18•51 E
16	Umeå	Swe.	63•48 N	20•29 E
55	Umhlatuzi (R.)	S.Afr.	28•47 S	31•17 E
6	Umiat	Ak.	69•20 N	152•28 W
55	Umkomaas	S.Afr.	30•12 S	30•48 E
53	Umm Durmān (Omdurman)	Sud.	15•45 N	32•30 E
54	Umniati (R.)	Rh.	17•08 S	29•11 E
54	Umtali	Rh.	18•49 S	32•39 E
55	Umtata	S.Afr.	31•36 S	28•47 E
55	Umtentweni	S.Afr.	30•41 S	30•29 E
55	Umzimkulu	S.Afr.	30•12 S	29•53 E
55	Umzinto	S.Afr.	30•19 S	30•41 E
20	Una (R.)	Yugo.	44•38 N	16•10 E
49	Unare (R.)	Ven.	9•45 N	65•12 W
49	Unare, Laguna de	Ven.	10•07 N	65•23 W
34	Unayzah	Sau.Ar.	25•50 N	44•02 E
48	Uncía	Bol.	18•28 S	66•32 W
53	Undo	Eth.	6•37 N	38•29 E
11	Ungava B.	Can.	59•46 N	67•18 W
11	Ungava (Pen.)	Can.	59•55 N	74•00 W
50	União da Vitória	Braz.	26•17 S	51•13 W
20	Unije (I.)	Yugo.	44•39 N	14•10 E
8	Union City	In.	40•10 N	85•00 W
8	Union City	Mi.	42•00 N	85•10 W
9	Union City	Pa.	41•50 N	79•50 W
9	Uniontown	Pa.	39•55 N	79•45 W
41	Unisan	Phil.	13•50 N	121•59 E
6	Unitas, Mts.	U.S.	40•35 N	111•00 W
32	United Arab Emirates	Asia	24•00 N	54•00 E
16	United Kingdom	Eur.	56•30 N	1•40 W
5	United States	N.A.	38•00 N	110•00 W
8	Universal	In.	39•35 N	87•30 W
22	Unst (I.)	Scot.	60•50 N	1•24 W
31	Unye	Tur.	41•00 N	37•10 E
30	Unzha (R.)	Sov.Un.	57•45 N	44•10 E
51	Upanda, Sierra do	Ang.	13•15 S	14•15 E
48	Upata	Ven.	7•58 N	62•27 W
54	Upington	S.Afr.	28•25 S	21•15 E
40	Upper Kapuas Mts.	Mala.	1•45 N	112•06 E
8	Upper Sandusky	Oh.	40•50 N	83•20 W
51	Upper Volta	Afr.	11•46 N	3•18 E
24	Uppsala	Swe.	59•53 N	17•39 E
31	Ural (R.)	Sov.Un.	49•50 N	51•30 E
28	Urals (Mts.)	Sov.Un.	56•28 N	58•13 E
31	Ural'sk	Sov.Un.	51•15 N	51•10 E
10	Uranium City	Can.	59•34 N	108•59 W
8	Urbana	Il.	40•10 N	88•15 W
8	Urbana	Oh.	40•05 N	83•50 W
20	Urbino	It.	43•43 N	12•37 E
31	Urda	Sov.Un.	48•50 N	47•30 E
41	Urdaneta	Phil.	15•59 N	120•34 E
28	Urdzhar	Sov.Un.	47•28 N	82•00 E
31	Urfa	Tur.	37•20 N	38•45 E
15	Urgench	Sov.Un.	41•32 N	60•33 E
21	Urla	Tur.	38•20 N	26•44 E
48	Urrao	Col.	6•19 N	76•11 W
48	Urubamba (R.)	Peru	11•48 S	72•34 W
50	Uruguaiana	Braz.	29•45 S	57•00 W
47	Uruguay	S.A.	32•45 S	56•00 W
50	Uruguay, Rio	Braz.	27•05 S	55•15 W
36	Urungu R.	China	46•31 N	87•44 E
31	Uryupinsk	Sov.Un.	50•50 N	42•00 E
21	Urziceni	Rom.	44•45 N	26•42 E
30	Usa (R.)	Sov.Un.	66•00 N	58•20 E
31	Uşak	Tur.	39•50 N	29•15 E
54	Usakos	Namibia	22•00 S	15•40 E
50	Ushuaia	Arg.	54•46 S	68•24 W
31	Üsküdar	Tur.	40•55 N	29•00 E
31	Usman	Sov.Un.	52•03 N	39•40 E
50	Uspallata P.	Arg.-Chile	32•47 S	70•08 W
37	Ussuri (R.)	China	46•30 N	133•56 E
29	Ussuriysk	Sov.Un.	43•48 N	132•09 E
29	Ust'-Bol'sheretsk	Sov.Un.	52•41 N	157•00 E
20	Ustica, I. di	It.	38•43 N	12•11 E
26	Ústí nad Labem	Czech.	50•39 N	14•02 E
26	Ustka	Pol.	54•34 N	16•52 E
29	Ust'-Kamchatsk	Sov.Un.	56•13 N	162•18 E
30	Ust'-Kulom	Sov.Un.	61•38 N	54•00 E
29	Ust'-Maya	Sov.Un.	60•33 N	134•43 E
29	Ust' Olenëk	Sov.Un.	72•52 N	120•15 E
28	Ust' Port	Sov.Un.	69•20 N	83•41 E
30	Ust'-Tsil'ma	Sov.Un.	65•25 N	52•10 E
29	Ust'-Tyrma	Sov.Un.	50•27 N	131•17 E
28	Ust'-Urt, Plato	Sov.Un.	44•03 N	54•58 E
30	Ustyuzhna	Sov.Un.	58•49 N	36•19 E
6	Utah	U.S.	39•25 N	112•40 W
25	Utena	Sov.Un.	55•32 N	25•40 E
55	Utete	Tan.	8•05 S	38•47 E
9	Utica	NY	43•05 N	75•10 W
17	Utiel	Sp.	39•34 N	1•13 W
23	Utrecht	Neth.	52•05 N	5•06 E
17	Utrera	Sp.	37•12 N	5•48 W
24	Utsira (I.)	Nor.	59•21 N	4•50 E
37	Utsunomiya	Jap.	36•35 N	139•52 E
40	Uttaradit	Thai.	17•47 N	100•10 E
35	Uttar Pradesh (State)	India	27•00 N	80•00 E
13	Utuado	P.R.	18•16 N	66•40 W
25	Uusikaupunki (Nystad)	Fin.	60•48 N	21•24 E
54	Uvira	Zaire	3•28 S	29•03 E
55	Uvongo Beach	S.Afr.	30•49 S	30•23 E
48	Uyuni	Bol.	20•28 S	66•45 W
48	Uyuni, Salar de (Salt Flat)	Bol.	20•58 S	67•09 W
28	Uzbek S.S.R.	Sov.Un.	42•42 N	60•00 E
31	Uzen, Bol'shoy (R.)	Sov.Un.	49•50 N	49•35 E
27	Uzhgorod	Sov.Un.	48•38 N	22•18 E
21	Uzunköpru	Tur.	41•17 N	26•42 E
	V			
54	Vaal (R.)	S.Afr.	28•15 S	24•30 E
25	Vaasa	Fin.	63•06 N	21•39 E
27	Vác	Hung.	47•46 N	19•10 E
16	Vadsö	Nor.	70•08 N	29•52 E
24	Vadstena	Swe.	58•27 N	14•53 E
26	Vaduz	Liech.	47•10 N	9•32 E
30	Vaga (R.)	Sov.Un.	61•55 N	42•30 E
24	Vågsöy (I.)	Nor.	61•58 N	4•44 E
27	Vah R.	Czech.	48•07 N	17•52 E
21	Valachia	Rom.	44•45 N	24•17 E
30	Valdai Hills	Sov.Un.	57•50 N	32•35 E
30	Valday (Valdai)	Sov.Un.	57•58 N	33•13 E
25	Valdemārpils	Sov.Un.	57•22 N	22•34 E
17	Valdepeñas	Sp.	38•46 N	3•22 W
50	Valdés, Pen.	Arg.	42•15 S	63•15 W
50	Valdivia	Chile	39•47 S	73•13 W
48	Valdivia	Col.	7•10 N	75•26 W
11	Val-d'Or	Can.	48•03 N	77•50 W
49	Valença	Braz.	13•43 S	38•58 W
17	Valence-sur-Rhône	Fr.	44•56 N	4•54 E
17	Valencia	Sp.	39•26 N	0•23 W
49	Valencia	Ven.	10•11 N	68•00 W
22	Valencia (I.)	Ire.	51•55 N	10•26 W
49	Valencia, Lago de	Ven.	10•11 N	67•45 W
23	Valenciennes	Fr.	50•24 N	3•36 E
6	Valentine	Ne.	42•52 N	100•34 W
48	Valera	Ven.	9•12 N	70•45 W
55	Valhalla	S.Afr.	25•49 S	28•09 E
21	Valjevo	Yugo.	44•17 N	19•57 E
17	Valladolid	Sp.	41•41 N	4•41 W
48	Valle (Dept.)	Col.	4•03 N	76•13 W
49	Valle de Guanape	Ven.	9•54 N	65•41 W
48	Valle de la Pascua	Ven.	9•12 N	65•08 W
48	Valledupar	Col.	10•13 N	73•39 W
48	Valle Grande	Bol.	18•27 S	64•03 W
6	Vallejo	Ca.	38•06 N	122•15 W
50	Vallenar	Chile	28•39 S	70•52 W
18	Valletta	Malta	35•50 N	14•29 E
6	Valley City	ND	46•55 N	97•59 W
9	Valleyfield	Can.	45•16 N	74•09 W
11	Valleyfield	Can.	45•05 N	74•00 W
20	Valli di Comàcchio (L.)	It.	44•38 N	12•15 E
25	Valmiera	Sov.Un.	57•34 N	25•54 E
50	Valparaíso	Chile	33•02 S	71•32 W
8	Valparaiso	In.	41•25 N	87•05 W
41	Vals, Tandjung (C.)	Indon.	8•30 S	137•15 E
54	Valsbaai (False Bay)	S.Afr.	34•14 S	18•35 E
31	Valuyki	Sov.Un.	50•14 N	38•04 E
31	Van	Tur.	38•04 N	43•10 E
8	Vanceburg	Ky.	38•35 N	83•20 W
10	Vancouver	Can.	49•16 N	123•06 W
6	Vancouver	Wa.	45•37 N	122•40 W
10	Vancouver I.	Can.	49•50 N	125•05 W
8	Vandalia	Il.	39•00 N	89•00 W
10	Vanderhoof	Can.	54•01 N	124•01 W
44	Van Diemen, C.	Austl.	11•05 S	130•15 E
44	Van Diemen G.	Austl.	11•50 S	131•30 E
12	Vanegas	Mex.	23•54 N	100•54 W
24	Vänern (L.)	Swe.	58•52 N	13•17 E
24	Vänersborg	Swe.	58•24 N	12•15 E
55	Vanga	Ken.	4•38 S	39•10 E
31	Van Gölü (L.)	Tur.	38•45 N	43•00 E
8	Van Lear	Ky.	37•45 N	82•50 W
17	Vannes	Fr.	47•42 N	2•46 W
41	Van Rees (Mtn.)	Indon.	2•30 S	138•45 E
25	Vantaan (R.)	Fin.	60•25 N	24•43 E
8	Van Wert	Oh.	40•50 N	84•35 W
24	Vara	Swe.	58•17 N	12•55 E
20	Varaklāni	Sov.Un.	56•38 N	26•46 E
20	Varallo	It.	45•44 N	8•14 E
35	Vārānasi (Benares)	India	25•25 N	83•00 E
16	Varanger Fd.	Nor.	70•05 N	30•53 E
20	Varano, Lago di	It.	41•52 N	15•55 E
20	Varaždin	Yugo.	46•17 N	16•20 E
20	Varazze	It.	44•23 N	8•34 E
24	Varberg	Swe.	57•06 N	12•16 E
21	Vardar (R.)	Yugo.	41•40 N	21•50 E
24	Varde	Den.	55•39 N	8•28 E
16	Vardö	Nor.	70•23 N	30•43 E
40	Varella, C.	Viet.	12•58 N	109•50 E
25	Varena	Sov.Un.	54•16 N	24•35 E
21	Varěs	Yugo.	44•10 N	18•20 E
20	Varese	It.	45•45 N	8•49 E
49	Varginha	Braz.	21•33 S	45•25 W
25	Varkaus	Fin.	62•19 N	27•51 E
21	Varna	Bul.	43•14 N	27•58 E
24	Värnamo	Swe.	57•11 N	13•45 E
26	Varnsdorf	Czech.	50•54 N	14•36 E
30	Vashka (R.)	Sov.Un.	63•20 N	47•50 E
31	Vasil'kov	Sov.Un.	50•10 N	30•22 E
27	Vaslui	Rom.	46•39 N	27•49 E
8	Vassar	Mi.	43•25 N	83•35 W
50	Vassouras	Braz.	22•25 S	43•40 W
24	Västanfors	Swe.	59•59 N	15•49 E
24	Västerås	Swe.	59•39 N	16•30 E
24	Väster-dalälven (R.)	Swe.	61•06 N	13•10 E
24	Västervik	Swe.	57•45 N	16•35 E
20	Vasto	It.	42•06 N	12•42 E
20	Vatican City	Eur.	41•54 N	12•22 E
20	Vaticano, C.	It.	38•38 N	15•52 E
16	Vatnajökull	Ice.	64•34 N	16•41 W
55	Vatomandry	Mad.	18•53 S	48•13 E
27	Vatra Dornei	Rom.	47•22 N	25•20 E
24	Vättern (L.)	Swe.	58•15 N	14•24 E
9	Vandreuil	Can.	45•24 N	74•02 W
48	Vaupés (R.)	Col.	1•18 N	71•14 W
24	Vaxholm	Swe.	59•26 N	18•19 E
24	Växjo	Swe.	56•53 N	14•46 E
30	Vaygach (I)	Sov.Un.	70•00 N	59•00 E
49	Veadeiros, Chapadas dos (Mts.)	Braz.	15•20 S	48•43 W
24	Veblungsnares	Nor.	62•33 N	7•46 E
21	Vedea (R.)	Rom.	44•25 N	24•45 E
8	Veedersburg	In.	40•05 N	87•15 W
16	Vegen (I.)	Nor.	65•38 N	10•51 E
24	Vejle	Den.	55•41 N	9•29 E
20	Velebit (Mts.)	Yugo.	44•25 N	15•23 E
17	Vélez-Málaga	Sp.	36•48 N	4•05 W
20	Velika Kapela (Mts.)	Yugo.	45•03 N	15•20 E
21	Velika Morava (R.)	Yugo.	44•20 N	21•10 E
21	Velika Tŭrnovo	Bul.	43•06 N	25•38 E
27	Velikiy Bychkov	Sov.Un.	47•59 N	24•01 E
28	Velikiye Luki	Sov.Un.	56•19 N	30•32 E
30	Velikiy Ustyug	Sov.Un.	60•45 N	46•38 E
30	Velizh	Sov.Un.	55•37 N	31•11 E
26	Velke Meziříčí	Czech.	49•21 N	16•01 E
45	Vella (I.)	Sol.Is.	8•00 S	156•42 E
20	Velletri	It.	41•42 N	12•48 E
35	Vellore	India	12•57 N	79•09 E
30	Vel'sk	Sov.Un.	61•00 N	42•18 E
48	Venadillo	Col.	4•43 N	74•55 W
50	Venado Tuerto	Arg.	33•28 S	61•47 W
20	Veneto	It.	45•58 N	11•24 E
30	Venëv	Sov.Un.	54•19 N	38•14 E
20	Venezia (Venice)	It.	45•25 N	12•18 E
20	Venezia, Golfo di	It.	45•23 N	13•00 E
47	Venezuela	S.A.	8•00 N	65•00 W
48	Venezuela, Golfo de	Ven.	11•34 N	71•02 W
25	Venta (R.)	Sov.Un.	57•05 N	21•45 E
50	Ventana, Sierra de la	Arg.	38•00 S	63•00 W
20	Ventimiglia	It.	43•46 N	7•37 E
9	Ventnor	NJ	39•20 N	74•25 W
25	Ventspils	Sov.Un.	57•24 N	21•41 E
48	Ventuari (R.)	Ven.	4•47 N	65•56 W
50	Vera	Arg.	29•22 S	60•09 W
12	Vera Cruz	Mex.	20•30 N	97•15 W
12	Veracruz	Mex.	19•13 N	96•07 W
35	Verāval	India	20•59 N	70•49 E
20	Vercelli	It.	45•18 N	8•27 E
41	Verde (I.)	Phil.	13•34 N	121•11 E
41	Verde Island Pass.	Phil.	13•36 N	120•39 E
26	Verden	F.R.G.	52•55 N	9•15 E
29	Verkhne-Kamchatsk	Sov.Un.	54•42 N	158•41 E
30	Verkhne Ural'sk	Sov.Un.	53•53 N	59•15 E
29	Verkhnyaya Tunguska (Angara) (R.)	Sov.Un.	58•13 N	97•00 E
29	Verkhoyansk	Sov.Un.	67•43 N	133•33 E
29	Verkhoyanskiy Khrebet (Mts.)	Sov.Un.	67•45 N	128•00 E
10	Vermilion	Can.	53•22 N	110•51 W
8	Vermilion (R.)	Il.	41•05 N	89•00 W
7	Vermont	U.S.	43•50 N	72•50 W
54	Verneuk Pan (L.)	S.Afr.	30•10 S	21•46 E
10	Vernon	Can.	50•18 N	119•15 W
8	Vernon	In.	39•00 N	85•40 W
21	Véroia	Grc.	40•30 N	22•13 E
20	Verona	It.	45•28 N	11•02 E
17	Versailles	Fr.	48•48 N	2•07 E
8	Versailles	Ky.	38•05 N	84•45 W
52	Vert, Cap	Senegal	14•43 N	17•30 W
55	Verulam	S.Afr.	29•39 S	31•08 E
23	Verviers	Bel.	50•35 N	5•57 E
25	Vesijärvi (L.)	Fin.	61•09 N	25•10 E
16	Vester Aalen (Is.)	Nor.	68•54 N	14•03 E
16	Vestfjord	Nor.	67•33 N	12•59 E
16	Vestmannaeyjar	Ice.	63•12 N	20•17 W
20	Vesuvio (Mtn.)	It.	40•35 N	14•26 E
27	Veszprem	Hung.	47•05 N	17•53 E
27	Vesztö	Hung.	46•55 N	21•18 E
24	Vetlanda	Swe.	57•26 N	15•05 E
30	Vetluga	Sov.Un.	57•50 N	45•42 E
30	Vetluga (R.)	Sov.Un.	56•50 N	45•50 E
21	Vetovo	Bul.	43•42 N	26•18 E
21	Vetren	Bul.	42•16 N	24•04 E
8	Vevay	In.	38•45 N	85•05 W
48	Viacha	Bol.	16•43 S	68•16 W
20	Viadana	It.	44•55 N	10•30 E
49	Viana	Braz.	3•09 S	44•44 W
18	Viana do Castélo	Port.	41•41 N	8•45 W
40	Viangchan	Laos	18•07 N	102•33 E
20	Viareggio	It.	43•52 N	10•14 E
24	Viborg	Den.	56•27 N	9•22 E
20	Vibo Valentia	It.	38•47 N	16•06 E
50	Vicente López	Arg.	34•15 S	58•29 W
20	Vicenza	It.	45•33 N	11•33 E
30	Vichuga	Sov.Un.	57•13 N	41•58 E
17	Vichy	Fr.	46•06 N	3•28 E
8	Vicksburg	Mi.	42•10 N	85•30 W
7	Vicksburg	Ms.	32•20 N	90•50 W
50	Victoria	Arg.	32•36 S	60•09 W
10	Victoria	Can.	48•26 N	123•23 W
50	Victoria	Chile	38•15 S	72•16 W
39	Victoria	Hong Kong	22•10 N	114•18 E
48	Victoria	Col.	5•19 N	74•54 W
41	Victoria	Phil.	15•34 N	120•41 E
45	Victoria (State)	Austl.	36•46 S	143•15 E
51	Victoria (L.)	Afr.	0•50 S	32•50 E
44	Victoria (R.)	Austl.	17•25 S	130•50 E
36	Victoria, Mt.	Bur.	21•26 N	93•59 E
47	Victoria, Mt.	Pap.N.Gui.	9•35 S	147•45 E
54	Victoria Falls	Rh.	17•55 S	25•51 E
10	Victoria I.	Can.	70•13 N	107•45 W
67	Victoria Land	Ant.	75•00 S	160•00 E
44	Victoria River Downs	Austl.	16•30 S	131•10 E
10	Victoria Str.	Can.	69•10 N	100•58 W
11	Victoriaville	Can.	46•04 N	71•59 W
54	Victoria West	S.Afr.	31•25 S	23•10 E
21	Vidin	Bul.	44•00 N	22•53 E
50	Viedma	Arg.	40•55 S	63•03 W
50	Viedma (L.)	Arg.	49•40 S	72•35 W
17	Vienne	Fr.	45•31 N	4•54 E
13	Vieques	P.R.	18•09 N	65•27 W
13	Vieques (I.)	P.R.	18•05 N	65•28 W
26	Vierwaldstätter See (L.)	Switz.	46•54 N	8•36 E
17	Vierzon	Fr.	47•14 N	2•04 E
20	Vieste	It.	41•52 N	16•10 E
40	Vietnam	Asia	18•00 N	107•00 E
41	Vigan	Phil.	17•36 N	120•22 E
20	Vigevano	It.	45•18 N	8•52 E
16	Vigo	Sp.	42•18 N	8•42 W
25	Vihti	Fin.	60•27 N	24•18 E
35	Vijayawāda	India	16•31 N	80•37 E
21	Vijosë (R.)	Alb.	40•15 N	20•30 E
16	Vik	Ice.	63•22 N	18•58 W
24	Vik	Nor.	61•06 N	6•35 E
45	Vila	New Hebr.	18•00 S	168•30 E
54	Vila de Manica	Moz.	18•48 S	32•49 E

Page	Name	Region	Lat.	Long.
54	Vilanculos	Moz.	22•03 S	35•13 E
18	Vila Real	Port.	41•18 N	7•48 W
16	Vilhelmina	Swe.	64•37 N	16•30 E
25	Viljandi	Sov.Un.	58•24 N	25•34 E
25	Vilkaviškis	Sov.Un.	54•40 N	23•08 E
25	Vilkija	Sov.Un.	55•04 N	23•30 E
31	Vilkovo	Sov.Un.	45•24 N	29•36 E
50	Villa Angela	Arg.	27•31 S	60•42 W
50	Villa Ballester	Arg.	34•18 S	58•33 W
48	Villa Bella	Bol.	10•25 S	65•22 W
26	Villach	Aus.	46•38 N	13•50 E
20	Villacidro	It.	39•28 N	8•41 E
52	Villa Cisneros	W.Sah.	23•45 N	16•04 W
49	Villa de Cura	Ven.	10•03 N	67•29 W
50	Villa Dolores	Arg.	31•50 S	65•05 W
20	Villafranca	It.	45•22 N	10•53 E
8	Villa Grove	Il.	39•55 N	88•15 W
50	Villaguay	Arg.	31•47 S	58•53 W
50	Villa Hayes	Par.	25•07 S	57•31 W
12	Villahermosa	Mex.	17•59 N	92•56 W
12	Villaldama	Mex.	26•30 N	100•26 W
50	Villa María	Arg.	32•17 S	63•08 W
50	Villa Mercedes	Arg.	33•38 S	65•16 W
48	Villa Montes	Bol.	21•13 S	63•26 W
48	Villanueva	Col.	10•44 N	73•08 W
17	Villanueva de la Serena	Sp.	38•59 N	5•56 W
17	Villanueva y Geltrú	Sp.	41•13 N	1•44 E
50	Villarrica	Par.	25•55 S	56•23 W
17	Villarrobledo	Sp.	39•15 N	2•37 W
48	Villavicencio	Col.	4•09 N	73•38 W
48	Villavieja	Col.	3•13 N	75•13 W
50	Villazón	Bol.	22•02 S	65•42 W
18	Villena	Sp.	38•37 N	0•52 W
48	Villeta	Col.	5•02 N	74•29 W
17	Villeurbanne	Fr.	45•43 N	4•55 E
26	Villingen	F.R.G.	48•04 N	8•28 E
25	Vilnius (Wilno)	Sov.Un.	54•40 N	25•26 E
25	Vilppula	Fin.	62•01 N	24•24 E
29	Vilyuy (R.)	Sov.Un.	65•22 N	108•45 E
29	Vilyuysk	Sov.Un.	63•41 N	121•47 E
24	Vimmerby	Swe.	57•41 N	15•51 E
26	Vimperk	Czech.	49•04 N	13•41 E
50	Viña del Mar	Chile	33•00 S	71•33 W
8	Vincennes	In.	38•40 N	87•30 W
16	Vindelälven (R.)	Swe.	65•02 N	18•30 E
16	Vindeln	Swe.	64•10 N	19•52 E
35	Vindhya Ra.	India	22•30 N	75•50 E
9	Vineland	NJ	39•30 N	75•00 W
39	Vinh	Viet.	18•38 N	105•42 E
21	Vinkovci	Yugo.	45•17 N	18•47 E
28	Vinnitsa	Sov.Un.	49•13 N	28•31 E
67	Vinson Massif (Mtn.)	Ant.	77•40 S	87•00 W
39	Virac	Phil.	13•38 N	124•20 E
25	Virbalis	Sov.Un.	54•38 N	22•55 E
10	Virden	Can.	49•51 N	101•55 W
7	Virginia	U.S.	37•00 N	80•45 W
9	Virginia Beach	Va.	36•50 N	75•58 W
13	Virgin Is.	N.A.	18•15 N	64•00 W
25	Virmo	Fin.	60•41 N	21•58 E
20	Virovitica	Yugo.	45•50 N	17•24 E
21	Virpazar	Yugo.	42•16 N	19•06 E
25	Virrat	Fin.	62•15 N	23•45 E
24	Virserum	Swe.	57•22 N	15•35 E
20	Vis	Yugo.	43•03 N	16•11 E
20	Vis (I.)	Yugo.	43•00 N	16•10 E
20	Visa, Mt.	It.	45•42 N	7•08 E
24	Visby	Swe.	57•39 N	18•19 E
5	Viscount Mellville Sound	Can.	74•80 N	110•00 W
21	Višegrad	Yugo.	43•45 N	19•19 E
35	Vishākhapatnam	India	17•48 N	83•21 E
54	Vishoek	S.Afr.(In.)	34•13 S	18•26 E
24	Viskan (R.)	Swe.	57•20 N	12•25 E
21	Visoko	Yugo.	43•59 N	18•10 E
21	Vistonís	Grc.	40•58 N	25•12 E
21	Vitanovac	Yugo.	43•44 N	20•50 E
28	Vitebsk	Sov.Un.	55•12 N	30•16 E
20	Viterbo	It.	42•24 N	12•08 E
29	Vitim	Sov.Un.	59•22 N	112•43 E
29	Vitim (R.)	Sov.Un.	56•12 N	115•30 E
49	Vitória	Braz.	20•09 S	40•17 W
17	Vitoria	Sp.	42•43 N	2•43 W
49	Vitória de Conquista	Braz.	14•51 S	40•44 W
17	Vittoria	It.	37•01 N	14•31 E
20	Vittorio	It.	45•59 N	12•17 E
41	Vitu Is.	Pap.N.Gui.	4•45 S	149•50 E
21	Vize	Tur.	41•34 N	27•46 E
35	Vizianagaram	India	18•10 N	83•29 E
30	Vladimir	Sov.Un.	56•08 N	40•24 E
27	Vladimir-Volynskiy	Sov.Un.	50•50 N	24•20 E
29	Vladivostok	Sov.Un.	43•06 N	131•47 E
21	Vlasenica	Yugo.	44•11 N	18•58 E
21	Vlasotinci	Yugo.	42•58 N	22•08 E
23	Vlieland (I.)	Neth.	53•19 N	4•55 E
23	Vlissingen	Neth.	51•30 N	3•34 E
21	Vlorë (Valona)	Alb.	40•28 N	19•31 E
26	Vltana R.	Czech.	49•24 N	14•18 E
30	Vodl (L.)	Sov.Un.	62•20 N	37•20 E
54	Voël (R.)	S.Afr.	32•52 S	25•12 E
20	Voghera	It.	44•58 N	9•02 E
55	Vohémar	Mad.	13•35 S	50•05 E
21	Voiviis (L.)	Grc.	39•34 N	22•50 E
48	Volcán Misti	Peru	16•04 S	71•20 W
31	Volga (R.)	Sov.Un.	47•30 N	46•20 E
31	Volgograd (Stalingrad)	Sov.Un.	48•40 N	42•20 E
31	Volgogradskoye (Res.)	Sov.Un.	51•10 N	45•10 E
30	Volkhov (R.)	Sov.Un.	58•45 N	31•40 E
27	Volkovysk	Sov.Un.	53•11 N	24•29 E
28	Vologda	Sov.Un.	59•12 N	39•52 E
21	Vólos	Grc.	39•23 N	22•56 E
31	Vol'sk	Sov.Un.	52•10 N	47•00 E
52	Volta, L.	Ghana	7•10 N	0•30 W
52	Volta Noire (Black Volta) (R.)	Afr.	10•30 N	2•55 W
49	Volta Redonda	Braz.	22•32 S	44•05 W
20	Volterra	It.	43•22 N	10•51 E
20	Voltri	It.	44•25 N	8•45 E
20	Volturno (R.)	It.	41•12 N	14•20 E
55	Voortekkerhoogte	S.Afr.	25•48 S	28•10 E
16	Vopnafjördhur	Ice.	65•43 N	14•58 W
24	Vordingborg	Den.	55•10 N	11•55 E
21	Voríai (Is.)	Grc.	39•12 N	24•03 E
21	Vorios Evvikós Kólpos (G.)	Grc.	38•48 N	23•02 E
30	Vorkuta	Sov.Un.	67•28 N	63•40 E
25	Vormsi (I.)	Sov.Un.	59•06 N	23•05 E
31	Vorona (R.)	Sov.Un.	51•50 N	42•00 E
30	Voron'ya (R.)	Sov.Un.	68•20 N	35•20 E
31	Voronezh	Sov.Un.	51•39 N	39•11 E
27	Voronovo	Sov.Un.	54•07 N	25•16 E
31	Voroshilovgrad	Sov.Un.	48•34 N	39•18 E
30	Võru	Sov.Un.	57•50 N	26•58 E
24	Voss	Nor.	60•40 N	6•24 E
30	Votkinsk	Sov.Un.	57•00 N	54•00 E
30	Votkinskoye Vdkhr (Res.)	Sov.Un.	57•30 N	55•00 E
24	Voxna älv (R.)	Swe.	61•30 N	15•24 E
30	Vozhe (L.)	Sov.Un.	60•40 N	39•00 E
31	Voznesensk	Sov.Un.	47•34 N	31•22 E
28	Vrangelya (Wrangel) (I.)	Sov.Un.	71•25 N	173•38 E
21	Vranje	Yugo.	42•33 N	21•55 E
21	Vratsa	Bul.	43•12 N	23•31 E
21	Vrbas	Yugo.	45•34 N	19•43 E
20	Vrbas (R.)	Yugo.	44•25 N	17•17 E
21	Vršac	Yugo.	45•08 N	21•18 E
27	Vrutky	Czech.	49•09 N	18•55 E
54	Vryburg	S.Afr.	26•55 S	29•45 E
54	Vryheid	S.Afr.	27•43 S	30•58 E
27	Vsetín	Czech.	49•21 N	18•01 E
21	Vukovar	Yugo.	45•20 N	19•00 E
8	Vulcan	Mi.	45•45 N	87•50 W
20	Vulcano (I.)	It.	38•23 N	15•00 E
21	Vŭlchedrŭm	Bul.	43•43 N	23•29 E
25	Vyartsilya	Sov.Un.	62•10 N	30•40 E
30	Vyatka (R.)	Sov.Un.	58•25 N	51•25 E
30	Vyaz'ma	Sov.Un.	55•12 N	34•17 E
30	Vyazniki	Sov.Un.	56•10 N	42•10 E
25	Vyborg	Sov.Un.	60•43 N	28•46 E
30	Vychegda (R.)	Sov.Un.	61•40 N	48•00 E
30	Vym (R.)	Sov.Un.	63•15 N	51•20 E
30	Vyshniy Volochëk	Sov.Un.	57•34 N	34•35 E
26	Vyskov	Czech.	49•17 N	16•58 E
26	Vysoké Mýto	Czech.	49•58 N	16•07 E
30	Vytegra	Sov.Un.	61•00 N	36•20 E
30	Vyur	Sov.Un.	57•55 N	27•00 E
	W			
23	Waal (R.)	Neth.	51•46 N	5•00 E
8	Wabash	In.	40•45 N	85•50 W
8	Wabash (R.)	Il.-In.	38•00 N	88•00 W
27	Wabrzeźno	Pol.	53•17 N	18•59 E
6	Waco	Tx.	31•35 N	97•06 W
23	Waddenzee (Sea)	Neth.	53•00 N	4•50 E
33	Wādī Mūsā	Jordan	30•19 N	35•29 E
53	Wad Madani	Sud.	14•27 N	33•31 E
27	Wadowice	Pol.	49•53 N	19•31 E
8	Wadsworth	Oh.	41•01 N	81•44 W
11	Wager B.	Can.	65•48 N	88•19 W
46	Wagga Wagga	Austl.	35•01 S	147•30 E
27	Wagrowiec	Pol.	52•47 N	17•14 E
6	Wahiawa	Hi.	21•30 N	158•03 W
27	Waidhofen	Aus.	47•58 N	14•46 E
45	Waikato (R.)	NZ	38•00 S	175•47 E
46	Waikerie	Austl.	34•15 S	140•00 E
6	Wailuku	Hi.	20•55 N	156•30 W
35	Wainganga (R.)	India	20•24 N	79•41 E
40	Waingapu	Indon.	9•32 S	120•00 E
6	Waipahu	Hi.	21•20 N	158•02 W
45	Wakatipu (R.)	NZ	44•24 S	169•00 E
37	Wakayama	Jap.	34•14 N	135•11 E
42	Wake (I.)	Oceania	19•25 N	167•00 E
22	Wakefield	Eng.	53•41 N	1•25 W
37	Wakkanai	Jap.	45•19 N	141•43 E
54	Wakkerstroom	S.Afr.	27•19 S	30•04 E
26	Walbrzych	Pol.	50•46 N	16•16 E
22	Wales	U.K.	52•12 N	3•40 W
26	Walez	Pol.	53•16 N	16•30 E
46	Walgett	Austl.	30•00 S	148•10 E
67	Walgreen Coast	Ant.	73•00 N	110•00 W
46	Wallaroo	Austl.	33•52 S	137•45 E
6	Walla Walla	Wa.	46•03 N	118•20 W
53	Wallel, Tulu (Mt.)	Eth.	9•00 N	34•52 E
9	Wallingford	Vt.	43•30 N	72•55 W
42	Wallis Is.	Oceania	13•00 S	176•10 E
22	Walney (C.)	Eng.	54•04 N	3•13 W
22	Walsall	Eng.	52•35 N	1•58 W
9	Walton	NY	42•10 N	75•05 W
54	Walvis Bay	S.Afr.	22•50 S	14•30 E
55	Wami (R.)	Tan.	6•31 S	37•17 E
46	Wandoan	Austl.	26•09 S	149•51 E
45	Wanganui	NZ	39•53 N	175•01 E
46	Wangaratta	Austl.	36•23 N	146•18 E
26	Wangeroog I.	F.R.G.	53•49 N	7•57 E
39	Wanhsien	China	30•48 N	108•22 E
54	Wankie	Rh.	18•22 S	26•29 E
39	Wantsai	China	28•05 N	114•25 E
46	Waodoan	Austl.	26•12 S	149•52 E
8	Wapakoneta	Oh.	40•35 N	84•10 W
9	Wappingers Falls	NY	41•35 N	73•55 W
35	Warangal	India	18•03 N	79•45 E
44	Warburton (R.)	Austl.	27•30 S	138•45 E
33	Wardān, Wādī (R.)	Egypt	29•22 N	33•00 E
35	Wardha	India	20•46 N	78•42 E
8	War Eagle	WV	37•30 N	81•50 W
26	Waren	F.R.G.	53•32 N	12•43 E
46	Warialda	Austl.	29•32 S	150•34 E
54	Warmbad	Namibia	28•25 S	18•45 E
24	Warnemünde	G.D.R.	54•11 N	12•04 E
26	Warnow R.	G.D.R.	53•51 N	11•55 E
46	Warracknabeal	Austl.	36•20 S	142•28 E
46	Warrangamba Res.	Austl.	33•40 S	150•00 E
45	Warrego (R.)	Austl.	27•13 S	145•58 E
8	Warren	In.	40•40 N	85•25 W
8	Warren	Mi.	42•33 N	83•03 W
8	Warren	Oh.	41•15 N	80•50 W
9	Warren	Pa.	41•50 N	79•10 W
9	Warrenton	Va.	38•45 N	77•50 W
52	Warri	Nig.	5•33 N	5•43 E
46	Warrnambool	Austl.	36•20 S	142•28 E
45	Warrumbungle Ra.	Austl.	31•18 S	150•00 E
8	Warsaw	In.	41•15 N	85•50 W
9	Warsaw	NY	42•45 N	78•10 W
27	Warsaw	Pol.	52•15 N	21•05 E
26	Warta R.	Pol.	52•35 N	15•07 E
55	Wartburg	S.Afr.	29•26 S	30•39 E
46	Warwick	Austl.	28•05 S	152•10 E
22	Warwick	Eng.	52•19 N	1•46 W
9	Warwick	RI	41•42 N	71•27 W
6	Wasatch Ra.	U.S.	39•10 N	111•30 W
55	Wasbank	S.Afr.	28•27 S	30•09 E
23	Wash, The	Eng.	53•00 N	0•20 E
9	Washington	DC	38•50 N	77•00 W
8	Washington	In.	38•40 N	87•10 W
6	Washington	U.S.	47•30 N	121•10 W
9	Washington, Mt.	NH	44•15 N	71•15 W
8	Washington Court House	Oh.	39•30 N	83•25 W
27	Wasilkow	Pol.	53•12 N	23•13 E
13	Water (I.)	Vir.Is.	18•20 N	64•57 W
9	Waterbury	Ct.	41•30 N	73•00 W
22	Waterford	Ire.	52•20 N	7•03 W
9	Waterloo	Can.	45•25 N	72•30 W
7	Waterloo	Ia.	42•30 N	92•22 W
9	Waterloo	NY	42•55 N	76•50 W
9	Watertown	NY	44•00 N	75•55 W
6	Watertown	SD	44•53 N	97•07 W
8	Watertown	Wi.	43•13 N	88•40 W
9	Watervliet	NY	42•45 N	73•54 W
53	Watsa	Zire	3•03 N	29•32 E
8	Watseka	Il.	40•45 N	87•45 W
10	Watson Lake	Can.	60•18 N	128•50 W
8	Waukesha	Wi.	43•01 N	88•13 W
8	Wausau	Wi.	44•58 N	89•40 W
8	Wauseon	Oh.	41•30 N	84•10 W
8	Wautoma	Wi.	4•04 N	89•11 W
8	Wauwatosa	Wi.	43•03 N	88•00 W
23	Waveney (R.)	Eng.	52•27 N	1•17 E
55	Waverly	S.Afr.	31•54 S	26•29 E
53	Wāw	Sud.	7•41 N	28•00 E
53	Wāw al-Kabīr	Libya	25•23 N	16•52 E
8	Wawasee (L.)	In.	41•25 N	85•45 W
7	Waycross	Ga.	31•11 N	82•24 W
9	Waynesboro	Pa.	39•45 N	77•35 W
9	Waynesboro	Va.	38•05 N	78•50 W
9	Waynesburg	Pa.	39•55 N	80•10 W
22	Weald, The	Eng.	50•58 N	0•15 W
9	Webster Springs	WV	38•30 N	80•20 W
67	Weddell Sea	Ant.	73•00 S	45•00 W
55	Weenen	S.Afr.	28•52 S	30•05 E
23	Weert	Neth.	51•16 N	5•39 E
27	Wegorzewo	Pol.	54•14 N	21•46 E
27	Wegrow	Pol.	52•23 N	22•02 E
38	Weich'ang	China	41•50 N	118•00 E
38	Weihai	China	37•30 N	122•05 E
38	Wei Ho (R.)	China	34•00 N	108•10 E
36	Weihsi	China	27•27 N	99•30 E
26	Weilheim	F.R.G.	47•50 N	11•06 E
26	Weimar	G.D.R.	50•59 N	11•20 E
38	Weinan	China	34•32 N	109•40 E
45	Weipa	Austl.	12•25 S	141•54 E
8	Weirton	WV	40•25 N	80•35 W
38	Weishih	China	34•23 N	114•12 E
26	Weissenburg	F.R.G.	49•04 N	11•20 E
26	Weissenfels	G.D.R.	51•13 N	11•58 E
27	Wejherowo	Pol.	54•36 N	18•15 E
46	Welford	Austl.	25•08 S	144•43 E
54	Welkom	S.Afr.	27•57 S	26•45 E
9	Welland	Can.	42•59 N	79•13 W
22	Welland (R.)	Eng.	52•38 N	0•40 W
44	Wellesley Is.	Austl.	16•15 S	139•25 E
46	Wellington	Austl.	32•40 S	148•50 E
45	Wellington	N.Z.	41•15 S	174•45 E
8	Wellington	Oh.	41•10 N	82•10 W
50	Wellington (I.)	Chile	49•30 S	76•30 W
44	Wells	Austl.	26•35 S	123•40 E
8	Wells	Mi.	45•50 N	87•00 W
9	Wellsboro	Pa.	41•45 N	77•15 W
8	Wellsburg	WV	40•10 N	80•40 W
8	Wellston	Oh.	39•05 N	82•30 W
9	Wellsville	NY	42•10 N	78•00 W
8	Wellsville	Oh.	40•35 N	80•40 W
26	Wels	Aus.	48•10 N	14•01 E
22	Welshpool	Wales	52•44 N	3•10 W
39	Wench'ang	China	19•32 N	110•42 E
39	Wenchou	China	28•00 N	120•40 E
38	Wench'üan (Halunrshan)	China	47•10 N	120•00 E
53	Wendo	Eth.	6•37 N	38•29 E
39	Wenshan	China	23•20 N	104•15 E
36	Wensu (Aksu)	China	41•45 N	80•30 E
23	Wensum (R.)	Eng.	52•45 N	1•08 E
46	Wentworth	Austl.	24•03 S	141•53 E
54	Wepener	S.Afr.	29•43 S	27•04 E
53	Were Ilu	Eth.	10•39 N	39•21 E
26	Werra R.	F.R.G.	51•16 N	9•54 E
26	Wertach R.	F.R.G.	48•12 N	10•40 E
23	Wesel	F.R.G.	51•39 N	6•37 E
26	Weser R.	F.R.G.	53•08 N	8•35 E
44	Wessel (Is.)	Austl.	11•45 S	136•25 E
12	West, Mt.	C.Z.	9•10 N	79•52 W
8	West Allis	Wi.	43•01 N	88•01 W
8	West Bend	Wi.	43•25 N	88•13 W
35	West Bengal	India	23•30 N	87•30 E
26	West Berlin	F.R.G.	52•31 N	13•20 E
8	West Branch	Mi.	44•15 N	84•10 W
22	West Bromwich	Eng.	52•32 N	1•59 W
9	Westbrook	Me.	43•41 N	70•23 W
44	West Cape Howe (C.)	Austl.	35•15 S	117•30 E
9	West Chester	Pa.	39•57 N	75•36 W
9	Westerly	RI	41•25 N	71•50 W
26	Western Alps (Mts.)	Switz.-Fr.	46•19 N	7•03 E
44	Western Australia (State)	Austl.	24•15 S	121•30 E
22	Western Downs	Eng.	50•50 N	2•25 W
35	Western Ghāts (Mts.)	India	17•35 N	74•00 E
9	Western Port	Md.	39•30 N	79•00 W
51	Western Sahara	Afr.	23•05 N	15•33 W
42	Western Samoa	Oceania	14•30 S	172•00 W
28	Western Siberian Lowland	Sov.Un.	63•37 N	72•45 E
8	Westerville	Oh.	40•10 N	83•00 W
26	Westerwald (For.)	F.R.G.	50•35 N	7•45 E
9	Westfield	Ma.	42•05 N	72•45 W
22	West Ham	Eng.	51•30 N	0•00
9	West Hartford	Ct.	41•45 N	72•45 W
13	West Indies	N.A.	19•00 N	78•30 W
8	West Lafayette	In.	40•25 N	86•55 W
9	Westminster	Md.	39•40 N	76•55 W
8	Weston	WV	39•00 N	80•30 W
22	Weston-super-Mare	Eng.	51•23 N	3•00 W
9	West Point	NY	41•23 N	73•58 W
9	West Point	Va.	37•25 N	76•50 W
22	Westport	Ire.	53•44 N	9•36 W
22	Westray (I.)	Scot.	59•19 N	3•05 W
23	West Schelde (R.)	Neth.	51•25 N	3•30 E
8	West Terre Haute	In.	39•30 N	87•30 W
8	Westville	Il.	40•00 N	87•40 W
7	West Virginia	U.S.	39•00 N	80•50 W
46	West Wyalong	Austl.	34•00 S	147•20 E
41	Wetar (I.)	Indon.	7•34 S	126•00 E
10	Wetaskiwin	Can.	52•58 N	113•22 W
41	Wewak	Pap.N.Gui.	3•19 S	143•30 E
22	Wexford	Ire.	52•20 N	6•30 W
53	Weyib (R.)	Eth.	6•25 N	41•21 E
22	Weymouth	Eng.	50•37 N	2•34 W
9	Weymouth	Ma.	42•44 N	70•57 W
22	Wharfe (R.)	Eng.	54•01 N	1•53 W
6	Wheeler Pk.	Nv.	38•58 N	114•15 W
8	Wheeling	WV	40•05 N	80•45 W
11	Whitby	Can.	43•50 N	79•00 W
7	White (R.)	Ar.	34•32 N	91•11 W
8	White (R.)	In.	39•15 N	86•45 W
9	White (R.)	Vt.	43•45 N	72•35 W
11	White B.	Can.	50•00 N	56•30 W
8	White Cloud	Mi.	43•35 N	85•45 W
10	Whitecourt	Can.	54•09 N	115•41 W
9	Whitefield	NH	44•20 N	71•35 W
8	Whitehall	Mi.	43•20 N	86•20 W
9	Whitehall	NY	43•30 N	73•25 W
22	Whitehaven	Eng.	54•35 N	3•30 W
10	Whitehorse	Can.	60•39 N	135•01 W
9	White Mts.	NH	42•20 N	71•05 W
53	White Nile (R.)	Sud.	14•00 N	32•35 E
10	White P.	Ak.-Can.	59•35 N	135•03 W
9	White Plains	NY	41•02 N	73•47 W
8	White R., East Fork	In.	38•45 N	86•20 W
30	White Sea	Sov.Un.	66•00 N	40•00 E
55	White Umfolzi (R.)	S.Afr.	28•12 S	30•55 E

Page	Name	Region	Lat.	Long.
8	Whitewater	Wi.	42•49 N	88•40 W
22	Whitham (R.)	Eng.	53•08 N	0•15 W
8	Whiting	In.	41•41 N	87•30 W
6	Whitney, Mt.	Ca.	36•34 N	118•18 W
45	Whitsunday (I.)	Austl.	20•16 S	149•00 E
55	Whittlesea	S.Afr.	32•11 S	26•51 E
46	Whyalla	Austl.	33•00 S	137•32 E
8	Wichita	Ks.	37•42 N	97•21 W
6	Wichita Falls	Tx.	33•54 N	98•29 W
6	Wichita Mts.	Ok.	34•48 N	98•43 W
22	Wick	Scot.	58•25 N	3•05 W
22	Wicklow Mts.	Ire.	52•49 N	6•20 W
9	Wiconisco	Pa.	43•35 N	76•45 W
8	Widen	WV	38•25 N	80•55 W
26	Wieden	F.R.G.	49•41 N	12•09 E
27	Wieliczka	Pol.	49•58 N	20•06 E
27	Wieluń	Pol.	51•13 N	18•33 E
26	Wien (Vienna)	Aus.	48•13 N	16•22 E
26	Wiener Neustadt	Aus.	47•48 N	16•15 E
27	Wieprz, R.	Pol.	51•25 N	22•45 E
26	Wiesbaden	F.R.G.	50•05 N	8•15 E
22	Wigan	Eng.	53•33 N	2•37 W
22	Wight, Isle of	Eng.	50•44 N	1•17 W
46	Wilcannia	Austl.	31•30 S	143•30 E
26	Wild Spitze Pk.	Aus.	46•49 N	10•50 E
9	Wildwood	NJ	39•00 N	74•50 W
45	Wilhelm, Mt.	Pap.N.Gui.	5•58 S	144•58 E
49	Wilhelmina Gebergte (Mts.)	Sur.	4•30 N	57•00 W
26	Wilhelmshaven	F.R.G.	53•30 N	8•10 E
9	Wilkes-Barre	Pa.	41•15 N	75•50 W
67	Wilkes Land	Ant.	71•00 S	126•00 E
10	Wilkie	Can.	52•25 N	108•43 W
6	Willamette R.	Or.	44•15 N	123•13 W
8	Willard	Oh.	41•00 N	82•50 W
48	Willemstad	Neth. Antilles	12•12 N	68•58 W
22	Willesden	Eng.	51•31 N	0•17 W
44	William Creek	Austl.	28•45 S	136•20 E
9	Williamsburg	Va.	37•15 N	76•41 W
8	Williamson	WV	37•40 N	82•15 W
9	Williamsport	Md.	39•35 N	77•45 W
9	Williamsport	Pa.	41•15 N	77•05 W
8	Williamstown	WV	39•20 N	81•30 W
9	Willimantic	Ct.	41•40 N	72•10 W
45	Willis Is.	Austl.	16•15 S	150•30 E
6	Williston	ND	48•08 N	103•38 W
8	Willoughby	Oh.	41•39 N	81•25 W
54	Willowmore	S.Afr.	33•15 S	23•37 E
55	Willowvale	S.Afr.	32•17 S	28•32 E
8	Wilmette	Il.	42•04 N	87•42 W
46	Wilmington	Austl.	32•39 S	138•07 E
9	Wilmington	De.	39•45 N	75•33 W
8	Wilmington	Oh.	39•20 N	83•50 W
8	Wilmore	Ky.	37•50 N	84•35 W
46	Wilson's Prom.	Austl.	39•05 S	146•50 E
44	Wiluna	Austl.	26•35 S	120•25 E
8	Winamac	In.	41•05 N	86•40 W
22	Winchester	Eng.	51•04 N	1•20 W
8	Winchester	In.	40•10 N	84•50 W
8	Winchester	Ky.	38•00 N	84•15 W
9	Winchester	NH	42•45 N	72•25 W
9	Winchester	Va.	39•10 N	78•10 W
9	Windber	Pa.	40•15 N	78•45 W
6	Wind Cave Natl. Park	SD	43•36 N	103•53 W
22	Windermere	Eng.	54•25 N	2•59 W
9	Windham	Ct.	41•45 N	72•05 W
54	Windhoek	Namibia	22•05 S	17•10 E
46	Windora	Austl.	25•15 S	142•50 E
6	Wind River Ra.	Wy.	43•19 N	109•47 W
8	Windsor	Can.	42•19 N	83•00 W
11	Windsor	Can.	44•59 N	64•08 W
11	Windsor	Can.	48•57 N	55•40 W
22	Windsor	Eng.	51•27 N	0•37 W
13	Windward Is.	N.A.	12•45 N	61•40 W
11	Winisk (R.)	Can.	54•30 N	86•30 W
8	Winnebago, L.	Wi.	44•09 N	88•10 W
6	Winnemucca	Nv.	40•59 N	117•43 W
8	Winnetka	Il.	42•07 N	87•44 W
10	Winnipeg	Can.	49•53 N	97•09 W
10	Winnipeg, L.	Can.	52•00 N	97•00 W
10	Winnipeg (R.)	Can.	52•20 N	95•54 W
10	Winnipegosis	Can.	51•39 N	99•56 W
10	Winnipegosis (L.)	Can.	52•30 N	100•00 W
9	Winnipesaukee (L.)	NH	43•40 N	71•20 W
7	Winona	Mn.	44•03 N	91•40 W
9	Winooski	Vt.	44•30 N	73•10 W
7	Winston-Salem	NC	36•05 N	80•15 W
55	Winterberge (Mts.)	S.Afr.	32•18 S	26•25 E
26	Winterthur	Switz.	47•30 N	8•32 E
55	Winterton	S.Afr.	28•51 S	29•33 E
9	Winthrop	Me.	44•19 N	70•00 W
45	Winton	Austl.	22•17 S	143•08 E
7	Wisconsin	U.S.	44•30 N	91•00 W
8	Wisconsin (R.)	Wi.	43•14 N	90•34 W
8	Wisconsin Dells	Wi.	43•38 N	89•46 W
8	Wisconsin Rapids	Wi.	44•24 N	89•50 W
27	Wisla (Vistula) R.	Pol.	52•48 N	19•02 E
27	Wisloka R	Pol.	49•55 N	21•26 E
49	Wismar	Guy.	5•58 N	58•15 W
26	Wismar	G.D.R.	53•53 N	11•28 E
26	Wissembourg	Fr.	49•03 N	7•58 E
55	Witberg (Mtn.)	S.Afr.	30•32 S	27•18 E
22	Witham (R.)	Eng.	53•11 N	0•20 W
8	Witt	Il.	39•10 N	89•15 W
26	Wittenberg	G.D.R.	51•53 N	12•40 E
26	Wittenberge	G.D.R.	52•59 N	11•45 E
26	Wittlich	F.R.G.	49•58 N	6•54 E
55	Witu	Ken.	2•18 S	40•28 E
55	Witwatersberg (Mts.)	S.Afr.	25•58 S	27•53 E
27	Wkra R.	Pol.	52•40 N	20•35 E
27	Wloclawek	Pol.	52•38 N	19•08 E
27	Wlodawa	Pol.	51•33 N	23•33 E
27	Wloszczowa	Pol.	50•51 N	19•58 E
9	Wolf (I.)	Can.	44•10 N	76•25 W
8	Wolf (R.)	Wi.	45•14 N	88•45 W
26	Wolfenbüttel	F.R.G.	52•10 N	10•32 E
26	Wolfsburg	F.R.G.	52•30 N	10•37 E
26	Wolgast	G.D.R.	54•04 N	13•46 E
55	Wolhuterskop	S.Afr.	25•41 S	27•40 E
10	Wollaston (L.)	Can.	58•15 N	103•20 W
10	Wollaston Pen.	Can.	70•00 N	115•00 W
46	Wollongong	Austl.	34•26 S	151•05 E
27	Wolomin	Pol.	52•19 N	21•17 E
22	Wolverhampton	Eng.	52•35 N	2•07 W
38	Wŏnsan	Kor.	39•08 N	127•24 E
46	Wonthaggi	Austl.	38•45 S	145•42 E
10	Wood Buffalo Natl. Park	Can.	59•50 N	118•53 W
41	Woodlark (I.)	Pap.N.Gui.	9•07 S	152•00 E
44	Woodroffe, Mt.	Austl.	26•05 S	132•00 E
44	Woods (L.)	Austl.	18•00 S	133•18 E
7	Woods, L. of the	Can.-Mn.	49•25 N	93•25 W
8	Woodsfield	Oh.	39•45 N	81•10 W
8	Woodstock	Il.	42•20 N	88•29 W
9	Woodstock	Va.	38•55 N	78•25 W
9	Woodsville	NH	44•10 N	72•00 W
46	Woomera	Austl.	31•15 S	136•43 E
8	Wooster	Oh.	40•50 N	81•55 W
22	Worcester	Eng.	52•09 N	2•14 W
9	Worcester	Ma.	42•16 N	71•49 W
54	Worcester	S.Afr.	33•35 S	19•31 E
22	Workington	Eng.	54•40 N	3•30 W
26	Worms	F.R.G.	49•37 N	8•22 E
22	Worthing	Eng.	50•48 N	0•29 W
8	Worthington	In.	39•05 N	87•00 W
41	Wowoni (I.)	Indon.	4•05 S	123•45 E
22	Wrath, C.	Scot.	58•34 N	5•01 W
45	Wreck Rfs.	Austl.	22•00 S	155•52 E
22	Wrexham	Wales	53•03 N	3•00 W
27	Wroclaw (Breslau)	Pol.	51•07 N	17•10 E
27	Wrzesnia	Pol.	52•19 N	17•33 E
39	Wuch'ang	China	30•32 N	114•25 E
38	Wuch'ang	China	44•59 N	127•00 E
38	Wuch'ing	China	39•32 N	116•51 E
39	Wu Chin Shan	China	18•48 N	109•30 E
39	Wuchou (Tsangwu)	China	23•32 N	111•25 E
39	Wuhan	China	30•30 N	114•15 E
39	Wuhsing	China	30•38 N	120•10 E
39	Wuhu	China	31•22 N	118•22 E
39	Wui Shan (Mts.)	China	26•38 N	116•35 E
40	Wu Liang Shan (Mts.)	China	23•07 N	100•45 E
36	Wulunuch'i (Urunchi)	China	43•49 N	87•43 E
39	Wup'ing	China	25•05 N	116•01 E
26	Wuppertal	F.R.G.	51•16 N	7•14 E
39	Wu R.	China	27•30 N	108•00 E
26	Würm See (L.)	F.R.G.	47•58 N	11•30 E
26	Würzburg	F.R.G.	49•48 N	9•57 E
26	Wurzen	G.D.R.	51•22 N	12•45 E
36	Wushih (Uch Turfan)	China	41•13 N	79•08 E
36	Wusu (Kweitun)	China	44•28 N	84•07 E
39	Wuwei	China	31•19 N	117•53 E
37	Wuyün	China	48•51 N	130•06 E
8	Wyandotte	Mi.	42•12 N	83•10 W
54	Wynberg	S.Afr.	34•00 S	18•28 E
44	Wyndham	Austl.	15•30 S	128•15 E
10	Wynyard	Can.	51•47 N	104•10 W
6	Wyoming	U.S.	42•50 N	108•30 W
6	Wyoming Ra.	Wy.	42•43 N	110•35 W
26	Wysokie Mazowieckie	Pol.	52•55 N	22•42 E
26	Wyszkow	Pol.	52•35 N	21•29 E

X

Page	Name	Region	Lat.	Long.
21	Xanthi	Grc.	41•08 N	24•53 E
54	Xau, L.	Bots.	21•15 S	24•38 E
8	Xenia	Oh.	39•40 N	83•55 W
49	Xingú (R.)	Braz.	6•20 S	52•34 W

Y

Page	Name	Region	Lat.	Long.
39	Yann	China	30•00 N	103•20 E
27	Yablonitskiy Pereval (P.)	Sov.Un.	48•20 N	24•25 E
29	Yablonovyy Khrebet (Mts.)	Sov.Un.	51•15 N	111•30 E
50	Yacuiba	Arg.	22•02 S	63•44 W
52	Yafran	Libya	31•57 N	12•04 E
39	Yaihsien	China	18•20 N	109•10 E
6	Yakima	Wa.	46•35 N	120•30 W
37	Yaku (I.)	Jap.	30•15 N	130•41 E
29	Yakut A.S.S.R.	Sov.Un.	65•21 N	117•13 E
29	Yakutsk	Sov.Un.	62•13 N	129•49 E
8	Yale	Mi.	43•05 N	82•45 W
53	Yalinga	Cen.Afr.Emp.	6•56 N	23•22 E
31	Yalta	Sov.Un.	44•29 N	34•12 E
38	Yalu (Amnok) (R.)	China-Kor.	41•20 N	126•35 E
36	Yalung Chiang (R.)	China	32•29 N	98•41 E
28	Yalutorovsk	Sov.Un.	56•42 N	66•32 E
37	Yamagata	Jap.	38•12 N	140•24 E
28	Yamal, P-ov (Pen.)	Sov.Un.	71•15 N	70•00 E
34	Yambi, Mesa de	Col.	1•55 N	71•45 W
41	Yamdena (I.)	Indon.	7•23 S	130•30 E
36	Yamdrog Tsho (L.)	China	29•11 N	91•26 E
36	Yamethin	Bur.	20•14 N	96•27 E
46	Yamma Yamma, L.	Austl.	26•15 S	141•30 E
29	Yamsk	Sov.Un.	59•41 N	154•09 E
35	Yamuna (R.)	India	26•50 N	80•10 E
29	Yana (R.)	Sov.Un.	69•42 N	135•45 E
46	Yanac	Austl.	36•10 S	141.30 E
35	Yanam	India	16•48 N	82•15 E
34	Yanbu'	Sau.Ar.	23•57 N	38•02 E
39	Yangchiang	China	21•51 N	111•58 E
37	Yangchou	China	32•24 N	119•24 E
39	Yangch'un	China	22•08 N	111•48 E
38	Yangkochuang	China	40•10 N	116•48 E
37	Yangtze (R.)	China	30•30 N	117•25 E
38	Yangyang	Kor.	38•02 N	128•38 E
6	Yankton	SD	42•51 N	97•24 W
53	Yao	Chad	13•00 N	17•38 E
52	Yaoundé	Cam.	3•52 N	11•31 E
42	Yap (I)	Pac.Is.Trust Ter.	11•00 N	138•00 E
12	Yaqui (R.)	Mex.	28•15 N	109•40 W
49	Yaracuy (State)	Ven.	10•10 N	68•31 W
46	Yaraka	Austl.	24•50 S	144•08 E
30	Yaransk	Sov.Un.	57•18 N	48•05 E
53	Yarda (Well)	Chad	18•29 N	19•13 E
11	Yarmouth	Can.	43•50 N	66•07 W
28	Yaroslavl	Sov.Un.	57•57 N	39•54 E
30	Yarra-to (L.)	Sov.Un.	68•30 N	71•30 E
30	Yartsevo	Sov.Un.	55•04 N	32•38 E
29	Yartsevo	Sov.Un.	60•13 N	89•52 E
48	Yarumal	Col.	6•57 N	75•24 W
27	Yasel'da R.	Sov.Un.	53•13 N	25•53 E
27	Yasinya	Sov.Un.	48•17 N	24•21 E
10	Yathkyed (L.)	Can.	62•41 N	98•00 W
27	Yavorvo	Sov.Un.	49•56 N	23•24 E
34	Yazd	Iran	31•59 N	54•03 E
7	Yazoo (R.)	Ms.	32•32 N	90•40 E
40	Ye	Bur.	15•13 N	97•52 E
30	Yegor'yevsk	Sov.Un.	55•23 N	38•59 E
36	Yehch'eng (Karghalik)	China	37•30 N	79•26 E
30	Yelabuga	Sov.Un.	55•50 N	52•18 E
31	Yelan	Sov.Un.	50•50 N	44•00 E
28	Yelets	Sov.Un.	52•35 N	38•28 E
29	Yelizavety, Mys (C.)	Sov.Un.	54•28 N	142•59 E
22	Yell (I.)	Scot.	60•35 N	1•27 W
10	Yellowknife	Can.	62•29 N	114•38 W
38	Yellow Sea	Asia	35•20 N	122•15 E
6	Yellowstone L.	Wy.	44•27 N	110•03 W
6	Yellowstone Natl. Park	Wy.	44•45 N	110•35 W
6	Yellowstone R.	Mt.	46•28 N	105•39 W
32	Yemen	Asia	15•45 N	44•30 E
32	Yemen, People's Democratic Republic of	Asia	14•45 N	46•45 E
30	Yemetsk	Sov.Un.	63•28 N	41•28 E
38	Yenan	China	36•35 N	109•32 E
36	Yenan (Fushih)	China	36•46 N	109•15 E
35	Yenangyaung	Bur.	20•27 N	94•59 E
38	Yencheng	China	33•38 N	113•59 E
38	Yenchi	China	42•55 N	129•35 E
52	Yendi	Ghana	9•26 N	0•01 W
21	Yenice (R.)	Tur.	41•10 N	33•00 E
28	Yenisey (R.)	Sov.Un.	67•48 N	87•15 E
29	Yeniseysk	Sov.Un.	58•27 N	90•28 E
38	Yenshan	China	38•05 N	117•15 E
38	Yenshou	China	45•25 N	128•43 E
38	Yent'ai (Chefoo)	China	37•32 N	121•22 E
44	Yeo (I.)	Austl.	28•15 S	124•00 E
31	Yerevan	Sov.Un.	40•10 N	44•30 E
22	Yerington	Nv.	38•59 N	119•10 W
30	Yermak (I.)	Sov.Un.	66•30 N	71•30 E
17	Yeu, Île d'	Fr.	46•43 N	2•45 W
31	Yevpatoriya	Sov.Un.	45•13 N	33•22 E
31	Yeysk	Sov.Un.	46•41 N	38•13 E
21	Yiannitsá	Grc.	40•47 N	22•26 E
38	Yinch'uan (Ninghsia)	China	38•22 N	106•22 E
36	Yingchisha	China	39•01 N	75•29 E
38	Yingk'ou	China	40•35 N	122•10 E
38	Yin Shan (Mtn.)	China	40•50 N	110•30 E
21	Yioúra (I.)	Grc.	37•52 N	24•42 E
21	Yíthion	Grc.	36•50 N	22•37 E
39	Yog Pt.	Phil.	14•00 N	124•30 E
40	Yogyakarta	Indon.	7•50 S	110•20 E
37	Yokohama	Jap.	35•57 N	139•40 E
52	Yola	Nig.	9•13 N	12•27 E
48	Yolombó	Col.	6•37 N	74•59 W
38	Yŏngdŏk	Kor.	36•28 N	129•25 E
38	Yŏnghŭng	Kor.	39•31 N	127•11 E
9	Yonkers	NY	40•57 N	73•54 W
44	York	Austl.	32•00 S	117•00 E
22	York	Eng.	53•58 N	1•10 W
9	York	Pa.	40•00 N	76•40 W
45	York, C.	Austl.	10•45 S	142•35 E
5	York, Kap (C.)	Grnld.	75•30 N	73•00 W
46	Yorketown	Austl.	35•00 S	137•28 E
46	York Pen	Austl.	34•24 S	137•20 E
22	Yorkshire Wolds (Hills)	Eng.	54•00 N	0•35 W
10	Yorkton	Can.	51•13 N	102•28 W
42	Yoron (I.)	Jap.	26•48 N	128•40 E
6	Yosemite Natl. Park	Ca.	38•30 N	119•36 W
30	Yoshkar-Ola	Sov.Un.	56•35 N	48•05 E
22	Youghal B.	Ire.	51•52 N	7•46 W
22	Youhal	Ire.	51•58 N	7•57 W
46	Young	Austl.	34•15 S	148•18 E
8	Youngstown	Oh.	41•05 N	80•40 W
31	Yozgat	Tur.	39•50 N	34•50 E
8	Ypsilanti	Mi.	42•15 N	83•37 W
26	Ystad	Swe.	55•29 N	13•28 E
26	Ytre Solund (I.)	Nor.	61•01 N	4•25 E
33	Yu'alliq, Jabal (Mts.)	Egypt	30•12 N	33•42 E
39	Yüan (R.)	China	28•50 N	110•50 E
39	Yüanan	China	31•08 N	111•28 E
39	Yüanling	China	28•30 N	110•18 E
38	Yüanshih	China	37•45 N	114•32 E
52	Yuby, C.	Mor.	28•01 N	13•21 W
12	Yucatan	Mex.	20•45 N	89•00 W
12	Yucatán Chan	Mex.	22•30 N	87•00 W
39	Yu Chiang (R.)	China	23•55 N	106•50 E
38	Yuch'eng	China	36•55 N	116•39 E
39	Yüehyang	China	92•25 N	113•05 E
30	Yug (R.)	Sov.Un.	59•50 N	45•55 E
14	Yugoslavia	Eur.	44•48 N	17•29 E
38	Yühsien	China	39•40 N	114•38 E
10	Yukon (Ter.)	Can.	63•16 N	135•30 W
6	Yukon R.	Ak.-Can.	62•10 N	143•00 W
39	Yülin	China	22•38 N	110•10 E
38	Yülin	China	38•18 N	109•45 E
8	Yuma	Az.	32•40 N	114•40 W
36	Yümen	China	40•14 N	96•56 E
38	Yünch'eng	China	35•00 N	110•40 E
39	Yüngan	China	26•00 N	117•22 E
38	Yungch'ing	China	39•18 N	116•27 E
39	Yungshun	China	29•05 N	109•58 E
38	Yungting Ho	China	40•25 N	115•00 E
39	Yünhsiao	China	24•00 N	117•20 E
38	Yünhsien	China	32•50 N	110•55 E
36	Yunnan (Prov.)	China	24•23 N	101•03 E
36	Yünnan Plat	China	26•03 N	101•26 E
48	Yurimaguas	Peru	5•59 S	76•12 W
30	Yur'yevets	Sov.Un.	57•15 N	43•08 E
39	Yüshan	China	28•42 N	118•20 E
38	Yüshu	China	44•58 N	126•32 E
36	Yütien (Keriya)	China	36•55 N	81•39 E
38	Yut'ien	China	39•54 N	117•45 E
50	Yuty	Par.	26•45 S	56•13 W
38	Yützu	China	37•32 N	112•40 E
30	Yuzha	Sov.Un.	56•38 N	42•20 E
29	Yuzhno-Sakhalinsk	Sov.Un.	47•11 N	143•04 E
26	Yverdon	Switz.	46•46 N	6•35 E

Z

Page	Name	Region	Lat.	Long.
18	Za R.	Mor.	34•19 N	2•23 W
23	Zaandam	Neth.	52•25 N	4•49 E
26	Zabkowice	Pol.	50•35 N	16•48 E
27	Zabrze	Pol.	50•18 N	18•48 E
12	Zacatecas (State)	Mex.	24•00 N	102•45 W
20	Zadar	Yugo.	44•08 N	15•16 E
26	Zagan	Pol.	51•34 N	15•32 E
25	Žagare	Sov.Un.	56•21 N	23•14 E
52	Zaghouan	Tun.	36•30 N	10•04 E
21	Zagorá	Grc.	39•29 N	23•04 E
20	Zagreb	Yugo.	45•50 N	15•58 E
34	Zagro Mts.	Iran	33•30 N	46•30 E
34	Zāhedān	Iran	29•37 N	60•31 E
33	Zahlah	Leb.	33•50 N	35•54 E
51	Zaire	Afr.	1•00 S	22•15 E
53	Zaire (Congo) (R.)	Afr.	1•10 N	18•25 E
21	Zaječar	Yugo.	43•54 N	22•16 E
21	Zákinthos	Grc.	37•48 N	20•55 E
21	Zákinthos (Zante) (I.)	Grc.	37•45 N	20•32 E
27	Zakopane	Pol.	49•18 N	19•57 E
26	Zalaegerszeg	Hung.	46•50 N	16•50 E
27	Zalău	Rom.	47•11 N	23•06 E
27	Zalew Wiślany (B.)	Pol.	54•22 N	19•39 E
53	Zaftan	Libya	28•20 N	19•40 E
54	Zambezi (R.)	Afr.	15•45 S	33•15 E
51	Zambia	Afr.	14•23 S	24•15 E
40	Zamboanga	Phil.	6•58 N	122•02 E
27	Zambrów	Pol.	52•59 N	22•17 E
12	Zamora	Mex.	19•59 N	102•16 W
18	Zamora	Sp.	41•32 N	5•43 W
27	Zamość	Pol.	50•42 N	23•17 E
8	Zanesville	Oh.	39•55 N	82•00 W
34	Zanjān	Iran	36•26 N	48•24 E
55	Zanzibar	Tan.	6•10 S	39•11 E
55	Zanzibar (I.)	Tan.	6•20 S	39•37 E
52	Zaouia el Kahla	Alg.	28•06 N	6•34 E
30	Zapadnaya Dvina (R.)	Sov.Un.	55•30 N	28•27 E
50	Zapala	Arg.	38•53 S	70•02 W
25	Zaporoshskoye			

ABCDEFGHIJ–RM–79